GW00676582

# Handbook of
# Proteolytic Enzymes

The cover design is based on the tertiary structure of a complex between the protease of the human immunodeficiency virus and one of the inhibitors that are proving to be valuable drugs in the fight against the virus.

# Handbook of Proteolytic Enzymes

*Edited by*

## Alan J Barrett
MRC Peptidase Laboratory
The Babraham Institute
Cambridge
UK

## Neil D Rawlings
MRC Peptidase Laboratory
The Babraham Institute
Cambridge
UK

## J Fred Woessner
Department of Biochemistry and Molecular Biology
University of Miami School of Medicine
Miami
USA

**Academic Press**

SAN DIEGO   LONDON   BOSTON   NEW YORK
SYDNEY   TOKYO   TORONTO

Copyight © 1998 by ACADEMIC PRESS

Academic Press
24–28 Oil Road, London NW1 7DX, UK
hp://www.hbuk.co.uk/ap/

Academic Press
525 B Street, Suite 190, San Diego, California 92101-4493, USA
ltp://www.apnet.com

IBN 0-12-079370-9

**Library of Congrss Cataloging-in-Publication Data**

Handbook of proteolytic enzmes / edited by Alan J. Barrett, Neil D. Rawlings, F. Fred Woessne, Jr.
    p.  cm.
   Includes index.
   ISBN 0-12-079370-9 (alk. paper)
    1. Proteolytic enzymes–Hadbooks, manuals, etc.   I. Barrett, Alan J.   II. Rawlings. Neil L   III. Woessner, J.F.
OP609.P78H36 1998
572'.76–dc21                                                                          98-23079
                                                                                        CIP

Typeset in 9.5/11pt Times by Laser Words, Madras, India
Printed in Great Britain by The Bath Press, Bath

98 99 00 01 02 03 BP 9 8 7 6 5 4 3 2 1

# Contents

# METALLOPEPTIDASES                                                            **987**

# 334 Introduction: metallopeptidases and their clans                         **989**

# UNCLASSIFIED PEPTIDASES      1551

## 543 Introduction: peptidases of unknown catalytic type      1553

# Contributors

*Chapter numbers are shown in parenthesis. Full addresses of the contributors are to be found at the end of each chapter.*

Hideki Adachi (501)
Osao Adachi (523)
Ibrahim M. Adham (27)
Patrizia Aducci (111)
David A. Agard (82)
Alexey A. Agranovsky (237)
Kyunghye Ahn (363, 364)
Yoshinori Akiyama (516)
Sean M. Amberg (88)
Norma W. Andrews (127)
Stuart M. Arfin (476, 477)
Gérard J. Arlaud (43, 44)
Nathan N. Aronson, Jr (168)
Yasuhisa Asano (146, 147)
Paolo Ascenzi (111)
Marina T. Assakura (440, 446)
David S. Auld (451, 452)
Francesc X. Avilés (455)
William M. Awad, Jr (490)
Lilia M. Babé (24)
S. Paul Bajaj (52)
Susan C. Baker (229)
Marcus D. Ballinger (94)
Alan J. Barrett (6, 86, 255, 267, 343, 371, 372, 531)
Isolda P. Baskova (552)
Karl Bauer (340, 527, 529)
Ulrich Baumann (386)
Wolfgang Baumeister (160, 169, 170)
Ann Beaumont (351, 352)
J. David Becherer (449)
Robert Belas (388)
Teruhiko Beppu (298)
Ernst M. Bergmann (240)
Neil A. Bersani (471)
Ch. Betzel (95)
Robert J. Beynon (351, 352)
Gabriele Bierbaum (109, 259)
Joseph G. Bieth (11, 12, 13, 15)
Jón B. Bjarnason (416, 418, 420, 423, 430, 441, 443)
Roy A. Black (449)
August Böck (561)
Albert Bolhuis (154)
Judith S. Bond (406, 407, 413)
Frank Bordusa (257)
Javier Bordallo (484, 496)
Jacques Bouvier (383)
Barry J. Bowman (289)
Ralph A. Bradshaw (476, 477)
Hans-Peter Braun (469)
Klaus Breddam (79, 80, 132)
Paul J. Brindley (254)
Anne Broadhurst (165)
Keith Brocklehurst (197, 221)

Peter E. Brodelius (279)
Johanna Broer (353)
Dieter Brömme (212)
Sierd Bron (154)
Murray Broom (550)
Leena Bruckner-Tuderman (414, 415)
Barbara A. Burleigh (127)
Michael J. Butler (114, 140, 346, 382)
David J. Buttle (188, 189, 190, 196)
Sandrine Cadel (348)
Javier Caldentey (557)
Daniel R. Caprioglio (344)
Juan Carbonell (199)
Ruth E. Carter (492)
Francis J. Castellino (59)
Joseph J. Catanese (421)
George H. Caughey (18, 20)
Niamh X. Cawley (304)
Tim E. Cawston (389)
Juan José Cazzulo (203)
Manuel Cercós (199)
Trinad Chakraborty (353)
Julie Chao (29, 30, 31, 32, 33, 34, 35, 36, 37, 38, 39, 40, 41, 61)
Marie-Pierre Chapot-Chartier (216)
Jinq-May Chen (371)
Valérie Chesneau (468)
Bernard Chevrier (491)
Jean-François Chich (136)
Kwan Yong Choi (244)
Michel Chrétien (118, 119, 120, 122, 123)
Chin Ha Chung (171, 178, 184)
Christine Clamagirand (540, 541)
Steven Clarke (502)
Paul P. Cleary (100)
Philippe Clauziat (268)
D. Coates (360)
Paul Cohen (348, 468, 540, 541)
D. Collen (58)
Ivan E. Collier (402)
Robert W. Colman (50)
Gregory E. Conner (277)
Andreas Conzelmann (256)
Tim Coolbear (99)
Graham H. Coombs (204)
Antony Cooper (569)
Jonathan B. Cooper (296)
Max. D. Cooper (339)
Maria C. Cordeiro (279)
Pierre Corvol (359)
Joseph T. Coyle (492)
Charles S. Craik (3, 7)
John W.M. Creemers (117)

Rosario Cueva (496)
Susan Daenke (314)
Bjorn Dahlbäck (56)
Ross E. Dalbey (153)
John Dalton (254)
John B. Dame (288)
Pam M. Dando (343, 372)
Roy M. Daniel (96, 97, 331)
Keld Danø (57)
Paul L. Darke (163)
Robert DeLotto (64, 65)
George N. DeMartino (173)
Johan A. den Boon (234)
Christophe d'Enfert (373, 374)
Mark R. Denison (246)
Lakshmi A. Devi (542)
Jan Maarten van Dijl (154)
Iztok Dolenc (172, 214)
Michelle L.L. Donnelly (566)
Hugues D'Orchymont (491)
Adam Dubin (224, 538)
Edward G. Dudley (545)
Ben M. Dunn (297, 310, 311, 319)
Bruno Dupuy (560)
Maria Conceição Duque-Magalhães (536)
Don R. Durham (355)
Torbjörn Egelrud (26)
Vincent Ellis (57)
Françoise Emily-Fenouil (400)
Wolfgang Engel (27)
Ervin G. Erdös (137, 459)
Herbert J. Evans (444)
Kurt M. Fenster (217)
Bent Foltmann (273, 274)
Simon J. Foster (464)
Thierry Foulon (348, 468)
Stephen I. Foundling (300, 301, 302)
Jay W. Fox (416, 418, 420, 423, 430, 441, 443)
Hudson H. Freeze (207)
Jean-Marie Frère (145, 150)
Teryl K. Frey (231)
Lloyd D. Fricker (458, 461)
Christopher J. Froelich (22, 23)
Konomi Fujimura-Kamada (366)
J. Fukushima (367)
Robert S. Fuller (115, 305)
Christian Gache (400)
John Galivan (264)
David Gani (566)
David J. Garfinkel (321)
Jonathan D. Gary (502)
G.M. Gaucher (108)
Christian Ghiglione (400)
Jean-Marie Ghuysen (144, 463)
C. Anne Gibson (246)
Paul Glynn (448)
Eiichi Gohda (261)
Alfred L. Goldberg (178, 184)
Daniel E. Goldberg (286, 287)
Gregory I. Goldberg (402)
Alexander E. Gorbalenya (92, 234, 235, 236)

Stuart G. Gordon (262)
Friedrich Götz (369)
László Gráf (8)
John S. Graham (398)
Antonio Granell (199)
Jonathan Green (307, 308)
Marvin J. Grubman (226)
Bernard Guillemain (315)
Micheline Guinand (265, 457, 553)
José Maria Gutiérrez (438)
Jesper Z. Haeggström (347)
James H. Hageman (98)
Marie-Luise Hagmann (539)
Sherin Halfon (3)
Charles H. Halsted (493)
Andrew J. Hamilton (179)
Theodor Hanekamp (517)
Ikuko Hara-Nishimura (253)
Minoru Harada (520)
Rod J. Hay (179)
Dirk F. Hendriks (453)
Bernhard Henrich (345, 376, 486, 487)
Andrés Hernández-Arana (198)
Paula Meese Hicks (558)
Bradley I. Hillman (233)
B. Yukihiro Hiraoka (534)
Theodor Hofmann (295)
Thomas Hohn (322)
John R. Hoidal (17)
Joachim-Volker Höltje (143, 555, 556)
Vivian Y.H. Hook (263)
Nigel M. Hooper (500, 522)
Christopher J. Howe (155)
Dennis E. Hruby (471)
H.-C. Huang (171)
Oumaïma Ibrahim-Granet (373, 374)
Eiji Ichishima (292, 294, 349, 512)
Yoshitaka Ikeda (174)
Yukio Ikehara (128, 129)
Tadashi Inagami (284, 285)
R. Elwyn Isaac (360)
Grazia Isaya (377)
Shin-ichi Ishii (361)
Koreaki Ito (125, 516)
Sadaaki Iwanaga (66, 67, 68, 69, 116, 425, 427, 433)
Mel C. Jackson (528)
Ralph W. Jack (109, 259)
John J. Jeffrey (391)
Weiping Jiang (413)
David A. Johnson (19)
Gary D. Johnson (406, 407)
John E. Johnson (324)
Stephen Albert Johnston (215)
Elizabeth W. Jones (104)
Leemor Joshua-Tor (215)
Carine Joudiou (541)
Ray Jupp (164, 165, 166)
Tsutomu Kabashima (139)
Shuichi Kaminogawa (498, 499, 504)
Yoshiyuki Kamio (518)
Chen-Chen Kan (162)

Makoto Kaneda (110)
Sung Gyun Kang (4)
Mary Kania (169, 170)
Shun-ichiro Kawabata (66, 67, 68, 69, 116, 425, 427, 433)
John Kay (164, 166, 275)
Kenneth C. Keiler (157)
Robert M. Kelly (558)
Susanna R. Keller (342)
Michael A. Kerr (46, 47)
Jukka Kervinen (278)
Efrat Kessler (357, 411, 506, 507)
Mogens Kilian (87)
Do-Hyung Kim (244)
In Seop Kim (4)
Yukio Kimura (535)
Heidrun Kirschke (210, 211, 213)
Ana Kitazono (139)
Naomi Kitamura (63)
Jürgen R. Klein (345, 376, 486, 487)
Hideyuki Kobayashi (299)
Jan Kok (365)
Pappachan E. Kolattukudy (105, 514)
Harald Kolmar (83)
Jan Konvalinka (320)
Alexander A. Kortt (101)
Hans-Georg Kräusslich (320)
Lawrence F. Kress (421)
Takashi Kumazaki (361)
Kotoku Kurachi (60)
Imelda J. Lambkin (179)
Gayle K. Lamppa (470)
Bernard F. Le Bonniec (55)
Kye Joon Lee (4)
Jonathan Leis (313)
Jo Ann C. Leong (564)
Thierry Lepage (400)
Stephen H. Leppla (511)
Ivan A. Lessard (503)
Guy Lhomond (400)
H.R. Lijnen (58)
Shie-Jea Lin (103)
Xinli Lin (332)
William N. Lipscomb (473)
John W. Little (152)
Mark O. Lively (156)
Y. Peng Loh (304, 306)
George P. Lomonossoff (242)
Yvan Looze (191)
Deshun Lu (14)
Andrei N. Lupas (169, 170, 172)
Vivian L. MacKay (303)
Richard E. Mains (121)
Kauko K. Mäkinen (354)
Fajga Mandelbaum (440, 446)
Walter F. Mangel (176)
James M. Manning (130)
Francis S. Markland, Jr (70, 71, 72, 417, 442)
Takeharu Masaki (85)
Sususmu Maruyama (532)
Carla Mason (564)
José Matos (481)

Lynn M. Matrisian (394)
Hiroshi Matsuzawa (103)
Yoshihiro Matsuda (384)
Michael R. Maurizi (177)
Majambu Mbikay (120)
Dewey G. McCafferty (503)
J. Ken McDonald (138, 182, 183, 266, 530, 546, 547, 548)
Gerard McGeehan (398)
Ultan G. McKeon (269)
David McMillan (166)
Robert P. Mecham (515)
Darshini P. Mehta (207)
Alan Mellors (505)
Roger G. Melton (483)
Armelle Ménard (315)
Robert Ménard (187)
Luis Menéndez-Arias (312, 317, 318)
Gregor Meyers (245)
Susan Michaelis (366)
Wojciech P. Mielicki (262)
Igor Mierau (365)
Charles G. Miller (375, 485, 544, 562)
John Mills (164, 165, 166)
Michel-Yves Mistou (216)
Yoshio Misumi (128, 129)
Shigehiko Mizutani (341)
William L. Mock (478)
Véronique Monnet (378, 481)
Cesare Montecucco (509, 510)
Kazuyuki Morihara (387)
James H. Morrissey (53)
John S. Mort (209)
Uffe H. Mortensen (132)
Hidemasa Motoshima (498, 499, 504)
Jerry C. Mottram (204, 205)
Kazuo Murakami (284, 285)
Kimiko Murakami-Murofushi (293)
Sawao Murao (290, 328, 329)
Gillian Murphy (401)
Tatsushi Muta (66, 67, 68, 69)
Hideaki Nagase (393)
Rajesh R. Naik (104)
Hiroshi Nakazato (341)
Yukio Nakamura (284, 285)
Toshiaki Nikai (426, 428, 431, 435)
Makoto Nishiyama (298)
Gordon W. Niven (495)
Shigemi Norioka (5)
Michael J. North (205)
Donald L. Nuss (228)
Richard J. Obiso, Jr (403)
Brendan O'Connor (269)
Kohei Oda (326, 330)
Douglas H. Ohlendorf (81)
Dennis E. Ohman (357, 507)
Junji Ohnishi (62)
Kenji Okuda (367)
Kjeld Olesen (132)
Stephen Oroszlan (312, 317, 318)
Michael Ovadia (438)
Jeremy C.L. Packer (155)

Mark J.I. Paine (429, 432)
Salomé M. Pais (279)
Himadri B. Pakrasi (158)
Kirk L. Parkin (217)
Pyong Woo Park (515)
Sudhir Paul (180)
Duanqing Pei (397)
Alan D. Pemberton (21)
Iva Pichová (316)
Michael Pieper (390)
Lothaire Pinck (243)
Euclides M.V. Pires (280, 281)
Hubert Pirkle (70, 71)
Henry C. Pitot (261)
Andrew G. Plaut (508)
Jan Potempa (224, 258, 538)
Knud Poulsen (87)
Annik Prat (468)
David G. Pritchard (379)
Alexey V. Pshezhetsky (133)
Mohammad Abul Qasim (77, 78)
Stephen D. Rader (82)
Robert R. Rando (260)
Oscar D. Ratnoff (49)
Neil D. Rawlings (563)
Karen E. Reed (567)
Antonia P. Reichl (440, 446)
Julian R. Reid (99)
Kenneth B.M. Reid (45)
S. James Remington (134)
Wolfgang J. Rettig (131)
Mohamed Rholam (540)
Charles M. Rice (88, 89, 567)
Todd W. Ridky (313)
Howard Riezman (256)
Lori A. Rinckel (321)
Alison Ritchie (165)
Janine Robert-Baudouy (268)
R. Michael Roberts (307, 308)
Michael Robinson (165)
Adela Rodriguez-Romero (198)
John C. Rogers (200)
Markus Rohrwild (171)
Philip J. Rosenthal (201)
J. Rosing (73)
Cesare Rossi (245)
Richard A. Roth (466, 467)
Andrew D. Rowan (192, 193, 194, 195)
Mikhail N. Rozanov (232)
Tillmann Rümenapf (565)
Martin D. Ryan (566)
Thomas J. Ryan (264)
J. Evan Sadler (14)
W. Saenger (106)
Hans-Georg Sahl (109, 259)
Fumio Sakiyama (5, 85)
Guy S. Salvesen (16,22,23)
Eladio F. Sanchez (424)
QingXiang Amy Sang (392)
Krishnan Sankaran (333)
Yoshikiyo Sasagawa (518)

Kimiyuki Satoh (159)
Robert T. Sauer (157)
Jennifer J. Schiller (229)
Manfred Schlösser (27)
Patrick M. Schlievert (81)
Brian F. Schmidt (24)
Udo K. Schmitz (469)
Henning Scholze (202)
Erika Seemüller (172)
Nabil G. Seidah (118, 119, 120, 122, 123)
Motoharu Seiki (399)
Giorgio Semenza (338)
Robert M. Senior (395)
Solange M.T. Serrano (74, 440, 446)
Peter Setlow (559)
M. Shanks (242)
Stephen D. Shapiro (395)
Lei Shen (56)
Roger F. Sherwood (483)
Hitoshi Shimoi (84)
Yun-Bo Shi (392)
William H. Simmons (135, 479, 480)
Robert B. Sim (48)
Alison Singleton (196)
Tatiana D. Sirakova (105, 514)
Tim Skern (239, 241)
Randal Skidgel (459, 460)
Jeffrey Slack (206)
Mark J. Smyth (25)
Eric J. Snijder (92, 234, 235, 236)
Hiroyuki Sorimachi (218, 219, 220)
Eric B. Springman (454)
Robert Stark (565)
James L. Steele (217, 545)
Henning R. Stennicke (79, 80)
Johan Stenflo (54)
Valentin M. Stepanov (456)
Erwin E. Sterchi (408)
Frank Steven (75)
Kenneth J. Stevenson (108)
David J. Stewart (101)
Raymond J. St. Leger (107)
Walter Stöcker (405)
Stuart R. Stone (55)
Andrew C. Storer (187)
Ellen G. Strauss (230)
James H. Strauss (230)
Norbert Sträter (473)
Paz Suárez-Rendueles (484, 496)
Kyoko Suefuji (103)
Fumiaki Suzuki (284, 285)
Koichi Suzuki (218, 219, 220)
Eugene D. Sverdlov (552)
Stephen Swenson (72, 417, 442)
Pal B. Szecsi (273, 274)
László Szilágyi (8, 9)
Muhamed-Kheir Taha (560)
Kenji Takahashi (291, 327)
Takayuki Takahashi (62)
Koji Takio (537)
Amy Tam (366)

Tomohiro Tamura (160)
Fulong Tan (137)
Nget Hong Tan (434)
Jordan Tang (272, 276, 332)
Martha M. Tanizaki (439)
Nayuki Taniguchi (174)
Egbert Tannich (202)
Anthony L. Tarentino (412)
Peter J. Tatnell (275)
Hiroki Tatsumi (513)
Norbert Tautz (91)
Markus F. Templin (143, 555, 556)
Gonzales Thierry (268)
Heinz-Jürgen Thiel (91, 245, 565)
Nicole M. Thielens (43, 44)
Nancy A. Thornberry (248, 249)
Peter E. Thorsness (517)
Natalie J. Tigue (164)
Harold Tjalsma (154)
Diana L. Toledo (176)
Birgitta Tomkinson (113)
Akito Tomomura (10)
Fiorella Tonello (510)
H.S. Toogood (96, 97, 331)
Paolo Tortora (112, 494)
József Tözsér (312, 317, 318)
James Travis (224, 258, 538)
Harald Tschesche (390)
Christopher A. Tsu (7)
Masafumi Tsujimoto (341)
Daisuke Tsuru (126)
Anthony T. Tu (419, 422)
Boris Turk (214)
Vito Turk (214)
Anthony J. Turner (337, 362)
Veli-Jukka Uitto (102)
Dirk Ullmann (257)
Wim J.M. Van de Ven (117)
Marco Vanoni (112, 494)
Harold E. Van Wart (368)
Gregory M. Vath (81)
Josep Vendrell (455)

Gerard Venema (154)
István Venekei (8)
Jeanmarie Verchot (90, 227)
John E. Volanakis (42)
Fred W. Wagner (436)
Kenneth W. Walker (476, 477)
Stephen J. Walker (156)
Christopher T. Walsh (503)
Peter N. Walsh (51)
Jiyang Wang (339)
Alfred L.M. Wassenaar (92, 235, 236)
Aaron B. Watts (197, 221)
Joseph M. Weber (251)
Daniel R. Webster (549)
Jürgen Wehland (353)
Stephen J. Weiss (397)
James A. Wells (94)
Judith M. White (447)
Gene L. Wijffels (208)
Sherwin Wilk (524, 533, 551)
Jean-Marc Wilkin (142, 148)
Keith D. Wilkinson (222, 223)
Tracy D. Wilkins (403)
Tracy A. Williams (359)
Jakob R. Winther (283)
J. Fred Woessner (149, 282, 358, 381, 396, 521, 525, 526)
Dieter H. Wolf (488)
Tyra G. Wolfsberg (447)
David O. Wood (474)
Gerry D. Wright (554)
Sancai Xie (307, 308)
Kenjiro Yamagami (409, 410)
Yoshio Yamakawa (437, 445)
Koichiro Yamamoto (356)
Noriko Yasuda (535)
Shigeki Yasumasu (409, 410)
Toshimasa Yasuhara (489)
Soon Ji Yoo (171)
Tadashi Yoshimoto (125, 139)
Ludmila L. Zavalova (552)
Nada Zein (568)

# *Preface*

Ever since the discovery of pepsin in the late eighteenth century there has been continuing investigation of the chemistry and activities of the proteolytic enzymes, or peptidases. But recent years have seen a remarkable acceleration of the pace of this research, fuelled by numerous practical applications in biotechnology, and the realization that the peptidases are major therapeutic targets. A striking example of the link between peptidases and drug design is the basis of the cover design of the *Hankbook*. This depicts the structure of retropepsin, the essential processing peptidase of the human immunodeficiency virus, in complex with a potent inhibitor of the class that are proving to be effective drugs against the virus (drawn from Brookhaven Protein Databank entry 1SBG).

The many ways in which proteolytic enzymes impinge on the health and welfare of mankind have made it essential for biological scientists in many fields to have ready access to data on peptidases, but the sheer numbers of these enzymes pose a problem. Analysis of complete sequences of several genomes has shown about 2% of all gene products to be peptidases, indicating that this is one of the larger functional groups of proteins. The great number of known peptidases creates many practical problems for those needing to work with them. For example, it is difficult to know how one peptidase can be distinguished from another and referred to unambiguously, and how a scientist can tell when he/she has discovered a novel peptidase. It is precisely this kind of question that the present *Handbook* is designed to answer. The rapid expansion of the field of proteolytic enzymes is bringing into the field new investigators who will find a comprehensive reference book a particularly valuable resource.

The present volume has grown out of a long-standing interest of the editors in preparing readily accessible compilations of data on peptidases. Its genesis can be traced back to the 2nd International Symposium on Intracellular Protein Catabolism held in Ljubljana, Slovenia, in 1975. Two of the editors (AJB and JFW) recognized that there was considerable confusion in the field concerning the various intracellular proteolytic enzymes and how they might be distinguished. A chapter was prepared with the collaboration of I. Kregar and V. Turk entitled 'Present knowledge of proteolytic enzymes and their inhibitors', and published in *Intracellular Protein Catabolism II* (V. Turk & N. Marks eds, Plenum Press, New York, 1977). A grand total of 23 intracellular enzymes were tabulated! This modest beginning was followed by *Mammalian Proteases: a Glossary and Bibliography* by AJB with J.K. McDonald in two volumes (Academic Press, London, 1980, 1986). Here, 173 peptidases were described in concise summaries supplemented by extensive bibliographies.

These two books have long been out of print and the number of known peptidases has risen steadily, so that the total across all kinds of organisms now exceeds 500. This growth has been reflected in a short doubling-time of the literature, so that publications on proteolytic enzymes now approach 8000 a year. Those responsible for subsection 3.4 of the EC List (the enzyme nomenclature of the International Union of Biochemistry and Molecular Biology) that deals with peptidases have been striving to keep up with this flood and have succeeded in including nearly 300 peptidases to date. Although the EC List has considerable value, notably as a source of unambiguous approved names, it has significant limitations. Only enzymes that have been subject to rigorous enzymological characterization can be included, and the endopeptidases are allocated to just four mechanistic classes, in which they are listed in random order.

During recent years, two of the editors (AJB, NDR) have been developing an alternative approach to the classification of peptidases that takes advantage of the new wealth of structural information. In this scheme the enzymes are allocated to clans and families within each major mechanistic class. The system is based on genetic relationships among the enzymes and has been presented in two recent volumes of *Methods in Enzymology* (244 & 248, Academic Press, San Diego, 1994, 1995). But this treatment was not designed to be comprehensive, and is not in convenient form for rapid reference. In contrast, it is the hope of the editors of the present *Handbook* to provide a convenient classification of all the known peptidases that meet minimal criteria for inclusion (see the Introduction). This *Handbook* provides a ready reference to the 500 or so peptidases known to date, but it also provides a framework for the addition of the many more proteolytic enzymes that may be expected to emerge during the coming years of intensive genome research. It is the hope of the editors that this work will be readily accessible to the multitude of workers in the field, so that it is truly a 'hand' book. To this end, we have arranged with the Publishers that a CD-ROM be made available in addition to the printed bound volume.

The chapters on individual peptidases have been contributed by over 500 expert authors who have worked hard and well to describe their favorite peptidases in the concise and strict format required for the *Handbook*. For practical reasons, the editors have had to assume full responsibility for the final editing and proofreading of each chapter, and sometimes have made substantive changes without providing the authors with much opportunity for rebuttal. It cannot be expected that every author will be in complete agreement with the way in which his or her work is presented, but we trust that they will understand that any apparent slights were unintended and are merely the consequences of our effort to produce a systematic and consistent overview of the peptidases. We thank them all for their good work.

The editors thank their secretaries, Desi O'Rourke and Michelle Gonzalez, for their skilled assistance in the task of soliciting, collecting and collating the many manuscripts for this project. We are particularly grateful for the assistance of the staff at Academic Press, London, including Susan Lord, Sarah Stafford, Roopa Baliga and Emma Parkinson, and to Harriet Stewart-Jones for meticulous copy-editing and Ian Ross for expert proof-reading. Finally, we express our heartfelt, personal appreciation to Jinq-May Chen, Cheow-Yong Rawlings and Nina Woessner, who generously tolerated our preoccupation with the *Handbook* over much of the past two years.

AJB, NDR, JFW
February 1998

# Introduction

## Terminology

The scientists who work on the large and important group of enzymes that hydrolyze peptide bonds currently allow themselves a great deal of freedom in the terms they use for their objects of study. The effect of this is that there are commonly several names in use for much the same thing. This is seen when one looks for a collective word for all of these enzymes. They are commonly termed *proteases*, *proteinases* and *peptidases*, as well as *proteolytic enzymes*. Historically, these terms had slightly different meanings (Barrett & McDonald, 1986), but these are now forgotten by many. The editors felt that *proteolytic enzymes* was perhaps the most generally understood term in the current usage and therefore adopted this for the title of the *Handbook*. The reader should note, however, that even this is not unambiguous, since many of the enzymes that hydrolyze peptide bonds (and are included in the *Handbook*) do not act on proteins directly.

In our editing of the present volume, we wished to encourage movement towards more rational and systematic terminology in the study of proteolytic enzymes. Fortunately, sound and authoritative recommendations are available in the form of the EC List. This is still named after the Enzyme Commission that compiled the first editions (Webb, 1993), but has now been curated by the Nomenclature Committee of the International Union of Biochemistry and Molecular Biology (NC-IUBMB) for many years. The EC List was last printed in full as *Enzyme Nomenclature 1992* (NC-IUBMB, 1992), but the part dealing with peptidases has subsequently been amended by regular supplements and can be found in its revised form on the World Wide Web (WWW) at http://www.chem.qmw.ac.uk/iubmb/enzyme/index.html. The EC List recommends the term *peptidase* as the general term for all the enzymes that hydrolyze peptide bonds, and the present editors support this recommendation. Such terms as protease and proteinase will be found to occur frequently in the *Handbook*, but peptidase is the term that is preferred by the editors and is used in the editorial chapters that introduce the major groups of these enzymes. Most peptidases are either *exopeptidases* cleaving one or a few amino acids from the N- or C-terminus, or *endopeptidases* that act internally in polypeptide chains. The EC List also provides terms for subtypes of exopeptidases and endopeptidases. The exopeptidases that act at a free N-terminus liberate a single amino acid residue (*aminopeptidases*) or a dipeptide or a tripeptide (*dipeptidyl-peptidases* and *tripeptidyl-peptidases*). Those acting at a free C-terminus liberate a single residue (*carboxypeptidases*) or a dipeptide (*peptidyl-dipeptidases*). Other exopeptidases are specific for dipeptides (*dipeptidases*) or remove terminal residues that are substituted, cyclized or linked by isopeptide bonds (peptide linkages other than those of $\alpha$-carboxyl to $\alpha$-amino groups) (*omega peptidases*). The endopeptidases are divided on the basis of catalytic mechanism into *serine endopeptidases*, *cysteine endopeptidases*, *aspartic endopeptidases* and *metalloendopeptidases*. The term *oligopeptidase* is used to refer to endopeptidases that act optimally on substrates smaller than proteins.

The muddled situation that exists in the language currently used in reference to peptidases is seen at its worst in the naming of the individual enzymes. It is all too common for scientists working on a single enzyme in different laboratories to use quite different names for it, even in their published work. Such a liberal practice must come as a surprise to scientists from other disciplines – it is as if biologists were to use their own local names for organisms rather then the scientific binomials!

Two reasons for the present relaxed attitude to the naming of peptidases are not hard to see. One is that the number of known peptidases is commonly perceived as small, so that confusion seems unlikely. And secondly, the naming of peptidases is notoriously difficult, so that it has never been possible to derive simple, objective rules for forming names that can be applied by anyone, yielding the same result. A consequence of this is that all of the names may tend to look arbitrary, none being better than another.

Neither of these reasons is adequate to justify the present situation. There are indeed fewer peptidases so far recorded than living organisms, say, but the reader of the present volume will soon realize that they are quite numerous. Even if each of the peptidases known today were referred to by only one name, there would be quite enough to tax any but the complete specialist. And although it is indeed difficult to name peptidases, there is no doubt that there are good names and bad ones. Some of the names in common use are very poorly suited to the functions that a name should serve, and there is every reason to encourage general use of one of the better names. The situation in which most peptidases have several names in common use can only be an obstacle to the progress of this field of study. Vast amounts of data about these enzymes are now being published in printed form and in computer databases, but the accessing of that information is made more difficult and less efficient when multiple names are used, especially when some of the names are ambiguous, being applied to several different enzymes. And as the number of known peptidases rises into the thousands, multiple names must be a recipe for chaos.

In assembling the present volume, we have attempted to encourage the rationalization of terminology that is needed in the field, and again we support the recommendations of the EC List. Although we have not insisted that only the names recommended in the EC List be used, we have encouraged their use, and have inserted cross-references to them where necessary. We have also invited each author of a chapter in the *Handbook* to start with a section on the history and naming of the enzyme.

## Scope of the Handbook

The *Handbook* contains information on almost every known peptidase. Amongst these there are formal entries on all the peptidases that are included in the current EC List, but it should be noted that a few of these now seem unlikely to represent distinct enzymes and may disappear in later revisions of the List (as is explained in the text). There are also chapters on many peptidases that have not yet been included in the EC List, as well as accounts of some gene products that are shown by their amino acid sequences to be close relatives of known peptidases but have not yet been demonstrated directly to be peptidases in their own right. Also included for their interest are a few entities that show peptidase-like activity, but would not normally be thought of as peptidases proper. Amongst these are a number of self-processing proteins, an antibody molecule, and a histone-splitting chromoprotein. By use of these broad criteria, we assembled a list of over 500 peptidases for inclusion in the *Handbook*, and invited a similar number of expert authors to write about them.

## Classification

A fundamental aspect of the *Handbook* is the way in which it is organized. The peptidases are grouped on the basis of primary and tertiary structures into families and clans, and these are further grouped by catalytic mechanism. The classification used in the *Handbook* is therefore very different from that adopted in the EC List (see above), where the peptidases, which form subclass 3.4 of all enzymes, are divided into 13 sub-subclasses. The sub-subclasses are not further divided, and the peptidases are listed in arbitrary order within each of them. The molecular structures and evolutionary relationships that are of key importance in the present *Handbook* are not taken into account in the EC classification.

The system of classification that we have employed in the *Handbook* is one that was introduced by Rawlings & Barrett (1993), and has subsequently been further developed by these authors. In this scheme, a *family* of peptidases is a group in which every member shows a statistically significant relationship in amino acid sequence to at least one other member of the family in the part of the molecule that is responsible for peptidase activity. Strict statistical criteria are applied so that we can be confident that any two peptidases that are placed in the same family have evolved from a common ancestor and thus are homologous by the definition of Reeck *et al.* (1987). The restriction of the comparison to the catalytically active part of the molecule is an important one, since many peptidases are chimeric proteins containing additional, nonpeptidase domains that are shared with other groups of proteins. We do not consider the relationships of these domains to be directly relevant to the classification of the peptidase, and for this reason, the peptidase families do not correspond closely to families of proteins recognized in other systems, such as the PIR protein sequence database (Barker *et al.*, 1990). A few of the peptidase families contain two or more rather distinct groups of peptidases (shown by a deep divergence in the dendrogram: see below), and for these, subfamilies are recognized.

Each peptidase family is named with a letter denoting the catalytic type (S, T, C, A, M or U, for serine, threonine, cysteine, aspartic, metallo- or unknown), followed by an arbitrarily assigned number. When a family disappears, usually because it is merged with another, the family name is not reused, and for this reason there are interruptions in the numerical sequences of families that are of no current significance.

*Clan* is the term used to describe a group of families the members of which have evolved from a single ancestral protein, but have diverged so far that we can no longer prove their relationship by comparison of the primary structures (Rawlings & Barrett, 1993; Barrett & Rawlings, 1995). The clearest kind of evidence for clan-level relationship between families is similarity in three-dimensional structures, but the arrangement of catalytic residues in the polypeptide chains and limited similarities in amino acid sequence around the catalytic amino acids can also be revealing. The name of a clan is formed from the letter for the catalytic type (as for families) followed by an arbitrary second capital letter. If a clan disappears, the name is not reused. Not all families can yet be assigned to clans. When a formal clan assignment is needed for these it is given as SX (for a serine peptidase family), or CX, AX, etc.

*Catalytic type* depends upon the chemical nature of the groups responsible for catalysis in a way that can be traced back to Hartley (1960) and was adapted for the EC List in 1972. The major catalytic types are Serine, Cysteine, Aspartic, Metallo and as yet Unclassified, and these initial letters can be seen on the right-hand margin of the pages of the present volume, to provide easy access to the relevant sections of the book. Generally, the use of these catalytic types as the top level of the hierarchical classification of peptidases works well, but there are a few anomalies. Most notably, the serine peptidases of clan SA undoubtedly share a common origin with the cysteine peptidases of clan CB (see Chapter 238). It should also be mentioned here that for practical convenience a few enzymes such as the proteasome in which threonine rather than serine forms the nucleophile of catalysis have been placed in the 'serine' section, and the 'aspartic' section contains several acid-acting endopeptidases that may possibly contain catalytic glutamic residues.

The *Handbook* contains many links to the **MEROPS** database. This is a WWW database of information on peptidases (curated by NDR) set up as a means to respond to the constant state of development of the system of families and clans of peptidases that is driven by the flow of new data for primary and tertiary structures. New releases of the database appear as needed, and we hope that readers will obtain updating information from it after publication of the book. The CD-ROM version of the *Handbook* contains a hypertext link to the MEROPS database on the WWW from each chapter. Anyone wishing to browse the MEROPS file for a chapter without the CD can do so by use of a URL in the form http://www.bi.bbsrc.ac.uk/Merops/HPE[chapter].htm, in which [chapter] is the chapter number in the *Handbook* (without the brackets). Each peptidase has a unique identifier (ID) in MEROPS. The ID is constructed of two parts, the family name (e.g. 'S01') and an arbitrary three-digit number for the individual peptidase within the family, the two parts being separated by a decimal point. The minority of peptidases for which sequences are not yet available cannot be assigned to families, and for these, provisional MEROPS IDs are formed as follows. As usual, the first character is a letter for the catalytic type, but this is always followed by a figure 9, and the third character is a letter indicating the type of peptidase activity: A, aminopeptidase; B, dipeptidase; C, dipeptidyl- or tripeptidyl-peptidase; D, peptidyl-dipeptidase; E, carboxypeptidase; F, omega peptidase, and G, endopeptidase. The second part of the ID is formed as usual. The ID also can be used for direct access to the relevant data file in MEROPS. The URL is exactly as described for the chapter links, but HPE[chapter] is replaced by the seven-character MEROPS ID modified by changing the decimal point to 'p'. Thus, the file for acrosin (ID S01.223) is http://www.bi.bbsrc.ac.uk/Merops/S01p223.htm.

## Editors' Introductory Chapters

Since the *Handbook* is organized according to the hierarchical classification: catalytic type, clan and family, families do not necessarily appear in numerical order. Text on any given family can be located by use of the subject index, but additionally, each sequence of chapters on peptidases from a single clan or large family is preceded by an introduction from the editors. The introductory chapter typically starts with a databanks table (see below) that gives an overview of the families within the clan and their constituent peptidases. Data are not repeated from the individual chapters, but cross-referenced. Sequence database accession numbers are, however, given for peptidases or putative peptidases that are not the subjects of separate chapters, including peptidase homologs that are known as sequences but have not yet been characterized biochemically.

Alignments of amino acid sequences have been prepared for most families. A preliminary alignment was constructed by use of the PILEUP program of the GCG package (Genetics Computer Group, 1994). This was improved manually, and when a structural alignment based on three-dimensional structure was available in the literature, the preliminary alignment was edited to match this. For brevity, the alignments are not presented in full, but are shown in one of two ways. For some families, the chimeric nature of the proteins is of interest, and 'domain diagrams' have been produced in which the arrangement of domains of various types, location of active-site residues, and disulfide loops are depicted schematically. On the CD-ROM only, there are also partial alignment diagrams in which key regions of the sequences are picked out, typically to show the conserved sequence motifs around the functionally important amino acid residues.

Dendrograms are shown for many families. These are based on the alignment of the amino acid sequences of the peptidase units for the family. That is to say, the alignment as described above was trimmed to exclude all portions of the sequences that are not part of the peptidase unit, including in some cases inserts within the peptidase unit. For some very large families, only a selection of peptidases was included. A difference matrix of percentage sequence identities was computed from the alignment and this was used to compute the tree. The Fitch–Margoliash algorithm (Fitch & Margoliash, 1967) with contemporary tips was used as implemented in the KITSCH program of the PHYLIP package (Felsenstein, 1989). The trees are used here to give a graphic depiction of the distances between the structures of peptidases within a family and are not intended to represent an accurate reconstruction of the evolutionary history of the family; statistical sampling has therefore not been performed. The trees are drawn so that the longest branches are uppermost. The x-axis is calibrated in PAM (percentage accepted mutations) calculated from the percentage identity at the amino acid level as described by Dayhoff *et al.* (1978).

At least one Richardson diagram depicting the protein tertiary structure has been constructed for every peptidase family for which a structure has been published. Each of these (in the style of Richardson, 1985) has been constructed from a PDB entry, and where possible structures have been selected that include a bound small molecule inhibitor or substrate. The diagrams were constructed by use of a series of programs. First, the RASMOL program (Sayle & Milner-White, 1995) was used to orient the molecule so that all the active-site residues are visible (and to place molecules with similar structures in the same orientation). For the book, the MOLSCRIPT program (Kraulis, 1991) was used to generate the images in grayscale. For the CD-ROM, the MOLSCRIPT program was then used to generate an input file for the RENDER program (Bacon & Anderson, 1988) of the RASTER3D package (Merritt & Murphy, 1994). We have shown helices as (red) coils, sheets as (green) arrows, and coils and turns as (cyan) wires. Active-site residues and bound inhibitors and substrates are shown in ball-and-stick representation, and metal ions as Corey–Pauling–Koltun spheres. In some cases, the PDB entries have been edited so that only one subunit of a multimeric structure is shown.

## Chapters on Individual Peptidases

A series of chapters on related peptidases is typically ordered to bring together enzymes that are shown as being closely related in the dendrogram for the family. Each chapter on an individual peptidase starts with a databanks table compiled by NDR. The first line of this gives the peptidase classification, i.e. the name of the clan and family and the MEROPS identifier (ID) of the peptidase. The peptidase classification line is followed by the EC number, the American Tissue Culture Collection (ATCC) clone numbers, and the Chemical Abstracts Service (CAS) Registry number.

The body of the databanks table contains the primary sequence database accession numbers, arranged alphabetically by scientific binomial of the organisms from which sequences are known. The organism name is printed for each distinct gene product. Accession numbers have been collected from the SwissProt protein sequence database (SW) (Bairoch & Apweiler, 1996) (http://expasy.hcuge.ch/sprot/sprot-top.html), the PIR database (Barker *et al.*, 1990) (http://www-nbrf.georgetown.edu/pir/), the EMBL nucleic acid sequence database (Kahn & Cameron, 1990) (http://www.ebi.ac.uk/ebi_docs/embl_db/embl_db.html), and occasionally the GenPept database. Of these databases, only SwissProt is nonredundant. Accession numbers in the EMBL database are the same as those in the GenBank (Benson *et al.*, 1998) (http://www.ncbi.nlm.nih.gov/Web/Genbank/index.html) and DDBJ (http://www.ddbj.nig.ac.jp/) databases. The entries for the EMBL database have been arranged in two columns for clarity. One column is headed 'cDNA' and refers to EMBL entries containing only the sequence of the peptidase mRNA (or the gene in the case of a peptidase from a bacterium). The second column, headed 'genomic' includes complete genes, exons, introns, gene promoters, and portions of genomes. The accession number may be followed by a brief comment describing the database entry.

A Brookhaven Protein Data Bank table is provided for those peptidases for which the coordinates of a three-dimensional structure have been deposited in the Protein Data Bank (PDB) (Bernstein *et al.*, 1977) (http://www.pdb.bnl.gov/). The entries in this table are arranged by scientific binomial and then alphabetical order of the PDB accession number. All structures relating to the peptidase in question are included in the table together with the computed resolution of each structure (in Angstrom units) and a brief description.

The remainder of each chapter has been prepared by the author(s) credited at the end of the chapter, and the chapters have a uniform structure. Following the databanks table there is the author's text describing the Name and History, Activity and Specificity, Structural Chemistry, Preparation, Biological Aspects, Distinguishing Features, Related Peptidases, Further Reading and References.

In the descriptions of the specificity of the peptidases, the symbol '+' has frequently been used to mark the bond that is hydrolyzed, the scissile bond, in the formulae of substrate molecules. Underlying the visible text on the CD-ROM there is encoded an invisible version of the same information that can be searched (see below).

The terminology used in describing the specificity of peptidases depends on a model in which the catalytic site is considered to be flanked on one or both sides by specificity subsites, each able to accommodate the side chain of a single amino acid residue. These sites are numbered from the catalytic site, S1...Sn towards the N-terminus of the substrate, and S1'...Sn' towards the C-terminus. The residues they accommodate are numbered P1...Pn, and P1'...Pn', respectively, as follows (the catalytic site of the enzyme being marked '*'):

Substrate:   – P3 – P2 – P1 + P1' – P2' – P3' –
Enzyme:    – S3 – S2 – S1 * S1' – S2' – S3' –

This scheme is essentially as described by Berger & Schechter (1970), but slightly simplified in that we print the numbers of subsites and the side chains they accommodate on the line rather than subscript, and place the prime signs after the numbers.

## Appendices

The sequence of 569 chapters is followed by two appendices: Appendix 1 provides the common English name (if any) and the type of organism for each of the species from which a peptidase sequence is listed in the book, and Appendix 2 provides address details for many of the suppliers of materials mentioned.

## CD-ROM

The CD-ROM version of the *Handbook* contains the full text with a number of enhancements. There are many hypertext links, including links through the WWW to the MEROPS database. The figures are in color, and the text is fully searchable, so that any term not in the printed Index should be easily located. These may include authors' names, words in titles or journal names for the more than 15 000 references cited throughout the text. Also, there is a specially designed search function for the scissile bonds in substrate structures that allows the reader to locate a peptidase with a given specificity. There are facilities for the insertion of bookmarks, and for copying and pasting the text. All of this is described more fully in the leaflet that accompanies the CD.

## Feedback to the Editors

The editors would be pleased to hear of any necessary corrections, additions, or information on new enzymes. Whether or not there will be a further edition of the *Handbook*, such information may be used to correct and update the MEROPS database and the EC List. Please address your comments to any of the editors at his email address.

## References

Bacon, D. & Anderson, W.F. (1988) A fast algorithm for rendering space-filling molecule pictures. *J. Mol. Graph.* **6**, 219–220.
Bairoch, A. & Apweiler, R. (1996) The SWISS-PROT protein-sequence data bank and its new supplement TREMBL. *Nucleic Acids Res.* **24**, 21–25.
Barker, W.C., George, D.G. & Hunt, L.T. (1990) Protein sequence database. *Methods Enzymol.* **183**, 31–49.
Barrett, A.J. & McDonald, J.K. (1986) Nomenclature: protease, proteinase and peptidase. *Biochem. J.* **237**, 935.
Barrett, A.J. & Rawlings, N.D. (1995) Families and clans of serine peptidases. *Arch. Biochem. Biophys.* **318**, 247–250.
Benson, D.A., Boguski, M.S., Lipman, D.J., Ostell, J. & Ouelette, B.F. (1998) GenBank. *Nucleic Acids Res.* **26**, 1–7.

Berger, A. & Schechter, I. (1970) Mapping the active site of papain with the aid of peptide substrates and inhibitors. *Philos. Trans. R. Soc. Lond. [Biol.]* **257**, 249–264.

Bernstein, F.C., Koetzle, T.F., Williams, G.J.B., Meyer, E.F., Jr, Brice, M.D., Rodgers, J.R., Kennard, O., Shimanouchi, T. & Tasumi, M. (1977) The protein data bank: a computer-based archival file for macromolecular structures. *J. Mol. Biol.* **112**, 535–542.

Dayhoff, M.O., Schwartz, R.M. & Orcutt, B.C. (1978) A model of evolutionary change in proteins. In: *Atlas of Protein Sequence and Structure* (Dayhoff, M.O., ed.). Washington DC: National Biomedical Research Foundation, pp. 345–362.

Felsenstein, J. (1989) PHYLIP – phylogeny inference package. *Cladistics* **5**, 164–166.

Fitch, W.M. & Margoliash, E. (1967) A method based on mutation distances as estimated from cytochrome *c* sequences is of general applicability. *Science* **155**, 281–284.

Genetics Computer Group (1994) *Program Manual for the Wisconsin Package, Version 8, September 1994*, University of Madison, Wisconsin.

Hartley, B.S. (1960) Proteolytic enzymes. *Annu. Rev. Biochem.* **29**, 45–72.

Kahn, P. & Cameron, G. (1990) EMBL data library. *Methods Enzymol.* **183**, 23–31.

Kraulis, P.J. (1991) *MOLSCRIPT*: a program to produce both detailed and schematic plots of protein structures. *J. Appl. Cryst.* **24**, 946–950.

Merritt, E.A. & Murphy, M.P. (1994) RASTER3D version 2.0 – a program for photorealistic molecular graphics. *Acta Crystallogr. D* **50**, 869–873.

NC-IUBMB (Nomenclature Committee of the International Union of Biochemistry and Molecular Biology) (1992) *Enzyme Nomenclature 1992. Recommendations of the Nomenclature Committee of the International Union of Biochemistry and Molecular Biology on the Nomenclature and Classification of Enzymes*. Orlando, FL: Academic Press.

Rawlings, N.D. & Barrett, A.J. (1993) Evolutionary families of peptidases. *Biochem. J.* **290**, 205–218.

Reeck, G.R., de Haën, C., Teller, D.C., Doolittle, R.F., Fitch, W.M., Dickerson, R.E., Chambon, P., McLachlan, A.D., Margoliash, E., Jukes, T.H. & Zuckerkandl, E. (1987) 'Homology' in proteins and nucleic acids: a terminology muddle and a way out of it. *Cell* **50**, 667.

Richardson, J.S. (1985) Schematic drawings of protein structures. *Methods Enzymol.* **115**, 359–380.

Sayle, R.A. & Milner-White, E.J. (1995) RASMOL: biomolecular graphics for all. *Trends Biochem. Sci.* **20**, 374.

Webb, E.C. (1993) Enzyme nomenclature: a personal perspective. *FASEB J.* **7**, 1192–1194.

**Alan J Barrett**
alan.barrett@bbsrc.ac.uk

**Neil D Rawlings**
neil.rawlings@bbsrc.ac.uk

**J. Fred Woessner**
fwoessne@mednet.med.miami.edu

# *Abbreviations*

| | |
|---|---|
| Abz- | *o*-aminobenzoyl |
| Ac- | acetyl |
| AEBSF | 4-(2-aminoethyl)benzenesulfonyl fluoride |
| Boc- | *t*-butyloxycarbonyl |
| Bz- | *N*-benzoyl |
| -CH$_2$Cl | chloromethane |
| -CHN$_2$ | diazomethane |
| Cpp | *N*-[1-(*RS*)-carboxy-3-phenylpropyl]- |
| Cya | cysteic |
| DCI | 3,4-dichloroisocoumarin |
| DFP | diisopropylfluorophosphate |
| DAN | diazoacetylnorleucine methyl ester |
| Dnp | 2,4-dinitrophenyl |
| Dns | 5-dimethylaminonaphthyl-1-sulfonyl |
| DMSO | dimethylsulfoxide |
| DTT | dithiothreitol |
| E-64 | L-3-carboxy-2,3-*trans*-epoxypropionyl-leucylamido(4-guanidino)butane |
| EDDnp | ethylenediamine dinitrophenyl |
| EDTA | ethylenediaminetetra-acetate |
| EGTA | [ethylenebis(oxonitrilo)]tetra-acetate |
| EPNP | 1,2-epoxy-3(p-nitrophenoxy)propane |
| FA- | furanacryloyl |
| FPLC | fast protein liquid chromatography |
| Glp | pyroglutamyl |
| Glt | glutaryl |
| H-kininogen | high molecular mass kininogen |
| HCO- | formyl |
| HPLC | high-performance liquid chromatography |

| | |
|---|---|
| IgA, IgG, IgM | immunoglobulins |
| ME | mercaptoethanol |
| Mca | (7-methoxycoumarin-4-yl)acetyl |
| Mcc | 7-methoxycoumarin-3-carboxylyl- |
| MHC | major histocompatibility complex |
| -NHMec | 7-(4-methyl)coumarylamide |
| -NHNap | 2-naphthylamide |
| -NHPhNO$_2$ | 4-nitroanilide |
| NMR | nuclear magnetic resonance |
| Nph | 4-nitrophenylalanine |
| -OEt | ethyl ester |
| -OMe | methyl ester |
| -OPhNO$_2$ | 4-nitrophenyl ester |
| pAb | 4-aminobenzoate |
| PAGE | polyacrylamide gel electrophoresis |
| PCMB | *p*-chloromercuribenzoate |
| pepstatin | isovaleryl-L-valyl-L-valyl-4-amino-3-hydroxy-6-methylheptanoyl-L-alanyl-4-amino-3-hydroxy-6-methylheptanoic acid |
| Pip | pipecolyl |
| PMSF | phenylmethanesulfonyl fluoride |
| PCR | polymerase chain reaction |
| Pz- | $N^\alpha$-(4-phenylazo)benzyloxycarbonyl |
| -SBzl | thiobenzyl ester |
| SDS | sodium dodecyl sulfate |
| Suc- | 3-carboxypropionyl |
| Tos- | toluene-*p*-sulfonyl- (tosyl-) |
| Z- | benzoylcarbonyl |

# SERINE AND THREONINE PEPTIDASES

# 1. Introduction: serine peptidases and their clans

Five catalytic types of peptidases can now be recognized, in which serine, threonine, cysteine, aspartic or metallo groups play primary roles in catalysis. The serine, threonine and cysteine peptidases are catalytically very different from the aspartic and metallopeptidases in that the nucleophile of the catalytic site is part of an amino acid, whereas it is an activated water molecule in the other groups. One consequence of this is that acyl enzyme intermediates can be formed only in the reactions of the Ser/Thr/Cys peptidases, and only these peptidases can readily act as transferases.

Peptidases in which the catalytic mechanism depends upon the hydroxyl group of a serine residue acting as the nucleophile that attacks the peptide bond are termed serine peptidases. For practical convenience, the few known families of threonine-dependent peptidases are grouped with the serine peptidases in this volume.

There are in total about 40 families of serine- and threonine-type peptidases that we distinguish on the basis of comparisons of the amino acid sequences. We have been able to group most of these families into seven clans by comparing the tertiary structures and the order of the catalytic residues in the sequence (summarized in Table 1.1). The catalytic machinery usually involves, in addition to the serine that carries the nucleophile, a proton donor (or general base). In clans SA, SB, SC and SH, the proton donor is a histidine residue, and there is a catalytic triad because a third residue is required, probably for orientation of the imidazolium ring of the histidine. This is usually an aspartate, but is another histidine in clan SH. In clans SE and SF, a lysine residue has the role of proton donor, and a third catalytic residue is not required. In clan SF, there are some peptidases that have a Ser/His catalytic dyad. The seventh clan, TA, contains endopeptidases in which the N-terminal residue is the nucleophile, and includes the threonine-type endopeptidases.

In *clan SA* (Chapters 2 and 76), the order of the catalytic triad is His, Asp, Ser, and the tertiary structure consists mainly of $\beta$ sheet. There is a two-domain structure, with each domain containing a $\beta$ barrel, and the active-site cleft is between the domains. All the proteolytic enzymes in the clan are endopeptidases. Examples are known from RNA viruses, bacteria and eukaryotes. There are no known examples from archaea. The fold is shared with clan CB (Chapter 238) of cysteine-type endopeptidases, in which the nucleophile is replaced with a cysteine.

In *clan SB* (Chapter 93), the order of the catalytic triad is Asp, His, Ser. The tertiary structure includes both $\alpha$ helices and $\beta$ sheets. The clan includes both endopeptidases and exopeptidases. Peptidases are known from bacteria, archaea and eukaryotes; there is a single representative from a DNA virus.

In *clan SC* (Chapter 124), the order of the catalytic triad is Ser, Asp, His. The peptidases in this clan share a tertiary structure with many other hydrolases, including acetylcholinesterase, lipases and haloalkane dehalogenase. The clan includes endopeptidases and exopeptidases, of which the endopeptidases are oligopeptidases, and the exopeptidases are aminopeptidases and carboxypeptidases. Examples are known from bacteria, archaea and eukaryotes, but not from viruses.

In *clan SE* (Chapter 141), the catalytic dyad is Ser, Lys occurring within the motif Ser-Xaa-Xbb-Lys. The peptidases in the clan are involved in the biosynthesis, turnover and lysis of bacterial cell walls and commonly act on D-alanyl bonds. Many of the peptidases are penicillin-binding proteins, and the tertiary fold is shared with class C $\beta$-lactamases. The clan contains mainly carboxypeptidases, but does include an aminopeptidase and endopeptidases that cleave bacterial cell wall cross-links. All the known examples are from bacteria.

In *clan SF* (Chapter 151), the order of the catalytic dyad is Ser, Lys or Ser, His. The catalytic residues are more widely spaced than in clan SE, and the tertiary structures show no similarity. The clan includes only endopeptidases. Examples are known from bacteriophages, bacteria, archaea and eukaryotes.

In *clan SH* (Chapter 161), the order of the catalytic triad is His, Ser, His. Again, the tertiary structure shows no similarity to that from any other protein. All known members of the clan are endopeptidases from DNA viruses, and are involved in virus prohead assembly.

*Clan TA* (Chapter 167) contains enzymes in which the nucleophile is at the N-terminus. The clan includes a number of enzymes whose only proteolytic activity is self-activation. The nucleophile can be serine, threonine or cysteine. The other residues in the active site, if any, are not well characterized: in the proteasome a lysine may act as a general base, but among the self-processing, nonproteolytic enzymes, the N-terminal amino group may provide the proton during hydrolysis. All peptidases in the clan are endopeptidases, and examples are known from bacteria, archaea and eukaryotes, but not from viruses.

A few families of serine peptidases cannot yet be assigned to clans, as is described in Chapter 175.

*Table 1.1* The clans of serine peptidases

| Clan SA | Catalytic residues: His, Asp, Ser | Fold: Double β-barrel |
|---|---|---|
| Family | S1 | Trypsin |
| | S2 | Streptogrisin A |
| | S3 | Togavirin |
| | S6 | IgA1-specific serine endopeptidase |
| | S7 | Flavivirin |
| | S29 | Hepatitis C polyprotein peptidase |
| | S30 | Helper component proteinase |
| | S31 | Pestivirus NS2-3/NS3 serine peptidase |
| | S32 | Arterivirus serine endopeptidase |
| **Clan SB** | **Catalytic residues: Asp, His, Ser** | **Fold: Parallel β-sheet** |
| Family | S8 | Subtilisin |
| **Clan SC** | **Catalytic residues: Ser, Asp, His** | **Fold: α,β Hydrolase** |
| Family | S9 | Prolyl oligopeptidase |
| | S10 | Carboxypeptidase C |
| | S15 | X-Pro dipeptidyl-peptidase |
| | S28 | Lysosomal Pro-X carboxypeptidase |
| | S33 | Prolyl aminopeptidase |
| | S37 | *Streptomyces* PS-10 peptidase |
| **Clan SE** | **Catalytic residues: Ser, Lys** | **Fold: Helices and α + β sandwich** |
| Family | S11 | Penicillin-binding protein 5 |
| | S12 | *Streptomyces* R61 D-Ala-D-Ala carboxypeptidase |
| | S13 | Penicillin-binding protein 4 |
| **Clan SF** | **Catalytic residues: Ser, Lys (His)** | **Fold: Single β barrel** |
| Family | S24 | LexA repressor |
| | S26 | Signal peptidase 1 |
| | S41 | TSP protease |
| | S44 | Tricorn protease |
| **Clan SH** | **Catalytic residues: His, Ser, His** | **Fold: All β** |
| Family | S21 | Cytomegalovirus assemblin |
| **Clan TA** | **Catalytic residues: Thr** | **Fold: α,β,β,α Sandwich** |
| Family | T1 | Proteasome |
| | S42 | γ-Glutamyl transpeptidase |
| | – | Ntn-hydrolases |
| **Other families** | | |
| Family | S14 | Endopeptidase Clp |
| | S16 | Endopeptidase La |
| | S18 | Omptin |
| | S19 | Cell wall-associated endopeptidase of *Trichophyton* |
| | S34 | HflA endopeptidase |
| | S38 | *Treponema* chymotrypsin-like endopeptidase |
| | S39 | Cocksfoot mottle virus endopeptidase |
| | S43 | Porin D |

# 2. *Introduction: family S1 of trypsin (clan SA)*

*Databanks*

MEROPS ID: S01

| Species | SwissProt | PIR | EMBL (cDNA) | EMBL (genomic) |
|---|---|---|---|---|
| **Family S1** | | | | |
| Achelase (Chapter 3) | | | | |
| Acrosin (Chapter 27) | | | | |
| Ancrod (Chapter 70) | | | | |
| Apolipoprotein A | | | | |
|   *Homo sapiens* | P08519 | S00657 | X06290 | – |
|   *Macaca mulatta* | P14417 | A32869 | J04635 | – |
| Batroxobin (Chapter 70) | | | | |
| Bilineobin | | | | |
|   *Agkistrodon bilineatus* | – | A45368 | – | – |
| Bothrombin (Chapter 70) | | | | |
| Brachyurin (Chapter 7) | | | | |
| Caldecrin (Chapter 10) | | | | |
| Cathepsin G (Chapter 16) | | | | |
| Cerastobin | | | | |
|   *Cerastes vipera* | P18692 | – | – | – |
| Cercarial elastase | | | | |
|   *Schistosoma mansoni* | P12546 | A28942 | J03946 | – |
| Chymase I (Chapter 18) | | | | |
| Chymase II (Chapter 18) | | | | |
| Chymotrypsin (Chapter 8) | | | | |
| Clotting enzyme, *Limulus* (Chapter 68) | | | | |
| Coagulation factor B, *Limulus* (Chapter 67) | | | | |
| Coagulation factor C, *Limulus* (Chapter 66) | | | | |
| Coagulation factor G, *Limulus* (Chapter 69) | | | | |
| Coagulation factor V-activating proteinase (Russell's viper) (Chapter 72) | | | | |
| Coagulation factor VIIa (Chapter 53) | | | | |
| Coagulation factor IXa (Chapter 52) | | | | |
| Coagulation factor Xa (Chapter 54) | | | | |
| Coagulation factor XIa (Chapter 51) | | | | |
| Coagulation factor XIIa (Chapter 49) | | | | |
| Cocoonase (Chapter 3) | | | | |
| Complement component C2 (Chapter 46) | | | | |
| Complement component B (Chapter 47) | | | | |
| Complement component C1r (Chapter 43) | | | | |
| Complement component C1s (Chapter 44) | | | | |
| Complement factor D (Chapter 42) | | | | |
| Complement factor I (Chapter 48) | | | | |
| Crotalase (Chapter 70) | | | | |
| Duodenase I (Chapter 21) | | | | |
| Easter protein (Chapter 65) | | | | |
| Enteropeptidase (Chapter 14) | | | | |
| Epidermal growth factor-binding protein (Chapter 34) | | | | |
| Flavoxobin (Chapter 70) | | | | |
| Fragmentin | | | | |
|   *Rattus norvegicus* | P18291 | A43520 | M34097 | – |
| Gilatoxin | | | | |
|   *Heloderma horridum* | P43685 | A48598 | – | – |
| Granzyme A (Chapter 22) | | | | |
| Granzyme B (Chapter 23) | | | | |

*continued overleaf*

| Species | SwissProt | PIR | EMBL (cDNA) | EMBL (genomic) |
|---|---|---|---|---|
| **Family S1** (*continued*) | | | | |
| Granzyme C | | | | |
|   *Mus musculus* | P08882 | B28952 | X12822 | M22527: complete gene |
| | | C26944 | | |
| | | S03004 | | |
|   *Rattus norvegicus* | – | – | U57062 | – |
| Granzyme D | | | | |
|   *Mus musculus* | P11033 | A36172 | J03255 | X56990: complete gene |
| | | D26944 | | |
| | | S24941 | | |
| Granzyme E | | | | |
|   *Mus musculus* | P08884 | B36172 | X12821 | X56988: complete gene |
| | | E26944 | X14093 | |
| | | S01006 | | |
| | | S06177 | | |
| | | S24939 | | |
| Granzyme F | | | | |
|   *Mus musculus* | P08883 | S01007 | – | X56989: complete gene |
| Granzyme G | | | | |
|   *Mus musculus* | P13366 | A33412 | J02872 | – |
| | | S06176 | | |
| Granzyme H | | | | |
|   *Homo sapiens* | P20718 | A32692 | J02907 | M72150: complete gene |
| | | I54748 | M36118 | M57888: complete gene |
| | | I59619 | | |
| Granzyme K (Chapter 24) | | | | |
| Granzyme M (Chapter 25) | | | | |
| Granzyme-like protein I | | | | |
|   *Rattus norvegicus* | Q06605 | – | X66693 | – |
| Hepatocyte growth factor-activating endopeptidase (Chapter 63) | | | | |
| Hepsin (Chapter 60) | | | | |
| House dust mite allergen | | | | |
|   *Dermatophagoides farinae* | P49275 | – | D63858 | – |
|   *Dermatophagoides farinae* | P49276 | – | – | – |
|   *Dermatophagoides pteronyssinus* | P39675 | – | U11719 | – |
|   *Dermatophagoides pteronyssinus* | P49277 | – | – | – |
| Hypodermin A | | | | |
|   *Hypoderma lineatum* | P35587 | A21590 | X74303 | – |
| Hypodermin B | | | | |
|   *Hypoderma lineatum* | P35588 | A20190 | L24915 | – |
| | | | X74304 | |
|   *Hypoderma lineatum* | – | – | X74305 | – |
| Hypodermin C (Chapter 8) | | | | |
| Leukocyte elastase (Chapter 15) | | | | |
| Mannose-binding protein-associated serine protease (Chapter 45) | | | | |
| Mastin (Chapter 20) | | | | |
| Myeloblastin (Chapter 17) | | | | |
| Natural killer cell protease 7 | | | | |
|   *Rattus norvegicus* | – | – | U57063 | – |
| γ-Nerve growth factor subunit (Chapter 35) | | | | |
| Pancreatic elastase (Chapter 11) | | | | |
| Pancreatic elastase II (Chapter 12) | | | | |
| Pancreatic endopeptidase E (Chapter 13) | | | | |
| Plasma kallikrein (Chapter 50) | | | | |
| Plasmin (Chapter 59) | | | | |
| Prostasin (Chapter 61) | | | | |

| Species | SwissProt | PIR | EMBL (cDNA) | EMBL (genomic) |
|---|---|---|---|---|
| **Family S1** (*continued*) | | | | |
| Protein C (activated) (Chapter 56) | | | | |
| γ-Renin (Chapter 33) | | | | |
| Semenogelase (Chapter 31) | | | | |
| Snake maternal effect gene product (*Drosophila*) (Chapter 64) | | | | |
| Stratum corneum chymotryptic enzyme (Chapter 26) | | | | |
| Stubble gene product (Chapter 64) | | | | |
| Submandibular endopeptidase A (Chapter 32) | | | | |
| Submaxillary gland endopeptidase K (Chapter 38) | | | | |
| t-Plasminogen activator (Chapter 58) | | | | |
| Thrombin (Chapter 55) | | | | |
| Tissue kallikrein (Chapter 29) | | | | |
| Tonin (Chapter 39) | | | | |
| Trypsin (cationic, anionic) (Chapter 3) | | | | |
| Trypsin (*Saccharopolyspora*) (Chapter 5) | | | | |
| Trypsin (*Streptomyces*) (Chapter 4) | | | | |
| Tryptase (Chapter 19) | | | | |
| u-Plasminogen activator (Chapter 57) | | | | |
| Venombin B (Taiwan habu snake) | | | | |
| *Trimeresurus mucrosquamatus* | – | A38940 JC2479 | – | – |
| *Trimeresurus okinavensis* | P20005 | PX0042 | – | – |
| Others (Chapter 74) | | | | |
| Unknown | – | – | M27347 | – |
| *Anopheles gambiae* | – | – | U21917 | – |
| *Anopheles gambiae* | – | – | Z49813 | – |
| *Anopheles gambiae* | – | – | Z49814 | – |
| *Anopheles gambiae* | – | – | Z49815 | – |
| *Anopheles gambiae* | – | – | Z49832 | – |
| *Anopheles gambiae* | – | – | Z49833 | – |
| *Botryllus schlosseri* | – | – | X75016 | – |
| *Haematobia irritans* | – | – | U09794 | – |
| *Herdmania momus* | – | – | U63517 | – |
| *Homo sapiens* | P26927 | A40331 | M74178 | U37055: complete gene |
| *Homo sapiens* | P34168 | S38295 | – | – |
| *Homo sapiens* | P40313 | – | X71877 | – |
| *Homo sapiens* | – | – | U62801 | – |
| *Homo sapiens* | – | – | X75363 | – |
| *Mus musculus* | P26928 | – | M74181 | M74180: complete gene |
| *Mus musculus* | Q08048 | – | D10212 D10213 | – |
| *Porphyra purpurea* | – | – | U08840 | – |
| *Rattus norvegicus* | P17945 | A35644 | D90102 M32987 X54400 | – |
| *Xenopus laevis* | – | A40242 | – | – |

*Family S1* is the largest family of peptidases, both in terms of the number of sequenced proteins, and in the number of recognizably different peptidase activities. All the peptidases in family S1 are endopeptidases, although the family also includes a number of proteins for which peptidase activity is unknown or unlikely, usually because catalytic residues have been replaced. As in all members of clan SA, the endopeptidases in family S1 have a catalytic triad consisting of His, Asp and Ser in that order in the sequence. An

alignment showing regions around these residues is shown in Alignment 2.1 (see CD-ROM).

Most of these peptidases enter the secretory pathway and have an N-terminal signal peptide, and all of them are synthesized as precursors with an N-terminal extension that acts as a propeptide, requiring proteolytic cleavage to generate the active enzyme (Figure 2.1). The length of this extension varies enormously, with the smallest being two residues in human cathepsin G (Chapter 16), and the larger

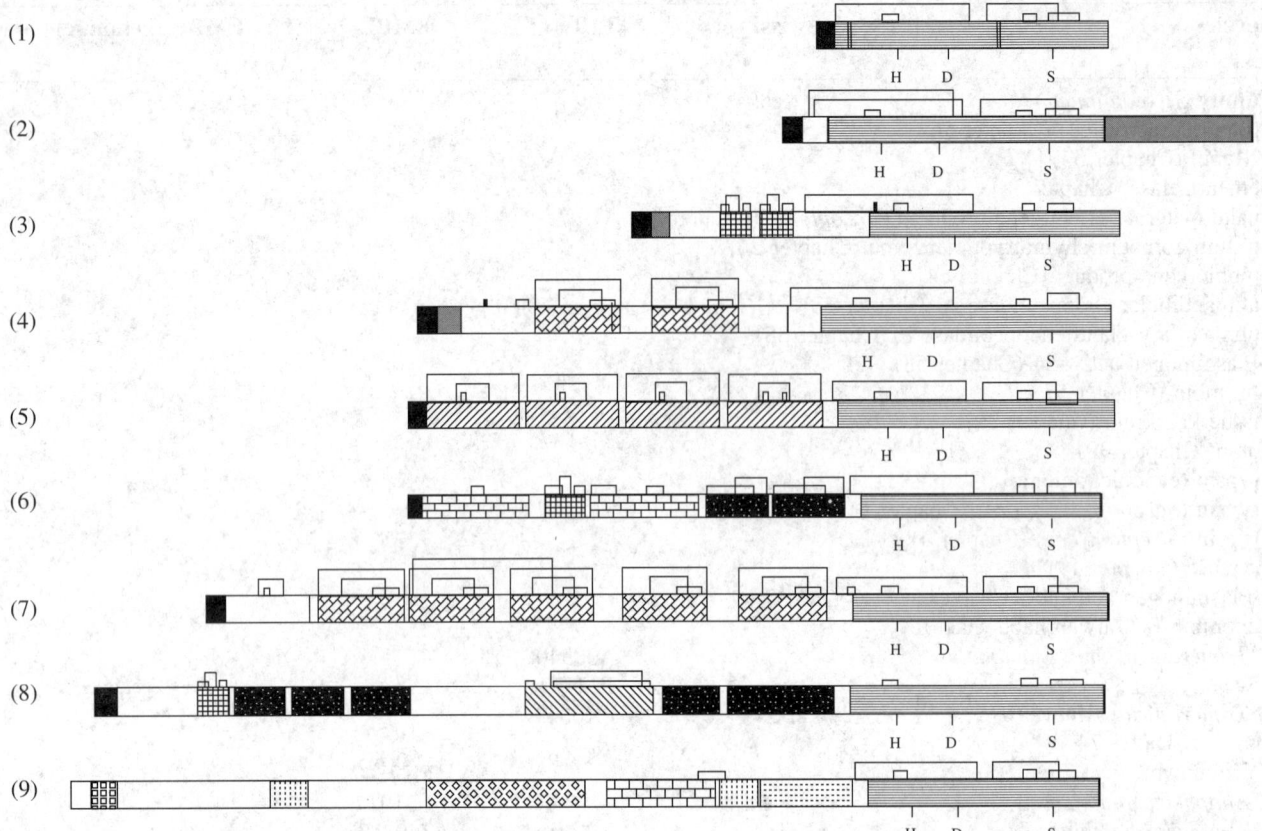

*Figure 2.1* Domain structures for selected endopeptidases from family S1. endopeptidase sequences are shown as rectangles. The length of the rectangle is a representation of the sequence length. Disulfide bridges are shown as open boxes above the sequence rectangle. The positions of the catalytic triad residues are shown below the sequence rectangles, and all structures are aligned at the catalytic serine. Horizontal lines are processing sites. Structures are arranged so that the simplest is at the top, and the most complex at the bottom. Shaded boxes within the sequence rectangle represent different structural domains. *Key to domains* (from top of figure to bottom): black, signal peptide; gray, N- or C-terminal propeptide; horizontal stripes, catalytic domain; checks, epidermal growth factor-like domain; diagonal bricks, kringle; left-to-right diagonal lines, apple domain; horizontal bricks, C1r-like module; black with white spots, sushi domain; left to right diagonal lines, lecithin-like domain; squares, transmembrane domain; short vertical lines, cysteine-rich domain; diamonds, MAM domain; short horizontal lines, CUB domain. *Key to sequences*: (1) *Homo sapiens* chymotrypsinogen B, (2) *H. sapiens* proacrosin, (3) *H. sapiens* coagulation factor X, (4) *H. sapiens* prothrombin, (5) *H. sapiens* plasma prokallikrein, (6) *H. sapiens* complement component C1r, (7) *H. sapiens* plasminogen, (8) *Tachypleus tridentatus* clotting factor C precursor, (9) *Sus scrofa* proenteropeptidase.

ones containing hundreds of amino acids. The larger extensions, found for example in the blood coagulation factors and complement components, commonly remain attached by disulfide bonds to the peptidase domain after proteolytic activation, and the extension is referred to as the 'heavy chain' of the heterodimeric active enzyme. The N-terminal extension may be mosaic in nature, containing domains that have had quite separate evolutionary origins from the peptidase domains. The acquisition of these additional domains is thought to have been the result of exon shuffling during evolution (Patthy, 1990). This was made possible by the presence of a phase 1 intron–exon junction near the 5′ end of the peptidase domain in the ancestral gene, which is retained in many present-day genes for peptidases of family S1.

Additional domains that occur in the N-terminal extensions of endopeptidases from family S1 are listed in Table 2.1, and several of these can be recognized from their characteristic patterns of disulfide bonds as shown in Figure 2.1. Apple domains are known only from peptidase family S1, and kringles are found mainly in these proteins, but also occur in some receptor tyrosine kinases. Other domains are found in a variety of other proteins, and some of these domains occur in other, unrelated peptidases; for example, the C1r-like repeats also occur in procollagen C-endopeptidase (Chapter 411), and the MAM domain occurs in meprins A (Chapter 406) and B (Chapter 407).

Tertiary structures have been solved for a number of peptidases of family S1, including not only simple proteins such as chymotrypsin, trypsin and pancreatic elastase, but also endopeptidases with more complex heavy chains. Figures 2.2–2.5 show representations of the structures for chymotrypsin, thrombin and coagulation factors IXa and Xa. The chymotrypsin fold shows a two-domain structure, with the active-site cleft between the domains. Each domain consists of an open-ended $\beta$ barrel, and the two barrels are

*Table 2.1*   Numbers of additional domains in the mosaic proteins of S1 family

| Protein | Domain | | | | | | | | |
|---|---|---|---|---|---|---|---|---|---|
| | **Kringle** | **Sushi** | **Apple** | **EGF-like** | **Finger** | **Ca²⁺-binding** | **C1r-like** | **Lectin** | **MAM** |
| Plasminogen | 5 | – | – | – | – | – | – | – | – |
| Apolipoprotein (a) | 37 | – | – | – | – | – | – | – | – |
| Hepatocyte growth factor | 4 | – | – | – | – | – | – | – | – |
| Coagulation factor XII | 1 | 1 | – | 2 | 1 | – | – | – | – |
| Prothrombin | 2 | – | – | – | – | 1 | – | – | – |
| t-Plasminogen activator | 2 | – | – | 1 | 1 | – | – | – | – |
| u-Plasminogen activator | 1 | – | – | 1 | – | – | – | – | – |
| Protein C | – | – | – | 2 | – | 1 | – | – | – |
| Coagulation factor VII | – | – | – | 2 | – | 1 | – | – | – |
| Coagulation factor IX | – | – | – | 2 | – | 1 | – | – | – |
| Coagulation factor X | – | – | – | 2 | – | 1 | – | – | – |
| Complement factor B | – | 3 | – | – | – | – | – | – | – |
| Complement factor 2 | – | 3 | – | – | – | – | – | – | – |
| Complement subcomponent C1r | – | 2 | – | 1 | – | – | 1 | – | – |
| Complement subcomponent C1s | – | 2 | – | 1 | – | – | 1 | – | – |
| Haptoglobin | – | 1 | – | – | – | – | – | – | – |
| Limulus clotting factor C | – | 5 | – | 1 | – | – | – | 1 | – |
| Plasma kallikrein | – | – | 4 | – | – | – | – | – | – |
| Coagulation factor XI | – | – | 4 | – | – | – | – | – | – |
| Enteropeptidase | – | – | – | – | – | – | 1 | – | 1 |

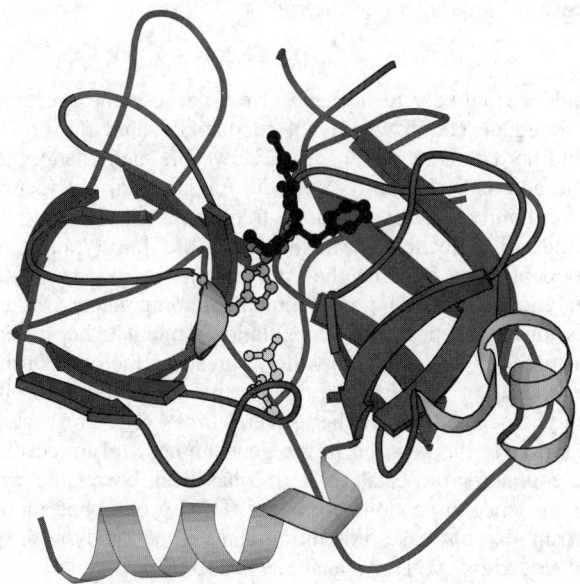

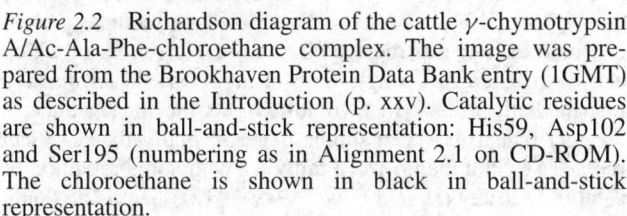

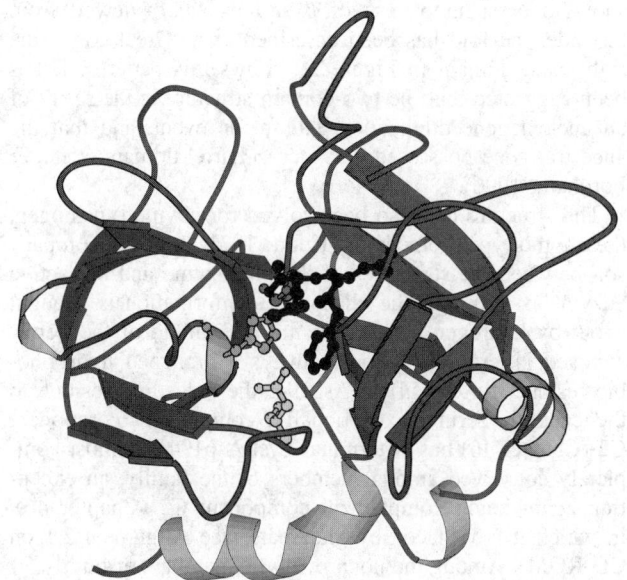

*Figure 2.2*   Richardson diagram of the cattle γ-chymotrypsin A/Ac-Ala-Phe-chloroethane complex. The image was prepared from the Brookhaven Protein Data Bank entry (1GMT) as described in the Introduction (p. xxv). Catalytic residues are shown in ball-and-stick representation: His59, Asp102 and Ser195 (numbering as in Alignment 2.1 on CD-ROM). The chloroethane is shown in black in ball-and-stick representation.

*Figure 2.3*   Richardson diagram of the human thrombin/ D-Phe-Pro-Arg-chloromethane complex. The image was prepared from the Brookhaven Protein Data Bank entry (1ABJ) as described in the Introduction (p. xxv). Catalytic residues are shown in ball-and-stick representation: His59, Asp102 and Ser195 (numbering as in Alignment 2.1 on CD-ROM). The chloromethane is shown in black in ball-and-stick representation.

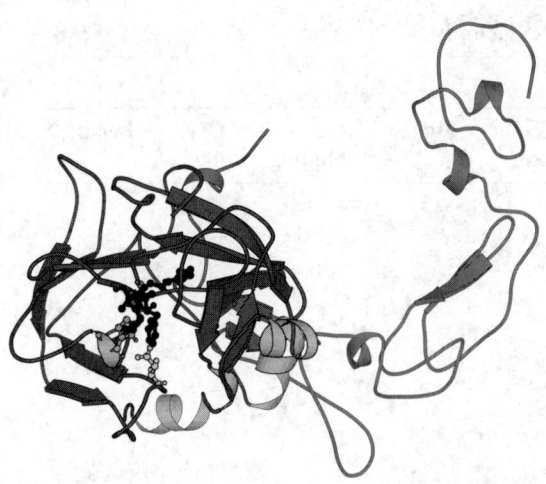

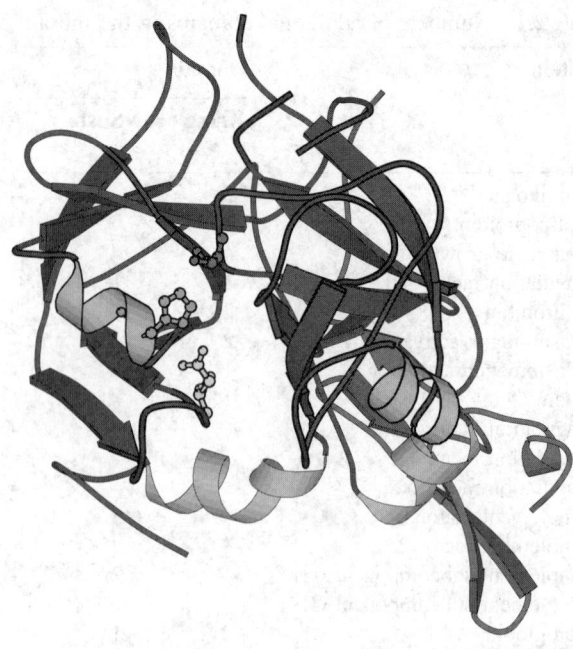

*Figure 2.4*   Richardson diagram of the pig coagulation factor IXa/D-Phe-Pro-Arg complex. The image was prepared from the Brookhaven Protein Data Bank entry (1PFX) as described in the Introduction (p. xxv). Catalytic residues are shown in ball-and-stick representation: His59, Asp102 and Ser195 (numbering as in Alignment 2.1 on CD-ROM). The N-terminal heavy chain is shown in gray (on the right), and the D-Phe-Pro-Arg peptide is shown in black in ball-and-stick representation.

*Figure 2.5*   Richardson diagram of human coagulation factor Xa. The image was prepared from the Brookhaven Protein Data Bank entry (1HCG) as described in the Introduction (p. xxv). Catalytic residues are shown in ball-and-stick representation: His59, Asp102 and Ser195 (numbering as in Alignment 2.1 on CD-ROM). The N-terminal light chain is shown in gray (again on the right).

arranged at right-angles to each other. Because a $\beta$ barrel shows a crossing pattern of $\beta$ strands when viewed from the side, the fold has been described as a 'Greek key' (the right-hand domain in Figure 2.2 shows this pattern). It has been suggested that the two-domain structure is the result of an ancient gene duplication and fusion event, and that the ancestral gene possessed only one $\beta$-barrel domain (Lesk & Fordham, 1996).

The structure has also been solved for chymotrypsinogen, the chymotrypsin precursor (Kraut, 1971), and the comparison between the structures of the proenzyme and the active enzyme is striking. The change in conformation is brought about by hydrogen bonding of the free amino of the newly exposed N-terminus (almost always a branched hydrophobic residue, Ile or Val) to Asp194, the preceding residue to the catalytic serine; exceptionally, complement component C2 (Chapter 46) has N-terminal Lys. Asp194 is almost completely conserved among members of the family, an exception again being complement component C2 (Chapter 46), in which it is replaced by glutamate (see Alignment 2.1 on CD-ROM). Among the nonproteolytic members of family S1 is component III of carboxypeptidase A, which is inactive because the N-terminus of the mature protein is aspartate, and the conformational shift does not occur (Venot *et al.*, 1986).

There has been much speculation about the nucleotide triplet encoding the catalytic serine. In the universal genetic code, there are six triplets that code for serine, and these fall naturally into two groups between which a single nucleotide replacement will lead to a change in the amino acid. These nucleotides are TCA, TCG, TCC, TCT (the 'TCX group'); AGC and AGT (the 'AGX group'). All six codons have been found coding for the catalytic serine in some member of family S1. It might have been thought that an evolutionary change from a codon of one group to a codon from the other

would be unlikely to have occurred, because the presumed intermediate codon would not encode Ser, and the protein could not be catalytically active. However, such changes are seen, and do not seem to represent fundamental divergences in the family. The evolutionary tree presented in Figure 2.6 includes a selection of members of the family, and the endopeptidases utilizing the AGX group of codons for the catalytic serine (including complement components C1r and C1s, thrombin, protein C, coagulation factor IX, hepsin and plasmin) can be seen to be widely spread on the tree. During the course of evolution, it seems that the coding for the catalytic serine has switched several times. The presence of an intron at the position in the gene immediately preceding the codon for the catalytic serine has been correlated with the presence of codons from the TCX group, but mouse acrosin also possesses this intron and yet the catalytic serine is encoded by AGT (Watanabe *et al.*, 1991).

The switch between types of Ser codon could possibly have occurred via a catalytically active intermediate protein in which the Ser was replaced by Cys or Thr (either of which can be reached in a single base change), or via an inactive intermediate protein. In favor of Cys as the intermediate is the fact that it has been shown experimentally that Cys can replace Ser in a member of family S1 with some retention of catalytic activity (Higaki *et al.*, 1989). Also, there is no doubt that the cysteine endopeptidases of clan CB had a common origin with the serine peptidases of clan SA (see Chapter 1), so a change in one direction or the other certainly occurred at least once in evolution. The possibility that the hypothetical intermediate peptidase contained Thr as its nucleophile is

PAM score

(1) Complement factor B (*Danio rerio*)
(2) Complement component C2 (*Homo sapiens*)
(3) Complement factor B (*Homo sapiens*)
(4) Cercarial elastase (*Schistosoma mansoni*)
(5) SNAKE protein (*Drosophila melanogaster*)
(6) Hypodermin C (*Hypoderma lineatum*)
(7) Complement component C1r (*Homo sapiens*)
(8) Complement component C1s (*Homo sapiens*)
(9) EASTER protein (*Drosophila melanogaster*)
(10) Brachyurin (*Uca pugilator*)
(11) Trypsin (*Pleuronectes platessa*)
(12) Trypsin (*Squalus acanthias*)
(13) Trypsin (*Xenopus laevis*)
(14) Trypsin I (*Homo sapiens*)
(15) Trypsin II (*Homo sapiens*)
(16) Trypsin IV (*Homo sapiens*)
(17) Trypsin III (*Homo sapiens*)
(18) Stratum corneum chymotryptic enzyme (*Homo sapiens*)
(19) Tissue kallikrein 10 (*Rattus norvegicus*)
(20) Tonin (*Rattus norvegicus*)
(21) Granzyme E (*Rattus norvegicus*)
(22) 7S nerve growth factor gamma chain (*Mus musculus*)
(23) Epidermal growth factor-binding protein type 3 (*Mus musculus*)
(24) Submandibular tissue kallikrein (*Mus musculus*)
(25) Semenogelase (*Homo sapiens*)
(26) Tissue kallikrein 1 (*Homo sapiens*)
(27) Glutoxin (*Helioderma horridum*)
(28) Factor X-activating alpha (*Daboia russelli*)
(29) Flavoxobin (*Trimeresurus flavoviridis*)
(30) Plasminogen activator (*Trimeresurus stejnegeri*)
(31) Plasminogen activator (*Trimeresurus stejnegeri*)
(32) Venombin A (*Lachesis muta*)
(33) Bothrombin (*Bothrops jararaca*)
(34) Batroxobin (*Bothrops atrox*)
(35) Fibrinogenolytic proteinase (*Trimeresurus macrosquamatus*)
(36) Ancrod (*Agkistrodon contortrix*)
(37) Myeloblastin (*Homo sapiens*)
(38) Neutrophil elastase (*Homo sapiens*)
(39) Complement factor D (*Homo sapiens*)
(40) Granzyme M (*Homo sapiens*)
(41) Cathepsin G (*Homo sapiens*)
(42) Duodenase I (*Bos taurus*)
(43) Granzyme C (*Mus musculus*)
(44) Granzyme B (*Homo sapiens*)
(45) Granzyme H (*Homo sapiens*)
(46) Chymase 1 (*Homo sapiens*)
(47) Chymase 1 (*Mus musculus*)
(48) Granzyme A (*Homo sapiens*)
(49) Granzyme K (*Homo sapiens*)
(50) Mannose-binding protein associated serine endopeptidase (*Homo sapiens*)
(51) Complement factor I (*Homo sapiens*)
(52) Complement factor I (*Xenopus laevis*)
(53) Limulus clotting factor C (*Tachypleus tridentatus*)
(54) Coagulation factor VII (*Homo sapiens*)
(55) Coagulation factor IX (*Homo sapiens*)
(56) Coagulation factor X (*Homo sapiens*)
(57) Thrombin (*Homo sapiens*)
(58) Protein C (*Homo sapiens*)
(59) Serine protease Hepsin 1 (*Hordeumia mornu*)
(60) Hepatocyte growth factor activator (*Homo sapiens*)
(61) Coagulation factor XII (*Homo sapiens*)
(62) t-Plasminogen activator (*Homo sapiens*)
(63) u-Plasminogen activator (*Homo sapiens*)
(64) Acrosin (*Homo sapiens*)
(65) Pancreatic elastase A (*Gadus morhua*)
(66) Pancreatic elastase 3a (*Homo sapiens*)
(67) Pancreatic elastase 3b (*Homo sapiens*)
(68) Pancreatic elastase 1 (*Rattus norvegicus*)
(69) Pancreatic elastase 2a (*Homo sapiens*)
(70) Pancreatic elastase 2b (*Homo sapiens*)
(71) Chymotrypsin-like protease (*Homo sapiens*)
(72) Chymotrypsinogen B (*Homo sapiens*)
(73) Chymotrypsin-like protease ctrl-1 (*Homo sapiens*)
(74) Mastin (*Canis familiaris*)
(75) Tryptase 2 (*Homo sapiens*)
(76) Tryptase 1 (*Homo sapiens*)
(77) Plasmin (*Homo sapiens*)
(78) Apolipoprotein (a) (*Homo sapiens*)
(79) Prostasin (*Homo sapiens*)
(80) Plasma kallikrein (*Homo sapiens*)
(81) Coagulation factor XI (*Homo sapiens*)
(82) Hepsin (*Homo sapiens*)
(83) Serine protease (*Homo sapiens*)
(84) NUDEL protein (*Drosophila melanogaster*)
(85) Limulus coagulation factor C beta subunit (*Tachypleus tridentatus*)
(86) Limulus coagulation factor B (*Tachypleus tridentatus*)
(87) Limulus proclotting enzyme (*Tachypleus tridentatus*)
(88) Trypsin (*Saccharopolyspora erythraea*)
(89) Trypsin (*Streptomyces griseus*)
(90) Serine proteinase sprC (*Streptomyces griseus*)
(91) Vitellin-degrading proteinase (*Bombyx mori*)
(92) Hypodermin A (*Hypoderma lineatum*)
(93) Trypsin alpha (*Drosophila melanogaster*)
(94) Faecal allergen Der P3 (*Dermatophagoides pteronyssinus*)
(95) Trypsin (*Fusarium oxysporum*)
(96) Trypsin (*Astacus fluviatilis*)

*Figure 2.6*   Evolutionary tree for selected members of family S1. The tree was prepared as described in the Introduction (p. xxv).

perhaps supported by the fact that in clan TA, Cys, Ser and Thr nucleophiles are all found (see Chapter 167). The proteasome (see Chapter 167) and its relatives are the only known threonine-type peptidases, but there are homologs of trypsin in which the catalytic Ser is replaced by Thr. These are mainly fragments of sequence from insects (see Alignment 2.1 on CD-ROM), and it not known whether they represent active peptidases. The intermediate protein may have been catalytically inactive, and regained activity as a result of subsequent mutations, and there are examples of homologs lacking the catalytic nucleophile in family S1 as well as several other families of serine peptidases.

### References

Higaki, J.N., Evnin, L.B. & Craik, C.S. (1989) Introduction of a cysteine protease active site into trypsin. *Biochemistry* **28**, 9256–9263.

Kraut, J. (1971) Chymotrypsinogen: X-ray structure. In: *The Enzymes*, vol. 3 (Boyer, P.D., ed.). New York: Academic Press, pp. 165–183.

Lesk, A.M. & Fordham, W.D. (1996) Conservation and variability in the structures of serine proteinases of the chymotrypsin family. *J. Mol. Biol.* **258**, 501–537.

Patthy, L. (1990) Evolutionary assembly of blood coagulation proteins. *Semin. Thromb. Hemost.* **16**, 245–259.

Venot, N., Sciaky, M., Puigserver, A., Desnuelle, P. & Laurent, G. (1986) Amino acid sequence and disulfide bridges of subunit III, a defective endopeptidase present in the bovine pancreatic 6 S procarboxypeptidase A complex. *Eur. J. Biochem.* **157**, 91–99.

Watanabe, K., Baba, T., Kashiwabara, S., Okamoto, A. & Arai, Y. (1991) Structure and organization of the mouse acrosin gene. *J. Biochem. (Tokyo)* **109**, 828–833.

# 3. Trypsin

## *Databanks*

*Peptidase classification: clan SA, family S1, MEROPS ID: S01.151*
*NC-IUBMB enzyme classification: EC 3.4.21.4*
*Chemical Abstracts Service registry number: 9002-07-7*
*Databank codes:*

| Species | Form | SwissProt | PIR | EMBL (cDNA) | EMBL (genomic) |
|---|---|---|---|---|---|
| Trypsin | | | | | |
| *Aedes aegypti* | 3A1 | P29786 | S19890 | X64362 | – |
| *Aedes aegypti* | 5G1 | P29787 | S19891 | X64363 | – |
|  | – | | | M77814 | |
|  | fragment | – | B61143 | – | – |
| *Anisakis simplex* | – | – | A32946 | – | – |
| *Anopheles gambiae* | 1 | P35035 | S35339 | Z18889 | – |
|  | | | S35412 | Z18889 | – |
|  | | | S40065 | Z18889 | – |
| *Anopheles gambiae* | 2 | P35036 | S35340 | Z18890 | – |
|  | | | S35413 | Z18890 | – |
| *Anopheles gambiae* | 3 | P35037 | S40007 | Z22930 | – |
| *Anopheles gambiae* | 4 | P35038 | S40005 | Z22930 | – |
| *Anopheles gambiae* | 5 | P35039 | S40004 | Z22930 | – |
| *Anopheles gambiae* | 6 | P35040 | S40003 | Z22930 | – |
| *Anopheles gambiae* | 7 | P35041 | S40006 | Z22930 | – |
| *Anopheles gambiae* | fragment | – | – | U50474 | – |
| *Anopheles quadrimaculatus* | fragment | – | A61143 | – | – |
| *Astacus fluviatilis* | – | P00765 | A00951 | – | – |
| *Astacus leptodactylus* | fragment | – | A61327 | – | – |
| *Balaenoptera acutorostrata* | fragment | – | A61328 | – | – |
| *Bos taurus* | cationic | P00760 | A00946 | D38507 | – |
|  | | | A90164 | | |
|  | | | S08774 | | |

| Species | Form | SwissProt | PIR | EMBL (cDNA) | EMBL (genomic) |
|---|---|---|---|---|---|
| Trypsin (*continued*) | | | | | |
| *Bos taurus* | anionic | – | S13813 | X54703 | – |
| *Canis familiaris* | cationic | P06871 | B26273 | M11590 | – |
| *Canis familiaris* | anionic | P06872 | A26273 | M11589 | – |
| *Choristoneura fumiferana* | CFT-1 | P35042 | – | L04749 U12917 | – |
| *Choristoneura fumiferana* | – | – | A56597 | – | – |
| *Cochliobolus carbonum* | – | – | – | U39500 | – |
| *Culex pipiens* | I | – | – | U65411 | – |
| *Culex pipiens* | II | – | – | U65412 | – |
| *Dermasterias imbricata* | 1 | – | A61334 | – | – |
| *Dermasterias imbricata* | 2 | – | B61334 | – | – |
| *Dissostichus mawsoni* | – | – | – | U58945 | – |
| *Drosophila erecta* | $\alpha$ | P54624 | – | U40653 | – |
| *Drosophila erecta* | $\beta$ | P54625 | – | U40653 | – |
| *Drosophila erecta* | $\delta/\gamma$ | P54626 | – | U40653 | – |
| *Drosophila erecta* | $\varepsilon$ | P54627 | – | U40653 | – |
| *Drosophila erecta* | $\theta$ | P54628 | – | U40653 | – |
| *Drosophila erecta* | $\eta$ | P54629 | – | U40653 | – |
| *Drosophila erecta* | $\zeta$ | P54630 | – | U40653 | – |
| *Drosophila melanogaster* | $\alpha$ | P04814 | A23493 | X02989 | – |
| *Drosophila melanogaster* | $\beta$ | P35004 | – | M96373 U04853 | – |
| *Drosophila melanogaster* | $\gamma$ | P42277 | – | U04853 | – |
| *Drosophila melanogaster* | $\delta$ | P42276 | – | U04853 | – |
| *Drosophila melanogaster* | $\varepsilon$ | P35005 | – | M96372 | – |
| *Drosophila melanogaster* | $\zeta$ | P42280 | – | U04853 | – |
| *Drosophila melanogaster* | $\eta$ | P42279 | – | U04853 | – |
| *Drosophila melanogaster* | $\theta$ | P42278 | – | U04853 | – |
| *Drosophila melanogaster* | $\iota$ | P52905 | – | U41476 | – |
| *Drosophila melanogaster* | try29F.2 | – | – | U28641 | – |
| *Drosophila melanogaster* | – | – | – | U09798 | – |
| *Fusarium oxysporum* | – | P35049 | – | S63827 | – |
| *Gadus morhua* | fragment | P16049 | S03570 | – | – |
| *Gadus morhua* | I | – | S39047 | S66913 U47819 X75998 X76886 | – |
| *Gadus morhua* | X | – | S39048 | X76887 | – |
| *Gallus gallus* | P1 | – | S55065 | U15155 | – |
| *Gallus gallus* | 1-P38 | – | – | U15156 | – |
| *Gallus gallus* | 2-P29 | – | S55066 | U15157 | – |
| *Haematobia irritans* | psp8 | – | S35208 S42696 | Z22567 | – |
| *Haematobia irritans* | fragment | – | – | U09801 | – |
| *Homo sapiens* | I | P07477 | A25852 A43988 B61066 | M22612 | U70137: exon 3 |
| *Homo sapiens* | II | P07478 | A61066 B25852 B43988 | M27602 | – |
| *Homo sapiens* | III | P15951 | S12764 | D45417 X15505 | – |
| *Homo sapiens* | IVA | P35030 | S33496 S37538 | X71345 | X72781: a-form |
| *Homo sapiens* | fragment | – | S50020 | – | – |

*continued overleaf*

| Species | Form | SwissProt | PIR | EMBL (cDNA) | EMBL (genomic) |
|---------|------|-----------|-----|-------------|----------------|
| Trypsin (*continued*) | | | | | |
| *Homo sapiens* | fragment | – | S50021 | – | – |
| *Homo sapiens* | fragment | – | S50022 | – | – |
| *Homo sapiens* | fragment | – | S50023 | – | – |
| *Lucilia cuprina* | α-4 | P35044 | – | U07693 U07684 | L15632 |
| *Lucilia cuprina* | α-3 | P35043 | – | – | L15632 |
| *Lucilia cuprina* | sbcard11 | – | – | U07685 | – |
| *Lumbricus rubellus* | FI | – | PN0653 PN0654 PN0656 | – | – |
| *Lumbricus rubellus* | FII | – | PN0655 | – | – |
| *Lumbricus rubellus* | FIII | – | PN0657 PN0658 | – | – |
| *Manduca sexta* | alkaline A | P35045 | – | L16805 | – |
| *Manduca sexta* | alkaline B | P35046 | – | L16806 | – |
| *Manduca sexta* | alkaline C | P35047 | – | L16807 | – |
| *Mus musculus* | – | P07146 | B25528 | X04574 | X04577: 5′ region X04578: 5′ region X04579: 5′ region X04580: 5′ region |
| *Paranotothenia magellanica* | – | – | S49489 | X82223 | – |
| *Penaeus monodon* | – | P35050 | S11537 | – | – |
| *Penaeus vannamei* | – | – | S54146 | X86369 | – |
| *Pleuronectes platessa* | – | P35034 | S31384 | X56744 | – |
| *Protopterus aethiopicus* | fragment | P35051 | A27719 | – | – |
| *Protopterus aethiopicus* | fragment | – | A61331 | – | – |
| *Rana esculenta* | fragment | – | A61333 | – | – |
| *Rattus norvegicus* | I, anionic | P00762 | A00948 B22657 | V01273 | J00778: complete gene |
| *Rattus norvegicus* | II, anionic | P00763 | A00949 A22657 | V01274 | K03042: 5′ flank K03043: 5′ flank K03044: 5′ flank L00130: exons 1–3 L00131: exons 4–5 and complete CDS |
| *Rattus norvegicus* | III, cationic | P08426 | A27547 | M16624 | – |
| *Rattus norvegicus* | IV | P12788 | S05494 | X15679 | – |
| *Rattus norvegicus* | Va | P32821 | JQ1471 S23784 | X59012 | – |
| *Rattus norvegicus* | Vb | P32822 | JQ1472 | X59013 | – |
| *Saccharopolyspora erythraea* | – | P24664 | JC4170 | D30760 | – |
| *Salmo salar* | I | P35031 | S31775 S31776 S31777 | X70071 X70072 X70075 | – |
| *Salmo salar* | II | P35032 | S31778 | X70073 | – |
| *Salmo salar* | III | P35033 | S31779 | X70074 | – |
| *Sarcophaga bullata* | – | P51588 | – | X94691 | – |
| *Simulium vittatum* | – | P35048 | – | L08428 | U22454: 5′ end |
| *Squalus acanthias* | – | P00764 | A00950 B27719 | – | – |
| *Streptomyces exfoliatus* | – | P80420 | – | – | – |
| *Streptomyces fradiae* | – | – | – | D16687 | – |

| Species | Form | SwissProt | PIR | EMBL (cDNA) | EMBL (genomic) |
|---|---|---|---|---|---|
| Trypsin (*continued*) | | | | | |
| *Streptomyces glaucescens* | – | – | – | U13770 | – |
| *Streptomyces griseus* | – | P00775 | A00962 | M64471 | – |
| | | | JQ1302 | | |
| *Streptomyces griseus* | – | – | – | X75952 | – |
| *Sus scrofa* | – | P00761 | A00947 | – | – |
| | | | A90368 | | |
| | | | A90641 | | |
| *Takifugu rubripes* | – | – | – | U25747 | – |
| *Xenopus laevis* | – | P19799 | A35871 | X53458 | – |
| | | | S12117 | | |
| *Xenopus laevis* | – | – | – | U72330 | – |
| **Related peptidases** | | | | | |
| Achelase | | | | | |
| *Lonomia achelous* | I | P23604 | S17537 | – | – |
| *Lonomia achelous* | II | P23605 | S17680 | – | – |
| Cocoonase | | | | | |
| *Limulus polyphemus* | | P35586 | – | – | – |
| Hypodermin | | | | | |
| *Hypoderma lineatum* | A | P35587 | A21590 | X74303 | – |
| *Hypoderma lineatum* | B | P35588 | A20190 | L24915 | – |
| | – | | | X74304 | |
| | – | | | X74305 | |
| Vitellin-degrading protease | | | | | |
| *Bombyx mori* | – | Q07943 | S32794 | D16232 | D16233: complete gene |

Brookhaven Protein Data Bank three-dimensional structures:

| Species | ID | Resolution | Notes |
|---|---|---|---|
| *Bos taurus* | 1GBT | 2 | guanidinobenzoylated at Ser195 |
| | 1NTP | 1.8 | modified $\beta$-trypsin |
| | 1PPC | 1.8 | complex with NAPAP |
| | 1PPE | 2 | complex with squash trypsin inhibitor |
| | 1PPH | 1.9 | complex with 3-TAPAP |
| | 1SMF | 2.1 | complex with Bowman–Birk inhibitor |
| | 1TAB | 2.3 | complex with adzuki bean Bowman–Birk inhibitor |
| | 1TGB | 1.8 | precursor |
| | 1TGC | 1.8 | precursor, in 50% methanol |
| | 1TGN | 1.65 | precursor |
| | 1TGS | 1.8 | complex with pig secretory trypsin inhibitor |
| | 1TGT | 1.7 | precursor at 173 degrees |
| | 1TLD | 1.5 | $\beta$-trypsin, orthorhombic at pH 5.3 |
| | 1TNG | 1.8 | complex with aminomethylcyclohexane |
| | 1TNH | 1.8 | complex with 4-fluorobenzylamine |
| | 1TNI | 1.9 | complex with 4-phenylbutylamine |
| | 1TNJ | 1.8 | complex with 2-phenylethylamine |
| | 1TNK | 1.8 | complex with 3-phenylpropylamine |
| | 1TNL | 1.9 | complex with tranylcypromine |
| | 1TPA | 1.9 | anhydro-trypsin; complex with pancreatic trypsin inhibitor |
| | 1TPO | 1.7 | $\beta$-trypsin at pH 5.0 |
| | 1TPP | 1.4 | $\beta$-trypsin; complex with *p*-amidino-phenyl pyruvate |
| | 1TPS | 1.9 | complex with inhibitor A90720A |
| | 1TYN | 2 | complex with cyclotheonamide A |
| | 2PTC | 1.9 | complex with pancreatic trypsin inhibitor |
| | 2PTN | 1.55 | orthorhombic |
| | 2TGA | 1.8 | precursor |

*continued overleaf*

| Species | ID | Resolution | Notes |
|---|---|---|---|
| | 2TGD | 2.1 | precursor; complex with diisopropylphosphoryl |
| | 2TGP | 1.9 | precursor; complex with pancreatic trypsin inhibitor |
| | 2TGT | 1.7 | precursor at 103 degrees |
| | 2TLD | 2.6 | complex with modified *Streptomyces* subtilisin inhibitor |
| | 2TPI | 2.1 | precursor; complex with pancreatic trypsin inhibitor |
| | 3PTB | 1.7 | $\beta$-trypsin; complex with benzamidine |
| | 3PTN | 1.7 | trigonal |
| | 3TPI | 1.9 | precursor; complex with pancreatic trypsin inhibitor |
| | 4PTP | 1.34 | $\beta$-trypsin; complex with diisopropylphosphoryl |
| | 4TPI | 2.2 | precursor; complex with pancreatic trypsin inhibitor Arg15 analog |
| *Fusarium oxysporum* | 1TRY | 1.55 | |
| *Rattus norvegicus* | 1BRA | 2.2 | complex with benzamidine |
| | 1BRB | 2.1 | variant, complex with bovine pancreatic trypsin inhibitor variant |
| | 1BRC | 2.1 | variant, complex with amyloid $\beta$-protein precursor inhibitor domain |
| | 1SLU | 1.8 | Asn143His mutant, complex with Ala86His ecotin |
| | 1SLV | 2.3 | Asn143His mutant, complex with Ala86His ecotin |
| | 1SLW | 2 | Asn143His mutant, complex with Ala86His ecotin |
| | 1SLX | 2.2 | Asn143His mutant, complex with Ala86His ecotin |
| | 1TRM | 2.3 | Asp102Asn mutant; complex with benzamidine |
| | 2TRM | 2.8 | Asp102Asn; complex with benzamidine at pH 8 |
| *Salmo salar* | 1BIT | 1.83 | complex with benzamidine inhibitor |
| | 1TBS | 1.8 | complex with benzamidine |
| *Streptomyces griseus* | 1SGT | 1.7 | |
| *Sus scrofa* | 1EPT | 1.8 | |
| | 1MCT | 1.6 | complex with bitter gourd inhibitor |

## Name and History

Trypsin was first described and named in 1876 by Kühne as the proteolytic activity in pancreatic secretions (Kühne, 1876; Neurath & Zwilling, 1986). Kühne differentiated this activity from that of pepsin by the higher optimal pH of trypsin. As separation and characterization of the individual pancreatic proteases was achieved, the name *trypsin* became associated with the proteolytic activity which cleaved peptide bonds C-terminal to Arg or Lys. The ready availability of trypsin from the pancreas of cattle allowed the enzyme to be purified by crystallization in 1931 (Northrop & Kunitz, 1931).

## Activity and Specificity

Trypsin can be seen as a prototype of the serine endopeptidases of family S1, and much of the fundamental knowledge about the family has been derived from the study of this enzyme (Perona & Craik, 1995). Trypsin strongly prefers to cleave amide substrates following P1 Arg or Lys residues. The preference for these basic side chains is reflected by relative values for catalytic efficiency ($k_{cat}/K_m$) at least $10^5$ greater than for other natural amino acids. The preference for Arg over Lys is 2- to 10-fold (Craik *et al.*, 1985). However, discrimination among ester substrates is much less strict. Secondary binding sites on both sides of the scissile bond play only a very minor role in the determination of substrate specificity, although occupancy of these sites does contribute to catalytic efficiency (Corey *et al.*, 1992; Schellenberger *et al.*, 1994). In small synthetic substrates, formation of an acyl enzyme intermediate is usually the rate-determining step for the cleavage of amide bonds by trypsin, whereas hydrolysis of this intermediate is the rate-determining step for ester cleavage. The binding of the substrate influences not only $K_m$ but also $k_{cat}$. In fact, the acylation rate with a substrate is a major specificity determinant (Hedstrom *et al.*, 1994b), and this is one reason why trypsin is much more promiscuous with ester substrates than with peptides. With protein substrates, the binding event may be the rate-limiting step.

Typical spectroscopic assay substrates include Bz-Arg-OEt, Tos-Arg-OMe, Bz-Arg-NHPhNO₂, Suc-Ala-Ala-Pro-Arg↓NHPhNO₂, and Z-Gly-Pro-Arg↓NHMec (Zimmerman *et al.*, 1977). The Pro commonly utilized in the P2 position of these substrates helps to align the substrate for productive binding, since positioning of this residue in either the P3 or P1 positions is strongly disfavored. A commonly used peptide substrate is the insulin B chain (Corey *et al.*, 1992), substrate cleavage being monitored by HPLC. A gel overlay assay has also been developed that can detect subnanogram amounts of trypsin (Vasquez *et al.*, 1989). Active-site titrations are usually accomplished with *p*-nitrophenyl-*p*'-guanidinobenzoate (Chase & Shaw, 1967) or 4-methylumbelliferyl-*p*-guanidinobenzoate (Jameson *et al.*, 1973).

The pH optimum of trypsin is approximately 8, although this varies slightly with species. The reaction buffer is required to contain moderate amounts (20 mM) of $CaCl_2$ for maximal activity and stability of the protease. Under these conditions, the catalytic efficiency ($k_{cat}/K_m$) of insulin B chain cleavage by rat trypsin is 18 min⁻¹ μM⁻¹, whereas the value for the smaller substrate, Z-Gly-Pro-Arg-NHMec, is 210 min⁻¹ μM⁻¹ (Corey *et al.*, 1992).

Trypsin is stable for extended periods of time as a lyophilized powder, or in solution at pH 3 (in which it is largely inactive). The enzyme from natural sources is available from many commercial suppliers at various degrees of purity. Tos-Phe-CH$_2$Cl-treatment of the enzyme minimizes contamination by active chymotrypsin. Trypsin is also available from Sigma (see Appendix 2 for full names and addresses of suppliers) bound to agarose and to polyacrylamide for easy separation from tryptic digests. Recombinant forms of trypsin have also been made (see below).

Many general serine protease inhibitors (i.e. PMSF, DFP, DCI) inhibit trypsin, but a greater specificity for enzymes with trypsin-like specificity is shown by leupeptin, benzamidine, Tos-Lys-CH$_2$Cl, and APMSF. Protein inhibitors include ecotin, soybean trypsin inhibitor, aprotinin, $\alpha_2$-macroglobulin, and $\alpha_1$-proteinase inhibitor. Trypsin can also be reversibly denatured by urea (Higaki & Light, 1985). The extinction coefficient at 280 nm is 15.4 for rat anionic trypsin and 14.6 for cattle cationic trypsin.

## Structural Chemistry

The three-dimensional structure of cattle trypsin was determined in 1974 independently by two groups (Huber *et al.*, 1974; Stroud *et al.*, 1974) and this structure has become the prototype for the S1 family of proteases. Structural analyses of eukaryotic and prokaryotic members of this family have revealed a common three-dimensional structure (Delbaere *et al.*, 1975; Bode *et al.*, 1983).

The tertiary structures of the enzymes belonging to the S1 family are strongly conserved, and this is seen very clearly for trypsin. Although the primary structures of trypsins can vary substantially, the folds are closely similar. For example, the cattle cationic and rat anionic trypsin backbones have a root-mean-square (r.m.s.) deviation of only 0.29 Å (Sprang *et al.*, 1987) and there is an r.m.s. deviation of only about 1 Å between the hydrophobic cores of the prokaryotic trypsin of *S. griseus* and rat anionic trypsin. The usual numbering system of the residues follows that of the homologous protease chymotrypsin (Chapter 8). Chymotrypsin has a nearly identical tertiary structure, but is less than 50% identical in primary structure, and differs from trypsin in substrate specificity. The positions of key residues, such as those of the catalytic triad, are identical in the two proteases. Active trypsin consists of a single polypeptide chain, in which the catalytic residues bridge two $\beta$-barrel domains. Other forms (termed $\alpha$, $\gamma$ and $\pi$), in which the polypeptide backbone has been clipped, also possess varying degrees of activity (Higaki & Light, 1985). As in other serine proteases of the family, the most important catalytic residues are those of the Asp/His/Ser triad. Ser195 acts as a nucleophile in the cleavage reaction, producing an acyl enzyme intermediate. The catalytic triad is no longer thought to act as a 'charge-relay system' as was once widely assumed, but instead His57 is thought to act as a general base (Kossiakoff & Spencer, 1981). Asp102 is believed to stabilize the correct tautomer of His57, and to provide compensation for the developing positive charge during the catalytic reaction. Replacement of this aspartic acid with asparagine results in an enzyme that is approximately 10$^4$-fold less active than the wild-type enzyme (Craik *et al.*, 1987), while relocation of Asp102 to position 214 yields a protease which retains approximately 0.5% of the wild-type activity on peptide substrates. Replacement of either His57 or Ser195 with alanine results in a protease which is 10$^5$- to 10$^6$-fold less active than

the wild-type enzyme (Corey & Craik, 1992). The residual activity in the His57Ala and Ser195Ala mutants is believed to be due to other structural features in the protease which help to stabilize the tetrahedral intermediate.

Work with recombinant trypsin has addressed the role of particular amino acid residues through site-directed mutagenesis (Sprang *et al.*, 1987; McGrath *et al.*, 1989; Corey & Craik, 1992; Corey *et al.*, 1992). While replacement of the active-site His57 by Ala reduced the activity of the enzyme by four orders of magnitude, significant 'substrate-assisted catalysis' could be observed with peptide substrates in which the function of the catalytic His could be fulfilled by the substrate (Corey *et al.*, 1995). In addition, it could be shown that many parts of the trypsin molecule contribute to the specific recognition of substrate (Craik *et al.*, 1985; Gráf *et al.*, 1987; Hedstrom *et al.*, 1992, 1994a; Perona *et al.*, 1993, 1995). Subsequent studies sought to alter the activity and substrate specificity of the enzyme. For instance, His+ substrate specificity has been engineered into the subsites of trypsin by creating metal-binding sites that bridge the substrate and enzyme (Willett *et al.*, 1996). Engineered metal-binding sites were also shown to be useful in the reversible regulation of trypsin activity: several variants that involve the active site histidine in metal binding were effective at allowing the reversible inhibition of trypsin with submicromolar concentrations of transition metal ions (Higaki *et al.*, 1990; Brinen *et al.*, 1996; Halfon & Craik, 1996).

All naturally occurring trypsins are synthesized as proenzymes. The mammalian propeptide (usually a hexapeptide) contains the consensus sequence for cleavage by enteropeptidase (Chapter 14), -(Asp)$_4$-Lys-preceding the mature N-terminal sequence Ile16-Val17-Gly18-Gly19-(chymotrypsinogen numbering). Cleavage of the propeptide results in disruption of a His40 to Asp194 hydrogen bond, and this is followed by rotation of Asp194 so that it can interact with the new N-terminus at Ile16. This conformational change completes formation of the oxyanion hole (comprising backbone amides of Gly193 and Ser195) and the binding pockets (Fehlhammer *et al.*, 1977; Bode *et al.*, 1978). Stabilization of these new conformations in the so-called activation domain is due principally to hydrophobic interactions of the Ile16 side chain (Hedstrom *et al.*, 1996). Only small adjustments in position are made in other regions (including the catalytic residues) of the protein during the activation process.

The substrate forms an antiparallel $\beta$ sheet with the protein-binding site. The substrate specificity is primarily determined by the Asp189 side chain which lies at the bottom of the S1 binding pocket. Substitution of this residue with Ser results in a 10$^5$-fold decrease in $k_{cat}/K_m$ for Arg/Lys substrates (with most of the decrease coming from a lowered $k_{cat}$), without providing significant activity towards substrates with hydrophobic side chains (Gráf *et al.*, 1987, 1988). Removal of the side chain at position 189 in the Asp189Gly variant permits binding of a well-ordered acetate ion in a similar position (Perona *et al.*, 1994). High concentrations of acetate increase the catalytic efficiency of the variant enzyme by 300-fold, demonstrating that the negative charge at the base of the trypsin specificity pocket may be provided by a noncovalently bound ligand.

In the case of Arg-containing substrates, a direct interaction occurs between the substrate guanidinium group and the carboxyl group of Asp189, but for Lys side chains, the contact is mediated by a water molecule. These specific interactions can be selectively altered by introducing Ala at positions 216 and 226 (normally Gly) in rat and cattle trypsin. This in turn dramatically alters the Arg/Lys specificity of the resulting enzymes (Craik *et al*., 1985). A disulfide bridge between Cys191 and Cys220, and a loop comprising residues 214–220 also contribute to the structure of the specificity pocket. Numerous experiments have shown that trypsin surface loops comprising amino acids 185–193 (loop 1) and amino acids 217–224 (loop 2) strongly influence the specificity of the enzyme, even though they do not directly contact the substrate (Perona *et al*., 1995; Perona & Craik, 1995).

Other structural features of trypsin include a high-affinity calcium-binding site, which is required for stability of the enzyme; autodegradation quickly occurs in its absence. This site is formed by the loop Glu70–Glu80 (Bode & Schwager, 1975). The protein has six completely conserved disulfide bonds, at positions 15–145, 33–49, 117–218, 124–191, 156–170, and 181–205. The 'autolysis' loop, comprising residues 143–151, is very flexible in both trypsin and trypsinogen. Cleavage of this loop at Lys145 yields the $\alpha$-trypsin form which retains some catalytic activity. This and other clipped forms are present in most preparations of trypsin (Higaki & Light, 1985). Molecular mass values are approximately 25 kDa, while the pI values can vary widely for forms of trypsin, both cationic and anionic forms existing in many species. Trypsin and its zymogen form, trypsinogen, contain no post-translational modifications aside from the proteolytic processing required for activity.

## Preparation

Trypsin is readily available from many sources. Cattle pancreatic trypsin is commercially available from Sigma, Boehringer Mannheim, Worthington and Fluka. However, it should be noted that trypsin from commercial sources is often contaminated with other pancreatic enzymes. Recombinant trypsin has been expressed in many different systems, including *Escherichia coli* (Higaki *et al*., 1989; Vasquez *et al*., 1989; Evnin *et al*., 1990), *Saccharomyces cerevisiae* (Hedstrom *et al*., 1992), and *Pichia pastoris* (Halfon & Craik, 1996). The mature form has been successfully expressed in *E. coli* (Vasquez *et al*., 1989). Purification from these systems often involves affinity chromatography on immobilized benzamidine or aprotinin.

Genetic selections have been established for isolating trypsins from libraries of variants. An *in vivo* selection in bacteria with a dynamic range of five orders of magnitude has been used to isolate trypsins with altered substrate specificities (Evnin *et al*., 1990; Perona *et al*., 1993). Recently, trypsin was displayed on bacteriophage, permitting an *in vitro* selection of trypsin activity (Corey *et al*., 1993).

## Biological Aspects

Trypsin is one of several digestive enzymes secreted into the intestine of animals. It is found in all animals, including insects, fish and mammals. In bovine pancreatic secretions,

it represents approximately 15% of the digestive enzymes (Keller *et al*., 1958). Trypsin is synthesized as a preproenzyme by the acinar cells of the pancreas and is stored as the proenzyme trypsinogen in secretory granules. Following release into the gut, trypsinogen is activated by enteropeptidase (Chapter 14), or by trypsin itself. Once activated, the enzyme is responsible for the activation of the proenzymes of all the other digestive enzymes such as chymotrypsin (see Chapter 8) and elastase (see Chapters 11 and 12), and contributes to the digestion of consumed protein.

Numerous genes encoding both anionic and cationic forms of trypsin are present in most animal species, together with the corresponding proteins. The individual proteins are expressed at differing levels. For instance, anionic and cationic trypsins have been isolated from human (Guy *et al*., 1978), cattle (Louvard & Puigserver, 1974), dog (Ohlsson & Tegner, 1973) and rat (Brodrick *et al*., 1980). In general, the percentage sequence identity of cationic trypsins of different organisms is closer than to the anionic forms in the same species. In most species, one cationic form predominates, while two anionic trypsins are present at much lower levels (less than 10% of the total). In rats, however, the anionic variant is the major form (Fletcher *et al*., 1987). The reason for the presence of these multiple forms is not known. Since the various forms have distinct amino acid sequences and the differences are distributed throughout the protein, they are clearly products of separate genes. In the adult rat, there are at least ten genes which encode trypsinogen (Craik *et al*., 1984). Approximately 2–5% of the total adult rat pancreatic mRNA encodes anionic trypsin (MacDonald *et al*., 1982). There are eight human genes, of which five are transcribed to RNA, two are pseudogenes, and one is a relic gene (Rowen *et al*., 1996). The human trypsinogen genes are intercalated in two pieces within the human $\beta$-T cell receptor locus on chromosome 7q35 (Honey *et al*., 1984a; Rowen *et al*., 1996). At the 5′-end of the locus lie the three nonfunctional trypsinogen genes, while other genes, including those coding for the known trypsinogen isoenzymes, lie approximately 500 kb 3′-terminal to these, near the opposite end of the locus. These corresponding genes have also been found in the TCR locus (Whitcomb *et al*., 1996) in the mouse on chromosome 6 (Honey *et al*., 1984b) and have been identified in the chicken (Wang *et al*., 1995). The rat genes corresponding to the predominant trypsin forms have been shown to contain four introns (Craik *et al*., 1984). Although these introns are positioned at the same sites within the genes, the introns have no homology across variants. The intron–exon junctional amino acids of trypsin and related serine proteases map to the surface of the enzyme and provide a model for evolution of the structure and function of the enzymes through 'junctional sliding' (Craik *et al*., 1982, 1983).

A recent report (Whitcomb *et al*., 1996) has suggested that hereditary pancreatitis may be associated with an Arg117His mutation in the human cationic trypsin which prevents autolysis at this site. The consequent stability of the mutated enzyme could result in increased proteolytic activity that could damage the pancreas. Some reports have indicated expression of very low levels of trypsin in nonpancreatic tissues on the basis of PCR analysis (Wiegand *et al*., 1993; Wang *et al*., 1995).

## Distinguishing Features

The form of trypsin present in higher animals has very little sequence identity with microbial trypsins, but is similar structurally. In particular, *Streptomyces griseus* trypsin has a very similar fold to these proteases (Read & James, 1988). Other microbial proteases are more distantly related, having shorter amino acid sequences and corresponding surface loops. For instance, $\alpha$-lytic protease (Chapter 82) has the same fold as trypsin, but differs greatly in many structural aspects (James, 1976). Many other serine proteases such as the kallikreins, elastase and chymotrypsin comprise the trypsin family and these are very similar structurally and catalytically to trypsin, but differ in other key aspects such as substrate preferences. Therefore, while the fold of this class of enzymes is similar among all its members, the activities and functions can vary greatly (Chapter 2).

In addition to the conservation of the three-dimensional fold in all known trypsins, the catalytic triad is entirely conserved, and amino acids flanking these residues are also conserved. Degenerate oligonucleotide probes have been designed for the sequences flanking the active site Ser and His residues, and used in PCR to isolate and identify trypsins and related serine proteases from organisms of diverse phyla (Sakanari *et al.*, 1989).

## Related Enzymes

Trypsin-like enzymes are present in many different organisms. Bacterial trypsins, such as those from *Streptomyces* species are dealt with elsewhere (Chapters 4 and 5). The trypsin from the fungus *Fusarium oxysporum* is equally (approximately 45%) identical to the bacterial and mammalian trypsins, and all trypsins are thought to have evolved from a common ancestor (Hewett-Emmett *et al.*, 1981). While functional parts of the proteins are strongly conserved, other regions are less so (Rypniewski *et al.*, 1993). For instance, the *Fusarium oxysporum* trypsin lacks a calcium-binding site, and has a propeptide that lacks homology to the mammalian trypsins. The optimal activity of this enzyme is at 40°C and pH 9.5.

## Further Reading

The reader is referred to the articles of Hedstrom (1996) and Perona & Craik (1995).

## Acknowledgements

This work was supported by a grant from the National Science Foundation MCB-9604379 (to C.S.C.) and post-doctoral NRSA Fellowship GM15884 (to S.H.).

## References

Bode, W. & Schwager, P. (1975) The refined crystal structure of bovine $\beta$-trypsin at 1.8 Å resolution. *J. Mol. Biol.* **98**, 693–717.

Bode, W., Chen, Z., Bartels, K., Kutzbach, C., Schmidt-Kastner, G. & Bartunik, H. (1983) refined 2 Å X-ray crystal structure of porcine pancreatic kallikrein A, a specific trypsin-like serine proteinase. Crystallization, structure determination, crystallographic refinement, structure and its comparison with bovine trypsin. *J. Mol. Biol.* **164**, 237–282.

Bode, W., Schwager, P. & Huber, R. (1978) The transition of bovine trypsinogen to a trypsin-like state upon strong ligand binding. The refined crystal structures of the bovine trypsinogen-pancreatic trypsin inhibitor complex and of its ternary complex with Ile-Val at 1.9 Å resolution. *J. Mol. Biol.* **118**, 99–112.

Brinen, L.S., Willett, W.S., Craik, C.S. & Fletterick, R.J. (1996) X-ray structures of a designed binding site in trypsin show metal-dependent geometry. *Biochemistry* **35**, 5999–6009.

Brodrick, J.W., Largman, C., Geokas, M.C., O'Rourke, M. & Ray, S.B. (1980) Clearance of circulating anionic and cationic pancreatic trypsinogens in the rat. *Am. J. Physiol.* **239**, G511–515.

Chase, T., Jr. & Shaw, E. (1967) *p*-Nitrophenyl-*p'*-guanidino-benzoate HCl: a new active site titrant for trypsin. *Biochem. Biophys. Res. Commun.* **29**, 508–514.

Corey, D.R. & Craik, C.S. (1992) An investigation into the minimum requirements for peptide hydrolysis by mutation of the catalytic triad of trypsin. *J. Am. Chem. Soc.* **114**, 1784–1790.

Corey, D.R., McGrath, M.E., Vasquez, J.R., Fletterick, R.J. & Craik, C.S. (1992) An alternate geometry for the catalytic triad of serine proteases. *J. Am. Chem. Soc.* **114**, 4905–4907.

Corey, D.R., Shiau, A.K., Yang, Q., Janowski, B.A. & Craik, C.S. (1993) Trypsin display on the surface of bacteriophage. *Gene* **128**, 129–134.

Corey, D.R., Willett, W.S., Coombs, G.S. & Craik, C.S. (1995) Trypsin specificity increased through substrate-assisted catalysis. *Biochemistry* **34**, 11521–11527.

Craik, C.S., Sprang, S., Fletterick, R. & Rutter, W.J. (1982) Intron-exon splices junctions map at protein surfaces. *Nature* **299**, 180–182.

Craik, C.S., Rutter, W.J. & Fletterick, R. (1983) Splice junctions: association with variation in protein structure. *Science* **220**, 1125–1129.

Craik, C.S., Choo, Q.L., Swift, G.H., Quinto, C., MacDonald, R.J. & Rutter, W.J. (1984) Structure of two related rat pancreatic trypsin genes. *J. Biol. Chem.* **259**, 14255–14264.

Craik, C.S., Largman, C., Fletcher, T., Barr, P., Fletterick, R. & Rutter, W.J. (1985) Redesigning trypsin: alteration of substrate specificity, catalytic activity and protein conformation. *Science* **228**, 291–297.

Craik, C.S., Roczniak, S., Largman, C. & Rutter, W.J. (1987) The catalytic role of the active site aspartic acid in serine proteases. *Science* **237**, 909–913.

Delbaere, L.T., Hutcheon, W.L., James, M.N. & Thiessen, W.E. (1975) Tertiary structural differences between microbial serine proteases and pancreatic serine enzymes. *Nature* **257**, 758–763.

Evnin, L.B., Vasquez, J.R. & Craik, C.S. (1990) Substrate specificity of trypsin investigated by using a genetic selection. *Proc. Natl Acad. Sci. USA* **87**, 6659–6663.

Fehlhammer, H., Bode, W. & Huber, R. (1977) Crystal structure of bovine trypsinogen at 1.8 Å resolution. *J. Mol. Biol.* **111**, 415–438.

Fletcher, T.S., Alhadeff, M., Craik, C.S. & Largman, C. (1987) Isolation and characterization of a cDNA encoding rat cationic trypsinogen. *Biochemistry* **26**, 3081–3086.

Gráf, L, Craik, C.S., Patthy, A., Roczniak, S., Fletterick, R.J. & Rutter, W.J. (1987) Selective alteration of substrate specificity by replacement of aspartic acid-89 with lysine in the binding pocket of trypsin. *Biochemistry* **26**, 2616–2623.

Gráf, L., Jancso, A., Szilágyi, L., Hegyi, G., Pinter, K., Naray-Szabó, G., Hepp, J., Medzihradszky, K. & Rutter, W.J. (1988)

Electrostatic complementarity within the substrate-binding pocket of trypsin. *Proc. Natl Acad. Sci. USA* **85**, 4961–4965.

Guy, O., Lombardo, D., Bartelt, D.C., Amic, J. & Figarella, C. (1978) Two human trypsinogens. Purification, molecular properties, and N-terminal sequences. *Biochemistry* **17**, 1669–1675.

Halfon, S. & Craik, C.S. (1996) Regulation of proteolytic activity by engineered metal binding loops. *J. Am. Chem. Soc.* **118**, 1227–1228.

Hedstrom, L. (1996) Trypsin: a case study in the structural determinants of enzyme specificity. *Biol. Chem.* **377**, 465–470.

Hedstrom, L., Szilagyi, L. & Rutter, W. (1992) Converting trypsin to chymotrypsin. *Science* **255**, 1249–1253.

Hedstrom, L., Perona, J.J. & Rutter, W.J. (1994a) Converting trypsin to chymotrypsin: residue 172 is a substrate specificity determinant. *Biochemistry* **33**, 8757–8763.

Hedstrom, L., Farr-Jones, S., Kettner, C.A. & Rutter, W.J. (1994b) Converting trypsin to chymotrypsin: ground-state binding does not determine substrate specificity. *Biochemistry* **33**, 8764–8769.

Hedstrom, L., Lin, T.Y. & Fast, W. (1996) Hydrophobic interactions control zymogen activation in the trypsin family of serine proteases. *Biochemistry* **35**, 4515–4523.

Hewett-Emmett, D., Czelusniak, J. & Goodman, M. (1981) The evolutionary relationship of the enzymes involved in blood coagulation and hemostasis. *Ann. N.Y. Acad. Sci.* **370**, 511–527.

Higaki, J.N. & Light, A. (1985) The identification of neotrypsinogens in samples of bovine trypsinogen. *Anal. Biochem.* **148**, 111–120.

Higaki, J.N., Evnin, L.B. & Craik, C.S. (1989) Introduction of a cysteine protease active site into trypsin. *Biochemistry* **29**, 8582–8586.

Higaki, J.N., Haymore, B.L., Chen, S., Fletterick, R. & Craik, C.S. (1990) Regulation of serine protease activity by an engineered metal switch. *Biochemistry* **29**, 8582–8586.

Honey, N.K., Sakaguchi, A.Y., Quinto, C., MacDonald, R.J., Bell, G.I., Craik, C.S., Rutter, W.J. & Naylor, S.L. (1984a) Chromosomal assignments of human genes for serine proteases trypsin, chymotrypsin B, and elastase. *Somat. Cell Mol. Genet.* **10**, 369–376.

Honey, N.K., Sakaguchi, A.Y., Lalley, P.A., Quinto, C., MacDonald, R.J., Craik, C.S., Bell, G.I., Rutter, W.J. & Naylor, S.L. (1984b) Chromosomal assignments of genes for trypsin, chymotrypsin B, and elastase in mouse. *Somat. Cell Mol. Genet.* **10**, 77–83.

Huber, R., Kukla, D., Bode, W., Schwager, P., Bartels, K., Deisenhofer, J. & Steigemann, W. (1974) Structure of the complex formed by bovine trypsin and bovine pancreatic trypsin inhibitor. II. Crystallographic refinement at 1.9 Å resolution. *J. Mol. Biol.* **89**, 73–101.

James, M.N.G. (1976) Relationship between the structures and activities of some microbial serine proteases. II. Comparison of the tertiary structures of microbial and pancreatic serine proteases. In: *Proteolysis and Physiological Regulation* (Ribbons, D.W. & Brew, K., eds). New York: Academic Press, pp. 125–142.

Jameson, G.W., Roberts, D.V., Adams, R.W., Kyle, W.S.A. & Elmore, D.T. (1973) Determination of the operational molarity of solutions of bovine α-chymotrypsin, trypsin, thrombin and factor Xa by spectrofluorimetric titration. *Biochem. J.* **131**, 107–117.

Keller, P., Cohen, E. & Neurath, H. (1958) The proteins of bovine pancreatic juice. *J. Biol. Chem.* **233**, 344–349.

Kossiakoff, A.A. & Spencer, S.A. (1981) Direct determination of the protonation states of aspartic acid-102 and histidine-57 in the tetrahedral intermediate of the serine proteases: neutron structure of trypsin. *Biochemistry* **20**, 6462–6474.

Kühne, W.F (1876) Über die Verdauung der Eiweisstoffe durch den Pankreassaft. *Virchows Arch.* **39**, 130 (reproduced by Gutfreund, H. (1976) *FEBS Lett.* **62**(suppl.), E1–E12).

Louvard, M.N. & Puigserver, A. (1974) On bovine and porcine anionic trypsinogens. *Biochim. Biophys. Acta* **371**, 177–185.

MacDonald, R.J., Stary, S.J. & Swift, G.H. (1982) Two similar but nonallelic rat pancreatic trypsinogens. Nucleotide sequences of the cloned cDNAs. *J. Biol. Chem.* **257**, 9724–9732.

McGrath, M.E., Wilke, M.E., Higaki, J.N., Craik, C.S. & Fletterick, R.J. (1989) Crystal structures of two engineered thiol trypsins. *Biochemistry* **28**, 9264–9270.

Neurath, H. & Zwilling, R. (1986) Willy Kühne und die Anfänge der Enzymologie. In: *Semper Apertus*. Berlin: Springer-Verlag, pp. 361–374.

Northrop, J.H. & Kunitz, M. (1931) Isolation of protein crystals possessing tryptic activity. *Science* **73**, 262–263.

Ohlsson, K. & Tegner, H. (1973) Experimental pancreatitis in the dog. Demonstration of trypsin in ascitic fluid, lymph and plasma. *Scand. J. Gastroenterol.* **8**, 129–133.

Perona, J.J. & Craik, C.S. (1995) Structural basis of substrate specificity in the serine proteases. *Protein Sci.* **4**, 337–360.

Perona, J.J., Tsu, C.A., Craik, C.S. & Fletterick, R.J. (1993) Crystal structures of rat anionic trypsin complexed with the protein inhibitors APPI and BPTI. *J. Mol. Biol.* **230**, 919–933.

Perona, J., Hedstrom, L., Wagner, R., Rutter, W., Craik, C.S. & Fletterick, R. (1994) Exogenous acetate reconstitutes the enzymatic activity of trypsin Asp189Ser. *Biochemistry* **33**, 3252–3259.

Perona, J.J., Hedstrom, L., Rutter, W.J. & Fletterick, R.J. (1995) Structural origins of substrate discrimination in trypsin and chymotrypsin. *Biochemistry* **34**, 1489–1499.

Read, R.J. & James, M.N.G. (1988) Refined crystal structure of *Streptomyces griseus* trypsin at 1.7 Å resolution. *J. Mol. Biol.* **200**, 523–551.

Rowen, L., Koop, B.F. & Hood, L. (1996) The complete 685-kilobase DNA sequence of the human β-T cell receptor locus. *Science* **272**, 1755–1762.

Rypniewski, W.R., Hastrup, S., Betzel, C., Dauter, M., Dauter, Z., Papendorf, G., Branner, S. & Wilson, K.S. (1993) The sequence and X-ray structure of the trypsin from *Fusarium oxysporum*. *Protein Eng.* **6**, 341–348.

Sakanari, J., Staunton, C.E., Eakin, A.E., Craik, C.S. & McKerrow, J.H. (1989) Serine proteases from nematode and protozoan parasites: Isolation of sequence homologous using generic molecular probes. *Proc. Natl Acad. Sci. USA* **86**, 4863–4867.

Schellenberger, V., Turck, C.W. & Rutter, W.J. (1994) Role of the S′ subsites in serine protease catalysis. Active-site mapping of rat chymotrypsin, rat trypsin, α-lytic protease, and cercarial protease from *Schistosoma mansoni*. *Biochemistry* **33**, 4251–4257.

Sprang, S., Standing, T., Fletterick, R.J., Stroud, R.M., Finer-Moore, J., Xuong, N.H., Hamlin, R., Rutter, W.J. & Craik, C.S. (1987) The three-dimensional structure of Asn102 mutant of trypsin: role of Asp102 in serine protease catalysis. *Science* **237**, 905–909.

Stroud, R.M., Kay, L.M. & Dickerson, R.E. (1974) The structure of bovine trypsin: electron density maps of the inhibited enzyme at 5 Å and at 2.7 Å resolution. *J. Mol. Biol.* **83**, 185–208.

Vasquez, J.R., Evnin, L.B., Higaki, J.N. & Craik, C.S. (1989) An expression system for trypsin. *J. Cell Biochem.* **39**, 265–276.

Wang, K., Gan, L. & Hood, L. (1995) Isolation and characterization of the chicken trypsinogen gene family. *Biochem. J.* **307**, 471–479.

Whitcomb, D.C., Gorry, M.C., Preston, R.A., Furey, W., Sossenheimer, M.J., Ulrich, C.D., Martin, S.P., Gates, Jr., L.K., Amann, S.T., Toskes, P.P., Liddle, R., McGrath, K., Uomo, G., Post, J.C. & Ehrlich, G.D. (1996) Hereditary pancreatitus is caused by a mutation in the cationic trypsinogen gene. *Nature Genet.* **14**, 141–145.

Wiegand, U., Corbach, S., Minn, A., Kang, J. & Muller-Hill, B. (1993) Cloning of the cDNA encoding human brain trypsinogen and characterization of its product. *Gene* **136**, 167–175.

Willett, W.S., Brinen, L.S., Fletterick, R.J. & Craik, C.S. (1996) Delocalizing trypsin specificity with metal activation. *Biochemistry* **35**, 5992–5998.

Zimmerman, M., Ashe, B., Yurewicz, E.C. & Patel, G. (1977) Sensitive assays for trypsin, elastase, and chymotrypsin using new fluorogenic substrates. *Anal. Biochem.* **78**, 47–51.

*Sherin Halfon*
*Department of Pharmaceutical Chemistry,*
*University of California, San Francisco,*
*San Francisco, CA 94301-0446, USA*

*Charles S. Craik*
*Department of Pharmaceutical Chemistry,*
*University of California, San Francisco,*
*San Francisco, CA 94301-0446, USA*
*Email: craik@cgl.ucsf.edu*

# 4. Trypsin (Streptomyces exfoliatus and S. albidoflavus)

## Databanks

*Peptidase classification: clan SA, family S1, MEROPS ID: S01.101*
*NC-IUBMB enzyme classification: none*
*Databank codes:*

| Species | SwissProt | PIR | EMBL (cDNA) | EMBL (genomic) |
|---|---|---|---|---|
| *Streptomyces exfoliatus* | P80420 | – | – | – |
| *Streptomyces fradiae* | – | – | D16687 | – |
| *Streptomyces glaucescens* | – | – | U13770 | – |
| *Streptomyces griseus* | P00775 | A00962 JQ1302 | M64471 | – |
| *Streptomyces griseus* | – | – | X75952 | – |

Brookhaven Protein Data Bank three-dimensional structures:

| Species | ID | Resolution | Notes |
|---|---|---|---|
| *Streptomyces griseus* | 1SGT | 1.7 | |

## Name and History

Trypsin-like proteases are widely distributed in streptomycetes (Kim *et al.*, 1995). These enzymes are commonly described as **'trypsins'**, and we shall also use this term here, for convenience, but it is important to note that there are significant differences between these and the pancreatic trypsins of mammals (see Chapter 3). Trypsins hydrolyze synthetic peptide substrates having arginine or lysine at the P1 position, and are inhibited by typical trypsin inhibitors such as Tos-Lys-CH$_2$Cl, DFP, PMSF, APMSF, leupeptin, soybean trypsin inhibitor, aprotinin, $\alpha_2$-macroglobulin, and $\alpha_1$-antitrypsin (Kang *et al.*, 1995b; Kim & Lee, 1995a).

## Activity and Specificity

Trypsins of *Streptomyces exfoliatus* and *S. albidoflavus* hydrolyze a broad range of native proteins such as bovine serum albumin, egg albumin, Hammarsten casein, collagen, hemoglobin and lysozyme. The $K_m$ and the $V$ values of *S. exfoliatus* trypsin with Bz-Arg-NHPhNO$_2$ as substrate were 160 μM and 48.1 μM min$^{-1}$ mg$^{-1}$, respectively. The optimum pH and temperature of *S. exfoliatus* trypsin were 7.5 and 35°C (Kim & Lee, 1996). The $K_m$ and the $V$ values with Bz-Arg-NHPhNO$_2$ of *S. albidoflavus* trypsin were 139 μM and 10 nmol min$^{-1}$ mg$^{-1}$, respectively. The optimum pH and temperature of *S. albidoflavus* trypsin were 10 and 40°C, respectively (Kang *et al.*, 1995a). Trypsins of *S. exfoliatus* and *S. albidoflavus* were competitively inhibited by leupeptin, and the inhibition constants were 0.015 μM and 0.0031 μM, respectively.

### Structural Chemistry

The molecular masses of *S. exfoliatus* and *S. albidoflavus* trypsins were estimated to be 31.8 kDa and 32.0 kDa, respectively. The N-terminal amino acid sequence of *S. exfoliatus* trypsin was RVGGTXAAQGNFPFQQXLSM.

### Biological Aspects

The production of trypsin was found to be related to morphological differentiation in streptomycetes. Production of the enzyme started just before the onset of aerial mycelium formation, and rapidly increased during aerial mycelium growth. Addition of the inhibitors leupeptin or TLCK inhibited aerial mycelium formation. In addition, *bld* mutants of *S. exfoliatus* and *S. albidoflavus*, defective in aerial mycelium formation, did not produce trypsin at all (Kang *et al.*, 1995a; Kim *et al.*, 1995; Kim & Lee, 1995a, 1996).

The production of trypsin in *S. exfoliatus* and *S. albidoflavus* is induced under conditions of limiting carbon, nitrogen and phosphate, in which aerial spore formation in solid culture, and submerged spore formation in submerged culture, occur (Kang *et al.*, 1995a; Kim & Lee, 1995b,c).

*S. exfoliatus* has been shown to produce leupeptin, leupeptin-inactivating enzyme (LIE) and trypsin, in sequence. The activity of trypsin is inhibited by the endogenous leupeptin, but this can be inactivated by LIE (Kim & Lee, 1995a,c, 1996). We have proposed that under biological conditions, leupeptin is inactivated by LIE when trypsin activity is required, and that trypsin may play a major role in the metabolism of substrate mycelial or nongrowing mycelial proteins for supporting the growth of aerial mycelium on surface cultures and tip growth in submerged cultures.

### References

Kang, S.G., Kim, I.S., Rho, Y.T. & Lee, K.J. (1995a) Production dynamics of extracellular proteases accompanying morphological differentiation of *Streptomyces albidoflavus* SMF301. *Microbiology* **141**, 3095–3103.

Kang, S.G., Kim, I.S., Ryu, J.G., Rho, Y.T. & Lee, K.J. (1995b) Purification and characterization of trypsin-like protease and metalloprotease in *Streptomyces albidoflavus* SMF301. *J. Microbiol.* **33**, 307–314.

Kim, I.S. & Lee, K.J. (1995a) Physiological roles of leupeptin and extracellular proteases in mycelium development of *Streptomyces exfoliatus* SMF13. *Microbiology* **141**, 1017–1025.

Kim, I.S. & Lee, K.J. (1995b) Regulation of production of leupeptin, leupeptin inactivation enzyme and trypsin like protease in *Streptomyces exfoliatus* SMF13. *J. Ferment. Bioeng.* **80**, 434–439.

Kim, I.S. & Lee, K.J. (1995c) Nutritional regulation of morphological and physiological differentiation on surface culture of *Streptomyces exfoliatus* SMF13. *J. Microbiol. Biotechnol.* **5**, 200–205.

Kim, I.S. & Lee, K.J. (1996) Trypsin like protease in *Streptomyces exfoliatus* SMF13, as a potential agent for mycelium differentiation. *Microbiology* **142**, 1797–1806.

Kim, I.S., Kang, S.G. & Lee, K.J. (1995) Physiological importance of trypsin like protease during morphological differentiation of *Streptomyces* spp. *J. Microbiol.* **33**, 315–321.

*Kye Joon Lee*
*Department of Microbiology,*
*Research Centre for*
*Molecular Microbiology,*
*Seoul National University,*
*151-742, Seoul, Korea*
*Email: lkj12345@plaza.snu.ac.kr*

*Sung Gyun Kang*
*Department of Microbiology,*
*Research Centre for*
*Molecular Microbiology,*
*Seoul National University,*
*151-742, Seoul, Korea*

*In Seop Kim*
*Department of Microbiology,*
*Research Centre for*
*Molecular Microbiology,*
*Seoul National University,*
*151-742, Seoul, Korea*

# 5. *Trypsin (Streptomyces erythraeus)*

### Databanks

*Peptidase classification: clan SA, family S1, MEROPS ID: S01.102*
*NC-IUBMB enzyme classification: none*

*Databank codes:*

| Species | SwissProt | PIR | EMBL (cDNA) | EMBL (genomic) |
|---|---|---|---|---|
| *Streptomyces erythraeus* | P24664 | JC4170 | D30760 | – |

## Name and History

*Streptomyces erythraeus* is one of the *Streptomyces* species that secrete trypsin-like proteases extracellularly (Yoshida *et al.*, 1971b; Inoue *et al.*, 1972). The enzyme produced by *S. erythraeus* is called **Streptomyces erythraeus trypsin (SET)**.

## Activity and Specificity

SET splits the peptide bonds of oxidized insulin B chain in accordance with the cleavage specificity of trypsin (Yoshida *et al.*, 1971b). The esterase activity for Bz-L-Arg-OEt is comparable with that of cattle trypsin (Yoshida *et al.*, 1971b), but the amidase activity for Bz-L-Arg-NHPhNO$_2$ is 200-fold greater than that of cattle trypsin, as a result of a much lower $K_m$ rather than a higher $k_{cat}$.

SET is more resistant to autolysis than cattle trypsin because of the absence of an arginine or lysine residue on the 'autolysis loop' (Yamane *et al.*, 1991).

SET is inhibited by Tos-Lys-CH$_2$Cl and Tos-Arg-CH$_2$Cl (Yoshida *et al.*, 1973). Although Tos-Arg-CH$_2$Cl inactivates cattle trypsin 3.4 times more rapidly than does Tos-Lys-CH$_2$Cl, there is little difference in the reactivity of SET with the two chloromethanes. SET is also inhibited by ovomucoids from the eggs of Japanese quail and chicken (Nagata & Yoshida, 1983, 1984).

## Structural Chemistry

The pI of SET is around 4.0 (Yoshida *et al.*, 1971b). CD spectra indicate a poorly ordered structure (Inoue *et al.*, 1972).

The amino acid sequence of SET, originally determined by protein sequencing (Miyamoto *et al.*, 1979), was revised in the light of the nucleotide sequence of the cloned SET gene (Nagamine-Natsuka *et al.*, 1995). SET consists of 227 amino acid residues with three disulfide bonds and shows 33% sequence identity with cattle trypsin (Nagamine-Natsuka *et al.*, 1995). The catalytic triad is composed of His42, Asp88 and Ser179. The p$K_a$ of His42 is normal (6.7), and is not perturbed by phenylmethanesulfonylation of Ser179 (Sakiyama & Kawata, 1983).

The tertiary structure of SET was elucidated by X-ray crystal structure analysis at 2.7 Å resolution. The structure showed that SET folds essentially as cattle trypsin, and the spatial arrangement of the catalytic triad is similar. The remarkable structural difference is that His42, one of the residues comprising the catalytic triad, is located in a short stretch of 3$_{10}$-helix formed from Ala41 to Thr44 in SET.

## Preparation

SET was isolated from the culture medium of *S. erythraeus* by acrinol precipitation, batchwise DEAE-cellulose treatment, and DEAE-cellulose and DEAE-Sephadex chromatographies (Yoshida *et al.*, 1971a), and further purified to homogeneity by QAE-Sephadex chromatography (Yoshida *et al.*, 1971b).

A high expression system for recombinant SET was constructed in *S. lividans*, and the enzyme was purified as the proenzyme (proSET) (Nagamine-Natsuka *et al.*, 1995). The proSET was activated by controlled chymotryptic digestion, and mature SET was purified by affinity chromatography on a soybean trypsin inhibitor-Sepharose 4B column with a yield of 16 mg liter$^{-1}$ culture broth (Nagamine-Natsuka *et al.*, 1995).

## Biological Aspects

The SET gene is composed of an 816 bp open reading frame encoding a 272 residue precursor protein (Nagamine-Natsuka *et al.*, 1995). The precusor has a 42 residue N-terminal pre-propeptide, in which the first 30 residues from the initiator methionine have a typical signal sequence for *Streptomyces* and the remaining 12 residues are thought to comprise a propeptide. The cleavage of the Phe(−1)-Ile1 bond was essential for tryptic activity in proSET. However, no chymotrypsin-like enzyme that could be responsible for the maturation of native proSET is yet known in *S. erythraeus*. The propeptide (TPAPPSDDLGTF) for SET contains part of the consensus sequence (PXDDDDK) for trypsinogen. The SET gene terminates at Arg230, but the major components of isolated native and recombinant SETs lack Ser228-Arg230 and Ala229-Arg230, respectively. In the SET gene, five sets of promoter-like sequences (two *E. coli*-type and three *Bacillus ctc*-type sequences) were detected in the A/T-rich upstream region, though it is unclear whether all or parts of these structures actually function as promoters in the expression of the SET gene. The physiological function of SET is not known.

## Distinguishing Features

No information is available as to whether polyclonal antisera against SET are specific for this protein or cross-reactive to cattle trypsin or other *Streptomyces* trypsins. The best way to distinguish SET from other trypsins is to perform amino acid analysis or N-terminal sequence analysis of the purified protein.

## Further Reading

The paper of Nagamine-Natsuka *et al.* (1995) gives further information about the *S. erythraeus* trypsin.

## References

Inoue, H., Sasaki, A. & Yoshida, N. (1972) An anionic trypsin-like enzyme from *Streptomyces erythraeus*: characterization. *Biochim. Biophys. Acta* **284**, 451–460.

Miyamoto, K., Matsuo, H. & Narita, K. (1979) Micro-circumstances analysis around the active site of *Streptomyces erythraeus* trypsin and its primary structure. In: Abstract of 30th Protein Structure Conference, pp. 77–80 (in Japanese).

Nagamine-Natsuka, Y., Norioka, S. & Sakiyama, F. (1995) Molecular cloning, nucleotide sequence, and expression of the gene encoding a trypsin-like protease from *Streptomyces erythraeus*. *J. Biochem.* **118**, 338–346.

Nagata, K. & Yoshida, N. (1983) Interaction between trypsin-like enzyme from *Streptomyces erythraeus* and Japanese quail ovomucoid. *J. Biochem.* **93**, 909–919.

Nagata, K. & Yoshida, N. (1984) Interaction between trypsin-like enzyme from *Streptomyces erythraeus* and chicken ovomucoid. *J. Biochem.* **96**, 1041–1049.

Sakiyama, F. & Kawata, Y. (1983) NMR titration studies of histidine 57 and the [methylene-13C]PMS group in the phenyl-

methanesulfonyl (PMS) derivative of *Streptomyces erythraeus* trypsin. *J. Biochem.* **94**, 1661–1669.

Yamane, T., Kobuke, M., Tsutsui, H., Toida, T., Suzuki, A., Ashida, T., Kawata, Y. & Sakiyama, F. (1991) Crystal structure of *Streptomyces erythraeus* trypsin at 2.7 Å resolution. *J. Biochem.* **110**, 945–950.

Yoshida, N., Inoue, H., Sasaki, A. & Otsuka, H. (1971a) Ribonuclease from *Streptomyces erythraeus*: purification and properties. *Biochim. Biophys. Acta* **228**, 636–647.

Yoshida, N., Sasaki, A. & Inoue, H. (1971b) An anionic trypsin-like enzyme from *Streptomyces erythraeus*. *FEBS Lett.* **15**, 129–132.

Yoshida, N., Sasaki, A. & Inouye, K. (1973) Active site trypsin-like from *Streptomyces erythraeus*: specific inactivation by new chloromethyl ketones derived from $N^\alpha$-dinitrophenyl-L-lysine and $N^\alpha$-tosyl-L-arginine. *Biochim. Biophys. Acta* **321**, 615–623.

*Shigemi Norioka*
*Institute for Protein Research, Osaka University,*
*3-2 Yamadaoka, Suita, Osaka 565, Japan*
*Email: norioka@protein.osaka-u.ac.jp*

*Fumio Sakiyama*
*Institute for Protein Research, Osaka University,*
*3-2 Yamadaoka, Suita, Osaka 565, Japan*
*Email: sakiyama@protein.osaka-u.ac.jp*

# 6. Metridin

## Databanks

*Peptidase classification: clan SA, family S1, MEROPS ID: S9G.034*
*NC-IUBMB enzyme classification: EC 3.4.21.3*
*Databank codes: no sequence data available*

## Name and History

Gibson & Dixon (1969) described a set of three chymotrypsin-like serine endopeptidases from the sea anemone *Metridium senile*, terming the enzymes protease A, protease B and protease C. **Protease A** was characterized in particular detail, and was included in the EC nomenclature in 1972 under the name **Metridium protease A**. The name **metridin** was recommended by NC-IUBMB in 1992.

## Activity and Specificity

Metridin hydrolyzes several blocked esters of aromatic amino acids with a specificity that can be described as chymotrypsin-like, but is distinctly different from that of cattle α-chymotrypsin (Table 6.1). Notably, Trp was much less favorable in S1 for metridin than for chymotrypsin. Ac-Tyr+OEt was the assay substrate used in the work of Gibson & Dixon (1969).

The action of metridin on glucagon showed cleavage of peptide bonds generally paralleling that by cattle α-chymotrypsin. The major points of hydrolysis of glucagon

Table 6.1   Action of metridin on ester substrates, as compared to that of α-chymotrypsin

| Substrate | Metridin | | | α-Chymotrypsin | | |
|---|---|---|---|---|---|---|
| | $k_{cat}$ | $K_m$ | $k_{cat}/K_m$ | $k_{cat}$ | $K_m$ | $k_{cat}/K_m$ |
| Ac-Tyr-OEt | 10.5 | 1.3 | 8.1 | 3.34 | 0.84 | 4.0 |
| Ac-Tyr-OMe | 8.8 | 0.125 | 70 | 3.70 | 0.55 | 6.7 |
| Ac-Phe-OMe | 5.2 | 0.20 | 26 | 0.88 | 1.39 | 0.63 |
| Ac-Trp-OMe | 0.16 | 0.19 | 0.84 | 0.95 | 0.08 | 11.9 |

Data are from Gibson & Dixon (1969).
The chymotrypsin was obtained commercially, and may possibly not have been fully pure and active.

were as follows: His-Ser-Gln-Gly-Thr-Phe+Thr-Ser-Asp-Tyr+ Ser-Lys-Tyr+Leu-Asp-Ser-Arg-Arg-Ala-Gln-Asp-Phe+Val-Gln-Trp-Leu+Met-Asn-Thr. All of these bonds were also found to be cleaved by cattle chymotrypsin, but the -Trp-Leu- bond also cleaved by chymotrypsin was not hydrolyzed by metridin, consistent with the low activity of the sea anemone enzyme on the ester of tryptophan.

Metridin was inhibited by DFP, Tos-Phe-CH$_2$Cl and indole ($K_i$ $1.4 \times 10^{-4}$ M), but not by Tos-Lys-CH$_2$Cl, benzamidine or soybean trypsin inhibitor. The latter inhibitors were effective against a trypsin-like enzyme present in the crude enzyme preparations, however.

## Structural Chemistry

Metridin is a cationic protein similar in molecular size to chymotrypsin. It is formed from a proenzyme that is activated by a *Metridium* enzyme that has characteristics of mammalian trypsins, being inhibited by Tos-Lys-CH$_2$Cl, benzamidine and soybean trypsin inhibitor. However, the prometridin could not be activated by cattle trypsin itself.

Reaction of metridin with [$^{32}$P]DFP, followed by partial acid hydrolysis, led to the formation of a labeled peptide identified as Asp-Ser*-Gly, the active-site peptide also found in chymotrypsin. This led Gibson & Dixon (1969) to the conclusion that metridin is a member of the trypsin/chymotrypsin family of serine endopeptidases.

Amino acid analysis of metridin by Stevenson *et al*. (1974) was also consistent with the view that metridin is a chymotrypsin-like enzyme. Both proteins contained about 240 amino acid residues, including 10 residues of Cys. However, there were also important differences; metridin contained far more proline (30 rather than 9 residues), and was glycosylated, as indicated by the presence of hexosamines. Also, there were apparently six residues of 2-aminoethylphosphonic acid, an unusual amino acid that had previously been found in another species of sea anemone.

## Preparation

The homogenate of the organism was allowed to stand for 3–5 h at 0°C for activation of metridin, and an acetone powder was then prepared. The powder was extracted with 2 M LiCl$_2$ at pH 8, and the soluble proteins were fractionated with acetone. The preparation was then applied to a column of DEAE-Sephadex at pH 8.6 through which metridin passed unadsorbed, being a cationic protein. The purification was completed by two steps of chromatography on the cation-exchange resin Bio-Rex 70 at pH 6.0 in the presence of 5 mM indole. The purified metridin ran as a single band in PAGE at pH 4.

## Biological Aspects

*Metridium senile* is a representative of the phylum Coelenterata, the members of which were amongst the earliest metazoans to appear, and its proteins are accordingly of evolutionary interest. These organisms illustrate a significant stage in the phylogeny of the digestive process, the appearance of the simplest digestive organ, a sac with a single opening, serving alternately as mouth and anus. Sea anemones have been known since ancient times to be carnivorous, and powerful proteolytic activity has been found associated with the gastric filaments of the animals. As described by Gibson & Dixon (1969), an electron microscopic study revealed the presence of dense granules in the cells of the gastric filaments. These granules were very similar in appearance to the zymogen granules of the mammalian pancreas. Taken together with the evidence described above that metridin is stored as a proenzyme that requires activation by a trypsin-like enzyme, it seems that the physiological system for handling of the digestive serine proteinases was already well advanced at the time of divergence of the coelenterates and higher animals.

## References

Gibson, D. & Dixon, G.H. (1969) Chymotrypsin-like proteases from the sea anemone, *Metridium senile*. *Nature* **222**, 753–756.

Stevenson, K.J., Gibson, D. & Dixon, G.H. (1974) Amino acid analyses of chymotrypsin-like proteases from the sea anemone (*Metridium senile*). *Can. J. Biochem.* **52**, 93–100.

*Alan J. Barrett*
*MRC Peptidase Laboratory,*
*The Babraham Institute,*
*Cambridge CB2 4AT, UK*
*Email: alan.barrett@bbsrc.ac.uk*

# 7. *Brachyurins*

## Databanks

*Peptidase classification: clan SA, family S1, MEROPS ID: S01.122*
*NC-IUBMB enzyme classification: EC 3.4.21.32*

*Databank codes:*

| Species | Form | SwissProt | PIR | EMBL (cDNA) | EMBL (genomic) |
|---|---|---|---|---|---|
| Brachyurin | | | | | |
| *Uca pugilator* | I | P00771 | A00958 | U49931 | – |
| Chymotrypsin | | | | | |
| *Penaeus monodon* | I | – | S61558 | – | – |
| *Penaeus monodon* | II | P35002 | S18356 | – | – |
| *Penaeus vannamei* | I | Q00871 | S22075 | X66415 | – |
| | | | S29239 | | |
| *Penaeus vannamei* | II | P36178 | – | – | – |
| Trypsin | | | | | |
| *Astacus fluviatilis* | – | P00765 | A00951 | – | – |
| *Astacus leptodactylus* | – | – | A61327 | – | – |
| *Penaeus monodon* | – | P35050 | S11537 | – | – |
| *Penaeus vannamei* | – | – | S54146 | X86369 | – |
| Others | | | | | |
| *Chionoecetes opilio* | 25 kDa II | P34153 | – | – | – |
| *Chionoecetes opilio* | 35 kDa II | P34154 | – | – | – |
| *Chionoecetes opilio* | 36 kDa | P34155 | – | – | – |
| *Paralithodes camtschatica* | 28 kDa | P20731 | A34817 | – | – |
| *Paralithodes camtschatica* | 36 kDa A | P20732 | B34817 | – | – |
| *Paralithodes camtschatica* | 36 kDa B | P20733 | C34817 | – | – |
| *Paralithodes camtschatica* | 36 kDa C | P20734 | D34817 | – | – |

## Name and History

**Brachyurin** was a term recommended by NC-IUBMB in 1992 for a distinctive type of serine endopeptidase that had been found in crabs and their relatives, the decapod crustacea. The name is reminiscent of the brachyura, the phylogenetic subgrouping of 'true' crabs, a common source of these enzymes (Brusca & Brusca, 1990). It has now become clear, however, that this is a diverse group of enzymes, which need to be further subdivided.

Brachyurins were first discovered in the late 1960s (DeVillez & Buschlen, 1967; Eisen & Jefferey, 1969). Early examples of the family include fiddler crab collagenase I (Eisen *et al*., 1973), crayfish trypsin (Zwilling *et al*., 1969) and shrimp trypsin (Gates & Travis, 1969). Other serine proteases have been isolated from krill (Turkiewicz *et al*., 1991), crab (Grant *et al*., 1983; Klimova *et al*., 1990, 1991), crayfish (Titani *et al*., 1983), shrimp (Lu *et al*., 1990; Tsai *et al*., 1986, 1991; Van Wormhoudt *et al*., 1992; Klein *et al*., 1996) and lobster (Johnston *et al*., 1995). Collagen cleavage is a distinctive characteristic of many of these enzymes. The study of brachyurins has contributed significantly to our understanding of the evolution of serine protease structure and function. Comparative structural and functional analysis indicates that these invertebrate enzymes are closely related to, yet distinct from, their vertebrate and bacterial cousins (Zwilling *et al*., 1975; Titani *et al*., 1983; Tsu *et al*., 1994).

## Activity and Specificity

There are at least three types of brachyurins: (Ia) Broadly specific, with activities similar to those of trypsin (Arg+, Lys+), chymotrypsin (Phe+, Leu+) and elastase (Ala+, Leu+) on synthetic substrates (Grant & Eisen, 1980;

Turkiewicz *et al*., 1991; Tsu *et al*., 1994; Tsu & Craik, 1996); (Ib) broadly specific, but with less trypsin-like activity (Van Wormhoudt *et al*., 1992); and (II) strict trypsin-like specificity (Arg+, Lys+) (Grant *et al*., 1983; Lu *et al*., 1990). Crab collagenase I serves as the paradigm for the type I brachyurins. Substrate structure–activity relationships show that the broad S1 specificity profile of crab collagenase I is novel, as directly compared to trypsin, chymotrypsin or elastase (Tsu & Craik, 1996). In particular, cleavage of Gln- and Leu-containing substrates has been noted for type I brachyurins by several groups (Grant *et al*., 1980; Tsai *et al*., 1991; Tsu *et al*., 1994). Acidic and $\beta$-branched residues are poor substrates. The range in P1 residue discrimination is $10^5$-fold in $k_{cat}/K_m$ (Tsu & Craik, 1996). The S1′ site of these enzymes may possess significant specificity for Arg and Lys and lesser specificity towards hydrophobic residues, while D-Ala, Pro, Asp and Glu are discriminated against (Tsu *et al*., 1994). The range in P1′ residue discrimination is greater than 300-fold, as measured in an acyl transfer reaction (Tsu *et al*., 1994). The S1′ site appears to play an important role in collagen recognition (see below).

Enzymes of types Ia and Ib, such as crab collagenase I and the various shrimp and krill chymotrypsins, have been shown to possess a strong proteolytic activity against native collagen (Welgus *et al*., 1982; Tsai *et al*., 1991; Turkiewicz *et al*., 1991; Van Wormhoudt *et al*., 1992). The initial cleavage event is similar to the 3/4:1/4 length cleavage pattern shown by vertebrate interstitial collagenase (Bornstein & Traub, 1979; Birkedal-Hansen *et al*., 1993; see also Chapter 389). The 3/4:1/4 cleavage is considered to be the signature of a 'true' collagenase, as compared to the nonspecific degradation of collagen by some bacterial proteases. Crab collagenase I cleaves type I collagen at six sites (two each: Gln+Arg, Arg+Gly and Leu+Xaa); five of the six are within

4–17 residues of the metallocollagenase sites (Tsu *et al.*, 1994). The metallocollagenase cleavages are located between residues 775 and 776 of collagen (Gly┼Ile or Gly┼Leu) (Bornstein & Traub, 1979). The similarity of 3/4:1/4 cleavage of collagen by serine and metalloproteases suggests convergent evolution of function. The 3/4:1/4 region of collagen may possess unusual structural features that render it an attractive target for collagenases (Bornstein & Traub, 1979). It is noteworthy that while the activity of type I brachyurins towards synthetic substrates is broad, cleavage of collagen is quite specific. This implies that secondary interactions play a major role in macromolecular substrate recognition by brachyurins (see below). Collagenolytic activity has also been noted for several brachyurin trypsins, but cleavage sites have not yet been identified (Grant *et al.*, 1983; Chen *et al.*, 1991).

The pH optimum of brachyurin is in the weakly alkaline range. A buffer that yields high levels of activity for brachyurin and other serine proteases is 50 mM Tris–HCl, 100 mM NaCl, 20 mM CaCl$_2$, pH 8.0, 25°C (Tsu & Craik, 1996). Sensitive substrates include Suc-Ala-Ala-Pro-(Phe, Leu or Arg)-NHPhNO$_2$ or Suc-Ala-Ala-Pro-(Phe or Arg)-thiobenzyl ester. Single-residue amides are poor substrates for type I brachyurins, but may be useful for brachyurin trypsins. Single-residue esters can be used in either spectroscopic or activity gel assays (Turkiewicz *et al.*, 1991; Tsu *et al.*, 1994). Assays of crude extracts can be complicated by the presence of other serine proteases. The copurification of tryptic, chymotryptic and collagenolytic activities corresponding to a single protein is a hallmark of the type I brachyurins (Eisen *et al.*, 1973; Turkiewicz *et al.*, 1991; Van Wormhoudt *et al.*, 1992).

Brachyurins are inhibited by many small molecule inhibitors (PMSF, DFP) and by protein inhibitors (soybean trypsin inhibitor, bovine pancreatic trypsin inhibitor) of family S1 enzymes. However, inhibition by specific small molecules (Tos-Lys-CH$_2$Cl, Tos-Phe-CH$_2$Cl, etc.) is inconsistent for the type I brachyurins, while brachyurin trypsins generally exhibit trypsin-like inhibition (Tos-Lys-CH$_2$Cl) (Grant & Eisen, 1980; Zwilling & Neurath, 1981; Grant *et al.*, 1983; Tsai *et al.*, 1991; Turkiewicz *et al.*, 1991; Van Wormhoudt *et al.*, 1992).

## Structural Chemistry

As a whole, brachyurins demonstrate a high degree of structural similarity (25–36 kDa, 35–40% amino acid sequence identity) to other members of the trypsin family of serine proteases. However, these invertebrate enzymes have fewer disulfide bonds than their vertebrate cousins (Grant *et al.*, 1980; Titani *et al.*, 1983; Sellos & Van Wormhoudt, 1992), and perhaps for this reason, they are particularly unstable at low pH (Eisen *et al.*, 1973; Tsai *et al.*, 1991). Another distinctive feature is their exceptionally acidic pI values (Eisen *et al.*, 1973; Lu *et al.*, 1990).

At the time of writing, the mature amino acid sequence is known for five brachyurins: crab collagenase I (Grant *et al.*, 1980; Tsu & Craik, 1996), crayfish trypsin (Titani *et al.*, 1983), shrimp chymotrypsins I and II (Sellos & Van Wormhoudt, 1992), and shrimp trypsin (Klein *et al.*, 1996). All of these enzymes except crayfish trypsin are cloned,

and thus provide additional information about the signal and propeptides. Numerous N-terminal sequences have also been reported (see the Databanks table). Most of these partial sequences align well with one or more of the complete sequences, allowing tentative categorization.

The major types of brachyurins can be distinguished on the basis of sequence similarity. The type Ia and Ib brachyurins are very similar (75–95% identical residues), as are the trypsins (75%). However, the sequence identity between these two groupings is no greater than that with other members of family S1 (35%). Although type Ia and Ib brachyurins appear virtually identical in sequence, several changes in the active site cause significant differences in substrate specificity. Type Ia brachyurins such as crab collagenase I and shrimp chymotrypsin II possess significant activity towards basic substrates (Arg and Lys in P1) in addition to activity towards hydrophobic or neutral polar substrates (Phe, Leu, Ala and Gln in P1) (Van Wormhoudt *et al.*, 1992; Tsu & Craik, 1996). The trypsin-like activity is likely due to the presence of a negative charge, the side-chain of Asp226 (chymotrypsinogen numbering), in the S1 binding pocket. Type Ib enzymes such as shrimp chymotrypsin I possess drastically reduced activity toward Arg substrates (40-fold as compared to shrimp chymotrypsin II), while retaining the other aspects of type Ia substrate specificity (Van Wormhoudt *et al.*, 1992). Accordingly, shrimp chymotrypsin I lacks a negative charge in the S1 site, possessing Ala rather than Asp at position 226 (Sellos & Van Wormhoudt, 1992). Mutagenesis studies removing Asp226 in crab collagenase I support this observation (Tsu *et al.*, 1997).

Comparative sequence analysis of brachyurin, bacterial and vertebrate trypsins provides insight into the evolution of this serine protease family (Titani *et al.*, 1983). Conservation in the primary (S1) binding pocket of all trypsins is quite strict, including Asp189, which interacts with the Arg or Lys (P1) substrate side chain. On closer inspection, brachyurin trypsin shares more sequence identity with vertebrate trypsins than with its bacterial counterpart, while important structural features such as disulfide bonds are conserved between the brachyurin and bacterial enzymes. This suggests that brachyurin trypsin occupies an intermediate evolutionary position between bacterial and vertebrate trypsins.

Comparison of the propeptides of these enzymes serves to further delineate the group, as they are of variable length and share little identity. Crab collagenase I, shrimp chymotrypsins I and II, and shrimp trypsin possess propeptides that are longer (29–30 residues) than those of the vertebrate proteases (8–15 residues). The purpose of these large activation domains is unclear, as they are not required for heterologous expression of vertebrate proteases. The activation site of procollagenase, Val-Lys-Ser-Ser-Arg┼Ile-Val-Gly-Gly, is more similar to those of chymotrypsinogen, Ser-Gly-Leu-Ser-Arg┼Ile-Val-Val-Gly, and proelastase, Glu-Thr-Asn-Ala-Arg┼Val-Val-Gly-Gly, which are activated by trypsin, than that of trypsinogen, Asp-Asp-Asp-Asp-Lys┼Ile-Val-Gly-Gly, which is activated by enteropeptidase (Light & Janska, 1989). Interestingly, the activation site of shrimp trypsin, Arg-Gly-Leu-Asn-Lys┼Ile-Val-Gly-Gly, also lacks an enteropeptidase site. This suggests that the vertebrate digestive protease activation cascade has diverged significantly from that of crustaceans. Brachyurins may self-activate or other trypsin-like proteases in the hepatopancreas may perform this function.

The X-ray crystal structure of crab collagenase I has recently been solved to 2.5 Å in complex with the *E. coli* serine protease inhibitor ecotin (Perona *et al.*, 1997; Tsu *et al.*, 1997). Ecotin inhibits a wide range of proteases such as trypsin, chymotrypsin and elastase, by forming a tetrameric 2 ecotin:2 protease complex. As expected from the primary sequence, the tertiary structure of crab collagenase I is globally similar to that of rat trypsin and other vertebrate enzymes (Perona & Craik, 1995; Perona *et al.*, 1997). The conservation of the double $\beta$ barrel core is strict, and the positioning of the catalytic triad (His57, Asp102, Ser195) is identical to that seen in other serine protease–protein inhibitor structures. The remarkable specificity of crab collagenase I for peptidyl and collagen substrates appears to arise from the subtle modification of several surface loops in the active site. These changes affect both the primary (S1) and secondary (S7–S4') substrate-binding sites.

As in trypsin and chymotrypsin, the S1 site of collagenase is formed from the juxtaposition of three $\beta$ strands and is shaped as a deep cylindrical cavity. The negatively charged carboxylate of Asp226 extends across the base of the site, and is well positioned to interact with the positively charged side chains of basic substrates. In both collagenase and trypsin, position 226 is directly across from position 189. The pocket is not as deep as that of trypsin since Asp226 is located closer to the catalytic residues than is Asp189 of trypsin. Its outer lip, formed from amino acids 214–220, is shaped quite differently owing to the insertion of two residues relative to trypsin, after the conserved Gly216. This elongates the mouth of the pocket and provides additional volume. Molecular modeling suggests that the lip of the crab collagenase I binding pocket forms a binding site for hydrophobic P1 residues that is distinct from that for basic and polar P1 residues (Asp226) (Tsu *et al.*, 1997). This observation is corroborated by site-directed mutagenesis of recombinant collagenase. Removal of the negative charge at position 226 (Asp226Gly), and its subsequent restoration at position 189 (Gly189Asp) dramatically reduces and recovers specificity for basic residues (10–100-fold in $k_{cat}/K_m$), while only slightly affecting that for hydrophobic residues (up to 5-fold) (Tsu *et al.*, 1997).

In addition to the S1 site, at least one additional enzyme subsite possesses some amino acid preference. By using a nucleophilic acyl transfer protocol, in which mixtures of peptides compete with water and each other for attack on the acyl enzyme, a significant preference of 20–30-fold was found in favor of Arg and Lys (as compared to Ala) at position P1' (Tsu *et al.*, 1994). Hydrophobic side-chains are also favored at this site, although to a lesser degree. The crystal structure shows that the S1' site forms a shallow hydrophobic cavity arising from the juxtaposition of two surface loops at residues 57–62 and 34–42. Asp60 likely provides specificity for basic residues (Tsu *et al.*, 1997).

The rates of hydrolysis of crab collagenase I and related brachyurins with protein and oligopeptide substrates are similar to those of the vertebrate enzymes, but their activities on small molecule substrates are much less (Grant & Eisen, 1980; Lu *et al.*, 1990; Tsai *et al.*, 1991). This implies that brachyurins may depend more heavily on secondary substrate-binding sites for catalysis than do trypsin or chymotrypsin, for example (Tsu & Craik, 1996; Tsu *et al.*, 1997). This behavior is similar to that of elastase (Tsu & Craik, 1996). The X-ray structure of crab collagenase I clearly illustrates the degree of secondary binding utilized by this enzyme. Binding of the primary inhibitory loop of ecotin to collagenase encompasses over 11 amino acid residues, from P7 to P4'. The equivalent trypsin-ecotin structure shows binding of only the P3–P2' residues (Perona *et al.*, 1997).

A unique feature of crab collagenase I relative to other well-characterized serine proteases is its ability to cleave a triple helical collagen substrate. Clues to the structural basis of this specificity are revealed by the extended nature of the ecotin primary binding site interactions. No other chymotrypsin-like serine protease is known to possess so large a binding interface. Numerous main-chain/main-chain interactions between collagenase and the primary inhibitory loop of ecotin suggest a binding site optimized to accommodate the high Gly and Pro content of collagen. Crab collagenase I appears to accommodate the steric interactions of the collagen triple helix via a well-defined groove on the leaving-group (S') side of the active site. An equivalent groove is not found in other serine proteases. The remarkable correlation of P1–P1' substrate specificity, collagen cleavage and three-dimensional structure of crab collagenase I suggests an enzyme highly adapted for collagen proteolysis (Tsu *et al.*, 1994, 1997; Perona *et al.*, 1997). One may expect similar adaptation in the other type I brachyurins from shrimp, crab and krill. The mechanisms of the reported collagenolytic activity of brachyurin trypsins are not yet well understood.

## Preparation

Brachyurins can be readily purified from whole organisms, digestive fluid or the dissected hepatopancreas. Anion-exchange (DEAE, Q) chromatography is a universal purification step, which takes advantage of the highly acidic pI of brachyurins (Grant *et al.*, 1981; Zwilling & Neurath, 1981; Turkiewicz *et al.*, 1991; Van Wormhoudt *et al.*, 1992). Gel filtration, affinity chromatography on immobilized inhibitors and reverse-phase HPLC have also proven useful (Zwilling & Neurath, 1981; Tsu *et al.*, 1994; Van Wormhoudt *et al.*, 1992). The proenzyme of crab collagenase I has been expressed heterologously in *Saccharomyces cerevisiae* (Tsu & Craik, 1996). The proenzyme is secreted into the medium and then partially purified by DEAE chromatography. Cleavage of the propeptide by agarose-immobilized trypsin releases the mature, active form of the enzyme. The mature collagenase is purified to homogeneity by a final aprotinin affinity chromatography step.

## Biological Aspects

The biologic role of brachyurins is primarily digestive. These enzymes are secreted by the hepatopancreas into the crustacean digestive tract. The unique collagenolytic activity exhibited by these serine proteases likely reflects the diet of these organisms. For example, the common fiddler crab, *Uca pugilator*, inhabits a scavenger niche in the sandy intertidal zone along the eastern seaboard of the United States (Miller, 1961; Williams, 1984). This creature is a deposit feeder, scouring sand grains with specialized mouthparts to recover particles of food (Miller, 1961; Warner, 1977). The diet of the fiddler crab contains mostly plant detritus, as well as some animal matter. This may include debris from marine animals such as fish and molluscs, as well as nematodes. These sources contain large amounts of collagen. It was therefore hypothesized that the digestive enzymes of the fiddler crab may be naturally adapted to degrade this protein, which is generally extremely resistant to proteases (Eisen & Jefferey, 1969). The resulting characterization yielded crab collagenase I, a paradigm of the brachyurin family.

## Distinguishing Features

In summary, some brachyurins are distinguished by their unusually broad substrate specificities, while others are strictly trypsin-like. Brachyurins are inhibited by many small molecule and proteinaceous serine protease inhibitors. However, inhibition by specific small molecules is inconsistent for the type I brachyurins, while brachyurin trypsins generally exhibit trypsin-like inhibition (Grant & Eisen, 1980; Zwilling & Neurath, 1981; Grant *et al.*, 1983; Tsai *et al.*, 1991; Turkiewicz *et al.*, 1991; Van Wormhoudt *et al.*, 1992). Many brachyurins possess significant collagenolytic activity. All brachyurins isolated to date have very acidic pI values. The low isoelectric point of these enzymes relative to other proteins is frequently taken advantage of in their purification.

The abundant natural source, the potential for producing recombinant forms and the unique proteolytic properties of brachyurins will certainly continue to provide insight into the structure and function of this remarkable class of proteases. In addition, the highly specific nature of these enzymes on macromolecular substrates may prove to be particularly useful in future industrial and medical applications.

## Further Reading

The reader is directed the articles of Tsu *et al*. (1994) and Perona & Craik (1995).

## Acknowledgements

This work was supported by a grant from the National Science Foundation MCB-9604379 (to C.S.C.).

## References

Birkedal-Hansen, H., Moore, W., Bodden, M., Windsor, L., Birkedal-Hansen, B., DeCarlo, A. & Engler, J. (1993) Matrix metalloproteinases: a review. *Crit. Rev. Oral Biol. Med.* **4**, 1197–1250.

Bornstein, P. & Traub, W. (1979) The chemistry and biology of collagen. In: *The Proteins*, 3rd edn. New York: Academic Press, pp. 411–632.

Brusca, R.C. & Brusca, G.J. (1990) *Invertebrates*. Sunderland, MA: Sinauer Associates, pp. 595–666.

Chen, Y.L., Lu, P.J. & Tsai, I.H. (1991) Collagenolytic activity of crustacean midgut serine proteases: comparison with the bacterial and mammalian enzymes. *Comp. Biochem. Physiol. [B]* **100**, 763–768.

DeVillez, E.J. & Buschlen, K. (1967) Survey of a tryptic digestive enzyme in various species of crustacea. *Comp. Biochem. Physiol.* **21**, 541–546.

Eisen, A.Z. & Jeffrey, J.J. (1969) An extractable collagenase from crustacean hepatopancreas. *Biochim. Biophys. Acta* **191**, 517–526.

Eisen, A.Z., Henderson, K.O., Jeffrey, J.J. & Bradshaw, R.A. (1973) A collagenolytic protease from the hepatopancreas of the fiddler crab, *Uca pugilator*, purification and properties. *Biochemistry* **12**, 1814–1822.

Gates, R.J. & Travis, J. (1969) Isolation and comparative properties of shrimp trypsin. *Biochemistry* **8**, 4483–4489.

Grant, G.A. & Eisen, A.Z. (1980) Substrate specificity of the collagenolytic serine protease from *Uca pugilator*: studies with non-collagenous substrates. *Biochemistry* **19**, 6089–6095.

Grant, G.A., Henderson, K.O., Eisen, A.Z. & Bradshaw, R.A. (1980) Amino acid sequence of a collagenolytic protease from the hepatopancreas of the fiddler crab, *Uca pugilator*. *Biochemistry* **19**, 4653–4659.

Grant, G.A., Eisen, A.Z. & Bradshaw, R.A. (1981) Collagenolytic protease from fiddler crab (*Uca pugilator*). *Methods Enzymol.* **80**, 722–734.

Grant, G.A., Sacchettini, J.C. & Welgus, H.G. (1983) A collagenolytic serine protease with trypsin-like specificity from the fiddler crab *Uca pugilator*. *Biochemistry* **22**, 2228–2233.

Johnston, D., Hermans, J.M. & Yellowlees, D. (1995) Isolation and characterization of a trypsin from the slipper lobster, *Thenus orientalis* (Lund). *Arch. Biochem. Biophys.* **324**, 35–40.

Klein, B., Le, M.G., Sellos, D. & Van Wormhoudt, A. (1996) Molecular cloning and sequencing of trypsin cDNAs from *Penaeus vannamei* (Crustacea, Decapoda): use in assessing gene expression during the moult cycle. *Int. J. Biochem. Cell Biol.* **28**, 551–563.

Klimova, O.A., Borukhov, S.I., Solovyeva, N.I., Balaevskaya, T.O. & Strongin, A. (1990) The isolation and properties of collagenolytic proteases from crab hepatopancreas. *Biochem. Biophys. Res. Commun.* **166**, 1411–1420.

Klimova, O.A., Vedishcheva, I. & Strongin, A. (1991) Isolation and characteristics of collagenolytic enzymes from the hepatopancreas of the crab *Chionoecetes opilio*. *Dokl. Akad. Nauk. SSSR* **317**, 482–484.

Light, A. & Janska, H. (1989) Enterokinase (enteropeptidase) comparative aspects. *Trends Biochem. Sci.* **14**, 110–112.

Lu, P.J., Liu, H.C. & Tsai, I.H. (1990) The midgut trypsins of shrimp (*Penaeus monodon*). High efficiency toward native protein substrates including collagens. *Biol. Chem. Hoppe-Seyler* **371**, 851–859.

Miller, D.C. (1961) The feeding mechanism of fiddler crabs, with ecological considerations of feeding adaptations. *Zoologica* **46**, 89–101.

Perona, J.J. & Craik, C.S. (1995) Structural basis of substrate specificity in the serine proteases. *Protein Sci.* **4**, 337–360.

Perona, J.J., Tsu, C.A., Craik, C.S. & Fletterick, R.J. (1997) Crystal structure of an ecotin-collagenase complex suggests a model for recognition and cleavage of the collagen triple helix. *Biochemistry* **36**, 5381–5392.

Sellos, D. & Van Wormhoudt, A. (1992) Molecular cloning of a cDNA that encodes a serine protease with chymotryptic and collagenolytic activities in the hepatopancreas of the shrimp *Penaeus vanameii*. (Crustacea, Decapoda). *FEBS Lett.* **309**, 219–224.

Titani, K., Sasagawa, T., Woodbury, R.G., Ericsson, L.H., Dorsam, H., Kraemer, M., Neurath, H. & Zwilling, R. (1983) Amino acid sequence of crayfish *(Astacus fluviatilis)* trypsin If. *Biochemistry* **22**, 1459–1465.

Tsai, I.H., Chuang, K.I. & Chuang, J.L. (1986) Chymotrypsins in digestive tracts of crustacean decapods (shrimps). *Comp. Biochem. Physiol.* **85B**, 235–239.

Tsai, I.H., Lu, P.J. & Chuang, J.L. (1991) The midgut chymotrypsins of shrimps *(Penaeus monodon, Penaeus japonicus and Penaeus penicillatus)*. *Biochim. Biophys. Acta* **1080**, 59–67.

Tsu, C.A. & Craik, C.S. (1996) Substrate recognition by recombinant serine collagenase 1 from *Uca pugilator*. *J. Biol. Chem.* **271**, 11563–11570.

Tsu, C.A., Perona, J.J., Schellenberger, V., Turck, C.W. & Craik, C.S. (1994) The substrate specificity of *Uca pugilator* collagenolytic serine protease 1 correlates with the bovine type I collagen cleavage sites. *J. Biol. Chem.* **269**, 19565–19572.

Tsu, C.A., Perona, J.J., Fletterick, R.J. & Craik, C.S. (1997)

Structural basis for broad substrate specificity of the fiddler crab collagenolytic serine protease 1. *Biochemistry* **36**, 5393–5401.

Turkiewicz, M., Galas, E. & Kalinowska, H. (1991) Collagenolytic serine protease from *Euphausia superba* dana (antarctic krill). *Comp. Biochem. Physiol.* **99B**, 359–371.

Van Wormhoudt, A., Le Chevalier, P. & Sellos, D. (1992) Purification, biochemical characterization and N-terminal sequence of a serine-protease with chymotrypsic and collagenolytic activities in a tropical shrimp, *Penaeus vannamei* (Crustacea, Decapoda). *Comp. Biochem. Physiol. [B]* **103**, 675–680.

Warner, G. (1977) *The Biology of Crabs*. London: Paul Elek.

Welgus, H.G., Grant, G.A., Jeffrey, J.J. & Eisen, A.Z. (1982) Substrate specificity of the collagenolytic serine protease from *Uca pugilator*: studies with collagenous substrates. *Biochemistry* **21**, 5183–5189.

Williams, A. (1984) *Shrimps, Lobsters and Crabs of the Atlantic Coast of the Eastern United States, Maine to Florida*. Washington DC: Smithsonian Institution Press.

Zwilling, R. & Neurath, H. (1981) Invertebrate proteases. *Methods Enzymol.* **80**, 633–664.

Zwilling, R., Pfleiderer, G., Sonneborn, H., Kraft, V. & Stucky, I. (1969) The evolution of the endopeptidases – V. Common and different traits of bovine and crayfish trypsin. *Comp. Biochem. Physiol.* **28**, 1275–1287.

Zwilling, R., Neurath, H., Ericsoon, L.H. & Enfield, D.L. (1975) The amino-terminal sequence of an invertebrate trypsin (crayfish *Astacus leptodactylus*): Homology with other serine proteases. *FEBS Lett.* **60**, 247–249.

*Christopher A. Tsu*
*Department of Pharmaceutical Chemistry,*
*University of California,*
*San Francisco, CA 94143-0446, USA*

*Charles S. Craik*
*Department of Pharmaceutical Chemistry,*
*University of California,*
*San Francisco, CA 94143-0446, USA*
*Email: craik@cgl.ucsf.edu*

# 8. Chymotrypsin

## Databanks

*Peptidase classification: clan SA, family S1, MEROPS ID: S01.152*
*NC-IUBMB enzyme classification: EC 3.4.21.1*
*Chemical Abstracts Service registry number: 9004-07-3*
*Databank codes:*

| Species | SwissProt | PIR | EMBL (cDNA) | EMBL (genomic) |
|---|---|---|---|---|
| Chymotrypsin | | | | |
| *Aedes aegypti* | – | – | U56423 | – |
| *Anopheles gambiae* | – | – | Z18887 | Z32645: complete gene |
| *Anopheles gambiae* | – | – | Z18888 | – |
| *Arenicola marina* | – | – | X95078 | – |

| Species | SwissProt | PIR | EMBL (cDNA) | EMBL (genomic) |
|---------|-----------|-----|-------------|----------------|
| Chymotrypsin (*continued*) | | | | |
| *Bos taurus* | P00766 | A00952 | – | – |
| | | A90235 | | |
| | | A93158 | | |
| | | S29650 | | |
| *Bos taurus* | P00767 | A00953 | – | – |
| *Canis familiaris* | P04813 | A21195 | K01173 | – |
| *Gadus morhua* | P47796 | S43163 | X78490 | – |
| | | S47537 | | |
| *Gadus morhua* | P80646 | – | – | – |
| *Haliotis rufescens* | P35003 | S32750 | X71438 | – |
| | | S35585 | | |
| *Homo sapiens* | P17538 | A31299 | M24400 | – |
| *Lucilia cuprina* | – | | U03760 | – |
| *Manduca sexta* | – | | L34168 | – |
| *Penaeus monodon* | – | S61558 | – | – |
| *Penaeus monodon* | P35002 | S18356 | – | – |
| *Penaeus vannamei* | Q00871 | S22075 | X66415 | – |
| | | | S29239 | |
| *Penaeus vannamei* | P36178 | – | – | – |
| *Rattus norvegicus* | P07338 | A22658 | K02298 | – |
| *Vespa crabro* | P00769 | A00955 | – | – |
| *Vespa orientalis* | P00768 | A00954 | | |
| Hypodermin C | | | | |
| *Hypoderma lineatum* | P08897 | A27802 | X74306 | – |

Brookhaven Protein Data Bank three-dimensional structures:

| Species | ID | Resolution | Notes |
|---------|-----|-----------|-------|
| *Bos taurus* | 1ACB | 2 | complex with eglin C |
| | 1CGI | 2.3 | complex with human pancreatic secretory trypsin inhibitor variant 3 |
| | 1CGJ | 2.3 | complex with human pancreatic secretory trypsin inhibitor variant 4 |
| | 1CHG | 2.5 | precursor |
| | 1CHO | 1.8 | complex with turkey ovomucoid third domain |
| | 1GCD | 1.9 | complex with diethyl phosphoryl |
| | 1GCT | 1.6 | $\gamma$-chymotrypsin A |
| | 1GHA | 2.2 | in 4% isopropanol |
| | 1GHB | 2 | complex with *N*-acetyl-D-tryptophan |
| | 1GMC | 2.2 | $\gamma$-chymotrypsin |
| | 1GMD | 2.2 | $\gamma$-chymotrypsin |
| | 1GMH | 2.1 | complex with diisopropylphosphorofluoridate |
| | 2CGA | 1.8 | precursor |
| | 2CHA | 2 | $\alpha$-chymotrypsin A |
| | 2GCH | 1.9 | $\gamma$-chymotrypsin A |
| | 2GCT | 1.8 | $\gamma$-chymotrypsin A; pH 2.0 |
| | 2GMT[a] | 1.8 | $\gamma$-chymotrypsin A; complex with Ac-Ala-Phe-chloroethane |
| | 3GCH | 1.9 | $\gamma$-chymotrypsin B; complex with *trans-O*-hydroxy-$\alpha$-methyl cinnamate |
| | 3GCT | 1.6 | $\gamma$-chymotrypsin A; pH 10.5 |
| | 4CHA | 1.68 | $\alpha$-chymotrypsin |
| | 4GCH | 1.9 | $\gamma$-chymotrypsin; complex with diethylamino-*O*-hydroxy-$\alpha$-methyl |
| | 5CHA | 1.67 | $\alpha$-chymotrypsin A |
| | 6CHA | 1.8 | $\alpha$-chymotrypsin A; complex with phenylethane boronic acid |
| | 6GCH | 2.1 | $\gamma$-chymotrypsin; complex with Ac-Phe-trifluoromethane |
| | 7GCH | 1.8 | $\gamma$-chymotrypsin; complex with Ac-Leu-Phe-trifluoromethane |
| | 8GCH | 1.6 | $\gamma$-chymotrypsin; complex with Gly-Ala-Trp |

[a]See Chapter 2 for an image constructed from these coordinates.

## Name and History

It has been known for a long time that cattle pancreatic juice possesses the property of digesting proteins (Kühne, 1867). Kühne suggested that this property of the juice was due to the presence of an 'unorganized ferment' or enzyme that he named 'trypsin'. Subsequent studies revealed that the pancreatic juice contains more than just one proteinase, and the name trypsin was retained to designate the one that was thought to be the most important (Northrop & Kunitz, 1932). Historic experiments reviewed by Northrop *et al*. (1939) showed that the freshly secreted pancreatic juice contains proteinases in inactive (proenzyme) forms and that these proenzymes undergo activation upon addition of extracts of the small intestine, or upon standing in slightly acid solution. **Chymotrypsin** was identified as the second major proteinase component of the pancreatic juice (see Northrop *et al*., 1939). Cattle **chymotrypsinogen A**, the precursor form of **chymotrypsin A**, was first crystallized by Kunitz & Northrop (1935). A second crystalline cattle chymotrypsinogen, named **chymotrypsinogen B**, which on activation leads to a crystallizable **chymotrypsin B**, was obtained by Laskowski (1955). Complete amino acid sequences of both forms, chymotrypsinogen A (Hartley, 1964; Meloun *et al*., 1966) and chymotrypsinogen B (Smillie *et al*., 1968) were determined, revealing an 80% sequence identity between the two proteins. They are present in the cattle pancreas in equal quantities, and have similar, though not identical, activation and activity properties (see below). Chymotrypsin cDNAs with apparent sequence similarities to either cattle chymotrypsin A (e.g. human chymotrypsin: Tomita *et al*., 1989) or to cattle chymotrypsin B (e.g. rat chymotrypsin: Bell *et al*., 1984) have been cloned from a variety of species.

Studies on the activation mechanisms of chymotrypsinogen A and on the characterization of a series of molecular forms of the active proteinase have a long and interesting history.

In theory, tryptic cleavage of a single peptide bond, the Arg15┤Ile16 bond of the proenzyme, is sufficient to activate chymotrypsinogen A (or B). The activation mechanism of the proenzyme, however, turned out to be a much more complex one, in fact a combination of activating (by trypsin action) and autolytic events (by chymotrypsin action). Northrop *et al*. (1939) and Jacobsen (1947) activated chymotrypsinogen A with chymotrypsinogen to trypsin ratios of $10^4$ and 30, respectively, giving rise to chymotrypsins of the 'α-type' by slow activation, or the 'γ-type' by fast activation, respectively (Figure 8.1). The α- and γ-chymotrypsins have identical primary structures: the three polypeptide chains of residues 1–13 (the activation peptide), residues 16–147 and residues 149–246 are held together with disulfide bridges. These two forms of chymotrypsin A, however, were shown to have somewhat different enzymatic properties and stabilities (Desnuelle, 1960), and X-ray structures (Cohen *et al*., 1981). Intermediate forms of the two activation mechanisms have the chain compositions as follows: residues 1–146 and 149–246 (neochymotrypsinogen) ['slow' activation], residues 1–15 and 16–246 (π-chymotrypsin), residues 1–13 and 16–246 (δ-chymotrypsin) ['fast' activation] (Neurath, 1957; Desnuelle, 1960).

During the past decades chymotrypsin and related members of the pancreatic serine proteinase family have become favorite models with which to investigate many aspects of enzymology, including the molecular mechanisms of proenzyme activation (Huber & Bode, 1978) and substrate-specific catalysis (Bergmann & Fruton, 1941; Gráf, 1995; Perona & Craik, 1995; Venekei *et al*., 1996a).

## Activity and Specificity

The specificity of chymotrypsin for hydrolysis of peptide bonds formed by the carboxyl groups of Tyr, Phe, Trp

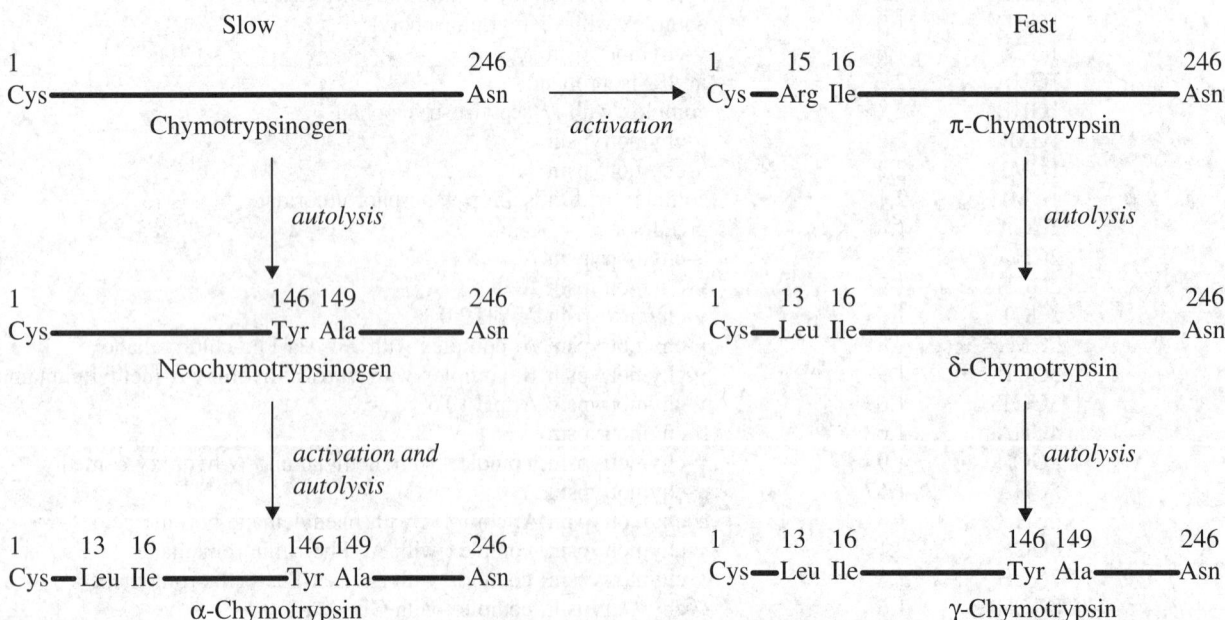

*Figure 8.1* 'Slow' and 'fast' activation of chymotrypsinogen.

and Leu has been recognized for some time (Bergmann & Fruton, 1941; Neurath, 1957; Desnuelle, 1960). As early as the late 1930s, Bergmann & Fruton (1941) realized that chymotrypsin (and trypsin), which were thus far considered as 'proteinases' specifically splitting large proteins, were actually able to cleave 'internal' peptide bonds of short peptides also. Bergmann & Fruton (1941) also found that various ester and amide derivatives of the chymotrypsin-specific amino acids were also hydrolyzed effectively by the enzyme. Chymotrypsin A, either the $\alpha$ or the $\gamma$ form, was investigated most extensively, but some comparative activity and specificity studies were also carried out with other variants. It was agreed that chymotrypsin $A_\pi$ is the most active form, and that further autolysis results in a gradual decrease of the catalytic activity (Neurath, 1957; Desnuelle, 1960). It was also noted that chymotrypsin B, in sharp contrast with chymotrypsin A, splits acyl-tryptophan esters very slowly (Keller et al., 1958). Indeed, our recent comparison of the substrate specificities of cattle chymotrypsin A and recombinant rat chymotrypsin B has revealed that the former is about 100-fold more active on the substrate Suc-Ala-Ala-Pro-Trp-NHMec (see below) than rat chymotrypsin B (P. Hudáky, G. Kaslik, L. Szilágyi, I. Venekei & L. Gráf, unpublished results).

The hydrolysis of amide and ester substrates by chymotrypsin is a three-step process in which an enzyme–substrate complex and an acyl enzyme intermediate are formed (Bender, 1962) (Figure 8.2). The first evidence for this mechanism was reported by Hartley & Kilby (1954) who observed a rapid initial burst in the liberation of $p$-nitrophenol when chymotrypsin was mixed with excess $p$-nitrophenyl acetate or $p$-nitrophenyl ethyl carbonate. They postulated that initially the ester rapidly acylated the enzyme in a mole-to-mole ratio, and that the rate of subsequent substrate turnover was limited by the slow hydrolytic deacylation of the enzyme. The existence of the acyl enzyme intermediate was ultimately proven by the isolation and crystallization of several stable forms such as indolylacryloyl-chymotrypsin (Henderson, 1970), tosyl-chymotrypsin (2CHA; Henderson et al., 1972) and two photoreversible cinnamoyl-chymotrypsins (Stoddard et al., 1990a). This later work is especially interesting because, due to the special structure of the bound inhibitor, light-induced cis–trans isomerization increases the rate of deacylation by several orders of magnitude. Photoirradiaton of the inhibited chymotrypsin crystals triggers deacylation, so that the process can be directly studied by X-ray crystallography (Stoddard et al., 1990b). Furthermore, the formation of acyl enzyme intermediates in the pathway of amide hydrolysis was also deduced by nucleophile partitioning experiments (Fastrez & Fersht, 1973). Recently, careful analysis of the X-ray structure of $\gamma$-chymotrypsin has revealed that this form is an acyl enzyme complex of $\alpha$-chymotrypsin with its autolysis product (Dixon & Matthews, 1989: structure 1GCT; Dixon et al., 1991: structures 2GCT, 3GCT; Harel et al., 1991: structure 8GCH). In hexane, the tetrahedral intermediate of the reaction was also observed (Yennawar et al., 1994: 1GCM).

Besides their theoretical importance, burst substrates offer a simple and convenient way to determine the concentration of active enzyme by active site titration. Initially, various $p$-nitrophenyl esters were used for this purpose, but more recently, substances with fluorogenic leaving groups, such as 4-methyl-umbelliferyl $p$-($N,N,N$-triethylammonium) cinnamate (Jameson et al., 1973) have been developed, which provide a significantly higher sensitivity (20–50 pmol). A recently described active-site titration of chymotrypsin with $\alpha_2$-macroglobulin and HPLC may estimate proteinase activity with as little as 5–10 pmol enzyme (Osada et al., 1992).

In the course of sequence analysis studies, chymotryptic digests of many peptides and proteins have been examined in detail. In addition to hydrolysis at aromatic amino acids and leucine, hydrolysis of bonds formed by asparagine, cysteic acid, glutamine, glycine, histidine, isoleucine, lysine, serine, threonine and valine have also been found. Cleavage at these sites was not extensive, however, and depended strongly on the P1' residue (Hill, 1965, and references therein). Chymotrypsin does not hydrolyze bonds formed to the imino group of proline in proteins. Schellenberger et al. (1991) performed quantitative structure/activity analysis of all available quantitative data on the chymotrypsin-catalyzed hydrolysis of amino acid and short peptide substrates. The substrates in the database span a range from the P5 to P3' positions. It was found that parameters for the P5, P4 and P3' subsites could be omitted from the calculations, which implied that interactions at these sites did not contribute significantly to the catalytic efficiency of chymotrypsin. The calculated P1 contributions to $\log(k_{cat}/K_m)$ gave linear correlation to the molar refraction of the P1 side chain. (Molar refraction is the measure of the volume and polarizability of a substituent.) Substrate specificity mapping using all 400 possible dipeptides in membrane-bound form gave similar results (Duan & Laursen, 1994).

The most common and convenient analytical technique for activity determination is spectrophotometry with $p$-nitroaniline as leaving group. Commercially available substrates include Suc-Phe-NHPhNO$_2$ (Schechter & Berger, 1967) and Suc-Ala-Ala-Pro-Phe-NHPhNO$_2$ (DelMar et al., 1979). Greater sensitivity can be achieved either by a fluorogenic leaving group (e.g. Suc-Ala-Ala-Pro-Phe-NHMec), with highly reactive thiobenzyl esters (Suc-Ala-Ala-Pro-Phe-SBzl; Harper et al., 1981) or with bioluminescence substrates like 6-($N$-acetyl-L-phenylalanyl)-aminoluciferin (Monsees et al., 1994).

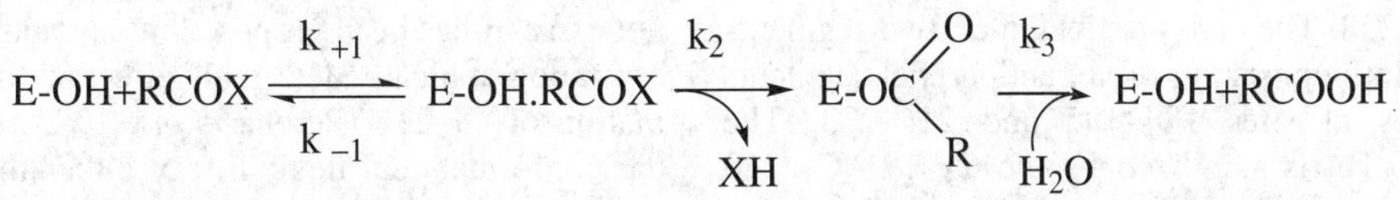

$$\text{E-OH+RCOX} \underset{k_{-1}}{\overset{k_{+1}}{\rightleftharpoons}} \text{E-OH.RCOX} \overset{k_2}{\underset{\text{XH}}{\longrightarrow}} \text{E-OC}\overset{\text{O}}{\underset{\text{R}}{<}} \overset{k_3}{\underset{\text{H}_2\text{O}}{\longrightarrow}} \text{E-OH+RCOOH}$$

Figure 8.2 Kinetic scheme of chymotrypsin action. In a rapid equilibrium process ($K_s = k_{+1}/k_{-1}$) the enzyme forms a Michaelis complex with an amide or ester substrate which is cleaved in the next step (characterized by rate constant $k_2$) to form an acyl-enzyme intermediate and an amide or alcohol. The rate constant of the hydrolysis of the acyl-enzyme is $k_3$.

Proline is commonly present at the P2 position in commercially available synthetic substrates (e.g. Suc-Ala-Ala-Pro-Phe-NHPhNO$_2$). Depending on pH and temperature, about 85% of the Ala-Pro bond is *trans*, providing a good substrate for chymotrypsin, whereas the *cis* form is not a substrate. The activity of peptidylprolyl *cis–trans* isomerase can therefore be measured in a coupled assay with chymotrypsin (Fisher *et al.*, 1984; Lang & Schmid, 1988).

## Structural Chemistry

Three amino acid residues, His57, Asp102 and Ser195, the catalytic triad, are essential for peptide bond cleavage. They are located at the entrance of a substrate-binding pocket and their conservative arrangement is stabilized by hydrogen bonds. These three amino acids are highly conserved in the sequences of the peptidases of family S1 (Chapter 2). A serine at position 214 is also highly conserved in the family, being present in all but three of almost 200 homologous bacterial and mammalian serine proteinases (Marquart *et al.*, 1983; Ohara *et al.*, 1989; Bazan & Fletterick, 1990). Ser214 is hydrogen bonded to Asp102 and to the main-chain nitrogen atom of the scissile bond in the substrate. This residue contributes significantly to the polar environment that stabilizes the charge of the buried Asp102. A mutagenesis study performed on trypsin supported the importance of the Ser214 function. This result and the invariance of this residue throughout the family led to the proposal that this residue might be considered as a fourth member of the catalytic triad (McGrath *et al.*, 1992). The peptide amido groups of Gly193 and Ser195, in the structural unit called the oxyanion hole, have important hydrogen bond donating interactions with the carbonyl group of the scissile peptide bond. These interactions are indispensable for catalysis since they orient the scissile bond NH to His57 and Ser195 (Blow, 1976), initiating the formation of the tetrahedral intermediate. In the course of this process they dissipate the negative charge developing on the scissile bond carbonyl oxygen (Rühlmann *et al.*, 1973; Huber & Bode, 1978).

The interactions between the S$n$,..S2, S1, S1′, S2′,..S$n$′ sites of the enzyme and the P$n$,..P2, P1, P1′, P2′,..P$n$′ amino acid residues of the substrate ensure the precise alignment of the substrate to the catalytic triad and the oxyanion hole, and thereby the selectivity of catalysis. The contributions of these subsite interactions to the substrate discrimination, as also discussed in section **Activity and Specificity**, is different (Schellenberger *et al.*, 1991). The primary specificity determinant is the P1–S1 interaction. S1 is a deep pocket-like structure and, in contrast to the situation in most enzymes, the subsite is bounded primarily by main-chain atoms. Amino acids in the pocket that are closer than 5 Å to substrate atoms are at sites 189–195, 214–220 and 225–228. They are distributed on two segments of a $\beta$-barrel structure consisting of six antiparallel segments and on two loops at sites 184–195 and 216–223. The structure of the pocket is stabilized by the Cys191–Cys220 disulfide bond and by a surrounding network of 11 hydrogen bonds between the amino acid residues that turn out of the pocket and some others in the two loops. Three

water molecules that are located in the pocket close to its entrance join loosely to this hydrogen bonding network. On substrate binding, these water molecules are expelled by the (aromatic) P1 residue that by making van der Waals contacts becomes sandwiched between peptide bonds 190–191–192 and 215–216 (when P1 = tryptophan; Steitz *et al.*, 1969), or 191–192 and 215–216 (when P1 = phenylalanine; Brady *et al.*, 1990). The Ala226Gly substitution in rat chymotrypsin (a B-type enzyme, see section **Name and History**) resulted in a mutant with chymotrypsin A-like substrate specificity (P. Hudáky, G. Kaslik, L. Szilágyi, I. Venekei & L. Gráf, unpublished results). Thus, the substrate specificity difference between chymotrypsins A and B (see section **Activity and Specificity**), a 100-fold higher activity of the former on P1 = tryptophan substrates, might be due to the size difference of the side chain of residue 226.

Though the X-ray structures of many chymotrypsins and chymotrypsin–inhibitor complexes are known, recent mutagenesis studies clearly indicate that the mechanisms by which chymotrypsin and trypsin achieve substrate-specific catalysis are far from being fully understood. The specificity subsites of chymotrypsin have essentially the same geometry as those of trypsin and elastase. The comparison of the structures of the trypsin and chymotrypsin S1 sites revealed only one significant difference, at amino acid position 189, suggesting a simple mechanism for the substrate discrimination by these two enzymes (Steitz *et al.*, 1969): Asp189 in trypsin provides charge compensation for the positively charged P1 residue of the substrate, and increases the polarity of the pocket. In contrast, in chymotrypsin, where Ser replaces Asp189, the substrate-binding pocket is relatively nonpolar (Dorovskaya *et al.*, 1972). However, exchange of the amino acids at this site by directed mutagenesis of trypsin and chymotrypsin failed to interchange the specificities of the two proteinases. The trypsin mutant Asp189Ser turned out to be a poor, nonspecific proteinase (Gráf *et al.*, 1988), the tryptic specificity of which, however, could be rescued by acetate ions (Perona *et al.*, 1994). The reverse mutant, Ser189Asp chymotrypsin, remained a chymotrypsin-like proteinase (Venekei *et al.*, 1996b), that has a reduced pH optimum of pH 6.0 relative to the value of pH 9.0 for the native enzyme (G. Kaslik, I. Venekei, L. Szilágyi, G. Szabó, W. Rutter & L. Gráf, unpublished results). In addition to rational modifications in the substrate-binding pocket, the substitutions of 15 other amino acids, some at sites distant from the substrate-binding cavity, were required to convert the specificity of trypsin into that of chymotrypsin (Hedstrom *et al.*, 1992, 1994a,b; Perona *et al.*, 1995). In contrast, substitutions in chymotrypsin in the reverse direction resulted only in nonspecific enzymes of very low activity (Venekei *et al.*, 1996b). It is clear from these studies that remote elements of the enzyme structure strongly influence substrate specificity, though the mechanism(s) by which they do so are not clear. It has been proposed that in addition to some X-ray crystallographically detectable effect on the backbone conformation of Gly216 (Perona *et al.*, 1995) they may modulate the conformational flexibility of either the substrate-binding site or the catalytic apparatus or both (Gráf, 1995; Perona & Craik, 1995). This conclusion is consistent with earlier results from proton magnetic resonance pH titration studies of the active site residue His57, which indicated that trypsin and trypsinogen have more rigid structures than chymotrypsin and chymotrypsinogen (for review see Markley, 1979). On the basis of such a dynamic model of the mechanism of substrate-specific binding and catalysis by this class of enzymes, it is not entirely surprising that the specific changes that were used for the trypsin to chymotrypsin conversion cited above do not work in reverse.

It will be clear from the above discussion that the trypsin- and chymotrypsin-specific networks of interactions involved in substrate-specific catalysis remain to be defined. A candidate for such a function is the so called 'activation domain' of these proteases (Gráf, 1995). Four adjacent peptide segments that are involved in the formation of the substrate-binding pocket are differently organized and more flexible in the proenzyme, rendering it inactive. Cleavage of the Arg15-Ile16 bond by trypsin during activation, and the concomitant formation of a salt bridge between the liberated $NH_2$ group of Ile16 and the carboxylate group of Asp194, initiates mutual conformational changes in the four neighboring segments (the 'activation domain': Ile16 to Gly19, Gly142 to Asp153, Gly184 to Asp194 and Gly216 to Ser223) to yield the active form of the enzyme (Sigler *et al.*, 1968; Wang *et al.*, 1985). Though the activation process for trypsin has features in common with that of chymotrypsin, the two mechanisms are considerably different (Huber & Bode, 1978). Just as the activation processes of chymotrypsin and trypsin differ, the resultant mutually stabilizing interactions between the four segments in the two enzymes are also different. It follows from this that conformational plasticities of the activation domains of chymotrypsin and trypsin, as opposed to the static structures, should also be different (Gráf, 1995). In this regard it is worth mentioning that the elimination of the disulfide bond Cys191-Cys220 from trypsin and chymotrypsin (constituent of the activation domains and the substrate-binding pockets of both enzymes) has different effects on their substrate specificities (Várallyay *et al.*, 1997).

## Preparation

The purification from natural sources usually consists of two major steps. It is the proenzyme which is prepared first by acidic extraction, ammonium sulfate precipitation and crystallization (Laskowski, 1955), or by acetone precipitation and ion-exchange chromatography (Wilcox, 1970). Then, after the activation of the proenzyme, the different forms of chymotrypsin can be purified by ion-exchange (Wilcox, 1970) or affinity chromatography (Schellenberger *et al.*, 1993; Venekei *et al.*, 1996b). For the preparation of cloned chymotrypsin, the expression, purification and activation of chymotrypsinogen is also the most convenient route. Rat chymotrypsinogen expressed in a yeast system is secreted into the culture medium on a scale of milligrams per liter (Schellenberger *et al.*, 1993; Venekei *et al.*, 1996b).

## Biological Aspects

The proteolytic enzymes of the digestive tract, including chymotrypsin, trypsin (see Chapter 3) and elastase (see Chap-

ters 11 and 12), are produced in inactive forms by the acinar cells of the pancreas, and they are carried as such by the pancreatic juice into the duodenum where they are activated. The initial step of a complicated activation process is the activation of trypsinogen by enteropeptidase (Chapter 14). All the proenzymes in the digestive tract, chymotrypsinogen, proelastase and procarboxypeptidase A, are then activated by trypsin. Like the activation of these proteases, their breakdown and inactivation are also regulated by limited proteolysis. In the case of chymotrypsin(ogen) at least two kinds of proteolytic activities are involved in the activation/inactivation of the enzyme: one from trypsin and the other from the accumulating chymotrypsin (see section **Name and History** also). Under physiological conditions there may be sufficient trypsin to activate chymotrypsinogen rapidly enough to bypass the autolysis of chymotrypsin and the breakdown of chymotrypsinogen by chymotrypsin. As discussed earlier (see section **Name and History**), 'fast' activation of chymotrypsinogen (Jacobsen, 1947), the process we expect in the duodenum, leads to the formation of $\gamma$-type chymotrypsin. Autolysis of chymotrypsin and its possible degradation by other proteases represents one of the physiologic mechanisms for the inactivation of chymotrypsin in the small intestine. The other mechanism to regulate the activity of serine proteases is their inhibition by pancreatic protease inhibitors and serpins. It was recently suggested from our laboratory (Kaslik *et al.*, 1995, 1997) and another laboratory (Plotnick *et al.*, 1996) that serpins, when covalently bound to serine proteases, convert them into an inactive, loose structure that serves as a 'conformational trap' of the enzyme, preventing catalytic deacylation. It is also suggested that this trap mechanism could be general for inhibitory serpins and that it may facilitate the degradation of the target proteases *in vivo*.

## Distinguishing Features

Pancreatic serine proteases are expressed as proenzymes with N-terminal propeptides of different lengths. These propeptides, unlike those of subtilisin and $\alpha$-lytic protease, which are inhibitory for the correctly folded enzymes (Abita *et al.*, 1969), prevent correct folding of the substrate-binding pocket and the oxyanion hole of the pancreatic proteases (see section **Structural Chemistry**). After the proenzymes reach the duodenum, where their activity is required, the propeptides are clipped off by enteropeptidase (for trypsinogen) or by trypsin (for chymotrypsinogen and proelastase). The chymotrypsin structure is different from that of trypsin in that its 15 amino acid propeptide remains linked to the enzyme through a disulfide bridge between Cys1 and Cys122. To explore the structural and functional significance of this disulfide bond, chymotrypsinogen mutants lacking the bridge, wild-type chymotrypsinogen, a chymotrypsin/trypsin propeptide chimera and wild-type trypsinogen were expressed in yeast and thoroughly characterized (Venekei *et al.*, 1996a). The conclusion of this study is that the disulfide bridge Cys1-Cys122 of chymotrypsinogen rather than the propeptide sequence itself plays a crucial role in keeping the proenzyme stable against nonspecific activation. The disulfide-linked propeptide in the active enzyme, however, does not seem to affect the activity and specificity of chymotrypsin. A comparison of proenzyme stabilities showed that the trypsinogen propeptide is about 10 times more effective than the chymotrypsinogen propeptide in preventing nonspecific proenzyme activation during heterologous expression and secretion from yeast (Venekei *et al.*, 1996a).

A curious feature of $\alpha$-chymotrypsin, in which it differs from $\gamma$-chymotrypsin and trypsin, is its ability to undergo self-association (Timasheff, 1969, and references therein). Sigler *et al.* (1968) suggested that the presence of a dyad axis of symmetry between $\alpha$-chymotrypsin molecules in crystals is related to the ability of these molecules to dimerize in solution. Based on the pH dependence of the dimerization of $\alpha$-chymotrypsin below pH 5.5, it was proposed that the most important intermolecular interactions are between Tyr146 and His57 (Aune & Timasheff, 1971). In this regard it is interesting to note that the mutation of a single residue of rat trypsin, Asp189 to Ser (Gráf *et al.*, 1988, see section **Structural Chemistry**), leads to significant changes of the X-ray structure of the enzyme uncomplexed with inhibitor. These include the appearance of chymotrypsin-like structural features of some parts of its activation domain, and (perhaps as a consequence) dimerization of the mutant molecules in a similar fashion to those of $\alpha$-chymotrypsin (E. Szabó, Z. Böcskei, G. Náray-Szabó & L. Gráf, unpublished results).

## Further Reading

Particularly recommended are the publications of Kühne (1867), Northrop *et al.* (1939) and Perona & Craik (1995).

## References

Abita, J.P., Delaage, M., Lazdunski, M. & Savrada, J. (1969) *Eur. J. Biochem.* **8**, 314–324.

Aune, K.C. & Timasheff, S.N. (1971) Dimerization of $\alpha$-chymotrypsin. I. pH dependence in the acid region. *Biochemistry* **10**, 1609–1616.

Bazan, J.F. & Fletterick, R.J. (1990) Structural and catalytic models of trypsin-like viral proteases. *Semin. Virol.* **1**, 311–322.

Bell, G.I., Quinto, C., Quiroga, M., Valenzuela, P., Craik, C.S. & Rutter, W.J. (1984) Isolation and sequence of a rat chymotrypsin B gene. *J. Biol. Chem.* **259**, 14265–14270.

Bender, M.M. (1962) The mechanism of $\alpha$-chymotrypsin catalyzed hydrolyses. *J. Am. Chem. Soc.* **84**, 2582–2590.

Bergmann, M. & Fruton, J.S. (1941) The specificity of proteinases. *Adv. Enzymol.* **1**, 63–98.

Blow, D.M. (1976) Structure and mechanism of chymotrypsin. *Acc. Chem. Res.* **9**, 145–152.

Brady, K., Wei, A., Ringe, D. & Abeles, R.H. (1990) Structure of chymotrypsin-trifluoromethyl ketone inhibitor complexes: comparison of slowly and rapidly equilibrating inhibitors. *Biochemistry* **29**, 7600–7607.

Cohen, G.H., Silverton, E.W. & Davies, D.R. (1981) Refined crystal structure of $\gamma$-chymotrypsin at 1.9 Å resolution. *J. Mol. Biol.* **148**, 449–479.

Del Mar, E.G., Largman, C., Brodrick, J.W. & Geokas, M.C. (1979) A sensitive new substrate for chymotrypsin. *Anal. Biochem.* **99**, 316–320.

Desnuelle, P. (1960) Chymotrypsin. In: *The Enzymes*, vol. 4 (Boyer, P.D., Lardy, H. & Myrback, K., eds). New York: Academic Press, pp. 93–118.

Dixon, M.M. & Matthews, B.W. (1989) Is $\gamma$-chymotrypsin a tetrapeptide acyl-enzyme adduct of $\alpha$-chymotrypsin? *Biochemistry* **28**, 7033–7038.

Dixon, M.M., Brennan, R.G. & Matthews, B.W. (1991) Structure of $\gamma$-chymotrypsin in the range pH 2.0 to pH 10.5 suggests that $\gamma$-chymotrypsin is a covalent acyl-adduct at low pH. *Int. J. Biol. Macromol.* **13**, 86–96.

Dorovskaya, V.N., Varfolomeyev, S.D., Kazanskaya, N.F., Klyosov, A.A. & Martinek, K. (1972) The influence of the geometric properties of the active centre on the specificity of $\alpha$-chymotrypsin catalysis *FEBS Lett.* **23**, 122.

Duan, Y. & Laursen, R.A. (1994) Proteinase substrate specificity mapping using membrane-bound peptides. *Anal. Biochem.* **216**, 431–438.

Erlanger, B.F. & Edel, F. (1964) The utilization of a specific chromogenic inactivator in an 'all or none' assay for chymotrypsin. *Biochemistry* **3**, 346–349.

Fastrez, J. & Fersht, A.R. (1973) Demonstration of the acyl-enzyme mechanism for the hydrolysis of peptides and anilides by chymotrypsin. *Biochemistry* **12**, 2025–2034.

Fisher, G., Bang, H., Berger, E. & Schellenberger, A. (1984) Conformational specificity of chymotrypsin toward proline containing substrates. *Biochem. Biophys. Acta* **791**, 87–97.

Gráf, L. (1995) Structural basis of serine protease action: the fourth dimension. In: *Natural Sciences and Human Thought* (Zwilling, R., ed.). Berlin, Heidelberg: Springer-Verlag, pp. 139–148.

Gráf, L., Jancsó, Á., Szilágyi, L., Hegyi, Gy., Pintér, K., Náray-Szabó, G., Hepp, J., Medzihradszky, K. & Rutter, W.J. (1988) Electrostatic complementarity within the substrate-binding pocket of trypsin. *Proc. Natl Acad. Sci. USA* **85**, 4961–4965.

Harel, M., Tsu, C.T., Frolow, F., Silman, I. & Sussman, J.L. (1991) $\gamma$-Chymotrypsin is a complex of $\alpha$-chymotrypsin with its own autolysis products. *Biochemistry* **30**, 5217–5225.

Harper, J.W., Ramirez, G., & Powers, J.C. (1981) Reaction of peptide thiobenzyl esters with mammalian chymotrypsinlike enzymes. A sensitive assay method. *Anal. Biochem.* **118**, 382–387.

Hartley, B.S. (1964) Amino-acid sequence of bovine chymotrypsinogen A. *Nature* **201**, 1284–1287.

Hartley, B.S. & Kilby, B.A. (1954) The reaction of p-nitrophenyl esters with chymotrypsin and insulin. *Biochem. J.* **56**, 288–297.

Hedstrom, L., Szilágyi, L. & Rutter, W.J. (1992) Converting trypsin to chymotrypsin: The role of surface loops. *Science* **255**, 1249–1253.

Hedstrom, L., Perona, J.J. & Rutter, W.J. (1994a) Residue 172 is a substrate specificity determinant. *Biochemistry* **33**, 8757–8763.

Hedstrom, L., Farr Jones, S., Kettner, C.A. & Rutter, W.J. (1994b) Converting trypsin to chymotrypsin: ground state binding does not determine substrate specificity. *Biochemistry* **33**, 8764–8769.

Henderson, R. (1970) Structure of crystalline $\alpha$-chymotrypsin. IV. The structure of indoleacryloyl-$\alpha$-chymotrypsin and its relevance to the hydrolytic mechanism of the enzyme. *J. Mol. Biol.* **54**, 341–354.

Henderson, R., Wright, C.S., Hess, G.P. & Blow, D.M. (1972) $\alpha$-Chymotrypsin: what can we learn about catalysis from X-ray diffraction? *Cold Spring Harb. Symp. Quant. Biol.* **36**, 63–70.

Hill, R.L. (1965) Hydrolysis of proteins. In: *Advances in Protein Chemistry*, vol. 20 (Anfinsen, C.B., Anson, M.L., Edsall, J.T. & Richards, F.M., eds). New York: Academic Press, pp. 68–74.

Huber, R. & Bode, W. (1978) Structural basis of the activation and action of trypsin. *Acc. Chem. Res.* **11**, 114–122.

Jacobsen, C.F. (1947) *C. R. trav. lab. Carlsberg Ser. chim.* **25**, 325.

Jameson, G.W., Roberts, D.V., Adams, R.W., Kyle, W.S.A. & Elmore, D.T. (1973) Determination of the operational molarity of solutions of bovine $\alpha$-chymotrypsin, trypsin, thrombin and factor Xa by spectrofluorimetric titration. *Biochem. J.* **131**, 107–117.

Kaslik, G., Patthy, A., Bálint, M. & Gráf, L. (1995) Trypsin complexed with $\alpha_1$-proteinase inhibitor has an increased structural flexibility. *FEBS Lett.* **370**, 179–183.

Kaslik, G., Kardos, J., Szabó, E., Szilágyi, L., Závodszky, P., Westler, W.M., Markley, J.L. & Gráf, L. (1997) Effects of serpin binding on the target proteinase: global stabilization, localized increased structural flexibility, and conserved hydrogen bonding at the active site. *Biochemistry* **36**, 5455–5464.

Keil, B. (1992) *Specificity of Proteolysis*. Springer-Verlag, Berlin.

Keller, P.J., Cohen, E. & Neurath, H. (1958) The proteins of bovine pancreatic juice. *J. Biol. Chem.* **233**, 344–349.

Kühne, W. (1867) Über die Verdauung der Eiweisstoffe durch den Pankreassaft. *Virchows Arch.* **39**, 130 (reproduced by Gutfreund, H. (1976) *FEBS Lett.* **62**(suppl.), E1–E12).

Kunitz, M. & Northrop, J.H., (1935) Crystalline chymotrypsin and chymotrypsinogen. I. Isolation, crystallization, and general properties of a new proteolytic enzyme and its precursor. *J. Gen. Physiol.* **18**, 433.

Lang, K. & Schmid, F.X. (1988) Protein-disulphide isomerase and prolyl isomerase act differently and independently as catalysts of protein folding. *Nature* **331**, 453–455.

Laskowski, M. (1955) Chymotrypsinogens and chymotrypsins. *Methods Enzymol.* **11**, 8–26.

Markley, J.L. (1979) Catalytic groups of serine proteinases. NMR investigations. In: *Magnetic Resonance Studies in Biology* (Shulman, R.G., ed.). New York: Academic Press, pp. 397–461.

Marquart, M., Walter, J., Deisenhofer, J., Bode, W. & Huber, R. (1983) *Acta Crystallogr.* **B39**, 480–490.

McGrath, M., Vásquez, J.R., Craik, C.S., Yang, A.S., Honig, B. & Fletterick, R.J (1992) Perturbing the polar environment of Asp 102 in trypsin: Consequences of replacing conserved Ser214. *Biochemistry* **31**, 3059–3064.

Meloun, B., Kluh, I., Kostka, V., Morávek, L., Prušik, Z., Vaneček, J., Keil, B. & Šorm, F. (1966) Covalent structure of bovine chymotrypsinogen A. *Biochim. Biophys. Acta* **130**, 543–545.

Monsees, T., Miska, W. & Geiger, R. (1994) Synthesis and characterization of a bioluminogenic substrate for $\alpha$-chymotrypsin. *Anal. Biochem.* **221**, 329–334.

Neurath, H. (1957) The activation of zymogens. In: *Advances in Protein Chemistry*, vol. 12 (Anfinsen, C.B., Anson, M.L., Bailey, K. & Edsall, J.T., eds). New York: Academic Press, pp. 319–386.

Northrop, J.H. & Kunitz, M. (1932) Crystalline trypsin. I. Isolation and tests of purity. *J. Gen. Physiol.* **16**, 267.

Northrop, J.H., Kunitz, M. & Herriott, R.M. (1939) *Crystalline Enzymes*. New York: Columbia University Press.

Ohara, T., Makino, K., Shinagawa, H., Nakata, A., Norioka, S. & Sakiyama, S. (1989) Cloning, nucleotide sequence, and expression of Achromobacter proteinase I gene. *J. Biol. Chem.* **264**, 20625–20631.

Osada, T., Ookata, K., Athauda, S.B., Takahashi, K. & Ikai, A. (1992) The active site titration of proteinases by using $\alpha_2$-macroglobulin and high-performance liquid chromatography. *Anal. Biochem.* **207**, 76–79.

Perona, J.J. & Craik, C.S. (1995) Structural basis of substrate specificity in the serine proteinases *Protein Sci.* **4**, 337–360.

Perona, J.J., Hedstrom, L., Wagner, R.L., Rutter, W.J., Craik, C.S. & Fletterick, R.J. (1994) Exogenous acetate reconstitutes the enzymatic activity of Asp189Ser trypsin. *Biochemistry* **33**, 3252–3259.

Perona, J.J., Hedstrom, L., Rutter, W.J. & Fletterick, R.J. (1995) Structural origins of substrate discrimination in trypsin and chymotrypsin. *Biochemistry* **34**, 1489–1499.

Plotnick, M.I., Mayne, L., Schechter, N.M. & Rubin, H. (1996) Distortion of the active site of chymotrypsin complexed with serpin. *Biochemistry* **35**, 7586–7590.

Rühlmann, A., Kukla, D., Schwager, P., Bartles, K. & Huber, R. (1973) Structure of the complex formed by bovine trypsin and bovine pancreatic trypsin inhibitor. *J. Mol. Biol.* **77**, 417–436.

Schechter, I. & Berger, A. (1967) On the size of the active site in proteases I. Papain. *Biochem. Biophys. Res. Commun.* **27**, 157–162.

Schellenberger, V., Braune, K., Hofman, H.-J. & Jakubke, H.-D. (1991) The specificity of chymotrypsin. A statistical analysis of hydrolysis data. *Eur. J. Biochem.* **199**, 623–636.

Schellenberger, V., Truck, C.W., Hedstrom, L. & Rutter, W.J. (1993) Mapping the SN subsites of serine proteinases using acyl transfer to mixtures of peptide nucleophiles. *Biochemistry* **32**, 4349–4353.

Sigler, P.B., Blow, D.M., Matthews, B.W. & Henderson, R. (1968) Structure of crystalline $\alpha$-chymotrypsin II. *J. Mol. Biol.* **35**, 143–164.

Smillie, L.B., Furka, A., Nagabhushan, N., Stevenson, K.J. & Parkes, C.O. (1968) Structure of chymotrypsinogen B compared with chymotrypsinogen A and trypsinogen. *Nature* **218**, 343–346.

Steitz, T.A., Henderson, R. & Blow, D.M. (1969) Structure of crystalline $\alpha$-chymotrypsin. *J. Mol. Biol.* **46**, 337–348.

Stoddard, B.L., Bruhnke, J., Porter, N., Ringe, D. & Petsko, G.A. (1990a) Structure and activity of two photoreversible cinnamates bound to chymotrypsin. *Biochemistry* **29**, 4871–4879.

Stoddard, B.L., Bruhnke, J., Koenig, P., Porter, N., Ringe, D. & Petsko, G.A. (1990b) Photolysis and deacylation of inhibited chymotrypsin. *Biochemistry* **29**, 8042–8051.

Timasheff, S.N. (1969) On the mechanism of $\alpha$-chymotrypsin dimerization. *Arch. Biochem. Biophys.* **132**, 165–169.

Tomita, N., Izumoto, Y., Horii, A., Doi, S., Yokouchi, H., Ogawa, M., Mori, T. & Matsubara, K. (1989) Molecular cloning and nucleotide sequence of human pancreatic prechymotrypsinogen cDNA. *Biochem. Biophys. Res. Commun.* **158**, 569–575.

Várallyay, I., Lengyel, Z., Gráf, L. & Szilágyi, L. (1997) The role of disulfide bond C191-C220 in trypsin and chymotrypsin. *Biochem. Biophys. Res. Commun.* **230**, 592–596.

Venekei, I., Gráf, L. & Rutter, W.J. (1996a) Expression of rat chymotrypsinogen in yeast: a study on the structural and functional significance of the chymotrypsinogen propeptide. *FEBS Lett.* **379**, 139–142.

Venekei, I., Szilágyi, L., Gráf, L. & Rutter, W.J. (1996b) Attempts to convert chymotrypsin to trypsin. *FEBS Lett.* **383**, 133–138.

Wang, D., Bode, W. & Huber, R. (1985) Bovine chymotrypsinogen A: X-ray crystal structure analysis and refinement of a new crystal form at 1.8 Å resolution. *J. Mol. Biol.* **185**, 595–624.

Wilcox, P.E. (1970) Chymotrypsinogens – Chymotrypsins. *Methods Enzymol.* **19**, 64–112.

Yennawar, N.H., Yennawar, H.P. & Farber, G.K. (1994) X-ray crystal structure of γ-chymotrypsin in hexane. *Biochemistry* **33**, 7326–7336.

**László Gráf**
*Department of Biochemistry,*
*University Eötvös Loránd,*
*Budapest, Hungary*
*Email: graf@ludens.elte.hu*

**László Szilágyi**
*Department of Biochemistry,*
*University Eötvös Loránd,*
*Budapest, Hungary*
*Email: szilagyi@ludens.elte.hu*

**István Venekei**
*Department of Biochemistry,*
*University Eötvös Loránd,*
*Budapest, Hungary*
*Email: venekei@ludens.elte.hu*

# 9. *Chymotrypsin C*

## Databanks

*Peptidase classification: clan SA, family S1, MEROPS ID: S01.157*
*NC-IUBMB enzyme classification: EC 3.4.21.2*
*Databank codes: The Brookhaven Protein Data Bank record 1PYT contains the complete amino acid sequence as well as coordinates for the three-dimensional structure of the cattle chymotrypsin C (see below).*

## Name and History

It was found in the 1960s that pig pancreas contained three different proenzymes, each of which can be activated by trypsin to give an enzyme with high esterolytic activity against Ac-Tyr-OEt. The first two, chymotrypsinogen A and chymotrypsinogen B, are described in Chapter 8, whereas the third form, *chymotrypsinogen C*, has been given an independent EC code number because its activity, stability and partial amino acid sequence display marked differences from the better-known chymotrypsins (Folk & Schirmer, 1965; McConnell & Gjessing, 1966; Gratecos *et al.*, 1969). Similar enzymatic activity was also described in cattle pancreatic extract (Brown *et al.*, 1961). Interestingly, in ruminants the proform of this enzyme was found to be associated with procarboxypeptidase in the binary complex procarboxypeptidase A-S5 (Brown *et al.*, 1963b), or with procarboxypeptidase and proproteinase E in the ternary complex procarboxypeptidase A-S6 (Brown *et al.*, 1963a; Peanasky *et al.*, 1969; Puigserver *et al.*, 1972; Chapus *et al.*, 1988). The tendency to form complexes with other proteinases is a characteristic of the pig enzyme as well, for it was shown that the elastase-associated acidic endopeptidase isolated from pig pancreas is identical with chymotrypsin C (Thomson & Dennis, 1976).

Recently, two chymotrypsins (ChT1 and ChT2) were identified from pyloric ceca of Atlantic cod. ChT1 exhibited broader cleavage specificity compared to ChT2 on the oxidized B chain of cattle insulin as substrate. This difference is reminiscent of the difference between chymotrypsin C and chymotrypsin A in mammals (Raae *et al.*, 1995).

## Activity and Specificity

Both pig and cattle chymotrypsins C are characterized by high activity for the hydrolysis of leucyl bonds. For example, the value of $k_{cat}$ for the hydrolysis of the ester substrate Bz-Leu-OEt by pig chymotrypsin C is $25.4\,\mathrm{s}^{-1}$, whereas under identical conditions $k_{cat}$ is $0.07\,\mathrm{s}^{-1}$ and $0.11\,\mathrm{s}^{-1}$ for cattle chymotrypsins A and B, respectively (Wilcox, 1970). On the other hand, the $k_{cat}$ of chymotrypsin C with synthetic tyrosyl and phenylalanyl substrates is comparable to that of chymotrypsin A, although $K_m$ values are usually higher for chymotrypsin C (Keil-Dlouha *et al.*, 1972). The preference for leucyl bonds is also seen on peptide substrates of natural origin. For example both pig and cattle chymotrypsin C readily cleave oxidized insulin B chain at three leucyl bonds (Leu6┼CysO₃H, Leu15┼Tyr, Leu17┼Val) that are resistant to chymotrypsin A, while phenylalanyl and tyrosyl bonds are cleaved somewhat slower by chymotrypsin C than by chymotrypsin A (Folk & Cole, 1965; Keil-Dlouha *et al.*, 1972). In this respect, chymotrypsin C more closely resembles cathepsin G than chymotrypsin A (Blow & Barrett, 1977). In addition to the above data, cleavages of natural peptides (oxidized insulin A and B chains, glucagon, oxidized oxytocin) at tryptophanyl, methionyl, glutaminyl and asparaginyl bonds have also been reported (Folk & Cole, 1965).

The distinctive specificity of chymotrypsin C is also observed in reactions with chloromethane inhibitors. Tos-Leu-CH₂Cl reacts rapidly with chymotrypsin C, whereas Tos-Phe-CH₂Cl, the typical affinity label for chymotrypsins A and B, proved to be a less effective inhibitor (Tobita & Folk, 1967).

## Structural Chemistry

The first cleavage during the tryptic activation of chymotrypsinogen C results in a disulfide-linked short peptide containing 13 residues instead of the 15 characteristic of other chymotrypsins. Sequence determination showed that the two missing residues correspond to a deletion at positions 12 and 13 (Peanasky *et al.*, 1969). Further studies revealed that the N-terminus liberated upon tryptic cleavage is a Val rather then Ile. The process of activation and autolytic cleavage in autolyzed pancreas gives rise to several forms similar to chymotrypsins A$_\alpha$ and A$_\delta$, respectively (Iio-Akama *et al.*, 1985).

The sequence peculiarity (i.e., Val being the N-terminal residue of the active enzyme) certainly does not explain the distinctive enzymatic properties of chymotrypsin C. It was demonstrated recently that the activity and substrate specificity of rat anionic trypsin remained unchanged when Ile16 was replaced by Val (Hedstrom *et al.*, 1996).

Quite recently, the X-ray structure of the cattle pancreatic ternary complex formed by chymotrypsinogen C with procarboxypeptidase A (see Chapter 451) and proendopeptidase E (see Chapter 13) has been solved (Gomis-Rüth *et al.*, 1995). The Brookhaven Protein Data Bank entry (1PYT) contains the complete amino acid sequence of cattle chymotrypsin C, consisting of 251 residues. Compared to chymotrypsin A it contains several insertions also present in the pancreatic elastases. The sequence of cattle chymotrypsin C is significantly closer to cattle pancreatic elastase (56% identity) or pig pancreatic elastase (54%) than to cattle chymotrypsinogen A (41%). The disulfide bridge pattern is chymotrypsin-like, however. Alignment 9.1 (see CD-ROM) shows an alignment of the sequence with related sequences, including that of rat caldecrin (Chapter 10).

The X-ray structure offers a possible explanation of the distinctive substrate specificity of chymotrypsin C. At two critical positions in the substrate-binding pocket, namely at 216 and 226, the residues are Gly and Thr, respectively. This results in a structure at this part of the molecule that is intermediate between those of chymotrypsin A (Gly216 and Gly226) and pancreatic elastase (Val216 and Thr226). The side chain of Thr226 in chymotrypsin C protrudes into the substrate-binding pocket and molds it to accommodate medium-sized residues. It is pertinent to note here that recent results from our laboratory show that a Gly to Ala substitution at position 226 in chymotrypsin B is the major reason for the low activity of this enzyme on tryptophanyl substrates (see Chapter 8; P. Hudáky, G. Kaslik, L. Szilágyi, I. Venekei & L. Gráf, unpublished results).

## Preparation

Chymotrypsinogen C has been prepared from frozen and lyophilized pancreatic juice of young pigs by chromatography on DEAE-cellulose (pH 6.0), gel filtration on Sephadex G-100, and further ion-exchange chromatography on CM-Sephadex (pH 6.0). The yield was about 60–70 mg of chymotrypsinogen C from 15 g of lyophilized pancreatic juice (Gratecos *et al.*, 1969). The active enzyme has also been isolated from the acetone powder of autolyzed pig pancreas by extracting it with water followed by ammonium sulfate fractionation. The pellet obtained at 0.6 saturation was dialysed and subjected to two chromatographic runs on DEAE-cellulose (pH 8). The yield of twice-chromatographed chymotrypsinogen C was 100–150 mg from 100 g of acetone powder (Folk, 1970).

Cattle chymotrypsinogen C (subunit II) can be isolated by the alkaline dissociation of procarboxypeptidase A-S6 isolated from the acetone powder of cattle pancreas (Puigserver *et al.*, 1972). Higher yields can be achieved after reversible modification of lysyl groups in the complex by 2,3-dimethyl maleic anhydride (Puigserver & Desnuelle, 1975; Kerfelec *et al.*, 1984).

## Biological Aspects

Besides its wide substrate specificity, the potential to form a ternary complex certainly makes chymotrypsin C an important component of the ruminant digestive system. Several hypotheses have been put forward for the biological function of the heterotrimer complex, including modulation of the activity of the subunits, proper coordination of the appearance of proteolytic activities in the duodenum, and the protection of procarboxypeptidase A against inactivation by the acidic gastric juice (Uren & Neurath, 1972; Puigserver & Desnuelle, 1977; Michon *et al.*, 1991; Avilés *et al.*, 1993). The recent X-ray structure of the ternary complex of procarboxypeptidase A with proendopeptidase E and chymotrypsin C shows that the two latter enzymes are associated with the long, extended activation segment of the procarboxypeptidase, thereby protecting it against tryptic attack. In contrast, the activation sites of both serine proteinase proenzymes are exposed on the surface of the complex and are presumably susceptible to tryptic cleavage. Activation of these enzymes by trypsin might destabilize the ternary complex and facilitate the access of trypsin to the activation sites of procarboxypeptidase (Gomis-Rüth *et al.*, 1995).

## Related Peptidases

The amino acid sequence of cattle chymotrypsin C is so close to that of rat caldecrin (Chapter 10) as to suggest that the two may be species variants of a single protein.

## References

Avilés, F.X., Vendrell, J., Guasch, A., Coll, M. & Huber, R. (1993) Advances in metallo-procarboxypeptidases. Emerging details on the inhibition mechanism and on the activation process. *Eur. J. Biochem.* **211**, 381–389.

Blow, A.M.J. & Barrett, A.J. (1977) Action of human cathepsin G on the oxidized B chain of insulin. *Biochem. J.* **161**, 17–19.

Brown, J.R., Cox, D.L., Greenshields, R.N., Walsh, H., Yamasaki, M. & Neurath, H. (1961) The chemical structure and enzymatic functions of bovine pancreatic procarboxypeptidase A. *Proc. Natl Acad. Sci. USA* **47**, 1554–1562.

Brown, J.R., Greenshields, R.N., Yamasaki, M. & Neurath, H. (1963a) The subunit structure of bovine procarboxypeptidase A-S6. Chemical properties and enzymatic activities of the products of molecular disaggregation. *Biochemistry* **2**, 867–876.

Brown, J.R., Yamasaki, M. & Neurath, H. (1963b) A new form of bovine pancreatic procarboxypeptidase A. *Biochemistry* **2**, 877–886.

Chapus, C., Puigserver, A. & Kerfelec, B. (1988) The bovine pro-carboxypeptidase A-S6 ternary complex: a rare case of a secreted protein complex. *Biochimie* **70**, 1143–1151.

Folk, J.E. (1970) Chymotrypsin C (porcine pancreas). *Methods Enzymol.* **19**, 109–112.

Folk, J.E. & Cole, J. (1965) Chymotrypsin C. II. Enzymatic specificity toward several polypeptides. *J. Biol. Chem.* **240**, 193–197.

Folk, J.E. & Schirmer, W.E. (1965) Chymotrypsin C. I. Isolation of the zymogen and the active enzyme: Preliminary structure and specificity studies. *J. Biol. Chem.* **240**, 181–192.

Gomis-Rüth, F.X., Gómez, M., Bode, W., Huber, R. & Avilés, F.X. (1995) The three-dimensional structure of the native ternary complex of bovine pancreatic procarboxypeptidase A with proproteinase E and chymotrypsinogen C. *EMBO J.* **14**, 4387–4394.

Gratecos, D., Guy, O., Rovery, M. & Desnuelle, P. (1969) On the two anionic chymotrypsinogens of bovine pancreas. *Biochim. Biophys. Acta* **181**, 82–92.

Hedstrom, L., Lin, T.-Y. & Fast, W. (1996) Hydrophobic interaction in the trypsin family of serine proteases. *Biochemistry* **35**, 4515–4523.

Iio-Akama, K., Sasamoto, H., Miyazawa, K., Miura, S., & Tobita, T. (1985) Active forms of chymotrypsin C isolated from autolyzed porcine pancreas glands. *Biochim. Biophys. Acta* **831**, 249–256.

Keil-Dlouha, V., Puigserver, A., Marie, A. & Keil, B. (1972) On subunit II of bovine procarboxypeptidase A. Enzymatic specificity after tryptic activation. *Biochim. Biophys. Acta* **276**, 531–535.

Kerfelec, B., Chapus, C. & Puigserver, A. (1984) Two step dissociation of bovine 6S procarboxypeptidase A by dimethylation. *Biochem. Biophys. Res. Commun.* **121**, 162–167.

McConnell, B. & Gjessing, E.C. (1966) Isolation and crystallization of an esteroproteolytic zymogen from porcine pancreas. *J. Biol. Chem.* **241**, 573–579.

Michon, T., Granon, S., Sauve, P. & Chapus, C. (1991) The activation peptide of procarboxypeptidase A is the keystone of the bovine procarboxypeptidase A-S6 ternary complex. *Biochem. Biophys. Res. Commun.* **181**, 449–455.

Peanasky, R.J., Gratecos, D., Baratti, J. & Rovery, M. (1969) Mode of activation of the N-terminal sequence of subunit II in bovine procarboxypeptidase A and of porcine chymotrypsinogen C. *Biochim. Biophys. Acta* **181**, 82–92.

Puigserver, A. & Desnuelle, P. (1975) Dissociation of bovine 6S procarboxypeptidase A by reversible condensation with 2,3-dimethyl maleic anhydride: application to the partial characterization of subunit III. *Proc. Natl Acad. Sci. USA* **72**, 2442–2445.

Puigserver, A. & Desnuelle, P. (1977) Reconstitution of bovine procarboxypeptidase A-S6 from the free subunits. *Biochemistry* **16**, 2497–2501.

Puigserver, A., Vaugoyeau, G. & Desnuelle, P. (1972) On subunit II of bovine procarboxypeptidase A. Properties after alkaline dissociation. *Biochim. Biophys. Acta* **276**, 519–530.

Raae, A.J., Fledsrud, R. & Sletten, K. (1995) Chymotrypsin isoenzymes in Atlantic cod; differences in kinetics and substrate specificity. *Comp. Biochem. Physiol. B Biochem. Mol. Biol.* **112**, 393–398.

Thomson, A. & Dennis, I.S. (1976) The identity of the elastase-associated acidic endopeptidase and chymotrypsin C from porcine pancreas. *Biochim. Biophys. Acta* **429**, 581–590.

Tobita, T. & Folk, J.E. (1967) Chymotrypsin C. III. Sequence of amino acids around an essential histidine residue. *Biochim. Biophys. Acta* **147**, 15–25.

Uren, J.R. & Neurath, H. (1972) Mechanism of the activation of bovine procarboxypeptidase A S5. Alteration in primary and quaternary structure. *Biochemistry* **11**, 4483–4492.

Wilcox, P.E. (1970) Chymotrypsinogens – chymotrypsins. *Methods Enzymol.* **19**, 64–108.

*László Szilágyi*
*Department of Biochemistry,*
*University Eötvös Loránd,*
*Budapest, Hungary*
*Email: szilagyi@ludens.elte.hu*

# 10. Caldecrin

### Databanks

*Peptidase classification: clan SA, family S1, MEROPS ID: S01.157*
*NC-IUBMB enzyme classification: none*
*Databank codes:*

| Species | SwissProt | PIR | EMBL (cDNA) | EMBL (genomic) |
|---|---|---|---|---|
| *Rattus norvegicus* | P55091 | – | S80379 X59014 | – |

## Name and History

A reduction in serum calcium (hypocalcemia) occurs in patients with acute pancreatitis. The serum calcium-decreasing factor was purified from porcine and rat pancreas (Tomomura *et al.*, 1992, 1995) and was given the name *caldecrin* to denote this serum **cal**cium-**decr**easing activity. The purified caldecrin was found to be a chymotrypsin-type serine protease. The caldecrin cDNAs were generated from rat and human pancreas RNA (Tomomura *et al.*, 1995, 1996). The sequence of rat caldecrin cDNA is, with the exception of its central region, almost identical to that of a cDNA described as that of *elastase IV*, which was identified by cloning with PCR from rat pancreas (Kang *et al.*, 1992). The amino acid sequences of fragments derived from purified rat caldecrin were identical to those deduced from rat caldecrin cDNA, but different from those deduced from the elastase IV, especially in the central region (Tomomura *et al.*, 1995). The difference in the central region can be largely accounted for by a minor change in nucleotide sequence that alters the reading frame. The putative elastase IV protein has not been investigated. The central region of rat caldecrin shows high homology with that of purified pig caldecrin and with the amino acid sequences deduced from human caldecrin cDNAs (Tomomura *et al.*, 1995, 1996).

## Activity and Specificity

Assay of this protease activity is conducted with azocasein and Suc-Ala-Ala-Pro-Phe-NHPhNO$_2$ (Tomomura *et al.*, 1992, 1993). These substrates are available from Sigma (see Appendix 2 for full names and addresses of suppliers). The pH optimum of the enzyme is about 9.0. The protease activity of caldecrin is inhibited by DFP, PMSF, chymostatin and soybean trypsin inhibitor, but not by APMSF (Tomomura *et al.*, 1992). Assay of the serum calcium-decreasing activity is made by intravenous injection of the caldecrin into mice and subsequent measurement of the serum calcium concentration (Tomomura *et al.*, 1992).

## Structural Chemistry

The primary structure of the rat caldecrin cDNA is considered to be that of a member of the elastase family (Tomomura *et al.*, 1995). Two homologous human caldecrin cDNA clones were isolated, the structures of which were identical except for one base change and the corresponding amino acid residue substitution (Tomomura *et al.*, 1996). Caldecrin is a single-chain protein of 268 amino acids (about 30 kDa) with a signal peptide of 16 amino acids and a propeptide (activation peptide) of 13 amino acids. It is converted into a fully active enzyme when the peptide bond between Arg(−1) and Val1 is cleaved by trypsin. The cysteines form the five disulfide bridges characteristic of chymotrypsin, including a bridge between the N-terminal Cys(−13) of the activation peptide and Cys112 of the mature enzyme. The C-terminal three amino residues of the activation peptide are lost after activation. The pI for the porcine enzyme is 4.5 (Tomomura *et al.*, 1992).

## Preparation

Caldecrin has been found only in the pancreas (Tomomura *et al.*, 1995). The recombinant rat and human enzymes have been expressed in Sf9 cells by baculovirus and purified from the culture medium (Tomomura *et al.*, 1995, 1996).

## Biological Aspects

The gene structure and regulation of gene expression of caldecrin have not been reported. Caldecrin administered to mice at a dose of 100 mg kg$^{-1}$ body weight decreases serum calcium by 15–20% by 4 h postinjection. Procaldecrin does not have the serum calcium-decreasing activity or the protease activity. Caldecrin activated by trypsin treatment expresses the serum calcium-decreasing activity. The serum calcium-decreasing activity of activated caldecrin is not affected by treatment with PMSF (Tomomura *et al.*, 1992, 1993, 1995). The catalytic amino acid mutants of recombinant caldecrin retain the serum calcium-decreasing activity (Tomomura *et al.*, 1995). Caldecrin decreases the serum hydroxyproline as well as serum calcium level (Tomomura *et al.*, 1992), and inhibits the bone resorption of cultured fetal mouse long bone stimulated by parathyroid hormone (Tomomura *et al.*, 1992). The serum calcium-decreasing activity of the enzyme might result from the suppression of bone resorption through an as yet unknown mechanism.

Nothing is known of the normal function and substrates of the enzyme or of the enzyme level in the serum of patients with acute pancreatitis.

## Distinguishing Features

Polyclonal antiserum against the enzyme from porcine pancreas cross-reacts with the rat and human enzymes (Tomomura *et al.*, 1993, 1995, 1996), but is not commercially available.

## Related Peptidases

The amino acid sequence of cattle chymotrypsin C (Chapter 9) is so close to that of rat caldecrin as to suggest that the two may be species variants of a single protein.

## References

Kang, J., Wiegand, U. & Muller-Hill, B. (1992) Identification of cDNAs encoding two novel rat pancreatic serine proteases. *Gene (Amst.)* **110**, 181–187.

Tomomura, A., Fukushige, T., Noda, T., Noikura, T. & Saheki, T. (1992) Serum calcium-decreasing factor (caldecrin) from porcine pancreas has proteolytic activity which has no clear connection with the calcium decrease. *FEBS Lett.* **301**, 277–281.

Tomomura, A., Fukushige, T., Tomomura, M., Noikura, T., Nishii, Y. & Saheki, T. (1993) Caldecrin proform requires trypsin activation for the acquisition of serum calcium-decreasing activity. *FEBS Lett.* **335**, 213–216.

Tomomura, A., Tomomura, M., Fukushige, T., Akiyama, M., Kubota, N., Kumaki, K., Nishii, Y., Noikura, T. & Saeki, T. (1995)

Molecular cloning and expression of serum calcium-decreasing factor (caldecrin). *J. Biol. Chem.* **270**, 30315–30321.
Tomomura, A., Akiyama, M., Itoh, H., Yoshino, I., Tomomura, M.,

Nishii, Y., Noikura, T. & Saheki, T. (1996) Molecular cloning and expression of human caldecrin. *FEBS Lett.* **386**, 26–28.

*Akito Tomomura*
*Department of Biochemistry,*
*Meikai University School of Dentistry,*
*1–1 Keyakidai, Sakado, Saitama 350–02 Japan*
*Email: atomomu@dent.meikai.ac.jp*

# 11. Pancreatic elastase

## Databanks

*Peptidase classification: clan SA, family S1, MEROPS ID: S01.153*
*NC-IUBMB enzyme classification: EC 3.4.21.36*
*Chemical Abstracts Service registry number: 9004-06-2*
*Databank codes:*

| Species | SwissProt | PIR | EMBL (cDNA) | EMBL (genomic) |
|---|---|---|---|---|
| *Bos taurus* | – | – | M80838 | – |
| *Gadus morhua* | P32197 | S33787 | – | – |
| *Homo sapiens* | P11423 | B29934 | – | – |
| *Rattus norvegicus* | P00773 | A00960 | V01234 | L00112: exons 1 and 2 |
| | | | | L00113: exons 3 and 4 |
| | | | | L00114: exon 5 |
| | | | | L00115: exon 6 |
| | | | | L00116: exon 7 |
| | | | | L00117: exon 8 and complete CDS |
| *Sus scrofa* | P00772 | JS0013 | X04036 | – |

Brookhaven Protein Data Bank three-dimensional structures:

| Species | ID | Resolution | Notes |
|---|---|---|---|
| *Sus scrofa* | 1EAS | 1.8 | complex with 3-[[(methylamino)sulfonyl]amino]-2-oxo-6-phenyl-N-[3,3,3-trifluoro-1-(1-methylethyl)-2-oxopropyl]-1(2H)-pyridine acetamide |
| | 1EAT | 2 | complex with 2-[5-methanesulfonylamino-2-(4-aminophenyl-6-oxo-1,6-dihydro-1-pyrimidinyl]-N-(3,3,3-trifluoro-1-isopropyl-2-oxopropyl) acetamide |
| | 1EAU | 2 | complex with 2-[5-amino-6-oxo-2-(2-thienyl)-1,6-dihydropyrimidin-1-yl)-N-[3,3-difluoro-1-isopropyl-2-oxo-3-(N-(2-morpholinoethyl)carbamoyl)propyl] acetamide |
| | 1ELA | 1.8 | complex with trifluoroacetyl-Lys-Pro-isopropylanilide |
| | 1ELB | 2.1 | complex with trifluoroacetyl-Lys-Leu-isopropylanilide |
| | 1ELC | 1.75 | complex with trifluoroacetyl-Phe-*p*-isopropylanilide |
| | 1ELD | 1.8 | complex with trifluoroacetyl-Phe-Ala-*p*-trifluoromethylanilide |
| | 1ELE | 1.8 | complex with trifluoroacetyl-Val-Ala-*p*-trifluoromethylanilide |
| | 1ELF | 1.7 | complex with N-(*tert*-butoxycarbonyl-Ala-Ala)-O-(*p*-nitrobenzoyl) hydroxylamine |
| | 1ELG | 1.65 | complex with N-(*tert*-butoxycarbonyl-Ala-Ala)-O-(*p*-nitrobenzoyl) hydroxylamine at pH 5 |
| | 1ESA | 1.65 | low-temperature form (−45°C) |
| | 1ESB | 2.3 | complex with Z-Ala-*p*-nitrophenol ester |
| | 1EST | 2.5 | Tos-elastase |

| Species | ID | Resolution | Notes |
|---|---|---|---|
| | 1FLE | 1.9 | complex with elafin |
| | 1INC | 1.94 | complex with benzoxazinone inhibitor |
| | 1JIM | 2.31 | complex with 3-methoxy-4-chloro-7-aminoisocoumarin inhibitor |
| | 2EST | 2.5 | complex with trifluoroacetyl-Lys-Ala-$p$-trifluoromethylphenylanilide |
| | 3EST | 1.65 | |
| | 4EST | 1.78 | complex with Ac-Ala-Pro-Val-difluoro-$N$-phenylethylacetamide |
| | 5EST | 2.09 | complex with carbobenzoxy-Ala-Ile boronic acid |
| | 6EST | 1.8 | crystallized in 10% DMF |
| | 7EST | 1.8 | complex with trifluoroacetyl-Leu-Ala-$p$-trifluoromethylphenylanilide |
| | 8EST | 1.78 | complex with guanidinium isocoumarin |
| | 9EST | 1.9 | complex with guanidinium isocoumarin |

## Name and History

The presence of elastinolytic activity in pancreatic extracts was described more than a century ago. However, this activity was thought for a long time to be due to trypsin, and it was not until the report of Baló & Banga (1949) that the presence of a separate *pancreatic elastase (PE)* was established. Elastases are defined by their ability to release soluble peptides from insoluble elastin fibers by a proteolytic process. Elastinolysis is a *sine qua non* for naming a given proteinase as an elastase. Pure pig pancreatic elastase was first obtained by Shotton (1970). A second elastinolytic proteinase was discovered in pig pancreas a few years later (Ardelt, 1974), and was termed *pancreatic elastase II*. The first-discovered enzyme, which had been referred to as *pancreatic elastase* or *pancreatopeptidase E*, was subsequently termed *pancreatic elastase I (PE I)*. The occurrence of two separate elastinolytic enzymes in pig pancreas is exceptional, as most animal species have only one pancreatic elastase, either of type I or of type II.

## Activity and Specificity

Pancreatic elastases are serine proteinases of the chymotrypsin (S1) family whose catalytic site is composed of three hydrogen-bonded amino acid residues, Asp102, His57 and Ser195 (chymotrypsinogen numbering), which form the so-called charge-relay system or catalytic triad. Peptide bond hydrolysis occurs in several steps. The initial reversible formation of an adsorption (Michaelis) complex between the serine proteinase and its substrate is followed by a nucleophilic attack by the $\gamma$-oxygen of Ser195 on the peptide bond with subsequent formation of an acyl enzyme intermediate and release of the first product, the C-terminal part of the substrate. The covalent intermediate is then hydrolyzed in the deacylation step, which regenerates active enzyme and releases the carboxyl group of the second product (Bieth, 1986). Recently, the acyl enzyme intermediate was directly observed by crystallographic cryoenzymology. The substrate Z-Ala-NHPhNO$_2$ was allowed to flow through a single crystal of pig PE I at $-26°$C. The acyl enzyme intermediate was trapped at $-55°$C before collection of crystallographic data. Electron density for the intermediate was clearly visible (Ding *et al.*, 1994).

Most studies on the substrate-binding site of PE (i.e. the site responsible for its specificity) have been done on the pig enzyme. Its action on the A and B chains of oxidized insulin shows a marked primary specificity for nonbulky residues such as Ala, Ser, Gly and Val in P1 (Bieth, 1978). Studies with acyl-tetrapeptide-$p$-nitroanilide substrates confirm the P1 Ala specificity (Largman, 1983). Dog PE I exhibits the same specificity, but the rat enzyme cleaves Ala+ and Tyr+ bonds equally rapidly (Largman, 1983). The elastases isolated from a number of fish pancreases also have a P1 Ala specificity. The specificity of pig PE I for small residues is in full agreement with the tertiary structure of its substrate-binding site (Bode *et al.*, 1989). Crystallographic analysis of enzyme–inhibitor complexes shows, however, that an Ile residue also binds tightly to the primary specificity pocket of pig PE I (Takahashi *et al.*, 1989).

The catalytic activity of pig PE I increases considerably with the chain length of the synthetic substrates. For example, the $k_{cat}/K_m$ value for the hydrolysis of Suc-(Ala)$_n$-$p$-nitroanilide is 4, 12 000 and 170 000 M$^{-1}$ s$^{-1}$, respectively, for the series $n = 2$, 3 and 4 (Bieth, 1986). Occupancy of subsite S4 thus appears to be of critical importance for efficient catalysis. The sensitivity of pig PE I to peptide chain elongation is much more pronounced than that of any other related proteinase. The substrate-binding site of pig PE I extends over 29 Å and is composed of eight subsites, S5–S3′. The specificity of each of the individual subsites has been thoroughly investigated (Bieth, 1978). The crystallographic structure of the complex of pig PE I with a hexapeptide shows the nature of the subsites on both sides of the scissile bond and mimics the binding of a natural peptide to the enzyme (Meyer *et al.*, 1988). The amino acid residues that are likely to form these subsites have been discussed by Bode *et al.* (1989). The catalytic activity of rat PE I also strongly increases with the chain length of the substrate. Again, occupancy of subsite S4 is of critical importance as it leads to a 500-fold increase in $k_{cat}/K_m$. The rat enzyme apparently has one more S subsite than the pig enzyme (Bieth *et al.*, 1989).

The catalytic activities of pig and rat PE I may be regulated by secondary enzyme–substrate interactions (i.e. by interactions remote from the S1–P1 binding). For instance, occupancy of subsite S4 by a succinyl group very strongly increases the catalytic rate constant $k_{cat}$ (Bieth *et al.*, 1989). The increase in $k_{cat}$ may result either from an indirect, P4-induced tightening of the S1–P1 binding, with resultant optimal closeness of the scissile bond of the substrate to the serine residue of the catalytic triad, or from a P4-induced enhancement of the nucleophilicity of O$\gamma$ of Ser195 of the catalytic site. Activation or inhibition of catalysis by excess substrate, as observed in the course of pig or rat PE I-catalyzed hydrolysis of short $p$-nitroanilide substrates, is another example of regulation of activity by secondary enzyme–substrate interactions. This phenomenon is best explained by assuming that at high substrate concentration two molecules of substrate bind to the active site, one favoring or impairing the hydrolysis of the other (Bieth *et al.*, 1989). $Ca^{2+}$ binds strongly to pig PE I, but this interaction does not influence the catalytic properties of the enzyme. Calcium and gadolinium ions bind pig PE I with similar affinities ($K_d = 45\,\mu M$ and $20\,\mu M$, respectively). A fluorine NMR investigation of the ternary complex of pig PE I, gadolinium ion and trifluoroacetyl-trialanine, an elastase inhibitor, showed that the metal ion and the fluorine atoms are distant by 20 Å. From this it was concluded that the calcium-binding site of pig PE I involves Glu70 and Glu80 (Dimicoli & Bieth, 1977). The calcium-binding site of pig PE I has also been investigated by terbium luminescence due to an energy transfer between a tryptophan residue close to the bound terbium ion. Moreover, circularly polarized terbium luminescence measurements allowed the detection of conformational changes in the elastase molecule following pH variation, or reaction with $\alpha_1$-proteinase inhibitor or $\alpha_2$-macroglobulin (Duportail *et al.*, 1980). Organic solvents partially inhibit the pig PE I-catalyzed hydrolysis of synthetic substrates (Bieth, 1978).

Elastases have the unique property of solubilizing fibrous elastin, a structural protein of lung, artery, skin and ligament, the insolubility of which is due both to a large number of nonpolar amino acid residues and to special cross-linking residues (desmosine, isodesmosine and lysinonorleucine). Elastin is rich in Gly (32%), Ala (26%) and Val (13%), the amino acid residues for which PE I shows a P1 specificity. Pig PE I mainly cleaves Ala┼Ala and Ala┼Gly bonds in elastin, and it is among the most potent elastinolytic proteinases, with a rate of elastinolysis 20-fold higher than that of human leukocyte elastase (Bieth, 1986).

Assay of pure pig PE I is best done with the commercially available synthetic substrate (from Bachem) (see Appendix 2 for full names and addresses of suppliers), Suc-Ala-Ala-Ala┼NHPhNO$_2$ ($k_{cat}/K_m = 14\,000\,M^{-1}\,s^{-1}$), which allows 10 nM enzyme to be determined accurately. This substrate may also be used to assay the elastases from dog, rat and fish pancreas. Elastinolytic assays of crude pancreatic extracts may be done with $^3$H-labeled elastin, or with rhodamine-elastin, which is less sensitive but commercially available (Elastin Products Co.). Pancreatic elastases have pH optima close to neutrality (Bieth, 1978).

Pancreatic elastases are inhibited by nonspecific irreversible serine proteinase inhibitors such as DFP and PMSF, which are commercially available from many sources (e.g. Sigma). More specific irreversible inhibitors include acyl-Ala-Ala-Pro-Ala-chloromethane (from Enzyme Systems Products) and DCI (ICN). Elastatinal, a low molecular mass reversible aldehyde inhibitor from microorganisms (Sigma) inhibits PE I of various origins with $K_i$ values ranging from 0.1 to 0.3 $\mu M$ (Bieth, 1986). The possible pathological role of human leukocyte elastase has led to the synthesis of inhibitors, most of which have also been tested on pig PE I. These compounds belong to a variety of chemical classes including peptide-based inhibitors such as trifluoroacetylated peptides (De la Sierra *et al.*, 1990), and heterocyclic and nonheterocyclic compounds (for a review see Edwards &

Bernstein, 1994). Some of these molecules are very potent. For example, the boronic acid derivative, MeOSuc-Ala-Ala-Pro-NH-CH-(CH(CH$_3$)$_2$)-B(OH)$_2$ inhibits pig PE I with a $K_i$ of 0.2 nM (Kettner & Shenvi, 1984).

Pancreatic elastases are also inhibited by naturally occurring or engineered protein proteinase inhibitors. Human plasma $\alpha_1$-proteinase inhibitor (53 kDa), a member of the serpin superfamily, and $\alpha_2$-macroglobulin (725 kDa), are commercially available (Athens Research & Technology) irreversible inhibitors whose second-order association rate constant $k_{ass}$ for pig PE I are $1 \times 10^5\,M^{-1}\,s^{-1}$ (Beatty *et al.*, 1980) and $4 \times 10^6\,M^{-1}\,s^{-1}$ (Bieth & Meyer, 1984), respectively. Rat plasma contains four proteins able to bind rat PE I, namely $\alpha_1$-proteinase inhibitor, $\alpha_2$-macroglobulin, $\alpha_2$-acute phase protein and $\alpha_1$-inhibitor$_3$ (Gauthier *et al.*, 1978). Elafin, a 6 kDa human leukocyte elastase inhibitor present in human keratinocytes or bronchial secretions also inhibits pig PE I with a $K_i$ of 1 nM (Wiedow *et al.*, 1990). Other proteins with moderate potencies against pig PE I include dog submandibular gland, chicken egg white, turkey ovomucoid and inhibitors from *Ascaris lumbricoides*, leech (eglin c), soybean and potato (Bode *et al.*, 1989).

## Structural Chemistry

Pig PE I is a 26 kDa protein with a pI of 9.5. It is composed of a single peptide chain of 240 amino acid residues with four disulfide bridges, and exhibits a high degree of sequence identity with other pancreatic proteinases (McDonald *et al.*, 1982). The structures of many forms of free or inhibitor-bound pig PE I have been determined crystallographically to atomic resolution (Meyer *et al.*, 1986, 1988; Bode *et al.*, 1989; De la Sierra *et al.*, 1990). Free pig PE I yields good crystals with an open binding site into which small inhibitors may easily be soaked. Like those of other serine proteinases in family S1, the polypeptide chain of pig PE I consists of two interacting antiparallel $\beta$ barrel cylindrical domains, with a small proportion of helices. The two domains form a crevice which contains the catalytic triad. The substrate-binding site lies across this crevice and comprises parts of the two domains. Pig PE I is topologically very similar to human leukocyte elastase (see Chapter 15), especially in its active-site region. It contains several internal water channels which might play a role in catalysis (Meyer *et al.*, 1988). Reaction of a pig PE I crystal with the synthetic substrate Ac-Ala-Pro-Ala-NHPhNO$_2$ led to the finding that two molecules of the product Ac-Ala-Pro-Ala are bound per elastase molecule. Both of these peptides are bound backwards with respect to the classical antiparallel pleated-sheet binding mode (Meyer *et al.*, 1986). Trifluoroacetyl-Leu-Ala-*p*-trifluoromethylanilide also binds in a nonconventional way to the active site of pig PE I. While the trifluoroacyl group binds at S1, Leu-Ala form a parallel $\beta$ sheet association with the 214–216 loop of the enzyme (De la Sierra *et al.*, 1990). Pig PE I has also been crystallized in complex with human elafin (Tsunemi *et al.*, 1993). The enzyme is remarkably stable at pH 7–8, but undergoes rapid inactivation at pH 3 as a result of a conformational change (Bieth, 1986). Rat PE I is a 27 kDa protein with a pI of 9.6 (Largman, 1983). Its amino acid sequence shows 84% identity with that of pig PE I (McDonald *et al.*, 1982).

## Preparation

Pig PE I is best isolated from a commercial pancreatic extract ('Trypsin 1-300', Calbiochem) by the method of Shotton (1970) which includes extraction, ammonium sulfate precipitation, dialysis, batch adsorption of impurities on DEAE-Sephadex, and crystallization. The procedure takes 2 weeks, is easy to carry out, and yields very pure enzyme. Pig PE I is also commercially available from different sources (Elastin Products Co., Sigma). Unfortunately, these manufacturers also sell a twice-crystallized aqueous suspension of pig PE I which is very impure and contains acidic endopeptidase, chymotrypsin, insulysin and carboxypeptidases in addition to elastase (Bieth, 1986). Rat pancreatic proelastase I has been purified from pancreatic tissue by a combination of ion-exchange and affinity chromatography (Largman, 1983). On the other hand, active rat PE I has been isolated from pancreatic juice using two cation-exchange chromatographic steps (Bieth *et al.*, 1989).

## Biological Aspects

Elastase activity has been found in human, rat, cow, horse, dog, cat, guinea pig, chicken and fish pancreas (Bieth, 1986). Rat PE I has a distinct elastinolytic activity (Largman, 1983) and is structurally related to pig PE I (McDonald *et al.*, 1982). The human pancreas does not synthesize an elastase similar to pig PE I (Shen *et al.*, 1987), because the human gene corresponding to the gene of rat pancreatic elastase I is silent (Kawashima *et al.*, 1992). The enzyme, which is sometimes called 'human pancreatic elastase I' (Murata *et al.*, 1983; Wendorf *et al.*, 1991), is in fact protease E (Shen *et al.*, 1987), now called pancreatic endopeptidase E (Chapter 13). The other pancreatic 'elastases' either do not solubilize elastin or have not been sequenced.

Like other pancreatic proteinases, elastases are synthesized as inactive precursors called proelastases which are stored in the acinar cells of the pancreas. Trypsin-induced activation takes place in the duodenum. Pig pancreatic proelastase I (Gertler & Birk, 1970) and rat pancreatic proelastase I (Largman, 1983) have been isolated and investigated. The zymogens of dog and African lung fish pancreatic elastases have also been studied (De Haën & Gertler, 1974; Geokas *et al.*, 1980). It is not known, however, whether the active enzymes corresponding to the latter zymogens are related to pig PE I. The rat pancreatic preproelastase I gene consists of seven introns and eight exons, two of which encode separately the pre and pro sequences (Swift *et al.*, 1984). The transcriptional enhancer of this gene contains an A element directing tissue-specific expression (Rose *et al.*, 1994).

The proteolytic action of pancreatic elastases is not restricted to elastin: they attack any soluble protein that contains appropriate, surface-exposed amino acid sequences. For instance, pig PE I readily hydrolyzes tropoelastin, the soluble precursor of elastin (Christner *et al.*, 1978), and is as powerful as trypsin or chymotrypsin in cleaving soluble proteins like casein or hemoglobin. Also, pig PE I inactivates $\alpha_1$-antichymotrypsin by proteolytic cleavage (Morii & Travis, 1983) and hydrolyzes plasma apolipoprotein AII (Byrne *et al.*, 1984) and lung connective tissue proteoglycans (van de

Lest *et al.*, 1995). Glutaryl-(Ala)$_3$-ethylamide, an inhibitor of rat PE I, has been used in the treatment of acute pancreatitis in the rat (Fric *et al.*, 1992). Pig PE I has been used for many years as a model enzyme for the induction of lung emphysema in animals (Janoff, 1985).

## References

Ardelt, W. (1974) Partial purification and properties of porcine pancreatic elastase II. *Biochim. Biophys. Acta* **341**, 257–265.

Baló, J. & Banga, I. (1949) Elastase and elastase inhibitor. *Nature Lond.* **164**, 491.

Beatty, K., Bieth, J.G. & Travis J. (1980) Kinetics of association of serine proteinases with native and oxidized $\alpha_1$-proteinase inhibitor and with $\alpha_1$-antichymotrypsin. *J. Biol. Chem.* **255**, 3931–3934.

Bieth, J.G. (1978) Elastases: structure, function and pathological role. In: *Frontiers of Matrix Biology*, vol. 6 (Robert, L., Collin-Lapinet, G.M. & Bieth, J.G., eds). Basel: Karger, pp. 1–82.

Bieth, J.G. (1986) Elastases: catalytic and biological properties. In: *Biology of Extra-Cellular Matrix*, vol. 1: *Regulation of Matrix Accumulation* (Mecham, R.P., ed.). New York: Academic Press, pp. 217–320.

Bieth, J.G. & Meyer, J.F. (1984) Temperature and pH dependence of the association rate constant of elastase with $\alpha_2$-macroglobulin. *J. Biol. Chem.* **259**, 8904–8906.

Bieth, J.G., Dirrig, S., Jung, M.L., Boudier, C., Papamichael, E., Sakarellos, C. & Dimicoli, J.L. (1989) Investigation of the active center of rat pancreatic elastase. *Biochim. Biophys. Acta* **994**, 64–74.

Bode, W., Meyer, E., Jr. & Powers, J.C. (1989). Human leukocyte and porcine pancreatic elastase: X-ray crystal structures, mechanism, substrate specificity, and mechanism-based inhibitors. *Biochemistry* **28**, 1951–1963.

Byrne, R.E., Placek, D., Gordon, J. & Scanu, A. (1984) The enzyme that cleaves apolipoprotein A-II upon *in vitro* incubation of human plasma high-density lipoprotein-3 with blood polymorphonuclear cells is an elastase. *J. Biol. Chem.* **259**, 14537–14544.

Christner, P., Weinbaum, G., Sloan, B. & Rosenbloom, J. (1978) Degradation of tropoelastin by proteases. *Anal. Biochem.* **88**, 682–688.

De Haën, C. & Gertler, A. (1974) Isolation and amino-terminal sequence analysis of two dissimular pancreatic proelastases from the African lungfish. *Biochemistry* **13**, 2673–2677.

de la Sierra, I.L., Papamichael, E., Sakarellos, C., Dimicoli, J.L. & Prange, T. (1990) Interactions of the peptide CF$_3$-Leu-Ala-NH-C$_6$H$_4$-CF$_3$ (TFLA) with porcine pancreatic elastase. *J. Mol. Recognit.* **3**, 36–44.

Dimicoli, J.L. & Bieth, J.G. (1977) Location of the calcium ion binding site in porcine pancreatic elastase using a lanthanide ion probe. *Biochemistry* **16**, 5532–5537.

Ding, X., Rasmussen, B.F., Petsko, G.A. & Ringe, D. (1994) Direct structural observation of an acyl-enzyme intermediate in the hydrolysis of an ester substrate by elastase. *Biochemistry* **33**, 9285–9293.

Duportail, G., Lefèvre, J.F., Lestienne, P., Dimicoli, J.L. & Bieth, J.G. (1980) Binding of terbium to porcine pancreatic elastase. Ligand-induced changes in the stability, the maximum luminescence intensity and the circularly polarized luminescence spectrum of the complex. *Biochemistry* **19**, 1377–1382.

Edwards, P.D. & Bernstein, P.R. (1994) Synthetic inhibitors of elastase. *Med. Res. Rev.* **14**, 127–194.

Fric, P., Slaby, J., Kasafirek, E., Kocna, P. & Marek, J. (1992) Effective peritoneal therapy of acute pancreatitis in the rat with glutaryl-trialanine-ethylamide: a novel inhibitor of pancreatic elastase. *Gut* **33**, 701–706.

Gauthier, F., Genell, S., Mouray, H. & Ohlsson, K. (1978) The *in vitro* interactions of rat pancreatic elastase and normal and inflammatory rat serum. *Biochim. Biophys. Acta* **526**, 218–226.

Geokas, M.C., Largman, C., Brodrick, J.W. & Fassett, M. (1980) Molecular forms of immunoreactive pancreatic elastase in canine pancreatic and peripheral blood. *Am. J. Physiol.* **238**, G238–G246.

Gertler, A. & Birk, Y. (1970) Isolation and characterization of porcine proelastase. *Eur. J. Biochem.* **12**, 170–176.

Janoff, A. (1985) Elastases and emphysema. Current assessment of the protease-antiprotease hypothesis. *Am. Rev. Respir. Dis.* **132**, 417–433.

Kawashima, I., Tani, T., Mita-Honjo, K., Shimoda-Takano, K., Ohmine, T., Furukawa, H. & Takigushi, Y. (1992) Genomic organization of the human homologue of the rat pancreatic elastase I gene. *DNA* **2**, 303–312.

Kettner, C.A. & Shenvi, A.B. (1984) Inhibition of the serine proteases leukocyte elastase, pancreatic elastase, cathepsin G, and chymotrypsin by peptide boronic acids. *J. Biol. Chem.* **259**, 15106–15114.

Largman, C. (1983) Isolation and characterization of rat pancreatic elastase. *Biochemistry* **22**, 3763–3770.

McDonald, R.J., Swift, G.H., Quinto, C., Swain, W., Pictet, R.L., Nikovits, W. & Rutter, W.J. (1982) Primary structure of two distinct rat pancreatic preproelastases determined by sequence analysis of the complete cloned messenger ribonucleic acid sequences. *Biochemistry* **21**, 1453–1463.

Meyer, E.F., Radhakrishnan, R., Cole, G.M. & Presta, L.G. (1986) Structure of the product complex of acetyl-Ala-Pro-Ala with porcine pancreatic elastase at 1.65 Å resolution. *J. Mol. Biol.* **189**, 533–539.

Meyer, E.F., Cole, G., Radhakrishnan, R. & Epp, O. (1988) Structure of native porcine pancreatic elastase at 1.65 Å resolution. *Acta Cystallogr. B* **44**, 26–38.

Morii, M. & Travis, J. (1983) Amino acid sequence at the reactive site of human $\alpha_1$-antichymotrypsin. *J. Biol. Chem.* **258**, 12749–12752.

Murata, A., Ogawa, M., Fujimoto, K.I., Kitahara, T. & Matsuda, Y. (1983) Radioimmunoassay of human pancreatic elastase 1. *In vitro* interaction of human pancreatic elastase 1 with serum protease inhibitors. *Enzyme* **30**, 29–37.

Rose, S.D., Kruse, F., Swift, G.H., MacDonald, R.J. & Hammer, R.E. (1994) A single element of the elastase I enhancer is sufficient to direct transcription selectively to the pancreas and gut. *Mol. Cell. Biol.* **14**, 2048–2057.

Shen, W.F., Fletcher, T.S. & Largman, C. (1987) Primary structure of human pancreatic protease E determined by sequence analysis of the cloned mRNA. *Biochemistry* **26**, 3447–3452.

Shotton, D.M. (1970) Elastases. *Methods Enzymol.* **19**, 113–140.

Swift, G.H., Craik, C.S., Stary, S.J., Quinot, C., Lahaie, R.G., Rutter, W.J. & MacDonald, R.J. (1984) Structure of the two related elastase genes expressed in the rat pancreas. *J. Biol. Chem.* **259**, 14271–14278.

Takahashi, L., Radhakrishnan, R., Rosenfield, R., Meyer, E. & Trainor, D. (1989) Crystal structure of the covalent complex formed by a peptidyl $\alpha,\alpha$-difluoro-$\beta$-keto amide with porcine pancreatic elastase at 1.78-Å resolution. *J. Am. Chem. Soc.* **111**, 3368–3374.

Tsunemi, M., Matsuura, Y., Sakakibara, S. & Katsube, Y. (1993) Crystallization of a complex between an elastase-specific inhibitor elafin and porcine pancreatic elastase. *J. Mol. Biol.* **232**, 310–311.

van de Lest, C.H., Versteeg, E.M., Veerkamp, J.H. & van Kuppevelt, T.H. (1995) Digestion of proteoglycans in porcine pancreatic elastase-induced emphysema in rats. *Eur. Respir. J.* **8**, 238–245.

Wendorf, P., Linder, D., Sziegoleit, A. & Geyer, R. (1991) Carbohydrate structure of human pancreatic elastase 1. *Biochem. J.* **278**, 505–514.

Wiedow, O., Schroder, J., Gregory, H., Young, J. & Christophers, E. (1990) Elafin: an elastase-specific inhibitor of human skin. Purification, characterization and complete amino acid sequence. *J. Biol. Chem.* **265**, 14791–14795.

*Joseph G. Bieth*
*INSERM U 392,*
*Faculté de Pharmacie,*
*74 route du Rhin, F-67400 Illkirch, France*
*Email: jgbieth@pharma.u-strasbg.fr*

# 12. *Pancreatic elastase II*

### Databanks

*Peptidase classification: clan SA, family S1, MEROPS ID: S01.155*
*NC-IUBMB enzyme classification: EC 3.4.21.71*

*Databank codes:*

| Species | Form | SwissProt | PIR | EMBL (cDNA) | EMBL (genomic) |
|---|---|---|---|---|---|
| *Gadus morhua* | A | – | – | U57055 | – |
| *Homo sapiens* | A | P08217 | B26823 | D00236 M16631 M16652 | – |
| *Homo sapiens* | B | P08218 | C26823 | J03516 M16653 M18692 | – |
| *Mus musculus* | – | P05208 | A25528 | X04573 | – |
| *Rattus norvegicus* | – | P00774 | A00961 | V01233 | L00118: exons 1 and 2 L00119: exon 3 L00120: exon 4 L00121: exon 5 L00122: exon 6 L00123: exon 7 L00124: exon 8 and complete CDS |
| *Sus scrofa* | – | P08419 | A26823 | D00237 M16651 | – |

## Name, History and Occurrence

Elastases are defined by their ability to release soluble peptides from insoluble elastin fibers by a proteolytic process called elastinolysis. Elastinolytic activity is a *sine qua non* for naming a given proteinase an elastase. The pig pancreas contains two genomically and structurally distinct elastinolytic enzymes called elastase I (see Chapter 11) and elastase II (Kawashima *et al.*, 1987). Other animal species express only one pancreatic elastase either of type I or of type II. To date, ***pancreatic elastase II (PE II)*** has been found only in pig and human pancreas (Ohlsson & Olsson, 1976; Gertler *et al.*, 1977).

## Enzymatic and Structural Properties

PE II is a serine proteinase of the chymotrypsin (S1) family, whose catalytic site is composed of Asp102, His57 and Ser195 (chymotrypsinogen numbering), the charge-relay system which cleaves peptide bonds via formation of an acyl enzyme intermediate. Studies with synthetic substrates have shown that both the pig and human forms of PE II have chymotrypsin-like primary substrate specificity, in which the P1 preference is for a medium to large amino acid residue, such as Leu, Met, Phe or Tyr (Gertler *et al.*, 1977; Del Mar *et al.*, 1980). When solubilizing elastin, pig PE II hydrolyzes the bonds formed between Leu, Phe or Tyr (in P1) and Gly or Ala (in P1′), whereas pig pancreatic elastase I cleaves Ala-Ala and Ala-Gly linkages (Gertler *et al.*, 1977). Human PE II is best assayed with glt-Ala-Ala-Pro-Leu+NHPhNO$_2$ ($k_{cat}/K_m = 6500\,\mathrm{M^{-1}\,s^{-1}}$; Del Mar *et al.*, 1980), a commercially available substrate (Bachem) (see Appendix 2 for full names and addresses of suppliers).

PE II is inactivated by nonspecific irreversible serine proteinase inhibitors such as DFP and PMSF (Sigma). Pig PE II is irreversibly inhibited by acyl-peptidyl chloromethanes with a Phe residue at the P1 position (Gertler *et al.*, 1977). Human PE II forms irreversible complexes with three human plasma protein proteinase inhibitors, namely $\alpha_1$-proteinase inhibitor ($k_{ass} = 8 \times 10^5\,\mathrm{M^{-1}\,s^{-1}}$), $\alpha_1$-antichymotrypsin ($k_{ass} = 9 \times 10^5\,\mathrm{M^{-1}\,s^{-1}}$) and $\alpha_2$-macroglobulin ($k_{ass} = 6 \times 10^6\,\mathrm{M^{-1}\,s^{-1}}$) (Laurent & Bieth, 1989, and references therein). Interestingly, proelastase II, the inactive zymogen of human PE II also forms a stable complex with $\alpha_1$-proteinase inhibitor (Largman *et al.*, 1979). Human PE II is reversibly inhibited by mucus proteinase inhibitor ($K_i = 67\,\mathrm{nM}$) and eglin c ($K_i = 0.37\,\mathrm{nM}$) (Faller *et al.*, 1990, and references therein).

The human and pig forms of PE II are 26 kDa cationic proteins. The former is a single-chain protein (Ohlsson & Olsson, 1976), while the latter is probably formed of two chains held together by a disulfide bond (Gertler *et al.*, 1977). The primary structures of their precursors were elucidated by molecular cloning and cDNA sequence analysis (Kawashima *et al.*, 1987). They are synthesized as preproenzymes of 269 amino acid residues, including predicted signal and activation peptides of 16 and 12 residues, respectively. Human PE II is synthesized as two isoenzymes called IIA and IIB which share 90% sequence identity. Pig PE II and human PE IIA share 84% sequence identity, whereas pig pancreatic elastase I and PE II share only 60% identity. The S1 subsite of pig pancreatic elastase I comprises residues Val216 and Thr226, which partially occlude it and hinder the binding of bulky P1 residues. In contrast, pig and human PE II have Gly216 and Ser226, which explains their ability to hydrolyze substrates with bulky P1 residues.

## Biological Aspects

Blot hybridization analysis shows that the pancreas is the only organ containing PE II mRNAs (Kawashima *et al.*, 1987). Human PE II cleaves a number of human plasma proteins including complement C3, kininogen and fibronectin (Hakansson & Ohlsson, 1992). This enzyme is able to induce acute pancreatitis in the animal. In addition, histological and enzymatic studies suggest that it is responsible for the destruction of elastic tissue of intrapancreatic blood vessels in the course of acute hemorrhagic pancreatitis (Geokas *et al.*, 1968). Atherosclerosis is characterized by a fragmentation of elastin fibers within the media of blood vessels, suggesting a pathogenic role of elastinolytic enzymes such as human PE II (discussed by Bieth, 1986).

## References

Bieth, J.G. (1986) Elastases: Catalytic and biological properties. In: *Biology of Extracellular Matrix*, vol. 1: *Regulation of Matrix Accumulation* (Mecham. R.P., ed.). New York: Academic Press, pp. 217–320.

Del Mar, E.G., Largman, C., Brodrick, J.W., Fassett, M. & Geokas, M.G. (1980) Substrate specificity of human pancreatic elastase 2. *Biochemistry* **19**, 468–472.

Faller, B., Dirrig, S., Rabaud, M. & Bieth, J.G. (1990) Kinetics of the inhibition of human pancreatic elastase by recombinant eglin c. *Biochem. J.* **270**, 639–644.

Geokas, M.C., Rinderknecht, H., Swansson, V. & Haverback, B.J. (1968) The role of elastase in acute hemorrhagic pancreatitis in man. *Lab. Invest.* **19**, 235–239.

Gertler, A., Weiss, Y. & Burstein, Y. (1977) Purification and characterization of porcine elastase II and investigation of its elastolytic specificity. *Biochemistry* **16**, 2709–2716.

Hakansson, H.O. & Ohlsson, K. (1992) Influence of plasma proteinase inhibitor and the secretory leucocyte proteinase inhibitor on pancreatic elastase-induced degradation of some plasma proteins. *Gastroenterol. Jpn* **27**, 652–656.

Kawashima, I., Tani, T., Shimoda, K. & Takiguchi, Y. (1987) Characterization of pancreatic elastase II cDNAs: two elastase II mRNAs are expressed in human pancreas. *DNA* **6**, 163–172.

Largman, C., Brodrick, J.W., Geokas, M.C., Sischo, W.M. & Johnson, J.H. (1979) Formation of a stable complex between human proelastase 2 and human $\alpha_1$-proteinase inhibitor. *J. Biol. Chem.* **254**, 8516–8523.

Laurent, P. & Bieth, J.G. (1989) Kinetics of the inhibition of free and elastin-bound human pancreatic elastase by $\alpha_1$-proteinase inhibitor and $\alpha_2$-macroglobulin. *Biochim. Biophys. Acta* **994**, 285–288.

Ohlsson, K. & Olsson, A.S. (1976) Purification and partial characterization of human pancreatic elastase. *Z. Physiol. Chem.* **357**, S.1153–1161.

*Joseph G. Bieth*
*INSERM U 392,*
*Faculté de Pharmacie,*
*74 route du Rhin, F-67400 Illkirch, France*
*Email: jgbieth@pharma.u-strasbg.fr*

# 13. *Pancreatic endopeptidase E*

## Databanks

*Peptidase classification: clan SA, family S1, MEROPS ID: S01.154*
*NC-IUBMB enzyme classification: EC 3.4.21.70*
*Chemical Abstracts Service registry number: 68073-27-8*
*Databank codes:*

| Species | Form | SwissProt | PIR | EMBL (cDNA) | EMBL (genomic) |
|---------|------|-----------|-----|-------------|----------------|
| *Homo sapiens* | A | P08861 | B29934 | M16630 M18692 | – |
| *Homo sapiens* | B | P09093 | A29934 | D00306 | M18693: exon 1 M18694: exon 2 M18695: exon 3 M18696: exon 4 M18697: exon 5 M18698: exon 6 M18699: exon 7 M18700: exon 8 and complete CDS |
| *Sus scrofa* | – | – | B33257 | – | – |

## Name, History and Occurrence

Mallory & Travis (1975) discovered a new proteinase in human pancreatic tissue and named it *protease E*. The enzyme was able to hydrolyze synthetic substrates of pig pancreatic elastase I (Chapter 11), but did not solubilize fibrous elastin to a significant extent. In addition, protease E differed from pancreatic elastase I in its anionic character. Protease E, now named *pancreatic endopeptidase E (PEE)* has also been called *proteinase E, elastase III, cholesterol-binding pancreatic proteinase, pancreatic protein P35, protein G32* and *subunit III of the procarboxypeptidase A S6 complex* (Figarella, 1992). PEE occurs in human, pig, cattle, sheep and goat pancreas as a trypsin-activatable zymogen (Figarella, 1992). Human PEE is sometimes misnamed *human pancreatic elastase I* (Sziegoleit *et al.*, 1985), an enzyme that probably does not exist (Shen *et al.*, 1987) (see Chapter 11), or *human pancreatic elastase III*, which is misleading, since it is not an elastase.

## Enzymatic and Structural Properties

As a serine proteinase of family S1, PEE has a catalytic site composed of Asp102, His57 and Ser195 (chymotrypsin numbering), the charge-relay system which cleaves peptide bonds via formation of an acyl enzyme intermediate. PEE is specific for nonbulky hydrophobic amino acid residues such as Ala or Val in P1 (Mallory & Travis, 1975; Kobayashi *et al.*, 1981) and has a pH optimum of 7.7–9.5 (Mallory & Travis, 1975). It hydrolyzes Ac-Ala-Ala-Ala-OMe and Suc-Ala-Ala-Ala-NHPhNO$_2$, two substrates of pancreatic elastase I, but does not solubilize fibrous elastin to a significant extent (Mallory & Travis, 1975). PEE may be conveniently assayed with Suc-Ala-Ala-Ala-NHPhNO$_2$ (Bachem) (see Appendix 2 for full names and addresses of suppliers) as substrate. It is inactivated by a series of acyl-peptidyl chloromethanes with Ala in P1, and by plasma $\alpha_1$-proteinase inhibitor (Mallory & Travis, 1975).

PEE is a 35 kDa glycoprotein (Guy-Crotte *et al.*, 1988) with a pI of 4.8 (Kobayashi *et al.*, 1981). Shen *et al.* (1987) deduced the amino acid sequence of human PEE from the nucleotide sequence of a cDNA clone. The amino acid residues forming the primary substrate-binding site of the enzyme showed a high degree of identity to those of pig pancreatic elastase I, explaining the specificity of PEE for alanine-containing substrates, but also, two new Cys residues not present in other serine proteinases were identified at positions 117 and 120. Tani *et al.* (1988) isolated two different, but closely related human pancreatic cDNAs encoding forms of PEE (misleadingly termed elastase III A and elastase III B), with 93% amino acid sequence identity. Analysis of the cloned genes revealed the presence of seven introns and eight exons, exons 1 and 2 coding for the signal peptide and the propeptide, respectively. The crystal structure of subunit III (a truncated form of PEE) has been solved for the cattle and pig species variants (Pignol *et al.*, 1994, 1995).

## Biological Aspects

Human PEE is also a cholesterol- and deoxycholate-binding protein (Sziegoleit *et al.*, 1985). Human pancreatic juice contains a 35 kDa glycoprotein called protein P35 whose N-terminal sequence is identical to that of PEE but lacks the first two Val residues. This truncated PEE was also found to be associated with procarboxypeptidase A in human pancreatic juice (Moulard *et al.*, 1989). *In vitro* experiments suggest that truncated PEE results from cleavage of the Val13↓Asn14 bond of proPEE by active PEE (Avilés *et al.*, 1989). Although inactive on Suc-Ala-Ala-Ala-NHPhNO$_2$, truncated human PEE slowly turns over the smaller substrate, Ac-OPhNO$_2$ (*p*-nitrophenylacetate), a behavior that resembles that of the zymogens of pancreatic serine proteinases.

Ruminant truncated PEE, also called subunit III, forms a ternary complex with procarboxypeptidase A (subunit I) and chymotrypsinogen C (subunit II) (Gomis-Rüth *et al.*, 1995). The exact nature of subunit III has been a mystery for three decades (Figarella, 1992). The truncation of ruminant PEE involves a deletion of the N-terminal Val-Val dipeptide, and is caused by the action of active PEE on proPEE, as is the case for the human enzyme (Pascual *et al.*, 1990). Like human truncated PEE, it lacks significant enzymatic activity. Recent crystallographic data confirm that the substrate-binding site of truncated cattle PEE has a zymogen-like conformation that prevents easy substrate binding (Pignol *et al.*, 1994).

## References

Avilés, F.X., Pascual, R., Salva, M., Bonicel, J. & Puigserver, A. (1989) Generation of a subunit III-like protein by autolysis of human and porcine proproteinase in a binary complex with procarboxypeptidase A. *Biochem. Biophys. Res. Commun.* **163**, 1191–1196.

Figarella, C. (1992) What is human pancreatic proelastase 1? *Int. J. Pancreatol.* **11**, 213–215.

Gomis-Rüth, F.X., Gómez, M., Bode, W., Huber, R. & Avilés, F.X. (1995) The three-dimensional structure of the native ternary complex of bovine pancreatic procarboxypeptidase A with proproteinase E and chymotrypsinogen C. *EMBO J.* **14**, 4387–4394.

Guy-Crotte, O., Barthe, C., Basso, D., Fournett, B. & Figarella, C. (1988) Characterization of two glycoproteins of human pancreatic juice: P35, a truncated protease E and P19, precursor of protein X. *Biochem. Biophys. Res. Commun.* **156**, 318–322.

Kobayashi, Y., Kobayashi, R. & Hirs, C.H.W. (1981) Identification of zymogen E in a complex with bovine procarboxypeptidase A. *J. Biol. Chem.* **256**, 2466–2470.

Mallory, P.A. & Travis, J. (1975) Human pancreatic enzymes: purification and characterization of a nonelastolytic enzyme, protease E, resembling elastase. *Biochemistry* **14**, 722–730.

Moulard, M., Kerfelec, B., Mallet, B. & Chapus, C. (1989) Identification of a procarboxypeptidase A-truncated protease E binary complex in human pancreatic juice. *FEBS Lett.* **250**, 166–170.

Pascual, R., Vendrell, J., Avilés, F.X., Bonicel, J., Wicker, C. & Puigserver, A. (1990) Autolysis of proproteinase E in bovine procarboxypeptidase A ternary complex gives rise to subunit III. *FEBS Lett.* **277**, 37–41.

Pignol, D., Gaboriaud, C., Michon, T., Kerfelec, B., Chapus, C. & Fontecilla-Camps, J.C. (1994) Crystal structure of bovine procarboxypeptidase A-S6 subunit III, a highly structured truncated zymogen E. *EMBO J.* **13**, 1763–1771.

Pignol, D., Granon, S., Chapus, C. & Fontecilla-Camps, J.C. (1995) Crystallographic study of a cleaved, non-activatable form of porcine zymogen E. *J. Mol. Biol.* **252**, 20–24.

Shen, W.F., Fletcher, T.S. & Largman, C. (1987) Primary structure of human pancreatic protease E determined by sequence analysis of the cloned mRNA. *Biochemistry* **26**, 3447–3452.

Sziegoleit, A., Linder, D., Schlütter, M., Ogawa, M., Nishibe, S. & Fujimoto, K. (1985) Studies on the specificity of the cholesterol-binding pancreatic proteinase and identification as

human pancreatic elastase 1. *Eur. J. Biochem.* **151**, 595–599.

Tani, T., Ohsumi, J., Mita, K. & Takigushi, Y. (1988) Identification of a novel class of elastase isozyme, human pancreatic elastase III, by cDNA and genomic gene cloning. *J. Biol. Chem.* **263**, 1231–1239.

*Joseph G. Bieth*
*INSERM U 392,*
*Faculté de Pharmacie,*
*74 route du Rhin, F-67400 Illkirch, France*
*Email: jgbieth@pharma.u-strasbg.fr*

# 14. Enteropeptidase

## Databanks

*Peptidase classification: clan SA, family S1, MEROPS ID: S01.156*
*NC-IUBMB enzyme classification: EC 3.4.21.9*
*Chemical Abstracts Service registry number: 9014-74-8*
*Databank codes:*

| Species | SwissProt | PIR | EMBL (cDNA) | EMBL (genomic) |
| --- | --- | --- | --- | --- |
| *Bos taurus* | P98072 | A43090 | L19663 U09859 | – |
| *Homo sapiens* | P98073 | A56318 | U09860 | – |
| *Sus scrofa* | P98074 | – | D30799 | |

## Name and History

Enteropeptidase was discovered in the laboratory of I.P. Pavlov, who was awarded the 1904 Nobel Prize in Medicine or Physiology for his studies of gastrointestinal physiology. Digestive enzymes within the pancreas were known to be inert compared to their potent activity within the intestine, but the basis of this difference was unknown. In 1899, Pavlov's student, N.P. Schepowalnikov, demonstrated that canine duodenal secretions dramatically stimulated the digestive activity of pancreatic enzymes, especially trypsinogen. The active principle in small intestine was heat labile and was attributed to a special enzyme in the intestine that could activate other enzymes ('a ferment of ferments'). Pavlov named it *enterokinase* (Schepowalnikov, 1900; Pavlov, 1902). Whether enterokinase was an enzyme or a cofactor remained controversial until Kunitz proved that the activation of trypsinogen by enterokinase was catalytic (Kunitz, 1939). In the 1950s, cattle trypsinogen was shown to be activated autocatalytically by cleavage of an N-terminal hexapeptide, Val-Asp-Asp-Asp-Asp-Lys6+Ile (Davie & Neurath, 1955) and enterokinase was shown to cleave the same bond (Yamashina, 1956). The more precise

IUBMB name *enteropeptidase* has been in existence since 1970. However, the original name 'enterokinase' has a long history and remains in common use.

## Activity and Specificity

Enteropeptidase cleaves internal bonds of peptides and proteins at sites that closely resemble the activation cleavage site of cattle trypsinogen: Asp-Asp-Asp-Asp-Lys+Ile. The $(Asp)_4$-Lys activation peptide sequence is highly conserved among vertebrate trypsinogens (Bricteux-Gregoire *et al.*, 1972). Most of the few exceptions have Glu instead of Asp at P2, P3 or P4; a minor rat trypsinogen has Arg at P1 and Asn at P3. Trypsinogens from some fish and from *Xenopus laevis* have Ile, Leu, Ala or Phe at P5. Limited data with small synthetic peptides indicate that cleavage requires two acidic residues at P2 and P3 in addition to Lys at P1; additional acidic residues at P4 and P5 markedly increased the affinity for enteropeptidase (Maroux *et al.*, 1971; Light & Janska, 1989). In contrast to the stringent requirements for the P1–P5 positions, enteropeptidase shows little dependence on the structure of the peptide on the C-terminal side of the scissile bond (Maroux *et al*., 1971). This remarkable specificity has made enteropeptidase a useful reagent for the separation of recombinant fusion proteins that are joined by engineered enteropeptidase cleavage sites (Hopp *et al*., 1988).

Enteropeptidase is active between pH 6 and 9, with a broad optimum at about pH 8 (Kunitz, 1939; Maroux *et al*., 1971; Magee *et al*., 1981). Activity toward trypsinogen is inhibited by increasing ionic strength, due mainly to an increase in $K_m$ (Baratti & Maroux, 1976). Calcium ions at 4–10 mM stabilize enteropeptidase to heat denaturation (Barns & Elmslie, 1974; Anderson *et al*., 1977) and also modestly increase its activity (Baratti *et al*., 1973a; Grant & Hermon-Taylor, 1979; Magee *et al*., 1981). The calcium-dependent increase in trypsinogen activation may be due in part to the binding of calcium ions by the acidic trypsinogen activation peptide (Abita *et al*., 1969).

The (Asp)$_4$-Lys┼Ile activation cleavage site of trypsinogen is the preferred substrate for enteropeptidase, but the multiple Asp residues markedly inhibit cleavage of trypsinogen by trypsin (Abita *et al*., 1969). For example, toward cattle trypsinogen the catalytic efficiency of enteropeptidase is 12 000-fold (pig) (Maroux *et al*., 1971) or 34 000-fold (cattle) (Anderson *et al*., 1977) greater than that of cattle trypsin. This reciprocal specificity protects trypsinogen against autoactivation, and promotes activation by enteropeptidase in the gut.

Several methods are used to assay trypsinogen activation by enteropeptidase; the trypsin product may be quantitated by cleavage of Bz-Arg-OEt (Baratti *et al*., 1973a), or trapped as a complex with chicken ovomucoid (Janska & Light, 1989). A relatively specific synthetic enteropeptidase substrate, Gly-(Asp)$_4$-Lys┼NHNap, is commercially available (Bachem) (see Appendix 2 for full names and addresses of suppliers) and cleavage can be assayed fluorometrically (LaVallie *et al*., 1993) or spectrophotometrically (Grant & Hermon-Taylor, 1979). This substrate can also be used for the histochemical identification of enteropeptidase in tissue sections (Lojda & Gossrau, 1983).

Enteropeptidase is not inhibited by chicken ovomucoid, soybean trypsin inhibitor or limabean trypsin inhibitor (Maroux *et al*., 1971; Liepnieks & Light, 1979; Green, 1983). It is inhibited by Tos-Lys-CH$_2$Cl, DFP, *p*-aminobenzamidine ($K_i$ 9.5 μM) and benzamidine ($K_i$ 22 μM) (Baratti & Maroux, 1976). Cattle enteropeptidase is inhibited by cattle pancreatic trypsin inhibitor (Liepnieks & Light, 1979), whereas pig and human enteropeptidase are resistant (Maroux *et al*., 1971;

Light & Janska, 1989). An enteropeptidase inhibitor (60 kDa) was purified from kidney bean (Jacob *et al*., 1983).

## Structural Chemistry

The N-terminal sequence of the cattle enteropeptidase light chain was determined by Edman degradation, and indicated that enteropeptidase is homologous to the trypsin-like serine proteases (family S1) (Light & Janska, 1991). The sequences of cattle (LaVallie *et al*., 1993; Kitamoto *et al*., 1994), human (Kitamoto *et al*., 1994, 1995), pig (Matsushima *et al*., 1994) and rat enteropeptidase (Yahagi *et al*., 1996) were determined indirectly by cDNA cloning. The human gene is on chromosome 21q21 (Kitamoto *et al*., 1995).

Enteropeptidase is synthesized as a single-chain zymogen (Figure 14.1). The primary translation product for cattle enteropeptidase consists of 1035 residues with a predicted mass of 114.9 kDa (Kitamoto *et al*., 1994). The observed apparent mass of about 160 kDa is consistent with the reported carbohydrate content of 30–40%, with equal amounts of neutral and amino sugars (Anderson *et al*., 1977; Liepnieks & Light, 1979). The activation cleavage site after Lys800 separates the heavy and light chains of mature cattle enteropeptidase. There are 17 potential *N*-linked glycosylation sites in the heavy chain and three in the light chain; most of these are conserved in other species. Digestion with peptide: *N*-glycosidase F reduces the apparent mass of both chains (Kitamoto *et al*., 1994).

The heavy chain has a hydrophobic segment near the N-terminus that could serve as a signal peptide or a transmembrane anchor. The heavy chain also contains structural motifs (Figure 14.1) found in other protein families (Kitamoto *et al*., 1994; Matsushima *et al*., 1994). These include a mucin-like domain that may be heavily *O*-glycosylated; two low-density lipoprotein receptor cysteine-rich repeats; two repeats found in the complement serine proteases C1r and C1s (see Chapters 43 and 44); a domain named 'MAM' for motifs found in the metalloprotease 'meprin' (Chapter 406), the *X. laevis* neuronal protein 'A5', and protein tyrosine phosphatase μ ('mu'); and a macrophage scavenger receptor cysteine-rich repeat.

The heavy chain influences the specificity of enteropeptidase. Native enteropeptidase is resistant to soybean trypsin inhibitor, but the isolated light chain is sensitive whether

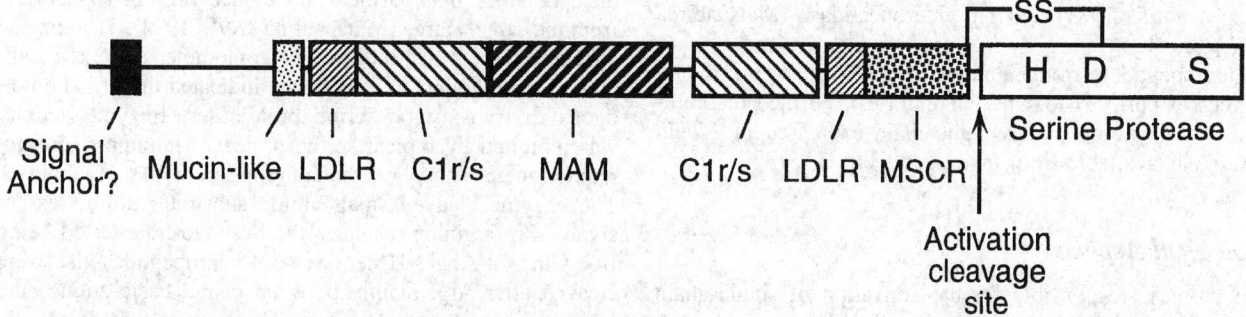

*Figure 14.1*  Schematic structure of enteropeptidase. The active site His (H), Asp (D), and Ser (S) of the serine protease domain are indicated. Locations are shown of the activation cleavage site between the heavy and light chains (arrow) and the predicted disulfide bond that connects them. The labeled domains of enteropeptidase are discussed in the text. (Adapted with permission from Kitamoto *et al*., 1995.)

prepared by limited reduction of the natural protein (Light & Fonseca, 1984) or by mutagenesis and expression in COS cells (LaVallie *et al*., 1993). Native enteropeptidase and the isolated light chain have similar activity toward Gly-(Asp)$_4$-Lys-NHNap, but the isolated light chain has markedly decreased activity toward trypsinogen (LaVallie *et al*., 1993). A similar selective defect in the recognition of trypsinogen can be produced in two-chain enteropeptidase by heating (Barns & Elmslie, 1974; Anderson *et al*., 1977) or by acetylation (Baratti & Maroux, 1976). This behavior suggests that the catalytic center and one or more secondary substrate-binding sites are necessary for optimal recognition of trypsinogen.

The enteropeptidase light chain is a typical trypsin-like serine protease; its active site residues, disulfide bonds and S1 subsite can be predicted by analogy with those of chymotrypsin. A disulfide bond is predicted to link the heavy and light chains. The acidic residues of preferred substrates may interact with a cluster of four basic amino acid residues in a surface loop that is adjacent to the catalytic center. These basic residues are conserved among enteropeptidases from different species, but are not found in other peptidases of family S1 (LaVallie *et al*., 1993; Kitamoto *et al*., 1994; Matsushima *et al*., 1994).

The enteropeptidase heavy chain is postulated to mediate association with intestinal brush border membranes (Fonseca & Light, 1983a), and this function is consistent with the presence of a hydrophobic domain near the N-terminus. Purified pig enteropeptidase was shown to lack this potential transmembrane domain, however, suggesting that another mechanism could be responsible for membrane attachment (Matsushima *et al*., 1994).

## Preparation

Enteropeptidase has been purified from human (Magee *et al*., 1981), cattle (Anderson *et al*., 1977; Liepnieks & Light, 1979; Fonseca & Light, 1983b), pig (Baratti *et al*., 1973b; Matsushima *et al*., 1994), and ostrich duodenal mucosa or fluid (Naude *et al*., 1993). The purification procedures generally include ion-exchange chromatography and affinity chromatography on either *p*-aminobenzamidime or pancreatic trypsin inhibitor. In most species, purified enteropeptidase appears to be a disulfide-linked heterodimer with a heavy chain of 82–140 kDa and a light chain of 35–62 kDa. However, heterotrimeric structures have been suggested for human (Magee *et al*., 1981) and pig enteropeptidase (Matsushima *et al*., 1994).

Enteropeptidase from cattle intestine is available in several degrees of purity (Biozyme Laboratories). Purified recombinant cattle enteropeptidase light chain expressed in *Pichia pastoris* is available from Invitrogen Corp.

## Biological Aspects

The protease responsible for the activation of single-chain proenteropeptidase has not been identified. The sequence surrounding the activation cleavage site is Ile-Thr-Pro-Lys+Ile (human), Val-Ser-Pro-Lys+Ile (cattle, pig), or Val-Gly-Pro-Lys+Ile (rat). These are very poor sites for autoactivation but excellent sites for cleavage by trypsin and many related proteases. If trypsin were the enteropeptidase activator *in vivo*, this would appear to invoke a closed cycle and would raise the logical question of how a trypsin-enteropeptidase cycle could be initiated or sustained.

Enteropeptidase mRNA expression is limited to the proximal small intestine (Kitamoto *et al*., 1995), and the protein is found in enterocytes of duodenum and proximal jejunum (Hermon-Taylor *et al*., 1977; Lojda & Gossrau, 1983; Miyoshi *et al*., 1990). Upon secretion from the pancreas into the duodenum, trypsinogen encounters enteropeptidase and is activated. Trypsin then cleaves and activates other pancreatic serine protease zymogens (chymotrypsinogen and proelastases), metalloprotease zymogens (procarboxypeptidases) and prolipases. By means of this simple two-step cascade, the destructive activity of these digestive hydrolases is confined to the lumen of the intestine. The physiological importance of this pathway is demonstrated by the severe intestinal malabsorption caused by congenital deficiency of enteropeptidase (Hadorn *et al*., 1969; Haworth *et al*., 1971). This condition can be life threatening, but responds to oral supplementation with pancreatic extract.

In animal models, injection of enteropeptidase into the pancreatic duct system activates trypsinogen within the pancreas and causes a fatal hemorrhagic pancreatitis. A similar reflux of enteropeptidase into the pancreatic duct is proposed to cause the acute necrotizing pancreatis that sometimes occurs in patients with gallstones or alcoholism (Grant, 1986).

## Distinguishing Features

Enteropeptidase is distinguished easily from related serine proteases by its localization in the microvillus membrane of duodenal enterocytes, by an unusually high degree of glycosylation, and by its extraordinary specificity for substrates that resemble the acidic activation peptide of trypsinogen. A monoclonal antibody to human enteropeptidase has been described (Miyoshi *et al*., 1990).

## Related Peptidases

A distinct membrane-bound serine proteinase was purified from pig small intestinal mucosa (Tsuchiya *et al*., 1994) by the laboratory that purified and cloned pig enteropeptidase from the same source (Matsushima *et al*., 1994). The enzyme appeared to be a disulfide-linked homodimer of 32 kDa subunits. Partial amino acid sequence indicated that it is homologous to trypsin-type serine proteinases. Enzyme activity was inhibited by typical serine proteases inhibitors, did not require metal ions, and had an optimum pH of about 9. The enzyme cleaved small amide substrates almost exclusively after arginine residues, the best substrate tested being Boc-Gln-Ala-Arg+NHMec. Several neuropeptides also were cleaved after Arg, mainly between paired basic amino acid residues, Arg+Arg or Arg+Lys. The enzyme was essentially inactive toward large proteins. This Arg-specific, trypsin-like endopeptidase was proposed to be involved in the metabolism of certain gastrointestinal neuropeptides or peptide hormones (Tsuchiya *et al.*, 1994).

## Further Reading

A concise review of the substrate requirements of enteropeptidase has been provided by Light & Janska (1989).

## References

Abita, J.P., Delaage, M., Lazdunski, M. & Savrda, J. (1969) The mechanism of activation of trypsinogen. The role of the four N-terminal aspartyl residues. *Eur. J. Biochem.* **8**, 314–324.

Anderson, L.E., Walsh, K.A. & Neurath, H. (1977) Bovine enterokinase. Purification, specificity, and some molecular properties. *Biochemistry* **16**, 3354–3360.

Baratti, J. & Maroux, S. (1976) On the catalytic and binding sites of porcine enteropeptidase. *Biochim. Biophys. Acta* **452**, 488–496.

Baratti, J., Maroux, S. & Louvard, D. (1973a) Effect of ionic strength and calcium ions on the activation of trypsinogen by enterokinase. *Biochim. Biophys. Acta* **321**, 632–638.

Baratti, J., Maroux, S., Louvard, D. & Desnuelle, P. (1973b) On porcine enterokinase. Further purification and some molecular properties. *Biochim. Biophys. Acta* **315**, 147–161.

Barns, R.J. & Elmslie, R.G. (1974) The active site of porcine enteropeptidase. Selective inactivation of the peptidase activity. *Biochim. Biophys. Acta* **350**, 495–498.

Bricteux-Gregoire, S., Schyns, R. & Florkin, M. (1972) Phylogeny of trypsinogen activation peptides. *Comp. Biochem. Physiol.* **42B**, 23–39.

Davie, E.W. & Neurath, H. (1955) Identification of a peptide released during autocatalytic activation of trypsinogen. *J. Biol. Chem.* **212**, 515–529.

Fonseca, P. & Light, A. (1983a) Incorporation of bovine enterokinase in reconstituted soybean phospholipid vesicles. *J. Biol. Chem.* **258**, 3069–3074.

Fonseca, P. & Light, P. (1983b) The purification and characterization of bovine enterokinase from membrane fragments in the duodenal mucosal fluid. *J. Biol. Chem.* **258**, 14516–14520.

Grant, D. (1986) Acute necrotizing pancreatitis – a role for enterokinase. *Int. J. Pancreatol.* **1**, 167–183.

Grant, D.A.W. & Hermon-Taylor, J. (1979) Hydrolysis of artificial substrates by enterokinase and trypsin and the development of a sensitive specific assay for enterokinase in serum. *Biochim. Biophys. Acta* **567**, 207–215.

Green, G.M. (1983) Use of trypsin inhibitors to prevent autoactivation in the assay of enterokinase. *Clin. Chim. Acta* **134**, 363–367.

Hadorn, B., Tarlow, M.J., Lloyd, J.K. & Wolff, O.H. (1969) Intestinal enterokinase deficiency. *Lancet* **i**, 812–813.

Haworth, J.C., Gourley, B., Hadorn, B. & Sumida, C. (1971) Malabsorption and growth failure due to intestinal enterokinase deficiency. *J. Pediatr.* **78**, 481–490.

Hermon-Taylor, J., Perrin, J., Grant, D.A.W., Appleyard, A. & Magee, A.I. (1977) Immunofluorescent localization of enterokinase in human small intestine. *Gut* **18**, 259–265.

Hopp, T.P., Prickett, K.S., Price, V.L., Libby, R.T., March, C.J., Cerretti, D.P., Urdal, D.L. & Conlon, P.J. (1988) A short polypeptide marker sequence useful for recombinant protein identification and purification. *Bio/Technology* **6**, 1204–1210.

Jacob, R.T., Bhat, P.G. & Pattabiraman, T.N. (1983) Isolation and characterization of a specific enterokinase inhibitor from kidney bean *(Phaseolus vulgaris)*. *Biochem. J.* **209**, 91–97.

Janska, H. & Light, A. (1989) A direct high-performance liquid chromatography assay of the enzymatic activity of enterokinase (enteropeptidase). *Anal. Biochem.* **176**, 132–136.

Kitamoto, Y., Yuan, X., Wu, Q., McCourt, D.W. & Sadler, J.E. (1994) Enterokinase, the initiator of intestinal digestion, is a mosaic protease composed of a distinctive assortment of domains. *Proc. Natl Acad. Sci. USA* **91**, 7588–7592.

Kitamoto, Y., Veile, R.A., Donis-Keller, H. & Sadler, J.E. (1995) cDNA sequence and chromosomal localization of human enterokinase, the proteolytic activator of trypsinogen. *Biochemistry* **34**, 4562–4568.

Kunitz, M. (1939) Formation of trypsin from crystalline trypsinogen by means of enterokinase. *J. Gen. Physiol.* **22**, 429–446.

LaVallie, E.R., Rehemtulla, A., Racie, L.A., DiBlasio, E.A., Ferenz, C., Grant, K.L., Light, A. & McCoy, J.M. (1993) Cloning and functional expression of a cDNA encoding the catalytic subunit of bovine enterokinase. *J. Biol. Chem.* **268**, 23311–23317.

Liepnieks, J.J. & Light, A. (1979) The preparation and properties of bovine enterokinase. *J. Biol. Chem.* **254**, 1677–1683.

Light, A. & Fonseca, P. (1984) The preparation and properties of the catalytic subunit of bovine enterokinase. *J. Biol. Chem.* **259**, 13195–13198.

Light, A. & Janska, H. (1989) Enterokinase (enteropeptidase): comparative aspects. *Trends Biochem. Sci.* **14**, 110–112.

Light, A. & Janska, H. (1991) The amino-terminal sequence of the catalytic subunit of bovine enterokinase. *J. Protein Chem.* **10**, 475–480.

Lojda, Z. & Gossrau, R. (1983) Histochemical demonstration of enteropeptidase activity. New method with a synthetic substrate and its comparison with the trypsinogen procedure. *Histochemistry* **78**, 251–270.

Magee, A.I., Grant, D.A.W. & Hermon-Taylor, J. (1981) Further studies on the subunit structure and oligosaccharide moiety of human enterokinase. *Clin. Chim. Acta* **115**, 241–254.

Maroux, S., Baratti, J. & Desnuelle, P. (1971) Purification and specificity of porcine enterokinase. *J. Biol. Chem.* **246**, 5031–5039.

Matsushima, M., Ichinose, M., Yahagi, N., Kakei, N., Tsukada, S., Miki, K., Kurokawa, K., Tashiro, K., Shiokawa, K., Shinomiya, K., Umeyama, H., Inoue, H., Takahashi, T. & Takahashi, K. (1994) Structural characterization of porcine enteropeptidase. *J. Biol. Chem.* **269**, 19976–19982.

Miyoshi, Y., Onishi, T., Sano, T. & Komi, N. (1990) Monoclonal antibody against human enterokinase and immunohistochemical localization of the enzyme. *Gastroenterol. Jpn* **25**, 320–327.

Naude, R.J., Da Silva, D., Edge, W. & Oelofsen, W. (1993) The isolation and partial characterization of enterokinase from ostrich *(Struthio camelus)* duodenal mucosa. *Comp. Biochem. Physiol.* **105B**, 591–595.

Pavlov, I.P. (1902) *The Work of the Digestive Glands*, 1st edn (trans. Thompson, W.H.). London: Charles Griffin & Co.

Schepowalnikov, N.P. (1900) Die Physiologie des Darmsaftes. *Jahresbericht über die Fortschritte der Thierchemie* **29**, 378–380.

Tsuchiya, Y., Takahashi, T., Sakurai, Y., Iwamatsu, A. & Takahashi, K. (1994) Purification and characterization of a novel membrane-bound arginine-specific serine proteinase from porcine intestinal mucosa. *J. Biol. Chem.* **269**, 32985–32991.

Yahagi, N., Ichinose, M., Matsushima, M., Matsubara, Y., Miki, K., Kurokawa, K., Fukamachi, H., Tashiro, K., Shiokawa, K., Kageyama, T., Takahashi, T., Inoue, H. & Takahashi, K. (1996) Complementary DNA cloning and sequencing of rat enteropepti-

dase and tissue distribution of its mRNA. *Biochem. Biophys. Res. Commun.* **219**, 806–812.
Yamashina, I. (1956) The action of enterokinase on trypsinogen, *Acta Chem. Scand.* **10**, 739–743.

*Deshun Lu*
Departments of Medicine and of Biochemistry and
Molecular Biophysics,
Washington University School of Medicine,
660 South Euclid Avenue, Box 8022,
St. Louis, Missouri 63110, USA

*J. Evan Sadler*
Departments of Medicine and of Biochemistry and
Molecular Biophysics,
Washington University School of Medicine,
660 South Euclid Avenue, Box 8022,
St. Louis, Missouri 63110, USA
Email: esadler@imgate.wustl.edu

# 15. Leukocyte elastase

## Databanks

Peptidase classification: clan SA, family S1, MEROPS ID: S01.131
NC-IUBMB enzyme classification: EC 3.4.21.37
Databank codes:

| Species | SwissProt | PIR | EMBL (cDNA) | EMBL (genomic) |
| --- | --- | --- | --- | --- |
| *Equus caballus* | P37357 | S44461 | – | – |
| *Equus caballus* | P37358 | S44462 | – | – |
| *Homo sapiens* | P08246 | A31976 | D00187 | M20199: exon 1 |
| | | | J03545 | M20200: exon 2 |
| | | | M27783 | M20201: exon 3 |
| | | | M34379 | M20202: exon 4 |
| | | | X05875 | M20203: exon 5 and complete CDS |
| | | | | Y00477: complete gene |

Brookhaven Protein Data Bank three-dimensional structures:

| Species | ID | Resolution | Notes |
| --- | --- | --- | --- |
| *Homo sapiens* | 1HNE | 1.84 | complex with MeOSuc-Ala-Ala-Pro-Ala chloromethane |
| | 1PPF | 1.8 | complex with turkey ovomucoid third domain |
| | 1PPG | 2.3 | complex with MeOSuc-Ala-Ala-Pro-Val chloromethylacetone |

## Name and History

Although **leukoprotease** activity was described early in this century, **human leukocyte elastase (human LE)** was only identified in 1968 (Janoff & Scherer, 1968). The term *elastase* describes an enzyme capable of the proteolytic release of soluble peptides from insoluble elastin, and such activity is a *sine qua non* for defining a proteinase as an elastase. The term **leukocyte** is not really precise enough in the present context, because elastase is not present in all white blood cells, but mainly in polymorphonuclear leukocytes.

**Polymorphonuclear leukocyte elastase** or **neutrophil elastase** might be better names.

## Activity and Specificity

Leukocyte elastase belongs to the chymotrypsin family of serine proteinases (S1). The pH optimum of these enzymes is generally close to neutrality, and the catalytic site is composed of three hydrogen-bonded amino acid residues, His57, Asp102 and Ser195 (in chymotrypsinogen numbering), which form the so-called charge-relay system or catalytic triad. Peptide bond hydrolysis occurs in several steps. The initial reversible formation of an adsorption (Michaelis) complex between the serine proteinase and its substrate is followed by a nucleophilic attack of the $\gamma$-oxygen of Ser195 on the peptide bond with subsequent formation of an acyl enzyme intermediate and release of the amine portion of the substrate. This covalent intermediate is then hydrolyzed in the deacylation step, which regenerates active enzyme and releases the carboxyl moiety of the substrate (Powers & Harper, 1986).

The substrate-binding site of human LE, i.e. the site responsible for its specificity, has been investigated with insulin, synthetic substrates, and by X-ray crystallography. Human LE preferentially cleaves the oxidized insulin B chain with Val at P1, but also hydrolyzes bonds with Ala, Ser or Cys in P1 (Blow, 1976). The preference of Val over Ala has been confirmed with model substrates. The catalytic activity of elastase increases with the chain length of the synthetic substrates, suggesting an extended substrate-binding site (Lestienne & Bieth, 1980). X-Ray crystallography of the complex between elastase and the third domain of ovomucoid, a protein proteinase inhibitor, shows that the substrate-binding site of human LE is composed of eight subsites, and that subsite S1 can accommodate Leu, in addition to Val or Ala. The amino acid residues that are likely to form the specificity subsites of human LE have been discussed by Bode *et al.* (1989).

The catalytic activity of human LE may be regulated by secondary enzyme–substrate interactions, i.e. by interactions remote from the S1–P1 binding. Thus, activity is affected by the substrate chain length, and also by the nature of amino acid residues in positions other than P1 (Yasutake & Powers, 1981). The catalytic activity of human LE is enhanced by ionic strength and by hydrophobic solvents (Lestienne & Bieth, 1980). Another example of regulation of activity by secondary enzyme–substrate interactions is the activation of catalysis by excess substrate that is observed with some synthetic substrates (Lestienne & Bieth, 1980). This phenomenon is best explained by assuming that at high substrate concentrations, two molecules of substrate bind to the active site, one in a productive way, and the other in a nonproductive mode that favors the hydrolysis of the productively bound molecule. For moderately sensitive substrates such as Suc-Ala-Ala-Val⧧NHPhNO$_2$, only one proton is involved in catalysis, whereas for good substrates such as MeOSuc-Ala-Ala-Pro-Val⧧NHPhNO$_2$, the two protons of the charge-relay system are transferred during catalysis. The catalytic site and the substrate-binding site are therefore intimately related (Stein, 1983).

Unlike most proteinases, elastases are able to solubilize fibrous elastin, an important extracellular matrix protein that has the unique property of elastic recoil, and plays a mechanical function in lungs, arteries, skin and ligaments. The insolubility of elastin is due to a high content of nonpolar amino acid residues, and to special cross-linking residues (desmosine, isodesmosine, lysinonorleucine). This flexible protein is rich in valine (13%) and alanine (26%), the amino acid residues for which human LE shows a P1 specificity. This may partially explain its ability to solubilize elastin. There is, however, no strict relationship between the catalytic power of elastases on model substrates and their elastinolytic activity

because (a) the enzyme must bind to the insoluble substrate and, when bound, must be able to perform surface proteolysis, i.e. it must be bound productively, (b) elastinolysis may generate soluble peptides with different molecular masses; as a consequence, a potent enzyme may cleave many more peptide bonds per unit time than a poor one does, but the latter may generate much larger peptides than the former and hence be a more effective elastase (Kagan *et al.*, 1972). Leukocyte elastases are significantly less active than pancreatic elastases (Chapters 11 and 12). The activity of human LE varies with the species of origin of the elastin substrate, and the method by which it has been prepared (Bieth, 1986). Ionic strength and elastin-binding ligands such as cathepsin G, platelet factor 4 and poly-lysine greatly increase the rate of elastin solubilization (Lonky & Wohl, 1983). Human LE is thought to interact with many negatively charged sites on the elastin surface to form nonproductive complexes, i.e. complexes in which the enzyme cannot cleave peptides bonds. Lysine-rich ligands such as platelet factor 4 are believed to occupy these negatively charged, nonproductive binding sites, and thus increase the rate of elastinolysis (Lonky & Wohl, 1983).

Pure human LE is best assayed with synthetic substrates, the most convenient being MeOSuc-Ala-Ala-Pro-Val⧧NHPhNO$_2$ (Bachem) (see Appendix 2 for full names and addresses of suppliers) ($k_{cat}/K_m = 120\,000\,M^{-1}\,s^{-1}$) which allows 5 nM enzyme to be determined accurately. MeOSuc-Ala-Ala-Pro-Val⧧SBzl (Enzyme Systems Products) is more sensitive ($k_{cat}/K_m = 5\,600\,000\,M^{-1}\,s^{-1}$), but the assay medium must include a thiol-reactive agent to detect the product of the enzymatic cleavage. Suc-Ala-Ala-Ala⧧NHPhNO$_2$ (Bachem) requires more enzyme (100 nM), but is less expensive than the two other substrates. These compounds may also be used to assay leukocyte elastases from other species. The elastinolytic activity of leukocyte elastases may be determined with [3]H-labeled elastin, or with the less sensitive fluorescent derivative, rhodamine-elastin (Elastin Products Co.) (Bieth, 1986).

Leukocyte elastases are inhibited by nonspecific irreversible serine proteinase inhibitors such as DFP and PMSF (from Sigma). More specific, commercially available irreversible inhibitors including MeOSuc-Ala-Ala-Pro-Val-CH$_2$Cl (from Bachem) and DCI (from ICN) (Powers & Harper, 1986) may be used to characterize leukocyte elastase. Elastatinal, a low molecular mass reversible inhibitor from microorganisms, may be purchased from Sigma.

Because of the possible involvement of human LE in inflammatory diseases (see below), an intense effort has been directed towards the synthesis of inhibitors of potential therapeutic usefulness. This has led to a wealth of reversible and irreversible inhibitors that belong to different chemical classes (peptide-based compounds, heterocyclic and nonheterocyclic inhibitors) (Powers & Harper, 1986; Edwards & Bernstein, 1994). Some of these molecules are extremely potent. For instance, recently synthesized compounds reversibly inhibit human LE with $K_i$ values in the picomolar range. On the other hand, the most potent azetidinone and isocoumarin derivatives inhibit irreversibly, with second-order rate constants close to $10^6\,M^{-1}\,s^{-1}$ (Edwards & Bernstein, 1994).

Human LE is also inhibited by several naturally occurring or engineered protein proteinase inhibitors. Human plasma $\alpha_1$-proteinase inhibitor (53 kDa), a member of the serpin superfamily, and $\alpha_2$-macroglobulin (725 kDa) are commercially available (Athens Research & Technology), fast-acting irreversible elastase inhibitors whose second-order association rate constants $k_{ass}$ are greater than $10^7\,M^{-1}\,s^{-1}$ (Barrett & Salvesen, 1986). There is a genetically determined polymorphism of $\alpha_1$-proteinase inhibitor called the Pi system (Crystal, 1995). A large number of Pi variants have been found: PiMM is the normal phenotype, whereas PiZZ, PiSS and PiNulNul are phenotypes giving rise to lowered plasma levels of $\alpha_1$-proteinase inhibitor. The 11.7 kDa human mucus proteinase inhibitor, also called secretory leukoprotease inhibitor (SLPI), inhibits elastase reversibly, with a $K_i$ of $4 \times 10^{-11}$ M (Faller *et al.*, 1992). Elafin, a 6 kDa protein from human keratinocytes or bronchial secretions is also a potent reversible inhibitor ($K_i = 2 \times 10^{-10}$ M) (Ying & Simon, 1993). The cytosolic fractions from human and horse neutrophils contain potent irreversible leukocyte elastase inhibitors that belong to the serpin family (Thomas *et al.*, 1991). Thrombospondin, a platelet protein, inhibits reversibly with a $K_i$ of $6 \times 10^{-8}$ M (Hogg *et al.*, 1993). High-affinity reversible inhibitors of nonhuman origin include dog submandibulary gland inhibitor, eglin c from *Hirudo medicinalis*, guamerin from *Hirudo nipponia* (Jung *et al.*, 1995), chemically engineered aprotinin (Trasylol), turkey ovomucoid, and soybean trypsin inhibitor AA (for reviews see Barrett & Salvesen, 1986; Bode *et al.*, 1989). Heparin and other glycosaminoglycans are also reversible inhibitors (Frommherz *et al.*, 1991).

## Structural Chemistry

Human LE is formed of a single peptide chain of 218 amino acid residues and four disulfide bridges. Its amino acid sequence was confirmed by cDNA sequencing (Takahashi *et al.*, 1988). It shows 30–40% sequence identity with other elastinolytic or nonelastinolytic serine proteinases (Bode *et al.*, 1989). Its tertiary structure (Navia *et al.*, 1989) is similar to that of other serine proteinases of the chymotrypsin family, and consists of two interacting antiparallel $\beta$-barrel cylindrical domains and a small proportion of helices. The two domains form a crevice which contains the catalytic site. The substrate-binding site lies across this crevice and includes parts of the two domains. The structure of human LE differs from that of the related pancreatic serine proteinases in the presence of several loop segments with unique conformations that are mainly arranged around the substrate-binding site. Human LE contains 19 Arg residues, 18 of which form clusters on the surface of this basic protein (pI > 9). It is a 30 kDa glycoprotein, containing 20% of neutral sugars with only small amounts of sialic acid (Baugh & Travis, 1976). There are two glycosylation sites, at Asn109 and Asn159 (Navia *et al.*, 1989).

## Preparation

Human LE, the most studied leukocyte elastase, is best isolated from cystic fibrosis purulent sputum using extraction, affinity chromatography on Sepharose-aprotinin (aprotinin, distributed commercially as Trasylol, is the basic pancreatic trypsin inhibitor) and chromatography on CM-Sephadex (Martodam *et al.*, 1979). Starting with 500 ml of sputum, one gets 50–150 mg of pure active enzyme within about 10 days. This procedure is derived from that of Baugh & Travis (1976), who isolated elastase from blood leukocytes. Their method may be used successfully if no sputum is available. The isolation of LE from other species usually follows a purification scheme close to that of Baugh & Travis (1976).

## Biological Aspects

Leukocyte elastase has been demonstrated in a number of animals including dogs, rats, rabbits, horses and hamsters. Dog leukocyte elastase (Delshammar & Ohlsson, 1976) is composed of a single polypeptide chain with a molecular mass of 24 kDa. It is a cationic protein with no Lys or Tyr residues. It is not a glycoprotein, unlike its human counterpart, but exhibits clear-cut elastinolytic activity. Partially purified rat leukocyte elastase has substrate and inhibitor specificities close to those of the human enzyme, but it has not been shown to solubilize elastin. In contrast, true elastinolytic activity was demonstrated in polymorphonuclear leukocytes from rabbits, hamsters and horses, although the elastase content of rabbit and hamster leukocytes is much lower than that of human leukocytes (Bieth, 1986). Horse leukocytes were recently shown to contain two elastinolytic proteinases with considerable sequence similarities (Dubin *et al.*, 1994).

Human LE is primarily located in the azurophil granules of polymorphonuclear leukocytes, but has also been detected in the nuclear membrane, the Golgi complex, the endoplasmic reticulum and the mitochondria of these cells. The human LE concentration of polymorphonuclear leukocytes is very high (3 µg enzyme per $10^6$ cells) (Liou & Campbell, 1995). Promyelocytes and more differentiated myeloid cells are the only cells that contain elastase. Monocytes have low levels (about 6%) of this enzyme while the promonocyte-like cell line U-937 synthesizes large quantities of this proteinase (Murphy & Reynolds, 1993). Human LE is synthesized during the promyelocytic and promonocytic stages of differentiation of leukocytes and monocytes, respectively. Mature leukocytes do not synthesize elastase, but are capable of rapidly mobilizing azurophil granules to the cell surface in response to varied stimuli (Liou & Campbell, 1995). When monocytes differentiate into macrophages, they gradually lose human LE, but develop an elastinolytic metalloproteinase (Chapter 395). Alveolar macrophages internalize human LE through receptor-mediated endocytosis (Bieth, 1986). Human LE is bound in part to the plasma membranes of monocytes (Allen & Tracy, 1995) and polymorphonuclear leukocytes (Owen *et al.*, 1995).

The human LE gene consists of five exons and four introns. The coding exons predict a 29 residue N-terminal propeptide, a 218 residue mature elastase protein, and a 20 residue C-terminal extension. The 5′-flanking region of the gene includes typical TATA, CAAT and GC sequences (Takahashi *et al.*, 1988). The gene is located on chromosome 19pter (Zimmer *et al.*, 1992). Elastase, which is fully active in the azurophil granules, is transiently present in the cell as a zymogen and loses its N- and C-terminal extensions within 90 min after the onset of its biosynthesis (Salvesen & Enghild, 1990).

In addition to elastin, elastase cleaves many proteins with important biological functions, including collagens of type I,

II, III, IV, VIII, IX, X and XI (Pipoly & Crouch, 1987; Gadher *et al.*, 1988; Kittelberger *et al.*, 1992). It also acts on other matrix proteins including fibronectin, laminin (Heck *et al.*, 1990), and cartilage proteoglycan (Janusz & Doherty, 1991). Human LE also indirectly favors the breakdown of matrix proteins by activating procollagenase, prostromelysin and pro-gelatinase (Rice & Banda, 1995, and references therein). On the other hand, human LE cleaves immunoglobulin G, immunoglobulin M, complement C3 and C5 and inactivates coagulation factor VII (Chapter 53), VIII, X (Chapter 54), XII (Chapter 49) and XIII (Turkington, 1991; Anderssen *et al.*, 1993) as well as protein C, another coagulation proteinase (Eckle *et al.*, 1991). Other coagulation proteins are cleaved but not inactivated. For instance, human LE activates factor V (Allen & Tracy, 1995), transforms plasminogen (Chapter 59) into 'mini-plasminogen' (Machovich & Owen, 1989), cleaves fibrinogen and transforms thrombin (Chapter 55) into a proteinase with decreased fibrinogen-clotting and platelet-stimulatory activity (Brower *et al.*, 1987). In addition, human LE inactivates a number of endogenous protein inhibitors such as tissue factor pathway inhibitor (Higushi *et al.*, 1992), antithrombin in complex with heparin (Jordan *et al.*, 1989), $\alpha_2$-antiplasmin, C1-inactivator, $\alpha_1$-antichymotrypsin (Bieth, 1986) and tissue inhibitor of metalloproteinases (Okada *et al.*, 1988). It also destroys the biological activity of tumor necrosis factor $\alpha$ (TNF$\alpha$) (van Kessel *et al.*, 1991), activates prorenin (Takada *et al.*, 1987), and cleaves surfactant proteins (Liau *et al.*, 1996) and platelet glycoprotein Ib (Aziz *et al.*, 1995).

The major physiological function of LE is probably to digest bacteria and immune complexes phagocytosed by the polymorphonuclear leukocyte. It is likely that human LE also plays a role in leukocyte migration from blood to tissue. Elastase activity may also be required after acute connective tissue injury, such as wounds, in the initial stage of tissue repair (Murphy & Reynolds, 1993). Polymorphonuclear leukocyte apoptosis is favored by human LE and other proteinases (Trevani *et al.*, 1996).

Human LE is thought to play a pathological role in lung emphysema, cystic fibrosis, the adult respiratory distress syndrome (ARDS), rheumatoid arthritis and infectious diseases. Emphysema is characterized by destructive changes of the alveolar walls with resultant limitation of airflow. The proteinase–antiproteinase hypothesis of emphysema holds that alveolar structures are normally protected from elastase attack by local antielastases (mainly $\alpha_1$-proteinase inhibitor), and that emphysema may occur if the elastase burden increases and/or if the antielastase protection is weakened (Crystal, 1996). Smoking, the major cause of the disease, increases the human LE burden of the lungs and oxidatively impairs the antielastase function of $\alpha_1$-proteinase inhibitor by a direct or a myeloperoxidase-mediated mechanism, in which Met358, the P1 residue of $\alpha_1$-proteinase inhibitor, is oxidized to methionine sulfoxide, which renders the inhibitor very inefficient (Bieth, 1986). It has been confirmed that the genetically engineered Met358Val inhibitor mutant, which resists oxidation, is much more efficient in protecting the extracellular matrix against neutrophil-mediated proteolysis than is the natural inhibitor (Crystal, 1996). Emphysema occurring in patients with hereditary $\alpha_1$-proteinase inhibitor deficiency (PiNulNul or PiZZ phenotype) is much less frequent but much easier to rationalize than smoke-induced emphysema since the lung

inhibitor concentration is very low (Snider, 1993; Crystal, 1996). Damiano *et al.* (1986) have shown that human LE is associated with elastin fibers in emphysematous lungs, which strengthens the view that it plays a pathogenic role in the disease.

Cystic fibrosis is a hereditary disease caused by a dysfunctioning of a chloride channel protein. Its major clinical manifestations are in the airways, where viscous mucus accumulates and favors microbial (*Pseudomonas aeruginosa*) infections and massive recruitment of neutrophils. Phagocytosis and cell death permanently release human LE which attacks proteins in an unimpaired way since the local antielastase potential is overcome (Birrer *et al.*, 1994). As a matter of fact, cystic fibrosis secretions contain large amounts of active elastase (Martodam *et al.*, 1979). In addition to its action on lung matrix proteins, human LE impairs the phagocytic function of neutrophils and is cytotoxic for bronchial epithelial cells, thus favoring bacterial colonization (Amitani *et al.*, 1991).

Most patients with ARDS have high elastase activity in their bronchoalveolar lavage fluids, while the functional activity of $\alpha_1$-proteinase inhibitor is considerably decreased. This combination of circumstances explains the lung damage that is seen in the disease (Snider, 1993). In rheumatoid arthritis there is extensive proteolytic degradation of the articular cartilage whose components, collagen and proteoglycan, are degraded by human LE *in vitro* (Barrett, 1994; Murphy & Reynolds, 1993). However, human LE does not overcome the local elastase inhibitory capacity, because synovial fluids have no free elastase activity, and possess active $\alpha_1$-proteinase inhibitor (Bieth, 1986). In addition, immunoglobulin A-bound $\alpha_1$-proteinase inhibitor, which is present in synovial fluids, is highly active on elastase (Adam & Bieth, 1996). Cartilage proteolysis therefore occurs despite the presence of active inhibitor.

Severe septicemia leads to leukocyte recruitment, phagocytosis, human LE release and partial destruction of plasma antithrombin III, prothombin, factor XIII, plasminogen, $\alpha_2$-antiplasmin and complement component C3. Human LE has also been detected in pleural empyema and in patients with acute bacterial pneumonia. Human LE is therefore an aggravating factor of infections (Snider, 1993).

Human LE cannot be considered as the sole proteinase important in the pathogenesis of the diseases mentioned above: other proteinases from leukocytes, macrophages and microorganisms may also participate in proteolysis. There is also an important question: how can LE-mediated tissue destruction occur in emphysema and arthritis despite the presence of active $\alpha_1$-proteinase inhibitor? Several hypotheses may be put forward. First, proteolysis in the presence of an inhibitor may be the normal consequence of competitive inhibition. The principles of classical enzymology predict that if an enzyme is faced with a substrate and an inhibitor it will partition between these two ligands according to their concentrations and the kinetic constants describing their binding to the enzyme. As a consequence, breakdown of substrate in the course of the inhibition of an enzyme must be considered as an inescapable event. The delay time of inhibition $d(t)$, i.e. the time required to inhibit a proteinase in the absence of substrate, attempts to predict the magnitude of this undesirable event: $d(t) = 5/k_{ass}[I]_o$, where $[I]_o$ is the *in vivo* inhibitor concentration, $k_{ass}$ is the second-order inhibition rate constant. If $d(t)^2 \leq 1$ s, proteolysis is probably small (Bieth, 1984). The delay time of human LE inhibition may vary within a given organ or tissue due to uneven concentrations of active $\alpha_1$-proteinase inhibitor, i.e. variable $[I]_o$, or the local presence of ligands (e.g. glycosaminoglycans) that decreases $k_{ass}$ (Frommherz *et al*., 1991). Hence, significant proteolysis may occur at some loci despite an overall excess of inhibitor over enzyme. Second, proteolysis may occur at sites where polymorphonuclear leukocytes degranulate, since the human LE concentration in the azurophil granules (5 mM) is three orders of magnitude higher than the inhibitor concentration. Thus, during a certain time interval, proteolysis will occur in a totally unimpaired way (Liou & Campbell, 1995). Third, the inhibition of human LE by $\alpha_1$-proteinase inhibitor is severely impaired if the enzyme is bound to elastin (Snider, 1993) or to leukocytes (Owen *et al*., 1995), or if the leukocyte is in close contact with an insoluble matrix protein (Campbell & Campbell, 1988; Rice & Weiss, 1990).

The indications of involvement of human LE in all these diseases has prompted therapeutic research based on elastase inhibitors. Plasma-derived $\alpha_1$-proteinase inhibitor is presently used in patients with hereditary inhibitor deficiency (Crystal, 1996). Attempts have been made to use recombinant mucus proteinase inhibitor in cystic fibrosis (Birrer *et al*., 1994). Also, some of the synthetic elastase inhibitors that have been designed in the last two decades are very potent and orally active molecules (Edwards & Bernstein, 1994) which might be used in medicine in the near future.

## Related Peptidases

Medullasin, an inflammatory proteinase isolated from human bone marrow cells (Ikai *et al*., 1989) and human spleen fibrinolytic proteinase (Okada *et al*., 1984) are similar to if not identical with human LE. Human polymorphonuclear leukocytes contain another elastinolytic serine proteinase, proteinase 3 or myeloblastin (Chapter 17), whose molecular, enzymatic and pathogenic properties are close to those of elastase, although it is undoubtedly a distinct enzyme (Hoidal *et al*., 1994).

## References
Adam, C. & Bieth, J.G. (1996) Inhibition of neutrophil elastase by the $\alpha_1$-proteinase inhibitor-immunoglobulin A complex. *FEBS Lett.* **385**, 201–204.

Allen, D.H. & Tracy, P.B. (1995) Human coagulation factor V is activated to the functional cofactor by elastase and cathepsin G expressed at the monocyte surface. *J. Biol. Chem.* **270**, 1408–1415.

Amitani, R., Wilson, R., Rutman, A., Read, R., Ward, C., Burnett, D., Stockley, R.A. & Cole, P.J. (1991) Effects of human neutrophil elastase and *Pseudomonas aeruginosa* proteinases on human respiratory epithelium. *Am. J. Respir. Cell. Mol. Biol.* **4**, 26–32.

Anderssen, T., Halvorsen, H., Bajaj, S.P. & Østerud, B. (1993) Human leukocyte elastase and cathepsin G inactivate factor VII by limited proteolysis. *Thromb. Haemost.* **70**, 414–417.

Aziz, K.A., Cawley, J.C., Kamiguti, A.S. & Zuzel, M. (1995) Degradation of platelet glycoprotein Ib by elastase released from primed neutrophils. *Br. J. Haemostol.* **91**, 46–54.

Barrett, A.J. (1994) The possible role of neutrophil proteinases in damage to articular cartilage. *Agents Actions* **43**, 194–201.

Barrett, A.J. & Salvesen, G. (eds) (1986) *Proteinase Inhibitors*. Amsterdam: Elsevier Science Publishers.

Baugh, R.J. & Travis, J. (1976) Human leukocyte granule elastase: rapid isolation and characterization. *Biochemistry* **15**, 836–841.

Bieth, J.G. (1984) *In vivo* significance of kinetic constants of protein proteinase inhibitors. *Biochem. Med.* **32**, 387–397.

Bieth, J.G. (1986) Elastases: catalytic and biological properties. In: *Biology of Extra-Cellular Matrix*, vol. 1: *Regulation of Matrix Accumulation* (Mecham, R.P., ed.). New York: Academic Press, pp. 217–320.

Birrer, P., McElvaney, N.G., Rüdeberg, A., Wirz Sommer, C., Liechti-Gallati, S., Kraemer, R., Hubbard, R. & Crystal, R.G. (1994) Protease-antiprotease imbalance in the lungs of children with cystic fibrosis. *Am. J. Respir. Crit. Care Med.* **150**, 207–213.

Blow, A.M.J. (1977) Action of human lysosomal elastase on the oxidized B chain of insulin. *Biochem. J.* **161**, 13–16.

Blow, D.M. (1976) Structure and mechanism of chymotrypsin. *Acc. Chem. Res.* **9**, 145–152.

Bode, W., Meyer, E., Jr. & Powers, J.C. (1989). Human leukocyte and porcine pancreatic elastase: X-ray crystal structures, mechanism, substrate specificity, and mechanism-based inhibitors. *Biochemistry* **28**, 1951–1963.

Brower, M.S., Walz, D.A., Gary, K.E. & Fenton II, J.W. (1987) Human neutrophil elastase alters human $\alpha$-thrombin function: limited proteolysis near the $\gamma$-cleavage site results in decreased fibrinogen clotting and platelet-stimulatory activity. *Blood* **69**, 813–819.

Campbell, E.J. & Campbell, M.A. (1988) Pericellular proteolysis by neutrophils in the presence of proteinase inhibitors: effects of substrate opsonization. *J. Cell. Biol.* **106**, 667–676.

Crystal, R.G. (1996) *Alpha1-Antitrypsin Deficiency. Biology – Pathogenesis – Clinical Manifestations – Therapy.* New York: Marcel Dekker.

Damiano, V.V., Tsang, A., Kucich, U., Abrams, W.R., Rosenbloom, J., Kimbel, P., Fallahnejad, M. & Weinbaum, G. (1986) Immunolocalization of elastase in human emphysematous lungs. *J. Clin. Invest.* **78**, 482–493.

Delshammar, M. & Ohlsson, K. (1976) Isolation of partial characterization of elastase from dog granulocytes. *Eur. J. Biochem.* **69**, 125–131.

Dubin, A., Potempa, J. & Travis, J. (1994) Structural and functional characterization of elastases from horse neutrophils. *Biochem. J.* **300**, 401–406.

Eckle, I., Seitz, R., Egbring, R., Kolb, G. & Havemann, K. (1991) Protein C degradation *in vitro* by neutrophil elastase. *Biol. Chem. Hoppe-Seyler* **372**, 1007–1013.

Edwards, P.D. & Bernstein, P.R. (1994) Synthetic inhibitors of elastase. *Med. Res. Rev.* **14**, 127–194.

Faller, B., Mely, Y., Gérard, D. & Bieth, J.G. (1992) Heparin-induced conformational change and activation of mucus proteinase inhibitor. *Biochemistry* **31**: 8285–8290.

Frommherz, K., Faller, B. & Bieth, J.G. (1991) Heparin strongly decreases the rate of inhibition of neutrophil elastase by $\alpha_1$-proteinase inhibitor. *J. Biol. Chem.* **266**, 15356–15362.

Gadher, S.J., Eyre, D.R., Duance, V.C., Wotton, S.F., Heck, L.W., Schmid, T.M. & Woolley, D.E. (1988) Susceptibility of cartilage collagens type II, IX, X, and XI to human synovial collagenase and neutrophil elastase. *Eur. J. Biochem.* **175**, 1–7.

Heck, L.W., Blackburn, W.D., Irwin, M.H. & Abrahamson, D.R. (1990) Degradation of basement membrane laminin by human neutrophil elastase and cathepsin G. *Am. J. Pathol.* **136**, 1267–1274.

Higushi, D.A., Wun, T.Z., Likert, K.M. & Broze Jr., G.J. (1992) The effect of leukocyte elastase on tissue factor pathway inhibitor. *Blood* **79**, 1712–1719.

Hogg, P.J., Owensby, D.A., Mosher, D.F., Misenheimer, T.M. & Chesterman, C.N. (1993) Thrombospondin is a tight-binding competitive inhibitor of neutrophil elastase. *J. Biol. Chem.* **268**, 7139–7146.

Hoidal, J.R., Rao, N.V. & Gray, B. (1994) Myeloblastin: Leukocyte proteinase 3. *Methods Enzymol.* **244**, 61–67.

Ikai, A., Nakashima, M. & Aoki, Y. (1989) Inhibition of inflammatory proteinase, medullasin, by $\alpha_2$-macroglobulin and ovomacroglobulin. *Biochem. Biophys. Res. Commun.* **158**, 831–836.

Janoff, A. & Scherer, J. (1968) Mediators of inflammation in leukocyte lysosomes. IX. Elastinolytic activity in granules of human polymorphonuclear leukocytes. *J. Exp. Med.* **128**, 1137–1140.

Janusz, M.J. & Doherty, N.S. (1991) Degradation of cartilage matrix proteoglycan by human neutrophils involves both elastase and cathepsin G. *J. Immunol.* **146**, 3922–3928.

Jordan, R.E., Nelson, R.M., Kilpatrick, J., Newgren, J.O., Esmon, P.C. & Fournel, M.A. (1989) Inactivation of human antithrombin by neutrophil elastase. *J. Biol. Chem.* **264**, 10493–10500.

Jung, H.I., Kim, S.I., Ha, K.S., Joe, C.O. & Kang, K.W. (1995) Isolation and characterization of guamerin, a new human leukocyte elastase inhibitor from *Hirudo nipponia*. *J. Biol. Chem.* **270**, 13879–13884.

Kagan, H.M., Crombie, G.D., Jordan, R.E., Lewis, W. & Franzblau, C. (1972) Proteolysis of elastin–ligand complexes. Stimulation of elastase digestion of insoluble elastin by sodium dodecyl sulfate. *Biochemistry* **11**, 3412–3418.

Kittelberger, R., Neale, T.J., Francky, K.T., Grennhill, N.S. & Gibson, G.J. (1992) Cleavage of type VIII collagen by human neutrophil elastase. *Biochim. Biophys. Acta* **1139**, 295–299.

Lestienne, P. & Bieth, J.G. (1980) Activation of human leukocyte elastase activity by excess substrate, ionic strength and hydrophobic solvents. *J. Biol. Chem.* **255**, 9289–9294.

Liau, D.F., Yin, N.X., Huang, J. & Ryan, S.F. (1996) Effects of human polymorphonuclear leukocyte elastase upon surfactant proteins *in vitro*. *Biochim. Biophys. Acta* **1302**, 117–128.

Liou, T.G. & Campbell, E.J. (1995) Nonisotropic enzyme-inhibitor interactions: a novel nonoxidative mechanism for quantum proteolysis by human neutrophils. *Biochemistry* **34**, 16171–16177.

Lonky, S.A. & Wohl, H. (1983) Regulation of elastolysis of insoluble elastin by human leukocyte elastase: stimulation by lysine-rich ligands, anionic detergents and ionic strength. *Biochemistry* **22**, 3714–3720.

Machovich, R. & Owen, W.G. (1989) An elastase-dependent pathway of plasminogen activation. *Biochemistry* **28**, 4517–4522.

Martodam, R.R., Baugh, R.J., Twumasi, D.Y. & Liener, I.E. (1979) A rapid procedure for the large-scale purification of elastase and cathepsin G from human sputum. *Prep. Biochem.* **9**, 15–31.

Murphy, G. & Reynolds, J.J. (1993) Extracellular matrix degradation. In: *Connective Tissue and its Heritable Disorders* (Royce, P.M. & Steinmann, B., eds). New York: Wiley-Liss, pp. 287–316.

Navia, M.A., McKeever, B.M., Springer, J.P., Lin, T.Y., Williams, H.R., Fluder, E.M., Dorn, C.P. and Hoogsteen, K. (1989) Structure of human neutrophil elastase in complex with a peptide chloromethyl ketone inhibitor at 1.84-Å resolution. *Proc. Natl Acad. Sci. USA* **86**, 7–11.

Okada, Y., Tsuda, Y., Nagamatsu, Y. & Okamoto, U. (1984) Synthesis of peptide inhibitors of human spleen fibrinolytic proteinase (SFP) and human leukocyte elastase-like proteinase (ELP). *Int. J. Pept. Protein Res.* **24**, 347–358.

Okada, Y., Watanabe, S., Natanishi, I., Kishi, J.I., Hayatawa, T., Watorek, W., Travis, J. & Nagase, H. (1988) Inactivation of tissue inhibitor of metalloproteinases by neutrophil elastase and other serine proteinases. *FEBS Lett.* **229**, 157–160.

Owen, C.A., Campbell, M.A., Sannes, P.L., Boukedes, S.S. & Campbell, E.J. (1995) Cell surface-bound elastase and cathepsin G on human neutrophils: a novel, non-oxidative mechanism by which neutrophils focus and preserve catalytic activity of serine proteinases. *J. Cell. Biol.* **131**, 775–789.

Pipoly, D.J. & Crouch, E.C. (1987) Degradation of native type IV procollagen by human neutrophil elastase. Implications for leukocyte-mediated degradation of basement membranes. *Biochemistry* **26**, 5748–5754.

Powers, J.C. & Harper, J.W. (1986) Inhibitors of serine proteinases. In: *Proteinase Inhibitors* (Barrett, A.J. & Salvesen, G., eds). Amsterdam: Elsevier Science Publishers, pp. 55–152.

Rice, A. & Banda, M.J. (1995) Neutrophil elastase processing of gelatinase A is mediated by extracellular matrix. *Biochemistry* **34**, 9249–9256.

Rice, W. & Weiss, S. (1990) Regulation of proteolysis at the neutrophil-substrate interface by secretory leukoprotease inhibitor. *Science* **249**, 178–181.

Salvesen, G. & Enghild, J.J. (1990) An unusual specificity in the activation of neutrophil serine proteinase zymogens. *Biochemistry* **29**, 5304–5308.

Snider, G.L. (1993) Emphysema: the first two centuries – and beyond. A historical overview, with suggestions for future research: part 1. *Am. Rev. Respir. Dis.* **146**: 1334–1344.

Stein, R.L. (1983) Catalysis by human leukocyte elastase: substrate structural dependence of rate-limiting proteolytic catalysis and operation of the charge relay system. *J. Am. Chem. Soc.* **105**, 5111–5116.

Takada, Y., Maruta, H., Wagner, B., Fa, X.G., James, H.L. & Erdös, E.G. (1987) Activation of human prorenin by neutrophil elastase. *J. Clin. Endocrin. Metab.* **65**, 1225–1230.

Takahashi, H., Nukiwa, T., Yoshimura, K., Quick, C.D., States, D.J., Holmes, M.D., Whang-Peng, J., Knutsen, T. & Crystal, R.G. (1988) Structure of the human neutrophil elastase gene. *J. Biol. Chem.* **263**, 14739–14747.

Thomas, R.M., Nauseef, W.M., Iyer, S.S., Peterson, M.W., Stone, P.J. & Clark, R.A. (1991) A cytosolic inhibitor of human neutrophil elastase and cathepsin G. *J. Leukocyte Biol.* **50**, 568–579.

Trevani, A.S., Andonegui, G., Giordano, M., Nociari, M., Fontan, P., Dran, G. & Geffner, J.R. (1996) Neutrophil apoptosis induced by proteolytic enzymes. *Lab. Invest.* **74**, 711–721.

Turkington, P. (1991) Degradation of human factor X by human polymorphonuclear leucocyte cathepsin G and elastase. *Haemostasis* **21**, 111–116.

Van Kessel, K.P.M., Van Strijp, J.A.G. & Verhoef, J. (1991) Inactivation of recombinant human tumor necrosis factor-$\alpha$ by proteolytic enzymes released from stimulated human neutrophils. *J. Immunol.* **147**, 3662–3668.

Yasutake, A. & Powers, J.C. (1981) Reactivity of human leukocyte elastase and porcine pancreatic elastase toward peptide 4-nitroanilides containing model desmosine residues. Evidence that human leukocyte elastase is selective for cross-linked regions of elastin. *Biochemistry* **20**, 3675–3679.

Ying, Q.L. & Simon, S.R. (1993) Kinetics of the inhibition of human leukocyte elastase by elafin, a 6-kilodalton elastase-specific inhibitor from human skin. *Biochemistry* **32**, 1866–1874.

Zimmer, M., Medcalf, R.L., Fink, T.M., Mattmann, C., Lichter, P. & Jenne, D.E. (1992) Three human elastase-like genes coordinately expressed in the myelomonocyte lineage are organized as a single genetic locus on 19pter. *Proc. Natl Acad. Sci. USA* **89**, 8215–8219.

***Joseph G. Bieth***
*INSERM U 392,*
*Faculté de Pharmacie,*
*74 route du Rhin, F-67400 Illkirch, France*
*Email: jgbieth@pharma.u-strasbg.fr*

# 16. Cathepsin G

## Databanks

*Peptidase classification: clan SA, family S1, MEROPS ID: S01.133*
*NC-IUBMB enzyme classification: EC 3.4.21.20*
*Chemical Abstracts Service registry number: 56645-49-9*
*Databank codes:*

| Species | SwissProt | PIR | EMBL (cDNA) | EMBL (genomic) |
|---|---|---|---|---|
| *Homo sapiens* | P08311 | A05307 | M16117 | J04990: complete gene |
| | | A27122 | | M59717: 5′ flank |
| | | A32627 | | |
| | | A37115 | | |
| | | A46471 | | |
| | | A90031 | | |
| | | S44427 | | |
| *Mus musculus* | P28293 | A48932 | M96801 | – |
| | | S23170 | X70057 | |
| | | S40162 | X78544 | |
| *Rattus norvegicus* | P17977 | S10608 | – | – |

Brookhaven Protein Data Bank three-dimensional structures:

| Species | ID | Resolution | Notes |
|---|---|---|---|
| *Homo sapiens* | 1CGH | – | complex with Suc-Val-Pro-Phe$^P$-(OPh)$_2$ inhibitor |

## Name and History

A human neutrophil-derived proteolytic activity specific for synthetic substrates of chymotrypsin was reported by Dewald *et al*. (1975), and Rindler-Ludwig & Braunsteiner (1975). The activity was later purified and shown to consist of a single-chain serine proteinase (Baugh & Travis, 1976; Starkey & Barrett, 1976a,b). The name *cathepsin G* was given by Starkey & Barrett (1976b) who purified the proteinase from spleen, and its location in neutrophils was sometimes signified by preceding the name with 'neutrophil' or 'polymorphonuclear leukocyte', according to personal preference. The prefix is now dropped, and the proteinase is simply known as cathepsin G. Sequence databases classify cathepsin G as a 'vimentin-specific protease'. This should not be construed to mean a role for the enzyme in vimentin processing since cathepsin G is unlikely to see vimentin (a cytosolic protein) under normal conditions.

## Activity and Specificity

Optimal pH on most substrates is 8.0. Activity of the purified enzyme is reportedly stimulated by 0.5 M MgCl$_2$ and inhibited by 0.5 M $CaCl_2$. Partially purified preparations are stimulated by 0.5 M KCl. The reasons for variability in activity with salt concentration are unknown, but may reflect the highly basic enzyme's binding to surfaces during purification or assay. The rules governing the specificity of cathepsin G have been studied extensively with oligopeptide nitroanilides and chloromethanes (Nakajima et al., 1979; Powers et al., 1985; Szabó et al., 1986). The best described synthetic substrates and inhibitors have Phe at P1, but the rates of hydrolysis or inhibition are surprisingly low compared to related proteinases. The early classification as an enzyme with chymotrypsin-like specificity (preferring Tyr and Phe at P1) was revised following the establishment of cathepsin G's protein sequence (Salvesen et al., 1987). Residue 226, governing the P1 preference of the close homolog rat mast cell chymase (Remington et al., 1988), was found to be Glu, implying that the enzyme could have specificity for basic side-chains at P1. For a related discussion of the influence of residue 226, see Chapter 23 on granzyme B. It was later confirmed that cathepsin G has similar specificity constants for synthetic substrates containing Lys, Phe or Trp at P1, though very little activity was shown for substrates containing Arg in P1 (Powers et al., 1989). Cathepsin G is thus unusual among the peptidases of the chymotrypsin family in having a somewhat dual P1 specificity. Extended specificities of P2–P4 have been examined, to demonstrate a certain consensus typical of chymotrypsin family members. As with most serine proteinases, optimal substrate occupation of prime side subsites has not been examined in detail.

Assays are most conveniently made with peptide nitroanilides, thioesters or aminomethyl coumarins containing Phe in the P1 position. Suc-Ala-Ala-Pro-Phe$\downarrow$NHPhNO$_2$ is a useful substrate for most purposes, though the aminomethyl coumarin derivative is more sensitive.

Inhibitors consist of the usual serine proteinase-specific reagents, though DFP and PMSF react so slowly as to be almost useless. DCI is the most effective general inhibitor, and inhibitors that take advantage of the specificity of the enzyme include Phe-Gly-Phe-CH$_2$Cl. Protein inhibitors include the physiologic regulator $\alpha_1$-antichymotrypsin, as well as $\alpha_1$-proteinase inhibitor, eglin C and limabean trypsin inhibitor.

## Structural Chemistry

Cathepsin G is a single-chain glycoprotein of about 28.5 kDa, pI about 12 (the high pI makes accurate measurements unreliable, so this is a rough estimate). The protein contains one potential N-glycosylation site and three disulfides (Salvesen et al., 1987). Though otherwise a typical chymotrypsin family member, cathepsin G does not contain the highly conserved 191–220 disulfide present in most members of the family. This disulfide, which is near the active site, is also absent from granzyme B and chymase. The genes for all three of these structurally similar enzymes are located at human chromosomal locus 14q11.2.

## Preparation

Natural sources of the enzyme are restricted to neutrophils, neutrophil precursors in bone marrow, and tissues that contain them (such as spleen). Preparation of pure enzyme is readily achieved by extraction of neutrophil granules, affinity chromatography on immobilized aprotinin, and cation-exchange chromatography to remove traces of the main contaminant, neutrophil elastase (Baugh & Travis, 1976). It is estimated that a human neutrophil contains about 3 pg of cathepsin G.

## Biological Aspects

The human azurophil granule is rich in serine proteinases of the chymotrypsin family, including cathepsin G, neutrophil elastase (Chapter 15), myeloblastin (Chapter 17) and the inactive homolog azurocidin. The latter has lost activity due to replacement of catalytic residues. It is presumed, therefore, that the biological activity of cathepsin G is linked to the function of the azurophil granule. This granule fuses with phagosomes during engulfment of microbes and is largely responsible for their killing and digestion. It is assumed that during normal phagocytosis by neutrophils, cathepsin G participates in the proteolytic degradation of the engulfed particles. Under extreme circumstances (e.g. stimulation by secretagogues, or during 'frustrated phagocytosis' of very large particles), the granule and its constituents can be released to the extracellular milieu. Here the proteinase can participate in local destruction of connective tissue proteins and, if degranulation takes place in the circulation, destruction of plasma proteins. The activity of cathepsin G is tightly regulated in the circulation, and possibly in inflamed extravascular sites, by its controlling serpin, $\alpha_1$-antichymotrypsin. This serpin is a major acute phase reactant, and a function of its upregulation during inflammatory episodes is presumed to be the inactivation of excess cathepsin G released from activated neutrophils.

Cathepsin G demonstrates significant bactericidal activity that is not dependent on its proteolytic activity. The sites that contain at least part of this activity have been mapped to short stretches of primary sequence on the surface of the molecule (Bangalore et al., 1990), though part of the activity is also probably due to the extremely basic nature of the protein. It is not clear that the bactericidal activity demonstrated in vitro has biological significance, since the assay conditions are usually hypo-osmotic compared to those expected in vivo (Gabay et al., 1989).

Other potential biological functions for cathepsin G may include effects on platelets (Selak et al., 1988), and proteolysis of certain blood coagulation cofactors (Schmidt et al., 1975; Turkington, 1991).

## Distinguishing Features

An activity that hydrolyzes Suc-Ala-Ala-Pro-Phe-NHPhNO$_2$, and is inhibited by $\alpha_1$-proteinase inhibitor or Phe-Gly-Phe-CHCl$_2$ is almost certainly due to cathepsin G. If the crude sample contains mast cells, then this activity could also be due to chymases. Antisera against the human enzyme are available from Athens Research & Technology (see Appendix 2 for full names and addresses of suppliers).

## References

Bangalore, N., Travis, J., Onunka, V.C., Pohl, J. & Shafer, W.M. (1990) Identification of the primary antimicrobial domains in human neutrophil cathepsin G. J. Biol. Chem. 265, 13584–13588.

Baugh, R. & Travis, J. (1976) Human leukocyte granule elastase: Rapid isolation and characterization. *Biochemistry* **15**, 836–841.

Dewald, B., Rindler-Ludwig, R., Bretz, U. & Baggiolini, M. (1975) Subcellular localization and heterogeneity of neutral proteases in neutrophilic polynuclear leukocytes. *J. Exp. Med.* **141**, 709–723.

Gabay, J.E., Scott, R.W., Campenelli, D., Griffith, J., Wilde, C., Marra, M.N., Seeger, M. & Nathan, C.F. (1989) Antibiotic proteins of human polymorphonuclear leukocytes. *Proc. Natl Acad. Sci. USA* **86**, 5610–5614.

Nakajima, K., Powers, J.C., Ashe, B.M. & Zimmerman, M. (1979) Mapping the extended substrate binding site of cathepsin G and human leukocyte elastase. *J. Biol. Chem.* **254**, 4027–4032.

Powers, J.C., Tanaka, T., Harper, J.W., Minematsu, Y., Barker, L., Lincoln, D. & Crumley, K.V. (1985) Mammalian chymotrypsin-like enzymes. Comparative reactivities of rat mast cell proteases, human and dog skin chymases, and human cathepsin G with peptide 4-nitroanilide substrates and with peptide chloromethyl ketone and sulfonyl fluoride inhibitors. *Biochemistry* **24**, 2048–2058.

Powers, J.C., Kam, C.-M., Narasimhan, L., Oleksyszyn, J., Hernandez, M. & Ueda, T. (1989) Mechanism-based isocoumarin inhibitors for serine proteases: use of active site structure and substrate specificity in inhibitor design. *J. Cell. Biochem.* **39**, 33–46.

Remington, J.R., Woodbury, R.G., Reynolds, R.A., Matthews, B.W. & Neurath, H. (1988) The structure of rat mast cell protease II at 1.9-Å resolution. *Biochemistry* **27**, 8097–8105.

Rindler-Ludwig, R. & Braunsteiner, H. (1975) Cationic proteins from human neutrophil granulocytes (evidence for their chymotrypsin-like properties). *Biochim. Biophys. Acta* **379**, 606–617.

Salvesen, G., Farley, D., Shuman, J., Pryzbyla, A., Reilly, C. & Travis, J. (1987) Molecular cloning of human cathepsin G: structural similarity to mast cell and cytotoxic lymphocyte proteinases. *Biochemistry* **26**, 2289–2293.

Schmidt, W., Egbring, R. & Havemann, K. (1975) Effect of elastase-like and chymotrypsin-like neutral proteases from human granulocytes on isolated clotting factors. *Thromb. Res.* **6**, 315–326.

Selak, M.A., Chignard, M. & Smith, J.B. (1988) Cathepsin G is a strong platelet agonist released by neutrophils. *Biochem. J.* **251**, 293–299.

Starkey, P.M. & Barrett, A.J. (1976a) Neutral proteinases of human spleen. Purification and criteria for homogeneity of elastase and cathepsin G. *Biochem. J.* **155**, 255–263.

Starkey, P.M. & Barrett, A.J. (1976b) Human cathepsin G: catalytic and immunological properties. *Biochem. J.* **155**, 273–278.

Szabó, G.C., Tözsér, J., Aurell, L. & Elödi, P. (1986) Mapping of the substrate binding site of human leukocyte chymotrypsin (cathepsin G) using tripeptidyl-*p*-nitroanilide substrates. *Acta Biochim. Biophys. Hung.* **21**, 349–362.

Turkington, P.T. (1991) Cathepsin G can produce a Gla-domainless protein C *in vitro*. *Thromb. Res.* **63**, 399–406.

*Guy S. Salvesen*
*The Burnham Institute,*
*10901 North Torrey Pines Road,*
*San Diego, CA 92037, USA*
*Email: gsalvesen@ljcrf.edu*

# 17. *Myeloblastin*

*Peptidase classification: clan SA, family S1, MEROPS ID: S01.134*
*NC-IUBMB enzyme classification: EC 3.4.21.76*
*Databank codes:*

| Species | SwissProt | PIR | EMBL (cDNA) | EMBL (genomic) |
|---|---|---|---|---|
| Myeloblastin | | | | |
| *Homo sapiens* | P24158 | A33751 | X55668 | M96839: exon 5 |
| | | A43983 | X56132 | M97911: exons 1–5 |
| | | A45080 | | X56606: 5′ end |
| | | A61176 | | |
| | | B46268 | | |
| | | JH0331 | | |
| | | S11091 | | |
| *Mus musculus* | – | – | U43525 | – |
| Proteinase 4 | | | | |
| *Homo sapiens* | P18078 | – | – | – |

## Name and History

*Myeloblastin* is the term initially coined by Bories and colleagues (Bories *et al*., 1989) to describe a serine proteinase identified in promyelocytic leukemia cells that was felt to be important in myeloid cell proliferation and differentiation. Downregulation of myeloblastin by antisense oligonucleotides causes growth arrest of HL-60 promyelocytic leukemia cells and differentiation to monocyte-like cells. It had been characterized previously as an emphysema-producing proteinase and named independently as *proteinase 3* (*PR-3*: Baggiolini *et al*., 1978; Kao *et al*., 1988) and subsequently as a protein with microbicidal activity, *p29b* (Campanelli *et al*., 1990).

## Activity and Specificity

Myeloblastin has an elastase-like specificity for small aliphatic residues (Ala, Val, Ser, Met) at the P1 and P1' sites (Rao *et al*., 1991). Assay of activity can be done with Boc-Ala-OPhNO$_2$, or the tetrapeptide substrate, MeOSuc-Ala-Ala-Pro-Val-NHPhNO$_2$. The pH optimum is about pH 7 (Kao *et al*., 1988).

Myeloblastin is inhibited by $\alpha_1$-proteinase inhibitor ($k_a = 8.1 \times 10^6\,\mathrm{M^{-1}\,s^{-1}}$) and $\alpha_2$-macroglobulin ($k_a = 1.1 \times 10^7\,\mathrm{M^{-1}\,s^{-1}}$), and elafin is also a potent inhibitor (Wiedow *et al*., 1991). In contrast to human leukocyte elastase (Chapter 15) and cathepsin G (Chapter 16), myeloblastin is not inhibited by secretory leukoprotease inhibitor (SLPI), a major serine proteinase inhibitor in the upper airways of humans (Rao *et al*., 1993).

## Structural Chemistry

The mature protein consists of 222 amino acids plus two carbohydrate chains at *N*-linked glycosylation sites. The elastase-like specificity of myeloblastin is consistent with its high level of sequence similarity to elastase at substrate-binding sites (Rao *et al*., 1991). The crystal structure of myeloblastin has been recently solved by molecular replacement using the leukocyte elastase structure (Fujinaga *et al*., 1996). These studies demonstrated that, in general, the substrate-binding sites, S4–S3', are more polar than comparable sites in leukocyte elastase. The preference of myeloblastin for small aliphatic residues at the P1 position of a substrate is explained by the Val to Ile substitution at position 190 when compared to the leukocyte elastase structure.

## Preparation

Myeloblastin was initially purified from polymorphonuclear leukocyte granule extracts by dye-ligand affinity chromatography on Matrex Gel Orange A followed by cation-exchange chromatography on Bio-Rex 70 cation-exchange resin (Kao *et al*., 1988). The recovery of myeloblastin in this method varied considerably because the Matrex Gel Orange A varied between batches. Subsequently, an HPLC/FPLC-based strategy was described (Rasmussen *et al*., 1990). More recently, recombinant myeloblastin and its zymogen have been expressed in baculovirus/insect cell systems (Fujinaga

*et al*., 1996; Witko-Sarsat *et al*., 1996) or human mast cell lines (Specks *et al*., 1996).

## Biological Aspects

The gene structure of myeloblastin has been elucidated (Sturrock *et al*., 1992). The gene spans approximately 6.5 kb and consists of five exons and four introns. The genomic organization is similar to that of the other serine proteinases expressed in hematopoietic cells. Each residue of the catalytic triad of myeloblastin is located on a separate exon, and the positions of the residues within the exons are similar to those in leukocyte elastase and cathepsin G. The gene for myeloblastin is located in a serine proteinase gene cluster localized on chromosome 19p13.3 (Zimmer *et al*., 1992; Sturrock *et al*., 1993), within 3 kb of leukocyte elastase and 8 kb of azurocidin (an homologous protein with bactericidal but no proteolytic activity).

Recent studies examining the human promoter indicate that the first 200 bp of the transcription initiation site is sufficient to give maximal and myeloid-specific expression of a reporter gene (Sturrock *et al*., 1996). Within this region two elements have been identified (−101 and −190) that confer the majority of activity. The element at −101 is a binding site for the PU.1 transcription factor. PU.1, a member of the *ets* family of transcription factors, is expressed only in cells of myeloid and B-cell lineage and is involved in multiple steps of myeloid development. The element at −190 has a core sequence of CCCCGCCC (CG element). The CG element binds a nuclear protein of approximately 40 kDa that has not yet been fully characterized. Expression of the myeloblastin gene is restricted to the promyelocytic stage of differentiation. Maturation of promyelocytic cells results in an inhibition of myeloblastin gene expression and a reduction in nuclear protein binding to the PU.1 and CG elements. The data suggest that there is cooperative interaction between the PU.1 and CG elements in conferring tissue- and developmental-specific expression of myeloblastin. Similar elements occur in the leukocyte elastase and cathepsin G promoters, although their importance is currently unknown. Leukocyte elastase and cathepsin G are coordinately expressed with myeloblastin during azurophil development, suggesting similar transcriptional control.

Myeloblastin is stored as an active enzyme within the azurophil granules of polymorphonuclear leukocytes, but is synthesized as a zymogen. The biosynthesis and processing of myeloblastin have been described (Rao *et al*., 1996). Myeloblastin is synthesized as a 256 amino acid preproenzyme. The processing of myeloblastin involves three proteolytic cleavages, two on the N-terminal side and one on the C-terminal side. The initial N-terminal cleavage results in the removal of a 25 amino acid endoplasmic reticulum-targeting signal sequence. In additional processing, a dipeptide propiece is removed from the N-terminal side and a seven amino acid peptide from the C-terminal side, leaving the 222 amino acid mature form that contains two carbohydrate chains at *N*-linked glycosylation sites. The removal of the activation dipeptide occurs in a post-Golgi organelle, and the enzyme responsible for the removal is a cysteine proteinase, identical or similar to dipeptidyl-peptidase I (Chapter 214).

Physiologically, myeloblastin may assist in the killing and digestion of microbes (Gabay & Almeida, 1993) and in the movement of polymorphonuclear leukocytes through the basement membrane at sites of inflammation. Myeloblastin degrades a variety of matrix proteins *in vitro* including fibronectin, laminin, vitronectin and collagen type IV (Rao *et al*., 1991). It shows no or minimal activity against interstitial collagens types I and III. Recently myeloblastin has been shown to be active against a number of nonmatrix substrates. It cleaves the nuclear factor-$\kappa$B subunit (Franzoso *et al*., 1994), a transcription factor involved in the regulation of a large number of genes. The ability of myeloblastin to activate tumor necrosis factor $\alpha$ (TNF$\alpha$) by cleaving the TNF$\alpha$ precursor to generate a bioactive form (Robache-Gallea *et al*., 1995) and potently activate latent transforming growth factor $\beta$ (TGF$\beta$) (Csernok *et al*., 1996) suggests proinflammatory actions of the enzyme that are independent of its ability to degrade extracellular matrix proteins. In addition, myeloblastin can enhance interleukin 8 (IL-8) production by endothelial cells *in vitro* (Berger *et al*., 1996), a further example of its potential proinflammatory capacity. It also degrades the 28 kDa mammalian heat shock protein (Spector *et al*., 1995) previously linked to the differentiation of normal and neoplastic cells.

Myeloblastin has been associated with a number of human diseases. Most importantly, it is the target antigen of the cytoplasmic pattern of antineutrophil cytoplasmic autoantibodies (c-ANCA) detected in the circulation of patients with Wegener's granulomatosis (Ludemann *et al*., 1990). Circulating antibodies to myeloblastin increase prior to vascular inflammation in patients with Wegener's disease and monitoring of their levels is an important part of both diagnosis and treatment (reviewed in Gross *et al*., 1995). Growing evidence supports the hypothesis that antibodies to myeloblastin are more than an epiphenomenon, and that interaction of myeloblastin with these autoantibodies is intimately involved in the pathogenesis of the disease. As a unifying construct, it has been proposed that proinflammatory cytokines such as TNF$\alpha$ and IL-8 induce surface expression of myeloblastin on polymorphonuclear leukocytes. The antimyeloblastin antibodies then activate the primed polymorphonuclear leukocytes by binding to cell surface-associated myeloblastin and engaging the low-affinity receptor for the Fc fragment of IgG (Porges *et al*., 1994). It is suggested that this activates the polymorphonuclear leukocytes to produce reactive oxygen species and release granule proteins (Falk *et al*., 1990), which causes endothelial cell injury (Savage *et al*., 1992). A role for cell-mediated immunity in the pathogenesis of antimyeloblastin-associated vasculitides has also been suggested. Brouwer and colleagues recently demonstrated that blood lymphocytes of patients with Wegener's granulomatosis respond vigorously to highly purified myeloblastin (Brouwer *et al*., 1994).

Animal models have indicated a potential role for myeloblastin in pulmonary emphysema (Kao *et al*., 1988). Hamsters given a single intratracheal instillation of myeloblastin (0.1 mg) develop dramatic emphysematous lesions comparable with those developed following instillation of a similar quantity of leukocyte elastase. The mechanism by which myeloblastin produces emphysema appears to relate directly to its ability to degrade elastin. Myeloblastin is only the second enzyme purified from human phagocytes that has been demonstrated to cause experimental emphysema.

Myeloblastin has also been implicated in the growth and differentiation of human leukemic cells (Bories *et al*., 1989). Downregulation of myeloblastin expression by antisense oligonucleotides inhibits proliferation and induces differentiation of HL-60-derived promyelocyte-like leukemia cells. Whether proteolytic activity of myeloblastin is required for its apparent proliferative effects has not been determined.

## Distinguishing Features

Myeloblastin can be distinguished from the other polymorphonuclear leukocyte neutral serine proteinases by its inhibitory profile in that secretory leukocyte protease inhibitor, a potent inhibitor of leukocyte elastase and cathepsin G, does not inhibit myeloblastin (Rao *et al*., 1993). Polyclonal antisera and monoclonal antibodies against the myeloblastin that do not cross-react with leukocyte elastase or cathepsin G have been described (Kao *et al*., 1988; Calafat *et al*., 1990; Rao *et al*., 1996), but are not commercially available at the time of writing.

## Related Peptidase

Myeloblastin is similar to but distinct from the two other azurophil granule serine proteinases, leukocyte elastase and cathepsin G. It resides in a gene cluster on chromosome 19p13.3 separated by 8 kb from the 3′ end of the azurocidin gene and by 3 kb from the 5′ end of the leukocyte elastase gene (Zimmer *et al*., 1992). Azurocidin is a proteolytically inactive homolog because of Ser to Gly and His to Ser substitutions in the catalytic triad (Almeida *et al*., 1991). Using the phase and position of the introns as criteria, myeloblastin, along with the other serine proteinases of hematopoietic origin, including the lymphocyte granzymes and mast cell chymases, has been placed in the sixth class of serine proteinases (Jenne *et al*., 1991).

## Further Reading

For a review, see Hoidal *et al*. (1995).

## References

Almeida, R.P., Melchior, M., Campanelli, D., Nathan, C. & Gabay, J.E. (1991) Complementary DNA sequence of human neutrophil azurocidin, an antibiotic with extensive homology to serine proteases. *Biochem. Biophys. Res. Commun.* **177**, 688–695.

Baggiolini, M., Bretz, U. & Dewald, B. (1978) The polymorphonuclear leukocyte. *Agents Actions* **8**, 3–10.

Berger, S.P., Seelen, M.A., Hiemstra, P.S., Gerritsma, J.S., Heemskerk, E., van der Woude, F.J. & Daha, M.R. (1996) Proteinase 3, the major autoantigen of Wegener's granulomatosis, enhances IL-8 production by endothelial cells *in vitro*. *J. Am. Soc. Nephrol.* **7**, 694–701.

Bories, D., Raynal, M.-C., Solomon, D.H., Darzynkiewicz, Z. & Cayre, Y.E. (1989) Downregulation of a serine protease, myeloblastin, causes growth arrest and differentiation of promyelocytic leukemia cells. *Cell* **59**, 959.

Brouwer, E., Stegeman, C.A., Huitema, M.G., Limburg, P.C. & Kallenberg, C.G. (1994) T cell reactivity to proteinase 3 and myeloperoxidase in patients with Wegener's granulomatosis (WG). *Clin. Exp. Immunol.* **98**, 448–453.

Calafat, J., Goldschmeding, R., Pingeling, P.L., Janssen, H. & van der Schoot, C.E. (1990) In situ localization by double-labeling immunoelectron microscopy of anti-neutrophil cytoplasmic autoantibodies in neutrophils and monocytes. *Blood* **75**, 242–250.

Campanelli, D., Melchior, M., Fu, Y., Nakata, M., Shuman, H., Nathan, C. & Gabay, J.E. (1990) Cloning of cDNA for proteinase 3: a serine protease, antibiotic, and autoantigen from human neutrophils. *J. Exp. Med.* **172**, 1709–1715.

Csernok, E., Szymkowiak, C.H., Mistry, N., Daha, M.R., Gross, W.L. & Kekow, J. (1996) Transforming growth factor-beta (TGF-beta) expression and interaction with proteinase 3 (PR3) in anti-neutrophil cytoplasmic antibody (ANCA)-associated vasculitis. *Clin. Exp. Immunol.* **105**, 104–111.

Falk, R.J., Terrell, R.S., Charles, L.A. & Jennette, J.C. (1990) Anti-neutrophil cytoplasmic autoantibodies induce neutrophils to degranulate and produce oxygen radicals *in vitro. Proc. Natl Acad. Sci. USA* **87**, 4115–4119.

Franzoso, G., Biswas, P., Poli, G., Carlson, L.M., Brown, K.D., Tomita-Yamaguchi, M., Fauci, A.S. & Seibenlist, U.K. (1994) A family of serine proteases expressed exclusively in myelo-monocytic cells specifically processes in nuclear factor-κB subunit p65 in vitro and may impair human immunodeficiency virus replication in these cells. *J. Exp. Med.* **180**, 1445–1456.

Fujinaga, M., Chernaia, M.M., Halenbeck, R., Koths, K. & James, M.N.G. (1996) The crystal structure of PR3, a neutrophil serine proteinase antigen of Wegener's granulomatosis antibodies. *J. Mol. Biol.* **261**, 267–278.

Gabay, J.E. & Almeida, R.P. (1993) Antibiotic peptides and serine protease homologs in human polymorphonuclear leukocytes: defensins and azurocidin. *Curr. Opin. Immunol.* **5**, 97–102.

Gross, W.L., Csernok, E. & Helmchen, U. (1995) Antineutrophil cytoplasmic autoantibodies, autoantigens, and systemic vasculitis. *APMIS* **103**, 81–97.

Hoidal, J.R., Rao, N.V. & Gray, B. (1995) Myeloblastin: leukocyte proteinase 3. *Methods Enzymol.* **244**, 61–67.

Jenne, D.E., Zimmer, M., Garcia-Sanz, J.A., Tschopp, J. & Lichter, P. (1991) Genomic organization and subchromosomal in situ localization of the murine granzyme F, a serine protease expressed in CD8+ T cells. *J. Immunol.* **147**, 1045–1052.

Kao, R.C., Wehner, N.G., Skubitz, K.M., Gray, B.H. & Hoidal, J.R. (1988) A distinct human polymorphonuclear leukocyte proteinase that produces emphysema in hamsters. *J. Clin. Invest.* **82**, 1963–1973.

Ludemann, J., Utecht, B. & Gross, W.L. (1990) Anti-neutrophil cytoplasm antibodies in Wegener's granulomatosis recognize an elastinolytic enzyme. *J. Exp. Med.* **171**, 357–362.

Porges, A.J., Redecha, P.B., Kimberly, W.T., Csernok, E., Gross, W.L. & Kimberly, R.P. (1994) Anti-neutrophil cytoplasmic antibodies engage and activate human neutrophils via Fc-γRIIa. *J. Immunol.* **153**, 1271–1280.

Rao, N.V., Wehner, N.G., Marshall, B.C., Gray, W.R., Gray, B.H. & Hoidal, J.R. (1991) Characterization of proteinase-3 (PR-3), a neutrophil serine proteinase. *J. Biol. Chem.* **266**, 9540–9548.

Rao, N.V., Marshall, B.C., Gray, B.H. & Hoidal, J.R. (1993) Interaction of secretory leukocyte protease inhibitor with proteinase-3. *Am. J. Respir. Cell. Mol. Biol.* **8**, 612–616.

Rao, N.V., Rao, G.V., Marshall, B.C. & Hoidal, J.R. (1996) Biosynthesis and processing of proteinase 3 in U937 cells. *J. Biol. Chem.* **271**, 2972–2978.

Rasmussen, N., Sjolin, C., Isakkson, B., Bygren, P. & Weislander, J. (1990) An ELISA for the detection of anti-neutrophil cytoplasm antibodies (ANCA). *J. Immunol. Methods* **127**, 139–145.

Robache-Gallea, S., Morand, V., Bruneau, J.M., Schoot, B., Tagat, E., Realo, E., Chouaib, S. & Roman-Roman, S. (1995) *In vitro* processing of human tumor necrosis factor-alpha. *J. Biol. Chem.* **270**, 23688–23692.

Savage, C.O.S., Pottinger, B.E. & Gaskin, G. (1992) Autoantibodies developing to myeloperoxidase and proteinase 3 in systemic vasculitis stimulate neutrophil cytotoxicity toward cultured endothelial cells. *Am. J. Pathol.* **141**, 335–342.

Specks, U., Fass, D.N., Fautsch, M.P., Hummel, A.M. & Viss, M.A. (1996) Recombinant human proteinase 3, the Wegener's autoantigen, expressed in HMC-1 cells is enzymatically active and recognized by c-ANCA. *FEBS* **390**, 265–270.

Spector, N.L., Hardy, L., Ryan, C., Miller, W.H., Jr., Humes, J.L., Nadler, L.M. & Luedke, E. (1995) 28-kDa mammalian heat shock protein, a novel substrate of a growth regulatory protease involved in differentiation of human leukemia cells. *J. Biol. Chem.* **270**, 1003–1006.

Sturrock, A.B., Franklin, K.F., Rao, G., Marshall, B.C., Rebentisch, M.B., Lemons, R.S. & Hoidal, J.R. (1992) Structure, chromosomal assignment, and expression of the gene for proteinase-3. *J. Biol. Chem.* **267**, 21193–21199.

Sturrock, A.B., Espinosa, R., III, Hoidal, J.R. & Le Beau, M.M. (1993) Localization of the gene encoding proteinase-3 (the Wegener's granulomatosis autoantigen) to human chromosome band 19p13.3. *Cytogenet. Cell. Genet.* **64**, 33–34.

Sturrock, A., Franklin, K.F. & Hoidal, J.R. (1996) Human proteinase-3 expression is regulated by PU.1 in conjunction with a cytidine-rich element. *J. Biol. Chem.* **271**, 32392–32402.

Wiedow, O., Lüdemann, J. & Utecht, B. (1991) Elafin is a potent inhibitor of proteinase 3. *Biochem. Biophys. Res. Commun.* **174**, 6–10.

Witko-Sarsat, V., Halbwachs-Mecarelli, L., Almeida, R.P., Nusbaum, P., Melchior, M., Jumaleddine, G., Lesavre, P., Descamps-Latscha, B. & Gabay, J.E. (1996) Characterization of a recombinant proteinase 3, the autoantigen in Wegener's granulomatosis and its reactivity with anti-neutrophil cytoplasmic autoantibodies. *FEBS Lett.* **382**, 130–136.

Zimmer, M., Medcalf, R.L., Fink, T.M., Mattmann, C., Lichter, P. & Jenne, D.E. (1992) Three human elastase-like genes coordinately expressed in the myelomonocyte lineage are organized as a single genetic locus on 19pter. *Proc. Natl Acad. Sci. USA* **89**, 8215–8219.

***John R. Hoidal***
*Department of Internal Medicine, Division of Respiratory, Critical Care and Occupational Medicine,*
*University of Utah Health Sciences Center,*
*Salt Lake City, Utah 84132, USA*
*Email: jhoidal@msscc.med.utah.edu*

# 18. Chymase

## Databanks

*Peptidase classification: clan SA, family S1, MEROPS ID: S01.140*
*NC-IUBMB enzyme classification: EC 3.4.21.39*
*Chemical Abstracts Service registry number: 97501-92-3*
*Databank codes:*

| Species | SwissProt | PIR | EMBL (cDNA) | EMBL (genomic) |
|---|---|---|---|---|
| **Chymase 1** | | | | |
| *Canis familiaris* | P21842 | A35842 | J02904 | – |
| | | | Z28392 | |
| *Homo sapiens* | P23946 | A40967 | M69136 | M64269: complete gene |
| | | | S61334 | M69137: complete gene |
| | | | | X59072: midsection |
| *Meriones unguiculatus* | P50340 | – | D45173 | – |
| *Mus musculus* | P11034 | A46504 | S44609 | X68803: complete gene |
| *Papio hamadryas* | P52195 | – | U38521 | U38463: complete gene |
| *Rattus norvegicus* | P09650 | JC2125 | S69206 | – |
| **Chymase 2** | | | | |
| *Meriones unguiculatus* | P50341 | – | D45174 | – |
| *Mus musculus* | P15119 | A34910 | J05177 | L08486: complete gene |
| | | A46721 | | |
| *Ovis aries* | – | – | Y08133 | – |
| *Rattus norvegicus* | P00770 | A00957 | J02712 | – |
| | | A29548 | | |
| **Chymase 3** | | | | |
| *Rattus norvegicus* | P50339 | – | D38495 | – |
| **Chymase 4** | | | | |
| *Mus musculus* | P21812 | A38678 | M55617 | M55616: complete gene |
| | | B35646 | X68804 | |
| | | B38678 | | |
| | | C38678 | | |
| **Chymase 5** | | | | |
| *Mus musculus* | P21844 | S26043 | M68898 | M73760: unannotated |
| | | | M73759 | |
| | | | X68805 | |
| **Chymase-L** | | | | |
| *Mus musculus* | P43430 | – | X78545 | – |
| **Others** | | | | |
| *Mus musculus* | P21843 | A35646 | – | – |
| | | A46741 | | |
| *Mus musculus* | Q00356 | A38678 | M57401 | |
| *Rattus norvegicus* | Q06606 | – | X68657 | – |

Brookhaven Protein Data Bank three-dimensional structures:

| Species | ID | Resolution | Notes |
|---|---|---|---|
| **Chymase 2** | | | |
| *Rattus norvegicus* | 3RP2 | 1.9 | |

## Name and History

The name *chymase* was proposed by Lagunoff & Benditt (1963) to denote an enzyme similar to pancreatic chymotrypsin (Chapter 8) that had been first detected in mast cells 10 years previously (Gomori, 1953). In more recent usage, the term refers to a group of chymotryptic serine proteases expressed in mast cell secretory granules. Some use the term to describe natural killer (NK) and cytolytic T cell chymotrypsin-like proteases, but these are better considered as granzymes (Chapter 23).

Chymases can be bewildering in number and variety, and their nomenclature tends to increase confusion. The first purified chymases, from mast cells of rat thyroid (Pastan & Almqvist, 1966) or peritoneum (Kawiak *et al.*, 1971), were termed *chymase* or *chymotrypsin-like protease*. Similar enzymes from muscle and small intestine were termed *skeletal muscle protease* and *group-specific protease*, respectively. When found to originate from mast cells, these were termed *rat mast cell protease I* and *rat mast cell protease II* (Woodbury & Neurath, 1980), and they are here termed *rat chymases 1* and 2. Related dog and human enzymes were labeled by tissue of origin (*skin neutral protease, lung chymotrypsin-like protease, heart chymase*, etc.). These enzymes are produced mainly or exclusively by mast cells, and for the sake of simplicity, are here all termed *chymase*. In mice, there are five or more chymases (Springman & Serafin, 1995). These have been named according to the mast cell or tissue in which they are found (e.g. *mucosal or intestinal mast cell protease*) or numbered in order of discovery (*mouse mast cell protease (MMCP)-1*, *-2*, etc.). We follow the numbers of Reynolds *et al.* (1990), who refer to *MMCP-1* to *MMCP-5*, here simplified to *mouse chymases 1–5*. MMCP-6 and -7 are tryptases (Chapter 19), not chymases. An additional mRNA sequence deposited in databases is termed MMCP-8, but it is not yet clear whether its product is a chymase. Also, a chymase-like gene (*MMCP-L*), here termed *chymase-L*, was described; there is no evidence of expression in normal mice, although low mRNA levels are found in *Nippostrongylus*-infected animals (Serafin *et al.*, 1991).

In recent years, a picture of structure–activity and phylogenetic relationships among chymases has emerged (Caughey, 1995). Based on structural similarities, chymases fall into one of two groups, α or β (Chandrasekharan *et al.*, 1996) (see Figure 18.1). Of mouse chymases, all but chymase 5 are β, as are rat 1 and 2, but human and dog are α. It is a mystery why mice have many β-chymases while humans apparently have none. Evidence that β genes have exchanged genetic information suggests that they have evolved by repeated gene conversions or unequal crossing mechanisms (Huang & Hellman, 1994). The chymase-L gene is a possible master gene for such exchanges.

## Activity and Specificity

Chymase specificity is chymotrypsin-like: protein and peptide targets are cleaved on the carbonyl side of aromatic residues, with the order of P1 preference Phe > Tyr ≫ Trp. Subsite specificity and catalytic efficiency vary among chymases, which also hydrolyze esters of amino acids (e.g. Bz-Tyr-OEt) and naphthols (e.g. the chloroacetyl

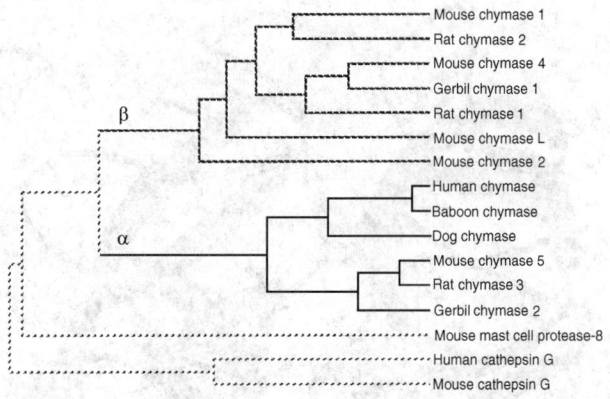

*Figure 18.1* Dendrogram of mast cell chymases and closely related proteases. This figure was generated using the multiple sequence alignment and tree programs implemented in GeneWorks. The lengths of the horizontal lines are proportional to the percentage of amino acid differences between the sequences of the prepro form of pairs or groups of enzymes. Two major branches of the chymase tree (α and β) are labeled. Note that cathepsin G and mouse mast cell protease 8 follow branches that are distinct from those of α and β chymases.

ester of naphthol-3-hydroxy-2-naphthoic acid *o*-toluidide (naphthol-ASD)). Chymases are active at neutral to alkaline pH, and high ionic strength increases activity. Available substrates include Suc-Ala-Ala-Pro-Phe↓NHPhNO$_2$ and Suc-Phe-Pro-Phe↓NHPhNO$_2$, which also are hydrolyzed by cathepsin G (Chapter 16) (Yoshida *et al.*, 1980). No truly specific substrate is known. Soluble chymases are inhibited by serpins, e.g. α$_1$-antichymotrypsin (Schechter *et al.*, 1989). However, some chymases (notably rat 1) are released as insoluble complexes resisting large inhibitors (Le Trong *et al.*, 1987). Other useful inactivators include soybean trypsin inhibitor and chymostatin.

## Structural Chemistry

Chymases are synthesized as proenzymes, which are activated by removal of a two-residue propeptide. They vary in primary structure to a surprising degree; mouse chymases are only 50–76% identical (Springman & Serafin, 1995). Mature chymases ($M_r$ 26 000–32 000) may be *N*-glycosylated at one or more sites; as an extreme example, mouse chymase 1 occurs in five glycoforms, which vary in kinetic properties (Miller *et al.*, 1995). Other chymases (e.g. rat 1 and 2) lack consensus sites and are not glycosylated. All α-chymases, including human (Caughey *et al.*, 1991), are known or predicted to be glycosylated. At least one chymase (rat 2) is modified at the C-terminus by removal of several amino acids whose presence is predicted from the cDNA (Benfey *et al.*, 1987) but are absent from the mature protein (Woodbury *et al.*, 1978). Compared to most other family S1 serine proteases, chymases have few cysteine residues (seven in α-chymases; six in β-chymases) and few disulfide bonds (three). As predicted from the crystal structure of rat chymase 2 (Remington *et al.*, 1988) (see Figure 18.2), the secondary structure of chymase is primarily β sheet, with the substrate-binding site formed by a cleft between twin β-barrel domains. Highly cationic

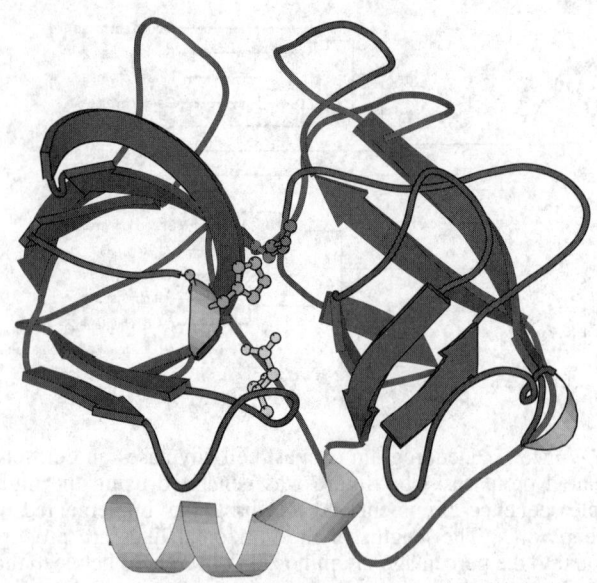

*Figure 18.2* Richardson diagram of rat chymase 2. The image was prepared from the Brookhaven protein databank entry (3RP2) as described in the Introduction (p. xxv). Catalytic residues are shown in ball-and-stick representation: His57, Asp102, and Ser195 (numbering as in Alignment 2.1 on the CD-ROM).

chymases (dog, human and rat 1) are released in complex with heparin proteoglycans. Nearly neutral chymases (rat 2 and mouse 1) have low affinity for heparin, and are highly soluble and diffusible upon release (Miller *et al.*, 1995). Thus, the physical properties of chymases vary considerably.

## Preparation

Chymases can be purified from many sources, reflecting the wide species and tissue distribution of mast cells. Species from which chymases have been purified include human, baboon, dog, sheep, pig, rat, mouse and hamster; cDNAs have been cloned from human, baboon, dog, rat, mouse, and gerbil (see the Databanks table). In a given mammal, mast cells differ in the kinds and amounts of chymases they produce. Mouse chymase 1 and rat 2, which are closely related products of mucosal mast cells, are abundant in helminth-infected animals, in which the number of intestinal mast cells rises dramatically (Miller *et al.*, 1995). Rat chymase 1 has been purified from peritoneal connective tissue mast cells, which are tedious to collect in large numbers. The same enzyme can be purified from cruder homogenates of skeletal muscle, tongue or skin, in which connective tissue mast cells abound. Mastocytomas are a good source of the dog enzyme. In humans, the best source is skin (Schechter *et al.*, 1983), a tissue rich in chymase-containing mast cells, but gut mucosa and lung parenchyma are poor sources (in contrast to the situation in rodents). Recently, recombinant human chymase was expressed in bacterial and mammalian cells (Urata *et al.*, 1993; Wang *et al.*, 1995), which may yield useful amounts of enzyme. Because chymases diverge in properties such as charge, solubility, glycosylation and heparin binding, strategies for purification vary.

## Biological Aspects

Regulation of chymase gene expression is complex, especially in mice, where the products of many chymase genes in different mast cells are controlled by various normal and pathologic stimuli. Studies with cells cultured from mouse bone marrow suggest the importance of kit ligand and interleukins 3, 9 and 10 in modulating steady-state levels of chymase mRNA (Springman & Serafin, 1995). In rats, regions in the 5′ flank of the chymase 2 gene regulate transcription (Benfey *et al.*, 1987). The relevance of these regions to regulation of other chymases with different patterns of expression is not yet clear. Other genes (see the Databanks table) have been described. Chymase genes of both humans (Caughey *et al.*, 1993) and mice (Gurish *et al.*, 1993) cluster with other granule-associated serine protease genes transcribed in lymphocytes and myelomonocytes. This clustering raises the possibility of higher-level orchestration of hematopoietic protease gene expression.

Mature chymases are packaged in secretory granules together with histamine, tryptases and other mediators. Activation of prochymases occurs intracellularly by the action of dipeptidyl-peptidase I (Chapter 214), which removes an acidic N-terminal dipeptide (Dikov *et al.*, 1994; Murakami *et al.*, 1995). Although chymases mature within mast cells, their activity in granules is limited by low pH; they become fully active when released into the neutral pH environment outside the cell. Mast cell degranulation, e.g. by antigen-bound IgE or substance P, releases chymase. The importance of chymases to organ development, tissue homeostasis and host defense is unknown. However, many lines of evidence support extracellular roles for these enzymes. This evidence, reviewed elsewhere (Caughey, 1995), suggests that chymases destroy extracellular matrix proteins, activate matrix metalloproteinases, potentiate plasma leakage, stimulate submucosal gland cell secretion, inactivate inflammatory neuropeptides, generate extravascular angiotensin II, catabolize lipoproteins, and control complement-mediated inflammation.

## Distinguishing Features

Common structural features of the chymases include an acidic activation dipeptide, a conserved octapeptide near the N-terminus of the mature enzyme (Caughey *et al.*, 1988), and the existence of only three disulfide bonds in the catalytic domain. Chymase genes examined to date are identical in the number, phase and placement of introns. Of those chymases examined as enzymes, all prefer substrates with P1 aromatic residues. A distinguishing feature of some chymases is selective hydrolysis of angiotensin I to II, a potent vasoconstrictor (Reilly *et al.*, 1982; Urata *et al.*, 1990). Thus, the $\alpha/\beta$ division of chymases may be functionally important. A useful distinction between chymase and cathepsin G is the resistance of chymase to inhibition by aprotinin.

## Related Peptidases

Granzyme B (Chapter 23), cathepsin G (Chapter 16) and human granzyme H have gene structures and propeptides similar to those of chymases, and are grouped with them in the mouse and human genomes (Caughey *et al.*, 1993; Gurish *et al.*, 1993). Several other granule-associated leukocyte serine proteases, including elastase (Chapter 15), are phylogenetically similar and share an activation pathway (Caughey, 1994).

## Further Reading

Two books that broadly cover chymase chemistry and biology are those of Caughey (1995) and Schwartz (1990).

## References

Benfey, P.N., Yin, F.H. & Leder, P. (1987) Cloning of the mast cell protease, RMCP II. Evidence for cell-specific expression and a multigene family. *J. Biol. Chem.* **262**, 5377–5384.

Caughey, G.H. (1994) Serine proteases of mast cell and leukocyte granules: a league of their own. *Am. J. Respir. Crit. Care Med.* **150**, S138–S142.

Caughey, G.H. (1995) Mast cell chymases and tryptases: phylogeny, family relations, and biogenesis. In: *Mast Cell Proteases in Immunology and Biology* (Caughey, G.H., ed.). New York: Marcel Dekker, pp. 305–329.

Caughey, G.H., Viro, N.F., Lazarus, S.C. & Nadel, J.A. (1988) Purification and characterization of dog mastocytoma chymase: identification of an octapeptide conserved in chymotryptic leukocyte proteinases. *Biochim. Biophys. Acta* **952**, 142–149.

Caughey, G.H., Zerweck, E.H. & Vanderslice, P. (1991) Structure, chromosomal assignment, and deduced amino acid sequence of a human gene for mast cell chymase. *J. Biol. Chem.* **266**, 12956–12963.

Caughey, G.H., Schaumberg, T.H., Zerweck, E.H., Butterfield, J.H., Hanson, R.D., Silverman, G.A. & Ley, T.J. (1993) The human mast cell chymase gene (CMA1): mapping to the cathepsin G/granzyme gene cluster and lineage-restricted expression. *Genomics* **15**, 614–620.

Chandrasekharan, U.M., Sanker, S., Glynias, M.J., Karnik, S.S. & Husain, A. (1996) Angiotensin II-forming activity in a reconstructed ancestral chymase. *Science* **271**, 502–505.

Dikov, M.M., Springman, E.B., Yeola, S. & Serafin, W.E. (1994) Processing of procarboxypeptidase A and other zymogens in murine mast cells. *J. Biol. Chem.* **269**, 25987–25904.

Gomori, G. (1953) Chloroacyl esters as histochemical substrates. *J. Histochem. Cytochem.* **1**, 469–470.

Gurish, M.F., Nadeau, J.H., Johnson, K.R., McNeil, H.P., Grattan, K.M., Austen, K.F. and Stevens, R.L. (1993) A closely linked complex of mouse mast cell-specific chymase genes on chromosome 14. *J. Biol. Chem.* **268**, 11372–11379.

Huang, R. & Hellman, L. (1994) Genes for mast-cell serine protease and their molecular evolution. *Immunogenetics* **40**, 397–414.

Kawiak, J., Vensel, W.H., Komender, J. & Barnard, E.A. (1971) Non-pancreatic proteases of the chymotrypsin family. I. A chymotrypsinlike protease from rat mast cells. *Biochim. Biophys. Acta* **235**, 172–187.

Lagunoff, D. & Benditt, E.P. (1963) Proteolytic enzymes of mast cells. *Ann. N.Y. Acad. Sci.* **103**, 185–198.

Le Trong, H., Neurath, H. & Woodbury, R.G. (1987) Substrate specificity of the chymotrypsin-like protease in secretory granules isolated from rat mast cells. *Proc. Natl Acad. Sci. USA* **84**, 364–367.

Miller, H.R.P., Huntley, J.F. & Newlands, G.F.J. (1995) Mast cell chymases in helminthosis and hypersensitivity. In: *Mast Cell Proteases in Immunology and Biology* (Caughey, G.H., ed.). New York: Marcel Dekker, pp. 305–329.

Murakami, M., Karnik, S.S. & Husain, A. (1995) Human prochymase activation. A novel role for heparin in zymogen processing. *J. Biol. Chem.* **270**, 2218–2223.

Pastan, I. & Almqvist, S. (1966) Purification and properties of a mast cell protease. *J. Biol. Chem.* **241**, 5090–5094.

Reilly, C.F., Tewksbury, D.A., Schechter, N.B. & Travis, J. (1982) Rapid conversion of angiotensin I to angiotensin II by neutrophil and mast cell proteinases. *J. Biol. Chem.* **257**, 8619–8622.

Remington, S.J., Woodbury, R.G., Reynolds, R.A., Matthews, B.W. & Neurath, H. (1988) The structure of rat mast cell protease II at 1.9-Å resolution. *Biochemistry* **27**, 8097–8145.

Reynolds, D.S., Stevens, R.L., Lane, W.S., Carr, M.H., Austen, K.F. & Serafin, W.E. (1990) Different mouse mast cell populations express various combinations of at least six distinct mast cell serine proteases. *Proc. Natl Acad. Sci. USA* **87**, 3230–3234.

Schechter, N.M., Fraki, J.E., Geesin, J.C. & Lazarus, G.S. (1983) Human skin chymotryptic proteinase. Isolation and relation to cathepsin G and rat mast cell proteinase I. *J. Biol. Chem.* **258**, 2973–2978.

Schechter, N.M., Sprows, J.L., Schoenberger, O.L., Lazarus, G.S., Cooperman, B.S. & Rubin, H. (1989) Reaction of human skin chymotrypsin-like proteinase chymase with plasma proteinase inhibitors. *J. Biol. Chem.* **264**, 21308–21315.

Schwartz, L.B. (1990) *Neutral Proteases of Mast Cells*. Basel: S. Karger AG.

Serafin, W.E., Sullivan, T.P., Conder, G.A., Ebrahimi, A., Marcham, P., Johnson, S.S., Austen, K.F., Reynolds, D.S. (1991) Cloning of the cDNA and gene for mouse mast cell protease 4. Demonstration of its late transcription in mast cell subclasses and analysis of its homology to subclass-specific neutral proteases of the mouse and rat. *J. Biol. Chem.* **266**, 1934–1941.

Springman, E.B. & Serafin, W.E. (1995) Secretory endo- and exopeptidases of mouse mast cells: structure, genetics and regulation of expression. In: *Mast Cell Proteases in Immunology and Biology* (Caughey, G.H., ed.). New York: Marcel Dekker, pp. 169–201.

Urata, H., Kinoshita, A., Misono, K.S., Bumpus, F.M. & Husain, A. (1990) Identification of a highly specific chymase as the major angiotensin II-forming enzyme in the human heart. *J. Biol. Chem.* **265**, 22348–22357.

Urata, H., Karnik, S.S., Graham, R.M. & Husain, A. (1993) Dipeptide processing activates recombinant human prochymase. *J. Biol. Chem.* **268**, 24318–24322.

Wang, Z., Rubin, H. & Schechter, N.M. (1995) Production of active recombinant human chymase from a construct containing the enterokinase cleavage site of trypsinogen in place of the native propeptide sequence. *Biol. Chem. Hoppe Seyler* **376**, 681–684.

Woodbury, R.G. & Neurath, H. (1980) Structure, specificity and localization of the serine proteases of connective tissue. *FEBS Lett.* **114**, 189–195.

Woodbury, R.G., Katunuma, N., Kobayashi, K., Titani, K. & Neurath, H. (1978) Covalent structure of a group-specific protease from rat small intestine. *Biochemistry* **17**, 811–819.

Yoshida, N., Everitt, M.T., Neurath, H., Woodbury, R.G. & Powers, J.C. (1980) Substrate specificity of two chymotrypsin-like proteases from rat mast cells. Studies with peptide 4-nitroanilides and comparison with cathepsin G. *Biochemistry* **19**, 5799–5804.

*George H. Caughey*
*Department of Medicine and Cardiovascular Research Institute,*
*University of California at San Francisco,*
*San Francisco, CA 94143-0911, USA*
*Email: ghc@itsa.ucsf.edu*

# 19. Tryptase

## Databanks

*Peptidase classification: clan SA, family S1, MEROPS ID: S01.143*
*NC-IUBMB enzyme classification: EC 3.4.21.59*
*Chemical Abstracts Service registry number: 97501-93-4*
*Databank codes:*

| Species | SwissProt | PIR | EMBL (cDNA) | EMBL (genomic) |
|---|---|---|---|---|
| Tryptase 1 | | | | |
| *Bos taurus* | – | – | X94982 | – |
| *Canis familiaris* | P15944 | A32410 | M24664 | – |
| *Homo sapiens* | P15157 | A35863 | M30038 | M33494: complete gene |
| | | A45754 | | |
| | | C35863 | | |
| *Meriones unguiculatus* | P50342 | – | D31789 | |
| *Mus musculus* | P21845 | A38654 | L31853 | M57625: complete gene |
| | | D35646 | M57626 | |
| *Rattus norvegicus* | P50343 | JC4171 | D38455 | – |
| Tryptase 2 | | | | |
| *Homo sapiens* | P20231 | A37193 | M33492 | – |
| | | B35863 | M37488 | |
| | | | S55551 | |
| Tryptase 3 | | | | |
| *Homo sapiens* | – | – | M33493 | |
| Others | | | | |
| *Mus musculus* | Q02844 | A47246 | L00654 | – |
| *Rattus norvegicus* | P27436 | S21275 | – | – |

Brookhaven Protein Data Bank three-dimensional structures:

| Species | ID | Resolution | Notes |
|---|---|---|---|
| Tryptase 1 | | | |
| *Homo sapiens* | 1AAO | – | theoretical model |
| | 1TRN | 2.2 | complex with DFP |

## Name and History

A trypsin-like activity was first discovered in mast cells by Glenner & Cohen (1960), using the substrate $N^\alpha$-benzoyl-arginine 2-naphthylamide as a histochemical stain. This activity was subsequently named ***tryptase*** by Lagunoff & Benditt (1963) and has become of increasing interest due to the broad distribution of mast cells in various human and animal tissues, and the enzyme's potential involvement in various normal and pathologic processes. Tryptase is stored in the dense cytoplasmic granules of mast cells and accounts for 23% of the total cellular protein of human mast cells (Schwartz *et al.*, 1981). Some nonmast cell trypsin-like proteases such as lymphocyte granzyme A (Chapter 22) (Poe *et al.*, 1988) have on occasion been referred to as 'tryptases'. Fortunately, the name tryptase is now almost exclusively used to refer to mast cell serine proteases with trypsin-like specificities.

There are multiple human variants of tryptase, such as tryptases $\alpha$ and $\beta$ cloned from human lung (Miller *et al.*, 1989, 1990) and tryptases 1, 2 and 3 cloned from human skin (Vanderslice *et al.*, 1990). Most biochemical studies have used enzyme isolated from human lung or skin and while the purified enzymes have not been correlated with the cloned variants it appears that the $\alpha$ variant is most likely the material isolated from lung (Sakai *et al.*, 1996b). Although human tryptase has been studied in greatest detail, dog, rat, mouse, gerbil and cattle mast cells also contain species variants of the enzyme.

## *Activity and Specificity*

Tryptase cleaves protein and peptide substrates on the carboxyl side of Arg and Lys residues and substrates normally used for trypsin have been employed. Specificity studies using a series 4-nitroanilide and thioester synthetic peptide substrates indicated some differences in human lung tryptase as compared to human skin tryptase (Tanaka *et al.*, 1983), with Pro in P3 decreasing the activity of both enzymes. A subsequent comparison of skin and lung tryptase showed no differences regarding substrate specificity using four substrates (Harvima *et al.*, 1988a). Tryptase isolated from human pituitary mast cells preferred substrates with Lys or Arg in P2 and preferentially cleaved prohormones with Pro in P3 or P4 (Cromlish *et al.*, 1987). Rat skin tryptase preferred Pro in P2 for 4-nitroanilides and Arg in P2 with 7-amino-4-methylcoumarin substrates (Braganza & Simmons, 1991). Tryptases lack the general proteolytic ability to hydrolyze protein substrates such as casein and albumin. Protein substrates cleaved by human tryptases (apparently at specific sites) include H-kininogen (Maier *et al.*, 1983), complement C3 to C3a (Schwartz *et al.*, 1983), fibrinogen (Schwartz *et al.*, 1985) and vasoactive intestinal peptide (Caughey *et al.*, 1988; Tam & Caughey, 1990). Additional substrates include proatrial natriuretic factor (Imada *et al.*, 1987; Proctor *et al.*, 1991), calcitonin gene-related peptide (Tam & Caughey, 1990), kinetensin (Goldstein *et al.*, 1991) and prothrombin (Chapter 55) (converted to thrombin) (Dietze *et al.*, 1990; Kido *et al.*, 1985a). Tryptase also activates prourokinase (Chapter 57) via cleavage at the normal activation site (Stack & Johnson, 1994) and prostromelysin (Chapter 393) is activated via cleavage at an unidentified site (Gruber *et al.*, 1989), but tryptase does not directly activate procollagenase (Johnson & Cawston, 1985). H-Kininogen is selectively cleaved between Arg431 and Asp432 (Little & Johnson, 1995), but cleavage sites in other substrates have not been determined.

Maximal activity occurs in the pH range 7–8 (Smith *et al.*, 1984). Salts inhibit activity on synthetic substrates (Smith *et al.*, 1984), with even 0.1 M KCl being inhibitory at low substrate concentrations (Harvima *et al.*, 1988b), but

inhibition can be reversed by higher substrate concentrations (Alter *et al.*, 1987). The concentration of active sites can be determined with the titrant 4-methylumbelliferyl *p*-guanidinobenzoate hydrochloride (MUGB; Jameson *et al.*, 1973). Typical substrates include Z-Lys-SBzl (Little & Johnson, 1995), Z-Gly-Pro-Arg-NHPhNO$_2$ (Schecter *et al.*, 1995) and Tos-Gly-Pro-Lys-NHPhNO$_2$ (Schwartz, 1994a). Assays based on the cleavage of vasoactive intestinal peptide (Delaria & Muller, 1996) and the activation of protease zymogens (Dietze *et al.*, 1990) also have been used.

A distinguishing feature is resistance to inhibition by most natural protein inhibitors of serine proteinases, such as $\alpha_1$-proteinase inhibitor, antithrombin III, $\alpha_2$-macroglobulin and soybean trypsin inhibitor (Smith *et al.*, 1984; Alter *et al.*, 1990). Low molecular weight compounds, such as DFP, PMSF, benzamidine and tosyl-lysyl chloromethane are effective inhibitors (Smith *et al.*, 1984). Although cattle pancreatic trypsin inhibitor (aprotinin), which comes from cattle lung mast cells (Fritz *et al.*, 1979), does not inhibit human lung tryptase (Smith *et al.*, 1984), it does inhibit dog (Caughey *et al.*, 1987), rat (Braganza & Simmons, 1991) and cattle tryptases (Fiorucci *et al.*, 1995). Several synthetic benzamidine derivatives are inhibitory (Sturzebecher *et al.*, 1992; Caughey *et al.*, 1993) and another synthetic inhibitor reportedly functions in sheep (Clark *et al.*, 1995). Trypstatin, a Kunitz-type inhibitor of rat tryptase, was isolated from rat peritoneal mast cells (Kido *et al.*, 1988) and was subsequently shown to be a fragment of the inter-$\alpha$-inhibitor light chain that had been taken into mast cells (Itoh *et al.*, 1994). A Kazal-type inhibitor from the leech has been isolated, cloned and expressed (Auerswald *et al.*, 1994; Sommerhoff *et al.*, 1994).

## *Structural Chemistry*

Tryptases are approximately 40% identical in sequence to trypsin and chymotrypsin, having an additional 21 amino acids for a total of 245 residues and core protein molecular weights of 27 500 (Johnson & Barton, 1992). Diffuse banding on SDS-PAGE in the $M_r$ range from 30 to 35 kDa is due in part to glycosylation (Cromlish *et al.*, 1987; Little & Johnson, 1995). Active enzyme exists as a tetramer of approximately 135 kDa (Schwartz *et al.*, 1983), with the enzyme activity (Smith *et al.*, 1984) and the tetrameric structure being stabilized by tightly bound heparin proteoglycan (Schwartz & Bradford, 1986). The transition from tetramer to inactive monomer in the absence of heparin have been the subject of several reports (Schwartz *et al.*, 1990; Schechter *et al.*, 1993, 1995). In the absence of heparin, activity decays with a half-life of 2 min (Schechter *et al.*, 1995) to an 'inactive' intermediate that can be reactivated by heparin, which in the absence of heparin transitions with a half-life of 20 min to less active monomers and then to irreversibly inactivated monomers (Addington & Johnson, 1996). Molecular modeling positioned two loops on either side of the active site, suggesting an explanation for the limited substrate specificity and resistance to inhibition by natural proteinase inhibitors (Johnson & Barton, 1992). Tryptases have pI values of about 6.0, are relatively rich in His, and contain a Pro-rich sequence (Johnson & Barton, 1992), but the functions of these features are unclear. The high $A_{1\%,280}$ value of 28.9 is due the presence of nine Trp residues (Smith *et al.*, 1994). Human tryptase binds heparin very tightly, requiring 0.8 M NaCl for elution from heparin columns at pH 7.6 (Schwartz & Bradford, 1986), but mouse tryptase (MMCP-7) only binds heparin at the granule pH of 5.5 via a cluster of positively charged histidine residues (Matsumoto *et al.*, 1995).

Human tryptases $\beta$ and 2 have identical catalytic domains and one putative *N*-linked glycosylation site, whereas tryptases $\alpha$, 1 and 3 have two such sites. Tryptases $\beta$, 1, 2 and 3 are more than 98% identical in amino acid sequence, in contrast to $\alpha$, which is only 92% identical with the others.

### Preparation

Human tryptase was first purified from human lung mast cells (Schwartz *et al.*, 1981) and then from human lung tissue (Smith *et al.*, 1984), but other sources include skin (Harvima *et al.*, 1988a), pituitary (Cromlish *et al.*, 1987) and a human mast cell line (Butterfield *et al.*, 1990). Recombinant human $\beta$-tryptase has been expressed as a precursor in a baculovirus/insect cell system (Sakai *et al.*, 1996a), and was autocatalytically processed to a proenzyme form prior to activation by dipeptidyl-peptidase I (Chapter 214) in the presence of dextran sulfate, yielding fully active tetrameric enzyme (Sakai *et al.*, 1996b). Tryptases have also been prepared from dog mastocytoma cells and tissues (Caughey *et al.*, 1987; Schecter *et al.*, 1988), rat peritoneal mast cells (Kido *et al.*, 1985b; Muramatu *et al.*, 1988), rat skin (Braganza & Simmons, 1991), cattle liver capsula (Fiorucci *et al.*, 1992) and guinea pig lung (McEuen *et al.*, 1996).

### Biological Aspects

Tryptases are found in all human mast cells (Irani *et al.*, 1986) and provide a much better marker of mast cell activation than does histamine (Schwartz *et al.*, 1987; Schwartz, 1994b). The predominant mast cell of the skin contains both tryptase (35 pg per cell) and chymase (Chapter 18), whereas lung mast cells primarily have only tryptase (10 pg per cell) (Schwartz, 1994b). Tryptases $\alpha$ and $\beta$ have been cloned from human lung and localized to chromosome 16 (Miller *et al.*, 1989, 1990) and tryptases 1, 2 and 3 have been cloned from human skin (Vanderslice *et al.*, 1990). The basophil cell line KU812 expresses only $\beta$-tryptase (Blom & Hellman, 1993), and the Mono Mac 6 and U-937 cell lines express tryptases $\alpha$ and $\beta$, respectively (Huang *et al.*, 1993). Tryptase has been cloned from a dog mastocytoma cDNA library, along with a dog mast cell protease 3 (Vanderslice *et al.*, 1989), an oligomeric relative of the tryptases (Raymond *et al.*, 1995). Two tryptases, designated mouse mast cell proteases 6 and 7, have been cloned (Reynolds *et al.*, 1991; McNeil *et al.*, 1992). Tryptases also have been cloned from rat (Ide *et al.*, 1995), gerbil (Murakumo *et al.*, 1995) and cattle mast cells (Pallaoro *et al.*, 1996).

Due to the absence of endogenous inhibitors in humans, tryptase released from mast cells may function in normal and pathologic situations and several substrates have been identified via *in vitro* studies. These include: inactivation of fibrinogen as a substrate for thrombin (Schwartz *et al.*, 1985), destruction of the procoagulant function of H-kininogen (Maier *et al.*, 1983), generation of C3a from complement C3 (Schwartz *et al.*, 1983), cleavage of fibronectin (Lohi

*et al.*, 1992), degradation of the vasoactive intestinal peptide (Caughey *et al.*, 1988) and degradation of calcitonin gene-related peptide (Walls *et al.*, 1992), activation of stromelysin (Chapter 393) (Gruber *et al.*, 1989), activation of prourokinase (Chapter 57) (Stack & Johnson, 1994) and activation of prekallikrein (Imamura *et al.*, 1996). Tryptase is mitogenic for fibroblasts (Hartmann *et al.*, 1992), smooth muscle cells (Brown *et al.*, 1995), and for epithelial cells (Cairns & Walls, 1996).

### Further Reading

Tryptase has been reviewed by Schwartz (1994a).

### References

Addington, A.K. & Johnson, D.A. (1996) Inactivation of human lung tryptase: evidence for a re-activatable tetrameric intermediate and active monomers. *Biochemistry* **35**, 13511–13518.

Alter, S.C., Metcalfe, D.D., Bradford, T.R. & Schwartz, L.B. (1987) Regulation of human mast cell tryptase. Effects of enzyme concentration, ionic strength and the structure and negative charge density of polysaccharides. *Biochem. J.* **248**, 821–827.

Alter, S.C., Kramps, J.A., Janoff, A. & Schwartz, L.B. (1990) Interactions of human mast cell tryptase with biological protease inhibitors. *Arch. Biochem. Biophys.* **276**, 26–31.

Auerswald, E.A., Morenweiser, R., Sommerhoff, C.P., Piechotka, G.P., Eckerskorn, C., Gurtler, L.G. & Fritz, H. (1994) Recombinant leech-derived tryptase inhibitor: construction, production, protein chemical characterization and inhibition of HIV-1 replication. *Biol. Chem. Hoppe-Seyler* **375**, 695–703.

Blom, T. & Hellman, L. (1993) Characterization of a tryptase mRNA expressed in the human basophil cell line KU812. *Scand. J. Immunol.* **37**, 203–208.

Braganza, V.J. & Simmons, W.H. (1991) Tryptase from rat skin: purification and properties. *Biochemistry* **30**, 4997–5007.

Brown, J.K., Tyler, C.L., Jones, C.A., Ruoss, S.J., Hartmann, T. & Caughey, G.H. (1995) Tryptase, the dominant secretory granular protein in human mast cells, is a potent mitogen for cultured dog tracheal smooth muscle cells. *Am. J. Respir. Cell. Mol. Biol.* **13**, 227–236.

Butterfield, J.H., Weiler, D.A., Hunt, L.W., Wynn, S.R. & Roche, P.C. (1990) Purification of tryptase from a human mast cell line. *J. Leukoc. Biol.* **47**, 409–419.

Cairns, J.A. & Walls, A.F. (1996) Mast cell tryptase is a mitogen for epithelial cells. Stimulation of IL-8 production and intercellular adhesion molecule-1 expression. *J. Immunol.* **156**, 275–283.

Castells, M.C., Irani, A.M. & Schwartz, L.B. (1987) Evaluation of human peripheral blood leukocytes for mast cell tryptase. *J. Immunol.* **138**, 2184–2189.

Caughey, G.H., Viro, N.F., Ramachandran, J., Lazarus, S.C., Borson, D.B. & Nadel, J.A. (1987) Dog mastocytoma tryptase: affinity purification, characterization, and amino-terminal sequence. *Arch. Biochem. Biophys.* **258**, 555–563.

Caughey, G.H., Leidig, F., Viro, N.F. & Nadel, J.A. (1988) Substance P and vasoactive intestinal peptide degradation by mast cell tryptase and chymase. *J. Pharmacol. Exp. Ther.* **244**, 133–137.

Caughey, G.H., Raymond, W.W., Bacci, E. Lombardy, R.L. & Tidwell, R.R. (1993) Bis(5-amidino-2-benzimidazolyl)methane and related amidines are potent, reversible inhibitors of mast cell tryptases. *J. Pharmacol. Exp. Ther.* **264**, 676–682.

Clark, J.M., Abraham, W.M., Fishman, C.E., Forteza, R., Ahmed, A., Cortes, A., Warne, R.L., Moore, W.R. & Tanaka, R.D. (1995) Tryptase inhibitors block allergen-induced airway and inflammatory responses in allergic sheep. *Am. J. Respir. Crit. Care Med.* **152**, 2076–2083.

Cromlish, J.A., Seidah, N.G., Marcinkiewicz, M., Hamelin, J., Johnson, D.A. & Chretien, M. (1987) Human pituitary·tryptase: molecular forms, NH2-terminal sequence, immunocytochemical localization, and specificity with prohormone and fluorogenic substrates. *J. Biol. Chem.* **262**, 1363–1373.

Delaria, K. & Muller, D. (1996) High-performance liquid chromatographic assay for tryptase based on the hydrolysis of dansyl-vasoactive intestinal peptide. *Anal. Biochem.* **236**, 74–81.

Dietze, S.C., Auerswald, E.A. & Fritz, H. (1990) A new, highly sensitive enzymic assay for human tryptase and its use for identification of tryptase inhibitors. *Biol. Chem. Hoppe-Seyler* **371**(suppl.), 65–73.

Fiorucci, L., Erba, F. & Ascoli, F. (1992) Bovine tryptase: purification and characterization. *Biol. Chem. Hoppe-Seyler* **373**, 483–490.

Fiorucci, L., Erba, F., Falasca, L., Dini, L. & Ascoli, F. (1995) Localization and interaction of bovine pancreatic trypsin inhibitor and tryptase in the granules of bovine mast cells. *Biochim. Biophys. Acta* **1243**, 407–413.

Fritz, H., Kruck, J., Russe, I. & Liebich, H.G. (1979) Immunofluorescence studies indicate that the basic trypsin-kallikrein-inhibitor of bovine organs (Trasylol) originates from mast cells. *Z. Physiol. Chem.* **360**, 437–444.

Glenner, G.G. & Cohen, L.A. (1960) Histochemical demonstration of a species-specific trypsin-like enzyme in mast cells. *Nature* **185**, 846–847.

Goldstein, S.M., Leong, J. & Bunnett, N.W. (1991) Human mast cell proteases hydrolyze neurotensin, kinetensin and Leu[5]-enkephalin[1]. *Peptides* **12**, 995–1000.

Gruber, B.L., Marchese, M.J., Suzuki, K., Schwartz, L.B., Okada, Y., Nagase, H. & Ramamurthy, N.S. (1989) Synovial procollagenase activation by human mast cell tryptase dependence upon matrix metalloproteinase 3 activation. *J. Clin. Invest.* **84**, 1657–1662.

Hartmann, T., Ruoss, S.J., Raymond, W.W., Seuwen, K. & Caughey, G.H. (1992) Human tryptase as a potent, cell-specific mitogen: role of signaling pathways in synergistic responses. *Am. J. Physiol.* **262**, L528–534.

Harvima, I.T., Harvima, R.J., Eloranta, T.O. & Fräki, J.E. (1988a) The allosteric effect of salt on human mast cell tryptase. *Biochim. Biophys. Acta* **956**, 133–139.

Harvima, I.T., Schechter, N.M., Harvima, R.J. & Fräki, J.E. (1988b) Human skin tryptase: purification, partial characterization and comparison with human lung tryptase. *Biochim. Biophys. Acta* **957**, 71–80.

Huang, R., Abrink, M., Gobl, A.E., Nilsson, G., Aveskogh, M., Larsson, L.G., Nilsson, K. & Hellman, L. (1993) Expression of a mast cell tryptase in the human monocytic cell lines U-937 and Mono Mac 6. *Scand. J. Immunol.* **38**, 359–367.

Ide, H., Itoh, H., Tomita, M., Murakumo, Y., Kobayashi, T., Maruyama, H., Osada, Y. & Nawa, Y. (1995) cDNA sequencing and expression of rat mast cell tryptase. *J. Biochem. (Tokyo)* **118**, 210–215.

Imada, T., Takayanagi, R. & Inagami, T. (1987) Identification of a peptidase which processes atrial natriuretic factor precursor to its active form with 28 amino acid residues in particulate fractions of rat atrial homogenate. *Biochem. Biophys. Res. Commun.* **143**, 587–592.

Imamura, T., Dubin, A., Moore, W., Tanaka, R. & Travis, J. (1996) Induction of vascular permeability enhancement by human tryptase: dependence on activation of prekallikrein and direct release of bradykinin from kininogens. *Lab. Invest.* **74**, 861–870.

Irani, A.A., Schechter, N.M., Craig, S.S., DeBlois, G. & Schwartz, L.B. (1986) Two types of human mast cells that have distinct neutral protease compositions. *Proc. Natl Acad. Sci. USA* **83**, 4464–4468.

Itoh, H., Ide, H., Ishikawa, N. & Nawa, Y. (1994) Mast cell protease inhibitor, trypstatin, is a fragment of inter-α-trypsin inhibitor light chain. *J. Biol. Chem.* **269**, 3818–3822.

Jameson, G.W., Roberts, D.V., Adams, R.W., Kyle, W.S.A. & Elmore, D.T. (1973) Determination of the operational molarity of solutions of bovine α-chymotrypsin, trypsin, thrombin and Factor Xa by spectrofluorimetric titration. *Biochem. J.* **131**, 107–117.

Johnson, D.A. & Barton, G.J. (1992) Mast cell tryptases: examination of unusual characteristics by multiple sequence alignment and molecular modelling. *Protein Sci.* **1**, 370–377.

Johnson, D.A. & Cawston, T.E. (1985) Human lung mast cell tryptase fails to activate procollagenase or degrade proteoglycan. *Biochem. Biophys. Res. Commun.* **132**, 453–459.

Kido, H., Fukusen, N., Katunuma, N., Morita, T. & Iwanaga, S. (1985a) Tryptase from rat mast cells converts bovine prothrombin to thrombin. *Biochem. Biophys. Res. Commun.* **132**, 613–619.

Kido, H., Fukusen, N. & Katunuma, N. (1985b) Chymotrypsin- and trypsin-type serine proteases in rat mast cells: properties and functions. *Arch. Biochem. Biophys.* **239**, 436–443.

Kido, H., Yokogoshi, Y. & Katunuma, N. (1988) Kunitz-type protease inhibitor found in rat mast cells. Purification, properties, and amino acid sequence. *J. Biol. Chem.* **263**, 18104–18107.

Lagunoff, D. & Benditt, E.P. (1963) Proteolytic enzymes of mast cells. *Ann. NY Acad. Sci.* **103**, 185–198.

Little, S.S. & Johnson, D.A. (1995) Human mast cell tryptase isoforms: separation and examination of substrate-specificity differences. *Biochem. J.* **307**, 341–346.

Lohi, J., Harvima, I. & Keski-Oja, J. (1992) Pericellular substrates of human mast cell tryptase: 72,000 dalton gelatinase and fibronectin. *J. Cell. Biochem.* **50**, 337–349.

Maier, M., Spragg, J. & Schwartz, L.B. (1983) Inactivation of human high molecular weight kininogen by human mast cell tryptase. *J. Immunol.* **130**, 2352–2356.

Matsumoto, R., Sali, A., Ghildyal, N., Karplus, M. & Stevens, R.L. (1995) Packaging of proteases and proteoglycans in the granules of mast cells and other hematopoietic cells. A cluster of histidines on mouse mast cell protease 7 regulates its binding to heparin serglycin proteoglycans. *J. Biol. Chem.* **270**, 19524–19531.

McEuen, A.R., He, S., Brander, M.L. & Walls, A.F. (1996) Guinea pig lung tryptase. Localisation to mast cells and characterisation of the partially purified enzyme. *Biochem. Pharmacol.* **52**, 331–340.

McNeil, P.H., Reynolds, D.S., Schiller, V., Ghildyal, N., Gurley, D.S., Austen, K.F. & Stevens, R.L. (1992) Isolation, characterization, and transcription of the gene encoding mouse mast cell protease 7. *Proc. Natl Acad. Sci. USA* **89**, 11174–11178.

Miller, J.S., Westin, E.H. & Schwartz, L.B. (1989) Cloning and characterization of complementary DNA for human tryptase. *J. Clin. Invest.* **84**, 1188–1195.

Miller, J.S., Moxley, G. & Schwartz, L.B. (1990) Cloning and characterization of a second complementary DNA for human tryptase. *J. Clin. Invest.* **86**, 864–870.

Murakumo, Y., Ide, H., Itoh, H., Tomita, M., Kobayashi, T., Maruyama, H., Horii, Y. & Nawa, Y. (1995) Cloning of the cDNA encoding mast cell tryptase of Mongolian gerbil, *Meriones unguic-ulatus*, and its preferential expression in the intestinal mucosa. *Biochem. J.* **309**, 921–926.

Muramatu, M., Itoh, T., Takei, M. & Endo, K. (1988) Tryptase in rat mast cells: properties and inhibition by various inhibitors in comparison with chymase. *Biol. Chem. Hoppe-Seyler* **369**, 617–625.

Pallaoro, M., Gambacurta, A., Fiorucci, L., Mignogna, G., Barra, D. & Ascoli, F. (1996) cDNA cloning and primary structure of tryptase from bovine mast cells, and evidence for the expression of bovine pancreatic trypsin inhibitor mRNA in the same cells. *Eur. J. Biochem.* **237**, 100–105.

Poe, M., Bennett, C.D., Biddison, W.E., Blake, J.T., Norton, G.P., Rodkey, J.A., Sigal, N.H., Turner, R.V., Wu, J.K. & Zweerink, H.J. (1988) Human cytotoxic lymphocyte tryptase. Its purification from granules and the characterization of inhibitor and substrate speci-ficity. *J. Biol. Chem.* **263**, 13215–13222.

Proctor, G.B., Chan, K-M., Garrett, J.R., & Smith, R.E. (1991) Pro-teinase activities in bovine atrium and the possible role of mast cell tryptase in the processing of atrial natriuretic factor (ANF). *Comp. Biochem. Physiol.* **99B**, 839–844.

Raymond, W.W., Tam, E.K., Blount, J.L. & Caughey, G.H. (1995) Purification and characterization of dog mast cell protease-3, an oligomeric relative of tryptases. *J. Biol. Chem.* **270**, 13164–13170.

Reynolds, D., Gurley, D.S., Austen, K.F. & Serafin, W.E. (1991) Cloning of the cDNA and gene of mouse mast cell protease-6. *J. Biol. Chem.* **266**, 3847–3853.

Sakai, K., Long, S.D., Pettit, D.A.D. Cabral, G.A. & Schwartz, L.B. (1996a) Expression and purification of recombinant human tryptase in a baculovirus system. *Protein Exp. Purif.* **7**, 67–73.

Sakai, K., Ren, S. & Schwartz, L.B. (1996b) A novel heparin-dependent processing pathway for human tryptase. Autocatalysis followed by activation with dipeptidyl-peptidase I. *J. Clin. Invest.* **97**, 988–995.

Schechter, N.M., Slavin, D., Fetter, R.D., Lazarus, G.S. & Fraki, J.E. (1988) Purification and identification of two serine class pro-teinases from dog mast biochemically and immunologically sim-ilar to human proteinases tryptase and chymase. *Arch. Biochem. Biophys.* **262**, 232–244.

Schechter, N.M., Eng, G.Y. & McCaslin, D.R. (1993) Human skin tryptase: kinetic characterization of its spontaneous inactivation. *Biochemistry* **32**, 2617–2625.

Schechter, N.M., Eng, G.Y., Selwood, T. & McCaslin, D.R. (1995) Structural changes associated with the spontaneous inactiva-tion of the serine proteinase human tryptase. *Biochemistry* **34**, 10628–10638.

Schwartz, L.B. (1994a) Tryptase: a mast cell serine protease. *Methods Enzymol.* **244**, 88–100.

Schwartz, L.B. (1994b) Tryptase: a clinical indicator of mast cell-dependent events. *Allergy Proc.* **15**, 119–123.

Schwartz, L.B. & Bradford, T.R. (1986) Regulation of tryptase from human lung mast cells by heparin. Stabilization of the active tetramer. *J. Biol. Chem.* **261**, 7372–7379.

Schwartz, L.B., Lewis, R.A. & Austen, K.F. (1981) Tryptase from human pulmonary mast cells. Purification and characterization. *J. Biol. Chem.* **256**, 11939–11943.

Schwartz, L.B., Kawahara, M.S., Hugli, T.E., Vik, D., Fearon, D.T. & Austen, K.F. (1983) Generation of C3a anaphylatoxin from human C3 by human mast cell tryptase. *J. Immunol.* **130**, 1891–1895.

Schwartz, L.B., Bradford, T.R., Littman, B.H. & Wintroub, B.U. (1985) The fibrinogenolytic activity of purified tryptase from human lung mast cells. *J. Immunol.* **135**, 2762–2767.

Schwartz, L.B., Metcalfe, D.D., Miller, J.S., Earl, H. & Sullivan, T. (1987) Tryptase levels as an indicator of mast-cell activation in systemic anaphylaxis and mastocytosis. *N. Engl. J. Med.* **316**, 1622–1626.

Schwartz, L.B., Bradford, T.R., Lee, D.C. & Chlebowski, J.F. (1990) Immunologic and physicochemical evidence for confor-mational changes occurring on conversion of human mast cell tryptase from active tetramer to inactive monomer. *J. Immunol.* **144**, 2304–2311.

Smith, T.J., Hougland, M.W. & Johnson, D.A. (1984) Human lung tryptase. Purification and characterization. *J. Biol. Chem.* **259**, 11046–11051.

Sommerhoff, C.P., Söllner, C., Mentele, R., Piechottka, G.P., Auer-swald, E.A. & Fritz, H. (1994) A Kazal-type inhibitor of human mast cell tryptase: isolation from the medical leech *Hirudo medici-nalis*, characterization, and sequence analysis. *Biol. Chem. Hoppe-Seyler* **375**, 685–694.

Stack, M.S. & Johnson, D.A. (1994) Human mast cell tryptase activates single-chain urinary-type plasminogen activator (pro-urokinase). *J. Biol. Chem.* **269**, 9416–9419.

Sturzebecher, J., Prasa, D. & Sommerhoff, C.P. (1992) Inhibition of human mast cell tryptase by benzamidine derivatives. *Biol. Chem. Hoppe-Seyler* **373**, 1025–1030.

Tam, E.K. & Caughey, G.H. (1990) Degradation of airway neuropeptides by human lung tryptase. *Am. J. Respir. Cell. Mol. Biol.* **3**, 27–32.

Tanaka, T., McRae, B.J., Cho, K., Cook, R., Fräki, J.E., John-son, D.A. & Powers, J.C. (1983) Mammalian tissue trypsin-like enzymes. Comparative reactivities of human skin tryptase, human lung tryptase, and bovine trypsin with peptide 4-nitroanilide and thioester substrates. *J. Biol. Chem.* **258**, 13552–13557.

Vanderslice, P., Craik, C.S., Nadel, J.A. & Caughey, G.H. (1989) Molecular cloning of dog mast cell tryptase and a related protease: structural evidence of a unique mode of serine protease activation. *Biochemistry* **28**, 4148–4155.

Vanderslice, P., Ballinger, S.M., Tam, E.K., Goldstein, S.M., Craik, C.S. & Caughey, G.H. (1990) Human mast cell tryptase: multiple cDNAs and genes reveal a multigene serine protease family. *Proc. Natl Acad. Sci. USA* **87**, 3811–3815.

Walls, A.F., Brain, S.D., Desai, A., Jose, P.J., Hawkings, E., Church, M.K. & Williams, T.J. (1992) Human mast cell tryptase attenuates the vasodilator activity of calcitonin gene-related pep-tide. *Biochem. Pharmacol.* **43**, 1243–1248.

***David A. Johnson***
*Department of Biochemistry and Molecular Biology,*
*J.H. Quillen College of Medicine, East Tennessee State University,*
*Johnson City, TN 37614-0581, USA*
*Email: johnsoda@etsuserv.etsu-tn.edu*

# 20. Mastin

## Databanks

*Peptidase classification: clan SA, family S1, MEROPS ID: S01.145*
*NC-IUBMB enzyme classification: none*
*Databank codes:*

| Species | SwissProt | PIR | EMBL | EMBL (genomic) |
|---|---|---|---|---|
| *Canis familiaris* | P19236 | B32410 | M24664 | – |
| | | | M24665 | |

## Name and History

*Mastin* is a mast cell protease. Its existence was predicted by an orphan dog mastocytoma cDNA (Vanderslice *et al.*, 1989). The anticipated product was termed *dog mastocytoma protease*. Later, when identified and localized to mast cells (Yezzi *et al.*, 1994), it was termed *dog mast cell protease 3* to distinguish it from chymase (Chapter 18) and tryptase (Chapter 19). Here it is termed *mastin*, in reference to its cell of origin.

## Activity and Specificity

Mastin is a trypsin-like serine endoprotease, hydrolyzing peptide bonds on the C-terminal side of basic residues, with Arg preferred to Lys. It hydrolyzes peptidyl 4-nitroanilides (with subsite preferences distinct from those of tryptase) but is not a general protease; for example, it has little caseinolytic activity. Like tryptase, it resists inactivation by plasma serpins but is sensitive to small inhibitors, such as leupeptin and aromatic amidines (Raymond *et al.*, 1995).

## Structural Chemistry and Preparation

Mastin cDNA predicts an activation peptide similar to those of tryptases, with which it may share a mode of activation (Vanderslice *et al.*, 1989). However, the protein is only about 48% identical to tryptases. Its 250 residue catalytic domain is one of the largest of serine proteases. Four of its 12 cysteines are not found in tryptases and do not correspond to those in other serine proteases. It is tetrameric, with half of the subunits disulfide-linked. Mastin has been purified from dog mastocytomas by affinity chromatography on immobilized benzamidine and heparin (Raymond *et al.*, 1995).

## Biological Aspects

The biological roles of mastin are unknown. *In vitro*, it cleaves calcitonin gene-related peptide and gelatin (Raymond *et al.*, 1995). It is activated intracellularly and released extracellularly during exocytosis. Mastin has been identified only in the dog.

## Distinguishing Features and Related Peptidases

Mastin's oligomerization, subunit size, inhibitor resistance and mast cell origin distinguish it from serine proteases other than tryptases. It differs from tryptase (its closest relative: Caughey, 1995) in the formation of intersubunit disulfide links, stability in the absence of heparin and substrate preferences.

## References

Caughey, G.H. (1995) Mast cell chymases and tryptases: phylogeny, family relations, and biogenesis. In: *Mast Cell Proteases in Immunology and Biology* (Caughey, G.H., ed.). New York: Marcel Dekker, pp. 305–329.

Raymond, W.W., Tam, E.K., Blount, J.L. and Caughey, G.H. (1995) Purification and characterization of dog mast cell protease-3, an oligomeric relative of tryptases. *J. Biol. Chem.* **270**, 13164–13170.

Vanderslice, P., Craik, C.S., Nadel, J.A. and Caughey, G.H. (1989) Molecular cloning of dog mast cell tryptase and a related protease: structural evidence of a unique mode of serine protease activation. *Biochemistry* **28**, 4148–4155.

Yezzi, M.J., Hsieh, I.E. and Caughey, G.H. (1994) Mast cell and neutrophil expression of dog mast cell proteinase-3, a novel tryptase-related serine proteinase. *J. Immunol.* **152**, 3064–3072.

*George H. Caughey*
*Department of Medicine and Cardiovascular Research Institute,*
*University of California at San Francisco,*
*San Francisco, CA 94143-0911, USA*
*Email: ghc@itsa.ucsf.edu*

# 21. Cattle duodenase and sheep mast cell proteinase 1

## Databanks

*Peptidase classification: clan SA, family S1, MEROPS ID: S01.142*
*NC-IUBMB enzyme classification: none*
*Databank codes:*

| Species | SwissProt | PIR | EMBL (cDNA) | EMBL (genomic) |
|---|---|---|---|---|
| Duodenase | | | | |
|   *Bos taurus* | P80219 | – | – | – |
| Mast cell proteinase 1 | | | | |
|   *Ovis aries* | P80931 | – | Y14654 | – |
| Mast cell proteinase 2 | | | | |
|   *Ovis aries* | – | – | Y08133 | – |
| Mast cell proteinase 3 | | | | |
|   *Ovis aries* | – | – | Y13462 | – |

## Name and History

A proteinase exhibiting chymotrypsin-like properties was purified by Huntley *et al.* (1986) from sheep gastric mucosa and isolated mucosal mast cells, and classified as a serine endopeptidase (Knox & Huntley, 1988). It was therefore named *sheep mast cell proteinase*, and subsequently *sheep mast cell proteinase 1 (sMCP-1)*, following the cloning of another putative chymase from sheep mast cell cDNA (sMCP-2; EMBL Y08133; S. McAleese, unpublished). An enzyme with specificity for substrates of both chymotrypsin and trypsin was subsequently isolated from cattle duodenal mucosa, and termed *duodenase* (Zamolodchikova *et al.*, 1995a). Further characterization of sMCP-1 indicated that like duodenase, it possesses restricted trypsin-like properties in addition to chymase-like characteristics (Pemberton *et al.*, 1997a). Comparison of the N-terminal sequence of sMCP-1 (Miller *et al.*, 1995) with the full amino acid sequence of duodenase (Zamolodchikova *et al.*, 1995b) reveals 92% identity over the first 24 residues, suggesting that they are species variants of the same enzyme.

## Activity and Specificity

Peptide and protein substrates are hydrolyzed at basic and also hydrophobic P1 residues: cleavages after Lys, Arg, Phe, Tyr and Leu have been noted for both duodenase and sMCP-1 (Zamolodchikova *et al.*, 1995a; Mirgorodskaya *et al.*, 1996; Pemberton *et al.*, 1997a). sMCP-1 shows a marked preference for Lys over Arg. Effective chromogenic substrates for sMCP-1 include Tos-Gly-Pro-Lys+NHPhNO₂, Z-Lys+SBzl and Suc-Phe-Leu-Phe+SBzl. Hydrolysis of proteins is quite restricted, and incubation with bovine serum albumin yields only a small number of cleavage products.

The pH optima of duodenase and sMCP-1 are 7.9–8.2, and 7.6–8.0, respectively (Knox *et al.*, 1986; Zamolodchikova *et al.*, 1995a).

Native inhibitors of sMCP-1 include $\alpha_2$-macroglobulin and $\alpha_1$-proteinase inhibitor, although inhibition by the latter is rather slow ($k_{on} = 1 \times 10^3\,\mathrm{M}^{-1}\,\mathrm{s}^{-1}$) (Pemberton *et al.*, 1997b). Both sMCP-1 and duodenase are effectively inhibited by soybean trypsin inhibitor.

## Structural Chemistry

Bovine duodenase is a single-chain glycoprotein of 29 kDa, and pI of 10. Similarly, sMCP-1 is a very basic protein of about 28 kDa. Duodenase contains the residues Asn189 and Asp226 (chymotrypsinogen numbering) lining the S1 substrate-binding pocket (Zamolodchikova *et al.*, 1995b). The presence of the acidic Asp residue presumably permits binding of substrates with basic P1 amino acids, whereas the uncharged Asn residue may assist with binding of hydrophobic side groups. Cathepsin G (Chapter 16) has a similar primary structure, with uncharged Ala189 and acidic Glu226, and the crystal structure reveals a compartmentalized substrate-binding pocket capable of binding both basic and hydrophobic amino acids (Hof *et al.*, 1996).

## Preparation

Natural or experimental parasitism by gastric nematodes causes massive mucosal mast cell hyperplasia, making the gastric mucosa of parasitized animals the tissue of choice for preparation of sMCP-1 (Miller *et al.*, 1995). The typical yield of sMCP-1 is approximately $100\,\mu\mathrm{g}\,\mathrm{g}^{-1}$ tissue. Duodenase has been isolated in comparable yields from cattle duodenal mucosa (Zamolodchikova *et al.*, 1995a).

## Biological Aspects

Sheep mast cell proteinase 1 has been immunolocalized to mucosal mast cell granules (Huntley *et al.*, 1986). The cellular origin of duodenase has not been demonstrated, but it is very likely to be the cattle homolog of sMCP-1, and similarly stored in mast cell granules.

Expulsion of parasites from the gastric mucosa is associated with mast cell degranulation, release of stored sMCP-1 into the gut lumen, and increased mucosal permeability (Miller *et al.*, 1995), but the causes of the increased permeability are not fully understood.

Fibrinogen is probably a natural substrate of sMCP-1, since sMCP-1 added to sheep plasma containing excess inhibitors substantially degrades fibrinogen before it is inhibited (Pemberton *et al.*, 1997b). Studies with cattle fibrinogen show that the β chain is hydrolyzed at P1 Lys28, releasing an N-terminal peptide that includes the thrombin-cleavage site (Pemberton *et al.*, 1997a).

In addition, sMCP-1 is mitogenic for cattle pulmonary artery fibroblasts, in common with trypsin, and unlike chymotrypsin (Pemberton *et al.*, 1997b). These effects suggest that sMCP-1 may play an important role in the processes of tissue remodeling following mast cell degranulation.

## Distinguishing Features and Related Peptidases

Duodenase and sMCP-1 seem to be species variants of the same enzyme. However, a minor form of duodenase (form II) exists with small differences in amino acid composition (Zamolodchikova *et al.*, 1995a). This form has similar properties to the major form, but they can be separated by chromatography on immobilized soybean trypsin inhibitor. A minor (<10%) form of sMCP-1 also exists (A. Pemberton, unpublished results), with similar properties to the major form, but binding less strongly to cation-exchange media during purification. A proteinase with very similar N-terminal sequence and substrate specificity to duodenase and sMCP-1 has been isolated from goat gastric mucosa (C. MacIldowie, personal communication). Homologs of duodenase have not been described from nonruminants.

## References

Hof, P., Mayr, I., Huber, R., Korzus, E., Potempa, J., Travis, J.,

Powers, J.C. & Bode, W. (1996) The 1.8 Å crystal structure of human cathepsin G in complex with Suc-Val-Pro-Phe^P-(OPh)₂: a Janus-faced proteinase with two opposite specificities. *EMBO J.* **15**, 5481–5491.

Huntley, J.F., Gibson, S., Knox, D. & Miller, H.R.P. (1986) The isolation and purification of a proteinase with chymotrypsin-like properties from ovine mucosal mast cells. *Int. J. Biochem.* **18**, 673–682.

Knox, D.P. & Huntley, J.F. (1988) Classification of sheep abomasal mucosal mast cell proteinase as a serine endopeptidase (EC 3.4.21) *Int. J. Biochem.* **20**, 193–195.

Knox, D.P., Gibson, S. & Huntley, J.F. (1986) The catalytic properties of a proteinase isolated from sheep abomasal mucosal mast cells. *Int. J. Biochem.* **18**, 961–964.

Miller, H.R.P., Huntley, J.F. & Newlands, G.F.J. (1995) Mast cell chymases in helminthosis and hypersensitivity. In: *Mast Cell Proteases in Immunology and Biology* (Caughey, G.H., ed.). New York: Marcel Dekker, pp. 203–235.

Mirgorodskaya, O., Kazanina, G., Mirgorodskaya, E., Vorotyntseva, T., Zamolodchikova, T. & Alexandrov, S. (1996) A comparative study of the specificity of melittin hydrolysis by duodenase, trypsin and plasmin. *Protein Pept. Lett.* **3**, 315–320.

Pemberton, A.D., Huntley, J.F. & Miller, H.R.P. (1997a) Sheep mast cell proteinase-1: characterization as a member of a new class of dual-specific ruminant chymases. *Biochem. J.* **321**, 665–670.

Pemberton, A.D., Belham, C.M., Huntley, J.F., Plevin, R. & Miller, H.R.P. (1997b) Sheep mast cell proteinase-1, a serine proteinase with both tryptase- and chymase-like properties, is inhibited by plasma proteinase inhibitors and is mitogenic for bovine pulmonary artery fibroblasts. *Biochem. J.* **323**, 719–725.

Zamolodchikova, T.S., Vorotyntseva, T.I. & Antonov, V.K. (1995a) Duodenase, a new serine protease of unusual specificity from bovine duodenal mucosa. Purification and properties. *Eur. J. Biochem.* **227**, 866–872.

Zamolodchikova, T.S., Vorotyntseva, T.I., Nazimov, I.V. & Grishina, G.A. (1995b) Duodenase, a new serine protease of unusual specificity from bovine duodenal mucusa. Primary structure of the enzyme. *Eur. J. Biochem.* **227**, 873–879.

***Alan D. Pemberton***
*Department of Veterinary Clinical Studies,*
*University of Edinburgh, Veterinary Field Station,*
*Easter Bush, Roslin, Midlothian EH25 9RG, UK*
*Email: alan.pemberton@ed.ac.uk*

# 22. Granzyme A

## Databanks

*Peptidase classification: clan SA, family S1, MEROPS ID: S01.135*
*NC-IUBMB enzyme classification: EC 3.4.21.78*

*Databank codes:*

| Species | SwissProt | PIR | EMBL (cDNA) | EMBL (genomic) |
|---|---|---|---|---|
| *Homo sapiens* | P12544 | A28943 | M18737 | – |
| | | A30525 | | |
| | | A30526 | | |
| | | A31372 | | |
| *Mus musculus* | P11032 | A26944 | X14799 | L01426: exon 1 |
| | | A27640 | | L01427: exon 2 |
| | | A45061 | | L01428: exon 3 |
| | | A47590 | | L01429: exon 4 and complete CDS |
| | | S10085 | | |
| | | S26184 | | |

## Name and History

Pasternack & Eisen (1985) first described a serine esterase with trypsin-like activity in cloned cytotoxic T lymphocytes (CTLs). Following this discovery, several serine esterases designated **granzyme A** (Masson *et al.*, 1986), **T cell-specific proteinase 1** (Fruth *et al.*, 1987), **serine esterase 1** (Young *et al.*, 1986), **cytotoxic cell protease 1** (Redmond *et al.*, 1987) or simply **serine esterase** (Pasternack *et al.*, 1986) were identified in the cytolytic granules from CTL and natural killer (NK) cells. Masson & Tschopp (1987) purified to homogeneity a family of serine proteases present in cytolytic granules of mouse CTLs, coining the term **granzyme A** (as well as defining granzymes B–F). Granzyme A is present in the highest amounts in these granules. Subsequently Poe *et al.* (1988) isolated granzyme A (CTL Q31) from human CTLs and Hameed *et al.* (1988) purified the protease from lymphokine-activated killer (LAK) cells as **granzyme 1**.

The identification and structural analysis of granzyme A were greatly facilitated by the application of cDNA subtraction library techniques. In murine CTLs, granzyme A is encoded by the cDNA called H factor (Gershenfeld & Weissman, 1986) or CTLA-1 (Brunet *et al.*, 1987). Human granzyme A is encoded by the HuHF cDNA clone isolated by Gershenfeld *et al.* (1988).

## Activity and Specificity

Preferential cleavage: Arg┼Xaa > Lys┼Xaa. The pH optimum of the enzyme on most substrates is about 8.0, though one report indicated a very high optimum (11.0) for a nitroanilide substrate (Simon & Kramer, 1994). Subsite mapping has been performed to yield data suggesting that granzyme A possesses a primary specificity for Arg residues, but Lys is also tolerated in the S1 position. Both mouse and human forms hydrolyze Arg- and Lys-containing thioesters very efficiently with $k_{cat}/K_m$ of $10^4$–$10^5$ $M^{-1}$ $s^{-1}$. PMSF, aprotinin, DFP, benzamidine and DCI rapidly inhibit granzyme A, whereas Tos-Lys-CH$_2$Cl and Tos-Phe-CH$_2$Cl have no or little effect. Though less efficiently cleaved than thioester substrates, Tos-Gly-Pro-Arg┼NHPhNO$_2$ provides a convenient substrate with which to measure granzyme A activity. Protein inhibitors include soybean trypsin inhibitor, aprotinin, antithrombin III and protease nexin 1. Each subunit of the dimer interacts with one molecule of antithrombin III or nexin 1.

## Structural Chemistry

Granzyme A has an unusual structure among serine proteinases in that it consists of a disulfide-linked homodimer of 60 kDa linked via Cys93 (chymotrypsinogen numbering). Like other members of the family, the protein is synthesized as a preproenzyme, and delivered to granules via the Golgi. It is not clear whether activation requires removal of a dipeptide Glu-Arg (mouse) or Glu-Lys (human) or a longer propeptide, and it is therefore not known whether activation is achieved by dipeptidyl-peptidase I (Chapter 214) as is the case with other granule-associated serine proteinases such as cathepsin G (Chapter 16), neutrophil elastase (Chapter 15) and granzyme B (Chapter 23). Granzyme A has been localized to chromosome 13 (p13.3) in mice (Baker *et al.*, 1994) and 5 (5q11–q12) in humans (Fink *et al.*, 1993). All granzyme genes studied to date share a common exon–intron organization, with five exons separated by introns situated at homologous positions. Exon 1 codes for the signal peptide and the first nucleotide of the propeptide, while exons 2, 3 and 5 contain the codons for the catalytic site residues His57, Asp102 and Ser195 (chymotrypsinogen numbering).

## Preparation

Natural sources of the enzyme are restricted to cytotoxic cells including CTLs, NK cells, and large granular lymphocytes. Granzyme A is contained within the lytic granules of these cells, which are analogous to the azurophil granules of neutrophils. Normally these cells must be cytokine-activated to produce sufficient enzyme for purification, but even then very small quantities are obtained. A fairly economical method to isolate natural granzyme A entails generation of interleukin 2-activated lymphocytes followed by granule isolation/extraction and cation-exchange chromatography. Granzymes A and B, however, elute at about 490 and 590 mM NaCl, respectively. Precautions must be taken to ensure that the eluted granzyme A is not contaminated with granzyme B.

Recently, recombinant human granzyme A zymogen was expressed in several eukaryotic cell lines (HepG2, Jurkat and COS-1) after infection with a recombinant vaccinia virus containing full-length granzyme A cDNA (Kummer *et al.*, 1996). Immunoblot analysis of cell lysates showed that all infected cells produced a disulfide-linked homodimer of molecular mass identical to that of the natural granzyme A. Infected HepG2 cells produced the largest amount of this protease (approximately 160 times more than LAK cells). The recombinant protein contained only high mannose-type oligosaccharides, as did the natural protein. Although infected HepG2 and COS cells contained high granzyme A antigen levels, lysates from these cells did not show any granzyme A proteolytic activity. However, the inactive proenzyme could be converted into active granzyme A by incubation with dipeptidyl-peptidase I (Chapter 214).

## Biological Aspects

The biological function of granzyme A remains enigmatic. Shi *et al.* (1992) demonstrated that rat granzyme A, when added to target cells together with perforin, induced apoptotic nuclear changes, and on this basis they proposed a role for the protein in granule-mediated apoptosis. This proposal has been supported by the results of reconstitution and ablation studies, which support a minor role for granzyme A in cell-mediated cytotoxicity, although a less important one than is played by granzyme B. Transfection studies in normally nonlytic basophils (Shiver *et al.*, 1992; Nakajima & Henkart, 1994) demonstrate some enhanced cytolysis, though the basophil line had less activity than CTLs directed against the same targets (Nakajima *et al.*, 1995a,b). The content of murine granzyme A in cytotoxic cells has been reduced by transfection with antisense mRNA. Although the transfectants displayed markedly reduced levels of granzyme A message, lytic activity against nucleated target cells was reduced by no more than 50% (Talento *et al.*, 1992). Most notably, generation of granzyme A-deficient mice through homologous recombination has been reported (Ebnet *et al.*, 1995). Their *in vitro-* and *ex vivo*-derived CTLs and NK cells are indistinguishable from those of normal mice in their activity in causing membrane disruption, apoptosis and DNA fragmentation in target cells. Furthermore, granzyme A-deficient mice readily recover from both lymphocytic choriomeningitis virus and *Listeria monocytogenes* infections, and eradicate syngeneic tumors with kinetics similar to the wild-type strain.

Although granzyme A plays a role in cell-mediated cytotoxicity, other host defense functions may be initiated by the granzyme after secretion from cytotoxic cells. It has been reported that the granzyme inactivates the reverse transcriptase from the Moloney murine leukemia virus and that pretreatment of the intact virus with the protease leads to cleavage of the envelope protein GP70, resulting in loss of infectivity (Simon *et al.*, 1987). Recently, granzyme A was reported to activate the tethered ligand thrombin receptor on neuronal cells (Suidan *et al.*, 1994). However, we have shown that the granzyme does not cleave the cloned receptor expressed in baculovirus (Sower *et al.*, 1996a). Granzyme A has been reported to induce the synthesis of IL-6 and IL-8 in monocytes and fibroblast lines (Sower *et al.*, 1996a,b). Taking these results together with those of the study that showed that the protease processed pre-IL-1 to a biologically active form (Falcieri *et al.*, 1994), it is tempting to speculate that the granzyme may be designed to activate a cytokine cascade during granule-mediated cytotoxicity.

Physiologic substrates for granzyme A, which may play a role in vascular migration of cytotoxic cells, include fibronectin (Simon *et al.*, 1988), type IV collagen (Simon *et al.*, 1991) and proteoglycans (Vettel *et al.*, 1991). In a pathologic light, granzyme A has been found to cleave myelin basic protein (Vanguri *et al.*, 1993) and various skeletal muscle proteins (dystrophin, myosin and nebulin) (Nakamura *et al.*, 1993). However, relevant catalytic efficiencies have not been described in any of these studies.

## Distinguishing Features

Distinguishing granzyme A from granzyme K (Chapter 24) is difficult, and one could probably expect only to determine a combination of the two in crude samples. Monoclonal IgG antibodies directed against linear epitopes of granzyme A have been reported (Kummer *et al.*, 1993).

## Further Reading

For more information, see the compilation by Sitkovsky & Henkart (1993).

## References

Baker, E., Sayers, T.J., Sutherland, G.R. & Smyth, M.J. (1994) The genes encoding NK cell granule serine proteases, human tryptase-2 (*TRYP2*) and human granzyme A (*HFSP*), both map to chromosome 5q11–q12 and define a new locus for cytotoxic lymphocyte granule tryptases. *Immunogenetics* **40**, 235–237.

Brunet, J.-F., Denizot, F., Suzan, M., Haas, W., Mercia-Huerta, J.-M., Berke, G., Luciani, M.-F. & Goldstein, P. (1987) CTLA-1 and CTLA-3 serine protease transcripts are detected mostly in cytotoxic cells, but not only and not always. *J. Immunol.* **138**, 4102–4108.

Ebnet, K., Hausmann, M., Lehmann-Grube, F., Mullbacher, A., Kopf, M., Lamers, M. & Simon, M.M. (1995) Granzyme A-deficient mice retain potent cell-mediated cytotoxicity. *EMBO J.* **14**, 4230–4239.

Falcieri, E., Zamai, L., Santi, S., Cinti, C., Gobbi, P., Bosco, D., Cataldi, A., Betts, C. & Vitale, M. (1994) The behaviour of nuclear domains in the course of apoptosis. *Histochemistry* **102**, 221–231.

Fink, T.M., Lichter, P., Wekerle, H., Zimmer, M. & Jenne, D.E. (1993) The human granzyme A (HFSP, CTLA3) gene maps to 5q11–q12 and defines a new locus of the serine protease superfamily. *Genomics* **18**, 401–403.

Fruth, U., Singaglia, F., Schlesier, M., Kilgus, J., Kramer, M. & Simon, M.M. (1987) A novel serine protease (HuTSP) isolated from a cloned human CD8+ cytolytic T cell line is expressed and secreted by activated CD4+ and CD8+ cells. *Eur. J. Immunol.* **17**, 1625–1633.

Gershenfeld, H.K. & Weissman, I.L. (1986) Cloning of a cDNA for a T cell-specific serine protease from a cytotoxic T lymphocyte. *Science* **232**, 854–858.

Gershenfeld, H.K., Hershberger, R., Shows, R. & Weissmann, I. (1988) Cloning and chromosomal assignment of a human cDNA encoding a T cell- and natural killer cell-specific trypsin-like serine protease. *Proc. Natl Acad. Sci. USA* **85**, 1184–1189.

Hameed, A., Lowrey, D., Lichtenheld, M.G. & Podack, E.R. (1988) Characterization of three serine proteases isolated from human IL-2 activated killer cells. *J. Immunol.* **141**, 3142–3147.

Kummer, J.A., Kamp, A.M.M., van Katwijk, J.P., Brakenhoff, K., Radosevic, A.M., van Leeuwen, J., Borst, C.L., Verweij, C.L. &

Hack, C.E. (1993) Production and characterization of monoclonal antibodies raised against recombinant human granzymes A and B showing cross reactions with the natural proteins. *J. Immunol. Methods* **163**, 77–83.

Kummer, J.A., Kamp, A.M., Citarella, F., Horrevoets, A.J. & Hack, C.E. (1996) Expression of human recombinant granzyme A zymogen and its activation by the cysteine proteinase cathepsin C. *J. Biol. Chem.* **271**, 9281–9286.

Masson, D. & Tschopp, J. (1987) A family of serine proteases in lytic granules of cytolytic T cells. *Cell* **49**, 679–685.

Masson, D., Nabholz, M., Estrade, C. & Tschopp, J. (1986) Granules of cytolytic T-lymphocytes contain two serine esterases. *EMBO J.* **7**, 1595–1600.

Nakajima, H. & Henkart, P.A. (1994) Cytotoxic lymphocyte granzymes trigger a target cell internal disintegration pathway leading to cytolysis and DNA breakdown. *J. Immunol.* **152**, 1057–1063.

Nakajima, H., Park, H.L. & Henkart, P.A. (1995a) Synergistic roles of granzymes A and B in mediating target cell death by rat basophilic leukemia mast cell tumors also expressing cytolysin/perforin. *J. Exp. Med.* **181**, 1037–1046.

Nakajima, H., Golstein, P. & Henkart, P.A. (1995b) The target cell nucleus is not required for cell-mediated granzyme- or Fas-based cytotoxicity. *J. Exp. Med.* **181**, 1905–1909.

Nakamura, K., Arahata, K., Ishiura, S., Osame, M. & Sugita, H. (1993) Degradative activity of granzyme A on skeletal muscle proteins *in vitro*: a possible molecular mechanism for muscle fiber damage in polymyositis. *Neuromusc. Disord.* **3**, 303–310.

Pasternack, M.S. & Eisen, H.N. (1985) A novel serine protease expressed by cytotoxic T lymphocytes. *Nature* **313**, 743–746.

Pasternack, M.S., Verret, C.R., Liu, M.A. & Eisen, H.N. (1986) Serine esterase in cytolytic T lymphocytes. *Nature* **322**, 740–743.

Poe, M., Bennett, C., Biddison, W.E., Norton, G., Rodkey, J., Sigal, N.H., Turner, R., Wu, J.K. & Zweerink, H.J. (1988) Human cytotoxic lymphocyte tryptase: its purification from granules and the characterization of inhibitor and substrate specificity. *J. Biol. Chem.* **263**, 13215–13222.

Redmond, M.J., Letellier, M., Parker, J.M.R., Lobe, C., Havele, C., Paetkau, V. & Bleackley, R.C. (1987) A serine protease (CCP1) is sequestered in the cytoplasmic granules of cytotoxic T lymphocytes. *J. Immunol.* **139**, 3184–3188.

Shi, L., Kam, C.-M., Powers, J.C., Aebersold, R. & Greenberg, A.H. (1992) Purification of three cytotoxic lymphocyte granule serine proteases that induce apoptosis through distinct substrate and target cell interactions. *J. Exp. Med.* **176**, 1521–1529.

Shiver, J.W., Su, L. & Henkart, P.A. (1992) Cytotoxicity with target DNA breakdown by rat basophilic leukemia cells expressing both cytolysin and granzyme A. *Cell* **71**, 315–322.

Simon, H.G. & Kramer, M.D. (1994) Granzyme A. *Methods Enzymol.* **244**, 69–79.

Simon, H.G., Fruth, U., Kramer, M.D. & Simon, M.M. (1987) A secretable serine proteinase with highly restricted specificity from cytolytic T cells inactivates retro-virus associated reverse transcriptase. *FEBS Lett.* **223**, 352–358.

Simon, M.M., Prester, M., Nerz, G., Kramer, M. & Furth, U. (1988) Degradation of fibronectin by mouse CTL trypsin-like protease TSP-1. *Biol. Chem.* **369**, 107–112.

Simon, M.M., Kramer, M., Prester, M. & Gay, S. (1991) Mouse T-cell associated serine proteinase 1 degrades collagen type IV: A structural basis for the migration of lymphocytes through vascular basement membranes. *Immunology* **73**, 117–119.

Sitkovsky, M.V. & Henkart, P. (eds) (1993) *Cytotoxic Cells: Recognition, Effector Function, Generation and Methods*. Boston: Birkhäuser.

Sower, L.E., Froelich, C.J., Allegretto, N., Rose, P.M., Hanna, W.L. & Klimpel, G.R. (1996a) Extracellular activities of human granzyme A. Monocyte activation by granzyme A versus α-thrombin. *J. Immunol.* **156**, 2585–2590.

Sower, L.E., Klimpel, G.R., Hanna, W. & Froelich, C.J. (1996b) Extracellular activities of human granzymes 1. Granzyme A induces IL6 and IL8 production in fibroblast and epithelial cell lines. *Cell. Immunol.* **171**, 159–163.

Suidan, H.S., Bouvier, J., Schaerer, E., Stone, S.R., Monard, D. & Tschopp, J. (1994) Granzyme A released upon stimulation of cytotoxic cells activates the thrombin receptor on neuronal cells and astrocytes. *Proc. Natl Acad. Sci. USA* **91**, 8112–8116.

Talento, A., Nguyen, M., Law, S., Wu, J.K., Poe, M., Blake, J.T., Patel, M., Wu, T.-J., Manyak, C.L., Silberklang, M., Mark, G., Springer, M., Sigal, N.H., Weissman, I.L., Bleackley, R.C., Podack, E.R., Tykocinski, M.L. & Koo, G.C. (1992) Transfection of mouse cytotoxic T lymphocyte with an antisense granzyme A vector reduces lytic activity. *J. Immunol.* **149**, 4009–4015.

Vanguri, P., Lee, E., Henkart, P.A. & Shin, M.L. (1993) Hydrolysis of myelin basic protein in myelin membranes by granzymes of large granular lymphocytes. *J. Immunol.* **150**, 2431–2439.

Vettel, U., Bar-Shavit, R., Simon, M.M., Brunner, G., Vlodavsky, I. & Kramer, M.D. (1991) Coordinate secretion and functional synergism of T-cell associated serine protease-1 and endoglycosidase(s) of activated T cells. *Eur. J. Immunol.* **21**, 2247–2251.

Young, J.D.-E., Leong, G. & Liu, C.-C. (1986) Isolation and characterization of a serine protease from cytolytic T cell granules. *Cell* **47**, 183–188.

*Christopher J. Froelich*
*Evanston Hospital,*
*Northwestern University,*
*2650 Ridege Ave., Evanston,*
*IL 60201, USA*

*Guy S. Salvesen*
*The Burnham Institute,*
*10901 North Torrey Pines Road,*
*San Diego, CA 92037, USA*
*Email: gsalvesen@ljcrf.edu*

# 23. Granzyme B

## Databanks

*Peptidase classification: clan SA, family S1, MEROPS ID: S01.136*
*NC-IUBMB enzyme classification: EC 3.4.21.79*
*Databank codes:*

| Species | SwissProt | PIR | EMBL (cDNA) | EMBL (genomic) |
|---|---|---|---|---|
| *Homo sapiens* | P10144 | A28659 | J03189 | J03072: complete gene |
| | | A31405 | J04071 | |
| | | A32168 | M17016 | |
| | | A61021 | M28879 | |
| | | I56092 | | |
| | | I71986 | | |
| | | JH0094 | | |
| *Mus musculus* | P04187 | A00956 | M12302 | M22526: complete gene |
| | | A28952 | X04072 | |
| | | A93382 | | |
| | | A94288 | | |
| | | B26944 | | |

Brookhaven Protein Data Bank three-dimensional structures:

| Species | ID | Resolution | Notes |
|---|---|---|---|
| *Mus musculus* | 2CP1 | – | theoretical model |

## Name and History

Early work on the mechanism of cytolysis of target cells by cytotoxic lymphocytes demonstrated the requirement for proteolytic activity. The activity went undescribed until subtractive cloning techniques and isolation of components of cytotoxic cell lytic granules showed that they contain several proteinases of the chymotrypsin family. The peptidases forming the series of chymotrypsin homologs discovered in the lytic granules of murine cytotoxic T lymphocytes were termed **granzymes** by Tschopp and colleagues. The enzyme known as **granzyme B** corresponded to the previously identified cDNA encoding CCP1. Other names for this enzyme include **CGL-1**, **HLP**, **CSP-B**, **SECT** and **granzyme 2** in humans, **CTLA-1** in mouse and **RNKP-1** and **fragmentin 1** in rat. The description of fragmentin 1 is particularly noteworthy because it provided the first direct evidence for the biological role of granzyme B. The protein was isolated on the basis of its ability to cause apoptotic damage to target cells when added in the presence of the pore-forming protein perforin.

## Activity and Specificity

Preferential cleavage:  Asp┼Xaa ≫ Asn┼Xaa > Met┼Xaa = Ser┼Xaa. Optimal pH is 7.5. Granzyme B is currently unique among the mammalian serine proteinases in its preference for Asp in P1. This is explained primarily by residue 226 (chymotrypsinogen numbering), governing the P1 preference

of the close homolog, rat mast cell chymase (Remington *et al.*, 1988), which in granzyme B is Arg (Caputo *et al.*, 1994). For a related discussion of the influence of residue 226, see the chapter on cathepsin G (Chapter 16). Little work on the extended specificities of P2–P4 has been reported and, as with most proteinases of family S1, optimal substrate occupation of prime-side subsites has not been examined in detail. Most work on substrate specificity has focused on proteins that may be activated during apoptosis (see below).

Assays are most conveniently made with the chromogenic substrate Suc-Ala-Ala-Pro-Asp┼NHPhNO$_2$, or slightly less conveniently, though with slightly greater sensitivity, with the thioester, Boc-Ala-Ala-Asp┼SBzl. A number of protein inhibitors have been reported to inhibit granzyme B very weakly, most notably $\alpha_1$-proteinase inhibitor (Poe *et al.*, 1991). However, we have not been able to reproduce this latter result (W. Hanna, C. Froelich & G. Salvesen, unpublished results). The cowpox virus serpin, CrmA, an inhibitor of some Asp-specific members of the caspase family, inhibits granzyme B ($k_{ass} = 2.9 \times 10^5\,\mathrm{M^{-1}\,s^{-1}}$; Quan *et al.*, 1995), and a serpin known as PI9, endogenous to some human cells, seems to be a physiologic inhibitor with $k_{ass}$ of $1.6 \times 10^6\,\mathrm{M^{-1}\,s^{-1}}$ (Sun *et al.*, 1996).

## Structural Chemistry

Granzyme B is a single-chain glycoprotein of about 28.5 kDa, pI about 12 (the high pI makes accurate measurements unreliable, so this is a rough estimate). The predicted human protein sequence contains two potential $N$-glycosylation sites, three disulfides, and one unpaired Cys (Cys88, chymotrypsin numbering). Though otherwise a typical chymotrypsin family member, granzyme B does not contain the highly conserved 191–220 disulfide present in most members of the family. The absence of this disulfide, located near the active site, is a defining characteristic of the group of granule serine proteinases, located at chromosomal locus 14q11.2 in humans, that also includes cathepsin G and chymase (Dahl *et al.*, 1990; Caughey *et al.*, 1993).

## Preparation

Natural sources of the enzyme are restricted to cytotoxic cells, including cytotoxic T-lymphocytes (CTLs), natural killer (NK) cells and large granular lymphocytes. Granzyme B is contained within the lytic granules of these cells, which are analogous to the azurophil granules of neutrophils. Normally these cells must be cytokine-activated to produce sufficient enzyme for purification, but even then very small quantities are obtained. Consequently, the enzyme is most readily purified from the YT natural killer cell line, by a combination of cell lysis, granule extraction and cation-exchange chromatography. A typical yield from $10^9$ cells is about 1 mg of enzyme (Hanna *et al.*, 1993). Alternately, granzyme B may be isolated from cellular extracts by immunoaffinity chromatography (Trapani *et al.*, 1993). Human granzyme B is available from Enzyme Systems Products (see Appendix 2 for full names and addresses of suppliers).

## Biological Aspects

### Role of Granzyme B in Cytotoxic Cell Granule-Mediated Apoptosis

The pivotal role of granzyme B in the granule secretion model of apoptosis was first demonstrated by Shi *et al.* (1992). Using rat granzyme B, these authors showed that the simultaneous addition of granzyme B and sublytic perforin to targets induced apoptosis. The paradigm was verified by experiments with granzyme B knockout mice (Heusel *et al.*, 1994; Shresta *et al.*, 1995), whose cytotoxic cells were defective in killing of canonical targets. An important clue that led to the clarification of the function of granzyme B is its preference for cleavage of peptide bonds after Asp residues. Some of the caspases appear to function at the execution phase of apoptosis. Their zymogens are processed to form active heterodimeric enzymes by cleavage at specific Asp residues. Thus granzyme B appears to be an ideal candidate designed to activate the caspases. Although the prototypic member of the family caspase-1 (Chapter 248) is not activated by granzyme B (Darmon *et al.*, 1994; Quan *et al.*, 1996), most of the others are. Thus it is likely that granzyme B achieves its apoptotic function by engaging pre-existing death pathways via activation of participating caspases. Which caspases are activated *in vivo* is yet to be determined in detail, though preliminary evidence comes from Darmon *et al.* (1995) who demonstrated that the precursor of caspase-3 (Chapter 249) is processed in murine targets treated with CTLs, and Chinnaiyan *et al.* (1996), who demonstrated that target cells exposed to granzyme B and perforin contained processed caspase-7 and caspase-3.

### Other Potential Biologic Roles of Granzyme B

Although granzyme B clearly plays a role in granule-mediated apoptosis, other potential physiological or pathological functions cannot be excluded. Granzyme B has been identified by the sandwich immunoassay technique in the synovial fluid of patients with rheumatoid arthritis (Tak *et al.*, 1996). Since granzyme B cleaves aggrecan, the resident proteoglycan of cartilage (Froelich *et al.*, 1993), it is possible that granzyme B could contribute to joint damage in rheumatoid arthritis.

## Distinguishing Features

Any activity that cleaves Suc-Ala-Ala-Pro-Asp+NHPhNO$_2$ is likely to be due to granzyme B. This can be further substantiated by including iodoacetamide in the incubation of crude samples that contain caspases, since these enzymes can hydrolyze the substrate at slow rates. Monoclonal IgG antibodies (mAbs) directed against linear epitopes (Hameed *et al.*, 1991; Kummer *et al.*, 1993) and an IgM antibody against a conformational epitope (Trapani *et al.*, 1993) of granzyme B have been reported. The latter may be purchased from Kamiya Inc. Further paired mAbs against distinct conformational epitopes of granzyme B have been reported that will allow the immunochemical measurement of extracellular granzyme B *in vitro* and *in vivo* (Tak *et al.*, 1996).

## Further Reading

For more information, see Sitkovsky & Henkart (1993).

## References

Caputo, A., James, M.N., Powers, J.C., Hudig, D. & Bleackley, R.C. (1994) Conversion of the substrate specificity of mouse proteinase granzyme B. *Nat. Struct. Biol.* **1**, 364–367.

Caughey, G.H., Schaumberg, T.H., Zerweck, E.H., Butterfield, J.H., Hanson, R.D., Silverman, G.A. & Ley, T.J. (1993) The human mast cell chymase gene (CMA1): mapping to the cathepsin G/granzyme gene cluster and lineage-restricted expression. *Genomics* **15**, 614–620.

Chinnaiyan, A.M., Orth, K., Hanna, W.L., Duan, H.J., Poirier, G.G., Froelich, C.J. & Dixit, V.M. (1996) Cytotoxic T cell-derived granzyme B activates the apoptotic protease ICE-LAP3. *Curr. Biol.* **6**, 897–899.

Dahl, C.A., Bach, F.H., Chan, W., Huebner, K., Russo, G., Croce, C.M., Herfurth, T. & Cairns, J.S. (1990) Isolation of a cDNA clone encoding a novel form of granzyme B from human NK cells and mapping to chromosome 14. *Hum. Genet.* **84**, 465–470.

Darmon, A.J., Ehrman, N., Caputo, A., Fujinaga, J. & Bleackley, R.C. (1994) The cytotoxic T cell proteinase granzyme B does not activate interleukin-1b-converting enzyme. *J. Biol. Chem.* **269**, 32043–32046.

Darmon, A.J., Nicholson, D.W. & Bleackley, R.C. (1995) Activation of the apoptotic protease CPP32 by cytotoxic T-cell-derived granzyme B. *Nature* **377**, 446–468.

Froelich, C.J., Zhang, X., Turbov, J., Hudig, D., Winkler, U. & Hanna, W.L. (1993) Human granzyme B degrades aggrecan proteoglycan in matrix synthesized by chondrocytes. *J. Immunol.* **151**, 7161–7169.

Hameed, A., Truong, L.D., Price, V., Krähenbühl, O. & Tschopp, J. (1991) Immunohistochemical localization of granzyme B antigen in cytotoxic cells in human tissues. *Am. J. Pathol.* **138**, 1069–1075.

Hanna, W.L., Zhang, X., Turbov, J., Winkler, U., Hudig, D. & Froelich, C.J. (1993) Rapid purification of cationic granule proteases: Application to human granzymes. *Protein Exp. Purif.* **4**, 398–402.

Heusel, J.W., Wesselschmidt, R.L., Shresta, S., Russell, J.H. & Ley, T.J. (1994) Cytotoxic lymphocytes require granzyme B for the rapid induction of DNA fragmentation and apoptosis in allogeneic target cells. *Cell* **76**, 977–987.

Kummer, J.A., Kamp, A.M.M., van Katwijk, J.P., Brakenhoff, K., Radosevic, A.M., van Leeuwen, J., Borst, C.L., Verweij, C.L. & Hack, C.E. (1993) Production and characterization of monoclonal antibodies raised against recombinant human granzymes A and B showing cross-reactions with the natural proteins. *J. Immunol. Methods* **163**, 77–83.

Poe, M., Blake, J.T., Boulton, D.A., Gammon, M., Sigal, N.H., Wu, J.K. & Zweerink, H.J. (1991) Human cytotoxic lymphocyte granzyme B. Its purification from granules and the characterization of substrate and inhibitor specificity. *J. Biol. Chem.* **266**, 98–103.

Quan, L.T., Caputo, A., Bleackley, R.C., Pickup, D.J. & Salvesen, G.S. (1995) Granzyme B is inhibited by the cowpox virus serpin cytokine response modifier A. *J. Biol. Chem.* **270**, 10377–10379.

Quan, L.T., Tewari, M., O'Rourke, K., Dixit, V., Snipas, S.J.,

Poirier, G.G., Ray, C., Pickup, D.J. & Salvesen, G.S. (1996) Proteolytic activation of the cell death protease Yama/CPP32 by granzyme B. *Proc. Natl Acad. Sci. USA* **93**, 1972–1976.

Remington, J.R., Woodbury, R.G., Reynolds, R.A., Matthews, B.W. & Neurath, H. (1988) The structure of rat mast cell protease II at 1.9-Å resolution. *Biochemistry* **27**, 8097–8105.

Shi, L., Kam, C.-M., Powers, J.C., Aebersold, R. & Greenberg, A.H. (1992) Purification of three cytotoxic lymphocyte granule serine proteases that induce apoptosis through distinct substrate and target cell interactions. *J. Exp. Med.* **176**, 1521–1529.

Shresta, S., MacIvor, D.M., Heusel, J.W., Russell, J.H. & Ley, T.J. (1995) Natural killer and lymphokine-activated killer cells require granzyme B for the rapid induction of apoptosis in susceptible target cells. *Proc. Natl Acad. Sci. USA* **92**, 5679–5683.

Sitkovsky, M.V. & Henkart, P. (eds) (1993) *Cytotoxic Cells: Recognition, Effector Function, Generation and Methods.* Boston: Birkhäuser.

Sun, J., Bird, C.H., Sutton, V., McDonald, L., Coughlin, P.B., De Jong, T.A., Trapani, J.A. & Bird, P.I. (1996) A cytosolic granzyme B inhibitor related to the viral apoptotic regulator cytokine response modifier A is present in cytotoxic lymphocytes. *J. Biol. Chem.* **271**, 27802–27809.

Tak, P.P., Kummer, J.A., Breedveld, F.C., Froelich, C.J. & Hack, C.E. (1996) Granzyme B is elevated in the synovial fluids and plasma of patients with rheumatoid arthritis. *Arthritis Rheum.* in the press (abstr.).

Trapani, J.A., Browne, K.A., Dawson, M. & Smyth, M.J. (1993) Immunopurification of functional Asp-ase (natural killer cell granzyme B) using a monoclonal antibody. *Biochem. Biophys. Res. Commun.* **195**, 910–920.

*Guy S. Salvesen*
*The Burnham Institute,*
*10901 North Torrey Pines Road,*
*San Diego, CA 92037, USA*
*Email: gsalvesen@ljcrf.edu*

*Christopher J. Froelich*
*Evanston Hospital,*
*Northwestern University,*
*2650 Ridege Ave.,*
*Evanston, IL 60201, USA*

# 24. Granzyme K

## Databanks

*Peptidase classification: clan SA, family S1, MEROPS ID: S01.146*
*NC-IUBMB enzyme classification: none*
*Databank codes:*

| Species | SwissProt | PIR | EMBL (cDNA) | EMBL (genomic) |
|---------|-----------|-----|-------------|----------------|
| *Homo sapiens* | P49863 | – | U26174 U35237 | – |
| *Rattus norvegicus* | P49864 | – | L19694 | – |

## Name and History

**Human granzyme K** was initially isolated from the granules of interleukin 2 (IL-2)-activated peripheral blood mononuclear cells (Hameed *et al.*, 1988). Since it was the third esterase characterized from these granules, it was originally designated as **granzyme 3**. Similar to granzyme A (Chapter 22), but not as abundant in the granules, granzyme K was purified by following its ability to cleave the synthetic substrate Z-Lys-SBzl. Thus, it has been described as a **tryptase** since it has trypsin-like P1 specificity (Sayers *et al.*, 1994), but this name is best reserved for the enzyme from mast cells (see Chapter 19).

The human granzyme K gene has been cloned and sequenced from a cDNA library constructed from human ascitic tissue (designated **granzyme 3**; Przetak *et al.*, 1995) and from natural killer (NK) cells (designated **HNK-Tryp-2**; Sayers *et al.*, 1996). The rat homolog of human granzyme K, designated **RNK-Tryp-2**, was cloned from cDNA constructed from NK leukemia cells (Sayers *et al.*, 1994).

## Activity and Specificity

Granzyme K esterase activity is readily detected with either Z-Lys┼SBzl or Z-Arg┼SBzl as substrate, release of thiobenzoate being monitored with Ellman's reagent (Hameed *et al.*, 1988; Odake *et al.*, 1991; Hanna *et al.*, 1993). The reaction mixture consists of 0.5 mM Ellman's reagent [5,5'-dithio-*bis*-(2-nitrobenzoic acid)], 0.5 mM CaCl$_2$, 0.5 mM MgCl$_2$, 50 mM HEPES, pH 7.4, 0.5 mM substrate, 5% DMSO. The reactions are carried out at room temperature, and activity is determined as rate of substrate hydrolysis by measuring increase in absorbance at 406 nm over time.

The recombinant enzyme exhibits $k_{cat}/K_m$ ratios of $2.8 \times 10^4\,M^{-1}\,s^{-1}$ for Z-Lys-SBzl substrate and $1.6 \times 10^4\,M^{-1}\,s^{-1}$ for Z-Arg-SBzl substrate (L.M. Babé & B.F. Schmidt, unpublished results). This preference of approximately 2-fold for Lys over Arg at position P1 is the inverse preference of granzyme A (Odake *et al.*, 1991). Measuring the initial rate of Z-Lys-SBzl hydrolysis, it was determined that the enzyme (recombinant or of granular origin) is most active near pH 7.5 (L.M. Babé & B.F. Schmidt, unpublished results; Hameed *et al.*, 1988).

To determine true peptidase activity, a 13 amino acid peptide (Cys-Gly-Tyr-Gly-Pro-Lys┼Lys-Lys-Arg┼Lys-Val-Gly-Gly) was incubated with recombinant human granzyme K. The parent peptide and hydrolysis products were fractionation by reverse-phase HPLC and then identified by electrospray mass spectroscopy. The peptide was cleaved at the two bonds indicated, C-terminal to Lys6 and Arg9, with comparable rates (L.M. Babé & B.F. Schmidt, unpublished results).

Inhibition of the proteolytic activity of recombinant granzyme K was monitored by measuring the decrease in the rate of hydrolysis of Z-Lys-SBzl at pH 7.4, following a 1 h preincubation with each potential inhibitor (L.M. Babé & B.F. Schmidt, unpublished results). Potent inhibition was observed with DFP (100 μM), DCI (100 μM), PMSF (2 mM), Tos-Lys-CH$_2$Cl (100 μM) and aprotinin (15 μM). Partial inhibition (50%) was observed with a high concentration of leupeptin (200 μM). No inhibition was observed upon incubation with Tos-Phe-CH$_2$Cl (50 μM), soybean trypsin inhibitor (1 μM), $\alpha_1$-proteinase inhibitor (1 μM), chymostatin (250 μM), turkey ovomucoid (150 μg ml$^{-1}$), ecotin (100 μg ml$^{-1}$), E64 (10 μM), 1,10-phenanthroline (1 mM), EDTA (10 mM) or pepstatin (1 μM).

## Structural Chemistry

The primary structure of human granzyme K has been deduced from the cDNA sequence, which codes for a 26 kDa mature serine protease with 75% identity to its rat homolog (RNK-Tryp-2) and 45% identity to human granzyme A (Przetak *et al.*, 1995; Sayers *et al.*, 1996). By amino acid sequence comparisons, granzyme K would be expected to have four disulfide bridges with no free cysteine residues. The gene codes for a typical signal sequence with most likely a short propeptide (10 amino acids or less) (Przetak *et al.*, 1995; Sayers *et al.*, 1996). There is an Asp residue at the bottom of the putative S1 binding pocket, consistent with its preference for lysine or arginine at P1. There are no consensus sequences for *N*-linked glycosylation sites (Przetak *et al.*, 1995; Sayers *et al.*, 1996).

The human granzyme K purified from granules has reduced and unreduced relative molecular mass values of 25 000 and 28 000, respectively, as judged by SDS-PAGE (Hameed *et al.*, 1988). The recombinant enzyme appears as a mature enzyme of $M_r$ 28 000 as judged by SDS-PAGE under reducing conditions (L.M. Babé & B.F. Schmidt, unpublished results). These values are in agreement with that of 25 849 Da deduced from the nucleotide sequence (Sayers *et al.*, 1996). The predicted pI is 10.24 (Sayers *et al.*, 1996).

## Preparation

Human granzyme K can be purified from the granules of peripheral blood mononuclear cells using a Mono-S FPLC column as the major purification step (Hameed *et al.*, 1988; Hanna *et al.*, 1993). About 20 mg of purified protein can be obtained from 10$^{10}$ cells.

Due to the difficulty in obtaining pure enzyme, active human granzyme K has recently been produced recombinantly with *Bacillus subtilis* as host. The granzyme K is secreted in its enzymatically mature form into the bacterial culture supernatant. About 200 mg of the recombinant enzyme can be purified to near homogeneity from 1 liter of culture supernatant. This is achieved by NaCl gradient fractionation on a 10 ml Mono-S column (L.M. Babé & B.F. Schmidt, unpublished results) in an adaptation of the protocol used to purify the enzyme from mononuclear granules (Hameed *et al.*, 1988; Hanna *et al.*, 1993).

## Biological Aspects

Human granzyme K mRNA can be detected in natural killer (NK) and T cells with an increase in transcripts generally found after T cell stimulation (e.g. with concanavalin A or IL-2) (Sayers *et al.*, 1996). High levels of granzyme K transcripts were found in tissues known to be rich in cytotoxic lymphocytes, such as lung, spleen and thymus (Przetak *et al.*, 1995; Sayers *et al.*, 1996). The human granzyme K gene has been mapped to the same locus on chromosome 5 as the granzyme A gene (5q11–q12) (Baker *et al.*, 1994).

Granzymes appear to play a role in the destruction of target cells (including tumor cells, virally infected cells and transplant tissue cells) by cytotoxic T lymphocytes (CTLs), NK cells or lymphokine-activated killer (LAK) cells. Currently, the function(s) or natural substrate(s) of granzyme K are unknown.

## Further Reading

Review articles that give fuller information on cell killing and roles of the granzymes include those by Haddad *et al*. (1993), Simon *et al*. (1993), Henkart (1994) and Berke (1995).

## References

Baker, E., Sayers, T.J., Sutherland, G.R. & Smyth, M.J. (1994) The genes encoding NK cell granule serine proteases, human tryptase-2 (TRYP2) and human granzyme A (HFSP), both map to chromosome 5q11–q12 and define a new locus for cytotoxic lymphocyte granule tryptases. *Immunogenetics* **40**, 235–237.

Berke, G. (1995) The CTL's kiss of death. *Cell* **81**, 9–12.

Haddad, P., Jenne, D.E., Krähenbühl, O. & Tschopp, J. (1993) Structure and possible functions of lymphocyte granzymes. In: *Cytotoxic Cells: Recognition, Effector Function, Generation and Methods* (Sikovsky, M. & Henkart, P., eds). Boston: Birkhäuser, pp. 251–262.

Hameed, A., Lowrey, D.M., Lichtenheld, M. & Podack, E.R. (1988) Characterization of three serine esterases isolated from human IL-2 activated killer cells. *J. Immunol.* **141**, 3142–3147.

Hanna, W.L., Zhang, X., Turbov, J., Winkler, U., Hudig, D. & Froelich, C.J (1993) Rapid purification of cationic granule proteases: application to human granzymes. *Protein Exp. Purif.* **4**, 398–404.

Henkart, P.A. (1994) Lymphocyte-mediated cytotoxicity: two pathways and multiple effector molecules. *Immunity* **1**, 343–346.

Odake, S., Kam, C.-M., Narasimhan, L., Poe, M., Blake, J.T., Krähenbühl, O., Tschopp, J. & Powers, J.C. (1991) Human and murine cytotoxic T lymphocyte serine proteases: subsite mapping with peptide thioester substrates and inhibition of enzyme activity and cytolysis by isocoumarins. *Biochemistry* **30**, 2217–2227.

Przetak, M.M., Yoast, S. & Schmidt, B.F. (1995) Cloning of cDNA for human granzyme 3. *FEBS Lett.* **364**, 268–271.

Sayers, T.J., Wiltrout, T.A., Smyth, M.J., Ottaway, K.S., Pilaro, A.M., Sowder, R., Henderson, L.E., Sprenger, H. & Lloyd, A.R. (1994) Purification and cloning of a novel serine protease, RNK-Tryp-2, from the granules of rat NK cell leukemia. *J. Immunol.* **152**, 2289–2297.

Sayers, T.J., Lloyd, A.R., McVicar, D.W., O'Connor, M.D., Kelly, J.M., Carter, C.R.D., Wiltrout, T.A., Wiltrout, R.H. & Smyth, M.J. (1996) Cloning and expression of a second human natural killer cell granule tryptase, HNK-Tryp-2/granzyme 3. *J. Leukoc. Biol.* **59**, 763–768.

Simon, M.M., Ebnet, K. & Kramer, M.D. (1993) Molecular analysis and possible pleiotropic function(s) of the T cell-specific serine proteinase-1 (TSP-1). In: *Cytotoxic Cells: Recognition, Effector Function, Generation and Methods* (Sikovsky, M. & Henkart, P., eds). Boston: Birkhäuser, pp. 278–294.

*Lilia M. Babé*
*Arris Pharmaceutical Corp.,*
*180 Kimball Way,*
*South San Francisco, CA 94080, USA*
*Email: babe@arris.com*

*Brian F. Schmidt*
*FibroGen, Inc.,*
*260 Littlefield Ave,*
*South San Francisco, CA 94080, USA*
*Email: bschmidt@fibrogen.com*

# 25. Granzyme M

## Databanks

*Peptidase classification: clan SA, family S1, MEROPS ID: S01.139*
*NC-IUBMB enzyme classification: none*
*Databank codes:*

| Species | SwissProt | PIR | EMBL (cDNA) | EMBL (genomic) |
|---|---|---|---|---|
| Homo sapiens | P51124 | – | L23134 | L36922: exon 1 L36936: exons 2–5 |
| Mas musculus | – | – | L76741 | – |
| Rattus norvegicus | Q03238 | A45161 | L05175 | – |

## Name and History

Granzymes are cytotoxic lymphocyte-specific granule-associated serine proteases that make up a small subgroup of the chymotrypsin-like serine protease family (S1). Until recently, granzymes A–H had been described with granzymes C–G isolated only from the mouse (Smyth & Trapani, 1995). The development of new thiobenzyl ester synthetic substrates (Boc-Ala-Ala-Met-SBzl; Hudig *et al.*, 1991) led to the discovery of rat Met-ase-1 as a new member of the granzyme subfamily (Smyth *et al.*, 1992). The name **Met-ase-1** was employed to describe this granzyme's primary serine protease activity and subsequently the cDNAs for rat (Smyth *et al.*, 1992), human (Smyth *et al.*, 1993) and mouse Met-ase-1 (Smyth *et al.*, 1996) were isolated. To fall in line with the nomenclature of granzyme family members (granzymes A–H), the gene for human Met-ase-1 was given the symbol *GZMM*, and the protein was named **granzyme M** (Pilat *et al.*, 1994). The gene for mouse granzyme M is known as *LMet1*, for lymphocyte Met-ase-1 (Thia *et al.*, 1995).

## Activity and Specificity

Data are available only for cleavage of thiobenzyl ester peptide substrates of up to four amino acid residues. The rules governing specificity remain unclear, with the exception that cleavage is favored by unbranched hydrophobic residues (e.g. Met, Nle, Leu) at P1 (Smyth *et al.*, 1992, 1995a; Kelly *et al.*, 1996). Assays with the thiobenzyl ester substrate (Boc-Ala-Ala-Met-SBzl) are described by Smyth *et al.* (1992). The peptidyl chloromethane, Boc-Ala-Ala-Met-CH$_2$Cl, is inhibitory. The substrate Boc-Ala-Ala-Met-SBzl may be available from Enzyme Systems Products (see Appendix 2 for full names and addresses of suppliers).

## Structural Chemistry

Granzyme M is a single-chain protein of about 26 kDa (protein core), predicted pI 10.4, containing four disulfide bonds (like α-chymotrypsin) (Smyth *et al.*, 1992). Like other granzymes, mature granzyme M has the N-terminus Ile-Ile-Gly-Gly- and shares many conserved sequences with other granzymes, particularly around the catalytic triad residues, His57, Asp102, Ser195 (chymotrypsinogen numbering).

## Preparation

Only rat natural killer cell leukemia (RNK-16) cytotoxic granules have been used to yield homogeneous preparations of granzyme M (30-fold purification) (Smyth *et al.*, 1992). The recombinant mouse and human granzyme M enzymes have also been expressed in the monkey kidney cell line COS-7, but were not purified to homogeneity (Smyth *et al.*, 1995a; Kelly *et al.*, 1996).

## Biological Aspects

Granzyme M is present in at least humans, rats and mice. It has only been detected in purified natural killer (NK) cells and NK cell lines, and not in resting or activated T-cell populations (unlike all other granzymes) (Smyth *et al.*, 1992, 1993). Expression in gd$^+$ T cells and thymus has not been rigorously evaluated. In particular, in human NK cells, Met-ase activity and granzyme M cDNA appear to be restricted to CD3$^-$CD56$^+$ large granular lymphocytes rather than CD3$^-$ small agranular lymphocytes (Smyth *et al.*, 1995b). Mouse granzyme M RNA was detected only in IL-2-activated spleen cells and CD3$^-$NK1.1$^+$ mouse large granular lymphocyte cell lines (Kelly *et al.*, 1996). In NK cell homogenates, the activity is predominantly in cytotoxic granules (Smyth *et al.*, 1993).

A great deal is known about the human, mouse and rat genes for granzyme M. Granzyme M genes encompass approximately 5–6 kb of genomic DNA. The genes encode granzyme M on five exons, with four intervening introns of approximately 2.9 kb, 0.8 kb, 0.1 kb and 0.4 kb (Smyth *et al.*, 1995c). The unusual position of intron 1 at residue (−7) in granzyme M is shared with the neutrophil elastase-like genes, azurocidin (*AZU*), myeloblastin (*PR3*) (Chapter 17) and neutrophil elastase (*NE*) (Chapter 15) (Zimmer *et al.*, 1992). The substrate specificity site is encoded within the fourth exon, as in other chymotrypsin-like serine proteases (Smyth *et al.*, 1995c). The evolution of a distinct granzyme M subfamily has also been suggested by chromosomal gene mapping studies. The human granzyme M gene (*GZMM*) is located on chromosome 19p13.3 (Baker *et al.*, 1994; Pilat *et al.*, 1994) and the equivalent mouse *LMet1* gene is located on a syntenic region of chromosome 10C (Pilat *et al.*, 1994; Thia *et al.*, 1995). Interestingly, both genes are closely linked to the same cluster of neutrophil elastase-like serine proteases (Zimmer *et al.*, 1992).

Glycosylation of native granzyme M (Smyth *et al.*, 1992) and recombinant granzyme M (Smyth *et al.*, 1996) up to an estimated 35 kDa has been observed. Granzymes are produced as prepropeptides. As predicted by the von Heijne consensus (von Heijne, 1986), expression of recombinant human and mouse granzyme M suggests that an activation pro-hexapeptide normally regulates processing of this granzyme. Two key residues of human granzyme M, Lys179 (equivalent to Met192 in chymotrypsinogen) and Ser201 (Gly216 in chymotrypsinogen) appear to restrict the preference of this granzyme for long narrow hydrophobic amino acids in the P1 (Smyth *et al.*, 1996). The biological activity of granzyme M is unknown.

## Distinguishing Features

In mammalian NK cell cytotoxic granules, any enzyme hydrolyzing Boc-Ala-Ala-Met-SBzl, but not Suc-Phe-Leu-Phe-SBzl, is likely to be granzyme M. A monoclonal antibody specific for human granzyme M has been described (Smyth *et al.*, 1995a), but is not commercially available.

## References

Baker, E., Sutherland, G.R. & Smyth, M.J. (1994) The gene encoding a human natural killer cell granule serine protease, Met-ase 1, maps to chromosome 19p13.3. *Immunogenetics* **39**, 294–295.

Hudig, D., Allison, N.J., Pickett, T.M., Winkler, U., Kam, C-M. & Powers, J.C. (1991) The function of lymphocyte proteases. Inhibition and restoration of granule-mediated lysis with isocoumarin serine protease inhibitors. *J. Immunol.* **147**, 1360–1368.

Kelly, J.M., O'Connor, M.D., Hulett, M.D., Thia, K.Y.T. & Smyth, M.J. (1996) Cloning and expression of the recombinant mouse natural killer cell granzyme Met-ase-1. *Immunogenetics* **44**, 340–350.

Pilat, D., Fink, T., Obermaier-Skrobanek, B., Zimmer, M., Wekerle, H., Lichter, P. & Jenne, D.E. (1994) The human Met-ase gene (GZMM): structure, sequence, and close physical linkage to the serine protease gene cluster on 19p13.3. *Genomics* **24**, 445–480.

Smyth, M.J. & Trapani, J.A. (1995) Granzymes – exogenous proteinases that induce target cell apoptosis. *Immunol. Today* **16**, 202–206.

Smyth, M.J., Wiltrout, T., Trapani, J.A., Ottaway, K.S., Sowder, R., Henderson, L., Powers, J. & Sayers, T.J. (1992) Purification and cloning of a novel serine protease, RNK-Met-1, from the granules of a rat natural killer cell leukemia. *J. Biol. Chem.* **267**, 24418–24425.

Smyth, M.J., Sayers, T.J., Wiltrout, T., Powers, J.C. & Trapani, J.A. (1993) Met-ase: cloning and distinct chromosomal location of a serine protease preferentially expressed in natural killer cells. *J. Immunol.* **151**, 6195–6205.

Smyth, M.J., O'Connor, M.D., Kelly, J.M., Ganesvaran, P., Thia, K.Y.T. & Trapani, J.A. (1995a) Expression of recombinant human Met-ase-1: a NK cell-specific granzyme. *Biochem. Biophys. Res. Commun.* **217**, 675–683.

Smyth, M.J., Browne, K.A., Kinnear, B.F., Trapani, J.A. & Warren, H. (1995b) Distinct granzyme expression in human CD3⁻ CD56⁺ large granular- and CD3⁻ CD56⁺ small high density-lymphocytes displaying non-MHC-restricted cytolytic activity. *J. Leukoc. Biol.* **57**, 88–93.

Smyth, M.J., Hulett, M.D., Thia, K.Y.T., Sayers, T.J., Carter, C.R.D., Young, H.A. & Trapani, J.A. (1995c) Cloning and characterization of a novel NK-cell specific serine protease gene and its functional 5′-flanking sequences. *Immunogenetics* **42**, 101–111.

Smyth, M.J., O'Connor, M.D., Trapani, J.A., Kershaw, M.H. & Brinkworth, R.I. (1996) A novel substrate binding pocket interaction restricts the specificity of the human natural killer cell-specific serine protease, Met-ase-1. *J. Immunol.* **156**, 4174–4181.

Thia, K.Y.T., Jenkins, N.A., Gilbert, D.J., Copeland, N.G. & Smyth, M.J. (1995) The natural killer cell serine protease gene, *Lmet1*, maps to mouse chromosome 10. *Immunogenetics* **41**, 47–49.

von Heijne, G. (1986) A new method for predicting signal sequence cleavage sites. *Nucleic Acids Res.* **14**, 4683–4690.

Zimmer, M., Medcalf, R.L., Fink, T.M., Mattmann, C., Lichter, P. & Jenne, D.E. (1992) Three human elastase-like genes coordinately expressed in the myelomonocytic lineage are organized as a single genetic locus on 19pter. *Proc. Natl Acad. Sci. USA* **89**, 8215–8219.

***Mark J. Smyth***
*Cellular Cytotoxicity Laboratory,*
*Austin Research Institute,*
*Studley Road, Heidelberg, 3084, Victoria, Australia*
*Email: mj_smyth@muwayf.unimelb.edu.au*

# 26. *Stratum corneum chymotryptic enzyme*

## Databanks

*Peptidase classification: clan SA, family S1, MEROPS ID: S01.300*
*NC-IUBMB enzyme classification: none*
Databank codes:

| Species | SwissProt | PIR | EMBL (cDNA) | EMBL (genomic) |
| --- | --- | --- | --- | --- |
| *Homo sapiens* | P49862 | – | L33404 | – |

## Name and History

Stratum corneum chymotryptic enzyme (SCCE) was discovered as a result of studies on the mechanisms of cell cohesion and cell shedding (desquamation) in the outermost layer of human skin, i.e. in the stratum corneum of the epidermis. The stratum corneum forms the physicochemical barrier between the interior of the body and the surroundings. The mechanical integrity of the stratum corneum is dependent on a strong cohesion between its cellular building blocks, the corneocytes. The stratum corneum is continuously produced in the process of terminal differentiation of keratinocytes. For the maintenance of a constant thickness of the stratum corneum its *de novo* production must be balanced by desquamation, i.e. shedding of cells as a result of loss of intercellular cohesion. The question was how this could take place in a regulated way in a tissue consisting of 'dead' cells. It was found that when a piece of stratum corneum from the heel was incubated in a simple buffer, the tissue apparently contained a mechanism which was 'programmed' to cause a release of cells from the surface that had faced outwards *in vivo*. This '*in vitro* desquamation' could be inhibited by aprotinin, and was thus apparently dependent on the activity of a serine proteinase (Lundström & Egelrud, 1988).

The inhibitor profile of this process (Lundström & Egelrud, 1990) and of the concomitant degradation of the desmosomal protein desmoglein I (Lundström & Egelrud, 1989) was similar to that of a protease which could be extracted from dissociated corneocytes (Egelrud & Lundström, 1991). This enzyme catalyzed the cleavage of a chymotrypsin substrate, and could be inhibited by chymostatin (an inhibitor of chymotrypsin), but not by leupeptin (an inhibitor of trypsin). It was found to be generally distributed in human stratum corneum (Lundström & Egelrud, 1991), and was therefore named ***stratum corneum chymotryptic enzyme (SCCE)***.

## Activity and Specificity

SCCE readily cleaves heat-denatured casein used as substrate in zymograms (Egelrud & Lundström, 1991). It also degrades the $\alpha$ chain of native human fibrinogen (Egelrud, 1992). The only low molecular mass peptide substrate known so far is MeOSuc-Arg-Pro-Tyr↓NHPhNO$_2$. SCCE is essentially inactive against the typical chymotrypsin substrate Suc-Ala-Ala-Pro-Phe-NHPhNO$_2$ (Egelrud, 1993). In the B chain of bovine insulin, the cleavage sites are Phe-Val-Asn-Gln-His-Leu↓Cys(SO$_3$)-Gly-Ser-His-Leu-Val-Glu-Ala-Leu-Tyr↓Leu-Val-Cys(SO$_3$)-Gly-Glu-Arg-Gly-Phe-Phe↓Tyr↓Thr-Pro-Lys-Ala (Skytt *et al.*, 1995).

The pH optimum of SCCE is between 7 and 8, but significant activity is found also at pH 5.5. SCCE is inhibited by PMSF, aprotinin and soybean trypsin inhibitor. It is also inhibited by chymostatin, but the concentration of this inhibitor needed to cause 50% inhibition of SCCE is about three orders of magnitude higher than for bovine chymotrypsin (Chapter 8) or human cathepsin G (Chapter 16). The activity of SCCE against MeOSuc-Arg-Pro-Tyr-NHPhNO$_2$ is sensitive to zinc ions, 20–30 μM causing 50% inhibition (Egelrud & Lundström, 1991; Egelrud, 1993).

## Structural Chemistry

SCCE is a single-chain protein of about 28 kDa, pI > 9. Its primary structure suggests five disulfide bonds and one potential *N*-glycosylation site. The native as well as the recombinant enzyme expressed in mammalian cells appear to be present as unglycosylated as well as glycosylated forms. When expressed in C127 cells, recombinant SCCE is produced as an inactive precursor with a propeptide consisting of seven amino acid residues. The propeptide can be released by means of trypsin cleavage, which yields catalytically active SCCE. From the cDNA, a signal peptide of 22 amino acid residues can be deduced. The deduced amino acid sequence of SCCE contains all the components of the conserved catalytic triad of mammalian serine proteases. A unique structural feature of SCCE is its Asn residue at the position deduced to be located at the bottom of the S1 specificity subsite (Hansson *et al.*, 1994).

## Preparation

SCCE from extracts of dissociated corneocytes from human plantar stratum corneum can be purified to near homogeneity by affinity chromatography on immobilized soybean trypsin inhibitor (Egelrud, 1993). Large amounts of recombinant proSCCE can be purified from the growth medium of C127 cells transfected with an expression vector containing the SCCE-cDNA (Hansson *et al.*, 1994).

## Biological Aspects

Although there is reasonably good evidence that SCCE is responsible for the degradation of intercellular cohesive structures, and thus involved in the desquamation-like process in plantar stratum corneum *in vitro*, it has not yet been firmly established that SCCE is involved in desquamation *in vivo*. Evidence in favor of this physiologic function is the immuno-electron microscopic observation that SCCE is secreted to the extracellular space during the transition between viable and cornified epidermal layers, and that the enzyme is often found in association with desmosomes in the stratum corneum extracellular space (Sondell *et al.*, 1995).

When mRNA from a large variety of human sources was analyzed, abundant expression of SCCE was found only in the skin. Low expression at the RNA level was found also in brain and kidney. By means of immunohistochemistry, a close correlation between SCCE expression and the presence of a cornified layer in squamous epithelia has been found. In addition to interfollicular epidermis, the inner root sheet of the hair follicle, the luminal part of the sebaceous duct, and the cornified epithelium of the hard palate also express SCCE. In the normally noncornified squamous epithelium of the human bucca, SCCE appears not to be expressed (Sondell *et al.*, 1994; Ekholm *et al.*, 1995). In processes of the oral mucosa where there is pathologic cornification of the epithelium, a concomitant expression of SCCE at sites where it is normally absent takes place (Sondell *et al.*, 1996). All these findings can be taken as evidence that wherever there is a stratum corneum there is a need to desquamate, and thus for an enzyme such as SCCE. Another interpretation, however, may be that SCCE is a protease that is involved in the cornification process, i.e. the apoptosis-like process by which differentiated epithelial cells lose all cellular organelles except the keratin filaments, and become surrounded by a cell envelope consisting of cross-linked proteins.

There may possibly be a link between SCCE and inflammation in the epidermis. Although keratinocytes do not contain active caspase-1 (Chapter 248), epidermis has been shown to contain biologically active interleukin 1$\beta$ (IL-1$\beta$) (Nylander-Lundqvist *et al.*, 1996). In a manner similar to chymotrypsin, SCCE can convert the IL-1$\beta$ precursor to biologically active IL-1$\beta$ *in vitro* (Nylander-Lundqvist & Egelrud, 1997).

A possible physiological inhibitor of SCCE in the epidermis is antileukoprotease, a low molecular mass protein inhibitor of a number of serine proteases (Franzke *et al.*, 1996).

## Distinguishing Features

SCCE differs from other chymotrypsin-like serine proteases in its inhibition profile and activity towards peptide substrates (Egelrud, 1993). SCCE-specific monoclonal (Sondell *et al.*, 1994) and polyclonal (Sondell *et al.*, 1996) antibodies have been raised.

## Further Reading

For a review of the potential role of protein structures and proteolytic enzymes in cell cohesion and desquamation, see Egelrud *et al.* (1996).

## References

Egelrud, T. (1992) Stratum corneum chymotryptic enzyme: evidence of its location in the stratum corneum intercellular space. *Eur. J. Dermatol.* **2**, 50–55.

Egelrud, T. (1993) Purification and preliminary characterization of stratum corneum chymotryptic enzyme: a proteinase that may be involved in desquamation. *J. Invest. Dermatol.* **101**, 200–204.

Egelrud, T. & Lundström, A. (1991) A chymotrypsin-like proteinase that may be involved in desquamation in plantar stratum corneum. *Arch. Dermatol. Res.* **283**, 108–112.

Egelrud, T., Lundström, A. & Sondell, B. (1996) Stratum corneum cell cohesion and desquamation in maintenance of the skin barrier. In: *Dermatotoxicology*, 5th edn (Marzulli, F.N. & Maibach, H.I., eds). Washington DC: Taylor & Francis, pp. 19–27.

Ekholm, E., Sondell, B., Jonsson, M., Leigh, I. & Egelrud, T. (1995) Expression of stratum corneum chymotryptic enzyme in normal sebaceous glands. *J. Invest. Dermatol.* **105**, 477 (abstr.).

Franzke, C.-W., Baici, A.B., Bartels, J., Christophers, E. & Wiedow, O. (1996) Antileukoprotease inhibits stratum corneum chymotryptic enzyme. *J. Biol. Chem.* **271**, 21886–21890.

Hansson, L., Stromqvist, M., Backman, A., Wallbrandt, P., Carlstein, A. & Egelrud, T. (1994) Cloning, expression, and characterization of stratum corneum chymotryptic enzyme. A skin-specific human serine proteinase. *J. Biol. Chem.* **269**, 19420–19426.

Lundström, A. & Egelrud, T. (1988) Cell shedding from human plantar skin *in vitro*: evidence of its dependence on endogenous proteolysis. *J. Invest. Dermatol.* **91**, 340–343.

Lundström, A. & Egelrud, T. (1989) Evidence that cell shedding form plantar skin in vitro involves endogenous proteolysis of the desmosomal protein desmoglein I. *J. Invest. Dermatol.* **94**, 216–220.

Lundström, A. & Egelrud, T. (1990) Cell shedding from human plantar skin in vitro: evidence that two different types of protein structures are degraded by a chymotrypsin-like enzyme. *Arch. Dermatol. Res.* **282**, 234–237.

Lundström, A. & Egelrud, T. (1991) Stratum corneum chymotryptic enzyme: a proteinase which may be generally present in the stratum corneum and with a possible involvement in desquamation. *Acta Derm. Venereol. (Stockh.)* **71**, 471–474.

Nylander-Lundqvist, E. & Egelrud, T. (1997) Formation of active IL-1$\beta$ from pro-IL1$\beta$ catalyzed by stratum corneum chymotryptic enzyme. *Acta Derm. Venereol. (Stockh.),* **77**, 203–206.

Nylander-Lundqvist, E., Bäck, O. & Egelrud, T. (1996) IL-1-beta activation in human epidermis. *J. Immunol.* **157**, 1699–1704.

Skytt, A., Stromqvist, M. & Egelrud, T. (1995) Primary substrate specificity of recombinant human stratum corneum chymotryptic enzyme. *Biochem. Biophys. Res. Commun.* **211**, 586–589.

Sondell, B., Thornell, L.E., Stigbrand, T. & Egelrud, T. (1994) Immunolocalization of stratum corneum chymotryptic enzyme in human skin and oral epithelium with monoclonal antibodies: evidence of a proteinase specifically expressed in keratinizing squamous epithelia. *J. Histochem. Cytochem.* **42**, 459–465.

Sondell, B., Thornell, L.E. & Egelrud, T. (1995) Evidence that stratum corneum chymotryptic enzyme is transported to the stratum corneum extracellular space via lamellar bodies. *J. Invest. Dermatol.* **104**, 819–823.

Sondell, B., Dyberg, P., Anneroth, G.K.B., Östman, P.-O. & Egelrud, T. (1996) Association between expression of stratum corneum chymotryptic enzyme and pathological keratinization in human oral mucosa. *Acta Derm. Venereol. (Stockh.)* **76**, 177–181.

*Torbjörn Egelrud*
*Department of Dermatology,*
*University Hospital,*
*S-901 85 Umeå, Sweden*
*Email: torbjorn.egelrud@dermven.umu.se*

# 27. Acrosin

## Databanks

*Peptidase classification: clan SA, family S1, MEROPS ID: S01.223*
*NC-IUBMB enzyme classification: EC 3.4.21.10*
*Chemical Abstracts Service registry number: 9068-57-9*
*Databank codes:*

| Species | SwissProt | PIR | EMBL (cDNA) | EMBL (genomic) |
|---|---|---|---|---|
| *Bos taurus* | – | – | – | X68212: complete gene |
| *Capra hircus* | P10626 | S02175 | – | – |
| | | S02176 | | |
| *Cavia porcellus* | – | S29599 | Z12153 | – |
| *Homo sapiens* | P10323 | A61022 | Y00970 | M77378: 5′ end |
| | | S03330 | | M77379: exons 2 and 3 |
| | | S11674 | | M77380: exon 4 |
| | | S12063 | | M77381: 3′ end |
| | | S23499 | | X54017: exon 1 |
| | | | | X54018: exons 2 and 3 |
| | | | | X54019: exon 4 |
| | | | | X54020: exon 5 |
| | | | | X66188: exon 1 |
| *Mus musculus* | P23578 | A37344 | D00754 | M96426: exon 1 |
| | | A55283 | M85170 | M96427: exon 2 |
| | | JX0138 | X52466 | M96428: exon 3 |
| | | JX0172 | | M96429: exon 4 |
| | | | | M96430: exon 5 and complete CDS |
| | | | | S64500: promoter segment 1 |
| | | | | S66243: promoter segment 2 |
| | | | | S66245: promoter segment 3 |
| *Oryctolagus cuniculus* | P48038 | S47538 | U05204 | – |
| *Rattus norvegicus* | P29293 | A56620 | X59254 | X58550: complete gene |
| | | S18407 | | |
| | | S30037 | | |
| *Sus scrofa* | P08001 | A34170 | J04950 | X58549: complete gene |
| | | S02428 | X14844 | |
| | | S02780 | | |
| | | S04940 | | |
| | | S08994 | | |
| | | S10695 | | |
| | | S12968 | | |
| | | S16657 | | |

## Name and History

Yamane (1935) described the activity of rabbit sperm extract in dissolving the investment layers of oocytes, and later, the discovery of a trypsin-like enzyme in acrosomal preparations (Srivastava *et al.*, 1965; Waldschmidt *et al.*, 1966) led to the view that this enzyme might be responsible for sperm penetration through the zona pellucida (Stambaugh & Buckley, 1968). Partial amino acid sequence analysis of the boar acrosin confirmed that the enzyme belongs to the chymotrypsin (S1) family of serine peptidases (Fock-Nüzel *et al.*, 1984). The name *acrosin* was applied to the trypsin-like protease of the acrosome of the sperm head.

## Activity and Specificity

Acrosin preferentially splits Arg+ and Lys+ bonds in natural and synthetic substrates (Schiessler *et al.*, 1975), and the enzyme is routinely assayed by its amidase activity on Bz-Arg-NHPhNO$_2$ (Schleuning & Fritz, 1976). The pH optimum is about 8.0, and the assay system typically includes calcium (50 mM) and a detergent such as Triton X-100 (0.01%).

Acrosin is inhibited by the typical inhibitors of trypsin-like serine proteinases, DFP, PMSF and *p*-nitrophenyl-*p*′-guanidinobenzoate (Polakoski & McRorie, 1973; Brown & Hartree, 1978).

## Structural Chemistry

Amino acid sequences of human (Baba *et al*., 1989b; Adham *et al*., 1990), pig (Adham *et al*., 1989b; Baba *et al*., 1989a), cattle (Adham *et al*., 1996), mouse (Klemm *et al*., 1990), rat (Klemm *et al*., 1991), rabbit (Richardson & O'Rand, 1994) and guinea pig proacrosin have been deduced from cDNA and gene sequences. The enzyme is synthesized as a preproenzyme, consisting of a signal peptide, a propeptide, the catalytic domain and a C-terminal domain. There is the typical catalytic triad of Asp, His, Ser that is found in all peptidases of family S1. The C-terminal domain of proacrosin is extremely rich in proline, and has not been found in any other serine proteinase. A cross-species comparison of the proacrosin sequences reveals that the catalytic domain is highly conserved, but the C-terminal domain is very variable (Adham *et al*., 1996). The mature acrosin is comprised of two chains, a light chain and a heavy chain, which are linked by two disulfide bonds (Fock-Nüzel *et al*., 1984). Acrosin is glycosylated (Fock-Nüzel *et al*., 1984; Mukerji & Meizel, 1979), and the sequence contains two *N*-linked glycosylation sites that are conserved in all characterized mammalian acrosins. The pI for the boar enzyme is 10.5.

## Preparation

Acrosin is found in spermatozoa of vertebrates, and has been purified from human (Tobias & Schumacher, 1977), pig (Polakoski & Parrish, 1977), guinea pig (Hardy *et al*., 1987), hamster (Meizel & Mukerji, 1976), mouse (Brown, 1983), rabbit (Mukerji & Meizel, 1979) and ram (Harrison & Brown, 1979).

Purification of proacrosin is done in the presence of trypsin inhibitors and at low pH, in order to prevent autoactivation. In a modified version of the procedure described by Schleuning & Fritz (1976), boar sperm is extracted with 0.3 M acetic acid, 50 mM NaCl and NaN₃, and the enzyme is purified by Sephadex G100 gel filtration, phenyl-Sepharose hydrophobic chromatography, CM-Sephadex C-50 ion-exchange chromatography, and finally, affinity chromatography, to give a 160-fold purification (Müller-Esterl & Fritz, 1981).

## Biological Aspects

The enzyme is stored in the sperm acrosome as its proenzyme, proacrosin, which is activated to the mature enzyme during the acrosome reaction (Polakoski & Parrish, 1977). The 53–55 kDa proteolytically inactive proacrosin forms are converted by autoactivation to active enzyme with molecular masses of 49, 34 and 25 kDa, designated $\alpha$-, $\beta$- and $\gamma$-acrosin, respectively (Polakoski & Parrish, 1977; Parrish & Polakoski, 1978). Proacrosin is converted to the mature enzyme by two processes: the cleavage of the peptide bond Arg23+Val24 that leads to the formation of the light and heavy chains (Fock-Nüzel *et al*., 1984) and the removal of 85 residues at the C-terminus (Baba *et al*., 1989a; Zelezna *et al*., 1989).

The proacrosin genes of human (Keime *et al*., 1990), boar and cattle (Adham *et al*., 1996), mouse (Kremling *et al*., 1991b; Watanabe *et al*., 1991) and rat (Kremling *et al*., 1991a) have been cloned and characterized. Alignment of the intron–exon structures of the proacrosin genes with those of other serine proteases reveals that the coding sequence of mammalian proacrosin is distributed in five exons, and the splice junction types are identical to those of the exons encoding the catalytic domains of other serine proteinases (Adham *et al*., 1996). The proacrosin gene (*Acr*) has been assigned to human chromosome 22q13-qter (Adham *et al*., 1989a), to mouse chromosome 15E (Kremling *et al*., 1991b), to rat chromosome 7 (Adham *et al*., 1991), to cattle chromosome 5q35 (Friedl *et al*., 1994) and to pig chromosome 5p15 (Rettenberger *et al*., 1995). Proacrosin gene expression is first detectable in the diploid phase of spermatogenesis (Kashiwabara *et al*., 1990; Kremling *et al*., 1991b), while translation occurs in haploid spermatids (Flörke-Gerloff *et al*., 1983; Mansouri *et al*., 1983). The results of experiments with transgenic mice indicate that the translational regulation of proacrosin is mediated by nucleotide sequences in the 5′-untranslated region (Nayernia *et al*., 1994).

Acrosin has been presumed to be involved in the recognition and binding of the sperm to the zona pellucida of the ovum (Saling, 1981; Jones *et al*., 1988) and in the penetration of the sperm through the zona pellucida (McRorie & Williams, 1974; Yanagimachi, 1981). The function of acrosin has been examined *in vivo* by generating mice carrying a mutation at the acrosin locus (*Acr*) generated by targeted disruption of the gene in embryonic stem cells (Baba *et al*., 1994; Adham *et al*., 1997). Mice homozygous for the disrupted acrosin gene do not exhibit defects in fertility. However, results obtained from *in vitro* fertilization revealed that the spermatozoa of mutant mice are slower to penetrate the zona pellucida, and thereby have a selective disadvantage when in competition with the spermatozoa of wild-type mice (Adham *et al*., 1997).

## Distinguishing Features

Polyclonal antisera for the enzyme from various species have been described (Flörke-Gerloff *et al*., 1983; Arboleda & Gerton, 1988), but are not commercially available at the time of writing.

## Further Reading

Acrosin is reviewed in Urch (1991).

## References

Adham, I.M., Grzeschik, K.-H., Geurts van Kessel, A.H.M. & Engel, W. (1989a) The gene encoding the human preproacrosin (ACR) maps to the q13-qter region on chromosome 22. *Hum. Genet.* **84**, 59–62.

Adham, I.M., Klemm, U., Maier, W.-M., Hoyer-Fender, S., Tsaousidou, S. & Engel, W. (1989b) Molecular cloning of preproacrosin and analysis of its expression pattern in spermatogenesis. *Eur. J. Biochem.* **182**, 563–568.

Adham, I.M., Klemm, U., Maier, W.-M. & Engel, W. (1990) Molecular cloning of human preproacrosin cDNA. *Hum. Genet.* **84**, 125–128.

Adham, I.M., Szpirer, C., Kremling, H., Keime, S., Szpirer, J., Levan, G. & Engel, W. (1991) Chromosomal assignment of four rat genes coding for the spermatid-specific proteins proacrosin (ACR), transition proteins 1 (TNP1) and 2 (TNP2), and protamine 1 (PRM1). *Cytogenet. Cell Genet.* **57**, 47–50.

Adham, I.M., Kremling, H., Nieter, S., Zimmermann, S., Hummel, M., Schroeter, U. & Engel, W. (1996) The structures of the bovine and porcine proacrosin genes and their conservation among mammals. *Biol. Chem. Hoppe-Seyler* **377**, 261–265.

Adham, I.M., Nayernia, K. & Engel, W. (1997) Spermatozoa lacking acrosin protein show delayed fertilization. *Mol. Reprod. Dev.* **46**, 370–376.

Arboleda, C.E. & Gerton, G.L. (1988) Proacrosin/acrosin during guinea pig spermatogenesis. *Dev. Biol.* **125**, 217–225.

Baba, T., Kashiwabara, S., Watanabe, K., Itoh, H., Michikawa, Y., Kimura, K., Takada, M., Fukamizu, A. & Arai, Y. (1989a) Activation and maturation mechanisms of boar acrosin zymogen based on the deduced primary structure. *J. Biol. Chem.* **264**, 11920–11927.

Baba, T., Watanabe, K., Kashiwabara, S. & Arai, Y. (1989b) Primary structure of human proacrosin deduced from its cDNA sequence. *FEBS Lett.* **244**, 296–300.

Baba, T., Azuma, S., Kashiwabara, S. & Toyoda, Y. (1994) Sperm from mice carrying a targeted mutation of the acrosin gene can penetrate the oocyte zona pellucida and effect fertilization. *J. Biol. Chem.* **269**, 31845–31849.

Brown, C.R. (1983) Purification of mouse acrosin, its activation from proacrosin and effect on homologous egg investments. *J. Reprod. Fertil.* **69**, 289–295.

Brown, C.R. & Hartree, E.F. (1978) Studies on ram acrosin. Activation of proacrosin accompanying the isolation of acrosin from spermatozoa, and purification of the enzyme by affinity chromatography. *Biochem. J.* **175**, 227–238.

Flörke-Gerloff, S., Töpfer-Petersen, E., Müller-Esterl, W., Schill, W.-B. & Engel, W. (1983) Acrosin and the acrosome in human spermatogenesis. *Hum. Genet.* **65**, 61–67.

Fock-Nüzel, R., Lottspeich, F., Henschen, A. & Müller-Esterl, W. (1984) Boar acrosin is a two chain molecule. Isolation and primary structure of the light chain; homology with the pro-part of other serine proteinase. *Eur. J. Biochem.* **141**, 441–446.

Friedl, R., Adham, I.M. & Rottmann, O.J. (1994) Mapping of the gene encoding bovine preproacrosin (ACR) to chromosome BTA5 region q35. *Mammalian Genome* **5**, 830–831.

Hardy, D.M., Wild, G.C. & Tung, K.S.K. (1987) Purification and initial characterization of proacrosins from guinea pig testes and epididymal spermatozoa. *Biol. Reprod.* **37**, 189–199.

Harrison, R.A.P. & Brown, C.R. (1979) The zymogen form of acrosin in testicular, epididymal, and ejaculated spermatozoa from ram. *Gamete Res.* **2**, 75–87.

Jones, R., Brown, C.R. & Lancaster, R.T. (1988) Carbohydrate binding properties of boar sperm proacrosin and assessment of its role in sperm–egg recognition and adhesion during fertilization. *Development* **102**, 781–792.

Kashiwabara, S., Arai, Y., Kodaira, K. & Baba, T. (1990) Acrosin biosynthesis in meiotic and postmeiotic spermatogenic cells. *Biochem. Biophys. Res. Commun.* **173**, 240–245.

Keime, S., Adham, I.M. & Engel, W. (1990) Nucleotide sequence and exon–intron organization of the human proacrosin gene. *Eur. J. Biochem.* **190**, 195–200.

Klemm, U., Maier, W.M., Tsaousidou, S., Adham, I.M., Willison, K. & Engel, W. (1990) Mouse preproacrosin: cDNA sequence, primary structure and postmeiotic expression in spermatogenesis. *Differentiation* **42**, 160–166.

Klemm, U., Flake, A. & Engel, W. (1991) Rat sperm acrosin: cDNA sequence, derived primary structure and phylogenetic origin. *Biochim. Biophys. Acta* **1090**, 270–272.

Kremling, H., Flake, A., Adham, I.M., Radtke, J. & Engel, W. (1991a) Exon–intron structure and nucleotide sequence of the rat proacrosin gene. *DNA Sequence* **2**, 57–60.

Kremling, H., Keime, S., Wilhelm, K., Adham, I.M., Hameister, H. & Engel, W. (1991b) Mouse proacrosin gene: nucleotide sequence, diploid expression and chromosomal localization. *Genomics* **11**, 828–834.

Mansouri, A., Phi-van, L., Geithe, H.P. & Engel, W. (1983) Proacrosin/acrosin activity during spermiohistogenesis of the bull. *Differentiation* **24**, 149–152.

McRorie, R.A. & Williams, W.L. (1974) Biochemistry of mammalian fertilization. *Annu. Rev. Biochem.* **43**, 777–803.

Meizel, S. & Mukerji, S.K. (1976) Biochemical studies of proacrosin and acrosin from hamster cauda epididymal spermatozoa. *Biol. Reprod.* **14**, 444–450.

Mukerji, S.K. & Meizel, S. (1979) Rabbit testis proacrosin. Purification, molecular weight estimation and amino acid and carbohydrate composition of the molecule. *J. Biol. Chem.* **254**, 11721–11728.

Müller-Esterl, W. & Fritz, H. (1981) Sperm acrosin. *Methods Enzymol.* **80**, 621–632.

Nayernia, K., Nieter, S., Kremling, H., Oberwinkler, H. & Engel, W. (1994) Functional and molecular characterization of the transcriptional regulatory region of the proacrosin gene. *J. Biol. Chem.* **269**, 32181–32186.

Parrish, R.F. & Polakoski, K.L. (1978) Boar $M_a$-acrosin. Purification and characterization of the initial active enzyme resulting from the conversion of boar proacrosin to acrosin. *J. Biol. Chem.* **253**, 8428–8432.

Polakoski, K.L. & McRorie, R.A. (1973) Boar acrosin. II. Classification, inhibition and specificity studies of a proteinase from sperm acrosomes. *J. Biol. Chem.* **248**, 8183–8188.

Polakoski, K.L. & Parrish, R.F. (1977) Boar proacrosin. Purification and preliminary activation studies of proacrosin isolated from ejaculated boar sperm. *J. Biol. Chem.* **252**, 1888–1894.

Rettenberger, G., Adham, I.M., Engel, W., Klett, C. & Hameister, H. (1995) Assignment of the porcine acrosin gene, ACR, to chromosome 5p15 by fluorescence in situ hybridization (FISH). *Mammalian Genome* **6**, 60–61.

Richardson, R.T. & O'Rand, M.G. (1994) Cloning and sequencing of cDNAs for rabbit preproacrosin and a noval preproacrosin-related cDNA. *Biochim. Biophys. Acta* **1219**, 215–218.

Saling, P.M. (1981) Involvement of trypsin-like activity in binding of mouse spermatozoa to zona pellucida. *Proc. Natl Acad. Sci. USA* **78**, 6231–6235.

Schiessler, H., Schleuning, W.D. & Fritz, H. (1975) Cleavage specifity of boar acrosin on polypeptide substrates, ribonuclease and insulin B-chain. *Z. Physiol. Chem.* **356**, 1931–1936.

Schleuning, W.-D., & Fritz, H. (1976) Sperm acrosin. *Methods Enzymol.* **45**, 330–342.

Srivastava, P.N., Adams, C.E. & Hartree, I.F. (1965) Enzymatic action of acrosomal preparations on the rabbit ovum in vitro. *J. Reprod. Fertil.* **10**, 61–67.

Stambaugh, R. & Buckley, J. (1968) Zona pellucida dissolution enzymes of the rabbit sperm head. *Science* **161**, 585–586.

Tobias, P.S. & Schumacher, G.F.B. (1977) Observation of two proacrosin in extracts of human spermatozoa. *Biochem. Biophys. Res. Commun.* **74**, 434–439.

Urch, U.A. (1991) Biochemistry and function of acrosin. In: *Elements of Mammalian Fertilization. Basic Concepts*, vol. 1 (Wassarman, P.M., ed.). Boston: CRC Press, pp. 233–248.

Waldschmidt, M., Karg, H. & Hoffmann, B. (1966) Untersuchungen über die tryptische Enzymaktivität in Geschlechtssekreten von Bullen. *Zuchthygiene* **1**, 15–21.

Watanabe, K., Baba, T., Kashiwabara, S., Okamoto, A. & Arai, Y. (1991) Structure and organization of the mouse acrosin gene. *J. Biochem.* **109**, 828–833.

Yamane, J. (1935) Kausal-analytische Studien über die Befruchtung des Kanincheneies. *Cytologia* **6**, 233–255.

Yanagimachi, R. (1981) Mechanisms of fertilization of mammals. In: *Fertilization and Embryonic Development in vitro* (Mastroianni, L.J. & Biggers, J.D., eds). New York: Plenum Press, pp. 81–182.

Zelezna, B., Cechova, D. & Henschen, A. (1989) Isolation of the boar sperm acrosin peptide, released during the conversion of α-form into β-form. *Biol. Chem. Hoppe-Seyler* **370**, 323–327.

**S**

*Ibrahim M. Adham*
*Institut für Humangenetik,*
*Gosslerstrasse 12d,*
*37073 Göttingen, Germany*

*Manfred Schlösser*
*Institut für Humangenetik,*
*Gosslerstrasse 12d,*
*37073 Göttingen, Germany*
*Email: Mschloe1@gwdg.de*

*Wolfgang Engel*
*Institut für Humangenetik,*
*Gosslerstrasse 12d,*
*37073 Göttingen, Germany*
*Email: Wengel@gwdg.de*

# 28. Introduction: tissue kallikrein and its relatives

## Databanks

| Species | Gene | SwissProt | PIR | EMBL (cDNA) | EMBL (genomic) |
|---|---|---|---|---|---|
| *Canis familiaris* | – | P09582 | A30981 A37938 | Y00751 | M63669: complete gene |
| *Canis familiaris* | – | – | S28195 | – | – |
| *Canis familiaris* | *dklk-2* | – | S45303 | X75479 | – |
| *Cavia porcellus* | – | P12323 | A27207 | – | – |
| *Cavia porcellus* | – | P12322 | A18671 | – | – |
| *Homo sapiens* | *klk1* | P06870 | A04628 A23587 A24696 A28678 A60248 JX0040 S05642 | M12706 M25629 X13561 | L10038: complete gene M18157: complete gene M33105: exon 1 M33106: exon 2 M33107: exon 3 M33108: exon 4 M33109: exon 5 and complete CDS |
| *Homo sapiens* | *klk2* | P20151 | A29586 B24696 | S39329 | M18156: 5′ flank |
| *Macaca fascicularis* | – | Q07276 | S33772 | L10039 | – |
| *Mus musculus* | *klk-1* | P15947 | A25606 | M13498 | – |
| *Mus musculus* | *klk-2* | – | A05308 | – | – |
| *Mus musculus* | *klk-3* | P00756 | A91005 | X01389 | X01798: 5′ flanking region X01799: exons 2–5 and 3′ flanking region |
| *Mus musculus* | *klk-4* | P00757 | B91005 | M11434 | X01800: 5′ flanking region X01801: exons 2–5 and 3′ flanking region |
| *Mus musculus* | *klk-5* | P15945 | I70019 S06305 | Y00500 | – |
| *Mus musculus* | *klk-6* | P00755 | A00941 | – | J00390: complete gene and pseudogene V00829: complete gene |
| *Mus musculus* | *klk-8* | P07628 | A24378 I70011 | – | M18607: exon 3 |
| *Mus musculus* | *klk-9* | P15949 | A27120 A29745 C29746 I70015 | M17962 | M17983: exon 1 M17984: exon 2 M17985: exons 3–5 M18588: exon 2 M18608: exon 3 |
| *Mus musculus* | *klk-11* | P15946 | I70023 S01971 | – | M18610: exon 3 X13215: exon 1 |
| *Mus musculus* | *klk-13* | P36368 | A41020 B29746 PC2014 S18674 | X58628 | – |
| *Mus musculus* | *klk-14* | – | I70030 | – | M18613: exon 3 |
| *Mus musculus* | *klk-16* | P04071 | A28062 I70032 I70033 | – | M18615: exon 3 |
| *Mus musculus* | *klk-21* | – | I70037 | – | M18617: exon 3 |

S

| Species | Gene | SwissProt | PIR | EMBL (cDNA) | EMBL (genomic) |
|---|---|---|---|---|---|
| *Mus musculus* | *klk-22* | P15948 | A29746 I70039 | – | M17977: exon 1 M17978: exon 1 M17979: exons 3–5 M18598: exon 2 M18618: exon 3 |
| *Mus musculus* | *klk-24* | – | I70017 | – | M18619: exon 3 |
| *Mus musculus* | *klk-26* | P36369 | A00940 S19310 | K01831 V00828 X63327 | – |
| *Mus musculus* | *mK1* | – | PC2013 | – | – |
| *Mus musculus* | – | – | A00939 | – | – |
| *Papio hamadryas* | *klk-1* | – | – | L43121 | – |
| *Praomys natalensis* | – | P32824 | S15686 | X17352 | – |
| *Praomys natalensis* | – | – | S15685 S28196 | X17351 | – |
| *Rattus norvegicus* | *klk-1* | P00758 | A00944 A23863 A25137 A33359 A41429 JX0073 | – | M23874: exon 1 M23875: exon 2 M23876: exons 3–5 |
| *Rattus norvegicus* | *klk-2* | P00759 | B33359 | M11565 | M23877: exon 1 M23878: exons 2–5 M26533: complete gene |
| *Rattus norvegicus* | *klk-3* | P15950 | B23863 B32340 | M26534 | – |
| *Rattus norvegicus* | *klk-7* | P36373 | A31136 B41429 D41429 S09315 S10698 S10699 | – | M19647: complete gene |
| *Rattus norvegicus* | *klk-8* | P36374 | A34079 S10700 | M27215 M27216 | M27217: exons 3–5 |
| *Rattus norvegicus* | *klk-9* | P07647 | A23710 B23710 D23863 S19302 S46212 | M11566 | – |
| *Rattus norvegicus* | *klk-10* | P36375 | A18966 | S48142 A38356 A44284 A56784 C18966 C41429 | – |
| *Rattus norvegicus* | *klk-12* | P36376 | B31136 | – | M19648: exons 3–5 M22922: exon 1 |
| *Sus scrofa* | – | P00752 | A00938 A92895 | – | – |

Brookhaven Protein Data Bank three-dimensional structures:

| Species | Gene | ID | Resolution | Notes |
|---|---|---|---|---|
| *Rattus norvegicus* | *klk-2* | 1TON | 1.8 | |
| *Sus scrofa* | – | 2KAI | 2.5 | complex with cattle pancreatic trypsin inhibitor |
| | | 2PKA | 2.05 | |

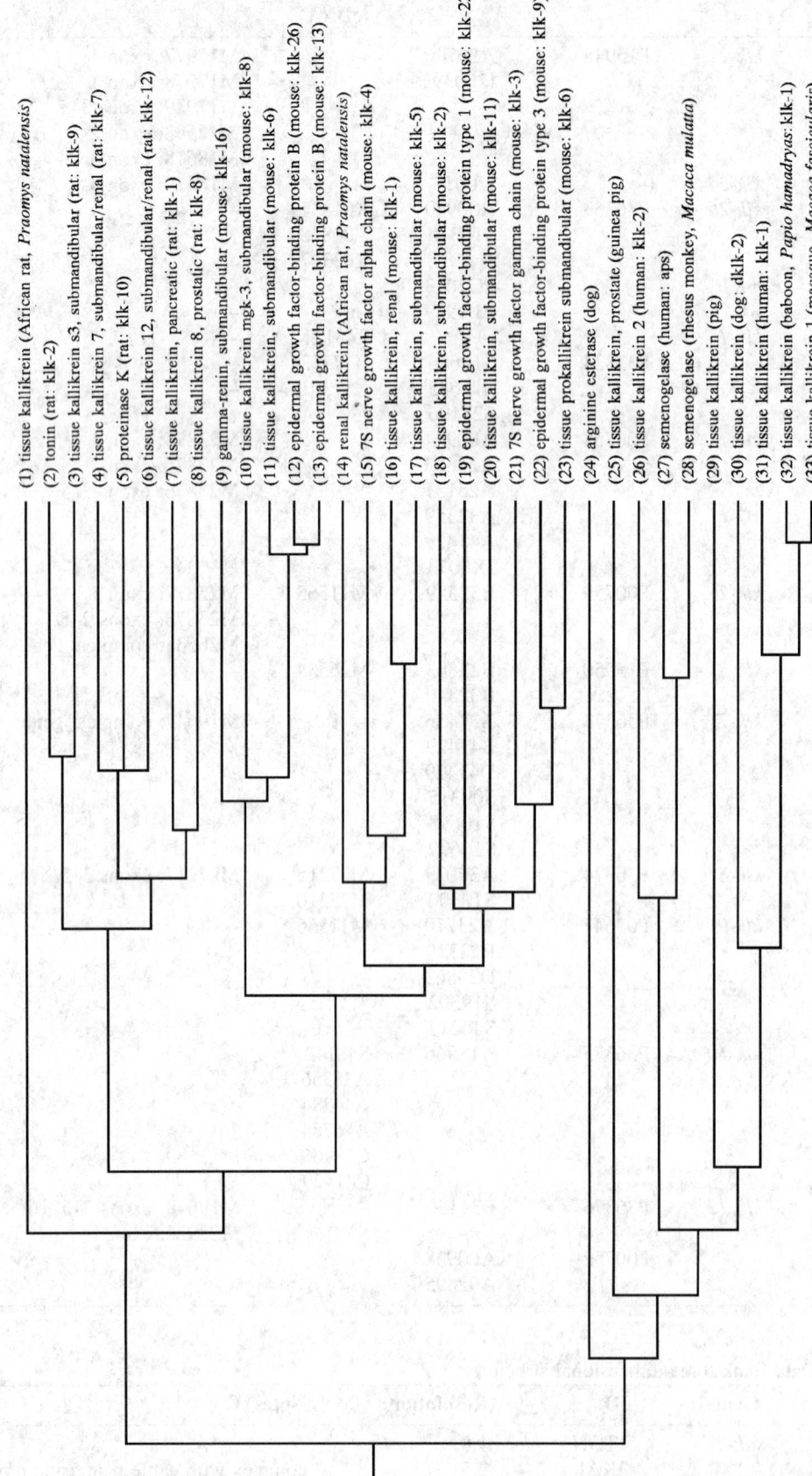

(1) tissue kallikrein (African rat, *Praomys natalensis*)
(2) tonin (rat: klk-2)
(3) tissue kallikrein s3, submandibular (rat: klk-9)
(4) tissue kallikrein 7, submandibular/renal (rat: klk-7)
(5) proteinase K (rat: klk-10)
(6) tissue kallikrein 12, submandibular/renal (rat: klk-12)
(7) tissue kallikrein, pancreatic (rat: klk-1)
(8) tissue kallikrein 8, prostatic (rat: klk-8)
(9) gamma-renin, submandibular (mouse: klk-16)
(10) tissue kallikrein mgk-3, submandibular (mouse: klk-8)
(11) tissue kallikrein, submandibular (mouse: klk-6)
(12) epidermal growth factor-binding protein B (mouse: klk-26)
(13) epidermal growth factor-binding protein B (mouse: klk-13)
(14) renal kallikrein (African rat, *Praomys natalensis*)
(15) 7S nerve growth factor alpha chain (mouse: klk-4)
(16) tissue kallikrein, renal (mouse: klk-1)
(17) tissue kallikrein, submandibular (mouse: klk-5)
(18) tissue kallikrein, submandibular (mouse: klk-2)
(19) epidermal growth factor-binding protein type 1 (mouse: klk-22)
(20) tissue kallikrein, submandibular (mouse: klk-11)
(21) 7S nerve growth factor gamma chain (mouse: klk-3)
(22) epidermal growth factor-binding protein type 3 (mouse: klk-9)
(23) tissue prokallikrein submandibular (mouse: klk-6)
(24) arginine esterase (dog)
(25) tissue kallikrein, prostate (guinea pig)
(26) tissue kallikrein 2 (human: klk-2)
(27) semenogelase (human: aps)
(28) semenogelase (rhesus monkey, *Macaca mulatta*)
(29) tissue kallikrein (pig)
(30) tissue kallikrein (dog: dklk-2)
(31) tissue kallikrein (human: klk-1)
(32) tissue kallikrein (baboon, *Papio hamadryas*: klk-1)
(33) tissue kallikrein 1 (macaque, *Macaca fascicularis*)

*Figure 28.1*   Evolutionary tree for the relatives of tissue kallikrein. The tree was prepared as described in the Introduction (p. xxv).

A large number of proteins closely similar in structure and/or catalytic activity to tissue kallikrein have been characterized, and the interrelationships of these can be confusing. One way of looking at this population of peptidases is through comparison of amino acid sequences (at least for the subset of enzymes that have so far been sequenced). Thus, the known sequences can be aligned, and the degrees of similarity in amino acid sequence can then be depicted graphically in an evolutionary tree (Figure 28.1). It can be seen that amongst the Muridae, the rats and mice appear to have given rise to their own branch of these sequences, and that within this, rat and mouse have each produced a subgroup, the members of which are structurally closer to each other than to proteins found in other species. This is consistent with very rapid divergence of these proteins since the separation of the species. Further evidence in support of this is the finding that all three of the human tissue kallikrein genes are closely associated on chromosome 19 (Riegman *et al.*, 1992), and all of the mouse homologs are located on chromosome 7 (Howles *et al.*, 1984).

Quite distinct from these groups in the typical rodent species is a third major group containing the three human tissue kallikrein-like enzymes, together with all of the sequenced homologs from other primates, dog, pig and guinea pig. Possibly the enzymes of this group may prove to be more typical of tissue kallikreins in animals generally. Of course, one must not assume that two peptidases that are closely similar in overall structure will necessarily be close in catalytic activity. The need for this caveat is demonstrated in the present set of enzymes, for instance, by the difference in substrate specificity between human tissue kallikrein and semenogelase, which may be primarily due to a single amino acid mutation (see Chapter 31).

In this *Handbook*, Chapters 29, 30 and 31 describe the three human enzymes that may be representative of tissue kallikreins in animals generally. Examples of the diverse kallikreins of mouse are described in Chapters 32, 33, 34, 35 and 36. Variants of kallikrein from rat are dealt with in Chapters 37, 38 and 39, and finally, the putatively primitive enzymes from mole (Chapter 40) and fish (Chapter 41) are described.

## References

Howles, P.N., Dickinson, D.P., DiCaprio, L.L., Woodworth-Gutai, M. & Gross, K.W. (1984) Use of a cDNA recombinant for the γ-subunit of mouse nerve growth factor to localize members of this multigene family near the TAM-1 locus on chromosome 7. *Nucleic Acids Res.* **12**, 2791–2805.

Riegman, P.H., Vlietstra, R.J., Suurmeijer, L., Cleutjens, C.B. & Trapman, J. (1992) Characterization of the human kallikrein locus. *Genomics* **14**, 6–11.

# 29. Human tissue kallikrein

## Databanks

*Peptidase classification: clan SA, family S1, MEROPS ID: S01.160*
*NC-IUBMB enzyme classification: EC 3.4.21.35*
*Chemical Abstracts Service registry number: 9001-01-8*
*Databank codes:*

| Species | Gene | SwissProt | PIR | EMBL (cDNA) | EMBL (genomic) |
|---|---|---|---|---|---|
| *Homo sapiens* | *klk-1* | P06870 | A04628 | M12706 | L10038: complete gene |
| | | | A23587 | M25629 | M18157: complete gene |
| | | | A24696 | X13561 | M33105: exon 1 |
| | | | A28678 | | M33106: exon 2 |
| | | | A60248 | | M33107: exon 3 |
| | | | JX0040 | | M33108: exon 4 |
| | | | S05642 | | M33109: exon 5 and complete CDS |

## Name and History

Kallikrein was discovered as a substance in human urine exhibiting hypotensive properties (Frey, 1926; Frey & Kraut, 1928). A similar substance was detected in the pancreas and was given the name **kallikrein** (Gr. *kallikreas*, pancreas) (Kraut *et al.*, 1930). This enzyme was also named **glandular kallikrein, true tissue kallikrein, pancreatic kallikrein, renal kallikrein** and **urinary kallikrein**. Its counterpart has been described in other species including monkey, dog, rabbit, rat, mouse, frog, newt, mole and fish.

## Activity and Specificity

Human tissue kallikrein releases lysyl-bradykinin from kininogens, whereas rat and mouse tissue kallikreins release bradykinin (Kato *et al.*, 1987; Hosoi *et al.*, 1994). Arg is favored in position P1. One cause of the narrow specificity of tissue kallikrein is its pronounced secondary specificity for a bulky, hydrophobic amino acid residue in the P2 position.

The pH optimum is about 8.5–9.0. Assays are most conveniently made with Tos-Arg-OMe by measuring the arginine esterase activity, and with chromogenic substrates (Bz-Pro-Phe-Arg┼NHPhNO$_2$ and DL-Val-Leu-Arg┼NHPhNO$_2$) and fluorogenic substrates (Pro-Phe-Arg┼NHMec and Val-Leu-Arg┼NHMec). Tissue kallikrein activity can also be measured as kininogenase activity with a kinin radioimmunoassay (Shimamoto *et al.*, 1978).

The activity of tissue kallikrein is inhibited by a number of serine proteinase inhibitors of the serpin family, including kallistatin, protein C inhibitor and $\alpha_1$-proteinase inhibitor (Geiger *et al.*, 1981; Zhou *et al.*, 1992; Ecke *et al.*, 1992). The activity of tissue kallikrein is inhibited by aprotinin, PMSF, benzamidine and leupeptin, and weakly inhibited by aprotinin (Trasylol), but not affected by soybean trypsin inhibitor, limabean trypsin inhibitor or ovomucoid trypsin inhibitor.

The esterolytic substrate is available from Sigma (see Appendix 2 for full names and addresses of suppliers), the chromogenic substrates are available from Kabi. The fluorescence substrates are available from Enzyme Systems Products.

## Structural Chemistry

Human tissue kallikrein is an acidic glycoprotein, variably and extensively (~20% of molecular mass) glycosylated. The human tissue kallikrein is synthesized as a zymogen (prokallikrein) with an attached 17 amino acid signal peptide preceding a seven amino acid propeptide that must be cleaved to activate the enzyme (Fukushima *et al.*, 1985). Human prokallikrein can be activated by thermolysin, trypsin and human plasma kallikrein *in vitro*, but the *in vivo* activating enzyme is unknown.

The molecular mass of human tissue kallikrein is 46 kDa and its isoelectric point is 3.9–4.0. Optimum pH values were found to be 8.5–9.0 in both the esterolytic and kininogenase assays. The 1 mg ml$^{-1}$ solution of human tissue kallikrein has an absorbance of 1.5 at 280 nm with a light path of 1 cm.

## Preparation

Essentially homogeneous preparations of human tissue kallikrein have been made from urine (Hial *et al.*, 1974; Shimamoto *et al.*, 1980; Takahashi *et al.*, 1988; Girolami *et al.*, 1989) and submaxillary glands (Fujimoto *et al.*, 1990). The preparation of recombinant human prokallikrein from recombinant baculovirus-infected insect cells has been described by Angermann *et al.* (1992). The yield of purified recombinant protein was 5 mg liter$^{-1}$ of insect cell culture.

## Biological Aspects

The complete amino acid sequence of human urinary prokallikrein has been determined by amino acid sequence analysis of peptide fragments obtained from chemical and enzymologic cleavages of kallikrein and alignment of peptides derived from different proteolytic cleavages (Takahashi *et al.*, 1988; Kellermann *et al.*, 1989; Lu *et al.*, 1989). The gene encoding human tissue kallikrein has been isolated and characterized (Fukushima *et al.*, 1985; Lin *et al.*, 1993).

Tissue kallikrein has been identified in mammals, amphibians and lower vertebrates (Bhoola *et al.*, 1992; Richards *et al.*, 1995, 1996a,b). Tissue kallikrein is present in cells of most tissues, but is particularly abundant in pancreas, kidney, urine and saliva. It has been localized to the epithelial or secretory cells of various ducts, including salivary, sweat, pancreatic, prostatic and intestinal ducts, and the distal nephron. In addition, it has been found in neutrophils, the colonic mucous cells, the trachea, nasal mucosa and the anterior pituitary.

The best-known physiological function of tissue kallikrein is the cleavage of L-kininogen to release the vasoactive kinin peptide (bradykinin or lysyl-bradykinin). Kinins binds to their receptors at target organs and exert a broad array of biological activities including vasodilation, blood pressure reduction, smooth muscle relaxation and contraction, pain induction and inflammation (Bhoola *et al.*, 1992; Madeddu, 1993; Lintz *et al.*, 1995).

The tissue kallikrein–kinin system has been implicated in the pathogenesis of hypertension. Kallikrein excretion in urine was decreased in patients with essential hypertension and increased by sodium depletion (Margolius *et al.*, 1971, 1974a,b). Utah family pedigree studies indicate that a dominant allele expressed as high urinary kallikrein excretion may be associated with a decreased risk of essential hypertension (Berry *et al.*, 1989). By employing molecular genetic approaches, a direct link between alteration of tissue kallikrein gene expression and blood pressure regulation has been demonstrated. Transgenic mice overexpressing human tissue kallikrein under the control of the metallothionein metal response element (MRE) or albumin gene enhancer/promoter are hypotensive throughout their lifespan (Wang *et al.*, 1994; Song *et al.*, 1996). In adult spontaneously hypertensive rats (SHR), somatic gene delivery of human tissue kallikrein by systemic and targeted approaches caused a sustained reduction of blood pressure for up to 8 weeks (Wang *et al.*, 1995; Xiong *et al.*, 1995; Chao *et al.*, 1996, 1997). These findings support the notion that tissue kallikrein plays an important physiological role in maintaining low blood pressure.

## Related Peptidases and Distinguishing Features

The human tissue kallikrein gene subfamily seems to be composed of just three genes, those of tissue kallikrein, human kallikrein hK2 (Chapter 30), and semenogelase (or prostate-specific antigen) (Chapter 31) (Clements, 1994).

Polyclonal and monoclonal antisera against human tissue kallikrein have been described (Shimanoto *et al.*, 1980; Bagshaw *et al.*, 1984; Chao *et al.*, 1985; Kizuki *et al*, 1989), but are not commercially available at the time of writing.

## References

Angermann, A., Rahn, H.P., Hektor, T., Fertig, G. & Kemme, M. (1992) Purification and characterization of human salivary-gland prokallikrein from recombinant baculovirus-infected insect cells. *Eur. J. Biochem.* **206**, 225–233.

Bagshaw, A.F., Bhoola, K.D., Lemon, M.J. & Whicher, J.T. (1984) Development and characterization of a radioimmunoassay to measure human tissue kallikrein in biological fluids. *J. Endocrinol.* **101**, 173–179.

Berry, T.D., Hasstedt, S.J., Hunt, S.C., Wu, L.L., Smith, J.B., Ash, K.O., Kuida, H. & Williams, R.R. (1989) A gene for high urinary kallikrein may protect against hypertension in Utah kindreds. *Hypertension* **13**, 3–8.

Bhoola, K.D., Figueroa, C.D. & Worthy, K. (1992) Bioregulation of kinins: kallikreins, kininogens, and kininases. *Pharmacol. Rev.* **44**, 1–80.

Chao, J., Chao, L., Tillman, D.M., Woodley, C.M. & Margolius, H.S. (1985) Characterization of monoclonal and polyclonal antibodies to human tissue kallikrein. *Hypertension* **7**, 931–937.

Chao, J., Jin, L., Chen, L.M., Chen, V.C. & Chao, L. (1996) Systemic and portal vein delivery of human kallikrein gene reduces blood pressure in hypertensive rats. *Hum. Gene Ther.* **7**, 901–911.

Chao, J., Yang, R., Jin, L., Lin, K.F. & Chao, L. (1997) Kallikrein gene therapy: a new strategy for hypertensive diseases. *Immunopharmacology* **36**, 229–236.

Clements, J.A. (1994) The human kallikrein gene family: a diversity of expression and function. *Mol. Cell. Endocrinol.* **99**, C1–C6.

Ecke, S., Geiger, M., Resch, I., Jerabek, I., Sting, L., Maier, M. & Binder, B.R. (1992) Inhibition of tissue kallikrein by protein C inhibitor. Evidence for identity of protein C inhibitor with the kallikrein binding protein. *J. Biol. Chem.* **267**, 7048–7052.

Frey, E.K. (1926) Zusammenhänge zwischen Herzabeit und Nierentätigkeit. *Arch. Klin. Chir.* **142**, 663.

Frey, E.K. & Kraut, H. (1928) Ein neues Kreislaufhormon und seine Wirkung. *Arch. Exp. Pathol. Pharmakol.* **133**, 1.

Fujimoto, Y., Suzuki, C., Watanabe, Y., Matsuda, Y. & Akihama, S. (1990) Purification and characterization of a kallikrein from human submaxillary glands. *Biochem. Med. Metab. Biol.* **44**, 218–227.

Fukushima, D., Kitamura, N. & Nakanishi, S. (1985) Nucleotide sequence of cloned cDNA for human pancreatic kallikrein. *Biochemistry* **24**, 8037–8043.

Geiger, R., Stuckstedte, U., Clausnitzer, B. & Fritz, H. (1981) Progressive inhibition of human glandular (urinary) kallikrein by human serum and identification of the progressive antikallikrein as alpha 1-antitrypsin (alpha 1-protease inhibitor). *Z. Physiol. Chem.* **362**, 317–325.

Girolami, J.P., Pecher, C., Bascands, J.L., Cabos, G., Adam, A. & Suc, J.M. (1989) Purification of human active urinary kallikrein: comparative inhibition studies of kininogenase and amidolytic activities. *Prep. Biochem.* **19**, 75–90.

Hial, V., Diniz, C.R. & Mares-Guia, M. (1974) Purification and properties of a human urinary kallikrein (kininogenase). *Biochemistry* **13**, 4311–4318.

Hosoi, K., Tsunasawa, S., Kurihara, K., Aoyama, H., Ueha, T., Murai, T. & Sakiyama, F. (1994) Identification of mK1, a true tissue (glandular) kallikrein of mouse submandibular gland: tissue distribution and a comparison of kinin-releasing activity with other submandibular kallikreins. *J. Biochem.* **115**: 137–143.

Kato, H., Nakawishi, E., Enjyoi, K., Hayashi, I., Oh-Ish, S. & Iwanaga, S. (1987) Characterization of serine proteases isolated from rat submaxillary gland: with special reference to the degradation of rat kininogen by these enzymes. *J. Biochem.* **102**, 1389–1404.

Kellermann, J., Lottspeich, F., Geiger, R. & Deutzmann, R. (1989) Human urinary kallikrein: amino acid sequence and carbohydrate attachment sites. *Adv. Exp. Med. Biol.* **247A**, 519–525.

Kizuki, K., Suzuki, S., Aoki, K., Kamada, M., Ikekita, M., Inaba, T. & Moriya, H. (1989) Enzyme-linked immunosorbent assays for human tissue kallikrein and analysis of immunoreactive kallikrein in the plasma by them. *Adv. Exp. Med. Biol.* **247B**, 211–216.

Kraut, H., Frey, E.K. & Werle, E. (1930) Der Nachweis eines Kreislaufhormones in der Pankreasdrüs. *Z. Physiol. Chem.* **189**, 97.

Lin, F.K., Lin, C.H., Chou, C.C., Chen, K., Lu, H.S., Bacheller, W., Herrera, C., Jones, T., Chao, J. & Chao, L. (1993) Molecular cloning and sequence analysis of the monkey and human tissue kallikrein genes. *Biochim. Biophys. Acta* **1173**, 325–328.

Lintz, W., Wiemer, G., Gohlke, P., Unger, T. & Scholkens, B.A. (1995) Contribution of kinins to the cardiovascular actions of angiotensin-converting enzyme inhibitors. *Pharmacol. Rev.* **47**, 25–49.

Lu, H.S., Lin, F.K., Chao, L. & Chao, J. (1989) Human urinary kallikrein. Complete amino acid sequences and sites of glycosylation. *Int. J. Peptide Protein Res.* **33**, 237–249.

Madeddu, P. (1993) Receptor antagonists of bradykinin: a new tool to study the cardiovascular effects of endogenous kinins. *Pharmacol. Rev.* **28**, 107–128.

Margolius, H.S., Geller, R., de Jong, W., Pisano, J.J. & Sjoerdsma, A. (1971) Altered urinary kallikrein excretion in human hypertension. *Lancet* **2**, 1063–1065.

Margolius, H.S., Horwitz, D., Geller, R.G., Alexander, R.W., Gill, J.R., Pisano, J.J. & Keiser, H.R. (1974a) Urinary kallikrein excretion in normal man. Relationships to sodium intake and sodium-retaining steroids. *Circ. Res.* **35**, 812–819.

Margolius, H.S., Horwitz, D., Pisano, J.J. & Keiser, H.R. (1974b) Urinary kallikrein excretion in hypertensive man. Relationships to sodium intake and sodium-retaining steroids. *Circ. Res.* **35**, 820–825.

Richards, G.P., Chao, J., Chung, P. & Chao, L. (1995) Purification and characterization of tissue kallikrein-like proteinases from the black sea bass (*Centropristis striata*) and the southern frog (*Rana berlandieri*). *Comp. Biochem. Physiol. [C]* **111**, 69–82.

Richards, G.P., Chao, J. & Chao, L. (1996a) Tissue kallikreins in evolutionarily diverse vertebrates. *Immunopharmacology* **32**, 94–95.

Richards, G.P., Zintz, C., Chao, J. & Chao, L. (1996b) Purification and characterization of salivary kallikrein from an insectivore (*Scalopus aquaticus*) – substrate specificities, immunoreactivity, and kinetic analyses. *Arch. Biochem. Biophys.* **329**, 104–112.

Shimamoto, K., Ando, T., Nakao, T., Tanaka, S., Sakuma, M. & Miyahara, M. (1978) A sensitive radioimmunoassay method for urinary kinins in man. *J. Lab. Clin. Med.* **91**, 721–728.

Shimamoto, K., Chao, J. & Margolius, H.S. (1980) The radioimmunoassay of human urinary kallikrein and comparisons with kallikrein activity measurements. *J. Clin. Endocrinol. Metab.* **51**, 840–848.

Song, Q., Chao, J. & Chao, L. (1996) Liver-targeted expression of human tissue kallikrein induces hypotension in transgenic mice. *Clin. Exp. Hypertens.* **18**, 975–993.

Takahashi, S., Irie, A. & Miyake, Y. (1988) Primary structure of human urinary prokallikrein. *J. Biochem.* **104**, 22–29.

Wang, C., Chao, L. & Chao, J. (1995) Direct gene delivery of human tissue kallikrein reduces blood pressure in spontaneously hypertensive rats. *J. Clin. Invest.* **95**, 1710–1716.

Wang, J., Xiong, W., Yang, Z., Davis, T., Dewey, M.J., Chao, J. & Chao, L. (1994). Human tissue kallikrein induces hypotension in transgenic mice. *Hypertension* **23**, 236–243.

Xiong, W., Chao, J. & Chao, L. (1995) Muscle delivery of human tissue kallikrein gene reduces blood pressure in spontaneously hypertensive rats. *Hypertension* **25**, 715–719.

Zhou, X.G., Chao, L. & Chao, J. (1992) Kallistatin: a novel human tissue kallikrein inhibitor. Purification, characterization, and reactive center sequence. *J. Biol. Chem.* **267**, 25873–25880.

*Julie Chao*
*Department of Biochemistry and Molecular Biology,*
*Medical University of South Carolina,*
*Charleston, SC 29425-2211, USA*
*Email: chaoj@musc.edu*

# 30. Human tissue kallikrein hK2

## Databanks

*Peptidase classification: clan SA, family S1, MEROPS ID: S01.161*
*NC-IUBMB enzyme classification: EC 3.4.21.35*
*Chemical Abstracts Service registry number: 9001-01-8*
*Databank codes:*

| Species | Gene | SwissProt | PIR | EMBL (cDNA) | EMBL (genomic) |
|---|---|---|---|---|---|
| *Homo sapiens* | *klk-2* | P20151 | A29586 B24696 | S39329 | M18156: 5′ flank |

## Name and History

The gene encoding kallikrein hK2 (initially termed **human glandular kallikrein 1, hGK-1**) was first isolated from a human genomic library by use of a mouse tissue kallikrein cDNA probe (Schedlich *et al.*, 1987). Kallikrein hK2 has also been termed **hKK-3** (Fukushima *et al.*, 1985; Schedlich *et al.*, 1987; Ropers & Pericak-Vance, 1990), but the name **kallikrein hK2** was agreed upon at the Kinin '91 Munich Symposium (Berg *et al.*, 1992). The counterpart of hK2 in other species has not been described.

## Activity and Specificity

It has not yet been possible to characterize the peptidase activity of human kallikrein hK2 directly, but the deduced amino acid sequence is very similar to that of semenogelase (Chapter 31). However, semenogelase lacks an Asp residue that is essential for kallikrein-specific cleavage, whereas this is present in hK2. The covalent complex of hK2 with a serpin, protein C inhibitor, has been isolated (Deperthes *et al.*, 1995).

## Structural Chemistry

The human kallikrein hK2 gene (*klk2*) encodes a unique preproprotein of 261 amino acids. Mature, active hK2 and semenogelase are both 237 residue proteinases. The sequence of the mature hK2 is 66% identical to that of human tissue kallikrein, and 78% identical to semenogelase. Molecular modeling shows that the structures of hK2 and semenogelase are closely similar. The molecular mass of naturally occurring hK2 is 22 kDa, as determined by immunoblotting of samples from human prostate gland and seminal plasma (Saedi *et al.*, 1995).

## Preparation

Overexpression of the entire hK2 preprotein has been achieved in *E. coli*, and the protein was purified to near homogeneity by preparative SDS-PAGE (Saedi *et al.*, 1995). Kallikrein hK2 has not been isolated from natural sources.

## Biological Aspects

A northern blot analysis with a synthetic oligonucleotide primer complementary to human semenogelase mRNA showed that the human kallikrein hK2 gene, *klk2*, is expressed at a high level in the human prostate gland (Chapdelaine *et al.*, 1988). *In situ* hybridization with a specific oligonucleotide probe that can differentiate hK2 mRNA from that of semenogelase demonstrated that the hK2 mRNA is located in the prostatic epithelium. Thus, it could be concluded that the mRNAs for hK2 and semenogelase are colocalized in the prostatic epithelia (Young *et al.*, 1992). The expression of both genes in the prostate is under androgen control (Murtha *et al.*, 1993). Levels of hK2 mRNA were increased significantly by mibolerone and dihydrotestosterone, but not by dexamethasone or diethylstilbestrol (Young *et al.*, 1992). In LNCaP prostatic cancer cells, epidermal growth factor was found to reduce the cellular secretion of hK2 (Henttu & Vihko, 1993).

Although specific assays are not yet available for the activity of hK2, the striking similarities of hK2 to semenogelase, including selective expression in the prostate, suggest that hK2 may also prove to be a useful clinical marker in relation to prostate cancer. Kallikrein hK2 may facilitate the growth and spread of cancerous prostatic cells.

The detection of the complex of kallikrein hK2 with protein C inhibitor in seminal plasma with monoclonal antibodies suggests that protein C inhibitor may regulate the activity of hK2 in seminal plasma (Deperthes *et al.*, 1995).

Genes for three human tissue kallikreins are clustered together on chromosome 19q13.2–q13.4 (Qin *et al.*, 1991). These are the genes for hK2 (*klk2*), semenogelase (*aps*) and tissue kallikrein itself (*klk1*). The *aps* and *klk2* genes are separated by 12 kb, and the *klk1* and *aps* genes by 31 kb (Riegman *et al.*, 1989, 1992).

## Distinguishing Features

Polyclonal and monoclonal antisera against the recombinant hK2-fusion protein have been described (Deperthes *et al.*, 1995; Saedi *et al.*, 1995), but are not commercially available at the time of writing.

## References

Berg, T., Bradshaw, R.A., Carretero, O.A., Chao, J., Chao, L., Clements, J.A., Fahnestock, M., Fritz, H., Gauthier, F., MacDonald, R.J., Margolius, H.S., Morris, B.J., Richards, R.I. & Scicli, A.G. (1992) A common nomenclature for members of the tissue (glandular) kallikrein gene families. *Agents Action* **38I**, 19–25.

Chapdelaine, P., Paradis, G., Tremblay, R.R. & Dube, J.Y. (1988) High level of expression in the prostate of a human glandular kallikrein mRNA related to prostate-specific antigen. *FEBS Lett.* **236**, 205–208.

Deperthes, D., Chapdelaine, P., Tremblay, R.R., Brunet, C., Berton, J., Hebert, J., Lazure, C. & Dube, J.Y. (1995) Isolation of prostatic kallikrein hK2, also known as hGK-1, in human seminal plasma. *Biochim. Biophys. Acta* **1245**, 311–316.

Fukushima, D., Kitamura, N. & Nakanishi, S. (1985) Nucleotide sequence of cloned cDNA for human pancreatic kallikrein. *Biochemistry* **24**, 8037–8043.

Henttu, P. & Vihko, P. (1993) Growth factor regulation of gene expression in the human prostatic carcinoma cell line LNCaP. *Cancer Res.* **53**, 1051–1058.

Murtha, P., Tindall, D.J. & Young, C.Y. (1993) Androgen induction of a human prostate-specific kallikrein, hKLK2: characterization of an androgen response element in the 5' promoter region of the gene. *Biochemistry* **32**, 6459–6464.

Qin, H., Kemp, J., Yip, M.Y., Lam-Po-Tang, P.R. & Morris, B.J. (1991) Localization of human glandular kallikrein-1 gene to chromosome 19q13.3–13.4 by *in situ* hybridization. *Hum. Hered.* **41**, 222–226.

Riegman, P.H., Vlietstra, R.J., Klaassen, P., van der Korput, J.A., Geurts van Kessel, A., Romijn, J.C. & Trapman, J. (1989) The prostate-specific antigen gene and the human glandular kallikrein-1 gene are tandemly located on chromosome 19. *FEBS Lett.* **247**, 123–126.

Riegman, P.H., Vlietstra, R.J., Suurmeijer, L., Cleutjens, C.B. & Trapman, J. (1992) Characterization of the human kallikrein locus. *Genomics* **14**, 6–11.

Ropers, H.H. & Pericak-Vance, M.A. (1990) Report of the committee on the genetic constitution of chromosome 19. *Cytogenet. Cell Genet.* **55**, 218–228.

Saedi, M.S., Cass, M.M., Goel, A.S., Grauer, L., Hogen, K.L., Okaneya, T., Griffin, B.Y., Klee, G.G., Young, C.Y. & Tindall, D.J. (1995) Overexpression of a human prostate-specific glandular kallikrein, hK2, in *E. coli* and generation of antibodies. *Mol. Cell. Endocrinol.* **109**, 237–241.

Schedlich, L.J., Bennetts, B.H. & Morris, B.J. (1987) Primary structure of a human glandular kallikrein gene. *DNA* **6**, 429–437.

Young, C.Y., Andrews, P.E., Montgomery, B.T. & Tindall, D.J. (1992) Tissue-specific and hormonal regulation of human prostate-specific glandular kallikrein. *Biochemistry* **31**, 818–824.

*Julie Chao*
*Department of Biochemistry and Molecular Biology,*
*Medical University of South Carolina,*
*Charleston, SC 29425-2211, USA*
*Email: chaoj@musc.edu*

# 31. Semenogelase or prostate-specific antigen

## Databanks

*Peptidase classification: clan SA, family S1, MEROPS ID: S01.162*
*NC-IUBMB enzyme classification: EC 3.4.21.77*
*Databank codes:*

| Species | SwissProt | PIR | EMBL (cDNA) | EMBL (genomic) |
|---|---|---|---|---|
| *Homo sapiens* | P07288 | A32297 | U17040 X05332 X07730 | M24543: complete gene M27274: complete gene X13940: exon 1 X13941: exon 2 X13942: exon 3 X13943: exon 4 X13944: exon 5 X14810: complete gene |
| *Macaca mulatta* | P33619 | S34239 S35711 | X73560 | – |

## Name and History

A protein generally termed **prostate-specific antigen (PSA)** has been shown to be a serine proteinase with chymotrypsin-like activities (Ban *et al.*, 1984). The same protein has also been identified as the **insulin-like growth factor-binding protein 3 proteinase (IGFBP-3 proteinase)** of seminal plasma (Cohen *et al.*, 1992). There is at least one other 'prostate-specific' peptidase (see Chapter 492), and taking account of a possible biological function of the protein, NC-IUBMB have recommended the name **semenogelase** for this enzyme.

Counterparts of semenogelase in rhesus monkey, dog and guinea pig have been described (Dunbar & Bradshaw, 1985, 1987; Chapdelaine *et al.*, 1991; Gauthier *et al.*, 1993). Using Southern blot analyses, genes related to human semenogelase have been detected in several primate species, including chimpanzee, orangutan, gorilla, macaque and rhesus monkey, but not in other mammalian species, including rabbit, cow, pig, dog, rat or mouse (Karr *et al.*, 1995).

## Activity and Specificity

Semenogelase is a serine proteinase manifesting restricted chymotrypsin-like activity. A specific activity of $9.21\,\mu M\,min^{-1}\,mg^{-1}$ of semenogelase was determined with the chromogenic substrate Suc-Ala-Ala-Pro-Phe$+$NHPhNO$_2$, $K_m$ and $k_{cat}$ values being 15.3 mM and $0.075\,s^{-1}$, respectively (Watt *et al.*, 1986).

At pH 7.8, semenogelase hydrolyzed insulin A and B chains, recombinant interleukin 2, and to a lesser extent, gelatin, myoglobin, ovalbumin and fibrinogen. The P1 residues recognized in the cleavage sites in these proteins were either hydrophobic (Tyr, Leu, Val, Phe) or basic (His, Lys, Arg). Semenogelase converts the single-chain proform of urokinase-type plasminogen activator (Chapter 57) to an active two-chain form *in vitro* (Yoshida *et al.*, 1995).

Proteinase inhibitors such as PMSF, Tos-Phe-CH$_2$Cl, aprotinin, leupeptin, soybean trypsin inhibitor, as well as Zn$^{2+}$ and spermidine, were found to be effective inhibitors of semenogelase (Watt *et al.*, 1986). Semenogelase was stable at $-20°C$ for 6 months.

Semenogelase interacts with at least two protein inhibitors of the serpin family: $\alpha_1$-proteinase inhibitor and protein C inhibitor, both *in vitro* and in human semen (Christensson & Lilja, 1994). A small amount of semenogelase is complexed to protein C inhibitor in semen. Semenogelase complexed to $\alpha_1$-proteinase inhibitor constitutes the predominant molecular form of serum semenogelase, although complex formation is slow between the purified proteins *in vitro*. Semenogelase also forms stable complexes with $\alpha_2$-macroglobulin *in vitro*, but the *in vivo* significance of this is presently unclear.

## Structural Chemistry

The mature form of semenogelase is a single-chain glycoprotein of 237 amino acids, with a calculated $M_r$ of 26 496. Semenogelase is synthesized as a proenzyme with extensive similarity to those of tissue kallikreins, but the mechanism of activation of the proenzyme has not been defined. An *N*-linked carbohydrate side-chain is predicted at Asn45, and *O*-linked carbohydrate side-chains are possibly attached to Ser69, Thr70 and Ser71.

Two heparin-binding sites are predicted to exist on the surface of the semenogelase molecule, and these could explain the relatively rapid complex formation with protein C inhibitor as compared to $\alpha_1$-proteinase inhibitor (Villoutreix et al., 1996).

The 1 mg ml$^{-1}$ solution of semenogelase has an absorbance of 1.61 at 280 nm with a light path of 1 cm. The molecular mass of semenogelase is about 33 kDa in SDS-PAGE under reducing or nonreducing conditions, and its pI is 6.9–7.2 (Zhang et al., 1995).

## Preparation

Essentially homogeneous preparations of semenogelase have been prepared from human seminal fluid (Ban et al., 1984; Zhang et al., 1995). Semenogelase is selectively expressed in prostate, the mRNA being localized within the glandular epithelium (Young et al., 1991).

## Biological Aspects

Semenogelase is one of the most abundant prostate-derived proteins in the seminal fluid. In semen, approximately two-thirds of the semenogelase molecules are enzymatically active, and the remainder are inactive as a result of internal proteolytic cleavage. Semenogelase is mainly responsible for gel dissolution in freshly ejaculated semen by proteolysis of the major gel-forming proteins semenogelins I and II and fibronectin.

In vitro studies suggest that the proteolytic activity of semenogelase may play a role in the growth and spread of cancerous prostatic cells. In patients with carcinoma of the prostate, the serum semenogelase level increases, and analysis of this level is used both for diagnosing and monitoring patients with carcinoma of the prostate.

Synthesis of semenogelase mRNA is induced by the natural androgen dihydrotestosterone, as well as by mibolerone, but not by the synthetic glucocorticoid dexamethasone or the synthetic estrogen diethylstilbestrol (Young et al., 1991). Regulation of expression of the gene for semenogelase (like that for human kallikrein hK2) is mediated not only by androgens, however, but also by a number of autocrine and paracrine factors, suggesting that the control mechanisms for expression of these genes are complex and multifaceted. Such factors may also be integral to the growth and differentiation of prostate cells. In LNCaP prostatic cancer cells, epidermal growth factor was found to reduce the secretion of semenogelase (Henttu & Vihko, 1993), and 13-cis-retinoic acid inhibited not only secretion, but expression at both mRNA and protein levels (Dahiya et al., 1994).

In situ hybridization and Southern blot analysis of a human × mouse somatic cell hybrid panel, mapped the human semenogelase gene to 19q13, where it occurs in a cluster with the two closely related genes for tissue kallikrein (klk1) (Chapter 29) and human kallikrein hK2 (klk2) (Chapter 30) (Sutherland et al., 1988; Riegman et al., 1992; Clements, 1994).

## Distinguishing Features

Semenogelase is a tissue-specific serine proteinase similar in structure to the trypsin-like tissue kallikreins, but distinguished by additional chymotrypsin-like activity. It is produced by prostate cells, and provides an excellent serum marker for prostatic cancer.

Polyclonal and monoclonal antisera against semenogelase have been described (Chu et al., 1989; Chen et al., 1992). Some of the semenogelase monoclonal antisera cross-react with identical affinities to recombinant hK2 (Lovgren et al., 1995). Commercial assay kits for measuring semenogelase in serum are available (TANDEM-R PSA, two-site monoclonal radioimmunometric assay, Hybritech, Inc.; IRMA-Count PSA, Diagnostic Products Corp.; ELSA-PSA and PROS-CHECK PSA Polyclonal Radioimmunoassay, Yang Laboratories (see Appendix 2 for full names and addresses of suppliers).

## Further Reading

Excellent reviews on semenogelase have been written by Lilja (1993) and Malm & Lilja (1995).

## References

Ban, Y., Wang, M.C., Watt, W.K., Loor, R. & Chu, T.M. (1984) The proteolytic activity of human prostate-specific antigen. Biochem. Biophys. Res. Commun. **123**, 482–488.

Chapdelaine, P., Gauthier, E., Ho-Kim, M.A., Bissonnette, L., Tremblay, R.R. & Dube, J.Y. (1991) Characterization and expression of the prostatic arginine esterase gene, a canine glandular kallikrein. DNA Cell Biol. **10**, 49–59.

Chen, K.W., Tsai, Y.L., Chen, W.S. & Lee, C.Y. (1992) Enzyme immunoassay for prostate-specific antigen and its diagnostic application in prostate cancer. J. Formosan Med. Assoc. **91**, 1039–1043.

Christensson, A. & Lilja, H. (1994) Complex formation between protein C inhibitor and prostate-specific antigen in vitro and in human semen. Eur. J. Biochem. **220**, 45–53.

Chu, T.M., Kawinski, E., Hibi, N., Croghan, G., Wiley, J., Killian, C.S. & Corral, D. (1989) Prostate-specific antigenic domain of human prostate specific antigen identified with monoclonal antibodies. J. Urol. **141**, 152–156.

Clements, J.A. (1994) The human kallikrein gene family: a diversity of expression and function. Mol. Cell. Endocrinol. **99**, C1–C6.

Cohen, P., Graves, H.C., Peehl, D.M., Kamarei, M., Giudice, L.C. & Rosenfeld, R.G. (1992) Prostate-specific antigen (PSA) is an insulin-like growth factor binding protein-3 protease found in seminal plasma. J. Clin. Endocrinol. Metab. **75**, 1046–1053.

Dahiya, R., Park, H.D., Cusick, J., Vessella, R.L., Fournier, G. & Narayan, P. (1994) Inhibition of tumorigenic potential and prostate-specific antigen expression in LNCaP human prostate cancer cell line by 13-cis-retinoic acid. Int. J. Cancer **59**, 126–132.

Dunbar, J.C. & Bradshaw, R.A. (1985) Nerve growth factor biosynthesis: isolation and characterization of a guinea pig prostate kallikrein. J. Cell. Biochem. **29**, 309–319.

Dunbar, J.C. & Bradshaw, R.A. (1987) Amino acid sequence of guinea pig prostate kallikrein. Biochemistry **26**, 3471–3478.

Gauthier, E.R., Chapdelaine, P., Tremblay, R.R. & Dube, J.Y. (1993) Characterization of rhesus monkey prostate specific antigen cDNA. Biochim. Biophys. Acta **1174**, 207–210.

Henttu, P. & Vihko, P. (1993) Growth factor regulation of gene expression in the human prostatic carcinoma cell line LNCaP. Cancer Res. **53**, 1051–1058.

Karr, J.F., Kantor, J.A., Hand, P.H., Eggensperger, D.L. & Schlom, J. (1995) The presence of prostate-specific antigen-related genes in primates and the expression of recombinant human prostate-specific antigen in a transfected murine cell line. *Cancer Res.* **55**, 2455–2462.

Lilja, H. (1993) Structure, function, and regulation of the enzyme activity of prostate-specific antigen. *World J. Urol.* **11**, 188–191.

Lovgren, J., Piironen, T., Overmo, C., Dowell, B., Karp, M., Pettersson, K., Lilja, H. & Lundwall, A. (1995) Production of recombinant PSA and HK2 and analysis of their immunologic cross-reactivity. *Biochem. Biophys. Res. Commun.* **213**, 888–895.

Malm, J. & Lilja, H. (1995) Biochemistry of prostate-specific antigen, PSA. *Scand. J. Clin. Lab. Invest.* **221**, 15–22.

Riegman, P.H., Vlietstra, R.J., Suurmeijer, L., Cleutjens, C.B. & Trapman, J. (1992) Characterization of the human kallikrein locus. *Genomics* **14**, 6–11.

Sutherland, G.R., Baker, E., Hyland, V.J., Callen, D.F., Close, J.A., Tregear, G.W., Evans, B.A. & Richards, R.I. (1988) Human prostate-specific antigen (APS) is a member of the glandular kallikrein gene family at 19q13. *Cytogenet. Cell Genet.* **48**, 205–207.

Villoutreix, B.O., Lilja, H., Pettersson, K., Lovgren, T. & Teleman, O. (1996) Structural investigation of the alpha-1-antichymotrypsin–prostate-specific antigen complex by comparative model building. *Protein Sci.* **5**, 836–851.

Watt, K.W., Lee, P.J., M'Timkulu, T., Chan, W.P. & Loor, R. (1986) Human prostate-specific antigen: structural and functional similarity with serine proteases. *Proc. Natl Acad. Sci. USA* **83**, 3166–3170.

Yoshida, E., Ohmura, S., Sugiki, M., Maruyama, M. & Mihara, H. (1995) Prostate-specific antigen activates single-chain urokinase-type plasminogen activator. *Int. J. Cancer* **63**, 863–865.

Young, C.Y., Montgomery, B.T., Andrews, P.E., Qui, S.D., Bilhartz, D.L. & Tindall, D.J. (1991) Hormonal regulation of prostate-specific antigen messenger RNA in human prostatic adenocarcinoma cell line LNCaP. *Cancer Res.* **51**, 3748–3752.

Zhang, W.M., Leinonen, J., Kalkkinen, N., Dowell, B. & Stenman, U.H. (1995) Purification and characterization of different molecular forms of prostate-specific antigen in human seminal fluid. *Clin. Chem.* **41**, 1567–1573.

*Julie Chao*
*Department of Biochemistry and Molecular Biology,*
*Medical University of South Carolina,*
*Charleston, SC 29425-2211, USA*
*Email: chaoj@musc.edu*

# 32. *Mouse proteinase A*

## Databanks

*Peptidase classification: clan SA, family S1, MEROPS ID: S01.164*
*NC-IUBMB enzyme classification: none*
*Chemical Abstracts Service registry number: 82047-85-6*
*Databank codes:*

| Species | SwissProt | PIR | EMBL (cDNA) | EMBL (genomic) |
| --- | --- | --- | --- | --- |
| *Mus musculus* | P00755 | A00941 | – | J00390: complete gene and pseudogene<br>V00829: complete gene |

## Name and History

**Proteinase A**, so far known only from mouse, has also been termed **β-nerve growth factor endopeptidase** (Hosoi *et al.*, 1983). Proteinase A and true tissue kallikrein are very closely related enzymes isolated from mouse submandibular gland. It is postulated that proteinase A is a member of a larger family of similar enzymes, all of which are involved in the processing of precursors to polypeptide hormones and growth factors.

## Activity and Specificity

Proteinase A has significant kininogenase activity toward purified mouse L-kininogen, which is comparable to the kininogenase activities of mouse true tissue kallikrein from other tissues (Bothwell *et al.*, 1979). The enzyme hydrolyzes Bz-Arg-NHPhNO$_2$, and activity on this substrate is slightly increased by zinc ions. The pH optimum for hydrolysis of the arginine nitroanilide is 8.5, whereas that for action on β-nerve growth factor is 6.0. Proteinase A is inhibited by pancreatic trypsin inhibitor but not soybean trypsin inhibitor (Wilson & Shooter, 1979).

## Structural Chemistry

The molecular mass of proteinase A is 26 kDa and the pI is 5.8 (Tanaka *et al*., 1984).

## Preparation

The essentially pure preparation of proteinase A from mouse submaxillary gland was described by Wilson & Shooter (1979).

## Biological Aspects

The immunoreactive proteinase A was localized in the cytoplasm of granular convoluted tubular cells in the mouse submandibular gland (Tanaka *et al*., 1984).

The quantity of proteinase A in the submandibular gland was reduced in hypophysectomized mice, and this reduction was reversed by injections of androgen, progesterone, glucocorticoid or thyroid hormones. Thyroid hormones acted synergistically with glucocorticoids or progesterone to increase the level of proteinase A in the submandibular gland, but androgen and thyroid hormones had additive effects (Hosoi *et al*., 1992). Estrogen was not synergistic with thyroid hormone, and it inhibited the effect of glucocorticoid; it also partially blocked the action of androgen on induction of proteinase activity. Therefore, progesterone, androgens, thyroid hormone or glucocorticoid hormones are capable of regulating expression of proteinase A, and their actions can be modulated by other pituitary-dependent hormones. In addition, estrogen does not regulate proteinase A expression, and it has an inhibitory effect on the inductive action of other hormones (Maruyama *et al*., 1993).

During postnatal development, proteinase A appears on the 22nd day of life. At maturity, the activity of proteinase A is higher in males than in females (Hosoi *et al*., 1990).

## Distinguishing Features

Polyclonal antiserum against proteinase A has been produced that does not cross-react with proteinase F (Tanaka *et al*., 1984).

## References

Bothwell, M.A., Wilson, W.H. & Shooter, E.M. (1979) The relationship between glandular kallikrein and growth factor-processing proteases of mouse submaxillary gland. *J. Biol. Chem.* **254**, 7287–7294.

Hosoi, K., Kamiyama, S., Atsumi, T., Nemoto, A., Tanaka, I. & Ueha, T. (1983) Characterization of two esteroproteases from the male mouse submandibular gland. *Arch. Oral Biol.* **28**, 5–11.

Hosoi, K., Ueha, T., Fukuuchi, H., Kohno, M. & Takahashi, T. (1990) Reversal of relative proteinase F activity and onset of androgen-dependent proteinases in the submandibular gland of postnatal mice. *Biochem. Int.* **22**, 179–187.

Hosoi, K., Maruyama, S., Ueha, T., Sato, S. & Gresik, E.W. (1992) Additive and/or synergistic effects of 5 alpha-dihydrotestosterone, dexamethasone, and triiodo-L-thyronine on induction of proteinases and epidermal growth factor in the submandibular gland of hypophysectomized mice. *Endocrinology* **130**, 1044–1055.

Maruyama, S., Hosoi, K., Ueha, T., Tajima, M., Sato, S. & Gresik, E.W. (1993) Effects of female hormones and 3,5,3'-triiodothyronine or dexamethasone on induction of epidermal growth factor and proteinases F, D, A, and P in the submandibular glands of hypophysectomized male mice. *Endocrinology* **133**, 1051–1060.

Tanaka, S., Hosoi, K., Tanaka, I., Kumegawa, M. & Ueha, T. (1984) Immunocytochemical study of proteinase F in the mouse submandibular gland. *J. Histochem. Cytochem.* **32**, 585–592.

Wilson, W.H. & Shooter, E.M. (1979) Structural modification of the NH₂ terminus of nerve growth factor. Purification and characterization of beta-nerve growth factor endopeptidase. *J. Biol. Chem.* **254**, 6002–6009.

*Julie Chao*
*Department of Biochemistry and Molecular Biology,*
*Medical University of South Carolina,*
*Charleston, SC 29425-2211, USA*
*Email: chaoj@musc.edu*

# 33. γ-Renin

## Databanks

*Peptidase classification: clan SA, family S1, MEROPS ID: S01.163*
*NC-IUBMB enzyme classification: EC 3.4.21.54*
*Chemical Abstracts Service registry number: 85270-20-8*

*Databank codes:*

| Species | SwissProt | PIR | EMBL (cDNA) | EMBL (genomic) |
|---|---|---|---|---|
| *Mus musculus* | P04071 | A28062 I70032 I70033 | J03877 | M18615: exon 3 |

### Name and History

Mouse *γ-renin* is obtained from male mouse submandibular gland (Poe *et al*., 1983). On the basis of molecular mass, arginyl esterase activity, inhibition profile and partial amino acid sequence, γ-renin has been included in a highly homologous group of mouse glandular or tissue kallikreins (see Chapter 28).

### Activity and Specificity

Unlike most of the kallikrein family members, which cleave at a basic residue, γ-renin cleaves at a Leu-Leu bond in a similar fashion to renin (Poe *et al*., 1983). Like renin, γ-renin cleaves synthetic renin substrate tetradecapeptide, Asp-Arg-Val-Tyr-Ile-His-Pro-Phe-His-Leu┼Leu-Tyr-Tyr-Ser, to give angiotensin I (Poe *et al*., 1983). Like γ-NGF, γ-renin can hydrolyze $Bz\text{-}Arg\text{-}NHPhNO_2$ at an appreciable rate, and is inhibited by benzamidine and DFP.

### Structural Chemistry

γ-Renin is synthesized as a larger precursor, from which presumably a leader sequence of 17 amino acids and a short propeptide are subsequently cleaved in a manner analogous to other glandular kallikreins (Drinkwater *et al*., 1988). SDS-PAGE of γ-renin under nonreducing conditions showed a band of 26 kDa, but under reducing conditions, no band larger than 10 kDa was detected (Poe *et al*., 1983).

### Preparation

Essentially homogeneous γ-renin has been prepared from male mouse submandibular gland (Poe *et al*., 1983). The $1\,\mathrm{mg\,ml^{-1}}$ solution of γ-renin has an absorbance of 0.75 at 280 nm with a light path of 1 cm.

### Biological Aspects

The gene (*mGK-16* or *KLK-16*) and a cDNA encoding γ-renin have been identified and sequenced (Drinkwater *et al*., 1988). The mRNA of γ-renin was detected in mouse submandibular gland at much higher levels in the male tissues (Drinkwater *et al*., 1988).

Based on the fact that γ-renin cleaves synthetic renin substrate tetradecapeptides to release angiotensin I, it has been proposed that it may participate in the renin–angiotensin system.

### References

Drinkwater, C.C., Evans, B.A. & Richards, R.I. (1988) Sequence and expression of mouse gamma-renin. *J. Biol. Chem.* **263**, 8565–8568.

Poe, M., Wu, J.K., Florance, J.R., Rodkey, J.A., Bennett, C.D. & Hoogsteen, K. (1983) Purification and properties of renin and gamma-renin from the mouse submaxillary gland. *J. Biol. Chem.* **258**, 2209–2216.

*Julie Chao*
*Department of Biochemistry and Molecular Biology,*
*Medical University of South Carolina,*
*Charleston, SC 29425-2211, USA*
*Email: chaoj@musc.edu*

# 34. Epidermal growth factor-binding protein

### Databanks

*Peptidase classification: clan SA, family S1, MEROPS ID: S01.169*
*NC-IUBMB enzyme classification: none*

*Databank codes:*

| Species | Gene | SwissProt | PIR | EMBL (cDNA) | EMBL (genomic) |
|---------|------|-----------|-----|-------------|----------------|
| *Mus musculus* | *klk-9* | P15949 | A27120 A29745 C29746 I70015 | M17962 | M17983: exon 1 M17984: exon 2 M17985: exons 3–5 M18588: exon 2 M18608: exon 3 |
| *Mus musculus* | *klk-13* | P36368 | A41020 B29746 PC2014 S18674 | X58628 | – |
| *Mus musculus* | *klk-22* | P15948 | A29746 I70039 | – | M17977: exon 1 M17978: exon 1 M17979: exons 3–5 M18598: exon 2 M18618: exon 3 |
| *Mus musculus* | *klk-26* | P36369 | A00940 S19310 | K01831 V00828 X63327 | – |

## Name and History

Epidermal growth factor (EGF) may be isolated from the mouse submandibular gland as a stable, high molecular weight complex containing arginine-specific esterase activity (Taylor *et al*., 1970). *Epidermal growth factor-binding protein (EGF-BP)*, discovered as a member of the mouse glandular kallikrein family, complexes with the biosynthetic precursor to EGF, and converts it to EGF (Taylor *et al*., 1974; Frey *et al*., 1979; Hirata & Orth, 1979). EGF-BP was also named as *high molecular weight epidermal growth factor (HMW-EGF), epidermal growth factor urogastrone* and *mGK-9* (Richards *et al*., 1982).

## Activity and Specificity

EGF-BP is the specific intracellular processing enzyme that converts the precursor of EGF. In so doing, it forms the stable high molecular weight complex of EGF (Frey *et al*., 1979; Hirata & Orth, 1979). EGF-BP also cleaves mouse L-kininogen, releasing kinin peptide, and it has plasminogen activator activity (Hiramatsu *et al*., 1982).

EGF-BP is easily assayed with Tos-Arg-OMe, Bz-Arg-OEt or D-Val-Leu-Arg-NHPhNO$_2$, and its optimum pH is 8.5.

## Structural Chemistry

Under reducing conditions, EGF-BP is seen as a two-chain protein of 7 kDa and 10 kDa (Server & Shooter, 1976).

## Preparation

Essentially homogeneous preparations of EGF-BP have been obtained from mouse submandibular gland (Taylor *et al*., 1974: Frey *et al*., 1979; Hirata & Orth, 1979).

EGF-BP has been expressed in *E. coli*; the system allows rapid isolation of EGF-BP with yields of approximately 1 mg of purified protein from 10 g of initial cell paste (Blaber *et al*., 1990).

## Biological Aspects

EGF-BP forms SDS-stable complexes with the proteinase inhibitor protease nexin 2 (the amyloid-precursor protein: Van Nostrand *et al*., 1989), and its biological activity may be regulated by protease nexin 2 (Kim *et al*., 1992). There is little evidence of the existence of other natural inhibitors.

The mouse kallikrein family members are located on mouse chromosome 7 (Howles *et al*., 1984).

## Distinguishing Features

EGF-BP, nerve growth factor $\gamma$ subunit (Chapter 35) and $\beta$-nerve growth factor endopeptidase (Chapter 32) have similar amino acid composition, molecular mass and immunological properties. Polyclonal antisera against EGF-BP have been described (Bothwell *et al*., 1979), but are not commercially available at the time of writing.

## References

Blaber, M., Isackson, P.J. & Bradshaw, R.A. (1990) The characterization of recombinant mouse glandular kallikreins from *E. coli*. *Proteins* **7**, 280–290.

Bothwell, M.A., Wilson, W.H. & Shooter, E.M. (1979) The relationship between glandular kallikrein and growth factor-processing proteases of mouse submaxillary gland. *J. Biol. Chem.* **254**, 7287–7294.

Frey, P., Forand, R., Maciag, T. & Shooter, E.M. (1979) The biosynthetic precursor of epidermal growth factor and the mechanism of its processing. *Proc. Natl Acad. Sci. USA* **76**, 6294–6298.

Hiramatsu, M., Hatakeyama, K., Kumegawa, M. & Minami, N. (1982) Plasminogen activator activity in the subunits of mouse submandibular gland nerve-growth factor and epidermal growth factor. *Arch. Oral Biol.* **27**, 517–518.

Hirata, Y. & Orth, D.N. (1979) Conversion of high molecular weight human epidermal growth factor (hEGF)/urogastrone (UG) to small molecular weight hEGF/UG by mouse EGF-associated arginine esterase. *J. Clin. Endocrinol. Metab.* **49**, 481–483.

Howles, P.N., Dickinson, D.P., DiCaprio, L.L., Woodworth-Gutai, M. & Gross, K.W. (1984) Use of a cDNA recombinant for the gamma-subunit of mouse nerve growth factor to localize members of this multigene family near the TAM-1 locus on chromosome 7. *Nucleic Acids Res.* **12**, 2791–2805.

Kim, T., Choi, B.H., Choe, W., Kim, R.C., Van Nostrand, W., Wagner, S. & Cunningham, D. (1992) Expression of protease nexin-II in human dorsal root ganglia. A correlative immunocytochemical and *in situ* hybridization study. *Mol. Chem. Neuropathol.* **16**, 225–239.

Richards, R.I., Catanzaro, D.F., Mason, A.J., Morris, B.J., Baxter, J.D. & Shrine, J. (1982) Mouse glandular kallikrein genes. *J. Biol. Chem.* **257**, 2758–2762.

Server, A.C. & Shooter, E.M. (1976) Comparison of the arginine esteropeptidases associated with the nerve and epidermal growth factor. *J. Biol. Chem.* **251**, 165–173.

Server, A.C., Sutter, A. & Shooter, E.M. (1976) Modification of the epidermal growth factor affecting the stability of its high molecular weight complex. *J. Biol. Chem.* **251**, 1188–1196.

Taylor, J.M., Cohen, S. & Mitchell, W.M. (1970) Epidermal growth factor: high and low molecular weight forms. *Proc. Natl Acad. Sci. USA* **67**, 164–171.

Taylor, J.M., Mitchell, W.M. & Cohen, S. (1974) Characterization of the high molecular weight form of epidermal growth factor. *J. Biol. Chem.* **249**, 3198–3203.

Van Nostrand, W.E., Wagner, S.L., Suzuki, M., Choi, B.H., Farrow, J.S., Geddes, J.W., Cotman, C.W. & Cunningham, D.D. (1989) Protease nexin-II, a potent antichymotrypsin, shows identity to amyloid beta-protein precursor. *Nature* **341**, 546–549.

*Julie Chao*
*Department of Biochemistry and Molecular Biology,*
*Medical University of South Carolina,*
*Charleston, SC 29425-2211, USA*
*Email: chaoj@musc.edu*

# 35. *Mouse γ-nerve growth factor*

## Databanks

Peptidase classification: clan SA, family S1, MEROPS ID: S01.170
NC-IUBMB enzyme classification: none
Databank codes:

| Species | SwissProt | PIR | EMBL (cDNA) | EMBL (genomic) |
|---|---|---|---|---|
| *Mus musculus* | P00756 | A91005 | X01389 | X01798: 5′ flanking region<br>X01799: exons 2–5 and 3′ flanking region |

## Name and History

The β subunit of 7S nerve growth factor forms a stable high molecular weight complex with the γ subunit, or *γ-nerve growth factor (γ-NGF)* in the mouse submandibular gland (Varon *et al.*, 1968).

## Activity and Specificity

Mouse γ-NGF cleaves D-Val-Leu-Arg-NHPhNO$_2$ and Bz-Arg-NHPhNO$_2$. Its pH optimum is 8.5. γ-NGF was shown to activate prorenin (Chapter 284) from human amniotic fluid (Morris *et al.*, 1981), and specifically cleaves the Phe┼His bond of a synthetic renin substrate. The enzyme therefore exhibits rat tonin-like activity, generating angiotensin II (Uddin & Beg, 1995).

## Structural Chemistry

The molecular mass of γ-NGF is 26.2 kDa, and polypeptide chains of 16.4 kDa, 11.4 kDa, 9.4 kDa and 6.8 kDa have been detected in SDS-PAGE (Server & Shooter, 1976). Mouse γ-nerve growth factor belongs to the kallikrein family of serine proteinases (see Chapter 28).

## Preparation

Essentially homogeneous preparations of γ-NGF have been purified from mouse submaxillary glands and saliva (Wilson

& Shooter, 1979; Petrides & Shooter, 1986). The expression of γ-NGF in *E. coli* allows rapid isolation of γ-NGF with a yield of approximately 1 mg of purified protein from 10 g of initial cell paste (Blaber *et al.*, 1990).

## Biological Aspects

Since other mouse kallikreins do not cleave the Phe-His bond, γ-NGF may play a regulatory role in the generation of angiotensin II (Uddin & Beg, 1995).

The genes for γ-NGF and closely related serine proteinases are located on mouse chromosome 7 (Howles *et al.*, 1984).

Protease nexin 2, a potent chymotrypsin inhibitor, forms SDS-stable complexes with γ-NGF (as well as epidermal growth factor-binding protein) (Kim *et al.*, 1992).

C1 Inhibitor inhibits the hydrolysis of Bz-Arg-NHPhNO$_2$ by γ-NGF, forming a 1:1 molecular complex (Faulmann *et al.*, 1987).

## Distinguishing Features

γ-NGF, epidermal growth factor-binding protein (Chapter 34) and β-nerve growth factor endopeptidase (Chapter 32) have similar amino acid compositions, molecular masses and immunological properties. Polyclonal antisera against γ-NGF have been described (Bothwell *et al.*, 1979), but are not commercially available at the time of writing.

## References

Blaber, M., Isackson, P.J. & Bradshaw, R.A. (1990) The characterization of recombinant mouse glandular kallikreins from *E. coli*. *Proteins* **7**, 280–290.

Bothwell, M.A., Wilson, W.H. & Shooter, E.M. (1979) The relationship between glandular kallikrein and growth factor-processing proteases of mouse submaxillary gland. *J. Biol. Chem.* **254**, 7287–7294.

Faulmann, E.L., Young, M. & Boyle, M.D. (1987) Inactivation of the proteolytic activity of mouse nerve growth factor by human C1 (activated)-inhibitor. *J. Immunol.* **138**, 4336–4340.

Howles, P.N., Dickinson, D.P., DiCaprio, L.L., Woodworth-Gutai, M. & Gross, K.W. (1984) Use of a cDNA recombinant for the gamma-subunit of mouse nerve growth factor to localize members of this multigene family near the TAM-1 locus on chromosome 7. *Nucleic Acids Res.* **12**, 2791–2805.

Kim, T., Choi, B.H., Choe, W., Kim, R.C., Van Nostrand, W., Wagner, S. & Cunningham, D. (1992) Expression of protease nexin-II in human dorsal root ganglia. A correlative immunocytochemical and *in situ* hybridization study. *Mol. Chem. Neuropathol.* **16**, 225–239.

Morris, B.J., Catanzaro, D.F. & De Zwart, R.T. (1981) Evidence that the arginine esteropeptidase (gamma) subunit of nerve growth factor can activate inactive renin. *Neurosci. Lett.* **24**, 87–92.

Petrides, P.E. & Shooter, E.M. (1986) Rapid isolation of the 7S-nerve growth factor complex and its subunits from murine submaxillary glands and saliva. *J. Neurochem.* **46**, 721–725

Server, A.C. & Shooter, E.M. (1976) Comparison of the arginine esteropeptidases associated with the nerve and epidermal growth factors. *J. Biol. Chem.* **251**, 161–173.

Uddin, M. & Beg, O.U. (1995) Specific cleavage of synthetic renin substrate by mouse gamma-nerve growth factor. *J. Protein Chem.* **14**: 621–625.

Varon, S., Normura, J. & Shooter, E.M. (1968) Reversible dissociation of the mouse nerve growth factor protein into different subunits. *Biochemistry* **7**, 1296–1303.

Wilson, W.H. & Shooter, E.M. (1979) Structural modification of the NH$_2$ terminus of nerve growth factor. Purification and characterization of beta-nerve growth factor endopeptidase. *J. Biol. Chem.* **254**, 6002–6009.

*Julie Chao*
*Department of Biochemistry and Molecular Biology,*
*Medical University of South Carolina,*
*Charleston, SC 29425-2211, USA*
*Email: chaoj@musc.edu*

# 36. *Mouse proteinase F*

## Databanks

*Peptidase classification: clan SA, family S1, MEROPS ID: S01.167*
*NC-IUBMB enzyme classification: none*
*Databank codes: see Chapter 28*

| Species | Gene | SwissProt | PIR | EMBL (cDNA) | EMBL (genomic) |
|---|---|---|---|---|---|
| *Mus musculus* | mK1 | – | PC2013 | – | |

## Name and History

*Proteinase F* was discovered as an esterase and proteinase in the mouse submandibular glands (Hosoi *et al.*, 1983). Significant amounts of proteinase F, but not other members of the mouse tissue kallikrein gene family, are present in the urine, pancreas and digestive organs, as well as in the salivary glands. On the basis of its kinin-releasing activity and tissue distribution, proteinase F is confirmed as a mouse true kallikrein which is identical to *mK1* and *tissue/renal kallikrein* (Hosoi *et al.*, 1994).

## Activity and Specificity

Proteinase F has the strongest kininogenase activity for both L- and H-kininogens, relative to other members of the mouse tissue kallikrein gene family. Like rat tissue kallikrein (Chapter 37), proteinase F specifically releases bradykinin from the two kininogen substrates, whereas human tissue kallikrein (Chapter 29) releases lysyl-bradykinin from the kininogens (Kato *et al.*, 1987; Hosoi *et al.*, 1994).

The pH optimum is about 9.0. Activity is strongly inhibited by $Cu^{2+}$ and $Hg^{2+}$, and is also inhibited by aprotinin, PMSF, iodoacetamide, leupeptin, antipain and benzamidine. However, the enzyme is not inhibited by trypsin inhibitors from pancreas, soybean or ovomucoid, or by Tos-Lys-$CH_2$Cl, Tos-Phe-$CH_2$Cl, or $\varepsilon$-amino-*n*-caproic acid.

## Structural Chemistry

Proteinase F is a protein of 27.6 kDa, with two subunits of 10 kDa and 18 kDa. A partial sequence of proteinase F was described by Hosoi *et al.* (1994), and the pI of the protein is 4.6.

## Preparation

Essentially homogeneous preparations of proteinase F have been prepared from mouse submandibular gland (Hosoi *et al.*, 1983).

## Biological Aspects

The enzyme is present in mammals and lower vertebrates. It is found in cells of most tissues, but is particularly abundant in submandibular glands (Hosoi *et al.*, 1994). Significant amounts of proteinase F, but not other members of the mouse tissue kallikrein family, are present in the urine, pancreas and digestive organs, as well as in the salivary glands. Like related proteinases in the submandibular gland of the mouse, proteinase F is immunologically detected in secretory granules of granular convoluted tubular cells (Hosoi *et al.*, 1984; Tanaka *et al.*, 1984).

There is no obvious hydrophobic segment that might serve as a signal peptide or a membrane-spanning domain, and there is no evidence of a proteolytically activatable proenzyme. The gene locus of proteinase F is on chromosome 7, with the other mouse kallikreins.

Proteinase F is present in the submandibular glands of female mice more abundantly than in those of males, and the amount increases in males following castration. Thus,

proteinase F appears to be affected by male hormones *in vivo* (Hosoi *et al.*, 1983). In contrast to proteinases A, D or P, proteinase F content in male (but not in female mice) is increased by gonadectomy and decreased by the injection of various androgens (Hosoi *et al.*, 1984). In females, the enzyme activity is decreased by hypophysectomy and increased by dihydrotestosterone administration. In males, on the other hand, it is increased by hypophysectomy and suppressed by triiodothyronine or triiodothyronine plus steroid hormones (Hosoi *et al.*, 1992).

In a recent study, it was found that progesterone alone slightly increased levels of proteinase F. This inductive action was augmented when progesterone was given with triiodothyronine, but blocked when it was given with dexamethasone. 17-$\beta$-Estradiol alone not only failed to induce any of the isozymes, but even further reduced levels of proteinase F. It also decreased the inductive effects of triiodothyronine on these isozymes, but with dexamethasone completely blocked induction of proteinases D, A and P. 17-$\beta$-Estradiol plus dihydrotestosterone suppressed proteinase F levels and only induced proteinase A (Maruyama *et al.*, 1993).

During postnatal development, proteinase F appeared on the 15th day after birth and increased in both sexes, but its percentage ratio to total activity decreased markedly with time because of the rapid increase of other proteinases. Proteinase F is appreciably different from the other four proteinases in its developmental pattern as well as in its responsiveness to sex hormones (Hosoi *et al.*, 1990).

## Distinguishing Features

Polyclonal antisera against proteinase F have been described (Hosoi *et al.*, 1983). Although the structures of the proteins are similar, antisera against proteinase F react only weakly against the epidermal growth factor-binding protein (Chapter 34). The antisera are not commercially available at the time of writing.

## References

Hosoi, K., Tanaka, I., Ishii, Y. & Ueha, T. (1983) A new esteroproteinase (proteinase F) from the submandibular glands of female mice. *Biochim. Biophys. Acta* **756**, 163–170.

Hosoi, K., Tanaka, I., Murai, T. & Ueha, T. (1984) Inhibitory effect of androgen on the synthesis of proteinase F in the male mouse submandibular gland. *J. Endocrinol.* **100**, 253–262.

Hosoi, K., Ueha, T., Fukuuchi, H., Kohno, M. & Takahashi, T. (1990) Reversal of relative proteinase F activity and onset of androgen-dependent proteinases in the submandibular gland of postnatal mice. *Biochem. Int.* **22**, 179–187.

Hosoi, K., Maruyama, S., Ueha, T., Sato, S. & Gresik, E.W. (1992) Additive and/or synergistic effects of 5 alpha-dihydrotestosterone, dexamethasone, and triiodo-L-thyronine on induction of proteinases and epidermal growth factor in the submandibular gland of hypophysectomized mice. *Endocrinology* **130**, 1044–1055.

Hosoi, K., Tsunasawa, S., Kurihara, K., Aoyama, H., Ueha, T., Murai, T. & Sakiyama, F. (1994) Identification of mK1, a true tissue (glandular) kallikrein of mouse submandibular gland: tissue distribution and a comparison of kinin-releasing activity with other submandibular kallikreins. *J. Biochem.* **115**, 137–143.

Kato, H., Nakawishi, E., Enjyoi, K., Hayashi, I., Oh-Ish, S. & Iwanaga, S. (1987) Characterization of serine proteases isolated from rat submaxillary gland: with special reference to the degradation of rat kininogen by these enzymes. *J. Biochem.* **102**, 1389–1404.

Maruyama, S., Hosoi, K., Ueha, T., Tajima, M., Sato, S. & Gresik, E.W. (1993) Effects of female hormones and 3,5,3'-triiodothyronine or dexamethasone on induction of epidermal

growth factor and proteinases F, D, A, and P in the submandibular glands of hypophysectomized male mice. *Endocrinology* **133**, 1051–1060.

Tanaka, S., Hosoi, K., Tanaka, I., Kumegawa, M. & Ueha, T. (1984) Immunocytochemical study of proteinase F in the mouse submandibular gland. *J. Histochem. Cytochem.* **32**, 585–592.

*Julie Chao*
*Department of Biochemistry and Molecular Biology,*
*Medical University of South Carolina,*
*Charleston, SC 29425-2211, USA*
*Email: chaoj@musc.edu*

# 37. Rat tissue kallikrein

## Databanks

*Peptidase classification: clan SA, family S1, MEROPS ID: S01.405*
*NC-IUBMB enzyme classification: EC 3.4.21.35*
*Chemical Abstracts Service registry number: 9001-01-8*
*Databank codes (see also Chapter 28):*

| Species | Gene | SwissProt | PIR | EMBL (cDNA) | EMBL (genomic) |
|---|---|---|---|---|---|
| *Rattus norvegicus* | *klk-1* | P00758 | A00944 A23863 A25137 A33359 A41429 JX0073 | – | M23874: exon 1 M23875: exon 2 M23876: exons 3–5 |
| Other tissue kallikreins | | | | | |
| *Rattus norvegicus* | *klk-3* | P15950 | B23863 B32340 | M26534 | – |
| *Rattus norvegicus* | *klk-7* | P36373 | A31136 B41429 D41429 S09315 S10698 S10699 | – | M19647: complete gene |
| *Rattus norvegicus* | *klk-8* | P36374 | A34079 S10700 | M27215 M27216 | M27217: exons 3–5 |
| *Rattus norvegicus* | *klk-9* | P07647 | A23710 B23710 D23863 S19302 S46212 | M11566 | – |
| *Rattus norvegicus* | *klk-12* | P36376 | B31136 | – | M19648: exons 3–5 M22922: exon 1 |

## Name and History

Rat *tissue kallikrein* was isolated from urine and pancreas (Nustad & Pierce, 1974; Hojima *et al.*, 1977; Chao & Margolius, 1979). It is also known as *glandular kallikrein, true tissue kallikrein, urinary kallikrein, renal kallikrein* and *pancreatic kallikrein* (Bhoola *et al.*, 1992).

## Activity and Specificity

Rat tissue kallikrein specifically cleaves bradykinin from kininogens, whereas human tissue kallikrein releases lysyl-bradykinin from kininogens (Kato *et al.*, 1987; Hosoi *et al.*, 1994). *In vitro*, tissue kallikrein has been shown to process a wide variety of peptide precursors including prorenin, proinsulin, atrial natriuretic peptide, apolipoprotein B100, plasminogen and angiotensinogen.

Assays of rat tissue kallikrein are most conveniently made with the Tos-Arg-OMe as substrate, in an arginine esterase reaction. Rat tissue kallikrein hydrolyzes the fluorescent substrate D-Val-Leu-Arg∤NHMec, as well as the chromogenic substrates Bz-Pro-Phe-Arg∤NHPhNO$_2$, D-Val-Leu-Arg∤NHPhNO$_2$ and L-Val-Leu-Arg∤NHPhNO$_2$. Arg is favored in position P1. In contrast to tissue kallikreins from other mammals, rat tissue kallikrein accommodates Pro at the P2' position (Chagas *et al.*, 1992). A hydrophobic side-chain in the P2 position substantially facilitates substrate hydrolysis (Wang *et al.*, 1992). An extended interaction site in kallikreins was demonstrated by using peptides of increasing length and with different amino acids in positions P5 and P6 (Brillard-Bourdet *et al.*, 1995). The measurement of tissue kallikrein activity can also be made by measuring kininogenase activity with a kinin radioimmunoassay (Shimamoto *et al.*, 1978). The pH optimum is about 8.5–9.0.

The activity of rat tissue kallikrein is inhibited by a number of serine proteinase inhibitors (serpins) including rat kallikrein-binding protein and rat $\alpha_1$-proteinase inhibitor (Chao *et al.*, 1990; Serveau *et al.*, 1992). Rat kallikrein-binding protein (RKBP) is a serine proteinase inhibitor of the serpin family that forms SDS-stable complexes with active kallikrein (Chao *et al.*, 1990). The activity of tissue kallikrein is inhibited by aprotinin, PMSF, benzamidine and leupeptin, but not affected by soybean trypsin inhibitor, limabean trypsin inhibitor and ovomucoid trypsin inhibitor.

The esterolytic substrate is available from Sigma (see Appendix 2 for full names and addresses of suppliers), the chromogenic substrates from Kabi, and the fluorescence substrates from Enzyme Systems Products.

## Structural Chemistry

Rat tissue kallikrein is an acidic, single-chain glycoprotein of about 38 kDa with a pI of 3.8–4.2. The 1 mg ml$^{-1}$ solution of rat tissue kallikrein has an absorbance of 1.5 at 280 nm with a light path of 1 cm.

Rat tissue kallikrein cDNA encodes a preproenzyme of 265 amino acids, including a proposed secretory prepeptide of 17 amino acids and a putative activation peptide of 11 amino acids (Swift *et al.*, 1982). Although several proteinases are capable of activating prokallikrein *in vitro*, the *in vivo* activating enzyme is unknown. The gene encoding the rat tissue kallikrein has been isolated and characterized (Inoue *et al.*, 1989).

## Preparation

Essentially homogeneous preparations of rat tissue kallikrein have been prepared from rat salivary gland, kidney, heart, intestine, vascular tissue and urine (Nustad & Pierce, 1974; Hojima *et al.*, 1977; Porcelli *et al.*, 1978; Chao & Margolius, 1979; Uchida *et al.*, 1979; Moriwaki *et al.*, 1980; Nolly *et al.*, 1985; Takaoka *et al.*, 1986; El-Thaher *et al.*, 1990; Xiong *et al.*, 1990). The purified rat pancreatic enzyme is synthesized as a proenzyme (prokallikrein) with a propeptide that must be removed to activate the enzyme.

The preparation of recombinant rat tissue kallikrein in *E. coli* and yeast expression systems has been described by Wang *et al.* (1991). The recombinant kallikrein from yeast is inactive.

## Biological Aspects

Rat tissue kallikrein mRNA, which encodes an enzyme with true kallikrein activity, is present in a wide range of tissues, with high levels in the submandibular gland, pancreas and kidney.

In the rat pancreas, kallikrein was found at the level of the rough endoplasmic reticulum (RER), Golgi cisternae, condensing vacuoles and zymogen granules of the pancreatic acinar cells as well as in the acinar lumen, suggesting that kallikrein is synthesized in the RER, processed through the Golgi apparatus, and packed in the zymogen granules before being released into the acinar lumen (Bendayan & Orstavik, 1982).

The circulating tissue kallikrein is cleared rapidly from the circulation of the rat, probably in the form of a complex with a plasma proteinase inhibitor. These complexes are removed mainly by the liver and to a lesser extent by the kidney (Rabito *et al.*, 1985).

## Distinguishing Features

Polyclonal and monoclonal antibodies against rat tissue kallikrein have been described (Nustad & Pierce, 1974; Savoy-Moore *et al.*, 1986; Chao & Chao, 1987).

## Further Reading

For a review, see Bhoola *et al.* (1992).

## References

Bendayan, M. & Orstavik, T.B. (1982) Immunocytochemical localization of kallikrein in the rat exocrine pancreas. *J. Histochem. Cytochem.* **30**, 58–66.

Bhoola, K.D., Figueroa, D.D. & Worthy, K. (1992) Bioregulation of kinins: Kallikreins, kininogens, and kininases. *Pharmacol. Rev.* **44**, 1–80.

Brillard-Bourdet, M., Moreau, T. & Gauthier, F. (1995) Substrate specificity of tissue kallikreins: importance of an extended interaction site. *Biochim. Biophys. Acta* **1246**, 47–52.

S

Chagas, J.R., Hirata, I.Y., Juliano, M.A., Xiong, W., Wang, C., Chao, J., Juliano, L. & Prado, E.S. (1992) Substrate specificities of tissue kallikrein and T-kininogenase: Their possible role in kininogen processing. *Biochemistry* **31**, 4969–4974.

Chao, J. & Chao, L. (1987) Identification and expression of kallikrein gene family in rat submandibular and prostate glands using monoclonal antibodies as specific probe. *Biochim. Biophys. Acta* **910**, 233–239.

Chao, J. & Margolius, H.S. (1979) Isozymes of rat urinary kallikrein. *Biochem. Pharmacol.* **28**, 2071–2079.

Chao, J., Chai, K.X., Chen, L.M., Xiong, W., Chao, S., Woodley-Miller, C., Wang, L.X., Lu, H.S. & Chao, L. (1990) Tissue kallikrein-binding protein is a serpin. I. Purification, characterization, and distribution in normotensive and spontaneously hypertensive rats. *J. Biol. Chem.* **265**, 16394–16401.

El-Thaher, T.S., Saed, G.M. & Bailey, G.S. (1990) A simple and rapid purification of kallikrein from rat submandibular gland. *Biochim. Biophys. Acta* **1034**, 157–161.

Hojima, Y., Yamashita, M., Ochi, N., Moriwaki, C. & Moriya, H. (1977) Isolation and some properties of dog and rat pancreatic kallikreins. *J. Biochem.* **81**, 599–610.

Hosoi, K., Tsunasawa, S., Kurihara, K., Aoyama, H., Ueha, T., Murai, T. & Sakiyama, F. (1994) Identification of mK1, a true tissue (glandular) kallikrein of mouse submandibular gland: tissue distribution and a comparison of kinin-releasing activity with other submandibular kallikreins. *J. Biochem.* **115**: 137–143.

Inoue, H., Fukui, K. & Miyake, Y. (1989) Identification and structure of the rat true tissue kallikrein gene expressed in the kidney. *J. Biochem.* **105**, 834–840.

Kato, H., Nakawishi, E., Enjyoi, K., Hayashi, I., Oh-Ish, S. & Iwanaga, S. (1987) Characterization of serine proteases isolated from rat submaxillary gland: with special reference to the degradation of rat kininogen by these enzymes. *J. Biochem.* **102**, 1389–1404.

Moriwaki, C., Fujimori, H., Toyono, Y. & Nagai, T. (1980) Studies of kallikreins. V. Purification and characterization of rat intestinal kallikrein. *Chem. Pharm. Bull.* **28**, 3612–3620.

Nolly, H., Scicli, A.G., Scicli, G. & Carretero, O.A. (1985) Characterization of a kininogenase from rat vascular tissue resembling tissue kallikrein. *Circ. Res.* **56**, 816–821.

Nustad, K. & Pierce, J.V. (1974) Purification of rat urinary kallikreins and their specific antibody. *Biochemistry* **13**, 2312–2319.

Porcelli, G., Marini-Bettolo, G.B., Croxatto, H.R., Iorio, M.D. & Micotti, G. (1978) Purification of renal rat kallikrein and chemical relations with urinary rat kallikrein. *Ital. J. Biochem.* **27**, 201–210.

Rabito, S.F., Seto, M., Maitra, S.R. & Carretero, O.A. (1985) Clearance and metabolism of glandular kallikrein in the rat. *Am. J. Physiol.* **248**, E664–E668.

Savoy-Moore, R.T., Khullar, M., Swartz, K., Scicli, A.G. & Carretero, O.A. (1986) Characterization of monoclonal antibodies against rat glandular kallikrein. *J. Immunol. Methods* **88**, 45–51.

Serveau, C., Moreau, T., Zhou, G.X., El Moujahed, A., Chao, J. & Gauthier, F. (1992) Inhibition of rat tissue kallikrein gene family members by rat kallikrein-binding protein and alpha 1-proteinase inhibitor. *FEBS Lett.* **309**, 405–408.

Shimamoto, K., Ando, T., Nakao, T., Tanaka, S., Sakuma, M., Miyahara, M. (1978) A sensitive radioimmunoassay method for urinary kinins in man. *J. Lab. Clin. Med.* **91**, 721–728.

Swift, G.H., Dagorn, J.C., Ashley, P.L., Cummings, S.W. & MacDonald, R.J. (1982) Rat pancreatic kallikrein mRNA: nucleotide sequence and amino acid sequence of the encoded preproenzyme. *Proc. Natl Acad. Sci. USA* **79**, 7263–7267.

Takaoka, M., Okamura, H., Iwamoto, T. & Morimoto, S. (1986) Purification of inactive kallikrein from rat urine. *Adv. Exp. Med. Biol.* **198A**, 339–345.

Uchida, K., Yokoshima, A., Niinobe, M., Kato, H. & Fujii, S. (1979) Rat stomach kallikrein: its purification and properties. *Adv. Exp. Med. Biol.* **120A**, 291–303.

Wang, J., Chao, J. & Chao, L. (1991) Purification and characterization of recombinant tissue kallikrein from *Escherichia coli* and yeast. *Biochem. J.* **276**, 63–71.

Wang, J., Chao, J., Juliano, L. & Chao, L. (1992) Comparative studies on P2 specificity of wild-type rat tissue kallikrein, Y99H:W215G mutant and tonin. *Agents Actions* **38**, 59–65.

Xiong, W., Chen, L.M., Woodley-Miller, C., Simson, J.A. & Chao, J. (1990) Identification, purification, and localization of tissue kallikrein in rat heart. *Biochem. J.* **267**, 639–646.

*Julie Chao*
*Department of Biochemistry and Molecular Biology,*
*Medical University of South Carolina,*
*Charleston, SC 29425-2211, USA*
*Email: chaoj@musc.edu*

# 38. *Salivary gland proteinase K*

## Databanks

*Peptidase classification: clan SA, family S1, MEROPS ID: S01.165*
*NC-IUBMB enzyme classification: none*

*Databank codes:*

| Species | SwissProt | PIR | EMBL (cDNA) | EMBL (genomic) |
|---|---|---|---|---|
| *Rattus norvegicus* | P36375 | A18966 A38356 A44284 A56784 C18966 C41429 | S48142 | – |

## Name and History

An enzyme responsible for releasing T-kinin (Ile-Ser-bradykinin) from T-kininogen was isolated and characterized from rat submandibular gland extracts (Barlas *et al*., 1987; Gutman *et al*., 1988), and termed **proteinase K**. The same enzyme has been known as **antigen γ, T-kininogenase, proteinase B, kallikrein k10, esterase B** and **endopeptidase K** (Barlas *et al*., 1987; Gutman *et al*., 1988; Xiong *et al*., 1990; Berg *et al*., 1991). It should be noted that there is a completely different serine proteinase, of the subtilisin family (S8), that is commonly known as proteinase K, and recommended by NC-IUBMB to be called endopeptidase K (Chapter 106).

## Activity and Specificity

Proteinase K shows a preference for cleaving Arg-X bonds in small molecule substrates, and releases T-kinin from T-kininogen. It also forms SDS- and heat-stable complexes with the purified recombinant rat kallikrein-binding protein *in vitro* (Ma *et al*., 1993). The enzyme (as antigen γ) acted on T-kininogen with a $K_m$ of $29 \pm 4 \mu M$ and a $k_{cat}/K_m$ of $140 M^{-1} s^{-1}$. With H- and L-kininogens, 7.4 and 10 μg of kinin $h^{-1}$ $mg^{-1}$, respectively, were released (Berg *et al*., 1991).

Proteinase K (as T-kininogenase) was shown to differ from tissue kallikrein in its interactions at subsites S2, S1′ and S2′. Taking advantage of these differences, selective quenched fluorescence substrates have been designed, containing the Abz (*o*-aminobenzoyl) group as fluorophore, and EDDnp (2,4-dinitrophenyl-ethylenendiamine) as quencher. Abz-Phe-Arg+Ser-Arg-EDDnp, with Arg at P2′, is a good substrate for tissue kallikreins from horse, pig and rat, but not for proteinase K. In contrast, Abz-Phe-Arg+Leu-Val-EDDnp and Abz-Phe-Arg+Leu-Val-Arg-EDDnp (the T-kininogen sequence) are good substrates for proteinase K, but not for tissue kallikrein. Arg in P1′, P2′ or P3′ lowers the $K_m$ values of proteinase K, whereas Val at P2′ increases $k_{cat}$ (Chagas *et al*., 1992). Proteinase K cleaves specifically after Arg residues, like true tissue kallikrein, but differs from tissue kallikrein in being able to accommodate either polar or nonpolar residues at P2.

Unusually for a serine proteinase, the activity of proteinase K is increased 10-fold by the presence of a thiol compound such as DTT. Inhibitors included leupeptin, aprotinin, Tos-Lys-CH$_2$Cl and soybean trypsin inhibitor. Pepstatin and PMSF did not inhibit the enzyme (Xiong *et al*., 1990; Gutman *et al*., 1991).

The pH optimum is about 8.0 (Xiong *et al*., 1990). Assays are most conveniently made with the intramolecularly quenched fluorogenic peptide substrates Abz-Phe-Arg+Leu-Val-EDDnp and Abz-Phe-Arg+Leu-Val-Arg-EDDnp (Chagas *et al*., 1992).

## Structural Chemistry

Proteinase K has two chains of about 21 kDa and 6 kDa, a pI of 4.3, and contains disulfide bonds (Xiong *et al*., 1990). The sequence of the cDNA from a rat salivary gland cDNA library predicts a mature protein of 235 amino acid residues and a propeptide (Ma *et al*., 1992).

## Preparation

Natural sources that have yielded essentially homogeneous preparations of proteinase K include rat and mouse submandibular gland and cattle erythrocyte membranes (Barlas *et al*., 1989; Xiong *et al*., 1990; Gutman *et al*., 1991; Ferreira *et al*., 1994).

## Biological Aspects

No change in the mean blood pressure was observed following injection of proteinase K into the circulation of normal rats, even when the amount of enzyme injected was up to 10 times that required for tissue kallikrein to induce a significant fall in blood pressure (Gutman *et al*., 1988).

## Distinguishing Features

Proteinase K is microheterogeneous, as a result of variable glycosylation of its N-terminal light chain, and variable processing at its kallikrein loop. The enzymatic properties of the molecular variants towards synthetic fluorogenic substrates are not significantly different, however (Gutman *et al*., 1991). Polyclonal antisera against the enzyme have been described (Xiong *et al*., 1990), but are not commercially available at the time of writing.

## References

Barlas, A., Gao, X.X. & Greenbaum, L.M. (1987) Isolation of a thiol-activated T-kininogenase from the rat submandibular gland. *FEBS Lett.* **218**, 266–270.

Barlas, A., Gao, X. & Greenbaum, L.M. (1989) Identification of thiol-activated T-kininogenases in the rat and mouse submandibular glands. *Adv. Exp. Med. Biol.* **247B**, 293–295.

Berg, T., Wassdal, I., Mindroiu, T., Sletten, K., Scicli, G., Carretero, O.A. & Scicli, A.G. (1991) T-kininogenase activity of the rat submandibular gland is predominantly due to the kallikrein-like serine protease antigen gamma. *Biochem. J.* **280**, 19–25.

Chagas, J.R., Hirata, I.Y., Juliano, M.A., Xiong, W., Wang, C., Chao, J., Juliano, L. & Prado, E.S. (1992) Substrate specificities of tissue kallikrein and T-kininogenase: their possible role in kininogen processing. *Biochemistry* **31**, 4969–4974.

Ferreira, L.A., Bergamasco, M. & Henriques, O.B. (1994) Isolation and properties of a T-kininogenase from bovine erythrocyte membranes. *J. Protein Chem.* **13**, 547–552.

Gutman, N., Moreau, T., Alhenc-Gelas, F., Baussant, T., Elmoujahed, A., Akpona, S. & Gauthier, F. (1988) T-kinin release from T-kininogen by rat-submaxillary-gland endopeptidase K. *Eur. J. Biochem.* **171**, 577–582.

Gutman, N., Elmoujahed, A., Brillard, M., Du Sorbier, B.M. & Gauthier, F. (1991) Microheterogeneity of rat submaxillary gland kallikrein k10, a member of the kallikrein family. *Eur. J. Biochem.* **197**, 425–429.

Ma, J.X., Chao, J. & Chao, L. (1992) Molecular cloning and characterization of rKlk10, a cDNA encoding T-kininogenase from rat submandibular gland and kidney. *Biochemistry* **31**, 10922–10928.

Ma, J.X., Chao, L., Zhou, G. & Chao, J. (1993) Expression and characterization of rat kallikrein-binding protein in *Escherichia coli. Biochem. J.* **292**, 825–832.

Xiong, W., Chen, L.M. & Chao, J. (1990) Purification and characterization of a kallikrein-like T-kininogenase. *J. Biol. Chem.* **265**: 2822–2827.

*Julie Chao*
*Department of Biochemistry and Molecular Biology,*
*Medical University of South Carolina,*
*Charleston, SC 29425-2211, USA*
*Email: chaoj@musc.edu*

# 39. Tonin

## Databanks

*Peptidase classification: clan SA, family S1, MEROPS ID: S01.172*
*NC-IUBMB enzyme classification: none*
*Databank codes:*

| Species | Gene | SwissProt | PIR | EMBL (cDNA) | EMBL (genomic) |
|---------|------|-----------|-----|-------------|----------------|
| *Rattus norvegicus* | *klk-2* | P00759 | B33359 | M11565 | M23877: exon 1<br>M23878: exons 2–5<br>M26533: complete gene |

Brookhaven Protein Data Bank three-dimensional structures:

| Species | ID | Resolution |
|---------|-----|-----------|
| *Rattus norvegicus* | 1TON | 1.8 |

## Name and History

**Tonin** was discovered as a rat kallikrein-related proteinase which forms angiotensin II directly from angiotensinogen, from a tetradecapeptide substrate or from angiotensin I (Genest *et al.*, 1975; Cemassieux *et al.*, 1976). Tonin is present at high concentration in the rat submandibular gland. Tonin-like activity is also exhibited by mouse γ-nerve growth factor (γ-NGF) (Chapter 35), an enzyme purified from mouse submandibular gland, and a highly active angiotensin-producing enzyme from dog serum (Haas *et al.*, 1989; Gualberto *et al.*, 1992; Uddin & Beg, 1995), but the exact counterpart of tonin in other species has not been described.

## Activity and Specificity

The enzymatic properties of tonin are very similar to those of rat prostatic kallikrein (S3, rk9) (Chapter 37) (Wang *et al.*, 1992) and differ significantly from those of other rat tissue kallikrein gene family members. Unlike other kallikrein-related proteinases, tonin has primarily chymotrypsin-like specificity, specifically cleaving Phe+ bonds in angiotensinogen and angiotensin I to yield angiotensin II (Cemassieux *et al.*, 1976). The kininogenase activity of tonin has been demonstrated (Ikeda & Arakawa, 1984). It cleaves on the carboxyl side of Arg residues and hydrolyzes Z-Val-Lys-Lys-Arg+7-amino-4-trifluoromethylcoumarin but not D-Val-Leu-Arg-7-amino-4-trifluoromethylcoumarin. The $K_m$ values of tonin for Z-Val-Lys-Lys-Arg-7-amino-4-trifluoromethylcoumarin were found to be approximately $20 \mu M$ (Shori *et al.*, 1992). Tonin hydrolyzes Bz-Arg-OEt, Bz-Arg-OMe, Tos-Arg-OMe, $Bz-Arg-NHPhNO_2$ and other small synthetic substrates (Thibault & Genest, 1981). Tonin levels can be measured by a direct radioimmunoassay (Shih *et al.*, 1986).

The pH optimum is about 8.0 (Ikeda & Arakawa, 1984). In contrast to tissue kallikrein, tonin activity is inhibited by soybean trypsin inhibitor but not by aprotinin (Shori *et al.*, 1992). Tonin is inhibited by PMSF at high concentrations (greater than 10 mM) (Thibault & Genest, 1981). Assays are most conveniently made with Z-Val-Lys-Lys-Arg-7-amino-4-trifluoromethylcoumarin available from Enzyme Systems Products (see Appendix 2 for full names and addresses of suppliers). Tonin prefers a prolyl residue in position P2 of the substrate and does not accommodate bulky and hydrophobic residues at that position, as do most of the other kallikrein-related proteinases. Residues in positions P5–P8 are essential for substrate binding and the specificity of tonin (Moreau *et al.*, 1992). Subsites accommodating residues on the prime side of the scissile bond were also shown to be important in determining the overall substrate specificity of this proteinase; tonin shows a preference for hydrophobic residues in P2′.

## Structural Chemistry

The tonin gene encodes a preprotonin of 259 amino acids (Shai *et al.*, 1989). The active enzyme consists of 235 amino acids and is preceded by a deduced signal peptide of 17 amino acids and a propeptide of 7 amino acids (Shai *et al.*, 1989). The deduced molecular mass of tonin is 25 658. Two *N*-glycosylation sites are located at positions 82 and 165. The tonin precursor was identified in rat salivary gland, pancreas, kidney and brain from cell-free translation products in a wheatgerm system (Woodley-Miller *et al.*, 1987). The $1 \text{ mg ml}^{-1}$ solution of tonin has an absorbance of 1.0 at 280 nm with a light path of 1 cm.

The X-ray crystallographic structure of tonin at 1.8 Å resolution was reported by Fujinaga & James (1987). The differences in the structures of tonin and tissue kallikrein are concentrated in several loop regions, and these are probably responsible for the differences in catalytic activity.

Tonin is a single-chain protein of about 32 kDa which contains disulfide bonds and has a pI value of 6.15–6.20 (El-Thaher & Bailey, 1993). Purified tonin was very stable when stored in buffers of low pH or when incubated at high temperatures in a neutral solution.

## Preparation

Essentially homogeneous preparations of tonin have been prepared from rat salivary gland (Cemassieux *et al.*, 1976; Ikeda *et al.*, 1981; Ørstavik *et al.*, 1982; Johansen *et al.*, 1987).

## Biological Aspects

Immunoreactive tonin was localized in the cytoplasm of granular convoluted tubular cells and on the apical surface of striated duct cells and collecting duct cells of the submandibular gland. In the parotid and sublingual glands, which lack granular cells, tonin was only found on the apical surface of striated duct and collecting duct cells. In the kidney, immunoreactive tonin was found only associated with cells of the distal convoluted tubules (Ledoux *et al.*, 1982).

The rat salivary gland secretes tonin into the saliva. Secretion of tonin from rat submandibular glands upon sympathetic stimulation is mediated through stimulation of $\alpha$-adrenoreceptors (Maitra *et al.*, 1986). The *in vitro* secretion of immunoreactive tonin from rat submandibular gland, initiated by activation of $\alpha$-adrenergic receptors, involves a mechanism dependent not on cAMP, but on the influx of extracellular $Ca^{2+}$ (Hirata *et al.*, 1986). $\beta$-Adrenergic stimulation enhances both tonin release into the saliva and tonin synthesis in the submaxillary gland, and these effects might be mediated by cAMP. Isoproterenol increases tonin release in perfused rat submaxillary gland *in vitro* (Garcia *et al.*, 1983). The biological significance of tonin in the salivary gland and saliva remains obscure.

Tonin circulates as a complex with $\alpha_1$-macroglobulin (the rat equivalent of human $\alpha_2$-macroglobulin), which, once formed, is quickly eliminated from the bloodstream by tissue uptake and is metabolized. The kidney is the major site of metabolism (Morris & Cheng, 1982).

Tonin of rat submandibular gland contracts rat uterus independently of addition of the substrate (Feitosa *et al.*, 1989). Tonin is capable of activating rat urinary prokallikrein and prorenin *in vitro* (Lis *et al.*, 1990). Peptide substrates reproducing the rat proatrial natriuretic peptide (proANP) sequence are rapidly and specifically cleaved by tonin. The biological significance of tonin in proANP processing remains to be established (Moreau *et al.*, 1995).

Intracerebroventricular administration of tonin stimulated water intake and increased blood pressure in the rat through local and direct generation of angiotensin II in the central nervous system (Kondo *et al.*, 1980).

The gene (*RSKG-5*) encoding for tonin has been characterized (Ashley & MacDonald, 1985; Shai *et al.*, 1989). It has five exons and contains the variant CCAAA and TTTAAA boxes in the 5′ region and an AATAAA polyadenylation signal at the 3′ end.

In salivary gland, the levels of immunoreactive tonin increase with age in male rats (Shih *et al.*, 1986). The biological activity of tonin is induced by androgen in salivary gland and is suppressed by estrogen (Clements *et al.*, 1990). The expression of tonin is regulated by ACTH, prolactin and growth hormone (Boucher *et al.*, 1977; Shih *et al.*, 1986).

## Distinguishing Features

Polyclonal antisera and monoclonal antibodies against tonin have been described (Gutkowska *et al.*, 1978; Ørstavik *et al.*, 1982; Shih *et al*, 1986; Chao & Chao, 1987; Woodley-Miller *et al.*, 1987), but are not commercially available at the time of writing.

# References

Ashley, P.L. & MacDonald, R.J. (1985) Tissue-specific expression of kallikrein-related genes in the rat. *Biochemistry* **24**, 4520–4527.

Boucher, R., Chretien, M. & Genest, J. (1977) Dependence of tonin activity in rat submaxillary gland on growth hormone and testosterone. *Am. J. Physiol.* **232**, E522–E525.

Cemassieux, S., Boucher, R., Crise, C. & Genest, J. (1976) Purification and characterization of tonin. *Can. J. Biochem.* **54**, 788–795.

Chao, J. & Chao, L. (1987) Identification and expression of kallikrein gene family in rat submandibular and prostate glands using monoclonal antibodies as specific probes. *Biochim. Biophys. Acta* **910**, 233–239.

Clements, J.A., Matheson, B.A., MacDonald, R.J. & Funder, J.W. (1990) Oestrogen administration and the expression of the kallikrein gene family in the rat submandibular gland. *J. Steroid Biochem.* **35**, 55–60.

El-Thaher, T., & Bailey, G.S. (1993) Rapid and convenient purification of tonin from rat submandibular gland. Comparison of tonin and tissue kallikrein in their interactions with inhibitors. *Int. J. Peptides Protein Res.* **41**, 196–200.

Feitosa, M.H., Pesquero, J.L., Ferreira, M.A., Oliveira, G.M., Rogana, E. & Beraldo, W.T. (1989) Tonin and kallikrein-kinin system. *Adv. Exp. Med. Biol.* **247A**, 573–580.

Fujinaga, M. & James, M.N. (1987) Rat submaxillary gland serine protease, tonin. Structure solution and refinement at 1.8 Å resolution. *J. Mol. Biol.* **195**, 373–396.

Garcia, R., Thibault, G., Gutkowska, J. & Genest, J. (1983) Release of tonin and of kallikrein by perfused rat submaxillary gland. *Am. J. Physiol.* **244**, R228–R234.

Genest, J., Nowaczynski, W., Boucher, R., Kuchel, O. & Rojo-Ortega, J.M. (1975) Aldosterone and renin in essential hypertension. *Can. Med. Assoc. J.* **113**, 421–431.

Gualberto, M.P., Nunes, R.L., Beraldo, W.T. & Pesquero, J.L. (1992) Tonin-like activity present in the human submandibular gland. *Agents Actions* **38**(part 1), 392–400.

Gutkowska, J., Boucher, R., Demassieux, S., Garcia, R. & Genest, J. (1978) A direct radioimmunoassay for tonin. *Can. J. Biochem.* **56**, 769–773.

Haas, E., Lewis, L., Koshy, T.J., Varde, A.U., Renerts, L. & Bagai, R.C. (1989) Angiotensin II-producing enzyme III from acidified serum of nephrectomized dogs. *Am. J. Hypertens.* **2**, 696–707.

Hirata, Y., Tomita, M., Fujita, T. & Ikeda, M. (1986) *In vitro* secretion of immunoreactive tonin from dispersed rat submandibular gland cells. *Hypertension* **8**, 883–889.

Ikeda, M. & Arakawa, K. (1981) Purification of tonin by chromatography using soybean trypsin inhibitor coupled CH-sepharose 4B. *Jpn Circ. J.* **45**, 1083–1089.

Ikeda, M. & Arakawa, K. (1984) Kininogenase activity of tonin. *Hypertension* **6**, 222–228.

Ikeda, M., Gutkowska, J., Thibault, G., Boucher, R. & Genest, J. (1981) Purification of tonin by affinity chromatography. *Hypertension* **3**, 81–86.

Johansen, L., Bergundhaugen, H. & Berg, T. (1987) Rapid purification of tonin, esterase B, antigen psi and kallikrein from rat submandibular gland by fast protein liquid chromatography. *J. Chromatogr.* **387**, 347–359.

Kondo, K., Garcia, R., Boucher, R. & Genest, J. (1980) Effects of intracerebroventricular administration of tonin on water intake and blood pressure in the rat. *Brain Res.* **200**, 437–441.

Ledoux, S., Gutkowska, J., Garcia, R., Thibault, G., Cantin, M. & Genest, J. (1982) Immunohistochemical localization of tonin in rat salivary glands and kidney. *Histochemistry* **76**, 329–339.

Lis, M., Kamada, M., Furuhata, N., Yamaguchi, T., Ikekita, M., Kizuki, K. & Moriya, H. (1990) Observation of tissue prokallikrein activation by some serine proteases, arginine esterases in rat submandibular gland. *Biochem. Biophys. Res. Commun.* **166**, 231–237.

Maitra, S.R., Rabito, S.F. & Carretero, O.A. (1986) Release of kallikrein and tonin from the rat submandibular gland. *Adv. Exp. Med. Biol.* **198A**, 247–254.

Moreau, T., Brillard-Bourdet, M., Bouhnik, J. & Gauthier, F. (1992) Protein products of the rat kallikrein gene family. Substrate specificities of kallikrein rK2 (tonin) and kallikrein rK9. *J. Biol. Chem.* **267**, 10045–10051.

Moreau, T., Brillard-Bourdet, M., Chagas, J. & Gauthier, F. (1995) Pro-rat atrial natriuretic peptide-mimicking peptides as substrates for rat kallikreins rK2 (tonin) and rK9. *Biochim. Biophys. Acta* **1249**, 168–172.

Morris, B.J. & Cheng, E.S. (1982) Clearance and metabolism of [125]I-labeled tonin in the rat. *Endocrinology* **111**, 1462–1468.

Ørstavik, T.B., Carretero, O.A., Hayashi, H., Scicli, G.A. & Johansen, L. (1982) Immunohistochemical localization of tonin and its relation to kallikrein in rat salivary glands. *J. Histochem. Cytochem.* **30**, 1123–1129.

Shai, S.Y., Chao, J. & Chao, L. (1989) Characterization of genes encoding rat tonin and a kallikrein-like serine protease. *Biochemistry* **28**, 5334–5343.

Shih, H.C., Chao, L. & Chao, J. (1986) Age and hormonal dependence of tonin levels in rat submandibular gland as determined by a new direct radioimmunoassay. *Biochem. J.* **238**, 145–149.

Shori, D.K., Proctor, G.B., Chao, J., Chan, K.M. & Garrett, J.R. (1992) New specific assays for tonin and tissue kallikrein activities in rat submandibular glands. Assays reveal differences in the effects of sympathetic and parasympathetic stimulation on proteinases in saliva. *Biochem. Pharmacol.* **43**, 1209–1217.

Thibault, G. & Genest, J. (1981) Tonin, an esteroprotease from rat submandibular glands. *Biochim. Biophys. Acta* **660**, 23–29.

Uddin, M. & Beg, O.U. (1995) Specific cleavage of synthetic renin substrate by mouse gamma-nerve growth factor. *J. Protein Chem.* **14**, 621–625.

Wang, C., Tang, C.Q., Zhou, G.X., Chao, L. & Chao, J. (1992) Biochemical characterization and substrate specificity of rat prostate kallikrein (S3): comparison with tissue kallikrein, tonin and T-kininogenase. *Biochim. Biophys. Acta* **1121**, 309–316.

Woodley-Miller, C., Chao, J. & Chao, L. (1987) Identification of tonin in brain and exocrine tissues and in the cell-free translation products encoded by the mRNA of these tissues. *Biochem. J.* **248**, 477–481.

*Julie Chao*
*Department of Biochemistry and Molecular Biology,*
*Medical University of South Carolina,*
*Charleston, SC 29425-2211, USA*
*Email: chaoj@musc.edu*

# 40. *Mole salivary kallikrein*

## Databanks

*Peptidase classification: clan SA, family S1, MEROPS ID: S9G.66*
*NC-IUBMB enzyme classification: none*
*Databank codes: no sequence data available*

## Name and History

This kallikrein was recently isolated from salivary gland of a primitive insectivore, the Eastern Atlantic mole (*Scalopus aquaticus*). Insectivores are among the most primitive mammals known and kallikreins from these species may provide clues regarding the evolution of tissue kallikreins and other serine proteinases.

## Activity and Specificity

Mole kallikrein cleaves on the carboxyl side of arginyl residues, and minimally at lysyl residues; Phe is preferred over Pro at the P3 position. Mole kallikrein cleaves D-Val-Leu-Arg↓NHMec and D-Phe-Phe-Arg↓NHMec with a high affinity and specificity. L-Kininogens from mammals are processed, with the release of vasoactive kinin peptides (Richards *et al*., 1996a,b). The pH optimum was determined as pH 9.0, but the enzyme has strong activity over a broad range of pH (Richards *et al*., 1996b).

## Structural Chemistry and Preparation

Mole kallikrein is a single polypeptide chain of 30 kDa and has a pI value of 5.3 (Richards *et al*., 1996b). An essentially pure preparation of mole salivary kallikrein was obtained by anion-exchange perfusion chromatography as described by Richards *et al*. (1996b).

## Biological Aspects

The gene encoding mole kallikrein has not been identified, but genomic Southern blot analysis with human and rat tissue kallikrein cDNA probes identified several weak bands in mole genomic DNA.

The biological functions of salivary kallikreins are uncertain, but they probably serve as processing enzymes to maintain the normal flora of the oral cavity.

## Distinguishing Features

Antisera against human tissue kallikrein cross-reacted weakly with mole kallikrein, but antisera against rat tissue kallikrein did not. A 1 mg ml$^{-1}$ solution of mole kallikrein has an absorbance of 1.4 at 280 nm with a light path of 1 cm, compared to 1.5 for both human and rat tissue kallikreins (Richards *et al*., 1996a,b).

### References
Richards, G.P., Chao, J. & Chao, L. (1996a) Tissue kallikreins in evolutionary diverse vertebrates. *Immunopharmacology* **32**, 94–95.
Richards, G.P., Zintz, C., Chao, J. & Chao, L. (1996b) Purification and characterization of salivary kallikrein from an insectivore (*Scalopus aquaticus*): substrate specificities, immunoreactivity, and kinetic analyses. *Arch. Biochem. Biophys.* **329**, 104–112.

*Julie Chao*
*Department of Biochemistry and Molecular Biology,*
*Medical University of South Carolina,*
*Charleston, SC 29425-2211, USA*
*Email: chaoj@musc.edu*

# 41. *Fish muscle prokallikrein*

## Databanks

*Peptidase classification: clan SA, family S1, MEROPS ID: S9G.65*
*NC-IUBMB enzyme classification: none*
*Databank codes: no sequence data available*

## Name and History

**Fish prokallikrein** was initially identified in skeletal muscle of the black sea bass (*Centropristis striata*) by its weak immunoreactivity with human tissue kallikrein antiserum.

## Activity and Specificity

Bass kallikrein is synthesized as a zymogen and can be activated by incubation with trypsin. A dimer of 72 kDa is also enzymatically active after trypsin activation. Upon activation, bass kallikrein cleaves L-kininogen and releases vasoactive kinin peptide (Richards *et al.*, 1996, 1997b). Bass kallikrein exhibited trypsin-like activities towards synthetic substrates D-Phe-Phe-Arg+NHMec and D-Val-Leu-Arg+NHMec. Like tissue kallikreins of other species, bass kallikrein is inhibited by aprotinin, benzamidine and PMSF, but not by elastatinal, soybean trypsin inhibitor or limabean trypsin inhibitor. The pH optimum for action on the synthetic substrates is 9.0 (Richards *et al.*, 1997b).

## Structural Chemistry

Bass kallikrein is a single-chain polypeptide of 36 kDa as determined by SDS-PAGE under reducing conditions. The enzyme readily complexes to form dimers, tetrameters and aggregates. The pI value is 4.95–5.15 (Richards *et al.*, 1997a,b).

## Preparation

The essentially pure preparation of bass prokallikrein from black sea bass skeletal muscle was obtained by anion-exchange perfusion chromatography and reversed-phase HPLC as described by Richards *et al.* (1997b).

## Biological Aspects

Immunoreactive bass kallikrein was identified by western blot analyses in bass heart, skeletal muscle, spleen, swimbladder, gill, kidney and liver, but not in the pyloric cecum or plasma. The gene encoding for bass prokallikrein has not been identified.

## Distinguishing Features

Polyclonal antisera against bass prokallikrein has been produced (Richards *et al.*, 1997a,b). Immunoblots revealed 36, 72 and 144 kDa proteins. The latter are believed to be dimers and tetramers of bass prokallikrein.

Polyclonal antisera against human tissue kallikrein cross-react weakly with bass kallikrein, but polyclonal antisera against rat tissue kallikrein do not.

Trypsin-activated bass kallikrein preferentially cleaves D-Phe-Phe-Arg-NHMec over D-Val-Leu-Arg-NHMec and D-Pro-Phe-Arg-NHMec (Richards *et al.*, 1997b).

### References

Richards, G.P., Chao, J. & Chao, L. (1996) Tissue kallikrein in evolutionarily diverse vertebrates. *Immunopharmacology* **32**, 94–95.

Richards, G.P., Chao, J. & Chao, L. (1997a) Distribution of tissue kallikreins in lower vertebrates: potential physiological roles for fish kallikreins. *Comp. Biochem. Physiol.* **118C**, 49–57.

Richards, G.P., Liang, Y.M., Chao, J. & Chao, L. (1997b) Purification, characterization and activation of fish muscle prokallikrein. *Comp. Biochem. Physiol.* **118C**, 39–48.

*Julie Chao*
*Department of Biochemistry and Molecular Biology,*
*Medical University of South Carolina,*
*Charleston, SC 29425-2211, USA*
*Email: chaoj@musc.edu*

# 42. Complement factor D

## Databanks

*Peptidase classification: clan SA, family S1, MEROPS ID: S01.191*
*NC-IUBMB enzyme classification: EC 3.4.21.46*
*Chemical Abstracts Service registry number: 37213-56-2*

*Databank codes:*

| Species | SwissProt | PIR | EMBL (cDNA) | EMBL (genomic) |
|---------|-----------|-----|-------------|-----------------|
| *Homo sapiens* | P00746 | A40197 | M84526 | – |
| *Mus musculus* | P03953 | C25952 | M11768 | M13386: complete gene<br>X04673: complete gene |
| *Rattus norvegicus* | P32038 | S19275 | M92059 | – |
| *Rattus norvegicus* | – | – | S73894 | – |
| *Sus scrofa* | P51779 | S54115 | U29948<br>Z49058 | – |

Brookhaven Protein Data Bank three-dimensional structures:

| Species | ID | Resolution | Notes |
|---------|-----|-----------|-------|
| *Homo sapiens* | 1DST | 2 | mutant with enhanced catalytic activity |
|  | 1DSU | 2 |  |

## Name and History

*Factor D* was initially described under the names *C3 pro-activator convertase (C3PAse)* (Müller-Eberhard & Götze, 1972) and *glycine-rich β-glycoprotein convertase (GBGase)* (Rosen & Alper, 1972) as a serum enzyme necessary for the formation of the C3 convertase in the alternative pathway of complement activation. Shortly thereafter, factor D was shown to be necessary for the formation of C3 convertase by mixtures of purified factor B and cobra venom factor (CoVF), the C3b homolog of cobra venom (Hunsicker *et al.*, 1973). It was further shown that factor D, also termed *properdin factor D*, could catalyze the formation of a C3-cleaving enzyme by interacting with factor B and red cell-bound C3b (Fearon *et al.*, 1974). The serine protease nature of the enzyme was established by the demonstration of its susceptibility to DFP (Fearon *et al.*, 1974). Vogt and his group then showed that formation of a complex between factor B and either CoVF (Vogt *et al.*, 1974) or C3b (Vogt *et al.*, 1975) is necessary for cleavage of factor B by factor D. The enzyme was rediscovered in mice as a serine protease, called *adipsin*, encoded by an adipocyte-specific mRNA (Cook *et al.*, 1985). The primary structure and the functional similarity to human factor D identified adipsin with mouse factor D (Rosen *et al.*, 1989). However, the name adipsin has persisted in the literature, probably because it is mainly produced by adipocytes and also because it is deficient in animal models of obesity and therefore thought to play a role in energy metabolism (Flier *et al.*, 1987).

## Activity and Specificity

Factor D hydrolyzes the Arg233┤Lys234 bond of its single known natural substrate, factor B, only in the context of a $Mg^{2+}$-dependent complex of factor B with C3b, C3 with a hydrolyzed thioester bond ($C3_{H_2O}$), or CoVF. It also hydrolyzes with low catalytic efficiency a small number of thioester substrates (Kam *et al.*, 1987). Both Z-Arg-SBzl and Z-Lys-SBzl (Kim *et al.*, 1994) are reactive with factor D. Of dipeptides with an Arg at the P1 site only those containing Arg, Val or Lys in P2 are hydrolyzed at measurable rates. Extension of peptide thioesters to include Gln at P3 and P4,

which are also found in factor B, results in complete loss of reactivity. Also, tripeptide thioesters containing four other amino acids at P3 (Gly, Glu, Lys, Phe) are not reactive with factor D. In addition to this high degree of specificity, factor D exhibits extremely low reactivity with peptide thioesters. Compared to C1s, its functional homolog in the classical pathway, the $k_{cat}/K_m$ of factor D is 2–3 orders of magnitude lower. Assays of thioester hydrolysis are performed at pH 7.5 as described (Kam *et al.*, 1987) by using Ellman's reagent as chromogen (Kim *et al.*, 1994). Z-Lys-SBzl is available from Calbiochem (see Appendix 2 for full names and addresses of suppliers). More specific are hemolytic assays which are described in detail in Volanakis *et al.* (1993).

Factor D is inhibited by DFP and APMSF. Isocoumarins substituted with basic groups are also inhibitory, but the $K_{obs}/[I]$ values are five orders of magnitude lower than those obtained with trypsin (Kam *et al.*, 1992). Nafamostat mesilate (6-amidino-2-napthyl *p*-guanidinobenzoate dimethylsulfonate) also inhibits factor D (Inagi *et al.*, 1994).

## Structural Chemistry

The single polypeptide chain of factor D consists of 228 amino acid residues with a calculated $M_r$ of 24 376 (Johnson *et al.*, 1984; Niemann *et al.*, 1984; White *et al.*, 1992). The primary structure corresponds to that of an active pancreatic serine protease and no structural zymogen is present in the blood (Lesavre & Müller-Eberhard, 1978). There are no potential *N*-glycosylation sites and no carbohydrate is present in purified factor D (Tomana *et al.*, 1985). Four disulfide bridges that are highly conserved among serine proteases are also present in factor D. The high-resolution crystal structure of factor D shows that it has conserved the characteristic chymotrypsin structural fold (Narayana *et al.*, 1994). However, key catalytic and substrate-binding residues exhibit unique conformations incompatible with expression of efficient catalytic activity. These atypical structural features include residues of the catalytic triad, the substrate specificity pocket, and the nonspecific substrate-binding site of factor D. Two molecules, A and B, related by a noncrystallographic 2-fold axis are present in the triclinic unit cell. In molecule

A, the carboxyl of Asp102 (chymotrypsinogen numbering) is pointed away from His57 and is freely accessible to the solvent. In molecule B, the imidazole of His57 is oriented away from Ser195, having assumed the energetically favored *trans*-conformation. Neither orientation is compatible with a functional catalytic triad. In addition, in both molecules, Asp189 at the bottom of the specificity pocket forms a salt bridge with Arg218. This salt link would probably restrict access of the guanidinium of the P1 Arg233 of factor B to the negative charge of Asp189. Finally, the loop formed by residues 214–219 of factor D, which forms one of the walls of the specificity pocket, is substantially raised towards the solvent, resulting in considerable narrowing of the pocket. A detailed presentation of the atypical structural features of factor D and their possible functional correlates is given in the review by Volanakis & Narayana (1996).

## Preparation

Factor D can be purified from human serum, but due to its low serum concentration, procedures are laborious and of low yield (Reid *et al*., 1981). A simpler, higher yield method for obtaining relative large amounts of factor D utilizes urine from patients with Fanconi's syndrome as starting material (Volanakis & Macon, 1987). The method has been adapted for purifying the protein from peritoneal dialysis fluid (Catana & Schifferli, 1991). Factor D has been expressed in CHO cells and purified from the media in milligram amounts (Kim *et al*., 1995). Expression in insect cells by using a baculovirus vector yields the zymogen form of the enzyme, which can be converted to active factor D by catalytic amounts of trypsin (Yamauchi *et al*., 1994). This has been attributed to the absence from insect cells of a trypsin-like processing enzyme, which in mammalian cells cleaves off the activation peptide before secretion.

## Biological Aspects

The concentration of factor D in blood, $1.8 \pm 0.4 \,\mu g \,ml^{-1}$ (Barnum *et al*., 1984) is the lowest of any complement protein and makes it the limiting enzyme in the activation sequence of the alternative pathway (Lesavre & Müller-Eberhard, 1978). The low concentration of factor D in blood is maintained by an extremely high fractional catabolic rate estimated at about 60% per hour (Pascual *et al*., 1988). Because of its small size, factor D is filtered at high rates through the glomerular membrane and is reabsorbed and catabolized by the proximal epithelial cells (Sanders *et al*., 1986). Thus, in renal insufficiency serum levels increase by as much as 10-fold, and in renal tubular dysfunction, such as Fanconi's syndrome, large amounts of the protein are secreted in the urine (Volanakis *et al*., 1985). The high catabolic rates of factor D imply correspondingly high synthetic rates. In that regard it is interesting that the main synthetic site for blood factor D is the adipocyte (Cook *et al*., 1985; White *et al*., 1992). Blood monocytes/macrophages (Whaley, 1980) and brain astrocytes (Barnum *et al*., 1992) also synthesize factor D, which may be important for complement-related functions at local sites.

The atypical conformation of the active center of factor D explains its low reactivity with synthetic substrates. However,

the proteolytic activity of the enzyme during activation of the alternative pathway is comparable to that of other complement enzymes. This paradox can be adequately explained by the proposal (Volanakis & Narayana, 1996) that the natural substrate, i.e. the C3bB complex, induces the conformational changes necessary for the realignment of the active center residues of factor D. Native factor B and small peptide esters cannot induce these conformational changes apparently because they lack crucial contacts with the enzyme. A corollary to this hypothesis is that, following cleavage of C3b-bound factor B, the active center of factor D reverts to its resting-state inactive conformation. This mechanism provides for the regulation of the proteolytic activity of factor D in the absence of a circulating zymogen. Thus, the need for a zymogen-activating enzyme and for a serpin-type inhibitor are obviated.

## Further Reading

See Volanakis & Narayana (1996) for a comprehensive review of the structural and functional aspects of the enzyme factor D.

## References

Barnum, S.R., Niemann, M.A., Kearney, J.F. & Volanakis, J.E. (1984) Quantitation of complement factor D in human serum by a solid-phase radioimmunoassay. *J. Immunol. Methods* **67**, 303–309.

Barnum, S.R., Ishii, Y., Agrawal, A. & Volanakis, J.E. (1992) Production and interferon-$\gamma$-mediated regulation of complement component C2 and factors B and D by the astroglioma cell line U105-MG. *Biochem. J.* **287**, 595–601.

Catana, E. & Schifferli, J.A. (1991) Purification of human complement factor D from the peritoneal fluid of patients on chronic ambulatory peritoneal dialysis. *J. Immunol. Methods* **138**, 265–271.

Cook, K.S., Groves, D.L., Min, H.Y. & Spiegelman, B.M. (1985) A developmentally regulated mRNA from 3T3 adipocytes encodes a novel serine protease homologue. *Proc. Natl Acad. Sci. USA* **82**, 6480–6484.

Fearon, D.T., Austen, K.F. & Ruddy, S. (1973) Formation of a hemolytically active cellular intermediate by the interaction between properdin factors B and D and the activated third component of complement. *J. Exp. Med.* **138**, 1305–1313.

Fearon, D.T., Austen, K.F. & Ruddy, S. (1974) Properdin factor D: Characterization of its active site and isolation of the precursor form. *J. Exp. Med.* **139**, 355–366.

Flier, J.S., Cook, K.S., Usher, P. & Spiegelman, B.M. (1987) Severely impaired adipsin expression in genetic and acquired obesity. *Science* **237**, 405–408.

Hunsicker, L.G., Ruddy, S. & Austen, K.F. (1973) Alternate complement pathway: factors involved in cobra venom factor (CoVF) activation of the third component of complement (C3). *J. Immunol.* **110**, 128–138.

Inagi, R., Miyata, T., Oda, O., Maeda, K. & Inoue, K. (1994) Evaluation of the proteolytic activity of factor D accumulated as an active serine protease in patients with chronic renal failure. *Nephron* **66**, 285–290.

Johnson, D.M.A., Gagnon, J. & Reid, K.B.M. (1984) Amino acid sequence of human factor D of the complement system. Similarity in sequence between factor D and proteases of non-plasma origin. *FEBS Lett.* **166**, 347–351.

Kam, C.-M., McRae, B.J., Harper, J.W., Niemann, M.A., Volanakis, J.E. & Powers, J.C. (1987) Human complement proteins D, C2, and B. Active site mapping with peptide thioester substrates. *J. Biol. Chem.* **262**, 3444–3451.

Kam, C.-M., Oglesby, T.J., Pangburn, M.K., Volanakis, J.E. & Powers, J.C. (1992) Substituted isocoumarins as inhibitors of complement serine proteases. *J. Immunol.* **149**, 163–168.

Kim, S., Narayana, S.V.L. & Volanakis, J.E. (1994) Mutational analysis of the substrate binding site of human complement factor D. *Biochemistry* **33**, 14393–14399.

Kim, S., Narayana, S.V.L. & Volanakis, J.E. (1995) Catalytic role of a surface loop of the complement serine protease factor D. *J. Immunol.* **154**, 6073–6079.

Lesavre, P.H. & Müller-Eberhard, H.J. (1978) Mechanism of action of factor D of the alternative complement pathway. *J. Exp. Med.* **148**, 1498–1509.

Müller-Eberhard, H.J. & Götze, O. (1972) C3 proactivator convertase and its mode of action. *J. Exp. Med.* **135**, 1003–1009.

Narayana, S.V.L., Carson, M., El-Kabbani, O., Kilpatrick, J.M., Moore, D., Chen, X., Bugg, C.E., Volanakis, J.E. & DeLucas, L.J. (1994) Structure of human factor D a complement system protein at 2.0 Å resolution. *J. Mol. Biol.* **235**, 695–708.

Niemann, M.A., Bhown, A.S., Bennett, J.C. & Volanakis, J.E. (1984) Amino acid sequence of human D of the alternative complement pathway. *Biochemistry* **23**, 2482–2486.

Pascual, M., Steiger, G., Estreicher, J., Macon, K., Volanakis, J.E. & Schifferli, J.A. (1988) Metabolism of complement factor D in renal failure. *Kidney Int.* **34**, 529–536.

Reid, K.B.M., Johnson, D.M.A., Gagnon, J. & Prohaska, R. (1981) Preparation of human factor D of the alternative pathway of complement. *Methods Enzymol.* **80**, 134–143.

Rosen, F.S. & Alper, C.A. (1972) An enzyme in the alternate pathway to C3 activation and its activation by a protein in normal serum. *J. Clin. Invest.* **51**, 80a.

Rosen, B.S., Cook, K.S., Yaglom, J., Groves, D.L., Volanakis, J.E., Damm, D., White, T. & Spiegelman, B.M. (1989) Adipsin and complement factor D activity: An immune-related disorder in obesity. *Science* **244**, 1483–1487.

Sanders, P.W., Volanakis, J.E., Rostand, S.G. & Galla, J.H. (1986) Human complement protein D catabolism by the rat kidney. *J. Clin. Invest.* **77**, 1299–1304.

Tomana, M., Niemann, M., Garner, C. & Volanakis, J.E. (1985) Carbohydrate composition of the second, third, and fifth components and factor B and D of human complement. *Mol. Immunol.* **22**, 107–111.

Vogt, W., Dieminger, L., Lynen, R. & Schmidt, G. (1974) Formation and composition of the complex with cobra venom factor that cleaves the third component of complement. *Z. Physiol. Chem.* **355**, 171–183.

Vogt, W., Schmidt, G., Dieminger, L. & Lynen, R. (1975) Formation and composition of the C3 activating enzyme complex of the properdin system. Sequential assembly of its components on solid-phase trypsin-agarose. *Z. Immun.-Forsch.* **149**, 440–455.

Volanakis, J.E. & Macon, K.J. (1987) Isolation of complement factor D from urine of patients with Fanconi's syndrome. *Anal. Biochem.* **163**, 242–246.

Volanakis, J.E. & Narayana, S.V.L. (1996) Complement factor D, a novel serine protease. *Protein Sci.* **5**, 553–564.

Volanakis, J.E., Barnum, S.R., Giddens, M. & Galla, J.H. (1985) Renal filtration and catabolism of complement protein D. *N. Engl. J. Med.* **312**, 395–399.

Volanakis, J.E., Barnum, S.R. & Kilpatrick, J.M. (1993) Purification and properties of human factor D. *Methods Enzymol.* **223**, 82–97.

Whaley, K. (1980) Biosynthesis of the complement components and the regulatory proteins of the alternative complement pathway by human peripheral blood monocytes. *J. Exp. Med.* **151**, 501–516.

White, R.T., Damm, D., Hancock, N., Rosen, B.S., Lowell, B.B., Usher, P., Flier, J.S. & Spiegelman, B.M. (1992) Human adipsin is identical to complement factor D and is expressed at high levels in adipose tissue. *J. Biol. Chem.* **267**, 9210–9213.

Yamauchi, Y., Stevens, J.W., Macon, K.J. & Volanakis, J.E. (1994) Recombinant and native zymogen forms of human factor D. *J. Immunol.* **152**, 3645–3653.

*John E. Volanakis*
*Division of Clinical Immunology and Rheumatology,*
*University of Alabama School of Medicine,*
*UAB, THT 437, Birmingham, AL 35294-0006, USA*
*Email: jvolanakis@medinfo.dom.uab.edu*

# 43. Complement component C1r

## Databanks

*Peptidase classification: clan SA, family S1, MEROPS ID: S01.192*
*NC-IUBMB enzyme classification: EC 3.4.21.41*
*Chemical Abstracts Service registry number: 80295-69-8*

*Databank codes:*

| Species | SwissProt | PIR | EMBL (cDNA) | EMBL (genomic) |
|---|---|---|---|---|
| *Homo sapiens* | P00736 | A24170 | M14058 X04701 | – |

## Name and History

It was shown originally by Lepow *et al.* (1963) that human C1, the factor responsible for the initiation of the classical pathway of complement, could be resolved by ion-exchange chromatography into three distinct components, designated C1q, C1r, and C1s. That C1r is the protease responsible for the activation of C1s was later demonstrated by Naff & Ratnoff (1968). The activated form of C1r is commonly indicated by an overbar (C$\overline{1}$r).

## Activity and Specificity

C1r shows very narrow trypsin-like proteolytic specificity restricted to a few arginyl bonds in its two natural protein substrates, C1s and itself. Thus, C1r autoactivation and C1s cleavage by activated C1r both involve cleavage of a single Arg┼Ile bond in the sequences Gln-Arg-Gln-Arg┼Ile-Ile-Gly-Gly and Glu-Lys-Gln-Arg┼Ile-Ile-Gly-Gly, respectively (Arlaud & Thielens, 1993). Purified activated C1r also undergoes autolytic proteolysis involving cleavage of two Arg┼Gly bonds (Arlaud *et al.*, 1987).

Esterolytic activity can be assayed on a restricted number of peptide esters containing a P1 Arg or Lys residue, including Ac-Gly-Lys┼OMe and Z-Gly-Arg┼SBzl (McRae *et al.*, 1981; Arlaud & Thielens, 1993). Both substrates are cleaved by other serine proteases too, and therefore are suitable only for the purified enzyme. The pH optimum is 8.0–8.5 (Naff & Ratnoff, 1968).

The only known physiological inhibitor of activated C1r is C1 inhibitor (Davis *et al.*, 1993). DFP, PMSF and *p*-nitrophenyl-*p'*-guanidinobenzoate are also potent irreversible inhibitors (Arlaud & Thielens, 1993).

## Structural Chemistry

Human C1r is a noncovalent homodimer of $M_r$ 172 600 and the pI is 4.9. Each monomer (688 amino acids) is a single chain in the proenzyme and is split on activation, through cleavage at Arg446-Ile447, into two disulfide-linked chains, A and B. The N-terminal A chain is homologous to the corresponding chains of C1s (Chapter 44) and mannose-binding protein-associated serine protease (Chapter 45), being subdivided into five structural modules (I–V) (Figure 43.1). Modules I and III belong to the CUB family (**c**omplement C1r/C1s, **u**egf, **b**one morphogenetic protein). Module II contains *erythro-β*-hydroxyasparagine (position 150) and belongs to a subfamily of epidermal growth factor (EGF)-like modules known to participate in calcium binding. Modules IV and V belong to the complement control protein (CCP) family. The B chain is homologous to type I serine proteases, but lacks the histidine loop disulfide bridge (Arlaud & Gagnon, 1983; Leytus *et al.*, 1986; Journet & Tosi, 1986; Arlaud *et al.*, 1987). Each C1r monomer contains 13 (1 interchain and 12 intrachain) disulfide bridges, and four complex-type Asn-linked carbohydrates located in both the A chain (Asn108 and Asn204) and the B chain (Asn497 and Asn564). A polymorphic site (Ser/Leu) is present at position 135 (Arlaud *et al.*, 1987). The N-terminal α region contains one high-affinity calcium-binding site (Thielens *et al.*, 1990) and mediates assembly of the calcium-dependent C1s–C1r–C1r–C1s tetramer, the catalytic subunit of C1.

## Preparation

Available methods for the purification of human C1r involve: (a) selective isolation of the whole C1 complex from human serum by affinity-based procedures utilizing IgG-Sepharose or insoluble antibody–antigen aggregates; and (b) C1r purification by ion-exchange and gel-filtration chromatography. Isolation of the proenzyme is carried out in the presence of protease inhibitors (Arlaud & Thielens, 1993). Human C1r has been expressed in a baculovirus–insect cell system (Gal *et al.*, 1989).

## Biological Aspects

C1r is the protease responsible for activation of the C1 complex, an intramolecular, two-step process involving C1r

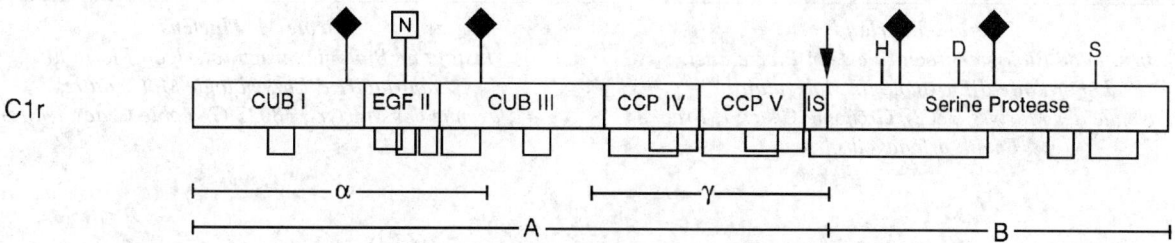

*Figure 43.1*   The modular structure of human C1r. The relative positions of the disulfide bridges, Asn-linked carbohydrates, and of the His, Asp and Ser residues of the catalytic triad are indicated. N, the Asn residue that undergoes post-translational hydroxylation; α, γ, fragments generated by limited proteolysis; IS, intermediary segment. The arrow shows the Arg-Ile bond cleaved on activation.

autoactivation, then C1s activation by active C1r. The average plasma concentration of C1r in human is $34\,\mathrm{mg\,liter}^{-1}$ $(0.2\,\mu\mathrm{M})$ (Cooper, 1985). C1r is synthesized mainly in hepatocytes and other cells such as monocytes and endothelial cells (Colten & Strunk, 1993). The genes coding for human C1r and C1s are located in region 12p13 (Van Cong *et al.*, 1988), and arranged in a tail-to-tail orientation (Kusumoto *et al.*, 1988). The serine protease domain of human C1r is encoded by a single exon (Tosi *et al.*, 1989). Partial or complete genetic deficiencies of C1r are correlated with lupus erythematosus and renal disease (Colten, 1993). C1r has been purified from plasma of rabbit and rat, and is probably present in all mammals and in a number of nonmammalian vertebrate species ranging from sharks to birds (Dodds & Petry, 1993).

## Distinguishing Features

A particular feature of C1r lies in its ability to undergo intramolecular autocatalytic activation. Autoactivation of purified proenzyme C1r is blocked in the presence of DFP and calcium ions (Thielens *et al.*, 1994).

## Further Reading

See Arlaud & Thielens (1993) for a full review of C1r.

## References

Arlaud, G.J. & Gagnon, J. (1983) Complete amino acid sequence of the catalytic chain of human complement subcomponent C1r. *Biochemistry* **22**, 1758–1764.

Arlaud, G.J. & Thielens, N.M. (1993) Human complement serine proteases C1r and C1s and their proenzymes. *Methods Enzymol.* **223**, 61–82.

Arlaud, G.J., Willis, A.C. & Gagnon, J. (1987) Complete amino acid sequence of the A chain of human complement-classical-pathway enzyme C1r. *Biochem. J.* **241**, 711–720.

Colten, H.R. (1993) Deficiencies of the first component of complement (C1): an update. *Behring Inst. Mitt.* **93**, 287–291.

Colten, H.R. & Strunk, R.C. (1993) Extrahepatic synthesis of complement. In: *Complement in Health and Disease*, 2nd edn (Whaley, K., Loos, M. & Weiler, J.M., eds). Dordrecht: Kluwer Academic, pp. 127–158.

Cooper, N.R. (1985) The classical complement pathway: activation and regulation of the first complement component. *Adv. Immunol.* **37**, 151–216.

Davis, A.E., Aulak, K.S., Zahedi, K., Bissler, J.J. & Harrison, R.A. (1993) C1 Inhibitor. *Methods Enzymol.* **223**, 97–120.

Dodds, A.W. & Petry, F. (1993) The phylogeny and evolution of the first component of complement, C1. *Behring Inst. Mitt.* **93**, 87–102.

Gal, P., Sarvari, M., Szilagyi, K., Zavodszky, P. & Schumaker, V.N. (1989) Expression of hemolytically active human complement component C1r proenzyme in insect cells using a baculovirus vector. *Complement Inflamm.* **6**, 433–441.

Journet, A. & Tosi, M. (1986) Cloning and sequencing of full-length cDNA encoding the precursor of human complement component C1r. *Biochem. J.* **240**, 783–787.

Kusumoto, H., Hirosawa, S., Salier, J.P., Hagen, F.S. & Kurachi, K. (1988) Human genes for complement components C1r and C1s in a close tail-to-tail arrangement. *Proc. Natl Acad. Sci. USA* **85**, 7307–7311.

Lepow, I.H., Naff, G.B., Todd, E.W., Pensky, J. & Hinz, C.F. (1963) Chromatographic resolution of the first component of human complement into three activities. *J. Exp. Med.* **117**, 983–1008.

Leytus, S.P., Kurachi, K., Sakariassen, K.S. & Davie, E.W. (1986) Nucleotide sequence of the cDNA coding for human complement C1r. *Biochemistry* **25**, 4855–4863.

McRae, B.J., Lin, T.-Y. & Powers, J.C. (1981) Mapping the substrate binding site of human C1r and C1s with peptide thioesters. *J. Biol. Chem.* **256**, 12362–12366.

Naff, G.B. & Ratnoff, O.D. (1968) The enzymatic nature of C1r. Conversion of C1s to C1 esterase and digestion of amino acid esters by C1r. *J. Exp. Med.* **128**, 571–593.

Thielens, N.M., Aude, C.A., Lacroix, M.B., Gagnon, J. & Arlaud, G.J. (1990) Calcium-binding properties and calcium-dependent interactions of the isolated NH2-terminal $\alpha$ fragments of human complement proteases C1r and C1s. *J. Biol. Chem.* **265**, 14469–14475.

Thielens, N.M., Illy, I., Bally, I.M. & Arlaud, G.J. (1994) Activation of human complement serine proteinase C1r is down-regulated by a calcium-dependent intramolecular control that is released in the C1 complex through a signal transmitted by C1q. *Biochem. J.* **301**, 509–516.

Tosi, M., Duponchel, C. & Meo, T. (1989) Complement genes C1r and C1s feature an intronless serine protease domain closely related to haptoglobin. *J. Mol. Biol.* **208**, 709–714.

Van Cong, N., Tosi, M., Gross, M.S., Cohen-Haguenauer, O., Jegou-Foubert, C., de Tand, M.F., Meo, T. & Frézal, J. (1988) Assignment of the complement serine protease genes C1r and C1s to chromosome 12 region 12p13. *Hum. Genet.* **78**, 363–368.

*Gérard J. Arlaud*
Institut de Biologie Structurale Jean-Pierre Ebel,
Laboratoire d'Enzymologie Moléculaire,
41, avenue des Martyrs, 38027 Grenoble Cedex 1, France
Email: arlaud@ibs.fr

*Nicole M. Thielens*
Institut de Biologie Structurale Jean-Pierre Ebel,
Laboratoire d'Enzymologie Moléculaire,
41, avenue des Martyrs, 38027 Grenoble Cedex 1, France

# 44. *Complement component C1s*

## Databanks

*Peptidase classification: clan SA, family S1, MEROPS ID: S01.193*
*NC-IUBMB enzyme classification: EC 3.4.21.42*
*Chemical Abstracts Service registry number: 80295-70-1*
*Databank codes:*

| Species | SwissProt | PIR | EMBL (cDNA) | EMBL (genomic) |
|---|---|---|---|---|
| *Homo sapiens* | P09871 | A40496 | J04080 | – |
| | | S00224 | M18767 | |
| | | | X06596 | |
| *Mesocricetus auratus* | P15156 | S05008 | X16160 | – |

## Name and History

Lepow *et al.* (1963) showed originally that human C1, the factor responsible for the triggering of the classical pathway of complement, could be resolved by ion-exchange chromatography into three distinct components, which they named C1q, C1r and C1s. With the isolation by Haines & Lepow (1964) of the activated form of C1s (originally called *C1 esterase*), and the demonstration of its ability to cleave C2 and C4, the natural substrates of C1, it became clear that C1s was the protease responsible for the enzymic activity of C1. The activated form of C1s is commonly indicated by an overbar (C1̄s).

## Activity and Specificity

C1s exhibits trypsin-like proteolytic activity restricted to the cleavage of arginyl bonds in its physiological substrates C2 and C4, in the sequences Ser-Leu-Gly-Arg↓Lys-Ile-Gln-Ile and Gly-Leu-Gln-Arg↓Ala-Leu-Glu-Ile, respectively. C1s was also shown to cleave noncomplement protein substrates such as $\beta_2$-microglobulin and the heavy chain of the major histocompatibility complex (MHC) class I antigens (Arlaud & Thielens, 1993). C1s hydrolyzes a wide range of synthetic substrates containing an Arg, Lys or, unexpectedly, a Tyr

residue at the P1 position. Esterolytic activity can be assayed on Tos-Arg↓OMe, Z-Tyr↓OPhNO₂, and Z-Gly-Arg↓SBzl (McRae *et al.*, 1981; Arlaud & Thielens, 1993). All of these substrates are cleaved by other proteases also, and therefore must be used only for purified C1s. The pH optimum is 7.0–8.0 (Haines & Lepow, 1964).

C1 inhibitor is the only known physiological inhibitor of activated C1s (Davis *et al.*, 1993). DFP, PMSF and *p*-nitrophenyl-*p*′-guanidinobenzoate are also potent irreversible inhibitors (Arlaud & Thielens, 1993).

## Structural Chemistry

Isolated C1s dimerizes in the presence of calcium ions, but preferentially associates with C1r to form calcium-dependent C1s–C1r–C1r–C1s tetramers. Monomeric human C1s (673 amino acids) is single chain in the proenzyme form and is split on activation, through cleavage at Arg422-Ile423, into two disulfide-linked chains A and B. The N-terminal A chain exhibits a five-module structure homologous to that of C1r (Chapter 43) and mannose-binding protein-associated serine protease (Chapter 45) (Figure 44.1), and comprises two CUB modules (**c**omplement C1r/C1s, **u**egf, **b**one morphogenetic protein), a single epidermal growth factor (EGF)-like module, and two complement control protein (CCP) modules.

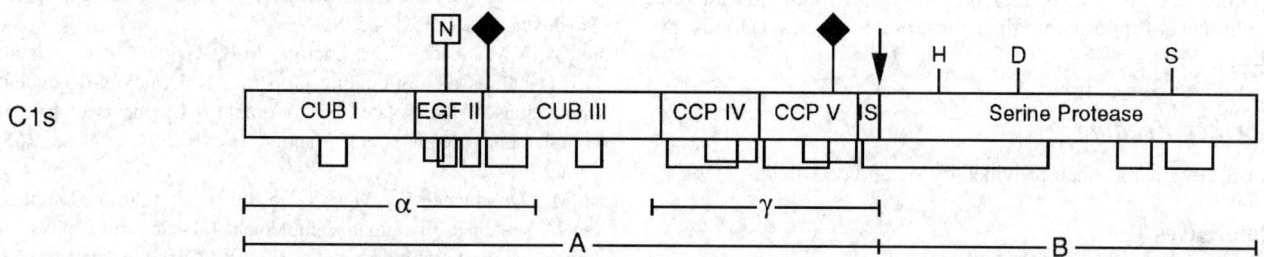

**Figure 44.1**   The modular structure of human C1s. The relative positions of the disulfide bridges, Asn-linked carbohydrates, and of the His, Asp and Ser residues of the catalytic triad are indicated. N, the Asn residue that undergoes partial post-translational hydroxylation; $\alpha$, $\gamma$, fragments generated by limited proteolysis; IS, intermediary segment. The arrow shows the Arg-Ile bond cleaved on activation.

Asn134 in the EGF module is partially modified to *erythro-β*-hydroxyasparagine. The B chain is homologous to type I serine proteases, but lacks the histidine loop disulfide bridge. C1s contains 13 (1 interchain and 12 intrachain) disulfide bridges (Tosi *et al.*, 1987; Mackinnon *et al.*, 1987; Thielens *et al.*, 1990a). Human C1s contains two complex-type Asn-linked carbohydrates, both located in the A chain (Asn159 and Asn391). Asn391 is occupied by either biantennary or tri-antennary species, giving rise to three types of C1s molecules of $M_r$ 79 320, 79 970 and 80 130 (average $M_r$ 79 800) (Petillot *et al.*, 1995). The pI is 4.5. The N-terminal $\alpha$ region contains one high-affinity calcium-binding site (Thielens *et al.*, 1990b). A three-dimensional model of the catalytic region of human C1s is available (Rossi *et al.*, 1995).

## Preparation

Current methods for the purification of human C1s involve: (a) isolation of the whole C1 complex from human serum by affinity-based procedures utilizing IgG-Sepharose or insoluble antibody–antigen aggregates; and (b) C1s purification by ion-exchange chromatography. Isolation of the proenzyme requires the use of protease inhibitors (Arlaud & Thielens, 1993). Human C1s has been expressed in a baculovirus–insect cell system and the recombinant protein shown to retain the functional properties of serum C1s (Luo *et al.*, 1992). Hamster C1s has been expressed in mouse fibroblasts transformed by polyoma virus (Toyoguchi *et al.*, 1995).

## Biological Aspects

C1s is the protease responsible for the cleavage by activated C1 of its natural substrates C4 and C2, a process which generates C4b2a, the protease that cleaves complement component C3. The average plasma concentration of C1s is 31 mg liter$^{-1}$ (0.2 mM) (Cooper, 1985). C1s is synthesized mainly in hepatocytes, as well as other cells including monocytes and endothelial cells (Colten & Strunk, 1993). The genes coding for human C1r and C1s are located in region 12p13 and arranged in a close tail-to-tail orientation (Van Cong *et al.*, 1988; Kusumoto *et al.*, 1988). The serine protease domain of human C1s is encoded by a single exon (Tosi *et al.*, 1989). Known cases of partial or complete genetic deficiency of C1s are usually associated with a C1r deficiency and correlated with lupus erythematosus and renal disease (Colten, 1993). C1s has been purified from rabbit, bovine and rat sera, and from malignant hamster fibroblasts. C1-like activity is probably present in all mammals and in a number of nonmammalian vertebrate species (Dodds & Petry, 1993).

## Further Reading

A full review has been provided by Arlaud & Thielens (1993).

## References

Arlaud, G.J. & Thielens, N.M. (1993) Human complement serine proteases C1r and C1s and their proenzymes. *Methods Enzymol.* **223**, 61–82.

Colten, H.R. (1993) Deficiencies of the first component of complement (C1): an update. *Behring Inst. Mitt.* **93**, 287–291.

Colten, H.R. & Strunk, R.C. (1993) Extrahepatic synthesis of complement. In: *Complement in Health and Disease*, 2nd edn (Whaley, K., Loos, M. & Weiler, J.M., eds). Dordrecht: Kluwer Academic, pp. 127–158.

Cooper, N.R. (1985) The classical complement pathway: activation and regulation of the first complement component. *Adv. Immunol.* **37**, 151–216.

Davis, A.E., Aulak, K.S., Zahedi, K., Bissler, J.J. & Harrison, R.A. (1993) C1 Inhibitor. *Methods Enzymol.* **223**, 97–120.

Dodds, A.W. & Petry, F. (1993) The phylogeny and evolution of the first component of complement, C1. *Behring Inst. Mitt.* **93**, 87–102.

Haines, A.L. & Lepow, I.H. (1964) Studies on human C1 esterase. I. Purification and enzymatic properties. *J. Immunol.* **92**, 456–467.

Kusumoto, H., Hirosawa, S., Salier, J.P., Hagen, F.S. & Kurachi, K. (1988) Human genes for complement components C1r and C1s in a close tail-to-tail arrangement. *Proc. Natl Acad. Sci. USA* **85**, 7307–7311.

Lepow, I.H., Naff, G.B., Todd, E.W., Pensky, J. & Hinz, C.F. (1963) Chromatographic resolution of the first component of human complement into three activities. *J. Exp. Med.* **117**, 983–1008.

Luo, C., Thielens, N.M., Gagnon, J., Gal, P., Sarvari, M., Tseng, Y., Tosi, M., Zavodszky, P., Arlaud, G.J. & Schumaker, V.N. (1992) Recombinant human complement subcomponent C1s lacking $\beta$-hydroxyasparagine, sialic acid, and one of its two carbohydrate chains still reassembles with C1q and C1r to form a functional C1 complex. *Biochemistry* **31**, 4254–4262.

Mackinnon, C.M., Carter, P.E., Smyth, S.J., Dunbar, B. & Fothergill, J.E. (1987) Molecular cloning of cDNA for human complement component C1s. The complete amino acid sequence. *Eur. J. Biochem.* **169**, 547–553.

McRae, B.J., Lin, T.-Y. & Powers, J.C. (1981) Mapping the substrate binding site of human C1r and C1s with peptide thioesters. *J. Biol. Chem.* **256**, 12362–12366.

Petillot, Y., Thibault, P., Thielens, N.M., Rossi, V., Lacroix, M., Coddeville, B., Spik, G., Schumaker, V.N., Gagnon, J. & Arlaud, G.J. (1995) Analysis of the *N*-linked oligosaccharides of human C1s using electrospray ionisation mass spectrometry. *FEBS Lett.* **358**, 323–328.

Rossi, V., Gaboriaud, C., Lacroix, M., Ulrich, J., Fontecilla-Camps, J.C., Gagnon, J. & Arlaud, G.J. (1995) Structure of the catalytic region of human complement protease C1s: study by chemical cross-linking and three-dimensional homology modeling. *Biochemistry* **34**, 7311–7321.

Thielens, N.M., van Dorsselaer, A., Gagnon, J. & Arlaud, G.J. (1990a) Chemical and functional characterization of a fragment of C1s containing the epidermal growth factor homology region. *Biochemistry* **29**, 3570–3578.

Thielens, N.M., Aude, C.A., Lacroix, M.B., Gagnon, J. & Arlaud, G.J. (1990b) Calcium-binding properties and calcium-dependent interactions of the isolated NH$_2$-terminal $\alpha$ fragments of human complement proteases C1r and C1s. *J. Biol. Chem.* **265**, 14469–14475.

Tosi, M., Duponchel, C., Meo, T. & Julier, C. (1987) Complete cDNA sequence of human complement C1s and close physical linkage of the homologous genes C1s and C1r. *Biochemistry* **26**, 8516–8524.

Tosi, M., Duponchel, C. & Meo, T. (1989) Complement genes C1r and C1s feature an intronless serine protease domain closely related to haptoglobin. *J. Mol. Biol.* **208**, 709–714.

Toyoguchi, T., Yamaguchi, K., Imajoh-Ohmi, S., Kato, N., Kuso-noki, M., Kageyama, H., Sakiyama, S., Nagasawa, S., Moriya, H. & Sakiyama, H. (1995) Purification and characterization of recombinant hamster tissue complement C1s. *Biochim. Biophys. Acta* **1250**, 90–96.

Van Cong, N., Tosi, M., Gross, M.S., Cohen-Haguenauer, O., Jegou-Foubert, C., de Tand, M.F., Meo, T. & Frézal, J. (1988) Assignment of the complement serine protease genes C1r and C1s to chromosome 12 region 12p13. *Hum. Genet.* **78**, 363–368.

**Gérard J. Arlaud**
*Institut de Biologie Structurale Jean-Pierre Ebel,*
*Laboratoire d'Enzymologie Moléculaire,*
*41, avenue des Martyrs, 38027 Grenoble Cedex 1, France*
*Email: arlaud@ibs.fr*

**Nicole M. Thielens**
*Institut de Biologie Structurale Jean-Pierre Ebel,*
*Laboratoire d'Enzymologie Moléculaire,*
*41, avenue des Martyrs, 38027 Grenoble Cedex 1, France*

# 45. *Mannose-binding protein-associated serine endopeptidases*

## Databanks

Peptidase classification: clan SA, family S1, MEROPS IDs: S01.198, S01.229
NC-IUBMB enzyme classification: none
Databank codes:

| Species | SwissProt | PIR | EMBL (cDNA) | EMBL (genomic) |
|---|---|---|---|---|
| MASP-1 | | | | |
| *Homo sapiens* | P48740 | – | D17525 D28593 | – |
| *Mus musculus* | P98064 | – | D16492 | – |
| MASP-2 | | | | |
| *Homo sapiens* | – | – | Y09926 | – |

## Name and History

Two ***mannose-binding protein-associated serine endopepti-dases*** (***MASP-1*** and ***MASP-2***) are now known to be associated in human serum with the lectin mannose-binding protein (MBP). Mannose-binding protein is also called 'mannan-binding protein' or 'mannan-binding lectin' (MBL) in the recent literature. The MBP–MASP complex plays an important role in innate immunity by virtue of the binding of MBP to carbohydrate structures present on a wide variety of microorganisms. The interaction of MBP with these specific arrays of carbohydrate structures brings about the activation of the MASP proenzymes which, in turn, activate the classical pathway of complement by cleaving the complement components C4 and C2 to form the C3 convertase C4b2b (Kawasaki *et al*., 1989; Matsushita & Fujita, 1992; Ji *et al*., 1993) (Chapter 46). The MBP–MASP proenzyme complex was, until very recently, considered to contain only one type of protease (MASP-1), but it is now clear that there

is a second distinct protease (MASP-2) associated with MBP (Thiel *et al*., 1996; Vorup-Jensen *et al*., 1996). MASP-1 is also called the ***P-100 protease*** component of the serum Ra-reactive factor, which is now recognized as being a complex composed of MBP plus MASP (Matsushita *et al*., 1992; Ji *et al*., 1993). The ability of the activated MASP, within the MBP–MASP complex, to act on the complement components C4 and C2 in a manner apparently identical to that of the C1s enzyme (Chapter 44) within the C1q–(C1r)$_2$–(C1s)$_2$ complex of the classical pathway of complement, suggests that there is a MBP–MASP-1–MASP-2 complex which is functionally analogous to the C1q–(C1r)$_2$–(C1s)$_2$ complex. However, to date, there has been no conclusive demonstration of such a complex, and the possibility that there are distinct MBP–MASP-1 and MBP–MASP-2 complexes cannot be excluded. The C1q–(C1r)$_2$–(C1s)$_2$ complex is activated by the interaction of C1q with the Fc regions of antibody IgG or IgM present in immune complexes. This brings about the autoactivation of the C1r proenzyme (Chapter 43) which, in turn, activates the C1s proenzyme which then acts on complement components C4 and C2. It has been speculated that MASP-1 and MASP-2 play analogous roles to C1r and C1s, respectively (Thiel *et al.*, 1996; Vorup-Jensen *et al.*, 1996).

## Activity and Specificity

Since it was not recognized until 1995 that there are two forms of MASP, it seems likely that all the activity and specificity studies carried out on MASP preparations were performed using a mixture of both MASP-1 and MASP-2. Therefore, unless stated otherwise, any activity/specificity which is described here should be regarded as possibly being a property of either MASP-1 and/or MASP-2, especially as some of the functional studies were carried out using an MBP–MASP complex rather than purified MASP (Ogata *et al.*, 1995). Both human MASP (Matsushita & Fujita, 1992) and mouse MASP (Ogata *et al.*, 1995) show proteolytic activity towards the complement components C4 and C2. As judged by the size of the fragments of C4 generated, and the functional C3 convertase activity generated, it seems likely that the activated MASP splits C4 and C2 in an identical manner to that carried out by activated C1s, i.e. at a single arginyl bond (Arg76$\dagger$Ala77) within the $\alpha$ chain of C4 and at a single arginyl bond (Arg223$\dagger$Lys224) within the proenzyme chain of C2. It has also been reported that the mouse MASP (in the form of the mouse MBP–MASP complex designated Ra-reactive factor) can, unlike C1s, split the $\alpha$ chain of complement component C3 to yield the biologically active fragments C3a and C3b (Ogata *et al.*, 1995). If this were to take place in the human system it would require the splitting of a single arginyl bond (Arg77$\dagger$Ser78) within the $\alpha$ chain of C3. The activated MASP, like activated C1s, is unable to split complement component C5.

## Structural Chemistry

The cDNA-derived amino acid sequences of human and mouse MASP-1 (Sato *et al.*, 1994; Takada *et al.*, 1993; Takayama *et al.*, 1994) and human MASP-2 (Vorup-Jensen *et al.*, 1996) indicate that both these proteases are serine peptidases having the expected triad of His, Asp and Ser residues within their putative catalytic domains. The coding sequences of human MASP-1 and MASP-2 encode polypeptide chains of 699 and 686 amino acid residues, respectively, which include leader peptides of 19 residues and 15 residues, respectively. When the leader peptides are omitted, the calculated molecular masses of MASP-1 and MASP-2 are 76 976 Da and 74 153 Da, respectively. The MASP-2 sequence does not contain any potential *N*-linked glycosylation sites, whereas the MASP-1 sequence contains four such sites. MASP-1 shows 44% sequence identity to MASP-2 and both sequences show approximately 37% sequence identity to the complement enzymes C1r and C1s.

Both MASP-1 and MASP-2 are predicted to consist of six distinct domains arranged as found in C1r and C1s (see Figs 43.1, 44.1), i.e. (I) an N-terminal C1r/s-like (or CUB) domain; (II) an epidermal growth factor (EGF)-like domain; (III) a second C1r/s-like domain; (IV and V) two complement control protein (CCP) domains; and (VI) the serine protease

domain. The single-chain proenzyme MASP-1 is activated (like proenzyme C1r and C1s) by cleavage of an Arg-Ile bond located between the second CCP domain (domain V) and the serine protease domain (domain VI). Proenzyme MASP-2 is also considered to be activated in a similar fashion. It is not yet known which proteases activate the proenzyme forms of MASP-1 and MASP-2. The activation of MASP-1 and MASP-2 leads to the formation, in each case, of H and L chains of approximately 43 kDa and 25 kDa, with the L chain corresponding to the entire, C-terminal serine protease domain. All the cysteine residues present in MASP-2, C1r and C1s align with equivalent residues in MASP-1, however MASP-1 has two cysteine residues (at positions 465 and 481 in the L chain) which are not found in the MASP-2 (Vorup-Jensen *et al.*, 1996), C1r and C1s (Arlaud & Gagnon, 1981). These two cysteine residues in MASP-1 are in the expected positions used to form the 'histidine-loop' disulfide bridge as found in trypsin and chymotrypsin. This suggests that MASP-2, C1r and C1s may have evolved, by gene duplication and divergence, from MASP-1.

## Preparation

Human MASP-1 was isolated by affinity chromatography, together with MBP, by running human serum on a column of mannan-Sepharose in the presence of calcium. It was then separated from MBP by passing the MBP–MASP preparation through an anti-MBP column in the presence of EDTA (Matsushita & Fujita, 1992). In another procedure, human plasma was used in the preparation of both MASP-1 and MASP-2 (Thiel *et al.*, 1996). The plasma was diluted with buffer containing EDTA and enzyme inhibitors and precleared by applying to Sepharose 2B-CL and then mannan-Sepharose. Then a thrombin inhibitor (D-Phe-Pro-Arg-CH$_2$Cl) and CaCl$_2$ was added to the precleared plasma. The recalcified plasma was passed through Sepharose 2B-CL and mannan-Sepharose and the proteins, which bound in a calcium-dependent fashion, were eluted with EDTA-containing buffer. The preparation was finally purified by application to an *N*-acetylglucosamine-Sepharose affinity column in the presence of calcium followed by elution with an EDTA-containing buffer. The preparation contained MBP, MASP-1 and MASP-2 as judged by use of specific antisera and gel filtration on Superose 6B-CL (Thiel *et al.*, 1996). It has been found that MASP-1, which is tightly bound to MBP even in the absence of Ca$^{2+}$, can be released from the complex by treatment at pH 5.0 and separated from MBP by gel filtration at the same pH (Tan *et al.*, 1996).

## Biological Aspects

Northern blot analysis indicates that liver is the major source of MASP-1 and MASP-2 mRNA. A major MASP-1 transcript was seen at 4.8 kb and a minor one at 3.4 kb, both of which are expressed in human fetal liver. For MASP-2 mRNA, three signals were consistently deleted at 2.6 kb, 1.4 kb and 1 kb on northern blot analysis of fetal liver mRNA. Initial Southern blotting data indicate that the three species of mRNA of MASP-2 are transcripts of a single structural gene and that they are possibly generated by alternative splicing (W. Schwaeble, unpublished results). No signals for

MASP-1 or MASP-2 mRNA could be detected in a variety of other tissues by northern blotting. The human MASP-1 gene has been mapped to the 3q27–28 region of the long arm of chromosome 3 and the mouse MASP-1 gene has been mapped to the 16B2–B3 region of chromosome 16 (Takada *et al*., 1995). The role of the MBP–MASP complex in innate immunity is mediated via the calcium-dependent binding of the C-type lectin domains (present in the MBP molecule) to carbohydrate structures found on yeast, bacteria, viruses and fungi. This recognition phase brings about the activation of the proenzyme MASP which then mimics the action of the activated C1s within the C1q–(C1r)$_2$–(C1s)$_2$ complex by cleaving C4 and C2 to form the C3 convertase C4b2b. This allows deposition of C4b and C3b on target pathogens and thus promotes killing and clearance through phagocytosis. Although levels of MASP have not yet been reported in normal and disease states, it is known that deficiency of MBP leads to a tendency to frequent infections in childhood (Super *et al*., 1989; Garred *et al*., 1995) and a decreased resistance to HIV infection (Nielsen *et al*., 1995; Garred *et al*., 1996). Thus, deficiency, or nonutilization, of MASP could have an adverse effect on an individual's ability to mount immediate, nonantibody-dependent defense against certain pathogens.

## Distinguishing Features

It should be emphasized that although MASP-1 and MASP-2 have been fully characterized at the cDNA level and found to show approximately 50% sequence identity, it is not yet clear if they are both associated with MBP within the same complex or whether they are independently associated with MBP in serum. Preparations of MASP have been shown to split, and activate, the complement components C4, C2 and C3. These activities have been attributed to MASP-1. However, a recent study suggests that MASP-2, and not MASP-1, may be responsible for C4 activation (Thiel *et al*., 1996; Vorup-Jensen *et al*., 1996). Until highly purified preparations of proenzyme MASP-1 and MASP-2 are isolated from plasma, or produced by recombinant DNA techniques, it will be difficult to determine how the proenzyme forms of MASP-1 and MASP-2 are activated and to firmly establish which of the complement components (C4, C2 and C3) are activated by MASP-1 and/or MASP-2. A study of the substrate specificities of the murine MBP–MASP preparation (in the form of a preparation of the Ra-reactive factor from mouse serum) showed that the mouse protease complex cleaved the C4 α chain with specific activities 20- to 100-fold greater than either human or murine C1s (Ogata *et al*., 1995). The murine MBP–MASP preparation was also shown to cleave the C3 α chain but at a lower efficiency than its cleavage of the C4 α chain. Under the same conditions, C1s showed no cleavage of the C3 α chain. The MBP–MASP preparation also showed different reactivities against synthetic substrates on comparison with the activated mouse C1q–(C1r)$_2$–(C1s)$_2$ complex and activated human C1s (Ogata *et al*., 1995).

## Further Reading

Matsushita & Fujita (1996) have provided a review of the cDNA cloning of MASP-1 and the enzymatic properties of MASP preparations, which may include MASP-2 as well as MASP-1.

## References

Arlaud, G.J. & Gagnon, J. (1981) C1r and C1s subcomponents of human complement: two serine proteases lacking the 'histidine-loop' disulphide bridge. *Biosci. Rep.* **1**, 779–784.

Garred, P., Madsen, H.O., Hofman, B. & Svejgaard, A. (1995) Increased frequency of homozygosity of abnormal mannan-binding-protein alleles in patients with suspected immunodeficiency. *Lancet* **346**, 941–943.

Garred, P., Madsen, H.O., Balslev, U., Hofmann, B., Gerstoft, J. & Svejgaard, A. (1996) Variant alleles of mannan-binding protein are associated with susceptibility to infection with HIV and influence the progression of AIDS. *Mol. Immunol.* **33** (suppl. 1), 8.

Ji, Y.H., Fujita, T., Hatsuse, H., Takahashi, A., Matsushita, M. & Kawakami, M. (1993) Activation of the C4 and C2 components of complement by a proteinase in serum bactericidal factor, Ra reactive factor. *J. Immunol.* **150**, 571–578.

Kawasaki, N., Kawasaki, T. & Yamashina, I. (1989) A serum lectin (mannan-binding protein) has complement-dependent bactericidal activity. *J. Biochem. (Tokyo)* **106**, 483–489.

Matsushita, M. & Fujita, T. (1992) Activation of the classical complement pathway by mannose-binding protein in association with a novel C1s-like serine protease. *J. Exp. Med.* **176**, 1497–1502.

Matsushita, M. & Fujita, T. (1996) MASP (MBP-associated serine protease). In: *Collectins and Innate Immunity* (Ezekowitz, R.A.B., Sastry, K. & Reid, K.B.M., eds). Austin, TX: R.G. Landes Co.

Matsushita, M., Takahashi, A., Hatsuse, H., Kawakami, M. & Fujita, T. (1992) Human mannose-binding protein is identical to a component of Ra-reactive factor. *Biochem. Biophys. Res. Commun.* **183**, 645–651.

Nielsen, S.L., Andersen, P.L., Koch, C., Jensenius, J.C. & Thiel, S. (1995) The level of the serum opsonin, mannan-binding protein in HIV-1 antibody-positive patients. *Clin. Exp. Immunol.* **100**, 219–222.

Ogata, R.T., Low, P.J. & Kawakami, M. (1995) Substrate specificities of the protease of mouse serum Ra-reactive factor. *J. Immunol.* **154**, 2351–2357.

Sato, T., Endo, Y., Matsushita, M. & Fujita, T. (1994) Molecular characterisation of a novel serine protease involved in activation of the complement system by mannose-binding protein. *Int. Immunol.* **6**, 665–669.

Super, M., Thiel, S., Lu, J., Levinsky, R.J. & Turner, M.W. (1989) Association of low levels of mannan-binding protein with a common defect of opsonisation. *Lancet* **2**, 1236–1239.

Takada, F., Takayama, Y., Hatsuse, H. & Kawakami, M. (1993) A new member of the C1s family of complement proteins found in a bactericidal factor, Ra-reactive factor, in human serum. *Biochem. Biophys. Res. Commun.* **196**, 1003–1009.

Takada, F., Seki, N., Matsuda, Y.I., Takayama, Y. & Kawakami, M. (1995) Localisation of the genes for the 11-kDa complement-activating components of Ra-reactive factor (CRARF and Crarf) to human 3q27-q28 and mouse 16B2-B3. *Genomics* **25**, 757–759.

. Takayama, Y., Takada, F., Takahashi, A. & Kawakami, M. (1994) A 100-kDa protein in the C4-activating component of Ra-reactive factor is a new serine protease having module organization similar to C1r and C1s. *J. Immunol.* **152**, 2308–2316.

Tan, S.M., Chung, M.C.M., Kon, O.L., Thiel, S., Lee, S.H. & Lu, J. (1996) Improvements on the purification of mannan-binding lectin

(MBL) and demonstration of its Ca$^{2+}$-independent association with a C1s-like serine protease. *Biochem. J.* **319**, 329–332.

Thiel, S., Vorup-Jensen, T., Laursen, S.B., Willis, A., Reid, K.B.M., Hansen, S. & Jensenius, J.C. (1996) Identification of a new mannan-binding lectin associated serine protease (MASP-2). *Mol. Immunol.* **33**(suppl. 1), 91.

Vorup-Jensen, T., Stover, C., Poulsen, K., Laursen, S.B., Eggleton, P., Reid, K.B.M., Willis, A.C., Schwaeble, W., Lu, J., Holmskov, U., Jensenius, J.C. & Thiel, S. (1996) Cloning of cDNA encoding a human MASP-like protein (MASP-2). *Mol. Immunol.* **33**(suppl. 1), 81.

*Kenneth B.M. Reid*
*MRC Immunochemistry, Department of Biochemistry,*
*University of Oxford,*
*South Parks Road, Oxford OX1 3QU, UK*
*Email: kbmreid@bioch.ox.ac.uk*

# 46. Complement component C2 and the classical pathway C3/C5 convertase

## Databanks

*Peptidase classification: clan SA, family S1, MEROPS ID: S01.194*
*NC-IUBMB enzyme classification: EC 3.4.21.43*
*Chemical Abstracts Service registry number: 56626-15-4*
*Databank codes:*

| Species | SwissProt | PIR | EMBL (cDNA) | EMBL (genomic) |
|---|---|---|---|---|
| Component C2 | | | | |
| *Homo sapiens* | P06681 | A25971 | X04481 | M15082: 5′ end |
| *Mus musculus* | P21180 | A38876 | M57891 | M60563: exon 1 |
| | | | | M60564: exon 2 |
| | | | | M60565: exon 3 |
| | | | | M60566: exon 4 |
| | | | | M60567: exon 5 |
| | | | | M60568: exon 6 |
| | | | | M60569: exon 7 |
| | | | | M60570: exon 8 |
| | | | | M60571: exon 9 |
| | | | | M60572: exon 10 |
| | | | | M60573: exon 11 |
| | | | | M60574: exon 12 |
| | | | | M60575: exon 13 |
| | | | | M60576: exon 15 |
| | | | | M60577: exon 16 |
| | | | | M60578: exon 17 |
| | | | | M60579: exon 18 and complete CDS |
| | | | | M60605: exon 14 (long) |

## Name and History

The classical pathway of complement activation was first identified at the end of the last century as a major effector system of the immune response, causing the lysis of antibody-coated microorganisms or triggering their ingestion by phagocytic cells. It is an enzyme cascade which, upon activation, leads to the assembly of a complex proteolytic enzyme, the *C3 convertase*, which cleaves C3, the central protein of the complement system. Proteolytic cleavage of C3 initiates the many important effector functions of the system.

The classical pathway C3 convertase is usually assembled following activation of C1 by antibody–antigen aggregates (immune complexes) containing IgM or IgG. Upon activation by immune complexes, the classical pathway C3 convertase is assembled from C2 and C4 following cleavage of these proteins into C2a and C2b and C4a and C4b by the C1s subcomponent (Chapter 44) of the activated C1 complex. There are other mechanisms of assembly of the convertase that do not require antibody. For example, a plasma serine proteinase called MASP (Chapter 45) when complexed to mannan-binding protein bound to bacterial or yeast cell walls, will activate C2 and C4 without the need for antibodies or C1 (Ji *et al.*, 1993).

The convertase cleaves C3 into a large fragment, C3b, and a small fragment, C3a. Immediately after cleavage, the C3b moiety is able to bind covalently to nucleophilic groups on the proteins and carbohydrates on the surface of the microorganism or immune complex which is activating the pathway. The covalent binding reaction is catalyzed by a buried thioester in C3 that becomes externalized upon proteolytic cleavage. When C3b molecules are attached covalently to C4b in the convertase, the specificity of the enzyme is altered and the enzyme becomes the C5 convertase. C3 and C5 are homologous proteins, each cleaved at a single Arg-Xaa bond by its respective convertase. The active site of the convertase is contained within the C2a fragment of the C2 component (Polley & Müller-Eberhard, 1968).

## Activity and Specificity

When the classical pathway C3 convertase is assembled, C4 is cleaved by C1s with the release of a small peptide, C4a (8 kDa), to form C4b (198 kDa), and C2 is cleaved to C2a (74 kDa) and C2b (34 kDa). For the generation of C3 convertase activity, C4 cleavage must precede C2 cleavage. The enzyme is not generated by the addition of C4 to a previously incubated mixture of C2 and C1s, but is generated by the addition of C2 to a previously incubated mixture of C4 and C1s (Nagasawa & Stroud, 1977; Kerr, 1980).

C4b and C2 form a $Mg^{2+}$-dependent complex, and it is the cleavage of C2 in this complex that results in C3 convertase activity. The interaction of C4b and C2 is mediated primarily by the C2b part of the C2 molecule (Kerr, 1980; Oglesby *et al.*, 1988). C2a is the catalytic subunit of C3 convertase. Assembly of the C3 convertase is very inefficient in free solution. It is much more efficient on cell surfaces or on immune complexes where the C4b that is formed can bind covalently via its thioester adjacent to the activated C1 complex that is initiating the activation of the classical pathway. When C2 binds to the C4b it is correctly oriented for cleavage by the C1s.

The C3 convertase, once assembled, is extremely unstable. In solution, the half-life of the enzyme at 37°C is less than 1 min. The decay of activity reflects the release of C2a from the C4b–C2b–C2a complex. The soluble classical pathway C3 convertase is more stable when assembled in the presence of $Ni^{2+}$ rather than $Mg^{2+}$ (Villiers *et al.*, 1985). The C3 convertase can also be stabilized by prior treatment of C2 with low concentrations of iodine, which increases the

affinity of C2a for C4b (Polley & Müller-Eberhard, 1967). The mechanism involves the oxidation of the thiol group of Cys241, suggesting that the site of interaction of C4b with C2a includes the residues around this cysteine (Parkes *et al.*, 1983; Kerr & Parkes, 1984; Horiuchi *et al.*, 1991). Thiol-blocking reagents destroy the hemolytic activity of C2. Iodine treatment is widely used to stabilize the convertase. It has greatly facilitated study of human C2. The C3 convertase prepared using guinea pig C2 is much more stable than that containing human C2 (Kerr & Gagnon, 1982).

The C3 convertase cleaves C3 at a single Arg┼Ser bond to give C3a (9 kDa) and C3b (185 kDa). In the presence of excess C3b, C5 is also a substrate for the enzyme. C2a is the active subunit of this C5 convertase. The C5 convertase of the classical complement pathway is thus a protein complex consisting of C4b, C2a and C3b (Ebanks & Isenman, 1995). Within this complex, C3b binds to C4b via an ester linkage to Ser1217 of C4b (Kim *et al.*, 1992; Kozono *et al.*, 1990). Although probably not physiological, it has been shown that C4b dimer, when complexed with C2, expresses C3/C5 convertase activity without participation of C3 (Masaki *et al.*, 1991).

In addition to its inherent instability, the activity of the C3 convertase is further controlled by several serum or cell membrane proteins that increase the rate of decay of the enzyme. Decay-accelerating factor (DAF) is a widely distributed membrane protein that increases the rate of dissociation of C2a from the complex (Fujita *et al.*, 1987). C4b-binding protein is a serum protein which dissociates the convertase and acts as a cofactor for factor I (Chapter 48), a proteinase that cleaves and inactivates C4b. Membrane cofactor protein (MCP) serves a similar function on the surfaces of cells of many types, protecting the cells of the host from the potentially lethal effects of the complement system (Reid *et al.*, 1986).

C2a in the C3 convertase, (C2b)–C4b–C2a, cleaves C3 at the single Arg77┼Ser78 bond and in the C5 convertase, (C2b)–C4b–C3b–C2a, cleaves C5 at the single Arg74┼Leu75 bond. No other natural substrates have been identified. Isolated C2a is unable to cleave either C3 or C5. Both C2 and C2a have been shown to possess weak esterase activity against certain arginine and lysine esters such as Ac-Gly-Lys-OMe. The $K_m$ for the ester is 18 mM. Cleavage of C2 by C1s, but not oxidation by iodine, increases enzymatic activity slightly. Ac-Gly-Lys-OMe competitively inhibits the cleavage of C3 by assembled convertase (Cooper, 1975). The specificities of C2 and C2a have been studied with a series of peptide thioester substrates. They had comparable reactivities and hydrolyzed peptides containing Leu-Ala-Arg and Leu-Gly-Arg, which have the same sequence as the cleavage sites of C3 and C5. The best substrates for C2 and C2a were Z-Gly-Leu-Ala-Arg┼SBzl and Z-Leu-Gly-Leu-Ala-Arg┼SBzl, respectively. Cattle trypsin hydrolyzed these thioester substrates with $k_{cat}/K_m$ values approximately a 1000-fold higher than the complement enzymes (Kam *et al.*, 1987). Peptides similar to the sequences around the scissile Arg-Ser bond of C3 only weakly inhibit the cleavage of C3 by the convertase, suggesting that the binding requirements of the C3 convertase are complex (Peake *et al.*, 1990). C2 has been reported to be inhibited by DFP (Medicus *et al.*, 1976), but this has not been confirmed by others.

## Structural Chemistry

Component C2 (and factor B, the homologous protein of the alternative pathway – see Chapter 47) have been shown to be serine proteinases by sequence analysis. They are mechanistically similar to other members of the trypsin (S1) family, but are activated by a mechanism that differs from that which is general in the family. The primary structure of human C2 has been deduced from cDNA (Bentley & Porter, 1984; Horiuchi *et al*., 1989). The protein (107 kDa) has 732 amino acids. It contains multiple structural domains: a 15 amino acid leader peptide, three short consensus repeats (SCRs), a von Willebrand factor type A-like domain and a serine proteinase domain (Bentley, 1986). In the electron microscope, C2 appears as a three-domain structure (Smith *et al*., 1984).

The serine proteinase domains of C2 and factor B have been modeled with reference to nine serine proteinases of known crystal structure and four other complement serine proteinases: C1r, C1s, factor I and factor D (Perkins & Smith, 1993). All sequence insertions and deletions were readily located at the protein surface. The internal location of disulfide bridges and the surface location of putative glycosylation sites were compatible with the predicted structures. Carbohydrate compositions but not their detailed structures have been determined (Tomana *et al*., 1985).

## Preparation

C2 is one of the least abundant of the complement components in human plasma, being present at a concentration of $15–20\,\mathrm{mg\,liter^{-1}}$. This and the extreme susceptibility of C2 to proteolytic digestion have in the past hampered its purification and molecular characterization. However, purification schemes which yield milligram quantities of human and guinea pig C2 have been described (Kerr, 1981; Kerr & Gagnon, 1982). C2 is often used in a functionally pure form.

## Biological Aspects

C2 from human serum is polymorphic, as judged by electrophoretic mobility. The two common alleles have gene frequencies of 0.96 and 0.04 (Campbell, 1987). C2 deficiency is the most common of the human complement deficiencies, and it is transmitted as an autosomal recessive trait. Many C2-deficient individuals are healthy, but most suffer from systemic lupus erythematosus (SLE)-like 'immune complex diseases', possibly reflecting the importance of the classical pathway C3 convertase in the handling of immune complexes (Whaley *et al*., 1992). The single gene coding for human C2 is located on the short arm of chromosome 6 (6p21.3) within the class III region of the MHC, adjacent to the gene coding for factor B and the genes for the two isotypes of C4 (Campbell *et al*., 1986; Ishii *et al*., 1993). The C2 gene consists of 18 exons and spans about 18 kb of DNA. Introns vary in length between 83 bp and 4.4 kb and all intron–exon boundaries follow the AG/GT consensus rule for splicing. Exon 1 encodes 270 bp of the 5'-untranslated region of C2 and 15 amino acids of the leader peptide. The three SCRs of C2b are encoded by the single exons 2, 3 and 4, respectively. The von Willebrand factor type A-like domain of C2a is encoded

by exons 6–10, and the serine proteinase domain by exons 11–18. Exons 5 and 15 appear to be unique to the C2 and factor B genes.

C4 is encoded by two genes in humans. The products C4A and C4B are both highly polymorphic with probably more than 50 allotypes. The variants have differing biological activities. In general, C4A allotypes of human C4 show one-fourth to one-third the hemolytic activity of C4B allotypes. The lack of hemolytic activity of one allotype, C4A6, suggests that the C5-binding site on C4b involves Arg458 of the C4 $\beta$ chain (Ebanks *et al*., 1992). C2 gene expression has been shown in several tissues and cell lines *in vitro* (Colten & Strunk, 1993) and is regulated by cytokines, mainly interferon $\gamma$, and also by glucocorticoids (Lappin & Whaley, 1989; Barnum *et al*., 1992). Alternatively spliced transcripts of the human C2 gene have been identified in cell lines. In all cell types, alternatively spliced transcripts represented a substantial fraction of the total C2 message, although relative amounts of individual C2 mRNA species appeared to vary with cell type. These findings raise the possibility that alternative splicing may provide a mechanism for tissue-specific C2 gene expression by regulating the message stability, translational efficiency or efficiency of post-translational modification and secretion (Cheng & Volanakis, 1994; Akama *et al*., 1995).

Mouse genomic and cDNA clones for C2 have been isolated. They suggest a protein with high identity to human C2 (74% amino acid identity). Like the human C2 gene, the mouse C2 gene consists of 18 exons with similar intron–exon organization. The mouse C2 gene (20 kb) is more than three times the size of that for factor B (6 kb) as a result of the presence of large intronic segments separating the exons encoding the N-terminal binding and central (von Willebrand factor) domains. Evidence from cDNA clones shows that the two C2 transcripts are generated by an alternative splice at the donor site of exon 14 producing long and short C2 mRNA species that differ by 21 base pairs. This 21 bp sequence encodes an amino acid sequence (Gly636-Ser-Thr-Cys-Lys-Asp-His642) that forms the binding pocket of the serine proteinase domain (Ishikawa *et al*., 1990).

Serum C2 levels are characteristically low in a number of diseases associated with immune complex deposition (Morgan, 1990; Whaley *et al*., 1992). They are also very low in patients with acquired or hereditary C1-inhibitor deficiencies. A small, poorly characterized fragment of C2 may be released by the action of C1s and plasmin on C2 in such patients. The fragment, termed C2 kinin, causes angioedema (Donaldson *et al*., 1977). *In vitro* the peptide could be obtained on cleavage of C2 with C1s and plasmin (Strang *et al*., 1988). A synthetic peptide derived from the C-terminus of C2b (residues 207–223 of C2) has been reported to enhance vascular permeability in guinea pig and human skin, and to induce contraction of estrous rat uterus (Cholin *et al*., 1989). This finding remains controversial, as other work has shown that C2 is probably not the source of the kinin-like activity generated in hereditary angioedema plasma (Smith & Kerr, 1985; Shoemaker *et al*., 1994).

## Distinguishing Features

Monoclonal and polyclonal antibodies recognizing C2 are available commercially. C2 is usually detected by hemolytic assay using functionally pure components or genetically deficient sera (Kerr, 1994).

## Related Peptidases

Complement components C2 and factor B are novel types of serine proteinase that are encoded by single adjacent loci in the MHC on human chromosome 6. The two proteins share 39% identity (50% similarity) in amino acid sequence. The catalytic chains, C2a (509 residues) and Bb (505 residues), show homology in their C-terminal domains to the catalytic polypeptides of other serine proteinases of family S1. The noncatalytic chains, C2b (223 residues) and Ba (234 residues), both contain three tandem repeats of approximately 60 amino acids each that are homologous to the repeats in complement proteins such as C4b-binding protein and factor H, and also in noncomplement proteins such as $\beta_2$-glycoprotein I. Molecular mapping and DNA sequence analysis has shown that the factor B gene is 6 kb in length and contains 18 exons, while the C2 gene is 18 kb in length; 425 bp separates the 3' end of the C2 gene from the 5' end of the factor B gene. C2 and factor B are polymorphic, and structural variants have been analyzed (Bentley & Campbell, 1986).

## Further Reading

The history of the initial analysis of C2, its biochemistry and molecular biology, clinical and laboratory aspects have been reviewed extensively by Ross (1986), Law & Reid (1995), Whaley (1992), Morgan (1990) and Kerr (1994).

## Comment on Nomenclature

The nomenclature of the fragments of C2 has been the subject of some confusion. The larger fragment was initially called C2a and the smaller C2b. For a brief period the nomenclature was reversed to bring the proteins in line with all other complement proteins for which the smaller fragments were given the suffix 'a'. This change was not widely accepted and the older nomenclature is now used (WHO-IUIS Nomenclature, 1992, 1993; Hauptmann *et al.*, 1990).

## References

Akama, H., Johnson, C.A.C. & Colten, H.R. (1995) Human complement protein C2: alternative splicing generates templates for secreted and intracellular C2 proteins. *J. Biol. Chem.* **270**, 2674–2680.

Barnum, S.R., Ishii, Y., Agrawal, A. & Volanakis, J.E. (1992) Production and interferon-γ-mediated regulation of complement component C2 and factors B and D by the astroglioma cell line U105-MG. *Biochem. J.* **287**, 595–602.

Bentley, D.R. (1986) Primary structure of human complement component C2: homology to two unrelated protein families. *Biochem. J.* **239**, 339–346.

Bentley, D.R. & Campbell, R.D. (1986) C2 and factor B: structure and genetics. *Biochem. Soc. Symp.* **51**, 7–18.

Bentley, D.R. & Porter, R.R. (1984) Isolation of cDNA clones for human complement component C2. *Proc. Natl Acad. Sci. USA*

**81**, 1212–1215.

Campbell, R.D. (1987) The molecular genetics and polymorphism of C2 and factor B. *Br. Med. Bull.* **43**, 37–49.

Campbell, R.D., Carroll, M.C. & Porter, R.R. (1986) The molecular genetics of components of complement. *Adv. Immunol.* **38**, 203–244.

Cheng, J. & Volanakis, J.E. (1994) Alternatively spliced transcripts of the human complement C2 gene. *J. Immunol.* **152**, 1774–1779.

Cholin, S., Gerard, N.P., Strang, C.J. & Davis, A.E. III. (1989) Biologic activity of a C2-derived peptide. Demonstration of a specific interaction with guinea pig lung tissues. *J. Immunol.* **142**, 2401–2404.

Colten, H.R. & Strunk, R.C. (1993) Extrahepatic complement synthesis. In: *Complement in Health and Disease*, 2nd edn (Whaley, K., Loos, M. & Weiler, J., eds). Amsterdam: Kluwer Publications, pp. 127–158.

Cooper, N.R. (1975) Enzymatic activity of the second component of complement. *Biochemistry* **24**, 4245–4251.

Donaldson, V.H., Rosen, F.S. & Bing, D.H. (1977) Role of the second component of complement (C2) and plasmin in kinin release in hereditary angioneurotic oedema (H.A.N.E.) plasma. *Trans. Assoc. Am. Physicians* **90**, 174–183.

Ebanks, R.O. & Isenman, D.E. (1995) Evidence for the involvement of Arginine 462 and the flanking sequence of human C4b chain in mediating C5 binding to the C4b subcomponent of the classical pathway C5 convertase. *J. Immunol.* **154**, 2808–2820.

Ebanks, R.O., Jaikaran, A.S., Carroll, M.C., Anderson, M.J., Campbell, R.D. & Isenman, D.E. (1992) A single arginine to tryptophan interchange at beta-chain residue 458 of human complement component C4 accounts for the defect in classical pathway C5 convertase activity of allotype C4A6. Implications for the location of a C5 binding site in C4. *J. Immunol.* **148**, 2803–2811.

Fujita, T., Inoue, T., Ogawa, K., Iida, K. & Tamura, N. (1987) The mechanism of action of decay-accelerating factor (DAF). DAF inhibits the assembly of C3 convertases by dissociating C2a and Bb. *J. Exp. Med.* **166**, 1221–1228.

Hauptmann, G., John, I. Uring-Lambert, B. & Arnold, D. (1990) C2 nomenclature statement. *Complement Inflamm.* **7**, 252–254.

Horiuchi, T., Macon, K.J., Kidd, V.J. & Volanakis, J.E. (1989) cDNA cloning and expression of human complement component C2. *J. Immunol.* **142**, 2105–2111.

Horiuchi, T., Macon, K.J., Engler, J.A. & Volanakis, J.E. (1991) Site-directed mutagenesis of the region around Cys-241 of complement component C2: evidence for a C4b binding site. *J. Immunol.* **147**, 584–589.

Ishii, Y., Zhu, Z.B., Macon, K.J. & Volanakis, J.E. (1993) Structure of the human C2 gene. *J. Immunol.* **151**, 170–174.

Ishikawa, N., Nonaka, M., Wetsel, R.A. & Colten, H.R. (1990) Murine complement C2 and factor B: genomic and cDNA cloning reveals different mechanisms for multiple transcripts of C2 and B. *J. Biol. Chem.* **265**, 19040–19046.

Ji, Y.H., Fujita, T., Hatsuse, H., Takahashi, A., Matsushita, M. & Kawakami, M. (1993) Activation of the C4 and C2 components of complement by a proteinase in serum bactericidal factor, Ra reactive factor. *J. Immunol.* **150**, 571–578.

Kam, C.M., McRae, B.J., Harper, J.W., Niemann, M.A., Volanakis, J.E. & Powers, J.C. (1987) Human complement proteins D, C2, and B. Active site mapping with peptide thioester substrates. *J. Biol. Chem.* **262**, 3444–3451.

Kerr, M.A. (1980) The human complement system: assembly of the classical pathway C3 convertase. *Biochem. J.* **189**, 173–181.

Kerr, M.A. (1981) The second component of human complement. *Methods Enzymol.* **80C**, 54–64.

Kerr, M.A. (1994) Complement. In: *LABFAX Immunochemistry* (Kerr, M.A. & Thorpe, R., eds). Oxford: Bios Scientific, pp. 211–233.

Kerr, M.A. & Gagnon, J. (1982) The purification and properties of the second component of guinea-pig complement. *Biochem. J.* **205**, 59–67.

Kerr, M.A. & Parkes, C. (1984) The effects of iodine and thiol-blocking reagents on complement component C2 and on the assembly of the classical-pathway C3 convertase. *Biochemical Journal* **219**, 391–399.

Kim, Y.U., Carroll, M.C., Isenman, D.E., Nonaka, M., Pramoonjago, P., Takeda, J., Inoue, K. & Kinoshita, T. (1992) Covalent binding of C3b to C4b within the classical complement pathway C5 convertase. Determination of amino acid residues involved in ester linkage formation. *J. Biol. Chem.* **267**, 4171–4176.

Kozono, H., Kinoshita, T., Kim, Y.U., Takata-Kozono, Y., Tsunasawa, S., Sakiyama, F., Takeda, J., Hong, K. & Inoue, K. (1990) Localization of the covalent C3b-binding site on C4b within the complement classical pathway C5 convertase, C4b2a3b. *J. Biol. Chem.* **265**, 14444–14449.

Lappin, D.F. & Whaley, K. (1989) Modulation of complement gene expression by glucocorticoids. *Biochem. J.* **280**, 117–124.

Law, S.K.A. & Reid, K.B.M. (1995) *Complement; In Focus*, 2nd edn. Oxford: IRL Press.

Masaki, T., Matsumoto, M., Yasuda, R., Levine, R.P., Kitamura, H. & Seya, T. (1991) A covalent dimer of complement C4b serves as a subunit of a novel C5 convertase that involves no C3 derivatives. *J. Immunol.* **147**, 927–932.

Medicus, R.G., Gotze, O. & Müller-Eberhard, H.J. (1976) The serine protease nature of the C3 and C5 convertases of the classical and alternative complement pathways. *Scand. J. Immunol.* **5**, 1049–1055.

Morgan, B.P. (1990) *Complement: Clinical Aspects and Relevance to Disease*. London: Academic Press.

Nagasawa, S. & Stroud, R.M. (1977) Cleavage of C2 by C1s into the antigenically distinct fragments C2a and C2b: demonstration of binding of C2b to C4b. *Proc. Natl Acad. Sci. USA* **74**, 2998–3001.

Oglesby, T.J., Accavitti, M.A. & Volanakis, J.E. (1988) Evidence for a C4b binding site on the C2b domain of C2. *J. Immunol.* **141**, 926–931.

Parkes, C., Gagnon, J. & Kerr, M. (1983) The reaction of iodine and thiol-blocking reagents with human complement components C2 and factor B. Purification and N-terminal amino acid sequence of a peptide from C2a containing a free thiol group. *Biochem. J.* **213**, 201–209.

Peake, P., Szelke, M., Jones, D.M., Singleton, A., Sueiras-Diaz, J. & Lachmann, P.J. (1990) Peptide inhibitors of C3 breakdown. *Clin. Exp. Immunol.* **79**, 454–458.

Perkins, S.J. & Smith, K.F. (1993) Identity of the putative serine-proteinase fold in proteins of the complement system with nine relevant crystal structures. *Biochem. J.* **295**, 109–114.

Polley, M.J. & Müller-Eberhard, H.J. (1967) Enhancement of the hemolytic activity of the second component of human complement by oxidation. *J. Exp. Med.* **126**, 1013–1025.

Polley, M.J. & Müller-Eberhard, H.J. (1968) The second component of human complement: its isolation, fragmentation by C'1 esterase, and incorporation into C'3 convertase. *J. Exp. Med.* **128**, 533–551.

Reid, K.B.M., Bentley, D.R., Campbell, D.R., Chung, L.P., Sim, R.B., Kristensen, T. & Tack, B.F. (1986) Complement proteins which interact with C3b or C4b. A superfamily of structurally related proteins. *Immunol. Today* **7**, 230–234.

Ross, G.D. (1986) *Immunobiology of the Complement System*. Orlando: Academic Press.

Shoemaker, L.R., Schurman, S.J., Donaldson, V.H. & Davis, A.E. III (1994) Hereditary angioneurotic oedema: characterisation of plasma kinin and vascular permeability enhancing activities. *Clin. Exp. Immunol.* **95**, 22–28.

Smith, C.A., Vogel, C-W. & Müller-Eberhard, H.J. (1984) MHC Class III products: an electron microscope study of the C3 convertases of human complement. *J. Exp. Med.* **159**, 324–332.

Smith, M.A. & Kerr, M.A. (1985) Cleavage of the second component of complement by plasma proteases: implications in hereditary C1-inhibitor deficiency. *Immunology* **56**, 561–570.

Strang, C.J., Cholin, S., Spragg, J., Davis A.E. III, Schneeberger, E.E., Donaldson, V.H. & Rosen, F.S. (1988) Angioedema induced by a peptide derived from complement component C2. *J. Exp. Med.* **168**, 1685–1698.

Tomana, M., Niemann, M., Garner, C. & Volanakis, J.E. (1985). Carbohydrate composition of the second, third and fourth components and factors B and D of human complement. *Mol. Immunol.* **22**, 107–111.

Villiers, M.B., Thielens, N.M. & Colomb, M.G. (1985) Soluble C3 proconvertase and convertase of the classical pathway of human complement. Conditions of stabilization *in vitro*. *Biochem. J.* **226**, 429–436.

Whaley, K., Loos, M. & Weiler, J. (1992) *Complement in Health and Disease*, 2nd edn. Amsterdam: Kluwer Publications.

WHO-IUIS Nomenclature Sub-Committee (1993) Nomenclature for human complement component C2. *J. Immunol. Methods* **163**, 1–2.

WHO-IUIS Nomenclature Sub-Committee (1992) Nomenclature for human complement component C2. *Bull. World Health Organ.* **70**, 527–530.

*Michael A Kerr*
*Department of Molecular and Cellular Pathology,*
*University of Dundee,*
*Ninewells Hospital Medical School,*
*Dundee DD1 9SY, Scotland*
*Email: makerr@ninewells.dundee.ac.uk*

# 47. Factor B and the alternative pathway C3/C5 convertase

## Databanks

*Peptidase classification: clan SA, family S1, MEROPS ID: S01.196*
*NC-IUBMB enzyme classification: EC 3.4.21.47*
*Chemical Abstracts Service registry number: 80295-67-6*
*Databank codes:*

| Species | SwissProt | PIR | EMBL (cDNA) | EMBL (genomic) |
|---|---|---|---|---|
| *Danio rerio* | – | – | U34662 | – |
| *Gallus gallus*[a] | – | – | – | – |
| *Homo sapiens* | P00751 | S34075 | J00126 | J00125: exons 1–11 |
| | | | J00185 | M15082: 5′ end |
| | | | J00186 | |
| | | | K01566 | |
| | | | L15702 | |
| | | | S67310 | |
| | | | X00284 | |
| | | | X72875 | |
| *Lampetra japonica* | – | I50807 | D13568 | – |
| *Mus musculus* | P04186 | A38875 | K01496 | M60629: exon 1 |
| | | | K01497 | M60630: exon 2 |
| | | | K01498 | M60631: exon 3 |
| | | | M57890 | M60632: exon 4 |
| | | | | M60633: exon 5 |
| | | | | M60634: exon 6 |
| | | | | M60635: exon 7 |
| | | | | M60636: exon 8 |
| | | | | M60637: exon 9 |
| | | | | M60638: exon 10 |
| | | | | M60639: exon 11 |
| | | | | M60640: exon 12 |
| | | | | M60641: exon 13 |
| | | | | M60642: exon 14 |
| | | | | M60643: exon 15 |
| | | | | M60644: exon 16 |
| | | | | M60645: exon 17 |
| | | | | M60646: exon 18 and complete CDS |
| *Oryzias latipes* | – | – | D84063 | – |
| *Sus scrofa* | – | A53274 | – | – |
| | – | B53274 | – | – |
| *Xenopus laevis* | – | – | D29796 | – |
| | | | D49373 | |

[a]The sequence of chicken complement factor B is deposited in GenPept only; the accession number is 425644.

## Name and History

The alternative pathway of complement activation was discovered in the 1950s (see Pillemer *et al*., 1954; Ross, 1986). The components of the pathway were identified during the following decades and subsequently characterized at the molecular level (reviewed by Pangburn & Müller-Eberhard, 1984). Its importance lies in the fact that it is activated not only by antibody–antigen aggregates but also by direct contact with the surfaces of many bacteria, yeasts and some viruses without the need for antibody. The alternative pathway is therefore part of the innate immune system, the first line of defense against infection (Figure 47.1).

Although they use different proteins, the alternative pathway and the classical pathway lead to the assembly of C3

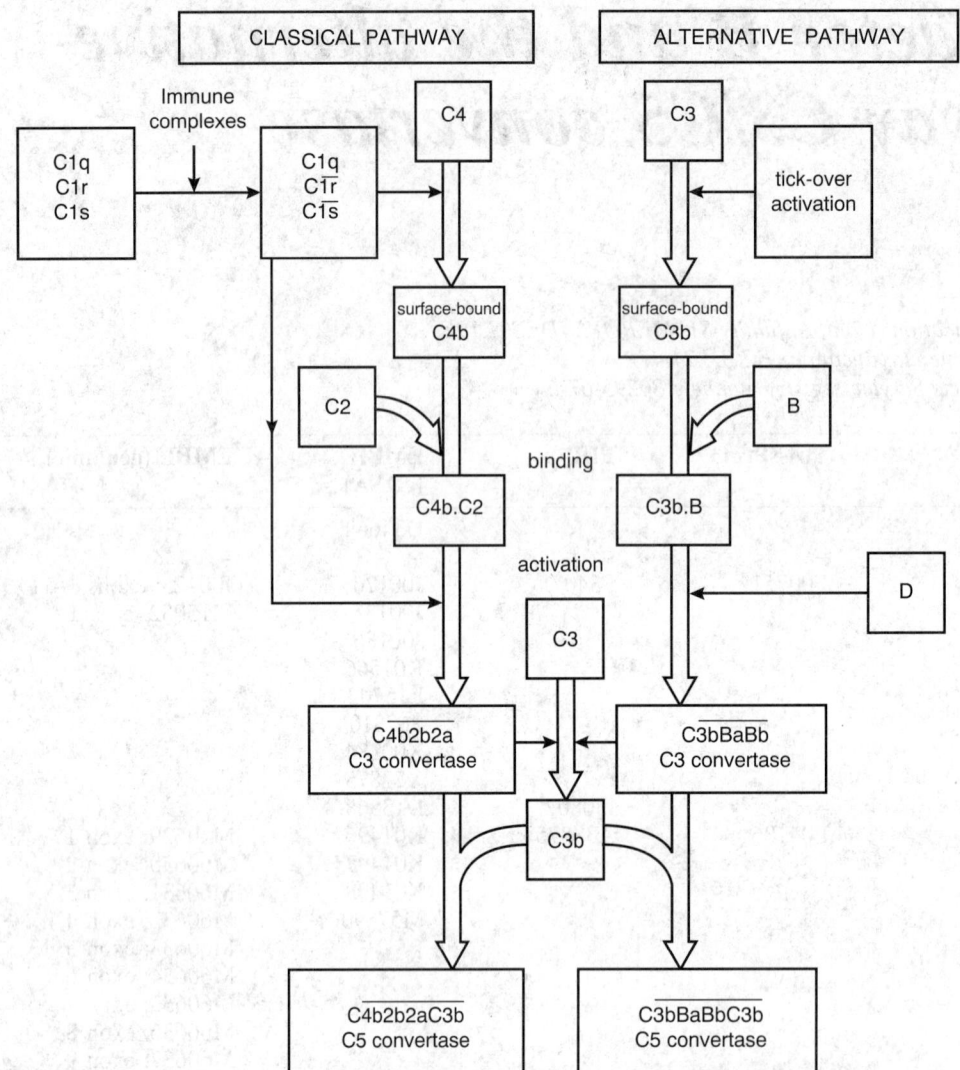

*Figure 47.1*  Analogous action of the classical and alternative pathways. (Modified with permission from Walport, 1996.)

convertases with very similar properties. Upon interaction with excess C3b, both C3 convertases become C5 convertases. **Factor B** was defined by early studies on the alternative pathway as a heat-labile protein necessary for the cleavage of C3. The identification of factor B as a specific protein came from the study of a serum protein termed **C3 proactivator** which was able to complex with a protein from cobra venom to produce a C3-cleaving enzyme. Factor B and C3 proactivator were shown to be the same protein, and furthermore, were subsequently shown to be identical to a human serum protein that had been called **glycine-rich β-glycoprotein** (Boenisch & Alper, 1970; Alper *et al*., 1973).

### Activity and Specificity

The alternative pathway C3 convertase cleaves complement component C3, a 185 kDa glycoprotein, to yield components C3a and C3b. The cleavage occurs as Arg77┼Ser of C3, and Arg77 becomes the essential C-terminal residue of the bioactive polypeptide, C3a. The alternative pathway C3 convertase is a complex of factor Bb with C3b, and is formed

by the action of factor D (Chapter 42) on factor B while factor B is bound in a $Mg^{2+}$-dependent complex with C3b. The two fragments of factor B are Ba (30 kDa) and Bb (63 kDa), of which Bb is the catalytic subunit of the alternative pathway C3 and C5 convertases.

It will be apparent that the unusual situation exists that C3b is a product of the action of the alternative pathway C3 convertase, but also is itself a component of the enzyme! This allows for rapid amplification of the activity of the alternative pathway, since cleavage of C3 leads rapidly to the formation of more convertase. It does raise the question of how the activation of the pathway can start, however. Although C3b is essential for cleavage of factor B, it can be substituted in the initial activation of the pathway by a form of C3 that is structurally very similar to C3b, but has not had C3a removed (Pangburn *et al*., 1981). This is formed by spontaneous hydrolysis of the thioester that is part of the structure of C3. The hydrolysis occurs at a rate of 0.2–0.4% per hour in buffer, but faster in the presence of nucleophiles or chaotropic agents. Unlike native C3, this form of C3 is capable of generating fluid-phase C3 convertase with factors B, D and P, and is cleaved by factor I (Chapter 48) in the presence of factor H (Pangburn *et al.*, 1981). Activation of the alternative pathway of complement thus commences with the formation of an initial fluid-phase C3 convertase containing the uncleaved, C3b-like form of C3. *In vitro*, the convertase can also be made using cobra venom factor, a protein now identified as cobra C3b (Alper & Balavitch, 1976; Vogel & Müller-Eberhard, 1982).

*In vitro*, the alternative pathway C3 convertase is extremely unstable, the activity decaying spontaneously with a half-life of minutes as Bb is released from the complex. The *in vitro* stability of the convertase is increased by the presence of $Ni^{2+}$ instead of $Mg^{2+}$ ions (Fishelson *et al.*, 1984). In serum, there are several factors that can increase or decrease the stability of the convertase. The decay is retarded by the binding of properdin (Fearon & Austen, 1975), and there is also stabilization by autoantibodies called nephritic factors that are associated with membrano-proliferative glomerulonephritis and other diseases (Daha *et al.*, 1976; Daha, 1986). Specific serum or cell-bound control proteins that increase the rate of dissociation of the C3b,Bb complex cause the alternative pathway C3 convertase to decay even faster (Fujita *et al.*, 1987). Of these, decay-accelerating factor (DAF), a protein found on the surface of many cell types, binds primarily to the Bb subunits, whereas CR1, a complement receptor found on some leukocytes and other cells, and factor H, a serum protein, interact primarily with the C3b (Pangburn, 1986, 1990). Attachment of C3b to activators of the alternative pathway of complement results in a decrease in regulatory activity of factor H.

The C3 convertase cleaves C3 at the same Arg⫫Ser bond as the classical pathway C3 convertase. In the presence of excess C3b, factor Bb binds noncovalently to a covalently linked C3b dimer, and becomes also the catalytic subunit of the alternative pathway C5 convertase (Isenman *et al.*, 1980; Kinoshita *et al.*, 1988). The C5 convertases of both pathways have the same specificity, cleaving an Arg⫫Leu bond.

Isolated Bb is unable to cleave C3 or C5. There is therefore considerable structural and functional similarity between the enzymes of the alternative and classical pathways, factor B being equivalent to C2, and C3b to C4b. Both Bb and C2a hydrolyze ester substrates, but are proteolytic only in the complexes. One major difference is in the formation of the alternative and classical activators: C2 is cleaved by C1s even in the absence of C4b, but factor B is cleaved by factor D only when the factor B is complexed to C3b. The binding site for C3b on intact B is located on the Ba portion of the molecule (Pryzdial & Isenman, 1987; Ueda *et al.*, 1987).

The kinetics of C3 cleavage by the convertase have been determined using the Michaelis–Menten equation, yielding $k_{cat}$ $1.8\,s^{-1}$, $K_m$ $5.9 \times 10^{-6}$ M and turnover number $107\,min^{-1}$. The $t_{0.5}$ of the enzyme was determined to be $90\,s$, and $k_{cat}/K_m$ $31 \times 10^4\,M^{-1}\,s^{-1}$ at physiological pH, ionic strength and temperature (Pangburn & Müller-Eberhard, 1986). No natural substrates are known besides C3 and C5. Bb in isolation has been shown to possess esterase activity against certain arginine and lysine esters, principally Ac-Gly-Lys⫫OMe (Cooper, 1971). The specificity and reactivity of factors B and Bb have been determined with a series of peptide thioester substrates.

Factor B was catalytically less active than factor D or C2. The best substrate for factor B was Z-Lys-Ar⫫SBzl with a $k_{cat}/K_m$ value of $1370\,M^{-1}\,s^{-1}$. The catalytic fragment of B, Bb, had higher activity toward these peptide thioester substrates. The best substrate for Bb was Z-Gly-Leu-Ala-Arg⫫SBzl with a $k_{cat}/K_m$ similar to that of C2a and 10-fold higher than the value for factor B. Both C2a and Bb were considerably more reactive against C3-like than C5-like substrates. Bovine trypsin hydrolyzed the thioester substrates with $k_{cat}/K_m$ values approximately 1000-fold higher than the complement enzymes (Kam *et al.*, 1987).

Factor B has been reported to be inhibited by DFP (Medicus *et al.*, 1976), but this has not been confirmed by others. Confirmation of the serine proteinase nature of factor B came from the deduced amino acid sequence. Although the catalytic Bb chain has twice the moleculzar mass of other serine proteinases of family S1, the C-terminal part of the molecule shows clear homology, especially around the catalytic triad residues. There is no characteristic N-terminal sequence like those of other serine proteinases that are important in the activity of these enzymes.

## Structural Chemistry

Factor B is a single polypeptide chain of 86–93 kDa as judged from SDS-PAGE. This molecular mass has been confirmed by gel filtration, sedimentation equilibrium, and sedimentation velocity measurements. The sedimentation coefficient is 5.9–6.2 S; partial specific volume $0.72\,ml\,g^{-1}$; Stokes radius 40 Å; diffusion constant ($D_{20,w}$) 5.4; and frictional ratio 1.28 (Lesavre *et al.*, 1979; Kerr, 1981). Carbohydrate analyses have been carried out (Tomana *et al.*, 1985). It has also been shown that glucose can be covalently attached to Lys266 of purified human complement factor B (Niemann *et al.*, 1991).

Following cleavage of factor B by factor D in the presence of C3b, the two fragments, Ba of 30 kDa and Bb of 63 kDa, are not linked by disulfide bonds, and can be separated by gel filtration or by ion-exchange chromatography without the need for denaturing agents. They represent separate domains within factor B, Ba being the N-terminal part of the molecule (Lesavre *et al.*, 1979; Kerr, 1979). The molecular architecture of the cobra venom factor (CoVF)-dependent C3/C5 convertase of human complement has been deduced from electron microscopy of the purified bimolecular complex (CoVF,Bb) and its isolated subunits, CoVF and Bb. CoVF appeared as an irregularly shaped cylindrical structure, with approximate dimensions of $137 \times 82$ Å (length × diameter). The zymogen factor B appeared globular with a diameter of about 80 Å and its image suggested a bipartite structure of two compact, closely associated regions, one about twice as large as the other. Bb, the larger, catalytic site-bearing cleavage fragment of factor B, revealed two globular domains, each 40 Å in diameter, connected by a short linker region about 10 Å long and equally thick. In the CoVF,Bb complex, Bb was attached to one end of the CoVF molecule through only one of its two domains, and in an orientation making the long axes of both molecules approximately orthogonal (Smith *et al.*, 1982).

The site of cleavage of factor B by factor D in the presence of C3b is an Arg-Lys bond at a site containing three basic residues, Lys-Arg⊹Lys. This is homologous to the site of cleavage of C2 by C1s. Although factor B is cleaved by factor D only in the presence of C3b, it is cleaved at the same site in the absence of C3b by trypsin, showing that the site is accessible in the uncomplexed molecule (Lesavre *et al.*, 1979; Kerr, 1979).

Circular-dichroism spectra suggested that 17% of the polypeptide chain of factor B is $\alpha$ helix, but none of the helix is in Ba (Lesavre *et al.*, 1979). The serine proteinase domain of factor B has been modeled with reference to nine homologous serine proteinases of known crystal structure and four other complement serine proteinases, C1r, C1s, factor I and factor D. All sequence insertions and deletions were readily located at the protein surface. The internal location of disulfide bridges and the surface location of putative glycosylation sites were compatible with the predicted structures (Perkins & Smith, 1993).

Sequence analysis of factor B shows it to consist of a leader sequence of 25 amino acids followed by the non-catalytic, Ba chain (234 residues) containing three tandem repeats of approximately 60 amino acids each that are homologous to the repeats in other complement proteins and some noncomplement proteins such as $\beta_2$-glycoprotein I and the selectins. Bb (505 residues) contains a von Willebrand factor-like domain and the serine proteinase domain (Mole *et al.*, 1984; Bentley & Campbell, 1986).

## Preparation

Factor B is present in normal human serum at 120–330 μg ml$^{-1}$. It is an acute phase reactant rising in concentration in inflammation. A number of purification schemes for human and guinea pig factor B have been described (e.g. Lesavre *et al.*, 1979; Kerr, 1981).

## Biological Aspects

Factor B from human serum is polymorphic as judged by electrophoretic mobility. The two common alleles, originally called F and S, have gene frequencies of 0.28 and 0.72 in the Caucasian population. There are seven rarer alleles. The single structural gene for factor B is in the class III region of the major histocompatibility complex, adjacent to the structural gene for C2 (reviewed by Bentley & Campbell, 1986; Rittner & Schneider, 1988). No well-documented cases of genetic deficiency of factor B have been reported. Full-length cDNA clones encoding human factor B have been isolated, sequenced and expressed (Campbell & Porter, 1983; Schwaeble *et al.*, 1993). The deduced amino acid sequence comprises 25 residues of a putative leader peptide and 739 residues of the mature polypeptide chain. The F allele of factor B and an S allele-like Q7R mutant obtained by site-directed mutagenesis had similar specific hemolytic activities when expressed (Horiuchi *et al.*, 1993).

Mouse genomic and cDNA clones for factor B show great similarity to the human gene products (85% amino acid identity). Like the human Bf gene, mouse Bf and C2 each consist of 18 exons with similar intron-exon organizations. Tissue-specific, multiple factor B transcripts are generated by alternative transcriptional initiation. It thus seems

that different molecular mechanisms account for the multiple forms of C2 and factor B mRNA. The structures of the different transcripts predict differences in function and expression of the respective gene products (Ishikawa *et al.*, 1990). cDNAs for factor B have been cloned from a number of other species including chicken, *Xenopus* and lamprey (Kjalke *et al.*, 1993).

Biosynthesis of mammalian factor B occurs primarily in the liver (reviewed by Colten & Dowton, 1986) but also in monocytes. The rate of synthesis is controlled by cytokines such as interleukin 1 (IL-1) and interferon $\gamma$ (IFN$\gamma$) and by lipopolysaccharides. Factor B is a typical acute phase protein whose concentration rises 2- to 3-fold in severe infections or other acute phase responses. Factor B levels are decreased in some infections and in glomerulonephritis, as a result of activation of the alternative pathway (Kerr, 1994); the fragments Ba and Bb can be detected in the serum under such circumstances.

## Distinguishing Features

Monoclonal and polyclonal antibodies recognizing factor B are widely available. Factor B is readily detected by immunological assays or by hemolytic assay. Commercial kits are available to detect the factor B cleavage products and therefore alternative pathway activation (Kerr, 1994).

## Related Proteinases

Complement components C2 and factor B are distinctive members of peptidase family S1 that are encoded by single loci in the major histocompatibility complex on human chromosome 6. The two proteins share 39% identity in amino acid sequence, or 50% taking into account conservative amino acid replacements. Molecular mapping and DNA sequence analysis have shown that the factor B gene is 6 kb in length and contains 18 exons, whereas the C2 gene is 18 kb in length. Just 425 bp separate the 3′ end of the C2 gene from the 5′ end of the factor B gene.

## Further Reading

The general properties of factor B, its biochemistry and molecular biology, clinical and laboratory aspects have been reviewed extensively by Pangburn & Müller-Eberhard (1984), Bentley & Campbell (1986), Morgan (1990), Whaley *et al.* (1993), Law & Reid (1995) and Kerr (1994).

## References

Alper, C.A. & Balavitch, D. (1976) Cobra venom factor: evidence for its being altered cobra C3 (the third component of complement). *Science* **191**, 1275–1276.

Alper, C.A., Goodkofsky, I. & Lepow, I.H. (1973) The relationship of glycine-rich $\beta$-glycoprotein to factor B in the properdin system and to the cobra factor-binding protein of human serum. *J. Exp. Med.* **137**, 424–437.

Bentley, D.R. & Campbell, R.D. (1986) C2 and factor B; structure and genetics. *Biochem. Soc. Symp.* **51**, 7–18.

Boenisch, T. & Alper, C.A. (1970) Isolation and properties of a glycine-rich beta-glycoprotein of human serum. *Biochim. Biophys. Acta* **221**, 529–535.

Campbell, R.D. & Porter, R.R. (1983) Molecular cloning and characterisation of the gene coding for complement protein factor B. *Proc. Natl Acad. Sci. USA* **80**, 4464–4467.

Colten, H.R. & Dowton, S.B. (1986) Regulation of complement gene expression. *Biochem. Soc. Symp.* **51**, 37–49.

Cooper, N.R. (1971) Esterolytic activity of complement factor B. *Prog. Immunol.* **1**, 567–577.

Daha, M.R. (1986) Autoantibodies to complement components. In: *Complement in Health and Disease* (Whaley, K., ed.). Lancaster: MTP Press, pp. 185–199.

Daha, M.R, Fearon, D.T. & Austen, K.F. (1976) C3 nephritic factor (C3Nef) stabilization of fluid phase and cell bound alternative pathway convertase *J. Immunol.* **116**, 1–7.

Fearon, D.T. & Austen, K.F. (1975) Properdin: binding to C3b and stabilization of the C3b-dependent C3 convertase. *J. Exp. Med.* **142**, 856–863.

Fishelson, Z., Pangburn, M.K. & Müller-Eberhard, H.J. (1984) Characterization of the initial C3 convertase of the alternative pathway of human complement. *J. Immunol.* **132**, 1430–1434.

Fujita, T., Inoue, T., Ogawa, K., Iida, K. & Tamura, N. (1987) The mechanism of action of decay-accelerating factor (DAF). DAF inhibits the assembly of C3 convertases by dissociating C2a and Bb. *J. Exp. Med.* **166**, 1221–1228.

Horiuchi, T., Kim, S., Matsumoto, M., Watanabe, I., Fujita, S. & Volanakis, J.E. (1993) Human complement factor B: cDNA cloning, nucleotide sequencing, phenotypic conversion by site-directed mutagenesis and expression. *Mol. Immunol.* **30**, 1587–1592.

Isenman, D.E., Podack, E.R. & Cooper, N.R. (1980) The interaction of C5 with C3b in free solution: a sufficient condition for cleavage by a fluid phase C3/C5 convertase. *J. Immunol.* **124**, 326–331.

Ishikawa, N., Nonaka, M., Wetsel, R.A. & Colten, H.R. (1990) Murine complement C2 and factor B genomic and cDNA cloning reveals different mechanisms for multiple transcripts of C2 and B. *J. Biol. Chem.* **265**, 19040–19046.

Kam, C.M., McRae, B.J., Harper, J.W., Niemann, M.A., Volanakis, J.E. & Powers, J.C. (1987) Human complement proteins D, C2, and B. Active site mapping with peptide thioester substrates. *J. Biol. Chem.* **262**, 3444–3451.

Kato, Y., Salter-Cid, L., Flajnik, M.F., Kasahara, M., Namikawa, C., Sasaki, M. & Nonaka, M. (1994) Isolation of *Xenopus* complement Factor B complementary cDNA and linkage to the frog MHC. *J. Immunol.* **153**, 4546–4554.

Kerr, M.A. (1979) Limited proteolysis of C2 and Factor B. *Biochem. J.* **183**, 615–622.

Kerr, M.A. (1981) Human factor B. *Methods Enzymol.* **80C**, 102–112.

Kerr, M.A. (1994) Complement. In: *LABFAX Immunochemistry* (Kerr, M.A. & Thorpe, R., eds). Oxford: Bios Scientific, pp. 211–233.

Kinoshita, T., Takata, Y., Kozono, H., Takeda, J., Hong, K.S. & Inoue, K. (1988) C5 convertase of the alternative complement pathway: covalent linkage between two C3b molecules within the trimolecular complex enzyme. *J. Immunol.* **41**, 3895–3901.

Kjalke, M., Welinder, K.G. & Koch, C. (1993) Structural analysis of chicken factor B-like protease and comparison with mammalian complement proteins factor B and C2. *J. Immunol.* **151**, 4147–4152.

Law, S.K.A. & Reid, K.B.M. (1995) Complement. In: *Focus*, 2nd edn. Oxford: IRL Press.

Lesavre, P.H., Hugli, T.E., Esser, A.F. & Müller-Eberhard, H.J. (1979) The alternative pathway C3/C5 convertase: chemical basis of factor B activation. *J. Immunol.* **123(2)**, 529–534.

Medicus, R.G., Götze, O. & Müller-Eberhard, H.J. (1976) The serine protease nature of the C3 and C5 convertases of the classical and alternative complement pathways. *Scand. J. Immunol.* **5**, 1049–1055.

Mole, J.E., Anderson, J.K., Davison, E.A. & Woods, D.E. (1984) Complete primary structure for the zymogen of human complement factor B. *J. Biol. Chem.* **259**, 3407–3412.

Morgan, B.P. (1990) *Complement: Clinical Aspects and Relevance to Disease.* London: Academic Press.

Niemann, M.A., Bhown, A.S. & Miller, E.J. (1991) The principal site of glycation of human complement factor B. *Biochem. J.* **274**, 473–480.

Pangburn, M.K. (1986) Differences between the binding sites of the complement regulatory proteins DAF, CR1, and factor H on C3 convertases. *J. Immunol.* **136**, 2216–2221.

Pangburn, M.K. (1990) Reduced activity of DAF on complement enzymes bound to alternative pathway activators. Similarity with factor H. *Immunology* **71**, 598–600.

Pangburn, M.K. & Müller-Eberhard, H.J. (1984) The alternative pathway of complement. *Springer Semin. Immunopathol.* **7**, 163–192.

Pangburn, M.K. & Müller-Eberhard, H.J. (1986) The C3 convertase of the alternative pathway of human complement. Enzymic properties of the bimolecular proteinase. *Biochem. J.* **235**, 723–730.

Pangburn, M.K., Schreiber, R.D. & Müller-Eberhard, H.J. (1981) Formation of the initial C3 convertase of the alternative complement pathway. Acquisition of C3b-like activities by spontaneous hydrolysis of the putative thioester in native C3. *J. Exp. Med.* **154**, 856–867.

Perkins, S.J. & Smith, K.F. (1993) Identity of the putative serine-proteinase fold in proteins of the complement system with nine relevant crystal structures. *Biochem. J.* **295**, 109–114.

Pillemer, L., Blum, L., Lepow, K.H., Ross, O.A., Todd, E.W. & Wardlaw, A.C. (1954) The properdin system and immunity. Demonstration and isolation of a new serum protein, properdin and its role in immune phenomena. *Science* **120**, 279–288.

Pryzdial, E.L.G. & Isenman, D.E. (1987) Alternative complement pathway activation fragment Ba binds to C3b: evidence that formation of the B.C3b complex involves two discrete points of contact. *J. Biol. Chem.* **262**, 1519–1526.

Rittner, C. & Schneider, P.M. (1988) Genetics and polymorphism of complement components. In: *The Complement System* (Rother, K. & Till, G.O., eds). Berlin: Springer, pp. 80–135.

Ross, G.D. (1986) *Immunobiology of the Complement System.* New York: Academic Press.

Schwaeble, W., Luttig, B., Sokolowski, T., Estaller, C., Weiss, E.H., Meyer zum Buschenfelde, K.H., Whaley, K. & Dippold, W. (1993) Human complement factor B: functional properties of a recombinant zymogen of the alternative activation pathway convertase. *Immunobiology* **188**, 221–232.

Smith, C.A,. Vogel, C.W. & Müller-Eberhard, H.J. (1982) Ultrastructure of cobra venom factor-dependent C3/C5 convertase and its zymogen, factor B of human complement. *J. Biol. Chem.* **257**, 9879–9882.

Tomana, M., Niemann, M., Garner, C. & Volanakis, J.E. (1985). Carbohydrate composition of the second, third and fourth components and factors B and D of human complement. *Mol. Immunol.* **22**, 107–111.

Ueda, A., Kearney, J.F., Roux, K.H. & Volanakis, J.E. (1987) Probing functional sites on complement protein B with monoclonal antibodies. Evidence for C3b-binding sites on Ba. *J. Immunol.* **138**, 1143–1149.

Vogel, C.W. & Müller-Eberhard, H.J. (1982) The cobra venom factor-dependent C3 convertase of human complement. A kinetic and thermodynamic analysis of a protease acting on its natural high molecular weight substrate. *J. Biol. Chem.* **257**, 8292–8299.

Walport, M.J. (1996) Complement. In: *Immunology*, 4th edn (Roitt, I., Brostoff, J. & Male, D., eds). London: Mosby, pp. 13.1–13.17.

Whaley, K., Loos, M. & Weiler, J. (1993) *Complement in Health and Disease*, 2nd edn. Amsterdam: Kluwer Publications.

WHO-IUIS Nomenclature Subcommittee (1993) Nomenclature for human factor B. *J. Immunol. Methods* **163**, 9–11.

***Michael A Kerr***
*Department of Molecular and Cellular Pathology,*
*University of Dundee,*
*Ninewells Hospital Medical School,*
*Dundee DD1 9SY, Scotland*
*Email: makerr@ninewells.dundee.ac.uk*

# 48. Complement factor I

## Databanks

*Peptidase classification: clan SA, family S1, MEROPS ID: S01.199*
*NC-IUBMB enzyme classification: EC 3.4.21.45*
*Chemical Abstracts Service registry number: 80295-66-5*
*Databank codes:*

| Species | SwissProt | PIR | EMBL (cDNA) | EMBL (genomic) |
|---|---|---|---|---|
| *Homo sapiens* | P05156 | A29154 | J02770 Y00318 | X78594: 5′UTR |
| *Mus musculus* | – | – | U47810 | – |
| *Xenopus laevis* | – | S15468 | X59959 | – |

## Name and History

***Complement factor I*** was first described and partially characterized from guinea pig and human serum as an enzyme involved in the physiological degradation of the major plasma complement protein C3 (Tamura & Nelson, 1967; Lachmann & Müller-Eberhard, 1968). In the 1970s it was also shown to act in the physiological breakdown of the human complement protein C4, a homolog of C3. It was previously named ***conglutinogen-activating factor (KAF)***, because its action on C3 exposed a binding site for the cattle protein, conglutinin. Other former names include ***C3b inactivator***, ***C3b INA***, and ***C3b/C4b*** inactivator. By the early 1980s, the action of factor I was demonstrated more precisely to be the cleavage of the complement protein fragment C3b at two sites to form iC3b, and cleavage of C4b at two sites to form C4c plus C4d (Law *et al*., 1979; Press & Gagnon, 1981). It was realized that highly purified factor I had poor activity on these substrates, and that a cofactor protein was required for efficient cleavage of the substrates. Complement factor H (formerly called β-1H globulin) was identified as the major cofactor in human serum

for the factor I-mediated cleavage of C3b (Whaley & Ruddy, 1976), and C4b-binding protein (C4bp) as the major cofactor for breakdown of C4b (Fujita *et al*., 1978). C4bp can also act as a cofactor for C3b degradation, but not in physiological conditions (Nagasawa & Stroud, 1980; Sim *et al*., 1981). Subsequently, two cell surface proteins present on blood cells and other tissues were shown to act as cofactors for factor I. These are CR1 (complement receptor type 1: CD35) (Fearon, 1980) and MCP (membrane cofactor protein: CD46) (Seya *et al*., 1986). CR1 and MCP act as cofactors for the factor I-mediated cleavage of both C3b and C4b (Fearon, 1980; Seya & Atkinson, 1989; Masaki *et al*., 1992) . These four cofactor proteins (factor H, C4bp, MCP and CR1) are homologous to each other, and are made up of multiple CCP domains (also called short consensus repeat (SCR) or sushi domains); they have no known enzymic activity. The cofactors are all encoded on human chromosome 1q31–32 in a region called the RCA (regulation of complement activation) cluster (Reid *et al*., 1986).

S

## *Activity and Specificity*

The only known activities of factor I are to cleave the complement proteins C3b and C4b. It has no action on C5b, or other members of the $\alpha_2$-macroglobulin family, which are homologs of C3b and C4b. The physiological substrates of factor I are produced only when the complement system is activated (for review, see Sim *et al*., 1993b). C3 and C4 are abundant plasma proteins. C4 is cleaved during complement activation by the serine protease C1s, releasing a 9 kDa fragment, C4a (Chapter 44). The remainder of C4, termed C4b (190 kDa), can be cleaved by factor I in the presence of one of the cofactor proteins C4bp, MCP or CR1. Cleavage occurs at two positions (Arg950+Thr951 and Arg1330+Asn1331) to excise a 45 kDa fragment named C4d (Press & Gagnon, 1981). The remainder of C4b is termed C4c. Similarly, C3 is activated by the serine proteases called C3 convertases (C4b,2a or C3b,Bb), releasing a 9 kDa fragment C3a (Chapters 46 and 47). The rest of the molecule, C3b (180 kDa), is a substrate for factor I. In the presence of one of the cofactor proteins, factor H, MCP or CR1, factor I cleaves two Arg+Ser bonds (positions 1303–1304 and 1320–1321), excising a small fragment C3f (17 residues) (Harrison & Lachmann, 1980; Davis & Harrison, 1981; Sim *et al*., 1981). The remainder of C3b is termed iC3b (or in older literature, C3bi). iC3b undergoes further slow cleavage in blood to form fragments C3c (140 kDa) and C3dg (37 kDa). This cleavage may perhaps be mediated by factor I (cleaving an Arg-Glu bond at position 954–955), but only CR1 appears to act as a cofactor for this cleavage (Seya & Nagasawa, 1985) . It is also possible that this cleavage is mediated not by factor I, but by leukocyte elastase or other proteases (Law *et al*., 1979; Harrison & Lachmann, 1980; Davis & Harrison, 1981; Malhotra & Sim, 1984; Jepsen *et al*., 1986).

The complement proteins C3 and C4 each contain an internal thioester. In circulation in the blood, the thioester undergoes hydrolysis (or nucleophilic attack by species such as ammonia) at a very low rate. This produces species of C3 or C4 which, although not cleaved by an activating protease to remove the C3a or C4a fragments, behave like C3b or C4b. These species are also substrates for factor I (Parkes *et al*., 1981; von Zabern *et al*., 1982). These substrates do not have a defined nomenclature, and are called by many different names, such as C3i, C4i, C3u, C4u, C3($H_2O$), C3($NH_3$).

The requirement for a cofactor protein for factor I-mediated proteolysis appears to be absolute, i.e. there is no proteolysis in the absence of a cofactor. The mechanism however requires further investigation. In the case of C3b cleavage by factor I in the presence of factor H, it is clear that both factor I and factor H bind independently and weakly to C3b (DiScipio, 1992; Soames & Sim, 1995, 1997). Factor I and factor H also have a weak direct interaction (Soames & Sim, 1997). The three-way interaction does reinforce the low affinity of factor I for C3b, and this alone may explain the effect of the cofactor. However the cofactor may act by altering the conformation of the substrate, or of the factor I, or may orient the substrate and the enzyme.

Isolated factor I has negligible activity against synthetic substrates such as peptide esters and amides. It is not inhibited by common serine protease inhibitors such as DFP, PMSF or Pefabloc-SC (4-[2-aminoethyl]benzenesulfonylfluoride). Nor is it inhibited by any of the plasma protease inhibitors, or by any of the common protein protease inhibitors such as soybean trypsin inhibitor, aprotinin, etc. (summarized by Crossley, 1981). Activity of factor I is susceptible to mild reduction (1–2 mM DTT). Tests of synthetic substrates and inhibitors have nearly all been done in the absence of a cofactor protein, so it is not clearly established whether the presence of the cofactor protein might induce the enzyme to take up an active conformation. There is one report that factor I can be inhibited by DFP if C3b, but not factor H, is present during the incubation with DFP (Nilsson-Ekdahl *et al*., 1990).

There are few known inhibitors of the action of factor I plus a cofactor. The cleavage of C3b in the presence of factors H and I is inhibited by low concentrations (0.05–0.1 mM) of $Zn^{2+}$ ions (Day & Sim, 1986; Jepsen *et al*., 1990), but this is likely to be an effect on the cofactor protein, not on factor I. The drug suramin (Antrypol) also inhibits breakdown of C3b, but it is not clear on which protein it acts, and it has several effects on the complement system (Lachmann & Hobart, 1978). K-76 monocarboxylic acid, a sesquiterpene derived from fungal cultures, also inhibits C3b breakdown, possibly by direct action on factor I, but it also has several other effects on complement system proteins (Hong *et al*., 1979; Miyazaki *et al*., 1980, 1984). A recent brief report indicates that novel compounds designed as thrombin inhibitors inhibit factor I (Knabb *et al*., 1996).

The pH optimum for cleavage of C3b in the presence of factor H is below pH 5.5 (Sim *et al*., 1981; Sim & Sim, 1983). When MCP or CR1 are used as cofactors, the pH optimum is pH 6 or 7.5–8.0, respectively (Seya *et al*., 1990). It is likely, therefore, that the pH optimum is determined by features other than the direct interaction with the susceptible bonds in the substrates.

The activity of factor I is species specific, and it is the interaction of factor I with factor H that confers this species specificity (see, e.g., Okada *et al*., 1993). For example, human C3b can be cleaved by human factor I plus human factor H, or by mouse factor I plus mouse factor H, but if the factor I and H are not from the same species (e.g. human C3b with mouse factor I plus human factor H), the rate of reaction is generally very low. This feature has not been investigated in detail for the other cofactors or with C4b as substrate.

Factor I activity is usually determined by following the cleavage of the substrate C3b to iC3b. This can be done by SDS-PAGE to observe the extent of cleavage of C3b (Sim *et al*., 1981; Sim & Sim, 1983), or by ELISA-type assays, detecting exposure of neoepitopes on iC3b (Vetvicka *et al*., 1993).

## *Structural Chemistry*

The cDNA sequence of human factor I encodes a 583 residue single preproprotein (Catterall *et al*., 1987; Goldberger *et al*., 1987). The secreted form of human factor I is a glycoprotein of about 88 kDa consisting of a 50 kDa heavy chain disulfide-linked to a 38 kDa light chain. The carbohydrate content is 27%. Each chain has three potential sites for attachment of *N*-linked oligosaccharides, and it is likely that all of these sites are occupied. Factor I exhibits considerable microheterogeneity in isoelectric focusing, showing more than 20 isoforms with pI values in the range 4.5–6.0 (Sim *et al.*, 1993a). Removal of sialic acid reduces the complexity and alters the pI to about 7.5. The light chain (243 amino acids) is a serine protease domain of the trypsin family. Its sequence is most similar to the serine protease domains of tissue plasminogen activator and urokinase. These serine proteases have an extra pair of cysteine residues (Cys355 and Cys426 in factor I) that are presumably disulfide bonded. The heavy chain (318 amino acids) is made up of four domains, and is derived from the N-terminal domain of the precursor polypeptide (Sim *et al.*, 1993b). The first domain (residues 23–89) is homologous to regions in complement proteins C6 and C7, and is called an I/C6/C7 module or FIM (factor I module). Next is a CD5 module, about 100 residues long, followed by two LDL-receptor modules each about 40 amino acids long. The cDNA sequences of *Xenopus laevis* factor I (Kunnath-Muglia *et al.*, 1993), mouse factor I (Minta *et al.*, 1996) and a partial cDNA sequence corresponding to a section of chicken factor I heavy chain (Minta *et al.*, 1996) have been determined. From these data, it appears that a short region of the heavy chain differs markedly between species, and this region may be associated with the species specificity discussed above.

The shape of the factor I molecule has been determined by electron microscopy and low-angle scattering techniques (DiScipio, 1992; Perkins *et al.*, 1993). It has a compact structure of maximum length about 13 nm.

## Preparation

Factor I is present in human plasma at about 40 mg liter$^{-1}$. Several procedures for purification by conventional precipitation and chromatography techniques are available (Fearon, 1977; Crossley, 1981). It can most conveniently be purified using monoclonal antibody affinity chromatography with the antibody MRC OX21 (available from the European Collection of Animal Cell Cultures, accession no. 91061417) (see Appendix 2 for full names and addresses of suppliers), followed by a gel-filtration step (Hsiung *et al.*, 1982; Sim *et al.*, 1993b).

## Biological Aspects

An activity like that of factor I plus a cofactor is detectable in a wide range of vertebrate sera. Human factor I is synthesized mostly in liver hepatocytes, although other tissues and cultured cells synthesize small amounts (e.g. glioma cells, lymphocytes, blood vessel endothelium: Gasque *et al.*, 1992; Julen *et al.*, 1992; Vetvicka *et al.*, 1993). Factor I circulates in its two-chain form, and there is no evidence that a single-chain precursor or proenzyme enters the circulation in normal circumstances. Factor I is a major regulatory protein of the complement system. It acts by controlling the turnover of the major complement protein C3, by destroying the C3b subunit of the enzyme C3b,Bb, which activates C3. It also converts C3b, attached to the surface of complement-activating particles, such as immune complexes or microorganisms, into the major opsonin, iC3b, thereby enhancing destruction of such particles by phagocytosis.

Deficiency of factor I is rare in humans, about 20 pedigrees having been reported (Sim *et al.*, 1993a; Vyse *et al.*, 1994b).

Lack of factor I leads to uncontrolled consumption of C3 and complement factor B, so that circulating C3 falls to negligible levels, and affected individuals are susceptible to recurrent bacterial infection (Rasmussen *et al.*, 1990; Floret *et al.*, 1991). Details of factor I deficiency are summarized in the OMIM database, entry 217030.

The gene symbol for factor I is *IF*. The gene is located on human chromosome 4q25 (Shiang *et al.*, 1989; Sim *et al.*, 1993a) and has been characterized by Vyse *et al.* (1994a). The human gene is about 63 kb long, with 13 exons. The mouse factor I gene has been localized to chromosome 3 (Minta *et al.*, 1996).

## Distinguishing Features

Factor I produces a characteristic pattern of limited proteolysis of the α chains of C3b and C4b (Harrison & Lachmann, 1980; Sim *et al.*, 1981; Press & Gagnon 1981). The activity of factor I plus a cofactor can be detected and monitored in many vertebrate sera using radioiodinated human C3b or C4b as substrate (Sim *et al.*, 1981; Kaidoh & Gigli, 1987). Polyclonal antisera against human factor I are widely available, as are several monoclonal antibodies. These do not generally react across species.

## Further Reading

Further information on factor I may be found in Sim *et al.* (1993b) and Vyse *et al.* (1994b).

## References

Catterall, C.F., Lyons, A., Sim, R.B., Day, A.J. & Harris, T.J.R. (1987) Characterisation of the primary amino acid sequence of human complement control protein Factor I from an analysis of cDNA clones. *Biochem. J.* **242**, 849–856.

Crossley, L.G. (1981) C3b-inactivator and beta-1H. *Methods Enzymol.* **80**, 112–124.

Davis, A.E. & Harrison, R.A. (1981) Structural characterization of factor I-mediated cleavage of the third component of complement. *Biochemistry* **21**, 5745–5749.

Day, A.J. & Sim, R.B. (1986) Inhibitory effect of $Zn^{2+}$ ions on the degradation of the complement activation fragment C3b. *Biochem. Soc. Trans.* **14**, 73–74.

DiScipio, R.G. (1992) Ultrastructures and interactions of complement Factor H and Factor I. *J. Immunol.* **149**, 2592–2599.

Fearon, D.T. (1977) Purification of C3b-inactivator and demonstration of its two polypeptide chain structure. *J. Immunol.* **119**, 1248–1252.

Fearon, D.T. (1980) Identification of the membrane glycoprotein that is the C3b receptor of the human erythrocyte, polymorphonuclear leukocyte, B lymphocyte, and monocyte. *J. Exp. Med.* **152**, 20–30.

Floret, D., Stamm, D. & Ponard, D. (1991) Increased susceptibility to infection in children with congenital deficiency of factor I. *Pediatr. Infect. Dis. J.* **10**, 615–618.

Fujita, T., Gigli, I. & Nussenzweig, V. (1978) Human C4-binding protein. II. Role in proteolysis of C4b by C3b-inactivator. *J. Exp. Med.* **148**, 1044–1051.

Gasque, P., Julen, N., Ischenko, A.M., Picot, C., Mauger, C., Chauzy, C., Ripoche, J. & Fontaine, M. (1992) Expression of complement components by glioma cell-lines. *J. Immunol.* **149**, 1381–1387.

Goldberger, G., Bruns, G.A.P., Rits, M., Edge, M.D. & Kwiatkowski, D.J. (1987) Human complement factor I: analysis of cDNA-derived structure and assignment of its gene to chromosome 4. *J. Biol. Chem.* **262**, 10065–10071.

Harrison, R.A. & Lachmann, P.J. (1980) The physiological breakdown of the third component of human complement. *Mol. Immunol.* **17**, 9–20.

Hong, K., Kinoshita, T., Miyazaki, W., Izawa, T. & Inoue, K. (1979) An anticomplementary agent, K-76 monocarboxylic acid: its site and mechanism of inhibition of the complement activation cascade. *J. Immunol.* **122**, 2418–2423.

Hsiung, L.-M., Barclay, A.N., Brandon, M.R., Sim, E. & Porter, R.R. (1982) Purification of human C3b-inactivator by monoclonal antibody affinity chromatography. *Biochem. J.* **203**, 293–298.

Jepsen, H.H., Svehag, S.-E., Jensenius, J.-C. & Sim, R.B. (1986) Release of immune complexes bound to erythrocyte complement receptor (CR1) with particular reference to the role of Factor I. *Scand. J. Immunol.* **24**, 205–213.

Jepsen, H.H., Teisner, B. & Svehag, S.-E. (1990) Zinc ions inhibit factor I-mediated release of CR1-bound immune complexes and degradation of cell-bound C3b and C4b. *Scand. J. Immunol.* **31**, 397–403.

Julen, N., Dauchel, H., Lemercier, C., Sim, R.B., Fontaine, M. & Ripoche, J. (1992) *In vitro* biosynthesis of complement Factor I by human endothelial cells. *Eur. J. Immunol.* **22**, 213–217.

Kaidoh, T. & Gigli, I. (1987) Phylogeny of C3b/C4b-cleaving activity: similar fragmentation patterns of human C4b and C3b produced by lower animals. *J. Immunol.* **139**, 194–201.

Knabb, R.M., Luettgen, J.M., Leamy, A.W., Barbera, F.A., Kettner, C.A. & Pangburn, M.K. (1996) Acute toxicity of synthetic thrombin inhibitors caused by inhibition of complement factor I. *Circulation* **94** (suppl.), 4069 (abstr.).

Kunnath-Muglia, L.M., Chang, G.C., Sim, R.B., Day A.J. & Ezekowitz, R.A.B. (1993) Characterisation of *Xenopus laevis* complement Factor I structure – conservation of modular structure except for an unusual insert not present in human Factor I. *Mol. Immunol.* **30**, 1249–1256.

Lachmann, P.J. & Hobart, M.J. (1978) Complement technology. In: *Handbook of Experimental Immunology*, vol. 1, 3rd edn (Weir, D.M., ed.). Oxford: Blackwell.

Lachmann, P.J. & Müller-Eberhard, H.J. (1968) The demonstration in human serum of conglutinogen-activating factor and its effect on the third component of complement. *J. Immunol.* **100**, 691–698.

Law, S.K., Fearon, D.T. & Levine, R.P. (1979) Action of the C3b-inactivator on the cell-bound C3b. *J. Immunol.* **122**, 759–765.

Masaki, T., Matsumoto, M., Nakanishi, I., Yasuda, R. & Seya, T. (1992) Factor I-dependent inactivation of human complement C4b by C3b/C4b receptor (CR1, CD35) and membrane cofactor protein (MCP, CD46). *J. Biochem.* **111**, 573–578.

Malhotra, V. & Sim, R.B. (1984) Role of complement receptor CR1 in the breakdown of soluble and zymosan-bound C3b. *Biochem. Soc. Trans.* **12**, 781–782.

Minta, J.O., Wong, M.J., Kozak, C.A., Kunnath-Muglia, L.M. & Goldberger, G. (1996) cDNA cloning, sequencing and chromosomal assignment of the gene for mouse complement factor I: identification of a species-specific divergent segment in factor I. *Mol. Immunol.* **33**, 101–112.

Miyazaki, W., Tamaoka, H., Shinohara, M., Kaise, H., Izawa, T., Nakano, Y., Kinoshita, T., Hong, K. & Inoue, K. (1980) A complement inhibitor produced by *Stachybotrys complementi*, nov. sp.

K-76, a new species of fungi imperfecti. *Microbiol. Immunol.* **24**, 1091–1108.

Miyazaki, W., Izawa, T., Nakano, Y., Shinohara, M., Hing, K., Kinoshita, T. & Inoue, K. (1984) Effects of K-76 monocarboxylic acid, an anticomplementary agent, on various in vivo immunological reactions and on experimental glomerulonephritis. *Complement* **1**, 134–146.

Nagasawa, S. & Stroud, R.M. (1980) Purification and characterization of a macromolecular weight cofactor for C3b-inactivator, C4bC3bINA-cofactor, of human plasma. *Mol. Immunol.* **17**, 1365–1372.

Nilsson-Ekdahl, K., Nilsson, U.R. & Nilsson, B. (1990) Inhibition of factor I by diisopropylfluorophosphate. *J. Immunol.* **144**, 4269–4274.

Okada, M., Kojima, A., Takano, H., Harada, Y., Nonaka, M., Nonaka, M., Kinoshita, T., Seya, T. & Natsuume-Sakai, S. (1993) Functional properties of the allotypes of mouse Factor H. Difference in compatibility of each allotype with human Factor I. *Mol. Immunol.* **30**, 841–848.

Parkes, C., DiScipio, R.G., Kerr, M.A. & Prohaska, R. (1981) The separation of functionally-distinct forms of human complement C3. *Biochem. J.* **193**, 963–970.

Perkins, S.J., Smith, K.F. & Sim, R.B. (1993) Molecular modelling of the domain structure of factor I of human complement by X-ray and neutron solution scattering. *Biochem. J.* **295**, 101–108.

Press, E.M. & Gagnon, J. (1981) Human complement component C4: structural studies of the fragments derived from cleavage with C3b-inactivator. *Biochem. J.* **199**, 351–357.

Rasmussen, J.M., Teisner, B., Brandt, J., Brandslund, J & Gry, H. (1990) Metabolism of C3 and factor B in patients with congenital Factor I deficiency. *J. Clin. Lab. Immunol.* **31**, 59–67.

Reid, K.B.M., Bentley, D.R., Campbell, R.D., Chung, L.P., Sim, R.B., Kristensen, T. & Tack, B.F. (1986) Complement system proteins which interact with C3b or C4b: a superfamily of structurally related proteins. *Immunol. Today* **7**, 230–234.

Seya, T. & Nagasawa, S. (1985) Limited proteolysis of complement protein C3b by regulatory enzyme C3b inactivator. Isolation and characterisation of a biologically active fragment, C3dg. *J. Biochem.* **97**, 373–382.

Seya, T. & Atkinson, J.P. (1989) Functional properties of membrane cofactor protein of complement. *Biochem. J.* **264**, 581–588.

Seya, T., Turner, J.R. & Atkinson, J.P. (1986) Purification and characterization of a membrane protein (gp45–70) that is a cofactor for cleavage of C3b and C4b. *J. Exp. Med.* **163**, 837–855.

Seya, T., Okada, M., Nishino, H. & Atkinson, J.P. (1990) Regulation of proteolytic activity of complement factor I by pH: CR1 and membrane cofactor protein (MCP) have different pH optima for factor I-mediated cleavage of C3b. *J. Biochem.* **107**, 310–315.

Shiang, R., Murray, J.C., Morton, C.C., Buetow, K.H., Wasmuth, J.J., Olney, A.H., Sanger, W.G. & Goldberger, G. (1989) Mapping of the human complement factor I gene to 4q25. *Genomics* **4**, 82–86.

Sim, E. & Sim, R.B. (1983) Enzymic assay of C3b receptor on intact cells and solubilized cells. *Biochem. J.* **210**, 567–576.

Sim, E., Wood, A.B., Hsiung, L.-M. & Sim, R.B. (1981) Pattern of degradation of human complement fragment C3b. *FEBS Lett.* **132**, 55–60.

Sim, R.B., Kölble, K., McAleer, M.A., Dominguez, O. & Dee, V.M. (1993a) Genetics and deficiencies of the soluble regulatory proteins of the complement system. *Int. Rev. Immunol.* **10**, 65–86.

Sim, R.B., Day, A.J., Moffatt, B.E. & Fontaine, M. (1993b) Complement factor I and its cofactors in control of the complement system convertase enzymes. *Methods Enzymol.* **223**, 13–35.

Soames, C.J. & Sim, R.B. (1995) An investigation of the interaction between human complement factor H and C3b. *Biochem. Soc. Trans.* **22**, 53s.

Soames, C.J. & Sim, R.B. (1997) The interactions between human complement factor H, factor I and C3b. *Biochem. J.* **326**, 553–561.

Tamura, N. & Nelson, R.A. (1967) Three naturally-occurring inhibitors of components of complement in guinea pig and rabbit serum. *J. Immunol.* **99**, 582–589.

Vetvicka, V., Reed, W., Hoover, M.L. & Ross, G.D. (1993) Complement Factor H and Factor I synthesised by B cell lines generate a growth factor activity from C3. *J. Immunol.* **150**, 4052–4060.

von Zabern, I., Bloom, E.L.. Chu, V. & Gigli, I. (1982) The fourth component of human complement treated with amines or chaotropes or frozen-thawed: interaction with C4-binding protein and C3b/C4b inactivator. *J. Immunol.* **128**, 1433–1438.

Vyse, T.J., Bates, G.P., Walport, M.J. & Morley, B.J. (1994a) The organisation of the human complement factor I gene (IF). *Genomics* **24**, 90–98.

Vyse, T.J., Spath, P., Davies, K.A., Morley, B.J. Philippe, P., Athanassiou, P., Giles, C.M. & Walport, M.J. (1994b) Hereditary complement Factor I deficiency. *Q. J. Med.* **87**, 385–401.

Whaley, K. & Ruddy, S. (1976) Modulation of the alternative complement pathway by beta-1H. *J. Exp. Med.* **144**, 1147–1155.

*Robert B. Sim*
*MRC Immunochemistry Unit, Department of Biochemistry,*
*Oxford University,*
*South Parks Road, Oxford OX1 3QU, UK*
*Email: rbsim@bioch.ox.ac.uk*

# 49. Coagulation factor XIIa

## Databanks

*Peptidase classification: clan SA, family S1, MEROPS ID: S01.211*
*NC-IUBMB enzyme classification: EC 3.4.21.38*
*Databank codes:*

| Species | SwissProt | PIR | EMBL (cDNA) | EMBL (genomic) |
|---|---|---|---|---|
| *Bos taurus* | P98140 | – | S70164 | – |
| *Cavia porcellus* | Q04962 | – | X68615 | – |
| *Homo sapiens* | P00748 | A29411 | M11723 | M17464: exon 1 |
|  |  |  | M13147 | M17465: exon 2 |
|  |  |  | M31315 | M17466: exons 3–14 and complete CDS |

## Name and History

In 1955, a plasma protein was found to be lacking from the circulation of one John Hageman (Ratnoff *et al*., 1968). This protein has subsequently been known as Hageman factor or coagulation factor XII. Factor XII is the proenzyme of *factor XIIa*, which is the first component of the intrinsic pathway of coagulation. As such, it does not require proteolytic activation by another endopeptidase, but self-activates, under the conditions that initiate this pathway. This happens when blood comes in contact with 'foreign' surfaces, or with polyanions in solution.

## Activity and Specificity

The proenzyme factor XII is activated either autolytically or by the action of plasma kallikrein (Chapter 50). *In vitro*, factor XII becomes activated when exposed to a negatively charged surface, or a similarly charged polymer in solution. Effective agents are glass, kaolin, diatomaceous earth, dextran sulfate and ellagic acid, as well as biological molecules such as chondroitin sulfate and sulfatides (Rosing *et al*., 1985). This form of activation is thought to model the physiological activation that occurs when a vessel is damaged, and blood comes into contact with the subendothelial basement membrane, although the identity of the physiological activator in that situation is not certain. Full activation of factor XII occurs only when the proenzyme is in contact with the negatively charged 'surface' in the presence of high H-kininogen and kallikrein. H-kininogen interacts with all of the early components in the intrinsic pathway, and has an important controlling influence. The molecular interactions that occur at this stage in the intrinsic pathway of coagulation are fully described in Chapters 50 and 51. The positively charged polymer, polybrene, inhibits contact activation so as to stabilize factor XII *in vitro*.

Once activated, factor XIIa will itself activate the precursors of factor XIa (Chapter 51) and plasma kallikrein (Chapter 50). Kallikrein can then mediate further proteolysis of factor XII to form more factor XIIa or factor XIIf (Colman, 1984; Tans & Rosing, 1987). Factor XIIa is also able to activate factor VII (Seligsohn *et al.*, 1979), prorenin (Gordon *et al.*, 1983) and complement component C1 (Ghebrehiwet *et al.*, 1981, 1983). Factor XIIf is not a coagulant molecule, since it is only a weak activator of factor XI, but it may have other significant activities.

The coagulant activity of factor XIIa can be determined by its correction of the abnormal clotting time of human plasma deficient in factor XII (Ratnoff & Colopy, 1955; Pixley & Colman, 1993). Spectrophotometric assays may be made with D-Pro-Phe-Arg-NHPhNO$_2$ (available from Kabi) (see Appendix 2 for full names and addresses of suppliers) as substrate (Pixley & Colman, 1993).

The major physiological inhibitor of factor XIIa is C1 inhibitor (Forbes *et al.*, 1970; Pixley & Colman, 1993), but the enzyme is significantly protected against inhibition while it remains bound to the activating surface. There is also inhibition by antithrombin III, which is dramatically enhanced by heparin (Fujikawa, 1988), and by $\alpha_2$-plasmin inhibitor. Less physiologically, factor XIIa is potently inhibited by ecotin, with a $K_i$ of 90 pM (Ulmer *et al.*, 1995). Limabean trypsin inhibitor is also effective, and an inhibitor from popcorn is quite selective for factor XIIa (Hojima *et al.*, 1980; Kambhu *et al.*, 1985; Pedersen *et al.*, 1994). Soybean trypsin inhibitor and ovomucoid do not inhibit factor XIIa. DFP, PMSF and benzamidine are inhibitory.

## Structural Chemistry

The proenzyme, factor XII, is a single-chain glycoprotein composed of 596 amino acids (human) and 16.8% carbohydrate, with a molecular mass of 76–80 kDa. The activation cleavage occurs at the Arg353$\dashv$Val354 bond within a disulfide loop, so that the N-terminal heavy chain (48 kDa) and the catalytically active light chain (28 kDa) remain covalently associated (Fujikawa & McMullen, 1983; McMullen & Fujikawa, 1985). The heavy chain contains several domains found in other mosaic proteins of the trypsin family (as described in Chapter 2). These include one kringle domain, one sushi domain, two EGF-like domains and a 'finger' domain. There is a heavily glycosylated, proline-rich sequence on the N-terminal side of the activation site. Species differences exist between human, cattle and guinea pig variants of factor XII in the amino acid sequences near the site of the activation cleavage, and it has been suggested that these account for

differences in the activation processes (Shibuya *et al.*, 1994; Yamamoto *et al.*, 1996).

## Preparation

Procedures for the purification of factor XII from human, cattle and pig plasma have been fully described (Fujikawa *et al.*, 1977; Silverberg & Kaplan, 1988; Fujikawa, 1988; Pixley & Colman, 1993; Mashiko *et al.*, 1996). In recent years, factor XII has generally been isolated either by zinc chelate chromatography or by immunoaffinity chromatography on an immobilized monoclonal antibody B7C9. The isolation of two distinct forms of pig factor XII has been described (Mashiko *et al.*, 1996). Factor XII is caused to autoactivate to factor XIIa by exposure to dextran sulfate, and factor XIIf can be isolated as a byproduct from factor XIIa preparations (Silverberg & Kaplan, 1988; Pixley & Colman, 1993).

## Biological Aspects

Factor XII is synthesized in the liver by hepatocytes (Saito *et al.*, 1983; Gordon *et al.*, 1990). The functional deficiency of factor XII, Hageman trait, is detected *in vitro* by the finding of a prolonged partial thromboplastin time, but *in vivo*, with rare exceptions, no obvious hemorrhagic tendency is observed in those who lack factor XII (Silverberg & Kaplan, 1988). Several individuals with Hageman trait, including the proband Mr Hageman, have sustained major thrombotic episodes, however. The normal plasma concentration is about 30 μg ml$^{-1}$ in human plasma and 6 μg ml$^{-1}$ in cattle plasma (Fujikawa, 1988). The gene encoding human factor XII is located at 5q33–qter (Que & Davie, 1986; Royle *et al.*, 1988); the gene is approximately 12 kb in length, and contains 14 exons (Cool & MacGillivray, 1987).

Both factor XII and factor XIIa enhance proliferation of cultured human hepatoma cells (Schmeidler-Sapiro *et al.*, 1991). It may therefore be significant that factor XII is very similar in amino acid sequence to an activator of hepatocyte growth factor (Miyazawa *et al.*, 1993) (see Chapter 63).

## Distinguishing Features

The activity of factor XIIa may be distinguished from that of plasma kallikrein by its resistance to inhibition by soybean trypsin inhibitor (50 μg ml$^{-1}$). A commercially available monoclonal antibody against human factor XII is described by Pixley & Colman (1993).

## Related Peptidases

Factor XII has structural similarities to several other coagulation factors described elsewhere in the present volume. The structural similarity of factor XII to an activator of hepatocyte growth factor (Chapter 63) was mentioned above.

## Further Reading

Practical aspects of the study of factor XIIa are well reviewed by Fujikawa (1988), Kluft (1988), Pixley & Colman (1993) and Silverberg & Kaplan (1988).

## References

Colman, R.W. (1984) Surface-mediated defense reactions. The plasma contact activation system. *J. Clin. Invest.* **73**, 1249–1253.

Cool, D.E. & MacGillivray, R.T.A. (1987) Characterization of the human blood coagulation factor XII gene. Intron/exon gene organization and analysis of the 5′-flanking region. *J. Biol. Chem.* **262**, 13662–13673.

Forbes, C., Pensky, J. & Ratnoff, O.D. (1970) Inhibition of activated Hageman factor and activated plasma thromboplastin antecedent by purified serum C1 inactivator. *J. Lab. Clin. Med.* **76**, 809–815.

Fujikawa, K. (1988) Bovine Hageman factor and its fragments. *Methods Enzymol.* **163**, 54–68.

Fujikawa, K. & McMullen, B.A. (1983) Amino acid sequence of human $\beta$-factor XIIa. *J. Biol. Chem.* **258**, 10924–10933.

Fujikawa, K., Kurachi, K. & Davie, E.W. (1977) Characterization of bovine factor XII$_a$ (activated Hageman factor). *Biochemistry* **16**, 4182–4188.

Ghebrehiwet, B., Randazzo, B.P., Dunn, J.T. Silverberg, M. & Kaplan, A.D. (1983) Mechanisms of activation of the classical pathway of complement by Hageman factor fragment. *J. Clin. Invest.* **71**, 1450–1456.

Ghebrehiwet, B., Silverberg, M. & Kaplan, A.P. (1981) Activation of the classical pathway of complement by Hageman factor fragment. *J. Exp. Med.* **153**, 665–676.

Gordon, E.M., Douglas, J. & Ratnoff, O.D. (1983) Influence of augmented Hageman factor (factor XII) titers on the cryoactivation of plasma prorenin in women using oral contraceptive agents. *J. Clin. Invest.* **72**, 1833–1838.

Gordon, E.M., Gallagher, C.A., Johnson, T.R., Blossey, B.K. & Ilan, J. (1990) Hepatocytes express blood coagulation factor XII (Hageman factor). *J. Lab. Clin. Med.* **115**, 463–469.

Hojima, Y., Pierce, J.V. & Pisano, J.J. (1980) Hageman factor fragment inhibitor in corn seeds: purification and characterization. *Thromb. Res.* **20**, 149–162.

Kambhu, S.A., Ratnoff, O.D. & Everson, B. (1985) Inhibition of Hageman factor (factor XII) by popcorn inhibitor. *J. Lab. Clin. Med.* **105**, 625–628.

Kluft, C. (1988) Synthetic substrates for the assay of prekallikrein and factor XII. *Methods Enzymol.* **163**, 170–179.

Mashiko, H., Kato, K., Fujii, K. & Takahashi, H. (1996) Studies on factor XII in porcine plasma: purification and its conversion to activated form by porcine plasma kallikrein. *Biochim. Biophys. Acta* **1296**, 198–206.

McMullen, B.A. & Fujikawa, K. (1985) Amino acid sequence of the heavy chain of human $\alpha$-factor XIIa (activated Hageman factor). *J. Biol. Chem.* **260**, 5328–5341.

Miyazawa, K., Shimomura, T., Kitamura, A., Kondo, J., Morimoto, Y. & Kitamura, N. (1993) Molecular cloning and sequence analysis of the cDNA for a human serine protease responsible for activation of hepatocyte growth factor. Structural similarity of the protease precursor to blood coagulation factor XII. *J. Biol. Chem.* **268**, 10024–10028.

Pedersen, L.C., Yee, V.C., von Dassow, G., Hazeghazam, M., Reeck, G.R., Stenkamp, R.E. & Teller, D.C. (1994) The corn inhibitor of blood coagulation factor XIIa. Crystallization and preliminary crystallographic analysis. *J. Mol. Biol.* **236**, 385–387.

Pixley, R.A. & Colman, R.W. (1993) Factor XII: Hageman factor. *Methods Enzymol.* **222**, 51–65.

Que, B.G. & Davie, E.W. (1986) Characterization of a cDNA coding for human factor XII (Hageman factor). *Biochemistry* **25**, 1525–1528.

Ratnoff, O.D. & Colopy, J.E. (1955) A familial hemorrhagic trait associated with a deficiency of a clot-promoting fraction of plasma. *J. Clin. Invest.* **34**, 603–613.

Ratnoff, O.D., Busse, R.J.J. & Sheon, R.P. (1968) The demise of John Hageman. *N. Engl. J. Med.* **279**, 760–761.

Rosing, J., Tans, G. & Griffin, J.H. (1985) Surface-dependent activation of human factor XII (Hageman factor) by kallikrein and its light chain. *Eur. J. Biochem.* **151**, 531–538.

Royle, N.J., Nigli, M., Cool, D., MacGillivray, R.T. & Hamerton, J.L. (1988) Structural gene encoding human factor XII is located at 5q33-qter. *Somatic Cell Mol. Genet.* **14**, 217–221.

Saito, H., Hamilton, S.M., Tavill, A.S. *et al.* (1983) Synthesis and release of Hageman factor (factor XII) by the isolated perfused rat liver. *J. Clin. Invest.* **72**, 948–954.

Schmeidler-Sapiro, K.T., Ratnoff, O.D. & Gordon, E.M. (1991) Mitogenic effects of coagulation factor XII and factor XIIa on HepG2 cells. *Proc. Natl Acad. Sci. USA* **88**, 4382–4385.

Seligsohn, U., Østerud, B., Brown, S.F., Griffin, J.H. & Rapaport, S.I. (1979) Activation of human factor VII in plasma and in purified systems. Roles of activated factor IX, kallikrein, and activated factor XII. *J. Clin. Invest.* **64**, 1056–1065.

Shibuya, Y., Semba, U., Okabe, H., Kambara, T. & Yamamoto, T. (1994) Primary structure of bovine Hageman factor (blood coagulation factor XII): comparison with human and guinea pig molecules. *Biochim. Biophys. Acta* **1206**, 63–70.

Silverberg, M. & Kaplan, A.P. (1988) Human Hageman factor and its fragments. *Methods Enzymol.* **163**, 68–80.

Tans, G. & Rosing, J. (1987) Structural and functional characterization of factor XII. *Semin. Thromb. Hemost.* **13**, 1–14.

Ulmer, J.S., Lindquist, R.N., Dennis, M.S. & Lazarus, R.A. (1995) Ecotin is a potent inhibitor of the contact system proteases factor XIIa and plasma kallikrein. *FEBS Lett.* **365**, 159–163.

Yamamoto, T., Shibuya, Y. & Semba, U. (1996) Species differences in amino acid sequences of Hageman factor and prekallikrein at region around scissile bond in activation. *Immunopharmacology* **32**, 34–38.

*Oscar D. Ratnoff*
*Department of Medicine,*
*University Hospitals of Cleveland,*
*Case Western Reserve University,*
*Cleveland, OH 44106, USA*

# 50. *Plasma prekallikrein and kallikrein*

## Databanks

*Peptidase classification: clan SA, family S1, MEROPS ID: S01.212*
*NC-IUBMB enzyme classification: EC 3.4.21.34*
*Chemical Abstracts Service registry number: 9001-01-8*
*Databank codes:*

| Species | SwissProt | PIR | EMBL (cDNA) | EMBL (genomic) |
|---|---|---|---|---|
| *Homo sapiens* | P03952 | A00921 | M13143 | – |
| | | A37939 | | |
| *Mus musculus* | P26262 | A36557 | M58588 | – |
| *Rattus norvegicus* | P14272 | A33320 | M30282 | M62345: exon 1 |
| | | A36557 | M58590 | M62346: exon 3 |
| | | A39180 | | M62347: exon 4 |
| | | I53041 | | M62348: exon 5 |
| | | S06851 | | M62349: exon 6 |
| | | S06852 | | M62350: exon 7 |
| | | | | M62351: exon 8 |
| | | | | M62352: exons 9 and 10 |
| | | | | M62353: exon 11 |
| | | | | M62354: exon 12 |
| | | | | M62355: exon 13 |
| | | | | M62356: exon 14 |
| | | | | M62357: exon 15 and complete CDS |
| | | | | M62358: exon 2 |
| | | | | M62359: intron 2 |

## Name and History

The name **kallikrein** (Gr. *kallikros*, pancreas) was first used for an enzyme from the pancreas that cleaved kininogen to yield bradykinin, or a bradykinin-like peptide with similar pharmacologic activity (Chapter 29). The plasma protein that was also capable of cleaving kininogen, **plasma kallikrein**, was discovered by Hathaway *et al.* (1965) who described a previously unrecognized coagulation protein that was affected in an hereditary abnormality termed Fletcher trait. This is characterized by a prolonged partial thromboplastin time which shortens to normal if the plasma is exposed for a prolonged period to a clot-promoting surface such as kaolin before addition of calcium ions, and these characteristics suggested an indirect or feedback role for the protein in the coagulation cascade. The protein was also designated as **Fletcher factor** after the surname of the affected family. The occurrence in four siblings – offspring of a consanguineous union – suggested autosomal recessive inheritance (Abildgaard & Harrison, 1974).

Wuepper (1973) showed that the missing protein in Fletcher trait was prekallikrein, since the purified protein corrected not only the coagulation defect but also abnormal surface-activated kinin formation and fibrinolysis seen in Fletcher trait. Affected individuals do not have a hemostatic disorder. Although the proenzyme should strictly be designated *prokallikrein*, the name **prekallikrein (PK)** has persisted because of long usage.

## Activity and Specificity

Plasma PK is a proenzyme and must be converted to the active enzyme by activated coagulation factor XII (Chapter 49). In normal plasma, the reactions are initiated by exposure to an anionic surface which alters the conformation of the proenzyme, factor XII, allowing it to autoactivate to factor XIIa, and then to convert PK to plasma kallikrein. Plasma kallikrein is an endopeptidase that cleaves at P1 Arg (preferred) and Lys bonds. The purified enzyme was first assayed by its ability to cleave the synthetic substrate Tos-Arg-OMe (sometimes abbreviated TAME, in the literature) (Webster & Pierce, 1961), and this assay was modified to measure PK in plasma (Mason & Colman, 1971). More recently, the synthetic substrate Bz-Pro-Phe-Arg+NHPhNO$_2$, available from Chromogenix (see Appendix 2 for full names and addresses of suppliers), has been used to measure both PK (DeLa Cadena *et al.*, 1987) and kallikrein. The pH optimum is 7.4, and there is no metal ion requirement.

Assays to measure correction of the abnormal partial thromboplastin time, or the release of bradykinin from plasma or purified kininogens, are less accurate and more laborious.

PK can be measured immunologically by a sandwich-type ELISA using two monoclonal antibodies against different epitopes (Page *et al.*, 1994a). The enzyme is a serine protease in family S1 is inhibited by DFP. Commercially available macromolecular inhibitors include soybean trypsin inhibitor and aprotinin.

## Structural Chemistry

Analysis of the cDNA for PK has allowed determination of the amino acid sequence (Figure 50.1) (Chung *et al.*, 1986). The leader sequence of 19 amino acids contains the typical hydrophobic amino acids needed for translocation across the rough endoplasmic reticulum in the secretion process. The cDNA codes for a mature protein of 619 amino

acids and shows a 58% identity with coagulation factor XI (Chapter 51). The conversion of PK to kallikrein is catalyzed on a surface by factor XIIa augmented by H-kininogen, or in the fluid phase by factor XII fragment (Wuepper & Cochrane, 1972). A single bond (Arg371-Ile372) is split, generating a heavy chain of 371 amino acids (33 kDa). The heavy chain, which is the N-terminal end, contains four tandem repeats of 90 or 91 amino acids, which result from two successive gene duplications. The location of the disulfide bonds (McMullen *et al.*, 1991) determines the presence of unique 'apple' domains, homologous to those of factor XI. Each of the four repeats has six conserved Cys residues. The fourth repeat contains two additional Cys residues, forming an additional disulfide bridge. Repeats 1 and 3, and repeats 2 and 4, are most similar to each other. Repeats 2 and 4

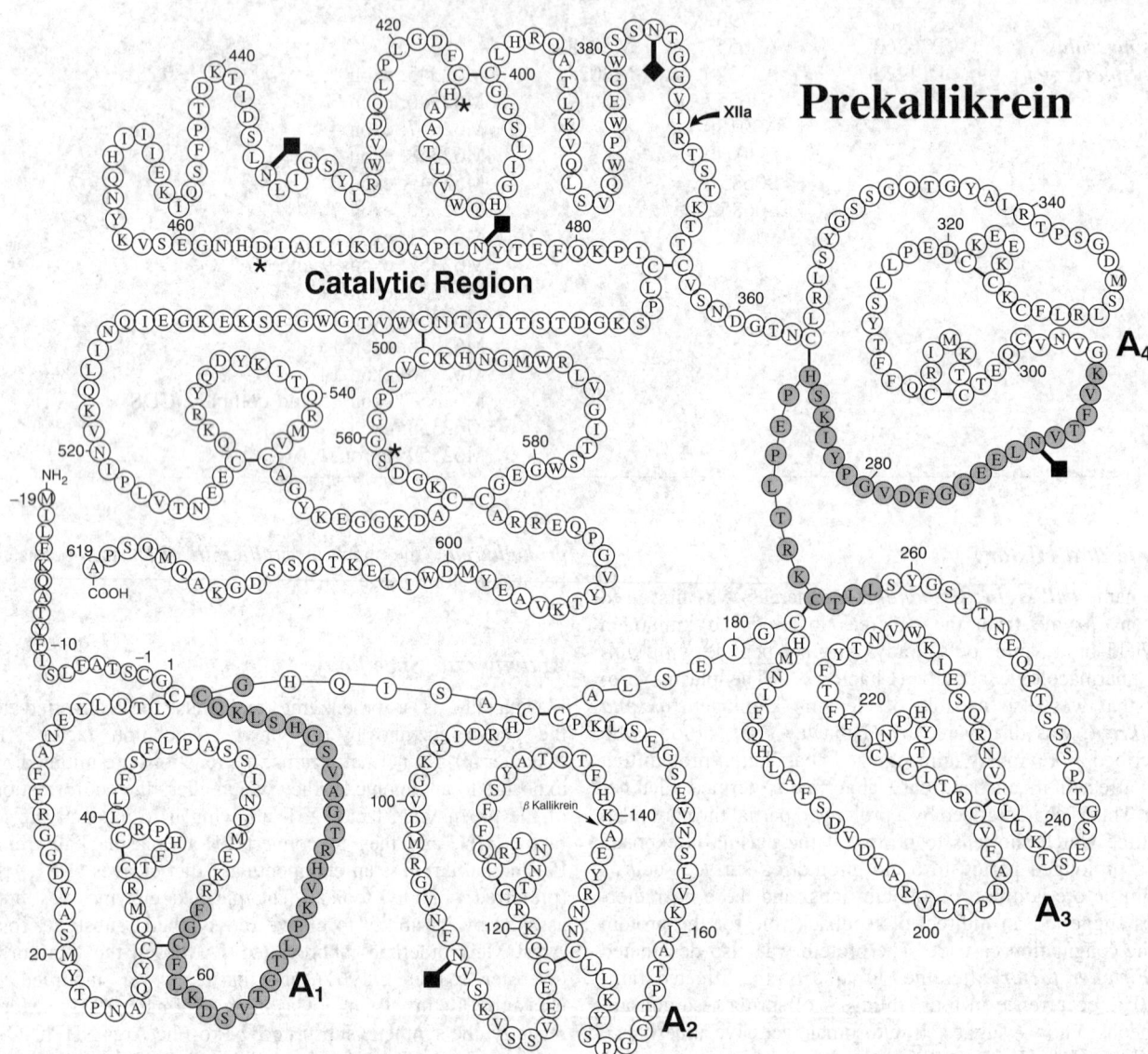

*Figure 50.1*   Primary structure of human PK. Location of the disulfide bonds in human prekallikrein. The four apple domains in the N-terminal portion of the molecule are labeled $A_1$, $A_2$, $A_3$ and $A_4$. The Arg371-Ile372 bond that is cleaved by factor XIIa during the activation reaction is marked by a solid curved arrow. The starred residues (His415, Asp464 and Ser559) are members of the catalytic triad characteristic of serine proteases. The Asn residues marked by solid diamonds (residues 108, 289, 377, 434 and 475) are attachment sites for carbohydrate chains. The N- and C-terminal residues are identified by $NH_2$ and COOH, respectively. The shaded residues are the H-kininogen binding peptides in $A_1$ and $A_4$.

each have a carbohydrate chain linked to the homologous Asn residues 108 and 289. A short connecting region of nine amino acids follows repeat 4, ending at the cleavage site between the light and heavy chains, Arg371$+$Ile372. Following cleavage, the enzyme consists of two chains linked by a single disulfide bridge. Kallikrein catalyzes an autolytic cleavage at Lys140$+$Ala141, forming $\beta$-kallikrein (Page & Colman, 1991). This enzyme, with a single cleavage in the heavy chain, exhibits decreased coagulant activity, a diminished rate of cleavage of H-kininogen, and decreased ability to stimulate neutrophils. The H-kininogen binding sequences of PK reside in the two domains apple 1 (Phe56–Gly86) and apple 4 (Lys266–Cys295) (Page *et al.*, 1994b). In addition, studies with a monoclonal antibody to the heavy chain region indicate that it contains a site recognized by factor XIIa; this is within the same 28 kDa fragment (residues 141–371), but the two subdomains are separate (Page & Colman, 1991).

The light chain contains the catalytic triad His415, Asp464 and Ser559. The isolated light chain retains the ability to activate factor XII in solution and is also able to cleave H-kininogen, albeit at a slower rate. The molecule contains 15% carbohydrate and five potential sites for *N*-linked glycosylation. Heterogeneity of the carbohydrate attachment sites provides the basis for two light chains of 36 and 33 kDa. The light chain of kallikrein is the site for reactions with protease inhibitors (Schapira *et al.*, 1982d). In plasma, kallikrein is rapidly inactivated by two inhibitors, $\alpha_2$-macroglobulin and C1 inhibitor (Schapira *et al.*, 1982b). C1 inhibitor forms a 1:1 stoichiometric complex with kallikrein (Gigli *et al.*, 1970), resulting in complete loss of proteolytic and amidolytic activity, while the kallikrein–$\alpha_2$-macroglobulin complex maintains 25% of its amidolytic activity. Although H-kininogen and C1 inhibitor bind to different regions of the kallikrein molecule, it has been observed that H-kininogen and its light chain protect kallikrein from inhibition by C1 inhibitor (Schapira *et al.*, 1981) and other plasma protease inhibitors (Schapira *et al.*, 1982c). This observation is compatible with substrate-protection of an enzyme from an active site-directed inhibitor, since H-kininogen is a substrate of kallikrein.

## Preparation

Plasma PK is synthesized in the liver, but mRNA has also been detected in the human and rat kidney, adrenal and placenta (Ciechanowicz *et al.*, 1993). PK is a fast $\gamma$-globulin with a pI between 8.5 and 9.0. It circulates in blood with an estimated concentration of 35–50 µg ml$^{-1}$. Approximately 75% circulates bound to H-kininogen (Mandle *et al.*, 1976), and only 25% is free. The molecular mass of human PK, as assessed by gel filtration, is approximately 100 kDa. By SDS-PAGE, PK consists of two components of $M_r$ 85 000 and 88 000 that probably differ in degree of glycosylation. PK can be purified by column chromatography (Scott *et al.*, 1979) or by an immunoaffinity method (Page & Colman, 1991).

## Biological Aspects

### Gene Structure

Although the nature of the gene for human PK remains to be determined, the gene structure and chromosomal localization

for rat PK have been described (Beaubien *et al.*, 1991). Like that of human factor XI, the rat plasma PK gene is composed of 15 exons and 14 introns. The sequence identity between the human and rat proteins is 75%. These data, together with the results of chromosomal localization for the human factor XI and rat PK genes (both chromosome 4), clearly point to a gene duplication event from a common ancestor of both PK and factor XI about 270 million years ago (Veloso *et al.*, 1987).

### Surface Activation Reactions

The majority of plasma PK exists in bimolecular complexes with H-kininogen. The kininogen can be activated so as to augment its endothelial cell surface binding, and this activation brings more PK to the surface (Figure 50.2). Factor XII initiates the contact system by diffusing to the surface and undergoing autoactivation. On the surface, factor XIIa can cleave PK to kallikrein. Kallikrein can promote reciprocal activation, generating additional factor XIIa from factor XII. The process of reciprocal activation is several orders of magnitude faster than autoactivation. The importance of plasma kallikrein in accelerating surface-bound activation of factor XII was shown by the observation that in normal plasma the binding of factor XII to the surface was rapidly followed by fragmentation of the molecule, whereas in PK-deficient plasma, the binding of factor XII to the surface occurred normally, but cleavage and activation were retarded. Kallikrein, once formed, may also diffuse off the surface to further hydrolyze its substrate, H-kininogen. As a result of kallikrein cleavage of H-kininogen, bradykinin is released. Thus, the net effect of the initial contact reactions is to release bradykinin, and position complexes of activated H-kininogen with PK and activated H-kininogen with factor XI on the surface in close proximity to factor XIIa. Factor XIIa acts on both of these surface-bound complexes to form the surface-bound complexes of activated H-kininogen with kallikrein and activated H-kininogen with factor XIa. Factor XIa inactivates the activated H-kininogen and is released to convert factor IX to factor IXa, and enhance blood coagulation.

### Fluid Phase Reactions

Kallikrein complexed with activated H-kininogen is not tightly bound and is released from the surface into the

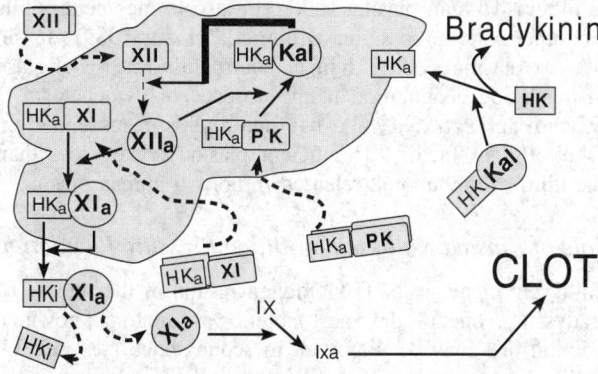

*Figure 50.2* Initiation of contact activation. Rectangles are zymogens, or procofactors. Circles are active enzymes. Irregular background is anionic or cell surface.

fluid phase. Besides H-kininogen, fluid-phase kallikrein has a variety of substrates. These include factor XII (Chapter 49), prourokinase (Chapter 57), prorenin (Chapter 284), and complement component C1 (Chapters 43 and 44). Besides activating additional factor XII, kallikrein can convert the factor XIIa into factor XIIf (Hageman factor fragment), which retains the active site of factor XIIa while losing the surface-binding domain. Factor XIIf is a potent fluid-phase activator of PK. In addition, activated H-kininogen complexed with either PK or factor XI can penetrate a barrier of fibrinogen to bind to an anionic surface, where it optimally positions the proenzyme for activation by adjacent factor XIIa molecules.

### Role of PK in Fibrinolysis

Plasma kallikrein plays an important role in fibrinolysis. Colman (1969) showed that plasma kallikrein can cleave plasminogen to plasmin (Chapter 59), but it requires stoichiometric amounts and is a very slow reaction. More important is the ability of kallikrein to cleave prourokinase to urokinase in purified systems (Ichinose *et al.*, 1986), plasma (Hauert *et al.*, 1989) and on the endothelial cell surface (Lin *et al.*, 1996). Since both prourokinase and plasminogen bind to endothelial cells, kallikrein may be the initiating enzyme in cell-associated plasmin formation.

### Role of Kallikrein in Neutrophil Stimulation

Human plasma kallikrein (the active species), but not PK (the proenzyme), is chemotactic for neutrophils (Kaplan *et al.*, 1972). Exposure of neutrophils to concentrations of human plasma kallikrein capable of eliciting chemotaxis increases aerobic glycolysis and activity of the hexose-monophosphate shunt (Goetzl & Austen, 1974). In the presence of calcium and magnesium ions, neutrophils aggregate in response to human plasma kallikrein. This interaction is associated with a stimulation of the respiratory burst in neutrophils, as indicated by an increase in oxygen uptake (Schapira *et al.*, 1982a). Human plasma kallikrein also induces neutrophils to release human neutrophil elastase (Chapter 15) from their azurophilic granules (Wachtfogel *et al.*, 1983). In the absence of plasma, washed neutrophils must be primed with cytochalasin B (a substance that causes the membranes of the azurophilic granules to fuse with the cellular membrane) for degranulation, thus facilitating the release of elastase in response to kallikrein. Human plasma kallikrein also primes neutrophils for superoxide production (Zimmerli *et al.*, 1989). In an *in vitro* plasma system, human neutrophils release elastase during blood coagulation in the absence of cytochalasin B (Wachtfogel *et al.*, 1983), but neutrophils resuspended in either PK- or factor XII-deficient plasma release less than one-third of the amount released in normal human plasma.

### Role of Plasma Kallikrein in Blood Pressure Regulation

Kallikrein cleavage of H-kininogen results in the release of bradykinin, one of the most potent vasodilators known, a reaction that may be important in septic shock (see below). Kallikrein can also activate partially purified plasma prorenin which, theoretically, could raise the blood pressure, but this reaction does not occur in normal plasma (Purdon *et al.*, 1985).

### Congenital Deficiency of PK (Fletcher Trait)

Since human PK deficiency was first identified (Hathaway *et al.*, 1965), it has been reported in black Americans, but studies with American, Mediterranean, Italian, Austrian and Nigerian populations suggest that the deficiency occurs generally. PK deficiency, like that of factor XII, is an autosomal-recessive trait with no apparent associated clinical bleeding disorder (Hattersley & Hayse, 1970). However, patients with this deficiency have experienced myocardial infarction, thromboembolism or multiple cerebral thromboses (Currembhoy *et al.*, 1976), confirming the lack of protection from arterial thrombotic disease. Additionally, surface-mediated fibrinolysis and increased vascular permeability are also defective in plasma of these patients (Saito *et al.*, 1974). Immunoreactive, PK-cross-reacting material (CRM) has been detected in 5 of 18 patients of Mediterranean extraction who lack functional PK activity (Saito *et al.*, 1981). Reduced levels of PK antigen have been detected in a family from Long Beach, California, whose members lack PK activity (Bouma *et al.*, 1986).

### Acquired Deficiency of PK

A deficiency of PK may result from depressed biosynthesis in chronic hepatocellular insufficiency such as that seen in cirrhosis (Wong *et al.*, 1972). A rare cause of PK deficiency would be that due to an autoantibody which can either inhibit the activity of kallikrein or block the activation of PK. We have recently reported the first case of the latter in which a patient with an elevated antinuclear antibody titer also manifested an antibody to the heavy chain of PK which prolonged the activated partial thromboplastin time (Page *et al.*, 1994a). A further cause of acquired PK deficiency is activation of the contact system, with consequent depletion of PK. Decreased PK levels have been reported in hereditary angioedema, septic shock, carcinoid syndrome, allergic reactions, postgastrectomy syndrome, nephrotic syndrome, polycythemia vera, Rocky Mountain spotted fever, adult respiratory distress syndrome and cardiopulmonary bypass. Examples of these are briefly discussed in the following sections.

## Hereditary Angioedema

Hereditary angioedema is due to decreased functional levels of C1 inhibitor, a major inhibitor of plasma kallikrein. During acute episodes, but not during the quiescent phases, functional activity and antigen levels of PK decrease (Schapira *et al.*, 1983) and kallikrein-$\alpha_2$-macroglobulin levels increase (Kaufman *et al.*, 1991). Blisters formed during acute episodes contain high levels of kallikrein (Curd *et al.*, 1980), which cleaves H-kininogen and gives rise to bradykinin, accounting for the increase in vascular permeability.

## Septicemia and Septic Shock

Bradykinin has been implicatd in the pathogenesis of septic shock through its ability to lower blood pressure in normal humans and animals. The early peripheral vascular changes accompanying gram-negative bacteremia and endotoxic shock are similar to those of bradykinin infusion, including arteriolar dilatation and venular constriction. Infusion of

*E. coli* endotoxin in normal volunteers (DeLa Cadena *et al.*, 1993), as well as naturally occurring sepsis (Mason *et al.*, 1970), leads to a decrease in functional levels of PK and H-kininogen in plasma, and an increase in plasma levels of kallikrein–$\alpha_2$-macroglobulin complexes (Kaufman *et al.*, 1991). Activation of the kallikrein–kinin system in septic shock is probably mediated by factor XIIa or factor XII fragments. In a baboon model of sepsis, an antibody to the factor XII light chain partially prevents the hypotension and prolongs survival (Pixley *et al.*, 1993).

The sequence of events of kinin production in human septic shock includes several stages. Initially, factor XII is activated by the lipid A component of endotoxin *in vivo*, which results in a decrease in the plasma concentration of the proenzyme and formation of factor XIIa and factor XII fragments. These enzymes then activate PK, causing a decrease in its plasma level. Kallikrein then cleaves H-kininogen to release bradykinin with depletion of plasma H-kininogen.

Other infectious diseases, such as typhoid fever (Colman *et al.*, 1978) and Rocky Mountain spotted fever (Rao *et al.*, 1988), also show decrease of PK. Adult respiratory distress syndrome associated with trauma or with sepsis shows decreases in PK functional activity in excess of antigen activity (Carvalho *et al.*, 1988).

## Allergic Diseases

Contact activation, with decreased circulating PK, has been detected in a wide variety of allergic syndromes including anaphylactic reactions due to insect sting, hypersensitivity (Ratnoff & Nossel, 1983), and contrast media reactions (Zuran *et al.*, 1988). Asthmatics have increased kallikrein–$\alpha_2$-macroglobulin complex concentrations.

## Cardiopulmonary Bypass

During cardiopulmonary bypass there is extensive contact between blood anticoagulated with heparin and synthetic surfaces of the extracorporeal circuit, triggering a chemical and cellular 'whole body inflammatory response'. Extracorporeal circulation has been associated with both qualitative and quantitative alterations of the contact systems. Release of elastase from neutrophil azurophilic granules and lactoferrin from neutrophil secondary granules occurs during clinical cardiopulmonary bypass and during simulated extracorporeal circulation (Wachtfogel *et al.*, 1987). Although no gross abnormalities in PK levels could be shown in either clinical cardiopulmonary bypass or simulated extracorporeal circulation, a significant increase in kallikrein–C1 inhibitor complex formation was noted in simulated extracorporeal circulation (Wachtfogel *et al.*, 1989). Because kallikrein can induce neutrophil degranulation, this result raises the possibility that contact activation contributes to the neutrophil activation seen during cardiopulmonary bypass. Aprotinin has been shown to inhibit kallikrein activation and neutrophil activation (Wachtfogel *et al.*, 1993a).

## *Distinguishing Features*

Enzymes that cleave kininogens (kininogenases) historically have been designated 'kallikreins'. Tissue kallikreins (Chapters 28–41) are a group of serine proteases similar in size to trypsin, lacking the unique heavy chain of plasma kallikrein. Tissue kallikreins generally cleave kininogens at a Met-Lys bond to yield Lys-bradykinin, whereas plasma kallikrein cleaves at a Lys-Arg bond to yield bradykinin. Coagulation factor XI is 60% identical in amino acid sequence to plasma kallikrein, but it cleaves H-kininogen to decrease its cofactor activity, whereas hydrolysis by plasma kallikrein produces the activated H-kininogen cofactor.

An enzyme termed follipsin that has been isolated from the follicular fluid of pig ovary (Chapter 62) is structurally similar, or possibly identical, to plasma kallikrein.

## *Further Reading*

A review is given in Wachtfogel *et al.* (1993b).

## References

Abildgaard, C.F. & Harrison, J. (1974) Fletcher factor deficiency: family study and detection. *Blood* **43**, 641.

Beaubien, G., Rosinski-Chupin, I., Mattei, M.G., Mbikay, M., Chretien, M. & Seidah, N.G. (1991) Gene structure and chromosomal localization of plasma kallikrein. *Biochemistry* **30**, 1628–1635.

Bouma, B.N., Kerbiriou, D.M., Baker, J. & Griffin, J.H. (1986) Characterization of a variant prekallikrein, prekallikrein Long Beach, from a family with mixed cross-reacting material-positive and cross-reacting material-negative prekallikrein deficiency. *J. Clin. Invest.* **78**, 170.

Carvalho, A.C., DeMarinis, S., Scott, C.F., Silver, L.D., Schmaier, A.H. & Colman, R.W. (1988) Activation of the contact system of plasma proteolysis in the adult respiratory distress syndrome. *J. Lab. Clin. Med.* **112**, 270–277.

Chung, D.W., Fujikawa, K., McMullen, B.A. & Davie, E.W. (1986) Human plasma prekallikrein, a zymogen to a serine protease that contains four tandem repeats. *Biochemistry* **25**, 2410–2417.

Ciechanowicz, A., Bader, M., Wagner, J. & Ganten, D. (1993) Extra-hepatic transcription of plasma prekallikrein gene in human and rat tissues. *Biochem. Biophys. Res. Commun.* **197**, 1370–1376.

Colman, R.W. (1969) Activation of plasminogen by human plasma kallikrein. *Biochem. Biophys. Res. Commun.* **35**, 273–279.

Colman, R.W., Edelman, R., Scott, C.F. & Gilman, R.H. (1978) Plasma kallikrein activation and inhibition during typhoid fever. *J. Clin. Invest.* **61**, 287–296.

Curd, J.G., Prograis, L.J. Jr. & Cochrane, C.G. (1980) Detection of active kallikrein in induced blister fluids of hereditary angioedema patients. *J. Exp. Med.* **152**, 742–747.

Currembhoy, Z., Vinciguerra, V., Palakarougs, P., Kulansky, P. & Degnan, T.J. (1976) Fletcher factor deficiency and myocardial infarction. *Am. J. Clin. Pathol.* **65**, 970.

De La Cadena, R.A., Scott, C.F. & Colman, R.W. (1987) Evaluation of a microassay for human plasma prekallikrein. *J. Lab. Clin. Med.* **109**, 601–607.

De La Cadena, R.A., Suffredini, A.F., Page, J.D., Pixley, R.A., Kaufman, N., Parrillo, J.E. & Colman, R.W. (1993) Activation of the kallikrein-kinin system after endotoxin administration to normal human volunteers. *Blood* **81**, 3313–3317.

Gigli, I., Mason, J.W., Colman, R.W. & Austen, K.F. (1970) Interaction of plasma kallikrein with the C1 inhibitor. *J. Immunol.* **104**, 574–581.

Goetzl, E.J. & Austen, K.F. (1974) Stimulation of human neutrophil leukocyte aerobic glucose metabolism by purified chemotactic factors. *J. Clin. Invest.* **53**, 591–599.

Hathaway, W.E., Belhasen, L.P. & Hathaway, H.S. (1965) Evidence for a new plasma thromboplastin factor. I. Case report, coagulation studies and physicochemical properties. *Blood* **26**, 521.

Hattersley, W.E. & Hayse, D. (1970) Fletcher factor deficiency: a report of three unrelated cases. *Br. J. Haematol.* **18**, 411–415.

Hauert, J., Nicoloso, G., Schleuning, W.D., Bachmann, F. & Schapira, M. (1989) Plasminogen activators in dextran sulfate-activated euglobulin functions: a molecular analysis of factor XII and prekallikrein-dependent fibrinolysis. *Blood* **73**, 994–999.

Ichinose, A., Fujikawa, K. & Suyama, T. (1986) The activation of prourokinase by plasma kallikrein and its inactivation by thrombin. *J. Biol. Chem.* **261**, 3486–3489.

Kaplan, A.P., Kay, A.B. & Austen, K.F. (1972) A prealbumin activator of prekallikrein. III. Appearance of chemotactic activity for human neutrophils by the conversion of human prekallikrein to kallikrein. *J. Exp. Med.* **135**, 81–97.

Kaufman, N., Page, J.D., Pixley, R.A., Schein, R., Schmaier, A.H. & Colman, R.W. (1991) $\alpha_2$-Macroglobulin-kallikrein complexes detect contact system activation in hereditary angioedema and human sepsis. *Blood* **77**, 2660–2667.

Lin, Y., Harris, R.B., Yan, W., McCrae, K.R. & Colman, R.W. (1996) Inhibition of prourokinase-activated fibrinolysis on endothelial cell surfaces by high molecular weight kininogen peptides. *FASEB J.* **10**, A1414 (abstr.).

Mandle, R., Jr., Colman, R.W. & Kaplan, A.P. (1976) Identification of prekallikrein and high molecular weight kininogen as a complex in human plasma. *Proc. Natl Acad. Sci. USA* **73**, 4179–4183.

Mason, J.W., Kleeberg, U., Dolan, P. & Colman, R.W. (1970) Plasma kallikrein and Hageman factor in Gram-negative bacteremia. *Ann. Intern. Med.* **73**, 545–551.

McMullen, B.A., Fujikawa, K. & Davie, E.W. (1991) Location of the disulfide bonds in human plasma prekallikrein: the presence of four novel Apple domains in the amino-terminal portion of the molecule. *Biochemistry* **30**, 2050–2056.

Page, J.D. & Colman, R.W. (1991) Localization of distinct functional domains on prekallikrein for interaction with both high molecular weight kininogen and activated factor XII in a 28-kDa fragment (amino acids 141–371). *J. Biol. Chem.* **266**, 8143–8148.

Page, J.D., DeLa Cadena, R.A., Humphries, J.E. & Colman, R.W. (1994a) An autoantibody to human plasma prekallikrein blocks activation of the contact system. *Br. J. Haematol.* **87**, 81–86.

Page, J.D., You, J.L., Harris, R.B. & Colman, R.W. (1994b) Localization of the binding site on plasma kallikrein for high molecular weight kininogen to both apple 1 and apple 4 domains of the heavy chain. *Arch. Biochem. Biophys.* **314**, 159–164.

Pixley, R.A., DeLa Cadena, R.A., Page, J.D., Kaufman, N., Wyshock, E.G., Chang, A., Taylor, F.B., Jr. & Colman, R.W. (1993) The contact system contributes to hypotension but not disseminated intravascular coagulation in lethal bacteremia: in vivo use of a monoclonal anti-factor XII antibody to block contact activation in baboons. *J. Clin. Invest.* **91**, 61–68.

Purdon, A.D., Schapira, M., DeAgostini, A. & Colman, R.W. (1985) Plasma kallikrein and prorenin in patients with hereditary angioedema. *J. Lab. Clin. Med.* **105**, 694–699.

Rao, A.K., Schapira, M., Clements, M.L., Niewiarowski, S., Budzynski, A.Z., Schmaier, A.H., Harpel, P.C., Blackwelder, W.C., Scherrer, J.R. & Sobel, E. (1988) A prospective study of platelets and plasma proteolytic systems during the early stages of Rocky Mountain spotted fever. *N. Engl. J. Med.* **318**, 1021–1028.

Ratnoff, O.D. & Nossel, H.L. (1983) Wasp sting anaphylaxis. *Blood* **61**, 132.

Saito, H., Ratnoff, O.D. & Donaldson, V.H. (1974) Defective activation of clotting, fibrinolysis and performance-enhancing systems in human Fletcher trait plasma. *Circ. Res.* **34**, 641.

Saito, H., Goodnough, L.T., Soria, J., Soria, C., Aznan, J. & Espana, F. (1981) Heterogeneity of human plasma prekallikrein deficiency (Fletcher trait): evidence that five of 18 cases are positive for crossreacting material. *N. Engl. J. Med.* **305**, 910.

Schapira, M., Scott, C.F. & Colman, R.W. (1981) Protection of human plasma kallikrein from inactivation by C1 inhibitor and other protease inhibitors. The role of high molecular weight kininogen. *Biochemistry* **20**, 2738–2743.

Schapira, M., Despland, E., Scott, C.F., Boxer, L.A. & Colman, R.W. (1982a) Purified human plasma kallikrein aggregates human blood neutrophils. *J. Clin. Invest.* **69**, 1199–1202.

Schapira, M., Scott, C.F. & Colman, R.W. (1982b) Contribution of plasma protease inhibitors to the inactivation of kallikrein in plasma. *J. Clin. Invest.* **69**, 462–468.

Schapira, M., Scott, C.F., James, A., Silver, L.D., Kueppers, F., James, H.L. & Colman, R.W. (1982c) High molecular weight kininogen or its light chain protects human plasma kallikrein from inactivation by plasma protease inhibitors. *Biochemistry* **21**, 567–572.

Schapira, M., Silver, L.D., Scott, C.F. & Colman, R.W. (1982d) New and rapid functional assay for C1 inhibitor in human plasma. *Blood* **59**, 719–724.

Schapira, M., Silver, L.D., Scott, C.F., Schmaier, A.H., Prograis, L.J., Curd, J.G. & Colman, R.W. (1983) Prekallikrein activation and high molecular weight kininogen consumption in hereditary angioedema. *N. Engl. J. Med.* **308**, 1050–1053.

Scott, C.F., Liu, C.Y. & Colman, R.W. (1979) Human plasma prekallikrein: a rapid high-yield method for purification. *Eur. J. Biochem.* **100**, 77–83.

Veloso, D., Silver, L.D., Hahn, S. & Colman, R.W. (1987) A monoclonal anti-human plasma prekallikrein antibody that inhibits activation of prekallikrein by factor XIIa on a surface. *Blood* **70**, 1053–1062.

Wachtfogel, Y.T., Kucich, U., James, H.L., Scott, C.F., Schapira, M., Zimmerman, M., Cohen, A.B. & Colman, R.W. (1983) Human plasma kallikrein releases neutrophil elastase during blood coagulation. *J. Clin. Invest.* **72**, 1672–1677.

Wachtfogel, Y.T., Kucich, U., Greenplate, J., Gluszko, P., Abrams, W., Weinbaum, G., Wenger, R.K., Rucinski, B., Niewiarowski, S., Edmunds, L.H. & Colman, R.W. (1987) Human neutrophil degranulation during extracorporeal circulation. *Blood* **69**, 324–330.

Wachtfogel, Y.T., Harpel, P.C., Edmunds, L.H., Jr. & Colman, R.W. (1989) Formation of C1s-C1-inhibitor, kallikrein-C1-inhibitor, and plasmin-alpha 2-plasmin-inhibitor complexes during cardiopulmonary bypass. *Blood* **73**, 468–471.

Wachtfogel, Y.T., Kucich, U., Hack, C.E., Gluszko, P., Niewiarowski, S., Colman, R.W. & Edmunds, L.H., Jr. (1993a) Aprotinin inhibits the contact, neutrophil and platelet activation systems during simulated extracorporeal perfusion. *J. Thorac. Cardiovasc. Surg.* **106**, 1–10.

Wachtfogel, Y.T., DeLa Cadena, R.A. & Colman, R.W. (1993b) Structural biology, cellular interactions and pathophysiology of the contact system. *Thromb. Res.* **72**, 1–21.

Webster, M.E. & Pierce, J.V. (1961) Action of kallikrein on synthetic substrates. *Proc. Soc. Exp. Biol. Med.* **107**, 1564–1583.

Wong, P., Colman, R.W., Talamo, R.C. & Babior, B.M. (1972) Kallikrein bradykinin system in chronic alcoholic liver disease. *Ann. Intern. Med.* **77**, 205–209.

Wuepper, K.D. (1973) Prekallikrein deficiency in man. *J. Exp. Med.* **138**, 1345.

Wuepper, K.D. & Cochrane, C.G. (1972) Plasma prekallikrein: isolation, characterization, and mechanism of activation. *J. Exp.*

*Med.* **135**, 1–20.

Zimmerli, W., Huber, I., Bouma, B.N. & Lammle, B. (1989) Purified human plasma kallikrein does not stimulate but primes neutrophils for superoxide production. *Thromb. Haemost.* **62**, 1121–1125.

Zuran, B.L., Christiansen, S.C., Sugimoto, S.L. & Long, D.L. (1988) Radiographic contact-mediated reactions associated with cleavage of high molecular weight kininogen in plasma. *J. Allergy Clin. Immunol.* **81**, 218A.

*Robert W. Colman*
*Sol Sherry Thrombosis Research Center,*
*Temple University School of Medicine,*
*3400 North Broad Street, Philadelphia, PA 19140, USA*
*Email: Colman-R@VM.Temple.edu*

# 51. Coagulation factor XIa

## Databanks

*Peptidase classification: clan SA, family S1, MEROPS ID: S01.213*
*NC-IUBMB enzyme classification: EC 3.4.21.27*
*Chemical Abstracts Service registry number: 37203-61-5*
*Databank codes:*

| Species | SwissProt | PIR | EMBL (cDNA) | EMBL (genomic) |
| --- | --- | --- | --- | --- |
| *Homo sapiens* | P03951 | A27431 | M13142 | M18295: exon 1 |
| | | | | M18296: exon 2 |
| | | | | M18297: exon 3 |
| | | | | M18298: exon 4 |
| | | | | M18299: exon 5 |
| | | | | M18300: exon 6 |
| | | | | M18301: exon 7 |
| | | | | M18302: exons 8–10 |
| | | | | M18303: exon 11 |
| | | | | M18304: exon 12 |
| | | | | M18306: exon 14 |
| | | | | M18307: exon 15 |
| | | | | M19417: exon 13 |

## Name and History

*Coagulation factor XI* (formerly known as *plasma thromboplastin antecedent, PTA)* was first identified by its functional deficiency in the plasma of patients with abnormal bleeding (Rosenthal *et al.*, 1953), especially in patients of Ashkenazi Jewish descent (Rapaport *et al.*, 1961; Leiba *et al.*, 1965). The protein was isolated from cattle (Koide *et al.*, 1977) and then human (Bouma & Griffin, 1977) plasma, and characterized in regard to biochemical properties (Bouma & Griffin, 1977; Koide *et al.*, 1977), interactions with other factors in the contact phase of blood coagulation (Ratnoff *et al.*, 1961; Kurachi & Davie, 1977; Thompson *et al.*, 1977; Kurachi *et al.*, 1980), capacity to activate factor IX (Fujikawa *et al.*, 1974; DiScipio *et al.*, 1978; Sinha *et al.*, 1987), and importance for clinical hemostasis (Rosenthal *et al.*, 1953, 1955; Rapaport *et al.*, 1961; Leiba *et al.*, 1965; Ragni *et al.*, 1985).

## Activity and Specificity

Factor XIa, once formed by limited proteolytic cleavage of factor XI, recognizes the proenzyme factor IX, a single-chain 57 kDa glycoprotein containing 17% carbohydrate, as its normal physiological substrate, and converts it to factor IXa (Fujikawa *et al.*, 1974; DiScipio *et al.*, 1978; Sinha *et al.*, 1987) (Chapter 52). Factor IXa can in turn cleave and activate factor X (Chapter 54), which completes the generation of thrombin (Chapter 55) by the intrinsic pathway of blood coagulation.

The scissile bonds in factor IX cleaved by factor XIa (Arg145$+$Ala146 and Arg180$+$Val181) are the same as those cleaved by factor VIIa (Chapter 53) in the presence of tissue factor (Fujikawa *et al.*, 1974; Østerud & Rapaport, 1977; DiScipio *et al.*, 1978). In addition, factor XIa has been demonstrated to cleave H-kininogen slowly, but the physiological relevance of this activity is not clear (Scott *et al.*, 1985).

In addition to the coagulation proteins, factor XIa hydrolyzes a variety of oligopeptide chromogenic substrates, including Glp-Pro-Arg-NHPhNO$_2$, which can be used for assay (Scott *et al.*, 1984). The specific activity of factor XIa against this substrate is 1.05 μmol of substrate hydrolyzed per min per mg of enzyme. In addition, factor XIa acts on a variety of peptide thioester substrates, and can be active-site titrated with the sensitive fluorescein mono-*p*-guanidinobenzoate. Finally, the fluorometric substrate Boc-Glu-Ala-Arg$+$NHMec provides assays about 1000-fold more sensitive than conventional chromogenic assays (Scandura *et al.*, 1997).

The most potent physiological inhibitors of factor XIa in the plasma are $\alpha_1$-proteinase inhibitor and the antithrombin III–heparin complex, whereas C1 inhibitor, $\alpha_2$-plasmin inhibitor and $\alpha_2$-macroglobulin are less effective (Scott *et al.*, 1982; Beeler *et al.*, 1986). A very potent reversible inhibitor of factor XIa is protease nexin 2. This reversible, slow, tight-binding inhibitor is released from human platelets, and is a truncated form of the Alzheimer's amyloid β-protein precursor (APP), which contains a Kunitz-like protease inhibitor domain (Bush *et al.*, 1990; Smith *et al.*, 1990; Van Nostrand *et al.*, 1990; Scandura *et al.*, 1997). Unphysiological inhibitors of factor XIa include DFP, soybean trypsin inhibitor and aprotinin, whereas limabean trypsin inhibitor and ovomucoid do not inhibit (Walsh, 1992).

## Structural Chemistry

The sequence and domain structure of factor XI have been deduced by sequencing of cDNA from human liver (Fujikawa *et al.*, 1986) and more recently from mouse liver (Gailani *et al.*, 1997). Plasma factor XI consists of a disulfide-linked homodimer (143 kDa) including 5% carbohydrate (0.6% hexose, 2.7% *N*-acetylhexosamine and 1.7% *N*-acetylneuraminic acid) (Bouma & Griffin, 1977; Koide *et al.*, 1977).

Each polypeptide chain of factor XI contains 607 amino acids, and the proenzyme can be cleaved and activated by three biologically relevant proteases, factor XIIa (Chapter 49), factor XIa and thrombin (Bouma & Griffin, 1977; Gailani & Broze, 1991; Naito & Fujikawa, 1991), as well as by trypsin. The scissile bond cleaved by all four enzymes is Arg369$+$Ile370, yielding a heavy chain (369 amino acids) and an active-site-containing light chain (238 amino acids) (Kurachi & Davie, 1977; Kurachi *et al.*, 1980; Fujikawa *et al.*, 1986).

The N-terminal heavy chain comprises four tandem repeat sequences termed apple domains (90–91 amino acids each). The amino acid sequences of these are 23–34% identical, and the entire apple domain-containing heavy chain region of factor XI is 58% identical to a similar region in plasma prekallikrein (Fujikawa *et al.*, 1986) (Chapter 50). Otherwise, no significant homologies exist.

After proteolytic cleavage, the C-terminal light chain contains the new N-terminal sequence, Ile-Val-Gly, typical of serine proteases in family S1 (Fujikawa *et al.*, 1986), together with a catalytic triad consisting of His413, Asp462 and Ser557, and a typical trypsin-like substrate-binding pocket (Fujikawa *et al.*, 1986; McMullen *et al.*, 1991). No specific tertiary structural information exists, since the factor XI structure has not been solved either by X-ray crystallography or NMR, but molecular modeling with energy minimization has resulted in putative structures of all four apple domains, which have led to predictions of protein-binding sites for H-kininogen (Baglia *et al.*, 1989, 1990, 1992; Seaman *et al.*, 1994), thrombin (Baglia & Walsh, 1996), factor IX (Baglia *et al.*, 1991), platelet receptor (Baglia *et al.*, 1995) and factor XIIa (Baglia *et al.*, 1993). Both the apple 2 domain (Baglia *et al.*, 1991) and the apple 3 domain (Sun & Gailani, 1996) have been suggested as putative binding sites for factor IX within factor XIa. Finally, there is strong evidence that the two monomers of factor XI are associated through tight, non-covalent binding of the apple 4 domains to one another, in addition to a disulfide bond between adjacent cysteine residues at position 321 (Meijers *et al.*, 1992). Glycosylation of the protein has been suggested to occur at the potential *N*-linked glycosylation sites Asn72, Asn108, Asn335, Asn432 and Asn473 (Fujikawa *et al.*, 1986). Thirty-six Cys residues are present in each monomer, including 6 or 7 in each of the apple domains and 10 in the catalytic domain (Fujikawa *et al.*, 1986; McMullen *et al.*, 1991). The physical properties of factor XI include $M_r$ about 143 000; sedimentation coefficient 6.96 S (sucrose gradient); pI value 8.9–9.1; and extinction coefficient 13.4 (280 nm, 1%, 1 cm).

## Preparation

Factor XI can be purified from human, cattle or other mammalian plasma by monoclonal antibody affinity chromatography (Sinha *et al.*, 1985) or physicochemical methods (Koide *et al.*, 1977; Bouma & Griffin, 1977). It is also commercially available. Both human (Fujikawa *et al.*, 1986) and mouse (Gailani *et al.*, 1997) factor XI have been cloned and expressed in baby hamster kidney cells (Meijers *et al.*, 1992) or human fetal kidney fibroblasts (293 cells) (Gailani *et al.*, 1997).

## Biological Aspects

Biosynthesis of factor XI occurs in the liver (Schmaier *et al.*, 1987; Walsh, 1992), although recent evidence of extrahepatic synthesis has yet to be evaluated (Gailani *et al*., 1997). Factor XI is secreted into the plasma as a proenzyme that circulates in a complex with H-kininogen (Thompson *et al*., 1977), which is required for the binding of factor XI to negatively charged surfaces (Wiggins *et al*., 1977) (Chapter 50). Factor XI participates in the contact phase of blood coagulation that requires the presence of anionic surfaces for optimal surface-mediated activation *in vitro* by the enzyme factor XIIa (Ratnoff *et al*., 1961; Griffin & Cochrane, 1976; Meier *et al*., 1977; Thompson *et al*., 1977; Wiggins *et al*., 1977) (Chapter 49).

Deficiencies of plasma proteins involved in the 'contact phase' of blood coagulation (factor XII, prekallikrein and H-kininogen) are not associated with hemostatic deficiency (Schmaier *et al*., 1987), but factor XI deficiency produces abnormal bleeding complications in at least 50% of affected individuals (Ragni *et al*., 1985). In view of this, an alternative mechanism to the contact system has recently been proposed for the activation of factor XI in normal hemostasis (Gailani & Broze, 1991; Naito & Fujikawa, 1991). This is suggested to involve proteases such as thrombin and factor XIa itself that are independent of the contact phase of blood coagulation. Factor XI can bind to the surface of activated platelets (Greengard *et al*., 1986) via structures exposed on the surface of the apple 3 domain (Baglia *et al*., 1995). Optimal rates of limited proteolytic activation of factor XI by thrombin can occur on the platelet surface (Baglia & Walsh, 1998), so this may be the physiological locus for factor XI activation either via the contact phase (Walsh & Griffin, 1981) or via thrombin-mediated activation (Baglia & Walsh, 1998).

The gene for factor XI is located on the distal end of the long arm of chromosome 4 (4q35) and contains 23 kb of nucleotide sequence with 14 introns and 15 exons (Asakai *et al*., 1987). Exon I codes for the 5′ untranslated region, exon II for the signal peptide, exons III–X for the four apple domains and exons XI–XV for the catalytic domain (Asakai *et al*., 1987).

Factor XI deficiency is inherited as an autosomal recessive trait characterized by low levels of circulating plasma factor XI antigen and activity in homozygotes, and intermediate levels of both antigen and activity in heterozygotes (Asakai *et al*., 1989, 1991; Ragni *et al*., 1985). Factor XI deficiency is estimated to occur with a high gene frequency in Ashkenazi Jews (1 in 90 individuals in Israel) and with very low frequency in non-Jewish ethnic groups (1 in 1 million) (Seligsohn, 1978). Analysis of cDNA sequencing data in patients with factor XI deficiency has identified several independent mutations in the factor XI gene giving rise to a variety of abnormalities including an abnormal mRNA splicing defect (type I), a nonsense mutation converting Glu117 to a stop codon that causes premature termination of translation (type II), and a missense mutation resulting in the substitution of Phe283 by Leu in the protein (type III) (Asakai *et al*., 1989, 1991). Several other specific abnormalities have been demonstrated in factor XI-deficient patients.

## Distinguishing Features and Related Peptidases

As is described above, factor XIa is an extremely specific endopeptidase, not difficult to distinguish from others.

A unique and structurally distinct form of factor XI associated with platelet membranes has been described (Lipscomb & Walsh, 1979; Tuszynski *et al*., 1982; Hsu *et al*., 1998). The only other protein with significant sequence similarity to factor XI is prekallikrein (Chapter 50), the amino acid sequence of which is 58% identical to that of factor XI (Fujikawa *et al*., 1986). Both monoclonal and polyclonal antibodies to factor XI have been described (Sinha *et al*., 1985).

## Further Reading

For a review, see Walsh (1992).

## References

Asakai, R., Davie, E.W. & Chung, D.W. (1987) Organization of the gene for human factor XI. *Biochemistry* **26**, 7221–7228.

Asakai, R., Chung, D.W., Ratnoff, O.D. & Davie, E.W. (1989) Factor XI (plasma thromboplastin antecedent) deficiency in Ashkenazi Jews in a bleeding disorder that can result from three types of point mutations. *Proc. Natl Acad. Sci. USA* **86**, 7667–7671.

Asakai, R., Chung, D.W., Davie, E.W. & Seligsohn, U. (1991) Factor XI deficiency in Ashkenazi Jews in Israel. *N. Engl. J. Med.* **325**, 153–158.

Baglia, F.A. & Walsh, P.N. (1996) A binding site for thrombin in the Apple 1 domain of factor XI. *J. Biol. Chem.* **271**, 3652–3658.

Baglia, F.A. & Walsh, P.N. (1997) Prothrombin is a cofactor for the binding of factor XI to the platelet surface and for platelet-mediated factor-XI activation by thrombin. *Biochemistry* **37**, 2271–2281.

Baglia, F.A., Sinha, D. & Walsh, P.N. (1989) Functional domains in the heavy chain region of factor XI: a high molecular weight kininogen-binding site and a substrate-binding site for factor IX. *Blood* **74**, 244–251.

Baglia, F.A., Jameson, B.A. & Walsh, P.N. (1990) Localization of the high molecular weight kininogen binding site in the heavy chain of human factor XI to amino acids Phe-56 through Ser-86. *J. Biol. Chem.* **265**, 4149–4154.

Baglia, F.A., Jameson, B.A. & Walsh, P.N. (1991) Identification and chemical synthesis of a substrate-binding site for factor IX on coagulation factor XIa. *J. Biol. Chem.* **266**, 24190–24197.

Baglia, F.A., Jameson, B.A. & Walsh, P.N. (1992) Fine mapping of the high molecular weight kininogen binding site on blood coagulation factor XI through the use of rationally designed synthetic analogs. *J. Biol. Chem.* **267**, 4247–4252.

Baglia, F.A., Jameson, B.A. & Walsh, P.N. (1993) Identification and characterization of a binding site for factor XIIa in the Apple 4 domain of coagulation factor XI. *J. Biol. Chem.* **268**, 3838–3844.

Baglia, F.A., Jameson, B.A. & Walsh, P.N. (1995) Identification and characterization of a binding site for platelets in the Apple 3 domain of coagulation factor XI. *J. Biol. Chem.* **270**, 6734–6740.

Beeler, D.L., Marcum, J.A., Schiffman, S. & Rosenberg, R.D. (1986) Interaction of factor XIa and antithrombin in the presence and absence of heparin. *Blood* **67**, 1488–1492.

Bouma, B.N. & Griffin, J.H. (1977) Human blood coagulation factor XI: purification properties and mechanisms of activation by activated factor XII. *J. Biol. Chem.* **252**, 6432–6437.

Bush, A.I., Martins, R.N., Rumble, B., Moir, R., Fuller, S., Milward, E., Currie, J., Ames, D., Weidemann, A., Fischer, P., Multhaup, G., Beyreuther, K. & Masters, C.L. (1990) The amyloid precursor protein of Alzheimer's disease is released by human platelets. *J. Biol. Chem.* **265**, 15977–15982.

DiScipio, R.G., Kurachi, K. & Davie, E.W. (1978) Activation of human factor IX (Christmas factor). *J. Clin. Invest.* **61**, 1528–1538.

Fujikawa, K., Legaz, M.E., Kato, H. & Davie, E.W. (1974) The mechanism of activation of bovine factor IX (Christmas factor) by bovine factor XIa (activated plasma thromboplastin antecedent). *Biochemistry* **13**, 4508–4516.

Fujikawa, K., Chung, D.W., Hendrickson, L.E. & Davie, E.W. (1986) Amino acid sequence of human factor XI, a blood coagulation factor with four tandem repeats that are highly homologous with plasma prekallikrein. *Biochemistry* **25**, 2417–2424.

Gailani, D. & Broze, G.J. Jr. (1991) Factor XI activation in a revised model of blood coagulation. *Science* **253**, 909–912.

Gailani, D., Sun, M.-F. & Sun, Y. (1997) A comparison of murine and human factor XI. *Blood* **90**, 1055–1064.

Greengard, J.S., Heeb, M.J., Ersdal, E., Walsh, P.N. & Griffin, J.H. (1986) Binding of coagulation factor XI to washed human platelets. *Biochemistry* **25**, 3884–3890.

Griffin J.H. & Cochrane C.G. (1976) Mechanisms for the involvement of high molecular weight kininogen in surface-dependent reactions of Hageman factor. *Proc. Natl Acad. Sci. USA* **73**, 2554–2558.

Hsu, T.-C., Shore, S.K., Seshsmma, T., Bagasra, O. & Walsh, P.N. (1998) Molecular cloning of platelet factor XI, an alternative splicing product of the plasma factor XI gene. *J. Biol. Chem.*, in the press.

Koide, T., Kato, H. & Davie, E.W. (1977) Isolation and characterization of bovine factor XI (plasma thromboplastin antecedent). *Biochemistry* **16**, 2279–2286.

Kurachi, K. & Davie, E.W. (1977) Activation of human factor XI (plasma thromboplastin antecedent) by factor XIIa (activated Hageman factor). *Biochemistry* **16**, 5831–5839.

Kurachi, K., Fujikawa, K. & Davie, E.W. (1980) Mechanism of activation of bovine factor XI by factor XII and factor XIIa. Biochemistry **19**, 1330–1338.

Leiba, H., Ramot, B. & Many, A. (1965) Heredity and coagulation studies in ten families with factor XI (plasma thromboplastin antecedent) deficiency. *Br. J. Haemat.* **11**, 654–665.

Lipscomb, M.S. & Walsh, P.N. (1979) Human platelets and factor XI. Localization in platelet membranes of factor-XI-like activity and its functional distinction from plasma factor XI. *J. Clin. Invest.* **63**, 1006–1014.

McMullen, B.A., Fujikawa, K. & Davie, E.W. (1991) Location of the disulfide bonds in human coagulation factor XI: the presence of tandem apple domains. *Biochemistry* **30**, 2056–2060.

Meier, H.L., Pierce, J.V., Colman, R.W. & Kaplan, A.P. (1977) Activation and function of human Hageman factor: the role of high molecular weight kininogen and prekallikrein. *J. Clin. Invest.* **60**, 18–31.

Meijers, J.C., Mulvihill, E.R., Davie, E.W. & Chung, D.W. (1992) Apple four in human blood coagulation factor XI mediates dimer formation. *Biochemistry* **31**, 4680–4684

Naito, K. & Fujikawa, K. (1991) Activation of human blood coagulation factor XI independent of factor XII. Factor XI is activated by thrombin and factor XIa in the presence of negatively charged surfaces. *J. Biol. Chem.* **266**, 7353–7358.

Østerud, B. & Rapaport, S.I. (1977) Activation of factor IX by the reaction product of tissue factor and factor VII: additional pathway for initiating blood coagulation. *Proc. Natl Acad. Sci. USA* **74**, 5260–5264.

Ragni, M.V., Sinha, D., Seaman, F., Lewis, J.H., Spero, J.A. & Walsh, P.N. (1985) Comparison of bleeding tendency, factor XI coagulant activity and factor XI antigen in twenty-five factor XI deficient kindreds. *Blood* **65**, 719–724.

Rapaport, S.I., Proctor, R.R., Patch, M.J. & Yettra, M. (1961) The mode of inheritance of PTA deficiency: Evidence for the existence of major PTA deficiency and minor PTA deficiency. *Blood* **18**, 149–165.

Ratnoff, O.D., Davie, E.W. & Mallett, D.L. (1961) Studies on the action of Hageman factor: evidence that activated Hageman factor in turn activates plasma thromboplastin antecedent. *J. Clin. Invest.* **40**, 803–819.

Rosenthal, R.L., Dreskin, O.H. & Rosenthal, N. (1953) New hemophilia-like disease caused by deficiency of a third plasma thromboplastin factor. *Proc. Soc. Exp. Biol. Med.* **82**, 171–174.

Rosenthal, R.L., Dreskin, O.H. & Rosenthal, N. (1955) Plasma thromboplastin antecedent (PTA) deficiency: clinical coagulation, therapeutic and hereditary aspects of a new hemophilia-like disease. *Blood* **10**, 120.

Scandura, J.M., Zhang, Y., Van Nostrand, W.E. & Walsh, P.N. (1997) Progress curve analysis of the kinetics with which blood coagulation factor XIa is inhibited by protease nexin-2. *Biochemistry* **36**, 412–420.

Schmaier, A.H., Silverberg, M., Kaplan, A.P. & Colman, R.W. (1987) Contact activation and its abnormalities. In: *Hemostasis and Thrombosis* (Colman, R.W., Hirsh, J., Marder, V.J. & Salzman, E.W., eds). Philadelphia: J.B. Lippincott, pp. 18–38.

Scott, C.F., Schapira, M., James, H.L., Cohen, A.B. & Colman, R.W. (1982) Inactivation of factor XIa by plasma protease inhibitors. Predominant role of α1-protease inhibitor and protective effect of high molecular weight kininogen. *J. Clin. Invest.* **69**, 844–852.

Scott, C.F., Sinha, D., Seaman, F.S., Walsh, P.N. & Colman, R.W. (1984) Amidolytic assay of human factor XI in plasma. Comparison with a coagulant assay and a new rapid radioimmunoassay. *Blood* **63**, 42–50.

Scott, C.F., Purdon, D.A., Silver, L.D. & Colman, R.W. (1985) Cleavage of high molecular weight kininogen (HMWK) by plasma factor XIa. *J. Biol. Chem.* **260**, 10856–10863.

Seaman, F.S., Baglia, F.A., Gurr, J.A., Jameson, B.A. & Walsh, P.N. (1994) Binding of high-molecular-mass kininogen to the Apple 1 domain of factor XI is mediated in part by Val[64] and Ile[77]. *Biochem. J.* **304**, 715–721.

Seligsohn, U. (1978) High gene frequency of factor XI (PTA) deficiency in Ashkenazi Jews. *Blood* **51**, 1223–1228.

Sinha, D., Koshy, A., Seaman, F.S. & Walsh, P.N. (1985) Functional characterization of human blood coagulation factor XIa using hybridoma antibodies. *J. Biol. Chem.* **260**, 10714–10719.

Sinha, D., Seaman, F.S. & Walsh, P.N. (1987) Role of calcium ions and the heavy chain of factor XIa in the activation of human coagulation factor IX. *Biochemistry* **26**, 3768–3775.

Smith, R.P., Higuchi, D.A. & Broze, G.J. Jr. (1990) Platelet coagulation factor XIa-inhibitor, a form of Alzheimer amyloid precursor protein. *Science* **248**, 1126–1128.

Sun, Y. & Gailani, D. (1996) Identification of a factor IX binding site on the third apple domain of activated factor XI. *J. Biol. Chem.* **271**, 29023–29028.

Thompson, R.E., Mandle, R. Jr. & Kaplan, A.P. (1977) Association of factor XI and high molecular weight kininogen in human plasma. *J. Clin. Invest.* **60**, 1376–1380.

Tuszynski, G.P., Bevaqua, S.J., Schmaier, A.H., Colman, R.W. & Walsh, P.N. (1982) Factor XI antigen and activity in human platelets. *Blood* **59**, 1148–1156.

Van Nostrand, W.E., Schmaier, A.H., Farrow, J.S. & Cunning-ham, D.D. (1990) Protease nexin-II (amyloid β-protein precursor): A platelet α-granule protein. *Science* **248**, 745–748.

Walsh, P.N. (1992) Factor XI: A Renaissance. *Semin. Hematol.* **29**, 189–201.

Walsh, P.N. & Griffin, J.H. (1981) Contributions of human platelets to the proteolytic activation of blood coagulation factors XII and XI. *Blood* **57**, 106–118.

Wiggins, R.C., Bouma, B.N., Cochrane, C.G. & Griffin, J.H. (1977) Role of high-molecular-weight kininogen in surface-binding and activation of coagulation factor XI and prekallikrein. *Proc. Natl Acad. Sci. USA* **74**, 4636–4640.

*Peter N. Walsh*
*Sol Sherry Thrombosis Research Center,*
*Temple University School of Medicine,*
*3400 North Broad Street, Philadelphia, PA 19140, USA*
*Email: pnw@astro.ocis.temple.edu*

# 52. Coagulation factor IXa

## Databanks

*Peptidase classification: clan SA, family S1, MEROPS ID: S01.214*
*NC-IUBMB enzyme classification: EC 3.4.21.22*
*ATCC entries: 79837 (2.8 kb of human IX cDNA)*
*Chemical Abstracts Service registry number: 9001-28-9 (factor IX)*
*Databank codes:*

| Species | SwissProt | PIR | EMBL (cDNA) | EMBL (genomic) |
|---|---|---|---|---|
| *Bos taurus* | P00741 | A14757 | A22490 J00007 | – |
| *Canis familiaris* | P19540 | A30351 | M21757 M33826 | – |
| *Cavia porcellus* | P16295 | – | M26237 | – |
| *Homo sapiens* | P00740 | A00922 | A22493 J00136 J00137 M11309 M35672 | K02048: exon 1 K02049: exons 2 and 3 K02051: exon 5 K02052: exon 6 K02053: exons 7 and 8 K02402: complete gene M19063: exon 8 S66752: 3′ end; genomic mutant S68634: Strasbourg 2 mutant; exon 2 |
| *Mus musculus* | P16294 | JQ0419 | M23109 M26236 | – |
| *Oryctolagus cuniculus* | P16292 | – | M26234 | – |
| *Ovis aries* | P16291 | – | M26233 | – |
| *Rattus norvegicus* | P16296 | – | M26247 | – |
| *Sus scrofa* | P16293 | – | M26235 | – |

Brookhaven Protein Data Bank three-dimensional structures:

| Species | ID | Resolution | Notes |
|---|---|---|---|
| *Homo sapiens* | 1CFH | – | NMR; residues 1–47 |
| | 1EDM | 1.5 | EGF-like domain |
| | 1IXA | – | NMR; EGF-like module |
| | 1MGX | – | NMR; $\alpha$-carbon backbone |
| *Sus scrofa* | 1PFX | 3.0 | complex with D-Phe-Pro-Arg-CH$_2$Cl |

## Name and History

The identification of **coagulation factor IX** (proenzyme form of factor IXa) as a substance required for normal blood coagulation was first established by Pavlovsky (1947), who reported that a mixture of blood from two unrelated hemophiliacs clotted normally. Based upon this landmark observation and subsequent studies (Biggs *et al.*, 1952), hemophilia was divided into two conditions. Hemophilia A, the commoner condition, had also been called classic hemophilia or factor VIII deficiency, and hemophilia B, the rarer form, had been known as plasma thromboplastin component (PTC) deficiency, Christmas disease or factor IX deficiency. Since then, much has been learned about the molecular and structural biology of factor IX, which is the name currently in use for the hemophilia B protein.

## Activity and Specificity

Factor IXa is an endopeptidase requiring Arg at the P1 position (Spitzer *et al.*, 1990). The rules governing the specificity for other positions remain unclear since a satisfactory synthetic substrate has as yet not been found; this may be due to an incompletely formed specificity pocket in factor IXa as compared to other serine proteases (Brandstetter *et al.*, 1995). The S2 site of factor IXa is relatively open and can accommodate Pro, Thr, Phe, Val or Trp found as P2 residues in natural or synthetic substrates (Brandstetter *et al.*, 1995). The enzyme cleaves Z-Trp-Arg-SBzl (Link & Castellino, 1983), Z-Glu-Gly-Arg-NHPhNO$_2$ (Cho *et al.*, 1984) and CH$_3$SO$_2$-D-Leu-Gly-Arg-NHPhNO$_2$ (Lenting *et al.*, 1995) at rates which are too low for general use; the CH$_3$SO$_2$-D-Leu-Gly-Arg-NHPhNO$_2$ substrate is available from American Diagnostica (see Appendix 2 for full names and addresses of suppliers). Factor IXa is resistant to inhibition by DFP (van Dieijen *et al*, 1981; Spitzer *et al.*, 1990), but it can be inhibited by Dns-Glu-Gly-Arg-CH$_2$Cl (Lollar & Fass, 1984) and D-Phe-Pro-Arg-CH$_2$Cl (Brandstetter *et al.*, 1995) upon incubation for hours; both the inhibitors are available from Calbiochem/Novabiochem. The pH optimum of the protease is about 7.5 and the assay system typically includes some NaCl ($\sim$0.15 M). A known natural inhibitor of factor IXa is antithrombin and this reaction is accelerated by heparin several-fold (DiScipio *et al.*, 1978). Antithrombin can be purchased from Chromogenix, Diagnostica Stago, Haematologic Technologies Inc. or Enzyme Research.

## Structural Chemistry

Factor IXa is composed of two chains held together by a single disulfide bond (Brandstetter *et al.*, 1995). The N-terminus of the light chain of human factor IXa contains 12 $\gamma$-carboxyglutamic acid (Gla) residues and represents the Gla domain (residues 1–40); the Gla domain is followed by a few aromatic residues (hydrophobic stack; residues 41–46), two epidermal growth factor (EGF)-like domains (EGF1 residues 47–84, EGF2 residues 85–127) and a short stretch of residues 128–145 (Yoshitake *et al.*, 1985). The heavy chain (residues 181–415) contains the serine protease domain essential for catalysis (Yoshitake *et al.*, 1985). The Gla domain possesses several weak- to intermediate-affinity Ca$^{2+}$-binding sites (Freedman *et al.*, 1995), whereas the protease and EGF1 domains each contain one high-affinity Ca$^{2+}$-binding site (Bajaj *et al.*, 1992; Rao *et al.*, 1995). The activation peptide (residues 146–180) that is removed upon conversion of factor IX to factor IXa, and the EGF1 domain, contain several N- and O-linked carbohydrate chains (Nishimura *et al.*, 1989, 1992; Harris *et al.*, 1993; Agarwala *et al.*, 1994). The Gla, EGF2 and protease domains appear not to contain carbohydrate (Bajaj & Birktoft, 1993). NMR coordinates of the Ca$^{2+}$-loaded form of the Gla domain (code 1MGX), X-ray coordinates of the calcium-bound form of the EGF1 domain (code 1EDM), and the X-ray coordinates of the protease (in the absence of Ca$^{2+}$) and EGF2 domains (code 1PFX) are available from the Brookhaven Protein Data Bank. According to Brandstetter *et al.* (1995), factor IXa has the characteristic appearance of a tulip, with the Gla domain representing the bulb, the two EGF domains the bent stalk, and the C-terminal protease domain the flower.

## Preparation

Isolation procedures to obtain homogeneous preparations of cattle (Fujikawa & Davie, 1976), human (Miletich *et al.*, 1981; Bajaj & Birktoft, 1993), pig (Lollar & Fass, 1984) and dog factor IX (Sugahara *et al.*, 1996) have been reported. Human and cattle factor IX can be commercially obtained from Haematologic Technologies Inc. or Enzyme Research. A commercially available vector (pRc/CMV from Invitrogen Corp.) and human factor IX cDNA clone (ATCC clone 79837) can be used to express recombinant human factor IX in 293 kidney cells; the recombinant protein purified using this system is fully carboxylated and active (Zhong *et al.*, 1994). The use of cattle papillomavirus vector and mouse C127 cells have also yielded satisfactory preparations of recombinant human factor IX (Lin *et al.*, 1990).

## Biological Aspects

Factor IX is a plasma protein that participates in the middle phase of the intrinsic as well as the extrinsic coagulation cascade. Absence or reduced activity of factor IX causes an

X-linked bleeding disorder commonly known as hemophilia B. The gene for factor IX consists of eight exons and seven introns, is approximately 34 kb long, and is located on the long arm of the X-chromosome at band Xq27.1 (Anson *et al.*, 1984; Kurachi & Kurachi, 1995; Yoshitake *et al.*, 1985). The sizes of the seven introns vary from about 200 bp to about 1000 bp and the gene is expressed in hepatocytes in a highly tissue-specific manner; both the NF-1 and LF-A1 binding elements have been identified in the 5′ end region of the gene (Kurachi & Kurachi, 1995). The positions of the introns in the factor IX gene are essentially identical to those of three other homologous proteins, namely factor VII (Chapter 53), factor X (Chapter 54) and protein C (Chapter 56). The factor IX gene lacks a typical TATA box sequence (Yoshitake *et al.*, 1985); however, based upon the functional analysis of 5′ flanking region, the TCAAAT sequence at nt −181 has been assigned as the TATA box for this gene (Kurachi & Kurachi, 1995). Sequences in the first intron also enhance the level of factor IX expression as compared to the constructs without the intron sequence. An interesting hemophilia B phenotype (hemophilia B Leyden) has been described; the factor IX gene of all families affected with this phenotype have mutations that span nt −40 to nt +20 in the 5′ end region of the gene. The molecular mechanism responsible for this phenotype has been extensively studied and the reader is referred to a recent review for further information (Kurachi & Kurachi, 1995).

Human factor IX is synthesized as a precursor molecule of 461 amino acids (Anson *et al.*, 1984; Yoshitake *et al.*, 1985). The first 46 residues contain the hydrophobic signal and the hydrophilic propeptide sequence. During biosynthesis, the protein undergoes several post-translational modifications which include glycosylation at Ser53, Ser61, Asn157, Asn167, Thr159 and Thr169, $\gamma$-carboxylation of the first 12 Glu residues, partial hydroxylation of Asp64, and removal of the signal sequence (Yoshitake *et al.*, 1985; Nishimura *et al.*, 1989, 1992; Harris *et al.*, 1993; Agarwala *et al.*, 1994). The resulting mature protein of 415 amino acids (57 kDa) is the proenzyme of factor IXa and contains 17% carbohydrate by weight (DiScipio *et al.*, 1978). During coagulation, factor IX can be activated by factor XIa/$Ca^{2+}$, and by factor VIIa/$Ca^{2+}$/tissue factor (Østerud & Rapaport, 1977). Activation by either enzyme occurs in two steps (Bajaj & Birktoft, 1993; DiScipio *et al.*, 1978). In the first step, the Arg145∤Ala146 bond is cleaved, which yields a two-chain disulfide-linked inactive intermediate called factor IX$\alpha$. In the second step, which is also the rate-limiting step, factor IX$\alpha$ is cleaved at the Arg180∤Val181 bond to yield an active serine protease called factor IX$\alpha\beta$ (or simply factor IXa) and an activation peptide of residues 146–180. Factor IX can also be activated by a protease from Russell's viper venom. This activation involves only cleavage of the Arg180∤Val181 bond, giving rise to a two-chain molecule, factor IXa$\alpha$, without the release of activation peptide (Fujikawa & Davie, 1976). Factor IX$\alpha$ can also be generated by factor Xa/$Ca^{2+}$/phospholipid (Lawson & Mann, 1991). The EGF1 domain of factor IX is required for its activation by factor VIIa/tissue factor but not by factor XIa (Zhong *et al.*, 1994).

Factor IXa proteolytically converts factor X to factor Xa in the coagulation cascade; this reaction requires three known cofactors, namely, $Ca^{2+}$, phospholipid and factor VIIIa (van Dieijen *et al.*, 1981; Bajaj & Birktoft, 1993). The Gla domain

mediates phospholipid binding to the protein (Freedman *et al.*, 1996) and the protease domain appears to provide primary specificity for binding to factor VIIIa (Astermark *et al.*, 1994; Bajaj *et al.*, 1985). The EGF2 domain may also contribute to the binding of factor IXa to factor VIIIa (Lin *et al.*, 1990). The EGF1 domain of factor IXa is necessary for properly aligning the protease domain (and possibly the EGF2 domain) in its interaction with factor VIIIa (Lenting *et al.*, 1996; Lin *et al.*, 1990). A database of point mutations and short additions and deletions, which cause hemophilia B, has been published (Giannelli *et al.*, 1996). Factor IXa, in addition to activating factor X in the clotting cascade, may also activate factor VII (Masys *et al.*, 1982); this reaction requires $Ca^{2+}$ and phospholipid.

## Distinguishing Features

The most reliable method to distinguish factor IX/IXa from other proteases is to measure its activity in a coagulant-based assay (Bajaj & Birktoft, 1993; Fujikawa & Davie, 1976). Preincubation of factor IX/IXa with monospecific polyclonal antibodies should result in >90% loss of coagulant activity. Monospecific antibodies to human factor IX can be obtained from American Diagnostica, Haematologic Technologies Inc. or Enzyme Research.

## Further Reading

The article of Bajaj & Birktoft (1993) describes isolation of the protein from human plasma and recombinant cultures, preparation of [³H]sialyl and ¹²⁵I-labeled tyrosyl factor IX, and the basic concepts involved in building a model structure of a domain using as template the experimentally determined three-dimensional structure of a homologous domain. Brandstetter *et al.* (1995) describe the three-dimensional structure of the protease (in the absence of $Ca^{2+}$) and EGF2 domains as well as the modeled structures of the EGF1 and Gla domains of pig factor IXa; an explanation of why factor IXa has low estrolytic/amidolytic activity towards synthetic substrates is also provided. Naturally occurring point mutations that cause hemophilia B are also discussed. Finally, Kurachi & Kurachi (1995) describe the biology and gene regulation of factor IX.

## References

Agarwala, K.L., Kawabata, S., Takao, T., Murata, H., Shimonishi, Y., Nishimura, H. & Iwanaga, S. (1994) Activation peptide of human factor IX has oligosaccharides *O*-glycosidically linked to threonine residues at 159 and 169. *Biochemistry* **33**, 5167–5171.

Anson, D.S., Choo, K.H., Rees, D.J.G., Giannelli, F., Gould, K., Huddleston, J.A. & Brownlee, G.G. (1984) The gene structure of human anti-hemophilic factor IX. *EMBO J.* **3**, 1053–1060.

Astermark, J., Hogg, P.J. & Stenflo, J. (1994) The $\gamma$-carboxyglutamic acid and epidermal growth factor-like modules of factor IX$\alpha\beta$. Effects on the serine protease module and factor X activation. *J. Biol. Chem.* **269**, 3682–3689.

Bajaj, S.P. & Birktoft, J.J. (1993) Human factor IX and factor IXa. *Methods Enzymol.* **222**, 96–128.

Bajaj, S.P., Rapaport, S.I. & Maki, S.L. (1985) A monoclonal antibody to factor IX that inhibits the factor VIII: Ca potentiation of factor X activation. *J. Biol. Chem.* **260**, 11574–11580.

Bajaj, S.P., Sabharwal, A.K., Gorka, J. & Birktoft, J.J. (1992) Antibody-probed conformational transitions in the protease domain of human factor IX upon calcium binding and zymogen activation: putative high-affinity $Ca^{2+}$-binding site in the protease domain. *Proc. Natl Acad. Sci. USA* **89**, 152–156.

Biggs, R., Douglas, A.S., Macfarlane, R.G., Dacie, J.V., Pitney, W.R., Merskey, C. & O'Brien, J.R. (1952) Christmas disease, a condition previously mistaken for hemophilia. *BMJ* **2**, 1378–1387.

Brandstetter, H., Bauer, M., Huber, R., Lollar, P. & Bode, W. (1995) X-ray structure of clotting factor IXa: active site and module structure related to Xase activity and hemophilia B. *Proc. Natl Acad. Sci. USA* **92**, 9796–9800.

Cho, K., Tanaka, T., Cook, R.R., Kisiel, W., Fujikawa, K., Kurachi, K. & Powers, J.C. (1984) Active-site mapping of bovine and human blood coagulation serine proteases using synthetic peptide 4-nitroanilide and thio ester substrates. *Biochemistry* **23**, 644–650.

DiScipio, R.G., Kurachi, K. & Davie, E.W. (1978) Activation of human factor IX (Christmas factor). *J. Clin. Invest.* **61**, 1528–1538.

Freedman, S.J., Furie, B.S., Furie, B. & Baleja, J.D. (1995) Structure of the calcium ion-bound gamma-carboxyglutamic acid-rich domain of factor IX. *Biochemistry* **34**, 12126–12137.

Freedman, S.J., Blostein, M.D., Baleja, J.D., Jacobs, M., Furie, B.C. & Furie, B. (1996) Identification of the phospholipid binding site in the vitamin K-dependent blood coagulation protein factor IX. *J. Biol. Chem.* **271**, 16227–16236.

Fujikawa, K. & Davie, E.W. (1976) Bovine factor IX. *Methods Enzymol.* **45**, 74–83.

Giannelli, F., Green, P.M., Sommer, S.S., Poon, M.-C., Ludwig, M., Schwaab, R., Reitsma, P.H., Goossens, M., Yosioka, A. & Brownlee, G.G. (1996) Haemophilia B (sixth edition): a database of point mutations and short additions and deletions. *Nucleic Acids Res.* **24**, 103–118.

Harris, R.J., van Halbeek, H., Glushka, J., Basa, L.J., Ling, V.T., Smith, K.J. & Spellman, M.W. (1993) Identification and structural analysis of the tetrasaccharide NeuAc alpha (266) Gal beta (164) GlcNAc beta (163) Fuc alpha 160-linked to serine 61 of human factor IX. *Biochemistry* **32**, 6539–6547.

Kurachi, K. & Kurachi, S. (1995) Regulatory mechanism of the factor IX gene. *Thromb. Haemost.* **73**, 333–339.

Lawson, J.H. & Mann, K.G. (1991) Cooperative activation of human factor IX by the human extrinsic pathway of blood coagulation. *J. Biol. Chem.* **266**, 11317–11327.

Lenting, P.J., ter Maat, H., Clijsters, P.P.F.M., Donath, M.-J.S.H., van Mourik, J.A. & Mertens, K. (1995) Cleavage at arginine 145 in human blood coagulation factor IX converts the zymogen into a factor VIII binding enzyme. *J. Biol. Chem.* **270**, 14884–14890.

Lenting, P.J., Christophe, O.D., ter Maat, H., Rees, J.G. & Mertens, K. (1996) $Ca^{2+}$-binding to the first epidermal growth factor-like domain of human blood coagulation factor IX promotes enzyme activity and factor VIII light chain binding. *J. Biol. Chem.* **271**, 25332–25337.

Lin, S.-W., Smith, K.G., Welsch, D. & Stafford, D.W. (1990) Expression and characterization of human factor IX and factor IX-factor X chimeras in mouse 127 cells. *J. Biol. Chem.* **265**, 144–150.

Link, R.P. & Castellino, F.J. (1983) Kinetic properties of bovine blood coagulation factors IXaα and IXaβ toward synthetic substrates. *Biochemistry* **22**, 4033–4041.

Lollar, P. & Fass, D.N. (1984) Inhibition of activated porcine factor IX by dansyl-glutamyl-glycyl-arginyl-chloromethylketone. *Arch. Biochem. Biophys.* **233**, 438–446.

Masys, D.R., Bajaj, S.P. & Rapaport, S.I. (1982) Activation of human factor VII by activated factors IX and X. *Blood* **60**, 1143–1150.

Miletich, J.P., Broze, G.J., Jr. & Majerus, P.W. (1981) Purification of human coagulation factors II, IX, and X using sulfated dextran beads. *Methods Enzymol.* **80**, 221–228.

Nishimura, H., Kawabata, S., Kisiel, W., Hase, S., Ikenaka, T., Takao, T., Shimonishi, Y. & Iwanaga, S. (1989) Identification of a disaccharide (Xyl-Glc) and a trisaccharide ($Xyl_2$-Glc) O-glycosidically linked to a serine residue in the first epidermal growth factor-like domain of human factors VII and IX and protein Z and bovine protein Z. *J. Biol. Chem.* **264**, 20320–20325.

Nishimura, H., Takao, T., Hase, S., Shimonishi, Y. & Iwanaga, S. (1992) Human factor IX has a tetrasaccharide O-glycosidically linked to serine 61 through the fucose residue. *J. Biol. Chem.* **267**, 17520–17525.

Østerud, B. & Rapaport, S.I. (1977) Activation of factor IX by the reaction product of tissue factor and factor VII: additional pathway for initiating blood coagulation. *Proc. Natl Acad. Sci. USA* **74**, 5260–5264.

Pavlovsky, A. (1947) Contribution to pathogenesis of hemophilia. *Blood* **2**, 185–191.

Rao, Z., Handford, P., Mayhew, M., Knott, V., Brownlee, G.G. & Stuart, D. (1995) The structure of a $Ca^{2+}$-binding epidermal growth factor-like domain: its role in protein–protein interactions. *Cell* **82**, 131–141.

Spitzer, S.G., Warn-Cramer, B.J., Kasper, C.K. & Bajaj, S.P. (1990) Replacement of isoleucine-397 by threonine in the clotting proteinase factor IXa (Los Angeles and Long Beach variants) affects macromolecular catalysis but not L-tosylarginine methyl ester hydrolysis. *Biochem. J.* **265**, 219–225.

Sugahara, Y., Catalfamo, J., Brooks, M., Hitomi, E., Bajaj, S.P. & Kurachi, K. (1996) Isolation and characterization of canine factor IX. *Thromb. Haemost.* **75**, 450–455.

van Dieijen, G., Tans, G., Rosing, J. & Hemker, H.C. (1981) The role of phospholipid and factor VIIIa in the activation of bovine factor X. *J. Biol. Chem.* **256**, 3433–3442.

Yoshitake, S., Schach, B.G., Foster, D.C., Davie, E.W. & Kurachi, K. (1985) Nucleotide sequence of the gene for human factor IX (antihemophilic factor B). *Biochemistry* **24**, 3736–3750.

Zhong, D., Smith, K.J., Birktoft, J.J. & Bajaj, S.P. (1994) First epidermal growth factor-like domain of human blood coagulation factor IX is required for its activation by factor VIIa/tissue factor but not by factor XIa. *Proc. Natl Acad. Sci. USA* **91**, 3574–3578.

*S. Paul Bajaj*
*Departments of Internal Medicine, Pathology and Biochemistry,*
*Saint Louis University School of Medicine,*
*3635 Vista Avenue at Grand Boulevard,*
*PO Box 15250, St. Louis, MO 63110-0250, USA*
*Email: BajajPS@wpogate.slu.edu*

# 53. *Coagulation factor VIIa*

## Databanks

*Peptidase classification: clan SA, family S1, MEROPS ID: S01.215*
*NC-IUBMB enzyme classification: EC 3.4.21.21*
*ATCC entries: 59790, 59791 (cDNA)*
*Chemical Abstracts Service registry number: 65312-43-8*
*Databank codes:*

| Species | SwissProt | PIR | EMBL (cDNA) | EMBL (genomic) |
|---|---|---|---|---|
| *Bos taurus* | P22457 | A31979 | – | – |
| *Canis familiaris* | – | – | D21213 | – |
| *Homo sapiens* | P08709 | A28322 | M13232 | J02933: complete gene |
| *Macaca mulatta* | – | I84615 | D21212 | |
| *Mus musculus* | – | – | U44795 | U66079: complete gene |
| *Oryctolagus cuniculus* | P98139 | – | S56300 | – |
| | | | U77477 | – |

Brookhaven Protein Data Bank three-dimensional structures:

| Species | ID | Resolution | Notes |
|---|---|---|---|
| *Homo sapiens* | 1DAN | 2 | complex with soluble recombinant tissue factor |

## Name and History

The name **coagulation factor VIIa** refers to the active serine proteinase, while coagulation factor VII refers to the inert precursor (proenzyme or zymogen). Factor VII/VIIa has had other names but they have not been used for decades (Wright, 1959). The proteolytic activity of factor VIIa is enhanced by many orders of magnitude upon binding to its protein cofactor, tissue factor. The resulting complex is a two-subunit enzyme, with factor VIIa as the catalytic subunit and tissue factor as the positive-acting regulatory subunit. Tissue factor, initially identified because tissue extracts shortened the clotting time of plasma, is also known as thromboplastin, coagulation factor III, and CD142 (see Bach, 1988, for historical review).

## Activity and Specificity

Factor VIIa recognizes three known protein substrates: coagulation factors VII (Chapter 53), IX (Chapter 52) and X (Chapter 54) (all proenzymes), of which factor X is the most intensively studied. Factor VIIa has little enzymatic activity unless it is bound to tissue factor, and for full activity, tissue factor must be incorporated into phospholipid vesicles containing anionic phospholipids (typically, 20% phosphatidylserine, 80% phosphatidylcholine) (reviewed by Carson & Brozna, 1993). The factor VIIa–tissue factor complex cleaves a single peptide bond in factor X (Arg152┼Ile153), two bonds in factor IX (Arg145┼Ala146 and Arg180┼Val181), and a single bond in factor VII (Arg194┼Ile195). The enzymatic activity of factor VIIa–tissue factor using factor X as the substrate is typically measured in two stages: (a) activation of factor X by limited proteolysis and (b) measurement of newly generated factor

Xa enzymatic activity. Since factor VIIa activity is dependent upon calcium ions (1–5 mM is optimal), the first stage is quenched by addition of EDTA and nonionic detergent. Factor Xa is quantified by measuring the hydrolysis of a chromogenic substrate (see, for example, Neuenschwander & Morrissey, 1994).

Factor VIIa catalyzes the hydrolysis of several commercially available chromogenic or fluorogenic tripeptidyl substrates (Neuenschwander *et al.*, 1993; Butenas *et al.*, 1993). These reactions are independent of phospholipid but enhanced by tissue factor (although to a lesser degree than for macromolecular substrates). Factor VIIa prefers Arg or Lys in the P1 position, with Arg being most preferred. The pH optimum for hydrolysis of tripeptidyl *p*-nitroanilide substrates is 7.6 for factor VIIa and 8.5 for factor VIIa–tissue factor (Neuenschwander *et al.*, 1993).

The natural inhibitors of factor VIIa–tissue factor are tissue factor pathway inhibitor (TFPI) and antithrombin (see, for example, Broze *et al.*, 1993). TFPI must complex with factor Xa before it can effectively inhibit factor VIIa–tissue factor, and antithrombin requires heparin. Bovine pancreatic trypsin inhibitor (or aprotinin) is a relatively weak inhibitor of factor VIIa–tissue factor, but site-directed mutagenesis of it and related proteins have led to the development of much more effective inhibitors of this enzyme (Dennis & Lazarus, 1994; Stassen *et al.*, 1995). Benzamidine and calcium chelators are reversible inhibitors, whereas PMSF and Phe-Phe-Arg-CH$_2$Cl are irreversible inhibitors.

## Structural Chemistry

Mature factor VII is a single-chain 50 kDa protein made up of modules or domains homologous to a variety of other proteins (Hagen *et al*., 1986). At the N-terminus is the Gla domain, so called because it contains ten γ-carboxyglutamic acid (Gla) residues. Next are two modules homologous to epidermal growth factor (EGF domains), followed by the catalytic or serine proteinase domain, which is homologous to trypsin and chymotrypsin. The Gla domain binds multiple calcium ions, and the EGF1 and proteinase domains each bind a calcium ion. Factor VII contains *N*-linked and *O*-linked carbohydrate and 12 disulfide bonds. It is converted to factor VIIa by cleavage of an Arg-Ile bond in the connecting region between EGF2 and the serine proteinase domain, yielding a 22 kDa light chain and a 26 kDa heavy chain (which remain disulfide-bonded together). Factor VIIa is closely related to coagulation factors IXa and Xa, and to the anticoagulant proteinase, protein C.

Tissue factor, an integral membrane protein, is a member of the cytokine receptor superfamily. The mature protein is a single polypeptide chain with an extracellular domain consisting of two fibronectin type III domains, a single membrane-spanning domain and a short cytoplasmic tail (see Martin *et al*., 1995, for review). The protein contains *N*-linked carbohydrate and two disulfide bonds, and a Cys residue in the cytoplasmic domain is thioester-linked to a fatty acid. The X-ray crystallographic structure of a complex of active site-blocked factor VIIa and a subtilisin-cleaved form of the isolated extracellular domain of tissue factor has been solved (Banner *et al*., 1996). Some database entries for human tissue factor are: SwissProt, TF_HUMAN; PIR, A28320; EMBL, J02931 (cDNA) and J02846 (genomic); Brookhaven Protein Data Bank entries, 2HFT and 1BOY; and Chemical Abstracts Service registry number 9035-58-9.

## Preparation

Factor VII has been purified 100 000-fold from human plasma (Broze & Majerus, 1980). Recombinant human factor VII and factor VIIa have been expressed in mammalian cells and purified from the culture supernatant (Thim *et al*., 1988). Factor VII and VIIa (recombinant and plasma-derived) are available commercially from several suppliers. Tissue factor has been purified 53 000-fold from human brain (Broze *et al*., 1986). Recombinant tissue factor has been produced and purified from *E. coli* and mammalian cell cultures, as has the isolated extracellular domain (soluble tissue factor) (Paborsky *et al*., 1989; Rehemtulla *et al*., 1991); both are available from several commercial suppliers.

## Biological Aspects

Factor VII is synthesized by the liver and circulates in plasma as 99% factor VII and about 1% factor VIIa. Tissue factor is an integral membrane protein present on a variety of nonvascular cell types and in atherosclerotic plaques (Drake *et al*., 1989; Wilcox *et al*., 1989). When exposed to plasma (e.g. following vascular injury or plaque rupture), tissue factor binds factor VII with high affinity, which is then rapidly converted to factor VIIa via hydrolysis of one peptide bond. The proteinase responsible for this *in vivo* is unclear, although *in vitro* the reaction can be catalyzed by thrombin and factors IXa, Xa, XIIa and VIIa. The result is a membrane-bound, two-subunit enzyme consisting of a 1:1 complex

of factor VIIa and tissue factor. It is the first enzyme in the blood coagulation cascade, proteolytically activating two serine proteinase proenzymes, factor IX and factor X, to their enzymatically active forms. Factor VIIa and tissue factor appear to be widely distributed in vertebrates (Lewis, 1996).

The gene for human factor VII has been sequenced. Individuals with very low factor VII levels have a severe bleeding tendency, and genetic defects have been mapped to the nucleotide level (reviewed by Tuddenham *et al*., 1995). Humans with hereditary deficiency in tissue factor have not been reported, but knocking out the gene in mice leads to embryonic lethality (Bugge *et al*., 1996). Site-directed mutagenesis of both factor VIIa and tissue factor have led to insights into the structure–function relationship of these two proteins (reviewed by Martin *et al*., 1995). Elevated levels of factor VII have been implicated as a risk factor for heart disease, and specific inhibitors of factor VIIa or tissue factor are currently under investigation as novel antithrombotic agents.

The gene for human tissue factor has been sequenced and the locus is at 1p21–p22. Tissue factor gene expression can be induced in certain cell types (such as fibroblasts, monocytes and vascular endothelial cells) by a variety of cytokines and inflammatory mediators, which may be important in inflammation and many thrombotic disorders. Transcriptional regulatory elements in the tissue factor gene have been identified and studied (reviewed by Mackman, 1995).

## Distinguishing Features

Factor VII or VIIa corrects the prolonged clotting time of factor VII-deficient plasma and requires tissue factor for function. Similarly, factor VIIa activates factors IX and X by limited proteolysis in a tissue factor-dependent manner. Monoclonal and polyclonal antibodies to both factor VIIa and tissue factor have been reported and are available from a variety of commercial suppliers.

## Further Reading

For a review of current thinking about how the blood-clotting system is activated and controlled, see Luchtman-Jones & Broze (1995).

## References

Bach, R. (1988) Initiation of coagulation by tissue factor. *CRC Crit. Rev. Biochem.* **23**, 339–368.
Banner, D.W., D'Arcy, A., Chène, C., Winkler, F.K., Guha, A., Konigsberg, W.H., Nemerson, Y. & Kirchhofer, D. (1996) The crystal structure of the complex of blood coagulation factor VIIa with soluble tissue factor. *Nature* **380**, 41–46.
Broze, G.J., Jr. & Majerus, P.W. (1980) Purification and properties of human coagulation factor VII. *J. Biol. Chem.* **255**, 1242–1247.
Broze, G.J., Jr., Leykam, J.E., Schwartz, B.D. & Miletich, J.P. (1986) Purification of human brain tissue factor. *J. Biol. Chem.* **260**, 10917–10920.
Broze, G.J., Jr., Likert, K. & Higuchi, D. (1993) Inhibition of factor VIIa/tissue factor by antithrombin III and tissue factor pathway inhibitor. *Blood* **82**, 1679–1680.
Bugge, T.H., Xiao, Q., Kombrinck, K.W., Flick, M.J., Holmbäck, K., Danton, H.J.S., Colbert, M.C., Witte, D.P., Fujikawa, K., Davie, E.W. & Degen, J.L. (1996) Fatal embryonic bleeding events in

mice lacking tissue factor, the cell-associated initiator of blood coagulation. *Proc. Natl Acad. Sci. USA* **93**, 6258–6263.

Butenas, S., Ribarik, N. & Mann, K.G. (1993) Synthetic substrates for human factor VIIa and factor VIIa-tissue factor. *Biochemistry* **32**, 6531–6538.

Carson, S.D. & Brozna, J.P. (1993) The role of tissue factor in the production of thrombin. *Blood Coagul. Fibrinol.* **4**, 281–292.

Dennis, M.S. & Lazarus, R.A. (1994) Kunitz domain inhibitors of tissue factor–factor VIIa. II. Potent and specific inhibitors by competitive phage selection. *J. Biol. Chem.* **269**, 22137–22144.

Drake, T.A., Morrissey, J.H. & Edgington, T.S. (1989) Selective cellular expression of tissue factor in human tissues: implications for disorders of hemostasis and thrombosis. *Am. J. Pathol.* **134**, 1087–1097.

Hagen, F.S., Gray, C.L., O'Hara, P., Grant, F.J., Saari, G.C., Woodbury, R.G., Hart, C.E., Insley, M., Kisiel, W., Kurachi, K. & Davie, E.W. (1986) Characterization of a cDNA coding for human factor VII. *Proc. Natl Acad. Sci. USA* **83**, 2412–2416.

Lewis, J.H. (1996) *Comparative Hemostasis in Vertebrates*. New York: Plenum Press.

Luchtman-Jones, L. & Broze, G.J., Jr. (1995). The current status of coagulation. *Ann. Med.* **27**, 47–52.

Mackman, N. (1995) Regulation of the tissue factor gene. *FASEB J.* **9**, 883–889.

Martin, D.M.A., Boys, C.W.G. & Ruf, W. (1995) Tissue factor: Molecular recognition and cofactor function. *FASEB J.* **9**, 852–859.

Neuenschwander, P.F. & Morrissey, J.H. (1994) Roles of the membrane-interactive regions of factor VIIa and tissue factor. The factor VIIa Gla domain is dispensable for binding to tissue factor but important for activation of factor X. *J. Biol. Chem.* **269**, 8007–8013.

Neuenschwander, P.F., Branam, D.E. & Morrissey, J.H. (1993) Importance of substrate composition, pH and other variables on tissue factor enhancement of factor VIIa activity. *Thromb. Haemost.* **70**, 970–977.

Paborsky, L.R., Tate, K.M., Harris, R.J., Yansura, D.G., Band, L., McCray, G., Gorman, C.M., O'Brien, D.P., Chang, J.Y., Swartz, J.R., Fung, V.P., Thomas, J.N. & Vehar, G.A. (1989) Purification of recombinant human tissue factor. *Biochemistry* **28**, 8072–8077.

Rehemtulla, A., Pepe, M. & Edgington, T.S. (1991) High level expression of recombinant human tissue factor in Chinese hamster ovary cells as a human thromboplastin. *Thromb. Haemost.* **65**, 521–527.

Stassen, J.M., Lambeir, A.-M., Matthyssens, G., Ripka, W.C., Nystrom, A., Sixma, J.J. & Vermylen, J. (1995) Characterization of a novel series of aprotinin-derived anticoagulants. II. Comparative antithrombotic effects on primary thrombus formation in vivo. *Thromb. Haemost.* **74**, 646–654.

Thim, L., Bjoern, S., Christensen, M., Nicolaisen, E.M., Lund-Hansen, T., Pedersen, A.H. & Hedner, U. (1988) Amino acid sequence and posttranslational modifications of human factor VIIa from plasma and transfected baby hamster kidney cells. *Biochemistry* **27**, 7785–7793.

Tuddenham, E.G.D., Pemberton, S. & Cooper, D.N. (1995) Inherited factor VII deficiency: Genetics and molecular pathology. *Thromb. Haemost.* **74**, 313–321.

Wilcox, J.N., Smith, K.M., Schwartz, S.M. & Gordon, D. (1989) Localization of tissue factor in the normal vessel wall and in the atherosclerotic plaque. *Proc. Natl Acad. Sci. USA* **86**, 2839–2843.

Wright, I.S. (1959) Nomenclature of blood clotting factors: Four factors, their characterization, and international number. *JAMA* **170**, 325–328.

*James H. Morrissey*
*Cardiovascular Biology Research Program,*
*Oklahoma Medical Research Foundation,*
*825 NE 13th Street, Oklahoma City, OK 73104, USA*
*Email: morrisseyj@omrf.uokhsc.edu*

# 54. *Coagulation factor X*

## Databanks

*Peptidase classification: clan SA, family S1, MEROPS ID: S01.216*
*NC-IUBMB enzyme classification: EC 3.4.21.6*
*Chemical Abstracts Service registry number: 9002-05-5*
*Databank codes:*

| Species | SwissProt | PIR | EMBL (cDNA) | EMBL (genomic) |
|---|---|---|---|---|
| *Bos taurus* | P00743 | A22867 | X00673 | – |
| *Gallus gallus* | P25155 | S15838 | D00844 | – |
| *Homo sapiens* | P00742 | A24478 | K03194 | L00390: exon 1 |

*continued overleaf*

| Species | SwissProt | PIR | EMBL (cDNA) | EMBL (genomic) |
|---------|-----------|-----|-------------|----------------|
| | | | M22613 | L00391: exon 2 |
| | | | M57285 | L00392: exon 3 |
| | | | | L00393: exon 4 |
| | | | | L00394: exon 5 |
| | | | | L00395: exon 6 |
| | | | | L00396: exon 7 |
| | | | | L29433: exon 8 |

Brookhaven Protein Data Bank three-dimensional structures:

| Species | ID | Resolution | Notes |
|---------|-----|-----------|-------|
| *Homo sapiens* | 1HCG | 2.2 | factor Xa |

## Name and History

In the mid-1950s it was becoming evident that a coagulation factor different from factors VII and IX was required for the activation of prothrombin to thrombin (Hougie *et al*., 1957; Graham *et al*., 1957). In 1957 it was found that the *in vitro* coagulation properties of blood samples from certain patients thought to lack factor VII (then called proconvertin), could be normalized by mixing the samples. This indicated that the disease was not a homogeneous entity, but could be divided into two groups. The new 'factor' was called **Stuart's factor** or **Prower's factor** after the names of two of the first patients identified with this unique bleeding disorder (Duckert *et al*., 1955). The factor was later designated by the Roman numeral X. **Factor X** was found to require vitamin K for normal biosynthesis, and treatment of patients with vitamin K antagonists such as dicoumarol or warfarin resulted in a decrease in the biological activity of factor X (Hougie *et al*., 1957). Factor X was first purified to homogeneity from cattle plasma and then from human plasma (Jackson *et al*., 1968; Fujikawa *et al*., 1972b; DiScipio *et al*., 1977b). Factor X is the proenzyme of a serine protease with trypsin-like specificity. The factor X structure has been determined by protein and cDNA sequencing, and the factor X gene has been characterized (Leytus *et al*., 1986). Several postribosomal modifications have been identified, including the vitamin K-dependent $\gamma$-carboxyglutamic acid (Gla). The three-dimensional structure of human factor X (lacking residues 1–45) has been determined by X-ray crystallography (Padmanabhan *et al*., 1993; Brandstetter *et al*., 1996).

## Activity and Specificity

Factor X is the proenzyme of a serine protease. The N-terminal Gla domain, which binds 8–10 calcium ions, is required for interaction with phosphatidylserine-containing membranes (e.g. platelet microparticles) (Furie & Furie, 1988; Mann *et al*., 1990; Davie *et al*., 1991; James, 1994; Stenflo & Dahlbäck, 1994). In the presence of calcium ions, factor Xa forms a phospholipid-bound, macromolecular complex with a cofactor, factor Va. The complex activates prothrombin (Chapter 55) by cleaving two peptide bonds. Factor Va is derived from factor V by limited proteolysis mediated by thrombin or factor Xa. Membrane-bound factor Va binds

factor Xa with high affinity ($K_d \sim 1$ nmol liter$^{-1}$), but has no measurable affinity for the proenzyme, factor X (Miletich *et al*., 1977; Mann *et al*., 1990; Jackson, 1994). The affinity between factor Xa and phospholipid in the absence of factor Va is about 1000-fold lower than with factor Va ($K_d \sim 1$ mmol liter$^{-1}$). In these reactions, the phospholipid can be regarded as a means of reducing the 'dimensionality' of the reaction to two dimensions, and hence increasing the likelihood of a productive collision between factor Xa and membrane-bound factor Va. Factor Xa in the macromolecular complex activates prothrombin $\sim 10^6$-fold more rapidly than does free factor Xa. Studies of this reaction have demonstrated that the phospholipid reduces the $K_m$ for prothrombin, whereas factor Va increases the $V$ (Mann *et al*., 1990). The biological membranes that interact with the enzymes and proenzymes (factors Xa/X), as well as with the cofactor (factor Va), require phospholipid vesicles rich in phosphatidylserine or microparticles derived from activated platelets (Sims *et al*., 1989; Jackson, 1994).

Activation of prothrombin, a single-chain molecule, requires the cleavage of two peptide bonds and may thus proceed through two pathways (Mann *et al*., 1990; Jackson, 1994). Cleavage of the Arg320-Ile321 bond yields meizothrombin (Rosing *et al*., 1986). Subsequent cleavage of the Arg271-Thr272 bond yields the so-called fragment 1-2 and thrombin. Alternatively, cleavage of the peptide bond Arg284┼Thr285 (mediated by thrombin) yields fragment 1-2-3 and prethrombin 2, whereas cleavage of the Arg271┼Thr272 peptide bond yields fragment 1-2 and prethrombin 2 with a 13 residues long N-terminal extension. Prethrombin 2 is enzymatically inactive. Cleavage of the Arg320-Ile321 bond in prethrombin 2 yields thrombin, a molecule (32 kDa) of two chains held together by a single disulfide bond. Meizothrombin has amidolytic activity, but is inactive against fibrinogen, although it readily activates protein C (Chapter 56). The pathway of prothrombin activation in which meizothrombin is an intermediate is generally thought to be physiologically more relevant than the prethrombin 2 pathway, but this is still debated. Factor Xa is an Arg-specific serine protease that is related to trypsin, but has much narrower substrate specificity (Padmanabhan *et al*., 1993; Brandstetter *et al*., 1996). Factor Xa can also activate factors V and VIII, and factor VII (Chapter 53).

Several low molecular mass synthetic substrates are available for factor Xa, including methoxycarbonyl-cyclo-hexylglycyl-Gly-Arg+NHPhNO$_2$ (Spectrozyme Xa), methane-sulfonyl-D-Leu-Gly-Arg+NHPhNO$_2$ (CBS 31.39), and Bz-Ile-Glu-(piperidine amide)-Gly-Arg+NHPhNO$_2$ (S2337).

Factor Xa can be inhibited by antithrombin III, $\alpha_1$-proteinase inhibitor and soybean trypsin inhibitor. The rate of inactivation by antithrombin III is greatly increased (several thousand-fold) by heparin (Bourin & Lindahl, 1993; Broze & Tollefsen, 1994). Factor Xa is also inactivated by the tissue factor pathway inhibitor (TFPI). TFPI consists of three domains in tandem that are homologous to the Kunitz-type protease inhibitors (Broze *et al.*, 1990; Rapaport, 1991). The second Kunitz-type domain interacts with factor Xa. This complex then interacts with the tissue factor–factor VIIa complex to form an inactive ternary complex. Low molecular mass inhibitors of factor Xa are Dns-Ile-Glu-Gly-Arg-CH$_2$Cl and DFP, although the latter is a poor inhibitor.

## Structural Chemistry

The primary structure of human factor X has been determined by sequencing at the amino acid, cDNA and genomic levels (McMullen *et al.*, 1983; Fung *et al.*, 1984; Leytus *et al.*, 1986). The sequences of bovine and chicken factor X have also been determined. Human factor X contains 488 amino acids including a prepro sequence of 40 amino acids. The propeptide serves as a recognition site for the vitamin K-dependent carboxylase. The molecular mass of the mature protein is 58.9 kDa (sedimentation equilibrium). It consists of a light chain with 139 amino acids, and a heavy chain with 306 amino acids. The protein is synthesized as a single polypeptide chain with the sequence Arg-Arg-Lys-Arg connecting the two chains. The mature protein has an N-terminal Gla domain that is followed by two EGF-like domains. The N-terminal part of the heavy chain is occupied by an activation peptide that is 52 amino acids long, whereas the heavy chain of the active enzyme contains 254 amino acids. The structure of the calcium-free form of factor Xa, lacking the 45 N-terminal amino acids (the Gla domain), has been determined by X-ray crystallography to 2.2 Å resolution (Padmanabhan *et al.*, 1993). The overall folding of the catalytic domain is similar to those of chymotrypsin and thrombin. His42, Asp88 and Ser185 of the activated heavy chain constitute the catalytic triad. The C-terminal EGF-like domain has a typical fold, and is in close contact with the catalytic domain. On the other hand, the N-terminal domain could not be localized in the structure, presumably because of its mobility. In the Gla domain, Glu residues 6, 7, 14, 16, 19, 20, 25, 26, 29, 32 and 39 are carboxylated to Gla. The structure of the Gla domain in prothrombin has been determined by X-ray crystallography (Soriano Garcia *et al.*, 1992). The Gla residues were found to ligate six calcium ions in the interior of the domain. The homologous Gla domain from factor X has been modeled on the prothrombin domain (Sunnerhagen *et al.*, 1995). Moreover, the structure of the Gla domain linked to the N-terminal EGF domain was determined in the absence of calcium by NMR spectroscopy. In the calcium-free structure the Gla residues were found to be exposed to the solution, whereas three N-terminal hydrophobic residues were in the interior of the domain. The structure made it evident that the calcium ions pull the Gla residues to the interior of the domain and thus provide the energy required to expose the hydrophobic side chains in solution. The structure, together with results of site-directed mutagenesis studies, established that the interaction between Gla domains and phospholipid membranes is essentially hydrophobic (Arni *et al.*, 1994; Zhang & Castellino 1994; Sunnerhagen *et al.*, 1995). In the first EGF-like domain, the Asp residue in position 63 is hydroxylated to $\beta$-hydroxyaspartic acid (Drakenberg *et al.*, 1983; Fernlund & Stenflo, 1983; Sugo *et al.*, 1984). The three-dimensional structure of the N-terminal EGF-like domain has been determined both in the presence of calcium and in its absence. The calcium structure revealed a calcium-binding site that is unique to EGF-like domains. Moreover, the structure showed that the calcium ion locks the Gla domain in the proper position relative to the EGF domain (Sunnerhagen *et al.*, 1996).

## Preparation

Factor X can be purified to homogeneity from plasma by conventional methods. As with other vitamin K-dependent coagulation factors, an initial step is used that is based on the affinity of these proteins for insoluble salts of divalent cations, such as barium citrate or sulfate. This is followed by ammonium sulfate precipitation and anion-exchange chromatography. These procedures are sufficient to obtain essentially homogeneous cattle factor X (Jackson *et al.*, 1968; Fujikawa *et al.*, 1972b). Purification of human factor X usually requires additional chromatography steps (DiScipio *et al.*, 1977b; Jackson, 1994). Numerous procedures have been developed for the expression of factor IX in eukaryotic cells, but no such method appears to have been developed for factor X, presumably because factor X deficiency is very rare and is not an important clinical problem (James, 1994).

## Biological Aspects

The factor X gene is found on chromosome 13, position q34 adjacent to the factor VII gene. It spans about 22 kb and is made up of seven introns and eight exons (Leytus *et al.*, 1986). Exon I encodes the signal peptide, exon II the propeptide/Gla domain, exon III the C-terminal part of the Gla domain, the so-called aromatic amino acid stack, exons IV and V the EGF-like domains, exon VI the activation peptide region and exons VII and VIII the catalytic serine protease part. Factor X is synthesized in the liver as a single-chain protein with an N-terminal signal peptide that is followed by a propeptide (Furie & Furie, 1988). The propeptide is the recognition signal for the vitamin K-dependent carboxylase that carboxylates the N-terminal Glu residues to Gla. Asp63 in the first EGF-like domain is hydroxylated to *erythro*-$\beta$-hydroxyaspartic acid by a dioxygenase (Stenflo *et al.*, 1989; Wang *et al.*, 1991). There are two *N*-linked carbohydrate side chains in the activation peptide region (Leytus *et al.*, 1986). Prior to secretion, a single peptide bond is cleaved to yield two peptide chains that are linked by a single disulfide bond. This cleavage is followed by removal of the three C-terminal basic amino acids (Arg-Lys-Arg; positions 140–142) in plasma by carboxypeptidase U (Chapter 435).

Activation of factor X to a serine protease proceeds by cleavage of two peptide bonds. Amidolytic activity is obtained after cleavage of an Arg-Ile peptide bond in the heavy chain with release of a 52 residue activation peptide (factor Xaα). A second, autocatalytic, peptide bond cleavage in the C-terminal part of the heavy chain yields factor Xaβ (Jesty *et al.*, 1974). The biological activities of the two forms of factor Xa are identical. Activation of factor X can follow two pathways. Factor X may be activated by factor VIIa in complex with tissue factor (TF), an integral membrane protein (Rapaport & Rao, 1995); this pathway is called the 'extrinsic pathway' and is responsible for the initiation of coagulation. Alternatively, factor X is activated by a membrane-bound macromolecular complex that consists of factor IXa bound to its cofactor, factor VIIIa; this pathway is called the 'intrinsic pathway' (see also Chapter 52). Assembly of this complex also requires calcium ions. The rate of activation of factor X by the intact macromolecular complex is about $10^6$-fold more rapid than its activation by factor IXa alone (Mann *et al.*, 1990). The physiological importance of this pathway is illustrated by the fact that hereditary deficiency of factor IX is associated with a severe bleeding disorder (hemophilia B). Both bovine and human factor X can be activated by an enzyme present in Russell's viper venom (Fujikawa *et al.*, 1972a; DiScipio *et al.*, 1977a) (Chapter 433).

Two cell surface receptors have been identified for factor X/Xa. One of these receptors, Mac-1, is expressed on monocytes. It binds factor X with high affinity ($K_d \sim 30$ nM), but has no affinity for factor Xa (Altieri & Edgington, 1988; Altieri *et al.*, 1988). ADP appears to induce high-affinity binding of factor X to the receptor. Bound factor X then appears to be rapidly activated by limited proteolysis, with subsequent activation of blood coagulation. Recently an endothelial cell receptor (EPR-1) for factor Xa was identified (Nicholson *et al.*, 1996). This receptor does not bind the proenzyme, factor X. Binding of factor Xa requires the active site, and results in the formation of specific protease–receptor complex. Formation of this complex results in an increase in [$^3$H]thymidine uptake by the cells. It has also been demonstrated that factor Xa induces calcium oscillations in certain cells (Camerer *et al.*, 1996).

## Distinguishing Features

Factor X is a liver-synthesized proenzyme of a serine protease that requires vitamin K for normal biosynthesis. The protein has the same domain structure as coagulation factors VII and IX, and protein C (Chapter 56), and the four proteins have pronounced similarities in amino acid sequence. All of these proteins are synthesized as single polypeptide chains, but factor X and protein C are cleaved into two-chain structures prior to secretion. The proteins are easily distinguished by SDS-PAGE under reducing conditions. Monoclonal and polyclonal antibodies that distinguish the proteins are commercially available from several sources.

## Further Reading

For a full review, see James (1994).

## References

Altieri, D.C. & Edgington, T.S. (1988) The saturable high affinity association of factor X to ADP-stimulated monocytes defines a novel function of the Mac-1 receptor. *J. Biol. Chem.* **263**, 7007–7015.

Altieri, D.C., Morissey, J.H. & Edgington, T.S. (1988) Adhesive receptor Mac-1 coordinates the activation of factor X on stimulated cells of monocytic myeloid differentiation: an alternative initiation of the coagulation protease cascade. *Proc. Natl Acad. Sci. USA* **85**, 7462–7466.

Arni, R.K., Padmanabhan, K., Padmanabkan, K.P., Wu, T.S. & Tulinsky, A. (1994) Structure of non-covalent complex of prothrombin kringle 2 with PPACK-thrombin. *Chem. Phys. Lipids* **67/68**, 59–66.

Bourin, M.C. & Lindahl, U. (1993) Glycosaminoglycans and the regulation of blood coagulation. *Biochem. J.* **289**, 313–330.

Brandstetter, H., Kühne, A., Bode, W., Huber, H.R., von der Saal, W., Wirthensohn, K. & Engh, R.A. (1996) X-ray structure of active site-inhibited clotting factor Xa. Implications for drug design and substrate recognition. *J. Biol. Chem.* **271**, 29988–29992.

Broze, G.J. & Tollefsen, D.M. (1994) Regulation of blood coagulation by protease inhibitors. In: *The Molecular Basis of Blood Diseases* (Stamatoyannopoulos, G., Nienhuis, A.W., Majerus, P.W. & Varmus, H., eds). Philadelphia: W.B. Saunders, pp. 629–656.

Broze, G.J., Girard, T.J. & Novotny, W.F. (1990) Regulation of coagulation by a multivalent Kunitz-type inhibitor. *Biochemistry* **29**, 7539–7546.

Camerer, E., Røftingen, J.-A., Iversen, J.G. & Prydz, H. (1996) Coagulation factors VII and X induce $Ca^{2+}$ oscillations in Madin-Darby canine kidney cells only when proteolytically active. *J. Biol. Chem.* **271**, 29034–29042.

Davie, E.W., Fujikawa, K. & Kisiel, W. (1991) The coagulation cascade: initiation, maintenance, and regulation. *Biochemistry* **30**, 10363–10370.

DiScipio, R.G., Hermodson, M.A. & Davie, E.W. (1977a) Activation of human factor X (Stuart factor) by a protease from Russell's viper venom. *Biochemistry* **16**, 5253–5260.

DiScipio, R.G., Hermodson, M.A., Yates, S.G. & Davie, E.W. (1977b) A comparison of human prothrombin, factor IX, factor X, and protein S. *Biochemistry* **16**, 698–706.

Drakenberg, T., Fernlund, P., Roepstorff, P. & Stenflo, J. (1983) β-Hydroxyaspartic acid in vitamin K-dependent protein C. *Proc. Natl Acad. Sci. USA* **80**, 1802–1806.

Duckert, F., Flückiger, P., Matter, M. & Koller, F. (1955) Clotting factor X physiologic and physico-chemical properties. *Proc. Soc. Exp. Biol. Med.* **90**, 17–22.

Fernlund, P. & Stenflo, J. (1983) β-Hydroxyaspartic acid in vitamin K-dependent proteins. *J. Biol. Chem.* **258**, 12509–12512.

Fujikawa, K., Legaz, M.E. & Davie, E.W. (1972a) Bovine factor X₁ (Stuart factor). Mechanism of activation by a protease from Russell's viper venom. *Biochemistry* **11**, 4892–4898.

Fujikawa, K., Legaz, M.E. & Davie, E.W. (1972b) Bovine factor X₁ and X₂ (Stuart factor). Isolation and characterization. *Biochemistry* **11**, 4882–4891.

Fung, M.R., Campbell, R.M. & MacGillivary, R.T.A. (1984) Blood coagulation factor X mRNA encodes a single polypeptide chain containing a prepro leader sequence. *Nucleic Acids Res.* **12**, 4481–4492.

Furie, B. & Furie, B.C. (1988) The molecular basis of blood coagulation. *Cell* **53**, 505–518.

Graham, J.B., Barrow, E.M. & Hougie, C. (1957) Stuart clotting defect. II. Genetic aspects of a 'new' hemorrhagic state. *J. Clin. Invest.* **36**, 497–503.

Hougie, C., Barrow, E.M. & Graham, J.B. (1957) Stuart clotting defect. I. Segregation of an hereditary hemorrhagic state from the heterogenous group heretofore called 'stable factor' (SPCA, proconvertin, factor VII) deficiency. *J. Clin. Invest.* **36**, 485–496.

Jackson, C.M. (1994). Physiology and biochemistry of prothrombin. In: *Haemostasis and Thrombosis* (Bloom, A.L., Forbes, C.D., Thomas, D.P. & Tuddenham, E.G.D., eds). Edinburgh: Churchill Livingstone, pp. 397–438.

Jackson, C.M., Johnson, T.F. & Hanahan, D.J. (1968) Studies on bovine factor X. I. Large-scale purification of the bovine plasma protein possessing factor X activity. *Biochemistry* **7**, 4492–4505.

James, H.L. (1994) Physiology and biochemistry of factor X. In: *Haemostasis and Thrombosis*, 3rd edn (Bloom, A.L., Forbes, C.D., Thomas, D.P. & Tuddenham, E.G.D., eds). Edinburgh: Churchill Livingstone, pp. 439–464.

Jesty, J., Spencer, A.K. & Nemerson, Y. (1974) The mechanism of activation of factor X. Kinetic control of alternative pathways leading to the formation of activated factor X. *J. Biol. Chem.* **249**, 5614–5622.

Leytus, S.P., Foster, D.C., Kurachi, K. & Davie, E.W. (1986) Gene for human factor X: A blood coagulation factor whose gene organization is essentially identical with that of factor IX and protein C. *Biochemistry* **25**, 5098–5102.

Mann, K.G., Nesheim, M.E., Church, W.R., Haley, P. & Krishnaswamy, S. (1990) Surface-dependent reactions of the vitamin K-dependent enzyme complexes. *Blood* **76**, 1–16.

McMullen, B.A., Fujikawa, K., Kisiel, W. Sasagawa, T., Howald, D.M., Kwa, E.Y. & Weinstein, B. (1983) Complete amino acid sequence of the light chain of human blood coagulation factor X: Evidence for identification of residue 63 as β-hydroxyaspartic acid. *Biochemistry* **22**, 2875–2884.

Miletich, J.P., Jackson, C.M. & Majerus, P.W. (1977). Interaction of coagulation factor Xa with human platelets. *Proc. Natl Acad. Sci. USA* **74**, 4033–4036.

Nicholson, A.C., Nachman, R.L., Alteri, D.C., Summers, B.D., Ruf, W., Edgington, T.S. & Hajjar, D.P. (1996) Effector cell protease receptor-1 is a vascular receptor for coagulation factor Xa. *J. Biol. Chem.* **271**, 28407–28413.

Padmanabhan, K., Padmanabhan, K.P., Park, C.H., Bode, W., Huber, R., Blankenship, D.T., Cardin, A.D. & Kisiel, W. (1993) Structure of human Des(1–45) factor Xa at 2.2 Å resolution.

*J. Mol. Biol.* **232**, 947–966.

Rapaport, S.I. (1991) The extrinsic pathway inhibitor: a regulator of tissue factor-dependent blood coagulation. *Thromb. Haemost.* **66**, 6–15.

Rapaport, S.I. & Rao, V.M. (1995) The tissue factor pathway: how it has become a 'prima ballerina'. *Thromb. Haemost.* **74**, 7–17.

Rosing, J., Zwaal, R.F.A. & Tans, G. (1986) Formation of meizothrombin as intermediate in factor Xa-catalyzed prothrombin activation. *J. Biol. Chem.* **261**, 4224–4228.

Sims, P.J., Wiedmer, T., Esmon, C.T., Weiss, H.J. & Shattil, S.J. (1989) Assembly of the platelet prothrombin complex is linked to vesiculation of the platelet plasma membrane. *J. Biol. Chem.* **264**, 17049–17057.

Soriano Garcia, M.W., Padmanabhan, K., de Vos, A.M. & Tublinsky, A. (1992) The $Ca^{2+}$ ion and membrane binding structure of the Gla-domain of $Ca^{2+}$-prothrombin fragment 1. *Biochemistry* **31**, 2554–2566.

Stenflo, J. & Dahlbäck, B. (1994) Vitamin K-dependent proteins. In: *The Molecular Basis of Blood Diseases* (Stamatoyannopoulos, G., Nienhuis, A.W., Majerus, P.W. & Varmus, H., eds). Philadephia: W.B. Saunders, pp. 565–598.

Stenflo, J., Holme, E., Lindstedt, S., Chandramouli, N., Tsai Huang, L.H., Tam, J.P. & Merrifield, R.B. (1989) Hydroxylation of aspartic acid in domains homologous to the epidermal growth factor precursor is catalyzed by a 2-oxoglutarate-dependent dioxygenase. *Proc. Natl Acad. Sci. USA* **86**, 444–447.

Sugo, T., Fernlund, P. & Stenflo, J. (1984) *Erythro*-β-hydroxyaspartic acid in bovine factor IX and factor X. *FEBS Lett.* **165**, 102–106.

Sunnerhagen, M., Forsén, S., Hoffrén, A.-M., Drahenberg, T., Teleman, O. & Stenflo, J. (1995) Structure of the $Ca^{2+}$ free GLA domain sheds light on membrane binding of blood coagulation proteins. *Nature Struct. Biol.* **2**, 504–509.

Sunnerhagen, M., Glenn, A.O., Stenflo, J., Forsén, S., Drakenberg, T. & Trewhella, J. (1996) The relative orientation of Gla and EGF domains in coagulation factor X is altered by $Ca^{2+}$ binding to the first EGF domain. A combined NMR-small angle X-ray scattering study. *Biochemistry* **35**, 11547–11559.

Wang, Q., Van Dusen, W.J., Petroski, C.J., Garsky, V.M., Stern, A.M. & Friedman, P.A. (1991) Bovine liver aspartyl beta-hydroxylase. Purification and characterization. *J. Biol. Chem.* **266**, 14004–14010.

Zhang, L. & Castellino, F.J. (1994) The binding energy of human coagulation protein C to acidic phospholipid vesicles contains a major contribution from leucine 5 in the gamma-carboxyglutamic acid domain. *J. Biol. Chem.* **269**, 3590–3595.

***Johan Stenflo***
*Lund University,*
*Department of Clinical Chemistry,*
*University Hospital, Malmö,*
*Malmö, Sweden*
*Email: johan.stenflo@klkemi.mas.lu.se*

# 55. *Thrombin*

## *Databanks*

*Peptidase classification: clan SA, family S1, MEROPS ID: S01.217*
*NC-IUBMB enzyme classification: EC 3.4.21.5*
*Chemical Abstracts Service registry number: 9002-04-4*
*Databank codes:*

| Species | SwissProt | PIR | EMBL (cDNA) | EMBL (genomic) |
|---|---|---|---|---|
| *Acipenser transmontanus* | – | H42696 | M81399 | – |
| *Bos taurus* | P00735 | A00915 | J00041 | – |
| | | A37552 | V00135 | |
| | | S02537 | | |
| *Cynops pyrrhogaster* | – | F42696 | M81395 | |
| *Eptatretus stoutii* | – | I42696 | M81393 | – |
| *Gallus gallus* | – | D42696 | M81391 | – |
| *Gecko gecko* | – | E42696 | M81392 | – |
| *Homo sapiens* | P00734 | A00914 | J00307 | M17262: complete gene |
| | | A29351 | M33031 | |
| | | A37549 | V00595 | |
| | | A37550 | | |
| | | B00914 | | |
| *Mus musculus* | P19221 | A35827 | M81394 | – |
| | | A42696 | X52308 | |
| | | S12081 | | |
| *Oncorhynchus mykiss* | – | G42696 | M81398 | |
| *Oryctolagus cuniculus* | – | C42696 | M81396 | – |
| *Rattus norvegicus* | P18292 | A60576 | M81397 | – |
| | | B42696 | X52835 | |
| | | S10511 | | |
| *Tachypleus gigas* | – | – | S77063 | – |
| *Tachypleus gigas* | – | – | S77064 | – |

Brookhaven Protein Data Bank three-dimensional structures:

| Species | ID | Resolution | Notes |
|---|---|---|---|
| *Bos taurus* | 1BBR | 2.3 | complex with fibrinopeptide $\alpha$ |
| | 1ETR | 2.2 | $\varepsilon$-thrombin; noncovalent complex with MQPA |
| | 1ETS | 2.3 | $\varepsilon$-thrombin; noncovalent complex with NAPAP |
| | 1ETT | 2.5 | $\varepsilon$-thrombin; noncovalent complex with 4-TAPAP |
| | 1HRT | 2.8 | $\alpha$-thrombin; complex with hirudin variant 1 |
| | 1TBQ | 3.1 | complex with rhodniin inhibitor |
| | 1TBR | 2.6 | complex with rhodniin inhibitor |
| | 2PF1 | 2.2 | prothrombin fragment 1 |
| | 2PF2 | 2.2 | prothrombin fragment 1; complex with calcium |
| | 2SPT | 2.5 | prothrombin fragment 1; complex with strontium |
| *Homo sapiens* | 1ABI | 2.3 | complex with hirulog 3 |
| | 1ABJ[a] | 2.4 | complex with D-Phe-Pro-Arg-chloromethane |
| | 1AHT | 1.6 | $\alpha$-thrombin; complex with *p*-amidino-Phe-pyruvate |
| | 1DWB | 3.16 | $\alpha$-thrombin; complex with des-Asp55 hirudin and benzamidine |
| | 1DWC | 3 | $\alpha$-thrombin; complex with des-Asp55 hirudin and MD-805 |
| | 1DWD | 3 | $\alpha$-thrombin; complex with des-Asp55 hirudin and NAPAP |

| Species | ID | Resolution | Notes |
|---|---|---|---|
| | 1DWE | 3 | $\alpha$-thrombin; complex with des-Asp55 hirudin and D-Phe-Pro-Arg-chloromethane |
| | 1FPC | 2.3 | $\alpha$-thrombin; complex with hirugen and Dns-Arg-*N*-(3-ethyl-1,5-pentanediyl)amide |
| | 1FPH | 2.5 | $\alpha$-thrombin; complex with hirudin and fibrinopeptide A |
| | 1HAG | 2.3 | prethrombin; complex with hirugen |
| | 1HAH | 2.3 | $\alpha$-thrombin; complex with hirugen |
| | 1HAI | 2.4 | $\alpha$-thrombin; complex with D-Phe-Pro-Arg-chloromethane |
| | 1HAO | 2.8 | $\alpha$-thrombin; complex with 15-mer oligonucleotide |
| | 1HAP | 2.8 | $\alpha$-thrombin; complex with 15-mer oligonucleotide |
| | 1HBT | 2 | $\alpha$-thrombin; complex with a peptidyl pyridinium methyl ketone |
| | 1HDT | 2.6 | $\alpha$-thrombin; complex with hirugen |
| | 1HGT | 2.2 | $\alpha$-thrombin; complex with hirugen |
| | 1HLT | 3 | $\alpha$-thrombin; complex with thrombomodulin and D-Phe-Pro-Arg-chloromethane |
| | 1HRT | 2.8 | $\alpha$-thrombin; complex with hirudin variant 1 |
| | 1HUT | 2.9 | $\alpha$-thrombin; complex with DNA and D-Phe-Pro-Arg-chloromethane |
| | 1IHS | 2 | $\alpha$-thrombin; complex with hirutonin-2 |
| | 1IHT | 2.1 | $\alpha$-thrombin; complex with hirutonin-6 |
| | 1NRN | 3.1 | $\alpha$-thrombin; noncovalent complex with receptor-based peptide NRS |
| | 1NRO | 3.1 | $\alpha$-thrombin; noncovalent complex with receptor-based peptide NRP |
| | 1NRP | 3 | $\alpha$-thrombin; noncovalent complex with receptor-based peptide NR′S |
| | 1NRQ | 3.5 | $\alpha$-thrombin; noncovalent complex with receptor-based peptide D-FPR′S |
| | 1NRR | 2.4 | $\alpha$-thrombin; noncovalent complex with receptor-based peptide XA and D-Phe-Pro-Arg-chloromethane |
| | 1NRS | 2.4 | $\alpha$-thrombin; noncovalent complex with receptor-based peptide NRP and hirugen |
| | 1PPB | 1.92 | $\alpha$-thrombin; complex with D-Phe-Pro-Arg-chloromethane |
| | 1THR | 2.3 | $\alpha$-thrombin; complex with hirullin |
| | 1THS | 2.2 | $\alpha$-thrombin; complex with MDL-28050 |
| | 1TMB | 2.3 | $\alpha$-thrombin; complex with hirugen and cyclotheonamide A |
| | 1TMT | 2.2 | $\alpha$-thrombin; complex with CGP 50,856 |
| | 1TMU | 2.5 | $\alpha$-thrombin; complex with hirudin fragment and D-Phe-Pro-Arg-chloromethane |
| | 2HAT | 2.5 | $\alpha$-thrombin; complex with hirugen and fibrinopeptide A mimic |
| | 2HGT | 2.2 | $\alpha$-thrombin; complex with hirulog 1 |
| | 2HNT | 2.5 | $\gamma$-thrombin |
| | 2HPP | 3.3 | $\alpha$-thrombin; complex with D-Phe-Pro-Arg-chloromethane and bovine prothrombin |
| | 2HPQ | 3.3 | $\alpha$-thrombin; complex with D-Phe-Pro-Arg-chloromethane and bovine prothrombin |
| | 3HAT | 2.5 | $\alpha$-thrombin; complex with hirugen and fibrinopeptide A mimic |
| | 3HTC | 2.3 | $\alpha$-thrombin; complex with recombinant hirudin variant 2 |
| | 4HTC | 2.3 | $\alpha$-thrombin; complex with recombinant hirudin variant 2 |

[a]See Chapter 2 for an image constructed from these coordinates.

## Name and History

The name ***thrombin*** derives from 'thrombus', since thrombin is the protease capable of producing a thrombus or blood clot, and is the last protease in the coagulation cascade. Thrombin is derived from its proenzyme or zymogen form (prothrombin or coagulation factor II) through cleavage by factor Xa (Chapter 54) in the presence of the cofactor Va, $Ca^{2+}$ and a phospholipid surface (Krishnaswamy *et al*., 1993). Thrombin is also known in the blood coagulation nomenclature as ***factor IIa***. Its other alternative name, ***fibrinogenase***, also derives from its activity in the coagulation cascade; thrombin cleaves fibrinogen to yield fibrin monomers which polymerize to form the basis of the blood clot. Prothrombin and thrombin were the subject of many protein chemical studies in the period 1960–1980 (Lundblad *et al*., 1976; Mann, 1976). In the early 1980s, the nucleotide sequence for prothrombin was determined (Degen *et al*., 1983). In 1989, the tertiary structure of thrombin was elucidated (Bode *et al*., 1989) and this led to a number of groups focusing on thrombin as a target for drug development (Tapparelli *et al*., 1993). Many studies in the period 1970–1990 demonstrated that thrombin had diverse effects on cells (Chen & Buchanan, 1975; Fenton,

1981). The molecular basis of many of these effects was elucidated in 1991 with the identification of a proteolytically activated receptor for thrombin (Vu *et al.*, 1991a).

## Activity and Specificity

The P1 residue of substrates and inhibitors of thrombin is almost always arginine, and the specificity is generally trypsin-like, but much more restricted. The restriction of specificity results from (a) partial occlusion of the active site, restricting access of macromolecular substrates, (b) a greater selectivity for P3, P2, P2' and P3' residues, and (c) the use of extended interaction areas, termed exosites, distant from the active site (Bode *et al.*, 1989, 1992).

The most abundant natural substrate of thrombin is fibrinogen. Thrombin triggers clot formation by releasing two oligopeptides (fibrinopeptides A and B) from the N-termini of the a and b chains of fibrinogen (Higgins *et al.*, 1983). The activity of thrombin is also important in stabilizing the fibrin clot; it activates factor XIII, which in turn cross-links the fibrin monomers (Lorand *et al.*, 1993). Thrombin amplifies its own production by activating essential cofactors in the coagulation cascade (factors V and VIII; Davie *et al.*, 1991) and by catalyzing platelet activation, which provides the surface necessary for the assembly of the protease–cofactor complexes of the cascade. However, not all of the activities of thrombin are procoagulant. When thrombin is bound to thrombomodulin at the surface of the endothelial cells, most of its procoagulant activities are inhibited, and it becomes an efficient activator of the protease proenzyme, protein C (Chapter 56), which in turn degrades factors Va and VIIa (Chapter 53) to shut down the coagulation cascade (Esmon, 1993).

Thrombin generated during blood coagulation is rapidly neutralized by the serpins antithrombin and heparin cofactor II (Olson & Björk, 1992). The most powerful inhibitors of thrombin, however, have been isolated from blood-sucking animals, including hirudin from the medicinal leech (Bagdy *et al.*, 1976), and rhodniin from the insect *Rhodnius prolixus* (Friedrich *et al.*, 1993).

The preferred sequence for interaction with the active site of thrombin has been established in a number of studies with synthetic substrates and inhibitors. Arg and Pro, respectively, are the preferred P1 and P2 residues. Phe or Val are the best L-amino acids in the P3 position, but a D-amino acid with an aromatic side chain is optimal (Kettner & Shaw, 1981; Lottenberg *et al.*, 1983). The aromatic side chain of the D-amino acid in P3 binds a site known as the aryl-binding site that would normally accommodate the P4 residue (Bode *et al.*, 1992). Phe is preferred in P2', and Arg or Lys in P3' (Le Bonniec *et al.*, 1996). Overall, the most favorable P3–P3' sequence for a thrombin cleavage site would be Val-Pro-Arg+Ser-Phe-Arg. Negatively charged residues hamper catalysis at any position between P3 and P3'; Thr in P2, and Pro in P1' also retard cleavage (Le Bonniec *et al.*, 1996).

The most important of the interaction areas outside the active site is the anion-binding exosite (also known as anion-binding exosite 1 and fibrinogen-recognition exosite). This is a positively charged groove on the surface of the molecule that binds negatively charged sequences in substrates such as fibrinogen (Binnie & Lord, 1993), in the thrombin receptor (Vu *et al.*, 1991b; Mathews *et al.*, 1994b), and in inhibitors such as hirudin (Stone *et al.*, 1987) and heparin cofactor II (Rogers *et al.*, 1992; Sheehan *et al.*, 1993). The cofactor thrombomodulin also binds to this site, and the interaction helps to overcome the unfavorable effects of the aspartates in the P3 and P3' positions of the substrate protein C (Le Bonniec & Esmon, 1991; Le Bonniec *et al.*, 1991; Esmon, 1993). Thrombomodulin also allosterically modulates to a small degree the activity of thrombin with synthetic tripeptidyl chromogenic substrates (Hofsteenge *et al.*, 1986). The binding of other ligands to the fibrinogen-recognition exosite has similar allosteric effects on the active site of thrombin (e.g. Dennis *et al.*, 1990; Hortin & Trimpe, 1991; Naski *et al.*, 1990). In addition, thrombin contains an exosite that binds heparin (also known as anion-binding exosite 2); heparin binds to this site and to one on antithrombin to accelerate greatly the rate of inhibition of thrombin by this serpin (Sheehan *et al.*, 1993; Gan *et al.*, 1994). Sodium ions bind to a site distinct from the two exosites and regulate the activity of thrombin on synthetic and natural substrates (e.g. Wells & Di Cera, 1992; De Cristofaro *et al.*, 1996). When $Na^+$ is bound, thrombin cleaves fibrinogen and many synthetic substrates more rapidly, and hence the $Na^+$–thrombin complex has been termed the 'fast' form (Wells & Di Cera, 1992).

The activity of thrombin can readily be assayed by use of commercially available chromogenic substrates (e.g. D-Phe-Pip-Arg+$NHPhNO_2$, S-2238, from Chromogenix or Tos-Gly-Pro-Arg+$NHPhNO_2$, Chromozym-TH, from Boehringer-Mannheim) (see Appendix 2 for full names and addresses of suppliers). Thrombin is most active at a pH of about 8 in the presence of at least 0.1 M NaCl. The chromogenic substrates are not strictly specific for thrombin, being hydrolyzed by other serine proteases also. Even with purified thrombin, attention should be given to the possibility of contamination by the degradation products, $\beta$- and $\gamma$-thrombins, that are produced by autolysis (Fenton *et al.*, 1977). The relative amounts of $\alpha$-, $\beta$- and $\gamma$-derivatives can be determined by PAGE, and contaminating $\beta$- and $\gamma$-thrombin lead to lower fibrinogen-clotting activity (Fenton *et al.*, 1977). The fibrinogen-recognition exosites of the $\beta$- and $\gamma$-derivatives are nonfunctional, and these derivatives have been used in many studies to examine the importance of the fibrinogen-recognition exosite to the different activities of thrombin (e.g. Fenton, 1981; Hofsteenge *et al.*, 1988).

Thrombin has a tendency to precipitate and to adsorb to glass and plastic surfaces, especially at low enzyme concentrations or low ionic strength; these effects can be minimized by performing assays in the presence of poly(ethylene glycol) or bovine serum albumin (Lottenberg *et al.*, 1983).

## Structural Chemistry

Two numbering systems are used for the amino acid sequence of thrombin in the literature. In one, the residues are numbered sequentially from the N-terminus of the B chain. The second system is based on chymotrypsinogen numbering, with topographically equivalent residues in thrombin being given the same number, and insertions in thrombin with respect to the chymotrypsinogen sequence being denoted by letters (Bode *et al*., 1989). The chymotrypsinogen numbering system is used in all structural studies and for this reason is preferable.

Thrombin ($\alpha$-thrombin) consists of two chains and has an apparent molecular mass of about 36.5 kDa; its pI value is 7.5. The catalytic domain (B chain) is homologous to the catalytic domains of chymotrypsin and trypsin; the A chain has no known function. The A chain (36 residues in human thrombin) is linked by a single disulfide bond to the B chain of 259 amino acids. Within the B chain, six additional Cys residues are engaged in three internal disulfide bridges. The B chain contains the residues forming the charge stabilizing system (residues His43, Asp99 and Ser205 from the N-terminus of the B chain, or His57, Asp102 and Ser195 in the chymotrypsinogen numbering system). The B chain of thrombin is glycosylated on Asn60f.

Factor Xa cleaves prothrombin at two specific arginines to generate an active enzyme with two polypeptide chains (Mann, 1976; Krishnaswamy *et al*., 1993). A single cleavage of prothrombin at Arg330 (the activating cleavage site) results in the formation of a transient intermediate (meizothrombin) which catalyzes some, but not all, of thrombin's reactions (Krishnaswamy *et al*., 1993). Human thrombin also removes autocatalytically the N-terminal 13 residues of the A chain to yield the mature 36 residue chain.

Numerous crystallographic structures of thrombin have been determined. The initial structure of the D-Phe-Pro-Arg-CH$_2$–thrombin complex provided great insight into the structural basis for thrombin's specificity (Bode *et al*., 1989, 1992). The binding of the D-Phe-Pro-Arg-CH$_2$ group delineates the substrate-binding regions for residues on the N-terminal side of the scissile bond. Access to the active site is partially blocked by a loop containing nine residues that represents an insertion into the thrombin sequence compared to that of chymotrypsin. The structures of the thrombin exosites were also elucidated in these initial studies. The fibrinogen-recognition exosite and the heparin-binding site consist of positively charged surface regions. Electrostatic calculations demonstrate that positive fields project from these two exosites at the poles of the molecule, whereas the active site between the two poles has a negative potential (Bode *et al*., 1992). The nature of interactions with the fibrinogen-recognition exosite has been further refined in complexes with hirudin (Rydel *et al*., 1991), hirudin analogs (Skrzypczak-Jankun *et al*., 1991), and fragments of thrombomodulin (Mathews *et al*., 1994a,b) and the thrombin receptor (Mathews *et al*., 1994a,b). Active-site interactions have also been further explored in structures with substrate analogs (Martin *et al*., 1992; Stubbs *et al*., 1992) and synthetic inhibitors (e.g. Brandstetter *et al*., 1992; Chen *et al*., 1995).

## Preparation

Thrombin can be prepared from blood plasma. The preparation involves three main steps: isolation of prothrombin from plasma, activation to thrombin, and purification of the mature enzyme from the activation mixture. Most protocols for prothrombin purification use a combination of adsorption to a barium salt followed by anion-exchange chromatography (e.g. Miletich *et al*., 1980). These protocols are also suitable for the simultaneous purification of the other major plasma vitamin K-dependent coagulation factors (factor IX (Chapter 52), factor X (Chapter 54) and protein C (Chapter 56)). Prothrombin is relatively abundant in plasma (0.1 mg ml$^{-1}$ or 1.3 $\mu$M) and consequently prothrombin can be isolated in good yield. Activation of prothrombin is most conveniently achieved with a prothrombin activator from a snake venom (e.g. Walker *et al*., 1980) (Chapter 73). Thrombin can then be purified from other activation components by cation-exchange chromatography (Lundblad *et al*., 1976; Fenton *et al*., 1977).

Plasma-derived thrombin preparations are available from a number of manufacturers (Worthington, Sigma). The specific activities of commercial preparations are, however, quite variable, and the purity is not always satisfactory. Commercial thrombin preparations often contain the autolytic derivatives of thrombin that do not clot fibrinogen.

The prothrombin molecule has three domains in addition to the thrombin domain: a Gla domain and two kringle domains. The Gla domain derives its name from the fact that the first ten glutamates of the prothrombin sequence are post-translationally modified to $\gamma$-carboxyglutamate (Gla) residues (Stenflo *et al*., 1974). The nonthrombin domains have been deleted in several systems used for recombinant expression of prothrombin. Prothrombin or prethrombin 1 (a shorter proenzyme of thrombin lacking the Gla and first kringle domains) have been expressed in mammalian cells (Falkner *et al*., 1992; Le Bonniec *et al*., 1991), and insect cells (T. Myles & S.R. Stone, unpublished results). Prethrombin 2 (a single-chain proenzyme consisting of the A and B chains of thrombin without other proenzyme sequences) has been expressed in *E. coli*, successfully refolded, and activated to thrombin (DiBella *et al*., 1995). The preparation of molecular chimeras containing one fragment expressed in *E. coli* together with one from plasma-derived thrombin has also been reported (Gan *et al*., 1991).

## Biological Aspects

Prothrombin is found in every vertebrate, and cDNA sequences are available for the proenzymes from a wide variety of species (see the Databanks table). Prothrombin deficiency occurs very rarely, and few dysfunctional thrombins have been characterized (e.g. Henriksen & Mann, 1989; Miyata *et al*., 1992). It appears that complete prothrombin deficiency may be lethal. Thrombin plays a central role in both hemostasis and thrombosis; low activity leads to bleeding, whereas excess activity contributes to thrombus formation. The importance of the enzyme in thrombosis is evidenced by the fact that thrombin inhibitors have been effective in preventing thrombosis in both animal models and clinical trials (Harker *et al*., 1995).

Prothrombin is activated after initiation of the coagulation cascade at a site of vascular injury. Injury to the vessel exposes tissue factor which complexes with factor VII (Chapter 53) to form an active protease–cofactor complex. After initiation by this complex, the cascade proceeds through further steps involving the conversion of proenzymes to active proteases, and each of these steps involves the assembly of protease–cofactor complexes on the surface of platelets. Exposure of the subendothelial cell matrix at the site of vascular injury also leads to activation of platelets. Upon activation, the properties of the platelet surface membranes change to facilitate the assembly of protease–cofactor complexes (Davie *et al.*, 1991). Through the action of this cascade of proenzyme activation, there is a dramatic amplification of the signal; picomolar concentrations of tissue factor can theoretically lead to micromolar concentrations of thrombin (Lawson *et al.*, 1994). The amount of thrombin produced is regulated at the level of proenzyme activation rather than by transcriptional control of the prothrombin gene. Once it is formed, interactions at the thrombin exosites are important for the control of its activity. Binding of thrombomodulin to the fibrinogen-recognition exosite converts thrombin into an efficient activator of the anticoagulant protease protein C with consequent inhibition of thrombin production (Esmon, 1993). Binding of heparin to another exosite is required for efficient inactivation of thrombin by antithrombin (Olson & Björk, 1992). As noted above, the reaction of thrombin with many of its substrates also involves the fibrinogen-recognition exosite. Thus, thrombin can be designated an allosteric enzyme, since interactions at exosites modulate activity.

In addition to its activity on the soluble components of the coagulation cascade, thrombin also induces responses from a variety of cells. For example, thrombin causes platelets to aggregate, is mitogenic for a number of cell types, and modulates neurite growth (Grand *et al.*, 1996). Many of these responses are mediated through a G protein-coupled receptor that thrombin activates by a unique mechanism. Thrombin cleaves the extracellular domain of the receptor, and the newly created N-terminus acts as a tethered ligand for the receptor. Interactions with both the active site and the fibrinogen-recognition exosite are required for efficient activation of the receptor (Vu *et al.*, 1991a,b). Some cellular responses to thrombin (e.g. neutrophil and monocyte chemotaxis), do not seem to involve this proteolytically activated receptor, however, in that they are induced by inactive thrombin (Bar Shavit *et al.*, 1983).

### Distinguishing Features

Thrombin is the only known mammalian protease capable of clotting fibrinogen. It is also the only protease that is inhibited by hirudin (Stubbs & Bode, 1993). Polyclonal antibodies that recognize human prothrombin are commercially available (Dako).

### Further Reading

There is a vast literature on thrombin and it has been possible only to mention a small fraction of it. The article by Stubbs & Bode (1993) gives a good introduction to the relationships between the many activities of thrombin and its structure, and the review by Grand *et al.* (1996) outlines the cellular effects of thrombin.

### References

Bagdy, D., Barabas, E., Graf, L., Petersen, T.E. & Magnusson, S. (1976) Hirudin. *Methods Enzymol.* **45**, 669–678.

Bar Shavit, R., Kahn, A., Wilner, G.D. & Fenton, J.W. II (1983) Monocyte chemotaxis: stimulation by specific exosite region in thrombin. *Science* **220**, 728–731.

Binnie, C.G. & Lord, S.T. (1993) The fibrinogen sequences that interact with thrombin. *Blood* **81**, 3186–3192.

Bode, W., Mayr, I., Baumann, U., Huber, R., Stone, S.R. & Hofsteenge, J. (1989) The refined 1.9-Å crystal structure of human α-thrombin: interaction with D-Phe-Pro-Arg chloromethylketone and significance of the Tyr-Pro-Pro-Trp insertion segment. *EMBO J.* **8**, 3467–3475.

Bode, W., Turk, D. & Karshikov, A. (1992) The refined 1.9-Å X-ray crystal structure of D-Phe-Pro-Arg chloromethylketone-inhibited human α-thrombin: structure analysis, overall structure, electrostatic properties, detailed active-site geometry, and structure-function relationships. *Protein Sci.* **1**, 426–471.

Brandstetter, H., Turk, D., Hoeffken, H.W., Grosse, D., Sturzebecher, J., Martin, P.D., Edwards, B.F. & Bode, W. (1992) Refined 2.3 Å X-ray crystal structure of bovine thrombin complexes formed with the benzamidine and arginine-based thrombin inhibitors NAPAP, 4-TAPAP and MQPA. A starting point for improving antithrombotics. *J. Mol. Biol.* **226**, 1085–1099.

Chen, L.B. & Buchanan, J.M. (1975) Mitogenic activity of blood components. I. Thrombin and prothrombin. *Proc. Natl Acad. Sci. USA* **72**, 131–135.

Chen, Z., Li, Y., Mulichak, A.M., Lewis, S.D. & Shafer, J.A. (1995) Crystal structure of human alpha-thrombin complexed with hirugen and *p*-amidinophenylpyruvate at 1.6 Å resolution. *Arch. Biochem. Biophys.* **322**, 198–203.

Davie, E.W., Fujikawa, K. & Kisiel, W. (1991) The coagulation cascade: initiation, maintenance, and regulation. *Biochemistry* **30**, 10363–10370.

De Cristofaro, R., Picozzi, M., Morosetti, R. & Landolfi, R. (1996) Effect of sodium on the energetics of thrombin–thrombomodulin interaction and its relevance for protein C hydrolysis. *J. Mol. Biol.* **258**, 190–200.

Degen, S.J., MacGillivray, R.T. & Davie, E.W. (1983) Characterization of the complementary deoxyribonucleic acid and gene coding for human prothrombin. *Biochemistry* **22**, 2087–2097.

Dennis, S., Wallace, A., Hofsteenge, J. & Stone, S.R. (1990) Use of fragments of hirudin to investigate thrombin–hirudin interaction. *Eur. J. Biochem.* **188**, 61–66.

DiBella, E.E., Maurer, M.C. & Scheraga, H.A. (1995) Expression and folding of recombinant bovine prethrombin-2 and its activation to thrombin. *J. Biol. Chem.* **270**, 163–169.

Esmon, C.T. (1993) Molecular events that control the protein C anticoagulant pathway. *Thromb. Haemost.* **70**, 29–35.

Falkner, F.G., Turecek, P.L., MacGillivray, R.T., Bodemer, W., Scheiflinger, F., Kandels, S., Mitterer, A., Kistner, O., Barrett, N., Eibl, J. & Dorner, F. (1992) High level expression of active human prothrombin in a vaccinia virus expression system. *Thromb. Haemost.* **68**, 119–124.

Fenton, J.W. II (1981) Thrombin specificity. *Annals N.Y. Acad. Sci.* **370**, 468–495.

Fenton, J.W. II, Fasco, M.J. & Stackrow, A.B. (1977) Human thrombins. Production, evaluation, and properties of α-thrombin. *J. Biol. Chem.* **252**, 3587–3598.

Friedrich, T., Kroger, B., Bialojan, S., Lemaire, H.G., Hoffken, H.W., Reuschenbach, P., Otte, M. & Dodt, J. (1993) A Kazal-type inhibitor with thrombin specificity from *Rhodnius prolixus*. *J. Biol. Chem.* **268**, 16216–16222.

Gan, Z.R., Lewis, S.D., Stone, J.R. & Shafer, J.A. (1991) Reconstitution of catalytically competent human ζ-thrombin by combination of ζ-thrombin residues A1–36 and B1–148 and an *Escherichia coli* expressed polypeptide corresponding to ζ-thrombin residues B149–259. *Biochemistry* **30**, 11694–11699.

Gan, Z.R., Li, Y., Chen, Z., Lewis, S.D. & Shafer, J.A. (1994) Identification of basic amino acid residues in thrombin essential for heparin-catalyzed inactivation by antithrombin III. *J. Biol. Chem.* **269**, 1301–1305.

Grand, R.J., Turnell, A.S. & Grabham, P.W. (1996) Cellular consequences of thrombin-receptor activation. *Biochem. J.* **313**, 353–368.

Harker, L.A., Hanson, S.R. & Kelly, A.B. (1995) Antithrombotic benefits and hemorrhagic risks of direct thrombin antagonists. *Thromb. Haemost.* **74**, 464–472.

Henriksen, R.A. & Mann, K.G. (1989) Substitution of valine for glycine-558 in the congenital dysthrombin thrombin Quick II alters primary substrate specificity. *Biochemistry* **28**, 2078–2082.

Higgins, D.L., Lewis, S.D. & Shafer, J.A. (1983) Steady state kinetic parameters for the thrombin-catalyzed conversion of human fibrinogen to fibrin. *J. Biol. Chem.* **258**, 9276–9282.

Hofsteenge, J., Taguchi, H. & Stone, S.R. (1986) Effect of thrombomodulin on the kinetics of the interaction of thrombin with substrates and inhibitors. *Biochem. J.* **237**, 243–251.

Hofsteenge, J., Braun, P.J. & Stone, S.R. (1988) Enzymatic properties of proteolytic derivatives of human α-thrombin. *Biochemistry* **27**, 2144–2151.

Hortin, G.L. & Trimpe, B.L. (1991) Allosteric changes in thrombin's activity produced by peptides corresponding to segments of natural inhibitors and substrates. *J. Biol. Chem.* **266**, 6866–6871.

Kettner, C. & Shaw, E. (1981) Inactivation of trypsin-like enzymes with peptides of arginine chloromethyl ketone. *Methods Enzymol.* **80**, 826–842.

Krishnaswamy, S., Nesheim, M.E., Pryzdial, E.L. & Mann, K.G. (1993) Assembly of prothrombinase complex. *Methods Enzymol.* **222**, 260–280.

Lawson, J.H., Kalafatis, M., Stram, S. & Mann, K.G. (1994) A model for the tissue factor pathway to thrombin. I. An empirical study. *J. Biol. Chem.* **269**, 23357–23366.

Le Bonniec, B.F. & Esmon, C.T. (1991) Glu-192 → Gln substitution in thrombin mimics the catalytic switch induced by thrombomodulin. *Proc. Natl Acad. Sci. USA* **88**, 7371–7375.

Le Bonniec, B.F., MacGillivray, R.T. & Esmon, C.T. (1991) Thrombin Glu-39 restricts the P$_3'$ specificity to nonacidic residues. *J. Biol. Chem.* **266**, 13796–13803.

Le Bonniec, B.F., Myles, T., Johnson, T., Knight, C.G., Tapparelli, C. & Stone, S.R. (1996) Characterization of the P$_2'$ and P$_3'$ specificities of thrombin using fluorescence-quenched substrates and mapping of the subsites by mutagenesis. *Biochemistry* **35**, 7114–7122.

Lorand, L., Jeong, J.M., Radek, J.T. & Wilson, J. (1993) Human plasma factor XIII: subunit interactions and activation of zymogen. *Methods Enzymol.* **222**, 22–35.

Lottenberg, R., Hall, J.A., Blinder, M., Binder, E.P. & Jackson, C.M. (1983) The action of thrombin on peptide *p*-nitroanilide substrates. Substrate selectivity and examination of hydrolysis under different reaction conditions. *Biochim. Biophys. Acta* **742**, 539–557.

Lundblad, R.L., Kingdon, H.S. & Mann, K.G. (1976) Thrombin. *Methods Enzymol.* **45**, 156–176.

Mann, K.G. (1976) Prothrombin. *Methods Enzymol.* **45**, 123–156.

Martin, P.D., Robertson, W., Turk, D., Huber, R., Bode, W. & Edwards, B.F. (1992) The structure of residues 7–16 of the A α-chain of human fibrinogen bound to bovine thrombin at 2.3-Å resolution. *J. Biol. Chem.* **267**, 7911–7920.

Mathews, I.I., Padmanabhan, K.P., Tulinsky, A. & Sadler, J.E. (1994a) Structure of a nonadecapeptide of the fifth EGF domain of thrombomodulin complexed with thrombin. *Biochemistry* **33**, 13547–13552.

Mathews, I.I., Padmanabhan, K.P., Ganesh, V., Tulinsky, A., Ishii, M., Chen, J., Turck, C.W., Coughlin, S.R. & Fenton, J.W. II (1994b) Crystallographic structures of thrombin complexed with thrombin receptor peptides: existence of expected and novel binding modes. *Biochemistry* **33**, 3266–3279.

Miletich, J.P., Broze, G.J., Jr. & Majerus, P.W. (1980) The synthesis of sulfated dextran beads for isolation of human plasma coagulation factors II, IX, and X. *Anal. Biochem.* **105**, 304–310.

Miyata, T., Aruga, R., Umeyama, H., Bezeaud, A., Guillin, M.C. & Iwanaga, S. (1992) Prothrombin Salakta: substitution of glutamic acid-466 by alanine reduces the fibrinogen clotting activity and the esterase activity. *Biochemistry* **31**, 7457–7462.

Naski, M.C., Fenton, J.W. II, Maraganore, J.M., Olson, S.T. & Shafer, J.A. (1990) The COOH-terminal domain of hirudin. An exosite-directed competitive inhibitor of the action of alpha-thrombin on fibrinogen. *J. Biol. Chem.* **265**, 13484–13489.

Olson, S.T. & Björk, I. (1992). Regulation of thrombin by antithrombin and heparin cofactor II. In: *Thrombin. Structure and Function* (Berliner, L.J., ed.). New York: Plenum, pp. 159–217.

Rogers, S.J., Pratt, C.W., Whinna, H.C. & Church, F.C. (1992) Role of thrombin exosites in inhibition by heparin cofactor II. *J. Biol. Chem.* **267**, 3613–3617.

Rydel, T.J., Tulinsky, A., Bode, W. & Huber, R. (1991) Refined structure of the hirudin–thrombin complex. *J. Mol. Biol.* **221**, 583–601.

Sheehan, J.P., Wu, Q., Tollefsen, D.M. & Sadler, J.E. (1993) Mutagenesis of thrombin selectively modulates inhibition by serpins heparin cofactor II and antithrombin III. Interaction with the anion-binding exosite determines heparin cofactor II specificity. *J. Biol. Chem.* **268**, 3639–3645.

Skrzypczak-Jankun, E., Carperos, V.E., Ravichandran, K.G., Tulinsky, A., Westbrook, M. & Maraganore, J.M. (1991) Structure of the hirugen and hirulog 1 complexes of α-thrombin. *J. Mol. Biol.* **221**, 1379–1393.

Stenflo, J., Ferlund, P., Egan, W. & Roepstorff, P. (1974) Vitamin K dependent modifications of glutamic acid residues in prothrombin. *Proc. Acad. Natl Sci. USA* **71**, 2730–2733.

Stone, S.R., Braun, P.J. & Hofsteenge, J. (1987) Identification of regions of α-thrombin involved in its interaction with hirudin. *Biochemistry* **26**, 4617–4624.

Stubbs, M.T. & Bode, W. (1993) A player of many parts: the spotlight falls on thrombin's structure. *Thromb. Res.* **69**, 1–58.

Stubbs, M.T., Oschkinat, H., Mayr, I., Huber, R., Angliker, H., Stone, S.R. & Bode, W. (1992) The interaction of thrombin with fibrinogen. A structural basis for its specificity. *Eur. J. Biochem.* **206**, 187–195.

Tapparelli, C., Metternich, R., Ehrhardt, C. & Cook, N.S. (1993) Synthetic low-molecular weight thrombin inhibitors: molecular design and pharmacological profile. *Trends Pharmacol. Sci.* **14**, 366–376.

Vu, T.K., Hung, D.T., Wheaton, V.I. & Coughlin, S.R. (1991a) Molecular cloning of a functional thrombin receptor reveals a novel proteolytic mechanism of receptor activation. *Cell* **64**, 1057–1068.

Vu, T.K., Wheaton, V.I., Hung, D.T., Charo, I. & Coughlin, S.R. (1991b) Domains specifying thrombin-receptor interaction. *Nature* **353**, 674–677.

Walker, F.J., Owen, W.G. & Esmon, C.T. (1980) Characterization of the prothrombin activator from the venom of *Oxyuranus scutellatus scutellatus* (Taipan venom). *Biochemistry* **19**, 1020–1023.

Wells, C.M. & Di Cera, E. (1992) Thrombin is a Na$^+$-activated enzyme. *Biochemistry* **31**, 11721–11730.

***Stuart R. Stone***
*Department of Biochemistry and Molecular Biology,*
*Monash University,*
*Clayton, Victoria 3168, Australia*

***Bernard F. Le Bonniec***
*INSERM U428,*
*Faculté de Pharmacie, Université Paris V,*
*4 Av. de l'Observatoire, 75270 Paris Cedex 06, France*

# 56. Protein C

## Databanks

*Peptidase classification: clan SA, family S1, MEROPS ID: S01.218*
*NC-IUBMB enzyme classification: EC 3.4.21.69*
*Chemical Abstracts Service registry number: 42617-41-4*
*Databank codes:*

| Species | SwissProt | PIR | EMBL (cDNA) | EMBL (genomic) |
|---------|-----------|-----|-------------|----------------|
| *Bos taurus* | P00745 | A26250 | – | – |
| *Canis familiaris* | – | – | D43751 | – |
| *Capra hircus* | – | – | D43752 | – |
| *Equus caballus* | – | – | D43753 | – |
| *Felis catus* | – | – | D43750 | – |
| *Homo sapiens* | P04070 | A22331 | K02059 | M11228: complete gene<br>M12682: exon 1<br>M12683: exon 2<br>M12684: exon 3<br>M12685: exons 4–6<br>M12686: exon 7<br>M12687: exon 8<br>M12712: exon 9 |
| *Macaca mulatta* | – | – | D43754 | – |
| *Mus musculus* | P33587 | JX0210 | D10445 | – |
| *Rattus norvegicus* | P31394 | S18994 | X64336 | – |

Brookhaven Protein Data Bank three-dimensional structures:

| Species | ID | Resolution | Notes |
|---------|-----|-----------|-------|
| *Homo sapiens* | 1AUT | 2.8 | complex with inhibitor |
| | 1PCU | – | theoretical model |

## Name and History

In 1960, an activity inhibitory to blood coagulation was found in thrombin-treated prothrombin complex, and termed ***autoprothrombin IIA*** (Mammen *et al.*, 1960). It was originally believed to be derived from prothrombin, but in 1972 it was shown that the precursor protein of autoprothrombin IIA was distinct from prothrombin (Marcianik, 1972).

Autoprothombin IIA was shown to inhibit factor Xa-mediated activation of prothrombin, and to facilitate fibrinolysis. Hyde *et al.* (1974) demonstrated that rabbits generated anticoagulant activity *in vivo* in response to intravenous injection of thrombin, and suggested that the inhibitory activity was identical to autoprothrombin IIA. In 1976, Stenflo purified a new vitamin K-dependent protein from bovine plasma and named it *protein C*, since it was the third-eluted peak from an ion-exchange column. Protein C was shown to be the proenzyme of a serine protease, and to be activated by thrombin. *Activated protein C (APC)* was rapidly confirmed to be identical to autoprothrombin IIA (Seegers *et al.*, 1976). Human protein C was purified by Kisiel (1979).

## Activity and Specificity

Protein C is the proenzyme of a serine protease, and Kisiel (1979) showed that it can be converted to its active form by thrombin (Chapter 55). Thrombin alone is a relatively poor activator, but when the thrombin is bound to an endothelial membrane protein, thrombomodulin, it becomes very efficient in the activation of protein C (Esmon & Owen, 1981). Thrombomodulin modulates the thrombin function and upon binding to thrombomodulin, thrombin loses its procoagulant properties. The process of activation of protein C is influenced by $Ca^{2+}$, which decreases the rate of activation by thrombin alone, but increases the rate for the thrombin–thrombomodulin complex.

Activated protein C cleaves several peptide bonds in the heavy chain of factor VIIIa, destroying the factor VIIIa activity. These bonds are: Arg336+Met337, Arg562+Gly563 and Arg740+Ser741 (Fay *et al.*, 1991). Activated protein C also inactivates factor Va by cleaving the peptide bonds Arg306+Asn307, Arg506+Gly507 and Arg679+Lys680 (Kalafatis *et al.*, 1994). The optimum pH is 7.5 for inactivation of factor Va and 6.5 for that of factor VIIIa.

Protein C can be activated by thrombin alone, by the thrombin–thrombomodulin complex, or by Protac, an activator from the venom of *Agkistrodon contortrix contortrix* (Kisiel *et al.*, 1976; Kisiel, 1979; Esmon & Owen, 1981; Klein & Walker, 1986). The function of activated protein C can be measured through the anticoagulant effect of the protein in a clotting assay, i.e. the 'activated partial thromboplastin time' reaction. In this assay, activated protein C inhibits the generation of thrombin by cleaving and inactivating factor Va and factor VIIIa in the plasma (Francis & Patch, 1983).

In the laboratory, activated protein C can be quantified by its hydrolysis of synthetic, nitroanilide substrates. These include D-Lys($\gamma$-Z)-Pro-Arg-NHPhNO$_2$ (Spectrozyme PCa; Francis & Seyfert, 1987), Glu-Pro-Arg-NHPhNO$_2$ (S-2366; Sala *et al.*, 1984) and D-Phe-Pip-Arg-NHPhNO$_2$ (S-2238; Kisiel, 1979). The release of nitroaniline is followed at 405 nm. Immunological assays are also used to measure the concentration of protein C (Comp *et al.*, 1984).

Activated protein C has three plasma inhibitors: the protein C inhibitor (PCI, also called plasminogen activator inhibitor 3, PAI-3), $\alpha_1$-proteinase inhibitor and $\alpha_2$-macroglobulin (Dahlbäck & Stenflo, 1994; Esmon, 1995).

## Structural Chemistry

The primary structures of the heavy and light chains of cattle protein C were determined by amino acid sequencing (Fernlund & Stenflo, 1982; Stenflo & Fernlund, 1982). Protein C cDNA has been isolated and sequenced from human, cattle, mouse and rat (Long *et al.*, 1984; Beckmann *et al.*, 1985; Okafuji *et al.*, 1992; Tada *et al.*, 1992). The human protein C

precursor contains 461 amino acid residues including a 42 residue prepropeptide. The mature plasma protein C (62 kDa) contains a light chain (21 kDa, 155 amino acids) and a heavy chain (41 kDa, 262 amino acids) held together by a disulfide bond. The light chain of protein C consists of several separate domains including a $\gamma$-carboxyglutamic acid-containing (Gla) domain and two epidermal growth factor homologous (EGF-like) domains, and the heavy chain forms the serine protease domain. Protein C is converted to activated protein C by the thrombin cleavage, which releases 12 amino acids from the N-terminus of the heavy chain (Kisiel, 1979; Stenflo, 1988). The three-dimensional structure of protein C has been determined with X-ray crystallography (Mather *et al.*, 1996).

## Preparation

Protein C can be isolated from citrated blood plasma by conventional methods including barium citrate adsorption and elution, ammonium sulfate precipitation, ion-exchange chromatography, dextran sulfate agarose chromatography and preparative electrophoresis (Stenflo, 1976; Kisiel, 1979; Suzuki *et al.*, 1983). Monoclonal antibodies have allowed simple purification by immunoaffinity chromatography (Laurell *et al.*, 1985; Stearns *et al.*, 1988). Recombinant protein C has been produced in cell lines (Grinnell *et al.*, 1987; Yan *et al.*, 1990) and also in transgenic animals (Velander *et al.*, 1992).

## Biological Aspects

The protein C gene is located on chromosome 2, position q14–21, it spans about 11.2 kb and the gene is made up of nine exons and eight introns (Plutzky *et al.*, 1986). Protein C is mainly synthesized in the liver as a single-chain protein containing a signal peptide and a prepropeptide (Long *et al.*, 1984; Beckmann *et al.*, 1985). It is also synthesized in the male reproductive tract in cells such as the Leydig cells of the testis, in the excretory epithelium of epididymis, and in some of the epithelial cells of the prostate glands (He *et al.*, 1995). After synthesis and prior to secretion, it undergoes proteolytic modification during which 42 amino acids constituting the signal peptide and the propeptide are removed. In addition, an endoproteolytic internal cleavage at Arg157, followed by carboxypeptidase-mediated removal of two amino acids, Lys156 and Arg157, converts the single-chain precursor into a two-chain molecule composed of a light chain and a heavy chain (Stenflo, 1988). The concentration of human plasma protein C is about 3–5 mg liter$^{-1}$ and the half-life is 6–8 h (Kisiel, 1979).

Protein C is converted to the active form by the thrombin–thrombomodulin complex on the surface of endothelial cells. Upon cleavage, a 12 amino acid peptide is released from the N-terminus of the heavy chain (Esmon & Owen, 1981). Activated protein C regulates the coagulation pathway by inhibiting the generation of thrombin through its selective proteolytic inactivation of coagulant cofactors factor Va and VIIIa (Dahlbäck & Stenflo, 1994). The anticoagulant activity of activated protein C is potentiated by its cofactors, protein S and intact factor V (Walker, 1981; Dahlbäck & Hildebrand, 1994; Shen & Dahlbäck, 1994). A receptor for protein C/activated protein C has recently been reported to be located on the surface of endothelial cells. It is suggested that the protein C receptor plays a role in regulation of the function of activated protein C in the inflammatory response (Esmon & Fukudome, 1995).

Inherited deficiency of protein C is associated with thromboembolic disease. The homozygous deficiency of protein C (less than 5% of normal plasma level) results in a severe thrombosis in the microcirculation in the neonatal period and if untreated, patients often die of a condition termed neonatal purpura fulminans. Heterozygous protein C deficiency increases the risk factor for venous thrombosis (Dahlbäck & Stenflo, 1994). Activated protein C resistance, a recently discovered major cause of familial thrombophilia is associated with a single-point mutation in the factor V gene, Arg506 is mutated to Gln. Arg506 forms one of the three cleavage sites for activated protein C, and the rate of inactivation of mutated factor Va is decreased compared to that of normal factor Va. Thus, the mutation leads to an increase in the activity of the coagulation system due to increased thrombin generation. As a result, patients with activated protein C resistance have a hypercoagulant state and increased risk of thrombosis. The Arg506Gln mutated factor V (also known as factor V Leiden) is confined to the Caucasian population and prevalence of 2–15% has been reported, the higher numbers representing the Scandinavian population (Bertina *et al*., 1994; Zöller & Dahlbäck, 1994; Dahlbäck, 1995; Rees *et al*., 1995).

## Distinguishing Features

Protein C is a vitamin K-dependent proenzyme of a serine protease composed of multiple domains. The amino acid sequence and domain structure are similar to those of the other vitamin K-dependent serine protease precursors, factor VII (identity 41%), factor IX (identity 38%) and factor X (identity 39%) (see Chapters 52, 53 and 54). Factor VII is a single-chain glycoprotein in plasma, of 50 kDa. Factor IX is a single-chain glycoprotein of 55 kDa. Factor X is a two-chain protein of 58 kDa containing light chain (16.2 kDa) and heavy chain (42 kDa). Under reducing conditions, protein C is easily distinguished from factor VII and IX in SDS-PAGE. Monoclonal and polyclonal antibodies against protein C distinguish protein C from factor X and the other vitamin K-dependent proteins. Rabbit polyclonal antibody against human protein C is available from Dako (see Appendix 2 for full names and addresses of suppliers), and monoclonal antibodies may be obtained from several commercial sources.

## Further Reading

For reviews, see Dahlbäck (1995) and Esmon & Fukudome (1995).

## References

Beckmann, R.J., Schmidt, R.J., Santerre, R.F., Plutzky, J., Crabtree, G.R. & Long, G.L. (1985) The structure and evolution of a 461 amino acid human protein C precursor and its messenger RNA, based upon the DNA sequence of cloned human liver cDNAs. *Nucleic Acids Res.* **13**, 5233–5247.

Bertina, R.M., Koeleman, B.P.C., Koster, T., Rosendaal, F.R., Dirven, R.J., de Ronde, H., van der Velden, P.A. & Reitsma, P.H. (1994) Mutation in blood coagulation factor V associated with resistance to activated protein C. *Nature* **369**, 64–67.

Comp, P.C., Nixon, R.R. & Esmon, C.T. (1984) Determination of functional levels of protein C, an antithrombotic protein, using thrombin–thrombomodulin complex. *Blood* **63**, 15–21.

Dahlbäck, B. (1995) The protein C anticoagulant system: inherited defects as basis for venous thrombosis. *Thromb. Res.* **77**, 1–43.

Dahlbäck, B. & Hildebrand, B. (1994) Inherited resistance to activated protein C is corrected by anticoagulant cofactor activity found to be a property of factor V. *Proc. Natl Acad. Sci. USA* **91**, 1396–1400.

Dahlbäck, B. & Stenflo, J. (1994) The protein C anticoagulant system. In: *The Molecular Basis of Blood Diseases* (Stamatoyannopoulos, G., Nienhuis, A.W., Majerus, P.W. & Varmus, H., eds). Philadelphia: Saunders, pp. 599–627.

Esmon, C.T. (1995) Thrombomodulin as a model of molecular mechanisms that modulate protease specificity and function at the vessel surface. *FASEB. J.* **9**, 946–955.

Esmon, C.T. & Fukudome, K. (1995) Cellular regulation of the protein C pathway. *Semin. Cell Biol.* **6**: 259–268.

Esmon, C.T. & Owen, W.G. (1981) Identification of an endothelial cell cofactor for thrombin-catalyzed activation of protein C. *Proc. Natl Acad. Sci. USA* **78**, 2249–2252.

Fay, P.J., Smudzin, T.M. & Walker, F.J. (1991) Activated protein C-catalyzed inactivation of factor VIII and VIIIa. Identification of cleavage sites and correlation of proteolysis with cofactor activity. *J. Biol. Chem.* **266**, 20139–20145.

Fernlund, P. & Stenflo, J. (1982) Amino acid sequence of the light chain of bovine protein C. *J. Biol. Chem.* **257**, 12170–12179.

Francis, R.B & Patch, M.J. (1983) A functional assay for protein C in human plasma. *Thromb. Res.* **32**, 605–613.

Francis, R.B. & Seyfert, U. (1987) Rapid amidolytic assay of protein C in whole plasma using an activator from the venom of *Agkistrodon contortrix*. *Am. J. Clin. Pathol.* **87**, 619–625.

Grinnell, B.W., Berg, D.T., Walls, J. & Yan, S.B. (1987) Transactivated expression of full gamma-carboxylated recombinant human protein C, an antithrombotic factor. *Bio/Technology* **5**, 1189–1192.

He, X., Shen, L., Bjartell, A., Malm, J., Lilia, H. & Dahlbäck, B. (1995) The gene encoding vitamin K-dependent anticoagulant protein C is expressed in human male reproductive tissues. *J. Histochem. Cytochem.* **43**, 563–570.

Hyde, E., Wetmore, R. & Gurewich, V. (1974) Isolation and characterization of an in vivo thrombin-induced anticoagulant activity. *Scand. J. Haematol.* **13**, 121–128.

Kalafatis, M., Matthew, D.R. & Mann, K.G. (1994) The mechanism of inactivation of human factor V and human factor Va by activated protein C. *J. Biol. Chem.* **269**, 31869–31880.

Kisiel, W. (1979) Human plasma protein C. Isolation, characterization and mechanism of activation by α-thrombin. *J. Clin. Invest.* **64**, 761–769.

Kisiel, W., Ericsson, L.H. & Davie, E.W. (1976) Proteolytic activation of protein C from bovine plasma. *Biochemistry* **15**, 4893–4900.

Klein, J.D. & Walker, F.J. (1986) Purification of a protein C activator from the venom of the southern copperhead snake (*Agkistrodon contortrix contortrix*). *Biochemistry* **15**, 4175–4179.

Laurell, M., Ikeda, K., Lindgren, S. & Stenflo, J. (1985) Characterization of monoclonal antibodies against human protein C specific for the calcium ion-induced conformation or for the activation peptide region. *FEBS Lett.* **191**, 75–81.

Long, G.L., Belagaje, R.M. & MacGillivray, R.T. (1984) Cloning and sequencing of liver cDNA coding for bovine protein C. *Proc. Natl Acad. Sci. USA* **81**, 5653–5656.

Mammen, E.F., Thomas, W.R. & Seegers, W.H. (1960) Activation of purified prothrombin to autoprothrombin II (platelet cofactor II or autoprothrombin IIA). *Thromb. Diath. Haemorrhag.* **5**, 218–249.

Marcianik, E. (1972) Inhibitor of human blood coagulation elicited by thrombin. *J. Lab. Clin. Med.* **79**, 924–934.

Mather, T., Oganessyan, V., Hof, P., Huber, R., Foundling, S., Esmon, C.T. & Bode, W. (1996) The 2.8 Å crystal structure of Gla-domainless activated protein C. *EMBO J.* **15**, 6822–6831.

Okafuji, T., Maekawa, K., Nawa, K. & Marumoto, Y. (1992) The cDNA cloning and mRNA expression of rat protein C. *Biochim. Biophys. Acta* **1131**, 329–332.

Plutzky, J., Hoskins, J.A., Long, J.L. & Crabtree, G.R. (1986) Evolution and organization of the human protein C gene. *Proc. Natl Acad. Sci. USA* **83**, 546–550.

Rees, D.C., Cox, M. & Clegg, J.B. (1995) World distribution of factor V Leiden. *Lancet* **346**, 1133–1134.

Sala, N., Owen, W.G. & Collen, D. (1984) A functional assay of protein C in human plasma. *Blood* **63**, 671–675.

Seegers, W.H., Novoa, E., Henry, R.L. & Hassouna, H.I. (1976) Relationship of 'new' vitamin K-dependent protein C and 'old' autoprothrombin II-a. *Thromb. Res.* **8**, 543–552.

Shen, L. & Dahlbäck, B. (1994) Factor V and protein S as synergistic cofactors to activated protein C in degradation of factor VIIIa. *J. Biol. Chem.* **269**, 18735–18738.

Stearns, D.J, Kurosawa, S, Sims, P.J, Esmon, N.L. & Esmon, C.T. (1988) The interaction of a $Ca^{2+}$-dependent monoclonal antibody with the protein C activation peptide region. Evidence for obligatory $Ca^{2+}$ binding to both antigen and antibody. *J. Biol. Chem.* **263**, 826–832.

Stenflo, J. (1976) A new vitamin K-dependent protein. *J. Biol. Chem.* **251**, 355–363

Stenflo, J. (1988) The biochemistry of protein C. In: *Protein C and Related Proteins* (Bertina, R.M., ed.). Edinburgh: Churchill Livingstone, pp. 21–54.

Stenflo, J. & Fernlund, P. (1982) Amino acid sequence of the heavy chain of bovine protein C. *J. Biol. Chem.* **257**, 12180–12190.

Suzuki, K., Stenflo, J., Dahlbäck, B. & Teodorsson, B. (1983) Inactivation of human coagulation factor V by activated protein C. *J. Biol. Chem.* **258**, 1914–1920.

Tada, N., Sato, M., Tsujimura, A., Iwase, R. & Hashimoto, G.T. (1992) Isolation and characterization of a mouse protein C cDNA. *J. Biochem. (Tokyo)* **111**, 491–495.

Velander, W.H., Johnson, J.L., Page, R.L., Russell, C.G., Subramanian, A. & Wilkins, T.D. (1992) High-level expression of a heterologous protein in the milk of transgenic swine using the cDNA encoding human protein C. *Proc. Natl Acad. Sci. USA* **89**, 12003–12007.

Walker, F.J. (1981) Regulation of bovine activated protein C by protein S. *Thromb. Res.* **22**, 321–327.

Yan, S.B., Razzno, P., Chao, Y.B., Walls, J.D., Berg, D.T., McClure, D.B. & Grinnell, B.W. (1990) Characterization and novel purification of recombinant human protein C from three mammalian cell lines. *Bio/Technology* **8**, 655–661.

Zöller, B. & Dahlbäck, B. (1994) Linkage between inherited resistance to activated protein C and factor V gene mutation in venous thrombosis. *Lancet* **343**, 1536–1538.

*Lei Shen*
*Department of Clinical Chemistry,*
*University Hospital, Malmö, Lund University,*
*S 205 02, Malmö, Sweden*

*Björn Dahlbäck*
*Department of Clinical Chemistry,*
*University Hospital, Malmö, Lund University,*
*S 205 02, Malmö, Sweden*
*Email: bjorn.dahlback@klkemi.mas.lu.se*

# 57. *u-Plasminogen activator*

## Databanks

*Peptidase classification: clan SA, family S1, MEROPS ID: S01.231*
*NC-IUBMB enzyme classification: EC 3.4.21.73*
*Chemical Abstracts Service registry number: 9039-53-6*
*Databank codes:*

| Species | SwissProt | PIR | EMBL (cDNA) | EMBL (genomic) |
|---|---|---|---|---|
| Bos taurus | Q05589 | JN0560 | L03546 X85801 | – |
| Gallus gallus | P15120 | A35005 | J05187 | J05188: complete gene |

*continued overleaf*

| Species | SwissProt | PIR | EMBL (cDNA) | EMBL (genomic) |
|---|---|---|---|---|
| *Homo sapiens* | P00749 | A00931 | A18397 | K02286: 3′ end |
| | | A35689 | A21571 | K03027: exons 6 and 7 |
| | | A37561 | D00244 | X02419: exons 1–9 |
| | | A37562 | D11143 | X12641: promoter |
| | | A37563 | K03226 | |
| | | A37564 | M15476 | |
| | | JT0102 | | |
| *Mus musculus* | P06869 | A24615 | X02389 | M17922: complete gene |
| | | A29420 | X52971 | |
| *Papio cynocephalus* | P16227 | S08651 | X51935 | – |
| | | S14687 | | |
| *Rattus norvegicus* | P29598 | I60186 | X63434 | X66907: exon 4 |
| | | S18932 | X65651 | |
| | | S24604 | | |
| *Sus scrofa* | P04185 | A00932 | X02724 | L27481: 5′ flank |
| | | | X51428 | X01648: complete gene |
| | | | | X74381: unannotated |

Brookhaven Protein Data Bank three-dimensional structures:

| Species | ID | Resolution | Notes |
|---|---|---|---|
| *Homo sapiens* | 1KDU | – | NMR; kringle domain |
| | 1LMW | 2.5 | peptidase domain; complex with Glu-Gly-Arg-chloromethane |
| | 1URK | – | NMR; N-terminal fragment |

## Name and History

The fibrinolytic activity of human urine, known since the end of the nineteenth century, was shown to be due to the presence of a plasminogen activator in the early 1950s (Williams, 1951). The term **urokinase** was first used by Sobel *et al.* (1952), although the enzyme is now more frequently termed **urinary** or **urokinase-type plasminogen activator (uPA)**, and NC-IUBMB has recommended **u-plasminogen activator**. The enzyme was first purified to homogeneity from urine as a protein of 54 kDa, although an enzymatically active form of 31 kDa was also detected which later proved to be a proteolytic degradation product (Gunzler *et al.*, 1982) and is termed LMW uPA. The development of anti-uPA antibodies led to the detection of the enzyme in human blood plasma, its purification to homogeneity from this source and its discrimination from tissue-type t-plasminogen activator (tPA) (Chapter 58) (Wun *et al.*, 1982). The single-chain proenzyme form of uPA was first purified from the conditioned medium of mouse sarcoma virus-transformed 3T3 cells (Skriver *et al.*, 1982) and later from the conditioned medium of human tumor cell lines (Nielsen *et al.*, 1982). The proenzyme, which could be activated by plasmin, was shown not to incorporate DFP and to be unreactive with plasminogen and synthetic substrates. This form of uPA is variously termed pro-uPA, prourokinase or scu-PA (single-chain urokinase-like plasminogen activator). The latter terminology, together with tc-uPA (two-chain uPA) for the activated enzyme, was adopted by the International Committee on Thrombosis and Haemostasis in 1986 on the premise that the proenzyme possessed high intrinsic catalytic activity.

## Activity and Specificity

uPA has an extremely limited substrate specificity, cleaving the sequence Cys-Pro-Gly-Arg560+Val561-Val-Gly-Gly-Cys that constitutes a small disulfide-bridged loop in plasminogen. Synthetic peptide substrates based on this sequence are available commercially both as chromogenic substrates, e.g. Glp-Gly-Arg+NHPhNO$_2$ (Chromogenix) and fluorogenic substrates, e.g. Glt-Gly-Arg+NHMec (Bachem) (see Appendix 2 for full names and addresses of suppliers). These substrates cannot discriminate between the activities of uPA and tPA, however. Peptides with Lys at P1 are poor substrates for uPA, although it can hydrolyze various esters, e.g. Z-Lys-SBzl (Lottenberg *et al.*, 1981).

uPA can be assayed fibrinolytically and quantified against standardized reference preparations with a specific activity of approximately 100 000 IU mg$^{-1}$ (Gaffney & Heath, 1990). These units have replaced the previously used Ploug and CTA units, which are approximately equivalent to 0.96 and 0.67 IU, respectively. It can also be assayed in two-stage plasminogen-activation assays with plasmin-specific peptide substrates, or directly by use of the specific peptide substrates described above (Lottenberg *et al.*, 1981) (Table 57.1). The plasminogen-activating activity of uPA can also be assayed when it is bound to its specific cellular receptor (Ellis *et al.*, 1993). uPA has a pH optimum of 8.4. Assays of plasminogen activation are strongly affected by Cl$^-$ and other anions (I$^-$ > SCN$^-$ > Cl$^-$ > IO$_3^-$ > HCOO$^-$ > F$^-$ > OAc$^-$), and lysine analogs due to conformational effects on plasminogen. The effects are reciprocal, with Cl$^-$ acting as a negative regulator ($K_i$ 10 mM) and

*Table 57.1*   Kinetic constants for action of uPA on plasminogen and synthetic substrates

| Substrate | Effector | $K_m$ ($\mu$M) | $k_{cat}$ (s$^{-1}$) | $k_{cat}/K_m$ ($\mu$M$^{-1}$s$^{-1}$) |
|---|---|---|---|---|
| Plasminogen | – | 20 | 1.80 | 0.090[a] |
| | uPA receptor (soluble) | 20 | 1.53 | 0.077[b] |
| | uPA receptor (cell surface) | 0.11 | 0.22 | 2.23[a] |
| | Monoclonal antibody | 0.10 | 0.45 | 4.32[c] |
| Glp-Gly-Arg-NHPhNO$_2$ | – | 55 | 20 | 0.32[d] |
| Z-Lys-SBzl | – | 30 | 60 | 2.0[d] |

[a]Ellis *et al*. (1991); [b]Ellis (1996); [c]Ellis & Danø (1993b); [d]Lottenberg *et al*. (1981).

*Table 57.2*   Kinetic constants for the inhibition of uPA[a]

| | PAI-1[b] | PAI-2[b] | PAI-3 | PN-1 | Amiloride[c] |
|---|---|---|---|---|---|
| uPA | $7.9 \times 10^6$ | $5.3 \times 10^5$ | $8 \times 10^3$ | $2 \times 10^5$ | $7 \times 10^{-6}$ M |
| uPA-uPA receptor (cell surface) | $4.5 \times 10^6$ | $3.3 \times 10^5$ | – | – | – |

[a]Second-order association constant, M$^{-1}$ s$^{-1}$; [b]Data from Ellis *et al*. (1990); [c]Competitive inhibitor, $K_i$.

6-aminohexanoic acid as a positive regulator ($K_a$ 5 mM) (Urano *et al*., 1987). In purified systems it is essential to include a detergent such as Tween 80 (0.01%) to prevent losses by adsorption.

Physiological inhibitors of uPA are limited to the specific plasminogen activator inhibitors types 1 and 2 (PAI-1 and PAI-2), protease nexin (PN-1), protein C inhibitor (PAI-3) and to some extent, $\alpha_2$-macroglobulin (Table 57.2). uPA is poorly inhibited by most other naturally occurring inhibitors. The diuretic drug amiloride (3,5-diamino-$N$-(aminoiminomethyl)-6-chloropyrazinecarboxamide) has been widely used as a specific competitive inhibitor of uPA as it does not inhibit tPA appreciably (Vassalli & Belin, 1987). Other selective synthetic inhibitors include 4-substituted benzo[*b*]thiophene-2-carboxamidines, the most potent of which has a $K_i$ of 0.16 $\mu$M (Towle *et al*., 1993). uPA is slowly inactivated by covalent inhibitors such as DFP and Glu-Gly-Arg-CH$_2$Cl.

## Structural Chemistry

uPA is a mosaic protein with a modular structure similar to other serine proteases of the coagulation and fibrinolytic systems, consisting of an epidermal growth factor (EGF)-like, a kringle and a serine protease domain. The kringle and the serine protease domains are separated by an unusually long (16 residue) linker region. The proenzyme comprises 411 amino acid residues after removal of a 20 residue signal peptide. uPA is post-translationally modified by glycosylation at Asn302 and fucosylation at Thr18, with no

other modifications detected by mass spectrometry (Buko *et al*., 1991). Phosphorylation of unidentified Ser, Thr and Tyr residues has been detected (Barlati *et al*., 1995). Proenzyme activation occurs at Lys158┼Ile159, giving rise to two polypeptides bridged by Cys148-Cys279. In commercial preparations purified from urine, the C-terminus of the A chain is often degraded to Phe157 or Arg156. uPA has a pI of 8.6.

The EGF-like module of uPA is the sole determinant of uPA binding to its specific cellular receptor, uPA-receptor (Appella *et al*., 1987). It has a structure closely resembling those of transforming growth factor $\beta$ (TGF$\beta$) and EGF itself (Hansen *et al*., 1994). The most notable difference in the structures is a seven residue $\omega$ loop (Asn22-Ile28) connecting the two $\beta$ strands in the EGF-like module of uPA compared to the type I $\beta$ bend in the others. Tyr24 within this loop has been unequivocally demonstrated to be involved in the interaction with uPA receptor (Ploug *et al*., 1995). The kringle module of uPA does not contain a lysine-binding site, as present in tPA and plasminogen kringles. Part of the kringle anionic center is replaced by Arg-Arg-Arg110, which results in a polyanion-binding site that can bind heparin with a $K_d$ of 85 $\mu$M (Stephens *et al*., 1992; Li *et al*., 1994). LMW-uPA is formed by cleavage of Glu143┼Leu144, which has been shown to be catalyzed by matrilysin (Chapter 396) in kidney cell cultures (Marcotte *et al*., 1992). The residual A chain is termed ATF (**a**mino-**t**erminal **f**ragment), although this is most often produced *in vitro* by 'autoproteolysis' (which may possibly be due to contaminating protease activity, rather than uPA itself) or plasmin cleavage of Lys135┼Lys136. 'Autoproteolysis' can also lead to cleavage of Lys23┼Tyr24 with loss of receptor-binding function.

The X-ray crystallographic structure of the catalytic domain of uPA in complex with Glu-Gly-Arg-CH$_2$Cl has been determined at 2.5 Å (Spraggon *et al*., 1995). The solution structure of ATF has been solved by heteronuclear NMR (Hansen *et al*., 1994). No interdomain constraints in uPA have been detected by NMR, demonstrating an exceptionally high degree of interdomain flexibility.

## Preparation

uPA in its activated two-chain form is prepared commercially from urine, e.g. Ukidan (Serono), or the conditioned medium of long-term kidney cell cultures as LMW-uPA, e.g. Abbokinase (Abbot Laboratories). Pro-uPA is usually purified from the conditioned media of the kidney cell line TCL-598, or neoplastic cells, e.g. CALU-3 (Stump *et al*., 1986) and HT-1080 (Petersen *et al*., 1988). The recombinant enzyme can be expressed in *E. coli* (Winkler & Blaber, 1986), yeast (Melnick *et al*., 1990) or mammalian cells such as CHO (Avgerinos *et al*., 1990). All these forms of the enzyme appear to be functionally equivalent. Recombinant pro-uPA is produced commercially as Saruplase (Grünenthal). Pro-uPA has been purified on fibrin-Celite in the presence of Zn$^{2+}$ (Husain, 1993), Zn$^{2+}$-chelate, *p*-aminobenzamidine-Sepharose (Skriver *et al*., 1982) and immunoaffinity matrices (Nielsen *et al*., 1982). *p*-Aminobenzamidine-Sepharose can also be used to specifically adsorb activated uPA from pro-uPA preparations (Melnick *et al*., 1990).

## Biological Aspects

### Genetics

The uPA gene is 6.4 kb in size, made up of 11 exons and 10 introns, located on chromosome 10 in humans, and the sequence identity between mammalian species is high (Degen *et al.*, 1987). Transcription of the gene produces a single mRNA form of 2.5 kb. uPA gene expression is upregulated by a wide variety of cytokines, growth factors and hormones in many cells in a cell type-specific manner.

### Substrates

Despite the highly restricted substrate specificity of the plasminogen activators, uPA has been demonstrated to hydrolyze a number of other proteins *in vitro* including fibronectin, fibrinogen, diphtheria toxin and possibly uPA itself. These are poor substrates for uPA and are probably of no biological significance. Interestingly uPA's own cellular receptor has been shown to be a relatively good substrate, specific cleavage (at a site resembling the activation sequence of cattle plasminogen) leading to its inactivation, which has been speculated to act as a regulatory mechanism *in vivo* (Høyer-Hansen *et al.*, 1992). *In vitro*, uPA can also activate the noncatalytic plasminogen homolog, hepatocyte growth factor (Naldini *et al.*, 1992).

### Proenzyme Activation

Proenzyme activation occurs by a single cleavage at Lys158+Ile159 catalyzed by plasmin (Chapter 59) (Gunzler *et al.*, 1982) and a variety of other trypsin-like serine proteases, including trypsin (Chapter 3), plasma kallikrein (Chapter 50) and cathepsin G (Chapter 16), and also thermolysin (Chapter 351) and cathepsin B (Chapter 209). In the presence of plasminogen, activation leads to a reciprocal proenzyme-activation system and a subsequent amplification of plasmin generation (Lijnen *et al.*, 1986; Ellis *et al.*, 1987). Arg156+Phe157 can be cleaved by thrombin (Chapter 55), leading to an inactive two-chain molecule that is very poorly activated by plasmin, but can be activated by dipeptidyl-peptidase I (Chapter 214) (Nauland & Rijken, 1994). Cleavage at Ile159+Ile160 by leukocyte elastase (Chapter 15) leads to an inactive two-chain molecule that cannot subsequently be activated.

### Proenzyme Activity

The sensitivity to plasmin activation means that the determination of the intrinsic proteolytic activity of pro-uPA is not a trivial problem, and this has led to much controversy over the level and potential biological significance of the activity. Early studies with natural pro-uPA suggested that it had little or no activity towards plasminogen, peptide substrates, plasminogen activator inhibitors or DFP. However Collen and coworkers (1986), using a recombinant protein expressed in *E. coli*, claimed that pro-uPA had a higher proteolytic activity than the activated enzyme and that this was due to a remarkably low $K_m$ for plasminogen, 0.2 μM compared to the 20–50 μM observed in most laboratories with the activated enzyme (Collen *et al.*, 1986; Ellis *et al.*, 1987; Christensen, 1988). In these experiments, a plasmin-specific

substrate was used at 4-fold $K_m$ in order to competitively suppress plasmin activity. The approximately 20% plasmin activity remaining under these conditions is enough however to initiate pro-uPA activation and interfere with the kinetic analysis, as initial rate conditions cannot be presumed to apply (Ellis, 1996). The low $K_m$, but not the high activity, was also reported for natural forms of the proenzyme (Stump *et al.*, 1986). These data have not been corroborated by other laboratories, using a variety of other strategies to suppress the effect of feedback activation, including detection of plasmin by fluorescein-labeled bovine pancreatic trypsin inhibitor (BPTI) and cleavage of [125]I-labeled plasminogen in the presence of BPTI and other inhibitors (Ellis *et al.*, 1987; Pannell & Gurewich, 1987; Petersen *et al.*, 1988; Husain, 1991). These studies all show that pro-uPA has a much lower activity than uPA, and with highly purified pro-uPA and plasminogen preparations (i.e. essentially intact proenzymes) the intrinsic activity of pro-uPA was less than 0.4% (Petersen *et al.*, 1988). Some workers have claimed that pro-uPA is completely inactive, but the validity of such a conclusion is circumscribed by the sensitivity and detection limit of the assay used. Experiments with nonplasmin activatable recombinant mutants with the P1 Lys158 replaced by Glu, Gly, Met or Val have confirmed the low activity of pro-uPA (Nelles *et al.*, 1987). With Lys158Glu mutant pro-uPA and Ser740Ala plasminogen, an intrinsic activity of 0.2% was obtained (Lijnen *et al.*, 1990). However some caution is needed in interpreting data obtained with these mutants, because of the possibility of conformational rearrangements affecting the active site and the potential for activation by other trace-contaminating proteases. Pro-uPA has been shown specifically to incorporate [3H]DFP in a manner consistent with a low intrinsic catalytic activity (Manchanda & Schwartz, 1991). PAI-1 has also been reported to inhibit pro-uPA, although with a mechanism differing from that for the activated protease (Manchanda & Schwartz, 1995; Ellis, 1996).

It has been suggested that the intrinsic activity of pro-uPA has a significant role in initiating plasminogen activation both in fibrinolysis by therapeutically administered pro-uPA and in physiological cell surface plasminogen activation (see below). Although there is no clear consensus over the absolute level of this intrinsic activity, an *in vitro* mechanistic model of the cell surface plasminogen activation has demonstrated that in theory any finite degree of pro-uPA intrinsic activity is sufficient to rapidly initiate efficient plasmin generation (Ellis & Danø, 1993b).

### Fibrinolysis

In contrast to t-plasminogen activator uPA does not bind to fibrin, due to the lack of Lys-binding function of its kringle module. However pro-uPA, but not uPA, is considered to have fibrin specificity *in vivo*, i.e. to promote plasmin cleavage of fibrin rather than fibrinogen. Native pro-uPA has an *in vitro* fibrinolytic potency 2000-fold higher than that of recombinant Lys158Glu pro-uPA, but only 2- to 5-fold lower than uPA, suggesting significant activation of pro-uPA at the fibrin surface (Lijnen *et al.*, 1989). In a rabbit model of jugular vein thrombosis, however, Lys158Glu and other nonplasmin-activatable mutants of pro-uPA were only 5-fold less potent than pro-uPA (Lijnen & Collen, 1991).

Although it is difficult to draw firm conclusions regarding mechanism from observations in such a complex milieu, these data tend to suggest that therapeutic levels of pro-uPA have a significant intrinsic plasminogen-activating potential *in vivo*. Fibrin fragment E-2 has been reported to specifically promote plasminogen activation by pro-uPA *in vitro* with little or no effect on uPA or tPA (Liu & Gurewich, 1992), and similar effects have been observed with immobilized fibrin (Fleury *et al.*, 1993). The mechanism underlying these effects is unknown, but possibly relates to specific effects on plasminogen conformation.

*uPA Receptor*

The outstanding biological feature of uPA is its high affinity binding ($K_d$ 0.5 nM) to a specific glycolipid-anchored cell surface receptor, uPA receptor, which has a major role in regulating the catalytic function of uPA (Ellis *et al.*, 1995). The binding is mediated solely by the EGF-like domain of uPA,

and all forms of uPA containing this domain, i.e. EGF-like domain, ATF, uPA and pro-uPA, bind equally well. Binding to cellular uPA receptor causes a large increase in plasmin generation through a large reduction in the apparent $K_m$ for plasminogen activation (see Table 57.1) (Ellis *et al.*, 1991) and an increased efficiency of the reciprocal activation of pro-uPA (Ellis *et al.*, 1989). By contrast, binding to soluble forms of uPA receptor (either recombinant truncated or solubilized wild-type protein) has no direct potentiating influence on the activities of either uPA or pro-uPA, but rather causes slight reductions in the rates of plasminogen activation and plasmin-catalyzed pro-uPA activation, presumably due to some degree of steric hindrance (Ellis, 1996; Ellis *et al.*, 1991). The differential effects of cell-associated versus soluble uPA receptor are accounted for by the concurrent cellular binding of plasminogen (Ellis *et al.*, 1989, 1991). This leads to formation of a kinetically favored activation complex (Ellis & Danø, 1993b) which may involve non-active site interactions between a kringle Lys-binding site

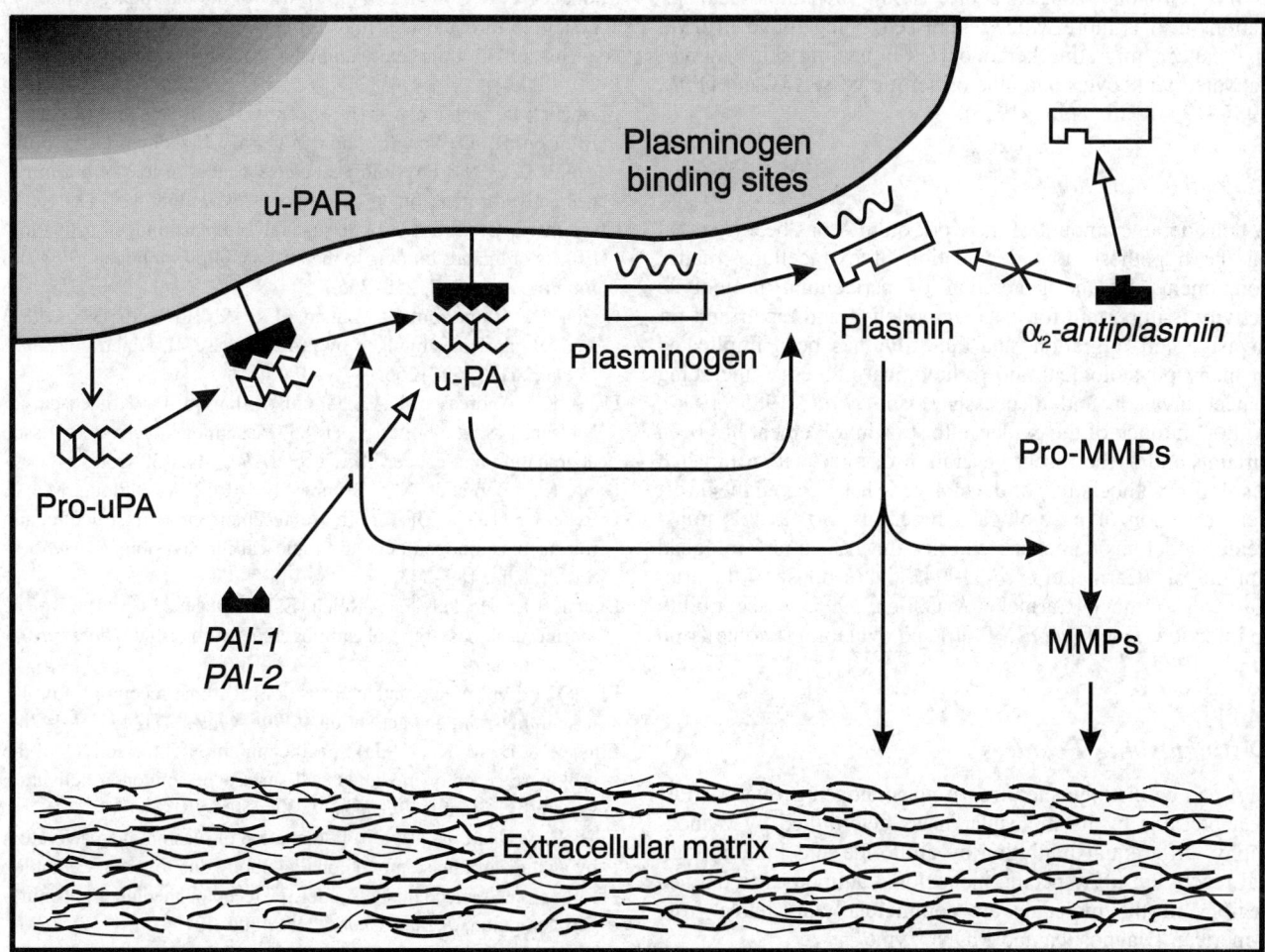

*Figure 57.1* Cell surface plasminogen activation. The components of the uPA receptor-mediated plasminogen activation system are shown, demonstrating that the principal reactions occur with membrane-bound components and make this a completely cell-associated system. uPA receptor-bound uPA preferentially activates cell-bound plasminogen, and cell-bound plasmin can activate receptor-bound pro-uPA. The plasmin generated remains active while cell bound, due to protection from inhibition by $\alpha_2$-antiplasmin, but is rapidly inactivated once dissociated from the cell surface. Receptor-bound uPA, however, is fully accessible to the plasminogen activator inhibitors PAI-1 and PAI-2, which can therefore either modulate or terminate the system. Plasmin generated by this system is thought to have a role in the activation of certain matrix metalloproteinases (MMPs).

in plasminogen and the uPA A chain C-terminal sequence Gly149–Lys158 (Ellis *et al*., 1996). Plasmin generated by this system is protected from inhibition by $\alpha_2$-antiplasmin while it remains cell associated, thus further favoring uPA receptor-mediated plasminogen activation (Figure 57.1) (Ellis *et al*., 1991).

## Distribution

Although uPA is expressed by a wide variety of cells in culture, its expression by normal mammalian tissues *in vivo* is limited, including proximal and distal kidney tubules, subepithelial fibroblasts in the gastrointestinal tract, bladder urothelium, ovarian theca and granulosa cells, and in the mouse, seminiferous epithelium (Larsson *et al*., 1984; Kristensen, 1992). Synthesis in the kidney tubules appears to be the source of uPA in urine. uPA is notably absent from normal endothelium (Kristensen *et al*., 1984). It is present in normal blood plasma at a concentration of approximately 20 pM (Grøndahl-Hansen *et al*., 1988). In pathological situations uPA is more widely expressed, being found in invasive cancer, migrating keratinocytes in healing skin wounds, activated leukocytes and atherosclerotic vessels (Danø *et al*., 1985, 1994; Ellis *et al*., 1995).

## Biological Functions

A considerable amount of the work on uPA has been focused on the hypothesis that degradation of extracellular matrix components by the generation of pericellular proteolytic activity is important for tissue remodeling and repair and for invasive cell migration, and thus uPA has been implicated in many physiological and pathological processes including cancer invasion and metastasis (Danø *et al*., 1985, 1994). Although much of the evidence for this involvement has been circumstantial, the recent generation of mice with a targeted disruption ('knockout') of the uPA gene has allowed the direct demonstration of its biological functions. So far this transgenic model has demonstrated roles for uPA in physiological fibrinolysis (Carmeliet *et al*., 1994), the response of the arterial wall to injury (Carmeliet & Collen, 1995), susceptibility to infection (Gyetko *et al*., 1996) and ovulation (Leonardsson *et al*., 1995).

## Distinguishing Features

uPA can be distinguished from t-plasminogen activator (tPA) (Chapter 58) by its molecular mass determined by zymography, differential inhibition by amiloride and immunological reactivity. uPA-specific monoclonal antibodies to defined regions of the molecule are commercially available from American Diagnostica and MONOzyme.

## Further Reading

An extensive review of the biological function of uPA is that of Danø *et al*. (1985), and the role of uPA and its receptor in relation to cancer has been considered by Danø *et al*. (1994). Ellis *et al*. (1995) provide a general review of the functions of uPA and its receptor, and Lijnen & Collen (1991) discuss the fibrinolytic activity of uPA and recombinant mutants.

## References

Appella, E., Robinson, E.A., Ullrich, S.J., Stoppelli, M.P., Corti, A., Cassani, G. & Blasi, F. (1987) The receptor-binding sequence of urokinase. A biological function for the growth-factor module of proteases. *J. Biol. Chem.* **262**, 4437–4440.

Avgerinos, G.C., Drapeau, D., Socolow, J.S., Mao, J.I., Hsiao, K. & Broeze, R.J. (1990) Spin filter perfusion system for high density cell culture: production of recombinant urinary type plasminogen activator in CHO cells. *Bio/Technology* **8**, 54–58.

Barlati, S., De Petro, G., Bona, C., Paracini, F. & Tonelli, M. (1995) Phosphorylation of human plasminogen activators and plasminogen. *FEBS Lett.* **363**, 170–174.

Buko, A.M., Kentzer, E.J., Petros, A., Menon, G., Zuiderweg, E.R.P. & Sarin, V.K. (1991) Characterization of a posttranslational fucosylation in the growth factor domain of urinary plasminogen activator. *Proc. Natl Acad. Sci. USA* **88**, 3992–3996.

Carmeliet, P. & Collen, D. (1995) Role of the plasminogen/plasmin system in thrombosis, hemostasis, restenosis and atherosclerosis: evaluation in transgenic animals. *Trends Cardiovasc. Med.* **5**, 117–122.

Carmeliet, P., Schoonjans, L., Kieckens, L., Ream, B., Degen, J., Bronson, R., De Vos, R., van den Oord, J.J., Collen, D. & Mulligan, R.C. (1994) Physiological consequences of loss of plasminogen activator gene function in mice. *Nature* **368**, 419–424.

Christensen, U. (1988) Urokinase-catalysed plasminogen activation. Effects of ligands binding to the AH-site of plasminogen. *Biochim. Biophys. Acta* **957**, 258–265.

Collen, D., Zamarron, C., Lijnen, H.R. & Hoylaerts, M. (1986) Activation of plasminogen by pro-urokinase. II. Kinetics. *J. Biol. Chem.* **261**, 1259–1266.

Danø, K., Andreasen, P.A., Grøndahl-Hansen, J., Kristensen, P., Nielsen, L.S. & Skriver, L. (1985) Plasminogen activators, tissue degradation and cancer. *Adv. Cancer Res.* **44**, 139–266.

Danø, K., Behrendt, N., Brünner, N., Ellis, V., Ploug, M. & Pyke, C. (1994) The urokinase receptor: protein structure and role in plasminogen activation and cancer invasion. *Fibrinolysis* **8** (suppl. 1), 189–203.

Degen, S.J., Heckel, J.L., Reich, E. & Degen, J.L. (1987) The murine urokinase-type plasminogen activator gene. *Biochemistry* **26**, 8270–8279.

Ellis, V. (1996) Functional analysis of the cellular receptor for urokinase in plasminogen activation. *J. Biol. Chem.* **271**, 14779–14784.

Ellis, V. & Danø, K. (1993a) Specific inhibition of the activity of the urokinase receptor-mediated cell-surface plasminogen activation system by suramin. *Biochem. J.* **296**, 505–510.

Ellis, V. & Danø, K. (1993b) Potentiation of plasminogen activation by an anti-urokinase monoclonal antibody due to ternary complex formation. A mechanistic model for receptor-mediated plasminogen activation. *J. Biol. Chem.* **268**, 4806–4813.

Ellis, V., Scully, M.F. & Kakkar, V.V. (1987) Plasminogen activation by single-chain urokinase in functional isolation. A kinetic study. *J. Biol. Chem.* **262**, 14998–15003.

Ellis, V., Scully, M.F. & Kakkar, V.V. (1989) Plasminogen activation initiated by single-chain urokinase-type plasminogen activator. Potentiation by U937 monocytes. *J. Biol. Chem.* **264**, 2185–2188.

Ellis, V., Wun, T.-C., Behrendt, N., Rønne, E. & Danø, K. (1990) Inhibition of receptor-bound urokinase by plasminogen-activator inhibitors. *J. Biol. Chem.* **265**, 9904–9908.

Ellis, V., Behrendt, N. & Danø, K. (1991) Plasminogen activation by receptor-bound urokinase. A kinetic study with both cell-associated and isolated receptor. *J. Biol. Chem.* **266**, 12752–12758.

Ellis, V., Behrendt, N. & Danø, K. (1993) Cellular receptor for urokinase-type plasminogen activator: function in cell-surface proteolysis. *Methods Enzymol.* **223**, 223–234.

Ellis, V., Ploug, M., Plesner, T. & Danø, K. (1995) Gene expression and function of the cellular receptor for u-PA (u-PAR). In: *Fibrinolysis in Disease: Molecular and Hemovascular Aspects of Fibrinolysis* (Glas-Greenwalt, P., ed.). Boca Raton, FL: CRC Press, pp. 30–42.

Ellis, V., Whawell, S.A. & Deadman, J. (1996) Direct interactions between uPA and plasminogen are involved in assembly of uPAR-mediated plasminogen activation complexes. *Fibrinolysis* **10** (suppl. 3), 65 (abstr.).

Fleury, V., Lijnen, H.R. & Angles-Cano, E. (1993) Mechanism of the enhanced intrinsic activity of single-chain urokinase-type plasminogen activator during ongoing fibrinolysis. *J. Biol. Chem.* **268**, 18554–18559.

Gaffney, P.J. & Heath, A.B. (1990) A collaborative study to establish a standard for high molecular weight urinary-type plasminogen activator (HMW/u-PA). *Thromb. Haemost.* **64**, 398–401.

Grøndahl-Hansen, J., Agerlin, N., Munkholm-Larsen, P., Bach, F., Nielsen, L.S., Dombernowsky, P. & Danø, K. (1988) Sensitive and specific enzyme-linked immunosorbent assay for urokinase-type plasminogen activator and its application to plasma from patients with breast cancer. *J. Lab. Clin. Med.* **111**, 42–51.

Günzler, W.A., Steffens, G.J., Ötting, F., Buse, G. & Flohé, L. (1982) Structural relationship between human high and low molecular mass urokinase. *Z. Physiol. Chem.* **363**, 133–141.

Gyetko, M.R., Chen, G.H., McDonald, R.A., Goodman, R., Huffnagle, G.B., Wilkinson, C.C., Fuller, J.A. & Toews, G.B. (1996) Urokinase is required for the pulmonary inflammatory response to *Cryptococcus neoformans* – a murine transgenic model. *J. Clin. Invest.* **97**, 1818–1826.

Hansen, A.P., Petros, A.M., Meadows, R.P., Nettesheim, D.G., Mazar, A.P., Olejniczak, E.T., Xu, R.X., Pederson, T.M., Henkin, J. & Fesik, S.W. (1994) Solution structure of the amino-terminal fragment of urokinase-type plasminogen activator. *Biochemistry* **33**, 4847–4864.

Høyer-Hansen, G., Rønne, E., Solberg, H., Behrendt, N., Ploug, M., Lund, L.R., Ellis, V. & Danø, K. (1992) Urokinase plasminogen activator cleaves its cell surface receptor releasing the ligand-binding domain. *J. Biol. Chem.* **267**, 18224–18229.

Husain, S.S. (1991) Single-chain urokinase-type plasminogen activator does not possess measurable intrinsic amidolytic or plasminogen activator activities. *Biochemistry* **30**, 5797–5805.

Husain, S.S. (1993) Fibrin affinity of urokinase-type plasminogen activator. Evidence that $Zn^{2+}$ mediates strong and specific interaction of single-chain urokinase with fibrin. *J. Biol. Chem.* **268**, 8574–8579.

Kristensen, P. (1992) Localization of components from the plasminogen activation system in mammalian tissues. *APMIS Suppl.* **29** (vol. 100), 1–27.

Kristensen, P., Larsson, L.I., Nielsen, L.S., Grøndahl-Hansen, J., Andreasen, P.A. & Danø, K. (1984) Human endothelial cells contain one type of plasminogen activator. *FEBS Lett.* **168**, 33–37.

Larsson, L.I., Skriver, L., Nielsen, L.S., Grøndahl-Hansen, J., Kristensen, P. & Danø, K. (1984) Distribution of urokinase-type plasminogen activator immunoreactivity in the mouse. *J. Cell Biol.* **98**, 894–903.

Leonardsson, G., Peng, X.R., Liu, K., Nordstrom, L., Carmeliet, P., Mulligan, R., Collen, D. & Ny, T. (1995) Ovulation efficiency is reduced in mice that lack plasminogen activator gene function: functional redundancy among physiological plasminogen activators. *Proc. Natl Acad. Sci. USA* **92**, 12446–12450.

Li, X., Bokman, A.M., Llinas, M., Smith, R.A. & Dobson, C.M. (1994) Solution structure of the kringle domain from urokinase-type plasminogen activator. *J. Mol. Biol.* **235**, 1548–1559.

Lijnen, H.R. & Collen, D. (1991) Strategies for the improvement of thrombolytic agents. *Thromb. Haemost.* **66**, 88–110.

Lijnen, H.R., Zamarron, C., Blaber, M., Winkler, M.E. & Collen, D. (1986) Activation of plasminogen by pro-urokinase. I. Mechanism. *J. Biol. Chem.* **261**, 1253–1258.

Lijnen, H.R., Van Hoef, B., De Cock, F. & Collen, D. (1989) The mechanism of plasminogen activation and fibrin dissolution by single chain urokinase-type plasminogen activator in a plasma milieu *in vitro*. *Blood* **73**, 1864–1872.

Lijnen, H.R., Van Hoef, B., Nelles, L. & Collen, D. (1990) Plasminogen activation with single-chain urokinase-type plasminogen activator (scu-PA). Studies with active site mutagenized plasminogen ($Ser^{740}$ > Ala) and plasmin-resistant scu-PA ($Lys^{158}$ > Glu). *J. Biol. Chem.* **265**, 5232–5236.

Liu, J.N. & Gurewich, V. (1992) Fragment E-2 from fibrin substantially enhances pro-urokinase-induced glu-plasminogen activation. A kinetic study using the plasmin-resistant mutant pro-urokinase Ala-158-rpro-UK. *Biochemistry* **31**, 6311–6317.

Lottenberg, R., Christensen, U., Jackson, C.M. & Coleman, P.L. (1981) Assay of coagulation proteases. *Methods Enzymol.* **80**, 341–361.

Manchanda, N. & Schwartz, B.S. (1991) Single chain urokinase. Augmentation of enzymatic activity upon binding to monocytes. *J. Biol. Chem.* **266**, 14580–14584.

Manchanda, N. & Schwartz, B.S. (1995) Interaction of single-chain urokinase and plasminogen activator inhibitor type 1. *J. Biol. Chem.* **270**, 20032–20035.

Marcotte, P.A., Kozan, I.M., Dorwin, S.A. & Ryan, J.M. (1992) The matrix metalloproteinase pump-1 catalyzes formation of low molecular weight (pro)urokinase in cultures of normal human kidney cells. *J. Biol. Chem.* **267**, 13803–13806.

Melnick, M.M., Turner, B.G., Puma, P., Price-Tillotson, B., Salvato, K.A., Dumais, D.R., Moir, D.T., Broeze, R.J., Averginos, G.C. (1990) Characterization of a nonglycosylated single chain urinary plasminogen activator secreted from yeast. *J. Biol. Chem.* **265**, 801–807.

Naldini, L., Tamagnone, L., Vigna, E., Sachs, M., Hartmann, G., Birchmeier, W., Daikuhara, Y., Tsubouchi, H., Blasi, F. & Comoglio, P.M. (1992) Extracellular proteolytic cleavage by urokinase is required for activation of hepatocyte growth factor/scatter factor. *EMBO J.* **11**, 4825–4833.

Nauland, U. & Rijken, D.C. (1994) Activation of thrombin-inactivated single-chain urokinase-type plasminogen activator by dipeptidyl peptidase I (cathepsin C). *Eur. J. Biochem.* **223**, 497–501.

Nelles, L., Lijnen, H.R., Collen, D. & Holmes, W.E. (1987) Characterization of recombinant human single chain urokinase-type plasminogen activator mutants produced by site-specific mutagenesis of lysine 158. *J. Biol. Chem.* **262**, 5682–5689.

Nielsen, L.S., Hansen, J.G., Skriver, L., Wilson, E.L., Kaltoft, K., Zeuthen, J. & Danø, K. (1982) Purification of zymogen to plasminogen activator from human glioblastoma cells by affinity chromatography with monoclonal antibody. *Biochemistry* **21**, 6410–6415.

Pannell, R. & Gurewich, V. (1987) Activation of plasminogen by single-chain urokinase or by two-chain urokinase – a demonstration that single-chain urokinase has a low catalytic activity. *Blood* **69**, 22–26.

Petersen, L.C., Lund, L.R., Nielsen, L.S., Danø, K. & Skriver, L. (1988) One-chain urokinase-type plasminogen activator from human sarcoma cells is a proenzyme with little or no intrinsic activity. *J. Biol. Chem.* **263**, 11189–11195.

Ploug, M., Rahbek-Nielsen, H., Ellis, V., Roepstorff, P. & Danø, K. (1995) Chemical modification of the urokinase-type plasminogen activator and its receptor using tetranitromethane. Evidence for the involvement of specific tyrosine residues in both molecules during receptor-ligand interaction. *Biochemistry* **34**, 12524–12534.

Skriver, L., Nielsen, L.S., Stephens, R. & Danø, K. (1982) Plasminogen activator released as inactive proenzyme from murine cells transformed by sarcoma virus. *Eur. J. Biochem.* **124**, 409–414.

Sobel, G.W., Mohler, S.R., Jones, N.W., Dowdy, A.B.C. & Guest, M.M. (1952) Urokinase: an activator of plasma profibrinolysin extracted from urine. *Am. J. Physiol.* **171**, 768–769.

Spraggon, G., Phillips, C., Nowak, U.K., Ponting, C.P., Saunders, D., Dobson, C.M., Stuart, D.I. & Jones, E.Y. (1995) The crystal structure of the catalytic domain of human urokinase-type plasminogen activator. *Structure* **3**, 681–691.

Stephens, R.W., Bokman, A.M., Myöhänen, H.T., Reisberg, T.,

Tapiovaara, H., Pedersen, N., Grøndahl-Hansen, J., Llinás, M. & Vaheri, A. (1992) Heparin binding to the urokinase kringle domain. *Biochemistry* **31**, 7572–7579.

Stump, D.C., Lijnen, H.R. & Collen, D. (1986) Purification and characterization of single-chain urokinase-type plasminogen activator from human cell cultures. *J. Biol. Chem.* **261**, 1274–1278.

Towle, M.J., Lee, A., Maduakor, E.C., Schwartz, C.E., Bridges, A.J. & Littlefield, B.A. (1993) Inhibition of urokinase by 4-substituted benzo[$\beta$]thiophene-2-carboxamidines: an important new class of selective synthetic urokinase inhibitor. *Cancer Res.* **53**, 2553–2559.

Urano, T., Sator de Serrano, V., Chibber, B.A. & Castellino, F.J. (1987) The control of the urokinase-catalyzed activation of human glutamic acid 1-plasminogen by positive and negative effectors. *J. Biol. Chem.* **262**, 15959–15964.

Vassalli, J.D. & Belin, D. (1987) Amiloride selectively inhibits the urokinase-type plasminogen activator. *FEBS Lett.* **214**, 187–191.

Williams, J.R.B. (1951) The fibrinolytic activity of urine. *Br. J. Exp. Pathol.* **32**, 530–537.

Winkler, M.E. & Blaber, M. (1986) Purification and characterization of recombinant single-chain urokinase produced in *Escherichia coli. Biochemistry* **25**, 4041–4045.

Wun, T.C., Schleuning, W.D. & Reich, E. (1982) Isolation and characterization of urokinase from human plasma. *J. Biol. Chem.* **257**, 3276–3283.

*Vincent Ellis*
*Proteolysis Laboratory,*
*Thrombosis Research Institute,*
*Manresa Road, London SW1 6LR, UK*

*Keld Danø*
*Finsen Laboratory,*
*Rigshospitalet,*
*Strandboulevarden 49, 2100 Copenhagen Ø, Denmark*

# 58. t-Plasminogen activator

## Databanks

*Peptidase classification: clan SA, family S1, MEROPS ID: S01.232*
*NC-IUBMB enzyme classification: EC 3.4.21.68*
*Chemical Abstracts Service registry number: 39639-23-9*
*Databank codes:*

| Species | SwissProt | PIR | EMBL (cDNA) | EMBL (genomic) |
|---|---|---|---|---|
| *Bos taurus* | – | – | X85800 | – |
| *Desmodus rotundus* | P49150 | JS0600 | M63990 | – |
| *Desmodus rotundus* | P98119 | JS0597 | M63986 | – |
| | | | M63987 | |
| *Desmodus rotundus* | P98121 | JS0599 | M63989 | – |
| *Gallus gallus* | – | – | U31988 | – |
| *Homo sapiens* | P00750 | A23529 | A01465 | K03021: complete gene |
| | | A91343 | A07197 | L00140: exon 1 |
| | | A93293 | D01096 | L00141: exon 2 |
| | | A93951 | M15518 | L00142: exon 3 |

| Species | SwissProt | PIR | EMBL (cDNA) | EMBL (genomic) |
|---|---|---|---|---|
| | | A94004 | M18182 | L00143: exon 4 |
| | | JT0562 | U63828 | L00144: exon 5 |
| | | S02125 | V00570 | L00145: exon 6 |
| | | | X02901 | L00146: exon 7 |
| | | | X07393 | L00147: exon 8 |
| | | | X13097 | L00148: exon 9 |
| | | | | L00149: exon 10 |
| | | | | L00150: exon 11 |
| | | | | L00151: exon 12 |
| | | | | L00152: exon 13 |
| | | | | L00153: exon 14 and complete CDS |
| | | | | M11888: exon 1 |
| | | | | M11889: exon 2 |
| | | | | M11890: exon 3 |
| | | | | S77144: 3′ end |
| *Megaderma lyra* | – | A34369 | J05082 | – |
| *Mus musculus* | P11214 | A29941 | J03520 | M26065: 5′ flank |
| | | S48205 | | |
| | | S48207 | | |
| *Rattus norvegicus* | P19637 | A31597 | A19618 | M31184: intron A |
| | | A35029 | M23697 | M31185: exon 2 |
| | | | | M31186: exon 3 |
| | | | | M31187: exon 4 |
| | | | | M31188: exon 5 |
| | | | | M31189: exon 6 |
| | | | | M31190: exon 7 |
| | | | | M31191: exon 8 |
| | | | | M31192: exon 9 |
| | | | | M31193: exon 10 |
| | | | | M31194: exon 11 |
| | | | | M31195: exon 12 |
| | | | | M31196: exon 13 |
| | | | | M31197: exon 14 |
| | | | | S73569: exon 1, promoter |

Brookhaven Protein Data Bank three-dimensional structures:

| Species | ID | Resolution | Notes |
|---|---|---|---|
| *Homo sapiens* | 1PK2 | | NMR; kringle 2 domain |
| | 1PML | 2.38 | kringle 2 |
| | 1TPK | 2.4 | kringle 2 |
| | 1TPM | | NMR; fibrin-binding finger domain |
| | 1TPN | | NMR; fibrin-binding finger domain |

## Name and History

The fibrinolytic system (Figure 58.1) comprises an inactive proenzyme, plasminogen, which can be converted to the active enzyme, plasmin, which in turn degrades fibrin into soluble fibrin degradation products. Two physiological plasminogen activators have been identified: the tissue-type activator, recommended by NC-IUBMB to be called **t-plasminogen activator (t-PA)** and the urokinase-type, u-plasminogen activator (u-PA) described in Chapter 57. Inhibition of the fibrinolytic system may occur either at the level of the plasminogen activator (PA), by specific plasminogen activator inhibitors (mainly PAI-1), or at the level of plasmin, by specific plasmin inhibitors (mainly $\alpha_2$-antiplasmin) (Collen & Lijnen, 1991).

As early as 1947 (Astrup & Permin, 1947) it was reported that animal tissues contain an agent that can activate plasminogen; this factor was originally called fibrinokinase. There have since been many further reports of the purification and characterization of tissue plasminogen activators from various sources, including pig heart and ovaries, and human postmortem vascular perfusates and postexercise blood. The first highly purified form of human t-PA was obtained from

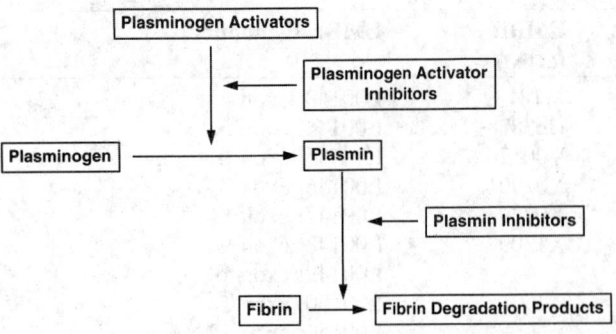

*Figure 58.1*   Schematic representation of the fibrinolytic system. Plasminogen activators convert plasminogen to plasmin, which degrades fibrin. Inhibition occurs either by plasminogen activator inhibitors or by plasmin inhibitors.

uterine tissue (Rijken *et al.*, 1979). Using an antiserum raised against uterine plasminogen activator, it has been shown that tissue plasminogen activator, vascular plasminogen activator and blood plasminogen activator are immunologically identical, but different from u-PA (Rijken *et al.*, 1980). The plasminogen activator found in blood represents vascular t-PA that has been released mainly from endothelial cells.

## Activity and Specificity

t-PA converts plasminogen to the two-chain serine proteinase plasmin (Chapter 59) by hydrolysis of a single Arg561+Val562 peptide bond. In synthetic peptide substrates, Arg residues in the P1 position are favored (e.g. H-D-Ile-Pro-Arg+NHPhNO$_2$). The enzymatic activity of t-PA can be determined in a parabolic rate assay, in which the enzyme is incubated with excess plasminogen (e.g. in 0.1 M sodium phosphate buffer, pH 7.4, containing 0.01% Tween 80) in the presence of fibrin fragments as stimulating cofactor; the concentration of generated plasmin is measured with a chromogenic substrate (Ranby *et al.*, 1982). This assay can also be applied to plasma following neutralization of plasma proteinase inhibitors (Verheijen *et al.*, 1982). By comparison with the International Reference Preparation for t-PA (WHO 2nd IS 86/670), a specific activity of 500 000 IU mg$^{-1}$ has been assigned to the purified protein (Gaffney & Curtis, 1987).

The activation of plasminogen by t-PA, both in the presence and the absence of fibrin, follows Michaelis–Menten kinetics, with a Michaelis constant ($K_m$) of 65 µM and a catalytic rate constant ($k_{cat}$) of 0.05 s$^{-1}$ in the absence of fibrin, whereas in the presence of fibrin the $K_m$ decreases to 0.16 µM with only a minor change in $k_{cat}$ (Hoylaerts *et al.*, 1982). Although different kinetic parameters have been reported, most authors agree that the activation rate of plasminogen increases several hundred-fold in the presence of fibrin. Thus, t-PA is a relatively poor enzyme in the absence of fibrin, but the presence of fibrin strikingly enhances its activity. The kinetic data support a mechanism in which fibrin provides a surface to which t-PA and plasminogen adsorb in a sequential and ordered way, yielding a cyclic ternary complex. Fibrin essentially increases the local plasminogen concentration by creating an additional interaction between t-PA and its substrate. The high affinity of t-PA for plasminogen in the presence of fibrin thus allows efficient activation on the fibrin clot, while plasminogen activation by t-PA in plasma is a comparatively inefficient process (Collen, 1980; Lijnen *et al.*, 1994).

Plasminogen activator inhibitor 1 (PAI-1) is the main physiological inhibitor. It forms irreversible, inactive, stoichiometric complexes with t-PA with a very high second-order rate constant ($>10^7$ M$^{-1}$ s$^{-1}$). PAI-1 is a single-chain glycoprotein with $M_r$ 52 000, consisting of 379 amino acids; its reactive-site peptide bond consists of Arg346-Met347. Synthetic t-PA inhibitors include tripeptide chloromethanes (e.g. D-Phe-Pro-Arg-CH$_2$Cl).

## Structural Chemistry

Human t-PA was first isolated as a single-chain serine proteinase with $M_r$ about 70 000 (pI = 7 to 8), consisting of 527 amino acids with Ser as the N-terminal amino acid (Pennica *et al.*, 1983) (Figure 58.2). It was subsequently shown that native t-PA contains an N-terminal extension of three amino acids, but in general the initial numbering system has been maintained. The molecule has 17 disulfide bonds and an additional free Cys at position 83. Limited hydrolysis of the Arg275-Ile276 peptide bond by plasmin converts t-PA to a two-chain molecule held together by one interchain disulfide bond. The t-PA molecule contains four domains: (1) an N-terminal region of 47 residues (residues 4–50) (F-domain) which is homologous with the finger domains mediating the fibrin affinity of fibronectin; (2) residues 50–87 (E-domain) which are homologous with epidermal growth factor; (3) two kringle regions (residues 87–176, K$_1$-domain, and 176–262, K$_2$-domain), which share a high degree of homology with the five kringles of plasminogen, and (4) a serine proteinase region (residues 276–527, P-domain) with the active-site residues His322, Asp371 and Ser478 (Pennica *et al.*, 1983). The distinct domains in t-PA are involved in several functions, including its binding to fibrin (mainly via F- and K$_2$-domains), rapid clearance *in vivo* with an initial half-life of 6 min in humans (mediated via F- and/or E-domains) and enzymatic activity (P-domain).

Physicochemical characterization of t-PA has indicated that the molecule is ellipsoidal and relatively compact (Margossian *et al.*, 1993), with the individual domains folded within the molecule yielding a globular structure, which is stabilized by strong interactions between the proteinase domain and the F- and/or E-domains (Novokhatny *et al.*, 1991). NMR studies on the solution structure of the F-domain of t-PA support this model (Downing *et al.*, 1992). In contrast to the single-chain precursor form of most serine proteinases, single-chain t-PA is enzymatically active. It has been postulated that the activity of single-chain t-PA would involve an equilibrium between an active and a zymogenic conformation, which would be shifted to the active conformation upon substrate binding (Nienaber *et al.*, 1992). The t-PA molecule comprises three potential *N*-glycosylation sites, at Asn117 (K$_1$), Asn184 (K$_2$) and Asn448 (P) (Pennica *et al.*, 1983). t-PA preparations usually contain a mixture of variant I (with all three sites glycosylated) and variant II (lacking carbohydrate at Asn184).

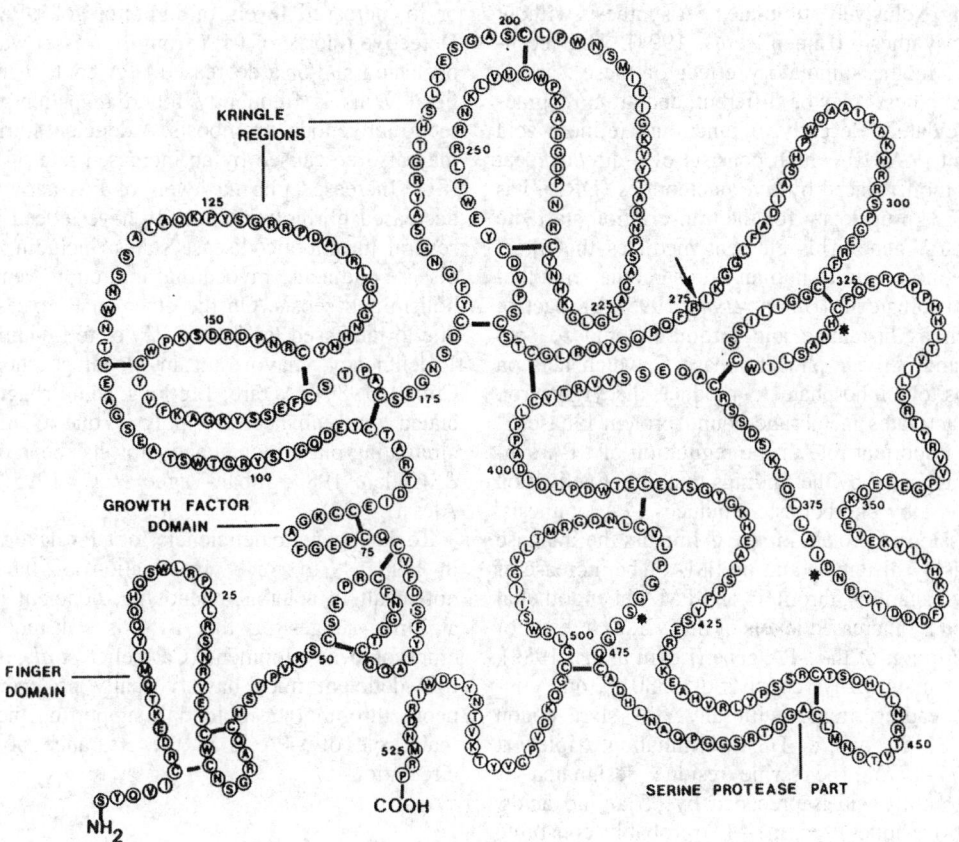

*Figure 58.2*   Schematic representation of the primary structure of t-PA. The amino acids are represented by their single letter symbols and black bars indicate disulfide bonds. The active-site residues His322, Asp371 and Ser478 are marked by asterisks. The arrow indicates the cleavage site for conversion of single-chain to two-chain t-PA.

## Preparation

t-PA has been purified from the culture fluid of a stable human melanoma cell line (Bowes, RPMI-7272), in sufficient amounts to study its biochemical and biological properties (Rijken & Collen, 1981). Purification methods include immunoadsorption and chromatography on zinc chelate, lysine-Sepharose or fibrin-Sepharose.

t-PA has subsequently been produced by recombinant DNA technology (rt-PA). The cDNA of human t-PA has been cloned and expressed in *E. coli* and in mammalian cell systems (Pennica *et al.*, 1983). More efficient expression was obtained in mammalian cells, yielding a properly processed and glycosylated molecule. This rt-PA was shown to be indistinguishable from the natural activator isolated from human melanoma cell cultures, with respect to biochemical properties, turnover *in vivo*, and specific thrombolytic activity (Collen *et al.*, 1984). The generation of CHO (Chinese hamster ovary) cells capable of producing single-chain human t-PA has allowed the development of large-scale tissue culture fermentation and purification procedures, yielding rt-PA for commercial purposes (Activase®, Genentech Inc.; Actilyse®, Boehringer Ingelheim). Biologically active t-PA has also been obtained by expression of cDNA in *Aspergillus nidulans* (Upshall *et al.*, 1987) and in mouse C127 cells (Reddy *et al.*, 1987).

## Biological Aspects

### Physiological Aspects

t-PA antigen levels (active plus inactive forms) in human plasma at rest (measured by ELISA) range between 3 and $7 \mu g\, liter^{-1}$. This level increases about 3-fold following exhaustive physical exercise, venous occlusion or infusion of 1-deamino-8-D-arginine vasopressin (DDAVP) (Rijken *et al.*, 1983; Holvoet *et al.*, 1985). Using an ELISA based on a specific monoclonal antibody, detectable levels of free single-chain t-PA (about $4 \mu g\, liter^{-1}$) were found in plasma of about 60% of healthy subjects (Holvoet *et al.*, 1987).

Vascular endothelial cells synthesize and secrete t-PA into the circulating blood. Stimulation of vascular endothelium by venous occlusion, infusion of DDAVP or epinephrine and physical exercise result in a rapid release (within minutes) of t-PA. This reponse is too rapid to represent increased synthesis and may reflect release from cellular storage pools, although such a storage pool has not been conclusively identified. A variety of agents have been shown to increase the synthesis of t-PA by cultured endothelial cells, including thrombin, histamine, butyrate, phorbol myristate acetate (PMA), basic fibroblast growth factor, activated protein C, butanol and alcohol derivatives, and retinoids. However, only histamine, butyrate or a combination of cAMP with protein kinase C agonists exclusively stimulate t-PA synthesis without affecting PAI-1 synthesis (Lijnen *et al*., 1994). The mechanisms involved in the stimulatory effect of these various agents on t-PA synthesis may be different, and are now gradually being elucidated. Recently, a functional retinoic acid response element (RARE), which consists of a direct repeat of the GGTCA motif spaced by five nucleotides (DR5), has been localized 7.3 kb upstream for the transcription start site of the human t-PA gene. This element mediates the direct regulation by retinoic acid in human fibrosarcoma, endothelial and neuroblastoma cells (Bulens *et al*., 1995). Vasoactive substances, such as histamine and thrombin, bind to specific receptors and activate phospholipase C which acts on phosphatidylinositol biphosphate to produce diacylglycerol. Diacylglycerol activates membrane-bound protein kinase C, which plays an important role in the regulation of t-PA synthesis. This is suggested by the findings that direct activation of protein kinase C by phorbol esters induces t-PA synthesis, whereas suppression of protein kinase C impairs the increase in t-PA synthesis by histamine and by PMA. The increase of t-PA induced by histamine, thrombin and PMA in endothelial cells is paralleled by increased levels of mRNA, as a result of enhanced transcription of the t-PA gene (Levin *et al*., 1989).

The cDNA of mature t-PA consists of 2530 bp, and contains a single reading frame with an ATG start codon at nucleotides 85–87 and a TGA termination triplet at nucleotides 1771–1775. The serine residue designated as the N-terminal amino acid is preceded by 35 amino acids, 20–23 of which (residues −35 to −13) probably constitute a hydrophobic signal peptide involved in the secretion of t-PA. The remaining hydrophobic amino acids immediately preceding the start of mature t-PA (residues −14 to −1) may constitute a 'pro' sequence similar to that found for serum albumin (Pennica *et al*., 1983). The human t-PA gene has been localized to chromosome 8 (8p12), and more than 36 kb of its sequence have been determined. The t-PA gene consists of 14 exons, and the intron–exon organization suggests that the gene has evolved by 'exon shuffling', whereby the distinct structural domains are each encoded by a single exon or by adjacent exons (Lijnen & Collen, 1991). The proximal promoter sequences in the human t-PA gene containing typical TATA and CAAT boxes and potential recognition sequences for transcription factors (e.g. AP1, NF1, SP1, AP2) have been identified (Feng *et al*., 1990; Kooistra *et al*., 1991). Consensus sequences of a cAMP-responsive element and of an AP2-binding site have been identified, which may have a cooperative effect on constitutive t-PA gene expression (Medcalf *et al*., 1990). Allelic dimorphism has been observed in the human t-PA gene as a result of an Alu insertion/deletion event which occurred early in evolution (Ludwig *et al*., 1992).

## Pathophysiological Aspects

The importance of the fibrinolytic system is demonstrated by the association between abnormal fibrinolysis and a tendency toward bleeding or thrombosis. Impairment of fibrinolysis represents a commonly observed hemostatic abnormality associated with thrombosis. It may be due to a defective synthesis and/or release of t-PA from the vessel wall, to a deficiency or functional defect in the plasminogen molecule or to increased levels of inhibitors of t-PA or of plasmin. Defective release of t-PA from the vessel wall during venous occlusion and/or a decreased t-PA content in walls of superficial veins is frequently found in patients with idiopathic recurrent venous thrombosis. A deficient fibrinolytic response may also be caused by an increased rate of neutralization of t-PA. Increased plasma levels of PAI activity resulting in a decreased fibrinolytic capacity have indeed been reported in several thrombotic disease states, including coronary artery disease and acute myocardial infarction, venous thromboembolism and sepsis. On the other hand, excessive fibrinolysis due to increased levels of t-PA, or to $\alpha_2$-antiplasmin or PAI-1 deficiency, may result in bleeding tendency (Lijnen & Collen, 1989). A rare, life-long hemorrhagic disorder associated with enhanced fibrinolysis, due to increased levels of circulating plasminogen activator, has been described (Lijnen & Collen, 1989; Juhan-Vague *et al*., 1995; Juhan-Vague & Alessi, 1996).

To date, genetic deficiencies of t-PA have not been reported in humans. In mice, inactivation of the t-PA gene did not result in a lethal phenotype; deficient mice are normal at birth, suggesting that t-PA is not required for normal embryonic development (Carmeliet *et al*., 1994). However, t-PA-deficient mice have virtually no endogenous spontaneous thrombolytic potential, supporting the notion that the main role of t-PA is in the clearance of fibrin from the circulation.

## Therapeutic Aspects

The fibrin specificity of t-PA has stimulated great interest in its use for thrombolytic therapy in patients with thromboembolic disease. Numerous clinical trials have compared the thrombolytic properties of rt-PA with those of other agents (de Bono, 1995; Verstraete, 1995), culminating in the GUSTO trial and its angiographic substudy (GUSTO Investigators, 1993; GUSTO Angiographic Investigators, 1993), which conclusively established the potential and limitations of rt-PA for thrombolytic therapy in patients with acute myocardial infarction. Several approaches have been followed to further improve the thrombolytic properties of rt-PA, essentially by enhancing its plasminogen-activating potency or its fibrin specificity and by reducing its plasma clearance. These include the construction of mutants and variants (with deletion or substitution of specific amino acids, of entire domains or of carbohydrate moieties) of chimeric molecules (e.g. with domains of u-PA) and of antibody-targeted t-PA (by coupling t-PA to fibrin-specific monoclonal antibodies) (Lijnen & Collen, 1991). These studies have yielded important insights into the structure–function relationships of t-PA and have led to the development of a few selected mutants which are now being evaluated in clinical trials. Such potentially new thrombolytic agents include TNK-rt-PA (with Thr103 substituted with Asn, introducing a new glycosylation site; with Asn117 substituted with Gln, eliminating a high mannose glycosylation site; and with the sequence Lys296-His-Arg-Arg299 replaced by Ala, conferring increased zymogenicity and resistance to PAI-1), rt-PA E6010 (with Lys84 substituted with Ser), and reteplase (rt-PA consisting only of the $K_2$- and P-domains) (Verstraete *et al*., 1995).

## Distinguishing Features

In contrast to other proenzyme forms of serine proteinases, t-PA is an active enzyme in its single-chain form. Proteolytic conversion by hydrolysis of the Arg275-Ile276 peptide bond yields a two-chain molecule with higher activity towards chromogenic substrates but comparable activity towards the natural substrate plasminogen.

t-PA is a poor enzyme in the absence of fibrin, but its activity is dramatically stimulated in the presence of a fibrin cofactor. This fibrin specificity has formed the basis for its development as a thrombolytic agent for the treatment of patients with thromboembolic disease.

## Related Peptidases

Different molecular forms of a plasminogen activator have been purified from the saliva of the vampire bat *Desmodus rotundus* (DSPA) and were subsequently prepared by recombinant DNA technology. Two high molecular weight forms (DSPA $\alpha$1 with $M_r$ 43 000 and DSPA $\alpha$2 with $M_r$ 39 000) exhibit about 85% homology with human t-PA but lack the $K_2$-domain and the plasmin cleavage site for conversion to a two-chain form. Both forms have a specific activity equal to or higher than that of t-PA, a relative resistance to PAI-1 and exhibit a greatly enhanced stimulation by fibrin. Two smaller forms lacking the F-domain (DSPA $\beta$) or the F- and E-domains (DSPA $\gamma$) have also been reported (Gardell *et al.*, 1989; Verstraete *et al.*, 1995).

## Further Reading

The article of Collen (1980) establishes the mechanism of the fibrin specificity of t-PA, and the first satisfactory purification procedure of t-PA from cell cultures is reported by Rijken & Collen (1981). Hoylaerts *et al.* (1982) provide a quantitative analysis of the stimulating effect of fibrin on the plasminogen-activating potential of t-PA. The successful cloning and expression of t-PA by recombinant DNA technology were achieved by Pennica *et al.* (1983), thereby opening the perspective of the availability of large amounts of rt-PA for clinical trials. The comparative clinical trial in over 40 000 patients with acute myocardial infarction, which established the superiority of rt-PA over streptokinase, is reported by the GUSTO Investigators (1993).

## References

Astrup, T. & Permin, P.M. (1947) Fibrinolysis in animal organism. *Nature* **159**, 681–682.

Bulens, F., Ibanez-Tallon, I., Van Acker, P., De Vriese, A., Nelles, L., Belayew, A. & Collen, D. (1995) Retinoic acid induction of human tissue-type plasminogen activator (t-PA) gene expression via a direct repeat element (DR5) located at −7 kilobases. *J. Biol. Chem.* **270**, 7167–7175.

Carmeliet, P., Schoonjans, L., Kieckens, L., Ream, B., Degen, J., Bronson, R., De Vos, R., van den Oord, J.J., Collen, D. & Mulligan, R. (1994) Physiological consequences of loss of plasminogen activator gene function in mice. *Nature* **368**, 419–424.

Collen, D. (1980) On the regulation and control of fibrinolysis. *Thromb. Haemost.* **43**, 77–89.

Collen, D. & Lijnen, H.R. (1991) Basic and clinical aspects of fibrinolysis and thrombolysis. *Blood* **78**, 3114–3124.

Collen, D., Stassen, J.M., Marafino, B.J., Builder, S., De Cock, F., Ogez, J., Tajiri, D., Pennica, D., Bennett, W.F. & Salwa, J. (1984) Biological properties of human tissue-type plasminogen activator obtained by expression of recombinant DNA in mammalian cells. *J. Pharmacol. Exp. Ther.* **231**, 146–152.

de Bono, D.P. (1995) Thrombolytic therapy of acute myocardial infarction. *Baillière's Clin. Haematol.* **8**, 403–412.

Downing, A.K., Driscoll, P.C., Harvey, T.S., Dudgeon, T.J., Smith, B.O., Baron, M. & Campbell, I.D. (1992) Solution structure of the fibrin binding finger domain of tissue-type plasminogen activator determined by 1H nuclear magnetic resonance. *J. Mol. Biol.* **225**, 821–833.

Feng, P., Ohlsson, M. & Ny, T. (1990) The structure of the TATA-less rat tissue-type plasminogen activator gene. Species-specific sequence divergences in the promoter predict differences in regulation of gene expression. *J. Biol. Chem.* **265**, 2022–2027.

Gaffney, P.J. & Curtis, A.D. (1987) A collaborative study to establish the 2nd International Standard for Tissue Plasminogen Activator (t-PA). *Thromb. Haemost.* **58**, 1085–1087.

Gardell, S.J., Duong, L.T., Diehl, R.E., York, J.D., Hare, T.R., Register, R.B., Jacobs, J.W., Dixon, R.A. & Friedman, P.A. (1989) Isolation, characterizaton, and cDNA cloning of a vampire bat salivary plasminogen activator. *J. Biol. Chem.* **264**, 17947–17952.

GUSTO Angiographic Investigators (1993) The effects of tissue plasminogen activator, streptokinase, or both on coronary-artery patency, ventricular function, and survival after acute myocardial infarction. *N. Engl. J. Med.* **329**, 1615–1622.

GUSTO Investigators (1993) An international randomized trial comparing four thrombolytic strategies for acute myocardial infarction. *N. Engl. J. Med.* **329**, 673–682.

Holvoet, P., Cleemput, H. & Collen, D. (1985) Assay of human tissue-type plasminogen activator (t-PA) with an enzyme-linked immunosorbent assay (ELISA) based on three murine monoclonal antibodies to t-PA. *Thromb. Haemost.* **54**, 684–687.

Holvoet, P., Boes, J. & Collen, D. (1987) Measurement of free, one-chain tissue-type plasminogen activator in human plasma with an enzyme-linked immunosorbent assay based on an active site-specific murine monoclonal antibody. *Blood* **69**, 284–289.

Hoylaerts, M., Rijken, D.C., Lijnen, H.R. & Collen, D. (1982) Kinetics of the activation of plasminogen by human tissue plasminogen activator. Role of fibrin. *J. Biol. Chem.* **257**, 2912–2919.

Juhan-Vague, I. & Alessi, M.-C. (1996) Fibrinolysis and risk of coronary artery disease. *Fibrinolysis* **10**, 127–136.

Juhan-Vague, I., Alessi, M.-C. & Declerck, P.J. (1995) Pathophysiology of fibrinolysis. *Baillière's Clin. Haematol.* **8**, 329–343.

Kooistra, T., Bosma, P.J., Toet, K., Cohen, C.H., Griffioen, M., van den Berg, E., le Clercq, L. & van Hinsbergh, V.W. (1991) Role of protein kinase C and cyclic adenosine monophosphate in the regulation of tissue-type plasminogen activator, plasminogen activator inhibitor-1, and platelet-derived growth factor mRNA levels in human endothelial cells. Possible involvement of proto-oncogenes c-jun and c-fos. *Arterioscler. Thromb.* **11**, 1042–1052.

Levin, E.G., Marotti, K.R. & Santell, L. (1989) Protein kinase C and the stimulation of tissue plasminogen activator release from human endothelial cells. Dependence on the elevation of messenger RNA. *J. Biol. Chem.* **264**, 16030–16036.

Lijnen, H.R. & Collen, D. (1989) Congenital and acquired deficiencies of components of the fibrinolytic system and their relation to bleeding or thrombosis. *Fibrinolysis* **3**, 67–77.

Lijnen, H.R. & Collen, D. (1991) Strategies for the improvement of thrombolytic agents. Thromb. Haemost. **66**, 88–110.

Lijnen, H.R., Bachmann, F., Collen, D., Ellis, V., Pannekoek, H., Rijken, D.C. & Thorsen S. (1994) Mechanisms of plasminogen activation. J. Intern. Med. **236**, 415–424.

Ludwig, M., Wohn, K.D., Schleuning, W.D. & Olek, K. (1992) Allelic dimorphism in the human tissue-type plasminogen activator (tPA) gene as a result of an Alu insertion/deletion event. Hum. Genet. **88**, 388–392.

Margossian, S.S., Slayter, H.S., Kaczmarek, E. & McDonagh, J. (1993) Physical characterization of recombinant tissue plasminogen activator. Biochim. Biophys. Acta **1163**, 250–256.

Medcalf, R.L., Ruegg, M. & Schleuning, W.D. (1990) A DNA motif related to the cAMP-responsive element and an exon-located activator protein-2 binding site in the human tissue-type plasminogen activator gene promoter cooperate in basal expression and convey activation by phorbol ester and cAMP. J. Biol. Chem. **265**: 14618–14626.

Nienaber, V.L., Young, S.L., Birktoft, J.J., Higgins, D.L. & Berliner, L.J. (1992) Conformational similarities between one-chain and two-chain tissue plasminogen activator (t-PA): implications to the activation mechanism of one-chain t-PA. Biochemistry **31**, 3852–3861.

Novokhatny, V.V., Ingham, K.C. & Medved, L.V. (1991) Domain structure and domain–domain interactions of recombinant tissue plasminogen activator. J. Biol. Chem. **266**, 12994–13002.

Pennica, D., Holmes, W.E., Kohr, W.J., Harkins, R.N., Vehar, G.A., Ward, C.A., Bennett, W.F., Yelverton, E., Seeburg, P.H., Heyneker, H.L., Goeddel, D.V. & Collen, D. (1983) Cloning and expression of human tissue-type plasminogen activator cDNA in E. coli. Nature **301**, 214–221.

Ranby, M., Norrman, B. & Wallen, P. (1982) A sensitive assay for tissue plasminogen activator. Thromb. Res. **27**, 743–749.

Reddy, V.B., Garramone, A.J., Sasak, H., Wei, C.M., Watkins, P., Galli, J. & Hsiung, N. (1987) Expression of human uterine tissue-type plasminogen activator in mouse cells using BPV vectors. J. Mol. Biol. **6**, 461–472.

Rijken, D.C. & Collen, D. (1981) Purification and characterization of the plasminogen activator secreted by human melanoma cells in culture. J. Biol. Chem. **256**, 7035–7041.

Rijken, D.C., Wijngaards, G., Zaal-de Jong, M. & Welbergen, J. (1979) Purification and partial characterization of plasminogen activator from human uterine tissue. Biochim. Biophys. Acta **580**, 140–153.

Rijken, D.C., Wijngaards, G. & Welbergen, J. (1980) Relationship between tissue plasminogen activator and the activators in blood and vascular wall. Thromb. Res. **18**, 815–830.

Rijken, D.C., Juhan-Vague, I., De Cock, F. & Collen, D. (1983) Measurement of human tissue-type plasminogen activator by a two-site immunoradiometric assay. J. Lab. Clin. Med. **101**, 274–284.

Upshall, A., Kumar, A.A., Bailey, M.C., Parker, M.D. & Favreau, M.A. (1987) Secretion of active human tissue plasminogen activator from the filamentous fungus Aspergillus-nidulans. Bio/Technology **5**, 1301–1304.

Verheijen, J.H., Mullaart, E., Chang, G.T.G., Kluft, C. & Wijngaards, G. (1982) A simple, sensitive spectrophotometric assay for extrinsic (tissue-type) plasminogen activator applicable to measurement in plasma. Thromb. Haemost. **48**, 266–269.

Verstraete, M. (1995) Thrombolytic therapy of non-cardiac disorders. Baillière's Clin. Haematol. **8**, 413–424.

Verstraete, M., Lijnen, H.R. & Collen, D. (1995) Thrombolytic agents in development. Drugs **50**, 29–42.

**H.R. Lijnen**
Center for Molecular and Vascular Biology,
University of Leuven, Campus Gasthuisberg O & N,
Herestraat 49, B-3000 Leuven, Belgium
Email: roger.lijnen@med.kuleuven.ac.be

**D. Collen**
Center for Molecular and Vascular Biology,
University of Leuven, Campus Gasthuisberg O & N,
Herestraat 49, B-3000 Leuven, Belgium
Email: desire.collen@med.kuleuven.ac.be

# 59. Plasmin

## Databanks

*Peptidase classification: clan SA, family S1, MEROPS ID: S01.233*
*NC-IUBMB enzyme classification: EC 3.4.21.7*
*ATCC entries: cell lines MB-8064 (Hep3B epithelial cells producing plasminogen) and MB-8065 (HepG2 epithelial cells producing plasminogen)*
*Chemical Abstract Service registry number: plasmin, 9001-90-5; plasminogen, 9001-91-6*

*Databank codes:*

| Species | SwissProt | PIR | EMBL (cDNA) | EMBL (genomic) |
|---------|-----------|-----|-------------|----------------|
| *Bos taurus* | P06868 | S45046 | K02935 X79402 | – |
| *Canis familiaris* | P80009 | – | – | – |
| *Capra hircus* | – | C61545 | – | |
| *Equus caballus* | P80010 | A61545 S17527 | – | – |
| *Erinaceus europaeus* | – | – | U33171 | – |
| *Gallus gallus* | – | A60140 | – | – |
| *Homo sapiens* | P00747 | A35229 | A22096 K02922 M74220 X05199 | K02921: exon in 4th kringle M33272: exon 1 M33273: intron 1 M33274: exon 2 M33275: exon 3 M33276: intron 3 M33277: intron 3 M33278: exon 4 M33279: exon 5 M33280: exon 6 M33281: intron F M33282: exon 7 M33283: exon 8 M33284: exon 9 M33285: exon 10 M33286: exon 11 M33287: exon 12 M33288: exon 13 M33289: exon 14 M33290: exon 15 M34272: exon 16 M34274: intron 17 M34275: exon 18 M34276: exon 19 and complete CDS M62890: exon 1 U07744: promoter |
| *Macaca mulatta* | P12545 | B30848 | J04697 | – |
| *Mus musculus* | P20918 | A38514 | J04766 | – |
| *Oryctolagus cuniculus* | – | D61545 | – | – |
| *Ovis aries* | – | B61545 S28200 | – | – |
| *Petromyzon marinus* | P33574 | S33879 | – | – |
| *Rattus norvegicus* | Q01177 | A40522 | M62832 | – |
| *Sus scrofa* | P06867 | A25834 S03733 | – | – |

Brookhaven Protein Data Bank three-dimensional structures:

| Species | ID | Resolution | Notes |
|---------|-----|-----------|-------|
| *Homo sapiens* | 1CEA | 2.1 | kringle 1 complex with ε-aminocaproic acid |
| | 1CEB | 2.1 | kringle 1 complex with *trans*-4-aminomethylcyclohexane-carboxylic acid |
| | 1PK4 | 1.9 | kringle 4 |
| | 1PKR | 2.48 | kringle 1 |
| | 1PMK | 2.25 | kringle 4 |
| | 2PK4 | 2.25 | kringle 4 complex with ε-aminocaproic acid |

## Name and History

As early as 1902, it was shown that the 'press juices' from various organs possessed a fibrinolytic activity (Conradi, 1902), and the active agent was called *fibrinolysin*. A thorough investigation of the clot lysis activities of a variety of animal tissues was subsequently carried out and the more general presence of such an activity was observed (Fleischer & Loeb, 1915). Fibrinolysis was then shown to develop in blood plasma after its treatment with chloroform, or by addition of culture filtrates of some strains of $\beta$-hemolytic streptococci (Tillet & Garner, 1933). In these early papers, the fibrinolysins to which reference is made were in reality plasminogen activators. This point was more fully appreciated in 1941, when it was argued that a human plasma factor, 'lysin factor', was necessary for fibrin digestion by bacterial culture filtrates (Milstone, 1941).

The next seminal events occurred in 1944 (Kaplan, 1944) and 1945 (Christensen, 1945; Christensen & MacLeod, 1945), when it was concluded that the fibrin lytic activities induced by chloroform or by streptococcal filtrates were proteolytic reactions, and that Milstone's 'lysin factor' was a plasma proenzyme that was converted to the functional proteolytic enzyme. The proenzyme, 'lysin factor', was then termed *profibrinolysin* or *plasminogen*, and the enzyme responsible for clot lysis was named *fibrinolysin*, or *plasmin*. This latter name was adopted in order to avoid the confusion with the bacterial fibrinolysin, which is a plasminogen activator, since named streptokinase. The tissue fibrinolysins originally referred to, also being plasminogen activators, were originally named fibrokinases (Astrup & Permin, 1947), and are now called tissue-type plasminogen activators (Chapter 58).

## Activity and Specificity

Plasmin catalyzes cleavages of Lys+X and Arg+X bonds, with a specificity similar to that of trypsin. However, plasmin is a much less efficient enzyme than trypsin, and cleaves only some of these bonds in proteins. Of the simple amino acid substrates, esters and amides of arginine and lysine are cleaved with relative rates affected by the inherent susceptibility of the leaving group to hydrolysis. With these substrates, potentiometric and colorimetric assays were developed (Castellino & Powell, 1981). More modern plasmin assays with small substrates, such as peptides that contain C-terminal lysine and arginine amides or esters, take advantage of $p$-nitrophenolate and $p$-nitroanilide leaving groups. These are assayed by the visible absorbancy (405 nm) of the released moiety. A comprehensive list of kinetic properties of synthetic substrates for plasmin has been published (Lottenberg *et al.*, 1981; Robbins *et al.*, 1981). Similarly, amide- and ester-based synthetic substrates possessing a variety of fluorescent leaving groups have been designed to provide assays for plasmin with much higher sensitivities (Pierzchala *et al.*, 1979; Lottenberg *et al.*, 1981). Single-turnover substrates that are derivatives of $p$-nitrophenylguanidinobenzoate (Chase & Shaw, 1969) or 4-methylumbelliferylguanidinobenzoate (Jameson *et al.*, 1973) are very effective spectrophotometric or fluorometric titrants of plasmin (Sodetz & Castellino, 1972; Sodetz *et al.*, 1976).

A wide variety of proteins are cleaved by plasmin at a very limited number of Arg-X and Lys-X bonds (*vide infra*). Plasmin assays based on hydrolysis of casein (Robbins & Summaria, 1970) and fibrin (Urano *et al.*, 1989) have been established.

Most inhibitors of trypsin also inhibit plasmin, as do general serine protease inhibitors. DFP and Tos-Lys-CH$_2$Cl are irreversible inhibitors (Summaria *et al.*, 1967; Groskopf *et al.*, 1969), and benzamidine (Markwardt *et al.*, 1968) and its derivatives (Walsmann, 1977; Andrews *et al.*, 1978), as well as other aromatic amidines (Ferroni *et al.*, 1988) and tetra-amidines (Ferroni *et al.*, 1986), function as competitive, reversible inhibitors of plasmin. Leupeptin (R-Leu-Leu-argininal), and its synthetic analogs (Saino *et al.*, 1988), in addition to naturally occurring fungal derivatives, R-Val-Val-argininal, R-Ile-Ile-argininal and R-Thr-Thr-argininal, are also excellent competitive inhibitors of plasmin (Chi *et al.*, 1989).

Among the protein inhibitors, the Kunitz-type, serpin, soybean and limabean trypsin inhibitors inactivate the various molecular forms of plasmin (Vogel *et al.*, 1968; Sakurama, 1984; Steiner *et al.*, 1987; Ascenzi *et al.*, 1990). $\alpha_2$-Macroglobulin forms stoichiometric complexes with plasmin and inhibits access to the active site of only large molecular mass substrates and inhibitors (Cummings & Castellino, 1984). The physiologically relevant, fast-acting inhibitor of plasmin in plasma is $\alpha_2$-antiplasmin (Wiman & Collen, 1977).

The different molecular forms of plasmin interact with the same substrates and inhibitors, but to different extents (Robbins *et al.*, 1981). A major difference is noted, in comparison to plasmin, in the enzymic specificity of the streptokinase–plasmin complex. While the activity of plasmin on small synthetic substrates is largely unaffected in the complex, the nonspecific protease activity is dramatically reduced. Importantly, plasmin alone does not catalyze activation of plasminogen, but the streptokinase–plasmin complex, although possessing diminished general proteolytic activity, serves as a very efficient plasminogen activator (Schick & Castellino, 1974).

## Structural Chemistry

The mature molecule of human plasminogen consists of 791 amino acid residues in a single polypeptide chain (Robbins *et al.*, 1967; Groskopf *et al.*, 1969; Wiman, 1973, 1977; Sottrup-Jensen *et al.*, 1978; Forsgren *et al.*, 1987). Variants of plasminogen in other species are very similar, showing high conservation of the protein sequence (Brunisholz *et al.*, 1981; Brunisholz & Rickli, 1981; Gyenes & Patthy, 1985; Marti *et al.*, 1985; Schaller *et al.*, 1985, 1987, 1989, 1991, 1992; Schaller & Rickli, 1988). No unusual amino acids have been found in the human protein, although recent evidence suggests that post-translational phosphorylation occurs on amino acid side-chains (Barlati *et al.*, 1995). Two major glycoforms of the protein have been discovered in several species; these are separable on affinity chromatography columns using specific ligands for plasminogen (Brockway & Castellino, 1972; Sodetz *et al.*, 1972). In the case of the human protein, one glycoform contains only *O*-linked sialylated trisaccharide on Thr346, and the second possesses this same *O*-linked glycan as well as an Asn289-linked biantennary, bisialylated oligosaccharide on the only *N*-linked consensus sequence in the protein (Hayes & Castellino, 1979a–c). Recent papers have shown that Ser248 bears an *O*-linked trisaccharide in one glycoform (Pirie-Shepherd *et al.*, 1997) and that Ser578 is a major phosphorylation locus in both of these human plasminogen forms (Wang *et al.*, 1997). Regarding glycosylation patterns, similarities have been found in other species of plasminogen (Marti *et al.*, 1988). A number of isoelectric forms have been observed for each glycoform (Wallen & Wiman, 1970, 1972; Summaria *et al.*, 1972), some of which are attributable to variable sialylation (Sodetz *et al.*, 1972; Siefring & Castellino, 1974; Pirie-Shepherd *et al.*, 1995) and others that may be due to polymorphisms in the protein sequence in normal populations (Raum *et al.*, 1980; Nishigaki & Omoto, 1982; Malinowski *et al.*, 1984; Forsgren *et al.*, 1987; Murray *et al.*, 1987; Petersen *et al.*, 1990; Nevo *et al.*, 1992; Yiping *et al.*, 1993). No amino acid sequence differences have been identified between the plasminogen glycoforms, including the area surrounding their *N*-linked glycosylation consensus sequences (Powell & Castellino, 1983).

Plasmin is a two-chain serine protease that is derived from plasminogen. Two peptide bonds in plasminogen are cleaved to form plasmin, one at Arg561+Val562 (original human plasminogen numbering), catalyzed by plasminogen activators (Robbins *et al.*, 1967), and a second, Lys77+Lys78, with secondary cleavages within the 77 residue peptide (Wallen, 1978), catalyzed only by the plasmin formed during the activation (Sodetz & Castellino, 1975; Violand & Castellino, 1976). The peptide chains are linked by two disulfide bonds, Cys548-Cys666 and Cys558-Cys566 (Wiman, 1977; Sottrup-Jensen *et al.*, 1978). The heavy, noncatalytic chain of plasmin consists of residues Lys78-Arg561, and is responsible for regulation of plasmin activity via interactions with various effector molecules (for a recent review, see Castellino, 1995). This polypeptide is composed of five homologous kringle domains, each of about 80 amino acid residues, and with three disulfide bridges. These are located at residues C84–C162 (kringle 1), C166–C243 (kringle 2), C256–C333 (kringle 3), C358–C435 (kringle 4), and C462–C541 (kringle 5), with interconnecting stretches of peptides (Sottrup-Jensen *et al.*, 1978). The kringle 1 module is preceded by the above-mentioned 77 residue activation peptide (Wallen & Wiman, 1970) that is usually cleaved from the final plasmin. However, this latter step is not necessary for expression of plasmin activity. The light chain of plasmin is homologous to serine proteases such as trypsin and pancreatic elastase (Wiman, 1977), and is responsible for the catalytic activities of the enzyme. In human plasmin, partial amino acid sequences surrounding the active-site His (Robbins *et al.*, 1973) and Ser (Groskopf *et al.*, 1969) residues, and subsequent alignments of the entire plasmin protease chain sequences (Wiman, 1977; Forsgren *et al.*, 1987), show the catalytic triad at His603, Asp646 and Ser741. The three-dimensional structure of the protease chain has not been determined, but structures for the isolated kringle 1 and 4 domains in the absence and presence of certain ligands have been published (Mulichak & Tulinsky, 1990; Mulichak *et al.*, 1991; Wu *et al.*, 1991, 1994; Mathews *et al.*, 1996). Other forms of active plasmin exist and are produced by proteolytic

cleavages and concomitant peptide release from the heavy chain. These include (beginning with the amino acid residue of the heavy chain): Glu1-plasmin (Violand & Castellino, 1976) and Lys78-plasmin (Robbins *et al.*, 1967) (both containing all five plasminogen kringles), Val355-plasmin (Powell & Castellino, 1981) (kringles 4, 5 and the protease chain), Val442-plasmin (Sottrup-Jensen *et al.*, 1978; Powell & Castellino, 1980) (miniplasmin, containing kringle 5 and the protease chain), and Pro551-plasmin (Wang *et al.*, 1995; Wang & Reich, 1995) (microplasmin, containing none of the kringles). Their properties are described in the references cited.

## *Preparation*

Plasminogen is readily isolated from plasma by affinity chromatography on lysine-Sepharose (Deutsch & Mertz, 1970). The two glycoforms of plasminogen can be efficiently separated by gradient elution of plasminogen from the column (Brockway & Castellino, 1972). An excellent expression system of the cDNA of plasminogen in baculovirus-infected lepidopteran cells has been identified (Whitefleet-Smith *et al.*, 1989).

Plasmin is readily prepared from plasminogen by activation with u-plasminogen activator (Chapter 57). The exact activation and isolation conditions have been described for Glu1-plasmin, Lys78-plasmin and Val442-plasmin, as well as for isolation of the intact protease chain of plasmin (Robbins *et al.*, 1981). Similarly, the procedures for isolation of Val355-plasmin (Powell & Castellino, 1981) and Pro551-plasmin (Wang *et al.*, 1995) have been published.

## *Biological Aspects*

Plasminogen is synthesized in the liver and secreted into the plasma. The cDNA (Forsgren *et al.*, 1987) and gene (Petersen *et al.*, 1990) for human plasminogen have been cloned and sequenced. The cap site of the plasminogen mRNA has been mapped to within 161 bases upstream of the initiation ATG sequence (Malgaretti *et al.*, 1990). The gene for human plasminogen encompasses 52.5 kb and has been mapped to chromosome 6q26–27 (Murray *et al.*, 1987), along with a family of related genes and pseudogenes, including that for apolipoprotein (a) (Malgaretti *et al.*, 1990). The coding sequence of plasminogen includes a 57 bp signal sequence and a total of 2373 nucleotides for the mature protein. The gene consists of a total of 19 exons, which range in size from 75 to 387 bp, with 18 introns of type I, type II and type O. Each kringle module is encoded by two exons, and three exons compose the protease domain, each coding for a peptide stretch containing one of the active-site residues. The coding of the different domains of the protein by separate exons argues strongly for exon shuffling in the evolution of the group of kringle-containing proteins.

Whereas several different proteases are able to activate plasminogen *in vitro*, including factor XIIa (Chapter 49) (Goldsmith *et al.*, 1978), factor XIa (Chapter 51) (Mandle & Kaplan, 1979), plasma kallikrein (Chapter 50) (Colman, 1980), and even trypsin (Chapter 3) (Kocholaty *et al.*, 1952), the most widely studied *in vivo* plasminogen activators

S

are t-plasminogen activator (Chapter 58) and u-plasminogen activator (Chapter 57), which are clinically employed for thrombolytic therapy. Bacterial plasminogen activators (streptokinase and staphylokinase), which function in that capacity only via stoichiometric complexes with plasminogen or plasmin (Castellino, 1983) are also important because of their *in vivo* thrombolytic potential. These activators, to various degrees, activate plasminogen in solution, when bound to a variety of cell types, and when bound to fibrin. Each activator generates a somewhat different form of plasmin.

Plasmin is a key enzyme in blood clot lysis, and its major physiological substrates are fibrinogen and fibrin. Based on phenotypes of heterozygous patients with type I or type II defects in plasminogen, in whom the clinical histories involved venous thromboses, it is clear that clot lysis is indeed an important function of plasmin (Miyato *et al.*, 1982; Soria *et al.*, 1983; Miyata *et al.*, 1984; Manabe & Matsuda, 1985; Ichinose *et al.*, 1991; Azuma *et al.*, 1993). The phenotype of homozygous plasminogen-deficient mice shows not only severe thrombosis (Bugge *et al.*, 1995; Ploplis *et al.*, 1995), but also impaired wound healing (Romer *et al.*, 1996) and reduced fertility (Ploplis *et al.*, 1995). Plasmin appears to be centrally involved in other physiological and pathological processes also. It is believed to function in processes where cell movement is essential, such as macrophage invasion in inflammation (Unkeless *et al.*, 1973; Wohlwend *et al.*, 1987), mammary cell involution after lactation (Ossowski *et al.*, 1979), breakdown of the follicular wall for ovulation (Reich *et al.*, 1985), trophoblast invasion into the endometrium during embryogenesis (Strickland *et al.*, 1976; Sappino *et al.*, 1989), angiogenesis (Gross *et al.*, 1983), and keratinocyte accumulation after wound healing (Morioka *et al.*, 1991). In addition, plasmin has been strongly implicated as an important mediator in pathological processes of cell migration that are involved in tumor cell growth, invasion of surrounding tissues, and metastasis (Ossowski & Reich, 1983; Ossowski, 1992). An important role for plasmin in these latter processes is supported by the ability of cell-bound plasmin to directly degrade extracellular matrix proteins (Mullin & Rohrlich, 1983), such as proteoglycans (Edmonds-Alt *et al.*, 1980), fibronectin (Jilek, 1977), laminin (Schlechte *et al.*, 1989) and type IV collagen (MacKay *et al.*, 1990). Plasmin may also be indirectly responsible for lysis of matrix proteins through activation of metalloprotease proenzymes, such as prostromelysin (Chapter 393) and procollagenase (Chapter 389) (Stricklin *et al.*, 1977; He *et al.*, 1989). The degradation of the extracellular matrix facilitates cell migration into surrounding areas.

Plasmin also functions in the conversion of latent cell-associated transforming growth factor $\beta_1$ (TGF$\beta_1$) to its active form, and via this route is involved in regulation of the inflammatory response (Khalil *et al.*, 1996). In addition, plasmin plays a major role as an extracellular or pericellular protease. In these cases, plasmin may be active in pathways such as proenzyme conversions in the classic (Agostoni *et al.*, 1994) and alternate (Brade *et al.*, 1974) complement pathways, in the contact phase of blood coagulation (Cochrane *et al.*, 1974), in the proinsulin to insulin conversion (Virgi *et al.*, 1980), in bradykinin generation from kininogen (Habal *et al.*, 1976), and in proteolytic activation/destruction of other plasma proteins, such as $\gamma$-globulin (Janeway *et al.*, 1968),

coagulation factor V (Lee & Mann, 1989), coagulation factor VIII (Denson & Redman, 1980), and anticoagulation protein C (Chapter 56) (Varadi *et al.*, 1994).

Many of the biologically relevant protein–protein interactions involving plasminogen and plasmin involve anchoring of these latter proteins through lysine-binding sites in the kringle domains. For example, digestion of virgin fibrin clots by plasmin, exposing C-terminal lysine residues, enhances binding of plasminogen and plasmin to these clots, and thus stimulates clot lysis (Christensen, 1985; Tran-Thang *et al.*, 1986). By releasing these terminal lysine residues carboxypeptidase U-type enzymes may interfere with clot dissolution (Bajzar *et al.*, 1995). Other proteins also interact with plasmin if they contain C-terminal lysine residues (Hortin *et al.*, 1988; Hamanoue *et al.*, 1994), or even internal lysine $\varepsilon$-amino groups with carboxylate groups of Asp and Glu side-chains appropriately positioned to mimic C-terminal Lys or $\varepsilon$-aminocaproic acid (Sugiyama *et al.*, 1988). The kringle domains that are known to bind these $\omega$-amino acid ligands, and that are probably involved in these interactions, are kringle 1 (Menhart *et al.*, 1991), kringle 2 (Marti *et al.*, 1994), kringle 4 (McCance *et al.*, 1994) and kringle 5 (Novokhatny *et al.*, 1989; McCance *et al.*, 1994). In addition, binding of $\omega$-amino acids such as lysine to kringles greatly increases the rate of activation of Glu1-plasminogen, through induction of a very large conformational change in the molecule (Menhart *et al.*, 1995; McCance & Castellino, 1995).

## Distinguishing Features

Most of the enzymatic activities of plasmin are also catalyzed to some degree by other closely related enzymes, such as trypsin, or enzymes of the coagulation or fibrinolysis systems. *In vitro*, lysis of fibrin clots is also catalyzed by trypsin, although not in as specific a fashion as by plasmin. The ability of human plasmin to form complexes with human-host-derived streptokinases (Castellino *et al.*, 1976) and staphylokinases (Collen *et al.*, 1993), and, as such, to serve to activate plasminogens from a variety of species, is a well-defined property of human plasmin. Similar complexes may form with other animal species of plasmin and their respective host-derived bacterial plasminogen activators. Human plasminogen and plasmin can be distinguished from other proteins by polyclonal antibodies to human plasminogen prepared in the usual fashion, or purchased commercially (Enzyme Research, Hematologic Technologies) (see Appendix 2 for full names and addresses of suppliers). An excellent library of monoclonal antibodies that recognize specific domains of human plasminogen and plasmin has been generated (Ploplis *et al.*, 1982; Cole & Castellino, 1984; Cummings & Castellino, 1985). Some of these monoclonal antibodies are commercially available (Enzyme Research).

## Further Reading

More complete reviews on the topics of plasminogen and plasmin structure, plasminogen activation, and plasmin chemistry have been written (Castellino, 1981, 1983, 1984, 1995;

Collen & Lijnen, 1986; Castellino *et al.*, 1988), and an excellent review on plasminogen-deficiencies is that of Robbins (1992).

## References

Agostoni, A., Gardinali, M., Frangi, D., Cafaro, C., Conciato, L., Sponzilli, C., Salvioni, A., Cugno, M. & Cicardi, M. (1994) Activation of complement and kinin systems after thrombolytic therapy in patients with acute myocardial infarction: a comparison between streptokinase and recombinant tissue-type plasminogen activator. *Circulation* **90**, 2666–2670.

Andrews, J.M., Roman, D.P. & Bing, D.H. (1978) Inhibition of four human serine proteases by substituted benzamidines. *J. Med. Chem.* **21**, 1202–1207.

Ascenzi, P., Amiconi, G., Bolognesi, M., Menegatti, E. & Guarneri, M. (1990) Binding of the bovine basic pancreatic trypsin inhibitor (Kunitz) to human Glu1-, Lys78-, Val442-, and Val561-plasmin: a comparative study. *Biochim. Biophys. Acta* **1040**, 134–136.

Astrup, T. & Permin, P.N. (1947) Fibrinolysis in the animal organism. *Nature* **159**, 681–700.

Azuma, H., Uno, Y., Shigekiyo, T. & Saito, S. (1993) Congenital plasminogen deficiency caused by a Ser572 to Pro mutation. *Blood* **82**, 475–480.

Bajzar, L., Manuel, R. & Nesheim, M.E. (1995) Purification and characterization of TAFI, a thrombin-activatable fibrinolysis inhibitor. *J. Biol. Chem.* **270**, 14477–14484.

Barlati, S., DePetro, G., Bona, C., Paracini, F. & Tonelli, M. (1995) Phosphorylation of human plasminogen activators and plasminogen. *FEBS Lett.* **363**, 170–174.

Brade, V., Nicholson, A., Bitter-Suermann, D. & Hadding, V. (1974) Formation of the C-3 cleaving properdin enzyme on zymosen. *J. Immunol.* **113**, 1735–1743.

Brockway, W.J. & Castellino, F.J. (1972) Measurement of the binding of antifibrinolytic amino acids to various plasminogens. *Arch. Biochem. Biophys.* **151**, 194–199.

Brunisholz, R.A. & Rickli, E.E. (1981) Primary structure of porcine plasminogen. Isolation and characterization of CNBr-fragments and their alignment within the polypeptide chain. *Eur. J. Biochem.* **119**, 15–52.

Brunisholz, R.A., Lerch, P.G., Schaller, J., Rickli, E.E., Lergier, W., Manneberg, M. & Gillessen, D. (1981) Comparison of the primary structure of the N-terminal CNBr fragments of human, bovine and porcine plasminogen. *Eur. J. Biochem.* **114**, 465–470.

Bugge, T.H., Flick, M.J., Daugherty, C.C. & Degen, J.L. (1995) Plasminogen deficiency causes severe thrombosis but is compatible with development and reproduction. *Gene Dev.* **9**, 794–807.

Castellino, F.J. (1981) Recent advances in the chemistry of the fibrinolytic system. *Chem. Rev.* **81**, 431–446.

Castellino, F.J. (1983) Plasminogen activators. *Bioscience* **33**, 647–650.

Castellino, F.J. (1984) Biochemistry of human plasminogen. *Semin. Thromb. Haemost.* **10**, 18–23.

Castellino, F.J. (1995) Plasminogen. In: *Molecular Basis of Thrombosis and Hemostasis* (High, K.A. & Roberts, H.R., eds). New York: Marcel Dekker, pp. 495–515.

Castellino, F.J. & Powell, J.R. (1981) Human plasminogen. *Methods Enzymol.* **80**, 365–378.

Castellino, F.J., Sodetz, J.M., Brockway, W.J. & Siefring, G.E. (1976) Streptokinase. *Methods Enzymol.* **45**, 244–257.

Castellino, F.J., Urano, T., De Serrano, V., Morris, J.P. & Chibber, B.A.K. (1988) Control of human plasminogen activation. *Haemostasis* **18**, 15–23.

Chase, T. & Shaw, E. (1969) Comparison of the esterase activities of trypsin, plasmin, and thrombin on guanidino esters. Titration of the enzymes. *Biochemistry* **8**, 2212–2224.

Chi, C.-W., Liu, H.-Z., Chibber, B.A.K. & Castellino, F.J. (1989) Inhibition of enzymic activity of blood coagulation and fibrinolytic serine proteases by a new leupeptin-like inhibitor, and its structural analogues, isolated from *Streptomyces griseus. J. Antibiot.* **42**, 1506–1512.

Christensen, L.R. (1945) Streptococcal fibrinolysis: proteolytic reaction due to serum enzyme activated by streptococcal fibrinolysin. *J. Gen. Physiol.* **28**, 363–383.

Christensen, L.R. & Macleod, C.M. (1945) Proteolytic enzyme of serum: characterization, activation, and reaction with inhibitors. *J. Gen. Physiol.* **28**, 559–583.

Christensen, U. (1985) C-terminal lysine residues of fibrinogen fragments essential for binding to plasminogen. *FEBS Lett.* **182**, 43–46.

Cochrane, C.G., Revak, S.D. & Wuepper, W.G. (1974) Activation of Hageman factor in solid and fluid phases. *J. Exp. Med.* **138**, 1564–1583.

Cole, K.R. & Castellino, F.J. (1984) The binding of antifibrinolytic amino acids to kringle 4-containing fragments of plasminogen. *Arch. Biochem. Biophys.* **229**, 568–575.

Collen, D. & Lijnen, H.R. (1986) The fibrinolytic system in man. *CRC Crit. Rev. Oncol. Hematol.* **4**, 249–301.

Collen, D., Schlott, B., Engelborghs, Y., Vanhoef, B., Hartmann, M., Lijnen, H.R. & Behnke, D. (1993) On the mechanism of the activation of human plasminogen by recombinant staphylokinase. *J. Biol. Chem.* **268**, 8284–8289.

Colman, R.W. (1980) Activation of plasminogen by human plasma kallikrein. *Biochem. Biophys. Res. Commun.* **35**, 273–279.

Conradi, H. (1902) Uber die beziehungen der autolyse zur blutgerrinnung. *Beitr. z. Chem. Phys. u. Path.* **i**, 136–142.

Cummings, H.S. & Castellino, F.J. (1984) Interaction of human plasmin with human $\alpha_2$-macroglobulin. *Biochemistry* **23**, 105–111.

Cummings, H.S. & Castellino, F.J. (1985) A monoclonal antibody to the $\varepsilon$-aminocaproic acid binding site on the kringle 4 region of human plasminogen that accelerates the activation of Glu1-plasminogen by urokinase. *Arch. Biochem. Biophys.* **236**, 612–618.

Denson, K.W. & Redman, C.W. (1980) The destruction of factor VIII clotting activity by thrombin and plasmin. *Thromb. Res.* **18**, 547–549.

Deutsch, D.G. & Mertz, E.T. (1970) Plasminogen: purification from human plasma by affinity chromatography. *Science* **170**, 1095–1096.

Edmonds-Alt, X., Quisquater, E. & Vaes, G. (1980) Proteoglycan- and fibrin-degrading neutral proteinase activities of Lewis lung carcinoma cells. *Eur. J. Cancer* **16**, 1257–1261.

Ferroni, R., Menegatti, E., Orlandini, P., Guarneri, M., Taddeo, U., Bolognesi, M., Ascenzi, P., Bertollini, A. & Amiconi, G. (1986) Aromatic tetra-amidines: antiproteolytic and antiesterolytic activities towards serine proteinases involved in blood coagulation and clot lysis. *Farmaco* **41**, 464–470.

Ferroni, R., Menegatti, E., Guarneri, M., Taddeo, U., Ascenzi, P. & Amiconi, G. (1988) Aromatic amidines: inhibitory effect on purified plasma serine proteinases, blood coagulation and platelet aggregation. *Farmaco* **43**, 5–13.

Fleischer, M.S. & Loeb, L. (1915) On tissue fibrinolysis. *J. Biol. Chem.* **21**, 477–501.

Forsgren, M., Råden, B., Israelsson, M., Larsson, K. & Hedín, L.-O. (1987) Molecular cloning and characterization of a full-length cDNA clone for human plasminogen. *FEBS Lett.* **213**, 254–260.

Goldsmith, G.H., Saito, H. & Ratnoff, O.D. (1978) The activation of plasminogen by Hageman factor (factor XII) and Hageman factor fragments. *J. Clin. Invest.* **62**, 54–60.

Groskopf, W.R., Hsieh, B., Summaria, L. & Robbins, K.C. (1969) Studies on the active center of human plasmin. Partial amino acid sequence of a peptide containing the active center serine residue. *J. Biol. Chem.* **244**, 359–365.

Gross, J.L., Moscatelli, D. & Rifkin, D.B. (1983) Increased capillary endothelial cell protease activity in response to angiogenic stimuli *in vitro. Proc. Natl Acad. Sci. USA* **80**, 2623–2627.

Gyenes, M. & Patthy, L. (1985) The kringle 4 domain of chicken plasminogen. *Biochim. Biophys. Acta* **832**, 326–330.

Habal, F.M., Burrowes, C.E. & Movat, H.Z. (1976) Generation of kinin by plasma kallikrein and plasmin and the effect of $\alpha$1-antitrypsin and antithrombin III on the kininogenases. *Adv. Exp. Med. Biol.* **70**, 23–36.

Hamanoue, M., Takemoto, N., Hattori, T., Kato, K. & Kohsaka, S. (1994) Plasminogen binds specifically to $\alpha$-enolase on rat neuronal plasma membrane. *J. Neurochem.* **63**, 2048–2057.

Hayes, M.L. & Castellino, F.J. (1979a) Carbohydrate of human plasminogen variants. I. Carbohydrate composition and glycopeptide isolation and characterization. *J. Biol. Chem.* **254**, 8768–8771.

Hayes, M.L. & Castellino, F.J. (1979b) Carbohydrate of human plasminogen variants. II. Structure of the asparagine-linked oligosaccharide unit. *J. Biol. Chem.* **254**, 8772–8776.

Hayes, M.L. & Castellino, F.J. (1979c) Carbohydrate of human plasminogen variants. III. Structure of the *O*-glycosidically-linked oligosaccharide unit. *J. Biol. Chem.* **254**, 8777–8780.

He, C., Wilhelm, S.M., Pentland, A.P., Marmer, B.L., Grant, G.A., Eisen, A.Z. & Goldberg, G.I. (1989) Tissue cooperation in a proteolytic cascade activating human interstitial collagenase. *Proc. Natl Acad. Sci. USA* **86**, 2632–2636.

Hortin, G.L., Gibson, B.L. & Fok, K.F. (1988) Alpha 2-antiplasmin's carboxy-terminal lysine residue is a major site of interaction with plasmin. *Biochem. Biophys. Res. Commun.* **155**, 591–596.

Ichinose, A., Espling, E.E., Takamatsu, J., Saito, H., Shinmyozu, K., Maruyama, I., Petersen, T.E. & Davie, E.W. (1991) Two types of abnormal genes for plasminogen in families with a predisposition for thrombosis. *Proc. Natl Acad. Sci. USA* **88**, 115–119.

Jameson, G.W., Roberts, D.V., Adams, R.W., Kyle, W.S.A. & Elmore, D.T. (1973) Determination of the operational molarity of solutions of bovine $\alpha$-chymotrypsin, trypsin, and factor Xa by spectrofluorometric titration. *Biochem. J.* **131**, 107–117.

Janeway, C.A., Merler, E., Rosen, F.S., Salmon, S. & Crain, J.O. (1968) Intravenous $\gamma$-globulin: Mechanism of gamma-globulin fragments in normal and agamma-globulinemic persons. *N. Engl. J. Med.* **278**, 919–923.

Jilek, F. (1977) Cold-insoluble globulin III. Cyanogen bromide and plasminolysis fragments containing a label introduced by transamidation. *Z. Physiol. Chem.* **358**, 1165–1168.

Kaplan, M.H. (1944) Nature and role of the lytic factor in hemolytic streptococcal fibrinolysis. *Proc. Soc. Exp. Biol. Med.* **57**, 40–43.

Khalil, N., Corne, S., Whitman, C. & Yacyshyn, H. (1996) Plasmin regulates the activation of cell-associated latent TGF-$\beta$1 secreted by rat alveolar macrophages after *in vivo* bleomycin injury. *Am. J. Respir. Cell Mol. Biol.* **15**, 252–259.

Kocholaty, W., Ellis, W. & Jensen, H. (1952) Activation of plasminogen by trypsin and plasmin. *Blood* **7**, 882–889.

Lee, C.D. & Mann, K.G. (1989) Activation/inactivation of human factor V by plasmin. *Blood* **73**, 185–190.

Lottenberg, R., Christensen, U., Jackson, C.M. & Coleman, P.L. (1981) Assay of coagulation proteases using peptide chromogenic and fluorogenic substrates. *Methods Enzymol.* **80**, 341–361.

MacKay, A.R., Corbitt, R.H., Hartzler, J.L. & Thorgeirsson, U.P. (1990) Basement membrane type IV collagen degradation: evidence for the involvement of a proteolytic cascade independent of metalloproteinases. *Cancer Res.* **50**, 5997–6001.

Malgaretti, N., Bruno, L., Pontoglio, M., Candiani, G., Meroni, G., Ottolenghi, S. & Taramelli, R. (1990) Definition of the transcription initiation site of human plasminogen gene in liver and non hepatic cell lines. *Biochem. Biophys. Res. Commun.* **173**, 1013–1018.

Malinowski, D.P., Sadler, J.E. & Davie, E.W. (1984) Characterization of a complementary DNA coding for human and bovine plasminogen. *Biochemistry* **23**, 4243–4250.

Manabe, S.-I. & Matsuda, M. (1985) Homozygous protein C deficiency combined with heterozygous dysplasminogenemia found in a 21-year-old thrombophilic male. *Thromb. Res.* **39**, 333–341.

Mandle, R.J. & Kaplan, A.P. (1979) Generation of fibrinolytic activity by the interaction of activated factor XI and plasminogen. *Blood* **54**, 850–861.

Markwardt, F., Landmann, H. & Walsmann, P. (1968) Comparative studies on the inhibition of trypsin, plasmin, and thrombin by derivatives of benzylamine and benzamidine. *Eur. J. Biochem.* **6**, 502–506.

Marti, D., Schaller, J., Ochensberger, B. & Rickli, E.E. (1994) Expression, purification and characterization of the recombinant kringle 2 and kringle 3 domains of human plasminogen and analysis of their binding affinity for $\omega$-aminocarboxylic acids. *Eur. J. Biochem.* **219**, 455–462.

Marti, T., Schaller, J. & Rickli, E.E. (1985) Determination of the complete amino-acid sequence of porcine miniplasminogen. *Eur. J. Biochem.* **149**, 279–285.

Marti, T., Schaller, J., Rickli, E.E., Schmid, K., Kamerling, J.P., Gerwig, G.J., van Halbeek, H. & Vliegenhart, J.F. (1988) The N- and O-linked carbohydrate chains of human, bovine and porcine plasminogen. Species specificity in relation to sialylation and fucosylation patterns. *Eur. J. Biochem.* **173**, 57–63.

Mathews, I.I., Vanderhoff-Hanaver, P., Castellino, F.J. & Tulinsky, A. (1996) Crystal structures of the recombinant kringle 1 domain of human plasminogen in complexes with the ligands $\varepsilon$-aminocaproic acid and *trans*-4-(aminomethyl)cyclohexane-1-carboxylic acid. *Biochemistry* **35**, 2567–2576.

McCance, S.G. & Castellino, F.J. (1995) Contributions of individual kringle domains toward maintenance of the chloride-induced tight conformation of human Glu1-plasminogen. *Biochemistry* **34**, 9581–9586.

McCance, S.G., Menhart, N. & Castellino, F.J. (1994) Amino acid residues of the kringle-4 and kringle-5 domains of human plasminogen that stabilize their interactions with $\Omega$-amino acid ligands. *J. Biol. Chem.* **269**, 32405–32410.

Menhart, N., Sehl, L.C., Kelley, R.F. & Castellino, F.J. (1991) Construction, expression and purification of recombinant kringle 1 of human plasminogen and analysis of its interaction with $\omega$-amino acids. *Biochemistry* **30**, 1948–1957.

Menhart, N., Hoover, G.J., McCance, S.G. & Castellino, F.J. (1995) Roles of individual kringle domains in the functioning of positive and negative effectors of human plasminogen activation. *Biochemistry* **34**, 1482–1488.

Milstone, H. (1941) Factor in normal blood which participates in hemolytic streptococcal fibrinolysis. *J. Immunol.* **42**, 109–116.

Miyata, T., Iwanaga, S., Sakata, Y., Aoki, N., Takamatsu, J. & Kamiya, T. (1984) Plasminogens Tochigi II and Nagoya: Two additional molecular defects with Ala600→Thr replacement found in plasmin light chain variants. *J. Biochem.* **96**, 277–287.

Miyata, T., Iwanaga, S., Sayata, Y. & Aoki, N. (1982) Plasminogen Tochigi: Inactive plasmin resulting from replacement of Ala600 by Thr at the active site. *Proc. Natl Acad. Sci. USA* **79**, 6132–6136.

Morioka, S., Lazarus, G.S., Baird, J.L. & Jensen, P.J. (1991) Migrating keratinocytes express urokinase-type plasminogen activator. *J. Invest. Dermatol.* **88**, 418–423.

Mulichak, A.M. & Tulinsky, A. (1990) Structure of the lysine-fibrin subsite of human plasminogen kringle 4. *Blood Coagul. Fibrinolysis* **1**, 673–679.

Mulichak, A.M., Tulinsky, A. & Ravichandran, K.G. (1991) Crystal and molecular structure of human plasminogen kringle 4 refined at 1.9-Å resolution. *Biochemistry* **30**, 10576–10588.

Mullin, D.E. & Rohrlich, S.T. (1983) The role of proteinases in cellular invasiveness. *Biochim. Biophys. Acta* **695**, 177–214.

Murray, J.C., Buetow, K.H., Donovan, M., Hornung, S., Motulsky, A.G., Disteche, C., Dyer, K., Swisshelm, K., Anderson, J., Giblet, E., Sadler, E., Eddy, R. & Shows, T.B. (1987) Linkage disequilibrium of plasminogen polymorphisms and assignment of the gene to human chromosome 6q26–6q27. *Am. J. Hum. Genet.* **40**, 338–350.

Nevo, S., Cleve, H., Koller, A., Eigel, E., Patutschnick, W., Kanaaneh, H. & Joel, A. (1992) Serum protein polymorphisms in Arab Moslems and Druze of Israel – BF, F13B, AHSG, GC, PLG, PI, and TF. *Hum. Biol.* **64**, 587–603.

Nishigaki, T. & Omoto, K. (1982) Genetic polymorphism of human plasminogen in Japanese: correspondence of alleles thus far reported in Japanese and difference of activity among phenotypes. *Jap. J. Hum. Genet.* **27**, 341–348.

Novokhatny, V.V., Matsuka, Y.V. & Kudinov, S.A. (1989) Analysis of ligand binding to kringles 4 and 5 fragments of human plasminogen. *Thromb. Res.* **53**, 243–252.

Ossowski, L. (1992) Invasion of connective tissue by human carcinoma cell lines. Requirement for urokinase, urokinase receptor, and interstitial collagenase. *Cancer Res.* **52**, 6754–6760.

Ossowski, L. & Reich, E. (1983) Antibodies to plasminogen activator inhibit human tumor metastasis. *Cell* **35**, 611–619.

Ossowski, L., Biegel, D. & Reich, E. (1979) Mammary plasminogen activator: Correlation with involution hormonal modulation and comparison between normal and neoplastic tissue. *Cell* **16**, 929–940.

Petersen, T.E., Martzen, M.R., Ichinose, A. & Davie, E.W. (1990) Characterization of the gene for human plasminogen, a key proenzyme in the fibrinolytic system. *J. Biol. Chem.* **265**, 6104–6111.

Pierzchala, P.A., Dorn, C.P. & Zimmerman, M. (1979) A new fluorogenic substrate for plasmin. *Biochem. J.* **183**, 555–559.

Pirie-Shepherd, S.R., Jett, E.A., Andon, N.L. & Pizzo, S.V. (1995) Sialic acid content of plasminogen 2 glycoforms as a regulator of fibrinolytic activity – isolation, carbohydrate analysis, and kinetic characterization of six glycoforms of plasminogen. *J. Biol. Chem.* **270**, 5877–5881.

Pirie-Sheperd, S.R., Stevens, R.D., Andon, N.L., Enghild, J.J. & Pizzo, S.V. (1997) Evidence for a novel *O*-linked sialylated trisaccharide on Ser-248 of human plasminogen 2. *J. Biol. Chem.* **272**, 7408–7411.

Ploplis, V.A., Cummings, H.S. & Castellino, F.J. (1982) Monoclonal antibodies to discrete regions of human Glu1-plasminogen. *Biochemistry* **21**, 5891–5897.

Ploplis, V.A., Carmeliet, P., Vazirzadeh, S., Van Vlaenderen, I., Moons, L., Plow, E.F. & Collen, D. (1995) Effects of disruption of the plasminogen gene on thrombosis, growth, and health in mice. *Circulation* **92**, 2585–2593.

Powell, J.R. & Castellino, F.J. (1980) Activation of neo-plasminogen-Val442 by urokinase and streptokinase and a kinetic characterization of neo-plasmin-Val442. *J. Biol. Chem.* **255**, 5329–5335.

Powell, J.R. & Castellino, F.J. (1981) Isolation of human Val354-plasminogen as an elastolytic fragment of human Glu1-plasminogen. *Biochem. Biophys. Res. Commun.* **102**, 46–52.

Powell, J.R. & Castellino, F.J. (1983) Amino acid sequence analysis of the Asn288 region of the carbohydrate variants of human plasminogen. *Biochemistry* **22**, 923–927.

Raum, D., Marcus, A. & Alper, C.A. (1980) Genetic polymorphism of human plasminogen. *Am. J. Hum. Genet.* **32**, 681–689.

Reich, R., Miskin, R. & Tsafriri, A. (1985) Follicular plasminogen activator: involvement in ovulation. *Endocrinology* **116**, 516–521.

Robbins, K.C. (1992) Dysplasminogenemias. *Prog. Cardiovasc. Dis.* **34**, 295–308.

Robbins, K.C. & Summaria, L. (1970) Human plasminogen and plasmin. *Methods Enzymol.* **19**, 184–199.

Robbins, K.C., Summaria, L., Hsieh, B. & Shah, R.J. (1967) The peptide chains of human plasmin. Mechanism of activation of human plasminogen to plasmin. *J. Biol. Chem.* **242**, 2333–2342.

Robbins, K.C., Bernabe, P., Arzadon, L. & Summaria, L. (1973) The primary structure of human plasminogen. II. The histidine loop of human plasmin: light (B) chain active center histidine sequence. *J. Biol. Chem.* **248**, 1631–1633.

Robbins, K.C., Summaria, L. & Wohl, R.C. (1981) Human plasmin. *Methods Enzymol.* **80**, 379–408.

Romer, J., Bugge, T.H., Pyke, C., Lund, L.R., Flick, M.J., Degen, J.L. & Dano, K. (1996) Impaired wound healing in mice with a disrupted plasminogen gene. *Nature Med.* **2**, 287–292.

Saino, T., Someno, T., Ishii, S., Aoyagi, T. & Umezawa, H. (1988) Protease-inhibitory activities of leupeptin analogues. *J. Antibiot.* **41**, 220–225.

Sakurama, S. (1984) The interaction of human α1-antitrypsin with human plasmin. *Hokkaido Igaku Zasshi* **59**, 48–58.

Sappino, A.P., Huarte, J., Belin, D. & Vassalli, J.D. (1989) Plasminogen activators in tissue remodeling and invasion: mRNA localization in mouse ovaries and implanting embryos. *J. Cell Biol.* **109**, 2471–2479.

Schaller, J. & Rickli, E.E. (1988) Structural aspects of the plasminogen of various species. *Enzyme* **40**, 63–69.

Schaller, J., Moser, P.W., Dannegger-Muller, G.A.K., Rosselet, S.J., Kampfer, U. & Rickli, E.E. (1985) Complete amino acid sequence of bovine plasminogen. Comparison with human plasminogen. *Eur. J. Biochem.* **149**, 267–278.

Schaller, J., Marti, T., Rosselet, S.J., Kampfer, U. & Rickli, E.E. (1987) Amino acid sequence of the heavy chain of porcine plasmin. Comparison of the carbohydrate attachment sites with the human and bovine species. *Fibrinolysis* **1**, 91–102.

Schaller, J., Straub, C., Kampfer, U. & Rickli, E.E. (1989) Complete amino acid sequence of canine miniplasminogen. *Protein Seq. Data Anal.* **2**, 445–450.

Schaller, J., Straub, C., Kampfer, U. & Rickli, E.E. (1991) Complete amino acid sequence of equine miniplasminogen. *Protein Seq. Data Anal.* **4**, 69–74.

Schaller, J., Straub, C., Kampfer, U. & Rickli, E.E. (1992) Complete amino acid sequence of ovine miniplasminogen. *Protein Seq. Data Anal.* **5**, 21–25.

Schick, L.A. & Castellino, F.J. (1974) Direct evidence for the generation of an active site in the plasminogen moiety of the streptokinase-human plasminogen activator complex. *Biochem. Biophys. Res. Commun.* **57**, 47–54.

Schlechte, W., Murano, G. & Boyd, D. (1989) Examination of the role of the urokinase receptor in human colon cancer mediated laminin degradation. *Cancer Res.* **49**, 6064–6069.

Siefring, G.E. & Castellino, F.J. (1974) The role of sialic acid in the distinct properties of the isozymes of rabbit plasminogen. *J. Biol. Chem.* **249**, 7742–7746.

Sodetz, J.M. & Castellino, F.J. (1972) A comparison of steady- and presteady-state kinetics of bovine and human plasmins. *Biochemistry* **11**, 3167–3171.

Sodetz, J.M. & Castellino, F.J. (1975) The mechanism of activation of rabbit plasminogen by urokinase. *J. Biol. Chem.* **250**, 3041–3049.

Sodetz, J.M., Brockway, W.J. & Castellino, F.J. (1972) Multiplicity of rabbit plasminogen. Physical characterization. *Biochemistry* **11**, 4451–4458.

Sodetz, J.M., Violand, B.N. & Castellino, F.J. (1976) A kinetic characterization of the rabbit plasmin isozymes. *Arch. Biochem. Biophys.* **174**, 209–215.

Soria, J., Soria, C., Bertrand, O., Dunn, F., Drouet, L. & Caen, J. (1983) Plasminogen Paris I: congenital abnormal plasminogen and its incidence in thrombosis. *Thromb. Res.* **32**, 229–238.

Sottrup-Jensen, L., Claeys, H., Zajdel, M., Petersen, T.E. & Magnusson, S. (1978) The primary structure of human plasminogen: isolation of two lysine-binding fragments and one 'mini' plasminogen (MW, 38000) by elastase-catalyzed-specific limited proteolysis. *Prog. Chem. Fibrinolysis Thrombolysis* **3**, 191–209.

Steiner, J.P, Migliorini, M. & Strickland, D.K. (1987) Characterization of the reaction of plasmin with $\alpha_2$-macroglobulin: effect of antifibrinolytic agents. *Biochemistry* **26**, 8487–8495.

Strickland, S., Reich, E. & Sherman, M.Z. (1976) Plasminogen activator in early embryogenesis: Enzyme production by trophoblast and parietal endotherm. *Cell* **9**, 231–240.

Stricklin, G.P., Bauer, E.A., Jeffrey, J.J. & Eisen, A. (1977) Human skin collagenase: Isolation of precursor and active forms from both fibroblast and organ cultures. *Biochemistry* **16**, 1607–1615.

Sugiyama, N., Sasaki, T., Iwamoto, M. & Abiko, Y. (1988) Binding site of $\alpha$2-plasmin inhibitor to plasminogen. *Biochim. Biophys. Acta* **952**, 1–7.

Summaria, L., Hsieh, B., Groskopf, W.R., Robbins, K.C. & Barlow, G.H. (1967) The isolation and characterization of the S-carboxymethyl 8 (light) chain derivative of human plasmin. *J. Biol. Chem.* **242**, 5046–5052.

Summaria, L., Arzadon, L., Bernabe, P. & Robbins, K.C. (1972) Studies on the isolation of the multiple molecular forms of human plasminogen and plasmin by isoelectric focusing methods. *J. Biol. Chem.* **247**, 6757–6762.

Tillet, W.S. & Garner, R.L. (1933) The fibrinolytic activity of hemolytic *streptococci*. *J. Exp. Med.* **58**, 485–502.

Tran-Thang, C., Kruithof, E.K.O., Atkinson, J. & Bachmann, F. (1986) High-affinity binding sites for human Glu-plasminogen unveiled by limited plasmic digestion of human fibrin. *Eur. J. Biochem.* **160**, 599–604.

Unkeless, J.C., Tobia, A., Ossowski, L., Quigley, J.P., Rifkin, D.B. & Reich, E. (1973) An enzymatic function associated with transformations of fibroblasts by oncogenic viruses. *J. Exp. Med.* **137**, 85–111.

Urano, S., Metzger, A.R. & Castellino, F.J. (1989) Plasmin-mediated fibrinolysis by variant recombinant tissue plasminogen activators. *Proc. Natl Acad. Sci. USA* **86**, 2568–2571.

Varadi, K., Philapitsch, A., Santa, T. & Schwarz, H.P. (1994) Activation and inactivation of human protein C by plasmin. *Thromb. Haemost.* **71**, 615–621.

Violand, B.N. & Castellino, F.J. (1976) Mechanism of urokinase-catalyzed activation of human plasminogen. *J. Biol. Chem.* **251**, 3906–3912.

Virgi, M.A.G., Vassalli, J.D., Estensen, R.D. & Reich, E. (1980) Plasminogen activator of islets of Langerhans: modulation by glucose and correlation with insulin production. *Proc. Natl Acad. Sci. USA* **77**, 875–879.

Vogel, R., Trautschold, I. & Werle, E. (1968) *Natural Proteinase Inhibitors*. New York: Academic Press.

Wallen, P. (1978) Chemistry of plasminogen and plasminogen activation. *Prog. Chem. Fibrinolysis Thrombolysis* **3**, 167–181.

Wallen, P. & Wiman, B. (1970) Characterization of human plasminogen I. On the relationship between different molecular forms of plasminogen demonstrated in plasma and found in purified preparations. *Biochim. Biophys. Acta* **221**, 20–30.

Wallen, P. & Wiman, B. (1972) Characterization of human plasminogen. II. Separation and partial characterization of different molecular forms of human plasminogen. *Biochim. Biophys. Acta* **257**, 122–134.

Walsmann, P. (1977) Inhibition of serine proteinases by benzamidine derivatives. *Acta Biol. Med. Ger.* **36**, 1931–1937.

Wang, H., Prorok, M., Bretthauer, R.K. & Castellino, F.J. (1997) Serine-578 I a major phosphorylation locus in human plasma plasminogen. *Biochemistry* **36**, 8100–8106.

Wang, J.Y. & Reich, E. (1995) Structure and function of microplasminogen. 2. Determinants of activation by urokinase and by the bacterial activator streptokinase. *Protein Sci.* **4**, 1768–1779.

Wang, J.Y., Brdar, B. & Reich, E. (1995) Structure and function of microplasminogen. 1. Methionine shuffling, chemical proteolysis, and proenzyme activation. *Protein Sci.* **4**, 1758–1767.

Whitefleet-Smith, J., Rosen, E., McLinden, J., Ploplis, V.A., Fraser, M.J., Tomlinson, J.E., McLean, J.W. & Castellino, F.J. (1989) Expression of human plasminogen cDNA in a baculovirus-infected insect cell system. *Arch. Biochem. Biophys.* **271**, 390–399.

Wiman, B. (1973) Primary structure of peptides released during activation of human plasminogen by urokinase. *Eur. J. Biochem.* **39**, 1–9.

Wiman, B. (1977) Primary structure of the $\beta$ (light) chain of human plasmin. *Eur. J. Biochem.* **76**, 129–137.

Wiman, B. & Collen, D. (1977) Purification and characterization of human antiplasmin. The fast acting plasmin inhibitor in plasma. *Eur. J. Biochem.* **78**, 19–26.

Wohlwend, A., Belin, D. & Vassali, J.-D. (1987) Plasminogen activator-specific inhibitors produced by human monocytes/macrophages. *J. Exp. Med.* **165**, 320–339.

Wu, T.-P., Padmanabhan, K., Tulinsky, A. & Mulichak, A.M. (1991) The refined structure of the ε-aminocaproic acid complex of human plasminogen kringle 4. *Biochemistry* **30**, 10589–10594.

Wu, T.P., Padmanabhan, K.P. & Tulinsky, A. (1994) The structure of recombinant plasminogen kringle 1 and the fibrin binding site.

*Blood Coagul. Fibrinolysis* **5**, 157–166.

Yiping, H., Qing, G. & Meiyun, W. (1993) Genetic polymorphism of human plasminogen (PLG) in a Chinese population. *Eur. J. Immunogenet.* **20**, 91–94.

*Francis J. Castellino*
*Department of Chemistry and Biochemistry,*
*University of Notre Dame,*
*Notre Dame, IN 46556, USA*
*Email: castellino.1@nd.edu*

# 60. Hepsin

## Databanks

*Peptidase classification: clan SA, family S1, MEROPS ID: S01.224*
*NC-IUBMB enzyme classification: none*
*Databank codes:*

| Species | SwissProt | PIR | EMBL (cDNA) | EMBL (genomic) |
|---|---|---|---|---|
| *Homo sapiens* | P05981 | S00845 | X07002 X07732 | – |
| *Rattus norvegicus* | Q05511 | S33777 | X70900 | – |

## Name and History

Hepsin was originally identified from a human liver cDNA library by screening with an oligonucleotide probe designed to the sequence Met-Phe-Cys-Ala-Gly that is highly conserved among serine proteases of family S1 (Leytus *et al*., 1988; Kurachi *et al*., 1994). Because of its liver origin, the name *hepsin* was given to the enzyme.

## Activity and Specificity

Hepsin has all the structural features of a trypsin-like protease (Leytus *et al*., 1988; Kurachi *et al*., 1994). These include the residues His203, Asp257 and Ser353 required to form the active-site triad upon limited proteolysis at Arg162+Ile163 in the conserved sequence, Arg-Ile-Val-Gly-Gly. During its purification, hepsin is partially activated by an unknown protease or proteases, and trypsin treatment *in vitro* significantly increases hepsin activity (Torres-Rosado *et al*., 1993). Synthetic substrates including Bz-Leu-Ser-Arg+NHPhNO$_2$, Bz-Ile-Glu-Phe-Ser-

Arg+NHPhNO$_2$ and Bz-Phe-Val-Arg+NHPhNO$_2$ are cleaved by activated hepsin (Tsuji *et al*., 1991; Torres-Rosado *et al*., 1993). Its detailed substrate specificity is not known, however.

Overexpressed hepsin at the BHK cell surface can apparently activate factor VII, but not factor X, factor IX, prothrombin or protein C (Kazama *et al*., 1995). This hepsin activity is substantially increased by incubation of the cells with trypsin. Thus, cell surface hepsin may have processing activity with high specificity for its potential protein substrates. It is not known whether or not factor VII is among its natural substrates, however.

## Structural Chemistry

The hepsin mRNA is approximately 1.8 kb in size (Leytus *et al*., 1988), and the cDNA sequence indicates that hepsin is synthesized as a single polypeptide chain of 417 amino acid residues including the first Met. It lacks a typical signal peptide sequence, but contains an internal hydrophobic sequence (residues 18–44) near its N-terminus. There is also a linking region and a typical serine protease catalytic domain at the C-terminus (residues 163–417). The sequence Arg+Ile-Val-Gly-Gly, identifiable as the proteolytic activation site by analogy to other proenzymes in family S1, is present at residues 162–166. Upon activation, hepsin is thought to be converted to a two-chain form, consisting of a light chain (the N-terminal half up to residue 163) and a heavy chain (the remainder of the C-terminal half, containing the catalytic subunit) linked by a disulfide bond (Tsuji *et al.*, 1991; Kazama *et al.*, 1995; Zhukov *et al.*, 1997). The internal hydrophobic sequence near the N-terminus is responsible for the plasma membrane association of hepsin in a type II membrane protein topology (Tsuji *et al.*, 1991) (Figure 60.1).

The sequence predicted from a rat hepsin cDNA (Farley *et al.*, 1993) is composed of 416 amino acid residues, lacking the Glu3 residue of human hepsin. Rat hepsin has an overall similarity of 88.7% with human hepsin.

The overall similarity of the heavy chain to those of chymotrypsin, trypsin and the catalytic subunit of plasmin is in the range 39–43%. The light chain has no significant similarity to other known protein sequences, except the typical hydrophobic sequence responsible for membrane spanning. The catalytic subunit has all the disulfide bonds expected of a serine protease of the trypsin family, but also has one additional cysteine residue in the catalytic subunit (Cys372), which may function to bind unidentified small ligands, but not to form homodimers or heterodimers with any other sizable proteins. There is no experimental evidence for dimerization.

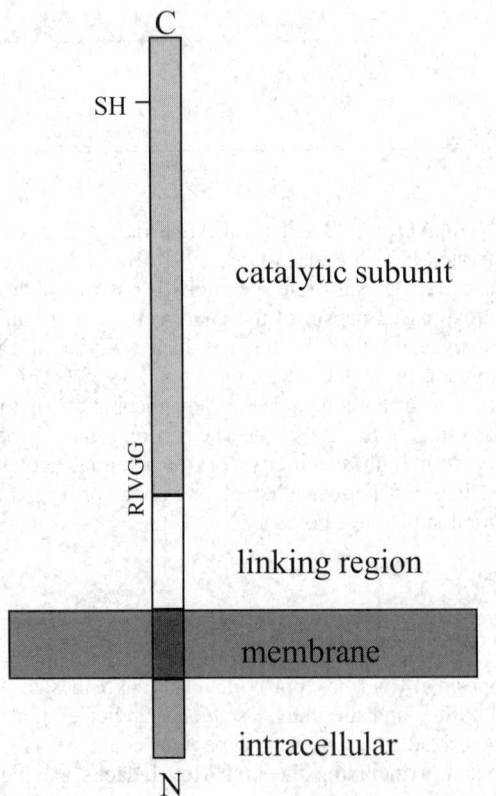

*Figure 60.1*  Hepsin in the type II topology at the cell surface.

Hepsin contains at least one potential *N*-linked glycosylation site at residue 112, but no information is currently available about post-translational modifications, or secondary or tertiary structures.

## Preparation

Hepsin was isolated from a detergent-solubilized membrane fraction of HepG2 cells (Torres-Rosado *et al.*, 1993). It was also isolated from rat liver microsomal membranes by Zhukov *et al.* (1993, 1997).

## Biological Aspects

Hepsin is expressed in the liver at a high level, but is also ubiquitously expressed at much lower levels (10% or less than that in liver) in many other tissues such as lung, kidney and skeletal muscle (Tsuji *et al.*, 1991). Hepsin is produced by cultured cells including hepatoma cells (HepG2 and Alexander cells), BHK cells and breast cancer cells (Kurachi *et al.*, 1994). Importantly, hepsin is present not only in the plasma membrane fraction, but also in the nuclear, mitochondrial and microsomal fractions. Hepsin is not detectable in the cytosolic fraction, nor in the culture medium of the HepG2 and BHK cells tested.

Hepsin plays a critical role in the growth of some mammalian cells such as hepatoma cells (Torres-Rosado *et al.*, 1993). The presence of accessible hepsin (catalytic subunit) on the cell surface is essential for cell proliferation, as tested by binding of a hepsin-specific antibody. Suppression of hepsin biosynthesis with hepsin-specific antisense oligonucleotides further confirmed this, and demonstrated other possible roles of hepsin, including the maintenance of cell morphology. Hepsin may play a critical role in organogenesis and tissue remodeling of, for instance, skin, liver and skeletal muscle (Torres-Rosado *et al.*, 1993), though the underlying molecular mechanisms have yet to be determined. Overexpressed hepsin at the cell surface can activate factor VII (Kazama *et al.*, 1995), although it is not known whether factor VII is among the natural substrates for hepsin.

Based on these findings, hepsin apparently has multiple functions in organogenesis, tissue rearrangement, maintenance of structural integrity of some mammalian cells, and possibly initiation of hemostasis at the cell surface.

The human hepsin gene is located on chromosome 19 at q13.1 (Tsuji *et al.*, 1991). No hepsin deficiency in human or animals has been reported to date.

## Distinguishing Features

Although hepsin has similarities to other serine proteases of the trypsin family, it is clearly distinguished by its structure (Kurachi *et al.*, 1994). Polyclonal antibodies have been produced against synthetic peptides (Torres-Rosado *et al.*, 1993) and a chimeric hepsin protein (Kazama *et al.*, 1995), but are not yet commercially available.

## References

Farley, D., Reymond, F. & Nick, H. (1993) Cloning and sequence analysis of rat hepsin, a cell surface serine proteinase. *Biochim. Biophys. Acta* **1173**, 350–352.

Kazama, Y., Hamamoto, T., Foster, D.C. & Kisiel, W. (1995) Hepsin, a putative membrane-associated serine protease, activates human factor VII and initiates a pathway of blood coagulation on the cell surface leading to thrombin formation. *J. Biol. Chem.* **270**, 66–72.

Kurachi, K., Torres-Rosado, A. & Tsuji, A. (1994) Hepsin. *Methods Enzymol.* **244**, 100–114.

Leytus, S.P., Loeb, K.R., Hagen, F.S., Kurachi, K. & Davie, E.W. (1988) A novel trypsin-like serine protease (hepsin) with a putative transmembrane domain expressed by human liver and hepatoma cells. *Biochemistry* **27**, 1067–1074.

Torres-Rosado, A., O'Shea, K.S., Tsuji, A., Chou, S.-H. & Kurachi, K. (1993) Hepsin, a putative cell-surface protease, is required for mammalian cell growth. *Proc. Natl Acad. Sci. USA* **90**, 7181–7185.

Tsuji, A., Torres-Rosado, A., Arai, T., Le Beau, M.M., Lemons, R.S., Chou, S.-H. & Kurachi, K. (1991) Hepsin, a cell membrane-associated protease. Characterization, tissue distribution, and gene localization. *J. Biol. Chem.* **266**, 16948–16953.

Zhukov, A., Werlinder, V. & Ingelman-Sundberg, M. (1993) Purification and characterization of two membrane bound serine proteinases from rat liver microsomes active in degradation of cytochrome P450. *Biochem. Biophys. Res. Commun.* **197**, 221–228.

Zhukov, A., Hellman, U. & Ingelman-Sundberg, M. (1997) Purification and characterization of hepsin from rat liver microsomes. *Biochim. Biophys. Acta* **1337**, 85–95.

*Kotoku Kurachi*
*Department of Human Genetics,*
*University of Michigan Medical School,*
*Ann Arbor, MI 48109-0618, USA*
*Email: kkurachi@umich.edu*

# 61. *Prostasin*

## Databanks

*Peptidase classification: clan SA, family S1, MEROPS ID: S01.159*
*NC-IUBMB enzyme classification: none*
*Databank codes:*

| Species | SwissProt | PIR | EMBL (cDNA) | EMBL (genomic) |
|---|---|---|---|---|
| *Homo sapiens* | – | – | L41351 | U33446: complete gene |

## Name and History

The name **prostasin** was introduced to denote a trypsin-like enzyme from the human prostate gland that was first discovered in seminal fluid (Yu *et al.*, 1994), but is now known to be widely distributed in human tissues. Prostasin has not yet been reported from other species.

## Activity and Specificity

Prostatin preferentially cleaves Arg∤ bonds in artificial substrates, and assays are conveniently made with synthetic fluorogenic substrates, D-Pro-Phe-Arg∤NHMec, D-Val-Leu-Arg∤NHMec, D-Phe-Phe-Arg∤NHMec and Z-Gly-Pro-Arg∤-NHMec (all available from Enzyme Systems Products) (see Appendix 2 for full names and addresses of suppliers). The pH optimum is about 9.0 and the activity is inhibited by aprotinin, benzamidine, antipain and leupeptin (Yu *et al.*, 1994).

## Structural Chemistry

Prostasin is synthesized as a preproenzyme of 343 amino acids. During translocation into the endoplasmic reticulum, a cleavage occurs to remove the 32 amino acid signal peptide generating proprostasin, which is then activated by a specific cleavage between Arg12∤Ile13 to produce an active form. The active prostasin consists of two chains held together by an interchain disulfide bond (Yu *et al.*, 1995). Prostasin is a single-chain protein of 40 kDa, which contains disulfide bonds and has a pI value of 4.5–4.8 (Yu *et al.*, 1994). A 1 mg ml$^{-1}$ solution of prostasin has an absorbance of 1.63 at 280 nm with a light path of 1 cm. The deduced amino acid sequence of prostasin is 34–42% identical to those of human plasma kallikrein, coagulation factor XI, $\beta$-tryptase, hepsin, plasminogen and acrosin (Yu *et al.*, 1995). The existence of a putative transmembrane domain at the C-terminus of prostasin suggests that the enzyme may be membrane-associated.

## Preparation

Essentially homogeneous prostasin has been prepared from seminal fluid, with 1960-fold purification (Yu *et al*., 1994).

## Biological Aspects

Prostasin is present in cells of most human tissues, but the levels identified in the prostate gland and in seminal fluid were over 20-fold higher than in any other tissues examined. Prostasin was localized to the epithelial cells of the human prostate gland, but was also found in the mucous cells of the salivary gland. In the kidney, prostasin was identified in the proximal and distal tubular cells. In addition, it has been found in bronchi, lung, colon, pancreas, kidney, liver, and male and female urine (Yu *et al*., 1994). Whether the enzyme is present in other organisms is not known.

A full-length prostasin gene, denoted *PRSS8*, has been isolated, together with its flanking regions. The 7 kb sequence of the gene showed a 1.4 kb 5′-flanking region, the 4.4 kb *PRSS8* open reading frame, and a 1.2 kb 3′-flanking region. The gene consists of six exons and five introns, as judged by comparison with its cDNA sequence. The gene locus of human prostasin is on chromosome 16 (Yu *et al*., 1996).

## Distinguishing Features

Polyclonal antisera against the enzyme and the C-terminal peptide (Pro262-Gln-Thr-Gln-Glu-Ser-Gln-Pro-Asp-Ser-Asn272) have been described (Yu *et al*., 1994, 1995), but are not commercially available at the time of writing.

## References

Yu, J.X., Chao, L. & Chao, J. (1994) Prostasin is a novel human serine proteinase from seminal fluid. Purification, tissue distribution, and localization in prostate gland. *J. Biol. Chem.* **269**, 18843–18848.

Yu, J.X., Chao, L. & Chao, J. (1995) Molecular cloning, tissue-specific expression, and cellular localization of human prostasin mRNA. *J. Biol. Chem.* **270**, 13483–13489.

Yu, J.X., Chao, L., Ward, D.C. & Chao, J. (1996) Structure and chromosomal localization of the human prostasin (PRSS8) gene. *Genomics* **32**, 334–340.

*Julie Chao*
*Department of Biochemistry and Molecular Biology,*
*Medical University of South Carolina,*
*Charleston, SC 29425-2211, USA*
*Email: chaoj@musc.edu*

# 62. Follipsin

## Databanks

*Peptidase classification: clan SA, family S1, MEROPS ID: S01.227*
*Enzyme classification: none*
*Databank codes: no sequence data available*

## Name and History

Follicular fluid from pig ovary was found to contain a high activity hydrolyzing Arg-containing 4-methylcoumaryl-7-amide (NHMec) substrates (Takahashi *et al*., 1992). Subsequent purification and characterization of the enzyme primarily responsible for the activity revealed that it is a serine protease apparently distinct from any other enzyme thus far reported (Hamabata *et al*., 1994). The enzyme, which is a trypsin-like protease present in the ovarian follicular fluid, was given the name *follipsin* (Hamabata *et al*., 1994).

## Activity and Specificity

Follipsin cleaves synthetic and peptide substrates only by an endopeptidase activity at amide and peptide bonds of arginine, indicating a strict preference for arginine at P1. The enzyme shows no apparent preference for residues at P2 with synthetic substrates, but for peptide substrates, hydrophobic residues are favored in this position (Ohnishi *et al*., 1995). Specificity for amino acids in other positions is not yet clear.

Assays are conveniently performed with -Arg-NHMec substrates, and Boc-Gln-Arg-Arg↓NHMec is a good substrate for the enzyme (Hamabata *et al.*, 1994; Ohnishi *et al.*, 1995). Activity detected in the crude follicular fluid of the pig ovary with this substrate is largely due to follipsin. It should be noted, however, that the substrate can be cleaved by many proteases with trypsin-like specificity. The enzyme shows virtually no activity on native proteins such as casein, BSA or histone, but the single-chain form of human t-plasminogen activator (Chapter 58) was recently shown to be a good substrate for the enzyme *in vitro* (Ohnishi *et al.*, 1995). The pH optimum of the protease is about 8.0. Follipsin is inhibited by DFP, benzamidine, leupeptin and antipain (Hamabata *et al.*, 1994), and soybean trypsin inhibitor and aprotinin also inhibit strongly (Ohnishi *et al.*, 1995).

## Structural Chemistry

The follipsin molecule consists of two covalently associated polypeptide chains of 45 kDa and 32 kDa. The N-terminal amino acid sequences of the two chains are very similar to those of the heavy and light chains, respectively, of human plasma kallikrein and human factor XIa (Chapters 50 and 51). A single-chain precursor form of follipsin is present in the follicular fluid of the pig ovary, and the purified precursor is readily converted to its two-chain active form by treatment with trypsin. The fluid also contains a proteolytic enzyme capable of activating the proenzyme (unpublished results of the authors).

## Preparation

Follipsin has been purified 12 400-fold from pig follicular fluid (Hamabata *et al.*, 1994).

## Biological Aspects

Follipsin is shown immunohistochemically to be located in the follicular fluid and stroma cells of the pig ovary (Hamabata *et al.*, 1994). The activity in the follicular fluid increases several-fold during follicular maturation (Takahashi *et al.*, 1992), so it seems probable that the enzyme is involved in the proteolytic events associated with follicular maturation and/or ovulation. Ohnishi *et al.* (1995) have recently reported that follipsin activates single-chain tissue-type plasminogen activator, converting it to its two-chain form. Based on this finding, those authors propose that the enzyme plays a role in triggering the plasminogen activator/plasmin system that is generally thought to contribute to the rupture of the follicle during ovulation (Beers, 1975; Tsafriri & Dekel, 1994; Leonardsson *et al.*, 1995). However, definitive evidence for this has not yet been obtained.

## Distinguishing Features

It is obvious from immunological and partial structural studies that follipsin is closely related to plasma kallikrein. Although existing evidence indicates that they are distinct enzymes, further studies including cDNA cloning study are necessary to confirm this. Polyclonal antisera against pig follipsin have been raised in mice (Hamabata *et al.*, 1994).

## Further Reading

Follipsin has been reviewed by Takahashi & Ohnishi (1995).

## References

Beers, W.H. (1975) Follicular plasminogen and plasminogen activator and the effect of plasmin on ovarian follicular wall. *Cell* **6**, 379–386.

Hamabata, T., Okimura, H., Yokoyama, N., Takahashi, T. & Takahashi, K. (1994) Purification, characterization, and localization of follipsin, a novel serine proteinase from the fluid of porcine ovarian follicles. *J. Biol. Chem.* **269**, 17899–17904.

Leonardsson, G., Peng, X.-R., Liu, K., Nordström, L., Carmeliet, P., Mulligan, R., Collen, D. & Ny, T. (1995) Ovulation efficiency is reduced in mice that lack plasminogen activator gene function: Functional redundancy among physiological plasminogen activators. *Proc. Natl Acad. Sci. USA.* **92**, 12446–12450.

Ohnishi, J., Kihara, T., Hamabata, T., Takahashi, K. & Takahashi, T. (1995) Cleavage specificity of porcine follipsin. *J. Biol. Chem.* **270**, 19391–19394.

Takahashi, T. & Ohnishi, J. (1995) Molecular mechanism of follicle rupture during ovulation. *Zool. Sci.* **12**, 359–365.

Takahashi, T., Tsuchiya, Y., Tamanoue, Y., Mori, T., Kawashima, S. & Takahashi, K. (1992) Occurrence of a novel 350-kDa serine proteinase in the fluid of porcine ovarian follicles and its increase during their maturation. *Zool. Sci.* **9**, 343–347.

Tsafriri, A. & Dekel, N. (1994) Molecular mechanism in ovulation. In: *Molecular Biology of the Female Reproductive System* (Findlay, J.K., ed.). San Diego: Academic Press, pp. 207–258.

*Takayuki Takahashi*
*Division of Biological Sciences,*
*Graduate School of Science,*
*Hokkaido University,*
*Sapporo 060, Japan*
*Email: ttakaha@bio.hokudai.ac.jp*

*Junji Ohnishi*
*Division of Biological Sciences,*
*Graduate School of Science,*
*Hokkaido University,*
*Sapporo 060, Japan*
*Email: ohnishi@bio.hokudai.ac.jp*

# 63. Hepatocyte growth factor activator

## Databanks

*Peptidase classification: clan SA, family S1, MEROPS ID: S01.228*
*NC-IUBMB enzyme classification: none*
*Databank codes:*

| Species | SwissProt | PIR | EMBL (cDNA) | EMBL (genomic) |
|---|---|---|---|---|
| *Homo sapiens* | Q04756 | A46688 | D14012 | – |

## Name and History

A protease that cleaves the bond Arg494┼Val495 in the single-chain precursor of human hepatocyte growth factor (HGF) to generate the active form of HGF was discovered and purified from fetal bovine serum (Shimomura *et al*., 1992) and from human serum (Miyazawa *et al*., 1993). Because the protease had a strong HGF-activating activity, it was named *hepatocyte growth factor activator (HGF activator)* (Miyazawa *et al*., 1993).

## Activity and Specificity

The assay of activity is done with single-chain HGF as substrate (Shimomura *et al*., 1993). Single-chain HGF can be prepared from serum-free conditioned medium of HGF-expressing cells as described (Shimomura *et al*., 1992).

Activity is stimulated by the presence of dextran sulfate. Optimal NaCl and dextran sulfate concentrations are around 120 mM and 50 $\mu$g ml$^{-1}$, respectively (Shimomura *et al*., 1995). Activity is inhibited by various serine protease inhibitors such as nafamostat mesilate, leupeptin and aprotinin (Shimomura *et al*., 1992).

## Structural Chemistry

HGF activator purified from human serum is a protein of 34 kDa, and consists of two chains held together by a disulfide bond. The nucleotide sequence of HGF activator cDNA shows that human HGF activator is derived from the C-terminal region of a precursor of 655 amino acids by proteolytic cleavage of the bonds Arg372┼Val373 and Arg407┼Ile408, and that the precursor consists of multiple domains homologous to those in blood coagulation factor XII (Chapter 49). These domains comprise a type II fibronectin homology region, two epidermal growth factor domains, a type I fibronectin homology region, a kringle domain and the catalytic domain (Miyazawa *et al*., 1993).

## Preparation

The active form of HGF activator was purified from fetal bovine serum (Shimomura *et al*., 1992), from human serum (Miyazawa *et al*., 1993) and from rat serum (Miyazawa *et al*., 1996). The yield was about 40 $\mu$g from 500 ml of human serum (Miyazawa *et al*., 1993). The zymogen form of HGF activator was purified from human plasma and the yield was about 70 $\mu$g from 190 ml of human plasma (Shimomura *et al*., 1993).

## Biological Aspects

HGF activator is produced and secreted by the liver (Miyazawa *et al*., 1993), and normally circulates in the blood as an inactive zymogen (Shimomura *et al*., 1993). In response to hepatic or renal injury, the zymogen is activated exclusively in the injured tissue by limited cleavage of the bond Arg407┼Ile408 (Miyazawa *et al*., 1996). The active enzyme acquires a strong affinity for heparin, and thus can associate with cell surface heparin-like molecules. This localized enzyme activity is responsible for the activation of HGF in injured tissues (Miyazawa *et al*., 1996). Thus, HGF activator appears to be a key regulator of the activity of HGF in the injured tissues.

The HGF activator zymogen is efficiently cleaved *in vitro* by thrombin at the bond Arg407┼Ile408 in the presence of negatively charged substances. An activity in human serum that activates the zymogen was eliminated by an antibody against thrombin. Thus, thrombin is most likely the enzyme that converts the zymogen to the active form *in vivo* (Shimomura *et al*., 1993).

## Distinguishing Features

HGF activator is homologous to coagulation factor XIIa. Factor XIIa has an ability to activate HGF, but the specific activity of HGF activator in HGF conversion is much higher than that of factor XIIa (Shimomura *et al*., 1995).

Monoclonal antibodies against human HGF activator have been prepared (Shimomura *et al*., 1993, 1995), and one of them efficiently inhibits the HGF-converting activity of the enzyme (Shimomura *et al*., 1995; Miyazawa *et al*., 1996).

## References

Miyazawa, K., Shimomura, T., Kitamura, A., Kondo, J., Morimoto, Y. & Kitamura, N. (1993) Molecular cloning and sequence analysis of the cDNA for a human serine protease responsible for activation of hepatocyte growth factor Y. Structural similarity of the protease precursor to blood coagulation factor XII. *J. Biol. Chem.* **268**, 10024–10028.

Miyazawa, K., Shimomura, T. & Kitamura, N. (1996) Activation of hepatocyte growth factor in the injured tissues is mediated by hepatocyte growth factor activator. *J. Biol. Chem.* **271**, 3615–3618.

Shimomura, T., Ochiai, M., Kondo, J. & Morimoto, Y. (1992) A novel protease obtained from FBS-containing culture supernatant, that processes single chain form hepatocyte growth factor to two chain form in serum-free culture. *Cytotechnology* **8**, 219–229.

Shimomura, T., Kondo, J., Ochiai, M., Naka, D., Miyazawa, K., Morimoto, Y. & Kitamura, N. (1993) Activation of the zymogen of hepatocyte growth factor activator by thrombin. *J. Biol. Chem.* **268**, 22927–22932.

Shimomura, T., Miyazawa, K., Komiyama, Y., Hiraoka, H., Naka, D., Morimoto, Y. & Kitamura, N. (1995) Activation of hepatocyte growth factor by two homologous proteases, blood coagulation factor XIIa and hepatocyte growth factor activator. *Eur. J. Biochem.* **229**, 257–261.

*Naomi Kitamura*
*Department of Life Science,*
*Faculty of Bioscience & Biotechnology,*
*Tokyo Institute of Technology,*
*4259 Nagatsuta-cho, Midori-ku, Yokohama 226, Japan*
*Email: nkitamur@bio.titech.ac.jp*

# 64. Snake protease of Drosophila

## Databanks

Peptidase classification: clan SA, family S1, MEROPS ID: S01.200
NC-IUBMB enzyme classification: none
Databank codes:

| Species | SwissProt | PIR | EMBL (cDNA) | EMBL (genomic) |
|---|---|---|---|---|
| Snake protein | | | | |
| *Drosophila melanogaster* | P05049 | A24702 | X04513 | – |
| **Related peptidases** | | | | |
| Easter protein | | | | |
| *Drosophila melanogaster* | P13582 | A30100 A32727 | J03154 | – |
| Nudel protein | | | | |
| *Drosophila melanogaster* | P98159 | – | U29153 | – |
| Stubble protein | | | | |
| *Drosophila melanogaster* | Q05319 | – | L11451 | – |

## Name and History

**Snake protease** is the product of the maternally required *snake* gene of *Drosophila melanogaster* (Anderson & Nuesslein-Volhard, 1984a,b). Embryos produced by females homozygous for strong mutant alleles of *snake* produce 'dorsalized' embryos consisting of hollow tubes of dorsal ectoderm, hence the name *snake*. The *snake* gene was isolated in genetic screens for maternally required loci which alter the specification of dorsal–ventral cell fates during embryogenesis (Anderson & Nuesslein-Volhard, 1984a,b). Subsequent analysis indicates that Snake is a component of an extracellular signal transduction pathway. Molecular cloning of the gene suggested that the protein product was a secreted serine protease proenzyme (DeLotto & Spierer, 1986). It is conventional in *Drosophila* nomenclature that the protein product of a gene adopts the gene name.

## Activity and Specificity

Snake is a component of a sequential protease activation cascade that propagates a spatially restricted signal within the perivitelline space of the syncytial blastoderm embryo. Genetic epistasis data positioned Snake upstream of another serine protease proenzyme, Easter (Chapter 65). Comparative amino acid sequence alignments suggested that the substrate specificity of Snake would be similar to that of trypsin and

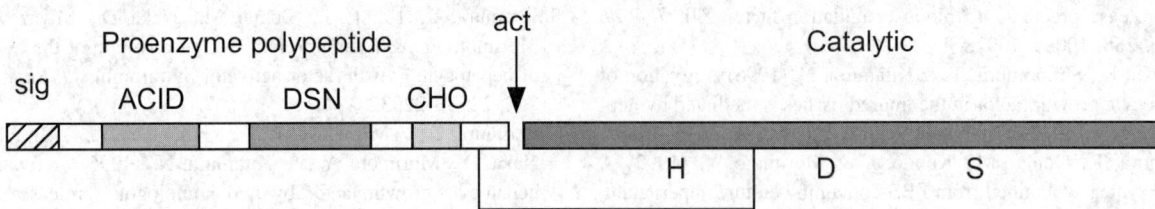

*Figure 64.1*    Diagrammatic structure of the Snake protease molecule.

compatible with activation of the Easter proenzyme. Coexpression studies confirm that the active form of Snake specifically hydrolyzes Easter at its activation peptide, consistent with Snake being the direct activator of Easter proenzyme (R. DeLotto, unpublished results). To date no synthetic substrates have been described for *in vitro* activity assay. A biological assay has been developed, consisting of microinjection of the protease into the perivitelline space to rescue the *snake* mutant phenotype (Stein & Nuesslein-Volhard, 1992; Smith *et al.*, 1995).

### Structural Chemistry

The structure of the primary translation product is illustrated in Figure 64.1. Secreted recombinant Snake has an $M_r$ of 52 000–54 000 as estimated by SDS-PAGE (Smith *et al.*, 1995). The catalytic chain has an $M_r$ of 29 000 and is covalently associated with the proenzyme polypeptide by a disulfide bridge. Within the proenzyme polypeptide three functional domains have been defined, an acidic region (ACID) which may be involved in stability or localization of the enzyme, a disulfide knot (DSN) domain which may function as protein–protein interaction and intramolecular inhibition of the catalytic chain, and a glycosylated region that has *O*-linked carbohydrates. The acidic region is not strictly required for biological activity, but the DSN and CHO regions are (Smith *et al.*, 1994). Smith & DeLotto (1992) showed that the proenzyme polypeptide has a similar domain structure to Easter and *Tachypleus tridentatus* proclotting enzyme (Chapter 68).

### Preparation

Due to the narrow time period of expression and the small physical size of *Drosophila melanogaster* embryos, Snake has not been purified from embryonic sources. However, recombinant forms of the enzyme have been produced from the molecularly cloned cDNA using the baculovirus expression system, and purified (Smith *et al.*, 1995). While the proenzyme form is relatively stable, an isolated active catalytic chain degrades rapidly in solution, suggesting a role of the proenzyme polypeptide in stabilization of the active protease catalytic chain. Snake binds to benzamidine-Sepharose and immobilized soybean trypsin inhibitor, and benzamidine-Sepharose affinity chromatography has been used in its purification.

### Biological Aspects

The current view of the dorsal–ventral signaling pathway suggests that positional information is laid down in the vitelline membrane of the egg during oogenesis (Roth, 1994). After

fertilization, a protease cascade is initiated in the perivitelline space of the embryo, resulting in the sequential activation of three serine protease proenzymes, gastrulation defective (Konrad & Marsh, 1990), Snake and Easter. The end-product of this cascade is the ventrally restricted processing of the Spaetzle precursor, resulting in a graded distribution of ligand with a local maximum on the ventral side of the embryo. The protease cascade in which Snake participates is involved in controlling or generating the graded spatial distribution of processed ligand in the perivitelline space. Thus, Snake represents a serine protease which is directly involved in spatially organizing the body plan of a multicellular animal.

### Related Peptidases

Several other components of the dorsal–ventral signaling pathway probably are proteases. For example, *nudel*, a somatically required component of the dorsal–ventral patterning system has been molecularly cloned and the gene was found to encode an unusual 315 kDa mosaic protein with a significant serine protease catalytic chain homology in the center (Hong & Hashimoto, 1995). The *gastrulation defective* gene has been molecularly cloned and the protein found to have serine protease catalytic chain homology at its C-terminus (Konrad & Marsh, 1990). Further downstream in dorsal–ventral axis formation, the zygotically required gene *tolloid* was found to encode a putative metalloprotease with homology to human bone morphogenetic protein 1 (Chapter 413) (Shimell *et al.*, 1991).

Other pathways in *Drosophila* such as those mediating imaginal disc morphogenesis appear to involve the function of serine proteases as well as metalloproteases. The *stubble/stubbloid* gene is required during imaginal disc development and it encodes a transmembrane serine protease (Appel *et al.*, 1993). Serine protease activities have been detected using zymography in explanted imaginal discs undergoing evagination (Pino-Heiss & Schubiger, 1989).

### Further Reading

The general role of proteases in patterning of *Drosophila* has been reviewed (Hecht & Anserson, 1992). The protease cascade leading to the ventrally restricted production of processed Spaetzle protein has also been reviewed (Roth, 1994).

### References

Anderson, K. & Nuesslein-Volhard, C. (1984a) Genetic analysis of dorsal–ventral embryonic pattern in *Drosophila*. In: *Pattern Formation, A Primer in Developmental Biology* (Malakinski, G. & Bryant, P., eds). New York: Macmillan, pp. 269–289.

Anderson, K. & Nuesslein-Volhard, C. (1984b) Information for the dorsal–ventral pattern of the *Drosophila* embryo is stored as maternal mRNA. *Nature* **311**, 223–227.

Appel, L., Prout, M., Abu-Shumays, R., Hammonds, A., Garbe, J., Fristrom, D. & Fristrom, J. (1993) The Drosophila Stubble-stubbloid gene encodes an apparent transmembrane serine protease required for epithelial morphogenesis. *Proc. Natl Acad. Sci. USA* **90**, 4937–4941.

DeLotto, R. & Spierer, P. (1986) A gene required for the specification of dorsal–ventral pattern in *Drosophila* appears to encode a serine protease. *Nature* **323**, 688–692.

Hecht, P. & Anserson, K. (1992) Extracellular proteases and embryonic pattern formation. *Trends Cell Biol.* **2**, 197–202.

Hong, C. & Hashimoto, C. (1995) An unusual mosaic protein with a protease domain encoded by the *nudel* gene is involved in defining dorsoventral polarity in *Drosophila*. *Cell* **82**, 785–794.

Konrad, K. & Marsh, L. (1990) The gastrulation defective (gd) gene displays homology to serine proteases. Paper presented at the 31st Annual Drosophila Research Conference, The Genetics Society of America, Bethesda, MD, pp. 43–44 (abstr.).

Pino-Heiss, S. & Schubiger, G. (1989) Extracellular protease production by *Drosophila* imaginal discs. *Dev. Biol.* **132**, 282–291.

Roth, S. (1994) Proteolytic generation of a morphogen. *Curr. Biol.* **4**, 755–757.

Shimell, M., Ferguson, E., Childs, R. & O'Conner, M. (1991) The *Drosophila* dorsal–ventral patterning gene tolloid is related to human bone morphogenetic protein 1. *Cell* **67**, 469–481.

Smith, C. & DeLotto, R. (1992) A common domain within the proenzyme regions of the *Drosophila* Snake and Easter proteins and *Tachypleus* proclotting enzyme defines a new subfamily of serine proteases. *Protein Sci.* **1**, 1225–1226.

Smith, C., Giordano, H. & DeLotto, R. (1994) Mutational analysis of the Drosophila *snake* protease: An essential role for domains within the proenzyme polypeptide chain. *Genetics* **136**, 1355–1365.

Smith, C., Giordano, H., Schwartz, M. & DeLotto, R. (1995) Spatial regulation of *Drosophila* Snake protease activity in the generation of dorsal–ventral polarity. *Development* **121**, 4127–4135.

Stein, D. & Nuesslein-Volhard, C. (1992) Multiple extracellular activities in *Drosophila* egg perivitelline fluid are required for establishment of embryonic dorsal–ventral polarity. *Cell* **68**, 429–440.

*Robert DeLotto*
*Department of Genetics,*
*University of Copenhagen,*
*Øster Farimagsgade 2A, DK-1353 Copenhagen K, Denmark*
*Email: rdelotto@biobase.dk*

# 65. *Easter protease of Drosophila*

## Databanks

*Peptidase classification: clan SA, family S1, MEROPS ID: S01.201*
*NC-IUBMB enzyme classification: none*
*Databank codes:*

| Species | SwissProt | PIR | EMBL (cDNA) | EMBL (genomic) |
|---|---|---|---|---|
| **Easter protein** | | | | |
| *Drosophila melanogaster* | P13582 | A30100 A32727 | J03154 | – |
| **Related peptidases** | | | | |
| Nudel protein | | | | |
| *Drosophila melanogaster* | P98159 | – | U29153 | – |
| Snake protein | | | | |
| *Drosophila melanogaster* | P05049 | A24702 | X04513 | – |
| Stubble protein | | | | |
| *Drosophila melanogaster* | Q05319 | – | L11451 | – |

## Name and History

***Easter protease*** is the product of the maternal effect *easter* gene of *Drosophila melanogaster*. Embryos produced by females homozygous for strong mutant alleles of *easter* produce 'dorsalized' embryos consisting of hollow tubes of dorsal ectoderm (Anderson & Nuesslein-Volhard, 1984). The *easter* gene was isolated in genetic screens for maternal effect genes required for the correct specification of dorsal–ventral cell fates during embryonic development. From genetic and molecular biological data it has become evident that Easter is an extracellular component of a signal transduction pathway. It is conventional in *Drosophila* nomenclature that the protein product of a gene adopts the gene name.

## Activity and Specificity

Easter is a component of a sequential protease activation cascade that propagates a spatially restricted signal within the 'extracellular' perivitelline space of the syncytial blastoderm embryo. Genetic epistasis experiments place Easter before Spaetzle and after Snake (Chapter 64) in the hierarchy of functions (Chasen *et al.*, 1992). Spaetzle is a secreted, cysteine-rich protein that appears to be a precursor form of the ligand for the Toll receptor (Hashimoto *et al.*, 1988; Morisato & Anderson, 1994). That Easter may directly process Spaetzle has been recently tested by coexpression studies, and it was found that an activated form of Easter specifically cleaves Spaetzle at a single Arg+Val bond generating a 106 amino acid C-terminal fragment with the characteristics of a neurotrophin-like growth factor (R. DeLotto, unpublished results). A biological assay consisting of microinjection into the perivitelline space and phenotypic rescue of the mutant phenotype has been described (Stein & Nuesslein-Volhard, 1992).

## Structural Chemistry

The *easter* gene was molecularly cloned and deduced by amino acid sequence alignment to encode a serine protease proenzyme with structural similarity to Snake protease (Chasen & Anderson, 1989). The structure of the Easter protease is shown in Figure 65.1. The primary translation product has a signal peptide which directs secretion of the proenzyme form. It was observed that Easter is structurally similar to *Tachypleus tridentatus* proclotting enzyme (Chapter 68) and Snake within the proenzyme polypeptide region (Gay & Keith, 1992; Smith & DeLotto, 1992). Easter shares with Snake a disulfide knot motif (DSN) and, as in proclotting enzyme, the proenzyme polypeptide and catalytic chain are joined by a disulfide bridge. The $M_r$ of the proenzyme is 50 000 and that of the catalytic chain is 32 000, as estimated

by SDS-PAGE (R. DeLotto, unpublished results). Mutant alleles have been described that either lateralize or ventralize cell fates (Jin & Anderson, 1990). From modeling studies, these alterations map to surface residues of the catalytic chain, suggesting the existence of an exosite that may regulate the activity of the protease catalytic chain.

## Preparation

Attempts to detect Easter activity in embryonic extracts have so far been unfruitful. An active form of Easter has been detected and partially purified from *Drosophila* hemolymph, but the biological role of the protease in hemolymph is at present unknown (Hecht & Anserson, 1992). Recombinant forms of Easter have been generated using the baculovirus expression system previously applied to Snake, and they are biologically active when microinjected into the perivitelline space (R. DeLotto, unpublished results; Smith *et al.*, 1995).

## Biological Aspects

The current view of the dorsal–ventral signaling pathway suggests that a protease cascade initiated in the perivitelline space of the embryo results in the sequential activation of three serine protease proenzymes, gastrulation defective (Konrad & Marsh, 1990), Snake and Easter. The end-product of this cascade is the ventrally restricted processing of Spaetzle precursor resulting in a graded distribution of ligand with a local maximum on the ventral side of the embryo. The protease cascade in which Easter participates is involved in either controlling or generating the graded spatial distribution of processed ligand in the perivitelline space.

## Related Peptidases

Several other components of the dorsal–ventral pathway probably involve protease function. For example, *nudel*, a somatically required component of the dorsal–ventral patterning system has been molecularly cloned and the gene was found to encode an unusual 315 kDa mosaic protein with a significant serine protease catalytic chain homology in the center (Hong & Hashimoto, 1995). The *gastrulation defective* gene has been molecularly cloned and the protein found to have serine protease catalytic chain homology at its C-terminus (Konrad & Marsh, 1990). Further downstream, the zygotically required gene *tolloid* was found to encode a putative metalloprotease with homology to human bone morphogenetic protein 1 (Chapter 413) (Shimell *et al.*, 1991).

Other pathways in *Drosophila*, such as those mediating imaginal disc morphogenesis, involve the function of

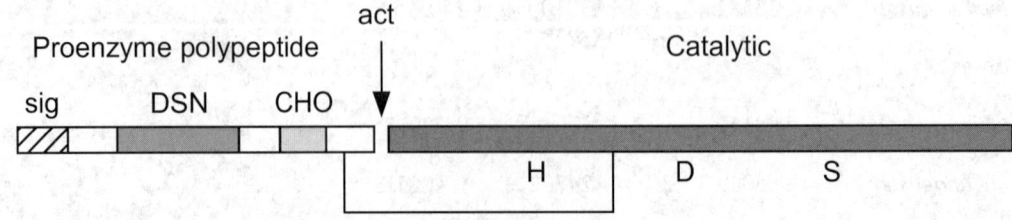

*Figure 65.1* Diagrammatic structure of the Easter protease molecule.

serine proteases and metalloproteases. For example, the *stubble/stubbloid* gene is required during imaginal disc development, and it encodes a transmembrane serine protease (Appel *et al*., 1993). Serine protease activities have been detected by zymography and implicated in imaginal disc morphogenesis (Pino-Heiss & Schubiger, 1989).

### Further Reading

The general role of proteases in patterning of *Drosophila* has been reviewed (Hecht & Anserson, 1992). The protease cascade leading to the ventrally restricted production of processed Spaetzle protein has been reviewed (Roth, 1994).

### References

Anderson, K. & Nuesslein-Volhard, C. (1984) Information for the dorsal–ventral pattern of the *Drosophila* embryo is stored as maternal mRNA. *Nature* **311**, 223–227.

Appel, L., Prout, M., Abu-Shumays, R., Hammonds, A., Garbe, J., Fristrom, D. & Fristrom, J. (1993) The *Drosophila Stubble-stubbloid* gene encodes an apparent transmembrane serine protease required for epithelial morphogenesis. *Proc. Natl Acad. Sci. USA* **90**, 4937–4941.

Chasen, R. & Anderson, K.V. (1989) The role of *easter*, an apparent serine protease, in organizing the dorsal–ventral axis of the *Drosophila* embryo. *Cell* **56**, 391–400.

Chasen, R., Jin, Y. & Anderson, K. (1992) Activation of the easter zymogen is regulated by five other genes to define dorsal–ventral polarity in the *Drosophila* embryo. *Development* **115**, 607–616.

Gay, N. & Keith, F. (1992) Regulation of translation and proteolysis during the development of embryonic dorso-ventral polarity in *Drosophila*. Homology of Easter proteinase with Limulus proclotting enzyme and translational activation of Toll receptor synthesis. *Biochim. Biophys. Acta* **1132**, 290–296.

Hashimoto, C., Hudson, K. & Anderson, K. (1988) The *Toll* gene of *Drosophila*, required for dorsal–ventral embryonic polarity, appears to encode a transmembrane protein. *Cell* **52**, 269–279.

Hecht, P. & Anserson, K. (1992) Extracellular proteases and embryonic pattern formation. *Trends Cell Biol.* **2**, 197–202.

Hong, C. & Hashimoto, C. (1995) An unusual mosaic protein with a protease domain encoded by the *nudel* gene is involved in defining dorsoventral polarity in *Drosophila*. *Cell* **82**, 785–794.

Jin, Y. & Anderson, K. (1990) Dominant and recessive alleles of the *Drosophila easter* gene are point mutations at conserved sites in the serine protease catalytic domain. *Cell* **60**, 873–881.

Konrad, K. & Marsh, L. (1990) The gastrulation defective (gd) gene displays homology to serine proteases. Paper presented at the 31st Annual Drosophila Research Conference, The Genetics Society of America, Bethesda, MD, pp. 43–44 (abstr.).

Morisato, D. & Anderson, K. (1994) The *spaetzle* gene encodes a component of the extracellular signalling pathway establishing the dorsal–ventral pattern of the *Drosophila embryo*. *Cell* **76**, 677–688.

Pino-Heiss, S. & Schubiger, G. (1989) Extracellular protease production by *Drosophila* imaginal discs. *Dev. Biol.* **132**, 282–291.

Roth, S. (1994) Proteolytic generation of a morphogen. *Curr. Biol.* **4**, 755–757.

Shimell, M., Ferguson, E., Childs, R. & O'Conner, M. (1991) The *Drosophila* dorsal–ventral patterning gene tolloid is related to human bone morphogenetic protein 1. *Cell* **67**, 469–481.

Smith, C. & DeLotto, R. (1992) A common domain within the proenzyme regions of the *Drosophila* snake and easter proteins and *Tachypleus* proclotting enzyme defines a new subfamily of serine proteases. *Protein Sci.* **1**, 1225–1226.

Smith, C., Giordano, H., Schwartz, M. & DeLotto, R. (1995) Spatial regulation of *Drosophila* Snake protease activity in the generation of dorsal–ventral polarity. *Development* **121**, 4127–4135.

Stein, D. & Nuesslein-Volhard, C. (1992) Multiple extracellular activities in *Drosophila* egg perivitelline fluid are required for establishment of embryonic dorsal–ventral polarity. *Cell* **68**, 429–440.

*Robert DeLotto*
*Department of Genetics,*
*University of Copenhagen,*
*Øster Farimagsgade 2A, DK-1353 Copenhagen K, Denmark*
*Email: rdelotto@biobase.dk*

# 66. *Limulus coagulation factor C*

## Databanks

*Peptidase classification: clan SA, family S1, MEROPS ID: S01.219*
*NC-IUBMB enzyme classification: EC 3.4.21.84*
*Databank codes:*

| Species | SwissProt | PIR | EMBL (cDNA) | EMBL (genomic) |
|---|---|---|---|---|
| *Tachypleus tridentatus* | P28175 | A38738 | D90271 | – |

## Name and History

The horseshoe crab (or limulus) hemolymph is known to be very sensitive to the lipopolysaccharide (LPS) located in the outer membranes of gram-negative bacteria (Iwanaga, 1993; Muta & Iwanaga, 1996; Kawabata *et al.*, 1996). A trace amount of LPS activates the hemocytes, causing them to release LPS-sensitive coagulation factors and antimicrobial substances by degranulation. Among the proteins released from the cells is an LPS-sensitive serine protease proenzyme, ***limulus coagulation factor C***, that is autocatalytically converted to its active form, factor $\overline{C}$ (Nakamura *et al.*, 1986b; Muta *et al.*, 1991). The active factor $\overline{C}$ in turn activates proenzyme factor B (Chapter 67) to factor $\overline{B}$ (Nakamura *et al.*, 1986a; Muta *et al.*, 1993), which then activates proclotting enzyme (Chapter 68) to clotting enzyme (Nakamura *et al.*, 1985; Muta *et al.*, 1990). The resulting clotting enzyme acts on coagulogen, causing the formation of an insoluble coagulin gel (Nakamura *et al.*, 1976; Miyata *et al.*, 1984). In addition to the LPS-mediated coagulation pathway, the hemocyte lysate also responds to $\beta$-1,3-glucans, which are among the major cell wall components of fungi. A serine protease proenzyme, factor G (Chapter 69), is directly activated in the presence of $\beta$-1,3-glucan, and its active form converts proclotting enzyme to clotting enzyme, resulting also in coagulin gel formation (Seki *et al.*, 1994; Muta *et al.*, 1995). The limulus hemolymph coagulation system is summarized diagrammatically in Figure 66.1.

## Activity and Specificity

Factor C is an LPS-sensitive serine protease proenzyme. Purified factor C is a mixture of a single-chain form (123 kDa)

and a two-chain form composed of an H chain (80 kDa) and an L chain (43 kDa) (Nakamura *et al.*, 1986a). In the presence of LPS, both forms of factor C are autocatalytically activated to an active form, factor $\overline{C}$ composed of three chains, the H chain, an A chain (7.9 kDa) and a B chain (34 kDa). Factor $\overline{C}$ selectively cleaves the Arg103$\downarrow$Ser104 and Ile124$\downarrow$Ile125 bonds in limulus clotting factor B to form factor $\overline{B}$, and it has a similar substrate specificity to $\alpha$-thrombin (Chapter 55), efficiently hydrolyzing Boc-Val-Pro-Arg$\downarrow$NHPhNO$_2$ and Boc-Val-Pro-Arg$\downarrow$NHMec.

A serpin, LICI-1, isolated from horseshoe crab hemocytes, specifically inhibits the activity of factor $\overline{C}$ (Miura *et al.*, 1994). Factor $\overline{C}$ is also inhibited by antithrombin III, but not $\alpha_2$-plasmin inhibitor. DFP is a potent inhibitor (Nakamura *et al.*, 1986a), and benzamidine (>20 mM) and leupeptin (>0.2 mM) are inhibitory at these high concentrations.

## Structural Chemistry

Factor C is structurally related to mammalian complement factors, and it is a novel mosaic protein containing five sushi domains, one EGF-like domain, one C-type lectin-like domain and one serine protease domain (Muta *et al.*, 1991). In addition to these domains, one Cys-rich and one Pro-rich region are located in the N-terminal and the C-terminal portions of the H chain, respectively. The factor C-derived B chain shows the highest sequence identity (36.7%) with human $\alpha$-thrombin, consistent with the substrate specificity mentioned above. The sushi domain, which is also called the SCR (short consensus repeat) or $\beta_2$-GP-I ($\beta_2$-glycoprotein I)-like domain, was named after the traditional Japanese dish, because of a resemblance between this and its schematic

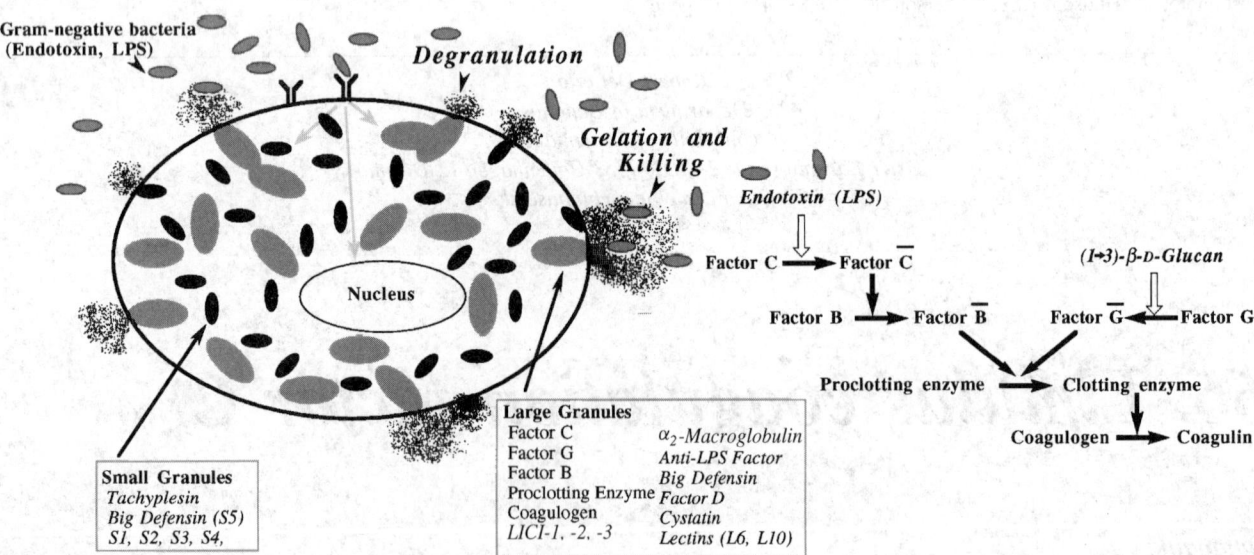

*Figure 66.1* Defense system and coagulation cascade in horseshoe crab hemocytes. The hemocyte detects lipopolysaccharide on gram-negative bacteria and initiates exocytosis of the large and small granules. The coagulation factors released are activated by the lipopolysaccharide or $\beta$-1,3-glucan (overbars indicate the activated forms), and this causes coagulation of the hemolymph. As a result, the pathogen cells are agglutinated by lectins and subsequently killed by antibacterial substances, such as anti-LPS factor, tachyplesin and big defensin. The large granules also contain protease inhibitors (printed in italics), such as $\alpha_2$-macroglobulin, serpins (LICIs, limulus intracellular coagulation inhibitors), cystatin and a pseudo-serine protease with antimicrobial activities, factor D.

domain structure built around two disulfide bonds (Ichinose *et al.*, 1990). Factor C was the first example of a protein from an invertebrate animal that has complement protein-like structure. The EGF-like domain of factor C is most closely related to that in the laminin B2 chain. The lectin-like domain of factor C shows similarity to the so-called C-type lectin, and no serine protease has previously been found to include this type of domain.

## Preparation

Factor C has been highly purified from the hemocyte lysate of *Tachypleus tridentatus* (Nakamura *et al.*, 1986).

## Biological Aspects

The activation site in the L chain of limulus factor C (between the A and B chains) is a Phe+Ile bond (Tokunaga *et al.*, 1987; Muta *et al.*, 1991), which differs from those in most serine protease proenzymes. Consistent with this, factor C is activated by chymotrypsin (Chapter 8) or chymase, a serine protease in mast cells with a chymotrypsin-like substrate specificity (Chapter 18), but not by trypsin (Chapter 3) (Tokunaga *et al.*, 1991). The B chain of factor $\bar{C}$ constitutes a serine protease domain with a DFP-sensitive active serine residue. LPS binds to the H chain, however (Nakamura *et al.*, 1988a). Factor C is also activated by lipid A, which has a core structure of LPS and is important for the expression of the endotoxin activity of LPS (Nakamura *et al.*, 1988b). Acidic phospholipids such as phosphatidylinositol, phosphatidylserine and cardiolipin, which have structures somewhat similar to that of lipid A, are also capable of activating the proenzyme factor C. The negatively charged surfaces that activate factor XII (Chapter 49) and the intrinsic pathway of the mammalian clotting cascade, such as kaolin, celite, glass, ellagic acid, sulfatide, amylose sulfate and dextran sulfate or sulfatides, do not activate factor C at all. The activation of factor C by LPS is dependent on the concentration of LPS, and excessively high concentrations inhibit activation.

## Distinguishing Features

A polyclonal antiserum and several monoclonal antibodies against factor C have been described (Tokunaga *et al.*, 1987; Miura *et al.*, 1992), but are not commercially available.

## Further Reading

An extensive review of all the clotting enzymes of horseshoe crab has been provided by Muta & Iwanaga (1996).

## References

Ichinose, A., Bottenus, R.E. & Davie, E.W. (1990) Structure of transglutaminases. *J. Biol. Chem.* **265**, 13411–13414.

Iwanaga, S. (1993) The limulus clotting reaction. *Curr. Opin. Immunol.* **5**, 74–82.

Kawabata, S., Muta, T. & Iwanaga, S. (1996) Clotting cascade and defense molecule found in hemolymph of horseshoe crab. In: *Invertebrate Immunology* (Söderhäll, G. *et al.*, eds). Fair Haven: SOS Publications, pp. 255–284.

Miura, Y., Tokunaga, F., Miyata, T., Moriyasu, M., Yoshikawa, K.

& Iwanaga, S. (1992) Preparation and properties of monoclonal antibodies against lipopolysaccharide-sensitive serine protease zymogen, factor C, from horseshoe crab (*Tachypleus tridentatus*) hemocytes. *J. Biochem. (Tokyo)* **112**, 476–481.

Miura, Y., Kawabata, S. & Iwanaga, S. (1994) A limulus intracellular coagulation inhibitor with characteristics of the serpin superfamily: Purification, characterization, and cDNA cloning. *J. Biol. Chem.* **269**, 542–547.

Miyata, T., Hiranaga, M., Umezu, M. & Iwanaga, S. (1984) Amino acid sequence of the coagulogen from *Limulus polyphemus* hemocytes. *J. Biol. Chem.* **259**, 8924–8933.

Muta, T. & Iwanaga, S. (1996) The role of hemolymph coagulation in innate immunity. *Curr. Opin. Immunol.* **8**, 41–47.

Muta, T., Hashimoto, R., Miyata, T., Nishimura, H., Toh, Y. & Iwanaga, S. (1990) Proclotting enzyme from horseshoe crab hemocytes: cDNA cloning, disulfide locations, and subcellular localization. *J. Biol. Chem.* **265**, 22426–22433.

Muta, T., Miyata, T., Misumi, Y., Tokunaga, F., Nakamura, T., Toh, Y., Ikehara, Y. & Iwanaga, S. (1991) Limulus factor C: An endotoxin-sensitive serine protease zymogen with a mosaic structure of complement-like, epidermal growth factor-like, and lectin-like domains. *J. Biol. Chem.* **266**, 6554–6561.

Muta, T., Oda, T. & Iwanaga, S. (1993) Horseshoe crab coagulation factor B: A unique serine protease zymogen activated by cleavage of an Ile-Ile bond. *J. Biol. Chem.* **268**, 21384–21388.

Muta, T., Seki, N., Takaki, Y., Hashimoto, R., Oda, T. Iwanaga, A., Tokunaga, F. & Iwanaga, S. (1995) Purified horseshoe crab factor G: Reconstitution and characterization of the $(1\rightarrow3)$-$\beta$-D-glucan-sensitive serine protease cascade. *J. Biol. Chem.* **270**, 892–897.

Nakamura, S., Iwanaga, S., Harada, T. & Niwa, M. (1976) A clottable protein (coagulogen) from amoebocyte lysate of Japanese horseshoe crab (*Tachypleus tridentatus*): its isolation and biochemical properties. *J. Biochem. (Tokyo)* **80**, 1011–1021.

Nakamura, T., Morita, T. & Iwanaga, S. (1985) Intracellular proclotting enzyme in limulus (*Tachypleus tridentatus*) hemocytes: its purification and properties. *J. Biochem. (Tokyo)* **97**, 1561–1574.

Nakamura, T., Horiuchi, T., Morita, T. & Iwanaga, S. (1986a) Purification and properties of intracellular clotting factor, factor B, from horseshoe crab (*Tachypleus tridentatus*) hemocytes. *J. Biochem. (Tokyo)* **99**, 847–857.

Nakamura, T., Morita, T. & Iwanaga, S. (1986b) Lipopolysaccharide-sensitive serine-protease zymogen (factor C) found in *Limulus* hemocytes: Isolation and characterization. *Eur. J. Biochem.* **154**, 511–521.

Nakamura, T., Tokunaga, F., Morita, T. & Iwanaga, S. (1988a) Interaction between lipopolysaccharide and intracellular serine protease zymogen, factor C, from horseshoe crab (*Tachypleus tridentatus*) hemocytes. *J. Biochem. (Tokyo)* **103**, 370–374.

Nakamura, T., Tokunaga, F., Morita, T., Iwanaga, S., Kusumoto, S., Shiba, T., Kobayashi, T. & Inoue, K. (1988b) Intracellular serine-protease zymogen, factor C, from horseshoe crab hemocytes: its activation by synthetic lipid A analogues and acidic phospholipids. *Eur. J. Biochem.* **176**, 89–94.

Seki, N., Muta, T., Oda, T., Iwaki, D., Kuma, K., Miyata, T. & Iwanaga, S. (1994) Horseshoe crab $(1,3)$-$\beta$-D-glucan-sensitive coagulation factor G: a serine protease zymogen heterodimer with similarities to $\beta$-glucan-binding proteins. *J. Biol. Chem.* **269**, 1370–1374.

Tokunaga, F., Miyata, T., Nakamura, T., Morita, T., Kuma, K., Miyata, T. & Iwanaga, S. (1987) Lipopolysaccharide-sensitive serine-protease zymogen (factor C) of horseshoe crab hemocytes:

identification and alignment of proteolytic fragments produced during the activation show that it is a novel type of serine protease. *Eur. J. Biochem.* **167**, 405–416.

Tokunaga, F., Nakajima, H. & Iwanaga, S. (1991) Further studies on

lipopolysaccharide-sensitive serine protease zymogen (factor C): Its isolation from *Limulus polyphemus* hemocytes and identification as an intracellular zymogen activated by alpha-chymotrypsin, not by trypsin. *J. Biochem. (Tokyo)* **109**, 150–157.

*Shun-ichiro Kawabata*
Department of Biology,
Faculty of Science, Kyushu University,
Fukuoka, 812-81 Japan
Email: skawascb@mbox.nc.kyushu-u.ac.jp

*Tatsushi Muta*
Department of Biology,
Faculty of Science, Kyushu University,
Fukuoka, 812-81 Japan

*Sadaaki Iwanaga*
Department of Biology,
Faculty of Science, Kyushu University,
Fukuoka, 812-81 Japan

# 67. *Limulus coagulation factor B*

## Databanks

*Peptidase classification: clan SA, family S1, MEROPS ID: S01.220*
*NC-IUBMB enzyme classification: EC 3.4.21.85*
*Databank codes:*

| Species | SwissProt | PIR | EMBL (cDNA) | EMBL (genomic) |
|---|---|---|---|---|
| *Tachypleus tridentatus* | – | – | D14701 | – |

## Name and History

*Limulus coagulation factor B* is a component of the serine protease cascade that leads to coagulation of the horseshoe crab hemolymph. The system and its components are introduced in the chapter on limulus coagulation factor C (Chapter 66).

## Activity and Specificity

The activation of proenzyme factor B by factor C̄ involves limited proteolysis of the H chain (32 kDa) to yield the active factor B̄. The H chain in factor B̄ incorporates DFP, indicating that this chain contains the active serine residue. Factor B̄ selectively cleaves the Arg98╪Ile99 bond in horseshoe crab proclotting enzyme to form the active clotting enzyme. The cleavage site occurs in the sequence Thr-Thr-Thr-Thr-Arg╪Ile. The occurrence of threonine in the P2 position of the cleavage site is consistent with the substrate specificity of factor B̄ (Nakamura *et al.*, 1986). With synthetic substrates, this protease has a preference for an arginine residue and a hydroxyamino acid at the P1 and the P2 sites, respectively, so that preferred substrates are Boc-Met-Thr-Arg╪NHMec, Bz-Thr-Thr-Arg╪NHMec, and Bz-Ser-Thr-Arg╪NHMec.

Factor B̄ is inhibited by $\alpha_2$-plasmin inhibitor, and is very sensitive to DFP and benzamidine (Nakamura *et al.*, 1986).

## Structural Chemistry

Factor B is a 64 kDa glycoprotein identified as a serine protease proenzyme whose active form (factor B̄) activates proclotting enzyme to clotting enzyme (Nakamura *et al.*, 1986). The cDNA sequence for factor B indicates that the proenzyme is homologous to the proclotting enzyme (Chapter 68) not only in the serine protease domain but also in the N-terminal clip domain (Muta *et al.*, 1993). The clip domain has been also found in *Drosophila* Snake and Easter protease precursors (see Chapters 64 and 65, respectively). Snake and Easter are both indispensable proteins for the normal embryonic development of *Drosophila* (Chasan & Anderson, 1989; DeLotto & Spierer, 1986). Although the significance of this domain is not clear yet, its presence in the *Drosophila* proteins strongly suggests the existence of protease cascade systems in insects similar to that found in horseshoe crab. Among the mammalian serine proteases, human plasma prekallikrein (Chapter 50), which is homologous to factor XI, shows the highest sequence similarity to limulus factor B. Factor B differs from proclotting enzyme in that factor B has an insertion in the N-terminus of the H chain; this is released upon activation as a peptide of 21 amino acid residues.

## Preparation

Factor B has been highly purified from the hemocyte lysate of *Tachypleus tridentatus* (Nakamura *et al.*, 1986).

## Biological Aspects

The activation of most of the serine protease proenzymes in the mammalian coagulation and complement systems occurs by the cleavage of Arg+Ile or Arg+Val bonds. In contrast, the activation of factor B by factor C depends upon cleavage of an Ile+Ile bond. Because of this requirement, factor B is activated neither by trypsin (Chapter 3) nor by chymotrypsin (Chapter 8). This unusual activation site should prevent factor B from being activated by clotting enzyme, a typical trypsin-type enzyme, in a positive feedback manner. So far, factor $\overline{C}$ is the only known activator of factor B.

## Further Reading

An extensive review of all the clotting enzymes of horseshoe crab has been provided by Muta & Iwanaga (1996).

## References

Chasan, R. & Anderson, K.V. (1989) The role of *easter*, an apparent serine protease, in organizing the dorsal-ventral pattern of the *Drosophila* embryo. *Cell* **56**, 391–400.

DeLotto, R. & Spierer, P. (1986) A gene required for the specification of dorsal-ventral pattern in *Drosophila* appears to encode a serine protease. *Nature* **323**, 688–692.

Muta, T. & Iwanaga, S. (1996) The role of hemolymph coagulation in innate immunity. *Curr. Opin. Immunol.* **8**, 41–47.

Muta, T., Oda, T. & Iwanaga, S. (1993) Horseshoe crab coagulation factor B: A unique serine protease zymogen activated by cleavage of an Ile-Ile bond. *J. Biol. Chem.* **268**, 21384–21388.

Nakamura, T., Horiuchi, T., Morita, T. & Iwanaga, S. (1986) Purification and properties of intracellular clotting factor, factor B, from horseshoe crab (*Tachypleus tridentatus*) hemocytes. *J. Biochem. (Tokyo)* **99**, 847–857.

**Shun-ichiro Kawabata**
*Department of Biology,
Faculty of Science, Kyushu University,
Fukuoka, 812-81 Japan
Email: skawascb@mbox.nc.kyushu-u.ac.jp*

**Tatsushi Muta**
*Department of Biology,
Faculty of Science, Kyushu University,
Fukuoka, 812-81 Japan*

**Sadaaki Iwanaga**
*Department of Biology,
Faculty of Science, Kyushu University,
Fukuoka, 812-81 Japan*

# 68. *Limulus clotting enzyme*

## Databanks

*Peptidase classification: clan SA, family S1, MEROPS ID: S01.221*
*NC-IUBMB enzyme classification: EC 3.4.21.86*
*Databank codes:*

| Species | SwissProt | PIR | EMBL (cDNA) | EMBL (genomic) |
|---|---|---|---|---|
| *Tachypleus tridentatus* | P21902 | A23689 | M58366 | – |

## Name and History

Limulus clotting enzyme is a component of the serine protease cascade that leads to coagulation of the horseshoe crab hemolymph. The system and its components are introduced in the chapter on limulus coagulation factor C (Chapter 66).

Synthetic chromogenic substrates having a Gly-Arg+ sequence at the P2 and P1 sites, such as Tos-Ile-Glu-Gly-Arg+NHPhNO$_2$ and Boc-Leu-Gly-Arg+NHPhNO$_2$, were developed based on the conserved sequence of the site cleaved during the transformation of coagulogen to coagulin in the coagulation of horseshoe crab hemolymph (Iwanaga *et al.*, 1978). By use of these substrates, an active serine protease, ***limulus clotting enzyme***, was discovered in the hemocyte lysate activated by bacterial lipopolysaccharide (LPS), and the enzyme and its proenzyme were purified (Nakamura *et al.*, 1982, 1985).

## Activity and Specificity

The activation of proclotting enzyme is catalyzed by factor $\overline{B}$. A higher concentration of trypsin also activates the proenzyme *in vitro*. The resulting clotting enzyme selectively cleaves the Arg18+Thr19 and Arg47+Gly48 bonds in coagulogen to form coagulin. The clotting enzyme also catalyzes the conversion of bovine prothrombin (Chapter 55) to $\alpha$-thrombin, showing its similar substrate specificity to the mammalian coagulation factor Xa.

The clotting enzyme is inhibited by antithrombin III, $\alpha_2$-plasmin inhibitor, soybean trypsin inhibitor and DFP. The amidase activity of the clotting enzyme is efficiently inhibited by a horseshoe crab serpin LICI-2 (Miura *et al*., 1995), but not by LICI-1 (Miura *et al*., 1994).

### Structural Chemistry

Proclotting enzyme is a single-chain glycoprotein of 54 kDa (Nakamura *et al*., 1985; Muta *et al*., 1990). Upon activation, the single-chain proenzyme is cleaved to yield a light (L, 25 kDa) and a heavy (H, 31 kDa) chain held together by a disulfide bond. The C-terminal H chain is a typical serine protease domain with sequence similarity to the mammalian coagulation factors IXa (34.5% identity) and Xa (34.1% identity) (see Chapters 52 and 54). The N-terminal L chain with pyroglutamic acid at its N-terminus contains a small compact domain with three disulfide bonds, called the clip domain, followed by a region with six *O*-linked carbohydrate chains.

### Preparation

Proclotting enzyme has been highly purified from the hemocyte lysate of *Tachypleus tridentatus* (Nakamura *et al*., 1985).

### Biological Aspects

The clip domain has also been found in limulus coagulation factor B and in *Drosophila* Snake and Easter protease precursors. Easter and Snake are both indispensable proteins for the normal embryonic development in *Drosophila* (Chasan & Anderson, 1989; DeLotto & Spierer, 1986) (see Chapters 64 and 65). Although the significance of this domain is not yet clear, its presence also in the *Drosophila* proteins strongly suggests the existence of protease cascade systems similar to those found in the horseshoe crab in other animals too.

### Distinguishing Features

A polyclonal antiserum against proclotting enzyme has been described (Muta *et al*., 1990), but is not commercially available.

### Further Reading

An extensive review of all the clotting enzymes of horseshoe crab has been provided by Muta & Iwanaga (1996).

### References

Chasan, R. & Anderson, K.V. (1989) The role of *easter*, an apparent serine protease, in organizing the dorsal–ventral pattern of the *Drosophila* embryo. *Cell* **56**, 391–400.

DeLotto, R. & Spierer, P. (1986) A gene required for the specification of dorsal–ventral pattern in *Drosophila* appears to encode a serine protease. *Nature* **323**, 688–692.

Iwanaga, S., Morita, T., Harada, T., Nakamura, S., Niwa, M., Takada, K., Kimura, T. & Sakakibara, S. (1978) Chromogenic substrates for horseshoe crab clotting enzyme: its application for the assay of bacterial endotoxin. *Haemostasis* **7**, 183–188.

Miura, Y., Kawabata, S. & Iwanaga, S. (1994) A limulus intracellular coagulation inhibitor with characteristics of the serpin superfamily: purification, characterization, and cDNA cloning. *J. Biol. Chem.* **269**, 542–547.

Miura, Y., Kawabata, S., Wakamiya, Y., Nakamura, T. & Iwanaga, S. (1995) A limulus intracellular coagulation inhibitor type 2: purification, characterization, cDNA cloning, and tissue localization. *J. Biol. Chem.* **270**, 558–565.

Muta, T. & Iwanaga, S. (1996) The role of hemolymph coagulation in innate immunity. *Curr. Opin. Immunol.* **8**, 41–47.

Muta, T., Hashimoto, R., Miyata, T., Nishimura, H., Toh, Y. & Iwanaga, S. (1990) Proclotting enzyme from horseshoe crab hemocytes: cDNA cloning, disulfide locations, and subcellular localization. *J. Biol. Chem.* **265**, 22426–22433.

Nakamura, S., Morita, T., Harada-Suzuki, T., Iwanaga, S., Takahashi, K. & Niwa, M. (1982) A clottable enzyme associated with the hemolymph coagulation system of horseshoe crab (*Tachypleus tridentatus*): its purification and characterization. *J. Biochem. (Tokyo)* **92**, 781–792.

Nakamura, T., Morita, T. & Iwanaga, S (1985) Intracellular proclotting enzyme in limulus (*Tachypleus tridentatus*) hemocytes: its purification and properties. *J. Biochem. (Tokyo)* **97**, 1561–1574.

*Shun-ichiro Kawabata*
*Department of Biology,*
*Faculty of Science, Kyushu University,*
*Fukuoka, 812-81, Japan*
*Email: skawascb@mbox.nc.kyushu-u.ac.jp*

*Tatsushi Muta*
*Department of Biology,*
*Faculty of Science, Kyushu University,*
*Fukuoka, 812-81, Japan*

*Sadaaki Iwanaga*
*Department of Biology,*
*Faculty of Science, Kyushu University,*
*Fukuoka, 812-81, Japan*

# 69. *Limulus coagulation factor G*

### Databanks

*Peptidase classification: clan SA, family S1, MEROPS ID: S01.222*
*NC-IUBMB enzyme classification: none*

*Databank codes:*

| Species | SwissProt | PIR | EMBL (cDNA) | EMBL (genomic) |
|---------|-----------|-----|-------------|----------------|
| α subunit | | | | |
| *Tachypleus tridentatus* | – | – | D16622 | – |
| β subunit (peptidase domain) | | | | |
| *Tachypleus tridentatus* | – | – | D16623 | – |

## Name and History

*Limulus coagulation factor G* is a component of the serine protease cascade that leads to coagulation of the horseshoe crab hemolymph. The system and its components are introduced in the chapter on limulus coagulation factor C (Chapter 66).

In the course of the diagnostic application of the limulus test, it was pointed out that positive reactions are observed with plasma of some patients even in the absence of bacterial lipopolysaccharide (LPS) (Pearson *et al.*, 1984). Since some of these patients were suffering from fungus infection or were undergoing hemodialysis with cellulosic dialyzers, this pseudo-positive reaction had been suggested to be at least in part caused by β-1,3-glucans. Eventually, the β-1,3-glucan-sensitive protease was found in the hemocyte lysate (Morita *et al.*, 1981; Kakinuma *et al.*, 1981) and purified, and termed *factor G* (Seki *et al.*, 1994; Muta *et al.*, 1995).

## Activity and Specificity

The proenzyme factor G is a heterodimeric protein composed of two noncovalently associated subunits, α (72 kDa) and β (37 kDa). The two subunits are derived from separate mRNA species and thus are encoded by different genes (Seki *et al.*, 1994). The purified factor G is autocatalytically activated in the presence of various glucans containing β-1,3 linkages from different origins, but it is not activated by LPS, sulfatides or cholesterol sulfates (Muta *et al.*, 1995). The most effective activators are linear β-1,3-glucans, such as curdlan and paramylon. The 72 kDa subunit α is converted to a 55 kDa fragment and a 17 kDa fragment, and the 37 kDa subunit β is converted to a 34 kDa fragment. In both subunits, arginyl bonds are cleaved, consistent with the substrate specificity of factor $\overline{\text{G}}$. Factor $\overline{\text{G}}$ selectively activates limulus proclotting enzyme, probably through cleavage of the Arg98↓Ile99 bond in the proclotting enzyme. Factor $\overline{\text{G}}$ efficiently hydrolyzes Boc-Glu(OBzl)-Gly-Arg↓NHMec, Boc-Ser(OBzl)-Ala-Arg↓NHMec and Boc-Met-Thr-Arg↓NHMec (Muta *et al.*, 1995).

Factor $\overline{\text{G}}$ is strongly inhibited by α₂-plasmin inhibitor, DFP and leupeptin (Muta *et al.*, 1995). The amidase activity of factor $\overline{\text{G}}$ is efficiently inhibited by a horseshoe crab serpin LICI-3 (Agarwala *et al.*, 1996), but not by LICI-1 (Miura *et al.*, 1994).

## Structural Chemistry

Subunit β shows the strongest sequence similarity to horseshoe crab factor B (40.5% identity) (Seki *et al.*, 1994), but also shows strong similarity to proclotting enzyme (37.7%

identity) (Chapter 68), suggesting that the horseshoe crab clotting factors, proclotting enzyme, factor B and factor G subunit β, may have evolved from a common origin. Subunit α is a new type of mosaic protein with intriguing features. The N-terminal portion of subunit α (Pro4–Ala236) shows sequence similarity to the C-terminal portion of β-(1,3)-glucanase A1 (EC 3.2.1.39) from *Bacillus circulans*, some bacterial lichenases, and the gene product of *meri-5* from *Arabidopsis thaliana*, which contains a glucanase catalytic domain. A dot matrix plot indicates that subunit α contains two types of internal repeat. In the central portion of the subunit, there are three tandem repeats of 47 amino acids showing similarity to repeats found in xylanase A (EC 3.2.1.8) from *Streptomyces lividans*. They are also seen in the yeast-lytic endopeptidase of *Rarobacter* (Chapter 84). β-(1,3)-glucanase from *Oerskovia xanthineolytica*, and ricin B chain. The C-terminal portion of subunit α has two long repeats (126 amino acids in length) with 91.3% identity, homologous to the N-terminal portion of xylanase Z (EC 3.2.1.8) from *Clostridium thermocellum* and xylanase D from *Bacillus polymyxa*.

## Preparation

Factor G has been highly purified from the hemocyte lysate of *Tachypleus tridentatus* (Muta *et al.*, 1995).

## Biological Aspects

As little as 1 ng of curdlan significantly activates the proenzyme factor G. Branching of the linear chain with β-1,4- or β-1,6 linkages appears to reduce the factor G-activating activity. Shorter oligosaccharides containing 2–6 glucose residues do not activate factor G at all. When the concentration of factor G is 10-fold higher or lower, the optimum concentration of curdlan for the activation shifts approximately 10-fold, implying that the molar ratio of factor G to β-1,3-glucan is important in the proenzyme activation. Upon activation of factor G, both subunits α and β undergo limited proteolysis (Muta *et al.*, 1995). A longer incubation with β-1,3-glucan causes a fragmentation of the 55 kDa fragment to a 46 kDa fragment, concomitant with loss of the amidase activity. Although the activation of factor G is associated with limited proteolysis at arginyl bonds, it is not activated by factor $\overline{\text{G}}$, factor $\overline{\text{B}}$ or clotting enzyme; nor can it be activated *in vitro* by digestive proteases such as trypsin or chymotrypsin.

## Further Reading

An extensive review of all the clotting enzymes of horseshoe crab has been provided by Muta & Iwanaga (1996).

## References

Agarwala, K.L., Kawabata, S., Miura, Y., Kuroki, Y. & Iwanaga, S. (1996) Limulus intracellular coagulation inhibitor type 3: purification, characterization, cDNA cloning, and tissue localization. *J. Biol. Chem.* **271**, 23768–23774.

Kakinuma, A., Asano, T., Torii, H. & Sugino, Y. (1981) Gelation of *Limulus* amoebocyte lysate by an antitumor (1 → 3)-β-D-glucan. *Biochem. Biophys. Res. Commun.* **101**, 434–439.

Miura, Y., Kawabata, S. & Iwanaga, S. (1994) A limulus intracellular coagulation inhibitor with characteristics of the serpin superfamily: purification, characterization, and cDNA cloning. *J. Biol. Chem.* **269**, 542–547.

Morita, T., Tanaka, S., Nakamura, T. & Iwanaga, S. (1981) A new (1 → 3)-β-D-glucan-mediated coagulation pathway found in limulus amebocytes. *FEBS Lett.* **129**, 318–321.

Muta, T. & Iwanaga, S. (1996) The role of hemolymph coagulation in innate immunity. *Curr. Opin. Immunol.* **8**, 41–47.

Muta, T., Seki, N., Takaki, Y., Hashimoto, R., Oda, T., Iwanaga, A.,

Tokunaga, F. & Iwanaga, S. (1995) Purified horseshoe crab factor G: Reconstitution and characterization of the (1 → 3)-β-D-glucan-sensitive serine protease cascade. *J. Biol. Chem.* **270**, 892–897.

Nakamura, T., Horiuchi, T., Morita, T. & Iwanaga, S. (1986) Purification and properties of intracellular clotting factor, factor B, from horseshoe crab (*Tachypleus tridentatus*) hemocytes. *J. Biochem. (Tokyo)* **99**, 847–857.

Pearson, F.C., Bohon, J., Lee, W., Bruszer, G., Sagona, M., Dawe, R., Jakubowski, G., Morrison, D. & Dinarello, C. (1984) Comparison of chemical analyses of hollow-fiber dialyzer extracts. *Artif. Organs* **8**, 291–298.

Seki, N., Muta, T., Oda, T., Iwaki, D., Kuma, K., Miyata, T. & Iwanaga, S. (1994) Horseshoe crab (1,3)-β-D-glucan-sensitive coagulation factor G: a serine protease zymogen heterodimer with similarities to β-glucan-binding proteins. *J. Biol. Chem.* **269**, 1370–1374.

*Shun-ichiro Kawabata*
Department of Biology,
Faculty of Science, Kyushu University,
Fukuoka, 812-81, Japan
Email: skawascb@mbox.nc.kyushu-u.ac.jp

*Tatsushi Muta*
Department of Biology,
Faculty of Science, Kyushu University,
Fukuoka, 812-81, Japan

*Sadaaki Iwanaga*
Department of Biology,
Faculty of Science, Kyushu University,
Fukuoka, 812-81, Japan

# 70. Venombin A

## Databanks

*Peptidase classification: clan SA, family S1, MEROPS ID: S01.175*
*NC-IUBMB enzyme classification: EC 3.4.21.74*
*Databank codes:*

| Species | SwissProt | PIR | EMBL (cDNA) | EMBL (genomic) |
|---|---|---|---|---|
| Ancrod | | | | |
|   *Agkistrodon bilineatus* | P33588 | A60489 | – | – |
|   *Agkistrodon contortrix* | P09872 | A60468 | – | – |
|   *Calloselasma rhodostoma* | P47797 | S36783 | L07308 | – |
| Batroxobin | | | | |
|   *Bothrops atrox* | P04971 | A28169 | J02684 | M20890: exon 1 |
| | | | | M20891: exon 2 |
| | | | | M20892: exon 3 |
| | | | | M20893: exon 4 |
| | | | | M20894: exon 5 and complete CDS |
| | | | | X12747: complete gene |
| Crotalase | | | | |
|   *Crotalus atrox* | – | A37002 | – | – |
| **Related peptidases** | | | | |
| Bothrombin | | | | |
|   *Bothrops jararaca* | – | A54361 | – | – |

| Species | SwissProt | PIR | EMBL (cDNA) | EMBL (genomic) |
|---|---|---|---|---|
| **Related peptidases** (*continued*) | | | | |
| Flavoxobin | | | | |
| *Trimeresurus flavoviridis* | P05620 | A41456 | – | – |
| Others | | | | |
| *Lachesis muta* | P33589 | A32415<br>S35689 | – | – |

S

## *Name and History*

Fontana was probably the first to show that snake venoms have the ability to clot whole animal blood and to cause circulating blood to become incoagulable (Fontana, 1787). Approximately 100 years later, studies were carried out with venom from the eastern diamondback rattlesnake (*Crotalus adamanteus*) indicating that the venom destroyed the coagulability of animal blood. The globulin fraction of the venom was found to be responsible for this action (Mitchell & Reichert, 1886). In 1937 it was shown that the venom of *C. adamanteus*, as well as other venoms, acted directly on fibrinogen to coagulate plasma, and that this action was not dependent on calcium (Eagle, 1937). Paradoxically, in several cases of envenomation of humans by *C. adamanteus*, the blood became incoagulable (Andrews, 1960; Weiss *et al*., 1969). Thus, it appears that *in vitro* the venom coagulates blood, whereas *in vivo* the venom acts as an anticoagulant. It was reported that crude venom from *C. adamanteus* acts like thrombin in converting fibrinogen to fibrin and that there is no effect on factors II, VII or X. Unlike thrombin, however, the venom does not aggregate platelets (Weiss *et al*., 1969). The enzyme responsible for this action was purified from *C. adamanteus* venom and named ***crotalase*** to reflect its origin from a *Crotalus* species (Markland and Damus, 1971). The enzyme is available commercially under the names ***Crotalase*** or ***Defibrizyme***.

Incoagulability of the blood of an animal following envenomation occurs with many snakes of the family Crotalidae. This activity was noted following envenomation by the Malayan pit viper (*Calloselasma rhodostoma*, formerly known as *Agkistrodon rhodostoma*) (Reid *et al*., 1963) and the enzyme responsible for this action was isolated and named ***ancrod*** (Esnouf & Tunnah, 1967). Ancrod is the generic term adopted by the World Health Organization for the coagulant enzyme isolated from *Calloselasma rhodostoma*. The enzyme is available commercially as ***Ancrod*** or ***Arvin***. A coagulant enzyme was also isolated from the venom of *Bothrops atrox*, a pit viper found in Central and South America (Stocker & Egberg, 1973). ***Batroxobin*** is the generic term adopted by the World Health Organization for the coagulant enzyme isolated from the venom of *B. atrox* or its subspecies. The enzyme is available commercially as ***Defibrase*** or ***Reptilase***.

The name ***venombin A*** recommended by NC-IUBMB for EC 3.4.21.74 embraces this set of three well-characterized endopeptidases. They have many similarities, including the cleavage of the Arg16+Gly17 bond in the fibrinogen A$\alpha$ chain, but also have distinctive properties. *In vivo*, the venombin A enzymes act as benign defibrinogenating agents to remove fibrinogen from the blood. In the test tube they act to form a fibrin clot. Thrombin-like, or defibrinogenating, snake venom enzymes have been the topic of several review articles over the years (Aronson, 1976; Markland & Pirkle, 1977; Stocker *et al*., 1982; Bell, 1988, 1990; Pirkle & Theodor, 1990, 1998). In the following paragraphs, the properties of the three venom enzyme will be described in some detail.

## *Activity and Specificity*

Members of this group of enzymes have been referred to as thrombin-like enzymes, in view of their ability to clot purified fibrinogen solutions or blood plasma (Aronson, 1976). One feature common to all three venom enzymes is cleavage of the Arg16+Gly17 peptide bond in the A$\alpha$ chain of fibrinogen, leading to the release of fibrinopeptides A, AY and AP, and the conversion of fibrinogen to a fibrin clot (Ewart *et al*., 1970; Holleman & Coen, 1970). The A, AP and AY peptides are all related; the AP peptide is phosphorylated at Ser3 and the AY peptide is one residue shorter than fibrinopeptide A at the N-terminus. The venom enzymes differ from thrombin in a number of ways, however, one of which is that they only cleave fibrinopeptide A and not fibrinopeptide B from fibrinogen. Further, the venom enzymes do not activate factor XIII and did not cause platelet aggregation. Early reports indicated that batroxobin purified from *B. atrox moojeni* activated factor XIII. However, it was shown with affinity-purified enzyme that the factor XIII-activating activity had been due to a contaminant (Holleman & Weiss, 1976). Thus, reports of factor XIII-activating activities in thrombin-like enzymes purified from the *B. atrox* subspecies appear in all cases to be due to contaminants. There are species-related differences in the rates of coagulation of fibrinogen by the venom coagulant enzymes which may be due to species differences in fibrinogen or to the susceptibility of the venom enzymes to inhibition by plasma proteinase inhibitors. Thus, batroxobin clots human fibrinogen about 10-fold more rapidly than rabbit fibrinogen (Stocker & Barlow, 1976).

Ancrod also differs from thrombin in that on prolonged incubation with fibrinogen the venom enzyme completely degrades the A$\alpha$ chains, forming a product with a molecular mass of ~40 kDa (Mattock & Esnouf, 1971; Pizzo *et al*., 1972). This means that peptides totaling 31 kDa in molecular mass are cleaved from the A$\alpha$ chains. Thrombin does not catalyze such a conversion. Batroxobin also does not degrade the A$\alpha$ chains of fibrinogen (Mattock & Esnouf, 1971). By contrast, crotalase degrades the B$\beta$ chains on prolonged incubation (Markland & Pirkle, 1977). Interestingly, crotalase cleaves a thrombin-susceptible bond in prothrombin without generating thrombin activity (Pirkle *et al*., 1976). Both $Ca^{2+}$

and phospholipid are required for this activity, probably to induce a conformational change in prothrombin. Neither ancrod nor batroxobin possess this activity.

Ancrod has esterase activity with small basic ester substrates such as Bz-Arg-OEt and Bz-Lys-OPhNO$_2$ (Hatton, 1973). After studying a series of Z-amino acid *p*-nitrophenyl esters, Ascenzi *et al*. (1985) concluded that the arginine ester showed the most favorable kinetic characteristics and is probably the substrate of choice for ancrod. These findings were confirmed by others who found a very strict specificity for arginine esters and the *N*-acylated derivatives (Exner & Koppel, 1972). From a study of substrate kinetics, it was concluded that ancrod showed catalytic properties similar to those of tissue kallikrein (Chapter 29) in the hydrolysis of small synthetic substrates (Ascenzi *et al*., 1986). The pH optimum for ancrod with arginine esters is ~8.0 (Collins & Jones, 1972; Ascenzi *et al*., 1985). Coagulant and esterase activities of ancrod are inhibited simultaneously by PMSF (Hatton, 1973). Batroxobin (Stocker & Barlow, 1976) and crotalase (Markland, 1976) also possess esterase activity with small peptide or basic amino acid esters. Optimum pH for hydrolysis of basic amino acid ester substrates by crotalase is above 8.0.

Ancrod is inhibited by small serine proteinase inhibitors such as DFP and PMSF. Titration with the active-site titrant *p*-nitrophenyl, *p'*-guanidinobenzoate indicates one active site per mole (Collins & Jones, 1972). The enzyme is also inhibited by the histidine-reactive agent and substrate analog *N*-α-nitrobenzyloxycarbonyl-L-Arg-CH$_2$Cl (Collins & Jones, 1974). Batroxobin is also inhibited by DFP (Stocker & Barlow, 1976) and the histidine-reactive inhibitor D-Phe-Pro-Arg-CH$_2$Cl (Stürzebecher *et al*., 1986). It is not inhibited by thrombin inhibitors, such as heparin or hirudin, nor by proteinase inhibitors such as soybean trypsin inhibitor (Stocker & Barlow, 1976). Batroxobin is inhibited by competitive, reversible inhibitors of the benzamidine type (Stürzebecher *et al*., 1986). Crotalase is similarly inactivated by serine proteinase inhibitors, such as DFP, but it is inhibited poorly by Tos-Lys-CH$_2$Cl (Markland & Damus, 1971). Interestingly, crotalase is inhibited rapidly by the specific inhibitor of plasma kallikrein (Chapter 50), Pro-Phe-Arg-CH$_2$Cl (Markland *et al*., 1981). Ancrod (Pitney & Regoeczi, 1970) and batroxobin (Egberg, 1972) are inactivated by α$_2$-macroglobulin. In view of the similarities in structure of the venombins, it is assumed that crotalase is likewise inhibited by α$_2$-macroglobulin.

Crotalase has been reported to exhibit kallikrein-like activity, on the basis of substrate cleaving preference (Markland *et al*., 1982), amino acid sequence homology (Pirkle *et al*., 1981), ability to liberate kinin from high molecular mass kininogen (Markland *et al*., 1982), and the profile of inhibition by peptidyl chloromethanes (Markland *et al*., 1981). The significance of the potential hypotensive action of crotalase is unknown at present. There have been reports in the literature of kinin-releasing enzymes in crotalid venoms, and the enzymes responsible for kallikrein-like activity have been purified from several rattlesnake species (Bjarnason *et al*., 1983; Komori *et al*., 1988; Mori & Sugihara, 1989) and shown to have N-terminal sequence homology with crotalase. Although ancrod and batroxobin have not been reported to possess kallikrein-like activity, Ascenzi *et al*. (1986) reported that ancrod possessed basic ester substrate hydrolyzing activity similar to that of pig pancreatic kallikrein. It was reported that both ancrod (Au *et al*., 1993) and batroxobin (Itoh *et al*., 1987) exhibit a high degree of sequence identity to mammalian serine proteinases including pancreatic kallikrein and thrombin. Further, the exon–intron organization of the batroxobin gene suggests that batroxobin belongs to the tissue kallikrein group (Itoh *et al*., 1988).

Assay for enzymatic activity of the venombin A enzymes can easily be performed using the hydrolysis of *N*-α-substituted arginine esters at pH 8.0 (Markland, 1976; Nolan *et al*., 1976). Coagulant activity can be measured by adding different dilutions of the enzyme to a standard control blood plasma or a standard fibrinogen solution and measuring the clotting time. By comparing these values to similarly measured clotting times for a standard thrombin solution of known activity, the activity of the venom enzyme in thrombin equivalent units can conveniently be obtained (Nolan *et al*., 1976).

## Structural Chemistry

The venombin A enzymes are all single-chain glycoproteins with molecular masses of approximately 35 kDa. The carbohydrate content varies considerably between the different enzymes, and this is most likely responsible for the differences in the masses of the different enzymes. Thus, different forms of batroxobin have been isolated from *Bothrops atrox* ($M_r$ 41 600, 10.2% neutral carbohydrate), and *B. a. moojeni* ($M_r$ 35 800, 5.8% neutral sugar) (Stocker & Barlow, 1976; Stocker & Meier, 1988). A separate report indicated that the enzyme from *B. a. moojeni* has a molecular mass of 29 000 and contains 27% carbohydrate including sialic acid (Holleman & Weiss, 1976). The enzyme from *B. a. moojeni* has an isoelectric point at pH 6.6 and is a more effective defibrinogenating agent *in vivo* (Stocker & Meier, 1988). Ancrod ($M_r$ 35 400) contains 36% carbohydrate and the isoelectric point varies from 4.2 to 6.2 for several different electrophoretic forms (Nolan *et al*., 1976). Crotalase ($M_r$ 32 700) was originally reported to contain 5.4% carbohydrate (Markland & Damus, 1971), but a more recent determination using high-performance anion-exchange chromatography with pulsed amperometric detection after trifluoroacetic acid hydrolysis indicated 8.3% carbohydrate content (H. Pirkle, unpublished result). Microheterogeneity of the venombin A enzymes has been observed and is most likely due to varying sialic acid content. Treatment of crotalase with neuraminidase resulted in the conversion of five isoelectric focusing bands into one, with virtually no loss of enzymatic activity; the isoelectric point of the desialylated enzyme is pH 4.6 (Bajwa & Markland, 1979). There appears to be some variability in the sialic acid content from preparation to preparation of crotalase, perhaps due to the conditions of purification (Bajwa & Markland, 1979). Microheterogeneity of ancrod (Hatton, 1973) and batroxobin (Holleman & Weiss, 1976) has also been observed, possibly due to the variability in sialic acid content, and loss of batroxobin activity was reported following neuraminidase treatment (Stocker & Barlow, 1976). Although the role of the carbohydrate in the venombin A enzymes is unknown at present, total removal of the carbohydrate from batroxobin led to considerable loss of enzymatic activity (Tanaka *et al*., 1992).

The primary structure of each of the three enzymes has been determined, and their extensive sequence similarity is shown in Alignment 70.1 (see CD-ROM). The structure of crotalase has been determined recently (Pirkle *et al.*, 1981, 1996) and reveals a single *N*-glycosylation site in the 237 amino acid sequence. The 12 cysteines are most likely involved in six disulfide bonds, since early investigations (Markland & Damus, 1971) indicated that there were no free sulfhydryl groups. The amino acid sequence of ancrod indicates that it contains 234 amino acids and the arrangement of the cysteine residues is identical to that in crotalase (Burkhart *et al.*, 1992). Glycosylation of all five *N*-glycosylation sites in ancrod was suspected on the basis of the blank sequencer cycles at each of the *N*-glycosylation sites. The cDNA sequence for ancrod was also determined (Au *et al.*, 1993) and there are several differences in the derived amino acid sequence from that determined by protein sequencing.

Batroxobin contains 231 amino acids and its 12 cysteine residues can be aligned with those in crotalase and ancrod (Itoh *et al.*, 1987). There are two potential glycosylation sites in batroxobin. There is general agreement on the positioning of the catalytic triad (using the ancrod numbering) of His43, Asp88 and Ser182 on the basis of alignment with mammalian serine proteinases of family S1.

The structures of the carbohydrate side-chains of ancrod have been determined recently (Pfeiffer *et al.*, 1992). The five glycosylation sites on the enzyme have been identified as Asn residues 23, 79, 99, 148 and 229 (Pfeiffer *et al.*, 1993). There are partially truncated di-, tri- and tetra-antennary complex type *N*-glycans with fucose α-1,6 residues at the innermost *N*-acetylglucosamine and solely α-2,3-linked sialic acid substituents. Many of the structures determined represent novel glycoprotein-*N*-glycan structures. All of the five *N*-glycosylation sites of ancrod are substituted by carbohydrate and each site displays a characteristic pattern of complex glycans which carry an α-fucosyl residue at C6 of *N*-acetylglucosamine-1 with, α,3-linked mono-, di-, tri- and tetrasialylated glycans (Pfeiffer *et al.*, 1993).

The structure of the *N*-linked oligosaccharide chain of batroxobin has also been reported recently (Tanaka *et al.*, 1992; Lochnit & Geyer, 1995). The potential *N*-glycosylation site is at Asn146 or Asn225 in the batroxobin structure (Itoh *et al.*, 1987). Although it was originally reported that there was but a single oligosaccharide chain attached to batroxobin (Tanaka *et al.*, 1992), recent studies suggest that there is some heterogeneity in the complex carbohydrate attached (Lochnit & Geyer, 1995). Most of the glycan chains are core-fucosylated at C6 of the innermost *N*-acetylglucosamine. In comparison with ancrod, the *N*-glycosylation site in batroxobin bears primarily diantennary complex-type oligosaccharides, carrying exclusively *N*-acetylgalactosamine-β,4-*N*-acetyl-glucosamine antennae. Thus, there is a novel type of glycoprotein *N*-glycan in batroxobin. In summary, the carbohydrate side-chains in batroxobin are significantly different from those in ancrod. There is only one glycosylation site in crotalase, Asn81 (Pirkle *et el.*, 1996), but the structure of the carbohydrate side-chain in crotalase has not been determined.

Crystallization of crotalase from phosphate solutions has been reported (Pirkle *et al.*, 1991), but no crystallographic structure has yet been determined for any of the venombin A enzymes.

The properties of the three enzymes are compared in Table 70.1.

*Table 70.1*   Comparison of characteristics of venombin A enzymes

| Property | Enzyme | | |
| --- | --- | --- | --- |
| | Ancrod[a] | Batroxobin[b] | Crotalase[c] |
| Venom source | *Calloselasma rhodostoma* | *Bothrops atrox (moojeni)* | *Crotalus adamanteus* |
| Molecular mass | 35 400 | 36 000 | 32 700 |
| Isoelectric point | pI 4.2–6.2 | pI 6.6 | Acidic, pI 4.6 after desialylation |
| Carbohydrate content | 36% | 5.8% | 8.3% |
| Fibrinogen Aα chain bond cleavage specificity | Arg16-Gly17 | Arg16-Gly17 | Arg16-Gly17 |
| Additional fibrinogen degradation | Aα chain degradation | None | Bβ chain degradation |
| Small molecule inhibitors | DFP | DFP | DFP |
| Protein inhibitors | $\alpha_2$-Macroglobulin | $\alpha_2$-Macroglobulin | $\alpha_2$-Macroglobulin (assumed) |
| Esterase activity | Small peptide or basic amino acid esters | Small peptide or basic amino acid esters | Small peptide or basic amino acid esters |
| Active-site residue | Serine | Serine | Serine |
| N-terminal residue | Valine | Valine | Valine |
| Number of amino acids | 234 | 231 | 237 |
| Stability | Stable | Stable | Stable |

[a]Data from Collins & Jones (1974), Nolan *et al.* (1976), Burkhart *et al.* (1992), Pfeiffer *et al.* (1992, 1993).
[b]Data from Stocker & Barlow (1976), Itoh *et al.* (1987, 1988), Tanaka *et al.* (1992), Lochnit & Geyer (1995).
[c]Data from Markland & Damus (1971), Markland (1976), Bajwa & Markland (1979), Pirkle *et al.* (1996).

## Preparation

Ancrod has been purified from the venom of the Malayan pit viper (*Calloselasma rhodostoma*) by a two-step procedure involving affinity chromatography on 4-aminobutylguanidine (agmatine)-agarose followed by molecular sieve chromatography on Sephadex G-100 (Nolan *et al.*, 1976). This method is superior to the original method which employed anion-exchange chromatography (TEAE-cellulose) as the first step and either Sephadex G-100 (Esnouf & Tunnah, 1967) or IRC-50 cation-exchange chromatography (Hatton, 1973) as the second step. Ancrod is readily available in the UK, Europe and Canada. The registered trade name of the enzyme is Arvin (Twyford Laboratories) (see Appendix 2 for full names and addresses of suppliers). The enzyme is also known as Arwin (the registered trade name for Twyford Pharmaceutical). Ancrod is prepared commercially by Abbott Laboratories as Venacil. In the USA, ancrod can also be obtained from Knoll Laboratories. Ancrod cDNA has been cloned from the venom glands of *Calloselasma rhodostoma*, but the enzyme has not been expressed (Au *et al.*, 1993).

Batroxobin is found in the venom of the South American pit viper *Bothrops atrox*, and in several subspecies including *B. a. moojeni*, and *B. a. asper*. The enzyme has been purified by a two-step procedure involving anion-exchange chromatography on DEAE-Sephadex A-50 followed by molecular sieve chromatography on Sephadex G-100 (Stocker & Barlow, 1976). An affinity chromatography procedure utilizing *p*-aminobenzamidine, linked to Sepharose 4B through a spacer of diaminodipropylaminosuccinate was also developed and combined with a second step which involved isoelectric focusing (Holleman & Weiss, 1976). The enzyme purified from *B. a. moojeni* venom by affinity chromatography only was devoid of factor XIII-activating activity. The enzyme is available from Pentapharm as Reptilase, and is also known as Defibrase. Batroxobin has been cloned from the venom gland cDNA library of *B. a. moojeni* (Itoh *et al.*, 1987). The cDNA has been expressed in *E. coli* and the protein isolated as an insoluble fusion protein that could be cleaved with thrombin to generate denatured batroxobin. The recombinant protein was refolded to generate an active coagulant enzyme (Maeda *et al.*, 1991).

Crotalase has been prepared from venom of *Crotalus adamanteus* (eastern diamondback rattlesnake). Initially, a five-step procedure was developed including molecular sieve chromatography on Sephadex G-100, anion-exchange chromatography on DEAE-cellulose, hydroxyapatite chromatography, rechromatography on Sephadex G-100 and rechromatography on DEAE-cellulose (Markland & Damus, 1971). Another procedure was developed subsequently that is less time consuming and provides an enzyme with higher activity than the original method. This procedure involves gel filtration on Sephadex G-100, anion-exchange chromatography on DEAE-cellulose, affinity chromatography on benzamidine-Sepharose (in which benzamidine is attached to Sepharose through a six carbon spacer), and cation-exchange chromatography on SP-Sephadex (Bajwa & Markland, 1979). Crotalase is available commercially from Sigma, or as Defibrizyme from Research Plus. The gene for crotalase has not yet been cloned.

## Biological Aspects

The gene for batroxobin has been isolated from the venom gland of *Bothrops atrox moojeni* (Itoh *et al.*, 1988). The gene covers 8 kb and has five exons and four introns. Batroxobin is synthesized as a preproenzyme with an 18 residue prepeptide and a six residue propeptide. The mature protein is encoded by four exons and the three active-site residues, His41, Asp86 and Ser178, are encoded by three separate exons. The exon–intron organization of the batroxobin gene differs from that of the mammalian prothrombin gene, but is quite similar to the organization of the trypsin and kallikrein genes. This suggests that batroxobin, being from snake venom gland which is believed to have been derived from the submaxillary gland, belongs to the tissue kallikrein group.

Ancrod is synthesized as a preproenzyme with a putative secretory peptide of 18 amino acids and a propeptide of six amino acids (Au *et al.*, 1993) similar to that in batroxobin. There are no data available concerning the genes for ancrod or crotalase.

Ancrod and batroxobin have been used as defibrinogenating agents for a number of clinical conditions including deep vein thrombosis, myocardial infarction, pulmonary embolus, central retinal vein occlusion, peripheral vascular disease, acute ischemic stroke, angina pectoris, glomerulonephritis, priapism, sickle cell crises and renal transplant rejection (Bell, 1988, 1990; Eschenfelder, 1996; Furukawa & Ishimaru, 1990; Stocker, 1988; Pollak *et al.*, 1990; Soutar & Ginsberg, 1993). Crotalase has not been used in humans. Ancrod (Cercek *et al.*, 1987) and batroxobin (Tomaru *et al.*, 1988) have been used separately in combination with thrombolytic agents in dog models of arterial thrombosis. Batroxobin acted to prevent coronary artery reocclusion, while ancrod enhanced the effect of thrombolytic agents in a carotid arterial thrombosis model, probably by depleting fibrinogen and preventing propagation of existing thrombi. Ancrod was also shown to decrease cyclic flow variations and to cause thrombolysis in a coronary artery model in the dog (Apprill *et al.*, 1987).

The venombin A enzymes are all serine proteinases that rapidly cleave fibrinopeptide A from fibrinogen, forming a fibrin clot *in vivo*. Unlike thrombin, the venombins A do not activate factor XIII, other coagulation factors or platelets (Aronson, 1976; Markland & Damus, 1971; Soutar & Ginsberg, 1993). It would appear that the mechanism whereby the venombin A enzymes produce a benign state of defibrinogenation involves the formation of an abnormal clot, distinguishable by the fact that it is soluble in 5 M urea or 1% monochloroacetic acid. This is in contrast to thrombin clots which are cross-linked by factor XIII and are insoluble in these solvents. Within minutes of administration of the venom enzymes *in vivo* there is a depression in the fibrinogen concentration in plasma and within an hour or two the fibrinogen levels become very low and remain so. Fibrinogen can be maintained at the depressed level by repeated infusions on a daily or twice daily basis (Bell, 1990). The low levels have been maintained in some patients for up to 7 weeks. Upon termination of venom enzyme infusion there is a slow recovery of the fibrinogen levels. There is a dramatic increase in the levels of fibrin(ogen) degradation products in the plasma upon venom-induced defibrinogenation. These degradation products appear to be derived by plasmin digestion of the non-crosslinked fibrin clot, which is highly susceptible to the action of the fibrinolytic enzyme. It has been reported that ancrod (Soszka *et al.*, 1985) and batroxobin (Klöcking *et al.*, 1987) induce the release of plasminogen activator from vascular endothelial cells, thereby activating the fibrinolytic response. However, a recent study in six humans indicated that although there is a massive rise in fibrin(ogen) degradation products and depletion of plasminogen following ancrod infusion, there is no rise in tissue or urokinase-type plasminogen activator levels (Prentice *et al.*, 1993). These authors reported that fibrinopeptide A was removed from fibrinogen to produce soluble fibrin which was then removed from the circulation with no evidence of increase in plasminogen activator levels, but with significant activation and consumption of plasminogen and depletion of $\alpha_2$-antiplasmin. Further, the fibrin degradation products formed were identical to those formed by plasmin degradation of soluble (non-crosslinked) fibrin, suggesting that plasminogen activation by the fibrinolytic system accounts for the degradation of fibrin.

If the venom defibrinogenating enzymes are used for a long time or if there is a necessity for repeated usage, there is a loss of enzyme activity due to the development of immunological resistance by the patient (Pitney *et al.*, 1969; Thomson *et al.*, 1977). Interestingly, there is no immunological cross-reaction between ancrod and batroxobin as shown by studies with specific antibodies (Barlow *et al.*, 1973). A test has been developed to determine whether antibodies to ancrod or batroxobin are present in human serum (Stocker & Yeh, 1975). Although crotalase has not been used in humans, there is a similar depression of fibrinogen with concomitant increase in fibrin(ogen) degradation products in animals following intravenous administration of the enzyme (Damus *et al.*, 1972).

The importance of the fibrinolytic system in the action of ancrod has been demonstrated by the appropriate use of inhibitors of fibrinolysis in animals following ancrod infusion. This treatment led to significant clot formation *in vivo* and death. A beneficial effect of ancrod infusion is the decrease in blood viscosity due to fibrinogen depletion. This improves the blood flow characteristics and may aid the therapeutic effectiveness of ancrod (Ehrly, 1976).

The danger of therapy with the defibrinogenating enzymes consists of bleeding. However, the side-effects and complications can be controlled by knowledge of contra-indications, careful control of dosage and daily monitoring of fibrinogen levels (Latallo, 1983; Bell, 1990; Soutar & Ginsberg, 1993).

## Distinguishing Features

The ability of these enzymes to release fibrinopeptide A from fibrinogen without releasing fibrinopeptide B, activating factor XIII or inducing platelet aggregation, distinguishes these enzymes from most other fibrinogen-clotting venom enzymes and thrombin. Antibodies that recognize ancrod but not batroxobin and vice versa have been prepared, indicating that there are immunologically unique epitopes in each enzyme (Barlow *et al.*, 1973). Specific antibodies to crotalase have not yet been reported.

## Related Peptidases

Defibrinogenating enzymes with properties similar to those of the enzymes in the venombin A group have been reported in a number of snake species from the pit viper family including members of the *Agkistrodon, Bothrops, Crotalus, Lachesis* and *Trimeresurus* genera. Similar enzymes have also been found in snakes from the true viper family and there is an example of a snake in the Colubridae family possessing a fibrinogen-clotting enzyme (Pirkle & Theodor, 1990). There are a smaller number of snake venom fibrinogen-clotting enzymes that preferentially release fibrinopeptide B (Herzig *et al.*, 1970), or release fibrinopeptides A and B simultaneously and activate factor XIII (Chapter 71).

## Further Reading

An excellent description of each of these enzymes is found in *Methods in Enzymology*: for ancrod, an article by Nolan *et al.* (1976), for batroxobin, an article by Stocker & Barlow (1976), and for crotalase, an article by Markland (1976). A review covering these enzymes has been published recently (Pirkle & Theodor, 1998).

## References

Andrews, C.E. (1960) The treatment of diamondback rattlesnake bite *(Crotalus adamanteus). Arch. Surg.* **81**, 699–705.

Apprill, P.G., Ashton, J., Guerrero, J., Glas-Greenwalt, P., Buja, L.M. & Willerson, J.T. (1987) Ancrod decreases the frequency of cyclic flow variations and causes thrombolysis following acute coronary thrombosis. *Am. Heart J.* **113**, 898–906.

Aronson, D.L. (1976) Comparison of the actions of thrombin and the thrombin-like venom enzymes ancrod and batroxobin. *Thromb. Haemost.* **36**, 9–13.

Ascenzi, P., Betollini, A., Bolognesi, M., Guarneri, M., Menegatti, E. & Amiconi, G. (1985) Catalytic properties of ancrod, the thrombin-like proteinase from the Malayan pit viper *(Agkistrodon rhodostoma)* venom. *Biochim. Biophys. Acta* **829**, 415–423.

Ascenzi, P., Betollini, A., Bolognesi, M., Guarneri, M., Menegatti, E. & Amiconi, G. (1986) Primary specificity of ancrod, the coagulating serine proteinase from the Malayan pit viper *(Agkistrodon rhodostoma)* venom. *Biochim. Biophys. Acta* **871**, 225–228.

Au, L.C., Lin, S.B., Chou, J.S., Teh, G.W., Chang, K.J. & Shih, C.M. (1993) Molecular cloning and sequence analysis of the cDNA for ancrod, a thrombin-like enzyme from the venom of *Calloselasma rhodostoma*. *Biochem. J.* **294**, 387–390.

Bajwa, S.S. & Markland, F.S. (1979) A new method for purification of the thrombin-like enzyme from the venom of the eastern diamondback rattlesnake. *Thrombos. Res.* **16**, 11–23.

Barlow, G.H., Lewis, J.L., Finley, R., Martin, D. and Stocker, K. (1973) Immunochemical identification of ancrod (A38414) and reptilase (defibrase). *Thromb. Res.* **2**, 17–22.

Bell, W.R. (1988) Clinical trials with ancrod. In: *Hemostasis and Animal Venoms* (Pirkle, H. & Markland. F.S., eds). New York: Marcel Dekker, pp. 541–551.

Bell, W.R. (1990) Defibrinogenating enzymes. In: *Hemostasis and Thrombosis* (Colman, R.W., Hirsh, J., Marder, V.J. & Salzman, E.W., eds). Philadelphia: J.B. Lippincott, pp. 886–900.

Bjarnason, J.B., Barsh, A., Direnzo, G.S., Campbell, R. & Fox, J.W. (1983) Kallikrein-like enzymes from *Crotalus atrox* venom. *J. Biol. Chem.* **208**, 12566–12573.

Burkhart, W., Smith, G.F.H., Su, J.L., Parikh, I. & LeVine, H. (1992) Amino acid sequence determination of Ancrod, the thrombin-like α-fibrinogenase from the venom of *Akistrodon rhodostoma. FEBS Lett.* **297**, 297–301.

Cercek, B., Lew, A.S. Hod, H., Yano, J., Lewis, B., Reddy, K.N.N. & Ganz, W. (1987) Ancrod enhances the thrombolytic effect of streptokinase and urokinase. *Thromb. Res.* **47**, 417–426.

Collins, J.P. & Jones, J.G. (1972) Studies on the active site IRC-50 Arvin, the purified coagulant enzyme from *Agkistrodon rhodostoma* venom. *Eur. J. Biochem.* **26**, 510–517.

Collins, J.P. & Jones, J.G. (1974) Identification of serine and histidine as essential amino-acid residues in the coagulant enzyme ancrod. *Eur. J. Biochem.* **42**, 81–87.

Damus, P.S., Markland, F.S., Jr., Davidson, T.M. & Shandley, J.D. (1972) A purified procoagulant enzyme from the venom of the eastern diamondback rattlesnake *(Crotalus adamanteus)*: In vivo and in vitro studies. *J. Lab. Clin. Med.* **79**, 906–923.

Eagle, H. (1937) The coagulation of blood by snake venoms and its physiological significance. *J. Exp. Med.* **65**, 613–639.

Egberg, N. (1972) On interaction of serum proteins with thrombin-like enzyme from *Bothrops atrox* venom. *Thromb. Res.* **1**, 637–640.

Ehrly, A.M. (1976) Improvement of the flow properties of blood: a new therapeutical approach in occlusive arterial disease. *Angiology* **27**, 188–196.

Eschenfelder, V. (1996) Ancrod as an antithrombotic and thrombolytic agent. In: *Advances in Anticoagulant, Antithrombotic and Thrombolytic Therapeutics* (Zavoico, G., ed.). Southborough: IBC Biomedical Library Series, pp. 6.8.1–6.8.21.

Esnouf, M.P. & Tunnah, G.W. (1967) The isolation and properties of the thrombin-like activity from *Ancistrodon rhodostoma* venom. *Br. J. Haematol.* **13**, 581–590.

Ewart, M.R., Hatton, M.W.C., Basford, J.M. & Dodgson, K.S. (1970) The proteolytic action of arvin on human fibrinogen. *Biochem. J.* **118**, 603–609.

Exner, T. & Koppel, J.L. (1972) Observations concerning the substrate specificity of Arvin. *Biochim. Biophys. Acta* **258**, 825–829.

Fontana, F. (1787) In: *Treatise on the Venom of the Viper*, vol. 1. London.

Furukawa, K. & Ishimaru, S. (1990) Use of thrombin-like snake venom enzymes in the treatment of vascular occlusive diseases. In: *Medical Use of Snake Venom Proteins* (Stocker, K.F., ed.). Boca Raton, FL: CRC Press, pp. 161–173.

Hatton, M.W.C. (1973) Studies on the coagulant enzyme from *Agkistrodon rhodostoma* venom. Isolation and some properties of the enzyme. *Biochem. J.* **131**, 799–807.

Herzig, R.H., Ratnoff, O.D. & Shainoff, J.R. (1970) Studies on a procoagulant fraction of southern copperhead snake venom: The preferential release of fibrinopeptide B. *J. Lab. Clin. Med.* **76**, 451–465.

Holleman, W.H. & Coen, L.J. (1970) Characterization of peptides released from human fibrinogen by Arvin. *Biochim. Biophys. Acta* **200**, 587–589.

Holleman, W.H. & Weiss, L.J. (1976) The thrombin-like enzyme from *Bothrops atrox* snake venom. Properties of the enzyme purified by affinity chromatography on *p*-aminobenzamidine-substituted agarose. *J. Biol. Chem.* **251**, 1663–1669.

Itoh, N., Tanaki, N., Mihasi, S. & Yamashina, I. (1987) Molecular cloning and sequence analysis of cDNA for batroxobin, a thrombin-like snake venom enzyme. *J. Biol. Chem.* **262**, 3132–3135.

Itoh, N., Tanaka, N., Funakoshi, I., Kawasaki, T., Mihashi, S. & Yamashina, I. (1988) Organization of the gene for batroxobin, a thrombin-like snake venom enzyme. *J. Biol. Chem.* **263**, 7628–7631.

Klöcking, H.P., Hoffman, A. & Markwardt, F. (1987) Release of plasminogen activator by batroxobin. *Haemostasis* **17**, 235–237.

Komori, Y., Nikai, T. & Sugihara, H. (1988) Biochemical and physiological studies on a kallikrein-like enzyme from the venom of *Crotalus viridis viridis* (Prairie rattlesnake). *Biochim. Biophys. Acta* **967**, 92–102.

Latallo, Z.S. (1983) Retrospective study on complications and adverse effects of treatment with thrombin-like enzymes – a multicentre trial. *Thromb. Haemost.* **50**, 604–609.

Lochnit, G. & Geyer, R. (1995) Carbohydrate structure analysis of batroxobin, a thrombin-like serine protease from *Bothrops moojeni* venom. *Eur. J. Biochem.* **228**, 805–816.

Maeda, M., Satoh, S., Suzuki, S., Niwa, M., Itoh, N. & Yamashina, I. (1991) Expression of cDNA for batroxobin, a thrombin-like snake venom enzyme. *J. Biochem.* **109**, 632–637.

Markland, F.S. (1976) Crotalase. *Methods Enzymol.* **45**, 223–236.

Markland, F.S. & Damus, P.S. (1971) Purification and properties of a thrombin-like enzyme from the venom of *Crotalus adamanteus* (eastern diamondback rattlesnake). *J. Biol. Chem.* **246**, 6460–6473.

Markland, F.S. & Pirkle, H. (1977) Thrombin-like enzyme from the venom of *Crotalus adamanteus* (eastern diamondback rattlesnake). *Thromb. Res.* **10**, 487–494.

Markland, F.S., Kettner, C., Shaw, E. & Bajwa, S.S. (1981) The inhibition of crotalase, a thrombin-like snake venom enzyme by several peptide chloromethyl ketone derivatives. *Biochem. Biophys. Res. Commun.* **102**, 1302–1309.

Markland, F.S., Kettner, C., Schiffman, S., Shaw, E., Bajwa, S.S., Reddy, K.N.N., Kirakossian, H., Patkos, G.B., Theodor, I. & Pirkle, H. (1982) Kallikrein-like activity of crotalase, a snake venom enzyme that clots fibrinogen. *Proc. Natl Acad. Sci. USA* **79**, 1688–1692.

Mattock, O. & Esnouf, M.P. (1971) Differences in the structure of human fibrin formed by the action of arvin, reptilase and thrombin. *Nature New Biol.* **233**, 277–279.

Mitchell, S.W. & Reichert, E.T. (1886) Researches upon the venoms of poisonous serpents. *Smithsonian Contributions to Knowledge* **26**, 1–157.

Mori, N. & Sugihara, H. (1989) Characterization of kallikrein-like enzyme from *Crotalus ruber ruber* (red rattlesnake venom). *Int. J. Biochem.* **21**, 83–90.

Nolan, C., Hall, L.S. & Barlow, G.H. (1976) Ancrod, the coagulating enzyme from Malayan pit viper *(Agkistrodon rhodostoma)* venom. *Methods Enzymol.* **45**, 205–213.

Pfeiffer, G., Dabrowski, U., Dabrowski, J., Stirm, S., Strube, KH. & Geyer, R. (1992) Carbohydrate structure of a thrombin-like serine protease from *Agkistrodon rhodostoma*. Structure elucidation of oligosaccharides by methylation analysis, liquid secondary-ion mass spectrometry and proton magnetic resonance. *Eur. J. Biochem.* **205**, 961–978.

Pfeiffer, G., Linder, D., Strube, K.H. & Geyer, R. (1993) Glycosylation of the thrombin-like serine protease ancrod from *Agkistrodon rhodostoma* venom. Oligosaccharide substitution pattern at each N-glycosylation site. *Glycoconjugate J.* **10**, 240–246.

Pirkle, H. & Theodor, I. (1990) Thrombin-like venom enzymes: structure and function. *Adv. Exp. Med. Biol.* **281**, 165–175.

Pirkle, H. & Theodor, I. (1998) Thrombin-like enzymes. In: *The Enzymology of Snake Venoms* (Bailey, G.S., ed.). Fort Collins: Alaken (in press).

Pirkle, H., Markland, F.S. & Theodor, I. (1976) Thrombin-like enzymes of snake venoms: action of prothrombin. *Thromb. Res.* **8**, 619–627.

Pirkle, H., Markland, F.S., Theodor, I., Baumgartner, R., Bajwa, S.S. & Kirakossian, H. (1981) The primary structure of crotalase, a thrombin-like venom enzyme, exhibits closer homology to kallikrein than to other serine proteases. *Biochem. Biophys. Res. Commun.* **99**, 715–721.

Pirkle, H., Theodor, I., Henschen, A., Kreiglstein, K., Martin, P. & Edwards, B.F.P. (1991) Structural studies on crotalase, a thrombin-like snake venom enzyme. *Thromb. Haemost.* **65**, 954.

Pirkle, H., Theodor, I. & Henschen, A. (1996) Crotalase, a fibrinogen-clotting venom enzyme: Primary structure and evidence for lack of a fibrinogen recognition exosite homologous to that of thrombin. *Haemostasis* **26**(suppl. 2), 452.

Pitney, W.R. & Regoeczi, E. (1970) Inactivation of 'Arvin' by plasma proteins. *Br. J. Haematol.* **19**, 67–81.

Pitney, W.R., Bray, C., Holt, P.J.L. & Bolton, G. (1969) Acquired resistance to treatment with Arvin. *Lancet* **1**, 79–81.

Pizzo, S.V., Schwartz, M.L., Hill, R.L. & Mckee, P.A. (1972) Mechanism of ancrod anticoagulation: a direct proteolytic effect on fibrin. *J. Clin. Invest.* **51**, 2841–2850.

Pollack, V.E., Glas-Greenwalt, P., Olinger, C.P., Wadhwa, N.K. & Myre, S.A. (1990) Ancrod causes rapid thrombolysis in patients with acute stroke. *Am. J. Med. Sci.* **299**, 319–325.

Prentice, C.R.M., Hampton, K.K., Grant, P.J., Nelson, S.R., Nieuwenhuizen, W. & Gaffeney, P.J. (1993) The fibrinolytic response to ancrod therapy: characterization of fibrinogen and fibrin degradation products. *Br. J. Haematol.* **83**, 276–281.

Reid, H.A., Chan, K.E. & Thean, P.C. (1963) Prolonged coagulation defect (defibrinogenation syndrome) in Malayan viper bite. *Lancet* **1**, 621–626.

Soszka, T., Kirschbaum, N.E., Stewart, G.J. & Budzynski, A.Z. (1985) Direct effect of fibrinogen-clotting enzymes on plasminogen activator secretion from human endothelial cells. *Thromb. Haemost.* **54**, 164.

Soutar, R.L. & Ginsberg, J.S. (1993) Anticoagulant therapy with ancrod. *Clin. Rev. Oncol./Hematol.* **15**, 23–33.

Stocker, K.F. (1988) Clinical trials with batroxobin. In: *Hemostasis and Animal Venoms* (Pirkle, H. & Markland, F.S., eds). New York: Marcel Dekker, pp. 525–540.

Stocker, K. & Barlow, G.H. (1976) The coagulant enzyme from *Bothrops atrox* venom (Batroxobin). *Methods Enzymol.* **45**, 214–223.

Stocker, K and Egberg, N. (1973) Reptilase as a defibrinogenating agent. *Thromb. Diath. Haemorrh.* **suppl. 54**, 361–370.

Stocker, K.F. & Meier, K. (1988) Thrombin-like snake venom enzymes. In: *Hemostasis and Animal Venoms* (Pirkle, H. & Markland, F.S., eds). New York: Marcel Dekker, pp. 67–84.

Stocker, K. & Yeh, H. (1975) A simple and sensitive test for the detection of inhibitors of Defibrase® and Arwin® in serum. *Thromb. Res.* **6**, 189–194.

Stocker, K., Fischer, H. & Meier, J. (1982) Thrombin-like snake venom proteinases. *Toxicon* **20**, 265–273.

Stürzebecher, J., Stürzebecher, U. & Markwardt, F. (1986) Inhibition of batroxobin, a serine proteinase from *Bothrops* snake venom by derivatives of benzamidine. *Toxicon* **24**, 585–595.

Tanaka, N., Nakada, H., Itoh, N., Mizuno, Y., Takanishi, M., Kawasaki, T., Tate, S., Inagaki, F. & Yamashina, I. (1992) Novel structure of the *N*-acetylgalactosamine containing *N*-glycosidic carbohydrate chain of batroxobin, a thrombin-like snake venom enzyme. *J. Biochem.* **112**, 68–74.

Thomson, N.C., Hutcheon, A.W. & Dagg, J.H. (1977) Multiple courses of ancrod (Arvin) therapy. *BMJ* **1**, 508.

Tomaru, T., Uchida, Y., Sonoki, H. & Sugimoto, T. (1988) Preventive effects of batroxobin on experimental canine coronary thrombosis. *Clin. Cardiol.* **11**, 223–230.

Weiss, H.J., Allan, S., Davidson, E. & Kochwa, S. (1969) Afibrinogenemia in man following the bite of a rattlesnake (*Crotalus adamanteus*). *Am. J. Med.* **47**, 625–634.

*Francis S. Markland, Jr.*
*Department of Biochemistry and Molecular Biology,*
*University of Southern California, School of Medicine,*
*Cancer Research Laboratory 106,*
*1303 N. Mission Road, Los Angeles, CA 90033, USA*
*Email: markland@hsc.usc.edu*

*Hubert Pirkle*
*Department of Pathology,*
*Medical Sciences I, University of California,*
*Irvine, CA 92697, USA*
*Email: hcpirkle@uci.edu*

# 71. *Venombin AB*

## Databanks

*Peptidase classification: presumably clan SA, family S1, MEROPS ID: S9G.026*
*NC-IUBMB enzyme classification: EC 3.4.21.55*
*Chemical Abstracts Service registry number: 104003-74-9*
*Databank codes: no sequence data available*

## Name and History

Gaffney *et al*. (1973) were the first to note that a fibrinogen-clotting fraction from the venom of *Bitis gabonica* released both fibrinopeptides A and B from fibrinogen and also activated factor XIII. Partially purified preparations of the coagulant enzyme were studied by Marsh & Whaler (1974) and by Viljoen *et al*. (1979). Pirkle *et al*. (1986) purified the enzyme essentially to homogeneity and described several of its physicochemical and coagulant properties. The name *gabonase* was proposed by Marsh & Whaler (1984). The name *venombin AB* has been recommended for any snake venom serine endopeptidase that releases both fibrinopeptides.

## Activity and Specificity

The coagulant action of gabonase is triggered primarily by the cleavage of the Arg+Gly bond nearest to the N-terminus of the A$\alpha$ chain of fibrinogen, releasing fibrinopeptide A. Fibrinopeptide B is released more slowly by cleavage of another Arg+Gly bond nearest to the N-terminus of the B$\beta$ chain. No other cleavages could be detected by PAGE of reduced gabonase fibrin, nor did gabonase hydrolyze any peptide bonds of insulin, glucagon or the S peptide of ribonuclease (Pirkle *et al*., 1986). This enzyme, like thrombin (Chapter 55), activates factor XIII added to fibrinogen-clotting mixtures, resulting in the formation of $\gamma$ chain dimers and $\alpha$ chain polymers. Gabonase exhibits strong Tos-Arg-OMe esterase activity, but hydrolyzes tripeptide nitroanilide derivatives weakly or not at all (Pirkle *et al*., 1986).

The coagulant activity of gabonase can be assayed from the clotting time of purified fibrinogen, but substrate inhibition occurs at fibrinogen concentrations of $3\,mg\,ml^{-1}$ or higher. Using bovine thrombin as a standard, gabonase has a specific activity of 45 NIH thrombin-equivalent units per mg protein (Pirkle *et al*., 1986). The activity of gabonase is stabilized by calcium ion (Pirkle *et al*., 1986).

Gabonase is inactivated by PMSF and by Tos-Lys-CH$_2$Cl, but not by hirudin or heparin (Pirkle *et al*., 1986).

## Structural Chemistry

The N-terminal sequence of gabonase is VVGGAECK-IDGHRCLALLY (Pirkle *et al*., 1986). Its $M_r$ is 30 600 (SDS-PAGE). Since gabonase has no free sulfhydryl groups, the Cys residues found by sequence analysis and by amino acid analysis are presumed to be disulfide bonded. Its carbohydrate composition is sialic acid 5.2%, mannose 4.5%, galactose 3.3% and glucosamine 7.6%. Other physical properties are an extinction coefficient (1%, 1 cm) of 9.6 and a pI of 5.3.

## Preparation

Whole venom from *Bitis gabonica* is readily available commercially (e.g. Sigma, Latoxan) (see Appendix 2 for full names and addresses of suppliers). The fibrinogen-clotting enzyme, which constitutes about 1.5% of the whole venom protein, can be isolated by gel filtration followed by DEAE-cellulose chromatography (Pirkle *et al*., 1986).

## Biological Aspects

Gabonase is one of many closely related fibrinogen-clotting enzymes found in the venoms of several pit viper genera (*Agkistrodon, Bothrops, Crotalus, Lachesis, Trimeresurus*), some true vipers (*Bitis, Cerastes*), and the colubrid, *Dispholidus typus* (Pirkle & Theodor, 1997). Enzymes of this type are used medically as defibrinating anticoagulant agents in the treatment mainly of thrombotic disorders.

## Distinguishing Features

The N-terminal amino acid sequence of gabonase (Pirkle *et al*., 1986) is unique among the 14 or so such sequences of fibrinogen-clotting venom enzymes that have been published to date (Pirkle & Theodor, 1998). As noted below, the fibrinopeptide-releasing characteristics of gabonase and its capacity to activate factor XIII distinguish it from most other fibrinogen-clotting venom enzymes.

## Related Peptidases

Most fibrinogen-clotting venom enzymes studied so far act selectively to release fibrinopeptide A from fibrinogen, while a smaller number preferentially release fibrinopeptide B (reviewed by Pirkle & Stocker, 1991). Apart from gabonase, the only other well-studied venom enzyme that displays noteworthy releasing activity for both fibrinopeptide A and fibrinopeptide B is okinaxobin II from *Trimeresurus okinavensis*, which exhibits a time-course of fibrinopeptide release that closely resembles the action of bovine thrombin (Nose *et al*., 1994). Whether okinaxobin II, like gabonase and thrombin, activates factor XIII was not reported.

## Further Reading

Pirkle *et al*. (1986) give the most comprehensive description of gabonase. The biochemistry of fibrinogen-clotting enzymes from snake venoms in general is reviewed by Pirkle & Theodor (1998).

## References

Gaffney, P.J., Marsh, N.A. & Whaler, B.C. (1973) A coagulant enzyme from Gaboon-viper venom: Some aspects of its mode of action. *Biochem. Soc. Trans.* **1**, 1208–1209.

Marsh, N.A. & Whaler, B.C. (1974) Separation and partial characterization of a coagulant enzyme from *Bitis gabonica* venom. *Br. J. Haematol.* **26**, 295–306.

Marsh, N.A. & Whaler, B.C. (1984) The Gaboon viper (*Bitis gabonica*): Its biology, venom components and toxinology. *Toxicon* **22**, 669–694.

Nose, T., Shimohigashi, Y., Hattori, S., Kihara, H. & Ohno, M. (1994) Purification and characterization of a coagulant enzyme, okinaxobin II, from *Trimeresurus okinavensis* (himehabu snake) venom which releases fibrinopeptides A and B. *Toxicon* **32**, 1509–1520.

Pirkle, H. & Stocker, K. (1991) Thrombin-like enzymes from snake venoms: An inventory. *Thromb. Haemost.* **65**, 444–450.

Pirkle, H. & Theodor, I. (1998) Thrombin-like enzymes. In: *The Enzymology of Snake Venoms* (Bailey, G.S., ed.). Fort Collins: Alaken (in press).

Pirkle, H., Theodor, I., Miyada, D. & Simmons, G. (1986) Thrombin-like enzyme from the venom of *Bitis gabonica*. Purification, properties, and coagulant actions. *J. Biol. Chem.* **261**, 8830–8835.

Viljoen, C.C., Meehan, C.M. & Botes, D.P. (1979) Separation of *Bitis gabonica* (Gaboon adder) venom arginine esterases into kinin-releasing, clotting and fibrinolytic factors. *Toxicon* **17**, 145–154.

*Hubert Pirkle*
*Department of Pathology,*
*Medical Sciences I, University of California,*
*Irvine, CA 92697, USA*
*Email: hcpirkle@uci.edu*

*Francis S. Markland, Jr*
*Department of Biochemistry and Molecular Biology,*
*University of Southern California School of Medicine,*
*1303 North Mission Road, Los Angeles, CA 90033, USA*
*Email: markland@hsc.usc.edu*

# 72. *Russell's viper venom factor V activator*

## Databanks

*Peptidase classification: clan SA, family S1, MEROPS ID: S01.184*
*NC-IUBMB enzyme classification: EC 3.4.21.95*
*Chemical Abstracts Service registry number: 79393-92-3*
*Databank codes:*

| Species | Form | SwissProt | PIR | EMBL (cDNA) | EMBL (genomic) |
|---------|------|-----------|-----|-------------|----------------|
| *Daboia russellii* | $\alpha$ | P18964 | A32121 | – | – |
| *Daboia russellii* | $\gamma$ | P18965 | B32121 | – | – |

## Name and History

The venom of Russell's viper (*Daboia russellii*, formerly *Vipera russelli*) contains a serine proteinase that has the property of significantly increasing the coagulant activity of factor V (Hjort, 1957). An increase in blood coagulability caused by Russell's viper venom (RVV) was first described by Macfarlane & Barnett (1934), and a protein that acted on factor X was isolated from RVV in the early 1960s (Esnouf & Williams, 1962) (Chapter 433). At about the same time, RVV was identified to have a calcium-independent effect, increasing the activity of factor V (Hjort, 1957). Although RVV has a profound effect on both factors X and V, the effector molecules were thought to be different. Schiffman *et al.* (1969) described the separation of an activator of factor V by size-exclusion chromatography, and reported that the proteinase that acted on factor X had an apparent molecular mass of 145 kDa, whereas the molecule that acted on factor V was much smaller, on the order of 20 kDa (Schiffman *et al.*, 1969). Subsequent characterization of the **Russell's viper venom factor V activator (RVV-V)** showed it to be a 29 kDa glycoprotein containing 6% carbohydrate (Kisiel & Canfield, 1981). The amino acid sequences of three isoforms of RVV-V have been determined (Tokunaga *et al.*, 1988);

each consists of 236 amino acids and possesses an *N*-linked oligosaccharide near the C-terminus.

## Activity and Specificity

Factor V is a critical component of the hemostatic system, since its cleavage promotes blood coagulation (see Chapter 54). The endogenous proteinase responsible for this cleavage is thrombin (Chapter 55). Factor V has three domains, A, B and C, which are cleaved by thrombin to yield active factor V (Va) (Jenny *et al.*, 1987; Keller *et al.*, 1995). Upon cleavage of factor V, coagulant activity increases 25-fold (Kane & Majerus, 1981). Activation with RVV-V cleaves factor V at a single locus, Arg1545$\downarrow$Ser1546, yielding two products, one of 303 kDa and a doublet with a molecular mass of 91–95 kDa (Kane & Davie, 1986; Keller *et al.*, 1995). The doublet band at 91–95 kDa is identical to a band formed also by the action of thrombin, and the increase in coagulant activity correlates with the appearance of this band (Kane & Majerus, 1981). Treatment of factor V with thrombin after it has been exposed to RVV-V yields no further cleavage of the 91–95 kDa band, but additional bands arise from cleavage of the 303 kDa fragment. This indicates that

RVV-V shares a single cleavage site with thrombin (Keller *et al.*, 1995). Many hemostatic activator proteins found in snake venom act on more than one target molecule, such as RVV-X (Chapter 433) which acts on both factors X and IX. By comparison, RVV-V activates only factor V (Smith & Hanahan, 1976). RVV-V was first identified as a serine proteinase through its sensitivity to inhibition by DFP (Esmon & Jackson, 1973).

RVV-V activity is measured *in vitro* by the ability to activate factor V, and activated factor V (Va) is then assayed by its ability to correct the clotting time of human plasma deficient in factor V. The correction in clotting time is proportional to the amount of factor Va present and thus the effectiveness of the activator in converting factor V to Va (Kisiel & Canfield, 1981).

## Structural Chemistry

RVV-V is a glycoprotein with a single *N*-linked oligosaccharide chain attached at Asn229. Three isoforms of RVV-V have been identified and the two predominant species have been sequenced (Tokunaga *et al.*, 1988). The two sequenced isoforms, RVV-V$\alpha$ and RVV-V$\gamma$, share an overall sequence identity of 98%, differing at just six amino acids, and each of the replacements can be accounted for by a single point mutation in the nucleotide sequence (Tokunaga *et al.*, 1988). RVV-V contains 12 cysteines, and six disulfide bonds are postulated on the basis of the homology with other members of the chymotrypsin family. Of the six disulfide bonds one, Cys76-Cys234, is unique to snake venom serine proteinases, being found also in the snake venom proteins batroxobin (Chapter 70) and flavoxobin (Chapter 70), but not in thrombin or any other normal physiologic proteinase. A major difference between RVV-V and other trypsin-like serine proteinases occurs in the highly conserved region close to the catalytic site that contains Ser214 of chymotrypsin (Chapter 8). The sequence Ser214-Trp215-Gly216 (chymotrypsinogen numbering) is replaced by Ala197-Gly198-Gly199 (RVV-V numbering). Few other members of the chymotrypsin family (S1) have been found to lack the Ser214-Trp215-Gly216 sequence.

## Preparation

Preparation of RVV-V from crude venom is by a two-step chromatographic procedure described by Kisiel & Canfield (1981). In the first step, a Sephadex G-150 molecular sieve column separates the crude venom into three peaks. The peaks containing RVV are then run on an SP-Sephadex C-50 cation-exchange column at pH 5.0. Most of the protein is not adsorbed at the starting concentration of 0.4 M NaCl, and RVV-V is then eluted at 0.6 M NaCl. The expected yield of RVV-V from 1 g of crude venom is 25 mg.

## Biological Aspects

Envenomation by *D. russellii* is one of the major causes of snake bite mortality in South-East Asia. It has been reported that Russell's viper is responsible for more than 1000 deaths annually in Burma (Aung-Khin, 1980). After envenomation, patients develop disseminated intravascular coagulation

which can be associated with spontaneous bleeding, since the blood will not coagulate normally (Myint-Lewin *et al.*, 1985). Hemostatic changes observed after envenomation are described by Than-Than *et al.* (1988), the most notable being the loss of coagulability, resulting from the depletion of fibrinogen, factors V, X, Xa and protein C. RVV-V contributes to the coagulopathy by its depletion of factor V.

## Distinguishing Features

RVV-V is as efficient as $\alpha$-thrombin in the activation of factor V and more specific in its cleavage of the protein. $\alpha$-Thrombin cleaves factor V at three loci, whereas RVV-V cleaves at only one site. It is apparent from activity assays that the activation of factor V is completed by this single cleavage. By contrast, Keller *et al.* (1995) have determined that multiple cleavages of factor V by thrombin are necessary for full activation of factor V. It is believed that the two additional thrombin cleavages change the conformation of factor V, allowing thrombin access to the crucial cleavage site at Arg1545; this is apparently not necessary for RVV-V. RVV-V does not have the multiple cleavage requirement since the single cleavage at the terminus of the B region of factor V is sufficient for generation of activated factor Va activity.

## Related Peptidases

An additional proteinase, thrombocytin, from the venom of *Bothrops atrox* has been reported to have an effect on factor V, but this enzyme is less well characterized. The enzyme is a 36 kDa, single-chain glycoprotein containing 5.6% carbohydrate (Kirby *et al.*, 1979). This protein has been determined to affect factor XIII as well as factor V, a property not observed with RVV-V. Thrombocytin is purified from *B. atrox* venom by a method similar to that described above for RVV-V. The conditions of preparation are identical, only the salt concentration in the eluting buffer for the SP-Sephadex column being changed. Thrombocytin elutes from the SP column at approximately 1 M NaCl. The yield of thrombocytin is 8–10 mg g$^{-1}$ crude venom, or approximately one-third to one-half the amount obtained in the purification of RVV-V. Thrombocytin has been determined to be a classical serine proteinase through its inhibition by standard serine proteinase inhibitors, and the presence of Ser and His in the active site (Niewiarowski *et al.*, 1979).

## Further Reading

For a review, see Kisiel & Canfield (1981).

## References

Aung-Khin (1980) The problem of snakebites in Burma. *Snake* **12**, 125–127.

Esmon, C.T. & Jackson, C.M. (1973) The factor V activating enzyme of Russell's viper venom. *Thromb. Res.* **2**, 509–524.

Esnouf, M.P. & Williams, W.J. (1962) The isolation and purification of a bovine-plasma protein which is a substrate for the coagulant fraction of Russell's viper venom. *Biochem. J.* **84**, 660–665.

Hjort, P.F. (1957) Intermediate reactions in the coagulation of blood with tissue thromboplastin: convertin, accelerin, prothrombinase. *Scand. J. Clin. Invest.* **9**(suppl. 27), 7–183.

Jenny, R.J., Pittman, D.D., Toole, J.J., Kriz, R.W., Aldape, R.A., Hewick, R.M., Kaufman, R.J. & Mann, K.G. (1987) Complete cDNA and derived amino acid sequence of human factor V. *Proc. Natl Acad. Sci. USA* **84**, 4846–4850.

Kane, W.H. & Davie, E.W. (1986) Cloning of a cDNA coding for human factor V, a blood coagulation factor homologus to factor VIII and ceruloplasmin. *Proc. Natl Acad. Sci. USA* **83**, 6800–6804.

Kane, W.H. & Majerus, P.W. (1981) Purification and characterization of human coagulation factor V. *J. Biol. Chem.* **256**, 1002–1007.

Keller, F.G., Ortel, T.L., Quinn-Allen, M. & Kane, W.H. (1995) Thrombin-catalyzed activation of recombinant human factor V. *Biochemistry* **34**, 4118–4124.

Kirby E.P., Niewiarowski, S., Stocker, K., Kettner, C., Shaw, E. & Brudzynski, T.M. (1979) Thrombocytin, a serine protease from *Bothrops atrox* venom. 1. Purification and characterization of the enzyme. *Biochemistry* **18**, 3564–3570.

Kisiel, W. & Canfield, W.M. (1981) Snake venom proteases that activate blood coagulation factor V. *Methods Enzymol.* **80**, 275–285.

Macfarlane, R.G. & Barnett, B. (1934) The haemostatic possibilities of snake venom. *Lancet* **ii**, 985–987.

Myint-Lewin, Warrell, D.A., Philips, R.E., Tin-Nu-Swe, Tun-Pe & Maung-Maung-Lay (1985) Bites by Russel's viper (*Vipera russelli siamensis*) in Burma: haemostatic, vascular and renal disturbances and response to treatment. *Lancet* **ii**, 1259–1264.

Niewiarowski, S., Kirby E.P., Brudzynski, T.M. & Stocker, K. (1979) Thrombocytin, a serine protease from *Bothrops atrox* venom. 2. Interaction with platelets and plasma-clotting factors. *Biochemistry* **18**, 3570–3577.

Schiffman, S., Theodor, I. & Rapaport, S.I. (1969) Separation from Russell's viper venom of one fraction reacting with factor X and another reacting with factor V. *Biochemistry* **8**, 1397–1405.

Smith, C.M. & Hanahan, D.J. (1976) The activation of factor V by factor Xa or $\alpha$-chymotrypsin and comparison with thrombin and RVV-V action. An improved factor V isolation procedure. *Biochemistry* **15**, 1830–1838.

Than-Than, Hutton, R.A., Myint-Lwin, Khin-Ei-Han, Soe-Soe, Tin-Nu-Swe, Phillips, R.E. & Warrell, D.A. (1988) Haemostatic disturbances in patients bitten by Russell's viper (*Vipera russelli siamensis*) in Burma. *Br. J. Haemotol.* **69**, 513–520.

Tokunaga, F., Nagasawa, K., Tamura, S., Miyata, T., Iwanagawa, S. & Kisiel, W. (1988) The factor V-activating enzyme (RVV-V) from Russell's viper venom: identification of isoproteins RVV-V$\alpha$, -V$\beta$, and -V$\gamma$ and their complete amino acid sequences *J. Biol. Chem.* **263**, 17471–1748.

*Stephen Swenson*
*Department of Biochemistry and Molecular Biology,*
*University of Southern California, School of Medicine,*
*Cancer Research Laboratory 104,*
*1303 N. Mission Road, Los Angeles, CA 90033, USA*
*Email: sswenson@hsc.usc.edu*

*Francis S. Markland, Jr*
*Department of Biochemistry and Molecular Biology,*
*University of Southern California, School of Medicine,*
*Cancer Research Laboratory 106,*
*1303 N. Mission Road, Los Angeles, CA 90033, USA*
*Email: markland@hsc.usc.edu*

# 73. *Scutelarin*

## Databanks

*Peptidase classification: presumably clan SA, family S1, MEROPS ID: S9G.027*
*NC-IUBMB enzyme classification: 3.4.21.60*
*Databank codes: no sequence data available*

## Name and History

The presence of procoagulant activities in snake venoms that interfere with the blood circulation was recognized many centuries ago (Helmont, 1648; Fontana, 1781). Martin (1893) observed that the venom from *Pseudechis porphyriacus*, an Australian snake belonging to the family Elapidae, coagulated blood. The procoagulant activity present in the venom of Australian Elapidae was later shown to be associated with an enzyme that converted the blood coagulation factor prothrombin into thrombin (Jobin & Esnouf, 1966; Denson, 1969). The venom prothrombin activators from several elapid species share many properties with the physiological prothrombin activator, factor Xa, in that their activities in prothrombin activation are greatly stimulated by the presence of phospholipids, $Ca^{2+}$ ions and blood coagulation factor Va (Jobin & Esnouf, 1966; Rosing & Tans, 1992). The venom from *Oxyuranus scutellatus*, also called *Oxyuranus scutellatus scutellatus* or taipan snake, was, however, only stimulated by phospholipids and not by the nonenzymatic protein cofactor Va (Pirkle *et al.*, 1972). The purified prothrombin activator from *Oxyuranus scutellatus*, presently referred to as **scutelarin** (Pirkle & Marsh, 1991), is a multisubunit protein (Walker *et al.*, 1980) that consists of a factor Xa-like serine protease domain and a factor Va-like cofactor domain (Speijer *et al.*, 1986).

## Activity and Specificity

Scutelarin is a serine protease that converts the blood coagulation factor prothrombin into thrombin (Chapter 55) by limited proteolysis (Speijer *et al.*, 1986). Although the exact cleavage sites have not been determined, it is likely that scutelarin activates prothrombin by cleavage of peptide bonds that are also susceptible to proteolysis by the physiological prothrombin activator factor Xa (Chapter 54), i.e. the Arg271┼Thr272 and Arg320┼Ile321 bonds in human prothrombin, and the Arg274┼Thr275 and Arg323┼Ile324 bonds in bovine prothrombin. The serine protease domain of scutelarin can also activate human blood coagulation factor VII (Chapter 53) (Nagaki *et al.*, 1992). Walker *et al.* (1980) and Speijer *et al.* (1986) have shown that scutelarin also cleaves the arginyl bonds in such small oligopeptide chromogenic substrates as S2222 (Bz-Ile-Glu-Gly-Arg┼NHPhNO$_2$; Chromogenix), S2237 (Bz-Ile-Glu-(piperidyl)-Gly-Arg┼NHPhNO$_2$; Chromogenix) and pefachrome FXa (Ac-D-CHA-Gly-Arg┼NHPhNO$_2$; Pentapharm) (see Appendix 2 for full names and addresses of suppliers).

The activity of scutelarin can be assayed either by measuring thrombin generation in a reaction system that contains prothrombin and $Ca^{2+}$ ions in the presence or absence of negatively charged phospholipid vesicles, or by measuring its amidolytic activity on chromogenic substrates (Speijer *et al.*, 1986). However, care should be taken in determining the activity with chromogenic substrates, since these are also cleaved by many other serine proteases. The pH optimum for cleavage of S2222 by scutelarin is about 8.0 (Walker *et al.*, 1980).

The activity of scutelarin on prothrombin is greatly enhanced by the presence of $Ca^{2+}$ ions and phospholipid vesicles (Denson, 1969; Pirkle *et al.*, 1972; Walker *et al.*, 1980; Speijer *et al.*, 1986). The stimulation by phospholipid vesicles is due to the fact that scutelarin and prothrombin both bind to phospholipids, a condition that promotes their interaction (see Kalafatis *et al.*, 1994, for a discussion of the mode of action of phospholipid surfaces in blood coagulation factor activation).

Incubation of purified scutelarin at high salt concentrations (Walker *et al.*, 1980) or with chaotropic reagents (Speijer *et al.*, 1986) results in a loss of its ability to activate prothrombin, but not of its amidolytic activity toward chromogenic substrates. The loss of prothrombin-converting activity appears to be due to dissociation of the factor Va-like cofactor domain from the serine protease domain of scutelarin (Speijer *et al.*, 1986). The serine protease domain of scutelarin can be isolated and has a low activity on prothrombin.

The activity of scutelarin on prothrombin and chromogenic substrates is inhibited by the serine protease inhibitors benzamidine, soybean trypsin inhibitor and Dns-Glu-Gly-Arg-CH$_2$Cl (Walker *et al.*, 1980; Speijer *et al.*, 1986). The enzyme is not inhibited by DFP or PMSF (Owen & Jackson, 1973; Walker *et al.*, 1980).

## Structural Chemistry

The $M_r$ of scutelarin is 300 000. It consists of subunits of 110 kDa, 80 kDa and two disulfide-linked polypeptides of 30 kDa. One or both of the 30 kDa subunits covalently binds Dns-Glu-Gly-Arg-CH$_2$Cl. This indicates that these subunits contain the active site (Speijer *et al.*, 1986). HPLC analysis has shown that scutelarin contains 5–10 $\gamma$-carboxyglutamic acid residues per mole (Speijer *et al.*, 1986), and these are probably involved in its interaction with phospholipids.

The dimeric, 58 kDa, factor Xa-like serine protease domain of scutelarin can be separated from its factor Va-like cofactor domain. The prothrombin-converting activity of the isolated factor Xa-like domain is greatly reduced and can be restored by reconstitution with mammalian factor Va (Speijer *et al.*, 1986).

## Preparation

Scutelarin is purified from the venom of *Oxyuranus scutellatus* (available from Sigma, Latoxan). The crude venom contains a considerable amount of prothrombin activator, which was purified to homogeneity with a recovery of 30% and a purification factor of 18 (Speijer *et al.*, 1986).

## Biological Aspects

*Oxyuranus scutellatus* is one of the most dangerous venomous snakes of Australia, and individuals bitten become seriously ill, and commonly die if left untreated. The venom is strongly neurotoxic, and it is not clear to what extent the scutelarin present in the venom contributes to the clinical symptoms and the mortality (Sutherland, 1983).

The crude venom of *Oxyuranus scutellatus* and several related species can be used as a diagnostic tool in clinical assays of the prothrombin concentration in plasma (Denson *et al.*, 1971; Kirchhof *et al.*, 1978) and in the screening of patients for the presence of lupus anticoagulants (Triplett *et al.*, 1993).

## Distinguishing Features

Scutelarin shares many functional properties with the mammalian factor Xa and with prothrombin activators present in venoms from several other Australian Elapidae (*Notechis* species, *Cryptophis nigrescens*, *Hoplocephalus stephensii*, *Pseudechis porphyriacus* and *Tropidechis carinatus*) (Rosing & Tans, 1991, 1992). These are all serine proteases that activate prothrombin, the prothrombin-converting activity of which is considerably enhanced by negatively charged phospholipids plus $Ca^{2+}$ ions, and by factor Va. In contrast to factor Xa and these elapid venom activators, scutelarin does not require factor Va to activate thrombin.

## Related Peptidases

Prothrombin activators that strongly resemble scutelarin are present in the venoms from related subspecies such as *Oxyuranus microlepidotus*, *Pseudonaja textilis textilis*, *Pseudonaja affinis* and *Pseudonaja nuchalis* (Speijer *et al.*, 1986; Rosing & Tans, 1991).

## Further Reading

A review of various aspects of snake venom prothrombin activators is that of Rosing *et al.* (1988).

## References

Denson, K.W.E. (1969) Coagulant and anticoagulant actions of snake venoms. *Toxicon* **7**, 5–11.

Denson, K.W.E., Borrett, R. & Biggs, R. (1971) The specific assay of prothrombin using the Taipan snake venom. *Br. J. Haematol.* **21**, 219–226.

Fontana, F. (1781) Traité sur le venin de la vipère, Florence.

Helmont, F.M. van (1648) Discussion on the opinion that the virulence of viper venom is due to the animal being angry. *Orthus Medicinae*, Amsterdam.

Jobin, F. & Esnouf, M.P. (1966) Coagulant activity of tiger snake (*Notechis scutarus scutarus*) venom. *Nature* **211**, 873–875.

Kalafatis, M., Swords, N.A., Rand, M.D. & Mann, K.G. (1994) Membrane-dependent reactions in blood coagulation: role of vitamin K-dependent enzyme complexes. *Biochim. Biophys. Acta* **1227**, 113–129.

Kirchhof, B.R.J., Vermeer, C. & Hemker, H.C. (1978) The determination of prothrombin using synthetic chromogenic substrates; choice of a suitable activator. *Thromb. Res.* **13**, 219–232.

Martin, C.J. (1893) On some effects upon the blood produced by injection of the Australian black snake (*Pseudechis porphyriacus*). *J. Physiol. (Lond.)* **15**, 380.

Nagaki, T., Lin, P. & Kisiel, W. (1992) Activation of human factor VII by the prothrombin activator from the venom of *Oxyuranus scutellatus* (Taipan snake). *Thromb. Res.* **65**, 105–116.

Owen, W.G. & Jackson, C.M. (1973) Activation of prothrombin with *Oxyuranus scutellatus scutellatus* (Taipan snake) venom. *Thromb. Res.* **3**, 705–714.

Pirkle, H. & Marsh, N. (1991) Nomenclature of exogenous hemostatic factors. *Thromb. Haemost.* **66**, 264.

Pirkle, H., McIntosh, M., Theodor, I. & Vernon, S. (1972) Activation of prothrombin with Taipan snake venom. *Thromb. Res.* **1**, 559–568.

Rosing, J. & Tans, G. (1991) Inventory of exogenous prothrombin activators. Report for the Subcommittee on Nomenclature of Exogenous Hemostatic Factors of the Scientific and Standardization Committee of the International Society on Thrombosis and Haemostasis. *Thromb. Haemost.* **65**, 627–630.

Rosing, J. & Tans, G. (1992) Structural and functional properties of snake venom prothrombin activators. *Toxicon* **30**, 1515–1527.

Rosing, J., Zwaal, R.F.A. & Tans, G. (1988) Snake venom prothrombin activators. In: *Hemostasis and Animal Venoms* (Pirkle, H. & Markland, Jr., F.S., eds). New York: Marcel Dekker, pp. 3–27.

Speijer, H., Govers-Riemslag, J.W.P., Zwaal, R.F.A. & Rosing, J. (1986) Prothrombin activation by an activator from the venom of *Oxyuranus scutellatus* (Taipan snake). *J. Biol. Chem.* **261**, 13258–13267.

Sutherland, S.K. (1983) *Australian Animal Toxins.* Australia: Oxford University Press.

Triplett, D.A., Stocker, K.F., Unger, G.A. & Barna, L.K. (1993) The textarin/ecarin ratio: a confirmatory test for lupus anticoagulants. *Thromb. Haemost.* **70**, 925–931.

Walker, F.J., Owen, W.G. & Esmon, C.T. (1980) Characterization of the prothrombin activator from the venom of *Oxyuranus scutellatus scutellatus* (taipan venom). *Biochemistry* **19**, 1020–1023.

*J. Rosing*
*Department of Biochemistry,*
*Cardiovascular Research Institute Maastricht,*
*Maastricht University,*
*PO Box 616, 6200 MD Maastricht, The Netherlands*
*Email: j.rosing@biochem.unimaas.nl*

# 74. Platelet-aggregating endopeptidase of Bothrops jararaca venom

## Databanks

*Peptidase classification: clan SA, family S1, MEROPS ID: S01.180*
*NC-IUBMB enzyme classification: none*
*Databank codes: The amino acid sequence was reported by Serrano et al. (1995), but is not yet to be found in the public databases.*

## Name and History

Serine endopeptidases with platelet-aggregating activity are present in snake venoms of crotalid and viperid species (Hutton and Warrel, 1993). *Platelet-aggregating endopeptidase of Bothrops jararaca venom (PA-BJ)* is a platelet-aggregating enzyme isolated from the venom of *Bothrops jararaca* and is the first for which the amino acid sequence has been elucidated (Serrano *et al.*, 1995).

## Activity and Specificity

PA-BJ cleaves *p*-nitroaniline from several synthetic substrates containing Arg or Lys at the scissile bond. For instance, $k_{cat}/K_m$ values for the substrates D-Phe-Pip-Arg↓NHPhNO$_2$ and Tos-Gly-Pro-Arg↓NHPhNO$_2$ are 85 and 31 mM$^{-1}$ s$^{-1}$. Both the platelet-aggregating and amidase activities of PA-BJ are inhibited by PMSF (1 mM). The benzamidine derivative *N*-[(2-naphthylsulfonyl)-glycyl]-4-amidinophenylalanine piperidide inhibits the amidase activity of PA-BJ with a $K_i$ value of 15 μM.

## Structural Chemistry

PA-BJ is a single-chain, basic glycoprotein with an $M_r$ of 30 000 estimated by SDS-PAGE. It is composed of 232 amino acid residues and contains one *N*- and one *O*-glycosidically linked carbohydrate moiety at residues Asn20 and Ser23. Sequence comparison to other venom serine endopeptidases of family S1 shows significant homology, mainly in regions around the catalytic triad and conserved cysteine residues.

## Preparation

The enzyme has been isolated from the venom of the snake *Bothrops jararaca* by saturation with ammonium sulfate followed by chromatography on DEAE-Sephacel, SP-Sephadex C-50 and Mono-S.

## Biological Aspects

PA-BJ promotes 95% platelet aggregation in platelet-rich plasma at a concentration of 10$^{-8}$ M. On washed platelet suspensions it causes 40% aggregation at a concentration of 10$^{-7}$ M.

## Distinguishing Features

PA-BJ resembles thrombin (Chapter 55) in its high platelet-aggregating activity. However, PA-BJ lacks the fibrinogen-clotting activity shown by thrombin.

## References

Hutton, R.A. & Warrel, D.A. (1993) Action of snake venom components on the haemostatic system. *Blood Reviews* **7**, 176–189.

Serrano, S.M.T., Mentele, R., Sampaio, C.A.M. & Fink, E. (1995) Purification, characterization, and amino acid sequence of a serine proteinase, PA-BJ, with platelet-aggregating activity from the venom of *Bothrops jararaca*. *Biochemistry* **34**, 7186–7193.

*Solange M. T. Serrano*
*Laboratório de Bioquímica e Biofísica,*
*Instituto Butantan,*
*Av. Vital Brasil 1500, CEP 05503-900, São Paulo, Brazil*

# 75. Guanidinobenzoatase

## Databanks

*Peptidase classification: clan SX, family S99, MEROPS ID: S9G.013*
*NC-IUBMB enzyme classification: none*
*Databank codes: no sequence data available*

## Name and History

Inhibition of a trypsin-like protease by an active-site titrant such as 4-nitrophenyl-*p*-guanidinobenzoate (NPGB) or 4-methylumbelliferyl-*p*-guanidinobenzoate (MUGB) is normally expected to be instantaneous and to be accompanied by the release of one molecule of product for each molecule of enzyme inhibited. It was observed that the instantaneous inhibition of a tumor cell surface protease by both NPGB and MUGB was accompanied by the continuous production of nitrophenol (yellow) and methylumbelliferone (blue fluorescence). This unexpected side-reaction led to studies on an enzyme referred to as *guanidinobenzoatase (GB)* which had the specific ability to cleave NPGB and MUGB as conventional substrates (Steven & Al-Ahmad,

1983). Particular attention has been given to the study of the possible significance of the enzyme to tumor cells.

## Activity and Specificity

GB was initially purified from the fluid surrounding mouse Ehrlich ascites tumor cells employing affinity techniques (Steven *et al*., 1985) and assaying with MUGB. Intact tumour cells possess cell surface receptors for GB from which GB can be dissociated using a fibrin overlay technique (Steven *et al*., 1993a) and the GB subsequently purified by affinity techniques (Steven *et al*., 1985). GB is similar to tissue-type plasminogen activator (t-PA) (Chapter 58) in its ability to bind to fibrils of fibrin (Steven *et al*., 1993a), its inhibition by plasminogen activator inhibitor (PAI-1) (Steven *et al*., 1995b) and its molecular mass of 70 kDa (Steven *et al*., 1986) in the monomeric form. GB was demonstrated to cleave the peptide Gly-Arg+Gly-Asp (Steven *et al*., 1986) which is part of the link hexapeptide region of fibronectin (Pierschbacher & Ruoslahti, 1984). More recently a tetrameric form of GB was demonstrated to degrade gelatin (Poustis-Delpont *et al*., 1994).

Monomeric GB (70 kDa) has been studied both in solution and also immobilized on cell surfaces and synthetic polysaccharides in more detail than the tetrameric form. The active center of the GB monomer has a high affinity for arginyl peptides and substituted guanidino-derivatives but no affinity for lysyl-derivatives (Steven *et al*., 1985). A number of $\alpha$-$N$-substituted agmatines and guanidino-derivatives have been designed as active site-directed fluorescent inhibitors of GB. These fluorescent molecules have played a useful role in the location of cells possessing active GB on their surfaces in pathological sections (Steven *et al*., 1987) as well as cells of cytological interest in cervical smears (Steven *et al*., 1993c, 1995a).

The ability of the fluorescent probes to locate active cell surface GB in sections of pathological tissue led to the realization that GB could also exist as a latent form, an enzyme–inhibitor complex (Steven & Griffin, 1988) and that different protein inhibitors of GB could be artificially exchanged at the cell surface (Steven *et al*., 1993b). The cytoplasm of tumor cells contained a protein which could specifically recognize and inhibit the GB on the surface of the cells from which the cytoplasmic inhibitor was prepared. This led to the concept of isoenzymic forms of GB which were also cell specific (Steven *et al*., 1992).

The above studies with monomeric GB were confirmed by Poustis-Delpont *et al*. (1992), who demonstrate that the affinity-purified monomer of GB was similar in function but immunologically distinct from plasminogen activators. This French group purified GB from human kidney tumor cell surface membranes and demonstrated that this enzyme was uniquely associated with human kidney carcinoma cells. Later studies (Poustis-Delpont *et al*., 1994) demonstrated a remarkable property of this purified GB in solution. They were able to demonstrate a reversible monomer-to-tetramer transition with a marked change in biological behavior. The tetramer behaved as a classical trypsin-like protease in the presence of NPGB; i.e. complete and instantaneous inhibition took place in an assay with gelatin as substrate. Dissociation of the tetramer to the monomer was accompanied by the

regain of the enzyme's ability to cleave NPGB continuously as though this was a true substrate rather than an inhibitor. It was also observed that proteolytic activity towards gelatin was lost by the monomer. These studies by Poustis-Delpont *et al*. (1994) help to explain the unexpected cleavage of NPGB and MUGB by the monomeric form of GB first studied by Steven & Al-Ahmab (1983), with release of massive amounts of nitrophenol and methylumbelliferone. No role has been defined for the tetrameric form of GB.

Since all the studies by the Manchester group involving kinetic analysis of the enzyme (a) in the solution, (b) linked to agmatine-Sepharose, and (c) on cell surfaces have employed MUGB or NPGB as substrate, we assume that the results obtained from these analyses apply to the monomeric form of GB, as the tetramer is totally inhibited by NPGB and MUGB (Poustis-Delpont *et al*., 1992).

## Structural Chemistry

Monomeric GB is a single-chain protein of approximately 70 kDa (Steven *et al*., 1986) and this is unaffected by reducing agents. A tetrameric form has been demonstrated by Poustis-Delpont *et al*. (1994).

## Preparation

GB has been purified from tumor tissue (Poustis-Delpont *et al*., 1992), frozen tumor sections (Steven *et al*., 1995b) and from ascitic fluid (Steven *et al*., 1985). In order to obtain a specific GB isoenzymic form associated with the surface of a defined type of tumor cell it is recommended that cultured cells be used. The washed, cultured tumor cells are coated with an overlay of fibrin fibrils; the fibrin–GB complex is dissociated with 9% (w/v) NaCl solution; the soluble fraction obtained is then used to isolate GB by an agmatine-Sepharose affinity technique. A modification of this fibrin overlay technique can be used to isolate the cytoplasmic inhibitor obtained from lysed cultured tumor cells (Steven *et al*., 1993c). It is also possible to purify GB as a unique cell specific protein from human kidney carcinoma tissue employing a physical and chromatographic separation procedure (Poustis-Delpont *et al*., 1992).

## Biological Aspects

GB is located on tumor cell surfaces and some other normal migratory cells. Cell specific isoenzymic forms of GB can be dissociated from the cell surface receptor protein by fibrin fibril overlays leaving the cell-bound functional receptor protein able to bind different isoenzymic forms of GB (Steven *et al*., 1995b). The selective inhibition of isoenzymic forms of cell surface GB by the corresponding cell cytoplasmic inhibitor suggest that this mechanism may play a significant regulatory role. These interactions may be observed at the cell surface on microscope slides employing fluorescence or by kinetic studies in solution with MUGB as substrate (Steven *et al*., 1993b). The observation that both the GB and its inhibitor can be detached from the cell surface receptor protein by fibrin fibrils suggests that the GB on malignant cell surfaces may have some role in the spread of such cells through the blood vessels.

## Distinguishing Features

The GB monomer cleaves MUGB and NPGB as true substrates rather than being inhibited by these active-site titrants. By contrast, the tetramer of GB behaves as a trypsin-like enzyme and is completely inhibited by these active-site titrants. All tumor cells so far examined possess specific cell surface GB isoenzymes which are selectively recognized and inhibited by corresponding cytoplasmic inhibitor proteins.

## References

Pierschbacher, M.D. & Ruoslahti, E. (1984) Cell attachment activity of fibronectin can be duplicated by small synthetic fragments of the molecule. *Nature* **309**, 30–33.

Poustis-Delpont, C., Descompes, R., Auberger, P., Delque-Bayer, P., Sudaka, P. & Rossi, B. (1992) Characterization and purification of a guanidinobenzoatase – a possible marker for renal carcinoma. *Cancer Res.* **52**, 3622–3628.

Poustis-Delpont, C., Thoan, S., Auberger, P., Gerardi-Laffin, C., Sudaka, P. & Rossi, B. (1994) Monomeric 55 kDa guanidinobenzoatase switches to a serine protease activity upon tetramerization. *J. Biol. Chem.* **269**, 14666–14671.

Steven, F.S. & Al-Ahmad, R.K. (1983) Evidence for an enzyme which cleaves the guanidinobenzoate moiety from active site titrants designed to inhibit and quantify trypsin. *Eur. J. Biochem.* **130**, 335–339.

Steven, F.S. & Griffin, M.M. (1988) Inhibitors of guanidinobenzoatase and their possible role in migration. *Biol. Chem. Hoppe-Seyler* **369**(suppl.), 137–143.

Steven, F.S., Griffin, M.M. & Al-Ahmad, R.K. (1985) The design of fluorescent probes which bind to the active centre of guanidinobenzoatase. Application to the location of cells possessing this enzyme. *Eur. J. Biochem.* **149**, 35–40.

Steven, F.S., Griffin, M.M., Wong, T.L.H. & Itzhaki, S. (1986) Evidence for inhibitors of the cell surface protease guanidinobenzoatase. *J. Enzym. Inhib.* **1**, 127–137.

Steven, F.S., Griffin, M.M., Wong, T.L.H., Jackson, H. & Barnett, F. (1987) Fluorescent inhibitors of a cell surface protease used to locate leukaemia cells in kidney sections. *J. Enzym. Inhib.* **1**, 203–213.

Steven, F.S., Griffin, M.M., Blakey, D.C., Talbot, I.C., Hanski, C. & Bell, J. (1992) Further evidence for different iso-enzymic forms of a cell surface protease, guanidinobenzoatase associated with tumour cells. *Anticancer Res.* **12**, 2159–2164.

Steven, F.S., Anees, M., Myers, J. & Hasleton, P. (1993a) Association and dissociation of a protease and its inhibitor on the surface of lung squamous cell carcinoma cells. *Anticancer Res.* **13**, 1063–1068.

Steven, F.S., Anees, M, Talbot, I.C., Blakey, D.C. & Hasleton, P (1993b) The interaction of protein inhibitors with tumour proteases studied in solution and immobilised on cell surfaces in frozen sections. *Anticancer Res.* **13**, 2003–2010.

Steven, F.S., Johnson, J. & Eason, P. (1993c) A protein inhibitor extracted from Hela cells which recognises a cell surface protease on preneoplastic cells obtained from cervical smears. *Anticancer Res.* **13**, 1059–1062.

Steven, F.S., Desai, M. Davis, J. Steadman, Y. McClure, J., Eason, P., Palcic, B. & Anderson, G. (1995a) Correlation of cell surface fluorescence with conventional PAP analysis of cells of cytological interest obtained from cervical scrapes. *Anticancer Res.* **15**, 1521–1525.

Steven, F.S., Anees, M. & Booth, N.A. (1995b) Selectivity of the plasminogen activator inhibitor (PAI-1) for the iso-enzyme of guanidinobenzoatase on the surface of colonic carcinoma cells. *Anticancer Res.* **15**, 205–210.

*Frank Steven*
*Division of Biochemistry,*
*School of Biological Sciences, University of Manchester,*
*Manchester M13 9PT, UK*

# 76. Introduction: other families of clan SA

## Databanks

*MEROPS ID: SA*

| Species | SwissProt | PIR | EMBL (cDNA) | EMBL (genomic) |
|---|---|---|---|---|
| **Family S2** | | | | |
| Epidermolytic toxin (Chapter 79) | | | | |
| Glutamyl endopeptidase (Chapter 79) | | | | |
| Glutamyl endopeptidase II (*Streptomyces*) (Chapter 80) | | | | |
| Lysyl endopeptidase (*Achromobacter*) (Chapter 85) | | | | |
| α-Lytic endopeptidase (*Lysobacter*) (Chapter 82) | | | | |
| Osteoblast serine protease | | | | |
|    *Homo sapiens* | – | – | D87258 | – |
|    *Mus musculus* | – | – | W33649 | – |
| Protease Do (Chapter 83) | | | | |
| SAM-P20 protein | | | | |
|    *Streptomyces albogriseolus* | – | – | D29744 | – |
| SFase protein | | | | |
|    *Streptomyces fradiae* | P41140 | – | S68947 | |
| Streptogrisin A (Chapter 78) | | | | |
| Streptogrisin B (Chapter 77) | | | | |
| Streptogrisin C | | | | |
|    *Streptomyces griseus* | P52320 | – | L29018 | – |
| Streptogrisin D | | | | |
|    *Streptomyces griseus* | P52321 | – | L29019 | – |
| Yeast-lytic endopeptidase (*Rarobacter*) (Chapter 84) | | | | |
| Others | | | | |
|    *Arabidopsis thaliana* | – | – | T44399 | – |
|    *Homo sapiens* | – | – | T80106 | – |
|    *Homo sapiens* | – | – | T80727 | – |
|    *Saccharomyces cerevisiae* | P53920 | – | – | Z69382: chromosome XIV segment<br>Z71399: chromosome XIV ORF YNL123w |
|    *Streptomyces lividans* | – | – | S79442 | – |
|    *Streptomyces* sp. | – | S34672 | – | – |
|    *Streptomyces* sp. | – | S34673 | – | – |
| **Family S3** | | | | |
| Togavirin (Chapter 86) | | | | |
| **Family S6** | | | | |
| IgA1-specific serine-type prolyl endopeptidase (*Neisseria, Haemophilus*) (Chapter 87) | | | | |
| hap protein | | | | |
|    *Haemophilus influenzae* | P44596 | – | U32710 | – |
|    *Haemophilus influenzae* | P45387 | – | U11024 | – |
| sepA protein | | | | |
|    *Shigella flexneri* | – | – | – | – |
| tsh protein | | | | |
|    *Escherichia coli* | – | – | L27423 | – |
| **Family S7** | | | | |
| Flavivirin (Chapter 88) | | | | |
| **Family S29** | | | | |
| Hepatitis C virus NS3 polyprotein peptidase (Chapter 89) | | | | |

*continued overleaf*

| Species | SwissProt | PIR | EMBL (cDNA) | EMBL (genomic) |
|---|---|---|---|---|

**Family S30**
Potyvirus P1 proteinase (Chapter 90)
**Family S31**
Pestivirus NS3 polyprotein peptidase (Chapter 91)
**Family S32**
Equine arteritis virus serine endopeptidase (Chapter 92)

Brookhaven Protein Data Bank three-dimensional structures:

| Species | ID | Resolution | Notes |
|---|---|---|---|
| **Family S2** | | | |
| Sfase2 | | | |
| *Streptomyces fradiae* | 2SFA | 1.6 | |

Structures are also known for streptogrisin B (Chapter 77), streptogrisin A (Chapter 78), glutamyl endopeptidase II (Chapter 80), exfoliative toxin A (Chapter 81), α-lytic endopeptidase (Chapter 82) and lysyl endopeptidase (Chapter 85).

Besides family S1, clan SA includes eight other families, as listed in the table. The peptidases in these families share several characteristics with chymotrypsin. All possess a catalytic triad consisting of serine, aspartate and histidine residues, and these are arranged within the sequence in the order: His, Asp, Ser. There is some conservation of glycine residues around the catalytic serine, in the motif Gly-Xaa-Ser-Gly, in which Xaa can be Asp, Asn, Ser, Thr or Trp. All the peptidases in clan SA are endopeptidases. Alignment 76.1 (see CD-ROM) shows the conservation of sequence around the active-site residues.

*Family S2* contains mainly bacterial endopeptidases that are either secreted or are membrane bound. The sequences are quite diverse, and the family can be divided into four subfamilies, each of which had at one time been considered a separate family until linking sequences were discovered. Subfamily S2A includes streptogrisins A and B, α-lytic endopeptidase and glutamyl endopeptidase II. Tertiary structures have been solved for all of these endopeptidases, and all show similarities to the structure of chymotrypsin, namely two β barrels at right-angles to each other with the active-site cleft between them. Figure 76.1 shows the tertiary structure of streptogrisin A.

Subfamily S2B includes glutamyl endopeptidase I, and is known only from Gram-positive bacteria. No tertiary structure has yet been determined for any member of this subfamily.

Subfamily S2C includes the membrane-bound, bacterial enzyme protease Do and a number of poorly characterized homologs. This subfamily includes the only nonbacterial members of the family: a protease Do homolog is known from human osteoblasts, a number of mouse-expressed sequence tags have been deposited in the databases, and there is a hypothetical protein from *Saccharomyces cerevisiae*. A protease Do homolog is present on the strand complementary to that for the gene for a 57 kDa immunogenic protein in *Chlamydia trachomatis* (EMBL: M31119). Because the reading frames overlap, and there are frameshifts in the open reading frame for the protease Do homolog, it is possible that the gene for the immunogenic protein has been misidentified.

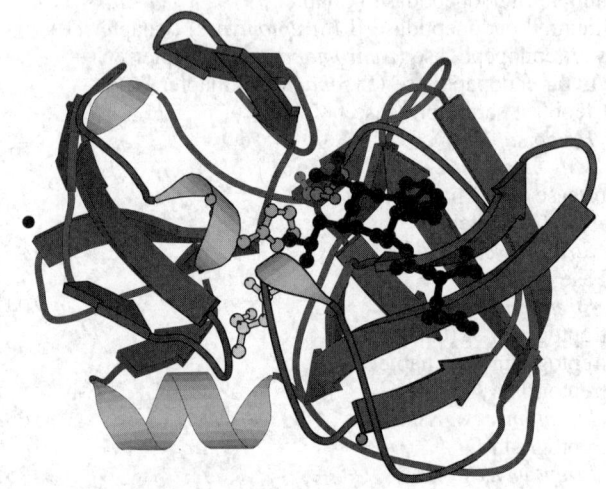

*Figure 76.1* Richardson diagram of the *Streptomyces griseus* streptogrisin A/chymostatin A complex. The image was prepared from the Brookhaven Protein Data Bank entry (1SGC) as described in the Introduction (p. xxv). Catalytic residues are shown in ball-and-stick representation: His59, Asp102 and Ser195 (numbering as in Alignment 76.1). Chymostatin A is shown in ball-and-stick representation.

Presumably, protease Do, being membrane bound, would be a likely immunogen. This subfamily includes a 47 kDa protein from *Rickettsia tsutsugamushi* that is probably not proteolytic, because all three members of the catalytic triad have been replaced. Protease Do is a mosaic protein with a C-terminal domain that shows some similarity to a tight-junction protein from dog (EMBL: L27152).

Subfamily S2D includes lysyl endopeptidase, known only from *Achromobacter*. The crystal structure for this enzyme has been determined, and the structure is similar to that of chymotrypsin, as can be seen in Figure 76.2.

The lysyl endopeptidase precursor is a mosaic protein that includes a C-terminal domain necessary for secretion through the cell membrane, and is homologous to similar secretion domains in a variety of other proteins from *Vibrio* and its close relatives, including peptidases in families S8, M4 and M28.

*Family S3* includes togavirin, the core protein C of togaviruses, which possesses only one catalytic activity, that

*Figure 76.2* Richardson diagram of the *Achromobacter lyticus* lysyl endopeptidase/$N^{\alpha}$-*p*-tosyl-lysine chloromethane hydrochloride complex. The image was prepared from the Brookhaven Protein Data Bank entry (1ARC) as described in the Introduction (p. xxv). Catalytic residues are shown in ball-and-stick representation: His59, Asp102 and Ser195 (numbering as in Alignment 76.1). The chloromethane hydrochloride is shown in ball-and-stick representation.

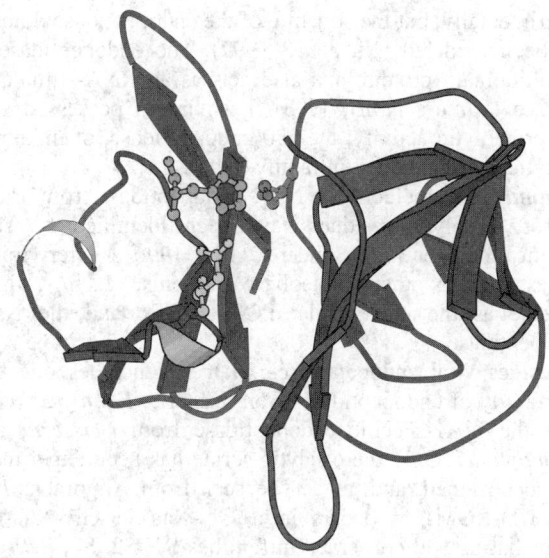

*Figure 76.3* Richardson diagram of Sindbis virus togavirin. The image was prepared from the Brookhaven Protein Data Bank entry (2SNV) as described in the Introduction (p. xxv). Catalytic residues are shown in ball-and-stick representation: His59, Asp102 and Ser195 (numbering as in Alignment 76.1).

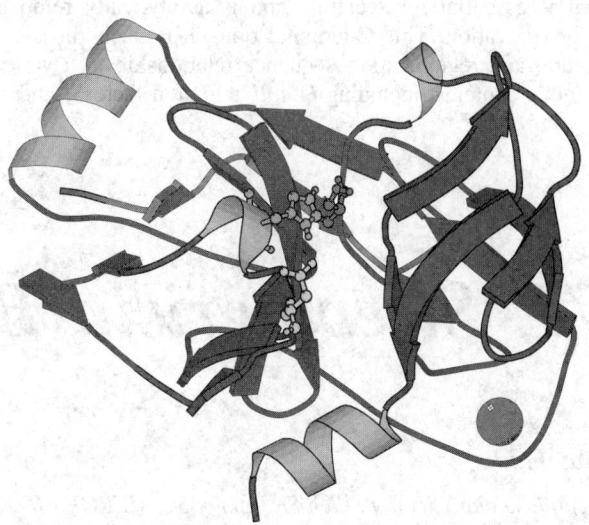

*Figure 76.4* Richardson diagram of the human hepatitis C virus NS3 endopeptidase/NS4A cofactor peptide complex. The image was prepared as described in the Introduction (p. xxv) from co-ordinates kindly provided by Dr J. L. Kim. The structural zinc ion is shown as a CPK sphere in grey. Catalytic residues are shown in ball-and-stick representation: His59, Asp102 and Ser195 (numbering as in Alignment 76.1 on the CD-ROM).

of releasing itself from the N-terminus of the polyprotein-containing structural proteins. Once released, togavirin has no further catalytic activity. The tertiary structure has been solved, and a representation is shown in Figure 76.3. Togaviruses encode two polyproteins, each possessing a processing endopeptidase, and the nonstructural polyprotein-processing enzyme is a cysteine endopeptidase in family C9.

*Family S29* includes a polyprotein-processing NS3 endopeptidase from the hepatitis C virus. Once again, the three-dimensional structure is known, and shows similarities to chymotrypsin. Unusually for a member of clan SA, the NS3 endopeptidase includes a structural zinc-binding site (Love *et al.*, 1996). A representation of the structure is shown in Figure 76.4. The NS3 endopeptidase is a mosaic protein with a C-terminal helicase domain that is homologous to helicases from related viruses. The hepatitis C polyprotein is processed by two endopeptidases, of which the second is of uncertain catalytic type, in family U39.

Besides S3 and S29, there are a number of other viral polyprotein-processing endopeptidases that are included in families within clan SA, either because the endopeptidases have been shown to have a His, Asp, Ser active site by site-directed mutagenesis, or they are assumed to possess one. Sequence comparisons around the proposed catalytic residues are shown in Alignment 76.1 (see CD-ROM).

*Family S7* includes flavirin from flaviviruses such as yellow fever virus, and the catalytic residues have been identified (Chambers *et al.*, 1990). No proteolytic activities have been described for other flavivirus proteins.

*Family S30* includes one of the three processing endopeptidases from potyviruses, the 35 kDa endopeptidase. Catalytic residues have been identified by site-specific mutagenesis in the tobacco etch virus (Verchot *et al.*, 1992). The other endopeptidases from potyviruses are the cysteine-type NIa endopeptidase in family C4 and HC-Pro endopeptidase in family C6.

*Family S31* includes the pestivirus NS2–3/NS3 or p80 endopeptidase. Catalytic residues have not been identified experimentally, but the structure of the endopeptidase domain has been modeled (Xu *et al.*, 1997). The endopeptidase is a multidomain protein, and also acts as an RNA-stimulated NTPase (Tamura *et al.*, 1993). Pestiviruses possess a second proteolytic activity, the p20 endopeptidase, of unknown catalytic type, included in family U44.

*Family S32* includes the Nsp4 endopeptidase from arteriviruses. Catalytic residues have been identified by site-specific mutagenesis (Snijder *et al.*, 1996). Arteriviruses possess three other proteolytic proteins, each with a cysteine as the nucleophile: PCPα, PCPβ and the Nsp2 endopeptidase.

Besides viral endopeptidases, clan SA includes one further family of endopeptidases from bacteria. *Family S6* contains the IgA1-specific endopeptidase from *Neisseria* and *Haemophilus*. Only the catalytic serine has been experimentally determined, and, as can be seen from Alignment 76.1 (see CD-ROM), it occurs in a Gly-Xaa-Ser-Gly-Xaa-Pro motif that is well-conserved in families S1, S2, S7, S29 and S30. Because of the conserved motif, family S6 is included in clan SA. Alignment 76.1 also shows our predictions for the catalytic His and Asp residues, which are the only His and Asp residues completely conserved in the family that occur in the order His, Asp, Ser. The IgA1 endopeptidase precursor is a large, multidomain protein with a large C-terminal domain that is essential for secretion and is autolytically removed upon activation. This C-terminal domain is itself mosaic in nature, and regions show sequence relationship to a variety of other proteins, including *E. coli* initiation factor 2 and rat

plectin. Multiple forms of the IgA1 endopeptidase are present in the *Haemophilus influenzae* genome.

## References

Chambers, T.J., Weir, R.C., Grakoui, A., McCourt, D.W., Bazan, J.F., Fletterick, R.J. & Rice, C.M. (1990) Evidence that the N-terminal domain of nonstructural protein NS3 from yellow fever virus is a serine protease responsible for site-specific cleavages in the viral polyprotein. *Proc. Natl Acad. Sci. USA* **87**, 8898–8902.

Love, R.A., Parge, H.E., Wickersham, J.A., Hostomsky, Z., Habuka, N., Moomaw, E.W., Adachi, T. & Hostomska, Z. (1996) The crystal structure of hepatitis C virus NS3 proteinase reveals a trypsin-like fold and a structural zinc binding site. *Cell* **87**, 331–342.

Snijder, E.J., Wassenaar, A.L.M., Van Dinten, L.C., Spaan, W.J.M. & Gorbalenya, A.E. (1996) The arterivirus Nsp4 protease is the protoype of a novel group of chymotrypsin-like enzymes, the 3C-like serine proteases. *J. Biol. Chem.* **271**, 4864–4871.

Tamura, J.K., Warrener, P. & Collett, M.S. (1993) RNA-stimulated NTPase activity associated with the p80 protein of the pestivirus bovine diarrhea virus. *Virology* **193**, 1–10.

Verchot, J., Gerndon, K.L. & Carrington, J.C. (1992) Mutational analysis of the tobacco etch potyviral 35-kDa proteinase: identification of essential residues and requirements for autoproteolysis. *Virology* **190**, 298–306.

Xu, J., Mendez, E., Caron, P.R., Lin, C., Murcko, M.A., Collett, M.S. & Rice, C.M. (1997) Bovine viral diarrhea virus NS3 serine proteinase: polyprotein cleavage sites, cofactor requirements, and molecular model of an enzyme essential for pestivirus replication. *J. Virol.* **71**, 5312–5322.

# 77. *Streptogrisin B*

## *Databanks*

*Peptidase classification: clan SA, family S2, MEROPS ID: S02.003*
*NC-IUBMB enzyme classification: EC 3.4.21.81*
*Databank codes:*

| Species | SwissProt | PIR | EMBL (cDNA) | EMBL (genomic) |
|---|---|---|---|---|
| *Streptomyces griseus* | P00777 | A00964 | A24973 | – |
| | | B26974 | M17104 | |
| *Streptomyces lividans* | – | – | L44109 | – |

Brookhaven Protein Data Bank three-dimensional structures:

| Species | ID | Resolution | Notes |
|---|---|---|---|
| *Streptomyces griseus* | 1SGP | 1.4 | complex with turkey ovomucoid third domain |
| | 1SGQ | 1.9 | complex with turkey ovomucoid third domain Gly18 variant |
| | 1SGR | 1.9 | complex with turkey ovomucoid third domain Leu18 variant |
| | 3SGB | 1.8 | complex with turkey ovomucoid third domain |
| | 4SGB | 2.1 | complex with potato inhibitor PCI-1 |

## Name and History

The first detailed investigation of the proteinases from the culture filtrate of *Streptomyces griseus* was done by Nomoto *et al.* (1960) who obtained what they thought to be a homogeneous preparation of a proteinase and named it *Streptomyces griseus* protease G. This preparation showed broad endo- and exopeptidase activities, and later was shown to contain several proteinases (Hiramatsu & Ouchi, 1963; Narahashi & Yanagita, 1967; Löfqvist & Sjöberg, 1971). During the period between 1968 and 1973 several laboratories worked on the fractionation of proteinases from the culture filtrate of *S. griseus* strain K-1 which had become available commercially as Pronase. Each of these labs showed the presence of at least three serine proteinases with chymotrypsin- or trypsin-like activities and named them according to their observations. The proteinases named as **PNPA hydrolase II** (Wählby, 1969), **pronase elastase** (Trop & Birk, 1970), **elastase like enzyme III** (Gertler & Trop, 1971), **Streptomyces griseus protease B (SGPB)** (Jurášek *et al.*, 1971), **serine endopeptidase** (Awad *et al.*, 1972), **alkaline proteinase C** (Narahashi & Yoda, 1973), **guanidine stable chymoelastase** (Siegel & Awad, 1973), and **Streptomyces griseus protease 1 (SGP1)** (Bauer, 1977) all refer to the same enzyme. The name most commonly used for the enzyme in recent times has been **Streptomyces griseus protease B**, but the term recommended by NC-IUBMB, **streptogrisin B**, is used here.

## Activity and Specificity

Streptogrisin B is a serine proteinase with broad substrate specificity. Its specificity has been studied by use of protein and peptide substrates (Narahashi & Yoda, 1973; Bauer, 1978; Sidhu *et al.*, 1995) as well as by studying its interaction with inhibitors (Gertler, 1974; James *et al.*, 1980; Bigler *et al.*, 1993; Lu *et al.*, 1997). The specificity of streptogrisin B can most simply be described as 'chymotrypsin-like'; it readily cleaves peptide bonds on the carboxyl side of Phe, Tyr, Leu and Met. There are, however, some differences between the specificity of streptogrisin B and α-chymotrypsin (Narahashi & Yoda, 1973; Bauer, 1978). Hydrolysis of substrates having different P1 amino acids reveals that Phe and Tyr are optimal P1 residues followed by Met, Leu, Trp, Ala, Val and Gly (Bauer, 1978; Sidhu *et al.*, 1995). Progressive increase in chain length from a single amino acid amide to a heptapeptide amide greatly increases the efficiency of substrate hydrolysis by the enzyme. The difference in $k_{cat}/K_m$ between Ac-Phe⊣NH$_2$ and Ac-Pro-Ala-Pro-Phe⊣Ala-Ala-Ala-NH$_2$ is of the order of $10^7$. Further increase in chain length produces insignificant changes in $k_{cat}/K_m$. The conclusion that streptogrisin B acts best on extended peptides is also supported by a study of inhibition by a series of peptidyl chloromethanes of different chain length (Gertler, 1974). Tos-Phe-CH$_2$Cl, a well known inhibitor of α-chymotrypsin, did not inhibit streptogrisin B, whereas a tripeptidyl chloromethane, Ac-Gly-Leu-Phe-CH$_2$Cl, was a potent inhibitor.

The interaction of streptogrisin B with protein proteinase inhibitors provides another way to study the specificity of this enzyme. In this interaction, the P6–P3′ (or P4′) residues (and a few secondary contacts) of the inhibitor make numerous contacts with the enzyme to form a stable complex (Read *et al.*, 1983; Greenblatt *et al.*, 1989). Bigler *et al.* (1993) and Lu *et al.* (1997) have measured association equilibrium constants of 25 P1 variants (20 coded and 5 noncoded) of turkey ovomucoid third domain (OMTKY3) with six serine proteinases including streptogrisin B. This study provides a quantitative measure of P1 specificity. Whereas in substrates the best P1 residues are Phe and Tyr, the best P1 amino acids in OMTKY3 are Cys and Leu. The P1 Phe and Tyr mutants of OMTKY3 also bind strongly, but are about 10 times weaker than the P1 Cys and Leu variants. The discrepancy is presumably due to the rigid binding loop in inhibitors and an unfavorable χ angle which aromatic amino acid side chains have to acquire to bind to the S1 pocket of the enzyme (Bateman *et al.*, 1996). A complete correlation of the binding of P1 residues with their binding and hydrolysis parameters in substrates is not possible, as substrate hydrolysis data are available only for a few amino acids. However, a limited comparison reveals that except for Phe and Tyr, the order of P1 preference for substrates and inhibitors is the same (Leu > Trp > Ala > Val > Gly). The complete preference order for position P1 is as follows (residues that have similar $K_a$ values being bracketed): (Cys, Leu) > Met > Phe > (Tyr, His) > Trp > Gln > (Ala, Val) > (Lys, Thr) > (Arg, Asn) > Ser > Ile > Gly > Asp > Glu ≫ Pro.

The primary specificity pocket, S1, of streptogrisin B is hydrophobic, and during association of inhibitors or substrates with streptogrisin B, the P1 side chain binds to this pocket. As shown by Huang *et al.* (1995), the strength of binding of P1 side chains is linearly dependent upon their buried hydrophobic surface area for the homologous aliphatic series of amino acids. This linear relationship breaks down (Lu *et al.*, 1997) when applied to aromatic amino acids for the reasons stated above. Because of the hydrophobic nature of the S1 pocket, polar and charged side chains bind poorly. Charged groups at P1 undergo large p$K$ shifts in complex with streptogrisin B (Qasim *et al.*, 1995). The binding of unionized forms of these groups is comparable to that of their hydrophobic counterparts.

The other contact positions for which complete specificity requirements are available are the P4, P2 and P3′ positions (Lu, 1994; Lu *et al.*, 1996; Ranjbar *et al.*, 1996). All 20 variants of OMTKY3 for these positions are available, and their $K_a$ values have been measured with streptogrisin B. The order of preference for these positions is as follows:

P4: (Tyr, Trp) > Phe > (Val, His, Ile) > (Asn, Ala, Cys, Met, Leu) > (Thr, Gln, Asp, Pro, Glu) > Ser > Arg > (Gly, Lys)

P2: Thr > (Ile, Arg) > (Met, Lys) > Ser > (Gln, Trp) > (Asn, Tyr) > (Leu, His) > (Val, Pro) > Phe > Glu > (Ala, Asp) > Gly

P3′: Lys > (Arg, Val, Gln, Ala, Met) > (Phe, Leu, Ile, Ser, Tyr, Trp, Asn) > (His, Thr, Cys) > (Asp, Glu) > Gly ≫ Pro

At present, our understanding of the role of other contact positions is relatively limited. Substrate hydrolysis data with different P1′ residues suggest that Phe is best followed by Leu, Tyr, Ala or Gly (Bauer, 1978).

Streptogrisin B shows maximal activity between pH 7.5 and 9.5 (Gertler, 1974). The activity can be assayed most conveniently by using peptide *p*-nitroanilide substrates, and following the release of the leaving group spectrophotometrically. A variety of commercially available chromogenic and fluorogenic substrates such as Suc-Ala-Ala-Pro-Phe↓NHPhNO$_2$, Suc-Ala-Ala-Pro-Leu↓NHPhNO$_2$, Suc-Ala-Pro-Ala↓NHPhNO$_2$, Suc-Ala-Ala-Pro-Ala↓NHPhNO$_2$, Suc-Gly-Gly-Phe↓NHPhNO$_2$, and Suc-Ala-Ala-Pro-Phe↓NHMec can be used over a broad range of enzyme concentration. The enzyme does not require metal ions for its activity, but stability is increased by the presence of Ca$^{2+}$ (Siegel *et al.*, 1972), so that it is advisable to include 5–20 mM Ca$^{2+}$ in the assay buffer. Streptogrisin B also shows surface inactivation, particularly at low concentrations (<10 nM), but the addition of a mild detergent such as Triton X-100 (0.005%) solves this problem (Empie & Laskowski, 1982; Wynn, 1990). The enzyme is inhibited by DFP (Wählby, 1969), *p*-iodobenzenesulfonyl fluoride, *p*-methoxy *m*-chloromercuribenzenesulfonyl fluoride, *p*-chloromercuribenzenesulfonyl fluoride (Codding *et al.*, 1974), Ac-Gly-Leu-Phe-CH$_2$Cl (Gertler, 1974), OMTKY3 (Wynn, 1990; Lu, 1994), potato inhibitor I (Greenblatt *et al.*, 1989), *Streptomyces* subtilisin inhibitor (Christensen *et al.*, 1985) and eglin c (Qasim *et al.*, 1996).

## Structural Chemistry

Mature streptogrisin B consists of a single polypeptide chain of 185 amino acid residues. Its complete amino acid sequence was determined by Jurášek *et al.* (1974). More recently, the complete sequence of the streptogrisin B gene (*str*B) has also been determined (Henderson *et al.*, 1987). Except for differences at positions 65, 82 and 186 (chymotrypsinogen numbering) the two sequences are identical.

The molecular mass of streptogrisin B is 18 500 and its pI is 6.89. Sequence comparison with serine proteinases of the chymotrypsin family shows about 20% sequence identity with α-chymotrypsin, bovine trypsin and porcine pancreatic elastase, and 60% identity with its sister enzyme, streptogrisin A (Chapter 78) (James *et al.*, 1978).

X-ray crystallographic structures of streptogrisin B alone (Delbaere *et al.*, 1975) and in complex with a tripeptidyl chloromethane (James *et al.*, 1980) and with protein proteinase inhibitors, OMTKY3 (Fujinaga *et al.*, 1982; Read *et al.*, 1983) and potato chymotrypsin inhibitor (PCI-1) (Greenblatt *et al.*, 1989) have been determined. In addition, X-ray structures of streptogrisin B with several P1 mutants of OMTKY3 have been determined (Huang *et al.*, 1995; Bateman *et al.*, 1996). A comprehensive project involving structure determination of the complexes of streptogrisin B with all 20 P1 mutants of OMTKY3 is underway in the laboratory of Professor M.N.G. James at Alberta. These structural studies provide a detailed insight into the molecular architecture and catalytic mechanism of the enzyme. The polypeptide chain is organized into two hydrophobic cores with the active-site residues lying at their junction. Despite low sequence identity between streptogrisin B and the enzymes of family S1, the tertiary structures are closely similar, as is the topology of the catalytic triad (His57, Asp102, Ser195, in chymotrypsinogen numbering).

In the complexes of streptogrisin B with OMTKY3 and PCI-1, about a dozen amino acid residues of the inhibitor make a total of about 105 close contacts with 17 (19 in the case of PCI-1) residues of the enzyme. The P1 side chain alone makes 29 (in OMTKY3) of these contacts. The other contact residues of the inhibitor (mainly P6 to P3′ or P4′) are involved in a battery of other noncovalent interactions including several hydrogen bonds. The backbone amide and carbonyl groups of P3–P1 and P2′ residues are involved in the formation of antiparallel hydrogen bonds with the backbone carbonyl oxygen and amide hydrogen of Gly216-Ser214 and Phe41, respectively, of streptogrisin B. In addition, the P1 carbonyl oxygen occupies the 'oxyanion hole' and is involved in hydrogen bond formation with the NHs of Gly193 and Ser195. The carbonyl carbon is 2.7–2.8 Å away from O$^\gamma$ of Ser195 – too close to be involved in favorable van der Waals interactions and too far to be considered a covalent bond. Since inhibitors are known to bind to serine proteinases in the manner of substrates (Laskowski & Kato, 1980), the structure of an inhibitor–streptogrisin B complex provides invaluable information about the mechanism of substrate hydrolysis. The scissile peptide bond in the inhibitor remains planar and does not undergo hydrolysis because of the rigidity of the inhibitor binding loop, and because of several restraining interactions around this bond that make formation of a tetrahedral intermediate energetically highly unfavorable. The overall global conformation of the enzyme and inhibitor remains unchanged upon complex formation, but there are significant localized conformational changes, particularly in the reactive-site region, that facilitate hydrolysis of susceptible peptide bonds in substrates. Most importantly, the side chains of Ser195 and His57 move in such a way that the hydrogen bond between Ser O$^\gamma$ and His N$^{\varepsilon2}$ is shortened from 3.03 Å in the free enzyme to 2.55 Å in the inhibitor–enzyme complex (Fujinaga *et al.*, 1982). This in turn makes Ser195 strongly nucleophilic, facilitating the transfer of a proton via His57 N$^{\varepsilon2}$ to the peptide NH of the peptide bond undergoing hydrolysis.

## Preparation

The biological source of streptogrisin B is the culture filtrate of *S. griseus* strain K-1. This is a mixture of enzymes and proteins, and is commercially available as Pronase (Sigma, Calbiochem and Boehringer Mannheim) (see Appendix 2 for full names and addresses of suppliers). The most commonly used purification procedure is that described by Jurášek *et al.* (1971). A slightly modified procedure that works well and reproducibly consists of the following steps:

1. Pronase (∼2 g per 100 ml) is dissolved in 0.05 M sodium acetate buffer, pH 5.0, containing 0.01 M CaCl$_2$, and run on CM-Sepharose CL-6B in this buffer. Unbound proteins are removed by washing with the starting buffer, and bound proteins are then eluted with a linear gradient from 0 to 0.3 M NaCl in the starting buffer. Appearance of streptogrisin A and B peaks can be monitored by measuring activity against Suc-Gly-Gly-Phe↓NHPhNO$_2$.

S

2.  Further purification to homogeneity can be achieved by size-exclusion chromatography on a Sephacryl S-100 column followed by ion-exchange chromatography on Q-Sepharose at pH 9 with a linear gradient from 0 to 0.3 M NaCl.

The gene of streptogrisin B has also been cloned and expressed in *B. subtilis* and *E. coli* (Sidhu *et al*., 1994; Sidhu & Borgford, 1996), and the expressed streptogrisin B has been purified to homogeneity.

## Biological Aspects

The gene structure of streptogrisin B has been determined (Henderson *et al*., 1987). It codes for a preproenzyme, the pre- and propeptides being 38 and 76 amino acid residues, respectively. As in many other bacterial serine proteinases, the pre and pro regions are responsible for the secretion of the enzyme across the membrane and for its correct folding. After synthesis, the pre region is cleaved off by a signal peptidase. The peptide bond connecting the propeptide to the mature enzyme is cleaved by a self-processing event, the hydrolysis of a Leu┼Ile bond. In a recent study (Sidhu & Borgford, 1996) a novel screening strategy has been used to engineer desired substrate specificity in streptogrisin B mutants. The amino acid at the pro/mature junction was mutated, together with other amino acids that form the primary specificity pocket of the enzyme. The specificity of the mutant enzymes is thus constrained by the pro/mature junction, and only those mutants that are capable of cleaving the pro/mature junction produce active, mature enzyme.

## Distinguishing Features

The proteinase that is most similar to streptogrisin B is streptogrisin A (Chapter 78), but there are some important differences between the two. Streptogrisin B is an unusually stable enzyme and shows full enzymatic activity in 6 M guanidine hydrochloride and 8 M urea (Siegel *et al*., 1972). Streptogrisin A, like most of the other serine proteinases, is rapidly denatured under these conditions. Streptogrisin B also shows unusual stability to acidic pH, and its enzymatic activity is measurable down to pH 2.5 (Qasim *et al*., 1995). Even more striking is its ability to convert virgin inhibitor (reactive-site peptide bond intact) into modified inhibitor (reactive-site peptide bond cleaved) at pH values as low as 1 (Ardelt & Laskowski, 1991). Streptogrisin A, on the other hand, shows rapid irreversible inactivation below pH 3.5 (M. Ranjbar & M. Laskowski, Jr, unpublished results).

## Related Peptidases

In addition to streptogrisin A, the gene structures of two further related proteinases from *S. griseus* have been determined (Sidhu *et al*., 1994, 1995). The expressed enzymes have been designated *Streptomyces griseus* protease C (streptogrisin C) and *Streptomyces griseus* protease D (streptogrisin D). Both of these enzymes are very similar to streptogrisins A and B in their enzymatic properties and amino acid sequence, but they differ in molecular mass and quaternary structure. Streptogrisin C contains an additional domain at its C-terminus that is homologous to the chitin-binding domain of chitinases from *Bacillus circulans*. Streptogrisin D on the other hand exists as a stable dimer. Another enzyme that shows strong sequence similarity to streptogrisin B and is also very similar to it in its tertiary structure is *Streptomyces fradiae* protease 2 (SFase 2) (Kitadokoro *et al*., 1994). The extracellular serine protease, SAM-P20, from a subtilisin inhibitor-deficient mutant of *S. albogriseolus* S-3253 (Taguchi *et al*., 1995) also shows strong sequence similarity to streptogrisins A and B.

## Acknowledgements

My thanks go to Professor M. Laskowski Jr, who introduced me to serine proteinases and their inhibitors. Supported by NIH grant GM 10831 to Professor M. Laskowski, Jr.

## Further Reading

Three articles that will provide detailed information about structural aspects, molecular biological aspects and specificity correlations between streptogrisin B and other serine proteinases are those of Huang *et al*. (1995), Sidhu & Borgford (1996), and Lu *et al*. (1997).

## References

Ardelt, W. & Laskowski, M. Jr. (1991) Effect of single amino acid replacements on the thermodynamics of the reactive site peptide bond hydrolysis in ovomucoid third domain. *J. Mol. Biol.* **220**, 1041–1053.

Awad, M. Jr., Soto, A.R., Siegel, S., Skiba, W.E., Bernstrom, G.G. & Ochoa, M.S. (1972) The proteolytic enzymes of the K-1 strain of *Streptomyces griseus* obtained from a commercial preparation (Pronase). I. Purification of four serine peptidases. *J. Biol. Chem.* **247**, 4144–4154.

Bateman, K.S., Huang, K., Lu, W., Anderson, S., Laskowski, M. Jr. & James, M.N.G. (1996) X-ray crystal structures of SGPB in complex with three aromatic P₁ variants of OMTKY3. Poster presented at Keystone symposium on Proteolytic Enzymes and Inhibitors in Biology and Medicine.

Bauer, C.-A. (1977) Purity and pH dependence of *Streptomyces griseus* protease 1. *Acta Chem. Scand.* **B31**, 637–640.

Bauer, C.-A. (1978) Active centers of *Streptomyces griseus* protease 1, *Streptomyces griseus* protease 3 and α-chymotrypsin: enzyme–substrate interactions. *Biochemistry* **17**, 375–380.

Bigler, T.L., Lu, W., Park, S.J., Tashiro, M., Wieczorek, M., Wynn, R. & Laskowski, M. Jr. (1993) Binding of amino acid side chains to preformed cavities: interaction of serine proteinases with turkey ovomucoid third domains with coded and noncoded P₁ residues. *Protein Sci.* **2**, 786–799.

Christensen, U., Ishida, S., Ishii, S.-I., Mitsui, Y., Iitaka, Y., McClarin, J. & Langridge, R. (1985) Interaction of *Streptomyces* subtilisin inhibitor with *Streptomyces griseus* protease A and B. Enzyme kinetic and computer simulation studies. *J. Biochem.* **98**, 1263–1274.

Codding, P.W., Delbaere, L.T.J., Hayakawa, K., Hutcheon, W.L.B., James, M.N.G. & Jurášek, L. (1974) The 4.5 Å resolution structure of a bacterial serine protease from *Streptomyces griseus*. *Can. J. Biochem.* **52**, 208–220.

Delbaere, L.T.J., Hutcheon, W.L.B., James, M.N.G. & Thiessen, W.E. (1975) Tertiary structural differences between microbial serine proteases and pancreatic serine enzymes. *Nature* **257**, 758–763.

Empie, M.W. & Laskowski, M. Jr. (1982) Thermodynamics and kinetics of single residue replacements in avian ovomucoid third domains: effect on inhibitor interactions with serine proteinases. *Biochemistry* **21**, 2274–2284.

Fujinaga, M., Read, R.J., Sielecki, A., Ardelt, W., Laskowski, M. Jr. & James, M.N.G. (1982) Refined crystal structure of the molecular complex of *Streptomyces griseus* protease B, a serine protease, with the third domain of the ovomucoid inhibitor from turkey. *Proc. Natl Acad. Sci. USA* **79**, 4868–4872.

Gertler, A. (1974) Inhibition of *Streptomyces griseus* protease B by peptide chloromethyl ketones: partial mapping of the binding site and identification of the reactive site residue. *FEBS Lett.* **43**, 81–85.

Gertler, A. & Trop, M. (1971) The elastase like enzymes from *Streptomyces griseus* (Pronase). Isolation and partial characterization. *Eur. J. Biochem.* **19**, 90–96.

Greenblatt, H.M., Ryan, C.A. & James, M.N.G. (1989) Structure of the complex of *Streptomyces griseus* proteinase B and polypeptide chymotrypsin inhibitor 1 from Russet Burbank potato tubers at 2.1 Å resolution. *J. Mol. Biol.* **205**, 201–228.

Henderson, G., Krygsman, P., Liu, C.J., Davey, C.C. & Malek, L.T. (1987) Characterization and structure of genes for proteases A and B from *Streptomyces griseus*. *J. Bacteriol.* **169**, 3778–3784.

Hiramatsu, A. & Ouchi, T. (1963) On the proteolytic enzymes from the commercial protease preparation of *Streptomyces griseus* (Pronase P). *J. Biochem.* **54**, 462–464.

Huang, K., Lu, W., Anderson, S., Laskowski, M. Jr & James, M.N.G. (1995) Water molecules participate in proteinase–inhibitor interactions: crystal structures of Leu18, Ala18, and Gly18 variants of turkey ovomucoid third domain complexed with *Streptomyces griseus* protease B. *Protein Sci.* **4**, 1985–1997.

James, M.N.G., Delbaere, L.T.J. & Brayer, G.D. (1978) Amino acid sequence alignment of bacterial and mammalian pancreatic serine proteinases based on topological equivalences. *Can. J. Biochem.* **56**, 396–402.

James, M.N.G., Brayer, G.D., Delbaere, L.T.J. & Sielecki, A.R. (1980) Crystal structure studies and inhibition kinetics of tripeptide chloromethyl ketone inhibitors with *Streptomyces griseus* protease B. *J. Mol. Biol.* **139**, 423–438.

Jurášek, L., Johnson, P., Olafson, K.W. & Smellie, L.B. (1971) An improved fractionation system for pronase on CM-Sephadex. *Can. J. Biochem.* **49**, 1196–1201.

Jurášek, L., Carpenter, M.R., Smillie, L.B., Gertler, A., Levy, S. & Ericsson. L.H. (1974) Amino acid sequence of *Streptomyces griseus* protease B, a major component of pronase. *Biochem. Biophys. Res. Commun.* **61**, 1095–1100.

Kitadokoro, K., Tsuzuki, H., Nakamura, E., Sato, T. & Teraoka, H. (1994) Purification, characterization, primary structure, crystallization and preliminary crystallographic study of a serine proteinase from *Streptomyces fradiae* ATCC 14544. *Eur. J. Biochem.* **220**, 55–61.

Laskowski, M. Jr & Kato, I. (1980) Protein inhibitors of proteinases. *Annu. Rev. Biochem.* **49**, 593–626.

Löfqvist, B. & Sjöberg, L-S. (1971) Studies of the heterogeneity of *Streptomyces griseus* protease. I. Polyacrylamide gel electrophoresis of commercial pronase P derived from *Streptomyces griseus* K-1. *Acta Chem. Scand.* **25**, 1663–1678.

Lu, S., Zhang, W., Qasim, S., Rauch, R., Ryan, K., Anderson, S. & Laskowski, M. Jr. (1996) The effect of $P_4$ residues upon the association of Standard Mechanism canonical protein inhibitors of serine proteinases with their cognate enzymes. *Protein Sci.* **5**(suppl. 1), 129.

Lu, W. (1994) Energetics of the interaction of ovomucoid third domain variants with different serine proteinases. PhD thesis, Purdue University.

Lu, W., Apostol, I., Qasim, M.A., Warne, N., Wynn, R., Zhang, W.L., Anderson, S., Chiang, Y.W., Rothberg, I., Ryan, K. & Laskowski, M. Jr. (1997) Binding of amino acid side chains to $S_1$ cavities of serine proteinases. *J. Mol. Biol.* **266**, 441–461.

Narahashi, Y. & Yanagita, M. (1967) Studies on proteolytic enzymes (pronase) of *Streptomyces griseus* K-1. I. Nature and properties of the proteolytic enzyme system. *J. Biochem.* **62**, 633–641.

Narahashi, Y. & Yoda, K. (1973) Studies on proteolytic enzymes (pronase) of *Streptomyces griseus* K-I. III. Purification and proteolytic specificity of alkaline proteinase C from pronase. *J. Biochem.* **73**, 831–841.

Nomoto, M., Narahashi, Y. & Murakami, M. (1960) A proteolytic enzyme of *Streptomyces griseus*. VII. Substrate specificity of *Streptomyces griseus* protease. *J. Biochem.* **48**, 906–918.

Qasim, M.A., Ranjbar, M., Wynn, R., Anderson, S. & Laskowski, M. Jr. (1995) Ionizable $P_1$ residues in serine proteinase inhibitor undergo pK shifts on complex formation. *J. Biol. Chem.* **270**, 27419–27422.

Qasim, M.A., Wynn, R., Saunders, S., Ganz, P.J., Bateman, K., James, M.N.G. & Laskowski, M. Jr. (1996) Interframe additivity comparisons of the 7 P1 eglin c and turkey ovomucoid third domain with six serine proteinases. *Protein Sci.* **5**(suppl. 1), 129.

Ranjbar, M., Wynn, R., Qasim, S., Kiri, A.J., Anderson, S. & Laskowski, M. Jr. (1996) The effect of P2 residue on the association of Standard Mechanism protein inhibitors with their cognate serine proteinases. *Protein Sci.* **5**(suppl. 1), 128.

Read, R.J., Fujinaga, M., Sielecki, A.R. & James, M.N.G. (1983) Structure of the complex of *Streptomyces griseus* protease B and third domain of the turkey ovomucoid inhibitor at 1.8 Å resolution. *Biochemistry* **22**, 4426–4433.

Sidhu, S.S. & Borgford, T.J. (1996) Selection of *Streptomyces griseus* protease B mutants with desired alterations in primary specificity using a library screening strategy. *J. Mol. Biol.* **257**, 233–245.

Sidhu, S.S., Kalman, G.B., Willis, L.G. & Borgford, T.J. (1994) *Streptomyces griseus* protease C. A novel enzyme of the chymotrypsin superfamily. *J. Biol. Chem.* **269**, 20167–20171.

Sidhu, S.S., Kalmer, G.B., Willis, L.G. & Borgford, T.J. (1995) Protease evolution in *Streptomyces griseus*. Discovery of a novel dimeric enzyme. *J. Biol. Chem.* **270**, 7594–7600.

Siegel, S. & Awad, W.M. Jr. (1973) The proteolytic enzymes of the K-1 strain of *Streptomyces griseus* obtained from a commercial preparation (Pronase). IV. Structure function studies of the two smallest serine endopeptidases, stabilization by glycerol during reaction with acetic anhydride. *J. Biol. Chem.* **248**, 3233–3240.

Siegel, S., Brady, A.H. & Awad, W.M. Jr (1972) The proteolytic enzymes of the K-1 strain of *Streptomyces griseus* obtained from a commercial preparation (Pronase) II. The activity of a serine enzyme in 6 M guanidinium chloride. *J. Biol. Chem.* **247**, 4155–4159.

Taguchi, S., Odaka, A., Watanabe, Y. & Momose, H. (1995) Molecular characterization of a gene encoding extracellular serine protease isolated from a subtilisin inhibitor deficient mutant of *Streptomyces albogriseolus* S-3253. *Appl. Environ. Microbiol.* **61**, 180–186.

Trop, M. & Birk, Y. (1970) The specificity of proteinases from *Streptomyces griseus* (pronase). *Biochem. J.* **116**, 19–25.

Wählby, S. (1969) Studies on *Streptomyces griseus* protease. III. Purification of two DFP reacting enzymes. *Biochim. Biophys.*

*Acta* **185**, 178–185.

Wynn, R. (1990) Design of a specific human leukocyte elastase inhibitor based on ovomucoid third domain. PhD thesis, Purdue University.

**M. Abul Qasim**
*Department of Chemistry,*
*1393 Brown Building,*
*Purdue University,*
*West Lafayette, IN 47907, USA*
*Email: qasim@omni.cc.purdue.edu*

# 78. *Streptogrisin A*

## Databanks

*Peptidase classification: clan SA, family S2, MEROPS ID: S02.001*
*NC-IUBMB enzyme classification: EC 3.4.21.80*
*Databank codes:*

| Species | SwissProt | PIR | EMBL (cDNA) | EMBL (genomic) |
|---|---|---|---|---|
| *Streptomyces griseus* | P00776 | A00963 | A24972 | – |
| | | A26974 | M17103 | |

Brookhaven Protein Data Bank three-dimensional structures:

| Species | ID | Resolution | Notes |
|---|---|---|---|
| *Streptomyces griseus* | 1SGC[a] | 1.8 | complex with chymostatin |
| | 2SGA | 1.5 | |
| | 3SGA | 1.8 | complex with Ac-Pro-Ala-Pro-Phe-aldehyde |
| | 4SGA | 1.8 | complex with Ac-Pro-Ala-Pro-Phe |
| | 5SGA | 1.8 | complex with Ac-Pro-Ala-Pro-Tyr |

[a]See Chapter 76 for an image constructed from these coordinates.

## Name and History

The discovery of streptogrisin A runs parallel to that of streptogrisin B (Chapter 77). During the late 1960s and early 1970s, workers in several laboratories purified this streptogrisin from Pronase (commercial source of the culture filtrate from *Streptomyces griseus*) and coined their own names. The names given to the enzyme included: *PNPA hydrolase I* (Wählby, 1969), *alkaline protease A* (Narahashi, 1970), *elastase like enzyme II* (Gertler & Trop, 1971), **Streptomyces griseus protease A** (Jurášek *et al.*, 1971), **Streptomyces griseus protease 3** (Löfqvist & Sjöberg, 1971) and *lysine-free chymoelastase* (Siegel & Awad, 1973). Here, the name *streptogrisin A* recommended by NC-IUBMB will be used.

## Activity and Specificity

Streptogrisin A is a serine endopeptidase with broad substrate specificity. Substrate hydrolysis data (Bauer *et al.*, 1976a,b) indicate that the order of preference for P1 residue is Phe > Tyr > Leu > Val > Ala > Gly. The enzyme has an extended substrate-binding site and requires the presence of P4, P3, P2, P1, P1′, P2′ and P3′ residues for maximal activity; the cleavage of single amino acid amides is very sluggish. The association equilibrium constants of the enzyme with 20 P1 variants of turkey ovomucoid third domain (OMTKY3) correlate well with those obtained with streptogrisin B (correlation coefficient 0.96), $\alpha$-chymotrypsin (correlation coefficient 0.89), and subtilisin Carlsberg (correlation coefficient 0.87) (Lu *et al.*, 1997), but poorly with the two elastases porcine pancreatic elastase and human leukocyte elastase (correlation coefficients 0.23 and 0.35). In general, streptogrisin A is very similar to streptogrisin B in structure and specificity, but it is more active, and its binding constants with OMTKY3 mutants are generally 3–10 times stronger than obtained with streptogrisin B. The P1 mutants with basic residues are exceptional, as they bind a little better to streptogrisin B than to streptogrisin A. Site-directed mutants of OMTKY3 involving replacements at P4, P2 and P3′ have also been prepared, and association equilibrium constants of the P4 and P3′ variants have been measured with this enzyme. The overall preference pattern is generally the same as that obtained for streptogrisin B (Chapter 77). Other known inhibitors of streptogrisin A are: *Streptomyces* subtilisin inhibitor (Christensen *et al.*, 1985), eglin c (Qasim *et al.*, 1996) and various aldehyde inhibitors (Brayer *et al.*, 1979; Tomkinson *et al.*, 1992). The enzyme can be assayed with the same substrates and under the same conditions as described for streptogrisin B.

## Structural Chemistry

Mature streptogrisin A contains 181 amino acids in a single polypeptide chain. The amino acid sequence has been determined chemically (Johnson & Smellie, 1974) and has also been deduced from the nucleotide sequence of the gene (Henderson *et al.*, 1987). The X-ray crystallographic structure of streptogrisin A alone and in complex with aldehyde inhibitors Ac-Pro-Ala-Pro-Phe-H and chymostatin has been determined (Sielecki *et al.*, 1979; Brayer *et al.*, 1979; Delbaere & Brayer, 1985). The overall fold of the polypeptide chain is similar to that of streptogrisin B, and nearly 85% of the backbone α-carbon atoms are topologically equivalent in the two enzymes (James *et al.*, 1978). The sequences around the members of the catalytic triad are identical in the two enzymes. The aldehyde inhibitors form stable hemiacetal adducts that are analogous to the tetrahedral transition intermediates formed during substrate hydrolysis.

## Preparation

The biological source of streptogrisin A is the culture filtrate of *S. griseus*, and purification of the enzyme involves the same steps as described for the streptogrisin B. The two enzymes are separated in the CM-Sepharose CL-6B column chromatography. Purification to homogeneity can be achieved by size-exclusion chromatography on Sephacryl S-100 followed by ion-exchange chromatography on Q-Sepharose.

## Biological Aspects

The structure of the gene encoding the preproenzyme has been determined. The pre and pro regions are 38 and 78 residues long respectively. Formation of the mature active enzyme involves cleavage of the pre region by a signal peptidase, followed by removal of the pro region in an autocatalytic proteolysis step (Henderson *et al.*, 1987).

## Distinguishing Features

Streptogrisin A is the smallest of the six known serine proteinases of *S. griseus*. In contrast to its closely related partner streptogrisin B, it is readily denatured by high concentrations of guanidine hydrochloride and urea. It is also irreversibly denatured below pH 3.5 (M. Ranjbar & M. Laskowski, Jr, unpublished results).

## Related Peptidases

Streptogrisins C and D from *S. griseus* (Sidhu *et al.*, 1994, 1995), *Streptomyces fradiae* protease 2 from *S. fradiae* (Kitadokoro *et al.*, 1994), and SAM-P20 (Taguchi *et al.*, 1995) from a subtilisin inhibitor-deficient mutant of *S. albogriseus* are structurally related to streptogrisin A.

## Acknowledgements

Supported by NIH grant GM 10831 to Professor M. Laskowski, Jr.

## Further Reading

The articles of Brayer *et al.* (1979) and Delbaere & Brayer (1985) give detailed structural information on streptogrisin A, and Lu *et al.* (1997) provides information about specificity correlations with other serine proteinases.

## References

Bauer, C.-A., Thompson, R.C. & Blout, E.R. (1976a) The active centers of *Streptomyces griseus* protease 3 and α-chymotrypsin: enzyme–substrate interactions remote from the scissile peptide bond. *Biochemistry* **15**, 1291–1295.

Bauer, C.-A., Thompson, R.C. & Blout, E.R. (1976b) The active centers of *Streptomyces griseus* protease 3, α-chymotrypsin, and elastase: enzyme–substrate interactions close to the scissile peptide bond. *Biochemistry* **15**, 1296–1299.

Brayer, G.D., Delbaere, L.T.J., James, M.N.G., Bauer, C.-A. & Thompson, R.C. (1979) Crystallographic and kinetic investigations of the covalent complex formed by a specific tetrapeptide aldehyde and the serine protease from *Streptomyces griseus*. *Proc. Natl Acad. Sci. USA* **76**, 96–100.

Christensen, U., Ishida, S., Ishii, S.-I., Mitsui, Y., Iitaka, Y., McClarin, J. & Langridge, R. (1985) Interaction of *Streptomyces* subtilisin inhibitor with *Streptomyces griseus* protease A and B. Enzyme kinetic and computer simulation studies. *J. Biochem.* **98**, 1263–1274.

Delbaere, L.T.J. & Brayer, G.D. (1985) The 1.8 Å structure of the complex between chymostatin and *Streptomyces griseus* protease A. A model for serine protease catalytic tetrahedral intermediates. *J. Mol. Biol.* **183**, 89–103.

Gertler, A. & Trop, M. (1971) The elastase like enzymes from *Streptomyces griseus* (Pronase). Isolation and partial characterization. *Eur. J. Biochem.* **19**, 90–96.

Henderson, G., Krygsman, P., Liu, C.J., Davey, C.C. & Malek, L.T. (1987) Characterization and structure of genes for proteases A and B from *Streptomyces griseus*. *J. Bacteriol.* **169**, 3778–3784.

James, M.N.G., Delbaere, L.T.J. & Brayer, G.D. (1978) Amino acid sequence alignment of bacterial and mammalian pancreatic serine proteinases based on topological equivalences. *Can. J. Biochem.* **56**, 396–402.

Johnson, P. & Smellie, L.B. (1974) The amino acid sequence and predicted structure of *Streptomyces griseus* protease A. *FEBS Lett.* **47**, 1–6.

Jurášek, L., Johnson, P., Olafson, K.W. & Smellie, L.B. (1971) An improved fractionation system for pronase on CM-Sephadex. *Can. J. Biochem.* **49**, 1196–1201.

Kitadokoro, K., Tsuzuki, H., Nakamura, E., Sato, T. & Teraoka, H. (1994) Purification, characterization, primary structure, crystallization and preliminary crystallographic study of a serine proteinase from *Streptomyces fradiae* ATCC 14544. *Eur. J. Biochem.* **220**, 55–61.

Löfqvist, B. & Sjöberg, L.-S. (1971) Studies of the heterogeneity of *Streptomyces griseus* protease. I. Polyacrylamide gel electrophoresis of commercial pronase P derived from *Streptomyces griseus* K 1. *Acta Chem. Scand.* **25**, 1663–1678.

Lu, W., Apostol, I., Qasim, M.A., Warne, N., Wynn, R., Zhang, W.L., Anderson, S., Chiang, Y.W., Rothberg, I., Ryan, K. & Laskowski, M. Jr. (1997) Binding of amino acid side chains to $S_1$ cavities of serine proteinases. *J. Mol. Biol.* **266**, 441–461.

Narahashi, Y. (1970) Pronase. *Methods Enzymol.* **19**, 651–664.

Qasim, M.A., Wynn, R., Saunders, S., Ganz, P.J., Bateman, K., James, M.N.G. & Laskowski, M. Jr. (1996) Interframe additivity comparisons of the 7 P1 eglin c and turkey ovomucoid third domain with six serine proteinases. *Protein Sci.* **5**(suppl. 1), 129.

Sidhu, S.S., Kalman, G.B., Willis, L.G. & Borgford, T.J. (1994) *Streptomyces griseus* protease C. A novel enzyme of the chymotrypsin superfamily. *J. Biol. Chem.* **269**, 20167–20171.

Sidhu, S.S., Kalmer, G.B., Willis, L.G. & Borgford, T.J. (1995) Protease evolution in *Streptomyces griseus*. Discovery of a novel dimeric enzyme. *J. Biol. Chem.* **270**, 7594–7600.

Siegel, S. & Awad, W.M. Jr. (1973) The proteolytic enzymes of the K-1 strain of *Streptomyces griseus* obtained from a commercial preparation (Pronase). IV. Structure function studies of the two smallest serine endopeptidases, stabilization by glycerol during reaction with acetic anhydride. *J. Biol. Chem.* **248**, 3233–3240.

Sielecki, A.R., Hendrickson, W.A., Broughton, C.G., Delbaere, L.T.J., Brayer, G.D. & James, M.N.G. (1979) Protein structure refinement: *Streptomyces griseus* serine protease A at 1.8 Å resolution. *J. Mol. Biol.* **134**, 781–804.

Taguchi, S., Odaka, A., Watanabe, Y. & Momose, H. (1995) Molecular characterization of a gene encoding extracellular serine protease isolated from a subtilisin inhibitor deficient mutant of *Streptomyces albogriseolus* S-3253. *Appl. Environ. Microbiol.* **61**, 180–186.

Tomkinson, N.P., Galpins, I.J. & Beynon, R.J. (1992) Synthetic analogues of chymostatin: inhibition of chymotrypsin and *Streptomyces griseus* protease A. *Biochem. J.* **286**, 475–480.

Wählby, S. (1969) Studies on *Streptomyces griseus* protease. III. Purification of two DFP reacting enzymes. *Biochim. Biophys. Acta* **185**, 178–185.

*M. Abul Qasim*
*Department of Chemistry,*
*1393 Brown Building,*
*Purdue University,*
*West Lafayette, IN 47907, USA*
*Email: qasim@omni.cc.purdue.edu*

# 79. *Glutamyl endopeptidase I*

## *Databanks*

*Peptidase classification: clan SA, family S2, MEROPS ID: S02.021*
*NC-IUBMB enzyme classification: EC 3.4.21.19*
*Chemical Abstracts Service registry number: 137010-42-5*
*Databank codes:*

| Species | Form | SwissProt | PIR | EMBL (cDNA) | EMBL (genomic) |
|---|---|---|---|---|---|
| Glutamyl endopeptidase | | | | | |
| *Bacillus licheniformis* | Blase | P80057 | A45134 | D10060 | – |
| *Bacillus subtilis* | MPR | P39790 | A35122 | L10505 M22916 | – |
| *Staphylococcus aureus* | V8 | P04188 | A00966 A26812 | D00730 Y00356 | – |
| Epidermolytic toxin | | | | | |
| *Staphylococcus aureus* | A | P09331 | A26680 | M17347 M17357 | – |
| *Staphylococcus aureus* | B | P09332 | A26050 B26680 | M17348 | – |

## Name and History

Proteases that specifically cleave after negatively charged residues were first described by Drapeau and coworkers (Drapeau *et al.*, 1972; Houmard & Drapeau, 1972), and a number of Glu-specific serine proteases have been isolated from bacteria (Birktoft & Breddam, 1994). The bacterial glutamyl endopeptidases may be divided into three groups according to the source organisms and sequence relationships: a staphylococcal group, a *Bacillus* group and a *Streptomyces* group. The staphylococcal and *Bacillus* groups are closer to each other than to the *Streptomyces* group, and are the subject of the present chapter. These enzymes are considered by NC-IUBMB to be variants of **glutamyl endopeptidase I**. The best characterized of these are from *Staphylococcus aureus (GluV8)* (Drapeau *et al.*, 1972; Houmard & Drapeau, 1972) and *Bacillus licheniformis (GluBL)* (Kakudo *et al.*, 1992a; Svendsen & Breddam, 1992), but there are also related enzymes from *Bacillus subtilis (GluBS)* (Rufo *et al.*, 1990; Sloma *et al.*, 1990) and *Staphylococcus aureus* ATCC12600 *(SPase)* (Yoshikawa *et al.*, 1992).

## Activity and Specificity

Common to the forms of glutamyl endopeptidase I isolated from species of *Staphylococcus* and *Bacillus* is a dominant preference for Glu in the P1 position, the ratio of $k_{cat}/K_m$ for Glu to Asp ranging from approximately 1000 with GluBL to 5300 with GluV8 (Breddam & Meldal, 1992). Otherwise, these enzymes are rather unspecific, although the other subsites exhibit some preferences (Table 79.1).

Forms of glutamyl endopeptidase I, especially GluV8, have found extensive use in the specific fragmentation of proteins prior to amino acid sequencing. The procedure for the use of GluV8 for this purpose usually specifies the use of ammonium bicarbonate buffers, but recent studies have demonstrated that the rate of hydrolysis is 10-fold higher in phosphate buffer, because of the inhibitory effect of bicarbonate, and that the change of buffer does not affect the specificity (Sørensen *et al.*, 1991).

The glutamyl endopeptidases are not irreversibly inhibited by PMSF (Drapeau *et al.*, 1972; Yoshida *et al.*, 1988) and as a consequence the enzyme from *B. subtilis* (GluBS) appears under the name 'metalloprotease' (MPR) in the literature (Rufo *et al.*, 1990; Sloma *et al.*, 1990). However, the sequence identity with GluBL (approx. 52%) combined with the preference for Glu in P1 unambiguously establishes it as one of the glutamyl endopeptidases. Members of both

groups of glutamyl endopeptidases have been found to be irreversibly inhibited by DFP as well as Boc-Leu-Glu-CH$_2$Cl (Drapeau *et al.*, 1972; Kakudo *et al.*, 1992a; Svendsen & Breddam, 1992). However, in contrast to glutamyl endopeptidase II (see Chapter 80), GluV8 is not inhibited by any known protein inhibitors, including the turkey ovomucoid third domain Leu18Glu mutant (Komiyama *et al.*, 1991), crmA (Komiyama *et al.*, 1996) or $\alpha_1$-proteinase inhibitor (Potempa *et al.*, 1986).

The pH dependence of the hydrolysis of fluorogenic substrates with Glu in the P1 position by GluV8 protease shows a bell-shaped profile with an optimum at pH 7.2 (Sørensen *et al.*, 1991). The activity of GluV8 depends on two ionizable groups with apparent $pK_a$ values of 5.8 and 8.4. With chromogenic substrates, i.e. *p*-nitroanilides, the pH optimum was at pH 7 with Asp in P1 and at pH 8 with Glu (Nagata *et al.*, 1991). The pH dependence of GluBL has been found to be quite similar to that of GluV8, although optimal activity is seen at a slightly higher pH (Svendsen & Breddam, 1992).

A number of the glutamyl endopeptidases have been found to be activated by Ca$^{2+}$ (Kakudo *et al.*, 1992a; Svendsen & Breddam, 1992). The GluBL protease has this characteristic, and is very sensitive to the presence of EDTA in the reaction mixture (Rufo *et al.*, 1990). The epidermolytic toxins A and B from *S. aureus* (see Chapter 81) contain a putative Ca$^{2+}$-binding site and the toxicity of these proteins depends on the presence of calcium (Dancer *et al.*, 1990). The role of calcium ions has not yet been established, but it is probably structural as in relatives of subtilisin (Bajorath *et al.*, 1988, 1989; Gros *et al.*, 1991; Betzel *et al.*, 1992; Goddette *et al.*, 1996).

Like the majority of serine proteases, the glutamyl endopeptidases may catalyze transacylation reactions. This has been demonstrated with GluBL in the semisynthesis of the superpotent analog of human growth hormone-releasing factor, [desNH$_2$Tyr$^1$,D-Ala$^2$,Ala$^{15}$]-GRF(1–29)-NH$_2$ (Bongers *et al.*, 1994). Furthermore, it has been demonstrated that GluBL may become saturated with nucleophile and that the yield of the aminolysis reaction under such conditions is less than 100%, indicating that the hydrolysis reaction may proceed even when nucleophile/leaving group is bound within the active site (Rolland-Fulcrand & Breddam, 1993). Thus, hydrolysis appears to proceed via a random ordered bi-bi mechanism rather than a ping-pong mechanism.

## Structural Chemistry

The molecular sizes of the forms of glutamyl endopeptidase I differ significantly, i.e. the mature forms of GluV8, SPase and GluBL contain 268, 289 and 222 amino acids, respectively (Carmona & Gray, 1987; Kakudo *et al.*, 1992a; Svendsen & Breddam, 1992). The primary structures are also widely different, although GluBL exhibits 26% identity to the 220 N-terminal amino acids of GluV8 (Svendsen & Breddam, 1992). Furthermore, the GluV8 protease differs by having a tail of 50 amino acid residues consisting of a unique 12-fold repeat of the tripeptide Pro-Asn/Asp-Asn (one tripeptide has Glu in the third position) (Carmona & Gray, 1987). The SPase protease, which is very similar to GluV8 (91.5% identity), contains seven additional tripeptide repeats (Yoshikawa *et al.*, 1992). GluV8, SPase and GluBL are all synthesized as preproenzymes with prepro segments of 69, 70 and 94

*Table 79.1* Summary of the substrate preferences of the glutamyl endopeptidases

| Enzyme | Preference | P4 | P3 | P2 | P1 | P1' | P2' |
|--------|-----------|------|------|------|------|------|------|
| GluV8 | Preferred | Asp | Val | Phe | Glu | Phe | Ala |
| | Poorly accepted | none | Pro | none | Ala, Phe, Asp | Asp, Pro | Pro |
| GluBL | Preferred | Asp | Ala | Phe | Glu | Val | Phe |
| | Poorly accepted | none | none | none | Ala, Phe, Asp | Asp, Pro | Pro |

Adapted from Breddam & Meldal (1992).

amino acid residues, respectively, each containing a signal peptide and a propeptide (Carmona & Gray, 1987; Kakudo *et al.*, 1992b). The proenzyme is activated by cleavage of either an Asn+Val (GluV8) or a Lys+Ser bond (GluBL) and thus, the activation appears to be controlled by another endogenous protease. The role of the C-terminal tail of the GluV8 protease has not yet been established, and recently it has been found that neither the prepro sequence nor the C-terminal segment are required for enzymatic activity and protein folding (Yabuta *et al.*, 1995). Although the three-dimensional structure has not yet been solved for any of the members of this group of enzymes, some insight into their molecular action may be gained from modeling studies. Such studies carried out on GluV8 and GluBL suggest that these enzymes utilize the same type of interactions as GluSGP (Chapter 80) for recognition of the P1 glutamyl residue, i.e. His213 and Thr190 are believed to interact with the $\gamma$-carboxylate group (Barbosa *et al.*, 1996). In the epidermolytic toxins (Chapter 81), an additional residue, Lys216, is suggested to play a role in recognition of the P1 substituent (Barbosa *et al.*, 1996).

## Preparation

The GluV8 protease is the only form of glutamyl endopeptidase I that is readily available from commercial sources (ICN, Miles Laboratories, Worthington and Sigma) (see Appendix 2 for full names and addresses of suppliers). Although not presently commercially available, GluBL and an acidic amino acid-specific protease from *B. subtilis* can easily be purified in large amounts from commercially available sources, i.e. Alcalase and Protease Type XVI (Sigma), respectively (Niidome *et al.*, 1990; Birktoft & Breddam, 1994). Systems for overexpression of several of the glutamyl endopeptidases have been described, e.g. for GluBL in *B. subtilis* (Kakudo *et al.*, 1992a; Matsumoto *et al.*, 1995), for SPase in *B. subtilis* (Kakudo *et al.*, 1992b) and for GluV8 in *E. coli* (Yabuta *et al.*, 1995).

## Related Peptidases

The epidermolytic toxins A and B from *S. aureus* induce staphylococcal scalded skin syndrome in newborn infants (Dancer *et al.*, 1990). In 1990 it was discovered that these toxins were homologs of GluV8 and that they contain a catalytic triad of the Ser-His-Asp type (Dancer *et al.*, 1990). The importance of the catalytic apparatus in epidermolysis has been confirmed by the fact that replacement of the catalytic serine with a glycine (Redpath *et al.*, 1991) or cysteine (Prevost *et al.*, 1991) as well as chemical modification by DFP (Dancer *et al.*, 1990) leads to a significant reduction in epidermolytic activity. A similar effect is observed when the toxins are treated with EDTA, consistent with the presence of a putative $Ca^{2+}$-binding site in the primary sequence (Dancer *et al.*, 1990). Furthermore, it has been demonstrated that the toxins exhibit esterolytic activity towards Glu-OPh (Bailey & Redpath, 1992). Despite these observations, it has not been established conclusively that proteolytic activity of these toxins is the reason for their toxicity. It has been suggested that the epidermolytic toxins also possess a phospholipase C-like activity (Wiley & Rogolsky, 1985) and recent modeling

studies suggest that the phosphonyl group of the phosphoester may fit into the active site in place of a peptide, and that the negative charge of the phosphonyl group may be stabilized by His213 and Lys216 (Barbosa *et al.*, 1996). It is then believed that the phosphoester bond is cleaved by nucleophilic attack by the catalytic serine in a manner similar to the alkaline phosphatases (Kim & Wyckoff, 1991; Barbosa *et al.*, 1996).

## Further Reading

For a review, see Birktoft & Breddam (1994).

## References

Bailey, C.J. & Redpath, M.B. (1992) The esterolytic activity of epidermolytic toxins. *Biochem. J.* **284**, 177–180.

Bajorath, J., Hinrichs, W. & Saenger, W. (1988) The enzymatic activity of proteinase K is controlled by calcium. *Eur. J. Biochem.* **176**, 441–447.

Bajorath, J., Raghunathan, S., Hinrichs, W. & Saenger, W. (1989) Long-range structural changes in proteinase K triggered by calcium ion removal. *Nature* **337**, 481–484.

Barbosa, J.A.R.G., Saldanha, J. & Garratt, R.C. (1996) Novel features of serine protease active sites and specificity pockets: sequence analysis and modelling studies of glutamate-specific endopeptidases and epidermolytic toxins. *Protein Eng.* **9**, 591–601.

Betzel, C., Klupsch, S., Papendorf, G., Hastrup, S., Branner, S. & Wilson, K.S. (1992) Crystal structure of the alkaline proteinase Savinase from *Bacillus lentus* at 1.4 Å resolution. *J. Mol. Biol.* **223**, 427–445.

Birktoft, J.J. & Breddam, K. (1994) Glutamyl endopeptidases. *Methods Enzymol.* **244**, 114–126.

Bongers, J., Liu, W., Lambros, T., Breddam, K., Campbell, R.M., Felix, A.M. & Heimer, E.P. (1994) Peptide synthesis catalyzed by the Glu/Asp-specific endopeptidase – influence of the ester leaving group of the acyl donor on yield and catalytic efficiency. *Int. J. Pept. Protein Res.* **44**, 123–129.

Breddam, K. & Meldal, M. (1992) Substrate preference of glutamic-acid-specific endopeptidases assessed by synthetic peptide substrates based on intramolecular fluorescence quenching. *Eur. J. Biochem.* **206**, 103–107.

Carmona, C. & Gray, G.L. (1987) Nucleotide sequence of the serine protease gene of *Staphylococcus aureus*, strain V8. *Nucleic Acids Res.* **15**, 6757.

Dancer, S.J., Garratt, R., Saldanha, J., Jhoti, H. & Evans, R. (1990) The epidermolytic toxins are serine proteases. *FEBS Lett.* **268**, 129–132.

Drapeau, G.R., Boily, Y. & Houmard, J. (1972) Purification and properties of an extracellular protease of *Staphylococcus aureus*. *J. Biol. Chem.* **247**, 6720–6726.

Goddette, D.W., Paech, C., Yang, S.S., Mielenz, J.R., Bystroff, C., Wilke, M.E. & Fletterick, R.J. (1996) The crystal structure of the *Bacillus lentus* alkaline protease, subtilisin BL, at 1.4 Å resolution. *J. Mol. Biol.* **228**, 580–595.

Gros, P., Kalk, K.H. & Hol, W.G. (1991) Calcium binding to thermitase. Crystallographic studies of thermitase at 0, 5, and 100 mM calcium. *J. Biol. Chem.* **266**, 2953–2961.

Houmard, J. & Drapeau, G.R. (1972) Staphylococcal protease: a proteolytic enzyme specific for glutamoyl bonds. *Proc. Natl Acad. Sci. USA* **69**, 3506–3509.

Kakudo, S., Kikuchi, N., Kitadokoro, K., Fujiwara, T., Nakamura, E., Okamoto, H., Shin, M., Tamaki, M., Teraoka, H., Tsuzuki, H. &

Yoshida, N. (1992a) Purification, characterization, cloning, and expression of a glutamic acid-specific protease from *Bacillus licheniformis* ATCC 14580. *J. Biol. Chem.* **267**, 23782–23788.

Kakudo, S., Yoshikawa, K., Tamaki, M., Nakamura, E. & Teraoka, H. (1992b) Secretory expression of a glutamic-acid-specific endopeptidase (SPase) from *Staphylococcus aureus* ATCC12600 in *Bacillus subtilis*. *Appl. Microbiol. Biotechnol.* **38**, 226–233.

Kim, E.E. & Wyckoff, H.W. (1991) Reaction mechanism of alkaline phosphatase based on crystal structure. Two-metal ion catalysis. *J. Mol. Biol.* **218**, 449–464.

Komiyama, T., Bigler, T.L., Yoshida, N., Noda, K. & Laskowski, M. (1991) Replacement of P$_1$ Leu[18] by Glu[18] in the reactive site of turkey ovomucoid third domain converts it into a strong inhibitor of Glu-specific *Streptomyces griseus* proteinase (GluSGP). *J. Biol. Chem.* **266**, 10727–10730.

Komiyama, T., Quan, L.T. & Salvesen, G.S. (1996) Inhibition of cysteine and serine proteinases by the cowpox virus serpin CrmA. In: *Intracellular Protein Catabolism* (Suzuki, K. & Bond, J.S., eds). New York: Plenum Press, pp. 173–176.

Matsumoto, K., Mitsushima, K., Tamaki, M., Iwamoto, H., Kakudo, S., Okamoto, H., Tsuzuki, H. & Teraoka, H. (1995) High-level production of recombinant glutamic acid-specific protease from *Bacillus licheniformis* in *Bacillus subtilis* expression system. *J. Ferment. Bioeng.* **79**, 23–27.

Nagata, K., Yoshida, N., Ogata, F., Araki, H. & Noda, K. (1991) Subsite mapping of an acidic amino acid-specific endopeptidase from *Streptomyces griseus*, GluSGP, and protease V8. *J. Biochem. (Tokyo)* **110**, 859–862.

Niidome, T., Yoshida, N., Ogata, F., Ito, A. & Noda, K. (1990) Purification and characterization of an acidic amino acid-specific endopeptidase of *Bacillus subtilis* obtained from a commercial preparation (Protease Type XVI, Sigma). *J. Biochem. (Tokyo)* **108**, 965–970.

Potempa, J., Watorek, W. & Travis, J. (1986) The inactivation of human plasma $\alpha_1$-proteinase inhibitor by proteinases from *Staphylococcus aureus*. *J. Biol. Chem.* **261**, 14330–14334.

Prevost, G., Rifal, S., Chaix, M.L. & Piemont, Y. (1991) Functional evidence that the Ser-195 residue of staphylococcal exfoliative toxin A is essential for biological activity. *Infect. Immun.* **59**, 3337–3339.

Redpath, M.B., Foster, T.J. & Bailey, C.J. (1991) The role of the serine protease active site in the mode of action of epidermolytic toxin of *Staphylococcus aureus*. *FEMS Lett.* **81**, 151–156.

Rolland-Fulcrand, V. & Breddam, K. (1993) The use of a glutamic acid specific endopeptidase in peptide synthesis. *Biocatalysis* **7**, 75–82.

Rufo, G.A.J., Sullivan, B.J., Sloma, A. & Pero, J. (1990) Isolation and characterization of a novel extracellular metalloprotease from *Bacillus subtilis*. *J. Bacteriol.* **172**, 1019–1023.

Sloma, A., Rudolph, C.F., Rufo, G.A.J., Sullivan, B.J., Theriault, K.A., Ally, D. & Pero, J. (1990) Gene encoding a novel extracellular metalloprotease in *Bacillus subtilis*. *J. Bacteriol.* **172**, 1024–1029.

Sørensen, S.B., Sørensen, T.L. & Breddam, K. (1991) Fragmentation of proteins by *S. aureus* strain V8 protease. Ammonium bicarbonate strongly inhibits the enzyme but does not improve the selectivity for glutamic acid. *FEBS Lett.* **294**, 195–197.

Svendsen, I. & Breddam, K. (1992) Isolation and amino acid sequence of a glutamic acid specific endopeptidase from *Bacillus licheniformis*. *Eur. J. Biochem.* **204**, 165–171.

Wiley, B.B. & Rogolsky, M.S. (1985) Manipulation of the extra-chromosomal genetic determinants for exfoliative toxin B and bacteriocin R1 synthesis in phage group II *Staphylococus aureus*. In: *The Staphylococci*, Zbl. Bakt. Suppl. 14 (Jeljaszewicz, J., ed.). New York: Gustav Fisher Verlag, pp. 295–300.

Yabuta, M., Ochi, N. & Ohsuye, K. (1995) Hyperproduction of a recombinant fusion protein of *Staphylococcus aureus* V8 protease in *Escherichia coli* and its processing by OmpT protease to release an active V8 protease derivative. *Appl. Microbiol. Biotechnol.* **44**, 118–125.

Yoshida, N., Tsuruyama, S., Nagata, K., Hirayama, K., Noda, K. & Makisumi, S. (1988) Purification and characterization of an acidic amino acid specific endopeptidase of *Streptomyces griseus* obtained from a commercial preparation (Pronase). *J. Biochem. (Tokyo)* **104**, 451–456.

Yoshikawa, K., Tsuzuki, H., Fujiwara, T., Nakamura, E., Iwamoto, H., Matsumoto, K., Shin, M., Yoshida, N. & Teraoka, H. (1992) Purification, characterization and gene cloning of a novel glutamic acid-specific endopeptidase from *Staphylococcus aureus* ATCC 12600. *Biochim. Biophys. Acta* **1121**, 221–228.

*Henning R. Stennicke*
*Department of Chemistry, Carlsberg Laboratory,*
*Gamle Carlsberg Vej 10, DK-2500 Valby,*
*Copenhagen, Denmark*
*Email: carlprot@biobase.dk*

*Klaus Breddam*
*Department of Chemistry, Carlsberg Laboratory,*
*Gamle Carlsberg Vej 10, DK-2500 Valby,*
*Copenhagen, Denmark*

# 80. Glutamyl endopeptidase II

## Databanks

*Peptidase classification: clan SA, family S2, MEROPS ID: S02.012*
*NC-IUBMB enzyme classification: EC 3.4.21.82*

*Databank codes:*

| Species | SwissProt | PIR | EMBL (cDNA) | EMBL (genomic) |
|---|---|---|---|---|
| *Streptomyces fradiae* | Q03424 | S33321 | D12470 | – |
| *Streptomyces griseus* | – | S18322 S37460 | L28762 | – |

Brookhaven Protein Data Bank three-dimensional structures:

| Species | ID | Resolution | Notes |
|---|---|---|---|
| *Streptomyces griseus* | 1HPG | 1.5 | complex with tetrapeptide ligand |

## Name and History

*Glutamyl endopeptidase II* is represented by two enzymes isolated from *Streptomyces griseus (GluSGP)* (Yoshida *et al.*, 1988; Svendsen *et al.*, 1991) and *Streptomyces fradiae* ATCC 14544 *(GluSF)* (Kitadokoro *et al.*, 1993), respectively. They are closely similar in primary structure (82%) (Kitadokoro *et al.*, 1993). The other bacterial glutamyl endopeptidases, from various strains of *Bacillus* and *Staphylococcus*, are discussed in Chapter 79.

## Activity and Specificity

Glutamyl endopeptidase II is characterized by a less dominant preference for Glu in the P1 position than the glutamyl endopeptidases discussed in Chapter 79, i.e. the Glu to Asp $k_{cat}/K_m$ ratio is only ~10 with GluSF (Kitadokoro *et al.*, 1993) and ~100 with GluSGP, as compared with 5300 for GluV8 (Breddam & Meldal, 1992). With respect to other parts of the substrate, the *Streptomyces* enzymes are rather unspecific, although certain preferences exist (Table 80.1).

Both GluSGP and GluSF have been found to be irreversibly inhibited by DFP and Boc-Phe-Leu-Glu-CH$_2$Cl (Yoshida *et al.*, 1988; Kitadokoro *et al.*, 1993). However, they are not efficiently inhibited by PMSF (Yoshida *et al.*, 1988). Chloromethane derivatives of glutamic and aspartic acid-containing peptides have also been found to inhibit GluSGP (Birktoft & Breddam, 1994).

The pH dependence of $k_{cat}/K_m$ for glutamyl endopeptidase II (GluSGP) has also been thoroughly investigated and it has been found to differ from that of glutamyl endopeptidase I in that activity does not approach zero even at high pH (Stennicke *et al.*, 1996). This indicates that His213, which is involved in the substrate binding, remains capable of donating a hydrogen bond to the substrate even at high pH.

*Table 80.1*  Summary of the substrate preferences of GluSGP

| Enzyme | Preference | P4 | P3 | P2 | P1 | P1' | P2' |
|---|---|---|---|---|---|---|---|
| GluSGP | Preferred | Asp | Val | Pro | Glu | Arg | Ala |
|  | Poorly accepted | Arg | Pro | Asp | Ala, Phe, Asp | Asp, Pro | Pro |

Adapted from Breddam & Meldal (1992).

Furthermore, the pH dependence of $K_m$ for the hydrolysis of Suc-Ala-Ala-Pro-Glu+NHPhNO$_2$ showed that the protonated form of Glu was not accepted as a substrate. This is in agreement with the ability of GluSGP to distinguish between Glu and Gln in the P1 position (Stennicke *et al.*, 1996).

A number of protein inhibitors of GluSGP have been described, e.g. wild-type and mutant turkey ovomucoid third domain, with the Leu18Glu mutant being the most potent ($K_i$ 0.02 nM) (Komiyama *et al.*, 1991), the GluSGP propeptide ($K_i$ 0.5 nM) (H.S. Stennicke, unpublished results), human $\alpha_1$-proteinase inhibitor ($\alpha$1-PI) (Nagata *et al.*, 1991) and an inhibitor from the seeds of bitter gourd ($K_i$ 70 nM) (Ogata *et al.*, 1991). Furthermore, GluSGP has been found to form an SDS-stable complex with the caspase-1 inhibitor, CMA ($K_a > 10^7$ M) (Komiyama *et al.*, 1996).

## Structural Chemistry

With only 187 amino acid residues, mature glutamyl endopeptidase II (GluSGP) is significantly smaller than glutamyl endopeptidase I, and is very different in sequence (Svendsen *et al.*, 1991; Sidhu *et al.*, 1993). It is much closer in sequence to streptogrisins A, B and D (Chapters 77 and 78), also from *S. griseus* (Svendsen *et al.*, 1991; Sidhu *et al.*, 1993), as well as SFase II from *S. fradiae* (Kitadokoro *et al.*, 1993, 1994), SAL and SALO from *S. lividans* (Binnie *et al.*, 1995; Taguchi, 1995), SAM-P20 from *S. albogriseolus* (Taguchi *et al.*, 1995), and the well-characterized $\alpha$-lytic protease (Chapter 82) from *Lysobacter enzymogenes* (Svendsen *et al.*, 1991; Sidhu *et al.*, 1993). Like $\alpha$-lytic protease, both GluSF and GluSGP are synthesized as preproenzymes with large propeptides (139 amino acid residues) (Sidhu *et al.*, 1993), and in both cases, the mature enzyme is generated from the proenzyme by cleavage of a Glu+Val bond (Svendsen *et al.*, 1991; Kitadokoro *et al.*, 1993; Sidhu *et al.*, 1993). This cleavage has been demonstrated to be autocatalytic (Stennicke *et al.*, 1996), again in contrast to the activation of glutamyl endopeptidase I.

GluSGP is the only glutamyl endopeptidase that has thus far been crystallized, and the three-dimensional structure has been solved at 1.5 Å resolution (Nienaber *et al.*, 1993). The overall fold of GluSGP closely resembles those of streptogrisin A (Chapter 78) (Sielecki *et al.*, 1979) and streptogrisin B (Chapter 77) (Read *et al.*, 1983), as well as $\alpha$-lytic protease (Chapter 82) (Fujinaga *et al*., 1985), but it also resembles the pancreatic-type serine proteases (Birktoft & Blow, 1972; Kraut, 1977). Like trypsin and related enzymes it is composed of two $\beta$ barrel cylindrical structures and a C-terminal $\alpha$ helix. The first $\beta$ barrel contains His57 and Asp102 of the catalytic triad (chymotrypsinogen numbering), and the second contains Ser195 of the catalytic triad and the residues constituting the S1 specificity pocket. The crystal structure of GluSGP with Boc-Ala-Ala-Pro-Glu bound in the active site revealed that the mode of recognition of Glu differs from that seen with other charge-specific proteases (Nienaber *et al*., 1993). Thus, no obvious counter charge for the Glu is found in GluSGP, in contrast to the Asp-specific caspase-1 (Chapter 248), where there are two Arg residues (Wilson *et al*., 1994). In GluSGP the side chain of the P1 Glu residue interacts with a histidine residue (His213) and two serine residues (Ser192 and Ser216). Furthermore, His213 appears to interact with His199 which in turn interacts with His228. However, only His199 plays a significant role in substrate recognition, by orienting the imidazole side chain of His213 relative to the P1 glutamate (Stennicke *et al*., 1996). Among the serines, it has been demonstrated that only Ser192 is very important for substrate recognition, Ser216 playing only a minor role in ground state stabilization (Stennicke *et al*., 1996).

## Preparation

Although not presently commercially available, glutamyl endopeptidase II can easily be purified from the commercially available Pronase (Yoshida *et al*., 1988; Svendsen *et al*., 1991). Systems for overexpression of GluSGP in *B. subtilis* (Sidhu *et al*., 1993; Stennicke *et al*., 1996) and *S. lividans* (Suzuki *et al*., 1994) have been described.

## Further Reading

For a review, see Birktoft & Breddam (1994).

## References

Binnie, C., Liao, L., Walczyk, E. & Malek, L.T. (1996) Isolation and characterization of a gene encoding a chymotrypsin-like serine protease from *Streptomyces lividans* 66. *Can. J. Microbiol.* **42**, 284–288.

Birktoft, J.J. & Blow, D.M. (1972) Structure of crystalline $\alpha$-chymotrypsin. V. The atomic structure of tosyl-$\alpha$-chymotrypsin at 2 Å resolution. *J. Mol. Biol.* **68**, 187–240.

Birktoft, J.J. & Breddam, K. (1994) Glutamyl endopeptidases. *Methods Enzymol.* **244**, 114–126.

Breddam, K. & Meldal, M. (1992) Substrate preference of glutamic-acid-specific endopeptidases assessed by synthetic peptide substrates based on intramolecular fluorescence quenching. *Eur. J. Biochem.* **206**, 103–107.

Fujinaga, M., Delbaere, L.T., Brayer, G.D. & James, M.N. (1985) Refined structure of alpha-lytic protease at 1.7 Å resolution. Analysis of hydrogen bonding and solvent structure. *J. Mol. Biol.* **184**, 479–502.

Kitadokoro, K., Nakamura, E., Tamaki, M., Horii, T., Okamoto, H., Shin, M., Sato, T., Fujiwara, T., Tsuzuki, H., Yoshida, N. & Teraoka, H. (1993) Purification, characterization and molecular

cloning of an acidic amino-acid specific proteinase from *Streptomyces fradiae* ATCC 14544. *Biochim. Biophys. Acta* **1163**, 149–157.

Kitadokoro, K., Tsuzuki, H., Okamoto, H. & Sato, T. (1994) Crystal structure analysis of a serine proteinase from *Streptomyces fradiae* at 0.16-nm resolution and molecular modeling of an acidic-amino-acid-specific proteinase. *Eur. J. Biochem.* **224**, 735–742.

Komiyama, T., Bigler, T.L., Yoshida, N., Noda, K. & Laskowski, M. (1991) Replacement of $P_1$ Leu[18] by Glu[18] in the reactive site of turkey ovomucoid third domain converts it into a strong inhibitor of Glu-specific *Streptomyces griseus* proteinase (GluSGP). *J. Biol. Chem.* **266**, 10727–10730.

Komiyama, T., Quan, L.T. & Salvesen, G.S. (1996) Inhibition of cystein and serine proteinases by the cowpox virus serpin CrmA. In: *Intracellular Protein Catabolism* (Suzuki, K. & Bond, J.S., eds). New York: Plenum Press, pp. 173–176.

Kraut, J. (1977) Serine proteases: structure and mechanism of catalysis. *Annu. Rev. Biochem.* **46**, 331–358.

Nagata, K., Yoshida, N., Ogata, F., Araki, H. & Noda, K. (1991) Subsite mapping of an acidic amino acid-specific endopeptidase from *Streptomyces griseus*, GluSGP, and protease V8. *J. Biochem. (Tokyo)* **110**, 859–862.

Nienaber, V.L., Breddam, K. & Birktoft, J.J. (1993) A glutamic acid specific serine protease utilizes a novel histidine triad in substrate binding. *Biochemistry* **32**, 11469–11475.

Ogata, F., Miyata, T., Fujii, N., Yoshida, N., Noda, K., Makisumi, S. & Ito, A. (1991) Purification and amino acid sequence of a bitter gourd inhibitor against acidic amino acid-specific endopeptidase of *Streptomyces griseus*. *J. Biol. Chem.* **266**, 16715–16721.

Read, R.J., Fujinaga, M., Sielecki, A.R. & James, M.N. (1983) Structure of the complex of *Streptomyces griseus* protease B and the third domain of turkey ovomucoid inhibitor at 1.8-Å resolution. *Biochemistry* **22**, 4420–4433.

Sidhu, S.S., Kalmar, G.B. & Borgford, T.J. (1993) Characterization of the gene encoding the glutamic-acid-specific protease of *Streptomyces griseus*. *Biochem. Cell Biol.* **71**, 454–461.

Sielecki, A.R., Hendrickson, W.A., Broughton, C.G., Delbaere, L.T., Brayer, G.D. & James, M.N. (1979) Protein structure refinement: *Streptomyces griseus* serine protease A at 1.8-Å resolution. *J. Mol. Biol.* **134**, 781–804.

Stennicke, H.R., Birktoft, J.J. & Breddam, K. (1996) Characterization of the $S_1$ binding site glutamic acid specific protease from *Streptomyces griseus*. *Protein Sci.* **5**, 2266–2275.

Suzuki, Y., Yabuta, M. & Ohsuye, K. (1994) Cloning and expression of the gene encoding the glutamic acid-specific protease of *Streptomyces griseus* ATCC10137. *Gene* **150**, 149–151.

Svendsen, I., Jensen, M.R. & Breddam, K. (1991) The primary structure of the glutamic acid-specific protease of *Streptomyces griseus*. *FEBS Lett.* **292**, 165–167.

Taguchi, H. (1995) Molecular cloning and sequence analysis of a DNA sequence encoding an extracellular serine protease from *Streptomyces lividans* 66. *Genebank* Accession number: D50081.

Taguchi, S., Odaka, A., Watanabe, Y. & Momose, H. (1995) Molecular characterization of a gene encoding extracellular serine protease isolated from a subtilisin inhibitor-deficient mutant of *Streptomyces albogriseolus* S-3253. *Appl. Environ. Microbiol.* **61**, 180–186.

Wilson, K.P., Black, J.-A.F., Thomson, J.A., Kim, E.E., Griffith, J.P., Navia, M.A., Murcko, M.A., Chambers, S.P., Aldape, R.A., Raybuck, S.A. & Livingston, D.J. (1994) Structure and mechanism of interleukin-1$\beta$ converting enzyme. *Nature* **370**, 270–275.

Yoshida, N., Tsuruyama, S., Nagata, K., Hirayama, K., Noda, K. & Makisumi, S. (1988) Purification and characterization of an acidic amino acid specific endopeptidase of *Streptomyces griseus*

obtained from a commercial preparation (Pronase). *J. Biochem. (Tokyo)* **104**, 451–456.

*Henning R. Stennicke*
*Department of Chemistry, Carlsberg Laboratory,*
*Gamle Carlsberg Vej 10, DK-2500 Valby,*
*Copenhagen, Denmark*
*Email: carlprot@biobase.dk*

*Klaus Breddam*
*Department of Chemistry, Carlsberg Laboratory,*
*Gamle Carlsberg Vej 10, DK-2500 Valby,*
*Copenhagen, Denmark*

# 81. *Exfoliative toxin A*

## Databanks

*Peptidase classification: clan SA, family S2, MEROPS ID: S02.022*
*NC-IUBMB enzyme classification: none*
*Databank codes:*

| Species | SwissProt | PIR | EMBL (cDNA) | EMBL (genomic) |
|---|---|---|---|---|
| Exfoliative toxin A | | | | |
| *Staphylococcus aureus* | P09331 | A26680 | M17347 M17357 | – |
| Exfoliative toxin B | | | | |
| *Staphylococcus aureus* | P09332 | A26050 B26680 | M17348 | – |

Brookhaven Protein Data Bank three-dimensional structures:

| Species | ID | Resolution |
|---|---|---|
| Exfoliative toxin A | | |
| *Staphylococcus aureus* | 1EXF 2EXF | |

## Name and History

***Exfoliative toxin A (ETA)*** was first purified and identified as the causative agent in staphylococcal scalded skin syndrome in 1972 (Melish *et al.*, 1972). Scalded skin syndrome involves sloughing of the skin that may encompass one-half or more of the total epidermis, hence the name of the toxin. Another closely related toxin from *Staphylococcus aureus* with similar biological effects was purified in 1974 and subsequently named exfoliative toxin B (ETB) (Kondo *et al.*, 1974). Other synonyms for the exfoliative toxins include ***exfoliatins, epidermolysins*** and ***epidermolytic toxins***.

## Activity and Specificity

Purified ETA does not demonstrate proteolytic activity *in vitro*, and protease inhibitors do not appear to inhibit biological activity completely. Nevertheless, data suggesting that ETA is a serine protease include the following observations: (a) ETA is 25% identical with staphylococcal V8 protease (glutamyl endopeptidase I) (Chapter 79); (b) application of DFP and PMSF delay the onset of exfoliation (Dancer *et al.*, 1990); (c) radiolabeled DFP reacts, although incompletely, with Ser195 (Bailey *et al.*, 1995); (d) mutation of any residue in the proposed catalytic triad, i.e. His72, Asp102 or Ser195 (ETA numbering), abolishes exfoliation (Redpath *et al.*, 1991; Prévost *et al.*, 1991, 1992; Vath *et al.*, 1997); and (e) the three-dimensional structure of ETA shows a serine protease fold (Vath *et al.*, 1997).

Exfoliative toxin activity is assayed by subcutaneous injection into newborn mice. Gentle stroking of the neck gives rise to permanent wrinkling of the skin referred to as a positive Nikolsky sign. A positive Nikolsky sign is obtained using microgram quantities of ETA 30 min to 3 h after injection.

## Structural Chemistry

The mature protein of ETA is 242 amino acids after its 38 residue signal sequence is cleaved, and has a molecular mass of 30 kDa. The crystal structure of ETA has recently been reported (Vath *et al.*, 1997). ETA has a fold that is very similar to that of the chymotrypsin family of serine proteases with His72, Asp102 and Ser195 forming the presumed catalytic triad. ETA appears to be more closely related structurally to the mammalian serine proteases than are their somewhat smaller microbial counterparts. An additional N-terminal domain which includes a 15 residue charged α helix is a unique feature of ETA and may be important for biological function. Based on the conformation of residues in the S1 binding pocket, including Thr190 and His210 (190 and 213 respectively in chymotrypsinogen numbering), ETA appears to recognize glutamic acid in the P1 position analogous to glutamyl endopeptidase II (Chapter 80) (Nienaber *et al.*, 1993). Unlike glutamyl endopeptidase II, which apparently utilizes a triad of histidine residues to neutralize the charge of glutamic acid, Lys213 (216 in chymotrypsin) of ETA appears to provide the counter ion. Reports have suggested that exfoliation by ETA involves a metal ion, and it has been proposed that the ETA molecule contains a calcium-binding site (Dancer *et al.*, 1990), but no such site was apparent in the crystal structure.

The reason why purified ETA does not demonstrate protease activity may be that the oxyanion hole is obstructed by the main-chain carbonyl oxygen of Pro192 (Figure 81.1). The conformation of the peptide bond between Pro192 and Gly193 is flipped approximately 180° relative to known active serine proteases for which structures are available. The carbonyl oxygen of Pro192 forms two hydrogen bonds with Ser195 (one main chain and one side chain). This presumably inactive conformation is further stabilized by a hydrogen bond between the side chain of Asp164 (144 in chymotrypsinogen) of the adjacent loop (loop D, residues 163–168 in ETA) and the main-chain nitrogen of Gly193. The conformation of loop D may then control whether ETA is able to bind substrate and is active. The binding of a specific receptor possibly involving the N-terminal domain may regulate activity by its effect on the conformation of loop D.

## Preparation

ETA was cloned and expressed in *Escherichia coli* (Lee *et al.*, 1987; O'Toole & Foster, 1987; Sakurai *et al.*, 1988). ETA was also expressed in overproducing strains of *S. aureus* isolated from affected patients (Melish *et al.*, 1972; Dancer *et al.*, 1990; Bailey *et al.*, 1995) as well as in recombinant vectors containing the gene in *S. aureus* (Prévost *et al.*, 1991; Redpath *et al.*, 1991; Vath *et al.*, 1997). ETA has been purified using isoelectric focusing (Bailey *et al.*, 1995;

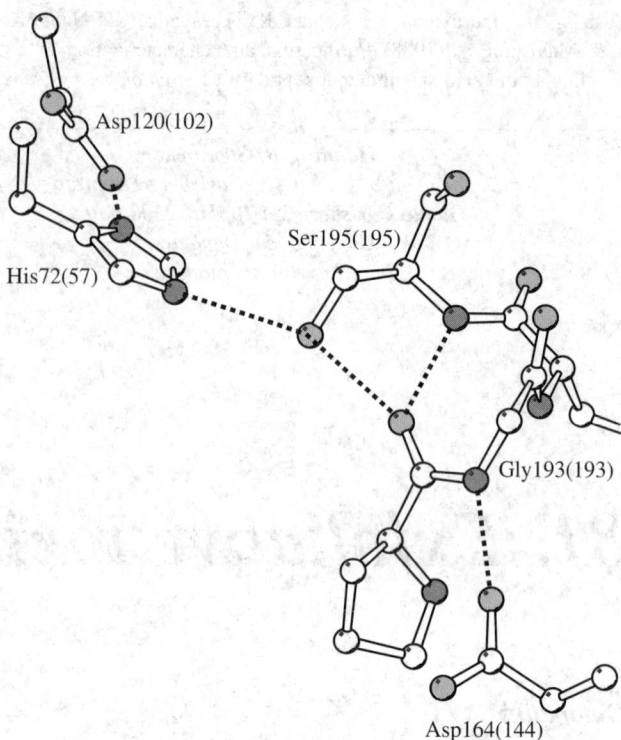

*Figure 81.1* The catalytic site of exfoliative toxin A, identifying the carbonyl oxygen of Pro192 that blocks access to the oxyanion hole. Oxygen atoms are in light grey, nitrogen atoms in dark grey, and carbon atoms in white. Hydrogen bonds are indicated as dashed lines. Numbering is based on that of ETA, with that of chymotrypsinogen in parentheses.

Vath *et al.*, 1997) and standard ion-exchange chromatography (Dancer *et al.*, 1990; Prévost *et al.*, 1991; Redpath *et al.*, 1991).

## Biological Aspects

ETA and ETB are the causative agents in staphylococcal scalded skin syndrome. This disease primarily affects neonates and is characterized by a specific intra-epidermal splitting of layers of the skin. The gene encoding ETA is chromosomal while the gene encoding ETB is located on a plasmid. In addition to exfoliation, ETA has also demonstrated superantigenic activity. A superantigen has the ability to stimulate the expansion of T lymphocytes containing a particular Vβ region in their T cell receptors (Marrack & Kappler, 1990). ETA has been shown to expand human Vβ2$^+$ T cells (Marrack & Kappler, 1990). The exfoliation caused by ETA appears to depend on protease activity while the superantigenic activity does not, such that mutation of Ser195 to a cysteine abolishes exfoliation but not T lymphocyte proliferation (Vath *et al.*, 1997). (There is further discussion of the mechanism of the biological activity of the exfoliative toxins in Chapter 79 – *editors*.)

## Distinguishing Features

Polyclonal antibodies specific for ETA or ETB produced in rabbits do not cross-react, but will neutralize the toxicity of their respective toxins. Purified ETA is available through Toxin Technology (see Appendix 2 for full names and addresses of suppliers).

## Related Peptidases

The mature protein of ETB is 246 amino acids after cleavage of its 31 residue signal sequence. ETB is 40% identical with ETA including the putative catalytic site.

## Further Reading

Recent review articles are those of Gemmell (1995) and Iandolo & Chapes (1997).

## References

Bailey, C.J., Lockhart, B.P., Redpath, M.B. & Smith, T.P. (1995) The epidermolytic (exfoliative) toxins of *Staphylococcus aureus*. *Med. Microbiol. Immunol.* **184**, 53–61.

Dancer, S.J., Garratt, R., Saldanha, J., Jhoti, H. & Evans, R. (1990) The epidermolytic toxins are serine proteases. *FEBS Lett.* **268**, 129–132.

Gemmell, C.G. (1995) Staphylococcal scalded skin syndrome. *J. Med. Microbiol.* **43**, 318–327.

Iandolo, J.J. & Chapes, S.K. (1997) The exfoliative toxins of *Staphylococcus aureus*. In: *Superantigens: Molecular Biology, Immunology, and Relevance to Human Disease* (Leung, D.Y.M., Huber, B.T. & Schlievert, P.M., eds). New York: Marcel Dekker, pp. 231–255.

Kondo, I., Sakurai, S. & Sarai, Y. (1974) New type of exfoliatin obtained from staphylococcal strains, belonging to phage groups other than group II, isolated form patients with impetigo and Ritter's disease. *Infect. Immun.* **10**, 851–861.

Lee, C.Y., Schmidt, J.J., Johnson-Winegar, A.D., Spero, L. & Iandolo, J.J. (1987) Sequence determination and comparison of the

exfoliative toxin A and toxin B genes from *Staphylococcus aureus*. *J. Bacteriol.* **169**, 3904–3909.

Marrack, P. & Kappler, J. (1990) The staphylococcal enterotoxins and their relatives. *Science* **248**, 705–711.

Melish, M.F., Glasgow, L.A. & Turner, M.D. (1972) The staphylococcal scalded-skin syndrome: isolation and partial characterization of the exfoliative toxin. *J. Infect. Dis.* **125**, 129–140.

Nienaber, V.L., Breddam, K. & Birktoft, J.J. (1993) A glutamic acid specific serine protease utilizes a novel histidine triad in substrate binding. *Biochemistry* **32**, 11469–11475.

O'Toole, P.W. & Foster, T.J. (1987) Nucleotide sequence of the epidermolytic toxin A gene of *Staphylococcus aureus*. *J. Bacteriol.* **169**, 3910–3915.

Prévost, G., Rifai, S., Chaix, M.L. & Piémont, Y. (1991) Functional evidence that the Ser-195 residue of staphylococcal exfoliative toxin A is essential for biological activity. *Infect. Immun.* **59**, 3337–3339.

Prévost, G., Rifai, S., Chaix, M.L., Meyer S. & Piémont, Y. (1992) Is the His72, Asp120, Ser195 triad constitutive of the catalytic site of staphylococcal exfoliative toxin A? In: *Bacterial Protein Toxins* (Witholt, B., Alouf, J.E., Boulnois, G.J., Cossart, P., Dijkstra, B.W., Falmagne, P., Fehrenbach, F.J., Freer, J., Niemann, H., Rappuoli, R., Wadstrom, T., eds). Stuttgart: Fischer, pp. 488–489.

Redpath, M.B., Foster, T.J. & Bailey, C.J. (1991) The role of the serine protease active site in the mode of action of epidermolytic toxin of *Staphylococcus aureus*. *FEMS Microbiol. Lett.* **81**, 151–156.

Sakurai, S., Suzaki, H. & Kondo, I. (1988) DNA sequencing of the eta gene coding for staphylococcal exfoliatin toxin serotype A. *J. Gen. Microbiol.* **134**, 711–717.

Vath, G.M., Earhart, C.A., Rago, J.V., Kim, M.H., Bohach, G.A., Schlievert, P.M & Ohlendorf, D.H. (1997) The structure of the superantigen exfoliative toxin A suggests a novel regulation as a serine protease. *Biochemistry* **36**, 1559–1566.

*Gregory M. Vath*
Department of Biochemistry,
University of Minnesota
Medical School,
4-225 Millard Hall,
435 Delaware Street S.E.,
Minneapolis, MN 55455, USA
Email: VATH@dcmit.med.umn.edu

*Patrick M. Schlievert*
Department of Microbiology,
University of Minnesota
Medical School,
4-225 Millard Hall,
435 Delaware Street S.E.,
Minneapolis, MN 55455, USA

*Douglas H. Ohlendorf*
Department of Biochemistry,
University of Minnesota
Medical School,
4-225 Millard Hall,
435 Delaware Street S.E.,
Minneapolis, MN 55455, USA
Email: ohlen@dccc.med.umn.edu

# 82. α-Lytic protease

## Databanks

*Peptidase classification: clan SA, family S2, MEROPS ID: S02.014*
*NC-IUBMB enzyme classification: EC 3.4.21.12*
*Chemical Abstracts Service registry number: 37288-76-9*

*Databank codes:*

| Species | SwissProt | PIR | EMBL (cDNA) | EMBL (genomic) |
|---|---|---|---|---|
| *Achromobacter lyticus* | P27459 | – | – | – |
| *Lysobacter enzymogenes* | P00778 | A31772 | J04052 M22763 | – |

Brookhaven Protein Data Bank three-dimensional structures:

| Species | ID | Resolution | Notes |
|---|---|---|---|
| *Lysobacter enzymogenes* | 1GBA | 2.15 | Met190Ala and Gly216Ala mutant |
| | 1GBB | 2.15 | Met190Ala and Gly216Ala mutant; complex with MeOSuc-Ala-Ala-Pro-Ala |
| | 1GBC | 2.2 | Met190Ala and Gly216Ala mutant; complex with MeOSuc-Ala-Ala-Pro-Leu |
| | 1GBD | 2.2 | Met190Ala and Gly216Ala mutant; complex with MeOSuc-Ala-Ala-Pro-Phe |
| | 1GBE | 2.3 | Met190Ala and Gly216Leu mutant |
| | 1GBF | 2.15 | Met190Ala and Gly216Leu mutant; complex with MeOSuc-Ala-Ala-Pro-Ala |
| | 1GBH | 2.2 | Met190Ala and Gly216Leu mutant; complex with MeOSuc-Ala-Ala-Pro-Leu |
| | 1GBI | 2.3 | Met190Ala and Gly216Leu mutant; complex with MeOSuc-Ala-Ala-Pro-Phe |
| | 1GBJ | 2 | Met190Ala mutant |
| | 1GBK | 2.15 | Met190Ala mutant; complex with MeOSuc-Ala-Ala-Pro-Ala boronic acid |
| | 1GBL | 2.15 | Met190Ala mutant; complex with MeOSuc-Ala-Ala-Pro-Leu boronic acid |
| | 1GBM | 2.28 | Met190Ala mutant; complex with MeOSuc-Ala-Ala-Pro-Phe boronic acid |
| | 1P01 | 2 | complex with Boc-Ala-Pro-Val boronic acid |
| | 1P02 | 2 | complex with MeOSuc-Ala-Ala-Pro-Ala boronic acid |
| | 1P03 | 2.15 | complex with MeOSuc-Ala-Ala-Pro-Val boronic acid |
| | 1P04 | 2.55 | complex with MeOSuc-Ala-Ala-Pro-Ile boronic acid |
| | 1P05 | 2.1 | complex with MeOSuc-Ala-Ala-Pro-Nle boronic acid |
| | 1P06 | 2.34 | complex with MeOSuc-Ala-Ala-Pro-Phe boronic acid |
| | 1P09 | 2.2 | Met213Ala mutant |
| | 1P10 | 2.25 | Met213Ala mutant; complex with MeOSuc-Ala-Ala-Pro-Val boronic acid |
| | 1P11 | 1.93 | complex with $N$-[(2$S$)-2-(phenoxy(1-$R$-($N$-T-Boc-L-Ala-Pro)-1-amino-2-methyl-propyl)-phosphinyloxy)-propanoyl]-L-Ala methyl ester |
| | 1P12 | 1.9 | complex with $N$-[(2$S$)-2-(phenoxy(1-$R$-($N$-T-Boc-oxycarbonyl-L-Ala-Pro)-1-amino-2-methylpropyl)-phosphinyloxy)-propanoyl]-L-Ala methyl ester |
| | 1TAL | 1.5 | single structure model |
| | 2ALP | 1.7 | |
| | 2LPR | 2.25 | Met192Ala mutant; complex with MeOSuc-Ala-Ala-Pro-Val boronic acid |
| | 3LPR | 2.15 | Met213Ala mutant; complex with MeOSuc-Ala-Ala-Pro-Nle boronic acid |
| | 5LPR | 2.13 | Met213Ala mutant; complex with MeOSuc-Ala-Ala-Pro-Ala boronic acid |
| | 6LPR | 2.1 | Met213Ala mutant; complex with MeOSuc-Ala-Ala-Pro-Nle boronic acid |
| | 7LPR | 2.05 | Met213Ala mutant; complex with MeOSuc-Ala-Ala-Pro-Leu boronic acid |
| | 8LPR | 2.25 | Met213Ala mutant; complex with MeOSuc-Ala-Ala-Pro-Phe boronic acid |
| | 9LPR | 2.2 | complex with MeOSuc-Ala-Ala-Pro-Leu boronic acid |

## Name and History

*α-Lytic protease (αLP)* was discovered during investigations of *Lysobacter enzymogenes*, a soil bacterium with the unusual property of strongly attracting nematodes (Katznelson *et al.*, 1964). *L. enzymogenes* was initially classified incorrectly as Myxobacter 495, genus *Sorangium*, and early papers on αLP refer to it as coming from either Myxobacter 495 or from *Sorangium* sp. *L. enzymogenes* is capable of lysing a variety of soil microorganisms including yeast, nematodes and bacteria (Gillespie & Cook, 1965). Two extracellular proteases were found to be responsible for most of the lytic activity, and were designated the *α-* and *β*-lytic proteases (Whitaker, 1965). Inhibition with DFP indicated that αLP is a serine protease, and peptide sequencing of the DFP-labeled, acid-digested fragments showed homology to the pancreatic serine proteases, making it their first-known bacterial homolog (Whitaker & Roy, 1967). The three-dimensional structure of αLP was determined by Brayer *et al.* (1979).

## Activity and Specificity

αLP is an endopeptidase that has a strong preference for cleaving after small, hydrophobic residues (Kaplan & Whitaker, 1969). While the dominant specificity determinant is at the P1 position, where Ala is favored, Pro is preferred at the P2 position, and Ala at the P1' position. According to Bauer *et al.* (1981), up to six substrate residues (P4–P2') can contribute to binding, but more recent evidence suggests that favorable interactions extend as far as P4' (Schellenberger *et al.*, 1994). The pH optimum for catalytic activity is 8.5 (Gillespie & Cook, 1965).

Activity assays have utilized the natural mucopeptide substrates (Tsai *et al.*, 1965), protein substrates (Gillespie & Cook, 1965), peptide esters (Kaplan & Whitaker, 1969), and peptide amides (Bauer *et al.*, 1981). The most convenient assays, however, are performed using peptide-*p*-nitroanilide chromogenic substrates such as Ac-Ala-Pro-Ala╂NHPhNO₂ (Hunkapiller *et al.*, 1976). A rapid mini-purification and kinetic characterization for screening large numbers of mutants has been described (Mace *et al.*, 1995). Fluorogenic peptide-aminomethylcoumarin substrates such as MeO-Suc-Ala-Ala-Pro-Ala╂NHMec have also been used in activity assays (Kettner *et al.*, 1988).

αLP is inhibited by standard serine protease inhibitors, including peptide aldehydes, ketones and sulfonyl fluorides. PMSF works poorly compared to 4-(2-aminoethyl)benzene-sulfonyl fluoride (Rader, 1996), presumably due to the small S1 binding pocket of αLP. αLP is also unusual in not being stably inhibited by dichloroisocoumarin, a potent inhibitor and active site titrant of other serine proteases (Harper *et al.*, 1985). A (terpyridine)platinum chromophore has been used in spectroscopic studies. Surprisingly, it is only a weak inhibitor despite being bound to the active-site histidine (Brothers & Kostic, 1990). Peptide-boronic acid inhibitors are extremely effective and have been used extensively in the structural characterization of αLP specificity (Bone *et al.*, 1987; Kettner *et al.*, 1988) (see also below). αLP is also very strongly inhibited by its 166 residue pro region, which inhibits by direct steric occlusion of the active site (Baker *et al.*, 1992a; Sohl *et al.*, 1997). Unlike several

homologous serine proteases, αLP is not inhibited by ecotin or aprotinin.

The reaction mechanism of αLP has been studied in detail. The enzyme has been used as a model system for serine proteases because its single active-site His36(57) makes it particularly amenable to spectroscopic analysis (Hunkapiller *et al.*, 1973, 1976; Bachovchin *et al.*, 1981). (Note: Amino acid numbering is sequential, with homology numbering according to Fujinaga *et al.* (1985) in parentheses.) Crystal structures of αLP with high-affinity peptide boronic acid inhibitors, and kinetic studies with the corresponding substrates, demonstrated convincingly that αLP provides a template for the tetrahedral transition state or nearby tetrahedral intermediates (Bone *et al.*, 1987; Mace & Agard, 1995). Furthermore, these crystal structures have shown that His36(57) is well positioned to extract the proton from the active-site Ser143(195), and is ideally positioned to donate a proton to the leaving group, thus allowing it to contribute to both acylation and deacylation. While the structures of the peptide boronic acid complexes most closely resemble the deacylation transition state, structural studies using a peptide phosphonate reveal features of the acylation transition state or nearby intermediates. Surprisingly, in these experiments the conformations of active-site residues and P1–P4 residues are identical. These studies also indicate that a negative charge on the substrate/inhibitor in the 'transition state' is required for the His to remain in the active site (Bone *et al.*, 1991b). NMR studies have confirmed the presence of a strong hydrogen bond between His36(57) and Ser143(195), supporting the idea that His36(57) polarizes the serine hydroxyl, increasing its nucleophilicity (Bachovchin, 1986). Solid-state NMR of αLP crystals demonstrated that this hydrogen bond had not been observed in crystal structures because the histidine p$K_a$ is increased to 7.9 due to the high concentration of sulfate anions in the crystals (Smith *et al.*, 1989). Interestingly, crystal structures of complexes indicate that peptide boronic acid inhibitors corresponding to poor substrates (with P1 = Phe or D-Val) form unusual trigonal covalent complexes with Ser143(195) in which the His36(57) N$^{\varepsilon 2}$ also makes a dative bond directly to the boron (Bone *et al.*, 1989a). Similar interactions between the His and the boron have been detected by NMR (Bachovchin *et al.*, 1988).

Although the specificity at the P1 position is determined primarily by the hydrophobic side chains that form the S1 binding pocket, Met138(190), Met158(213), and Val163(218) (Brayer *et al.*, 1979), mutagenetic studies have shown that specificity is actually a distributed property of the enzyme, and residues up to 21 Å from the active-site Ser contribute to specificity (Mace *et al.*, 1995). Furthermore, the crucial importance of protein dynamics (the available dynamic modes and ensembles of conformations) to the determination of specificity has been revealed by binding-site mutations (Bone *et al.*, 1989a,b; Mace & Agard, 1995; Mace *et al.*, 1995) and structures of complexes with peptide boronic acid inhibitors (Bone *et al.*, 1989b, 1991a; Mace & Agard, 1995). Multiple-conformation refinement of the structure at 120 K indicates the existence of conformational substrates in residues around the active site and supports the view that substrate binding makes use of compatible enzyme conformations selected out of a pre-existing ensemble (Rader & Agard, 1997). This is also corroborated by NMR studies of protein dynamics (J. Davis, unpublished results; S. Rader, unpublished results).

Activity has been increased by certain mutations at the active site, e.g. Met138(190)Ala (Bone *et al.*, 1989b), and extensive mutagenetic studies have resulted in a variety of altered specificity profiles and rates (Bone *et al.*, 1989b; Epstein & Abeles, 1992; Graham *et al.*, 1993; Mace & Agard, 1995; Mace *et al.*, 1995).

## Structural Chemistry

Mature αLP is a single chain of 198 residues with three sequential disulfide bonds (Olson *et al.*, 1970). Its amino acid composition is unusual in that it has a high alanine content (24 residues) and a skewed ratio of arginine (12) to lysine (2) (Jurášek & Whitaker, 1967). It has a molecular mass of 19.8 kDa, and a pI of 11, and it is formed by proteolytic cleavage of a 397 residue preproenzyme (Silen *et al.*, 1988). The initial crystal structure revealed a very similar fold to the pancreatic serine proteases (Brayer *et al.*, 1979) (Figure 82.1). Subsequently, the structure refined at 1.7 Å suggested that the structural rigidity, manifested in the low B factors (mean B factor = 14.3 Å$^2$) and high thermal stability (Kaplan *et al.*, 1970), is due to extensive hydrogen bonding (Fujinaga *et al.*, 1985). Finally, low-temperature (120 K) crystallography reveals an unusually small thermal expansion coefficient which is consistent with a very rigid structure (Rader & Agard, 1997). Many crystal structures of mutants and complexes with inhibitors (both boronate and phosphonate), as well as the low-temperature structure, are available from the Brookhaven Protein Data Bank. Structures of the pro region and the complex between the pro region and αLP have recently been determined in the authors' laboratory.

The complete assignment of αLP's backbone resonances by NMR has been made (J. Davis, unpublished results), and

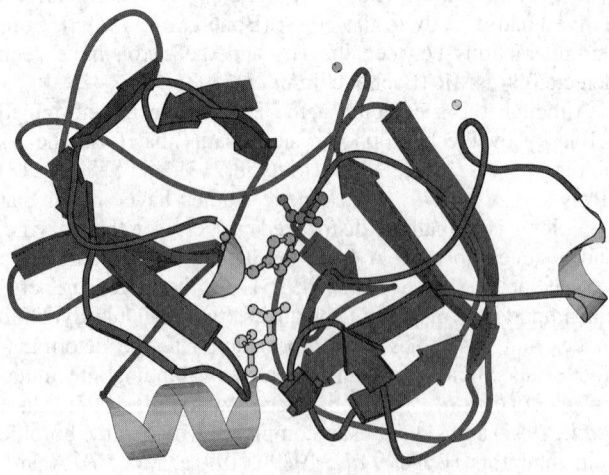

*Figure 82.1* Richardson diagram of *Lysobacter enzymogenes* α-lytic protease. The image was prepared from the Brookhaven Protein Data Bank entry (2ALP) as described in the Introduction (p. xxv). Catalytic residues are shown in ball-and-stick representation: His59, Asp102 and Ser195 (numbering as in Alignment 76.1, see CD-ROM).

the solution structure is in progress (D.A. Agard, unpublished results). Backbone dynamics of the free enzyme and of the inhibitory complex with the peptide boronic acid Boc-Ala-Pro-boroVal have been determined by solution NMR (Davis *et al.*, 1998).

## Preparation

α-Lytic protease is easily purified by cation-exchange chromatography from the culture medium of *L. enzymogenes* (Hunkapiller *et al.*, 1973), routinely yielding 250 mg liter$^{-1}$ when grown in rich medium or 100 mg liter$^{-1}$ of isotopically labeled material when grown in defined media for NMR (Davis *et al.*, 1998). Several expression systems have been employed for heterologous production in *E. coli* (Silen & Agard, 1989; Epstein & Abeles, 1992; Graham *et al.*, 1993; Mace *et al.*, 1995). Protease production is absolutely dependent on the coexpression of the pro region, either in the precursor or as a separate polypeptide chain. *In vivo* formation of the three disulfide bonds requires that αLP and its pro region be targeted into the periplasmic space. The highest yields of wild-type and mutant αLP from *E. coli* (up to 50 mg liter$^{-1}$) have been obtained from the proprotease precursor with an isopropyl β-D-thiogalactopyronoside (IPTG)-inducible T7 $P_{A1}$ early-gene promoter and the signal sequence from PhoA (Mace *et al.*, 1995). Although the time-course of production is slow, 3–5 days, mature protease is secreted into the growth medium, considerably simplifying the purification (Silen *et al.*, 1989). The protein can be purified from the culture supernatant to a single band on a Coomassie-stained gel either with a mini-purification for rapid screening of mutants (Mace *et al.*, 1995) or on a large scale for biochemical or crystallographic purposes (Mace & Agard, 1995).

Although αLP is readily produced, expression of active protease is temperature sensitive; there is no detectable activity above ∼30°C, and optimal levels are obtained in *E. coli* at temperatures ranging from 12°C to 22°C. This effect is presumably due to the low thermal stability of the pro region (see below).

Inactive or misprocessed mutants must be expressed with a 'trans' construct, in which the pro region is expressed as a separate polypeptide chain from the mature domain, because the processing that separates the pro region from the protease is autolytic (Mace *et al.*, 1995; Silen *et al.*, 1989). Otherwise, misfolded precursor accumulates embedded in the outer membrane (Fujishige *et al.*, 1992).

## Biological Aspects

α-Lytic protease is secreted extracellularly by *L. enzymogenes*, a Gram-negative, Canadian soil bacterium, resulting in the lysis of other common soil organisms, including bacteria, fungi and nematodes. Its lytic effect on a variety of soil organisms has been tabulated (Gillespie & Cook, 1965) and the mechanism of bacterial lysis by mucopeptide degradation has also been investigated (Tsai *et al.*, 1965). Nevertheless, the means by which *L. enzymogenes* attracts nematodes is not understood.

The signal sequence and pro region of αLP are both unusually large (Silen *et al.*, 1988), but nothing is known about the secretory pathway in *Lysobacter*. Experiments in *E. coli* indicate that the precursor is translocated into the periplasmic space in a signal sequence-dependent manner. Once in the periplasm, it is rapidly folded and processed. Extracellular secretion through the outer membrane via the extended general secretion pathway occurs rapidly in *Lysobacter* ($t_{1/2} = 3$ min) but considerably more slowly in *E. coli* ($t_{1/2} = 4$ h) (Fujishige-Boggs & Agard, 1996).

The maturation and folding of the preproenzyme have been studied extensively in *E. coli*. The 166 residue pro region is absolutely required for the production of active protease. The pro region can be present either in *cis*, as part of the same polypeptide, or in *trans*, expressed as a separate polypeptide (Silen & Agard, 1989). Cleavage between the pro region and the protease is required for protease secretion across the outer membrane in *E. coli* and the cleavage reaction has been shown to require α-lytic protease activity. Production of inactive mutants such as Ser143(195)Ala results in the unprocessed precursor being trapped in the outer membrane. However, production of the same active-site mutant when the pro region is supplied in *trans* results in normal export through the outer membrane (Fujishige *et al.*, 1992; Silen & Agard, 1989). The location of the pro region C-terminus in the protease active site of the αLP–pro region complex, and comparison to the proenzyme structure of subtilisin, suggest that this reaction is intramolecular (Sohl *et al.*, 1997).

*In vitro* experiments have shown that the pro region has at least two roles: it is a very strong inhibitor of αLP, and it is both necessary and sufficient for catalysis of protease folding (Baker *et al.*, 1992a). Refolding the protease in the absence of the pro region results in the formation of a stable conformation referred to as the I state. The I state has an expanded radius of gyration according to gel filtration, native-like secondary structure as measured by circular dichroism, and no observable tertiary structure (Baker *et al.*, 1992b). Thus the I state has all the characteristics of a stable 'molten globule'. The I state and the folded mature domain do not interconvert on a time-scale of weeks under folding conditions, but addition of pro region at any time results in the rapid recovery of αLP activity (Baker *et al.*, 1992b). Therefore, αLP folding is under kinetic, not thermodynamic, control (Baker & Agard, 1994). Based on kinetic and energetic analyses, it is clear that the pro region acts much like an enzyme to catalyze the folding of the protease domain by dramatically lowering a rate-limiting barrier (Baker *et al.*, 1992a; Sohl & Agard, 1995).

It is not clear why pro regions have evolved, but it has been noted that all secreted bacterial proteases have them (Baker *et al.*, 1993). Similarly, eukaryotic proteases that require pro regions are often found in harsh environments (e.g. lysosomes). It is hypothesized that the pro region-assisted folding allows αLP to attain an unusually stable conformation necessary for function in the extracellular milieu. Unlike most proteins, which achieve stability through thermodynamics, αLP has achieved stability through kinetics. In this model, the biological 'pressure' to prevent the protein from unfolding on a relevant time-scale requires the development of a high barrier to unfolding, which also implies a high barrier to folding. Furthermore, the high unfolding barrier (highly cooperative unfolding) presumably decreases breathing fluctuations which would allow destruction through proteolysis. The problem with a kinetically controlled system is that the large barrier to folding necessitates use of a folding catalyst such as the pro region. Postfolding degradation of the pro region ensures that the unfolding reaction will not be catalyzed.

## Distinguishing Features

α-Lytic protease has a single His residue which is located at the active site. It also has a larger pro region than the *S. griseus* proteases other than the glutamate endopeptidase of *Streptomyces griseus* ATCC 10137 (Suzuki *et al.*, 1994), or subtilisin (Silen *et al.*, 1988). It is not irreversibly inhibited by the active-site titrant dichloroisocoumarin (Harper *et al.*, 1985). A large combination of structural, biochemical and mutant data makes this an ideal system in which to explore function and folding.

## Further Reading

Bone *et al.* (1989a,b) describe structure and activity studies of mutants of αLP with broad specificity, and Baker *et al.* (1992b) report on the demonstration of kinetic control of the folding reaction. Baker & Agard (1994) review folding pathways of proteins that are controlled kinetically, and Sohl & Agard (1995) give evidence for metastability of the native state of αLP, and discuss kinetic stability in proteins.

## References

Bachovchin, W.W. (1986) 15N NMR spectroscopy of hydrogen-bonding interactions in the active site of serine proteases: evidence for a moving histidine mechanism. *Biochemistry* **25**, 7751–7759.

Bachovchin, W.W., Kaiser, R., Richards, J.H. & Roberts, J.D. (1981) Catalytic mechanism of serine proteases: re-examination of the pH dependence of the histidyl 1J13C2-H coupling constant in the catalytic triad of α-lytic protease. *Proc. Natl Acad. Sci. USA* **78**, 7323–7326.

Bachovchin, W.W., Wong, W.Y., Farr-Jones, S., Shenvi, A.B. & Kettner, C.A. (1988) Nitrogen-15 NMR spectroscopy of the catalytic-triad histidine of a serine protease in peptide boronic acid inhibitor complexes. *Biochemistry* **27**, 7689–7697.

Baker, D., Silen, J.L. & Agard, D.A. (1992a) Protease pro region required for folding is a potent inhibitor of the mature enzyme. *Proteins* **12**, 339–344.

Baker, D., Sohl, J.L. & Agard, D.A. (1992b) A protein-folding reaction under kinetic control [see comments]. *Nature* **356**, 263–265.

Baker, D., Shiau, A.K. & Agard, D.A. (1993) The role of pro regions in protein folding. *Curr. Opin. Cell Biol.* **5**, 966–970.

Baker, D. & Agard, D.A. (1994) Kinetics versus thermodynamics in protein folding. *Biochemistry* **33**, 7505–7509.

Bauer, C.A., Brayer, G.D., Sielecki, A.R. & James, M.N. (1981) Active site of α-lytic protease: enzyme–substrate interactions. *Eur. J. Biochem.* **120**, 289–294.

Bone, R., Shenvi, A.B., Kettner, C.A. & Agard, D.A. (1987) Serine protease mechanism: structure of an inhibitory complex of α-lytic protease and a tightly bound peptide boronic acid. *Biochemistry* **26**, 7609–7614.

Bone, R., Frank, D., Kettner, C.A. & Agard, D.A. (1989a) Structural analysis of specificity: α-lytic protease complexes with analogues of reaction intermediates. *Biochemistry* **28**, 7600–7609.

Bone, R., Silen, J.L. & Agard, D.A. (1989b) Structural plasticity broadens the specificity of an engineered protease. *Nature* **339**, 191–195.

Bone, R., Fujishige, A., Kettner, C.A. & Agard, D.A. (1991a) Structural basis for broad specificity in α-lytic protease mutants. *Biochemistry* **30**, 10388–10398.

Bone, R., Sampson, N.S., Bartlett, P.A. & Agard, D.A. (1991b) Crystal structures of α-lytic protease complexes with irreversibly bound phosphonate esters. *Biochemistry* **30**, 2263–2272.

Brayer, G.D., Delbaere, L.T. & James, M.N. (1979) Molecular structure of the α-lytic protease from Myxobacter 495 at 2.8 Angstroms resolution. *J. Mol. Biol.* **131**, 743–775.

Brothers, H.M. & Kostic, N.M. (1990) Catalytic activity of the serine proteases α-chymotrypsin and α-lytic protease tagged at the active site with a (terpyridine)platinum(II) chromophore. *Biochemistry* **29**, 7468–7474.

Davis, J., Agard, D.A., Handel, T.M. & Basus, V.J. (1998) Alterations in chemical shifts and exchange broadening upon peptide boronic acid inhibitor binding to α-lytic protease. *J. Biomolecular NMR* **10**, 21–27.

Epstein, D.M. & Abeles, R.H. (1992) Role of serine 214 and tyrosine 171, components of the S2 subsite of α-lytic protease, in catalysis. *Biochemistry* **31**, 11216–11223.

Fujinaga, M., Delbaere, L.T., Brayer, G.D. & James, M.N. (1985) Refined structure of α-lytic protease at 1.7 Å resolution. Analysis of hydrogen bonding and solvent structure. *J. Mol. Biol.* **184**, 479–502.

Fujishige, A., Smith, K.R., Silen, J.L. & Agard, D.A. (1992) Correct folding of α-lytic protease is required for its extracellular secretion from *Escherichia coli*. *J. Cell Biol.* **118**, 33–42.

Fujishige-Boggs, A. & Agard, D.A. (1996) Bacterial extracellular secretion: transport of α-lytic protease across the outer membrane of *Escherichia coli*. In: *Membrane Protein Transport*, vol. 3 (Rothman, S.S., ed.). Greenwich, CT: JAI Press, pp. 165–179.

Gillespie, D.C. & Cook, F.D. (1965) Extracellular enzymes from strains of *Sorangium*. *Can. J. Microbiol.* **11**, 109–118.

Graham, L.D., Haggett, K.D., Jennings, P.A., Le Brocque, D.S., Whittaker, R.G. & Schober, P.A. (1993) Random mutagenesis of the substrate-binding site of a serine protease can generate enzymes with increased activities and altered primary specificities. *Biochemistry* **32**, 6250–6258.

Harper, J.W., Hemmi, K. & Powers, J.C. (1985) Reaction of serine proteases with substituted isocoumarins: discovery of 3,4-dichloroisocoumarin, a new general mechanism based serine protease inhibitor. *Biochemistry* **24**, 1831–1841.

Hunkapiller, M.W., Smallcombe, S.H., Whitaker, D.R. & Richards, J.H. (1973) Carbon nuclear magnetic resonance studies of the histidine residue in α-lytic protease. Implications for the catalytic mechanism of serine proteases. *Biochemistry* **12**, 4732–4743.

Hunkapiller, M.W., Forgac, M.D. & Richards, J.H. (1976) Mechanism of action of serine proteases: tetrahedral intermediate and concerted proton transfer. *Biochemistry* **15**, 5581–5588.

Jurášek, L. & Whitaker, D.R. (1967) Amino acid and metal composition of the α- and β-lytic proteases of *Sorangium* sp. *Can. J. Biochem.* **45**, 917–927.

Kaplan, H. & Whitaker, D.R. (1969) Kinetic properties of the α-lytic protease of *Sorangium* sp., a bacterial homologue of the pancreatopeptidases. *Can. J. Biochem.* **47**, 305–316.

Kaplan, H., Symonds, V.B., Dugas, H. & Whitaker, D.R. (1970) A comparison of properties of the α-lytic protease of *Sorangium* sp. and porcine elastase. *Can. J. Biochem.* **48**, 649–658.

Katznelson, H.K., Gillespie, D.C. & Cook, F.D. (1964) Studies on the relationships between nematodes and other soil microorganisms. III. Lytic action of soil myxobacters on certain species of nematodes. *Can. J. Microbiol.* **10**, 699.

Kettner, C.A., Bone, R., Agard, D.A. & Bachovchin, W.W. (1988) Kinetic properties of the binding of α-lytic protease to peptide boronic acids. *Biochemistry* **27**, 7682–7688.

Mace, J.E. & Agard, D.A. (1995) Kinetic and structural characterization of mutations of glycine 216 in α-lytic protease: a new target for engineering substrate specificity. *J. Mol. Biol.* **254**, 720–736.

Mace, J.E., Wilk, B.J. & Agard, D.A. (1995) Functional linkage between the active site of α-lytic protease and distant regions of structure: scanning alanine mutagenesis of a surface loop affects activity and substrate specificity. *J. Mol. Biol.* **251**, 116–134.

Olson, M.O., Nagabhushan, N., Dzwiniel, M., Smillie, L.B. & Whitaker, D.R. (1970) Primary structure of α-lytic protease: a bacterial homologue of the pancreatic serine proteases. *Nature* **228**, 438–442.

Rader, S.D. (1996) Protein folding and enzyme function: X-ray crystallographic studies of α-lytic protease. PhD dissertation, Graduate Group in Biophysics, University of California, San Francisco.

Rader, S.D. & Agard, D.A. (1997) Conformational substrates in enzyme mechanism: The 120 K structure of α-lytic protease at 1.5 Å resolution. *Protein Sci.* **6**, 1375–1386.

Schellenberger, V., Turck, C.W. & Rutter, W.J. (1994) Role of the S′ subsites in serine protease catalysis. Active-site mapping of rat chymotrypsin, rat trypsin, α-lytic protease, and cercarial protease from *Schistosoma mansoni*. *Biochemistry* **33**, 4251–4257.

Silen, J.L. & Agard, D.A. (1989) The α-lytic protease pro region does not require a physical linkage to activate the protease domain in vivo. *Nature* **341**, 462–464.

Silen, J.L., McGrath, C.N., Smith, K.R. & Agard, D.A. (1988) Molecular analysis of the gene encoding α-lytic protease: evidence for a preproenzyme. *Gene* **69**, 237–244.

Silen, J.L., Frank, D., Fujishige, A., Bone, R. & Agard, D.A. (1989) Analysis of prepro-α-lytic protease expression in *Escherichia coli* reveals that the pro region is required for activity. *J. Bacteriol.* **171**, 1320–1325.

Smith, S.O., Farr-Jones, S., Griffin, R.G. & Bachovchin, W.W. (1989) Crystal versus solution structures of enzymes: NMR spectroscopy of a crystalline serine protease. *Science* **244**, 961–964.

Sohl, J.L. & Agard, D.A. (1995) Alpha-lytic protease: dynamic stability via kinetic control. In: *Intramolecular Chaperones and Protein Folding* (Shinde, U. & Inouye, M., eds). Austin, TX: R.G. Landes, pp. 61–79.

Sohl, J.L., Shiau, A.K., Rader, S.D., Wilk, B. & Agard, D.A. (1997) Inhibition of α-lytic protease by pro region C-terminal steric occlusion of the active site. *Biochemistry* **36**, 3894–3902.

Suzuki, Y., Yabuta, M. & Ohsuye, K. (1994) Cloning and expression of the gene encoding the glutamic acid-specific protease of *Streptomyces griseus* ATCC10137. *Gene* **150**, 149–151.

Tsai, C.S., Whitaker, D.R., Jurasek, L. & Gillespie, D.C. (1965) Lytic enzymes of *Sorangium* sp. Action of the α- and β-lytic proteases on two bacterial mucopeptides. *Can. J. Biochem.* **43**, 1971–1983.

Whitaker, D.R. (1965) Lytic enzymes of *Sorangium* sp. Isolation and enzymatic properties of the α- and β-lytic proteases. *Can. J. Biochem.* **43**, 1935–1954.

Whitaker, D.R. & Roy, C. (1967) Concerning the nature of the α- and β-lytic proteases of *Sorangium* sp. *Can. J. Biochem.* **45**, 911–916.

***Stephen D. Rader***
*Howard Hughes Medical Institute and
Graduate Group in Biophysics,
University of California, San Francisco,
San Francisco, CA 94143-0448, USA*

***David A. Agard***
*Howard Hughes Medical Institute and
Graduate Group in Biophysics,
University of California, San Francisco,
San Francisco, CA 94143-0448, USA
Email: agard@msg.ucsf.edu*

# 83. DegP or protease Do

## Databanks

*Peptidase classification: clan SA, family S2, MEROPS ID: S02.031*
*NC-IUBMB enzyme classification: none*
*Databank encodes:*

| Species | SwissProt | PIR | EMBL (cDNA) | EMBL (genomic) |
|---------|-----------|-----|-------------|----------------|
| Protease Do | | | | |
| *Azotobacter vinelandii* | – | – | U30799 | – |
| *Bacillus subtilis* | P39668 | – | L22006 | – |
| *Bartonella henselae* | P54925 | – | L20127 | – |
| *Brucella abortus* | – | I40060 | L09274 | – |
| *Campylobacter jejuni* | – | – | U27271 X82628 | – |
| *Chlamydia trachomatis* | – | – | M31119 | – |
| *Escherichia coli* | P09376 | B35993 S01899 S45229 | M29955 M31772 M36536 X12457 | D26562: chromosome 2.4–4.1′ |
| *Mycobacterium leprae* | – | – | – | U15180: cosmid B1756 |
| *Mycobacterium paratuberculosis* | – | S47170 | Z23092 | – |
| *Pseudomonas aeruginosa* | – | – | U32853 | – |
| *Rhizobium meliloti* | – | – | U31512 | – |
| *Rhodobacter capsulatus* | – | – | Y11304 | – |
| *Rickettsia typhi* | – | – | D78346 | – |
| *Rickettsia tsutsugamushi* | – | – | L11697 L31933 L31934 | – |
| *Salmonella typhimurium* | P26982 | S15337 | X54548 | – |
| *Synechocystis* sp. | – | – | – | D90905: genome section 7 of 27 |
| *Yersinia enterocolitica* | – | – | D78376 X94153 | – |
| DegQ protein | | | | |
| *Escherichia coli* | – | – | M24777 | – |
| *Escherichia coli* | P39099 | – | U15661 | – |
| *Haemophilus influenzae* | P45129 | – | U32805 | – |
| *Pseudomonas aeruginosa* | – | – | U29172 | – |
| *Synechocystis* sp. | – | – | – | D90900: genome section 2 of 27 |

*continued overleaf*

| Species | SwissProt | PIR | EMBL (cDNA) | EMBL (genomic) |
|---|---|---|---|---|
| DegS protein | | | | |
| *Escherichia coli* | P31137 | – | U15661 | – |
| *Haemophilus influenzae* | P44947 | – | U32775 | – |
| *Synechocystis* sp. | – | – | – | D90911: genome section 13 of 27 |

Note: the *Chlamydia* homolog is on the complementary strand to the gene that is included and translated in M31119.

## Name and History

During the early 1980s, nine distinct endopeptidases were isolated from *Escherichia coli*, and named Do, Re, Mi, Fa, So, La, Ti, Pi and Ci (Chung, 1993). ***Protease Do*** was characterized as an oligomeric serine protease with an unusually high molecular mass of about 500 kDa (Swamy *et al.*, 1983), and was later found to be the product of a gene termed *degP* and *htrA* (Seol *et al.*, 1991). The *degP* function was originally identified by use of a genetic screen for *E. coli* mutants that fail to degrade an unstable chimeric protein located in the inner membrane (Strauch & Beckwith, 1988). A number of other ordinarily unstable cell envelope proteins have also been shown to be stabilized by the *degP* mutation. The *degP* gene codes for a proteolytically active periplasmic protein (Strauch *et al.*, 1989). Hence, it was concluded that the product of the *degP* gene, **DegP**, plays a major role in the **deg**radation of proteins exported beyond the cytoplasm.

DegP is a heat-shock protein and was identified independently as a function needed for growth of *E. coli* at higher temperatures. Georgopoulos and coworkers cloned an *E. coli* gene called **htrA** (**h**igh **t**emperature **r**equirement) which proved to be identical with *degP*. *htrA⁻* mutants were shown to undergo a block in macromolecular synthesis, and eventually lysed at temperatures above 42°C (Lipinska *et al.*, 1989, 1990).

## Activity and Specificity

DegP efficiently degrades β-casein *in vitro*, which is largely unstructured in solution (Lipinska *et al.*, 1990), and small proteins like the phage P22 Arc repressor and the N-terminal domain of λ repressor, which are only slightly more stable than the denatured states (Kolmar *et al.*, 1996). The native forms of proteins such as bovine serum albumin, ovalbumin and carbonic anhydrase are not cleaved, and nor are any of 20 synthetic peptides routinely used as protease substrates (Lipinska *et al.*, 1990). At the level of primary sequence, only Val+Xaa and Ile+Xaa peptide bonds appear to be efficiently cleaved, suggesting a preference for a β-branched side chain at the P1 position. However, several Val-Xaa and Ile-Xaa bonds are not cleaved in the small proteins mentioned above, indicating cleavage determinants in addition to the presence of a nonpolar β-branched residue. These additional determinants could involve sequence or structural features (Kolmar *et al.*, 1996).

Proteolytic activity of DegP increases rapidly with temperature in the range 37–55°C with the highest value at 55°C (Skórko-Glonek *et al.*, 1995). A drastic decrease in DegP activity observed at 60°C coincided with the initial phase of DegP denaturation between 55 and 60°C. DegP proteolytic activity is largely independent of the pH of the reaction in the range pH 4.8–10 (Lipinska *et al.*, 1990). The enzyme is potently inhibited by DFP but not by PMSF (Lipinska *et al.*, 1990).

## Structural Chemistry

DegP is a homo-oligomeric protein with a monomer molecular mass of about 48 kDa. It is produced from a 51 kDa precursor protein by removal of a 26 amino acid signal peptide (Lipinska *et al.*, 1990). Based on sequence conservation and mutagenesis experiments, DegP seems to contain the catalytic triad found in other serine proteases of clan SA, consisting of His105, Asp135 and Ser210 (Skórko-Glonek *et al.*, 1995; Waller & Sauer, 1996). In gel-filtration experiments, DegP protease eluted in either of two forms which chromatographed at apparent molecular masses of approximately 300 kDa or 500 kDa (Swamy *et al.*, 1983). Molecular cross-linking studies resulted in formation of covalently linked oligomers ranging from dimers to dodecamers (Kolmar *et al.*, 1996).

## Preparation

DegP has been purified from *E. coli* bacterial cultures carrying a multicopy plasmid containing the *degP* gene under transcriptional control of the plasmid-borne $P_{trc}$ promoter (Waller & Sauer, 1995; Kolmar *et al.*, 1996), or, alternatively, under T7 promoter control (Skórko-Glonek *et al.*, 1995). The protein accumulated in the periplasmic space and could be purified by standard methods.

## Biological Aspects

DegP is required for survival of *E. coli* at elevated temperatures, and mutations in the *degP* gene result in decreased degradation of chimeric membrane and periplasmic proteins (Strauch & Beckwith, 1988; Strauch *et al.*, 1989; Lipinska *et al.*, 1989). The temperature sensitivity may reflect an accumulation at high temperatures of misfolded proteins that interferes with essential periplasmic functions. Hence, the main function of DegP is most likely the removal of misfolded membrane and periplasmic proteins or proteins that are not properly processed. Supporting evidence to the view that DegP recognizes the non-native states of proteins comes from several findings, as follows. (a) Thermally aggregated, endogenous proteins of *E. coli* that can be isolated from heat-shocked cells are efficiently degraded by DegP *in vitro* (Laskowska *et al.*, 1997). (b) DegP is able to cleave aliphatic, β-branched residues, which are typically buried in the hydrophobic core of most proteins. In model substrates, these peptide bonds were generally inaccessible in the native three-dimensional structures (Kolmar *et al.*, 1996). (c) DegP efficiently degrades β-casein, which is largely unstructured in solution, but fails to cleave the native forms of the proteins analyzed so far (Lipinska *et al.*, 1990; Skórko-Glonek *et al.*, 1995). (d) DegP has been shown to cleave a chimeric β-lactamase protein only after its disulfide bonds are broken (Kolmar *et al.*, 1996). This is similar to the archaebacterial 20S proteasome (Chapter 169) which consists of two seven-subunit rings with the active sites located in an inner channel of the complex. Only unfolded proteins devoid of disulfide bonds appear capable of entering the channel, thereby preventing nonspecific proteolysis of folded proteins (Wenzel & Baumeister, 1995). Since DegP is also a large oligomer, geometric restrictions on access to the active site may well be involved in the substrate selectivity of this protease too.

Unlike the cytoplasmic heat-shock proteins, DegP is not transcriptionally regulated by the classical heat-shock regulon coordinated by $\sigma^{32}$. Rather, the *degP* gene is regulated by an alternate heat-shock factor, $\sigma^E$ (Lipinska *et al.*, 1989). RNA polymerase containing $\sigma^E$ is thought to be specifically involved in coordinating responses to extracytoplasmic stress by directing transcription of envelope stress proteins like DegP. In addition, *degP* transcription is partially dependent on the action of the *cpx* two-component signal transduction pathway. The CxpA inner membrane sensor responds to the alterations in the physiology of the bacterial envelope and communicates this information to its response regulator, CpxR, which then activates *degP* transcription by working in concert with RNA polymerase containing the $\sigma^E$ subunit (Danese *et al.*, 1995).

## Related Peptidases

Several groups have recently identified two other proteases of *E. coli*, DegQ (HhoA) and DegS (HhoB), which are homologous to DegP (Bass *et al.*, 1996; Waller & Sauer, 1996). DegQ is of similar size, and displays approximately 60% sequence identity to DegP. It is also indistinguishable in substrate specificity towards several protein substrates *in vitro* (Kolmar *et al.*, 1996). The similarities in structure and properties between the two proteases, taken together with their common subcellular localization in the periplasm, and the finding that overproduction of DegQ rescues the temperature-sensitive growth defect of a *degP⁻* strain (Waller & Sauer, 1996), suggest that they are essentially interchangeable. In contrast with *degP*, transcription from the *degQ* gene is not heat inducible, however, and the *degQ* gene is not important for normal *E. coli* growth. Furthermore, strains containing deletions in the *degQ* gene show no obvious phenotype. In contrast with DegQ, the DegS protein, which is likely to be located on the periplasmic side of the cytoplasmic membrane, cannot substitute for DegP, but is required for normal bacterial growth (Waller & Sauer, 1996).

Searches of the sequence databases reveal a rapidly growing family of *degP* (*htrA*) homologous genes in a variety of other bacteria including *Salmonella typhimurium*, *Yersinia enterocolitica*, *Haemophilus influenzae*, *Brucella abortus*, *Bartonella henselae*, *Campylobacter jejuni*, *Pseudomonas aeruginosa*, *Rhodobacter capsulatus*, *Rhizobium meliloti*, *Synechocystis* sp., *Rickettsia tsutsugamushi* and *Mycobacterium tuberculosis*. Some of these species are facultative intracellular pathogens. Several DegP/DegQ homologs have been implicated in the pathogenic virulence of these organisms. For instance, the *S. typhimurium degP* homolog was initially isolated as a gene required for full virulence of this organism in mice (Johnson *et al.*, 1991). A similar role of *degP* and *degQ*-like genes in *Brucella abortus* virulence has been observed within the first and second weeks of infection (Tatum *et al.*, 1994). Likewise, the *Pseudomonas aeruginosa degP*-homologous gene *algW* is required for expression of mucoidy and survival at elevated temperatures (Boucher *et al.*, 1996). Since *degP* mutants are more susceptible to heat and oxidative stress, they are presumably less viable in the host tissue and more easily killed. Nevertheless, a direct contribution of the DegP protease homologs to the pathogenesis of the various bacterial infections cannot be excluded.

## References

Bass, S., Gu, Q. & Christen, A. (1996) Multicopy suppressors of Prc mutant *Escherichia coli* include two HtrA (DegP) protease homologs (HhoAB), DksA, and a truncated RlpA. *J. Bacteriol.* **178**, 1154–1161.

Boucher, J.C., Martinez-Salazar, J., Schurr, M.J., Mudd, M.H., Yu, H. & Deretic, V. (1996) Two distinct loci affecting conversion to mucoidy in *Pseudomonas aeruginosa* in cystic fibrosis encode homologs of the serine protease HtrA. *J. Bacteriol.* **178**, 511–523.

Chung, C.H. (1993) Proteases in *Escherichia coli*. *Science* **262**, 372–374.

Danese, P.N., Snyder, W.B., Cosma, C.L., Davis, L.J.B. & Silhavy, T.J. (1995) The Cpx two-component signal transduction pathway of *Escherichia coli* regulates transcription of the gene specifying the stress-inducible periplasmic protease, DegP. *Genes Dev.* **9**, 387–398.

Johnson, K.I., Dougan, C.G., Pickard, D., O'Gaora, P., Costa, G., Ali, T., Miller, I. & Hormaeche, C. (1991) The role of stress-response protein in *Salmonella typhimurium* virulence. *Mol. Microbiol.* **5**, 401–407.

Kolmar, H., Waller, P.R. & Sauer, R.T. (1996) The DegP and DegQ periplasmic endoproteases of *Escherichia coli*: specificity for cleavage sites and substrate conformation. *J. Bacteriol.* **178**, 5925–5929.

Laskowska, E., Kuczynska-Wisnik, D., Skórko-Glonek, J. & Tylor, A. (1997) Degradation by proteases Lon, Clp and HtrA, of *Escherichia coli* proteins aggregated *in vivo* by heat shock; HtrA protease action *in vivo* and *in vitro*. *Mol. Microbiol.* **22**, 555–571.

Lipinska, B., Fayet, O., Baird, L. & Georgopoulos, C. (1989) Identification, characterization, and mapping of the gene, whose product is essential for bacterial growth only at elevated temperatures. *J. Bacteriol.* **171**, 1574–1584.

Lipinska, B., Zylicz, M. & Georgopoulos, C. (1990) The HtrA (DegP) protein, essential for *Escherichia coli* survival at high temperatures, is an endopeptidase. *J. Bacteriol.* **172**, 1791–1797.

Seol, J.H., Woo, S.K., Jung, E.M., Yoo, S.J., Lee, C.S., Kim, K.J., Tanaka, K., Ichihara, A., Ha, D.B. & Chung, C.H. (1991) Protease Do is essential for survival of *Escherichia coli* at high temperatures: its identity with the *htrA* gene product. *Biochem. Biophys. Res. Commun.* **176**, 730–736.

Skórko-Glonek, J., Krzewski, K., Lipinska, B., Bertoli, E. & Tanfani, F. (1995) Comparison of the structure of wild-type HtrA heat shock protease and mutant HtrA proteins. *J. Biol. Chem.* **270**, 1140–1146.

Strauch, K.L. & Beckwith, J. (1988). An *Escherichia coli* mutation preventing degradation of abnormal periplasmic proteins. *Proc. Natl Acad. Sci. USA* **85**, 1576–1580.

Strauch, K.L., Johnson, K. & Beckwith, J. (1989) Characterization of *degP*, a gene required for proteolysis in the cell envelope and essential for growth of *Escherichia coli* at high temperatures. *J. Bacteriol.* **171**, 2689–2696.

Swamy, K.H.S., Chung, C.H. & Goldberg, A. (1983) Isolation and characterization of protease Do from *Escherichia coli*, a large serine protease containing multiple subunits. *Arch. Biochem. Biophys.* **224**, 543–554.

Tatum, F.M., Cheville, N.F. & Morfitt, D. (1994) Cloning, characterization and construction of *htrA* and *htrA*-like mutants of *Brucella abortus* and their survival in BALB/c mice. *Microb. Pathog.* **17**, 23–36.

Waller, P.R.H. & Sauer, R.T. (1996) Characterization of *degQ* and *degS*, *Escherichia coli* genes encoding homologs of the DegP protease. *J. Bacteriol.* **178**, 1146–1153.

Wenzel, T. & Baumeister, W. (1995) Conformational constraints in protein degradation by the 20S proteasome. *Struct. Biol.* **2**, 199–204.

*Harald Kolmar*
*Institut für Molekulare Genetik,*
*Georg-August-Universität Göttingen,*
*Grisebachstrasse 8, D-37077 Göttingen, Germany*
*Email: hkolmar@Uni-MolGen.gwdg.de*

# 84. Yeast-lytic endopeptidase (Rarobacter)

## Databanks

*Peptidase classification: SA, family S2, MEROPS ID: S02.035*
*NC-IUBMB enzyme classification: none*
*Databank codes:*

| Species | SwissProt | PIR | EMBL (cDNA) | EMBL (genomic) |
|---|---|---|---|---|
| *Rarobacter faecitabidus* | Q05308 | A45053 | D10753 | – |

## Name and History

A yeast-lytic endopeptidase was first isolated from *Oerskovia xanthineolytica*. It is a serine endopeptidase that has an affinity for Sephadex gel and lyses yeast cells cooperatively with β-1,3-glucanase (Obata *et al.*, 1977). A commercial yeast-lytic enzyme (Zymolyase) also contains a yeast-lytic endopeptidase as a major component (Kitamura, 1982). **Yeast-lytic endopeptidase** (*Rarobacter*) is referred to as **RPI**, an acronym for **Rarobacter faecitabidus protease I** (Shimoi *et al.*, 1992). RPI is the major yeast-lytic endopeptidase of *R. faecitabidus* and is well characterized in its structure and function (Shimoi *et al.*, 1992).

## Activity and Specificity

The substrate specificity of RPI is similar to that of elastase (Shimoi & Tadenuma, 1991). Amongst amino acid nitrophenyl ester substrates, RPI preferentially hydrolyzes the ester of alanine. It also efficiently hydrolyzes Suc-Ala-Pro-Ala+NHPhNO$_2$, the specific synthetic substrate for pancreatic elastase (Chapter 11). With oxidized insulin B chain, the enzyme hydrolyzes almost exclusively the peptide bond between Val18+Cys(SO$_3$)19 at an early stage in the reaction, and thereafter it partially hydrolyzes Val12+Glu13, Ala14+Leu15 and Leu15+Tyr16 (Shimoi & Tadenuma, 1991).

Assays are most conveniently made with the chromogenic substrate, Suc-Ala-Pro-Ala-NHPhNO$_2$ as described by Shimoi *et al.* (1992). Yeast-lytic activity of the protease is assayed with yeast cells as substrate (Shimoi *et al.*, 1991). The pH optimum is about 9 for both proteolytic and yeast-lytic activity (Shimoi *et al.*, 1991). Activity is strongly inhibited by DFP and PMSF (Shimoi *et al.*, 1991).

## Structural Chemistry

The gene cloning of RPI revealed that it contained an open reading frame encoding a 525 amino acid protein (Shimoi *et al.*, 1992). Prepro-RPI consists of three domains: (a) an N-terminal prepro domain that is not found in the mature form of RPI, (b) a protease domain homologous to α-lytic protease (Chapter 82), and (c) a C-terminal domain homologous to the C-terminal part of *O. xanthineolytica* β-1,3-glucanase and the

N-terminal part of ricin B chain, a lectin isolated from the castor bean. The mature enzyme has a mass of 33 855 Da. The C-terminal domain participates in the mannose-binding activity and is required for yeast-lytic activity. Mutant RPI, in which the C-terminal domain was truncated by a site-directed mutagenesis, lacked both mannose-binding and yeast-lytic activity, although the protease activity was not affected.

### Preparation

RPI is purified from culture supernatant of *R. faecitabidus* (Shimoi *et al.*, 1991). The enzyme is readily separated from other proteases by gel filtration (Shimoi *et al.*, 1991). The recombinant enzyme has been expressed in *E. coli* under the *lacZ* promoter as an extracellular protein with concomitant cell lysis (Shimoi *et al.*, 1992). This protein can be easily purified by affinity chromatography on immobilized ovoinhibitor (Shimoi *et al.*, 1995).

### Biological Aspects

The natural substrate for RPI is mannoproteins of the yeast cell wall (Shimoi *et al.*, 1995). *R. faecitabidus* is a yeast-lytic bacterium originally isolated from a wastewater treatment system. It adheres to and agglutinates yeast cells, then lyses them to utilize the decomposed cells as nutrients (Yamamoto *et al.*, 1988). *R. faecitabidus* secretes one $\beta$-1,3-glucanase and two proteases (Shimoi *et al.*, 1991). The $\beta$-1,3-glucanase cannot lyse yeast cells without the aid of the proteases, but the two proteases alone are able to lyse them. The $\beta$-1,3-glucanase combined with one of the proteases completely and synergistically lyses yeast cells (Shimoi *et al.*, 1991). RPI has a lectin-like affinity for mannose and is adsorbed by yeast cells (Shimoi & Tadenuma, 1991). The yeast cell wall contains many species of mannoproteins (Klis, 1994) which cover the surface of cell and are the natural substrates for RPI (Shimoi *et al.*, 1995). RPI specifically recognizes the mannose chains of mannoproteins and cleaves peptide bonds in their vicinity. RPI was used for the solubilization and identification of yeast cell wall mannoproteins (Shimoi *et al.*, 1995).

### Distinguishing Features

RPI has elastase-like substrate specificity and lectin-like affinity for mannose (Shimoi & Tadenuma, 1991); it is a chimera with an N-terminal serine protease domain and a C-terminal mannose-binding domain (Shimoi *et al.*, 1992). Rabbit polyclonal antibodies have been produced and used in western blotting analysis (Shimoi *et al.*, 1992).

### Further Reading

The reader is referred to the paper of Shimoi *et al.* (1992).

### References

Kitamura, K. (1982) Re-examination of Zymolyase purification. *Agric. Biol. Chem.* **46**, 963–969.

Klis, F. M. (1994) Cell wall assembly in yeast. *Yeast* **10**, 851–869.

Obata, T., Iwata, H. & Namba, Y. (1977) Proteolytic enzyme from *Oerskovia* sp. lysing viable yeast cells. *Agric. Biol. Chem.* **41**, 2387–2394.

Shimoi, H. & Tadenuma, M. (1991) Characterization of *Rarobacter faecitabidus* protease I, a yeast-lytic serine protease having mannose-binding activity. *J. Biochem. (Tokyo)* **110**, 608–613.

Shimoi, H., Muranaka, Y., Sato, S., Saito, K. & Tadenuma, M. (1991) Purification of the enzymes responsible for the lysis of yeast cells by *Rarobacter faecitabidus*. *Agric. Biol. Chem.* **55**, 371–378.

Shimoi, H., Iimura, Y., Obata, T. & Tadenuma, M. (1992) Molecular structure of *Rarobacter faecitabidus* protease I. A yeast-lytic serine protease having mannose-binding activity. *J. Biol. Chem.* **267**, 25189–25195.

Shimoi, H., Iimura, Y. & Obata, T. (1995) Molecular cloning of CWP1: a gene encoding a *Saccharomyces cerevisiae* cell wall protein solubilized with *Rarobacter faecitabidus* protease I. *J. Biochem. (Tokyo)* **118**, 302–311.

Yamamoto, N., Sato, S., Saito, K., Hasuo, T., Tadenuma, M., Suzuki, K., Tamaoka, J. & Komagata, K. (1988) *Rarobacter faecitabidus* gen. nov., sp. nov., a yeast-lysing coryneform bacteria. *Int. J. Syst. Bacteriol.* **38**, 7–11.

*Hitoshi Shimoi*
*National Research Institute of Brewing,*
*3-7-1, Kagamiyama, Higashihiroshima, 739 Japan*
*Email: simoi@nrib.go.jp*

# 85. Lysyl endopeptidase

### Databanks

*Peptidase classification: clan SA, family S2, MEROPS ID: S02.041*
*NC-IUBMB enzyme classification: EC 3.4.21.50*

*ATCC entries: Lysobacter enzymogenes strain ATCC 29487*
*Chemical Abstracts Service registry number: 123175-82-6*
*Databank codes:*

| Species | SwissProt | PIR | EMBL (cDNA) | EMBL (genomic) |
|---------|-----------|-----|-------------|----------------|
| *Achromobacter lyticus* | P15636 | A32687 | D23664 | – |
| | | A32960 | J05128 | |

Brookhaven Protein Data Bank three-dimensional structures:

| Species | ID | Resolution | Notes |
|---------|-----|------------|-------|
| *Achromobacter lyticus* | 1ARB | 1.2 | |
| | 1ARC[a] | 2 | complex Tos-Lys-CH$_2$Cl |

[a]See Chapter 76 for an image constructed from these coordinates.

## Name and History

*Achromobacter lyticus* M497-1, lysing gram-negative bacterial cell walls, secretes three proteases which were named *Achromobacter* proteases I, II and III (Masaki *et al.*, 1978). The name ***Achromobacter protease I (API)***, has been widely used for the major component, which specifically hydrolyzes lysyl bonds (Masaki *et al.*, 1981a). No further information is available about the two other minor proteases. The name recommended by NC-IUBMB is lysyl endopeptidase.

## Activity and Specificity

API is active at pH 8.5–10.5 and the maximal specific activity is an order of magnitude higher than that of bovine trypsin (Chapter 3). This is due to a higher $k_{cat}$ for the esterase activity optimal around pH 8 and a lower $K_m$ for the amidase activity optimal at pH 9–9.5 (Masaki *et al.*, 1978). All lysyl bonds in the denatured protein are usually cleaved when incubated at a molar ratio of substrate to enzyme of 200–400 to 1 in 0.1 M Tris-HCl buffer, pH 9, for 6 h at 37°C (Tsunasawa *et al.*, 1987). The Lys╀Pro bond can be hydrolyzed, but at a reduced rate. Also, a lysyl bond is hydrolyzed slowly when it is that of an unblocked, N-terminal residue, or when the lysine is preceded by a basic amino acid, or followed by an acidic one (Tsunasawa *et al.*, 1987). Three substrate-binding subsites, S3, S2 and S1, exist on the amino side of the scissile lysyl bond (Sakiyama *et al.*, 1990). Transpeptidation potency for lysyl bond formation is higher than that of bovine trypsin (Morihara *et al.*, 1986). API activity is stable to exposure to 4 M urea or 0.1% SDS at pH 8 for 20 min at 30°C (Masaki *et al.*, 1984), but is lost in aqueous formic acid. The enzyme is stable in the range pH 4.0–10.0 and active up to 40°C. Lysinal derivatives are strong, reversible inhibitors of API (Masaki *et al.*, 1992). Tos-Lys-CH$_2$Cl, DFP, PMSF, *n*-amylamine and *n*-butylamine, and Zn$^{2+}$ are also strong inhibitors (Masaki *et al.*, 1981a, 1984).

The lysine specificity is generally strict, but hydrolysis of Arg╀Ser (Yonetsu *et al.*, 1986) or Arg╀Ala (Kitagawa *et al.*, 1986) bonds has been reported. The *S*-aminoethylcysteinyl peptide bond is also sensitive to hydrolysis by API (Kawata *et al.*, 1988; Masaki *et al.*, 1994).

## Structural Chemistry

API (pI 6.9) is a single-chain protein consisting of 268 residues with three disulfide bonds (Tsunasawa *et al.*, 1989). The catalytic triad is composed of His57, Asp113 and Ser194, and substrate-binding subsites are located at His210, Gly211 and Gly212. Asp225 primarily determines the lysine specificity and loss of the negative charge inactivates the enzyme (Norioka *et al.*, 1994). When compared with bovine trypsin, API bears extensions at N- and C-termini which may be associated with the peptidase activity. The protease is synthesized as an inactive preproprotein of 653 residues which, in addition to the mature protein, includes a 20 residue signal peptide and the 185 residue propeptide at the N-terminus and a 180 residue Ser/Thr-rich sequence at the C-terminus (Ohara *et al.*, 1989). Activation is probably autocatalytic. The three-dimensional structure resembles that of bovine trypsin, although the sequence identity is only 16% (Y. Kitagawa *et al.*, unpublished results). The folds of the peptide chain in the active site are nearly superimposable for the two proteins, but the nature and geometry of the side chains of key residues, apart from the catalytic triad, are different. The critical difference is seen at the His210 side chain, which is located at a position that allows it to interact with both Asp113 and Trp169. Whether these structural characteristics are related to the particular enzymatic properties of API is unknown. The positive charge of the side chain of Tos-Lys-CH$_2$Cl directly interacts with the negative charge of Asp225, displacing a bound water molecule upon binding to API (Y. Kitagawa *et al.*, unpublished results).

## Preparation

The active enzyme is prepared through five steps from culture broth of *Achromobacter lyticus* M497-1 (50% yield from acetone powder) (Masaki *et al.*, 1981b; Sakiyama & Masaki, 1994). The recombinant protein has also been expressed in *E. coli*. It is secreted to the periplasm as the mature form, and may be purified by affinity chromatography (Ohara *et al.*, 1989). Lysyl endopeptidase is commercially available as such or as *Achromobacter* protease I (Wako) (see Appendix 2 for full names and addresses of suppliers). Strain ATCC 29487 of *Lysobacter enzymogenes* available from ATCC also produces the enzyme (Ohara *et al*., 1991).

## Biological Aspects

Possible digestive roles have been considered, but no definite information is available on the biological functions of this extracellular protease.

## Related Peptidases

Lysyl endopeptidases are produced by *Lysobacter enzymogenes* (Jekel *et al*., 1983) and *Pseudomonas aeruginosa* (Elliott & Cohen, 1986). The *Lysobacter* enzyme is probably API. The *Pseudomonas* enzyme is less active than API, and no structural information is available for either of these enzymes.

## References

Elliott, B.W. & Cohen, C. (1986) Isolation and characterization of a lysine-specific protease from *Pseudomonas aeruginosa*. *J. Biol. Chem.* **261**, 11259–11265.

Jekel, P.A., Weijer, W.J. & Beintema, J.J. (1983) Use of endoproteinase Lys-C from *Lysobacter enzymogenes*. *Anal. Biochem.* **134**, 347–354.

Kawata, Y., Sakiyama, F. & Tamaoki, H. (1988) Amino acid sequence of ribonuclease T2 from *Aspergillus oryzae*. *Eur. J. Biochem.* **176**, 683–697.

Kitagawa, Y., Tsunasawa, S., Tanaka, N., Katsube, Y., Sakiyama, F. & Asada, K. (1986) Amino acid sequence of copper, zinc-superoxide dismutase from spinach leaves. *J. Biochem.* **99**, 1289–1298.

Masaki, T. & Soejima, M. (1985) Actions of *Achromobacter* protease I on some zymogens and dimethylcasein. *Agric. Biol. Chem.* **49**, 1867–1868.

Masaki, T., Nakmura, K., Isono, M. & Soejima, M. (1978) A new proteolytic enzyme from *Achromobacter lyticus* M497-1. *Agric. Biol. Chem.* **47**, 1443–1445.

Masaki, T., Funahashi, T., Nakamura, K. & Soejima, M. (1981a) Studies on a new proteolytic enzyme from *Achromobacter lyticus* M497-1. II. Specificity and inhibition studies of *Achromobacter* protease I. *Biochim. Biophys. Acta* **660**, 51–55.

Masaki, T., Tanabe, M., Nakamura, K. & Soejima, M. (1981b) Studies on a new proteolytic enzyme from *Achromobacter lyticus* M497-1. I. Purification and some enzymatic properties. *Biochim.* *Biophys. Acta* **660**, 44–50.

Masaki, T., Fujihashi, T. & Soejima, M. (1984) Effect of various inhibitors on the activity of *Achromobacter* protease I (in Japanese). *Nippon Nogeikagaku Kaishi* **58**, 865–870.

Masaki, T., Tanaka, T., Tsunasawa, S., Sakiyama, F. & Soejima, M. (1992) Inhibition of *Achromobacter* protease I by lysinal derivatives. *Biosci. Biotech. Biochem.* **56**, 1604–1607.

Masaki, T., Takiya, T., Tsunasawa, S., Kuwahara, S., Sakiyama, F. & Soejima, M. (1994) Hydrolysis of S-2-aminoethylcysteinyl peptide bond by *Achromobacter* protease I. *Biosci. Biotech. Biochem.* **58**, 215–216.

Morihara, K., Ueno, Y. & Sakina, K. (1986) Influence of temperature on the enzyme semisynthesis of human insulin by coupling and transpeptidation methods. *Biochem. J.* **240**, 803–810.

Norioka, S., Ohata, S., Ohara, T., Lim, S.-I. & Sakiyama, F. (1994) Identification of three catalytic triad constituents and Asp225 essential for function of lysine-specific serine protease, *Achromobacter* protease I. *J. Biol. Chem.* **269**, 17025–17029.

Ohara, T., Makino, K., Shinagawa, H., Nakata, A., Norioka, S. & Sakiyama, F. (1989) Cloning, nucleotide sequence and expression of *Achromobacter* protease I gene. *J. Biol. Chem.* **264**, 20625–20631.

Ohara, T., Yamamoto, A., Shinagawa, H., Nakata, A., Norioka, S. & Sakiyama, F. (1991) Primary structures of a lysylendopeptidase from *Lysobacter enzymogenes* and *Achromobacter* protease I are identical. Abstracts of Fifth Symposium of the Protein Society, Baltimore, MD, p. 120 (T77).

Sakiyama, F. & Masaki, T. (1994) Lysyl endopeptidase from *Achromobacter* lyticus. *Methods Enzymol.* **244**, 126–137.

Sakiyama, F., Suzuki, M., Yamamoto, A., Aimoto, S., Norioka, S., Masaki, T. & Soejima, M. (1990) Mapping of substrate binding subsites of a lysine-specific serine protease, *Achromobacter* protease I. *J. Protein Chem.* **9**, 297–298.

Tsunasawa, S., Sugihara, A., Masaki, T., Sakiyama, F., Takeda, Y., Miwatani, T. & Narita, K. (1987) Amino acid sequence of thermostable direct hemolysin produced by *Vibrio parahaemolyticus*. *J. Biochem.* **101**, 111–121.

Tsunasawa, S., Masaki, T., Hirose, M., Soejima, M. & Sakiyama, F. (1989) The primary structure and structural characteristics of *Achromobacter* protease I, a lysine-specific serine protease. *J. Biol. Chem.* **264**, 3832–3839.

Yonetsu, T., Higuchi, K., Tsunasawa, S., Takagi, S., Sakiyama, F. & Takeda, T. (1986) High homology is present in the primary structure between murine senile amyloid protein and human apolipoprotein A-II. *FEBS Lett.* **203**, 149–152.

*Fumio Sakiyama*
*Institute for Protein Research,*
*Osaka University,*
*Suita, Osaka 565, Japan*
*Email: sakiyama@protein.osaka-u.ac.jp*

*Takeharu Masaki*
*Faculty of Agriculture,*
*Ibaraki University,*
*Ami-machi, Ibaraki 300-03, Japan*

# 86. *Togavirin*

## Databanks

*Peptidase classification: clan SA, family S3, MEROPS ID: S03.001*
*NC-IUBMB enzyme classification: EC 3.4.21.90*
*Databank codes:*

| Species | SwissProt | PIR | EMBL (genomic) |
|---|---|---|---|
| Aura virus | – | – | S78478: complete genome |
| Barmah Forest virus | – | – | U73745: complete genome |
| Eastern equine encephalitis virus | P08768 | A26816 | X05816: region of polyprotein gene |
| Eastern equine encephalitis virus | P27284 | A39992 | M69094: structural polyprotein |
| O'Nyong-nyong virus | P22056 | B34680 | M20303: complete genome |
| Ross River virus | P08491 | A31833 | K00046: complete polyprotein gene |
| Ross River virus | P13890 | B28605 | M20162: complete genome |
| Semliki forest virus & O'Nyong-nyong virus | P03315 | A93861 | X04129: complete genome |
| Sindbis virus | P03316 | A03916 | V01403: structural polyprotein |
| | | B03916 | |
| Sindbis virus | P27285 | B39991 | M69205: complete genome |
| Sindbis-like virus | – | – | U38304: complete genome |
| Sindbis-like virus | – | – | U38305: complete genome |
| Venezuelan equine encephalitis virus | P05674 | A27871 | X04368: complete polyprotein gene |
| Venezuelan equine encephalitis virus | P09592 | B31467 | J04332: complete genome |
| | | | M14937: complete polyprotein gene |
| Venezuelan equine encephalitis virus | P36329 | D44213 | L00930: complete genome |
| Venezuelan equine encephalitis virus | P36330 | JQ1978 | L04598: complete polyprotein gene |
| Venezuelan equine encephalitis virus | P36331 | JQ1979 | L04599: structural polyprotein |
| Venezuelan equine encephalitis virus | P36332 | B44213 | L04653: complete genome |
| Western equine encephalitis virus | P13897 | A35587 | J03854: 3' end of genome |

Brookhaven Protein Data Bank three-dimensional structures:

| Species | ID | Resolution | Notes |
|---|---|---|---|
| Sindbis virus | 1SNW | 3 | |
| | 2SNV[a] | 2.8 | |

[a]See Chapter 76 for an image constructed from these coordinates.

## Name and History

Ross River, Semliki Forest and Sindbis viruses are togaviruses, members of the Togaviridae, genus *Alphavirus*. They are single-stranded, enveloped RNA viruses that are vertebrate pathogens transmitted by arthropods. They can cause encephalitis, arthritis, fever and rash in mammals. In the particle of the Sindbis virus ($M_r$ $40 \times 10^6$), the genomic RNA is surrounded by an icosahedral nucleocapsid. The nucleocapsid core is surrounded by a lipid bilayer, and this is penetrated by 80 glycoprotein spikes that carry the recognition sites for the host cell receptor. The genome encodes two polyproteins, p130 and p270. Polyprotein p270 contains a cysteine endopeptidase Sindbis virus NsP2 endopeptidase (Chapter 230), whereas p130 contains the serine endopeptidase. Polyprotein p130 also contains structural proteins for the nucleocapsid core. The serine endopeptidase is located at the N-terminus of the p130 polyprotein. The enzyme has generally been termed **Sindbis virus core protein (SCP)** in the literature, but the name **togavirin** has been recommended by NC-IUBMB to include the homologous peptidases of related viruses also.

## Activity and Specificity

Sindbis togavirin is located at the N-terminus of the polyprotein, from which it cleaves itself in a *cis* intramolecular event, by hydrolysis of the bond Trp264+Ser265. After the cleavage, the new, C-terminal Trp residue remains in the P1 subsite of the enzyme, blocking the active site, so that once the core protein is released, it retains no detected peptidase activity. No other substrate is known for the enzyme.

## Structural Chemistry

Mutagenetic and crystallographic studies have identified a catalytic triad in Sindbis togavirin, composed of His141,

Asp163 and Ser215 (Hahn & Strauss, 1990; Tong *et al.*, 1993). The X-ray crystallography has shown that the structure of the enzyme is very similar to those of chymotrypsin (Chapter 8) and α-lytic endopeptidase (Chapter 82) (Tong *et al.*, 1993; Choi *et al.*, 1996). The structure of togavirin also shows marked similarities to those of the picornains (Chapter 239). Deletions in the loop regions on the surface of the protein account for the smaller size of the togavirin molecule (151 residues) as compared to chymotrypsin (236 residues) (Tong *et al.*, 1993), and permits the *cis* autocatalytic cleavage of the viral polyprotein to produce the capsid protein. Amongst the loops that are deleted are those that form the P2 and P3 substrate-binding pockets in chymotrypsin.

Despite the evident similarity of the fold of togavirin to that of chymotrypsin, there are also some significant differences. There are no disulfide bridges in togavirin, and the 'methionine loop' and C-terminal helix are missing (the C-terminus being truncated by the autoproteolysis). The Ser214 residue (chymotrypsinogen numbering) that is conserved in family S1, and hydrogen bonds to the side chain of the aspartate, Asp163, in the catalytic triad, is a nonhydrogen-bonding leucine in togavirin. And Asp163 itself, which is buried in endopeptidases of the chymotrypsin family, is exposed to solvent in togavirin.

The active-site serine is encoded by AGX codons in both Sindbis virus and Semliki Forest virus (Rawlings & Barrett, 1994).

The apparently very efficient blocking of the catalytic site of togavirin by its own C-terminus, despite the fact that there are no prime-side interactions to stabilize the binding, may be accounted for by the fact that the interaction is an intramolecular one, and thus energetically favored.

## Preparation

Togavirin has been isolated from the virus particles (Choi *et al.*, 1991), and expressed in a number of forms in *E. coli* (Choi *et al.*, 1996). Purification of the recombinant protein was by ammonium sulfate fractionation, Sephadex G-50 gel filtration, and cation-exchange chromatography. The yield was about 10 mg of the enzyme per liter of bacterial culture (Choi *et al.*, 1996).

## Biological Aspects

Most of the single-stranded RNA viruses encode endopeptidases that are distantly related to chymotrypsin; these are either serine peptidases or cysteine peptidases. But togavirin is unusual amongst these enzymes in that the protein remains as part of the structure of the mature virus particle, in a catalytically inactive form.

## Related Peptidases

The togavirins of other species of togavirus (see the Databanks table) have not been studied in detail, but amino acid sequence comparisons show that all of them retain the Trp264 residue that is the point of cleavage, and the great majority have the catalytic triad residues conserved (see Chapter 76). This suggests that they are likely to be catalytically active, and to function in a similar way to the Sindbis virus enzyme. An exception is the aura virus protein, for which the published sequence does not show conservation of the His and Asp residues of the catalytic triad.

## Further Reading

An excellent commentary on the discovery that the Sindbis virus core protein is a peptidase was provided by Ringe & Petsko (1991).

## References

Choi, H.-K., Tong, L., Minor, W., Dumas, P., Boege, U., Rossmann, M.G. & Wengler, G. (1991) Structure of Sindbis virus core protein reveals a chymotrypsin-like serine proteinase and the organization of the virion. *Nature* **354**, 37–43.

Choi, H.K., Lee, S., Zhang, Y.P., McKinney, B.R., Wengler, G., Rossmann, M.G. & Kuhn, R.J. (1996) Structural analysis of Sindbis virus capsid mutants involving assembly and catalysis. *J. Mol. Biol.* **262**, 151–167.

Hahn, C.S. & Strauss, J.H. (1990) Site-directed mutagenesis of the proposed catalytic amino acids of the Sindbis virus capsid protein autoprotease. *J. Virol.* **64**, 3069–3073.

Rawlings, N.D. & Barrett, A.J. (1994) Families of serine peptidases. *Methods Enzymol.* **244**, 19–61.

Ringe, D. & Petsko, G.A. (1991) Viral proteases: molecular metamorphosis. *Nature* **354**, 22–23.

Tong, L., Wengler, G. & Rossmann, M.G. (1993) Refined structure of Sindbis virus core protein and comparison with other chymotrypsin-like serine proteinase structures. *J. Mol. Biol.* **230**, 228–247.

***Alan J. Barrett***
*MRC Peptidase Laboratory,*
*The Babraham Institute,*
*Cambridgeshire CB2 4AT, UK*
*Email: alan.barrett@bbsrc.ac.uk*

# 87. IgA1-specific serine endopeptidase

## Databanks

*Peptidase classification: clan SA, family S6, MEROPS ID: S06.001*
*NC-IUBMB enzyme classification: EC 3.4.21.72*
*Chemical Abstracts Service registry number: 55127-02-1*
*Databank codes:*

| Species | Strain/clone | SwissProt | PIR | EMBL (cDNA) | EMBL (genomic) |
|---|---|---|---|---|---|
| *Haemophilus influenzae* | BPF | – | – | X86103 | – |
| | Da66 | – | – | X82467 | – |
| | HK61 | P45386 | – | M87491 | – |
| | HK284 | – | – | X82487 | – |
| | HK368 | P42782 | – | M87492 X64357 | – |
| | HK393 | P45385 | – | M87490 | – |
| | HK635 | – | – | X82488 | – |
| | HK715 | P45384 | – | M87489 | – |
| | serotype d | P44969 | – | X59800 | U32779: section of genome |
| *Neisseria gonorrhoeae* | MS11 | P09790 | A26039 | X04835 | – |
| *Neisseria meningitidis* | HF13 | – | – | X82474 Z21614 | |

## Name and History

Microbial proteolytic activity that cleaves human serum and secretory immunoglobulin A (IgA) in the hinge region was first detected in the human gastrointestinal tract (Mehta *et al.*, 1973). The enzyme responsible was termed **IgA protease** (Plaut *et al.*, 1974) and this activity has since been demonstrated in a small number of bacteria associated with humans. The enzyme is also called **IgA1 protease** and **Igase**, and the precursor of the enzyme has been termed **Iga** after the *iga* gene that encodes it. Expression of IgA1 protease activity by bacteria represents a striking example of convergent evolution, since the IgA1 proteases produced by different bacterial genera belong to different classes of proteinases, serine or metallo (see also Chapter 508). The present chapter deals with the IgA1 proteases of the serine peptidase class produced by *Neisseria gonorrhoeae*, *Neisseria meningitidis* and *Haemophilus influenzae* (Plaut *et al.*, 1975; Kilian *et al.*, 1979; Male, 1979). The name recommended by NC-IUBMB is **IgA-specific serine endopeptidase**, but it is important to note that the enzyme is selective for IgA1, and does not act on IgA2 (see below).

## Activity and Specificity

Bacterial IgA1 proteases cleave human IgA1 in a part of the hinge region that is absent from IgA2 (reviewed in Plaut, 1983; Mulks, 1985; Kilian & Reinholdt, 1986). The typical assay for IgA1 protease activity uses as substrate human IgA of the IgA1 subclass purified from serum of patients with multiple myeloma, or secretory IgA from human colostrum (Mestecky & Kilian, 1985). After incubation, the Fc and Fab fragments

resulting from the action of the enzyme are detected by either immunoelectrophoresis developed with antiserum to human IgA (Mehta *et al.*, 1973), SDS-PAGE, or an ELISA system (Reinholdt, 1996). An overlay method has also been developed to screen bacterial colonies or phage plaques for IgA protease activity (Gilbert & Plaut, 1983). It should be noted that normal human serum and secretions may contain antibodies inhibitory to the IgA1 protease (Gilbert *et al.*, 1983).

Bacterial IgA1 proteases are highly sequence-specific prolyl endopeptidases (i.e. hydrolyzing bonds on the carboxyl side of proline, or prolyl bonds). The exact site of cleavage in the hinge region of human IgA1 differs between IgA1 proteases from different strains of bacteria, however. For the three species mentioned above, each isolate produces one of two distinct types of IgA1 protease. Within the sequence Pro227-Thr-Pro-Ser-Pro-Ser-Thr-Pro-Pro-Thr-Pro-Ser-Pro239 present in the IgA1 heavy chain molecule, the *H. influenzae* type 1 IgA1 protease cleaves at Pro231+Ser232, whereas the gonococcal and meningococcal type 1 enzymes cleave at Pro237+Ser238. The type 2 enzymes of all three species hydrolyze the Pro235+Thr236 bond. The respective Ser residues are *O*-glycosylated. Synthetic peptides covering the IgA1 hinge region are not cleaved. No natural substrate other than IgA1 from humans, chimpanzees and gorillas has been detected, apart from the autocatalytic sites in the preproteases. The amino acid sequence around these autocatalytic cleavage sites in a type 2 IgA1 protease from *N. gonorrhoeae* are similar but not identical to that of the IgA1 hinge region. The three sites may be termed **A**, **B** and **C**, in which the sequences are: **A**: -Val-Lys-Pro+Ser-Pro-Ala-Ala-Asn-; **B**: -Val-Val-Ala-Pro-Pro+Ser-Pro-Gln-Ala-Asn-, and **C**: -Leu-Pro-Arg-Pro-Pro+Ala-Pro-Val-Phe-Ser-. Decapeptides based on the sequences around autocatalytic sites B and C, but not A, are cleaved by the gonococcal type 2 IgA1 protease with a catalytic efficiency ($k_{cat}/K_m$) about 10% of that reported for intact IgA1 (Wood & Burton, 1991). A consensus sequence of the cleavage specificity of the gonococcal type 2 IgA1 protease is Xaa-Pro┼Xbb-Pro, where Xaa stands for Pro (or rarely for Pro in combination with Ala, Gly or Thr) and Xbb stands for Thr, Ser or Ala (Pohlner et al., 1992).

The region in the IgA1 protease that determines the cleavage site in the IgA1 hinge region resides within 124 amino acids in the N-terminal part of the protein (Grundy et al., 1990). The size of this region is characteristic for each of the different cleavage types (Lomholt et al., 1995). In the human IgA1 substrate, cleavage appears to depend on structures C-terminal to the cleavage site in the hinge region (Lomholt et al., 1995).

## Structural Chemistry

The amino acid sequences deduced from the iga genes encoding IgA1 proteases have revealed that the H. influenzae, N. gonorrhoeae and N. meningitidis IgA1 proteases are closely related (Pohlner et al., 1987; Poulsen et al., 1989; Lomholt et al., 1995). Presence of the conserved sequence Gly-Asp-**Ser**-Gly-Ser-Pro-Leu and inhibition of the enzyme activity by peptide boronic acids (Bachovchin et al., 1990) indicate that these IgA1 proteases are distantly related to the chymotrypsin–trypsin family of serine proteases, family S1, and thus in clan SA (see Chapter 2). Elimination of activity by site-directed mutagenesis of the putative active-site Ser to Thr confirms the serine peptidase nature of the enzymes (Poulsen et al., 1992).

Expression of iga genes in E. coli has revealed that each of the IgA1 proteases in H. influenzae, N. gonorrhoeae and N. meningitidis is synthesized as a large precursor of approximately 170 kDa with four distinct functional domains: (a) an N-terminal signal peptide; (b) the secreted IgA1 protease of approximately 110 kDa; (c) autoproteolytic cleavage sites that are similar in sequence to the target sites in human IgA1, and which in some Neisseria strains surround an α-peptide that is cleaved off upon secretion, and (d) a C-terminal β domain that forms a pore structure in the outer membrane and thereby facilitates secretion of the protease domain (Pohlner et al., 1987; Poulsen et al., 1989; Lomholdt et al., 1995). The secreted IgA1 protease domain contains two conserved cysteine residues. The α-peptide found in a gonococcal IgA1 protease precursor contains a functional eukaryotic nuclear location signal as well as a potential DNA-binding motif (Pohlner et al., 1995), but the significance of this peptide in vivo is unknown.

## Preparation

IgA1 protease is produced constitutively during exponential growth and secreted into the culture medium, from which it can be purified chromatographically (Simpson et al., 1988; Blake & Eastby, 1991). Production of IgA1 protease may be enhanced by iron starvation (Simpson et al., 1988). Recombinant E. coli harboring the iga genes also secrete the IgA1 protease (Pohlner et al., 1992; Poulsen et al., 1992). A gonococcal IgA1 protease is commercially available from

Boehringer Mannheim (see Appendix 2 for full names and addresses of suppliers).

## Biological Aspects and Biotechnological Applications

The IgA1 proteases have been shown to be active in vivo. The convergent evolution strongly suggests that IgA1-cleaving activity is essential to these bacteria. However, the biological significance of the IgA1 proteases is incompletely understood (Kilian et al., 1996). By cleaving human IgA1, these enzymes interfere with the protective functions of the principal mediator of specific immunity on mucosal surfaces. Cleavage in the hinge region of IgA1 abolishes the Fc-mediated secondary effector functions, while the Fab fragments retain antigen-binding capacity and therefore may mask epitopes on the bacteria. Presumably, the IgA1 proteases are important for the ability of the bacteria to colonize human mucosal surfaces in the presence of secretory IgA antibodies. By use of neutralizing antibodies raised in rabbits, several serologically distinct versions, termed inhibition types, of the IgA1 protease have been identified in H. influenzae, and this variation appears to serve immune escape purposes by allowing a succession of clones of the same species to colonize one host (Lomholt et al., 1993). The three leading causes of bacterial meningitis, H. influenzae, N. meningitidis and Streptococcus pneumoniae, all produce functionally identical IgA1 proteases, suggesting that this activity is a virulence factor associated with this particular invasive disease. The IgA1 protease of S. pneumoniae is a metalloproteinase described elsewhere (see Chapter 508).

Because of the sequence specificity of the IgA1 proteases they have potential applications for in vitro processing of recombinant proteins (Pohlner et al., 1992). In addition, the β domain of the IgA1 protease precursor may be used to direct secretion or surface exposure of recombinant proteins.

## Related Peptidases

A small number of bacteria associated with humans produce IgA1 proteases belonging to proteinase classes other than the serine proteinases (Plaut, 1983; Mulks, 1985; Kilian & Reinholdt, 1986). The serine endopeptidase-type IgA1 proteases described here belong to a family of proteins (peptidase family S6) that includes Tsh from E. coli, SepA from Shigella flexneri, and Hap from H. influenzae (see Chapter 76). These other proteins are used by a diverse group of Gram-negative bacteria for colonization and invasion. This may suggest additional functions for the IgA1 proteases.

## References

Bachovchin, W.W., Plaut, A.G., Flentke, G.R., Lynch, M. & Kettner, C.A. (1990) Inhibition of IgA1 proteinases from Neisseria gonorrhoeae and Haemophilus influenzae by peptide prolyl boronic acids. J. Biol. Chem. **265**, 3738–3743.

Blake, M.S. & Eastby, C. (1991) Studies on the gonococcal IgA1 protease II. Improved methods of enzyme purification and production of monoclonal antibodies to the enzyme. J. Immunol. Methods **144**, 215–221.

Gilbert, J.V. & Plaut, A.G. (1983) Detection of IgA protease activity among multiple bacterial colonies. J. Immunol. Methods **57**, 247–251.

Gilbert, J.V., Plaut, A.G., Longmaid, B. & Lamm, M.E. (1983) Inhibition of microbial IgA proteases by human secretory IgA and serum. *Mol. Immunol.* **20**, 1039–1049.

Grundy, F.J., Plaut, A.G. & Wright, A. (1990) Localization of the cleavage site specificity determinant of *Haemophilus influenzae* immunoglobulin A1 protease genes. *Infect. Immun.* **58**, 320–331.

Kilian, M. & Reinholdt, J. (1986) Interference with IgA defence mechanisms by extracellular bacterial enzymes. In: *Medical Microbiology*, vol. 5 (Easmon, C.S.F. & Jeljaszewics, J., eds). London: Academic Press, pp. 173–208.

Kilian, M, Mestecky, J & Schrohenloher, R.E. (1979) Pathogenic species of the genus *Haemophilus and Streptococcus pneumoniae* produce immunoglobulin A1 protease. *Infect. Immun.* **26**, 143–149.

Kilian, M., Reinholdt, J., Lomholt, H., Poulsen, K. & Frandsen, E.V.G. (1996) Biological significance of IgA1 proteases in bacterial colonization and pathogenesis: critical evaluation of experimental evidence. *APMIS* **104**, 321–338.

Lomholt, H., van Alphen, L. & Kilian, M. (1993) Antigenic variation of immunoglobulin A1 proteases among sequential isolates of *Haemophilus influenzae* from healthy children and patients with chronic obstructive pulmonary disease. *Infect. Immun.* **61**, 4575–4581.

Lomholt, H., Poulsen, K. & Kilian, M. (1995) Comparative characterization of the *iga* gene encoding IgA1 protease in *Neisseria meningitidis, Neisseria gonorrhoeae* and *Haemophilus influenzae*. *Mol. Microbiol.* **15**, 495–506.

Male, C.J. (1979) Immunoglobulin A1 protease production by *Haemophilus influenzae* and *Streptococcus pneumoniae*. *Infect. Immun.* **26**, 254–261.

Mehta, S.K., Plaut, A.G., Calvanico, N.J. & Tomasi, T.B. (1973) Human immunoglobulin A: production of an Fc fragment by an enteric microbial proteolytic enzyme. *J. Immunol.* **111**, 1274–1276.

Mestecky, J. & Kilian, M. (1985) Immunoglobulin A. *Methods Enzymol.* **116**, 37–75.

Mulks, M.H. (1985) Microbial IgA proteases. In: *Bacterial Enzymes and Virulence* (Holder, I.A., ed.). Boca Raton, FL: CRC Press, pp. 81–104.

Plaut, A.G. (1983) The IgA1 proteases of pathogenic bacteria. *Annu. Rev. Microbiol.* **37**, 603–622.

Plaut, A.G., Genco, R.J. & Tomasi, T.B. (1974) Isolation of an enzyme from *Streptococcus sanguis* which specifically cleaves IgA. *J. Immunol.* **113**, 289–291.

Plaut, A.G., Gilbert, J.V., Artenstein, M.S. & Capra, J.D. (1975) *Neisseria gonorrhoeae* and *Neisseria meningitidis*: extracellular enzyme cleaves human immunoglobulin A. *Science* **190**, 1103–1105.

Pohlner, J., Halter. R., Beyreuther, K. & Meyer, T.F. (1987) Gene structure and extracellular secretion of *Neisseria gonorrhoeae* IgA protease. *Nature* **325**, 458–462.

Pohlner, J., Klauser, T., Kuttler, E. & Halter, R. (1992) Sequence-specific cleavage of protein fusions using a recombinant *Neisseria* type 2 IgA protease. *Biotechnology* **10**, 799–804.

Pohlner, J., Langenberg, U., Wolk, U., Beck, S.C. & Meyer, T.F. (1995) Uptake and nuclear transport of *Neisseria* IgA1 protease-associated α-proteins in human cells. *Mol. Microbiol.* **17**, 1073–1083.

Poulsen, K., Brandt, J., Hjorth, J.P., Thorgersen, H.C. & Kilian, M. (1989) Cloning and sequencing of the immunoglobulin A1 protease gene (*iga*) of *Haemophilus influenzae* serotype b. *Infect. Immun.* **57**, 3097–3105.

Poulsen, K., Reinholdt, J. & Kilian, M. (1992) A comparative genetic study of serologically distinct *Haemophilus influenzae* type 1 immunoglobulin A1 proteases. *J. Bacteriol.* **174**, 2913–2921.

Reinholdt, J. (1996) A method for titration of inhibiting antibodies to bacterial immunoglobulin A1 proteases in human serum and secretions. *J. Immunol. Methods* **191**, 39–48.

Simpson, D.A., Hausinger, R.P. & Mulks, M.H. (1988) Purification, characterization, and comparison of the immunoglobulin A1 proteases of *Neisseria gonorrhoeae*. *J. Bacteriol.* **170**, 1866–1873.

Wood, S.G. & Burton, J. (1991) Synthetic peptide substrates for the immunoglobulin A1 protease from *Neisseria gonorrhoeae* (type 2). *Infect. Immun.* **59**, 1818–1822.

*Knud Poulsen*
*Department of Medical Microbiology and Immunology,*
*The Bartholin Building,*
*University of Aarhus,*
*DK-8000 Aarhus C, Denmark*

*Mogens Kilian*
*Department of Medical Microbiology and Immunology,*
*The Bartholin Building,*
*University of Aarhus,*
*DK-8000 Aarhus C, Denmark*
*Email: mogens.kilian@svfcd.aau.dk*

# 88. Flavivirin

## Databanks

*Peptidase classification: clan SA, family S7, MEROPS ID: S07.001*
*NC-IUBMB enzyme classification: EC 3.4.21.91*

*ATCC entries: Several virus isolates are available.*
*Databank codes:*

| Species | SwissProt | PIR | EMBL (genomic) |
|---|---|---|---|
| Dengue virus type 2 | P07564 | A94346 | M20558: complete genome |
| Dengue virus type 2 | P12823 | A29972 | M19197: complete genome |
| Dengue virus type 4 | P09866 | A94352 | M14931: complete genome |
| Japanese encephalitis virus | P19110 | – | D90194: complete genome |
| Kunjin virus | P14335 | A28697 | D00246: complete genome |
| Murray Valley encephalitis virus | P05769 | A24635 | X03467: 5′ end of genome |
| Tick-borne encephalitis virus | P07720 | A24055 | – |
| | | A33776 | |
| Tick-borne encephalitis virus | P14336 | – | U27495: complete genome |
| West Nile virus | P06935 | A25256 | M12294: complete genome |
| Yellow fever virus | P03314 | A03914 | K02749: complete genome |
| Yellow fever virus | P19901 | S07757 | X15062: complete genome; unannotated |

## Name and History

*Flavivirin* is the name recommended by NC-IUBMB for the NS2B-3 endopeptidase of the *Flavivirus* genus, that comprises at least 68 known members of the Flaviviridae family (see Rice, 1996). Flaviviruses contain a positive-strand RNA molecule that encodes a single long open reading frame; the expressed polyprotein is processed by several distinct enzymatic activities. The structural proteins (C, prM and E) are processed from the N-terminal quarter of the polyprotein, while the nonstructural proteins (NS1, NS2A, NS2B, NS3, NS4A, NS4B and NS5) are generated from the remainder. The N-terminal third of NS3 was predicted to be a trypsin-like protease based upon sequence homology (Bazan & Fletterick, 1989, 1990; Gorbalenya *et al.*, 1989a), and this was borne out by mutagenesis of the catalytic triad (yellow fever NS3 numbering): His53, Asp77 and Ser138 (Chambers *et al.*, 1990b; Pugachev *et al.*, 1993; Wengler *et al.*, 1991; Zhang *et al.*, 1992). Between the flavivirus groups for which sequence data have been obtained, the catalytic domain (N-terminal third) of NS3 exhibits between 30 and 55% amino acid identity. NS3 alone has not been shown to exhibit proteolytic activity; instead, flavivirin appears to be a complex of NS3 (the catalytic domain) and a requisite cofactor, NS2B.

## Activity and Specificity

Flavivirin mediates the cleavages that generate the N-termini of nonstructural proteins NS2B, NS3, NS4A and NS5. In addition, flavivirin also mediates cleavages within the C-terminal regions of the capsid (Amberg *et al.*, 1994; Yamshchikov & Compans, 1994) and NS4A (Lin *et al.*, 1993a). Analysis of these cleavage sites has identified a substrate motif. Flavivirin cleaves on the C-terminal side of a pair of basic amino acids, and on the N-terminal side of an amino acid with a short side chain. Arg is usually found at the P1 site, although Lys can also be found in this position. Lys and Arg are equally represented at the P2 site, although Gln is sometimes observed. The P1′ residue is most commonly Ser or Gly, less commonly Ala or Thr. Mutagenesis experiments have verified the importance of this motif as well as the high tolerance to Gln substitution at the P2 site (Lin *et al.*, 1993b; Nestorowicz *et al.*, 1994; Chambers *et al.*, 1995). However,

partial cleavage products can still be detected with a range of nonconservative substitutions in the P1 position of the 3/4A and 4B/5 sites, the P1′ position of the 2B/3 site, and in the P2 position of the 2B/3 and 4B/5 sites. This suggests that additional substrate determinants also play an important role in specificity. Substitutions in the P3 and P4 positions demonstrate minor or no effects on cleavage efficiency (Nestorowicz *et al.*, 1994; Chambers *et al.*, 1995). Mutations within the putative substrate-binding pocket were found to have significant effects on cleavage order and efficiency (Preugschat *et al.*, 1991). The flavivirus polyprotein contains several flavivirin substrate motif sequences for which the corresponding cleavages have not been detected, again emphasizing the significance of the context of the substrate. However, additional flavivirin cleavage sites have been identified within NS2A (Nestorowicz *et al.*, 1994) and NS3 (Arias *et al.*, 1993).

## Structural Chemistry

The NS3 protein is located in the middle of the flavivirus polyprotein and represents about 18% of the entire polyprotein (68–70 kDa). The N-terminal third of NS3 is essential for proteolytic activity, while the remaining two-thirds of NS3 are dispensable (Chambers *et al.*, 1990b; Preugschat *et al.*, 1990; Falgout *et al.*, 1991; Wengler *et al.*, 1991). The C-terminal two-thirds contain an RNA triphosphatase activity (Warrener *et al.*, 1993; Wengler & Wengler, 1991) and a motif for the DEAD family of RNA helicases (Gorbalenya *et al.*, 1989b). Flavivirin is found as a nonionic detergent-stable complex of NS2B and NS3 (Arias *et al.*, 1993; Chambers *et al.*, 1993). Although it is associated with membranes (Wengler *et al.*, 1990), NS3 has no significant hydrophobic regions. NS3 membrane localization is likely to be due to its interactions with NS2B. NS2B is a small membrane-bound protein (14 kDa) that is not particularly well conserved among flaviviruses. A more conserved, highly charged central domain of NS2B is critical for interacting with NS3 (Chambers *et al.*, 1993; Falgout *et al.*, 1993). This domain is flanked by hydrophobic stretches that are likely to be responsible for the membrane localization of NS2B. The region critical for activating NS3 flavivirin has been mapped to 40 amino acids for dengue 4 (Falgout *et al.*, 1993). Flavivirin cleavages occur in the cytoplasm, so the current model proposes that the central, hydrophilic domain of NS2B is exposed on the cytoplasmic side of the endoplasmic reticulum, where it is accessible for interactions with NS3. Within the flavivirus polyprotein, NS2B is located adjacent to the N-terminus of NS3; the NS2B termini are generated by rapid autocatalytic flavivirin cleavages at the 2A/2B and 2B/3 sites (Chambers *et al.*, 1990b; Preugschat *et al.*, 1990). Mutations at the 2B/3 site which inactivate cleavage clearly do not inactivate the enzyme, since cleavages at other sites still occur (Chambers *et al.*, 1995). The 3/4A cleavage event, which generates the C-terminus of NS3, also occurs primarily in *cis* (Chambers *et al.*, 1991), although inefficient cleavage at this site allows for a fairly stable NS3-4A species in infected cells (Chambers *et al.*, 1990a). Efficient *in vivo trans* cleavage has been reported within the C-terminus of NS4A (Lin *et al.*, 1993a) and at the 2B/3 (Chambers *et al.*, 1991) and 4B/5 sites (Chambers *et al.*, 1991; Zhang *et al.*, 1992).

## Preparation

Flavivirin has not yet been purified. Expression of the non-structural region of the polyprotein by cell-free translation in a rabbit reticulocyte lysate system demonstrates rapid *cis* cleavage at the NS2A/2B and NS2B/3 sites, even in the absence of membranes (Chambers *et al.*, 1990b; Preugschat *et al.*, 1990). An *in vitro* assay for flavivirin has been reported using Triton X-100-solubilized lysates of infected cells as an enzyme source (Amberg *et al.*, 1994). The substrate for this assay is a cell-free translation product containing the dibasic site from the C-terminus of the capsid protein. Activity was found to segregate with the membrane fraction, and was found to be dependent on the addition of detergent. Lysates of cells infected with vaccinia recombinants expressing NS2B and NS3, either as a polyprotein or as individual recombinants, are also competent to cleave this substrate (Amberg *et al.*, 1994; Yamshchikov & Compans, 1994). Cotranslation of flavivirin and the capsid substrate in a cell-free translation system has also been used to demonstrate activity (Yamshchikov & Compans, 1994).

## Biological Aspects

The role of flavivirin, or any viral protease, is to process polyproteins to mature proteins, and assembly of the viral RNA replication complex is likely to depend on flavivirin-mediated processing of the nonstructural proteins. Processing intermediates may have functions distinct from their terminal products such that flavivirin could play an important regulatory role. It has been demonstrated that active flavivirin is essential for viral replication (Chambers *et al.*, 1990b), although the precise stage in the replication cycle that is inhibited has not been identified; in addition, mutations at the 2A/2B (Nestorowicz *et al.*, 1994) and 2B/3 (Chambers *et al.*, 1995) substrate sites that block cleavage are also lethal for plaque formation, as are mutations within NS2B (Chambers *et al.*, 1993). An interesting feature of the polyprotein processing scheme is the flavivirin-mediated processing within the C-terminal portions of the C (capsid) and NS4A proteins; these cleavages appear to be required for subsequent signal peptidase (Chapter 156) cleavage, which occur 14–23 amino acids C-terminal to the dibasic site, to generate the

N-termini of their respective downstream proteins, prM and NS4B. Thus, flavivirin generates the C-terminus of the capsid protein (Nowak *et al.*, 1989; Speight & Westaway, 1989), and probably NS4A as well. It has been demonstrated that proper processing of a translocated C-prM species, in the absence of flavivirin, can be restored by trypsin digestion of the cytoplasmic domain (the C protein), which allows signal peptidase cleavage to occur post-translationally on the luminal side of the membrane to release the N-terminus of prM (Stocks & Lobigs, 1995). Presumably flavivirin functions similarly by removing the capsid protein from the polyprotein. The prM glycoprotein is a structural component of the immature virion and a precursor to the M protein found in the mature virion envelope. Extracellular release of structural proteins prM and E is dependent on either active flavivirin, or genetic removal of the C protein coding region, suggesting that flavivirin plays a critical role in virus assembly (Pincus *et al.*, 1992; Lobigs, 1993; Sato *et al.*, 1993; Yamshchikov & Compans, 1993, 1995). As mentioned previously, alternative cleavages within the NS2A and NS3 proteins have been identified. Infectious virus was not recovered from transcripts in which the internal NS2A site was inactivated by site-directed mutagenesis, suggesting a functional role in viral replication (Nestorowicz *et al.*, 1994). The biological significance of the alternative cleavage site within NS3 has not been determined. The importance of uncleaved NS3-4A in infected cells is also unclear; possibly the distribution of NS3 versus NS3-4A might have some regulatory function within the replication complex. Cellular targets for flavivirin have not yet been identified.

## Distinguishing Features

The two other genera of the Flaviviridae family, the hepatitis C viruses (Chapter 89) and the pestiviruses (Chapter 91), also encode endopeptidases of clan SA in the same region of the polyprotein. The overall processing schemes of these viruses are similar, the serine protease-mediating cleavages being within the nonstructural region of the polyprotein. The flavivirus genus is unique, however, in the two cleavages that occur 14–23 residues upstream of signal peptidase cleavages (at the C-termini of capsid and NS4A). While NS2B is an obligatory cofactor for the flavivirus NS3, the NS3 proteases of hepatitis C virus and pestivirus utilize NS4A as a cofactor that is necessary for some, but not all, cleavages. Substrate specificity also differs; flavivirus substrates generally have basic residues in the P1 and P2 positions, hepatitis C virus substrates have a Cys or Thr residue in the P1 position, and pestiviruses prefer Leu in the P1 location. Finally, the *Flavivirus* genus is the only one of the three in which the serine protease activity of NS3 mediates the cleavage that generates its own N-terminus (see Chapter 567).

## Acknowledgement

We would like to thank Brett Lindenbach for his help with *Flavivirus* sequence comparisons.

## Further Reading

Further information is to be found in Bazan & Fletterick (1990) and Rice (1996).

## References

Amberg, S.M., Nestorowicz, A., McCourt, D.W. & Rice, C.M. (1994) NS2B-3 proteinase-mediated processing in the yellow fever virus structural region: *In vitro* and *in vivo* studies. *J. Virol.* **68**, 3794–3802.

Arias, C.F., Preugschat, F. & Strauss, J.H. (1993) Dengue 2 virus NS2B and NS3 form a stable complex that can cleave NS3 within the helicase domain. *Virology* **193**, 888–899.

Bazan, J.F. & Fletterick, R.J. (1989) Detection of a trypsin-like serine protease domain in flaviviruses and pestiviruses. *Virology* **171**, 637–639.

Bazan, J.F. & Fletterick, R.J. (1990) Structural and catalytic models of trypsin-like viral proteases. *Semin. Virol.* **1**, 311–322.

Chambers, T.J., McCourt, D.W. & Rice, C.M. (1990a) Production of yellow fever virus proteins in infected cells: identification of discrete polyprotein species and analysis of cleavage kinetics using region-specific polyclonal antisera. *Virology* **177**, 159–174.

Chambers, T.J., Weir, R.C., Grakoui, A., McCourt, D.W., Bazan, J.F., Fletterick, R.J. & Rice, C.M. (1990b) Evidence that the N-terminal domain of nonstructural protein NS3 from yellow fever virus is a serine protease responsible for site-specific cleavages in the viral polyprotein. *Proc. Natl Acad. Sci. USA* **87**, 8898–8902.

Chambers, T.J., Grakoui, A. & Rice, C.M. (1991) Processing of the yellow fever virus nonstructural polyprotein: a catalytically active NS3 proteinase domain and NS2B are required for cleavages at dibasic sites. *J. Virol.* **65**, 6042–6050.

Chambers, T.J., Nestorowicz, A., Amberg, S.M. & Rice, C.M. (1993) Mutagenesis of the yellow fever virus NS2B protein: effects on proteolytic processing, NS2B-NS3 complex formation, and viral replication. *J. Virol.* **67**, 6797–6807.

Chambers, T.J., Nestorowicz, A. & Rice, C.M. (1995) Mutagenesis of the yellow fever virus NS2B/3 cleavage site: determinants of cleavage site specificity and effects on polyprotein processing and viral replication. *J. Virol.* **69**, 1600–1605.

Falgout, B., Pethel, M., Zhang, Y.-M. & Lai, C.-J. (1991) Both nonstructural proteins NS2B and NS3 are required for the proteolytic processing of Dengue virus nonstructural proteins. *J. Virol.* **65**, 2467–2475.

Falgout, B., Miller, R.H. & Lai, C.-J. (1993) Deletion analysis of Dengue virus type 4 nonstructural protein NS2B: identification of a domain required for NS2B-NS3 proteinase activity. *J. Virol.* **67**, 2034–2042.

Gorbalenya, A.E., Donchenko, A.P., Koonin, E.V. & Blinov, V.M. (1989a) N-terminal domains of putative helicases of flavi- and pestiviruses may be serine proteases. *Nucleic Acids Res.* **17**, 3889–3897.

Gorbalenya, A.E., Koonin, E.V., Donchenko, A.P. & Blinov, V.M. (1989b) Two related superfamilies of putative helicases involved in replication, recombination, repair and expression of DNA and RNA genomes. *Nucleic Acids Res.* **17**, 4713–4729.

Lin, C., Amberg, S.M., Chambers, T.J. & Rice, C.M. (1993a) Cleavage at a novel site in the NS4A region by the yellow fever virus NS2B-3 proteinase is a prerequisite for processing at the downstream 4A/4B signalase site. *J. Virol.* **67**, 2327–2335.

Lin, C., Chambers, T.J. & Rice, C.M. (1993b) Mutagenesis of conserved residues at the yellow fever virus 3/4A and 4B/5 dibasic cleavage sites: effects on cleavage efficiency and polyprotein processing. *Virology* **192**, 596–604.

Lobigs, M. (1993) Flavivirus premembrane protein cleavage and spike heterodimer secretion requires the function of the viral proteinase NS3. *Proc. Natl Acad. Sci. USA* **90**, 6218–6222.

Nestorowicz, A., Chambers, T.J. & Rice, C.M. (1994) Mutagenesis of the yellow fever virus NS2A/2B cleavage site: effects on proteolytic processing, viral replication and evidence for alternative processing of the NS2A protein. *Virology* **199**, 114–123.

Nowak, T., Färber, P.M., Wengler, G. & Wengler, G. (1989) Analyses of the terminal sequences of West Nile virus structural proteins and of the *in vitro* translation of these proteins allow the proposal of a complete scheme of the proteolytic cleavages involved in their synthesis. *Virology* **169**, 365–376.

Pincus, S., Mason, P.W., Konishi, E., Fonseca, B.A.L., Shope, R.E., Rice, C.M. & Paoletti, E. (1992) Recombinant vaccinia virus producing the prM and E proteins of yellow fever virus protects mice from lethal yellow fever encephalitis. *Virology* **187**, 290–297.

Preugschat, F., Yao, C.-W. & Strauss, J.H. (1990) In vitro processing of dengue virus type 2 nonstructural proteins NS2A, NS2B, and NS3. *J. Virol.* **64**, 4364–4374.

Preugschat, F., Lenches, E.M. & Strauss, J.H. (1991) Flavivirus enzyme–substrate interactions studied with chimeric proteinases: identification of an intragenic locus important for substrate recognition. *J. Virol.* **65**, 4749–4758.

Pugachev, K.V., Nomokonova, N.Y., Dobrikova, E.Y. & Wolf, Y.I. (1993) Site-directed mutagenesis of the tick-borne encephalitis virus NS3 gene reveals the putative serine protease domain of the NS3 protein. *FEBS Lett.* **328**, 115–118.

Rice, C.M. (1996) *Flaviviridae*: The viruses and their replication. In: *Fields Virology* (Fields, B.N., Knipe, D.M. & Howley, P.M., eds). New York: Raven Press, pp. 931–960.

Sato, T., Takamura, C., Yasuda, A., Miyamoto, M., Kamogawa, K. & Yasui, K. (1993) High-level expression of the Japanese encephalitis virus E protein by recombinant vaccinia virus and enhancement of its extracellular release by the NS3 gene product. *Virology* **192**, 483–490.

Speight, G. & Westaway, E.G. (1989) Carboxy-terminal analysis of nine proteins specified by the flavivirus Kunjin: Evidence that only the intracellular core protein is truncated. *J. Gen. Virol.* **70**, 2209–2214.

Stocks, C.E. & Lobigs, M. (1995) Posttranslational signal peptidase cleavage at the flavivirus C-prM junction *in vitro*. *J. Virol.* **69**, 8123–8126.

Warrener, P., Tamura, J.K. & Collett, M.S. (1993) An RNA-stimulated NTPase activity associated with yellow fever virus NS3 protein expressed in bacteria. *J. Virol.* **67**, 989–996.

Wengler, G. & Wengler, G. (1991) The carboxy-terminal part of the NS3 protein of the West Nile flavivirus can be isolated as a soluble protein after proteolytic cleavage and represents an RNA-stimulated NTPase. *Virology* **184**, 707–715.

Wengler, G., Wengler, G., Nowak, T. & Castle, E. (1990) Description of a procedure which allows isolation of viral nonstructural proteins from BHK vertebrate cells infected with the West Nile flavivirus in a state which allows their direct chemical characterization. *Virology* **177**, 795–801.

Wengler, G., Czaya, G., Färber, P.M. & Hegemann, J.H. (1991) *In vitro* synthesis of West Nile virus proteins indicates that the amino-terminal segment of the NS3 protein contains the active centre of the protease which cleaves the viral polyprotein after multiple basic amino acids. *J. Gen. Virol.* **72**, 851–858.

Yamshchikov, V.F. & Compans, R.W. (1993) Regulation of the late events in flavivirus protein processing and maturation. *Virology* **192**, 38–51.

Yamshchikov, V.F. & Compans, R.W. (1994) Processing of the intracellular form of the west Nile virus capsid protein by the viral NS2B-NS3 protease: an *in vitro* study. *J. Virol.* **68**, 5765–5771.

Yamshchikov, V.F. & Compans, R.W. (1995) Formation of the flavivirus envelope: role of the viral NS2B-NS3 protease. *J. Virol.* **69**, 1995–2003.

Zhang, L., Mohan, P.M. & Padmanabhan, R. (1992) Processing and localization of Dengue virus type 2 polyprotein precursor NS3-NS4A-NS4B-NS5. *J. Virol.* **66**, 7549–7554.

**Sean M. Amberg**
*Department of Molecular Microbiology,*
*Washington University School of Medicine,*
*660 South Euclid Avenue, St. Louis, MO 63110, USA*

**Charles M. Rice**
*Department of Molecular Microbiology,*
*Washington University School of Medicine,*
*660 South Euclid Avenue, St. Louis, MO 63110, USA*
*Email: Rice@borcim.wustl.edi*

# 89. *Hepatitis C virus polyprotein peptidase*

## Databanks

Peptidase classification: clan SA, family S29, MEROPS ID: S29.001
NC-IUBMB enzyme classification: none
ATCC entries: catalog no. 68275: plasmid clone of HCV-1 encompassing polyprotein residues 946–1630 (approximate location of the NS3 serine protein catalytic domain: residues 1027–1207) fused to superoxide dismutase
Databank codes:

| Species | Isolate | SwissProt | PIR | EMBL (genomic) |
|---|---|---|---|---|
| Hepatitis C virus type 2c | BEBE1 | – | – | D50409: complete genome |
| Hepatitis C virus | BK | P26663 | A38465 | M58335: complete genome |
| Hepatitis C virus | H | P27958 | A36814 | M67463: complete genome |
| Hepatitis C virus type 1c | HC-G9 | – | – | D14853: complete genome |
| Hepatitis C virus type 2a | HC-J6 | P26660 | JQ1303 | D00944: complete genome |
| Hepatitis C virus type 2b | HC-J8 | P26661 | A40250 | D10988: complete genome |
| Hepatitis C virus | HC-JT | Q00269 | A45573 | D11168: complete genome |
| Hepatitis C virus type 1a[a] | HCV-1 | P26664 | A39166 | M62321: complete genome |
| Hepatitis C virus type 1b[a] | HCV-J | P26662 | A39253 | D90208: complete genome |
| Hepatitis C virus type 11a or 6 | JK046 | – | – | D63822: complete genome |
| Hepatitis C virus type 3a[a] | NZL1 | – | – | D17763: complete genome |
| Hepatitis C virus | Taiwan | P29846 | A40244 | M84754: 5' end of genome |
| Hepatitis C virus type 3b | HCV-Tr | – | – | D49374: complete genome |
| Hepatitis G virus | R10291 | – | – | U45966: complete genome |
| Hepatitis GB virus A | | – | – | U22303: polyprotein gene |
| Hepatitis GB virus B | | – | – | U22304: complete genome |
| Hepatitis GB virus C | | – | – | U36380: polyprotein gene |

[a]Additional cloned isolates are available for these genotypes; see Bukh *et al*. (1995) for a discussion of HCV genotypes, and quasispecies and Tokita *et al*. (1996) for a more complete listing of isolate designations and accession numbers. Isolate JK046 is classified as genotype 11a by Tokita *et al*. (1996), but as genotype 6 by Simmonds *et al*. (1996).

Brookhaven Protein Data Bank three-dimensional structures:

| Species | ID | Resolution | Notes |
|---|---|---|---|
| hepatitis C Virus | 1A1Q | | monomer; ON HOLD |
| | 1A1R | | complex with NS4A peptide; ON HOLD |
| | 1A1V | | helicase domain; ON HOLD till 17 Dec 1998 |

## Name and History

The family Flaviviridae contains positive-stranded RNA viruses of three genera: the classical flaviviruses such as yellow fever virus (see Chapter 88), the animal pestiviruses such as bovine viral diarrhea virus (BVDV) (see Chapter 91), and the hepatitis C virus (HCV) group, provisionally named hepaciviruses, that are the subject of the present chapter. For all of these viruses, viral proteins are produced by co- and post-translational proteolytic processing of a single, long polyprotein translation product (reviewed in Rice, 1996). Both cellular proteinases such as signal peptidase (Chapter 156) and viral proteinases are involved. In 1989, two groups reported the discovery of a serine proteinase-like motif in flavivirus and pestivirus polyproteins (Bazan & Fletterick, 1989, 1990; Gorbalenya et al., 1989). As sequences became available, this observation was extended to HCV (Chambers et al., 1990a; Miller & Purcell, 1990) and the unclassified HCV-related viruses GBV-A, GBV-B and GBV-C/HGV (see below). For HCV, the predicted serine proteinase catalytic domain is approximately 180 residues in length, lies between polyprotein residues 1027 and 1207 (genotype 1a numbering), and makes up the N-terminal one-third of the 70 kDa nonstructural protein 3 (NS3). Site-directed mutagenesis of predicted catalytic triad residues (His1083, Asp1107 and Ser1165) has implicated this activity in cleavage at four sites in the polyprotein (Bartenschlager et al., 1993; Eckart et al., 1993; Grakoui et al., 1993; Hijikata et al., 1993a; Manabe et al., 1994; Tomei et al., 1993). In addition to the catalytic domain, the 54 residue NS4A 'cofactor' protein leads to dramatically enhanced cleavage at some processing sites (Bartenschlager et al., 1994; Failla et al., 1994; Lin et al., 1994; Tanji et al., 1995). Names commonly used to describe the **hepatitis C virus polyprotein peptidase** include **NS3 serine proteinase, NS3-4A serine proteinase complex**, and **Cpro-2**.

## Activity and Specificity

The four NS3 serine proteinase-dependent cleavage sites (3/4A, 4A/4B, 4B/5A and 5A/5B) were defined by N-terminal sequence analyses of NS4A, NS4B, NS5A and NS5B (Grakoui et al., 1993; Komoda et al., 1994b; Leinbach et al., 1994; Pizzi et al., 1994; Hirowatari et al., 1995b). Conserved features of the cleavage sites include an acidic residue (Asp or Glu) at the P6 position, a Cys or Thr residue at the P1 position, and a Ser or Ala residue at the P1′ position. Site-directed mutagenesis of residues flanking these cleavage sites verified the preference for Cys at P1 and for amino acids with small side chains at the P1′ position (Kolykhalov et al., 1994; Komoda et al., 1994a; Leinbach et al., 1994; Bartenschlager et al., 1995). The 3/4A site, which is believed to be cleaved efficiently in cis, is least sensitive to mutagenesis and contains a suboptimal Thr at P1 (Komoda et al., 1994a). The Cys P1 preference is also consistent with the S1 specificity pocket in molecular models (Pizzi et al., 1994) and recently determined X-ray structures (Kim et al., 1996; Love et al., 1996). Relatively large hydrophobic residues form the shallow, hydrophobic S1 pocket. Phe1180 is located at the bottom of the pocket in an appropriate position to make favorable contacts with the Cys P1 residue. Mutations that increase the size of the S1 pocket allow cleavage of substrates with a bulkier Phe residue at P1 (Failla et al., 1996). The acidic residue at P6 is not required for cleavage (Kolykhalov et al., 1994; Leinbach et al., 1994; Bartenschlager et al., 1995), although effects of mutations at this position have been observed in inefficient trans-cleavage assays (Komoda et al., 1994a). Additional cleavage-site determinants and the NS4A cofactor can influence processing efficiency. For instance, using decapeptide substrates (P6–P4′) and purified

proteinase domain, the relative order of cleavage efficiency ($k_{cat}/K_m$) observed is 5A/5B > 4A/4B ≫ 4B/5A (Steinkuhler et al., 1996b). Stoichiometric amounts of synthetic NS4A cofactor peptide increase the rate of cleavage of all substrates, but the greatest effect (100-fold) is seen for the 4B-5A decapeptide and the least effect (3-fold) for the 5A/5B decapeptide (Steinkuhler et al., 1996b). These results mimic those of in vivo, trans-processing studies, where NS4A is required for cleavage at the 4B/5A site but not for cleavage at the 5A/5B site (Bartenschlager et al., 1994; Failla et al., 1994; Lin et al., 1994; Tanji et al., 1995). The effect of NS4A appears to be primarily to enhance substrate binding affinity (Steinkuhler et al., 1996b).

Conditions for optimal activity and inhibitor profiles vary among reports, possibly depending upon the subregion expressed, genotype or isolate-specific sequence differences, the degree of purity, and the substrates utilized for processing assays. For the genotype 1b enzyme purified from E. coli (Steinkuhler et al., 1996b), the pH optimum is between 8.0 and 8.5 in reactions containing 50% glycerol, 2% CHAPS, and 10 mM DTT. The enzyme is inhibited by salt and destabilized by high pH, oxidizing conditions and high temperature. $k_{cat}/K_m$ values are low and vary between 0.4 and 650 M$^{-1}$ s$^{-1}$, depending upon the peptide substrate and the presence of NS4A cofactor peptide. By incorporating an ester bond between the P1 and P1′ residues, $k_{cat}/K_m$ values can be increased more than 100-fold, allowing assay of subnanomolar proteinase concentrations (Bianchi et al., 1996) and development of continuous activity assays based on resonance energy transfer (Taliani et al., 1996).

Some preparations of the enzyme show inhibition by high concentrations of serine proteinase inhibitors (D'Souza et al., 1995; Hahm et al., 1995; Hong et al., 1996; Lin & Rice, 1995; Shoji et al., 1995; Mori et al., 1996), by copper (Hahm et al., 1995; Han et al., 1995), and slight inhibition by EDTA (Lin & Rice, 1995). Much effort is being devoted to developing specific and potent HCV serine proteinase inhibitors. Besides in vitro biochemical assays for inhibitor screening and evaluation, cell-based assays using reporter genes have been developed for yeast (Song et al., 1996) and mammalian cells (Hirowatari et al., 1995a). Since HCV replication in cell culture is inefficient, engineered chimeric RNA viruses that depend upon the HCV NS3 proteinase for propagation (Filocamo et al., 1997; Hahm et al., 1996) may prove useful for isolating and characterizing inhibitor-resistant proteinase variants.

## Structural Chemistry

In normal polyprotein processing, cleavage at the 2/3 site by a second HCV-encoded autoproteinase (see Chapter 567) produces the N-terminus of NS3 (residue 1027). Cis cleavage at the 3/4A site is believed to occur in conjunction with the assembly of the NS3-4A proteinase complex. Regions implicated in NS3-NS4A complex formation and NS4A-dependent proteinase activation map to the N-terminal portion of the NS3 serine proteinase catalytic domain (Bartenschlager et al., 1994, 1995; Failla et al., 1995; Satoh et al., 1995; Koch et al., 1996) and a central region of the 54 residue NS4A protein (Lin et al., 1995; Tanji et al., 1995; Butkiewicz et al., 1996; Shimizu et al., 1996; Tomei et al., 1996). The

X-ray structure of the serine proteinase domain complexed with a synthetic NS4A cofactor peptide has been solved at 2.5 Å resolution (Kim *et al.*, 1996). This work reveals a two-domain chymotrypsin-like fold and shows that NS4A forms an integral part of the N-terminal domain comprising one $\beta$ strand of an eight-stranded twisted $\beta$ sheet. The C-terminal domain consists of a six-stranded $\beta$ barrel typical of the chymotrypsin family, followed by a structurally conserved $\alpha$ helix (Kim *et al.*, 1996; Love *et al.*, 1996). This domain is stabilized by a single tetrahedrally coordinated structural zinc ion (DeFrancesco *et al.*, 1996; Kim *et al.*, 1996; Love *et al.*, 1996) similar to that observed in the picornavirus 2A proteinases (see Chapter 241).

## Preparation

The serine proteinase from authentic HCV-infected cells has not been characterized. All studies to date have used cDNA clones to express the enzyme by cell-free translation (Bouffard *et al.*, 1995; Hahm *et al.*, 1995; Lin and Rice, 1995), or in *E. coli* (D'Souza *et al.*, 1995; Kakiuchi *et al.*, 1995; Shoji *et al.*, 1995; Mori *et al.*, 1996; Steinkuhler *et al.*, 1996b), insect (Overton *et al.*, 1995; Steinkuhler *et al.*, 1996a; Suzuki *et al.*, 1995) or mammalian (Hong *et al.*, 1996) systems. Soluble active enzyme from the *E. coli* and recombinant baculovirus-infected insect cell systems has been purified in milligram quantities (e.g. Kim *et al.*, 1996; Love *et al.*, 1996; Steinkuhler *et al.*, 1996a,b). To avoid problems with the hydrophobic N-terminus of the NS4A cofactor, synthetic peptides encompassing the NS4A central domain are often used to form the activated serine proteinase complex.

## Biological Aspects

Processing of the nonstructural (NS) region of the HCV polyprotein is believed to be essential for the assembly and function of the viral RNA replication machinery. In the case of the classical flaviviruses and pestiviruses, inactivation of the NS3 serine proteinase abrogates detectable virus replication (see Chambers *et al.*, 1990b; Xu *et al.*, 1997). Moreover, even mutations inhibiting or blocking cleavage at single processing sites can be lethal (Nestorowicz *et al.*, 1994; Chambers *et al.*, 1995). The HCV serine proteinase mediates four cleavages in the NS region important for liberating five NS proteins. During and after processing, the resulting proteins are believed to remain associated with each other and with host components, such as cytoplasmic cellular membranes, to form the RNA replication complex. Although far from being well understood, an early step in this assembly process probably involves *cis* cleavage at the 3/4A site coincident with formation of the NS3-4A proteinase complex. This complex may remain membrane associated, perhaps via the hydrophobic N-terminus of NS4A (Hijikata *et al.*, 1993b; Tanji *et al.*, 1995). The highly conserved C-terminal portion of NS3 contains both RNA-stimulated NTPase and RNA helicase activities which presumably function in some aspect of RNA replication (reviewed in Rice, 1996). The other three cleavages, which can occur in *trans*, produce mature NS4A, NS4B, NS5A and the NS5B RNA-dependent RNA polymerase. The precise details of the way in which these cleavage events modulate HCV RNA replication remain to be determined. In addition to its role in HCV polyprotein processing, the ability of the NS3 serine proteinase to function in *trans* raises the intriguing possibility that it may also cleave cellular proteins. To date, however, cellular substrates of the enzyme have not been reported.

## Distinguishing Features

The subunit structure and cleavage-site specificity of the HCV NS3 serine proteinase differ from those observed for the classical flaviviruses (see Chapter 88). The flavivirus enzyme consists of a complex of the upstream NS2B protein and NS3. NS2B is absolutely required for cleavage at all serine proteinase-dependent sites, which include one site in the structural region, an autocatalytic cleavage at the 2B/3 site, and at least four other sites in the NS region. In contrast to the HCV enzyme, the flavivirus enzyme typically cleaves after dibasic residues. The situation for the pestiviruses, as determined for BVDV, is much more similar to HCV (see Xu *et al.*, 1997, and Chapter 91). The pestivirus serine proteinase is responsible for four downstream cleavages and requires the 64 residue pestivirus NS4A protein for cleavage at some of these sites (Xu *et al.*, 1997). In contrast to HCV, cleavage sites for the pestiviral enzyme contain Leu at P1, and molecular models predict a narrow, hydrophobic S1 specificity pocket. Much less is known about the enzymes encoded by the HCV-related GB viral agents (Simons *et al.*, 1995b) and GBV-C/HGV (Simons *et al.*, 1995a; Linnen *et al.*, 1996). Alignment of the GBV-B and HCV polyproteins suggests a similar processing scheme and cleavage-site specificity (Muerhoff *et al.*, 1995). Cleavage-site preferences for the GBV-A and GBV-C/HGV enzymes have not been reported. Thus, although the serine proteinases of the Flaviviridae and related viruses are likely to have similar structures and functions in virus replication, they are poorly conserved at the sequence level and often differ in their cleavage-site specificity.

## Further Reading

Fuller reviews of several aspects are to be found in Bazan & Fletterick (1990), Houghton (1996), and Rice (1996).

## References

Bartenschlager, R., Ahlborn-Laake, L., Mous, J. & Jacobsen, H. (1993) Nonstructural protein 3 of the hepatitis C virus encodes a serine-type proteinase required for cleavage at the NS3/4 and NS4/5 junctions. *J. Virol.* **67**, 3835–3844.

Bartenschlager, R., Ahlborn-Laake, L., Mous, J. & Jacobsen, H. (1994) Kinetic and structural analyses of hepatitis C virus polyprotein processing. *J. Virol.* **68**, 5045–5055.

Bartenschlager, R., Lohmann, V., Wilkinson, T. & Koch, J.O. (1995) Complex formation between the NS3 serine-type proteinase of the hepatitis C virus and NS4A and its importance for polyprotein maturation. *J. Virol.* **69**, 7519–7528.

Bazan, J.F. & Fletterick, R.J. (1989) Detection of a trypsin-like serine protease domain in flaviviruses and pestiviruses. *Virology* **171**, 637–639.

Bazan, J.F. & Fletterick, R.J. (1990) Structural and catalytic models of trypsin-like viral proteases. *Semin. Virol.* **1**, 311–322.

Bianchi, E., Steinkuhler, C., Taliani, M., Urbani, A., Francesco, R.D. & Pessi, A. (1996) Synthetic depsipeptide substrates for the assay of human hepatitis C virus protease. *Anal. Biochem.* **237**, 239–244.

Bouffard, P., Bartenschlager, R., Ahlbornlaake, L., Mous, J., Roberts, N. & Jacobsen, H. (1995) An in vitro assay for hepatitis C virus NS3 serine proteinase. *Virology* **209**, 52–59.

Bukh, J., Miller, R.H. & Purcell, R.H. (1995) Genetic heterogeneity of hepatitis C virus: Quasispecies and genotypes. *Semin. Liver Dis.* **15**, 41–63.

Butkiewicz, N.J., Wendel, M., Zhang, R.M., Jubin, R., Pichardo, J., Smith, E.B., Hart, A.M., Ingram, R., Durkin, J., Mui, P.W., Murray, R.G., Ramanathan, L. & Dasmahapatra, B. (1996) Enhancement of hepatitis C virus NS3 proteinase activity by association with NS4A-specific synthetic peptides: identification of sequence and critical residues of NS4A for the cofactor activity. *Virology* **225**, 328–338.

Chambers, T.J., Hahn, C.S., Galler, R. & Rice, C.M. (1990a) Flavivirus genome organization, expression and replication. *Annu. Rev. Microbiol.* **44**, 649–688.

Chambers, T.J., Weir, R.C., Grakoui, A., McCourt, D.W., Bazan, J.F., Fletterick, R.J. & Rice, C.M. (1990b) Evidence that the N-terminal domain of nonstructural protein NS3 from yellow fever virus is a serine protease responsible for site-specific cleavages in the viral polyprotein. *Proc. Natl Acad. Sci. USA* **87**, 8898–8902.

Chambers, T.J., Nestorowicz, A. & Rice, C.M. (1995) Mutagenesis of the yellow fever virus NS2B/3 cleavage site: determinants of cleavage site specificity and effects on polyprotein processing and viral replication. *J. Virol.* **69**, 1600–1605.

D'Souza, E.D.A., Grace, K., Sangar, D.V., Rowlands, D.J. & Clarke, B.E. (1995) In vitro cleavage of hepatitis C virus polyprotein substrates by purified recombinant NS3 protease. *J. Gen. Virol.* **76**, 1729–1736.

DeFrancesco, R., Urbani, A., Nardi, M.C., Tomei, L., Steinkuhler, C. & Tramontano, A. (1996) A zinc binding site in viral serine proteinases. *Biochemistry* **35**, 13282–13287.

Eckart, M.R., Selby, M., Masiarz, F., Lee, C., Berger, K., Crawford, K., Kuo, C., Kuo, G., Houghton, M. & Choo, Q.-L. (1993) The hepatitis C virus encodes a serine protease involved in processing of the putative nonstructural proteins from the viral polyprotein precursor. *Biochem. Biophys. Res. Commun.* **192**, 399–406.

Failla, C., Tomei, L. & DeFrancesco, R. (1994) Both NS3 and NS4A are required for proteolytic processing of hepatitis C virus nonstructural proteins. *J. Virol.* **68**, 3753–3760.

Failla, C., Tomei, L. & DeFrancesco, R. (1995) An amino-terminal domain of the hepatitis C virus NS3 protease is essential for interaction with NS4A. *J. Virol.* **69**, 1769–1777.

Failla, C.M., Pizzi, E., DeFrancesco, R. & Tramontano, A. (1996) Redesigning the substrate specificity of the hepatitis C virus NS protease. *Folding & Design* **1**, 35–42.

Filocamo, G., Pacini, L. & Migliaccio, G. (1997) Chimeric Sindbis virus dependent upon the NS3 protease of hepatitis C virus. *J. Virol.* **71**, 1417–1427.

Gorbalenya, A.E., Donchenko, A.P., Koonin, E.V. & Blinov, V.M. (1989) N-terminal domains of putative helicases of flavi- and pestiviruses may be serine proteases. *Nucleic Acids Res.* **17**, 3889–3897.

Grakoui, A., McCourt, D.W., Wychowski, C., Feinstone, S.M. & Rice, C.M. (1993) Characterization of the hepatitis C virus-encoded serine proteinase: Determination of proteinase-dependent polyprotein cleavage sites. *J. Virol.* **67**, 2832–2843.

Hahm, B., Han, D.S., Back, S.H., Song, O.K., Cho, M.J., Kim, C.J., Shimotohno, K. & Jang, S.K. (1995) NS3–4A of hepatitis C virus is a chymotrypsin-like protease. *J. Virol.* **69**, 2534–2539.

Hahm, B., Back, S.H., Lee, T.G., Wimmer, E. & Jang, S.K. (1996) Generation of a novel poliovirus with a requirement of hepatitis C virus protease activity. *Virology* **226**, 318–326.

Han, D.S., Hahm, B., Rho, H.-M. & Jang, S.K. (1995) Identification of the protease domain in NS3 of hepatitis virus. *J. Gen. Virol.* **76**, 985–993.

Hijikata, M., Mizushima, H., Akagi, T., Mori, S., Kakiuchi, N., Kato, N., Tanaka, T., Kimura, K. & Shimotohno, K. (1993a) Two distinct proteinase activities required for the processing of a putative nonstructural precursor protein of hepatitis C virus. *J. Virol.* **67**, 4665–4675.

Hijikata, M., Mizushima, H., Tanji, Y., Komoda, Y., Hirowatari, Y., Akagi, T., Kato, N., Kimura, K. & Shimotohno, K. (1993b) Proteolytic processing and membrane association of putative nonstructural proteins of hepatitis C virus. *Proc. Natl Acad. Sci. USA* **90**, 10773–10777.

Hirowatari, Y., Hijikata, M. & Shimotohno, K. (1995a) A novel method for analysis of viral proteinase activity encoded by hepatitis C virus in cultured cells. *Anal. Biochem.* **225**, 113–120.

Hirowatari, Y., Hijikata, M., Tanji, Y. & Shimotohno, K. (1995b) Expression and processing of putative nonstructural proteins of hepatitis C virus in insect cells using baculovirus vector. *Virus Res.* **35**, 43–61.

Hong, Z., Ferrari, E., Wright-Minogue, J., Chase, R., Risano, C., Seelig, G., Lee, C.G. & Kwong, A.D. (1996) Enzymatic characterization of hepatitis C virus NS3/4A complexes expressed in mammalian cells by using the herpes simplex virus amplicon system. *J. Virol.* **70**, 4261–4268.

Houghton, M. (1996) Hepatitis C viruses. In: *Fields Virology* (Fields, B.N., Knipe, D.M. & Howley, P.M., eds). New York: Raven Press, pp. 1035–1058.

Kakiuchi, N., Hijikata, M., Komoda, Y., Tanji, Y., Hirowatari, Y. & Shimotohno, K. (1995) Bacterial expression and analysis of cleavage activity of HCV serine proteinase using recombinant and synthetic substrate. *Biochem. Biophys. Res. Commun.* **210**, 1059–1065.

Kim, J.L., Morgenstern, K.A., Lin, C., Fox, T., Dwyer, M.D., Landro, J.A., Chambers, SP., Markland, W., Lepre, C.A., O'Malley, E.T., Harbeson, S.L., Rice, C.M., Murcko, M.A., Caron, P.R. & Thomson, J.A. (1996) Crystal structure of the hepatitis C virus NS3 protease domain complexed with a synthetic NS4A cofactor peptide. *Cell* **87**, 343–355.

Koch, J.O., Lohmann, V., Herian, U. & Bartenschlager, R. (1996) In vitro studies on the activation of the hepatitis C virus NS3 proteinase by the NS4A cofactor. *Virology* **221**, 54–66.

Kolykhalov, A.A., Agapov, E.V. & Rice, C.M. (1994) Specificity of the hepatitis C virus serine proteinase: Effects of substitutions at the 3/4A, 4A/4B, 4B/5A & 5A/5B cleavage sites on polyprotein processing. *J. Virol.* **68**, 7525–7533.

Komoda, Y., Hijikata, M., Sato, S., Asabe, S.-I., Kimura, K. & Shimotohno, K. (1994a) Substrate requirements of Hepatitis C virus proteinase for intermolecular polypeptide cleavage in *Escherichia coli. J. Virol.* **68**, 7351–7357.

Komoda, Y., Hijikata, M., Tanji, Y., Hirowatari, Y., Mizushima, H., Kimura, K. & Shimotohno, K. (1994b) Processing of hepatitis C viral polyprotein in *Escherichia coli. Gene* **145**, 221–226.

Leinbach, S.S., Bhat, R.A., Xia, S.-M., Hum, W.-T., Stauffer, B., Davis, A.R., Hung, P.P. & Mizutani, S. (1994) Substrate specificity of the NS3 serine proteinase of hepatitis C virus as determined by mutagenesis at the NS3/NS4A junction. *Virology* **204**, 163–169.

Lin, C. & Rice, C.M. (1995) The hepatitis C virus NS3 serine proteinase and NS4A cofactor: establishment of a cell-free trans-processing assay. *Proc. Natl Acad. Sci. USA* **92**, 7622–7626.

Lin, C., Prágai, B., Grakoui, A., Xu, J. & Rice, C.M. (1994) The hepatitis C virus NS3 serine proteinase: *trans* processing requirements and cleavage kinetics. *J. Virol.* **68**, 8147–8157.

Lin, C., Thomson, J.A. & Rice, C.M. (1995) A central region in the hepatitis C virus NS4A protein allows formation of an active NS3-NS4A serine proteinase complex *in vivo* and *in vitro*. *J. Virol.* **69**, 4373–4380.

Linnen, J., Wages, J., Zhangkeck, Z.Y., Fry, K.E., Krawczynski, K.Z., Alter, H., Koonin, E., Gallagher, M., Alter, M., Hadziyannis, S., Karayiannis, P., Fung, K., Nakatsuji, Y., Shih, J., Young, L., Piatak, M., Hoover, C., Fernandez, J., Chen, S., Zou, J.C., Morris, T., Hyams, K.C., Ismay, S., Lifson, J.D., Hess, G. & Kim, J.P. (1996) Molecular cloning and disease association of hepatitis G virus: a transfusion-transmissible agent. *Science* **271**, 505–508.

Love, R.A., Parge, H., Wickersham, J.A., Hostomsky, Z., Habuka, N., Moomaw, E.W., Adachi, T. & Hostomska, Z. (1996) The crystal structure of hepatitis C virus NS3 proteinase reveals a trypsin-like fold and a structural zinc binding site. *Cell* **87**, 331–342.

Manabe, S., Fuke, I., Tanishita, O., Kaji, C., Gomi, Y., Yoshida, S., Mori, C., Takamizawa, A., Yoshida, I. & Okayama, H. (1994) Production of nonstructural proteins of hepatitis C virus requires a putative viral protease encoded by NS3. *Virology* **198**, 636–644.

Miller, R.H. & Purcell, R.H. (1990) Hepatitis C virus shares amino acid sequence similarity with pestiviruses and flaviviruses as well as members of two plant virus supergroups. *Proc. Natl Acad. Sci. USA* **87**, 2057–2061.

Mori, A., Yamada, K., Kimura, J., Koide, T., Yuasa, S., Yamada, E. & Miyamura, T. (1996) Enzymatic characterization of purified NS3 serine proteinase of hepatitis C virus expressed in *Escherichia coli*. *FEBS Lett.* **378**, 37–42.

Muerhoff, A.S., Leary, T.P., Simons, J.N., Pilotmatias, T.J., Dawson, G.J., Erker, J.C., Chalmers, M.L., Schlauder, G.G., Desai, S.M. & Mushahwar, I.K. (1995) Genomic organization of GB viruses A and B: two new members of the *Flaviviridae* associated with GB agent hepatitis. *J. Virol.* **69**, 5621–5630.

Nestorowicz, A., Chambers, T.J. & Rice, C.M. (1994) Mutagenesis of the yellow fever virus NS2A/2B cleavage site: Effects on proteolytic processing, viral replication and evidence for alternative processing of the NS2A protein. *Virology* **199**, 114–123.

Overton, H., McMillan, D., Gillespie, F. & Mills, J. (1995) Recombinant baculovirus-expressed NS3 proteinase of hepatitis C virus shows activity in cell-based and in vitro assays. *J. Gen. Virol.* **76**, 3009–3019.

Pizzi, E., Tramontano, A., Tomei, L., La Monica, N., Failla, C., Sardana, M., Wood, T. & DeFrancesco, R. (1994) Molecular-model of the specificity pocket of the hepatitis C virus protease: Implications for substrate recognition. *Proc. Natl Acad. Sci. USA* **91**, 888–892.

Rice, C.M. (1996) *Flaviviridae*: The viruses and their replication. In: *Fields Virology* (Fields, B.N., Knipe, D.M. & Howley, P.M., eds). New York: Raven Press, pp. 931–960.

Satoh, S., Tanji, Y., Hijikata, M., Kimura, K. & Shimotohno, K. (1995) The N-terminal region of hepatitis C virus nonstructural protein 3 (NS3) is essential for stable complex formation with NS4A. *J. Virol.* **69**, 4255–4260.

Shimizu, Y., Yamaji, K., Masuho, Y., Yokota, T., Inoue, H., Sudo, K., Satoh, S. & Shimotohno, K. (1996) Identification of the sequence on NS4A required for enhanced cleavage of the NS5A/5B site by hepatitis C virus NS3 protease. *J. Virol.* **70**, 127–132.

Shoji, I., Suzuki, T., Chieda, S., Sato, M., Harada, T., Chiba, T., Matsuura, Y. & Miyamura, T. (1995) Proteolytic activity of NS3 serine proteinase of hepatitis C virus efficiently expressed in *Escherichia coli*. *Hepatology* **22**, 1648–1655.

Simmonds, P., Mellor, J., Sakuldamrongpanich, T., Nuchaprayoon, C., Tanprasert, S., Holmes, E.C. & Smith, D.B. (1996) Evolutionary analysis of variants of hepatitis C virus found in South-East Asia: comparison with classifications based upon sequence similarity. *J. Gen. Virol.* **77**, 3013–3024.

Simons, J.N., Leary, T.P., Dawson, G.J., Pilotmatias, T.J., Muerhoff, A.S., Schlauder, G.G., Desai, S.M. & Mushahwar, I.K. (1995a) Isolation of novel virus-like sequences associated with human hepatitis. *Nature Med.* **1**, 564–569.

Simons, J.N., Pilotmatias, T.J., Leary, T.P., Dawson, G.J., Desai, S.M., Schlauder, G.G., Muerhoff, A.S., Erker, J.C., Buijk, S.L., Chalmers, M.L., Vansant, C.L. & Mushahwar, I.K. (1995b) Identification of two flavivirus-like genomes in the GB hepatitis agent. *Proc. Natl Acad. Sci. USA* **92**, 3401–3405.

Song, O.K., Cho, O.H., Hahm, B. & Jang, S.K. (1996) Development of an in vivo assay system suitable for screening inhibitors of hepatitis C viral protease. *Molecules Cells* **6**, 183–189.

Steinkühler, C., Tomei, L. & DeFrancesco, R. (1996a) In vitro activity of hepatitis C virus protease NS3 purified from recombinant baculovirus-infected Sf9 cells. *J. Biol. Chem.* **271**, 6367–6373.

Steinkuhler, C., Urbani, A., Tomei, L., Biasiol, G., Sardana, M., Bianchi, E., Pessi, A. & DeFrancesco, R. (1996b) Activity of purified hepatitis C virus protease NS3 on peptide substrates. *J. Virol.* **70**, 6694–6700.

Suzuki, T., Sato, M., Chieda, S., Shoji, I., Harada, T., Yamakawa, Y., Watabe, S., Matsuura, Y. & Miyamura, T. (1995) *In vivo* and *in vitro* trans-cleavage activity of hepatitis C virus serine proteinase expressed by recombinant baculoviruses. *J. Gen. Virol.* **76**, 3021–3029.

Taliani, M., Bianchi, E., Narjes, F., Fossatelli, M., Urbani, A., Steinkuhler, C., DeFrancesco, R. & Pessi, A. (1996) A continuous assay of hepatitis C virus protease based on resonance energy transfer depsipeptide substrates. *Anal. Biochem.* **240**, 60–67.

Tanji, Y., Hijikata, M., Satoh, S., Kaneko, T. & Shimotohno, K. (1995) Hepatitis C virus-encoded nonstructural protein NS4A has versatile functions in viral protein processing. *J. Virol.* **69**, 1575–1581.

Tokita, H., Okamoto, H., Iizuka, H., Kishimoto, J., Tsuda, F., Lesmana, L.A., Miyakawa, Y. & Mayumi, M. (1996) Hepatitis C virus variants from Jakarta, Indonesia classifiable into novel genotypes in the second (2e and 2f), tenth (10a) and eleventh (11a) genetic groups. *J. Gen. Virol.* **77**, 293–301.

Tomei, L., Failla, C., Santolini, E., DeFrancesco, R. & La Monica, N. (1993) NS3 is a serine protease required for processing of hepatitis C virus polyprotein. *J. Virol.* **67**, 4017–4026.

Tomei, L., Failla, C., Vitale, R.L., Bianchi, E. & DeFrancesco, R. (1996) A central hydrophobic domain of the hepatitis C virus NS4A protein is necessary and sufficient for the activation of the NS3 protease. *J. Gen. Virol.* **77**, 1065–1070.

Xu, J., Mendez, E.M., Caron, P., Lin, C., Murcko, M., Collett, M.S. & Rice, C.M. (1997) Bovine viral diarrhea virus NS3 serine proteinase: polyprotein cleavage sites, cofactor requirements, and molecular model of an enzyme required for pestivirus replication. *J. Virol.* **71**, 5312–5322.

*Charles M. Rice*
*Department of Microbiology,*
*Washington University School of Medicine,*
*Box 8230, South Euclid Avenue, St. Louis, MO 63110-1093, USA*
*Email: Rice@borcim.wustl.edu*

# 90. *Potyvirus P1 proteinase*

## Databanks

*Peptidase classification: clan SA, family S30, MEROPS ID: S30.001*
*NC-IUBMB enzyme classification: none*
*ATCC entries: PVAS-69 (TEV antisera); 45036 (molecular cloned TEV antisera)*
*Databank codes:*

| Species | SwissProt | PIR | EMBL (genomic) |
|---|---|---|---|
| Johnson grass mosaic virus | – | – | Z26920: complete genome |
| Papaya ringspot virus | Q01901 | JQ1899 | S46722: complete genome |
| | | | X67673: complete genome |
| Peanut stripe virus | – | – | U05771: complete genome |
| Plum pox potyvirus | P13529 | JA0078 | D00298: 3′ end |
| | | S06929 | X16415: complete genome |
| Potato virus Y | P18247 | JS0166 | D00441: complete genome |
| | | | U33454: midsection |
| | | | X12456: complete genome |
| Tobacco etch virus | P04517 | A04207 | L38714: complete genome |
| | | | M11216 |
| | | | M11458: complete genome |
| | | | M15239: complete genome |
| Tobacco vein mottling virus | P09814 | A23647 | U38621: complete genome |
| | | | X04083: complete genome |

## Name and History

The RNA genome of potyviruses encodes a single large polyprotein that is proteolytically processed by three viral proteinases to produce mature products (Reichmann *et al.*, 1992). The *P1 proteinase* lies near the N-terminus of the viral polyprotein and is required for processing a site at its own C-terminus. Prior to characterization of its activity, the P1 protein was named primarily for its molecular weight or was identified as the *NT protein* (N-terminal protein).

Processing between P1 and the helper component proteinase (HC-Pro) (Chapter 227) was first discovered in transgenic plants expressing truncated fragments of tobacco etch potyviral (TEV) polyproteins (Carrington *et al.*, 1990). Processed polypeptides were detected in the absence of a functional HC-Pro proteinase, suggesting that an alternative enzyme was responsible for cleavage events at this site. P1 proteinase activity was further characterized by *in vitro* translation of tobacco etch virus and tobacco vein mottling virus (TVMV) transcripts in the presence of wheatgerm extracts (Mavankal & Rhoads, 1991; Verchot *et al.*, 1991, 1992). The C-terminal half of P1 functions as a proteolytic domain containing three conserved amino acid residues, His214, Asp223 and Ser256, which resemble the active-site residues of known serine-type proteinases. The catalytic residue Ser256 lies within a conserved Gly-Ser-Ser-Gly motif similar to the Gly-Xaa-Ser-Gly active-site motif of known chymotrypsin-like proteinases (Verchot *et al.*, 1992).

## Activity and Specificity

The P1 proteinase cotranslationally cleaves within a Phe(Tyr)↓Ser dipeptide located at its C-terminus (Mavankal & Rhoads, 1991; Verchot *et al.*, 1992). While deletion analyses revealed that sequences C-terminal to the P1′ position are not essential for substrate recognition, the contribution of sequences lying N-terminal to the P1 position has not yet been determined. With an *in vitro* assay of P1 proteolytic activity, processing of substrate molecules in *trans* could not be detected, suggesting that the autoproteolytic activity of P1 occurs primarily by an intramolecular, *cis* mechanism. Proteolysis may involve a plant-derived cofactor, since processed products were not detected after translation in the presence of rabbit reticulocyte lysates, but were detected after translation in the presence of a wheatgerm extract and by western analysis of infected or transgenic plants. The alternative interpretation that an inhibitor is present in the rabbit reticulocyte lysate is less likely, since processed products were detected after translation in a mixed rabbit reticulocyte lysate and wheatgerm extract system (Verchot *et al.*, 1992).

## Structural Chemistry

Based on the compact linear arrangement of the proteinase active-site residues, the P1 proteinase is suggested to most closely resemble the Sindbis capsid proteinase togavirin (Chapter 86). Both are chymotrypsin-like enzymes and both are autoproteolytically released from the viral polyprotein by a *cis*-acting mechanism. A model for P1 activity has been proposed based on the mechanism described from the crystallographic data of the Sindbis virus capsid protein (Choi *et al.*, 1991). After processing of the scissile bond, the C-terminal residue stays within the substrate-binding pocket, inhibiting further processing activities of this enzyme. Therefore, processing can only occur once and can only occur in *cis*.

## Preparation

The purification of the P1 proteins derived from several potyviruses including TVMV, zucchini yellow mosaic virus, and pea seedborne mosaic virus by overexpression in *E. coli* has been described (Rodriguez-Cerezo & Shaw, 1991; Albrechtsen & Borkhardt, 1992; Wisler *et al.*, 1995). In most instances the P1 protein was recovered from the insoluble fraction on an SDS-PAGE gel and used for preparing antisera. Brantley & Hunt (1993) fused P1 to the GST (glutathione-*S*-transferase) tag and after overexpression of the protein in *E. coli*, recovered the fusion protein from the soluble fraction by chromatography on glutathione-Sepharose 4B.

## Biological Aspects

The P1 proteinase is one of the most variable proteins in the potyviral genomes and can range in size from a 29 kDa protein in TVMV to a protein of 63 kDa encoded by the papaya ringspot virus genome. While the active site lies within a conserved domain near the C-terminus of the protein, the N-terminal extension varies among potyviruses in sequence and length. Since virus replication and intercellular movement, as well as

P1 proteolysis, are unaffected by deletion of the N-terminal extension, no function has yet been ascribed to these sequences. Deletion of the entire P1 protein was shown to decrease viral replication to a level that is 1% of wild-type virus. Replication was shown to be restored when P1 was supplied in *trans*, indicating that P1 has additional functions beyond its role in proteolysis (Verchot & Carrington, 1995b). One experiment, in which the cleavage site recognized by P1 was functionally replaced by one recognized by the potyviral NIa proteinase (Chapter 244), showed that proteolytic separation of P1 and HC-Pro is essential for viral infectivity, but not specifically the proteinase activity of P1 (Verchot & Carrington, 1995a). From these studies one can conclude that while the first function of P1 identified is its proteinase activity, its primary contribution to the viral life cycle has yet to be discovered.

## References

Albrechtsen, M. & Borkhardt, B. (1992) Detection of a 45-kDa protein derived from the N-terminus of the pea seedborne mosaic potyvirus polyprotein in vivo and in vitro. *Virus Genes* **8**, 7–13.

Brantley, J.D. & Hunt, A.G. (1993) The N-terminal protein of the polyprotein encoded by the potyvirus tobacco vein mottling virus is an RNA-binding protein. *J. Gen. Virol.* **74**, 1157–1162.

Carrington, J.C., Freed, D.D. & Oh, C.-S. (1990) Expression of potyviral polyproteins in transgenic plants reveals three proteolytic activities required for complete processing. *EMBO* **9**, 1347–1353.

Choi, H.-K., Tong, L., Minor, W., Dumas, P., Boege, U., Grossman, M.G. & Wengler, G. (1991) Structure of Sindbis virus core protein reveals a chymotrypsin-like serine proteinase and the organization of the virion. *Nature* **354**, 37–43.

Mavankal, G. & Rhoads, R.E. (1991) In vitro cleavage at or near the N-terminus of the helper component protein in the tobacco vein mottling virus polyprotein. *Virology* **185**, 721–731.

Reichmann, J.L., Lain, S. & Garcia, J.A. (1992) Highlights and prospects of potyvirus molecular biology. *J. Gen. Virol.* **73**, 1–16.

Rodriguez-Cerezo, E. & Shaw, J.G. (1991) Two newly detected nonstructural viral proteins in potyvirus-infected cells. *Virology* **185**, 572–579.

Verchot, J. & Carrington, J.C. (1995a) Debilitation of plant potyvirus infectivity by P1 proteinase-inactivating mutations and restoration by second-site modifications. *J. Virol.* **69**, 1582–1590.

Verchot, J. & Carrington, J.C. (1995b) Evidence that the potyvirus P1 proteinase functions in *trans* as an accessory factor for genome amplification. *J. Virol.* **69**, 3668–3674.

Verchot, J., Koonin, E.V. & Carrington, J.C. (1991) The 35-kDa protein from the N-terminus of the potyviral polyprotein functions as a third virus-encoded proteinase. *Virology* **185**, 527–535.

Verchot, J., Herndon, K.L. & Carrington, J.C. (1992) Mutational analysis of the tobacco etch potyviral 35-kDa proteinase: Identification of essential residues and requirements for autoproteolysis. *Virology* **190**, 298–306.

Wisler, G.C., Purcifull, D.E. & Hiebert, E. (1995) Characterization of the P1 protein and coding region of the zucchini yellow mosaic virus. *J. Gen. Virol.* **76**, 37–45.

*Jeanmarie Verchot*
*The Sainsbury Laboratory,*
*Norwich Research Park,*
*Colney, Norwich NR47UH, UK*
*Email: verchot@bbsrc.ac.uk*

# 91. *Pestivirus NS2-3/NS3 serine peptidase*

## Databanks

*Peptidase classification: clan SA, family S31, MEROPS ID: S31.001*
*NC-IUBMB enzyme classification: none*
*Databank codes:*

| Species | SwissProt | PIR | EMBL (cDNA) | EMBL (genomic) |
|---|---|---|---|---|
| Border disease virus | – | – | – | U00892: mid-section |
| Cattle (bovine) viral diarrhea virus | P19711 | A29198 | M62430 | M31182: complete genome |
| Cattle (bovine) viral diarrhea virus | Q01499 | A44217 | – | M96751: complete polyprotein |
| Cattle (bovine) viral diarrhea virus | – | – | – | U18059: complete polyprotein |
| Hog cholera virus | P19712 | A34037 | – | J04358: complete genome |
| Hog cholera virus | P21530 | A35317 | – | M31768: complete genome |

## Name and History

The pestiviruses, hepatitis C viruses (HCV) and flaviviruses are grouped together in the family Flaviviridae. The genus *Pestivirus* encompasses bovine viral diarrhea virus (BVDV), classical swine fever virus (CSFV) and border disease virus of sheep (BDV). The single-stranded RNA genome of pestiviruses contains one long open reading frame (ORF) with about 4000 codons. The putative polyprotein is cleaved co- and post-translationally by host cell- and virus-encoded proteases. The existence of a chymotrypsin- or trypsin-like serine protease in the N-terminal domain of the pestiviral non-structural protein NS3 (p80) was predicted by comparative amino acid pattern analyses (Bazan & Fletterick, 1989; Gorbalenya *et al.*, 1989). Subsequently, transient expression studies in the T7 vaccinia virus system (Wiskerchen & Collett, 1991; Tautz *et al.*, 1994, 1996) and expression via baculovirus recombinants (Petric *et al.*, 1992) provided experimental evidence for this assumption. The studies revealed that NS2-3 (p125) as well as NS3 are proteolytically active serine proteases. The serine protease domain residing in NS2-3/NS3 is essential for processing of the polyprotein downstream of NS2-3 and, with respect to genomic localization, function and amino acid motifs is homologous to the NS3 serine proteases of flaviviruses (Chapter 88) and HCV (Chapter 89).

The existence of unprocessed NS2-3 as a mature viral protein is unique for pestiviruses, within the family Flaviviridae. Whereas uncleaved NS2-3 is apparently expressed by all pestiviruses, certain pestivirus isolates have evolved mechanisms for the release of NS3 from the polyprotein (Meyers *et al.*, 1992; Tautz *et al.*, 1993, 1994, 1996; Meyers & Thiel, 1995). Interestingly, high-level expression of NS3 is correlated with cytopathogenicity of the respective viruses.

## Activity and Specificity

Processing of the pestiviral polyprotein downstream of NS2-3 is mediated by the NS2-3/NS3 serine protease. The cleavage generating the C-terminus of NS2-3/NS3 is believed to occur only in *cis*. The cleavages at the NS4A-4B, NS4B-5A and NS5A-5B sites can be mediated in *trans* (Wiskerchen & Collett, 1991).

Cleavage at the NS5A-5B site by the NS2-3/NS3 serine protease requires a cofactor, most likely derived from the NS4A-B region of the polyprotein (Wiskerchen & Collett, 1991). The exact nature of the hypothesized cofactor and its role in processing of the pestiviral polyprotein have yet to be defined. Interestingly, in the HCV system NS4A represents a cofactor of the NS3 protease which is essential for cleavage at the NS3-4A, NS4A-B and NS4B-5A sites (Failla *et al.*, 1994, 1995; Lin *et al.*, 1994; Bartenschlager *et al.*, 1995); (see also Chapter 89); NS4A was found to be not essential for cleavage at the NS5A-5B site.

NS2-3 remains uncleaved in cells infected with non-cytopathogenic (noncp) BVDV. It was postulated that in cytopathogenic (cp) BVDV NADL-infected cells the NS2-3/NS3 serine protease also mediates NS2-3 cleavage; however, the data concerning this point are controversial since other studies carried out in the baculovirus system by the same group led to different results (Wiskerchen & Collett, 1991; Petric *et al.*, 1992). In the case of other cytopathogenic BVDV strains it has been demonstrated that the NS2-3/NS3 serine protease is not involved in processing of NS2-3 (Tautz *et al.*, 1993, 1996).

The cleavage sites acted on by the NS2-3/NS3 serine protease have not been identified. The corresponding enzyme of flaviviruses cleaves after two amino acids with basic side chains (P2 and P1) which are usually followed by amino acids with small side chains (P1'). According to a model proposed by Bazan & Fletterick (1990) an Asp residue is positioned at the bottom of the substrate-binding pocket (corresponding to Asp189 in chymotrypsin) and coordinates the positively charged amino acid at P1. In the case of HCV the NS3 serine protease-dependent cleavage sites show a consensus sequence for P6 (Asp or Glu), P1 (Thr or Ser) and P1′ (Ser or Ala). The amino acid suggested to reside at the bottom of the substrate-binding pocket is Ser, which might interact with the Thr/Cys residues found at P1. In the pestiviral NS2-3/NS3 serine protease a Lys residue is found at the bottom of the hypothetical substrate-binding pocket. Interestingly, studies with a trypsin derivative carrying an Asn to Lys mutation at position 189 showed that this modified enzyme has a substrate preference for Leu or Phe in P1 (Graf *et al*., 1987). Accordingly, it is possible that the amino acid sequence at the NS3-dependent cleavage sites in the pestiviral polyprotein might be similar to the ones identified for HCV.

## Structural Chemistry

According to the model of Bazan & Fletterick (1989) the catalytic triad of the pestiviral enzyme consists of His1658, Asp1695 and Ser1752 (numbers according to BVDV SD-1; Deng & Brock, 1992). Gorbalenya and coworkers (1989) proposed a similar model; however, they postulated the involvement of Asp1686. So far, an essential role for the activity of the enzyme was experimentally proven only for Ser1752 (Wiskerchen & Collett, 1991; Tautz *et al*., 1996). For NS2-3 neither the N- nor the C-terminus has been determined. It is assumed that the C-termini of NS2-3 and NS3 are identical. Recent studies on cp BVDV (Meyers *et al*., 1992; Tautz *et al*., 1993, 1994) and cp CSFV strains (Meyers & Thiel, 1995) strongly suggested that Gly1590 represents the N-terminus of NS3 in these cases even though direct protein sequencing data are not available. Studies in the T7 vaccinia virus system showed that NS3 starting with Gly1590 is catalytically active, at least with respect to generation of NS4A (Tautz *et al*., 1994). N-terminally elongated NS3 molecules as well as NS2-3 have also been found to be proteolytically active (Wiskerchen & Collett, 1991; Tautz *et al*., 1996). Mapping of the catalytic serine protease suggested that more than 430 amino acids of the N-terminal region of NS3 (BVDV NADL) are necessary for an active enzyme (Wiskerchen & Collett, 1991); however, protease activity was measured with respect to the NS2-3 site which is probably not processed by the NS3 serine protease. Thus, the minimal active serine protease domain remains to be determined. For flaviviruses and HCV the minimal length of a catalytically active NS3 serine protease is about 180 amino acids (Preugschat *et al*., 1990; Falgout *et al*., 1991; Failla *et al*., 1995). Because of the homology between these enzymes a similar result is expected for NS3 from all family members.

## Preparation

Enzymatically active proteases have been obtained by expression in the T7 vaccinia virus system and the baculovirus system (Wiskerchen & Collett, 1991; Petric *et al*., 1992; Tautz *et al*., 1994, 1996). Studies carried out with purified NS2-3 or NS3 serine proteases have not been published.

## Biological Aspects

Based on the knowledge about viruses with similar genome organization it can be assumed that cleavages performed by the NS3 serine peptidase are essential for the replication of pestiviruses. Polyprotein cleavage kinetics as studied by pulse chase experiments with BVDV NADL-infected cells showed that cleavages at the NS3-4A and NS4B-5A sites are very rapid; cleavages at the NS4A-B and NS5A-B sites occur in a delayed fashion since respective precursor molecules were detected (Collett *et al*., 1991). Accordingly, processing apparently occurs in a regulated manner. For other RNA viruses it has been demonstrated that temporal regulation of polyprotein processing has an essential role in virus replication (Lemm *et al*., 1994). In this context it will be interesting to compare the enzymatic activities of the NS2-3 and NS3 serine proteases. Any differences may be of biological significance since high level expression of NS3 is linked to cytopathogenicity of pestiviruses. Moreover, the presence of cytopathogenic BVDV in animals persistently infected with noncp BVDV causes a lethal disease (Thiel *et al*., 1996).

## Distinguishing Features

The expression of uncleaved NS2-3 by pestiviruses is a unique feature within the family Flaviviridae. The viruses of the other genera of this virus family show efficient processing of NS2-3. Experiments with one member of the genus *Flavivirus* imply that NS2-3 cleavage is essential for viability of this virus (Chambers *et al*., 1995).

With respect to cofactor requirements the pestiviral protease might resemble the HCV protease; the latter depends on NS4A for enzymatic activity. However, for the pestiviral enzyme the cofactor and its role in proteolytic processing have yet to be defined. Furthermore, the available data show that the NS2 region is not essential for the proteolytic function of the pestiviral NS2-3/NS3 serine protease (Tautz *et al*., 1994). This is in contrast to the situation in the flavivirus system where NS2B is an essential cofactor of the NS3 serine protease (Rice & Strauss, 1990). Thus pestiviral and HCV NS3 serine protease seem to be more closely related to each other than to the corresponding enzyme of flaviviruses. Monoclonal antibodies reactive to NS2-3/NS3 have been generated (Corapi *et al*., 1990).

## Further Reading

The characterization of the NS3 endopeptidase has been described in detail by Wiskerchen & Collett (1991).

## References

Bartenschlager, R., Lohmann, V., Wilkinson, T. & Koch, J.O. (1995) Complex formation between the NS3 serine-type proteinase of hepatitis C virus and NS4A and its importance for polyprotein maturation. *J. Virol.* **69**, 7519–7528.

Bazan, J.F. & Fletterick, R.J. (1989) Detection of a trypsin-like serine protease domain in flaviviruses and pestiviruses. *Virology* **171**, 637–639.

Bazan, J.F. & Fletterick, R.J. (1990) Structural and catalytic models of trypsin-like viral proteases. *Semin. Virol.* **1**, 311–322.

Chambers, T.J., Nestorowicz, A. & Rice, C.M. (1995) Mutagenesis of the yellow fever virus NS2B/3 cleavage site: determinants of cleavage site specificity and effects on polyprotein processing and viral replication. *J. Virol.* **69**, 1600–1605.

Collett, M.S., Wiskerchen, M.A., Welniak, E. & Belzer, S.K. (1991) Bovine viral diarrhea virus genomic organization. *Arch. Virol.* **3**, 19–27.

Corapi, W.V., Donis, R.O. & Dubovi, E.J. (1990) Characterization of a panel of monoclonal antibodies and their use in the study of the antigenic diversity of bovine viral diarrhea virus. *Am. J. Vet. Res.* **51**, 1388–1394.

Deng, R. & Brock, K.V. (1992) Molecular cloning and nucleotide sequence of a pestivirus genome, noncytopathogenic bovine viral diarrhea virus strain SD-1. *Virology* **191**, 867–879.

Failla, C., Tomei, L. & de Francesco, R. (1994) Both NS3 and NS4A are required for proteolytic processing of hepatitis C virus nonstructural proteins. *J. Virol.* **68**, 3753–3760.

Failla, C., Tomei, L. & de Francesco, R. (1995) An amino-terminal domain of the hepatitis C virus NS3 protease is essential for interaction with NS4A. *J. Virol.* **69**, 1769–1777.

Falgout, B., Pethel, M., Zhang, Y.-M. & Lai, C.-J. (1991) Both nonstructural proteins NS2B and NS3 are required for the proteolytic processing of dengue virus nonstructural proteins. *J. Virol.* **65**, 2467–2475.

Gorbalenya, A.E., Donchenko, A.P., Koonin, E.V. & Blinov, V.M. (1989) N-terminal domains of putative helicases of flavi- and pestiviruses may be serine proteases. *Nucleic Acids Res.* **17**, 3889–3897.

Graf, L., Craik, C.S., Patthy, A., Roczniak, S., Fletterick, R.J. & Rutter, W.J. (1987) Selective alteration of substrate specificity by replacement of aspartic acid-189 with lysine in the binding pocket of trypsin. *Biochemistry* **26**, 2616–2622.

Lemm, J.A., Rümenapf, T., Strauss, E.G., Strauss, J.H. & Rice, C.M. (1994) Polypeptide requirements for assembly of functional sindbis virus replication complexes: a model for the temporal regulation of minus- and plus-strand RNA synthesis. *EMBO J.* **13**, 2925–2934.

Lin, C., Pragai, B.M., Grakoui, A., Xu, J. & Rice, C.M. (1994) Hepatitis C virus NS3 serine proteinase: trans-cleavage requirements and processing kinetics. *J. Virol.* **68**, 8147–8157.

Meyers, G. & Thiel, H.-J. (1995) Cytopathogenicity of classical swine fever virus caused by defective interfering particels. *J. Virol.* **69**, 3683–3689.

Meyers, G., Tautz, N., Stark, R., Brownlie, J., Dubovi, E.J., Collett, M.S. & Thiel, H.-J. (1992) Rearrangement of viral sequences in cytopathogenic pestiviruses. *Virology* **191**, 368–386.

Petric, M., Yolken, R.H., Dubovi, E.J., Wiskerchen, M. & Collett, M.S. (1992) Baculovirus expression of pestivirus non-structural proteins. *J. Gen. Virol.* **73**, 1867–1871.

Preugschat, F., Yao, C.-W. & Strauss, J.H. (1990) In vitro processing of dengue virus type 2 nonstructural proteins NS2A, NS2B and NS3. *J. Virol.* **64**, 4364–4374.

Rice, C.M. & Strauss, J.H. (1990) Production of flavivirus polypeptides by proteolytic processing. *Semin. Virol.* **1**, 357–367.

Tautz, N., Meyers, G. & Thiel, H.-J. (1993) Processing of polyubiquitin in the polyprotein of an RNA virus. *Virology* **197**, 74–85.

Tautz, N., Thiel, H.-J., Dubovi, E.J. & Meyers, G. (1994) Pathogenesis of mucosal disease: a cytopathogenic pestivirus generated by internal deletion. *J. Virol.* **68**, 3289–3297.

Tautz, N., Meyers, G., Stark, R., Dubovi, E.J. & Thiel, H.-J. (1996) Cytopathogenicity of a pestivirus correlated with a 27-nucleotide insertion. *J. Virol.* **70**, 7851–7858.

Thiel, H.-J., Plagemann, P.G.W. & Moennig, V. (1996) Pestiviruses. In: *Fields Virology*, 3rd edn, vol. 1 (Fields, B.N., Knipe, D.M. & Howley, P.M., eds). Philadelphia, New York: Lippincott-Raven, pp. 1059–1073.

Wiskerchen, M.A. & Collett, M.S. (1991) Pestivirus gene expression: protein p80 of bovine viral diarrhea virus is a proteinase involved in polyprotein processing. *Virology* **184**, 341–350.

*Norbert Tautz*
*Institut für Virologie,*
*Fachbereich Veterinärmedizin (FB 18),*
*Justus-Liebig-Universität Giessen,*
*Frankfurter Strasse 107, D-35392 Giessen, Germany*
*Email: norbert.tautz@vetmed.uni-giessen.de*

*Heinz-Jürgen Thiel*
*Institut für Virologie,*
*Fachbereich Veterinärmedizin (FB 18),*
*Justus-Liebig-Universität Giessen,*
*Frankfurter Strasse 107, D-35392 Giessen, Germany*
*Email: heinz-juergen.thiel@vetmed.uni-giessen.de*

# 92. *Arterivirus serine endopeptidase*

## Databanks

*Peptidase classification: clan SA, family S32, MEROPS ID: S32.001*
*NC-IUBMB enzyme classification: none*

*Databank codes:*

| Species | SwissProt[a] | PIR | EMBL (genomic)[b] |
|---|---|---|---|
| Equine arteritis virus | P19811 | A39925 | X52277: 5′ end |
| | | S10158 | X53459: complete genome |
| Lactate-dehydrogenase-elevating virus | – | – | U15146: complete genome |
| Lelystad virus | Q04561 | A36861 | L04493: 3′ end |
| | | A45392 | M96262: 3′ end |

[a] Accession numbers refer to replicase polyprotein precursors of which the proteinase is a part.
[b] Accession numbers refer to full-length genomic sequences.

## Name and History

Arteriviruses are enveloped, positive-stranded RNA viruses that contain a polycistronic genome (12–15 kb) (Den Boon *et al.*, 1991; Godeny *et al.*, 1993; Meulenberg *et al.*, 1993; Snijder & Spaan, 1995). Their replicase proteins are expressed from the open reading frames (ORFs) 1a and 1b that encode two large precursor proteins: the ORF1a protein (187–260 kDa) and, due to ribosomal frameshifting, the ORF1ab protein (345–422 kDa). Both precursors are processed extensively by three or four ORF1a-encoded endopeptidases (Snijder *et al.*, 1992, 1994, 1995, 1996; Den Boon *et al.*, 1995; Van Dinten *et al.*, 1996). The equine arteritis virus (EAV) serine proteinase (SP) domain is located in nonstructural protein 4 (Nsp4), a 21 kDa cleavage product from the central region of the ORF1a protein. The SP was first identified and linked to cellular chymotrypsin-like proteinases by sequence analysis of the ORF1a protein of the arterivirus prototype, equine arteritis virus.

## Activity and Specificity

The activity of the EAV SP domain has been detected *in vivo*, both in infected cells and in transient expression systems upon expression of appropriate parts of the ORF1a polyprotein (Snijder *et al.*, 1994, 1996). In addition to the mature Nsp4, an Nsp34 processing intermediate with a long half-life is synthesized. The Nsp4 SP is now known to mediate (at least) five cleavages at Glu⧸Ser or Glu⧸Gly sites in the C-terminal half of the ORF1a protein (Snijder *et al.*, 1996). Three additional sites in the ORF1b protein are probably cleaved by the EAV SP (Van Dinten *et al.*, 1996). These eight sites are conserved in arteriviruses and can be described by the formula Glu⧸(Gly/Ser/Ala/Lys). The EAV Nsp4 SP domain has been shown to be able to cleave a separately expressed ORF1b polyprotein in *trans*. However, it is likely that the two sites flanking the proteinase in the ORF1a polyprotein (the Nsp3/Nsp4 and Nsp4/Nsp5 junctions) are cleaved in *cis*.

## Structural Chemistry

The EAV SP is a 204 residue proteolytic enzyme that displays strong sequence similarities (Figure 92.1) with chymotrypsin-like enzymes belonging to two different groups. The SP utilizes the canonical His-Asp-Ser catalytic triad found in classical chymotrypsin-like proteinases. On the other hand, its putative substrate-binding region contains Thr and His residues which are conserved in the so-called viral 3C-like cysteine peptidases (Chapter 239) (Snijder *et al.*, 1996) and which determine their specificity for Gln(Glu)⧸Gly(Ala/Ser) cleavage sites.

The replacement of the members of the predicted catalytic triad (His1103, Asp1129 and Ser1184; here and hereafter polyprotein numbering is used) confirmed their indispensability (Snijder *et al.*, 1996). The Ser1184 to Cys mutant did

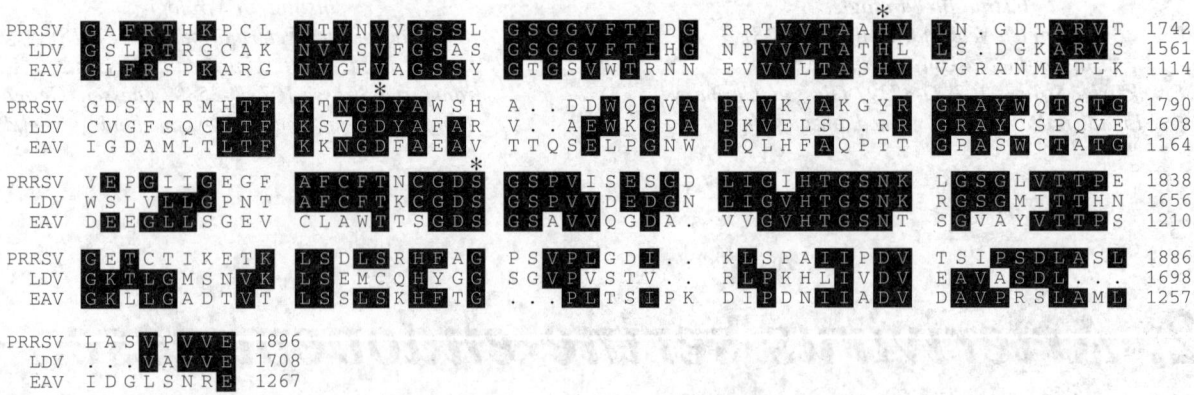

*Figure 92.1* Alignment of the Nsp4 SP domains of the arteriviruses pig reproductive and respiratory syndrome virus (PRRSV), lactate-dehydrogenase-elevating virus (LDV), and equine arteritis virus (EAV). The alignment was produced with the ClustalW program and presented in the PrettyBox format derived from the GCG package. Black indicates identical residues. The percentages of identical residues in the various protein pairs are: PRRSV versus LDV, 47%; LDV versus EAV, 36%; EAV versus PRRSV, 38%. The respective N-terminal and C-terminal boundaries of the protease domains in this alignment are the Nsp3/Nsp4 and Nsp4/Nsp5 cleavage sites (EAV) or putative cleavage sites (PRRSV and LDV).

not show any proteolytic activity. However, the replacement of Asp1129 by Glu, a substitution which can also be found in certain 3C-like cysteine peptidases, was partially tolerated. The putative role of Thr1179 and His1199 in substrate recognition was supported by replacements at these positions and by the results of mutagenesis of the predicted cleavage sites (Snijder *et al*., 1996).

## Biological Aspects

The extensive proteolytic processing of the replicase polyprotein is assumed to play a key role in the replication of arteriviruses. The Nsp4 SP is the main proteinase responsible for the production of the majority of replicative proteins from the ORF1a and ORF1ab proteins (Snijder *et al*., 1994, 1996; Van Dinten *et al*., 1996). The activity of the Nsp4 SP domain is strongly influenced by other processing steps. Just downstream of the SP region, the EAV Nsp5/Nsp6 site is cleaved by an unknown proteinase that may be both host encoded and membrane associated. This processing step is closely coupled (in time and space) to the first cleavage carried out by the SP at the Nsp4/Nsp5 junction and depends on the presence of cleaved Nsp2 (Chapter 236). In the absence of cleaved Nsp2, the Nsp5/Nsp6 and Nsp4/Nsp5 junctions are not cleaved, which leads to increased processing of the C-terminal part of the ORF1a protein along an alternative pathway.

## Distinguishing Features

The SP displays properties of two different groups of chymotrypsin-like enzymes: it utilizes the His-Asp-Ser catalytic triad found in classical chymotrypsin-like proteinases (clan SA), and its putative substrate-binding region contains Thr and His residues which are typical of the picornains (peptidase family C3, a major subset of clan CB). Thus, it can be considered as the prototype of a novel subgroup, the 3C-like serine proteinases (Snijder *et al*., 1996). Similar proteinases were predicted to be encoded by plant sobemoviruses, luteoviruses and pea enation mottle virus, and animal astroviruses.

Polyclonal peptide antisera against three regions of EAV Nsp4 have been raised in rabbits and are available from the authors for research purposes on request (Snijder *et al*., 1994).

## Further Reading

For further information, the reader is referred to Snider *et al*. (1996).

## References

Den Boon, J.A., Snijder, E.J., Chirnside, E.D., de Vries, A.A.F., Horzinek, M.C. & Spaan, W.J.M. (1991) Equine arteritis virus is not a togavirus but belongs to the coronaviruslike superfamily. *J. Virol.* **65**, 2910–2920.

Den Boon, J.A., Faaberg, K.S., Meulenberg, J.J.M., Wassenaar, A.L.M., Plagemann, P.G.W., Gorbalenya, A.E. & Snijder, E.J. (1995) Processing and evolution of the N-terminal region of the arterivirus replicase ORF1a protein: identification of two papainlike cysteine proteases. *J. Virol.* **69**, 4500–4505.

Godeny, E.K., Chen, L., Kumar, S.N., Methven, S.L., Koonin, E.V. & Brinton, M.A. (1993) Complete genomic sequence and phylogenetic analysis of the lactate dehydrogenase-elevating virus. *Virology* **194**, 585–596.

Meulenberg, J.J.M., Hulst, M.M., de Meijer, E.J., Moonen, P.L.J.M., den Besten, A., de Kluyver, E.P., Wensvoort, G. & Moormann, R.J.M. (1993) Lelystad virus, the causative agent of porcine epidemic abortion and respiratory syndrome (PEARS), is related to LDV and EAV. *Virology* **192**, 62–72.

Snijder, E.J. & Spaan, W.J.M. (1995) The coronaviruslike superfamily. In: *The Coronaviridae* (Siddell, S.G., ed.). New York: Plenum Press, pp. 239–255.

Snijder, E.J., Wassenaar, A.L.M. & Spaan, W.J.M. (1992) The 5′ end of the equine arteritis virus replicase gene encodes a papainlike cysteine protease. *J. Virol.* **66**, 7040–7048.

Snijder, E.J., Wassenaar, A.L.M. & Spaan, W.J.M. (1994) Proteolytic processing of the replicase ORF1a protein of equine arteritis virus. *J. Virol.* **68**, 5755–5764.

Snijder, E.J., Wassenaar, A.L.M., Spaan, W.J.M. & Gorbalenya, A.E. (1995) The arterivirus Nsp2 protease. An unusual cysteine protease with primary structure similarities to both papain-like and chymotrypsin-like proteases. *J. Biol. Chem.* **270**, 16671–16676.

Snijder, E.J., Wassenaar, A.L.M., van Dinten, L.C., Spaan, W.J.M. & Gorbalenya, A.E. (1996) The arterivirus nsp4 protease is the prototype of a novel group of chymotrypsin-like enzymes, the 3C-like serine proteases. *J. Biol. Chem.* **271**, 4864–4871.

Van Dinten, L.C., Wassenaar, A.L.M., Gorbalenya, A.E., Spaan, W.J.M. & Snijder, E.J. (1996) Processing of the equine arteritis virus replicase ORF1B protein: identification of cleavage products containing the putative viral polymerase and helicase domains. *J. Virol.* **70**, 6625–6633.

*Eric J. Snijder*
*Department of Virology, Institute of*
*Medical Microbiology,*
*Leiden University,*
*PO Box 9600, 2300 RC Leiden,*
*The Netherlands*
*Email: snijder@virology.azl.nl*

*Alfred L.M. Wassenaar*
*Department of Virology, Institute of*
*Medical Microbiology,*
*Leiden University,*
*PO Box 9600, 2300 RC Leiden,*
*The Netherlands*

*Alexander E. Gorbalenya*
*M.P. Chumakov Institute of*
*Poliomyelitis and Viral*
*Encephalitides,*
*Russian Academy of Medical*
*Sciences, 142782 Moscow*
*Region, and A.N. Belozersky Institute*
*of Physico-Chemical Biology,*
*Moscow State University,*
*119899 Moscow, Russia*

# 93. Introduction: clan SB containing the subtilisin family

## *Databanks*

MEROPS ID: SB

| Species | SwissProt | PIR | EMBL (cDNA) | EMBL (genomic) |
|---|---|---|---|---|
| **Family S8** | | | | |
| Acidic serine proteinase (V5) | | | | |
|   *Dichelobacter nodosus* | – | – | L38395 | – |
| Aqualysin (Chapter 103) | | | | |
| Bacillopeptidase F (Chapter 98) | | | | |
| Basic serine proteinase, *Dichelobacter* (Chapter 101) | | | | |
| C5a peptidase (*Streptococcus*) (Chapter 100) | | | | |
| Cerevisin (Chapter 104) | | | | |
| Cucumisin (Chapter 110) | | | | |
| Cuticle-degrading endopeptidase (*Metarhizium*, etc.) (Chapter 107) | | | | |
| Endopeptidase K (Chapter 106) | | | | |
| Furin (Chapter 117) | | | | |
| Halolysin (Chapter 112) | | | | |
| Intracellular serine proteinase | | | | |
|   *Bacillus polymyxa* | P29139 | C41335 | D00862 | – |
|   *Bacillus* sp. | P29140 | S27501 | D10730 | – |
|   *Bacillus subtilis* | P11018 | – | M13760 | – |
|   *Thermoactinomyces* sp. | – | – | D87557 | – |
| Kexin (Chapter 115) | | | | |
| Kexin, *Limulus* (Chapter 116) | | | | |
| Lactocepin (Chapter 99) | | | | |
| Mesentericopeptidase | | | | |
|   *Bacillus mesentericus* | P07518 | A23624 | – | – |
| Nisin leader peptidase NisP (*Lactococcus*) PepP (Chapter 109) | | | | |
| Oryzin (Chapter 105) | | | | |
| PACE4 (Chapter 121) | | | | |
| PrcA protease | | | | |
|   *Anabaena variabilis* | P23916 | A38816<br>S13335<br>S29552 | X56955 | – |
| Proprotein convertase 1 (Chapter 118) | | | | |
| Proprotein convertase 2 (Chapter 119) | | | | |
| Proprotein convertase 4 (Chapter 120) | | | | |
| Proprotein convertase 5 (Chapter 122) | | | | |
| Proprotein convertase 7 (Chapter 123) | | | | |
| Protease A, *Vibrio* | | | | |
|   *Vibrio alginolyticus* | P16588 | JS0173 | M25499 | – |
| Proteinase T | | | | |
|   *Tritirachium album* | P20015 | JQ0380 | M54900 | M54901: 3′ end |
| Pyrolysin (*Pyrococcus*) (Chapter 112) | | | | |

| Species | SwissProt | PIR | EMBL (cDNA) | EMBL (genomic) |
|---|---|---|---|---|
| **Family S8** (*continued*) | | | | |
| Savinase | | | | |
| *Bacillus lentus* | P29600 | – | – | – |
| *Bacillus lentus* | P41362 | JC1244 | D13157 | – |
| Subtilisin (Chapter 94) | | | | |
| Subtilisin homolog, *Bacillus* strain Ak.1 (Chapter 97) | | | | |
| Subtilisin homolog, *Thermus* strain Rt41A (Chapter 96) | | | | |
| Thermitase (Chapter 95) | | | | |
| Thermomycolin (Chapter 108) | | | | |
| Tripeptidyl-peptidase II (Chapter 113) | | | | |
| Tripeptidyl-peptidase S (*Streptomyces lividans*) (Chapter 114) | | | | |
| Others | | | | |
| *Acremonium chrysogenum* | P29118 | JU0332 PN0129 | – | D00923: complete gene |
| *Aeromonas salmonicida* | P31339 | S26691 | X67043 | – |
| *Alteromonas* sp. | – | – | D38600 | – |
| *Arabidopsis thaliana* | – | – | X85974 | – |
| *Bacillus* sp. | P20724 | A33973 B33973 | M28537 | – |
| *Bacillus* sp. | P28842 | S23407 | X62369 | – |
| *Bacillus* sp. | P41363 | A48373 JS0714 | D13158 | – |
| *Bacillus* sp. | – | – | D29688 | – |
| *Bacillus* sp. | – | – | S50880 | – |
| *Bacillus subtilis* | P16396 | A31386 S11504 S20207 S39670 | X53307 | – |
| *Bacillus subtilis* | P29141 | A41341 | M76590 | X73124: genomic region 325–333 |
| *Bacillus subtilis* | P54423 | – | U58981 | – |
| *Dictyostelium discoideum* | P54683 | – | U20432 | – |
| *Homo sapiens* | – | – | D42053 | – |
| Ictalurid herpesvirus 1 | Q00139 | C36791 | M75136 | – |
| *Ophiostoma piceae* | – | S58795 | – | – |
| *Paecilomyces lilacinus* | – | – | L29262 | – |
| *Pasteurella haemolytica* | P31631 | A43608 | S61890 | – |
| *Pseudomonas* sp. | – | – | U36429 | – |
| *Saccharomyces cerevisiae* | P25036 | – | M77197 | – |
| *Saccharomyces cerevisiae* | P25381 | S19458 | – | X59720: complete chromosome III |
| *Schizosaccharomyces pombe* | P40903 | S45493 | D14063 | – |
| *Serratia marcescens* | P09489 | A29840 | M13469 | – |
| *Serratia marcescens* | P29805 | S19882 | X59719 | – |
| *Serratia marcescens* | – | – | D78380 | – |
| *Streptomyces coelicolor* | – | – | U33176 | – |
| *Treponema denticola* | – | – | D83264 | – |
| *Trichoderma harzianum* | Q03420 | S32905 S34787 | M87516 | M87518: complete gene |
| *Tritirachium album* | P23653 | S11985 | X56116 | – |
| *Vibrio metschnikovii* | – | – | Z28354 | – |
| *Yarrowia lipolytica* | P09230 | A26955 A31563 | M17741 | M23353: complete gene |

Clan SB contains only the single subtilisin family, S8, but the family is a large and diverse one. It includes peptidases from bacteria, archaea and eukaryotes, and one distinctive characteristic of the family is the order of catalytic residues in the linear sequence, which is Asp, His, Ser and different to that in other clans. Most of the members of clan SB are secreted endopeptidases, although a few are intracellular, and the family includes one exopeptidase, tripeptidyl-peptidase II (Chapter 113).

A curious feature of family S8 is that the catalytic serine is always encoded by TCX triplets (where X can be any base), whereas in all other families of serine peptidases AGC and AGT codons are used in addition to TCX. Sequence conservation within family S8 around the catalytic residues is shown in Alignment 93.1 (see CD-ROM). Note that the catalytic aspartate commonly occurs in an Asp-Thr/Ser-Gly motif, which is also found in family A1 and has been used in the past to determine membership of the pepsin family A1.

Tertiary structures have been solved for several members of the family, and Figure 93.1 shows the structure for subtilisin (Chapter 94). The structure differs significantly from that of chymotrypsin (shown in Chapter 2) in containing $\alpha$ as well as $\beta$ elements.

Because of large divergences among the sequences, the family can be divided into at least three subfamilies. Subfamily S8A includes the bacterial and archaean endopeptidases, as well as some fungal (such as endopeptidase K; Chapter 106) and plant endopeptidases (such as cucumisin; Chapter 110). Relationships between the sequences in subfamily S8A are shown in the evolution tree (Figure 93.2). Subfamily S8B contains eukaryote endopeptidases that process certain pro-hormones at pairs of basic residues, and include the yeast enzyme kexin (Chapter 115), which processes the $\alpha$-mating factor precursor, and the animal enzyme furin (Chapter 117), which processes the precursors of albumin, complement component C3 and von Willebrand factor. An endopeptidase from the cyanobacterium *Anabaena variabilis* is also a member of this subfamily. Relationships between the sequences in subfamily S8B are shown in the evolutionary tree (Figure 93.3). Subfamily S8C includes one of the few exopeptidases in the family, tripeptidyl-peptidase II, which is a mammalian, cytosolic enzyme. There has been some controversy surrounding identification of the catalytic aspartate in this exopeptidase, and there appears to be a large insert in the catalytic domain. As can be seen from Alignment 93.1, membership of a subfamily can be determined from the conservation of motifs around the catalytic residues. Subfamilies S8A and S8C possess Thr(or Ser)139, His172 and Met324, whereas in subfamily S8B the residues are Asp139, Arg172 and Ala324. Members of subfamily S8C are the only ones to possess Asn140, Ser175 and Cys330.

Family S8 includes a number of mosaic proteins. There are many peptidases from the bacterium *Vibrio* and its close relatives which require a special C-terminal domain for secretion that is proteolytically removed upon extracellular activation. This domain is found in members of family S8, and homologous domains are found in families S2 (*Achromobacter* lysyl

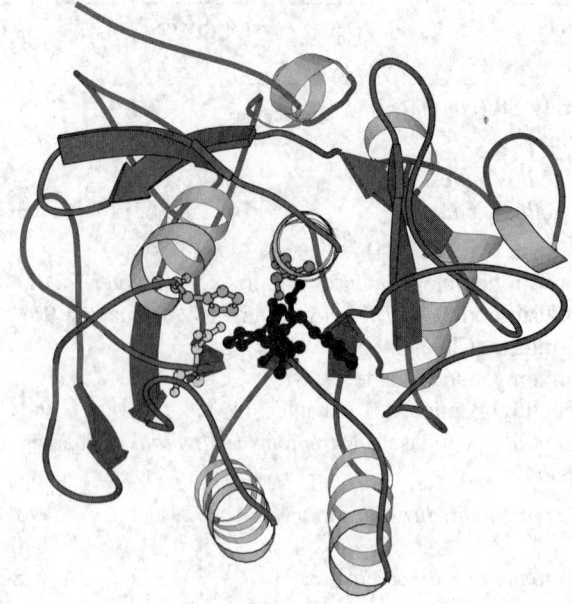

*Figure 93.1*  Richardson diagram of the *Bacillus licheniformis* subtilisin Carlsberg/*N*-tert-butoxycarbonyl-Ala-Pro-Phe-*O*-benzoyl hydroxylamine complex. The image was prepared from the Brookhaven Protein Data Bank entry (1SCN) as described in the Introduction (p. xxv). Catalytic residues are shown in ball-and-stick representation: Asp138, His169 and Ser323 (Alignment 93.1 numbering). The hydroxylamine is shown in ball-and-stick representation in black.

endopeptidase; Chapter 85), M4 (vibriolysin; Chapter 355) and M28 (*Vibrio* aminopeptidase; Chapter 491). There are large inserts in the catalytic domain between the active site His and Ser residues in lactocepin; Chapter 99, but these inserts have not been found in other proteins. Many members of subfamily S8B have C-terminal extensions, which may contain transmembrane regions (for example, furin), cysteine-rich regions (for example, proprotein convertase 5; Chapter 122), and a domain known as the P-domain. The P-domain is homologous to a C-terminal extension of the subtilisin homolog from *Aeromonas salmonicida*, and also an uncharacterized protein from the fungus *Pneumocystis carinii* (EMBL: D31909).

There is a single representative of this family in viruses. The Channel catfish (ictalurid) herpesvirus is a double-stranded DNA virus, and the gene 47 encodes a subtilisin homolog. This homolog appears to be derived from a horizontal transfer of genes, because it is part of a large insert in the viral genome not found in other herpesviruses (Davison, 1992).

## Reference

Davison, A.J. (1992) Channel catfish virus: a new type of herpesvirus. *Virology* **186**, 9–14.

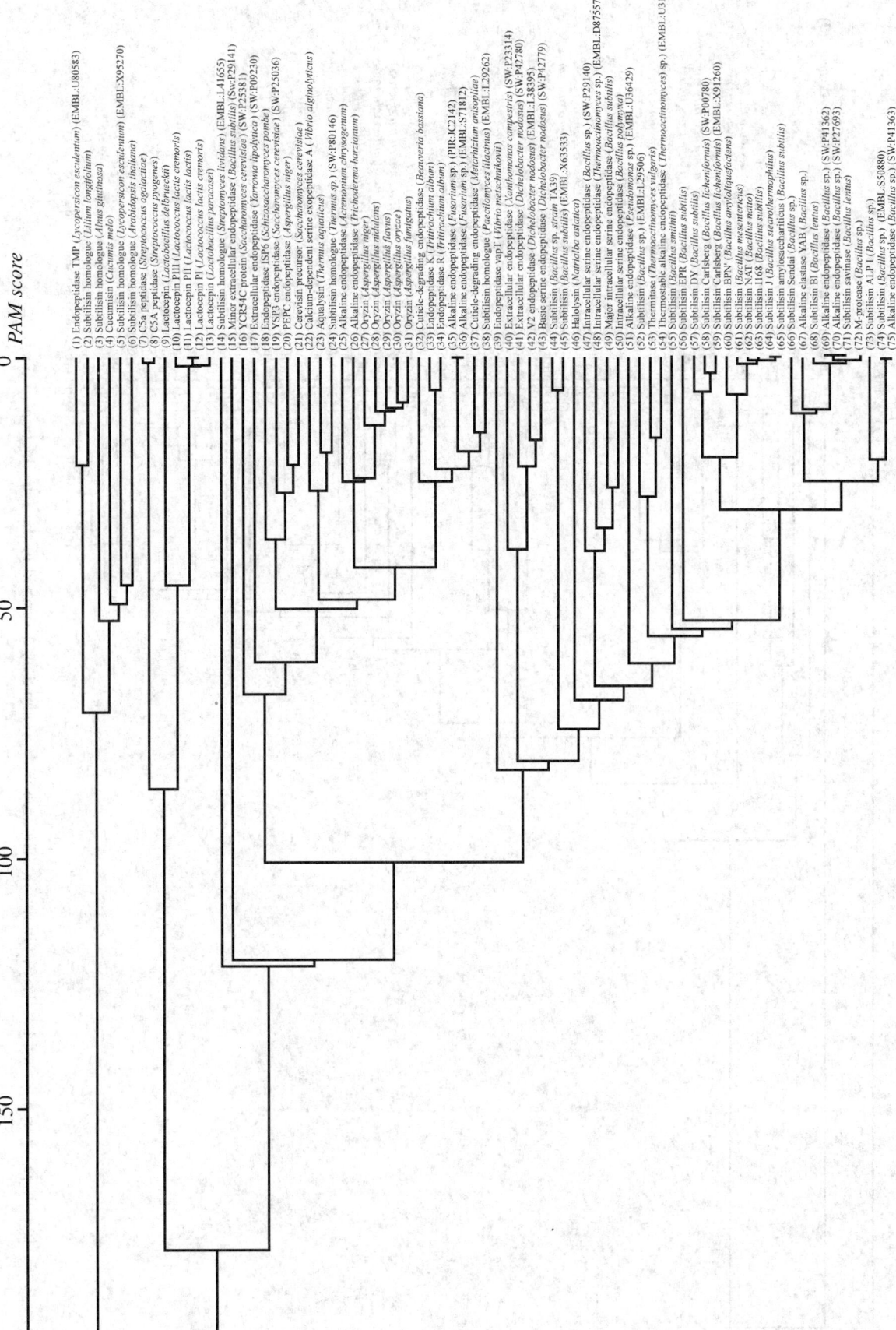

*Figure* 93.2   Evolutionary tree for subfamily A, family S8. The tree was prepared as described in the Introduction (p. xxv).

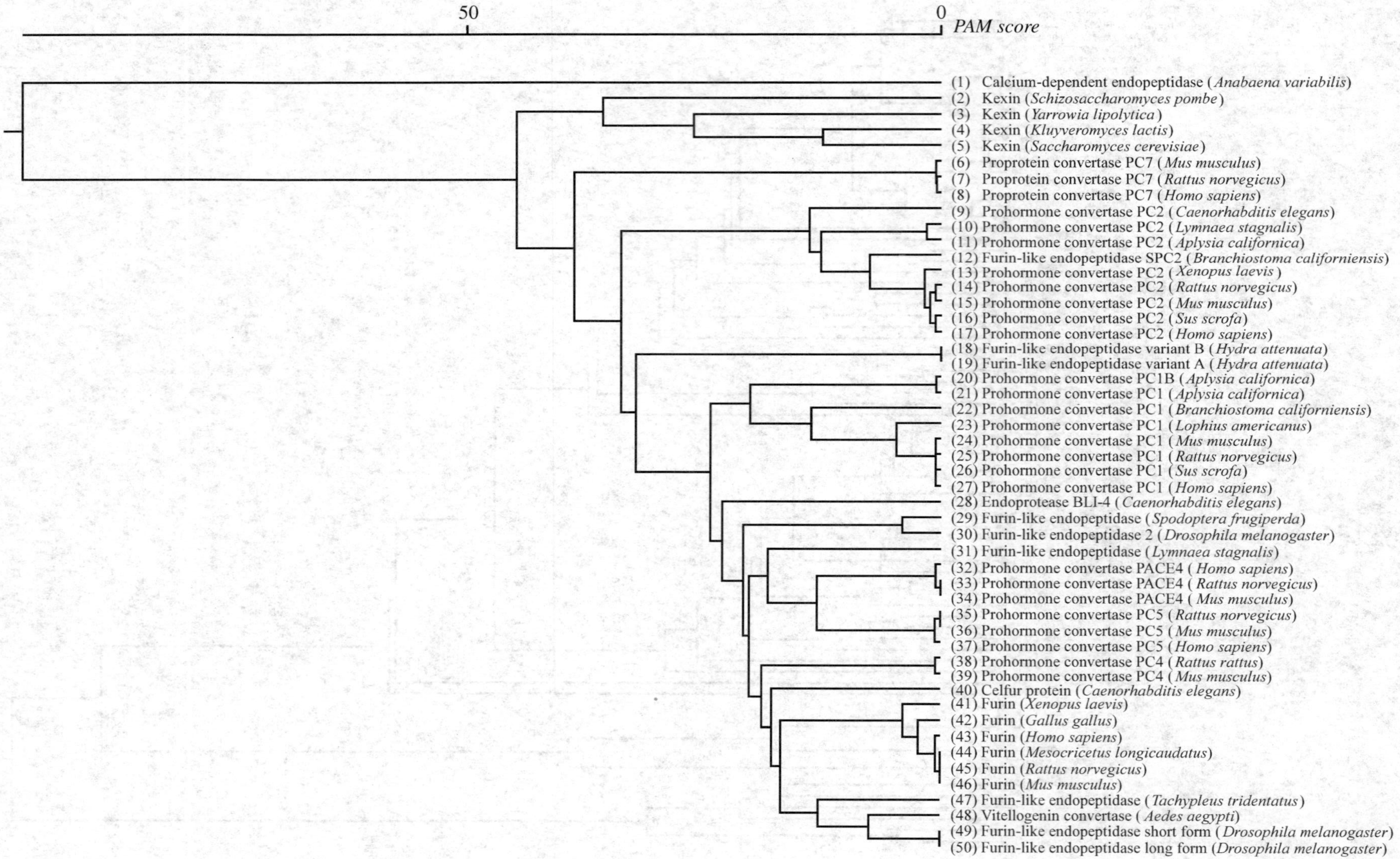

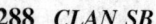

*Figure 93.3*   Evolutionary tree for subfamily B, family S8. The tree was prepared as described in the Introduction (p. xxv).

# 94. *Subtilisin*

## *Databanks*

*Peptidase classification: clan SB, family S8, MEROPS ID: S08.001*
*NC-IUBMB enzyme classification: EC 3.4.21.62*
*Chemical Abstracts Service registry number: 9014-01-1*
*Databank codes:*

| Species | Gene | SwissProt | PIR | EMBL (cDNA) | EMBL (genomic) |
|---|---|---|---|---|---|
| *Bacillus alcalophilus* | – | P27693 | A49778 | M65086 | – |
| *Bacillus amyloliquefaciens* | *apr* | P00782 | A00970 | K02496 | – |
|  |  | A92033 | X00165 |  |  |
|  |  | A93495 |  |  |  |
|  |  | B25415 |  |  |  |
| Subtilisin lentus |  |  |  |  |  |
| *Bacillus lentus* |  | P29599 | – | – | – |
| *Bacillus licheniformis* | *apr* | P00780 | A00968 | X03341 | – |
|  |  | A24111 |  |  |  |
|  |  | JC1085 |  |  |  |
| *Bacillus licheniformis* | *subC* | – | – | X91260 | – |
| *Bacillus natto* | *aprN* | P35835 | JC2036 | D25319 | – |
|  |  | JH0778 | S51909 |  |  |
|  |  | JS0517 |  |  |  |
|  |  | JS0601 |  |  |  |
| *Bacillus smithii* | *apr* | – | – | L24202 | – |
| *Bacillus* sp. | *aprQ* | – | – | D29736 | – |
| *Bacillus* sp. KSM-K16 |  | Q99405 | – | – | – |
| *Bacillus stearothermophilus* | *aprJ* | P29142 | JQ1487 | M64743 | – |
| *Bacillus subtilis* strain DY | *apr* | P00781 | A00969 | – | – |
| *Bacillus subtilis* var. amylosacchariticus | *apr* | P00783 | A41448 | D00264 | – |
| *Bacillus subtilis* | *aprE* | P04189 | A00972 | K01443 | – |
|  |  | A26116 | K01988 |  |  |
|  |  | I39779 | M16639 |  |  |
| *Bacillus subtilis* |  | – | – | X63533 | – |

Brookhaven Protein Data Bank three-dimensional structures:

| Species | ID | Resolution | Notes |
|---|---|---|---|
| *Bacillus amyloliquefaciens* | 1S01 | 1.7 | mutant |
|  | 1SBN | 2.1 | complex with eglin C Leu45Arg |
|  | 1SBT | 2.5 |  |
|  | 1SIB | 2.4 | complex with eglin C Arg53Lys |
|  | 1SPB | 2 | pro segment; complex with mutant subtilisin BPN |
|  | 1ST2 | 2 | peroxide inactivated |
|  | 1SUB | 1.75 | mutant Asn218Ser, Ser221Cys |
|  | 1SUC | 1.8 | mutant Met50Phe, Tyr217Lys, Asn218Ser, Ser221Cys, 75–83 deleted |
|  | 1SUD | 1.9 | mutant Met50Phe, Tyr217Lys, Asn218Ser, Ser221Cys |
|  | 2SBT | 2.8 |  |
|  | 2SIC | 1.8 | complex with *Streptomyces* subtilisin inhibitor |
|  | 2SNI | 2.1 | complex with chymotrypsin inhibitor 2 |
|  | 2ST1 | 1.8 | recombinant |
|  | 3SIC | 1.8 | complex with *Streptomyces* subtilisin inhibitor Met73Lys |
|  | 5SIC | 2.2 | complex with *Streptomyces* subtilisin inhibitor Met70Gly, Met73Lys |

*continued overleaf*

| Species | ID | Resolution | Notes |
|---|---|---|---|
| *Bacillus* sp. KSM-K16 | 1MPT | 2.4 | |
| *Bacillus subtilis* | 1CIS | | complex with barley chymotrypsin inhibitor 2 |
| | 1CSE | 1.2 | complex with eglin C |
| | 1SBC | 2.5 | |
| | 1SCA | 2 | |
| | 1SCB | 2.3 | acetonitrile-soaked crystals |
| | 1SCD | 2.3 | |
| | 1SCN[a] | 1.9 | complex with Boc-Ala-Pro-Phe-*O*-benzoyl hydroxylamine |
| *Bacillus subtilis* | 1SEL | 2 | Ser221selenocysteine mutant |
| *Bacillus subtilis* | 2SEC | 1.8 | complex with *N*-acetyl eglin C |

[a]See Chapter 93 for an image constructed from these coordinates.

## Name and History

The name *subtilisin* derives from the name of the bacterial species, *Bacillus subtilis*, from which the enzyme was first isolated (Ottensen & Svensen, 1970; Markland & Smith, 1971). Subtilisins are extracellular serine endopeptidases, and the term now encompasses the enzymes isolated from related *Bacillus* species as well. The first enzyme of this type, *subtilisin Carlsberg* from *B. licheniformis* (also known as *subtilisin A, subtilopeptidase A, Alcalase Novo*), was discovered by Linderstrøm-Lang and Ottensen while they were studying the conversion of ovalbumin to plakalbumin (Linderstrøm-Lang & Ottensen, 1947). Many other subtilisins are now known and the nomenclature is confusing, especially with the advent of engineered commercial variants. Subtilisin from *B. amyloliquefaciens* (also known as *subtilisin BPN′, Nagarse, subtilisin B, subtilopeptidase B, subtilopeptidase C* and *bacterial proteinase Novo*) was independently isolated by several groups (Hagihara, 1954; Ottensen & Spector, 1960), and simultaneously given several different names (Robertus *et al.*, 1971). Other species variants that are prominent in biochemical literature include those from *B. lentus* (*Subtilisin 147* or *Esperase*), *B. alcalophilus* PB92 (*maxacal*; an N85S variant of this enzyme, expressed in *B. lentus*, is known as *Subtilisin 309* or *Savinase*), and *B. subtilis* strain 168 (*Subtilisin I168, Subtilisin E*). For a complete listing see the reviews of Siezen (Siezen *et al.*, 1991; Siezen, 1996).

The three-dimensional, X-ray crystallographic structure of subtilisin BPN′ was first elucidated by Wright and coworkers in 1969 (Wright *et al.*, 1969) and found to show a fold distinct from the pancreatic serine protease, chymotrypsin (Matthews *et al.*, 1967). A large family (S8) of homologous proteins is now known (Chapter 93), and the members of the family have sometimes been termed 'subtilases' (Siezen, 1996). This family extends into eukaryotes, including the kexin subfamily of polybasic-specific proprotein convertases (see Chapter 115 and following chapters).

Subtilisins have been studied extensively because of their usefulness as additives to detergents. Enzymologic studies have been crucial to understanding the mechanism of serine protease action (Markland & Smith, 1971; Kraut, 1977; Polgár, 1987). Subtilisin BPN′ was first cloned in 1983 (Wells *et al.*, 1983), and subsequently has served as an excellent system for protein engineering (Wells & Estell, 1988).

## Activity and Specificity

Subtilisins are serine proteases that use a catalytic triad composed of Asp32, His64 and Ser221 (subtilisin BPN′ numbering). Site-directed mutagenesis has been used to assess the relative importance of each of these residues to catalysis (Carter & Wells, 1988). The reaction mechanism has been studied in detail and is described elsewhere (Polgár, 1987). The rate-limiting step is acylation for amide bond hydrolysis and deacylation for ester hydrolysis (Markland & Smith, 1971). The requirement for a deprotonated histidine gives rise to the basic pH dependence, with the $pK_a$ of $\sim 7$ (Phillip *et al.*, 1979), but it is possible to alter this by protein engineering (Russell & Fersht, 1987). Subtilisins show good stability in the pH range 7–10, with the *B. lentus* and *B. alcalophilus* varieties being particularly stable to alkali (van der Laan *et al.*, 1991; Betzel *et al.*, 1992). Assays are typically performed at pH 8.2–8.6, 25°C, with the chromogenic substrate Suc-Ala-Ala-Pro-Phe$\downarrow$NHPhNO$_2$ (Estell *et al.*, 1986), which is dissolved in DMSO or dimethyl formamide. Two millimolar CaCl$^{2+}$ and 0.1 M KCl are often included (Phillip & Bender, 1983), as subtilisin has two Ca$^{2+}$ sites that stabilize its structure (Pantoliano *et al.*, 1988). The kinetic parameters for this substrate vary depending on the enzyme species: values of $k_{cat}$ range from 30 to 200 s$^{-1}$ and $K_m$ from 0.13 to 1.2 mM, but $k_{cat}/K_m$ values are relatively constant at $10^5$–$10^6$ M$^{-1}$ s$^{-1}$.

A substantial effort has been made to enhance the stability of subtilisins to extremes of temperature (Pantoliano *et al.*, 1989) and nonaqueous conditions (Wong *et al.*, 1990; Chen & Arnold, 1993). Significant improvements in both respects have resulted from directed evolution strategies (Bryan, 1992; You & Arnold, 1994).

Subtilisins are covalently inactivated by standard serine protease inhibitors such as PMSF and DFP, and also by peptidyl-halomethanes (Morihara & Oka, 1970; Poulos *et al.*, 1976) and peptidyl-boronic acids (Matthews *et al.*, 1975). They are strongly inhibited by many natural inhibitor proteins or domains such as turkey ovomucoid third domains (Empie & Laskowski, 1982), bacitracins (Pfeffer *et al.*, 1991), eglins (Seemüller *et al.*, 1986), *Streptomyces* subtilisin inhibitor (Laskowski & Kato, 1980), and inhibitors from barley and potato (Svendsen *et al.*, 1982).

Subtilisins are relatively nonspecific proteases. The subsite preferences for subtilisin BPN′ and savinase on both the N- and C-terminal sides of the scissile bond have been mapped by the comprehensive work of Grøn and coworkers (Grøn *et al.*, 1992, 1996), who used full-length, internally quenched sets of fluorogenic substrates having restrictions on the site of cleavage imposed by appropriate design of the parent peptides. Surveys of the N-terminal subsite specificity have been performed with chromogenic substrates (Phillip & Bender, 1983). The subsites having the most prominent cavities and strongest preferences are S1 and S4. In S1, large, non-$\beta$-branched hydrophobic side chains are preferred (Estell *et al.*, 1986; Grøn *et al.*, 1992), although basic side chains can be accepted in an alternate binding mode (Robertus *et al.*, 1972; Poulos *et al.*, 1976). The S4 subsite strongly prefers hydrophobic side chains (Grøn *et al.*, 1992; Rheinnecker *et al.*, 1993). Nonadditivity is generally observed with regard to the preferences at other sites, depending on the residues in the P4 and P1 positions (Grøn & Breddam, 1992). Virtually any amino acid is accepted at P3 since the side chain at this position is directed into the solvent. The S2 subsite is smaller, accommodating alanine particularly well (Grøn *et al.*, 1992; Ballinger *et al.*, 1995).

Subtilisin has proved to be very amenable to alteration of substrate specificity by protein engineering. These efforts, which include changing the size (Estell *et al.*, 1986; Teplyakov *et al.*, 1992; Sørensen *et al.*, 1993; Rheinnecker *et al.*, 1994) and charge (Russell & Fersht, 1987; Wells *et al.*, 1987; Ballinger *et al.*, 1995) preferences, particularly at S1 and S4 subsites, have been reviewed (Perona & Craik, 1995). In a particularly successful specificity design, a requirement for histidine at the P2 or P1′ position in substrates has been introduced by a H64A mutation in subtilisin BPN′ (Carter & Wells, 1987). This variant and its activity-enhanced form, Genenase I (Carter *et al.*, 1989, 1991), operate by 'substrate-assisted catalysis', in which the histidine from the substrate effectively substitutes as a general base for the missing enzymic His64. This mutant has sufficient specificity to be useful for site-specific proteolysis of fusion proteins (Forsberg *et al.*, 1992). Multiple subsite modification has also produced mutants with multibasic specificities similar to the related eukaryotic hormone-processing enzymes kexin (Chapter 115) and furin (Chapter 117) (Ballinger *et al.*, 1995, 1996).

Two types of subtilisin variants have been produced with novel activities by mutating or chemically modifying the nucleophilic Ser221. A Ser221Cys mutation causes aminolysis of an acyl enzyme intermediate to be favored over hydrolysis (Nakatsuka *et al.*, 1987). The additional substitution, Pro225Ala, yields a double mutant ('subtiligase') that can be used to ligate esterified peptides site-specifically to the N-termini of proteins or peptides in aqueous solution and in high yield (Abrahmsen *et al.*, 1991; Chang *et al.*, 1994). This makes the enzyme an effective tool for blockwise peptide or protein synthesis, and has been employed to create fully synthetic ribonuclease A variants containing unnatural amino acids (Jackson *et al.*, 1994). Chemical modification of Ser221 with PMSF and hydrogen selenide yields a selenocysteine-containing variant ('selenosubtilisin'). Under anaerobic conditions it acts as a superb peptide ligase (Wu & Hilvert, 1989), but in aerobic conditions the selenium becomes oxidized and the enzyme has peroxidase activity (Wu & Hilvert, 1990; Bell *et al.*, 1993).

## Structural Chemistry

Subtilisins are monomeric proteins of 268–275 amino acid residues, and 26.9–27.5 kDa (Markland & Smith, 1971; Siezen *et al.*, 1991). The pI varies from 7.8 for subtilisin BPN′ to ~11 for *B. lentus* subtilisins (Betzel *et al.*, 1996). They are initially expressed as preproenzymes, with the pro segment typically containing 77 amino acid residues (Power *et al.*, 1986; Siezen *et al.*, 1995). The pro-segment is required for correct folding of the protease (Strausberg *et al.*, 1993; Shinde & Inouye, 1994). No Cys residues are present in the protein. Several crystal structures are available for subtilisins including subtilisin BPN′ (Wright *et al.*, 1969; McPhalen & James, 1988; Takeuchi *et al.*, 1991), subtilisin Carlsberg (Bode *et al.*, 1986; McPhalen & James, 1988), subtilisin 309 (Betzel *et al.*, 1992), and *Bacillus alcalophilus* subtilisin PB92 (van der Laan *et al.*, 1992). The subtilisin structures are highly superimposable, with an r.m.s. deviation of 0.4–0.65 Å in the 194 amino acids that make up the structural core (Siezen *et al.*, 1991). Most are enzyme–inhibitor complexes that are useful for defining the mode of substrate binding, and the specificity determinants. In addition, a structure of the complex between mature subtilisin BPN′ and its pro segment has recently been determined, which illustrates the intramolecular nature of the propeptide cleavage, and the basis of the chaperone activity of the propeptide (Gallagher *et al.*, 1995).

The protease fold consists of a central seven-stranded parallel $\beta$ sheet surrounded by nine $\alpha$ helices, with two additional strands of antiparallel $\beta$ sheet near the C-terminus (see Chapter 93). The catalytic triad is arranged near the surface with Ser221 and His64 at the ends of adjacent $\alpha$ helices, and Asp32 lying at the end of a strand of the central $\beta$ sheet. The oxyanion hole, which stabilizes the negative charge of the tetrahedral intermediate, is comprised of the Asn155 side chain amide group, the backbone amide of Ser221 (Robertus *et al.*, 1972; Poulos *et al.*, 1976), and the Thr220 side chain hydroxyl group (Braxton & Wells, 1991). The enzyme is capable of binding a nine residue stretch of the substrate or inhibitor, from S6 to S3′ (McPhalen & James, 1988). The N-terminal portion (P4–P1) of the substrate backbone is bound as the central strand of an antiparallel $\beta$ sheet, between residues 100 and 102 on one side and 125–127 on the other. The S1 subsite is a large open cleft that has residues 125–127 on one side, 153–156 on the other, and the 165–170 loop at the base. The hydrophobic S4 pocket is comprised primarily of the side chains from Tyr104, Ile107 and Leu126, with Leu135 slightly more distant. The P3 side-chain is directed out into the solvent. P2–S2 contacts include Thr33, Asn62, His64, Leu96 and Gly100; the P1′ residue contacts His64, Asn155, Asn218 and Ser221; and P2′ contacts Asn155, Phe189 and Asn218 (subtilisin Carlsberg residue numbers: McPhalen & James, 1988).

The enzyme has one low-affinity and one high-affinity calcium-binding site, although a calcium-independent form has been engineered (Gallagher *et al.*, 1993). The high-affinity calcium-binding site is a surface loop nearly opposite the catalytic site. It is comprised of a circle of three backbone carbonyls (75, 79, 81), and the side-chain oxygen of Asn77, and is additionally coordinated on either side by side-chain oxygens from Gln2 and Asp41. The low-affinity site is formed by backbone carbonyls from Ala168, Tyr170, Val173, the side chain of Glu194, and two water molecules (McPhalen & James, 1988).

## Preparation

Many forms of subtilisin are commercially available. Subtilisin can be prepared in large quantities from bacterial culture medium (yields $>100 \, mg \, liter^{-1}$ of shake-flask culture, and much larger amounts in an optimized fermenter; Wells *et al*., 1983). The enzyme is secreted and represents a significant proportion of the protein in crude supernatants, thus making purification relatively simple. Several isolation procedures have been described (Carter & Wells, 1988; Grøn *et al*., 1992; Rheinnecker *et al*., 1993; Ballinger *et al*., 1995). A straightforward procedure used in this laboratory has involved ethanol precipitation and ammonium sulfate precipitation steps, followed by cation-exchange chromatography.

Subtilisin has also been expressed in *Escherichia coli* (Ikemura *et al*., 1987). Inclusion bodies are produced, and a refolding step is applied to yield the active enzyme (Li & Inouye, 1994; Shinde & Inouye, 1994).

Catalytically inactive variants of subtilisin may be expressed and processed either using the *E. coli* system or by expressing in *B. subtilis* using a coculture of wild-type-like 'helper' subtilisin. In the latter case, an additional Ser24Cys substitution has been incorporated into the inactive mutant to allow for covalent chromatography on a thiol resin and separation from the helper enzyme (Carter & Wells, 1988).

## Biological Aspects

*Bacillus* species secrete subtilisins into the extracellular space together with neutral metalloproteases (Chapter 351). The enzymes were at one time suggested to be involved in sporulation (Hoch, 1976; Piggot & Coote, 1976; Priest, 1977) or cell wall turnover (Jolliffe *et al*., 1980), but the lack of morphological changes associated with deletion of either gene indicated that they probably serve only nutritional functions (Yang *et al*., 1984).

## Distinguishing Features

Subtilisins from *Bacillus* spp. have many close relatives in eukaryotic cells (Siezen *et al*., 1991). Bacterial subtilisins are characterized by their substrate specificity, high expression in secreted form, and alkaline stability.

## Further Reading

Polgár (1987) has reviewed mechanistic aspects of serine protease activity, and Wells & Estell (1988) provide a full account of the protein engineering of subtilisins. Perona & Craik (1995) review the natural and variant substrate specificities in the proteases of the subtilisin and chymotrypsin families.

## References

Abrahmsen, L., Tom, J., Burnier, J., Butcher, K.A., Kossiakoff, A. & Wells, J.A. (1991) Engineering subtilisin and its substrates for efficient ligation of peptide bonds in aqueous solution. *Biochemistry* **30**, 4151–4159.

Ballinger, M.D., Tom, J. & Wells, J.A. (1995) Designing subtilisin BPN′ to cleave substrates containing dibasic residues. *Biochemistry* **34**, 13312–13319.

Ballinger, M.D., Tom, J. & Wells, J.A. (1996) Furilisin: a subtilisin BPN′ variant engineered for cleaving tribasic substrates. *Biochemistry* **35**, 13579–13585.

Bell, I.M., Fisher, M.L., Wu, Z.P. & Hilvert, D. (1993) Peroxide dependence of the semisynthetic enzyme selenosubtilisin. *Biochemistry* **32**, 13969–13973.

Betzel, C., Klupsch, S., Papendorf, G., Hastrup, S., Branner, S. & Wilson, K.S. (1992) Crystal structure of the alkaline proteinase savinase from *Bacillus lentus* at 1.4 Å resolution. *J. Mol. Biol.* **223**, 427–445.

Betzel, C., Klupsch, S., Branner, S. & Wilson, K.S. (1996). Crystal structure from the alkaline proteases savinase and esperase from *Bacillus lentus*. In: *Subtilisin Enzymes: Practical Protein Engineering* (Bott, R. & Betzel, C., eds). New York: Plenum, pp. 49–61.

Bode, W., Papamokos, E., Musil, D., Seemüller, U. & Fritz, H. (1986) Refined 1.2 Å crystal structure of the complex formed between subtilisin Carlsberg and the inhibitor eglin C. Molecular structure of eglin and its detailed interaction with subtilisin. *EMBO J.* **5**, 813–818.

Braxton, S. & Wells, J.A. (1991) Importance of a distal hydrogen bonding group in stabilizing the transition state of subtilisin BPN′. *J. Biol. Chem.* **266**, 11797–11800.

Bryan, P. (1992). Engineering dramatic increases in the stability of subtilisin. In: *Stability of Protein Pharmaceuticals Part B: In Vivo Pathways of Degradation and Strategies for Protein Stabilization* (Ahern, T.J. & Manning, M.C., eds). New York: Plenum, pp. 147–181.

Carter, P. & Wells, J.A. (1987) Engineering enzyme specificity by 'substrate-assisted catalysis'. *Science* **237**, 394–399.

Carter, P. & Wells, J. (1988) Dissecting the catalytic triad of a serine protease. *Nature* **322**, 546–568.

Carter, P., Nilsson, B., Burnier, J.P., Burdick, D. & Wells, J.A. (1989) Engineering subtilisin BPN′ for site-specific proteolysis. *Proteins: Struct. Funct. Genet.* **6**, 240–248.

Carter, P., Abrahmsen, L. & Wells, J.A. (1991) Probing the mechanism and improving the rate of substrate-assisted catalysis in subtilisin BPN′. *Biochemistry* **30**, 6142–6148.

Chang, T.K., Jackson, D.Y., Burnier, J.B. & Wells, J.A. (1994) Subtiligase: a tool for semisynthesis of proteins. *Proc. Natl Acad. Sci. USA* **91**, 12544–12548.

Chen, K. & Arnold, F.H. (1993) Tuning the activity of an enzyme for unusual environments: sequential random mutagenesis of subtilisin E for catalysis in dimethylformamide. *Proc. Natl Acad. Sci. USA* **90**, 5618–5622.

Empie, M.W. & Laskowski, M. (1982) Thermodynamics and kinetics of single residue replacements in avian ovomucoid third domains: effect on inhibitor interactions with serine proteinases. *Biochemistry* **21**, 2274–2284.

Estell, D.A., Graycar, T.P., Miller, J.V., Powers, D.B., Burnier, J.P., Ng, P.G. & Wells, J.A. (1986) Probing steric and hydrophobic effects on enzyme–substrate interactions by protein engineering. *Science* **233**, 659–663.

Forsberg, G., Baastrup, B., Rondahl, H., Holmgren, E., Pohl, G., Hartmanis, M. & Lake, M. (1992) An evaluation of different enzymatic cleavage methods for recombinant fusion proteins, applied on des(1–3)insulin-like growth factor I. *J. Protein Chem.* **11**, 201–211.

Gallagher, T., Bryan, P. & Gilliland, G.L. (1993) Calcium-independent subtilisin by design. *Proteins: Struct. Funct. Genet.* **16**, 205–213.

Gallagher, T., Gilliland, G., Wang, L. & Bryan, P. (1995) The prosegment-subtilisin BPN' complex: crystal structure of a specific 'foldase'. *Structure* **3**, 907–914.

Grøn, H. & Breddam, K. (1992) Interdependency of binding subsites in subtilisin. *Biochemistry* **31**, 8967–8971.

Grøn, H., Meldal, M. & Breddam, K. (1992) Extensive comparison of the substrate preferences of two subtilisins as determined with peptide substrates which are based on the principle of intramolecular quenching. *Biochemistry* **31**, 6011–6018.

Grøn, H., Bech, L.M., Sorensen, S.B., Meldal, M. & Breddam, K. (1996) Studies of binding sites in the subtilisin from *Bacillus lentus* by means of site directed mutagenesis and kinetic investigations. In: *Subtilisin Enzymes: Practical Protein Engineering* (Bott, R. & Betzel, C., eds). New York: Plenum, pp. 105–112.

Hagihara, B. (1954) *Ann. Rep. Sci. Works. Fac. Sci. Osaka Univ.* **2**, 35.

Hoch, J.A. (1976) Genetics of bacterial sporulation. *Adv. Genet.* **18**, 69–98.

Ikemura, H., Takagi, H. & Inouye, M. (1987) Requirement of prosequence for the production of active subtilisin E in *Escherichia coli*. *J. Biol. Chem.* **262**, 7859.

Jackson, D.Y., Burnier, J., Quan, C., Stanley, M., Tom, J. & Wells, J.A. (1994) A designed peptide ligase for total synthesis of ribonuclease A with unnatural catalytic residues. *Science* **266**, 243–247.

Jolliffe, L.K., Doyle, R.J. & Streips, U.N. (1980) Extracellular proteases modify cell wall turnover in *Bacillus subtilis*. *J. Bacteriol.* **141**, 1199–1208.

Kraut, J. (1977) Serine proteases: structure and mechanism of catalysis. *Annu. Rev. Biochem.* **46**, 331–358.

Laskowski, M. & Kato, I. (1980) Protein inhibitors of proteinases. *Annu. Rev. Biochem.* **149**, 593–626.

Li, Y. & Inouye, M. (1994) Autoprocessing of prothiolsubtilisin E in which active-site serine 221 is altered to cysteine. *J. Biol. Chem.* **269**, 4169–4174.

Linderstrøm-Lang, K. & Ottensen, M. (1947) A new protein from ovalbumin. *Nature* **159**, 807.

Markland, F.S. & Smith, E.L. (1971). Subtilisins: primary structure, chemical and physical properties. In: *The Enzymes*, vol. 3 (Boyer, P.D., ed.). London: Academic Press, pp. 561–608.

Matthews, B.W., Sigler, P.B., Henderson, R. & Blow, D.M. (1967) Three-dimensional structure of tosyl-α-chymotrypsin. *Nature* **214**, 652–656.

Matthews, D.A., Alden, R.A., Birktoft, J.J., Freer, S.T. & Kraut, J. (1975) X-ray crystallographic study of boronic acid adducts with subtilisin BPN' (Novo). *J. Biol. Chem.* **250**, 7120–7126.

McPhalen, C.A. & James, N.G. (1988) Structural comparison of two serine proteinase-protein inhibitor complexes: eglin-C-subtilisin Carlsberg and CI-2-subtilisin Novo. *Biochemistry* **27**, 6582–6598.

Morihara, K. & Oka, T. (1970) Subtilisin BPN': inactivation by chloromethyl ketone derivatives of peptide substrates. *Arch. Biochem. Biophys.* **138**, 526–531.

Nakatsuka, T., Sasaki, T. & Kaiser, E.T. (1987) Peptide segment coupling catalyzed by the semisynthetic enzyme thiolsubtilisin. *J. Am. Chem. Soc.* **109**, 3808–3810.

Ottensen, M. & Spector, A. (1960) *C. R. Trav. Lab. Carlsberg* **32**, 63.

Ottensen, M. & Svensen, I. (1970) The subtilisins. *Methods Enzymol.* **19**, 199–215.

Pantoliano, M.W., Whitlow, M., Wood, J.F., Rollence, M.L., Finzel, B.C., Gilliland, G.L., Poulos, T.L. & Bryan, P.N. (1988) The engineering of binding affinity at metal binding sites for the stabilization of proteins: subtilisin as a test case. *Biochemistry* **27**, 8311–8317.

Pantoliano, M.W., Whitlow, M., Wood, J.F., Dodd, S.W., Hardman, K.D., Rollence, M.L. & Bryan, P.N. (1989) Large increases in general stability for subtilisin BPN' through incremental changes in the free energy of unfolding. *Biochemistry* **28**, 7205–7213.

Perona, J.J. & Craik, C.S. (1995) Structural basis of substrate specificity in the serine proteases. *Protein Sci.* **4**, 337–360.

Pfeffer, S., Hohne, W., Branner, S., Wilson, K. & Betzel, C. (1991) X-ray structure of the antibiotic bacitracin A. *FEBS Lett.* **285**, 115–119.

Phillip, M. & Bender, M.L. (1983) Kinetics of subtilisin and thiolsubtilisin. *Mol. Cell. Biochem.* **51G**, 5–32.

Phillip, M., Tsai, L.-H. & Bender, M.L. (1979) Comparison of the kinetic specificity of subtilisin and thiolsubtilisin toward *n*-alkyl *p*-nitrobenzyl esters. *Biochemistry* **18**, 3769–3773.

Piggot, P.J. & Coote, J.G. (1976) Genetic aspects of bacterial endospore formation. *Bacteriol. Rev.* **40**, 908–962.

Polgár, L. (1987) Structure and function of serine proteases. In: *Mechanisms of Protease Action* (Neuberger, A. & Brocklehurst, K., eds). Boca Raton, FL: CRC Press.

Poulos, T.L., Alden, R.A., Freer, S.T., Birktoft, J.J. & Kraut, J. (1976) Polypeptide halomethyl ketones bind to serine proteases as analogs of the tetrahedral intermediate. *J. Biol. Chem.* **251**, 1097–1103.

Power, S.D., Adams, R.M. & Wells, J.A. (1986) Secretion and autoproteolytic maturation of subtilisin. *Proc. Natl Acad. Sci. USA* **83**, 3096–3100.

Priest, F.G. (1977) Extracellular enzyme synthesis in the genus *Bacillus*. *Bacteriol. Rev.* **41**, 711–753.

Rheinnecker, M., Baker, G., Eder, J. & Fersht, A.R. (1993) Engineering a novel specificity in subtilisin BPN'. *Biochemistry* **32**, 1199–1203.

Rheinnecker, M., Eder, J., Pandey, P.S. & Fersht, A.R. (1994) Variants of subtilisin BPN' with altered specificity profiles. *Biochemistry* **33**, 221–225.

Robertus, J.D., Alden, R.A. & Kraut, J. (1971) On the identity of subtilisins BPN' and Novo. *Biochem. Biophys. Res. Commun.* **42**, 334–339.

Robertus, J.D., Alden, R.A., Birktoft, J.J., Kraut, J., Powers, J.C. & Wilcox, P.E. (1972) An X-ray crystallographic study of the binding of peptide chloromethyl ketone inhibitors to subtilisin BPN'. *Biochemistry* **11**, 2439–2449.

Russell, A.J. & Fersht, A.R. (1987) Rational modification of enzyme catalysis by engineering surface charge. *Nature* **328**, 496–500.

Seemüller, U., Dodt, J., Fink, E. & Fritz, H. (1986) Eglins. In: *Proteinase Inhibitors* (Barrett, A.J. & Salvesen, G., eds). Amsterdam: Elsevier, pp. 347–355.

Shinde, U. & Inouye, M. (1994) The structural and functional organization of intramolecular chaperones: the N-terminal propeptides which mediate protein folding. *J. Biochem.* **115**, 629–636.

Siezen, R.J. (1996) Subtilases: subtilisin-like serine proteases. In: *Subtilisin Enzymes: Practical Protein Engineering* (Bott, R. & Betzel, C., eds). New York: Plenum, pp. 75–93.

Siezen, R.J., de Vos, W.M., Leunissen, A.M. & Dijkstra, B.W. (1991) Homology modeling and protein engineering strategy of subtilases, the family of subtilisin-like serine proteases. *Protein Eng.* **4**, 719–737.

Siezen, R.J., Leunissen, J.A.M. & Shinde, U. (1995). Homology analysis of the propeptides of subtilisin-like serine proteases (subtilases). In: *Intramolecular Chaperones and Protein Folding* (Shinde, U. & Inouye, M., eds). New York: Springer, pp. 233–256.

Sørensen, S.B., Bech, L.M., Meldal, M. & Breddam, K. (1993) Mutational replacements of the amino acid residues forming the hydrophobic S4 binding pocket of subtilisin 309 from *Bacillus lentus*. *Biochemistry* **32**, 8994–8999.

Strausberg, S., Alexander, P., Wang, L., Schwartz, F. & Bryan, P. (1993) Catalysis of a protein folding reaction: Thermodynamic and kinetic analysis of subtilisin BPN′ interactions with its propeptide fragment. *Biochemistry* **32**, 8112–8119.

Svendsen, I., Boisen, S. & Hejgaard, J. (1982) Amino acid sequence of serine protease inhibitor CI-1 from barley. Homology with inhibitor CI-2, potato inhibitor I, and leech eglin. *Carlsberg Res. Commun.* **47**, 45–53.

Takeuchi, Y., Satow, Y., Nakamura, K.T. & Mitsui, Y. (1991) Refined crystal structure of the complex of subtilisin BPN′ and *Streptomyces* subtilisin inhibitor at 1.8 Å resolution. *J. Mol. Biol.* **221**, 309–325.

Teplyakov, A.V., van der Laan, J.C., Lammers, A.A., Kelders, H., Kalk, K.H., Misset, O., Mulleners, L.J.S.M. & Dijkstra, B.W. (1992) Protein engineering of the high-alkaline serine protease PB92 from *Bacillus alcalophilus*: functional and structural consequences of mutation at the S4 substrate binding pocket. *Protein Eng.* **5**, 413–420.

van der Laan, J.C., Gerritse, G., Mulleners, L.J.S.M., Van der Hoek, R.A.C. & Quax, W.J. (1991) Cloning, characterization, and multiple chromosomal integration of a *Bacillus* alkaline protease gene. *Appl. Environ. Microbiol.* **57**, 901.

van der Laan, J.C., Teplyakov, A.V., Kelders, H., Kalk, K.H., Misset, O., Mulleners, L.J.S.M. & Dijkstra, B.W. (1992) Crystal structure of the high-alkaline serine protease PB92 from *Bacillus alcalophilus*. *Protein Eng.* **5**, 405–411.

Wells, J.A. & Estell, D.A. (1988) Subtilisin B an enzyme designed to be engineered. *Trends Biochem. Sci.* **13**, 291–297.

Wells, J.A., Ferrari, E., Henner, D.J. & Chen, E.Y. (1983) Cloning, sequencing, and secretion of *Bacillus amyloliquefaciens* subtilisin in *Bacillus subtilis*. *Nucleic Acids Res.* **11**, 7911–7925.

Wells, J.A., Powers, D.B., Bott, R.R., Graycar, T.P. & Estell, D.A. (1987) Designing substrate specificity by protein engineering of electrostatic interactions. *Proc. Natl Acad. Sci. USA* **84**, 1219–1223.

Wong, C.-H., Chen, S.-T., Hennen, W.J., Bibbs, J.A., Wang, Y.-F., Liu, J.L.-C., Pantoliano, M.W., Whitlow, M. & Bryan, P.N. (1990) Enzymes in organic synthesis: use of subtilisin and a highly stable mutant derived from multiple site-specific mutations. *J. Am. Chem. Soc.* **112**, 945–953.

Wright, C.S., Alden, R.A. & Kraut, J. (1969) Structure of subtilisin BPN′ at 2.5 Å resolution. *Nature* **221**, 235–242.

Wu, Z.-P. & Hilvert, D. (1989) Conversion of a protease into an acyl transferase: selenolsubtilisin. *J. Am. Chem. Soc.* **111**, 4513–4514.

Wu, Z.-P. & Hilvert, D. (1990) Selenosubtilisin as a glutathione peroxidase mimic. *J. Am. Chem. Soc.* **112**, 5647–5648.

Yang, M.Y., Ferrari, E. & Henner, D.J. (1984) Cloning of the neutral protease gene of *Bacillus subtilis* and the use of the cloned gene to create an in vitro-derived deletion mutation. *J. Bacteriol.* **160**, 15–21.

You, L. & Arnold, F.H. (1994) Directed evolution of subtilisin E in *Bacillus subtilis* to enhance total activity in aqueous dimethylformamide. *Protein Eng.* **9**, 77–83.

*Marcus D. Ballinger*
*Department of Protein Engineering,*
*Genentech, Incorporated,*
*460 Pt. San Bruno Blvd.,*
*South San Francisco, CA 94080, USA*
*Email: marcus@gene.com*

*James A. Wells*
*Department of Protein Engineering,*
*Genentech, Incorporated,*
*460 Pt. San Bruno Blvd.,*
*South San Francisco, CA 94080, USA*
*Email: jaw@gene.com*

# 95. Thermitase

## Databanks

*Peptidase classification: clan SB, family S8, MEROPS ID: S08.007*
*NC-IUBMB enzyme classification: EC 3.4.21.66*
*Chemical Abstracts Service registry number: 69772-87-8*
*Databank codes:*

| Species | SwissProt | PIR | EMBL (cDNA) | EMBL (genomic) |
|---|---|---|---|---|
| *Thermoactinomyces* sp. | – | – | U31759 | – |
| *Thermoactinomyces vulgaris* | P04072 | A00973 | – | – |

Brookhaven Protein Data Bank three-dimensional structures:

| Species | ID | Resolution | Notes |
|---|---|---|---|
| *Thermoactinomyces vulgaris* | 1TEC | 2.2 | complex with eglin C |
| | 1THM | 1.37 | |
| | 2TEC | 1.98 | complex with eglin C |
| | 3TEC | 2 | complex with eglin C |

**S**

## Name and History

**Thermitase** is an extracellular, thermostable serine proteinase of the thermophilic microorganism *Thermoactinomyces vulgaris* and was first isolated by Frömmel *et al*. (1978). The enzyme is one of five proteolytic secreted components of the culture filtrate, but is responsible for 70–80% of the total activity and is therefore the dominating protease (Kleine & Rothe, 1977; Meloun *et al*., 1985). The enzyme is a member of the subtilisin family (Hausdorf *et al*., 1980), a well-characterized family of serine proteinases that have been extensively studied because of their considerable industrial importance in the food industry and as protein-degrading additives in washing powders (Siezen *et al*., 1991) (see Chapter 93). They are also a target of protein engineering studies carried out to provide insight into the mechanism and specificity of enzyme catalysis (Wells & Estell, 1988) (see Chapter 94). Thermitase is more stable against thermal denaturation and proteolytic degradation than most members of the family (Frömmel & Höhne, 1981).

## Activity and Specificity

Like other subtilisins, thermitase shows proteinase, esterase and amidase activities. It hydrolyzes soluble proteins and is also capable of efficient degradation of insoluble proteins including elastin and collagen (Kleine, 1982). The catalytic triad is formed by the residues Asp38, His71 and Ser225. In general, the active site and substrate-binding site can be described as a surface channel or crevice capable of accommodating at least six amino acid residues (P4–P2′) of a polypeptide substrate or inhibitor. The crystallographic analysis of the structure of the thermitase–eglin complex (Dauter *et al*., 1988; Gros *et al*., 1989) provided direct evidence for the participation of particular residues. The N-terminal, P4–P1, part of the substrate lines up between the extended enzyme backbone segments 108–110 and 133–135, forming the central strand of a three-stranded antiparallel $\beta$ sheet. The C-terminal or leaving portion, P1′–P3′, of the substrate appears to be held less tightly as it runs along the enzyme backbone segment 224–226 (Teplyakov *et al*., 1990). The substrate-binding region shows a preference for hydrophobic side chains, particular Phe, at the P1 site; the first substrate site on the N-terminal side (Brömme *et al*., 1986). This site is large and hydrophobic in thermitase as in other subtilisins, which explains the broad specificity of these enzymes with the preference for aromatic or large nonpolar substrate residues. Thermitase, as well as other subtilisins, has additional specificity sites on the C-terminal side of the scissile bond. The specificity shows a preference for small residues like Gly and Ala at the P1′ site and shows slower hydrolysis for large residues such as Phe or Leu, and no hydrolysis for

Pro in the P1′ site (Dauter *et al*., 1988). Depending on the size of the substrate, thermitase shows maximal activity at temperatures ranging from 60°C for Ac-Ala-Ala-Ala+OMe, or 65°C for Ac- or Suc-Ala-Ala-Ala+NHPhNO₂, up to 85°C for azocasein as substrate. Like other subtilisins, thermitase excludes Pro residues from S1 and S3, but accepts them in S2 and S4 (Brömme *et al*., 1986). An example of cleavage position is Ac-Pro-Ala-Pro-Phe+Ala-NH₂.

The pH profile shows a single but rather broad optimum lying between 7.5 and 9.5. Activity is stimulated by low concentrations of thiol compounds (e.g. 0.1 mM DTT, 2 mM 2-ME), but inhibited by higher concentrations (Kleine, 1982).

In addition to several protein inhibitors known from microorganisms, animals and plants (Laskowski & Kato, 1980), short peptidyl chloromethanes such as Z-Ala-Ala-Phe-CH₂Cl are irreversible inhibitors, and peptidyl methanes such as Z-Ala-Ala-Phe-CH₃ are reversible inhibitors (Baudys *et al*., 1983; Brömme & Fittkau, 1985; Ermer *et al*., 1990; Fittkau *et al*., 1986). The free cysteine in thermitase is placed 'below' the functional His71, with an S–N$^{\varepsilon 2}$ distance of about 4 Å (Hansen *et al*., 1982) and appears to be practically inaccessible to the solvent. However, HgCl₂ inhibits thermitase through its binding to the free SH group. Some flexibility in this region of the protein is required to allow HgCl₂ access to the cysteine thiol group (Dauter *et al*., 1988).

## Structural Chemistry

Thermitase consists of a single polypeptide chain of 279 amino acids (Meloun *et al*., 1985), with calculated $M_r$ of 28 369. It belongs to the subtilisin family and shows many similarities to the other members with respect to the structure and the enzymatic properties. The native structure solved by X-ray crystallography at 1.4 Å resolution included two Ca²⁺ ions (Teplyakov *et al*., 1990). In addition, the structure of the complex with the inhibitor eglin-c was analyzed in detail independently by two groups (Dauter *et al*., 1988; Gros *et al*., 1989, 1992). The overall appearance of thermitase is that of a half sphere, with the active site located on the flat side which is 40 Å in diameter. There is no clear indication of domains that would subdivide the globular structure. The tertiary structure of the molecule is composed of a very twisted, eight-stranded parallel $\beta$ sheet and a central helix with the active serine at the N-terminal site. The core is further decorated by five $\alpha$ helices, four of which are aligned antiparallel to the $\beta$ strands and three short antiparallel $\beta$ sheets.

Enzymes of the subtilisin family require generally Ca²⁺ for maximal stability toward thermal- or urea-induced denaturation and autolysis, and for optimal activity (Voordouw *et al*., 1976; Frömmel & Höhne, 1981; Bajorath *et al*., 1988). Binding of Ca²⁺ ions at specific sites in external loops increases the stability by reducing the flexibility of the molecule and hence the denaturation and/or autolysis rate. For thermitase, biochemical studies suggest that there are two tightly bound $Ca^{2+}$ ions, which remain attached during extraction, and one more weakly bound calcium with a dissociation constant of about $10^{-4}$ M at 25°C, pH 3.5–7.5 (Frömmel & Höhne, 1981; Briedigkeit & Frömmel, 1989). The removal of the weakly bound $Ca^{2+}$ slightly reduces the activity of the enzyme, and decreases significantly its stability to denaturation and autolysis. It is impossible to remove all three $Ca^{2+}$ ions from thermitase without destroying the three-dimensional structure. Therefore, it can be concluded that the binding of $Ca^{2+}$ is the main structural basis for thermostability.

## Preparation

As an extracellular enzyme with a pI value of 9.1, thermitase was isolated from concentrated culture filtrate of *Thermo-actinomyces vulgaris* by adsorption chromatography, and found to be essentially homogeneous, as judged by isoelectric focusing (Behnke *et al*., 1978; Frömmel *et al*., 1978).

## Biological Aspects

The physiological function of thermitase is to digest external protein substrates to membrane-permeable peptides that are further degraded by the intracellular peptidases of the bacterial cell. In the case of the thermophilic microorganism *Thermoactinomyces vulgaris*, several proteolytic enzymes are secreted into the surrounding medium during the logarithmic phase of growth, whereas the level of intracellular proteases constantly decreases (Schalinatus *et al*., 1980).

## Related Peptidases

Thermitase belongs to a subfamily of subtilisins containing a single free sulfhydryl group in the chain close to the active site (Meloun *et al*., 1985). Proteinase K (Chapter 106) is the only other member of this group for which the three-dimensional structure is known (Betzel *et al*., 1988), but the degree of sequence similarity between the two enzymes and the presence of the cysteine residue at the same site confirms that they are closely related (Betzel *et al*., 1990).

## Further Reading

Fuller accounts of thermitase and the other members of the subtilisin family are to be found in Bott & Betzel (1996) and Wells *et al*. (1987).

## References

Bajorath, J., Hinrichs, W. & Saenger, W. (1988) The enzymatic activity of proteinase K is controlled by calcium. *Eur. J. Biochem.* **176**, 441–447.

Baudys, M., Kostka, V., Hausdorf, G., Fittkau, S. & Hohne, W.E. (1983) Amino acid sequence of the tryptic SH-peptide of thermitase, a thermostable serine proteinase from *Thermoactinomyces vulgaris*. Relation to the subtilisins. *Int. J. Pept. Protein Res.* **22**, 66–72.

Behnke, U., Schalinatus, E., Ruttloff, H., Höhne, W.E. & Frömmel, C. (1978) Charakterisierung einer Protease aus *Thermoactinomyces vulgaris* (Thermitase) 1. Untersuchung zur Reinigung der Thermitase. *Acta Biol. Med. Germ.* **37**, 1185–1192.

Betzel, Ch., Pal, G. & Saenger, W. (1988) Three-dimensional structure of proteinase K at 0.15 nm resolution. *Eur. J. Biochem.* **178**, 155–171.

Betzel, Ch., Teplyakov, A.V., Harutyunyan, E.H., Saenger, W. & Wilson, K.S. (1990) Thermitase and proteinase K: a comparison of the refined three-dimensional structures of the native enzymes. *Protein Eng.* **3**, 161–172.

Bott, R. & Betzel, Ch. (eds) (1996) *Subtilisin Enzymes, Practical Protein Engineering. Advances in Experimental Medicine and Biology*, vol. 379. New York: Plenum Press.

Briedigkeit, L. & Frömmel, C. (1989) Calcium ion binding by thermitase. *FEBS Lett.* **253**, 83–87.

Brömme, D. & Fittkau, S. (1985) A new substrate and two inhibitors applicable for thermitase, subtilisin BPN′ and α-chymotrypsin. Comparison of kinetic parameters with customary substrates and inhibitors. *Biomed. Biochim. Acta* **44**, 1089–1094.

Brömme, D., Peters, K., Fink, S. & Fittkau, S. (1986) Enzyme–substrate interactions in the hydrolysis of peptide substrate by thermitase, subtilisin BPN′ and proteinase K. *Arch. Biochem. Biophys.* **244**, 439–446.

Dauter, Z., Betzel, Ch., Höhne, W.E., Ingelmann, M. & Wilson, K.S. (1988) Crystal structure of a complex between thermitase from *Thermoactinomyces vulgaris* and the leech inhibitor eglin. *FEBS Lett.* **236**, 171–178.

Ermer, A., Baumann, H., Steude, G., Peters, K., Fittkau, S., Dolaschka, P. & Genov, N. (1990) Peptidyl diazomethyl ketones are inhibitors of subtilisin-type serine proteases. *J. Enzym. Inhib.* **4**, 35–42.

Fittkau, S., Pauli, D., Bauaravong, Ph. & Damerau, W. (1986) Thermitase – eine thermostabile Serinprotease. VIII. Kinetische und ESR-Unteruschungen der Wechselwirkungen des Enzyms mit markierten Peptidmethylketonen. *Biomed. Biochim. Acta* **45**, 877–886.

Frömmel, C. & Höhne, W.E. (1981) Influence of calcium binding on the thermal stability of 'Thermitase', a serine protease from *Thermoactinomyces vulgaris*. *Biochim. Biophys. Acta* **670**, 25–31.

Frömmel, C., Hausdorf, G., Höhne, W.E., Behnke, U. & Ruttloff, H. (1978) Charakterisierung einer Protease aus *Thermoactinomyces vulgaris* (Thermitase). 2. Einschritt-Feinreinigung und proteinchemische Charakterisierung. *Acta Biol. Med. Germ.* **37**, 1193–1204.

Gros, P., Betzel, Ch., Dauter, Z., Wilson, K.S. & Hol, W.G.J. (1989) Molecular dynamics refinement of a thermitase-eglin-c complex at 1.98 Å resolution and comparison of two crystal forms that differ in calcium content. *J. Mol. Biol.* **210**, 347–367.

Gros, P., Teplyakov, A.V. & Hol, W.G.J. (1992) Effects of eglin-c binding to thermitase: three-dimensional structure comparison of native thermitase and thermitase eglin-c complexes. *Proteins* **12**, 63–74.

Hansen, G., Frömmel, C., Hausdorf, G. & Bauer, S. (1982) Thermitase, eine themostabile Serin-Protease aus *Thermoactinomyces vulgaris*. Wechselwirkungen zwischen aktiven Zentrum und SH Gruppe des Enzyms. *Acta Biol. Med. Germ.* **41**, 137–144.

Hausdorf, G., Krüger, K. & Höhne, W. (1980) Thermitase, a thermostable serine protease from *Thermoactinomyces vulgaris*. Classification as a subtilisin-type protease. *Int. J. Pept. Protein Res.* **15**, 420–429.

Kleine, R. (1982) Properties of thermitase, a thermostable serine protease from *Thermoactinomyces vulgaris. Acta Biol. Med. Germ.* **41**, 89–102.

Kleine, R. & Rothe, U. (1977) Isolierung, Kristallisation und teilweise Charakterisierung einer kationischen Protease aus *Thermoactinomyces vulgaris. Acta Biol. Med. Germ.* K27–K31.

Laskowski, M. & Kato, I. (1980) Protein inhibitors of proteinases. *Annu. Rev. Biochem.* **49**, 593–626.

Meloun, B., Baudys, M., Kostka, V., Hausdorf, G., Frömmel, C. & Höhne, W.E. (1985) Complete primary structure of thermitase from *Thermoactinomyces vulgaris* and its structural features related to the subtilisin-type proteinases. *FEBS Lett.* **183**, 195–200.

Schalinatus, E., Ruttloff, H. & Behnke, U. (1980) 2nd Symposium der sozialistischen Länder für Biotechnologie, Leipzig, Poster.

Siezen, R.J., de Vos, W.M., Leunissen, J.A.M. & Dijkstra, B.W. (1991) Homology modelling and protein engineering strategy of subtilases, the family of subtilisin-like serine proteinases. *Protein Eng.* **4**, 719–737.

Teplyakov, A., Kuranova, I.P., Harutyunyan, E.H., Vainshtein, B.K., Frömmel, C., Höhne, W.E. & Wilson, K.S. (1990) Crystal structure of thermitase at 1.4 Å resolution. *J. Mol. Biol.* **214**, 261–279.

Voordouw, G., Milo, C. & Roche, R.S. (1976) Role of bound calcium ions in thermostable, proteolytic enzymes. Separation of intrinsic and calcium ion contributions to the kinetic thermal stability. *Biochemistry* **15**, 3716–3724.

Wells, J.A. & Estell, D.A. (1988) Subtilisin – an enzyme designed to be engineered. *Trends Biochem. Sci.* **13**, 291–297.

Wells, J.A., Powers, D.B., Bott, R., Katz, B.A., Ultsch, M.H., Kossiakoff, A., Power, S., Adams, R.M., Heyneker, H., Cunningham, B.C., Miller, J.V., Graycar, T.P. & Estell, D. (1987) Protein engineering of subtilisin. *Protein Eng.* **1**, 279–287.

*Ch. Betzel*
*Arbeitsgruppe für Makromolekulare Strukturanalyse,*
*Institut für Biochemie & Lebensmittelchemie,*
*Institut für Physiologische Chemie,*
*Universität Hamburg,*
*c/o DESY, Geb 22a, Notkestrasse 85, 22603 Hamburg, Germany*
*Email: BETZEL@embl-hamburg.de*

# 96. *Thermus strain Rt41A protease*

## Databanks

*Peptidase classification: clan SB, family S8, MEROPS ID: S08.008*
*NC-IUBMB enzyme classification: none*
*Databank codes:*

| Species | SwissProt | PIR | EMBL (cDNA) | EMBL (genomic) |
|---|---|---|---|---|
| *Thermus* sp. | P80146 | – | U17342 | – |

## Name and History

*Thermus* strain Rt41A is a thermophilic eubacterium isolated from a geothermal hot pool in New Zealand (Peek *et al.*, 1992). This organism, like many other *Thermus* isolates (Daniel *et al.*, 1995), produces an extracellular protease. The **Rt41A protease** is a thermostable alkaline protease that is known commercially as **Pretaq**™, being used in the preparation of DNA (McHale *et al.*, 1991) and mRNA (Fung & Fung, 1991) prior to amplification by PCR.

## Activity and Specificity

Activity has been determined by the detection of trichloroacetic acid-soluble peptides from the cleavage of azocasein and with a variety of nitroanilide and nitrophenyl ester substrates (Peek *et al.*, 1992).

Rt41A protease is almost completely inhibited by PMSF, and partially inhibited by DFP, chymostatin and Z-Phe-CH₂Cl, but not by phosphoramidon or 1,10-phenanthroline. It also displays reduced activity in the presence of EDTA and

*Table 96.1*  Kinetic parameters for hydrolysis of nitroanilide and nitrophenyl ester substrates by Rt41A protease

| Substrate | $K_m$ (mM) | $k_{cat}$ (s$^{-1}$) | $k_{cat}/K_m$(s$^{-1}$ M$^{-1}$) |
|---|---|---|---|
| Suc-Ala-Ala-Pro-Phe-NHPhNO$_2$ | 2.5 | 508 | 203 200 |
| Suc-Ala-Ala-Pro-Leu-NHPhNO$_2$ | 3.6 | 69 | 19 167 |
| Suc-Ala-Ala-Ala-NHPhNO$_2$ | 13.4 | 15 | 1 119 |
| Z-Trp-OPhNO$_2$ | 0.3 | 3.0 | 10 133 |
| Z-Gly-OPhNO$_2$ | 3.4 | 33.0 | 9 686 |
| Z-Ala-OPhNO$_2$ | 5.0 | 38.0 | 7 600 |
| Z-Phe-OPhNO$_2$ | 0.9 | 6.7 | 7 389 |

From Peek *et al.* (1992).

EGTA. This apparent inhibition is due to a loss of stabilizing $Ca^{2+}$ ions rather than the removal of a catalytic $Zn^{2+}$ ion (Peek *et al.*, 1992).

Rt41A protease is thermostable, showing no loss of activity after 24 h at 70°C, and a half-life at 90°C of 20 min. In the absence of $Ca^{2+}$ ions, the half-life of the protease is only 3 min at 70°C, however. Thermal denaturation is the cause of loss of activity in this case, with a rate constant of $4 \times 10^{-3}$ s$^{-1}$. Rt41A is stabilized against thermal denaturation at 10 µM $Ca^{2+}$ (high-affinity $Ca^{2+}$ binding) and against autolysis at 5 mM $Ca^{2+}$ (low-affinity $Ca^{2+}$ binding) (Peek *et al.*, 1992; Wilson *et al.*, 1994b).

The protease showed substrate inhibition with azocasein and casein above concentrations of 0.02% (w/v) with an apparent $K_i$ of 0.6% (w/v) of azocasein.

Table 96.1 shows kinetic parameters for the hydrolysis of synthetic substrates by Rt41A protease. It can be seen that small aliphatic or aromatic amino acids are preferred at the P1 position, and the substrates with aromatic P1 amino acids generally had the lowest $K_m$ values. All of the amino acids tested in the P1 position led to hydrolysis except isoleucine, indicating a broad substrate specificity. The best substrate identified was Suc-Ala-Ala-Pro-Phe+NHPhNO$_2$, which is cleaved by other proteases in the subtilisin family, such as subtilisin BPN′ (Chapter 94).

The major cleavage site of oxidized insulin B chain was Leu15+Tyr16, the bond being cleaved completely within 15 min. The secondary and tertiary sites were Gln4+His5 and Phe24+Phe25, respectively. Rt41A appears to have some preference for amino acids containing aromatic groups at the P1′ site of the scissile bond.

## Structural Chemistry

Rt41A protease has a molecular mass of 32.5 kDa and a very high pI of about 10.5. It contains two disulfide bonds per enzyme molecule (Peek *et al.*, 1992), and a carbohydrate content of 0.7% by mass as glucose equivalents.

The Rt41A protease gene was sequenced, and showed that the protease is a member of the subtilisin family; the sequence was particularly close to those of aqualysin I (Chapter 103) (70%) and exoprotease A from *Vibrio alginolyticus* (Munro *et al.*, 1995).

## Preparation

*Thermus* st. Rt41A was cultivated aerobically in a 600 liter fermenter at 70°C (Peek *et al.*, 1992). The supernatant was made 80% saturation with ammonium sulfate, and the precipitated protease was desalted and concentrated before being purified by chromatography on S-Sepharose FF, and two successive steps on a Mono-S column (Peek *et al.*, 1992).

The protease has also been expressed in *E. coli* as a glutathione-*S*-transferase (GST) fusion protein in an inactive form. Activation of the protease was by heat treatment of the fusion protein at 70°C, resulting in the cleavage of the GST protein (Munro *et al.*, 1995).

## Biological Aspects

Being extracellular, the Rt41A protease is thought to have a nutritional role, and the broad substrate specificity is consistent with this.

## Distinguishing Features and Related Peptidases

Table 96.2 summarizes some characteristics of *Thermus* serine proteases. They are a broad group of proteases of molecular masses between 20 and 33 kDa, pI between 8.5 and 10.5 and pH optima in the range 7–10.4. They are all more or less thermostable, the most stable being the *Thermus* st. Rt4A2 protease (Freeman *et al.*, 1993). Some of the proteases have distinctive features, such as the chelator resistance of the *Thermus* st. Rt4A2 protease (Freeman *et al.*, 1993), and the dependence on high ionic strength to maintain stability of caldolase (Saravani *et al.*, 1989).

The specificity of hydrolysis of nitrophenyl esters by the Rt41A protease is similar to those of caldolase (Saravani *et al.*, 1989) and aqualysin I (Chapter 103) (Matsuzawa *et al.*, 1988). For example, both caldolase and aqualysin I showed the highest activity towards the alanine ester, and Rt41A protease has the highest $k_{cat}$ with this substrate. This cleavage pattern of oxidized insulin B chain for Rt41A protease is similar to other serine proteases such as aqualysin I (Matsuzawa *et al.*, 1988) and proteinase K (Chapter 106) (Kraus *et al.*, 1976).

*Table 96.2*  Some properties of *Thermus* serine proteases

| Source | Mass (kDa) | pI | $pH_{opt}$ | Stability |
|---|---|---|---|---|
| *Thermus* st. Rt41A[a] | 32.5 | ~10.5 | 8.0 | $t_{1/2}$ 20 min, 90°C |
| *Thermus* st. Rt4A2[b] | 31.6 | 10.25 | 9.0 | $t_{1/2}$ 1.5 h, 90°C |
| *Thermus* st. Rt$_6$[c] | 27 | 8.5 | 7.7–8.8 | $t_{1/2}$ 6 h, 85°C |
| *Thermus* st. Tok$_3$ (Caldolase)[d] | 25 | 8.9 | 9.5 | $t_{1/2}$ 45 min, 90°C |
| *Thermus* st. T351 (Caldolysin)[e] | 21 | 8.5 | 8.0 | $t_{1/2}$ 1 h, 90°C |
| *Thermus aquaticus* st. YT-1 I[f] | 28.5 | >9–10 | 10.4 | 40% after 30 min at 90°C |
| *Thermus aquaticus* st. YT-1 II[g] | – | – | 7.0 | – |
| *Thermus caldophilus*[h] | 31 | – | 7.8 | $t_{1/2}$ 2 h, 80°C |

Data from [a]Peek *et al.* (1992); [b]Freeman *et al.* (1993); [c]Cowan *et al.* (1987); [d]Saravani *et al.* (1989); [e]Cowan & Daniel (1982); [f]Matsuzawa *et al.* (1983); [g]Matsuzawa *et al.* (1988); [h]Taguchi *et al.* (1983).

## Further Reading

The general properties of the *Thermus* st. Rt41A protease are reviewed by Peek *et al.* (1992), and an overview of the *Thermus* proteases is provided by Daniel *et al.* (1995). The applications of Rt41A protease have been described by Wilson *et al.* (1992, 1994a,b) and Peek *et al.* (1990).

## References

Cowan, D.A. & Daniel, R.M. (1982) Purification and some properties of an extracellular protease (caldolysin) from an extreme thermophile. *Biochim. Biophys. Acta* **705**, 293–305.

Cowan, D.A., Daniel, R.M. & Morgan, H.W. (1987) A comparison of extracellular serine proteases from four strains of *Thermus aquaticus*. *FEMS Microbiol. Lett.* **43**, 155–159.

Daniel, R.M., Toogood, H.S. & Bergquist, P.L. (1995) Thermostable proteases. *Biotechnol. Genet. Eng. Rev.* **13**, 51–100.

Freeman, S.-A., Peek, K., Prescott, M. & Daniel, R.M. (1993) Characterisation of a chelator-resistant protease from *Thermus* st. Rt4A2. *Biochem. J.* **295**, 463–469.

Fung, M.-C. & Fung, K.Y.-M. (1991) PCR Amplification of messenger RNA directly from a crude cell lysate prepared by thermophilic protease digestion. *Nucleic Acids Res.* **19**, 4300.

Kraus, E., Klitz, H.H. & Fempert, U.F. (1976) Specificity of proteinase K against oxidized insulin-B chain. *Hoppe-Seyler's Z. Physiol. Chem.* **357**, 233–237.

McHale, R.H., Stapleton, P.M. & Bergquist, P.L. (1991) Rapid preparation of blood and tissue samples for polymerase chain reaction. *Biotechniques* **10**, 20–22.

Matsuzawa, H., Hamaoki, M. & Ohta, T. (1983) Production of thermophilic extracellular proteases (aqualysins I and II) by *Thermus aquaticus* YT-1, an extreme thermophile. *Agric. Biol. Chem.* **47**, 25–28.

Matsuzawa, H., Tokugawa, K., Hamaoki, M., Mizoguchi, M., Taguchi, H., Terada, I., Kwon, S.-T. & Ohta, T. (1988) Purification and characterisation of aqualysin I (a thermophilic alkaline serine protease) produced by *Thermus aquaticus* YT-1. *Eur. J. Biochem.* **171**, 441–447.

Munro, G.K.L., McHale, R.H., Reeves, R.A., Saul, D.J. & Bergquist, P.L. (1995) A gene encoding a thermophilic alkaline serine proteinase from *Thermus* sp. strain Rt41A and its expression in *Escherichia coli*. *Microbiology* **141**, 1731–1738.

Peek, K., Janssen, P.H., Morgan, H.W. & Daniel, R.M. (1990) Enhancement of *in vitro* stability and recovery of a proteinase produced by an extreme thermophile. In: *Fermentation Technologies: Industrial Applications* (Pak-Lam Yu, ed.). London: Elsevier Applied Science, pp. 97–102.

Peek, K., Daniel, R.M., Monk, C., Parker, L. & Coolbear, T. (1992) Purification and characterisation of a thermostable protease isolated from *Thermus* sp. strain Rt41A. *Eur. J. Biochem.* **207**, 1035–1044.

Saravani, G.-A., Cowan, D.A., Daniel, R.M. & Morgan, H.W. (1989) Caldolase, a chelator-insensitive extracellular serine proteinase from a *Thermus* spp. *Biochem. J.* **262**, 409–416.

Taguchi, H., Hamaoki, M., Matsuzawa, H. & Ohta, T. (1983) Heat-stable extracellular proteolytic enzyme produced by *Thermus caldophilus* strain GK24, an extremely thermophilic bacterium. *J. Biochem.* **93**, 7–13.

Wilson, S.-A., Young, O., Coolbear, T. & Daniel, R.M. (1992) The use of proteases from extreme thermophiles for meat tenderising. *Meat Sci.* **32**, 92–103.

Wilson, S.-A., Daniel, R.M. & Peek, K. (1994a) Peptide synthesis with a protease from the extremely thermophilic microorganism *Thermus* Rt41A. *Biotechnol. Bioeng.* **43**, 1–10.

Wilson, S.-A., Peek, K. & Daniel, R.M. (1994b) The immobilisation of a protease from the extremely thermophilic organism *Thermus* Rt41A. *Biotechnol. Bioeng.* **43**, 225–231.

**H.S. Toogood**
*Thermophile Research Unit,*
*University of Waikato,*
*Private Bag 3105, Hamilton, New Zealand*
*Email: Biolsec3@Waikato.ac.nz*

**R.M. Daniel**
*Thermophile Research Unit,*
*University of Waikato,*
*Private Bag 3105, Hamilton, New Zealand*

# 97. Bacillus strain AK.1 protease

## Databanks

*Peptidase classification: clan SB, family S8, MEROPS ID: S08.009*
*NC-IUBMB enzyme classification: none*
Databank codes:

| Species | SwissProt | PIR | EMBL (cDNA) | EMBL (genomic) |
|---|---|---|---|---|
| *Bacillus* sp. | – | – | L29506 | – |

## Name and History

*Bacillus* strain AK.1 is a thermophilic bacterium that produces an extracellular serine protease (Peek *et al*., 1993). The organism was discovered as a contaminant on an agar plate containing a thermophilic *Bacillus* species.

## Activity and Specificity

Activity of AK.1 protease was determined by the detection of trichloroacetic acid-soluble peptides from azocasein and with a variety of nitroanilide and nitrophenyl ester substrates (Peek *et al*., 1993).

The protease was sensitive to inhibition by PMSF (1 mM), inhibition being complete after 30 min at room temperature (Peek *et al*., 1993). The protease was also inhibited slightly (10%) by iodoacetate at 100 μM. It was insensitive to Tos-Lys-CH$_2$Cl, Tos-Phe-CH$_2$Cl and pepstatin. EDTA destabilized the protease by the removal of stabilizing Ca$^{2+}$ ions (Peek *et al*., 1993). Low Ca$^{2+}$ concentrations (<10 μM) do not increase stability to heat, but higher Ca$^{2+}$ concentrations do cause a substantial increase in thermostability. Maximal thermostability is at a Ca$^{2+}$ concentration >5 mM; this results in an increase in the half-life at 70°C from 2.5 min (without Ca$^{2+}$) to 18 days (H.S. Toogood, unpublished results). The relationship between the rate of autolytic loss of activity and Ca$^{2+}$ ion concentration is not linear, suggesting that there may be more than one type of Ca$^{2+}$-binding site in the protease (H.S. Toogood, unpublished results). While Ca$^{2+}$ has a dramatic effect on thermostability, other metal ions have less or no effect. Small effects were seen with 5 mM Sr$^{2+}$ and 50 μM Cd$^{2+}$, but none of the lanthanides tested had any significant stabilizing effect (H.S. Toogood, unpublished results).

Table 97.1 shows the specific activities of AK.1 protease towards a variety of peptide substrates. It can be seen that the protease has maximal activity towards the substrate Suc-Ala-Ala-Pro-Phe+NHPhNO$_2$, a good substrate of subtilisin-like enzymes generally. The kinetic parameters for this substrate were $K_m$ 1.125 mM and $V$ 595 μmol min$^{-1}$ mg$^{-1}$.

The major cleavage site of oxidized insulin B chain by AK.1 protease is Leu15+Tyr16 (Peek *et al*., 1993). Other major hydrolysis sites are Gln4+His5 and Glu13+Ala14. This indicates that negatively charged side chains can be accommodated in the P1 site quite readily (Peek *et al*., 1993).

## Structural Chemistry

AK.1 protease has a molecular mass of 36.9 kDa and a pI of 4.0. The gene encoding the protease was cloned and sequenced, and shows that the enzyme is a member of the subtilisin family (MacIver *et al*., 1994). This sequence showed 68% and 43% sequence similarity to thermitase (Chapter 95) and to subtilisin from *B. licheniformis* (Chapter 94), respectively (Daniel *et al*., 1995). The AK.1 protease contains two Cys residues, but no disulfide bond is expected as the residues are adjacent (Peek *et al*., 1993).

*Table 97.1* Substrates cleaved by AK.1 protease

| Substrate | Units[a] | Temperature (°C) | Specific activity (U mg$^{-1}$) |
|---|---|---|---|
| Suc-Ala-Ala-Pro-Phe+NHPhNO$_2$ | 1 | 75 | 313.79 |
| Suc-Ala-Ala-Ala+NHPhNO$_2$ | 1 | 75 | 16.89 |
| Ala-Ala+NHPhNO$_2$ | 1 | 75 | 0.75 |
| Ala-Ala-Phe+NHPhNO$_2$ | 1 | 40 | 0.18 |
| Suc-Tyr-Leu-Val+NHPhNO$_2$ | 1 | 75 | 0.17 |
| Z-Gly+OPhNO$_2$ | 2 | 40 | 5.30 |
| Z-Leu+OPhNO$_2$ | 2 | 40 | 1.38 |
| Z-Nle+OPhNO$_2$ | 2 | 40 | 0.45 |
| Z-Ala+OPhNO$_2$ | 2 | 40 | 0.42 |

[a]Unit 1 = μmol *p*-nitroaniline min$^{-1}$ mg$^{-1}$; unit 2 = μmol *p*-nitrophenol min$^{-1}$ mg$^{-1}$.

## Preparation

*Bacillus* st. AK.1 was cultivated aerobically in batch cultures for study of the natural enzyme (Peek *et al.*, 1993), but most work has been done with the recombinant protease. The gene coding for the protease was cloned, sequenced and expressed in *E. coli* (MacIver *et al.*, 1994). The gene was obtained by amplifying a region using primers for the active-site His and Ser residues. The gene was added to a *lac* operon promoter, which was later found to have mutated, with the result that the protease production was constitutive. The protein was shown to be produced in a prepro form, and was activated by heat treatment at 70°C for 2 h in the presence of $Ca^{2+}$ ions.

For the preparation of the recombinant enzyme, *E. coli* clone PB5517, expressing the recombinant proteinase from *Bacillus* st. AK.1, was grown aerobically in batch cultures at 37°C (Peek *et al.*, 1993). The supernatants were desalted and concentrated, and the protease was purified by passing the extract through phenyl-Sepharose and Mono-Q columns (Peek *et al.*, 1993).

## Biological Aspects

As this protease is produced extracellularly in the native organism, its role is presumably in nutrition.

## Distinguishing Features

Thermitase is also dependent on $Ca^{2+}$ for thermostability, and Table 97.2 shows a comparison of the half-lives at 80°C of thermitase and AK.1 protease in the presence and absence of $Ca^{2+}$. This shows that although thermitase is intrinsically more thermostable, AK.1 protease is significantly more stabilized by $Ca^{2+}$. This difference in $Ca^{2+}$-stabilizing ability may indicate a difference in the number of $Ca^{2+}$-binding sites, and their affinity.

Other subtilisins such as thermitase and subtilisin BPN′ cleave oxidized insulin B chain in a similar manner to AK.1 protease, but do not cleave at Glu13+Ala14 (Brömme &

*Table 97.2*  Half-lives of AK.1 protease and thermitase at 80°C in the presence and absence of $Ca^{2+}$

| Enzyme | Half-life at 80°C | |
|---|---|---|
| | $Ca^{2+}$-free | With $Ca^{2+}$ |
| AK.1 protease[a] | <1 min | 15 h |
| Thermitase[b] | 9 min | 18.9 min |

[a]H.S. Toogood & R.M. Daniel, unpublished results.
[b]From Frömmel *et al.* (1980).

Kleine, 1984). Overall, AK.1 protease appears to have a more restricted substrate specificity than many other subtilisins.

## Further Reading

A fuller account of the protease is given by Peek *et al.* (1993).

## References

Brömme, D. & Kleine, R. (1984) Substrate specificity of thermitase, a thermostable serine proteinase from *Thermoactinomyces vulgaris*. *Curr. Microbiol.* **11**, 93–100.
Daniel, R.M., Toogood, H.S. & Bergquist, P.L. (1995) Thermostable proteases. *Biotechnol. Genet. Eng. Rev.* **13**, 51–100.
Frömmel, C., Höhne, W.E. & Hansen, G. (1980) Influences on the stability of thermitase, a thermostable protease from *Thermoactinomyces vulgaris*. *Studia Biophys.* **79**, 153–154.
MacIver, B., McHale, R.H., Saul, D.J. & Bergquist, P.L. (1994) Cloning and sequencing of a serine proteinase gene from a thermophilic *Bacillus* species and its expression in *Escherichia coli*. *Appl. Environ. Microbiol.* **60**, 3981–3988.
Peek, K.P., Veitch, D.P., Prescott, M., Daniel, R.M., MacIver, B. & Bergquist, P.L. (1993) Some characteristics of a proteinase from thermophilic *Bacillus* sp. expressed in *Escherichia coli*: comparison with the native enzyme and its processing in *E. coli* and in vitro. *Appl. Environ. Microbiol.* **59**, 1168–1175.

**H.S. Toogood**
*Thermophile Research Unit,*
*University of Waikato,*
*Private Bag 3105, Hamilton, New Zealand*
*Email: Biolsec3@Waikato.ac.nz*

**R.M. Daniel**
*Thermophile Research Unit,*
*University of Waikato,*
*Private Bag 3105, Hamilton, New Zealand*

# 98. Bacillopeptidase F

## Databanks

*Peptidase classification: clan SB, family S8, MEROPS ID: S08.017*
*NC-IUBMB enzyme classification: none*

*Databank codes:*

| Species | SwissProt | PIR | EMBL (cDNA) | EMBL (genomic) |
|---|---|---|---|---|
| *Bacillus subtilis* | P16397 | A36734 | D44498 J05400 M29035 | – |

## Name and History

Bacillopeptidase F is a serine endopeptidase with an acidic isoelectric point; it was first isolated from the medium of stationary-phase cultures of *Bacillus subtilis* cells and called **acidic protease** (Boyer & Carlton, 1968). A similar activity has been called **enzyme C** (Prestidge *et al.*, 1971), **esterase** (Millet, 1970; Mäntsälä & Zalkin, 1980), and **enzyme II** (Mamas & Millet, 1975). Hageman & Carlton (1970) proposed the name **bacillopeptidase F (bpF)**. The protein is encoded by the gene designated *bpf* or *bpr* (Pero & Sloma, 1993).

## Activity and Specificity

Bacillopeptidase F effectively hydrolyzes casein (Boyer & Carlton, 1968), *N,N*-dimethylated casein (Hageman & Carlton, 1970) and azocasein (Roitsch & Hageman, 1983). Boyer & Carlton (1968) reported a pH optimum of 8.1 on casein using a continuous ultraviolet spectral-difference assay, but Roitsch & Hageman (1983) found the pH optimum to be very broad and between 9 and 10 by use of an assay measuring release of acid-soluble peptides (Roitsch, 1978). BpF hydrolyzes *N,N*-dimethylated casein at a $V$ one-sixth that of subtilisin, but it has an apparent $K_m$ of 0.32 mg ml$^{-1}$ versus 0.62 mg ml$^{-1}$ for subtilisin. In addition, bpF can hydrolyze *N,N*-dimethylated protamine, hide powder azure (Hageman & Carlton, 1970), gelatin (Wu *et al.*, 1990) and azocollagen (Sloma *et al.*, 1990).

bpF digests casein very differently than does the subtilisin of the same organism (Hageman & Carlton, 1970). Following a 5 h digest of 30 mg of casein, only 12 major peptides can be detected, whereas subtilisins yield 50. Peptides with unique C-termini of Val, Ile, Gly and Pro are produced, and 17 µmol of 14 different free amino acids are released (subtilisins yield <0.1 µmol). Kato *et al.* (1992) reported cleavage specificities against the oxidized B chain of insulin consistent with these.

bpF exhibits relatively high rates of hydrolysis of amino acid esters (Boyer & Carlton, 1968). Relative to subtilisin, bpF has a $V$ for hydrolysis of Bz-Tyr╀OEt and Bz-Arg╀OEt relative to that for casein of 7166 and 93, respectively (Hageman & Carlton, 1970). bpF also hydrolyzes Tyr╀NHMec (Kato *et al.*, 1992), and Boc-Glu-α-OPh (Sloma *et al.*, 1990).

Inhibitors of bpF include PMSF, DFP (Boyer & Carlton, 1968), chymostatin, elastinal and antipain (Kato *et al.*, 1992), soybean trypsin inhibitor (partial), and egg white trypsin inhibitor (partial) (Wu *et al.*, 1990). EDTA, DTT and several metal ions do not inhibit (Boyer & Carlton, 1968).

## Structural Chemistry

Estimates of molecular masses for bpF obtained by hydrodynamic methods have been: 36 kDa (Carlton & Boyer, 1968), 33 kDa (pI 4.4), 50 kDa (pI 5.4) (Roitsch & Hageman, 1983) and 88 kDa (pI 3.9) (Kato *et al.*, 1992). Values obtained by SDS-PAGE have been 65 and 74 kDa (Roitsch & Hageman, 1983), 48, 50, 68 and 80 kDa (pI 5.4) (Wu *et al.*, 1990), and 47 kDa (pI 4.5) (Sloma *et al.*, 1990).

The gene for bpF codes for 1433 amino acids (Sloma *et al.*, 1990). Sequences of the N-termini of homogeneous isolated forms of 50 kDa, 80 kDa and 90 kDa (SDS-PAGE) revealed that a prepro sequence of 191 or 194 residues is removed from the N-terminus by cleavage at Lys╀Ala bonds in the sequence Asn184-Met-Lys-Lys-Ala-Gln-Lys╀Ala-Ile-Lys╀Ala-Thr-Asp-Gly-Val-Glu (Wu *et al.*, 1990; Sloma *et al.*, 1990; Kato *et al.*, 1992; Pero & Sloma, 1993). The mature protein has a catalytic triad typical of peptidase family S8, at Asp33, His80 and Ser258 (Wu *et al.*, 1990). C-terminal processing also occurs. Roitsch & Hageman (1983) showed that bpF is susceptible to autolysis without detectable loss in catalytic activity and Wu *et al.* (1990) have directly shown that 800 amino acids from the C-terminus of the 1433 can be eliminated without altering the activity of bpF. Hydrolysis at Lys521-Ala522 (preproprotein numbering) would yield a protein of 30 kDa. Such processing may be autocatalytic since bpF can cleave on the carboxyl side of Lys and Arg (Hageman & Carlton, 1970).

## Preparation

bpF has been purified from sporulating cultures of *B. subtilis* by a variety of standard methods. A purification of 150-fold with a yield of 10% has been found to provide homogeneous product (Roitsch & Hageman, 1983; Sloma *et al.*, 1990).

## Biological Aspects

No biological function can be assigned to bpF, which accounts for a few per cent of total proteinase activity in typical cultures (Boyer & Carlton, 1968). Genetic elimination of bpF activity caused no detected deficiency in growth or sporulation (Sloma *et al.*, 1990; Wu *et al.*, 1990). In soil environments, bpF is presumed to contribute to the nutrition of *B. subtilis* cells.

## Distinguishing Features

Although the deduced amino acid sequence of bpF has many homologies with the four other serine proteinases of *B. subtilis* (Pero & Sloma, 1993), bpF has several unique features. The encoded sequence is larger than those of the other serine proteinases. bpF has an acidic isoelectric point, a relatively high ratio of esterolytic to proteolytic activity, a rather broad specificity for amino acid substrates and releases free amino acids from intact casein (Boyer & Carlton, 1968; Hageman

& Carlton, 1970). The large C-terminus of bpF might be involved in anchoring the enzyme to the bacterial cell wall or membrane in the soil (Mäntsälä & Zalkin, 1980).

### References

Boyer, H.W. & Carlton, B.C. (1968) Production of two proteolytic enzymes by a transformable strain of *Bacillus subtilis. Arch. Biochem. Biophys.* **128**, 422–455.

Hageman, J.H. & Carlton, B.C. (1970) An enzymatic and immunological comparison of two proteases from a transformable strain of *Bacillus subtilis* with the 'subtilisins'. *Arch. Biochem. Biophys.* **139**, 67–69.

Kato, T., Yamagata, Y., Arai, T. & Ichishima, E. (1992) Purification of a new extracellular 90-kDa serine proteinase with isoelectric point of 3.9 from *Bacillus subtilis* (natto) and elucidation of its distinct mode of action. *Biosci. Biotechnol. Biochem.* **96**, 1166–1168.

Mamas, S. & Millet, J. (1975) Purification et propriétiés d'une estérase excrétée pedant la sporulation de *Bacillus subtilis. Biochimie* **57**, 9–16.

Mäntsälä, P. & Zalkin, H. (1980) Extracellular and membrane-bound proteases of *Bacillus subtilis. J. Bacteriol.* **141**, 493–501.

Millet, J. (1970) Characterization of proteinases excreted by *Bacillus subtilis* Marburg strain during sporulation. *J. Appl. Bacteriol.* **23**, 207–219.

Pero, J. & Sloma, A. (1993) Proteases. In: Bacillus subtilis *and Other Gram-Positive Bacteria* (Sonenshein, A.L. *et al.*, eds). Washington, DC: American Society of Microbiology, pp. 939–952.

Prestige, L., Gage, V. & Spizizen, J. (1971). Protease activities during the course of sporulation in *Bacillus subtilis. J. Bacteriol.* **107**, 815–823.

Roitsch, C.A. (1978) Purification and characterization of protease F from *Bacillus subtilis*. PhD dissertation, NMSU.

Roitsch, C.A. & Hageman, J.H. (1983) Bacillopeptidase F: two forms of a glycoprotein serine protease from *Bacillus subtilis* 168. *J. Bacteriol.* **155**, 145–152.

Sloma, A., Rufo, Jr., G.A., Rudolph, C.F., Sullivan, B.J., Theriault, K.A. & Pero, J. (1990) Bacillopeptidase F of *Bacillus subtilis*: purification of the protein and cloning of the gene. *J. Bacteriol.* **172**, 1470–1477. (Erratum, **172**, 5520–5521.)

Wu, X.-C., Nathoo, S., Pang, A.S.-H., Carne, T. & Wong, S.-L. (1990) Cloning, genetic organization, and characterization of a structural gene encoding bacillopeptidase F from *Bacillus subtilis. J. Biol. Chem.* **265**, 6845–6850.

*James H. Hageman*
*Department of Chemistry and Biochemistry,*
*New Mexico State University,*
*Las Cruces, NM 88003, USA*
*Email: jhageman@nmsu.edu*

# 99. Lactocepin: the cell envelope-associated endopeptidase of lactococci

*Peptidase classification: clan SB, family S8, MEROPS ID: S08.019*
*NC-IUBMB enzyme classification: EC 3.4.21.96*
*Databank codes:*

| Species | Type | SwissProt | PIR | EMBL (cDNA) | EMBL (genomic) |
|---|---|---|---|---|---|
| *Lactobacillus delbrueckii* | – | – | – | L48487 | – |
| *Lactobacillus paracasei* | – | Q02470 | B44858 | M83946 | – |
| *Lactococcus lactis cremoris* | PIII | C44858 | A32634 | A15841 | J04962 |
| *Lactococcus lactis cremoris* | PI | P16271 | A60460 B45764 | M24767 | |
| *Lactococcus lactis lactis* | PI/III | P15293 | A44833 B44833 S06997 S08082 | M85288 X14130 | – |

## Name and History

The ability of *Lactococcus lactis* subsp. *lactis* and, in particular, subsp. *cremoris* to ferment milk to produce lactic acid is the cornerstone of the cheese (and lactic casein) industry. In order to grow in milk, lactococci need to degrade caseins to obtain essential amino acids (for review, see Thomas & Pritchard, 1987). The hydrolysis of caseins by lactococcal enzymes also makes an essential contribution to flavor development in ripening cheese (Visser, 1993). A cell wall location for the proteinase responsible for the initial breakdown of casein was described by Thomas *et al.* (1974) for *lactis* strains and by Exterkate (1975) for *cremoris* strains. Since this time, knowledge of the organization and characteristics of the proteinase has progressed considerably. It is now accepted that the enzyme is associated with the cell envelope through a membrane anchor and a cell wall-spanning region and exists in several closely related, but catalytically distinct forms (for review, see Pritchard & Coolbear, 1993). For this reason the trivial name **lactocepin** is proposed (**lacto**coccal **c**ell **e**nvelope-associated **p**roteinase), using I, III and I/III to distinguish the general specificity types (type II originally thought to exist proved to be an artifact of temperature and pH; Visser *et al.*, 1986).

## Activity and Specificity

The activity of the lactocepins has generally been measured by use of caseins labeled with $^{125}$I (Thomas *et al.*, 1974), $^{14}$C (Mills & Thomas, 1978), azo dye (Exterkate, 1979), fluorescamine (Hugenholtz *et al.*, 1984), or fluorescein isothiocyanate (Reid *et al.*, 1991a). Assays, often undertaken at about pH 6.4, generally involve precipitation of undigested casein substrate with trichloroacetic acid followed by determination of the concentration of soluble labeled peptides.

Specificity studies on the lactocepins have concentrated on their ability to hydrolyze caseins, but other substrates such as hemoglobin (Monnet *et al.*, 1987) and oxidized B chain of insulin (Monnet *et al.*, 1992), have been studied for comparative purposes. Both lactocepin I and lactocepin III hydrolyze $\beta$- and $\kappa$-caseins, although with different bond specificities, while $\alpha_{s1}$-casein is hydrolyzed efficiently only by lactocepin III in simple buffered systems. Definite rules concerning specificity have not been established, although some preferences have been observed (Monnet *et al.*, 1987; Visser *et al.*, 1991). Large hydrophobic residues are often found in the P1 and P4 positions, and a Pro is favored in the P2 position. Glu and Ser residues often occupy the P1 and P1′ positions, respectively, relative to bonds hydrolyzed by lactocepin I.

Studies on the hydrolysis of the N-terminal fragment of $\alpha_{s1}$-casein comprising residues 1–23 have indicated that lactocepin I and lactocepin III prefer positively and negatively charged residues, respectively, to be N-terminally located relative to the scissile bond (Exterkate *et al.*, 1991). Further, charged residues located at least up to seven amino acids N-terminally from the scissile bond have been shown to have a marked influence on the rate of hydrolysis (Reid *et al.*, 1995).

Differences in the rates of cleavage of small chromogenic peptide substrates have been used to differentiate between lactocepin I and lactocepin III. The former readily hydrolyzes MeOSuc-Arg-Pro-Tyr+NHPhNO$_2$ (Chromogenix) (see Appendix 2 for full names and addresses of suppliers), whereas the latter only hydrolyzes this substrate efficiently in the presence of 10% (w/v) NaCl (Exterkate, 1990). In contrast, lactocepin III efficiently hydrolyzes Suc-Ala-Glu-Pro-Phe+NHPhNO$_2$ (Bachem), whereas lactocepin I has negligible activity against this substrate (Exterkate *et al.*, 1993). These substrates have been used recently in a survey of the different lactocepin specificity types from a range of lactococcal starters (Bruinenberg & Limsowtin, 1995).

In simple buffered systems, the patterns of at least the initial hydrolysis of $\alpha_{s1}$- (Monnet *et al.*, 1992; Reid *et al.*, 1991a), $\beta$- (Monnet *et al.*, 1986, 1989; Reid *et al.*, 1991b; Visser *et al.*, 1988, 1991) and $\kappa$-caseins (Monnet *et al.*, 1992; Reid *et al.*, 1994; Visser *et al.*, 1994) have been reasonably well determined, with a study using HPLC-MS yielding the most comprehensive data on $\beta$-casein degradation (Juillard *et al.*, 1995). However, there are indications that the specificity of the lactocepins is influenced by release from the cell surface (Exterkate & Alting, 1993) and by the cheese environment, i.e. pH and NaCl levels (at least on the $\alpha_{s1}$-casein (1–23) fragment and chromogenic peptide substrates; see above) and water activity ($a_w$). Studies currently under way using *in vitro* model systems of the cheese environment with regard to water activity have shown that the specificity of lactocepin I acquires characteristics of lactocepin III at an $a_w$ of 0.95 (the water activity of Cheddar-type cheese; J.R. Reid & T. Coolbear, unpublished results).

Genetic studies have identified a small number of amino acid residues in the primary structure of lactocepin that contribute to the specificity differences between the type I and III enzymes (Vos *et al.*, 1991; Siezen *et al.*, 1993). The most conclusive evidence implicates residues 138, 166, 747 and 748. In lactocepin I, residue 166 is negatively charged while the other three are uncharged. However, in lactocepin III, residue 166 is uncharged while the other three are positively charged. These differences are thought to result overall in a more positively charged substrate-binding region in lactocepin III compared with that in lactocepin I.

In one study it has been suggested that dephosphorylation of casein can lead to an increase in the initial rate of hydrolysis by lactocepin (Kyriakidis *et al.*, 1993), but work in our laboratory has shown that for lactocepin I from *Lactococcus lactis* subsp. *cremoris* H2 there is no difference in either the rate or specificity of hydrolysis (D.J. Body, J.R. Reid & T. Coolbear, unpublished results).

At low concentrations (1–5 mM), Ca$^{2+}$ both activates and stabilizes lactocepin, whereas at higher concentrations inhibition of activity occurs. Activity is partially inhibited by citrate, EDTA and 1,10-phenanthroline (presumably by Ca$^{2+}$ chelation, since Zn$^{2+}$ does not appear to be involved) and is almost completely inhibited by HgCl$_2$, CuSO$_4$, PMSF and DFP (Exterkate & de Veer, 1987, 1989; Monnet *et al.*, 1987; Laan & Konings, 1989; Coolbear *et al.*, 1992b).

## Structural Chemistry

From genetic studies, structural/functional relationships of different regions of the lactocepin molecule have been deduced (for a review, see Kok, 1990). The cell-associated form of lactocepin III was predicted to comprise 1775 amino acid residues (Vos *et al.*, 1989). Assuming the same processing mechanism, lactocepin I would comprise 1715 amino acid residues. The enzyme has close homology with other members of the subtilisin family (S8) of serine peptidases (Siezen *et al.*, 1991), particularly with regard to the Asp30, His94, Ser433 catalytic triad located in the N-terminal region of the molecule. The enzyme also possesses a long C-terminal domain which ends in a membrane anchor sequence (Vos *et al.*, 1989); deletion of the last 130 amino acids results in the release of the enzyme into the growth medium (Haandrikman *et al.*, 1989). Whether the lactocepin from *Lactococcus lactis* subsp. *cremoris* ML1, which is the only known lactococcal strain to release the proteinase into the growth medium (Hugenholtz *et al.*, 1984), is naturally truncated in this manner is not known. The majority of the C-terminal region has no established function, apart from the specificity determinants at positions 747 and 748 (Vos *et al.*, 1991).

Lactocepin is protected from autoproteolysis by $Ca^{2+}$ (which also stabilizes the enzyme when in a soluble form) and by the structure of the cell wall, but the exact manner in which this is achieved is unknown. The pronounced stabilizing effect of $Ca^{2+}$ is consistent with the prediction, based on the close homology with subtilisin homologs of known three-dimensional structure, that lactocepin possesses at least three $Ca^{2+}$-binding loops (Siezen *et al.*, 1991). Lactocepin is released from the cell surface of some strains when grown in M17 broth, most likely a result of autoproteolysis (Nissen-Meyer & Sletten, 1991). Once released from the cell surface, autoproteolysis can lead to the production of a series of active fragments which differ in both molecular mass and antigenicity (for a review, see Pritchard & Coolbear, 1993). The pattern of autoproteolytic degradation products is different in the presence and absence of $Ca^{2+}$ (Laan & Konings, 1991). From studies in our laboratory it is clear that there are numerous bonds within the proteinase molecule that are cleaved during the autoproteolytic process, with no definite trends with respect to specificity of action (R.J. Baldwin, J.R. Reid & T. Coolbear, unpublished results).

Purified lactocepin fragments of 145–150 kDa exist as dimers under nondenaturing conditions, while the 180 kDa fragment (see below) exists in a larger multimeric state under the same conditions (Nissen-Meyer & Sletten, 1991; Coolbear *et al.*, 1992b). The pI of purified lactocepin fragments is reported to range from 4.6 to 4.8 (Geis *et al.*, 1985; Exterkate & de Veer, 1989).

## Preparation

The preparation of lactocepin is generally preceded by induction of autoproteolytic release of the enzyme from the cell surface. Two distinct methods have been used to achieve this. The first involves incubating washed lactococcal cells in a $Ca^{2+}$-free buffer at 30°C (Mills & Thomas, 1978) and results in the release of active fragments from the cell surface ranging in size from 80 to 165 kDa, depending partly on the lactococcal strain used. The second method involves enzymatic disruption of the cell wall structure in an osmotically stabilized medium to prevent lysis of the spheroplasts. Phage lysin (Thomas *et al.*, 1974) and lysozyme (Coolbear *et al.*, 1992a,b) have been used for this purpose and in the latter case significant quantities of enzyme were released even in the presence of 50 mM $Ca^{2+}$. The size of the largest lactocepin fragment released by lysozyme was 180 kDa, which is close to the predicted size (based on the gene sequence; see below) for the intact, activated, cell-associated form of the enzyme, and indicates that when the cell wall is disrupted autoproteolysis can occur at a site adjacent to the C-terminal anchor region.

Following release from the cell surface, lactocepin can readily be purified using standard chromatographic techniques, particularly anion-exchange and gel-permeation chromatography. Due to continued proteolysis during purification, poor yields can be a problem. This can be partially overcome by using rapid purification strategies at low temperatures and by inclusion of $Ca^{2+}$.

## Biological Aspects

Complete nucleotide sequences have been determined for the lactocepin genes *(prtP)* from *Lactococcus lactis* subsp. *cremoris* strains Wg2 (Kok *et al.*, 1988) and SK11 (Vos *et al.*, 1989) and for the subsp. *lactis* strain NCDO 763 (Kiwaki *et al.*, 1989). The Wg2, SK11 and NCDO 763 lactocepins possess type I, type III and type I/III activities, respectively. The primary gene products for lactocepin I and lactocepin III have been shown to comprise 1902 (Kok *et al.*, 1988) and 1962 (Vos *et al.*, 1989) amino acid residues, respectively. The primary product of *prtP* is, therefore, approximately 200 kDa (Kok & de Vos, 1994) and is processed by removal of an N-terminal signal sequence of 33 amino acid residues and a pro sequence of 154 amino acid residues to give a catalytically active molecule of about 180 kDa. This processing is dependent on a membrane-bound lipoprotein which is thought to impose a particular conformation on the proenzyme which leads to autoproteolytic removal of the pro region (Haandrikman *et al.*, 1991). The gene for this maturation protein *(prtM)* is separated from *prtP* by an AT-rich promoter region. Production of lactocepin is repressed by low molecular mass peptides and stimulated by $Ca^{2+}$ (Exterkate, 1979, 1985). However, mechanisms regulating lactocepin production appear to be strain dependent (Bruinenberg *et al.*, 1992).

## Distinguishing Features

As already discussed, lactocepin is a member of the subtilisin family of serine proteases, distinguished from all others by an extended C-terminal domain. Only one other extracellular lactococcal proteinase, the nisin leader peptidase (Chapter 109), has been identified (van der Meer *et al.*, 1993). This enzyme, which also belongs to the subtilisin family and is attached to the cell surface via a C-terminal membrane anchor region, converts pronisin to active nisin. However, it is only found in nisin-producing strains and is distinguishable from lactocepin by its size (54 kDa) and its high specificity towards a positively charged residue in the P1 position (Siezen *et al.*, 1995).

Individual strains of lactococci produce only one form of lactocepin of type I, type III or intermediate specificity (the range of which can vary). Only one strain has been tentatively described as producing two separate forms of the enzyme (Reid *et al.*, 1994).

Rabbit polyclonal antibodies (Hugenholtz *et al.*, 1984) and mouse monoclonal antibodies (Laan *et al.*, 1988) against lactocepin from a number of lactococcal strains have been described, but are not available commercially.

## Related Peptidases

As well as exhibiting homology with other members of the subtilisin family of serine proteases, lactocepin also appears to be related to the cell wall-associated serine proteinases of *Lactobacillus casei* NCDO 151 (Holck & Næs, 1992) and HN14 (Kojic *et al.*, 1991). There are both similarities and differences in terms of the gene organization. The specificity of casein hydrolysis of these *Lactobacillus* proteinases is similar to that of lactocepin. The proteinase from *Lactobacillus helveticus* L89 is also thought to be genetically related to lactocepin (Martín-Hernández *et al.*, 1994), indicating that related proteinases are probably widespread throughout the *Lactobacillus* species (Bockelmann, 1995).

## Further Reading

A number of excellent reviews have been cited in the text which deal with both the biochemical and the genetic characterisation of lactocepin and its relevance to lactococcal growth and the role of the proteinase (and the other enzymes in the proteolytic system) to the cheese ripening process. The reader is referred specifically to the reviews of Thomas & Pritchard (1987), Kok (1990), Pritchard & Coolbear (1993), Visser (1993), Kok & de Vos (1994) and Bockelmann (1995).

## References

Bockelmann, W. (1995) The proteolytic system of starter and non-starter bacteria: components and their importance for cheese ripening. *Int. Dairy J.* **5**, 977–994.

Bruinenberg, P.G. & Limsowtin, G.K.Y. (1995) Diversity of proteolytic enzymes among lactococci. *Aust. J. Dairy Technol.* **50**, 47–50.

Bruinenberg, P.G., Vos, P. & de Vos, W.M. (1992) Proteinase production in *Lactococcus lactis* strains: regulation and effect on growth and acidification in milk. *Appl. Environ. Microbiol.* **58**, 78–84.

Coolbear, T., Holland, R. & Crow, V.L. (1992a) Parameters affecting the release of cell surface components and lysis of *Lactococcus lactis* subsp. *cremoris*. *Int. Dairy J.* **2**, 213–232.

Coolbear, T., Reid, J.R. & Pritchard, G.G. (1992b) Stability and

specificity of the cell wall-associated proteinase from *Lactococcus lactis* subsp. *cremoris* H2 released by treatment with lysozyme in the presence of calcium ions. *Appl. Environ. Microbiol.* **58**, 3263–3270.

Exterkate, F.A. (1975) An introductory study of the proteolytic system of *Streptococcus cremoris* strain HP. *Neth. Milk Dairy J.* **29**, 303–318.

Exterkate, F.A. (1979) Accumulation of proteinase in the cell wall of *Streptococcus cremoris* AM1 and regulation of its production. *Arch. Microbiol.* **120**, 247–254.

Exterkate, F.A. (1985) A dual-directed control of cell wall proteinase production in *Streptococcus cremoris* AM1: a possible mechanism of regulation during growth in milk. *J. Dairy Sci.* **68**, 562–571.

Exterkate, F.A. (1990) Differences in short peptide-substrate cleavage by two cell-envelope-located serine proteases of *Lactococcus lactis* subsp. *cremoris* are related to secondary binding specificity. *Appl. Environ. Microbiol.* **33**, 401–406.

Exterkate, F.A. & Alting, A.C. (1993) The conversion of the $\alpha_{s1}$-casein-(1–23)-fragment by the free and bound form of the cell-envelope proteinase of *Lactococcus lactis* subsp. *cremoris* under conditions prevailing in cheese. *Syst. Appl. Microbiol.* **16**, 18.

Exterkate, F.A. & de Veer, G.J.C.M. (1987) Complexity of the native cell wall proteinase of *Lactococcus lactis* subsp. *cremoris* HP and purification of the enzyme. *Syst. Appl. Microbiol.* **9**, 183–191.

Exterkate, F.A. & de Veer, G.J.C.M. (1989) Characterization of the cell wall proteinase $P_{III}$ of *Lactococcus lactis* subsp. *cremoris* strain AM1 and its relationship with the catalytically different cell wall proteinase $P_I/P_{II}$ of strain HP. *Syst. Appl. Microbiol.* **11**, 108–115.

Exterkate, F.A., Alting, A.C. & Slangen, C.J. (1991) Specificity of two genetically related cell-envelope proteinases of *Lactococcus lactis* subsp. *cremoris* towards $\alpha_{s1}$-casein-(1–23)-fragment. *Biochem. J.* **273**, 135–139.

Exterkate, F.A., Alting, A.C. & Bruinenberg, P.G. (1993) Diversity of cell envelope proteinase specificity among strains of *Lactococcus lactis* and its relationship to charge characteristics of the substrate-binding region. *Appl. Environ. Microbiol.* **59**, 3640–3647.

Geis, A., Bockelmann, W. & Teuber, M. (1985) Simultaneous extraction and purification of a cell wall-associated peptidase and β-casein specific protease from *Streptococcus cremoris* AC1. *Appl. Microbiol. Biotechnol.* **23**, 79–84.

Haandrikman, A.J., Kok, J., Laan, H., Soemitro, S., Ledeboer, A.M., Konings, W.N. & Venema, G. (1989) Identification of a gene required for the maturation of an extracellular lactococcal serine proteinase. *J. Bacteriol.* **171**, 2789–2794.

Haandrikman, A.J., Meesters, R., Laan, H., Konings, W.N., Kok, J. & Venema, G. (1991) Processing of the lactococcal extracellular serine proteinase. *Appl. Environ. Microbiol.* **57**, 1899–1904.

Holck, A. & Næs, H. (1992) Cloning, sequencing and expression of the gene encoding the cell-envelope-associated proteinase from *Lactobacillus paracasei* subsp. *paracasei* NCDO 151. *J. Gen. Microbiol.* **138**, 1353–1364.

Hugenholtz, J., Exterkate, F. & Konings, W.N. (1984) The proteolytic system of *Streptococcus cremoris* – An immunological analysis. *Appl. Environ. Microbiol.* **48**, 1105–1110.

Juillard, V., Laan, H., Kunji, E.R., Jeronimus-Stratingh, C.M., Bruins, A.P. & Konings, W.N. (1995) The extracellular $P_I$-type proteinase of *Lactococcus lactis* hydrolyzes β-casein into more than one hundred different oligopeptides. *J. Bacteriol.* **177**, 3472–3478.

Kiwaki, M., Ikemura, H., Shimizu-Kadota, M. & Hirashima, A. (1989) Molecular characterization of a cell wall-associated proteinase gene from *Streptococcus lactis* NCDO 763. *Mol. Microbiol.* **3**, 359–369.

Kojic, M., Fira, D., Banina, A. & Topisirovic, L. (1991) Characterisation of the cell wall-bound proteinase of *Lactobacillus casei* HN14. *Appl. Environ. Microbiol.* **57**, 1753–1757.

Kok, J. (1990) Genetics of the proteolytic system of lactic acid bacteria. *FEMS Microbiol. Rev.* **87**, 15–42.

Kok, J. & de Vos, W.M. (1994) The proteolytic system of lactic acid bacteria. In: *Genetics and Biotechnology of Lactic Acid Bacteria* (Gasson, M.J. & de Vos, W.M., eds). Glasgow: Blackie and Professional, pp. 169–210.

Kok, J., Leenhouts, K.J., Haandrikman, A.J., Ledeboer, A.M. & Venema, G. (1988) Nucleotide sequence of the cell wall proteinase gene of *Streptococcus cremoris* Wg2. *Appl. Environ. Microbiol.* **54**, 231–238.

Kyriakidis, S.M., Sakellaris, G. & Sotiroudis, T.G. (1993) Is protein phosphorylation a control mechanism for the degradation of caseins by lactic acid bacteria? The detection of an extracellular acid phosphatase activity. *Lett. Appl. Microbiol.* **16**, 295–298.

Laan, H. & Konings, W.N. (1989) Mechanism of proteinase release from *Lactococcus lactis* subsp. *cremoris*. *Appl. Environ. Microbiol.* **55**, 3101–3106.

Laan, H. & Konings, W.N. (1991) Autoproteolysis of the extracellular serine proteinase of *Lactococcus lactis* subsp. *cremoris* Wg2. *Appl. Environ. Microbiol.* **57**, 2586–2590.

Laan, H., Smid, E.J., De Leij, L., Schwander, E. & Konings, W.N. (1988) Monoclonal antibodies to the cell-wall-associated proteinase of *Lactococcus lactis* subsp. *cremoris* Wg2. *Appl. Environ. Microbiol.* **54**, 2250–2256.

Martín-Hernández, M.C., Alting, A.C. & Exterkate, F.A. (1994) Purification and characterisation of the mature, membrane-associated cell-envelope proteinase of *Lactobacillus helveticus* L89. *Appl. Microbiol. Biotechnol.* **40**, 828–834.

Mills, O.E. & Thomas, T.D. (1978) Release of cell wall-associated proteinase(s) from lactic streptococci. *N.Z. J. Dairy Sci.* **13**, 209–215.

Monnet, V., Le Bars, D. & Gripon, J.-C. (1986) Specificity of a cell wall proteinase from *Streptococcus lactis* NCDO763 towards bovine β-casein. *FEMS Microbiol. Lett.* **36**, 127–131.

Monnet, V., Le Bars, D. & Gripon, J.-C. (1987) Purification and characterization of a cell wall proteinase from *Streptococcus lactis* NCDO763. *J. Dairy Res.* **54**, 247–255.

Monnet, V., Bockelmann, W., Gripon, J.-C. & Teuber, M. (1989) Comparison of cell wall proteinases from *Lactococcus lactis* subsp. *cremoris* AC1 and *Lactococcus lactis* subsp. *lactis* NCDO 763. II. Specificity towards bovine β-casein. *Appl. Microbiol. Biotechnol.* **31**, 112–118.

Monnet, V., Ley, J.P. & Gonzalez, S. (1992) Substrate specificity of the cell envelope-located proteinase of *Lactococcus lactis* subsp. *lactis* NCDO 763. *Int. J. Biochem.* **24**, 707–718.

Nissen-Meyer, J. & Sletten, K. (1991) Purification and characterization of the free form of the lactococcal extracellular proteinase and its autoproteolytic cleavage products. *J. Gen. Microbiol.* **137**, 1611–1618.

Pritchard, G.G. & Coolbear, T. (1993) The physiology and biochemistry of the proteolytic system in lactic acid bacteria. *FEMS Microbiol. Rev.* **12**, 179–206.

Reid, J.R., Moore, C.H., Midwinter, G.G. & Pritchard, G.G. (1991a) Action of a cell wall proteinase from *Lactococcus lactis* subsp. *cremoris* SK11 on bovine $\alpha_{s1}$-casein. *Appl. Microbiol. Biotechnol.* **35**, 222–227.

Reid, J.R., Ng, K.H., Moore, C.H., Coolbear, T. & Pritchard, G.G. (1991b) Comparison of bovine β-casein hydrolysis by $P_I$ and $P_{III}$-type proteinases from *Lactococcus lactis* subsp. *cremoris*. *Appl. Microbiol. Biotechnol.* **36**, 344–351.

Reid, J.R., Coolbear, T., Pillidge, C.J. & Pritchard, G.G. (1994) Specificity of hydrolysis of bovine κ-casein by cell envelope-associated proteinases from *Lactococcus lactis* strains. *Appl. Environ. Microbiol.* **60**, 801–806.

Reid, J.R., Coolbear, T., Moore, C.H., Harding, D.R.K. & Pritchard, G.G. (1995) Involvement of enzyme–substrate charge interactions in the caseinolytic specificity of lactococcal cell envelope-associated proteinases. *Appl. Environ. Microbiol.* **61**, 3934–3939.

Siezen, R.J., de Vos, W.M., Leunissen, J.A.M. & Dijkstra, B.W. (1991) Homology modelling and protein engineering strategy of subtilases, the family of subtilisin-like serine proteases. *Protein Eng.* **4**, 719–737.

Siezen, R.J., Bruinenberg, P.G., Vos, P., van Alen-Boerrigter, I., Nijhuis, M., Alting, A.C., Exterkate, F.A. & de Vos, W.M. (1993) Engineering of the substrate-binding region of the subtilisin-like, cell envelope proteinase of *Lactococcus lactis*. *Protein Eng.* **6**, 927–937.

Siezen, R.J., Rollema, H.S., Kuipers, O.P. & de Vos, W.M. (1995) Homology modelling of the *Lactococcus lactis* leader peptidase NisP and its interaction with the precursor of the lantibiotic nisin. *Protein Eng.* **8**, 117–125

Thomas, T.D. & Pritchard, G.G. (1987) Proteolytic enzymes of dairy starter cultures. *FEMS Microbiol. Rev.* **46**, 245–268.

Thomas, T.D., Jarvis, B.D.W. & Skipper, N.A. (1974) Localization of proteinases near the cell surface of *Streptococcus lactis*. *J. Bacteriol.* **118**, 329–333.

Van de Meer, J.R., Polman, J., Beerthuyzen, M.M., Siezen, R.J., Kuipers, O.P. & de Vos, W.M. (1993) Characterization of the *Lactococcus lactis* nisin A operon genes *nisP*, encoding a subtilisin-like serine protease involved in precursor processing, and *nisR*, encoding a regulatory protein involved in nisin biosynthesis. *J. Bacteriol.* **175**, 2578–2588.

Visser, S. (1993) Proteolytic enzymes and their relationship to cheese ripening and flavour: an overview. *J. Dairy Sci.* **76**, 329–350.

Visser, S., Exterkate, F.A., Slangen, C.J. & de Veer, G.J.C.M. (1986) Comparative study of action of cell wall proteinases from various strains of *Streptococcus cremoris* on bovine $\alpha_{s1}$-, β- and κ-casein. *Appl. Environ. Microbiol.* **52**, 1162–1166.

Visser, S., Slangen, C.J., Exterkate, F.A. & de Veer, G.J.C.M. (1988) Action of a cell wall proteinase ($P_I$) from *Streptococcus cremoris* HP on bovine β-casein. *Appl. Microbiol. Biotechnol.* **29**, 61–66.

Visser, S., Robben, A.J.P.M. & Slangen, C.J. (1991) Specificity of a cell-envelope-located proteinase ($P_{III}$-type) from *Lactococcus lactis* subsp. *cremoris* AM1 in its action on bovine β-casein. *Appl. Microbiol. Biotechnol.* **35**, 477–483.

Visser, S., Slangen, C.J., Robben, A.J.P.M., van Dongen, W., Heerma, W. & Haverkamp, J. (1994) Action of a cell envelope proteinase ($CEP_{III}$-type) from *Lactococcus lactis* subsp. *cremoris* AM1 on bovine κ-casein. *Appl. Microbiol. Biotechnol.* **41**, 644–651.

Vos, P., Simons, G., Siezen, R.J. & De Vos, W.M. (1989) Primary structure and organization of the gene for a procaryotic cell envelope-located serine protease. *J. Biol. Chem.* **264**, 13579–13585.

Vos, P., Boerrigter, I.J., Buist, G., Haandrikman, A.J., Nijhuis, M., De Reuver, M.B., Siezen, R.J., Venema, G., de Vos, W.M. & Kok, J. (1991) Engineering of the *Lactococcus lactis* serine proteinase by construction of hybrid enzymes. *Protein Eng.* **4**, 479–484.

**Julian R. Reid**
*Biological and Nutritional Science Section,*
*New Zealand Dairy Research Institute,*
*Private Bag 11 029, Palmerston North, New Zealand*
*Email: julian.reid@nzdri.org.nz*

**Tim Coolbear**
*Biological and Nutritional Science Section,*
*New Zealand Dairy Research Institute,*
*Private Bag 11 029, Palmerston North, New Zealand*
*Email: tim.coolbear@nzdri.org.nz*

# 100. C5a peptidase

## Databanks

*Peptidase classification: clan SB, family S8, MEROPS ID: S08.020*
*NC-IUBMB enzyme classification: none*
*Chemical Abstracts Service registry number: 80295-54-1*
*Databank codes:*

| Species | Serotype | SwissProt | PIR | EMBL (cDNA) | EMBL (genomic) |
|---|---|---|---|---|---|
| *Streptococcus agalactiae* | Ia | – | – | U56908 | – |
| *Streptococcus pyogenes* | M12 | P15926 | A35066 | J05229 | – |
| | M49 | – | – | X78055 | – |

## Name and History

Virulent strains of *Streptococcus pyogenes* weakly activate the alternate complement pathway (Wexler *et al.*, 1983), an observation which led to the discovery of the *S. pyogenes* C5a peptidase. Incubation of M+ strains with zymosan-activated serum, a source of human C5a, rapidly reduced the capacity of that serum to attract polymorphonuclear leukocytes (PMNs), in an under-agarose assay for chemotaxis (Wexler *et al.*, 1983). The bacterial enzyme responsible for this activity, first named **streptococcal chemotactic factor inactivator (SCFI)**, was extracted from a serotype M49 strain of *S. pyogenes* by limited trypsin digestion, and purified by sequential ammonium sulfate precipitation, hydrophobic chromatography and anion-exchange chromatography (Wexler & Cleary, 1985). The enzyme was renamed **streptococcal C5a peptidase (SCP)** by O'Connor & Cleary (1986).

The *scpA* gene was cloned from a serotype M12 strain of *S. pyogenes* and shown to be present in many different serotypes of this species. Analysis of the nucleotide sequence revealed that SCP is a serine protease related to subtilisin (Chapter 94) (Chen & Cleary, 1990). The *scpA* gene was also cloned from serotype M49 streptococci and shown to be 98% similar to that from M12 streptococci (Podbielski *et al.*, 1995). The C5a peptidase from *S. agalactiae* was initially thought to be significantly different from SCP because it seemed smaller, and failed to bind antibody directed against SCP (Hill *et al.*, 1988; Bohnsack *et al.*, 1991). However, an *scp*-like gene was cloned and sequenced from *S. agalactiae*, and discovered to be 97–98% similar to those from *S. pyogenes* (Chmouryguina *et al.*, 1996). Insertion mutagenesis of the *scp* gene from *S. agalactia* has confirmed that the cloned *scp* gene encodes all C5a peptidase activity associated with this species (P. Cleary, unpublished results). The suggested abbreviated nomenclature for these peptidases is **SCP** followed by an upper-case letter to indicate the group or species of streptococci and a number to indicate the serotype from which the peptidase is derived. **SCPA12** is, therefore, derived from an M12 strain of group A streptococcus or *S. pyogenes*; whereas, **SCPB** is derived from group B streptococci or *S. agalactiae*.

## Activity and Specificity

SCPA was shown to be cell associated, bound to the surface of *S. pyogenes* (O'Connor & Cleary, 1986). Cell-bound activity and free enzyme are assayed by one of three methods, all indirect assays which assess residual C5a following pre-incubation with cells or protein extracts. Two assays have been used to quantitate the capacity of residual C5a in buffer or serum to induce a chemotactic response by purified PMNs. One assay uses an under-agarose format, in which PMNs migrate under agarose toward wells containing C5a (Nelson *et al.*, 1975). The second uses a micro-modified Boyden chamber in which PMNs migrate through a membrane (Neuroprobe) (see Appendix 2 for full names and addresses of suppliers) (Bohnsack *et al.*, 1993). The third assay method depends on the increased capacity of PMNs to bind to bovine serum albumin (BSA) immobilized on microtiter plate wells (Booth *et al.*, 1992). Either recombinant human C5a from Sigma or C5a in serum, produced by activation of the complement pathway by yeast zymosan, can be the source of C5a substrate for use in the assays.

SCP is a highly specific endopeptidase that was initially thought, on the basis of carboxypeptidase sequencing of the N-terminal peptide fragment of C5a, to hydrolyze human C5a at Lys68 in the primary PMN-binding site. Later, however, Bohnsack *et al.* (1991) showed that a C5a peptidase from *Streptococcus agalactiae* hydrolyzed C5a at His67+Lys68. Therefore, the cleavage position of SCP was re-examined, and determined to be at His67 rather than Lys68, by use of synthetic oligopeptide substrates that correspond to this region of C5a (Cleary *et al.*, 1992). A synthetic oligopeptide that corresponds to the C-terminal 20 amino acid residues of human C5a was readily hydrolyzed. An analog with Gln substituted for Lys68 was equally cleaved by the peptidase. SCP is unable to act upon this peptide when Pro is inserted between His67 and Lys68.

C5a is a product of proteolysis of the C5 serum complement protein that results from the activation of the complement pathway (see Chapters 47 and 48). The His-Lys bond that is cleaved in C5a is apparently inaccessible in intact C5, because it is resistant to cleavage by SCP. Six other proteins known to contain internal His residues are also very resistant to SCP. BSA is somewhat susceptible only after denaturation with urea and prolonged incubation with SCP (Wexler *et al.*, 1985). *Streptococcus* cell-associated SCPB was unable to destroy the chemotactic activity of rodent, zymosan-activated serum (Bohnsack *et al.*, 1993). In contrast, Ji *et al.* (1996) reported that mouse C5a was destroyed by SCPA. This contradiction may be explained by the source of PMN used by Bohnsack *et al.* (1993) in their chemotaxis assays.

## Structural Chemistry

Little attention has been given to the chemistry of the enzyme. The deduced sequence shows a typical signal peptide, and also the Gram-positive cell wall binding consensus sequence, Leu-Pro-Thr-Thr-Asn-Asp, near the C-terminus (Schneewind *et al.*, 1995). The enzyme has an overall globular structure without regions of extensive $\alpha$ helix, characteristic of other streptococcal surface proteins. Sequence alignment showed that regions of SCPA12 are very similar to the catalytic domain of subtilisin, an unexpected finding considering the narrow specificity of SCP compared to the broad specificity of subtilisin. Asp130, His194 and Ser512 of SCP correspond to the catalytic triad (Chen & Cleary, 1990). Siezen *et al.* (1991) suggested that residues 276–295 of SCPA aligned with 136–156 of subtilisin BPN' and that this region corresponds to the S1 cleft (Siezen *et al.*, 1991). SCP is much larger than subtilisin, having a long stretch of sequence between its more N-terminal catalytic triad and C-terminal membrane anchor.

## Preparation

SCP can be extracted from most strains of *S. pyogenes, S. agalactiae* and human isolates of group G streptococci (Cleary *et al.*, 1991). Enzyme can be extracted by digestion of exponential phase streptococci with Mutanolysin or Phage lysin (Cleary *et al.*, 1991). Peptidase activity is highly resistant to heat, 90% of the activity being retained after exposure to 100°C for 10 min. Moreover, activity is relatively stable in acetonitrile, permitting the use of HPLC reverse-phase chromatography for purification of the enzyme. Peptide fusions of SCPA and SCPB with the thrombin-binding site of glutathione-S-transferase have been constructed and the protein expressed from the Pharmacia pGex-4T-1, high-expression vector (Chmouryguina *et al.*, 1996). The fusion protein can then be purified by affinity chromatography on a glutathione column. Yields of up to 4 mg of highly pure protein are obtained using this system.

## Biological Aspects

The primary interest in the C5a peptidase revolves around its role in the virulence of pathogenic streptococci. In *S. pyogenes*, the *scpA* gene is linked to and coregulated with genes that encode other surface-associated virulence factors (Simpson *et al.*, 1990). Plasmid insertion mutagenesis and gene replacement were used to construct strains of *S. pyogenes* that lack peptidase activity. When compared to wild-type parental cultures, SCPA$^-$ streptococci are cleared more rapidly from subdermal sites of infection and from the nasopharynx of mice. Other experiments showed that SPCA influences the trafficking of streptococci from skin sites of infection to lymph nodes and the spleen (Ji *et al.*, 1996). More recent experiments have demonstrated that intranasal inoculation of mice with purified SCPA protects against colonization of the nasopharynx by wild-type SCPA$^+$ streptococci, raising the possibility that SCPA can be used as a vaccine to prevent streptococcal throat infections in human populations (P. Cleary, unpublished results).

## Further Reading

A good account of the place of the C5a peptidase in the subtilisin family has been provided by Siezen *et al.* (1991), and the paper of Ji *et al.* (1996) is a full discussion of the influence of the peptidase on streptococcal virulence.

## References

Bohnsack, J.F., Mollison, K.W., Buko, A.M., Ashworth, J.C. & Hill, H.R. (1991) Group B streptococci inactivate C5a by enzymatic cleavage at the carboxy terminus. *Biochem. J.* **273**, 635–640.

Bohnsack, J.F., Chang, J. & Hill, H.R. (1993) Restricted ability of group B streptococcal C5a-ase to inactivate C5a prepared from different animal species. *Infect. Immun.* **61**, 1421–1426.

Booth, S.A., Moody, C.E., Dahl, M.V., Herron, M.J. & Nelson, R.D. (1992) Dapsone suppresses integrin-mediated neutrophil adherence function. *J. Invest. Dermatol.* **98**, 135–140.

Chen, C. & Cleary, P. (1990) Complete nucleotide sequence of the streptococcal C5a peptidase gene of *Streptococcus pyogenes. J. Biol. Chem.* **265**, 3161–3167.

Chmouryguina, I., Suvorov, A., Ferrieri, P. & Cleary, P.P. (1996) Conservation of the C5a peptidase genes in group A and B streptococci. *Infect. Immun.* **64**, 2387–2390.

Cleary, P.P., Peterson, J. & Nelson, C. (1991) Virulent human strains of group G streptococci express a C5a peptidase enzyme similar to that produced by group A streptococci. *Infect. Immun.* **59**, 2305–2310.

Cleary, P., Prabu, U., Dale, J., Wexler, D. & Handley, J. (1992) Streptococcal C5a peptidase is a highly specific endopeptidase. *Infect. Immun.* **60**, 5219–5223.

Hill, H.R., Bohnsack, J.F., Morris, E.Z., Augustine, N.H., Parker, C.J., Cleary, P.P. & Wu, J.T. (1988) Group B streptococci inhibit the chemotactic activity of the fifth component of complement. *J. Immunol.* **141**, 3551–3556.

Ji, Y., McLandsborough, L., Kondagunta, A. & Cleary, P.P. (1996) C5a peptidase alters clearance and trafficking of group A streptococcus by infected mice. *Infect. Immun.* **64**, 503–510.

Nelson, R.D., Quie, P.G. & Simmons, R.L. (1975) Chemotaxis under agarose: a new and simple method for measuring chemotaxis and spontaneous migration of human polymorphonuclear leukocytes and monocytes. *J. Immunol.* **115**, 1650–1656.

O'Connor, S.P. & Cleary, P.P. (1986) Localization of the streptococcal C5a peptidase to the surface of group A streptococci. *Infect. Immun.* **53**, 432–434.

Podbielski, A., Flosdorff, A. & Weber-Heynemann, J. (1995) The group A streptococcal virR49 gene controls expression of four structural vir Regulon genes. *Infect. Immun.* **63**, 9–20.

Schneewind, O., Fowler, A. & Faull, K.F. (1995) Structure of the cell wall anchor of surface proteins in *Staphylococcus aureus*. *Science* **268**, 103–106.

Siezen, R.J., de Vos, W.M., Leunissen, J.A. & Dijkstra, B.W. (1991) Homology modelling and protein engineering strategy of subtilases, the family of subtilisin-like serine proteinases. *Protein Eng.* **4**, 719–737.

Simpson, W.J., LaPenta, D., Chen, C. & Cleary, P. (1990) Coregulation of type 12 M protein and streptococcal C5a peptidase genes in group A streptococci: evidence for a virulence regulon controlled by the *virR* locus. *J. Bacteriol.* **172**, 696–700.

Wexler, D.E. & Cleary, P.P. (1985) Purification and characteristics of the streptococcal chemotactic factor inactivator. *Infect. Immun.* **50**, 757–764.

Wexler, D.E., Nelson, R.D. & Cleary, P. (1983) Human neutrophil chemotactic response to group A streptococci: bacteria-mediated interference with complement-derived chemotactic factors. *Infect. Immun.* **39**, 757–764.

Wexler, D.E., Chenoweth, E.D. & Cleary, P.P. (1985) Mechanism of action of the group A streptococcal C5a inactivator. *Proc. Natl Acad. Sci. USA* **82**, 8144–8148.

*Paul P. Cleary*
*Department of Microbiology,*
*University of Minnesota,*
*Minneapolis, MN 55455, USA*

# 101. *Dichelobacter (sheep foot-rot) basic serine proteinase*

## Databanks

*Peptidase classification: clan SB, family S8, MEROPS ID: S08.022*
*NC-IUBMB enzyme classification: none*
*Databank codes:*

| Species | SwissProt | PIR | EMBL (cDNA) | EMBL (genomic) |
|---|---|---|---|---|
| *Dichelobacter nodosus* | P42779 | – | Z16080 | – |
| *Dichelobacter nodosus* | P42780 | – | L08175 | – |
| **Related peptidase** | | | | |
| *Xanthomonas campestris* | P23314 | S11890 | X51635 | – |

## Name and History

*Dichelobacter nodosus* (formerly *Bacteroides nodusus*; Dewhirst *et al.*, 1990), an anaerobic gram-negative bacterium, is the causative organism of sheep foot-rot (Thomas, 1962; Egerton *et al.*, 1969) and secretes a family of extracellular proteinases (Every, 1982; Kortt *et al.*, 1982) which may be responsible for the pathogenesis of the disease (Thomas, 1964). **Dichelobacter basic serine proteinase** was discovered in culture supernatant of virulent strain 198 of *D. nodosus* as the proteolytic activity binding to a cationic exchanger at high pH (8.6) (Lilley *et al.*, 1992). This proteinase was termed *Dichelobacter* basic proteinase because of its alkaline pI of >9.5. *D. nodosus* benign strains, which cause only a mild infection, secrete a distinct but closely related proteinase characterized by a lower pI of ~8.6, which is referred to as benign basic proteinase (Lilley *et al.*, 1995).

## Activity and Specificity

Basic proteinase hydrolyzes ester and amide substrates, and cleaves extensively native proteins such as bovine serum albumin, ovalbumin and lysozyme. Denatured proteins such as *S*-carboxymethylated hemoglobin and lysozyme, and peptides such as glucagon and oxidized insulin A and B chains are completely fragmented (Kortt *et al.*, 1994a).

The specificity of *Dichelobacter* basic proteinase has been determined using the A and B chains of oxidized cattle insulin (Kortt *et al.*, 1994a) and other peptide substrates. Basic proteinase cleaves at bonds with Ala, Asn, Leu, Tyr, Trp, Phe, Met, Ser, Thr, Gln, Arg and Lys residues at the P1 position. The specificity of *D. nodosus* basic proteinase is thus very broad with cleavage at hydrophobic, aromatic and hydrophilic residues as well as the basic residues Lys and Arg. No evidence of residue preferences at subsites P2, P3 and P4 was apparent (Kortt *et al.*, 1994a).

Assay of proteolytic activity is performed with hide powder azure as substrate (Kortt *et al.*, 1982) and esterolytic activity can be determined by a spectrophotometric rate assay with Bz-Tyr-OEt (Kortt *et al.*, 1994b). *Dichelobacter* basic proteinase requires an ionic strength of about 0.2 for solubility at pH 8.6; dialysis against low salt buffers (0.01 M Tris–HCl, 5 mM CaCl$_2$, pH 8.6) results in the quantitative precipitation of the proteinase as microcrystals (Kortt *et al.*, 1994a).

The pH optimum is about 8.0 and the enzyme is stable at 55°C for 30 min. Basic proteinase is completely inhibited by DFP, PMSF and peptidyl chloromethanes with Phe at P1 and Ala or Gly at P2, but is not inhibited by Tos-Phe-CH$_2$Cl. The chymotrypsin inhibitor, chymostatin, is only a weak inhibitor. 1,10-Phenanthroline has no inhibitory effect, but EDTA produces an irreversible loss of activity. SDS-PAGE and reverse-phase HPLC analyses have shown that removal of divalent ions from the molecule with EDTA leads to complete autolysis of the proteinase to small peptide fragments. Thus bivalent metal ions are essential for maintaining integrity of the proteinase structure and stabilizing the enzyme against autolysis and hydrolysis by other proteinases (Kortt *et al.*, 1994a).

## Structural Chemistry

*Dichelobacter* basic proteinase is a single-chain protein of 344 amino acid residues (37 kDa) with two disulfide bonds (Lilley *et al.*, 1992). It belongs to the subtilisin family of proteinases and is synthesized as a precursor composed of a signal peptide (21 residues), a propeptide (111 residues) and a mature proteinase region of 344 residues. The precursor also contains a C-terminal region (127) residues. The signal peptide, propeptide and C-terminal regions are removed on secretion and folding to produce the mature proteinase. *Dichelobacter* basic proteinase shows a strong conservation of sequence around the catalytic-site residues with other

members of the subtilisin family, and shows a remarkable sequence similarity to the serine proteinase of *Xanthomonas campestris* with respect to the length of the precursor segments, conservation of disulfide bridges and approximately 50% sequence identity of the mature proteinases.

## Preparation

Native *Dichelobacter* basic proteinase has been isolated as a homogeneous protein from the culture supernatant of six virulent strains of *D. nodosus* (Kortt *et al.*, 1994a). Recombinant basic proteinase has been expressed in *E. coli* at a yield of 25 mg of purified protein per liter (Vaughan *et al.*, 1994). Recombinant basic proteinase has also been expressed in two heterologous systems, *Bacillus subtilis* (Wang *et al.*, 1993) and *Corynebacterium glutamicum* (Billman-Jacobe *et al.*, 1995), and the expressed precursor in each system was correctly processed to yield enzyme indistinguishable from the native proteinase.

## Biological Aspects

*Dichelobacter* basic proteinase and the extracellular acidic proteinases (see below) contribute to the virulence of the organism by promoting tissue invasion (Thomas, 1964) and the severity of the disease (Egerton *et al.*, 1969; Stewart *et al.*, 1986). Furthermore, virulence of isolates of *D. nodosus* correlates with differences in properties of the extracellular proteinases and these differences have been used in diagnostic tests (Kortt *et al.*, 1983; Depiazzi *et al.*, 1991; Palmer, 1993) to distinguish between virulent and benign strains of the organism.

Diagnostic tests using monoclonal antibodies raised against the basic proteinase (V2) and acidic proteinases (B2) isoenzymes have been developed, and are able to distinguish between virulent and benign strains of *D. nodosus*.

The basic proteinase has also been used as an immunogen in a vaccine formulation and shown to protect sheep against foot-rot infections (Stewart *et al.*, 1990). The potential use of the basic proteinase as a cross-protective vaccine may overcome some difficulties associated with the production of vaccines composed of multiple pilus serogroups (Stewart *et al.*, 1990).

## Related Peptidases

*Dichelobacter nodosus* also secretes four to five 'acidic' proteinases with pI values ranging from 5.2 to 5.6 and apparent molecular mass of 40–42 kDa. Acidic proteinases from 'virulent' and 'benign' strains of *D. nodosus* have been purified to homogeneity and shown to represent two distinct types; isoenzymes V1–V3 from virulent strains were similar to each other and to B1–B4 from benign strains, but were clearly distinct from proteinases V5 and B5, which were identical (Kortt *et al.*, 1994b; Kortt & Stewart, 1994). The primary structure of proteinase V5 was determined by both protein sequence analysis (Kortt *et al.*, 1993) and cDNA sequencing of the gene (Riffkin *et al.*, 1993). The gene sequences of proteinases V2 and B2 were also determined (Riffkin *et al.*, 1995). Sequence comparisons revealed that proteinase V2 showed 69% sequence similarity with the basic proteinase and 73% similarity with proteinase V5; proteinase V5 showed 64% similarity with the basic proteinase.

## Distinguishing Features

*Dichelobacter* basic proteinase from virulent strains of *D. nodosus* is distinguished from other *D. nodosus* serine proteinases isolated from both virulent and benign strains by its basic pI and low solubility at low ionic strength.

## References

Billman-Jacobe, H., Wang L.-F., Kortt, A.A., Stewart D.J. & Radford, A. (1995) Expression and secretion of heterologous proteases by *Corynebacterium glutamicum*. *Appl. Environ. Microbiol.* **61**, 1610–1613.

Depiazzi, L.J., Richards, R.B., Henderson, J., Rood, J.I., Palmer, M. & Penhale, W.J. (1991) Characterisation of virulent and benign strains of *Bacteroides nodosus*. *Vet. Microbiol.* **26**, 151–160.

Dewhirst, F.E., Paster, B.J., LaFontaine, S. & Rood, J.I. (1990) Transfer of *Kingella indologenes* (Snell & Lapage, 1976) to the genus *Suttonella* gen. nov. as *Suttonella indologenes* comb. nov. as *Dichelobacter nodosus* comb. nov.; and assignment of the genera *Cardiobacterium*, *Dichelobacter* and *Suttonella* to *Cardiobacteriacae* fam. nov. in the gamma division of *Proteobacteria* on the basis of 16S rRNA sequence comparisons. *Int. J. System. Bacteriol.* **40**, 426–433.

Egerton, J.R., Roberts, D.S. & Parsonson, I.M. (1969) The aetiology and pathogenesis of ovine foot rot. A histological study of the bacterial invasion. *J. Comp. Pathol.* **79**, 207–216.

Every, D. (1982) Proteinase isoenzymes patterns of *Bacteroides nodosus*: distinction between ovine virulent isolates, ovine benign isolates and bovine isolates. *J. Gen. Microbiol.* **128**, 809–812.

Kortt, A.A. & Stewart, D.J. (1994) Properties of the extracellular acidic proteases of *Dichelobacter nodosus*. Stability and specificity of peptide bond cleavage. *Biochem. Mol. Biol. Int.* **34**, 1167–1176.

Kortt, A.A., O'Donnell, I.J., Stewart, D.J. & Clark, B.L. (1982) Activities and partial purification of extracellular proteases of *Bacteroides nodosus* from virulent and benign footrot. *Aust. J. Biol. Sci.* **35**, 481–489.

Kortt, A.A., Burns, J.E. & Stewart, D.J. (1983) Detection of the extracellular proteases of *Bacteroides nodosus* in polyacrylamide gels: a rapid method of distinguishing virulent and benign ovine isolates. *Res. Vet. Sci.* **35**, 171–174.

Kortt, A.A., Riffkin, M.C., Focareta, A. & Stewart, D.J. (1993) Amino acid sequences of extracellular acidic protease V5 of *Dichelobacter nodosus*, the causative organism of ovine footrot. *Biochem. Mol. Biol. Int.* **29**, 989–998.

Kortt, A.A., Caldwell, J.B., Lilley, G.G., Edwards, R., Vaughan, J.

& Stewart, D.J. (1994a) Characterisation of a basic serine proteinase (pI ~ 9.5) secreted by virulent strains of *Dichelobacter nodosus* and identification of a distinct, but closely related, proteinase secreted by benign strains. *Biochem. J.* **299**, 521–525.

Kortt, A.A., Burns, J.E., Vaughan, J.A. & Stewart, D.J. (1994b) Purification of the extracellular acidic proteases of *Dichelobacter nodosus*. *Biochem. Mol. Biol. Int.* **34**, 1157–1166.

Lilley, G.G., Stewart, D.J. & Kortt, A.A. (1992) Amino acid and cDNA sequences of an extracellular basic protease of *Dichelobacter nodosus* show that it is a member of the subtilisin family of proteases. *Eur. J. Biochem.* **210**, 13–21.

Lilley, G.G., Riffkin, M.C., Stewart, D.J. & Kortt A.A. (1995) Nucleotide and deduced sequence of extracellular, serine basic protease gene (*bprB*) from *Dichelobacter nodosus* strain 305: comparison with the basic protease gene (*bprV*) from virulent strain 198. *Biochem. Mol. Biol. Int.* **36**, 101–111.

Palmer, M.A. (1993) A gelatin gel test to detect activity and stability of proteases produced by *Dichelobacter (Bacteroides) nodosus*. *Vet. Microbiol.* **36**, 113–122.

Riffkin, M.C., Focareta, A., Edwards, R.D., Stewart, D. J. & Kortt A.A. (1993) Cloning, sequence and expression of the gene (*aprV5*) encoding serine acidic protease V5 from *Dichelobacter nodosus*. *Gene* **137**, 259–264.

Riffkin, M.C., Wang, L.-F., Kortt, A.A. & Stewart, D.J. (1995) A single amino acid change between the antigenically different extracellular serine proteases V2 and B2 from *Dichelobacter nodosus*. *Gene* **167**, 279–283.

Stewart, D.J., Peterson, J.E., Vaughan, J.A., Clark, B.L., Emery, D.L., Caldwell, J.B. & Kortt, A.A. (1986) The pathogenicity and cultural characteristics of virulent, intermediate and benign strains of *Bacteroides nodosus* causing ovine footrot. *Aust. Vet. J.* **63**, 317–326.

Stewart, D.J., Kortt, A.A. & Lilley, G.G. (1990) New approaches to footrot vaccination and diagnosis utilising the proteases of *Bacteroides nodosus*. In: *Advances in Veterinary Dermatology* (von Tsharner, C. & Halliwell, R.E.W., eds). London: Baillière Tindall, pp. 359–369.

Thomas, J.H. (1962) The bacteriology and histopathology of foot-rot in sheep. *Aust. J. Agric. Res.* **13**, 725–732.

Thomas, J.H. (1964) The pathogenesis of foot-rot in sheep with reference to proteases of *Fusiformis nodosus*. *Aust. J. Agric. Res.* **15**, 1001–1016.

Vaughan, P.R., Wang, L.-F., Stewart, D.J., Lilley, G.G. & Kortt, A.A. (1994) Expression in *Escherichia coli* of the extracellular basic protease from *Dichelobacter nodosus*. *Microbiology* **140**, 2093–2100.

Wang, L.-F., Kortt, A.A. & Stewart, D.J. (1993) Use of a Gram-signal peptide for protein secretion by Gram+ hosts: basic protease of *Dichelobacter nodosus* is produced and secreted by *Bacillus subtilis*. *Gene* **131**, 97–102.

*Alexander A. Kortt*
*CSIRO, Division of Biomolecular Engineering,*
*343 Royal Parade, Parkville,*
*3052 Victoria, Australia*
*Email: alexk@mel.dbe.csiro.au*

*David J. Stewart*
*CSIRO, Division of Animal Health,*
*PO Bag 24, Geelong,*
*3220 Victoria, Australia*

# 102. *Trepolisin*

## Databanks

*Peptidase classification: clan SX, family S8, MEROPS ID: S08.024, also S38.001*
*NC-IUBMB enzyme classification: none*
*ATCC entries: T. denticola strains 33520 and 35405*
*Databank codes:*

| Species | SwissProt | PIR | EMBL (cDNA) | EMBL (genomic) |
|---|---|---|---|---|
| Family S8, *prtP* gene | | | | |
| *Treponema denticola* | – | – | D83264 | – |
| Family S38, *prtB* gene | | | | |
| *Treponema denticola* | – | – | L25603 | – |

## Name and History

*Treponema denticola* is a spirochete bacterium associated with progression of several periodontal diseases. When the proteolytic activity of *T. denticola* was analyzed with a range of peptide and protein substrates a powerful proteinase was discovered (Uitto *et al.*, 1988). Substrate-gel zymography with a variety of proteins revealed a single band of proteolytic activity in the bacterial extracts. Because the enzyme degraded Suc-Ala-Ala-Pro-Phe┼NHPhNO$_2$, and was inhibited by chymotrypsin inhibitors it was called **chymotrypsin-like proteinase** (Uitto *et al.*, 1988), and more recently, **dentilysin** (Ishihara *et al.*, 1996). Because the proteinase has a broad substrate specificity, and may contribute to tissue invasion (Grenier *et al.*, 1990) and lysis of host cells (Uitto *et al.*, 1995) the name **trepolisin** is used here.

## Activity and Specificity

The proteins that are degraded by trepolisin include IgA, IgG, fibrinogen, transferrin, $\alpha_1$-proteinase inhibitor, serum albumin, casein, fibronectin, laminin, type IV collagen, and gelatin (Uitto *et al.*, 1988; Mäkinen *et al.*, 1995). Native type I collagen is not degraded by the enzyme. The enzyme also cleaves peptides, e.g. substance P and angiotensin (Mäkinen *et al.*, 1995). The proteins are in many cases degraded into fragments smaller than 12 kDa. In the cases of some proteins, such as type IV collagen and IgA, only a limited cleavage takes place (Uitto *et al.*, 1988; Grenier *et al.*, 1990). Phe is the preferred amino acid residue at the cleavage site. Leu, Pro and Tyr bonds may be cleaved at a slower rate (Uitto *et al.*, 1988; Mäkinen *et al.*, 1995). The proteinase is inhibited by serine proteinase inhibitors PMSF, Bowman–Birk soybean trypsin–chymotrypsin inhibitor and Tos-Phe-CH$_2$Cl. Tos-Lys-CH$_2$Cl, $\alpha_1$-proteinase inhibitor and pepstatin have no effect on the enzyme activity (Uitto *et al.*, 1988; Mäkinen *et al.*, 1995). In early work (Uitto *et al.*, 1988), it was found that cysteine and DTT increased activity while thiol-blocking reagents inhibited it, but this was not confirmed by other authors using a more extensively purified enzyme preparation (Mäkinen *et al.*, 1995). Calcium chelators increase the activity somewhat (Uitto *et al.*, 1988) or have no effect (Mäkinen *et al.*, 1995). The pH optimum of the enzyme is 7.5–8.0 (Uitto *et al.*, 1988).

## Structural Chemistry

Trepolisin is a 95 kDa protein rich in proline (Mäkinen *et al.*, 1995). Heating the enzyme at 100°C results in formation of three polypeptides. The sizes of these are variously reported as of 72, 27 and 23 kDa (Uitto *et al.*, 1988), 76, 39 and 32 kDa (Mäkinen *et al.*, 1995) and 72, 43 and 38 kDa (Ishihara *et al.*, 1996). Chemical modification of the enzyme by DFP, *N*-ethoxycarbonyl-2-ethoxy-1,2-dihydroquinoline or diethyl pyrocarbonate inactivate it, revealing that the activity depends on an active serine residue, an active carboxyl group, and an active imidazole group.

A 72 kDa *T. denticola* proteinase, the product of the *ptrP* gene, has recently been cloned and sequenced, and may well be a form of trepolisin; it is a homolog of subtilisin. The 'chymotrypsin-like' characteristics of trepolisin could also be described as 'subtilisin-like', and it seems probable that trepolisin is a member of the subtilisin family of serine proteases.

## Preparation

Trepolisin may be purified from cultured ATCC strains, e.g. 33520 and 35405. Also the clinical *T. denticola* isolates carry the proteinase. A rapid purification involves extraction of *T. denticola* cells from 4–5 day cultures by sonication in 0.05 M Tris–HCl buffer, pH 7.5 (with or without 0.1% Triton X-100). A 13 000 × *g* supernatant is subjected to preparative SDS-PAGE under nondenaturing conditions, using 12% gels, 6 mm in thickness. The enzyme band is localized by incubating a narrow strip of the gel on an agarose gel containing skimmed milk. The enzyme is transferred from the corresponding area of the gel by electroelution or extraction of the gel homogenate, followed by removal of SDS by extensive dialysis, pressure filtration or by chromatographic means (Uitto *et al.*, 1988). By performing the electrophoresis twice a relatively pure preparation is obtained as judged by silver staining. A more extensive purification of the 0.1% Triton X-100 *T. denticola* extracts has been reported to involve FPLC ion-exchange chromatography with Fractogel EMD TMAE 650 and Mono-Q HR columns, followed by gel-permeation chromatography using Superose 12 HR and Superose 6 HR. A 19% recovery and a 101-fold purification was reported (Mäkinen *et al.*, 1995).

## Biological Aspects

Trepolisin was localized to the outer membrane of the organism using immunogold electron microscopy (Grenier *et al.*, 1990). It therefore has a good potential to function in the spirochete invasion. Indeed, the enzyme appears important in migration of the organism through basement membrane (Grenier *et al.*, 1990). *T. denticola* exerts several cytopathic effects on epithelial cells, such as loss of cell contacts, degradation of pericellular fibronectin, cytoskeletal collapse, formation of surface blebs and intracellular vacuoles, and cell death (Uitto *et al.*, 1995). Trepolisin appears to play a part in these effects, possibly in concert with another outer membrane protein (Haapasalo *et al.*, 1992). *T. denticola* is able to release these proteins in a form of membrane vesicles that penetrate into epithelial cells (Uitto *et al.*, 1995). The enzyme is also able to trigger release of polymorphonuclear leukocyte granules and activate latent matrix metalloproteinases (Sorsa *et al.*, 1992; Ding *et al.*, 1996). Trepolisin may also serve in the adhesion of the organism. All the studied factors that inhibit the enzyme activity, i.e. heating, treatment at pH 3.2, treatment with enzyme inhibitors or specific anti-trepolisin antibodies also inhibit attachment of *T. denticola* to both hyaluronan (Haapasalo *et al.*, 1996) and epithelial cell surface (Leung *et al.*, 1996). All in all, trepolisin appears to be a key *T. denticola* virulence determinant associated with periodontal disease progression.

## Distinguishing Features and Related Peptidases

Trepolisin is the product of the *PrtP* gene of *Treponema denticola*, and is a cell membrane-associated protein. The products of the genes *prtA* and *prtB* of the organism have also been described as 'chymotrypsin-like' (Que & Kuramitsu, 1990; Arakawa & Kuramitsu, 1994), but are apparently unrelated to trepolisin. Both are probably intracellular proteases. Little is known about *prtA* (Que & Kuramitsu, 1990). The *prtB*

protein is about 30 kDa, and is unrelated to other known proteolytic enzymes (Arakawa & Kuramitsu, 1994), forming peptidase family S38.

The many homologs of trepolisin in the subtilisin family are the subjects of Chapter 93 and following chapters, and have also been reviewed by Siezen *et al.* (1991) and Siezen & Leunnisen (1997).

## References

Arakawa, S. & Kuramitsu, H.K. (1994) Cloning and sequence analysis of a chymotrypsinlike protease from *Treponema denticola*. *Infect. Immun.* **62**, 3424–3433.

Ding, Y., Uitto, V.-J., Haapasalo, M., Lounatmaa, K., Kontinen, Y.T., Salo, T., Grenier, D. & Sorsa, T. (1996) Membrane components of *Treponema denticola* trigger proteinase release from human polymorphonuclear leukocytes. *J. Dent. Res.* **75**, 1986–1993.

Grenier, D., Uitto, V.-J. & McBride, B. (1990) Cellular location of a *Treponema denticola* chymotrypsinlike protease and importance of the protease in migration through basement membrane. *Infect. Immun.* **58**, 347–351.

Haapasalo, M., Müller, K.-H., Uitto, V.-J., Leung, W. & McBride, B. (1992) Characterization, cloning, and binding properties of the major 53-kilodalton *Treponema denticola* surface antigen. *Infect. Immun.* **60**, 2058–2065.

Haapasalo, M., Hannam, P., McBride, B.C. & Uitto, V.-J. (1996) Hyaluronan, a possible ligand mediating *Treponema denticola* binding to periodontal tissue. *Oral Microbiol. Immunol.* **11**, 156–160.

Ishihara, K., Miura, T., Kuramitsu, H.K. & Okuda, K. (1996) Characterization of the *Treponema denticola prtP* gene encoding a prolyl-phenylalanine-specific protease (Dentilisin). *Infect. Immun.* **64**, 5178–5186.

Leung, W.K., Haapasalo, M., Uitto, V.-J., Hannam, P. & McBride, B.C. (1996) The surface protease of *Treponema denticola* may mediate attachment of the bacteria to epithelium. *Anaerobe* **2**, 39–46.

Mäkinen, P.-L., Mäkinen, K. & Syed, S. (1995) Role of the chymotrypsin-like membrane-associated proteinase from *Treponema denticola* ATCC 35405 in inactivation of bioactive peptides. *Infect. Immun.* **58**, 3567–3575.

Que, X.C. & Kuramitsu, H.K. (1990) Isolation and characterization of the *Treponema denticola prtA* gene coding for chymotrypsinlike protease activity and detection of a closely linked gene encoding PZ-PLGPA-hydrolyzing activity. *Infect. Immun.* **58**, 4099–4105.

Siezen, R.J. & Leunissen, J.A.M. (1997) Subtilases: the superfamily of subtilisin-like serine proteases. *Protein Sci.* **6**, 501–523.

Siezen, R.J., de Vos, W.M., Leunissen, J.A.M. & Dijkstra, B.W. (1991) Homology modelling and protein engineering strategy of subtilases, the family of subtilisin-like serine proteinases. *Protein Eng.* **4**, 719–737.

Sorsa, T., Ingman, T., Suomalainen, K., Haapasalo, M., Konttinen, Y., Lindy, O., Saari, H. & Uitto, V.-J. (1992) Identification of proteases from potent periodontopathogenic bacteria as activators of latent human neutrophil and fibroblast-type interstitial collagenases. *Infect. Immun.* **60**, 4491–4495.

Uitto, V.-J., Grenier, D., Chan, E.C.S. & McBride, B.C. (1988) Isolation of a chymotrypsin-like enzyme from *Treponema denticola*. *Infect. Immun.* **56**, 2717–2722.

Uitto, V.-J., Pan, Y.-M., Leung, W.K., Larjava, H., Ellen, R., Finlay, B.B. & McBride, B.C. (1995) Cytopathic effects of *Treponema denticola* chymotrypsin-like protease on migrating and stratified epithelial cells. *Infect. Immun.* **63**, 3401–3410.

*Veli-Jukka Uitto*
*Department of Oral Biology,*
*University of British Columbia,*
*2199 Wesbrook Mall, Vancouver, BC, Canada V6S 1J9*
*Email: jukka@unixg.ubc.ca*

# 103. Aqualysin I

## Databanks

*Peptidase classification: clan SB, family S8, MEROPS ID: S08.151*
*NC-IUBMB enzyme classification: none*
*Databank codes:*

| Species | SwissProt | PIR | EMBL (cDNA) | EMBL (genomic) |
|---|---|---|---|---|
| *Thermus aquaticus* | P08594 | A35742 | D90108 X07734 | – |

## Name and History

*Thermus aquaticus* YT-1, an extremely thermophilic bacterium, was found to produce at least two extracellular proteases. One of them, an alkaline serine protease, was named **aqualysin I** after its producer, *T. aquaticus* (Matsuzawa *et al.*, 1983).

## Activity and Specificity

Assays can be performed with Hammarsten casein at 70°C, pH 7.5–10.4, as described by Matsuzawa *et al.* (1988), or Suc-Ala-Ala-Pro-Phe⫯NHPhNO$_2$ at 40°C, pH 7.5 (Lee *et al.*, 1992).

The optimum pH for activity is around 10.0, and maximum proteolytic activity is observed at 80°C in the presence of 1 mM CaCl$_2$, which stabilizes the enzyme (Matsuzawa *et al.*, 1988). The sites of cleavage of oxidized insulin B chain by aqualysin I are not very specific and differ from the cleavage sites by proteinase K (Chapter 106), thermitase (Chapter 95) and subtilisin BPN′ (Chapter 94), although there are three common cleavage sites. Elastinolytic activity toward elastin-orcein has also been detected. Aqualysin I exhibits specificity toward ester of amino acids with small hydrophobic or aromatic residues in P1 (Matsuzawa *et al.*, 1988).

The enzyme activity is strongly inhibited by DFP, Z-Ala-Gly-Phe-CH$_2$Cl, and *Streptomyces* subtilisin inhibitor (Matsuzawa *et al.*, 1988). The N-terminal pro region of aqualysin I is a potent inhibitor of the enzyme (K. Suefuji, unpublished results), as found for subtilisins (Ohta *et al.*, 1991).

## Structural Chemistry

Aqualysin I comprises 281 amino acid residues and contains two disulfide bonds. Its calculated molecular mass is 28 350 Da. The pI of the enzyme is above 9–10 (Matsuzawa *et al.*, 1988; Kwon *et al.*, 1988a). The active enzyme contains two (or more) atoms of calcium per molecule, and one atom of calcium is tightly bound to the enzyme (S.-J. Lin, unpublished results). The enzyme shows high sequence homology with subtilisin-type serine proteases, 43% identity with proteinase K, and 37–39% identity with subtilisins (Kwon *et al.*, 1988b).

Aqualysin I is synthesized as a large precursor (calculated molecular mass, 53 910 Da) consisting of four domains: an N-terminal signal sequence (14 residues), an N-terminal pro sequence (113 residues), a protease domain, and a C-terminal pro sequence (105 residues) (Terada *et al.*, 1990). Processing of the N- and C-terminal pro sequences is accomplished through the proteolytic activity of aqualysin I itself (Terada *et al.*, 1990; Kurosaka *et al.*, 1996).

## Preparation

The purification of aqualysin I from the culture medium of *T. aquaticus* YT-1 can be performed as described by Matsuzawa *et al.* (1988). The purification of recombinant aqualysin I from *Escherichia coli* cells harboring an expression plasmid using the *tac* promoter can be performed by cation-exchange chromatography on SP-Sepharose and Mono-S columns (Lin *et al.*, 1997). Larger scale production of the recombinant enzyme can also be performed by the method of Sakamoto *et al.* (1996) using bacteriophage T7 RNA polymerase/promoter.

## Biological Aspects

In an *E. coli* expression system for the gene encoding the entire precursor of aqualysin I, a 38 kDa precursor protein with a C-terminal pro sequence is found in the membrane fraction of *E. coli* cells. Heat treatment at 65–70°C is required for the production of active aqualysin I and processing of the C-terminal pro sequence (Terada *et al.*, 1990).

In the production of the active enzyme, the N-terminal pro region functions as an intramolecular chaperone for the proper folding of the protease domain (Lee *et al.*, 1991), as in the cases of the N-terminal pro sequences of many other proteases. The N-terminal pro region of aqualysin I also noncovalently facilitates the production of the active enzyme as an intermolecular chaperone (Lee *et al.*, 1992), as shown in the cases of subtilisin E (Zhu *et al.*, 1989) and α-lytic protease (Silen & Agard, 1989) (Chapter 82).

On the other hand, the C-terminal pro region is not essential for the production of active aqualysin I in the *E. coli* expression system (Lee *et al.*, 1992). The C-terminal pro sequence is, however, involved in its translocation across the membranes in a *Thermus thermophilus* expression system (Lee *et al.*, 1994; D.-W. Kim, unpublished results). Furthermore, the C-terminal pro region prevents the protease domain from taking on a properly folded structure on its secretion by *Saccharomyces cerevisiae* (Kim *et al.*, 1997). These findings suggest that the C-terminal pro sequence of aqualysin I also functions as an intramolecular chaperone that stabilizes the partially unfolded structure of the protease domain, and thereby facilitates its secretion by *T. aquaticus* and *T. thermophilus*.

## References

Kim, D.-W., Lin, S.-J., Morita, S., Terada, I. & Matsuzawa, H. (1997) A carboxy-terminal pro-sequence of aqualysin I prevents proper folding of the protease domain on its secretion by *Saccharomyces cerevisiae*. *Biochem. Biophys. Res. Commun.* **231**, 535–539.

Kurosaka, K., Ohta, T. & Matsuzawa, H. (1996) A 38 kDa precursor protein of aqualysin I (a thermophilic subtilisin-type protease) with a C-terminal extended sequence: its purification and in vitro processing. *Mol. Microbiol.* **20**, 385–389.

Kwon, S.-T., Matsuzawa, H. & Ohta, T. (1988a) Determination of the disulfide bonds in aqualysin I (a thermophilic alkaline serine protease) of *Thermus aquaticus* YT-1. *J. Biochem.* **104**, 557–559.

Kwon, S.-T., Terada, I., Matsuzawa, H. & Ohta, T. (1988b) Nucleotide sequence of the gene for aqualysin I (a thermophilic alkaline serine protease) of *Thermus aquaticus* YT-1 and characteristics of the deduced primary structure of the enzyme. *Eur. J. Biochem.* **173**, 491–497.

Lee, Y.-C., Miyata, Y., Terada, I., Ohta, T. & Matsuzawa, H. (1991) Involvement of NH$_2$-terminal pro-sequence in the production of active aqualysin I (a thermophilic serine protease) in *Escherichia coli*. *Agric. Biol. Chem.* **55**, 3027–3032.

Lee, Y.-C., Ohta, T. & Matsuzawa, H. (1992) A non-covalent NH$_2$-terminal pro-region aids the production of active aqualysin I (a thermophilic protease) without the COOH-terminal pro-sequence in *Escherichia coli*. *FEMS Microbiol. Lett.* **92**, 73–78.

Lee, Y.-C., Koike, H., Taguchi, H., Ohta, T. & Matsuzawa, H. (1994) Requirement of a COOH-terminal pro-sequence for the extracellular secretion of aqualysin I (a thermophilic subtilisin-type protease) in *Thermus thermophilus*. *FEMS Microbiol. Lett.* **120**, 69–74.

Lin, S.-J., Kim, D.-W., Ryugo, Y., Wakagi, T. & Matsuzawa, H. (1997) Increase of the protease activity of aqualysin I, a thermostable serine protease, by replacing Asn219 near the catalytic residue Ser222. *Biosci. Biotech. Biochem.* **61**, 718–719.

Matsuzawa, H., Hamaoki, M. & Ohta, T. (1983) Production of thermophilic extracellular proteases (aqualysins I and II) by *Thermus aquaticus* YT-1, an extreme thermophile. *Agric. Biol. Chem.* **47**, 25–28.

Matsuzawa, H., Tokugawa, K., Hamaoki, M., Mizoguchi, M., Taguchi, H., Terada, I., Kwon, S.-T. & Ohta, T. (1988) Purification and characterization of aqualysin I (a thermophilic alkaline serine protease) produced by *Thermus aquaticus* YT-1. *Eur. J. Biochem.* **171**, 441–447.

Ohta, Y., Hojo, H., Aimoto, S., Kobayashi, T., Zhu, X., Jordan, F. & Inouye, M. (1991) Pro-peptide as an intramolecular chaperone: renaturation of denatured subtilisin E with a synthetic pro-peptide. *Mol. Microbiol.* **5**, 1507–1510.

Sakamoto, S., Terada, I., Lee, Y.-C., Uehara, K., Matsuzawa, H. & Iijima, M. (1996) Efficient production of *Thermus* protease aqualysin I in *Escherichia coli*: effects of cloned gene structure and two-stage culture. *Appl. Microbiol. Biotechnol.* **45**, 94–101.

Silen, J.L. & Agard, D.A. (1989) The α-lytic protease pro-region does not require a physical linkage to activate the protease domain *in vivo*. *Nature* **341**, 462–464.

Terada, I., Kwon, S.-T., Miyata, Y., Matsuzawa, H. & Ohta, T. (1990) Unique precursor structure of an extracellular protease, aqualysin I, with NH$_2$- and COOH-terminal pro-sequences and its processing in *Escherichia coli*. *J. Biol. Chem.* **265**, 6576–6581.

Zhu, X., Ohta, Y., Jordan, F. & Inouye, M. (1989) Pro-sequence of subtilisin can guide the refolding of denatured subtilisin in an intermolecular process. *Nature* **339**, 483–484.

**Kyoko Suefuji**
Department of Biotechnology,
The University of Tokyo,
1-1-1 Yayoi, Bunkyo-ku,
Tokyo 113, Japan

**Shie-Jea Lin**
Department of Biotechnology,
The University of Tokyo,
1-1-1 Yayoi, Bunkyo-ku,
Tokyo 113, Japan

**Hiroshi Matsuzawa**
Department of Biotechnology,
The University of Tokyo,
1-1-1 Yayoi, Bunkyo-ku,
Tokyo 113, Japan
Email: ahmatsu@hongo.ecc.u-tokyo.ac.jp

# 104. Cerevisin

## Databanks

*Peptidase classification: clan SB, family S8, MEROPS ID: S08.052*
*NC-IUBMB enzyme classification: EC 3.4.21.48*
*Chemical Abstracts Service registry number: 37288-81-6*
*Databank codes:*

| Species | SwissProt | PIR | EMBL (cDNA) | EMBL (genomic) |
|---|---|---|---|---|
| *Saccharomyces cerevisiae* | P09232 | A29358 | M18097 M90522 Z11859 | U18795: chromosome V cosmids |

## Name and History

Yeast autolysates were reported to contain two proteinases 'Hefetryptase' and 'Hefepepsin' with pH optima of 7.0 and 4.5, respectively (Dernby, 1917). The term 'proteinase' was used because the activities were able to hydrolyze proteins like hemoglobin, gelatin and casein. Later work confirmed the presence of two proteinases, called A and B (Lenney, 1956). Many groups purified and characterized ***proteinase B (PrB)*** (Lenney, 1956; Felix & Brouillet, 1966; Hata *et al.*, 1967; Cabib & Ulane, 1973; Schott & Holzer, 1974; Kominami *et al.*, 1981a; Huse *et al.*, 1982; Moehle *et al.*, 1987b). PrB is found within yeast vacuoles (Lenney *et al.*, 1974). PrB, ***tryptophan synthase-inactivating enzyme II*** and ***chitin synthetase-activating factor*** are one and the same enzyme (Cabib & Ulane, 1973; Schott & Holzer, 1974). The enzyme is also known as ***protease B, endoproteinase B*** and ***proteinase yscB*** (yeast *Saccharomyce-s cerevisiae*). The name ***cerevisin***, now recommended by IUBMB, derives from the species name, but is a trivial name intended to be applicable to the enzyme also when it is found in other organisms.

## Activity and Specificity

Cerevisin (PrB) is a soluble vacuolar endopeptidase with broad specificity. Esterolytic activity is in the order Z-Tyr-OPhNO$_2$ > Bz-Arg-OEt > Ac-Tyr-OEt; peptide amides are not hydrolyzed (Hata *et al.*, 1967; Ulane & Cabib, 1976; Kominami *et al.*, 1981b). The oxidized B chain of insulin was rapidly cleaved at Leu+Tyr and Phe+Phe bonds, but cleavage at five other sites occurred as well (Kominami *et al.*, 1981b).

Activity can be assayed by use of collagen derivatives like Hide Powder Azure or Azocoll (Ulane & Cabib, 1976; Jones, 1991a). The pH optimum for PrB is typically in the neutral range, in keeping with its location in the pH 6.2 vacuole (Hata *et al.*, 1967; Lenney *et al.*, 1974; Preston *et al.*, 1989).

PrB is inhibited noncompetitively by a 74 amino acid cytoplasmic inhibitor, IB2 (Schu *et al.*, 1991). PrB activity is cryptic in lysates because of its association with IB2.

Autolysis at pH 5 results in destruction of IB2, presumably by saccharopepsin, proteinase A (Chapter 283) (Hata *et al.*, 1967). The cryptic activity of PrB poses a considerable problem during purification of yeast proteins, since PrB and IB2 continue to dissociate throughout various purification procedures. The development and use of protease-deficient strains has proved helpful (Jones, 1991a). Inhibitors from other biological sources have also been reported (Afting *et al.*, 1978). PrB is inhibited by chymostatin, and also by mercurials, presumably because it contains a cysteine residue near the active-site histidine. It is inactivated by DFP and PMSF (Lenney & Dalbec, 1967; Fujishiro *et al.*, 1980; Kominami *et al.*, 1981b).

## Structural Chemistry

The *PRB1* gene encodes a 635 codon open reading frame (ORF) with a predicted (unglycosylated) $M_r$ of 69 621 and five potential sites for *N*-glycosylation, of which only Asn594 is glycosylated, however. The post-translational processing of the PrB precursor (Prb1p), involves four proteolytic cleavages, two N-terminal cleavages in the endoplasmic reticulum and two C-terminal cleavages in the vacuole (Figure 104.1). After removal of the signal sequence by signal peptidase, an intramolecular autocatalytic cleavage removes an additional N-terminal 260 amino acids. This N-terminal propeptide is an essential part of the structure; deletion results in the absence of further processing, and degradation of the precursor, unless the missing peptide is supplied in *trans*. The third cleavage, catalyzed by saccharopepsin, removes about 30 C-terminal amino acids. The fourth and final cleavage removes about 30 additional amino acids (including the *N*-linked glycosyl side-chain) and is catalyzed autocatalytically or by PrB (Mechler *et al.*, 1982; Moehle *et al.*, 1987b, 1989; Nebes & Jones, 1991; Hirsch *et al.*, 1992). This fourth cleavage yields active PrB. PrB can catalyze the fourth cleavage in *trans* even if the third cleavage has not occurred, leading to a phenomenon known as 'phenotypic lag' (Zubenko *et al.*, 1982).

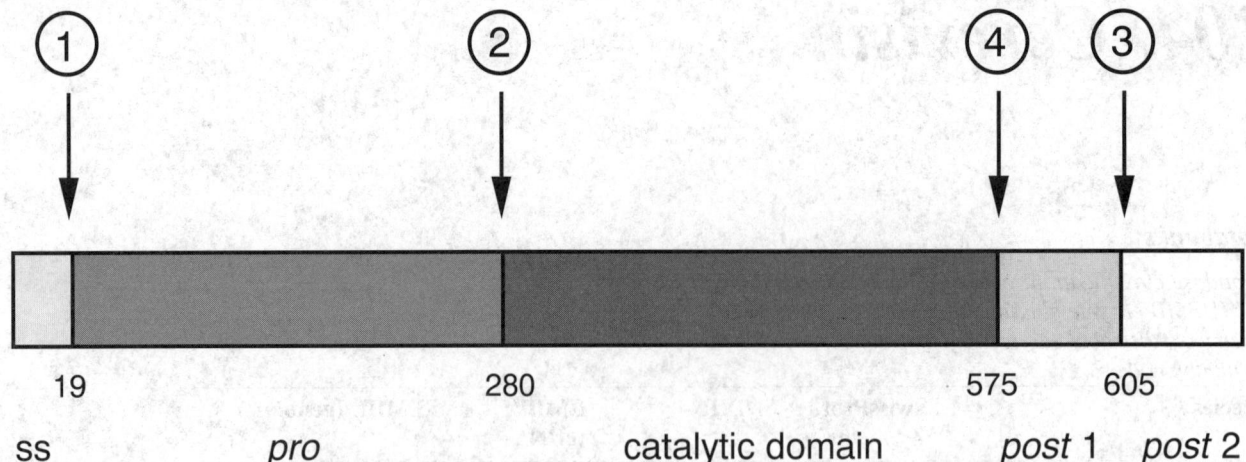

*Figure 104.1* The proteolytic processing of the cerevisin precursor (Prb1p). The arrows identify cleavages in the order of occurrence. Signal peptidase (1) and autocatalytic (2) cleavages occur in the endoplasmic reticulum. PrA (3) and PrB/autocatalytic (4) cleavages occur in the vacuole. PrB can catalyze cleavage 4 even if cleavage 3 has not occurred. Numbers indicate the amino acid position at which the cleavages occur, but all save 2 are approximated.

Mature active PrB is a single-chain protein of 31–33 kDa, with a pI of 6.3. PrB contains three Cys residues, one of which (Cys361) is near the active-site His357. It is not known whether Cys460 and Cys491 form a disulfide bond. PrB is a glycoprotein; it is presumed to carry *O*-linked mannosyl residues, since the single *N*-linked glycosyl side chain is removed during processing (Mechler *et al*., 1982; Moehle *et al*., 1987b, 1989; Nebes & Jones, 1991).

### Preparation

PrB was purified to homogeneity by many groups (Schott & Holzer, 1974; Ulane & Cabib, 1976; Kominami, *et al*., 1981a; Huse *et al*., 1982; Moehle *et al*., 1987b). Moehle *et al*. (1989) fused the full-length *PRB1* ORF to the *lac* promoter for expression in *E. coli*. The species variant of cerevisin from *Candida albicans* (ycaB) was purified to near homogeneity and showed properties similar to those of the *S. cerevisiae* enzyme (yscB) (Farley *et al*., 1986).

### Biological Aspects

Transcription of *PRB1* is repressed by nitrogen and glucose during the prediauxic log phase of growth. As cells growing on glucose exhaust the supply of glucose, transcription of *PRB1* is activated without the appearance of PrB activity. During the postdiauxic phase of logarithmic growth on ethanol, transcription of *PRB1* rises and PrB activity appears. PrB activity is maximal in stationary phase cells, being approximately 100-fold higher than in repressed cells. Growth in acetate-based medium or starvation for nitrogen also elicits high levels of PrB activity (Moehle *et al*., 1987a; Moehle & Jones, 1990; R. Naik, V. Nebes & E.W. Jones, unpublished results). Glucose repression of *PRB1* is partly effected by the Mig1p–Ssn6p–Tup1p repressor complex. Release from nitrogen repression is dependent

on the transcriptional activator Gln3p (R. Naik, V. Nebes & E.W. Jones, unpublished results). The $RAS2^{val19}$ mutation abrogates the postdiauxic derepression of *PRB1* (R. Naik & E.W. Jones, unpublished results).

Intracellularly, PrB plays a major role in the degradation of proteins and peptides during starvation for nitrogen or carbon sources, during autophagy, meiosis and sporulation (Zubenko & Jones, 1981; Teichert *et al*., 1989; Takeshige *et al*., 1992) and in the constitutive and/or environmentally triggered degradation of endocytosed plasma membrane proteins (Jones & Murdock, 1994; van den Hazel *et al*., 1996).

Precursors to several vacuolar hydrolases are processed, apparently by specific cleavages, by PrB. These include cleavages that generate the mature N-termini of saccharopepsin (PrA) (Chapter 283), carboxypeptidase Y (Chapter 132), Gly-X carboxypeptidase (Chapter 484), aminopeptidase I (Chapter 496), aminopeptidase Y (Chapter 489) and PrB itself, as well as the mature C-termini of yeast vacuolar alkaline phosphatase and PrB (Jones, 1991b; Jones & Murdock, 1994). An inspection of the sequences flanking the termini reveals no consensus, so the specificity is presumed to be defined by the tertiary structures of the precursors.

### Distinguishing Features

The most striking difference between PrB and other members of the subtilisin family is the presence of the N- and C-terminal peptide extensions that are removed during processing of the PrB precursor, Prb1p. The Prb1p N-terminal propeptide is very long: 260 amino acids. This peptide is required if active PrB is to be produced; it appears to catalyze a conformational change that allows C-terminal processing by PrA (Nebes & Jones, 1991). The N-terminal highly charged half of this peptide appears to modulate the rate of passage through the secretory pathway; the rest is required for the conformational change (V. Nebes & E.W. Jones, unpublished results). The prototypical bacterial subtilisins (Chapter 94) do not have C-terminal extensions. The C-terminal extension (*post* 1 and *post* 2) in Prb1p must be removed by one or two cleavages to give active PrB. The prohormone processing proteases of this family, like kexin (Chapter 115) and furin (Chapter 117), bear C-terminal P-domains that must be retained if N-terminal maturation is to occur (Gluschankof & Fuller, 1994). PrB, unlike many other subtilisins, contains a free Cys residue near the active site (Moehle *et al.*, 1987b) (but see also endopeptidase K, Chapter 106).

Polyclonal antisera were raised to the full-length Prb1p fused to *lacZ* and to the tetradecapeptide corresponding to the mature PrB N-terminus (Moehle *et al.*, 1989). Polyclonal antisera were also raised to purified PrB and to amino acids 83–259 of the N-terminal extension peptide (Mechler *et al.*, 1982; Hirsch *et al.*, 1992). None are commercially available.

## Further Reading

Jones & Murdock (1994) have provided an overview of the yeast proteinases and their functions, and van den Hazel *et al.* (1996) review the biosynthesis, trafficking and processing of proteases in yeast.

## References

Afting, E., Hinze, H. & Holzer, H. (1978) A new inhibitor protein from rat uterus against yeast proteinase B. *Z. Physiol. Chem.* **359**, 999–1003.

Cabib, E. & Ulane, R. (1973) Chitin synthetase activating factor from yeast, a protease. *Biochem. Biophys. Res. Commun.* **50**, 186–191.

Dernby, K.G. (1917) Studien uber die proteolytischen Enzyme der Hefe unde ihre Beziehung zu der Autolyse. *Biochem. Z.* **81**, 107–207.

Farley, P.C., Shepherd, M.G. & Sullivan, P.A. (1986). The purification of yeast proteinase B from *Candida albicans*. *Biochem. J.* **236**, 177–184.

Felix, F. & Brouillet, N. (1966) Purification and properties of two peptidases from brewers yeast. *Biochem. Biophys. Acta* **122**, 127–144.

Fujishiro, K., Sanada, Y., Tanaka, H. & Katunuma, N. (1980) Purification and characterization of yeast protease B. *J. Biochem.* **87**, 1321–1326.

Gluschankof, P. & Fuller, R.S. (1994) A C-terminal domain conserved in precursor processing proteases is required for intramolecular N-terminal maturation of pro-Kex2 protease. *EMBO* **13**, 2280–2288.

Hata, T., Hayashi, R. & Doi, E. (1967) Purification of yeast proteinases. *Agric. Biol. Chem.* **31**, 150–159.

Hirsch, H.H., Schiffer, H.H. & Wolf, D.H. (1992) Biogenesis of the yeast vacuole (lysosome). *Eur. J. Biochem.* **207**, 867–876.

Huse, G., Kopperschlaeger, G. & Hofmann, E. (1982) A new procedure for the purification of proteinase B from baker's yeast and interaction of the purified enzyme with a specific inhibitor. *Acta Biol. Med. Germ.* **41**, 991–1002.

Jones, E.W. (1991a) Tackling the protease problem in *Saccharomyces cerevisiae*. *Methods Enzymol.* **194**, 428–453.

Jones, E.W. (1991b) Three proteolytic systems in the yeast *Saccharomyces cerevisiae*. *J. Biol. Chem.* **266**, 7963–7966.

Jones, E.W. & Murdock, D.G. (1994) Proteolysis in the yeast vacuole. In: *Cellular Proteolytic Systems* (Ciechanover, A.J. & Schwartz, A.L., eds). New York: Wiley-Liss, pp. 115–134.

Kominami, E., Hoffschulte, H. & Holzer, H. (1981a) Purification and properties of proteinase B from yeast. *Biochim. Biophys. Acta* **661**, 124–135.

Kominami, E., Hoffschulte, H., Leuschel, L., Maier, K. & Holzer, H. (1981b) The substrate specificity of proteinase B from baker's yeast. *Biochim. Biophys. Acta* **661**, 136–141.

Lenney, J.F. (1956) A study of two yeast proteinases. *J. Biol. Chem.* **221**, 919–930.

Lenney, J.F. & Dalbec, J.M. (1967) Purification and properties of two proteinases from *Saccharomyces cerevisiae*. *Arch. Biochem. Biophys.* **120**, 42–48.

Lenney, J.P., Matile, A., Wiemken, M., Schellenberg, M. & Meyer, J. (1974) Activities and cellular localization of yeast proteinases and their inhibitors. *Biochem. Biophys. Res. Commun.* **60**, 1378–1383.

Mechler, B., Muller, M., Muller, H., Muessdoerffer, F. & Wolf, D. (1982) *In vivo* biosynthesis of the vacuolar proteinases A and B in the yeast *Saccharomyces cerevisiae*. *J. Biol. Chem.* **257**, 11203–11206.

Moehle, C.M. & Jones, E.W. (1990) Consequences of growth media, gene copy number, and regulatory mutations on the expression of the *PRB1* gene of *Saccharomyces cerevisiae*. *Genetics* **124**, 39–55.

Moehle, C.M., Aynardi, M.W., Kolodny, M.R., Park, F.J. & Jones, E.W. (1987a) Protease B of *Saccharomyces cerevisiae*: isolation and regulation of the *PRB1* structural gene. *Genetics* **115**, 255–263.

Moehle, C.M., Tizard, R., Lemmon, S.K., Smart, J. & Jones, E.W. (1987b) Protease B of the lysosomelike vacuole of the yeast *Saccharomyces cerevisiae* is homologous to the subtilisin family of serine proteases. *Mol. Cell. Biol.* **7**, 4390–4399.

Moehle, C.M., Dixon, C.K. & Jones, E.W. (1989) Processing pathway for protease B of *Saccharomyces cerevisiae*. *J. Cell. Biol.* **108**, 309–324.

Nebes, V.L. & Jones, E.W. (1991) Activation of the proteinase B precursor of the yeast *Saccharomyces cerevisiae* by autocatalysis and by an internal sequence. *J. Biol. Chem.* **266**, 22851–22857.

Preston, R.A., Murphy, R.F. & Jones, E.W. (1989) Assay of vacuolar pH in yeast and identification of acidification-defective mutants. *Proc. Natl Acad. Sci. USA* **86**, 7027–7031.

Schott, E.H. & Holzer, H. (1974) Purification and some properties of tryptophan synthase inactivase II from yeast. *Eur. J. Biochem.* **42**, 61–66.

Schu, P., Suarez Renduelles, P. & Wolf, D.H. (1991) The proteinase yscB inhibitor *(PBI2)* gene of yeast and studies on the function of its protein product. *Eur. J. Biochem.* **197**, 1–7.

Takeshige, K., Baba, M., Tsuboi, S., Noda, T. & Ohsumi, Y. (1992) Autophagy in yeast demonstrated with proteinase-deficient mutants and conditions for its induction. *J. Cell Biol.* **19**, 301–311.

Teichert, U., Mechler, B., Müller, H. & Wolf, D.H. (1989) Lysosomal (vacuolar) proteinases of yeast are essential catalysts for protein degradation, differentiation and cell survival. *J. Biol. Chem.* **264**, 16037–16045.

Ulane, R.E. & Cabib, E. (1976) The activating system of chitin synthetase from *Saccharomyces cerevisiae*. *J. Biol. Chem.* **251**, 3367–3374.

van den Hazel, H.B., Kielland-Brandt, M.C. & Winter, J.R. (1996) Biosynthesis and function of yeast vacuolar proteases. *Yeast* **12**, 1–6.

Zubenko, G.S. & Jones, E.W. (1981) Protein degradation, meiosis and sporulation in proteinase-deficient mutants of *Saccharomyces cerevisiae*. *Genetics* **97**, 45–64.

Zubenko, G.S., Park, F.J. & Jones, E.W. (1982) Genetic properties of mutations at the *PEP4* locus in *Saccharomyces cerevisiae*. *Genetics* **102**, 679–690.

*Rajesh R. Naik*
*Department of Biological Sciences,*
*Carnegie Mellon University,*
*Mellon Institute,*
*4400 Fifth Ave, Pittsburgh, PA 15213, USA*

*Elizabeth W. Jones*
*Department of Biological Sciences,*
*Carnegie Mellon University,*
*Mellon Institute,*
*4400 Fifth Ave, Pittsburgh, PA 15213, USA*
*Email: ej09@andrew.cmu.edu*

# 105. Oryzin

## Databanks

*Peptidase classification: clan SB, family S8, MEROPS ID: S08.053*
*NC-IUBMB enzyme classification: EC 3.4.21.63*
*Databank codes:*

| Species | SwissProt | PIR | EMBL (cDNA) | EMBL (genomic) |
|---|---|---|---|---|
| *Aspergillus flavus* | P35211 | – | – | L08473: complete gene |
| | | | | S67840: complete gene |
| *Aspergillus fumigatus* | P28296 | S22184 | M99165 | X66935: 5′ end |
| | | | M99420 | Z11580: complete gene |
| *Aspergillus niger* | P33295 | JU0146 | M96758 | – |
| *Aspergillus niger* | – | – | L19059 | – |
| *Aspergillus nidulans* | – | – | L31778 | – |
| *Aspergillus oryzae* | P12547 | JU0278 | D00350 | D10062: complete gene |
| | | | X17561 | X54726: 3′ end |

## Name and History

Alkaline proteinase activity that hydrolyzes a variety of protein substrates has been found in many fungal species, especially in *Aspergillus* (Reichard *et al.*, 1990; Monod *et al.*, 1991; Bouchara *et al.*, 1993; Kolattukudy *et al.*, 1993; Ramesh *et al.*, 1994). Purification and characterization indicate that the major alkaline proteinase is a 33 kDa serine proteinase of the subtilisin family. The enzyme has generally been termed **Aspergillus alkaline proteinase, elastinolytic proteinase** or **serine proteinase**, in the literature, but the name **oryzin** (EC 3.4.21.63) has now been recommended by NC-IUBMB, and is used here.

## Activity and Specificity

Oryzin is an alkaline serine proteinase that shows catalytic activity similar to that of subtilisin (Chapter 94). It can easily be assayed spectrophotometrically with Suc-Ala-Ala-Pro-Leu+NHPhNO$_2$ as substrate (Ramesh *et al.*, 1994). It readily hydrolyzes elastin and hydrolyzes collagen at less than half the rate obtained with elastin, but does not hydrolyze laminin at measurable rates. Other common protein substrates

hydrolyzed by chymotrypsin and subtilisin-like enzymes are also hydrolyzed. The pH optimum is at 8.0–9.0 with very low activity below pH 7.0. It is severely inhibited by active serine-directed reagents such as DFP and PMSF, and by the histidine-directed reagent diethylpyrocarbonate, but not by metal ion chelators or thiol-blocking reagents. Oryzin is also inhibited by potato chymotrypsin inhibitors 1 and 2 with $K_i$ values of $7.4 \times 10^{-9}$ M and $2.6 \times 10^{-8}$ M, respectively, by eglin C with a $K_i$ of $3.3 \times 10^{-7}$ M, by barley $\alpha$-amylase/subtilisin inhibitor with a $K_i$ of $7.7 \times 10^{-8}$ M, and by *Streptomyces* subtilisin inhibitor (SSI) with a $K_i$ of $10^{-9}$ M (Markaryan *et al.*, 1996a).

## Structural Chemistry

Oryzin consists of a single polypeptide of 33 kDa. In *A. flavus*, the primary translation product contains a 21 amino acid signal peptide and 101 amino acid propeptide (Ramesh *et al.*, 1994). Similar N-terminal regions are also present in *A. oryzae* (Tatsumi *et al.*, 1989) and *A. fumigatus (Sartoria fumigata)* (Jaton-Ogay *et al.*, 1992), and these are processed to generate the mature enzyme. In *A. niger* both termini are thought to be processed to generate the mature enzyme (Frederick *et al.*, 1993). The amino acid sequence of the mature enzyme from *A. flavus* has 83, 82, 61 and 63% identity to those of *A. fumigatus, A. oryzae, Acremonium chrysogenum* and *Trichoderma harzianum* proteinases, respectively (Ramesh *et al.*, 1994). The catalytic properties and the homologous sequences around the active Ser, His and Asp residues show that these serine proteinases belong to the subtilisin family (Jaton-Ogay *et al.*, 1992; Kolattukudy *et al.*, 1993; Markaryan *et al.*, 1996a). Based on the high degree of similarity to the subtilisin-like proteinase K (Chapter 106) from *Tritirachium album* (Betzel *et al.*, 1988), the fungal serine proteinase has been modeled. The SSI complex with oryzin has been modeled (Markaryan *et al.*, 1996a) after the known SSI complex with subtilisin BPN' (Takeuchi *et al.*, 1991) (Chapter 94). This modeling shows that the P1–P3 residues of SSI can interact with the S1–S3 site at residues 142–144 of oryzin. However, the S4–S6 site present in subtilisin and proteinase K that interacts with the P4–P6 residues of SSI is not present in oryzin, predicting that the SSI interaction would be slightly weaker, as is experimentally found.

## Preparation

Oryzin, being found in the culture fluid, can be purified at high yields from elastin-grown *A. fumigatus* culture filtrates by hydrophobic chromatography on a phenyl Sepharose column in 10% $(NH_4)_2SO_4$; a 10–0% gradient elutes the enzyme at 3.5% $(NH_4)_2SO_4$ (Kolattukudy *et al.*, 1993). Since the enzyme undergoes self-digestion it is stored as the $(NH_4)_2SO_4$ precipitate. Recombinant oryzin can be expressed in yeast *Saccharomyces cerevisiae* (Tatsumi *et al.*, 1989), and in osmophilic yeast *Zygosaccharomyces rouxii* to yield 300 mg of enzyme liter$^{-1}$ (Ogawa *et al.*, 1990).

## Biological Aspects

The cDNA and gene for oryzin have been cloned from several species (Tatsumi *et al.*, 1989; Chewevadhanarak *et al.*, 1991; Monod *et al.*, 1993; Tang *et al.*, 1993; Ramesh *et al.*, 1994). The genes from *A. fumigatus, A. flavus* and *A. oryzae* contain three introns, but that from *A. niger* has only one intron, corresponding to the position of the first intron in the other genes. The propeptide of the enzyme from *A. fumigatus*, expressed in *E. coli*, selectively and competitively inhibits the mature serine proteinase with a $K_i$ of 5 mM. This inhibition is highly specific; the propeptide is much less inhibitory to the very similar (83% identical) *A. flavus* enzyme, and does not inhibit subtilisin (Markaryan *et al.*, 1996b). *A. fumigatus* and *A. flavus* isolates that show high virulence on immunocompromised animals also show high levels of secretion of oryzin, and immunocytochemical studies show that the enzyme is secreted into the host tissue during invasion (Kolattukudy *et al.*, 1993). Disruption of the gene does not lead to lower virulence (Monod *et al.*, 1993; Tang *et al.*, 1993), but in view of the fact that disruption of the serine proteinase gene can cause a compensatory increase in production of other proteinases (Ramesh & Kolattukudy, 1996), the gene disruption studies should be interpreted with caution. Probably the production of all of the proteinases is under the control of master genes such as *areA*. Disruption of such genes lowers production of

all proteinases and virulence (Hensel *et al.*, 1995). An elastase-deficient *A. fumigatus* mutant that was found to have much lower virulence (Kolattukudy *et al.*, 1993) was recently found to be deficient not only in the production of oryzin but also in fungalysin 42 and fungal acid proteinase (Chapter 294). Potent inhibitors of oryzin protect immunocompromised animals against infection.

## Related Peptidases

Seaprose is a semialkaline proteinase from *A. melleus*. It is a 32 kDa single-chain serine proteinase with an optimal pH of 7.5–8.5 and an optimal temperature of 45°C. It has elastinolytic activity and preferentially cleaves substrates containing Leu or Phe residues in the P1 position (Tasaka *et al.*, 1980; Morihara *et al.*, 1984). It is not inhibited by most plasma inhibitors of proteinases (Eguchi & Yamamoto, 1988). Each molecule of $\alpha_2$-macroglobulin complexes two molecules of seaprose and inhibits them. $\alpha_1$-Proteinase inhibitor is inactivated by seaprose, but $\alpha_1$-antichymotrypsin forms a complex with seaprose and inhibits it (Korzus *et al.*, 1994). Seaprose, also known as onoprose, promelasum and prozime-10, has been tested for medical applications mostly as an anti-inflammatory agent over a long period of time. It can be absorbed from the gut and is thought to cause the release of glucocorticoids (Tasaka *et al.*, 1980). Anti-inflammatory effects on venous disease were observed (Bracale & Selvetella, 1996). Seaprose reduces sputum viscoelastic properties in chronic hypersecretory bronchitis (Braga *et al.*, 1990; Moretti *et al.*, 1993).

## References

Betzel, C., Pal, G.P. & Saenger, W. (1988) Synchrontron X-ray data collection and restrained least-squares refinement of the crystal structure of proteinase K at 1.5 Å resolution. *Eur. J. Biochem.* **178**, 155–171.

Bouchara, J.P., Larcher, G., Joubaud, F., Penn, P., Tronchin, G. & Chabasse, D. (1993) Extracellular fibrinogenolytic enzyme of *Aspergillus fumigatus*: substrate-dependent variations in the proteinase synthesis and characterization of the enzyme. *FEMS Immunol. Med. Microbiol.* **7**, 81–91.

Bracale, G. & Selvetella, L. (1996) Clinical study of the efficacy of and tolerance to seaprose S in inflammatory venous disease. Controlled study versus serratio-peptidase. *Minerva Cardioangiol.* **44**, 515–524.

Braga, P.C., Rampoldi, C., Ornaghi, A., Caminiti, G., Beghi, G. & Allegra, L. (1990) *In vitro* rheological assessment of mucolytic activity induced by Seaprose. *Pharmacol. Res.* **22**, 611–617.

Chewevadhanarak, S., Renno, D.V., Saunders, G. & Holt, G. (1991) Cloning and selective overexpression of an alkaline protease-encoding gene from *Aspergillus oryzae*. *Gene* **108**, 151–155.

Eguchi, M. & Yamamota, Y. (1988) Comparison of serum protein inhibitors from various mammals, chicken and silkworms against four proteases. *Comp. Biochem. Physiol. B* **91**, 625–630.

Frederick, G.D., Rombouts, P. & Buxton, F.P. (1993) Cloning and characterization of pepC, a gene encoding a serine proteinase from *Aspergillus niger*. *Gene* **125**, 57–64.

Hensel, M., Tang, C.M., Arst, Jr., H.N. & Holden, D.W. (1995) Regulation of fungal extracellular proteases and their role in mammalian pathogenesis. *Can. J. Bot.* **73** (suppl. 1), S1065–S1070.

Jaton-Ogay, K., Suter, M., Crameri, R., Falchetto, R., Faith, A. & Monod, M. (1992) Nucleotide sequence of a genomic and a cDNA clone encoding an extracellular alkaline proteinase of *Aspergillus fumigatus*. *FEMS Microbiol. Lett.* **92**, 163–168.

Kolattukudy, P.E., Lee, J.D., Rogers, L.M., Zimmerman, P., Ceselski, S., Fox, B., Stein, B. & Copelan, E.A. (1993) Evidence for possible involvement of an elastolytic serine protease in aspergillosis. *Infect. Immun.* **61**, 2357–2368.

Korzus, E., Luisetti, M. & Travis, J. (1994) Interactions of α-1-antichymotrypsin, α-1-proteinase inhibitor, and α-2-macroglobulin with the fungal enzyme, seaprose. *Biol. Chem. Hoppe-Seyler* **375**, 335–341.

Markaryan, A., Beall, C.J. & Kolattukudy, P.E. (1996a) Inhibition of *Aspergillus* serine proteinase by *Streptomyces* subtilisin inhibitor and high-level expression of this inhibitor in *Pichia pastoris*. *Biochem. Biophys. Res. Commun.* **220**, 372–376.

Markaryan, A., Lee, J.D., Sirakova, T.D. & Kolattukudy, P.E. (1996b) Specific inhibition of mature fungal serine and metalloproteinases by their propeptides. *J. Bacteriol.* **178**, 2211–2215.

Monod, M., Togni, G., Rahalison, L. & Frenk, E. (1991) Isolation and characterisation of an extracellular alkaline protease of *Aspergillus fumigatus*. *J. Med Microbiol.* **35**, 23–28.

Monod, M., Paris, S., Sanglard, D., Jaton-Ogay, K., Bille, J. & Latge, JP. (1993) Isolation and characterization of a secreted metalloprotease of *Aspergillus fumigatus*. *Infect. Immun.* **61**, 4099–4101.

Moretti, M., Bertoli, E., Bulgarelli, S., Testoni, C., Guffanti, E.E., Marchioni, C.F. & Braga, P.C. (1993) Effects of seaprose on sputum biochemical components in chronic bronchitic patients: a double-blind study vs placebo. *Int. J. Clin. Pharmacol. Res.* **13**, 275–280.

Morihara, K., Tsuzuki, H., Harada, M. & Iwata, T. (1984) Purification of human plasma α-1-proteinase inhibitor and its inactivation by *Pseudomonas aeruginosa* elastase. *J. Biochem. (Tokyo)* **95**, 795–804.

Ogawa, Y., Tatsumi, H., Murakami, S., Ishida, Y., Murakami, K., Masaki, A., Kawabe, H., Arimura, H., Nakano, E., Motai, H. et al. (1990) Secretion of *Aspergillus oryzae* alkaline protease in an osmophilic yeast, *Zygosaccharomyces rouxii*. *Agric. Biol. Chem.* **54**, 2521–2529.

Ramesh, M.V. & Kolattukudy, P.E. (1996) Disruption of the serine proteinase gene (*sep*) in *Aspergillus flavus* leads to a compensatory increase in the expression of a metalloproteinase gene (*mep20*). *Gene* **173**, 3899–3907.

Ramesh, M.V., Sirakova, T. & Kolattukudy, P.E. (1994) Isolation, characterization, and cloning of cDNA and the gene for an elastinolytic serine proteinase from *Aspergillus flavus*. *Infect. Immun.* **62**, 79–85.

Reichard, U., Buttner, S., Eiffert, H., Staib, F. & Ruchel, R. (1990) Purification and characterisation of an extracellular serine proteinase from *Aspergillus fumigatus* and its detection in tissue. *J. Med. Microbiol.* **33**, 243–251.

Takeuchi, Y., Satow, Y., Nakamura, K.T. & Mitsui, Y. (1991) Refined crystal structure of the complex of subtilisin BPN′ and *Streptomyces* subtilisin inhibitor at 1.8 Å resolution. *J. Mol. Biol.* **221**, 309–325.

Tang, C.M., Cohen, J., Krausz, T., Van Noorden, S. & Holden, D.W. (1993) The alkaline protease of *Aspergillus fumigatus* is not a virulence determinant in two murine models of invasive pulmonary aspergillosis. *Infect. Immun.* **61**, 1650–1656.

Tasaka, K., Meshi, T., Akagi, M., Kakimoto, M., Saito, R., Okada, I. & Maki, K. (1980) Anti-inflammatory activity of a proteolytic enzyme, Prozime-10. *Pharmacology* **21**, 43–52.

Tatsumi, H., Ogawa, Y., Murakami, S., Ishida, Y., Murakami, J., Masaki, A., Kawabe, H., Arimura, H., Nakano, E. & Motai, H. (1989) A full length cDNA clone for alkaline proteinase from *Aspergillus oryzae*: structural analysis and expression in *Saccharomyces cerevisiae*. *Mol. Gen. Genet.* **219**, 33–38.

**P.E. Kolattukudy**
*Ohio State University,*
*Department of Biochemistry and Medical Biochemistry,*
*The Neurobiotechnology Center,*
*206 Rightmire Hall, 1060 Carmack Drive,*
*Columbus, OH 43210, USA*
*Email: kolattukudy.2@osu.edu*

**Tatiana D. Sirakova**
*Ohio State University,*
*Department of Biochemistry and Medical Biochemistry,*
*The Neurobiotechnology Center,*
*206 Rightmire Hall, 1060 Carmack Drive,*
*Columbus, OH 43210, USA*
*Email: sirakova.1@osu.edu*

# 106. Proteinase K

## Databanks

*Peptidase classification: clan SB, family S8, MEROPS ID: S08.054*
*NC-IUBMB enzyme classification: EC 3.4.21.64*
*Chemical Abstracts Service registry number: 39450-01-6*

*Databank codes:*

| Species | SwissProt | PIR | EMBL (cDNA) | EMBL (genomic) |
|---|---|---|---|---|
| *Tritirachium album* | P06873 | A23188 A24541 S02142 | X14688 | X14689: complete gene |

Brookhaven Protein Data Bank three-dimensional structures:

| Species | ID | Resolution | Notes |
|---|---|---|---|
| *Tritirachium album* | 1PEK | 2.2 | complex with Ac-Pro-Ala-Pro-Phe-D-Ala-Ala-NH$_2$ |
| | 1PTK | 2.4 | complex with mercury |
| | 2PKC | 1.5 | calcium-free form |
| | 2PRK | 1.5 | |
| | 3PRK | 2.2 | complex with inhibitor MeOSuc-Ala-Ala-Pro-Ala-CH$_2$Cl |

## Name and History

The alkaline proteinase secreted into the culture medium by the mold *Tritirachium album* Limber (Ebeling *et al.*, 1974) is commonly known as ***proteinase K***. The designation 'K' was chosen to indicate that it can even hydrolyze native keratin.

## Activity and Specificity

Proteinase K is a typical member of the subtilisin family of proteinases (S8). The amino acid sequence has been derived by Edman degradation (Jany *et al.*, 1986) and from the gene sequence (Gunkel & Gassen, 1989); the polypeptide chain consists of 278 amino acids, with molecular mass 28 930, and the pI is 8.9 (Ebeling *et al.*, 1974). Proteinase K is an endopeptidase and carries at its active site the catalytic triad Asp39, His69, Ser224. Full enzymatic activity requires the presence of calcium ions (Ebeling *et al.*, 1974; Bajorath *et al.*, 1988). Maximal activity is in the pH range 7.5–12.0. The specific activity is about 300 U mg$^{-1}$, when one unit hydrolyzes 1 µmole of Ac-Tyr-OEt per min at pH 9.3 and 30°C in Tris-HCl buffer (Sweeney & Walker, 1993). Peptidyl chloromethanes (Betzel *et al.*, 1986, 1988a; Bajorath *et al.*, 1988; Wolf *et al.*, 1991) and Hg$^{2+}$ (Bagger *et al.*, 1991; Müller & Saenger, 1993; Saxena *et al.*, 1996) act as inhibitors, with 7% residual activity retained even with excess Hg$^{2+}$ (Müller & Saenger, 1993). Proteinase K suffers from autolysis at low concentrations ($\sim$0.01 mg ml$^{-1}$), but is much more stable at concentrations above 1.0 mg ml$^{-1}$ (Bajorath *et al.*, 1988; Sweeney & Walker, 1993).

Proteinase K is relatively unspecific. As shown by the cleavage of oxidized insulin B chain (Kraus *et al.*, 1976), it has a preference for aromatic and hydrophobic amino acids in P1, like other subtilisins. It does not hydrolyze synthetic substrates with Pro in the P1$'$ position (Morihara & Tsuzuki, 1975; Brömme *et al.*, 1986). Elongation of the substrate to the second C-terminal position (P2$'$) enhances the hydrolysis rate, as for the related enzymes subtilisin BPN$'$ (Chapter 94) and thermitase (Chapter 95). In the S1$'$ subsite, the specificity of these proteinases is very similar, with small residues like Ala and Gly favored, and peptide linkages at bulky residues like Phe and Leu being cleaved more slowly (Brömme *et al.*, 1986). Besides peptide bonds, proteinase K

also hydrolyzes esters (Borhan *et al.*, 1996). It can be used in nonaqueous solvents containing some water to synthesize peptides (Cerovsky & Martinek, 1989).

Several inhibitors of proteinase K have been isolated from wheat germ (Jany & Lederer, 1985; Pal *et al.*, 1994).

## Structural Chemistry

The crystal structure of proteinase K at 1.5 Å resolution revealed its structural similarity to subtilisin and thermitase, although the sequence identity between these enzymes is only about 35% (Pähler *et al.*, 1984; Betzel *et al.*, 1988b). These studies and the crystal structures of complexes of proteinase K with Z-Ala-Ala-CH$_2$Cl (Betzel *et al.*, 1986), with MeOSuc-Ala-Ala-Pro-Ala-CH$_2$Cl (Wolf *et al.*, 1991) and Z-Ala-Phe-CH$_2$Cl (Betzel *et al.*, 1988a) have illustrated the geometries of the catalytic site and the substrate recognition sites. The 'chloromethyl-ketone' groups attack the catalytic residues His69 and Ser224 to form two covalent links, one the His69 N$^\varepsilon$ methylene group, and the other the Ser224 O$^\gamma$ ketone carbon atom. The ketonic oxygen carries a negative charge and is hydrogen bonded to Asn161 N$^\delta$ and Ser224 NH forming the 'oxyanion hole'. The peptide parts of these inhibitors are in extended conformations and fill subsites S1–S5 of the substrate-recognition site. Their backbones hydrogen bond with strands 100–104 and 132–136 of proteinase K to form three-stranded antiparallel β-pleated sheets. The complex between proteinase K and inhibitor PKI3 from wheat germ shows at 2.5 Å resolution that the inhibitor interacts with the active site and causes distortion of the active Ser224 such that it is no longer part of the catalytic triad (Pal *et al.*, 1994).

Proteinase K binds two calcium ions, one tightly with a pK of $7.6 \times 10^{-8}$ M$^{-1}$, and the other so loosely that its dissociation constant could not be determined (Bajorath *et al.*, 1988). If the calcium ions are removed, a structural change is observed (Bajorath *et al.*, 1989; Müller *et al.*, 1994). Proteinase K contains five cysteines, of which four form two disulfide bonds and one, Cys72, is located close to His69 of the catalytic triad. This is where the inhibitor Hg$^{2+}$ binds, and inhibits by distorting the active-site geometry (Müller & Saenger, 1993). In the crystal structure of a complex between proteinase K, $Hg^{2+}$ and Ac-Pro-Ala-Pro-Phe-Pro-Ala-NH$_2$, the latter is cleaved into a tetramer, bound in the substrate-binding site, and a dimer bound in a site nearby and on the surface of the protein (Saxena *et al.*, 1996).

## Preparation

Proteinase K is secreted by *Tritirachium album* Limber if this mold is grown in the presence of keratin as the only source of nitrogen. It is isolated from the culture medium by precipitation with ammonium sulfate, DEAE-cellulose chromatography and crystallization (Ebeling *et al.*, 1974).

## Biological Aspects

The chromosomal gene of proteinase K isolated from *Tritirachium album* Limber as cDNA has been expressed in *Escherichia coli* (Gunkel & Gassen, 1989) as a preprotease with 384 amino acids; a fragment of 106 N-terminal amino acids is cleaved off to produce the mature enzyme. No mutational studies have been carried out on proteinase K. Since proteinase K remains active even at higher temperatures (<60°C), in the presence of urea, in 0.5% (w/v) SDS or 1% (w/v) Triton X, it is used in the degradation of proteins and in the preparation of DNA or RNA because no nuclease activity is retained (Sweeney & Walker, 1993; Goldenberger *et al.*, 1995).

## Distinguishing Features

Proteinase K contains five cysteines, in contrast to subtilisin (Chapter 94), which contains none. Proteinase K is more similar to thermitase (Chapter 95), which is also heat stable and contains one cysteine, Cys75, close to the active site, like Cys72 in proteinase K (Betzel *et al.*, 1990). Proteinase K resembles subtilisin in binding calcium, but in different positions on the molecule.

## Related Peptidases

The amino acid sequence and structure of proteinase K show its relationship to other members of the subtilisin family (Siezen *et al.*, 1991).

Fluorescence properties of native and photo-oxidized forms are described by Dolaska *et al.* (1992). If *Tritirachium album* Limber is not grown on keratin but on proteins like serum albumin, another proteinase, called proteinase R, is excreted (Samal *et al.*, 1990). This is heat stable and expressed as a preproteinase from which an N-terminal, 108 amino acids peptide is cleaved to produce mature enzyme with 279 amino acids. It is 87% homologous with proteinase K both on the nucleotide and on the amino acid level, contains five cysteines in identical positions, and probably has an identical three-dimensional structure.

## References

Bagger, S., Breddam, K. & Byberg, B.R. (1991) Binding of mercury (II) to protein thiol groups: a study of proteinase K and carboxypeptidase Y. *J. Inorg. Biochem.* **42**, 97–103.

Bajorath, J., Saenger, W. & Pal, G.P. (1988) Autolysis and inhibition of proteinase K, a subtilisin-related serine proteinase isolated from the fungus *Tritirachium album* Limber. *Biochim. Biophys.* **954**, 176–182.

Bajorath, J., Raghunathan, S., Hinrichs, W. & Saenger, W. (1989) Long-range structural changes in proteinase K triggered by calcium ion removal. *Nature* **337**, 481–484.

Betzel, C., Pal, G.P., Struck, M., Jany, K.-D. & Saenger, W. (1986) Active-site geometry of proteinase K. Crystallographic study of its complex with a dipeptide chloromethyl ketone inhibitor. *FEBS Lett.* **197**, 105–110.

Betzel, C., Bellemann, M., Pal, G.P., Bajorath, J., Saenger, W. & Wilson, K.S. (1988a) X-Ray and model-building studies on the specificity of the active site of proteinase K. *Proteins* **4**, 157–164.

Betzel, C., Pal, G.P. & Saenger, W. (1988b) Three-dimensional structure of proteinase K at 0.15 nm resolution. *Eur. J. Biochem.* **178**, 155–171.

Betzel, C., Teplyakov, A.V., Harutyunyan, E.H., Saenger, W. & Wilson (1990). Thermitase and proteinase K: a comparison of the refined three-dimensional structures of the native enzymes. *Protein Eng.* **3**, 161–172.

Borhan, B., Hammock, B., Seifert, J. & Wilson, B.W. (1996) Methyl and phenyl esters and thioesters of carboxylic acids as surrogate substrates for microassays of proteinase K esterase activity. *Fresenius' J. Anal. Chem.* **354**, 490–492.

Brömme, D., Peters, K., Fink, S. & Fittkau, S. (1986) Enzyme–substrate interactions in the hydrolysis of peptide substrates by thermitase, subtilisin BPN′ and proteinase K. *Arch. Biochem. Biophys.* **244**, 439–446.

Cerovsky, V. & Martinek, K. (1989) Amino acids and peptides. Part CCXII. Peptide synthesis catalyzed by native proteinase K in water-miscible organic solvents with low water content. *Collect. Czech. Chem. Commun.* **54**, 2027–2041.

Dolaska, P., Dimov, I., Genov, N., Svendsen, I., Wilson, K.S. & Betzel, C. (1992) Fluorescence properties of native and photo-oxidized proteinase K: the X-ray model in the region of the two tryptophans. *Biochim. Biophys. Acta* **1118**, 303–312.

Ebeling, W., Hennrich, N., Klockow, M., Metz, H., Orth, H.D. & Lang, H. (1974) Proteinase K from *Tritirachium album* Limber. *Eur. J. Biochem.* **47**, 91–97.

Goldenberger, D., Perschil, I., Ritzler, M. & Altwegg, M. (1995) A simple universal DNA extraction procedure using SDS and proteinase K is compatible with direct PCR amplification. *PCR Methods Appl.* **4**, 368–370.

Gunkel, E.A. & Gassen, H.-G. (1989) Proteinase K from *Tritirachium album* Limber. Characterization of the chromosomal gene and expression of the cDNA in *Escherichia coli. Eur. J. Biochem.* **179**, 185–194.

Jany, K.-D. & Lederer, G. (1985) Proteinase K – protein inhibition from wheat. *Biol. Chem. Hoppe-Seyler* **366**, 807–808.

Jany, K.-D., Lederer, G. & Mayer, B. (1986) Amino acid sequence of proteinase K from the mold *Tritirachium album* Limber. Proteinase K – a subtilisin-related enzyme with disulfide bonds. *FEBS Lett.* **199**, 139–144.

Kraus, E., Kiltz, H.H. & Femfert, U.F. (1976) The specificity of proteinase K against oxidized insulin B chain. *Z. Physiol. Chem.* **357**, 233–237.

Morihara, K. & Tsuzuki, H. (1975) Specificity of proteinase K from *Tritirachium album* Limber for synthetic peptides. *Agric. Biol. Chem.* **39**, 1489–1492.

Müller, A. & Saenger, W. (1993) Studies on the inhibitory action of mercury upon proteinase K. *J. Biol. Chem.* **268**, 26150–26154.

Müller, A., Hinrichs, W., Wolf, W.M. & Saenger, W. (1994) The structure of calcium-free proteinase K at 1.5 Å resolution. *J. Biol. Chem.* **269**, 23108–23115.

Pähler, A., Banerjee, A., Dattagupta, J.K., Fujiwara, T., Lindner, K., Pal, G.P., Suck, D., Weber, G. & Saenger W. (1984) Three-dimensional structure of fungal proteinase K reveals similarity to bacterial subtilisin. *EMBO J.* **6**, 1311–1314.

Pal, G.P., Kavounis, C.A., Jany, K.D. & Tsernoglou, D. (1994) The three-dimensional structure of the complex of proteinase K with its naturally occurring protein inhibitor, PKI3. *FEBS Lett.* **341**, 167–170.

Samal, B.B., Karan, B., Boone, T.C., Osslund, T.D., Chen, K.K. & Stabinsky, Y. (1990) Isolation and characterization of the gene encoding a novel, thermostable serine proteinase from the mould *Tritirachium album* Limber. *Mol. Microbiol.* **4**, 1789–1792.

Saxena, A.K., Singh, T.P., Peters, K., Fittkan, S., Bisanji, M., Wilson, K.S. & Betzel, Ch. (1996) Structure of a ternary complex of proteinase K, mercury and a substrate-analogue hexa-peptide at 2.2 Å resolution. *Proteins: Struct. Funct. Genet.* **25**, 195–201.

Siezen, R.J., de Vos, W.M., Leunissen, J.A.M. & Dijkstra, B.W. (1991) Homology modelling and protein engineering strategy of subtilases, the family of subtilisin-like serine proteinases. *Protein Eng.* **4**, 719–737.

Sweeney, P.J. & Walker, J.M. (1993) Proteinase K (EC 3.4.21.14). *Methods Mol. Biol.* **16**, 305–311.

Wolf, W.M., Bajorath, J., Müller, A., Raghunathan, S., Singh, T.P., Hinrichs, W. & Saenger, W. (1991) Inhibition of proteinase K by methoxysuccinyl-Ala-Ala-Pro-Ala-chloromethyl ketone. *J. Biol. Chem.* **266**, 17695–17699.

*W. Saenger*
*Institut für Kristallographie,*
*Freie Universität Berlin,*
*Takustrasse 6, D-14195 Berlin, Germany*
*Email: saenger@chemie.fu-berlin.de*

# 107. *Cuticle-degrading endopeptidase (Metarhizium)*

## Databanks

*Peptidase classification: clan SB, family S8, MEROPS ID: S08.056*
*NC-IUBMB enzyme classification: none*
*Databank codes:*

| Species | SwissProt | PIR | EMBL (cDNA) | EMBL (genomic) |
| --- | --- | --- | --- | --- |
| Arthrobotrys oligospora | – | – | – | X94121: complete gene |
| Beauveria bassiana | – | – | U16305 | – |
| Fusarium sp. | – | – | – | S71812: complete gene |
| Fusarium sp. | – | JC2142 | – | – |
| Metarhizium anisopliae | P29138 | S22387 | M73795 | – |
| | | S45442 | | |
| Verticillium chlamydosporium | P80406 | – | – | – |

## Name and History

The search for factors that are required for insect infection led to the discovery of a serine endopeptidase that facilitates degradation of the insect cuticle by fungal pathogens that directly penetrate the host surface. The enzyme was initially called a **chymoelastase** because it possessed a chymotrypsin-like primary specificity for hydrophobic residues and also degraded elastin (St. Leger *et al.*, 1987a). It was also referred to as **Pr1** to distinguish it from the trypsin-like *Pr2* enzymes produced by most entomopathogens (St. Leger *et al.*, 1987a,b). *Metarhizium anisopliae* Pr1 occurs as multiple isoenzymes identified on zymograms and by protein sequencing and cDNA cloning. These were referred to as **Pr1a, Pr1b** and **Pr1c** (St. Leger *et al.*, 1994; Joshi *et al.*, 1996). These enzymes are members of the subtilisin family (S8) and have been designated subtilisin-like Pr1 proteinases (St. Leger *et al.*, 1994).

## Activity and Specificity

Pr1 possesses a broad primary specificity for amino acids with a hydrophobic side chain (e.g. Phe, Met or Leu) but also possesses a secondary specificity for extended hydrophobic peptide chains, with the active site recognizing at least five subsite residues. This broad specificity accounts for its being a good general protease with activity against a range of proteins (casein, elastin, serum albumin, collagen) and insect cuticles (St. Leger *et al.*, 1987a). The positive charge (pI 8.5–10.2) of the Pr1 isoforms (St. Leger *et al.*, 1994) facilitates adsorption to negatively charged cuticle groups (carboxyl–hydroxyl complexes), which is a prerequisite for activity (Bidochka & Khachatourians, 1990; St. Leger *et al.*, 1986a). Consequently, variations in enzyme–substrate binding determine proteolytic efficiency against different cuticles (St. Leger *et al.*, 1991a). Only after adsorption does the active site come into contact with susceptible peptide bonds. Solubilized peptides are further degraded until a chain length of about five amino acids is obtained (St. Leger *et al.*, 1986b).

Assays are most conveniently made with the chromogenic substrate Suc-Ala-Ala-Pro-Phe$+$NHPhNO$_2$. This and other assays are described by St. Leger *et al.* (1994). pH optimum is about 8 (St. Leger *et al.*, 1987a). PMSF and DFP are inhibitory (St. Leger *et al.*, 1987a), and potent inhibition is shown by Suc-Ala-Ala-Pro-Phe-CH$_2$Cl (St. Leger *et al.*, 1994).

## Structural Chemistry

Pr1a is synthesized as a large precursor (40.3 kDa) containing an 18 amino acid signal peptide, an 89 amino acid propeptide, and the mature protein (28.6 kDa, pI 10.2) containing 281 amino acids. The active site contains an Asp, His, Ser catalytic triad, as expected for family S8, and the protein is probably stabilized by two disulfide bonds (St. Leger *et al.*, 1992).

## Preparation

A protocol has been published for the purification of pathogen proteases from culture filtrates by isoelectric focusing and affinity chromatography (St. Leger *et al.*, 1987a). Pr1 has been expressed in an insect cell line and purified from the media (Huang *et al.*, 1996).

## Biological Aspects

Pr1 production is under dual control of a general catabolite repression/derepression mechanism and a cuticle-specific induction mechanism (St. Leger *et al.*, 1988a; Paterson *et al.*, 1994). Studies on the expression of the Pr1 gene demonstrated that the 40.3 kDa primary translation product is produced within 45 min of nutrient deprivation and has a half-life of less than 5 min. The transit (processing) time between production of the translation product and the appearance of active Pr1 is only 7 min, demonstrating a very rapid export (secretion) of Pr1 to the external milieu (St. Leger *et al.*, 1991b).

Pr1 appears to be a pathogenicity determinant in *M. anisopliae* on the grounds of its ability to degrade cuticle, its high activity at the site of cuticle penetration (St. Leger *et al.*, 1989; Goettel *et al.*, 1989), and the results of experiments with specific inhibitors of Pr1 or polyclonal anti-Pr1 antibodies, which delayed mortality of host insects when applied to infected cuticles (St. Leger *et al.*, (1988b). The efficacy of *M. anisopliae* as a biological control agent is improved by transforming multiple copies of the Pr1 gene into the genome (St. Leger *et al.*, 1996).

## Distinguishing Features

Pr1a is one of the most basic members of the subtilisin family due to a group of positively charged residues on the surface of the protein. This functional domain facilitates binding to host cuticles and may account for Pr1a showing ~6-fold higher activity against cuticles than otherwise similar proteinases from nonpathogens (St. Leger *et al.*, 1992). Pr1b possesses several substitutions in the highly conserved sequences comprising the active sites of subtilisins. In particular the substitution of Thr220 by Ser is unique to Pr1b (Joshi *et al.*, 1996).

## Related Peptidases

Analogous cuticle-degrading peptidases with alkaline, neutral or acid pI values have been resolved in culture filtrates from *Beauveria bassiana*, *Verticillium lecanii* and *Aschersonia aleyrodis* (St. Leger *et al.*, 1987b). The cDNA for the *Beauveria* enzyme has been cloned (Joshi *et al.*, 1995).

## Further Reading

For a review, see St. Leger (1995).

## References

Bidochka, M.J. & Khachatourians, G.G. (1990) Basic proteases of entomopathogenic fungi differ in their absorption properties to insect cuticle. *J. Invert. Pathol.* **64**, 26–32.

Goettel, M.S., St. Leger, R.J., Rizzo, N.W., Staples, R.C. & Roberts, D.W. (1989) Ultrastructural localization of a cuticle-degrading protease produced by the entomopathogenic fungus *Metarhizium anisopliae* during penetration of host *(Manduca sexta)* cuticle. *J. Gen. Microbiol.* **135**, 2233–2239.

Huang, X.P., St. Leger, R.J., Davis, T.R., Joshi, L., Hughes, P.R. & Wood, H.A. (1996) Construction and characterization of a recombinant AcMNPV with a fungal gene encoding an insecticidal protease. In: *Abstracts Society of Invertebrate Pathology 29th Annual Meeting*, Cordoba, University of Cordoba, Spain, p. 38.

Joshi, L., St. Leger, R.J. & Bidochka, M.J. (1995) Cloning of a cuticle-degrading protease from the entomopathogenic fungus, *Beauveria bassiana. FEMS Microbiol. Lett.* **125**, 211–218.

Joshi, L., St. Leger, R.J. & Roberts, D.W. (1996) Isolation of a cDNA encoding a novel subtilisin-like protease (Pr1b) from the entomopathogenic fungus, *Metarhizium anisopliae* using differential display-RT-PCR. *Gene* **197**, 1–8.

Paterson, I.C., Charnley, A.K., Cooper, R.M. & Clarkson, J.M. (1994) Specific induction of a cuticle-degrading protease of the insect pathogenic fungus *Metarhizium anisopliae. Microbiology* **140**, 185–189.

St. Leger, R.J. (1995) The role of cuticle-degrading proteases in fungal pathogens of insects. *Can. J. Botany* (Special Edition: Proceedings of the Vth International Congress of Mycology) **73** (suppl. 1), S1119–S1125.

St. Leger, R.J., Charnley, A.K. & Cooper, R.M. (1986a) Cuticle-degrading enzymes of entomopathogenic fungi: mechanisms of interaction between pathogen enzymes and insect cuticle. *J. Invertebr. Pathol.* **47**, 295–302.

St. Leger, R.J., Cooper, R.M. & Charnley, A.K. (1986b) Cuticle-degrading enzymes of entomopathogenic fungi: cuticle degradation *in vitro* by enzymes from entomopathogens. *J. Invertebr. Pathol.* **47**, 167–177.

St. Leger, R.J., Charnley, A.K. & Cooper, R.M. (1987a) Characterization of cuticle-degrading proteases produced by the entomopathogen *Metarhizium anisopliae*. *Arch. Biochem. Biophys.* **253**, 221–232.

St. Leger, R.J., Cooper, R.M. & Charnley, A.K. (1987b) Distribution of chymoelastases and trypsin-like enzymes in five species of entomopathogenic deuteromycetes. *Arch. Biochem. Biophys.* **258**, 123–131.

St. Leger, R.J., Durrands, P.K., Cooper, R.M. & Charnley, A.K. (1988a) Regulation of production of proteolytic enzymes by the entomopathogenic fungus *Metarhizium anisopliae*. *Arch. Microbiol.* **150**, 413–416.

St. Leger, R.J., Durrands, P.K., Charnley, A.K. & Cooper, R.M. (1988b) Role of extracellular chymoelastase in the virulence of *Metarhizium anisopliae* for *Manduca sexta*. *J. Invertebr. Pathol.* **52**, 285–293.

St. Leger, R.J., Butt, T.M., Staples, R.C. & Roberts, D.W. (1989) Synthesis of proteins including a cuticle-degrading protease during differentiation of the entomopathogenic fungus *Metarhizium anisopliae*. *Exp. Mycol.* **13**, 253–262.

St. Leger, R.J., Charnley, A.K. & Cooper, R.M. (1991a) Kinetics of the digestion of insect cuticles by a protease (Pr1) from *Metarhizium anisopliae*. *J. Invertebr. Pathol.* **57**, 146–147.

St. Leger, R.J., Staples, R.C. & Roberts, D.W. (1991b) Changes in translatable mRNA species associated with nutrient deprivation and protease synthesis in *Metarhizium anisopliae*. *J. Gen. Microbiol.* **137**, 807–815.

St. Leger, R.J., Frank, D.C., Roberts, D.W. & Staples, R.C. (1992) Molecular cloning and regulatory analysis of the cuticle-degrading protease structural gene from the entomopathogenic fungus *Metarhizium anisopliae*. *Eur. J. Biochem.* **204**, 991–1001.

St. Leger, R.J., Bidochka, M.J. & Roberts, D.W. (1994) Isoforms of the cuticle-degrading Pr1 protease and production of a metalloproteinase by *Metarhizium anisopliae*. *Arch. Biochem. Biophys.* **313**, 1–7.

St. Leger, R.J., Joshi, L., Bidochka, M.J. & Roberts, D.W. (1996) Construction of an improved mycoinsecticide over-expressing a toxic protease. *Proc. Natl Acad. Sci. USA* **93**, 6349–6354.

*Raymond J. St. Leger*
*Boyce Thompson Institute at Cornell University,*
*Tower Road,*
*Ithaca, NY 14853, USA*
*Email: rs50@cornell.edu*

# 108. *Thermomycolin*

## Databanks

*Peptidase classification: clan SB, family S8, MEROPS ID: S08.057*
*NC-IUBMB enzyme classification: EC 3.4.21.65*
*Chemical Abstracts Service registry number: 52233-31-5*
*Databank codes:*

| Species | SwissProt | PIR | EMBL (cDNA) | EMBL (genomic) |
|---|---|---|---|---|
| *Malbranchea sulfurea* | P13858 | A24174 | – | – |

## Name and History

**Thermomycolin** was isolated as an extracellular alkaline endopeptidase that is produced by the thermophilic fungus *Malbranchea pulchella* var. *sulfurea* when grown in submerged cultures of 1–2% casein and salts (Ong & Gaucher, 1973). The proteinase was stabilized by $Ca^{2+}$, but stoichiometrically inhibited by DFP, thereby establishing it as a serine proteinase. The proteinase was originally named **thermomycolase.**

## Activity and Specificity

Proteolytic activity is most readily assayed by the spectrophotometric determination of the *p*-nitrophenol produced by the hydrolysis of Z-Gly-OPhNO$_2$ or Z-Ala-OPhNO$_2$ at pH 8. A titrimetric assay with Ac-Ala-Ala-Ala⊦OCH$_3$, and a colorimetric assay with Congo red-elastin as substrate have also been used. Proteinase activity exists between pH 4.5 and 10.5 with an optimum at 8.5. The pH stability is pronounced at pH 6.5–9.5 (30°C, 20 h) and is further enhanced by calcium ions (binding constant $5.0 \times 10^5 \, M^{-1}$). The half-life is 110 min at 73°C, pH 7.4 in 10 mM Ca$^{2+}$ (Voordouw & Roche, 1975; Gaucher & Stevenson, 1976). The specificity of the endopeptidase for small N-blocked *p*-nitrophenyl ester substrates is Ala > Tyr > Phe ≫ Gly ≫ Leu ≫ Val, but with the polypeptides glucagon and the oxidized A and B chains of insulin there is a broad specificity at 45°C and pH 7.0 with no marked selectivity for any particular P1 residue. Reducing the temperature to 28°C at pH 7.0, with only a 5 min digestion period, resulted in complete cleavage of a Gln-Trp bond in the sequence Phe-Val-Gln⊦Trp-Leu-Met of glucagon. Similarly selective cleavages within hydrophobic sequences were observed with the oxidized A and B chains of insulin during short digestion periods. There is an overall preference for four out of six amino acids surrounding the scissile bond to be nonpolar residues (Stevenson & Gaucher, 1975).

Thermomycolin is an elastase-like proteinase but with a much enhanced proteolysis rate compared to porcine elastase (Chapter 11). Moderate 'collagenase-like' activity was implied by the hydrolysis of the synthetic peptides Z-Gly-Pro-Leu⊦Gly-Pro and Z-Gly-Pro-Gly⊦Gly-Pro-Ala. The proteinase did not, however, liberate amino acids or peptides from keratin. Thermomycolin is strongly inhibited by mercuric, zinc and copper ions, as well as Z-Phe-CH$_2$Br and the standard serine peptidase inhibitors: DFP, PMSF, *p*-methylphenyl- and *p*-nitrophenyl-sulfonyl fluorides. Inhibitors of cysteine peptidases such as *p*-chloromercuribenzoate, iodoacetate and iodoacetamide were without effect. EDTA had no effect on the activity (as opposed to stability) of the enzyme, suggesting that thermomycolin is not a metalloproteinase.

## Structural Chemistry

$^{32}$P-labeled DFP inactivated thermomycolin stoichiometrically, and a serine residue was labeled in the active-site sequence -Leu-Ser-(Gly)-Thr-Ser*-Met-. Although the active-site sequence is typical for a member of the subtilisin family, thermomycolin possesses one disulfide bridge, which is exceptional. The N-terminal sequence of 28 residues (see the Databanks table) was reported by Gaucher & Stevenson (1976). Thermomycolin is a 325 residue, single-chain protein of 32–33 kDa, as determined by SDS-PAGE, sedimentation equilibrium and amino acid composition. The pI is 6.0, sedimentation coefficient ($s^0_{20,w}$) is $2.97 \pm 0.05$ S, diffusion coefficient ($D^0_{20,w}$) is $8.4 \times 10^{-7} \, cm^2 \, s^{-1}$, frictional ratio ($f/f_0$) is 1.09, the intrinsic viscosity is $3.05 \pm 0.05 \, ml \, g^{-1}$, and the partial specific volume is 0.736 (Voordouw *et al.*, 1974).

## Preparation

The thermophilic fungus *Malbranchea pulchella* var. *sulfurea* (Emerson no. 27) was cultured as described earlier (Cooney & Emerson, 1964; Ong & Gaucher, 1972, 1976; Stevenson & Gaucher, 1975). The culture medium was filtered, concentrated by vacuum evaporation and dialyzed. Color reduction and partial purification were achieved by a batchwise adsorption on DEAE-Sephadex. Thermomycolin was then purified by affinity chromatography on Sepharose-4-phenylbutylamine (Stevenson & Landman, 1971; Ong & Gaucher, 1976; Gaucher & Stevenson, 1976).

## Biological Aspects

A study of thermomycolin production media (Ong & Gaucher, 1972; Dunham, 1974) revealed that many sugars (1% ribose, glucose, galactose, sorbitol, sucrose) were repressive, whereas lactose was not. Except for methionine and tryptophan, most amino acids and dipeptides (0.3% concentration) repressed proteinase production; 0.3% NaNO$_2$, NaNO$_3$ and (NH$_4$)$_2$SO$_4$ were nonrepressive, whereas 3-indole pyruvate was stimulatory.

## Distinguishing Features

Thermomycolin is not as thermostable as the extracellular serine proteinases of thermophilic bacteria, but it is more stable than most fungal proteinases. This thermostability results from the conformational stability produced by the binding of a single calcium ion (Voordouw & Roche, 1975).

## References

Cooney, D.G. & Emerson, R. (1964) *Thermophilic Fungi*. San Francisco, CA: Freeman.

Dunham, M.J. (1974) A regulatory study of the extracellular thermostable alkaline serine protease from *Malbranchea pulchella* var. *sulfurea*. Masters thesis, University of Calgary.

Gaucher, G.M. & Stevenson, K.J. (1976) Thermomycolin. *Methods Enzymol.* **45**, 415–433.

Ong, P.S. & Gaucher, G.M. (1972) Production, purification and partial characterization of an extracellular protease from a thermophilic fungus. In: *Fermentation Technology Today, Proceedings of the IVth International Fermentation Symposium, Kyoto, Japan* (Terui, G., ed.). Osaka: Society of Fermentation Technology, pp. 271–278.

Ong, P.S. & Gaucher, G.M. (1973) Protease production by thermophilic fungi. *Can. J. Microbiol.* **19**, 129–133.

Ong, P.S. & Gaucher, G.M. (1976) Production, purification and characterization of thermomycolase, the extracellular serine protease of the thermophilic fungus *Malbranchea pulchella* var. *sulfurea. Can. J. Microbiol.* **22**, 165–176.

Stevenson, K.J. & Gaucher, G.M. (1975) The substrate specificity of thermomycolase, an extracellular serine proteinase from the thermophilic fungus *Malbranchea pulchella* var. *sulfurea. Biochem. J.* **151**, 527–542.

Stevenson, K.J. & Landman, A. (1971) The isolation of chymotrypsin-like enzymes by affinity chromatography using Sepharose-4-phenylbutylamine. *Can. J. Biochem.* **49**, 119–126.

Voordouw, G. & Roche, R.S. (1975) The role of bound calcium ions in thermostable, proteolytic enzymes. I. Studies on thermomycolase, the thermostable protease from the fungus *Malbranchea pulchella. Biochemistry* **14**, 4659–4666.

Voordouw, G., Gaucher, G.M. & Roche, R.S. (1974) Physiochemical

properties of thermomycolase, the thermostable, extracellular, serine protease of the fungus *Malbranchea pulchella. Can. J. Biochem.* **52**, 981–990.

*G.M. Gaucher*
*Department of Biological Sciences,*
*Division of Biochemistry,*
*The University of Calgary,*
*Calgary T2N 1NH, Alberta, Canada*
*Email: gaucher@acs.ucalgary.ca*

*Kenneth J. Stevenson*
*Department of Biological Sciences,*
*Division of Biochemistry,*
*The University of Calgary,*
*Calgary T2N 1NH, Alberta, Canada*

# *109. Lantibiotic leader peptidases*

## Databanks

*Peptidase classification: clan SB, family S8, MEROPS ID: S08.059, S08.060*
*NC-IUBMB enzyme classification: none*
*Databank codes:*

| Species | Genes | SwissProt | PIR | EMBL (cDNA) | EMBL (genomic) |
|---|---|---|---|---|---|
| *Enterococcus faecalis* | *cylA,cylB* | – | – | – | L37110: plasmid pAD1 cytotoxin genes |
| *Lactococcus lactis* | *nisP* | Q07596 | – | L11061 | – |
|  |  |  |  | Z22725 | – |
| *Lactococcus sake* | *lasP* | – | – | – | Z54312 |
| *Staphylococcus epidermidis* | *epiP* | P30199 | S23420 | X62386 | – |
| *Staphylococcus epidermidis* | *elkP,elkT* | – | – | – | X87412: *elk* operon |
| *Staphylococcus epidermidis* | *pepP* | – | – | Z49865 | – |

## Name and History

As early as 1969, a factor called **component A** was found to be necessary for activation of the *Enterococcus faecalis* cytolysin, now identified as a lantibiotic (Granato & Jackson, 1969; Sahl *et al*., 1995). Sequencing of the structural gene showed that component A is a serine protease of the subtilisin family which was renamed *CylA* (Segarra *et al*., 1991). Other lantibiotic leader peptidases were discovered during the sequencing of the respective lantibiotic biosynthetic gene clusters: *EpiP* of the epidermin biosynthetic gene cluster in *Staphylococcus epidermidis* Th 3298 (Schnell *et al*., 1992), *NisP* in nisin-producing strains of *Lactococcus lactis* (e.g. van der Meer *et al*., 1993), and *PepP* of the Pep5 gene cluster in *Staphylococcus epidermidis* 5 (Meyer *et al*., 1995). For *ElkP* of the epilancin K7 gene cluster, only the N-terminal sequence has been published (van de Kamp *et al*., 1995). The sequence of *LasP* which is involved in activation of lactocin S of *Lactococcus sake* can be found in the EMBL database (Z54312). It has been proposed that the term *CylP* be used in place of CylA, consistent with *LanP*, as a collective designation for lantibiotic leader peptidases (Sahl *et al*., 1995).

## Activity and Specificity

LanP proteins can be divided into two groups: in the cleavage sites of NisP (Ala-Ser-Pro-Arg↓Ile-Dhb), EpiP (Ala-Glu-Pro-Arg↓Ile-Ala), PepP (Leu-Glu-Pro-Gln↓Dhb-Ala) and ElkP (Leu-Ser-Pro-Gln↓Dha-Ala) (Dha: dehydroalanine; Dhb: dehydrobutyrine) P4 is a hydrophobic amino acid (Ala or Leu), P3 is a negatively charged or polar residue (Glu or Ser), P2 is always Pro, and P1 is a positively charged or polar amino acid (Arg or Gln). In contrast, the cleavage sites of CylP (Val-Gln-Ala-Glu↓Dhb-Dhb) and LasP (Met-Asn-Ala-Asp↓Dha-Dhb) contain a polar residue in position P3, a hydrophobic residue in position P2 and a negative charge in position P1. The residues in positions P1' and P2' are hydrophobic in both groups. Cleavage by EpiP is blocked if the Arg residue in P1 is exchanged for Gln (Geissler *et al*., 1996). Similarly, there is no activity of NisP with Gln or Asp in P1. In contrast, the presence of Val or Gly in P2 or exchanges in P1' (Trp) or P2' (Ser) did not inhibit NisP (van der Meer *et al*., 1994; Siezen *et al*., 1995).

EpiP is inhibited by DCI (1 mM) (Geissler *et al*., 1996).

## Structural Chemistry

EpiP is a 40.3 kDa protein (Geissler *et al*., 1996) and NisP is a 54 kDa protein, as expressed in *Escherichia coli* (van der Meer *et al*., 1993). The three-dimensional structure of the catalytic site of NisP has been modeled in homology to subtilisin BPN′ and thermitase (Siezen *et al*., 1995) (see Chapters 94 and 95).

## Preparation

The LanP proteins are expressed by several gram-positive bacteria during biosynthesis of the corresponding lantibiotics. EpiP was partially purified after overexpression in *Staphylococcus carnosus* using a HiTrap-Q column (Geissler *et al*., 1996). NisP was expressed in *E. coli* (van der Meer *et al*., 1993).

## Biological Aspects

The LanP proteins differ with respect to their location in the producer cell. PepP, LasP and ElkP do not possess a prepro sequence and are intracellular enzymes (Meyer *et al*., 1995). EpiP is synthesized as a preproenzyme with a 25 amino acid signal peptide and a 77 amino acid pro sequence (Geissler *et al*., 1996). NisP contains an N-terminal prepro sequence of 195 amino acids and a C-terminal spacer region of about 80 amino acids that is followed by a 30 amino acid membrane anchor sequence motif. This suggests that NisP is anchored to the cell membrane of *Lactococcus lactis* (van der Meer *et al*., 1993).

## References

Geissler, S., Götz, F. & Kupke, T. (1996) Serine protease EpiP from *Staphylococcus epidermidis* catalyzes the processing of the epidermin precursor peptide. *J. Bacteriol.* **178**, 284–288.

Granato, P.A. & Jackson, R.W. (1969) Bicomponent nature of lysin from *Streptococcus zymogenes*. *J. Bacteriol.* **100**, 856–868.

Meyer, C., Bierbaum, G., Heidrich, C., Reis, M., Shling, J., Iglesias-Wind, M.I., Kempter, C., Molitor, E. & Sahl, H.-G. (1995) Nucleotide sequence of the lantibiotic Pep5 biosynthetic gene cluster and functional analysis of PepP and PepC. Evidence for a role of PepC in thioether formation. *Eur. J. Biochem.* **232**, 478–489.

Sahl, H.-G., Jack, R.W. & Bierbaum, G. (1995) Biosynthesis and biological activities of lantibiotics with unique post-translational modifications. *Eur. J. Biochem.* **230**, 827–853.

Segarra, R.A., Booth, M.C., Morales, D.A., Huycke, M.M. & Gilmore, M.S. (1991) Molecular characterization of the *Enterococcus faecalis* cytolysin activator. *Infect. Immun.* **59**, 1239–1246.

Siezen, R.J., Rollema, H.S., Kuipers, O.P. & de Vos, W.M. (1995) Homology modelling of the *Lactococcus lactis* leader peptidase NisP and its interaction with the precursor of the lantibiotic nisin. *Protein Eng.* **8**, 117–125.

Schnell, N., Engelke, G., Augustin, J., Rosenstein, R., Ungermann, V., Götz, F. & Entian, K.-D. (1992) Analysis of genes involved in biosynthesis of lantibiotic epidermin. *Eur. J. Biochem.* **204**, 57–68.

van der Meer, J.R., Polman, J., Beerthuyzen, M.M., Siezen, R.J., Kuipers, O.P. & de Vos, W.M. (1993) Characterization of the *Lactococcus lactis* nisin A operon genes *nisP*, encoding a subtilisin-like serine protease involved in precursor processing, and *nisR*, encoding a regulatory protein involved in nisin biosynthesis. *J. Bacteriol.* **175**, 2578–2588.

van der Meer, J.R., Rollema, H.S., Siezen, R.J., Beerthuyzen, M.M., Kuipers, O.P. & de Vos, W.M. (1994) Influence of amino acid substitutions in the nisin leader peptide on biosynthesis and secretion of nisin by *Lactococcus lactis*. *J. Biol. Chem.* **269**, 3555–3562.

van de Kamp, M., van den Hooven, H.W., Konings, R.N.H., Hilbers, C.W., van de Ven, F.J.M., Bierbaum, G., Sahl, H.-G., Kuipers, O.P., Siezen, R.J. & de Vos, W.M. (1995) Elucidation of the primary structure of the lantibiotic epilancin K7: cloning and characterisation of the epilancin-K7-encoding gene and NMR analysis of mature epilancin K7. *Eur. J. Biochem.* **230**, 587–600.

*Gabriele Bierbaum*
*Institut für Medizinische*
*Mikrobiologie der Universität Bonn,*
*Sigmund-Freud-Strasse 25,*
*D-53105 Bonn, Germany*
*Email: gabi@mibi03.meb.uni-bonn.de*

*Ralph W. Jack*
*Institut für Medizinische*
*Mikrobiologie der Universität Bonn,*
*Sigmund-Freud-Strasse 25,*
*D-53105 Bonn, Germany*

*Hans-Georg Sahl*
*Institut für Medizinische*
*Mikrobiologie der Universität Bonn,*
*Sigmund-Freud-Strasse 25,*
*D-53105 Bonn, Germany*

# 110. Cucumisin

## Databanks

*Peptidase classification: clan SB, family S8, MEROPS ID: S08.092*
*NC-IUBMB enzyme classification: EC 3.4.21.25*
*Chemical Abstracts Service registry number: 82062-89-3*

*Databank codes:*

| Species | SwissProt | PIR | EMBL (cDNA) | EMBL (genomic) |
|---|---|---|---|---|
| *Alnus glutinosa* | – | – | X85975 | – |
| *Cucumis melo* | – | – | D32206 | – |
| *Lilium longifolium* | – | – | D21815 | – |
| *Lycopersicon esculentum* | – | – | – | U80583: complete gene |
| *Lycopersicon esculentum* | – | – | X95270 | – |

## Name and History

**Cucumisin** is a serine protease (Kaneda *et al*., 1984) isolated from the sarcocarp of melon fruit, *Cucumis melo* L. var. *Prince* (Kaneda & Tominaga, 1975; Yamagata *et al*., 1989; Uchikoba *et al*., 1995). Cucumisin-like proteases have now been purified from other cucurbitaceous plants, including white gourd (Kaneda & Tominaga, 1977), snake gourd (Kaneda *et al*., 1986a), fig-leaf gourd (Curotto *et al*., 1989; Dryjanski *et al*., 1990), yellow snake gourd (Uchikoba *et al*., 1990), honeydew melon (Uchikoba *et al*., 1993), and *Trichosanthes bracteata* (Kaneda & Uchikoba, 1994).

## Activity and Specificity

Good synthetic substrates are Glt-Ala-Ala-Pro-Leu┼NHPhNO$_2$ ($K_m$ 1.25 mM, $k_{cat}$ 5.9 s$^{-1}$) and Suc-Ala-Ala-Pro-Phe┼NHPhNO$_2$ ($K_m$ 1.00 mM, $k_{cat}$ 4.17 s$^{-1}$). The catalytic constant ($k_{cat}/K_m$) for Suc-Ala-Pro-Ala-NHPhNO$_2$ is 30-fold greater than that for Suc-Ala-Ala-Ala-NHPhNO$_2$ (Uchikoba *et al*., 1995). Assays can also be made with the peptide thiobenzyl esters (Kaneda *et al*., 1986b). The molarity of active enzyme is determined by titration with Ac-Ala-Ala-azaalanine┼OPhNO$_2$ (Kaneda *et al*., 1986c). The substrate specificity for oligopeptides and proteins is broad (Kaneda & Tominaga, 1975; Kaneda *et al*., 1995; Uchikoba *et al*., 1995). Inhibitors of chymotrypsin-family serine proteases such as soybean trypsin inhibitor, ovomucoid and aprotinin have no effect. $\alpha_2$-Macroglobulin shows 38% inhibition of the caseinolytic activity (Uchikoba *et al*., 1995). The enzyme is stable over a wide pH range (pH 4–11). The pH optimum is at alkaline pH with casein as substrate. The optimum temperature is 70°C at pH 7.1. Full caseinolytic activity is preserved in 8 M urea at pH 9.1, 50°C (Kaneda *et al*., 1995). Cucumisin immobilized on Sepharose is more stable against alkaline inactivation or heat than the soluble enzyme (Kaneda & Tominaga, 1987; Kaneda *et al*., 1988; Uchikoba & Kaneda, 1996).

## Structural Chemistry

From the chymotryptic digest of the DFP derivative of the enzyme, the octapeptide containing phosphorous was isolated, and its amino acid sequence was shown to be Asn-Ile-Ile-Ser-Gly-Thr-Ser(P)-Met. The four residues Gly-Thr-Ser(P)-Met are identical with those of subtilisin (Chapter 94) (Kanada *et al*., 1984).

Cucumisin modified with $^3$H-labeled Z-Ala-Ala-Pro-Phe-CH$_2$Cl was reduced and pyridylethylated, and then digested with trypsin. The radioactive peptide obtained contained 32 amino acid residues, and the sequence around the labeled residues was -Asp-Thr-Asn-Gly-His*-Gly-Thr-His-Thr-Ala-. This sequence is similar to that around the active-site His of the subtilisin family (Yonezawa *et al*., 1995). The sites of carbohydrate attachment are -Asn(sugar)-Ala-Ser- and -Asn(sugar)-Arg-Thr- (Kaneda *et al*., 1986d). The molecular mass of the enzyme was initially estimated to be about 50 kDa, and later corrected to 54 kDa (Yamagata *et al*., 1989). It was then shown that the 54 kDa enzyme is formed by limited autolysis of a 67 kDa form. When the enzyme was quickly purified from ripe melons, it was almost completely in the 67 kDa form, the 54 kDa form being negligible (Uchikoba *et al*., 1995). The substrate specificity and enzymatic stability of the 67 kDa enzyme are similar to those of the 54 kDa one.

The complete nucleotide sequence of a cucumisin cDNA (2552 nucleotides) has been determined and the corresponding amino acid sequence deduced (Yamagata *et al*., 1994). The open reading frame of the cDNA consists of 731 codons and encodes a large precursor (predicted molecular mass 78 815), consisting of four functional domains: a possible signal peptide (22 amino acid residues), an N-terminal pro sequence (88 residues), a 54 kDa protease domain (505 residues) that is the active enzyme domain of the 67 kDa cucumisin, and a 14 kDa C-terminal polypeptide (116 residues), which is cleaved off in the limited autolysis of the 67 kDa cucumisin. The protease domain of the cucumisin precursor contains highly conserved sequences around the catalytic triad amino acids Asp, His and Ser, and the substrate-binding site, characteristic of a peptidase of the subtilisin family.

## Preparation

In the melon fruit, the enzyme is localized to mesocarp and endocarp. Most of the proteins present in the pressed juice of melon are not bound to a cation exchanger, so cucumisin can be purified simply by CM-Sepharose column chromatography (Kaneda *et al*., 1995; Uchikoba *et al*., 1995).

## References

Curotto, E., Gonzalez, G., O'Reilly, S. & Tapia, G. (1989) Isolation and partial characterization of a protease from *Cucurbita ficifolia*. *FEBS* **243**, 363–365.

Dryjanski, M., Otlewski, J., Polanowski, A. & Wilusz, T. (1990) Serine proteinase from *Cucurbita ficifolia* seed; purification, properties, substrate specificity and action on native squash trypsin inhibitor (CMTII). *Biol. Chem. Hoppe-Seyler* **371**, 889–895.

Kaneda, M. & Tominaga, N. (1975) Isolation and characterization of a proteinase from the sarcocarp of melon fruit. *J. Biochem.* **78**, 1287–1296.

Kaneda, M. & Tominaga, N. (1977) Isolation and characterization of a proteinase from white gourd. *Phytochemistry* **16**, 345–346.

Kaneda, M. & Tominaga, N. (1987) Properties of a new plant serine protease cucumisin. *Agric. Biol. Chem.* **51**, 489–492.

Kaneda, M. & Uchikoba, T. (1994) Protease from the sarcocarp of *Trichosanthes bracteata*. *Phytochemistry* **35**, 583–586.

Kaneda, M., Ohmine, H., Yonezawa, H. & Tominaga, N. (1984) Amino acid around the reactive serine of cucumisin from melon fruit. *J. Biochem.* **95**, 825–829.

Kaneda, M., Sobue, A., Eida, S. & Tominaga, N. (1986a) Isolation and characterization of proteinase from the sarcocarp of snake-gourd fruit. *J. Biochem.* **99**, 569–577.

Kaneda, M., Minematsu, Y., Powers, J.C. & Tominaga, N. (1986b) Specificity of cucumisin in hydrolysis of peptide thiobenzyl ester. *Agric. Biol. Chem.* **50**, 1075–1076.

Kaneda, M., Minematsu, Y., Powers, J.C. & Tominaga, N. (1986c) Active site titration of the serine protease cucumisin from *Cucumis melo*. *Phytochemistry* **25**, 2407–2408.

Kaneda, M., Kamikubo, Y. & Tominaga, N. (1986d) Amino acid sequences of glycopeptides from cucumisin. *Agric. Biol. Chem.* **50**, 2413–2414.

Kaneda, M., Nisimura, S. & Tominaga, N. (1988) Caseinolytic activity of immobilized cucumisin. *J. Dairy Sci.* **71**, 1132–1134.

Kaneda, M., Yonezawa, H. & Uchikoba, T. (1995) Improved isolation, stability and substrate specificity of cucumisin, a plant serine endopeptidase. *Biotechnol. Appl. Biochem.* **22**, 215–222.

Uchikoba, T. & Kaneda, M. (1996) Milk-clotting activity of cucumisin, a plant serine protease from melon fruit. *Appl. Biochem. Biotechnol.* **56**, 325–330.

Uchikoba, T., Horita, H. & Kaneda, M. (1990) Proteases from the sarcocarp of yellow snake-gourd. *Phytochemistry* **29**, 1879–1881.

Uchikoba, T., Niidome, T., Sata, I. & Kaneda, M. (1993) Protease D from the sarcocarp of honeydew melon fruit. *Phytochemistry* **33**, 1005–1008.

Uchikoba, T., Yonezawa, H. & Kaneda, M. (1995) Cleavage specificity of cucumisin, a plant serine protease. *J. Biochem.* **117**, 1126–1130.

Yamagata, H., Ueno, S. & Iwasaki, T. (1989) Isolation and characterization of a possible native cucumisin from developing melon fruits and its limited autolysis to cucumisin. *Agric. Biol. Chem.* **53**, 1009–1017.

Yamagata, H., Masuzawa, T., Nagaoka, Y., Ohnishi, T. & Iwasaki, T. (1994) Cucumisin, a serine protease from melon fruit, shares structural homology with subtilisin and is generated from a large precursor. *J. Biol. Chem.* **269**, 32725–32731.

Yonezawa, H., Uchikoba, T. & Kaneda, M. (1995) Identification of the reactive histidine of cucumisin, a plant serine protease: modification with peptidyl chloromethyl ketone derivative of peptide substrate. *J. Biochem.* **118**, 917–920.

*Makoto Kaneda*
*Department of Chemistry,*
*Faculty of Science,*
*Kagoshima University,*
*Kagoshima 890, Japan*

# *111. Leucyl endopeptidase*

## Databanks

*Peptidase classification: clan SX, family S99, MEROPS ID: S9G.031*
*NC-IUBMB enzyme classification: EC 3.4.21.57*
*Databank codes: no sequence data available*

## Name and History

In the course of solubilization and purification of fusicoccin-binding sites present in microsomal fractions of spinach (*Spinacia oleracea*) leaves, some endogenous hydrolases responsible for the poor stability of the receptors were identified (Aducci *et al.*, 1982, 1984). Among them there was a serine proteinase displaying leucine-specific proteolytic activity. To reflect its primary specificity, the enzyme has been termed **leucyl endopeptidase**; other names are **plant Leu-proteinase** and **leucine-specific serine proteinase**

(Aducci *et al.*, 1986a,b). In the absence of sequence data, the enzyme cannot be assigned to a family, but it is likely to be in family S8 (p. 536).

## Activity and Specificity

Leucyl endopeptidase catalyzes the hydrolysis of proteins and small molecule substrates, and in the small molecule substrates, the cleavage preference is for leucyl bonds (Aducci *et al.*, 1986a,b).

*Table 111.1* Values of catalytic parameters for the leucyl endopeptidase-catalyzed hydrolysis of chromogenic substrates at 23°C

| Substrate | $K_m$ (μM) | $k_{cat}$ (s$^{-1}$) | $K_s$ (μM) | $k_{+2}$ (s$^{-1}$) | $k_{+3}$ (s$^{-1}$) |
|---|---|---|---|---|---|
| Bz-Gly-NHPhNO$_2$[a] | $8.9 \times 10^2$ | 1.2 | | | |
| Bz-Ala-NHPhNO$_2$[a] | $2.5 \times 10^2$ | 4.9 | | | |
| Bz-Val-NHPhNO$_2$[a] | $1.8 \times 10^2$ | 6.1 | | | |
| Bz-Leu-NHPhNO$_2$[a] | $5.0 \times 10^1$ | $2.1 \times 10^1$ | | | |
| Bz-Arg-NHPhNO$_2$[a] | $4.7 \times 10^2$ | 2.1 | | | |
| Bz-Lys-NHPhNO$_2$[a] | $4.5 \times 10^2$ | 1.8 | | | |
| Z-Gly-OPhNO$_2$[b] | $5.2 \times 10^2$ | $2.1 \times 10^1$ | $6.0 \times 10^2$ | $2.4 \times 10^1$ | $1.6 \times 10^2$ |
| Z-Ala-OPhNO$_2$[b] | $1.2 \times 10^2$ | $9.4 \times 10^1$ | $1.4 \times 10^2$ | $1.1 \times 10^2$ | $6.5 \times 10^2$ |
| Z-Val-OPhNO$_2$[b] | $6.6 \times 10^1$ | $1.4 \times 10^2$ | $7.6 \times 10^1$ | $1.6 \times 10^2$ | $1.1 \times 10^3$ |
| Z-Ile-OPhNO$_2$[b] | $6.5 \times 10^1$ | $1.4 \times 10^2$ | $7.4 \times 10^1$ | $1.6 \times 10^2$ | $1.1 \times 10^3$ |
| Z-Leu-OPhNO$_2$[b] | $1.8 \times 10^1$ | $4.3 \times 10^2$ | $2.1 \times 10^1$ | $5.0 \times 10^2$ | $3.0 \times 10^3$ |
| Z-Arg-OPhNO$_2$[b] | $2.8 \times 10^2$ | $4.3 \times 10^1$ | $3.2 \times 10^2$ | $5.0 \times 10^1$ | $3.3 \times 10^2$ |
| Z-Lys-OPhNO$_2$[b] | $2.6 \times 10^2$ | $4.5 \times 10^1$ | $3.0 \times 10^2$ | $5.1 \times 10^1$ | $3.6 \times 10^2$ |
| Azocasein[a] | $6.3 \times 10^2$ | $2.1 \times 10^{-4}$ | | | |
| Azocoll[a] | | | $k_{cat}/K_m = 6.0 \times 10^{-2}$ μM$^{-1}$ s$^{-1}$ | | |
| Remazol brilliant blue-hide powder[a] | | | $k_{cat}/K_m = 1.6 \times 10^{-2}$ μM$^{-1}$ s$^{-1}$ | | |

[a]pH 7.5, phosphate buffer, $I = 0.1$ M. From Aducci *et al*. (1986a).
[b]pH 8.0, phosphate buffer, $I = 0.1$ M. From Aducci *et al*. (1986b).

Leucyl endopeptidase action conforms to simple Michaelis–Menten kinetics, and pre-steady-state and steady-state data fit the minimal three-step catalytic mechanism of serine proteinases. The acylation step (i.e. $k_{+2}$) is rate limiting in the leucyl endopeptidase-catalyzed hydrolysis of *p*-nitrophenyl esters of $N^\alpha$-substituted L-amino acids (see Table 111.1) (Aducci *et al*., 1986a,b).

Bz-Leu┼NHPhNO$_2$ and Z-Leu┼OPhNO$_2$ are the best substrates for leucyl endopeptidase. Activity on *p*-nitroanilides and *p*-nitrophenyl esters of the $N^\alpha$-substituted L-amino acids was in the order: Leu > Ile ~ Val > Ala > Arg ~ Lys > Gly (see Table 111.1). Notably, *p*-nitroanilides and *p*-nitrophenyl esters of aromatic *N*-substituted L-amino acids are not hydrolyzed by leucyl endopeptidase (Aducci *et al*., 1986a,b).

Leucyl endopeptidase assays are most conveniently made with the chromogenic substrates Bz-Leu-NHPhNO$_2$ and Z-Leu-OPhNO$_2$. Simple calculations based on $k_{cat}$ values (see Table 111.1), indicate that the assays with Bz-Leu-NHPhNO$_2$ and Z-Leu-OPhNO$_2$, performed at the optimum pH (approximately 7.5; phosphate buffer, $I = 0.1$ M) and 23°C, allow the determination of a leucyl endopeptidase concentration as low as $10^{-9}$ M (Aducci *et al*., 1986a,b).

Tos-Leu-CH$_2$Cl, DFP and PMSF are the strongest inhibitors of leucyl endopeptidase (see Table 111.2). Consistent with the primary specificity of the enzyme, the inhibition behavior of the chloromethanes of $N^\alpha$-tosyl-L-amino acids

parallels the affinity of *p*-nitroanilides and p-nitrophenyl esters for the S1 specificity subsite of leucyl endopeptidase (see Tables 111.1 and 111.2 for comparison; Aducci *et al*., 1986a,b).

The activity of leucyl endopeptidase is not affected by Tos-Phe-CH$_2$Cl, the I-20K trypsin inhibitor from spinach

*Table 111.2* Inhibitory effect of chloromethane of $N^\alpha$-tosyl-L-amino acids, DFP and PMSF on the catalytic properties of leucyl endopeptidase, at pH 7.5, phosphate buffer, $I = 0.1$ M, and 23°C[a]

| Inhibitor[b] | Leucyl endopeptidase relative activity[c] (%) |
|---|---|
| None | 100 |
| Tos-Gly-CH$_2$Cl | 94 |
| Tos-Ala-CH$_2$Cl | 81 |
| Tos-Val-CH$_2$Cl | 68 |
| Tos-Leu-CH$_2$Cl | 0 |
| Tos-Arg-CH$_2$Cl | 87 |
| Tos-Lys-CH$_2$Cl | 83 |
| DFP | 0 |
| PMSF | 0 |

[a]From Aducci *et al*. (1986a).
[b]The inhibitor concentration was $1.0 \times 10^{-3}$ M.
[c]The enzyme activities are given as a per cent of the control.

leaves (Satoh *et al*., 1985), typical inhibitors of cysteine proteinases, chelating agents or cations (Aducci *et al*., 1986a).

### Structural Chemistry

The molecular mass of leucyl endopeptidase is $60 \pm 3$ kDa,
and pI is $4.8 \pm 0.1$ (Aducci *et al*., 1986a).

### Preparation

Fresh leaves of spinach (*Spinacia oleracea*) were obtained
from local markets. Leucyl endopeptidase from the crude
leaf extract was purified 281-fold by acetone precipitation
followed by chromatography on Sephadex G-100, DEAE
Bio-Gel A and hydroxyapatite Bio-Gel HPT. The protein
contaminants in the final preparation were less than 5%
(Aducci *et al*., 1986a).

### Biological Aspects

Leucyl endopeptidase activity has been detected in leaves,
roots and seeds of *Spinacia oleracea* (Aducci *et al*., 1986a),
but nothing is known of its function.

### Related Peptidases

Leucyl endopeptidase may correspond to the enzyme type 3
detected in the soluble fraction of spinach leaves by [$^3$H]DFP
affinity labeling (Satoh & Fujii, 1982), as discussed by Aducci
*et al*. (1986a).

### Further Reading

A full article on all aspects of the enzyme is that of Aducci
*et al*. (1986a), and a detailed study of the catalytic properties
is described by Aducci *et al*. (1986b).

### References

Aducci, P., Ballio, A. & Federico, R. (1982) Solubilization of
fusicoccin binding sites. In: *Plasmalemma and Tonoplast. Their
Function in the Plant Cell* (Marmè, D., Marrè, E. & Hertel, R.,
eds). Amsterdam: Elsevier Biomedical Press, pp. 279–284.

Aducci, P., Ballio, A., Fiorucci, L. & Simonetti, E. (1984) Inactivation of solubilized fusicoccin-binding sites by endogenous plant
hydrolases. *Planta* **160**, 422–427.

Aducci, P., Ascenzi, P., Pierini, M. & Ballio, A. (1986a) Purification and characterization of Leu-proteinase, the leucine specific
serine proteinase from spinach (*Spinacia oleracea* L.) leaves.
*Plant. Physiol.* **81**, 812–816.

Aducci, P., Ascenzi, P. & Ballio, A. (1986b) Esterolytic properties
of leucine-proteinase, the leucine-specific serine proteinase from
spinach (*Spinacia oleracea* L.) leaves. A steady-state and pre-
steady-state study. *Plant. Physiol.* **82**, 591–593.

Satoh, S. & Fujii, T. (1982) Detection and evaluation of serine
enzymes by [$^3$H]DFP affinity labeling in spinach plants. *Plant.
Cell Physiol.* **23**, 1383–1389.

Satoh, S., Satoh, E., Watanabe, T. & Fujii, T. (1985) Isolation and
characterization of a protease inhibitor from spinach leaves.
*Phytochemistry* **24**, 419–423.

*Patrizia Aducci*
*Department of Biology,*
*University of Rome 'Tor Vergata',*
*Via della Ricerca Scientifica, 00133 Rome, Italy*
*Email: aducci@tovvx1.ccd.utovrm.it*

*Paolo Ascenzi*
*Department of Biology,*
*Third University of Rome,*
*Viale Marconi 446, 00146 Rome, Italy*
*Email: segr-amm@bio.uniroma3.it*

# 112. Archaean serine proteases

### Databanks

Peptidase classification: clan SB, family S8, MEROPS ID: S08.096, S08.100–102, S08.105
NC-IUBMB enzyme classification: none
Databank codes:

| Species | SwissProt | PIR | EMBL (cDNA) |
|---|---|---|---|
| *Halobacterium mediterranei* | P28308 | – | – |
| *Halobacterium mediterranei* | – | – | D64073 |
| *Natrialba asiatica* | P29143 | A42605 | D10201 |
| *Pyrobaculum aerophilum* | – | – | S76079 |
| *Pyrococcus furiosus* | – | – | AA13597 |
| *Staphylothermus marinus* | – | – | U57968 |

## Name and History

Two proteinases from extreme halophiles (**halolysin 172P1** from *Natrialba asiatica* and **halolysin R4** from *Halobacterium* (formerly *Haloferax) mediterranei*) and **aerolysin** from *Pyrobaculum aerophilum*, a hyperthermophile belonging to the order of Thermoproteales, have been completely sequenced, and have been found to be members of the subtilisin family (S8). Some of the other enzymes have been characterized to some extent, but not yet sequenced. Obviously, this makes it impossible to establish reliable evolutionary relationships between them, or with their eubacterial counterparts. Nevertheless, some have been identified as serine proteases on the basis of their inhibitor sensitivities, and as endopeptidases on the basis of their catalytic behavior towards artificial or natural substrates.

## Activity and Specificity

The substrate specificity of these endopeptidases, as assessed with *p*-nitroanilides of *N*-blocked amino acids or peptides, is generally broad, with some preference for phenylalanine in the S1 subsite (see Table 112.1).

## Structural Chemistry

A schematic multiple alignment of the three sequenced enzymes with subtilisin BPN' (Chapter 94) and thermitase (Chapter 95) is shown in Figure 112.1. Each of the structures shows a catalytic core that is clearly related to those of the subtilisins. The halolysin and aerolysin sequences show a prepro region that is likely to play the same role as the prepropeptide in subtilisin, which allows folding into the active conformation, and is later autocatalytically cleaved. Both halolysins show long C-terminal extensions of 117 (halolysin R4) or 123 (halolysin 172P1) residues, and that of halolysin R4 has been shown to be essential for activity and stability.

Currently available knowledge about the biochemical properties of archaebacterial serine proteases is summarized in Table 112.1. On the whole, these data suggest that the endopeptidases that have been sequenced, i.e. the halophilic enzymes and aerolysin, are similar in structure and properties. Less can be said about unsequenced enzymes. Some resemblance may be recognized between *Desulfurococcus* archaelysin and *Sulfolobus* proteinase I, based on similar molecular masses, substrate specificity, pH and temperature optima, in spite of substantially different pI values. This fits well with the short evolutionary distance between *Sulfolobus* and *Desulfurococcus* (Woese, 1987). It is notable that these enzymes all display molecular masses substantially higher than those of their eubacterial counterparts. In the case of halolysin R4 at least this is accounted for by the presence of a C-terminal tail, which is likely to play a role in protein stability.

The characterization of the extracellular serine proteases from *Thermococcus* strains and from *Staphylothermus marinus* (Klingeberg *et al*., 1991), and intracellular ones from *Pyrococcus furiosus* (Blumentals *et al*., 1990), is only preliminary, and only the S66 proteinase from *P. furiosus* has yet even been purified to homogeneity. In the *Thermococcus* strains and in *S. marinus*, temperature optima range from 80°C to 95°C, and pH optima from 7 to 9, mostly with broad profiles.

Zymogram staining shows the presence of multiple bands with molecular masses ranging from 30 to 300 kDa, which might be partly due to aggregation and/or limited proteolysis.

Five different bands endowed with proteolytic activity and with molecular masses of 140, 125, 116, 102 and 66 kDa, have been detected by SDS-PAGE of *P. furiosus* crude extracts (Blumentals *et al*., 1990). The proteolytic activities increase up to 105°C, as assessed by azocasein hydrolysis. Particularly striking is the SDS-resistance of the 102 and 66 kDa proteases (designated as S102 and S66), which fully retain their activity after a 24 h preincubation at 98°C. Electrophoretic analyses also show that S66 undergoes C- or N-terminal cleavage, resulting in a progressively smaller protein. Also, it is likely to arise from a 200 kDa, catalytically inactive but immunologically cross-reactive precursor. Based on its sensitivity to PMSF and DFP, S66 has been identified as a serine protease, whereas no certain identification has been so far achieved for the other proteolytic activities. S66 also acts as an esterase, but it is unclear whether it is endowed with endo- or exopeptidase activity.

## Structural Features Responsible for Adaptations to Extreme Conditions

No general paradigm has emerged regarding mechanisms of protein adaptation to high temperature and/or high salt concentration. The only paper which has systematically addressed this issue in the case of archaebacterial serine proteases is that dealing with structural features related to thermostability in aerolysin (Völkl *et al*., 1994). Models of secondary and tertiary structure of the enzyme have been built on the basis of sequence alignment of this enzyme with other subtilisin-type proteases, which has allowed identification of sites possibly responsible for increased thermostability. In particular, Asp residues have been found at the N-terminus of surface helices, which might strengthen helix dipoles. Also, several Ala replacements have been detected at the beginning and end of surface helices, but it is unclear how they contribute to protein stability. In the case of *S. solfataricus* proteinase I, a role for intramolecular ionic interactions in thermostability is substantiated by the strong destabilizing effect of high NaCl concentrations (Burlini *et al*., 1992). In keeping with this hypothesis, a thorough investigation performed on a carboxypeptidase from the same microorganism also confirms the involvement of ionic interactions in kinetic thermal stability (Villa *et al*., 1993). This suggests that salt bridges might represent a general strategy of thermostabilization in *S. solfataricus*.

Halophilic enzymes are stable and active at high salt concentrations, and most of them are rapidly inactivated below 2 M NaCl. This property has been found to be related to a high content of acidic amino acids in these enzymes, although the structural reasons underlying this behavior are not understood (Kamekura *et al*., 1996). On the other hand, it has also been proposed that halophilicity of the serine proteinase halolysin R4 and its relatives may be supported by the presence of the C-terminal extension of about 120 residues, which is not found in subtilisin and related eubacterial serine proteases. In particular, the two Cys residues located in this extension seem to be essential for enzyme stability, as substantiated by characterization of wild-type protease and two cysteine mutants (Cys316Ser and Cys316Ser/Cys352Ser) with respect

*Table 112.1* Biochemical properties of archaebacterial serine proteinases

| Enzyme | Source | Localization | kDa | pI | Inhibitors | Preferred residue in subsite S1 | Optimum activity | | pH stability optimum | Known sequence | Reference |
|---|---|---|---|---|---|---|---|---|---|---|---|
| | | | | | | | pH | T (°C) | | | |
| Halolysin 172P1 | *Natrialba asiatica* | Extracellular | 41.96 | 3.68[a] | PMSF | F>Y, L | 10.7 | 70 (in 13.8% NaCl) | 5–11 (in 25% NaCl) | All | Kamekura & Seno (1989), Kamekura et al. (1992, 1996) |
| Halolysin R4 | *Halobacterium mediterranei* R4 | Extracellular | 41.26 | 4.08[a] | ND | L, Y, F | ND | ND[b] | ND | All | Kamekura & Seno (1993), Kamekura et al. (1996) |
| Unnamed | *Halobacterium mediterranei* | Extracellular | 41 | 7.5 | DFP, PMSF, ovomucoid | L>F, Y | 8–8.5 | 55 | 5.5–8 (in 4.5 M NaCl) | N-terminus | Stepanov et al. (1992) |
| Aerolysin | *Pyrobaculum aerofilum* | Extracellular | 33.37[a] | 6.57[a] | ND | ND | neutral to alkaline | >100 | ND | All | Völkl et al. (1994) |
| Archaelysin | *Desulfurococcus* strain | Extracellular | 52 | 8.7 | DFP, PMSF, chymostatin | Y, A, S, R | 7.2 | 95 | 8 | None | Cowan et al. (1987) |
| Proteinase I | *Sulfolobus solfataricus* | Intracellular | 54 × 2 | 5.6 | PMSF, TPCK chymostatin | Y>L>F | 7.5 | >90 | 7.5 | None | Burlini et al. (1992) |
| Proteinase II | *Sulfolobus solfataricus* | Intracellular | 32 | ND | PMSF, TPCK chymostatin | F | ND | 60 | ND | None | Fusi et al. (1991) |

ND, not done; TPCK, Tos-Phe-CH$_2$Cl.
[a]Deduced from primary structure.
[b]Requires 15% NaCl for optimal activity at pH 7.6, 37°C.

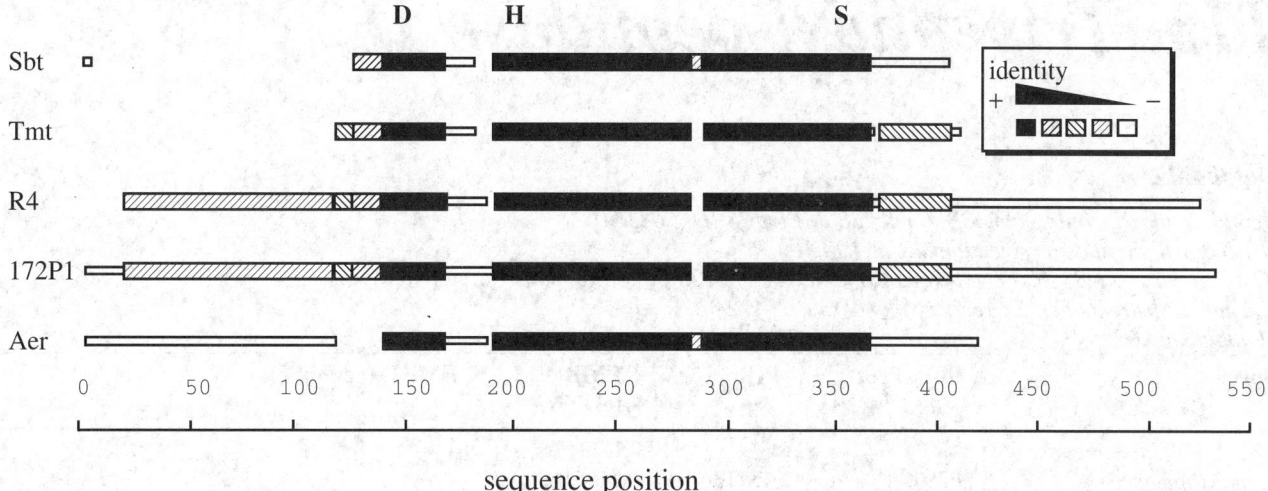

*Figure 112.1* Schematic multiple alignment of subtilisin BPN′ from *Bacillus subtilis* (Sbt), thermitase from *Thermoactinomyces vulgaris* (Tmt), halolysin R4 from *Halobacterium mediterranei* (R4), halolysin 172P1 from a *Natrialba asiatica* (172P1), and aerolysin from *Pyrobaculum aerophilum* (Aer). **D**, **H** and **S** indicate the locations of the catalytic triad residues, and the shading indicates the degree of similarity between the blocks.

to activity and stability (Kamekura *et al*., 1996). Experimental evidence suggests that decreased stability in hypotonic solutions of the mutants possibly results from disruption of potential disulfide bonds or from perturbation of calcium-binding site(s), as a consequence of cysteine replacement.

## Further Reading

The reader is referred to the paper of Völkl *et al*. (1994) for a detailed analysis of the sequence similarity between the archaebacterial aerolysin and several eubacterial serine proteinases.

## References

Blumentals, I.I., Robinson, A.S. & Kelly, R.M. (1990) Characterization of sodium dodecyl sulfate-resistant proteolytic activity in the hyperthermophilic archaebacterium *Pyrococcus furiosus*. *Appl. Environ. Microbiol.* **56**, 1992–1998.

Burlini, N., Magnani, P., Villa, A., Macchi, F., Tortora, P. & Guerritore, A. (1992) A heat-stable serine proteinase from the extreme thermophilic archaebacterium *Sulfolobus solfataricus*. *Biochim. Biophys. Acta* **1122**, 283–292.

Cowan, D.A., Smolenski, K.A., Daniel, R.M. & Morgan, H.W. (1987) An extremely thermostable extracellular proteinase from a strain of the archaebacterium *Desulfurococcus* growing at 88°C. *Biochem. J.* **247**, 121–133.

Fusi, P., Burlini, N., Villa, M., Tortora, P. & Guerritore, A. (1991) Intracellular proteases from the extreme thermophilic archaebacterium *Sulfolobus solfataricus*. *Experientia* **47**, 1057–1060.

Kamekura, M. & Seno, Y. (1989) A halophilic extracellular protease from a halophilic archaebacterium strain 172 P1. *Biochem. Cell.*

*Biol.* **68**, 352–359.

Kamekura, M. & Seno, Y. (1993) Partial sequence of the gene for a serine protease from a halophilic archaeum *Haloferax mediterranei* R4, and nucleotide sequences of 16S rRNA encoding genes from several halophilic archaea. *Experientia* **49**, 503–513.

Kamekura, M., Seno, Y., Holmes, M.L. & Dyall-Smith, M.L. (1992) Molecular cloning and sequencing of the gene for a halophilic alkaline serine protease (halolysin) from an unidentified halophilic archaea strain (172 P1) and expression of the gene in *Haloferax volcanii*. *J. Bacteriol.* **174**, 736–742.

Kamekura, M., Seno, Y. & Dyall-Smith, M.L. (1996) Halolysin R4, a serine proteinase from the halophilic archaeon *Haloferax mediterranei*; gene cloning, expression and structural studies. *Biochim. Biophys. Acta* **1294**, 159–167.

Klingeberg, M., Hashwa, F. & Antranikian, G. (1991) Properties of extremely thermostable proteases from hyperthermophilic bacteria. *Appl. Microbiol. Biotechnol.* **34**, 715–719.

Stepanov, V.M., Rudenskaya, G.N., Revina, L.P., Gryaznova, Y.B., Lysogorskaya, E.N., Filippova, I.Y. & Ivanova, I.I. (1992) A serine proteinase of an archaebacterium, *Halobacterium mediterranei*. A homologue of eubacterial subtilisins. *Biochem. J.* **285**, 281–286.

Villa, A., Zecca, L., Fusi, P., Colombo, S., Tedeschi. G. & Tortora, P. (1993) Structural features responsible for kinetic thermal stability of a carboxypeptidase from the archaebacterium *Sulfolobus solfataricus*. *Biochem. J.* **295**, 827–831.

Völkl, P., Markiewicz, P., Stetter, K.O. & Miller, J.I. (1994) The sequence of a subtilisin-type protease (aerolysin) from the hyperthermophilic archaeum *Pyrobaculum aerophilum* reveals sites important to thermostability. *Protein Sci.* **3**, 1329–1340.

Woese, C.R. (1987) Bacterial evolution. *Microbiol. Rev.* **51**, 221–271.

***Marco Vanoni***
*Dipartimento di Fisiologia e Biochimica generali,*
*Via Celoria 26,*
*I-20133 Milano, Italy*

***Paolo Tortora***
*Dipartimento di Fisiologia e Biochimica generali,*
*Via Celoria 26,*
*I-20133 Milano, Italy*
*Email: pator@imiucca.csi.unimi.it*

# 113. Tripeptidyl-peptidase II

## Databanks

*Peptidase classification: clan SB, family S8, MEROPS ID: S08.090*
*NC-IUBMB enzyme classification: EC 3.4.14.10*
*ATCC entries: 65988–65995 (various human clones); 63335 (3.86 kb mouse clone)*
*Chemical Abstracts Service registry number: 101149-94-4*
*Databank codes:*

| Species | SwissProt | PIR | EMBL (cDNA) | EMBL (genomic) |
|---|---|---|---|---|
| *Caenorhabditis elegans* | – | – | U23176 | – |
| *Homo sapiens* | P29144 | A37136 | M55169 | – |
| | | A39887 | M73047 | |
| | | S09033 | | |
| | | S54376 | | |
| | | S54393 | | |
| | | S54394 | | |
| *Mus musculus* | – | – | X81323 | – |
| *Rattus norvegicus* | – | – | U50194 | – |

## Name and History

**Tripeptidyl-peptidase II (TPP II)** was discovered during attempts to find a peptidase specific for phosphorylated sequences (Bålöw *et al.*, 1983). The peptide Arg-Arg-Ala+Ser($^{32}$P)-Val-Ala was used as a probe for the screening of rat liver extract, and an enzyme capable of cleaving the Ala-Ser($^{32}$P) bond was identified and purified. In the event the peptidase was not specific for phosphorylated sites, but could remove tripeptides sequentially from a free N-terminus. The enzyme was originally called **tripeptidyl aminopeptidase**, but was later given the systematic name **tripeptidyl peptidase II** (McDonald & Barrett, 1986). The hyphen was introduced in the 1992 edition of the IUBMB Enzyme Nomenclature.

## Activity and Specificity

TPP II removes tripeptides from a free N-terminus of peptides or chromogenic substrates. The pH optimum is 7.5, and the substrate specificity broad, except that Pro is not accepted in the P1 or P1' positions. There are few apparent similarities between the tripeptides removed, but the rate of cleavage of different peptide bonds varies more than 100-fold between different substrates (Bålöw *et al.*, 1983, 1986). The standard substrate used for measuring the enzyme activity has been the peptide Arg-Arg-Ala+Ser($^{32}$P)-Val-Ala, where binding of the product Ser($^{32}$P)-Val-Ala to the anion-exchanger QAE-Sephadex is measured (Bålöw *et al.*, 1983). However, the assay can be done more conveniently with the chromogenic substrate Ala-Ala-Phe+NHPhNO$_2$, available from Bachem (Tomkinson *et al.*, 1994) (see Appendix 2 for full names and addresses of suppliers). The enzyme is stabilized by 30% glycerol and 1 mM DTT, and inhibited by high concentrations of NaCl. TPP II is inhibited by serine protease inhibitors such as DFP and PMSF, and by some thiol-reactive compounds (Hg$^{2+}$, *N*-ethylmaleimide) but not others (iodoacetic acid, iodoacetamide, E64) (Bålöw *et al.*, 1986). Efficient specific inhibitors with $K_i$ values in the nanomolar range have been developed, e.g. a dehydroalanine-containing peptide (Tomkinson *et al.*, 1994) and a peptoid inhibitor, butabindide (Rose *et al.*, 1996). No natural inhibitors have been identified.

## Structural Chemistry

TPP II is an exceptionally large peptidase with a subunit $M_r$ of 138 000, as calculated from the amino acid sequence, and pI of 6.2. The subunits are assembled into a large oligomeric complex (Figure 113.1), giving it a native $M_r$ of about $4 \times 10^6$ (Bålöw *et al.*, 1983, 1986). We have shown that this complex is a prerequisite for enzymatic activity, since the enzyme loses activity upon dissociation (Macpherson *et al.*, 1987) (Figure 113.1). However, others claim that the subunits are active as monomers (Wilson *et al.*, 1993).

The active site is of the subtilisin type. This was first demonstrated after the active-site Ser residue was labeled with [$^3$H]DFP and the labeled peptide purified (Tomkinson *et al.*, 1987), and later confirmed when the sequence data were obtained. The cDNA encoding the human (Tomkinson, 1991; Tomkinson & Jonsson, 1991), mouse (Tomkinson, 1994) and rat (Rose *et al.*, 1996) enzymes have been isolated. The structure is well conserved between species, i.e. 96% of the amino acids are identical between human TPP II and the enzyme from mouse or rat. Apart from the subtilisin-like catalytic domain, no similarities to other proteins in the databases were detected (Tomkinson & Jonsson, 1991; Tomkinson, 1994), although an immunological similarity to fibronectin had been demonstrated (Tomkinson & Zetterqvist, 1990).

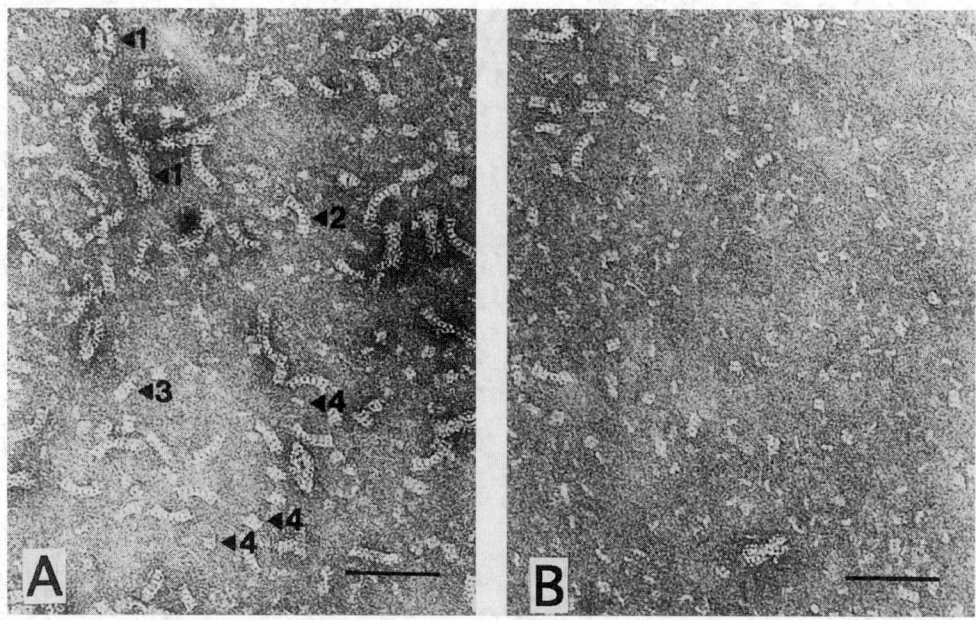

*Figure 113.1*   Electron micrographs of human TPP II negatively stained with ammonium molybdate before (A) and after (B) dialysis. Purified human TPP II was dialysed against 2 mM Tris buffer, pH 7.5, containing 3% glycerol and 0.5 mM ME for 4 h. The numbers and arrowheads indicate examples of different size classes. Bars indicate 100 nm. (Reprinted after modification from Macpherson *et al.* (1987), with permission.)

## Preparation

TPP II has been purified to apparent homogeneity from rat liver (4000-fold) and human red blood cells (80 000-fold) (Bålöw *et al.*, 1986), human brain (600-fold) (Wilson *et al.*, 1993) and rat brain (8600-fold) (Rose *et al.*, 1996).

## Biological Aspects

TPP II has been identified in a number of cells in different species (Bålöw *et al.*, 1986; Bålöw & Eriksson, 1987; Rose *et al.*, 1996; Tomkinson & Jonsson, 1991). Although initially identified as a cytosolic enzyme (Bålöw *et al.*, 1983), a membrane-associated variant, which constitutes 31% of the activity in brain, but only 4% of the activity in liver, has been recently described (Rose *et al.*, 1996). The widespread distribution, broad substrate specificity and conserved structure indicate a physiological role for the enzyme in the general intracellular protein turnover, where the formation of tripeptides is likely to facilitate the generation of free amino acids from oligopeptides. In addition, it was recently demonstrated that the membrane-associated variant of TPP II inactivates the neuropeptide cholecystokinin (Rose *et al.*, 1996). TPP II can cleave a number of different neuropeptides *in vitro* (Nyberg *et al.*, 1987; Wilson *et al.*, 1993; Rose *et al.*, 1996), which suggests a specific role for the membrane-associated variant in the degradation of neuropeptides.

The cDNA sequence contains no evident signal peptide or membrane-spanning domain, and Rose *et al.* (1996) suggested that TPP II is bound to the membrane through a glycosyl phosphatidylinositol anchor. Different variants of the mRNA encoding TPP II have been identified, i.e. mRNA with long and short untranslated 3′ ends, respectively (Tomkinson & Jonsson, 1991; Tomkinson, 1994) and a splicing variant

containing 39 bp extra, encoding 13 amino acids in the C-terminal part of the enzyme (Tomkinson, 1994). It remains to be seen whether the membrane-associated variant corresponds to one of these or a new mRNA variant.

The gene encoding human TPP II has been localized to 13q32–33 (Martinsson *et al.*, 1993), the mouse gene to chromosome 1 (Bermingham *et al.*, 1996) and the pig gene to chromosome 11 (Chowdhary *et al.*, 1993). No genetic diseases have yet been coupled to the gene in any of these species.

## Distinguishing Features

The combined action of other peptidases can be wrongly identified as tripeptidyl-peptidase activity, so that particular care must be taken if chromogenic substrates are used (Kenny & Ingram, 1988). In mammalian cells, the only other hitherto characterized tripeptidyl-peptidase is the lysosomal tripeptidyl-peptidase I (McDonald & Barrett, 1986) (Chapter 183). This enzyme is active at pH 4, at which TPP II is completely inactive. Antibodies against human TPP II do not recognize TPP I (Bålöw & Eriksson, 1987). Polyclonal chicken antibodies against human TPP II (Tomkinson & Zetterqvist, 1990) and rabbit antibodies against rat TPP II have been prepared (Rose *et al.*, 1996), but are not commercially available.

## Related Peptidases

At least three different tripeptidyl-peptidases have also been characterized in *Streptomyces lividans*. The amino acid sequences show that two of them are unrelated to TPP II (Binnie *et al.*, 1995; Butler *et al.*, 1995) (Chapter 140), whereas the third, tripeptidyl-peptidase S, has an active site of the subtilisin type (Chapter 114).

## Further Reading

The cloning of TPP II from mouse and identification of a splicing variant of the enzyme are described by Tomkinson (1994). The recent publication by Rose *et al.* (1996) describes the identification of TPP II as a cholecystokinin-inactivating enzyme, and the development of a potent inhibitor and its use for inhibition of the enzyme in rats.

## References

Bålöw, R.-M. & Eriksson, I. (1987) Tripeptidyl peptidase II in haemolysates and liver homogenates of various species. *Biochem. J.* **241**, 75–80.

Bålöw, R.-M., Ragnarsson, U. & Zetterqvist, Ö. (1983) Tripeptidyl aminopeptidase in the extralysosomal fraction of rat liver. *J. Biol. Chem.* **258**, 11622–11628.

Bålöw, R.-M., Tomkinson, B., Ragnarsson, U. & Zetterqvist, Ö. (1986) Purification, substrate specificity and classification of tripeptidyl peptidase II. *J. Biol. Chem.* **261**, 2409–2417.

Bermingham, N.A., McKay, T., Hoyle, J., Hernandez, D., Martin, J.E. & Fisher, E.M.C. (1996) The gene encoding tripeptidyl peptidase II maps to chromosome 1 in the mouse. *Mammalian Genome* **7**, 390.

Binnie, C., Butler, M.J., Aphale, J.S., Bourgault, R., DiZonno, M.A., Krygsman, P., Liao, L., Walcyk, E. & Malek, L.T. (1995) Isolation and characterization of two genes encoding proteases associated with the mycelium of *Streptomyces lividans* 66. *J. Bacteriol.* **177**, 6033–6040.

Butler, M.J., Binnie, C., DiZonno, M.A., Krygsman, P., Soltes, G.A., Soostmeyer, G., Walczyk, E. & Malek, L.T. (1995) Cloning and characterization of a gene encoding a secreted tripeptidyl amino-peptidase from *Streptomyces lividans* 66. *Appl. Environ. Microbiol.* **61**, 3145–3150.

Chowdhary, B.P., Johansson, M., Gu, F., Bräuner-Nielsen, P., Tomkinson, B., Andersson, L. & Gustavsson, I. (1993) Assignment of the linkage group EAM-TYRP2-TPP2 to chromosome 11 in pigs by *in situ* hybridization mapping of the TPP2 gene. *Chromosome Res.* **1**, 175–179.

Kenny, A.J. & Ingram, J. (1988) Is there a tripeptidyl peptidase in the renal brush-border membrane? *Biochem. J.* **255**, 373–376.

Macpherson, E., Tomkinson, B., Bålöw, R.-M., Höglund, S. & Zetterqvist, Ö. (1987) Supramolecular structure of tripeptidyl peptidase II from human erythrocytes as studied by electron microscopy, and its correlation to enzyme activity. *Biochem. J.* **248**, 259–263.

Martinsson, T., Vujic, M. & Tomkinson, B. (1993) Localization of the human tripeptidyl peptidase II gene (TPP2) to 13q32–33 by non-radioactive *in situ* hybridization and somatic cell hybrids. *Genomics* **17**, 493–495.

McDonald, J.K. & Barrett, A.J. (1986) *Mammalian Proteases*, vol. 2: *Exopeptidases*. London: Academic Press.

Nyberg, F., Bålöw, R.-M., Tomkinson, B. & Zetterqvist, Ö. (1987) Degradation of Leu- and Met-enkephalin and their C-terminal extensions by tripeptidyl peptidase II. In: *Protides of the Biological Fluids* (Peeters, H., ed.). Oxford: Pergamon Press, pp. 193–196.

Rose, C., Vargas, F., Facchinetti, P., Bourgeat, P., Bambal, R.B., Bishop, P.B., Chan, S.M.T., Moore, A.N.J., Ganellin, C.R. & Schwartz, J.-C. (1996) Characterization and inhibition of a cholecystokinin-inactivating serine peptidase. *Nature* **380**, 403–409.

Tomkinson, B. (1991) Nucleotide sequence of cDNA covering the N-terminus of human tripeptidyl peptidase II. *Biomed. Biochim. Acta* **50**, 727–729.

Tomkinson, B. (1994) Characterization of cDNA for murine tripeptidyl peptidase II reveals alternative splicing. *Biochem. J.* **304**, 517–523.

Tomkinson, B. & Jonsson, A.-K. (1991) Characterization of cDNA for human tripeptidyl peptidase II. The N-terminal part of the enzyme is similar to subtilisin. *Biochemistry* **30**, 168–174.

Tomkinson, B. & Zetterqvist, Ö. (1990) Immunological cross-reactivity between human tripeptidyl peptidase II and fibronectin. *Biochem. J.* **267**, 149–154.

Tomkinson, B., Wernstedt, C., Hellman, U. & Zetterqvist, Ö. (1987) Active site of tripeptidyl peptidase II from human erythrocytes is of the subtilisin-type. *Proc. Natl Acad. Sci. USA* **84**, 7508–7512.

Tomkinson, B., Grehn, L., Fransson, B. & Zetterqvist, Ö. (1994) Use of a dehydroalanine containing peptide as an efficient inhibitor of tripeptidyl peptidase II. *Arch. Biochem. Biophys.* **314**, 276–279.

Wilson, C., Gibson, A.M. & McDermott, J.R. (1993) Purification and characterization of tripeptidylpeptidase-II from post-mortem human brain. *Neurochem. Res.* **18**, 743–749.

*Birgitta Tomkinson*
*Department of Veterinary Medical Chemistry,*
*Swedish University of Agricultural Sciences,*
*Biomedical Center, Box 575,*
*S-751 23 Uppsala, Sweden*
*Email: Birgitta.Tomkinson@vmk.slu.se*

# 114. Tripeptidyl-peptidase S

## Databanks

*Peptidase classification: clan SB, family S8, MEROPS ID: S08.091*
*NC-IUBMB enzyme classification: none*

*Databank codes:*

| Species | SwissProt | PIR | EMBL (cDNA) | EMBL (genomic) |
|---|---|---|---|---|
| *Streptomyces lividans* | – | – | L41655 | – |

## Name and History

This peptidase was discovered during a search for minor secreted tripeptidyl-peptidase activities produced by *Streptomyces lividans* 66. After removal of the major secreted activity (TPP A) by recombinational deletion of the chromosomally encoded gene (Butler *et al.*, 1995), a low but detectable level of activity remained. The deletion strain was used as a host for expression cloning of an *S. lividans* genomic library by use of the chromogenic peptide substrate Ala-Pro-Ala-|-NHNap. Clones overexpressing activity were observed as dark red colonies against a faintly red background colony color in a direct agar plate substrate assay. Two genes homologous to *tppA* were isolated (*tppB* and *C*: Binnie *et al.*, 1995) (see Chapter 140) but in addition, another two genetically unrelated species were noticed. Nucleotide sequence determination revealed (Butler *et al.*, 1996) that one of these species encoded a secreted protease with significant homology to subtilisin BPN′ (particularly strong amino acid conservation being detected around the catalytic active-site residues, Asp, His and Ser). This protein was originally named *Ssp*, but is here redesignated *tripeptidyl-peptidase S (TPP S)* (for **tri**peptidyl-**p**eptidase **s**ubtilisin-like).

## Activity and Specificity

Preliminary characterization was carried out on crude fermentation both from a strain of *S. lividans* in which the *tppA* had been deleted and which carried the *tppS* gene on a multicopy plasmid. The relative rates of hydrolysis of chromogenic substrates by TPP S were compared to those for purified subtilisin BPN′ (Boehringer Mannheim) (see Appendix 2 for full names and addresses of suppliers). The observed enzyme specificities were quite different. TPP S was unable to hydrolyze complex substrates such as azocasein which were readily cleaved by subtilisin BPN′. Conversely, Ala-Pro-Ala-NHNap was hydrolyzed faster by TPP S. Substrates which did not contain a free amino group at the N-terminus were not cleaved by TPP S, but were attacked by subtilisin. Thus, TPP S is an N-terminal exopeptidase, not an endopeptidase. The activity was calcium dependent and inhibited by PMSF and EDTA

(but not 1,10-phenanthroline). This is consistent with the serine protease catalytic mechanism expected of a protein showing primary homology to subtilisin BPN′.

## Structural Chemistry

The protein was observed as a 45 kDa species on SDS-PAGE and its behavior in gel-filtration chromatography was consistent with a monomeric active form. The nucleotide sequence of the gene suggests that the protein is synthesized as an inactive precursor. This species is likely to be processed to the active form by another endopeptidase, since the active form of TPP S does not activate the proenzyme.

## Biological Aspects

TPP S appears to represent a minor secreted tripeptidyl-peptidase activity in *S. lividans*. It is a nonessential function, in that viable strains can be made in which both the *tppA* and *tppS* genes have been deleted. Such strains show a reduced ability to hydrolyze Ala-Pro-Ala-NHNap, although a small but significant amount of such activity remains, indicating the presence of other minor active species.

## References

Binnie, C., Butler, M.J., Aphale, J.S., Bourgault, R., DiZonno, M.A., Krygsman, P., Laio, L., Walczyk, E. & Malek, L.T. (1995) Isolation and characterisation of two genes encoding proteases associated with the mycelium of *Streptomyes lividans* 66. *J. Bacteriol.* **177**, 6033–6040.

Butler, M.J., Binnie, C., DiZonno, M.A., Krygsman, P., Soltes, G.A., Soostmeyer, G., Walczyk, E. & Malek, L.T. (1995) Cloning and characterization of a gene encoding a secreted tripeptidyl aminopeptidase from *Streptomyces lividans* 66. *Appl. Environ. Microbiol.* **61**, 3145–3150.

Butler, M.J., Aphale, J.S., Binnie, C., DiZonno, M.A., Krygsman, P., Soltes, G.A., Walczyk, E. & Malek, L.T. (1996) Cloning and analysis of a gene from *Streptomyces lividans* 66 encoding a novel secreted protease exhibiting homology to subtilisin BPN′. *Appl. Microbiol. Biotechnol.* **45**, 141–147.

**Michael J. Butler**
*Strangeways Research Laboratory,*
*Worts Causeway,*
*Cambridge CB1 4RN, UK*
*Email: mb@srl.mrc-lmb.cam.ac.uk*

# 115. Kexin

## Databanks

*Peptidase classification: clan SB, family S8, MEROPS ID: S08.070*
*NC-IUBMB enzyme classification: EC 3.4.21.61*
*Chemical Abstracts Service registry number: 99676-46-7*
*Databank codes:*

| Species | SwissProt | PIR | EMBL (cDNA) | EMBL (genomic) |
|---|---|---|---|---|
| *Kluyveromyces lactis* | P09231 | S01013 | X07038 | – |
| *Saccharomyces cerevisiae* | P13134 | A28931 | M22870 | – |
|  |  | C45108 | M24201 |  |
|  |  | S42157 |  |  |
|  |  | S63204 |  |  |
| *Schizosaccharomyces pombe* | Q09175 | – | X82435 | – |
| *Yarrowia lipolytica* | P42781 | – | L16238 | – |

## Name and History

In 1976, Leibowitz and Wickner reported that mutations in two *Saccharomyces cerevisiae* genes blocked secretion of killer toxin activity. Mutations in one of these genes, *KEX2* (**k**iller **ex**pression defective) also blocked production of the mating pheromone $\alpha$-factor (Leibowitz & Wickner, 1976). Characterization of $\alpha$-factor, killer toxin and intermediates in $\alpha$-factor biosynthesis, together with the cloning and sequence analysis of the genes encoding $\alpha$-factor and killer toxin precursors, suggested that both precursors undergo cleavage C-terminal to pairs of basic residues (Lys-Arg+ and Arg-Arg+) during their maturation (for review, see Fuller *et al.*, 1988). Biochemical analysis of *KEX2* mutant cells and cloning, sequence analysis and expression studies of the *KEX2* gene demonstrated that *KEX2* encoded a transmembrane serine protease of the subtilisin family (Julius *et al.*, 1984; Mizuno *et al.*, 1988, 1989; Fuller *et al.*, 1988, 1989a,b). In standard yeast genetic nomenclature, the *Saccharomyces KEX2* gene product is termed Kex2p or Kex2 protein (or protease). Kexin activity was designated ***protease YscF*** in one report (Achstetter & Wolf, 1985). The name ***kexin*** was recommended in 1992 by IUBMB. Homologous enzymes carrying out similar processing reactions have been identified in other fungi, including *Kluyveromyces lactis, Yarrowia lipolytica* and *Schizosaccharomyces pombe* (Tanguy-Rougeau *et al.*, 1988; Enderlin & Ogrydziak, 1994; Davey *et al.*, 1994). The predicted sequence of Kex2p led directly to the discovery of the mammalian proprotein processing proteases (Fuller *et al.*, 1989b; Smeekens & Steiner, 1990).

## Activity and Specificity

*S. cerevisiae* Kex2p has been more thoroughly characterized than other fungal kexins. Like subtilisin (Chapter 94), kexin is calcium dependent; however, kexin can be reactivated by micromolar $Ca^{2+}$ after inactivation by $Ca^{2+}$-chelators (Fuller *et al.*, 1989a). Like subtilisin, kexin exhibits a relatively simple pH profile, presumably reflecting titration of the catalytic His213, with an apparent $pK_a$ for $k_{cat}$ of $\sim 5.7$, $\sim 1.0\,pH$ unit more acidic than that of subtilisin (Brenner & Fuller, 1992). Resistance to standard serine protease inhibitors combined with sensitivity to heavy metals and thiol reagents led to the erroneous conclusion that kexin was a thiol protease (Julius *et al.*, 1984; Achstetter & Wolf, 1985; Fuller *et al.*, 1986; Mizuno *et al.*, 1987). Kexin is resistant to PMSF, Tos-Phe-CH$_2$Cl and Tos-Lys-CH$_2$Cl, and requires $>10\,mM$ DFP for inactivation. The enzyme is inactivated by heavy metals, including $Zn^{2+}$, $Cu^{2+}$ and $Hg^{2+}$, by *p*-chloromercuriphenyl sulfonate and by iodoacetate and iodoacetamide. However, homology to subtilisin, sensitivity to DFP, albeit only at high concentrations, and the effects of mutation of catalytic triad residues (His213, Ser385) and the oxyanion hole (Asn314) confirmed that kexin is a serine protease (Fuller *et al.*, 1989a,b; Germain *et al.*, 1992; Brenner *et al.*, 1993). The most specific inhibitors are dibasic peptidyl chloromethanes that yield $K_i$ values in the low nanomolar range and $k_2/K_i$ values of $\sim 10^7\,M^{-1}\,s^{-1}$ (Angliker *et al.*, 1993).

Peptide methylcoumarylamides and internally quenched fluorogenic peptides have been used in systematic analyses of kexin specificity (Brenner & Fuller, 1992; Rockwell *et al.*, 1997). Much like subtilisin, kexin interacts with P1, P2 and P4, but exhibits greater side-chain selectivity for each residue than do degradative subtilisins. The S1 subsite is highly selective for Arg. Decreases of 100- to 10 000-fold in $k_{cat}/K_m$ occur upon introducing Lys at P1. With citrulline at P1, $k_{cat}/K_m$ drops $>10^5$-fold, indicating that both charge and steric factors are important in P1 recognition (Rockwell *et al.*, 1997). With Arg at P1, $k_{cat}$ is generally in the range of $20–40\,s^{-1}$, while substitutions at P2 principally affect $K_m$. S2 selects for positive charge, exhibits neither exclusion of, nor interaction with, linear aliphatic side chains and exhibits negative interactions with bulky aromatic residues. No systematic analysis has been made of P'-side specificity subsites. However, the similarity between $k_{cat}/K_m$ values for the best tetrapeptide methycoumarylamides and internally quenched peptides indicates that positive interactions with P' side chains do not significantly contribute to specificity. The best substrate sequences for kexin are based on cleavage sites in pro-α-factor: Pro-Met-Tyr-Lys-Arg⊣Glu-Ala-Glu-Ala, with $k_{cat}/K_m$ values in the range of $2–5 \times 10^7 \, M^{-1} \, s^{-1}$. High $k_{cat}/K_m$ values do not indicate diffusion control, because kexin exhibits burst kinetics in cleavage of the best peptide methylcoumarylamide substrates (Brenner & Fuller, 1992). The rate of acylation of the enzyme is vastly in excess of the steady-state rate, reflecting either rate-limiting hydrolysis of the acyl enzyme or a slow conformational change.

## Structural Chemistry

In the endoplasmic reticulum, kexin undergoes signal peptide cleavage, intramolecular autoproteolytic cleavage of an N-terminal propeptide, and addition of Asn-linked and Ser/Thr-linked carbohydrate that is modified further in the Golgi (Wilcox & Fuller, 1991; Germain et al., 1992; Gluschankof & Fuller, 1994). After intramolecular cleavage at Lys⊣Arg110, kexin undergoes removal of N-terminal Leu-Pro and Val-Pro dipeptides by the *STE13*-encoded dipeptidyl-peptidase A (Chapter 129) (Brenner & Fuller, 1992). The full-length mature kexin polypeptide consists of 699 amino acids in several domains (see Figure 118.1). The N-terminal subtilisin domain is followed by the P-domain, which is also required for formation of an active enzyme (Gluschankof & Fuller, 1994), a Ser/Thr-rich domain (Fuller et al., 1989a), a single hydrophobic transmembrane domain and a C-terminal cytosolic tail. The apparent molecular mass of the protein, ∼130 000 Da, is greatly exaggerated by effects of the highly acidic cytosolic tail and the extensive O-glycosylation of the Ser/Thr-rich region (Fuller et al., 1989a).

## Preparation

The inability to purify full-length kexin intact, due to artifactual proteolysis after cell lysis (Fuller et al., 1989a), led instead to the purification and characterization of a genetically engineered, secreted, soluble enzyme consisting of the subtilisin and P-domains (Brenner & Fuller, 1992). This form of kexin can be secreted into yeast culture medium at concentrations of 10–30 mg liter$^{-1}$ and purified by concentration on an anion-exchange resin, vacuum dialysis and high-performance anion-exchange chromatography (Brenner et al., 1994).

## Biological Aspects

*Saccharomyces cerevisiae* kexin (Kex2p) processes several substrates *in vivo*, including pro-α-factor, pro-killer toxin, Hsp150p (Russo et al., 1992), a secreted exoglucanase precursor (Basco et al., 1990), and a variety of heterologous molecules (e.g. Brake, 1990). Other fungal kexins also cleave mating pheromone precursors and/or pro-killer toxins as well. The *Yarrowia lipolytica* kexin (Xpr6p) processes the precursor of an alkaline, extracellular protease (Enderlin & Ogrydziak, 1994). The existence of additional substrates for fungal kexins is indicated by the variety of phenotypes exhibited by mutant strains. *S. cerevisiae kex2* null mutants exhibit defective sporulation, cold-sensitive growth, hypersensitivity to several drugs and altered morphology (Leibowitz & Wickner, 1976; Martin & Young, 1989; Conklin et al., 1994; Komano & Fuller, 1995). *Y. lipolytica xpr1* mutants exhibit severe morphological abnormalities suggestive of cell wall defects (Enderlin & Ogrydziak, 1994). Mutation of the *S. pombe krp1* gene is lethal (Davey et al., 1994). Partial complementation of *kex2* mutant phenotypes led to the identification of the *YAP3* and *MKC7* genes that encode homologous, GPI-anchored cell surface aspartic proteases termed yapsin 1 (Chapter 304) and yapsin 2 (Chapter 305) (Egel-Mitani et al., 1990; Komano & Fuller, 1995). Synergistic effects of *kex2, yap3* and *mkc7* mutations suggest that there is at least a partial overlap in the physiological roles of kexin and the yapsins (Komano & Fuller, 1995).

Studies of the localization of *S. cerevisiae* kexin to the yeast *trans*-Golgi network (TGN) revealed the existence of a TGN-localization signal in the cytosolic tail (Wilcox et al., 1992). Clathrin heavy chain plays a role in TGN localization (Seeger & Payne, 1992), though possibly an indirect one (Redding et al., 1996a). Novel genes have been identified that are required for the signal-dependent localization of kexin (Redding et al., 1996b).

Perhaps of the greatest wider significance is that studies of the *KEX2* gene led directly to the identification of proprotein convertases in other eukaryotes, including mammals. These discoveries were foreshadowed by the demonstration that kexin could accurately process mammalian proproteins, including proalbumin and pro-opiomelanocortin (Bathurst et al., 1987; Thomas et al., 1988). Kexin, its mammalian homologs furin (Chapter 117), PACE4 (Chapter 121), PC1/3 (Chapter 118), PC2 (Chapter 119), PC4 (Chapter 120), PC5/6 (Chapter 122) (collectively reviewed in Rouille et al., 1995), LPC/PC8 (Chapter 123) (Bruzzaniti et al., 1996; Meerabux et al., 1996) and other similar eukaryotic enzymes define a distinct subfamily of subtilisins distinguished by: (a) physiological roles in biosynthetic, as opposed to degradative, proteolysis, (b) a higher degree of similarity within the subtilisin domain to one another than to subtilisin, (c) the presence of distinctive consensus sequences surrounding catalytic residues, (d) a high degree of selectivity for Arg at P1 and (e) the presence of the P-domain (also known as Homo B or the middle domain), which is not found in degradative subtilisins. Mutation of this domain blocks the intramolecular cleavage of the pro domain in both kexin and furin (Gluschankof & Fuller, 1994; Takahashi et al., 1995).

## Distinguishing Features

Kexin is distinguished as the prototype of a group of eukaryotic proprotein-processing enzymes. In assaying activity, kexin is distinguishable by its $Ca^{2+}$ requirement, sensitivity to heavy metals, resistance to generic serine protease inhibitors and by its pattern of specificities in cleaving peptide substrates and inactivation by peptidyl chloromethanes.

## Acknowledgement

The work of the author is supported by NIH grant GM39697.

## Further Reading

For a review, see Brenner *et al*. (1994).

## References

Achstetter, T. & Wolf, D.H. (1985) Hormone processing and membrane-bound proteinases in yeast. *EMBO J.* **4**, 173–177.

Angliker, H., Wikstrom, P., Shaw, E., Brenner, C. & Fuller, R.S. (1993) The synthesis of inhibitors for processing proteinases and their action on the Kex2 proteinase of yeast. *Biochem. J.* **293**, 75–81.

Basco, R.D., Giménez-Gallego, G. & Larriba, G. (1990) Processing of yeast exoglucanase (β-glucosidase) in a KEX2-dependent manner. *FEBS Lett.* **268**, 99–102.

Bathurst, I.C., Brennan, S.O., Carrell, R.W., Cousens, L.S., Brake, A.J. & Barr, P.J. (1987) Yeast KEX2 protease has the properties of a human proalbumin converting enzyme. *Science* **235**, 348–350.

Brake, A.J. (1990) Alpha-factor leader-directed secretion of heterologous proteins from yeast. *Methods Enzymol.* **185**, 408–421.

Brenner, C. & Fuller, R.S. (1992) Structural and enzymatic characterization of a purified prohormone-processing enzyme: secreted, soluble Kex2 protease. *Proc. Natl Acad. Sci. USA* **89**, 922–926.

Brenner, C., Bevan, A. & Fuller, R.S. (1993) One-step site-directed mutagenesis of the oxyanion hole of Kex2 protease. *Current Biol.* **3**, 498–506.

Brenner, C., Bevan, A. & Fuller, R.S. (1994) Biochemical and genetic methods for analyzing specificity and activity of a precursor-processing enzyme: yeast Kex2 protease, kexin. *Methods Enzymol.* **244**, 152–167.

Bruzzaniti, A., Goodge, K., Jay, P., Taviaux, S.A., Lam, M.H., Berta, P., Martin, T.J., Moseley, J.M. & Gillespie, M.T. (1996) PC8, a new member of the convertase family. *Biochem. J.* **314**, 727–731.

Conklin, D.S., Culbertson, M.R. & Kung, C. (1994) *Saccharomyces cerevisiae* mutants sensitive to the antimalarial and antiarrhythmic drug, quinidine. *FEMS Microbiol. Lett.* **119**, 221–227.

Davey, J., Davis, K., Imai, Y., Yamamoto, M. & Matthews, G. (1994) Isolation and characterization of krp, a dibasic endopeptidase required for cell viability in the fission yeast *Schizosaccharomyces pombe*. *EMBO J.* **13**, 5910–5921.

Enderlin, C.S. & Ogrydziak, D.M. (1994) Cloning, nucleotide sequence and functions of *XPR6*, which codes for a dibasic processing endoprotease from the yeast *Yarrowia lipolytica*. *Yeast* **10**, 67–79.

Egel-Mitani, M., Flygenring, H.P. & Hansen, M.T. (1990) A novel aspartyl protease allowing KEX2-independent *MFα* propheromone processing in yeast. *Yeast* **6**, 127–137.

Fuller, R.S., Brake, A.J. & Thorner, J. (1986) The yeast *KEX2* gene, required for processing prepro-α-factor, encodes a calcium-dependent endopeptidase that cleaves after Lys-Arg and Arg-Arg sequences. In: *Microbiology-1986* (Lieve, L., ed.). Washington, DC. American Society for Microbiology, pp. 273–278.

Fuller, R.S., Sterne, R.E. & Thorner, J. (1988) Enzymes required for yeast prohormone processing. *Annu. Rev. Physiol.* **50**, 345–362.

Fuller, R.S., Brake., A. & Thorner, J. (1989a) Yeast prohormone processing enzyme (*KEX2* gene product) is a Ca$^{2+}$-dependent serine protease. *Proc. Natl Acad. Sci. USA* **86**, 1434–1438.

Fuller, R.S., Brake, A.J. & Thorner, J. (1989b) Intracellular targeting and structural conservation of a prohormone-processing endoprotease. *Science* **246**, 482–486.

Germain, D., Dumas, F., Vernet, T., Bourbonnais, Y., Thomas, D.Y. & Boileau, G. (1992) The pro-region of the Kex2 endoprotease of *Saccharomyces cerevisiae* is removed by self-processing. *FEBS Lett.* **299**, 283–286.

Gluschankof, P. & Fuller, R.S. (1994) A C-terminal domain conserved in precursor processing proteases is required for intramolecular N-terminal maturation of pro-Kex2 protease. *EMBO J.* **13**, 2280–2288.

Julius, D., Brake, A., Blair, L., Kunisawa, R. & Thorner, J. (1984) Isolation of the putative structural gene for the lysine-arginine-cleaving endopeptidase required for processing of yeast prepro-alpha-factor. *Cell* **37**, 1075–1089.

Komano, H. & Fuller, R.S. (1995) Shared functions *in vivo* of a glycosyl-phosphatidylinositol-linked aspartyl protease, Mkc7, and the proprotein-processing protease Kex2 in yeast. *Proc. Natl Acad. Sci. USA* **92**, 10752–10756.

Leibowitz, M.J. & Wickner, R.B. (1976) A chromosomal gene required for killer plasmid expression, mating, and spore maturation in *Saccharomyces cerevisiae*. *Proc. Natl Acad. Sci. USA* **73**, 2061–2065.

Martin, C. & Young, R.A. (1989) *KEX2* mutations suppress RNA polymerase II mutants and alter the temperature range of yeast cell growth. *Mol. Cell. Biol.* **9**, 2341–2349.

Meerabux, J., Yaspo, M.L., Roebroek, A.J., Van de Ven, W.J., Lister, T.A. & Young, B.D. (1996) A new member of the proprotein convertase gene family (LPC) is located at a chromosome translocation breakpoint in lymphomas. *Cancer Res.* **56**, 448–451.

Mizuno, K., Nakamura, T., Takada, K., Sakakibara, S. & Matsuo, H. (1987) A membrane-bound, calcium-dependent protease in yeast alpha-cell cleaving on the carboxyl side of paired basic residues. *Biochem. Biophys. Res. Commun.* **144**, 807–814.

Mizuno, K., Nakamura, T., Ohshima, T., Tanaka, S. & Matsuo, H. (1988) Yeast *KEX2* gene encodes an endopeptidase homologous to subtilisin-like serine proteases. *Biochem. Biophys. Res. Commun.* **156**, 246–254.

Mizuno, K., Nakamura, T., Ohshima, T., Tanaka, S. & Matsuo, H. (1989) Characterization of *KEX2*-encoded endopeptidase from yeast *Saccharomyces cerevisiae*. *Biochem. Biophys. Res. Commun.* **159**, 305–311.

Redding, K., Seeger, M. Payne, G.S. & Fuller, R.S. (1996a) The effects of clathrin inactivation on localization of Kex2 protease are independent of the TGN-localization signal in the cytosolic tail of Kex2p. *Mol. Biol. Cell* **7**, 1667–1677.

Redding, K., Brickner, J.H., Marschall, L.G., Nichols, J.W. & Fuller, R.S. (1996b) Allele-specific suppression of a defective *trans*-Golgi network (TGN) localization signal in Kex2p identifies three genes involved in localization of TGN transmembrane proteins. *Mol. Cell. Biol.* **16**, 6208–6217.

Rockwell, N.C., Wang, G.T., Krafft, G.A. & Fuller, R.S. (1997) Internally consistent libraries of fluorogenic substrates demonstrate that Kex2 protease specificity is generated by multiple mechanisms. *Biochemistry* **36**, 1912–1917.

Rouille, Y., Duguay, S.J., Lund, K., Furuta, M., Gong, Q., Lipkind, G., Oliva, A.A., Jr, Chan, S.J. & Steiner, D.F. (1995) Proteolytic processing mechanisms in the biosynthesis of neuroendocrine peptides: the subtilisin-like proprotein convertases. *Front. Neuroendocrinol.* **16**, 322–361.

Russo, P., Kalkkinen, N., Sareneva, H., Paakkola, J. & Makarow, M. (1992) A heat shock gene from *Saccharomyces cerevisiae* encoding a secretory glycoprotein. *Proc. Natl Acad. Sci. USA* **89**, 3671–3675.

Seeger, M. & Payne, G.S. (1992) Selective and immediate effects of clathrin heavy chain mutations on Golgi membrane protein retention in *Saccharomyces cerevisiae. J. Cell Biol.* **118**, 531–540.

Smeekens, S.P. & Steiner, D.F. (1990) Identification of a human insulinoma cDNA encoding a novel mammalian protein structurally related to the yeast dibasic processing protease Kex2. *J. Biol. Chem.* **265**, 2997–3000.

Takahashi, S., Nakagawa, T., Kasai, K., Banno, T., Duguay, S.J., Van de Ven, W.J., Murakami, K. & Nakayama K. (1995) A second mutant allele of furin in the processing-incompetent cell line, LoVo. Evidence for involvement of the homo B domain in autocatalytic activation. *J. Biol. Chem.* **270**, 26565–26569.

Tanguy-Rougeau, C., Wesolowski-Louvel, M. & Fukuhara, H. (1988) The *Kluyveromyces lactis KEX1* gene encodes a subtilisin-type serine proteinase. *FEBS Lett.* **234**, 464–470.

Thomas, G., Thorne, B.A., Thomas, L., Allen, R.G., Hruby, D.E., Fuller, R. & Thorner, J. (1988) Yeast KEX2 endopeptidase correctly cleaves a neuroendocrine prohormone in mammalian cells. *Science* **241**, 226–230.

Wilcox, C.A. & Fuller, R.S. (1991) Posttranslational processing of the prohormone-cleaving Kex2 protease in the *Saccharomyces cerevisiae* secretory pathway. *J. Cell Biol.* **115**, 297–307.

Wilcox, C.A., Redding, K., Wright, R. & Fuller, R.S. (1992) Mutation of a tyrosine localization signal in the cytosolic tail of yeast Kex2 protease disrupts Golgi retention and results in default transport to the vacuole. *Mol. Biol. Cell* **3**, 1353–1371.

*Robert S. Fuller*
*Department of Biological Chemistry,*
*University of Michigan Medical School,*
*Ann Arbor, MI 48109-0606, USA*
*Email: bfuller@umich.edu*

# 116. *Limulus kexin*

## Databanks

*Peptidase classification: clan SB, family S8, MEROPS ID: S08.080*
*NC-IUBMB enzyme classification: none*
*Databank codes:*

| Species | SwissProt | PIR | EMBL (cDNA) | EMBL (genomic) |
|---|---|---|---|---|
| *Tachypleus tridentatus* | – | – | D83994 | – |
| **Related peptidases** | | | | |
| *Aedes aegypti* | – | – | L46373 | – |
| *Drosophila melanogaster* | P26016 | S17546 | X59384 | – |
| *Drosophila melanogaster* | P30430 | – | M81431 | – |
| *Drosophila melanogaster* | P30432 | A43434 | M94375 | – |

## Name and History

A number of kexin-like proteases (protein-convertase subfamily) with different substrate specificities are present in eukaryotic organisms, including mammals, insects, mollusks, cnidarians and protochordates. They are $Ca^{2+}$-dependent serine proteases with a subtilisin-like catalytic domain, responsible for proteolytic processing of proproteins at dibasic sites.

The hemolymph circulating in the chelicerata horseshoe crab contains granular hemocytes, comprising 99% of the total hemocytes. These are highly sensitive to bacterial endotoxins, which are lipopolysaccharides (Toh *et al*., 1991). The granular hemocytes of horseshoe crab store granule-specific proteins in two types of granules, large and small, which are released in response to external stimuli such as lipopolysaccharides and $Ca^{2+}$ ionophores (Muta & Iwanaga, 1996; Kawabata *et al*., 1996a). The granular components, which include serine protease zymogens and protease inhibitors participating in hemolymph coagulation, lectin-like proteins and antimicrobial substances, have been characterized and cloned. The cDNA sequence analyses have indicated that some of them are synthesized as preproproteins that have not only a signal peptide, but also an N-terminal propeptide that is linked to the mature protein through an -Arg-Xaa-Arg/Lys-Arg- motif, suggesting the presence of kexin-like processing proteases. In 1996, a new kexin-like protease (named *limulus kexin*) that is specifically expressed in the hemocytes was identified by PCR and cDNA cloning (Kawabata *et al*., 1996b).

## Structural Chemistry

The open reading frame of the cDNA encoded 752 amino acid residues of a preprotein containing a putative signal sequence of 27 residues, a propeptide of 112 residues, and a mature protein of 640 residues with a calculated $M_r$ of 71 040 (Kawabata *et al.*, 1996b). Amino acid residues functionally important for the catalytic triad, the oxyanion hole and a substrate-binding site are well conserved. The catalytic domain of limulus kexin exhibits striking sequence similarity to those of furins, in which *Drosophila* furin 1 is closest (79% identity). The degree of similarity to the mammalian proprotein convertases PC1 and PC2 (Chapters 118 and 119) is lower, however. A potential integrin-binding (Arg-Gly-Asp) sequence with unknown function in the kexin family is also conserved at the corresponding position. Furthermore, there is a Ser/Thr-rich domain containing 31% of these residues. This domain does not exist in proprotein convertases other than yeast kexin, being replaced in furins and PACE4 by a cysteine-rich domain.

## Biological Aspects

Northern blot analysis with poly(A)$^+$ RNAs prepared from hemocytes, heart, hepatopancreas, stomach, intestine and skeletal muscle detected the expression of limulus kexin only in hemocytes, with three different sizes of transcripts, 3.4 kb, 6.4 kb and 7.7 kb; no transcripts were detectable in other tissues (Kawabata *et al.*, 1996b). Limulus kexin has a hydrophobic segment at the C-terminus, and may therefore bind to a membrane, but it does not have a cytoplasmic tail as do the yeast kexin and furins. Since the cytoplasmic tail of mammalian furins is known to play an important role in their localization at the *trans*-Golgi network (Bosshart *et al.*, 1994; Chapman & Munro, 1994), the subcellular localization of limulus kexin may be different.

## References

Bosshart, H., Humphrey, J., Deignan, E., Davidson, J., Drazba, J., Yuan, L., Oorshot, V., Peters, P.J. & Bonifacino, J.S. (1994) The cytoplasmic domain mediates localization of furin to the *trans*-Golgi network en route to the endosomal/lysosomal system. *J. Cell Biol.* **126**, 1157–1172.

Chapman, R.E. & Munro, S. (1994) Retrieval of TGN proteins from the cell surface requires endosomal acidification. *EMBO J.* **13**, 2305–2312.

Kawabata, S., Muta, T. & Iwanaga, S. (1996a) Clotting cascade and defense molecule found in hemolymph of horseshoe crab. In: *Invertebrate Immunology* (Söderhäll, G. *et al.*, eds). Fair Haven: SOS Publications, pp. 255–284.

Kawabata, S., Saeki, K. & Iwanaga, S. (1996a) Limulus kexin: a new type of Kex2-like endoprotease specifically expressed in hemocytes of the horseshoe crab. *FEBS Lett.* **386**, 201–204.

Muta, T. & Iwanaga, S. (1996) Clotting and immune defense in Limulidae. In: *Progress in Molecular and Subcellular Biology*, vol. 15: *Invertebrate Immunology* (Rinkevich, B. & Müller, W.E.G., eds). Berlin, Heidelberg, New York: Springer-Verlag, pp. 154–189.

Toh, Y., Mizutani, A., Tokunaga, F., Muta, T. & Iwanaga, S. (1991) Morphology of the granular hemocytes of the Japanese horseshoe crab *Tachypleus tridentatus* and immunocytochemical localization of clotting factors and antimicrobial substances. *Cell Tissue Res.* **266**, 137–147.

*Shun-ichiro Kawabata*
*Department of Biology,*
*Faculty of Science,*
*Kyushu University,*
*Fukuoka, 812-81, Japan*
*Email: skawascb@mbox.nc.kyushu-u.ac.jp*

*Sadaaki Iwanaga*
*Department of Biology,*
*Faculty of Science,*
*Kyushu University,*
*Fukuoka, 812-81, Japan*

# 117. Furin

## Databanks

*Peptidase classification: clan SB, family S8, MEROPS ID: S08.071*
*NC-IUBMB enzyme classification: EC 3.4.21.75*
*ATCC entries: 79822, 79823 (4.10 kb of human furin), 63248 (2.75 kb of mouse furin)*
*Databank codes:*

| Species | SwissProt | PIR | EMBL (cDNA) | EMBL (genomic) |
|---|---|---|---|---|
| *Bos taurus* | – | S41191 | X75956 | – |
| *Caenorhabditis elegans* | – | – | U12682 | – |

| Species | SwissProt | PIR | EMBL (cDNA) | EMBL (genomic) |
|---------|-----------|-----|-------------|----------------|
| *Drosophila melanogaster* | P26016 | S17546 | X59384 | – |
| *Drosophila melanogaster* | P30430 | – | M81431 | – |
| *Drosophila melanogaster* | P30432 | A43434 | M94375 | – |
| *Gallus gallus* | – | – | Z68093 | – |
| *Homo sapiens* | P09958 | A24892 A38424 A39552 S08226 | X17094 | X15723: complete gene sequence |
| *Lymnaea stagnalis* | – | S43656 | S69833 | – |
| *Mesocricetus longicaudatus* | – | – | U20434 | – |
| *Mesocricetus longicaudatus* | – | – | U20435 | – |
| *Mesocricetus longicaudatus* | – | – | U20436 | – |
| *Mus musculus* | P23188 | A23679 | L26489 X54056 | – |
| *Mus musculus* | – | A23679 I49677 | X54056 | – |
| *Rattus norvegicus* | P23377 | S13106 | X55660 | – |
| *Spodoptera frugiperda* | – | – | Z68888 | – |
| *Xenopus laevis* | P29119 | A41627 B41627 | M80471 | – |

## Name and History

The name *furin* is derived from the name of its corresponding gene *FUR* (Van de Ven *et al*., 1990). This gene is located immediately upstream of the *FES/FPS* proto-oncogene (*FUR: FES/FPS* **u**pstream **r**egion), and was discovered serendipitously by Roebroek *et al*. (1986) during studies which were designed to molecularly characterize the human *FES/FPS* gene. Based on sequence homology, it was shown (Van den Ouweland *et al*., 1990) to be the mammalian homolog of the yeast endopeptidase kexin (Chapter 115) (Fuller *et al*., 1989). By some groups furin was given the acronym *PACE* (**p**aired basic **a**mino acid residue **c**leaving **e**nzyme) (Barr *et al*., 1991). When additional family members were cloned they were all called prohormone (or proprotein) convertases (e.g. PC1, PC2, etc.) with the exception of PACE4 (for review, see Steiner *et al*., 1992; Van de Ven *et al*., 1993; Seidah *et al*., 1994). Due to the fact that several convertases were cloned simultaneously by different groups, different names were given to the same enzyme (e.g. PC1 versus PC3, PC5 versus PC6 and PC7 versus LPC) (see Chapters 118, 122 and 123).

In an attempt to solve the problem of the confusing nomenclature, it was proposed to rename all mammalian convertases as SPCs (**s**ubtilisin-like **p**roprotein **c**onvertases) (Steiner *et al*., 1992). In this proposal, furin was given the name *SPC1*. However, this nomenclature has never been generally accepted.

## Activity and Specificity

The furin consensus cleavage site is C-terminal to the sequence motif -Arg-Xaa-Lys/Arg-Arg↓ (Hosaka *et al*., 1991). Alternative motifs in which the presence of a P6 Arg can compensate for the absence of a basic residue at either the P4 or P2 position have also been reported (Watanabe

*et al*., 1992). Under overexpression conditions, it has been found that substrates with the motifs -Arg-Xaa-Xaa-Arg↓ and -Arg/Lys-Arg↓ are also cleaved (Molloy *et al*., 1992; Creemers *et al*., 1993). However, cleavage does not occur at all such motifs within a protein. The residues at positions P2′, P3 and P5 and secondary protein structures like $\beta$ turns may provide additional determinants for endoproteolytic processing (Rholam *et al*., 1986; Siezen *et al*., 1994).

The pH optimum of furin is 7.0, but activities higher than 50% of maximal activity have been observed between pH 6.0 and 8.5 (Hatsuzawa *et al*., 1992; Molloy *et al*., 1992). Furin requires $Ca^{2+}$, with an optimal concentration of 1 mM. Furin activity can be completely inhibited by $Ca^{2+}$ chelators like EDTA and EGTA, by the heavy metal ions $Zn^{2+}$, $Hg^{2+}$ and $Cu^{2+}$, by the reducing agent DTT and by *p*-chloromercuribenzenesulfonic acid. Moderate inhibition was observed with the serine protease inhibitors DFP, PMSF and antipain (Hatsuzawa *et al*., 1992; Molloy *et al*., 1992). Tailor-made inhibitors are mutants of $\alpha_1$-proteinase inhibitor and peptidyl chloroalkanes containing the -Arg-Xaa-Lys/Arg-Arg↓ motif (Misumi *et al*., 1990; Hallenberger *et al*., 1992).

## Structural Chemistry

Furin is a type I transmembrane protein composed of a signal peptide, a propeptide terminating in an endoproteolytic cleavage site comprised of a cluster of basic residues, a subtilisin-like catalytic domain, a middle domain, and a cysteine-rich domain, followed by a C-terminal transmembrane anchor and a cytosolic tail (see Figure 118.1). Furin is sythesized as a precursor protein of about 100 kDa that is converted into a 94 kDa protein by autoproteolytic cleavage of the propeptide (Creemers *et al*., 1993). Furin is glycosylated and phosphorylated (Creemers *et al*., 1992; Jones *et al*., 1995).

The three-dimensional structure of the catalytic domain of furin and its interaction with substrates is predicted (Siezen *et al.*, 1994), based on homology modeling from the crystal structures of subtilisin BPN′ (Chapter 94) and thermitase (Chapter 95). In this model, two disulfide bonds and two $Ca^{2+}$-binding sites have been predicted. However, the most remarkable feature is the large number of acidic residues in and near the S1, S2 and S4 subsites of the substrate-binding region, which have been shown to be the main determinant of specificity for basic substrate segments (Creemers *et al.*, 1993; Siezen *et al.*, 1994).

## Preparation

The *FUR* gene is ubiquitously expressed at rather low levels (Van Duijnhoven *et al.*, 1992; Seidah *et al.*, 1994). For this reason, furin has mainly been purified from transfected cells. Recombinant soluble furin, lacking the transmembrane anchor, has been purified from culture medium (Hatsuzawa *et al.*, 1992; Molloy *et al.*, 1992). Purification of full-length furin from transfected cells has been described by Molloy *et al.* (1994) and Ayoubi *et al.* (1996).

## Biological Aspects

Expression of the *FUR* gene is directed by three alternative promoters, two of which are house-keeping (GC-rich), and one regulated (TATA-containing), suggesting that their differential use may be a mechanism to modulate levels of the enzyme (Ayoubi *et al.*, 1994). The TATA-containing promoter was found to be transactivated by transcription factor C/EBPb, but not C/EBPa or C/EBPd.

Furin is cotranslationally translocated into the lumen of the endoplasmic reticulum where its propeptide is cleaved. *N*-linked oligosaccharide chains are subsequently modified, and the protein takes up residence in the *trans*-Golgi network (TGN) (Creemers *et al.*, 1993, 1995, 1996; Molloy *et al.*, 1994; Vey *et al.*, 1994). The information required for the steady-state concentration of furin in the TGN and the targeted recycling from the cell surface is contained in its cytoplasmic tail, which contains several independent trafficking signals (Molloy *et al.*, 1994; Schäfer *et al.*, 1995; Jones *et al.*, 1995).

Furin has been shown to be able to process a variety of proproteins in the exocytotic pathway into their mature products, including growth factors, receptors, neurotrophic factors and serum factors. In addition, several coat proteins of enveloped viruses (e.g. gp160 of HIV: Hallenberger *et al.*, 1992) and secreted bacterial pathogens (e.g. anthrax toxin: Molloy *et al.*, 1992) are efficiently cleaved. All these data are obtained from *in vitro* studies, so it remains to be established whether or not these are physiological substrates of furin.

No disease has yet been associated with the absence or malfunctioning of furin. Preliminary data on the functional inactivation of *FUR* in mice by gene targeting suggest that the absence of furin is lethal early during embryogenesis (A.J.M. Roebroek and coworkers, unpublished results).

## Distinguishing Features

Distinguishing furin from other prohormone convertases on the basis of inhibitor profiles has not yet been described. However, specific antibodies have been produced by many groups, including a panel of monoclonal antibodies (Van Duijnhoven *et al.*, 1992). These monoclonal antibodies are directed against different domains of furin and with the exception of one monoclonal antibody which is specific for human furin, they cross-react with mouse, rat and cattle furins.

## Further Reading

An extensive review of the role of different convertases in the regulated and constitutive exocytic pathway is that of Halban & Irminger (1994).

## References

Ayoubi, T.A.Y., Creemers, J.W.M., Roebroek, A.J.M. & Van de Ven, W.J.M. (1994) Expression of the dibasic proprotein processing enzyme furin is directed by multiple promoters. *J. Biol. Chem.* **269**, 9298–9303.

Ayoubi, T.A.Y., Meulemans, S.M.P., Roebroek, A.J.M. & Van de Ven W.J.M. (1996) Production of recombinant proteins in Chinese hamster ovary cells overexpressing the subtilisin-like proprotein converting enzyme furin. *Mol. Biol. Rep.* **23**, 87–95.

Barr, P.J., Mason, O.B., Landsberg, K.E., Wong, P.A., Kiefer, M.C. & Brake A.J. (1991) cDNA and gene structure for a human subtilisin-like protease with cleavage specificity for paired basic amino acid residues. *DNA Cell Biol.* **10**, 319–328.

Creemers, J.W.M., Roebroek, A.J.M., Van den Ouweland, A.W.M., Van Duijnhoven, J.L.P. & Van de Ven, W.J.M. (1992) Cloning and functional expression of a 4.3 kbp mouse *FUR* cDNA: evidence for differential transcription. *Mol. Biol. (Life Sci. Adv.)* **11**, 127–138.

Creemers, J.W.M., Siezen, R.J., Roebroek, A.J.M., Ayoubi, T.A.Y., Huylebroeck, D. & Van de Ven, W.J.M. (1993) Modulation of furin-mediated proprotein processing by site-directed mutagenesis. *J. Biol. Chem.* **268**, 21826–21834.

Creemers, J.W.M., Vey, M., Schäfer, W., Ayoubi, T.A.Y., Roebroek, A.J.M., Klenk, H.-D., Garten, W. & Van de Ven, W.J.M. (1995) Endoproteolytic cleavage of its propeptide is a prerequisite for efficient transport of furin out of the endoplasmic reticulum. *J. Biol. Chem.* **270**, 2695–2702.

Creemers, J.W.M., Usac, E.F., Bright, N.A., Van de Loo, J.-W., Jansen, E., Van de Ven, W.J.M. & Hutton, J.C. (1996) Identification of a transferable sorting domain for the regulated pathway in the prohormone convertase PC2. *J. Biol. Chem.* **271**, 25284–25291.

Fuller, R.S., Brake, A.J. & Thorner, J. (1989) Intracellular targeting and structural conservation of a prohormone-processing endoprotease. *Science* **246**, 482–486.

Halban, P.A. & Irminger J.-C. (1994) Sorting and processing of secretory proteins. *Biochem. J.* **299**, 1–18.

Hallenberger, S., Bosch, V., Angliker, H., Shaw, E., Klenk, H.-D. & Garten, W. (1992) Inhibition of furin mediated cleavage activation of HIV-1 glycoprotein gp160. *Nature* **360**, 358–361.

Hatsuzawa, K., Nagahama, M., Takahashi, S., Takada, K., Murakami, K. & Nakayama, K. (1992) Purification and characterization of furin, a Kex2-like processing endoprotease, produced in Chinese hamster ovary cells. *J. Biol. Chem.* **267**, 16094–16099.

Hosaka, M., Nagahama, M., Kim, W.S., Watanabe, T., Hatsuzawa, K., Ikemizu, J., Murakami, K. & Nakayama, K. (1991) Arg-X-Lys/Arg-Arg motif as a signal for precursor cleavage catalyzed by furin within the constitutive secretory pathway. *J. Biol. Chem.* **266**, 12127–12130.

Jones, B.G., Thomas, L., Molloy, S.S., Thulin, C.D., Fry, M.D., Walsh, K.A. & Thomas, G. (1995) Intracellular trafficking of furin is modulated by the phosphorylation state of a casein kinase II site in its cytoplasmic tail. *EMBO J.* **14**, 5869–5883.

Misumi, Y., Ohkubo, K., Sohda, M., Takami, N., Oda, K. & Ikehara, Y. (1990) Intracellular processing of complement pro-C3 and proalbumin is inhibited by rat $\alpha_1$-protease inhibitor variant (Met$^{352}$ → Arg) in transfected cells. *Biochem. Biophys. Res. Commun.* **171**, 236–242.

Molloy, S.S., Bresnahan, P.A., Leppla, S.H., Klimpel, K.R. & Thomas, G. (1992) Human furin is a calcium-dependent serine endoprotease that recognizes the sequence Arg-X-X-Arg and efficiently cleaves anthrax toxin protective antigen. *J. Biol. Chem.* **267**, 16396–16402.

Molloy, S.S., Thomas, L., Van Slyke J.K., Stenberg, P.E. & Thomas, G. (1994) Intracellular trafficking and activation of the furin proprotein convertase: localization to the TGN and recycling from the cell surface. *EMBO J.* **13**, 18–33.

Rholam, M., Nicolas, P. & Cohen, P. (1986) Precursors of peptide hormones share common secondary structures forming features at the proteolytic sites. *FEBS Lett.* **207**, 1–6.

Roebroek, A.J.M., Schalken, J.A., Bussemakers, M.J.G., Van Heerikhuizen, H., Onnekink, C., Debruyne F.M.J., Bloemers, H.P.J. & Van de Ven, W.J.M. (1986) Characterization of human *FES/FPS* reveals a new transcription unit (*FUR*) in the immediately upstream region of the proto-oncogene. *Mol. Biol. Rep.* **11**, 117–125.

Schäfer, W., Stroh, A., Berghöfer, S., Seiler, J., Vey, M., Kern, M.F., Klenk, H.-D. & Garten, W. (1995) Two independent targeting signals in the cytoplasmic domain determine *trans*-Golgi network localization and endosomal trafficking of the proprotein convertase furin. *EMBO J.* **14**, 2424–2435.

Seidah, N.G., Chretien, M. & Day, R. (1994) The family of subtilisin/kexin like pro-protein and pro-hormone convertases: divergent or shared functions. *Biochimie* **76**, 197–209.

Siezen, R.J., Creemers, J.W.M. & Van de Ven, W.J.M. (1994) Subtilisin-like proprotein convertases: Homology modelling of the catalytic domain of human furin. *Eur. J. Biochem.* **222**, 255–266.

Steiner, D.F., Smeekens, S.P., Ohagi, S. & Chan, S.J. (1992) The new enzymology of precursor processing endoproteases. *J. Biol. Chem.* **267**, 23435–23438.

Van den Ouweland, A.M.W., Van Duijnhoven, J.L.P., Keizer, G.D., Dorssers, L.C.J & Van de Ven, W.J.M. (1990) Structural homology between the human *FUR* gene product and the subtilisin-like protease encoded by yeast *KEX2*. *Nucleic Acids Res.* **18**, 664.

Van de Ven, W.J.M., Voorberg, J., Fontijn, R., Pannekoek, H., Van den Ouweland, A.M.W. & Siezen, R.J. (1990) Furin is a subtilisin-like proprotein-processing enzyme in higher eukaryotes. *Mol. Biol. Rep.* **14**, 265–275.

Van de Ven, W.J.M., Van Duijnhoven, J.L.P. & Roebroek, A.J.M. (1993) Structure and function of eukaryotic proprotein processing enzymes of the subtilisin family of serine proteases. *Crit. Rev. Oncog.* **4**, 115–136.

Van Duijnhoven, J.L.P., Creemers, J.W.M., Kranenborg, M.G.C., Timmer, E.D.J., Groeneveld A., Van den Ouweland, A.M.W., Roebroek, A.J.M. & Van de Ven, W.J.M. (1992) Development and characterization of a panel of monoclonal antibodies against the novel subtilisin-like proprotein processing enzyme furin. *Hybridoma* **11**, 71–86.

Vey, M., Schäfer, W., Berghöfer, S., Klenk, H.-D. & Garten, W. (1994) Maturation of the *trans*-Golgi network protease furin: compartmentalization of propeptide removal, substrate cleavage, and COOH-terminal truncation. *J. Cell Biol.* **127**, 1829–1842.

Watanabe, T., Nakagawa, T., Ikemizu, J., Nagahama, M., Murakami, K. & Nakayama, K. (1992) Sequence requirements for precursor cleavage within the constitutive secretory pathway. *J. Biol. Chem.* **267**, 8270–8274.

*John W.M. Creemers*
*Laboratory for Molecular Oncology,*
*Center for Human Genetics,*
*University of Leuven and Flanders*
*Interuniversity Institute for Biotechnology,*
*Herestraat 49, B-3000 Leuven, Belgium*
*Email: john.creemers@med.kuleuven.ac.be*

*Wim J.M. Van de Ven*
*Laboratory for Molecular Oncology,*
*Center for Human Genetics,*
*University of Leuven and Flanders*
*Interuniversity Institute for Biotechnology,*
*Herestraat 49, B-3000 Leuven, Belgium*
*Email: wim.vandeven@med.kuleuven.ac.be*

# *118. Proprotein convertase 1*

## *Databanks*

*Peptidase classification: clan SB, family S8, MEROPS ID: S08.072*
*NC-IUBMB enzyme classification: EC 3.4.21.93*

*Databank codes:*

| Species | SwissProt | PIR | EMBL (cDNA) | EMBL (genomic) |
|---------|-----------|-----|-------------|----------------|
| *Aplysia californica* | – | – | L28767 | – |
| *Aplysia californica* | – | – | U40481 | – |
| *Branchiostoma californiensis* | – | – | U22052 | – |
| *Caenorhabditis elegans* | P51559 | – | L29438 | – |
| | | | L29439 | |
| | | | L29440 | |
| *Homo sapiens* | P29120 | I52991 | M90753 | U24128: promoter and 5′ flanking region |
| | | S21106 | X64810 | |
| *Hydra attenuata* | P29145 | – | M95932 | – |
| *Hydra attenuata* | P29146 | A46184 | M95931 | – |
| *Lophius americanus* | – | – | U01910 | |
| *Mus musculus* | P21662 | A35571 | M55668 | S74618: 5′ flanking region |
| | | A37951 | M58589 | |
| | | A39002 | M69196 | |
| | | A39604 | X57088 | |
| | | A46622 | | |
| | | JX0171 | | |
| | | S19165 | | |
| *Rattus norvegicus* | P28840 | A41556 | M76705 | – |
| | | S27361 | M83745 | |
| | | S36358 | | |
| *Sus scrofa* | – | – | U20545 | – |

## Name and History

In the early 1960s it was proposed that polypeptide hormones are first synthesized as inactive precursors that require specific cleavage following pairs of basic residues (such as Lys-Arg-, Arg-Arg-, Lys-Lys- and Arg-Lys-) in order to release the active hormone. Since then, this model has been extended to other precursor proteins, as it is also applicable to progrowth factors, proneurotrophic factors, hormonal receptors, adhesion molecules, retroviral surface glycoproteins, proenzymes and even certain protoxins. The search for the physiologically important processing enzymes, the proprotein convertases (PCs), was very laborious, and a number of laboratories, including our own, participated actively in the hunt (Seidah *et al.*, 1991a). The major breakthrough came in 1984, from the molecular identification of the convertase responsible for the activation of the yeast α-mating factor and killer toxin. The proteinase identified by genetic complementation of a *KEX2* mutant strain was found to be a subtilisin-like serine proteinase (Fuller *et al.*, 1988; Mizuno *et al.*, 1989), now called kexin (Chapter 115). The search for the mammalian counterpart of kexin took about five years, before it was realized through computer database searches for sequence identity to kexin that a partial human genomic sequence coding for a protein called furin had already been reported by Roebroek and colleagues in 1986 (Roebroek *et al.*, 1986). In the reported DNA sequence only the active-site Ser and the catalytically important Asn residue found in all subtilisin-like proteases were identified. The complete sequence of the 5′ end of the gene was completed in 1990 and it included the other two active-site residues, Asp and His (Van de Ven *et al.*, 1990). In 1989, the partial sequence of furin (from the catalytically important Asn up to the C-terminus) (Roebroek *et al.*,

1986) and the full sequence of kexin (Mizuno *et al.*, 1989) were known. Accordingly, based on the concept of sequence conservation around the active sites of serine proteinases, PCR applied to mRNA amplification (reverse transcriptase PCR, RT-PCR) allowed two laboratories to simultaneously isolate for the first time another mammalian homolog of kexin, known as *PC1* (Seidah *et al.*, 1990, 1991b) or *PC3* (Smeekens *et al.*, 1991) representing the first endocrine- and neuroendocrine-processing enzyme molecularly characterized in mammalian tissues. Thus, PCR amplification on cDNA obtained from mouse pituitary total RNA using oligonucleotides coding for the sequence around the catalytically important Asn and the active-site Ser of human furin, allowed the isolation of a 260 bp probe (Seidah *et al.*, 1990, 1991b). The screening of libraries from mouse pituitary and mouse insulinoma with this probe led to the isolation of the full-length cDNA clone coding for mouse PC1 (Seidah *et al.*, 1990, 1991b). This was the first of a cascade of events which led ultimately to the identification of seven different PCs (Figure 118.1). The name *proprotein convertase 1 (PC1)* has been recommended by IUBMB.

## Activity and Specificity

PC1 is a 753 amino acid enzyme usually responsible for the processing of precursors whose products are stored in dense-core secretory vesicles. These include mainly endocrine and neural polypeptide hormones, such as pro-opiomelanocortin (POMC), prosomatostatin, proenkephalin, proinsulin, prodynorphin and prothyrotropin-releasing hormone (proTRH). In general, PC1 cleaves after dibasic or monobasic sequences of general structures (Lys/Arg)-Arg↓ or

No. Amino Acids

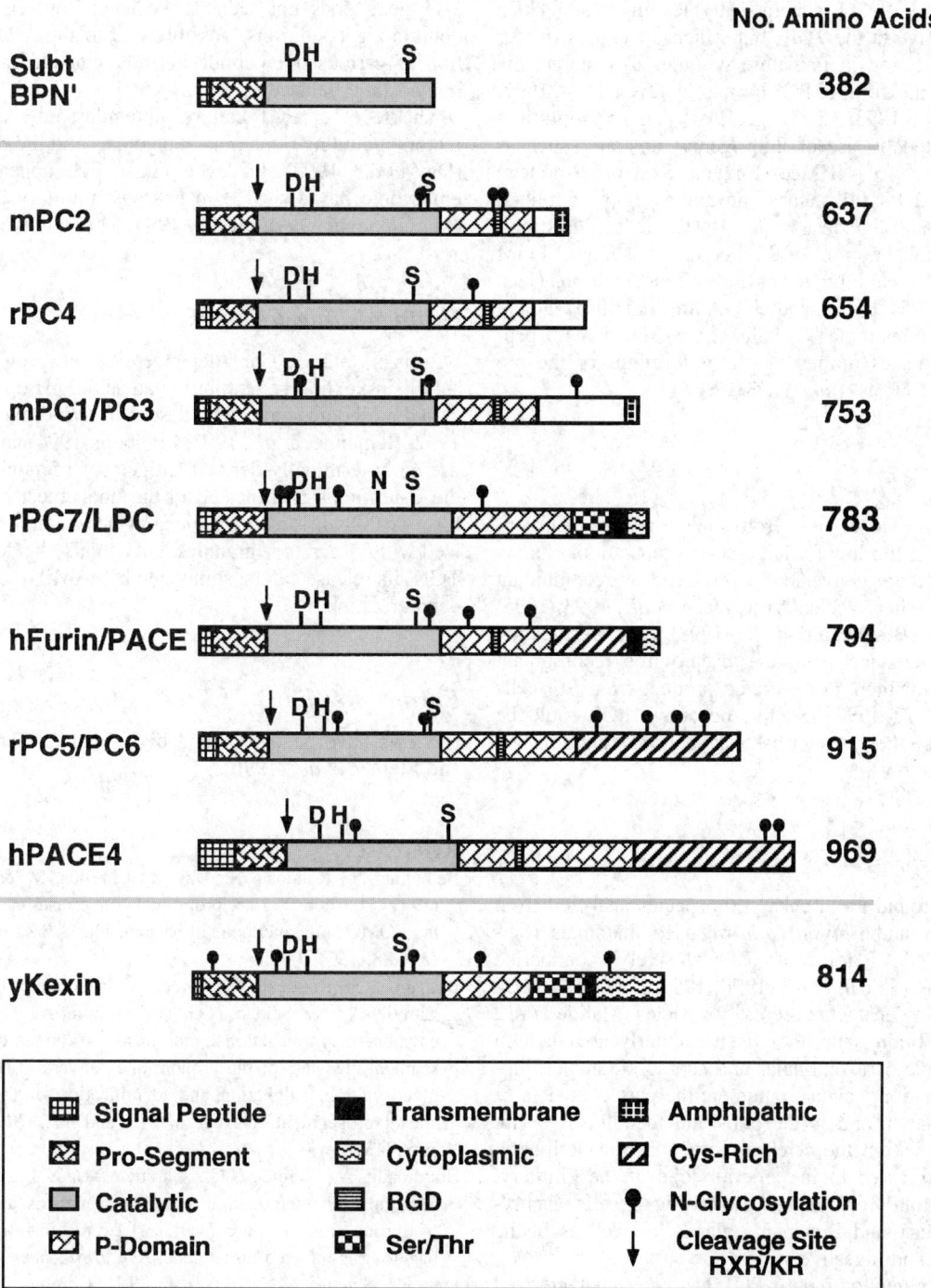

| | No. Amino Acids |
|---|---|
| Subt BPN' | 382 |
| mPC2 | 637 |
| rPC4 | 654 |
| mPC1/PC3 | 753 |
| rPC7/LPC | 783 |
| hFurin/PACE | 794 |
| rPC5/PC6 | 915 |
| hPACE4 | 969 |
| yKexin | 814 |

Key:

- Signal Peptide
- Pro-Segment
- Catalytic
- P-Domain
- Transmembrane
- Cytoplasmic
- RGD
- Ser/Thr
- Amphipathic
- Cys-Rich
- N-Glycosylation
- Cleavage Site RXR/KR

*Figure 118.1*  Schematic alignment of the mammalian proprotein convertases in comparison with subtilisin and kexin. *Key*: Subt BPN′, *Bacillus subtilis* subtilisin BPN′ (Chapter 94); mPC2, mouse proprotein convertase 2 (Chapter 119); rPC4, rat proprotein convertase 4 (Chapter 120); mPC1/PC3, mouse proprotein convertase 1; rPC7/LPC, proprotein convertase 7 (Chapter 123); hFurin/PACE, human furin (Chapter 117); rPC5/PC6, rat proprotein convertase 5 (Chapter 122); hPACE4, human proprotein convertase PACE4 (Chapter 121); yKexin, *Saccharomyces cerevisiae* kexin (Chapter 115).

(Arg/Lys)-(Xaa)$_n$-Arg↓, where $n = 2$, 4 or 6 (Seidah, 1995). Except in the case of proTRH, for which cleavage by PC1 leads to the production of the tripeptide TRH, PC1 generally acts early in the secretory pathway, usually within the trans-Golgi network (TGN) or immature secretory granules, and generates products of intermediate size that are then processed further by the second neural endocrine convertase PC2 (Chapter 119) (Benjannet *et al.*, 1991; Seidah *et al.*, 1994; Rouille *et al.*, 1995).

PC1 is a soluble, Ca$^{2+}$-dependent enzyme first synthesized as a zymogen (proPC1), which undergoes autocatalytic processing of its pro segment within the endoplasmic reticulum

(Benjannet *et al.*, 1993; Lindberg, 1994) leading to an 84 kDa form that is active in the TGN, but which undergoes further C-terminal truncation in immature granules to generate the most active 66 kDa form of PC1 (Vindrola & Lindberg, 1992; Benjannet *et al.*, 1993; Lindberg, 1994). The pH optimum of PC1 estimated by use of fluorogenic substrates such as Glp-Arg-Thr-Lys-Arg↓NHMec (Enzyme Systems Products) (see Appendix 2 for full names and addresses of suppliers) is between 5.5 and 6.0 (Jean *et al.*, 1993). Activity is stimulated by low concentrations of thiol compounds (e.g. 0.1 mM DTT, 1 mM ME), but inhibited at higher concentrations (Jean *et al.*, 1993, 1995). EDTA and EGTA are inhibitory. Potent inhibition (low nanomolar $K_i$ values) was shown with peptidyl chloromethanes mimicking the C-terminus of the pro segment of PC1 (Jean *et al.*, 1993).

## Preparation

In general, the yield of PC1 from endocrine and neuroendocrine tissues is too low for large-scale preparation. Human and mouse PC1 are conveniently produced as recombinant products from either vaccinia virus (Jean *et al.*, 1993, 1995) or baculovirus (Boudreault *et al.*, 1998). The purification of the mouse enzyme from the baculovirus recombinant yields about 1 mg liter$^{-1}$ of secreted product from Sf9 cells (Boudreault *et al.*, 1998). So far, no active PC1 could be generated from either yeast or bacterial cultures (unpublished results of the author).

## Biological Aspects

PC1 has been found in all eukaryotic species analyzed from *Hydra vulgaris* and *Aplysia californica* to mammals (Seidah *et al.*, 1997). It is found almost exclusively in endocrine and neural cells (Seidah *et al.*, 1990, 1991b) and its 66 kDa form is stored in dense-core secretory vesicles (Malide *et al.*, 1995). Within brain neurons, it is particularly abundant in the hypothalamic paraventricular and supraoptic nuclei, hippocampal formation, caudate putamen, the pars compacta of the mesencephalon, and in the pons and medulla. It is not very abundant within the cerebral cortex. In the periphery, it is highly expressed in the anterior lobe of the pituitary, thyroid and parathyroid glands, pancreatic $\beta$ cells, lachrymal, submaxillary and Bowman's glands, as well as in the cardiovascular and digestive systems.

The gene coding for mouse PC1 has been characterized (Ftouhi *et al.*, 1994) and it encodes 15 exons and 14 introns of various sizes, estimated to cover a 42 kb gene. The active-site residues Asp, His, Ser and the oxyanion hole Asn are each on separate exons. PC1 maps to human and mouse chromosomes 5q15–q21 and 13 (C1–C3 band), respectively (Seidah *et al.*, 1994). The promoter of PC1 is somewhat GC rich and does not contain TATA or CAAT boxes (Ftouhi *et al.*, 1994).

A human patient exhibiting a phenotype of obesity and diabetes has recently been identified as having an inactive PC1 gene; both alleles are affected by either a missense mutation or an early termination codon, both leading to inactive protein (O'Rahilly *et al.*, 1995). Thus, silencing of the PC1 gene does not seem to be lethal, but could result in debilitating conditions, possibly due to the reduced production of various polypeptide hormones needed to maintain a homeostatic balance.

The level of PC1 can be upregulated by a number of factors including dopamine antagonists and the thyroid status (Day *et al.*, 1992). In development, PC1 appears following embryonic day 13 (e13) in neurons, pituitary and pancreas (Marcinkiewicz *et al.*, 1993, 1994; Zheng *et al.*, 1994).

## Distinguishing Features

PC1 is a soluble, $Ca^{2+}$-dependent serine proteinase of the subtilisin-kexin type, exhibiting an acidic pH optimum. Polyclonal antibodies have been described (Vindrola & Lindberg, 1992; Benjannet *et al.*, 1993; Lindberg, 1994) and it is best to use an N-terminally derived antigen for immunization, since the C-terminus is truncated in the final, most active 66 kDa form of PC1. It is a secretable enzyme which is mostly localized within secretory granules, but can also be detected in the TGN. Its release can be stimulated by cAMP or by membrane depolarization.

## Further Reading

Reviews have been provided by Seidah & Chrétien (1994) and Steiner *et al.* (1996).

## References

Benjannet, S., Rondeau, N., Day, R., Chrétien, M. & Seidah, N.G. (1991) PC1 and PC2 are proprotein convertases capable of cleaving POMC at distinct pairs of basic residues. *Proc. Natl Acad. Sci. USA* **88**, 3564–3568.

Benjannet, S., Rondeau, N., Paquet, L., Boudreault, A., Lazure, C., Chrétien, M. & Seidah, N.G. (1993) Comparative biosynthesis, covalent post-translational modifications and efficiency of prosegment cleavage of the prohormone convertases PC1 and PC2: glycosylation, sulphation and identification of the intracellular site of prosegment cleavage of PC1 and PC2. *Biochem. J.* **294**, 735–743.

Boudreault, A., Seidah, N.G., Chrétien, M. & Lazure, C. (1998) Molecular characterization, enzymatic analysis and purification of murine prohormone convertase-1 expressed from recombinant baculovirus-infected insect cells. *Eur. J. Biochem.*, in the press.

Day, R., Schäfer, M.K.H., Watson, S.J., Chrétien, M. & Seidah, N.G. (1992) Distribution and regulation of the prohormone convertases PC1 and PC2 in the rat pituitary. *Mol. Endocrinol.* **6**, 485–497.

Ftouhi, N., Day, R., Mbikay, M., Chrétien, M. & Seidah, N.G. (1994) Gene organization of the mouse pro-hormone and proprotein convertase PC1. *DNA Cell Biol.* **13**, 395–407.

Fuller, R.S, Sterne, R.E. & Thorner, J. (1988) Enzymes required for yeast prohormone processing. *Annu. Rev. Biochem.* **50**, 345–362.

Jean, F., Basak, A., Rondeau, N., Benjannet, S., Hendy, G.N., Seidah, N.G., Chrétien, M. & Lazure, C. (1993) Enzymic characterization of murine and human prohormone convertase-1 (mPC1 and hPC1) expressed in mammalian GH4C1 cells. *Biochem. J.* **292**, 891–900.

Jean, F., Basak, A., Dimaio, J., Seidah, N.G. & Lazure, C. (1995) An internally quenched fluorogenic substrate of prohormone convertase 1 and furin leads to a potent prohormone convertase inhibitor. *Biochem. J.* **307**, 689–695.

Lindberg, I. (1994) Evidence for cleavage of the PC1/PC3 prosegment in the endoplasmic reticulum. *Mol. Cell. Neurosci.* **5**, 263–268.

Malide, D., Seidah, N.G., Chrétien, M. & Bendayan, M. (1995) Electron microscopic immunocytochemical evidence for the involvement of the convertases PC1 and PC2 in the processing of proinsulin in pancreatic β-cells. *J. Histochem. Cytochem.* **43**, 11–19.

Marcinkiewicz, M., Day, R., Seidah, N.G. & Chrétien, M. (1993) Ontogeny of the prohormone convertases PC1 and PC2 in the mouse hypophysis and their colocalization with corticotropin and α-melanotropin. *Proc. Natl Acad. Sci. USA* **90**, 4922–4926.

Marcinkiewicz, M., Ramla, D., Seidah, N.G. & Chrétien, M. (1994) Developmental expression of the prohormone convertases PC1 and PC2 in mouse pancreatic islets. *Endocrinology* **135**, 1651–1660.

Mizuno, K., Nakamura, T., Ohshima, T., Tanaka, S. & Matsuo, H. (1989) Characterization of kex2-encoded endopeptidase from yeast *Saccharomyces cerevisiae*. *Biochem. Biophys. Res. Commun.* **159**, 305–311.

O'Rahilly, S., Gray, H., Humphreys, P.J., Krook, A., Polonsky, K.S., White, A., Gibson, S., Taylor, K. & Carr, C. (1995) Brief report: impaired processing of prohormones associated with abnormalities of glucose homeostasis and adrenal function. *New Engl. J. Med.* **333**, 1386–1390.

Roebroek, A.J.M., Schalken, J.A., Leunissen, J.A.M., Onnekink, C., Bloemers, H.P.J. & Van de Ven, W.J.M. (1986) Evolutionary conserved close linkage of the c-fes/fps proto-oncogene and genetic sequences encoding a receptor-like protein. *EMBO J.* **5**, 2197–2202.

Rouille, Y., Duguay, S.J., Lund, K., Furuta, M., Gong, Q.M., Lipkind, G., Oliva, A.A., Chan, S.J. & Steiner, D.F. (1995) Proteolytic processing mechanisms in the biosynthesis of neuroendocrine peptides – the subtilisin-like proprotein convertases. *Front. Neuroendocrinol.* **16**, 322–361.

Seidah, N.G. (1995) The mammalian family of subtilisin/kexin-like proprotein convertases. In: *Intramolecular Chaperones and Protein Folding* (Shinde, U. & Inouye, M., eds). Austin, TX: R.G. Landes, pp. 181–203.

Seidah, N.G. & Chrétien, M. (1994) Pro-protein convertases of the subtilisin/kexin family. *Methods. Enzymol.* **244**, 175–188.

Seidah, N.G., Gaspar, L., Mion, P., Marcinkiewicz, M., Mbikay, M. & Chrétien, M. (1990) cDNA sequence of two distinct pituitary proteins homologous to Kex2 and furin gene products: tissue-specific mRNAs encoding candidates for pro-hormone processing proteinases. *DNA* **9**, 415–424.

Seidah, N.G., Day, R., Marcinkiewicz, M., Benjannet, S. & Chrétien, M. (1991a) Mammalian neural and endocrine pro-protein and pro-hormone convertases belonging to the subtilisin family of serine proteinases. *Enzyme* **45**, 271–284.

Seidah, N.G., Marcinkiewicz, M., Benjannet, S., Gaspar, L., Beaubien, G., Mattei, M.G., Lazure, C., Mbikay, M. & Chrétien, M. (1991b) Cloning and primary sequence of a mouse candidate pro-hormone convertase PC1 homologous to PC2, furin and Kex2: distinct chromosomal localization and mRNA distribution in brain and pituitary as compared to PC2. *Mol. Endocrinol.* **5**, 111–122.

Seidah, N.G., Day, R. & Chrétien, M. (1994) The family of subtilisin/kexin-like pro-protein convertases: divergent or shared functions. *Biochimie* **76**, 197–209.

Seidah, N.G., Day, R., Marcinkiewicz, M. & Chrétien, M. (1997) Precursor convertases: an evolutionary ancient, cell-specific, combinatorial mechanism yielding diverse bioactive peptides and proteins. *Ann. NY Acad. Sci.* (in press).

Smeekens, S.P., Avruch, A.S., LaMendola, J., Chan, S.J. & Steiner, D.F. (1991) Identification of a cDNA encoding a second putative prohormone convertase related to PC2 in AtT20 cells and islets of Langerhans. *Proc. Natl Acad. Sci. USA* **88**, 340–344.

Steiner, D.F., Rouille, Y., Gong, Q., Martin, S., Carroll, R. & Chan, S.J. (1996) The role of prohormone convertases in insulin biosynthesis: evidence for inherited defects in their action in man and experimental animals. *Diabetes Metab.* **22**, 94–104.

Van de Ven, W.J., Voorberg, J., Fontijn, R., Pannekoek, H., Van den Ouweland, A.M., Van Duijnhoven, H.L., Roebroek, A.J. & Siezen, R.J. (1990) Furin is a subtilisin-like proprotein processing enzyme in higher eukaryotes. *Mol. Biol. Rep.* **14**, 265–275.

Vindrola, O. & Lindberg, I. (1992) Biosynthesis of the prohormone convertase-mPC1 in AtT-20 Cells. *Mol. Endocrinol.* **6**, 1088–1094.

Zheng, M., Streck, R.D., Scott, R.E.M., Seidah, N.G. & Pintar, J.E. (1994) The developmental expression in rat of proteases furin, PC1, PC2, and carboxypeptidase E: implications for early maturation of proteolytic processing capacity. *J. Neurosci.* **14**, 4656–4673.

***Nabil G. Seidah***
*Laboratory of Biochemical Neuroendocrinology,*
*Clinical Research Institute of Montreal,*
*110 Pine Ave West, Montreal, QC, H2W 1R7 Canada*
*Email: seidahn@ipcmumontreal.ca*

***Michel Chrétien***
*Laboratory of Molecular Neuroendocrinology,*
*Clinical Research Institute of Montreal,*
*110 Pine Ave West, Montreal, QC, H2W 1R7 Canada*

# 119. *Proprotein convertase 2*

## Databanks

*Peptidase classification: clan SB, family S8, MEROPS ID: S08.073*
*NC-IUBMB enzyme classification: EC 3.4.21.94*
*Databank codes:*

| Species | SwissProt | PIR | EMBL (cDNA) | EMBL (genomic) |
|---|---|---|---|---|
| *Aplysia californica* | – | S40449 | L34741 | – |
| *Branchiostoma californiensis* | – | – | U22051 | – |
| *Caenorhabditis elegans* | – | – | U04995 | – |
| *Homo sapiens* | P16519 | A35062 | J05252 | M95960: exon 1 |
| | | A45382 | | M95961: exon 2 |
| | | | | M95962: exon 3 |
| | | | | M95963: exon 4 |
| | | | | M95964: exon 5 |
| | | | | M95965: exon 6 |
| | | | | M95966: exon 7 |
| | | | | M95967: exon 8 |
| | | | | M95968: exon 9 |
| | | | | M95969: exon 10 |
| | | | | M95970: exon 11 |
| | | | | M95971: exon 12 and complete CDS |
| | | | | S75955: promoter |
| *Lymnaea stagnalis* | – | S27270 | X68850 | |
| *Mus musculus* | P21661 | B35571 | M55669 | – |
| *Rattus norvegicus* | P28841 | A42751 | M76706 | – |
| | | B41556 | M83746 | |
| | | S27362 | | |
| | | S36359 | | |
| *Sus scrofa* | Q03333 | S29244 | X68603 | – |
| *Xenopus laevis* | – | S23118 | X66493 | – |

## Name and History

Since the early 1960s it has been proposed that polypeptide hormones are first synthesized as inactive precursors that require specific cleavage following pairs of basic residues (Lys-Arg-, Arg-Arg-, Lys-Lys- and Arg-Lys-) in order to release the active hormone. Since then this model has been extended to other precursors, such as progrowth factors, proneurotrophic factors, hormone receptors, adhesion molecules, retroviral surface glycoproteins, proenzymes, and even certain protoxins. The search for the physiologically important processing enzymes, termed proprotein convertases (PCs), was very laborious and a number of laboratories, including our own, participated actively in this hunt (Marcinkiewicz *et al*., 1994). The major breakthrough came in 1984, from the molecular identification of the convertase responsible for the activation of the yeast $\alpha$-mating factor and killer toxin. The proteinase identified by genetic complementation of a *KEX2* mutant strain was found to be a subtilisin-like serine proteinase (Fuller *et al*., 1988; Mizuno *et al*., 1989) and is now called kexin (Chapter 115). The search for the mammalian counterpart of kexin took about five years, before it was discovered by computer database searches for sequence identity to kexin that a partial human genomic sequence coding for a related protein called furin (Chapter 117) had already been reported by Roebroek and colleagues in 1986 (Roebroek *et al*., 1986). In the reported DNA sequence only the active-site Ser and the catalytically important Asn residue found in all subtilisin-like proteinases were identified.

In 1989, the partial sequence of furin (from the catalytically important Asn to the C-terminus) (Roebroek *et al*., 1986) and the full sequence of kexin (Mizuno *et al*., 1989) were known. Accordingly, based on the concept of sequence conservation around the active sites of serine proteinases, PCR applied to mRNA amplification (reverse transcriptase PCR, RT-PCR) allowed two laboratories to simultaneously isolate the second endocrine and neural mammalian homolog of kexin, termed *PC2*. PCR amplification of a cDNA synthesized from a human insulinoma total RNA using degenerate oligonucleotides coding for the consensus sequence surrounding the active-site Asp and His residues in kexin and related subtilisins gave a 150 bp probe. The latter was used to screen a human insulinoma library and to isolate a full-length cDNA coding for the novel convertase (Smeekens & Steiner, 1990).

Independently, PCR on cDNA obtained from mouse pituitary total RNA using oligonucleotides coding for the sequence around the catalytically important Asn and the active-site Ser of human furin allowed the isolation of a 260 bp probe. The screening of libraries from mouse pituitary and mouse insulinoma with this probe led to the isolation of full-length cDNA clones coding for mouse PC2 (Seidah *et al.*, 1990). This was the second case of a cascade of events which ultimately led to the identification of seven different proprotein convertases (see Figure 118.1). The name **proprotein convertase 2** has been recommended by IUBMB.

## Activity and Specificity

Mammalian PC2 is a 637–638 amino acid enzyme usually responsible for the processing of precursors whose peptide products are stored in dense-core secretory vesicles. These include mainly endocrine and neural polypeptide hormones, such as pro-opiomelanocortin, prosomatostatin, proenkephalin, proinsulin, prodynorphin and prothyrotropin-releasing hormone (proTRH). In general, PC2 cleaves after dibasic or monobasic sequences of general structures (Lys/Arg)-Arg↓ or (Arg/Lys)-(Xaa)$_n$-Arg↓, where $n=2$, 4 or 6 (Seidah, 1995). Except in the case of proTRH cleavage by protein convertase 1 (Chapter 118) leading to the production of the tripeptide TRH, in general PC2 is the convertase responsible for the generation of small bioactive peptides. PC2 acts late in the secretory pathway, usually within immature secretory granules, and generates peptide products from those initially produced by PC1. These include $\beta$-endorphin from $\beta$-lipotropin (Benjannet *et al.*, 1991), luteinizing hormone-releasing hormone (LHRH) from progonadotropin-releasing hormone (proGnRH) (Wetsel *et al.*, 1995), somatostatin 14 from prosomatostatin (Brakch *et al.*, 1995), [Met]$^5$- and [Leu]$^5$-enkephalins from proenkephalin (Breslin *et al.*, 1993), dynorphin(1–8) from prodynorphin (Day *et al.*, 1992), and insulin from (des-31,32)proinsulin generated by PC1 (Smeekens *et al.*, 1992).

PC2 is a soluble, $Ca^{2+}$-dependent enzyme (inhibitable by EDTA and EGTA) first synthesized as a proenzyme (proPC2), which undergoes autocatalytic activation to a 66 kDa form within either the *trans*-Golgi network (TGN) or immature secretory granules, and unlike PC1, PC2 does not undergo further C-terminal truncation (Benjannet *et al.*, 1993; Guest *et al.*, 1992; Matthews *et al.*, 1994). The pH optimum of PC2 with a fluorogenic substrate such as Glp-Glu-Arg-Thr-Lys-Arg↓NHMec (Enzyme Biosystems) (see Appendix 2 for full names and addresses of suppliers) is 5.5–6.0 (Lindberg *et al.*, 1992). Since PC2 is the only one of the seven known convertases that undergoes proenzyme processing and activation late along the secretory pathway, it was suspected that the proPC2 to PC2 conversion might be tightly regulated. Indeed, recent reports (Martens *et al.*, 1994; Seidah *et al.*, 1994; Zhu & Lindberg, 1995) clearly established that the proenzyme cleavage and activation of PC2 are under the control of the pan-neuronal polypeptide 7B2 discovered in 1982 (Hsi *et al.*, 1982; Seidah *et al.*, 1983; Seidah, 1995). The 186

amino acid precursor of 7B2 (pro7B2) contains two domains: (a) an N-terminal, 150 amino acid segment which facilitates the folding of proPC2, ultimately leading to a favorable rate of proPC2 to PC2 conversion (Benjannet *et al.*, 1995b; Zhu & Lindberg, 1995), and (b) a C-terminal, 31 amino acid CT-peptide which is a potent (nanomolar) inhibitor of PC2 (Martens *et al.*, 1994) and which is inactivated by a PC2 cleavage at a Lys-Lys↓ bond within immature granules (Zhu *et al.*, 1996). Recent studies suggest that pro7B2 binds to proPC2 within the endoplasmic reticulum (Benjannet *et al.*, 1995a), a binding requiring the Asp of the oxyanion hole of PC2 (Benjannet *et al.*, 1995b), and the complex traverses the secretory pathway towards the TGN, where pro7B2 is processed to 7B2 by a furin-like enzyme (Paquet *et al.*, 1994; Benjannet *et al.*, 1995b). The released CT-peptide and the N-terminal 150 amino acid 7B2 remain associated with proPC2 until the complex enters immature secretory granules and autocatalytic processing of proPC2 to PC2 occurs with the concomitant inactivation of the CT-peptide by PC2 and expression of the full PC2 activity within the confinement of the granules.

## Preparation

In general, the yield of PC2 from endocrine and neuroendocrine tissues is too low for large-scale preparations. Milligram amounts of active PC2 can conveniently be obtained from stably transfected Chinese hamster ovary cells (Shen *et al.*, 1993), but only when they also express the binding protein 7B2 (Lamango *et al.*, 1996). Vaccinia virus expression of PC2 in mammalian cell lines has so far yielded only low levels of active PC2, even when the 186 amino acid pro7B2 is coexpressed. However, recent data suggest that coexpression of only the 21 kDa N-terminal 150 amino acid segment of 7B2 may be an alternative procedure that could be applied to baculovirus production of active PC2 (N.G. Seidah, unpublished results).

## Biological Aspects

The enzyme PC2 is found in all eukaryotic species analyzed from *Lymnea stagnalis* and *Aplysia californica* to mammals (Seidah *et al.*, 1997). It is found almost exclusively in endocrine and neural cells (Day *et al.*, 1992; Schäfer *et al.*, 1993; Seidah *et al.*, 1990; Smeekens & Steiner, 1990) and the active 66 kDa form is stored in dense-core secretory vesicles (Malide *et al.*, 1995). It is very abundant in the central nervous system within neurons and is not expressed in glial cells. Within the brain, it is particularly abundant in the telencephalon including the hippocampus, cerebral cortex, diencephalon, especially within the thalamus and hypothalamus, mesencephalon and in pons and medulla. In the periphery it is highly expressed in the intermediate lobe of the pituitary, thyroid but not parathyroid, and in all endocrine pancreatic cells, as well as in the cardiovascular and digestive systems.

The gene coding for mouse PC2 has been characterized (Ohagi *et al.*, 1992) and it encodes 12 exons and 13 introns of various sizes, estimated to cover a >130 kb gene. The active-site residues Asp, His, Ser and the oxyanion hole Asp are on separate exons. The PC2 gene maps to human chromosome 20p11.1–p11.2 and mouse chromosome 2 (F3–H2 region), respectively (Ohagi *et al.*, 1992). The promoter of PC2 is very GC rich and does not contain TATA or CAAT boxes.

The level of PC2 can be upregulated by a number of factors including dopamine antagonists and the thyroid status (Day *et al.*, 1992). Developmental studies of PC2 revealed it to appear following embryonic day 13 (e13) in neurons, pituitary and pancreas (Marcinkiewicz *et al.*, 1993, 1994; Zheng *et al.*, 1994).

## Distinguishing Features

PC2 is a soluble, $Ca^{2+}$-dependent serine proteinase of the subtilisin-kexin type, exhibiting an acidic pH optimum. Polyclonal antibodies have been described (Guest *et al.*, 1992; Benjannet *et al.*, 1993; Matthews *et al.*, 1994). It is a secretable enzyme which is mostly localized within secretory granules, but can also be detected in the TGN. Its release can be stimulated by cAMP or by membrane depolarization.

## Further Reading

For reviews, see Rouille *et al.* (1995) and Seidah & Chrétien (1994).

## References

Benjannet, S., Rondeau, N., Day, R., Chrétien, M. & Seidah, N.G. (1991) PC1 and PC2 are proprotein convertases capable of cleaving POMC at distinct pairs of basic residues. *Proc. Natl Acad. Sci. USA* **88**, 3564–3568.

Benjannet, S., Rondeau, N., Paquet, L., Boudreault, A., Lazure, C., Chrétien, M. & Seidah, N.G. (1993) Comparative biosynthesis, covalent post-translational modifications and efficiency of prosegment cleavage of the prohormone convertases PC1 and PC2: glycosylation, sulphation and identification of the intracellular site of prosegment cleavage of PC1 and PC2. *Biochem. J.* **294**, 735–743.

Benjannet, S., Lusson, J., Hamelin, J., Savaria, D., Chrétien, M. & Seidah, N.G. (1995a) Structure–function studies on the biosynthesis and bioactivity of the precursor convertase PC2 and the formation of the PC2/7B2 complex. *FEBS Lett.* **362**, 151–155.

Benjannet, S., Savaria, D., Chrétien, M. & Seidah, N.G. (1995b) 7B2 is a specific intracellular binding protein of the prohormone convertase PC2. *J. Neurochem.* **64**, 2303–2311.

Brakch, N., Galanopoulou, A.S., Patel, Y.C., Boileau, G. & Seidah, N.G. (1995) Comparative proteolytic processing of rat prosomatostatin by the convertases PC1, PC2, furin, PACE4 and PC5 in constitutive and regulated secretory pathways. *FEBS Lett.* **362**, 143–146.

Breslin, M.B., Lindberg, I., Benjannet, S., Mathis, J.P., Lazure, C. & Seidah, N.G. (1993) Differential processing of proenkephalin by prohormone convertase-1(3) and convertase-2 and furin. *J. Biol. Chem.* **268**, 27084–27093.

Day, R., Schäfer, M.K.H., Watson, S.J., Chrétien, M. & Seidah, N.G. (1992) Distribution and regulation of the prohormone convertases PC1 and PC2 in the rat pituitary. *Mol. Endocrinol.* **6**, 485–497.

Fuller, R.S., Sterne, R.E. & Thorner, J. (1988) Enzymes required for yeast prohormone processing. *Annu. Rev. Biochem.* **50**, 345–362.

Guest, P.C., Arden, S.D., Bennett, D.L., Clark, A., Rutherford, N.G. & Hutton, J.C. (1992) The post-translational processing and intracellular sorting of PC2 in the islets of langerhans. *J. Biol. Chem.* **267**, 22401–22406.

Hsi, K.L., Seidah, N.G., De Serres, G. & Chrétien, M. (1982) Isolation and NH2-terminal sequence of a novel porcine anterior pituitary polypeptide. Homology to proinsulin, secretin and rous sarcoma virus transforming protein TVFV60. *FEBS Lett.* **147**, 261–266.

Lamango, N.S., Zhu, X.R. & Lindberg, I. (1996) Purification and enzymatic characterization of recombinant prohormone convertase 2 – stabilization of activity by 21 kDa 7B2. *Arch. Biochem. Biophys.* **330**, 238–250.

Lindberg, I., Lincoln, B. & Rhodes, C.J. (1992) Fluorometric assay of a calcium-dependent, paired-basic processing endopeptidase present in insulinoma granules. *Biochem. Biophys. Res. Commun.* **183**, 1–7.

Malide, D., Seidah, N.G., Chrétien, M. & Bendayan, M. (1995) Electron microscopic immunocytochemical evidence for the involvement of the convertases PC1 and PC2 in the processing of proinsulin in pancreatic $\beta$-cells. *J. Histochem. Cytochem.* **43**, 11–19.

Marcinkiewicz, M., Day, R., Seidah, N.G. & Chrétien, M. (1993) Ontogeny of the prohormone convertases PC1 and PC2 in the mouse hypophysis and their colocalization with corticotropin and alpha-melanotropin. *Proc. Natl Acad. Sci. USA* **90**, 4922–4926.

Marcinkiewicz, M., Ramla, D., Seidah, N.G. & Chrétien, M. (1994) Developmental expression of the prohormone convertases PC1 and PC2 in mouse pancreatic islets. *Endocrinology* **135**, 1651–1660.

Martens, G.J.M., Braks, J.A.M., Eib, D.W., Zhou, Y. & Lindberg, I. (1994) The neuroendocrine polypeptide 7B2 is an endogenous inhibitor of prohormone convertase PC2. *Proc. Natl Acad. Sci. USA* **91**, 5784–5787.

Matthews, G., Shennan, K.I.J., Seal, A.J., Taylor, N.A., Colman, A. & Docherty, K. (1994) Autocatalytic maturation of the prohormone convertase PC2. *J. Biol. Chem.* **269**, 588–592.

Mizuno, K., Nakamura, T., Ohshima, T., Tanaka, S. & Matsuo, H. (1989) Characterization of kex2-encoded endopeptidase from yeast saccharomyces cerevisiae. *Biochem. Biophys. Res. Commun.* **159**, 305–311.

Ohagi, S., Lamendola, J., Lebeau, M.M., Espinosa, R., Takeda, J., Smeekens, S.P., Chan, S.J. & Steiner, D.F. (1992) Identification and analysis of the gene encoding human PC2, a prohormone convertase expressed in neuroendocrine tissues. *Proc. Natl Acad. Sci. USA* **89**, 4977–4981.

Paquet, L., Bergeron, F., Boudreault, A., Seidah, N.G., Chrétien, M., Mbikay, M. & Lazure, C. (1994) The neuroendocrine precursor 7B2 is a sulfated protein proteolytically processed by a ubiquitous furin-like convertase. *J. Biol. Chem.* **269**, 19279–19285.

Roebroek, A.J.M., Schalken, J.A., Leunissen, J.A.M., Onnekink, C., Bloemers, H.P.J. & Van de Ven, W.J.M. (1986) Evolutionary conserved close linkage of the c-fes/fps proto-oncogene and genetic sequences encoding a receptor-like protein. *EMBO J.* **5**, 2197–2202.

Rouille, Y., Duguay, S.J., Lund, K., Furuta, M., Gong, Q.M., Lipkind, G., Oliva, A.A., Chan, S.J., & Steiner, D.F. (1995). Proteolytic processing mechanisms in the biosynthesis of neuroendocrine peptides – the subtilisin-like proprotein convertases. *Front. Neuroendocrinol.* **16**, 322–361.

Schäfer, M.K.H., Day, R., Cullinan, W.E., Chrétien, M., Seidah, N.G. & Watson, S.J. (1993) Gene expression of prohormone and proprotein convertases in the rat CNS – a comparative in situ hybridization analysis. *J. Neurosci.* **13**, 1258–1279.

Seidah, N.G. (1995) The mammalian family of subtilisin/kexin-like proprotein convertases. In: *Intramolecular Chaperones and Protein Folding* (Shinde, U. & Inouye, M., eds). Austin, TX: R.G. Landes, pp. 181–203.

Seidah, N.G. & Chrétien, M. (1994) Pro-protein convertases of the subtilisin/kexin family. *Methods Ezymol.* **244**, 175–188.

Seidah, N.G., Hsi, K.L., De Serres, G., Rochemont, J., Hamelin, J., Antakly, T., Cantin, M. & Chrétien M. (1983) Isolation and NH2-terminal sequence of a highly conserved human and porcine pituitary protein belonging to a new superfamily. Immunocytochemical localization in pars distalis and pars nervosa of the pituitary and in the supraoptic nucleus of the hypothalamus. *Arch. Biochem. Biophys.* **225**, 525–534.

Seidah, N.G., Gaspar, L., Mion, P., Marcinkiewicz, M., Mbikay, M. & Chrétien, M. (1990) cDNA sequence of two distinct pituitary proteins homologous to Kex2 and furin gene products: tissue-specific mRNAs encoding candidates for pro-hormone processing proteinases. *DNA* **9**, 415–424.

Seidah, N.G., Day, R. & Chrétien, M. (1994) The family of subtilisin/kexin-like pro-protein convertases: divergent or shared functions. *Biochimie* **76**, 197–209.

Seidah, N.G., Day, R., Marcinkiewicz, M. & Chrétien, M. (1997) Precursor convertases: an evolutionary ancient, cell-specific, combinatorial mechanism yielding diverse bioactive peptides and proteins. *Ann. NY Acad. Sci.* (in press).

Shen, F.S., Seidah, N.G. & Lindberg, I. (1993) Biosynthesis of the prohormone convertase PC2 in chinese hamster ovary cells and in rat insulinoma cells. *J. Biol. Chem.* **268**, 24910–24915.

Smeekens, S.P. & Steiner, D.F. (1990) Identification of a human insulinoma cDNA encoding a novel mammalian protein structurally related to the yeast dibasic processing protease Kex2. *J. Biol. Chem.* **265**, 2997–3000.

Smeekens, S.P., Montag, A.G., Thomas, G., Albiges-Rizo, C., Carroll, R., Benig, M., Phillips, L.A., Martin, S., Ohagi, S., Gardner, P., Swift, H.H. & Steiner, D.F. (1992) Proinsulin processing by the subtilisin-related proprotein convertases furin, PC2, and PC3. *Proc. Natl Acad. Sci. USA* **89**, 8822–8826.

Wetsel, W.C., Liposits, Z., Seidah, N.G. & Collins, S. (1995) Expression of candidate pro-GnRH processing enzymes in rat hypothalamus and an immortalized hypothalamic neuronal cell line. *Neuroendocrinology* **62**, 166–177.

Zheng, M., Streck, R.D., Scott, R.E.M., Seidah, N.G. & Pintar, J.E. (1994) The developmental expression in rat of proteases furin, PC1, PC2, and carboxypeptidase E: implications for early maturation of proteolytic processing capacity. *J. Neurosci.* **14**, 4656–4673.

Zhu, X. & Lindberg, I. (1995) 7B2 facilitates the maturation of propc2 in neuroendocrine cells and is required for the expression of enzymatic activity. *J. Cell Biol.* **129**, 1641–1650.

Zhu, X., Rouille, Y., Lamango, N.S., Steiner, D.F. & Lindberg, I. (1996) Internal cleavage of the inhibitory 7B2 carboxyl-terminal peptide by PC2: a potential mechanism for its inactivation. *Proc. Natl Acad. Sci. USA* **93**, 4919–4924.

*Nabil G. Seidah*
*Laboratory of Biochemical Neuroendocrinology,*
*Clinical Research Institute of Montreal,*
*110 Pine Ave West, Montreal, QC, H2W 1R7 Canada*
*Email: seidahn@ipcmumontreal.ca*

*Michel Chrétien*
*Laboratory of Molecular Neuroendocrinology,*
*Clinical Research Institute of Montreal,*
*110 Pine Ave West, Montreal, QC, H2W 1R7 Canada*

# 120. *Proprotein convertase 4*

## Databanks

Peptidase classification: clan SB, family S8, MEROPS ID: S08.074
NC-IUBMB enzyme classification: none
Databank codes:

| Species | SwissProt | PIR | EMBL (cDNA) | EMBL (genomic) |
| --- | --- | --- | --- | --- |
| *Mus musculus* | P29121 | A42151 | D01093 D45357 | L21210: promoter and exons 1 and 2 |
| *Rattus rattus* | – | A45357 | L14937 | – |

## Name and History

From the common observation that many peptide hormones are generated from larger polypeptide precursors through limited proteolysis at either specific single or paired basic residues, the existence of a class of endopeptidases dedicated to this type of processing was predicted many years ago (Seidah & Chrétien, 1992; Rouille *et al.*, 1995). It was the discovery in 1984 of yeast kexin (reviewed in Fuller *et al.*, 1988) (Chapter 115), responsible for the processing of pro-α-mating factor that ultimately led to the identification of three mammalian homologs within the period 1987–1991 (see Chapter 118). These $Ca^{2+}$-dependent serine proteinases belonging to the subtilisin family were called furin (Roebroek *et al.*, 1986) (Chapter 117), PC1 (Seidah *et al.*, 1990, 1991; also called PC3 in Smeekens *et al.*, 1991) (Chapter 118) and PC2 (Seidah *et al.*, 1990, 1991) (Chapter 119). The search for other proprotein convertases of the subfamily led two groups to identify and clone the cDNA coding for the fourth member, which was independently termed *proprotein convertase 4 (PC4)* by both groups (Nakayama *et al.*, 1992; Seidah *et al.*, 1992).

## Activity and Specificity

So far, very little is known about the natural substrates of PC4. In fact, vaccinia virus expression of PC4 with a number of substrates including proenkephalin and pro-opiomelanocortin did not allow us to identify any cleavages. From the site of zymogen activation of PC4, and in analogy to other convertases, it is predicted that PC4 will cleave precursors at the consensus PC sequence of (Lys/Arg)-Arg↓ or (Arg/Lys)-(Xaa)$_n$-Arg↓, where $n = 2$, 4 or 6 (Seidah *et al.*, 1992; Seidah, 1995).

## Structural Chemistry

The structure of PC4, not unlike that of PC2, is illustrated diagrammatically in Figure 118.1. Interestingly, the structure of the catalytic domain of PC4 is closer to that of furin than that of any other proprotein convertase (Seidah *et al.*, 1997).

## Preparation

We have been able to isolate small amounts of recombinant, active PC4 following overexpression of its vaccinia virus recombinant in the constitutively secreting green monkey kidney epithelial BSC40 cells (A. Bazak & N.G. Seidah, unpublished results). Our preliminary data suggest that PC4 is a soluble, $Ca^{2+}$-dependent enzyme, of about 52 kDa, exhibiting a neutral pH optimum, and that it can cleave fluorogenic substrates such as Glp-Arg-Thr-Lys-Arg↓NHMec.

## Biological Aspects

Analysis of the tissue expression of PC4 mRNA revealed it to be exclusively expressed within germ cells, in the testis (Nakayama *et al.*, 1992; Seidah *et al.*, 1992), but also within gonadotropin-activated ovarian cells (Seidah *et al.*, 1992). A number of C-terminally (Seidah *et al.*, 1992) and N-terminally (Mbikay *et al.*, 1994a) modified isoforms were predicted from the isolated cDNAs, probably arising from alternative splicing of the gene (Mbikay *et al.*, 1994a, 1995).

PC4 has only been detected and identified in rat and mouse testis, and it is not known whether invertebrates also express a PC4 ortholog. *In situ* hybridization histochemistry demonstrated that PC4 mRNA is detected with pachetene spermatocytes and round spermatids (Seidah *et al.*, 1992; Torii *et al.*, 1993; Mbikay *et al.*, 1994a). However, preliminary immunocytochemical data demonstrated that the PC4 protein is expressed at later stages of spermiogenesis and can even be detected in mature spermatozoa (Mbikay *et al.*, 1997). Developmental studies revealed that PC4 mRNA is only detected at the onset of spermiogenesis, at the pubertal stage (Seidah *et al.*, 1992).

Since nothing is yet known about the function of PC4, we have recently been successful in isolating mice deficient in PC4 expression by homologous recombination. These PC4 null mice exhibit impaired fertility, especially for the male. *In vitro*, PC4 deficiency appears to cause reduced fertilization as well as early embryonic death (Mbikay *et al.*, 1997). These data strongly suggest that PC4 will play an important role in fertility, and that functional redundancy mediated by any other PC is quite limited in the testis.

## Distinguishing Features

PC4 is a testicular enzyme that is not expressed in any other male tissue and which only appears postnatally in the rat between days 19 and 22, coinciding with the first stages of spermiogenesis. In the female, it seems that PC4 is expressed in the ovarian thecal and interstitial cells (M. Mbikay, unpublished results). It is secreted as a 52 kDa protein with a cleavage specificity for paired basic residues, similar to that of other PCs (Mbikay *et al.*, 1997). Its subcellular localization is not yet known.

The gene coding for mouse PC4 has been characterized (Mbikay *et al.*, 1994a, 1995) and it encodes 15 exons and 14 introns of various sizes, estimated to cover a 9.5 kb gene. The active-site residues Asp, His, Ser and the oxyanion hole Asn are each on separate exons. The gene maps to human and mouse chromosomes 19 and 10, respectively (Mbikay *et al.*, 1995). The promoter of PC1 is GC rich and carries three CCAAT boxes, but no TATA box (Mbikay *et al.*, 1994a).

## Further Reading

For reviews, see Seidah *et al.* (1994) and Mbikay *et al.* (1994b).

## References

Fuller, R.S., Sterne, R.E. & Thorner, J. (1988) Enzymes required for yeast prohormone processing. *Annu. Rev. Physiol.* **50**, 345–362.

Mbikay, M., Raffin-Samson, M.L., Tadros, H., Sirois, F., Seidah, N.G. & Chrétien, M. (1994a) Structure of the gene for the testis-specific proprotein convertase-4 and of its alternate messenger RNA isoforms. *Genomics* **20**, 231–237.

Mbikay, M., Raffin-Samsón, M.-L., Tadros, H., Sirois, F., Seidah, N.G. & Chrétien, M. (1994b) Testis-specific proprotein convertase 4: gene structure, optional exons, and mRNA isoforms. In: *Function of Somatic Cells in the Testis* (Barthe, A., ed.). New York: Springer-Verlag, Serono Symposia, pp. 388–399.

Mbikay, M., Seidah, N.G., Chrétien, M. & Simpson, E.M. (1995) Chromosomal assignment of the genes for proprotein convertases PC4, PC5, and PACE 4 in mouse and human. *Genomics* **26**, 123–129.

Mbikay, M., Tadros, H., Lerner, C.P., Chen, A., El-Alfy, M., Clermont, Y., Seidah, N.G., Chrétien, M. & Simpson, E.M. (1997) Impaired fertility in mice deficient for the testicular germ-cell protease PC4. *Proc. Natl Acad. Sci. USA* **94**, 6842–6846.

Nakayama, K., Kim, W.S., Torii, S., Hosaka, M., Nakagawa, T., Ikemizu, J., Baba, T. & Murakami, K. (1992) Identification of the fourth member of the mammalian endoprotease family homologous to the yeast kex2 protease – its testis-specific expression. *J. Biol. Chem.* **267**, 5897–5900.

Roebroek, A.J.M., Schalken, J.A., Leunissen, J.A.M., Onnekink, C., Bloemers, H.P.J. & Van de Ven, W.J.M. (1986) Evolutionary conserved close linkage of the c-fes/fps proto-oncogene and genetic sequences encoding a receptor-like protein. *EMBO J.* **5**, 2197–2202.

Rouille, Y., Duguay, S.J., Lund, K., Furuta, M., Gong, Q.M., Lipkind, G., Oliva, A.A., Chan, S.J. & Steiner, D.F. (1995) Proteolytic processing mechanisms in the biosynthesis of neuroendocrine peptides – the subtilisin-like proprotein convertases. *Front. Neuroendocrinol.* **16**, 322–361.

Seidah, N.G. (1995) The mammalian family of subtilisin/kexin-like proprotein convertases. In: *Intramolecular Chaperones and Protein Folding* (Shinde, U. & Inouye, M., eds). Austin, TX: R.G. Landes, pp. 181–203.

Seidah, N.G. & Chrétien, M. (1992) Proprotein and prohormone convertases of the subtilisin family – recent developments and future perspectives. *Trends Endocrinol. Metab.* **3**, 133–140.

Seidah, N.G., Gaspar, L., Mion, P., Marcinkiewicz, M., Mbikay, M. & Chrétien, M. (1990) cDNA sequence of two distinct pituitary proteins homologous to Kex2 and furin gene products: tissue-specific mRNAs encoding candidates for pro-hormone processing proteinases. *DNA* **9**, 415–424.

Seidah, N.G., Marcinkiewicz, M., Benjannet, S., Gaspar, L., Beaubien, G., Mattei, M.G., Lazure, C., Mbikay, M. & Chrétien, M. (1991) Cloning and primary sequence of a mouse candidate prohormone convertase PC1 homologous to PC2, furin and Kex2: distinct chromosomal localization and mRNA distribution in brain and pituitary as compared to PC2. *Mol. Endocrinol.* **5**, 111–122.

Seidah, N.G., Day, R., Hamelin, J., Gaspar, A., Collard, M.W. & Chrétien, M. (1992) Testicular expression of PC4 in the rat – molecular diversity of a novel germ cell-specific kex2/subtilisin-like proprotein convertase. *Mol. Endocrinol.* **6**, 1559–1570.

Seidah, N.G., Day, R. & Chrétien, M. (1994) The family of subtilisin/kexin-like pro-protein convertases: divergent or shared functions. *Biochimie* **76**, 197–209.

Seidah, N.G., Day, R., Marcinkiewicz, M. & Chrétien, M. (1997) Precursor convertases: an evolutionary ancient, cell-specific, combinatorial mechanism yielding diverse bioactive peptides and proteins. *Ann. NY Acad. Sci.* (in press).

Smeekens, S.P., Avruch, A.S., LaMendola, J., Chan, S.J. & Steiner, D.F. (1991) Identification of a cDNA encoding a second putative prohormone convertase related to PC2 in AtT20 cells and islets of Langerhans. *Proc. Natl Acad. Sci. USA* **88**, 340–344.

Torii, S., Yamagishi, T., Murakami, K. & Nakayama, K. (1993) Localization of Kex2-like processing endoproteases, furin and PC4 within mouse testis by *in situ* hybridization. *FEBS Lett.* **316**, 12–16.

***Nabil G. Seidah***
*Laboratory of Biochemical
Neuroendocrinology,
Clinical Research Institute of
Montreal,
110 Pine Ave West, Montreal, QC,
H2W 1R7 Canada
Email: seidahn@ipcmumontreal.ca*

***Majambu Mbikay***
*Laboratory of Molecular
Neuroendocrinology,
Clinical Research Institute of
Montreal,
110 Pine Ave West, Montreal, QC,
H2W 1R7 Canada*

***Michel Chrétien***
*Laboratory of Molecular
Neuroendocrinology,
Clinical Research Institute of
Montreal,
110 Pine Ave West, Montreal, QC,
H2W 1R7 Canada*

# 121. *Proprotein convertase PACE4*

## Databanks

*Peptidase classification: clan SB, family S8, MEROPS ID: S08.075*
*NC-IUBMB enzyme classification: none*
*Databank codes:*

| Species | Form | SwissProt | PIR | EMBL (cDNA) | EMBL (genomic) |
|---------|------|-----------|-----|-------------|----------------|
| *Homo sapiens* | A | P29122 | A39490 B39490 | M80482 | – |

*continued overleaf*

| Species | Form | SwissProt | PIR | EMBL (cDNA) | EMBL (genomic) |
|---------|------|-----------|-----|-------------|----------------|
| *Homo sapiens (continued)* | | | | | |
| | C | – | JC2191 | D28513 | – |
| | D | – | JC2192 | D28514 | – |
| *Mus musculus* | A | – | – | D50060 | – |
| *Rattus norvegicus* | pc7B | – | JC2345 | D31855 | – |
| *Rattus norvegicus* | A | – | I53282 | L31894 | – |

## Name and History

***Proprotein convertase PACE4*** was discovered in 1991, shortly after the discoveries of proprotein convertase 1 (Chapter 118) and proprotein convertase 2 (Chapter 119), and the realization that furin (Chapter 117) was also a member of the proprotein convertase subfamily of the subtilisin family (Kiefer *et al.*, 1991). In the original report on PACE4 (Kiefer *et al.*, 1991), it was noted that PACE4 had a very long signal peptide region, an observation borne out by the cloning of the rat and mouse variants of PACE4 (Hosaka *et al.*, 1994; Johnson *et al.*, 1994; Tsuji *et al.*, 1994a,b). The original report also stated that the distribution of PACE4 mRNA was rather uniform among many tissues, an observation that has been adjusted in light of the very high levels of PACE4 mRNA in anterior pituitary and cerebellum (Johnson *et al.*, 1994; Dong *et al.*, 1995; Zheng *et al.*, 1997). The term 'PACE' comes from **p**aired basic **a**mino acid **c**leaving **e**nzyme, and this enzyme was numbered 4 because it was fourth in the series after PC1, PC2 and furin. The 'PACE' nomenclature has not been widely adopted. In one publication, the term ***PC7*** was used for a proprotein convertase that can now be seen to have been PACE4 (subject to a frameshifting sequencing error) (Tsuji *et al.*, 1994b). This 'PC7' should not be confused with the proprotein convertase 7 (Chapter 123) (Seidah *et al.*, 1996), also termed PC8 and LPC (Bruzzaniti *et al.*, 1996; Meerabux *et al.*, 1996).

## Activity and Specificity

The enzymatic specificity of PACE4 has not been studied in as much detail as that of the other proprotein convertases. The data that are available have been obtained from three quite distinct systems: (a) recombinant enzyme preparation (not purified to homogeneity) in a test tube; (b) expression of PACE4 together with a potential protein substrate in cell lines with no large dense-core secretory vesicles or regulated pathway of secretion, and (c) expression of PACE4 together with a potential protein substrate in neuroendocrine cell lines that contain dense-core secretory vesicles and a regulated pathway of secretion. The results obtained from these different systems are not in simple agreement, although it is clear that the substrate specificity of PACE4 is more limited than that of furin, with which PACE4 is most often compared. PACE4 does not bind to pro7B2 in coexpression studies (Benjannet *et al.*, 1995), and does not seem to require this kind of chaperone inhibitor–activator. Consistent with this, PACE4 is activated in fibroblasts that lack 7B2 (Mains *et al.*, 1997).

Isolated PACE4 cleaves model substrates such as Boc-Arg-Val-Arg-Arg+NHMec to release the fluorescent aminomethyl-coumarin. In such an assay (Mains *et al.*, 1997), decanoyl-Arg-Val-Lys-Arg-CH$_2$Cl (10 µM) inactivates the enzyme as expected. PACE4 is much more sensitive to inhibition by leupeptin than is furin, but in contrast, is much less sensitive to DTT, EDTA, EGTA and MnCl$_2$ (Mains *et al.*, 1997). $\alpha_1$-Proteinase inhibitor-Portland and $\alpha_1$-proteinase inhibitor-Pittsburgh inhibit furin but have no effect on PACE4 activity (Rehemtulla *et al.*, 1993; Denault *et al.*, 1995; Mains *et al.*, 1997), further confirming that furin and PACE4 have distinct proteolytic activities.

In fibroblasts, when PACE4 is coexpressed with various potential protein substrates, the ability of PACE4 to cleave substrates is much more limited than that of furin. A large number of receptors and several zymogens, as well as growth factors and viral proteins, that are efficiently cleaved by furin under these circumstances are cleaved less well or not at all by PACE4 (Duguay *et al.*, 1997). An exception is the activation of the envelope glycoprotein gp160 of the human immunodeficiency virus (HIV), which is activated well by PACE4 but only poorly by furin (Inocencio *et al.*, 1997). When PACE4 is expressed with pro-von Willebrand's factor, cleavage occurs at the normal -Arg-Ser-Lys-Arg+ activation site, but when mutated forms of the substrate with the sequences -Arg-Ser-Lys-**Gly**- and -**Ala**-Ser-Lys-Arg- were tested, no cleavage was seen, and only poor cleavage was found with the -Arg-Ser-**Ala**-Arg- mutant (Creemers *et al.*, 1993). Thus PACE4 seems to need the consensus sequence Arg-Xaa-Lys/Arg-Arg+ for efficient cleavage of this substrate in a cellular environment.

In neuroendocrine cells, when PACE4 is coexpressed with various potential protein substrates, the pattern of cleavage depends on the intracellular routing of the substrates. Proopiomelanocortin, proglucagon, prosomatostatin and secretogranin are all normally trafficked to dense-core secretory vesicles, and none of these precursors is cleaved by coexpressed PACE4 in neuroendocrine cells (Brakch *et al.*, 1995; Mains *et al.*, 1997). This is consistent with the observation that PACE4 is not stored in dense-core secretory vesicles (as judged by immunocytochemical and pulse-chase analyses in primary anterior pituitary cells and in transfected neuroendocrine cells stably expressing PACE4) (Mains *et al.*, 1997). It is also consistent with the specificity for Arg-Xaa-Lys/Arg-Arg sites as noted above, since Arg-Xaa-Lys/Arg-Arg sites are not common among the prohormones tested. However, the *Aplysia* analog of PACE4 is found in dense-core secretory vesicles, so it is quite possible that PACE4 does contribute to prohormone processing under some circumstances (Chun *et al.*, 1994).

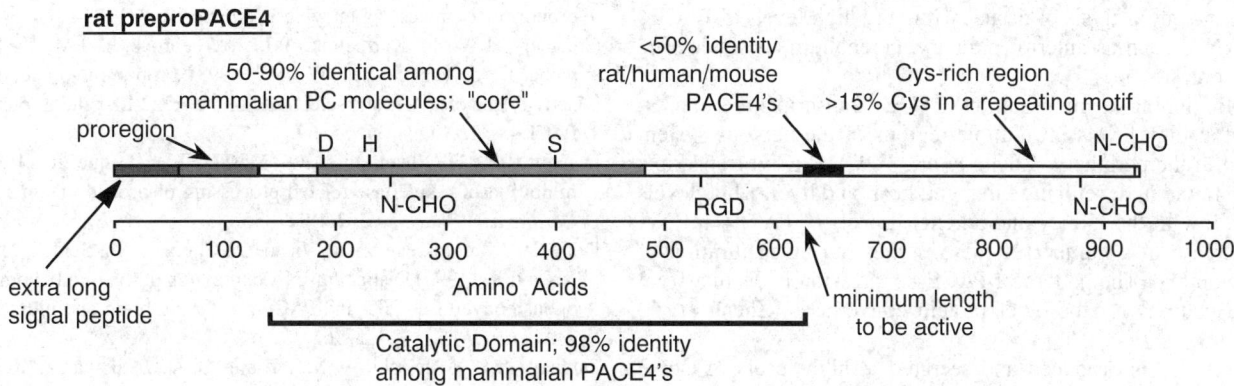

**rat preproPACE4**

*Figure 121.1*   Diagrammatic representation of the structure of the polypeptide chain of rat preproPACE4.

## Structural Chemistry

A schematic diagram of the structure of rat PACE4 is shown in Figure 121.1 (Mains *et al.*, 1997). Rat preproPACE4 has 937 amino acid residues ($M_r$ 104 053), while human (969 residues) and mouse (946 residues) preproPACE4 are slightly larger (Kiefer *et al.*, 1991; Hosaka *et al.*, 1994). The mature protein in each case is about 800 residues in length, since the bulk of the differences are in the signal peptides, which are all extremely long (45–63 residues, depending on species). Within the catalytic domain, the three mammalian sequences are ~98% identical at the amino acid level. All three enzymes have the Asp, His, Ser catalytic triad expected in the subtilisin family (Figure 121.1). There are three potential *N*-glycosylation sites (N-CHO in Figure 121.1) and one RGD potential integrin-binding site (Figure 121.1). C-Terminal to the highly conserved catalytic domain, there is a short stretch (solid black bar in Figure 121.1) where the PACE4 sequences are highly variable between species. After that point, each PACE4 has five copies of a 50 amino acid Cys-rich motif (>15% Cys) which may have a role in intracellular routing or membrane association.

A number of splicing variants and shorter clones have been reported; the version illustrated in Figure 121.1 is now often called PACE4A. In the original report on PACE4, a shorter (presumably inactive) variant called PACE4.1 was found in hEK-293 cells but in no other cell lines or tissue extracts. Naturally occurring forms called PACE4B, C, CS and D are progressively shorter, and only forms A, B and C are active (Nagamune *et al.*, 1995; Tsuji *et al.*, 1994a,b; Zhong *et al.*, 1996). PACE4s is an active, artificial mutant form of PACE4 which stops at amino acid 650 (~515 residues mature length, depending on the precise site of pro-region cleavage, which is not known) (Mains *et al.*, 1997); PACE4C, a natural variant, is 15 residues shorter than PACE4s and is also active (Zhong *et al.*, 1996).

## Preparation

Good sources for the preparation of recombinant PACE4 have been reported, but have not yet been widely used (Hubbard *et al.*, 1997; Mains *et al.*, 1997). Enzymatically active PACE4 is secreted by stably transfected fibroblast cell lines (hEK-293 and CHO) in which 1% of total protein synthesis is PACE4. Purifying the PACE4 or shortened variants from the medium usually involves concentrating the medium, ion-exchange, gel filtration and hydrophobic interaction chromatography, in a procedure similar to that developed for peptidylglycine $\alpha$-hydroxylating monooxygenase (Mains *et al.*, 1997).

## Biological Aspects

PACE4 is synthesized from the preproPACE4 molecule by at least two endoproteolytic cleavage steps. The very long signal peptide is cleaved during transport into the lumen of the endoplasmic reticulum (Johnson *et al.*, 1994). About an hour after the initial synthesis, the pro region is cleaved to liberate the active enzyme (Mains *et al.*, 1997). The presence of the Cys-rich C-terminal domain slows the pro region cleavage process (Mains *et al.*, 1997). In addition, some of the PACE4 synthesized in anterior pituitary cells and by stably transfected neuroendocrine cells is never recovered in a form recognized by the antisera, while most of the PACE4 produced in fibroblasts is secreted and recovered in the medium (Mains *et al.*, 1997). All indications from immunocytochemistry, stimulation of secretion with secretagogues, and analyses of cleavage of peptide products found in dense-core secretory vesicles are consistent with the proposal that PACE4 evades secretory vesicles even in neuroendocrine cells and is secreted rapidly and constitutively (Brakch *et al.*, 1995; Mains *et al.*, 1997).

PACE4 mRNA in the adult is found at the highest levels in the anterior pituitary, notably in endocrine cells making growth hormone, prolactin and ACTH (Johnson *et al.*, 1994). The level of expression of PACE4 mRNA in anterior pituitary increases with circulating thyroid hormone and decreases during thyroid deficiency, covering more than a 10-fold range (Johnson *et al.*, 1994). PACE4 is also very highly expressed in cerebellar Purkinje cells and in over a dozen subcortical nuclei (Dong *et al.*, 1995). PACE4 is produced in the insulin-producing $\beta$ cells of the endocrine pancreas, but not in the $\alpha$, $\delta$ or PP cells (Nagamune *et al.*, 1995). PACE4 is also found at substantial levels in cardiac myocytes (Johnson *et al.*, 1994). PACE4 is found at lower levels in a number of peripheral tissues such as liver, lung, kidney, spleen and gut. The original report of comparable levels of PACE4 mRNA in many tissues (Kiefer *et al.*, 1991) was probably colored by the lack of analysis of tissues with very high levels of PACE4 mRNA, such as anterior pituitary, cerebellum and endocrine pancreas.

Beginning at midgestation, PACE4 transcripts are expressed at high levels in multiple regions of the nervous system and in the periphery (Zheng *et al*., 1997). PACE4 transcripts are found in developing lung, gut, heart and liver, while levels are low in the kidney and adrenal. Levels of PACE4 mRNA decrease in a number of these tissues during maturation of the animal. The pattern of PACE4 expression is distinct from the patterns for the other protein convertases (Zheng *et al*., 1997).

PACE4 is produced and secreted at high levels by some invasive tumors, while closely related noninvasive tumors tend to express much less PACE4 (Hubbard *et al*., 1997). When tumor cells are stably transfected with a vector encoding PACE4, the cells produce and secrete PACE4 and adopt an invasive phenotype; tumor cells stably transfected with a control vector remain less invasive (Hubbard *et al*., 1997). Consistent with the role of secreted PACE4 in tumor invasiveness, the pattern of PACE4 expression in development suggests a role in the extracellular actions or activation of inductive signals (Zheng *et al*., 1997).

The human PACE4 gene is found in the *FES/FPS* region (within 5 Mb of the furin gene) of chromosome 15 at 15q25–qter, and the mouse gene is found at the *Pcsk6* locus on chromosome 7, only 13 cM from the *Pcsk3* (furin) locus (Kiefer *et al*., 1991; Mbikay *et al*., 1995).

## Distinguishing Features

Polyclonal antisera against the pro region cleavage site (Rehemtulla *et al*., 1993) and against a ~100 residue recombinant protein at the C-terminus of the enzymatic domain (Mains *et al*., 1997) have been described but are not commercially available. At the time of this writing, no monoclonal antibodies to PACE4 have been reported.

## Acknowledgement

This work was supported by the National Institute of Drug Abuse DA-00266.

## Further Reading

For a review, see Mains *et al*. (1997).

## References

Benjannet, S., Savaria, D., Chrétien, M. & Seidah, N.G. (1995) 7B2 is a specific intracellular binding protein of the prohormone convertase PC2. *J. Neurochem.* **64**, 2303–2311.

Brakch, N., Galanopoulou, A.S., Patel, Y.C., Boileau, G. & Seidah, N.G. (1995) Comparative proteolytic processing of rat prosomatostatin by the convertases PC1, PC2, furin, PACE4 and PC5 in constitutive and regulated pathways. *FEBS Lett.* **362**, 143–146.

Bruzzaniti, A., Goodge, K., Jay, P., Taviaux, S.A., Lam, M.H.C., Berta, P., Martin, T.J., Moseley, J.M. & Gillespie, M.T. (1996) PC8, a new member of the convertase family. *Biochem. J.* **314**, 727–731.

Chun, J.Y., Korner, J., Kreiner, T., Scheller, R.H. & Axel, R. (1994) The function and differential sorting of a family of *Aplysia*

prohormone processing enzymes. *Neuron* **12**, 831–844.

Creemers, J.W.M., Kormelink, P.J.G., Roebroek, A.J.M., Nakayama, K. & van de Ven, W.J.M. (1993) Proprotein processing activity and cleavage site selectivity of the Kex2-like endoprotease PACE4. *FEBS Lett.* **36**, 65–69.

Denault, J.B., D'Orleans-Juste, O., Masaki, T. & Leduc, R. (1995) Inhibition of convertase-related processing of proendothelin-1. *J. Cardiovasc. Pharmacol.* **26**, S47–S50.

Dong, W., Marcinkiewicz, M., Vieau, D., Chrétien, M., Seidah, N.G. & Day, R. (1995) Distinct mRNA expression of the highly homologous convertases PC5 and PACE4 in the rat brain and pituitary. *J. Neurosci.* **15**, 1778–1796.

Duguay, S.J., Milewski, W.M., Young, B.D., Nakayama, K. & Steiner, D.F. (1997) Processing of wild-type and mutant proinsulin-like growth factor-IA by subtilisin-related proprotein convertases. *J. Biol. Chem.* **272**, 6663–6670.

Hosaka, M., Murakami, K. & Nakayama, K. (1994) PACE4A is a ubiquitous endoprotease that has similar but not identical substrate specificity to other kex2-like processing endoproteases. *Biomed. Res.* **15**, 383–390.

Hubbard, F.C., Goodrow, T.L., Liu, S.C., Brilliant, M.H., Basset, P., Mains, R.E. & Klein-Szanto, A.J.P. (1997) Expression of PACE4 in chemically induced carcinomas is associated with spindle cell tumor conversion and increased invasive ability. *Cancer Res.* **57**, 5226–5231.

Inocencio, N.M., Sucic, J.F., Moehring, J.M., Spence, M.J. & Moehring, T.J. (1997) Endoprotease activities other than furin and PACE4 with a role in processing of HIV-1 gp-160 glycoproteins in CHO-K1 cells. *J. Biol. Chem.* **272**, 1344–1348.

Johnson, R.C., Darlington, D.N., Hand, T.A., Bloomquist, B.T. & Mains, R.E. (1994) PACE4: a subtilisin-like endoprotease prevalent in the anterior pituitary and regulated by thyroid status. *Endocrinology* **135**, 1178–1185.

Kiefer, M.C., Tucker, J.E., Joh, R., Landsberg, K.E., Saltman, D. & Barr, P.J. (1991) Identification of a second human subtilisin-like protease gene in the fes/fps region of chromosome 15. *DNA Cell Biol.* **10**, 757–769.

Mains, R.E., Berard, C.A., Denault, J.B., Zhou, A., Johnson, R.C. & Leduc, R. (1997) PACE4: a subtilisin-like endoprotease with unique properties. *Biochem. J.* **321**, 587–593.

Mbikay, M., Seidah, N.G., Chrétien, M. & Simpson, E.M. (1995) Chromosomal assignment of the genes for proprotein convertases PC4, PC5 and PACE4 in mouse and human. *Genomics* **26**, 123–129.

Meerabux, J., Yaspo, M.L., Roebroek, A.J.M., Van de Ven, W.J.M., Lister, A. & Young, B.D. (1996) A new member of the proprotein convertase gene family, LPC, is located at a chromosome translocation breakpoint in lymphomas. *Cancer Res.* **56**, 448–451.

Nagamune, H., Muramatsu, K., Akamatsu, T., Tamai, Y., Izumi, K., Tsuji, A. & Matsuda, Y. (1995) Distribution of the kexin family proteases in pancreatic islets: PACE4C is specifically expressed in B cells of pancreatic islets. *Endocrinology* **136**, 357–360.

Rehemtulla, A., Barr, P.J., Rhodes, C.J. & Kaufman, R.J. (1993) PACE4 is a member of the mammalian propeptidase family that has overlapping but not identical substrate specificity to PACE. *Biochemistry* **32**, 11586–11590.

Seidah, N.G., Hamelin, J., Mamarbachi, M., Dong, W., Tadros, H., Mbikay, M., Chrétien, M. & Day, R. (1996) cDNA structure, tissue distribution and chromosomal localization of rat PC7, a novel mammalian proprotein convertase closest to yeast kexin-like proteinases. *Proc. Natl Acad. Sci. USA* **93**, 3388–3393.

Tsuji, A., Higashine, K., Hine, C., Mori, K., Tamai, Y., Naga-mune, H. & Matsuda, Y. (1994a) Identification of novel cDNAs encoding human kexin-like protease, PACE4 isoforms. *Biochem. Biophys. Res. Commun.* **200**, 943–950.

Tsuji, A., Hine, C., Mori, K., Tamai, Y., Higashine, K., Naga-mune, H. & Matsuda, Y. (1994b) A novel member, PC7, of the mammalian kexin-like protease family: homology to PACE4A, its brain-specific expression and identification of isoforms. *Biochem. Biophys. Res. Commun.* **202**, 1452–1459.

Zheng, M., Seidah, N.G. & Pintar, J.E. (1997) The developmental expression in the rat CNS and peripheral tissues of proteases PC5 and PACE4 mRNAs: comparison with other proprotein processing enzymes. *Dev. Biol.* **181**, 268–283.

Zhong, M., Benjannet, S., Lazure, C., Munzer, S. & Seidah, N.G. (1996) Functional analysis of human PACE4-A and PACE4-C isoforms: identification of a new PACE4-CS isoform. *FEBS Lett.* **396**, 31–36.

*Richard E. Mains*
*Neuropeptide Laboratory, Departments of Neuroscience and Physiology,*
*The Johns Hopkins University School of Medicine,*
*Baltimore, MD 21205-2185, USA*
*Email: dick.mains@jhu.edu*

# 122. Proprotein convertase 5

## Databanks

*Peptidase classification: clan SB, family S8, MEROPS ID: S08.076*
*NC-IUBMB enzyme classification: none*
*Databank codes:*

| Species | Type | SwissProt | PIR | EMBL (cDNA) | EMBL (genomic) |
|---------|------|-----------|-----|-------------|----------------|
| *Homo sapiens* | PC5 | – | – | U49114 | – |
| *Homo sapiens* | PC6 | – | – | U56387 | – |
| *Mus musculus* | PC5/PC6-A | Q04592 | A48225 | D12619 | – |
| | | | JX0248 | L14932 | |
| *Mus musculus* | PC5/PC6-B | – | S34583 | D17583 | – |
| *Rattus norvegicus* | PC5-A | P41413 | B48225 | L14933 | – |
| *Rattus norvegicus* | PC5-B | – | – | U47014 | – |

## Name and History

Following the discovery of the mammalian kexin/subtilisin-like convertases furin (Chapter 117), PC1 (Chapter 118), PC2 (Chapter 119) and PC4 (Chapter 120), a new effort was made to detect other convertases in various tissues and cells. Using the PCR methodology as applied to mRNA isolated from rat and mouse testis, we were successful in isolating a novel convertase, which we termed *proprotein convertase 5 (PC5)* (Lusson *et al.*, 1993). We had also obtained a partial cDNA sequence of yet another enzyme from rat testis, which turned out to be the rat ortholog of human proprotein convertase PACE4 (Chapter 121), independently discovered by Kiefer *et al.* (1991). At about the same time, another group led by K. Nakayama in Japan, also succeeded in isolating a similar cDNA from mouse brain and called it *PC6* (Nakagawa *et al.*, 1993a), thinking that PACE4 would be the fifth member of the group. It turned out that PC5 and PC6 were identical, and the term *PC5/6* is often used (Seidah *et al.*, 1994). Alternative splicing of the mRNA from the single gene gives rise to two major forms of PC5, *PC5-A* and *PC5-B* (see below).

## Activity and Specificity

Both isoforms of PC5 exhibit neutral pH optima and require 0.2–1.0 mM $Ca^{2+}$ for maximal activity. Both isoforms can cleave a number of fluorogenic substrates of general formula Arg/Lys-(Xaa)$_n$-(Arg)$+$ where $n=2$, 4 or 6. Both isoforms are capable of processing the junction Val-Gln-**Arg**-Glu-**Lys**-**Arg**$+$Ala-Val-Gly-Leu between gp120 and gp41 within the HIV-1 gp160 precursor both *in vitro* (Decroly *et al.*, 1996) and *ex vivo* (Seidah *et al.*, 1994). This cleavage is inhibitable by the $\alpha_1$-proteinase inhibitor variant $\alpha$1-PDX (Anderson *et al.*, 1993) both *in vitro* (Decroly *et al.*, 1996) and *ex vivo* (Seidah *et al.*, 1994).

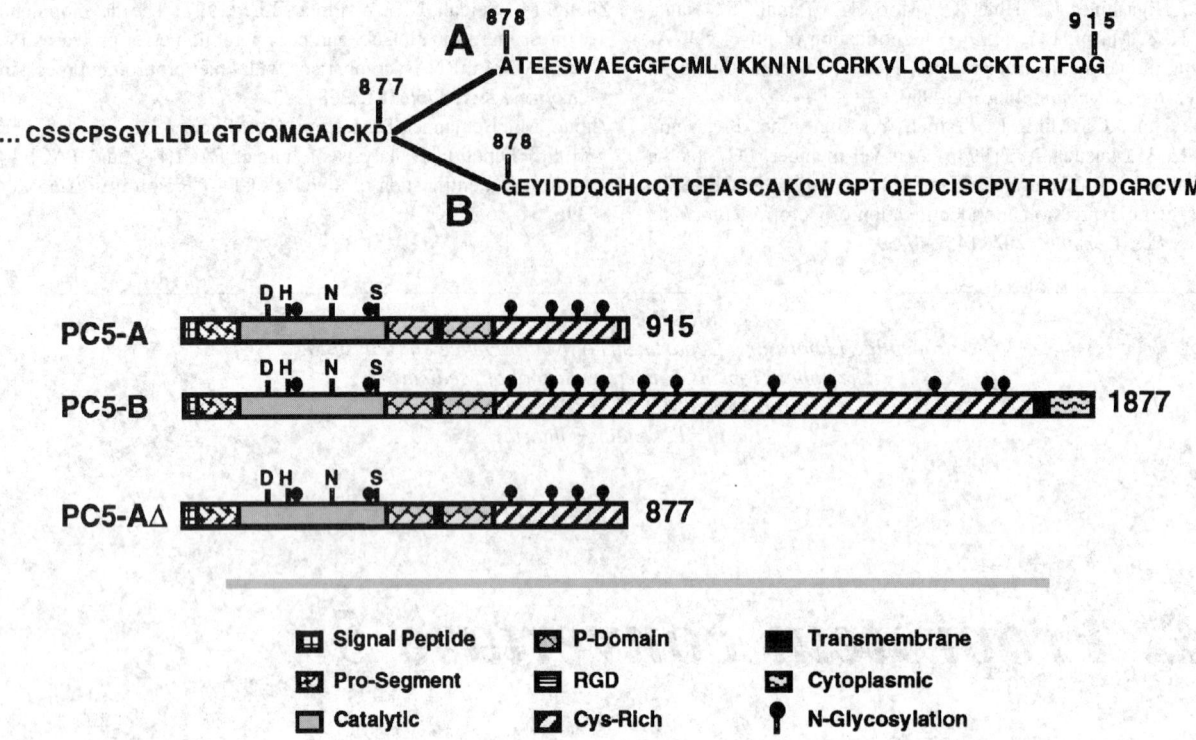

*Figure 122.1*   Schematic representation of mammalian proprotein convertases PC5-A and PC5-B. Amino acid length is given for each convertase. The sequence at the end of the C-terminal region common to PC5-A and PC5-B and that which is unique to each isoform is given at the top of the picture. The PC5-AΔ form in which the 38 amino acid C-terminal segment is deleted is emphasized.

## Structural Chemistry

Overexpression of a vaccinia virus recombinant of PC5-A or PC5-B allowed us to isolate from the medium of GH4C1 cells soluble forms of each isoform. The PC5-A isoform migrated with an apparent molecular mass of 117 kDa. Although the PC5-B form of the convertase was predicted to be membrane bound like furin (see below), soluble forms of both enzymes were released into the medium as a result of a cleavage N-terminal to the transmembrane domain by an as yet unidentified 'sheddase' or 'secretase' (Hooper *et al.*, 1997). The shed form of PC5-B exhibits an apparent mass on SDS-PAGE of 170 kDa (Anderson *et al.*, 1993; Decroly *et al.*, 1996).

Although PC5 exhibits an overall structure similar to the other known convertases (see Figure 118.1), its C-terminal segment contains five copies of a Cys-rich motif of general sequence: **Cys**-(Xaa)$_{9-16}$-**Cys**-(Xaa)$_2$-**Cys**-(Xaa)$_3$-**Cys**-(Xaa)$_2$-**Cys**-(Xaa)$_{5-7}$-**Cys**-(Xaa)$_2$-**Cys**-(Xaa)$_{8-15}$-**Cys**-(Xaa)$_3$. This motif was also found as two copies in furin and five copies in PACE4 (Seidah, 1995).

Analysis of the tissue expression of PC5 by northern blot (Lusson *et al.*, 1993; Nakagawa *et al.*, 1993a) revealed that multiple mRNA forms could be detected in a number of tissues including the adrenal cortex and gut (Lusson *et al.*, 1993). The major mRNA form (3.8 kb) was, however, the one coding for the first isolated 915 amino acid PC5. It was therefore not surprising to find that other PC5 isoforms exist, and the group of Nakayama soon reported the structure of the new C-terminally extended isoform termed PC5-B (or PC6-B) which was 1877 amino acids long, and was predicted to be a type I membrane-bound enzyme (Nakagawa *et al.*, 1993b) (Figure 122.1). This isoform, generated from a single PC5 gene (Mbikay *et al.*, 1995) presumably by alternative splicing of the PC5 heteronuclear RNA transcript, contains 22 repeats of the above mentioned Cys-rich motif (Nakagawa *et al.*, 1993b).

## Preparation

PC5 has been prepared as a soluble 117 kDa product isolated from the medium of GH4C1 cells infected with the vaccinia virus recombinant (Decroly *et al.*, 1996). It has been partially purified on a DEAE column (Decroly *et al.*, 1996). The yield from this procedure is about 0.1 mg liter$^{-1}$ of culture. A smaller, 69 kDa form of PC5 has also been detected on SDS-PAGE of a stable transfectant in AtT20 cells (De Bie *et al.*, 1996).

## Biological Aspects

PC5 has been reported to be widely expressed in both endocrine and nonendocrine tissues, with high expression in the adrenal cortex, the gut, in testicular Sertoli cells (Lusson *et al.*, 1993) and in endothelial cells (Beaubien *et al.*, 1995). These observations led to the analysis of the processing of human prorenin in the adrenals (Mercure *et al.*, 1996), pro-Mullerian inhibiting substance (proMIS) in Sertoli cells (Nachtigal & Ingraham, 1996), and the receptor tyrosine phosphatase R-PTP-m in endothelial cells (Campan *et al.*, 1996). Interestingly, in both proMIS (**Arg**-Gly-**Arg**-Ala-Gly-**Arg**+Ser-*Lys*) and R-PTP-m (Glu-Glu-**Arg**-Pro-**Arg**-**Arg**+Thr-*Lys*) the cleavage site contains a P2′ *Lys*, within the general sequence: **Arg**-Xaa-Xbb-**Arg**+(Ser/Thr)-*Lys*.

The human ortholog of the 915 amino acid PC5-A has been identified, and suggested to play a role in prorenin activation within the adrenal (Mercure *et al.*, 1996) or in HIV gp160 processing in CD4$^+$ cells (Decroly *et al.*, 1996; Miranda *et al.*, 1996; Vollenweider *et al.*, 1996).

From the limited data available on the specificity of PC5 and its physiological substrates, it is expected that this convertase will have a major impact on cellular proliferation and on sexual differentiation. The richness of PC5 within the digestive system also implies involvement in food metabolism. Interestingly, the two isoforms, PC5-A and PC5-B, which arise from alternative splicing of a single gene transcript, are not expected to act on similar precursors, since this C-terminal variation causes the two isoforms to either remain within the constitutive secretory pathway (PC5-B) or enter dense-core secretory vesicles (PC5-A) to process precursors destined to the regulated secretory pathway (De Bie *et al.*, 1996). Indeed, deletion of the unique C-terminal 38 amino acid segment in PC5-A (Figure 122.1) leads to its secretion from the constitutive secretory pathway, suggesting that it contains a powerful signal for sorting towards the granules (De Bie *et al.*, 1996).

The gene coding for PC5 is not characterized, but we do know that it maps to human and mouse chromosomes 9 and 19, respectively (Lusson *et al.*, 1993; Mbikay *et al.*, 1995). The role of the Cys-rich domain in PC5 is not yet understood, but it could well be implicated in protein-protein interactions either at the cell surface (for PC5-B) or intracellularly. PC5 exhibits a distinct distribution within the brain (Dong *et al.*, 1995) and it is expressed early during development (Zheng *et al.*, 1997); it could be involved in embryonic bone morphogenesis (Constam *et al.*, 1996; Zheng *et al.*, 1997).

## *Further Reading*

For a review, see Seidah *et al.* (1993).

## References

Anderson, E.D., Thomas, L., Hayflick, J.S. & Thomas, G. (1993) Inhibition of HIV-1 gp160-dependent membrane fusion by a furin-directed α-1-antitrypsin variant. *J. Biol. Chem.* **268**, 24887–24891.

Beaubien, G., Schafer, M.K.H., Weihe, E., Dong, W., Chrétien, M., Seidah, N.G. & Day, R. (1995) The distinct gene expression of the pro-hormone convertases in the rat heart suggests potential substrates. *Cell Tissue Res.* **279**, 539–549.

Campan, M., Yoshizumi, M., Seidah, N.G., Lee, M.E., Bianchi, C. & Haber, E. (1996) Increased proteolytic processing of protein tyrosine phosphatase-m in confluent vascular endothelial cells – the role of PC5, a member of the subtilisin family. *Biochemistry* **35**, 3797–3802.

Constam, D.B., Calfon, M. & Robertson, E.J. (1996) SPC4, SPC6, and the novel protease SPC7 are coexpressed with bone morphogenetic proteins at distinct sites during embryogenesis. *J. Cell Biol.* **134**, 181–191.

De Bie, I., Marcinkiewicz, M., Malide, D., Lazure, C., Nakayama, K., Bendayan, M. & Seidah, N.G. (1996) The isoforms of proprotein convertase PC5 are sorted to different subcellular compartments. *J. Cell Biol.* **135**, 1261–1275.

Decroly, E., Wouters, S., Dibello, C., Lazure, C., Ruysschaert, J.M. & Seidah, N.G. (1996) Identification of the paired basic convertases implicated in HIV gp160 processing based on in vitro assays and expression in CD4(+) cell lines. *J. Biol. Chem.* **271**, 30442–30450.

Dong, W., Marcinkiewicz, M., Vieau, D., Chrétien, M., Seidah, N.G. & Day, R. (1995) Distinct mRNA expression of the highly homologous convertases PC5 and PACE4 in the rat brain and pituitary. *J. Neurosci.* **15**, 1778–1796.

Hooper, N.M., Karran, E.H. & Turner, A.J. (1997) Membrane protein secretases. *Biochem. J.* **321**, 265–279.

Kiefer, M.C., Tucker, J.E., Joh, R., Landsberg, K.E., Saltman, D. & Barr, P.J. (1991) Identification of a second human subtilisin-like protease gene in the fes/fps region of chromosome 15. *DNA Cell Biol.* **10**, 757–769.

Lusson, J., Vieau, D., Hamelin, J., Day, R., Chrétien, M. & Seidah, N.G. (1993) cDNA structure of the mouse and rat subtilisin/kexin-like PC5 – a candidate proprotein convertase expressed in endocrine and nonendocrine cells. *Proc. Natl Acad. Sci. USA* **90**, 6691–6695.

Mbikay, M., Seidah, N.G., Chrétien, M. & Simpson, E.M. (1995) Chromosomal assignment of the genes for proprotein convertases PC4, PC5, and PACE 4 in mouse and human. *Genomics* **26**, 123–129.

Mercure, C., Jutras, I., Day, R., Seidah, N.G. & Reudelhuber, T.L. (1996) Prohormone convertase PC5 is a candidate processing enzyme for prorenin in the human adrenal cortex. *Hypertension* **28**, 840–846.

Miranda, L., Wolf, J., Pichuantes, S., Duke, R. & Franzusoff, A. (1996) Isolation of the human PC6 gene encoding the putative host protease for HIV-1 gp160 processing in CD4(+) T-lymphocytes. *Proc. Natl Acad. Sci. USA* **93**, 7695–7700.

Nachtigal, M.W. & Ingraham, H.A. (1996) Bioactivation of mullerian inhibiting substance during gonadal development by a kex2/subtilisin-like endoprotease. *Proc. Natl Acad. Sci. USA* **93**, 7711–7716.

Nakagawa, T., Hosaka, M., Torii, S., Watanabe, T., Murakami, K. & Nakayama, K. (1993a) Identification and functional expression of a new member of the mammalian kex2-like processing endoprotease family – its striking structural similarity to PACE4. *J. Biochem. (Tokyo)* **113**, 132–135.

Nakagawa, T., Murakami, K. & Nakayama, K. (1993b) Identification of an isoform with an extremely large Cys-rich region of PC6, a kex2-like processing endoprotease. *FEBS Lett.* **327**, 165–171.

Seidah, N.G. (1995) The mammalian family of subtilisin/kexin-like proprotein convertases. In: *Intramolecular Chaperones and Protein Folding* (Shinde, U. & Inouye, M., eds). Austin, TX: R.G. Landes, pp. 181–203.

Seidah, N.G., Day, R. & Chrétien, M. (1993) The family of pro-hormone and pro-protein convertases. *Biochem. Soc. Trans.* **21**, 685–691.

Seidah, N.G., Day, R. & Chrétien, M. (1994) The family of subtilisin/kexin-like pro-protein convertases: divergent or shared functions. *Biochimie* **76**, 197–209.

Vollenweider, F., Benjannet, S., Decroly, E., Savaria, D., Lazure, C., Thomas, G., Chrétien, M. & Seidah, N.G. (1996) Comparative cellular processing of the human immunodeficiency virus (HIV-1) envelope glycoprotein gp160 by the mammalian subtilisin/kexin-like convertases. *Biochem. J.* **314**, 521–532.

Zheng, M., Seidah, N.G. & Pintar, J.E. (1997) The developmental expression in the rat CNS and peripheral tissues of proteases PC5 and PACE4 mRNAs – comparison with other proprotein processing enzymes. *Dev. Biol.* **181**, 268–283.

**Nabil G. Seidah**
*Laboratory of Biochemical Neuroendocrinology,*
*Clinical Research Institute of Montreal,*
*110 Pine Ave West, Montreal, QC, H2W 1R7 Canada*
*Email: seidahn@ipcmumontreal.ca*

**Michel Chrétien**
*Laboratory of Molecular Neuroendocrinology,*
*Clinical Research Institute of Montreal,*
*110 Pine Ave West, Montreal, QC, H2W 1R7 Canada*

# 123. *Proprotein convertase 7*

## Databanks

Peptidase classification: clan SB, family S8, MEROPS ID: S08.077
NC-IUBMB enzyme classification: none
Databank codes:

| Species | SwissProt | PIR | EMBL (cDNA) | EMBL (genomic) |
|---|---|---|---|---|
| *Homo sapiens* | – | – | U33849 U40623 | – |
| *Mus musculus* | – | – | U48830 | – |
| *Rattus norvegicus* | – | – | U36580 | – |

## Name and History

The generation of bioactive proteins and peptides by limited proteolysis of inactive precursors is an evolutionarily ancient mechanism determining the level and duration of specific bioactivities. Processing of proproteins and prohormones often occurs at specific single or pairs of basic amino acids in precursors of neuropeptide hormones, peptide receptors, proteolytic enzymes, growth factors, cell adhesion molecules and cell surface glycoproteins of pathogenic species such as viruses and bacteria. The evolutionarily conserved enzymes responsible for this cleavage have now been molecularly and functionally characterized and shown to belong to a subfamily of serine proteinases of the subtilisin/kexin type. The seven known mammalian **p**roprotein **c**onvertases (PCs) have been named furin (also called PACE) (Chapter 117), PC1 (also called PC3) (Chapter 118), PC2 (Chapter 119), PC4 (Chapter 120), PACE4 (Chapter 121) and PC5 (also called PC6 or PC5/6) (Chapter 122) (Van de Ven *et al.*, 1993; Seidah *et al.*, 1994, 1997; Rouille *et al.*, 1995; Seidah, 1995) and the seventh one identified in 1996 was called **PC7** (Seidah *et al.*, 1996). PC7 has also been termed **LPC** (Meerabux *et al.*,1996), **PC8** (Bruzzaniti *et al.*, 1996), and **SPC7** (Constam *et al.*, 1996). [A form of PACE4 that seemed to be novel as the result of a sequencing error was termed PC7 by Tsuji *et al.* (1994), but this usage is now obsolete.]

Detailing the tissue expression, regulation, ontogeny and biological functions of the proprotein convertases and their mutants is starting to give us clues as to their specific roles in cells and whole animals, as well as their clinical importance. What is emerging is that the convertases play a pivotal role in many biological processes and that in a cell-specific combinatorial fashion they exhibit either complementary or specific functions. Interestingly, type I membrane-bound PC7 is the most ancestral member of the mammalian family, exhibiting the closest relationship to yeast kexin (Chapter 115) (Seidah *et al.*, 1996; see also Figure 93.3).

## Activity and Specificity

Biochemical and enzymatic characterization of the novel proprotein convertase rat PC7 (rPC7) was carried out using vaccinia virus recombinants overexpressed in mammalian BSC40 cells. ProPC7 is synthesized as a glycosylated proenzyme (101 kDa) and processed into mature rPC7 (89 kDa) in the endoplasmic reticulum. No endogenously produced soluble forms of this membrane-anchored protein were detected. A deletion mutant (65 kDa), truncated well beyond the expected C-terminal boundary of the P-domain (see Figure 118.1), produced soluble rPC7 in the culture medium. Enzymatic activity assays of rPC7 using fluorogenic substrates indicated that the enzyme exhibits a neutral pH optimum, and a $Ca^{2+}$ dependence, and cleavage specificity largely similar to that of furin, with a general recognition sequence of either Arg/Lys-Arg↓ or Arg/Lys-$(Xaa)_n$-(Arg)↓ where $n = 2$, 4 or 6 (Seidah, 1995; Seidah *et al.*, 1997). However, with some substrates, cleavage specificity more closely resembled that of yeast kexin, suggesting differential processing of proprotein substrates by this novel convertase. We examined the rPC7- and human furin-mediated cleavage of synthetic peptides containing the processing sites of three proteins known to colocalize *in situ* with rPC7. Whereas both enzymes correctly processed the proparathyroid hormone tridecapeptide **Lys**-Ser-Val-Lys-**Lys-Arg**↓Ser-Val-Ser-Glu-Ile-Gln-Leu (Hendy *et al.*, 1995) and the proPC4 heptadecapeptide Tyr-Glu-Thr-Leu-**Arg**-Arg-**Arg**-Val-**Lys-Arg**↓Ser-Leu-Val-Val-Pro-Thr-Asp (Seidah *et al.*, 1992), neither enzyme cleaved a hexadecapeptide His-Leu-Arg-Glu-Asp-Asp-His-His-Tyr-Ser-Val-**Arg**-Asn-Ser-Asp-Ser spanning the N-terminal cleavage site of proepidermal growth factor. Thus, rPC7 is an enzymatically functional subtilisin/kexin-like serine proteinase with a cleavage specificity resembling that of human furin (Munzer *et al.*, 1997).

In *ex vivo* expression systems, PC7 was shown to cleave the HIV-1 gp160 into gp120/gp41, but in contrast to furin, it did not process gp120 into gp77/gp53 (Meerabux *et al.*, 1997). It was also able to process pro7B2 into 7B2 (Benjannet *et al.*, 1997), and was inhibited by the $\alpha_1$-proteinase inhibitor variant α1-PDX (Anderson *et al.*, 1993) and was best able to cleave substrates within the environment of constitutively secreting cells (Benjannet *et al.*, 1997).

## Preparation

Biosynthetic analysis of full-length PC7 demonstrated that proPC7 is a membrane-bound proenzyme that is processed to PC7 within the endoplasmic reticulum (ER), possibly by an autocatalytic mechanism. Thus PC7 is like all other convertases, except PC2, in that it undergoes autocatalytic proenzyme processing within the ER (Seidah, 1995; Munzer *et al.*, 1997). Since PC7 is membrane bound (Munzer *et al.*, 1997), to isolate an active soluble form of PC7, we found it necessary to place a stop codon close to the N-terminus of the C-terminal transmembrane segment, called BTMD-PC7 (**b**efore **t**rans**m**embrane **d**omain) (Seidah *et al.*, 1996; Munzer *et al.*, 1997). Thus the soluble, *in vitro* active form produced spans the segment ending at Gly-Tyr-Ser622 (Seidah *et al.*, 1996; Munzer *et al.*, 1997). We thus obtained both vaccinia virus and baculovirus recombinants and showed that both systems yield active enzyme, in the range of 1–3 mg liter$^{-1}$.

## Biological Aspects

In adult rats both furin and PC7 are highly expressed in lymphoid-associated tissues such as thymus and spleen. In the spleen, furin is expressed more in the red pulp than in the white pulp, whereas the reverse is true for PC7. Thus, although these two class I convertases are widely expressed, the cells expressing the highest amount of these enzymes are not the same, suggesting that *in vivo* furin and PC7 are likely to process different precursors. However, the high expression of furin and PC7 in human T lymphocytes, including CD4$^+$ cells (Decroly *et al.*, 1997), suggests that both could participate in the activation of viral surface glycoproteins including the HIV gp160 type I glycoprotein.

Aside from their role in the activation of viral glycoproteins, the function of PCs in T lymphocytes and in the immune system in general remains to be elucidated. Growth factors, cytokines and cytokine receptors (e.g. tumor necrosis factor receptor) as well as integrins, whose activation is PC dependent, are synthesized in immune cells. The function of PC7 within the immune system remains to be fully uncovered. In that context, it is interesting to note that human PC7 (called lymphoma PC, or LPC) was originally discovered because of a translocation of a segment of chromosome 14 into chromosome 11 within the exact locus of PC7, and more specifically within the 3' end untranslated region of human PC7 (Meerabux *et al.*, 1996). The relationship of PC7 and lymphoid function is yet to be discovered, but it seems likely that perturbance of the levels of PC7 could lead to cancer of lymphatic tissues (Meerabux *et al.*, 1996). In a similar vein, it was suggested that PC7 (called PC8 in this case) could play an important role in the activation of parathyroid-related peptides (Bruzzaniti *et al.*, 1996), and hence could be important in cellular proliferation.

Finally, developmental studies on PC7 have revealed that it is widely expressed early during development, and that its expression in the adult becomes more limited and enriched in lymphoid-associated tissues (Munzer *et al.*, 1997; Seidah *et al.*, 1997). In adult rat brain, PC7 is mostly expressed in neurons, in contrast to the other widely expressed enzyme furin, which is found both in glia and neurons.

## Distinguishing Features

Polyclonal antibodies were produced against the segment Ala449-Ser-Tyr-Val-Ser-Pro-Met-Leu-Lys-Glu-Asn-Lys-Ala-Val-Pro-Arg-Ser465 located in the P-domain of rPC7 (see Figure 118.1). These antibodies were proven to work very well in immunoprecipitations, immunoblots and immunocytochemistry, suggesting this is a well-exposed antigen (Munzer *et al.*, 1997). The gene coding for PC7 is not yet characterized, but we do know that it maps to human and mouse chromosomes 11q23 and 9, respectively (Meerabux *et al.*, 1996; Seidah *et al.*, 1996).

## Further Reading

For a review, see Seidah *et al.* (1994).

## References

Anderson, E.D., Thomas, L., Hayflick, J.S. & Thomas, G. (1993) Inhibition of HIV-1 gp160-dependent membrane fusion by a furin-directed alpha 1-antitrypsin variant. *J. Biol. Chem.* **268**, 24887–24891.

Benjannet, S., Savaria, D., Laslop, A., Munzer, J.S., Chrétien, M., Marcinkiewicz, M. & Seidah, N.G. (1997) α1-Antitrypsin-Portland inhibits processing of precursors mediated by proprotein convertases within the constitutive secretory pathway. *J. Biol. Chem.* **272**, 26210–26218.

Bruzzaniti, A., Goodge, K., Jay, P., Taviaux, S.A., Lam, M.H.C., Berta, P., Martin, T.J., Moseley, J.M. & Gillespie, M.T. (1996)

PC8, a new member of the convertase family. *Biochem. J.* **314**, 727–731.

Constam, D.B., Calfon, M. & Robertson, E.J. (1996) SPC4, SPC6, and the novel protease SPC7 are coexpressed with bone morphogenetic proteins at distinct sites during embryogenesis. *J. Cell Biol.* **134**, 181–191.

Decroly, E., Benjannet, S., Savaria, D. & Seidah, N.G. (1997) Comparative functional role of PC7 and furin in the processing of the HIV envelope glycoprotein gp160. *FEBS Lett.* **405**, 68–72.

Hendy, G.N., Bennett, H.P.J., Gibbs, B.F., Lazure, C., Day, R. &. Seidah, N.G. (1995) Proparathyroid hormone is preferentially cleaved to parathyroid hormone by the prohormone convertase furin: a mass spectrometric study. *J. Biol. Chem.* **270**, 9517–9525.

Meerabux, J., Yaspo, M.L., Roebroek, A.J., Van de Ven, W.J., Lister, T.A. & Young, B.D. (1996) A new member of the proprotein convertase gene family (LPC) is located at a chromosome translocation breakpoint in lymphomas. *Cancer Res.* **56**, 448–451.

Munzer, J.S., Basak, A., Zhong, M., Mamarbachi, M., Hamelin, J., Savaria, D., Lazure, C., Benjannet, S., Chrétien, M. & Seidah, N.G. (1997) *In vitro* characterization of the novel proprotein convertase PC7. *J. Biol. Chem.* **272**, 19672–19681.

Rouille, Y., Duguay, S.J., Lund, K., Furuta, M., Gong, Q.M., Lipkind, G., Oliva, A.A., Chan, S.J. & Steiner, D.F. (1995) Proteolytic processing mechanisms in the biosynthesis of neuroendocrine peptides – the subtilisin-like proprotein convertases. *Front. Neuroendocrinol.* **16**, 322–361.

Seidah, N.G. (1995) The mammalian family of subtilisin/kexin-like proprotein convertases. In: *Intramolecular Chaperones and Protein Folding* (Shinde, U. & Inouye, M., eds). Austin, TX. R.G. Landes, pp. 181–203.

Seidah, N.G., Day, R., Hamelin, J., Gaspar, A., Collard, M.W. & Chrétien, M. (1992) Testicular expression of PC4 in the rat – molecular diversity of a novel germ cell-specific kex2/subtilisin-like proprotein convertase. *Mol. Endocrinol.* **6**, 1559–1570.

Seidah, N.G., Day, R. & Chrétien, M. (1994) The family of subtilisin/kexin-like pro-protein convertases: divergent or shared functions. *Biochimie* **76**, 197–209.

Seidah, N.G., Hamelin, J., Mamarbachi, M., Dong, W., Tadros, H., Mbikay, M., Chrétien, M. & Day, R. (1996) cDNA structure, tissue distribution, and chromosomal localization of rat PC7, a novel mammlian proprotein convertase closes to yeast kexin-like proteinases. *Proc. Natl Acad. Sci. USA* **93**, 3388–3393.

Seidah, N.G., Day, R., Marcinkiewicz, M. & Chrétien, M. (1997) Precursor convertases: an evolutionary ancient, cell-specific, combinatorial mechanism yielding diverse bioactive peptides and proteins. *Ann. NY Acad. Sci.* (in press).

Tsuji, A., Hine, C., Mori, K., Tamai, Y., Higashine, K., Nagamune, H. & Matsuda, Y. (1994) A novel member, PC7, of the mammalian kexin-like protease family: homology to PACE4A, its brain-specific expression and identification of isoforms. *Biochem. Biophys. Res. Commun.* **202**, 1452–1459.

Van de Ven, W.J.M., Roebroek, A.J.M. & Vanduijnhoven, H.L.P. (1993) Structure and function of eukaryotic proprotein processing enzymes of the subtilisin family of serine proteases. *Crit. Rev. Oncog.* **4**, 115–136.

***Nabil G. Seidah***
*Laboratory of Biochemical Neuroendocrinology,*
*Clinical Research Institute of Montreal,*
*110 Pine Ave West, Montreal, QC, H2W 1R7 Canada*
*Email: seidahn@ipcmumontreal.ca*

***Michel Chrétien***
*Laboratory of Molecular Neuroendocrinology,*
*Clinical Research Institute of Montreal,*
*110 Pine Ave West, Montreal, QC, H2W 1R7 Canada*

# 124. Introduction: clan SC containing peptidases with the α/β hydrolase fold

## Databanks

MEROPS ID: SC

| Species | SwissProt | PIR | EMBL (cDNA) | EMBL (genomic) |
|---|---|---|---|---|
| **Family S9** | | | | |
| Acylaminoacyl-peptidase (Chapter 130) | | | | |
| Dipeptidyl aminopeptidase A (Chapter 129) | | | | |
| Dipeptidyl aminopeptidase B (Chapter 129) | | | | |
| Dipeptidyl-peptidase IV (Chapter 128) | | | | |
| Fibroblast activation protein α subunit (Chapter 131) | | | | |
| Oligopeptidase B (Chapter 126) | | | | |
| Oligopeptidase B (*Trypanosoma*) (Chapter 127) | | | | |
| Prolyl oligopeptidase (Chapter 125) | | | | |
| **Family S10** | | M76427 | | |
| Carboxypeptidase C (Chapter 132) | | | | |
| Carboxypeptidase D (Chapter 134) | | | | |
| Carboxypeptidase S3 | | | | |
|    *Penicillium janthinellum* | – | S57294 | – | |
| Lysosomal carboxypeptidase A (Chapter 133) | | | | |
| Others | | | | |
|    *Absibia zychae* | – | – | D16519 | – |
|    *Aedes aegypti* | P42660 | – | L46594 | – |
| | | | M79452 | |
|    *Arabidopsis thaliana* | – | – | F13862 | – |
|    *Aspergillus saitoi* | P52719 | S55328 | D25288 | – |
| | | S55357 | | |
| | | S60126 | | |
|    *Caenorhabditis elegans* | P52714 | – | Z54342 | – |
|    *Caenorhabditis elegans* | P52717 | – | U23521 | – |
|    *Caenorhabditis elegans* | – | – | Z70203 | – |
|    *Naegleria fowleri* | P42661 | – | M88397 | – |
|    *Oryza sativa* | P52712 | – | D17587 | – |
|    *Saccharomyces cerevisiae* | P38109 | S46008 | Z36008 | X75891: part of chromosome II |
| | | S46581 | | |
|    *Schizosaccharomyces pombe* | P32825 | B42249 | D10199 | – |
|    *Trypanosoma brucei* | – | – | N99293 | – |
|    *Vigna radiata* | – | – | U49382 | – |
|    *Vigna radiata* | – | – | U49741 | – |
| **Family S15** | | | | |
| X-Pro dipeptidyl-peptidase (Chapter 136) | | | | |
| **Family S28** | | | | |
| Dipeptidyl-peptidase II (Chapter 138) | | | | |
| Lysosomal Pro-X carboxypeptidase (Chapter 137) | | | | |
| **Family S33** | | | | |
| Leucine aminopeptidase pepL | | | | |
|    *Lactobacillus delbrueckii* | – | – | Z34898 | – |
|    *Lactobacillus helveticus* | – | – | Z30709 | – |

*continued overleaf*

| Species | SwissProt | PIR | EMBL (cDNA) | EMBL (genomic) |
|---|---|---|---|---|
| **Family S33** (*continued*) | | | | |
| Prolyl aminopeptidase (Chapter 139) | | | | |
| Tripeptidyl-peptidases A, B, C (*Streptomyces lividans*) (Chapter 140) | | | | |
| **Family S37** | | | | |
| PS-10 peptidase | | | | |
| *Streptomyces lividans* | – | – | L46588 | – |

Clan SC contains peptidase families in which the catalytic triad has been identified as being in the order Ser, Asp, His. The clan includes the families S9, S10, S15, S28, S33 and S37, and the members of the clan are listed in the Databanks table. An alignment around the catalytic residues for selected members of this clan is shown in Alignment 124.1 (see CD-ROM).

Tertiary structures have been solved for families S10 and S33, although no peptidase structures are known for S33. The three-dimensional structure of wheat carboxypeptidase II is shown in Figure 124.1. The structure can be described as mainly parallel β sheets comprised of β/α/β units, and shows similarity to a variety of other proteins, including acetylcholinesterase, haloalkane dehalogenase, dienelactone hydrolase, bromoperoxidase A2, thioesterase, triacylglycerol lipase and hydroxynitrile lyase. This fold has been described as the 'α/β hydrolase fold' (even though not all members are hydrolases), and is unlike that of any other serine peptidase. Not all proteins with this fold have serine as a nucleophile, for example dienelactone hydrolase has cysteine, and haloalkane dehalogenase has an aspartate (Ollis *et al.*, 1992; Cygler *et al.*, 1993).

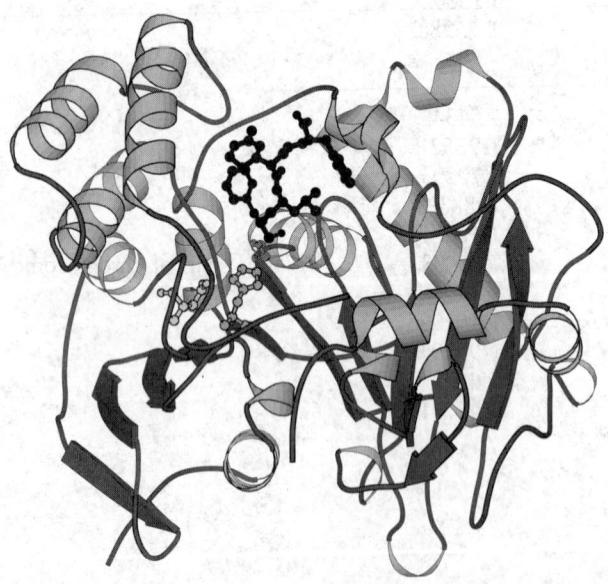

*Figure 124.1* Richardson diagram of the wheat serine carboxypeptidase II/chymostatin complex. The image was prepared from the Brookhaven Protein Data Bank entry (1BCS) as described in the Introduction (p. xxv). Catalytic residues are shown in ball-and-stick representation; Ser158, Asp340 and His392 (numbering as in Alignment 124.1). Chymostatin is shown in black in ball-and-stick representation.

*Family S9* includes prolyl oligopeptidase, acylaminoacyl-peptidase and dipeptidyl-peptidase IV. Although these peptidases are very divergent in structure, and different in catalytic activity, they all possess C-terminal catalytic domains showing significant sequence similarity (Rawlings *et al.*, 1991); each can be considered a representative of a subfamily. Subfamily S9A includes cytosolic oligopeptidases from bacteria, archaea and eukaryotes, such as prolyl oligopeptidase and oligopeptidase B. Subfamily S9B includes the cytoplasmic omega peptidase acylaminoacyl-peptidase, which is known only from eukaryotes. Subfamily S9C includes the membrane-bound dipeptidyl-peptidase IV, which is known from bacteria and eukaryotes. The catalytic residues have been identified only in dipeptidyl-peptidase IV (David *et al.*, 1993).

*Family S10* contains only carboxypeptidases, which are unique among serine peptidases in being active at acidic pH, and are mainly targeted to the lysosome in animals and to the vacuoles in fungi and plants. Examples are known only from eukaryotes. Alignment 124.1 shows an alignment of the sequences around the catalytic triad residues. Glu157, preceding the catalytic serine, is thought to be responsible for the acidic pH optimum of these carboxypeptidases (but it should be noted that a conserved Asp precedes the catalytic Ser in family S1, and does not have this effect there). The family includes carboxypeptidase C, which has a specificity for releasing C-terminal hydrophobic amino acids, and carboxypeptidase D, which has a preference for a basic residue in P1'. These specificities mirror those of the metallocarboxypeptidases in family M14 (see Chapter 450). Unusually, a gene from *Caenorhabditis elegans* is predicted to encode a protein with three copies of a serine carboxypeptidase domain (EMBL: Z70203). A full alignment of the amino acid sequences of the mature peptidases of family S10 has been published on the World Wide Web by Olesen & Breddam (1997).

Figure 124.2 is an evolutionary tree for family S10. It is notable that the distinction between carboxypeptidases C and D does not represent a single divergence on the tree, and carboxypeptidases that release basic amino acids are distributed in all three main branches. This situation is analogous to that of trypsin (Chapter 3) and other endopeptidases from family S1 (Chapter 2) that cleave after Lys and Arg residues, which are also derived not from one branch but from several. In family S1, this can be explained by mutation of a single residue within the S1 specificity site, and a similar explanation may also apply to family S10.

*Family S15* includes X-Pro dipeptidyl-peptidase, an intracellular enzyme known only from *Lactococcus* and *Lactobacillus*. The catalytic residues have not been determined biochemically, but as can be seen from Alignment 124.1, the

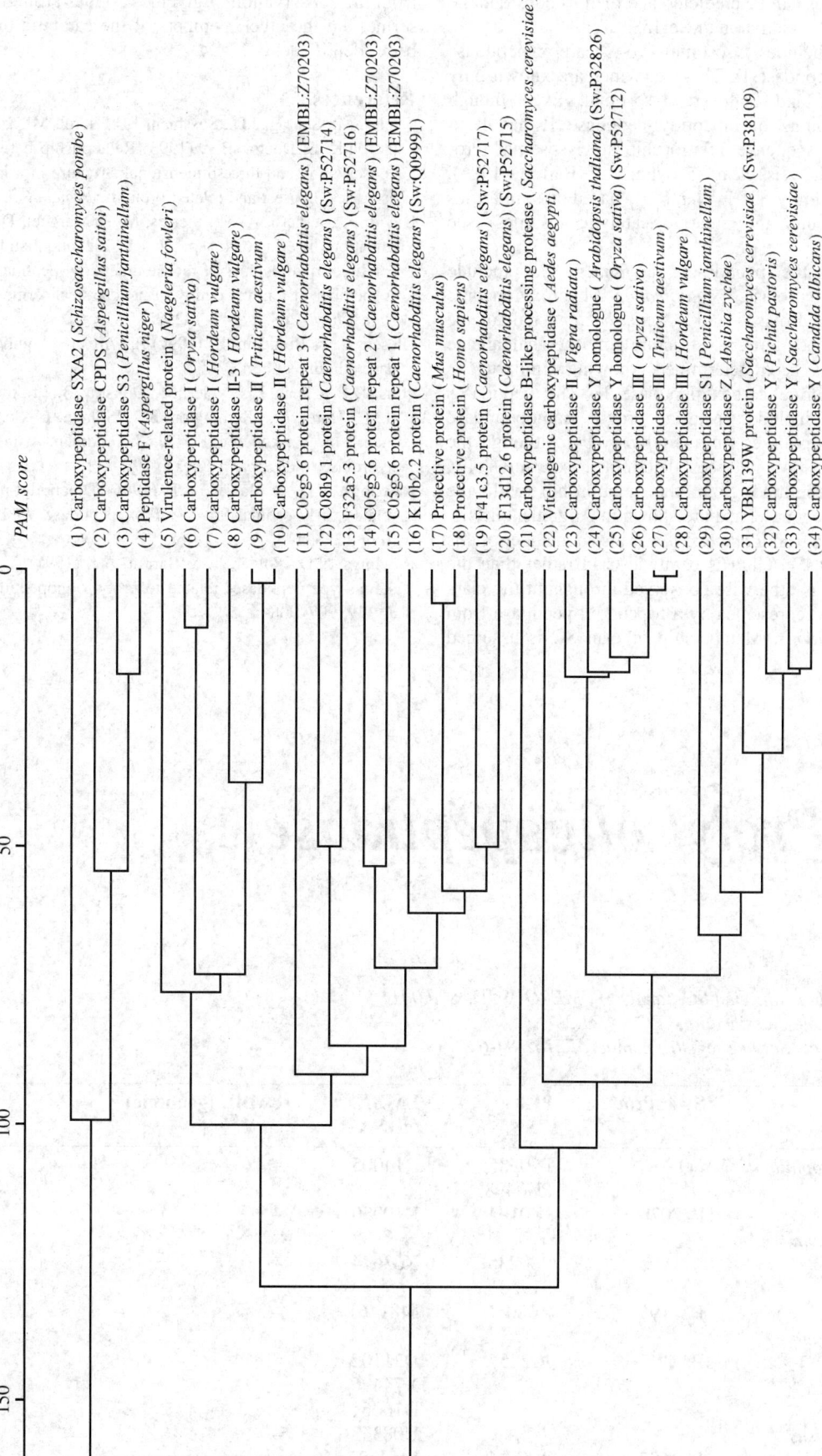

*PAM score*

(1) Carboxypeptidase SXA2 (*Schizosaccharomyces pombe*)
(2) Carboxypeptidase CPDS (*Aspergillus saitoi*)
(3) Carboxypeptidase S3 (*Penicillium janthinellum*)
(4) Peptidase F (*Aspergillus niger*)
(5) Virulence-related protein (*Naegleria fowleri*)
(6) Carboxypeptidase I (*Oryza sativa*)
(7) Carboxypeptidase I (*Hordeum vulgare*)
(8) Carboxypeptidase II-3 (*Hordeum vulgare*)
(9) Carboxypeptidase II (*Triticum aestivum*)
(10) Carboxypeptidase II (*Hordeum vulgare*)
(11) C05g5.6 protein repeat 3 (*Caenorhabditis elegans*) (EMBL:Z70203)
(12) C08h9.1 protein (*Caenorhabditis elegans*) (Sw:P52714)
(13) F32a5.3 protein (*Caenorhabditis elegans*) (Sw:P52716)
(14) C05g5.6 protein repeat 2 (*Caenorhabditis elegans*) (EMBL:Z70203)
(15) C05g5.6 protein repeat 1 (*Caenorhabditis elegans*) (EMBL:Z70203)
(16) K10b2.2 protein (*Caenorhabditis elegans*) (Sw:Q09991)
(17) Protective protein (*Mus musculus*)
(18) Protective protein (*Homo sapiens*)
(19) F41c3.5 protein (*Caenorhabditis elegans*) (Sw:P52717)
(20) F13d12.6 protein (*Caenorhabditis elegans*) (Sw:P52715)
(21) Carboxypeptidase B-like processing protease (*Saccharomyces cerevisiae*)
(22) Vitellogenic carboxypeptidase (*Aedes aegypti*)
(23) Carboxypeptidase II (*Vigna radiata*)
(24) Carboxypeptidase Y homologue (*Arabidopsis thaliana*) (Sw:P32826)
(25) Carboxypeptidase Y homologue (*Oryza sativa*) (Sw:P52712)
(26) Carboxypeptidase III (*Oryza sativa*)
(27) Carboxypeptidase III (*Triticum aestivum*)
(28) Carboxypeptidase III (*Hordeum vulgare*)
(29) Carboxypeptidase S1 (*Penicillium janthinellum*)
(30) Carboxypeptidase Z (*Absidia zychae*)
(31) YBR139W protein (*Saccharomyces cerevisiae*) (Sw:P38109)
(32) Carboxypeptidase Y (*Pichia pastoris*)
(33) Carboxypeptidase Y (*Saccharomyces cerevisiae*)
(34) Carboxypeptidase Y (*Candida albicans*)

*Figure* 124.2   Evolutionary tree for family S10. The tree was prepared as described in the Introduction (p. xxv).

S

active-site residues can be predicted and there is considerable conservation in the clan around Ser158.

*Family S28* includes lysosomal Pro-X carboxypeptidase and dipeptidyl-peptidase II. These enzymes are known only from eukaryotes and both are lysosomal. Even though only the N-terminus of dipeptidyl-peptidase II has been determined, the sequence relationship to lysosomal Pro-X carboxypeptidase is clear (Rawlings & Barrett, 1996). Membership of family S28 in clan SC is based on similarities in sequence around the putative catalytic residues (see Alignment 124.1).

Homologs of the peptidases in *family S33* include human epoxide hydrolase, *Pseudomonas* 2-hydroxymuconic semialdehyde hydrolase and *Xanthobacter* haloalkane dehalogenase. The tertiary structure of the haloalkane dehalogenase has been determined, showing similarity to that of wheat serine carboxypeptidase II. The peptidases of family S33 include bacterial prolyl aminopeptidase and tripeptidyl-peptidases from *Streptomyces*. Although the catalytic residues have not been determined experimentally, they can be identified from the relationship with the haloalkane dehalogenase, and are shown in Alignment 124.1. X-Pro dipeptidase from *Lactobacillus* is also a member of this family, but has a Ser/Glu/His catalytic triad rather than the Ser/Asp/His that is otherwise conserved throughout the clan.

*Family S37* is represented by a tripeptidyl-peptidase from *Streptomyces lividans*. Membership of clan SC is assumed from the conservation of glycine residues around the catalytic serine, but the other members of the catalytic triad have not been identified.

## References

Cygler, M., Schrag, J.D., Sussman, J.L., Harel, M., Silman, I., Gentry, M.K. & Doctor, B.P. (1993) Relationship between sequence conservation and three-dimensional structure in a large family of esterases, lipases, and related proteins. *Protein Sci.* **2**, 366–382.

David, F., Bernard, A.-M., Pierres, M. & Marguet, D. (1993) Identification of serine 624, aspartic acid 702, and histidine 734 as the catalytic triad residues of mouse dipeptidyl-peptidase IV (CD26). A member of a novel family of nonclassical serine hydrolases. *J. Biol. Chem.* **268**, 17247–17252.

Olesen, K. & Breddam, K. (1997) ftp://ftp.crc.dk/pub/carboxypeptidase/Alignment.htm.

Ollis, D.L., Cheah, E., Cygler, M., Dijkstra, B., Frolow, F., Franken, S.M., Harel, M., Remington, S.J., Silman, I., Schrag, J., Sussman, J.L., Verscheuren, K.H.G. & Goldman, A. (1992) The $\alpha/\beta$ hydrolase fold. *Protein Eng.* **5**, 197–211.

Rawlings, N.D. & Barrett, A.J. (1996) Dipeptidyl-peptidase II is related to lysosomal Pro-X carboxypeptidase. *Biochim. Biophys. Acta* **1298**, 1–3.

Rawlings, N.D., Polgár, L. & Barrett, A.J. (1991) A new family of serine-type peptidases related to prolyl oligopeptidase. *Biochem. J.* **279**, 907–908.

# 125. *Prolyl oligopeptidase*

## *Databanks*

*Peptidase classification: clan SC, family S9, MEROPS ID: S09.001*
*NC-IUBMB enzyme classification: EC 3.4.21.26*
*Chemical Abstracts Service registry number: 72162-84-6*
*Databank codes:*

| Species | SwissProt | PIR | EMBL (cDNA) | EMBL (genomic) |
|---|---|---|---|---|
| *Aeromonas hydrophila* | Q06903 | JN0585 PN0498 | D14005 | – |
| *Flavobacterium meningosepticum* | P27028 | JX0194 | D10980 | – |
| | | PS0235 S19201 | X63674 | |
| *Flavobacterium meningosepticum* | P27195 | A38086 | M81461 | – |
| *Homo sapiens* | P48147 | JC2257 | D21102 X74496 | – |
| *Oryza sativa* | – | – | D16061 | – |
| *Pyrococcus furiosus* | – | – | U08343 | – |
| *Sus scrofa* | P23687 | A37942 | M64227 | – |

## Name and History

*Prolyl oligopeptidase* was discovered in the human uterus as an oxytocin-degrading enzyme (Walter *et al.*, 1971). Since the enzyme showed a high specificity for proline residues, and hydrolyzed the peptide bond on their carboxyl side, it was originally named *post-proline cleaving enzyme*. During the period 1978–1983, several enzymes hydrolyzing the Pro-Xaa bonds of biologically active peptides (thyrotropin-releasing hormone, bradykinin, substance P, neurotensin, angiotensin II, luteinizing hormone-releasing hormone) were described. They were referred to as *post-proline endopeptidase, TRH deamidase* and *brain kinase B* or *endooligopeptidase B*. However, all of them were finally found to be identical to the post-proline cleaving enzyme.

The name post-proline endopeptidase was recommended by IUBMB in 1978, and then changed to *prolyl endopeptidase* in the supplement to Enzyme Nomenclature in 1981. Although prolyl oligopeptidase was recognized as a serine peptidase as early as 1977 (Yoshimoto *et al.*, 1977), it is commonly found to be activated by thiol comounds, and for a time the thiol-dependent activity was recognized by a separate EC number, EC 3.4.22.18. On the basis of the oligopeptidase nature of the reaction catalyzed, and the amino acid sequence homology with other oligopeptidases, Barrett & Rawlings (1992) proposed the name *prolyl oligopeptidase*.

## Activity and Specificity

Prolyl oligopeptidase activity is assayed with Z-Gly-Pro┼NHNap, Z-Gly-Pro┼NHPhNO$_2$ or Z-Gly-Pro┼NHMec (Yoshimoto *et al.*, 1979). The optimum pH for hydrolysis of these substrates is around 7–8. In many studies, activity has been found to be enhanced by thiol compounds, and these are commonly included in assay buffers.

The substrate specificity of prolyl oligopeptidase has been studied with a range of synthetic substrates (Walter & Yoshimoto, 1978; Yoshimoto *et al.*, 1978). The active site has been characterized with regard to the number of specificity subsites and their stereospecificity, and the $K_m$ and $k_{cat}$ values for hydrolysis of peptides of different lengths have also been determined. The studies on prolyl oligopeptidases from several sources indicated that the active site of the enzyme is composed of five binding subsites, S3, S2, S1, S1′ and S2′, showing high stereospecificity in the S2, S1 and S1′ subsites.

The enzyme also hydrolyzed -Xaa┼Ala- (Yoshimoto *et al.*, 1978), -Xaa┼(*N*-methyl)alanine- and -Xaa┼Ser- bonds (Nomura, 1986). This suggests that the enzyme also recognizes an amino acid side chain that fits in the space corresponding to the proline ring.

Potent inhibitors such as Z-Gly-Pro-CH$_2$Cl, Z-Pro-prolinal and Z-thioprolyl-thioprolinal (Wilk & Orlowski, 1983; Yoshimoto *et al.*, 1985) have been synthesized. Inhibitors from microorganisms (Aoyagi *et al.*, 1991) and intracellular inhibitors in porcine pancreas (Yoshimoto *et al.*, 1982) and rat brain (Soeda *et al.*, 1985) have also been found.

## Structural Chemistry

Prolyl oligopeptidase was identified as a serine enzyme by use of DFP and proline-containing chloromethanes (Yoshimoto *et al.*, 1977; Stone *et al.*, 1991). Nevertheless, the enzyme from several sources has been found to be activated by thiol compounds, and inhibited by PCMB. This is thought to be due to the presence of a thiol group near the catalytic serine residue. The enzyme is generally a soluble, 76–80 kDa-protein, but that from *Flavobacterium meningosepticum* has a signal peptide sequence and is expressed in the periplasm (Chevallier *et al.*, 1992).

The prolyl oligopeptidase genes already cloned are those from pig brain (Rennex *et al.*, 1991), *F. meningosepticum* (Yoshimoto *et al.*, 1991), *Aeromonas hydrophyla* (Kanatani *et al.*, 1993) and *Pyrococcus furiosus* (Robinson *et al.*, 1995).

Sequence similarities have shown that prolyl oligopeptidase, oligopeptidase B (Chapter 126), acylaminoacyl-peptidase (Chapter 130) and dipeptidyl-peptidase IV (Chapter 128) are all members of the prolyl oligopeptidase family (Barrett & Rawlings, 1992). Interestingly, the family is distantly related to carboxypeptidase P, serine carboxypeptidases and acetylcholine esterase (Tan *et al.*, 1993), which suggests that all these enzymes share the $\alpha/\beta$ hydrolase fold (see Chapter 124).

The active-site Ser and His residues were identified as Ser554 and His680, respectively, for the pig brain enzyme. Since the conservation of amino acid sequence in the prolyl oligopeptidase family is predominantly in the C-terminal half, which includes the catalytic triad, the enzyme may be composed of at least two domains. Limited proteolysis by trypsin cleaved the enzyme at Lys196┼Ser197, but the two fragments could not be separated by size-exclusion chromatography under nondenaturing conditions (Polgár & Patthy, 1992).

## Preparation

Prolyl oligopeptidase is widely present in microorganisms, plants and animals. The enzyme has been purified from various tissues of sheep, rat, rabbit, cattle, pig and human. The enzyme was also purified from carrot, mushroom, *Halocynthia roretzi* and *F. meningosepticum*. The enzyme from *F. meningosepticum* is commercially available.

## Biological Aspects

Prolyl oligopeptidase cleaves specifically the Pro┼Xaa bond in biologically active peptides, i.e. substance P, oxytocin, vasopressin, gonadoliberin, bradykinin, neurotensin, angiotensin II, thyrotropin-releasing hormone, luteinizing hormone-releasing hormone, α-melanocyte-stimulating hormone and dynorphin. However, the larger molecules of denatured casein, collagen, gastrin and adrenocorticotropin-releasing hormone are not hydrolyzed by the enzyme. Therefore, the wide distribution of prolyl oligopeptidase and the high rate of degradation of specific biologically active peptides suggest that this enzyme participates in the *in vivo* regulation of the action of these peptides. There has been much research interest in indications of roles for prolyl oligopeptidase in memory and other neural processes (Li *et al.*, 1996; Toide *et al.*, 1997). The human gene for prolyl oligopeptidase has been mapped to 6q22 (Goossens *et al.*, 1996).

## Further Reading

Fuller reviews of prolyl oligopeptidase are those of Walter *et al*. (1980), Yaron & Naider (1993) and Polgár (1995).

## References

Aoyagi, T., Nagai, M., Ogawa, K., Kojima, F., Okada, M., Ikeda, T., Hamada, M. & Takeuchi, T. (1991) Poststatin, a new inhibitor of prolyl endopeptidase, produced by *Streptomyces viridochromogenes* MH534–30F3. I. Taxonomy, production, isolation, physico-chemical properties and biological activities. *J. Antibiot.* **44**, 949–955.

Barrett, A.J. & Rawlings, N.D. (1992) Oligopeptidases, and the emergence of the prolyl oligopeptidase family. *Biol. Chem. Hoppe-Seyler* **373**, 353–360.

Chevallier, S., Goeltz, P., Thibault, P., Banville, D. & Gagnon, J. (1992) Characterization of a prolyl endopeptidase from *Flavobacterium meningosepticum*. Complete sequence and localization of the active-site serine. *J. Biol. Chem.* **267**, 8192–8199.

Goossens, F.J., Wauters, J.G., Vanhoof, G.C., Bossuyt, P.J., Schatteman, K.A., Loens, K. & Scharpé, S.L. (1996) Subregional mapping of the human lymphocyte prolyl oligopeptidase gene (PREP) to human chromosome 6q22. *Cytogenet. Cell Genet.* **74**, 99–101.

Kanatani, A., Yoshimoto, T., Kitazono, A., Kokubo, T. & Tsuru, D. (1993) Prolyl endopeptidase from *Aeromonas hydrophyla*: cloning, sequencing, and expression of the enzyme gene, and characterization of the expressed enzyme. *J. Biochem.* **113**, 790–796.

Li, J.R., Wilk, E. & Wilk, S. (1996) Inhibition of prolyl oligopeptidase by Fmoc-aminoacylpyrrolidine-2-nitriles. *J. Neurochem.* **66**, 2105–2112.

Nomura, K. (1986) Specificity of prolyl endopeptidase. *FEBS Lett.* **209**, 235–237.

Polgár, L. (1995) Prolyl oligopeptidases. *Methods Enzymol.* **248**, 188–200.

Polgár, L. & Patthy, A. (1992) Cleavage of the Lys196-Ser197 bond of prolyl oligopeptidase: enhanced catalytic activity for one of the two active enzyme forms. *Biochemistry* **31**, 10769–10773.

Rennex, D., Hemmings, B.A., Hofsteenge, J. & Stone, S.R. (1991) cDNA cloning of porcine brain prolyl endopeptidase and identification of the active-site seryl residue. *Biochemistry* **30**, 2195–2203.

Robinson, K.A., Bartley, D.A., Robb, F.T. & Schreier, H.J. (1995) A gene from the hyperthermophile *Pyrococcus furiosus* whose deduced product is homologous to members of the prolyl oligopeptidase family of proteases. *Gene* **152**, 103–106.

Soeda, S., Yamakawa, N.M., Ohyama, H.S. & Nagamatsu, A. (1985) An inhibitor of proline endopeptidase: purification from rat brain and characterization. *Chem. Pharm. Bull.* **33**, 2445–2451.

Stone, S.R., Rennex, D., Wikstrom, P., Shaw, E. & Hofsteenge, J. (1991) Inactivation of prolyl endopeptidase by a peptidylchloromethane. Kinetics of inactivation and identification of sites of modification. *Biochem. J.* **276**, 837–840.

Tan, F., Morris, P.W., Skidgel, R.A. & Erdös, E.G. (1993) Sequencing and cloning of human prolylcarboxypeptidase (Angiotensinase C): similarity to both serine carboxypeptidase and prolylendopeptidase families. *J. Biol. Chem.* **268**, 16631–16638.

Toide, K., Shinoda, M., Iwamoto, Y., Fujiwara, T., Okamiya, K. & Uemura, A. (1997) A novel prolyl endopeptidase inhibitor, JTP-4819, with potential for treating Alzheimer's disease. *Behav. Brain Res.* **83**, 147–151.

Walter, R. & Yoshimoto, T. (1978) Post-proline cleaving enzyme: kinetic studies of size and stereospecificity of its active site. *Biochemistry* **17**, 4139–4144.

Walter, R., Shlank, H., Glass, J.D., Schwartz, I.L. & Kerenyi, T.D. (1971) Leucylglycineamide released from oxytocin by human uterine enzyme. *Science* **173**, 827–829.

Walter, R., Simmons, W.H. & Yoshimoto, T. (1980) Proline specific endo- and exopeptidases. *Mol. Cell. Biochem.* **30**, 111–127.

Wilk, S. & Orlowski, M. (1983) Inhibition of rabbit brain prolyl endopeptidase by *n*-benzyloxycarbonyl-prolyl-prolinal, a transition state aldehyde inhibitor. *J. Neurochem.* **41**, 69–75.

Yaron, A. & Naider, F. (1993) Proline-dependent structural and biological properties of peptides and proteins. *CRC Crit. Rev. Biochem. Mol. Biol.* **28**, 31–81.

Yoshimoto, T., Orlowski, R. C. & Walter, R. (1977) Post-proline cleaving enzyme: identification as serine protease using active site specific inhibitors. *Biochemistry* **16**, 2942–2948.

Yoshimoto, T., Fischl, M., Orlowski, R.C. & Walter, R. (1978) Post-proline cleaving enzyme and post-proline dipeptidyl aminopeptidase: comparison of two peptidases with high specificity for proline residues. *J. Biol. Chem.* **253**, 3708–3716.

Yoshimoto, T., Ogita, K., Walter, R., Koida, M. & Tsuru, D. (1979) Post-proline cleaving enzyme: synthesis of a new fluorogenic substrate and distribution of the endopeptidase in rat tissues and body fluids of man. *Biochim. Biophys. Acta* **569**, 184–192.

Yoshimoto, T., Tsukumo, K., Takatsuka, N. & Tsuru, D. (1982) An inhibitor for post-proline cleaving enzyme: distribution and partial purification from porcine pancreas. *J. Pharm. Dyn.* **5**, 734–740.

Yoshimoto, T., Kawahara, K., Matsubara, F., Kado, K. & Tsuru, D. (1985) Comparison of inhibitory effects of prolinal-containing peptide derivatives on prolyl endopeptidase from bovine brain and *Flavobacterium*. *J. Biochem.* **98**, 975–979.

Yoshimoto, T., Kanatani, A., Shimoda, T., Inaoka, T., Kokubo, T. & Tsuru, D. (1991) Prolyl endopeptidase from *Flavobacterium meningosepticum*: cloning and sequencing of the enzyme gene. *J. Biochem.* **110**, 873–878.

*T. Yoshimoto*
*Department of Biotechnology,*
*Nagasaki University School of Pharmaceutical Sciences,*
*Nagasaki, 852 Japan*
*Email: t-yoshimoto@cc.nagasaki-u.ac.jp*

*K. Ito*
*Department of Biotechnology,*
*Nagasaki University School of Pharmaceutical Sciences,*
*Nagasaki, 852 Japan*
*Email: k-ito@cc.nagasaki-u.ac.jp*

# 126. Oligopeptidase B

## Databanks

*Peptidase classification: clan SC, family S9, MEROPS ID: S09.010*
*NC-IUBMB enzyme classification: EC 3.4.21.83*
*Databank codes:*

| Species | SwissProt | PIR | EMBL (cDNA) | EMBL (genomic) |
|---|---|---|---|---|
| *Escherichia coli* | P24555 | JQ1151 PQ0214 | D10976 | – |
| *Moraxella lacunata* | – | – | D38405 | – |

## Name and History

*Escherichia coli* produces a variety of intracellular proteolytic enzymes. More than 15 endopeptidases have been reported: proteases I to VII, and/or Do to Pi (Tsuru & Yoshimoto, 1987; Gottesman, 1996). *Protease II* has been purified from a strain of *E. coli* (Pacaud & Richaud, 1975) and characterized as a trypsin-like endopeptidase which is only slightly active toward high molecular mass proteins (Pacaud & Richaud, 1975; Pacaud, 1978; Kanatani *et al.*, 1991). This enzyme has now been designated *oligopeptidase B*, where 'B' refers to the preferential specificity of the enzyme toward basic amino acid residues. Similar enzymes have also been found in plant seeds (Nishikata, 1984), *Rhodococcus erythropolis* (Shannon *et al.*, 1982) and *Trypanosoma brucei* (Chapter 127) (Kornblatt *et al.*, 1992), suggesting that oligopeptidase B is widely distributed in nature, although the physicochemical properties of the enzyme differ somewhat between sources (Tsuru & Yoshimoto, 1994).

A mammalian enzyme once given the similar name 'endo-oligopeptidase B' is not the trypsin-like enzyme, but a proline-specific oligopeptidase that inactivates biologically active peptides containing a proline residue (Camargo *et al.*, 1979, 1984; Greene *et al.*, 1982). This enzyme is now called prolyl oligopeptidase (Rawlings *et al.*, 1991; Polgár, 1994) (see Chapter 125).

## Activity and Specificity

Oligopeptidase B catalyzes hydrolysis of synthetic substrates and oligopeptides exclusively at the carboxyl side of basic amino acid residues, even when the P1′ residue is proline. The optimal pH for activity is in the neutral to slightly alkaline region: 7.0 for *R. erythropolis* and 8–8.5 for *E. coli*, *T. brucei* and soybean seed. Good substrates are Bz-Arg┼OEt, Tos-Arg-OMe, Bz-Arg-NHPhNO$_2$ and Bz-Arg (or Lys)-NHNap in *E. coli* and soybean seed, Bz-Arg-NHMec and Bz-Arg-NHPhNO$_2$ for *T. brucei*, and Z-Phe-Arg-, Z-Arg- and Bz-Arg-NHMec for *R. erythropolis* (Tsuru & Yoshimoto, 1994). Oligopeptidase B is inhibited by DFP, antipain, leupeptin and Tos-Lys-CH$_2$Cl, but not affected by ovomucoid, or trypin inhibitors from lima- and soybeans. Aprotinin has been reported partially to inhibit the enzymes from soybean seed (Nishikata, 1984) and *T. brucei* (Kornblatt *et al.*, 1992).

Enzyme activity is assayed spectrophotometrically by measuring liberation of 2-naphthylamine from Bz-Arg-NHNap (Kanatani *et al.*, 1991; Tsuru & Yoshimoto, 1994), or fluorometrically by measuring liberation of 7-amino-4-methylcoumarin from Bz-Arg-NHMec (Shannon *et al.*, 1982).

## Structural Chemistry

Oligopeptidase B is a monomeric protein of about 82 kDa, except for the soybean seed enzyme (59 kDa), and pI values of enzymes from *E. coli* and *T. brucei* are near 5. Oligopeptidase B of *E. coli* is synthesized as an active form. The amino acid sequence of this enzyme has been established by protein sequencing and by nucleotide sequence analysis of the gene *prtB* (Kanatani *et al.*, 1991). The enzyme is composed of 707 amino acid residues with Met and Tyr as N- and C-termini, respectively. The active-site Ser residue is identified to be Ser532 by labeling with tritiated DFP (Kanatani *et al.*, 1991). The whole sequence of *E. coli* oligopeptidase B is quite different from those of enzymes in the trypsin and subtilisin families, and little similarity is seen with other serine proteases of *E. coli*. Surprisingly, however, the enzyme is 24–25% homologous to prolyl oligopeptidases of various origins (Yoshimoto *et al.*, 1991; Kanatani *et al.*, 1993; Tsuru & Yoshimoto, 1994). Thus, the enzyme has been classified in the prolyl oligopeptidase family (Rawlings & Barrett, 1994; Tsuru & Yoshimoto, 1994). From the sequence homology survey, the catalytic triad of *E. coli* oligopeptidase B is concluded to be Ser532, Asp617 and His652 (see Chapter 124). The primary structures of other oligopeptidases B have not yet been reported.

## Preparation

Oligopeptidase B has been purified from *E. coli* (Pacaud & Richaud, 1975; Pacaud, 1976), *Rhodococcus erythropolis* (Shannon *et al.*, 1982), soybean seed (Nishikata, 1984), and *Trypanosoma brucei* (Kornblatt *et al.*, 1992). The oligopeptidase B gene (*ptrB*) of *E. coli* was cloned and expressed in *E. coli* JM83, and the enzyme was easily purified from the transformant cells, which showed a 90-fold higher activity than the host cells, by DEAE-Toyopearl chromatography and gel filtration on Sephadex G-150 (Kanatani *et al.*, 1991).

## Biological Aspects and Distinguishing Features

Oligopeptidase B of *E. coli* is a periplasmic enzyme. Its biological role and the natural substrates remain unknown. The *E. coli prtB* gene is located at 1981–1984 kb on the physical map (Kanatani *et al.*, 1992). The enzyme is distinguished from mammalian trypsin and trypsin-like proteinases of microbial origin in that it is almost inert toward polypeptides larger than 6 kDa.

A unique trypsin-like peptidase from *E. coli* (protease In) has been described (Kato *et al.*, 1992). It is insensitive to DFP and inert toward Bz-Arg-NHPhNO$_2$, but is strongly inhibited by aprotinin. Omptin is another *E. coli* peptidase with trypsin-like specificity, but is a membrane-associated enzyme (Chapter 176).

## Further Reading

For reviews, see Tsuru & Yoshimoto (1994) and Rawlings & Barrett (1994).

## References

Camargo, A.C., Caldo, H. & Reis, M.L. (1979) Susceptibility of a peptide derived from bradykinin to hydrolysis by brain endo-oligopeptidases and pancreatic proteinases. *J. Biol. Chem.* **254**, 5304–5307.

Camargo, A.C., Almeida, M.L. & Emson, P.C. (1984) Involvement of endo-oligopeptidases A and B in the degradation of neurotensin by rabbit brain. *J. Neurochem.* **42**, 1758–1761.

Gottesman, S. (1996) Proteases and their targets in *Escherichia coli*. *Annu. Rev. Genet.* **30**, 465–506.

Greene, L.J., Spadaro, A.C., Martins, A.R., Perussi De Jesus, W.D. & Camargo, A.C. (1982) Brain endo-oligopeptidase B: a post-proline cleaving enzyme that inactivates angiotensin I and II. *Hypertension* **4**, 178–184.

Kanatani, A., Masuda, T., Shimoda, T., Misoka, F., Xu, S.L., Yoshimoto, T. & Tsuru, D. (1991) Protease II from *Escherichia coli*: sequencing and expression of the enzyme gene and characterization of the expressed enzyme. *J. Biochem. (Tokyo)* **110**, 315–320.

Kanatani, A., Yoshimoto, T., Nagai, H., Ito, K. & Tsuru, D. (1992) Location of the protease II gene (*ptrB*) on the physical map of the *Escherichia coli* chromosome. *J. Bacteriol.* **174**, 7881.

Kanatani, A., Yoshimoto, T., Kitazono, A., Kokubo, T. & Tsuru, D. (1993) Prolyl endopeptidase from *Aeromonas hydrophila*: cloning, sequencing, and expression of the enzyme gene, and characterization of the expressed enzyme. *J. Biochem. (Tokyo)* **113**, 790–796.

Kato, M., Irisawa, T., Ohtani, M. & Muramatu, M. (1992) purification and characterization of proteinase In, a trypsin-like proteinase, in *Escherichia coli*. *Eur. J. Biochem.* **210**, 1007–1014.

Kornblatt, M.J., Mpimbaza, G.W. & Lonsdale-Eccles, J.D. (1992) Characterization of an endopeptidase of *Trypanosoma brucei brucei*. *Arch. Biochem. Biophys.* **293**, 25–31.

Nishikata, M. (1984) Trypsin-like protease from soybean seeds: purification and some properties. *J. Biochem. (Tokyo)* **95**, 1169–1177.

Pacaud, M. & Richaud, C. (1975) Protease II from *Escherichia coli*: purification and characterization. *J. Biol. Chem.* **250**, 7771–7779.

Pacaud, M. (1976) Purification of protease II from *Escherichia coli* by affinity chromatography and separation of two enzyme species from cells harvested at late log phase. *Eur. J. Biochem.* **64**, 199–204.

Pacaud, M. (1978) Protease II from *Escherichia coli*: substrate specificity and kinetic parameters. *Eur. J. Biochem.* **82**, 439–451.

Polgár, L. (1994) Prolyl oligopeptidases. *Methods Enzymol.* **244**, 188–200.

Rawlings, N.D., Polgár, L. & Barrett, A.J. (1991) A new family of serine-type peptidases related to prolyl oligopeptidase. *Biochem. J.* **279**, 907–908.

Rawlings, N.D. & Barrett, A.J. (1994) Families of serine peptidases. *Methods Enzymol.* **244**, 19–61.

Shannon, J.D., Bond, J.S. & Bradley, S.G. (1982) Isolation and characterization of an intracellular serine protease from *Rhodococcus erythropolis*. *Arch. Biochem. Biophys.* **219**, 80–88.

Tsuru, D. & Yoshimoto, T. (1987) Microbial proteases. In: *Handbook of Microbiology*, 2nd edn, vol. VIII (Askin, A.L. & Lechevalier, H.A., eds). Boca Raton, FL: CRC Press, pp. 239–284.

Tsuru, D. & Yoshimoto, T. (1994) Oligopeptidase B: protease II from *Escherichia coli*. *Methods Enzymol.* **244**, 201–215.

Yoshimoto, T., Kanatani, A., Inaoka, T., Kokubo, T. & Tsuru, D. (1991) Cloning and sequencing of prolyl endopeptidase gene from *Flavobacterium meningosepticum*. *J. Biochem. (Tokyo)* **110**, 873–878.

*D. Tsuru*
*Department of Applied Microbiology and Biotechnology,*
*Kumamoto Institute of Technology,*
*Ikeda 4-22-1, Kumamoto 860, Japan*
*Email: tsuru@bio.kumamoto-it.ac.jp*

# 127. Oligopeptidase B of Trypanosoma

## Databanks

*Peptidase classification: clan SC, family S9, MEROPS ID: S09.011*
*NC-IUBMB enzyme classification: EC 3.4.21.83*

*Databank codes:*

| Species | SwissProt | PIR | EMBL (cDNA) | EMBL (genomic) |
|---|---|---|---|---|
| *Trypanosoma cruzi* | – | – | U69897 | – |

## Name and History

A peptidase activity with specificity for cleavage on the carboxyl side of basic amino acid residues was originally purified from lysates of *Trypanosoma cruzi* epimastigotes by Bongertz & Hungerer (1978). This enzyme was referred to as *T. cruzi* **alkaline peptidase** (Ashall, 1990) to distinguish it from the lysosomal cysteine proteinase cruzipain (Chapter 203) (Cazzulo *et al.*, 1989), the other major Arg-cleaving activity in these protozoan parasites. The names ***120 kDa alkaline proteinase*** (Santana *et al.*, 1992) and ***120 kDa alkaline peptidase*** (Burleigh & Andrews, 1995) were subsequently used since the peptidase was originally described as >200 kDa (Ashall, 1990). The cloning and sequencing of a full-length cDNA showed the enzyme to be a form of ***oligopeptidase B*** (Burleigh *et al.*, 1997).

## Activity and Specificity

*T. cruzi* oligopeptidase B is a serine endopeptidase (Burleigh *et al.*, 1997) that cleaves small synthetic peptide substrates containing Arg or Lys in the P1 position with preference for dibasic pairs (Ashall *et al.*, 1990), but exhibits little or no activity on polypeptide substrates (Ashall, 1990; Santana *et al.*, 1992; Burleigh & Andrews, 1995). The pH optimum is 8.0 (Santana *et al.*, 1992). Activity is inhibited by Tos-Lys-CH$_2$Cl, leupeptin, antipain, HgCl$_2$, PCMB, DFP (Ashall, 1990; Santana *et al.*, 1992; Burleigh & Andrews, 1995; Burleigh *et al.*, 1997) as well as arginyl fluoromethanes and chloromethanes such as Z-Phe-Arg-CH$_2$F and Z-Arg-Arg-CH$_2$Cl (Burleigh & Andrews, 1995).

## Structural Chemistry

Purified *T. cruzi* oligopeptidase B has been reported to be 120 kDa (Santana *et al.*, 1992; Burleigh & Andrews, 1995) and >200 kDa (Ashall, 1990). The nucleotide sequence of a full-length cDNA clone predicted a hydrophilic 81 kDa polypeptide with no signal sequence (Burleigh *et al.*, 1997). Expression of the full-length cDNA in *E. coli* revealed that the 81 kDa recombinant peptidase migrated at 120 kDa on SDS-PAGE gels under conditions known to preserve peptidase activity (Burleigh *et al.*, 1997). Based on amino acid composition, the predicted pI is 5.9, but the enzyme elutes from Mono-P chromatofocusing columns with an apparent pI of 4.8–5.0 (Santana *et al.*, 1992; Burleigh & Andrews, 1995). Gel-filtration elution profiles for the native and recombinant enzymes suggest that oligopeptidase B is a dimer (Burleigh *et al.*, 1997).

Oligopeptidase B contains the Gly-Xaa-**Ser**-Xaa-Gly-Gly-Zbb-Zbb consensus sequence (where Xaa is any amino acid, Zbb is a hydrophobic residue and **Ser** is the active-site serine) present in all members of the prolyl oligopeptidase family (Barrett & Rawlings, 1992). The *T. cruzi* enzyme is 34% and 30% identical in sequence to the *Moraxella lacunata* and *Escherichia coli* forms of oligopeptidase B (Burleigh *et al.*, 1997), all of which exhibit similar substrate specificity (Pacaud & Richaud, 1975; Yoshimoto *et al.*, 1995) (see Chapter 126). Less similarity (18–22% identity) was found between *T. cruzi* oligopeptidase B and prolyl oligopeptidase (Chapter 125), also in this family (Burleigh *et al.*, 1997).

## Preparation

*T. cruzi* oligopeptidase B is expressed in all life-cycle stages of *T. cruzi* (Burleigh *et al.*, 1997) and has been purified from epimastigotes using several chromatographic steps (Bongertz & Hungerer, 1978; Ashall, 1990; Santana *et al.*, 1992; Burleigh & Andrews, 1995). Epimastigotes are the most convenient source of oligopeptidase B since *in vitro* cultivation yields relatively large numbers of parasites. Expression of recombinant oligopeptidase B in *E. coli* as an N-terminal poly-histidine-tagged fusion protein resulted in the production of milligram quantities of active enzyme (Burleigh *et al.*, 1997). A similar peptidase activity has been demonstrated in other trypanosomatids (Ashall, 1990; Ashall *et al.*, 1990), but has not been fully characterized.

## Biological Aspects

Soluble extracts prepared from trypomastigotes, an invasive form of *T. cruzi*, induce intracellular free Ca$^{2+}$ transients in mammalian cells (Tardieux *et al.*, 1994; Burleigh & Andrews, 1995). The requirement for peptidase activity in the induction of Ca$^{2+}$ transients was demonstrated when signaling was blocked by addition of a subset of protease inhibitors. Based on the profile of effective inhibitors, oligopeptidase B was identified as the likely candidate for involvement in this pathway (Burleigh & Andrews, 1995). Direct evidence for the participation of oligopeptidase B in Ca$^{2+}$ signaling has been obtained recently using neutralizing antibodies generated to the active recombinant peptidase (Burleigh *et al.*, 1997). However, purified oligopeptidase B does not induce Ca$^{2+}$ transients in mammalian cells, and evidence suggests that this process requires an additional factor(s) (Burleigh & Andrews, 1995). It is likely that oligopeptidase B is a processing enzyme which generates an active Ca$^{2+}$ agonist for mammalian cells from a precursor expressed exclusively in invasive forms of the parasite. This putative substrate molecule has not yet been identified. In addition, since oligopeptidase B is a cytosolic enzyme expressed in all life-cycle stages of *T. cruzi*, it may fulfill a more general role in protein metabolism in these parasites.

## References

Ashall, F. (1990) Characterization of an alkaline peptidase of *Trypanosoma cruzi* and other trypanosomatids. *Mol. Biochem. Parasitol.* **38**, 77–87.

Ashall, F., Harris, D., Roberts, H., Healy, N. & Shaw, E. (1990) Substrate specificity and inhibitor sensitivity of a trypanosomatid alkaline peptidase. *Biochim. Biophys. Acta* **1035**, 293–299.

Barrett, A.J. & Rawlings, N.D. (1992) Oligopeptidases, and the emergence of the prolyl oligopeptidase family. *Biol. Chem. Hoppe-Seyler* **373**, 353–360.

Bongertz, V. & Hungerer, K.D. (1978) *Trypanosoma cruzi*: isolation and characterization of a protease. *Exp. Parasitol.* **45**, 8–18.

Burleigh, B. & Andrews, N.W. (1995) A 120 kDa alkaline peptidase from *Trypanosoma cruzi* is involved in the generation of a novel $Ca^{2+}$-signaling factor for mammalian cells. *J. Biol. Chem.* **270**, 5172–5180.

Burleigh, B.A., Caler, E.V., Webster, P. & Andrews, N.W. (1997) Cytosolic serine endopeptidase from *Trypanosoma cruzi* is required for the generation of $Ca^{2+}$ signaling in mammalian cells. *J. Cell Biol.* **136**, 609–620.

Cazzulo, J.J., Couso, R., Raimondi, A., Wernstedt, C. & Hellman, U. (1989) Further characterization and partial amino acid sequence of a cysteine proteinase from *Trypanosoma cruzi*. *Mol. Biochem. Parasitol.* **33**, 33–42.

Pacaud, M. & Richaud, C. (1975) Protease II from *Escherichia coli*. *J. Biol. Chem.* **250**, 7771–7779.

Santana, J.M., Grellier, P., Rodier, M.-H., Schrevel, J. & Teixeira, A. (1992) Purification and characterization of a new 120 kDa alkaline proteinase of *Trypanosoma cruzi*. *Biochem. Biophys. Res. Commun.* **187**, 1466–1473.

Tardieux, I., Nathanson, M.H. & Andrews, N.W. (1994) Role in host cell invasion of *Trypanosoma cruzi*-induced cytosolic free $Ca^{2+}$ transients. *J. Exp. Med.* **179**, 1017–1022.

Yoshimoto, T., Tabira, J., Kabashima, T., Inoue, S. & Ito, K. (1995) Protease II from *Moraxella lacunata*: cloning, sequencing, and expression of the enzyme gene, and crystallization of the expressed enzyme. *J. Biochem.* **117**, 654–660.

*Barbara A. Burleigh*
*Department of Cell Biology,*
*Yale University School of Medicine,*
*333 Cedar Street, New Haven, CN 06520-8002, USA*

*Norma W. Andrews*
*Department of Cell Biology,*
*Yale University School of Medicine,*
*333 Cedar Street, New Haven, CN 06520-8002, USA*
*Email: andrews_lab@quickmail.yale.edu*

# *128. Dipeptidyl-peptidase IV*

## Databanks

*Peptidase classification: clan SC, family S9, MEROPS ID: S09.003, S09.013*
*NC-IUBMB enzyme classification: EC 3.4.14.5*
*Chemical Abstracts Service registry number: 54249-88-6*
*Databank codes:*

| Species | SwissProt | PIR | EMBL (cDNA) | EMBL (genomic) |
|---|---|---|---|---|
| *Bos taurus* | P42659 | – | M76428 M76429 | – |
| *Flavobacterium meningosepticum* | – | – | D42121 | |
| *Homo sapiens* | P27487 | A42408 | M74777 | S79876: promoter |
| | | B42408 | M80536 | U13710: exon 1 |
| | | B61136 | X60708 | U13711: exon 2 |
| | | I56154 | | U13712: exon 3 |
| | | S24313 | | U13713: exon 4 |
| | | S59510 | | U13714: exon 5 |
| | | S59857 | | U13715: exon 6 |
| | | | | U13716: exon 7 |
| | | | | U13717: exon 8 |
| | | | | U13718: exon 9 |
| | | | | U13719: exon 10 |
| | | | | U13720: exon 11 |
| | | | | U13721: exon 12 |
| | | | | U13722: exon 13 |

| Species | SwissProt | PIR | EMBL (cDNA) | EMBL (genomic) |
|---|---|---|---|---|
| | | | | U13723: exon 14 |
| | | | | U13724: exon 15 |
| | | | | U13725: exon 16 |
| | | | | U13726: exon 17 |
| | | | | U13727: exon 18 |
| | | | | U13728: exon 19 |
| | | | | U13729: exon 20 |
| | | | | U13730: exon 21 |
| | | | | U13731: exon 22 |
| | | | | U13732: exon 23 |
| | | | | U13733: exon 24 |
| | | | | U13734: exon 25 |
| | | | | U13735: exon 26 and complete CDS |
| *Mus musculus* | P28843 | A46465 | X58384 | – |
| | | A56030 | | |
| | | S23752 | | |
| *Rattus norvegicus* | P14740 | A31781 | J04591 | |
| | | A33315 | | |
| | | A39914 | | |
| | | A42203 | | |
| | | A60730 | | |
| | | B33315 | | |
| | | S38949 | | |
| *Stenotrophomonas maltophilia* | – | – | D83263 | – |
| *Sus scrofa* | P22411 | S14746 | – | – |
| *Sus scrofa* | – | – | X73278 | – |

## Name and History

*Dipeptidyl-peptidase IV (DPP IV)* is a serine exopeptidase that cleaves Xaa-Pro+ dipeptides from the N-terminus of oligo- and polypeptides. It was first reported as *glycylproline naphthylamidase* by Hopsu-Havu & Glenner (1966), and has been named *dipeptidyl aminopeptidase IV* or *postproline dipeptidyl peptidase IV* in early work (Yoshimoto *et al.*, 1978). The enzyme was found to be abundant in kidney, small intestine, submaxillary gland and liver, from which it has been purified in a soluble form released by autolysis of microsomes (Oya *et al.*, 1974; Kenny *et al.*, 1976; Yoshimoto & Walter, 1977; Svensson *et al.*, 1978). Immunocytochemical studies revealed that the enzyme is localized in an apical domain of the plasma membrane in polarized epithelial cells (Fukasawa *et al.*, 1981a), accounting for its enrichment in the above mentioned tissues. The entire sequence of the enzyme molecule was first determined by cloning and sequencing the cDNA for rat DPP IV (Ogata *et al.*, 1989), allowing structural comparison with other proteins (Misumi *et al.*, 1992; Rawlings *et al.*, 1991). Both *CD26*, a surface marker involved in transduction of mitogenic signals in thymocytes and T lymphocytes (Ulmer *et al.*, 1990), and rat liver plasma membrane glycoprotein *gp110* were found to be identical to DPP IV (McCaughan *et al.*, 1990).

## Activity and Specificity

DPP IV is specific for a Pro residue at the penultimate position, and hydrolyzes on the carboxyl side of the Pro residue (Xaa-Pro+Xbb-). The Pro residue can be substituted by an Ala or a hydroxyproline, although the rates of hydrolysis for these substrates are much lower than that for the corresponding substrate containing Pro. The identity of the N-terminal residue of the substrate is not important for the enzyme activity (Yoshimoto *et al.*, 1978), although it must have a free amino group. DPP IV cannot hydrolyze substrates with Pro (McDonald *et al.*, 1970) or hydroxyproline in the third position (P1') (Oya *et al.*, 1974). DPP IV shows quite similar substrate specificity to that of the unrelated X-Pro dipeptidyl-peptidase of lactococci (Chapter 136) (Yoshpe-Besançon *et al.*, 1994).

Assay of activity can be done with Gly-Pro+NHPhNO$_2$ by measuring released *p*-nitroaniline photometrically (Nagatsu *et al.*, 1976). This synthetic substrate is available from Sigma and Peptide Institute (see Appendix 2 for full names and addresses of suppliers). One unit of activity is defined as the amount of activity which is capable of cleaving 1 $\mu$mol of substrate min$^{-1}$ at 37°C. DPP IV requires neither metals nor any cofactors for its activity. The pH optimum is about 7.8. DPP IV shows pH stability over a wide range (pH 5–10), is thermostable (up to 72°C), and is remarkably resistant to denaturation with 8 M urea (Yoshimoto *et al.*, 1978; Fukasawa *et al.*, 1981b).

DPP IV is very sensitive to DFP, but much less sensitive to other well-known serine protease inhibitors such as diethyl *p*-nitrophenylphosphate and PMSF (Kenny *et al.*, 1976). Ala-(or Pro)-boroPro dipeptide (Flentke *et al.*, 1991) and aminoacylpyrrolidine-2-nitriles (Li *et al.*, 1995) are specific and potent inhibitors of DPP IV, but are not yet commercially available. Heavy metals such as zinc, cadmium, mercury and lead inhibit the enzyme activity remarkably (Püschel *et al.*, 1982).

## Structural Chemistry

The entire amino acid sequences of DPP IV for human, rat and mouse have been determined by cDNA cloning and sequencing. Rat DPP IV contains 767 amino acids with a calculated molecular mass of 88 107 Da (Ogata *et al.*, 1989). It is a type II membrane protein whose N-terminal hydrophobic sequence represents an uncleavable signal peptide that also functions as a membrane-anchoring domain. Thus, DPP IV has a short cytoplasmic tail (six residues), a transmembrane domain (23 residues) and a long extracellular domain (738 residues). The extracellular domain contains glycosylation sites (mostly in the N-terminal half), a cysteine-rich region (C-terminal half) and the catalytic active site. The presence of eight potential *N*-linked glycosylation sites accounts for the difference in molecular mass between the predicted polypeptide (88 kDa) and the mature form (109 kDa). In the native state, DPP IV is present on the cell surface as a noncovalently linked homodimer (Macnair & Kenny, 1979). In the rat enzyme, the active-site Ser631 residue was identified by analysis of its chemical modification and site-directed mutagenesis (Ogata *et al.*, 1992), and is in the sequence Gly-Trp-Ser631-Tyr-Gly, which corresponds to the consensus motif Gly-Xaa-Ser-Xaa-Gly proposed for the active site of serine proteases generally (Brenner, 1988). It is of interest to note that a single substitution of either Gly residue in the motif results in retention of the newly synthesized enzyme in the endoplasmic reticulum, and rapid degradation (Tsuji *et al.*, 1992; Fujiwara *et al.*, 1992), suggesting that both residues are also essential for correct folding of the molecule and its transport to the cell surface. Other essential residues required for the catalytic triad of rat enzyme are Asp709 and His741 (David *et al.*, 1993). The sequential order of the catalytic triad residues (Ser/Asp/His) in DPP IV from all species is characteristic of clan SC (see Chapter 124).

The pI for the sheep DPP IV is 4.9 (Yoshimoto *et al.*, 1978) and that for the pig kidney is 5.2 (Fukasawa *et al.*, 1981b).

## Preparation

DPP IV is widely distributed in a variety of species and tissues, from which it has been prepared with various degrees of purity. Since it is most enriched in kidney, small intestine, submaxillary gland and placenta (Oya *et al.*, 1974; Svensson *et al.*, 1978; Macnair & Kenny, 1979; Fukasawa *et al.*, 1981a) and relatively enriched in liver (Elovson, 1980), the enzyme has been purified to homogeneity from these tissues. The intact membrane form is usually prepared by solubilizing the enzyme with Triton X-100, while a soluble form is obtained by autolysis of microsomes incubated at low pH or by digestion of membranes with papain (Macnair & Kenny, 1979) which cleaves a site adjacent to the transmembrane domain and releases the enzyme with a slightly smaller molecular mass (Ogata *et al.*, 1989). DPP IV was transiently expressed in COS-1 cells (Tsuji *et al.*, 1992), and stably transfected CHO cells (Hong *et al.*, 1989b), mouse fibroblast Lm (David *et al.*, 1993) and Jurkat cells (Tanaka *et al.*, 1993) have been established.

## Biological Aspects

The gene structure of human DPP IV has been elucidated by Abbott *et al.* (1994). The human DPP IV gene located at 2q24.3 spans approximately 70 kbp and contains 26 exons. The nucleotides that encode the active-site sequence (Gly-Xaa-Ser-Xaa-Gly) are split between two exons. This clearly distinguishes the genomic organization of DPP IV from that of the classical serine proteases (Rawlings *et al.*, 1991). The promoter region contains an unmethylated CpG island with several Sp1-binding sites and no consensus TATA box, which are characteristics of promoters found in housekeeping genes. In spite of these sequence features, the DPP IV promoter shows widely varying transcriptional activity which mirrors the enzyme activities in several cells (Böhm *et al.*, 1995). These characteristics of the promoter may play a role in expressing the DPP IV transcript ubiquitously, but at different levels; a high-level expression in kidney, small intestine and placenta, a moderate level in lung, spleen and liver, and a low level in heart, pancreas and brain (Hong *et al.*, 1989a; Abbott *et al.*, 1994).

DPP IV has been proposed to serve a number of different functions, though not all yet established. In kidney brush border membranes, the enzyme may be involved in a transport system specific for dipeptides and tripeptides (Tiruppathi *et al.*, 1990). DPP IV levels in rat small intestine can be increased by feeding a gelatin (high Pro-containing) diet (Suzuki *et al.*, 1993). These results suggest that DPP IV is one of the peptidases important in the digestion and assimilation of prolyl peptides. Studies of cell adhesion have suggested an involvement of DPP IV in cell–extracellular matrix interactions, especially with collagens (Johnson *et al.*, 1993). Based on experiments with monoclonal antibodies that inhibit the binding of DPP IV to collagen matrix, Löster *et al.* (1995) proposed that a putative binding site of the enzyme against collagens is the cysteine-rich domain distinct from the catalytic domain. In the immune system, DPP IV (CD26) is considered to be an activation factor on the surface of human T lymphocytes, and to play a key role in the regulation of proliferation and differentiation of the lymphocytes, including production of lymphokines (Reinhold *et al.*, 1994). Studies using specific inhibitors of the enzyme suggest that DPP IV contributes to T-cell activation through its dipeptidyl-peptidase activity (Flentke *et al.*, 1991). However, experiments in which the putative catalytic Ser residue of DPP IV was mutated to Ala, eliminating catalytic activity, suggested that the peptidase activity plays an important, but not absolute role in the enhancement of interleukin 2 (IL-2) production by Jurkat cells (Tanaka *et al.*, 1993).

## Distinguishing Features

The cellular localization and enzymatic properties of DPP IV are quite different from those of other dipeptidyl-peptidases (dipeptidyl-peptidases I, II and III) found in mammalian tissues (McDonald & Schwabe, 1977; McDonald & Barrett, 1986). Dipeptidyl-peptidase I (Chapter 214) and dipeptidyl-peptidase II (Chapter 138) are localized in lysosomes and require acidic pH (5.0–6.0) for maximal activity, while dipeptidyl-peptidase III (Chapter 182) is a cytosolic enzyme with an optimum pH near 9. Dipeptidyl-peptidase II hydrolyzes substrates with Pro or Ala in the penultimate position as does DPP IV, but neither dipeptidyl-peptidase I nor dipeptidyl-peptidase III can cleave the Xaa-Pro dipeptide from substrates. Dipeptidyl-peptidase I is a cysteine peptidase, while the others are serine peptidases.

Polyclonal antisera to rat liver (Ogata *et al.*, 1989; Hong *et al.*, 1989a) and pig kidney (Fukasawa *et al.*, 1981a) DPP IV have been produced in rabbits. Monoclonal anti-human DPP IV (Coulter, Eurogenetics), polyclonal anti-human DPP IV (Transformation Research) and polyclonal anti-pig DPP IV antibodies (Serva) are commercially available.

## Further Reading

A full description of the purification and assay methods for liver DPP IV has been provided by Ikehara *et al.* (1994).

## References

Abbott, C.A., Baker, E., Sutherland, G.R. & McCaughan, G.W. (1994) Genomic organization, exact localization, and tissue expression of the human CD26 (dipeptidyl peptidase IV) gene. *Immunogenetics* **40**, 331–338.

Böhm, S.K., Gum, J.R., Erickson, R.H., Hicks, J.W. & Kim, Y.S. (1995) Human dipeptidyl peptidase IV gene promoter: tissue-specific regulation from a TATA-less GC-rich sequence characteristic of a housekeeping gene promoter. *Biochem. J.* **311**, 835–843.

Brenner, S. (1988) The molecular evolution of genes and proteins: a tale of two serines. *Nature* **334**, 528–530.

David, F., Bernard, A.M., Pierres, M. & Marguet, D. (1993) Identification of serine 624, aspartic acid 702, and histidine 734 as the catalytic triad residues of mouse dipeptidyl-peptidase IV (CD26). A member of a novel family of nonclassical serine hydrolases. *J. Biol. Chem.* **268**, 17247–17252.

Elovson, J. (1980) Biogenesis of plasma membrane glycoproteins. *J. Biol. Chem.* **255**, 5807–5815.

Flentke, G.R., Munoz, E., Huber, B.T., Plaut, A.G., Kettner, C.A. & Bachovchin, W.W. (1991) Inhibition of dipeptidyl aminopeptidase IV (DP-IV) by Xaa-boroPro dipeptides and use of these inhibitors to examine the role of DP-IV in T-cell function. *Proc. Natl Acad. Sci. USA* **88**, 1556–1559.

Fujiwara, T., Tuji, E., Misumi, Y., Takami, N. & Ikehara, Y. (1992) Selective cell-surface expression of dipeptidyl peptidase IV with mutations at the active site sequence. *Biochem. Biophys. Res. Commun.* **185**, 776–784.

Fukasawa, K.M., Fukasawa, K., Sahara, N., Harada, M., Kondo, Y. & Nagatsu, I. (1981a) Immunohistochemical localization of dipeptidyl aminopeptidase IV in rat kidney, liver, and salivary glands. *J. Histochem. Cytochem.* **29**, 337–343.

Fukasawa, K.M., Fukasawa, K., Hiraoka, B.Y. & Harada, M. (1981b) Comparison of dipeptidyl peptidase IV prepared from pig liver and kidney. *Biochem. Biophys. Res. Commun.* **657**, 179–189.

Hong, W.J., Petell, J.K., Swank, D., Sanford, J., Hixson, D.C. & Doyle, D. (1989a) Expression of dipeptidyl peptidase IV in rat tissues is mainly regulated at the mRNA levels. *Exp. Cell Res.* **182**, 256–266.

Hong, W.J., Piazza, G.A., Hixson, D.C. & Doyle, D. (1989b) Expression of enzymatically active rat dipeptidyl peptidase IV in Chinese hamster ovary cells after transfection. *Biochemistry* **28**, 8474–8479.

Hopsu-Havu, V.K. & Glenner, G.G. (1966) A new dipeptide naphthylamidase hydrolyzing glycyl-prolyl-beta-naphthylamide.

Histochemie **7**, 197–201.

Ikehara, Y., Ogata, S. & Misumi, Y. (1994) Dipeptidyl-peptidase IV from rat liver. *Methods Enzymol.* **244**, 215–227.

Johnson, R.C., Zhu, D., Augustin-Voss, H.G. & Pauli, B.U. (1993) Lung endothelial dipeptidyl peptidase IV is an adhesion molecule for lung-metastatic rat breast and prostate carcinoma cells. *J. Cell Biol.* **121**, 1423–1432.

Kenny, A.J., Booth, A.G., George, S.G., Ingram, J., Kershaw, D., Wood, E.J. & Young, A.R. (1976) Dipeptidyl peptidase IV, a kidney brush-border serine peptidase. *Biochem. J.* **157**, 169–182.

Li, J., Wilk, E. & Wilk, S. (1995) Aminoacylpyrrolidine-2-nitriles: potent and stable inhibitors of dipeptidyl-peptidase IV (CD 26). *Arch. Biochem. Biophys.* **323**, 148–154.

Löster, K., Zeilinger, K., Schuppan, D. & Reutter, W. (1995) The cysteine-rich region of dipeptidyl peptidase IV (CD 26) is the collagen-binding site. *Biochem. Biophys. Res. Commun.* **217**, 341–348.

Macnair, D.C. & Kenny, A.J. (1979) Proteins of the kidney microvillar membrane. The amphipathic form of dipeptidyl peptidase IV. *Biochem. J.* **179**, 379–395.

McCaughan, G.W., Wickson, J.E., Creswick, P.F. & Gorrell, M.D. (1990) Identification of the bile canalicular cell surface molecule GP110 as the ectopeptidase dipeptidyl peptidase IV: an analysis by tissue distribution, purification and N-terminal amino acid sequence. *Hepatology* **11**, 534–544.

McDonald, J.K. & Barrett, A.J. (1986) *Mammalian Proteases: a Glossary and Bibliography*, vol. 2: *Exopeptidases*. London: Academic Press.

McDonald, J.K. & Schwabe, C. (1977) Intracellular exopeptidases. In: *Proteinases in Mammalian Cells and Tissues* (Barrett, A.J., ed.). Amsterdam: North-Holland Publishing, pp. 331–391.

McDonald, J.K., Zeitman, B.B. & Ellis, S. (1970) Leucine naphthylamide: an inappropriate substrate for the histochemical detection of cathepsins B and B'. *Nature* **225**, 1048–1049.

Misumi, Y., Hayashi, Y., Arakawa, F. & Ikehara, Y. (1992) Molecular cloning and sequence analysis of human dipeptidyl peptidase IV, a serine proteinase on the cell surface. *Biochim. Biophys. Acta* **1131**, 333–336.

Nagatsu, T., Hino, M, Fuyamada, H., Hayakawa, T. & Sakakibara, S. (1976) New chromogenic substrates for X-prolyl dipeptidyl-aminopeptidase. *Anal. Biochem.* **74**, 466–476.

Ogata, S., Misumi, Y. & Ikehara, Y. (1989) Primary structure of rat liver dipeptidyl peptidase IV deduced from its cDNA and identification of the $NH_2$-terminal signal sequence as the membrane-anchoring domain. *J. Biol. Chem.* **264**, 3596–3601.

Ogata, S., Misumi, Y., Tsuji, E., Takami, N., Oda, K. & Ikehara, Y. (1992) Identification of the active site residues in dipeptidyl peptidase IV by affinity labeling and site-directed mutagenesis. *Biochemistry* **31**, 2582–2587.

Oya, H., Harada, M. & Nagatsu, T. (1974) Peptidase activity of glycylprolyl beta-naphthylamidase from human submaxillary gland. *Arch. Oral Biol.* **19**, 489–491.

Püschel, G., Mentlein, R. & Heymann, E. (1982) Isolation and characterization of dipeptidyl peptidase IV from human placenta. *Eur. J. Biochem.* **126**, 359–365.

Rawlings, N.D., Polgár, L. & Barrett, A.J. (1991) A new family of serine peptidases related to prolyl oligopeptidase. *Biochem. J.* **279**, 907–911.

Reinhold, D., Bank, U., Bühling, F., Kähne, T., Kunt, D., Faust, J., Neubert, K. & Ansorge, S. (1994) Inhibitors of dipeptidyl peptidase

IV (DP IV, CD26) specifically suppress proliferation and modulate cytokine production of strongly CD26 expressing U937 cells. *Immunobiology* **192**, 121–136.

Suzuki, Y., Erickson, R.H., Sedlmayer, A., Chang, S.-K., Ikehara, Y. & Kim, Y.S. (1993) Dietary regulation of rat intestinal angiotensin-converting enzyme and dipeptidyl peptidase IV. *Am. J. Physiol.* **264**, G1153–G1159.

Svensson, B., Danielsen, M., Staun, M., Jeppesen, L., Norén, O. & Sjöström, H. (1978) An amphiphilic form of dipeptidyl peptidase IV from pig small-intestinal brush-border membrane. Purification by immunoadsorbent chromatography and some properties. *Eur. J. Biochem.* **90**, 489–498.

Tanaka, T., Kameoka, J., Yaron, A., Schlossman, S.F. & Morimoto, C. (1993) The costimulatory activity of the CD26 antigen requires dipeptidyl peptidase IV enzymatic activity. *Proc. Natl Acad. Sci. USA* **90**, 4586–4590.

Tiruppathi, C., Ganapathy, V. & Leibach, F.H. (1990) Evidence for tripeptide-proton symport in renal brush border membrane vesicles. Studies in a novel rat strain with a genetic absence of dipeptidyl peptidase IV. *J. Biol. Chem.* **265**, 2048–2053.

Tsuji, E., Misumi, Y., Fujiwara, T., Takami, N., Ogata, S. & Ikehara, Y. (1992) An active-site mutation (Gly633→Arg) of dipeptidyl peptidase IV causes its retention and rapid degradation in the endoplasmic reticulum. *Biochemistry* **31**, 11921–11927.

Ulmer, A.J., Mattern, T., Feller, A.C., Heymann, E. & Flad, H.-D. (1990) CD26 antigen is a surface dipeptidyl peptidase IV (DPP IV) as characterized by monoclonal antibodies clone TII-19-4-7 and 4EL1C7. *Scand. J. Immunol.* **31**, 429–435.

Yoshimoto, T. & Walter, R. (1977) Post-proline dipeptidyl aminopeptidase (dipeptidyl aminopeptidase IV) from lamb kidney, Purification and some enzymatic properties. *Biochim. Biophys. Acta* **485**, 391–401.

Yoshimoto, T., Fischl, M., Orlowski, R.C. & Walter, R. (1978) Post-proline cleaving enzyme and post-proline dipeptidyl aminopeptidase. Comparison of two peptidases with high specificity for proline residues. *J. Biol. Chem.* **253**, 3708–3716.

Yoshpe-Besançon, I., Gripon, J.C. & Ribadeau-Dumas, B. (1994) Xaa-Pro-dipeptidyl-aminopeptidase from *Lactococcus lactis* catalyses kinetically controlled synthesis of peptide bonds involving proline. *Biotechnol. Appl. Biochem.* **20**, 131–140.

*Yoshio Misumi*
*Department of Biochemistry,*
*Fukuoka University School of Medicine,*
*Jonan-ku, Fukuoka 814-80, Japan*
*Email: mm034023@cc.fukuoka-u.ac.jp*

*Yukio Ikehara*
*Department of Biochemistry,*
*Fukuoka University School of Medicine*
*Jonan-ku, Fukuoka 814-80, Japan*

# 129. Dipeptidyl-peptidases A and B

## Databanks

*Peptidase classification: clan SC, family S9, MEROPS ID: S09.005, S09.006*
*NC-IUBMB enzyme classification: none*
*Databank codes:*

| Species | SwissProt | PIR | EMBL (cDNA) | EMBL (genomic) |
| --- | --- | --- | --- | --- |
| Dipeptidyl aminopeptidase A | | | | |
| *Saccharomyces cerevisiae* | P33894 | A49737 | L21944 U08230 | – |
| Dipeptidyl aminopeptidase B | | | | |
| *Saccharomyces cerevisiae* | P18962 | A30107 | X15484 | – |
| Dipeptidyl-peptidase V | | | | |
| *Aspergillus fumigatus* | – | – | L48074 | – |

## Name and History

Suarez-Rendueles *et al*. (1981) discovered an activity hydrolyzing Ala-Pro+NHPhNO₂ and Gly-Pro+NHPhNO₂ in a particulate fraction from *Saccharomyces cerevisiae*. This sedimentable activity accounted for about 40% of the total activity of the cells. The membrane-associated activity was resolved by Julius *et al*. (1983) into a heat-stable component, termed *dipeptidyl aminopeptidase A* (or *yscIV*), and a heat-labile enzyme, termed *dipeptidyl aminopeptidase B* (or *yscV*). The term now recommended by NC-IUBMB for this type of peptidase activity is 'dipeptidyl-peptidase', but the more familiar 'dipeptidyl aminopeptidase' names will be used in the present text.

## Activity and Specificity

Dipeptidyl aminopeptidase A and dipeptidyl aminopeptidase B catalyze hydrolysis of Xaa-Pro+Y or Xaa-Ala+Y (in which Xaa is an amino acid, and Y is an amino acid, peptide, amide or ester). Both enzymes require the free amino group of the N-terminal residue, as is expected of dipeptidyl-peptidases. The apparent $K_m$ values for Ala-Pro-NHPhNO$_2$ are 0.4 mM for dipeptidyl aminopeptidase A and 0.06 mM for dipeptidyl aminopeptidase B (Bordallo *et al.*, 1984). The enzymes differ greatly in thermostability; dipeptidyl aminopeptidase A being stable at 60°C for 15 min, whereas dipeptidyl aminopeptidase B loses most of its activity under these conditions. Because of this large difference, the two dipeptidyl-peptidase activities in the membrane fraction can be differentially determined with the single substrate, Ala-Pro-NHPhNO$_2$ (Julius *et al.*, 1983).

## Structural Chemistry

The predicted sequence of dipeptidyl aminopeptidase A contains 931 amino acids (107 kDa), forming three domains: an N-terminal cytoplasmic domain (112 residues), a trans-membrane domain (36 residues) and a long luminal domain (782 residues) with two potential glycosylation sites and the predicted catalytic site (Anna-Arriola & Herskowitz, 1994). Mutational analysis has indicated that the motif Phe85-Xaa-Phe-Xaa-Asp89 in the cytoplasmic domain is responsible for retention of the molecule in the late Golgi (Nothwehr *et al.*, 1993).

The predicted amino acid sequence of dipeptidyl aminopeptidase B contains 841 amino acids (96 kDa). Eight potential *N*-glycosylation sites are found in the sequence, and the fully matured form of dipeptidyl aminopeptidase B is 120 kDa (Roberts *et al.*, 1989). The catalytic domains of both dipeptidyl-peptidases contain the conserved catalytic residues characteristic of family S9 (Chapter 124).

## Preparation

These enzymes are integral membrane proteins, and have not been extensively purified.

## Biological Aspects

Both dipeptidyl aminopeptidase A and dipeptidyl aminopeptidase B are type II membrane proteins; dipeptidyl aminopeptidase A is localized in the late Golgi compartment and dipeptidyl aminopeptidase B in the vacuolar membrane (Roberts *et al.*, 1989).

Dipeptidyl aminopeptidase A is encoded by the *STE13* gene, and plays a role in propheromone processing *in vivo*, converting the immature α-factor into the mature form by stepwise removal of Xaa-Ala dipeptides from the N-terminus of the pheromone, typically Glu-Ala+Glu-Ala+α-factor (Julius *et al.*, 1983). Achstetter (1989) reported that repeats of TGAAACA, the pheromone-responsive promoter element, were found in the 5′ flanking region of the *STE13* gene, and that the dipeptidyl aminopeptidase A mRNA level increases 2.5-fold after the induction of α-factor.

Dipeptidyl aminopeptidase B is a vacuolar membrane glycoprotein encoded by the *DAP2* gene (Roberts *et al.*, 1989);

natural substrates for the enzyme have not been identified, however.

## Distinguishing Characteristics

In the study of Achstetter *et al.* (1983), three cytosolic dipeptidyl-peptidases of differing specificity were resolved. One of these, DAP III, was active on Xaa-Pro-NHPhNO$_2$, but this enzyme exhibited a strong preference for Arg-Pro-NHPhNO$_2$ as substrate, whereas the membrane-associated dipeptidyl aminopeptidases A and B hydrolyze Arg-Pro-NHPhNO$_2$ at a much lower rate. Dipeptidyl aminopeptidase A and dipeptidyl aminopeptidase B, with similar substrate specificities, are distinguished by the much greater thermal stability of dipeptidyl aminopeptidase A (Julius *et al.*, 1983). Polyclonal antibodies against the luminal domain of dipeptidyl aminopeptidase A, and polyclonal antibodies against recombinant dipeptidyl aminopeptidase B expressed in *E. coli*, have been described (Roberts *et al.*, 1989).

## Related Peptidases

The specificities of dipeptidyl aminopeptidase A and dipeptidyl aminopeptidase B are similar to that of the mammalian dipeptidyl-peptidase IV (Chapter 128), also a member of peptidase family S9. A further member of family S9, also from a fungus, has recently been described (Beauvais *et al.*, 1997). This is dipeptidyl-peptidase V, secreted by *Aspergillus fumigatus*. In most respects this is structurally similar to the yeast dipeptidyl-peptidases, but the substrate specificity is different, and the enzyme is synthesized with a signal peptide.

## References

Achstetter, T. (1989) Regulation of α-factor production in *Saccharomyces cerevisiae. Mol. Cell. Biol.* **9**, 4507–4514.

Achstetter, T., Ehmann, C. & Wolf, D.H. (1983) Proteolysis in eucaryotic cells: aminopeptidases and dipeptidyl aminopeptidases of yeast revisited. *Arch. Biochem. Biophys.* **226**, 292–305.

Anna-Arriola, S.S. & Herskowitz, I. (1994) Isolation and DNA sequence of the STE13 gene encoding dipeptidyl aminopeptidase. *Yeast* **10**, 801–810.

Beauvais, A., Monod, M., Debeaupuis, J.P., Diaquin, M., Kobayashi, H. & Latgé, J.P. (1997) Biochemical and antigenic characterization of a new dipeptidyl-peptidase isolated from *Aspergillus fumigatus. J. Biol. Chem.* **272**, 6238–6244.

Bordallo, C., Schwencke, J. & Suarez-Rendueles, M. (1984) Localization of the thermosensitive X-prolyl dipeptidyl aminopeptidase in the vacuolar membrane of *Saccharomyces cerevisiae. FEBS Lett.* **173**, 199–203.

Julius, D., Blair, L., Brake, A., Sprague, G. & Thorner, J. (1983) Yeast alpha factor is processed from a larger precursor polypeptide: the essential role of a membrane-bound dipeptidyl aminopeptidase. *Cell* **32**, 839–852.

Nothwehr, S.F., Roberts, C.J. & Stevens, T.H. (1993) Membrane protein retention in the yeast Golgi apparatus: dipeptidylaminopeptidase A is retained by a cytoplasmic signal containing aromatic residues. *J. Cell Biol.* **121**, 1197–1209.

Roberts, C.J., Pohlig, G., Rothman, J.H. & Stevens, T.H. (1989) Structure, biosynthesis, and localization of dipeptidyl aminopeptidase B, an integral membrane glycoprotein of the yeast vacuole. *J. Cell Biol.* **108**, 1363–1373.

Suarez-Rendueles, M.P., Schwencke, J., Garcia-Alvarez, N. & Gas-

con, S. (1981) A new X-prolyl-dipeptidyl aminopeptidase from yeast associated with a particulate fraction. *FEBS Lett.* **131**, 296–300.

***Yoshio Misumi***
*Department of Biochemistry,*
*Fukuoka University School of Medicine,*
*Jonan-ku, Fukuoka 814-80, Japan*
*Email: mm034023@cc.fukuoka-u.ac.jp*

***Yukio Ikehara***
*Department of Biochemistry,*
*Fukuoka University School of Medicine,*
*Jonan-ku, Fukuoka 814-80, Japan*

# 130. *Acylaminoacyl-peptidase*

## Databanks

*Peptidase classification: clan SC, family S9, MEROPS ID: S09.004*
*NC-IUBMB enzyme classification: EC 3.4.19.1*
*Chemical Abstracts Service registry number: 73562-30-8*
*Databank codes:*

| Species | SwissProt | PIR | EMBL (cDNA) | EMBL (genomic) |
|---|---|---|---|---|
| *Caenorhabditis elegans* | P34422 | S44807 | L23648 | – |
| *Homo sapiens* | P13798 | A30145 A42257 | J03068 | – |
| *Oryctolagus cuniculus* | P25154 | – | – | – |
| *Rattus norvegicus* | P13676 | A33706 S07624 | J04733 | – |
| *Sus scrofa* | P19205 | JU0132 | X73277 | – |

## Name and History

*Acylaminoacyl-peptidase* (EC 3.4.19.1) has also been referred to by the names *acylpeptide hydrolase* (Gade & Brown, 1978; Jones & Manning, 1985; Kobayashi *et al*., 1989), *acylamino acid-releasing enzyme* (Tsunasawa *et al*., 1975; Mitta *et al*., 1989) and *acylaminoacyl peptide hydrolase* (Radhakrishna & Wold, 1989).

## Activity and Specificity

Acylaminoacyl peptidase catalyzes the removal of an *N*-acylated amino acid from a blocked peptide: Block-Xaa┼Xbb-Xcc.... The products of the reaction are the free acyl amino acid and a peptide with a free N-terminus shortened by one amino acid. The enzyme acts on a variety of peptides with different N-terminal acyl groups, including acetyl, chloroacetyl, formyl and carbamyl (Jones *et al*., 1986). The optimum length of the blocked peptide substrate is 2–3 amino acids, but larger substrates are also cleaved, at slower rates (Jones *et al*., 1991). For instance, the blocked 13 residue peptide α-melanocyte stimulating hormone (αMSH) is a substrate (Jones *et al*., 1986). On the other hand, N-terminally blocked

proteins are not substrates for the enzyme. The enzyme is active over a wide pH range, depending on the substrate used (Jones & Manning, 1985). For example, with Ac-Glu-$NHPhNO_2$, the pH optimum is around 6, but for Ac-Ala-$NHPhNO_2$, it is near pH 8.4. For most acetylated peptide substrates the pH optimum is in the range 7.3–7.6. Esters such as *p*-nitrophenyl propionate and α-naphthyl butyrate are also efficiently hydrolyzed by the enzyme.

## Structural Chemistry

The enzyme is composed of four identical subunits each containing 732 amino acids and a blocked N-terminus for a total molecular mass of about 300 kDa (Tsunasawa *et al*., 1975; Gade & Brown, 1978; Jones & Manning, 1985; Kobayashi *et al*., 1989) with an isoionic point of 4.1 (Scaloni *et al*., 1994). The enzyme is inhibited by several types of reagents including DFP, PCMB (Scaloni *et al*., 1992a), diethylpyrocarbonate and some heavy metals, such as $Hg^{2+}$, $Zn^{2+}$, and $Cd^{2+}$ (Gade & Brown, 1978). The inhibition by DFP is due to modification at Ser597 and Ac-Leu-$CH_2Cl$ inactivates the enzyme by reaction at the active-site His707 (Scaloni *et al*.,

1992a). These sites constitute two parts of the catalytic triad (Rawlings *et al*., 1991; Scaloni *et al*., 1992a). It is notable that the His residue of the catalytic triad is located closest to the C-terminus of the protein. Studies have been initiated on the structure of the enzyme, which has the features of α/β proteins (Feese *et al*., 1993).

## Preparation

The enzyme has been purified to homogeneity from human erythrocytes (Scaloni *et al*., 1992a; Jones *et al*., 1994), cattle liver (Gade & Brown, 1978) and rabbit muscle (Radhakrishna & Wold, 1989). However, activity has been detected in practically all tissues examined (Tsunasawa *et al*., 1975).

## Biological Aspects

The true biological function of this enzyme is not known. Although it does not act on acetylated protein substrates, it does hydrolyze small N-terminally blocked peptides, some of which are bioactive peptides. It could conceivably act on the exposed N-terminus of nascent polypeptide chains. In this regard, some tumor cells, in which 80% of all proteins are blocked at their N-termini, do not contain this enzyme (Scaloni *et al*., 1992b). In small-cell lung carcinoma cell lines, in which the locus on the short arm of chromosome 3 containing the gene encoding this enzyme undergoes deletions, the enzyme is not expressed (Scaloni *et al*., 1992b). In such cells an accumulation of unhydrolyzed acetylated peptide growth factors could lead to cell proliferation, but this hypothesis requires verification.

## References

Feese, M., Scaloni, A., Jones, W.M., Manning, J.M. & Remington, S.J. (1993) Crystallization and preliminary X-ray studies of human erythrocyte acylpeptide hydrolase. *J. Mol. Biol.* **223**, 546–549.

Gade, W. & Brown, J.L. (1978) Purification and partial characterization of α-*N*-acylpeptide hydrolase from bovine liver. *J. Biol. Chem.* **253**, 5012–5018.

Jones, W.M. & Manning, J.M. (1985) Acylpeptide hydrolase activity from erythrocytes. *Biochem. Biophys. Res. Commun.* **126**, 933–940.

Jones, W.M., Manning, L.R. & Manning, J.M. (1986) Enzymic cleavage of the blocked amino terminal residues of peptides. *Biochem. Biophys. Res. Commun.* **139**, 244–250.

Jones, W.M., Scaloni, A., Bossa, F., Popowicz, A.M., Schneewind, O. & Manning, J.M. (1991) Genetic relationship between acylpeptide hydrolase and acylase, two hydrolytic enzymes with similar binding but different catalytic specificities. *Proc. Natl Acad. Sci. USA* **88**, 2194–2198.

Jones, W.M., Scaloni, A. & Manning, J.M. (1994) Acylaminoacyl peptidase. *Methods Enzymol.* **244**, 227–231.

Kobayashi, K., Lin, L.-W., Yeadon, J.E., Klickstein, L.B. & Smith, J.A. (1989) Cloning and sequence analysis of a rat liver cDNA encoding acylpeptide hydrolase. *J. Biol. Chem.* **264**, 8892–8899.

Mitta, M., Asada, K., Uchimura, Y., Kimizuka, F., Kato, I., Sakiyama, F. & Tsunasawa, S. (1989) The primary structure of porcine liver acylamino acid-releasing enzyme deduced from cDNA sequences. *J. Biochem.* **106**, 548–551.

Radhakrishna, G. & Wold, F. (1989) Purification and characterization of an *N*-acylaminoacyl-peptide hydrolase from rabbit muscle. *J. Biol. Chem.* **264**, 11076–11081.

Rawlings, N.D., Polgar, L. & Barrett, A.J. (1991) A new family of serine-type peptidases related to prolyl oligopeptidase. *Biochem. J.* **279**, 907–908.

Scaloni, A., Jones, W.M., Barra, D., Pospischil, M., Sassa, S., Popowicz, A., Manning, L.R., Schneewind, O. & Manning, J.M. (1992a) Acylpeptide hydrolase: inhibitors and some active site residues of the human enzyme. *J. Biol. Chem.* **267**, 3811–3818.

Scaloni, A., Jones, W., Pospischil, M., Sassa, S., Schneewind, O., Popowicz, A.M., Bossa, F., Graziano, S.L. & Manning, J.M. (1992b) Deficiency of acylpeptide hydrolase in small-cell lung carcinoma cell lines. *J. Lab. Clin. Med.* **120**, 546–552.

Scaloni, A., Barra, D., Jones, W.M. & Manning, J.M. (1994) Human acylpeptide hydrolase. *J. Biol. Chem.* **269**, 15076–15084.

Tsunasawa, S., Narita, K. & Ogata, K. (1975) Purification and properties of acylamino acid-releasing enzyme from rat liver. *J. Biochem.* **77**, 89–102.

*James M. Manning*
*Department of Biology,*
*Northeastern University,*
*Boston, MA 02115, USA*
*Email: jmanning@lynx.neu.edu*

# 131. *Fibroblast activation protein* α

## Databanks

*Peptidase classification: clan SC, family S9, MEROPS ID: S09.007*
*NC-IUBMB enzyme classification: none*

*Databank codes:*

| Species | SwissProt | PIR | EMBL (cDNA) | EMBL (genomic) |
|---|---|---|---|---|
| Fibroblast activation protein α | | | | |
| *Homo sapiens* | – | – | U09278 | – |
| *Mus musculus* | – | – | Y10007 | – |
| *Xenopus laevis* | – | – | U41856 | – |
| DPP-X protein | | | | |
| *Homo sapiens* | P42658 | – | M96859 | |
| | | | M96860 | |
| *Rattus norvegicus* | P46101 | – | M76426 | |

## Name and History

The *fibroblast activation protein α (FAP)* was discovered with a monoclonal antibody, mAb F19, that was generated in the course of a serological survey of cell surface antigens expressed on cultured human fibroblasts, sarcomas and neuroectodermal tumor cells (Rettig *et al.*, 1986, 1987, 1988). This antibody was used to characterize the plasma membrane-associated 95 kDa FAPα glycoprotein (Rettig *et al.*, 1993, 1994), to isolate the FAP-encoding cDNA (Scanlan *et al.*, 1994), and to examine FAPα expression in a broad range of normal and neoplastic human tissues (Rettig *et al.*, 1988; Garin-Chesa *et al.*, 1990). The immunochemical studies also led to the detection of high molecular mass complexes, comprising FAPα multimers, and noncovalently linked, heteromeric complexes between FAPα and a distinct, FAPα-associated protein of 105 kDa designated as FAPβ (Rettig *et al.*, 1993). Subsequent investigations have shown that FAPβ is closely related or identical to the CD26 cell surface glycoprotein, which in its dimeric form is a dipeptidyl-peptidase, DPP IV (Hegen *et al.*, 1990) (see Chapter 128). However, FAP/CD26 heteromeric complexes are found only on the surface membrane of some cell types, notably cultured fibroblasts and melanocytes, whereas other cells express only FAPα (e.g. some sarcoma cell lines) or CD26 (e.g. kidney epithelial cells, activated T lymphocytes) or neither molecule (Figure 131.1B). Moreover, immunohistochemical tests with human tissues have revealed distinct and generally nonoverlapping expression patterns for FAPα and CD26 (Stein *et al.*, 1989; Garin-Chesa *et al.*, 1990; Scanlan *et al.*, 1994).

## Activity and Specificity

The natural substrates for FAPα in different antigen-expressing tissues have not yet been described. However, based on the similarity in structure between FAPα and DPP IV, dipeptidyl-peptidase substrates such as Gly-Pro⊣AFC or Ala-Pro⊣AFC (AFC: 7-amino-4-trifluoromethylcoumarin) have been used to assess FAPα activity (Rettig *et al.*, 1994). These substrates are commercially available, for instance through Enzyme Systems Products (see Appendix 2 for full names and addresses of suppliers).

## Structural Chemistry

FAPα homodimers and FAP/DPP IV heterodimers as well as higher molecular mass complexes of FAPα have been identified in cell extracts of cultured human fibroblasts,

melanocytes and sarcoma cells. When analyzed by immunoblotting of boiled extracts separated by SDS-PAGE, FAPα appears as a single protein species of about 95 kDa. Digestion of this 95 kDa glycoprotein with neuraminidase or endoglycosidase H, but not *O*-glycanase, reduces the apparent molecular size on SDS gels, and complete removal of all *N*-linked sugars with *N*-glycanase generates polypeptides of 75 kDa (Rettig *et al.*, 1993).

The *Mus musculus* FAPα cDNA has been cloned on the basis of its sequence similarity to human FAPα (Niedermeyer *et al.*, 1997). The predicted human and mouse FAPα polypeptides share 89% amino acid sequence identity and contain identical catalytic domains. There is evidence from mouse FAPα studies that differential mRNA splicing generates at least three distinct transcripts that differ in their extracellular portions adjacent to the transmembrane domain (Niedermeyer *et al.*, 1997), but the corresponding protein isoforms have not yet been characterized.

The FAP homolog in *Xenopus laevis* was discovered independently in the course of a differential gene expression survey of the thyroid hormone-induced tail resorption program during tadpole metamorphosis (Brown *et al.*, 1996). The human and *Xenopus* FAPα polypeptides show about 50% amino acid sequence identity.

## Preparation

Natural sources of human FAPα include cultured fibroblasts, melanocytes and certain sarcoma cell lines as well as embryonic tissues, granulation tissue of healing wounds and certain tumor tissues (Rettig *et al.*, 1986, 1987, 1988, 1994; Garin-Chesa *et al.*, 1990). Isolation of human FAPα has been accomplished with a combination of lectin and antibody affinity chromatography (Rettig *et al.*, 1994). The homologous proteins in mice and *Xenopus* have not yet been characterized, in part due to a lack of species cross-reactive antibodies.

## Biological Aspects

Although FAPα was discovered as a cell surface antigen of cultured normal fibroblasts, its expression pattern *in vivo*, as determined by immunohistochemistry with mAb F19 and other, second-generation antibodies against distinct FAPα epitopes (Rettig *et al.*, 1994), is highly restricted. In particular, resting fibroblasts in normal tissues lack detectable FAPα expression, and FAPα induction upon *in vitro* cell culture appears to be part of a more general activation program which fibroblasts assume when grown under artificial tissue culture

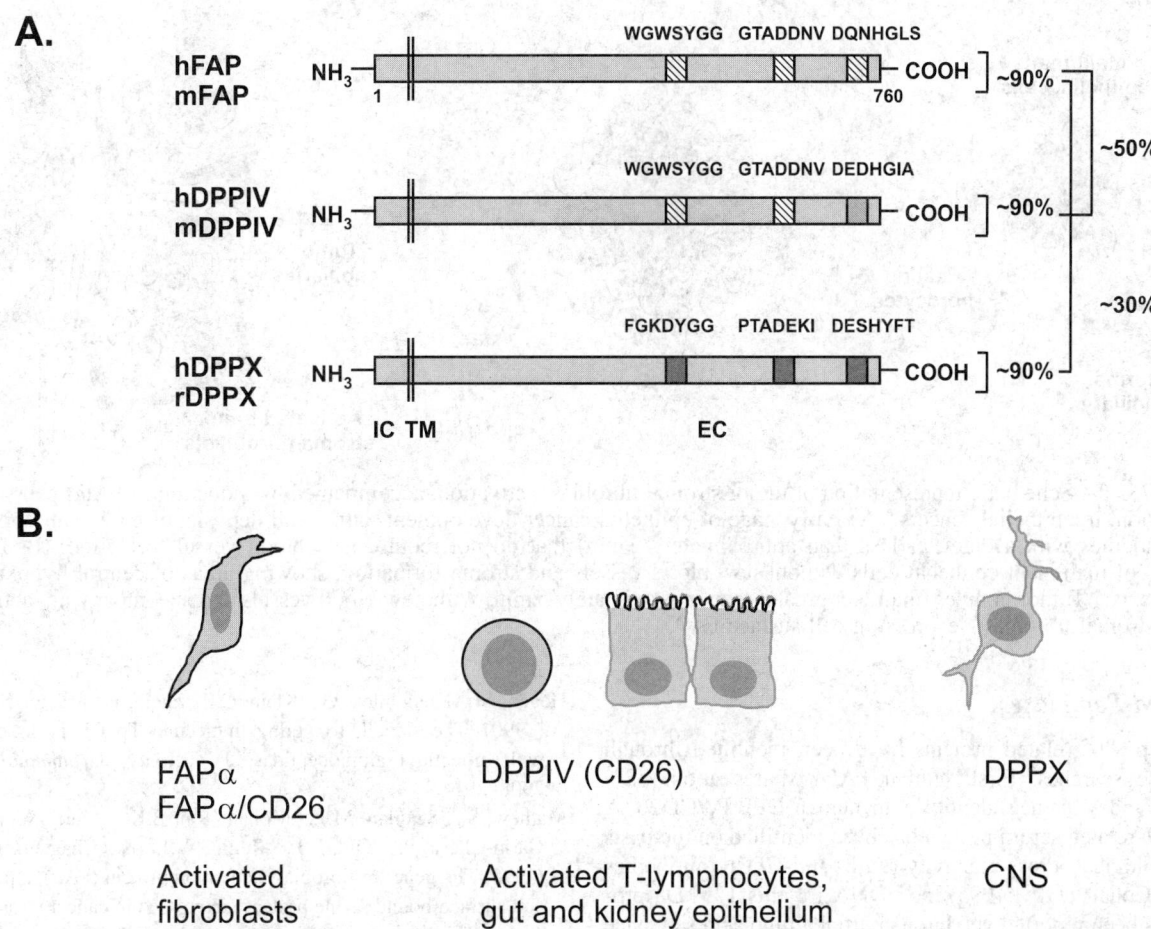

*Figure 131.1*  Distinguishing features of FAPα and related gene products. (A) Model of FAPα protein structure deduced from its cognate cDNA, and alignment with the predicted structures of DPP IV and DPP X. Putative catalytic domains are indicated (shaded boxes) with amino acid sequences above; percentage values to the right indicate amino acid sequence identity. *Abbreviations*: hFAP, human FAPα; mFAP, mouse FAPα; hDPPIV, human DPP IV; mDPPIV, mouse DPP IV; hDPPX, human DPP X; rDPPX, rat DPP X; IC, intracellular domain; TM, transmembrane domain; EC, extracellular domain. (B) Cell type- and domain-specific expression of FAP, DPP IV/CD26 and DPP X as derived from immunohistochemical staining (indicated by heavy lines for FAPα and DPP IV/CD26) and northern blot (DPP X) analyses.

conditions (Sappino *et al.*, 1990). Corresponding activation phenotypes have been postulated for the reactive fibroblasts found in the granulation tissue of healing wounds, certain chronic inflammatory lesions, the supporting stroma of malignant epithelial neoplasms, and normal embryonic tissues, and all of these fibroblasts also express FAPα (Garin-Chesa *et al.*, 1990; Rettig *et al.*, 1988; Scanlan *et al.*, 1994). Notably, the activated tumor stromal fibroblasts found in carcinomas of the breast, colorectum, lung, stomach, pancreas and esophagus show prominent FAPα expression (Figure 131.2). Since very few cell types in normal organs express this cell surface antigen, it has been possible to use the [131]I-labeled mAb F19 for selective immunological targeting in patients with metastatic colorectal cancers (Welt *et al.*, 1994), opening up the possibility of tumor stroma-directed approaches for cancer detection and therapy. The few normal human cell types known to express FAPα *in vivo* include fetal mesenchymal cells and a distinct subset of glucagon-producing endocrine cells (A cells) in the pancreatic islets. The biological role of FAPα

expression in these diverse tissue types and pathological processes is still unknown, but it is tempting to speculate about functions related to tissue remodeling and repair (Garin-Chesa *et al.*, 1990). Such a role would be consistent with the pattern of FAPα induction seen during *Xenopus* metamorphosis, which coincides with the induction of several extracellular matrix-degrading proteases (Brown *et al.*, 1996).

### Distinguishing Features

FAPα is readily distinguished as a unique protein by its primary structure, deduced from its cDNA and partial peptide sequence analysis (Figure 131.1A), and its pattern of reactivity with a panel of monoclonal antibodies (Rettig *et al.*, 1994). These antibodies identify several distinct epitopes of human FAP, but they do not react with FAPα in any other species tested and they do not bind to the structurally related DPP IV protein, for which the CD26 panel of antibodies has been defined (Stein *et al.*, 1989).

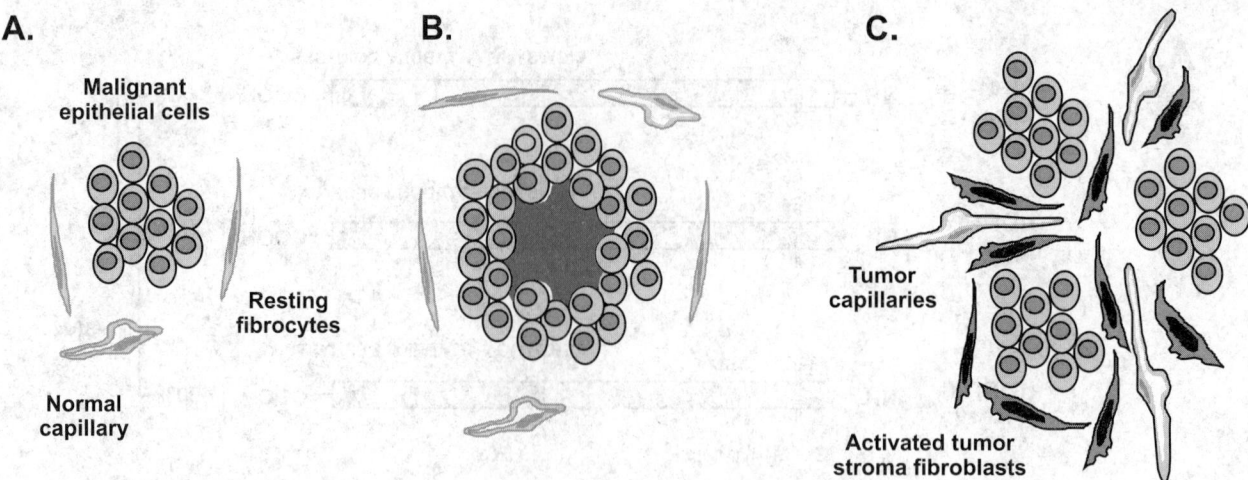

*Figure 131.2*   Schematic representation of tumor stromal fibroblast activation, accompanied by induction of FAPα cell surface expression, in epithelial cancers. (A) Early stage of epithelial cancer development with small deposits of malignant epithelial cells (nodules with a diameter less than approximately 2 mm) that do not require new blood vessel formation. (B) Larger nodules of malignant epithelial cells without new blood vessels and stroma formation, showing areas of central hypoxia and necrosis. (C) Tumor nodules (diameter greater than approximately 5 mm) with new blood vessels and accompanying, activated tumor stromal fibroblasts expressing cell surface FAP.

## Related Peptidases

Two types of related proteins have been identified through database searches. First, human FAPα shares about 52% amino acid sequence identity with human DPP IV/CD26. A DPP IV-related serum protein has been identified on the basis of serological cross-reactivity with anti-CD26 antibodies (Duke-Cohan *et al.*, 1995); no cDNA for this 175 kDa protein has been reported yet, but its partial amino acid sequence appears to be unrelated to either DPP IV/CD26 or FAP. Second, FAPα shares about 30% amino acid sequence identity with the predicted product of the dipeptidyl-peptidase X (DPP X) cDNA, which was isolated from a differential gene expression screen in brain tissues. The DPP X protein has not yet been characterized, but sequence analysis reveals that it lacks the characteristic Ser residue in the catalytic domain that is present both in FAPα and DPP IV (Wada *et al.*, 1992; Yokotani *et al.*, 1993). The FAPα and DPP IV gene loci reside on human chromosome 2q23 (Mathew *et al.*, 1995), suggesting that they may have arisen through gene duplication. By contrast, the DPP X gene locus has been mapped to human chromosome 7 (Yokotani *et al.*, 1993).

## References

Brown, D.D., Wang, Z., Furlow, J.D., Kanamori, A., Schwartz-man, R.A., Remo, B.F. & Pinder, A. (1996) The thyroid hormone-induced tail resorption program during *Xenopus laevis* metamorphosis. *Proc. Natl Acad. Sci. USA* **93**, 1924–1929.

Duke-Cohan, J.S., Morimoto, C., Rocker, J.A. & Schlossman, S.F. (1995) A novel form of dipeptidylpeptidase IV found in human serum. Isolation, characterization, and comparison with T lymphocyte membrane dipeptidylpeptidase IV (CD26). *J. Biol. Chem.* **270**, 14107–14114.

Garin-Chesa, P., Old, L.J. & Rettig, W.J. (1990) Cell surface glycoprotein of reactive stromal fibroblasts as a potential antibody-target in human epithelial cancers. *Proc. Natl Acad. Sci. USA* **87**, 7235–7239.

Hegen, M., Niedobitek, G., Klein, C.E., Stein, H. & Fleischer, B. (1990) The T cell triggering molecule Tp103 is associated with dipeptidyl aminopeptidase IV activity. *J. Immunol.* **144**, 2908–2914.

Mathew, S., Scanlan, M.J., Mohan Raj, B.K., Murty, V.V.V.S., Garin-Chesa, P., Old, L.J., Rettig, W.J. & Chaganti, R.S.K. (1995) The gene for fibroblast activation protein (FAP), a putative cell surface-bound serine protease expressed in cancer stroma and wound healing maps to chromosome band 2q23. *Genomics* **25**, 335–337.

Nidermeyer, J., Matthew, J., Scanlan, M.J., Garin-Chesa, J., Dai-ber, C., Fiebig, H.H., Old, L.J., Rettig, W.J. & Schnapp, A. (1997) Mouse fibroblast activation protein: molecular cloning, alternative splicing and expression in the reactive stroma of epithelial cancers. *Int. J. Cancer* **71**, 383–389.

Rettig, W.J., Garin-Chesa, P., Beresford, H.R., Feickert, H.J., Jennings, M.T., Cohen, J., Oettgen, H.F. & Old, L.J. (1986) Differential expression of cell surface antigens and glial fibrillary acidic protein in human astrocytoma subsets. *Cancer Res.* **46**, 6406–6412.

Rettig, W.J., Spengler, B.A., Garin-Chesa, P., Old, L.J. & Biedler, J.L. (1987) Coordinate changes in neuronal phenotype and surface antigen expression in human neuroblastoma cell variants. *Cancer Res.* **47**, 1383–1389.

Rettig, W.J., Garin-Chesa, P., Beresford, H.R., Oettgen, H.F., Melamed, M.R. & Old, L.J. (1988) Cell surface glycoproteins of human sarcomas: differential expression in normal and malignant tissues and cultured cells. *Proc. Natl Acad. Sci. USA* **85**, 3110–3114.

Rettig, W.J., Garin-Chesa, P., Healey, J.H., Su, S.L., Ozer, H.L., Schwab, M., Albino, A.P. & Old, L.J. (1993) Regulation and heteromeric structure of the fibroblast activation protein in normal and transformed cells of mesenchymal and neuroectodermal origin. *Cancer Res.* **53**, 3327–3335.

Rettig, W.J., Su, S.L., Fortunato, S.R., Scanlan, M.J., Mohan Raj, B.K., Garin-Chesa, P., Healey, J.H. & Old, L.J. (1994) Fibroblast activation protein: purification, epitope mapping, and induction by growth factors. *Int. J. Cancer* **58**, 385–392.

Sappino, A.P., Schhrich, W. & Gabbiani, G. (1990) Differentiation repertoire of fibroblastic cells: expression of cytoskeletal proteins as marker of phenotypic modulations. *Lab. Invest.* **63**, 144–161.

Scanlan, M.J., Mohan Raj, B.K., Calvo, B., Garin-Chesa, P., Sanz-Moncasi, M.P., Healey, J.H., Old, L.J. & Rettig, W.J. (1994) Molecular cloning of fibroblast activation protein, a member of the serine protease family selectively expressed in stromal fibroblasts of epithelial cancers. *Proc. Natl Acad. Sci. USA* **91**, 5657–5661.

Stein, H., Schwarting, R. & Niedobitek, G. (1989) Cluster report: CD26. In: *Leukocyte Typing IV: White Cell Differentiation Antigens* (Knapp, W., Dörken, B., Gilks, W.R., Rieber, E.P., Schmitt, R.E., Stein, H. & van dem Borne, A.E.G.K., eds). New York: Oxford University Press, pp. 412–415.

Wada, K., Yokotani, N., Hunter, C., Doi, K., Wenthold, R.J. & Shimasaki, S. (1992) Differential expression of two distinct forms of mRNA encoding members of a dipeptidyl aminopeptidase family. *Proc. Natl Acad. Sci. USA* **89**, 197–201.

Welt, S., Divgi, C.R., Scott, A.M., Garin-Chesa, P., Finn, R.D., Graham, M., Carswell, E.A., Cohen, A., Larson, S.M., Old, L.J. & Rettig, W.J. (1994) Antibody targeting in metastatic colon cancer: A phase I study of monoclonal antibody F19 against a cell surface protein of reactive tumor stromal fibroblasts. *J. Clin. Oncol.* **12**, 1193–1203.

Yokotani, N., Doi, K., Wenthold, R.J. & Wada, K. (1993) Non-conservation of a catalytic residue in a dipeptidyl peptidase IV-related protein encoded by a gene on human chromosome 7. *Hum. Mol. Genet.* **2**, 1037–1039.

*Wolfgang J. Rettig*
*Oncology Research,*
*Dr. Karl Thomae GmbH/Boehringer Ingelheim Pharma,*
*88400 Biberach an der Riss, Germany*
*Email: wolfgang.rettig@bid.de*

# 132. Carboxypeptidase C including carboxypeptidase Y

## Databanks

*Peptidase classification: clan SC, family S10, MEROPS ID: S10.001*
*NC-IUBMB enzyme classification: EC 3.4.16.5*
*Databank codes:*

| Species | SwissProt | PIR | EMBL (cDNA) | EMBL (genomic) |
|---|---|---|---|---|
| Carboxypeptidase I/Y | | | | |
| *Arabidopsis thaliana* | P32826 | – | M81130 | – |
| *Arabidopsis thaliana* | – | – | Z26528 | – |
| *Candida albicans* | P30574 | JC1380 | M95182 | – |
| *Hordeum vulgare* | P07519 | A25858 | J03897 | – |
| | | A29226 | | |
| | | B25858 | | |
| *Oryza sativa* | P37890 | S43516 | D17586 | – |
| *Pichia pastoris* | P52710 | – | X87987 | – |
| *Pisum sativum* | – | – | Z68130 | – |
| *Saccharomyces cerevisiae* | P00729 | A00909 | M15482 | X80836: chromosome XIII cosmid |
| | | A26597 | | |
| | | A90763 | | |
| | | A94609 | | |
| | | S47458 | | |
| Carboxypeptidase III | | | | |
| *Hordeum vulgare* | P21529 | A35275 | – | – |
| *Oryza sativa* | P37891 | S22530 | S40458 | D10985: complete gene |
| *Triticum aestivum* | P11515 | A29412 | – | J02817: complete gene |

Brookhaven Protein Data Bank three-dimensional structures:

| Species | ID | Resolution | Notes |
|---|---|---|---|
| *Saccharomyces cerevisiae* | 1YSC | 2.8 | |

## Name and History

***Carboxypeptidase C*** is a broad term for the subclass of serine carboxypeptidases that prefer hydrophobic residues in the P1' position. Within this subclass, carboxypeptidases MI and MIII (Breddam *et al.*, 1983; Breddam & Sørensen, 1987) accept almost exclusively hydrophobic residues in P1', whereas carboxypeptidase Y and carboxypeptidase I are more promiscuous (Breddam, 1986; Dal Degan *et al.*, 1992). Since carboxypeptidase Y is by far the best characterized variant of carboxypeptidase C it will be described in more detail. Due to sequence conservation of essential regions we expect that many of the characteristics described for carboxypeptidase Y will be either identical or similar to those of other enzymes within this subclass, however.

***Carboxypeptidase Y*** was originally termed ***proteinase C*** (Hata *et al.*, 1967). Subsequent work showed that the enzyme released amino acids from the C-terminus of peptides (Hayashi *et al.*, 1970) and that it could be inactivated by reaction of a unique Ser residue with DFP, thus defining the enzyme as a serine carboxypeptidase (Hayashi *et al.*, 1973a). This property of the enzyme has rendered it an important tool for C-terminal sequencing (Hayashi, 1977; Breddam & Ottesen, 1987). The preference for hydrophobic amino acids in the P1' position of the substrate classified carboxypeptidase Y as a C-type serine carboxypeptidase (see Remington & Breddam, 1994).

## Structural Chemistry

Mature carboxypeptidase Y is a 421 amino acid, single-chain enzyme with a pI of 3.6 and a mass of 64 kDa (Hayashi *et al.*, 1973b; Johansen *et al.*, 1976). The protein is glycosylated at four positions (Trimble & Maley, 1977; Hasilik & Tanner, 1978; Winther *et al.*, 1991). Some of the carbohydrate chains are phosphorylated and each carboxypeptidase Y molecule contains about 4–5 diesterified phosphates (Hashimoto *et al.*, 1981). The crystal structure has been solved at a resolution of 2.8 Å and the fold shows that the enzyme belongs to the $\alpha/\beta$ hydrolase family (Endrizzi *et al.*, 1994). The structure contains 14 $\alpha$ helices, 11 strands of mixed $\beta$ sheets, five disulfide bridges and a single free Cys residue. One of the carbohydrate chains is partially buried in the enzyme and is not removed by endoglycosidase H (Trimble & Maley, 1977; Endrizzi *et al.*, 1994).

## Activity and Specificity

### Catalytic Mechanism

The carboxypeptidases C contain a catalytic Ser, His, Asp triad (see Chapter 124), and there is also a trypsin-like oxyanion hole (Endrizzi *et al.*, 1994). The catalytic mechanism is therefore believed to be similar to that of the well-characterized serine endopeptidases of clans SA and SB. However, differences do exist. Firstly, the $k_{cat}$ values for the hydrolysis of peptide and ester substrates decrease only 3–5-fold as pH is lowered from 8 to 3.5 (Mortensen *et al.*, 1994a; Christensen, 1994). In contrast with serine endopeptidases of family S1, $k_{cat}$ decreases dramatically when pH is lowered below 7 (see Fersht, 1985) due to protonation of the essential His. Thus, in the carboxypeptidases either the $pK_a$ of the His is dramatically perturbed in the ES complex or alternatively, the mechanism at low pH may be different from the classical serine endopeptidase mechanism (for discussion see Stennicke *et al.*, 1996a). Second, the serine carboxypeptidases have the ability to interact with the C-terminal carboxylate group of a peptide substrate. These interactions are used solely to stabilize the transition state, lowering the activation energy by approximately 25 kJ mol$^{-1}$. This allows serine carboxypeptidases to cleave even short peptides, such as N-blocked dipeptides, at a very high rate (Mortensen *et al.*, 1994a; Stennicke *et al.*, 1996a). In addition to hydrolysis, carboxypeptidase Y is capable of catalyzing the reverse reaction, i.e. aminolysis, a feature that has been utilized in C-terminal modification/labeling of peptides and proteins (see Remington & Breddam, 1994). Furthermore, partitioning experiments have demonstrated that carboxypeptidase Y employs a random order bi-bi mechanism rather than a ping-pong mechanism (Mortensen *et al.*, 1994b).

### The Carboxylate-Binding Site

The specificities of both serine carboxypeptidases and metallo-carboxypeptidases (family M14) (Chapter 450) are determined by the requirement for binding of the C-terminal amino acid residue in a dead-end pocket (Christianson & Lipscomb, 1987; Endrizzi *et al.*, 1994). However, the negatively charged carboxylate group of the substrate is stabilized differently in the two. In the metallo-enzymes it forms a salt-bridge with an Arg (Riordan, 1973; Christianson & Lipscomb, 1987), whereas in carboxypeptidase Y it interacts with an extensive hydrogen-bond network consisting of Trp49, Asn51, Glu65 and Glu145 (Mortensen *et al.*, 1994a). This explains why the metallo-carboxypeptidases are sensitive to increasing salt concentrations whereas carboxypeptidase Y is not (Williams & Auld, 1986; Mortensen *et al.*, 1994a). The absence of a positive charge in serine carboxypeptidases may be the reason why they also exhibit peptidyl amino acid amide hydrolase activity (hereafter termed carboxyamidase activity) as well as esterase activity (see Breddam, 1986). The hydrogen-bond network is important for carboxypeptidase and carboxyamidase activities but not for esterase and amidase activities. Thus, efficient hydrolysis of peptides requires Asn51 and Glu145 (in the protonated form) each to donate one hydrogen bond to the carboxylate of the peptide substrate. The roles of Trp49 and Glu65 are to keep the two former residues in the proper orientation. When Glu145 is deprotonated, substrate binding is abolished, as a result of charge repulsion between the two negative charges. At low pH, when both the C-terminus of the peptide substrate and Glu145 are protonated, binding is adversely affected, since formation of optimal hydrogen bonds is prevented. Consequently, the pH profile of $K_m$ for peptide hydrolysis has a minimum value at pH 4 (Mortensen *et al.*, 1994a). The value of $k_{cat}$ varies 4–5-fold between pH 3 and 8.5, and the optimum condition for peptide hydrolysis is around pH 4–4.5. When peptide amides are degraded by carboxyamidase activity, the network acts slightly differently. The C-terminal carboxyamide $NH_2$ and $C=O$ groups of the substrate form hydrogen bonds to Glu145 (now in the charged form) and Asn51, respectively. Accordingly, $K_m$ does not increase significantly from pH 4.0 to 8.5, resulting in maximum $k_{cat}/K_m$ above pH 7 (Mortensen *et al.*, 1994c).

## The P1' Substrate Preference

Class C serine carboxypeptidases are defined as preferring hydrophobic residues in P1'. The crystal structure of carboxypeptidase Y revealed the structural basis for this specificity, showing that the large S1' pocket is mostly formed by hydrophobic residues: Thr60, Phe64, Tyr256, Leu272 and Met398 (Endrizzi *et al.*, 1994). In addition, two structural elements are found, a disulfide bridge (Cys56-Cys298) and a glutamic acid bridge (Glu65-Glu145) (Endrizzi *et al.*, 1994; Mortensen & Breddam, 1994). The S1' pocket accommodates small as well as large hydrophobic residues due to the flexible Met398 that modulates the size of the pocket upon substrate binding (Sørensen *et al.*, 1995). Even though carboxypeptidase Y shows a dominant preference for hydrophobic residues, it does accept basic and acidic leaving groups. However, Lys and Arg are strongly preferred over Asp and Glu due to favorable hydrophobic interactions between the long apolar part of the basic residues and the S1' binding site (Stennicke *et al.*, 1996b). Hydrophobic residues in the S1' pocket have been substituted with acidic residues in order to convert carboxypeptidase Y from a C-type into a D-type carboxypeptidase. The largest effects were obtained with the mutants Leu272Asp/Glu, which showed a marked decrease in the preference for hydrophobic residues, but only a slight increase in the specificity against basic residues, underlining the difficulty of engineering a charge specificity into a hydrophobic environment (Stennicke *et al.*, 1994).

## The P1 Substrate Preference

The S1 pocket, like the S1' pocket, is predominantly composed of hydrophobic residues (Tyr147, Leu178, Tyr185, Tyr188, Trp312, Ile340 and Cys341) resulting in a strong preference for hydrophobic amino acids at the P1 position (see Breddam, 1986; Endrizzi *et al.*, 1994). The discrimination against a basic amino acid in P1 is primarily due to the presence of Trp312, which prevents the positive charge on the side chain of the substrate from gaining access to bulk water (Olesen & Breddam, 1995). Mutated enzymes with a higher preference for basic residues in P1 have been obtained by rendering the S1 pocket more spacious as well as less hydrophobic (Olesen & Kielland-Brandt, 1993; Olesen *et al.*, 1994; Olesen & Breddam, 1995). Thus, the mutant enzyme Leu178Asp+Trp312Asp exhibits a Lys/Leu ratio which is 380 000-fold increased as compared to wild-type, and the overall P1 substrate preference of this enzyme resembles that of wheat carboxypeptidase D (Olesen & Breddam, 1995) (Chapter 134).

## The Extended Substrate Preference

The extended substrate preference of carboxypeptidase Y for the P2–P5 side chains has been investigated using fluorescent hexapeptides. For all substrate positions the enzyme exhibits a preference for hydrophobic side chains, with the highest discrimination observed around the scissile bond (Olesen *et al.*, 1996).

## Preparation

Efficient carboxypeptidase Y expression vectors are available, e.g. p72UG (Nielsen *et al.*, 1990) or pRA21 (Olesen *et al.*, 1994). Expression is most easily performed in a *vps1* yeast strain, as processed carboxypeptidase Y can be harvested directly from the growth media (Nielsen *et al.*, 1990). Purification is best performed as described by Johansen *et al.* (1976).

## Biological Aspects

The structural gene for carboxypeptidase Y, *PRC1*, has been cloned and sequenced, and it encodes a prepro form of the enzyme (Stevens *et al.*, 1986; Blachly-Dyson & Stevens, 1987; Valls *et al.*, 1987). Carboxypeptidase Y is an important marker for the study of membrane trafficking and sorting in yeast (see van den Hazel *et al.*, 1996). Accordingly, maturation and sorting of the enzyme have been extensively characterized. The 20 amino acid N-terminal leader sequence directs synthesis of procarboxypeptidase Y into the endoplasmic reticulum where the leader is removed (Blachly-Dyson & Stevens, 1987). The subsequent folding process is dependent on the presence of the C-terminal third of the pro region (Winther & Sørensen, 1991; Ramos *et al.*, 1994). This part of the propeptide is also sufficient to maintain the zymogen in an inactive form (Winther & Sørensen, 1991; Sørensen *et al.*, 1994). The four sites for *N*-linked glycosylation become core-glycosylated in the endoplasmic reticulum. During the passage through the Golgi apparatus further modification of the glycosyl residues takes place (Stevens *et al.*, 1982). These modifications influence the rate of intracellular transport of the protein, but not the vacuolar targeting (Winther *et al.*, 1991). A Gln-Arg-Pro-Leu motif within the pro region is recognized by the carboxypeptidase Y sorting receptor that directs the protein to the vacuole (Valls *et al.*, 1990; Marcusson *et al.*, 1994; van Voorst *et al.*, 1996). The final maturation of carboxypeptidase Y occurs in the vacuole where the pro region is cleaved off in a reaction dependent on proteinase A (saccharopepsin) (Chapter 283) and proteinase B (cerevisin) (Chapter 104) (van den Hazel *et al.*, 1992, 1996).

In addition to their catalytic activity, carboxypeptidases C may play structural roles. For example, one of these enzymes is essential for the correct assembly and function of a particle containing $\beta$-galactosidase and neuraminidase in the lysosomal compartment of human cells. Mutations in this enzyme, called 'human protective protein' or CarbL, can lead to genetic disorders such as galactosialidosis (Zhou *et al.*, 1991) (see Chapter 133).

## Related Peptidases

See Chapter 134 on carboxypeptidase D.

## Further Reading

Breddam (1986), Remington (1993) and Remington & Breddam (1994) provide further review articles on carboxypeptidase C.

## References

Blachly-Dyson E. & Stevens T.H. (1987) Yeast carboxypeptidase Y can be translocated and glycosylated without its amino-terminal signal sequence. *J. Cell. Biol.* **104**, 1183–1191.

Breddam, K. (1986) Serine carboxypeptidases. A review. *Carlsberg Res. Commun.* **51**, 83–128.

Breddam, K. & Ottesen, M. (1987) Determination of C-terminal sequences by digestion with serine carboxypeptidases: the influence of enzyme specificity. *Carlsberg Res. Commun.* **52**, 55–63.

Breddam, K. & Sørensen, S.B. (1987) Isolation of carboxypeptidase III from malted barley by affinity chromatography. *Carlsberg Res. Commun.* **52**, 275–283.

Breddam, K., Sørensen, S.B. & Ottesen, M. (1983) Isolation of carboxypeptidase from malted barley by affinity chromatography. *Carlsberg Res. Commun.* **48**, 217–230.

Christensen, U. (1994) Effects of pH on carboxypeptidase-Y-catalyzed hydrolysis and aminolysis reactions. *Eur. J. Biochem.* **220**, 149–153.

Christianson, D.W. & Lipscomb, W.N. (1987) Carboxypeptidase A: novel enzyme-substrate-product complex. *J. Am. Chem. Soc.* **109**, 5536–5538.

Dal Degan, F., Ribadeau-Dumas, B. & Breddam, K. (1992) Purification and characterization of two serine carboxypeptidases from *Aspergillus niger* and their use in C-terminal sequencing of proteins and peptide synthesis. *Appl. Environ. Microbiol.* **58**, 2144–2152.

Endrizzi, J.A., Breddam, K. & Remington, S.J. (1994) 2.8-Å Structure of yeast serine carboxypeptidase. *Biochemistry* **33**, 11106–11120.

Fersht, A. (1985) *Enzyme Structure and Mechanism*, 2nd edn. New York: W.H. Freeman and Co.

Hashimoto, C., Cohen, R.E., Zhang, W.-J. & Ballou, C. (1981) Carbohydrate chains on yeast carboxypeptidase Y are phosphorylated. *Proc. Natl Acad. Sci. USA* **78**, 2244–2248.

Hasilik, A. & Tanner W. (1978) Carbohydrate moiety of carboxypeptidase Y and perturbation of its biosynthesis. *Eur. J. Biochem.* **91**, 567–575.

Hata, T., Hayashi, R. & Doi, E. (1967) Purification of yeast proteinases. Part I. Fractionation and some properties of the proteinases. *Agric. Biol. Chem.* **31**, 150–159.

Hayashi, R. (1977) Carboxypeptidase Y in sequence determination of peptides. *Methods Enzymol.* **47**, 84–93.

Hayashi, R., Aibara, S. & Hata, T. (1970) A unique carboxypeptidase activity of yeast proteinase C. *Biochem. Biophys. Acta* **212**, 359–361.

Hayashi, R., Moore, S. & Stein, W.H. (1973a) Serine at the active center of yeast carboxypeptidase. *J. Biol. Chem.* **248**, 8366–8369.

Hayashi, R., Moore, S. & Stein, W.H. (1973b) Carboxypeptidase from yeast. Large scale preparation and the application to COOH-terminal analysis of peptides and proteins. *J. Biol. Chem.* **248**, 2296–2302.

Johansen, J.T., Breddam, K. & Ottesen, M. (1976) Isolation of carboxypeptidase Y by affinity chromatography. *Carlsberg Res. Commun.* **41**, 1–14.

Marcusson, E.G., Horazdovsky, B.F., Cereghino, J.L., Gharakhanian, E. & Emr, S.D. (1994) The sorting receptor for yeast vacuolar carboxypeptidase Y is encoded by the *VPS10* gene. *Cell* **77**, 579–586.

Mortensen, U.H. & Breddam, K. (1994) A conserved glutamic acid bridge in serine carboxypeptidases, belonging to the $\alpha/\beta$ hydrolase fold, acts as a pH-dependent protein-stabilizing element. *Protein Sci.* **3**, 838–842.

Mortensen, U.H., Remington, S.J. & Breddam, K. (1994a) Site-directed mutagenesis on (serine) carboxypeptidase Y. A hydrogen bond network stabilizes the transition state by interaction with the C-terminal carboxylate group of the substrate. *Biochemistry* **33**, 508–517.

Mortensen, U.H., Stennicke, H.R., Raaschou-Nielsen, M. & Breddam, K. (1994b) Mechanistic study on carboxypeptidase Y-catalyzed transacylation reactions. Mutationally altered enzymes for peptide synthesis. *J. Am. Chem. Soc.* **116**, 34–41.

Mortensen, U.H., Raaschou-Nielsen, M. & Breddam, K. (1994c) Recognition of C-terminal amide groups by (serine) carboxypeptidase Y investigated by site-directed mutagenesis. *J. Biol. Chem.* **22**, 15528–15532.

Nielsen, T.L, Holmberg, S. & Petersen, J.G.L. (1990) Regulated overproduction and secretion of yeast carboxypeptidase Y. *Appl. Microbiol. Biotechnol.* **33**, 307–312.

Olesen, K. & Breddam, K. (1995) Increase of the Lys/Leu substrate preference of carboxypeptidase Y by rational design based on known primary and tertiary structures of serine carboxypeptidases. *Biochemistry* **34**, 15689–15699.

Olesen, K. & Kielland-Brandt, M.C. (1993) Altering substrate preference of carboxypeptidase Y by a novel strategy of mutagenesis eliminating wild type background. *Protein Eng.* **6**, 409–415.

Olesen, K., Mortensen, U.H., Aasmul-Olsen, S., Kielland-Brandt, M.C. & Breddam, K. (1994) The activity of carboxypeptidase Y toward substrates with basic $P_1$ amino acid residues is drastically increased by mutational replacement of leucine 178. *Biochemistry* **33**, 11121–11126.

Olesen, K., Meldal, M. & Breddam, K. (1996) Extended subsite characterization of carboxypeptidase Y using substrates based on intramolecularly quenched fluorescence. *Protein Pept. Lett.* **3**, 67–74.

Ramos, C., Winther, J.R. & Kielland-Brandt, M.C. (1994) Requirement of the propeptide for in vivo formation of active yeast carboxypeptidase Y. *J. Biol. Chem.* **269**, 7006–7012.

Remington, S.J. (1993) Serine carboxypeptidases: a new and versatile family of enzymes. *Curr. Opin. Biotechnol.* **4**, 462–468.

Remington, S.J. & Breddam, K. (1994) Carboxypeptidases C and D. *Methods Enzymol.* **244**, 231–248.

Riordan, J.F. (1973) Functional arginyl residues in carboxypeptidase A. Modification with butanedione. *Biochemistry* **12**, 3915–3923.

Stennicke, H.R., Mortensen, U.H., Christensen, U., Remington, S.J. & Breddam, K. (1994) Effects of introduced aspartic and glutamic acid residues on the P1′ substrate specificity, pH dependence and stability of carboxypeptidase Y. *Protein Eng.* **7**, 911–916.

Stennicke, H.R., Mortensen, U.H., Christensen, U., Remington, S.J. & Breddam, K. (1996a) Studies on the hydrolytic properties of (serine) carboxypeptidase Y. *Biochemistry* **35**, 7131–7141.

Stennicke, H.R., Henriksen, D.B. & Breddam, K. (1996b) Evaluation of the significance of charge and length of side chain on the $P_1'$-$S_1'$ interactions in (serine) carboxypeptidase Y. *Protein Pept. Lett.* **3**, 75–80.

Stevens, T., Esmon, B. & Schekman, R. (1982) Early stages in the yeast secretory pathway are required for transport of carboxypeptidase Y to the vacuole. *Cell* **30**, 439–448.

Stevens, T.H., Rothman J.H., Payne, G.S. & Schekman, R. (1986) Gene dosage-dependent secretion of yeast vacuolar carboxypeptidase Y. *J. Cell Biol.* **102**, 1551–1557.

Sørensen, S.O., van den Hazel, H.B., Kielland-Brandt, M.C. & Winther, J.R. (1994) pH-Dependent processing of yeast procarboxypeptidase Y by proteinase A *in vivo* and *in vitro*. *Eur. J. Biochem.* **220**, 19–27.

Sørensen, S.B., Raaschou-Nielsen, M., Mortensen, U.H., Remington, S.J. & Breddam, K. (1995) Site-directed mutagenesis on (serine) carboxypeptidase Y from yeast. The significance of Thr60 and Met398 in hydrolysis and aminolysis reactions. *J. Am. Chem. Soc.* **117**, 5944–5950.

Trimble, R.B. & Maley, F. (1977) The use of endo-*β-N*-acetylglucosaminidase H in characterizing the structure and function of glycoproteins. *Biochem. Biophys. Res. Commun.* **78**, 935–944.

Valls, L.A., Hunter, C.P., Rothman, J.H. & Stevens, T.H. (1987) Protein sorting in yeast: the localization determinant of yeast vacuolar carboxypeptidase Y resides in the propeptide. *Cell* **48**, 887–897.

Valls, L.A., Winther, J.R. & Stevens, T.H. (1990) Yeast carboxypeptidase Y vacuolar targeting signal is defined by four propeptide amino acids. *J. Cell Biol.* **111**, 361–368.

van den Hazel, H.B., Kielland-Brandt, M.C. & Winther, J.R. (1992)

Autoactivation of proteinase A initiates activation of yeast vacuolar zymogens. *Eur. J. Biochem.* **207**, 277–283.

van den Hazel, H.B., Kielland-Brandt, M.C. & Winther, J.R. (1996) Biosynthesis and function of yeast vacuolar proteases. *Yeast* **12**, 1–16.

van Voorst, F., Kielland-Brandt, M.C. & Winther, J.R. (1996) Mutational analysis of the vacuolar sorting signal of procarboxypeptidase Y in yeast shows a low requirement for sequence conservation. *J. Biol. Chem.* **271**, 841–846.

Williams, A.C. & Auld, D.S. (1986) Kinetic analysis by stopped-flow radiationless energy transfer studies: effect of anions on the activity of carboxypeptidase A. *Biochemistry* **25**, 94–100.

Winther, J.R. & Sørensen, P. (1991) Propeptide of carboxypeptidase Y provides a chaperone-like function as well as inhibition of the enzymatic activity. *Proc. Natl Acad. Sci. USA* **88**, 9330–9334.

Winther J.R., Stevens, T.H. & Kielland-Brandt, M.C. (1991) Yeast carboxypeptidase Y requires glycosylation for efficient intracellular transport, but not for vacuolar sorting, *in vivo* stability, or activity. *Eur. J. Biochem.* **197**, 681–689.

Zhou, X.Y., Galjart, N.J., Willemsen, R., Gillemans, N., Galjaard, H. & d'Azzo, A. (1991) A mutation in a mild form of galactosialidosis impairs dimerization of the protective protein and renders it unstable. *EMBO J.* **10**, 4041–4048.

*Uffe H. Mortensen*
*Department of Chemistry,*
*Carlsberg Laboratory,*
*Gamle Carlsbergvej 10, DK-2500*
*Copenhagen Valby, Denmark*

*Kjeld Olesen*
*Department of Chemistry,*
*Carlsberg Laboratory,*
*Gamle Carlsbergvej 10, DK-2500*
*Copenhagen Valby, Denmark*

*Klaus Breddam*
*Department of Chemistry,*
*Carlsberg Laboratory,*
*Gamle Carlsbergvej 10, DK-2500*
*Copenhagen Valby, Denmark*

# 133. Lysosomal carboxypeptidase A

## Databanks

*Peptidase classification: clan SC, family S10, MEROPS ID: S10.002*
*NC-IUBMB enzyme classification: EC 3.4.16.5*
*Databank codes:*

| Species | SwissProt | PIR | EMBL (cDNA) | EMBL (genomic) |
|---|---|---|---|---|
| *Caenorhabditis elegans* | P52715 | – | – | Z49127: cosmid F13D12 |
| *Caenorhabditis elegans* | P52716 | – | – | U20864: cosmid F32A5 |
| *Caenorhabditis elegans* | Q09991 | – | – | U28730: cosmid K10B2 |
| *Homo sapiens* | P10619 | A31589 | M18453 M22960 | – |
| *Mus musculus* | P16675 | A35732 | J05261 | – |

Brookhaven Protein Data Bank three-dimensional structures:

| Species | ID | Resolution | Notes |
|---|---|---|---|
| *Homo sapiens* | 1IVY | 2.2 | Physiological dimer of precursor |

## Name and History

Lysosomal carboxypeptidase A was originally described under the name of **cathepsin I**, later **cathepsin A**, as an enzyme responsible for the hydrolysis of Z-Glu+Tyr in beef kidney and spleen extracts that had an acidic pH optimum and was not activated by thiol compounds (Fruton & Bergmann, 1939; Fruton *et al.*, 1941). Later, cathepsin A was shown to be a carboxypeptidase (Iodice *et al.*, 1966; Iodice, 1967). Highly purified preparations of cathepsin A were obtained from different mammalian tissues, and their biochemical properties were studied (Logunov & Orekhovich, 1972; Kawamura *et al.*, 1974, 1975, 1977; Matsuda & Misaka, 1974, 1975; Matsuda, 1976). Independently, lysosomal carboxypeptidase A was discovered as a component of a high molecular mass complex of human lysosomal β-galactosidase (EC 3.2.1.32) and *N*-acetyl-α-neuraminidase (EC 3.2.1.18), and named **protective protein** because of its ability to protect β-galactosidase and neuraminidase against rapid intralysosomal proteolytic degradation (d'Azzo *et al.*, 1982; Verheijen *et al.*, 1982, 1985). The cloning and sequencing of both human and mouse protective proteins demonstrated that they have high amino acid sequence homology with yeast serine carboxypeptidases (Galjart *et al.*, 1988, 1990). Then Tranchemontagne *et al.* (1990) demonstrated the carboxypeptidase activity of protective protein and named it **lysosomal carboxypeptidase** or **carboxypeptidase L**. The enzyme was rediscovered as a **deamidase** from human platelets, which showed deamidase and esterase activity at neutral pH and carboxypeptidase activity at acidic pH (Jackman *et al.*, 1990). The catalytic properties of deamidase resembled those of cathepsin A, and the N-terminal amino acid sequence was identical to that of protective protein. Finally the similarity of cathepsin A to protective protein was proven when protective protein was shown to be responsible for all activity against Z-Phe+Ala and Z-Glu+Tyr in human placenta tissue (Galjart *et al.*, 1990; Pshezhetsky & Potier, 1994).

## Activity and Specificity

Like the other serine carboxypeptidases (Remington, 1993), lysosomal carboxypeptidase A is a multifunctional enzyme that expresses deamidase and esterase activities at neutral pH (optimal pH is 7.0) and carboxypeptidase activity at acidic pH (optimal pH is 5.0–5.2) (Jackman *et al.*, 1990). The enzyme has a preference for substrates with hydrophobic amino acid residues in the P1′ position (Pshezhetsky *et al.*, 1995b) and therefore should be classified as a C-type carboxypeptidase (Remington & Breddam, 1994), but it also shows a high affinity for positively charged amino acid residues in the P1′ position. The recent modeling of the substrate binding in the S1′ subsite of lysosomal carboxypeptidase A as compared to that of wheat carboxypeptidase II and yeast carboxypeptidase Y (Elsliger *et al.*, 1996) provided a theoretical explanation for the observed substrate specificity. The Tyr247 residue in the enzyme active site provides the majority of the substrate-binding energy with some contributions from Met430 and Asn305, whereas Asp64 promotes the binding of substrates with positively charged residues in the P1′ position (Figure 133.1).

The esterase activity of lysosomal carboxypeptidase A is assayed with Bz-Tyr+OEt (Jackman *et al.*, 1990), and the carboxypeptidase activity with Z-Phe+Xaa dipeptides followed by fluorometric (Galjart *et al.*, 1991) or spectrophotometric (Tranchemontagne *et al.*, 1990) detection of the released amino acids. Z-Phe-Phe and Z-Phe-Leu are the most specific substrates (as judged by $k_{cat}/K_m$), and with Z-Phe-Ala the enzyme shows the highest activity ($k_{cat}$). Furylacryloyl-Phe-Phe (available from Bachem) (see Appendix 2 for full names and addresses of suppliers) can be used for the continuous spectrophotometric assay (Pshezhetsky *et al.*, 1995b). A fluorescent substrate, 5-dimethylaminonaphthalene-1-sulfonyl-D-Tyr-Val+NH$_2$ is useful for the HPLC-based determination of the deamidase activity of the enzyme (Chikuma *et al.*, 1996).

Lysosomal carboxypeptidase A is inhibited by PMSF, iodoacetamide, DFP, thiol reagents in high concentrations and heavy metals such as $Hg^{2+}$, $Ag^{2+}$ and $Cu^{2+}$ (Tranchemontagne *et al.*, 1990; Itoh *et al.*, 1993; Chikuma *et al.*, 1996). A specific inhibitor from potato has also been described (Worowsky, 1975). Pepstatin, leupeptin and phosphoramidon have no effect on the enzyme (Itoh *et al.*, 1993).

## Structural Chemistry

Lysosomal carboxypeptidase A contains two polypeptide chains of about 32 and 20 kDa which are linked by disulfide bonds. Both chains contain *N*-linked oligosaccharides, but only the oligosaccharide attached to the 32 kDa chain is mannose-6-phosphorylated (Morreau *et al.*, 1992).

The amino acid sequence of lysosomal carboxypeptidase A has about 30% identity with the other serine carboxypeptidases, yeast carboxypeptidase Y and wheat carboxypeptidase II (Galjart *et al.*, 1988; Elsliger & Potier, 1994). X-Ray atomic coordinates of the wheat enzyme were used to model the lysosomal carboxypeptidase A structure (Elsliger & Potier, 1994). Recently, the X-ray structure of the lysosomal carboxypeptidase A precursor, expressed in a baculovirus system, was determined with a 2.2–2.4 Å resolution (Rudenko *et al.*, 1995). The structure is similar to those of plant and yeast carboxypeptidases, and shows that lysosomal carboxypeptidase A belongs to a so-called α/β hydrolase family (Remington, 1993). The protein contains a core domain and a cap domain. The core domain consists of a central, 10-stranded β sheet which is flanked by 10 α helices and two small β strands on both sides. The cap domain consists of three α helices and a three-stranded mixed β sheet. The catalytic triad in the active site is formed by the residues Ser150, His429 and Asp372.

At acidic pH, the enzyme forms 95–98 kDa homodimers, of which the X-ray structure is also known (Rudenko *et al.*, 1995). The pI for the human enzyme is 5.4.

## Preparation

Lysosomal carboxypeptidase A is widely distributed in mammalian tissues with the highest expression in kidney, liver, lung (Satake *et al.*, 1994) and placenta (Pshezhetsky & Potier, 1994). These tissues are the most convenient sources, and homogeneous preparations of the enzyme can be obtained by the affinity purification of the β-galactosidase–lysosomal

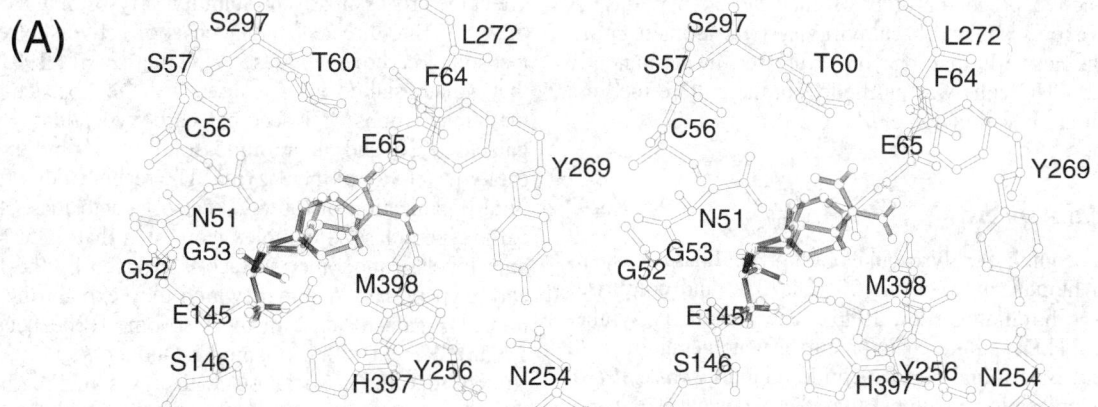

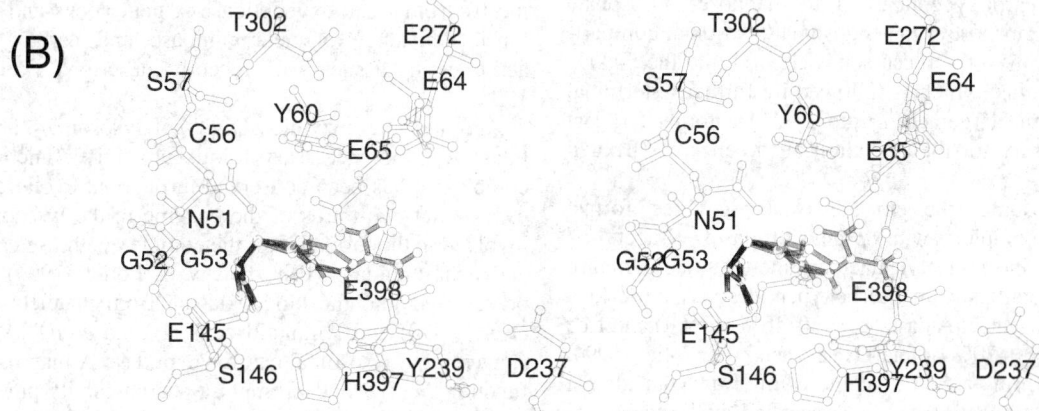

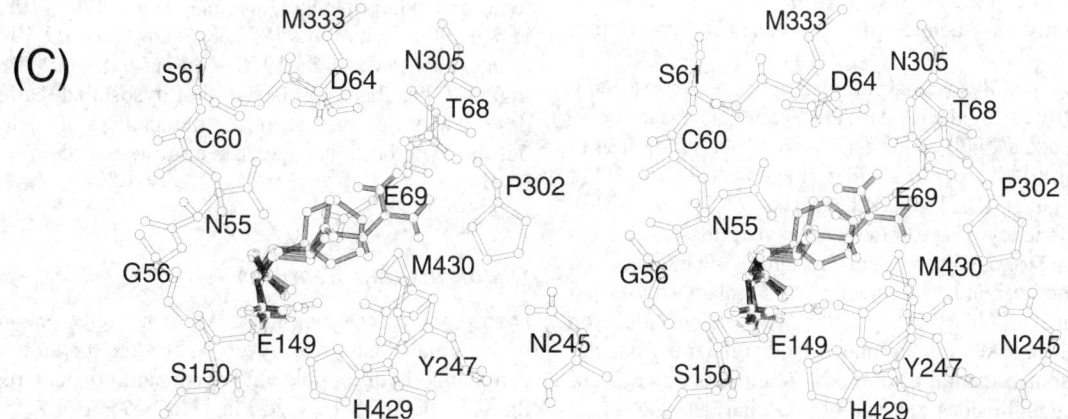

*Figure 133.1*  Stereo view of the modeled P1′ substrates (filled bonds) bound to the S1′ binding pockets (open bonds) of (A) yeast carboxypeptidaseY, (B) wheat carboxypeptidase II and (C) lysosomal carboxypeptidase A (Elsliger *et al.*, 1996).

carboxypeptidase A complex on the β-galactosidase-binding *p*-aminophenyl β-D-thiogalactopyranoside-agarose (available from Sigma), followed by the separation of the complex components at neutral pH (Pshezhetsky & Potier, 1993). Alternatively, one can use affinity chromatography on a

Phe-Leu-agarose column which directly binds the lysosomal carboxypeptidase A (Pshezhetsky & Potier, 1994). The purification factor is about 3700-fold. Homogeneous enzyme released from human platelets by thrombin has been purified 230-fold (Jackman *et al.*, 1990).

The inactive precursor of lysosomal carboxypeptidase A has been expressed in a baculovirus system (Bonten *et al.*, 1995). The active human enzyme expressed in permanently transfected CHO cells was purified from the culture medium on a microgram scale (Itoh *et al.*, 1993).

## Biological Aspects

The gene coding for lysosomal carboxypeptidase A spans 7.5 kb on human chromosome 20 (20q13.1) and comprises 15 exons (Shimmoto *et al.*, 1996). The protein is synthesized as a 54 kDa, single-chain precursor which contains a 28 amino acid N-terminal signal peptide (Galjart *et al.*, 1988). In the endoplasmic reticulum, after the cleavage of the signal peptide, the protein is folded and glycosylated at Asn117 and Asn305. In the late Golgi compartment the oligosaccharide chain at Asn117 gets a mannose-6-phosphate recognition signal (Morreau *et al.*, 1992) which mediates the binding of lysosomal carboxypeptidase A to mannose-6-phosphate receptors and transport to the endosomal/lysosomal compartment. In lysosomes, the precursor is cleaved into the 34 kDa and 20 kDa chains, which is followed by further C-terminal processing of the larger chain into a 32 kDa form. The last event is necessary for the activation of the enzyme (Bonten *et al.*, 1995).

In the lysosome, the enzyme exists in three forms: a 1.27 MDa complex with $\beta$-galactosidase, $N$-acetyl-$\alpha$-neuraminidase and $N$-acetylgalactosamine-6-sulfate sulfatase (about 1% of total lysosomal carboxypeptidase A), a 680 kDa complex with $\beta$-galactosidase only (30–40% of total) and free 98 kDa dimer (60–70% of total) (Pshezhetsky & Potier, 1994, 1996). These forms are in dynamic equilibrium, but almost all placental $\beta$-galactosidase is associated with lysosomal carboxypeptidase A in the 680 kDa complex. The association of lysosomal carboxypeptidase A and $\beta$-galactosidase was reported to occur already in the endoplasmic reticulum (Morreau *et al.*, 1992). However, *in vitro* the complex is stable only at acidic pH (Pshezhetsky & Potier, 1993).

Association with lysosomal carboxypeptidase A is essential for the stability of $\beta$-galactosidase and acetylgalactosamine-6-sulfate sulfatase as well as for activation of neuraminidase in the lysosome (d'Azzo *et al.*, 1982; Hoogeveen *et al.*, 1983; Van der Horst *et al.*, 1989; Pshezhetsky & Potier, 1996). Inherited deficiency of lysosomal carboxypeptidase A causes the lysosomal storage disorder galactosialidosis, characterized by a combined secondary deficiency of $\beta$-galactosidase and neuraminidase (d'Azzo *et al.*, 1995). The patients develop edema, ascites, skeletal dysplasia and bilateral macular cherry-red spots, mental and motor retardation and accumulate sialogangliosides and sialoligosaccharides due to the deficiency of neuraminidase (d'Azzo *et al.*, 1995). Recent modeling of lysosomal carboxypeptidase A based on the X-ray structure of the homologous wheat carboxypeptidase II (Elsliger & Potier, 1994) as well as transient expression studies of mutant enzyme from various galactosialidosis patients (Zhou *et al.*, 1991; Shimmoto *et al.*, 1993) suggested that most of the missense mutations in galactosialidosis patients

affect the processing or the stability of lysosomal carboxypeptidase A. Site-directed mutagenesis of active-site residues of lysosomal carboxypeptidase A (Galjart *et al.*, 1991) showed that galactosialidosis is caused by the absence of structural interactions of lysosomal carboxypeptidase A with $\beta$-galactosidase and neuraminidase, but not by the lack of catalytic activity of the enzyme. The recent study of the structural organization of the 680 kDa $\beta$-galactosidase–lysosomal carboxypeptidase A complex suggested that ∼35% of the $\beta$-galactosidase monomer surface is covered by the lysosomal carboxypeptidase A dimer, which may explain the stabilization of $\beta$-galactosidase in the lysosome (Pshezhetsky *et al.*, 1995a). Recent work (Okamura-Oho *et al.*, 1996) demonstrated that the association with lysosomal carboxypeptidase A is also necessary for the proper proteolytic processing of the C-terminal region of $\beta$-galactosidase precursor. Nothing is known about the biochemical basis of neuraminidase deficiency in galactosialidosis. A mouse model of galactosialidosis which is homozygous for a null mutation at the lysosomal carboxypeptidase A gene locus and has combined deficiency of neuraminidase and carboxypeptidase activities in tissues was recently described (Zhou *et al.*, 1995).

Although the enzymic activity of lysosomal carboxypeptidase A is not necessary for its protective function in the complex, it has been conserved throughout evolution. Moreover, about two-thirds of the enzyme in the lysosome is not involved in the formation of the complex with $\beta$-galactosidase and neuraminidase (Pshezhetsky & Potier, 1994), and can be secreted into the blood plasma from platelets (Jackman *et al.*, 1990) and lymphoid cells (Hanna *et al.*, 1994). This suggests that lysosomal carboxypeptidase A may have a dual function *in vivo* and prompts a search for its physiological peptide substrates. Several authors have shown that lysosomal carboxypeptidase A from platelets and lymphocytes hydrolyzes *in vitro* some regulatory peptides, i.e. endothelin, substance P, angiotensin I, Met-enkephalin-Arg6-Phe7, oxytocin and others (Jackman *et al.*, 1990, 1992, 1993; Hanna *et al.*, 1994; Itoh *et al.*, 1995; Kawamura *et al.*, 1977; Marks *et al.*, 1981; Matsuda 1976; Miller *et al.*, 1988). Accordingly, it has been proposed that lysosomal carboxypeptidase A may be involved in the metabolism of such peptides, but the hypothesis has never been tested at the physiological level.

## Distinguishing Features

Lysosomal carboxypeptidase A is the only enzyme found in mammals that hydrolyzes N-blocked peptide substrates containing hydrophobic and basic amino acid residues in the P1′ position (e.g. Z-Phe┼Phe, Z-Phe┼Leu) at acidic pH, and should therefore not be confused with the other mammalian carboxypeptidases, such as tissue carboxypeptidases A (Chapter 451) and B (Chapter 455) and lysosomal carboxypeptidase B (Chapter 267). Additionally, the enzyme can be distinguished by its complete inhibition in the presence of 5 mM PMSF or 50 mM DTT. Polyclonal rabbit antisera against the human enzyme or peptides selected from its amino acid sequence have been described (Galjart *et al*., 1988, 1991; Pshezhetsky & Potier, 1994; Satake *et al*., 1994; Bonten *et al*., 1995) but are not commercially available.

## Further Reading

The articles of d'Azzo *et al*. (1995) and Okamura-Oho *et al*. (1994) provide extensive reviews of all aspects of the enzyme.

## References

Bonten, E.J., Galjart, N.J., Willemsen, R., Usmany, M., Vlak, J.M. & d'Azzo, A. (1995) Lysosomal protective protein/cathepsin A. Role of the 'linker' domain in catalytic activation. *J. Biol. Chem.* **270**, 26441–26445.

Chikuma, T., Matsumoto, K., Furukawa, A., Nakayama, N., Yajima, R., Kato, T., Ishii, Y. & Tanaka, A. (1996) A fluorometric assay for measuring deamidase (lysosomal protective protein) using high-performance liquid chromatography. *Anal. Biochem.* **233**, 36–41.

d'Azzo, A., Hoogeveen, A., Reuser, A.J., Robinson, D. & Galjaard, H. (1982) Molecular defect in combined β-galactosidase and neuraminidase deficiency in man. *Proc. Natl Acad. Sci. USA* **79**, 4535–4539.

d'Azzo, A., Andria, G., Strisciuglio, P. & Galijaard, H. (1995) Galactosialidosis. In: *Metabolic and Molecular Bases of Inherited Disease* (Scriver, C.R., Beaudet, A.L, Sly, W.S. & Valle, D., eds). New York: McGraw-Hill, pp. 2835–2837.

Elsliger, M.-A. & Potier, M. (1994) Homologous modeling of the lysosomal protective protein/carboxypeptidase L: structural and functional implications of mutations identified in galactosialidosis patients. *Proteins* **18**, 81–93.

Elsliger, M.-A., Pshezhetsky, A.V., Vinogradova, M.V., Svedas, V.K. & Potier, M. (1996) Comparative modeling of substrate binding in the $S'_1$ subsite of serine carboxypeptidases from yeast, wheat and human. *Biochemistry* **35**, 14899–14909.

Fruton, J.S. & Bergmann, M. (1939) On the proteolytic enzymes of animal tissues. I. Beef spleen. *J. Biol. Chem.* **130**, 19-27.

Fruton, J.S., Irving, G.W., Jr. & Bergmann, M. (1941) On the proteolytic enzymes of animal tissues. II. The composite nature of beef spleen cathepsin. *J. Biol. Chem.* **138**, 249–262.

Galjart, N.J., Gillemans, N., Harris, A., Gijsbertus, T., van der Horst, J., Verheijen, F.W., Galjaard, H. & d'Azzo, A. (1988) Expression of cDNA encoding the human 'protective protein' associated with lysosomal β-galactosidase and neuraminidase: homology to yeast proteases. *Cell* **54**, 755–764.

Galjart, N.J., Gillemans, N., Meijer, D. & d'Azzo, A. (1990) Mouse 'protective protein'. cDNA cloning, sequence comparison, and expression. *J. Biol. Chem.* **265**, 4678–4684.

Galjart, N.J., Morreau, H., Willemsen, R., Gillemans, N., Bonten, E.J. & d'Azzo, A. (1991) Human lysosomal protective protein has cathepsin A-like activity distinct from its protection function. *J. Biol. Chem.* **266**, 14754–14762.

Hanna, W.L., Turbov, J.M., Jackman, H.L., Tan, F. & Froelich, C.J. (1994) Dominant chymotrypsin-like esterase activity in human lymphocyte granules is mediated by the serine carboxypeptidase called cathepsin A-like protective protein. *J. Immunol.* **153**, 4663–4672.

Hoogeveen, A.T., Verheijen, F.W. & Galjaard, H. (1983) The relation between human lysosomal β-galactosidase and its protective protein. *J. Biol. Chem.* **258**, 12143–12146.

Iodice, A.A. (1967) The carboxypeptidase nature of cathepsin A. *Arch. Biochem. Biophys.* **121**, 241–242.

Iodice, A.A., Leong, V. & Weinstock, I.M. (1966) Separation of cathepsins A and D of skeletal muscle. *Arch. Biochem. Biophys.* **117**, 477–486.

Itoh, K., Takiyama, N., Kase, R., Kondoh, K., Sano, A., Oshima, A., Sakuraba, H. & Suzuki, Y. (1993) Purification and characterization of human lysosomal protective protein expressed in stably transformed Chinese hamster ovary cells. *J. Biol. Chem.* **268**, 1180–1186.

Itoh, K., Kase, R., Shimmoto, M., Satake, H. & Suzuki, Y. (1995) Protective protein as an endogenous endothelin degradation enzyme in human tissues. *J. Biol. Chem.* **270**, 515–518.

Jackman, H.L., Tan, F.L., Tamei, H., Beurling-Harbury, C., Li, X.Y., Skidgel, R.A. & Erdos, E.G. (1990) A peptidase in human platelets that deamidates tachykinins. Probable identity with the lysosomal 'protective protein'. *J. Biol. Chem.* **265**, 11265–11272.

Jackman, H.L., Morris, P.W., Deddish, P.A., Skidgel, R.A. & Erdös, E.G. (1992) Inactivation of endothelin I by deamidase (lysosomal protective protein). *J. Biol. Chem.* **267**, 2872–2875.

Jackman, H.L., Morris, P.W., Rabito, S.F., Johansson, G.B., Skigel, R.A. & Erdös, E.G. (1993) Inactivation of endothelin-1 by an enzyme of the vascular endothelial cells. *Hypertension* **21**, 925–928.

Kawamura, Y., Matoba, T., Hata, T. & Doi, E. (1974) Purification and some properties of cathepsin A of large molecular size from pig kidney. *J. Biochem.* **76**, 915–924.

Kawamura, Y., Matoba, T., Hata, T. & Doi, E. (1975) Purification and some properties of cathepsin A of small molecular size from pig kidney. *J. Biochem.* **77**, 729–737.

Kawamura, Y., Matoba, T., Hata, T. & Doi, E. (1977) Substrate specificities of cathepsin A,L and A,S from pig kidney. *J. Biochem.* **81**, 435–441.

Logunov, A.I. & Orekhovich, V.N. (1972) Isolation and some properties of cathepsin A from bovine spleen. *Biochem. Biophys. Res. Commun.* **46**, 1161–1168.

Marks, N., Sachs, L., & Stern, F. (1981) Conversion of Met-enkephalin-Arg6-Phe7 by a purified brain carboxypeptidase (cathepsin A). *Peptides* **2**, 159–164.

Matsuda, K. (1976) Studies on cathepsins of rat liver lysosomes. III. Hydrolysis of peptides, and inactivation of angiotensin and bradykinin by cathepsin A. *J. Biochem.* **80**, 659–669.

Matsuda, K. & Misaka, E. (1974) Studies on cathepsins of rat liver lysosomes. I. Purification and multiple forms. *J. Biochem.* **76**, 639–649.

Matsuda, K. & Misaka, E. (1975) Studies on cathepsin of rat liver lysosomes. II. Comparative studies on multiple forms of cathepsin A. *J. Biochem.* **78**, 31–39.

Miller, J.J., Changaris, D.G. & Levy, R.S. (1988) Conversion of angiotensin I to angiotensin II by cathepsin A isoenzymes of porcine kidney. *Biochem. Biophys. Res. Commun.* **154**, 1122–1129.

Morreau, H., Galjart, N.J., Willemsen, R., Gillemans, N., Zhou, XY. & d'Azzo, A. (1992) Human lysosomal protective protein. Glycosylation, intracellular transport, and association with β-galactosidase in the endoplasmic reticulum. *J. Biol. Chem.* **267**, 17949–17956.

Okamura-Oho, Y., Zhang, S.Q. & Callahan, J.W. (1994) The biochemistry and clinical features of galactosialidosis. *Biochem. Biophys. Acta* **1225**, 244–254.

Okamura-Oho, Y., Zhand, S., Hilson, W., Hinek, A. & Callahan, J.W. (1996) Early proteolytic cleavage with loss of a C-terminal fragment underlies altered processing of the $\beta$-galactosidase precursor in galactosialidosis. *Biochem. J.* **313**, 787–794.

Pshezhetsky, A.V. & Potier, M. (1993) Stoichiometry of the human lysosomal carboxypeptidase-$\beta$-galactosidase complex. *Biochem. Biophys. Res. Commun.* **195**, 354–362.

Pshezhetsky, A.V. & Potier, M. (1994) Direct affinity purification and supramolecular organization of human lysosomal cathepsin A. *Arch. Biochem. Biophys.* **313**, 64–70.

Pshezhetsky, A.V. & Potier, M. (1996) Association of *N*-acetyl-galactosamine-6-sulfate sulfatase with the multienzyme lysosomal complex of $\beta$-galactosidase, cathepsin A and neuraminidase: possible implication for intralysosomal catabolism of keratan sulfate. *J. Biol. Chem.* **271**, 28359–28365.

Pshezhetsky, A.V., Elsliger, M.-A., Vinogradova, M.V. & Potier, M. (1995a) Human lysosomal $\beta$-galactosidase-cathepsin A complex: definition of the $\beta$-galactosidase-binding interface on cathepsin A. *Biochemistry* **34**, 2431–2440.

Pshezhetsky, A.V., Vinogradova, M.V., Elsliger, M.-A., El-Zein, F., Svedas, V.K. & Potier, M. (1995b) Continuous spectrophotometric assay of human lysosomal cathepsin A/protective protein in normal and galactosialidosis cells. *Anal. Biochem.* **230**, 303–307.

Remington, S.J. (1993) Serine carboxypeptidases: a new and versatile family of enzymes. *Curr. Opin. Biotechnol.* **4**, 462–468.

Remington, S.J. & Breddam, K. (1994) Carboxypeptidases C and D. *Methods Enzymol.* **244**, 231–248.

Rudenko, G., Bonten, E., d'Azzo, A. & Hol, W.G.J. (1995) Three-dimensional structure of the human protective protein: structure of the precursor form suggests a complex activation mechanism. *Structure* **3**, 1249–1259.

Satake, A., Itoh, K. Shimmoto, M., Saido, T.C., Sakuraba, H. & Suzuki, Y. (1994) Distribution of lysosomal protective protein in human tissues. *Biochem. Biophys. Res. Commun.* **205**, 38–43.

Shimmoto, M., Fukuhara, Y., Itoh, K., Oshima, A., Sakuraba, H. & Suzuki, Y. (1993) Protective protein gene mutations in galactosialidosis. *J. Clin. Invest.* **91**, 2393–2398.

Shimmoto, M., Nakahori, Y., Matsuhita, I., Shinka, T., Kuroki, Y., Itoh, K. & Sakuraba, H. (1996) A human protective protein gene partially overlaps the gene encoding phospholipid transfer protein on the complementary strand of DNA. *Biochem. Biophys. Res. Commun.* **220**, 802–806.

Tranchemontagne, J., Michaud, L. & Potier, M. (1990) Deficient lysosomal carboxypeptidase activity in galactosialidosis. *Biochem. Biophys. Res. Commun.* **168**, 22–29.

Van der Horst, G.J., Galjart, N.J., d'Azzo, A., Galjaard, H. & Verheijen, F.W. (1989) Identification and *in vitro* reconstitution of lysosomal neuraminidase from human placenta. *J. Biol. Chem.* **264**, 1317–1322.

Verheijen, F., Brossmer, R. & Galjaard, H. (1982) Purification of acid $\beta$-galactosidase and acid neuraminidase from bovine testis: Evidence for an enzyme complex. *Biochem. Biophys. Res. Commun.* **108**, 868–875.

Verheijen, F., Palmeri, S., Hoogeveen, A.T. & Galjaard, H. (1985) Human placental neuraminidase: activation, stabilization and association with $\beta$-galactosidase and its 'protective' protein. *Eur. J. Biochem.* **149**, 315–321.

Worowski, J. (1975) Inhibition of cathepsin A activity by the potato protease inhibitor. *Experientia* **31**, 637–638.

Zhou, X.Y., Galjart, N.J., Willemsen, R., Gillemans, N., Galjaard, H. & d'Azzo, A. (1991) A mutation in a mild form of galactosialidosis impairs dimerization of the protective protein and renders it unstable. *EMBO J.* **10**, 4041–4048.

Zhou, X.Y., Morreau, H., Rottier, R., Davis, D., Bonten, E., Gillemans, N., Wenger, D., Grosveld, F.G., Doherty, P., Suzuki, K., Grosveld, G. & d'Azzo, A. (1995) Mouse model for the lysosomal disorder galactosialidosis and correction of the phenotype with overexpressing erythroid precursor cells. *Genes Dev.* **9**, 2623–2634.

*Alexey V. Pshezhetsky*
*Service de Genetique Medicale,*
*Hôpital St-Justine, Universite de Montreal,*
*3175, Côte St-Catherine, Montreal, PQ, H3T 1C5, Canada*
*Email: alex@justine.umontreal.ca*

# 134. Carboxypeptidase D

## Databanks

*Peptidase classification: clan SC, family S10, MEROPS ID: S10.005*
*NC-IUBMB enzyme classification: EC 3.4.16.6*
*Chemical Abstracts Service registry number: 153967-26-1*

*Databank codes:*

| Species | SwissProt | PIR | EMBL (cDNA) | EMBL (genomic) |
|---------|-----------|-----|-------------|----------------|
| *Arabidopsis thaliana* | – | – | F14107 | – |
| *Aspergillus niger* | P52718 | S57907 | – | L33408: complete gene |
| | | | | X79541: complete gene |
| *Hordeum vulgare* | P08818 | A29640 | – | – |
| *Hordeum vulgare* | P52711 | – | X78877 | – |
| *Hordeum vulgare* | – | – | X78878 | – |
| *Penicillium janthinellum* | P34946 | S38953 | – | – |
| *Saccharomyces cerevisiae* | P09620 | A29651 | M17231 | – |
| *Triticum aestivum* | P08819 | A29639 | – | – |

Brookhaven Protein Data Bank three-dimensional structures:

| Species | ID | Resolution | Notes |
|---------|-----|-----------|-------|
| *Saccharomyces cerevisiae* | 1AC5 | 2.4 | |
| *Triticum aestivum* | 1BCR | 2.5 | complex with antipain |
| | 1BCS[a] | 2.1 | complex with chymostatin |
| | 1WHS | 2.3 | |
| | 1WHT | 2 | complex with L-benzylsuccinate |
| | 3SC2 | 2.2 | |

[a]See Chapter 124 for an image constructed from these coordinates.

## Name and History

Carboxypeptidases D comprise a subset of the serine carboxypeptidases. These are serine exopeptidases that release C-terminal amino acids. Originally, they were identified as carboxypeptidases with acidic pH optima (pH 4–5.5) and were named 'acid carboxypeptidases' (Zuber & Matile, 1968). This terminology tended to imply that the enzymes might utilize a mechanism similar to that of the aspartic peptidases, but it was later established that an essential serine is involved, so the enzymes were renamed 'serine carboxypeptidases' (Hayashi *et al.*, 1973). Based on the observation that the serine carboxypeptidases fall into two major groups in terms of substrate specificity (with some overlap), they have been further divided into two divergent classes, carboxypeptidase C (see Chapter 132) and *carboxypeptidase D*. Carboxypeptidase C exhibits a preference for bulky hydrophobic and aromatic side chains on either side of the scissile bond, whereas carboxypeptidase D exhibits a preference for basic groups. Both sets of enzymes have recently been reviewed (Breddam, 1986; Remington, 1993; Remington & Breddam, 1994). The best characterized member of the carboxypeptidase D group is wheat serine carboxypeptidase II, often abbreviated *CPW-II* or *carboxypeptidase WII* in the literature.

## Activity and Specificity

Serine carboxypeptidase activity is conveniently assayed by use of commercially available furylacryloyl-dipeptide substrates. The activity can be followed as decrease in absorbance at 340 nm, which is outside of the range that most proteins absorb. Alternatively, assays may be performed with Z- and Bz- groups, but with these compounds ultraviolet wavelengths must be employed. A general assay for carboxypeptidase D activity might be as follows: 965 μl 0.05 M acetate buffer, pH 4.5, 1 mM EDTA; 25 μl 8 mM furylacryloyl-Ala+Lys dissolved in methanol; 10 μl enzyme solution.

Carboxypeptidase WII shows a pronounced preference for basic amino acids in either the P1′ (leaving) or P1 positions, but peptides containing bulky hydrophobic or aromatic side chains are also efficiently hydrolyzed. Typical $k_{cat}$ and $K_m$ values are in the range of $10\,700\,min^{-1}$ and 0.39 mM for furylacryloyl-Phe-Leu and $8600\,min^{-1}$ and 0.17 mM for furylacryloyl-Ala-Lys, respectively (Breddam *et al.*, 1987). The enzyme will release ammonia from peptide amides, and alcohols from peptides with esterified C-termini. The *Saccharomyces cerevisiae KEX1* gene product is a vacuolar carboxypeptidase D that is essential for the maturation of peptide hormones, such as the $\alpha$-factor mating pheromone (Dmochowska *et al.*, 1987) and the active K1 killer toxin (Wickner & Leibowitz, 1976), by removal of C-terminal Lys-Arg sequences. It is highly specific for its peptide substrates ($K_m$ for $\alpha$-factor-Lys-Arg is 22 μM) and presumably has an extended binding site.

Carboxypeptidase II is inhibited by various compounds that mimic peptide substrates, such as benzylsuccinate ($K_i = 200$ μM: Bullock *et al.*, 1994) and by reagents that react covalently with the reactive serine, such as PMSF. More recently, inhibition by the bacterially produced peptide aldehydes antipain and chymostatin has been studied by X-ray crystallography (Bullock *et al.*, 1996); these also react covalently with the active-site serine. In that study, the inhibition constants were not determined. Whether there are proteins that specifically inhibit carboxypeptidases D is not known, but there are reports of a yeast protein that strongly inhibits yeast carboxypeptidase C (carboxypeptidase Y) (Lenney, 1975).

## Structural Chemistry

Carboxypeptidases D are glycoproteins, typically of 50–70 kDa. Although a number have been identified either by activity or sequence homology, four in particular have been characterized at various levels of detail (the enzyme from barley in the Databanks table is essentially identical to that of wheat). Carboxypeptidase WII is a two-chain enzyme (A: ~260 residues, B: ~160 residues) with two disulfide bridges covalently linking the two chains. Its biosynthesis has not been investigated, but the gene for a related enzyme from barley (carboxypeptidase MI: Doan & Fincher, 1988) has been cloned, and encodes a single polypeptide with a 50 amino acid linker, which is subsequently excised from the middle of the chain. Whether this processing is required for activity is not known. In contrast, the fungal enzymes from *Penicillium janthinellum* (carboxypeptidase S1) (Svendsen *et al.*, 1993) and *Aspergillus niger* (carboxypeptidase AII) (Dal Degan *et al.*, 1992) are single-chain enzymes, each of which shares about 50% sequence identity with carboxypeptidase WII. There is one additional disulfide bond in carboxypeptidase WII and several sites of glycosylation, three of which have been identified by the crystallographic studies. It is not known whether glycosylation or disulfide linkages are important for activity.

The gene for the more distantly related Kex1 enzyme has been cloned and the protein characterized (Dmochowska *et al.*, 1987; Cooper & Bussey, 1989). It codes for a 729 residue protein that contains a 22 residue signal peptide, a catalytic domain of 484 residues, a highly acidic domain that is probably unstructured, and a single membrane-spanning region from residues 619–637. Over all, Kex1 shares about 22% sequence identity with carboxypeptidase WII.

Only one crystal structure is known for a carboxypeptidase D, that of carboxypeptidase WII, and this was determined at 2.2 Å resolution (Liao & Remington, 1990; Liao *et al.*, 1992). The binding of several inhibitors has been investigated (Bullock *et al.*, 1994, 1996). A recombinant fragment expressing the entire catalytic domain of Kex1 has been crystallized (Shilton *et al.*, 1996), but as of this writing the structure has not been published.

Carboxypeptidase WII consists of an 11-stranded, mixed β sheet, the central six strands of which are parallel, surrounded on either side by 14 helices (Figure 134.1). The fold is common to a large family of lipases, esterases and hydrolases known as the 'α/β hydrolases' (Ollis *et al.*, 1992), all of which contain a catalytic triad located on conserved segments of the fold, consisting of nucleophile (Ser, Asp or Cys)–Acid (Glu or Asp)–His. The active site of carboxypeptidase WII is centrally located at the C-terminal end of strand 6, where the nucleophilic Ser146 is located. Asp338 and His397 complete the catalytic triad. The backbone amides of residues 53 and 147 constitute the 'oxyanion hole', which stabilizes the tetrahedral oxyanion that is formed in the transition state. Thus, carboxypeptidases D possess the full complement of catalytic groups found in the other serine peptidases, but the surprising discovery was recently made (Bullock *et al.*, 1996) that the arrangement of these groups is approximately the mirror image of the arrangement of catalytic groups in the families of trypsin (S1) and subtilisin (S8). This suggests that the stereochemical courses of the

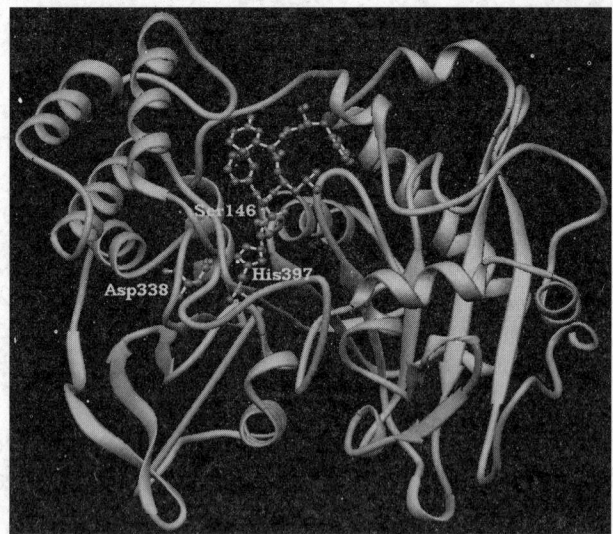

*Figure 134.1*   Schematic ribbon drawing of carboxypeptidase WII containing bound chymostatin. The active-site residues Asp338, His397 and Ser146 are labeled.

two enzymatic reactions are also approximate mirror images of each other.

Carboxypeptidases D also interact with their substrates in a manner unlike that of trypsin and subtilisin. In studying the binding of the tetrapeptide aldehydes antipain and chymostatin, Bullock *et al.* (1996) reported that hydrogen bonds to the peptide substrate backbone were made almost exclusively by side chains on the enzyme, and that other interactions were essentially van der Waals interactions with hydrophobic and aromatic groups. In contrast, both trypsin (Chapter 3) and subtilisin (Chapter 94) bind substrates by forming a short segment of antiparallel β sheet (Segal *et al.*, 1971) with the peptide backbone. The S1 binding site consists of a shallow depression lined by Val340, Val341, Phe215 and Leu216, while the S1' site is a cleft between the parallel aromatic rings of Tyr60 and Tyr239. Each binding site can accommodate either hydrophobic or basic side chains, as the ends of the pockets are exposed to solvent, or are terminated by acidic groups on the enzyme.

## Preparation

Carboxypeptidases D can conveniently be prepared in large quantities from cereal grains by affinity chromatography. Crude preparations can be loaded on to a column containing covalently bound benzylsuccinate ligands (which resemble a C-terminal phenylalanine), and enzyme eluted with benzylsuccinate. Alternatively, ion-exchange chromatography can be used to purify the enzymes. These procedures have been described in detail by Breddam *et al.* (1987). The enzyme does not seem to be available from commercial sources at present.

## Biological Aspects

Carboxypeptidases D have been found exclusively in eukaryotes, and they seem to be ubiquitous in all higher organisms. They may be extracellular, where their function would be to release free amino acids for nutrition, but when intracellular, they generally appear to be compartmentalized (Breddam, 1986), for example in the vacuole of yeast or the lysosomal compartment of human cells. They participate in general protein turnover, but also have more specific functions such as in the maturation of peptide hormones by removal of C-terminal basic residues. Kex1 is the paradigm for this activity, and has been shown to be capable of activating mammalian neuropeptides when heterologously expressed in mammalian cell cultures (Dmochowska *et al*., 1987; Cooper & Bussey, 1989).

## *Distinguishing Features*

A distinctive feature of a carboxypeptidase is its ability to recognize and bind the C-terminal carboxylate of the substrate. This is accomplished in a most unusual manner in serine carboxypeptidases, and the machinery is conserved in all known sequences. As shown by Liao *et al*. (1992), the enzyme recognizes the substrate carboxylate by forming hydrogen bonds with the carboxylate of Glu145, the side-chain amide of Asn51, and a backbone amide. In turn, Glu145 and Asn51 are linked by the carboxylate of Glu65, all buried in a very hydrophobic cavity on the enzyme. It has also been shown that at most one negative charge resides on the network, and further deprotonation of the network rapidly abolishes peptide substrate binding as the pH increases above ~6 (Mortensen *et al*., 1994). This network is also essential for a high level of peptidase activity (Stennicke *et al*., 1996). Although several possible explanations of the way in which this network participates in catalysis have been put forward (Christensen, 1994; Bullock *et al*., 1994, Stennicke *et al*., 1996), no completely satisfactory understanding of its function has been achieved, and further study will be required.

## *Further Reading*

The reader is directed to comprehensive (Breddam, 1986) and less extensive, but more recent (Remington & Breddam, 1994) reviews for additional information.

## References

Breddam, K. (1986) Serine carboxypeptidases. A review. *Carlsberg Res. Commun.* **51**, 83–128.

Breddam, K., Sorensen, S.B. & Svendsen, I. (1987) Primary structure and enzymatic properties of carboxypeptidase II from wheat bran. *Carlsberg Res. Commun.* **52**, 297–311.

Bullock, T.L., Branchaud, B. & Remington, S.J. (1994) Structure of the complex of L-benzylsuccinate with wheat serine carboxypeptidase II at 2.0 Å resolution. *Biochemistry* **33**, 11127–11134.

Bullock, T.L., Breddam, K. & Remington, S.J. (1996) Peptide aldehyde complexes with wheat serine carboxypeptidase. II: implications for the catalytic mechanism and substrate specificity. *J. Mol. Biol.* **255**, 714–725.

Cooper, A. & Bussey, H. (1989) Characterization of the yeast KEX1 gene product: a carboxypeptidase involved in processing secreted precursor proteins. *Mol. Cell. Biol.* **9**, 2706–2714.

Christensen, U. (1994) Effects of pH on carboxypeptidase Y catalyzed hydrolysis and aminolysis reactions. *Eur. J. Biochem.* **220**, 149–153.

Dal Degan, F.D., Ribadeau-Dumas, B. & Breddam, K. (1992) Purification and characterization of two serine carboxypeptidases from *Aspergillus niger* and their use in C-terminal sequencing of proteins and peptide synthesis. *Appl. Environ. Microbiol.* **58**, 2144–2152.

Doan, N.P. & Fincher, G.B. (1988) The A and B chains of carboxypeptidase I from germinated barley originate from a single precursor polypeptide. *J. Biol. Chem.* **263**, 11106–11110.

Dmochowska, A., Dignard, D., Henning, D., Thomas, D.Y. & Bussey, H. (1987) Yeast KEX1 gene encodes a putative protease with a carboxypeptidase B-like function involved in killer toxin and alpha-factor precursor processing. *Cell* **50**, 573–584.

Hayashi, R., Moore, S. & Stein, W.H. (1973) Serine at the active center of yeast carboxypeptidase. *J. Biol. Chem.* **248**, 8366–8369.

Lenney, J.F. (1975) Three yeast proteins that specifically inhibit yeast proteases A, B, and C. *J. Bacteriol.* **122**, 1265–1273.

Liao, D.-I. & Remington, S.J. (1990) Structure of wheat serine carboxypeptidase II at 3.5-Å resolution. A new class of serine proteinase. *J. Biol. Chem.* **265**, 6528–6531.

Liao, D.-I., Breddam, K., Sweet, R.M., Bullock, T. & Remington, S.J. (1992) Refined atomic model of wheat serine carboxypeptidase II at 2.2-Å resolution. *Biochemistry* **31**, 9796–9812.

Mortensen, U.H., Remington, S.J. & Breddam, K. (1994). Site-directed mutagenesis on (serine) carboxypeptidase Y. A hydrogen bond network stabilizes the transition state by interaction with the C-terminal carboxylate group of the substrate. *Biochemistry* **33**, 508–517.

Ollis, D.L., Cheah, E., Cygler, M., Dykstra, B., Frolow, F., Fraken, S., Harel, M., Remington, S.J., Silman, I., Schrag, J., Sussman, J. & Goldman, A. (1992) The alpha/beta hydrolase fold. *Protein Eng.* **5**, 197–211.

Remington, S.J. (1993) Serine carboxypeptidases: a new and versatile family of enzymes. *Curr. Opin. Biotech.* **4**, 462–468.

Remington, S.J. & Breddam, K. (1994) Carboxypeptidases C and D. *Methods Enzymol.* **244**, 231–248.

Shilton, B.H., Yunge, L., Tessier, D., Thomas, D.H. & Cygler, M. (1996) Crytallization of a soluble form of the Kex1p serine carboxypeptidase from *Saccharomyces cerevisiae*. *Protein Sci.* **5**, 395–397.

Segal, D.M., Cohen, G.H., Davies, D.R., Powers, J.C. & Wilcox, P.E. (1971) The stereochemistry of substrate binding to chymotrypsin A. *Cold Spring Harbor Symp. Quant. Biol.* **36**, 85–80.

Stennicke, H.R., Mortensen, U.H. & Breddam, K. (1996) Studies on the hydrolytic properties of (serine) carboxypeptidase Y. *Biochemistry* **35**, 7131–7141.

Svendsen, I., Hofmann, T., Endrizzi, J., Remington, S.J. & Breddam, K. (1993) The primary structure of carboxypeptidase S1 from *Penicillium janthinellum*. *FEBS Lett.* **333**, 39–43.

Wickner, R.B. & Leibowitz, M.J. (1976) Two chromosomal genes required for killing expression in killer strains of *Saccharomyces cerevisiae*. *Genetics* **82**, 420–442.

Zuber, H. & Matile, P.H. (1968) Acid carboxypeptidases: their occurrence in plants, intracellular distribution and possible function. *Z. Naturforsch.* **23b**, 663–665.

*S. James Remington*
*Institute of Molecular Biology,*
*University of Oregon,*
*Eugene, OR 97403-1229, USA*

# 135. Peptidyl-glycinamidase

## Databanks

*Peptidase classification: clan SX, family S99 (see p. 536), MEROPS ID: S9F.001*
*NC-IUBMB enzyme classification: EC 3.4.19.2*
*Databank codes: no sequence data available*

## Name and History

Several apparently distinct enzyme activities have been described which remove the C-terminal glycinamide from oxytocin and vasopressin (Walter & Simmons, 1977). These include (a) a cytosolic enzyme with chymotrypsin-like specificity which cleaves oxytocin but not vasopressin (Koida *et al*., 1971; Fruhaufová *et al*., 1973), (b) a membrane-bound peptidase which cleaves vasopressin (Nardacci *et al*., 1975), and (c) an enzyme from toad urinary bladder and skin which cleaves both oxytocin and vasopressin (Campbell *et al*., 1965; Glass *et al*., 1969). The following names have been used to refer to one or all of these activities: carboxyamidase, carboxamidopeptidase, antidiuretic hormone-inactivating enzyme, peptidyl aminoacylamidase and peptidyl-glycinamidase. Only the enzyme referred to as *carboxamidopeptidase* has been purified (from skin of *Bufo marinus*) and characterized in any detail (Simmons & Walter, 1980, 1981; Simmons, 1981), and the discussion below refers to this enzyme. The peptidase has been recommended to be termed *peptidyl-glycinamidase* (EC 3.4.19.2) by NC-IUBMB.

## Activity and Specificity

Peptidyl-glycinamidase can cleave oxytocin (-Leu8+Gly9-NH$_2$) and vasopressin (-Arg8+Gly9-NH$_2$) as well as other position 8 analogs of these hormones. However, the enzyme is not a strict peptidyl-glycinamidase, but can release other amino acid amides (or ammonia) from a variety of amidated peptides (e.g. substance P: -Leu10+Met11-NH$_2$) and can hydrolyze ester substrates such as Bz-Arg+OEt and Ac-Tyr+OEt. The pH optimum for these substrates is 7.0–8.5. Peptidyl-glycinamidase is also a carboxypeptidase and can release free amino acids from the C-terminus of nonamidated peptides at pH 5.5. The enzyme has a broad specificity for both the P1 and P1' residues. Activity toward all classes of substrates is inhibited by DFP, and the enzyme is also sensitive to thiol-blocking agents (e.g. PCMB).

## Structural Chemistry

Peptidyl-glycinamidase exists as a dimer in the presence of 1 M NaCl. Each subunit consists of two disulfide-bonded polypeptide chains of 28 kDa and 19 kDa. The active site is in the heavy chain, and the enzyme is a glycoprotein.

## Preparation

Peptidyl-glycinamidase has been purified 3800-fold to homogeneity from the skin of the toad (*Bufo marinus*) (Simmons & Walter, 1980).

## Distinguishing Features

Peptidyl-glycinamidase is very similar to members of the carboxypeptidase C family (S10) (EC 3.4.16.5) (Rawlings & Barrett, 1994; Remington & Breddam, 1994; see Chapter 124). In a direct comparison, peptidyl-glycinamidase hydrolyzed amidated peptides more readily than did yeast carboxypeptidase Y (Chapter 132).

## Biological Aspects

Peptidyl-glycinamidase is present in connective tissue of the toad urinary bladder (as opposed to the epithelium) and is present in high concentrations in toad skin. The similarity of the enzyme to carboxypeptidase C-like enzymes suggest that it may have a lysosomal localization, but this has not yet been demonstrated.

## References

Campbell, B.J., Thysen, B. & Chu, F.S. (1965) Peptidase catalyzed hydrolysis of antidiuretic hormone in toad bladder. *Life Sci.* **4**, 2129–2140.

Fruhaufová, L., Suska-Brezezinska, E., Barth, T. & Rychlik, I. (1973) Rat liver enzyme inactivating oxytocin and its deamino-carba analogues. *Coll. Czech. Chem. Commun.* **38**, 2793–2798.

Glass, J.D., Schwartz, I.L. & Walter, R. (1969) Enzymatic inactivation of peptide hormones possessing a C-terminal amide group. *Proc. Natl Acad. Sci. USA* **63**, 1426–1430.

Koida, M., Glass, J.D., Schwartz, I.L. & Walter, R. (1971) Mechanism of inactivation of oxytocin by rat kidney enzymes. *Endocrinology* **88**, 633–643.

Nardacci, N.J., Mukhopadhyay, S. & Campbell, B.J. (1975) Partial purification and characterization of the antidiuretic hormone-inactivating enzyme from renal plasma membranes. *Biochim. Biophys. Acta* **377**, 146–157.

Rawlings, N.D. & Barrett, A.J. (1994) Families of serine peptidases. *Methods Enzymol.* **244**, 19–61.

Remington, S.J. & Breddam, K. (1994) Carboxypeptidases C and D. *Methods Enzymol.* **244**, 231–248.

Simmons, W.H. (1981) Carboxamidopeptidase: substrate specificity and similarity to cathepsin A. *Fedn Proc.* **40**, 1440.

Simmons, W.H. & Walter, R. (1980) Carboxamidopeptidase: purification and characterization of a neurohypophyseal hormone inactivating peptidase from toad skin. *Biochemistry* **19**, 39–48.

Simmons, W.H. & Walter, R. (1981) Enzyme inactivation of oxytocin: properties of carboxamidopeptidase. In: *Neurohypophyseal Peptide Hormones and Other Biologically Active Peptides* (Schlesinger, D.H., ed.). New York: Elsevier North-Holland, pp. 151–165.

Walter, R. & Simmons, W.H. (1977) Metabolism of neurohypophyseal hormones: considerations from a molecular viewpoint. In: *Neurohypophysis* (Moses, A.M. & Share, L., eds). Basel: S. Karger, pp. 167–188.

***William H. Simmons***
*Department of Molecular and Cellular Biochemistry,*
*Loyola University Chicago Stritch School of Medicine,*
*2160 S. First Avenue, Maywood, IL 60153, USA*
*Email: WSIMMON@wpo.it.luc.edu*

# 136. X-Pro dipeptidyl-peptidase

## Databanks

*Peptidase classification: clan SC, family S15, MEROPS ID: S15.001*
*NC-IUBMB enzyme classification: EC 3.4.14.11*
*Databank codes:*

| Species | SwissProt | PIR | EMBL (cDNA) | EMBL (genomic) |
|---|---|---|---|---|
| *Lactobacillus delbrueckii* | P40334 | S32244 | Z14230 | – |
| *Lactobacillus helveticus* | – | – | U22900 | – |
| *Lactobacillus helveticus* | – | – | Z48236 | – |
| *Lactococcus lactis cremoris* | P22093 | A43747 | M58315 | – |
| *Lactococcus lactis lactis* | P22346 | A43748 | M35865 | – |

## Name and History

The presence of **X-Pro dipeptidyl-peptidase** activity in intracellular extracts of 21 strains of lactobacilli and lactococci was first detected by Casey & Meyer (1985) using Gly-Pro-NHMec as substrate. The enzyme, encoded by the *pepX* gene, was subsequently purified from several of these strains (Meyer & Jordi, 1987; Kiefer-Partsch *et al*., 1989; Booth *et al*., 1990; Khalid & Marth, 1990; Zevaco *et al*., 1990; Lloyd & Pritchard, 1991; Yan *et al*., 1991; Habibi-Nafaji & Lee, 1994; Vesanto *et al*., 1995). The *pepX* genes from *Lactococcus lactis* subsp. *lactis* and subsp. *cremoris* were cloned and sequenced simultaneously by Nardi *et al*. (1991) and Mayo *et al*. (1991), respectively. The genes from the lactobacilli were cloned by Meyer-Barton *et al*. (1993), Vesanto *et al*. (1995) and Yüksel & Steele (1996). The *pepX* gene was named according to the standard terminology for peptidases from lactic acid bacteria (Tan *et al*., 1993), and accordingly, the enzyme is commonly known as **PepX**; it is also sometimes called **PepXP**.

## Activity and Specificity

Dipeptides Xaa-Pro are cleaved from the N-termini of peptides (Zevaco *et al*., 1990; Lloyd & Pritchard, 1991), provided that the amino acids in the P2 and P1′ positions are not Pro (Lloyd & Pritchard, 1991; Habibi-Nafaji & Lee, 1994). The enzyme also releases, though less efficiently, Xaa-Ala and Xaa-Gly dipeptides (Khalid & Marth, 1990; Lloyd & Pritchard, 1991). The optimal pH lies in the range 6.0–9.0 (Meyer & Jordi, 1987; Kiefer-Partsch *et al*., 1989; Booth *et al*., 1990; Khalid & Marth, 1990; Zevaco *et al*., 1990; Lloyd & Pritchard, 1991; Yan *et al*., 1991; Habibi-Nafaji & Lee, 1994; Vesanto *et al*., 1995), and the optimal temperature is around 40°C (Meyer & Jordi, 1987; Kiefer-Partsch *et al*., 1989; Khalid & Marth, 1990; Zevaco *et al*., 1990; Yan *et al*., 1991; Habibi-Nafaji & Lee, 1994; Vesanto *et al*., 1995). PepX is a serine peptidase inhibited by DFP and PMSF. Some enzymes are inhibited by divalent cations such as $Hg^{2+}$ and $Cu^{2+}$ (Meyer & Jordi, 1987; Yan *et al*., 1991; Habibi-Nafaji & Lee, 1994; Vesanto *et al*., 1995).

## Structural Chemistry

The variants of PepX are generally dimeric enzymes (Meyer & Jordi, 1987; Kiefer-Partsch *et al.*, 1989; Booth *et al.*, 1990; Khalid & Marth, 1990; Zevaco *et al.*, 1990; Lloyd & Pritchard, 1991) without disulfide bridges. The molecular mass of the monomer is 72–90 kDa. One monomeric form (Atlan *et al.*, 1990) and one trimeric form (Miyakawa *et al.*, 1991) have been described. The pI of PepX is 4.5–4.7 (Meyer & Jordi, 1987; Habibi-Nafaji & Lee, 1994). The sequence surrounding the active-site serine was identified (Chich *et al.*, 1992), and differs from the known consensus sequences of other serine peptidases (see Chapter 124).

PepX from *Lactococcus lactis* has been crystallized (Chich *et al.*, 1995) and its three-dimensional structure is being resolved. The monomer is composed of two domains linked by a single polypeptide chain. The N-terminal domain (one-sixth of the molecule) is folded in α helices, whereas the other domain, which contains the active site, is mainly folded in β sheets, with a remarkable 10-stranded β sheet pattern (unpublished results of the author).

## Preparation

Strain variants of PepX have mainly been prepared from intracellular extracts of the lactic acid bacteria (Meyer & Jordi, 1987; Booth *et al.*, 1990; Khalid & Marth, 1990; Zevaco *et al.*, 1990; Lloyd & Pritchard, 1991; Yan *et al.*, 1991; Habibi-Nafaji & Lee, 1994; Vesanto *et al.*, 1995). The PepXs from *L. lactis* subsp. *lactis* (Chich *et al.*, 1995) and from *Lactobacillus helveticus* (Vesanto *et al.*, 1995) were prepared as recombinant enzymes from *E. coli*.

## Biological Aspects

It has been shown that PepXs are able to hydrolyze sequentially peptides derived from β-casein, including β-casomorphin (Zevaco *et al.*, 1990; Lloyd & Pritchard, 1991), and it is therefore believed that these enzymes play a role in the degradation of caseins. According to Atlan *et al.* (1990) and Yüksel & Steele (1996), PepX is involved in the casein-degradation pathway, providing essential amino acids to the lactobacilli. In contrast, Mayo *et al.* (1993) have shown that PepX is not essential for the growth of lactococci in milk, but affects the peptide composition of fermented milk products. The same observation was made by Mierau *et al.* (1996), who have studied the growth in milk of multiple-peptidase mutants of *Lactococcus lactis*. Thus, the roles of PepX in the proteolytic systems of lactobacilli and lactococci are probably different.

PepX has been used for kinetically controlled peptide synthesis of proline-containing peptides (Yoshpe-Besançon *et al.*, 1994; Houbart *et al.*, 1995).

## Distinguishing Features

No specific inhibitor is available for PepX. The catalytic activity is similar to that of dipeptidyl-peptidase IV (Chapter 128); the enzymes are both serine peptidases in clan SC, but differ enough in amino acid sequence to be placed in different families. As far as is known at present, PepX is confined to the lactic acid bacteria.

## References

Atlan, D., Laloi, P. & Portalier, R. (1990) X-prolyl-dipeptidyl aminopeptidase of *Lactobacillus delbrueckii* subsp. *bulgaricus*: characterization of the enzyme and isolation of deficient mutants. *Appl. Environ. Microbiol.* **56**, 2174–2179.

Booth, M., Fhaolain, I.N., Jennings, P.V. & O'Cuinn, G. (1990) Purification and characterization of a post-proline dipeptidyl aminopeptidase from *Streptococcus cremoris* AM2. *J. Dairy Res.* **57**, 89–99.

Casey, M.G. & Meyer, J. (1985) Presence of X-prolyl-dipeptidyl-peptidase in lactic acid bacteria. *J. Dairy Sci.* **68**, 3212–3215.

Chich, J.-F., Chapot-Chartier, M.-P., Ribadeau-Dumas, B. & Gripon, J.-C. (1992) Identification of the active site serine of the X-prolyl dipeptidyl aminopeptidase from *Lactococcus lactis*. *FEBS Lett.* **314**, 139–142.

Chich, J.-F., Rigolet, P., Nardi, M., Gripon, J.-C., Ribadeau-Dumas, L. & Brunie, S. (1995) Purification, crystallization and preliminary X-ray analysis of PepX, an X-prolyl dipeptidyl aminopeptidase from *Lactococcus lactis*. *Proteins* **23**, 278–281.

Habibi-Nafaji, M.B. & Lee, B.H. (1994) Purification and characterization of X-prolyl-dipeptidyl-aminopeptidase from *Lactobacillus casei* subsp. *casei* LLG. *Appl. Microbiol. Biotechnol.* **42**, 280–286.

Houbart, V., Ribadeau-Dumas, B. & Chich, J.-F. (1995) Synthesis of enterostatin-amide by the Xaa-prolyl dipeptidyl aminopeptidase from *Lactococcus lactis* subsp. lactis NCDO 763. *Biotechnol. Appl. Biochem.* **21**, 149–159.

Khalid, N.M. & Marth, E.H. (1990) Purification and partial characterization of a prolyl-dipeptidyl-aminopeptidase from *Lactobacillus helveticus* CNRZ 32. *Appl. Environ. Microbiol.* **56**, 381–388.

Kiefer-Partsch, B., Bockelmann, W., Geis, A. & Teuber, M. (1989) Purification of an X-prolyl-dipeptidyl aminopeptidase from *Lactococcus lactis* subsp. *cremoris*. *Appl. Microbiol. Biotechnol.* **31**, 75–78.

Lloyd, R.J. & Pritchard, G.G. (1991) Characterization of X-prolyl dipeptidyl aminopeptidase from *Lactococcus lactis* subsp. *lactis*. *J. Gen. Microbiol.* **137**, 49–55.

Mayo, B., Kok, J., Venema, K., Bockelmann, W., Teuber, M., Reinke, H. & Venema, G. (1991) Molecular cloning and sequence analysis of the X-prolyl dipeptidyl aminopeptidase gene from *Lactococcus lactis* subsp. *cremoris*. *Appl. Environ. Microbiol.* **57**, 38–44.

Mayo, B., Kok, J., Bockelmann, W., Haandrikman, A., Leenhouts, K.J. & Venema, G. (1993) Effect of X-prolyl dipeptidyl aminopeptidase deficiency on *Lactococcus lactis*. *Appl. Environ. Microbiol.* **59**, 2049–2055.

Meyer, J. & Jordi, R. (1987) Purification and characterization of X-prolyl-dipeptidyl-aminopeptidase from *Lactobacillus lactis* and from *Streptococcus thermophilus*. *J. Dairy Sci.* **70**, 738–745.

Meyer-Barton, E.C., Klein, J.R., Imam, M. & Plapp, R. (1993) Cloning and sequence analysis of the X-prolyl-dipeptidyl-aminopeptidase gene (pepX) from *Lactobacillus delbrueckii* spp. *lactis* DSM7290. *Appl. Microbiol. Biotechnol.* **40**, 82–89.

Mierau, I., Kunji, E.R.S., Leenhouts, K.J., Hellendorn, M.A., Haandrikman, A.J., Poolman, B. Konings, W.N., Venema, G. & Kok, J. (1996) Multiple peptidase mutants of *Lactococcus lactis* are severely impaired in their ability to grow in milk. *J. Bact.* **178**, 2794–2803.

Miyakawa, H., Kobayashi, S., Shimamura, S. & Tomita, M. (1991) Purification and characterization of an X-prolyl dipeptidyl aminopeptidase from *Lactobacillus delbrueckii* spp. *bulgaricus* LBU-147. *J. Dairy Sci.* **74**, 2375–2381.

Nardi, M., Chopin, M.-C., Chopin, A., Cals, M.-M. & Gripon, J.-C. (1991) Cloning and DNA sequence analysis of an X-prolyl dipeptidyl aminopeptidase gene from *Lactococcus lactis* subsp. *lactis* NCDO 763. *Appl. Environ. Microbiol.* **57**, 45–50.

Tan, P.S.T., Poolman, B. & Konings, W.N. (1993) Proteolytic enzymes of *Lactococcus lactis*. *J. Dairy Res.* **60**, 269–286.

Vesanto, E., Savijoki, K., Rantanen, T., Steele, J.L. & Palva, A. (1995) Molecular characterization, heterologous expression and purification of an X-prolyl-dipeptidyl aminopeptidase (*pepX*) gene from *Lactobacillus helveticus*. *Microbiology* **141**, 3067–3075.

Yan, T.-R., Lin, M.-Z., Lin, M.-J. & Sun, B.J. (1991) Purification and characterization of an X-prolyl-dipeptidyl-aminopeptidase from *Streptococcus cremoris* nTR. *J. Chinese Biochem. Soc.* **20**, 21–32.

Yoshpe-Besançon, I., Gripon, J.-C. & Ribadeau-Dumas, B. (1994) Xaa-Pro-dipeptidyl-aminopeptidase from *Lactococcus lactis* catalyses kinetically controlled synthesis of peptide bonds involving proline. *Biotechnol. Appl. Biochem.* **20**, 131–140.

Yüksel, G.Ü. & Steele, J.L. (1996) DNA sequence analysis, expression, distribution and physiological role of the Xaa-prolyldipeptidyl aminopeptidase gene from *Lactobacillus helveticus* CNRZ32. *Appl. Microbiol. Biotechnol.* **44**, 766–773.

Zevaco, C., Monnet, V. & Gripon, J.-C. (1990) Intracellular X-prolyl dipeptidyl peptidase from *Lactococcus lactis* subsp. *lactis*: purification and properties. *J. Appl. Bacteriol* **68**, 357–366.

*Jean-François Chich*
*I.N.R.A., Laboratoire de Biochimie et Structure des Protéines,*
*Unité Enzymologie, Domaine de Vilvert,*
*78352 Jouy-en-Josas Cedex, France*
*Email: chich@jouy.inra.fr*

# *137. Lysosomal Pro-X carboxypeptidase*

## Databanks

*Peptidase classification: clan SC, family S28, MEROPS ID: S28.001*
*NC-IUBMB enzyme classification: EC 3.4.16.2*
*Chemical Abstracts Service registry number: 9075-64-3*
*Databank codes:*

| Species | SwissProt | PIR | EMBL (cDNA) | EMBL (genomic) |
|---|---|---|---|---|
| *Arabidopsis thaliana* | – | – | T43129 | – |
| *Arabidopsis thaliana* | – | – | T43129 | – |
| *Caenorhabditis elegans* | P34676 | S44916 | L16621 | – |
| *Homo sapiens* | P42785 | A47352 | L13977 | – |
| *Mus musculus* | – | – | W49039 | – |

## Name and History

*Lysosomal Pro-X carboxypeptidase (prolylcarboxypeptidase, peptidyl prolylamino acid hydrolase)* was discovered when researchers studying the degradation of bradykinin noticed that a pig kidney extract unexpectedly cleaved (des-Arg9)-bradykinin at the -Pro7↓Phe8-OH bond (Yang *et al.*, 1968). Because angiotensins II and III [(des-Asp1) angiotensin II)] have the same C-terminus (-Pro-Phe) and are hydrolyzed by the enzyme, it was first named *angiotensinase C* (Yang *et al.*, 1968, 1970), but since the enzyme cleaves C-terminal amino acids linked to a penultimate Pro in several peptides, the name *prolylcarboxypeptidase (PCP)* was suggested and has been used since 1970 (Yang & Erdös, 1970). The name lysosomal Pro-X carboxypeptidase recommended by NC-IUBMB emphasizes the lysosomal location of the enzyme. We will refer to the enzyme here as *Pro-X carboxypeptidase*.

## Activity and Specificity

Pro-X carboxypeptidase cleaves C-terminal amino acids from peptides with the general structure R1-Pro↓R2, where R1 is either a blocking group, another protected amino acid, or a peptide, and R2 is an aromatic or aliphatic amino acid with a free carboxyl group (Yang & Erdös, 1970). With protected dipeptide substrates, the rate of hydrolysis of Z-Pro┼Ala, Z-Pro┼Val and Z-Pro┼Leu was several-fold higher than that of Z-Pro-Phe ($0.52 \mu mol\,min^{-1}\,mg^{-1}$). Serine or tyrosine were not released faster than phenylalanine, whereas the hydrolysis rate of Z-Pro-Gly was only 6% of that of Z-Pro-Phe. Although the bioactive peptides angiotensin II and angiotensin III have the same C-terminus as Z-Pro-Phe, angiotensins II and III were cleaved 4- and 13-fold faster than Z-Pro-Phe. The enzyme did not hydrolyze either Z-Pro-Pro or Z-Pro-hydroxyproline, or substrates of carboxypeptidases A (Bz-Gly-Phe), B (Bz-Gly-Arg) or C (Z-Phe-Pro) (Odya et al., 1978).

The optimal pH for hydrolysis of angiotensins II and III, and Z-Pro-Phe, is about 5, but at pH 7, the enzyme retains 20–50% of the maximal activity with the physiologically active substrates. Thus, like another serine carboxypeptidase, lysosomal carboxypeptidase A (Chapter 133) (Jackman et al., 1990), it has a broader pH optimum with oligopeptides than with shorter peptide substrates (Odya et al., 1978).

Pro-X carboxypeptidase activity can be determined by the release of $[^3H]$Ala from Z-Pro-$[^3H]$Ala as described by Skidgel et al. (1981); the substrate is synthesized from Z-Pro and $[^3H]$Ala, but is not yet commercially available.

As a serine peptidase, Pro-X carboxypeptidase is completely inhibited by 1 mM DFP and PMSF. Pepstatin inhibits the enzyme, but at a concentration of 0.3 mM. Z-Pro-prolinal, sometimes thought of as a specific noncompetitive inhibitor of prolyl oligopeptidase (Chapter 125) (Wilk, 1983), inhibits 50% of the hydrolysis of Z-Pro-$[^3H]$Ala (10 mM) by Pro-X carboxypeptidase at a concentration of $8 \times 10^{-7}$ M (Tan et al., 1993). The compounds p-chloromercuribenzoic acid and p-chloromercuriphenyl sulfonic acid ($10^{-3}$ M) inhibit only 15%. Other activators or inhibitors of lysosomal enzymes such as DTT, 2-mercaptoethanol, iodoacetic acid and iodoacetamide are ineffective at 1 mM concentration.

## Structural Chemistry

Pro-X carboxypeptidase is single-chain glycoprotein of 451 residues with a molecular mass of 58 kDa, containing six potential N-glycosylation sites (Tan et al., 1993). The native enzyme exists as a homodimer. The carbohydrate content is about 12%. The 45 amino acid sequence upstream from the N-terminal lysine of the mature protein contains a signal peptide and propeptide (Tan et al., 1993). The inactive proenzyme can be activated by chymotrypsin or trypsin (unpublished results of the author). Pro-X carboxypeptidase is a serine carboxypeptidase (Breddam, 1986). Although the overall sequence identity with other serine carboxypeptidases is low (10–18%), sequences around the residues of the putative catalytic triad (Ser134, Asp333, His411) are similar in both the serine carboxypeptidase family (e.g. deamidase or lysosomal protective protein, yeast carboxypeptidase Y, and KEX1 gene product) and the prolyl oligopeptidase family (Tan et al., 1993). Thus, Pro-X carboxypeptidase links these two families, suggesting an evolutionary relationship (Tan et al., 1993). Pro-X carboxypeptidase contains Ser or Thr repeated as the 26th residue 7 out of 9 times, with identical or similar amino acids in other positions in the repeats. The

KEX1 gene product contains a similar motif, with Ser or Thr as every 27th residue.

## Preparation

Pro-X carboxypeptidase was purified 1200–2000-fold from human kidney cortex (Odya et al., 1978, 1981). It has been expressed in High-5 insect cells by using recombinant baculovirus; the proenzyme was secreted into the medium and, after partial purification, was activated (unpublished results of the author).

## Biological Aspects

The enzyme has been found in the lysosomal fraction of human and pig kidney (Yang et al., 1968, 1970; Odya et al., 1978; Odya & Erdös, 1981). It has also been found in human granulocytes and lymphocytes (Kumamoto et al., 1981), urine (Sorrells & Erdös, 1972; Skidgel et al., 1981; Yang et al., 1970), synovial fluid (Sorrells & Erdös, 1972), fibroblasts (Kumamoto et al., 1981), and alveolar macrophages (Jackman et al., 1995). In addition, it occurs in organs of other animals such as rat (Kumamoto et al., 1981) and dog (Sorrells et al., 1972). Vascular endothelial cells are also rich in Pro-X carboxypeptidase (Skidgel et al., 1981).

A single 2.4 kb transcript of the enzyme has been found in the mRNAs of human brain, heart, placenta, lung, liver, skeletal muscle, kidney and pancreas. The lung, liver and placenta are relatively the richest in this mRNA, but the brain and heart also contain substantial amounts. The relative mRNA level in the kidney was somewhat lower than expected from the results of enzyme assays (Tan et al., 1993).

The presence of the enzyme in, and its release from, a variety of cells and organs, and the cleavage of the Pro-X bond that is usually resistant to carboxypeptidases, suggest several possible functions. For example, the localization of the enzyme in leukocytes, macrophages (Kumamoto et al., 1981; Jackman et al., 1995) and synovial fluid (Sorrells & Erdös, 1972) points to its possible participation in the breakdown of collagen fragments in inflammation. The enzyme is also highly concentrated in endothelial cells grown in culture, for example from the umbilical vessels (Skidgel et al., 1981). Here, it might inactivate the potent vasoconstrictor angiotensin II that can decrease the umbilical blood flow. One of the naturally occurring peptide substrates of the enzyme, (des-Arg9)-bradykinin, is a product of the action of kininase I-type enzymes (carboxypeptidases N or M, or deamidase) on bradykinin (Jackman et al., 1990; Erdös & Skidgel, 1996). (Des-Arg9)-bradykinin acts on a different receptor (B1) than intact bradykinin (B2), and endotoxin upregulates B1 receptors (Bhoola et al., 1992). Pro-X carboxypeptidase inactivates (des-Arg9)-bradykinin by cleaving the Pro7┼Phe8 bond. As the enzyme is released into blood during shock, it may modulate the activity of this peptide, for example, in septic shock (Sorrells et al., 1972; Sorrells & Erdös, 1972). The hydrolysis of the Pro7-Phe8 bond of angiotensin II negates most, but not all, of the effects of this peptide. Angiotensin(1–7) has a variety of actions; for example, it releases vasopressin, prostaglandins and catecholamines and is present in hypophysial-portal plasma and potentiates the actions of bradykinin in vivo (Ferrario, 1992; Handa et al.,

1996). The release of this heptapeptide and a tripeptide from angiotensin I by brain extracts was attributed mainly to prolyl oligopeptidase and to neprilysin (Chapter 362) (Gafford *et al*., 1983), based on inhibition of the former by Z-Proprolinal. Alternatively, the sequential release of Leu10 and His9 from angiotensin I by other enzymes such as deamidase or cathepsin A (Jackman *et al*., 1990), followed by the liberation of Phe8 by Pro-X carboxypeptidase, can also result in formation of angiotensin(1–7).

## Distinguishing Features

Pro-X carboxypeptidase can be distinguished from other carboxypeptidases by its specificity for the Pro┼Xaa bond, its acidic pH optimum and localization in lysosomes. It is also the only known member of the family S28 (Barrett, 1995; but see also Rawlings & Barrett, 1996). Polyclonal antiserum against the human enzyme was described (Kumamoto *et al*., 1981), but is not commercially available at the time of writing.

## References

Barrett, A.J. (1995) Classification of peptidases. *Methods Enzymol.* **244**, 1–15.

Bhoola, K.D., Figueroa, C.D. & Worthy, K. (1992) Bioregulation of kinins: kallikreins, kininogens and kininases. *Pharmacol. Rev.* **44**, 1–22.

Breddam, K. (1986) Serine carboxypeptidases. A review. *Carlsberg Res. Commun.* **51**, 83–128.

Erdös, E.G. & Skidgel, R.A.S. (1996) Metabolism of bradykinin by peptidases in health and disease. In: *The Kinin System.* Series: *Handbook of Immunopharmacology* (Farmer, S.G., ed.). London: Academic Press pp. 111–141.

Ferrario, C.M. (1992) Biological roles of angiotensin-(1–7). *Hypertens. Res.* **15**, 61–66.

Gafford, J.T., Skidgel, R.A., Erdös, E.G. & Hersh, L.B. (1983) Human kidney 'enkephalinase', a neutral metalloendo-peptidase that cleaves active peptides. *Biochemistry* **22**, 3265–3271.

Handa, R.K., Ferrario, C.M. & Strandhoy, J.W. (1996) Renal actions of angiotensin-(1–7): *in vivo* and *in vitro* studies. *Am. J. Physiol.* **39**, F141–F147.

Jackman, H.L., Tan, F., Tamei, H., Beurling-Harbury, C., Li, X-Y., Skidgel, R.A. & Erdös, E.G. (1990) A peptidase in human platelets that deamidates tachykinins: probable identity with the lysosomal 'protective protein'. *J. Biol. Chem.* **265**, 11265–11272.

Jackman, H.L., Tan, F., Schraufnagel, D., Dragovic, T., Dezsö, B., Becker, R.P. & Erdös, E.G. (1995) Plasma membrane-bound and lysosomal peptidases in human alveolar macrophages. *Am. J. Respir. Cell Mol. Biol.* **13**, 196–204.

Kumamoto, K., Stewart, T.A., Johnson, A.R. & Erdös, E.G. (1981) Prolylcarboxypeptidase (angiotensinase C) in human cultured cells. *J. Clin. Invest.* **67**, 210–215.

Odya, C.E. & Erdös, E.G. (1981) Human prolylcarboxypeptidase. *Methods Enzymol.* **80**, 460–466.

Odya, C.E., Marinkovic, D., Hammon, K.J., Stewart, T.A. & Erdös, E.G. (1978) Purification and properties of prolylcarboxypeptidase (angiotensinase C) from human kidney. *J. Biol. Chem.* **253**, 5927–5931.

Rawlings, N.D. & Barrett, A.J. (1996) Dipeptidyl-peptidase II is related to lysosomal Pro-X carboxypeptidase. *Biochim. Biophys. Acta* **1298**, 1–3.

Sorrells, K. & Erdös, E.G. (1972) Prolylcarboxypeptidase in biological fluids. In: *The Fundamental Mechanisms of Shock* (Hinshaw, L.B. & Cox, B.G., eds). New York: Plenum Press, pp. 393–397.

Sorrells, K., Erdös, E.G. & Massion, W.H. (1972) Effect of prostaglandin E1 on the pulmonary vascular response to endotoxin. *Proc. Soc. Exp. Biol. Med.* **140**, 310–313.

Skidgel, R.A., Wickstrom, E., Kumamoto, K. & Erdös, E.G. (1981) Rapid radioassay for prolylcarboxypeptidase (angiotensinase C). *Anal. Biochem.* **118**, 113–119.

Tan, F., Morris, P.W., Skidgel, R.A. & Erdös, E.G. (1993) Sequencing and cloning of human prolylcarboxypeptidase (angiotensinase C): similarity to both serine carboxypeptidase and prolylendopeptidase families. *J. Biol. Chem.* **268**, 16631–16638.

Wilk, S. (1983) Prolyl endopeptidase. *Life Sci.* **33**, 2149–2157.

Yang, H.Y.T. & Erdös, E.G. (1970) Prolylcarboxypeptidase: a recently described lysosomal enzyme. *Excerpta Medical International Congress: Immunopathology of Inflammation* **229**, 146–148.

Yang, H.Y.T., Erdös, E.G. & Chiang, T.S. (1968) New enzymatic route for the inactivation of angiotensin. *Nature* **218**, 1224–1226.

Yang, H.Y.T., Erdös, E.G., Chiang, T.S., Jenssen, T.A. & Rodgers, J.G. (1970) Characterization of an enzyme that inactivates angiotensin II (angiotensinase C). *Biochem. Pharmacol.* **19**, 1201–1211.

*Fulong Tan*
*Departments of Pharmacology and Anesthesiology,*
*University of Illinois College of Medicine at Chicago,*
*835 South Wolcott Avenue, M/C 868,*
*Chicago, IL 60612-7344, USA*

*Ervin G. Erdös*
*Departments of Pharmacology and Anesthesiology,*
*University of Illinois College of Medicine at Chicago,*
*835 South Wolcott Avenue, M/C 868,*
*Chicago, IL 60612-7344, USA*
*Email: EGErdos@uic.edu*

# 138. Dipeptidyl-peptidase II

## Databanks

*Peptidase classification: clan SC, family S28, MEROPS ID: S28.002*
*NC-IUBMB enzyme classification: EC 3.4.14.2*
*Databank codes: no sequence data available*

## Name and History

This enzyme was originally detected in a side fraction generated during the purification of dipeptidyl-peptidase I (EC 3.4.14.1) from cattle pituitary glands. Its activity was initially distinguished and characterized by use of dipeptidyl-NHNap substrates, and was thus referred to as *dipeptidyl arylamidase II* (McDonald *et al.*, 1966, 1968a). Following the characterization of the purified enzyme on peptide substrates, its name was changed to *dipeptidyl aminopeptidase II* (McDonald *et al.*, 1968b). The name *dipeptidyl peptidase II* was recommended by NC-IUB in 1978 when it was included as the second member of the sub-subclass (3.4.14) known systematically as the dipeptidylpeptide hydrolases. In 1992, IUBMB extended the sub-subclass to include the tripeptidylpeptide hydrolases and recommended the hyphenated name *dipeptidyl-peptidase II* (abbreviated here *DPP II*) for EC 3.4.14.2.

In an early paper by Stein *et al.* (1968), an acid carboxypeptidase isolated from cattle spleen was described that acted preferentially on tripeptides to release the C-terminal amino acid. They named the enzyme *carboxytripeptidase*, but its properties strongly suggest that it was the same enzyme as dipeptidyl-peptidase II. A similar enzyme purified from rat skin has also been characterized as a carboxytripeptidase (Hopsu-Havu *et al.*, 1970).

## Activity and Specificity

DPP II releases N-terminal dipeptides from oligopeptides, most especially tripeptides, and from 2-naphthylamides, amides and methyl esters of dipeptides, provided their N-termini are unsubstituted (McDonald *et al.*, 1968a,b). Virtually any residue may reside at the terminal (P2) position, although acidic ones are the least favorable. In general, Ala and Pro are the preferred residues in the P1 position.

Rates of hydrolysis of naphthylamides by cattle pituitary DPP II are highest for derivatives of Lys-Ala-NHNap and Lys-Pro-NHNap. Some relative rates are: Lys-Ala+ 100, Lys-Pro+ 100, Phe-Pro- 75, Arg-Pro- 62, Ala-Pro- 58, Arg-Ala- 30, Leu-Ala- 10, Ala-Ala- 7, and Gly-Pro- 5 (McDonald *et al.*, 1968b). A similar order of specificity is exhibited by DPP II derived from a cattle fibroblast-rich (dental pulp) connective tissue (McDonald & Schwabe, 1980). Some species show an altered order of specificity on naphthylamides. Relative rates for the enzyme from rat skin are Leu-Ala+ 117, Lys-Ala+ 100, Ala-Ala- 39, Gly-Pro- 29 (Hopsu-Havu *et al.*, 1970). Relative rates for a novel DPP II purified from pig ovaries are: Phe-Pro+ 700, Lys-Pro+

300, Arg-Pro- 180, Lys-Ala- 100, Arg-Ala- 73, Gly-Pro- 20 (Eisenhauer & McDonald, 1986). A similar marked preference for substrates containing a P1 prolyl residue is displayed by DPP II isolated from another pig source (seminal plasma). The rate on Gly-Pro-NHMec is double that for Lys-Ala-NHMec (Huang *et al.*, 1996).

Relative rates of hydrolysis for *p*-nitroanilides by rat kidney DPP II are: Ala-Ala- 100, Gly-Pro- 94, Lys-Pro- 66, and Glu-Pro- 60 (Fukasawa *et al.*, 1983).

Action on peptides by DPP II from the aforementioned sources is virtually limited to tripeptides, with rates usually highest on Ala-Ala-Ala. Relative rates on a wide range of tripeptides have been summarized and compared elsewhere (McDonald & Barrett, 1986). By way of contrast, a nonlysosomal, membrane-associated species of DPP II in rat brain ('DPP II-M') is capable of releasing X-Pro dipeptides from a range of oligopeptides at pH 5.5 (Mentlein & Struckhoff, 1989). Examples include Arg-Pro and Lys-Pro from substance P (an 11 residue peptide), and Tyr-Pro and Phe-Pro from casomorphin (a pentapeptide). Tripeptides such as Pro-Pro-Gly, His-Pro-Val and Gly-Hyp-Ala are also cleaved. In contrast to DPP IV (Chapter 128), a membrane-bound peptidase that is unable to cleave the Pro-Pro bond (Kato *et al.*, 1978; Püschel *et al.*, 1982), DPP II-M from rat brain cleaves this bond, as shown for Pro-Pro+Pro and for N-terminal fragments of bradykinin, e.g. Arg-Pro-Pro and Arg-Pro+Pro-Gly-Phe. However, action on intact bradykinin, a nine residue peptide, is trivial. X-Pro sequences at the N-termini of larger molecules, e.g. prolactin and aprotinin, are resistant to the action of DPP II-M (Mentlein & Struckhoff, 1989). A species of DPP II ('DPP II-S') present in a soluble fraction of rat brain has a substrate specificity that is similar to that of DPP II-M, and also to that reported for a dipeptidyl-peptidase purified from rat brain by Imai *et al.* (1983). Two dipeptidyl-peptidase species displaying postproline cleaving activity at acidic pH occur in the human cerebral cortex (Kato *et al.*, 1980). The catalytic activity of one species ('DPP-A') is characteristic of cattle pituitary DPP II, while that of the other ('DPP-B') is anomalous. The latter is capable of liberating Arg-Pro and Lys-Pro sequentially, at pH 5.2, from substance P – a characteristic that is shared with DPP II-M of rat brain.

Assays of activity may be performed at pH 5.5 with a fluorogenic substrate such as Lys-Ala+NHNap (McDonald *et al.*, 1968a) or Lys-Ala+NHMec (Nagatsu *et al.*, 1985). When DPP II is assayed in impure biological fluids, 1,10-phenanthroline (1 mM) can be employed to block aminopeptidase action without affecting DPP II activity (Nagatsu *et al.*, 1985; Rao *et al.*, 1990). *p*-Nitroanilide derivatives have been employed in colorimetric procedures (Imai *et al.*, 1983; Mentlein & Struckhoff, 1989). Assays may also be performed on a tripeptide substrate such as Ala-Ala-Ala, but at pH 4.5. The liberated C-terminal amino acid is detected colorimetrically with a trinitrobenzene sulfonic acid reagent (Okuyama & Satake, 1960) to which copper is added, greatly increasing sensitivity by preventing reaction with peptides (McDonald *et al.*, 1968b).

DPP II from all sources is sensitive to classical inhibitors of the serine catalytic class of peptidases, e.g. DFP, PMSF, and *p'*-nitrophenyl-*p*-guanidinobenzoate. It is uniquely sensitive to cations, especially ones of high molecular mass, e.g. Tris, puromycin and Hyamine 10-X, all of which inhibit competitively (McDonald *et al.*, 1968a). An active-site-directed inhibitor, Lys-Ala-CH$_2$Cl, provides specific, irreversible inhibition (McDonald & Schwabe, 1980). The aminopeptidase inhibitors bestatin and bacitracin are ineffective (Mentlein & Struckhoff, 1989).

## Structural Chemistry

The molecular mass of native DPP II purified from a variety of sources varies considerably. Values range from about 100 kDa for the enzyme in pig ovary (Eisenhauer & McDonald, 1986), human kidney (Sakai *et al.*, 1987), pig spleen (Lynn, 1991), and rat brain (Mentlein & Struckhoff, 1989) to 185 kDa for the enzyme in pig seminal plasma (Huang *et al.*, 1996). Except for the latter, which is trimeric, all are comprised of two identical protein chains that are noncovalently associated. The subunit structure of the 150 kDa enzyme from human kidney was not reported (Mantle, 1991). DPP II from the cattle pituitary gland (McDonald *et al.*, 1968b), rat kidney (Fukasawa *et al.*, 1983), and guinea pig testes (DiCarlantonio *et al.*, 1986) are representative, with $M_r$ values of about 130 000 and pI values of 4.8–5.0. The native enzyme is typically a glycoprotein containing mannose and glucosamine, but having little or no sialic acid (Eisenhauer & McDonald, 1986; Mentlein & Struckhoff, 1989; Lynn, 1991). The amino acid composition of the pig spleen enzyme has been reported (Lynn, 1991), as has the sequence of the first 41 residues (Huang *et al.*, 1996). On the basis of the N-terminal amino acid sequence, Rawlings & Barrett (1996) have shown that dipeptidyl-peptidase II is homologous with lysosomal Pro-X carboxypeptidase, and thus belongs to peptidase family S28.

## Preparation

As a lysosomal peptidase, DPP II can generally be prepared in acceptable yields from lysosome-rich tissues. Essentially homogeneous preparations have been obtained from the following sources, with the corresponding degree of purification as shown: cattle pituitary glands, 1000-fold (McDonald *et al.*, 1968b); rat brain, 2600-fold (Imai *et al.*, 1983); rat kidney, 3740-fold (Fukasawa *et al.*, 1983); pig ovary, 1400-fold (Eisenhauer & McDonald, 1986); rat brain, 7700-fold (Mentlein & Struckhoff, 1989); pig spleen, 6300-fold (Lynn, 1991); and pig seminal plasma, 1700-fold (Huang *et al.*, 1996). A method has been described by Sakai *et al.* (1987) for the rapid, 5000-fold purification of DPP II from human kidney.

Partially purified preparations have been obtained from rat skin (Hopsu-Havu *et al.*, 1970), human brain (Kato *et al.*, 1980), cattle dental pulp (McDonald & Schwabe, 1980), guinea pig testis (DiCarlantonio *et al.*, 1986), cattle adrenal medulla (Kecorius *et al.*, 1988), and human kidney (Mantle, 1991).

## Biological Aspects

The lysosomal distribution of DPP II in tissues (McDonald *et al.*, 1968a, 1971) and its manifest preferential action on tripeptides, with possible significant action on some small oligopeptides (Mentlein & Struckhoff, 1989), strongly support a role for DPP II in the terminal stages of intracellular protein degradation. Most tripeptides acted upon by DPP II are probably generated by tripeptidyl-peptidase I (Chapter 183), a lysosomal peptidase (McDonald *et al.*, 1985) of broad specificity that is capable of releasing tripeptides sequentially from large polypeptides (Doebber *et al.*, 1978). The complementary specificities displayed by DPP II and tripeptidyl-peptidase I provide the basis for a proposed synergistic action of these exopeptidases. The demonstrated ability of DPP II to complement the action of tripeptidyl-peptidase I in the breakdown of a model collagen α chain, poly Gly-Pro-Ala, supports such a proposition (McDonald *et al.*, 1985). Furthermore, in contrast to all other endo- and exopeptidases, DPP II possesses the unique ability to cleave penultimate Pro-Xaa, Xaa-Pro, Pro-Pro and Hyp-Xaa bonds (McDonald & Schwabe, 1980; Mentlein & Struckhoff, 1989). This action again speaks for a special contribution by DPP II to the breakdown of proline-rich proteins such as collagen, which is composed largely of repeating Gly-Pro-Xaa units ('collagen triplets').

The noteworthy complementarity of the specificities of dipeptidyl-peptidase I (Chapter 214) and DPP II may also contribute to a significant degree of synergism. Despite the broad specificity exhibited by dipeptidyl-peptidase I, this exopeptidase is unable to act on tripeptides (or polypeptides) in which Arg or Lys occurs at the N-terminus, or when Pro occurs at the P1 (penultimate) or P1' positions. These configurations, on the other hand, strongly favor action by DPP II. Conversely, dipeptides not removed by DPP II are generally cleaved by dipeptidyl-peptidase I (McDonald *et al.*, 1971, 1987).

Because dipeptides, in contrast to tripeptides, easily cross the lysosomal membrane (Lloyd & Foster, 1986), DPP II (together with dipeptidyl-peptidase I) probably plays a critical role in returning the products of proteolysis to the metabolic pool – a process that should meaningfully engage the terminal activities of lysosomal dipeptidase (McDonald *et al.*, 1972; McDonald & Barrett, 1986) and cytosolic dipeptidases such as the Pro-X (EC 3.4.13.8) and X-Pro (EC 3.4.13.9) dipeptidases (Chapters 525 and 478).

DPP II of brain is not only capable of releasing N-terminal X-Pro dipeptides from arylamide derivatives and tripeptides, but also from oligopeptides such as casomorphin (five residues) and substance P (11 residues) (Mentlein & Struckhoff, 1989). DPP II has been localized to specific neuronal cell populations of the rat brain, where, in contrast to other tissues, its distribution is primarily extralysosomal (Gorenstein *et al.*, 1981). But others have detected

DPP II in glial cells, where it shows a typical lysosomal localization (Koshiya *et al.*, 1984; Stevens *et al.*, 1987, 1988). In the brain, DPP II may serve to modulate and metabolize proline-rich neuropeptides, and fragments therefrom, that are resistant to attack by most other peptidases. Astrocytes grown in media conditioned by meningeal cells show an increased rate of proliferation and an associated increase in DPP II production that could also be elicited by cAMP. It is thought that DPP II may contribute to the catabolic processes of cellular differentiation (Struckhoff, 1993).

DPP II has a broad distribution in tissues. In the rat, highest levels occur in the thyroid gland where it has been localized by electron microscopy to the lysosomes of the secretory epithelial cells of the thyroid follicles (McDonald *et al.*, 1971). Its tissue distribution in the mouse and guinea pig has also been reported (Gossrau & Lojda, 1980). The epididymis of the guinea pig is an especially rich source. DPP II has been localized histochemically to the acrosomes of washed sperm from this source (McDonald *et al.*, 1987). DPP II displays a wide distribution in cattle reproductive tissues. That of seminal plasma is probably derived from the epididymis and seminal vesicle (Agrawal & Vanha-Perttula, 1986). In human seminal plasma, DPP II activity occurs in both soluble- and particle-bound forms (Vanha-Perttula, 1984).

DPP II is also prominent in the lysosomal elements of rat peritoneal (Sannes *et al.*, 1977) and alveolar (Sannes, 1983) macrophages, and of mast cells (Sannes *et al.*, 1979; Struckhoff & Heymann, 1986). Normally, DPP II is present in both T and B lymphocytes, whereas DPP IV (Chapter 128) is strictly T cell specific (Andrews *et al.*, 1985; Khalaf *et al.*, 1986). Extremely high levels of DPP II present in various human carcinoma cells (6- to 24-fold higher than in human fibroblasts) offer a possibly useful marker for malignancy (Komatsu *et al.*, 1987). In cancer patients, serum DPP II activity increases whereas that of DPP IV decreases. The resulting increase in the serum DPP II/DPP IV ratio may also be a useful diagnostic marker (Kojima *et al.*, 1987). This ratio is also elevated in the cerebrospinal fluid of patients with Parkinson's disease, primarily as a result of significant rises in the levels of DPP II activity, which may be derived from the brain (Hagihara *et al.*, 1987).

DPP II can be visualized histochemically in most cell types associated with bone and surrounding connective tissues, including osteocytes, osteoblasts, chondrocytes, chondroblasts, fibroblasts and macrophages. A lack of detectable DPP II in osteoclasts may reflect a higher rate of secretion from this cell type. Cells in the resting zone of the growth plate are intensely reactive for DPP II, but are only moderately reactive for dipeptidyl-peptidase I and cathepsin B (Sannes *et al.*, 1986).

## Distinguishing Features

Any hydrolytic activity detected on Lys-Ala-NHNap or Lys-Ala-NHMec at pH 5.5 in the presence of (1 mM) 1,10-phenanthroline is in all probability attributable to DPP II, especially if it is inhibited by large cations such as Tris and puromycin.

## References

Agrawal, Y. & Vanha-Perttula, T. (1986) Dipeptidyl peptidases in bovine reproductive organs and secretions. *Int. J. Androl.* **9**, 435–452.

Andrews, C., Crockard, C., San Miguel, J.F. & Catovsky, D. (1985) Dipeptidylaminopeptidase IV (DAP IV) in B- and T-cell leukaemias. *Clin. Lab. Haematol.* **7**, 359–369.

DiCarlantonio, G., Talbot, P. & Dudenhausen, E. (1986) Partial purification and characterization of dipeptidyl peptidase II (DPP II) from guinea pig testes. *Gamete Res.* **15**, 161–175.

Doebber, T.W., Givor, A.R. & Ellis, S. (1978) Identification of a tripeptidyl aminopeptidase in the anterior pituitary gland: effect of the chemical and biological properties of rat and bovine growth hormone. *Endocrinology* **103**, 1794–1804.

Eisenhauer, D.A. & McDonald, J.K. (1986) A novel dipeptidyl peptidase II from the porcine ovary. Purification and characterization of a lysosomal serine protease showing enhanced specificity for prolyl bonds. *J. Biol. Chem.* **261**, 8859–8865.

Fukasawa, K., Fukasawa, K.M., Hiraoka, B.Y. & Harada, M. (1983) Purification and properties of dipeptidyl peptidase II from rat kidney. *Biochim. Biophys. Acta* **754**, 6–11.

Gorenstein, C., Tran, V.T. & Snyder, S.H. (1981) Brain peptidase with a unique neuronal localization: the histochemical distribution of dipeptidyl-aminopeptidase II. *J. Neurosci.* **1**, 1096–1102.

Gossrau, R. & Lojda, Z. (1980) Study on dipeptidylpeptidase II (DPP II). *Histochemistry* **70**, 53–76.

Hagihara, M., Mihara, R., Togari, A. & Nagatsu, T. (1987) Dipeptidyl-aminopeptidase II in human cerebrospinal fluid: changes in patients with Parkinson's disease. *Biochem. Med. Metab. Biol.* **37**, 360–365.

Hopsu-Havu, V.K., Jansén, C.T. & Järvinen, M. (1970) Partial purification and characterization of an acid dipeptidyl naphthylamidase (carboxytripeptidase) of the rat skin. *Arch. Klin. Exp. Dermatol.* **236**, 282–296.

Huang, K., Takagaki, M., Kani, K. & Ohkubo, I. (1996) Dipeptidyl peptidase II from porcine seminal plasma: purification, characterization, and its homology to granzymes, cytotoxic cell proteinases (CCP 1–4). *Biochim. Biophys. Acta* **1290**, 149–156.

Imai, K., Hama, T. & Kato, T. (1983) Purification and properties of rat brain dipeptidyl aminopeptidase. *J. Biochem.* **93**, 431–437.

Kato, T., Nagatsu, T., Fukasawa, K., Harada, M., Nagatsu, I. & Sakakibara, S. (1978) Successive cleavage of N-terminal Arg[1]-Pro[2] and Lys[3]-Pro[4] from substance P but no release of Arg[1]-Pro[2] from bradykinin, by X-Pro dipeptidyl-aminopeptidase. *Biochim. Biophys. Acta* **525**, 417–422.

Kato, T., Hama, T. & Nagatsu, T. (1980) Separation of two dipeptidyl aminopeptidases in the human brain. *J. Neurochem.* **34**, 602–608.

Kecorius, E., Small, D.H., & Livett, B.G. (1988) Characterization of a dipeptidyl aminopeptidase from bovine adrenal medulla. *J. Neurochem.* **50**, 38–44.

Khalaf, M.R., Bevan, P.C. & Hayhoe, F.G.J. (1986) Comparative cytochemical study of dipeptidyl aminopeptidase (DAP) II and IV in normal and malignant haemic cells. *J. Clin. Pathol.* **39**, 891–896.

Kojima, K., Mihara, R., Sakai, T., Togari, A., Matsui, T., Shinpo, K., Fujita, K., Fukasawa, K., Hanada, M. & Nagatsu, T. (1987) Serum activities of dipeptidyl-aminopeptidase II and dipeptidyl-aminopeptidase IV in tumor-bearing animals and in cancer patients. *Biochem. Med. Metab. Biol.* **37**, 35–41.

Komatsu, M., Urade, M., Yamaoka, M., Fukasawa, K. & Harada, M. (1987) Alteration in dipeptidyl peptidase activities in cultured human carcinoma cells. *J. Natl Cancer Inst.* **78**, 863–868.

Koshiya, K., Kato, T., Tanaka, R. & Kato, T. (1984) Brain peptidases: their possible neuronal and glial localization. *Brain Res.* **324**, 261–270.

Lloyd, J.B. & Foster, S. (1986) The lysosomal membrane. *Trends Biochem. Sci.* **11**, 365–368.

Lynn, K.R. (1991) The isolation and some properties of dipeptidyl peptidases II and III from porcine spleen. *Int. J. Biochem.* **23**, 47–50.

Mantle, D. (1991) Characterization of dipeptidyl and tripeptidyl aminopeptidases in human kidney soluble fraction. *Clin. Chim. Acta* **196**, 135–142.

McDonald, J.K. & Barrett, A.J. (1986) *Mammalian Proteases: A Glossary and Bibliography*, vol. 2: *The Exopeptidases*. London: Academic Press, pp. 120–126.

McDonald, J.K. & Schwabe, C. (1980) Dipeptidyl peptidase II of bovine dental pulp. Initial demonstration and characterization as a fibroblastic, lysosomal peptidase of the serine class active on collagen-related peptides. *Biochim. Biophys. Acta* **616**, 68–81.

McDonald, J.K., Ellis, S. & Reilly, T.J. (1966) Properties of dipeptidyl arylamidase I of the pituitary. Chloride and sulfhydryl activation of seryltyrosyl-$\beta$-naphthylamide hydrolysis. *J. Biol. Chem.* **241**, 1494–1501.

McDonald, J.K., Reilly, T.J., Zeitman, B. & Ellis, S. (1968a) Dipeptidyl arylamidase II of the pituitary. Properties of lysylalanyl-$\beta$-naphthylamide hydrolysis: inhibition by cations, distribution in tissues, and subcellular localization. *J. Biol. Chem.* **243**, 2028–2037.

McDonald, J.K., Leibach, F.H., Grindeland, R.E. & Ellis, S. (1968b) Purification of dipeptidyl aminopeptidase II (dipeptidyl arylamidase II) of the anterior pituitary gland. Peptidase and dipeptide esterase activities. *J. Biol. Chem.* **243**, 4143–4150.

McDonald, J.K., Callahan, P.X., Ellis, S. & Smith, R.E. (1971) Polypeptide degradation by dipeptidyl aminopeptidase I (cathepsin C) and related peptidases. In: *Tissue Proteinases* (Barrett, A.J. & Dingle, J.T., eds). Amsterdam: North-Holland Publishing, pp. 69–107.

McDonald, J.K., Zeitman, B.B. & Ellis, S. (1972) Detection of a lysosomal carboxypeptidase and a lysosomal dipeptidase in highly-purified dipeptidyl aminopeptidase I (cathepsin C) and the elimination of their activities from preparations used to sequence peptides. *Biochem. Biophys. Res. Commun.* **46**, 62–70.

McDonald, J.K., Hoisington, A.R. & Eisenhauer, D.A. (1985) Partial purification and characterization of an ovarian tripeptidyl peptidase: a lysosomal exopeptidase that sequentially releases collagen-related (Gly-Pro-X) triplets. *Biochem. Biophys. Res. Commun.* **126**, 63–71.

McDonald, J.K., Schwabe, C. & Owers, N.O. (1987) Peptidases in connective tissue degradation and remodelling in reproductive and invasive tissues. In: *Cells, Membranes, and Disease, Including Renal (Methodological Surveys in Biochemistry and Analysis*, vol. 17) (Reid, E., Cook, G.M.W. & Luzio, J.P., eds). New York: Plenum Press, pp. 335–349.

Mentlein, R. & Struckhoff, G. (1989) Purification of two dipeptidyl aminopeptidases II from rat brain and their action on proline-containing neuropeptides. *J. Neurochem.* **52**, 1284–1293.

Nagatsu, T., Sakai, T., Kojima, K., Araki, E., Sakakibara, S., Fukasawa, K. & Harada, M. (1985) A sensitive and specific assay for dipeptidyl aminopeptidase II in serum and tissues by liquid chromatography-fluorometry. *Anal. Biochem.* **147**, 80–85.

Okuyama, T. & Satake, K. (1960) The preparation and properties of 2,4,6-trinitrophenylamino acids and peptides. *J. Biochem.* **47**, 454–466.

Püschel, G., Mentlein, R. & Haymann, E. (1982) Isolation and characterization of dipeptidyl peptidase IV from human placenta. *Eur. J. Biochem.* **126**, 359–365.

Rao, A.J., Hagihara, M., Nagatsu, T. & Yanagita, N. (1990) Presence of dipeptidyl peptidase II, dipeptidyl peptidase IV, and prolyl endopeptidase in effusion from patients with serous otitis media. *Biochem. Med. Metab. Biol.* **43**, 276–282.

Rawlings, N.D. & Barrett, A.J. (1996) Dipeptidyl-peptidase II is related to lysosomal Pro-X carboxypeptidase. *Biochim. Biophys. Acta* **1298**, 1–3.

Sakai, T., Kojima, K. & Nagatsu, T. (1987) Rapid chromatographic purification of dipeptidyl-aminopeptidase II from human kidney. *J. Chromatogr.* **416**, 131–137.

Sannes, P.L. (1983) Subcellular localization of dipeptidyl peptidases II and IV in rat and rabbit alveolar macrophages. *J. Histochem. Cytochem.* **31**, 684–690.

Sannes, P.L., McDonald, J.K. & Spicer, S.S. (1977) Dipeptidyl aminopeptidase II in rat peritoneal wash cells. Cytochemical localization and biochemical characterization. *Lab. Invest.* **37**, 243–253.

Sannes, P.L., McDonald, J.K., Allen, R.C. & Spicer, S.S. (1979) Cytochemical localization and biochemical characterization of dipeptidyl aminopeptidase II in macrophages and mast cells. *J. Histochem. Cytochem.* **27**, 1496–1498.

Sannes, P.L., Schofield, B.H. & McDonald, D.F. (1986) Histochemical localization of cathepsin B, dipeptidyl peptidase I, and dipeptidyl peptidase II in rat bone. *J. Histochem. Cytochem.* **34**, 983–988.

Stein, V.U., Weber, U. & Buddecke, E. (1968) Uber eine neue Carboxypeptidase (EC 3.4.2.?) aus der Milz. *Z. Physiol. Chem.* **349**, 472–484.

Stevens, B.R., Raizada, M., Sumners, C. & Fernandez, A. (1987) Dipeptidyl peptidase-II activity in cultured astroglial cells from neonatal rat brain. *Brain Res.* **406**, 113–117.

Stevens, B.R., Fernandez, A., Sumners, C. & Hearing, L. (1988) Neonatal rat brain astroglial dipeptidyl peptidase II activity regulation by cations and anions. *Neurosci. Lett.* **89**, 319–322.

Struckhoff, G. (1993) Dipeptidyl peptidase II in astrocytes of the rat brain. Meningeal cells increase enzymic activity in cultivated astrocytes. *Brain Res.* **620**, 49–57.

Struckhoff, G. & Heymann, E. (1986) Rat peritoneal mast cells release dipeptidyl peptidase II. *Biochem. J.* **236**, 215–219.

Vanha-Perttula, T. (1984) Studies on alanine aminopeptidase, dipeptidyl aminopeptidase I and II of the human seminal fluid and prostasomes. *Select. Top. Clin. Enzymol.* **2**, 545–564.

*J. Ken McDonald*
*Department of Biochemistry and Molecular Biology,*
*Medical University of South Carolina,*
*171 Ashley Avenue, Charleston, SC 29425, USA*

# 139. Prolyl aminopeptidase

## Databanks

Peptidase classification: clan SC, family S33, MEROPS ID: S33.001
NC-IUBMB enzyme classification: EC 3.4.11.5
Chemical Abstracts Service registry number: 9025-40-5
Databank codes:

| Species | SwissProt | PIR | EMBL (cDNA) | EMBL (genomic) |
|---|---|---|---|---|
| Aeromonas sobria | P46547 | – | D30714 | – |
| Bacillus coagulans | P46541 | A47038 | D11037 | – |
| Flavobacterium meningosepticum | – | | D83254 | – |
| Hafnia alvei | – | – | D61383 | – |
| Lactobacillus delbrueckii | P46542 | – | Z26948 | – |
| Lactobacillus delbrueckii | P46544 | S44282 | L10712 | – |
| Lactobacillus helveticus | P52278 | – | Z56283 | – |
| Mycoplasma genitalium | P47266 | – | – | U02229: random genomic clone xb12 U39680: complete genome section 2 |
| Neisseria gonorrhoeae | P42786 | S39592 | Z25461 | – |
| Salmonella typhimurium | – | – | X99945 | – |
| Xanthomonas campestris | P52279 | – | Z54150 | – |

## Name and History

This enzyme catalyzes the removal of N-terminal proline residues from peptides, and has been the object of many studies since the first report of the Escherichia coli enzyme (Sarid et al., 1959). However, further studies have shown that this activity might have been mistakenly characterized (Fanghanel et al., 1991). Originally named **proline iminopeptidase (PIP)**, it has also been called **proline aminopeptidase** and **Pro-X aminopeptidase**. The activity has been detected in a variety of organisms, e.g. human oral cavity microorganisms (Mäkinen, 1969), bacteria (Yoshimoto et al., 1983; Yoshimoto & Tsuru, 1985; Kitazono et al., 1992; Albertson & Koomey, 1993; Atlan et al., 1994), mushroom (Sattar et al., 1989), and apricot seeds (Ninomiya et al., 1982). Many microbial genes for the enzyme have already been cloned, but no eukaryotic prolyl aminopeptidase gene has yet been isolated, and even the presence of this activity in mammalian tissues is still controversial (Heymann & Peter, 1993).

## Activity and Specificity

Prolyl aminopeptidases are highly specific for proline at P1, but some of them will also accept a hydroxyproline residue. Small peptides are easily cleaved by most of the enzymes, and the highest activities have been observed for di- and tripeptides. Prolyl aminopeptidase had long been assumed to be a cysteine peptidase. However, the dependence of its catalytic activity on a serine residue, the conservation of this residue and its surroundings in all the sequences known to date (Kitazono et al., 1994b), and the uniform inhibition by DCI have conclusively characterized it as a serine enzyme. On the other hand, $Mn^{2+}$, initially described as an activator of the E. coli enzyme, inhibits most of the enzymes.

## Structural Chemistry

Prolyl aminopeptidase has been described as a monomeric or tetrameric enzyme. The monomeric enzymes (B. coagulans, L. delbrueckii, N. gonorrhoeae, F. meningosepticum) are 30 kDa proteins, with a strict specificity for terminal Pro. The tetrameric enzymes (A. sobria, H. alvei) are formed by subunits of around 50 kDa, with a broader specificity, releasing also terminal hydroxyproline (Kitazono et al., 1994a, 1996a). Ser101, and His267 for the B. coagulans prolyl aminopeptidase, and Ser146 for the A. sobria enzyme, have been shown by site-directed mutagenesis to be essential for activity (Kitazono et al., 1994a).

## Preparation

Most of the prolyl aminopeptidases hitherto purified are those from microorganisms. The bacterial enzymes are predominantly soluble, cytosolic enzymes that have been purified from cell-free extracts prepared after mechanical cell disruption. In addition, the purification of prolyl aminopeptidases from a mushroom, Lyophyllum cinerascens (Sattar et al., 1989), and apricot seeds (Ninomiya et al., 1982) has been reported.

## Distinguishing Features

In spite of the divergence in the two types of prolyl aminopeptidases, there is still a significant homology in their primary structures. The A. sobria and H. alvei enzymes share 45% of identical residues, while the B. coagulans, F. meningosepticum, N. gonorrhoeae, and L. delbrueckii prolyl aminopeptidases are 24–36% identical. When enzymes from the different groups are compared, the identity values decrease to 18–21%, but are still significant, and the alignment of all the sequences shows the conservation of the active-site Ser and His residues (Kitazono *et al.*, 1996b). Homology searches have shown that these prolyl aminopeptidase sequences might be evolutionary related to hydrolases of diverse specificity that belong to the $\alpha/\beta$ hydrolase fold family (Atlan *et al.*, 1994; Rawlings & Barrett, 1994; Kitazono *et al.*, 1996b).

## References

Albertson, N. & Koomey, M. (1993) Molecular cloning and characterization of a proline iminopeptidase gene from *Neisseria gonorrhoeae*. *Mol. Microb.* **9**, 1203–1211.

Atlan, D., Gilbert, C., Blanc, B. & Portalier, R. (1994) Cloning, sequencing and characterization of the *pepIP* gene encoding a proline iminopeptidase gene from *Lactobacillus delbrueckii* subsp. *bulgaricus* CNRZ397. *Microbiology* **140**, 527–535.

Fanghanel, S., Reissbrodt, R. & Giesecke, H. (1991) L-Proline aminopeptidase activity as a tool for identification and differentiation of *Serratia marcescens*, *Serratia liquefaciens* and *Hafnia alvei* strains. *Zbl. Bakt.* **275**, 11–15.

Heymann, E. & Peter, K. (1993) A note on the identity of porcine liver carboxylesterase and prolyl-$\beta$-naphthylamidase. *Biol. Chem. Hoppe-Seyler* **374**, 1033–1036.

Kitazono, A., Yoshimoto, T. & Tsuru, D. (1992) Cloning, sequencing, and high expression of the proline iminopeptidase gene from *Bacillus coagulans*. *J. Bacteriol.* **174**, 7919–7925.

Kitazono, A., Tsuru, D. & Yoshimoto, T. (1994a) Isolation and characterization of the prolyl aminopeptidase gene *(pap)* from *Aeromonas sobria*. Comparison with the *Bacillus coagulans* enzyme. *J. Biochem.* **116**, 818–825.

Kitazono, A., Ito, K. & Yoshimoto, T. (1994b) Prolyl aminopeptidase is not a sulfhydryl enzyme: Identification of the active serine residue by site-directed mutagenesis. *J. Biochem.* **116**, 943–945.

Kitazono, A., Kitano, A., Kabashima, T., Ito, K. & Yoshimoto, T. (1996a) Prolyl aminopeptidase is also present in *Enterobacteriaceae*: cloning and sequencing of the *Hafnia alvei* enzyme-gene and characterization of the expressed enzyme. *J. Biochem.* **119**, 468–474.

Kitazono, A., Kabashima, T., Huang, H., Ito, K. & Yoshimoto, T. (1996b) Prolyl aminopeptidase gene from *Flavobacterium meningosepticum*: Cloning, purification of the expressed enzyme and analysis of its sequence. *Arch. Biochem. Biophys.* **336**, 35–40.

Mäkinen, K. (1969) The proline iminopeptidases of the human oral cavity – partial purification and characterization. *Acta Chem. Scand.* **23**, 1409–1438.

Ninomiya, K., Kawatani, K., Tanaka, S., Kawata, S. & Makisumi, S. (1982) Purification and properties of a proline iminopeptidase from apricot seeds. *J. Biochem.* **92**, 413–421.

Rawlings, N.D. & Barrett, A.J. (1994) Families of serine peptidases. *Methods Enzymol.* **244**, 19–61.

Sarid, S., Berger, A. & Katchalski, E. (1959) Proline iminopeptidase. *J. Biol. Chem.* **234**, 1740–1744.

Sattar, A.K.M., Yoshimoto, T. & Tsuru, D. (1989) Purification and characterization of proline iminopeptidase from *Lyophyllum cinerascens*. *J. Ferment. Bioeng.* **68**, 178–182.

Yoshimoto, T. & Tsuru, D. (1985) Proline iminopeptidase from *Bacillus coagulans*. *J. Biochem.* **97**, 1477–1485.

Yoshimoto, T., Saeki, T. & Tsuru, D. (1983) Proline iminopeptidase from *Bacillus megaterium*: purification and characterization. *J. Biochem.* **93**, 469–477.

*Ana Kitazono*
*Department of Radiation Oncology,*
*Albert Einstein College of Medicine,*
*Bronx, NY 10461, USA*
*Email: kitazono@aecom.yu.edu*

*Tadashi Yoshimoto*
*Department of Biotechnology,*
*Nagasaki University School of*
*Pharmaceutical Sciences,*
*Nagasaki, 852 Japan*
*Email: t-yoshimoto@cc.nagasaki-u.ac.jp*

*Tsutomu Kabashima*
*Department of Biotechnology,*
*Nagasaki University School of*
*Pharmaceutical Sciences,*
*Nagasaki, 852 Japan*

# 140. Tripeptidyl-peptidases A, B and C

## Databanks

*Peptidase classification: clan SC, family S33, MEROPS ID: S33.002, S33.006, S33.007*
*NC-IUBMB enzyme classification: none*
*Databank codes:*

| Species | Type | SwissProt | PIR | EMBL (cDNA) | EMBL (genomic) |
|---|---|---|---|---|---|
| Streptomyces lividans | TPP-A | – | – | L27466 | – |
| Streptomyces lividans | TPP-B | – | – | L42758 | – |
| Streptomyces lividans | TPP-C | – | – | L42759 | – |

## Name and History

*Streptomyces lividans* has been used as a host strain for the production of a number of heterologous proteins (Chang & Chang, 1988; Malek *et al.*, 1990; Brawner *et al.*, 1991). The nonpathogenic nature of *Streptomyces* strains, combined with their widespread use in large-scale fermentation processes, stimulated interest in the development of *S. lividans* as a host capable of secreting the protein of interest directly into the culture medium. Purification of the secreted heterologous protein can be relatively simple, but the productivity of some processes was found to be limited by the presence of low levels of secreted host proteases (Aretz *et al.*, 1989). A specific problem with aminopeptidase-like activity was observed (Krieger *et al.*, 1994) when producing recombinant human granulocyte-macrophage colony-stimulating factor (rhGM-CSF). Biochemical purification and characterization of the peptidase degrading rhGM-CSF suggested that the responsible protease (initially christened *TAP*) required a free N-terminal amino group, and caused the loss of three amino acid residues from the N-terminus of the rhGM-CSF. The tripeptide chromogenic substrate Ala-Pro-Ala+NHPhNO$_2$ (modeled on the N-terminal sequence of rhGM-CSF) was hydrolyzed by the same column fractions as rhGM-CSF during purification, whereas the rate of hydrolysis of mono- and di-amino acid p-nitroanilides was much slower, suggesting that a single direct cleavage event had occurred to produce the '-3' form of rhGM-CSF. At the same time, the relevant gene was isolated (Butler *et al.*, 1995) by expression screening of genomic libraries of *S. lividans* with a different tripeptide p-nitroanilide substrate, Gly-Pro-Leu-NHPhNO$_2$. The deduced amino acid sequence was consistent with the expectation from the biochemical inhibition profile that the enzyme was a serine protease. For clarity and consistency this enzyme is renamed here as ***tripeptidyl-peptidase A (TPP A)***.

Recombinational deletion of the chromosomal gene eliminated the ability of the strain to hydrolyze Gly-Pro-Leu-NHPhNO$_2$, and reduced but did not eliminate the degrading activity against rhGM-CSF. The deletion strain was used (Binnie *et al.*, 1995) as a host for further expression screening of the *S. lividans* genomic library with Ala-Pro-Ala-NHPhNO$_2$ to identify three further chromosomally encoded proteins (SlpD, SlpE and Ssp) which are renamed here as tripeptidyl-peptidase B (TPP B), tripeptidyl-peptidase C (TPP C) and tripeptidyl-peptidase S (TPP S), respectively. TPP B and C were both shown to be homologs of TPP A, whereas TPP S (see Chapter 114) showed homology to subtilisin (Chapter 94).

## Activity and Specificity

TPP A activity was initially identified (Krieger *et al.*, 1994) by use of the rhGM-CSF degradation assay, and was characterized by the observation of a discrete series of faster migrating species of the factor on native PAGE analysis. The first band was shown by N-terminal sequence determination to represent a '-3' truncated form of rhGM-CSF. Using this assay to monitor biochemical purification of the TPP A protein from *S. lividans* fermentation broth, a purified preparation was obtained consisting of essentially one band of material migrating in SDS-PAGE at a position

consistent with a molecular mass of 55 kDa. Characterization of the enzyme was facilitated by the availability of strains of *S. lividans* carrying the *tppA* gene on a multicopy plasmid. The enzyme showed an absolute requirement for an unblocked N-terminus in peptide substrates, and was unable to hydrolyze mono-amino acid, di- or tripeptide substrates. The inhibition profile suggested that it was a serine protease, since it was only substantially inhibited by PMSF and peptidyl chloromethanes. The enzyme cleaved Ala-Pro-Ala-NHPhNO$_2$ 10-fold more rapidly than Ala-Pro-Met-NHPhNO$_2$ consistent with the observation that *S. lividans* strains producing rhGM-CSF showed more rapid degradation of their product than those producing recombinant human interleukin 3 (rhIL-3). The TPP A showed a broad pH activity spectrum, being active in the range pH 5.0–9.0, with maximal activity at pH 7.0–8.5. The Ala-Pro-Ala-NHPhNO$_2$ substrate is available from Bachem (see Appendix 2 for full names and addresses of suppliers), and can conveniently be used to assay the activity of streptomycete colonies on agar plates or in liquid culture.

The gene encoding the TPP B protein was identified by expression screening of a *S. lividans* genomic library with the substrate Ala-Pro-Ala-NHPhNO$_2$; its chromosomal locus was designated *slpD*. However, unlike the case of the *tppA* gene, no enzymic activity was observed in the liquid culture medium, although very significant activity was seen in the mycelium of colonies grown on agar medium. The gene encoding TPP C was similarly isolated using a longer peptide substrate designed to search for a protease activity that was observed to degrade rhGM-CSF to a '-4' form: Boc-Ala-Pro-Ala-Arg+Ser-Pro-Ala-NHPhNO$_2$. In this screen, TPP A activity was used in a linked assay to release the p-nitroaniline chromophore from the first product (Ser-Pro-Ala-NHPhNO$_2$). Surprisingly, clones representing the *tppB* gene were reisolated using this blocked substrate. Three other clones were also identified and shown to encode another chromosomal locus which was designated *slpE*. Once again, no secreted protease activity was observed. Nucleotide sequence determination of the *tppB* and *tppC* gene loci revealed the presence of TPP proteins which were clearly homologous to TPP A. The protein sequences translated from the DNA sequences showed hydrophobic N-terminal residues preceded by several positively charged residues. However, no consensus for a signal peptidase I (Chapter 153) cleavage site was detected in an analysis according to von Heijne (1989), but instead, putative signal peptidase II (Chapter 333) sequences were noticed. This was consistent with the possibility of the proteases being attached to the mycelial surfaces through lipids post-translationally added to the cysteine residues.

## Structural Chemistry

TPP A is a single-chain protein of 55 kDa, pI approximately 4.8, secreted directly into the medium of *S. lividans* cultures. It contains eight cysteine residues that are presumably disulfide bonded and contribute to its enzymic stability. Soluble forms of both TPP B and TPP C were produced by replacement of the nucleotides encoding the putative promoter/leader mRNA region and the lipoprotein signal sequence by sequences encoding the aminoglycoside phosphotransferase *(aph)* promoter and the cerevisin

(Chapter 104) signal peptide (Henderson *et al.*, 1987). A small spacer peptide was also introduced (Binnie *et al.*, 1995) between the cerevisin signal peptide cleavage site and the TPP B and C proteins. Secreted proteins of approximately 55 kDa were observed for both constructs. The Ala-Pro-Ala-Ala-Pro-Ala spacer peptides were not present in the TPP B protein, but were intact in the TPP C. The TPP B protein showed the ability to hydrolyze Ala-Pro-Ala-NHPhNO$_2$ whereas the TPP C protein did not. Neither protein was able to hydrolyze the blocked substrate Boc-Ala-Pro-Ala-Arg-Ser-Pro-Ala-NHPhNO$_2$.

## Preparation

Very small amounts of TPP A were isolated from the medium of *S. lividans* cultures. Purification and yield were greatly aided by the use of a strain transformed with the *tppA* gene cloned on a multicopy plasmid. With this strain, secreted enzyme was produced earlier in the fermentation and consequently amounts of contaminating host proteases (both low-level secreted proteases and intracellular proteases released by mycelial lysis) were very significantly reduced.

## Biological Aspects

Recombinational deletion of the *tppA* gene produced strains with significantly reduced TPP activity. However, recombinational deletion of the *tppB* gene was not possible, indicating that this function is essential to the viability of the strain. The *tppC* gene was dispensable since deletion strains were still able to grow under laboratory conditions. The enzymic activities of TPP B and C have not been fully characterized. The presence of all three *tpp* genes was confirmed in all of the strains of *Streptomyces* tested by the observation of bands representing homologous DNA in low stringency hybridization experiments. TPP A is likely to be an important part of the degradative capability of *Streptomyces* strains since it is such an efficient secreted peptidase. Together with endopeptidases, it will produce rapidly absorbable tripeptides from environmentally encountered proteins. The function of the mycelial-associated TPPs is less easy to envisage, but may be related to transport of small peptides into the mycelium or perhaps, more speculatively, in remodeling of cell wall proteoglycan components.

## Distinguishing Features

TPP A is observed as a 55 kDa protein in culture medium and can readily hydrolyze a wide range of tripeptide substrates and proteins. TPP B and C were inferred to be associated with the mycelium and not released into the culture medium. The *tppB* and *tppC* chromosomal loci appear to be essential for growth of *S. lividans* in the absence of complex nitrogen sources such as yeast extract.

## References

Aretz, W., Koller, K.-P. & Riess, G. (1989) Proteolytic enzymes from recombinant *Streptomyces lividans* TK24. *FEMS Microbiol. Lett.* **65**, 31–36.

Binnie, C., Butler, M.J., Aphale, J.S., Bourgault, R., DiZonno, M.A., Krygsman, P., Laio, L., Walczyk, E. & Malek, L.T. (1995) Isolation and characterisation of two genes encoding proteases associated with the mycelium of *Streptomyces lividans 66. J. Bacteriol.* **177**, 6033–6040.

Brawner, M., Poste, G., Rosenberg, M. & Westphaling, J. (1991) *Streptomyces*: a host for heterologous gene expression. *Curr. Opin. Biotechnol.* **2**, 674–681.

Butler, M.J., Binnie, C., DiZonno, M.A., Krygsman, P., Soltes, G.A., Soostmeyer, G., Walczyk, E. & Malek, L.T. (1995) Cloning and characterisation of a gene encoding a secreted tripeptidyl aminopeptidase from *Streptomyces lividans 66. Appl. Environ. Microbiol.* **61**, 3145–3150.

Chang, S.-Y. & Chang, S. (1988) Secretion of heterologous proteins in *Streptomyces lividans*. In: *Biology of Actinomycetes '88* (Okami, Y., Beppu, T. & Ogawara, H., eds). Tokyo: Japan Scientific Press, pp. 103–107.

Henderson, G., Krygsman, P., Lui, C.J., Davey, C.C. & Malek, L.T. (1987) Characterisation and structure of genes for proteases A and B from *Streptomyces griseus. J. Bacteriol.* **169**, 3778–3784.

Krieger, T., Bartfeld, D., Jenish, D. & Hadary, D. (1994) Purification and characterisation of a novel tripeptidyl aminopeptidase from *Streptomyces lividans 66. FEBS Lett.* **352**, 385–388.

Malek, L.T., Soostmeyer, G., Davey, C.C., Krygsman, P., Compton, J., Gray, J., Zimny, T. & Stewart, D. (1990) Secretion of granulocyte macrophage colony stimulating factor (GM-CSF) in *Streptomyces lividans. J. Cell Biochem.* **suppl. 14A**, 127.

von Heijne, G. (1989) The structure of signal peptides from bacterial lipoproteins. *Protein Eng.* **2**, 531–534.

*Michael J. Butler*
*Strangeways Research Laboratory,*
*Worts Causeway,*
*Cambridge CB1 4RN, UK*
*Email: m.butler@uea.ac.uk*

# *141. Introduction: clan SE containing serine-type D-Ala-D-Ala peptidases*

## *Databanks*

*MEROPS ID: SE*

| Species | SwissProt | PIR | EMBL (cDNA) | EMBL (genomic) |
|---|---|---|---|---|
| **Family S11** | | | | |
| D-Ala-D-Ala carboxypeptidase PBP5 (Chapter 142) | | | | |
| Murein-DD-endopeptidase PBP7 (Chapter 143) | | | | |
| *Streptomyces* K15 D-Ala-D-Ala transpeptidase (Chapter 144) | | | | |
| Others | | | | |
|    *Escherichia coli* | P33013 | – | – | U00009: chromosome sbcB region |
|    *Salmonella typhimurium* | P37604 | – | L31538 L32188 | – |
| **Family S12** | | | | |
| Alkaline D-peptidase (Chapter 147) | | | | |
| D-Stereospecific aminopeptidase (*Ochrobactrum*) (Chapter 146) | | | | |
| *Streptomyces* R61 D-Ala-D-Ala carboxypeptidase (Chapter 145) | | | | |
| **Family S13** | | | | |
| D-Ala-D-Ala carboxypeptidase (*Actinomadura* R39) (Chapter 150) | | | | |
| D-Ala-D-Ala peptidase PBP4 (Chapter 148) | | | | |

Clan SE contains families S11, S12 and S13, the enzymes of which are mostly involved in the biosynthesis and remodeling of bacterial cell walls. Both gram-positive and gram-negative bacterial cell walls are complex polymers of amino sugars and amino acids. Chains of alternating *N*-acetylglucosamine and *N*-acetylmuramic acid units are cross-linked to one another by short peptides. In *E. coli* the structure of the link peptide is L-alanyl-D-isoglutamyl-L-*meso*-diaminopimelyl-D-alanine, but the nature of the third amino acid in the chain varies with the bacterial species. These chains are cross-linked, usually between the carboxyl group of D-alanine and the free amino group of diaminopimelate. In the biosynthesis of the cell wall peptidoglycan the precursor has the four-residue structure above, but with an additional C-terminal D-alanine residue. The D-Ala-D-Ala transpeptidases and carboxypeptidases are involved in the metabolism of the cell wall components.

The clan is assembled on the basis of similarities revealed by hydrophobic cluster analysis (Palomeque-Messia *et al.*, 1991; Granier *et al.*, 1992) and conservation of motifs around the catalytic residues (Alignment 141.1) (see CD-ROM). It is also notable that there are many similarities in the catalytic activities and biological roles of the enzymes.

The catalytic mechanism involves Ser35, Lys38 and Ser96; Ser35 acting as the nucleophile, and Lys38 being the general base. The motif Ser35-Xaa-Xaa-Lys38 is conserved in the members of all three families. The motif Ser96-Xaa-Asn is conserved in most members of families S11 and S13, but in family S12 the serine is replaced by tyrosine. The third conserved motif is Lys214-Thr-Gly. These two motifs include

residues that form the sides of the active-site cavity (Englebert *et al.*, 1994; E. Fonzé, M. Vermeire, Y. Toth, M. Nguyen-Distèche, J.-M. Ghuysen & P. Charlier, unpublished results).

These enzymes are also known for being penicillin-binding proteins, and have been classified on this basis (Table 141.1). The D-Ala-D-Ala peptidases are known as low $M_r$ penicillin-binding proteins, and probably are related to the high $M_r$ penicillin-binding proteins. The antibiotic action of penicillin is due to its binding to the high $M_r$ penicillin-binding proteins and not the D-Ala-D-Ala peptidases. Enzymes that are capable of degrading penicillins and related antibiotics are $\beta$-lactamases, amongst which the class C $\beta$-lactamases are homologous to D-Ala-D-Ala peptidases (S12), and the class A $\beta$-lactamases are more distantly related, as is revealed by the three-dimensional structures (Oefner *et al.*, 1990). Some of the D-Ala-D-Ala peptidases (PBP-1A, PBP-1B and PBP-2) perform a transpeptidation reaction in which the peptidoglycan monomer minus the C-terminal D-Ala is transferred to an exogenous acceptor; the *Streptomyces* K15 peptidase performs this reaction exclusively, and does not hydrolyze peptide bonds (Ghuysen, 1991).

Three-dimensional structures are known for *Streptomyces* R61 D-Ala-D-Ala peptidase (Kelly *et al.*, 1985) and *Citrobacter* class C $\beta$-lactamase (Oefner *et al.*, 1990) (both from family S12), as well as class A $\beta$-lactamases from *Streptomyces* (Lamotte-Brasseur *et al.*, 1992) and *E. coli* (Strynadka *et al.*, 1992) (not members of the peptidase families but homologous at the clan level). The three-dimensional structure of *Streptomyces* R61 D-Ala-D-Ala peptidase is shown in Figure 141.1. It can be seen that the molecule is a two-lobed structure, with

*Table 141.1* Penicillin-binding proteins of *Escherichia coli*

| PBP | Family | Activity | Gene names |
|---|---|---|---|
| PBP-1A | – | transpeptidase | *mrcA, ponB* |
| PBP-1B | – | transpeptidase | *mrcB, ponB, pbpF* |
| PBP-2 | S12 | transpeptidase | *mrdA, pbpA* |
| PBP-3 | S11 | transpeptidase | *ftsI, pbpB* |
| PBP-4 | S13 | DD-carboxypeptidase; DD-endopeptidase | *dacB* |
| PBP-5 | S11 | DD-carboxypeptidase | *dacA, pfv* |
| PBP-6 | S11 | DD-carboxypeptidase | *dacC* |
| PBP-6B | S11 | DD-carboxypeptidase | *dacD, phsE* |
| PBP-7 | S11 | DD-endopeptidase | *pbpG* |
| – | – | ? | *pbpC* |

PBP-3 is a high $M_r$ PBP, as are PBP-2x from *Streptococcus pneumoniae* (S12) and PBP-3 of *Enterococcus hirae* (S12). PBP-1A and-1B are 'class A', the remainder 'class B'.

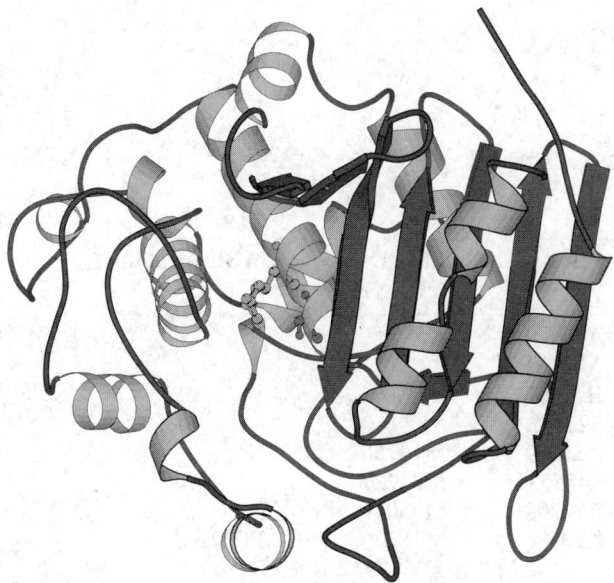

*Figure 141.1* Richardson diagram of the *Streptomyces* sp. R61 D-Ala-D-Ala carboxypeptidase. The image was prepared from the Brookhaven Protein Data Bank entry (3PTE) as described in the Introduction (p. xxv). Catalytic residues are shown in ball-and-stick representation: Ser35 and His38 (numbering as in Alignment 141.1).

the catalytic Ser and Lys residues on a helix that traverses the cleft between the lobes.

Families S11 and S13 include peptidases capable of degrading the D-Ala-D-Ala cross-links in the bacterial cell wall peptidoglycans, which have been described as 'endopeptidases'. A common profile for the D-Ala-D-Ala endopeptidases can be defined which excludes the D-Ala-D-Ala carboxypeptidases (see Chapter 143). Some of the enzymes in family S11 are partially inhibited by thiol-blocking reagents such as PCMB, and in the *Streptomyces* K15 peptidase there are two cysteine residues, one of which is close to the general base (Palomeque-Messia *et al.*, 1991).

Family S12 contains enzymes with the most diverse specificities. In addition to the *Streptomyces* R61 D-Ala-D-Ala peptidase and class C β-lactamases, there is a D-aminopeptidase from *Ochrobactrum* (the only aminopeptidase known to be specific for D-amino acids) (see Chapter 146), a lipolytic esterase from *Pseudomonas* (EMBL: M68491) and proteins from *Bacteroides nodosus* (FMDH_BACNO, FMDD_BACNO) that may be involved in the assembly of fimbriae, which are external appendages found on the surface of gram-negative bacteria and used for cell adhesion. The fimbrial subunits that are related to D-Ala-D-Ala peptidases are found only in some strains, and seem to represent a recent lateral transfer of genes. The function of these proteins is unknown, but the active-site motif **Ser**-Xaa-Xaa-Lys is retained.

The metallopeptidases of family M15 (clan MD) are also D-Ala-D-Ala carboxypeptidases, similar in specificity to serine peptidases in clan SE, but not catalyzing the transpeptidase reaction (presumably because the metallopeptidases do not form acyl enzyme intermediates) (see Chapter 462). Also, there are three enzymes that have activities similar to those of peptidases in clan SE, but are of unknown catalytic type, and show no sign of structural relationship to the enzymes of clan SE. These are VanY D-Ala-D-Ala carboxypeptidase (Chapter 554), penicillin-insensitive murein endopeptidase (Chapter 556), and phage murein endopeptidase (Chapter 557).

## References

Englebert, S., Charlier, P., Fonzé, E., Toth, Y., Vermeire, M., Van Beeumen, J., Grandchamps, J., Hoffmann, K., Leyh-Bouille, M., Nguyen-Distèche, M. & Ghuysen, J.-M. (1994) Crystallization and X-ray diffraction study of the *Streptomyces* K15 penicillin-binding DD-transpeptidase. *J. Mol. Biol.* **241**, 295–297.

Ghuysen, J.-M. (1991) Serine β-lactamases and penicillin-binding proteins. *Annu. Rev. Microbiol.* **45**, 37–67.

Granier, B., Duez, C., Lepage, S., Englebert, S., Dusart, J., Dideberg, O., Van Beeumen, J., Frère, J.-M. & Ghuysen, J.-M. (1992) Primary and predicted secondary structures of the *Actinomadura* R39 extracellular DD-peptidase, a penicillin-binding protein (PBP) related to the *Escherichia coli* PBP4. *Biochem. J.* **282**, 781–788.

Kelly, J.A., Knox, J.R., Moews, P.C., Hite, G.J., Bartolone, J.B., Zhao, H., Joris, B., Frère, J.-M. & Ghuysen, J.-M. (1985) 2.8-Å Structure of penicillin-sensitive D-alanyl carboxypeptidase-transpeptidase from *Streptomyces* R61 and complexes with β-lactams. *J. Biol. Chem.* **260**, 6449–6458.

Korat, B., Mottl, H. & Keck, W. (1991) Penicillin-binding protein 4 of *Escherichia coli*: molecular cloning of the dacB gene, controlled overexpression, and alterations in murein composition. *Mol. Microbiol.* **5**, 675–684.

Lamotte-Brasseur, J., Jacob-Dubuisson, F., Dive, G., Frère, J.-M. & Ghuysen, J.-M. (1992) *Streptomyces* albus G serine β-lactamase. Probing of the catalytic mechanism via molecular modelling of mutant enzymes. *Biochem. J.* **282**, 189–195.

Oefner, C., D'Arcy, A., Daly, J.J., Gubernator, K., Charnas, R.L., Heinze, I., Hubschwerlen, C. & Winkler, F.K. (1990) Refined crystal structure of β-lactamase from *Citrobacter freundii* indicates a mechanism for β-lactam hydrolysis. *Nature* **343**, 284–288.

Palomeque-Messia, P., Englebert, S., Leyh-Bouille, M., Nguyen-Distèche, M., Duez, C., Houba, S., Dideberg, O., Van Beeumen, J. & Ghuysen, J.-M. (1991) Amino acid sequence of the penicillin-binding protein/DD-peptidase of *Streptomyces* K15. Predicted secondary structures of the low-$M_r$ penicillin-binding proteins of class A. *Biochem. J.* **279**, 223–230.

Strynadka, N.C.J., Adachi, H., Jensen, S.E., Johns, K., Sielecki, A., Betzel, C., Sutoh, K. & James, M.N.G. (1992) Molecular structure of the acyl-enzyme intermediate in $\beta$-lactam hydrolysis at 1.7 Å resolution. *Nature* **359**, 700–705.

# 142. Penicillin-binding protein 5, a serine-type D-Ala-D-Ala carboxypeptidase

## Databanks

*Peptidase classification: clan SE, family S11, MEROPS ID: S11.001*
*NC-IUBMB enzyme classification: EC 3.4.16.4*
*Databank codes:*

| Species | Gene | SwissProt | PIR | EMBL (cDNA) | EMBL (genomic) |
|---|---|---|---|---|---|
| *Bacillus stearothermophilus* | *dacA* | Q05523 | A16374 A61335 JN0801 S26433 | X68587 | – |
| *Bacillus subtilis* | *dacA* | P08750 | A23307 | M13766 | – |
| *Bacillus subtilis* | *dacB* | P35150 | S45552 | M84227 | – |
| *Bacillus subtilis* | *dacF* | P38422 | B42708 | M85047 | – |
| *Escherichia coli* | *dacA* | P04287 | A28536 | M18276 X06479 | AE000168 D90703 |
| *Escherichia coli* | *dacC* | P08506 | B28536 | X06480 | – |
| *Haemophilus influenzae* | *dacA* | P44466 | – | U32688 | – |
| *Streptococcus pneumoniae* | *dacA* | – | – | M37688 | – |

## Name and History

The peptidase activity termed *serine-type D-Ala-D-Ala carboxypeptidase* by NC-IUBMB is expressed by several proteins, all in clan SE, but in families S11, S12 or S13 (see Chapter 141). *Penicillin-binding protein 5 (PBP-5)* is one of these.

Tamura *et al*. (1976) separated *Escherichia coli* D-alanine carboxypeptidase I activity into three fractions characterized as D-alanine carboxypeptidases IA, IB and IC. Enzyme IA was membrane associated and in fact contained two proteins which were identified as *penicillin-binding protein 5 (PBP-5)* and penicillin-binding protein 6 (PBP-6) (Spratt & Strominger, 1976). The genes encoding PBP-5 and PBP-6 were identified in 1988 as the *dacA* and *dacC* genes, respectively (Broome-Smith, 1988). Extensive study of the membrane

anchoring system (Pratt *et al*., 1986; Jackson & Pratt, 1987; Phoenix, 1990; Phoenix & Pratt, 1990, 1993; van der Linden *et al*., 1993; Harris *et al*., 1995) and various mutant analyses (Broome-Smith & Spratt, 1984; Nicholas *et al*., 1985; Nicholas & Strominger, 1988a,b; Malhotra & Nicholas, 1992; van der Linden *et al*., 1994) were later carried out.

## Activity and Specificity

PBP-5 catalyzes two reactions:

1) $\text{R-D-Ala-D-Ala} + H_2O \longrightarrow \text{R-D-Ala} + \text{D-Ala}$
   (DD-carboxypeptidation)
2) $\text{R-D-Ala-D-Ala} + NH_2\text{-R}' \longrightarrow \text{R-D-Ala-NH-R}' + \text{D-Ala}$
   (DD-transpeptidation)

where R-D-Ala-D-Ala and $NH_2$-R′ are the donor and acceptor substrate, respectively.

## Substrate Specificity

Four donor substrates have been utilized to characterize the DD-carboxypeptidase activity of PBP-5 (Tamura *et al.*, 1976; Amanuma & Strominger, 1980; Ferreira *et al.*, 1988): uridine diphosphate-*N*-acetylmuramoyl-L-alanyl-D-$\gamma$-glutamyl-L-lysyl-D-alanyl┼D-alanine (UDPMur-AEKAA), uridine diphosphate-*N*-acetylmuramoyl-L-alanyl-D-$\gamma$-glutamyl-*meso*-2,6-diaminopimelyl-D-alanyl┼D-alanine (UDPMur-AEA$_2$pm-AA), $N^{\alpha}N^{\epsilon}$-diacetyl-L-lysyl-D-alanyl┼D-alanine (Ac$_2$KAA), and $N^{\alpha}N^{\epsilon}$-diacetyl-L-lysyl-D-alanyl┼D-lactate (Ac$_2$KALa). The specific activities with UDPMur-AEA$_2$pmAA, Ac$_2$KAA and Ac$_2$KALa were respectively 95, 4.4 and 92 $\mu$mol min$^{-1}$ mg$^{-1}$ (Amanuma & Strominger, 1980). The increase of specific activity when D-alanine is replaced by D-lactate in the substrates containing the Ac$_2$KA moiety showed that acylation might be the rate-limiting step of the overall reaction. The steady-state parameters on the purified enzyme were only determined with Ac$_2$KAA at pH 8.0 ($k_{cat} = 1.5$ s$^{-1}$, $K_m = 14$ mM and $k_{cat}/K_m = 110$ M$^{-1}$ s$^{-1}$) (van der Linden *et al.*, 1992).

Glycine and hydroxylamine have been utilized as acceptors in combination with Ac$_2$KAA, Ac$_2$KALa and UDPMur-AEA$_2$pmAA as donors to probe the transpeptidation reaction. In particular, the influence of glycine on the utilization of Ac$_2$KAA or Ac$_2$KALa was examined. In both cases, the increase of the transpeptidation compensated for the decrease of the hydrolysis so that the overall utilization of the donor (release of D-alanine or D-lactate) remained almost constant. The concentrations of glycine at which the same amounts of transpeptidation and hydrolysis products were formed were 15 mM and 17 mM with Ac$_2$KAA and Ac$_2$KALa as donor substrates, respectively (Amanuma & Strominger, 1980). Hydroxylamine was shown to be a less potent acceptor than glycine with both donor substrates (Tamura *et al.*, 1976).

## Mechanism

In the hydrolysis reaction, the enzyme follows the classical acyl enzyme pathway of serine proteases:

$$E + S \xrightleftharpoons{K} ES \xrightarrow{k_{+2}} \overset{P_1}{ES^*} \xrightarrow{k_{+3}} E + P_2$$

where ES*, $P_1$ and $P_2$ are the acyl enzyme, the first leaving group and the hydrolysis products, respectively.

## Assay and Influence of Physicochemical Conditions

Hydrolysis and transpeptidation reactions were formerly quantified by high-voltage paper electrophoresis using radioactive substrates (Tamura *et al.*, 1976; Amanuma & Strominger, 1980). The carboxypeptidase activity is now monitored by a colorimetric, discontinuous assay of the released D-alanine with Ac$_2$KAA as substrate (Frère *et al.*, 1976).

The enzyme showed a broad pH optimum (pH 8.5–10) in 100 mM Tris–HCl buffer (Amanuma & Strominger, 1980).

Some buffers inhibited the enzyme activity: 100 mM sodium or potassium borate (pH 9–10), 100 mM sodium or potassium phosphate (pH 6–8), 100 mM sodium carbonate (pH 9–10). In addition, the following compounds were reported to inhibit the enzyme activity (Tamura *et al.*, 1976) (the inhibitor concentration and the percentage of inhibition are indicated in parentheses): EDTA (8 mM, 58%), MgCl$_2$ (7 mM, 50%), MnCl$_2$ (10 mM, 60%), CoCl$_2$ (10 mM, 44%) and ZnCl$_2$ (10 mM, 96%).

## Inhibitors

Antibiotics of the $\beta$-lactam family are efficient inactivators that acylate the active-site serine according to the pathway described above for the substrates, but no $P_1$ is formed due to the cyclic $\beta$-lactam structure, and deacylation is slow so that an inactive acyl enzyme accumulates. So far, four penicillins (benzylpenicillin, ampicillin, penicillin V, 6-aminopenicillanate), seven cephalosporins (cephalexin, cephaloridine, cephalosporins C and G, cephalothin, cephradin, cefoxitin) and one amidino-penicillin (mecillinam) have been found to decrease the binding of [$^{14}$C]benzylpenicillin to the membrane form of PBP-5 or to inhibit the D-alanine carboxypeptidase activity of the purified enzyme IA (Tamura *et al.*, 1976; Spratt, 1977). Except for cefoxitin and cephalothin, the penicillins are better inhibitors than the cephalosporins and mecillinam. Hence, the concentration of $\beta$-lactam required to inhibit 50% of the enzyme activity in 15 min ranged from 3 to 43 $\mu$g ml$^{-1}$ for penicillins, cephalothin and cefoxitin, 90 $\mu$g ml$^{-1}$ for 6-aminopenicillanate and cephalosporin C, to more than 500 $\mu$g ml$^{-1}$ for the other inhibitors tested. The acylation rates have only been determined with benzylpenicillin ($k_{+2}/K' = 12\,000$ M$^{-1}$ s$^{-1}$: van der Linden *et al.*, 1994). The deacylation rates of PBP-5 are relatively high for a penicillin-binding protein. With benzylpenicillin, the acyl enzyme half-life is about 5 min (compared to more than 120 min with most other PBPs) and PBP-5 has been characterized as a weak penicillinase (Spratt, 1977). However, with cefoxitin and cephradin, no significant deacylation was observed after 45 min (Spratt, 1977). Benzylpenicillin is degraded into a mixture of phenylacetylglycine and penicilloic acid (Ananuma & Strominger, 1984).

## Structural Chemistry

The mature protein consists of one 374 residues polypeptide chain ($M_r$ 41 337) synthesized as a preprotein which contains a 29 residue N-terminal signal sequence that is cleaved upon translocation of the protein to the periplasm (Pratt *et al.*, 1981). The membrane association of PBP-5 is mediated by an 18 residue C-terminal membrane anchor forming an amphiphilic $\alpha$ helix that is embedded in the lipidic bilayer (Pratt *et al.*, 1981, 1986; Jackson *et al.*, 1985). Truncation of that C-terminal anchor yielded five soluble forms of the protein termed PBP-5s$^{353+9}$ (Ferreira *et al.*, 1988), PBP-5s$^{353}$, PBP-5s$^{347}$, PBP-5s$^{275}$ and PBP-5s$^{260}$ (van der Linden *et al.*, 1992). All but the last form retained full enzyme activity. The soluble form PBP-5s$^{353+9}$ has been used in numerous studies. *E. coli* PBP-5 belongs to the superfamily of the penicillin-recognizing enzymes (PREs) which comprises the serine $\beta$-lactamases and the penicillin-sensitive DD-peptidases (Frère *et al*., 1992). Accordingly, it contains the three conserved elements found in all PREs: Ser44-Xaa-Xaa-Lys47, Ser110-Xaa-Asn112 and Lys213-Thr214-Gly215, where Ser44 is the active-site serine (van der Linden *et al*., 1994). Crystallization has been reported by Ferreira *et al*. (1988).

## Preparation

Purification of PBP-5 was formerly carried out from cell lysates of *E. coli* in three steps: membrane solubilization using Triton X-100, covalent penicillin affinity and CM-cellulose column chromatography (Amanuma & Strominger, 1980; Ferreira *et al*., 1988). Much higher yields (50 mg liter$^{-1}$) were later obtained with the gene cloned under the control of the Lambda $P_R$ promoter. Purification is now achieved by dye-affinity chromatography (van der Linden *et al*., 1992).

## Biological Aspects

*E. coli* PBP-5 and the other low molecular mass PBPs (PBP-4 and PBP-6) are responsible for most of the DD-carboxypeptidase activity of the bacterium and are thought to maintain a correct balance between the various precursors involved in peptidoglycan synthesis in such a way that cells can elongate or divide (Frère *et al*., 1992). However, its precise physiological role remains obscure. PBP-5 and PBP-6 seem to be dispensable. Strains in which the chromosome contains single or double deletions for PBP-5 and/or PBP-6 grow with normal morphology under laboratory conditions (Broome-Smith, 1988). Note that PBP-5 and 6 account for 85% of all the PBPs (Frère *et al*., 1992) and are present at about 1800 and 570 molecules per cell, respectively (Spratt, 1977). Since the PBP-5 activity is 3- to 4-fold higher than that of PBP-6, PBP-5 represents 90–93% of the PBP-5/6 DD-carboxypeptidase activity.

## Distinguishing Features

Among the penicillin-binding proteins, PBP-5 is remarkable for the high deacylation rates observed with penicillins.

## Related Enzymes

Amino acid sequence alignments with other PBPs or $\beta$-lactamases showed 32 and 64% identity with PBP-5 from *Bacillus subtilis* (Joris *et al*., 1988) and PBP-6 from *Escherichia coli* (Broome-Smith *et al*., 1988), respectively.

## Further Reading

Several reviews have been published that summarize the properties of the DD-peptidases and their structural relationships with $\beta$-lactamases: Waxman & Strominger (1983), Frère & Joris (1985), Frère *et al*. (1992) and Ghuysen (1997). Jamin *et al*. (1995) specifically analyze the transpeptidase reaction.

## References

Amanuma, H. & Strominger, J.L. (1980) Purification and properties of penicillin-binding protein 5 and 6 from *Escherichia coli. J. Biol. Chem.* **259**, 1294–1298.

Amanuma, H. & Strominger, J.L. (1984) Simultaneous release of penicilloic acid and phenylacetylglycine by penicillin-binding protein 5 and 6 of *Escherichia coli. J. Bacteriol. Chem.* **160**, 822–823.

Broome-Smith, J.K. & Spratt, B.G. (1984) An amino acid substitution that blocks the deacylation step in the enzyme mechanism of penicillin-binding protein 5 of *Escherichia coli. FEBS Lett.* **165**, 185–189.

Broome-Smith, J.K. (1988) Construction of a mutant of *Escherichia coli* that has deletions of both penicillin-binding protein 5 and 6 genes. *J. Gen. Microbiol.* **131**, 2115–2118.

Broome-Smith, J.K., Ionnidis, I., Edelman, A. & Spratt, B.G. (1988) Nucleotide sequences of the penicillin-binding protein 5 and 6 of *Escherichia coli. Nucleic Acids Res.* **16**, 1617.

Ferreira, L.C.S., Schwarz, U., Keck, W., Charlier, P., Dideberg, O. & Ghuysen, J.-M. (1988) Properties and crystallization of a genetically engineered, water soluble derivative of penicillin-binding protein 5 of *Escherichia coli* K12. *Eur. J. Biochem.* **171**, 11–16.

Frère, J.-M. & Joris, B. (1985) Penicillin-sensitive enzymes in peptidoglycan biosynthesis. *CRC Crit. Rev. Microbiol.* **11**, 299–396.

Frère, J.-M., Leyh-Bouille, M., Ghuysen, J.-M., Nieto, M. & Perkins, H.R. (1976) Exocellular DD-carboxypeptidases-transpeptidases from *Streptomyces. Methods Enzymol.* **45**, 610–636.

Frère, J.-M., Nguyen-Distèche, M., Coyette, J. & Joris, B. (1992) Mode of action: interaction with the penicillin-binding proteins. In: *The Chemistry of Beta-Lactams* (Page, M.I., ed.). London: Chapman & Hall, pp. 148–197.

Ghuysen, J.-M. (1997) Penicillin-binding proteins. Wall peptidoglycan assembly and resistance to penicillin: facts, doubts and hopes. *Int. J. Antimicrob. Agents* **8**, 45–60.

Harris, F., Chatfield, L. & Phoenix, D.A. (1995) The possible involvement of anionic phospholipids in the anchoring of penicillin-binding protein 5 to the inner membrane of *Escherichia coli. Biochem. Soc. Trans.* **23**, 32S.

Jackson, M.E. & Pratt, J.M. (1987) An 18 amino acid amphiphilic helix forms the membrane-anchoring domain of *Escherichia coli* penicillin-binding protein 5. *Mol. Microbiol.* **1**, 23–28.

Jackson, M.E. & Pratt, J.M. (1988) Analysis of the membrane-binding domain of penicillin-binding protein 5 of *Escherichia coli. Mol. Microbiol.* **2**, 563–568.

Jackson, M.E., Pratt, J.M., Stocker, N.G. & Holland, B.I. (1985) An inner membrane protein N-terminal sequence is able to promote efficient localisation of an outer membrane-protein in *Escherichia coli. EMBO J.* **271**, 399–406.

Jamin, M., Wilkin, J.-M. & Frère, J.-M. (1995) Bacterial transpeptidases and penicillin. *Essays Biochem.* **29**, 1–24.

Joris, B., Ghuysen, J.-M., Dive, G., Renard, R., Dideberg, O., Charlier, P., Frère, J.M., Kelly, J.A., Boyington, J.C., Moews, P.C. & Knox, J.R. (1988) The active-site serine penicillin-recognizing enzymes as members of the Streptomyces R61 DD-peptidase family. *Biochem. J.* **250**, 313–324.

Malhotra, K.T. & Nicholas, R.A. (1992) Substitution of lysine 213 with arginine in penicillin-binding protein 5 of *Escherichia coli* abolishes D-alanine carboxypeptidase activity without affecting penicillin binding. *J. Biol. Chem.* **267**, 11386–11391.

Nicholas, R.A. & Strominger, J.L. (1988a) Site-directed mutants of a soluble form of penicillin-binding protein 5 from *Escherichia coli* and their catalytic properties. *J. Biol. Chem.* **263**, 2034–2040.

Nicholas, R.A. & Strominger, J.L. (1988b) Relations between β-lactamases and penicillin-binding proteins: β-lactamase activity of penicillin-binding protein 5 from *Escherichia coli*. *Rev. Infect. Dis.* **10**, 733–738.

Nicholas, R.A., Ishino, F., Park, W., Matsuhashi, M. & Strominger, J.L. (1985) Purification and sequencing of the active site tryptic peptide from penicillin-binding protein 5 from the *dacA* mutant strain of *Escherichia coli* (TMRL 1222). *J. Biol. Chem.* **260**, 6394–6397.

Phoenix, D.A. (1990) Investigation into the structural features of *Escherichia coli* penicillin-binding protein 5. *Biochem. Soc. Trans.* **18**, 948–949.

Phoenix, D.A. & Pratt, J.M. (1990) pH-induced insertion of the amphiphilic alpha-helical anchor of *Escherichia coli* penicillin-binding protein 5. *Eur. J. Biochem.* **190**, 365–369.

Phoenix, D.A. & Pratt, J.M. (1993) Membrane interaction of *Escherichia coli* penicillin-binding protein 5 is modulated by the ectomembranous domain. *FEBS Lett.* **332**, 215–218.

Pratt, J.M., Holland, B. & Spratt, B.J. (1981) Precursor forms of penicillin-binding protein 5 and 6 of *E. coli* cytoplasmic membrane. *Nature* **293**, 307–309.

Pratt, J.M., Jackson, M.E. & Holland, I.B. (1986) The C-terminus of penicillin-binding protein 5 is essential for localization to the *E. coli* inner membrane. *EMBO J.* **5**, 2399–2405.

Spratt, B.J. (1977) Properties of the penicillin-binding proteins of *Escherichia coli* K12. *Eur. J. Biochem.* **72**, 341–352.

Spratt, B.G. & Strominger, J.L. (1976) Identification of the major penicillin-binding proteins of *Escherichia coli* as D-alanine carboxypeptidase IA. *J. Bacteriol.* **127**, 660–663.

Tamura, T., Yasuo, I. & Strominger, J.L. (1976) Purification to homogeneity and properties of two D-alanine carboxypeptidases from *Escherichia coli*. *J. Biol. Chem.* **251**, 414–423.

van der Linden, M.P.G., Mottl, H. & Keck, W. (1992) Cytoplasmic high-level expression of a soluble, enzymatically active form of the *Escherichia coli* penicillin-binding protein 5 and purification by dye chromatography. *Eur. J. Biochem.* **204**, 197–202.

van der Linden, M.P.G., de Haan, L. & Keck, W. (1993) Domain organization of penicillin-binding protein 5 from *Escherichia coli* analyzed by C-terminal truncation. *Biochem. J.* **289**, 593–598.

van der Linden, M.P.G., de Haan, L., Dideberg, O. & Keck, W. (1994) Site-directed mutagenesis of proposed active-site residues of penicillin-binding protein 5 from *Escherichia coli*. *Biochem. J.* **303**, 357–362.

Waxman, D.J. & Strominger, J.L. (1983) Penicillin-binding proteins and the mechanism of action of β-lactam antibiotics. *Annu. Rev. Biochem.* **52**, 825–869.

*Jean-Marc Wilkin*
*Center for Protein Engineering and Laboratory of Enzymology,*
*University of Liège,*
*Institut de Chemistry, B6 Sart Tilman,*
*B-4000 Liège, Belgium*
*Email: JM.Wilkin@ulg.ac.be*

# 143. Murein DD-endopeptidase PBP-7

## Databanks

*Peptidase classification: clan SE, family S11, MEROPS ID: S11.002*
*NC-IUBMB enzyme classification: none*
*Databank codes:*

| Species | SwissProt | PIR | EMBL (cDNA) | EMBL (genomic) |
|---|---|---|---|---|
| *Escherichia coli* | P33364 | – | – | U00007: 47–48 centrosome region of chromosome |
| *Haemophilus influenzae* | P44664 | – | U32720 | – |
| *Pseudomonas aeruginosa* | – | – | U62582 | |

## Name and History

Although they were described as early as the first publications on the penicillin-binding proteins (PBPs) of *Escherichia coli* (Spratt, 1975, 1977), the functions and enzymatic specificity of the low molecular mass PBP-7 and PBP-8 remained obscure, and the proteins were even suggested to represent degradation products of other PBPs (Barbas *et al.*, 1986). One of the problems with these two proteins was an inconsistency in their appearance in the PBP assay, which could not be explained (Rodriguez-Tebar *et al.*, 1985a). A direct relationship (two forms of the same protein) between PBP-7 and PBP-8 had been postulated (Rodriguez-Tebar *et al.*, 1985b), and eventually it was possible to show that PBP-8 originates from PBP-7 as a result of proteolytic degradation by omptin (Chapter 176) (Henderson *et al.*, 1994).

Being a protein that is able to bind penicillin covalently, PBP-7 was likely to be an enzyme that interacts with the DD-peptide bond present in the murein peptidoglycan of the bacterial cell wall (Rogers *et al.*, 1980; Ghuysen & Hakenbeck, 1994). The purification and characterization of the protein did indeed show that it is an autolytic enzyme, degrading whole murein sacculi of *E. coli* by splitting the peptide cross-bridges that interlink the glycan strands in the murein net (Romeis & Höltje, 1994a). Accordingly, it is termed *murein* DD-*endopeptidase PBP-7*.

## Activity and Specificity

PBP-7 (like its degradation product PBP-8) is a penicillin-sensitive DD-endopeptidase that accepts insoluble, high molecular mass murein sacculi as substrate. The bond hydrolyzed by the enzyme is the DD peptide bond between the $\varepsilon$-amino group of the *meso*-diaminopimelic acid of one peptide moiety and the $\alpha$-carboxyl group of the terminal D-alanine of a second peptide moiety (L-Ala-D-Glu-*meso*-diaminopimelic acid$+$D-Ala-*meso*-diaminopimelic acid-D-Glu-L-Ala); therefore this cleavage reverses the transpeptidation reaction of the synthetic PBPs (Izaki *et al.*, 1966). The bonds split include all characterized DD cross-links in dimeric, trimeric and tetrameric cross-links (for a complete discussion of the structure of the murein and of all possible DD cross-links, see Glauner *et al.*, 1988). Isolated, soluble, low molecular mass muropeptides such as lysozyme degradation products are not accepted as substrates. The second type of cross-bridge found in the murein, the LD bond between two *meso*-diaminopimelic acid residues of neighboring peptide moieties (Glauner *et al.*, 1988), is not hydrolyzed by this endopeptidase, so the enzyme is highly specific for DD peptide bonds. (As a contrasting example of an unspecific murein endopeptidase, see Chapter 556 on penicillin-insensitive murein endopeptidase.)

The enzyme has a pH optimum of about 6.2, and is most active at temperatures between 37 and 42°C. It is sensitive to high salt concentration so that 100 mM NaCl inhibits the enzyme by 80%.

Enzyme activity is easily measured by following the formation of soluble murein degradation products (muropeptides) from high molecular mass sacculi. A sensitive assay uses radioactively labeled murein sacculi as substrate (Romeis & Höltje, 1994a).

As expected of a penicillin-binding protein, the enzyme is inactivated by $\beta$-lactam antibiotics. Certain members of the penem class of antibiotics bind with high affinity to this PBP (Tuomanen & Schwartz, 1987).

## Structural Chemistry

The DD-endopeptidase PBP-7 is a protein of about 31.2 kDa and a pI of 10.2. The degradation product PBP-8 (29.8 kDa, pI 10.2) lacking the C-terminus, retains enzymatic activity. Being a soluble, periplasmic protein, the enzyme is synthesized as a preprotein, and the leader peptide is cleaved off after export into the periplasm. Omptin, an outer membrane protease, is responsible for the formation of PBP-8, which lacks the 17 C-terminal amino acids.

The similarity to other low molecular mass PBPs is high, with the known penicillin-binding DD-carboxypeptidases showing the highest similarity scores (Henderson *et al.*, 1995). The similarity to another group of DD-endopeptidases, penicillin-binding protein 4 (Chapter 148), is less obvious, but still a significant similarity around the active center can be found. Using the sequences of PBP-7 and PBP-4 from *E. coli* and the homologs from *H. influenzae* (Fleischmann *et al.*, 1995), a common profile can be defined, which only detects these enzymes. The characterized DD-carboxypeptidases do not contain this profile, so it is possible to detect the DD-endopeptidases selectively in database searches.

## Preparation

Different purification procedures for PBP-7/8 from *E. coli* have been described (Romeis & Höltje, 1994a; Henderson *et al.*, 1995). They yield essentially homogeneous preparations with a 1600- to 4000-fold enrichment of the protein. An expression system that uses a MalE-fusion has been used to prepare milligrams of the protein (G. Schiffer & J.-V. Höltje, unpublished results).

## Biological Aspects

The enzyme has been isolated from *E. coli* and characterized biochemically. The presence of homologous proteins could be shown in the gram-negative organisms *H. influenzae* (Fleischmann *et al.*, 1995) and *Pseudomonas aeruginosa*.

The gene *pbpG* could be mapped to 48 min on the *E. coli* chromosome and has been sequenced (Henderson *et al.*, 1995). A s70 promoter directly upstream of the coding sequence could be demonstrated and a terminator at the end of the gene has been identified (Hara *et al.*, 1996). The characterization of a multicopy suppressor of a *spr* mutation led also to the cloning of the structural gene for PBP-7, which in this work was designated *psv* (Hara *et al.*, 1996).

From experiments with the penem antibiotics Imipenem and CGP 31608, which could be shown to bind with high affinity to PBP-7 (Tuomanen & Schwartz, 1987), a role of this endopeptidase during growth and antibiotic-induced autolysis was proposed (Cozens *et al.*, 1989). The specific inhibition of PBP-7 has been claimed to cause lysis of nongrowing *E. coli* (Tuomanen & Tomasz, 1986). In apparent contradiction to this report, a deletion mutant in *pbpG* is viable and an obvious phenotype could not be demonstrated (Henderson *et al.*, 1995).

Murein DD-endopeptidase activity together with glycosylase activity has been postulated to be required during growth and division of the murein sacculus of *E. coli* (Höltje & Glauner, 1990; Höltje, 1995). Indeed, protein–protein interactions between PBP-7 and lytic transglycosylases could be demonstrated. Affinity chromatography of *E. coli* extracts on immobilized lytic transglycosylases, such as Slt70-Sepharose, showed an enrichment of PBP-8 together with the murein synthesizing enzymes PBP-1B and PBP-3 (Romeis & Höltje, 1994b; von Rechenberg *et al.*, 1996). Using PBP-7 as a specific ligand for affinity chromatography, similar results could be obtained (G. Schiffer & J.-V. Höltje, unpublished results). In addition, PBP-7/8 was found not only to stabilize but also to stimulate the enzymatic activity of Slt70 by a protein–protein interaction (Romeis & Höltje, 1994b). These results and studies on the metabolism of the murein sacculus (Glauner & Schwarz, 1988; Glauner & Höltje, 1990) suggest the existence of a multienzyme complex, consisting of murein synthases and murein hydrolases. The DD-endopeptidase PBP-7 has been proposed to be part of this hypothetical holoenzyme, which catalyzes the enlargement and division of the murein sacculus during growth of *E. coli* (Höltje, 1996).

## Distinguishing Features

Besides PBP-7, two other enzymes are known to split the DD bond in the cross-linked peptide side chains in the murein. Penicillin-binding protein 4 (Chapter 148), also a penicillin-sensitive peptidase, not only has DD-carboxypeptidase activity but also shows DD-endopeptidase activity. However, unlike PBP-7, this enzyme is able to act on isolated muropeptides. The other known murein endopeptidase, penicillin-insensitive murein endopeptidase (Chapter 556), is not only penicillin-insensitive, but also hydrolyzes the LD-type cross-links found in the murein sacculus of *E. coli* (Engel *et al.*, 1992) as well as the DD peptide bonds.

## Further Reading

Recommended reviews are those of Ghuysen & Hakenbeck (1994) and Rogers *et al.* (1980).

## References

Barbas, J.A., Diaz, J., Rodriguez-Tebar, A. & Vazquez, D. (1986) Specific location of penicillin-binding proteins within the cell envelope of *Escherichia coli*. *J. Bacteriol.* **165**, 269–275.

Cozens, R.M., Markiewicz, Z. & Tuomanen, E. (1989) Role of autolysins in the activities of imipenem and CGP 31608, a novel penem, against slowly growing bacteria. *Antimicrob. Agents Chemother.* **33**, 1819–1821.

Engel, H., van Leeuwen, A., Dijkstra, A. & Keck, W. (1992) Enzymatic preparation of 1,6-anhydro-muropeptides by immobilized murein hydrolases from *Escherichia coli* fused to staphylococcal protein A. *Appl. Microbiol. Biotechnol.* **37**, 772–783.

Fleischmann, R.D. *et al.* (1995) Whole-genome random sequencing and assembly of *Haemophilus influenzae* Rd. *Science* **269**, 496–512.

Ghuysen, J.-M. & Hakenbeck, R. (1994) Bacterial cell wall. In: *New Comprehensive Biochemistry*, vol. 27 (Neuberger, A. & van Deenen, L.L.M., eds). Amsterdam: Elsevier.

Glauner, B. & Höltje, J.-V. (1990) Growth pattern of the murein sacculus of *Escherichia coli*. *J. Biol. Chem.* **265**, 18988–18996.

Glauner, B. & Schwarz, U. (1988) Investigation of murein structure and metabolism by high pressure liquid chromatography. In: *Modern Microbiological Methods*, vol. III: *Bacterial Cell Surface Techniques* (Hancock, I.C. & Poxton, I.R., eds). Chichester: John Wiley & Sons, pp. 158–171.

Glauner, B., Höltje, J.-V. & Schwarz, U. (1988) The composition of the murein of *Escherichia coli*. *J. Biol. Chem.* **263**, 10088–10095.

Hara, H., Abe, N., Nakakouji, M., Nishimura, Y. & Horiuchi, K. (1996) Overproduction of penicillin-binding protein 7 suppresses thermosensitive growth defect at low osmolarity due to an spr mutation of *Escherichia coli*. *Microb. Drug Resistance* **2**, 63–72.

Henderson, T.A., Dombrosky, P.M. & Young, K.D. (1994) Artifactual processing of penicillin-binding proteins 7 and 1b by the OmpT protease of *Escherichia coli*. *J. Bacteriol.* **176**, 256–259.

Henderson, T.A., Templin, M. & Young, K.D. (1995) Identification and cloning of the gene encoding penicillin-binding protein 7 of *Escherichia coli*. *J. Bacteriol.* **177**, 2074–2079.

Höltje, J.-V. (1995) From growth to autolysis: the murein hydrolases in *Escherichia coli*. *Arch. Microbiol.* **164**, 243–254.

Höltje, J.-V. (1996) A hypothetical holoenzyme involved in the replication of the murein sacculus of *Escherichia coli*. *Microbiology* **142**, 1911–1918.

Höltje, J.-V. & Glauner, B. (1990) Structure and metabolism of the murein sacculus. *Res. Microbiol.* **141**, 75–89.

Izaki, K., Matsuhashi, M. & Strominger, J.L. (1966) Glycopeptide transpeptidase and D-alanine carboxypeptidase: penicillin-sensitive enzymatic reactions. *Proc. Natl Acad. Sci. USA* **55**, 656–663.

Rodriguez-Tebar, A., Barbas, J.A. & Vazquez, D. (1985a) Location of some proteins involved in peptidoglycan synthesis and cell division in the inner and outer membranes of *Escherichia coli*. *J. Bacteriol.* **161**, 243–248.

Rodriguez-Tebar, A., Prats, R., Diaz, J.V.A., Barbas, J.A. & Vazquez, D. (1985b) Penicillin-binding proteins and cell division in *Escherichia coli*. In: *Proceedings of the 16th FEBS Congress*, part B (Ovchinnikov, Y.A., ed.). Utrecht: VNU Science Press, pp. 551–563

Rogers, H.J., Perkins, H.R. & Ward, J.B. (1980) *Microbial Cell Walls and Membranes*. London: Chapman & Hall.

Romeis, T. & Höltje, J.-V. (1994a) Penicillin-binding protein 7/8 of *Escherichia coli* is a DD-endopeptidase. *Eur. J. Biochem.* **224**, 597–604.

Romeis, T. & Höltje, J.-V. (1994b) Specific interaction of penicillin-binding proteins 3 and 7/8 with the soluble lytic transglycosylase in *Escherichia coli*. *J. Biol. Chem.* **269**, 21603–21607.

Spratt, B.G. (1975) Distinct penicillin binding proteins involved in the division, elongation, and shape of *Escherichia coli* K12. *Proc. Natl Acad. Sci. USA* **72**, 2999–3003.

Spratt, B.G. (1977) Properties of the penicillin binding proteins of *Escherichia coli* K12. *Eur. J. Biochem.* **72**, 341–352.

Tuomanen, E. & Schwartz, J. (1987) Penicillin-binding protein 7 and its relationship to lysis of nongrowing *Escherichia coli. J. Bacteriol.* **169**, 4912–4915.

Tuomanen, E. & Tomasz, A. (1986) Induction of autolysis in non-growing *Escherichia coli. J. Bacteriol.* **167**, 1077–1080.

von Rechenberg, M., Ursinus, A. & Höltje, J.-V. (1996) Affinity chromatography as a means to study multi-enzyme-complexes involved in murein synthesis. *Microb. Drug Resistance* **2**, 155–158.

***Markus F. Templin***
*Abteilung Biochemie,*
*Max-Planck-Institut für Entwicklungsbiologie,*
*Spemannstrasse 35,*
*72076 Tübingen, Germany*
*Email: marcustemplin@tuebingen.mpg.de*

***Joachim-Volker Höltje***
*Abteilung Biochemie*
*Max-Planck-Institut für Entwicklungsbiologie*
*Spemannstrasse 35*
*72076 Tübingen, Germany*
*Email: joho@tuebingen.mpg.de*

# 144. *Streptomyces K15* D-Ala-D-Ala transpeptidase

## Databanks

*Peptidase classification: clan SE, family S11, MEROPS ID: S11.004*
*NC-IUBMB enzyme classification: none*
*Databank codes:*

| Species | SwissProt | PIR | EMBL (cDNA) | EMBL (genomic) |
|---|---|---|---|---|
| *Streptomyces* K15 | P39042 | S04638 S17674 | X59965 | – |

## Name and History

Serine-type acyl transferases that characteristically cleave the C-terminal peptide bond of acyl-D-alanyl-D-alanine peptides are involved in the assembly and metabolism of the bacterial cell wall peptidoglycan. They couple breaking the peptide bond and transfer of the acyl-D-alanyl moiety to the amino group of another peptide (DD-transpeptidase activity) or to water (DD-carboxypeptidase activity) via the formation of a serine ester-linked acyl-D-alanyl enzyme. With penicillin, they form a long-lived, serine ester-linked, penicilloyl enzyme, and thus behave as penicillin-binding proteins (PBPs) (Ghuysen, 1991).

Exocellular DD-carboxypeptidases and membrane-bound DD-transpeptidases of *Streptomyces* sp. were identified using $N^{\alpha}N^{\varepsilon}$-diacetyl($Ac_2$)-L-lysyl-D-alanyl-D-alanine as an analog of the natural carbonyl donor and glycylglycine as an analog of the natural amino acceptor in *Streptomyces* (Pollock *et al.*, 1972; Dusart *et al.*, 1973; Ghuysen *et al.*, 1973, 1974a,b; Marquet *et al.*, 1974). Membrane-bound DD-transpeptidases had peculiar properties (Dusart *et al.*, 1975, 1977; Leyh-Bouille *et al.*, 1977; Nguyen-Distèche *et al.*, 1977). Membrane preparations were enzymatically active in the frozen state at temperatures down to $-35°C$. The enzymes were extractible with *N*-cetyl-*N,N,N*-trimethylammonium bromide (Cetavlon) in an active form. Under certain conditions, the Cetavlon-extracted enzymes catalyzed the quantitative transpeptidation of the tripeptide donor in aqueous media. The **Streptomyces K15 D-Ala-D-Ala transpeptidase** was identified as a serine enzyme by active-site peptide mapping (Leyh-Bouille *et al.*, 1989).

## Activity and Specificity

In water, acylation of the essential serine of the serine D-Ala-D-Ala transpeptidase by the carbonyl donors $Ac_2$-L-lysyl-D-alanyl-(COR)-D-CH(CH$_3$)COOH (where R is NH, O or

S) produces the same peptidyl(Ac$_2$-L-lysyl-D-alanyl) enzyme, but the course of the reaction depends on the nature of the leaving group (Nguyen-Distèche *et al.*, 1986; Grandchamps *et al.*, 1995). In the case of the peptide, the released D-alanine is utilized as amino acceptor in the enzyme deacylation step, the peptide donor is continuously regenerated, and the enzyme is seemingly inert, although it turns over ($k_{cat} \approx 0.1 \, s^{-1}$; $K_m = 8 \, mM$). In the case of the ester and thioester, the released D-lactate and D-thiolactate cannot compete with water, the peptidyl enzyme undergoes hydrolytic breakdown, and the ester and thioester are hydrolyzed to completion.

In the presence of 2–5 mM glycylglycine, the enzyme-catalyzed conversion of the peptide, ester and thioester carbonyl donors is quantitatively channeled into the transpeptidated product Ac$_2$-L-lysyl-D-alanyl-glycylglycine. Kinetic studies suggest that in the case of the peptide donor, glycylglycine binds to the Michaelis complex and increases the rate of enzyme acylation by the donor. In the case of the ester and thioester donors, glycylglycine behaves as a simple alternative nucleophile at the level of the peptidyl enzyme.

Formation of the serine ester-linked acyl(Ac$_2$-L-lysyl-D-alanyl) enzyme implies that the carbonyl donor binds to the enzyme in a position that allows the proton of the serine $\gamma$OH to be abstracted, the activated $\ddot{O}^\gamma$ to attack the carbonyl carbon atom of the scissile bond, and the abstracted proton to be back-donated to the adjacent atom. Replacement of the penultimate D-alanine of the tripeptide Ac$_2$-L-lysyl-D-alanyl-D-alanine by D-leucine, glycine or L-alanine, and replacement of the C-terminal D-alanine by L-alanine abolishes donor activity. Replacement of the C-terminal D-alanine by D-leucine or glycine and shortening the neutral side chain of L-lysine results in a much decreased donor activity.

Aminolysis of the serine ester-linked acyl(Ac$_2$-L-lysyl-D-alanyl) enzyme implies that the amino acceptor binds to the enzyme in a position that allows the proton of the NH$_2$ group to be abstracted, the activated $\ddot{N}$H to attack the carbonyl carbon atom of the ester bond and the abstracted proton to be back-donated to the O$^\gamma$ atom of the serine residue. Of all the amino compounds tested, glycylglycine and glycyl-L-alanine have the highest acceptor activity. In comparison with D-alanine, *meso*-diaminopimelate, D-aminoadipate and glycine are acceptors of lower efficacy, and D-lactate and D-thiolactate have negligible activity.

### Active-site-directed β-Lactam Inactivators

The serine DD-transpeptidase is a PBP (Leyh-Bouille *et al.*, 1986). The value of the second-order rate constant of acylation by benzylpenicillin is $150 \, M^{-1} \, s^{-1}$. The benzylpenicilloyl enzyme undergoes hydrolytic breakdown at a very slow rate via two pathways. Hydrolysis of the serine ester linkage (first-order rate constant: $8 \times 10^{-5} \, s^{-1}$) causes the release of penicilloate which is the reaction product of β-lactamase action on penicillin. Hydrolytic attack on the C$_5$-C$_6$ bond (first-order rate constant: $1 \times 10^{-5} \, s^{-1}$) causes the formation of a new acyl(phenylacetylglycyl) enzyme that is rapidly hydrolyzed and/or aminolysed. Combining the two pathways, the enzyme turns over once every 10 000 s. In comparison with benzylpenicillin, cefoxitine is a better inactivating agent. The value of the second-order rate constant of enzyme acylation is

$860 \, M^{-1} \, s^{-1}$, and the enzyme turns over more slowly, once every 170 000 s.

### Preparation

The wild-type, Cetavlon-extracted, serine D-Ala-D-Ala transpeptidase was purified with a 8000-fold specific enrichment by Sephadex filtration and affinity chromatography on ampicillin-linked CH Sepharose 4B in the presence of Cetavlon (Nguyen-Distèche *et al.*, 1982). Gene overexpression in *Streptomyces lividans* (Palomeque-Messia *et al.*, 1991, 1992) resulted in the export of an appreciable amount of the synthesized enzyme (yield 4 mg enzyme liter$^{-1}$ of culture supernatant). The cloned enzyme was purified by a two-step procedure (acetone precipitation and dialysis) in the absence of detergent with a yield of 75%. The purified enzyme requires the presence of 0.5 M NaCl to remain soluble. It is indistinguishable from the Cetavlon-extracted enzyme.

### Structural Chemistry

The serine D-Ala-D-Ala transpeptidase is synthesized as a 291 amino acid residue precursor that contains a cleavable 29 amino acid signal peptide (Palomeque-Messia *et al.*, 1991). The mature 262 amino acid residue protein lacks transmembrane segments. It possesses two cysteine residues at positions 98 and 223, respectively, one of which (presumably Cys98) is accessible to PCMB, resulting in a drastic decrease of peptidase activity and penicillin-binding capacity (Leyh-Bouille *et al.*, 1987).

The serine D-Ala-D-Ala transpeptidase has the amino acid sequence and fold signature of the penicilloyl serine transferases superfamily (Ghuysen, 1991). It is a two-globular domain protein (Englebert *et al.*, 1994; E. Fonzé, M. Vermeire, Y. Toth, M. Nguyen-Distèche, J.-M. Ghuysen & P. Charlier, unpublished results). One domain is of the all-$\alpha$ type. The other domain is of the $\alpha/\beta$ type consisting of a five-stranded $\beta$ sheet that is covered by $\alpha$ helices on both faces. The active site, located between the two domains, is defined by three structural elements. A long $\alpha$ helix of the all-$\alpha$ domain containing motif Ser35-Thr-Thr-Lys (where Ser35 is the essential serine) is central to the cavity. One loop connecting two other $\alpha$ helices of the all-$\alpha$ domain and containing motif Ser96-Gly-Cys forms one side of the cavity. The other side is occupied by the Lys213-Thr-Gly-Thr motif-containing strand $\beta$3 of the five-stranded sheet.

An $\alpha$ carbon atom overlay shows that the three active-site-defining motifs Ser35-Thr-Thr-Lys/Ser70-Thr-Phe-Lys, Ser96-Gly-Cys/Ser130Asp-Asn and Lys213-Thr-Gly-Thr/Lys234-Ser-Gly-Ala of the serine D-Ala-D-Ala transpeptidase and the *Escherichia coli* TEM serine β-lactamase of class A superimpose almost exactly, and that polypeptide backbone segments of the two enzymes, representing 170 amino acid residues of the transpeptidase, superimpose with an average r.m.s. value of 2.3 Å. However, an X-ray structure-based alignment reveals that the two penicilloyl serine transferases have insignificant similarity in their amino acid sequences (15% identities). In comparison with the class A β-lactamases, the serine D-Ala-D-Ala transpeptidase lacks the N-terminal $\alpha$ helix, the bottom of its active site is constructed differently, and a four-stranded $\beta$ sheet exposed at the surface of the all-$\alpha$ domain may be the site of attachment of the protein to the plasma membrane.

There is more similarity between the serine D-Ala-D-Ala transpeptidase and the class A $\beta$-lactamases (group 1), or between the *Streptomyces* R61 serine D-Ala-D-Ala carboxypeptidase and the class C $\beta$-lactamases (group 2) than between the transpeptidase and the carboxypeptidase or between the class A and class C $\beta$-lactamases (E. Fonzé, M. Vermeire, Y. Toth, M. Nguyen-Distèche, J.-M. Ghuysen & P. Charlier, unpublished results). The active-site serine-containing $\alpha$ helices and the five-stranded $\beta$ sheets of the enzymes of groups 1 and 2 are superimposable, but the other $\alpha$ helices have different lengths and orientations. In addition, the active-site-defining loops are displaced, allowing the OH group of the serine residue of motif Ser-Gly-Cys/Ser-Asp-Asn (of group 1) and the OH group of the tyrosine residue of motif Tyr-Xaa-Asn (of group 2) to occur at equivalent positions with respect to the nucleophilic serine and strand $\beta 3$.

It is evident that the *Streptomyces* K15 serine D-Ala-D-Ala transpeptidase, the *Streptomyces* R61 serine D-Ala-D-Ala carboxypeptidase and the serine $\beta$-lactamases well illustrate the notion of a superfamily (or clan, in the context of the present volume) that includes proteins with no overall similarity in sequences or in biological functions, but nevertheless with similar three-dimensional structures and catalytic properties.

## Further Reading

Fuller, recent reviews are those of Ghuysen *et al.* (1996) and Ghuysen (1997).

## References

Dusart, J., Marquet, A., Ghuysen, J.-M., Frère, J.-M., Moreno, R., Leyh-Bouille, M., Johnson, K., Lucchi, Ch., Perkins, H.R. & Nieto, M. (1973) DD-Carboxypeptidase-transpeptidase and killing site of $\beta$-lactam antibiotics in *Streptomyces* strains R39, R61 and K11. *Antimicrob. Agents Chemother.* **3**, 181–187.

Dusart, J., Marquet, A., Ghuysen, J.-M. & Perkins, H.R. (1975) The catalytic activity and penicillin sensitivity in the liquid and frozen states of membrane-bound and detergent-solubilized transpeptidases of *Streptomyces* R61. *Eur. J. Biochem.* **56**, 57–65.

Dusart, J., Leyh-Bouille, M. & Ghuysen, J.-M. (1977) The peptidoglycan crosslinking enzyme system in *Streptomyces* strains R61, K15 and *rimosus*. Kinetic coefficients involved in the interactions of the membrane-bound transpeptidase with peptide substrates and $\beta$-lactam antibiotics. *Eur. J. Biochem.* **81**, 33–44.

Englebert, S., Charlier, P., Fonzé, E., Toth, Y., Vermeire, M., Van Beeumen, J., Grandchamps, J., Hoffmann, K., Leyh-Bouille, M., Nguyen-Distèche, M. & Ghuysen, J.-.M. (1994) Crystallization and X-ray diffraction study of the *Streptomyces* K15 penicillin-binding DD-transpeptidase. *J. Mol. Biol.* **241**, 295–297.

Ghuysen, J.-M. (1991) Serine $\beta$-lactamases and penicillin-binding proteins. *Annu. Rev. Microbiol.* **45**, 37–67.

Ghuysen, J.-M. (1997) Penicillin-binding proteins. Wall peptidoglycan assembly and resistance to penicillin. Facts, doubts and hopes. *Int. J. Antimicrob. Agents* **8**, 45–60.

Ghuysen, J.-M., Leyh-Bouille, M., Campbell, J.N., Moreno, R., Frère, J.-M., Duez, C., Nieto, M. & Perkins, H.R. (1973) Structure of the wall peptidoglycan of *Streptomyces* R39 and the specificity profile of its exocellular DD-carboxypeptidase-transpeptidase

for peptide acceptors. *Biochemistry* **12**, 1243–1251.

Ghuysen, J.-M., Leyh-Bouille, M., Frère, J.-M., Dusart, J., Marquet, A., Perkins, H.R. & Nieto, M. (1974a) The penicillin receptor in *Streptomyces*. *Ann. N.Y. Acad. Sci.* **235**, 236–266.

Ghuysen, J.-M., Reynolds, P.E., Perkins, H.R., Frère, J.-M. & Moreno, R. (1974b) Effects of donor and acceptor peptides on concomitant hydrolysis and transfer reactions catalysed by the exocellular DD-carboxypeptidase-transpeptidase from *Streptomyces* R39. *Biochemistry* **13**, 2539–2547.

Ghuysen, J.-M., Charlier, P., Coyette, J., Duez, C., Fonzé, E., Fraipont, C., Goffin, C., Joris, B. & Nguyen-Distèche, M. (1996) Penicillin and beyond. Evolution, protein fold, multimodular polypeptides and multiprotein complexes. *Microb. Drug Resistance: Mech. Epidemiol. Dis.* **2**, 163–175.

Grandchamps, J., Nguyen-Distèche, M., Damblon, C., Frère, J.-M. & Ghuysen, J.-M. (1995) The *Streptomyces* K15 active-site serine DD-transpeptidase. Specificity profile for peptide, thiolester and ester carbonyl donors and pathways of the transfer reactions. *Biochem. J.* **307**, 335–339.

Leyh-Bouille, M., Dusart, J., Nguyen-Distèche, M., Ghuysen, J.-M., Reynolds, P.E. & Perkins, H.R. (1977) The peptidoglycan crosslinking enzyme system in *Streptomyces* strains R61, K15 and *rimosus*. Exocellular, lysozyme-releasable and membrane-bound enzymes. *Eur. J. Biochem.* **81**, 19–28.

Leyh-Bouille, M., Nguyen-Distèche, M., Pirlot, S., Veithen, A., Bourguignon, C. & Ghuysen, J.-M. (1986) *Streptomyces* K15 DD-peptidase-catalysed reaction with suicide $\beta$-lactam carbonyl donors. *Biochem. J.* **235**, 177–182.

Leyh-Bouille, M., Nguyen-Distèche, M., Bellefroid-Bourguignon, C. & Ghuysen, J.-M. (1987) Effects of thiol reagents on the *Streptomyces* K15 DD-peptidase-catalysed reactions. *Biochem. J.* **241**, 893–897.

Leyh-Bouille, M., Van Beeumen, J., Renier-Pirlot, S., Joris, B., Nguyen-Distèche, M. & Ghuysen, J.-M. (1989) The *Streptomyces* K15 DD-peptidase/penicillin-binding protein. Active-site and sequence of the amino terminal region. *Biochem. J.* **260**, 601–604.

Marquet, A., Dusart, J., Ghuysen, J.-M. & Perkins, H.R. (1974) Membrane-bound transpeptidase and penicillin binding sites in *Streptomyces* R61. *Eur. J. Biochem.* **46**, 515–523.

Nguyen-Distèche, M., Frère, J.-M., Dusart, J., Leyh-Bouille, M., Ghuysen, J.-M., Pollock, J.J. & Iacono, V.J. (1977) The peptidoglycan crosslinking enzyme system in *Streptomyces* R61, K15 and rimosus. Immunological studies. *Eur. J. Biochem.* **81**, 29–32.

Nguyen-Distèche, M., Leyh-Bouille, M. & Ghuysen, J.-M. (1982) Isolation of the membrane-bound 26,000-Mr penicillin-binding protein of *Streptomyces* strain K15 in the form of a penicillin-sensitive D-alanyl-D-alanine-cleaving transpeptidase. *Biochem. J.* **207**, 109–115.

Nguyen-Distèche, M., Leyh-Bouille, M., Pirlot, S., Frère, J.-M. & Ghuysen, J.-M. (1986) *Streptomyces* K15 DD-peptidase-catalysed reactions with ester and amide carbonyl donors. *Biochem. J.* **235**, 167–176.

Palomeque-Messia, P., Englebert, S., Leyh-Bouille, M., Nguyen-Distèche, M., Duez, C., Houba, S., Dideberg, O., Van Beeumen, J. & Ghuysen, J.-M. (1991) Amino acid sequence of the penicillin-binding protein/DD-peptidase of *Streptomyces* K15. Predicted secondary structures of the low-Mr penicillin-binding proteins of class A. *Biochem. J.* **279**, 223–230.

Palomeque-Messia, P., Quittre, V., Leyh-Bouille, M., Nguyen-Distèche, M., Gershater, C.J.L., Dacey, I.K., Dusart, J., Van Beeumen, J. & Ghuysen, J.-M. (1992) Secretion by overexpression

and purification of the water-soluble *Streptomyces* K15 DD-transpeptidase/penicillin-binding protein. *Biochem. J.* **288**, 87–91.
Pollock, J.J., Ghuysen, J.-M., Linder, R., Salton, M.R.J., Perkins,

H.R., Nieto, M., Leyh-Bouille, M., Frère, J.-M. & Johnson, K. (1972) Transpeptidase activity of *Streptomyces* D-alanyl-D carboxypeptidases. *Proc. Natl Acad. Sci. USA* **69**, 662–666.

*Jean-Marie Ghuysen*
*Centre d'Ingénierie des Protéines,*
*Université de Liège,*
*Institut de Chimie, B6,*
*Sart Tilman, B-4000 Liège, Belgium*

# 145. *Streptomyces R61 D-Ala-D-Ala carboxypeptidase*

## Databanks

*Peptidase classification: clan SE, family S12, MEROPS ID: S12.001*
*NC-IUBMB enzyme classification: EC 3.4.16.4*
*Databank codes:*

| Species | SwissProt | PIR | EMBL (CDNA) | EMBL (genomic) |
|---|---|---|---|---|
| *Bacillus subtilis* | – | – | – | – |
| *Enterococcus hirae* | – | – | X69092 | – |
| *Streptomyces* sp. | P15555 | S00765 S11947 | X05109 | – |

Brookhaven Protein Data Bank three-dimensional structures:

| Species | ID | Resolution | Notes |
|---|---|---|---|
| *Streptomyces* sp. | 2PTE | | theoretical model |
| | 3PTE[a] | 1.6 | |

[a]See Chapter 141 for an image constructed from these coordinates.

## Name and History

After the identification of the peptidoglycan cross-linking transpeptidation as the penicillin-sensitive reaction in bacteria (Martin, 1963; Tipper & Strominger, 1965), the study of D-Ala-D-Ala carboxypeptidase activities became central to the understanding of the mode of action of $\beta$-lactam antibiotics. *Streptomyces* strain R61 was found to secrete such a penicillin-sensitive enzyme into the culture medium (Leyh-Bouille *et al.*, 1971). In contrast to most D-Ala-D-Ala carboxypeptidases, which are membrane bound, **Streptomyces D-Ala-D-Ala carboxypeptidase** was soluble, and this allowed a relatively easy purification, a detailed study of its kinetic properties and its crystallization. When supplied with suitable substrates, it was found to catalyze transpeptidation reactions (Pollock *et al.*, 1972; Perkins *et al.*, 1973; Frère *et al.*, 1973a) and, in consequence, is often

referred to as a **DD-peptidase**. It has been successfully used as a model protein for analyzing the interactions between penicillin-sensitive enzymes and $\beta$-lactam antibiotics. This resulted in the demonstration of the formation of a covalent bond between a serine residue and penicillin (Frère *et al.*, 1976a), a conclusion which was later extended to all the penicillin-sensitive enzymes.

## Activity and Specificity

The catalyzed reactions are:

R-D-Ala-D-Ala + $H_2O$ $\longrightarrow$ R-D-Ala + D-Ala
(carboxypeptidase)
R-D-Ala-D-Ala + R'-$NH_2$ $\longrightarrow$ R-D-Ala-NH-R' + D-Ala
(transpeptidase)

where R-D-Ala-D-Ala and R′-NH$_2$ are the donor and acceptor substrates, respectively.

## Substrate Specificity

The standard donor peptide, with which the highest activity has been recorded, is $N^\alpha N^\varepsilon$-diacetyl-D-Ala┤D-Ala ($k_{cat}$ = 50 s$^{-1}$; $k_{cat}/K_m$ = 4000 M$^{-1}$ s$^{-1}$: Frère *et al.*, 1973b). The C-terminal D-Ala residue can be replaced by Gly, D-Lys, D-Leu or D-Glu (10% of efficiency) and the penultimate D-Ala residue by D-Leu (1.5%) or Gly (0.15%). An L-Ala residue in either of these positions results in a complete loss of activity (Leyh-Bouille *et al.*, 1971). The structure of the R group is also very important. If R is acetyl, only 1% of efficiency is observed (Nieto *et al.*, 1973). *N*-Acetyl-L-Xaa- is favorable if Xaa contains a long aliphatic side chain (acetyl-L-Ala- is not better than acetyl). But if Xaa is Lys, the $\varepsilon$-amino group must be acylated: $N^\alpha$-acetyl-L-Lys exhibits 0.5% efficiency and $N^\alpha$-acetyl,$N^\varepsilon$-glycyl-L-Lys, more than 20%. Glycine, many D-amino acids and several Gly-L-Xaa dipeptides are good acceptors, whereas D-Ala-L-Ala and Gly-D-Ala are significantly less efficient (about 10% of Gly-L-Ala). Finally, D-Ala-D-Ala and L-Ala-L-Ala are very poor acceptors (1%) and L-amino acids are not recognized. A variety of other peptides and amines have also been analyzed (Perkins *et al.*, 1973). The best substrates are those which closely mimic the donor and acceptor parts of the peptides in the nascent *Streptomyces* peptidoglycan. Note that it is not strictly accurate to list this enzyme as a proteolytic enzyme, since it is completely inactive on proteins and derived peptides.

When supplied with compounds containing both a D-Ala-D-Ala C-terminus and an adequately positioned amino group, such as $N^\alpha$-acetyl,$N^\varepsilon$-glycyl-L-Lys-D-Ala-D-Ala, the enzyme catalyzes the formation of polymers (Zeiger *et al.*, 1975). The trimer is preferentially formed by reaction between a donor monomer and an acceptor dimer (Frère *et al.*, 1976b).

The enzyme also acts as an esterase and a thioesterase on compounds of general structure R-Xaa-Y-CH(R″)-COOH, where Y is O or S (Adam *et al.*, 1990) and Xaa is Gly or a D residue, preferentially D-Ala. These compounds are also utilized in transacylation reactions. The best reported substrate is Bz-D-Ala-S-CH$_2$-COOH ($k_{cat}$ = 50 s$^{-1}$; $k_{cat}/K_m$ = 430 000 M$^{-1}$ s$^{-1}$: Damblon *et al.*, 1995). The structural requirements for the R group and stereospecificity of the thioesters are somewhat modified when compared to those for the peptides (Damblon *et al.*, 1995).

## Mechanism

In the hydrolysis reaction, the enzyme follows the classical acyl enzyme pathway of serine proteases:

$$E + S \longleftrightarrow ES \longrightarrow ES^* + P1 \longrightarrow E + P2$$

where ES* is the acyl enzyme. With the peptides, acylation is the rate-limiting step and the acyl enzyme cannot be detected. By contrast, deacylation is strongly rate limiting with most thioesters (Jamin *et al.*, 1991).

The transpeptidation is not a simple partitioning of the acyl enzyme between water and the acceptor substrate. The latter binds to a specific site on the enzyme (Jamin *et al.*, 1991)

and it seems that multiple binding sites exist for both donor and acceptor (Frère *et al.*, 1973a; Jamin *et al.*, 1993).

## Assays and Influence of Physicochemical Conditions

Hydrolysis of peptides is monitored by a colorimetric, discontinuous assay of the released D-alanine (Frère *et al.*, 1976c), while that of thioesters can be followed directly at 260 nm (Adam *et al.*, 1990). Transpeptidation was formerly quantified by high-voltage paper electrophoresis (Frère *et al.*, 1973a), but HPLC techniques are now preferred (Jamin *et al.*, 1993).

The enzyme shows a broad pH optimum between 6 and 9 (Varetto *et al.*, 1987). It is inhibited by increasing ionic strength (Frère *et al.*, 1973a). Transpeptidation is favored by high pH, high and low acceptor and donor concentrations, respectively, and by a decrease in the water content of the medium (Frère *et al.*, 1973a; Jamin *et al.*, 1993).

## Inhibitors

Some peptide analogs of the donor substrates and an excess of some acceptors are inhibitory at high (millimolar) concentrations (Perkins *et al.*, 1973; Frère *et al.*, 1973a). Antibiotics of the $\beta$-lactam family are efficient transient inactivators that acylate the active-site serine according to the pathway described above for the substrates, but no P1 is formed (due to the $\beta$-lactam structure) and the deacylation step is very slow ($10^{-3}$–$10^{-6}$ s$^{-1}$), so that an inactive acyl enzyme accumulates (Frère *et al.*, 1975a,b). With some but not all $\beta$-lactams, the deacylation step does not yield the expected penicilloic acid, but first involves a rate-limiting cleavage of the C5–C6 bond of the antibiotic (Frère *et al.*, 1975c, 1976d).

## Structural Chemistry

The mature protein consists of one 349 residue polypeptide chain with a single disulfide bridge. The gene encodes a 406 residue precursor containing a 31 residue signal peptide and a 26 residue C-terminal extension (Duez *et al.*, 1987; Joris *et al.*, 1987) whose role remains obscure. When the cloned gene is expressed in *E. coli*, no periplasmic cleavage of this C-terminal extension occurs and the protein is catalytically inactive (Fanuel *et al.*, 1994).

The three-dimensional structure has been established by X-ray crystallography to a resolution of 0.16 nm (Kelly *et al.*, 1986, 1989; Kelly & Kuzin, 1995). The protein is composed of an all-$\alpha$ and an $\alpha/\beta$ domain, with the active site situated at the interface between the two. The crystals are enzymatically active (Kelly *et al.*, 1992).

The calculated $M_r$ is 37 389, in good agreement with the value (37 500) obtained by physicochemical methods. The pI is close to 4.8 (Frère *et al.*, 1973b).

## Preparation

The enzyme was initially extracted and purified from culture filtrates of *Streptomyces* R61 (Frère *et al.*, 1973b; Fossati *et al.*, 1978). Much higher yields were later obtained with the gene cloned in *Streptomyces lividans* (Duez *et al.*, 1987; Erpicum *et al.*, 1990) or *E. coli* (Fanuel *et al.*, 1994).

## Biological Aspects

No membrane-bound form of the R61 DD-peptidase has been found. Since it is unlikely that a protein secreted into the culture medium would have a role in peptidoglycan metabolism, the biological role of this enzyme remains mysterious. It could represent a primitive attempt to protect the cell against β-lactams produced by organisms living in the same environmental niches. The much more efficient class C β-lactamases, which exhibit vastly increased deacylation rates, may be evolved from a protein like this.

## Distinguishing Features

Among the penicillin-binding proteins, the R61 enzyme is remarkable for its high activity on simple peptides and thioesters and its ability to catalyze the formation of dimers and higher polymers.

## Related Enzymes

The *Streptomyces* K11 DD-peptidase (Leyh-Bouille *et al.*, 1972) exhibits very similar enzymatic and molecular characteristics. A search of the databanks reveals sequence similarities only with class C β-lactamases (Joris *et al.*, 1988). The three-dimensional structure is however similar to those of class A β-lactamases (Kelly *et al.*, 1986) and of PBP-2x of *Streptococcus pneumoniae* (Pares *et al.*, 1996), in addition to class C β-lactamases (Lobkovsky *et al.*, 1993). It seems likely that all active-site serine β-lactamases and penicillin-binding proteins derive from a common ancestor, and are therefore more or less distantly related.

## Further Reading

The properties of DD-peptidases and their structural and functional relationships with β-lactamases have been discussed in detail by Frère & Joris (1985), Joris *et al.* (1991), Ghuysen (1991), Frère *et al.* (1992) and Jamin *et al.* (1995).

## References

Adam, M., Damblon, C., Plaitin, B., Christiaens, L. & Frère, J.-M. (1990) Chromogenic depsipeptide substrates for β-lactamases and penicillin-sensitive DD-peptidases. *Biochem. J.* **270**, 525–529.

Damblon, C., Zhao, G.-H., Jamin, M., Ledent, P., Dubus, A., Vanhove, M., Raquet, X., Christiaens, L. & Frère, J.-M. (1995) Breakdown of the stereospecificity of DD-peptidases and β-lactamases with thiolester substrates. *Biochem. J.* **309**, 431–436.

Duez, C., Piron-Fraipont, C., Joris, B., Dusart, J., Urdea, M., Martial, J., Frère, J.-M. & Ghuysen, J.-M. (1987) Primary structure of the *Streptomyces* R61 extracellular DD-peptidase. *Eur. J. Biochem.* **162**, 509–518.

Erpicum, T., Granier, B., Delcour, M., Lenzini, M.V., Nguyen-Distèche, M., Dusart, J. & Frère, J.-M. (1990) Enzyme production by genetically engineered *Streptomyces* strains: influence of culture conditions. *Biotechnol. Bioeng.* **35**, 719–726.

Fanuel, L., Granier, B., Wilkin, J.-M., Bellefroid-Bourguignon, C., Joris, B., Knowles, J., Komives, E., Van Beeumen, J., Ghuysen, J.-M. & Frère, J.-M. (1994) The precursor of the *Streptomyces* R61 DD-peptidase containing a C-terminal extension is inactive. *FEBS Lett.* **351**, 49–52.

Fossati, P., Saint-Ghislain, M., Sicard, P.J., Frère, J.-M., Dusart, J., Klein, D. & Ghuysen, J.-M. (1978) Large scale preparation of purified exocellular DD-carboxypeptidase-transpeptidase of *Streptomyces* strain R61. *Biotechnol. Bioeng.* **20**, 577–587.

Frère, J.-M. & Joris, B. (1985) Penicillin-sensitive enzymes in peptidoglycan biosynthesis. *CRC Crit. Rev. Microbiol.* **11**, 299–396.

Frère, J.-M., Ghuysen, J.-M., Perkins, H.R. & Nieto, M. (1973a) Kinetics of concomitant transfer and hydrolysis reactions catalysed by the exocellular DD-carboxypeptidase transpeptidase of *Streptomyces* R61. *Biochem. J.* **135**, 483–492.

Frère, J.-M., Ghuysen, J.-M., Perkins, H.R. & Nieto, M. (1973b) Molecular weight and amino acid composition of the exocellular DD-carboxypeptidase-transpeptidase of *Streptomyces* R61. *Biochem. J.* **135**, 463–468.

Frère, J.-M., Ghuysen, J.-M. & Iwatsubo, M. (1975a) Kinetics of interaction between the exocellular DD-carboxypeptidase-transpeptidase from *Streptomyces* R61 and β-lactam antibiotics. A choice of models. *Eur. J. Biochem.* **57**, 343–351.

Frère, J.-M., Ghuysen, J.-M. & Perkins, H.R. (1975b) Interaction between the exocellular DD-carboxypeptidase-transpeptidase from *Streptomyces* R61, substrate and β-lactam antibiotics. A choice of models. *Eur. J. Biochem.* **57**, 353–359.

Frère, J.-M., Ghuysen, J.-M., Degelaen, J., Loffet, A. & Perkins, H.R. (1975c) Fragmentation of benzylpenicillin after interaction with the exocellular DD-carboxypeptidase-transpeptidases of *Streptomyces* R61 et R39. *Nature* **258**, 168–170.

Frère, J.-M., Duez, C., Ghuysen, J.-M. & Vandekerckhove, J. (1976a) Occurrence of a serine residue in the penicillin-binding site of the exocellular DD-carboxy-peptidase-transpeptidase from *Streptomyces* R61. *FEBS Lett.* **70**, 257–260.

Frère, J.-M., Ghuysen, J.-M., Zeiger, A.R. & Perkins, H.R. (1976b) The direction of trimer synthesis from the donor-acceptor substrate $N^{\alpha}$-(acetyl)-$N^{\varepsilon}$-(glycyl)-L-lysyl-D-alanyl-D-alanine by the exocellular DD-carboxypeptidase-transpeptidase of *Streptomyces* R61. *FEBS Lett.* **63**, 112–116.

Frère, J.-M., Leyh-Bouille, M., Ghuysen, J.-M., Nieto, M. & Perkins, H.R. (1976c) Exocellular DD-carboxypeptidases-transpeptidases from *Streptomyces*. *Methods Enzymol.* **45**, 610–636.

Frère, J.-M., Ghuysen, J.-M., Vanderhaeghe, H., Adriaens, P., Degelaen, J. & De Graeve, J. (1976d) The fate of the thiazolidine ring during fragmentation of penicillin by the exocellular DD-carboxypeptidase-transpeptidase of *Streptomyces* R61. *Nature* **260**, 451–454.

Frère, J.-M., Nguyen-Distèche, M., Coyette, J. & Joris, B. (1992) Mode of action: interaction with the penicillin binding proteins. In: *The Chemistry of Beta-lactams* (Page, M.I., ed.). Glasgow: Chapman & Hall, pp. 148–195.

Ghuysen, J.-M. (1991) Serine beta-lactamases and penicillin-binding proteins. *Annu. Rev. Microbiol.* **45**, 37–67.

Jamin, M., Adam, M., Damblon, C., Christiaens, L. & Frère, J.-M. (1991) Accumulation of acyl-enzyme in DD-peptidase-catalysed reactions with analogues of peptide substrates. *Biochem. J.* **280**, 499–506.

Jamin, M., Wilkin, J.-M. & Frère, J.-M. (1993) A new kinetic mechanism for the concomitant hydrolysis and transfer reactions catalysed by bacterial DD-peptidases. *Biochemistry* **32**, 7278–7285.

Jamin, M., Wilkin, J.-M. & Frère, J.-M. (1995) Bacterial DD-transpeptidases and penicillin. *Essays Biochem.* **29**, 1–24.

Joris, B., Jacques, P., Frère, J.-M., Ghuysen, J.-M. & Van Beeumen, J. (1987) Primary structure of *Streptomyces* R61 extracellular

DD-peptidase 2. Amino acid sequence data. *Eur. J. Biochem.* **162**, 519–524.

Joris, B., Ghuysen, J.-M., Dive, G., Renard, A., Dideberg, O., Charlier, P., Frère, J.-M., Kelly, J., Boyington, J., Moews, P. & Knox, J. (1988) The active-site-serine penicillin-recognizing enzymes as members of the *Streptomyces* R61 DD-peptidase family. *Biochem. J.* **250**, 313–324.

Joris, B., Ledent, P., Dideberg, O., Fonzé, E., Lamotte-Brasseur, J., Kelly, J.A., Ghuysen, J.-M. & Frère, J.-M. (1991) Comparison of the sequences of class-A beta-lactamases and of the secondary structure elements of penicillin-recognizing proteins. *Antimicrob. Agents Chemother.* **35**, 2294–2301.

Kelly, J.A. & Kuzin, A.P. (1995) The refined crystallographic structure of a DD-peptidase penicillin-target enzyme at 1.6 Å resolution. *J. Mol. Biol.* **254**, 223–236.

Kelly, J.A., Dideberg, O., Charlier, P., Wéry, J., Libert, M., Moews, P., Knox, J., Duez, C., Fraipont, C., Joris, B., Dusart, J., Frère, J.-M. & Ghuysen, J.-M. (1986) On the origin of bacterial resistance to penicillin: comparison of a *β*-lactamase and a penicillin target. *Science* **231**, 1429–1431.

Kelly, J.A., Knox, J.R., Zhao, H., Frère, J.-M. & Ghuysen, J.-M. (1989) Crystallographic mapping of *β*-lactams bound to a D-alanyl-D-alanine peptidase target enzyme. *J. Mol. Biol.* **209**, 281–295.

Kelly, J.A., Waley, S.G., Adam, M. & Frère, J.-M. (1992) Crystalline enzyme kinetics: activity of the *Streptomyces* R61 D-alanyl-D-alanine peptidase. *Biochim. Biophys. Acta* **1119**, 256–260.

Leyh-Bouille, M., Coyette, J., Ghuysen, J.-M., Idczak, J., Perkins, H.R. and Nieto, M. (1971) Penicillin-sensitive DD-carboxypeptidase from *Streptomyces* strain R61. *Biochemistry* **10**, 2163–2170.

Leyh-Bouille, M., Nakel, M., Frère, J.-M., Johnson, K., Ghuysen, J.-M., Nieto, M. & Perkins, H.R. (1972) Penicillin-sensitive DD-carboxypeptidases from *Streptomyces* strains R39 et K11. *Biochemistry* **11**, 1290–1298.

Lobkovsky, E., Moews, P.C., Hansong, L., Haiching, Z., Frère, J.-M. & Knox, J.R. (1993) Evolution of an enzyme activity: crystallographic structure at 2-Å resolution of cephalosporinase from the *ampC* gene of *Enterobacter cloacae* P99 and comparison with a class A penicillinase. *Proc. Natl Acad. Sci. USA* **90**, 11257–11261.

Martin, H.H. (1963) Composition of the mucopolymer of the cell wall of the unstable L-form of *Proteus mirabilis*. *J. Gen. Microbiol.* **36**, 441–450.

Nieto, M., Perkins, H.R., Leyh-Bouille, M., Frère, J.-M. & Ghuysen, J.-M. (1973) Peptide inhibitors of *Streptomyces* DD-carboxypeptidases. *Biochem. J.* **131**, 163–171.

Pares, S., Mouz, N., Pétillot, Y., Hakenbeck, R. & Dideberg, O. (1996) X-ray structure of *Streptococcus pneumoniae* PBP2x, a primary penicillin target. *Nature Struct. Biol.* **3**, 284–289.

Perkins, H.R., Nieto, M., Frère, J.-M., Leyh-Bouille, M. & Ghuysen, J.-M. (1973) *Streptomyces* DD-carboxypeptidases as transpeptidases: the specificity for amino compounds acting as carboxyl acceptors. *Biochem. J.* **131**, 707–718.

Pollock, J.J., Ghuysen, J.-M., Linder, R., Salton, M.R.J., Perkins, H.R., Nieto, M., Leyh-Bouille, M., Frère, J.-M. & Johnson, K. (1972) Transpeptidase activity of *Streptomyces* D-alanyl-D carboxypeptidases. *Proc. Natl Acad. Sci. USA* **69**, 662–666.

Tipper, D.J. & Strominger, J.L. (1965) Mechanism of action of penicillins: a proposal based on their structural similarity to acyl-D-alanyl-D-alanine. *Proc. Natl Acad. Sci. USA.* **54**, 1133–1141.

Varetto, L., Frère, J.-M., Nguyen-Distèche, M., Ghuysen, J.-M. & Houssier, C. (1987) The pH dependence of the active-site serine peptidase of *Streptomyces* R61. *Eur. J. Biochem.* **162**, 525–531.

Zeiger, A.R., Frère, J.-M. & Ghuysen, J.-M. (1975) A donor-acceptor substrate of the exocellular DD-carboxypeptidase-transpeptidase from *Streptomyces* R61. *FEBS Lett.* **52**, 221–225.

*Jean-Marie Frère*
*Department of Biochemistry,*
*University of Liège,*
*Institut of Chemistry (B6) Sart-Tilman,*
*B-4000 Liège 1, Belgium*

# *146. D-Stereospecific aminopeptidase*

## *Databanks*

*Peptidase classification: clan SE, family S12, MEROPS ID: S12.002*
*NC-IUBMB enzyme classification: EC 3.4.11.19*
*ATCC entries: 49237*
*Databank codes:*

| Species | SwissProt | PIR | EMBL (cDNA) | EMBL (genomic) |
|---------|-----------|-----|-------------|----------------|
| *Ochrobactrum anthropi* | – | A42209 | M84523 | – |

## Name and History

An enrichment culture in a medium containing a synthetic substrate D-Ala-NH$_2$ as the sole nitrogen source led to the isolation of a bacterial strain *Ochrobactrum anthropi* SCRC C1–38. The enzyme hydrolyzing D-Ala-NH$_2$ was purified to homogeneity from the cell-free extract of the strain and characterized (Asano *et al.*, 1989c). Since the enzyme exhibited a mode of action typical of aminopeptidases which liberate an N-terminal D-amino acid residue with a free amino group, the enzyme was named as *D-aminopeptidase*, and the name *D-stereospecific aminopeptidase* is recommended by NC-IUBMB. During the screening, an amidase specific toward aromatic D-amino acid amides was also found (Asano *et al.*, 1989a), although there was no associated peptidase activity. There has been no other report of an aminopeptidase catalyzing the stereospecific hydrolysis of D-amino acid-containing peptides, although the peptides do exist in nature. At present, the occurrence of D-aminopeptidase is apparently limited to *O. anthropi*.

## Activity and Specificity

The enzyme activity was assayed by the hydrolysis of D-Ala-NH$_2$ or by the liberation of *p*-nitroaniline from D-Ala *p*-nitroanilide at 405 nm as follows. A reaction mixture (1.0 ml) containing 5 mM D-Ala+NHPhNO$_2$, 100 mM Tris–HCl, pH 8.0, and the enzyme was monitored by the change in absorbance at 405 nm (Asano *et al.*, 1992). The substrate D-Ala-NHPhNO$_2$ is available from Bachem (see Appendix 2 for full names and addresses of suppliers).

The enzyme exhibited maximal activity at 45°C and pH 8.5. About 80% activity remained after incubation at 45°C in 0.1 M potassium phosphate, pH 8.0 for 10 min. No loss of activity was found between pH 7.0 and 10.0 after incubation at 30°C for 1 h in 0.05 M buffers of various pH values (Asano *et al.*, 1989c).

The enzyme acted on D-Ala-NH$_2$ (relative activity 100%, $K_m$ value 0.65 mM), Gly-NH$_2$ (44%, 22.3 mM), D-$\alpha$-aminobutyric acid amide (30%, 18.3 mM), D-Ser-NH$_2$ (29%, 27.0 mM), D-Ala 3-aminopentane amide (32%, 2.27 mM), D-Ala anilide (73%), D-Ala benzylamide (72%, 0.51 mM), D-Ala *p*-nitroanilide (96%, 0.51 mM), D-Ala *n*-butylamide (66%, 0.73 mM), D-Ala-OMe (75%), Gly-OMe (229%), D-Ala-Gly (95%, 0.98 mM), D-Ala-(Gly)$_2$ (45%, 0.37 mM), (D-Ala)$_2$ (21%, 10.2 mM), (D-Ala)$_3$ (92%, 0.57 mM), (D-Ala)$_4$ (89%, 0.32 mM), D-Ala-L-Ala (46%, 1.03 mM) and D-Ala-(L-Ala)$_2$ (100%, 0.65 mM), etc. (Asano *et al.*, 1989c). These results show that the enzyme has higher affinity toward peptide substrates than amino acid amides.

The compounds inactive as substrates (less than 0.05% the activity for D-Ala-NH$_2$) include most of the D-amino acid amides with more than five carbon atoms, most of the L-amino acid amides, L-amino acid methylesters, D- and L-$\alpha$-amino-$\varepsilon$-caprolactams, low molecular mass aliphatic amides, $\beta$-Ala-dipeptides, and N-protected D-Ala derivatives. The rate of hydrolysis of L-Ala-NH$_2$ was less than 0.01% that of D-Ala-NH$_2$. The enzyme showed neither endopeptidase nor carboxypeptidase activity, but exhibited a wide range of activities: (a) aminopeptidase activity which requires D-configuration at the N-terminus (except for Gly amides), (b) D-amino acylamidase and aryl-D-amino acylamidase activities, and (c) amino acid esterase activity. With amino acid esters as substrates, the enzyme showed low stereoselectivity, suggesting that these are not the true substrates of the enzyme. With the peptide substrates, the enzyme recognized the configuration of the N-terminal D-amino acid, and hydrolyzed not only the peptide bond between (D-Ala)$_2$, but also that of D-Ala-L-Ala.

## Effects of Metal Ions and Inhibitors

The enzyme activity was measured after the enzyme was preincubated at 30°C for 30 min with various compounds (at 1 mM unless otherwise noted). The enzyme activity was inhibited to 20–50% by Ca$^{2+}$, Ni$^{2+}$, Cd$^{2+}$, Cu$^{2+}$, Zn$^{2+}$, 5,5′-dithio-bis(2-nitrobenzoic acid), hydroxylamine (at 10 mM) and *N*-ethylmaleimide; and 25–100% by Ag$^+$, Hg$^{2+}$ (at 0.02 mM), and PCMB (at 0.074 mM). PMSF did not inhibit the enzyme at 1 mM. Although the effect of the inhibitors showed that activity is thiol-dependent, the enzyme was later shown to be a serine peptidase as described below.

The enzyme was significantly inhibited by $\beta$-lactam compounds, such as 6-aminopenicillanic acid (6-APA), 7-aminocephalosporanic acid (7-ACA), benzylpenicillin and ampicillin (Asano *et al.*, 1992). A time-dependent decrease in activity, reaching a plateau in 5–10 min, was observed with 6-APA and 7-ACA, whereas there was immediate inhibition by ampicillin and benzylpenicillin. The inhibited activities by 6-APA and 7-ACA were gradually restored in a few minutes by more than 10-fold dilution of the reaction mixture with 0.05 M Tris–HCl, pH 8.0, or by overnight dialysis against 10 mM potassium phosphate buffer, pH 7.0. When D-Ala-NHPhNO$_2$ was used as substrate, $N^{\alpha}N^{\varepsilon}$-diacetyl-L-Lys-(D-Ala)$_2$ acted as a competitive inhibitor with $K_i$ value of 2.4 mM, while 7-ACA ($K_i$ 2.0 mM), ampicillin ($K_i$ 5.7 mM), and 6-APA ($K_i$ 10 mM) were noncompetitive inhibitors. Benzylpenicillin acted as an uncompetitive inhibitor with relatively high $K_i$ value (19 mM). When (D-Ala)$_2$ was used as substrate, 7-ACA, 6-APA, ampicillin, *N*-Boc-6-APA, $N^{\alpha}N^{\varepsilon}$-diacetyl-L-Lys-(D-Ala)$_2$ and benzylpenicillin were all good competitive inhibitors of the enzyme with $K_i$ values ranging from 0.1 to 5.9 mM. When D-Ala oligomers were used as substrates, $N^{\alpha}N^{\varepsilon}$-diacetyl-L-Lys-(D-Ala)$_2$ showed competitive inhibition with lower $K_i$ values (2.4–5.3 mM).

## Structural Chemistry

The native enzyme is a homodimer, with $M_r$ approximately 122 000 by gel filtration. The $M_r$ of the subunits is 59 000 by SDS-PAGE (Asano *et al.*, 1989c). The *dap* gene coding for D-aminopeptidase consists of an open reading frame of 1560 nucleotides which specifies a protein of $M_r$ of 57 257 (Asano *et al.*, 1992). The enzyme has a pI value of 4.2.

The deduced primary structure of D-aminopeptidase shows similarity to that of *Streptomyces* R61 DD-carboxypeptidase (Chapter 145) (Duez *et al.*, 1987) (27% identical over 287 amino acids) and class C $\beta$-lactamase from *E. coli* (22% identical over 234 amino acids) (Jaurin & Grundström, 1981; Lindberg & Normark, 1986). The enzyme is also structurally related to class A $\beta$-lactamases (Ambler & Scott, 1978; Neugebauger, 1981), penicillin-binding proteins (Nakamura *et al.*, 1983; Broome-Smith *et al.*, 1985; Asoh *et al.*, 1986; Song *et al.*, 1987; Broome-Smith *et al.*, 1988; Spratt, 1988) and class D β-lactamases (Dale *et al.*, 1985, Huovinen *et al.*, 1988). The sequence Ser-Xaa-Xaa-Lys is perfectly conserved among this class of enzymes, including the peptidases of clan SE. With DD-carboxypeptidase and class C β-lactamases the sequences at the highly similar area, Ser-Val-Xaa-Lys-Xaa-Phe-Xaa-Ala-Xaa-Val-Leu-Leu (Duez *et al.*, 1987) and Ser-Val-Ser-Lys-Xaa-Phe-Tyr (Lindberg & Normark, 1986), respectively, are well conserved. Furthermore, some of the residues scattered around the conserved region are similar among the enzymes. The Met57 located four residues upstream from the Ser61 is found similarly in the DD-carboxypeptidase and β-lactamases as a hydrophobic residue Phe. A common observation between D-aminopeptidase and the β-lactamase is that the consensus sequence is located around 60 from the N-terminal of the enzymes (Joris *et al.*, 1986).

The inhibition by PCMB and other thiol-blocking reagents suggested that the enzyme was a cysteine peptidase (Asano *et al.*, 1989b). However, site-directed mutagenesis shows that the Ser61 and Lys64, located in the sequence Ser-Xaa-Xaa-Lys which is conserved in the penicillin-recognizing enzymes are essential for the D-aminopeptidase activity. The $V/K_m$ values for the Ser61Cys and Lys64Asn mutants were decreased to 0.26 and 0.008%, respectively, of those of the natural enzyme, although the $K_m$ values were hardly affected. The Ser61Gly mutant had a still lower activity; $10^5$-fold of that of the wild-type enzyme. On the other hand, the mutations at other sites (Met57 and Cys68) did not greatly affect activity. Mutations at Cys60, which is adjacent to the likely active center Ser61, had marked effects on the kinetics, showing that the residue is important in the substrate. Mutants Cys60Ser and Cys60Gly had slightly altered $V/K_m$ values. They were tolerant to inhibition by PCMB, which suggests that the inhibition by PCMB of the native enzyme is caused by steric hindrance resulting from mercaptide bond formation between Cys60 and PCMB.

A Gly155Ser mutant of D-aminopeptidase showed increased thermal stability (Asano & Yamaguchi, 1995).

## Preparation

The enzyme producer, *O. anthropi*, isolated from a soil sample, is a gram-negative, motile, obligatory aerobic, nonfermenting rod, now available from ATCC. The enzyme is intracellularly and constitutively formed. The enzyme was purified about 2800-fold with a 17% yield from the cell-free extracts by use of ammonium sulfate fractionation and column chromatography on DEAE-Toyopearl, butyl-Toyopearl, Sephacryl S-300, and DEAE-5PW (HPLC) (Asano *et al.*, 1989c).

The gene for the enzyme, located on the chromosome of *O. anthropi*, was cloned into a plasmid for expression in *E. coli*. The amount of the enzyme in the cell-free extract of the *E. coli* transformant was elevated about 3600-fold over the wild strain (Asano *et al.*, 1992).

## Biological and Biotechnological Aspects

Considering the high stereospecificity and rather narrow substrate specificity toward derivatives of low molecular mass D-amino acids, the physiological role of the enzyme may be to hydrolyze the dipeptide (D-Ala)$_2$ which is a product of (D-Ala)$_2$ ligase, a fragment of bacterial peptidoglycan composed of D-Ala and Gly, or to degrade D-Ala-Gly and (D-Ala)$_2$ synthesized in rice plants (Manabe, 1986).

The intact cells of *E. coli* transformant were used as a catalyst for the D-stereospecific hydrolysis of several racemic amino acid amide hydrochlorides (Asano *et al.*, 1991). Complete hydrolysis of D-Ala-NH$_2$ was achieved in a short time (4.5 h) from 5.0 M of racemic Ala-NH$_2$ HCl using the cells of *E. coli* transformant. The concentration of D-Ala reached up to 220 g liter$^{-1}$. The cells or the cell-free extract catalyzed the synthesis of D-α-amino butyric acid, D-Met, D-norvaline and D-norleucine from their amides in a similar manner. An enzyme-catalyzed D-stereospecific aminolysis of amino acid ester was also carried out. D-Amino acid *N*-alkylamides were stereoselectively synthesized by the immobilized enzyme with urethane prepolymer PU-6, in water-saturated butyl-acetate, benzene, and 1,1,1-trichloroethane, from racemic amino acid esters or amide, and nucleophiles. The acyl donor specificity was rather limited with methylester or amide of D-Ala, Gly, and D-α-aminobutyric acid. The acyl acceptor specificity was relatively wide with bulky, aromatic-containing and straight-chain amides (Asano *et al.*, 1989b; Kato *et al.*, 1989). The enzyme was also active in synthesizing D-Ala dimer and trimer in water-saturated toluene (Kato *et al.*, 1990).

## Distinguishing Features

D-Aminopeptidase is distinguished from the following enzymes. A carboxypeptidase-like D-peptidase was characterized in an actinomycete, although it is not strictly specific towards peptides containing D-amino acids (Sugie & Suzuki, 1986). An endopeptidase that acts D-stereospecifically upon peptides composed of aromatic D-amino acids named alkaline D-peptidase was discovered in *Bacillus cereus* (Asano *et al.*, 1996) (see Chapter 147).

## References

Ambler, R.P. & Scott, G.K. (1978) Partial amino acid sequence of penicillinase coded by *Escherichia coli* plasmid R6K. *Proc. Natl Acad. Sci. USA* **75**, 3732–3736.

Asano, Y. & Yamaguchi, K. (1995) Mutants of D-aminopeptidase with increased thermal stability. *J. Ferment. Bioeng.* **79**, 614–616.

Asano, Y., Mori, T., Hanamoto, S., Kato, Y. & Nakazawa, A. (1989a) A new D-stereospecific amino acid amidase from *Ochrobactrum anthropi. Biochem. Biophys. Res. Commun.* **162**, 470–474.

Asano, Y., Nakazawa, A., Kato, Y. & Kondo, K. (1989b) Isolierung einer D-stereospezifischen Aminopeptidase und ihre Anwendung als Katalysator in der organischen Synthese. *Angew. Chem.* **101**, 511–512.

Asano, Y., Nakazawa, A., Kato, Y. & Kondo, K. (1989c) Properties of a novel D-stereospecific aminopeptidase from *Ochrobactrum anthropi. J. Biol. Chem.* **264**, 14233–14239.

Asano, Y., Kishino, K., Yamada, A., Hanamoto, S. & Kondo, K. (1991) Plasmid-based, D-aminopeptidase-catalyzed synthesis of (R)-amino acids. *Recl. Trav. Chim. Pays-Bas* **110**, 206–208.

Asano, Y., Kato, Y., Nakazawa, A. & Kondo, K. (1992) Structural similarity of D-aminopeptidase to carboxypeptidase DD and β-lactamase. *Biochemistry* **31**, 2316–2328.

Asano, Y., Ito, H., Dairi, T. & Kato, Y. (1996) An alkaline D-stereospecific endopeptidase with β-lactamase activity from *Bacillus cereus*. *J. Biol. Chem.* **271**, 30256–30261.

Asoh, S., Matsuzawa, H., Ishino, F., Strominger, J.L., Matsuhashi, M. & Ohta, T. (1986) Nucleotide sequence of the *pbpA* gene and characteristics of the deduced amino acid sequence of penicillin-binding protein 2 of *Escherichia coli* K12. *Eur. J. Biochem.* **160**, 231–238.

Broome-Smith, J.K., Edelman, A., Yousif, S. & Spratt, B.G. (1985) The nucleotide sequences of the *ponA* and *ponB* genes encoding penicillin-binding proteins 1A and 1B of *Escherichia coli* K12. *Eur. J. Biochem.* **147**, 437–446.

Broome-Smith, J.K., Ioannidis, I., Edelman, A. & Spratt, B.G. (1988) Nucleotide sequences of the penicillin-binding protein 5 and 6 genes of *Escherichia coli*. *Nucleic Acids Res.* **16**, 1617.

Dale, J.W., Godwin, D., Mossakowska, D., Stephenson, P. & Wall, S. (1985) Sequence of the OXA2 β-lactamase: comparison with other penicillin-reactive enzymes. *FEBS Lett.* **191**, 39–44.

Duez, C., Piron-Fraipont, C., Joris, B., Dusart, J., Urdea, M.S., Martial, J.A., Frère, J.-M. & Ghuysen, J.-M. (1987) Primary structure of the *Streptomyces* R61 extracellular DD-peptidase. *Eur. J. Biochem.* **162**, 509–518.

Huovinen, P., Huovinen, S. & Jacoby, G.A. (1988) Sequence of PSE-2 β-lactamase. *Antimicrob. Agents Chemother.* **32**, 134–136.

Jaurin, B. & Grundström, T. (1981) *ampC* cephalosporinase of *Escherichia coli* K-12 has a different evolutionary origin from that of β-lactamases of the penicillinase type. *Proc. Natl Acad. Sci. USA* **78**, 4897–4901.

Joris, B., De Messer, F., Galleni, M., Masson, S., Dusart, J., Frère, J.-M., Van Beeumen, J., Bush, K. & Sykes, R. (1986) Properties of a class C β-lactamase from *Serratia marcescens*. *Biochem. J.* **239**, 581–586.

Kato, Y., Asano, Y., Nakazawa, A. & Kondo, K. (1989) First stereoselective synthesis of D-amino acid amide catalyzed by a novel aminopeptidase. *Tetrahedron* **45**, 5743–5754.

Kato, Y., Asano, Y., Nakazawa, A. & Kondo, K. (1990) Synthesis of D-alanine oligopeptides by D-aminopeptidase in non-aqueous media. *Biocatalysis* **3**, 207–215.

Lindberg, F. & Normark, S. (1986) Sequence of the *Citrobacter freundii* OS60 chromosomal *ampC* β-lactamase gene. *Eur. J. Biochem.* **156**, 441–445.

Manabe, H. (1986) Distribution of D-alanylglycine and related compounds in *Oryza* species. *Phytochemistry* **25**, 2233–2235.

Nakamura, M., Maruyama, I.N., Soma, M., Kato, J., Suzuki, H. & Horota, Y. (1983) On the process of cellular division of the gene for penicillin-binding protein 3. *Mol. Gen. Genet.* **191**, 1–9.

Neugebauer, K., Sprengel, R. & Schaller, H. (1981) Penicillinase from *Bacillus licheniformis*: nucleotide sequence of the gene and implications for the biosynthesis of a secretory protein in a gram-positive bacterium. *Nucleic Acids Res.* **9**, 2577–2588.

Song, M.D., Wachi, M., Doi, M., Ishino, F. & Matsuhashi, M. (1987) Evolution of an inducible penicillin-target protein in methicillin-resistant *Staphylococcus aureus* by gene fusion. *FEBS Lett.* **221**, 167–171.

Spratt, B.G. (1988) Hybrid penicillin-binding proteins in penicillin-resistant strains of *Neisseria gonorrhoeae*. *Nature* **332**, 173–176.

Sugie, M. & Suzuki, H. (1986) Purification and properties of a peptidase from *Nocardia orientalis* specific to D-amino acid peptides. *Agric. Biol. Chem.* **50**, 1397–1402.

*Yasuhisa Asano*
*Biotechnology Research Center,*
*Toyama Prefectural University,*
*Kosugi, Toyama 939-03 Japan*
*Email: asano@pu-toyama.ac.jp*

# 147. Alkaline D-peptidase

## Databanks

*Peptidase classification: clan SE, family S12, MEROPS ID: S12.003*
*NC-IUBMB enzyme classification: none*
*Databank codes:*

| Species | SwissProt | PIR | EMBL (cDNA) | EMBL (genomic) |
|---|---|---|---|---|
| *Bacillus cereus* | – | – | D86380 | – |

## Name and History

An enrichment culture in LB medium containing a synthetic substrate (D-Phe)$_4$ led to the isolation of a bacterial strain *Bacillus cereus* DF4-B. The extracellular peptidase hydrolyzing (D-Phe)$_4$ was purified and characterized (Asano *et al.*, 1997). The enzyme is an endopeptidase that acts D-stereospecifically upon peptides composed of aromatic D-amino acids, recognizing the D-configuration of the amino acid whose C-terminal peptide bond is hydrolyzed. The enzyme had an optimum pH at around 10.3. Thus, the enzyme was named **alkaline D-peptidase** (D-stereospecific peptide hydrolase). Although the enzyme showed poor $\beta$-lactamase activity towards ampicillin and penicillin G, no activities of DD-carboxypeptidase (Duez *et al.*, 1987) and D-aminopeptidase (Asano *et al.*, 1989, 1992) were detected. The enzyme can be categorized as one of the 'penicillin-recognizing enzymes' (Frère *et al.*, 1988).

## Activity and Specificity

Alkaline D-peptidase (ADP) activity was routinely assayed at 30°C by measuring the production of (D-Phe)$_2$ from (D-Phe)$_4$ by HPLC. The enzyme activity was maximal at 45°C. About 60% activity remained after incubation at 43°C in 0.1 M potassium phosphate buffer, pH 8.0, for 10 min. No activity was lost between pH 5.0 and 10.0 after incubation at 30°C for 1 h in 0.05 M buffers at various pH (Asano *et al.*, 1996).

The enzyme was active towards (D-Phe)$_4$ (relative activity 100%, $K_m$ value 0.398 mM) and (D-Phe)$_3$ (90%, 0.127 mM), forming (D-Phe)$_2$ and D-Phe. The enzyme is also active towards tripeptides L-Phe-(D-Phe)$_2$ (119%, 0.455 mM), L-Phe-D-Phe-L-Phe (28.1%, 1.63 mM), D-Tyr-(D-Phe)$_2$ (83.6%), (D-Phe)$_2$-D-Tyr (83.6%), and on Boc-(D-Phe)$_n$ ($n = 2$–$4$) forming Boc-D-Phe, (D-Phe)$_2$ and D-Phe. The enzyme had esterase activity towards D-Phe-OMe and (D-Phe)$_2$-OMe. The products from Boc-(D-Phe)$_3$ *tert*-butylester were Boc-D-Phe, D-Phe and D-Phe *tert*-butylester. A dimer was formed when D-Phe methylester and D-Phe-NH$_2$ were the substrates. The enzyme also showed poor $\beta$-lactamase activity towards ampicillin (8.9%, 73.1 mM) and penicillin G (9.7%, 48.9 mM). On the other hand, DD-carboxypeptidase (Duez *et al.*, 1987) and D-aminopeptidase (Asano *et al.*, 1989, 1992) activities were undetectable. The following compounds were inert as substrates: (L-Phe)$_4$, (L-Phe)$_3$, D-Phe-L-Phe-D-Phe, D-Phe-(L-Phe)$_2$, (L-Phe)$_2$-D-Phe, (L-Phe)$_2$, L-Phe-D-Phe, L-Phe-OMe, L-Phe-NH$_2$, Boc-(L-Phe)$_2$, Boc-(L-Phe)$_4$ methylester, D-Leu *p*-nitroanilide, D-Ala *p*-nitroanilide, D-phenylglycine-NH$_2$, (D-Ala)$_5$, (D-Ala)$_4$, (D-Ala)$_3$, (D-Ala)$_2$, (D-Val)$_3$, (D-Leu)$_2$, and DL-Ala-DL-Phe (Asano *et al.*, 1996).

## Effects of Metal Ions and Inhibitors

The enzyme activity was enhanced by Mg$^{2+}$ (138%), Mo$^{3+}$ (130%) and Ba$^{2+}$ (123%). When measured after incubating at 30°C for 30 min, the activity was inhibited to 94% by PMSF, 76% by Ag$^+$, 74% by Fe$^{2+}$ and 32% by Hg$^{2+}$ at 5 mM. These results were consistent with structural evidence that the enzyme is a serine peptidase (see below) (Asano *et al.*, 1996).

## Structural Chemistry

The gene coding for ADP (*adp*) was cloned into plasmid pUC118, and a 1164 bp open reading frame consisting of 388 codons with an $M_r$ of 42 033 was identified as the *adp* gene (Asano *et al.*, 1996). ADP would be synthesized with a signal peptide. The $M_r$ of the subunit calculated was about 36 000 by SDS-PAGE, 37 000 by HPLC, and 37 952 by mass spectrophotometry. The absorption of the purified enzyme in 0.01 M potassium phosphate buffer, pH 7.0, was maximal at 281 nm.

The deduced primary structure of ADP is similar to that of DD-carboxypeptidase from *Streptomyces* R61 (35.0% identical over 346 amino acids) (Duez *et al.*, 1987), penicillin-binding proteins (PBPs) from *Streptomyces (Nocardia) lactamdurans* (28.1% over 263 amino acids) (Coque *et al.*, 1993) and PBPs of *B. subtilis* (28.5% over 309 amino acids) (Popham & Setlow, 1993), and class C $\beta$-lactamases of *Serratia marcescens* (24.9% over 217 amino acids) (Joris *et al.*, 1985), class C $\beta$-lactamases of *Enterobacter cloacae* (25.1% over 191 amino acids) (Galleni *et al.*, 1988), fimbrial protein D from *Dichelobacter nodosus* (24.1% over 261 amino acids) (Hobbs *et al.*, 1991), D-aminopeptidase from *Ochrobactrum anthropi* (27.5% over 182 amino acids) (Asano *et al.*, 1992), and esterase from *Pseudomonas* sp. (30.5% over 154 amino acids) (McKay *et al.*, 1992). The sequence Ser-Xaa-Xaa-Lys is perfectly conserved among the 'penicillin-recognizing enzymes' (Frère *et al.*, 1988), which include PBPs, $\beta$-lactamases and D-aminopeptidase (Chapter 146).

## References

Asano, Y., Nakazawa, A., Kato, Y. & Kondo, K. (1989) Properties of a novel D-stereospecific aminopeptidase from *Ochrobactrum anthropi*. *J. Biol. Chem.* **264**, 14233–14239.

Asano, Y., Kato, Y., Nakazawa, A. & Kondo, K. (1992) Structural similarity of D-aminopeptidase to carboxypeptidase DD and $\beta$-lactamase. *Biochemistry* **31**, 2316–2328.

Asano, Y., Ito, H., Dairi, T. & Kato, Y. (1996) An alkaline D-stereospecific endopeptidase with $\beta$-lactamase activity from *Bacillus cereus*. *J. Biol. Chem.* **271**, 30256–30262.

Coque, J.J.R., Liras, P. & Martin, J.F. (1993) Genes for a $\beta$-lactamase, a penicillin-binding protein and a transmembrane protein are clustered with the cephamycin biosynthetic genes in *Nocardia lactamdurans*. *EMBO J.* **12**, 631–639.

Duez, C., Piron-Fraipont, C., Joris, B., Dusart, J., Urdea, M.S., Martial, J.A., Frère, J.-M. & Ghuysen, J.-M. (1987) Primary structure of the *Streptomyces* R61 extracellular DD-peptidase. *Eur. J. Biochem.* **162**, 509–518.

Frère, J.-M., Joris, B., Dideberg, O., Charlier, P. & Ghuysen, J.-M. (1988) Penicillin-recognizing enzymes. *Biochem. Soc. Trans.* **16**, 934–938.

Galleni, M., Lindberg, F., Normark, S., Cole, S., Honore, N., Joris, B. & Frère, J.-M. (1988) Sequence and comparative analysis of three *Enterobacter cloacae* ampC $\beta$-lactamase genes and their products. *Biochem. J.* **250**, 753–760.

Hobbs, M., Dalrymple, B.P., Cox, P.T., Livingstone, S.P., Delaney, S.F. & Mattick, J.S. (1991) Organization of the fimbrial gene region of *Bacteroides nodosus*: class I and class II strains. *Mol. Microbiol.* **5**, 543–560.

Joris, B., De Meester, F., Galleni, M., Reckinger, G., Coyette, J. & Frère, J.-M. (1985) The $\beta$-lactamase of *Enterobacter cloacae* P99.

Chemical properties, N-terminal sequence and interaction with 6 β-halogenopenicillanates. *Biochem. J.* **228**, 241–248.

McKay, D.B., Jennings, M.P., Godfrey, E.A., MacRae, I.C., Rogers, P.J. & Beacham, I.R. (1992) Molecular analysis of an esterase-encoding gene from a lipolytic psychrotrophic

pseudomonad. *J. Gen. Microbiol.* **138**, 701–708.

Popham, D.L. & Setlow, P. (1993) Cloning, nucleotide sequence, and regulation of the *Bacillus subtilis pbpE* operon, which codes for penicillin-binding protein 4* and an apparent amino acid racemase. *J. Bacteriol.* **175**, 2917–2925.

*Yasuhisa Asano*
*Biotechnology Research Center,*
*Toyama Prefectural University,*
*Kosugi, Toyama 939-03 Japan*
*Email: asano@pu-toyama.ac.jp*

# 148. Penicillin-binding protein 4

## Databanks

*Peptidase classification: clan SE, family S13, MEROPS ID: S13.001*
*NC-IUBMB enzyme classification: EC 3.4.16.4*
Databank codes:

| Species | SwissProt | PIR | EMBL (cDNA) | EMBL (genomic) |
|---|---|---|---|---|
| *Bacillus subtilis* | P39844 | – | Z34883 | – |
| *Escherichia coli* | P24228 | – | U01376 X59460 X60038 | U18997: chromosomal region 67.4–76.0' |
| *Haemophilus influenzae* | P45161 | – | – | U32812: complete genome section 127 of 163 |

## Name and History

In 1976, Tamura *et al.* separated *Escherichia coli* D-alanine carboxypeptidase I activity into three fractions characterized as D-alanine carboxypeptidases IA, IB and IC. Enzymes IB and IC were similar in their activity profiles, the main difference being their localization, in that after cell disruption, enzyme IC was soluble, but enzyme IB was solubilized from the membrane fraction by LiCl. In 1977, studies of mutants showed that IB and IC were identical, and corresponded to *penicillin-binding protein 4 (PBP-4)* which is encoded by the *dacB* gene situated at 68 min on the *E. coli* chromosome (Iwaya & Strominger, 1977; Matsuhashi *et al.*, 1977). Molecular cloning of the *dacB* gene and the overexpression of PBP-4 were achieved in 1991 (Korat *et al.*, 1991). A preliminary crystallographic analysis was carried out in 1995 (Thunnissen *et al.*, 1995).

## Activity and Specificity

Three reactions are catalyzed by PBP-4–

R-D-Ala-D-Ala + H₂O ⟶ R-D-Ala + D-Ala
(DD-carboxypeptidation)

R-D-Ala-D-Ala + NH₂-R′ ⟶ R-D-Ala-NH-R′ + D-Ala
(DD-transpeptidation)
R-D-Ala-D-Xaa + H₂O ⟶ R-D-Ala + D-Xaa
(so-called 'DD-endopeptidation')

where R-D-Ala-D-Ala and NH₂-R′ are the donor and acceptor substrate, respectively, and D-Xaa is the D-end of *meso*-2,6-diaminopimelic acid whose L-end is part of another peptide. Thus, the so-called 'endopeptidase' activity of PBP-4 hydrolyzes the D-alanyl-γ-*meso*-2,6-diaminopimelyl cross-bridge formed by the transpeptidase activity (Tamura *et al.*, 1976; Korat *et al.*, 1991). However, the name 'endopeptidase' is a misnomer, since the hydrolyzed peptide bond is always in α to a free carboxyl group on a D center so that the activity is actually a DD-carboxypeptidase activity (Frère & Joris, 1985).

## Substrate Specificity

Most of the following information originates from the study of purified preparations of enzymes IB and IC (Tamura *et al.*, 1976). Three donor substrates have been utilized to characterize the DD-carboxypeptidase activity of PBP-4: uridine diphosphate-*N*-acetylmuramoyl-L-alanyl-D-γ-glutamyl-L-lysyl-D-alanyl-D-alanine (UDPMur-AEKAA), uridine diphosphate-*N*-acetylmuramoyl-L-alanyl-D-γ-glutamyl-*meso*-2,6-diaminopimelyl-D-alanyl-D-alanine (UDPMur-AEA$_2$pmAA) and $N^\alpha N^\epsilon$-diacetyl-L-lysyl-D-alanyl-D-alanine (Ac$_2$KAA). Of these, UDPMur-AEA$_2$pmAA is the only one reported to be processed by the enzyme (specific activity 2–4 µmol min$^{-1}$ mg$^{-1}$, $K_m$ 1–2 mM).

Transpeptidation was studied using UDPMur-AEA$_2$pmAA and glycine as donor and acceptor substrates, respectively. An increase of the transpeptidation product accompanied by a decrease of the hydrolysis product was observed, so that the total donor utilization (release of D-alanine) remained constant up to 1.2 M glycine. The concentration of glycine necessary to yield equivalent amounts of hydrolysis and transpeptidation products was 600–700 mM. No transpeptidation was observed with Ac$_2$KAA and glycine.

Bis(*N*-acetylglucosamyl-*N*-acetylmuramoyl-L-alanyl-D-γ-glutamyl-*meso*-2,6-diaminopimelyl-D-alanine) was used to probe the so-called 'endopeptidase' activity, but no kinetic parameters were reported.

## Mechanism

In the hydrolysis reaction, the enzyme appears to follow the classical acyl enzyme pathway of serine proteases:

$$E + S \underset{K}{\overset{}{\rightleftharpoons}} ES \xrightarrow{k_{+2}} ES^* \xrightarrow{k_{+3}} E + P_2$$

with $P_1$ branching off above $ES^*$.

where ES*, P$_1$ and P$_2$ are the acyl enzyme, the first leaving group and the hydrolysis products, respectively.

## Assay and Influence of Physicochemical Conditions

Hydrolysis, 'endopeptidase' and transpeptidation reactions were quantified by high-voltage paper electrophoresis using radioactive substrates (Tamura *et al.*, 1976). Maximal activity was obtained in the pH range 8–9. The following compounds have been reported to influence the alanine carboxypeptidase activity of PBP-4 D: 8 mM EDTA (26–40% inhibition), 10 mM MgCl$_2$ (2.5–3.7-fold increase), 10 mM MnCl$_2$ (38–72% inhibition), 10 mM CoCl$_2$ (1.8–3.0-fold increase) and 10 mM ZnCl$_2$ (63–78% inhibition).

## Inhibitors

Antibiotics of the β-lactam family are efficient inactivators which acylate the active-site serine according to the pathway described above for the substrates, but no P$_1$ is formed due to the cyclic β-lactam structure and deacylation is slow so that an inactive acyl enzyme accumulates. So far, four penicillins (benzylpenicillin, ampicillin, penicillin V, 6-aminopenicillanate), five cephalosporins (cephalexin, cephaloridine, cephalosporin C and G, cephalothin) and one amidino-penicillin (mecillinam) have been found to decrease the binding of [$^{14}$C]benzylpenicillin to the membrane form of PBP-4 or to inhibit the D-alanine carboxypeptidase activity of the purified enzyme IB–IC (Tamura *et al.*,

1976; Spratt, 1977). The concentration of β-lactam required to inhibit 50% of the enzyme activity in 15 min ranged from 0.1 to 2 µg ml$^{-1}$ for the penicillin, 10–50 µg ml$^{-1}$ for the cephalosporins and more than 500 µg ml$^{-1}$ with mecillinam. Cephamycins (Matsuhashi *et al.*, 1982), cefoxitin, cephradin and penems like imipenem and CGP31608 (Hashizume *et al.*, 1984) also bind to the enzyme but no kinetic parameters were determined. Values for the acylation ($k_{+2}/K'$) and deacylation ($k_{+3}$) rates with benzylpenicillin were reported (Frère & Joris, 1985) as 7000 M$^{-1}$ s$^{-1}$ (37°C) and 0.4 × 10$^{-3}$ s$^{-1}$ (30°C), respectively. The breakdown of the benzylpenicilloyl–PBP-4 adduct generates benzylpenicilloic acid (Tamura *et al.*, 1976).

Finally, boronic acids have been described as competitive inhibitors of the penicillin-binding capacity of PBP-4 (Then *et al.*, 1994).

## Structural Chemistry

*E. coli* PBP-4 has a molecular mass of 49 568 Da and consists of 457 residues (Korat *et al.*, 1991). This periplasmic protein is loosely attached to the cytoplasmic membrane. When overexpressed, it can be detected both in the membrane and the soluble fractions obtained after cell disruption. The protein does not show any transmembrane or amphiphilic helices nor attachment site for a lipid anchor, and the anchoring mechanism remains unclear (Mottl *et al.*, 1991). *E. coli* PBP-4 belongs to the superfamily of the penicillin-recognizing enzymes (PREs) which comprises the serine β-lactamases and the penicillin-sensitive DD-peptidases (Frère *et al.*, 1992). Analysis of the amino acid sequence showed that PBP-4 possesses the three conserved elements Ser42-Xaa-Xaa-Lys45, Ser289-Xaa-Asn288 and Lys397-Thr398-Gly399 characteristic of the PREs with Ser42 as the active-site serine. Preliminary X-ray data (Thunnissen *et al.*, 1995) indicated that if one excepts a 188 residues stretch (from Asp59 to Ala246), the PBP-4 scaffolding is similar to that of the class A β-lactamases.

## Preparation

PBP-4 was formerly prepared from *E. coli* lysates in two steps. After cell disruption and membrane and debris removal, the protein was purified by dye-affinity chromatography (Cibacron Navyblue 2G-E), followed by an anion-exchange Fractogel TSK DEAE 650-S column (Mottl & Keck, 1991). A revised version of this protocol utilized covalent penicillin affinity instead of dye affinity (Thunnissen *et al.*, 1995).

## Biological Aspects

*E. coli* PBP-4 (110 molecules per cell; Spratt, 1977) has multiple biological functions. The enzyme is thought to be involved in the recycling and maturation of the peptidoglycan polymer. Its so-called 'endopeptidase' activity would be responsible for the hydrolysis of the peptidoglycan cross-links that are needed to allow the incorporation of new glycan strands into the murein sacculus (Goodell & Schwartz, 1983; Burman & Park, 1984). In addition, various studies indicate a possible participation of PBP-4 (with other endopeptidases such as the penicillin-insensitive murein endopeptidase MepA, Chapter 556) in the bacterial autolysis system (Hackenbeck & Messer, 1977; Iwaya & Stromginger, 1977; Kitano et al., 1980).

PBP-4, together with PBP-5 and -6, is responsible for the bacterial D-alanine carboxypeptidase activity that would maintain a correct balance between the various peptidoglycan precursors required for cell growth and division (Frère et al., 1992). From the number of molecules per cell and their catalytic activities (Tamura et al., 1976; Spratt, 1977), the ratio of the contribution of PBP-4 to that of PBP-5/6 in the total hydrolysis activity can be estimated to vary from 0.5 to 0.2, depending upon the substrate concentration.

## Distinguishing Features

E. coli PBP-4 is the only PBP for which boronic acids are competitive inhibitors (Then et al., 1990), but note that these compounds are also specific inhibitors of the active-site serine $\beta$-lactamases (Martin et al., 1994).

## Related Enzymes

Hydrophobic cluster analysis indicated that E. coli PBP-4 was similar to the Actinomadura R39 DD-peptidase (Chapter 150). In addition, providing that a large deletion of 188 amino acid was made in the two PBPs, the analysis revealed possible similarities in the polypeptide folding between the two PBPs and the known three-dimensional structure of the class A Streptomyces albus G $\beta$-lactamase (Granier et al., 1992). After the same deletion, sequence alignments of PBP-4 and three class A $\beta$-lactamases (TEM1, Bacillus licheniformis and Staphylococcus aureus) showed 21% identical residues and 51% similar residues (Thunnissen et al., 1995).

## Further Reading

The properties of DD-peptidases and their structural relationships with $\beta$-lactamases have been analyzed in detail by Waxman & Stromginger (1983), Frère & Joris (1985), Frère et al. (1992) and Ghuysen (1997). Jamin et al. (1995) focus specifically on the transpeptidase activity.

## References

Burman, L.G. & Park, J.T. (1984) Molecular model for elongation of the murein sacculus of Escherichia coli. Proc. Natl Acad. Sci. USA **81**, 1844–1848.

Frère, J.-M. & Joris, B. (1985) Penicillin-sensitive enzymes in peptidoglycan biosynthesis. CRC Crit. Rev. Microbiol. **11**, 299–396.

Frère, J.-M., Nguyen-Distèche, M., Coyette, J. & Joris, B. (1992) Mode of action; interaction with the penicillin-binding proteins. In: The Chemistry of Beta-Lactams (Page, M.I., ed.). London: Chapman & Hall, pp. 148–197.

Ghuysen, J.-M. (1997) Penicillin-binding proteins. Wall peptidoglycan assembly and resistance to penicillin: facts, doubts and hopes. Int. J. Antimicrob. Agents **8**, 45–60.

Goodell, W.E. & Schwarz, U. (1983) Cleavage and resynthesis of peptide crossbridges in Escherichia coli murein. J. Bacteriol. **156**, 136–140.

Granier, B., Duez, C., Lepage, S., Englebert, S., Dusart, J., Dideberg, O., van Beeumen, J., Frère, J.-M. & Ghuysen, J.-M. (1992) Primary and predicted secondary structures of the Actinomadura R39 extracellular DD-peptidase, a penicillin-binding protein (PBP) related to the Escherichia coli PBP4. Biochem. J. **282**, 781–788.

Hackenbeck, R. & Messer, W. (1977) Activity of murein hydrolases in synchronized cultures of Escherichia coli. J. Bacteriol. **129**, 1239–1244.

Hashizume, T., Ishino, F., Nagakawa, J.-I., Tamaki, S. & Matsuhashi, M. (1984) Studies on the mechanism of action of Imipenem (N-formimidoyl-thienamycin) in vitro: binding to the penicillin-binding proteins (PBPs) in Escherichia coli and Pseudomonas aeruginosa and inhibition of enzyme activities due to the PBPs in E. coli. J. Antibiotics **37**, 394–400.

Iwaya, M. & Stromginger, J.L. (1977) Simultaneous deletion of D-alanine carboxypeptidase IB-C and penicillin-binding component IV in a mutant of Escherichia coli K12. Proc. Natl Acad. Sci. USA **74**, 2980–2984.

Jamin, M., Wilkin, J.-M. & Frère, J.-M. (1995) Bacterial transpeptidases and penicillin. Essays Biochem. **29**, 1–24.

Kitano, K., Williamson, R. & Tomasz, A. (1980) Murein hydrolase defect in the $\beta$-lactam tolerant mutants of Escherichia coli. FEMS Microbiol. Lett. **7**, 133–136.

Korat, B., Mottl, H. & Keck, W. (1991) Penicillin-binding protein 4 of Escherichia coli, molecular cloning of the dacB gene, controlled overexpression, and alterations in murein composition. Mol. Microbiol. **5**, 675–684.

Martin, R., Gold, M. & Bryan Jones, J. (1994) Inhibition of R-TEM1 $\beta$-lactamase by boronic acids. Biorg. Med. Chem. **4**, 1229–1234.

Matsuhashi, M., Takagaki, Y., Maruyama, I.N., Tamaki, S., Nishimura, Y., Suzuki, H., Ogino, U. & Hirota, Y. (1977) Mutants of Escherichia coli lacking in highly penicillin-sensitive D-alanine carboxypeptidase activity. Proc. Natl Acad. Sci. USA **74**, 2976–2979.

Matsuhashi, M., Ishino, F., Tamaki, S., Nakajima-Iijima, S.T., Tomokia, S., Nakagawa, J.-L., Hirata, A, Spratt, B., Tsuruoka, T., Inouye, S. & Yamada, Y. (1982) Mechanism of action of $\beta$-lactam antibiotics. In: Trends in Antibiotic Research. Tokyo: Japan Antibiotic Research Association, pp. 99–114.

Mottl, H. & Keck, W. (1991) Purification of penicillin-binding protein 4 of Escherichia coli as a soluble protein by dye-affinity chromatography. Eur. J. Biochem. **200**, 767–773.

Mottl, H., Terpstra, P. & Keck, W. (1991) Penicillin-binding protein 4 of Escherichia coli shows a novel type of primary structure among penicillin-interacting proteins. FEMS Microbiol. Lett. **78**, 213–220.

Spratt, B.J. (1977) Properties of the penicillin-binding proteins of Escherichia coli K12. Eur. J. Biochem. **72**, 341–352.

Tamura, T., Yasuo, I. & Stromginger, J.L. (1976) Purification to homogeneity and properties of two D-alanine carboxypeptidases from Escherichia coli. J. Biol. Chem. **251**, 414–423.

Then, R.L., Hubschwerlen, C. & Charnas, R. (1994) Antagonism between some novel boronic acids and cephalosporins in inducible strains of Enterobacter cloacae. In: Fifty Years of Penicillin Application. History and Trends (Kleinkauf, H. & von Dôhren, H., eds). Czech Republic: PUBLIC Ltd, pp. 383–387.

Thunnissen, M.M.G.M., Fusetti, F., de Boer, B. & Dijkstra, B.W. (1995) Purification, crystallisation and preliminary X-ray analysis of penicillin-binding protein 4 from *Escherichia coli*, a protein related to class A β-lactamases. *J. Mol. Biol.* **247**, 149–153.

Waxman, D.J. & Strominger, J.L. (1983) Penicillin-binding proteins and the mechanism of action of β-lactam antibiotics. *Annu. Rev. Biochem.* **52**, 825–869.

**Jean-Marc Wilkin**
*Center for Protein Engineering and Laboratory of Enzymology,*
*University of Liège,*
*Institut de Chemistry, B6 Sart Tilman,*
*B-4000 Liège Belgium*
*Email: jmwilkin@ben.vub.zc.be*

# *149. Muramoyl pentapeptide carboxypeptidase*

## Databanks

*Peptidase classification: clan SE, family S13, MEROPS ID: S13.001*
*NC-IUBMB enzyme classification: currently EC 3.4.17.8 (inappropriate)*
*Databank codes: see Chapter 148*

## Name and History

This enzyme was purified 120-fold from a solubilized fraction of *Escherichia coli* cells and named **D-alanine carboxypeptidase I** by Izaki & Strominger (1968). It was rapidly incorporated into the Enzyme Commission list as EC 3.4.12.6 and carried through all subsequent editions. The number was changed in 1978 to EC 3.4.17.8, in the subclass of metallo-carboxypeptidases.

## Enzyme Properties

The enzyme is released from the particulate fraction of *E. coli* by prolonged (25 min) ultrasonication which releases 90% of the total activity. It has a pH optimum of 9 against the substrate uridine diphosphate-$N$-acetylmuramoyl-L-alanyl-D-$\gamma$-glutamyl-*meso*-2,6-diaminopimelyl-D-alanyl-D-alanine (UDPMur-AEA$_2$pmAA). It can cleave only the C-terminal Ala residue, but it also shows slight transpeptidase activity. The enzyme is activated by 5 mM $Mg^{2+}$ or $Ca^{2+}$ and to a lesser extent by $Zn^{2+}$, $Mn^{2+}$ and $Fe^{3+}$. This led to its earlier (mis)identification as a metallocarboxypeptidase. The enzyme is exquisitely sensitive to inhibition by penicillin ($2\,ng\,ml^{-1}$).

## Current Status

Suspicions are aroused by the penicillin inhibition that is now known to be typical of the penicillin-binding proteins (PBPs) that are serine enzymes. In fact, the history of this enzyme can be traced through further work from the Strominger lab. Tamura *et al*. (1976) purified the D-Ala carboxypeptidase I to homogeneity and found that it could be separated into three forms: IA, IB and IC. The first two forms were membrane bound and the third proved to be a soluble form of IB. Only the IB–C enzyme was highly sensitive to penicillin. The following year, Iwaya & Strominger (1977) used gene deletion analysis to demonstrate that the IB form was, in fact, PBP-4 (Chapter 148). Therefore, this carboxypeptidase should no longer be considered a metalloproteinase. It is the product of the *dacB* gene, the membrane-associated D-Ala-D-Ala carboxypeptidase of serine type (EC 3.4.16.4). The current EC listing should be deleted. The reader is referred to Chapter 148 for a full description of this enzyme.

## References

Iwaya, M. & Strominger, J.L. (1977) Simultaneous deletion of D-alanine carboxypeptidase IB-C and penicillin-binding component IV in a mutant of *Escherichia coli* K12. *Proc. Natl Acad. Sci. USA* **74**, 2980–2984.

Izaki, K. & Strominger, J.L. (1968) Biosynthesis of the peptidoglycan of bacterial cell walls. XIV. Purification and properties of two D-alanine carboxypeptidases from *Escherichia coli. J. Biol. Chem.* **243**, 3193–3201.

Tamura, Y., Imae, Y. & Strominger, J.L. (1976) Purification to homogeneity and properties of two D-alanine carboxypeptidases I from *Escherichia coli. J. Biol. Chem.* **251**, 414–423.

**J. Fred Woessner**
*Department of Biochemistry and Molecular Biology,*
*University of Miami School of Medicine,*
*Miami, FL 33101, USA*
*Email: fwoessne@mednet.med.miami.edu*

# 150. *Actinomadura R39* D-Ala-D-Ala carboxypeptidase

## Databanks

*Peptidase classification: clan SE, family S13, MEROPS ID: S13.002*
*NC-IUBMB enzyme classification: none*
*Databank codes:*

| Species | SwissProt | PIR | EMBL (cDNA) | EMBL (genomic) |
|---|---|---|---|---|
| *Actinomadura* R39 | P39045 | S22409 | X64790 | – |

## Name and History

After the discovery of the penicillin-sensitive *Streptomyces* R61 D-Ala-D-Ala carboxypeptidase (Leyh-Bouille *et al.*, 1971), a similar soluble enzyme was found in the culture medium of *Actinomadura* R39 (formerly *Streptomyces* R39) (Leyh-Bouille *et al.*, 1972). Like its R61 counterpart, it catalyzed transpeptidation reactions when supplied with adequate substrates (Pollock *et al.*, 1972; Perkins *et al.*, 1973; Ghuysen *et al.*, 1973, 1974). Inactivation by β-lactam antibiotics results from the formation of a covalent bond with the essential serine residue (Duez *et al.*, 1981).

## Activity and Specificity

The catalyzed reactions are:

R-D-Ala-D-Ala + $H_2O \longrightarrow$ R-D-Ala + D-Ala
(carboxypeptidase)
R-D-Ala-D-Ala + R′-NH$_2 \longrightarrow$ R-D-Ala-NH-R′ + D-Ala
(transpeptidase)

where R-D-Ala-D-Ala and R′-NH$_2$ are the donor and acceptor substrates, respectively.

## Substrate Specificity

The donor peptide with which the highest activity has been recorded is $N^\alpha$-acetyl-L-Lys-D-Ala$+$D-Ala ($k_{cat} = 18\,s^{-1}$, $k_{cat}/K_m = 60000\,M^{-1}\,s^{-1}$). Acetylation of the ε-amino group of L-Lys results in an 8-fold decrease in the $k_{cat}/K_m$ value. The C-terminal residue can be replaced by Gly (10% efficiency), D-Glu (100%) or D-Leu (75%). Replacement of the penultimate D-Ala residue by D-Leu, Gly or L-Ala yields substrates on which the enzyme is not significantly active (Leyh-Bouille *et al.*, 1972; Nieto *et al.*, 1973). The structure of the R group is also very important. If R is acetyl, only 1% of efficiency is observed (Nieto *et al.*, 1973). *N*-Acetyl-L-Xaa is favorable if Xaa contains a long aliphatic side chain (acetyl-L-Ala is not better than acetyl), such as L-Lys or *meso*-diaminopimelic acid (A$_2$pm).

The only good acceptors are glycine and D-amino acids, but the latter may have complex side chains. For instance, they may be derivatives of *meso*-A$_2$pm in which the amino and carboxyl groups on the D center are free while those on the L center are engaged in peptide bonds. The best substrates closely mimic the donor and acceptor parts of the peptides in the nascent *Actinomadura* peptidoglycan (Ghuysen *et al.*, 1973).

When supplied with compounds containing both a D-Ala-D-Ala C-terminus and an adequately positioned amino group, such as the 'natural' pentapeptide L-Ala-D-Glu-ε(L)-A₂pm-(L)-D-Ala-D-Ala, where A₂pm is the *meso* isomer, the enzyme catalyzes the formation of dimers (Ghuysen *et al.*, 1974).

The DD-peptidase also acts as an esterase and a thioesterase on compounds of general structure R-Xaa-Y-CH(R″)-COOH, where Y is O or S (Adam *et al.*, 1990) and Xaa is Gly or a D residue, preferentially D-Ala. These compounds are also utilized in transacylation reactions. Among these, the best reported substrate is Bz-D-Ala-S-CH₂-COOH ($k_{cat} = 6\,s^{-1}$, $k_{cat}/K_m = 330\,000\,M^{-1}\,s^{-1}$). The structural requirements for the R group and stereospecificity of the thioesters are somewhat modified when compared to those for the peptides (Damblon *et al.*, 1995).

## Mechanism

In the hydrolysis reaction, the enzyme follows the classical acyl enzyme pathway of serine proteases:

$$E + S \longleftrightarrow ES \rightarrow ES^* + P_1 \rightarrow E + P_2$$

where ES* is the acyl enzyme. With the peptide Ac₂-L-Lys-D-Ala-D-Ala, acylation is the rate-limiting step, and the acyl enzyme cannot be detected. By contrast, deacylation is strongly rate limiting with most thioesters (Jamin *et al.*, 1991).

## Assays and Influence of Physicochemical Conditions

The hydrolysis of peptides is monitored by a colorimetric, discontinuous assay of the released D-alanine (Frère *et al.*, 1976), while that of the thioesters can be followed directly at 260 nm (Adam *et al.*, 1990). Transpeptidation was formerly quantified by high-voltage paper electrophoresis (Pollock *et al.*, 1972), but HPLC techniques are now preferred (Jamin *et al.*, 1993).

The enzyme presents a broad optimum pH between 5.5 and 9.5. Transpeptidation is favored by high pH, high ionic strength and high and low acceptor and donor concentrations, respectively (Ghuysen *et al.*, 1973, and unpublished results of the author).

## Inhibitors

Some derivatives of lysine, such as $N^{\alpha}$-acetyl,$N^{\varepsilon}$-glycyl-L-Lys, act as specific inhibitors of the transpeptidation reaction at high (millimolar) concentrations (Perkins *et al.*, 1981). An excess (5 mM) of a complex acceptor, the 'Glu-amidated tetrapeptide' can completely inhibit both hydrolysis and transpeptidation reactions (Ghuysen *et al.*, 1973).

Antibiotics of the β-lactam family are very efficient transient inactivators that acylate the active-site serine according to the pathway described above for the substrates, but no $P_1$ is formed (due to the β-lactam structure) and the deacylation step is very slow ($10^{-3} - 10^{-6}\,s^{-1}$), so that an inactive acyl enzyme accumulates (Frère *et al.*, 1974b; Fuad *et al.*, 1976). The rate of acylation is consistently much larger than with the R61 enzyme. In fact, among the presently known enzymes, the R39 DD-peptidase is one of the most penicillin sensitive. With some but not all β-lactams, the deacylation step does not

yield the expected penicilloic acid but involves the cleavage of the C5–C6 bond of the antibiotic (Frère *et al.*, 1975).

## Structural Chemistry

The mature protein consists of one 489 residue polypeptide chain with one single disulfide bridge ($M_r = 50\,051$). The gene encodes a 538 residue precursor including a 49 residue signal peptide (Granier *et al.*, 1992). Incomplete cleavage of this signal peptide (before position −20) when the protein is produced by *Streptomyces lividans* yields an inactive enzyme.

Attempts to crystallize the protein have remained unsuccessful.

## Preparation

The enzyme was initially extracted and purified from culture filtrates of the original *Actinomadura* R39 strain (Frère *et al.*, 1974a). Much higher yields were later obtained when the cloned gene was reintroduced into the R39 strain itself by electroporation of the pDML15 plasmid (Granier *et al.*, 1992).

## Biological Aspects

As for the similar enzyme produced by *Streptomyces* R61 (Chapter 145), the physiological role of the R39 DD-peptidase remains obscure.

## Distinguishing Features

Among the penicillin-binding proteins (PBPs), the R39 enzyme is remarkable for its high activity on simple peptides and thioesters, its ability to catalyze the formation of dimers and its high sensitivity to most β-lactams. This latter property has been utilized in the design of a sensitive method for the assay of β-lactams in biological fluids (Frère *et al.*, 1980).

## Related Enzymes

The sequence of the enzyme can be aligned with that of *E. coli* PBP-4 (Chapter 148) with 19% of identical residues. The two proteins also share some enzymatic properties. The three conserved elements, characteristic of penicillin-recognizing enzymes (**Ser**-Xaa-Xaa-Lys, where **Ser** is the active-site serine, Ser-Xaa-Asn and Lys-Thr-Gly), also allow an alignment on the sequences of class A β-lactamases in which the prominent feature is a 170 residue insertion between the first and second elements of the two PBPs (Granier *et al.*, 1992).

## Further Reading

The properties of DD-peptidases and their structural and functional relationships with β-lactamases have been considered in detail by Frère & Joris (1985), Joris *et al.* (1991), Frère *et al.* (1992), Ghuysen (1991) and Jamin *et al.* (1995).

## References

Adam, M., Damblon, C., Plaitin, B., Christiaens, L. & Frère, J.M. (1990) Chromogenic depsipeptide substrates for β-lactamases and penicillin-sensitive DD-peptidases. *Biochem. J.* **270**, 525–529.

Damblon, C., Zhao, G.-H., Jamin, M., Ledent, P., Dubus, A., Vanhove, M., Raquet, X., Christiaens, L. & Frère, J.-M. (1995) Breakdown of the stereospecificity of DD-peptidases and β-lactamases with thiolester substrates. *Biochem. J.* **309**, 431–436.

Duez, C., Joris, B., Frère, J.-M. & Ghuysen, J.-M. (1981) The penicillin-binding site in the exocellular DD-carboxypeptidase-transpeptidase of *Actinomadura* R39. *Biochem. J.* **193**, 83–86.

Frère, J.-M. & Joris, B. (1985) Penicillin-sensitive enzymes in peptidoglycan biosynthesis. *CRC Crit. Rev. Microbiol.* **11**, 299–396.

Frère, J.-M., Moreno, R., Ghuysen, J.-M., Perkins, H.R., Dierickx, L. & Delcambe, L. (1974a) Molecular weight, amino acid composition and physicochemical properties of the exocellular DD-carboxypeptidase-transpeptidase of *Streptomyces* R39. *Biochem. J.* **143**, 233–240.

Frère, J.-M., Ghuysen, J.-M., Reynolds, P.E., Moreno, R. & Perkins, H.R. (1974b) Binding of β-lactam antibiotics to the exocellular DD-carboxypeptidase-transpeptidase of *Streptomyces* R39. *Biochem. J.* **143**, 241–249.

Frère, J.-M., Ghuysen, J.-M., Degelaen, J., Loffet, A. & Perkins, H.R. (1975) Fragmentation of benzylpenicillin after interaction with the exocellular DD-carboxypeptidase-transpeptidases of *Streptomyces* R61 et R39. *Nature* **258**, 168–170.

Frère, J.-M., Leyh-Bouille, M., Ghuysen, J.-M., Nieto, M. & Perkins, H.R. (1976) Exocellular DD-carboxypeptidases-transpeptidases from *Streptomyces*. *Methods Enzymol.* **45**, 610–636.

Frère, J.-M., Klein, D. & Ghuysen, J.-M. (1980) Enzymatic method for rapid and sensitive determination of beta-lactam antibiotics. *Antimicrob. Agents Chemother.* **18**, 506–510.

Frère, J.-M., Nguyen-Distèche, M., Coyette, J. & Joris, B. (1992) Mode of action: interaction with the penicillin binding proteins. In: *The Chemistry of Beta-Lactams* (Page, M.I., ed.). Glasgow: Chapman & Hall, pp. 148–197.

Fuad, N., Frère, J.-M., Ghuysen, J.-M., Duez, C. & Iwatsubo, M. (1976) Mode of interaction between β-lactam antibiotics and the exocellular DD-carboxypeptidase-transpeptidase from *Streptomyces* R39. *Biochem. J.* **155**, 623–629.

Ghuysen, J.-M. (1991) Serine β-lactamases and penicillin-binding proteins. *Annu. Rev. Microbiol.* **45**, 37–67.

Ghuysen, J.-M., Leyh-Bouille, M., Campbell, J.N., Moreno, R., Frère, J.-M., Duez, C., Nieto, M. & Perkins, H.R. (1973) Structure of the wall peptidoglycan of *Streptomyces* R39 and the specificity profile of its exocellular DD-carboxypeptidase-transpeptidase for peptide acceptors. *Biochemistry* **12**, 1243–1250.

Ghuysen, J.-M., Reynolds, P.E., Perkins, H.R., Frère, J.-M. & Moreno, R. (1974) Effects of donor and acceptor peptides on concomitant hydrolysis and transfer reactions catalysed by the exocellular DD-carboxypeptidase-transpeptidase from *Streptomyces* R39. *Biochemistry* **13**, 2539–2547.

Granier, B., Duez, C., Lepage, S., Englebert, S., Dusart, J., Dideberg, O., Van Beeumen, J., Frère, J.-M. & Ghuysen, J.-M. (1992) Primary and predicted secondary structures of the *Actinomadura* R39 extracellular DD-peptidase, a penicillin-binding protein (PBP) related to the *Escherischia coli* PBP4. *Biochem. J.* **282**, 781–788.

Jamin, M., Adam, M., Damblon, C., Christiaens, L. & Frère, J.-M. (1991) Accumulation of acyl-enzyme in DD-peptidase-catalysed reactions with analogues of peptide substrates. *Biochem. J.* **280**, 499–506.

Jamin, M., Wilkin, J.-M. & Frère, J.-M. (1993) A new kinetic mechanism for the concomitant hydrolysis and transfer reactions catalysed by bacterial DD-peptidases. *Biochemistry* **32**, 7278–7285.

Jamin, M., Wilkin, J.-M. & Frère, J.-M. (1995) Bacterial DD-transpeptidases and penicillin. *Essays Biochem.* **29**, 1–24.

Joris, B., Ledent, P., Dideberg, O., Fonze, E., Lamotte-Brasseur, J., Kelly, J.A., Ghuysen, J.-M. & Frère, J.-M. (1991) Comparison of the sequences of class-A beta-lactamases and of the secondary structure elements of penicillin-recognizing proteins. *Antimicrob. Agents Chemother.* **35**, 2294–2301.

Leyh-Bouille, M., Coyette, J., Ghuysen, J.-M., Idczak, J., Perkins, H.R. & Nieto, M. (1971) Penicillin-sensitive DD-carboxypeptidase from *Streptomyces* strain R61. *Biochemistry* **10**, 2163–2170.

Leyh-Bouille, M., Nakel, M., Frère, J.-M., Johnson, K., Ghuysen, J.-M., Nieto, M. & Perkins, H.R. (1972) Penicillin-sensitive DD-carboxypeptidases from *Streptomyces* strains R39 and K11. *Biochemistry* **11**, 1290–1298.

Nieto, M., Perkins, H.R., Leyh-Bouille, M., Frère, J.-M. & Ghuysen, J.-M. (1973) Peptide inhibitors of *Streptomyces* DD-carboxypeptidases. *Biochem. J.* **131**, 163–171.

Perkins, H.R., Nieto, M., Frère, J.-M., Leyh-Bouille, M. & Ghuysen, J.-M. (1973) *Streptomyces* DD-carboxypeptidases as transpeptidases: the specificity for amino compounds acting as carboxyl acceptors. *Biochem. J.* **131**, 707–718.

Perkins, H.R., Frère, J.-M. & Ghuysen, J.-M. (1981) Synthetic peptide inhibitors of transpeptidation by the exocellular DD-carboxypeptidase-transpeptidase from *Actinomadura* R39. FEBS Lett. **123**, 75–78.

Pollock, J.J., Ghuysen, J.-M., Linder, R., Salton, M.R.J., Perkins, H.R., Nieto, M., Leyh-Bouille, M., Frère, J.-M. & Johnson, K. (1972) Transpeptidase activity of *Streptomyces* D-alanyl-D carboxypeptidases. *Proc. Natl Acad. Sci. USA* **69**, 662–666.

***Jean-Marie Frère***
*Department of Biochemistry,*
*University of Liège,*
*Institut of Chemistry (B6) Sart-Tilman, B-4000 Liège 1, Belgium*
*Email: u217902@vm1.ulg.ac.be*

# 151. Introduction: clan SF containing peptidases with a Ser/Lys catalytic dyad

## Databanks

*MEROPS ID: SF*

| Species | SwissProt | PIR | EMBL (cDNA) | EMBL (genomic) |
|---|---|---|---|---|
| **Family S24** | | | | |
| Repressor LexA and bacterial homologs (Chapter 152) | | | | |
| **Family S26** | | | | |
| Mitochondrial/chloroplast leader peptidase (Chapter 155) | | | | |
| Signal peptidase I (Chapter 153) | | | | |
| Signal peptidase SipS (chromosomal) (Chapter 154) | | | | |
| Signal peptidase (eukaryote) (Chapter 156) | | | | |
| Others | | | | |
| *Agrobacterium tumefaciens* | P15595 | S15913 | X13981 | – |
| *Bacillus subtilis* | P54506 | – | – | D84432: genome section |
| *Clostridium perfringens* | – | – | X86488 | – |
| *Escherichia coli* | Q03450 | A45739 | M94366 | – |
| *Escherichia coli* | Q05119 | B45739 | M94367 | X54458: plasmid R751 transfer region |
| *Methanococcus jannaschii* | – | – | U67481 | – |
| **Family S41** | | | | |
| C-terminal processing enzyme (*Synechocystis*) (Chapter 158) | | | | |
| C-terminal processing protease (higher plant) (Chapter 159) | | | | |
| Tail-specific protease (Chapter 157) | | | | |
| **Family S44** | | | | |
| Tricorn protease (Chapter 160) | | | | |

Clan SF consists of families S24, S26, S41 and S44. All members of the clan have a catalytic mechanism comprising a Ser/Lys dyad except the mammalian signal peptidase, in which the catalytic lysine is replaced by histidine (Alignment 151.1) (see CD-ROM). The members of the clan are shown in the Databanks table. Although the enzymes in clan SF perform different biological functions, one striking similarity is the preference for Ala in P1.

The tertiary structure of only one member of the clan has so far been reported, that of the umuD′ protein from *Escherichia coli*. The umuD peptidase undergoes autocatalysis, and 31 amino acids are removed from the N-terminus to generate the inactive umuD′ protein. A representation of the umuD′ protein structure is shown in Figure 151.1, and it is clear from the separation of the N-terminus and the catalytic

dyad (50 Å) that a considerable conformational change has occurred. It is therefore possible that the fold of the uncleaved protein is very different. The umuD′ protein contains mainly β-structure and is therefore unrelated to the D-Ala-D-Ala carboxypeptidases of clan SE, which also possess a Ser/Lys catalytic dyad.

**Family S24** is confined entirely to bacteria and bacteriophages, and many members are known to be repressor proteins that undergo autocatalysis. The LexA protein from *E. coli* is involved in the control of DNA repair, and is activated in the presence of single-stranded DNA (see Chapter 152).

**Family S26** is found in all organisms. The peptidases in this family are responsible for removal of N-terminal peptides that target proteins for entry into the secretory path-

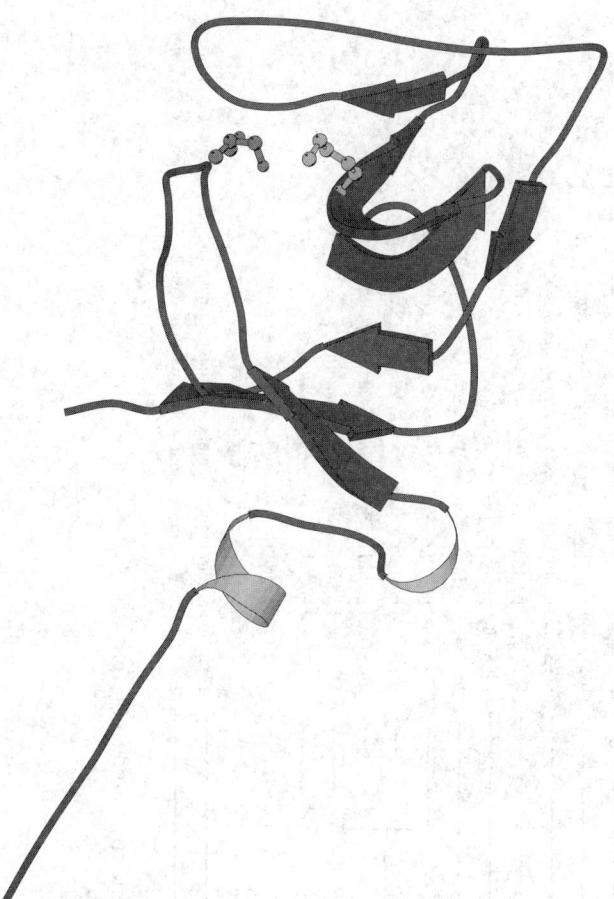

*Figure 151.1* Richardson diagram of *Salmonella typhimurium* UmuD′ protein. The image was prepared from the Brookhaven Protein Data Bank entry (1UMU) as described in the Introduction (p. xxv). Catalytic residues Ser35 and Lys38 are shown in ball-and-stick representation (numbering as in Alignment 151.1).

from a single ancestral protein. The other subunits of the eukaryotesingle peptidase fall into three families of proteins that show no sequence similarity to known peptidases. In the mitochondrial signal peptidase there are two subunits, both of which are catalytically active and members of family S26 (see Chapter 155).

Figure 151.2 is an evolutionary tree for family S26, and shows the deep divergence between the eukaryote microsomal signal peptidase subunits and the bacterial signal peptidase. The subunits of the mitochondrial inner membrane signal peptidase form a distinct branch that diverged from bacterial signal peptidase more recently, consistent with a horizontal transfer of genes from the bacterial endosymbiont ancestral to the mitochondrion. The archean signal peptidase is the oldest branch from the line leading to the microsomal signal peptidase. This line includes two bacterial homologs that are also derived from horizontal transfers, this time from eukaryote to prokaryote.

*Family S41* contains the tail-specific endopeptidase from *E. coli*, which removes nonpolar residues from the C-terminus of proteins (Silber *et al.*, 1992) and an endopeptidase from the photoreactive II center of plants and cyanobacteria that removes the C-terminal extension of D1 protein (Anbudurai *et al.*, 1994; Shestakov *et al.*, 1994). The peptidase domain of tail-specific endopeptidase is also homologous to a domain in the mammalian interphotoreceptor retinoid-binding protein; the catalytic serine is conserved, but the lysine is not, and the retinoid-binding protein is probably not a peptidase.

*Family S44* includes the catalytically active core protein of the tricorn endopeptidase, which is a multisubunit peptidase known only from archaea. The sequence of the core protein shows limited similarity to the tail-specific endopeptidase around the catalytic serine, with conservation of the serine in the core protein from both archaean species (Schneider & Hartl, 1996). However, the only lysine residues conserved between the two archaean species are very close to the proposed catalytic serine (see Figure 151.1).

way and are known as signal or leader peptidases. In most eukaryotes, there are separate signal peptidases for processing proteins synthesized in the cytoplasm and mitochondria, but in plants there is a third signal peptidase in the chloroplast. The eukaryote microsomal and mitochondrial signal peptidases are multisubunit complexes, the number of subunits varying with the species. In the microsomal signal peptidase, only one or two subunits are members of family S26, and these are presumed to be catalytic subunits, although they have not been identified biochemically. The subunits of the microsomal signal peptidase were originally placed in a separate family from the bacterial enzyme, but two bacterial sequences have provided links between the bacterial signal peptidases and eukaryote signal peptidases: namely homologs from *Clostridium perfringens* (EMBL: X86488) and *Bacillus subtilis* (Sw: P54506). This means that we are now sure that the bacterial and eukaryote enzymes evolved

## References

Anbudurai, P.R., Mor, T.S., Ohad, I., Shestakov, S.V. & Pakrasi, H.B. (1994) The *ctpA* gene encodes the C-terminal processing protease for the D1 protein of the photosystem II reaction center complex. *Proc. Natl Acad. Sci. USA* **91**, 8082–8086.

Schneider, C. & Hartl, F.U. (1996) Cell biology – hats off to the tricorn protease. *Science* **274**, 1323–1324.

Shestakov, S.V., Anbudurai, P.R., Stanbekova, G.E., Gadzhiev, A., Lind, L.K. & Pakrasi, H.B. (1994) Molecular cloning and characterization of the *ctpA* gene encoding a carboxyl-terminal processing protease. Analysis of a spontaneous photosystem II-deficient mutant strain of the cyanobacterium *Synechocystis* sp. PCC 6803. *J. Biol. Chem.* **269**, 19354–19359.

Silber, K.R., Keiler, K.C. & Sauer, R.T. (1992) Tsp: a tail-specific protease that selectively degrades proteins with nonpolar C termini. *Proc. Natl Acad. Sci. USA* **89**, 295–299.

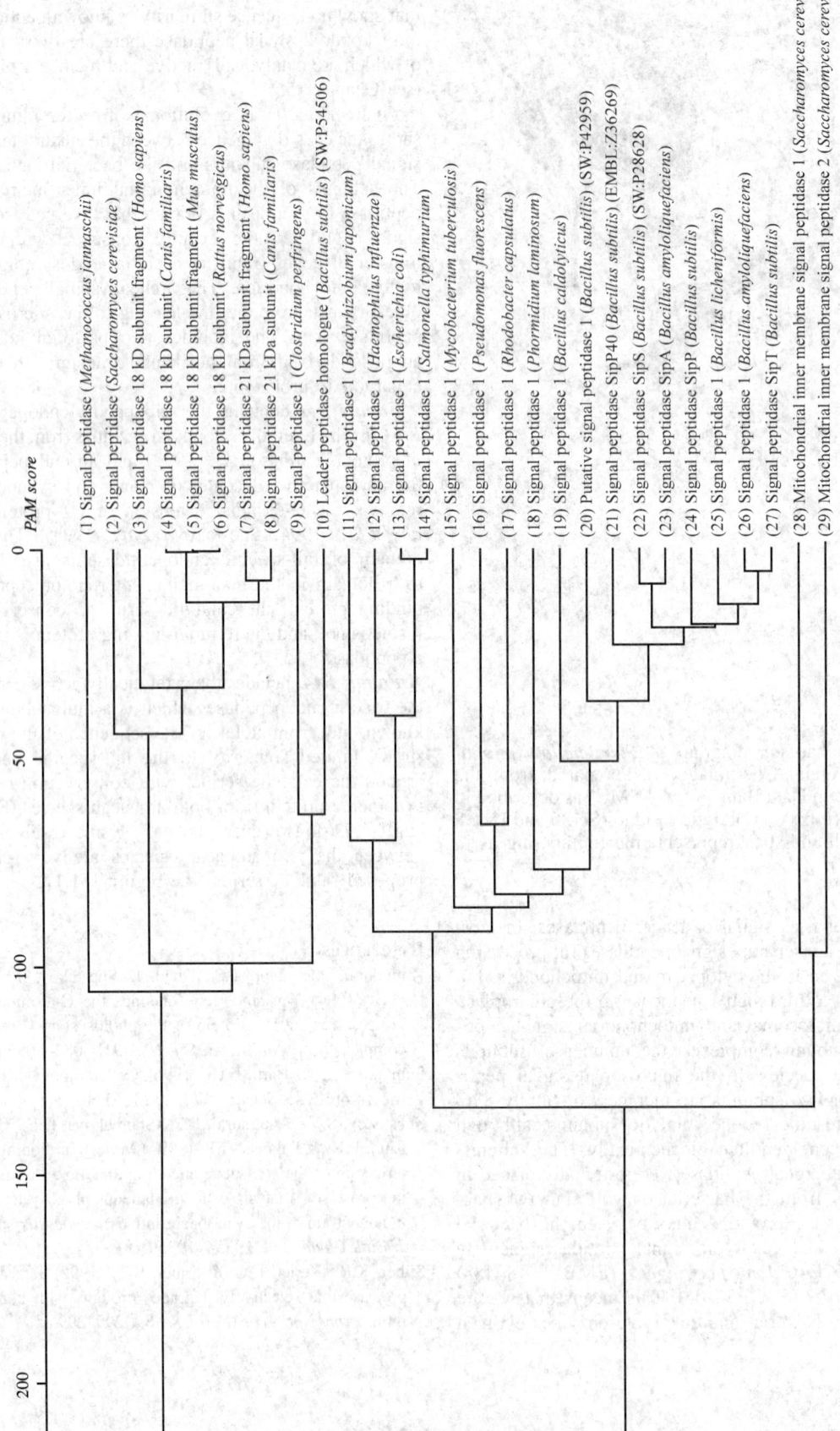

*Figure 151.2* Evolutionary tree for family S26. The tree was prepared as described in the Introduction (p. xxv).

# 152. *Repressor LexA*

## Databanks

*Peptidase classification: clan SF, family S24, MEROPS ID: S24.001*
*NC-IUBMB enzyme classification: EC 3.4.21.88*
*Databank codes:*

| Species | Gene | SwissProt | PIR | EMBL (cDNA) | EMBL (genomic) |
|---------|------|-----------|-----|-------------|----------------|
| *Bacillus subtilis* | *lexA* | P31080 | A41315 | M64684 | – |
|   bacteriophage lambda | | P03034 | A14086 | – | J02459: complete genome |
|   bacteriophage P22 | | P03035 | A91518 | V01153 | – |
| *Erwinia carotovora* | *lexA* | Q04596 | S30163 | X63189 | – |
| *Escherichia coli* | | – | ZWECBP | – | – |
| *Escherichia coli* | *lexA* | P03033 | A03569 | J01643 | U00006: chromosomal region from 89.2 to 92.8′ |
| | | | | L02362 | |
| *Escherichia coli* | *umuD* | P04153 | A03551 | – | M10107: umu operon |
| | | | | | M13387: umuDC operon |
| *Haemophilus influenzae* | *lexA* | P44858 | – | U32759 | – |
| *Mycobacterium leprae* | *lexA* | – | – | – | U00019: cosmid B2235 |
| *Providencia rettgeri* | *lexA* | Q07267 | S33693 | X70965 | – |
| *Pseudomonas aeruginosa* | *lexA* | P37452 | S30165 | X63018 | – |
| *Pseudomonas putida* | *lexA* | P37453 | S30164 | X63017 | – |
| *Salmonella typhimurium* | *impA* | P18641 | JQ0660 | – | X53528: imp operon |
| *Salmonella typhimurium* | *lexA* | P29831 | S30166 | M83220 | – |
| | | | S35496 | X63002 | |
| | | | | X63562 | |
| *Salmonella typhimurium* | *mucA* | P07376 | D23157 | X16596 | – |
| *Salmonella typhimurium* | *samA* | P23831 | A38176 | D90202 | – |
| *Salmonella typhimurium* | *umuD* | P22493 | A36713 | – | M57431: umuDC operon |

Brookhaven Protein Data Bank three-dimensional structures:

| Species | Gene | ID | Resolution | Notes |
|---------|------|-----|------------|-------|
| *Salmonella typhimurium* | *umuD* | 1UMU[a] | 2.5 | cleaved UmuD′ |

[a] See Chapter 151 for an image constructed from these coordinates.

## Name and History

The *lexA* gene of *Escherichia coli* was identified in a mutant strain that was sensitive to DNA-damaging treatments. The original name (*lex*) had a trivial meaning. It is now known that the mutant protein in this strain cannot undergo a specific cleavage reaction, and that this accounts for its change in phenotype. The specific cleavage reaction was originally discovered not in LexA but in a related protein, phage λ CI repressor. It had been known that this protein was inactivated when cells were treated with agents that induce the SOS regulatory system (see below). It was observed that CI repressor was cleaved into two discrete fragments (Roberts & Roberts, 1975). This reaction was then demonstrated *in vitro*, and found to require a ternary complex containing another protein, RecA, a nucleoside triphosphate such as ATP, and single-stranded DNA (Craig & Roberts, 1980; see also Kowalczykowski *et al.*, 1994). LexA was shown to

undergo a closely similar reaction (Little *et al.*, 1980; Horii *et al.*, 1981). These findings led to the view that RecA, also a central enzyme in homologous recombination, is a highly specific protease with a few repressors as substrates. This view was superseded by the finding that LexA and λ repressors can cleave themselves at elevated pH in an intramolecular reaction termed autodigestion (Little, 1984). The current view is that the chemistry of specific cleavage is carried out by groups in the protein itself, and that activated RecA facilitates this reaction in the cell (Little, 1993). Hence, this activity of RecA is termed co-protease to emphasize its indirect role in cleavage.

The term ***repressor LexA*** is recommended by NC-IUBMB rather than ***LexA repressor***, simply to avoid giving the impression that this protein might somehow repress LexA, but the term LexA repressor is the one that is in common use, and will be used herein.

## Activity and Specificity

LexA self-cleavage is stimulated by alkaline pH (Slilaty *et al.*, 1986), which leads to titration of a group with a $pK_a$ of about 10, believed to be Lys156. Autodigestion does not require metal ions, and its rate is largely independent of ionic strength (Slilaty *et al.*, 1986). LexA cleavage cuts a single bond in the protein, the Ala84⊥Gly85 bond near the center of the 202 amino acid protein. The same bond is cleaved in the RecA-mediated reaction. Related proteins also cut an Ala⊥Gly or Cys⊥Gly bond located in a homologous site.

## Structural Chemistry

LexA contains two structural and functional domains. The N-terminal domain binds DNA, and the C-terminal domain allows dimerization and contains the active site required for cleavage. Cleavage occurs in a hinge region between the two domains, separating them and inactivating the repressor function. It is believed that the active site for cleavage, like that in an enzyme, consists of a catalytic center that carries out the chemistry of bond breakage, and a binding pocket for binding the substrate and positioning it relative to the catalytic center. It is not clear whether the substrate is normally bound in the active site, or is free in solution.

The nucleophile attacking the peptide bond appears to be Ser119. This residue is completely conserved among all known self-cleaving proteins of this type (Little, 1996). If Ser119 is changed to Ala, cleavage is abolished (Slilaty & Little, 1987); the Cys119 variant has some activity. DFP reacts selectively with Ser119. This residue is probably activated by Lys156, another completely conserved residue; this is suggested by the pH rate profile, the effects of mutations on cleavage, and the altered pH profile of the Arg156 variant. A recent crystal structure (Peat *et al.*, 1996) of a homologous self-cleaving protein, UmuD′, shows that the side-chains of the corresponding Ser60 and Lys97 residues lie within hydrogen-bonding distance. This structure does not contain the cleavage site, since UmuD′ is the product of the cleavage reaction. These active-site residues lie on the surface of the protein, at the end of a groove that may represent the substrate-binding site.

## Preparation

LexA is made from overproducing strains containing a T7 promoter:LexA fusion. A purification scheme has been described (Little *et al.*, 1994).

## Biological Aspects

LexA cleavage is an early event in the induction of the *E. coli* SOS response to treatments that damage DNA or inhibit DNA replication (Little & Mount, 1982; Friedberg *et al.*, 1995).

These treatments are thought to produce single-stranded DNA (Sassanfar & Roberts, 1990), which activates RecA, which in turn mediates LexA cleavage. Cleavage of LexA inactivates its repressor function, leading to high levels of LexA-controlled genes, which are involved in DNA repair, mutagenesis, and other events promoting cell survival.

Several other prokaryotic proteins undergo a closely similar reaction. These include the CI repressors of various bacteriophages and a group of proteins involved in SOS mutagenesis, such as UmuD. In each case, both autodigestion and RecA-mediated cleavage can occur.

Specific cleavage of viral repressor proteins, typified by λCI repressor, can lead to a process termed prophage induction. Phage λ and related bacterial viruses, termed lambdoid phages, can exist in a so-called lysogenic state in which the viral DNA is integrated into the host genome; expression of genes required for lytic growth is blocked by the CI repressor. Induction of the SOS system leads to specific cleavage of this repressor, leading to expression of lytic genes and consequent growth of the virus. This change in pattern of gene expression is termed the 'genetic switch' (Ptashne, 1992).

Cleavage of LexA occurs at about a 50-fold greater rate than that of λCI, both *in vivo* and *in vitro*. This disparity is reflected in the biological roles of these two proteins. LexA controls a cellular response that needs to be rapid and decisive. By contrast, prophage induction is a last resort for the latent virus when its host is likely not to survive due to extensive DNA damage. Indeed, induction of λ prophage becomes efficient only at doses of DNA damage that kill a substantial fraction of the cells. Hence, it is likely that the rate of cleavage for each protein has evolved to fit its particular ecological niche.

A second major class of self-cleaving molecules is a group of proteins, typified by *E. coli* UmuD, involved in mutagenesis. Cleavage is also stimulated *in vivo* by RecA. Unlike the repressors, which are inactivated by cleavage, UmuD is activated for its role in SOS mutagenesis (Friedberg *et al.*, 1995). This process occurs on damaged DNA when the SOS system is activated, and it involves other proteins, including UmuC and RecA protein, and leads to mutations at the original site of damage. Its mechanism is poorly understood.

## Distinguishing Features

The most distinctive feature of LexA cleavage is its intramolecular nature (Little, 1993). Since the substrate and active site lie in the same molecule, the protein must be designed to self-cleave very slowly, but to be capable of large rate increases upon interaction with RecA. Evidence with rapidly cleaving LexA mutant proteins (termed Ind[S]) suggests that RecA stabilizes a reactive conformation of LexA (Roland *et al.*, 1992), but direct physical evidence is lacking.

LexA cleavage can also occur in an intermolecular reaction in which one molecule of LexA acts as an enzyme to cut other molecules (Kim & Little, 1993). The rate of this reaction is greatly increased when the enzyme or the substrate contains Ind[S] mutations. RecA protein does not appear to accelerate this reaction. Even with the best enzyme and substrate, this reaction appears to have a $K_m > 4$ mM, so that it is not yet possible to determine kinetic parameters. The C-terminal domain of λCI repressor can also act as an enzyme, attacking the LexA substrates about as rapidly as does the LexA C-terminal domain. This finding strongly suggests that the active site of the two proteins is strongly conserved, and that their differing rates of self-cleavage are controlled by differences in the substrate.

One final distinctive feature, shared by a growing group of other peptidases, is the Ser-Lys dyad mechanism by which LexA cleavage is apparently activated.

### Further Reading

The article of Little & Mount (1982) reviews the SOS regulatory system, and is still a relatively up-to-date summary of the major aspects, aside from the revised understanding of specific cleavage. Little (1993) describes LexA cleavage in more detail, and compares it with other self-processing biochemical reactions, and Friedberg *et al*. (1995) give an extensive discussion of the SOS response and SOS mutagenesis.

### References

Craig, N.L. & Roberts, J.W. (1980) *E. coli* recA protein-directed cleavage of phage λ repressor requires polynucleotide. *Nature* **283**, 26–30.

Friedberg, E.C., Walker, G.C. & Siede, W. (1995) *DNA Repair and Mutagenesis*. Washington, DC: American Society of Microbiology Press.

Horii, T., Ogawa, T., Nakatani, T., Hase, T., Matsubara, H. & Ogawa, H. (1981) Regulation of SOS functions: purification of *E. coli* LexA protein and determination of its specific site cleaved by the RecA protein. *Cell* **27**, 515–522.

Kim, B. & Little, J.W. (1993) LexA and λ CI repressors as enzymes: specific cleavage in an intermolecular reaction. *Cell* **73**, 1165–1173.

Kowalczykowski, S.C., Dixon, D.A., Eggleston, A.K., Lauder, S.D. & Rehrauer, W.M. (1994) Biochemistry of homologous recombination in *Escherichia coli*. *Microbiol. Rev.* **58**, 401–465.

Little, J.W. (1984) Autodigestion of lexA and phage lambda repressors. *Proc. Natl Acad. Sci. USA* **81**, 1375–1379.

Little, J.W. (1993) LexA cleavage and other self-processing reactions. *J. Bacteriol.* **175**, 4943–4950.

Little, J.W. (1996) The SOS regulatory system. In: *Regulation of Gene Expression in* Escherichia coli (Lin, E.C.C. & Lynch, A.S., eds). Austin, TX: R.C. Landes.

Little, J.W. & Mount, D.W. (1982) The SOS regulatory system of *Escherichia coli*. *Cell* **29**, 11–22.

Little, J.W., Edmiston, S.H., Pacelli, L.Z. & Mount, D.W. (1980) Cleavage of the *Escherichia coli lexA* protein by the recA protease. *Proc. Natl Acad. Sci. USA* **77**, 3225–3229.

Little, J.W., Kim, B., Roland, K.L., Smith, M.H., Lin, L.-L. & Slilaty, S.N. (1994) Cleavage of LexA repressor. *Methods Enzymol.* **244**, 266–284.

Peat, T.S., Frank, E.G., McDonald, J.P., Levine, A.S., Woodgate, R. & Hendrickson, W.A. (1996) Structure of the UmuD′ protein and its regulation in response to DNA damage. *Nature* **380**, 727–730.

Ptashne, M. (1992) *A Genetic Switch: Phage λ and Higher Organisms*, 2nd edn. Cambridge, MA: Cell Press and Blackwell Scientific.

Roberts, J.W. & Roberts, C.W. (1975) Proteolytic cleavage of bacteriophage lambda repressor in induction. *Proc. Natl Acad. Sci. USA* **72**, 147–151.

Roland, K.L., Smith, M.H., Rupley, J.A. & Little, J.W. (1992) *In vitro* analysis of mutant LexA proteins with an increased rate of specific cleavage. *J. Mol. Biol.* **228**, 395–408.

Sassanfar, M. & Roberts, J.W. (1990) Nature of the SOS-inducing signal in *Escherichia coli*. The involvement of DNA replication. *J. Mol. Biol.* **212**, 79–96.

Slilaty, S.N. & Little, J.W. (1987) Lysine-156 and serine-119 are required for LexA repressor cleavage: a possible mechanism. *Proc. Natl Acad. Sci. USA* **84**, 3987–3991.

Slilaty, S.N., Rupley, J.A. & Little, J.W. (1986) Intramolecular cleavage of LexA and phage lambda repressors: dependence of kinetics on repressor concentration, pH, temperature, and solvent. *Biochemistry* **25**, 6866–6875.

*John W. Little*
*Department of Biochemistry,*
*University of Arizona,*
*Tucson, AZ 85721, USA*
*Email: jlittle@biosci.arizona.edu*

# 153. *Signal peptidase I*

### Databanks

*Peptidase classification: clan SF, family S26, MEROPS ID: S26.001*
*NC-IUBMB enzyme classification: EC 3.4.21.89*

| Species | SwissProt | PIR | EMBL (cDNA) | EMBL (genomic) |
|---|---|---|---|---|
| *Bradyrhizobium japonicum* | – | – | U33883 | – |
| *Escherichia coli* | P00803 | A00998 | D64044 | – |
| | | S12563 | K00426 | |
| *Haemophilus influenzae* | P44454 | – | – | U32687: genome section 2 of 163 |
| *Mycobacterium tuberculosis* | Q10789 | – | Z74024 | – |
| *Pseudomonas fluorescens* | P26844 | S22414 | X56466 | – |
| *Rhodobacter capsulatus* | – | – | Z68305 | – |
| *Salmonella typhimurium* | P23697 | S12020 | X54933 | – |
| *Streptococcus pneumoniae* | – | – | U90721 | – |

## Name and History

In the early 1970s, Milstein and coworkers reported the identification of proteases that are involved in the processing of secreted proteins in eukaryotic cells (Milstein *et al.*, 1972). The secretory proteins were synthesized in a higher molecular mass form with N-terminal extension peptides, termed 'signal peptides', and consequently, the proteases were called signal peptidases. A few years later, Chang and coworkers showed that similar proteases exist in bacteria: *Escherichia coli* extracts were found to process correctly bacteriophage M13 procoat protein to the coat protein (Chang *et al.*, 1978). Exploiting this assay, the protease in *E. coli* was soon purified to homogeneity by Zwizinski & Wickner (1980) who named the protease **leader peptidase** because they referred to the signal peptide as a leader peptide. Therefore, in the literature, there are two names that are commonly used for the same protease: **signal peptidase I** (recommended by NC-IUBMB) and **leader peptidase**. The unrelated signal peptidase, lipoprotein signal peptidase or signal peptidase II, is described elsewhere (Chapter 333).

## Activity and Specificity

Signal peptidase activity is typically assayed by use of a cell-free system in which the mRNA for the precursor protein is translated in the presence of radioactive amino acids to produce labeled precursor protein (Ray *et al.*, 1986). Cleavage of the precursor protein is monitored by the production of radioactive mature protein, which is separated by PAGE and assayed by autoradiography.

Kinetic parameters of the *E. coli* signal peptidase I can conveniently be determined using a synthetic peptide Phe-Ser-Ala-Ser-Ala-Leu-Ala⫮Lys-Ile (Dev *et al.*, 1990) or a preprotein substrate, pro-OmpA nuclease A (Chatterjee *et al.*, 1995). This latter substrate is the best known for the *E. coli* signal peptidase I, with a $k_{cat}$ of $44 s^{-1}$ and $K_m$ of $19.2 \mu M$ (Tschantz *et al.*, 1995).

The rules governing specificity have been addressed using both preprotein substrates and peptide substrates. Cleavage occurs when preprotein substrates have small residues at the −1 (P1) or small uncharged or aliphatic residues at the −3 (P3) position relative to the cleavage site (Fikes *et al.*, 1990; Shen *et al.*, 1991). This cleavage requirement coincides with a common pattern of small residues at the −1 and −3 position within the preprotein signal peptide (von Heijne, 1983; Perlman & Halvorson, 1983). Similar studies with chemically synthesized peptides corresponding to the cleavage site of the precursor proteins point to the −1 and −3 positions as being critical for signal peptide cleavage (Dierstein & Wickner, 1986; Dev *et al.*, 1990; Kuo *et al.*, 1993).

The pH optimum for catalysis is around 9.0 (Kim *et al.*, 1995). Activity is inhibited by ionic detergents or by high concentrations of NaCl or $MgCl_2$ (Wolfe *et al.*, 1983b).

The *E. coli* signal peptidase I is not sensitive to standard inhibitors of the serine, cysteine, aspartic or metallopeptidases (Zwizinski *et al.*, 1981; Black *et al.*, 1992; Kuo *et al.*, 1993). However, it is inhibited by the leader peptide of the M13 procoat (Wickner *et al.*, 1987) and by the precursor maltose-binding protein with proline in the +1 position of the cleavage site (Barkocy-Gallagher & Bassford, 1992). To date, the best inhibitors reported are penem compounds (Allsop *et al.*, 1995).

Bacterial signal peptidase 1 is a mechanistically novel serine protease that is believed to use a lysine residue as the general base in much the same way as the proteins of the LexA family (Chapter 152). Site-directed mutagenesis has indicated that Ser90 (Sung & Dalbey, 1992) and Lys145 (Black, 1993; Tschantz *et al.*, 1993) are essential for the activity of the *E. coli* signal peptidase 1, but no essential His or Cys residues are present. The *Bacillus subtilis* signal peptidase 1 also has critical Ser and Lys residues as well as an important Asp (van Dijl *et al.*, 1995). Therefore, type 1 signal peptidases in bacteria most likely use a Ser/Lys dyad as the reactive center.

## Structural Chemistry

The amino acid sequence of signal peptidase 1 from *E. coli* was deduced from its gene sequence and indicates that the protein is made without a cleavable signal peptide (Wolfe *et al.*, 1983a). The N-terminus of the protein is blocked (Wolfe *et al.*, 1982), but the significance of this is not known. In *E. coli*, signal peptidase 1 spans the plasma membrane twice, with a large C-terminal domain containing the active site protruding into the periplasmic space (Wolfe *et al.*, 1983a; Moore & Miura, 1987; Whitley *et al.*, 1993).

The *E. coli* protein consists of a single polypeptide chain with a molecular mass of 35 988 Da, contains one disulfide bond, and has a pI of 6.9 (Wolfe *et al.*, 1983b). A soluble and catalytically active fragment of leader peptidase ($\Delta$2–75), lacking the membrane-spanning domain, has been engineered (Kuo *et al*., 1993). This fragment, although water soluble, still requires detergent or phospholipid for optimal activity (Tschantz *et al*., 1995). The soluble fragment has now been crystallized (Paetzel *et al*., 1995) and the structure is being solved by X-ray diffraction methods.

Except in the bacterium *Mycoplasma genitalium* (Fraser *et al*., 1995), a homologous type I signal peptidase has been discovered in all bacteria examined. In fact, it is believed that signal peptidases are part of a larger family of proteases present in prokaryotes and eukaryotes (van Dijl *et al*., 1992; Dalbey & von Heijne, 1992). Signal peptidases share sequence similarity from different organisms and have similar substrate specificity. Ser90, the likely catalytic residue in the *E. coli* signal peptidase 1, is conserved in the whole family. However, the critical Lys145 residue of the *E. coli* signal peptidase is conserved only in the bacterial and the mitochondrial IMP1/IMP2 signal peptidase (Chapter 155). The homologous subunits within the endoplasmic reticulum signal peptidase do not contain the conserved Lys; rather, they contain a His residue at this position (see Chapter 156).

All the type 1 signal peptidases in bacteria are integral membrane proteins. The protein is believed to have two transmembrane segments in *E. coli, S. typhimurium* and *P. fluorescens*, and three transmembrane segments in *H. influenzae*. The related, bacterial SipS signal peptidases of the grampositive *Bacillus* spp. contain one transmembrane segment (see Chapter 154). In the signal peptidase of the mitochondrial inner membrane, the IMP1 and IMP2 subunits are believed to span the membrane once.

## Preparation

The *E. coli* signal peptidase is purified essentially as described by Wolfe *et al*. (1982) except that the overproducing strain MC1061pRD8 (Tschantz & Dalbey, 1994) is used, in which the synthesis of signal peptidase is controlled by the arabinose operon (Dalbey & Wickner, 1985). Approximately 40 mg of pure leader peptidase can be obtained from 10 liters of bacterial cells. The recombinant soluble protein $\Delta$2–75 from *E. coli* can be purified on a scale of hundreds of milligrams (Paetzel *et al*., 1995). At this time, purification of signal peptidase 1 from other bacterial sources has not been reported.

## Biological Aspects

Signal peptidase 1 has been shown to be an essential enzyme in *E. coli* (Date, 1983; Dalbey & Wickner, 1985; Inada *et al*., 1989). The removal of signal peptides from exported proteins is an essential function for the cell because the uncleaved signal peptide acts as a membrane anchor (Dalbey & Wickner, 1985; Fikes and Bassford, 1987), and cleavage releases the secretory protein from the membrane. The removal of the signal peptide from exported proteins is not necessary for translocation of the exported protein across the membrane (Kuhn & Wickner, 1985) nor is signal peptide processing necessary for the activity of most exported proteins (Haugen & Heath, 1979; Ito, 1982).

Like many bacterial plasma membrane proteins, signal peptidase 1 does not contain a cleaved signal peptide; rather it contains uncleaved signal sequences. In *E. coli*, the second hydrophobic domain of signal peptidase 1 initiates translocation of the large C-terminal domain containing the active site (Dalbey & Wickner, 1987; Dalbey *et al*., 1987) and remains as a transmembrane anchor.

## Distinguishing Features

The type 1 signal peptidases in bacteria are not inhibited by classical serine protease inhibitors. They are mechanistically novel serine proteases that utilize a Ser/Lys catalytic dyad rather than the Ser/His/Asp catalytic triad found in the majority of serine peptidases (Black, 1993; Tschantz *et al*., 1993). As signal peptidase 1 is an essential enzyme in bacteria, the protein is being used as a drug target (Allsop *et al*., 1995). The forthcoming crystal structure of the water-soluble, catalytically active fragment of leader peptidase is expected to facilitate the design of novel antibacterial compounds.

## Further Reading

The reader is referred to articles in von Heijne (1994).

## References

Allsop, A.E., Brooks, G., Bruton, G., Coulton, S., Edwards, P.D., Hatton, I.K., Kaura, A.C., McLean, S.D., Pearson, N.D., Smale, T.C. & Southgate, R. (1995) Penem inhibitors of bacterial signal peptidase. *Bioorg. Med. Chem. Lett.* **5**, 443–448.

Barkocy-Gallagher, G.A. & Bassford, P.J., Jr. (1992) Synthesis of precursor maltose-binding protein with proline in the +1 position of the cleavage site interferes with the activity of *Escherichia coli* signal peptidase I. *J. Biol. Chem.* **267**, 1231–1238.

Black, M.T. (1993) Evidence that the catalytic activity of prokaryote leader peptidase depends upon the operation of a serine-lysine catalytic dyad. *J. Bacteriol.* **175**, 4957–4961.

Black, M.T., Munn, J.G.R. & Allsop, A.E. (1992) On the catalytic mechanism of prokaryotic leader peptidase I. *Biochem. J.* **282**, 539–543.

Chang, C.N., Blobel, G. & Model, P. (1978) Detection of prokaryotic signal peptidase in an *Escherichia coli* membrane fraction: endoproteolytic cleavage of nascent f1 pre-coat protein. *Proc. Natl Acad. Sci. USA* **75**, 361–365.

Chatterjee, S., Suciu, D., Dalbey, R.E., Kahn, P.C. & Inouye, M. (1995) Determination of $K_m$ and $k_{cat}$ for signal peptidase 1 using a full length secretory precursor, pro-OmpA-nuclease A. *J. Mol. Biol.* **245**, 311–314.

Dalbey, R.E. & von Heijne, G. (1992) Signal peptidase in prokaryotes and eukaryotes – a new protease family. *Trends Biochem. Sci.* **17**, 474–478.

Dalbey, R.E. & Wickner, W. (1985) Leader peptidase catalyzes the release of exported from the outer surface of the *Escherichia coli* plasma membrane. *J. Biol. Chem.* **260**, 15925–15931.

Dalbey, R.E. & Wickner, W. (1987) Leader peptidase of *Escherichia coli*: critical role of a small domain in membrane assembly. *Science* **235**, 783–787.

Dalbey, R.E., Kuhn, A. & Wickner, W. (1987) The internal signal sequence of *Escherichia coli* leader peptidase is necessary, but not sufficient, for its rapid membrane assembly. *J. Biol. Chem.* **262**, 13241–13245.

Date, T. (1983) Demonstration by a novel genetic technique that leader peptidase is an essential enzyme of *Escherichia coli. J. Bacteriol.* **154**, 76–83.

Dev, I.K., Ray, P.H. & Novak, P. (1990) Minimum substrate sequence for signal peptidase 1 of *Escherichia coli. J. Biol. Chem.* **265**, 20069–20072.

Dierstein, R. & Wickner, W. (1986) Requirement for substrate recognition by bacterial leader peptidase. *EMBO J.* **5**, 427–431.

Fikes, J.D. & Bassford, P.J., Jr. (1987) Export of unprocessed precursor maltose-binding protein to the periplasm of *Escherichia coli. J. Bacteriol.* **169**, 2352–2359.

Fikes, J.D., Barkocy-Gallagher, G.A., Klapper, D.G. & Bassford, P.J., Jr. (1990) Maturation of *Escherichia coli* maltose binding protein by signal peptidase 1. *J. Biol. Chem.* **265**, 3417–3423.

Fraser, C.M., Gocayne, J.D., White, O. *et al.* (1995) The minimal gene complement of *Mycoplasma genitalium. Science* **270**, 397–403.

Haugen, T.H. & Heath, E.C. (1979) De novo biosynthesis of an enzymatically active precursor form of bovine pancreatic RNAse. *Proc. Natl Acad. Sci. USA* **76**, 2689–2693.

Inada, T., Court, D.L., Ito, K. & Nakamura, Y. (1989) Conditionally lethal amber mutations in the leader peptidase gene of *Escherichia coli. J. Bacteriol.* **171**, 585–587.

Ito, K. (1982) Purification of the precursor form of maltose binding protein, a periplasmic protein of *Escherichia coli. J. Biol. Chem.* **257**, 9895–9897.

Kim, Y.-T., Muramatsu, T. & Takahashi, K. (1995) Leader peptidase from *Escherichia coli*: overexpression, characterization, and inactivation by modification of tryptophan residues 300 and 310 with *N*-bromosuccinimide. *J. Biochem.* **117**, 535–544.

Kuhn, A. & Wickner, W. (1985) Conserved residues of the leader peptide are essential for cleavage by leader peptidase. *J. Biol. Chem.* **260**, 15914–15918.

Kuo, D.W., Chan, H.K., Wilson, D.J., Griffin, P.R., Williams, H. & Knight, W.B. (1993) *Escherichia coli* leader peptidase: production of an active form lacking a requirement for detergent and development of peptide substrate. *Arch. Biochem. Biophys.* **303**, 274–280.

Milstein, C., Brownlee, G.G., Harrison, T.M. & Mathews, M.B. (1972) A possible precursor of immunoglobin light chains. *Nature (Lond.) New Biol.* **239**, 117–120.

Moore, K.E. & Miura, S. (1987) A small hydrophobic domain anchors leader peptidase to the cytoplasmic membrane of *Escherichia coli. J. Biol. Chem.* **262**, 8806–8813.

Paetzel, M., Chernaia, M., Strynadka, N., Tschantz, W., Cao, G., Dalbey, R.E. & James, M.N.G. (1995) Crystallization of a soluble, catalytically active form of *Escherichia coli* leader peptidase. *Proteins: Struct. Funct. Genet.* **23**, 122–125.

Perlman, D. & Halvorson, H.O. (1983). A putative signal peptidase recognition site and sequence in eukaryotic and prokaryotic signal peptides. *J. Mol. Biol.* **167**, 391–409.

Ray, P., Dev, I., MacGregor, C. & Bassford, P., Jr. (1986) Signal peptidases. *Curr. Top. Microbiol. Immunol.* **125**, 75–102.

Shen, L.M., Lee, J.-I., Cheng, S., Jutte, H., Kuhn, A. & Dalbey, R.E. (1991) Use of site-directed mutagenesis to define the limits of sequence variation tolerated for processing of the M13 procoat protein by the *Escherichia coli* leader peptidase. *Biochemistry* **30**, 11775–11781.

Sung, M. & Dalbey, R.E. (1992) Identification of potential active-site residues in the *Escherichia coli* leader peptidase. *J. Biol. Chem.* **267**, 13154–13159.

Tschantz, W.R. & Dalbey, R.E. (1994) Bacterial leader peptidase 1. *Methods Enzymol.* **244**, 285–301.

Tschantz, W.R., Sung, M., Delgado-Partin, V.M. & Dalbey, R.E. (1993) A serine and a lysine residue implicated in the catalytic mechanism of the *Escherichia coli* leader peptidase. *J. Biol. Chem.* **268**, 27349–27354.

Tschantz, W.R., Paetzel, M., Cao, G., Suciu, D., Inouye, M. & Dalbey, R.E. (1995) Characterization of a soluble, catalytically active form of *Escherichia coli* leader peptidase. Requirement of detergent or phospholipids for optimal activity. *Biochemistry* **34**, 3935–3941.

van Dijl, J.M., de Jong, A., Vehmaanpera, J., Venema, G. & Bron, S. (1992) Signal peptidase 1 of *Bacillus subtilis*: patterns of conserved amino acids in prokaryotic and eukaryotic type I signal peptidases. *EMBO J.* **11**, 2819–2828.

van Dijl., J.M., de Jong, A., Venema, G. & Bron, S. (1995) Identification of the potential active site of the signal peptidase Sips of *Bacillus subtilis. J. Biol. Chem.* **270**, 3511–3618.

von Heijne, G. (1983) Patterns of amino acids near signal-sequence cleavage sites. *Eur. J. Biochem.* **133**, 17–21.

von Heijne, G. (ed.) (1994) *Signal Peptidases*. Austin, TX: R.G. Landes.

Whitley, P., Nilsson, L. & von Heijne, G. (1993) Three-dimensional model for the membrane domain of *Escherichia coli* leader peptidase based on disulfide mapping. *Biochemistry* **32**, 8534–8539.

Wickner, W., Moore, K., Dibb, N., Geissert, D. & Rice, M. (1987) Inhibition of purified *Escherichia coli* leader peptidase by the leader (signal) peptide of bacteriophage M13 procoat. *J. Bacteriol.* **169**, 3821–3822.

Wolfe, P.B., Silver, P. & Wickner, W. (1982) The isolation of homogeneous leader peptidase from a strain of *Escherichia coli* which overproduces the enzyme. *J. Biol. Chem.* **257**, 7898–7902.

Wolfe, P.B., Wickner, W. & Goodman, J.M. (1983a) Sequence of the leader peptidase gene of *Escherichia coli* and the orientation of leader peptidase in the bacterial envelope. *J. Biol. Chem.* **258**, 12073–12080.

Wolfe, P.B., Zwizinski, C. & Wickner, W. (1983b) Purification and characterization of leader peptidase from *Escherichia coli. Methods Enzymol.* **97**, 40–46.

Zwizinski, C. & Wickner, W. (1980) Purification and characterization of leader (signal) peptidase from *Escherichia coli. J. Biol. Chem.* **255**, 7973–7977.

Zwizinski, C., Date, T. & Wickner, W. (1981) Leader peptidase is found in both the inner and outer membrane of *Escherichia coli. J. Biol. Chem.* **256**, 3593–3597.

*Ross E. Dalbey*
*Department of Chemistry,*
*Ohio State University,*
*100 West 18th Avenue, Columbus, OH 43210, USA*
*Email: rdalbey@chemistry.ohio-state.edu*

# 154. Signal peptidase SipS

## Databanks

*Peptidase classification: clan SF, family S26, MEROPS ID: S26.003*
*NC-IUBMB enzyme classification: included in EC 3.4.21.89*
*Databank codes:*

| Species | Gene | SwissProt | PIR | EMBL (cDNA) | EMBL (genomic) |
|---------|------|-----------|-----|-------------|----------------|
| *Bacillus amyloliquefaciens* | *sipS* | P41026 | S38885 | L26259 Z27458 | – |
| *Bacillus amyloliquefaciens* | *sipT* | P41025 | S45022 | Z33640 | – |
| *Bacillus caldolyticus* | *sipC* | P41027 | S41913 | L26257 Z27457 | – |
| *Bacillus licheniformis* | *sip* | P42668 | – | X75604 | – |
| *Bacillus subtilis* | *sipS* | P28628 | S23381 S45540 | Z11847 | L09228: multiple entry |
| *Bacillus subtilis* | *sipP* | P37943 | S38888 | L26258 Z27459 | U32379: complete plasmid pTA1015 |
| *Bacillus subtilis* | *sipP* | – | I40552 | U32378 Z36269 | – |
| *Bacillus subtilis* | *sipT* | – | – | U45883 | – |
| *Bacillus subtilis* | *sipU* | P42959 | – | D38161 | – |
| *Staphylococcus aureus* | *spsA* | – | – | U65000 | – |
| *Staphylococcus aureus* | *spsB* | – | – | U65000 | – |

## Name and History

Type I signal peptidases, also known as leader peptidases, remove signal peptides from secretory proteins, and were originally characterized from gram-negative bacteria (Chapter 153). **Signal peptidase SipS** (**si**gnal **p**eptidase of *Bacillus subtilis*) is the first characterized signal peptidase from a gram-positive bacterium (van Dijl *et al*., 1992). More recently, it was found that *B. subtilis* contains at least two other chromosomally encoded type I signal peptidases, denoted SipT and SipU (Akagawa *et al*., 1995; Bolhuis *et al*., 1996). In addition, some strains of *B. subtilis* contain endogenous plasmids (pTA1015 or pTA1040) specifying type I signal peptidases, denoted SipP (Meijer *et al*., 1995). Related signal peptidases have been identified in *Bacillus amyloliquefaciens, Bacillus caldolyticus* and *Bacillus licheniformis* (Hoang & Hofemeister, 1995; Meijer *et al*., 1995). The sequence of SipS allowed the identification of conserved domains in type I signal peptidases of bacteria, the mitochondrial inner membrane, and the endoplasmic reticular membrane (van Dijl *et al*., 1992; Dalbey, 1994).

## Activity and Specificity

Signal peptidase I cleavage regions in secretory precursor proteins are poorly conserved. In general, small, uncharged residues precede the site of cleavage, the residues at the −3 (P3) and −1 (P1) positions being most important (von Heijne, 1990). More specifically, signal peptidase cleavage regions in secretory preproteins of gram-positive bacteria are longer than those of other organisms (von Heijne & Abrahmsén, 1989), probably reflecting differences in the substrate specificity of the corresponding signal peptidases. Differences in the substrate specificities of the type I signal peptidases of *Escherichia coli* and *B. subtilis* were exploited for the identification of signal peptidase-encoding genes of the latter organism and related bacilli (van Dijl *et al*., 1992; Meijer *et al*, 1995).

A useful substrate in the analysis of SipS signal peptidase activity is the hybrid precursor pre(A13i)-β-lactamase, containing a randomly selected signal peptide from *B. subtilis*, modified by site-directed mutagenesis. *In vivo*, pre(A13i)-β-lactamase is processed by *Bacillus* signal peptidases, but not by the signal peptidase of *E. coli*, which can be visualized in a plate assay and in pulse-chase labeling assays, as described in van Dijl *et al*. (1992). An *in vitro* assay for *Bacillus* signal peptidases is described in Vehmaanperä *et al*. (1993).

Like signal peptidases from eukaryotes and *E. coli* (see Chapters 153 and 156), SipS and other known signal peptidases of bacilli are resistant to most general inhibitors of serine, cysteine, metallo and aspartic proteases (Vehmaanperä *et al*., 1993).

## Structural Chemistry

SipS and other known type I signal peptidases from bacilli are proteins of about 21 kDa, with an N-terminal, membrane-spanning domain. Site-directed mutagenesis of residues of

SipS that are conserved in all type I signal peptidases showed that Ser43, Lys83 and Asp153 are indispensable for activity, whereas Arg84 and Asp146 are likely to constitute conformational determinants (van Dijl *et al*., 1995). As proposed for the signal peptidase of *E. coli* (Chapter 153) (Dalbey, 1994), SipS may employ a Ser43–Lys83 catalytic dyad.

### Biological Aspects

The genes *sipS*, *sipT* and *sipU* have been mapped at 209, 126 and 39° of the *B. subtilis* chromosome, respectively (van Dijl *et al*., 1992; Akagawa *et al*., 1995, and unpublished results of the author). None of these genes is essential for cell viability, or protein secretion. Nevertheless, SipS is important for protein secretion, as cells lacking this enzyme secrete several (but not all) proteins at greatly reduced levels. These findings suggest that SipS, SipT and SipU have different, but overlapping, substrate specificities. The transcription of *sipS* is temporally controlled, being markedly increased in the postexponential growth phase, like the transcription of genes for most secretory proteins of *B. subtilis*. Apparently, *B. subtilis* can modify its capacity and specificity for protein secretion through temporally controlled expression of the *sipS* gene (Bolhuis *et al*., 1996).

### References

Akagawa, E., Kurita, K., Sugawara, T., Nakamura, K., Kasahara, Y., Ogasawara, N. & Yamane, K. (1995) Determination of a 17484 bp nucleotide sequence around the 39 degrees region of the *Bacillus subtilis* chromosome and similarity analysis of the products of putative ORFs. *Microbiology* **141**, 3241–3245.

Bolhuis, A., Sorokin, A., Azevedo, V., Ehrlich, S.D., Braun, P.G., de Jong, A., Venema, G., Bron, S. & van Dijl, J.M. (1996)
*Bacillus subtilis* can modulate its capacity and specificity for protein secretion through temporally controlled expression of the *sipS* gene for signal peptidase I. *Mol. Microbiol.* **22**, 605–618.

Dalbey, R.E. (1994) Bacterial signal peptidase I. In: *Signal Peptidases* (von Heijne, G., ed.). Austin, TX: R.G. Landes, pp. 5–15.

Hoang, V. & Hofemeister, J. (1995) *Bacillus amyloliquefaciens* possesses a second type I signal peptidase with extensive sequence similarity to other *Bacillus* signal peptidases. *Biochim. Biophys. Acta* **1269**, 64–68.

Meijer, W.J.J., de Jong, A., Wisman, G.B.A., Tjalsma, H., Venema, G., Bron, S. & van Dijl, J.M. (1995) The endogenous *Bacillus subtilis (natto)* plasmids pTA1015 and pTA1040 contain signal peptidase-encoding genes: identification of a new structural module on cryptic plasmids. *Mol. Microbiol.* **17**, 621–631.

van Dijl, J.M., de Jong, A., Vehmaanperä, J., Venema, G. & Bron, S. (1992) Signal peptidase I of *Bacillus subtilis*: patterns of conserved amino acids in prokaryotic and eukaryotic type I signal peptidases. *EMBO J.* **11**, 2819–2828.

van Dijl, J.M., de Jong, A., Venema, G. & Bron, S. (1995) Identification of the potential active site of the signal peptidase SipS of *Bacillus subtilis*: structural and functional similarities with LexA-like proteases. *J. Biol. Chem.* **270**, 3611–3618.

Vehmaanperä, J., Görner, A., Venema, G., Bron, S. & van Dijl, J.M. (1993) In vitro assay for the *Bacillus subtilis* signal peptidase SipS: systems for efficient in vitro transcription-translation and processing of precursors of secreted proteins. *FEMS Microbiol. Lett.* **114**, 207–214.

von Heijne, G. (1990) The signal peptide. *J. Membr. Biol.* **115**, 195–201.

von Heijne, G. & Abrahmsén, L. (1989) Species-specific variation in signal peptide design. Implications for protein secretion in foreign hosts. *FEBS Lett.* **244**, 439–446.

*Jan Maarten van Dijl*
*Department of Genetics,*
*Groningen Biomolecular Sciences and*
*Biotechnology Institute,*
*Kerklaan 30, 9751 NN Haren,*
*The Netherlands*
*Email: J.M.van.Dijl@Biol.RuG.NL*

*Albert Bolhuis*
*Department of Genetics,*
*Groningen Biomolecular Sciences and*
*Biotechnology Institute,*
*Kerklaan 30, 9751 NN Haren,*
*The Netherlands*

*Harold Tjalsma*
*Department of Genetics,*
*Groningen Biomolecular Sciences and*
*Biotechnology Institute,*
*Kerklaan 30, 9751 NN Haren,*
*The Netherlands*

*Gerard Venema*
*Department of Genetics,*
*Groningen Biomolecular Sciences and Biotechnology Institute,*
*Kerklaan 30, 9751 NN Haren, The Netherlands*

*Sierd Bron*
*Department of Genetics,*
*Groningen Biomolecular Sciences and Biotechnology Institute,*
*Kerklaan 30, 9751 NN Haren, The Netherlands*

# 155. Organellar signal peptidases

### Databanks

*Peptidase classification: clan SF, family S26, MEROPS ID: S26.002*
*NC-IUBMB enzyme classification: none*

*Databank codes:*

| Species | SwissProt | PIR | EMBL (cDNA) | EMBL (genomic) |
|---|---|---|---|---|
| Mitochondrial signal peptidase | | | | |
| *Homo sapiens* | – | – | R06820 | – |
| *Saccharomyces cerevisiae* | P28627 | S16817 | S55518 | – |
| *Saccharomyces cerevisiae* | P46972 | – | – | Z49213: chromosome XIII cosmid 9973 |
| Plastid signal peptidase | | | | |
| *Arabidopsis thaliana* | – | – | Y10477 | U93215: chromosome II BAC |
| *Phormidium laminosum* | – | – | X81990 | – |
| *Synechocystis* sp. | – | – | – | D90899: genome section 1 of 27 |
| *Synechocystis* sp. | – | – | – | D90904: genome section 6 of 27 |

## Name and History

Proteins destined for the mitochondrial intermembrane space are synthesized as precursors either in the mitochondrion (e.g. cytochrome oxidase subunit II), or in the cytoplasm. The cytoplasmically synthesized proteins carry bipartite signal peptides; the first part directs the precursor to the mitochondrion and is removed in the matrix by the mitochondrial matrix processing peptidase (see Chapter 469), whilst the second targets the protein to the intermembrane space (e.g. cytochrome $c_1$). The mature proteins are generated by proteolytic processing by the *inner membrane protease (IMP)* complex (Schneider *et al.*, 1991; Pratje *et al.*, 1994). Two genes, designated *imp*1 (*PET2858*) and *imp*2, have been identified in *Saccharomyces cerevisiae* that are each essential for the processing of a subset of mitochondrial precursors (Behrens *et al.*, 1991; Nunnari *et al.*, 1993). Both encode homologs of bacterial type 1 signal peptidases (Chapter 153).

A somewhat parallel system exists in the chloroplast. *Thylakoidal processing peptidase (TPP)* processes precursor proteins targeted to the thylakoid lumen (Hageman *et al.*, 1986; Robinson, 1994). Such proteins, with the exception of the chloroplast-encoded cytochrome *f*, are encoded by nuclear genes and targeted to the thylakoid lumen by bipartite signal sequences. The second, thylakoid-targeting domain resembles a bacterial signal peptide, and is cleaved by TPP to generate the mature protein. An *Arabidopsis thaliana* cDNA encoding a type 1 signal peptidase homolog which is targeted to the chloroplast has recently been isolated (B.K. Chaal, R.M. Mould, A.C. Barbrook, J.C. Gray & C.J. Howe, unpublished results).

## Activity and Specificity

The enzymes have been assayed only by their ability to cleave radiolabeled precursor proteins synthesized *in vitro*. Signal peptidases generally recognize a consensus sequence Pro/Gly-Xaa-Ala-Xaa-Ala↓ at the C-terminal end of the signal peptide. IMP1 and IMP2 each process a subset of mitochondrial precursors. IMP1 may have a preference for Asn in the P1 ('−1') position (Nunnari *et al.*, 1993). The properties of the enzymes have not been well characterized, and no inhibitors are known. The pea thylakoid enzyme has very similar specificity to *E. coli* leader peptidase (Halpin *et al.*, 1989). Mutants in the P1 ('−1') and P3 ('−3') positions of the TPP cleavage site of the precursor of the 33 kDa extrinsic protein of photosystem II have been examined (Shackleton & Robinson, 1991). Mutants with Ser or Gly in P1 are processed to a limited extent, but those with Lys, Leu, Glu or Thr are not processed. At P3, Val is processed normally, and the Glu, Leu and Lys mutants are processed to decreasing extents. The pH optimum for preplastocyanin processing by TPP is 6.5–7.0. The activity is moderately stimulated by EDTA, EGTA and 1,10-phenanthroline (Kirwin *et al.*, 1987; Musgrove *et al.*, 1989) and inhibited by the penem compound SB214357 (Barbrook *et al.*, 1996), an inhibitor of *E. coli* leader peptidase (Perry *et al.*, 1995).

## Structural Chemistry

A number of regions of high sequence similarity have been identified in members of the type 1 signal peptidase family (van Dijl *et al.*, 1992). On the basis of sequence comparison it is likely that Ser39 and Ser41 of IMP1 and IMP2 respectively are equivalent to the catalytic Ser90 of *E. coli* leader peptidase, and that Lys83 and Lys91 are equivalent to Lys145, which is also essential for activity. Both IMP1 and IMP2 are predicted to have catalytic domains in the intermembrane space and to be anchored in the inner membrane by single transmembrane domains. IMP1 and IMP2 copurify as a heterodimer (Schneider *et al.*, 1994) and can be cross-linked (Nunnari *et al.*, 1993), and it is possible that additional polypeptides are associated with the complex.

The chloroplast enzyme is synthesized as a precursor with a chloroplast-targeting domain at its N-terminus. By comparison with the *E. coli* enzyme, Ser184 and Lys237 of the *A. thaliana* chloroplast precursor are likely to be essential active-site residues. The topology of the mature protein is uncertain, especially since the maturation site cannot be predicted from the sequence, although it is likely to have a single transmembrane domain by analogy with the cyanobacterial enzymes to which it is highly similar. Native TPP is likely to be an oligomeric complex (Kirwin *et al.*, 1988).

## Preparation

A heterodimer of IMP1 and IMP2 has been partially purified from detergent-solubilized yeast mitochondria by ion-exchange chromatography, glycerol density gradient centrifugation and immunoaffinity chromatography (Schneider *et al.*, 1994).

The thylakoidal enzyme is more abundant in stromal lamellae, and can be selectively solubilized from washed, broken chloroplasts by Triton X-100 in the presence of 5 mM $MgCl_2$. The extracted activity is unstable, but has been partially purified by a combination of ion-exchange and hydroxyapatite chromatography (Kirwin *et al.*, 1987). A high molecular mass complex containing the activity can be partially resolved from other components of the stromal lamellar extract on sucrose density gradients (Kirwin *et al.*, 1988). The N-terminally His-tagged catalytic domain of the *A. thaliana* enzyme has been expressed in *E. coli* and shows activity against precursor proteins (B.K. Chaal, R.M. Mould, A.C. Barbrook, J.C. Gray & C.J. Howe, unpublished results).

## Biological Aspects

The mitochondrial inner membrane protease has been characterized only from yeast. The two catalytic subunits are specific for different precursor proteins; IMP1 processes cytochrome oxidase subunit II and cytochrome $b_2$, while IMP2 generates the mature cytochrome $c_1$. The products of two other nuclear genes are essential for IMP1 function *in vivo*, SOM1 and PET1402 (Bauer *et al.*, 1994; Esser *et al.*, 1996). The functions of these polypeptides are unknown and their association with the IMP1/2 complex has not been demonstrated.

Proteins are imported into the chloroplast thylakoid lumen by several mechanisms, requiring at least two or three distinct transport machineries (Robinson & Klösgen, 1994). The signal peptidase recognition sites of precursor proteins translocated by the various mechanisms are indistinguishable, but the occurrence of multiple forms of TPP is possible.

## Related Peptidases

The *A. thaliana* chloroplast enzyme is very similar in sequence to enzymes from the cyanobacteria *Phormidium laminosum* (Packer *et al.*, 1995) and *Synechocystis* sp. PCC6803 (Kaneko *et al.*, 1996). A TPP-like activity has been solubilized from both thylakoid and cytoplasmic membranes from *P. laminosum* and the activities are inhibited by SB214357 (Barbrook *et al.*, 1993, 1996). The thylakoidal enzyme has very similar specificity to the higher plant TPP (Wallace *et al.*, 1990). Two genes encoding leader peptidase homologs are found in the genome of *Synechocystis* sp. PCC6803, which has been completely sequenced. The functions of the two genes are at present unknown.

## Antibodies

Because of the very low level of sequence similarity between members of the type 1 signal peptidase family, antibodies raised against one do not cross-react well with others. Because of the low abundance of the protein, they can often not be detected immunochemically in crude preparations. Antibodies raised against a C-terminal peptide of IMP1 have been used (Schneider *et al.*, 1994) and an antibody against IMP1 has been reported elsewhere (Nunnari *et al.*, 1993).

## References

Barbrook, A.C., Packer, J.C.L. & Howe, C.J. (1993) Components of the protein translocation system in the thermophilic cyanobacterium *Phormidium laminosum*. *Biochem. Biophys. Res. Commun.* **197**, 874–877.

Barbrook, A.C., Packer, J.C.L. & Howe, C.J. (1996) Inhibition by penem of processing peptidases from cyanobacteria and chloroplast thylakoids. *FEBS Lett.* **398**, 198–200.

Bauer, M., Behrens, M., Esser, K., Michaelis, G. & Pratje, E. (1994) *PET1402*, a nuclear gene required for proteolytic processing of cytochrome oxidase subunit 2 in yeast. *Mol. Gen. Genet.* **245**, 272–278.

Behrens, M., Michaelis, G. & Pratje, E. (1991) Mitochondrial inner membrane protease 1 of *Saccharomyces cerevisiae* shows sequence similarity to the *Escherichia coli* leader peptidase. *Mol. Gen. Genet.* **228**, 167–176.

Esser, K., Pratje, E. & Michaelis, G. (1996) *SOM1*, a small new gene required for mitochondrial inner membrane peptidase function in *Saccharomyces cerevisiae. Mol. Gen. Genet.* **252**, 437–445.

Hageman, J., Robinson, C., Smeekens, S. & Weisbeek, P. (1986) A thylakoid processing peptidase is required for complete maturation of the lumen protein plastocyanin. *Nature* **324**, 567–569.

Halpin, C., Elderfield, P.D., James, H.E., Zimmermann, R., Dunbar, B. & Robinson, C. (1989) The reaction specificities of the thylakoidal processing peptidase and *Escherichia coli* leader peptidase are identical. *EMBO J.* **8**, 3917–3929.

Kaneko, T. *et al.* (1996) Sequence analysis of the genome of the unicellular cyanobacterium *Synechocystis* sp. PCC6803. II. Sequence determination of the entire genome and assignment of protein coding regions. *DNA Res.* **3**, 109–136.

Kirwin, P.M., Elderfield, P.D. & Robinson, C. (1987) Transport of proteins into chloroplasts: partial purification of a thylakoidal processing peptidase involved in plastocyanin biogenesis. *J. Biol. Chem.* **262**, 16386–16390.

Kirwin, P.M., Elderfield, P.D., Williams, R.S. & Robinson, C. (1988) Transport of proteins into chloroplasts: organization, orientation and lateral distribution of the plastocyanin processing peptidase in the thylakoid network. *J. Biol. Chem.* **263**, 18128–18132.

Musgrove, J.E., Elderfield, P.D. & Robinson, C. (1989) Endopeptidases in the stroma and thylakoids of pea chloroplasts. *Plant Physiol.* **90**, 1616–1620.

Nunnari, J., Fox, T.D. & Walter, P. (1993) A mitochondrial protease with two catalytic subunits of nonoverlapping specificities. *Science* **262**, 1997–2004.

Packer, J.C.L., André, D. & Howe, C.J. (1995) Cloning and sequence analysis of a signal peptidase I from the thermophilic cyanobacterium *Phormidium laminosum. Plant Mol. Biol.* **27**, 199–204.

Perry, C.R., Ashby, M.J. & Elsmere, S.A. (1995) Penems as research tools to investigate the activity of *E. coli* leader peptidase. *Biochem. Soc. Trans.* **23**, 548S.

Pratje, E., Esser, K. & Michaelis, G. (1994) The mitochondrial inner membrane peptidase. In: *Signal Peptidases* (von Heijne, G., ed.). Austin, TX: R.G. Landes, pp. 105–112.

Robinson, C. (1994) The chloroplast stromal and thylakoidal processing peptidases. In: *Signal Peptidases* (von Heijne, G., ed.). Austin, TX: R.G. Landes, pp. 113–123.

Robinson, C. & Klösgen, R.B. (1994) Targeting of proteins into and across the thylakoid membrane: a multitude of mechanisms. *Plant Mol. Biol.* **26**, 15–24.

Schneider, A., Behrens, M., Scherer, P., Pratje, E., Michaelis, G. & Schatz, G. (1991) Inner mitochondrial membrane protease I, an enzyme mediating intramitochondrial protein sorting in yeast. *EMBO J.* **10**, 247–254.

Schneider, A., Oppliger, W. & Jenö, P. (1994) Purified inner membrane protease I of yeast mitochondria is a heterodimer. *J. Biol. Chem.* **269**, 8635–8638.

Shackleton, J.B. & Robinson, C. (1991) Transport of proteins into chloroplasts: the thylakoidal processing peptidase is a signal-type peptidase with stringent substrate requirements at the −3 and −1 positions. *J. Biol. Chem.* **266**, 12152–12156.

van Dijl, J.M., de Jong, A., Vehmaanperä, J., Venema, G. & Bron, S. (1992) Signal peptidase I of *Bacillus subtilis*: patterns of conserved amino acids in prokaryotic and eukaryotic type I signal peptidases. *EMBO J.* **11**, 2819–2828.

Wallace, T.P., Robinson, C. & Howe, C.J. (1990) The reaction specificities of the pea and a cyanobacterial thylakoid processing peptidase are similar but not identical. *FEBS Lett.* **272**, 141–144.

*Jeremy C.L. Packer*
*Department of Biochemistry,*
*University of Cambridge, Tennis Court Road,*
*Cambridge CB2 1QW, UK*

*Christopher J. Howe*
*Department of Biochemistry,*
*University of Cambridge, Tennis Court Road,*
*Cambridge CB2 1QW, UK*
*Email: c.j.howe@bioc.cam.ac.uk*

# 156. *Signal peptidase (eukaryote)*

## Databanks

*Peptidase classification: clan SF, family S26, MEROPS ID: S26.010*
*NC-IUBMB enzyme classification: none*
*Databank codes:*

| Species | Subunit | SwissProt | PIR | EMBL (cDNA) | EMBL (genomic) |
|---|---|---|---|---|---|
| Putative catalytic components (Family S26) | | | | | |
| *Arabidopsis thaliana* | EST | – | – | H76216 | – |
| *Arabidopsis thaliana* | EST | – | – | H76629 | – |
| *Arabidopsis thaliana* | EST | – | – | H76949 | – |
| *Canis familiaris* | SPC21 | P13679 | A34229 | J05069 | – |
| *Canis familiaris* | SPC18 | P21378 | A35309 | J05466 | – |
| *Gallus gallus* | 19 kDa | – | – | – | – |
| *Homo sapiens* | EST | – | – | R11105 | – |
| *Homo sapiens* | EST | – | – | T81127 | – |
| *Homo sapiens* | EST | – | – | T84807 | – |
| *Mus musculus* | EST | – | – | W77326 | – |
| *Rattus norvegicus* | SPC18 | P42667 | – | L11319 | – |
| *Saccharomyces cerevisiae* | Sec11p | P15367 | A30184 S48484 | X07694 | Z47047: chromosome IX complete sequence |
| Other components (25 kDa subunit) | | | | | |
| *Canis familiaris* | SPC25 | – | A55012 | U12687 | – |
| *Homo sapiens* | 25 kDa | – | – | L38950 | – |
| *Saccharomyces cerevisiae* | Spc2p | – | – | – | Z46729: chromosome XIII cosmid 9958 |
| Other components (22 kDa subunit) | | | | | |
| *Caenorhabditis elegans* | glycoprotein | P34525 | S44854 | – | L14331: cosmid K12H4 |
| *Caenorhabditis elegans* | | – | – | M79663 | – |

*continued overleaf*

| Species | Subunit | SwissProt | PIR | EMBL (cDNA) | EMBL (genomic) |
|---|---|---|---|---|---|
| Other components (22 kDa subunit) *(continued)* | | | | | |
| *Canis familiaris* | SPC22/23 | P12280 | – | J04067 | – |
| *Drosophila melanogaster* | – | – | – | M32022 | – |
| *Gallus gallus* | gp23 | P28687 | – | X60795 | – |
| *Saccharomyces cerevisiae* | Spc3p | Q12133 | – | – | X94607: chromosome XII left arm |
| | | | | – | Z73238: chromosome XII ORF |
| *Schizosaccharomyces pombe* | glycoprotein | Q10259 | – | – | Z69728: chromosome I cosmid c56F8 |
| Other components (12 kDa subunit) | | | | | |
| *Homo sapiens* | 12 kDa | – | – | L38852 | – |
| *Saccharomyces cerevisiae* | Spc1p | P46965 | – | U26257 | Z49510 |

## Name and History

Microsomal *signal peptidase* is the endopeptidase responsible for the cleavage of precursor proteins targeted to the rough endoplasmic reticulum (ER) during biosynthesis and secretion. The enzyme was first identified in preparations of ER microsomes from mouse myeloma cells (Milstein *et al.*, 1972). Immunoglobulin light chain proteins synthesized in the absence of ER contained an N-terminal peptide extension not present on mature, secreted light chains. Synthesis in the presence of ER microsomes resulted in cleavage of the light chain precursor. The N-terminal peptide extension is now recognized as a feature common to most, but not all, secretory precursor proteins and many membrane proteins. It is known as a signal peptide because of its role in signaling the membrane transport of precursor proteins. Discovery of signal peptides led to the formulation of the signal hypothesis of peptide-mediated protein targeting and translocation (Blobel & Dobberstein, 1975a,b). Signal peptides are cleaved during translocation of the protein across the bilayer by a highly selective, membrane-bound peptidase. This signal peptidase is found in prokaryotic and eukaryotic cells as well as in the protein import pathways of mitochondria and chloroplasts (Dalbey *et al.*, 1997; see also Chapters 153, 154 and 155).

The first eukaryotic signal peptidase gene, *Sec11*, was identified genetically as an essential gene in *Saccharomyces cerevisiae* (Böhni *et al.*, 1988). The encoded protein is designated Sec11p (see Databanks table). Isolated yeast signal peptidase is associated with a complex of four membrane proteins (YaDeau *et al.*, 1991). cDNA clones encoding all four proteins have been characterized: a glycoprotein, Spc3p, 21 kDa (Fang *et al.*, 1997); Spc2p, 18 kDa (Mullins *et al.*, 1996); Sec11p, 17 kDa (Böhni *et al.*, 1988); and Spc1p, 11 kDa (Fang *et al.*, 1996).

Eukaryotic signal peptidase was first purified from dog pancreas ER as a complex of five membrane proteins (Evans *et al.*, 1986). In the current nomenclature, the subunits of the dog signal peptidase complex are abbreviated 'SPC' followed by a number corresponding to the apparent molecular mass indicated by SDS-PAGE (see Databanks table). The subunits are: SPC25 (Greenburg & Blobel, 1994); SPC22/23, a single glycoprotein that migrates as a doublet during electrophoresis (Shelness *et al.*, 1988); SPC21 (Greenburg *et al.*, 1989), SPC18 (Shelness & Blobel, 1990); and SPC12 (Fang *et al.*, 1996). SPC21 and SPC18 are homologous isoforms that are also homologous to yeast Sec11p.

Chicken oviduct signal peptidase is also an integral membrane protein complex but is purified as an active complex of only two proteins, a 23 kDa glycoprotein, named gp23, and a 19 kDa subunit, named p19 (Baker & Lively, 1987; Newsome *et al.*, 1992). Genetic evidence from other species suggests that chickens probably express homologs of SPC25 and SPC12 as well as a second homologous isoform of p19, but those proteins have not yet been observed in purified preparations of chicken signal peptidase. Only p19 and gp23 are required for chicken signal peptidase activity *in vitro*.

## Activity and Specificity

Signal peptidase is highly specific for cleavage of N-terminal secretory signal peptides. The natural substrate for signal peptidase is a nascent precursor protein molecule that is being elongated by a ribosome bound to the ER translocation site and is being actively transported into the ER lumen. Signal peptides range in length from 13 to 31 amino acids (Briggs & Gierasch, 1986; Nielsen *et al.*, 1997) and differ significantly in their amino acid sequences. While it appears that the primary substrates of signal peptidases are precursor proteins targeted to the ER, evidence in yeast indicates that the Sec11p subunit of the yeast signal peptidase complex plays an essential role in protein degradation (Mullins *et al.*, 1995). Therefore, microsomal signal peptidases may also cleave internal sites within proteins targeted for degradation (Klausner & Sitia, 1990; Yuk & Lodish, 1993).

No simple amino acid consensus sequence can be written to generalize the substrate determinants of signal peptides. However, all signal peptides have similar structural elements arranged into three domains: a basic N-terminus of 1–5 residues known as the N-region, a central hydrophobic core of

7–15 residues known as the H-region, and a C-terminal polar region with 3–7 residues that is positioned on the N-terminal side of the cleavage site, and known as the C-region (von Heijne, 1990). Amino acids at the P1 and P3 sites appear to be the primary determinants for signal peptidase cleavage, but extended subsite interactions also play a role in substrate recognition. (Note that the P1 and P3 sites are frequently referred to as the '−1' and '−3' sites in the signal peptidase literature.) Amino acid side chains at P1 and P3 must be small and nonpolar (Nielsen *et al.*, 1997). While the amino acid sequences of type I bacterial and eukaryotic signal peptidases are quite different, the enzymes have very similar substrate specificities.

The only substrates for signal peptidase assays *in vitro* are secretory precursor proteins or synthetic peptides with sequences based on known signal peptides. The short peptides are generally poor substrates that are hydrolyzed slowly and do not have easily detectable cleavage products (Caulfield *et al.*, 1989; Dev *et al.*, 1990). Small molecule substrates with readily detected leaving groups such as *p*-nitroanilides or thiobenzyl esters have not yet been discovered. The most commonly used assay is a post-translational cleavage method in which the enzyme is incubated with a radiolabeled precursor protein synthesized by cell-free translation of a secretory protein mRNA in the presence of a radiolabeled amino acid (Jackson, 1983). Cotranslational assays of signal peptidase activity are performed by including intact rough ER microsomes in the cell-free translation reaction mixture during synthesis of the precursor protein (Walter & Blobel, 1983). The reaction mixtures from both post- and cotranslational assays are analyzed by SDS-PAGE to separate the smaller molecular mass cleavage product from the larger precursor substrate. Reactants and products are detected by autoradiography of the dried electrophoresis gel, and conversion to product can be quantified by densitometry of the autoradiogram.

Several precursor proteins have been used in signal peptidase assays, including bovine preprolactin (Jackson, 1983) and human preplacental lactogen (Szczesna & Boime, 1976). The post-translational assay using radiolabeled precursor proteins is limited and is not sufficiently flexible to permit detailed kinetic analyses of signal peptidase. Only very small amounts of substrate, in the range of $10^{-15}$ moles, are typically produced. With this small amount of substrate it is not possible to vary the substrate concentration significantly.

An improved version of the signal peptidase assay has been developed using a substrate that is a chimeric precursor protein, called pONA, composed of the signal peptide of the *E. coli* outer membrane protein A (OmpA) and the mature enzyme domain of *Staphylococcus aureus* nuclease A (Chatterjee *et al.*, 1995). pONA can be expressed in *E. coli* and milligrams of the uncleaved precursor protein can be isolated for use as a signal peptidase substrate. It is an excellent substrate for both the bacterial (Chatterjee *et al.*, 1995) and eukaryotic signal peptidases (M.K. Nusier, D. Suciu, M. Inouye & M.O. Lively, unpublished results). The precursor and the cleavage product are separated by SDS-PAGE and detected by staining the proteins with a dye such as Coomassie blue.

Microsomal signal peptidase is resistant to inhibition by the general mechanistic classes of inhibitors of proteolytic enzymes. Early experiments with solubilized dog microsomal signal peptidase showed that the enzyme was not inhibited by *N*-ethylmaleimide, iodoacetamide, Tos-Phe-CH₂Cl, PMSF, EDTA, benzamidine, chymostatin, leupeptin, antipain, pepstatin or soybean trypsin inhibitor (Jackson & Blobel, 1980). Purified chicken oviduct signal peptidase is irreversibly inhibited by a specific peptide chloromethane derivative, MeOSuc-Ala-Ala-Pro-Val-CH₂Cl as well as certain peptidyl ($\alpha$-aminoalkyl) phosphonates (Nusier *et al.*, 1996). Inhibition by these compounds suggests that signal peptidase may employ a mechanism that involves an active-site Ser similar to the classical serine proteinase families. However, analysis of the amino acid sequences of the subunits, especially the Sec11 family subunits thought to contain the active site (see below), does not reveal any amino sequence similarity with the recognized protease families. Signal peptidase potentially functions with a novel endoproteolytic mechanism (Dalbey *et al.*, 1997; Dalbey & von Heijne, 1992; Paetzel & Dalbey, 1997).

## Structural Chemistry

Eukaryotic microsomal signal peptidase is a complex with up to five different integral membrane protein subunits. The stoichiometry of the subunits in the native enzyme complex in the ER has not yet been established. cDNAs encoding all of the identified dog, chicken and yeast signal peptidase subunits have now been cloned and the sequences currently in the sequence databases are identified in the Databanks table.

All of the signal peptidase subunits are integral membrane proteins. In the dog signal peptidase complex, SPC18, SPC21 and SPC22/23 are anchored in the ER by single N-terminal membrane-spanning domains, with the bulk of each protein exposed in the lumen of the ER (Shelness *et al.*, 1993). SPC12 and SPC25 span the bilayer twice, such that both the N- and C-termini are exposed in the cytoplasm (Kalies & Hartmann, 1996). Because of the high degree of sequence identity among the corresponding subunits in the hen oviduct and yeast signal peptidases, these signal peptidases are thought to have similar topologies.

The function and stoichiometry of the different signal peptidase subunits within the native enzyme complex are unknown. Genetic studies in *S. cerevisiae* have established that the Sec11p and Spc3p subunits are essential for viability of the yeast as well as for peptidase activity (Böhni *et al.*, 1988; Fang *et al.*, 1997; Meyer & Hartmann, 1997). Notably, the chicken homologs of Sec11p and Spc3p, which are p19 and gp23, respectively, are the only two proteins required for activity of the chicken signal peptidase complex (Baker & Lively, 1987). SPC12 and SPC25, the dog homologs of Spc1p and Spc2p, respectively, are not essential genes and are not required for signal peptidase activity (Mullins *et al.*, 1996). Homologs of the SPC25/Spc2p and SPC12/Spc1p proteins have not yet been identified in chickens but are likely to be present. Partial DNA sequences encoding human homologs of these proteins are found in the expressed sequence tag (EST) databases, suggesting that these genes are present in all eukaryotic species.

The subunits that comprise the Sec11 family (SPC18, SPC21, Sec11p and p19) are thought to contain the catalytic sites for signal peptide hydrolysis, but direct proof of this hypothesis has not yet been obtained. Except for the yeast *S. cerevisiae*, there are two homologous isoforms of the Sec11 proteins expressed in humans, monkeys, dogs, rats, frogs and chickens (Walker *et al.*, 1996). These subunits are hypothesized to contain the catalytic sites because they appear to be distantly related to the bacterial signal peptidases, as indicated by comparison of their amino acid sequences. The amino acid sequence surrounding the active-site Ser of *E. coli* type I signal peptidase can be aligned with a similar region containing a conserved Ser in the Sec11 family (Dalbey *et al.*, 1997) (see also Chapter 151). Significantly, the mitochondrial inner membrane peptidase (IMP) (Chapter 155) of *S. cerevisiae* is a heterodimer of homologous subunits that are also homologous to the Sec11 subunits. The IMP subunits have nonoverlapping substrate specificities for cleavage of mitochondrial precursor substrates (Nunnari *et al.*, 1993; Schneider *et al.*, 1994). The subunits of the microsomal signal peptidases related to Sec11p may have similar functions in the ER.

## Preparation

Microsomal signal peptidase is an integral membrane protein complex in the ER of all eukaryotic cells and can be obtained from any source from which ER microsomes can be isolated. The enzyme was first purified from dog pancreas ER microsomes as a complex of five subunits using a multistep protocol (Evans *et al.*, 1986). Chicken oviduct signal peptidase is isolated as a complex of the p19 and gp23 subunits from ER microsomes obtained from tubular gland cells of the oviduct magnum region of laying hens (Baker & Lively, 1987; Lively *et al.*, 1994). Purification of yeast signal peptidase yields a complex of four polypeptides (YaDeau & Blobel, 1989; YaDeau *et al.*, 1991). Because all of the subunits are integral membrane proteins with hydrophobic membrane-spanning domains, detergents must be used in buffers at all stages of the enzyme purification to solubilize the proteins. Useful detergents include Triton X-100, octylglucoside, deoxycholate, digitonin and other mild detergents. Expression of functional eukaryotic signal peptidase from cloned cDNAs has not yet been accomplished.

## Biological Aspects

Microsomal signal peptidase has been detected in virtually all cells and tissues examined including human ascites tumor cells (Szczesna & Boime, 1976), rat liver (Kaschnitz & Kreil, 1978), *Drosophilia melanogaster* embryos (Brennan *et al.*, 1980), chicken oviduct (Lively & Walsh, 1983), dog pancreas (Evans *et al.*, 1986) and yeast (YaDeau & Blobel, 1989). The enzyme is essential for life in yeast and thus presumably in all eukaryotes. There are examples of naturally occurring mutations within the signal peptides of secreted proteins that block cleavage by signal peptidase and result in human disease. A mutation that results in inefficient processing of the signal peptide of preprovasopressin appears to be the cause

of a form of familial central diabetes insipidus (Ito *et al.*, 1993; McLeod *et al.*, 1993). Similarly, a form of familial hypoparathyroidism is the result of a signal peptide mutation that blocks cleavage and secretion of preproparathyroid hormone (Arnold *et al.*, 1990; Karaplis *et al.*, 1995). A similar signal peptide mutation prevents secretion of human coagulation factor X (Chapter 54) and results in a severe form of hemophilia (Racchi *et al.*, 1993).

## Distinguishing Features

Microsomal signal peptidase is most readily distinguished by its localization as a membrane-bound protein complex in the ER and by its strict specificity for cleavage of precursor proteins. The enzyme is related to the inner mitochondrial processing peptidase (Nunnari *et al.*, 1993) and presumably to the thylakoid processing peptidase of chloroplasts (Robinson, 1994) (see Chapter 155).

## Further Reading

A collection of articles on signal peptidases is to be found in von Heijne (1994). A recent review is that of Dalbey *et al.* (1997).

## References

Arnold, A., Horst, S.A., Gardella, T.J., Baba, H., Levine, M.A. & Kronenberg, H.M. (1990) Mutation of the signal-encoding region of the preproparathyroid hormone gene in familial isolated hypoparathyroidism. *J. Clin. Invest.* **86**, 1084–1087.

Baker, R.K. & Lively, M.O. (1987) Purification and characterization of hen oviduct signal peptidase. *Biochemistry* **26**, 8561–8567.

Blobel, G. & Dobberstein, B. (1975a) Transfer of proteins across membranes. I. Presence of proteolytically processed and unprocessed nascent immunoglobulin light chains on membrane-bound ribosomes of murine myeloma. *J. Cell Biol.* **67**, 835–851.

Blobel, G. & Dobberstein, B. (1975b) Transfer of proteins across membranes. II. Reconstitution of functional rough microsomes from heterologous components. *J. Cell Biol.* **67**, 852–862.

Böhni, P.C., Deshaies, R.J. & Schekman, R.W. (1988) *SEC11* is required for signal peptide processing and yeast cell growth. *J. Cell Biol.* **106**, 1035–1042.

Brennan, M.D., Warren, T.G. & Mahowald, A.P. (1980) Signal peptides and signal peptidase in *Drosophila melanogaster. J. Cell Biol.* **87**, 516–520.

Briggs, M.S. & Gierasch, L.M. (1986) Molecular mechanisms of protein secretion: the role of the signal sequence. *Adv. Protein Chem.* **38**, 108–180.

Caulfield, M.P., Duong, L.T., Baker, R.K., Rosenblatt, M. & Lively, M.O. (1989) Synthetic substrate for eukaryotic signal peptidase. Cleavage of a synthetic peptide analog of the precursor of preproparathyroid hormone. *J. Biol. Chem.* **264**, 15813–15817.

Chatterjee, S., Suciu, D., Dalbey, R.E., Kahn, P.C. & Inouye, M. (1995) Determination of $k_m$ and $k_{cat}$ for signal peptidase 1 using a full length secretory precursor, pro-OmpA-nuclease A. *J. Mol. Biol.* **245**, 311–314.

Dalbey, R.E. & von Heijne, G. (1992) Signal peptidases in prokaryotes and eukaryotes – a new protease family. *Trends Biochem. Sci.* **17**, 474–478.

Dalbey, R.E., Lively, M.O., Bron, S. & van Dijl, J.M. (1997) The chemistry and enzymology of the type I signal peptidases. *Protein Sci.* **6**, 1129–1138.

Dev, I.K., Ray, P.H. & Novak, P. (1990) Minimum substrate sequence for signal peptidase I of *Escherichia coli. J. Biol. Chem.* **265**, 20069–20072.

Evans, E.A., Gilmore, R. & Blobel, G. (1986) Purification of microsomal signal peptidase as a complex. *Proc. Natl Acad. Sci. USA* **83**, 581–585.

Fang, H., Panzner, S., Mullins, C., Hartmann, E. & Green, N. (1996) The homologue of mammalian SPC12 is important for efficient signal peptidase activity in *Saccharomyces cerevisiae. J. Biol. Chem.* **271**, 16460–16465.

Fang, H., Mullins, C. & Green, N. (1997) In addition to SEC11, a newly identified gene, SPC3, is essential for signal peptidase activity in the yeast endoplasmic reticulum. *J. Biol. Chem.* **272**, 13152–13158.

Greenburg, G. & Blobel, G. (1994) cDNA-derived primary structure of the 25-kDa subunit of canine microsomal signal peptidase complex. *J. Biol. Chem.* **269**, 25354–25358.

Greenburg, G., Shelness, G.S. & Blobel, G. (1989) A subunit of mammalian signal peptidase is homologous to yeast SEC11 protein. *J. Biol. Chem.* **264**, 15762–15765.

Ito, M., Oiso, Y., Murase, T., Kondo, K., Saito, H., Chinzei, T., Racchi, M. & Lively, M.O. (1993) Possible involvement of inefficient cleavage of preprovasopressin by signal peptidase as a cause for familial central diabetes insipidus. *J. Clin. Invest.* **91**, 2565–2571.

Jackson, R.C. (1983) Quantitative assay for signal peptidase. *Methods Enzymol.* **96**, 784–794.

Jackson, R.C. & Blobel, G. (1980) Post-translational processing of full-length presecretory proteins with canine pancreatic signal peptidase. *Ann. NY Acad. Sci.* **343**, 391–404.

Kalies, K.-U. & Hartmann, E. (1996) Membrane topology of the 12- and the 25-kDa subunits of the mammalian signal peptidase complex. *J. Biol. Chem.* **271**, 3925–3929.

Karaplis, A.C., Lim, S.-K., Baba, H., Arnold, A. & Kronenberg, H.M. (1995) Inefficient membrane targeting, translocation, and proteolytic processing by signal peptidase of a mutant preproparathyroid hormone protein. *J. Biol. Chem.* **270**, 1629–1635.

Kaschnitz, R. & Kreil, G. (1978) Processing of prepromelittin by subcellular fractions from rat liver. *Biochem. Biophys. Res. Commun.* **83**, 901–907.

Klausner, R.D. & Sitia, R. (1990) Protein degradation in the endoplasmic reticulum. *Cell* **62**, 611–614.

Lively, M.O. & Walsh, K.A. (1983) Hen oviduct signal peptidase is an integral membrane protein. *J. Biol. Chem.* **258**, 9488–9495.

Lively, M.O., Newsome, A.L. & Nusier, M. (1994) Eukaryote microsomal signal peptidases. *Methods Enzymol.* **244**, 301–314.

McLeod, J.F., Kovács, L., Gaskill, M.B., Rittig, S., Bradley, G.S. & Robertson, G.L. (1993) Familial neurohypophyseal diabetes insipidus associated with a signal peptide mutation. *J. Clin. Endocrinol. Metab.* **77**, 599A–599G.

Meyer, H.-A. & Hartmann, E. (1997) The yeast SPC22/23 homolog Spc3p is essential for signal peptidase activity. *J. Biol. Chem.* **272**, 13159–13164.

Milstein, C., Brownlee, G.G., Harrison, T.M. & Mathews, M.B. (1972) A possible precursor of immunoglobulin light chains. *Nature New Biol.* **239**, 117–120.

Mullins, C., Lu, Y.Q., Campbell, A., Fang, H. & Green, N. (1995) A mutation affecting signal peptidase inhibits degradation of an abnormal membrane protein in *Saccharomyces cerevisiae. J. Biol. Chem.* **270**, 17139–17147.

Mullins, C., Meyer, H.-A., Hartmann, E., Green, N. & Fang, H. (1996) Structurally related Spc1p and Spc2p of yeast signal peptidase complex are functionally distinct. *J. Biol. Chem.* **271**, 29094–29099.

Newsome, A.L., McLean, J.W. & Lively, M.O. (1992) Molecular cloning of a cDNA encoding the glycoprotein of hen oviduct microsomal signal peptidase. *Biochem. J.* **282**, 447–452.

Nielsen, H., Engelbrecht, J., Brunak, S. & von Heijne, G. (1997) Identification of prokaryotic and eukaryotic signal peptides and prediction of cleavage sites. *Protein Eng.* **10**, 1–6.

Nunnari, J., Fox, T.D. & Walter, P. (1993) A mitochondrial protease with two catalytic subunits of nonoverlapping specificities. *Science* **262**, 1997–2004.

Nusier, M.K., Kam, C.-M., Powers, J.C. & Lively, M.O. (1996) Inhibition of microsomal signal peptidase by site-specific serine protease inhibitors. *FASEB J.* **10**, A1406.

Paetzel, M., Strynadka, N.C.J., Tschantz, W.R., Casareno, R., Bullinger, P.R. & Dalbey, R.E. (1997) Use of site-directed chemical modification to study an essential lysine in *Escherichia coli* leader peptidase. *J. Biol. Chem.* **272**, 9994–10003.

Racchi, M., Watzke, H.H., High, K.A. & Lively, M.O. (1993) Human coagulation factor X deficiency caused by a mutant signal peptide that blocks cleavage by signal peptidase but not targeting and translocation to the endoplasmic reticulum. *J. Biol. Chem.* **268**, 5735–5740.

Robinson, C. (1994) The chloroplast stromal and thylakoid processing peptidases. In: *Signal Peptidases* (von Heijne, G., ed.). Austin, TX: R.G. Landes, pp. 113–123.

Schneider, A., Opplinger, W. & Jeno, P. (1994) Purified inner membrane protease 1 of yeast mitochondria is a heterodimer. *J. Biol. Chem.* **269**, 8635–8638.

Shelness, G.S. & Blobel, G. (1990) Two subunits of the canine signal peptidase complex are homologous to yeast SEC11 protein. *J. Biol. Chem.* **265**, 9512–9519.

Shelness, G.S., Kanwar, Y.S. & Blobel, G. (1988) cDNA-derived primary structure of the glycoprotein component of canine microsomal signal peptidase complex. *J. Biol. Chem.* **263**, 17063–17070.

Shelness, G.S., Lin, L. & Nicchitta, C.V. (1993) Membrane topology and biogenesis of eukaryotic signal peptidase. *J. Biol. Chem.* **268**, 5201–5208.

Szczesna, E. & Boime, I. (1976) mRNA-dependent synthesis of authentic precursor of human placental lactogen: conversion to its mature form in ascites cell-free extracts. *Proc. Natl Acad. Sci. USA* **73**, 1179–1183.

von Heijne, G. (1990) The signal peptide. *Membr. Biol.* **115**, 195–201.

von Heijne, G. (1994) *Signal Peptidases*. Austin, TX: R.G. Landes.

Walker, S.J., Racchi, M., Langner, J., Shelness, G.S. & Lively, M.O. (1996) Demonstration of multiple isoforms of the 19 kDa subunit of hen oviduct signal peptidase in five species. *FASEB J.* **10**, A1406.

Walter, P. & Blobel, G. (1983) Preparation of microsomal membranes for co-translational protein translocation. *Methods Enzymol.* **96**, 84–93.

YaDeau, J.T. & Blobel, G. (1989) Solubilization and characterization of yeast signal peptidase. *J. Biol. Chem.* **264**, 2928–2934.

YaDeau, J.T., Klein, C. & Blobel, G. (1991) Yeast signal peptidase contains a glycoprotein and the *SEC11* gene product. *Proc. Natl Acad. Sci. USA* **88**, 517–521.

Yuk, M.H. & Lodish, H.F. (1993) Two pathways for the degradation of the H2 subunit of the asialoglycoprotein receptor in the endoplasmic reticulum. *J. Cell Biol.* **123**, 1735–1749.

*Stephen J. Walker*
Biochemistry Department,
The Bowman Gray School of Medicine of
Wake Forest University,
Medical Center Blvd., Winston-Salem, NC 27157, USA

*Mark O. Lively*
Biochemistry Department,
The Bowman Gray School of Medicine of
Wake Forest University,
Medical Center Blvd., Winston-Salem, NC 27157, USA
Email: lively@mgrp.bgsm.edu

# 157. Tsp protease

## Databanks

Peptidase classification: clan SF, family S41, MEROPS ID: S41.001
NC-IUBMB enzyme classification: none
Databank codes:

| Species | SwissProt | PIR | EMBL (cDNA) | EMBL (genomic) |
|---|---|---|---|---|
| *Bacillus subtilis* | – | – | X98341 | – |
| *Bartonella bacilliformis* | – | – | L37094 | – |
| *Escherichia coli* | P23865 | A41798 | D00674 | – |
| | | A42475 | M75634 | |
| *Haemophilus influenzae* | P45306 | F64135 | U32840 | – |
| *Neisseria gonorrhoeae* | – | – | U11547 | – |

## Name and History

Tsp protease was purified from *Escherichia coli* as an activity that degrades protein variants with nonpolar residues at their C-termini, but does not degrade otherwise identical variants with charged residues at the C-terminus (Silber *et al.*, 1992). Based on this substrate selectivity, the enzyme was named **Tsp** for **t**ail-**s**pecific **p**rotease. The protein has also been called **Prc** after the name of its gene. The gene for Tsp protease, *prc*, was first identified as the locus of a mutation that disrupts the C-terminal processing of penicillin-binding protein 3 (Hara *et al.*, 1991). The gene was named *prc* for **pr**ocessing involving **C**-terminal cleavage, and has also been called *tsp* (Silber *et al.*, 1992). Genes from other bacteria have been identified by sequence homology.

Tsp may also be identical to two previously isolated proteases, **protease Re** (Park *et al.*, 1988) and an enzyme that degrades oxidized glutamine synthetase (Roseman & Levine, 1987), but there is no sequence information on these enzymes.

## Activity and Specificity

Proteins and peptides with hydrophobic or nonpolar residues at their C-terminal positions are selectively degraded by Tsp. Ala is preferred at the C-terminus, Ala or Tyr at the penultimate position, and Ala or Leu at the third position from the end (Keiler & Sauer, 1996). The $\alpha$-carboxylate group appears to be required for cleavage, since amidation of this group protects peptides from cleavage by Tsp (Keiler *et al.*, 1995).

Although Tsp recognizes the C-terminal residues, it cleaves its substrates at internal peptide bonds, sometimes far from the C-terminus (Keiler *et al.*, 1995). The cleavage site specificity is broad, with Ala, Ser, Val (and to a lesser extent Ile, Leu, Lys or Arg) preferred at the P1 position, and these same residues plus Met, Tyr, or Trp at the P1′ position (Keiler *et al.*, 1995).

A Ser/Lys dyad mechanism similar to those proposed for LexA and the type I signal peptidases (Chapter 153) has been suggested for Tsp. Ser430 is the nucleophile for peptide bond hydrolysis, and Lys455 and Asp441 are also important for catalysis (Keiler & Sauer, 1995). The pH profile of activity shows a broad optimum from 5.0 to 8.5 (Keiler *et al*., 1995). Activity is stimulated by 1 mM $MnCl_2$, $CoCl_2$ or $CaCl_2$, and inhibited by 1 mM $FeCl_3$, $CuSO_4$ or $ZnSO_4$, but is not affected by typical inhibitors of serine, cysteine, metallo or aspartic proteases (Silber *et al*., 1992).

## Structural Chemistry

The mature Tsp protein is a 660 residue polypeptide of 75 kDa, which is a monomer in solution (Silber *et al*., 1992). Although unprocessed forms of the protein have not been observed, the gene contains a putative signal sequence of 22 residues (Hara *et al*., 1991; Silber *et al*., 1992).

## Preparation

Tsp is expressed at low levels from the chromosome of *E. coli* (Silber *et al*., 1992). The recombinant enzyme, however, has been purified from periplasmic fractions and whole-cell extracts of *E. coli* on the scale of tens of milligrams by use of a variety of expression systems (Keiler *et al*., 1995).

## Biological Aspects

The *prc* gene is located at 40.4′ on the *E. coli* chromosome (Hara *et al*., 1991). Deletion of *prc* results in temperature-sensitive growth under conditions of osmotic stress, a reduced heat-shock response, and leakage of periplasmic proteins (Hara *et al*., 1991). An increase in the sensitivity to antibiotics has also been observed in *prc*⁻ cells (Seoane *et al*., 1992). The gene contains a putative signal sequence, and the mature protein is localized to the periplasmic space of *E. coli* (Silber *et al*., 1992).

Tsp is a component of the ssrA RNA protein-tagging pathway for the removal of incorrectly synthesized proteins (Keiler *et al*., 1996). In this pathway, proteins synthesized from damaged RNA are tagged with a C-terminal peptide that targets them for degradation. Tsp degrades proteins tagged with this peptide (the C-terminal residues are Leu-Ala-Ala), and is responsible for a majority of the degradation of periplasmic proteins tagged via this pathway. In addition, Tsp is responsible for the C-terminal processing of penicillin-binding protein 3 (Hara *et al*., 1991), and potentiates the transport of long-chain fatty acids via the fadL pathway (Azizan & Black, 1994).

## Related Peptidases

Tsp shares sequence homology with the CtpA and CtpB proteases from cyanobacteria (Chapter 158) and plants (Chapter 159). These enzymes, which are responsible for the C-terminal processing of the D1 protein, share 50% similarity overall with Tsp, and over 90% identity in the active-site region, including conservation of the three active-site residues (Keiler & Sauer, 1995).

## References

Azizan, A. & Black, P. (1994) Use of transposon Tn*phoA* to identify genes for cell envelope proteins of *Escherichia coli* required for long-chain fatty acid transport: the periplasmic protein Tsp potentiates fatty acid transport. *J. Bacteriol.* **176**, 6653–6662.

Hara, H., Yamamoto, Y., Higashitani, A., Suzuki, H. & Nishimura, Y. (1991) Cloning, mapping, and characterization of the *Escherichia coli* prc gene, which is involved in C-terminal processing of penicillin-binding protein 3. *J. Bacteriol.* **173**, 4799–4813.

Keiler, K.C. & Sauer, R.T. (1995) Identification of active site residues of the Tsp protease. *J. Biol. Chem.* **270**, 28864–28868.

Keiler, K.C. & Sauer, R.T. (1996) Sequence determinants of C-terminal recognition by the Tsp protease. *J. Biol. Chem.* **271**, 2589–2593.

Keiler, K.C., Silber, K.S., Downard, K.M., Papayannopoulos, I.A., Biemann, K. & Sauer, R.T. (1995) C-terminal specific protein degradation: activity and substrate specificity of the Tsp protease. *Protein Sci.* **4**, 1507–1515.

Keiler, K.C., Waller, P.R.H. & Sauer, R.T. (1996) Role of a peptide tagging system in degradation of protein sythesized from damaged messenger RNA. *Science* **271**, 990–993.

Nagasawa, H., Sakagami, Y., Suzuki, A., Suzuki, H., Hara, H. & Hirota, Y. (1989) Determination of the cleavage site involved in C-terminal processing of penicillin-binding protein 3 of *Escherichia coli*. *J. Bacteriol.* **171**, 5890–5893.

Park, J.H., Lee, Y.S., Chung, C.H. & Goldberg, A.L. (1988) Purification and characterization of protease Re, a cytoplasmic endoprotease in *Escherichia coli*. *J. Bacteriol.* **170**, 921–926.

Roseman, J.E. & Levine, R.L. (1987) Purification of a protease from *Escherichia coli* with specificity for oxidized glutamine synthetase. *J. Biol. Chem.* **262**, 2101–2110.

Seoane, A., Sabbaj, A., McMurray, L.M. & Levy, S.B. (1992) Multiple antibiotic susceptibility associated with inactivation of the *prc* gene. *J. Bacteriol.* **174**, 7844–7847.

Silber, K.S., Keiler, K.C. & Sauer, R.T. (1992) Tsp: a tail-specific protease that selectively degrades proteins with nonpolar C termini. *Proc. Natl Acad. Sci. USA* **89**, 295–299.

*Kenneth C. Keiler*
*Department of Biology, 68-571,*
*Massachusetts Institute of Technology,*
*Cambridge, MA 02139, USA*
*Email: ken@rosa.mit.edu*

*Robert T. Sauer*
*Department of Biology, 68-571,*
*Massachusetts Institute of Technology,*
*Cambridge, MA 02139, USA*
*Email: bobsauer@mit.edu*

# 158. C-Terminal processing peptidase (Cyanobacteria)

## Databanks

*Peptidase classification: clan SF, family S41, MEROPS ID: S41.002*
*NC-IUBMB enzyme classification: none*
*Databank codes:*

| Species | SwissProt | PIR | EMBL (cDNA) | EMBL (genomic) |
|---|---|---|---|---|
| *Synechococcus* sp. | P42784 | S18125 | X63049 | – |
| *Synechocystis* sp. | – | – | X96490 | D90900: genome section 2 of 27 |
| *Synechocystis* sp. | – | A53964 | L25250 | D64000: genome section 19 of 27 |

## Name and History

Photosystem II (PSII), a multisubunit pigment–protein complex, is localized in the photosynthetic membranes of cyanobacteria, and chloroplasts of plants and algae (Pakrasi, 1995). Under physiological conditions, D1, one of the protein components of PSII, is turned over rapidly. Usually, D1 is synthesized in its precursor form (pD1), with a C-terminal extension. After the integration of pD1 in the membrane, this extension is cleaved off before the formation of an active Mn-cluster in PSII (Bowyer *et al.*, 1992; Nixon *et al.*, 1992). Analysis of a spontaneous PSII-deficient mutant strain of the unicellular cyanobacterium *Synechocystis* 6803 revealed that its D1 protein remains in the unprocessed pD1 form (Anbudurai *et al.*, 1994). Complementation analysis of this strain identified the gene *ctpA* (**C**-**t**erminal **p**rocessing), a mutation which eliminates the pD1 processing activity in this cyanobacterium (Shestakov *et al.*, 1994). Using this cyanobacterial gene as a hybridization probe, homologous *ctpA* cDNAs have been isolated from spinach and barley (Oelmüller *et al.*, 1996). The CtpA enzyme has been purified from spinach chloroplasts (Fujita *et al.*, 1995), and based on its N-terminal sequence, a *ctpA* cDNA clone has also been isolated from spinach (Chapter 159) (Inagaki *et al.*, 1996).

## Structural Chemistry

The cyanobacterial CtpA protein has 427 residues, with a 31 residues long N-terminal signal peptide (Shestakov *et al.*, 1994). The mature CtpA protein has a molecular mass of 43 kDa, with a pI value of 4.7. The spinach CtpA protein has been shown to be localized in the lumen of the photosynthetic membrane in chloroplasts (Oelmüller *et al.*, 1996). The *ctpA* gene is present in the nuclear genomes of plants. The precursor form of the spinach CtpA protein has an unusually long (150 residues) N-terminal bipartite transit peptide that is involved in the translocation of the protein across the envelope and photosynthetic membranes of chloroplasts. The sequence of the CtpA protein exhibits a high degree of similarity with that of Prc (or Tsp), a C-terminal processing protease in *Escherichia coli* (Chapter 157). The sequence of the *ctpA* gene was also determined in another cyanobacterium, *Synechococcus* 7002 (Brand *et al.*, 1992; Inagaki *et al.*, 1996). Recent determination of the sequence of the complete genome of the cyanobacterium *Synechocystis* 6803 has revealed the presence of two additional *ctpA*-like genes, *ctpB* (Sidoruk *et al.*, 1996) and *ctpC*. The exact functions of these gene products are currently unknown, although they are likely to be C-terminal peptidases.

## Biological Aspects

In oxygenic photosynthesis, the CtpA enzyme plays a crucial role during the biogenesis of the PSII complex. The four-Mn cluster in PSII is essential for the catalysis of light-induced oxidation of water and consequent evolution of molecular oxygen. It is thought that the C-terminus of the mature D1 protein in PSII acts as one of the ligands for the Mn-cluster (Nixon *et al.*, 1992). Thus, the processing of the C-terminal extension of pD1 is necessary for the biogenesis of an active PSII complex. However, a seeming paradox is the finding that an engineered cyanobacterial mutant strain in which the D1 protein is synthesized only in its mature form, can form normal PSII complexes (Nixon *et al.*, 1992). While it is possible that the C-terminal processing event of pD1 might offer some functional advantage to the photosynthetic organisms, the mechanisms of such a regulatory function of the CtpA enzyme remain to be determined.

## References

Anbudurai, P.R., Mor, T.S., Ohad, I., Shestakov, S.V. & Pakrasi, H.B. (1994) The *ctpA* gene encodes the carboxy-terminal processing protease for the D1 protein of photosystem II reaction center complex. *Proc. Natl Acad. Sci. USA* **91**, 8082–8086.

Bowyer, J.R., Packer, J.C.L., McCormack, B.A., Whitelegge, J.P., Robinson, C. & Taylor, M.A. (1992) Carboxyl-terminal processing of the D1 protein and photoactivation of water-splitting in photosystem II. *J. Biol. Chem.* **267**, 5424–5433.

Brand, S.N., Tan, X. & Widger, W.R. (1992) Cloning and sequencing of the *petBD* operon from the cyanobacterium *Synechococcus* sp. PCC 7002. *Plant Mol. Biol.* **20**, 481–491.

Fujita, S., Inagaki, N., Yamamoto, Y., Taguchi, F., Matsumoto, A. & Satoh, K. (1995) Identification of the carboxyl-terminal processing protease for the D1 precursor protein of the photosystem II reaction center of spinach. *Plant Cell Physiol.* **36**, 1169–1177.

Inagaki, N., Yamamoto, Y., Mori, H. & Satoh, K. (1996) Carboxyl-terminal processing protease for the D1 precursor protein: cloning and sequencing of the spinach cDNA. *Plant Mol. Biol.* **30**, 39–50.

Nixon, P.J., Trost, J.T. & Diner, B.A. (1992) Role of the carboxy terminus of polypeptide D1 in the assembly of a functional water-oxidizing manganese cluster in photosystem II of the cyanobacterium *Synechocystis* sp. PCC 6803: assembly requires a free carboxyl group at C-terminal position 344. *Biochemistry* **31**: 10859–10871.

Oelmüller, R., Herrmann, R.G. & Pakrasi, H.B. (1996) Molecular studies of CtpA, the carboxyl-terminal processing protease for the D1 protein of the photosystem II reaction center in higher plants. *J. Biol. Chem.* **271**, 21848–21852.

Pakrasi, H.B. (1995) Genetic analysis of the form and function of photosystem I and photosystem II. *Annu. Rev. Genet.* **29**, 755–776.

Shestakov, S.V., Anbudurai, P.R., Stanbekova, G.E., Gadzhiev, A., Lind, L.K. & Pakrasi, H.B. (1994) Molecular cloning and characterization of the *ctpA* gene encoding a carboxyl-terminal processing protease: analysis of a spontaneous photosystem II-deficient mutant strain of the cyanobacterium, *Synechocystis* sp. PCC 6803. *J. Biol. Chem.* **269**, 19354–19359.

Sidoruk, K.V., Shestopalov, V.I., Babykin, M.M. & Shestakov, S.V. (1996) Nucleotide sequence of a new member of the carboxyl-terminal processing protease family in the cyanobacterium *Synechocystis* sp. PCC 6803. *Plant Physiol.* **111**, 947.

*Himadri B. Pakrasi*
*Department of Biology,*
*Washington University,*
*St. Louis, MO 63130, USA*
*Email: Pakrasi@biodec.wustl.edu*

# 159. *C-Terminal processing peptidase (spinach)*

## Databanks

*Peptidase classification: clan SF, family S41, MEROPS ID: S41.003*
*NC-IUBMB enzyme classification: none*
*Databank codes:*

| Species | SwissProt | PIR | EMBL (cDNA) | EMBL (genomic) |
|---|---|---|---|---|
| *Hordeum vulgare* | – | – | X90929 | – |
| *Spinacia oleracea* | – | JH0263 | D50585 | – |
| *Spinacia oleracea* | – | – | X90558 | – |

## Name and History

The D1 protein, a subunit of photosystem II (PSII) reaction center, was shown to be synthesized as a precursor that is processed by cleavage at the C-terminus during assembly of the active PSII complex (Marder *et al.*, 1984). The cleavage site was identified as -Ala344⊦-, in spinach (Takahashi *et al.*, 1988). The protease involved in the processing was solubilized from *Pisum sativum* (Taylor *et al.*, 1988; Bowyer *et al.*, 1992) and *Spinacia oleracea* (Inagaki *et al.*, 1989), either by sonication or by Triton X-100 treatment of thylakoid membranes of chloroplasts, purified nearly in homogeneity (Fujita *et al.*, 1995), and the cDNA was sequenced (Inagaki *et al.*, 1996). A complementation analysis of a spontaneous photosynthesis-deficient mutant strain of a cyanobacterium, *Synechocystis* sp. PCC 6803, identified a gene (*ctpA*) involved in this process (Chapter 158) (Anbudurai *et al.*, 1994).

## Activity and Specificity

Three sorts of compounds can be used as substrate for the solubilized enzyme: (a) synthetic oligopeptides corresponding to the C-terminal sequence of the precursor D1 protein

($K_m$ 0.3 mM), (b) *in vitro* transcribed/translated precursor D1 protein (34 kDa), and (c) *in vivo* synthesized precursor D1 protein in thylakoid membranes (Fujita *et al*., 1995; Taguchi *et al*., 1995). The linkage Ala344┼Ala (higher plants) or the homologous Ala┼Ser (*Chlamydomonas reinhardtii* and some cyanobacteria) in the C-terminal region of protein is specifically cleaved by the enzyme. Among substitutions around the cleavage site on substrate, the replacement of Asp342 by Asn or Leu343 by Ala destroyed the ability of the oligopeptide to serve as a substrate, in *in vitro* experiments (Taguchi *et al*., 1995). A series of substitutions at Ala345 had marked effects on the value of $V$, without affecting the binding affinity as represented by $K_m$; the order of substitutions at the residue 345 in terms of their effects on $V$ was Ala, Ser, Phe, Cys > Gly > Val >> Pro = 0 (Taguchi *et al*., 1995). No specific inhibitor, except for substituted C-terminal oligopeptides of precursor D1 protein, could be found (Bowyer *et al*., 1992; Taguchi *et al*., 1995). The pH optimum is about 7.7.

## Structural Chemistry

The spinach enzyme consists of 389 amino acid residues and exhibits homology with two groups of proteins: (a) the deduced sequence of a protein proposed to be the C-terminal processing peptidase for precursor D1 protein in *Synechocystis* sp. PCC 6803, and (b) proteases for C-terminal cleavage identified in *E. coli* and *Bartonella bacilliformis* (Anbudurai *et al*., 1994; Inagaki *et al*., 1996) (Chapter 157). The pI value is 5.6. The enzyme is active in the monomeric state and is located in the luminal space of thylakoids in a soluble state or is loosely associated with the luminal surface (Bowyer *et al*., 1992; Fujita *et al*., 1995). The nucleotide sequence analysis of the cDNA demonstrated that spinach protein has an N-terminal extension of 150 amino acids (Inagaki *et al*., 1996). This part of the precursor protein could be divided into two domains, a chloroplast transit peptide and a signal for thylakoid translocation (Inagaki *et al*., 1996).

## Preparation

The spinach enzyme was purified from sonicated extracts of thylakoids, by a method that includes anion-exchange chromatography, hydroxyapatite, copper-chelating affinity and gel-filtration columns (Fujita *et al*., 1995). The enzyme can be expressed in *E. coli* in its active form and purified to homogeneity. The enzyme has been partially purified from *Scenedesmus* and *Pisum* (Bowyer *et al*., 1992).

## Biological Aspects

The D1 protein constitutes the core of the PSII reaction center (Nanba & Satoh, 1987), and the C-terminal processing of the precursor D1 protein is necessary for the assembly of the water-oxidizing machinery (Mn-cluster) (Nixon *et al*.,

1992). The C-terminal portion of the mature protein is highly conserved among species, whereas the C-terminal extension, which is cleaved off by the peptidase, is variable both in terms of chain length (8–16 amino acids) and in sequence. In *Euglena gracilis*, the C-terminal extension is absent from the gene (*psbA*).

## References

Anbudurai, P.R., Mor, T.S., Ohad, I., Shestakov, S.V. & Pakrasi, H.B. (1994) The *ctpA* gene encodes the C-terminal processing protease for the D1 protein of the photosystem II reaction center complex. *Proc. Natl Acad. Sci. USA* **91**, 8082–8086.

Bowyer, J.R., Packer, J.C.L., McCormack, B.A., Whitelegge, J.P., Robinson, C. & Taylor, M.A. (1992) Carboxyl-terminal processing of the D1 protein and photoactivation of water-splitting in photosystem II: partial purification and characterization of the processing enzyme from *Scenedesmus obliquus* and *Pisum sativum*. *J. Biol. Chem.* **267**, 5424–5433.

Fujita, S., Inagaki, N., Yamamoto, Y., Taguchi, F., Matsumoto, A. & Satoh, K. (1995) Identification of the carboxyl-terminal processing protease for the D1 precursor protein of the photosystem II reaction center of spinach. *Plant Cell Physiol.* **36**, 1169–1177.

Inagaki, N., Fujita, S. & Satoh, K. (1989) Solubilization and partial purification of a thylakoidal enzyme of spinach involved in the processing of D1 protein. *FEBS Lett.* **246**, 218–222.

Inagaki, N., Yamamoto, Y., Mori, H. & Satoh, K. (1996) Carboxyl-terminal processing protease for the D1 precursor protein: cloning and sequencing of the spinach cDNA. *Plant. Mol. Biol.* **30**, 39–50.

Marder, J.B., Goloubinoff, P. & Edelman, M. (1984) Molecular architecture of the rapidly metabolized 32-kilodalton protein of photosystem II: indications for COOH-terminal processing of a chloroplast membrane polypeptide. *J. Biol. Chem.* **259**, 3900–3908.

Nanba, O. & Satoh, K. (1987) Isolation of a photosystem II reaction center consisting of D-1 and D-2 polypeptides and cytochrome b-559. *Proc. Natl Acad. Sci. USA* **84**, 109–112.

Nixon, P.J., Trost, J.T. & Diner, B.A. (1992) Role of the carboxy terminus of polypeptide D1 in the assembly of a functional water-oxidizing manganese cluster in photosystem II of the cyanobacterium *Synechocystis* sp. PCC 6803: assembly requires a free carboxyl group at C-terminal position 344. *Biochemistry* **31**, 10859–10871.

Taguchi, F., Yamamoto, Y. & Satoh, K. (1995) Recognition of the structure around the site of cleavage by the carboxyl-terminal processing protease for D1 precursor protein of the photosystem II reaction center. *J. Biol. Chem.* **270**, 10711–10716.

Takahashi, M., Shiraishi, T. & Asada, K. (1988) COOH-terminal residues of D1 and 44-kDa CPa-2 at spinach photosystem II core complex. *FEBS Lett.* **240**, 6–8.

Taylor, M.A., Packer, J.C.L. & Bowyer, J.R. (1988) Processing of the D1 polypeptide of the photosystem II reaction centre and photoactivation of a low fluorescence mutant (LF-1) of *Scenedesmus obliquus*. *FEBS Lett.* **237**, 229–233.

*Kimiyuki Satoh*
*Department of Biology,*
*Okayama University,*
*Okayama 700, Japan*
*Email: kimiyuki@cc.okayama-u.ac.jp*

# 160. Tricorn protease

## Databanks

*Peptidase classification: clan SF, family S44, MEROPS ID: S44.001*
*NC-IUBMB enzyme classification: none*
*Databank codes:*

| Species | SwissProt | PIR | EMBL (cDNA) | EMBL (genomic) |
|---|---|---|---|---|
| *Sulfolobus solfataricus* | – | – | – | Y08256: 100 kbp DNA fragment |
| *Thermoplasma acidophilum* | – | – | U72850 | – |

## Name and History

Tricorn protease was first isolated from the archaeon *Thermoplasma acidophilum*, in which it appears to form the core of a modular proteolytic system (Tamura *et al*., 1996a). It was named **tricorn protease** after the French three-cornered hat, because of the peculiar triangular shape of the molecule.

## Activity and Specificity

Tricorn protease substrate specificity was characterized by means of synthetic fluorogenic peptides: the enzyme preferentially hydrolyzes trypsin substrates (e.g. Bz-Val-Gly-Arg+NHMec, Z-Ala-Arg-Arg+NHMec), but also some chymotrypsin substrates (e.g. Ala-Ala-Phe+NHMec). The peptidase activity of the tricorn protease is efficiently inhibited by Tos-Phe-CH$_2$Cl (IC$_{50}$ = 40 µM) and Tos-Lys-CH$_2$Cl. The temperature optimum ($\sim 65°$C) is slightly above the growth temperature of *T. acidophilum*, and the pH optimum is 8.5–8.8 (Tamura *et al*., 1996a).

## Structural Chemistry

Tricorn protease exists as a 720 kDa complex that is composed of six identical copies of a 120 kDa polypeptide. The primary structure of the 120 kDa protein does not show similarity to other known proteins, except for its C-terminal region (180 residues) where a weak but significant similarity to several C-terminal processing proteases of bacterial and eukaryotic origin is observed (Silber *et al*., 1992; Oelmüller *et al*., 1996) (Chapter 157). Of the catalytic dyad Ser and Lys residues that have been identified in the C-terminal processing protease of *Escherichia coli* (Keiler & Sauer, 1995), the tricorn core protease contains only the Ser, however.

Electron microscopy in conjunction with image analysis has shown that the tricorn protease complex is approximately 18 nm in diameter and approximately 10 nm in the direction of its 3-fold axis. The complex encloses a central cavity which is about 8 nm in diameter. The tricorn protease can further assemble into an icosahedral capsid structure that comprises 20 copies of the 720 kDa complex and is approximately 54 nm in diameter (Tamura *et al*., 1996a).

## Preparation

The tricorn protease has been isolated and purified to homogeneity (2300-fold) from *Thermoplasma acidophilum* cell extract. In molecular sieve chromatography, concentrated tricorn protease fractionated into peaks containing capsids and 720 kDa complex, respectively. The recombinant enzyme has also been expressed in *E. coli* as a soluble protein (Tamura *et al*., 1996a).

## Biological Aspects

Upon mixing tricorn protease with low molecular weight proteins referred to as factor 1 (F1) or factor 2 (F2), several different peptidase activities were generated or enhanced, including activity on Suc-Leu-Leu-Val-Tyr+NHMec and Boc-Leu-Arg-Arg+NHMec. The factors were purified from *T. acidophilum* cell extracts and characterized (Tamura *et al*., 1996a). F1 was found to be a 33 kDa protein with significant similarity to eubacterial prolyl aminopeptidases (Tamura *et al*.; 1996b) (Chapter 139). Experiments with inactive mutant recombinant prolyl aminopeptidases indicate that the additional activities of the complex between the tricorn core protease and prolyl aminopeptidase are contributed by prolyl aminopeptidase. Preliminary experiments with partially purified F2 suggests that it may also be an aminopeptidase. Although F1 and F2 both seem to be aminopeptidases, the new activities of their complexes with the tricorn protease include endopeptidase as well as aminopeptidase activities. These results suggest that tricorn protease acts as the core of a novel proteolytic system, and that upon interacting with several smaller proteins it displays multicatalytic activities.

## References

Keiler, K.C. & Sauer, R.T. (1995) Identification of active site residues of the Tsp protease. *J. Biol. Chem.* **270**, 28864–28868.

Oelmüller, R., Herrmann, R.G. & Pakrasi, H.B. (1996) Molecular studies of CtpA, the carboxyl-terminal processing protease for the D1 protein of the photosystem II reaction center in higher plants. *J. Biol. Chem.* **271**, 21848–21852.

Silber, K.R., Keiler, K.C. & Sauer, R.T. (1992) Tsp: a tail-specific protease that selectively degrades proteins with nonpolar C termini. *Proc. Natl Acad. Sci. USA* **89**, 295–299.

Tamura, T., Tamura, N., Cejka, Z., Hegerl, R., Lottspeich, F. & Baumeister, W. (1996a) Tricorn protease – the core of a modular proteolytic system. *Science* **274**, 1385–1389.

Tamura, T., Tamura, N., Lottspeich, F. & Baumeister, W. (1996b) Tricorn protease (TRI) interacting factor 1 from *Thermoplasma acidophilum* is a proline iminopeptidase. *FEBS Lett.* **398**, 101–105.

**Tomohiro Tamura**
*Max-Planck-Institute for Biochemistry,*
*D-82152 Martinsried bei München, Germany*
*Email: tamura@vms.biochem.mpg.de*

**Wolfgang Baumeister**
*Max-Planck-Institute for Biochemistry,*
*D-82152 Martinsried bei München, Germany*
*Email: tamura@vms.biochem.mpg.de*

# 161. Introduction: clan SH containing herpesvirus assemblins

## Databanks

*MEROPS ID: SH*

### Family S21

Assemblin (Chapter 163)

Assemblin homolog, cytomegalovirus (Chapter 162)

Assemblin homolog, Epstein–Barr virus (Chapter 165)

Assemblin, herpesvirus 6 (Chapter 164)

Assemblin, varicella-zoster (Chapter 166)

Clan SH includes only one family, S21, the members of which are shown in the Databanks table. The family contains the assemblins of herpesviruses. Assemblin is involved in the late stages of virion assembly, and breaks down the scaffold protein upon which the viral prohead is assembled. (A similar process occurs in the assembly of bacteriophage proheads, see Chapter 563.) The catalytic residues of the human cytomegalovirus assemblin are His63, Ser132 and His157, which is a combination not seen in serine peptidases of any other family. The alignment is shown in Alignment 161.1 (see CD-ROM). The tertiary structures of assemblins from human cytomegalovirus (Chen *et al.*, 1996; Qiu *et al.*, 1996; Shieh *et al.*, 1996; Tong *et al.*, 1996) and the varicella-zoster virus (Qiu *et al.*, 1997) have been determined, and the structures show no resemblance to structures of any other peptidase. The tertiary structure of human cytomegalovirus assemblin is shown in Figure 161.1. Unlike trypsin, assemblin does not possess a bilobed structure, but is composed of a single domain and is active as a dimer.

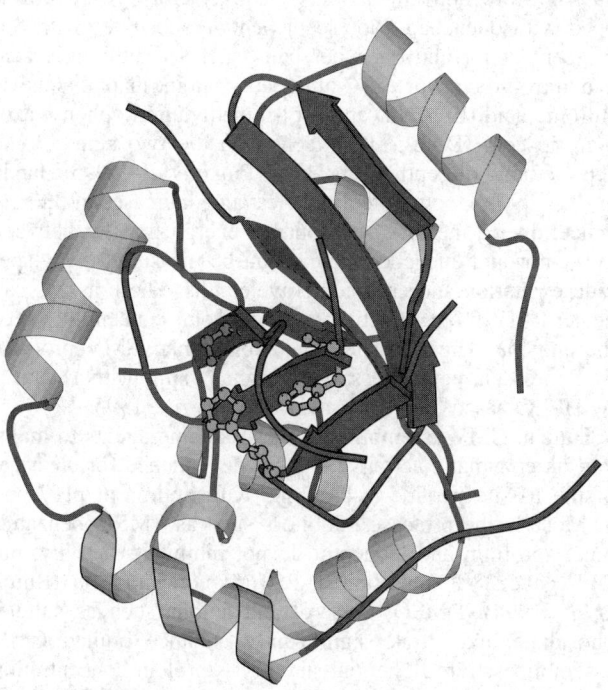

*Figure 161.1*   Richardson diagram of human cytomegalovirus assemblin. The image was prepared from the Brookhaven Protein Data Bank entry (1LAY) as described in the Introduction (p. xxv). Catalytic residues are shown in ball-and-stick representation: His63, Ser132 and His159 (numbering as in Alignment 161.1). Ser132 is visible in front of a $\beta$ strand close to the center of the image.

## References

Chen, P., Tsuge, H., Almassy, R.J., Gribskov, C.L., Katoh, S., Vanderpool, D.L., Margosiak, S.A., Pinko, C., Matthews, D.A. & Kan, C.C. (1996) Structure of the human cytomegalovirus protease catalytic domain reveals a novel serine protease fold and catalytic triad. *Cell* **86**, 835–843.

Qiu, X.Y., Culp, J.S., DiLella, A.G., Hellmig, B., Hoog, S.S., Janson, C.A., Smith, W.W. & Abdel-Meguid, S.S. (1996) Unique fold and active site in cytomegalovirus protease. *Nature* **383**, 275–279.

Qiu, X.Y., Janson, C.A., Culp, J.S., Richardson, B., Debouck, C., Smith, W.W. & Abdel-Meguid, S.S. (1997) Crystal structure of varicella-zoster virus protease. *Proc. Natl Acad. Sci. USA* **94**, 2874–2879.

Shieh, H.S., Kurumbail, R.G., Stevens, A.M., Stegeman, R.A., Sturman, E.J., Pak, J.Y., Wittwer, A.J., Palmier, M.O., Wiegand, R.C., Holwerda, B.C. & Stallings, W.C. (1996) Three dimensional structure of human cytomegalovirus protease. *Nature* **383**, 279–282.

Tong, L., Qian, C.G., Massariol, M.J., Bonneau, P.R., Cordingley, M.G. & Lagace, L. (1996) A new serine-protease fold revealed by the crystal structure of human cytomegalovirus protease. *Nature* **383**, 272–275.

# 162. *Cytomegalovirus assemblin*

## Databanks

*Peptidase classification: clan SH, family S21, MEROPS ID: S21.002*
*NC-IUBMB enzyme classification: none*
*Databank codes:*

| Species | SwissProt | PIR | EMBL (cDNA) | EMBL (genomic) |
|---|---|---|---|---|
| Human cytomegalovirus | P16753 | S09843 | X17403 | – |
| Simian cytomegalovirus | P16046 | A40414 | M64627 | – |

Brookhaven Protein Data Bank three-dimensional structures:

| Species | ID | Resolution | Notes |
|---|---|---|---|
| Human cytomegalovirus | 1LAY | 2.3 | Ala143Val mutant |

## Name and History

Cytomegalovirus (CMV) a betaherpesvirus, together with herpes simplex virus 1 and 2 (HSV-1, HSV-2) and Epstein–Barr virus (EBV) from the alpha and gamma herpes subfamilies, forms the Herpesviridae family. All herpesviruses encode a serine protease. It cleaves the assembly protein precursor which is a major component of the intermediate capsid and forms the scaffolding structure for capsid assembly. For human CMV, the 74 kDa full-length protease or assemblin precursor is encoded by the $U_L80$ open reading frame and is enzymatically active. The CMV assembly protein precursor is transcribed and translated independently from a gene nested inside, and 3'-coterminal with $U_L80$ (Welch *et al.*, 1991). Thus, the full-length human CMV protease and its protein substrate share an identical sequence of 373 amino acids.

Full-length protease cleaves the substrate and itself at the C-terminal maturation (M-) site at Ala643+Ser644. Subsequently, CMV protease-catalyzed autoproteolysis occurs at the release (R-) site (Ala256+Ser257) and releases the 28 kDa catalytic domain, the *cytomegalovirus assemblin*. This comprises amino acid residues 1–256 and retains full proteolytic activity. Only CMV proteases, unlike proteases from the alpha and gamma herpes subgroups, have additional internal (I-) cleavage(s) site within their catalytic domains. For human CMV protease, the two I-sites are at Ala143+Ala144 and Ala209+Ser210 (Jones *et al.*, 1994). Autoproteolytic cleavage at position 143 produces an active, two-chain catalytic domain which becomes dissociated only in high concentrations of urea (Holwerda *et al.*, 1994).

## Activity and Specificity

Human CMV protease cleaves at the M-, R- and I-sites. They share the consensus sequence Val-Xaa-Ala+Ala/Ser where P3 is usually Val, P1' is usually Ser and Xaa at P2 can be Asn, Asp, Gln, Lys or Ala (Gibson *et al.*, 1994). For efficient cleavage, peptide substrates require minimally amino acid residues spanning the P4–P4' of cleavage sites with the specificity of Ala for P1 and Ser for P1' (Sardana *et al.*, 1994). HPLC-based peptide assays were established with peptide substrates mimicking the M- or R-sites (Burck *et al.*, 1994). More convenient and sensitive assays were developed with quenched fluorescent peptide substrates with fluorescent donor and quencher pairs such as anthranilic acid and nitrotyrosine or 5-[(2-aminoethyl)amino]naphthalene-1-sulfonic acid (EDANS) and 4-(4-dimethylamino phenyl azo) benzoic acid (DABCYL) attached on the two sides of the scissile bond of peptide substrates (Pinko *et al.*, 1995; Handa *et al.*, 1995). Upon proteolytic cleavage, increases in fluorescence, due to the loss of resonance energy transfer between the donor and quencher groups, can be measured using specific excitation and emission wavelengths. These assays can detect CMV protease activity at nanomolar concentrations of the enzyme. The $k_{cat}$ determined for human CMV protease by fluorescent peptide assay is $22 \, min^{-1}$, similar to $18 \, min^{-1}$ by HPLC-based peptide assays (Pinko *et al.*, 1995).

Human CMV assemblin requires reducing agents to maintain its enzymatic activity. The pH dependence for cleavage of the M-site peptide is biphasic, with optima at pH 7 and 9. Most serine protease inhibitors such as PMSF, benzamidine, aprotinin and leupeptin do not inhibit its activity, but DFP gave 28% inhibition at $250 \, \mu M$ concentration (Burck *et al.*, 1994). DMSO, glycerol and anions such as sulfate, phosphate and citrate significantly enhance human CMV assemblin activity. Thermal stability is protein concentration dependent, and temperature optima are 27 and 33°C at 1 and $26.5 \, \mu M$ concentration, respectively (Margosiak *et al.*, 1996).

## Structural Chemistry

Human CMV assemblin forms an obligate, noncovalently-linked dimer with a $K_d$ of ~5 μM or 1.9 nM, in the absence or presence of 25% glycerol, respectively (Darke *et al.*, 1996; Margosiak *et al.*, 1996). It has an estimated pI value of 5.49. Each monomer contains five reduced cysteines, residues 84, 87, 138, 161 and 202, and no disulfide bonds. Cys84 is conserved among CMV proteases, and Cys161 is conserved among all known herpesvirus proteases. Alkylation or oxidation of cysteines leads to inactivation of the enzyme (Burck *et al*., 1994; Baum *et al*., 1996).

The X-ray structure of human CMV assemblin reveals that it is an $\alpha/\beta$ protein consisting of a central core comprising two $\beta$ sheets surrounded by eight $\alpha$ helices (Chen *et al*., 1996). The structure defines a new class of serine protease with respect to global fold topology. Two $\beta$ sheets are formed by amino acids within conserved sequence motifs.

Human CMV protease has a catalytic triad made of His63, Ser132 and His157. The substitution of His157 for aspartic acid found in His-Ser-Asp triads in most other serine proteases probably explains why human CMV assemblin has a $k_{cat}$ that is more than 100-fold lower than the $k_{cat}$ for trypsin and subtilisin-like proteases. The structure also shows that the absolutely conserved Cys161 is near the active site, so that alkylation or oxidation of Cys161 could either cause conformational changes around the active site or block substrate access to the active site. Formation of the active dimer is mediated primarily by burying an $\alpha$ helix of one protomer into a deep cleft in the protein surface of the other. Two active sites in the catalytic dimer are far apart and may function independently. For small peptide substrates, binding determinants appear to reside within a single subunit.

## Preparation

When the wild-type human CMV assemblin is expressed in *Escherichia coli*, formation of inclusion bodies and autoproteolysis at the I-sites during protein refold and purification make large-scale production difficult (Smith *et al*., 1994; Burck *et al*., 1994; Holwerda *et al*., 1994). However, using the T7 promoter and *E. coli* strain BL21(DE3) as expression host, I-site mutants such as Ala143Gln yielded active and soluble single-chain assemblin. Hundreds of milligrams of human CMV assemblin were purified from *E. coli* by hydrophobic interaction and ionic-exchange chromatography (Pinko *et al*., 1995).

## Biological Aspects

Proteolytic cleavage of assembly protein precursor is essential for dissolution of the capsid scaffold and packaging of the viral genome during capsid maturation (Liu & Roizman, 1991). Therefore, inhibitors of human CMV assemblin may potentially become antiviral agents for treating retinitis in patients with AIDS and CMV infection in transplant and immunocompromised patients.

Herpesvirus protease is expressed late in the viral replication cycle. It has been speculated that the assemblin precursor remains inactive in the cytoplasm, gets translocated into the nucleus via binding to the major capsid protein (MCP), and only becomes activated at the site of capsid assembly and maturation. For *in vivo* regulation of the CMV protease activity, multiple mechanisms have been proposed. Thus, it has been suggested (a) that

cytoplasmic enzyme exists mostly as monomer, and only becomes active in the nucleus or formed capsid scaffold where oligomerization of assemblin is favored (Margosiak *et al*., 1996), (b) that CMV protease is redox-regulated, and inactivation of cytoplasmic enzyme occurs via disulfide bond formation (Baum *et al*., 1996), and (c) that CMV protease activity can be enhanced by high local phosphate concentration in the nucleus during the late viral replication cycle (Darke *et al*., 1996). A recent study indicates that during HSV-1 viral replication, proteolytic processing of the assembly protein precursor occurs inside capsids. It also shows that the protease catalytic domain serves, in addition to its enzymatic role, another essential function for capsid maturation (Gao *et al*., 1996). To establish if this is also the case for CMV assemblin awaits further studies.

## Related Peptidases

Related peptidases are described in adjacent chapters of the present volume, and are introduced in Chapter 161.

## Further Reading

A review has been provided by Gibson *et al*. (1994).

## References

Baum, E.Z., Siegel, M.M., Bebernitz, G.A., Hulmes, J.D., Sridharan, L., Sun L., Tabei, K., Johnston, S.H., Wildey, M.J., Nygaard, J., Jones, T.R. & Gluzman, Y. (1996) Inhibition of human cytomegalovirus UL80 protease by specific intramolecular disulfide bond formation. *Biochemistry* **35**, 5838–5846.

Burck, P.J., Berg, D.H., Luk, T.P., Sassmannshausen, L.M., Wakulchik, M., Smith, D.P., Hsiung, H.M., Becker, G.W., Gibson, W. & Villarreal, E.C. (1994) Human cytomegalovirus maturational proteinase: expression in *Escherichia coli*, purification, and enzymatic characterization by using peptide substrate mimics of natural cleavage sites. *J. Virol.* **68**, 2937–2946.

Chen, P., Almassy, R.J., Gribskov, C.L., Katoh, S., Vanderpool, D.L., Margosiak, S.A., Pinko, C., Matthews, D.A. & Kan, C.-C. (1996) Structure of the human cytomegalovirus protease catalytic domain reveals a novel serine protease fold and catalytic triad. *Cell* **86**, 835–843.

Darke, P.L. Cole, J.L. Waxman, L., Hall, D.L. Sardana, M.K. & Kuo, L.C. (1996) Active human cytomegalovirus protease is a dimer. *J. Biol. Chem.* **271**, 7445–7449.

Gao, H.Q., Schiller, J.J. & Baker, S.C. (1996) Identification of the polymerase polyprotein products p72 and p65 of the murine coronavirus MHV-JHM. *Virus Res.* **45**, 101–109.

Gibson, W., Welch, A.R. & Hall, M.R.T. (1994) Assemblin, a herpes virus serine maturational proteinase and new molecular target for antiviral. *Perspect. Drug Discovery Design* **2**, 413–426.

Handa, B.K., Keech, E., Conway, E.A., Broadhurst, A. & Ritchie, A. (1995) Design and synthesis of a quenched fluorogenic peptide substrate for human cytomegalovirus proteinase. *Antiviral Chem. Chemother.* **6**, 255–261.

Holwerda, B.C., Wittwer, A.J., Duffin, K.L., Smith, C., Toth, M.V., Carr, L.S., Wiegand, R.C. & Bryant, M.L. (1994) Activity of two-chain recombinant human cytomegalovirus protease. *J. Biol. Chem.* **269**, 25911–25915.

Jones, T.R., Sun, L., Bebernitz, G.A., Muzithras, V.P., Kim, H.J., Johnston, S.H. & Baum, E.Z. (1994) Proteolytic activity of human cytomegalovirus UL80 protease cleavage site mutants. *J. Virol.* **68**, 3742–3752.

Liu, F. & Roizman, B. (1991) The herpes simplex virus 1 gene encoding a protease also contains within its coding domain the gene encoding the more abundant substrate. *J. Virol.* **65**, 5149–5156.

Margosiak, S.A., Vanderpool, D.L., Sisson, W., Pinko, C. & Kan, C.-C. (1996) Dimerization of the human cytomegalovirus protease: kinetic and biochemical characterization of the catalytic homodimer. *Biochemistry* **35**, 5300–5307.

Pinko, C., Margosiak, S.A., Vanderpool, D., Gutowski, J.C., Condon, B. & Kan, C.-C. (1995) Single-chain recombinant human cytomegalovirus protease. *J. Biol. Chem.* **270**, 23634–23640.

Sardana V.V., Wolfgang, C.A., Veloski, C.A., Long, W.J., LeGrow, K., Wolanski, B., Emini, E.A. & LaFemina R.L. (1994) Peptide substrate cleavage specificity of the human cytomegalovirus protease. *J. Biol. Chem.* **269**, 14337–14340.

Smith, M.C., Giordano, J., Cook, J.A., Wakulchik, M., Villarreal, E.C., Becker, G.W., Bemis, K., Labus, J. & Manetta, J.S. (1994) Purification and kinetic characterization of human cytomegalovirus assemblin. *Methods Enzymol.* **244**, 412–423.

Welch, A.R., McNally, L.M. & Gibson, W. (1991) Cytomegalovirus assembly protein nested gene family: four 3′p-coterminal transcripts encode four in-frame, overlapping proteins. *J. Virol.* **65**, 4091–4100.

*Chen-Chen Kan*
*Department of Molecular Biology and Biochemistry,*
*Agouron Pharmaceuticals, Inc.,*
*3565 General Atomics Court,*
*San Diego, CA 92121, USA*
*Email: kan@agouron.com*

# 163. *Herpesvirus assemblin*

## Databanks

*Peptidase classification: clan SH, family S21, MEROPS ID: S21.001*
*NC-IUBMB enzyme classification: none*
*Databank codes:*

| Species | SwissProt | PIR | EMBL (cDNA) | EMBL (genomic) |
|---|---|---|---|---|
| Cattle herpesvirus 1 | P54817 | – | U31809 | – |
| Equine herpesvirus type 1 | P28936 | I36798 | M86664 | – |
| Equine herpesvirus type 2 | P52369 | – | U20824 | – |
| Herpes simplex virus | P10210 | H30084 | X14112 | D10879: long unique region of genome |
| Herpes simplex virus | P52351 | – | U43400 | – |
| Herpes simplex virus type 2 | – | – | L37443 | – |
| Pseudorabies virus | – | – | X95710 | – |
| Herpesvirus saimiri | Q01002 | H36807 | X64346 | – |

## Name and History

A temperature-sensitive herpes simplex virus type 1 *(HSV-1)* gene that governs the correct proteolytic processing of certain herpesvirus proteins and that is essential for viral replication was identified in 1983 (Preston *et al.*, 1983). More recently, the protease gene has been specifically mutated and recombined into viral genomes to prove the essential nature of the catalytic activity (Gao *et al.*, 1994). The $U_L26$ gene of HSV-1 encodes the 635 amino acid protease precursor protein in which the N-terminal 247 amino acids comprise the catalytic domain (Liu & Roizman, 1991; Weinheimer *et al.*, 1993). The 3′ end of this coding region also contains the separately transcribed gene, $U_L26.5$, which codes for the 329 amino acid substrate, known as ICP35 (infected cell protein 35) or VP22 (virion protein 22). Due to its role as a scaffold protein during nucleocapsid assembly, ICP35 is known as 'the assembly protein', and hence the name *herpesvirus assemblin* has been given to the protease that processes it. Other designations for the precursor protease and its processed products are: *Pra*, the 635 amino acid protease; *Prb*, the precursor after a C-terminal cleavage (residues 1–610); *VP24* or *N$_o$*, the mature protease (assemblin, 1–247); *VP21* or *N$_b$*, the third product of the precursor (248–610), and *N$_a$* (248–635). Recent literature on the biochemistry of the mature herpesvirus proteases often omits these designations entirely, using descriptive terms (e.g. 'full-length protease precursor') to avoid confusion.

## Activity and Specificity

The two natural cleavage sites for the HSV-1 protease have the sequences Thr-Tyr-Leu-Gln-Ala+Ser-Glu-Lys-Phe-Lys and Ala-Leu-Val-Asn-Ala+Ser-Ser-Ala-Ala-His (DiIanni *et al.*, 1993a). The first of these corresponds to the cleavage site at the C-terminus of the catalytic domain and the second to the site within ICP35 (also contained within the C-terminus of the precursor protease). Peptides representing these sites are substrates, and a peptide encompassing the P5–P8′ residues gives optimal cleavage rates (DiIanni *et al.*, 1993b). Mutagenesis of the substrate proteins has been used to examine the sequence specificity of cleavage within the natural context (McCann & O'Boyle, 1994). A common motif of assemblin cleavage sites throughout the herpes group of viruses is Val/Leu-Xaa-Ala+Ser, where Xaa is a polar residue (Welch *et al.*, 1991). Assay of activity can be performed by cleavage of a suitable peptide substrate (available from Bachem) (see Appendix 2 for full names and addresses of suppliers) and HPLC analysis of the products. The pH optimum is 8 (DiIanni *et al.*, 1993b). Activity is very sensitive to solvent composition. Antichaotropic agents such as glycerol and polyvalent anions (e.g. phosphate, citrate) can increase the apparent specific activity as much as 200-fold (Hall & Darke, 1995; Yamanaka *et al.*, 1995). The activation is almost certainly due to the promotion of a dimeric form of the enzyme by the antichaotropes. The only active form of the related cytomegalovirus (CMV) protease (Chapter 162) is a dimer, and its activation by antichaotropes is known to be due to an increase in the dimer–monomer ratio (Darke *et al.*, 1996; Margosiak *et al.*, 1996). The dependence of activity upon solvent conditions, protein concentration, and time after dilution should therefore be carefully evaluated when comparing specific activities.

DFP is an inhibitor and was used to identify Ser129 as the active-site serine (DiIanni *et al.*, 1994). Other serine protease inhibitors such as Tos-Phe-CH$_2$Cl, Tos-Lys-CH$_2$Cl and PMSF are effective only at millimolar concentrations (Liu & Roizman, 1992).

## Structural Chemistry

The 247 amino acid sequence of the 27 kDa protein is 31% identical to that of the CMV protease. The three-dimensional structure of the HSV assemblin has yet to be determined, but it is expected to be similar to that of the CMV enzyme, which contains two closely associated subunits. The fold of the monomer is unique among serine proteases, as is the Ser/His/His catalytic triad of the active site (Chen *et al.*, 1996; Qui *et al.*, 1996; Shieh *et al.*, 1996; Tong *et al.*, 1996).

## Preparation

HSV-1 assemblin has been expressed in *Escherichia coli* for convenient purification on a scale of tens of milligrams. The enzyme was originally prepared as a glutathione-*S*-transferase fusion protein which exhibited appropriate catalytic specificity (Weinheimer *et al.*, 1993). A subsequent procedure to produce the mature nonfusion enzyme with higher specific activity employs the N-terminal 306 amino acids of the precursor (Darke *et al.*, 1994; Hall & Darke, 1995). This construct contains the catalytic domain followed by the 59 naturally occurring residues of the precursor. The soluble precursor is first partially purified by cation-exchange chromatography, then induced to autoprocess by the addition of citrate, an activator. In rechromatography exactly as before, the contaminants elute at the same position as in the first run, and the 247 amino acid mature enzyme elutes at a new position, essentially pure (Darke *et al.*, 1994).

## Biological Aspects

The herpes simplex virus nucleocapsid is assembled in the nucleus of infected cells with at least six viral proteins. An intermediate stage of the process involves construction of a scaffold using the assembly protein substrate, which is subsequently cleaved by the protease before DNA packaging (Gibson & Roizman, 1974). The herpesvirus proteases have been targeted for inhibitor development as virostatic agents, but significant inhibitors have yet to be described. The human herpesvirus 6 proteinase (Chapter 164) and Epstein–Barr virus proteinase (Chapter 165) have also been expressed and characterized (Donaghy & Jupp, 1995; Tigue *et al.*, 1996).

## Further Reading

An authoritative overview that places proteolytic processing in the context of HSV replication is that of Roizman & Sears 1995).

## References

Chen, P., Tsuge, H., Almassy, R.J., Gribskov, C.L., Katoh, S., Vanderpooi, D.L., Margosiak, S.A., Pinko, C., Matthews, D.A. & Kan, C.-C. (1996) Structure of the human cytomegalovirus protease catalytic domain reveals a novel serine protease fold and catalytic triad. *Cell* **86**, 835–843.

Darke, P.L., Chen, E., Hall, D.L., Sardana, M.K., Veloski, C.A., LaFemina, R.L., Shafer, J.A. & Kuo, L.C. (1994) Purification of active herpes simplex virus-1 protease expressed in *Escherichia coli*. *J. Biol. Chem.* **269**, 18708–18711.

Darke, P.L., Cole, J.L., Waxman, L., Hall, D.L., Sardana, M.K. & Kuo, L.C. (1996) Active human cytomegalovirus protease is a dimer. *J. Biol. Chem.* **271**, 7445–7449.

DiIanni, C.L., Drier, D.A., Deckman, I.C., McCann, P.J., III, Liu, F., Roizman, B., Colonno, R.J. & Cordingley, M.G. (1993a) Identification of the herpes simplex virus-1 protease cleavage sites by direct sequence analysis of autoproteolytic cleavage products. *J. Biol. Chem.* **268**, 2048–2051.

DiIanni, C.L., Mapelli, C., Drier, D.A., Tsao, J., Natarajan, S., Riexinger, D., Festin, S.M., Bolgar, M., Yamanaka, G., Weinheimer, S.P., Meyers, C.A., Colonno, R.J. & Cordingley, M.G.

(1993b) In vitro activity of the herpes simplex virus type 1 protease with peptide substrates. *J. Biol. Chem.* **268**, 25449–25454.

DiIanni, C.L., Stevens, J.T., Bolgar, M., O'Boyle, D.R., II, Weinheimer, S.P. & Colonno, R.J. (1994) Identification of the serine residue at the active site of the herpes simplex virus type 1 protease. *J. Biol. Chem.* **269**, 12672–12676.

Donaghy, G. & Jupp, R. (1995) Characterization of the Epstein-Barr virus proteinase and comparison with the human cytomegalovirus proteinase. *J. Virol.* **69**, 1265–1270.

Gao, M., Matusick-Kumar, L., Hurlburt, W., DiTusa, S.F., Newcomb, W.W., Brown, J.C., McCann, P.J., III, Deckman, I.C. & Colonno, R.J. (1994) The protease of herpes simplex virus type 1 is essential for functional capsid formation and viral growth. *J. Virol.* **68**, 3702–3712.

Gibson, G. & Roizman, B. (1974) Proteins specified by herpes simplex virus. X. Staining and radiolabeling properties of B capsid and virion proteins in polyacrylamide gels. *J. Virol.* **13**, 155–165.

Hall, D.L. & Darke, P.L. (1995) Activation of the herpes simplex virus type 1 protease. *J. Biol. Chem.* **270**, 22697–22700.

Liu, F. & Roizman, B. (1991) The herpes simplex virus 1 gene encoding a protease also contains within its coding domain the gene encoding the more abundant substrate. *J. Virol.* **65**, 5149–5156.

Liu, F. & Roizman, B. (1992) Differentiation of multiple domains in the herpes simplex virus 1 protease encoded by the UL26 gene. *Proc. Natl Acad. Sci. USA* **89**, 2076–2080.

Margosiak, S.A., Vanderpool, D.L., Sisson, W., Pinko, C. & Kan, C.-C. (1996) Dimerization of the human cytomegalovirus protease: kinetic and biochemical characterization of the catalytic homodimer. *Biochemistry* **35**, 5300–5307.

McCann, P.J., III, & O'Boyle, D.R., II (1994) Investigation of the specificity of the herpes simplex virus type 1 protease by point mutagenesis of the autoproteolysis sites. *J. Virol.* **68**, 526–529.

Preston, V.G., Coates, J.A.V. & Rixon, F.J. (1983) Identification and characterization of a herpes simplex virus gene product required for encapsidation of virus DNA. *J. Virol.* **45**, 1056–1064.

Qui, X., Culp, J.S., DiLella, A.G., Hellmig, B., Hoog, S.S., Janson, C.A., Smith, W.W. & Abdel-Meguid, S.S. (1996) Unique fold and active site in cytomegalovirus protease. *Nature* **383**, 275–279.

Roizman, B. & Sears, A. (1995) Herpes simplex viruses and their replication. In: *Fields Virology*, 3rd edn (Fields, B.N., Knipe, D.M. & Howly, P.M., eds). Philadelphia: Lippincott-Raven, pp. 2231–2295.

Shieh, H.-S., Kurumbail, R.G., Stevens, A.M., Stegeman, R.A., Sturman, E.J., Pak, J.Y., Wittwer, A.J., Palmier, M.O., Wiegand, R.C., Holwerda, B.C. & Stallings, W.C. (1996) Three-dimensional structure of human cytomegalovirus protease. *Nature* **383**, 279–282.

Tigue, N.J., Matharu, P.J., Roberts, N.A., Mills, J.S., Kay, J. & Jupp, R. (1996) Cloning, expression, and characterization of the proteinase from human herpesvirus 6. *J. Virol.* **70**, 4136–4141.

Tong, L., Qian, C., Massariol, M.-J., Bonneau, P.R., Cordingley, M.G. & Lagace, L. (1996) A new serine-protease fold revealed by the crystal structure of human cytomegalovirus protease. *Nature* **383**, 272–275.

Weinheimer, S.P., McCann, P.J., III, O'Boyle, D.R., II, Stevens, J.T., Boyd, B.A., Drier, D.A., Yamanaka, G.A., DiIanni, C.L., Deckman, I.C. & Cordingley, M.G. (1993) Autoproteolysis of herpes simplex virus type 1 protease releases an active catalytic domain found in intermediate capsid particles. *J. Virol.* **67**, 5813–5822.

Welch, A.R., Woods, A.S., McNeill, L.M., Cotter, R.J. & Gibson, W. (1991) A herpesvirus maturational proteinase assemblin: Identification of its gene putative active site domain and cleavage site. *Proc. Natl Acad. Sci. USA* **88**, 10792–10796.

Yamanaka, G., DiIanni, C.L., O'Boyle II, D.R., Stevens, J., Weinheimer, S.P., Deckman, I.C., Matusick-Kumar, L. & Colonno, R.J. (1995) Stimulation of the herpes simplex virus type I protease by antichaotropic salts. *J. Biol. Chem.* **270**, 30168–30172.

*Paul L. Darke*
*Department of Antiviral Research,*
*Merck Research Laboratories,*
*WP26-431, West Point, PA 19486, USA*
*Email: paul_darke@merck.com*

# 164. Human herpesvirus type 6 assemblin

## Databanks

*Peptidase classification: clan SH, family S21, MEROPS ID: S21.004*
*NC-IUBMB enzyme classification: none*

*Databank codes:*

| Species | SwissProt | PIR | EMBL (cDNA) | EMBL (genomic) |
|---|---|---|---|---|
| Human herpesvirus type 6 | P24433 | – | X87419 | M68963: multiple entry<br>X83413: complete genome |

## Name and History

Herpesvirus research has primarily focused on herpes simplex virus type 1 (HSV-1) to elucidate the characteristics of this family of viruses. Studies with temperature-sensitive mutants of HSV-1 demonstrated that the $U_L26$ gene product was involved in DNA packaging, although no proteinase activity was ascribed to the protein at that time (Rixon *et al.*, 1988). Subsequently, the $U_L26$ gene product as well as the homologous $U_L80$ gene product of human cytomegalovirus (HCMV) were shown to have proteolytic activity that was essential for viral replication (Liu & Roizman, 1991; Welch *et al.*, 1991; Gao *et al.*, 1994). Human herpesvirus type 6 (HHV-6) which also encodes a proteinase, belongs to the same subfamily (the betaherpesviruses) as HCMV (Gompels *et al.*, 1995). The name assemblin has been given to HCMV proteinase (Welch *et al.*, 1991), and a comparison of the properties of this enzyme with those of HHV-6 proteinase (Tigue *et al.*, 1996) leads us to feel that the name *herpesvirus type 6 assemblin* can be applied to this enzyme also.

## Activity and Specificity

HHV-6 assemblin is expressed as part of a larger polyprotein. Autolytic cleavage at a junction referred to as the release (R-) site results in the generation of mature proteinase (Tigue *et al.*, 1996). In addition, HHV-6 proteinase hydrolyzes a peptide bond at a second location close to the C-terminus of the viral assembly protein referred to as the maturation (M-) site. The synthetic peptide substrates, Suc-Arg-Arg-Tyr-Ile-Lys-Ala+Ser-Glu-Pro-Pro-Val-NH₂ and Suc-Arg-Arg-Ile-Leu-Asn-Ala+Ser-Leu-Ala-Pro-Glu-NH₂ based on the release and maturation sites respectively, are readily hydrolyzed by HHV-6 proteinase. The $K_m$ value obtained for the former (300 µM) is about 7-fold lower than that for the latter (Tigue *et al.*, 1996). Herpesvirus assemblins including HHV-6 proteinase exhibit a distinct preference for Ala and Ser in the P1 and P1′ positions respectively, although some other residues are tolerated (McCann *et al.*, 1994). The enzyme has a pH optimum of 8.5 and operates more effectively at 21°C than at 37°C (Tigue *et al.*, 1996). No naturally occurring or synthetic inhibitors have been reported for HHV-6 proteinase.

## Structural Chemistry

As a single-chain enzyme (29 kDa) HHV-6 assemblin differs from HCMV proteinase (which is a two-chain enzyme) but is presumed to form the obligate dimer that is necessary in the case of HCMV proteinase for activity (Darke *et al.*, 1996; Margosiak *et al.*, 1996). No structure is available for HHV-6 proteinase, but modeling on the structure solved for HCMV proteinase by X-ray crystallography indicates

that the active-site Ser115, His46 and His135 residues are conserved in the enzyme from HHV-6 (Chen *et al.*, 1996; Qiu *et al.*, 1996; Shieh *et al.*, 1996; Tong *et al.*, 1996). It is likely, therefore, to be a serine proteinase with the same fold as its counterpart from HCMV. The catalytic residues are different from the Ser/His/Asp catalytic triad of more classical serine proteinases, and the absence of the Asp residue may account for the relatively poor $k_{cat}$ values measured for HHV-6 proteinase acting on a variety of substrates (Tigue *et al.*, 1996). This is a common feature of herpesvirus proteinases.

## Preparation

No reports to date have described isolation of the naturally occurring proteinase from HHV-6 virions. Recombinant enzyme has been expressed in soluble form in *E. coli* and purified to homogeneity for characterization (Tigue *et al.*, 1996). Purified HHV-6 proteinase is stable at −70°C in 50 mM Tris–HCl, pH 8.0 containing 100 mM NaCl, 1 mM DTT, 1 mM EDTA and 20% glycerol.

## Biological Aspects

During childhood, infection with HHV-6 can result in the disease exanthem subitum (Lopez & Honess, 1990). Although there is no direct evidence, HHV-6 has been suggested to be a cofactor in the progression of HIV-infected patients into AIDS, and has also been associated with diseases such as chronic fatigue syndrome and myalgic encephalomyelitis (Wakefield *et al.*, 1988; Lusso *et al.*, 1991; Buchwald *et al.*, 1992).

The herpesvirus genome is housed within the capsid that is located at the center of the viral particle. By analogy with HSV-1, the icosahedral capsid is assembled from repeating blocks of viral proteins built on to a multisubunit scaffold that is composed of the proteinase in its precursor form and the assembly protein. Following an as yet undefined signal, the proteinase undergoes autolytic processing and the assembly protein is cleaved, presumably to facilitate removal of these scaffold proteins from the center of the capsid prior to or concomitant with packaging of the viral DNA.

## References

Buchwald, D., Cheney, P.R., Peterson, D.L., Henry, B., Wormsley, S.B., Geiger, A., Ablashi, D.V., Salahuddin, S.Z., Saxinger, C., Biddle, R., Kikinis, R., Jolesz, F.A. & Folks, T.A. (1992) A chronic illness characterized by fatigue, neurologic and immunologic disorders, and active human herpesvirus type 6 infection. *Ann. Intern. Med.* **116**, 103–113.

Chen, P., Tsuge, H., Almassy, R.J., Gribskov, C.L., Katoh, S., Vanderpool, D.L., Margosiak, S.A., Pinko, C. & Kan, C.C. (1996) Structure of the human cytomegalovirus protease catalytic domain reveals a novel serine protease fold and catalytic triad. *Cell* **86**, 835–843.

Darke, P.L., Cole, J.L., Waxman, L., Hall, D.L., Sardana, M.K. & Kuo, L.C. (1996) Active human cytomegalovirus protease is a dimer. *J. Biol. Chem.* **271**, 7445–7449.

Gao, M., Matusick-Kumar, L., Hurlburt, W., DiTusa, S.F., Newcomb, W.W., Brown, J.C., McCann, P.J., Deckman, I. & Colonno, R.J. (1994) The protease of herpes simplex virus 1 is essential for functional capsid formation and viral growth. *J. Virol.* **68**, 3702–3712.

Gompels, U.A., Nicholas, J., Lawrence, G., Jones, M., Thomson, B.J., Martin, M.E.D., Efstathiou, S., Craxton, M. & Macaulay, H.A. (1995) The DNA sequence of human herpesvirus-6: structure, coding content and genome evolution. *Virology* **209**, 29–51.

Liu, F.Y. & Roizman, B. (1991) The herpes simplex virus 1 gene encoding a protease also contains within its coding domain the gene encoding the more abundant substrate. *J. Virol.* **65**, 5149–5156.

Lopez, C. & Honess, R.W (1990) Human herpesvirus-6. In: *Virology* (Fields, B.N. & Knipe, D.M., eds). New York: Raven Press, pp. 2055–2062.

Lusso, P., De Maria, A., Malnati, M., Lori, F., De Rocco, S.E., Baleser, M. & Gallo, R.C. (1991) Induction of CD4 and susceptibility to HIV infection in human CD8[+] T lymphocytes by human herpesvirus 6. *Nature* **349**, 533–535.

Margosiak, S.A., Vanderpool, D.L., Sisson, W., Pinko, C. & Kan, C.C. (1996) Dimerisation of the human cytomegalovirus protease: kinetic and biochemical characterisation of the catalytic homodimer. *Biochemistry* **35**, 5300–5307.

McCann, P.J., O'Boyle, D.R. & Deckman, I.C. (1994) Investigation of the specificity of the herpes simplex virus type 1 protease by point mutagenesis of the autoproteolysis sites. *J. Virol.* **68**, 526–529.

Qiu, X., Culp, J.S., DiLella, A.G., Hellmig, B., Hoog, S.S., Janson, C.A., Smith, W.W. & Abdel-Meguid, S.S. (1996) Unique fold and active site in cytomegalovirus protease. *Nature* **383**, 275–279.

Rixon, F.J., Cross, A.M., Addison, C. & Preston, V.G. (1988) The products of herpes simplex virus type 1 gene U$_L$26 which are involved in DNA packaging are strongly associated with empty but not with full capsids. *J. Gen. Virol.* **69**, 2879–2891.

Shieh, H.-S., Kurumbail, R.G., Stevens, A.M., Stegeman, R.A., Sturman, E.J., Pak, J.Y., Wittwer, A.J., Palmier, M.O., Wiegand, R.C., Holwerda, B.C. & Stallings, W.C. (1996) Three-dimensional structure of human cytomegalovirus protease. *Nature* **383**, 279–282.

Tigue, N., Matharu, P.J., Roberts, N.A., Mills, J.S., Kay, J. & Jupp, R. (1996) Cloning, expression and characterisation of the proteinase from human herpesvirus-6. *J. Virol.* **70**, 4136–4141.

Tong, L., Qian, C., Massariol, M.-J., Bonneau, P., Cordingley, M.G. & Lagace, L. (1996) A new serine-protease fold revealed by the crystal structure of human cytomegalovirus proteinase. *Nature* **383**, 272–275.

Wakefield, D., Lloyd, A., Dwyer, J., Salahuddin, S.Z. & Ablashi, D.V. (1988) Human herpesvirus 6 and myalgic encephalomyelitis [letter]. *Lancet* **1**, 1059.

Welch, A.R., Woods, A.S., McNally, L.M., Cotter R.J. & Gibson, W. (1991) A herpesvirus maturational proteinase, assemblin: identification of its gene, putative active site domain, and cleavage site. *Proc. Natl Acad. Sci. USA* **88**, 10792–10796.

***Ray Jupp***
*Roche Research Center,*
*40 Broadwater Road,*
*Welwyn Garden City,*
*Herts AL7 3AY, UK*

***John Mills***
*Roche Research Center,*
*40 Broadwater Road,*
*Welwyn Garden City,*
*Herts AL7 3AY, UK*

***Natalie J. Tigue***
*School of Molecular and Medical Biosciences,*
*University of Wales, Cardiff,*
*Cardiff, CF1 3US, Wales, UK*

***John Kay***
*School of Molecular and Medical Biosciences,*
*University of Wales, Cardiff,*
*Cardiff, CF1 3US, Wales, UK*
*Email: kayj@cardiff.ac.uk*

# *165. Epstein–Barr virus assemblin*

## Databanks

*Peptidase classification: clan SH, family S21, MEROPS ID: S21.003*
*NC-IUBMB enzyme classification: none*

*Databank codes:*

| Species | SwissProt | PIR | EMBL (cDNA) | EMBL (genomic) |
|---------|-----------|-----|-------------|----------------|
| Epstein-Barr virus | P03234 | A03798 S33049 | V01555 | – |

## Name and History

Epstein–Barr virus (EBV) is a gammaherpesvirus and differs in some of its characteristics from the alpha and beta subfamilies of herpesviruses whose members include herpes simplex virus (HSV) and cytomegalovirus (CMV) respectively (Miller, 1990). An indication that EBV encodes a proteinase similar to those present in HSV and CMV came from alignment of the amino acid sequences that were deduced from the nucleotide sequences of the viral genomes (Welch *et al.*, 1991). Subsequently, the BVRF2 gene product of EBV was characterized and shown to have proteolytic activity (Donaghy & Jupp, 1995). This is the **Epstein–Barr virus assemblin**.

## Activity and Specificity

EBV proteinase hydrolyzes two kinds of scissile peptide bond, both located within its own precursor sequence. Cleavage at a junction referred to as the release (R-) site liberates the mature form of the proteinase, whilst cleavage at the second or maturation (M-) site results in the removal of a short C-terminal peptide from the original polyprotein (Donaghy & Jupp, 1995). Translation of a shorter 3′ coterminal mRNA yields the assembly protein, which completely lacks the proteinase domain but contains the maturation site (Lau *et al.*, 1993). Hydrolysis of the maturation site of this protein by recombinant EBV proteinase has also been characterized (Donaghy & Jupp, 1995). To further define the specificity of EBV proteinase, the hydrolysis of a number of synthetic peptides was measured and compared to the values obtained using human and mouse CMV proteinases (Table 165.1).

Although EBV proteinase readily hydrolyzed the peptide substrate mimicking its release site, solubility constraints prevented determination of the $V$ value and hence

the value of $k_{cat}$. By contrast, under these conditions no cleavage of this peptide was observed with mouse or human CMV proteinase. Moreover, the rate of hydrolysis of virtually every other synthetic substrate by EBV proteinase was at least equal to, if not considerably greater than, that obtained with mouse or human CMV proteinases. With a temperature and pH optimum close to 21°C and 9.0, respectively, EBV proteinase is similar to the other herpesvirus proteinases.

## Structural Chemistry

EBV proteinase (28 kDa) is a single-chain enzyme which on gel filtration migrates as a protein of $\sim 60$ kDa and is thus believed to form a dimer as previously shown for human CMV assemblin (Darke *et al.*, 1996; Margosiak *et al.*, 1996). There is no structural information available for EBV proteinase, but based on the recent X-ray crystal structure solved for human CMV proteinase (Chen *et al.*, 1996; Qiu *et al.*, 1996; Shieh *et al.*, 1996; Tong *et al.*, 1996) the active-site Ser116, His48 and His139 residues are conserved in the enzyme from EBV. Like its counterpart from human CMV, EBV proteinase is likely to be a serine proteinase with a fold distinct from those of classical serine proteinases that utilize a Ser/His/Asp catalytic triad.

## Preparation

Characterization of EBV proteinase has relied on the expression of this protein in heterologous systems to produce adequate quantities of protein for purification (Donaghy & Jupp, 1995). There are no reports of the purification of EBV proteinase from infected cells or virions.

*Table 165.1* Cleavage of synthetic peptide substrates by recombinant EBV, MCMV and HCMV proteinases

| Site[a] | Sequence[b] | HCMV | | MCMV | | EBV[c] | |
|---------|-------------|------|------|------|------|------|------|
| | | $K_m$ (mM) | $k_{cat}$ (min⁻¹) | $K_m$ (mM) | $k_{cat}$ (min⁻¹) | $K_m$ (mM) | $k_{cat}$ (min⁻¹) |
| HCMV-R | Suc-S-Y-V-K-A┼S-V-S-NH₂ | 9 ± 1 | 1 | 5 ± 0.6 | 1 | 1 ± 0.1 | 17 |
| EBV-R | Suc-Y-L-K-A┼S-D-A-NH₂ | – | 0 | – | 0 | N/S | N/S |
| EBV-M | Suc-K-K-L-V-Q-A┼S-A-S-NH₂ | 14 ± 1 | 1 | 5 ± 1 | 3 | 0.7 ± 0.1 | 7 |
| MCMV-M | Suc-L-V-N-A┼S-X-E-P-T-NH₂ | 11 ± 0.8 | 5 | 0.2 ± 0.06 | 3 | N/S | N/S |
| HCMV-M | Suc-V-V-N-A┼S-X-R-NH₂ | 4 ± 0.6 | 8 | 2 ± 0.5 | 4 | 15 ± 5 | 7 |

Incubations were performed at pH 8.0 (HCMV) or pH 9.0 (MCMV and EBV) and 21°C. Values for $k_{cat}$ were calculated from $V/[E]$ assuming that all of the proteinase present was in an active form.

[a] R and M denote the release and maturation sites respectively for each of the indicated viruses.

[b] X is 2-aminobutyric acid.

[c] N/S, not saturable due to insufficient substrate solubility.

## Biological Aspects

EBV is the cause of a latent, asymptomatic infection in a large percentage of the population. In young adults, however, primary infection can result in infectious mononucleosis (glandular fever) prior to the virus entering a latent state in which gene expression is restricted (Miller, 1990). Proteins expressed in the latent phase of the viral life cycle have been associated with a number of diseases including Burkitt's lymphoma, nasopharyngeal carcinoma and Hodgkin's disease (Miller, 1990; Ring, 1994). Periodically, EBV will reactivate from this latent state and enter into the lytic phase of its life cycle, in which the proteinase and the other structural proteins involved in the formation of the viral particles are produced.

By analogy with HSV, the precursor form of the proteinase and the assembly protein together form a scaffold on to which the other viral proteins that form the shell of the capsid coalesce. In order for the viral genome to be packaged into the capsid shell, the scaffold proteins have to be removed. This is facilitated by proteolysis of both the precursor proteinase and the assembly protein.

## References

Chen, P., Tsuge, H., Almassy, R.J., Gribskov, C.L., Katoh, S., Vanderpool, D.L., Margosiak, S.A., Pinko, C. & Kan, C.C. (1996) Structure of the human cytomegalovirus protease catalytic domain reveals a novel serine protease fold and catalytic triad. *Cell* **86**, 835–843.

Darke, P.L., Cole, J.L., Waxman, L., Hall, D.L., Sardana, M.K. & Kuo, L.C. (1996) Active human cytomegalovirus protease is a dimer. *J. Biol. Chem.* **271**, 7445–7449.

Donaghy, G. & Jupp, R. (1995) Characterisation of the Epstein–Barr virus proteinase and comparison with the human cytomegalovirus proteinase. *J. Virol.* **69**, 1265–1270.

Lau, R., Middeldorp. J. & Farrell, P.J. (1993) Epstein–Barr virus gene expression in oral hairy leukoplakia. *Virology* **195**, 463–474.

Margosiak, S.A., Vanderpool, D.L., Sisson, W., Pinko, C. & Kan, C.C. (1996) Dimerisation of the human cytomegalovirus protease: kinetic and biochemical characterisation of the catalytic homodimer. *Biochemistry* **35**, 5300–5307.

Miller, G. (1990) Epstein–Barr virus: biology, pathogenesis and medical aspects. In: *Virology* (Fields, B.N. & Knipe, D.M., eds). New York: Raven Press, pp. 1921–1958.

Qiu, X., Culp, J.S., DiLella, A.G., Hellmig, B., Hoog, S.S., Janson, C.A., Smith, W.W. & Abdel-Meguid, S.S. (1996) Unique fold and active site in cytomegalovirus protease. *Nature* **383**, 275–279.

Ring, C.J. (1994) The B cell immortalizing functions of Epstein–Barr virus. *J. Gen. Virol.* **75**, 1–13.

Shieh, H.-S., Kurumbail, R.G., Stevens, A.M., Stegeman, R.A., Sturman, E.J., Pak, J.Y., Wittwer, A.J., Palmier, M.O., Wiegand, R.C., Holwerda, B.C. & Stallings, W.C. (1996) Three-dimensional structure of human cytomegalovirus protease. *Nature* **383**, 279–282.

Tong, L., Qian, C., Massariol, M.-J., Bonneau, P., Cordingley, M.G. & Lagace, L. (1996) A new serine-protease fold revealed by the crystal structure of human cytomegalovirus proteinase. *Nature* **383**, 272–275.

Welch, A.R., Woods, A.S., McNally, L.M., Cotter, R.J. & Gibson, W. (1991) A herpesvirus maturational proteinase, assemblin: identification of its gene, putative active site domain, and cleavage site. *Proc. Natl Acad. Sci. USA* **88**, 10792–10796.

*Ray Jupp*
*Roche Research Center,*
*40 Broadwater Road,*
*Welwyn Garden City,*
*Herts AL7 3AY, UK*

*Alison Ritchie*
*Roche Research Center,*
*40 Broadwater Road,*
*Welwyn Garden City,*
*Herts AL7 3AY, UK*

*Michael Robinson*
*Roche Research Center,*
*40 Broadwater Road,*
*Welwyn Garden City,*
*Herts AL7 3AY, UK*

*Anne Broadhurst*
*Roche Research Center,*
*40 Broadwater Road,*
*Welwyn Garden City, Herts AL7 3AY, UK*

*John Mills*
*Roche Research Center,*
*40 Broadwater Road,*
*Welwyn Garden City, Herts AL7 3AY, UK*

# 166. Varicella-zoster virus assemblin

## Databanks

*Peptidase classification: clan SH, family S21, MEROPS ID: S21.005*
*NC-IUBMB enzyme classification: none*

*Databank codes:*

| Species | SwissProt | PIR | EMBL (cDNA) | EMBL (genomic) |
|---|---|---|---|---|
| Varicella-zoster virus | P09286 | G27214 | X04370 | – |

Brookhaven Protein Data Bank three-dimensional structures:

| Species | ID | Resolution | Notes |
|---|---|---|---|
| Varicella-zoster virus | 1VZV | | residues 10–236; Cys10Met mutant |

## Name and History

Varicella-zoster virus (VZV) is classified as an alphaherpesvirus. Considering it was first isolated in tissue culture some 40 years ago, the molecular biology of VZV has lagged far behind that of the other human herpesviruses. The main reason is the difficulty of obtaining cell-free virus. The complete genome of VZV has been sequenced and most of the open reading frames have been assigned functions by sequence homology with the analogous genes in herpes simplex virus type 1 (HSV-1) (Davison & Scott, 1986). The VZV 33 ORF has been reported to have homology with the proteinase-encoding genes of other herpesviruses, i.e. HSV-1 $U_L26$, human cytomegalovirus (HCMV) $U_L80$ and Epstein–Barr virus (EBV) BVRF2 (Gibson *et al.*, 1994) (see Chapters 161, 162, 163, 164 and 165). The product of VZV gene 33 when expressed in heterologous systems has proteolytic activity similar to that of HSV-1 proteinase (McMillan *et al.*, 1997), and this is the *varicella-zoster virus assemblin*.

## Activity and Specificity

Like the proteinases from all other herpesviruses studied to date, the VZV proteinase is released from a larger precursor by autocatalytic cleavage at a junction referred to as the release (R-) site. A second site recognized by the VZV proteinase, the maturation (M-) site, is located at the C-terminal end of both the proteinase precursor and the VZV assembly protein (reviewed in Gibson *et al.*, 1994). Proteolytic processing at the M-site generates the mature form of the assembly proteins. Autolytic cleavage of the VZV assemblin precursor was not observed when it was expressed in *Escherichia coli*, a result which deviates from the trend set by the other herpes proteinases, which all proved active from this expression system. The VZV proteinase was expressed in active form in insect cells by use of recombinant baculovirus, however (McMillan *et al.*, 1997). Recombinant VZV proteinase cleaves synthetic peptide substrates Ac-Asn-Ala-Val-Glu-Ala↓Ser-Ser-Lys-Ala-Pro-Leu-Ile-Gln-Arg-$NH_2$ and Ac-Val-Tyr-Leu-Gln-Ala↓Ser-Thr-Gly-Tyr-Gly-Leu-Ala-Arg-$NH_2$ modeled on the M- and R-sites respectively. Kinetics of hydrolysis have proved difficult to determine due to solubility constraints. Removing the P5 or P8′ residues from the M-site substrate results in a marked drop in the rate of hydrolysis of the substrates. Similar observations were made for the HSV-1 enzyme. In contrast, the proteinases of HCMV, human herpesvirus type 6 (HHV-6) and EBV readily hydrolyze peptide substrates spanning the P4–P4′ residues (DiIanni *et al.*, 1993; Burck *et al.*, 1994; Tigue *et al.*, 1996). VZV proteinase demonstrates optimum activity at 21°C.

## Structural Chemistry

Most structural information on herpes proteinases has been derived from studies on HCMV proteinase which is active as a dimer (Darke *et al.*, 1996; Margosiak *et al.*, 1996). The structure of the HCMV proteinase has been solved by X-ray crystallography, revealing a new class of serine proteinase with a novel fold and a catalytic triad consisting of His/Ser/His (Chapter 162) (Chen *et al.*, 1996; Qui *et al.*, 1996; Shieh *et al.*, 1996; Tong *et al.*, 1996). VZV proteinase contains each of the conserved domains revealed by amino acid sequence alignments of several herpesvirus proteinases. Residues important to secondary structure and dimer formation are located in the conserved domains, suggesting that VZV proteinase is structurally similar to HCMV proteinase and may also need to form a dimer to be active. Sequence alignments suggest a catalytic triad for the VZV proteinase consisting of His52, Ser120 and His139. Mutation of His52 to Ala inactivates the VZV enzyme. The X-ray crystallographic structure of VZV assemblin has been solved, but the coordinates are not public at the time of writing.

## Preparation

The low copy number of the proteinase in the VZV particle and the difficulty in preparing adequate quantities of VZV makes the virus an unsuitable source for purifying the proteinase. Active VZV proteinase has been overexpressed and purified from *E. coli* (McMillan *et al.*, 1997).

## Biological Aspects

Infection with VZV usually occurs in childhood and the primary infection produces the disease varicella (chicken pox). After primary infection, the virus persists in the sensory ganglia and reactivation later in life results in a second disease, herpes zoster (shingles). After the lesions associated with herpes zoster have healed, patients can suffer pain in the region where the lesions had formed, a condition known as postherpatic neuralgia.

The genome of all herpesviruses including VZV is contained within an icosahedral capsid. The role the herpes proteinase plays in the capsid assembly and maturation process has been extensively studied for HSV-1 (Gibson *et al.*, 1994).

The assembly protein and the proteinase precursor associate with the major capsid protein in the host cell cytoplasm. Following transport into the nucleus, these protein complexes interact via regions of the assembly protein sequence, resulting in the formation of the capsid shell. Completion of the capsid shell somehow triggers the activity of the proteinase which processes itself and the assembly proteins. The processed assembly proteins leave the capsid, allowing the viral genome to be packaged.

## References

Burck, P.J., Berg, D.H., Luk, T.P., Sassmannshausen, L.M., Wakulchik, M., Smith, D.P., Hsiung, H.M., Becker, G.W., Gibson, W. & Villarreal, E.C. (1994) Human cytomegalovirus maturational proteinase: expression in *Escherichia coli*, purification, and enzymatic characterization by using peptide substrate mimics of natural cleavage sites. *J. Virol.* **68**, 2937–2946.

Chen, P., Tsuge, H., Almassy, R.J., Gribskov, C.L., Katoh, S., Vanderpool, D.L., Margosiak, S.A., Pinko, C., Matthews, D.A. & Kan, C. (1996) Structure of the human cytomegalovirus protease catalytic domain reveals a novel serine protease fold and catalytic triad. *Cell* **86**, 835–843.

Darke, P.L., Cole, J.L., Waxman, L., Hall, D.L., Sardana, M.K. & Kuo, L.C. (1996). Active human cytomegalovirus protease is a dimer. *J. Biol. Chem.* **271**, 7445–7449.

Davison, A.J. & Scott, J.E. (1986) The complete DNA sequence of varicella-zoster virus. *J. Gen. Virol.* **67**, 1759–1816.

DiIanni, C.L., Mapelli, C., Drier, D.A., Tsao, J., Natarajan, S., Riexinger, D., Festin, S.M., Bolgar, M., Yamanaka, G., Weinheimer, S.P., Meyers, C.A., Colonno, R.J. & Cordingley, M.G. (1993) In vitro activity of the herpes simplex virus type 1 protease with peptide substrates. *J. Biol. Chem.* **268**, 25449–25454.

Gibson, W., Welch, A.R. & Ludford, M.J. (1994) Transient transfection assay of the herpesvirus maturational proteinase assemblin. *Methods Enzymol.* **244**, 399–411.

Gibson, W., Welch, A.R. & Hall, M.R.T. (1995) Assemblin, a herpes virus serine maturational proteinase and new molecular target for antivirals. *Perspect. Drug Discovery Design* **2**, 413–426.

Margosiak, S.A., Vanderpool, D.L., Sisson, W., Pinko, C. & Kan, C. (1996) Dimerization of the cytomegalovirus protease: Kinetic and biochemical characterization of the catalytic homodimer. *Biochemistry* **35**, 5330–5307.

McMillan, D.J., Kay, J. & Mills, J.S. (1997) Characterization of the proteinase specified by varicella-zoster virus gene 33. *J. Gen. Virol.* **78**, 2153–2157.

Qiu, X., Culp, J.S., Dilella, A.G., Hellmig, B., Hoog, S.S., Janson, C.A. Smith, W.W. & Abdel-Meguid, S.S. (1996). Unique fold and active site in cytomegalovirus protease. *Nature* **383**, 275–279.

Shieh, H., Kurumbail, R.G., Stevens, A.M., Stegeman, R.A., Sturman, E.J., Pak, J.Y., Wittwer, A.J., Palmier, M.O., Wiegand, R.C., Holwerda, B.C. & Stallings, W.C. (1996). Three-dimensional structure of human cytomegalovirus protease. *Nature* **383**, 279–282.

Tigue, N.J., Matharu, P.J., Roberts, N.A., Mills, J.S., Kay, J. & Jupp, R. (1996) Cloning, expression, and characterisation of the proteinase from human herpesvirus-6. *J. Virol.* **70**, 4136–4141.

Tong, L., Qian, C., Massariol, M., Bonneau, P.R., Cordingley, M.G. & Lagace, L. (1996) A new serine-protease fold revealed by the crystal structure of human cytomegalovirus protease. *Nature* **383**, 272–275.

*David McMillan*
*Roche Research Center,*
*40 Broadwater Road,*
*Welwyn Garden City,*
*Herts AL7 3AY, UK*

*Ray Jupp*
*Roche Research Center,*
*40 Broadwater Road,*
*Welwyn Garden City,*
*Herts AL7 3AY, UK*

*John Mills*
*Roche Research Center,*
*40 Broadwater Road,*
*Welwyn Garden City,*
*Herts AL7 3AY, UK*

*John Kay*
*School of Molecular and Medical Biosciences,*
*University of Wales, Cardiff,*
*Cardiff CF1 3US, Wales, UK*

# 167. Introduction: clan TA containing N-terminal nucleophile peptidases

## Databanks

*MEROPS ID: TA*

### Family T1
Archaean proteasome (Chapter 169)
Bacterial proteasome (Chapter 170)
HslVU protease (Chapter 171)
Eukaryotic proteasome (Chapter 172)
Ubiquitin-conjugate degrading enzyme (Chapter 173)

### Family S42
γ-Glutamyltransferase (Chapter 174)

Other self-processing N-terminal nucleophile amidohydrolases (Chapter 168 and the MEROPS database)

Clan TA contains the peptidases of the N-terminal nucleophile (Ntn) hydrolase class. These have been recognized only very recently, and knowledge remains in a state of flux. These enzymes are distantly related at the clan level, as revealed by the three-dimensional structures, but show very little conservation of amino acid sequence. As is explained in Chapter 168, there is reason to think that many of these enzymes autoactivate by an intramolecular proteolytic cleavage that liberates the N-terminal, nucleophilic amino acid residue that is primarily responsible for the hydrolytic activity subsequently expressed by the enzyme. If this is the case, then all of them may be classified as peptidases, even though each molecule may catalyze only one peptide bond hydrolysis. In general, these enzymes are not known to show further endopeptidase activity, after the initial autoactivation, but the important exception is the proteasome.

In many of the Ntn-hydrolases, the N-terminal nucleophile is indeed the N-terminal amino acid once the initiating methionine is removed. For these, there is no implication of peptidase activity, and they will not be considered further here. Others are synthesized as precursor proteins, however, and it is thought that they are processed autolytically. This kind of processing is similar to that of the serine endopeptidase togavirin (Chapter 86), which releases itself from a viral polyprotein and then has no further proteolytic activity, and repressor LexA (Chapter 152), which cleaves itself so as to destroy its repressor and proteolytic activity.

The Ntn-hydrolases (other than the proteasome) that are thought to autoactivate fall into at least three families (see also the Databanks table in Chapter 168 and the MEROPS database). Generally, the activation cleavage occurs internally in the polypeptide chain, so that the active enzyme is a two-chain protein in which the catalytic nucleophile is the N-terminal residue of the chain derived from the C-terminus of the precursor. The first of the three families contains aspartylglucosylaminase (EC 3.5.1.26), in which activation after removal of the signal peptide involves cleavage into a two-chain form that exposes a new N-terminal Thr that is the nucleophile. The second family includes bacterial penicillin amidases (EC 3.5.1.11) and a homolog from the protozoan *Naegleria fowleri*; once again, the activation cleavage exposes the nucleophile, but in this family it is Ser rather then Thr. The third family of Ntn-hydrolases includes amidophosphoribosyltransferases (EC 2.4.2.14) from eukaryotes and gram-positive bacteria. These are intracellular proteins with a hydrophilic propeptide, removal of which exposes an N-terminal Cys. Thus, rather like clan SA/CB (Chapters 1 and 238), clan TA includes endopeptidases with Ser as well as Cys as the reactive nucleophiles. The catalytic activities that appear as a result of the self-processing of these Ntn-hydrolases commonly involve action on the amide group of glutamine or asparagine, as is explained fully in Chapter 168, and these are not peptidase activities. But in the forms of the proteasome, endopeptidase activity is retained after the activation. The proteasome is exceptional among peptidases in its possession of a Thr residue as its nucleophile. A catalytic tetrad has been proposed consisting of the Thr1, Glu17, Lys33 and the N-terminal amine (Löwe *et al.*, 1995). In bacterial and eukaryote proteasomes Glu17 is replaced by Asp. Activation is by autocatalysis at the N-terminus of the subunit precursor. An alignment of residues around the catalytic threonine is shown in Alignment 167.1 (see CD-ROM).

In both eukaryotes and archaea, the proteasome is a multisubunit complex comprising four stacked rings each containing seven subunits. In eukaryotes, there are 14 different but homologous subunits, and the proteasome has several different catalytic activities (see Chapter 172). A simpler situation exists in archaea, with the proteasome containing just two different kinds of subunits and possessing only one catalytic activity (Chapter 169). In bacteria, the

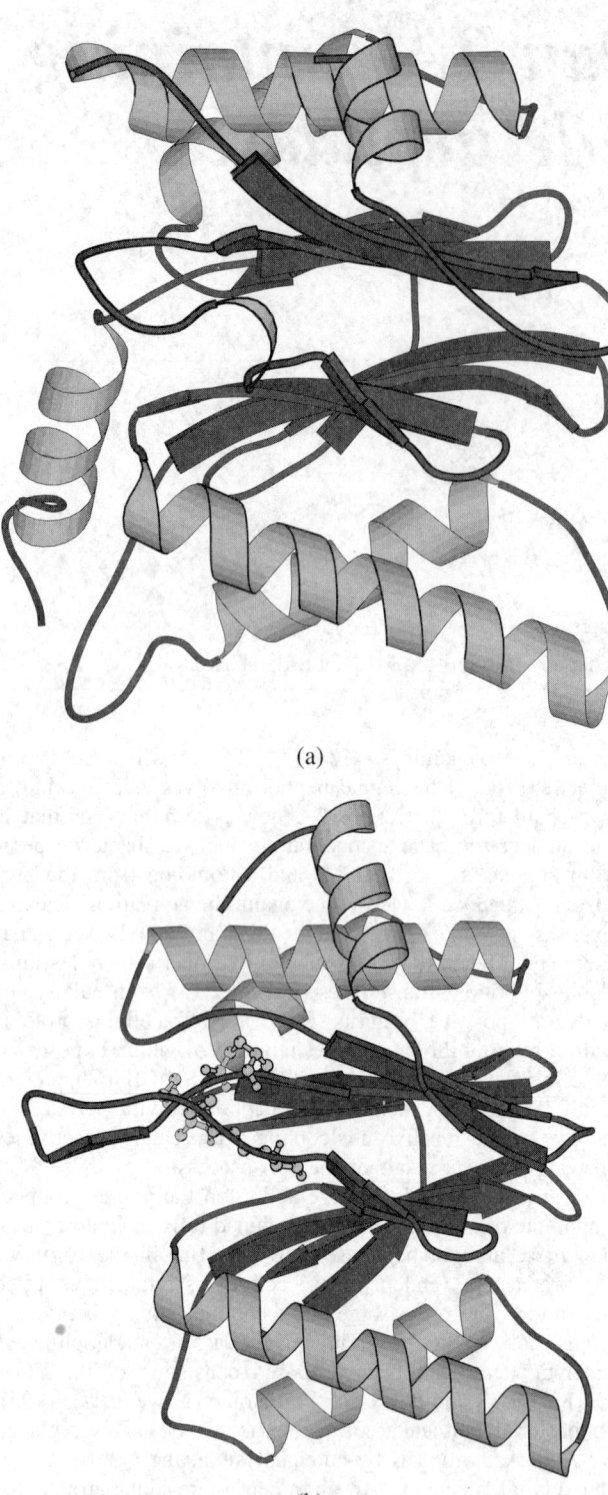

(a)

(b)

*Figure 167.1* Richardson diagram of the *Thermoplasma acidophilum* proteasome subunits (a) α-subunit, (b) β-subunit. The images were prepared from the Brookhaven Protein Data Bank entry (1PMA) as described in the Introduction (p. xxv). The catalytic Thr1, Glu17 and Lys33 are shown in ball-and-stick representation (numbering as in Alignment 167.1).

proteasome consists of two rings of six subunits, but there are only two different subunits, only one of which is related to the eukaryote and archaean proteasome subunits, the other being an ATPase (Chapter 170). The proteasome from *Rhodococcus* is an exception among bacteria, because both subunits are homologous (Zühl *et al.*, 1997). In all cases, only some of the subunits are catalytically active. The archaean proteasome has been crystallized and its tertiary structure has been resolved (Löwe *et al.*, 1995). Each ring contains only one type of subunit and the rings are stacked so that the catalytically active β subunits are in the middle of the four-ring sandwich, with the noncatalytic α subunits on the outside. Figure 167.1 shows a schematic diagram of the α and β subunits. The α subunits are longer at the N-terminus and are not proteolytically processed.

Although the tertiary structure shows no similarity to that of any other peptidase, the fold is typical of the Ntn-hydrolases. Structures are known for the N-terminal domains of class II glutamine amidotransferases, the catalytic domain of penicillin acylase and glycosylasparaginase.

The catalytically active subunits of bacterial proteasome and HslVU peptidases differ from those of other organisms in that proteolytic processing is unnecessary, and Thr1 is exposed once the initiating methionine is removed (see Alignment 167.1).

The eukaryote proteasome is thought to possess a similar organization of subunits to that of the archaean proteasome, with each middle ring of the sandwich including three catalytic components among the seven β subunits. The first and fourth rings are each thought to be composed of seven different α subunits. Replacements exist for some of the catalytic subunits, so that not only does the proteasome possess three different catalytic activities, but different proteasomes constructed with the alternative catalytic components have subtly different versions of these activities. The mouse β1i subunit (subunit δ) is an alternative to the MHC-linked β1 subunit, but is N-terminally truncated and is therefore probably not active (see Alignment 167.1).

A phylogenetic tree has been constructed from an alignment of all proteasome subunits (Figure 167.2). The tree shows a deep divergence between the α and β subunits, which is consistent with the occurrence of both subunits in bacteria, archaea and eukaryotes. The catalytically active components from all three superkingdoms are clustered together on the tree.

The eukaryote proteasome is also a component of the larger 26S proteasome (Chapter 173). This complex contains ATPases and other subunits that are unrelated to the proteasome.

## References

Löwe, J., Stock, D., Jap, B., Zwickl, P., Baumeister, W. & Huber, R. (1995) Crystal structure of the 20S proteasome from the archaeon *T. acidophilum* at 3.4 Å resolution. *Science* **268**, 533–539.

Zühl, F., Tamura, T., Dolenc, I., Cejka, Z., Nagy, I., De Mot, R. & Baumeister, W. (1997) Subunit topology of the *Rhodococcus* proteasome. *FEBS Lett.* **400**, 83–90.

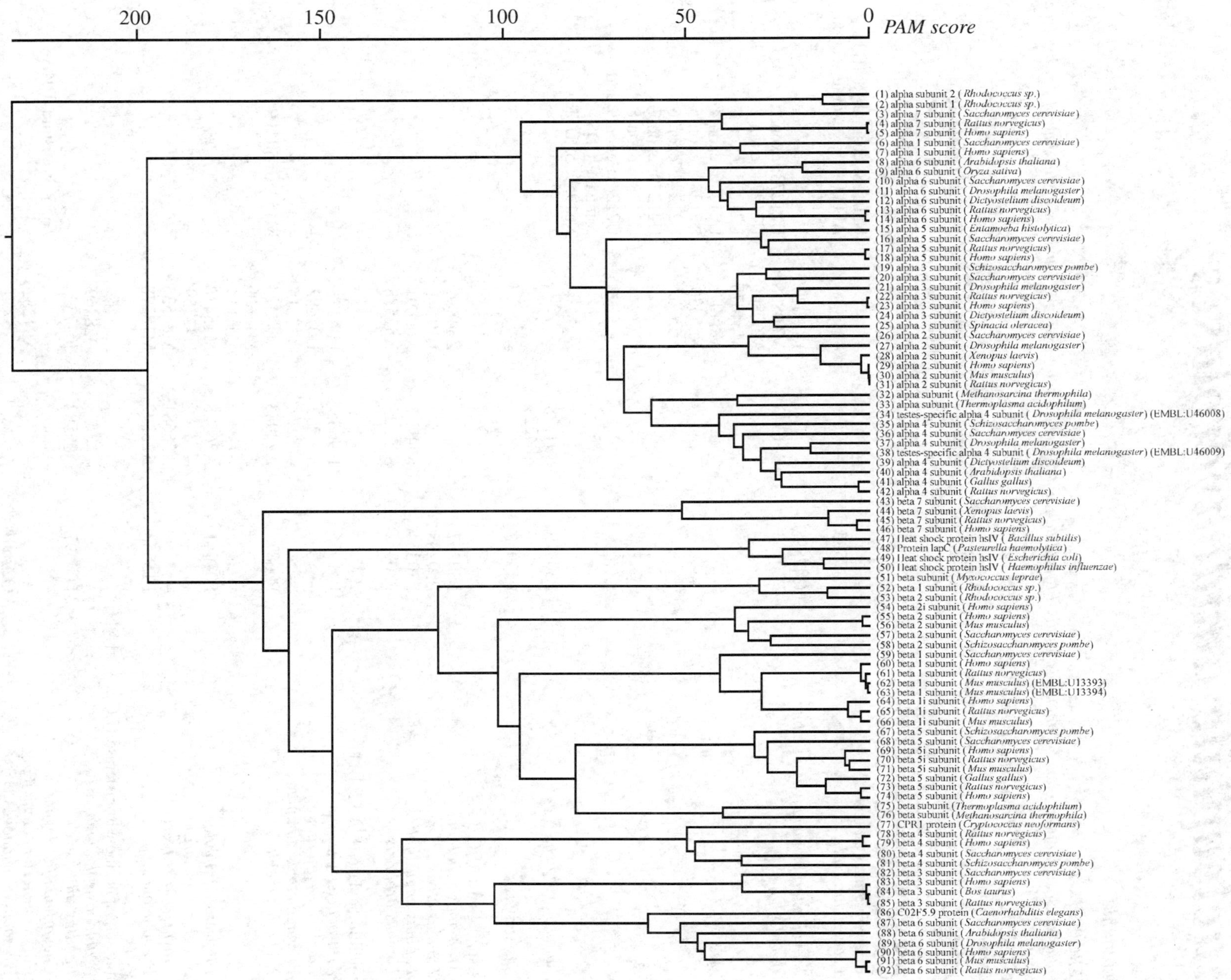

*Figure 167.2*    Evolutionary tree for family T1. The tree was prepared as described in the Introduction (p. xxv).

# 168. Self-processing N-terminal nucleophile amidohydrolases

## Databanks

*MEROPS ID: TA*
*Peptidase classification: clan TA, includes family T1, and additional families not assigned numbers at the time of writing (see Chapter 167)*
*NC-IUBMB enzyme classification: none*
*Databank codes:*

| Species | SwissProt | PIR | EMBL (cDNA) | EMBL (genomic) |
|---|---|---|---|---|
| Proteasome (Chapters 169 and 170) | | | | |
| Glycosylasparaginase (EC 3.5.1.26) | | | | |
| *Caenorhabditis elegans* | – | – | – | U50198: cosmid R04B3 |
| *Escherichia coli* | P37595 | – | – | AE000185: genome section 75 of 400 |
| *Flavobacterium meningosepticum* | Q47898 | – | U08028 | – |
| *Homo sapiens* | P20933 | S11343 | M60808 | U21273: exon 1 |
| | | | M60809 | U21274: exon 2 |
| | | | M64073 | U21275: exon 3 |
| | | | M64076 | U21276: exon 4 |
| | | | X55330 | U21277: exon 5 |
| | | | X55762 | U21278: exon 6 |
| | | | | U21279: exon 7 |
| | | | | U21280: exon 8 |
| | | | | U21281: exon 9 and complete CDS |
| | | | | X61959: 5′ UTR |
| *Rattus norvegicus* | P30919 | S04228 | – | – |
| | | S04229 | | |
| *Sus scrofa* | P30918 | – | – | – |
| L-Asparaginase (EC 3.5.1.1) | | | | |
| *Arabidopsis thaliana* | P50287 | – | Z34884 | – |
| *Lupinus albus* | P50288 | – | L19141 | – |
| *Lupinus angustifolius* | P30364 | S24757 | X60691 | – |
| *Lupinus arboreus* | P30362 | S22523 | X52588 | – |
| Penicillin acylase (EC 3.5.1.11) | | | | |
| *Arthrobacter viscosus* | P31956 | I39665 | L04471 | – |
| *Bacillus megaterium* | – | S49252 | U07682 | – |
| *Brevundimonas diminuta* | – | – | A17015 | – |
| *Escherichia coli* | P06875 | A23983 | M14424 | M15950: 5′UTR |
| | | | M17609 | |
| | | | X04114 | |
| *Kluyvera citrophila* | P07941 | A26528 | M15418 | – |
| *Naegleria fowleri* | – | – | U42759 | – |
| *Providencia rettgeri* | – | A56681 | M86533 | – |
| *Pseudomonas* sp. | P15558 | A28392 | M18278 | – |
| Amidophosphoribosyltransferase (EC 2.4.2.14) | | | | |
| *Arabidopsis thaliana* | – | – | D28869 | – |
| *Bacillus subtilis* | P00497 | A00582 | J02732 | – |
| *Caenorhabditis elegans* | – | – | – | Z35663: cosmid T04A8 |
| *Corynebacterium ammoniagenes* | – | – | X91252 | – |
| *Drosophila melanogaster* | Q27601 | – | L23759 | – |
| *Gallus gallus* | P28173 | A38337 | M60069 | L12533: 5′ end |
| *Glycine max* | P52418 | – | L23833 | |

| Species | SwissProt | PIR | EMBL (cDNA) | EMBL (genomic) |
|---|---|---|---|---|
| Amidophosphoribosyltransferase (EC 2.4.2.14) (*continued*) | | | | |
| *Homo sapiens* | Q06203 | A53342 | D13757 | – |
| | | JC1414 | U00238 | |
| | | | U00239 | |
| *Lactobacillus casei* | P35853 | PC1136 | M85265 | – |
| *Mycobacterium tuberculosis* | – | – | – | Z95618: cosmid SCY07H7 region A |
| *Rattus norvegicus* | P35433 | A46088 | D10853 | – |
| *Rhizobium etli* | – | – | U65392 | – |
| *Saccharomyces cerevisiae* | P04046 | A22642 | K02203 | M57633: 5′ UTR |
| | | | M74309 | Z49212: chromosome XIII cosmid 9952 |
| *Saccharomyces kluyveri* | – | – | U32992 | – |
| *Schizosaccharomyces pombe* | P41390 | S43526 | X72293 | – |
| *Synechococcus* sp. | Q55038 | – | U33211 | – |
| *Synechocystis* sp. | – | – | – | D64000: genome section 19 of 27 |
| *Vigna aconitifolia* | P52419 | – | L23834 | – |
| Acyl-coenzyme A:6-aminopenicillanic acid acyltransferase precursor | | | | |
| *Emericella nidulans* | P21133 | A36142 | – | M58293: complete gene |
| | | S09090 | | X53310: complete gene |
| | | S12169 | | |
| *Penicillium chrysogenum* | P15802 | JQ0118 | – | A15528: complete gene |
| | | S09089 | | A15359: complete gene |
| | | S09091 | | M31454: complete gene |

Brookhaven Protein Data Bank three-dimensional structures:

| Species | ID | Resolution | Notes |
|---|---|---|---|
| Glycosylasparaginase | | | |
| *Homo sapiens* | 1APY | 2.0 | |
| | 1APZ | 2.3 | complex with reaction product |
| Penicillin acylase | | | |
| *Escherichia coli* | 1PNK | 1.9 | |
| | 1PNL | 2.5 | |
| | 1PNM | 2.5 | |
| Amidophosphoribosyltransferase | | | |
| *Bacillus subtilis* | 1GPH | 3.0 | |

## Name and History

The name ***N-terminal-nucleophile (Ntn)-hydrolase*** was coined by Brannigan *et al.* (1995) who first noted distinct structural features of this enzyme group. So far, all examples are amidases (Aronson, 1996) and the term ***Ntn-amidohydrolase*** seems appropriate. Recognition of Ntn-amidohydrolases as a new enzyme class began with determination of the crystal structure of *Escherichia coli* penicillin amidohydrolase (Duggleby *et al.*, 1995). Earlier, the N-terminal serine of the $\beta$ subunit of this $\alpha/\beta$ heterodimeric enzyme had been shown to be the catalytic nucleophile by site-specific mutagenesis and inactivation with PMSF. Duggleby *et al.* (1995) noted that the nearest base available at the active site to polarize $O^\gamma$ of this Ser was its own free $\alpha$ amino group. It soon became evident that other enzymes had a similar unusual structure at their active sites, and shared the same reaction mechanism. It was noteworthy that the proteasome from the archeon *Thermoplasma acidophilum* (Löwe *et al.*, 1995) was a protein of similar fold that used the N-terminal Thr of some of its $\beta$

subunits as reactive nucleophile (Chapter 169). Next, it was realized that glutamine-phosphoribosylpyrophosphate (PRPP) amidotransferase (EC 2.4.2.14), the three-dimensional structure of which had previously been determined (Smith *et al.*, 1994), has a glutaminase domain with the Ntn-amidase structure. Crystal structures have now been solved for two further Ntn-amidohydrolases, human glycosylasparaginase (Oinonen *et al.*, 1995) and the glutaminase domain of glucosamine 6-phosphate synthase (Isupov *et al.*, 1996).

The justification for treating most of the Ntn-amidohydrolases as peptidases and including them in the present volume is the striking aspect that they undergo autoproteolysis of the nascent polypeptide chain to create the new Ntn-Cys, Thr or Ser residue necessary for catalytic activity.

## Activity and Specificity

All of the Ntn-hydrolases require a peptide bond cleavage of some kind to generate the catalytically active form of the enzyme, but they can be divided on this basis into three classes. In some of the enzymes, the autoproteolysis occurs internally in the polypeptide chain to produce a two-chain active enzyme, a heterodimer. Examples of this class are glycosylasparaginase (Figure 168.1), plant asparaginase, penicillin acylase, cephalosporin acylase, γ-glutamyltranspeptidase and acyl-CoA:isopenicillin *N*-acyltransferase. Members of the second group autocatalytically release just an N-terminal peptide (proteasome and some forms of glutamine PRPP aminotransferase). Finally, for the three Ntn-glutamine aminotransferases, the essential Ntn-nucleophilic Cys is penultimate to the initiation Met; these enzymes rely only on methionine aminopeptidase (Chapter 476) to expose their N-terminal nucleophiles, and are not known to have autoproteolytic activity.

The mechanism of the autoproteolysis that occurs in the first two groups listed above is just beginning to be studied. For both glycosylasparaginase and proteasome β subunits the same catalytic Thr of the active enzyme appears to play the role of nucleophile in the self-cleavage reaction prior to its α amino group being freed (Chen & Hochstrasser, 1996; Guan *et al.*, 1996; Schmidtke *et al.*, 1996; Seemüller *et al.*, 1996). Much less is understood about the other protein residues involved in autoproteolysis. It is hoped that site-specific mutagenesis of the Ntn-Thr in glycosylasparaginase will allow large-scale production of an inactive single-chain precursor

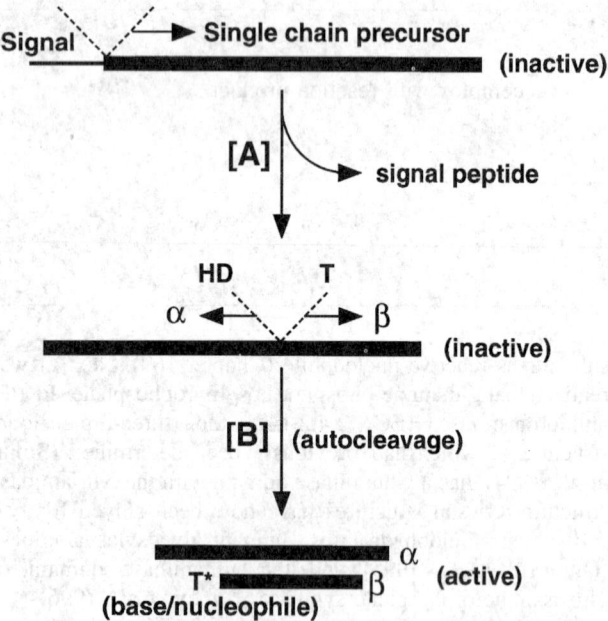

*Figure 168.1* Processing and autoproteolytic activation of glycosylasparaginase. (A) Signal peptidase cotranslationally removes the signal peptide from the inactive precursor. (B) The single-chain precursor is rapidly activated by self-cleavage at a highly conserved His-Asp+Thr sequence (Guan *et al.*, 1996; Liu *et al.*, 1996), and the newly exposed N-terminal Thr of the β subunit (T\*) is the active-site nucleophile. The free α-amino group of this Thr helps to polarize its O$^\gamma$ for reaction. Not shown is further nonspecific cleavage of a C-terminal peptide from the α subunit that occurs in the lysosome.

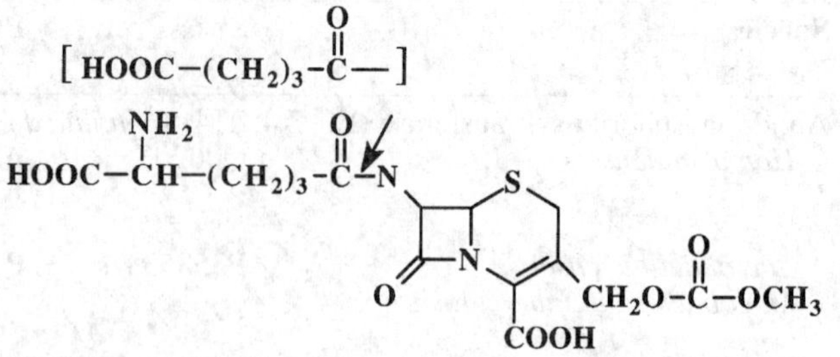

*Figure 168.2* Specificity of cephalosporin acylase for glutamate analogs. The arrow indicates the cleavage site. The glutaric (bracketed) and α-aminoadipic amides of cephalosporin are among the best substrates for this Ntn-amidohydrolase.

whose three-dimensional structure, once determined, can help explain this important process.

Once converted to their active forms, the enzymes of the Ntn group vary in the overall reaction that they catalyze. Thus, a number of them are aminotransferases that utilize an Ntn-glutaminase domain in transferring the ω amino group from glutamine to a second substrate. Included in this group of Ntn-aminotransferases are glutamine PRPP aminotransferase (Smith *et al.*, 1994), glucosamine 6-phosphate synthase (Isupov *et al.*, 1996) and asparagine synthase (Van Hecke & Schuster, 1989). Most Ntn-amidohydrolases show specificity for the glutamyl moiety of glutamine or glutamine analogs (Aronson, 1996). For example, cephalosporin acylase cleaves glutaric and D-α-aminoadipic acid amides of cephalosporin (Figure 168.2) (Matsuda *et al.*, 1987); glutaric acid is the dicarboxylate analog of glutamic acid lacking an α amino moiety, while α-amino adipic is the next higher homolog of glutamic acid.

Two other Ntn-amidohydrolases are plant asparaginase (Lough *et al.*, 1992) and glycosylasparaginase (Mononen *et al.*, 1993). Glycosylasparaginase hydrolyzes GlcNAc+Asn, the natural linkage structure between protein and carbohydrate in Asn-linked glycoproteins. The enzyme requires that both α amino and α carboxyl groups of the Asn residue be free, but there can be wide variability in the sugar component of the substrate, ranging from other monosaccharides (e.g. glucose or galactose) to complete biantennary oligosaccharide chains. Commercial artificial substrates for glycosylasparaginase include fluorometric L-Asp+β-NHMec (Mononen *et al.*, 1993), and colorimetric L-Asp+β-NHPhNO$_2$ (Tarentino & Plummer, 1993). The proteasome is the only Ntn-amidohydrolase known to cleave α peptide bonds as a peptidase while in its mature state, rather than catalyzing reactions of the side-chain amides of compounds of asparagine or glutamine.

Many active-site inhibitors have been found for various members of the Ntn group of enzymes. 5-Diazo-4-oxonorvaline (DONV) reacts with the N-terminal Thr in glycosylasparaginase. The reaction was first discovered by Tarentino & Maley (1968) working with the hen oviduct enzyme, and later was used by Kaartinen *et al.* (1991) to identify the active-site Thr of the β subunit. For the amidotransferase members, their glutaminase domain is similarly covalently modified by the next higher homolog, 6-diazo-5-oxonorleucine (DON). This inhibitor reacts covalently with the Cys residues at the N-termini of glucosamine

6-phosphate synthase and glutamine PRPP aminotransferase (Leriche *et al*., 1997). Lactacystin is a *Streptomyces* metabolite that covalently bonds to the nucleophilic N-terminal Thr of the β subunits of mammalian proteasomes (Fenteany *et al*., 1995) and the first reported crystal structure of the *T. acidophilum* proteasome has the inhibitor DCI bonded to these same Thr residues (Seemüller *et al*., 1995).

## Structural Chemistry

The structural features of Ntn-amidases define this group of enzymes as an evolutionary superfamily of proteins, termed clan TA in the peptidase classification. First, they all rely on the side-chain of an N-terminal amino acid as the source of the active-site nucleophile. Second, the free α amino group becomes the base that helps polarize the side-chain nucleophile, and third, the protein fold that creates this unusual active-site mechanism is found only in this class of enzyme. The common protein fold of Ntn-amidohydrolases is a characteristic four-layer structure, in which two central antiparallel β sheets lie between two layers of α helices, forming a so-called 'αββα sandwich' (Brannigan

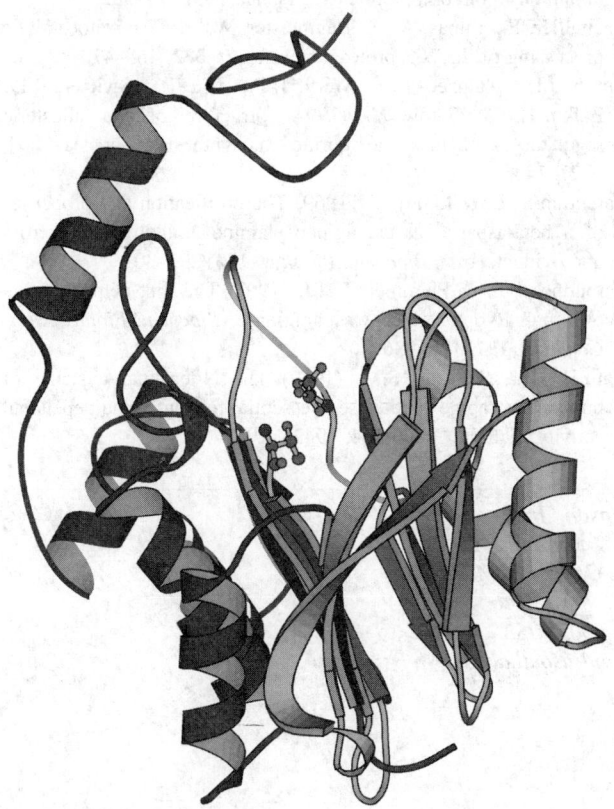

*Figure 168.3* Unique protein fold of Ntn-amidohydrolases. A MOLSCRIPT (Kraulis, 1991) view of the active-site region of human glycosylasparaginase (Oinonen *et al*., 1995). The Ntn-Thr and the product aspartic acid are rendered as ball-and-sticks. Note the overall αββα sandwich structure of the enzyme and the nature of the β sheets forming the active site. The left sheet is flat while the right sheet is twisted. These structural features are a hallmark of the Ntn-amidohydrolase superfamily (Reproduced from Brannigan *et al*., 1995 with permission from R.G. Landes & Co.).

*et al*., 1995). This fold is seen in a MOLSCRIPT representation of human glycosylasparaginase with its bound product, aspartic acid (Figure 168.3). Another hallmark of this superfamily is that the active site is located in a narrow pocket between the edges of the β sheets. The generalized catalytic cycle of these enzymes is (a) activation of an N-terminal Ser (or Thr or Cys) by its own α amino group, (b) nucleophilic attack on the amide carbon of a substrate (most often the side-chain carbonyl of a glutamine or asparagine compound), (c) stabilization of a tetrahedral intermediate in an oxyanion hole, (d) formation through general acid catalysis of a γ-glutamyl ester or thioester, or a β-aspartyl ester; and (e) deacylation of the enzyme by this same general base/acid mechanism.

## Biological Aspects

One of the most remarkable features of the Ntn-amidohydrolase family is its evolutionary history. How this common protein fold and distinctive mechanism employing a catalytic N-terminal nucleophile became so widely used in nature is most intriguing (Artymiuk, 1995). Many of the enzymes have extremely important roles in metabolism. For example, the three glutamine aminotransferases with Ntn-amidohydrolase domains catalyze controlling steps in purine nucleotide, amino acid and amino sugar synthesis. Glycosylasparaginase is essential for the degradation of Asn-linked glycoproteins, and its genetic deficiency is responsible for a prominent lysosomal storage disease in humans (Mononen *et al*., 1993). Discovery of further members of the Ntn-superfamily is likely; for example Brannigan *et al*. (1995) correctly predicted that the membrane enzyme γ-glutamyltranspeptidase (Chapter 174) is an Ntn-amidohydrolase.

## Further Reading

Brannigan *et al*. (1995) and Aronson (1996) give overviews of the Ntn-amidohydrolase field. The protein crystal structure papers are recommended (Duggleby *et al*., 1995; Isupov *et al*., 1996; Löwe *et al*., 1995; Oinonen *et al*., 1995 and Smith *et al*., 1994). Brannigan *et al*. (1995) suggested that protein self-splicing (Perler *et al*., 1997; Chapter 569) may be related to Ntn-amidohydrolases, and this possibility will be interesting to follow.

## References

Aronson, N.N., Jr (1996) Lysosomal glycosylasparaginase: a member of a family of amidases that employ a processed N-terminal threonine, serine or cysteine as a combined base-nucleophile catalyst. *Glycobiology* **6**, 669–675.

Artymiuk, P.J. (1995) A sting in the (N-terminal) tail. *Nature Struct. Biol.* **2**, 1035–1037.

Brannigan, J.A., Dodson, G., Duggleby, H.J., Moody, P.C.E., Smith, J.L., Tomchick, D.R. & Murzin, A.G. (1995) A protein catalytic framework with an N-terminal nucleophile is capable of self-activation. *Nature* **378**, 416–419.

Chen, P. & Hochstrasser, M. (1996) Autocatalytic subunit processing couples active site formation in the 20S proteasome to completion of assembly. *Cell* **86**, 961–972.

Duggleby, H.J., Tolley, S.P., Hill, C.P., Dodson, E.J., Dodson, G. &
Moody, P.C.E. (1995) Penicillin acylase has a single-amino-acid
catalytic center. *Nature* **373**, 264–268.

Fenteany, G., Standaert, R.F., Lane, W.S., Choi, S., Corey, E.J. &
Schreiber, S.L. (1995) Inhibition of proteasome activities and
subunit-specific amino terminal threonine modification by lactacystin. *Science* **268**, 726–731.

Guan, C., Cui, T., Rao, V., Liao, W., Benner, J., Lin, C-L. &
Comb, D. (1996) Activation of glycosylasparaginase: formation
of active N-terminal threonine by intramolecular proteolysis. *J.
Biol. Chem.* **271**, 1732–1737.

Isupov, M.J., Obmolova, G., Butterworth, S., Badet-Denisot, M.-A.,
Badet, B., Polikarpov, I., Littlechild, J.A. & Teplyakov, A. (1996)
Substrate binding is required for assembly of the active conformation of the catalytic site in Ntn amidotransferases: evidence
from the 1.8 Å crystal structure of the glutaminase domain of
glucosamine 6-phosphate synthase. *Structure* **4**, 801–810.

Kaartinen, V., Williams, J.C., Tomich, J., Yates, J.R., Hood, L.E. &
Mononen, I. (1991) Glycosasparaginase from human leukocytes:
inactivation and covalent modification with diazo-oxonorvaline.
*J. Biol. Chem.* **266**, 5860–5869.

Kraulis, P.J. (1991) MOLSCRIPT: a program to produce both
detailed and schematic plots of protein structures. *J. Appl. Cryst.*
**24**, 946–950.

Leriche, C., Badet-Denisot, M.A. & Badet, B. (1997) Affinity labeling of *Escherichia coli* glucosamine-6-phosphate synthase with a
fructose 6-phosphate analog – evidence for proximity between the
N-terminal cysteine and the fructose-6-phosphate-binding. *Eur. J.
Biochem.* **245**, 418–422.

Liu, Y., Dunn, G. & Aronson, N.N. Jr. (1996) Purification, biochemistry and molecular cloning of an insect glycosylasparaginase from
*Spodoptera frugiperda*. *Glycobiology* **6**, 527–536.

Lough, T.J., Reddington, B.D., Grant, M.R., Hill, D.F., Reynolds,
P.H.S. & Farnden, K.J.F. (1992) The isolation and characterization of a cDNA clone encoding L-asparaginase from developing
seeds of lupin (*Lupinus arboreus*). *Plant Mol. Biol.* **19**, 391–399.

Löwe, J., Stock, D., Jap, B., Zwickl, P., Baumeister, W. &
Huber, R. (1995) Crystal structure of the 20S proteasome from
the archaeon *T. acidophilum* at 3.4 Å resolution. *Science* **268**,
533–539.

Matsuda, A., Toma, K. & Komatsu, K.-I. (1987) Nucleotide
sequences of the genes for two distinct cephalosporin acylases
from a *Pseudomonas* strain. *J. Bacteriol.* **169**, 5821–5826.

Mononen, I., Fisher, K.J., Kaartinen, V. & Aronson, N.N. Jr. (1993)
Aspartylglycosaminuria: protein chemistry and molecular biology
of the most common lysosomal storage disorder of glycoprotein
degradation. *FASEB J.* **7**, 1247–1256.

Oinonen, C., Tikkanen, R., Rouvinen, J. & Peltonen, L. (1995)
Three-dimensional structure of human lysosomal aspartylglu-
cosaminidase. *Nature Struct. Biol.* **2**, 1102–1108.

Perler, F.B., Olsen, G.J. & Adam, E. (1997) Compilation and
analysis of intein sequences. *Nucleic Acids Res.* **25**, 1087–1093.

Schmidtke, G., Kraft, R., Kostka, S., Henklein, P., Frömmel, C.,
Löwe, J., Huber, R., Kloetzel, P.M. & Schmidt. M. (1996) Analysis of mammalian 20S proteasome biogenesis: the maturation of
β-subunits is an ordered two-step mechanism involving autocatalysis. *EMBO J.* **15**, 6887–6898.

Seemüller, E., Lupas, A., Stock, D., Löwe, J., Huber, R. &
Baumeister, W. (1995) Proteasome from *Thermoplasma acidophilum*: a threonine protease. *Science* **268**, 579–582.

Seemüller, E., Lupas, A. & Baumeister, W. (1996) Autocatalytic
processing of the 20S proteasome. *Nature* **382**, 468–470.

Smith, J.L., Zaluzec, E.J., Wery, J.-P., Nia, L., Switzer, R.L.,
Zalkin H. & Satow, Y. (1994) Structure of the allosteric
regulatory enzymes of purine biosynthesis. *Science* **264**,
1427–1433.

Tarentino, A.L. & Maley, F. (1969) The purification and properties
of a beta-aspartyl N-acetylglucosylamine amidohydrolase from
hen oviduct. *Arch. Biochem. Biophys.* **130**, 295–303.

Tarentino, A.L. & Plummer, T.H.J. (1993) The first demonstration
of a procaryoptic glycosylasparaginase. *Biochem. Biophys. Res.
Commun.* **197**, 179–186.

Van, H.G. & Schuster, S.M. (1989) The N-terminal cysteine of
human asparagine synthetase is essential for glutamine-dependent
activity. *J. Biol. Chem.* **264**, 19475–19477.

*Nathan N. Aronson, Jr*
*Department of Biochemistry & Molecular Biology,*
*University of South Alabama College of Medicine,*
*307 University Blvd., MSB 2146,*
*Mobile, AL 36688-0002, USA*
*Email: naronson@jaguar1.usouthal.edu*

# 169. Archaean proteasome

## Databanks

*Peptidase classification: clan TA, family T1, MEROPS ID: T01.002*
*NC-IUBMB enzyme classification: none*

*Databank codes:*

| Species | SwissProt | PIR | EMBL (cDNA) | EMBL (genomic) |
|---|---|---|---|---|
| *α* subunit (noncatalytic) | | | | |
|   *Methanosarcina thermophila* | – | – | U30483 | – |
|   *Methanococcus jannaschii* | Q60177 | G64373 | – | U67507: genome section 49 of 150 |
|   *Thermoplasma acidophilum* | P25156 | S21973 | X59507 | – |
| | | S55350 | | |
| | | S13920 | | |
| | | | | |
| *β* subunit (catalytic) | | | | |
|   *Methanosarcina thermophila* | – | – | U22157 | – |
|   *Methanococcus jannaschii* | Q58634 | D64454 | – | U67564: genome section 106 of 150 |
|   *Thermoplasma acidophilum* | P28061 | A42068 | M83674 | – |

Brookhaven Protein Data Bank three-dimensional structures:

| Species | ID | Resolution | Notes |
|---|---|---|---|
| *Thermoplasma acidophilum* | 1PMA | 3.4 | *α* and *β* subunits |

## Name and History

The search in prokaryotes for particles resembling the eukaryotic proteasome led to the discovery in *Thermoplasma acidophilum* of a proteolytic complex indistinguishable from eukaryotic 20S proteasomes by electron microscopy, but of much simpler subunit composition (Dahlmann *et al.*, 1989). In the initial description, the particle was called **multicatalytic proteinase (prosome)**, but subsequently, the name **proteasome**, which had been proposed for the homologous eukaryotic complex, was adopted; it was chosen to describe the proteolytic activity of the particle ('protea-') as well as its size and complexity ('-some'). In eukaryotes, the **20S proteasome** (named for its sedimentation coefficient of 20 S) forms the core protease of the much larger 26S proteasome complex. In prokaryotes, only 20S proteasomes have been discovered to date. Most data on the archaebacterial proteasome have been obtained from *Thermoplasma*, but recently proteasomes have also been isolated from *Methanosarcina thermophila* (Maupin-Furlow & Ferry, 1995) and from *Pyrococcus furiosus* (Bauer *et al.*, 1997).

## Activity and Specificity

The *Thermoplasma* proteasome has chymotryptic specificity when assayed with fluorogenic peptides. The temperature optimum is around 90°C and the pH optimum 8–9 (Dahlmann *et al.*, 1992). Activity assays are generally performed at 60°C, pH 7.5, with the peptide Suc-Leu-Leu-Val-Tyr⊣NHMec, which it degrades with a $K_m$ of 85 $\mu$M and a $k_{cat}$ of 0.03 s$^{-1}$ (Seemüller *et al.*, 1995a,b). Neither the specificity nor the catalytic efficiency can be extrapolated to protein substrates. For example, in the insulin B chain, the proteasome can cleave most peptide bonds and many of the preferred sites cannot be described in terms of a chymotryptic specificity (Wenzel *et al.*, 1994). Of the 29 different peptide bonds, seven are cleaved rapidly and 23 after prolonged incubation; the main cleavage sites are Gly⊣Ser, Glu⊣Ala and Val⊣Cys (E. Seemüller, unpublished results). Insulin B chain is also a good substrate for the proteasome and is

degraded with a $k_{cat}/K_m$ value of 25 mM$^{-1}$ s$^{-1}$ at 60°C, pH 7.5 (I. Dolenc, unpublished results). For the proteasome of *Pyrococcus furiosus*, a temperature optimum of 95°C and a pH optimum of 6.5 have been reported (Bauer *et al.*, 1997). Its substrate specificity against fluorogenic peptides is similar to that of the *Thermoplasma* enzyme, whereas the proteasome from *Methanosarcina* also displays significant peptidyl-glutamyl hydrolyzing activity (Maupin-Furlow & Ferry, 1995).

The *Thermoplasma* proteasome is a processive enzyme: after taking up a substrate protein, it degrades it entirely to peptides before attacking the next molecule (Goldberg *et al.*, 1997). At $V$, digestion of a 200 residue protein requires approximately 1–2 min. The peptides produced are mostly between 4 and 10 residues, which has led to the suggestion of a 'molecular ruler' in the complex (Wenzel *et al.*, 1994). Although neighboring active sites in the *Thermoplasma* proteasome are 2.8 nm apart (Löwe *et al.*, 1995), a distance that can be bridged by a hepta- or octapeptide in extended conformation, it is not yet clear that this is the structural basis for the ruler.

The *Thermoplasma* proteasome is moderately sensitive to DCI and very sensitive to peptide aldehydes such as Ac-Leu-Leu-norleucinal (Dahlmann *et al.*, 1992; Seemüller *et al.*, 1995b), but not to the eukaryotic proteasome inhibitor lactacystin or its $\beta$-lactone derivative (Goldberg *et al.*, 1997). Both DCI and Ac-Leu-Leu-norleucinal covalently modify O$^\gamma$ of Thr1 of the $\beta$ subunit, which is the catalytic nucleophile (Löwe *et al.*, 1995).

## Structural Chemistry

Archaebacterial proteasomes contain two different subunits, $\alpha$ and $\beta$. In the *Thermoplasma* proteasome, for which the crystal structure has been determined (Löwe *et al.*, 1995), the subunits are arranged in four rings with the $\alpha$ subunits in the outer and the $\beta$ subunits in the inner two rings. Collectively, they form an elongated cylinder of 14.8 nm length and 11.3 nm diameter (mass 673 kDa), which is traversed from end to end by a channel, and contains three large inner cavities separated by narrow constrictions. Access to the cavities is controlled by the $\alpha$ ring constrictions, whose diameter is 1.3 nm and which allow only entirely unfolded polypeptide chains to pass. Disulfide bonds or residual secondary structure elements (as are found for example in the molten globule state) are sufficient to prevent substrate uptake (Wenzel & Baumeister, 1995). The proteolytic active-site clefts are located in the $\beta$ subunits and open into the central cavity.

The $\alpha$ and $\beta$ subunits have related sequences and assume the same fold: a four-layer, $\alpha + \beta$ structure with two central antiparallel $\beta$ sheets, flanked on either side by $\alpha$ helices (Löwe *et al.*, 1995). The central $\beta$ sheets open at one end to form the active-site cleft. The main difference between $\alpha$ and $\beta$ subunits lies in an N-terminal extension of $\alpha$ subunits, which forms an $\alpha$ helix across the top of the central $\beta$ sandwich, thus filling the active-site cleft. In place of this helix, $\beta$ subunits contain a pro sequence (Zwickl *et al.*, 1992a), which is cleaved off during proteasome assembly to free the active site. The sequence properties of the *Thermoplasma* subunits allow the classification of eukaryotic subunits into an $\alpha$-type and a $\beta$-type group (Pühler *et al.*, 1994). The most highly conserved elements of the $\alpha$-type subunits are the N-terminal helix, an Arg-Pro-Xaa-Gly motif in the loop constricting the central channel, and a Gly-Xaa$_3$-Asp motif, which makes multiple interactions with the Arg-Pro-Xaa-Gly loop, both of its own and of the neighboring subunit. In $\beta$-type subunits, sequence similarity is concentrated in the N-terminal region that contains the active-site residues, in the Gly-Xaa$_3$-Asp motif, and in a Gly-Ser-Gly motif that is located at one side of the active-site cleft. In *Thermoplasma*, but not in *Methanosarcina* or in *Pyrococcus*, the $\alpha$ subunits are N-terminally blocked by an unknown modification.

The fold and catalytic mechanism of proteasome subunits identify them as members of the N-terminal nucleophile (Ntn)-hydrolase family (Brannigan *et al.*, 1995) (see Chapter 168). These enzymes have a 'single-residue' active site, in which both the catalytic nucleophile and the primary proton acceptor are located in the N-terminal residue. The nucleophilic attack is initiated when the free N-terminus strips the proton off the catalytic side-chain via an intermediate water molecule. The N-terminal residue may be serine (penicillin acylase), cysteine (glutamine PRPP amidotransferase) or threonine (the proteasome). In this respect, Ntn-hydrolases resemble inteins, the self-excision elements of protein splicing, but the reaction mechanism appears to be different (Xu & Perler, 1996) (Chapter 569). In all Ntn-hydrolases, the amide backbone group of a residue at the C-terminal end of strand $\beta 4$ (Gly47 in *Thermoplasma* $\beta$) is involved in forming the oxyanion hole. Aside from these general elements, the proteasome requires two further residues for activity, whose exact role remains to be clarified: Lys33 and Glu17 (Seemüller *et al.*, 1995b, 1996). These form a salt-bridge across the bottom of the active site and may either lower the p$K_a$ of the N-terminus by electrostatic effects, thus facilitating its function as a reversible proton acceptor, or may participate in the delocalization of the threonine side-chain proton by forming a charge-relay system (Löwe *et al.*, 1995).

## Preparation

Proteasomes have been purified from crude cell extracts of *T. acidophilum* (Dahlmann *et al.*, 1989), *M. thermophila* (Maupin-Furlow & Ferry, 1995), and *P. furiosus* (Bauer *et al.*, 1997). Recombinant *Thermoplasma* proteasomes and proteasome subunits have been purified with and without a His$_6$ tag from *Escherichia coli* expression systems (Zwickl *et al.*, 1992b; Seemüller *et al.*, 1995a).

## Biological Aspects

The occurrence of proteasomes in methanogenic, thermoacidophilic and hyperthermophilic branches of the archaebacteria suggest that the proteasome is universally distributed, at least among the Euryarchaeota. So far, its function and regulation in the archaean cell have not been studied.

Archaean $\beta$ subunits are synthesized in an inactive form containing a propeptide of 6–9 residues. This is much shorter than most eukaryotic and eubacterial proteasome propeptides. In *Thermoplasma*, the propeptide is cleaved off autocatalytically during assembly of the complex, and neither its presence nor its cleavage are required for proper assembly (Seemüller *et al.*, 1996). The processing only proceeds in the presence of $\alpha$ subunits and is relatively insensitive to the length and nature of the prosequence. The $\beta$ subunits do not assemble in the absence of $\alpha$ subunits, and there are indications that they do not fold correctly without the assistance of $\alpha$ subunits. The $\alpha$ subunits assemble by themselves into single and double seven-membered rings; this tendency to self-associate may be an important cause for the low yields of proteasomes that can be reconstituted *in vitro* from purified subunits.

In eukaryotes, the 20S proteasome interacts with the 19S caps, which contain a set of ATPases of the AAA family (see Chapters 172 and 173). No such complex has been identified in archaean to date, but an ATPase clearly related to the 19S ATPases has been found in the genome of *Methanococcus jannaschii*.

## Distinguishing Features

The *Thermoplasma* proteasome is activated 2- to 4-fold by low concentrations of SDS or guanidine HCl (Goldberg *et al.*, 1997) and is resistant to temperatures around 100°C, far in excess of the optimum growth temperature of the organism at 59°C. Antibodies against the $\alpha$ subunit of *Methanosarcina* are commercially available from Calbiochem (see Appendix 2 for full names and addresses of suppliers).

## Related Peptidases

The archaebacterial proteasome has eubacterial and eukaryotic homologs that are discussed elsewhere in the present volume (see Chapter 170 and Chapter 172). In addition, it is related to the simpler eubacterial protease HslV (see Chapters 170 and 171).

## References

Bauer, M.W., Bauer, S.H. & Kelly, R.M. (1997) Purification and characterization of a proteasome from the hyperthermophilic archaeon *Pyrococcus furiosus*. *Appl. Environ. Microbiol.* **63**, 1160–1164.

Brannigan, J.A., Dodson, G., Duggleby, H.J., Moody, P.C.E., Smith, J.L., Tomchick, D.R. & Murzin, A. (1995) A protein catalytic framework with an N-terminal nucleophile is capable of self-activation. *Nature* **378**, 416–419.

Dahlmann, B., Kopp, F., Kuehn, L., Niedel, B., Pfeifer, G., Hegerl, R. & Baumeister, W. (1989) The multicatalytic proteinase (prosome) is ubiquitous from eukaryotes to archaebacteria. *FEBS Lett.* **251**, 125–131.

Dahlmann, B., Kuehn, L., Grziwa, A., Zwickl, P. & Baumeister, W. (1992) Biochemical properties of the proteasome from *Thermoplasma acidophilum*. *Eur. J. Biochem.* **208**, 789–797.

Goldberg, A.L., Akopian, T.N., Kisselev, A.F. & Lee, D.H. (1997) Protein degradation by the proteasome and dissection of its in vivo importance with synthetic inhibitors. *Mol. Biol. Rep.* **24**, 69–75.

Löwe, J., Stock, D., Jap, B., Zwickl, P., Baumeister, W. & Huber, R. (1995) Crystal structure of the 20S proteasome from the Archaeon *Thermoplasma acidophilum* at 3.4 Å resolution. *Science* **28**, 533–539.

Maupin-Furlow, J.A. & Ferry, J.G. (1995) A proteasome from the methanogenic archaeon *Methanosarcina thermophila*. *J. Biol. Chem.* **270**, 28617–28622.

Pühler, G., Pitzer, F., Zwickl, P. & Baumeister, W. (1994) Proteasomes – multisubunit proteinases common to *Thermoplasma* and

eukaryotes. *System. Appl. Microbiol.* **16**, 734–741.

Seemüller, E., Lupas, A., Zuhl, F., Zwickl, P. & Baumeister, W. (1995a) The proteasome from *Thermoplasma acidophilum* is neither a cysteine nor a serine protease. *FEBS Lett.* **359**, 173–178.

Seemüller, E., Lupas, A., Stock, D., Löwe, J., Huber, R. & Baumeister, W. (1995b) Proteasome from *Thermoplasma acidophilum* – a threonine protease. *Science* **268**, 579–582.

Seemüller, E., Lupas, A. & Baumeister, W. (1996) Autocatalytic processing of the 20S proteasome. *Nature* **382**, 468–470.

Wenzel, T. & Baumeister, W. (1995) Conformational constraints in protein degradation by the 20S proteasome. *Nature Struct. Biol.* **2**, 199–204.

Wenzel, T., Eckerskorn, C., Lottspeich, F. & Baumeister, W. (1994) Existence of a molecular ruler in proteasomes suggested by analysis of degradation products. *FEBS Lett.* **349**, 205–209.

Xu, M.-Q. & Perler, F. (1996) The mechanism of protein splicing and its modulation by mutation. *EMBO J.* **15**, 5146–5153.

Zwickl, P., Grziwa, A., Puhler, G., Dahlmann, B., Lottspeich, F. & Baumeister, W. (1992a) Primary structure of the *Thermoplasma proteasome* and its implications for the structure, function, and evolution of the multicatalytic proteinase. *Biochemistry* **31**, 964–972.

Zwickl, P., Lottspeich, F. & Baumeister, W. (1992b) Expression of functional *Thermoplasma acidophilum* proteasomes in *Escherichia coli*. *FEBS Lett.* **312**, 157–160.

**Andrei Lupas**
*Molecular Structural Biology,*
*Max-Planck-Institute for Biochemistry,*
*D-82152 Martinsried, Germany*
*Email: lupas@vms.biochem.mpg.de*

**Mary Kania**
*Molecular Structural Biology,*
*Max-Planck-Institute for Biochemistry,*
*D-82152 Martinsried, Germany*

**Wolfgang Baumeister**
*Molecular Structural Biology,*
*Max-Planck-Institute for Biochemistry,*
*D-82152 Martinsried, Germany*
*Email: baumeist@alf.biochem.mpg.de*

# *170. Bacterial proteasome*

## Databanks

*Peptidase classification: clan TA, family T1, MEROPS ID: T01.005*
*NC-IUBMB enzyme classification: none*
Databank codes:

| Species | Type | SwissProt | PIR | EMBL (cDNA) | EMBL (genomic) |
|---|---|---|---|---|---|
| *α* subunit | | | | | |
| *Myxococcus leprae* | – | – | – | – | U00017: cosmid B2126 |
| *Rhodococcus* sp. | 1 | – | – | U26421 | – |
| *Rhodococcus* sp. | 2 | – | – | U26422 | – |
| *β* subunit | | | | | |
| *Myxococcus leprae* | – | – | – | – | U00017: cosmid B2126 |
| *Rhodococcus* sp. | 1 | – | – | U26421 | – |
| *Rhodococcus* sp. | 2 | – | – | U26422 | – |

## Name and History

Sequences related to eukaryotic proteasome β-type subunits were initially discovered during the *Escherichia coli* and *Mycobacterium leprae* genome projects. The *E. coli* sequence, HslV (ClpQ), had homologs in many other eubacteria and was found to form a complex with an ATPase of the Clp family, HslU (ClpY) (HslV and HslVU are discussed in the section on Related peptidases, below, and also Chapter 171). The *M. leprae* sequence, on the other hand, had no close homologs in other eubacteria until two operons with striking sequence similarity were discovered in the soil actinomycete *Rhodococcus erythropolis* NI86/21 (Tamura *et al.*, 1995). The proteasome of this organism was purified and shown to correspond structurally to the 20S proteasomes already described in eukaryotes and archaea. The name **proteasome** stems from the eukaryotic complex; it describes the proteolytic activity of the particle ('protea-') as well as its size and complexity ('-some'). In eukaryotes, the 20S proteasome (named for its sedimentation coefficient of 20 S) forms the core protease of the much larger 26S proteasome complex. In prokaryotes, only 20S proteasomes have been discovered so far.

## Activity and Specificity

The *Rhodococcus* proteasome has a chymotryptic specificity when assayed with fluorogenic peptides (Tamura *et al.*, 1995). The highest activity was measured against the peptide Suc-Leu-Leu-Val-Tyr+NHMec with a $K_m$ of 125 mM and a $k_{cat}$ of $0.1 \, s^{-1}$ at 37°C, pH 7.5 (Zühl *et al.*, 1997). Results obtained with the proteasomes of *Thermoplasma* and humans suggest that this specificity is not likely to be relevant to its activity against protein substrates, but this point has not been investigated yet.

## Structural Chemistry

The *Rhodococcus* proteasome is a complex of approximately 720 kDa and contains four kinds of subunits, of 28.3 kDa (pI 5.3), 27.8 kDa (pI 5.2), 24.2 kDa (pI 4.6), and 24.1 kDa (pI 4.6) (Tamura *et al.*, 1995; Zühl *et al.*, 1997). The two larger subunits, α1 and α2, belong to the α-type proteasome subunits and the two smaller ones, β1 and β2, to the β-type subunits. The two subunit types are evolutionarily related; they were defined in relation to the subunits of the *Thermoplasma* proteasome and their sequence properties are therefore discussed in Chapter 169. The two α-type subunits are proteolytically inactive and their N-termini are blocked by an unknown modification. The two β-type subunits express the proteolytic activity and are generated from precursor forms by cleavage of prosequences of 65 (β1) and 59 (β2) residues. The cleavage frees the amino group of Thr1, which is thought to act as the proton acceptor in the activation of Thr1 $O^\gamma$ prior to the nucleophilic attack.

Although not specifically determined for this complex, the three-dimensional fold of the subunits must be assumed to parallel that of archaean and eukaryotic proteasomes, for which crystal structures have been obtained. The size and shape of the complex is indistinguishable from that of eukaryotic and archaean proteasomes in electron micrographs (Tamura *et al.*, 1995) and an initial uncertainty on the number of subunits in the four rings forming the complex has now been resolved: as in all other proteasomes, the rings are seven-membered (Zühl *et al.*, 1997). Surprisingly, the two α-type and two β-type subunits appear to occur randomly in their respective rings.

## Preparation

The proteasome is a constitutive protein in *Rhodococcus*, but may be heat-shock-inducible in some actinomycetes. It has been purified 500-fold from *Rhodococcus* crude extracts (Tamura *et al.*, 1995). Recombinant proteasomes containing all possible combinations of the four subunits have also been obtained in $His_6$-tagged form from an *E. coli* expression system (Zühl *et al.*, 1997).

## Biological Aspects

20S proteasomes or their genes have so far been detected in several actinomycetes but in no other eubacteria. Instead, most eubacteria contain the related protease, HslV. The actinomycetes may have acquired the proteasome genes by horizontal transfer after separation from other gram-positive bacteria. If so, their assembly pathway and sensitivity to inhibitors indicate that the source was a eukaryote, rather than an archaean. The proteasome genes of *Rhodococcus erythropolis* NI86/21 are organized in two operons, each of which encodes one β and one α subunit in this order. Gene hybridization experiments show that other actinomycetes, including other *Rhodococcus* species, contain only one operon and that therefore four subunits are the exception rather than the rule (I. Nagy *et al.*, unpublished results). The two operons of *R. erythropolis* NI86/21 have a significantly different GC content, suggesting that they are the result of a recent gene transfer event rather than of duplication. In all operons sequenced to date, a short open reading frame of unknown function occurs prior to the genes for the β and α subunits. The three genes are translationally coupled.

*In vitro*, the *Rhodococcus* proteasome assembles via half-proteasome intermediates consisting of an α and a β subunit ring (F. Zühl *et al.*, unpublished results). The β subunits are still unprocessed at this stage and are activated by autocatalytic cleavage of their pro sequence during the association of half-proteasomes to full complexes. The pro sequences appear to play an active role in several steps of the assembly pathway prior to their cleavage. The assembly of proteasomes from any combination of *Rhodococcus* α and β subunits *in vitro* is essentially quantitative.

In eukaryotes, the 20S proteasome interacts with the 19S caps, which contain a set of ATPases of the AAA family (see Chapters 172 and 173). Although such a complex has not been identified in *Rhodococcus* yet, an ATPase of the AAA family is found in close proximity to the proteasome genes in *Rhodococcus* and *Mycobacterium*. The ATPase from *Rhodococcus* has been cloned and expressed in *E. coli*; it forms a cylindrical complex of two, six-membered rings (S. Wolf *et al.*, unpublished results). An interaction with the 20S proteasome has not been observed yet.

## Distinguishing Features

The *Rhodococcus* proteasome is resistant to low concentrations of SDS (0.05%), and this property has been exploited to detect its presence in crude extracts after fractionation on a sizing column (Tamura *et al.*, 1995).

## Related Peptidases

The eubacterial proteasome has archaean and eukaryotic homologs that are discussed in separate chapters (see Chapters 169 and 172). In addition, it is related to a heat-shock-inducible protease of simpler quaternary structure, HslV (ClpQ) (Lupas *et al.*, 1995), which is widespread in eubacteria and whose crystal structure has been determined recently. HslV is formed from only two stacked rings of $\beta$-type subunits, with an overall height of 8 nm and a diameter of 11 nm (Kessel *et al.*, 1996; Rohrwild *et al.*, 1997) and thus essentially corresponds to the two central rings of the proteasome; however, HslV is built of six-membered rings. HslV occurs in an operon with an ATPase of the Clp family, HslU (ClpY) (Chuang *et al.*, 1993). This arrangement parallels the operon structure of the ClpXP protease, which is also heat-shock inducible. Both HslU and HslV have been purified to apparent homogeneity (Rohrwild *et al.*, 1996; Shin *et al.*, 1996; Missisakas *et al.*, 1996). HslU forms six- and seven-membered rings in the presence of ATP or nonhydrolyzable analogs and binds to both ends of the HslV core protease (Rohrwild *et al.*, 1997). This association stimulates the proteolytic activity of HslV by two orders of magnitude in the presence of ATP but not in the presence of nonhydrolyzable analogs. Proteasomes and HslV (further described in Chapter 171) are the only currently known peptidase members of the N-terminal (Ntn-)hydrolase family (Brannigan *et al.*, 1995).

## Further Reading

For a review, see Lupas *et al.* (1997).

## References

Brannigan, J.A., Dodson, G., Duggleby, H.J., Moody, P.C.E., Smith, J.L., Tomchick, D.R. & Murzin, A. (1995) A protein catalytic framework with an N-terminal nucleophile is capable of self-activation. *Nature* **378**, 416–419.

Chuang, S.-E., Burland, V., Plunkett III, G., Daniels, D.L. & Blattner, F. (1993) Sequence analysis of four new heat-shock genes constituting the *hslTS/ibpAP* and *hslVU* operons in *Escherichia coli. Gene* **134**, 1–6.

Kessel, M., Wu, W., Gottesman, S., Kocsis, E., Steven, A.C. & Maurizi, M.R. (1996) Six-fold rotational symmetry of ClpQ, the *E. coli* homolog of the 20S proteasome, and its ATP-dependent activator, ClpY. *FEBS Lett.* **398**, 274–278.

Lupas, A., Zwickl, P. & Baumeister, W. (1995) Proteasome sequences in eubacteria. *Trends Biochem. Sci.* **19**, 533–534.

Lupas, A., Zühl, F., Tamura, T., Wolf, S., Nagy, I., De Mot, R. & Baumeister, W. (1997) Eubacterial proteasomes. *Mol. Biol. Rep.* **24**, 125–131.

Missiakas, D., Schwager, F., Betton, J.-M., Georgopoulos, C. & Raina, S. (1996) Identification and characterization of HslV HslU (ClpQ ClpY) proteins involved in overall proteolysis of misfolded proteins in *Escherichia coli. EMBO J.* **15**, 6899–6909.

Rohrwild, M., Coux, O., Huang, H.-C., Moerschell, R.P., Yoo, S.J., Seol, J.H., Chung, C.H. & Goldberg, A.L. (1996) HslV-HslU: a novel ATP-dependent protease complex in *Escherichia coli* related to the eukaryotic proteasome. *Proc. Natl Acad. Sci. USA* **93**, 5808–5813.

Rohrwild, M., Pfeifer, G., Santarius, U., Müller, S.A., Huang, H.-C., Engel, A., Baumeister, W. & Goldberg, A.L. (1997) The ATP-dependent HslVU protease from *Escherichia coli* is a four-ring structure resembling the proteasome. *Nature Struct. Biol.* **4**, 133–139.

Shin, D.H., Yoo, S.J., Shim, Y.K., Seol, J.H., Kang, M.-S. & Chung, C.H. (1996) Mutational analysis of the ATP-binding site in HslU, the ATPase component of HslVU protease in *Escherichia coli. FEBS Lett.* **398**, 151–154.

Tamura, T., Nagy, I., Lupas, A., Lottspeich, F., Cejka, Z., Schoofs, G., Tanaka, K., De Mot, R. & Baumeister, W. (1995) The first characterization of a eubacterial proteasome: the 20S complex of *Rhodococcus. Curr. Biol.* **5**, 766–774.

Zühl, F., Tamura, T., Dolenc, I., Cejka, Z., Nagy, I., De Mot, R. & Baumeister, W. (1997) Subunit topology of the *Rhodococcus* proteasome. *FEBS Lett.* **400**, 83–90.

*Andrei Lupas*
*Molecular Structural Biology,*
*Max-Planck-Institute for Biochemistry,*
*D-82152 Martinsried, Germany*
*Email: lupas@vms.biochem.mpg.de*

*Mary Kania*
*Molecular Structural Biology,*
*Max-Planck-Institute for Biochemistry,*
*D-82152 Martinsried, Germany*

*Wolfgang Baumeister*
*Molecular Structural Biology,*
*Max-Planck-Institute for Biochemistry,*
*D-82152 Martinsried, Germany*
*Email: baumeist@alf.biochem.mpg.de*

# 171. HslVU protease

## Databanks

*Peptidase classification: clan TA, family T1, MEROPS ID: T01.006*
*NC-IUBMB enzyme classification: none*
*Databank codes:*

| Species | SwissProt | PIR | EMBL (cDNA) | EMBL (genomic) |
|---|---|---|---|---|
| Proteolytic component, HslV | | | | |
| *Bacillus subtilis* | P39070 | – | Z33639 | – |
| *Borrelia burgdorferi* | – | – | L40503 | U43739: complete fesmid clone 31 |
| *Escherichia coli* | P31059 | – | – | L19201: chromosomal region from 87.2 to 89.2′ |
| *Haemophilus influenzae* | P43772 | – | U32731 | – |
| *Lactobacillus leichmannii* | – | – | X84261 | – |
| *Pasteurella haemolytica* | P49617 | – | M59210 | – |
| *Ralstonia eutropha*[a] | – | – | U09865 | – |
| ATPase, HslU | | | | |
| *Bacillus subtilis* | P39778 | S61495 S72310 | U13634 | – |
| *Borrelia burgdorferi* | – | – | – | U43739: complete fesmid clone 31 |
| *Escherichia coli* | P32168 | JT0761 S40874 | – | L19201: chromosomal region from 87.2 to 89.2′ |
| *Haemophilus influenzae* | P43773 | – | – | U32731: genome section 46 of 163 |
| *Helicobacter pylori* | – | – | – | AC000108: chromosomal fragment |
| *Lactobacillus leichmannii* | – | – | X84261 | – |
| *Pasteurella haemolytica* | P32180 | S27609 | – | M59210: arginine permease gene cluster |
| *Treponema denticola* | – | – | U78776 | – |

[a] *Ralstonia eutropha* is the new name for *Alicaligenes eutrophus*.

## Name and History

The *Escherichia coli* genome sequencing project has identified 26 new heat shock genes, termed *hsl* genes (Chuang & Blattner, 1993; Chuang *et al.*, 1993). Of these, the *hslVU* operon encodes the multimeric protease complex HslVU which consists of the proteolytic component HslV (subunit 19 kDa) of the **HslVU protease**, and the ATPase HslU (subunit 50 kDa). The *hslV* gene has strong similarity to *hslV* genes in several gram-positive and gram-negative bacteria and belongs to the β-type proteasome family (Lupas *et al.*, 1994). Proteasomes are large multicatalytic proteolytic complexes found in all eukaryotic cells, and simpler forms have been found in archaea and certain eubacteria (reviewed by Coux *et al.*, 1996, and see also Chapters 169 and 170). The 20S particle consists of four seven-membered rings whose subunits fall into two families: the α-type family forms the two outer rings, and the β-type family, which contains the proteolytic sites, comprise the two inner rings of the complex ($\alpha_7\beta_7\beta_7\alpha_7$). In eukaryotes, the 20S proteasome also exists as the central core of the 26S (2000 kDa) proteasome complex, which catalyzes the degradation of ubiquitin-conjugated proteins and contains an additional 19S complex, composed of regulatory proteins and six ATPases. HslU is a member of the HSP100/Clp family of proteins and contains only one ATP-binding domain per subunit (Schirmer *et al.*, 1996). HslU has therefore also been designated ClpY (Gottesman *et al.*, 1993), and the name **ClpQ** has been proposed for HslV (Gottesman *et al.*, 1995). However, the name ClpQ appears to be misleading, since HslV has no sequence similarity to the HSP100/Clp family or the ClpP peptidase (Chapter 177), and HslV and ClpP have very different proteolytic mechanisms.

## Activity and Specificity

The rapid hydrolysis of proteins and certain hydrophobic peptides by HslV requires ATP hydrolysis by HslU (Rohrwild *et al.*, 1996; Yoo *et al.*, 1996). The enzyme was initially characterized with fluorogenic peptide substrates and no endogenous substrates for HslVU have been reported. In the presence of ATP, the protease rapidly hydrolyzes the chymotrypsin substrate Z-Gly-Gly-Leu↓NHMec, and cleaves very slowly Suc-Ala-Ala-Phe↓NHMec and Suc-Leu-Leu-Val-Tyr↓NHMec, but does not hydrolyze Suc-Leu-Tyr-NHMec, Z-Gly-Pro-NHMec, Suc-Phe-Leu-PheNHNap or Glt-Ala-Ala-Ala-NHNapOMe.

No trypsin-like activity (against Boc-Leu-Arg-Arg-NHMec, Boc-Gly-Lys-Arg-NHMec, Bz-Phe-Val-Arg-NHMec, Z-Gly-Gly-Arg-NHMec, Z-Arg-NHMec) or peptidyl-glutamyl-peptidase activity (against Z-Leu-Leu-Glu-NHNapOMe, Ac-Tyr-Val-Ala-Asp-NHMec) was detected.

Purified HslVU has also been shown to degrade certain unfolded proteins such as insulin B chain, $\alpha$-, $\beta$- and $\kappa$-casein, and carboxymethylated lactalbumin, but showed little or no activity against $\gamma$-globulin, lysozyme or bovine serum albumin (Seol *et al.*, 1997). ATP stimulates peptidase activity up to 150-fold, whereas other nucleotide triphosphates (GTP, UTP), or nonhydrolyzable ATP analogs, ADP and AMP have no effect (Rohrwild *et al.*, 1996; Yoo *et al.*, 1996). ADP is an inhibitor of Z-Gly-Gly-Leu-NHMec hydrolysis and at equivalent concentrations, inhibits the effect of ATP (Rohrwild *et al.*, 1996). HslV stimulates ATP hydrolysis by HslU, and HslU stimulates peptide and protein hydrolysis by HslV, especially in the presence of ATP (Yoo *et al.*, 1996; Seol *et al.*, 1997). Therefore, each of the two components is likely to influence the conformation of the other within the active HslVU complex. Moreover, HslVU is both a protein-activated ATPase and an ATP-dependent protease, and these two functions are closely linked (Yoo *et al.*, 1996; Seol *et al.*, 1997). Recent evidence indicates that although ATP hydrolysis is normally essential for proteolysis, under certain conditions AMP-PNP can substitute for ATP in peptide and protein hydrolysis (Huang & Goldberg, 1997). Conditions for peptidase assays and ATP hydrolysis have been described previously (Rohrwild *et al.*, 1996; Yoo *et al.*, 1996).

Several types of inhibitors that block the mammalian proteasome also inhibit the HslVU protease, including the peptide aldehydes Z-Leu-Leu-norleucinal (MG132) and Ac-Leu-Leu-norleucinal (calpain inhibitor I). Activity is also inhibited by DCI and *N*-ethylmaleimide, but not by inhibitors of other classes of proteases (Rohrwild *et al.*, 1996). Lactacystin strongly inhibits protein degradation, but curiously it has little effect on the hydrolysis of small molecule substrates (Rohrwild *et al.*, 1996; Seol *et al.*, 1997). HslV therefore seems to have a threonine-dependent catalytic mechanism similar to that of proteasomes from archaea and eukarya. Recent mutagenesis studies indicate that replacement of the N-terminal Thr residue of HslV with a Val prevents peptidase and proteinase activity (S.J. Yoo, unpublished results).

## Structural Chemistry

Purified HslV behaves as a 250 kDa protein on gel filtration, and thus appears to be an oligomer of 12–14 subunits (Yoo *et al.*, 1996). Recent negative staining electron microscopy and image analysis revealed that HslV forms a dodecamer ($\sim$ 230 kDa) consisting of two stacked hexameric rings with a dyad across the equatorial plane of the molecule ($V_6V_6$) (Rohrwild *et al.*, 1997). Purified HslU has been shown to assemble in the presence of ATP into a large complex, which appeared to contain at least 8–10 subunits by gel filtration (Yoo *et al.*, 1996). Image analysis, cross-linking and scanning transmission electron microscopy (STEM) mass estimations indicate that HslU can exist both as hexameric ($\sim$ 296 kDa) and heptameric rings ($\sim$ 346 kDa) (Rohrwild *et al.*, 1997). HslV and HslU can be coimmunoprecipitated from *E. coli* extracts and thus appear to form a complex *in vivo* (Rohrwild *et al.*, 1996). In the presence of AMP-PNP purified HslV and HslU form cylindrical particles resembling 20S proteasomes or ClpAP in which the HslV dodecamer is flanked at each end by a HslU ring ($U_6V_6V_6U_6$ or $U_7V_6V_6U_7$) (Rohrwild *et al.*, 1997). The experimentally determined diameter for the HslU heptamer is almost identical to the diameter of the outer rings of the HslVU complex. Thus, the heptameric HslU species is probably the form that associates with HslV, although further evidence is necessary.

## Preparation

The purification of HslV and HslU has recently been detailed by Yoo and coworkers (Yoo *et al.*, 1996).

## Biological Aspects

The discovery of proteasome-related genes in eubacteria was surprising since proteasomes were thought to exist exclusively in eukaryotes and certain archaea (Dahlmann *et al.*, 1989). Today, DNA sequences similar to *hslV* have been reported in the purple bacteria *Alcaligenes eutrophus* (Hein & Steinbuchel, 1994), *Haemophilus influenzae* (Fleischmann *et al.*, 1995), *Pasteurella haemolytica* (Highlander *et al.*, 1993), *Pseudomonas aeruginosa* (Lupas *et al.*, 1994), the spirochete *Borrelia burgdorferi* (Genbank: U43739) and the gram-positive bacteria *Bacillus subtilis* (Slack *et al.*, 1995) and *Lactobacillus leichmannii* (Becker & Brendel, 1996). In all these organisms with the exception of *Alcaligenes eutrophus*, the *hslV* gene is accompanied by a gene encoding HslU. The *hslU* gene shows no sequence similarity to *hslV* or to genes encoding proteasome $\alpha$ and $\beta$ subunits. Thus far, no sequence resembling $\alpha$-type subunits, which comprise the outer rings of the 20S proteasome, have been found in *E. coli*. The function of the $\alpha$ rings in proteolysis is unknown, although these structures are essential for the processing, folding and assembly of the $\beta$ rings in *Thermoplasma* proteasomes (Zwickl *et al.*, 1994; Seemüller *et al.*, 1996) and for nuclear localization of proteasomes (Nederlof *et al.*, 1995). By contrast, HslV associates into stable oligomeric rings in the absence of other proteins (Yoo *et al.*, 1996; Rohrwild *et al.*, 1997).

In response to heat shock and other conditions that cause protein damage, cells increase the synthesis of a group of proteins known as heat shock proteins (HSPs) (Goldberg, 1992; Parsell & Lindquist, 1993), which function either as molecular chaperones in the refolding of the damaged polypeptides or as proteases which catalyze the degradation of such abnormal proteins (Goldberg, 1992; Parsell & Lindquist, 1993; Hayes & Dice, 1996). Among the *E. coli* proteases induced by heat shock are four ATP-dependent proteolytic complexes: protease La (Lon) (Goff *et al.*, 1984; Chapter 178), ClpAP/ClpXP (Ti) (Kroh & Simon, 1990; Wojtkowiak *et al.*, 1993; Chapter 177), FtsH

(HflB) (Tomoyasu *et al.*, 1995; Chapter 516), and the HslVU protease (Rohrwild *et al.*, 1996; Yoo *et al.*, 1996). The *hslVU* promoter perfectly matches the consensus of promoters recognized by $\sigma^{32}$, and the transcription of this operon increases upon heat shock (Chuang *et al.*, 1993). Moreover, it has been shown that HslV and HslU participate in the degradation of incomplete polypeptides induced by the incorporation of puromycin (a tRNA analog) which induces the premature release of polypeptide chains from ribosomes (Missiakas *et al.*, 1997). Therefore, it seems likely that HslV and HslU are involved in the degradation of highly abnormal polypeptides or proteins damaged during heat shock. In addition, there may be normal cell proteins that are substrates for this proteolytic enzyme. However, mutants lacking the HslVU protease are hypersensitive to puromycin and show a mucoid phenotype, but are otherwise capable of normal growth (Missiakas *et al.*, 1997). Therefore, other cytosolic proteases may have overlapping functions with the HslVU protease and can compensate for the loss of HslVU activity.

## References

Becker, J. & Brendel, M. (1996) Molecular characterization of the *xerC* gene of *Lactobacillus leichmannii* encoding a site-specific recombinase and two adjacent heat shock genes. *Curr. Microbiol.* **32**, 232–236.

Chuang, S.E. & Blattner, F.R. (1993) Characterization of twenty-six new heat shock genes of *Escherichia coli. J. Bacteriol.* **175**, 5242–5252.

Chuang, S.E., Burland, V., Plunkett, III, G., Daniels, D.L. & Blattner, F.R. (1993) Sequence analysis of four new heat-shock genes constituting the hslTS/ibpAB and hslVU operons in *Escherichia coli. Gene* **134**, 1–6.

Coux, O., Tanaka, K. & Goldberg, A.L. (1996) Structure and functions of the 20S and 26S proteasomes. *Annu. Rev. Biochem.* **65**, 801–847.

Dahlmann, B., Kopp, F., Kuehn, L., Niedel, B., Pfeifer, G., Hegerl, R. & Baumeister, W. (1989) The multicatalytic proteinase (prosome) is ubiquitous from eukaryotes to archaebacteria. *FEBS Lett.* **251**, 125–131.

Fleischmann, R.D., Adams, M.D., White, O., Clayton, R.A., Kirkness, E.F., Kerlavage, A.R., Bult, C.J., Tomb, J.F., Dougherty, B.A., Merrick, J.M. *et al.* (1995) Whole-genome random sequencing and assembly of *Haemophilus influenzae* Rd. *Science* **269**, 496–512.

Goff, S.A., Casson, L.P. & Goldberg, A.L. (1984) Heat shock regulatory gene htpR influences rates of protein degradation and expression of the lon gene in *Escherichia coli. Proc. Natl Acad. Sci. USA* **81**, 6647–6651.

Goldberg, A.L. (1992) The mechanism and functions of ATP-dependent proteases in bacterial and animal cells. *Eur. J. Biochem.* **203**, 9–23.

Gottesman, S., Clark, W.P., Crecy-Lagard, V. & Maurizi, M.R. (1993) ClpX, an alternative subunit for the ATP-dependent Clp protease of *Escherichia coli*. Sequence and *in vivo* activities. *J. Biol. Chem.* **268**, 22618–22626.

Gottesman, S., Wickner, S., Jubete, Y., Singh, S.K., Kessel, M. & Maurizi, M. (1995) Selective, energy-dependent proteolysis in *Escherichia coli. Cold Spring Harbor Symp. Quant. Biol.* **LX**, 533–548.

Hayes, S.A. & Dice, J.F. (1996) Roles of molecular chaperones in protein degradation. *J. Cell Biol.* **132**, 255–258.

Hein, S. & Steinbuchel, A. (1994) Biochemical and molecular characterization of the *Alcaligenes eutrophus* pyruvate dehydrogenase complex and identification of a new type of dihydrolipoamide dehydrogenase. *J. Bacteriol.* **176**, 4394–4408.

Highlander, S.K., Wickersham, E.A., Garza, O. & Weinstock, G.M. (1993) Expression of the *Pasteurella haemolytica* leukotoxin is inhibited by a locus that encodes an ATP-binding cassette homolog. *Infect. Immun.* **61**, 3942–3951.

Huang, H.C. & Goldberg, A.L. (1997) Proteolytic activity of the ATP-dependent protease HslVU can be uncoupled from ATP hydrolysis. *J. Biol. Chem.* **272**, 21364–21372.

Kroh, H.E. & Simon, L.D. (1990) The ClpP component of Clp protease is the $\sigma^{32}$-dependent heat shock protein F21.5. *J. Bacteriol.* **172**, 6026–6034.

Lupas, A., Zwickl, P. & Baumeister, W. (1994) Proteasome sequences in eubacteria. *Trends Biochem. Sci.* **19**, 533–534.

Missiakas, D., Schwager, F., Betton, J.-M., Georgopoulos, C. & Raina, S. (1997) Identification and characterization of HslV HslU (ClpQ ClpY) proteins involved in overall proteolysis of misfolded proteins in *Escherichia coli. EMBO J.* **15**, 6899–6909.

Nederlof, P.M., Wang, H.R. & Baumeister, W. (1995) Nuclear localization signals of human and *Thermoplasma* a subunits are functional *in vitro. Proc. Natl Acad. Sci. USA* **92**, 12060–12064.

Parsell, D.A. & Lindquist, S. (1993) The function of heat-shock proteins in stress tolerance: degradation and reactivation of damaged proteins. *Annu. Rev. Genet.* **27**, 437–496.

Rohrwild, M., Coux, O., Huang, H.C., Moerschell, R.P., Yoo, S.J., Seol, J.H., Chung, C.H. & Goldberg, A.L. (1996) HslV-HslU – a novel ATP-dependent protease complex in *Escherichia coli* related to the eukaryotic proteasome. *Proc. Natl Acad. Sci. USA* **93**, 5808–5813.

Rohrwild, M., Pfeifer, G., Santarius, U., Muller, S.A., Huang, H.C., Engel, A., Baumeister, W. & Goldberg, A.L. (1997) The ATP-dependent HslVU protease from *Escherichia coli* is a four-ring structure resembling the proteasome. *Nature Struct. Biol.* **4**, 133–139.

Schirmer, E.C., Glover, J.R., Singer, M.A. & Lindquist, S. (1996) HSP100/Clp proteins: a common mechanism explains diverse functions. *Trends Biochem. Sci.* **21**, 289–296.

Seemüller, E., Lupas, A. & Baumeister, W. (1996) Autocatalytic processing of the 20S proteasome. *Nature* **382**, 468–470.

Seol, J.H., Yoo, S.L., Shin, D.H., Shim, Y.K., Kang, M.S., Goldberg, A.L. & Chung, C.H. (1997) The HslVU Protease from *Escherichia coli* has both ATP-dependent proteinase and protein-activated ATPase activity. *Eur. J. Biochem.* **247**, 1143–1150.

Slack, F.J., Serror, P., Joyce, E. & Sonenshein, A.L. (1995) A gene required for nutritional repression of the *Bacillus subtilis* dipeptide permease operon. *Mol. Microbiol.* **15**, 689–702.

Tomoyasu, T., Gamer, J., Bukau, B., Kanemori, M., Mori, H., Rutman, A.J., Oppenheim, A. B., Yura, T., Yamanaka, K., Niki, H. *et al.* (1995) *Escherichia coli* FtsH is a membrane-bound, ATP-dependent protease which degrades the heat-shock transcription factor sigma 32. *EMBO J.* **14**, 2551–2560.

Wojtkowiak, D., Georgopoulos, C. & Zylicz, M. (1993) Isolation and characterization of ClpX, a new ATP-dependent specificity component of the Clp protease of *Escherichia coli. J. Biol. Chem.* **268**, 22609–22617.

Yoo, S.J., Seol, J.H., Shin, D.H., Rohrwild, M., Kang, M.S., Tanaka, K., Goldberg, A.L. & Chung, C.H. (1996) Purification and characterization of the heat shock proteins HslV and HslU that form a new ATP-dependent protease in *Escherichia coli. J. Biol.*

*Chem.* **271**, 14035–14040.

Zwickl, P., Kleinz, J. & Baumeister, W. (1994) Critical elements in proteasome assembly. *Nature Struct. Biol.* **1**, 765–770.

**Markus Rohrwild**
Department of Cell Biology,
Harvard Medical School,
240 Longwood Avenue,
Boston, MA 02115-5730, USA

**H.-C. Huang**
Department of Cell Biology,
Harvard Medical School,
240 Longwood Avenue,
Boston, MA 02115-5730, USA

**Alfred L. Goldberg**
Department of Cell Biology,
Harvard Medical School,
240 Longwood Avenue,
Boston, MA 02115-5730, USA
Email: agoldberg@bcmp.med.harvard.edu

**Soon Ji Yoo**
Department of Molecular Biology and Research
Center for Cell Differentiation,
College of Natural Sciences,
Seoul National University,
Seoul 151-742, Korea

**Chin Ha Chung**
Department of Molecular Biology and Research
Center for Cell Differentiation,
College of Natural Sciences,
Seoul National University,
Seoul 151-742, Korea

# 172. Eukaryotic 20S proteasome

## Databanks

Peptidase classification: clan TA, family T1, MEROPS ID: T01.001
NC-IUBMB enzyme classification: EC 3.4.99.46
Databank codes:

| Species | Gene | SwissProt | PIR | EMBL (cDNA) | EMBL (genomic) |
|---|---|---|---|---|---|
| α1 subunit (ι) | | | | | |
| Arabidopsis thaliana | – | – | – | Z27274 | – |
| Arabidopsis thaliana | – | – | – | Z29892 | – |
| Caenorhabditis elegans | – | – | – | T00439 | – |
| Homo sapiens | pros-27 | P34062 | PC2322 | X59417 | – |
| | | | S25413 | | |
| | | | S17523 | | |
| | | | S30274 | | |
| Rattus norvegicus | – | – | JX0230 | D10755 | – |
| Saccharomyces cerevisiae | prs2 | P21243 | S11658 | M55440 | – |
| | | | S12939 | M63641 | |
| | | | S31557 | X56732 | |
| | | | JQ0115 | Z72533 | |
| | | | A38770 | | |
| | | | S11199 | | |
| α2 subunit (C3) | | | | | |
| Drosophila melanogaster | pros-25 | P40301 | S36117 | X70304 | – |
| | | | A49550 | | |
| | | | S36116 | | |

*continued overleaf*

| Species | Gene | SwissProt | PIR | EMBL (cDNA) | EMBL (genomic) |
|---|---|---|---|---|---|
| **α2 subunit (C3) (*continued*)** | | | | | |
| *Homo sapiens* | *psC3* | P25787 | PC2325<br>S15970 | D00760 | – |
| *Mus musculus* | – | P49722 | – | X70303 | – |
| *Rattus norvegicus* | – | P17220 | A34535<br>A38800<br>S09742 | J02897 | – |
| *Saccharomyces cerevisiae* | *prs4* | P23639 | S12938<br><br>A38773<br>S49635 | X56731 | Z46660: chromosome XIII<br>cosmid |
| *Xenopus laevis* | – | P24495 | JH0421 | S51111 | – |
| **α3 subunit (C9)** | | | | | |
| *Dictyostelium discoideum* | *prdD* | P34119 | – | L22212 | – |
| *Drosophila melanogaster* | *pros-29* | P18053 | S10318 | X52319 | – |
| *Homo sapiens* | *psc9* | P25789 | PC2324<br>S15972 | D00763 | – |
| *Rattus norvegicus* | – | P21670 | A38766<br>S15066<br>JQ0653<br>A38765<br>S10566 | X55986 | – |
| *Saccharomyces cerevisiae* | *prs5* | P23638 | S12940<br>A38774<br>S64444 | M63851<br>X56730 | Z72920: chromosome VII ORF |
| *Sus scrofa* | – | – | – | X91847 | – |
| *Schizosaccharomyces pombe* | – | Q09682 | – | Z50112 | – |
| *Spinacia oleracea* | – | .P52427 | – | X96974 | – |
| **α4 subunit** | | | | | |
| *Arabidopsis thaliana* | – | P30186 | – | X66825 | – |
| *Dictyostelium discoideum* | *prdE* | P34120 | – | L22213 | – |
| *Drosophila melanogaster* | *pros28.1* | P22769 | JQ0681 | M57712 | X62286: complete gene |
| *Drosophila melanogaster* | – | – | – | U46008 | – |
| *Drosophila melanogaster* | – | – | – | U46009 | – |
| *Homo sapiens* | – | – | PC2326 | – | – |
| *Rattus norvegicus* | – | P48004 | – | D30804 | – |
| *Saccharomyces cerevisiae* | *pre6* | P40303 | B55904 | L34348 | Z74780: chromosome XV ORF |
| *Schizosaccharomyces pombe* | – | Q10329 | – | Z69909 | – |
| **α5 subunit (ζ)** | | | | | |
| *Drosophila melanogaster* | *prosma5* | Q95083 | – | U64721 | – |
| *Entamoeba histolyticum* | – | – | – | X99382 | – |
| *Homo sapiens* | – | P28066 | S25411<br>PC2315<br>S17521 | X61970 | – |
| *Rattus norvegicus* | – | P34064 | JX0229 | D10756 | – |
| *Saccharomyces cerevisiae* | *pup2* | P32379 | S64585<br>S55335<br>S26705 | X64918 | Z73038: chromosome VII ORF |
| *Schistosoma mansoni* | – | – | – | T14568 | – |
| **α6 subunit (C2, ν)** | | | | | |
| *Arabidopsis thaliana* | *psm30* | P34066 | S39900 | M98495 | – |
| *Dictyostelium discoideum* | *prtC* | – | – | – | U60168: complete gene |
| *Drosophila melanogaster* | *pros-35* | P12881 | A38761<br>S23450<br>S05507 | X15497 | X62285: complete gene |

| Species | Gene | SwissProt | PIR | EMBL (cDNA) | EMBL (genomic) |
|---|---|---|---|---|---|
| α6 subunit (C2, ν) (*continued*) | | | | | |
| *Homo sapiens* | *psC2* | P25786 | S17520 | M64992 | – |
| | | | S15897 | D00759 | |
| | | | PC2321 | | |
| | | | JC1445 | | |
| | | | S25410 | | |
| *Oryza sativa* | – | P52428 | – | D37886 | – |
| *Rattus norvegicus* | – | P18420 | S09741 | M29859 | – |
| | | | A38799 | D90265 | |
| | | | A32968 | | |
| *Saccharomyces cerevisiae* | *pre5* | P40302 | S47909 | L34347 | Z54141: chromosome XIII cosmid |
| | | | S59307 | | |
| | | | A55904 | | |
| α7 subunit (C8) | | | | | |
| *Caenorhabditis elegans* | *zk945.2* | Q09583 | – | Z48544 | – |
| *Homo sapiens* | *psc8* | P25788 | PC2317 | D00762 | – |
| | | | S15971 | | |
| *Rattus norvegicus* | – | P18422 | A30482 | D90258 | – |
| | | | A35894 | M58593 | |
| | | | B35894 | X55985 | |
| | | | S09743 | | |
| | | | JQ0652 | | |
| | | | S14004 | | |
| *Saccharomyces cerevisiae* | *prs1* | 21242 | A38769 | M55436 | Z75270: chromosome XV ORF |
| | | | S11182 | | |
| | | | S67274 | | |
| β1 subunit (δ) | | | | | |
| *Homo sapiens* | *psmb6* | P28072 | S17522 | D29012 | – |
| | | | B54589 | | |
| | | | S25412 | | |
| | | | S08188 | | |
| *Mus musculus* | – | – | – | U13393 | – |
| *Mus musculus* | – | – | – | U13394 | – |
| *Rattus norvegicus* | – | P28073 | JX0228 | D10754 | – |
| | | | S09086 | | |
| *Saccharomyces cerevisiae* | *pre3* | P38624 | S61338 | X78991 | X87611: chromosome X |
| | | | S61337 | X86020 | Z49276: chromosome X ORF |
| | | | S56771 | | |
| | | | S43669 | | |
| | | | S56167 | | |
| | | | S55187 | | |
| β1i subunit (C7) | | | | | |
| *Homo sapiens* | *lmp2* | P28065 | S27332 | S75169 | Z14982: complete gene, alternative splicing |
| | | | S17896 | U01025 | Z14977: complete gene |
| | | | B44324 | X62741 | X66401: complete gene |
| *Mus musculus* | *lmp2* | P28076 | A45744 | U22919 | L11613: complete gene |
| | | | | U22920 | |
| | | | | U22448 | |
| | | | | U22447 | |
| *Rattus norvegicus* | – | P28077 | JX0231 | D10757 | – |
| β2 subunit (Z) | | | | | |
| *Homo sapiens* | – | – | PC2319 | D38048 | – |
| *Mus musculus* | *psmb7* | – | – | D83585 | – |

*continued overleaf*

| Species | Gene | SwissProt | PIR | EMBL (cDNA) | EMBL (genomic) |
|---|---|---|---|---|---|
| **β2 subunit (Z) (continued)** | | | | | |
| Saccharomyces cerevisiae | pup1 | P25043 | S26996 | X61189 | Z75065: chromosome XV ORF |
| Schizosaccharomyces pombe | – | Q09841 | – | Z64354 | – |
| **β2i subunit (MECL-1)** | | | | | |
| Homo sapiens | mecl1 | P40306 | I38135 | – | X71874: complete gene |
| **β3 subunit (θ)** | | | | | |
| Bos taurus | – | P33672 | – | – | – |
| Homo sapiens | – | P49720 | S50148 | D26598 | – |
| Rattus norvegicus | – | P40112 | S40468 | D21800 | – |
| Saccharomyces cerevisiae | pup3 | P25451 | S27450 | M88470 | U18839: chromosome V cosmids |
| **β4 subunit (C7-I)** | | | | | |
| Caenorhabditis elegans | – | – | – | T00209 | – |
| Homo sapiens | – | P49721 | S50149 | D26599 | – |
| Rattus norvegicus | – | P40307 | – | D21799 | – |
| Saccharomyces cerevisiae | pre1 | P22141 | S50470 | X56812 | U18778: chromosome V cosmids |
| Schizosaccharomyces pombe | – | Q09720 | – | Z50113 | – |
| **β5 subunit (ε)** | | | | | |
| Arabidopsis thaliana | – | – | – | F13852 | – |
| Caenorhabditis elegans | – | – | – | T00391 | – |
| Gallus gallus | – | P34065 | S30539 | X57210 | – |
| Homo sapiens | psmb5 | P28074 | A54589 S08189 | D29011 S74378 | – |
| Rattus norvegicus | – | P28075 | S09087 | D45247 | – |
| Saccharomyces cerevisiae | pre2 | P30656 | A45411 | Z28348 M96667 S78566 X68662 | U32445: chromosome XVI cosmids |
| Schizosaccharomyces pombe | pts1 | P30655 | – | D13094 | – |
| **β5i subunit (C13)** | | | | | |
| Homo sapiens | lmp7 | P28062 | C44324 A44324 | U32862 U17497 U17496 X62598 U32863 | Z14982: alternatively spliced gene X66401: complete gene |
| Mus musculus | lmp7 | P28063 | – | U22035 U22033 U22034 U22032 X64449 U22031 | L11145: complete gene |
| Rattus norvegicus | – | P28064 | S21126 | D10729 | – |
| **β6 subunit (C5, γ)** | | | | | |
| Arabidopsis thaliana | – | P42742 | – | Z35020 X67338 X79806 | – |
| Caenorhabditis briggsae | – | – | – | R04379 | – |
| Caenorhabditis elegans | c02f5.9 | P34286 | S44611 | L14745 | – |
| Drosophila melanogaster | pros26 | P40304 | – | U00790 | – |
| Homo sapiens | psC5 | P20618 | S08187 S15973 | D00761 | – |
| Mus musculus | – | – | – | X80686 | – |
| Rattus norvegicus | – | P18421 | – | X52783 | – |

| Species | Gene | SwissProt | PIR | EMBL (cDNA) | EMBL (genomic) |
|---|---|---|---|---|---|
| β6 subunit (C5, γ) (*continued*) | | | | | |
| *Saccharomyces cerevisiae* | *prs3* | P23724 | S42436 | D00845 | X78214: chromosome II fragment |
| | | | | M34777 | Z35802: chromosome II ORF |
| *Zea mays* | – | – | – | T18278 | – |
| β7 subunit (β) | | | | | |
| *Homo sapiens* | – | P28070 | S05147 | D26600 | – |
| | | | S08186 | S71381 | |
| *Rattus norvegicus* | – | P34067 | S32507 | L17127 | |
| *Saccharomyces cerevisiae* | *pre4* | P30657 | A46610 | X68663 | D50617: chromosome VI |
| | | | | | D44597: chromosome VI cosmid |
| *Xenopus laevis* | – | P28024 | S17568 | X62709 | – |
| Others | | | | | |
| *Arabidopsis thaliana* | – | – | – | F13823 | – |
| *Caenorhabditis briggsae* | – | – | – | R04354 | – |
| *Medicago sativa* | – | – | – | Z71998 | – |

Note: The subunit names correspond to the systematic names for the respective subunits in the yeast crystal structure (Groll *et al.*, 1997). When viewed along the axis of pseudo-sevenfold symmetry, the subunits in the proximal rings are numbered counterclockwise relative to the axis of C2 symmetry. The names formerly used for these subunits are listed in Table 172.2.

## Name and History

The first description of proteasome particles resulted from an electron microscopic study of human erythrocyte lysates ('cylindrin': Harris, 1968), but the function of these particles remained unknown. From 1980 on, a growing number of reports described a large cellular protease with multicatalytic activities (Wilk & Orlowski, 1980; Hase *et al.*, 1980), but its identity with the cylindrin particles was not established until 1988 (Arrigo *et al.*, 1988; Falkenburg *et al.*, 1988; Baumeister *et al.*, 1988). Because of its ubiquity in eukaryotes, and central role in cellular proteolysis, the proteasome was independently discovered in many different systems. The more than 30 different names it accumulated reflect the problems which were encountered in determining its biochemical and cellular functions; these names are shown in Table 172.1. In 1988 it was established beyond doubt that the different names referred to the same particle and the name ***proteasome*** was proposed (Arrigo *et al.*, 1988) to indicate its proteolytic and particulate nature. From 1989 on it became clear that the proteasome functions as the proteolytic core of a larger ATP-dependent complex (Eytan *et al.*, 1989; Driscoll & Goldberg, 1990; Orino *et al.*, 1991; Peters *et al.*, 1991) (see Chapter 173). The two complexes are now referred to as the 20S and 26S proteasomes, in reference to their sedimentation coefficients of 20 and 26 S, respectively (although the sedimentation coefficient of the 26S proteasome is in fact about 30S: Yoshimura *et al.*, 1993). The 19S cap complexes, with which the 20S proteasome associates to form the 26S proteasome (Peters *et al.*, 1993) have themselves accumulated a large number of names, including CF-1 (Ganoth *et al.*, 1988), PA700 (Chu-Ping *et al.*, 1994), 20S ball (Hoffman *et al.*, 1992) and m particle (Udvardy, 1992). Besides the 19S cap complex, an 11S regulator (Dubiel *et al.*, 1992), also called PA28 (Chu-Ping *et al.*, 1992), can associate with the 20S proteasome (Gray *et al.*, 1994). Several other regulators of the 20S proteasome have been identified but remain poorly characterized (reviewed in Peters, 1994).

## Activity and Specificity

The proteasomes of different organisms may vary significantly in their catalytic properties and, to a lesser extent, also in their choice of preferred cleavage sites (reviewed in Rivett *et al.*, 1994). In addition, in mammalian proteasomes, the three catalytically active β-type subunits can be replaced by γ-interferon-inducible homologs, which are expressed even in the absence of stimulation at different levels in different tissues. It has therefore been impossible to isolate compositionally homogeneous proteasomes from mammalian cells so far.

When assayed with fluorogenic peptide substrates, eukaryotic proteasomes have three main cleavage specificities: after basic residues (trypsin-like), after large hydrophobic residues (chymotrypsin-like), and after acidic residues (peptidylglutamyl peptide hydrolyzing) (Wilk & Orlowski, 1980). Two additional specificities have been determined for mammalian proteasomes, involving cleavage after branched-chain amino acids (BrAAP activity) and between small neutral amino acids (SNAAP activity) (Cardozo *et al.*, 1992; Orlowski *et al.*, 1993). The main synthetic substrates used to assay proteasome activity are Suc-Leu-Leu-Val-Tyr↓NHMec, Suc-Ala-Ala-Phe↓NHMec, Boc-Leu-Ser-Thr-Arg↓NHMec, Boc-Leu-Arg-Arg↓NHMec, Z-Gly-Gly-Arg↓NHMec and Z-Leu-Leu-Glu↓NHNap (available from Sigma or Bachem) (see Appendix 2 for full names and addresses of suppliers). The $K_m$ values for these substrates are typically between 0.1 and 1 mM. The main nonsynthetic substrates used are oxidized insulin B chain and casein (available from Sigma). The pH optima are different for proteasome preparations from different organisms and tissues, and are also dependent

*Table 172.1*   History of names given to the 20S proteasome complex

| Name | Source | Reference |
|---|---|---|
| Cylindrin | Human erythrocytes | Harris (1968) |
| Small cytoplasmic particles | HeLa cells | Spohr et al. (1970) |
| Minute ring-shaped particles | Human adenocarcinoma | Narayan & Rounds (1973) |
| High molecular weight protease | Mouse liver | Rose et al. (1979) |
| Alkaline protease | Carp muscle | Hase et al. (1980) |
| Cation-sensitive neutral endopeptidase | Cattle pituitary | Wilk & Orlowski (1980) |
| High molecular weight enzyme | Human skeletal muscle | Hardy et al. (1981) |
| High molecular weight peptide hydrolase | Human erythrocytes | Edmunds & Pennington (1982) |
| Donut-shaped 'miniparticles' | Various human tissues | Domae et al. (1982) |
| Low molecular weight polypeptides (LMP) | Mouse macrophages | Monaco & McDevitt (1982) |
| High molecular weight cysteine proteinase II/III | Rat skeletal muscle | Dahlmann et al. (1983) |
| High molecular weight cysteine endopeptidase | Rat skeletal muscle | Ismail & Gevers (1983) |
| 22S Cylinder particle | *Xenopus laevis* ovaries | Kleinschmidt et al. (1983) |
| Proteinase YscE | Yeast | Achstetter et al. (1984) |
| Ring-shaped miniparticle | Various mammalian tissues | Dang (1984) |
| Prosome | Mouse and duck blood cells, HeLa cells | Schmidt et al. (1984) |
| Neutral endopeptidase | Cattle lens | Ray & Harris (1985) |
| 19S Ring-type particles | *Drosophila melanogaster* | Schuldt & Kloetzel (1985) |
| Ingensin | Pig skeletal muscle | Ishiura et al. (1985) |
| Alkaline protease | Mouse liver | Rivett (1985) |
| Multifunctional protease or LAMP (large alkaline multifunctional complex) | Rat liver and human erythrocytes | Tanaka et al. (1986) |
| Cylinder-shaped particle | *Xenopus laevis* ovaries | Castaño et al. (1986) |
| Macropain | Human erythrocytes | McGuire & DeMartino (1986) |
| 20S Protease | Rabbit reticulocytes | Hough et al. (1987) |
| 20S Particle | Yeast, wheat, Drosophila, chicken, rabbit, HeLa cells | Arrigo et al. (1987) |
| Endopeptidase-24.5 | Human lung | Zolfaghari et al. (1987) |
| Multicatalytic proteinase | Rat skeletal muscle | Dahlmann et al. (1988) |
| Multicatalytic proteinase complex | Cattle pituitary | Orlowski & Wilk (1988) |
| Large multifunctional protease complex | Various mammalian cells | Arrigo et al. (1988) |
| Proteasome | | Arrigo et al. (1988) |
| Multicatalytic endopeptidase complex | | NC-IUBMB (1992) |

on the substrate used. In general the optima cluster around pH 7–8, but may lie anywhere between 5 and 10 in individual cases.

The classification of cleavage specificity using residues in the P1 position is useful for the description of individual active sites in the 20S complex, but has little relevance to the cleavage specificities in nonsynthetic peptides and proteins. Analyses of preferred cleavage sites in substrates such as insulin B chain (Rivett, 1985; Dick et al., 1991; Takahashi et al., 1993; Ehring et al., 1996), human histone H3 (Ehring et al., 1996), ovalbumin (Dick et al., 1994), gonadotropin-releasing hormone (Leibovitz et al., 1995), bradykinin (Zolfaghari et al., 1987) or neurotensin (Cardozo et al., 1992) show that these frequently cannot be interpreted in terms of the cleavage specificities determined with synthetic peptides. Rather, residues in the P3 and P4 positions are likely to contribute to the selection of a cleavage site by the proteasome (Cardozo et al., 1994; Coux et al., 1996; Tsubuki et al., 1996; Groll et al., 1997). The P2 position is solvent-exposed and unconstrained by the active-site cleft (Groll et al., 1997), and is therefore unlikely to contribute

to the cleavage specificity. So far, nothing is known about the influence of the P'-side positions in determining preferred cleavage sites.

The proteasome is a processive enzyme: after taking up a protein, it degrades it entirely to peptides before attacking the next molecule (Dick et al., 1991; Akopian et al., 1997). The peptides produced are mostly between 4 and 10 residues long. In *Thermoplasma*, this feature has led to the proposal of a 'molecular ruler' (see Chapter 169). Although not widely studied so far, the presence of multiple active sites in one complex leads to cooperative effects in substrate degradation (Djaballah & Rivett, 1992; Stein et al., 1996).

Upon purification, proteasomes are generally in a latent state. Two natural activators have been described: the 19S cap complex (PA700), which stimulates the degradation of peptides and proteins in an ATP-dependent manner, and the 11S complex (PA28), which only stimulates the degradation of peptides and is ATP independent. Latent proteasomes can also be activated by a large number of different chemical agents, such as SDS, polylysine, some divalent cations, fatty acid derivatives (reviewed in Rivett et al., 1994), some synthetic peptides (Wilk & Chen, 1997), and by brief heating to 60°C (Mykles, 1989). The most commonly used agent is 0.03% SDS.

In the search for the nature of the proteasome active site, a very large number of protease inhibitors were tested, mostly with poor inhibitory results (reviewed in Rivett *et al.*, 1994) and affecting the different catalytic activities to varying extents. The most useful inhibitors proved to be certain peptide aldehydes that act as substrate analogs. Amongst these, chymostatin and Ac-Leu-Leu-norleucinal (calpain inhibitor I) mainly inhibit the chymotrypsin-like activity, and leupeptin mainly the trypsin-like activity, all with $K_i$ values around 1 $\mu$M (Wilk & Orlowski, 1983; Vinitsky *et al.*, 1992; Rock *et al.*, 1994). (These inhibitors are available from Bachem and Calbiochem.) In the yeast proteasome crystal structure, Ac-Leu-Leu-norleucinal is found covalently attached to the catalytic threonines of all three active $\beta$-type subunits (Groll *et al.*, 1997). The only known specific inhibitor of the proteasome is lactacystin, a *Streptomyces* metabolite (available from Calbiochem), which converts spontaneously to the $\beta$-lactone (Dick *et al.*, 1996) and attaches covalently in this form to the catalytic threonine of the subunit with chymotrypsin-like activity, $\beta$5 (Fenteany *et al.*, 1995; Groll *et al.*, 1997). Lactacystin inhibits all three catalytic activities

of the proteasome, albeit with different efficiency (Fenteany & Schreiber, 1996).

## Structural Chemistry

Eukaryotic proteasomes contain 14 different but related subunits, which can be divided into two groups by their similarity to the $\alpha$ and $\beta$ subunits of the *Thermoplasma* proteasome (Pühler *et al.*, 1994). In the crystal structure of the yeast proteasome (Groll *et al.*, 1997), the seven $\alpha$-type subunits form the outer rings and the seven $\beta$-type subunits the inner rings of the complex. Their arrangement within the rings has led to a systematic nomenclature based on their location relative to the axis of C2 symmetry (Groll *et al.*, 1997). Table 172.2 lists the new systematic names together with the former names in yeast and mammals. The $\beta$-type subunits $\beta$1i, $\beta$2i and $\beta$5i, which are present only in mammals, are the $\gamma$-interferon-inducible homologs of subunits $\beta$1, $\beta$2 and $\beta$5.

Collectively, the $\alpha$ and $\beta$ subunits form a cylinder of approximately 15 nm length and 11 nm diameter ($M_r$ approximately 700 kDa), which is traversed from end to end by a channel and contains three large inner cavities separated by narrow constrictions. In the *Thermoplasma* proteasome, access to the cavities is controlled by the $\alpha$ ring constrictions,

*Table 172.2*   Names and biophysical properties of yeast and human 20S proteasome subunits

| Subunit | Yeast names | Yeast | | Mammalian names | Human | |
|---|---|---|---|---|---|---|
| | | $M_r$ | pI | | $M_r$ | pI |
| $\alpha$1 | C7a, PRS2, Y8, PRC2, SCL1 | 28.0 | 6.05 | iota, PROS-27 | 27.4 | 5.97 |
| $\alpha$2 | PRS4, Y7, PRE8 | 27.2 | 5.54 | C3 | 25.9 | 7.29 |
| $\alpha$3 | PRS5, Y13, PRE9 | 28.7 | 4.96 | C9 | 29.5 | 7.69 |
| $\alpha$4 | PRE6 | 28.4 | 7.27 | XAPC-7, C6 | 27.9 | 8.69 |
| $\alpha$5 | PUP2, DOA5 | 28.6 | 4.58 | zeta | 26.4 | 4.59 |
| $\alpha$6 | PRE5 | 25.6 | 7.30 | C2, PROS-30 | 29.6 | 6.16 |
| $\alpha$7 | C1, PRS1, PRC1, PRE10 | 31.6 | 5.20 | C8 | 28.4 | 5.06 |
| $\beta$1*PGPH | PRE3 | 23.5 (21.5) | 5.26 (5.26) | Y, delta, N5 | 25.3 (21.9) | 4.65 (4.76) |
| $\beta$1i* | | | | LMP2, RING12, N7 | 23.3 (21.3) | 4.75 (4.62) |
| $\beta$2*tryp | PUP1 | 28.3 (25.2) | 6.60 (6.53) | Z, alpha, N1 | 30.0 (25.3) | 7.61 (5.54) |
| $\beta$2i* | | | | MECL-1, N2 | 28.9 (24.6) | 7.73 (6.07) |
| $\beta$3 | PUP3, RAD51-ORF1 | 22.6 | 4.93 | C10-II, theta | 22.9 | 6.15 |
| $\beta$4 | C11, PRE1 | 22.5 | 6.01 | C7-I | 22.8 | 6.61 |
| $\beta$5*chym | PRE2, PRG1, DOA3 | 31.6 (23.3) | 6.02 (6.19) | X, MB1, epsilon | 22.9 (22.5) | 8.67 (8.64) |
| $\beta$5i* | | | | LMP7E1, C1, C13 | 29.8 | 5.46 |
| | | | | LMP7E2, RING10, | 30.4 (22.7) | 7.18 (7.69) |
| $\beta$6 | PRS3, PTS1, PRE7 | 27.1 (25.1) | 6.11 (6.86) | C5, gamma, N4 | 26.5 (23.5) | 8.20 (8.18) |
| $\beta$7 | PRE4 | 29.4 (25.1) | 5.75 (5.84) | N3, PROS-26, beta | 29.2 (24.4) | 5.63 (5.34) |

The $M_r$ and pI values are for the longest translation products of the respective open reading frames. Data were obtained from sequence databanks, except those for human XAPC-7 ($\alpha$4) which were taken from Tanahashi *et al.* (1993). Values in brackets are for processed forms. The catalytically active subunits are marked with asterisks. The subunits carrying the peptidylglutamyl-peptide hydrolyzing activity (PGPH), chymotrypsin-like activity (chym) and trypsin-like activity (tryp) are so labeled.

whose diameter is 1.3 nm and which allow only entirely unfolded polypeptide chains to pass (see Chapter 169). In the yeast proteasome, the entrances to the channel are blocked by a highly conserved N-terminal sequence of the α-type subunits (Groll *et al*., 1997); in the *Thermoplasma* proteasome structure, this sequence is disordered and not visible in the density map. The location of the conserved sequence within the channel may provide a structural explanation for the observation that eukaryotic 20S proteasomes are purified in a latent state. Their activation by chaotropic agents such as SDS or heat may be due to a selective denaturation of the conserved sequences, resulting in an opening of the channel. In the yeast proteasome structure, several small openings are visible in the outer wall of the inner cavities, which had not been seen in the *Thermoplasma* proteasome structure. It seems conceivable that these are exit points for the cleavage products.

The α and β subunits have a four-layer α + β fold and belong to the N-terminal nucleophile (Ntn)-hydrolase family. These enzymes have a 'single-residue' active site, which is freed by the autocatalytic cleavage of a pro sequence. After processing, the nucleophilic attack is initiated when the free amino group of the N-terminal residue strips the proton off the side chain via an intermediate water molecule. In the proteasome, the catalytic residue is always threonine (the fold and mechanism of proteasome subunits was first determined for the *Thermoplasma* proteasome and is discussed in detail in Chapter 169). The α-type subunits contain a conserved N-terminal extension in place of a pro sequence, are not processed, and do not carry proteolytic activity. Their N-termini are blocked to Edman degradation; in several cases, the modification has been identified as an acetyl group. Of the seven different β-type subunits, only three contain the residues necessary for processing and activity, and the three main cleavage specificities of the proteasome map to these subunits (Heinemeyer *et al*., 1993; Enenkel *et al*., 1994; Fenteany *et al*., 1995). Of the other four, inactive subunits, two are processed at an intermediate site (presumably by neighboring active subunits), and two remain unprocessed (Groll *et al*., 1997).

## Preparation

The proteasome is ubiquitous in eukaryotes and has been isolated from a very large number of organisms and tissues. Proteasome concentration varies considerably among cell types and is greater in organs in which average rate of protein breakdown is high (e.g. liver) than in other tissues (Tanaka *et al*., 1986). A detailed description of purification procedures as well as an overview of sources used for purification are given in Rivett *et al*. (1994).

## Biological Aspects

The 20S proteasome associates with two 19S cap complexes to form the 26S proteasome. The 26S complex is present in all eukaryotes and functions as the central protease of the ubiquitin-dependent pathway of protein degradation (reviewed in Peters, 1994; Lupas *et al*., 1995; Coux *et al*., 1996). The function of the cap complexes is to recognize substrate proteins, generally after these have been ubiquitinated, and to present them in unfolded form to the 20S core in an ATP-dependent process. It is now clear that the 26S proteasome not only degrades proteins that are targeted by the ubiquitin-dependent pathway but also directly recognizes cellular substrates; it must therefore be considered the main agent of nonlysosomal protein degradation in the cell. It is involved in the turnover of many regulatory proteins, such as cell cycle control proteins (cyclin), transcriptional regulators (Gal4, I6B), and rate-limiting enzymes (ornithine decarboxylase). In addition, it participates in the bulk degradation of proteins during cell differentiation and apoptosis, and may therefore also play a role in the accelerated degradation of muscle tissue observed in certain diseases. In at least one case – that of NFκB – it processes a precursor protein rather than entirely degrading it. It remains to be seen whether this is an exceptional case.

Because of the focused size distribution of the peptides produced by the 20S core (typically 4–10 residues), the proteasome has acquired a further function in mammals: it provides the peptides for MHC class I antigen display (Rock *et al*., 1994). Correspondingly, the three active β-type subunits can be replaced by interferon γ-inducible homologs upon immune stimulation. This replacement alters the catalytic specificity of the proteasome and is thought to increase the yield of peptides suitable for presentation by MHC class I molecules (Gaczynska *et al*., 1994). The system is further modulated by the association of the 20S core with the 11S regulator, which is also interferon γ inducible and which activates the peptidolytic but not the proteolytic activity of the proteasome (Groettrup *et al*., 1996). It has recently been found that HIV-1 Tat protein inhibits the 20S proteasome and interferes with the formation of the 20S–11S complex (which appears to be involved in the immune response), but activates the 26S proteasome (which is required for transcriptional regulation) (Seeger *et al*., 1997). Tat is known to bind to some of the ATPase subunits of the 19S complex and thereby activate viral gene expression (Nelbock *et al*., 1990; Shibuya *et al*., 1992).

Like most proteases, proteasome β-type subunits are synthesized in an inactive form containing a propeptide. In order to prevent proteolytic damage to cellular proteins, processing is coupled to proteasome assembly. This insures that, once processed, the activity of β-type subunits is confined to the innermost cavity of a molecular microcompartment, well segregated from the cellular environment. In eukaryotes, pro sequences vary greatly in length and may be involved in several steps along the assembly pathway, although their functions are still largely unknown. They appear to allow subunits to find their correct position in the proteasome rings and may be subunit specific (Chen & Hochstrasser, 1996). Eukaryotic proteasomes assemble via 15S precursor particles, which may correspond to 'half-proteasomes' and which contain unprocessed β subunits (Frentzel *et al*., 1994). The eukaryotic assembly pathway resembles that of *Rhodococcus*, which is discussed in Chapter 170.

## Distinguishing Features

A wide palette of proteasome-related reagents, including proteins of the 20S, 19S and 11S complexes, antibodies, substrates and inhibitors are available from Affinity Research Products.

## Related Peptidases

The eukaryotic proteasome has archaean and eubacterial homologs, which are discussed elsewhere in the present volume. In addition, it is related to the simpler eubacterial protease HslV (see Chapters 170 and 171).

## References

Achstetter, T., Ehmann, C., Osaki, A. & Wolf, D.H. (1984) Proteolysis in eukaryotic cells. Proteinase yscE, a new peptidase. *J. Biol. Chem.* **259**, 13344–13448.

Akopian, T.N., Kisselev, A.F. & Goldberg, A.L. (1997) Processive degradation of proteins and other catalytic properties of the proteasome from *Thermoplasma acidophilum*. *J. Biol. Chem.* **272**, 1791–1798.

Arrigo, A.-P., Simon, M., Darlix, J.-L. & Spahr, P.-F. (1987) A 20S particle ubiquitous from yeast to human. *J. Mol. Evol.* **25**, 141–150.

Arrigo, A.P., Tanaka, K., Goldberg, A.L. & Welch, W.J. (1988) Identity of the 19S 'prosome' particle with the large multifunctional protease complex of mammalian cells (the proteasome). *Nature* **331**, 192–194.

Baumeister, W., Dahlmann, B., Hegerl, R., Kopp, F., Kuehn, L. & Pfeifer, G. (1988) Electron microscopy and image analysis of the multicatalytic proteinase. *FEBS Lett.* **241**, 239–245.

Cardozo, C., Vinitsky, A., Hidalgo, M.C., Michaud, C. & Orlowski, M. (1992) A 3,4-dichloroisocoumarin-resistant component of the multicatalytic proteinase complex. *Biochemistry* **31**, 7373–7380.

Cardozo, C., Vinitsky, A., Michaud, C. & Orlowski, M. (1994) Evidence that the nature of amino acid residues in the $P_3$ position directs substrates to distinct catalytic sites of the pituitary multicatalytic proteinase complex (proteasome). *Biochemistry* **33**, 6483–6489.

Castaño, J.G., Ornberg, R., Koster, J.G., Tobian, J.A. & Zasloff, M. (1986) Eukaryotic pre-tRNA 5′ processing nuclease: copurification with a complex cylindrical particle. *Cell* **46**, 377–387.

Chen, P. & Hochstrasser, M. (1996) Autocatalytic subunit processing couples active-site formation in the 20S proteasome to completion of assembly. *Cell* **86**, 961–972.

Chu-Ping, M., Slaughter, C.A. & DeMartino, G.N. (1992) Identification, purification and characterization of a protein activator (PA28) of the 20S proteasome. *J. Biol. Chem.* **267**, 10515–10523.

Chu-Ping, M., Vu, J.H., Proske R.J., Slaughter, C.A. & DeMartino, G.N. (1994) Identification, purification and characterization of a high-molecular weight, ATP-dependent activator (PA700) of the 20S proteasome. *J. Biol. Chem.* **269**, 3539–3547.

Coux, O., Tanaka, K., and Goldberg, A.L. (1996) Structure and functions of the 20S and 26S proteasomes. *Annu. Rev. Biochem.* **65**, 801–847.

Dahlmann, B., Kuehn, L. & Reinauer, H. (1983) Identification of three high molecular mass cysteine proteinases from rat skeletal muscle. *FEBS Lett.* **160**, 243–247.

Dahlmann, B., Kuehn, L., Ishiura, S., Tsukahara, T., Sugita, H., Tanaka, K., Rivett, A.J., Hough, R.F., Rechsteiner, M., Mykles, D.L., Fagan, M., Waxman, L., Ishii, S., Sasaki, M., Kloetzel, P.-M., Harris, H., Ray, K., Behal, F.J., DeMartino, G.N. & McGuire, M.J. (1988) The multicatalytic proteinase: a high-$M_r$ endopeptidase. *Biochem. J.* **255**, 750–751.

Dang, C.V. (1984) Identity of the ubiquitous eukaryotic ring-shaped miniparticle. *Cell Biol. Int. Rep.* **8**, 323–327.

Dick, L.R., Moomaw, C.R., DeMartino, G.N. & Slaughter, C.A. (1991) Degradation of oxidized insulin B chain by the multiproteinase complex macropain (proteasome). *Biochemistry* **30**, 2725–2734.

Dick, L.R., Aldrich, C., Jameson, S.C., Moomaw, C.R., Pramanik, B.C., Doyle, C.K., DeMartino, G.N., Bevan, M.J. Forman, J.M. & Slaughter, C.A. (1994) Proteolytic processing of ovalbumin and beta-galactosidase by the proteasome to yield antigenic peptides. *J. Immunol.* **152**, 3884–3894.

Dick, L.R., Cruikshank, A.A., Grenier, L., Melandri, F.D., Nunes, S.L. & Stein, R.L. (1996) Mechanistic studies on the inactivation of the proteasome by lactacystin. *J. Biol. Chem.* **271**, 7273–7276.

Djaballah, H. & Rivett, A.J. (1992) Peptidylglutamyl-peptide hydrolase activity of the multicatalytic proteinase complex: evidence for a new high-affinity site, analysis of cooperative kinetics, and the effect of manganese ions. *Biochemistry* **31**, 4133–4141.

Domae, N., Harmon, F.R., Busch, R.K., Spohn, W., Subrahmanyan, C.S. & Busch, H. (1982) Donut-shaped 'Miniparticles' in nuclei of human and rat cells. *Life Sci.* **30**, 469–477.

Driscoll, J. & Goldberg, A.L. (1990) The proteasome (multicatalytic protease) is a component of the 1500 kDa proteolytic complex which degrades ubiquitin-conjugated proteins. *J. Biol. Chem.* **265**, 4789–4792.

Dubiel, W., Pratt, G., Ferrell, K. & Rechsteiner, M. (1992) Purification of an 11S regulator of the multicatalytic proteases. *J. Biol. Chem.* **267**, 22369–22377.

Edmunds, T. & Pennington, R.J.T. (1982) A high-molecular weight peptide hydrolase in erythrocytes. *Int. J. Biochem.* **14**, 701–703.

Ehring, B., Meyer, T.H., Eckerskorn, C., Lottspeich, F. & Tampé, R. (1996) Effects of major-histocompatibility-complex-encoded subunits on the peptidase and proteolytic activities of human 20S proteasomes. *Eur. J. Biochem.* **235**, 404–415.

Enenkel, C., Lehmann, H., Kipper, J., Gückel, R., Hilt, W. & Wolf, D.H. (1994) *PRE3*, highly homologous to the human major histocompatibility complex-linked *LMP2 (RING12)* gene, codes for a yeast proteasome subunit necessary for the peptidylglutamyl-peptide hydrolyzing activity. *FEBS Lett.* **341**, 193–196.

Eytan, E., Ganoth, D., Armon, T. & Hershko, A. (1989) ATP-dependent incorporation of 20S protease into the 26S complex that degrades proteins conjugated to ubiquitin. *Proc. Natl Acad. Sci. USA* **86**, 7751–7755.

Falkenburg, P.-E., Haass, C., Kloetzel, P.-M., Niedel, B., Kopp, F., Kuehn, L. & Dahlmann, B. (1988) *Drosophila* small cytoplasmic 19S ribonucleoprotein is homologous to the rat multicatalytic proteinase. *Nature* **331**, 190–192.

Fenteany, G. & Schreiber, S.L. (1996) Specific inhibition of the chymotrypsin-like activity of the proteasome induces a bipolar morphology in neuroblastoma cells. *Chem. Biol.* **3**, 905–912.

Fenteany, G., Standaert, R.F., Lane, W.S., Choi, S., Corey, E.J. & Schreiber, S.L. (1995) Inhibition of proteasome activities and subunit-specific amino-terminal threonine modification by lactacystin. *Science* **268**, 726–731.

Frentzel, S., Pesold, H.B., Seelig, A. & Kloetzel, P.M. (1994) 20S proteasomes are assembled *via* distinct precursor complexes. Processing of LMP2 and LMP7 proproteins takes place in 13–16S preproteasome complexes. *J. Mol. Biol.* **236**, 975–981.

Gaczynska, M., Rock, K.L., Spies, T. & Goldberg, A.L. (1994) Peptidase activities of proteasomes are differentially regulated by the major histocompatibility complex-encoded genes for LMP2 and LMP7. *Proc. Natl Acad. Sci. USA* **91**, 9213–9217.

Ganoth, D., Leshinsky, E., Eytan, E. & Hershko, A. (1988) A multi-component system that degrades proteins conjugated to ubiquitin. *J. Biol. Chem.* **263**, 12412–12419.

Gray, C.W., Slaughter, C.A. & DeMartino, G.N. (1994) PA28 activator protein forms regulatory caps on proteasome stacked rings. *J. Mol. Biol.* **236**, 7–15.

Groettrup, M., Soza, A., Eggers, M., Kuehn, L., Dick, T.P., Schild, H., Rammensee, H.G., Koszinowski, U.H. & Kloetzel, P.M. (1996) A role for the proteasome regulator PA28a in antigen presentation. *Nature* **381**, 166–168.

Groll, M., Ditzel, L., Löwe, J., Stock, D., Bochtler M., Bartunik, H.D. & Huber, R. (1997) Structure of the 20S proteasome from yeast at 2.4 Å resolution. *Nature* **386**, 463–471.

Hardy, M.F., Mantle, D., Edmunds, T. & Pennington, R.J.T. (1981) A high-molecular-weight enzyme from skeletal muscle which hydrolyses chymotrypsin substrates. *Biochem. Soc. Trans.* **11**, 348–349.

Harris, J.R. (1968) Release of a macromolecular protein component from human erythrocyte ghosts. *Biochim. Biophys. Acta* **150**, 534–537.

Harris, J.R. (1988) Erythrocyte cylindrin: possible identity with the ubiquitous 20S high molecular weight protease complex and the prosome particle. *Indian J. Biochem. Biophys.* **25**, 459–466.

Hase, J., Kobashi, K., Nakai, N., Mitsui, K., Iwata, K. & Takadera, T. (1980) The quaternary structure of carp muscle alkaline protease. *Biochim. Biophys. Acta* **611**, 205–213.

Heinemeyer, W., Gruhler, A., Möhrle, V., Mahé, Y. & Wolf, D.H. (1993) *PRE2*, highly homologous to the human major histocompatibility complex-linked *RING10* gene, codes for a yeast proteasome subunit necessary for chymotryptic activity and degradation of ubiquitinated proteins. *J. Biol. Chem.* **268**, 5115–5120.

Hoffman, L. Pratt, G. & Rechsteiner, M. (1992) Multiple forms of the 20S multicatalytic and 26S ubiquitin-ATP-dependent proteases from rabbbit reticulocyte lysates. *J. Biol. Chem.* **267**, 22362–22368.

Hough, R., Pratt, G. & Rechsteiner, M. (1987) Purification of two high molecular weight proteases from rabbit reticulocyte lysate. *J. Biol. Chem.* **262**, 8303–8313.

Ismail, F. & Gevers, W. (1983) A high-molecular-weight cysteine endopeptidase from rat skeletal muscle. *Biochim. Biophys. Acta* **742**, 399–408.

Ishiura, S., Sano, M., Kamakura, K. & Sugita, H. (1985) Isolation of two forms of the high-molecular-mass serine protease, ingensin, from porcine skeletal muscle. *FEBS Lett.* **189**, 119–123.

Kleinschmidt, J.A., Hügle, B., Grund, C. & Franke, W.W. (1983) The 22S cylinder particles of *Xenopus laevis*. I. Biochemical and electron microscopic characterization. *Eur. J. Cell Biol.* **32**, 157–163.

Leibovitz, D., Koch, Y., Fridkin, M., Pitzer, F., Zwickl, P., Dantes, A., Baumeister, W. & Amsterdam, A. (1995) Archaebacterial and eukaryotic proteasomes prefer different sites in cleaving gonadotropin-releasing-hormone. *J. Biol. Chem.* **270**, 11029–11032.

Lupas, A., Zwickl, P., Wenzel, T., Seemüller, E. & Baumeister, W. (1995) Structure and function of the 20S proteasome and of its regulatory complexes. *Cold Spring Harbor Symp. Quant. Biol.* **60**, 515–524.

McGuire, M.J. & DeMartino, G.N. (1986) Purification and characterization of a high molecular weight proteinase (macropain) from human erythrocytes. *Biochim. Biophys. Acta* **873**, 279–289.

Monaco, J.J. & McDevitt, H.O. (1982) Identification of a fourth class of proteins linked to the murin major histocompatibility complex. *Proc. Natl Acad. Sci. USA* **79**, 3001–3005.

Mykles, D.L. (1989) Purification and characterization of a multicatalytic proteinase from crustacean muscle: comparison of latent and heat-activated forms. *Arch. Biochem. Biophys.* **274**, 216–228.

Narayan, K.S. & Rounds, D.E. (1973) Minute ring-shaped particles in cultured cells of malignant origin. *Nature New Biol.* **243**, 146–150.

NC-IUBMB (Nomenclature Committee of the International Union of Biochemistry and Molecular Biology) (1992) *Enzyme Nomenclature 1992. Recommendations of the Nomenclature Committee of the International Union of Biochemistry and Molecular Biology on the Nomenclature and Classification of Enzymes*. Orlando: Academic Press.

Nelbock, P., Dillon, P.J., Perkins, A. & Rosen, C.A. (1990) A cDNA for a protein that interacts with the human immunodeficiency virus Tat transactivator. *Science* **248**, 1650–1653.

Orino, E., Tanaka, K., Tamura, T., Sone, S., Ogura, T. & Ichihara, A. (1991) ATP-dependent reversible association of proteasomes with multiple protein components to form 26S complexes that degrade ubiquitinated proteins in human HL-60 cells. *FEBS Lett.* **248**, 206–210.

Orlowski, M. & Wilk, S. (1988) Multicatalytic proteinase complex or multicatalytic proteinase: a high $M_r$ endopeptidase. *Biochem. J.* **255**, 751.

Orlowski, M., Cardozo, C. & Michaud, C. (1993) Evidence for the presence of five distinct proteolytic components in the pituitary multicatalytic proteinase complex. Properties of two components cleaving bonds on the carboxyl side of branched chain and small neutral amino acids. *Biochemistry* **32**, 1563–1572.

Peters, J.-M. (1994) Proteasomes: protein degradation machines of the cell. *Trends Biochem. Sci.* **19**, 377–382.

Peters, J.-M., Harris, J.R. & Kleinschmidt, J.A. (1991) Ultrastructure of the ~26S complex containing the ~20S cylinder particle (multicatalytic proteinase/proteasome). *Eur. J. Cell Biol.* **56**, 422–432.

Peters, J.M., Cejka, Z., Harris, J.R., Kleinschmidt, J.A. & Baumeister, W. (1993) Structural features of the 26 S proteasome complex. *J. Mol. Biol.* **234**, 932–937.

Pühler, G., Pitzer, F., Zwickl, P. & Baumeister, W. (1994) Proteasomes – multisubunit proteinases common to *Thermoplasma* and eukaryotes. *Syst. Appl. Microbiol.* **16**, 734–741.

Ray, K. & Harris, H. (1985) Purification of neutral lens endopeptidase: close similarity to a neutral proteinase in pituitary. *Biochemistry* **82**, 7545–7549.

Rivett, A.J. (1985) Purification of a liver alkaline protease which degrades oxidatively modified glutamine synthetase. *J. Biol. Chem.* **260**, 12600–12606.

Rivett, A.J., Savory, P.J. & Djaballah H. (1994) Multicatalytic endopeptidase complex (proteasome). Methods Enzymol. **244**, 331–350.

Rock, K.L., Gramm, C., Rothstein, L., Clark, K., Stein, R., Dick, L., Hwang, D. & Goldberg, A.L. (1994) Inhibitors of the proteasome block the degradation of most cell proteins and the generation of peptides presented on MHC class I molecules. *Cell* **78**, 761–771.

Rose, I.A., Warms, J.V.B. & Hershko, A. (1979) A high-molecular-weight protease in liver cytosol. *J. Biol. Chem.* **254**, 8135–8138.

Schmidt, H.P., Akhayat, O., Martins De Sa, C., Puvion, F., Koehler, K. & Scherrer, K. (1984) The prosome: a ubiquitous morphologically distinct RNP particle associated with repressed mRNPs and containing ScRNA and a characteristic set of proteins. *EMBO J.* **3**, 29–34.

Schuldt, C. & Kloetzel, P.-M. (1985) Analysis of cytoplasmic 19 S ring-type particles in *Drosophila* which contain hsp 23 at normal growth temperature. *Dev. Biol.* **110**, 65–74.

Seeger, M., Ferrell, K., Frank, R & Dubiel, W. (1997) HIV-1 Tat inhibits the 20S proteasome and its 11S regulator-mediated activation. *J. Biol. Chem.* **272**, 8145–8148.

Shibuya, H., Irie, K., Ninomiya-Tsuji, J., Goebl, M., Taniguchi, T. & Matsumoto, K. (1992) New human gene encoding a positive modulator of HIV Tat-mediated transactivation. *Nature* **357**, 700–702.

Spohr, G., Granboulan, N., Morel, C. & Scherrer, K. (1970) Messenger RNA in HeLa cells: an investigation of free and polyribosome-bound cytoplasmic messenger ribonucleoprotein particles by kinetic labelling and electron microscopy. *Eur. J. Biochem.* **17**, 296–318.

Stein, R.L., Melandri, F. & Dick, L. (1996) Kinetic characterization of the chymotrytic activity of the 20S proteasome. *Biochemistry* **35**, 3899–3908.

Takahashi, R., Tokumoto, T., Ishikawa, K. & Takahashi, K. (1993) Cleavage specificity and inhibition profile of proteasome isolated from the cytosol of *Xenopus* oocyte. *J. Biochem. (Tokyo)* **113**, 225–228.

Tanahashi, N., Tsurumi, C, Tamura, T. & Tanaka, K. (1993) Molecular structures of 20S and 26S proteasomes. *Enzyme Protein* **47**, 241–251.

Tanaka, K., Ii, K., Ichihara, A., Waxman, L. & Goldberg, A.L. (1986) A high molecular weight protease in the cytosol of rat liver. *J. Biol. Chem.* **261**, 15197–15203.

Tsubuki, S., Saito, Y., Tomioka, M., Ito, H. & Kawashima, S. (1996) Differential inhibition of calpain and proteasome activities by peptidyl aldehydes of di-leucine and tri-leucine. *J. Biol. Chem.* **119**, 572–576.

Udvardy, A. (1992) Purification and characterization of a multiprotein component of the *Drosophila* 26S (1500 kDa) proteolytic complex. *J. Biol. Chem.* **268**, 9055–9062.

Vinitsky, A., Michaud, D., Powers, J.C. & Orlowski, M. (1992) Inhibition of the chymotrypsin-like activity of the pituitary multicatalytic proteinase complex. *Biochemistry* **31**, 9421–9428.

Wilk, S. & Chen, W.-E. (1997) Synthetic peptide-based activators of the proteasome. *Mol. Biol. Rep.* **24**, 119–124.

Wilk, S. & Orlowski, M. (1980) Cation-sensitive neutral endopeptidase: isolation and specificity of the bovine pituitary enzyme. *J. Neurochem.* **35**, 1172–1182.

Wilk, S. & Orlowski, M. (1983) Evidence that pituitary cation-sensitive neutral endopeptidase is a multicatalytic protease complex. *J. Neurochem.* **40**, 842–849.

Yoshimura, T., Kameyama, K., Takagi, T., Ikai, A., Tokunaga, F., Koide, T., Tanahashi, N., Tamura, T., Cejka, Z., Baumeister, W., Tanaka, K. & Ichihara, A. (1993) Molecular characterization of the '26S' proteasome complex from rat liver. *J. Struct. Biol.* **111**, 200–211.

Zolfaghari, R., Baker Jr., C.R.F., Canizaro, P.C., Amirgholami, A. & Behal, F.J. (1987) A high-molecular-mass neutral endopeptidase-24.5 from human lung. *Biochem. J.* **241**, 129–135.

**Erika Seemüller**
*Molecular Structural Biology,*
*Max-Planck-Institute for Biochemistry,*
*D-82152 Martinsried, Germany*
*Email: baumeist@alf.biochem.mpg.de*

**Istok Dolenc**
*Molecular Structural Biology,*
*Max-Planck-Institute for Biochemistry,*
*D-82152 Martinsried, Germany*

**Andrei Lupas**
*Molecular Structural Biology,*
*Max-Planck-Institute for Biochemistry,*
*D-82152 Martinsried, Germany*
*Email: Andrei_N_Lupas@sbphrd.com*

# 173. 26S Proteasome

## Databanks

*Peptidase classification: clan TA, family T1, MEROPS ID: T01.003*
*NC-IUBMB enzyme classification: none*
*Databank codes: The 26S proteasome consists of approximately 32 distinct gene products, about 18 of which represent subunits of the PA700 regulatory complex discussed in this chapter. Many but not all of these subunits have been cloned from yeast and/or mammalian sources. For clarity, accession numbers for these polypeptides are provided in Table 173.1, which also lists other features of the PA700 subunits. Databank codes for the subunits of the proteasome have been provided in Chapter 172.*

## Name and History

The **26S proteasome** is a 2100 kDa protease complex consisting of the 20S proteasome (also known as the multicatalytic proteinase, macropain and prosome; see Chapter 172) and at least one other multisubunit regulatory protein known as **PA700** (Ma *et al.*, 1994), **19S cap** (Peters *et al.*, 1994), **μ-particle** (Udvardy, 1993), **ball** (Hoffman *et al.*, 1992), and **ATPase complex** (Dubiel *et al.*, 1995). The

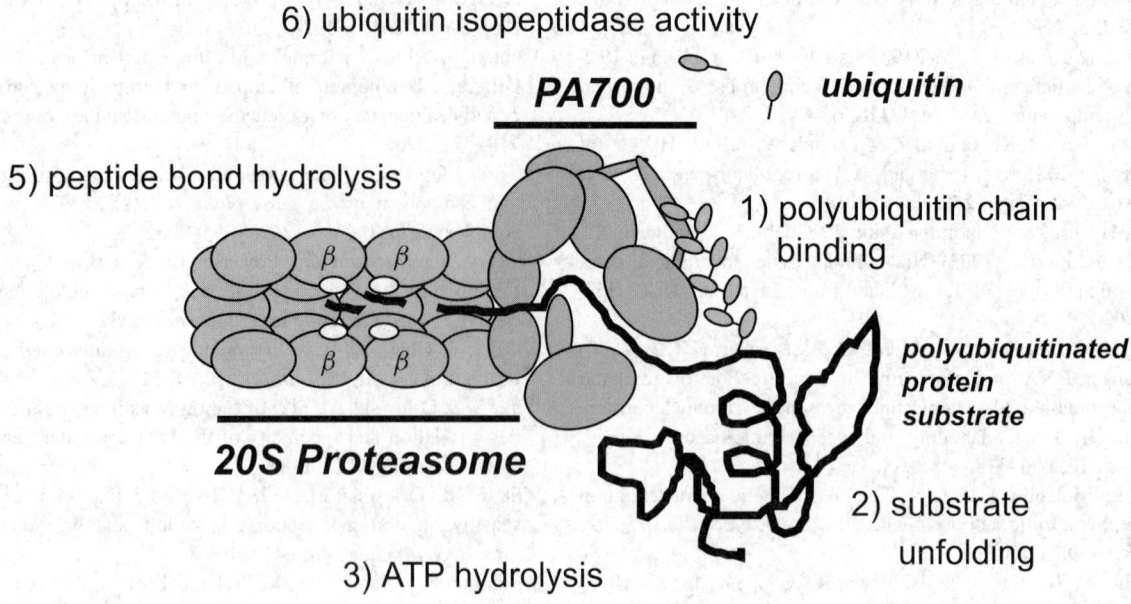

6) ubiquitin isopeptidase activity

**PA700**  *ubiquitin*

5) peptide bond hydrolysis

1) polyubiquitin chain binding

*polyubiquitinated protein substrate*

**20S Proteasome**

2) substrate unfolding

3) ATP hydrolysis

4) translocation of polypeptide chain

*Figure 173.1*   Structure and function of the 26S proteasome. The 26S proteasome is composed of two multisubunit complexes: the 20S proteasome and PA700. PA700 can bind to one (as shown) or both ends of the proteasome. The figure indicates known or likely functions of the 26S proteasome that result in proteolysis. The active catalytic sites of the proteasome are located on β subunits and face the interior of the central channel (drawn here in cross-section).

name 'proteasome' provides the connotation of a protease ('protea-') which is a large particle ('-some'). '26S' refers to the approximate sedimentation coefficient of the complex, 26 S. The 26S proteasome was first identified in extracts of rabbit reticulocytes by its ability to degrade ubiquitinated proteins in an ATP-dependent fashion (Hough *et al.*, 1987; Waxman *et al.*, 1987). Therefore, the enzyme has also been called **ubiquitin-conjugate degrading enzyme (UCDEN)** (Waxman *et al.*, 1987). However, this protease can degrade certain nonubiquitinated proteins and thus the latter name is probably inappropriate. The 26S proteasome nomenclature has achieved widespread use and continued usage of this nomenclature seems reasonable. Because a description of the 20S proteasome has been provided in Chapter 172, the present chapter will concentrate on the unique features of the 26S proteasome with particular emphasis on the PA700 regulatory complex. Table 173.1 summarizes the terminology that has been applied to the subunits of the PA700 complex.

### Activity and Specificity

The 20S proteasome subcomplex of the 26S proteasome accounts for all proteolytic activity of the 26S proteasome. There is no evidence that any of the regulatory polypeptides directly catalyzes the degradation of protein substrates. However, the PA700 regulatory complex displays other enzymatic activities that may be linked mechanistically to the degradation of protein substrates, such as ATPase activity (DeMartino *et al.*, 1994), isopeptidase activity (Lam *et al.*, 1997), and possibly a substrate protein unfolding activity (Rubin & Finley, 1995).

As described in Chapter 172, the eukaryotic 20S proteasome contains multiple catalytic sites with distinct specificities as defined by the hydrolysis of characteristic synthetic peptide substrates containing arginine, tyrosine or leucine, and glutamic acid, respectively, at the P1 cleavage position (Orlowski, 1990; Rivett, 1993; Coux *et al.*, 1996). Appropriate peptides are available with fluorescent or chromogenic reporter groups whose release can be monitored as a measure of protease activity. One of these substrates, Suc-Leu-Leu-Val-Tyr+NHMec, appears to be a specific substrate for the proteasome and therefore has been used extensively as an indicator of proteasome activity in unpurified extracts. The 26S proteasome cleaves the same small synthetic substrates as does the 20S proteasome but at much greater rates. Thus, the PA700 regulatory complex serves as an activator of the 20S proteasome's catalytic functions (Ma *et al.*, 1994). The hydrolysis of protein substrates by each form of the proteasome appears to be accomplished by the combined action of the catalytic sites that cleave peptide substrates described above, although little is known about the specificity of these sites in the degradation of proteins. In fact, there is evidence that little relationship exists between the specificity of a given catalytic site for a short peptide and the peptide bond(s) that the same site hydrolyzes in larger polypeptides (Dick *et al.*, 1991). It is also unknown whether PA700 alters the cleavage specificity of the 20S proteasome. Like the 20S proteasome, the 26S proteasome degrades most protein substrates slowly. However, in contrast to the 20S proteasome, the 26S proteasome degrades proteins rapidly and extensively if they are attached to a polyubiquitin chain (Hough *et al.*, 1987; Waxman *et al.*, 1987; Ugai *et al.*, 1993). Physiological

Table 173.1  Subunits of PA700, the regulatory complex of the 26S proteasome

| Subunits of PA700 (Homo sapiens) | | Also identified as | Accession number | Yeast homolog | Accession number | Comments/references |
|---|---|---|---|---|---|---|
| (a) | (b) | | | | | |
| **ATPase subunits** | | | | | | |
| p56 | S4 | | L02426 | MTS2/YTA5 | | Putative ATPase (Dubiel et al., 1992) |
| p50 | S6 | TBP1 | | YTA1 | | Putative ATPase (Nelbock et al., 1990; DeMartino et al., 1996) |
| p48 | S6 | TBP7 | | YTA2 | | Putative ATPase (Ohana et al., 1993; DeMartino et al., 1994; Dubiel et al., 1994) |
| p47 | S7 | MSS1 | | CIM5 | | Putative ATPase (Monacco, 1992a; Shibuya et al., 1992; Ghisiain et al., 1993) |
| p45 | S8 | Trip1 | D44467 | SUG1/CIM3 | | Putative ATPase (Swaffield et al., 1992; Ghisiain et al., 1993; Akiyama et al., 1995; Lee, et al., 1995) |
| p42 | | | D78275 | SUG2 | | Putative ATPase (DeMartino et al., 1996) |
| **Non-ATPase subunits** | | | | | | |
| p112 | S1 | TRAP2/55.1 | D44466 | SEN3 | | Identified as TNF-receptor interacting protein (DeMarini et al., 1995; Yokota et al., 1996; Boldin et al., 1995; Rothe et al., 1994) |
| p97 | S2 | | D78151 | NAS1/HRD2 | U10399 | Identified as gene required for HMG-COA reductase degradation (Hampton et al., 1996; Tsurumi et al., 1996) |
| p58 | S3 | P91A | D67025 | SUN2 | D78023 | Identified as suppressor of nin1 mutation (Gaczynska et al., 1996; Lurquin et al., 1989) |
| p55 | | | D31889 | SUN1 | D78022 | Deveraux et al., 1995 |
| p50.5 | S5b | | | | | |
| p44.5 | S5a | | | | | Polyubiquitin chain binding protein (Deveraux et al., 1994; Ferrell et al., 1996; Kominami et al., 1997) |
| p44 | S10 | | | | | |
| p40 | S12 | MOV34 | D50063 (Human) M64634 (Mouse) | | | Embryonic lethal mutation in mouse (Rothe et al., 1994; Tsurumi et al., 1995) |
| p39 | | | | | | Isopeptidase? (Lam et al., 1997) |
| p37 | | | | | | Isopeptidase? (Lam et al., 1997) |
| p31 | S14 | | D37047 | NIN1 | | Identified as cell cycle mutant (Kominami et al., 1995) |
| p28 | S15 | | | | | |

S

substrates of the 26S proteasome have been identified only in the past several years (Rechsteiner, 1991; Ciechanover, 1994; Hochstrasser, 1996) (see below), and therefore most biochemical characterization of the proteolytic properties of the enzyme have been achieved with model protein substrates such as [$^{125}$I]ubiquitin-lysozyme. The inconvenience and difficulty of preparing large amounts of biochemically defined ubiquitinated proteins has posed a significant obstacle to complete and rigorous biochemical and kinetic characterization of this activity. The degradation of ubiquitinated proteins by the 26S proteasome requires ATP and occurs with a pH optimum of 7.5–8.0. Most assays are based on quantitation of the rates of production of acid-soluble peptides. Although these peptides exhibit a range of sizes, they are typically from 5 to 10 residues in length and only limited amounts of free amino acids are produced. The novel nature of the threonine active site of the proteasome (see Chapter 172) confused the interpretation of many early inhibitor studies of the proteasome. Recently, potent inhibitors of the proteasome that affect both the 20S and 26S forms of the enzyme have been identified. Some inhibitors are peptide aldehydes developed by ProScript (see Appendix 2 for full names and addresses of suppliers). Furthermore, lactacystin, a *Streptomyces* metabolite (available from Calbiochem), is a potent and specific irreversible inhibitor of the proteasome (Fenteany *et al.*, 1995; Dick *et al.*, 1996). Importantly, this compound is cell permeant and should be an important reagent for elucidation of physiological functions of the proteasome.

Although the 26S proteasome shows a preference for ubiquitinated proteins as compared to nonubiquitinated counterparts, it can degrade certain nonubiquitinated proteins. The best characterized example is ornithine decarboxylase (ODC), whose degradation does not require ubiquitination, but does require ATP and the ODC inhibitor protein, antizyme (Murakami *et al.*, 1992; Tokunaga *et al.*, 1994; Elias *et al.*, 1995). The mechanism by which antizyme promotes the degradation of ODC is unclear, but it may serve a 'targeting' function in a manner analogous to polyubiquitin (Li *et al.*, 1996) (see below). Additional proteins that are degraded by the 26S proteasome by ubiquitin-independent mechanisms are likely to be identified. The specificity of the 26S proteasome for ubiquitinated proteins is probably related to the high-affinity binding of polyubiquitin chains to the PA700 regulatory complex, and one identified PA700 subunit displays this property (Deveraux *et al.*, 1994; Ferrell *et al.*, 1996) (see Table 173.1). However, genetic studies in yeast indicate that this subunit is not essential for the degradation of all ubiquitinated proteins (van Nocker *et al.*, 1996). Thus, PA700 may contain multiple polyubiquitin-binding proteins.

In addition to peptide bond hydrolysis, at least two other enzymatic activities are part of the 26S proteasome: ATPase activity and isopeptidase activity. ATPase activity serves at least two distinct functions, being required for (a) assembly of the 26S proteasome from 20S proteasome and PA700 subcomplexes (Eytan *et al.*, 1989; Hoffman *et al.*, 1992; Udvardy, 1993; Ma *et al.*, 1994), and (b) the degradation of protein substrates by the assembled complex (Eytan *et al.*, 1989; Hoffman *et al.*, 1992; Udvardy, 1993). The exact molecular basis for these effects of ATP is unknown. Because peptide bond hydrolysis *per se* should not be energy dependent, ATP hydrolysis may be utilized for unfolding protein

substrates and/or for translocation of the unfolded polypeptide substrate into the central channel of the proteasome (in a manner similar to polypeptide translocation across membranes). Finally, by its isopeptidase activity, PA700 disassembles polyubiquitin chains by cleaving ubiquitin monomers progressively from the distal end of the chain (Lam *et al.*, 1997). This activity may spare proteins from inappropriate degradation, thus serving an editing function. Alternatively, isopeptidase activity may be necessary for continuous translocation of the substrate into the narrow opening of the proteasome core. Much additional work will be required to determine the precise and coordinated function of all enzymatic activities for the degradation of protein substrates.

## Structural Chemistry

The 26S proteasome consists of two multimeric subcomplexes: the 20S proteasome and PA700. PA700 binds to one or both ends of the cylindrical 20S proteasome to form a characteristic complex which has been observed by electron microscopy (Yoshimura *et al.*, 1993; Coux *et al.*, 1996). The structural features of the 20S proteasome and its component subunits have been described in Chapter 172. PA700 contains approximately 18 subunits with molecular weights ranging from 20 to 112 kDa. The primary structures for most of these subunits, all of which are distinct gene products, have been determined in species from yeast to humans, and some progress has been achieved in relating specific functions of the complex to the structures of individual subunits. For example, six subunits are members of a large protein family characterized by a 200 amino acid domain that contains an ATP-binding motif (Dubiel *et al.*, 1992; Confalonieri & Duguet, 1995; DeMartino *et al.*, 1996). These subunits presumably participate in the ATP-dependent assembly of the 26S proteasome from 20S proteasome and PA700, and/or in the ATP-dependent degradation of protein substrates. It is unclear, however, whether each of the six subunits actually hydrolyzes ATP, and if so, whether they play similar or distinct roles in the various ATP-dependent functions of the complex. At least one subunit can bind polyubiquitin chains, and thus may serve as the polyubiquitin 'receptor' (Deveraux *et al.*, 1994; Kominami *et al.*, 1997). Another subunit, whose primary structure is not yet known, binds to ubiquitin aldehyde and can be affinity labeled with ubiquitin nitrile, indicating that it represents the isopeptidase of PA700 (Lam *et al.*, 1997). The genes for many PA700 subunits have been identified in yeast. Mutations in these genes result in pleiotropic phenotypes, including cell cycle arrest and abnormal accumulation of polyubiquitinated proteins (Ghisiain *et al.*, 1993; Gordon *et al.*, 1993; Kominami *et al.*, 1995; DeMarini *et al.*, 1995; Rubin *et al.*, 1996; Yokota *et al.*, 1996).

## Preparation

Preparations of the 26S proteasome have been reported from a variety of sources, including yeast, plants and mammals (Coux *et al.*, 1996). The preparation of large amounts of highly purified 26S proteasome has been difficult to achieve (although recent work has demonstrated significant improvements in these methodologies). In our judgement and experience, this difficulty may result from dissociation of the complex into 20S proteasome and PA700 subcomplexes during purification. ATP and glycerol (10–20%) appear to stabilize the 26S proteasome during purification. Alternatively, the 26S proteasome can be assembled from purified 20S proteasome and PA700 components, each of which can be obtained in large amounts and in high purity by conventional protein purification methods (McGuire *et al.*, 1989; Ma *et al.*, 1994). Assembly requires preincubation of proteasome and PA700 in the presence of ATP and is characterized by: (a) increased hydrolysis of synthetic substrates, (b) ATP-dependent degradation of ubiquitinated proteins, and (c) formation of a complex that has a significantly greater size than either of the individual proteins (as detected by faster sedimentation in density gradient centrifugation). The 26S proteasome is widely distributed and highly conserved, and therefore sources can be chosen for convenience and/or scientific interest. Because the 26S proteasome is composed of approximately 32 gene products, it is not suitable for preparation from expression of cloned genes and must be purified from cell and tissue sources. (In contrast, the 20S proteasome from archaea can be prepared from expressed proteins because it is the product of only two genes (Zwickl *et al.*, 1992); the existence of the 26S proteasome in archaea has not been demonstrated, however.) In yeast, expression of tagged subunits of the 26S proteasome has proven useful for rapid purification of the complex by affinity chromatography (Rubin *et al.*, 1996).

## Biological Aspects

The 26S proteasome plays an essential role in a surprisingly large number of diverse physiological functions. Some of these have been demonstrated directly for the protease, while others are inferred from the enzyme's role as the proteolytic component of the ubiquitin pathway. Inhibitor studies indicate that the proteasome is responsible for turnover of most cellular proteins in mammalian cells (Rock *et al.*, 1994) and of most normal short-lived proteins in yeast (Lee & Goldberg, 1996). It is also responsible for the selective degradation of proteins with abnormal structures and thus serves a critical quality control function that prevents the cellular accumulation of malfunctioning proteins. An unexpected extension of this latter role has been discovered recently during studies of the quality control mechanism of the endoplasmic reticulum (Kopito, 1997). Mutations in proteins that normally reside in the endoplasmic reticulum or transit the ER's secretory/sorting pathways, result in the selective and rapid degradation of these proteins (just as mutations in cytoplasmic proteins often result in their selective degradation). It had been assumed that this proteolysis occurred within the ER by a resident proteolytic system, even though there has been no strong evidence for such a system. Recent data indicate that mutant proteins (either proteins normally destined for the plasma membrane [Ward *et al.*, 1995], the ER membrane [Biederer *et al.*, 1997], or secretion [Werner *et al.*, 1996]) are 'reverse-translocated' across the ER membrane and degraded in the cytoplasm by the proteasome. Some of these proteins are ubiquitinated in the cytoplasm prior to their degradation (Wiertz *et al.*, 1996; Hughes *et al.*, 1997). Many of the molecular details of this process are not yet defined, including the possible role of the 26S proteasome in the process of 'reverse translocation' (Kopito, 1997).

As emphasized by Schimke (1970), even constitutive rapid turnover of cellular proteins has important implications for cellular regulation, but in addition to these housekeeping functions, the 26S proteasome participates in the regulation of numerous cellular processes by rapidly degrading regulatory proteins that are involved in them. Usually, such control proteins are repressors or obligatory forward activators of processes with which they are involved, and their proteolytic elimination affects these processes accordingly. Examples of these processes and their respective control proteins include: cell cycle progression by cyclins and cyclin control proteins (King *et al.*, 1996), transcription in mammals by NFκB (Palombella *et al.*, 1994), transcription in yeast by MATα2 repressor (Hochstrasser *et al.*, 1991), cell growth and oncogenesis by jun (Palombella *et al.*, 1994), mos (Isakasson *et al.*, 1996), and p53 (Scheffner *et al.*, 1990; Lam *et al.*, 1997). Studies in a number of laboratories also indicate that the 26S proteasome, via its role in the ubiquitin pathway, accounts for the increased proteolysis characteristic of a number of muscle wasting conditions (Mitch & Goldberg, 1996).

Finally, the proteasome is critically involved in the production of antigenic peptides for presentation by MHC class I complexes (Coux *et al.*, 1996; Rock *et al.*, 1994). Interestingly, antigenic peptides may be generated by a specific subpopulation of proteasomes that contain two or three subunits encoded in the major histocompatibility complex (Monacco, 1992b). These subunits are upregulated under conditions of increased antigen presentation, while their corresponding non-MHC-encoded homologs are downregulated (Akiyama *et al.*, 1994). Thus, cells appear to assemble a subpopulation of 'immunoproteasomes' containing subunits that alter proteasome specificity in a manner favoring production of specific antigenic peptides (Driscoll *et al.*, 1993). The exact role of the ubiquitin pathway in antigen presentation is under debate, and currently it is unclear whether any PA700 subunits are also controlled to play a role in antigen presentation. A distinct proteasome regulator, PA28, appears to regulate proteasome function in antigen presentation even though it has no obvious role in ubiquitin-dependent proteolysis (Realini *et al.*, 1994; Ahn *et al.*, 1995).

## Further Reading

The rapid, extensive and diverse recent progress on the proteasome and its functions have sparked a large number of excellent reviews. These include Coux *et al.* (1996), an extensive review of all major aspects of the 20S and 26S proteasomes, and Rubin & Finley (1995), a short review dealing with critical aspects of the structure and function of the 26S proteasome.

## References

Ahn, J., Tanahashi, N., Akiyama, K., Hisamatsu, H., Noda, C., Tanaka, K., Chung, C.H., Shibmara, N., Willy, P.J., Mott, J.D., Slaughter, C.A. & DeMartino, G.N. (1995) Primary structures of two homologous subunits of PA28, a γ-interferon-inducible protein activator of the 20S proteasome. *FEBS Lett.* **366**, 37–42.

Akiyama, K., Yokota, K., Kagawa, S., Shimbara, N., Tamura, T., Akioka, H., Nothwang, H.G., Noda, C., Tanaka, K. & Ichihara, A. (1994) cDNA cloning and interferon γ down-regulation of proteasomal subunits X and Y. *Science* **265**, 1231–1234.

Akiyama, K., Yokota, K., Kagawa, S., Shimbara, N., DeMartino, G.N., Slaughter, C.A., Noda, C. & Tanaka, K. (1995) cDNA cloning of a new putative ATPase subunit p45 of the human 26S proteasome, a homolog of yeast transcriptional factor Sug1p. *FEBS Lett.* **363**, 151–156.

Biederer, T., Volkwein, C. & Sommer, T. (1997) Degradation of subunits of the Sec61p complex, an integral component of the ER membrane, by the ubiquitin-proteasome pathway. *EMBO J.* **15**, 2069–2076.

Boldin, M.P., Mett, I.L. & Wallach, D. (1995) A protein related to a proteasomal subunit binds to the intracellular domain of the p55 TNF receptor upstream of its 'death domain'. *FEBS Lett.* **367**, 39–44.

Ciechanover, A. (1994) The ubiquitin-proteasome proteolytic pathway. *Cell* **79**, 13–21.

Confalonieri, F. & Duguet, M. (1995) A 200-amino acid ATPase module in search of a basic function. *BioEssays* **17**, 639–650.

Coux, O., Tanaka, K. & Goldberg, A.L. (1996) Structure and functions of the 20S and 26S proteasomes. *Annu. Rev. Biochem.* **65**, 801–847.

DeMarini, D.J., Papa, F.R., Swaminathan, S., Ursic, D., Rasmussen, T.P., Culbertson, M.R. & Hochstrasser, M. (1995) The yeast SEN3 gene encodes a regulatory subunit of the 26S proteasome complex required for ubiquitin-dependent protein degradation *in vivo*. *Mol. Cell. Biol.* **15**, 6311–6321.

DeMartino, G.N., Moomaw, C.R., Zagnitko, O.P., Proske, R.J., Ma, C., Afendis, S.J., Swaffield, J.C. & Slaughter, C.A. (1994) PA700, an ATP-dependent activator of the 20S proteasome, is an ATPase containing multiple members of a nucleotide-binding protein family. *J. Biol. Chem.* **269**, 20878–20884.

DeMartino, G.N., Proske, R.J., Moomaw, C.R., Strong, A.A., Song, X., Hisamatsu, H., Tanaka, K. & Slaughter, C.A. (1996) Identification, purification, and characterization of a PA700-dependent activator of the proteasome. *J. Biol. Chem.* **271**, 3112–3118.

Deveraux, Q., Ustrell, V., Pickart, C. & Rechsteiner, M. (1994) A 26S protease subunit that binds ubiquitin conjugates. *J. Biol. Chem.* **269**, 7059–7061.

Deveraux, Q., Jensen, C. & Rechsteiner, M. (1995) Molecular cloning and expression of a 26S protease subunit enriched in dileucine repeats. *J. Biol. Chem.* **270**, 23726–23729.

Dick, L.R., Moomaw, C.R., DeMartino, G.N. & Slaughter, C.A. (1991) Degradation of oxidized insulin B chain by the multiproteinase complex macropain (proteasome). *Biochemistry* **30**, 2725–2734.

Dick, L.R., Cruikshank, A.A., Grenier, L., Melandri, F.D., Nunes, S.L. & Stein, R.L. (1996) Mechanistic studies on the inactivation of the proteasome by lactacystin. *J. Biol. Chem.* **271**, 7273–7226.

Driscoll, J., Brown, M.G., Finley, D. & Monaco, J.J. (1993) MHC-linked LMP gene products specifically alter peptidase activities of the proteasome. *Nature* **365**, 262–264.

Dubiel, W., Ferrell, K., Pratt, G. & Rechsteiner, M. (1992) Subunit 4 of the 26S protease is a member of a novel eukaryotic ATPase family. *J. Biol. Chem.* **267**, 22699–22702.

Dubiel, W., Ferrell, K. & Rechsteiner, M. (1994) Tat-binding protein 7 is a subunit of the 26S protease. *Biol. Chem. Hoppe-Seyler* **375**, 237–240.

Dubiel, W., Ferrell, K. & Rechsteiner, M. (1995) Subunits of the regulatory complex of the 26S protease. *Mol. Biol. Rep.* **21**, 27–34.

Elias, S., Bercovich, B., Kahana, C., Coffino, P., Fischer, M., Hilt, W., Wolf, D.H. & Ciechanover, A. (1995) Degradation of ornithine decarboxylase by the mammalian and yeast 26S proteasome complexes requires all the components of the protease. *Eur. J. Biochem.* **229**, 276–283.

Eytan, E., Ganoth, D., Armon, T. & Hershko, A. (1989) ATP-dependent incorporation of 20S protease into the 26S complex that degrades proteins conjugated to ubiquitin. *Proc. Natl Acad. Sci. USA* **86**, 7751–7755.

Fenteany, G., Standaert, R.F., Lane, W.S., Choi, S., Corey, E.J. & Schreiber, S.L. (1995) Inhibition of proteasome activities and subunit-specific amino-terminal threonine modification by lactacystin. *Science* **268**, 726–731.

Ferrell, K., Deveraux, Q., van Nocker, S. & Rechsteiner, M. (1996) Molecular cloning and expression of a multiubiquitin chain binding subunit of the human 26S protease. *FEBS Lett.* **381**, 143–148.

Gaczynska, M., Goldberg, A.L., Tanaka, K., Hendil, K.B. & Rock, K.L. (1996) Proteasome subunits X and Y alter peptidase activities in opposite ways to the interferon-gamma-induced subunits LMP2 and LMP7. *J. Biol. Chem.* **271**, 17275–17280.

Ghislain, M., Udvardy, A. & Mann, C. (1993) *S. cerevisiae* 26S protease mutants arrest cell division in G2/metaphase. *Nature* **366**, 358–362.

Gordon, C., McGurk, G., Dillon, P., Rosen, C. & Hastle, N.D. (1993) Defective mitosis due to a mutation in the gene for a fission yeast 26S protease subunit. *Nature* **366**, 355–357.

Hampton, R.Y., Gardner, R.G. & Rine, J. (1996) Role of the 26S proteasome and *HRD* genes in the degradation of 3-hydroxy-3-methylglutaryl-CoA reductase, an integral endoplasmic reticulum membrane protein. *Mol. Biol. Cell* **7**, 2029–2044.

Hochstrasser, M. (1996) Ubiquitin-dependent protein degradation. *Annu. Rev. Genet.* **30**, 405–439.

Hochstrasser, M., Ellison, M.J., Chau, V. & Varshavsky, A. (1991) The short-lived MATα2 transcriptional regulator is ubiquitinated *in vivo*. *Proc Natl Acad. Sci. USA* **88**, 4606–4610.

Hoffman, L., Pratt, G. & Rechsteiner, M. (1992) Multiple forms of the 20S multicatalytic and the 26S ubiquitin-ATP-dependent proteases from rabbit reticulocyte lysate. *J. Biol. Chem.* **267**, 22362–22368.

Hough, R., Pratt, G. & Rechsteiner, M. (1987) Purification of two high molecular weight proteases from rabbit reticulocyte lysate. *J. Biol. Chem.* **262**, 8303–8313.

Hughes, E.A., Hammond, C. & Cresswell, P. (1997) Misfolded major histocompatibility complex class I heavy chains are translocated into the cytoplasm and degraded by the proteasome. *Proc. Natl Acad. Sci. USA* **94**, 1896–1901.

Isakasson, A., Musti, A.M. & Bohmann, D. (1996) Ubiquitin in signal transduction and cell transformation. *Biochim. Biophys. Acta* **1288**, F21–F29.

King, R.W., Deshaies, R.J., Peters, J. & Kirschner, M.W. (1996) How proteolysis drives the cell cycle. *Science* **274**, 1652–1658.

Kominami, K., DeMartino, G.N., Moomaw, C.R., Slaughter, C.A., Shimbara, N., Fujimuro, M., Yokosawa, H., Hisamatsu, H., Tanahashi, N., Shimizu, Y., Tanaka, K. & Toh-e, A. (1995) Nin1p, a regulatory subunit of the 26S proteasome, is necessary for activation of Cdc28p kinase of *Saccharomyces cerevisiae*. *EMBO J.* **14**, 3105–3115.

Kominami, K., Okura, N., Kawamura, M., DeMartino, G.N., Slaughter, C.A., Simbara, N., Chung, C.H., Fujimuro, M., Yokosawa, H., Shimizu, Y., Tanahashi, N., Tanaka, K. & Toh-e, A. (1997) Yeast counterparts of subunits S5a and p58 (S3) of the human 26S proteasome are encoded by two multicopy suppressors of *nin1-1*. *Mol. Biol. Cell* **8**, 171–187.

Kopito, R.R. (1997) ER quality control: the cytoplasmic connection. *Cell* **88**, 427–430.

Lam, Y.A., Xu, W., DeMartino, G.N. & Cohen, R.E. (1997) Editing of ubiquitin conjugates by an isopeptidase in the 26S proteasome. *Nature* **385**, 737–740.

Lee, D.H.L. & Goldberg, A.L. (1996) Selective inhibitors of the proteasome-dependent and vacuolar pathways of protein degradation in *Saccharomyces cerevisiae*. *J. Biol. Chem.* **271**, 27280–27284.

Lee, J.W., Ryan, F., Swaffield, J.C., Johnston, S.A. & Moore, D.D. (1995) Interaction of thyroid-hormone receptor with a conserved transcriptional mediator. *Nature* **374**, 91–94.

Li, X., Stebbens, B., Hoffman, L., Pratt, G., Rechsteiner, M. & Coffino, P. (1996) The N terminus of antizyme promotes degradation of heterologous proteins. *J. Biol. Chem.* **271**, 4441–4466.

Lurquin, C., Van Pel, A., Mariamé, B., Janssens, C., Reddehase, M.J., Lejeune, J. & Boon, T. (1989) Structure of the gene of tum-transplantation antigen P91A: the mutated exon encodes a peptide recognized with Ld by cytolytic T cells. *Cell* **58**, 293–303.

Ma, C., Vu, J.H., Proske, R.J., Slaughter, C.A. & DeMartino, G.N. (1994) Identification, purification & characterization of a high-molecular weight, ATP-dependent activator (PA700) of the 20S proteasome. *J. Biol. Chem.* **269**, 3539–3547.

McGuire, M.J., McCullough, M.L., Croall, D.E. & DeMartino, G.N. (1989) The high molecular weight multicatalytic proteinase, macropain, exists in a latent form in human erythrocytes. *Biochim. Biophys. Acta* **995**, 181–186.

Mitch, W.E. & Goldberg, A.L. (1996) Mechanisms of muscle wasting: the role of the ubiquitin-proteasome pathway. *New Engl. J. Med.* **335**, 1897–1905.

Monaco, J.J. (1992a) Pathways of antigen processing. *Immunol. Today* **13**, 173–179.

Monaco, J.J. (1992b) A molecular model of MHC class-I-restricted antigen processing. *Immunol. Today* **13**, 173–179.

Murakami, Y., Matsufuji, S., Kameji, T., Hayashi, S., Igarashi, K., Tamura, T., Tanaka, K. & Ichihara, A. (1992) Ornithine decarboxylase is degraded by the 26S proteasome without ubiquitination. *Nature* **360**, 597–599.

Nelbock, P., Dillon, P.J., Perkins, A. & Rosen, C.A. (1990) A cDNA for a protein that interacts with the human immunodeficiency virus Tat transactivator. *Science* **248**, 1650–1653.

Ohana, B., Moore, P.A., Ruben, S.M., Southgate, C.D., Green, M.R. & Rosen, C.A. (1993) The type 1 human immunodeficiency virus Tat binding protein is a transcriptional activator belonging to an additional family of evolutionarily conserved genes. *Proc. Natl Acad. Sci. USA* **90**, 138–142.

Orlowski, M. (1990) The multicatalytic proteinase complex, a major extralysosomal proteolytic system. *Biochemistry* **29**, 10289–10297.

Palombella, V.J., Rando, O.J., Goldberg, A.L. & Maniatis, T. (1994) The ubiquitin-proteasome pathway is required for processing the NF-κB1 precursor protein and the activation of NF-κB. *Cell* **78**, 773–785.

Peters, J., Franke, W.W. & Kleinschmidt, J.A. (1994) Distinct 19S and 20S subcomplexes of the 26S proteasome and their distribution in the nucleus and the cytoplasm. *J. Biol. Chem.* **269**, 7709–7718.

Realini, C., Dubiel, W., Pratt, G., Ferrell, K. & Rechsteiner, M. (1994) Molecular cloning and expression of a γ-interferon-inducible activator of the multicatalytic protease. *J. Biol. Chem.* **269**, 20727–20732.

Rechsteiner, M. (1991) Natural substrates of the ubiquitin proteolytic pathway. *Cell* **66**, 615–618.

Rivett, A.J. (1993) Proteasomes: multicatalytic proteinase complexes. *Biochem. J.* **291**, 1–10.

Rock, K.L., Gramm, C., Rothstein, L., Clark, K., Stein, R., Dick, L., Hwang, D. & Goldberg, A.L. (1994) Inhibitors of the proteasome block the degradation of most cell proteins and the generation of peptides presented on MHC class I molecules. *Cell* **78**, 761–771.

Rothe, M., Wong, S.C., Henzel, W.J. & Goeddel, D.V. (1994) A novel family of putative signal transducers associated with the cytoplasmic domain of the 75 a tumor necrosis factor receptor. *Cell* **78**, 681–692.

Rubin, D.M. & Finley, D. (1995) The proteasome: a protein-degrading organelle? *Curr. Biol.* **5**, 854–858.

Rubin, D.M., Coux, O., Wefes, I., Hengartner, C., Young, R.A., Goldberg, A.L. & Finley, D. (1996) Identification of the *gal4* suppressor *Sug1* as a subunit of the yeast 26S proteasome. *Nature* **379**, 655–658.

Scheffner, M., Werness, B.A., Huibregtse, J.M., Levine, A.J. & Howley, P.M. (1990) The E6 oncoprotein encoded by human papilomavirus types 16 and 18 promotes the degradation of p53. *Cell* **63**, 1129–1136.

Schimke, R.T. (1970) Regulation of protein degradation in mammalian tissues. In: *Mammalian Protein Metabolism* (Munro, H.N., ed.). New York: Academic Press, pp. 177–228.

Shibuya, H., Irie, K., Ninomiya-Tsuji, J., Goebll, M., Taniguchi, T. & Matsumoto, K. (1992) New human gene encoding a positive modulator of HIV tat-mediated transactivation. *Nature* **357**, 700–702.

Swaffield, J.C., Bromber, J.F. & Johnston, S.A. (1992) Alterations in a yeast protein resembling HIV tat-binding protein relieve requirement for an acidic activation domain in GAL4. *Nature* **357**, 698–700.

Tokunaga, F., Goto, T., Koide, T., Murakami, Y., Hayashi, S., Tamura, T., Tanaka, K. & Ichihara, A. (1994) ATP- and antizyme-dependent endoproteolysis of ornithine decarboxylase to oligopeptides by the 26S proteasome. *J. Biol. Chem.* **269**, 17382–17385.

Tsurumi, C., DeMartino, G.N., Slaughter, C.A., Shimbara, N. & Tanaka, K. (1995) cDNA cloning of p40, a regulatory subunit of the human 26S proteasome, and a homolog of the *Mov-34* gene product. *Biochem. Biophys. Res. Commun.* **210**, 600–608.

Tsurumi, C., Shimizu, Y., Saeki, M., Kato, S., DeMartino, G.N., Slaughter, C.A., Fujimuo, M., Yokasawa, H., Yamasaki, M., Hendil, K.B., Toh-e, A., Tanahashi, N. & Tanaka, K. (1996) cDNA cloning and functional analysis of the p97 subunit of the 26S proteasome, a polypeptide identical to the type-1 tumor-necrosis-factor-associated protein-2/55.11. *Eur. J. Biochem.* **239**, 912–921.

Udvardy, A. (1993) Purification and characterization of a multiprotein component of the *Drosophila* 26S (1500 kDa) proteolytic complex. *J. Biol. Chem.* **268**, 9055–9062.

Ugai, S., Tamura, T., Tanahashi, N., Takai, S., Komi, N., Chung, C.H., Tanaka, K. & Ichihara, A. (1993) Purification and characterization of the 26S proteasome complex catalyzing ATP-dependent breakdown of ubiquitin-ligated proteins from rat liver. *J. Biochem.* **113**, 754–768.

van Nocker, S., Sadis, S., Rubin, D.M., Glickman, M., Fu, H., Coux, O., Wefes, I., Finley, D. & Vierstra, R. (1996) The multiubiquitin-chain-binding protein Mcb1 is a component of the 26S proteasome in *Saccharomyces cerevisiae* and plays a nonessential, substrate-specific role in protein turnover. *Mol. Cell. Biol.* **16**, 6020–6028.

Ward, C.L., Omura, S. & Kopito, R.R. (1995) Degradation of CFTR by the ubiquitin-proteasome pathway. *Cell* **83**, 121–127.

Waxman, L., Fagan, J.M. & Goldberg, A.L. (1987) Demonstration of two distinct high molecular weight proteases in rabbit reticulocytes, one of which degrades ubiquitin conjugates. *J. Biol. Chem.* **262**, 2451–2457.

Werner, E.D., Brodsky, J.L. & McCracken, A.A. (1996) Proteasome-dependent endoplasmic reticulum-associated protein degradation: an unconventional route to a familiar fate. *Proc. Natl Acad. Sci USA* **93**, 13797–13801.

Wiertz, E.J.H.J., Tortorella, D., Bogyo, M., Yu, J., Mothes, W., Jones, T.R., Rapoport, T.A. & Ploegh, H.L. (1996) Sec61-mediated transfer of a membrane protein from the endoplasmic reticulum to the proteasome for destruction. *Nature* **384**, 432–438.

Yokota, K., Kagawa, S., Shimizu, Y., Akioka, H., Tsurumi, C., Noda, C., Fujimura, M., Yokosawa, H., Fujiwara, T., Takahashi, E., Ohba, M., Yamasaki, M., DeMartino, G.N., Slaughter, C.A., Toh-e, A. & Tanaka, K. (1996) cDNA cloning of p112, the largest regulatory subunit of the human 26S proteasome, and functional analysis of its yeast homologue, Sen3p. *Mol. Biol. Cell* **7**, 853–870.

Yoshimura, T., Kameyama, K., Takagi, T., Ikai, A., Tokunaga, F., Koide, T., Tanahashi, N., Tamura, T., Cejka, Z., Baumeister, W., Tanaka, K. & Ichihara, A. (1993) Molecular characterization of the '26S' proteasome complex from rat liver. *J. Struct. Biol.* **111**, 200–211.

Zwickl, P., Lottspeich, F. & Baumeister, W. (1992) Expression of functional *Thermoplasma acidophilum* proteasomes in *Escherichia coli. FEBS Lett.* **312**, 157–160.

*George N. DeMartino*
*Department of Physiology,*
*University of Texas Southwestern Medical Center at Dallas,*
*5323 Harry Hines Boulevard,*
*Dallas, TX 75235-9040, USA*
*Email: gdemar@mednet.swmed.edu*

# 174. γ-Glutamyl transpeptidase

## Databanks

*Peptidase classification: possibly clan TA, family S42, although there is no definite evidence of this. MEROPS ID: S42.001*
*NC-IUBMB enzyme classification: EC 2.3.2.2*
*Databank codes:*

| Species | Gene | SwissProt | PIR | EMBL (cDNA) | EMBL (genomic) |
|---|---|---|---|---|---|
| *Arabidopsis thaliana* | D22 | – | – | Z49240 | – |
| *Bacillus anthracis* | dep | – | S36209 | D14037 | – |
| *Bacillus subtilis* | ggt | P54422 | – | U49358 | – |
| *Escherichia coli* | ggt | P18956 | JV0028 | M28722 | – |
| *Homo sapiens* | ggt1 | P19440 | A31253 | J04131 J05235 M24087 M24903 | – |
| *Homo sapiens* | ggt2 | P36268 | A36742 | L20494 M30474 | M30475: exon – 4 M30476: exon – 3 M30477: exon – 2 M30478: exon – 1 M30479: last exon and complete CDS |
| *Homo sapiens* | ggt5 | P36269 | A41125 | M64099 | – |
| *Homo sapiens* | – | – | – | X98922 | – |
| *Mus musculus* | ggt | – | – | U30509 | – |
| *Pseudomonas* sp. | acyI | P15557 | B28392 | M18279 | – |

| Species | Gene | SwissProt | PIR | EMBL (cDNA) | EMBL (genomic) |
|---------|------|-----------|-----|-------------|----------------|
| *Pseudomonas* sp. | *acyI* | Q05053 | S27199 | X69020 | – |
| *Pseudomonas* sp. | *ggt* | P36267 | – | S63255 | – |
| *Rattus norvegicus* | *ggt* | P07314 | A05225 | M33822 X03518 X15443 | M57672: 5′ end |
| *Sus scrofa* | *ggt* | P20735 | S05532 | X16533 Z46916 Z46917 Z46922 | Z46918: exon |

## Name and History

*γ-Glutamyl transpeptidase* is also referred to as *γ-glutamyltransferase* (the name recommended by NC-IUBMB), and *γ-GTP*, *γ-GT* and *GGT* are abbreviations that are frequently used. The activity of the enzyme was first discovered in the pancreas, by which glutathione, L-γ-Glu-L-Cys-Gly, was degraded (Dakin & Dudley, 1913). Many years later, it was demonstrated that the enzyme catalyzes the initial step of glutathione breakdown (Hanes *et al.*, 1950), according to the following reaction: glutathione + amino acid → γ-glutamyl amino acid + cysteinylglycine.

The enzyme has been subsequently found in a number of animal tissues as well as in microorganisms and in plants. Although the enzyme has been classified as a transferase in the IUBMB enzyme nomenclature, it catalyzes not only the transfer of a γ-glutamyl group but also hydrolysis of a unique type of peptide bond formed by the γ-carboxyl group of glutamate and the α-amino group of cysteine or another amino acid (Tate & Meister, 1985). In addition, γ-glutamyl transpeptidase exhibits glutaminase activity and has been found to be identical to the enzyme formerly known as **phosphate-independent glutaminase** or **maleate-stimulated glutaminase** (Tate & Meister, 1975; Curthoys & Kuhlenschmidt, 1975).

## Activity and Specificity

Three kinds of reaction can be catalyzed by γ-glutamyl transpeptidase:

γ-glutamyl-X + Y → X + γ-glutamyl-Y *(transpeptidation)*

γ-glutamyl-X + γ-glutamyl-X → γ-(γ-glutamyl)-

glutamyl-X + X *(autotranspeptidation)*

γ-glutamyl-X → glutamate + X *(hydrolysis)*

In the transpeptidation reaction, γ-glutamyl transpeptidase catalyzes the transfer of a γ-glutamyl moiety of glutathione, its *S*-substituted derivatives or other γ-glutamyl compounds to a variety of acceptor substrates such as amino acids and dipeptides. Both L- and D-γ-glutamyl compounds are active as donor substrate (Tate & Meister, 1985). While the enzyme strictly recognizes the γ-glutamyl moiety of the donor, it exhibits relatively broad specificity for the leaving group of the donor, which corresponds to the Cys or Cys-Gly portion of glutathione. Glutathione analogs such as reduced

glutathione, L-γ-glutamyl-L-α-aminobutyrate, *S*-propanone-glutathione, *S*-acetamidoglutathione and *S*-acetophenone-glutathione are relatively good substrates, whereas oxidized glutathione is not so active. L-Isomers of neutral amino acids are good acceptors; cystine and glutamine being especially active (Thompson & Meister, 1976, 1977). Methionine, alanine and serine also can act as acceptors, but the branched-chain amino acids are rather poor acceptors, and proline, α-substituted amino acids and D-amino acids are inactive (Tate & Meister, 1974). Several dipeptides also are efficient acceptors, and some of these are more active than L-glutamine; the best dipeptide acceptors include Met-Gly, Gln-Gly, Ala-Gly, cystinyl-*bis*-Gly, Ser-Gly, Gly-Gly and Cys-Gly. However, high concentrations of amino acid and dipeptide acceptor substrates inhibit the enzyme to various extents, in competition with the γ-glutamyl donor (Allison, 1985).

Since L-γ-glutamyl donor substrates can also act as acceptors at sufficiently high concentrations, the γ-glutamyl moiety of the donor may be transferred to a second donor molecule. This reaction yields a γ-(γ-glutamyl)-glutamyl compound (autotranspeptidation). Water also serves as an acceptor of the γ-glutamyl moiety, so that the reaction can lead to hydrolysis of the γ-glutamyl bond (hydrolysis).

In the assay of transpeptidation activity, 2–5 mM *p*-nitroaniline derivatives such as L-γ-glutamyl-*p*-nitroanilide and L-γ-glutamyl-3-carboxy-4-nitroanilide (available from Sigma; see Appendix 2 for full names and addresses of suppliers) are most commonly used as donor substrate. The assay system typically includes 20 mM Gly-Gly as acceptor, and the pH optimum of the enzyme is about 7.5–9.0. These compounds are commercially available. On the other hand, in order to see the hydrolytic reaction, it is necessary to eliminate the autotranspeptidation by use of either a D-γ-glutamyl compound or a sufficiently low concentration of the L-γ-glutamyl donor (Thompson & Meister, 1976).

The enzyme is inhibited specifically and competitively by the L- and D-isomers of γ-glutamyl-(*O*-carboxyl)phenylhydrazide and by other similar compounds (Minato, 1979). The glutamine analogs, L-azaserine, 6-diazo-5-oxo-L-norleucine, and α-amino-3-chloro-4,5-dihydro-5-isoxazoleacetic acid (AT-125, NSC-16530, acivicin) are nonspecific but potent irreversible inhibitors (Tate & Meister, 1977; Allen *et al.*, 1980); these compounds are available from Sigma. The combination of serine and borate (serine–borate complex) is known to inhibit the enzyme by acting as a transition-state analog (Tate & Meister, 1978).

The reaction catalyzed by γ-glutamyl transpeptidase is believed to proceed via a γ-glutamyl enzyme intermediate, which is analogous to acyl enzymes found in serine class hydrolases. Although many studies have suggested that the enzyme employs a hydroxyl as a catalytic group involved in the formation of the intermediate (Stole *et al.*, 1994; Ikeda *et al.*, 1995a), such a catalytic residue has not yet been definitely identified.

## Structural Chemistry

Mammalian γ-glutamyl transpeptidases are membrane-bound glycoproteins with *N*-glycans. The enzyme is composed of two nonidentical subunits designated as the heavy and light subunits, that are encoded by a common mRNA and are derived from a single-chain precursor by proteolytic processing (Nash & Tate, 1982). The subunits associate in a noncovalent manner. The molecular masses of the peptide backbones of the heavy and light subunits are about 40 kDa and 20 kDa, respectively, but the masses of the mature subunits vary with glycosylation, in the ranges 46–60 kDa and 23–25 kDa. The *Escherichia coli* enzyme which has no sugar chains has been crystallized, and X-ray crystallographic analysis has been reported (Kumagai *et al.*, 1993; Sakai *et al.*, 1996). Carbohydrate structures are known for forms of the enzyme purified from rat and human kidney, hepatoma tissues and other tumor tissues, yolk sac and seminal fluid. The most interesting observation in terms of sugar chain structure is the detection of bisecting *N*-acetylglucosamine in the enzyme purified from ascites tumor cells (Yamashita *et al.*, 1983). The enzyme is anchored to the extracellular surface of the plasma membrane by the N-terminal transmembrane domain of the heavy subunit, and therefore is a typical type II membrane protein.

The enzyme is translated as a single-chain form and then is processed into the heavy and light subunits. The single-chain precursor is believed to be enzymatically inactive. When the subunits were dissociated by treatment with urea or SDS, the light subunit degraded the heavy subunit, and also albumin, suggesting that the light subunit has latent proteinase activity (Gardell & Tate, 1979). This finding and the exclusive labeling of the light subunit by the irreversible inhibitors (see above) provide evidence that the activity to hydrolyze a peptide bond is associated with the subunit. Therefore, the light subunit has been regarded as the catalytic one.

## Preparation

The enzyme is widely distributed in various mammalian tissues such as kidney, brain, liver, pancreas, intestine, ciliary body, epididymis, thyroid epithelium, bile duct epithelium, bronchial epithelium and seminal vesicles. The highest activity is found in kidney. Although normal liver displays very low activity, relatively high activity was detected in some hepatoma tissues. The enzyme has been purified from rat kidney, rat tumor tissues, human kidney, cattle kidney and various tissues. In these purification procedures, the enzyme was solubilized from cell membranes by a detergent such as Lubrol or Triton X-100, or alternatively by limited proteolysis with protease such as bromelain or papain (Hughey & Curthoys, 1976). The enzyme has been expressed in COS cells (Ikeda *et al.*, 1993) and V79 cells (Visvikis *et al.*, 1991) and also in baculovirus/insect systems (Ikeda *et al.*, 1995b), and the purified recombinant enzyme has been obtained with these expression systems.

## Biological Aspects

γ-Glutamyl transpeptidase is of critical importance in glutathione metabolism, which involves a γ-glutamyl cycle that includes synthesis and degradation of glutathione, a major reducing agent within cells (Meister & Larsson, 1995). A primary role of the enzyme would be to degrade glutathione into its constituent amino acids, acting in conjunction with a dipeptidase. γ-Glutamyl transpeptidase thereby facilitates recovery of the constituent amino acids from the extracellular fraction of glutathione into cells. In fact, inhibition or deficiency of the enzyme in kidney leads to severe glutathionuria. In particular, the formation of γ-glutamyl cystine resulting from transpeptidation and the subsequent uptake of the derivative into cells seem to be a significant salvage pathway of cyst(e)ine. Furthermore, the enzyme is also involved in metabolism of glutathione-*S*-conjugates of xenobiotics and endogenous substances.

The expression of the γ-glutamyl transpeptidase gene is tissue specific and developmentally regulated. It is also known that the expression during hepatocarcinogenesis is remarkably high. The gene was found to be a single copy in mouse and rat, whereas the human gene is a multiple copy one (reviewed by Lieberman *et al.*, 1995). The gene is transcribed by use of multiple promoters, and there are several RNAs with different 5′ ends, even though all the different transcripts would encode the same polypeptide of γ-glutamyl transpeptidase. The multipromoter system appears to serve for the tissue-specific expression of the enzyme, but DNA methylation is also implicated in the regulation of the tissue-specific and development-stage specific expression of γ-glutamyl transpeptidase.

In clinical chemistry, the enzyme is widely used for the diagnosis and monitoring of various liver diseases, including alcoholic hepatitis, cirrhosis and hepatocellular carcinoma, and also for pancreatic diseases, biliary diseases and various cancers. A number of assay kits for the enzyme activity are commercially available. Basic research questions remain, however. We still do not know whether the increased activity in serum is really associated with the increased level of the enzyme protein, and alcohol intake elevates the enzyme activity in serum, but the mechanism of this effect is still unknown.

## Distinguishing Features

A γ-glutamyl transpeptidase-related enzyme has been identified, and its cDNA has been cloned from human placenta (Heisterkamp *et al.*, 1991). The nucleotide sequence of the cDNA revealed 40% of amino acid sequence identity with human γ-glutamyl transpeptidase. This enzyme catalyzed hydrolysis of glutathione and cleavage of the γ-glutamyl moiety of glutathione-S-conjugate with leukotriene, similarly to γ-glutamyl transpeptidase, but interestingly, did not hydrolyze L-γ-glutamyl-p-nitroanilide. However, this related enzyme is not yet fully characterized.

Several studies have indicated that only portions of the γ-glutamyl transpeptidase sequence are translated in human tissues (Pawlak *et al*., 1990; Wetmore *et al*., 1993). These atypical translation events would lead to immunoreactive but enzymatically inactive proteins, and therefore might account for the discrepancy between the immunologically assessed level of the enzyme and the activity.

A monoclonal antibody against human γ-glutamyl transpeptidase that also reacts with mouse enzyme is commercially available (Cosmo Bio Co., Ltd.).

## Further Reading

For a review, see Taniguchi & Ikeda (1997).

## References

Allen, L., Meck, R. & Yunis, A. (1980) The inhibition of γ-glutamyl transpeptidase from human pancreatic carcinoma cells by (αS,5S)-α-amino-3-chloro-4,5-dihydro-5 isoxazole acetic acid (AT-125, NSC-16501). *Res. Commun. Chem. Pathol. Pharmacol.* **27**, 175–182.

Allison, R.D. (1985) γ-Glutamyl transpeptidase: kinetics and mechanism. *Methods Enzymol.* **113**, 419–437.

Curthoys, N.P. & Kuhlenschmidt, T. (1975) Phosphate-independent glutaminase from rat kidney. *J. Biol. Chem.* **250**, 2099–2105.

Dakin, H.D. & Dudley, H.W. (1913) Glyoxalase, part III. The distribution of the enzyme and its relation to the pancreas. *J. Biol. Chem.* **15**, 463–474.

Gardell, S.J. & Tate, S.S. (1979) Latent proteinase activity of γ-glutamyl transpeptidase light subunit. *J. Biol. Chem.* **254**, 4942–4945.

Hanes, C.S., Hird, F.J.R. & Isherwood, F.A. (1950) Synthesis of peptides in enzymic reactions involving glutathione. *Nature* **166**, 288–292.

Heisterkamp, N., Rajpert-De-Meyts, E., Uribe, L., Forman, H.J. & Groffen, J. (1991) Identification of a human γ-glutamyl cleaving enzyme related to, but distinct from, γ-glutamyl transpeptidase. *Proc. Natl Acad. Aci. USA* **88**, 6303–6307.

Hughey, R.P. & Curthoys, N.P. (1976) Comparison of the size and physical properties of γ-glutamyltranspeptidase purified from rat kidney following solubilization with papain or with Triton X-100. *J. Biol. Chem.* **251**, 8763–8770.

Ikeda, Y., Fujii, J. & Taniguchi, N. (1993) Significance of Arg-107 and Glu-108 in the catalytic mechanism of human γ-glutamyl transpeptidase: identification by site-directed mutagenesis. *J. Biol. Chem.* **268**, 3980–3985.

Ikeda, Y., Fujii, J., Anderson. M.E., Taniguchi, N. & Meister, A. (1995a) Involvement of Ser-451 and Ser-452 in the catalysis of human γ-glutamyl transpeptidase. *J. Biol. Chem.* **270**, 22223–22228.

Ikeda, Y., Fujii, J., Taniguchi, N. & Meister, A. (1995b) Expression of an active glycosylated human γ-glutamyl transpeptidase mutant that lacks a membrane anchor domain. *Proc. Natl Acad. Aci. USA*

**92**, 126–130.

Kumagai, H., Nohara, S., Suzuki, H., Hashimoto, W., Yamamoto, K., Sakai, H., Sakabe, K., Fukuyama, K. & Sakabe, N. (1993) Crystallization and preliminary X-ray analysis of γ-glutamyltranspeptidase. *J. Mol. Biol.* **234**, 1259–1262.

Lieberman, M.W., Barrios, R., Carter, B.Z., Habib, G.M., Lebovitz, R.M., Rajagopalan, S., Sepulveda, A.R., Shi, Z.-Z. & Wan, D.-F. (1995) γ-Glutamyl transpeptidase: what does the organization and expression of a multipromoter gene tell us about its function. *Am. J. Pathol.* **147**, 1175–1185.

Meister, A. & Larsson, A. (1995) Glutathione synthetase deficiency and other disorders of the γ-glutamyl cycle. In: *The Metabolic and Molecular Bases of Inherited Disease*, 7th edn, vol. I (Scriver, C.R., Baeudet, A.L., Sly, W.S. & Valle, D., eds). New York: McGraw-Hill, pp. 1461–1477.

Minato, S. (1979) Isolation of anthglutin, an inhibitor of γ-glutamyl transpeptidase. *Arch. Biochem. Biophys.* **192**, 235–240.

Nash, B. & Tate, S.S. (1982) Biosynthesis of rat renal γ-glutamyl transpeptidase: evidence for a common precursor of the two subunits. *J. Biol. Chem.* **257**, 585–588.

Pawlak, A., Cohen, E.H., Octave, J.N., Schweickhardt, R., Wu, S.J., Bulle, F., Chikhi, N., Baik, J.H., Siegrist, S. & Guellaen, G. (1990) An alternatively processed mRNA specific for γ-glutamyl transpeptidase in human tissues. *J. Biol. Chem.* **265**, 3256–3262.

Sakai, H., Sakabe, N., Sasaki, K., Hashimoto, W., Suzuki, H., Tachi, H., Kumagai, H. & Sakabe, K. (1996) A preliminary description of the crystal structure of γ-glutamyltranspeptidase from *E. coli* K-12. *J. Biochem.* **120**, 26–28.

Stole, E., Smith, T.K., Manning, J.M. & Meister, A. (1994) Interaction of γ-glutamyl transpeptidase with acivicin. *J. Biol. Chem.* **269**, 21435–21439.

Taniguchi, N. & Ikeda, Y. (1997) γ-Glutamyl transpeptidase: catalytic mechanism and gene expression. *Adv. Enzymol.* (in press).

Tate, S.S. & Meister, A. (1974) Interaction of γ-glutamyl transpeptidase with amino acids, dipeptides, and derivatives and analogs of glutathione. *J. Biol. Chem.* **249**, 7593–7602.

Tate, S.S. & Meister, A. (1975) Identity of maleate-stimulated 'glutaminase' with γ-glutamyl transpeptidase in rat kidney. *J. Biol. Chem.* **250**, 4619–4624.

Tate, S.S. & Meister, A. (1977) Affinity labeling of γ-glutamyl transpeptidase and location of the γ-glutamyl binding site on the light subunit. *Proc. Natl Acad. Sci. USA* **74**, 931–935.

Tate, S.S. & Meister, A. (1985) γ-Glutamyl transpeptidase from kidney. *Methods Enzymol.* **113**, 400–419.

Tate, S.S. & Meister, A. (1978) Serine–borate complex as a transition state inhibitor of γ-glutamyl transpeptidase. *Proc. Natl Acad. Sci. USA* **75**, 4806–4809.

Thompson, G.A. & Meister, A. (1976) Hydrolysis and transfer reactions catalyzed by γ-glutamyl transpeptidase, evidence for separate substrate sites and for high affinity of L-cystine. *Biochem. Biophys. Res. Commun.* **71**, 32–36.

Thompson, G.A. & Meister, A. (1977) Interrelationship between the binding sites for amino acids, dipeptides, and γ-glutamyl donors in γ-glutamyl transpeptidase. *J. Biol. Chem.* **252**, 6792–6798.

Visvikis, A., Thioudellet, C., Oster, T., Fournel-Gigleux, S., Wellman, M. & Siest, G. (1991) High-level expression of enzymatically active mature human γ-glutamyltransferase in transgenic V79 Chinese hamster cells. *Proc. Natl Acad. Aci. USA* **88**, 7361–7365.

Wetmore, L.A., Gerard, C. & Drazen, J.M. (1993) Human lung expresses unique γ-glutamyl transpeptidase transcripts. *Proc. Natl Acad. Aci. USA* **90**, 7461–7465.

Yamashita, K., Hitoi, A., Taniguchi, N., Yokosawa, N., Tsukada, Y. & Kobata, A. (1983) Comparative study of the sugar chains of γ-glutamyltranspeptidases purified from rat liver and rat AH-66 hepatoma cells. *Cancer Res.* **43**, 5059–5063.

***Nayuki Taniguchi***
*Department of Biochemistry,*
*Osaka University Medical School,*
*2–2 Yamadaokia, Suita, Osaka 565, Japan*
*Email: proftani@biochem.med.osaka-u.ac.jp*

***Yoshitaka Ikeda***
*Department of Biochemistry,*
*Osaka University Medical School,*
*2–2 Yamadaokia, Suita, Osaka 565, Japan*

# 175. Introduction: other families of serine peptidases

## Databanks

MEROPS ID: SX

| Species | SwissProt | PIR | EMBL (cDNA) | EMBL (genomic) |
|---|---|---|---|---|
| **Family S14**<br>Endopeptidase Clp (Chapter 177) | | | | |
| **Family S16**<br>Endopeptidase La (Chapter 178)<br>Others | | | | |
| *Haemophilus influenzae* | P43865 | – | – | U32812: complete genome section 127 of 163 |
| **Family S18**<br>Omptin (Chapter 176) | | | | |
| **Family S19**<br>Chymotrypsin-like protease (*Coccidioides*) (Chapter 179) | | | | |
| **Family S34**<br>Erythrocyte membrane band 7 protein | | | | |
| *Homo sapiens* | – | – | M81635 | – |
| HflA protease | | | | |
| *Escherichia coli* | P25661 | C43653 | U00005 | U14003: chromosome 92.8–00.1' |
| *Haemophilus influenzae* | – | – | L44796 | – |
| *Vibrio proteolyticus* | P40606 | – | U09005 | – |
| **Family S38**<br>Chymotrypsin-like protease (*Treponema denticola*) (Chapter 102) | | | | |
| **Family S39**<br>Cocksfoot mottle sobemovirus proteinase | – | – | Z48630 | |
| **Family S43**<br>Porin D | | | | |
| *Pseudomonas aeruginosa* | – | S34969 | D12711 | – |

There are eight families of serine peptidases for which neither the tertiary structures nor the order of catalytic residues are known. These cannot be assigned to clans, and are referred to as members of clan 'SX'. When crystal structures are solved, or when active-site residues have been identified in a member of one of these families, that family will be moved into an existing or a new clan.

*Family S14* includes the cytosolic, ATP-dependent endopeptidase Clp. This is an oligomeric complex composed of 18 subunits arranged in rings and stacked so that they resemble the proteasome in electron micrographs. There are only two different subunits, and the complex is composed of six ATP-binding subunits and 12 peptidase subunits. The ATP-binding subunits are unrelated to peptidases. There are at least two different but homologous ATP-binding subunits, known as ClpA and ClpB, and the complex contains only one

type. The peptidase domain is known as clpP, and without the ATP-binding domain, clpP acts only on oligopeptides. Only the catalytic serine has been identified (Maurizi *et al.*, 1994). Endopeptidase Clp is found in bacteria, but homologs are known from plant chloroplasts, *Caenorhabditis* and humans. The potato leaf roll luteovirus possesses a sequence at the 5' untranslated end of its RNA genome which is remarkably similar to that encoding part of clpP, including the catalytic serine (EMBL: D00530).

*Family S16* includes another bacterial, cytosolic ATP-dependent enzyme, endopeptidase La. Unlike endopeptidase Clp, endopeptidase La is a homotetramer and has both the ATP-binding domain and peptidase domain in one protein subunit. The catalytic serine (Ser679) has been identified by site-specific mutagenesis (Fischer & Glockshuber, 1993). Besides bacteria, the family includes representatives from eukaryotes, which are nuclear-encoded mitochondrial endopeptidases.

*Family S18* includes the bacterial outer membrane endopeptidase omptin, which is specific for cleavage between paired basic residues. A homolog from *Yersinia pestis* has coagulant and fibrinolytic activities, and is suggested to be involved in plague transmission (Sodeinde *et al.*, 1992).

*Family S19* includes cell wall-associated endopeptidases from fungi, including the human dermal pathogen *Trichophyton rubrum* and the soil fungus *Coccidiodes imitis* (Chapter 179).

*Family S38* contains an intracellular endopeptidase from the oral spirochete *Treponema denticola* that has been characterized as chymotrypsin-like from its inhibitor profile (Arakawa & Kuramitsu, 1994). It should be noted, however, that *Treponema* also contains a subtilisin homolog that could have similar activity (see Chapter 102).

*Family S34* includes a subunit of a poorly characterized *Escherichia coli* membrane endopeptidase complex. This complex consists of at least two subunits, the products of the *hflC* and *hflK* genes, and is known as the HflKC or HflA endopeptidase; it is believed to be responsible for the degradation of phage λ cII protein, cleavage of which prevents the phage entering the lysogeny pathway (Cheng *et al.*, 1988). The HflC protein sequence shows limited similarity to sequence around the catalytic serine of the clpP subunit of endopeptidase Clp, and is presumed to be a serine peptidase (Noble *et al.*, 1993). More recently, the HflKC endopeptidase has been implicated in the turnover of uncomplexed SecY protein. The SecY protein forms a complex with the SecA and SecE proteins that is involved in protein export and forms a channel permitting the translocation of proteins across the plasma membrane. Uncomplexed SecY protein is degraded by the FtsH complex (Chapter 516), and HflKC acts as an antagonist by making FtsH inactive (Kihara *et al.*, 1996). The HflC protein shows some similarity to human erythrocyte band 7 integral membrane protein, but this is not known to be a peptidase, and the relationship is probably one between mosaic proteins.

*Family S39* includes a putative polyprotein processing endopeptidase from sobemoviruses (Gorbalenya *et al.*, 1988).

*Family S43* includes porin D from *Pseudomonas aeruginosa*. This is an outer membrane protein that forms a pore permitting the passage of basic amino acids. Porin D has been shown to hydrolyze several synthetic peptides, and to be inhibited by DFP (Yoshihara *et al.*, 1996).

Chapter 180 is an account of UIPase, a monoclonal antibody with peptidase activity.

## References

Arakawa, S. & Kuramitsu, H.K. (1994) Cloning and sequence analysis of a chymotrypsinlike protease from *Treponema denticola*. *Infect. Immun.* **62**, 3424–3433.

Cheng, H.H., Muhlrad, P.J., Hoyt, M.A. & Echols, H. (1988) Cleavage of the cII protein of phage λ by purified HflA protease: control of the switch between lysis and lysogeny. *Proc. Natl Acad. Sci. USA* **85**, 7882–7886.

Fischer, H. & Glockshuber, R. (1993) ATP hydrolysis is not stoichiometrically linked with proteolysis in the ATP-dependent protease La from *Escherichia coli*. *J. Biol. Chem.* **268**, 22502–22507.

Gorbalenya, A.E., Koonin, E.V., Blinov, V.M. & Donchenko, A.P. (1988) Sobemovirus genome appears to encode a serine protease related to cysteine proteases of picornaviruses. *FEBS Lett.* **236**, 287–290.

Kihara, A., Akiyama, Y. & Ito, K. (1996) A protease complex in the *Escherichia coli* plasma membrane: HflKC (HflA) forms a complex with FtsH (HflB), regulating its proteolytic activity against SecY. *EMBO J.* **15**, 6122–6131.

Maurizi, M.R., Thompson, M.W., Singh, S.K. & Kim, S.-H. (1994) Endopeptidase Clp: the ATP-dependent Clp protease from *Escherichia coli*. *Methods Enzymol.* **244**, 314–331.

Noble, J.A., Innis, M.A., Koonin, E.V., Rudd, K.E., Banuett, F. & Herskowitz, I. (1993) The *Escherichia coli hflA* locus encodes a putative GTP-binding protein and two membrane proteins, one of which contains a protease-like domain. *Proc. Natl Acad. Sci. USA* **90**, 10866–10870.

Sodeinde, O.A., Subrahmanyam, Y.V.B.K., Stark, K., Quan, T., Bao, Y. & Goguen, J.D. (1992) A surface protease and the invasive character of plague. *Science* **258**, 1004–1007.

Yoshihara, E., Gotoh, N., Nishino, T. & Nakae, T. (1996) Protein D2 porin of the *Pseudomonas aeruginosa* outer membrane bears the protease activity. *FEBS Lett.* **394**, 179–182.

# *176. Omptin*

## *Databanks*

*Peptidase classification: clan SX, family S18, MEROPS ID: S18.001*
*NC-IUBMB enzyme classification: EC 3.4.21.87*
*Chemical Abstracts Service registry number: 150770-86-8*

*Databank codes:*

| Species | Gene | SwissProt | PIR | EMBL (cDNA) | EMBL (genomic) |
|---------|------|-----------|-----|-------------|----------------|
| *Escherichia coli* | *ompT* | P09169 | A31387 | M23630 | – |
| | | | S01751 | X06903 | |
| *Escherichia coli* | *ompP* | P34210 | S37473 | X74278 | – |
| *Salmonella typhimurium* | *pgtE* | P06185 | B28255 | M13923 | – |
| *Yersinia pestis* | *pla* | P17811 | A30916 | M27820 | – |
| | | | A42928 | X15136 | |

## Name and History

**Omptin** is present in many gram-negative bacteria. Its name is derived from the fact that it is an **o**uter **m**embrane **p**rotein (OMP) and the expression of its gene (*ompT*) is **t**emperature regulated (Manning & Reeves, 1977). Omptin was discovered and partially characterized as a proteinase activity in preparations of outer membranes from *Escherichia coli* that could activate the serum proenzyme plasminogen to the active proteinase plasmin (Chapter 59) (Leytus *et al*., 1981). Heretofore, the only known specific plasminogen activators were urokinase and tissue plasminogen activator (Stoppelli *et al*., 1986; Mangel, 1990). Omptin was later identified as the proteinase that cleaved recombinant T7 RNA polymerase during purification from *E. coli* (Grodberg & Dunn, 1988). Its gene was mapped to the *ompT* locus (Rupprecht *et al*., 1983; Grodberg & Dunn, 1988), and only *ompT*+ strains were shown to be able to cleave T7 RNA polymerase or activate plasminogen to plasmin (Grodberg & Dunn, 1988).

Omptin is the first member of a new class of proteinase; the sequence of its gene is not related to gene sequences of other proteinases in the databases (Grodberg *et al*., 1988). Omptin has an unusual specificity in that it cleaves between two basic amino acids. It is insensitive to most proteinase inhibitors. Because it remains active under extreme denaturing conditions, e.g. 8 M urea, it can cleave recombinant proteins expressed in *E. coli* as they are solubilized from inclusion bodies (White *et al*., 1995). Biologically, omptin has been associated with virulence factors in pathogenic *E. coli* isolated from urinary tract infections (Foxman *et al*., 1995). In *Yersinia pestis*, the organism that causes bubonic plague (Sodeinde *et al*., 1992), a homolog of omptin is a major virulence factor.

The Nomenclature Committee of the IUBMB has recommended that the name omptin be used in place of the earlier terms **OmpT** (Grodberg & Dunn, 1988), **protein a** or **3b** (Hollifield *et al*., 1978; Rupprecht *et al*., 1983), and **protease VII** (Sugimura & Nishihara, 1988). Omptin homologs are OmpP from *E. coli* K12 (Kaufmann *et al*., 1994), Pla from *Yersinia pestis* (Sodeinde & Goguen, 1989), and protein E from *Salmonella typhimurium* (Yu & Hong, 1986; Grodberg & Dunn, 1989) later renamed PrtA (Sodeinde & Goguen, 1989).

## Activity and Specificity

Omptin cleaves between two basic amino acid residues (Grodberg & Dunn, 1988; Mangel *et al*., 1994). It also cleaves between Arg and Met (Zhao & Somerville, 1993), and between Arg and Val (Sodeinde *et al*., 1992).

There are three assays for omptin. One assay monitors by SDS-PAGE the cleavage of T7 RNA polymerase to fragments of 80 and 20 kDa (Grodberg & Dunn, 1988; Mangel *et al*., 1994). This assay is not quantitative; there are many sites on T7 RNA polymerase that are cleaved by omptin, and some of them are cleaved early in an incubation, others late.

The plasminogen activator activity of omptin may be assayed by measuring with chromogenic or fluorogenic substrates the amount of plasmin formed upon incubating plasminogen with the enzyme (Leytus *et al*., 1981; Mangel *et al*., 1994). The pH optimum is 5.0. The rate of plasminogen activation increases as the temperature is raised from 20°C to 45°C. At 37°C the activity is 40% of what it is at 45°C. Activation of plasminogen by omptin is quite sensitive to ionic strength; the activity decreases by 40% at 0.04 M NaCl compared to no NaCl. The assay for the plasminogen activator activity of omptin is highly quantitative and sensitive. However, this assay is dependent on the homogeneity of a plasminogen preparation. There are several different forms of plasminogen and each may interact with omptin with different kinetic parameters, as they do with human urokinase (Chapter 57) (Peltz *et al*., 1982).

A new assay is based on the observations that omptin cleaves between basic amino acids and that aminopeptidases will not cleave after a blocked amino acid, so that the compounds Z-Arg+Arg-NHMec and Z-Ala+Lys-Arg-NHMec will not be cleaved by aminopeptidase M. In the presence of omptin, either substrate is cleaved to Arg-NHMec, and when Arg-NHMec is incubated with aminopeptidase M, the highly fluorescent cleavage product, aminomethylcoumarin is formed. This two-step procedure to measure the activity of omptin can be combined into a coupled assay or separated in an uncoupled assay. Omptin prefers Z-Ala-Lys-Arg-NHMec as a substrate compared to Z-Arg-Arg-NHMec. These synthetic substrates are available from Novabiochem (see Appendix 2 for full names and addresses of suppliers). When assaying with Z-Ala-Lys-Arg-NHMec and aminopeptidase M, the pH optimum is 6.0. Maximal activity is at 37°C, and ionic strength has little effect on activity up to a concentration of 170 mM NaCl. The activity is stimulated 2-fold by 10 mM octylglucoside. The assays with synthetic, fluorogenic substrates are 10-fold more sensitive than those measuring the rate of plasminogen activation.

The activity of omptin is not affected by most proteinase inhibitors. Known inhibitors include DFP (Leytus *et al*., 1981; Sugimura & Nishihara, 1988), $ZnCl_2$ (Sugimura & Nishihara, 1988; Mangel *et al*., 1994), benzamidine, $CuCl_2$, $FeSO_4$ (Sugimura & Nishihara, 1988), and arginine (Mangel *et al*., 1994).

## Structural Chemistry

Omptin is synthesized as a preprotein of 317 amino acids with a molecular mass of 35 562 Da (Grodberg *et al.*, 1988). During export to the outer membrane a signal sequence of 20 amino acids is removed, decreasing the molecular mass to 33 477 Da. The pI is calculated to be 5.62.

Although there is indirect evidence that omptin is a serine proteinase, there is no evidence at the sequence level for this. A feature of microbial and pancreatic serine proteinases of family S1 is conservation of sequence surrounding the catalytic histidine and serine residues, Thr-Ala-Gly-His-Cys and Gly-Asp-Ser-Gly-Gly (Delbaere *et al.*, 1975; see also Chapter 2). None of the three histidine residues or the 25 serine residues in omptin is flanked by a sequence that has strong homology with catalytic residues in other serine proteinases (Grodberg & Dunn, 1989). Searching current sequence databanks also failed to demonstrate any convincing regions of homology with any other known proteinases.

## Preparation

Omptin is found on the outer membrane of certain gram-negative bacteria such as *Escherichia coli* and *Yersinia pestis*. The entire *E. coli* omptin gene was cloned into the pET plasmid vector, thereby placing the gene under control of a T7 promoter and efficient translation signals from phage T7; the expression host was *E. coli* BL26(DE3) (Maniatis *et al.*, 1982; Grodberg & Dunn, 1988; Studier *et al.*, 1990; Mangel *et al.*, 1994). After induction by IPTG, the proteinase was purified from the outer membrane by sequential extraction with solutions containing $Mg^{2+}$, EDTA, and 40 mM octylglucoside. The remaining contaminants were removed by centrifugation after placing omptin in a boiling water bath for 45 s. Other purification procedures do not yield a pure protein (Sugimura & Nishihara, 1988; White *et al.*, 1995).

## Biological Aspects

The physiological role of omptin is unknown, and deletion of the *ompT* gene in *E. coli* has no discernible effect on the life cycle of the bacterium in the laboratory (Baneyx & Georgiou, 1990; Grodberg & Dunn, 1988). The role of omptin in protein processing and turnover in the inner and outer membrane is currently being investigated.

A homolog of omptin has been implicated as a virulence factor. *Yersinia pestis*, the organism that causes bubonic plague, contains a plasmid whose expression correlates with virulence (Ben-Gurion & Shafferman, 1981; Ferber & Brubaker, 1981). The virulence gene was named *pla* (for **pl**asminogen **a**ctivator activity) (Beesley *et al.*, 1967) and is 47% identical in amino acid sequence to omptin (Sodeinde & Goguen, 1989). Inactivation of the *pla* gene increased the median lethal dose, $LD_{50}$, for mice by a million-fold (Sodeinde *et al.*, 1992). Studies are being done to determine whether the expression of *ompT* correlates with the pathogenic potential of strains of *E. coli* that infect humans (Lundrigan & Webb, 1992). Omptin has been shown to be associated with virulence factors in pathogenic *E. coli* isolated from urinary tract infections (Foxman *et al.*, 1995).

Although both omptin and *pla* can activate plasminogen, it is not clear whether this activity is related to virulence. Both human plasminogen (Parkkinen & Korhonen, 1989; Ullberg *et al.*, 1990) and plasmin (Broeseker *et al.*, 1988; Lottenberg *et al.*, 1987, 1992) can bind specifically to the surface of bacteria, so an entire fibrinolytic enzyme system can assemble there, perhaps in ways similar to the assembly of plasminogen activation systems on the surface of eukaryotic cells and structures (Stoppelli *et al.*, 1986; Mangel, 1990). Cell surface-associated plasmin has been implicated in the invasion of host cells by pathogenic bacteria (Lottenberg *et al.*, 1994).

## Distinguishing Features

Omptin shows anomalous behavior when subjected to SDS-PAGE (Mangel *et al.*, 1994). When heated in a boiling water bath for 1.5 min in SDS-PAGE sample buffer containing 5% 2-ME and then fractionated by SDS-PAGE, omptin runs as a 42 kDa protein. However, if it is heated to less than 42°C for 5 min and then fractionated, it runs as a 28 kDa protein. When omptin in SDS-PAGE sample buffer, 5% 2-ME and either 41% ammonium sulfate or 375 mM arginine is heated in a boiling water bath for 1.5 min and then fractionated by SDS-PAGE, it runs not as a 42 kDa protein but as a 28 kDa protein. Removal of the ammonium sulfate or arginine by dialysis reverses the apparent decrease in molecular mass in that it runs as a 42 kDa protein.

If pure omptin is placed in SDS-PAGE sample buffer with 5% 2-ME, heated in a boiling water bath for 90 s, and left at room temperature overnight before SDS-PAGE, omptin migrates as 31, 18 and 16 kDa proteins. These seem to be autodigestion products. The 31 and 16 kDa bands have the same N-terminus as omptin whereas the 16 kDa band has the N-terminal sequence Asp-Asp-Ile-Gly indicating cleavage occurred between Arg157 and Asp158. The same three bands are irreversibly formed if omptin is left at room temperature at pH 5 in the presence of a detergent above its critical micelle concentration.

Most scientists who have worked with omptin have done so unknowingly; it cleaves their recombinant proteins. Among the many examples are the T7 RNA polymerase (Grodberg & Dunn, 1988), a recombinant human interferon $\gamma$ (Sugimura & Higashi, 1988), creatine kinase (White *et al.*, 1995), tryptophan synthetase (Zhao & Somerville, 1993), and penicillin-binding proteins (Henderson *et al.*, 1994). The T7 RNA polymerase is cleaved when cells expressing the recombinant enzyme are broken, exposing the soluble T7 RNA polymerase to omptin in the outer membrane (Grodberg & Dunn, 1988). Many recombinant proteins form inclusion bodies in bacterial expression systems. Active enzyme can sometimes be obtained by solubilizing inclusion bodies in 8 M urea and then refolding the protein, but the insoluble pellet that contains inclusion bodies also contains membrane fragments, and since omptin is active in 8 M urea, it can cleave the recombinant proteins as they are being solubilized (White *et al.*, 1995). Strains lacking omptin ($ompT^-$) have been constructed that have proven useful for the expression of recombinant proteins (Studier *et al.*, 1990; Baneyx & Georgiou, 1990).

## Related Peptidases

OmpP is 87% identical to omptin and has the same specificity for cleavage between two basic residues (Kaufman *et al.*, 1994). The E or prtA protein of *Salmonella typhimurium* is 48.2% identical to omptin (Grodberg & Dunn, 1989; Sodeinde & Goguen, 1989). The E protein can localize to the outer membrane of *E. coli* and can cleave T7 RNA polymerase (Grodberg & Dunn, 1988). The *pla* gene of *Yersinia pestis* is 47.5% identical to the *ompT* gene and the Pla protein is 71% identical to the E protein from *Salmonella typhimurium* (Sodeinde & Goguen, 1989). Pla in *Yersinia pestis* has been shown to promote bacterial cell adhesion to type IV collagen (Kienle *et al.*, 1992) and is required for virulence.

The proteins of the omptin family show some sequence similarity to those of the OmpX family. The OmpX proteins are outer membrane proteins of about 18 kDa, all of which are involved in virulence (Mecsas *et al.*, 1995). Of the 171 amino acid residues of OmpX from *E. coli*, 48 (28%) are identical to omptin. Other members of the OmpX family include *ail* of *Yersinia enterocolitica*, Rck and PagC of *Salmonella typhimurium*, Omp4 of *Serratia marcescens* (Guasch *et al.*, 1995), ompK17 of *Klebsiella pneumoniae*, and the *lom* gene product of bacteriophage λ (Barondess & Beckwith, 1990). The Ail protein of *Yersinia enterocolitica* mediates bacterial resistance to complement killing (Bliska & Falkow, 1992) and promotes invasion of epithelial cells (Miller & Falkow, 1988). *pagC* mutants are less able to survive within cultured macrophages and are 1000-fold less virulent in mice (Miller *et al.*, 1989).

## Acknowledgement

This research was supported by a grant to W.F.M. from the Office of Health and Environmental Research of the United States Department of Energy.

## Further Reading

The article of Mangel *et al.* (1994) contains a description of the cloning, expression and purification of omptin from *E. coli*. It also describes the different assays for the enzyme. Sodeinde *et al.* (1992) show that Pla is a virulence factor of *Yersinia pestis* and is homologous to omptin. Finally, Baneyx & Georgiou (1990) describe strain SF110 of *E. coli*, which has two genes involved in the degradation of secreted proteins, *ompT* and *degP*, inactivated. This strain should be useful for the expression of recombinant and/or secreted proteins in *E. coli*.

## References

Baneyx, F. & Georgiou, G. (1990) *In vivo* degradation of secreted fusion proteins by *Escherichia coli* outer membrane protease OmpT. *J. Bacteriol.* **172**, 491–494.

Barondess, J.J. & Beckwith, J. (1990) A bacterial virulence determinant encoded by lysogenic coliphage lambda. *Nature* **346**, 871–874.

Beesley, E.D., Brubaker, R.R., Jansen, W.A. & Surgalla, M.J. (1967) Pesticins. 3. Expression of coagulase and mechanisms of fibrinolysis. *J. Bacteriol.* **94**, 19–26.

Ben-Gurion, R. & Shafferman, A. (1981) Essential virulence determinants of different *Yersinia* species are carried on a common plasmid. *Plasmid* **5**, 183–187.

Bliska, J.B. & Falkow S. (1992) Bacterial resistance to complement killing mediated by the Ail protein of *Yersinia enterocolitica*. *Proc. Natl Acad. Sci. USA* **89**, 3561–3565.

Broeseker, T.A., Boyle, M.D.P. & Lottenberg, R. (1988) Characterization of the interaction of human plasmin with its specific receptor on a group A streptococcus. *Microb. Pathogen.* **5**, 19–27.

Delbaere, L.T.J., Hutcheon, W.L.B., James, M.N.G. & Thiessen, W.E. (1975) Tertiary structural differences between microbial serine proteases and pancreatic serine enzymes. *Nature* **257**, 758–763.

Ferber, D.M. & Brubaker, R.R. (1981) Plasmids in *Yersinia pestis*. *Infect. Immun.* **31**, 839–841.

Foxman, B., Zhang, L., Palin, K., Tallman, P. & Marrs, C.F. (1995) Bacterial virulence characteristics of *Escherichia coli* isolates from first-time urinary tract infection. *J. Infect. Dis.* **171**, 1514–1521.

Grodberg, J. & Dunn, J.J. (1988) OmpT encodes the *Escherichia coli* outer membrane protease that cleaves T7 RNA polymerase during purification. *J. Bacteriol.* **170**, 1245–1253.

Grodberg, J. & Dunn, J.J. (1989) Comparison of *Escherichia coli* K-12 outer membrane protease *ompT* and *Salmonella typhimurium* E protein. *J. Bacteriol.* **171**, 2903–2905.

Grodberg, J., Lundrigan, M.D., Toledo, D.L., Mangel, W.F. & Dunn, J.J. (1988) Complete nucleotide sequence and deduced amino acid sequence of the *OmpT* gene of *Escherichia coli* K-12. *Nucleic Acids Res.* **16**, 1209.

Guasch, J.F., Ferrer, S., Enfedaque, J., Viejo, M.B. & Regue, M. (1995) A 17 kDa outer-membrane protein (Omp4) from *Serratia marcescens* confers partial resistance to bacteriocin 28b when expressed in *Escherichia coli*. *Microbiology* **141**, 2535–2542.

Henderson, T.A., Dombrosky, P.M. & Young, K.D. (1994) Artifactual processing of penicillin-binding proteins 7 and 1b by the OmpT protease of *Escherichia coli*. *J. Bacteriol.* **176**, 256–259.

Hollifield, W.C., Jr, Fiss, E.H. & Nielands, J.B. (1978) Modification of a ferric enterobacter receptor protein from the outer membrane of *Escherichia coli*. *Biochem. Biophys. Res. Commun.* **83**, 739–746.

Kaufmann, A., Stierhof, Y.-D. & Henning, U. (1994) New outer membrane-associated protease of *Escherichia coli* K-12. *J. Bacteriol.* **176**, 359–367.

Kienle, Z., Emody, L., Svanborg, C. & O'Toole, P.W. (1992) Adhesive properties conferred by the plasminogen activator of *Yersinia pestis*. *J. Gen. Microbiol.* **138**, 1679–1687.

Leytus, S.P., Bowles, L.K., Konisky, J. & Mangel, W.F. (1981) Activation of plasminogen to plasmin by a protease associated with the outer membrane of *Escherichia coli*. *Proc. Natl Acad. Sci. USA* **78**, 1485–1489.

Lottenberg, R., Broder, C.C. & Boyle, M.P.D. (1987) Identification of a specific receptor for plasmin on group A streptococcus. *Infect. Immun.* **55**, 1914–1918.

Lottenberg, R., Broder, C.C., Boyle, M.D.P., Kain, S.J., Schroeder, B.L. & Curtiss III, R. (1992) Cloning, sequence analysis, and expression in *Escherichia coli* of a streptococcal plasmin receptor. *J. Bacteriol.* **174**, 5202–5210.

Lottenberg, R., Minning-Wenz, D. & Boyle, M.D.P. (1994) Capturing host plasmin(ogen) – a common mechanism for invasive pathogens. *Trends Microbiol.* **2**, 20–24.

Lundrigan, M.D. & Webb, R.M. (1992) Prevalence of OmpT among *Escherichia coli* isolates of human origin. *FEMS Microbiol. Lett.* **76**, 51–56.

Mangel, W.F. (1990) Enzyme systems. Better reception for urokinase. *Nature (Lond.)* **344**, 488–489.

Mangel, W.F, Toledo, D.L., Brown, M.T., Worzalla, K., Lee, M. & Dunn, J.J. (1994) Omptin: an *Escherichia coli* outer membrane proteinase that activates plasminogen. *Methods Enzymol.* **244**, 384–399.

Maniatis, T., Fritsch, E.F. & Sambrook, J. (1982) *Molecular Cloning: A Laboratory Manual.* Cold Spring Harbor, NY: Cold Spring Harbor Laboratory.

Manning, P. & Reeves, P. (1977) Outer membrane protein 3b of *Escherichia coli* K-12: effects of growth temperature on the amount of the protein and further characterization on acrylamide gels. *FEMS Microbiol. Lett.* **1**, 275–278.

Mecsas, J., Welch, R., Erickson, J.W. & Gross, C.A. (1995) Identification and characterization of an outer membrane protein, OmpX, in *Escherichia coli* that is homologous to a family of outer membrane proteins including Ail of *Yersinia enterocolitica. J. Bacteriol.* **177**, 799–804.

Miller, S.I., Kukral, A.M. & Mekalanos, J.J. (1989) A two component regulatory system (*pho*P and *pho*Q) controls *Salmonella typhimurium* virulence. *Proc. Natl Acad. Sci. USA* **86**, 5048–5058.

Miller, V.L. & Falkow, S. (1988) Evidence for two genetic loci in *Yersinia enterocolitica* that can promote invasion of epithelial cells. *Infect. Immun.* **56**, 1242–1248.

Parkkinen, J. & Korhonen, T.K. (1989) Binding of plasminogen to *Escherichia coli* adhesion proteins. *FEBS Lett.* **250**, 437–440.

Peltz, S.W., Hardt, T.A. & Mangel, W.F. (1982) Positive regulation of the activation of plasminogen by urokinase: differences in $K_m$ for (glutamic acid)-plasminogen and lysine-plasminogen and effect of certain alpha, omega-amino acids. *Biochemistry* **21**, 2798–2804.

Rupprecht, K.R., Gordon, G., Lundrigan, M., Gayda, R.C., Markovitz, A. & Earhart, C. (1983) *Ompt: Escherichia coli* K-12 structural gene for protein a (3b). *J. Bacteriol.* **153**, 1104–1106.

Sodeinde, O.A. & Goguen, J.D. (1989) Nucleotide sequence of the plasminogen activator gene of *Yersinia pestis*: relationship to *OmpT* of *Escherichia coli* and E gene of *Salmonella typhimurium. Infect. Immun.* **57**, 1517–1523.

Sodeinde, O.A., Subrahmanyam, Y.V.B.K., Stark, K., Quan, T., Bao, Y. & Goguen, J.D. (1992) A surface protease and the invasive character of plague. *Science* **258**, 1004–1007.

Stoppelli, M.P., Tacchetti, C., Cubellis, M.V., Corti, A., Hearing, V.J., Cassani, G., Appella, E. & Blasi, F. (1986) Autocrine saturation of pro-urokinase receptors on human A431 cells. *Cell* **45**, 675–684.

Studier, F.W., Rosenberg, A.H., Dunn, J.J. & Dubendorff, J.W. (1990) Use of T7 RNA polymerase to direct expression of cloned genes. *Methods Enzymol.* **185**, 60–89.

Sugimura, K. & Higashi, N. (1988) A novel outer-membrane-associated protease in *Escherichia coli. J. Bacteriol.* **170**, 3650–3654.

Sugimura, K. & Nishihara, T. (1988) Purification, characterization, and primary structure of *Escherichia coli* protease VII with specificity for paired basic residues: identity of protease VII and OmpT. *J. Bacteriol.* **170**, 5625–5632.

Ullberg, M., Kronvall, G., Karllson, I. & Wiman, B. (1990) Receptors for human plasminogen on gram-negative bacteria. *Infect. Immun.* **58**, 21–25.

White, C.B., Chen, Q., Kenyon, G.L. & Babbitt, P.C. (1995) A novel activity of OmpT. *J. Biol. Chem.* **270**, 12990–12994.

Yu, G.-Q. & Hong, J.-S. (1986) Identification and nucleotide sequence of the activator gene of the externally induced phosphoglycerate transport system of *Salmonella typhimurium. Gene* **45**, 51–57.

Zhao, G.-P. & Somerville, R.L. (1993) An amino acid switch (Gly$^{281}$ → Arg) within the 'hinge' region of the tryptophan synthetase *B* subunit creates a novel cleavage site for the OmpT protease and selectively diminishes affinity toward a specific monoclonal antibody. *J. Biol. Chem.* **268**, 14912–14920.

*Diana L. Toledo*
*Biology Department,*
*Brookhaven National Laboratory,*
*Upton, NY 11973-5000, USA*

*Walter F. Mangel*
*Biology Department,*
*Brookhaven National Laboratory,*
*Upton, NY 11973-5000, USA*
*Email: mangel@bnl.gov*

# 177. Endopeptidase Clp

## Databanks

*Peptidase classification: clan SX, family S14, MEROPS ID: S14.001*
*NC-IUBMB enzyme classification: EC 3.4.21.92*

*Databank codes:*

| Species | Gene | SwissProt | PIR | EMBL (cDNA) | EMBL (genomic) |
|---|---|---|---|---|---|
| **Bacterial endopeptidase subunit** | | | | | |
| *Bacillus subtilis* | *clpP* | P80244 | – | U59754 | – |
| *Campylobacter jejuni* | *clpP* | P54413 | – | X85954 | – |
| *Cyanophora paradoxa* | *clpP* | P48254 | – | U30821 | – |
| *Escherichia coli* | *clpP* | P19245 | B36575 | J05534 | – |
| *Haemophilus influenzae* | *clpP* | P43867 | – | – | U32754: complete genome section 69 of 163 |
| *Paracoccus denitrificans* | *clpP* | P54414 | – | U34346 | – |
| *Streptococcus salivarius* | *clpP* | P36398 | – | L07793 | – |
| *Synechococcus* sp. | *clpP* | P54415 | – | U16135 | – |
| *Synechocystis* sp. | *clpP* | P54416 | – | – | D64006: complete genome part 25 |
| **Mitochondrial endopeptidase** | | | | | |
| *Caenorhabditis elegans* | *clpP* | – | – | Z49073 | – |
| *Homo sapiens* | *clpP* | – | – | Z50853 | – |
| **Chloroplast endopeptidase** | | | | | |
| *Arabidopsis thaliana* | – | – | – | F13836 | – |
| *Arabidopsis thaliana* | – | – | – | T21974 | – |
| *Chlamydomonas eugametos* | *clpP* | P42379 | – | L29402 | – |
| *Chlamydomonas reinhardtii* | *clpP* | P42380 | – | L28803 | – |
| *Epifagus virginiana* | *clpP* | P30063 | – | – | M81884: complete chloroplast genome |
| *Lycopersicon esculentum* | *clpP* | – | – | L38581 | – |
| *Marchantia polymorpha* | *clpP* | P12208 | A05056 | X04465 | – |
| *Nicotiana otophora* | *clpP* | P12210 | – | U32397 | Z00044: complete chloroplast genome |
| *Oenothera organensis* | – | – | – | X55899 | – |
| *Oryza sativa* | *clpP* | P12209 | JQ0251 | X15901 | – |
| *Pinus contorta* | *clpP* | P36387 | S50763 | L28807 | – |
| *Pinus thunbergii* | *clpP* | P41609 | – | D17510 | – |
| *Triticum aestivum* | *clpP* | P24064 | S12408 | X54484 | – |
| *Zea mays* | *clpP* | P26567 | S19126 | X60548 | – |
| **ATP-binding subunits** | | | | | |
| *Arabidopsis thaliana* | *cd4B* | – | – | Z29026 | – |
| *Bacillus subtilis* | *clpC* | P37571 | – | D26185 X75930 | U02604: 180 kb region |
| *Bacteroides nodosus* | *clpB* | – | C35905 S15253 S15254 S18723 | M32229 | – |
| *Bos taurus* | *clpA* | – | – | L34677 | – |
| *Brassica napus* | *clpA* | P46523 | – | X75328 | – |
| *Escherichia coli* | *clpA* | P15716 | A35365 | M31045 | – |
| *Escherichia coli* | *clpB* | P03815 | A04440 D35905 S18736 | M29364 X57620 | – |
| *Lactococcus lactis* | *clpL* | Q06716 | – | X62333 | – |
| *Lycopersicon esculentum* | *cd4A* | P31541 | A35905 | M32603 | – |
| *Lycopersicon esculentum* | *cd4B* | P31542 | B35905 | M32604 | – |
| *Mycobacterium leprae* | *clpA* | – | C43601 S11163 | – | – |
| *Pisum sativum* | *clpA* | P35100 | S31164 | L09547 | – |
| *Rhodococcus erythropolis* | *clpA* | Q01357 | JC1175 | M76451 | – |
| *Rhodopseudomonas blastica* | *clpA* | P05444 | S04667 | – | – |
| *Trypanosoma brucei* | *clp* | P31543 | E35905 | M92325 | – |

## Name and History

***ClpAP*** was isolated from *E. coli lon* mutant cell extracts based on its ability to degrade casein in the presence of ATP (Katayama-Fujimura *et al.*, 1987; Hwang *et al.*, 1988). The enzyme is a complex of two separate proteins, ClpA and ClpP, which can be purified separately and reconstituted to obtain the active ATP-dependent protease (Maurizi, 1992; Kessel *et al.*, 1995). ClpAP is also referred to as ***protease Ti***, and is ***heat-shock protein F21.5***. ClpA and ClpP are readily separated from each other in the absence of a stabilizing ligand, ATP or a nonhydrolyzable analog of ATP, which accounts for the apparent instability of the active complex in dilute crude cell extracts and upon purification. The homonymous name ***Clp*** was originally intended to suggest the clipping of proteins into a great number of small peptides without the generation of free amino acids and was derived from its activity as a **c**aseinolytic **p**rotease (Katayama *et al.*, 1988). The name might be more usefully interpreted as **c**haperone-linked **p**rotease in light of its recently discovered mechanism of action (Wickner *et al.*, 1994), which will be discussed below. ClpAP should be distinguished from ClpXP (Gottesman *et al.*, 1993; Wojtkowiak *et al.*, 1993), which has distinct substrate specificity and is a complex of ClpX (a homolog of ClpA) and ClpP.

## Activity and Specificity

### Peptidase Activity of ClpP

The proteolytic active site of ClpAP resides in ClpP, which, in the absence of other proteins, has peptidase activity against short (<10 amino acids) peptides (Woo *et al.*, 1989; Thompson & Maurizi, 1994). Degradation of longer polypeptides and proteins requires the complex of ClpP and ClpA (Katayama-Fujimura *et al.*, 1987). Cleavage occurs preferentially following nonpolar residues, but significant rates of cleavage (5–15%) after polar and even charged residues have been observed with protein substrates (Thompson & Maurizi, 1994; Thompson *et al.*, 1994). The products of degradation by ClpP or ClpAP are peptides that range in size from 5 to 15 amino acids. Although cleavage of the terminal amino acid residue has been seen with some short peptide substrates, the enzyme does not possess significant exopeptidase activity.

## ATP-dependent Protein Degradation by ClpAP

The defining characteristic of protein degradation by ClpAP is its dependence on the binding of, and in most cases the hydrolysis of, ATP. Peptides longer than about 10 amino acids are neither substrates nor inhibitors of ClpP alone, apparently having limited access to the active sites. Formation of the ClpAP complex, which is promoted by ATP binding to ClpA (discussed below), expands the substrate range to include polypeptides or short proteins with little structure (Thompson & Maurizi, 1994). Cleavage of such substrates occurs in the absence of ATP hydrolysis, and the fastest turnover rates for peptide bond cleavage have been observed with such substrates (Table 177.1). Degradation of larger polypeptides and proteins occurs only with ClpAP and only with continuous ATP hydrolysis (Thompson *et al.*, 1994). Turnover rates for cleavage of peptide bonds in proteins are considerably lower than those seen with polypeptide substrates, presumably reflecting a rate-determining step involving unfolding or remodeling of the protein structure (Wickner *et al.*, 1994). Site-directed mutations in either of the ATPase sites of ClpA affect activity. The ATPase site in the N-terminal region is important in assembly of ClpA and of the ClpAP complex. Mutations in the ATPase site of the C-terminal region caused >90% loss of ATPase activity and of the ability to activate proteolysis.

## ATPase Activity and Chaperone Activity of ClpA

ClpA has a basal ATPase activity, cleaving ATP between the $\beta$ and $\gamma$ phosphates. The turnover number is about $70\ min^{-1}$ and is stimulated 20–40% in the presence of various protein and peptide substrates (Maurizi *et al.*, 1994). ATPase activity does not increase significantly when ClpP is present or when protein degradation occurs. ClpA interacts strongly with the phage P1 replication protein, RepA, and, in a reaction requiring ATP hydrolysis, will convert the inactive dimeric form of this protein into an active monomer (Wickner *et al.*, 1994). This structure-remodeling activity of ClpA duplicates exactly that carried out by the classical DnaK/DnaJ chaperone system and reflects the ability of ClpA to partially unfold protein substrates. When ClpP is present together with ClpA, the RepA protein is degraded, suggesting that the primary function of the unfolding activity of ClpA is to make protein substrates more accessible to the active sites of ClpP.

*Table 177.1* Reactions catalyzed by ClpP, ClpA and ClpAP

| Clp component | Substrate | Reaction | Turnover rate ($min^{-1}$) | ATP hydrolysis required? |
|---|---|---|---|---|
| ClpP | Short peptides (<10 amino acids) | Peptidase | 2 | No |
| ClpAP | Polypeptides (<40 amino acids) | Peptidase/proteolysis | 800 | No |
| ClpAP | Proteins | Proteolysis | 15 | Yes |
| ClpA | Proteins (RepA) | Chaperone/remodeling | n.a. | Yes |

n.a., not applicable.

## General Enzymatic Properties

The pH optimum for peptidase activity of ClpP is 7.0–7.5. Protein degradation has a broad optimum pH range between pH 7.5 and 9.5. ClpP is inactivated by DFP, but only slowly (Maurizi *et al*., 1990b). The enzyme is rapidly inactivated by Suc-Leu-Tyr-CH$_2$Cl (M.R. Maurizi & H.Y. Yong, unpublished results). Protein degradation by ClpAP requires ATP (or deoxyATP); no other nucleoside triphosphates can activate the enzyme (Hwang *et al*., 1988; Katayama *et al*., 1988). EDTA inhibits ClpA-dependent reactions when added in excess over Mg$^{2+}$. Peptidase activity of ClpP is easily measured with the fluorogenic peptide, Suc-Leu-Tyr$+$NHMec (Woo *et al*., 1989). Peptide cleavage by ClpAP in the presence of nonhydrolyzable analogs of ATP (ATP$\gamma$S or 5′-adenylyl imidodiphosphate) can be measured with Phe-Ala-Pro-His-Met$+$Ala-Leu-Val-Pro-Val (Thompson & Maurizi, 1994). Protein degradation is measured by release of trichloroacetic acid-soluble peptides from $\alpha$-casein that has been reductively methylated in the presence of [$^3$H]formaldehyde (Katayama-Fujimura *et al*., 1987; Maurizi *et al*., 1994). Protein degradation requires ATP and Mg$^{2+}$, and inhibition of ATP hydrolysis, for example by high concentrations of divalent anions, blocks protein degradation. Assays and methods for preparation of reagents are described elsewhere (Maurizi *et al*., 1994).

## Structural Chemistry

ClpP is synthesized *in vivo* as a 207 amino acid precursor that is processed by removal of the N-terminal 14 amino acids to produce the mature form of the protease (Maurizi *et al*., 1990a). Processing of ClpP appears to be autocatalytic, dependent on an active ClpP but not ClpA or ClpX (Maurizi *et al*., 1990b). ClpP subunits are organized into rings of seven, and the native protein has two such rings axially aligned and presumably isologously bonded (Kessel *et al*., 1995). The interface of the two rings produces an interior aqueous cavity to which there is limited access and which, by analogy with the archaean 20S proteasome (Löwe *et al*., 1995; see Chapter 169), may contain the proteolytic active sites.

ClpA is a polypeptide of 756 amino acids (Gottesman *et al*., 1990). Upon binding of ATP or a nonhydrolyzable analog of ATP, ClpA assembles into rings of six subunits (Kessel *et al*., 1995). The amino acid sequence of ClpA indicates that it is composed of tandem ATPase modules, which must have arisen by fusion of two nonhomologous ATPase genes. Side views of ClpA in the electron microscope show ClpA to be bilobed (Kessel *et al*., 1995). The lobes, which are connected by a region of relatively low mass density, appear to correspond to the two ATPase domains and form two structurally distinct rings. Hexameric ClpA has high affinity for ClpP, whereas the disassembled ClpA does not interact in a detectable way with ClpP. ClpA binds to the two exposed faces of the double ring of ClpP, producing a cylindrical molecule about 13 nm in diameter (slightly less in the center where ClpP is) and 45 nm long (Kessel *et al*., 1995) (Figure 177.1).

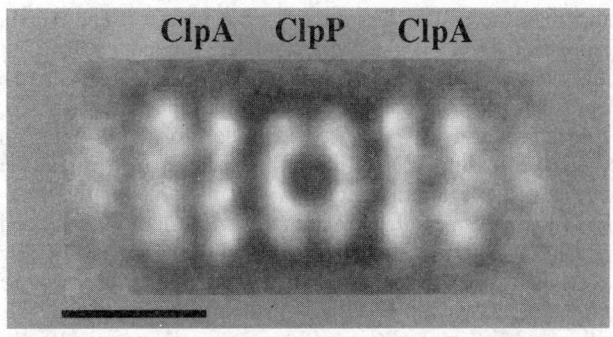

*Figure 177.1*  Organization of the ClpAP complex. Image of CplAP obtained by averaging negatively stained electron micrographs. The ClpAP complex was formed in the presence of ATP$\gamma$S. Scale bar = 10 nm.

## Preparation

ClpP and ClpA are found in moderate abundance in wild-type *E. coli* cells, constituting about 0.05% and 0.02% of total cellular protein, respectively. Both genes have been cloned into vectors that allow overexpression of the proteins. *E. coli* cells are tolerant of relatively high levels of ClpP (<10% of cell protein). About 0.5 g of ClpP can be purified from 20 g of cells expressing the protein under control of a T7 promoter (Maurizi *et al*., 1994). Cells do not grow well in the presence of high levels of ClpA (>2% of cell protein). For ClpA production, it is important to control the basal level of expression until cell density is sufficient to begin induction of synthesis. About 0.2 g of ClpA can be obtained from 25 g of cells expressing the protein under control of a p$_{tac}$ promoter (Maurizi *et al*., 1994). The purification procedures for ClpA and ClpP have been published (Maurizi *et al*., 1994).

## Biological Aspects

Neither ClpA nor ClpP is essential for growth of *E. coli* under any conditions tested. The *clpP* and *clpA* genes map to separate locations on the *E. coli* chromosome (Katayama *et al*., 1988; Maurizi *et al*., 1990a). *clpP*, in fact, is the first gene in a bi-cistronic operon also containing *clpX* (Maurizi *et al*., 1990a). In exponentially growing cells, *clpP* and *clpX* are transcribed in a single messenger RNA. The 5′-upstream region of *clpP/clpX* has a $\sigma^{32}$-dependent promoter and transcription is enhanced 5- to 10-fold upon heat shock (Kroh & Simon, 1990). ClpA transcription also increases at elevated temperatures, but ClpA is not under $\sigma^{32}$ control and does not appear to be coordinately regulated with ClpP (Katayama *et al*., 1988).

ClpP plays a relatively minor role in the degradation of abnormal proteins in *E. coli*, and the effects of mutations in either *clpP* or *clpA* are most easily seen in cells also carrying a *lon* mutation (Katayama *et al*., 1988). It appears that for the broad range of misfolded proteins in cells, degradative activities may be redundant. Only a single natural physiological substrate for ClpAP has been identified. An 81 amino acid protein, MazE, is degraded in wild-type cells and has an extended half-life in either *clpA* or *clpP* mutants (Aizenman *et al*., 1996). The MazE protein is required to neutralize the cell-killing activity of a protein called MazF. Expression of MazE and MazF is under stringent control, but little is known about their functions or the role of ClpAP-dependent degradation in modulating that function.

Several activities have been identified for ClpP acting in conjunction with ClpX *in vivo*. ClpXP helps to regulate the activity of the starvation sigma factor, $\sigma^S$, which is the product of the *rpoS* gene (Schweder *et al*., 1996). $\sigma^S$ is degraded by ClpXP and its levels are maintained at relatively low levels in exponentially growing cells and then increase in stationary phase cells when the $\sigma^S$ becomes resistant to degradation. ClpXP also degrades O protein (Gottesman *et al*., 1993; Wojtkowiak *et al*., 1993), the major replication protein of phage λ, and Phd, a protein involved in maintenance of the lysogenic form of phage P1 (Lehnherr & Yarmolinsky, 1995).

Clp proteins are highly conserved and have been found in a variety of prokaryotic and eukaryotic organisms. Genes homologous to ClpP have been found in plants, where the gene is contained within the chloroplast genome (Maurizi *et al*., 1990b). All chloroplast genomes sequenced to date contain coding regions homologous to ClpP. The plant ClpA homolog, referred to as ClpC, is encoded in the chromosome and includes a chloroplast transit sequence; the protein is found in the stroma (Shanklin *et al*., 1995). ClpP obtained from the cloned chloroplast gene had peptidase activity, and purified recombinant ClpC activated *E. coli* ClpP for casein degradation, suggesting that the plant enzymes constitute a functional ATP-dependent protease.

A human ClpP cDNA has been isolated and sequenced (Bross *et al*., 1995; Wang *et al*., 1997). Human ClpP is encoded in the nuclear chromosomal gene and imported into mitochondria. A recombinant form of the human ClpP has been purified after expression in *E. coli* and was shown to be similar in structure to the *E. coli* protein. The activity and function of human ClpP have not been characterized.

## Distinguishing Features

ClpP is structurally similar to the inner core of β subunits of the 20S proteasome, which is composed of two face-to-face rings with seven subunits each. ClpP however has no sequence homology to the proteasome. Their remarkable similarity in structure suggests that these major cytosolic proteases are designed to shield the proteolytic active sites from most soluble proteins in the cell. By restricting access to the active sites, damage to normal, functional cell proteins is avoided. Proteins are degraded only after they are screened by interaction with the regulatory, ATP-dependent components.

Recent data suggest that in addition to activating protein degradation *in vivo*, the ATPase components of the Clp proteases may have molecular chaperone activities. For example, ClpX is required for replication of Mu phage whereas ClpP is not, because ClpX disassembles the strand transfer complex of MuA and DNA (Levchenko *et al*., 1995; Kruklitis *et al*., 1996). Demonstration of chaperone activities *in vitro* for ClpA in activating RepA and of ClpX in disaggregating λ O protein suggests that ATP-dependent proteases may be considered bifunctional enzymes consisting of a chaperone linked to a protease. This chaperone activity may function in the holoenzyme to unfold or disassemble protein substrates,

which are then kinetically partitioned between refolding and degradation pathways (Gottesman *et al*., 1997).

## Further Reading

Maurizi *et al*. (1994) describe the purification, properties and assays for ClpAP. Gottesman *et al*. (1997) explore the topic of coupled chaperone and proteolytic activities.

## References

Aizenman, E., Engelberg-Kulka, H. & Glaser, G. (1996) An *Escherichia coli* chromosomal 'addiction module' regulated by ppGpp: a model for programmed bacterial cell death. *Proc. Natl Acad. Sci. USA* **93**, 6059–6063.

Bross, P., Andresen, B.S., Knudsen, I., Kruse, T.A. & Gregersen, N. (1995) Human ClpP protease: cDNA sequence, tissue-specific expression and chromosomal assignment of the gene. *FEBS Lett.* **377**, 249–252.

Gottesman, S., Clark, W.P. & Maurizi, M.R. (1990) The ATP-dependent Clp protease of *Escherichia coli*: sequence of *clpA* and identification of a Clp-specific substrate. *J. Biol. Chem.* **265**, 7886–7893.

Gottesman, S., Clark, W.P., Crecy-Lagard, V.D. & Maurizi, M.R. (1993) ClpX, an alternative subunit for the ATP-dependent Clp protease of *Escherichia coli*: sequence and *in vivo* activities. *J. Biol. Chem.* **268**, 22618–22626.

Gottesman, S., Wickner, S. & Maurizi, M.R. (1997) Protein quality control: triage by chaperones and proteases. *Genes Dev.* **11**, 815–823.

Hwang, B.J., Woo, K.M., Goldberg, A.L. & Chung, C.H. (1988) Protease Ti, a new ATP-dependent protease in *Escherichia coli* contains protein-activated ATPase and proteolytic functions in distinct subunits. *J. Biol. Chem.* **263**, 8727–8734.

Katayama, Y., Gottesman, S., Pumphrey, J., Rudikoff, S., Clark, W.P. & Maurizi, M.R. (1988) The two-component ATP-dependent Clp protease of *Escherichia coli*: purification, cloning, and mutational analysis of the ATP-binding component. *J. Biol. Chem.* **263**, 15226–15236.

Katayama-Fujimura, Y., Gottesman, S. & Maurizi, M.R. (1987) A multiple-component ATP-dependent protease from *Escherichia coli*. *J. Biol. Chem.* **262**, 4477–4485.

Kessel, M., Maurizi, M.R., Kim, B., Trus, B.L., Kocsis, E., Singh, S.K. & Steven, A.C. (1995) Homology in structural organization between *E. coli* ClpAP protease and the eukaryotic 26 S proteasome. *J. Mol. Biol.* **250**, 587–594.

Kroh, H.E. & Simon, L.E. (1990) The ClpP component of Clp protease is the $\sigma^{32}$-dependent heat shock protein F21.5. *J. Bacteriol.* **172**, 6026–6034.

Kruklitis, R., Welty, D.J. & Nakai, H. (1996) ClpX protein of *Escherichia coli* activates bacteriophage Mu transposase in the strand transfer complex for initiation of Mu DNA synthesis. *EMBO J.* **15**, 935–44.

Lehnherr, H. & Yarmolinsky, M.B. (1995) Addiction protein Phd of plasmid prophage P1 is a substrate of the ClpXP serine protease of *Escherichia coli*. *Proc. Natl Acad. Sci. USA* **92**, 3274–3277.

Levchenko, I., Luo, L. & Baker, T.A. (1995) Disassembly of the Mu transposase tetramer by the ClpX chaperone. *Genes Dev.* **9**, 2399–2408.

Löwe, J., Stock, D., Jap, B., Zwickl, P., Baumeister, W. & Huber, R. (1995) Crystal structure of the 20 S proteasome from the archaeon *T. acidophilum* at 3.4 Å resolution. *Science* **268**, 533–539.

Maurizi, M.R. (1992) Proteases and protein degradation in *Escherichia coli*. *Experientia* **48**, 178–201.

Maurizi, M.R., Clark, W.P., Katayama, Y., Rudikoff, S., Pumphrey, J., Bowers, B. & Gottesman, S. (1990a) Sequence and structure of Clp P, the proteolytic component of the ATP-dependent Clp protease of *Escherichia coli*. *J. Biol. Chem.* **265**, 12536–12545.

Maurizi, M.R., Clark, W.P., Kim, S.-H. & Gottesman, S. (1990b) ClpP represents a unique family of serine proteases. *J. Biol. Chem.* **265**, 12546–12552.

Maurizi, M.R., Thompson, M.W., Singh, S.K. & Kim, S.-H. (1994) Endopeptidase Clp: the ATP-dependent Clp protease from *Escherichia coli*. *Methods Enzymol.* **244**, 314–331.

Schweder, T., Lee, K.H., Lomovskaya, O. & Matin, A. (1996) Regulation of *Escherichia coli* starvation sigma factor ($\sigma^s$) by ClpXP protease. *J. Bacteriol.* **178**, 470–476.

Shanklin, J., DeWitt, N.D. & Flanagan, J.M. (1995) The stroma of higher plant plastids contain ClpP and ClpC, functional homologs of *Escherichia coli* ClpP and ClpA: an archetypal two-component ATP-dependent protease. *Plant Cell* **7**, 1713–1722.

Thompson, M.W. & Maurizi, M.R. (1994) Activity and specificity of *Escherichia coli* ClpAP protease in cleaving model peptide substrates. *J. Biol. Chem.* **269**, 18201–18208.

Thompson, M.W., Singh, S.K. & Maurizi, M.R. (1994) Processive degradation of proteins by the ATP-dependent Clp protease from *Escherichia coli*: requirement for the multiple array of active sites in ClpP but not ATP hydrolysis. *J. Biol. Chem.* **169**, 18209–18215.

Wang, N., Huang, N.-N., Gottesman, M.M. & Maurizi, M.R. (1997) Human mitochondria contain a homolog of ClpP, the proteolytic component of the *E. coli* ATP-dependent ClpAP protease. (in preparation).

Wickner, S., Gottesman, S., Skowyra, D., Hoskins, J., McKenney, K. & Maurizi, M.R. (1994) A molecular chaperone, ClpA, functions like DnaK and DnaJ. *Proc. Natl Acad. Sci. USA* **91**, 12218–12222.

Wojtkowiak, D., Georgopoulos, C. & Zylicz, M. (1993) Isolation and characterization of ClpX, a new ATP-dependent specificity component of the Clp protease of *Escherichia coli*. *J. Biol. Chem.* **268**, 22609–22617.

Woo, K.M., Chung, W.J., Ha, D.B., Goldberg, A.L. & Chung, C.H. (1989) Protease Ti from *Escherichia coli* requires ATP hydrolysis for protein breakdown but not for hydrolysis of small peptides. *J. Biol. Chem.* **264**, 2088–2091.

*Michael R. Maurizi*
*Laboratory of Cell Biology,*
*National Cancer Institute,*
*Building 37 Room 1B07,*
*Bethesda, MD 20892, USA*
*Email: mmaurizi@helix.nih.gov*

# 178. Endopeptidase La

## Databanks

*Peptidase classification: clan SX, family S16, MEROPS ID: S16.001*
*NC-IUBMB enzyme classification: EC 3.4.21.53*
*Chemical Abstracts Service registry number: 79818-35-2*
*Databank codes:*

| Species | Gene | SwissProt | PIR | EMBL (cDNA) | EMBL (genomic) |
|---|---|---|---|---|---|
| Eubacterial endopeptidase La | | | | | |
| *Azospirillum brasilense* | *lon* | – | – | U35611 | – |
| *Bacillus brevis* | *lon* | P36772 | B42375 I39874 JQ0901 | D00863 | – |
| *Bacillus subtilis* | *lon* | P37945 | I40421 S45101 | X76424 | – |
| *Bacillus subtilis* | *lon2* | P42425 | – | U18229 X76424 | – |

*continued overleaf*

| Species | Gene | SwissProt | PIR | EMBL (cDNA) | EMBL (genomic) |
|---|---|---|---|---|---|
| Eubacterial endopeptidase La (*continued*) | | | | | |
| *Borrelia burgdorferi* | *lon* | – | – | L77216 | – |
| *Caulobacter crescentus* | *lon* | P52977 | – | U56652 | – |
| *Erwinia amylovora* | *lon* | P46067 | S47270 | X77706 | – |
| *Escherichia coli* | *lon* | P08177 | A23101 | J03896 | – |
| | | | A31069 | L12349 | |
| | | | JN0303 | L20572 | |
| | | | S06368 | M10153 | |
| | | | | M38347 | |
| *Haemophilus influenzae* | *lon* | P43864 | A64070 | U32729 | – |
| *Mycoplasma genitalium* | *lon* | P47481 | – | U39702 | – |
| *Myxococcus xanthus* | *lonV* | P36773 | A49844 | D12923 | – |
| *Myxococcus xanthus* | *lonD* | P36774 | A36894 | D13204 | – |
| | | | A36895 | | |
| *Vibrio parahaemolyticus* | *lonS* | – | – | U66708 | |
| Mitochondrial endopeptidase La | | | | | |
| *Caenorhabditis elegans* | *lon* | – | – | Z36719 | – |
| *Homo sapiens* | – | P36776 | S57342 | U02389 | – |
| *Homo sapiens* | – | P36777 | S42366 | X74215 | – |
| | | | | X76040 | |
| *Saccharomyces cerevisiae* | *PIM1* | P36775 | S43938 | L28110 | Z35783: chromosome II ORF |
| | | | | X74544 | |
| *Schizosaccharomyces pombe* | *PIM1* | Q09769 | – | Z54285 | – |

## Name and History

In bacteria, as in eukaryotic cells, the degradation of most intracellular proteins requires ATP (Goldberg, 1992). Protease La was the first example discovered of an ATP-dependent proteolytic enzyme, and was isolated by Goldberg and coworkers in the course of studies to understand the enzymatic basis for the energy requirement for proteolysis in *E. coli* (Swamy & Goldberg, 1981). By systematic screening of different fractions of *E. coli* extracts, a number of novel proteinase activities were found, including one which degraded [³H]methyl globin and [³H]casein only in the presence of ATP (Swamy & Goldberg, 1981). This enzyme was subsequently purified and shown to be encoded by the *lon* gene (Chung & Goldberg, 1981; Charette *et al*., 1981), which had been implicated in the rapid degradation of many abnormal proteins.

*Protease La* catalyzes the initial ATP-requiring steps in the degradation of polypeptides with highly abnormal structure, as may result from nonsense or missense mutations, biosynthetic errors or intracellular denaturation (Kowit & Goldberg, 1977; Chung & Goldberg, 1981), as well as many short-lived regulatory proteins (Gottesman, 1996). Mutants with a reduced capacity for degrading abnormal proteins (initially called *deg*) were first isolated by Bukhari & Zipser (1973), who subsequently found them to map in the *lon* locus (Gottesman & Zipser, 1978), also called *capR* (Zehnbauer *et al*., 1981). Therefore, protease La has also been called the **Lon protease**.

## Activity and Specificity

This enzyme is not only an ATP-dependent endopeptidase, but also a protein-activated ATPase (Waxman & Goldberg,

1982). In the presence of ATP and $Mg^{2+}$, protease La rapidly hydrolyzes a variety of polypeptides (especially unfolded) processively to short peptides of 6–20 amino acids in length. Protease La also has intrinsic ATPase activity, and binds ATP very tightly to each subunit (Menon & Goldberg, 1987). Protein substrates (e.g. casein, globin or denatured albumin) allosterically stimulate ATP hydrolysis 2- to 4-fold, while nondegraded proteins (e.g. native albumin or hemoglobin) have no such effect on the ATPase (Waxman & Goldberg, 1986). Nonhydrolyzed ATP analogs cannot replace ATP in protein breakdown, but 5′-adenylyl imidodiphosphate does allow limited cleavage of some proteins (Goldberg *et al*., 1994; Van Melderen *et al*., 1996). ADP does not support proteolytic activity, and at intracellular levels is a potent inhibitor of the proteolytic activity (Goldberg & Waxman, 1985). In fact, La binds ADP more tightly than ATP, and the rate-limiting step in proteolysis seems to be the release of the bound ADP which is stimulated by the presence of a protein substrate (Menon & Goldberg, 1987).

Normally, about two ATP molecules are consumed for each bond cut in protein substrates, but ATP hydrolysis and proteolysis can be partially uncoupled by changes in the ionic medium (e.g. by increasing $Mg^{2+}$ concentration) (Menon & Goldberg, 1987). Also, inactivation of the proteolytic site by mutagenesis of the active-site serine residue (Fischer & Glockshuber, 1993) or treatment with DFP (Waxman & Goldberg, 1982) blocks proteolysis, but does not affect the ATPase activity or its stimulation by proteins. Although degradation of proteins requires ATP hydrolysis, the cleavage of short peptides occurs maximally with nonhydrolyzable ATP analogs. This difference between protein and peptide substrates led to mechanistic models in which cycles of

ATP binding and hydrolysis are linked to activation and inactivation of proteolytic sites so as to prevent inappropriate proteolysis *in vivo* (Goldberg, 1992).

Protease La also degrades certain hydrophobic peptide substrates, such as glutaryl-Ala-Ala-Phe┼NHNapOMe and Suc-Phe-Leu-Phe┼NHNapOMe, which are also substrates for chymotrypsin, but it does not cleave ester substrates of chymotrypsin or the characteristic peptide substrates of trypsin-like or elastase-like enzymes (Waxman & Goldberg, 1985). In addition, changing amino acids in the P2, P3 and P4 positions, or blocking the N-termini of fluorogenic peptides reduces the rate of cleavage. Analysis of cleavage sites in insulin and glucagon as well as the phage N protein also indicates that protease La prefers to cleave after hydrophobic and nonpolar residues (Maurizi, 1987). Even under optimal conditions, protease La has a much lower turnover number against these fluorogenic peptides or proteins than chymotrypsin, presumably because intracellular proteolysis is a slower, more selective process than extracellular digestion.

Protease La is sensitive to DFP and DCI (Waxman & Goldberg, 1985), which covalently modify the active-site serine. Certain hydrophobic peptidyl chloromethanes, which are substrate analogs (e.g. Z-Gly-Leu-Phe-$CH_2Cl$), can also rapidly inactivate protease La. However, the peptidyl chloromethane can inhibit only in the presence of ATP or a nonhydrolyzed ATP analog, which allows enzyme activation. By contrast, ADP completely blocks this inactivation reaction, just as it inhibits peptide hydrolysis (Waxman & Goldberg, 1985).

## Structural Chemistry

The cloning and sequencing of the *lon* gene from *E. coli* revealed that each subunit contains an ATP-binding site (Chin *et al.*, 1988) and a serine proteolytic site (Amerik *et al.*, 1991). The sequence lacks homology to any other serine protease or any other peptidase family. Protease La, with a predicted subunit size of 88 kDa (Chin *et al.*, 1988; Amerik *et al.*, 1991), migrates on polyacrylamide gels under denaturing conditions with an apparent molecular mass of 94 kDa (Chung & Goldberg, 1981). Gel filtration has not given a clear indication of its multimeric size. Early gel-filtration studies with Sephacryl S-300 columns or glycerol-gradient studies indicated that the enzyme from wild-type *E. coli* or encoded on the low-copy plasmid pJMC40 had a molecular mass of about 450 kDa, which suggested that protease La is a tetramer (Chung & Goldberg, 1981). However, more recent studies with strains carrying the high-copy plasmid pSG11 indicate a molecular mass of approximately 840 kDa, more consistent with an octameric structure (Goldberg *et al.*, 1994). Its sedimentation coefficient determined by velocity sedimentation with exponential sucrose gradients (15–30%, w/v) was 24 S. The Stokes radius, $R_s$, was determined from the elution pattern on Superose 6 to be 8.5–9 nm. From the sedimentation coefficient and Stokes radius, and assuming a partial specific volume of 0.73 $cm^3\,ml^{-1}$, a molecular mass of 840–900 kDa was calculated, which would suggest an octamer (or possibly a larger multimer). Possibly the discrepancy in the apparent size of the enzyme prepared from different strains may indicate a tendency of this enzyme to dimerize under certain conditions. On the other hand, the homologous ATP-dependent protease from rat liver mitochondria (Desautels & Goldberg, 1982) behaved as a 600 kDa multimer on gel filtration, which is most consistent with a hexameric structure.

## Preparation

Methods for assay and purification of protease La from *E. coli* have recently been described in detail by Goldberg *et al.* (1994) and for the related protease from mitochondria by Kuzela & Goldberg (1994).

## Biological Aspects

Enzymes closely homologous to protease La appear to be widespread in nature, although the catalytic properties of these other ATP-dependent proteases have not been studied. Proteases homologous to La are present in both gram-negative and gram-positive bacteria; in bacilli and mycobacteria, these enzymes are necessary for cell differentiation (Gill *et al.*, 1993; Tojo *et al.*, 1993). Genes encoding the mitochondrial protease La homologs have been cloned from human cells (Wang *et al.*, 1993) and yeast (Kutejova *et al.*, 1993; Van Dyck *et al.*, 1994; Suzuki *et al.*, 1994). The mitochondrial protease La is essential for respiration and can function as a molecular chaperone as well as an ATP-dependent protease. Comparison of the amino acid sequences of protease La from *E. coli* (Chin *et al.*, 1988; Amerik *et al.*, 1991), *Bacillus brevis* (Ito *et al.*, 1992), *Saccharomyces cerevisiae* (Suzuki *et al.*, 1994; Van Dyck *et al.*, 1994), and humans (Wang *et al.*, 1993) reveals identical amino acids in 25% and similar or identical amino acids in 53% of the positions in all four species. Conservation is highest in two regions: one containing the well-defined ATP-binding consensus motif and the other near the C-terminus containing the putative active-site serine residue.

*E. coli* carrying *lon* mutations show a variety of unusual phenotypic features, including mucoidy and increased sensitivity to DNA-damaging agents (Gottesman & Maurizi, 1992; Gottesman, 1996). These properties result from the inability of the *lon* mutants to degrade critical regulatory proteins such as SulA and RcsA rapidly. In wild-type cells, these proteins have half-lives of less than 2 min, but are stabilized 10- to 20-fold in *lon* mutants (Gottesman & Maurizi, 1992). In *lon* mutants, the degradation of abnormal polypeptides containing canavanine or puromycin occurs at 30–50% the rate in wild-type cells (Kowit & Goldberg, 1977), and the residual proteolysis seems to involve other ATP-dependent proteases (e.g. protease Ti (ClpAP)) (Chapter 177) (Katayama-Fujimura *et al.*, 1987; Hwang *et al.*, 1987) and protease HslVU (Chapter 171) (Rohrwild *et al.*, 1996). Thus, *lon* is not essential for normal growth, but is necessary for viability under certain harsh conditions, such as for recovery from DNA damage. On the other hand, cells carrying multiple copies of the *lon* gene, resulting in increased levels of protease La, hydrolyze both abnormal and certain normal proteins at increased rates (Goff & Goldberg, 1987). However, overproduction of this enzyme prevents normal growth, and mutations that inactivate the plasmid-encoded *lon* gene appear with high frequency (Goff & Goldberg, 1985).

Like the other ATP-dependent proteinases in *E. coli*, protease La is a heat-shock protein expressed under control of the *rpoH* operon (htpR) (Goff *et al.*, 1984; Grossman *et al.*,

1984; Goff & Goldberg, 1985). Consequently, the content of protease La and the cell's overall proteolytic capacity increase in wild-type cells during the heat-shock response, when cells accumulate large amounts of denatured proteins (Goff & Goldberg, 1985). Under such conditions, the ATP-dependent proteases and molecular chaperones are induced coordinately and catalyze the elimination of unfolded potentially toxic polypeptides (Sherman & Goldberg, 1992; Gottesman *et al*., 1995).

## Distinguishing Features

An intriguing feature of protease La is that it has a high affinity for DNA. In fact, Zehnbauer *et al*. (1981) originally isolated this protein as a DNA-binding protein, assuming it to be a regulator of gene expression. DNA, especially single-stranded DNA, activates the proteolytic activity and the protein-activated ATPase of protease La (Chung & Goldberg, 1982). mRNA and tRNA have no effect, although poly(dT), poly(rC), and poly(rU) stimulate casein hydrolysis provided the polynucleotide is of sufficient length (>(dT)10). A stoichiometric binding of protease La to plasmid DNA has been demonstrated and has been shown to increase the $V$ of the protease against casein. Moreover, these protein substrates promote the dissociation of the enzyme from DNA. The physiological significance of the effect of DNA remains unclear, although it appears likely that *in vivo* some protease La is associated with the bacterial chromosome, where rapid degradation of damaged polypeptides or regulatory proteins may be particularly important for homeostasis.

## Further Reading

Further information on protease La has been published by Goldberg (1992), Goldberg *et al*. (1994), and Gottesman (1996).

## References

Amerik, A.Y., Antonov, V.K., Gorbalenya, A.E., Kotova, S.A., Rotanova, T.V. & Shimbarevich, E.V. (1991) Site-directed mutagenesis of protease La. *FEBS Lett.* **287**, 211–214.

Bukhari, A.I. & Zipser, D. (1973) Mutants of *Escherichia coli* with a defect in the degradation of nonsense fragments. *Nature New Biol.* **243**, 238–241.

Charette, M.F., Henderson, G.W. & Markovitz, A. (1981) ATP-hydrolysis dependent activity of the lon (capR) protein of *E. coli* K12. *Proc. Natl Acad. Sci. USA* **78**, 4728–4732.

Chin, D.T., Goff, S.A., Webster, T., Smith, T. & Goldberg, A.L. (1988) Sequence of the *lon* gene in *Escherichia coli*: a heat shock gene which encodes the ATP-dependent protease La. *J. Biol. Chem.* **263**, 11718–11728.

Chung, C.H. & Goldberg, A.L. (1981) The product of the *lon* (capR) gene in *Escherichia coli* is the ATP-dependent protease, protease La. *Proc. Natl Acad. Sci. USA* **78**, 4931–4935.

Chung, C.H. & Goldberg, A.L. (1982) DNA stimulates ATP-dependent proteolysis and protein-dependent ATPase activity of protease La from *Escherichia coli*. *Proc. Natl Acad. Sci. USA* **79**, 795–799.

Desautels, M. & Goldberg, A.L. (1982) Demonstration of an ATP-dependent vanadate-sensitive endoprotease in the matrix of rat-liver mitochondria. *J. Biol. Chem.* **257**, 11673–11679.

Fischer, H. & Glockshuber, R. (1993) ATP hydrolysis is not stoichiometrically linked with proteolysis in the ATP-dependent protease La from *E. coli*. *J. Biol. Chem.* **268**, 22502–22506.

Gill, R.E., Korluk, M. & Benton, D. (1993) *Myxococcus xanthus* encodes an ATP-dependent protease which is required for developmental gene transcription and intercellular signaling. *J. Bacteriol.* **175**, 4538–4544.

Goff, S.A. & Goldberg, A.L. (1985) Production of abnormal proteins in *E. coli* stimulates transcription of *lon* and other heat shock genes. *Cell* **41**, 587–595.

Goff, S.A. & Goldberg, A.L. (1987) An increased content of protease La, the *lon* gene product, increases protein degradation and blocks growth in *Escherichia coli*. *J. Biol. Chem.* **262**, 4508–4515.

Goff, S.A., Casson, L.P. & Goldberg, A.L. (1984) Heat shock regulatory gene htpR influences rates of protein degradation and expression of the *lon* gene in *Escherichia coli*. *Proc. Natl Acad. Sci. USA* **81**, 6647–6651.

Goldberg, A.L. (1992) The mechanisms and functions of ATP-dependent proteases in bacterial and animal cells. *Eur. J. Biochem.* **203**, 9–23.

Goldberg, A.L. & Waxman, L. (1985) The role of ATP hydrolysis in the breakdown of proteins and peptides by protease La from *Escherichia coli*. *J. Biol. Chem.* **260**, 12029–12034.

Goldberg, A.L., Moerschell, R.P., Chung, C.H. & Maurizi, M.R. (1994) ATP-dependent protease La (Lon) from *Escherichia coli*. *Methods Enzymol.* **244**, 350–375.

Gottesman, S. (1996) Proteases and their targets in *Escherichia coli*. *Annu. Rev. Genet.* **30**, 465–506.

Gottesman, S. & Maurizi, M.R. (1992) Regulation by proteolysis: energy-dependent proteases and their targets. *Microbiol. Rev.* **56**, 592–621.

Gottesman, S. & Zipser, D. (1978) Deg phenotype of *Escherichia coli lon* mutants. *J. Bacteriol.* **133**, 744–851.

Gottesman, S., Wickner, S., Jubete, Y., Singh, S.K., Kessel, M. & Maurizi, M.R. (1995) Selective, energy-dependent proteolysis in *Escherichia coli*. *Cold Spring Harbor Symp. Quant. Biol.* **LX**, pp. 533–548.

Grossman, A.D., Erickson, J.W. & Gross, C.A. (1984) The htpR gene product of *E. coli* is a sigma factor for heat shock promoters. *Cell* **38**, 383–390.

Hwang, B.J., Park, W.J., Chung, C.H. & Goldberg, A.L. (1987) *Escherichia coli* contains a soluble ATP-dependent protease (Ti) distinct from protease La. *Proc. Natl Acad. Sci. USA* **84**, 5550–5554.

Ito, K., Udaka, S. & Yamagato, H. (1992) Cloning, characterization, and inactivation of the *Bacillus brevis lon* gene. *J. Bacteriol.* **174**, 2281–2287.

Katayama-Fujimura, Y., Gottesman, S. & Maurizi, M.R. (1987) A multi-component, ATP-dependent protease from *Escherichia coli*. *J. Biol. Chem.* **262**, 4477–4485.

Kowit, J.D. & Goldberg, A.L. (1977) Intermediate steps in the degradation of a specific abnormal protein in *Escherichia coli*. *J. Biol. Chem.* **252**, 8350–8357.

Kutejova, E., Durcova, G., Surovkova, E. & Kuzela, S. (1993) Yeast mitochondrial ATP-dependent protease: purification and comparison with the homologous rat enzyme and the bacterial ATP-dependent protease La. *FEBS Lett.* **329**, 47–50.

Kuzela, S. & Goldberg, A.L. (1994) The mitochondrial ATP-dependent protease from rat liver and yeast. *Methods Enzymol.* **244**, 376–383.

Maurizi, M.R. (1987) Degradation *in vitro* of bacteriophage lambda N protein by Lon protease from *Escherichia coli*. *J. Biol. Chem.* **262**, 2696–2703.

Maurizi, M.R., Trisler, P. & Gottesman, S. (1985) Insertional mutagenesis of the *lon* gene in *Escherichia coli: lon* is dispensable. *J. Bacteriol.* **164**, 1124–1135.

Menon, A.S. & Goldberg, A.L. (1987) Binding of nucleotides to the ATP-dependent protease La from *Escherichia coli*. *J. Biol. Chem.* **262**, 14921–14928.

Rohrwild, M., Coux, O., Huang, H.-C., Moerschell, R.P., Yoo, S.J., Seol, J.H., Chung, C.H. & Goldberg, A.L. (1996) HslV-HslU: a novel ATP-dependent protease complex in *Escherichia coli* related to the eukaryotic proteasome. *Proc. Natl Acad. Sci. USA* **93**, 5808–5813.

Sherman, M. & Goldberg, A.L. (1992) Involvement of the chaperonin dnaK in the rapid degradation of a mutant protein in *Escherichia coli*. *EMBO J.* **11**, 71–77.

Suzuki, C., Suda, K., Wang, N. & Schatz, G. (1994) Requirement for the yeast gene *lon* in intramitochondrial proteolysis and maintenance of respiration. *Science* **264**, 273–276.

Swamy, K.H.S. & Goldberg, A.L. (1981) *Escherichia coli* contains eight soluble proteolytic activities, one of which is ATP-dependent. *Nature* **292**, 652–654.

Tojo, N., Inouye, S. & Komano, J. (1993) The *lonD* gene is homologous to the *lon* gene encoding an ATP-dependent protease and is essential for the development of *Myxococcus xanthus*. *J. Bacteriol.*

**175**, 4545–4549.

Van Dyck, L., Pearce, D.A. & Sherman, F. (1994) PIM1 encodes a mitochondrial ATP-dependent protease that is required for mitochondrial function in the yeast *Saccharomyces cerevisiae*. *J. Biol. Chem.* **269**, 238–242.

Van Melderen, L., Thi, M.H.D., Lecchi, P., Gottesman, S., Couturier, M. & Maurizi, M.R. (1996) ATP-dependent degradation of CdcA by Lon protease. *J. Biol. Chem.* **271**, 27730–27738.

Wang, N., Gottesman, S., Willingham, M.C., Gottesman, M.M. & Maurizi, M.R. (1993) A human mitochondrial ATP-dependent protease that is highly homologous to bacterial Lon protease. *Proc. Natl Acad. Sci. USA* **90**, 11247–11251.

Waxman, L. & Goldberg, A.L. (1982) Protease La from *Escherichia coli* hydrolyzes ATP and proteins in a linked fashion. *Proc. Natl Acad. Sci. USA* **79**, 4883–4887.

Waxman, L. & Goldberg, A.L. (1985) Protease La, the lon gene product, cleaves specific fluorogenic peptides in an ATP-dependent reaction. *J. Biol. Chem.* **260**, 12022–12028.

Waxman, L. & Goldberg, A.L. (1986) Selectivity of intracellular proteolysis: protein substrates activate the ATP-dependent protease (La). *Science* **232**, 500–503.

Zehnbauer, B.A., Foley, E.C., Henderson, G.W. & Markovitz, A. (1981) Identification and purification of the lon+ (capR+) gene product, a DNA-binding protein. *Proc. Natl Acad. Sci. USA* **78**, 2043–2047.

*Chin Ha Chung*
*Department of Molecular Biology and Research*
*Center for Cell Differentiation,*
*College of Natural Sciences,*
*Seoul National University,*
*Seoul 151-742, Korea*
*Email: chchung@plaza.snu.ac.kr*

*Alfred L. Goldberg*
*Department of Cell Biology,*
*Harvard Medical School,*
*240 Longwood Avenue,*
*Boston, MA 02115, USA*
*Email: agoldber@bcmp.med.harvard.edu*

# 179. Cell-associated endopeptidase of Trichophyton

## Databanks

Peptidase classification: clan SX, family S19, MEROPS ID: S19.001
NC-IUBMB enzyme classification: none
Databank codes:

| Species | SwissProt | PIR | EMBL (cDNA) | EMBL (genomic) |
|---|---|---|---|---|
| *Coccidioides immitis* | P42783 | – | M81863 S77562 X63114 | – |

## Name and History

The fungus *Trichophyton rubrum* is a major dermal pathogen of humans, responsible for the majority of all recorded cases of dermatophytoses. The purification and characterization of a cell-associated proteinase from *T. rubrum* has recently been described (Lambkin *et al.*, 1996). Prior to this report studies have focused on *T. rubrum* extracellular proteinases (Apodaca & McKerrow, 1989a,b; Lambkin *et al.*, 1994; Sanyal *et al.*, 1985). Comparison of the characteristics of the cell-associated proteinase and the extracellular proteinases provides the first information on the relationship between *T. rubrum* proteinases detected within the cell and secreted proteinases, as well as elucidating their relationship with other fungal proteinases.

## Activity and Specificity

The cell-associated proteinase actively degrades azocasein, azocoll and azoalbumin while weaker activity is detected against keratin and elastin (Lambkin *et al.*, 1996). Multiple degradation products are detectable by SDS-PAGE after incubation with laminin and fibronectin substrates (Lambkin *et al.*, 1996). The enzyme is active over a broad pH range with optimal activity at pH 7.5 (Lambkin *et al.*, 1996). PMSF, chymostatin and $Zn^{2+}$ markedly inhibit enzyme activity, whereas EDTA and 1,10-phenanthroline inhibit the enzyme to a lesser extent (Lambkin *et al.*, 1996). All of these features are shared with the extracellular 235 kDa proteinase (Lambkin *et al.*, 1994).

## Structural Chemistry

The relative molecular mass of the proteinase is estimated as 37 000 on substrate SDS-PAGE gels and 34 000 on nonreduced SDS-PAGE gels (Lambkin *et al.*, 1996). Reduction of the protein produces nonproteolytic bands of $M_r$ 15 000, suggesting that the active enzyme exists as a disulfide-linked dimer, as do the *T. rubrum* extracellular $M_r$ 71 000 and $M_r$ 90 000 dimeric proteinases (Asahi *et al.*, 1985). The pI of the enzyme is 4.5 (Lambkin *et al.*, 1996). The N-terminal 20 amino acid sequence of the *T. rubrum* proteinase exhibits 50% homology with part of the deduced amino acid sequence of the *Coccidioides immitis* chymotrypsin-like serine proteinase (Lambkin *et al.*, 1996).

## Preparation

The enzyme has been purified to homogeneity from cytoplasmic antigen preparations of *T. rubrum* using liquid isoelectric focusing and gel-filtration FPLC (Lambkin *et al.*, 1996).

## Biological Aspects and Related Peptidases

Of the fungal serine proteinases so far characterized, most belong to the subtilisin family, but exceptions are a *Coccidioides immitis* 34 kDa wall-associated proteinase (Cole *et al.*, 1989) and an *Aspergillus niger* 21 kDa semialkaline aspergillopeptidase (Barthomeuf *et al.*, 1989). As was mentioned above, the *T. rubrum* enzyme is almost certainly homologous to the *C. immitis* chymotrypsin-like serine proteinase, and the *C. immitis* proteinase also resembles the *T. rubrum* enzyme in its low pI and the fact that it is a disulfide-linked dimer in the active state. The *A. niger* aspergillopeptidase exhibits a similar pI and a pH optimum of 7.8 (Barthomeuf *et al.*, 1989). Based on the biochemical similarities it appears that the cell-associated proteinase of *T. rubrum*, the *C. immitis* proteinase and the aspergillopeptidase constitute a unique family of fungal serine proteinases. It has been postulated that both the *C. immitis* and *A. niger* enzymes play a role in the pathogenesis of their respective diseases, and the *T. rubrum* enzyme may well fulfill a similar function.

## Distinguishing Features

Inhibition by both serine proteinase inhibitors and metalloproteinase inhibitors is an unusual feature of the proteinase. A similar inhibition profile has been observed for the *A. niger* aspergillopeptidase (Barthomeuf *et al.*, 1989).

## Further Reading

A fuller account of the *Trichophyton* enzyme has been given by Lambkin *et al.* (1996). The *Coccidioides* homolog is described by Yuan & Cole (1987), Cole *et al.* (1992), and Pan & Cole (1995).

## References

Apodaca, G. & McKerrow, J.H. (1989a) Purification and characterisation of a 27,000-$M_r$ extracellular proteinase from *Trichophyton rubrum*. *Infect. Immun.* **57**, 3072–3080.

Apodaca, G. & McKerrow, J.H. (1989b) Regulation of *Trichophyton rubrum* proteolytic activity. *Infect. Immun.* **57**, 3081–3090.

Asahi, M., Lindquist, R., Fukuyama, K., Apodaca, G., Epstein, W.L. & McKerrow, J.H. (1985) Purification and characterisation of major extracellular proteinases from *Trichophyton rubrum*. *Biochem. J.* **232**, 139–144.

Barthomeuf, C., Pourrat, H. & Pourrat, A. (1989) Properties of a new alkaline proteinase from *Aspergillus niger*. *Chem. Pharm. Bull.* **37**, 1333–1336.

Cole, G.T., Zhu, S., Pan, S., Yuan, L., Kruse, D. & Sun, S.H. (1989) Isolation of antigens with proteolytic activity from *Coccidioides immitis*. *Infect. Immun.* **57**, 1524–1534.

Cole, G.T., Zhu, S.W., Hsu, L.L., Kruse, D., Seshan, K.R. & Wang, F. (1992) Isolation and expression of a gene which encodes a wall-associated proteinase of *Coccidioides immitis*. *Infect. Immun.* **60**, 416–427.

Lambkin, I., Hamilton, A.J. & Hay, R.J. (1994) Partial purification and characterization of a 235,000-$M_r$ extracellular proteinase from *Trichophyton rubrum*. *Mycoses* **37**, 85–92.

Lambkin, I., Hamilton, A.J. & Hay, R.J. (1996) Purification and characterization of a novel 34,000-$M_r$ cell-associated proteinase from the dermatophyte *Trichophyton rubrum*. *FEMS Immunol. Med. Microbiol.* **13**, 131–140.

Pan, S. & Cole, G.T. (1995) Molecular and biochemical characterization of a *Coccidioides immitis*-specific antigen. *Infect. Immun.* **63**, 3994–4002.

Sanyal, A.K., Das, S.K. & Banerjee, A.B. (1985) Purification and partial characterisation of an extracellular proteinase from *Trichophyton rubrum. J. Med. Vet. Mycol.* **23**, 165–178.

Yuan, L. & Cole, G.T. (1987) Isolation and characterization of an extracellular proteinase of *Coccidioides immitis. Infect. Immun.* **55**, 1970–1978.

**Imelda J. Lambkin**
*Dermatology Unit,*
*Clinical Sciences Laboratory,*
*18th Floor Guys Tower,*
*Guys Hospital,*
*London SE1 9RT, UK*

**Andrew J. Hamilton**
*Dermatology Unit,*
*Clinical Sciences Laboratory,*
*18th Floor Guys Tower,*
*Guys Hospital,*
*London SE1 9RT, UK*

**Rod J. Hay**
*Dermatology Unit,*
*Clinical Sciences Laboratory,*
*18th Floor Guys Tower,*
*Guys Hospital,*
*London SE1 9RT, UK*

# 180. VIPase: antibody light chain hydrolyzing vasoactive intestinal peptide

## Databanks

*Peptidase classification: none; the protein is an immunoglobulin light chain, MEROPS ID: none*
*NC-IUBMB enzyme classification: none*
*ATCC entries: 63377*
*Databank codes:*

| Species | SwissProt | PIR | GenBank (cDNA) | EMBL (genomic) |
|---|---|---|---|---|
| *Mus musculus* | | A55491 | L34775 | |

## Name and History

The name **VIPase** is applied to the recombinant light chain subunit of an antibody to the neuropeptide vasoactive intestinal peptide (VIP) that has the property of hydrolyzing the peptide (Gao *et al.*, 1994). The name signifies a comparatively high level of substrate specificity. The light chain was obtained by immunization of mice with VIP, preparation of the light chain cDNA from hybridoma cells making a monoclonal antibody to VIP, and expression of the light chain cDNA in a bacterial host. Although this was the first example of a proteolytic light chain available in recombinant form, the ability of light chains and antibodies to catalyze chemical reactions is by no means an uncommon phenomenon. Polyclonal antibodies (Paul *et al.*, 1989; Suzuki *et al.*, 1992) and their light chain subunit (Sun *et al.*, 1991) have previously been described to express proteolytic activity, as have monoclonal light chains isolated from multiple myeloma patients (Matsuura *et al.*, 1994; Paul *et al.*, 1995). Furthermore, these proteins are capable of expressing other forms of catalytic activity, exemplified by observations of DNA cleavage by polyclonal antibody preparations (Shuster *et al.*, 1992) and

the peroxidase activity of a recombinant light chain (Takagi *et al.*, 1995).

## Activity and Specificity

The light chain cleaves four internal peptide bonds in VIP on the C-terminal side of basic residues. The amide bonds linking aminomethylcoumarin to Arg or Lys residues in short peptide-NHMec substrates are also cleaved.

Kinetic studies indicate that specificity for VIP derives from efficient ground state recognition. The $K_m$ for VIP is two or more orders of magnitude smaller than the $K_m$ for the nonhomologous substrates. Turnover numbers for VIP and the nonhomologous substrates are comparable (Gao *et al.*, 1994).

Changes in the activity and specificity of the light chain are found following treatment with conformational perturbants. Varying levels of proteolytic activity can be recovered upon renaturation of the reduced and alkylated form of the light chain prepared by chemical cleavage of the parent immunoglobulin G (IgG) molecule from a 6 M guanidine hydrochloride solution, depending on the concentration at

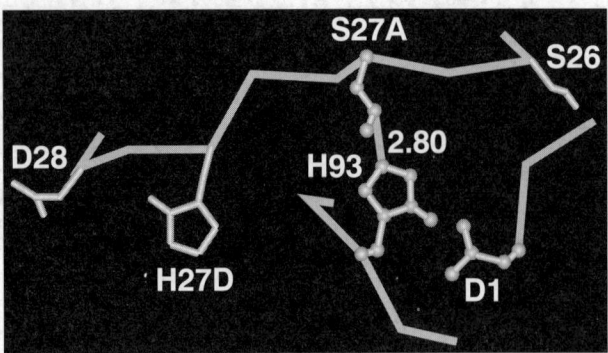

*Figure 180.1* Selected residues of anti-VIP light chain positioned by molecular modeling and confirmed by mutagenesis to serve as catalytic residues (Ser27A, His93, Asp1) and VIP ground state binding residues (Ser26, His27D/Asp28).

which the light chain is held during the renaturation step, and the inclusion of the substrate (VIP) in this step (Sun *et al.*, 1994). Loss of the basic residue cleavage preference is evident in the light chain subjected to denaturing and renaturing treatments.

Mechanistic similarities of the light chain to serine proteases are evident. Certain serine protease inhibitors, DFP and aprotinin, inhibit the proteolytic activity of the light chains, whereas inhibitors of other classes of proteases are without effect (Gao *et al.*, 1994). Molecular modeling has suggested that Asp1, Ser27a and His93 (see Kabat *et al.*, 1991 for numbering system) are present in a catalytic triad-like arrangement (Figure 180.1). Replacement of Ser27a or His93 by Ala by mutagenesis produces about 100-fold loss of turnover (Gao *et al.*, 1995), and replacement of Asp1 by Ala, a smaller loss of turnover (unpublished results of the author), implicating these amino acids as essential catalytic components.

Significant alterations in $K_{m}$ of the Ser27a, His93 and Asp1 mutants were not evident. In contrast, mutations at residues remote from the catalytic center (Ser26, His27d and Asp28) produced an increase in $K_{m}$, and consequently, an increase in turnover, as predicted by transition state theory (Fersht, 1985). These findings are consistent with the idea that the light chain contains a large site responsible for interaction with VIP, composed of a chemically reactive subsite and a VIP ground state binding subsite.

## Structural Chemistry

The recombinant light chain is a 27 kDa protein containing three framework regions and three complementarity determining regions. Among the known germline $V_{L}$ genes, the sequence of the anti-VIP light chain is most similar to the germline $V_{L}$ gene 2(70/3) (EMBL/GenBank: Z72384). The anti-VIP light chain contains the J 1 gene. Four sequence substitutions are evident with respect to the germline $V_{L}$/J genes (germline residues shown in parentheses): His27d(Asp), Thr27e(Ser), Ile34(Asn) and Gln96(His). In addition to the naturally encoded $V_{L}$ domain, the recombinant light chain contains a 10 residue c-*myc* tag and a six residue poly(His) tag at its C-terminus.

The light chain exists predominantly in monomer form in neutral aqueous solution. Small amounts of dimers of the light chain can be detected at concentrations greater than 1 μM.

## Preparation

cDNA encoding the light chain was prepared from the hybridoma cell line c23.5 by the reverse-transcriptase/PCR and cloned into pCantab5His6 vector. Light chain expressed in the periplasm of the transformed *E. coli* was purified by metal-affinity chromatography on a nickel-chelate column (Gao *et al.*, 1995). The expression level is 2–10 mg light chain liter$^{-1}$ bacterial culture. Additional rounds of nickel-chelate chromatography or ion-exchange and gel-filtration chromatographic procedures can be conducted to obtain electrophoretically homogeneous light chains.

## Biological Aspects

The proteolytic activity of this light chain validates previous observations of catalysis by autoantibodies. The impetus for these studies has been the idea that autoantibodies and antimicrobial antibodies express catalytic activities capable of participating in the pathogenesis of autoimmune disease and defense against infection. From a basic science perspective, catalytic antibodies offer the opportunity to study the development of specific ligand-binding and chemical-conversion activities in an organismal time scale, rather than the evolutionary time scale.

Better antibody catalysts than the anti-VIP light chain described here will undoubtedly be found for detailed examination of the biological importance of catalytic antibodies. Perhaps this light chain is significant only in that it is the first of such reagents available for the study of mechanism and biological function.

## Distinguishing Features

Compared to other proteolytic enzymes, the anti-VIP light chain is characterized by high specificity for the substrate to which it was raised, i.e. VIP. This feature is brought about by the presence of a large combining site, which includes the chemically reactive Ser-His-Asp triad and remote residues capable of recognizing residues in VIP far from the scissile bond.

## Further Reading

For reviews, see Paul (1996a,b).

## References

Fersht, A. (1985) *Enzyme Structure and Mechanism*, 2nd edn. New York: W.H. Freeman.

Gao, Q.-S., Sun, M., Tyutyulkova, S., Webster, D., Rees, A., Tramontano, A., Massey, R. & Paul, S. (1994) Molecular cloning of a proteolytic antibody light chain. *J. Biol. Chem.* **269**, 32389–32393.

Gao, Q.-S., Sun, M., Rees, A. & Paul, S. (1995) Site-directed mutagenesis of proteolytic antibody light chain. *J. Mol. Biol.* **253**, 658–664.

Kabat, E.A., Wu, T.T., Perry, H.M., Gottesman, K.S. & Foeller, C. (1991) *Sequences of Proteins of Immunological Interest*, 5th edn. Washington, D.C.: U.S. Department of Health and Human Services.

Matsuura, K., Yamamoto, K. & Sinohara, H. (1994) Amidase activity of human Bence Jones proteins. *Biochem. Biophys. Res. Commun.* **204**, 57–62.

Paul, S. (1996a) Proteolytic antibodies. *Isr. J. Chem.* **36**, 207–214.

Paul, S. (1996b) Natural catalytic antibodies. *Mol. Biotechnol.* **5**, 197–207.

Paul, S., Volle, D.J., Beach, C.M., Johnson, D.R., Powell, M.J. & Massey, R.J. (1989) Catalytic hydrolysis of vasoactive intestinal peptide by human autoantibody. *Science* **244**, 1158–1162.

Paul, S., Li, L., Kalaga, R., Wilkins-Stevens, P., Stevens, F.J. & Solomon, A. (1995) Natural catalytic antibodies: peptide hydrolyzing activities of Bence Jones proteins and VL fragment. *J. Biol. Chem.* **270**, 15257–15261.

Shuster, A.M., Gololobov, G.V., Kvashuk, O.A., Bogomolova, A.E., Smirnov, I.V. & Gabibov, A.G. (1992) DNA hydrolyzing autoantibodies. *Science* **256**, 665–667.

Sun, M., Mody, B., Eklund, S.H. & Paul, S. (1991) Vasoactive intestinal peptide hydrolysis by antibody light chains. *J. Biol. Chem.* **266**, 15571–15574.

Sun, M., Gao, Q.-S., Li, L. & Paul, S. (1994) Proteolytic activity of an antibody light chain. *J. Immunol.* **153**, 5121–5126.

Suzuki, H., Imanishi, H., Nakai, T. & Konishi, Y.K. (1992) Human autoantibodies that catalyze the hydrolysis of vasoactive intestinal polypeptide. *Biochemistry (Life Sci. Adv.)* **11**, 173–177.

Takagi, M., Kohda, K., Hamuro, T., Harada, A., Yamaguchi, H., Kamachi, M. & Imanaka, T. (1995) Thermostable peroxidase activity with a recombinant antibody L chain-porphyrin Fe(III) complex. *FEBS Lett.* **375**, 273–276.

***Sudhir Paul***
*University of Nebraska Medical Center,*
*Department of Anesthesiology,*
*600 South 42nd Street,*
*Omaha, NE 68198-6830, USA*
*Email: spaul@mail.unmc.edu*

# 181. Introduction: unsequenced serine peptidases

Of the serine peptidases that have been characterized in sufficient detail to merit inclusion in the present volume, there are less than 5% for which few or no sequence data are available so that the enzymes cannot be assigned to families. For several of the peptidases of unknown sequence, we can make judicious guesses about possible relationships, and the chapters describing these peptidases have been interpolated into appropriate locations in the preceding sequence of chapters. The three peptidases for which we have not made such guesses follow this Introduction. The unsequenced serine peptidases are as follows:

*Metridin* (Chapter 6). There are strong indirect lines of evidence, including the identification of an active site tripeptide, indicating that this sea anemone digestive enzyme is a relative of the pancreatic endopeptidases of higher animals. Accordingly, it has been grouped with the endopeptidases of family S1.

*Guanidinobenzoatase* (Chapter 75) has sufficient similarities to the plasminogen activators and tryptase also to justify a tentative assignment to family S1.

*Trepolisin* (Chapter 102) may well be the product of the *prtP* gene of *Treponema denticola*, as discussed in the chapter, and if so, it is a member of family S8, and that is where we have placed it.

*Leucyl endopeptidase* (Chapter 111) has also been placed with the members of the subtilisin family. There have been many reports of serine endopeptidase activity in plants, but few of the enzymes have been isolated and subjected to structural studies. The work on cucumisin proved that plants do indeed contain subtilisin-related endopeptidases, and the editors think it likely that many more such 'plant subtilisins' remain to be characterized. The spinach leucyl endopeptidase may very well be one of these.

*Peptidyl-glycinamidase* (Chapter 135) has several similarities to the serine carboxypeptidases, as discussed in the chapter, and has therefore been placed with these enzymes.

The three serine peptidases of unknown amino acid sequence that we can make no guesses about are *dipeptidyl-peptidase III*, *tripeptidyl-peptidase I* and *endopeptidase So*. Structural data on these interesting peptidases will be awaited with interest.

# 182. Dipeptidyl-peptidase III

## Databanks

*Peptidase classification: clan SX, family S99, MEROPS ID: S9C.002*
*NC-IUBMB enzyme classification: EC 3.4.14.4*
*Databank codes: no sequence data available*

## Name and History

The enzyme was discovered in extracts of cattle anterior pituitary glands when a range of aminoacyl and dipeptidyl 2-naphthylamides were employed as fluorogenic substrates to survey the spectrum of peptidases in that tissue (Ellis & Nuenke, 1967). Its distinctive action on a very restricted number of unsubstituted dipeptidyl-2-NHNap substrates resulted in its identification as the third in a series of *dipeptidyl arylamidases*. Based on results of specificity studies conducted with peptide substrates, its name was changed to *dipeptidyl aminopeptidase III* when it was included as the third member of a family of four such exopeptidases designated I through IV (McDonald *et al*., 1971). In 1984, IUB recommended the name *dipeptidyl peptidase III* and assigned an EC number (3.4.14.4). In 1992, IUBMB recommended the current hyphenated form of the name, *dipeptidyl-peptidase III*. The name is here abbreviated to *DPP III*. DPP III activity encountered by investigators employing physiologically active peptides as substrates has been reported under such names as *red cell angiotensinase* (which was believed to be an endopeptidase: Kokubu *et al*., 1969) and *enkephalinase B* (Gorenstein & Snyder, 1980).

## Activity and Specificity

Action on dipeptidyl-NHNap substrates is very restricted, with that of the cattle pituitary enzyme being virtually limited to Arg-Arg-NHNap (Ellis & Nuenke, 1967). Specificity determinants characteristically include a dipeptidyl moiety with a free N-terminus. Whereas the pituitary enzyme has no action on Leu-Arg┤NHNap, DPP III from rat skin is reported to have a significant (44%) level of activity (at pH 8.0) on that substrate relative to its rate on Arg-Arg-NHNap (Hopsu-Havu *et al.*, 1970). DPP III from human placenta shows a significant (30%) rate on Ala-Arg-NHNap relative to that on the Arg-Arg-derivative (Shimamori *et al.*, 1986). No action occurs on substrates employed for the assay of DPP I (Chapter 214), e.g. Ser-Tyr-, Gly-Phe- and Gly-Arg-NHNap, or DPP II (Chapter 138), e.g. Lys-Ala-NHNap, or DPP IV (Chapter 128), e.g. Gly-Pro-NHNap (McDonald & Barrett, 1986).

The enzyme removes N-terminal dipeptides sequentially from oligopeptides having unsubstituted N-termini. In contrast to its very restricted specificity on dipeptidyl arylamides, DPP III displays a relatively broad specificity on oligopeptides comprised of four or more residues. Most tripeptides are resistant to attack, as are large polypeptides and proteins, e.g. corticotropin, glucagon, and the S-peptide of ribonuclease (Ellis & Nuenke, 1967), as well as bovine serum albumin, $\beta$-lipotropin and proinsulin (Lee & Snyder, 1982). Some tetrapeptides are also resistant, e.g. $Glu_4$ and $Gly_4$ (Ellis & Nuenke, 1967). No action occurs on bonds involving a (P1 or P1') proline. Because DPP III has little action on tripeptides, oligopeptides with an odd number of residues are reduced to dipeptides and one tripeptide. Even-numbered oligopeptides are generally reduced completely to dipeptides. Examples of susceptible oligopeptides include $Ala_4$, $Ala_6$, $Lys_4$, $Phe_4$, Val-Leu┤Ser-Glu-Gly, $Asn^1$, $Val^5$angiotensin II (Asn-Arg┤Val-Tyr┤Val-His-Pro-Phe) (McDonald *et al.*, 1971), Leu-enkephalin (Tyr-Gly┤Gly-Phe-Leu) (Lee & Snyder, 1982), angiotensin III (Arg-Val┤Tyr-Ile┤His-Pro-Phe), Leu-Trp┤Met-Arg┤Phe-Ala, and Leu-Arg┤Arg-Ala-Ser-Leu-Gly (Abramic *et al.*, 1988).

Activity is assayed specifically on Arg-Arg-NHNap at pH 9.0 by a direct fluorometric procedure (Ellis & Nuenke, 1967), or colorimetrically by taking stopped aliquots of the reaction mixture for color development by diazotization of the liberated 2-naphthylamine (Shimamori *et al.*, 1986). A special assay for red cell DPP III is available (Jones & Kapralou, 1982). The enzyme from most sources hydrolyzes the assay substrate optimally at about pH 9.0, but the optimum can range from pH 8.0, as seen for DPP III from rat skin (Hopsu-Havu *et al.*, 1970), to pH 10, as reported for the enzyme from the slime mold *Dictyostelium discoideum* (Huang *et al.*, 1992). Substrate is usually employed at a concentration in the range of 0.05–0.5 mM since significant inhibition of activity occurs at higher levels (Ellis & Nuenke, 1967; Hopsu-Havu *et al.*, 1970). Activators are not generally required, but the purified enzyme (having been exposed to air) may benefit greatly from the addition of 2-ME (2 mM) (Ellis & Nuenke, 1967). Activity may also be enhanced by 0.1 M $Cl^-$ (Hopsu-Havu *et al.*, 1970). When assays are performed on impure preparations, bestatin ($10\,\mu g\,ml^{-1}$) may be employed to block aminopeptidase action without affecting DPP III

activity (Hazato *et al.*, 1984; Shimamori *et al.*, 1986). Tests for the hydrolysis of peptide substrates were originally made at pH 8.7 (Ellis & Nuenke, 1967), but more recent studies with physiologically active peptides have been in the range pH 6.8 (Kokubu *et al.*, 1969) to pH 7.5 (Shimamori *et al.*, 1988).

Inhibition (65%) by a 5 min exposure to 1 mM DFP has been shown for DPP III from both the cattle pituitary gland and rabbit red blood cells (McDonald & Schwabe, 1977). Inhibition by DFP ranging from 20% to 100% has been reported for DPP III derived from mammalian sources such as erythrocytes (Kokubu *et al.*, 1969; Abramic *et al.*, 1988), monkey brain (Hazato *et al.*, 1984), human eye lens (Swanson *et al.*, 1984), human placenta (Shimamori *et al.*, 1986), cattle adrenal medulla (Kecorius *et al.*, 1988), and human seminal plasma (Vanha-Perttula, 1988). DFP and PMSF have little or no effect on DPP III from yeast, a form of the enzyme that requires $Co^{2+}$ (Watanabe *et al.*, 1990). Ac-L-leucyl-L-argininal, a naturally occurring inhibitor of DPP III, has been isolated from the culture filtrate of a bacterium. Enzyme purified from rat pancreas is inhibited 50% by Ac-Leu-Arg-H at $0.04\,\mu g\,ml^{-1}$ (Nishikiori *et al.*, 1984). The activity of DPP III from these sources also appears to depend on a reduced sulfhydryl function. Purified (aerated) enzyme not only benefits from the addition of sulfhydryl compounds, but the latter also reverse the effects of sulfhydryl reagents such as PCMB (Ellis & Nuenke, 1967; Hopsu-Havu *et al.*, 1970). Whereas metals such as $Hg^{2+}$ and $Zn^{2+}$ are reported to be inhibitory (Hopsu-Havu *et al.*, 1970), it has also been found that $Zn^{2+}$ (Shimamori *et al.*, 1986) and $Co^{2+}$ (Shimamori *et al.*, 1988) are either required or stimulatory. A peptidase derived from *Saccharomyces cerevisiae* and characterized as DPP III is insensitive to DFP and requires $Co^{2+}$ for its activity (Watanabe *et al.*, 1990). Although chelating agents such as EDTA are widely reported to be inhibitory, and their effects to be reversed by metal ions such as $Zn^{2+}$ (Shimamori *et al.*, 1986), there is evidence that EDTA may be directly inhibitory. Peptidase and arylamidase activities inhibited by EDTA can be restored by 2-ME following the removal of EDTA by dialysis (Ellis & Nuenke, 1967).

## Structural Chemistry

Most estimates place $M_r$ at about 80 000, with values ranging from 68 100 (Kecorius *et al.*, 1988) to 100 000 (Hazato *et al.*, 1984). It is an acidic protein (pI 4.4–4.6) with a monomeric structure, as shown for DPP III from human erythrocytes (Abramic *et al.*, 1988). The purified enzyme is sensitive to freezing.

## Preparation

A five-step purification procedure has been employed to obtain apparently homogeneous enzyme in 35% yield from human erythrocytes (Abramic *et al.*, 1988). DPP III of apparent homogeneity has also been prepared from the cytosol (postmicrosomal) fraction of fresh human placenta homogenized in 0.25 M sucrose. A 9% yield was obtained with a six-step procedure (Shimamori *et al.*, 1986). Rat spleen, a more readily available tissue, is a rich source of DPP III (McDonald *et al.*, 1971) that has been employed for the partial purification of the enzyme (Nishikiori *et al.*, 1984).

An enzyme having properties that strongly resemble those of mammalian DPP III has been purified 91-fold from *S. cerevisiae* (Watanabe *et al.*, 1990).

## Biological Aspects

The relativity broad specificity of DPP III on peptides, together with its wide distribution in the cytosol of mammalian cells, including erythrocytes (McDonald & Schwabe, 1977), supports a general role for this peptidase in the terminal stages of intracellular protein degradation, and possibly red cell maturation. The activity of rat brain DPP III in the degradation of neuropeptides suggests that it may serve to modulate their levels in the brain.

During pregnancy, DPP III activity increases 11.6-fold in retroplacental serum (Shimamori *et al.*, 1986). It is believed that placental DPP III, like cystinyl aminopeptidase (EC 3.4.11.3; oxytocinase) (Chapter 341), is synthesized in placental cells and released into the maternal circulation. Since DPP III has a high affinity for angiotensin II, which is about 17-fold greater than for Leu-enkephalin (Lee & Snyder, 1982), it is quite possible that this peptidase contributes to the elevated level of plasma angiotensin-hydrolyzing activity that exists during pregnancy (Berger & Langhans, 1967), and may also serve to regulate the level of angiotensin in the cells of origin.

An enzyme characterized as DPP III in the slime mold *Dictyostelium discoideum* appears not to be developmentally associated, but rather associated with peptide breakdown throughout its life cycle, as indicated by its levels in cells grown on bacteria or in axenic media (Huang *et al.*, 1992).

## Distinguishing Features

An activity that releases 2-naphthylamine from Arg-Arg-NHNap at pH 8.5–9.0 in the presence of bestatin ($10\,\mu g\,ml^{-1}$) is most probably attributable to DPP III.

## References

Abramic, M., Zubanovic, M. & Vitale, L. (1988) Dipeptidyl peptidase III from human erythrocytes. *Biol. Chem. Hoppe-Seyler* **369**, 29–38.

Berger, M. & Langhans, J. (1967) Angiotensinase activity in pregnant and nonpregnant women. *Am. J. Obstet. Gynecol.* **98**, 215–228.

Ellis, S. & Nuenke, J.M. (1967) Dipeptidyl arylamidase III of the pituitary. Purification and characterization. *J. Biol. Chem.* **242**, 4623–4629.

Gorenstein, C. & Snyder, S.H. (1980) Enkephalinases. *Proc. R. Soc. [Biol.]* **210**, 123–132.

Hazato, T., Shimamura, M., Ichimura, A. & Katayama, T. (1984) Purification and characterization of two distinct dipeptidyl aminopeptidases in soluble fraction from monkey brain and their action on enkephalins. *J. Biochem.* **95**, 1265–1271.

Hopsu-Havu, V.K., Jansén, C.T. & Järvinen, M. (1970) Partial purification and characterization of an alkaline dipeptide naphthylamidase (Arg-Arg-NAase) of the rat skin. *Arch. Klin. Exp. Dermatol.* **236**, 267–281.

Huang, J., Kim, J., Ramamurthy, P. & Jones, T.H.D. (1992) The purification, specificity, and role of dipeptidyl peptidase III in *Dictyostelium discoideum. Exp. Mycol.* **16**, 102–109.

Jones, T.H.D. & Kapralou, A. (1982) A rapid assay for dipeptidyl aminopeptidase III in human erythrocytes. *Anal. Biochem.* **119**, 418–423.

Kecorius, E., Small, D.H. & Livett, B.G. (1988) Characterization of a dipeptidyl aminopeptidase from bovine adrenal medulla. *J. Neurochem.* **50**, 38–44.

Kokubu, T., Akutsu, H., Fujimoto, S., Ueda, E., Hiwada, K. & Yamamura, Y. (1969) Purification and properties of endopeptidase from rabbit red cells and its process of degradation of angiotensin. *Biochim. Biophys. Acta* **191**, 668–676.

Lee, C.-M. & Snyder, S.H. (1982) Dipeptidyl-aminopeptidase III of rat brain. Selective affinity for enkephalin and angiotensin. *J. Biol. Chem.* **257**, 12043–12050.

McDonald, J.K. & Barrett, A.J. (1986) Dipeptidyl peptidase III. *Mammalian Proteases: A Glossary and Bibliography*, vol. 2: *The Exopeptidases*. London: Academic Press, pp. 127–131.

McDonald, J.K. & Schwabe, C. (1977) Intracellular exopeptidases. In: *Proteinases in Mammalian Cells and Tissues* (Barrett, A.J., ed.). Amsterdam: North-Holland Publishing, pp. 311–391 (see pp. 361–364).

McDonald, J.K., Callahan, P.X., Ellis, S. & Smith, R.E. (1971) Polypeptide degradation by dipeptidyl aminopeptidase I (cathepsin C) and related peptidases. In: *Tissue Proteinases* (Barrett, A.J. & Dingle, J.T., eds). Amsterdam: North-Holland Publishing, pp. 69–107.

Nishikiori, T., Kawahara, F., Naganawa, H., Muraoka, Y., Aoyagi, T. & Umezawa, H. (1984) Production of acetyl-L-leucyl-L-argininal, inhibitor of dipeptidyl aminopeptidase III by bacteria. *J. Antibiot.* **37**, 680–681.

Shimamori, Y., Watanabe, Y. & Fujimoto, Y. (1986) Purification and characterization of dipeptidyl aminopeptidase III from human placenta. *Chem. Pharm. Bull.* **34**, 3333–3340.

Shimamori, Y., Watanabe, Y. & Fujimoto, Y. (1988) Human placental dipeptidyl aminopeptidase III: hydrolysis of enkephalins and its stimulation by cobaltous ion. *Biochem. Med. Metab. Biol.* **40**, 305–310.

Swanson, A.A., Davis, R.M. & McDonald, J.K. (1984) Dipeptidyl peptidase III of human cataractous lenses. Partial purification. *Curr. Eye Res.* **3**, 287–291.

Vanha-Perttula, T. (1988) Dipeptidyl peptidase III and alanyl aminopeptidase in the human seminal plasma: origin and biochemical properties. *Clin. Chim. Acta* **177**, 179–196.

Watanabe, Y., Kumagai, Y. & Fujimoto, Y. (1990) Presence of a dipeptidyl aminopeptidase III in *Saccharomyces cerevisiae. Chem. Pharm. Bull.* **38**, 246–248.

*J. Ken McDonald*
*Department of Biochemistry and Molecular Biology,*
*Medical University of South Carolina,*
*171 Ashley Avenue, Charleston, SC 29425, USA*

# 183. *Tripeptidyl-peptidase I*

## Databanks

*Peptidase classification: clan SX, family S99, MEROPS ID: S9C.003*
*NC-IUBMB enzyme classification: EC 3.4.14.9*
*Databank codes: no sequence data available*

## Name and History

The activity of tripeptidyl-peptidase I was first manifested as early as 1961 when neutral to mildly acidic conditions were employed for the purification of growth hormone from cattle anterior pituitary glands (Ellis, 1961). Under these conditions, the phenylalanyl residue at the N-terminus of one of the chains of this 42 kDa dimeric protein (Li & Ash, 1953) was removed. The newly formed methionyl N-terminus arose as a consequence of the proteolytic release of the three amino acid residues (Phe, Pro, Ala) that precede methionine at the N-terminus, but the peptidase(s) responsible for this remained to be identified. In 1978 an exopeptidase was discovered in the cattle pituitary gland that was capable of releasing intact Phe-Pro-Ala, at acidic pH, from the N-terminus of the phenylalanyl-monomer of cattle growth hormone (Doebber *et al.*, 1978). The enzyme was named *tripeptidyl aminopeptidase*. In a paper that dealt with a similar activity in the pig ovary (McDonald *et al.*, 1985), it was proposed that the enzyme be termed *tripeptidyl peptidase*, and more specifically *tripeptidyl peptidase I* to distinguish it from a cytosolic, tripeptide-releasing enzyme active at neutral pH (Bålöw *et al.*, 1983), to be named tripeptidyl peptidase II (Chapter 113). The name *tripeptidyl-peptidase I* (EC 3.4.14.9) was recommended by NC-IUBMB in 1992, and is here abbreviated *TPP I*.

## Activity and Specificity

TPP I releases tripeptides sequentially from the unsubstituted N-termini of oligopeptides and proteins (Doebber *et al.*, 1978). It also catalyzes the hydrolysis of arylamide bonds in unsubstituted tripeptidyl-NHNap, -NHMec and -NHPhNO$_2$ derivatives (McDonald *et al.*, 1985). Virtually any amino acid residue except proline may occupy the P1 and P1′ positions, although hydrophobic residues are generally preferred in the former. Poly(Gly-Pro-Ala-), a 14 kDa polypeptide that mimics the repeating Gly-Pro-Xaa sequence that is prominent in collagen α chains, is depolymerized to its constituent triplets (McDonald *et al.*, 1985). Tripeptides are also released from small peptides such as angiotensin III (seven residues) and des-¹Tyr-dynorphin (seven residues). But peptides having a proline residue in the P1 or P1′ position, e.g. bradykinin and substance P, are not attacked (Watanabe *et al.*, 1992).

The optimum pH for TPP I action varies from 4.0 to 5.0. Cattle TPP I acts optimally on cattle growth hormone at about pH 4.0, releasing 11 tripeptides in succession (Doebber *et al.*, 1978). Pig TPP I readily depolymerizes poly(Gly-Pro-Ala-) at pH 5.0 (McDonald *et al.*, 1985). On tripeptidyl arylamide assay substrates, pH optima range from 4.3 for nitroanilides to 5.0 for methylcoumarylamides (McDonald *et al.*, 1985).

The activity of TPP I can be determined fluorometrically with Phe-Pro-Ala+NHNap at pH 4.0, which requires a correction for the acid quenching of the 2-naphthylamine fluorescence (Doebber *et al.*, 1978), or Gly-Pro-Met+NHMec at pH 5.0 (McDonald *et al.*, 1985), or Ala-Ala-Phe+NHMec at pH 5.5 (Page *et al.*, 1993), neither of which requires a quenching correction. Colorimetric assays may also be made with *p*-nitroanilide derivatives (McDonald *et al.*, 1985; Watanabe *et al.*, 1992).

DFP is strongly inhibitory, as are the active-site-directed chloromethane inhibitors Gly-Pro-Met-CH$_2$Cl (McDonald *et al.*, 1985) and Ala-Ala-Phe-CH$_2$Cl (Watanabe *et al.*, 1992). Partial inhibition is seen with *p*-chloromercurisulfonate and DCI. Consistent with its serine catalytic character, TPP I is unaffected by pepstatin, E-64, leupeptin and EDTA (McDonald *et al.*, 1985; Page *et al.*, 1993).

## Structural Chemistry

TPP I is a single-chain protein. Estimates of its molecular mass have been reported for enzyme derived from the following sources: human osteoclastoma, 48 kDa (Page *et al.*, 1993), pig ovary, 55 kDa (McDonald *et al.*, 1985), and cattle pituitary gland, 57 kDa (Doebber *et al.*, 1978). Variants of TPP I from the first two sources are notably hydrophobic proteins that are extracted as aggregates of 700 kDa and 250 kDa, respectively. Whereas the aggregate from the pig ovary dissociates to monomers in 3 M urea–0.1% Brij-35 at pH 4.5, the aggregate from osteoclastoma cells is unaffected by 4 M urea–1% Triton X-100 at pH 5.5. No amino acid sequence data are available.

## Preparation

TPP I has been purified from specialized tissues, e.g. the pituitary gland (Doebber *et al.*, 1978), the pig ovary (McDonald *et al.*, 1985) and human osteoclastoma (Page *et al.*, 1993). However, in view of its wide distribution in the tissues of the rat, wherein spleen, liver and kidney rank among the richest sources (Ellis & Divor, 1979), it is to be expected that such lysosome-rich tissues could be employed for the preparation of TPP I. It has, in fact, been partially purified from rat liver. Whereas the solubilization of TPP I from tissue extracts usually requires the use of a detergent in an acidic buffer, the rat liver enzyme was released from a tritosome preparation by freeze–thaw fracture and sonication (Watanabe *et al.*, 1992). Due to its hydrophobic character, TPP I activity is usually lost in column chromatographic steps unless buffers contain a detergent such as Brij-35 (McDonald *et al.*, 1985; Watanabe *et al.*, 1992) or Triton X-100 (Page *et al.*, 1993). Additionally, only acidic buffers, e.g. pH 4.0–5.5, may be employed, because of the instability of TPP I in the neutral pH range.

## Biological Aspects

In view of its wide distribution in the tissues, and its lysosomal localization, TPP I undoubtedly contributes, in concert with other lysosomal peptidases, to intracellular protein degradation. Its ability to reduce large polypeptides to tripeptide fragments would be expected to play an important role in a degradative pathway that very likely depends on dipeptidyl-peptidase II (Chapter 138) for the further reduction of the liberated tripeptides to dipeptides and free amino acids. The permeability of the lysosomal membrane allows for the escape of such final products to the cytosol, and thus for the return of amino acids to the metabolic pool (Lloyd & Foster, 1986).

The ability of TPP I to depolymerize poly(Gly-Pro-Ala-), a synthetic model of the collagen $\alpha$ chain, supports the notion that this exopeptidase may play a special role in the terminal stages of intracellular collagen degradation (McDonald *et al.*, 1985). Studies by Page *et al.* (1993) provide support for such a proposition. Ala-Ala-Phe-$CH_2Cl$, a potent active-site-directed inhibitor of TPP I isolated from an osteoclastoma, was shown to inhibit osteoclastic bone resorption in an *in vitro* test system. The complementary action of TPP I and interstitial collagenase (EC 3.4.24.7) (Chapter 389) appears possible since the latter has been identified within the osteoclast and its extracellular resorption zone (Delaissé *et al.*, 1993). Lysosomal hydrolases are secreted into this zone by the osteoclast (Vaes, 1988).

## Distinguishing Features

At pH 4.5, TPP I catalyzes the release of an intact tripeptide, but not an *N*-acetylated tripeptide, from an arylamide derivative, e.g. Gly-Pro-Met-NHMec or Ala-Ala-Phe-NHMec.

A 55 kDa tripeptidyl-peptidase of the serine-catalytic class has been isolated from cell-free cultures of *Streptomyces lividans* 66, and is believed to be a major component of its secreted proteolytic activity. This prokaryotic enzyme hydrolyzes Ala-Phe-Ala$+$NPhNO$_2$ optimally at pH 8.0–9.0, a characteristic that more closely resembles that of tripeptidyl-peptidase II. The sequence of the first 15 amino acid residues has been reported for this prokaryotic tripeptidyl-peptidase (Krieger *et al.*, 1994) (see Chapter 140).

## References

Bålöw, R-M., Ragnarsson, U. & Zetterqvist, Ö. (1983) Tripeptidyl aminopeptidase in the extralysosomal fraction of rat liver. *J. Biol. Chem.* **258**, 11622–11628.

Delaissé, J.M., Eeckhout, Y., Neff, L., Francois-Gillet, Ch., Henriet, P., Su, Y., Vaes, G. & Baron, R. (1993) (Pro)collagenase (matrix metalloproteinase-1) is present in rodent osteoclasts and in the underlying bone-resorbing compartment. *J. Cell Sci.* **106**, 1071–1082.

Doebber, T.W., Divor, A.R. & Ellis, S. (1978) Identification of a tripeptidyl aminopeptidase in the anterior pituitary gland: effect on the chemical and biological properties of rat and bovine growth hormones. *Endocrinology* **103**, 1794–1804.

Ellis, S. (1961) Studies on the serial extraction of pituitary proteins. *Endocrinology* **69**, 554–570.

Ellis, S. & Divor, A.R. (1979) Specificity and lysosomal localization of tripeptidyl aminopeptidase. *Fedn Proc.* **38**, 832.

Krieger, T.J., Bartfeld, D., Jenish, D.L. & Hadary, D. (1994) Purification and characterization of a novel tripeptidyl aminopeptidase from *Streptomyces lividans* 66. *FEBS Lett.* **352**, 385–388.

Li, C.H. & Ash, L. (1953) The nitrogen terminal end groups of hypophyseal growth hormone. *J. Biol. Chem.* **203**, 419–424.

Lloyd, J.B. & Foster, S. (1986) The lysosomal membrane. *Trends Biochem. Sci.* **11**, 365–368.

McDonald, J.K., Hoisington, A.R. & Eisenhauer, D.A. (1985) Partial purification and characterization of an ovarian tripeptidyl peptidase: a lysosomal exopeptidase that sequentially releases collagen-related (Gly-Pro-X) triplets. *Biochem. Biophys. Res. Commun.* **126**, 63–71.

Page, A.E., Fuller, K., Chambers, T.J. & Warburton, M.J. (1993) Purification and characterization of a tripeptidyl peptidase I from human osteoclastomas: evidence for its role in bone resorption. *Arch. Biochem. Biophys.* **306**, 354–359.

Vaes, G. (1988) Cellular biology and biochemical mechanism of bone resorption. A review of recent developments on the formation, activation, and mode of action of osteoclasts. *Clin. Orthop.* **231**, 239–271.

Watanabe, Y., Kumagai, Y. & Fujimoto, Y. (1992) Acidic tripeptidyl aminopeptidase in rat liver tritosomes: partial purification and determination of its primary substrate specificity. *Biochem. Int.* **27**, 869–877.

*J. Ken McDonald*
*Department of Biochemistry and Molecular Biology,*
*Medical University of South Carolina,*
*171 Ashley Avenue, Charleston, SC 29425, USA*

# 184. Endopeptidase So

## Databanks

*Peptidase classification: clan SX, family S99, MEROPS ID: S9G.038*
*NC-IUBMB enzyme classification: EC 3.4.21.67*
*Databank codes: no sequence data available*

## Name and History

In studies aimed at identifying the proteolytic enzymes responsible for the ATP-dependent breakdown of proteins in *Escherichia coli*, Goldberg and coworkers systematically screened *E. coli* extracts for their capacity to degrade [$^3$H]methyl globin and [$^3$H]casein (Goldberg *et al.*, 1981; Swamy & Goldberg, 1981). Six such enzymes were found and named proteases Do, Re, Mi, Fa, So and La. *Endopeptidase So* was shown to be a cytosolic enzyme (Swamy & Goldberg, 1982) and was subsequently purified to homogeneity (Chung & Goldberg, 1983).

## Activity and Specificity

In addition to casein, denatured bovine serum albumin, glucagon and globin (Chung & Goldberg, 1983; Novak *et al.*, 1986), this enzyme hydrolyzes peptide substrates including Ac-Ala$_4$, but not Ac-Ala$_3$, Bz-Tyr-OEt or Ac-Phe-$\beta$-naphthylester. It is a serine protease inhibited by DFP and PMSF as well as by Tos-Phe-CH$_2$Cl, but not by Tos-Lys-CH$_2$Cl. Its activity is stabilized by Mg$^{2+}$.

## Structural Chemistry

Endopeptidase So behaves as a 150 kDa enzyme on gel filtration, and appears to be a homodimer composed of 77–83 kDa subunits. The enzyme has a pI of 6.4 (Chung & Goldberg, 1983; Novak *et al.*, 1986).

## Preparation

The enzyme can be purified to apparent homogeneity from soluble extracts of *E. coli* by conventional chromatographic procedures (Chung & Goldberg, 1983). No other cellular source of the enzyme is yet known.

## Biological Aspects

Multiple *in vivo* functions for this protease have been proposed. One possible role is in the selective elimination of oxidant-damaged polypeptides, whose rapid destruction can be an important protective mechanism for cells (Fucci *et al.*, 1983; Davies & Goldberg, 1987). Protease So degrades the oxidant-damaged form of *E. coli* glutamine synthetase (i.e. after damage induced by a hydroxyl-generating system composed of iron, ascorbate and oxygen) to acid-soluble peptides. The degradation of the oxidant-damaged glutamine synthetase is 5- to 10-fold faster than that of the native protein (Lee *et al.*, 1988). In addition, the enzyme *in vitro* rapidly hydrolyzes the signal peptides generated during the maturation of prolipoproteins by signal peptidase II (Chapter 333) (Novak *et al.*, 1986). Thus, the enzyme may well play a role in the elimination of signal peptides, whose accumulation could be toxic and interfere with the export of secretory proteins by competitively binding to the translocation machinery of the cell. Definitive information on its *in vivo* importance will require mutants lacking this enzyme.

## Further Reading

For a review, see Lee *et al.* (1988).

## References

Chung, C.H. & Goldberg, A.L. (1983) Purification and characterization of protease So, a cytoplasmic serine protease in *Escherichia coli*. *J. Bacteriol.* **154**, 231–238.

Davies, K.J.A. & Goldberg, A.L. (1987) Oxygen radicals stimulate intracellular proteolysis and lipid peroxidation by independent mechanism in erythrocytes. *J. Biol. Chem.* **262**, 8220–8226.

Fucci, L., Oliver, C.N., Coon, M.J. & Stadtman, E.R. (1983) Inactivation of key metabolic enzymes by mixed function oxidation reactions: possible implication in protein turnover and aging. *Proc. Natl Acad. Sci. USA* **80**, 1521–1525.

Goldberg, A.L., Swamy, K.H.S., Chung, C.H. & Larimore, F.S. (1981) Proteases in *Escherichia coli*. *Methods Enzymol.* **80**, 680–702.

Lee, Y.S., Park, S.C., Goldberg, A.L. & Chung, C.H. (1988) Protease So from *Escherichia coli* preferentially degrades oxidatively damaged glutamine synthetase. *J. Biol. Chem.* **263**, 6643–6646.

Novak, P., Ray, P.H. & Dev, I.K. (1986) Localization and purification of two enzymes from *Escherichia coli* capable of hydrolyzing a signal peptide. *J. Biol. Chem.* **261**, 420–427.

Swamy, K.H.S. & Goldberg, A.L. (1981) *Escherichia coli* contains eight soluble proteolytic activities, one of which is ATP-dependent. *Nature* **292**, 652–654.

Swamy, K.H.S. & Goldberg, A.L. (1982) Subcellular distribution of various proteases in *Escherichia coli. J. Bacteriol.* **149**, 1027–1033.

***Chin Ha Chung***
*Department of Molecular Biology and*
*Research Center for Cell Differentiation,*
*Seoul National University,*
*Seoul 151-742, Korea*
*Email: chchung@plaza.snu.ac.kr*

***Alfred L. Goldberg***
*Department of Cell Biology,*
*Harvard Medical School,*
*Boston, MA 02115, USA*
*Email: agoldberg@bcmp.med.harvard.edu*

# CYSTEINE

CYSTEINE

# 185. Introduction: cysteine peptidases and their clans

MEROPS ID: CA, CB, CC, CD, CE

Peptidases in which the nucleophile is the sulfhydryl group of a Cys residue are known as cysteine-type peptidases. The catalytic mechanism is similar to that of the serine-type peptidases in that a nucleophile and a proton donor/general base are required, and the proton donor in all cysteine peptidases in which it has been identified is a His residue, as in the majority of serine-type peptidases. Although there is evidence in some families that a third residue is required to orientate the imidazolium ring of the His (a role analogous to that of the essential aspartate, Asp102, seen in some serine peptidases), there are a number of families in which only a catalytic dyad is necessary.

The first clearly recognized cysteine peptidase was papain, and this forms the foundation of *clan CA* (Chapter 186). Crystal structures of papain and several closely related peptidases of family C1 have been determined, and the catalytic residues have been identified as Cys, His and Asn, occurring in that order in the sequence. A fourth residue, a Gln preceding the catalytic cysteine, is also essential for catalytic activity and is believed to help form the 'oxyanion hole'. The crystal structure of a ubiquitin C-terminal hydrolase of family C12 (see Figure 186.2) shows that it has a fold very similar to that of papain (see Figure 186.8), despite very little conservation of amino acid sequence, and there seems every reason to conclude that family C12 had a common origin with papain, and thus belongs in clan CA. No three-dimensional structures are available for peptidases of families C2, C10 or C19, but these are placed in clan CA on the basis of the Cys, His order of the catalytic dyad, and limited conservation of amino acids around the catalytic residues (see Alignment 186.1 on CD-ROM); there are also some similarities in properties. Like family C12, C19 contains ubiquitin C-terminal hydrolases. In family C12, Asn175 is replaced by Asp, and we predict a similar replacement for family C19, as well as the replacement of the equivalent of Gln19 by Asn.

A second group of cysteine peptidases that contain the Cys, His order of catalytic dyad is formed by the 'papain-like' endopeptidases of RNA viruses. The group is very diverse in sequence, and there is as yet no tertiary structure information for any member of it. These peptidases contain no residues that are obviously equivalent to the essential Gln19 and Asn175 of papain, and in the absence of substantial evidence that they share a common ancestor with papain, we place these peptidases in a separate clan, *clan CC* (Chapter 225). There is, however, nothing to exclude the possibility that when a crystal structure is determined it may indeed show the papain fold, and require the merging of clans CA and CC.

Three clans of cysteine peptidases contain the catalytic dyad in the His, Cys order in the sequence. The peptidases of *clan CB* (Chapter 238) emphasize the mechanistic similarities between cysteine- and serine-type peptidases, in that they have evolved from a common ancestor with chymotrypsin and the

many other peptidases of clan SA. This was originally suggested on the basis of amino acid sequence comparisons, and was clearly confirmed by the similarity in three-dimensional structure of the picornain 3C from hepatitis A virus (see Figure 238.1) to chymotrypsin (see Figure 2.2). The viral endopeptidase shares the same two $\beta$-barrel domains, and the folds are superimposable. The catalytic His is equivalent in position to His57 of chymotrypsin, and the Cys replaces Ser195. In rhinovirus picornain 3C there is even evidence that a catalytic triad exists, with a Glu in the position equivalent to Asp102 of chymotrypsin. There are several families of endopeptidases from RNA viruses that are also predicted to have the chymotrypsin fold, and all of these are placed in clan CB. In view of the evident common origin of the peptidases in clans SA and CB, these clans should, in principle, be merged, if it were not for the importance currently placed on the classification of peptidases primarily by catalytic type (but see the MEROPS database).

The two clans of viral cysteine endopeptidases contrast in several respects. In clan CC, the catalytic Cys is generally followed by a large hydrophobic residue, whereas in clan CB the active site Cys is followed by Gly (like the Ser in chymotrypsin). Other differences between clans CB and CC are the preference for the residue in the P1 site of the polyprotein substrate: in clan CC cleavage usually occurs at Gly+Gly bonds, but in clan CB it is commonly at Gln+Gly. The endopeptidases from clan CC generally perform only one cleavage, whereas those from clan CB tend to be general processing enzymes, cleaving several bonds in the polyprotein. Finally, a clan CB endopeptidase is usually positioned between the helicase and the RNA polymerase in the polyprotein, whereas a clan CC enzyme usually precedes both the helicase and the RNA polymerase.

One group of cysteine endopeptidases in which only a catalytic dyad appears to exist is the caspase family, C14. The tertiary structure (see Figure 247.1) shows a unique $\alpha/\beta$ fold unlike that of any other protein, with the catalytic residues in the order His, Cys in the sequence. The caspases form *clan CD* (Chapter 247).

The adenovirus endopeptidase (Chapter 251) is believed to have a catalytic mechanism similar to that of papain, involving four amino acids. However, these occur in the order His, Glu (or Asp), Gln, Cys, and the tertiary fold (see Figure 250.1) is different to that of papain or any other cysteine-type peptidase. For this reason, the single family (C5) is placed in *clan CE* (Chapter 250).

There are a number of other families of peptidases for which tertiary structures are unknown and too little is known about the catalytic machinery to permit any conclusions about which clan they may belong to. These are assigned to a catch-all *clan CX* (Chapter 252) for convenience.

# 186. Introduction: clan CA containing papain and its relatives

## Databanks

*MEROPS ID: CA*

| Species | SwissProt | PIR | EMBL (cDNA) | EMBL (genomic) |
|---|---|---|---|---|
| **Family C1** | | | | |
| Actinidain (Chapter 197) | | | | |
| Aleurain (Chapter 200) | | | | |
| Ananain (Chapter 194) | | | | |
| Asclepain (Chapter 198) | | | | |
| Bleomycin hydrolase (Chapter 215) | | | | |
| Calotropin | | | | |
| *Calotropis* sp. | P20728 | – | – | – |
| *Carica candamarcensis* endopeptidase (Chapter 191) | | | | |
| Caricain (Chapter 189) | | | | |
| Cathepsin B (Chapter 209) | | | | |
| Cathepsin H (Chapter 213) | | | | |
| Cathepsin K (Chapter 212) | | | | |
| Cathepsin L (Chapter 210) | | | | |
| Cathepsin O | | | | |
| *Homo sapiens* | P43234 | – | X77383 | – |
| Cathepsin S (Chapter 211) | | | | |
| Chymopapain (Chapter 188) | | | | |
| Comosain (Chapter 195) | | | | |
| Cruzipain (Chapter 203) | | | | |
| Der p1 proteinase | | | | |
| *Dermatophagoides farinae* | P16311 | A27634 | – | X65196: mid-section |
| *Dermatophagoides microceras* | P16312 | – | – | – |
| *Dermatophagoides pteronyssinus* | P08176 | C27634 JQ0337 | – | X65197: mid-section |
| *Euroglyphus maynei* | P25780 | S21864 | – | S52929: 3′ end X60073: mid-section |
| *Dictyostelium* cysteine proteinase 7 (Chapter 207) | | | | |
| Dipeptidyl-peptidase I (Chapter 214) | | | | |
| Falcipain (Chapter 201) | | | | |
| *Fasciola* cysteine endopeptidases (Chapter 208) | | | | |
| Ficain (Chapter 196) | | | | |
| Fruit bromelain (Chapter 193) | | | | |
| Glycyl endopeptidase (Chapter 190) | | | | |
| Histolysain (Chapter 202) | | | | |
| *Lactobacillus* PepG | – | – | Z71782 | – |
| *Lactococcus* PepC (Chapter 216) | | | | |
| *Leishmania* cysteine endopeptidases (Chapter 204) | | | | |
| Oligopeptidase E (Chapter 217) | | | | |
| Papain (Chapter 187) | | | | |
| Plant cysteine proteinases in germination and senescence (Chapter 199) | | | | |
| Stem bromelain (Chapter 192) | | | | |
| Tpr proteinase | | | | |
| *Porphyromonas gingivalis* | P25806 | S27608 | M84471 | – |
| Trichomonad and *Giardia* cysteine endopeptidases (Chapter 205) | | | | |
| V-cath proteinase (Chapter 206) | | | | |
| Other plant proteinases | | | | |
| *Alnus glutinosa* | – | – | U13940 | |
| *Arabidopsis thaliana* | – | – | Z17766 | |

| Species | SwissProt | PIR | EMBL (cDNA) | EMBL (genomic) |
|---|---|---|---|---|
| **Family C1** (*continued*) | | | | |
| *Arabidopsis thaliana* | – | – | Z17788 | – |
| *Arabidopsis thaliana* | – | – | Z18778 | – |
| *Arabidopsis thaliana* | P43295 | S46535 | X74359 | – |
| *Brassica napus* | – | – | U68221 | – |
| *Carica candamarcensis* | P32954 | – | – | – |
| *Carica candamarcensis* | P32955 | – | – | – |
| *Carica candamarcensis* | P32957 | – | – | – |
| *Carica papaya* | P05993 | B26074 | X03971 | – |
| *Cicer arietinum* | – | S49451 | X82011 | – |
| *Glycine max* | – | – | U71379 | – |
| *Glycine max* | – | – | U71380 | – |
| *Glycine max* | – | S44266 | Z32795 | – |
| | | S55923 | | |
| *Hemerocallis* sp. | – | – | U12637 | – |
| *Lycopersicon esculentum* | – | S24988 | Z14028 | – |
| *Mesembryanthemum crystallinum* | – | – | U30322 | – |
| *Nicotiana otophora* | – | – | U57824 | – |
| *Nicotiana otophora* | – | – | U57825 | – |
| *Nicotiana otophora* | – | – | X81995 | – |
| *Nicotiana otophora* | – | S24242 | Z13959 | – |
| *Nicotiana otophora* | – | S24246 | Z13964 | – |
| *Nicotiana otophora* | – | S30149 | Z13959 | – |
| *Nicotiana otophora* | – | S30150 | Z13964 | – |
| *Oryza sativa* | – | – | D76415 | – |
| *Oryza sativa* | – | S47434 | X80876 | – |
| *Phalaenopsis* sp. | – | – | U34747 | – |
| *Phaseolus vulgaris* | – | – | U52970 | – |
| *Phaseolus vulgaris* | P25803 | S16251 | X56753 | X63102: complete gene |
| *Pisum sativum* | – | – | U44947 | – |
| *Pisum sativum* | P25804 | S11862 | X54358 | – |
| *Pseudotsuga menziesii* | – | – | U41902 | – |
| | | | Z49765 | |
| *Ricinus communis* | – | – | T14997 | – |
| *Thaumatococcus daniellii* | P21381 | – | – | – |
| *Triticum aestivum* | – | – | U32430 | – |
| *Vicia faba* | – | – | U59465 | – |
| *Vicia sativa* | – | S49166 | Z34895 | – |
| *Vigna mungo* | P12412 | S05497 | X15732 | X51900: complete gene |
| | | S12581 | | |
| | | S20213 | | |
| | | S48684 | | |
| *Vigna radiata* | – | – | U49445 | – |
| *Zinnia elegans* | – | – | U19267 | – |
| Other protozoan proteinases | | | | |
| *Naegleria fowleri* | – | – | U42758 | – |
| *Paramecium tetraurelia* | – | – | X91756 | – |
| *Tetrahymena thermophila* | – | – | L03212 | – |
| Others | | | | |
| *Bos taurus* | P05689 | A29172 | X01809 | – |
| *Onchocerca volvulus* | – | – | U71150 | – |
| *Rattus norvegicus* | – | – | L14776 | – |
| *Urechis caupo* | – | – | U30877 | – |
| **Family C2** | | | | |
| Calpain p94 (Chapter 220) | | | | |
| m-Calpain (Chapter 219) | | | | |
| $\mu$-Calpain (Chapter 218) | | | | |
| nCL-2 calpain | | | | |
| *Rattus norvegicus* | – | – | D14478 | – |
| | | | D14479 | |
| | | | D14480 | |

*continued overleaf*

| Species | SwissProt | PIR | EMBL (cDNA) | EMBL (genomic) |
|---|---|---|---|---|
| **Family C2** (*continued*) | | | | |
| nCL-3 calpain | | | | |
| *Homo sapiens* | – | – | U94346 | – |
| *Mus musculus* | – | – | Y10656 | – |
| nCL-4 calpain | | | | |
| *Homo sapiens* | – | – | AF022799 | – |
| *Mus musculus* | – | – | U89513 | – |
| *Rattus norvegicus* | – | – | U89514 | – |
| palB protein | | | | |
| *Aspergillus nidulans* | – | – | Z54244 | – |
| Sol protein | | | | |
| *Drosophila melanogaster* | P27398 | A41146 | M64084 | – |
| Others | | | | |
| *Caenorhabditis elegans* | – | – | AF040660 | – |
| *Caenorhabditis elegans* | – | – | U12921 | – |
| *Caenorhabditis elegans* | – | – | U80838 | – |
| *Caenorhabditis elegans* | – | – | Z35663 | – |
| *Caenorhabditis elegans* | – | – | Z72515 | – |
| *Caenorhabditis elegans* | – | – | Z81083 | – |
| *Caenorhabditis elegans* | P34308 | S44749 | L25598 | – |
| *Drosophila melanogaster* | Q11002 | A55054 | X78555 Z46891 Z46892 | – |
| *Mus musculus* | P12815 | – | X149380 | – |
| *Schistosoma mansoni* | P27730 | A39343 | M67499 M74233 | – |
| *Trypanosoma brucei* | – | – | T26731 | |
| **Family C10** | | | | |
| Streptopain (Chapter 221) | | | | |
| **Family C12** | | | | |
| Ubiquitin C-terminal hydrolase PGP 9.5 (Chapter 222) | | | | |
| **Family C19** | | | | |
| Isopeptidase T (Chapter 223) | | | | |

Clan CA contains five families of cysteine peptidases: C1, C2, C10, C12 and C19. Tertiary structures are known for peptidases in families C1 and C12, and show that besides the Cys and His of the catalytic dyad, two other residues are important for catalysis (Alignment 186.1 on CD-ROM). These are a Gln (Gln19) that helps in the formation of the 'oxyanion hole', an electrophilic center that stabilizes the tetrahedral intermediate, and either an Asn or Asp, which is thought to orientate the imidazolium ring of the catalytic His. The order of these residues in the sequence is Gln, Cys, His, Asn/Asp. In family C19, only the catalytic Cys has been identified, and Alignment 186.1 shows our predictions for the other catalytic residues, including the suggestion that Gln19 can be replaced by Asn. The catalytic Cys is usually followed by an aromatic, hydrophobic amino acid, but a Gly occupies this position in some peptidases of family C2 and all those of family C12. Alignment 186.1 shows alignments around the catalytic residues for selected members of clan CA, and Figure 186.1 shows chain structures of selected peptidases from the clan.

Many peptidases of clan CA are inhibited by E-64 and by cystatins, but neither class of inhibitor is entirely selective for clan CA. Thus, E-64 also inhibits viral 'papain-like' cysteine endopeptidases of clan CC (see Chapter 225), and

the legumains in family C13 (Chapter 255) are also inhibited by cystatin (though not by E-64). Family C13 cannot yet be assigned to a clan, but there are no other indications that it would belong to clan CA.

*Family C1* includes endopeptidases from DNA viruses, protozoa, plants and animals, and exopeptidases from gram-positive bacteria, fungi and animals. Most of the endopeptidases enter the secretory pathway, though they may be targeted to vacuoles or lysosomes, whereas the exopeptidases are mainly cytosolic. The family includes the plant enzymes papain, chymopapain and actinidain, and the animal lysosomal enzymes cathepsins B, H and L. Figure 186.2 shows the three-dimensional structure of papain.

The distinction between exo- and endopeptidases is blurred for some members of family C1. Dipeptidyl-peptidase I (Chapter 214) acts principally as an exopeptidase, removing N-terminal dipeptides, but is a lysosomal enzyme and may have limited endopeptidase activity. The lysosomal cathepsins B and H (Chapters 209 and 213) are arguably more important for their exopeptidase activities than for the endopeptidase activities that both possess. Cathepsin B acts as a peptidyl-dipeptidase, releasing C-terminal dipeptides, and this activity is attributable to the existence of an extended loop that forms a cap to the active-site cleft, and carries a pair of His residues

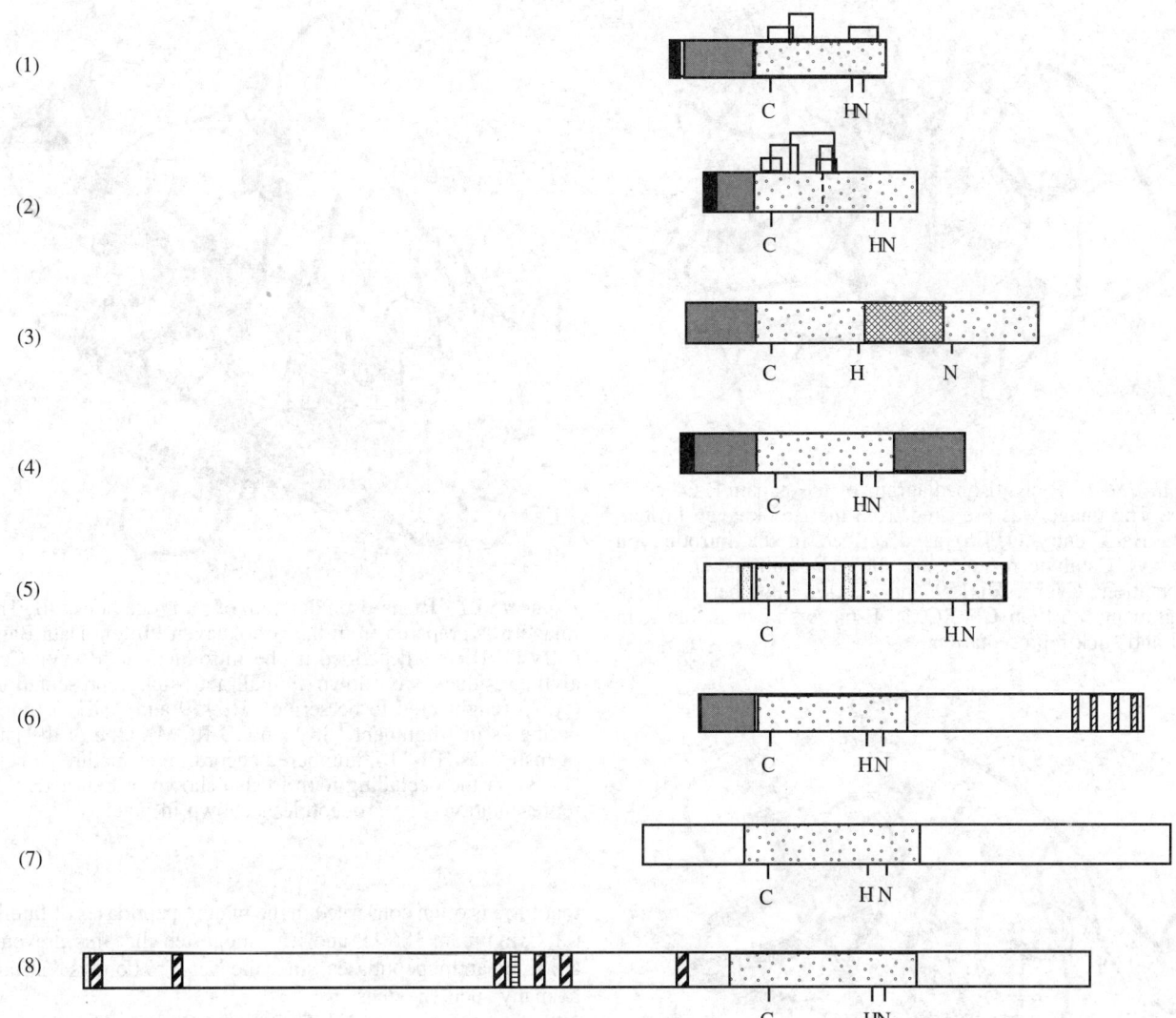

*Figure 186.1*   Domain structures for selected endopeptidases from clan CA. Endopeptidase sequences are shown as rectangles. The length of the rectangle is a representation of the sequence length. Disulfide bridges are shown as open boxes above the sequence rectangle. The positions of the catalytic triad residues are shown below the sequence rectangles, and all structures are aligned at the catalytic cysteine. Structures are arranged so that the simplest is at the top, and the most complex at the bottom. A dashed vertical line marks the position of His111 on the occluded loop of cathepsin B (2). Shaded boxes within the sequence rectangle represent different structural domains. *Key to domains*: black, signal peptide; gray, N- or C-terminal propeptide; dots, catalytic domain; cross-hatching, serine-rich domain; stiple, internal repeats; light hatching, calcium-binding EF-hand structure domain; dark hatching, zinc finger-like domains; horizontal stripes, opa repeat. *Key to sequences*: (1) papain, (2) *Homo sapiens* procathepsin B, (3) *Dictyostelium discoideum* cysteine proendopeptidase 7, (4) *Trypanosoma brucei* cruzipain, (5) *Saccharomyces cerevisiae* bleomycin hydrolase, (6) *H. sapiens* μ-calpain, (7) *Aspergillus nidulans* palB protein, (8) *Drosophila melanogaster* sol protein.

that are thought to bind the C-terminal carboxylate of the substrate (Figure 186.3).

The enzymes that enter the secretory pathway are synthesized as precursors, with N-terminal propeptides, as well as the signal peptides. Most members of family C1 have propeptides homologous to that of papain, but the propeptide of cathepsin B is much shorter and shows little sequence similarity to that of papain. However, the propeptides of both cathepsin B and papain are thought to act in the same way, blocking the active site by being bound in the reverse orientation to a substrate, as can be seen from the structure of procathepsin B (Figure 186.4). The propeptide of

dipeptidyl-peptidase I is different to, and longer, than those of cathepsin B or papain. Papain-like propeptides may be identified from the presence of the ERFNIN motif, in which some of the following residues (numbered according to the papain propeptide) are conserved: Glu64, Arg68, Phe72, Asn75, Asn83, Phe96, Asp98 and Glu103. The papain propeptide is homologous to proteins CTLA-2α and CTLA-2β of activated T cells, which have been shown to be cysteine endopeptidase inhibitors (Delaria *et al*., 1994).

Activation of peptidases of family C1 by removal of the N-terminal propeptide is sometimes followed by further processing. For cathepsins B, H and L, the mature enzyme consists

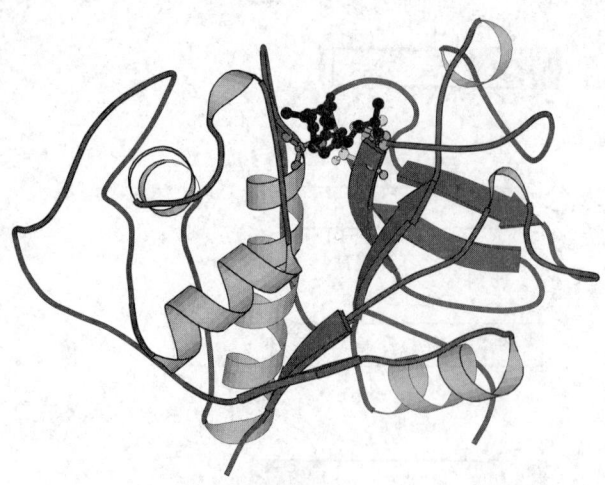

*Figure 186.2*   Richardson diagram of the papain/E-64 complex. The image was prepared from the Brookhaven Protein Data Bank entry (1PE6) as described in the Introduction (p. xxv). Catalytic residues are shown in ball-and-stick representation: Cys25, His159 and Asn175 (numbering as in Alignment 186.1 on CD-ROM). E-64 is shown in black in ball-and-stick representation.

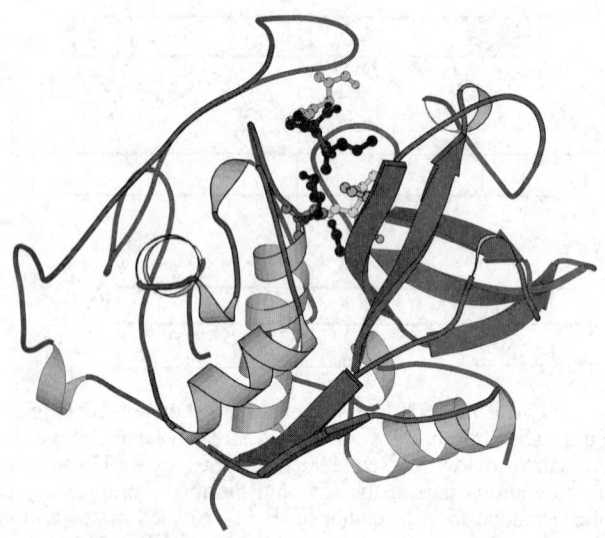

*Figure 186.3*   Richardson diagram of the human cathepsin B/CA030 complex. The image was prepared from the Brookhaven Protein Data Bank entry (1CSB) as described in the Introduction (p. xxv). Catalytic residues are shown in ball-and-stick representation: Cys25, His159 and Asn175 (numbering as in Alignment 186.1 on CD-ROM). One of the pair of histidines (His110, numbered according to mature cathepsin B) on the occluding loop is also shown in ball-and-stick representation. CA030 is shown in black in ball-and-stick representation.

of two chains, usually referred to as heavy and light chains. These are formed by proteolytic cleavage of the single-chain form of the molecule, with loss of a tri- or dipeptide between the chains. In cathepsin B, a C-terminal hexapeptide is also removed. In cathepsin H, after removal of the propeptide, the new N-terminus may be further trimmed, and distinct forms are known with different degrees of trimming. It is notable

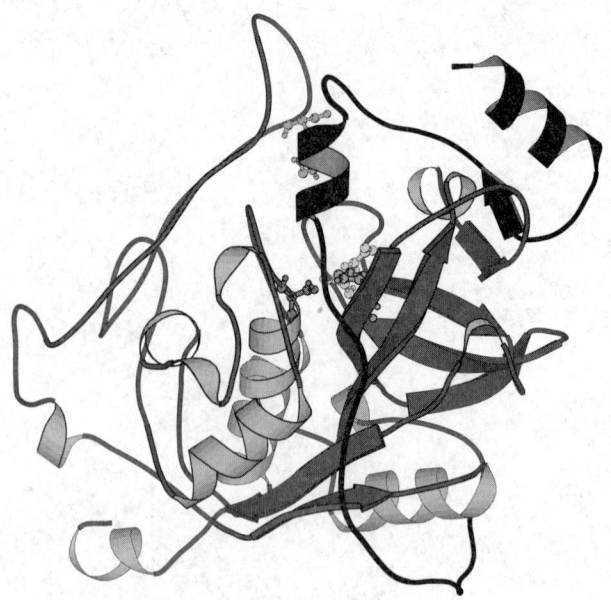

*Figure 186.4*   Richardson diagram of rat procathepsin B. The image was prepared from the Brookhaven Protein Data Bank entry (1MIR) as described in the Introduction (p. xxv). Catalytic residues are shown in ball-and-stick representation: Cys25 (engineered to be serine), His159 and Asn175 (numbering as in Alignment 186.1 on CD-ROM). One of the pair of histidines (His110, numbered according to mature cathepsin B) on the occluding loop is also shown in ball-and-stick representation. The propeptide is shown in black.

that Pro2 is often conserved in the mature peptidases of family C1 (Alignment 186.1), and it is suggested that this prevents attack by aminopeptidases, since the Xaa-Pro bond is resistant to many such enzymes.

Some peptidases of family C1 have C-terminal extensions relative to papain. Among the endopeptidases from plants, the extensions at the C-terminus of oryzains α and β, and cysteine endopeptidases from tomato (SW: P20721), *Arabidopsis thaliana* (SW: P43297), Douglas fir (EMBL: U41902), carnation (EMBL: U17135) and pea (EMBL: X66061) are homologous. The C-terminal extensions found in endopeptidases from *Trypanosoma* and *Leishmania* are unique in the family.

An evolutionary tree for selected members of family C1 is shown in Figure 186.5. It can be seen that the family is divided into two subfamilies. The larger of these (numbered 8–56 in Figure 186.5) consists of secreted and lysosomal enzymes. There is a clear division within the subfamily between the cathepsin B-like enzymes (numbered 8–19) and the papain-like enzymes (numbered 20–56). Among the cathepsin B-like enzymes are dipeptidyl-peptidase I, and the endopeptidases from *Giardia*. Included among the papain-like enzymes are cathepsins O, H, L, K and S. Cathepsin O is very divergent, whereas the others are more closely related. The close relationship between mammalian cathepsin H and plant aleurain is apparent.

The smaller subfamily (numbered 1–5 in Figure 186.5) contains several soluble, intracellular peptidases. These are aminopeptidases that commonly show selectivity for release

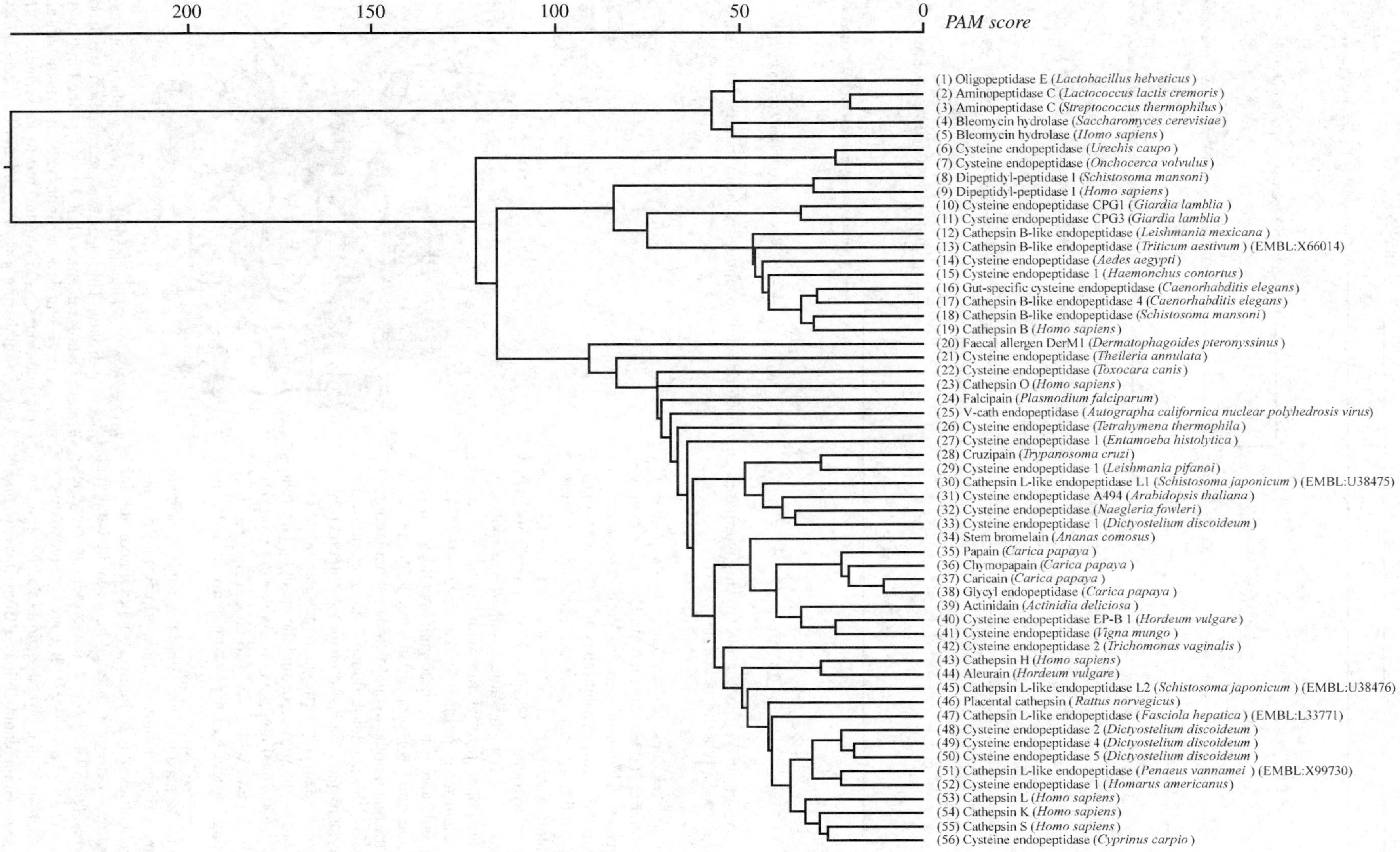

*Figure 186.5* Evolutionary tree for selected members of family C1. The tree was prepared as described in the Introduction (p. xxv).

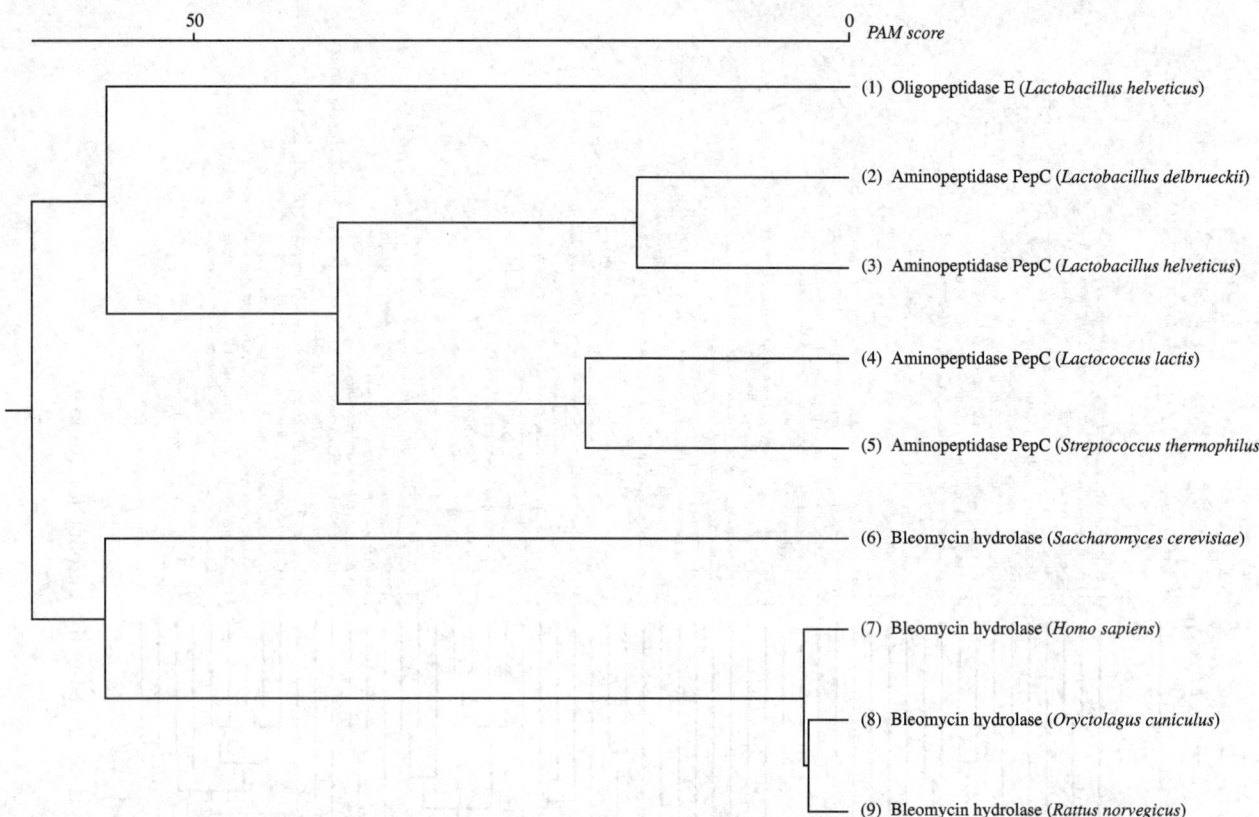

50                                                                  0  *PAM score*

(1) Oligopeptidase E (*Lactobacillus helveticus*)

(2) Aminopeptidase PepC (*Lactobacillus delbrueckii*)

(3) Aminopeptidase PepC (*Lactobacillus helveticus*)

(4) Aminopeptidase PepC (*Lactococcus lactis*)

(5) Aminopeptidase PepC (*Streptococcus thermophilus*)

(6) Bleomycin hydrolase (*Saccharomyces cerevisiae*)

(7) Bleomycin hydrolase (*Homo sapiens*)

(8) Bleomycin hydrolase (*Oryctolagus cuniculus*)

(9) Bleomycin hydrolase (*Rattus norvegicus*)

*Figure 186.6* Evolutionary tree for bleomycin hydrolases and bacterial aminopeptidases from family C1. The tree was prepared as described in the Introduction (p. xxv).

of N-terminal Arg residues, and include aminopeptidase C (PepC) from bacteria, and bleomycin hydrolase, from eukaryotes. The sequence relationships in this group are shown in greater detail in the second evolutionary tree (Figure 186.6).

Unlike the peptidases that enter the secretory pathway, the aminopeptidase C-like enzymes are oligomeric. The yeast bleomycin hydrolase (Chapter 215) is probably representative, and is a homohexamer, with the active sites arranged on the inner face of the central channel, in an arrangement reminiscent of that in the proteasome. Also, unlike papain and cathepsin B, these aminopeptidases do not contain disulfide bonds or propeptides. As is discussed in Chapter 215, the mature bleomycin hydrolase subunit consists of three domains, the peptidase domain, an oligomerization (or 'hook') domain, and a helical domain (see Figure 186.1). Half of the hook domain is N-terminal to the catalytic domain, but the other half is an insert preceding the catalytic His, relative to papain. The helical domain corresponds to two inserts in the catalytic domain with respect to papain. Figure 186.7 shows the tertiary structure of the yeast bleomycin hydrolase subunit.

Inserts relative to papain occur within the catalytic domain in other members of the family. In cathepsin B, the 'occluding loop' that carries the histidine residues important for peptidyl-dipeptidase activity is inserted between the catalytic Cys and His residues. This loop is not present in enzymes from *Giardia* (Ward *et al.*, 1997) that are the most divergent of the cathepsin B-like proteins and probably represent the primitive state. Some endopeptidases from *Dictyostelium*

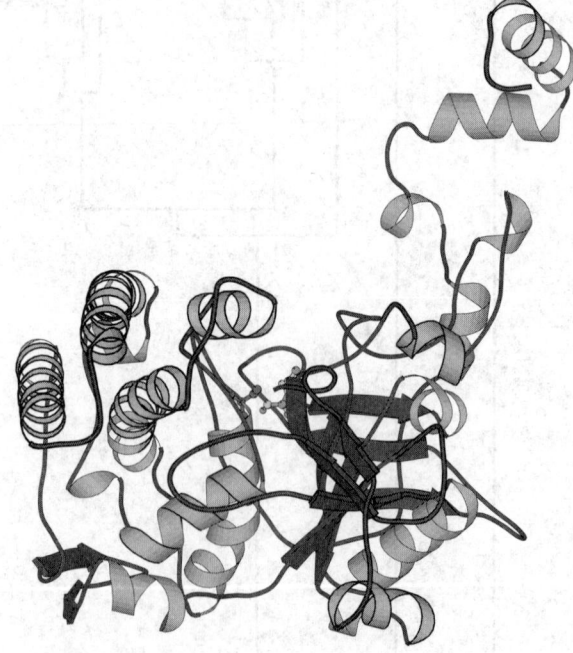

*Figure 186.7* Richardson diagram of a subunit of yeast bleomycin hydrolase. The image was prepared from the Brookhaven Protein Data Bank entry (1GCB) as described in the Introduction (p. xxv). Catalytic residues are shown in ball-and-stick representation: Cys25 (engineered to be serine), His159 and Asn175 (numbering as in Alignment 186.1 on CD-ROM).

possess glycine- and serine-rich inserts between His159 and Asn175 (papain numbering).

The specificity subsite that is dominant in most peptidases of family C1 is S2, which commonly displays a preference for occupation by a bulky hydrophobic side chain, and not a charged one. Unusually, the S2 subsite of cathepsin B readily accepts Arg, however, so that Z-Arg-Arg↓NHMec is a good and highly selective substrate for the enzyme. This distinctive specificity of cathepsin B can be explained in terms of the residue lying at the bottom of the S2 binding pocket, which in papain is Ser205, but in cathepsin B is Glu. Glu205 also occurs in cysteine endopeptidases from *Brassica* (SW: P25251), tomato (SW: P20721), barley (SW: P25249), *Plasmodium* (SW: P25805), the baculovirus of *Autographa* (SW: P25783), lobster (SW: P25784) and oryzain α (SW: P25776). Ser205 is replaced by Asp in endopeptidases from *Entamoeba* that show a preference for Arg in S2, and this replacement also occurs in stem bromelain and a lobster digestive endopeptidase (SW: P13277).

There are a number of proteins homologous to papain that lack peptidase activity because catalytic residues have been replaced. These include an oil-body-associated protein from soyabean (SW: P22895), in which the catalytic Cys is replaced by Gly, as well as a surface protective protein from *Plasmodium* (SW: P08676) and a protein from *Schistosoma japonicum* (EMBL: X70969) in which Cys25 has been replaced by Ser.

***Family C2*** contains the calpains. These are intracellular, multidomain proteins that may be involved in turnover of cytoskeletal proteins such as spectrin. Typically, a calpain is a heterodimer of a heavy chain, and a light chain.

The heavy chain is a mosaic protein, bearing not only the peptidase domain (domain II), but also a C-terminal domain (domain IV) with four calcium-binding EF-hand structures. The calcium-binding domain is homologous to similar domains in sorcin and the calpain light chain. The functions of heavy chain domains I and III are unknown, and the sequences show no relationship to those of other proteins.

In most mammalian cells there are two calpains with different calcium requirements for activity: μ-calpain (Chapter 218) and m-calpain (Chapter 219). Muscle cells possess a third calpain, calpain p94 (Chapter 220), that may be involved in the proteolysis of connectin and titin, and is lacking in limb girdle muscular dystrophy type 2A. Calpain p94 has a 48 residue insert in the catalytic domain between Cys25 and His159.

A *Drosophila* homolog of calpain possesses an insert in the C-terminal domain, and there is evidence that differential splicing can give rise to a form lacking the calcium-binding domain (Theopold *et al*., 1995). Another calpain homolog exists in *Drosophila* and is involved in the development of the small optic lobe. This putative endopeptidase is the product of the *sol* gene, and the protein is predicted to be multidomain, including six zinc fingers as well as a peptidase domain, but no calcium-binding domain (Delaney *et al*.,1991; Figure 186.1). Again, an alternatively spliced form exists, this time lacking the peptidase domain.

The calpain family is not restricted to animals: there is a sequence from *Aspergillus* (but none in the *Saccharomyces* genome), and an EST fragment from *Trypanosoma*, so the family appears to have developed late in protozoan evolution. The evolutionary tree (Figure 186.8) shows a

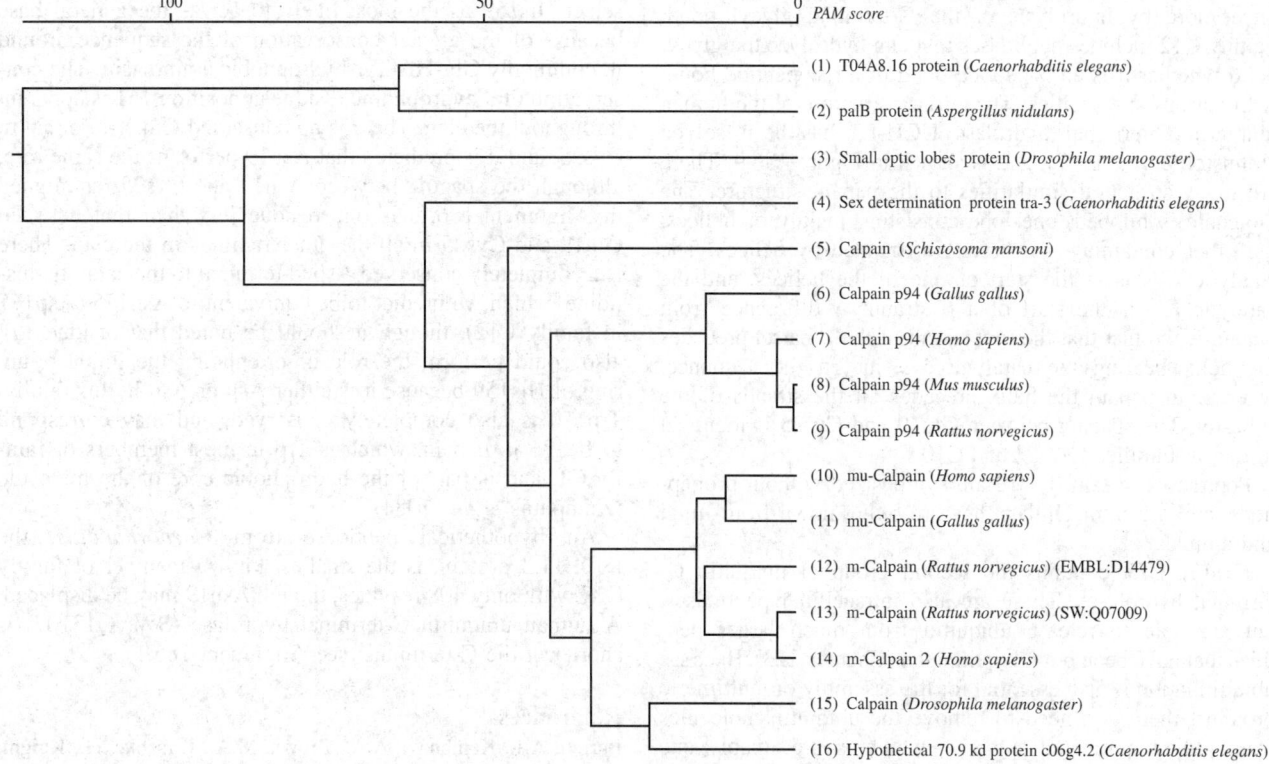

*Figure 186.8*  Evolutionary tree for family C2. The tree was prepared as described in the Introduction (p. xxv).

close relationship between the *Aspergillus* palB protein and a hypothetical protein from *Caenorhabditis*. Both of these are unusual among peptidases in clan CA in that Cys25 is followed by Ser rather than a hydrophobic residue (see Alignment 186.1 on CD-ROM). Both the *Aspergillus* and *Caenorhabditis* proteins are multidomain, but show only limited sequence similarity to domains I and III of human $\mu$-calpain. Neither protein possesses the calcium-binding domain IV. As can be seen from the Databanks table, there are human and mouse ESTs that show similarities to the *Aspergillus* and *Caenorhabditis* proteins.

**Family C10** contains streptopain (Chapter 221) and a few similar enzymes, all of which are from gram-positive bacteria. Streptopain is inactivated by E-64 much more slowly than papain, but has a similar specificity, with a preference for a hydrophobic residue at P2 (Kortt & Liu, 1973; Barrett *et al.*, 1982). In some members of the family, Asn175 is replaced by Asp. The prtT proteinase from *Porphyromonas* is a mosaic protein with a C-terminal hemagglutinin domain unrelated to those in gingipains R and K in family C25 but 98% identical to the C-terminal domain of hemin-regulated protein from *Porphyromonas* (EMBL: U54787).

**Family C12** is one of the two families of ubiquitin C-terminal hydrolases. Ubiquitin is a 76 amino acid protein that is attached to other proteins as a signal for degradation that is usually mediated by the proteasome. Ubiquitin is attached through the carboxyl group of its C-terminal glycine residue either to the N-terminus of another polypeptide or to the $\epsilon$-amino group of a Lys residue, which forms an isopeptide bond (Chapter 222). Ubiquitin may also be attached to another molecule of ubiquitin, and proteins targeted for degradation may be polyubiquitinated. In order for ubiquitin to be recycled there is a need for peptidases to liberate the free ubiquitin once more by hydrolysis of the C-terminal glycyl bond. Family C12 includes peptidases that can hydrolyze the glycyl bond whether it is an $\alpha$-peptide bond or an isopeptide bond, with various specificities. The tertiary structure of the human ubiquitin C-terminal hydrolase UCH-L3 has been solved (Johnston *et al.*, 1997), and is shown in Figure 186.9. There are many structural similarities to the papain structure. The molecule is bilobed, one lobe consisting mainly of helices, the other containing a $\beta$ barrel surrounded by helices. The catalytic Cys is at the start of one of the helices, and the catalytic His is the start of a $\beta$ strand. A difference from papain is the fact that the first strand of the $\beta$ barrel precedes the helix bearing the catalytic cysteine in the sequence, whereas in papain the helix precedes all the strands of the $\beta$ barrel. The spacing between Gln19 and Cys25 is identical to that in families C1, C2 and C10.

Peptidases in family C12 are synthesized without propeptides, and are intracellular; they are only known from fungi and animals.

**Family C19** contains the second group of ubiquitin C-terminal hydrolases. These are also intracellular peptidases, but are able to release ubiquitin from much larger peptides that have been polyubiquitinated (Chapter 223). Because ubiquitination is also essential for the assembly of multimeric proteins, there is a need to remove the ubiquitin molecules from polyubiquitinated subunits, which is presumably one of the functions of peptidases from family C19. Peptidases in family C19 have much more complicated structures than

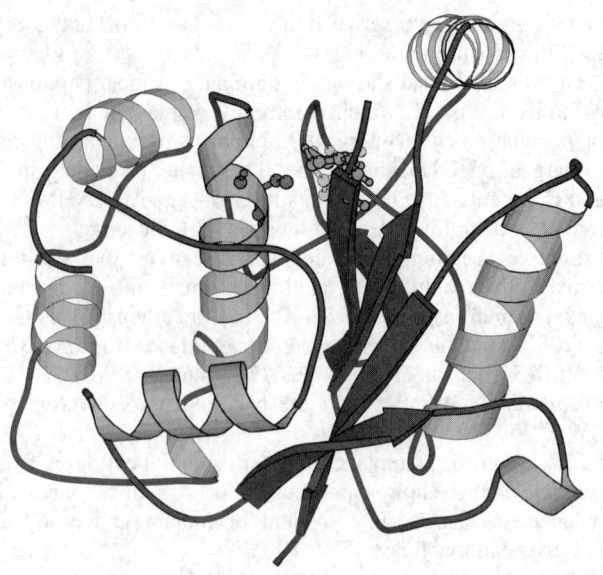

*Figure 186.9* Richardson diagram of human ubiquitin C-terminal hydrolase UCH-L3. Coordinates were kindly provided by Dr Keith Wilkinson. The image was prepared as described in the Introduction (p. xxv). Catalytic residues are shown in ball-and-stick representation: Cys25, His159 and Asp175 (numbering as in Alignment 186.1 on CD-ROM).

those in C12, and many are multidomain proteins. There is also a much greater variety, with *Saccharomyces* possessing at least 19 homologs. Only the catalytic Cys has been identified with certainty, and as can be seen from Alignment 186.1, there are two conserved histidines either of which may complete a catalytic dyad: His153 and His159. We would select His159 as the most likely to act as the general base because of the greater conservation of the sequence around it, commonly Gly-**His**-Tyr-hydrophobic amino acid. The conservation of a hydrophobic residue at position 161 is apparent throughout the clan. There is no conserved Gln N-terminal to Cys25, and it is predicted that Asn19 performs the same role, although the spacing between Asn19 and Cys25 (as aligned in Alignment 186.1) is one residue less than that between Gln19 and Cys25 in all the other families in the clan. There is a completely conserved Asp C-terminal to the catalytic histidine, which we predict to be equivalent to Asn175 (Asp175 in family C12), though it should be noted that residue 174 also could perform the role of orientating the imidazolium ring of His159 because it is either Asp or Asn in this family. Trp170 is also completely conserved, and may correspond to Ile171 in papain, which is Trp in most members of family C1, and is part of the hydrophobic core of the molecule (Kamphuis *et al.*, 1984).

An hypothetical peptidase from *Caenorhabditis*, the R10E11.3 protein, is the smallest known member of family C19, with only 408 residues, though Asn19 may be displaced. A human ubiquitin C-terminal hydrolase (SW: Q13107) is shorter at the C-terminus (see Alignment 186.1).

**References**

Barrett, A.J., Kembhavi, A.A., Brown, M.A., Kirschke, H., Knight, C.G., Tamai, M. & Hanada, K. (1982) L-*trans*-Epoxysuccinyl-leucylamido(4-guanidino)butane (E-64) and its analogues as inhibitors of cysteine proteinases including cathepsins B, H and L. *Biochem. J.* **201**, 189–198.

Delaney, S.J., Hayward, D.C., Barleben, F., Fischbach, K.F. & Miklos, G.L.G. (1991) Molecular cloning and analysis of small optic lobes, a structural brain gene of *Drosophila melanogaster*. *Proc. Natl Acad. Sci. USA* **88**, 7214–7218.

Delaria, K., Fiorentino, L., Wallace, L., Tamburini, P., Brownell, E. & Muller, D. (1994) Inhibition of cathepsin L-like cysteine proteases by cytotoxic T-lymphocyte antigen-2β. *J. Biol. Chem.* **269**, 25172–25177.

Johnston, S.C., Larsen, C.N., Cook, W.J., Wilkinson, K.D. & Hill, C.P. (1997) Crystal structure of a deubiquitinating enzyme (human UCH-L3) at 1.8 Å resolution. *EMBO J.* **16**, 3787–3796.

Kamphuis, I.G., Kalk, K.H., Swarte, M.B.A. & Drenth, J. (1984)

Structure of papain refined at 1.65 Å resolution. *J. Mol. Biol.* **179**, 233–256.

Kortt, A.A. & Liu, T.-Y. (1973) On the mechanism of action of streptococcal proteinase. II. Comparison of the kinetics of proteinase- and papain-catalyzed hydrolysis of *N*-acylamino acid esters. *Biochemistry* **12**, 328–337.

Theopold, U., Pintér, M., Daffre, S., Tryselius, Y., Friedrich, P., Nässel, D.R. & Hultmark, D. (1995) *CalpA*, a *Drosophila* calpain homolog specifically expressed in a small set of nerve, midgut, and blood cells. *Mol. Cell. Biol.* **15**, 824–834.

Ward, W., Alvarado, L., Rawlings, N.D., Engel, J., Franklin, C. & McKerrow, J.H. (1997) A primitive enzyme for a primitive cell: a protease required for excystation of *Giardia*. *Cell* **89**, 437–444.

# 187. Papain

## Databanks

*Peptidase classification: clan CA, family C1, MEROPS ID: C01.001*
*NC-IUBMB enzyme classification: EC 3.4.22.2*
*Chemical Abstracts Service registry number: 9001-73-4*
*Databank codes:*

| Species | SwissProt | PIR | EMBL (cDNA) | EMBL (genomic) |
|---|---|---|---|---|
| *Carica papaya* | P00784 | A00974 A26466 A37903 | M15203 | – |

Brookhaven Protein Data Bank three-dimensional structures:

| Species | ID | Resolution | Notes |
|---|---|---|---|
| *Carica papaya* | 1PAD | 2.8 | Cys25Ac-Ala-Ala-Phe-methylenyl Ala derivative |
| | 1PE6 | 2.1 | complex with E-64C |
| | 1PIP | 1.7 | complex with Suc-Gln-Vla-Val-Ala-Ala-pNA |
| | 1POP | 2.1 | complex with leupeptin |
| | 1PPD | 2 | hydroxyethylthiopapain |
| | 1PPN | 1.6 | Cys25 with bound atom |
| | 1PPP | 1.9 | complex with E-64C |
| | 1STF | 2.37 | complex with cystatin B Cys18Ser |
| | 2PAD | 2.8 | cysteinyl derivative of Cys25 |
| | 4PAD | 2.8 | Tos-methylenyl Lys derivative of Cys25 |
| | 5PAD | 2.8 | benzyloxycarbonyl-Gly-Phe-methylenyl Gly derivative |
| | 6PAD | 2.8 | benzyloxycarbonyl-Gly-Phe-methylenyl Gly derivative |
| | 9PAP | 1.65 | Cys25 oxidized |

## Name and History

*Papain* is the English translation of 'papaïne', the name given by Wurtz & Bouchut (1879) to a proteolytically active constituent in the latex of the tropical papaya fruit (*Carica papaya*). The crude dried latex contains a mixture of at least four cysteine proteinases and other enzymes (Brocklehurst *et al*., 1981) and initial preparations of 'papain' were contaminated by other enzymes. Improvements in purification procedures introduced by Kimmel & Smith (1954) (Brocklehurst *et al*., 1981) have allowed the isolation of pure papain. It must be noted that the term 'papain' is also applied to the commercial crude dried latex used in industry. Early work on papain, which is summarized in a number of reviews (e.g. Smith & Kimmel, 1960; Brocklehurst *et al*., 1981, 1987; Baker & Drenth, 1987), is responsible for major advances not only in the field of cysteine proteinases but also in enzymology generally. Papain is the most widely studied member of the cysteine proteinase class of enzymes. It was also the first cysteine proteinase to have its three-dimensional structure determined (Drenth *et al*., 1968). Consequently, this enzyme is considered to be the archetype of cysteine proteinases and accordingly, family C1 is referred to as the papain family of cysteine proteinases.

## Activity and Specificity

Papain exhibits endopeptidase, amidase and esterase activities (Glazer & Smith, 1971). The concentration of active papain can be obtained by titration with 5,5'-dithiobis(2-nitrobenzoic acid) as described by Ellman (1959), or more accurately through active-site titration with an irreversible inhibitor such as E-64 (Barrett *et al*., 1981). Schechter & Berger (1967) have shown that the active site of papain can be considered to consist of seven subsites (S1–S4 and S1'–S3'), each able to accommodate one amino acid residue of a substrate (P1–P4 and P1'–P3') located on the N-terminal and C-terminal sides of the site of cleavage, respectively. Preference is for substrates containing a bulky nonpolar side chain (e.g. Phe) at the P2 position (Berger & Schechter, 1970). The S1 subsite of papain is not as selective as the S2 subsite, but there is some preference for Arg and Lys at this position (Kimmel & Smith, 1957) while Val is not accepted (Baker & Drenth, 1987). Very little is known about the S' subsite specificities, which seem to be relatively broad (Ménard *et al*., 1993). Over all, papain can be considered to possess a fairly broad specificity. This is a consequence of the lack of any clearly defined residue selectivity other than S2 subsite preference and of the existence of an extended binding site for the substrate, involving interactions in many subsites of the enzyme. The requirements for interaction in a given subsite can, in the case of a polypeptide substrate, be over-ruled by interactions of the polypeptide chain with other subsites of the enzyme. This gives rise to numerous possible cleavage patterns. For this reason, most of the specificity requirements of cysteine proteinases have been defined by the use of small synthetic substrates instead of large protein substrates, and cleavage sites in proteins are difficult to predict.

## Structural Chemistry

Papain is a single-chain nonglycosylated polypeptide of 212 amino acids (23 429 Da) containing three disulfide bonds. The initial 2.8 Å resolution structure (Drenth *et al*., 1968) has been refined to 1.65 Å (Kamphuis *et al*., 1984). A number of structures are also available for papain complexes with ligands and inhibitors. The polypeptide chain is folded to form a globular protein with two interacting domains delimiting a cleft at the surface of the enzyme where substrates can bind. The protein is relatively basic, with a pI of 8.75. It is generally accepted that the active form of papain consists of a thiolate-imidazolium ion pair formed by the active-site residues Cys25 and His159 (Storer & Ménard, 1994).

## Preparation

The papaya latex from which papain can be purified is obtained by tapping the green unripe fruit of the plant and collecting the resulting exudate. The fresh latex is a milk-like liquid which rapidly coagulates. The latex is dried prior to further handling (Jones & Mercier, 1974). The purification of papain from spray-dried latex has been optimized by Baines & Brocklehurst (1979) to yield enzyme that is 80% active. The residual inactive (irreversibly oxidized) papain can be removed by affinity chromatography (Funk *et al*., 1979; Brocklehurst *et al*., 1985). Papain is most stable (for several months) when stored at 4°C in a reversibly inactivated form, e.g. mercuripapain (prepared by the addition of one equivalent of mercuric chloride) or 2-pyridyl papain disulfide (prepared using the reagent 2,2'-dipyridyl disulfide), both of which can be reactivated with thiol reagents such as 2-ME or DTT. When stored in its native state papain can lose as much as 50% of its activity within one week due to oxidation of the active-site thiol group. This oxidation can be only partially reversed by thiol reagents. The recombinant enzyme has been expressed as propapain using both the baculovirus/insect cell system (Vernet *et al*., 1990) and *Escherichia coli* (Taylor *et al*., 1992). In the former, the enzyme was secreted as a fully folded protein and in the latter it was produced intracellularly as an insoluble form requiring refolding. Final yields from both sources upon purification and processing to the mature enzyme were in the low milligram per liter range.

## Biological Aspects

Very little is known about the biology of papain. The enzyme is produced as an inactive precursor (Cohen *et al*., 1986; Vernet *et al*., 1990) and is located in the plant within the latex of the laticifer system. There is evidence that several papain-like enzymes appear in the laticifer system when the plants are a few months old, and that these enzymes are not detectable in other cell types (Smith *et al*., 1986; Yamamoto & Tabata, 1989). This suggests that differentiation into laticifers is required for papain production in papaya. The presence of the enzyme in the latex of the plant is probably the result of the natural breakdown of the laticifer cells to form the latex channels as occurs during maturation of the fruit (Roth & Clausnitzer, 1972). It is also known that the major proteinaceous components of the latex are cysteine proteinases of the papain family (Brocklehurst *et al.*, 1981). Based on this limited information it is not possible to clearly define the biological role of papain, but given the location and large quantity of the enzyme and its promiscuous specificity it is possible to speculate that it plays a protective role, guarding the plant against attack by pests such as insects and fungi.

## Distinguishing Features

The enzyme is stable and active under a wide range of conditions: from pH 4 to 10 and at temperatures up to 80°C (Glazer & Smith, 1971). It also retains activity in 8 M urea. Amongst the papaya cysteine proteinases, papain is the least basic protein and elutes last from an electrofocusing fractionation.

## Further Reading

Review articles include those of Baker & Drenth (1987), Brocklehurst *et al.* (1987) and Storer & Ménard (1994, 1996).

## References

Baines, B.S. & Brocklehurst, K. (1979) A necessary modification to the preparation of papain from any high-quality latex of *Carica papaya* and evidence for the structural integrity of the enzyme produced by traditional methods. *Biochem. J.* **177**, 541–548.

Baker, E.N. & Drenth, J. (1987) The thiol proteases: structure and mechanism. In: *Biological Macromolecules and Assemblies*, vol. 3: *Active Sites of Enzymes* (Jurnak, F.A. & McPherson, A., eds). New York: John Wiley, pp. 313–368.

Barrett, A.J., Kembhavi, A.A. & Hanada, K. (1981) E-64 [L-*trans*-epoxysuccinyl-leucyl-amido(4-guanidino)butane] and related epoxides as inhibitors of cysteine proteinases. *Acta Biol. Med. Ger.* **40**, 1513–1517.

Berger, A. & Schechter, I. (1970) Mapping the active site of papain with the aid of peptide substrates and inhibitors. *Philos. Trans. R. Soc. Lond. [Biol.]* **257**, 249–264.

Brocklehurst, K., Baines, B.S. & Kierstan, M.P.J. (1981) Papain and other constituents of carica papaya L. *Top. Enzyme Fermentation Biotechnol.* **5**, 262–335.

Brocklehurst, K., Carlsson, J. & Kierstan, M.P.J. (1985) Covalent chromatography in biochemistry and biotechnology. *Top. Enzyme Fermentation Biotechnol.* **10**, 146–188.

Brocklehurst, K., Willenbrock, F. & Salih, E. (1987) Cysteine proteinases. In: *Hydrolytic Enzymes* (Neuberger, A. & Brocklehurst, K., eds). Amsterdam: Elsevier Biomedical Press, pp. 39–158.

Cohen, L.W., Coghlan, V.M. & Dihel, L.C. (1986) Cloning and sequencing of papain-encoding cDNA. *Gene* **48**, 219–227.

Drenth, J., Jansonius, J.N., Koekoek, R., Swen, H.M. & Wolthers, B.G. (1968) Structure of papain. *Nature* **218**, 929–932.

Ellman, G.L. (1959) Tissue sulfhydryl groups. *Arch. Biochem. Biophys.* **82**, 70–77.

Funk, M.O., Nakagawa, Y., Skochdopole, J. & Kaiser, E.T. (1979) Affinity chromatographic purification of papain. *Int. J. Pept. Protein Res.* **13**, 296–303.

Glazer, A.N. & Smith, E.L. (1971) Papain and other plant sulfhydryl proteolytic enzymes. In: *The Enzymes*, vol. 3, 3rd edn (Boyer, P.D., ed.). New York: Academic Press, pp. 501–546.

Jones, J.G. & Mercier, P.L. (1974) Refined papain. *Process Biochem.* 21–24.

Kamphuis, I.G., Kalk, K.H., Swarte, M.B.A. & Drenth, J. (1984) Structure of papain refined at 1.65 Å resolution. *J. Mol. Biol.* **179**, 233–256.

Kimmel, J.R. & Smith, E.L. (1954) Crystalline papain. I. Preparation, specificity, and activation. *J. Biol. Chem.* **207**, 515–531.

Kimmel, J.R. & Smith, E.L. (1957) The properties of papain. *Adv. Enzymol.* **19**, 267–334.

Ménard, R., Carmona, E., Plouffe, C., Brömme, D., Konishi, Y., Lefebvre, J & Storer, A.C. (1993) The specificity of the S1' subsite of cysteine proteases. *FEBS Lett.* **328**, 107–110.

Roth, I. & Clausnitzer, I. (1972) Desarrollo y anatomia del fruto y de la semilla de *Carica papaya* L. (Lechosa). *Acta Bot. Venez.* **7**, 187–206.

Schechter, I. & Berger, A. (1967) On the size of the active site in proteases. I. Papain. *Biochem. Biophys. Res. Commun.* **27**, 157–162.

Smith, E.L. & Kimmel, J.R. (1960) Papain (with a section on ficin). *The Enzymes*, vol. 4, 2nd edn (Boyer, P.D., Lardy, H. & Myrback, K., eds). New York: Academic Press, pp. 133–173.

Smith, H., McKee, R.A. & Praekelt, U. (1986) The molecular biology of plant proteinases. *Spec. Publ. – R. Soc. Chem.* **63**, 196–207.

Storer, A.C. & Ménard, R. (1994) Catalytic mechanism in the papain family of cysteine peptidases. *Methods Enzymol.* **244**, 486–500.

Storer, A.C. & Ménard, R. (1996) Recent insights into cysteine protease specificity: lessons for drug design. *Perspect. Drug Discovery Design* **6**, 33–46.

Taylor, M.A.J., Pratt, K.A., Revell, D.F., Baker, K.C., Sumner, I.G. & Goodenough, P.W. (1992) Active papain renatured and processed from insoluble recombinant propapain expressed in *Escherichia coli*. *Protein Eng.* **5**, 455–459.

Vernet, T., Tessier, D.C., Richardson, C., Laliberte, F., Khouri, H.E., Bell, A.W., Storer, A.C. & Thomas, D.Y. (1990) Secretion of functional papain precursor from insect cells. *J. Biol. Chem.* **265**, 16661–16666.

Wurtz, A. & Bouchut, E. (1879) Sur le ferment digestif du Carica papaya. *Compt. Rend.* **89**, 425–430.

Yamamoto, H. & Tabata, M. (1989) Correlation of papain-like enzyme production with laticifer formation in somatic embryos of papaya. *Plant Cell Rep.* **8**, 251–254.

*Robert Ménard*
*Enzyme Engineering Group, Biotechnology*
*Research Institute,*
*National Research Council of Canada,*
*6100 Royalmount Avenue,*
*Montreal, Quebec H4P 2R2, Canada*
*Email: robert.menard@nrc.ca*

*Andrew C. Storer*
*Enzyme Engineering Group, Biotechnology*
*Research Institute,*
*National Research Council of Canada,*
*6100 Royalmount Avenue,*
*Montreal, Quebec H4P 2R2, Canada*
*Email: andrew.storer@nrc.ca*

# 188. *Chymopapain*

## Databanks

*Peptidase classification: clan CA, family C1, MEROPS ID: C01.002*
*NC-IUBMB enzyme classification: EC 3.4.22.6*
*Chemical Abstracts Service registry number: 9001-09-6*
*Databank codes:*

| Species | SwissProt | PIR | EMBL (cDNA) | EMBL (genomic) |
|---------|-----------|-----|-------------|----------------|
| *Carica papaya* | P14080 | S04222<br>S08285 | X97789 | – |

Brookhaven Protein Data Bank three-dimensional structures:

| Species | ID | Resolution | Notes |
|---------|-----|-----------|-------|
| *Carica papaya* | 1YAL | 1.7 | |

## Name and History

The original description of **chymopapain** was given by Jansen & Balls (1941), who obtained the enzyme in crystalline form from fresh papaya latex. They made use of the remarkable stability of the enzyme at acid pH and its high solubility in salt solutions. The name derives from the same root as that of chymotrypsin, the stem 'chyme' originally referring to a food (milk) in the process of being broken down by gastric secretion in the stomach. Thus, 'chymo-' was used for an enzyme having a higher ratio of milk-clotting to hemoglobin-digesting activity than the already familiar enzyme, trypsin in the case of chymotrypsin, or papain for chymopapain.

The first demonstration of the charge heterogeneity of chymopapain (see below) was provided by Kunimitsu & Yasunobu (1967) who described the separation by cation-exchange chromatography of **chymopapain B** from the earlier-eluting **chymopapain A**. The description of chymopapains A and B as early- and late-eluting forms on cation-exchange columns was confused by the introduction of another definition based on reactivity of the two forms with 2,2'-dipyridyldisulfide (Baines & Brocklehurst, 1982).

## Activity and Specificity

As with most endopeptidases in family C1, chymopapain accepts hydrophobic residues in S2 and S3. However, many other residues may also be accommodated. A large range of side chains are accepted by other subsites, only proline being excluded from S1', so that the specificity of chymopapain can be described as truly catholic. For instance, out of a total of 140 peptide bonds that comprise the $\alpha$-globin chain of manatee hemoglobin, 49 were cleaved by chymopapain. Of these, 29 were also hydrolyzed by caricain (Chapter 189) (Jacquet *et al.*, 1989a). The similarity of the specificity of chymopapain to those of caricain and papain (Chapter 187) is further illustrated by the cleavage patterns of the oxidized B

chain of insulin; all the bonds cleaved by chymopapain were also cleaved by papain, and chymopapain hydrolyzed six of the seven bonds cleaved by caricain (Johansen & Ottesen, 1968).

The pH optimum is broad, centered around 7, and largely controlled by groups with $pK_a$ values of 4.2 and 8.5. Reducing conditions are essential for enzyme activity, and are best achieved by the inclusion of low millimolar concentrations of cysteine in assay buffers. Bz-Arg+NHPhNO$_2$ is a convenient and inexpensive small-molecule spectrophotometric substrate for assay (Buttle & Barrett, 1984), $k_{cat}/K_m$ being $100\,M^{-1}\,s^{-1}$ at pH 6.8 and 40°C. More sensitive assays can be made by use of substrates with fluorescent leaving groups, such as Z-Phe-Arg+NHMec ($k_{cat}/K_m = 1.16 \times 10^6\,M^{-1}\,s^{-1}$, pH 6.8, 40°C). Although this value of $k_{cat}/K_m$ is very similar to that of papain for Z-Phe-Arg-NHMec, the individual values of both $K_m$ and $k_{cat}$ are much lower for chymopapain (Zucker *et al.*, 1985).

Like most members of family C1, chymopapain is potently inhibited by cystatins (Buttle *et al.*, 1986; Björk & Yli-nenjärvi, 1990), and inactivated by E-64. Both inhibitors interact stoichiometrically, allowing active-site titrations of the enzyme (Zucker *et al.*, 1985). Titrations of chymopapain have also been made with iodoacetate (Buttle *et al.*, 1990) and 2,2'-dipyridyldisulfide (Brocklehurst *et al.*, 1987).

## Structural Chemistry

The amino acid sequence of chymopapain deduced from the cDNA demonstrates that it is synthesized as a preproenzyme. The primary structure of the mature form of the enzyme has been elucidated (Jacquet *et al.*, 1989b) and is identical to the predicted sequence from the mRNA. The enzyme comprises a single nonglycosylated polypeptide chain of 218 amino acids, with an $M_r$ of 23 650. The sequence is 58% identical to that of papain, 65% to caricain and 70% to glycyl endopeptidase (Ritonja *et al.*, 1989). All seven cysteines are conserved amongst the papaya endopeptidases, and chymopapain retains the normal pattern of disulfide bonding. However, chymopapain has a novel eighth cysteine, Cys117 (papain numbering), accounting for the thiol content of >1 mol of thiol per mol protein, which has been observed by many investigators.

The $A_{1\%,280}$ is 18.3 (Robinson, 1975) giving a molar extinction coefficient of $4.26 \times 10^4 \, M^{-1} \, cm^{-1}$. Like the other papaya cysteine endopeptidases, chymopapain is a basic protein, the pI being in the region of 10.2–10.6 (Baines & Brocklehurst, 1982). For reasons that are still not entirely clear, chymopapain demonstrates charge heterogeneity, although there are no detected polymorphisms in the primary structure. As with the cysteine endopeptidases from pineapple stem (Napper *et al.*, 1994), some of the heterogeneity may be due to variation in the oxidation state of the active-site cysteine, exacerbated in the case of chymopapain by the presence of a second thiol group. The charge heterogeneity, and the fact that chymopapain preparations may be contaminated to varying degrees with glycyl endopeptidase (Chapter 190) has been at the root of the confusion over the number of different forms of chymopapain (discussed above). To date, however, there is no compelling evidence for the existence of more than one gene product.

## Preparation

The charge heterogeneity displayed by chymopapain, its similar physicochemical properties to the other papaya cysteine endopeptidases, and in many cases the absence of reliable criteria of purity, have conspired to make the purification of chymopapain difficult to achieve and to assess.

The discovery that chymopapain was more stable than glycyl endopeptidase at acidic pH now allows the preparation of chymopapain essentially free from this contaminant. The development of active site-directed affinity ligands that show some discrimination between the individual cysteine endopeptidases from papaya has also aided the purification (Buttle, 1994). The recommended method for the preparation of chymopapain consists of incubation of the soluble components of papaya latex at pH 1.5, centrifugation to remove denatured protein, cation-exchange chromatography, and application of the chymopapain-containing pool to a column of immobilized Ala-Phe-aldehyde semicarbazone. Fully activatable chymopapain is eluted from the affinity column by the addition of HgCl$_2$. Specific antibodies against the individual papaya cysteine endopeptidases can be used to determine the degree of contamination of purified chymopapain by the other endopeptidases. Such an analysis of chymopapain prepared as described above has shown levels of $\leq 0.02\%$ contamination by the other endopeptidases (Buttle *et al.*, 1990).

## Biological Aspects and Practical Uses

The biological functions of the papaya latex cysteine endopeptidases are not understood, but possibilities are discussed in Chapters 189 and 190. Whatever these functions may be, it is unlikely that they take place in the latex itself, where chymopapain represents 26–30% of total protein (Buttle *et al.*, 1990; Dando *et al.*, 1995).

Crude preparations of papaya latex, in which chymopapain is the most abundant proteolytic component, are used in a number of industrial applications, such as meat tenderization and food processing, and the dehairing of hide in the leather industry (Caygill, 1979).

For many years chymopapain has been used in chemonucleolysis; the treatment of herniated or prolapsed lumbar intervertebral disks. The enzyme is injected directly into the center of the affected disk, where it digests the proteoglycan component (Buttle *et al.*, 1986; Dekeyser *et al.*, 1995). The fragmented proteoglycan molecules diffuse from the disk, causing a reduction in hydrostatic pressure on the nerve root and thus an easing of pain and disability. The most common adverse reactions associated with chemonucleolysis are due to an allergic response (Dando *et al.*, 1995). Because of their allergenicity, preparations containing the papaya endopeptidases, particularly powders such as dried papaya latex, should be handled with care.

The licensing of chymopapain as a drug made it potentially attractive for other medical uses. One application has been found in the preparation of bone marrow cells for grafting. Cells are chosen by binding to specific antibodies on magnetic beads. The cells are recovered with a magnet, and chymopapain is used to digest the antibody and thus return the cells to a suspension ready for grafting (Fruehauf *et al.*, 1994).

## Distinguishing Features

Chymopapain shares many of its properties with two other cysteine endopeptidases from the same source, papain and caricain (Chapters 187 and 189). Differences in substrate specificity and inhibition profiles have been detected (Jacquet *et al.*, 1989a; Zucker *et al.*, 1985; Johansen & Ottesen, 1968; Björk & Ylinenjärvi, 1990) but are not remarkable. The enzymes react with 2,2'-dipyridyldisulfide at somewhat different rates (Baines & Brocklehurst, 1982) and the presence of a second thiol group in chymopapain can be detected. Antisera that discriminate completely between the different papaya cysteine endopeptidases can be prepared (Buttle *et al.*, 1990), and probably represent the best method for distinguishing between the papaya cysteine endopeptidases, but are not commercially available.

## References

Baines, B.S. & Brocklehurst, K. (1982) Isolation and characterization of the four major cysteine-proteinase components of the latex of *Carica papaya* L. Reactivity characteristics towards 2,2'-dipyridyldisulfide of the thiol groups of papain, chymopapains A and B, and papaya peptidase A. *J. Protein Chem.* **1**, 119–139.

Björk, I. & Ylinenjärvi, K. (1990) Interaction between chicken cystatin and the cysteine proteinases actinidin, chymopapain A, and ficin. *Biochemistry* **29**, 1770–1776.

Brocklehurst, K., Willenbrock, F. & Salih, E. (1987) Cysteine proteinases. In: *Hydrolytic Enzymes* (Neuberger, A. & Brocklehurst, K., eds). Amsterdam: Elsevier Science Publishers, pp. 39–158.

Buttle, D.J. (1994) Affinity chromatography of cysteine peptidases. *Methods Enzymol.* **244**, 639–648.

Buttle, D.J. & Barrett, A.J. (1984) Chymopapain. Chromatographic purification and immunological characterization. *Biochem. J.* **223**, 81–88.

Buttle, D.J., Abrahamson, M. & Barrett, A.J. (1986) The biochemistry of the action of chymopapain in relief of sciatica. *Spine* 11, 688–694.

Buttle, D.J., Dando, P.M., Coe, P.F., Sharp, S.L., Shepherd, S.T. & Barrett, A.J. (1990) The preparation of fully active chymopapain free of contaminating proteinases. *Biol. Chem. Hoppe-Seyler* 371, 1083–1088.

Caygill, J.C. (1979) Sulphydryl plant proteases. *Enzyme Microb. Technol.* 1, 233–242.

Dando, P.M., Sharp, S.L., Buttle, D.J. & Barrett, A.J. (1995) Immunoglobulin E antibodies to papaya proteinases and their relevance to chemonucleolysis. *Spine* 20, 981–985.

Dekeyser, P.M., Buttle, D.J., Devreese, B., van Beeumen, J., Demeester, J. & Lauwers, A. (1995) Kinetic constants for the hydrolysis of aggrecan by the papaya proteinases and their relevance for chemonucleolysis. *Arch. Biochem. Biophys.* 320, 375–379.

Fruehauf, S., Haas, R., Zeller, W.J. & Hunstein, W. (1994) CD34 selection for purging in multiple myeloma and analysis of CD34[+] B cell precursors. *Stem Cells* 12, 95–102.

Jacquet, A., Kleinschmidt, T., Dubois, T., Schnek, A.G., Looze, Y. & Braunitzer, G. (1989a) The thiol proteinases from the latex of *Carica papaya* L. IV. Proteolytic specificities of chymopapain and papaya proteinase Ω determined by digestion of α-globin chains. *Biol. Chem. Hoppe-Seyler* 370, 819–829.

Jacquet, A., Kleinschmidt, T., Schnek, A.G., Looze, Y. & Braunitzer, G. (1989b) The thiol proteinases from the latex of *Carica papaya* L. III. The primary structure of chymopapain. *Biol. Chem. Hoppe-Seyler* 370, 425–434.

Jansen, E.F. & Balls, A.K. (1941) Chymopapain: a new crystalline proteinase from papaya latex. *J. Biol. Chem.* 137, 459–460.

Johansen, J.T. & Ottesen, M. (1968) The proteolytic degradation of the B-chain of oxidized insulin by papain, chymopapain and papaya peptidase. *Comptes Rendus Lab. Carlsberg* 36, 265–283.

Kunimitsu, D.K. & Yasunobu, K.T. (1967) Chymopapain. IV. The chromatographic fractionation of partially purified chymopapain and the characterization of crystalline chymopapain B. *Biochim. Biophys. Acta* 139, 405–417.

Napper, A.D., Bennett, S.P., Borowski, M., Holdridge, M.B., Leonard, M.J.C., Rogers, E.E., Duan, Y., Laursen, R.A., Reinhold, B. & Shames, S.L. (1994) Purification and characterization of multiple forms of the pineapple-stem-derived cysteine proteinases ananain and comosain. *Biochem. J.* 301, 727–735.

Ritonja, A., Buttle, D.J., Rawlings, N.D., Turk, V. & Barrett, A.J. (1989) Papaya proteinase IV amino acid sequence. *FEBS Lett.* 258, 109–112.

Robinson, G.W. (1975) Isolation and characterization of papaya peptidase A from commercial chymopapain. *Biochemistry* 14, 3695–3700.

Zucker, S., Buttle, D.J., Nicklin, M.J.H. & Barrett, A.J. (1985) Proteolytic activities of papain, chymopapain and papaya proteinase III. *Biochim. Biophys. Acta* 828, 196–204.

***David J. Buttle***
*Department of Human Metabolism & Clinical Biochemistry,*
*University of Sheffield Medical School,*
*Sheffield S10 2RX, UK*
*Email: D.J.Buttle@sheffield.ac.uk*

# *189. Caricain*

## *Databanks*

*Peptidase classification: clan CA, family C1, MEROPS ID: C01.003*
*NC-IUBMB enzyme classification: EC 3.4.22.30*
*Chemical Abstracts Service registry number: 39307-22-7*
*Databank codes:*

| Species | SwissProt | PIR | EMBL (cDNA) | EMBL (genomic) |
|---|---|---|---|---|
| *Carica papaya* | P10056 | JN0633 | X51899 | – |
| | | JN0634 | X66060 | |
| | | S39367 | X69877 | |

Brookhaven Protein Data Bank three-dimensional structures:

| Species | ID | Resolution | Notes |
|---|---|---|---|
| *Carica papaya* | 1MEG | 2 | Asp185Glu mutant; complex with E-64 |
| | 1PPO | 1.8 | Cys25 with bound mercury |

## Name and History

The first description of this enzyme was provided by Schack (1967), who named it **papaya peptidase A**. The same enzyme has since been given a number of different names, including **papaya peptidase II** (Brocklehurst *et al.*, 1984), **papaya proteinase III** (Zucker *et al.*, 1985) and **papaya proteinase Ω** (Brocklehurst *et al.*, 1985). The name **caricain** was recommended by NC-IUBMB in 1992.

## Activity and Specificity

In common with most enzymes in family C1, caricain accepts hydrophobic amino acid residues in both S2 and S3. However, other residues are also accommodated in these subsites, including proline in S2, and lysine in S3 (Jacquet *et al.*, 1989). The specificities of three cysteine endopeptidases from papaya latex were found to be very similar. Caricain and chymopapain (Chapter 188) appeared to prefer an aliphatic to a hydrophobic residue at P2. The similarity in specificity of caricain and chymopapain was demonstrated by the fact that, of 44 peptide bonds in manatee hemoglobin cleaved by caricain, 29 were also cleaved by chymopapain (Jacquet *et al.*, 1989). An earlier study (Johansen & Ottesen, 1968) had highlighted the similarity in specificity of caricain, chymopapain and papain (Chapter 187). All seven bonds of the oxidized B chain of insulin that were hydrolyzed by caricain were also cleaved by papain, and six were hydrolyzed by chymopapain.

Caricain can be assayed with Bz-Arg┼NHPhNO$_2$, $k_{cat}/K_m$ being $187\,M^{-1}\,s^{-1}$ at pH 6.8 and 40°C. More sensitive substrates may employ a fluorometric leaving group, $k_{cat}/K_m$ for the hydrolysis of Z-Phe-Arg┼NHMec being $1.06 \times 10^6\,M^{-1}\,s^{-1}$ (pH 6.8, 40°C) (Zucker *et al.*, 1985). The enzyme exhibits a broad pH-activity profile, with the optimum near 7.0. About half-maximal activity is still achieved at pH values of about 5.3 and 8.3, and the profile is reported to be governed by at least three ionizing groups (Taylor *et al.*, 1994). The active-site sulfur requires reduction for catalytic competence, and this is best achieved by the inclusion of low millimolar concentrations of cysteine in assay buffers.

Caricain is inactivated by E-64, making the inhibitor a convenient active-site titrant (Zucker *et al.*, 1985), and it is inhibited by cystatins, $K_i$ for inhibition by papaya cystatin being 1.5 nM (Song *et al.*, 1995).

## Structural Chemistry

Caricain is synthesized as a preproenzyme. There is evidence at the mRNA level for polymorphism, two very similar clones being isolated, one of which contained a C-terminal extension (Revell *et al.*, 1993). The primary structure of the mature form of the enzyme has been determined (Dubois *et al.*, 1988a) and is as predicted from one of the cDNA sequences. The protein is 216 amino acids in length ($M_r$ 23 280), and is 68% identical in sequence to papain, 65% to chymopapain and 81% to glycyl endopeptidase (Chapter 190). The three disulfide bonds are conserved between all the papaya proteinases, and there is no evidence for glycosylation. Caricain is an extremely basic protein, with pI estimated to be 11.7 (Kaarsholm & Schack, 1983). The $A_{280,1\%}$ is reported to be

18.3 (Robinson, 1975), giving a molar extinction coefficient of $4.19 \times 10^4\,M^{-1}\,cm^{-1}$.

As with some other plant cysteine endopeptidases, caricain exhibits charge heterogeneity. This may be partly due to variation in the oxidization state of the active-site sulfur, as is the case with homologous enzymes from pineapple stem (Napper *et al.*, 1994), although genetic polymorphism (above) may also contribute.

The crystal structure of caricain has been solved to a resolution of 1.8 Å (Pickersgill *et al.*, 1991), and demonstrates main-chain conformation very similar to that of papain. Caricain has four amino acid residues (Ser169–Lys172) not present in papain, but it is papain that is exceptional at this point in the sequence, showing a deletion not seen in other members of the family. The architecture of the active site of caricain is very similar to that of papain.

## Preparation

The highly basic character of caricain makes it relatively easy to separate from the other papaya cysteine endopeptidases in cation-exchange chromatography of preparations of commercially available papaya latex. A sodium acetate gradient, pH 5.0, was first used successfully by Robinson (1975) and has since been adopted by others. Caricain is found in the latest-eluting protein peak. Due to the charge heterogeneity of caricain (above) the peak may not be symmetrical, but this does not necessarily indicate the presence of contaminants. Covalent chromatography on thiol-Sepharose allows isolation of fully active caricain from the material obtained by cation exchange (Dubois *et al.*, 1988b).

## Biological Aspects

Cysteine endopeptidases contribute 69–89% of total protein in papaya latex, caricain alone making up 14–26% (Buttle *et al.*, 1990; Dando *et al.*, 1995). The reasons for such high levels of potential proteolytic activity are not clear, but other constituents of the latex include lysozyme and chitinases, suggesting a defensive role against microbial pathogens and arthropods. The proteinases are present in quantities well in excess of those expected for catalysts, suggesting that they function once the latex has been released from the laticifers and diluted. Certain insects have developed sophisticated methods to avoid latex contamination (Dussourd & Eisner, 1987). An untested hypothesis is therefore that ingested papaya endopeptidases may be harmful to arthropods.

## Distinguishing Features

The enzymatic properties of caricain are very similar to those of two other papaya cysteine endopeptidases, papain and chymopapain, and thus cannot be used reliably to discriminate between these enzymes. The property of caricain that distinguishes it from almost all other proteins is its extreme basicity, due in part to a Lys/Arg excess over Asp/Glu of 19 residues.

## References

Brocklehurst, K., Carey, P.R., Lee, H.-H., Salih, E. & Storer, A.C. (1984) Comparative resonance Raman spectroscopic and kinetic studies of acyl-enzymes involving papain, actinidin and papaya peptidase II. *Biochem. J.* **223**, 649–657.

Brocklehurst, K., Salih, E., McKee, R. & Smith, H. (1985) Fresh non-fruit latex of *Carica papaya* contains papain, multiple forms of chymopapain A and papaya proteinase S. *Biochem. J.* **228**, 525–527.

Buttle, D.J., Dando, P.M., Coe, P.F., Sharp, S.L., Shepherd, S.T. & Barrett, A.J. (1990) The preparation of fully active chymopapain free of contaminating proteinases. *Biol. Chem. Hoppe-Seyler* **371**, 1083–1088.

Dando, P.M., Sharp, S.L., Buttle, D.J. & Barrett, A.J. (1995) Immunoglobulin E antibodies to papaya proteinases and their relevance to chemonucleolysis. *Spine* **20**, 981–985.

Dubois, T., Kleinschmidt, T., Schnek, A.G., Looze, Y. & Braunitzer, G. (1988a) The thiol proteinases from the latex of *Carica papaya* L. II. The primary structure of proteinase Ω. *Biol. Chem. Hoppe-Seyler* **369**, 741–754.

Dubois, T., Jacquet, A., Schnek, A.G. & Looze, Y. (1988b) The thiol proteinases from the latex of *Carica papaya* L. I. Fractionation, purification and preliminary characterization. *Biol. Chem. Hoppe-Seyler* **369**, 733–740.

Dussourd, D.E. & Eisner, T. (1987) Vein-cutting behavior: insect counterploy to the latex defense of plants. *Science* **237**, 898–901.

Jacquet, A., Kleinschmidt, T., Dubois, T., Schnek, A.G., Looze, Y. & Braunitzer, G. (1989) The thiol proteinases from the latex of *Carica papaya* L. IV. Proteolytic specificities of chymopapain and papaya proteinase Ω determined by digestion of α-globin chains. *Biol. Chem. Hoppe-Seyler* **370**, 819–829.

Johansen, J.T. & Ottesen, M. (1968) The proteolytic degradation of the B-chain of oxidized insulin by papain, chymopapain and papaya peptidase. *Comptes Rendus Lab. Carlsberg* **36**, 265–283.

Kaarsholm, N.C. & Schack, P. (1983) Characterization of papaya peptidase A as an enzyme of extreme basicity. *Acta Chem. Scand.* **B37**, 607–611.

Napper, A.D., Bennett, S.P., Borowski, M., Holdridge, M.B., Leonard, M.J.C., Rogers, E.E., Duan, Y., Laursen, R.A., Reinhold, B. & Shames, S.L. (1994) Purification and characterization of multiple forms of the pineapple-stem-derived cysteine proteinases ananain and comosain. *Biochem. J.* **301**, 727–735.

Pickersgill, R.W., Rizkallah, P., Harris, G.W. & Goodenough, P.W. (1991) Determination of the structure of papaya protease Ω. *Acta Crystallogr. B* **47**, 766–771.

Revell, D.F., Cummings, N.J., Baker, K.C., Collins, M.E., Taylor, M.A.J., Sumner, I.G., Pickersgill, R.W., Connerton, I.F. & Goodenough, P.W. (1993) Nucleotide sequence and expression in *Escherichia coli* of cDNAs encoding papaya proteinase Ω from *Carica papaya*. *Gene* **127**, 221–225.

Robinson, G.W. (1975) Isolation and characterization of papaya peptidase A from commercial chymopapain. *Biochemistry* **14**, 3695–3700.

Schack, P. (1967) Fractionation of proteolytic enzymes of dried papaya latex. Isolation and preliminary characterization of a new proteolytic enzyme. *Comptes Rendus Lab. Carlsberg* **36**, 67–83.

Song, I., Taylor, M., Baker, K. & Bateman, R.C. (1995) Inhibition of cysteine proteinases by *Carica papaya* cystatin produced in *Escherichia coli*. *Gene* **162**, 221–224.

Taylor, M.A.J., Baker, K.C., Connerton, I.F., Cummings, N.J., Harris, G.W., Henderson, I.M.J., Jones, S.T., Pickersgill, R.W., Sumner, I.G., Warwicker, J. & Goodenough, P.W. (1994) An unequivocal example of cysteine proteinase activity affected by multiple electrostatic interactions. *Protein Eng.* **7**, 1267–1276.

Zucker, S., Buttle, D.J., Nicklin, M.J.H. & Barrett, A.J. (1985) Proteolytic activities of papain, chymopapain and papaya proteinase III. *Biochim. Biophys. Acta* **828**, 196–204.

*David J. Buttle*
*Department of Human Metabolism & Clinical Biochemistry,*
*University of Sheffield Medical School,*
*Sheffield S10 2RX, UK*
*Email: D.J.Buttle@sheffield.ac.uk*

# 190. Glycyl endopeptidase

## Databanks

*Peptidase classification: clan CA, family C1, MEROPS ID: C01.004*
*NC-IUBMB enzyme classification: EC 3.4.22.25*
*Chemical Abstracts Service registry number: 149719-24-4*
*Databank codes:*

| Species | SwissProt | PIR | EMBL (cDNA) | EMBL (genomic) |
|---------|-----------|-----|-------------|----------------|
| *Carica papaya* | P05994 | A26074 | X03970 | – |
| | | S06837 | X78056 | |

Brookhaven Protein Data Bank three-dimensional structures:

| Species | ID | Resolution | Notes |
|---|---|---|---|
| *Carica papaya* | 1GEC | 2.1 | complex with benzyloxycarbonyl-Leu-Val-Gly methylene bound to Cys25 |

## Name and History

The assay of preparations of *Carica papaya* latex with blocked -Gly-OPhNO$_2$ substrates gave the first indication that the latex may contain an esterase with a markedly different substrate specificity to those of the other papaya endopeptidases, papain, chymopapain and caricain (Chapters 187, 188 and 189) (Lynn, 1979; Polgár, 1981). The novel fraction, which was called *papaya peptidase B*, hydrolyzed the glycine ester, but not Bz-Arg-NHPhNO$_2$ that is cleaved by the other three enzymes. Following purification and initial characterization, the enzyme was named *papaya proteinase IV* (Buttle *et al.*, 1989). Further characterization revealed the novel, restricted specificity for cleavage of glycyl amide and peptide bonds (Buttle *et al.*, 1990a), and the name *glycyl endopeptidase* was recommended by IUBMB in 1992. Despite this, the term *chymopapain M* has been used (Thomas *et al.*, 1994). This name is inappropriate in that glycyl endopeptidase shares no more sequence similarity to chymopapain than it does to the other cysteine endopeptidases from papaya, and the restricted specificity demonstrated by glycyl endopeptidase contrasts markedly with the catholic nature of the specificity of chymopapain.

## Activity and Specificity

The selectivity for glycyl bonds is apparent in the cleavage of proteins by glycyl endopeptidase. A compilation of known cleavage sites shows that of 42 bonds cleaved, 38 were glycyl bonds (Buttle, 1994; M. Abrahamson, personal communication). As with most enzymes in family C1, there is a preference for hydrophobic side chains at P2 and P3, particularly aliphatic ones. Small-molecule substrates include *N*-blocked esters and amides of glycine (Lynn, 1979; Polgár, 1981; Buttle *et al.*, 1990a; Thomas *et al.*, 1994). The strict selectivity for glycyl bonds is demonstrated by the 60-fold reduction in $k_{cat}$ resulting from the presence of the methyl side chain of Ala in P1 of an amide substrate (Buttle *et al.*, 1990a).

The pH optimum is broad, centered around 7 (Buttle, 1994) and governed primarily by groups with $pK_a$ values of 4.3, 5.6 and 8.7 (Thomas *et al.*, 1994). As expected for a proteinase dependent upon a thiolate anion for activity, reducing conditions are essential, and best obtained by the inclusion of low millimolar concentrations of cysteine in the assay buffer. Blocked -Gly-NHPhNO$_2$ or -NHMec substrates such as Boc-Ala-Ala-Gly┼NHPhNO$_2$ are conveniently used, and are available from Bachem (see Appendix 2 for full names and addresses of suppliers) (Buttle *et al.*, 1990a).

Glycyl endopeptidase binds to $\alpha_2$-macroglobulin, a process expected to inhibit its hydrolysis of proteins. However, the enzyme is not inhibited by most cystatins (Buttle *et al.*, 1990b). This property is not unique among the family C1 enzymes, both stem bromelain (Chapter 192) and fruit bromelain (Chapter 193) also being refractory to inhibition. However, glycyl endopeptidase also cleaves and inactivates some type II cystatins, such as chicken cystatin and human cystatin C (Buttle *et al.*, 1990b). This capacity appears to be unique among the family C1 enzymes. The one exception to the lack of inhibition by cystatins is moderate inhibition by papaya cystatin ($K_i = 3.44$ nM) (Song *et al.*, 1995), suggesting selection pressure for the control of glycyl endopeptidase activity in the plant. Among small-molecule inhibitors, the enzyme is inactivated by E-64. Rate constants for inactivation by peptidyl diazomethanes are greatly influenced according to the substrate specificity of the enzyme. Inactivation by iodoacetate and iodoacetamide is unusually slow (Buttle *et al.*, 1990b).

## Structural Chemistry

Glycyl endopeptidase is synthesized as a preproenzyme (Taylor *et al.*, 1995). The mature processed enzyme is a single polypeptide of $M_r$ 23 313 that is not glycosylated (Ritonja *et al.*, 1989). It shows a great deal of sequence similarity to the other cysteine endopeptidases from papaya latex, being 81% identical to caricain, 70% to chymopapain and 67% to papain. The crystal structure confirms the overall similarity to papain, the two structures being superimposable with a r.m.s. difference of 0.64 Å for 212 common C$\alpha$ positions (O'Hara *et al.*, 1995). The highly distinctive substrate specificity of glycyl endopeptidase stems largely from the substitutions of Gly23 and Gly65 (as found in papain and conserved in all other known members of family C1) by Glu and Arg, respectively. This dramatically alters the S1 subsite, as the side chains of Glu23 and Arg65 form a barrier across the binding pocket and sterically exclude residues with large side chains. Molecular modeling has suggested that the same substitutions are mainly responsible for the lack of inhibition by cystatins, the residues clashing with part of the N-terminal region of cystatin B, and Glu23 also being involved in a steric clash with the hairpin loop of the inhibitor (O'Hara *et al.*, 1995). It has recently been reported that crowding of the active-site cleft also prevents autocatalytic activation of the proenzyme (Baker *et al.*, 1996).

Independent confirmation of the importance of Glu23 and Arg65 to the restricted substrate specificity and inhibitor profile of glycyl endopeptidase has come from site-directed mutagenesis of the equivalent residues in cathepsin B (Chapter 209). Both the Gly27Glu and Gly73Arg mutants showed a drastic reduction in their ability to hydrolyze substrates with Arg in P1, although the kinetics of hydrolysis of substrates with Gly in P1 were much less affected. The rate of inactivation by E-64 was not affected by the Gly73Arg substitution, yet the affinity for cystatin C was decreased by a factor of 20 000 (Fox *et al*., 1995).

The pro regions of family C1 enzymes are inhibitors of their respective enzymes and often of other enzymes in the family. The propeptide of glycyl endopeptidase is no exception, $K_i$ for the inhibition of glycyl endopeptidase by its propeptide being $8.6 \times 10^{-7}$ M. Inhibition of the other papaya cysteine endopeptidases was even stronger (Taylor *et al*., 1995).

Glycyl endopeptidase is a basic protein with pI probably above 10. Its $A_{1\%,280}$ is 16.5 (Buttle *et al*., 1989), giving a molar extinction coefficient of $3.75 \times 10^4$ M$^{-1}$ cm$^{-1}$.

## Preparation

The similar physicochemical properties of the papaya cysteine endopeptidases means that they can be difficult to purify by traditional ion-exchange and size-exclusion techniques. This is particularly true for glycyl endopeptidase and chymopapain, which exhibit very similar charge properties over a range of pH values, and are also very similar in size. The first purification of glycyl endopeptidase awaited the discovery that this enzyme binds to the immobilized active site-directed affinity ligand, Gly-Phe-NHCH$_2$CN, much more tightly than does chymopapain (Buttle *et al*., 1989). The purification was later simplified by taking advantage of the fact that inactivation of glycyl endopeptidase by iodoacetate was much slower than of the other papaya enzymes (Buttle *et al*., 1990b), and in fact was not very different from the rate of reaction of iodoacetate with cysteine. The other papaya endopeptidases can therefore be selectively inactivated with iodoacetate prior to purification of glycyl endopeptidase on the affinity column (Buttle, 1994). Advantage can also be taken of the failure of glycyl endopeptidase to bind to Sepharose-glutathione-2-pyridyl disulfide, which binds the other papaya cysteine endopeptidases (Thomas *et al*., 1994), or of the selective failure of glycyl endopeptidase to bind to cystatin-Sepharose (Taylor *et al*., 1995).

Glycyl endopeptidase is a potent allergen (Dando *et al*., 1995). Care should therefore be taken when handling preparations (particularly powders) containing this enzyme.

## Biological Aspects and Practical Application

The biological functions of the papaya cysteine endopeptidases are not understood, but as pointed out in Chapter 189, they may include a defensive role against some plant pathogens and herbivorous arthropods. If this is the case, then it is tempting to speculate that glycyl endopeptidase may have evolved under evolutionary pressure for proteolytic activity that is resistant to the cystatins of pathogens or predatory insects. Glycyl endopeptidase is not inhibited by the cystatins, and its inactivation of some of them could allow the other endopeptidases to express their activity, providing *Carica papaya* with a useful defensive weapon. The concentration of glycyl endopeptidase in the latex, where it represents 23–28% of total protein (Buttle *et al*., 1990c; Dando *et al*., 1995), indeed suggests that it may act after escape of the latex from the laticifers and dilution.

Due to its restricted substrate specificity, glycyl endopeptidase is a helpful tool in the analysis of protein primary structure (Buttle, 1994). Cleavage of glycyl bonds occurs efficiently in the vicinity of disulfide bridges, making the enzyme very useful in the assignment of disulfide bonds in proteins (Bernard & Peanasky, 1993). Glycyl endopeptidase is currently not available from any of the major suppliers, but can be obtained from the author.

## Distinguishing Features

The major distinguishing features of glycyl endopeptidase undoubtedly relate to its novel substrate specificity, its lack of inhibition by the cystatins and its ability to inactivate some of these.

## Further Reading

A fuller account of glycyl endopeptidase has been provided by Buttle (1994), and O'Hara *et al*. (1995) have described how the solution of the crystal structure of the enzyme provides a rational structural basis for the understanding of its enzymatic properties.

## References

Baker, K.C., Taylor, M.A.J., Cummings, N.J., Tunon, M.A., Worboys, K.A. & Connerton, I.F. (1996) Autocatalytic processing of pro-papaya proteinase IV is prevented by crowding of the active-site cleft. *Protein Eng.* **9**, 525–529.

Bernard, V.D. & Peanasky, R.J. (1993) The serine protease inhibitor family from *Ascaris suum*: chemical determination of the five disulfide bonds. *Arch. Biochem. Biophys.* **303**, 367–376.

Buttle, D.J. (1994) Glycyl endopeptidase. *Methods Enzymol.* **244**, 539–555.

Buttle, D.J., Kembhavi, A.A., Sharp, S., Shute, R.E., Rich, D.H. & Barrett, A.J. (1989) Affinity purification of the novel cysteine proteinase papaya proteinase IV and papain from papaya latex. *Biochem. J.* **261**, 469–476.

Buttle, D.J., Ritonja, A., Pearl, L.H., Turk, V. & Barrett, A.J. (1990a) Selective cleavage of glycyl bonds by papaya proteinase IV. *FEBS Lett.* **260**, 195–197.

Buttle, D.J., Ritonja, A., Dando, P.M., Abrahamson, M., Shaw, E.N., Wikstrom, P., Turk, V. & Barrett, A.J. (1990b) Interaction of papaya proteinase IV with inhibitors. *FEBS Lett.* **262**, 58–60.

Buttle, D.J., Dando, P.M., Coe, P.F., Sharp, S.L., Shepherd, S.T. & Barrett, A.J. (1990c) The preparation of fully active chymopapain free of contaminating proteinases. *Biol. Chem. Hoppe-Seyler* **371**, 1083–1088.

Dando, P.M., Sharp, S.L., Buttle, D.J. & Barrett, A.J. (1995) Immunoglobulin E antibodies to papaya proteinases and their relevance to chemonucleolysis. *Spine* **20**, 981–985.

Fox, T., Mason, P., Storer, A.C. & Mort, J.S. (1995) Modification of S1 subsite specificity in the cysteine protease cathepsin B. *Protein Eng.* **8**, 53–57.

Lynn, K.R. (1979) A purification and some properties of two proteases from papaya latex. *Biochim. Biophys. Acta* **569**, 193–201.

O'Hara, B.P., Hemmings, A.M., Buttle, D.J. & Pearl, L.H. (1995) Crystal structure of glycyl endopeptidase from *Carica papaya*: a cysteine endopeptidase of unusual substrate specificity. *Biochemistry* **34**, 13190–13195.

Polgár, L. (1981) Isolation of highly active papaya peptidases A and B from commercial chymopapain. *Biochim. Biophys. Acta* **658**, 262–269.

Ritonja, A., Buttle, D.J., Rawlings, N.D., Turk, V. & Barrett, A.J. (1989) Papaya proteinase IV amino acid sequence. *FEBS Lett.* **258**, 109–112.

Song, I., Taylor, M., Baker, K. & Bateman, R.C. (1995) Inhibition of cysteine proteinases by *Carica papaya* cystatin produced in *Escherichia coli. Gene* **162**, 221–224.

Taylor, M.A.J., Baker, K.C., Briggs, G.S., Connerton, I.F., Cummings, N.J., Pratt, K.A., Revell, D.F., Freedman, R.B. & Goodenough, P.W. (1995) Recombinant pro-regions from papain and papaya proteinase IV are selective high affinity inhibitors of the mature papaya enzymes. *Protein Eng.* **8**, 59–62.

Thomas, M.P., Topham, C.M., Kowlessur, D., Mellor, G.W., Thomas, E.W., Whitford, D. & Brocklehurst, K. (1994) Structure of chymopapain M the late-eluted chymopapain deduced by comparative modelling techniques and active-centre characteristics determined by pH-dependent kinetics of catalysis and reactions with time-dependent inhibitors: the Cys-25/His-159 ion-pair is insufficient for catalytic competence in both chymopapain M and papain. *Biochem. J.* **300**, 805–820.

*David J. Buttle*
*Department of Human Metabolism & Clinical Biochemistry,*
*University of Sheffield Medical School,*
*Sheffield S10 2RX, UK*
*Email: D.J.Buttle@sheffield.ac.uk*

# 191. *Carica candamarcensis* endopeptidase

## Databanks

*Peptidase classification: clan CA, family C1, MEROPS ID: C01.020*
*NC-IUBMB enzyme classification: none*
*Databank codes:*

| Species | SwissProt | PIR | EMBL (cDNA) | EMBL (genomic) |
|---|---|---|---|---|
| *Carica candamarcensis* | P32956 | S46476 | S46476 | – |

## Name and History

*Carica candamarcensis* (syn. *Carica pubescens*), the source of this endopeptidase, belongs to the Caricaceae family (order Parietales). The so-called mountain papaya grows naturally at elevated altitudes (from 1500 to 2000 m) in various tropical regions of the world. The latex of the fruit is the source of at least four cysteine peptidases including the *Carica candamarcensis endopeptidase* described here (Walraevens *et al.*, 1993; Bravo *et al.*, 1994; Demoraes *et al.*, 1994).

## Activity and Specificity

Assays are conveniently made with synthetic substrates such as Bz-Arg┼NHPhNO$_2$, Bz-Pro-Phe-Arg┼NHPhNO$_2$ or Z-Phe-Arg┼NHMec after activation of the enzyme preparation by low concentrations of thiol compounds (e.g. 2.5 mM DTT). The pH optimum is about 6.8. The rules governing specificity have not been thoroughly investigated, but with synthetic substrates Phe residues tend to be favored in position P2. The free essential thiol function, located on Cys25 of the enzyme, is stoichiometrically titrated by iodoacetic acid or by E-64 (Walraevens *et al.*, 1993; Demoraes *et al.*, 1994; Jaziri *et al.*, 1994).

## Structural Chemistry

*Carica candamarcensis* endopeptidase is a glycoprotein containing 214 amino acid residues with the carbohydrate moiety bound to asparagine at position 44. The structure of the polypeptide chain is well-ordered (30% $\alpha$ helix, 27% $\beta$ sheet and 26% $\beta$ turn) and further stabilized by the presence of three disulfide bonds. Two free thiol functions are located at positions 25 and 196 in the chain. The enzyme shares 125 identical residues (59.5%) with papain, 142 (67.6%) with glycyl endopeptidase, 146 (69.5%) with caricain and 156 (74.3%) with chymopapain (the four cysteine peptidases isolated from *Carica papaya*) (Jaziri *et al.*, 1994).

## Preparation

Freshly collected latex of *C. candamarcensis* was the source of this endopeptidase. Latex was collected in the presence of methylmethanethiolsulfonate. The basic glycoprotein was then purified by a combination of cation-exchange chromatography and affinity chromatography on a column of concanavalin A-Sepharose (Walraevens *et al.*, 1993; Jaziri *et al.*, 1994).

## Biological Aspects

The enzyme is present in the latex that is contained in the laticifers that are to be found in all parts of the plant except the roots. The function of the endopeptidase is unknown, but it may contribute to the defenses of the plant against pathogens or predators, as has been suggested for chymopapain (Chapter 188).

## References

Bravo, L.M., Hermosilla, J. & Salas, C.E. (1994) A biochemical comparison between latex from *Carica candamarcensis* and *C. papaya. Braz. J. Med. Res.* **27**, 2831–2842.

Demoraes, M.G., Termignoni, C. & Salas, C. (1994) Biochemical characterization of a new cysteine endopeptidase from *Carica candamarcensis* L. *Plant Sci.* **102**, 11–18.

Jaziri, M., Kleinschmidt, T., Walraevens, V., Schneck, A.G. & Looze, Y. (1994) Primary structure of CC-III, the glycosylated cysteine proteinase from the latex of *Carica candamarcensis* Hook. *Biol. Chem. Hoppe-Seyler* **375**, 379–385.

Walraevens, V., Jaziri, M., Van Beeumen, J., Schneck, A.G., Kleinschmidt, T. & Looze, Y. (1993) Isolation and preliminary characterization of the cysteine proteinases of the latex of *Carica candamarcensis* Hook. *Biol. Chem. Hoppe-Seyler* **374**, 501–506.

*Yvan Looze*
*Protein Chemistry Department,*
*Faculty of Medicine,*
*Free University of Brussels,*
*808, route de Lennik, B-1070, Brussels, Belgium*
*Email: ylooze@resulb.ulb.ac.be*

# 192. Stem bromelain

*Peptidase classification: clan CA, family C1, MEROPS ID: C01.005*
*NC-IUBMB enzyme classification: EC 3.4.22.32*
*Chemical Abstracts Service registry number: 37189-34-7*
*Databank codes:*

| Species | SwissProt | PIR | EMBL (cDNA) | EMBL (genomic) |
|---|---|---|---|---|
| *Ananas comosus* | P14518 | S03964 | – | – |

## Name and History

All the endopeptidases of the pineapple plant (*Ananas comosus*) have generally been referred to as 'the bromelains', and indeed, the name 'bromelain' was originally applied to any protease from any member of the family Bromeliaceae (Heinicke, 1953). Following this definition, the names 'stem bromelain' and 'fruit bromelain' have been used to describe the major activities in the juice of pineapple stem and fruit, respectively. These enzymes were originally assigned separate systematic numbers (EC 3.4.4.24 and EC 3.4.4.25) IUBMB, then lumped together under EC 3.4.22.4, and separated again in 1992. Commercially available dried powder prepared from waste pineapple stem material is widely used in industry under the name 'bromelain' (Caygill, 1979), and this has further contributed to the muddled nomenclature.

There have been contradictory reports describing up to six different proteolytic components in the stem (reviewed by Murachi, 1970, and Rowan & Buttle, 1994). Recently, the situation has been clarified (Rowan *et al.*, 1990), and the major endopeptidase present in extracts of plant stem is to be termed ***stem bromelain***, whilst fruit bromelain is the major endopeptidase in the fruit (Chapter 193). Additional minor cysteine endopeptidases (ananain and comosain) have also been detected in the stem (Rowan *et al.*, 1988, 1990) (Chapters 194 and 195).

## Activity and Specificity

Stem bromelain represents almost 90% of the proteolytically active material present in pineapple stem extract (Rowan *et al.*, 1988). Surprisingly in view of its high proteolytic activity with protein substrates (Rowan *et al.*, 1990), it has been found to act efficiently only on Arg-Arg-containing synthetic substrates, as does comosain (Rowan *et al.*, 1988; Napper *et al.*, 1994). Z-Arg-Arg+NHMec is a convenient substrate that is known to be scarcely affected by ananain or fruit bromelain (Rowan *et al.*, 1988, 1990). A preference for polar amino acids in both the P1 and P1′ positions has been reported (Napper *et al.*, 1994). The pH optima with synthetic and protein substrates are broad, but most assays are performed near neutral pH (Rowan & Buttle, 1994). For full activity of the enzyme, the presence of a reducing reagent such as DTT or cysteine is required.

Unlike the other pineapple endopeptidases, stem bromelain exhibits unusual kinetics of inhibition with E-64 (Rowan *et al.*, 1988, 1990; Napper *et al.*, 1994), and is also atypical in its resistance to inhibition by chicken cystatin, a property it shares with fruit bromelain (Rowan *et al.*, 1990).

## Structural Chemistry

Stem bromelain is a glycosylated, single-chain protein of 24.5 kDa (Harrach *et al.*, 1995), a pI value of 9.55 (Murachi, 1970), an $A_{1\%, 280}$ of 20.1 (Murachi, 1970), containing seven cysteines (Napper *et al.*, 1994; Harrach *et al.*, 1995) and therefore most probably three disulfide bonds. For reasons unknown, it can exhibit charge heterogeneity which produces multiple chromatographic forms, although these are all immunologically identical (Rowan *et al.*, 1988). The complete amino acid sequence of stem bromelain has been deduced (Ritonja *et al.*, 1989) and shows it to be a member of the papain family. Comparison with N-terminal sequence data for the other pineapple endopeptidases demonstrate that it is distinct, however (Napper *et al.*, 1994).

## Preparation

Stem bromelain can be purified from commercially available dried pineapple stem powders on a milligram scale with over 80% activity, and is conveniently stored lyophilized after reversible inhibition by 2-hydroxyethyl disulfide (Rowan *et al.*, 1988).

## Biological Aspects

Why there should be several different but closely related endopeptidases in the pineapple plant is unknown, but the situation is seen also with other tropical fruits. It is possible that these enzymes may provide the plant with protection from parasites, pathogens and herbivores. Natural inhibitors of 'bromelain' that are present in high concentrations in the pineapple plant have been described (Perlstein & Kezdy, 1973), and these may regulate the activity of the cysteine endopeptidases of the plant *in vivo*.

## Distinguishing Features

The relatively weak inhibition by E-64, and lack of inhibition by chicken cystatin, distinguish stem bromelain from most other peptidases of family C1. Furthermore, stem bromelain has been shown to be immunologically distinct from both fruit bromelain (Rowan *et al.*, 1990) and ananain (Rowan *et al.*, 1988).

## Further Reading

Fuller reviews of the pineapple endopeptidases have been provided by Murachi (1970), Cooreman *et al.* (1976) and Rowan & Buttle (1994).

## References

Caygill, J.C. (1979) Sulphydryl plant proteases. *Enzyme Microb. Technol.* **1**, 233–242.

Cooreman, W.M., Scharpe, S., Demeester, J. & Lauwers, A. (1976) Bromelain, biochemical and pharmacological properties. *Pharm. Acta Helv.* **51**, 73–112.

Harrach, T., Eckert, K., Schulze-Forster, K., Nuck, R., Grunow, D. & Maurer, H.R. (1995) Isolation and partial characterization of basic proteinases from stem bromelain. *J. Protein Chem.* **14**, 41–52.

Heinicke, R.M. (1953) Complementary enzyme actions in the clotting of milk. *Science* **118**, 753–754.

Murachi, T. (1970) Bromelain enzymes. *Methods Enzymol.* **19**, 273–284.

Napper, A.D., Bennett, S.P., Borowski, M., Holdridge, M.B., Leonard, M.J.C., Rogers, E.E., Duan, Y., Laursen, R.A., Reinhold, B. & Shames, S.L. (1994) Purification and characterization of the pineapple stem-derived cysteine proteinases ananain and comosain. *Biochem. J.* **301**, 727–735.

Perlstein, S.H. & Kezdy, F.J. (1973) Isolation and characterization of a protease inhibitor from commercial stem bromelain acetone powder. *J. Supramol. Struct.* **1**, 249–254.

Ritonja, A., Rowan, A.D., Buttle, D.J., Rawlings, N.D., Turk, V. & Barrett, A.J. (1989) Stem bromelain: amino acid sequence and implications for weak binding of cystatin. *FEBS Lett.* **247**, 419–424.

Rowan, A.D. & Buttle, D.J. (1994) Pineapple cysteine endopeptidases. *Methods Enzymol.* **244**, 555–568.

Rowan, A.D., Buttle, D.J. & Barrett, A.J. (1988) Ananain: a novel cysteine proteinase found in pineapple stem. *Arch. Biochem. Biophys.* **267**, 262–270.

Rowan, A.D., Buttle, D.J. & Barrett, A.J. (1990) The cysteine proteinases of the pineapple plant. *Biochem. J.* **266**, 869–875.

*Andrew D. Rowan*
*Department of Rheumatology,*
*The Medical School,*
*Framlington Place, Newcastle upon Tyne NE2 4HH, UK*
*Email: a.d.rowan@ncl.ac.uk*

# 193. Fruit bromelain

## Databanks

*Peptidase classification: clan CA, family C1, MEROPS ID: C01.028*
*NC-IUBMB enzyme classification: EC 3.4.22.33*
*Databank codes: EMBL: D38534*

## Name and History

All the enzymes of the pineapple plant (*Ananas comosus*) have generally been referred to as 'the bromelains' with the major enzyme fraction found in the juice of the pineapple fruit being called **fruit bromelain**. Although once assigned its own separate systematic number in the EC system, it was later grouped together with stem bromelain before its recent reclassification. Heterogeneity of the proteinases present in pineapple fruit has long been suspected, with contradictory reports describing up to three different proteolytic components that may vary with geographical location (see Rowan & Buttle, 1994, and references therein). More recently, it has become clear that although trace amounts of stem bromelain are present (Rowan *et al.*, 1990), the major endopeptidase present in the juice of the pineapple fruit is fruit bromelain. Also classified under EC 3.4.22.33 is **pinguinain** (formerly EC 3.4.99.18), the major cysteine endopeptidase of the related bromeliad, *Bromelia pinguin* (commonly called rat or mouse pineapple).

## Activity and Specificity

Fruit bromelain constitutes 30–40% of the total fruit protein and represents almost 90% of the proteolytically active material of the pineapple fruit (Murachi, 1970). A convenient assay substrate is Bz-Phe-Val-Arg+NHMec, which is also hydrolyzed by pinguinain, but is scarcely affected by stem bromelain (Rowan *et al.*, 1990). Both enzymes require the presence of a reducing agent for full activity. Fruit bromelain has high proteolytic activity compared to stem bromelain (Rowan *et al.*, 1990), with broad pH optima for synthetic and protein substrates, although most assays are performed around neutral pH (Rowan & Buttle, 1994).

## Structural Chemistry

Fruit bromelain is a single-chain protein of approximately 25 kDa (Rowan *et al.*, 1990), with a pI of 4.6, and an $A_{1\%,280}$ of 19.2 (Ota *et al.*, 1964). The carbohydrate content is unclear (Ota *et al.*, 1985; Yamada *et al.*, 1976). The N-terminal sequence data are identical to those of stem bromelain (Yamada *et al.*, 1976). Pinguinain is reported to be a glycoprotein (Toro-Goyco *et al.*, 1968) of approximately 26 kDa (Rowan *et al.*, 1990), $A_{1\%,280}$ of 24.6 and pI of 6.5 (Toro-Goyco *et al.*, 1968). Both enzymes contain seven cysteines (Toro-Goyco *et al.*, 1968; Yamada *et al.*, 1976) and have been shown be be immunologically distinct from stem bromelain, ananain and one another (Rowan *et al.*, 1990).

## Preparation

Fruit bromelain can be affinity-purified from fresh pineapple fruit by use of an immobilized peptidylsemicarbazide on a milligram scale with over 85% activity (Rowan *et al.*, 1990). Pinguinain can be isolated from fresh fruit on a gram scale (Toro-Goyco *et al.*, 1968), and is similarly affinity-purified (Rowan *et al.*, 1990). Both enzymes can be stored lyophilized after reversible inhibition by mercuric chloride (Rowan *et al.*, 1990).

## Distinguishing Features

Fruit bromelain and pinguinain have been shown be be immunologically distinct from stem bromelain, ananain and one another (Rowan *et al.*, 1990). Fruit bromelain is atypical in its resistance to inhibition by chicken cystatin, whereas pinguinain is inhibited. Fruit bromelain, unlike stem bromelain, is scarcely able to hydrolyze Z-Arg-Arg-NHMec (Rowan *et al.*, 1990).

## Further Reading

Fuller reviews of fruit bromelain have been provided by Toro-Goyco *et al.* (1968), Ota *et al.* (1985) and Rowan & Buttle (1994).

## References

Murachi, T. (1970) Bromelain enzymes. *Methods Enzymol.* **18**, 273–284.
Ota, S., Moore, S. & Stein, W.H. (1964) Preparation and properties of purified stem and fruit bromelains. *Biochemistry* **3**, 180–185.
Ota, S., Muta, E., Katanita, Y. & Okamoto, Y. (1985) Reinvestigation of the fractionation and some properties of the proteolytically active components of stem and fruit bromelains. *J. Biochem. (Tokyo)* **98**, 219–228.
Rowan, A.D. & Buttle, D.J. (1994) Pineapple cysteine endopeptidases. *Methods Enzymol.* **244**, 555–568.
Rowan, A.D., Buttle, D.J. & Barrett, A.J. (1990) The cysteine proteinases of the pineapple plant. *Biochem. J.* **266**, 869–875.

Toro-Goyco, E., Maretzki, A. & Matos, M.L. (1968) Isolation, purification and partial characterization of pinguinain, the proteolytic enzyme from *Bromelia pinguin* L. *Arch. Biochem. Biophys.* **126**, 91–104.

Yamada, F., Takahashi, N. & Murachi, T. (1976) Purification and characterization of a proteinase from pineapple fruit, fruit bromelain A2. *J. Biochem. (Tokyo)* **79**, 1223–1234.

**Andrew D. Rowan**
*Department of Rheumatology,*
*The Medical School,*
*Framlington Place, Newcastle upon Tyne NE2 4HH, UK*
*Email: a.d.rowan@ncl.ac.uk*

# 194. Ananain

## Databanks

*Peptidase classification: clan CA, family C1, MEROPS ID: C01.026*
*NC-IUBMB enzyme classification: EC 3.4.22.31*
*Databank codes:*

| Species | SwissProt | PIR | EMBL (cDNA) | EMBL (genomic) |
|---|---|---|---|---|
| *Ananas comosus* | – | S46204 | – | – |

## Name and History

Early chromatographic fractionations of pineapple stem extract yielded up to six proteolytically active components, all typically referred to as 'bromelain' (Rowan & Buttle, 1994, and references therein). Two minor components of crude stem extract, termed glycine esterases A and B, were detected due to their markedly different activities against peptidyl nitrophenyl ester substrates compared to the major enzyme (stem bromelain) fractions (Silverstein, 1974). More recently, a commercial crystalline preparation of 'bromelain' was shown to contain components differing greatly in their susceptibility to inhibition by kininogen (Gounaris et al., 1984). Subsequently, a late-eluting chromatographic fraction of stem extract was shown to contain an enzyme quite distinct from stem bromelain, with activity towards Z-Phe-Arg+NHMec. This was named **ananain** from the botanical name of the pineapple plant, *Ananas comosus* (Rowan et al., 1988).

## Activity and Specificity

Ananain is the second most abundant endopeptidase of pineapple stem extract, comprising up to 5% of the total protein (Rowan et al., 1988; Napper et al., 1994). It was first isolated using Z-Phe-Arg-NHMec as substrate; this is known to be scarcely affected by stem bromelain (Rowan et al., 1988). For full activation of the enzyme the presence of a reducing agent such as DTT is required. A preference for polar amino acids in the P1′ position has been reported with a much broader specificity for the P1 site (Napper et al., 1994) and a preference for a hydrophobic side chain in P2 (Rowan et al., 1990a). The pH optimum with both synthetic and protein substrates is close to neutral. Ananain has high proteolytic activity, and useful colorimetric substrates include azocasein, azocoll and hide powder azure (Rowan & Buttle, 1994); this high activity has prompted the development of ananain as a debriding agent for burn injuries (Rowan et al., 1990b).

## Structural Chemistry

Ananain is a nonglycosylated, single-chain protein of 23.5 kDa (Rowan et al., 1988; Napper et al., 1994; Harrach et al., 1995), pI >10 (Rowan et al., 1988; Harrach et al., 1995), and an $A_{1\%,280}$ of 16.5 (Rowan et al., 1988). It contains seven cysteines (Napper et al., 1994) and therefore probably three disulfide bonds. N-terminal sequence data clearly demonstrate that ananain is distinct from the other pineapple endopeptidases (Rowan & Buttle, 1994).

## Preparation

Ananain can be purified from commercially available dried pineapple stem powders on a milligram scale with a 20% yield. The product is 70–95% active by titration with E-64 (Rowan et al., 1988; Napper et al., 1994). It is conveniently stored lyophilized after reversible inhibition by 2-hydroxyethyl disulfide (Rowan *et al.*, 1988, 1990a).

### Distinguishing Features

Ananain has been shown to be immunologically distinct from both stem and fruit bromelains (Rowan *et al.*, 1990a). It behaves similarly to papain in being efficiently inhibited by both chicken cystatin ($K_i = 1.1$ nM) and E-64 ($k_2 = 3 \times 10^5$ M$^{-1}$ s$^{-1}$), which clearly distinguishes it from stem bromelain (Rowan *et al.*, 1988).

### Further Reading

Reviews have been provided by Napper *et al.* (1994) and Rowan & Buttle (1994).

### References

Gounaris, A.D., Brown, M.A. & Barrett, A.J. (1984) Human plasma α-cysteine proteinase inhibitor. *Biochem. J.* **221**, 445–452.

Harrach, T., Eckert, K., Schulze-Forster, K., Nuck, R., Grunow, D. & Maurer, H.R. (1995) Isolation and partial characterization of basic proteinases from stem bromelain. *J. Protein Chem.* **14**, 41–52.

Napper, A.D., Bennett, S.P., Borowski, M., Holdridge, M.B., Leonard, M.J.C., Rogers, E.E., Duan, Y., Laursen, R.A., Reinhold, B. & Shames, S.L. (1994) Purification and characterization of the pineapple stem-derived cysteine proteinases ananain and comosain. *Biochem. J.* **301**, 727–735.

Rowan, A.D. & Buttle, D.J. (1994) Pineapple cysteine endopeptidases. *Methods Enzymol.* **244**, 555–568.

Rowan, A.D., Buttle, D.J. & Barrett, A.J. (1988) Ananain: a novel cysteine proteinase found in pineapple stem. *Arch. Biochem. Biophys.* **267**, 262–270.

Rowan, A.D., Buttle, D.J. & Barrett, A.J. (1990a) The cysteine proteinases of the pineapple plant. *Biochem. J.* **266**, 869–875.

Rowan, A.D., Christopher, C.W., Kelly, S.F., Buttle, D.J. & Ehrlich, H.P. (1990b) Debridement of experimental full-thickness skin burns of rats with enzyme fractions derived from pineapple stem. *Burns* **16**, 243–246.

Silverstein, R.M. (1974) The assay of the bromelains using *N*-CBZ-L-lysine-*p*-nitrophenyl ester as substrates. *Anal. Biochem.* **62**, 478–484.

*Andrew D. Rowan*
*Department of Rheumatology,*
*The Medical School,*
*Framlington Place, Newcastle upon Tyne NE2 4HH, UK*
*Email: a.d.rowan@ncl.ac.uk*

# 195. *Comosain*

### Databanks

*Peptidase classification: clan CA, family C1, MEROPS ID: C01.027*
*NC-IUBMB enzyme classification: none*
*Databank codes:*

| Species | SwissProt | PIR | EMBL (cDNA) | EMBL (genomic) |
|---|---|---|---|---|
| *Ananas comosus* | – | S46205 | – | – |

### Name and History

Until recently there has been considerable confusion as to the identity and number of proteolytically active components in crude pineapple stem extract (see Rowan & Buttle, 1994). A late-eluting protein fraction of stem extract was further purified and shown to contain two distinct enzymes (Rowan *et al.*, 1990a; Napper *et al.*, 1994). One of these was ananain (Chapter 194) while the other was termed *comosain*, the name being derived from *Ananas comosus* (Rowan *et al.*,

1990a). Comosain may well be one of the glycine esterases first described by Silverstein (1974).

### Activity and Specificity

Comosain is the least abundant endopeptidase of pineapple stem extract yet characterized, representing less than 1% of the total protein (Napper *et al.*, 1994). It was first identified because it copurified with ananain (Rowan *et al.*, 1990a,b), but it has activity against Z-Arg-Arg+NHMec which is a poor substrate for ananain (Rowan *et al.*, 1990a). Presence of a reducing agent is necessary for full activity. Comosain displays a preference for polar residues in both the P1 and P1′ substrate-binding sites (Napper *et al.*, 1994), is highly proteolytic, and like ananain has been used as a debriding agent for burn injuries (Rowan *et al.*, 1990b). It is inhibited by E-64 ($k_2 = 8900 \, \text{M}^{-1} \, \text{s}^{-1}$).

### Structural Chemistry

Comosain is a single-chain glycoprotein of 24.5 kDa, pI >10, containing seven cysteines (Napper *et al.*, 1994) and therefore probably three disulfide bonds. It has a similar carbohydrate composition to stem bromelain, but a markedly different amino acid composition (Napper *et al.*, 1994). N-terminal sequence data show that comosain is distinct from the other pineapple endopeptidases (Napper *et al.*, 1994; Rowan & Buttle, 1994).

### Preparation

Comosain can be purified from commercially available, acetone-precipitated, pineapple stem powders on a microgram scale with over 95% activity as judged by E-64 titration (Napper *et al.*, 1994). It is conveniently stored lyophilized

after reversible inhibition by 2-hydroxyethyl disulfide (Rowan *et al.*, 1990a; Napper *et al.*, 1994).

### Further Reading

A full account of comosain has been provided by Napper *et al.* (1994).

### References

Napper, A.D., Bennett, S.P., Borowski, M., Holdridge, M.B., Leonard, M.J.C., Rogers, E.E., Duan, Y., Laursen, R.A., Reinhold, B. & Shames, S.L. (1994) Purification and characterization of the pineapple stem-derived cysteine proteinases ananain and comosain. *Biochem. J.* **301**, 727–735.

Rowan, A.D. & Buttle, D.J. (1994) Pineapple cysteine endopeptidases. *Methods Enzymol.* **244**, 555–568.

Rowan, A.D., Buttle, D.J. & Barrett, A.J. (1990a) The cysteine proteinases of the pineapple plant. *Biochem. J.* **266**, 869–875.

Rowan, A.D., Christopher, C.W., Kelly, S.F., Buttle, D.J. & Ehrlich, H.P. (1990b) Debridement of experimental full-thickness skin burns of rats with enzyme fractions derived from pineapple stem. *Burns* **16**, 243–246.

Silverstein, R.M. (1974) The assay of the bromelains using *N*-CBZ-L-lysine-*p*-nitrophenyl ester as substrates. *Anal. Biochem.* **62**, 478–484.

*Andrew D. Rowan*
*Department of Rheumatology,*
*The Medical School,*
*Framlington Place, Newcastle upon Tyne NE2 4HH, UK*
*Email: a.d.rowan@ncl.ac.uk*

# 196. Ficain

### Databanks

*Peptidase classification: clan CA, family C1, MEROPS ID: C01.006*
*NC-IUBMB enzyme classification: EC 3.4.22.3*
*Chemical Abstracts Service registry number: 9001-33-6*
*Databank codes: no sequence data available*

### Name and History

It has been known for many years that the milky latex flowing from cuts of the stem, leaves and unripe fruit of species in the genus *Ficus* contains proteolytic activity. The name *ficin* was coined by Robbins (1930) for the purified white powder with antihelminthic activity that was obtained from any member of the genus. A crystalline preparation from an unnamed species was obtained by Walti (1938) and also named ficin. There are more than 1300 species of *Ficus*, many of which contain

proteolytic activity (Williams *et al.*, 1968), sometimes due to the presence of more than one proteinase (Sgarbieri *et al.*, 1964). The term ficin must therefore be regarded as generic. In 1992 IUBMB recommended the name *ficain* for the 'major proteolytic component of the latex of the fig, *Ficus glabrata*'. We suggest that this specific definition of ficain be retained, on the assumption that there is no doubt as to which is the major component of *F. glabrata* latex. Although the enzymes from the latex of the Mediterranean fig (*Ficus carica*) have also been the subject of study, in this chapter we restrict ourselves to a discussion of the major proteolytic component from the latex of *F. glabrata*.

## Activity and Specificity

The data available suggest that ficain is broadly 'papain-like' in terms of its specificity. Hydrolysis of the oxidized B chain of insulin illustrated some differences, however. Papain (Chapter 187) hydrolyzed ten bonds to a significant extent (Johansen & Ottesen, 1968), and ficain six (Englund *et al*., 1968), only three of which were amongst those cleaved by papain. Gly, Ser, Glu, Tyr and Phe were all accommodated in the S1 pocket of ficain. The subsite S2, generally regarded as the primary specificity determinant in papain-like enzymes, accepted cysteic acid, Val, Leu, Gly and Phe, suggesting that ficain has some specificity for hydrophobic side chains in this pocket.

Bz-Arg+NHPhNO$_2$ has traditionally been used as a small-molecule substrate for assay (Englund *et al*., 1968), but in our hands is very inefficiently cleaved by ficain. Z-Phe-Arg+NHPhNO$_2$ is a much more sensitive substrate for colorimetric assay, as is Bz-Phe-Val-Arg+NHPhNO$_2$, but the greater solubility of the former makes its use more convenient. Even greater sensitivity can be achieved with fluorometric substrates, such as Z-Phe-Arg+NHMec, $K_m$ being 300 µM (Björk *et al*., 1995), and Bz-Phe-Val-Arg+NHMec ($K_m$ 2.4 µM, pH 6.8, 37°C) (unpublished results of the author). The pH optimum of ficain is near 7.0, and the broad pH-activity profile is governed primarily by groups with p$K_a$ values of 4.46 and 8.37 (Whitaker, 1969). Ficain requires activation by reducing agents, and we find that the inclusion of 4 mM cysteine in assay buffers gives satisfactory results.

Ficain is inhibited extremely tightly by chicken egg white cystatin, the $K_d$ of $5 \times 10^{-14}$ M (Björk & Ylinenjärvi, 1990) being virtually identical to that of papain (Björk *et al*., 1989).

## Structural Chemistry

Remarkably little structural information about ficain is available. The enzyme is probably a single polypeptide. The amino acid compositions of ficain and papain are similar, although ficain, like chymopapain (Chapter 188) has an additional Cys residue (Englund *et al*., 1968). The amino acid sequences around the active-site Cys and His residues have been determined (Husain & Lowe, 1970) and found to show a great deal of similarity to those around the equivalent residues in papain. This information, together with the tight inhibition by chicken cystatin (above) indicates that ficain is a member of family C1.

## Preparation

Dried *Ficus glabrata* latex is commercially available, but before choosing a source it is wise to check that the origin of the latex is *F. glabrata*, as often material is described simply as 'from fig tree latex'. Ficain can be separated from most of the other components of the latex by cation-exchange chromatography, this enzyme being the latest-eluting peak with activity against Z-Phe-Arg-NHPhNO$_2$. Ficain elutes from the Resource S column (Pharmacia) (see Appendix 2 for full names and addresses of suppliers) in a sodium acetate gradient, pH 5.0, between 11.7 and 14.5 mS.

In order to separate activatable from irreversibly oxidized ficain, and thus obtain a product with high specific activity, it is necessary to follow cation-exchange chromatography with a form of active site-directed affinity chromatography. Thiol-specific ligands such as 2,2′-dipyridyl disulfide-glutathione (Malthouse & Brocklehurst, 1976) or aminophenylmercuric acetate (Anderson & Hall, 1974) have been successfully employed. We have obtained highly active ficain by chromatography on Gly-Phe-NHCH$_2$CN coupled to Sepharose by way of a linker (Buttle, 1994), with elution in 2,2′-dipyridyl disulfide (unpublished results of the author).

## Biological Aspects and Practical Uses

The functions of plant latex cysteine endopeptidases are not understood, but as discussed in Chapter 189 they may have a protective role against plant pathogens and herbivorous insects.

Proteolytic fractions from fig latex are used for unmasking antigens in serology (Mazda *et al*., 1995). The historical interest in 'ficin' originated from its ability to digest gastrointestinal nematodes. Despite more recent evidence that the enzyme(s) may be effective for this purpose (Hansson *et al*., 1986), ficain has not been adopted widely as a treatment for nematode infections.

## References

Anderson, C.D. & Hall, P.L. (1974) Purification of ficin by affinity chromatography. *Anal. Biochem.* **60**, 417–423.

Björk, I. & Ylinenjärvi, K. (1990) Interaction between chicken cystatin and the cysteine proteinases actinidin, chymopapain A, and ficin. *Biochemistry* **29**, 1770–1776.

Björk, I., Alriksson, E. & Ylinenjärvi, K. (1989) Kinetics of binding of chicken cystatin to papain. *Biochemistry* **28**, 1568–1573.

Björk, I., Brieditis, I. & Abrahamson, M. (1995) Probing the functional role of the N-terminal region of cystatins by equilibrium and kinetic studies of the binding of Gly-11 variants of recombinant human cystatin C to target proteinases. *Biochem. J.* **306**, 513–518.

Buttle, D.J. (1994) Affinity chromatography of cysteine peptidases. *Methods Enzymol.* **244**, 639–648.

Englund, P.T., King, T.P., Craig, L.C. & Walti, A. (1968) Studies on ficin. I. Its isolation and characterization. *Biochemistry* **7**, 163–175.

Hansson, A., Veliz, G., Naquira, C., Amren, M., Arroyo, M. & Arevalo, G. (1986) Preclinical and clinical studies with latex from *Ficus glabrata* HBK, a traditional intestinal anthelminthic in the Amazonian area. *J. Ethnopharmacol.* **17**, 105–138.

Husain, S.S. & Lowe, G. (1970) The amino acid sequence around the active-site cysteine and histidine residues, and the buried cysteine residue in ficin. *Biochem. J.* **117**, 333–340.

Johansen, J.T. & Ottesen, M. (1968) The proteolytic degradation of the B-chain of oxidized insulin by papain, chymopapain and papaya peptidase. *Comptes Rendus Lab. Carlsberg* **36**, 265–283.

Malthouse, J.P.G. & Brocklehurst, K. (1976) Preparation of fully active ficin from *Ficus glabrata* by covalent chromatography and characterization of its active centre by using 2,2′-dipyridyl disulphide as a reactivity probe. *Biochem. J.* **159**, 221–234.

Mazda, T., Makino, K., Yabe, R., Nakata, K., Fujisawa, K. & Ohshima, H. (1995) Use of standardized protease enzymes for antibody screening of blood donor samples with the microplate system autoanalyzer. *Transfusion Med.* **5**, 43–50.

Robbins, B.H. (1930) A proteolytic enzyme in ficin, the anthelmintic principle of leche de higueron. *J. Biol. Chem.* **87**, 251–257.

Sgarbieri, V.C., Gupte, S.M., Kramer, D.E. & Whitaker, J.R. (1964) *Ficus* enzymes. I. Separation of the proteolytic enzymes of

*Ficus carica* and *Ficus glabrata* latices. *J. Biol. Chem.* **239**, 2170–2177.

Walti, A. (1938) Crystalline ficin. *J. Am. Chem. Soc.* **60**, 493.

Whitaker, J.R. (1969) Papain- and ficin-catalysed reactions. Effect of pH on activity and conformation of ficin. *Biochemistry* **8**, 1896–1901.

Williams, D.C., Sgarbieri, V.C. & Whitaker, J.R. (1968) Proteolytic activity in the genus *Ficus*. *Plant Physiol.* **43**, 1083–1088.

*Alison Singleton*
*Departments of Chemistry, and Human*
*Metabolism and Clinical Biochemistry,*
*University of Sheffield,*
*Sheffield S10 2RX, UK*

*David J. Buttle*
*Departments of Chemistry, and Human*
*Metabolism and Clinical Biochemistry,*
*University of Sheffield,*
*Sheffield S10 2RX, UK*
*Email: D.J.Buttle@sheffield.ac.uk*

# 197. Actinidain

## Databanks

*Peptidase classification: clan CA, family C1, MEROPS ID: C01.007*
*NC-IUBMB enzyme classification: EC 3.4.22.14*
*Chemical Abstracts Service registry number: 39279-27-1*
*Databank codes:*

| Species | SwissProt | PIR | EMBL (cDNA) | EMBL (genomic) |
|---|---|---|---|---|
| *Actinidia deliciosa* | P00785 | S12618 | M21336 X13013 X13139 X16466 | – |

Brookhaven Protein Data Bank three-dimensional structures:

| Species | ID | Resolution | Notes |
|---|---|---|---|
| *Actinidia deliciosa* | 1AEC 2ACT | 1.86 1.7 | complex with E-64 |

## Name and History

*Actinidia chinensis* (Chinese gooseberry, kiwifruit or yang-tao) is native to China and was first described in 1847 from specimens that were sent to England. Arcus (1959) was the first to obtain a crude preparation of the cysteine proteinase and to describe some of its properties. It appears that he was prompted to search for an enzyme from *A. chinensis* by the fact well-known in New Zealand households that incorporation of the raw fruit into table jelly prevents it from setting. In 1970 McDowall began work on actinidain and by 1984 it was the subject of 18 papers that dealt with its purification and physical and molecular properties, specificity, catalytic characteristics, amino acid sequence, crystal structure and catalytic site characteristics as studied by reactivity probe and spectroscopic methods (Brocklehurst *et al.*, 1987 and references cited therein). More recently, attention has turned to the electrostatic fields within actinidain and structurally related enzymes to explain reactivity characteristics (Brocklehurst *et al.*, 1984, 1989; Pickersgill *et al.*, 1988; Reid *et al.*, 1997). The name **actinidain** is a slightly altered form of the previously used '*actinidin*', which was the anglicized version of *Actinidia*, the genus of the source plant. Actinidain has been referred to previously also as ***Actinidia anionic protease***.

## Activity and Specificity

Arcus (1959) characterized the proteolytic activity of extracts of *A. chinensis* with gelatin and hemoglobin as substrates, finding that the activity was enhanced by the presence of reagents such as cysteine, $Na_2S$, KCN and EDTA. The first study on a crystalline preparation from *A. chinensis* was by McDowall (1970), who demonstrated that it was the active component of the extracts studied by Arcus (1959). This preparation contained two closely eluted species which could not be well resolved by chromatography on DEAE-cellulose (McDowall, 1970). In a later study McDowall (1973) further purified his crystalline preparation by recycling gel chromatography, resolving it into an active and an inactive form. The active form was then derivatized as an inactive *S*-2-aminoethyl sulfenyl compound and resolved further by chromatography on DEAE-Sephadex A-25 into two components, termed $A_1$ and $A_2$, indicating their positions in the elution profile. McDowall used the $A_2$ fraction for specificity studies with the B chain of insulin as substrate. The pattern of hydrolysis was similar to that of papain (Chapter 187). Thus, for seven of the ten bonds hydrolyzed McDowall (1973) found that the P2 residue (Schechter & Berger, 1967, 1968; Berger & Schechter, 1970) was hydrophobic (Leu, Val or Phe, but not Tyr). For the other three bonds it was suggested that the main specificity determinant was the P1 residue. This is consistent with the findings of Boland & Hardman (1972) which indicated that a cationic residue was preferred at the P1 position together with an aromatic *N*-acyl blocking group, presumably interacting with the hydrophobic S2 subsite. Differences in specificity between papain and actinidain are revealed by comparison of values of $k_{cat}/K_m$ (Boland & Hardman, 1973; Baker *et al.*, 1980). For substrates with aromatic substituents approximately in the P2 position, values of $k_{cat}/K_m$ are 10–110 times smaller for actinidain than for papain whereas those containing aliphatic *N*-acyl groups have similar values of $k_{cat}/K_m$ for the two enzymes. This difference is accounted for by a key structural difference between the S2 subsites. In papain, Ser205 is located at the bottom of the S2 binding pocket and in actinidain the analogous residue is Met211 whose side chain extends across the pocket making it shorter.

Actinidain has been quantified by use of a variety of compounds. McDowall (1970) first employed both 5,5′-dithiobis-(2-nitrobenzoic acid) (DTNB) and 4,4′-dipyridyl disulfide (4PDS) as active-center titrants of free thiol in the enzyme. DTNB was also employed as an active-center titrant by Björk & Ylinenjärvi (1990), although they stated that it underestimated the thiol content because of its low reactivity. Brocklehurst and coworkers have used 2,2′-dipyridyl disulfide (2PDS) as a more effective thiol titrant (Brocklehurst *et al.*, 1981; see Brocklehurst, 1996a for a general review of enzyme active-center titration). Assays of enzymic activity have employed Tos-Gln┤OPhNO$_2$ (McDowall, 1970), Z-Lys┤OPhNO$_2$ (Boland & Hardman, 1972; Baker *et al.*, 1980; Carotti *et al.*, 1984), Bz-L-Arg┤NHPhNO$_2$ (Salih *et al.*, 1987; Topham *et al.*, 1991) and Ac-Phe-Gly┤NHPhNO$_2$ (Patel *et al.*, 1992; Reid *et al.*, 1997).

Actinidain is inhibited by the common cysteine proteinase inhibitors (e.g. $Hg^{2+}$, iodoacetate, aromatic disulfides such as DTNB and 2-pyridyl disulfides, E-64 and chicken cystatin: Brocklehurst *et al.*, 1987; Björk & Ylinenjärvi, 1990; Varughese *et al.*, 1992).

## Structural Chemistry

The primary structure of actinidain was first determined by Carne & Moore (1978) by examination of the tryptic peptides. More recently the amino acid sequence was deduced from the cDNA sequence by Podivinsky *et al.* (1989). Actinidain contains 220 amino acids ($M_r$ 23 500; Baker & Drenth, 1987), 17 of which were differently assigned in the cDNA sequence. Most of the changes are made from one hydrophobic residue to another of similar hydrophobicity, size and shape and, therefore, are not of great importance in terms of overall structure and catalytic activity of actinidain (see Brocklehurst *et al.*, 1997). The discrepancies were further examined by Varughese *et al.* (1992) using the crystal structure of actinidain derivatized by reaction with E-64. They favored the assignment of nine amino acid residues in agreement with the cDNA sequence, equivocal assignment of six, and four residues that could not be assigned according to either sequence. Actinidain has a pI of 3.1 (Baker & Drenth, 1987).

The crystal structure at 5.5 Å resolution (Baker, 1976) was available prior to knowledge of the amino acid sequence. A more precise structure was derived through extensive crystallographic refinement to a resolution of 1.7 Å (Baker & Dodson, 1980). The polypeptide is folded into a two-domain structure. The L-domain (domain I) consists of residues 19–115 and 214–220 and the R-domain (domain II) residues 1–18 and 116–213. The L-domain contains three helical regions, with the main helix running through the center of the molecule at the interface between the two domains. The R-domain forms a twisted antiparallel $\beta$-sheet barrel, with a hydrophobic interior. The two ends are then sealed by short helices at the molecular surface (Baker & Drenth, 1987). Each domain is made up of a hydrophobic core and the interface between the two domains, where the catalytic site is located, is mainly polar, being made up of hydrophilic side chains. These side chains can further interact with a network of eight buried water molecules. The structure is further stabilized by the presence of three disulfide bonds: Cys22-Cys65, Cys56-Cys98 and Cys156-Cys206.

The marked differences between actinidain and some other cysteine proteinases, notably papain, in the electrostatic environments to which the catalytic sites are exposed have been characterized by reactivity probe kinetics (Brocklehurst *et al.*, 1983, 1988, 1989; Salih & Brocklehurst, 1983; Salih *et al.*, 1987; Kowlessur *et al.*, 1989; Topham *et al.*, 1991) and by use of a spectroscopic reporter group (Brocklehurst *et al.*, 1984). The unusual specificity exhibited by actinidain towards covalent chromatography gels has its origin in the large negative potential that envelops the enzyme almost completely and extends out to about 8 Å from the mouth of the active-center cleft (Thomas *et al.*, 1995). Variation in the P2–S2 stereochemical selectivity of actinidain, ficain and papain has been reported (Patel *et al.*, 1992).

## Preparation

The first preparation of actinidain was carried out by McDowall (1970) who obtained a crystalline sample of the enzyme, but this contained three components. Two of the components, one active and one inactive, were detected by McDowall (1970), and the original active component was later further resolved into two active components by chromatography on DEAE-Sephadex A-25 (McDowall, 1973). More recently, Brocklehurst *et al*. (1981; see also Thomas *et al*., 1995) developed a rapid method for the isolation of fully active actinidain by covalent chromatography. Brocklehurst (1996b) has provided a review of covalent chromatography by thiol-disulfide interchange, the method used.

## Biological Aspects

The precise biological role of the plant cysteine proteinases remains unclear. They tend to have a broad specificity towards a variety of substrates, and may well protect the ripening fruit from attack by, for example, insects and fungi (Baker & Drenth, 1987). Due to their broad specificity there is also the possibility that they play a role in a variety of physiological processes (see Brocklehurst *et al*., 1987). Actinidain is found primarily in the fruit of *Actinidia chinensis*, with traces occurring in the leaves and apex tissues. Actinidain constitutes up to 50% of soluble fruit protein (Praekelt *et al*., 1988). As such, the plant cell itself must be protected in some way from the proteolytic activity. Actinidain is expressed as an inactive zymogen approximately 15 kDa larger than the native enzyme, with both N- and C-terminal extensions (Praekelt *et al*., 1988; Podivinsky *et al*., 1989). The mechanism of activation of the zymogen has yet to be determined.

## Further Reading

An extensive review of major types of cysteine proteinase including actinidain has been provided by Brocklehurst *et al*. (1987), and Brocklehurst *et al*. (1997) have written a comprehensive review of the mechanism of action of cysteine proteinases.

## References

Arcus, A.C. (1959) Proteolytic enzyme of *Actinidia chinensis*. *Biochim. Biophys. Acta* **33**, 242–243.
Baker, E.N. (1976) The structure of actinidin at 5.5 Å resolution. *J. Mol. Biol.* **101**, 185–196.
Baker, E.N. & Dodson, E.J. (1980) Crystallographic refinement of the structure of actinidin at 1.7 Å resolution by fast fourier least-squares methods. *Acta Crystallogr. A* **36**, 559–572.
Baker, E.N. & Drenth, J. (1987) The cysteine proteinases: structure and mechanism. In: *Biological Macromolecules and Assemblies*, vol. 3 (Jurnak, F. & McPherson, A., eds)., New York: John Wiley & Sons, pp. 313–368.
Baker, E.N., Boland, M.J., Calder, P.C. & Hardman, M.J. (1980) The specificity of actinidin and its relationship to the structure of the enzyme. *Biochim. Biophys. Acta* **616**, 30–34.
Berger, A. & Schechter, I. (1970) Mapping the active site of papain with the aid of peptide substrates and inhibitors. *Philos. Trans. R. Soc. Lond. [Biol.]* **257**, 249–264.
Björk, I. & Ylinenjärvi, K. (1990) Interaction between chicken cystatin and the cysteine proteinases actinidin, chymopapain A, and ficin. *Biochemistry* **29**, 1770–1776.
Boland, M.J. & Hardman, M.J. (1972) Kinetic studies on the cysteine proteinase from *Actinidia chinensis*. *FEBS Lett.* **27**, 282–284.
Boland, M.J. & Hardman, M.J. (1973) The actinidin-catalysed hydrolysis of *N*-α-benzyloxycarbonyl-L-lysine *p*-nitrophenyl ester. *Eur. J. Biochem.* **36**, 575–582.
Brocklehurst, K. (1996a) Active centre titration. In: *Enzymology Labfax* (Price, N.C., ed.). Oxford: Bios Scientific and San Diego: Academic Press, pp. 59–75.
Brocklehurst, K. (1996b) Covalent chromatography by thiol-disulfide interchange using solid-phase alkyl 2-pyridyl disulphides. In: *Proteins Labfax* (Price, N.C., ed.). Oxford: Bios Scientific and San Diego: Academic Press, pp. 65–71.
Brocklehurst, K., Baines, B.S. & Malthouse, J.P.G. (1981) Differences in the interactions of the catalytic groups of the active centres of actinidin and papain. *Biochem. J.* **197**, 739–746.
Brocklehurst, K., Willenbrock, S.J.F. & Salih, E. (1983) Effects of conformational selectivity and of overlapping kinetically influential ionizations on the characteristics of pH-dependent enzyme kinetics. Implications of free enzyme $pK_a$ variability in reactions of papain and for its catalytic mechanism. *Biochem. J.* **211**, 701–708.
Brocklehurst, K., Salih, E. & Lodwig, T.S. (1984). Differences between the electric fields of the catalytic sites of papain and actinidin by using the thiol-located nitrobenzofurazan label as a spectroscopic reporter group. *Biochem. J.* **220**, 609–612.
Brocklehurst, K., Willenbrock, F. & Salih, E. (1987) Cysteine proteinases. In: *Hydrolytic Enzymes* (Neuberger, A. & Brocklehurst, K., eds). Amsterdam: Elsevier, pp. 39–158.
Brocklehurst, K., Brocklehurst, S.M., Kowlessur, D., O'Driscoll, M., Patel, G., Salih, E., Templeton, W., Quigley, K., Thomas, E.W., Wharton, C.W., Willenbrock, F. & Szawelski, R. (1988) Supracrystallographic resolution of interactions contributing to enzyme catalysis by use of natural structural variants and reactivity-probe kinetics. *Biochem. J.* **256**, 543–555.
Brocklehurst, K., O'Driscoll, M., Kowlessur, D., Phillips, I.R., Templeton, W., Thomas, E.W., Topham, C.M. & Wharton, C.W. (1989) The interplay of electrostatic and binding interactions determining active centre chemistry and catalytic activity in actinidin and papain. *Biochem. J.* **257**, 309–312.
Brocklehurst, K., Watts, A.B., Patel, M., Verma, C. & Thomas, E.W. (1997) Cysteine proteinases. In: *Comprehensive Biological Catalysis: A Mechanistic Reference* (Sinnott, M.L., ed.). London: Academic Press, 1998, pp. 381–424.
Carne, A. & Moore, C.H. (1978) The amino acid sequence of the tryptic peptides from actinidin, a proteolytic enzyme from the fruit of *Actinidia chinensis*. *Biochem. J.* **173**, 73–83.
Carotti, A., Hansch, C., Mueller, M.M. & Blaney, J.M. (1984) Actinidin hydrolysis of substituted-phenyl hippurates: a quantitative structure-activity relationship and graphics comparison with hydrolysis by papain. *J. Med. Chem.* **27**, 1401–1405.
Kowlessur, D., O'Driscoll, M., Topham, C.M., Templeton, W., Thomas, E.W. & Brocklehurst, K. (1989) The interplay of electrostatic fields and binding interactions determining catalytic-site reactivity in actinidin. A possible origin of differences in the behaviour of actinidin and papain. *Biochem. J.* **259**, 443–452.
McDowall, M.A. (1970) Anionic proteinase from *Actinidia chinensis*. Preparation and properties of the crystalline enzyme. *Eur. J. Biochem.* **14**, 214–221.
McDowall, M.A. (1973) The action of proteinase A₂ of *Actinidia chinensis* on the B-chain of oxidized insulin. *Biochim. Biophys. Acta* **293**, 226–231.

Patel, M., Saleem, I., Mellor, G.W., Sreedharan, S., Templeton, W., Thomas, E.W., Thomas, M. & Brocklehurst, K. (1992) Variation in the P$_2$-S$_2$ stereochemical selectivity towards the enantiomeric *N*-acetylphenylalanylglycine-4-nitroanilides among the cysteine proteinases papain, ficin and actinidin. *Biochem. J.* **281**, 553–559.

Pickersgill, R.W., Goodenough, P.W., Sumner, I.G. & Collins, M.E. (1988) The electrostatic fields in the active-site clefts of actinidin and papain. *Biochem. J.* **254**, 235–238.

Podivinsky, E., Forster, R.L.S. & Gardner, R.C. (1989) Nucleotide sequence of actinidin, a kiwi fruit protease. *Nucleic Acids Res.* **17**, 8363.

Praekelt, U.M., McKee, R.A. & Smith, H. (1988) Molecular analysis of actinidin, the cysteine proteinase of *Actinidia chinensis*. *Plant Mol. Biol.* **10**, 193–202.

Reid, J.D., Pinitglang, S., Topham, C.M., Verma, C., Thomas, E.W. & Brocklehurst, K. (1997) Actinidin and chymopapain B provide variation in the common electrostatic environment of Glu50 in papain and caricain. *Biochem. Soc. Trans.* **25**, 89S.

Salih, E. & Brocklehurst, K. (1983) Investigation of the catalytic site of actinidin by using benzofuroxan as a reactivity probe with selectivity for the thiolate-imidazolium ion-pair systems of cysteine proteinases. Evidence that the reaction of the ion-pair of actinidin (p$K_I$ 3.0, p$K_{II}$ 9.6) is modulated by the state of ionization of a group associated with a molecular p$K_a$ of 5.5. *Biochem. J.* **213**, 713–718.

Salih, E., Malthouse, J.P.G., Kowlessur, D., Jarvis, M., O'Driscoll, M. & Brocklehurst, K. (1987) Differences in the chemical and catalytic characteristics of two crystallographically 'identical' enzyme catalytic sites. Characterization of actinidin and papain by a combination of pH-dependent substrate catalysis kinetics and reactivity probe kinetics and reactivity probe studies targeted on the catalytic-site thiol group and its immediate microenvironment. *Biochem. J.* **247**, 181–193.

Schechter, I. & Berger, A. (1967) On the size of the active site in proteases. I. Papain. *Biochem. Biophys. Res. Commun.* **27**, 157–162.

Schechter, I. & Berger, A. (1968) On the active site of proteases. III. Mapping the active site of papain. Specific peptide inhibitors of papain. *Biochem. Biophys. Res. Commun.* **32**, 898–912.

Thomas, M.P., Verma, C., Boyd, S.M. & Brocklehurst, K. (1995) The structural origins of the unusual specificities observed in the isolation of chymopapain M and actinidin by covalent chromatography and the lack of inhibition of chymopapain M by cystatin. *Biochem. J.* **306**, 39–46.

Topham, C.M., Salih, E., Frazao, C., Kowlessur, D., Overington, J.P., Thomas, M., Brocklehurst, S.M., Patel, M., Thomas, E.W. & Brocklehurst, K. (1991) Structure-function relationships in the cysteine proteinases actinidin, papain and papaya proteinase Ω. Three dimensional structure of papaya proteinase Ω deduced by knowledge-based modelling and active centre characteristics determined by two-hydronic-state reactivity probe kinetics and kinetics of catalysis. *Biochem. J.* **280**, 79–92.

Varughese, K., Su, Y., Cromwell, D., Hasnain, S. & Xuong, N.-H. (1992) Crystal structure of an actinidin-E64 complex. *Biochemistry* **31**, 5172–5176.

***Aaron B. Watts***
*Laboratory of Structural and Mechanistic Enzymology,*
*Department of Biochemistry,*
*Queen Mary and Westfield College,*
*University of London,*
*Mile End Road, London E1 4NS, UK*
*Email: ha5345@qmw.ac.uk*

***Keith Brocklehurst***
*Laboratory of Structural and Mechanistic Enzymology,*
*Department of Biochemistry,*
*Queen Mary and Westfield College,*
*University of London,*
*Mile End Road, London E1 4NS, UK*
*Email: kb1@qmw.ac.uk*

# 198. Asclepain

## Databanks

*Peptidase classification: clan CA, family C1, MEROPS ID: C01.008*
*NC-IUBMB enzyme classification: EC 3.4.22.7*
*Chemical Abstracts Service registry number: 37288-80-5*
*Databank codes: no sequence data available*

## Name and History

The presence of a proteolytic enzyme in the latex of milkweed (*Asclepias speciosa*) was first reported by Winnick *et al.* (1940), who named the enzyme **asclepain**. This name has also been used to describe the protease from *Asclepias syriaca* (Brockbank & Lynn, 1979); the enzyme obtained from *Asclepias glaucescens* has been named **asclepain g** to distinguish it from other related asclepains (Barragán *et al.*,

1985; Tablero *et al.*, 1991). Irrespective of the source, the proteolytic activity isolated from a single species is composed of multiple forms.

## Activity and Specificity

Asclepains require a reducing agent for maximal activity and hydrolyze a variety of ester, amide and peptide bonds (Winnick *et al.*, 1940; Brockbank & Lynn, 1979; Tablero *et al.*, 1991). When synthetic esters and amides are used as substrates, the multiple forms of the asclepains show only small differences in reactivity among them. Overall, they seem to have a lower preference for esters or amides of aromatic residues (Brockbank & Lynn, 1979), though these latter substrates might be hydrolyzed preferentially by some individual forms of asclepain g (Tablero *et al.*, 1991). Casein, hemoglobin and other proteins are good substrates for asclepain (Winnick *et al.*, 1940). Lynn *et al.* (1980a), studying the digestion of insulin B chain by asclepain $A_3$, concluded that this enzyme is even less specific than papain.

Activity can easily be assayed with casein (Robinson, 1975), *p*-nitrophenyl esters (Brockbank & Lynn, 1979) and Bz-L-Arg-OEt (Tablero *et al.*, 1991). Maximal activity is around pH 7.5 with casein (Brockbank & Lynn, 1979) and 8.0 with the arginine ester (Tablero *et al.*, 1991). At neutral pH, asclepains are completely inhibited by iodoacetate, sodium tetrathionate and phenyl mercuric acetate (Brockbank & Lynn, 1979; Tablero *et al.*, 1991).

## Structural Chemistry

All forms of asclepains studied so far have molecular masses of about 22 kDa (Brockbank & Lynn, 1979; Tablero *et al.*, 1991) as determined by SDS gel electrophoresis. Gel filtration under strongly denaturing and reducing conditions yields the same value, suggesting that these enzymes are single polypeptide chain molecules (Tablero *et al.*, 1991). The pI values are greater then 9.0 (Tablero *et al.*, 1991).

Lynn *et al.* (1980b) reported the N-terminal sequences to residue 21 for two molecular forms of asclepain (A and B) from *A. syriaca*. These were compared with the sequence of papain and clear homology was found (67% identity).

The fold of asclepain is also similar to that of papain and other cysteine proteinases, as judged from circular dichroism spectra (Figure 198.1) and the content of secondary structures (Tablero *et al.*, 1991).

## Preparation

Similar procedures have been used to purify asclepain from different species. The latex, collected from cut stems or petioles of the plant, is centrifuged. The aqueous layer containing the soluble protein is fractionated by gel filtration chromatography (Tablero *et al.*, 1991) or run on an organomercurial-agarose affinity column (Brockbank & Lynn, 1979). To separate the electrophoretically distinct molecular forms, cation-exchange chromatography can be used. It has been shown that the multiple molecular forms do not arise from autoproteolysis, but their relative proportions depend on the season in which the latex is collected (Tablero *et al.*, 1991).

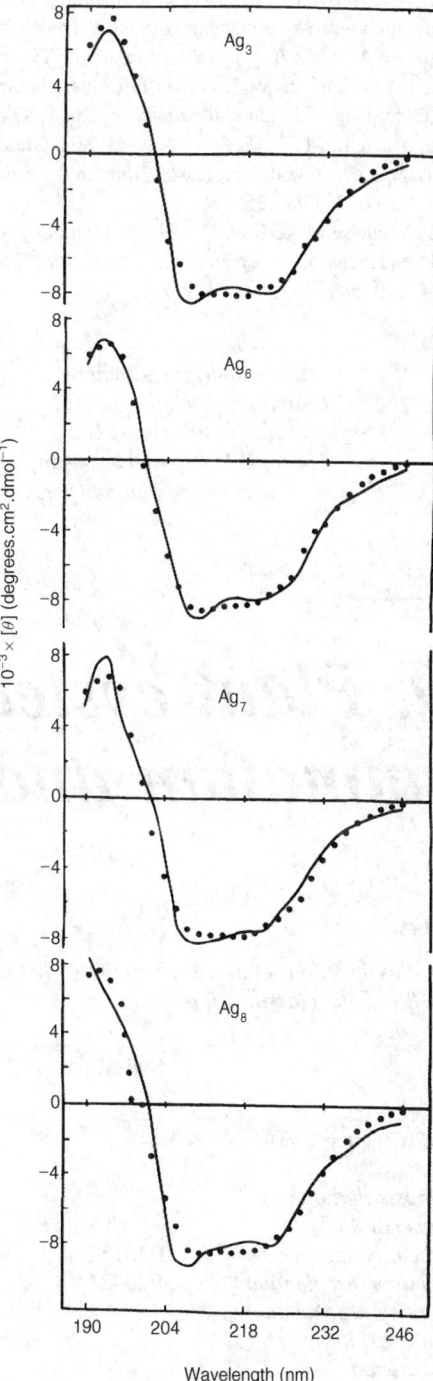

*Figure 198.1* Circular dichroism spectra of four isolated asclepain g forms. Spectra were recorded at pH 7.0 and 25°C; dots represent curves calculated from the results of the least-squares fitting to the basic spectra of Hennessey & Johnson (1981).

## References

Barragán, B.E., Hernández-Arana, A., Oliver, M., Castañeda-Agulló, M. & Del Castillo, L.M. (1986) Proteinases of Mexican plants. XII. Kinetic and conformational studies with asclepain g. *Rev. Latinoamer. Quim.* **16**, 158–160.

Brockbank, W.J. & Lynn, K.R. (1979) Purification and preliminary characterization of two asclepains from the latex of *Asclepias syriaca* L. (milkweed). *Biochim. Biophys. Acta* **578**, 13–22.

Hennessey, J.P. & Johnson W.C. (1981) Information content in the circular dichroism of proteins. *Biochemistry* **20**, 1085–1094.

Lynn, K.R., Yaguchi, M. & Clevette, N.A. (1980a) Multiple forms of the asclepains cysteinyl proteases from milkweed. *Biochim. Biophys. Acta* **612**, 119–125.

Lynn, K.R., Yaguchi, M. & Roy, C. (1980b) Homologies of the N-terminal sequences of asclepains and papain. *Biochim. Biophys. Acta* **624**, 579–580.

Robinson, G.W. (1975) Isolation and characterization of papaya peptidase A from commercial chymopapain. *Biochemistry* **14**, 3695–3700.

Tablero, M., Arreguín, R., Arreguín, B., Soriano, M., Sanchéz, R.I., Rodríguez-Romero, A. & Hernández-Arana, A. (1991) Purification and characterization of multiple forms of asclepain g from *Asclepias glaucescens* H.B.K. *Plant Sci.* **74**, 7–15.

Winnick, T., Davies, A.R. & Greenberg, D.M. (1940) Physicochemical properties of the proteolytic enzyme from the latex of the milkweed, *Asclepias speciosa* Torr. Some comparisons with other proteases. *J. Gen. Physiol.* **23**, 275–308.

*Adela Rodríguez-Romero*
*Instituto de Química, UNAM,*
*Circuito Exterior, CU,*
*México, D.F. 04510, Coyoacán*
*Email: adela@servidor.unam.mx*

*Andrés Hernández-Arana*
*Departamento de Química,*
*Universidad Autónoma Metropolitana-Iztapalapa,*
*México, D.F. 09340, Iztapalapa*
*Email: aha@xanun.uam.mx*

# 199. Plant cysteine proteinases in germination and senescence

## Databanks

*Peptidase classification: clan CA, family C1, MEROPS ID: C01.010, C01.029*
*NC-IUBMB enzyme classification: none*
Databank codes:

| Species | SwissProt | PIR | EMBL (cDNA) | EMBL (genomic) |
|---|---|---|---|---|
| **Legumain (Chapter 253)** | | | | |
| Senescence | | | | |
| *Arabidopsis thaliana* | – | – | U37336 | – |
| *Dianthus caryophyllus* | – | – | U17135 | – |
| *Hemerocallis* sp. | P43156 | – | X74406 | – |
| *Lycopersicon esculentum* | P20721 | JA0159 | M21444 | – |
| *Lycopersicon esculentum* | – | – | Z48736 | – |
| *Petunia hybrida* | – | – | U31094 | – |
| *Pisum sativum* | – | S24602 | X66061 | – |
| Germination | | | | |
| *Arabidopsis thaliana* | P43296 | JN0718 | – | D13042: complete gene |
| *Arabidopsis thaliana* | P43297 | JN0719 | – | D13043: complete gene |
| *Brassica napus* | P25251 | JQ1121 | – | – |
| *Cicer arietinum* | – | S31914 | X70375 | – |
| | | S46541 | | |
| *Hordeum vulgare* | P05167 | A25492 | – | X05167: complete gene |
| *Hordeum vulgare* | P25249 | JQ1111 | – | U19359: complete gene (no introns) |
| *Hordeum vulgare* | P25250 | JQ1110 | – | U19384: complete gene (no introns) |
| *Oryza sativa* | P25776 | JU0388 | D90406 | – |
| *Oryza sativa* | P25777 | JU0389 | D90407 | – |
| *Oryza sativa* | P25778 | JU0390 | D90408 | – |
| *Phaseolus vulgaris* | P25803 | S16251 | X56753 | X63102: complete gene |

| Species | SwissProt | PIR | EMBL (cDNA) | EMBL (genomic) |
|---|---|---|---|---|
| Germination (*continued*) | | | | |
| *Triticum aestivum* | – | – | X66012 | – |
| *Triticum aestivum* | – | – | X66014 | X66013: complete gene |
| *Triticum aestivum* | – | – | X66015 | – |
| *Triticum aestivum* | – | – | – | X66116: promoter |
| *Vicia sativa* | – | S42882 | Z30338 | – |
| *Vicia sativa* | – | S47312 | X75749 | – |
| *Vigna mungo* | P12412 | S05497 | X15732 | X51900: complete gene |
| | | S12581 | | |
| | | S20213 | | |
| | | S48684 | | |
| *Zea mays* | Q10716 | – | D45402 | – |
| | | | X82185 | |

## Name and History

Plant cysteine proteinases active in germination and senescence are major proteolytic enzymes induced in the plants for the mobilization of nitrogen from storage proteins in the first case or as part of a salvage mechanism in the latter. In both situations, digestion products are to be exported to the growing parts of the plant.

Most of the cysteine proteinases that are induced during germination or senescence have not been given names, but are referred to by short acronyms such as **EP-A, EP-B, SH-EP, CCP1, CCP2, MEP-1**, or by general names like cysteine endoprotease, cysteine proteinase, sulfhydryl proteinase and proteinase A. When specific names have been given they are commonly derived from the plant of origin; such names are **oryzain** (from the Latin name for rice, *Oryza*) and **vignain**, which was found in *Vigna aconitifolia*. In the absence of more satisfactory names, we shall here term the group of cysteine proteinases that are induced in germination **GerCP** (for **germination cysteine proteinases**), and those that are involved in senescence, **SenCP** (for **senescence cysteine proteinases**).

Information about most of the GerCP and SenCP is fragmentary, but all of them are characterized by small molecular size, generally acidic pI and pH optimum, broad substrate specificity, and sensitivity to cysteine proteinase inhibitors. Most of the GerCP and SenCP are relatives of papain, in peptidase family C1 (Chapter 186), but a few belong to the quite different family C13. These latter enzymes are forms of legumain (Chapter 253) (earlier called proteinase B) (Shutov & Vaintraub, 1987) and may be involved in the post-translational processing of other GerCP, SenCP, or other types of proproteins, or may simply participate in protein mobilization.

GerCP were first described from barley by Jacobsen & Varner (1967) and later studied mainly in monocots (wheat, barley, corn and rice) and in legumes (soyabean, *Vigna mungo*, *Phaseolus vulgaris*). SenCP have been characterized during senescence of leaves and flower organs in different plants: leaf senescence mainly in monocots (Huffaker, 1990; Feller & Fischer, 1994) and flower senescence in ephemeral and economically important plants (Jones *et al.*, 1995).

## Activity and Specificity

GerCP and SenCP activities have been assayed by using commercial proteins (hemoglobin, casein, azocasein) as substrates. Activity against possible endogenous substrates such as storage proteins for GerCP or Rubisco (D-ribulose 1,5-diphosphate carboxylase) for SenCP have also been determined. The few reported studies of cleavage-site specificity indicate that the enzymes behave as typical papain-family members with broad specificity, but sometimes preferring particular amino acids (e.g. glutamine for the barley MEP-1 form of GerCP) (Phillips & Wallace, 1989)

Although most SenCP and GerCP exhibit broadly acidic pH-activity profiles, there is a minor group that shows maximal activity at neutral pH. Both types of activity have been shown to increase during senescence and germination, although the temporal pattern of expression may differ between species of plants and systems.

Electrophoresis in substrate-containing gels has shown multiple isoforms of these enzymes, although in most cases, one or two of the forms represent more than 90% of the total cysteine proteinase activity (Enari & Mikola, 1967; Vercher *et al.*, 1989; Wrobel & Jones, 1992; Domínguez & Cejudo, 1995).

Activity is stimulated by a low concentration of thiol compounds and inhibited by typical cysteine proteinase inhibitors, as is expected for cysteine proteinases (Sundblom & Mikola, 1972; Csoma & Polgár, 1984; Hammerton & Ho, 1986). Cystatins able to block rice oryzains *in vitro* have been described from rice seeds (Kondo *et al.*, 1990) suggesting that *in vivo* these inhibitors may have a role in regulating GerCP protease activity.

The activity of the legumain-type enzymes is induced during both senescence (Alonso & Granell, 1995) and germination (Cornel & Plaxton, 1994; Okamoto & Minamikawa, 1995). These enzymes are specific for hydrolysis of asparaginyl bonds, as is described by Cornel & Plaxton (1994) and in Chapter 253.

## Structural Chemistry

GerCP and SenCP are synthesized as single-chain, monomeric preproenzymes with a signal peptide that is cleaved upon entering the secretory pathway. In some cases the proenzyme has been shown to be further processed through a number of intermediate forms of smaller $M_r$ to give the active enzyme (Mitsuhashi *et al.*, 1986; Mitsuhashi & Minamikawa, 1989; Koehler & Ho, 1990b; Marttila *et al.*, 1995). Processing sometimes seems to depend upon the activity of other, legumain-type cysteine proteinases that are coordinately expressed (Okamoto & Minamikawa, 1995). In a number of SenCP, the proenzyme contains a C-terminal extension that seems to be only partially removed during processing (Karrer *et al.*, 1993). For at least some of the GerCP, like EP-A and EP-B, the mature protein is a single-chain form (Koehler & Ho, 1990b), in contrast to the usual situation for mammalian cathepsins B, H and L, which are processed to heavy and light chains.

Except for the legumain-like and the cathepsin B-like types (Cejudo *et al.*, 1992a), all GerCP and SenCP have the Glu-Arg-Phe-Asn-Ile-Asn (or 'ERFNIN') motif in the pro sequence. The function of this motif is not known, but it has being suggested to block the active site in the proenzyme (Karrer *et al.*, 1993).

The mature forms of most GerCP and SenCP are not glycosylated, but exceptions have been described, like the GerCP MEP-1 (Phillips & Wallace, 1989). Some prepro forms may be also glycosylated (Koehler & Ho, 1990b). Some members contain vacuolar targeting signals in the pro sequence that may mediate their sorting to the vacuole while others are secreted to the extracellular space (Koehler & Ho, 1990b; Chrispeels & Raikhel, 1992; Domoto *et al.*, 1995; Marttila *et al.*, 1995; Drake *et al.*, 1996). Some of the SenCP contain an endoplasmic reticulum (ER) retention signal (Akasofu *et al.*, 1989; Tanaka *et al.*, 1991; Valpuesta *et al.*, 1995) and are probably residents of the ER or contained in the autophagic vesicles formed from the rough ER during senescence. The long C-terminal extensions of some SenCP are preceded by a conserved proline-rich motif similar to that which has been shown to mediate protein–protein interaction with the SH3 domains of intracellular signaling proteins (Yu *et al.*, 1994). At least part of this C-terminal region seems to be conserved in the mature protein (Karrer *et al.*, 1993).

In germinating winged-bean (*Psophocarpus tetragonolobus*), an acidic proteinase of the legumain family that has an affinity for $Ca^{2+}$ has been implicated in the mobilization of the 29 kDa storage protein of the bean (Usha & Singh, 1996).

## Preparation

Germinating seeds are a good source for the purification of GerCP, and similarly, senescing petals and ovaries, and yellowing leaves, are a good source of SenCP.

Purified enzyme preparations, homogeneous in analytical gel electrophoresis have been obtained from a number of sources of GerCP (Baumgartner & Chrispeels, 1977; Koehler & Ho, 1988, 1990a; Poulle & Jones, 1988; Papastoitsis & Wilson, 1991; Okamoto & Minamikawa, 1995) and for SenCP (Cercós & Carbonell, 1993).

Recombinant forms of GerCP (Holwerda *et al.*, 1990) and SenCP (Aghero *et al.*, 1996) have been obtained by overexpression in *Escherichia coli*; these constructs were not directed to obtaining active proteins and therefore no protease activity was determined.

## Biological Aspects

GerCP have been reported to carry out limited proteolysis, initiating the hydrolysis of seed storage proteins, and also to have a lysosomal role. The activity seems to be regulated both at the level of transcription (Koehler & Ho, 1990b; Cejudo *et al.*, 1992a; Yamauchi *et al.*, 1996) and at the post-translational level by limited proteolysis of an inactive precursor by an asparaginyl endopeptidase, or legumain, that is coordinately regulated (Okamoto & Minamikawa, 1995).

GerCP and SenCP can be specific for germinating seeds and senescent organs, not being expressed at all in other plant tissues, or they can be less specific, their levels nevertheless increasing dramatically during those processes (Watanabe *et al.*, 1991; Granell *et al.*, 1992; Smart *et al.*, 1995). In a number of cases, GerCP seem to be expressed during senescence of plant tissues (Drake *et al.*, 1996). Whether the mRNA detected is truly the product of the same gene, or a very closely related but distinct gene, remains to be determined. Some germination-specific cysteine proteinases are not only seed specific but expressed in particular cell types, e.g. aleurone and scutelar epithelial cells (Marttila *et al.*, 1993; Mikkonen *et al.*, 1996). During senescence, SenCP expression can also be highly regulated both spatially and temporally (Granell *et al.*, 1992).

Many GerCP are normally synthesized in one cell type to initiate the degradation of the storage proteins in other cell types; they are therefore secreted (Sundblom & Mikola, 1972). Other GerCP have been detected in Golgi, secretion vesicles (Marttila *et al.*, 1995) or vacuole (Holwerda *et al.*, 1990; Cejudo *et al.*, 1992b), and might serve a lysosomal-like role or digest peptide products produced by the secreted enzymes. The subcellular localization of SenCP extends to endoplasmic reticulum (Valpuesta *et al.*, 1995), vacuole, cell wall and extracellular space (Cercós *et al.*, 1993).

In some plant species at least, GerCP seem to be hormonally induced by gibberellins and repressed by abscisic acid, during seed germination. Abscisic acid is present in the dry seed and its concentration decreases at the onset of germination while gibberellins are synthesized by the embryo and act on the aleurone layer. In chickpea, ethylene, another plant growth regulator, seems to modulate GerCP expression (Cervantes *et al.*, 1994). In most plants, SenCP activity is regulated at the level of mRNA abundance by ethylene, during flower organ senescence (Jones *et al.*, 1995). The expression of SenCP during the senescence of ethylene-insensitive flowers has also been reported (Valpuesta *et al.*, 1995). In senescing leaves, the involvement of ethylene is more controversial (Hensel *et al.*, 1993; Drake *et al.*, 1996) and in some systems, a decrease in the levels of cytokinins (Tournaire *et al.*, 1996) or the balance between different growth regulators causes the increase in cysteine proteinase expression (Ye & Varner, 1993).

The specific expression of GerCP during germination is regulated at the level of transcription and is at least partially conserved through different species, as demonstrated by the ability of the promoter of GerCP from legumes to drive specific expression during germination in transgenic tobacco (Yamauchi *et al.*, 1996). The promoter region of GerCP appears also to contain the hormonal regulatory regions of the gene (Cejudo *et al.*, 1992a; Yamauchi *et al.*, 1996; Mikonnen *et al.*, 1996).

GerCP and SenCP are involved in the endoproteolytic digestion of storage or salvage proteins, perhaps to the point of extensive proteolysis to small peptides (Shutov & Vaintraub, 1987). In addition, it is possible that GerCP participate in the releasing of $\beta$-amylase from a protein-bound, less active form to a soluble, active form primarily responsible for the mobilization of sugars (Sopanen & Lauriere, 1989).

The accessibility and compartmentation, as well as the redox state, of the substrate proteins have been suggested to determine their susceptibilities to degradation by cysteine proteinases (Shutov & Vaintraub, 1987; Buchanan, 1991; Moreno *et al.*, 1995).

Evidence obtained with antibodies against SenCP and GerCP in western blots and with DNA probes in northern blots suggests that in a number of plants either the same or highly similar proteins and genes are expressed in both germination and senescence (Drake *et al.*, 1996). The different clusters obtained by sequence data analysis include members of both SenCP and GerCP, suggesting that specific types of proteases can be used in both situations.

## Distinguishing Features

Polyclonal antisera against purified or recombinant GerCP (Holwerda *et al.*, 1990; Koehler & Ho, 1990b; Marttila *et al.*, 1995) and SenCP (Cercós & Carbonell, 1993; Aghero *et al.*, 1996) have been described, but are not commercially available.

Specific patterns of degradation of presumed natural substrates could provide the best fingerprint for identification of these endopeptidases (Poulle & Jones, 1988; Torrent *et al.*, 1989; Barros & Larkins, 1990; Koehler & Ho, 1990b; Cercós *et al.*, 1992).

A subset of GerCP (Watanabe *et al.*, 1991) and SenCP (Schaffer & Fischer, 1988; Granell *et al.*, 1992; Drake *et al.*, 1996) contain a well-conserved C-terminus that is specific for them, only a cysteine proteinase from *Trypanosoma brucei* sharing 20% homology (Mottram *et al.*, 1989).

## References

Aghero, M.S., Granell, A. & Carbonell, J. (1996) Expression of thiol proteases decreases in tomato ovaries after fruit set induced by pollination or gibberellic acid. *Physiol. Plant.* **98**, 235–240.

Akasofu, H., Yamauchi, D., Mitsuhashi, W. & Minamikaya, T. (1989) Nucleotide sequence of cDNA for sulfhydryl-endopeptidase (SH-EP) from cotyledons of germinating *Vigna mungo* seeds. *Nucleic Acids Res.* **17**, 6733.

Alonso, J.M. & Granell, A. (1995) A putative vacuolar processing protease is regulated by ethylene and also during fruit ripening in citrus fruit. *Plant Physiol.* **109**, 541–547.

Baumgartner, B. & Chrispeels, M.J. (1977) Purification and characterization of vicilin peptidohydrolase, the major endopeptidase in the cotyledons of mung-bean seedlings. *Eur. J. Biochem.* **77**, 223–233.

Barros, E.G. & Larkins, B.A. (1990) Purification and characterization of zein-degrading proteases from endosperm of germinating maize seeds. *Plant Physiol.* **94**, 297–303.

Buchanan, B.B. (1991) Thioredoxin: an ubiquituous protein regulating photosynthesis functions in reduction of key seed proteins. *Cereals Foods World* **36**, 687.

Cejudo, F.J., Ghose, T.K., Stabel, P. & Baulcombe, D.C. (1992a) Analysis of the gibberellin-responsive promoter of a cathepsin B-like gene from wheat. *Plant Mol. Biol.* **20**, 849–856.

Cejudo, F.J., Murphy, G., Chinoy, C. & Baulcombe, D.C. (1992b) A gibberellin-regulated gene from wheat with sequence homology to cathepsin B of mammalian cells. *Plant J.* **2**, 937–948.

Cercós, M. & Carbonell, J. (1993) Purification and characterization of a thiol-protease induced during senescence of unpollinated ovaries of *Pisum sativum*. *Physiol. Plant.* **88**, 267–274.

Cercós, M., Carrasco, P., Granell, A. & Carbonell, J. (1992) Biosynthesis and degradation of Rubisco during ovary senescence and fruit development induced by gibberellic acid in *Pisum sativum*. *Physiol. Plant.* **85**, 476–482

Cercós, M., Harris, N. & Carbonell, J. (1993) Immunolocalization of a thiol-protease induced during the senescence of unpollinated pea ovaries. *Physiol. Plant.* **88**, 275–280.

Cervantes, E., Rodríguez, A. & Nicolás, G. (1994) Ethylene regulates the expression of a cysteine proteinase gene during germination of chickpea (*Cicer arietinum* L.). *Plant Mol. Biol.* **25**, 207–215.

Chrispeels, M.J. & Raikhel, N.V. (1992) Short peptide domains target proteins to plant vacuoles. *Cell* **68**, 613–616.

Cornel, F.A. & Plaxton, W.C. (1994) Characterization of asparaginyl endopeptidase activity in endosperm of developing and germinating castor oil seeds. *Physiol. Plant.* **91**, 599–604.

Csoma, C. & Polgár, L. (1984) Proteinase from germinating bean cotyledons. Evidence for involvement of a thiol group in catalysis. *Biochem. J.* **222**, 769–776.

Domínguez, F. & Cejudo, F.J. (1995) Pattern of endoproteolysis following wheat grain germination. *Physiol. Plant.* **95**, 253–259.

Domoto, C., Watanabe, H., Abe, M., Abe, K. & Arai, S. (1995) Isolation and characterization of two distinct cDNA clones encoding corn seed cysteine proteinases. *Biochim. Biophys. Acta* **1263**, 241–244.

Drake, R., John, I., Farrell, A., Cooper, W., Schuch, W. & Grierson, D. (1996) Isolation and analysis of cDNAs encoding tomato cysteine proteases expressed during leaf senescence. *Plant Mol. Biol.* **30**, 755–767.

Enari, T.-M. & Mikola, J. (1977) Peptidases in germinating barley grain: properties, localization and possible functions. In: *Peptide Transport and Hydrolysis*. CIBA Foundation Symposium, No. 50. Amsterdam: North-Holland Publishing, pp. 335–352.

Feller, U. & Fischer, A. (1994) Nitrogen metabolism in senescing leaves. *Crit. Rev. Plant Sci.* **13**, 241–273.

Granell, A., Harris, N., Pisabarro, A.G. & Carbonell, J. (1992) Temporal and spatial expression of a thiolprotease gene during pea ovary senescence, and its regulation by gibberellins. *Plant J.* **2**, 907–915.

Hammerton, R.W. & Ho, T.-H.D. (1986) Hormonal regulation of the development of protease and carboxypeptidase activities in barley aleurone layers. *Plant Physiol.* **80**, 692–697.

Hensel, L.L., Grbic, V., Baumgarten, D.A. & Bleecker, A.B. (1993) Developmental and age-related processes that influence the longevity and senescence of photosynthetic tissues in *Arabidopsis*. *Plant Cell* **5**, 553–564.

Holwerda, B.C., Galvin, N.J., Baranski, T.J. & Rogers, J.C. (1990) In vitro processing of aleurain, a barley vacuolar thiol protease. *Plant Cell* **2**, 1091–1106.

Huffaker, R.C. (1990) Proteolytic activity during senescence of plants. *New Phytol.* **116**, 199–231.

Jacobsen, J.V. & Varner, J.E. (1967) Gibberellic acid-induced synthesis of protease by isolated alurone layers of barley. *Plant Physiol.* **42**, 1596–1600.

Jones, M.L., Larsen, P.B. & Woodson, W.R. (1995) Ethylene-regulated expression of a carnation cysteine proteinase during flower petal senescence. *Plant Mol. Biol.* **28**, 505–512.

Karrer, K.M., Peiffer, S.L. & Di Tomas, M.E. (1993) Two distinct genes subfamilies within the family of cysteine protease genes. *Proc. Natl Acad. Sci. USA* **90**, 3063–3067.

Koehler, S. & Ho, T.-H.D. (1988) Purification and characterization of gibberellic acid-induced cysteine endoproteases in barley aleuron layers. *Plant Physiol.* **87**, 95–103.

Koehler, S. & Ho, T.-H.D. (1990a) A major gibberellic acid-induced barley aleurone cysteine proteinase which digest hordein. Purification and characterization. *Plant Physiol.* **94**, 251–258.

Koehler, S. & Ho, T.-H.D. (1990b) Hormonal regulation, processing and secretion of cysteine proteinases in barley aleurone layers. *Plant Cell* **2**, 769–783.

Kondo, H., Abe, K., Nishimura, I., Watanabe, H., Emori, Y. & Arai, S. (1990) Two distinct cystatin species in rice seeds with different specificities against cysteine proteinases. *J. Biol. Chem.* **265**, 15832–15837.

Marttila, S., Porali, I., Ho, T.-H.D. & Mikkonen, A. (1993) Expression of the 30 kD cysteine endoprotease B in germinating barley seeds. *Cell Biol. Int.* **2**, 205–212.

Marttila, S., Jones, B.L. & Mikkonen, A. (1995) Differential localization of two acid proteinases in germinating barley (*Hordeum vulgare*) seed. *Physiol Plant.* **93**, 317–327.

Mikkonen, A., Porali, I., Cercós, M. & Ho, T.-H.D. (1996) A major cysteine proteinase, EPB, in germinating barley seeds: structure of two intronless genes and regulation of expression. *Plant Mol. Biol.* **31**, 239–254.

Mitsuhashi, W. & Minamikawa, T. (1989) Synthesis and posttranslational activation of sulfhydryl endopeptidases in cotyledons of germinating *Vigna mungo* seed. *Plant Physiol.* **89**, 274–279.

Mitsuhashi, W., Koshiba, T. & Minamikawa, T. (1986) Separation and characterization of two endopeptidases from cotyledon of germinating *Vigna mungo* seeds. *Plant Physiol.* **80**, 628–634.

Moreno, J., Peñarrubia, L. & García-Ferris, C. (1995) The mechanism of redox regulation of ribulose-1,5-bisphosphate carboxylase/oxygenase turnover. A hypothesis. *Plant Physiol. Biochem.* **33**, 121–127.

Mottram, J.C., North, M.J., Barry, J.D. & Coombs, G.H. (1989) A cysteine proteinase cDNA from *Trypanosoma brucei* predicts an enzyme with an unusual C-terminal extension. *FEBS Lett.* **258**, 211–215.

Okamoto, T. & Minamikawa, T. (1995) Purification of a processing enzyme (VMPE-1) that is involved in post-translational processing of a plant cysteine endopeptidase (SH-EP). *Eur. J. Biochem.* **231**, 300–305.

Papastoitsis, G. & Wilson, K.A. (1991) Initiation of degradation of the soybean Kunitz and Bowman-Birk trypsin inhibitors by a cysteine protease. *Plant Physiol.* **96**, 1086–1092.

Phillips, H.A. & Wallace, W. (1989) A cysteine endopeptidase from barley malt which degrades hordein. *Phytochemistry* **28**, 3285–3290.

Poulle, M. & Jones, B.L. (1988) A proteinase from germinating barley. Purification and some physical properties of a 30 kD cysteine endoproteinase from green malt. *Plant Physiol.* **88**, 1454–1460.

Schaffer, M.A. & Fischer, R.L. (1988) Analysis of mRNAs that accumulate in response to low temperature identifies a thiol protease gene in tomato. *Plant Physiol.* **87**, 431–436.

Shutov, A.D. & Vaintraub, I.A. (1987) Degradation of storage proteins in germinating seeds. *Phytochemistry* **26**, 1557–1566.

Smart, C.M., Hosken, S.E., Thomas, H., Greaves, J.A., Blair, B.G. & Schuch, W. (1995) The timing of maize leaf senescence and characterisation of senescence-related cDNAs. *Physiol. Plant.* **93**, 673–682.

Sopanen, T. & Lauriere, C. (1989) Release and activity of bound α-amylase in a germinating barley grain. *Plant Physiol.* **89**, 244–249.

Sundblom, N.O. & Mikola, J. (1972) On the nature of the proteinases secreted by the aleurone layer of barley grain. *Physiol. Plant.* **27**, 281–284.

Tanaka, T., Yamauchi, D. & Minamikawa, T. (1991) Nucleotide sequence of cDNA for an endopeptidase (EP-C1) from pods of maturing *Phaseolus vulgaris* fruits. *Plant Mol. Biol.* **16**, 1083–1084.

Torrent, M., Geli, M.I. & Ludevid, M.D. (1989) Storage-protein hydrolysis and protein body breakdown in germinated *Zea mays* L. seeds. *Planta* **180**, 90–95.

Tournaire, C., Kushnir, S., Bauw, G., Inze, D., Delaserve, B.T. & Ranudin, J.P. (1996) A thiol protease and an anionic peroxidase are induced by lowering cytokinins during callus growth in petunia. *Plant Physiol.* **111**, 159–168.

Usha, R. & Singh, F. (1996) Proteases of germinating winged-bean (*Psophocarpus tetragonolobus*) seeds – purification and characterization of an acidic protease. *Biochem. J.* **313**, 423–429.

Valpuesta, V., Lange, N.E., Guerrero, C. & Reid, M.S. (1995) Up-regulation of a cysteine protease accompanies the ethylene-insensitive senescence of daylily (*Hemerocallis*) flowers. *Plant Mol. Biol.* **28**, 575–582.

Vercher, Y., Carrasco, P. & Carbonell, J. (1989) Biochemical and histochemical detection of endoproteolytic activites involved in ovary senescence or fruit development in *Pisum sativum. Physiol. Plant.* **76**, 405–411.

Watanabe, H., Abe, K., Emori, Y., Hosoyama, H. & Arai, S. (1991) Molecular cloning and gibberellin induced expression of multiple cysteine proteinases of rice seeds (oryzains). *J. Biol. Chem.* **266**, 16897–16902.

Wrobel, R. & Jones, B.L. (1992) Appearance of endoproteolytic enzymes during the germination of barley. *Plant Physiol.* **100**, 1508–1516.

Yamauchi, D., Terasaki, Y., Okamoto, T. & Minamikawa, T. (1996) Promoter regions of cysteine endopeptidase genes from legumes confer permination-specific expression in transgenic tobacco seeds. *Plant Mol. Biol.* **30**, 321–329.

Ye, Z.-H. & Varner, J.E. (1993) Gene expression patterns associated with *in vitro* tracheary formation in isolated single mesophyll cells of *Zinnia elegans. Plant Physiol.* **103**, 805–813.

Yu, H., Chen, J.K., Feng, S., Dalgarno, D.C., Brauer, A.W. & Schreiber, S.L. (1994) Structural basis for the binding of proline-rich polypeptides to SH3 domains. *Cell* **76**, 933–945.

**Antonio Granell**
*Departamento de Biología
del Desarrollo,
Instituto de Biología Molecular y
Celular de Plantas,
Universidad Politécnica de
Valencia-CSIC,
Camino de Vera s/n,
E-46022-Valencia, Spain*

**Manuel Cercós**
*Departamento de Biología
del Desarrollo,
Instituto de Biología Molecular y
Celular de Plantas,
Universidad Politécnica de
Valencia-CSIC,
Camino de Vera s/n,
E-46022-Valencia, Spain*

**Juan Carbonell**
*Departamento de Biología
del Desarrollo,
Instituto de Biología Molecular y
Celular de Plantas,
Universidad Politécnica de
Valencia-CSIC,
Camino de Vera s/n,
E-46022-Valencia, Spain
Email: jcarbon@ibmcp.upv.es*

# 200. Aleurain

## Databanks

*Peptidase classification: clan CA, family C1, MEROPS ID: C01.041*
*NC-IUBMB enzyme classification: none*
*Databank codes:*

| Species | SwissProt | PIR | EMBL (cDNA) | EMBL (genomic) |
|---|---|---|---|---|
| *Arabidopsis thaliana* | – | – | Z35026 | – |
| *Cicer arietinum* | – | – | X93220 | – |
| *Hordeum vulgare* | P05167 | A25492 | – | X05167: complete gene |
| *Lycopersicon esculentum* | – | – | Z48736 | – |
| *Nicotiana tabacum* | – | – | Y11003 | – |
| *Oryza sativa* | P25778 | JU0390 | D90408 | – |
| *Petunia hybrida* | – | – | U31094 | – |
| *Pisum sativum* | – | – | Z68291 | – |
| *Zea mays* | – | – | D45403 | – |
| | | | X99936 | |

## Name and History

The cDNA for **aleurain** was identified in a differential screen of a library prepared from gibberellin-treated barley aleurone layers; it hybridized more strongly to [32]P-labeled cDNA prepared from poly(A)-containing RNA from gibberellin-treated as compared to that from untreated aleurone cells. The sequence of the cDNA clone predicted a protein closely related to the mammalian lysosomal cysteine protease cathepsin H. The name of the barley gene (and its protein product) was derived from 'aleur-', indicating the cell type from which the cDNA was obtained, and '-ain', designating a cysteine protease (Rogers *et al*., 1985). Subsequently cDNAs representing aleurain homologs from other plants, such as **oryzain** from rice (Watanabe *et al*., 1991), **CCP2** from maize (Domoto *et al*., 1995), and **PeTh3** from petunia (Tournaire *et al*., 1996) have been cloned.

## Activity and Specificity

Aleurain is primarily an aminopeptidase with a pH optimum of about 5; it hydrolyzed three different aminopeptidase substrates, Arg+NHNap, Arg+NHMec and Leu+NHMec, with similar catalytic efficiency, but was 16-fold less efficient at hydrolyzing an *N*-blocked substrate analog, Bz-Arg-NHNap (Holwerda & Rogers, 1992). A more detailed analysis of $k_{cat}/K_m$ specificity constants for different substrates demonstrated a decrease in the order citrulline-NHMec > Arg-NHMec = Phe-NHMec >> Ala-NHMec (Rothe *et al.*, 1994). Those authors also observed a 75-fold decrease in the specificity constant when Bz-Arg-NHMec was used instead of Arg-NHMec (Rothe *et al.*, 1994). Although the enzyme showed optimal activity over a broad temperature range, 25–35°C, most of the activity was lost when it was incubated at 40°C (Holwerda & Rogers, 1992). Aleurain was inhibited to varying extents by compounds specific for cysteine proteases, E-64 being the most effective inhibitor (Holwerda & Rogers, 1992). Chicken cystatin inhibited aleurain competitively with $K_i$ of 133 nM (Rothe *et al.*, 1994).

## Structural Chemistry

Aleurain is synthesized as a 42 kDa proenzyme and proteolytically processed to the mature form in an acidified compartment of the plant cell secretory pathway (Holwerda *et al.*, 1990). A single N-terminal sequence, Ala-Xaa-Leu-Pro-Glu-Thr, was obtained from the purified 32 kDa aleurain polypeptide after denaturing electrophoresis and transfer to a polyvinylidene difluoride membrane (Holwerda & Rogers, 1992), while Rothe *et al.* (1994) obtained an N-terminal sequence of Leu-Pro-Glu-Thr-Lys-Asp for the 32 kDa polypeptide. Two smaller forms of 29 and 26 kDa were identified from cells incubated in the presence of tunicamycin to inhibit *N*-linked glycosylation, indicating that the mature protein has two *N*-linked oligosaccharide chains (Holwerda *et al.*, 1990). Most cysteine proteases of the papain family have six conserved Cys residues that form three intramolecular disulfide bonds in their mature protease domains, but aleurain and cathepsin H (Chapter 213) possess an additional pair of Cys residues not found in other members of this family. As in cathepsin H (Ritonja *et al.*, 1988), the N-terminally positioned extra Cys residue in aleurain is part of a minichain of eight amino acids linked through a disulfide bond with the other extra Cys residue near the C-terminus of the 32 kDa polypeptide (Holwerda & Rogers, 1993; Rothe *et al.*, 1994). In comparison to procathepsin H, proaleurain has 21 extra amino acids at the N-terminus of its propeptide (Holwerda & Rogers, 1993). These additional residues comprise the determinants responsible for targeting proaleurain to an acidified, lytic plant vacuole (Holwerda *et al.*, 1992) through their interaction with a sorting receptor protein (Kirsch *et al.*, 1994).

## Preparation

The protease domain of aleurain was expressed as a fusion protein in *Escherichia coli* to obtain denatured protein for antibody preparation (Holwerda *et al.*, 1990). Proaleurain was expressed in *Xenopus* oocytes, from which it was secreted in a properly folded form suitable as a substrate for processing enzymes (Holwerda *et al.*, 1990). Enzymatically active aleurain was purified >1000-fold to homogeneity from barley leaf tissue (Holwerda & Rogers, 1993; Rothe *et al.*, 1994).

## Biological Aspects

Processing of proaleurain *in vivo* involves two sequential steps: 'clipping', in which about 9 kDa is removed, and 'trimming', in which another 1 kDa is removed (Holwerda *et al.*, 1990). These steps were reproduced *in vitro* with extracts from barley aleurone cells, and the results indicated involvement of two separate enzymes: the clipping activity was not blocked by any inhibitor tested, but the trimming activity was blocked by E-64. Processing proceeded optimally at pH 5. Sorting of proaleurain in the secretory pathway to the vacuole is mediated by the amino acids SSSSFADSNPIR-PVTDRAAST, which comprise the N-terminus of the pro sequence (Holwerda *et al.*, 1992). A synthetic peptide with this sequence binds with high affinity to a type I transmembrane protein, identified as BP-80, that is highly enriched in pea clathrin-coated vesicles, and mutations of the sequence that prevent proper vacuolar sorting also prevent binding to BP-80 (Kirsch *et al.*, 1994). All evidence supports the concept that BP-80 is a receptor responsible for sorting proaleurain into a clathrin-coated vesicle pathway that carries it to the vacuole (Paris *et al.*, 1996). Aleurain was used in immunofluorescence experiments with barley root tip cells as a marker for a vacuolar compartment that has an acidic pH and contains other hydrolases. These studies demonstrated that the acidified, lytic vacuolar compartment containing aleurain is separate and distinct from the vacuolar compartment containing storage proteins (Paris *et al.*, 1996). Aleurain is a single-copy gene in the barley genome. Its close similarity to the mammalian lysosomal enzyme cathepsin H indicates that the two proteins have a function that will not tolerate further sequence divergence. The fact that two dicotyledonous plants, *Petunia hybrida* and *Arabidopsis thaliana*, and barley, a monocotyledonous plant, all have single highly conserved forms of aleurain further indicates that a very specific function of the protein must be important for normal plant growth and development. This function has yet to be clarified. Aleurain is expressed in root and shoot as well as aleurone tissue in barley, and the PeTh3 mRNA was found in all organs of *Petunia* plants (Tournaire *et al.*, 1996).

## Distinguishing Features

The presence of a minichain has been discussed above. Specific rabbit polyclonal antibodies have been raised to a portion of aleurain expressed in *E. coli* (Holwerda *et al.*, 1990; Paris *et al.*, 1996). Recently, aleurain-specific monoclonal antibodies have been obtained that also recognize the homologs in *Arabidopsis* and *Petunia* (S.W. Rogers & J.C. Rogers, unpublished results).

## Further Reading

A review has been published by Holwerda & Rogers (1993).

## References

Domoto, C., Watanabe, H., Abe, M., Abe, K. & Arai, S. (1995) Isolation and characterization of two distinct cDNA clones encoding corn seed cysteine proteinases. *Biochim. Biophys. Acta* **1263**, 241–244.

Holwerda, B.C. & Rogers, J.C. (1992) Purification and characterization of aleurain: a plant thiol protease functionally homologous to mammalian cathepsin H. *Plant Physiol.* **99**, 848–855.

Holwerda, B.C. & Rogers, J.C. (1993) Structure, functional properties and vacuolar targeting of the barley thiol protease, aleurain. *J. Exp. Bot.* **44** (suppl.), 321–339.

Holwerda, B.C., Galvin, N.J., Baranski, T.J. & Rogers, J.C. (1990) In vitro processing of aleurain, a barley vacuolar thiol protease. *Plant Cell* **2**, 1091–1106.

Holwerda, B.C., Padgett, H.S. & Rogers, J.C. (1992) Proaleurain vacuolar targeting is mediated by short contiguous peptide interactions. *Plant Cell* **4**, 307–318.

Kirsch, T., Paris, N., Butler, J.M., Beevers, L. & Rogers, J.C. (1994) Purification and initial characterization of a potential plant vacuolar targeting receptor. *Proc. Natl Acad. Sci. USA* **91**, 3403–3407.

Paris, N., Stanley, C.M., Jones, R.L. & Rogers, J.C. (1996) Plant cells contain two functionally distinct vacuolar compartments. *Cell* **85**, 563–572.

Ritonja, A., Popovič, T., Kotnik, M., Machleidt, W. & Turk, V. (1988) Amino acid sequences of the human kidney cathepsins H and L. *FEBS Lett.* **228**, 341–345.

Rogers, J.C., Dean, D. & Heck, G.R. (1985) Aleurain: a barley thiol protease closely related to mammalian cathepsin H. *Proc. Natl Acad. Sci. USA* **82**, 6512–6516.

Rothe, M., Zichner, A., Auerswald, E.A. & Dodt, J. (1994) Structure/function implications for the aminopeptidase specificity of aleurain. *Eur. J. Biochem.* **224**, 559–565.

Tournaire, C., Kushnir, L., Bauw, G., Inzé, D., Tessendier de la Serve, B. & Renaudin, J.-P. (1996) A thiol protease and an anionic peroxidase are induced by lowering cytokinins during callus growth in *Petunia. Plant Physiol.* **111**, 159–168.

Watanabe, H., Abe, K., Emori, Y., Hosoyama, H. & Arai, S. (1991) Molecular cloning and gibberellin-induced expression of multiple cysteine proteinases of rice seeds (Oryzains). *J. Biol. Chem.* **266**, 16897–16902.

*John C. Rogers*
*Biochemistry Department,*
*117 Schweitzer Hall, University of Missouri,*
*Columbia, MO 65203, USA*
*Email: bcjroger@muccmail.missouri.edu*

# 201. Falcipain

## Databanks

*Peptidase classification: clan CA, family C1, MEROPS ID: C01.077*
*NC-IUBMB enzyme classification: none*
*Databank codes:*

| Species | SwissProt | PIR | EMBL (cDNA) | EMBL (genomic) |
|---|---|---|---|---|
| *Plasmodium berghei* | – | – | U33420 | – |
| *Plasmodium cynomolgi* | – | – | U33422 | – |
| *Plasmodium falciparum* | P25805 | A45624 | M81341 | – |
| *Plasmodium fragile* | – | – | U33421 | – |
| *Plasmodium gallinaceum* | – | – | U33424 | – |
| *Plasmodium malariae* | – | – | U33425 | – |
| *Plasmodium ovale* | – | – | U33423 | – |
| *Plasmodium reichenowi* | – | – | U33426 | – |
| *Plasmodium vinckei* | P46102 | – | L08500 | – |
| *Plasmodium vivax* | P42666 | – | L26362 | – |
| *Theileria annulata* | P25781 | A45565 | M86659 | – |
| *Theileria parva* | P22497 | A36083 | – | M37791: complete gene M67476: 3′ end |

## Name and History

Malarial proteinase activity was noted in numerous older reports (McKerrow *et al.*, 1993), but a cysteine proteinase of malarial trophozoites was not clearly identified until 1987. At that time a single cysteine proteinase activity of trophozoites of the human malaria parasite *Plasmodium falciparum* was demonstrated in gelatin substrate PAGE (Rosenthal *et al.*, 1987). The enzyme was biochemically characterized as a papain-family proteinase and named **trophozoite cysteine proteinase** (Rosenthal *et al.*, 1988). Subsequently, a single-copy *P. falciparum* cysteine proteinase gene was characterized (Rosenthal & Nelson, 1992). When the heterologously expressed product of the cysteine proteinase gene was demonstrated to have similar biochemical properties to the native trophozoite cysteine proteinase (Salas *et al.*, 1995), it was concluded that the gene encoded that enzyme. As the predicted sequence of the protein and its biochemical features showed it to be a member of the papain family, it was renamed **falcipain**.

## Activity and Specificity

Studies of the activity of falcipain suggest that it is a fairly typical papain-family proteinase (Rosenthal *et al.*, 1988, 1989). The enzyme has an acidic pH optimum (pH 5.0–6.5), consistent with its action in the acidic malarial food vacuole. Falcipain activity is markedly stimulated by reducing agents including DTT, cysteine and glutathione. The enzyme cleaves the nonspecific substrates azocoll, gelatin and casein. With small substrates, the enzyme has a cathepsin L-like specificity, preferring Z-Phe-Arg+NHMec and Z-Val-Leu-Arg+NHMec over typical substrates of cathepsin B (Z-Arg-Arg-NHMec) and cathepsin H (Z-Arg-NHMec). As will be discussed below, the natural substrate of falcipain appears to be native or denatured hemoglobin.

Falcipain is inhibited by *N*-ethylmaleimide and the peptide cysteine proteinase inhibitors leupeptin, E-64 and chymostatin (Rosenthal, 1995; Rosenthal *et al.*, 1988). A number of peptidyl fluoromethanes (Rosenthal *et al.*, 1989, 1991, 1993) and vinyl sulfones (Rosenthal *et al.*, 1996) inhibit falcipain at picomolar to low nanomolar concentrations. For these compounds, preferred P2 and P1 substrate amino acids (in decreasing order of effectiveness) are Leu-homophenylalanine, Phe-Arg, Phe-homophenylalanine, Phe-Lys and Phe-Ala.

## Structural Chemistry

Translation of the falcipain gene predicts a 569 amino acid preproenzyme with a molecular mass of 66 800 (Rosenthal & Nelson, 1992). This size is unusually large for a papain-family proteinase. Alignment with papain and related enzymes identifies a presumptive cleavage site between Lys332 and Val333 to yield a 26.8 kDa mature proteinase that is 37% identical in sequence to human cathepsin L and 33% identical to papain. The mature form of falcipain is predicted to have four disulfide bonds. Molecular modeling led to the proposal of a tertiary structure for falcipain similar to that of other papain-family enzymes (Ring *et al.*, 1993). Genes encoding two falcipain analogs, from the human malaria parasite

*P. vivax* (Rosenthal *et al.*, 1994) and the murine parasite *P. vinckei* (Rosenthal, 1993) have also been sequenced. All of the malarial cysteine proteinases have unusually large pro sequences. The predicted mature proteinase sequences are highly conserved, with 71% identity between falcipain and its *P. vivax* analog and 56% identity between falcipain and its *P. vinckei* analog. These three malarial proteinases (and an additional seven proteinases from other malarial species for which partial sequences are available: Rosenthal, 1996) share the presence of a poorly conserved stretch of about 30 amino acids (amino acids 499–527 for falcipain) that is not present in most other papain-family enzymes.

## Preparation

Falcipain is routinely obtained from harvests of cultured malaria parasites (Rosenthal *et al.*, 1989). Trophozoite-infected erythrocytes are collected, erythrocyte membranes are selectively lysed with the detergent saponin, washed parasites are lysed by hypotonic freeze–thaw, and the supernatant containing falcipain is collected. An optimal, high-yield purification strategy for falcipain has not yet been developed. Enzyme purified by anion-exchange and gel-filtration (unpublished results of the author) or anion-exchange and hydroxyapatite chromatography (Gluzman *et al.*, 1994) has biochemical properties that are identical to the activity in the soluble fraction of parasite lysates when assayed with specific fluorogenic substrates. As purified material is obtained at low yield and is quite unstable, many assays have been performed with soluble lysates. Inhibitor results obtained with this material have correctly predicted antimalarial effects, suggesting that the only cysteine proteinase activity in the lysates is that of falcipain.

Enzymatically active falcipain has been expressed using a baculovirus system (Salas *et al.*, 1995). The expressed proteinase was purified using ion-exchange chromatography. Yields with the baculovirus expression system have been suboptimal, however. In addition, the enzyme appeared to be incompletely processed, and it differed somewhat from the native enzyme in the kinetics of cleavage of peptide substrates. Attempts to optimize expression with this and other systems are underway.

## Biological Aspects

Malaria, one of the most important infectious diseases in the world, is a febrile illness responsible for hundreds of millions of cases and over one million deaths each year. During the medically most important portion of their life cycle, malaria parasites reside within erythrocytes and degrade hemoglobin as a principal source of amino acids (Rosenthal & Meshnick, 1996). Falcipain appears to be a required malarial hemoglobinase. Hemoglobin is degraded in an acidic food vacuole. Three food vacuole proteinases have been identified, falcipain and the aspartic proteinases plasmepsins I (Chapter 286) and II (Chapter 287). A critical role for falcipain is suggested by the finding that cysteine proteinase inhibitors cause a specific morphological abnormality in cultured parasites whereby the food vacuole fills with undegraded globin (Rosenthal *et al*., 1988; Bailly *et al*., 1992; Rosenthal, 1995). Falcipain inhibitors that block parasite hemoglobin degradation also block parasite development (Rosenthal *et al*., 1991) and cure malaria in a murine model system (Rosenthal *et al*., 1993). For these reasons falcipain is considered a promising target for antimalarial chemotherapy.

The precise mechanism of hemoglobin degradation in the malarial food vacuole is not known (Rosenthal & Meshnick, 1996). Initial steps in the activation of falcipain or the plasmepsins are not characterized, though the processing of falcipain from a ∼40 kDa pro form to a 28 kDa mature form is inhibited by leupeptin, suggesting a requirement for autohydrolysis in the last step of falcipain activation. Regarding the sequence of events by which hemoglobin is degraded, cysteine proteinase inhibitors partially block the dissociation of the hemoglobin tetramer (Gamboa de Domínguez & Rosenthal, 1996) and strongly inhibit the hydrolysis of globin (Rosenthal, 1995; Rosenthal *et al*., 1991), suggesting principal roles for falcipain in the initial steps of hemoglobin degradation. Supporting this conclusion, falcipain cleaved native hemoglobin in an acidic, reducing environment predicted to approximate that in the food vacuole (Salas *et al*., 1995). However, in other studies, plasmepsin I was shown to readily cleave native hemoglobin at a hinge region of the molecule, plasmepsin II to prefer denatured hemoglobin as a substrate, and falcipain to cleave only denatured globin, suggesting an ordered pathway of hemoglobin catabolism initiated by plasmepsin I (Goldberg *et al*., 1991; Gluzman *et al*., 1994). In these *in vitro* studies, falcipain was shown to degrade α- and β-globin at five specific sites. It is unclear which *in vitro* studies best demonstrate the proteolytic events occurring *in vivo*. Thus, the precise roles of falcipain and the two plasmepsins in malarial hemoglobin degradation are not yet known, though it seems clear that all three enzymes participate in this process.

## Distinguishing Features

The activity of falcipain is, in general, very similar to that of cathepsin L and other related papain-family enzymes. A more precise analysis necessary to distinguish differences between these related enzymes has not yet been done. As will be discussed below, sequence comparisons of proteinases from multiple malarial species suggest that there are unique *Plasmodium*-specific sequences that are likely to provide specificity.

## Related Peptidases

Analogs of falcipain, the trophozoite proteinase of *P. falciparum*, have been identified by biochemical or molecular means from nine other species of malaria parasites. A proteinase of the murine malaria parasite *P. vinckei* has been studied biochemically. The proteinase is very similar to falcipain in terms of its substrate preference, pH optimum, and inhibitor sensitivity, though sensitivity to some highly effective falcipain inhibitors is significantly lower with the *P. vinckei* proteinase. The genes encoding two falcipain analogs, those of the human parasite *P. vivax* and the murine parasite *P. vinckei*, have been cloned and sequenced. Portions of genes encoding analogous cysteine proteinases of seven other malarial species, including parasites of humans,

other primates, rodents and birds, have been amplified and sequenced. The proteinases of different malarial species share important structural features, including >50% sequence identity in the predicted mature proteinase sequences, fairly strong conservation of sequences in the C-terminal portion of the pro sequences, and the presence of an unusual (for papain-family enzymes) ∼30 amino acid insert near the C-terminus of the proteins. Upon more detailed analysis in the light of the modeled structure of falcipain, a number of amino acids that are predicted to reside near the active site are highly conserved among the malarial proteinases but poorly conserved with other papain family proteinases (Rosenthal *et al*., 1994; Rosenthal, 1996). The *Plasmodium*-specific active-site sequences were presumably conserved to optimize efficiency of hemoglobin hydrolysis. The unique features of the malarial proteinases may allow the design of specific inhibitors of these enzymes.

Parasites of the genus *Theileria* are tickborne protozoans that parasitize cattle and cause economically important disease in the tropics. Papain-family proteinases have been cloned and sequenced from *Theileria parva* (Nene *et al*., 1990), the cause of East Coast Fever in Africa and *Theileria annulata* (Baylis *et al*., 1992), the cause of tropical theileriosis. The proteinase sequences are cathepsin L-like. Upon alignment/phylogeny analysis, they were found to be most similar in sequence to falcipain and related malarial proteinases (Berti & Storer, 1995). The *T. parva* proteinase was also studied biochemically. A cathepsin L-like activity was identified in parasite extracts. The activity is maximal against the substrate Z-Phe-Arg↓NHMec, has a pH optimum of 6.0, and is inhibited by E-64. These properties are all similar to those of falcipain and related enzymes.

## References

Bailly, E., Jambou, R., Savel, J. & Jaureguiberry, G. (1992) *Plasmodium falciparum*: differential sensitivity in vitro to E-64 (cysteine protease inhibitor) and pepstatin A (aspartyl protease inhibitor). *J. Protozool.* **39**, 593–599.

Baylis, H.A., Megson, A., Mottram, J.C. & Hall, R. (1992) Characterisation of a gene for a cysteine protease from *Theileria annulata. Mol. Biochem. Parasitol.* **54**, 105–107.

Berti, P.J. & Storer, A.C. (1995) Alignment/phylogeny of the papain superfamily of cysteine proteases. *J. Mol. Biol.* **246**, 273–283.

Gamboa de Domínguez, N.D. & Rosenthal, P.J. (1996) Cysteine proteinase inhibitors block early steps in hemoglobin degradation by cultured malaria parasites. *Blood* **87**, 4448–4454.

Gluzman, I.Y., Francis, S.E., Oksman, A., Smith, C.E., Duffin, K.L. & Goldberg, D.E. (1994) Order and specificity of the *Plasmodium falciparum* hemoglobin degradation pathway. *J. Clin. Invest.* **93**, 1602–1608.

Goldberg, D.E., Slater, A.F.G., Beavis, R., Chait, B., Cerami, A. & Henderson, G.B. (1991) Hemoglobin degradation in the human malaria pathogen *Plasmodium falciparum*: a catabolic pathway initiated by a specific aspartic protease. *J. Exp. Med.* **173**, 961–969.

McKerrow, J.H., Sun, E., Rosenthal, P.J. & Bouvier, J. (1993) The proteases and pathogenicity of parasitic protozoa. *Annu. Rev. Microbiol.* **47**, 821–853.

Nene, V., Gobright, E., Musoke, A.J. & Lonsdale-Eccles, J.D. (1990) A single exon codes for the enzyme domain of a protozoan cysteine protease. *J. Biol. Chem.* **265**, 18047–18050.

Ring, C.S., Sun, E., McKerrow, J.H., Lee, G.K., Rosenthal, P.J., Kuntz, I.D. & Cohen, F.E. (1993) Structure-based inhibitor design by using protein models for the development of antiparasitic agents. *Proc. Natl Acad. Sci. USA* **90**, 3583–3587.

Rosenthal, P.J. (1993) A *Plasmodium vinckei* cysteine proteinase shares unique features with its *Plasmodium falciparum* analogue. *Biochim. Biophys. Acta* **1173**, 91–93.

Rosenthal, P.J. (1995) *Plasmodium falciparum*: effects of proteinase inhibitors on globin hydrolysis by cultured malaria parasites. *Exp. Parasitol.* **80**, 272–281.

Rosenthal, P.J. (1996) Conservation of key amino acids among the cysteine proteinases of multiple malarial species. *Mol. Biochem. Parasitol.* **75**, 255–260.

Rosenthal, P.J. & Meshnick, S.R. (1996) Hemoglobin catabolism and iron utilization by malaria parasites. *Mol. Biochem. Parasitol.* **83**, 131–139.

Rosenthal, P.J. & Nelson, R.G. (1992) Isolation and characterization of a cysteine proteinase gene of *Plasmodium falciparum*. *Mol. Biochem. Parasitol.* **51**, 143–152.

Rosenthal, P.J., Kim, K., McKerrow, J.H. & Leech, J.H. (1987) Identification of three stage-specific proteinases of *Plasmodium falciparum*. *J. Exp. Med.* **166**, 816–821.

Rosenthal, P.J., McKerrow, J.H., Aikawa, M., Nagasawa, H. & Leech, J.H. (1988) A malarial cysteine proteinase is necessary for hemoglobin degradation by *Plasmodium falciparum*. *J. Clin.* *Invest.* **82**, 1560–1566.

Rosenthal, P.J., McKerrow, J.H., Rasnick, D. & Leech, J.H. (1989) *Plasmodium falciparum*: inhibitors of lysosomal cysteine proteinases inhibit a trophozoite proteinase and block parasite development. *Mol. Biochem. Parasitol.* **35**, 177–184.

Rosenthal, P.J., Wollish, W.S., Palmer, J.T. & Rasnick, D. (1991) Antimalarial effects of peptide inhibitors of a *Plasmodium falciparum* cysteine proteinase. *J. Clin. Invest.* **88**, 1467–1472.

Rosenthal, P.J., Lee, G.K. & Smith, R.E. (1993) Inhibition of a *Plasmodium vinckei* cysteine proteinase cures murine malaria. *J. Clin. Invest.* **91**, 1052–1056.

Rosenthal, P.J., Ring, C.S., Chen, X. & Cohen, F.E. (1994) Characterization of a *Plasmodium vivax* cysteine proteinase gene identifies uniquely conserved amino acids that may mediate the substrate specificity of malarial hemoglobinases. *J. Mol. Biol.* **241**, 312–316.

Rosenthal, P.J., Olson, J.E., Lee, G.K., Palmer, J.T., Klaus, J.L. & Rasnick, D. (1996) Antimalarial effects of vinyl sulfone cysteine proteinase inhibitors. *Antimicrob. Agents Chemother.* **40**, 1600–1603.

Salas, F., Fichmann, J., Lee, G.K., Scott, M.D. & Rosenthal, P.J. (1995) Functional expression of falcipain, a *Plasmodium falciparum* cysteine proteinase, supports its role as a malarial hemoglobinase. *Infect. Immun.* **63**, 2120–2125.

*Philip J. Rosenthal*
*Department of Medicine,*
*San Francisco General Hospital and University of California, San Francisco,*
*San Francisco, CA 94143, USA*
*Email: rosnthl@itsa.ucsf.edu*

# 202. *Histolysain and other Entamoeba cysteine endopeptidases*

## Databanks

*Peptidase classification: clan CA, family C1, MEROPS ID: C01.050*
*NC-IUBMB enzyme classification: EC 3.4.22.35*
*Chemical Abstract Service registry number: 92228-52-9*
*Databank codes:*

| Species | SwissProt | PIR | EMBL (cDNA) | EMBL (genomic) |
|---|---|---|---|---|
| Amoebapain (CP1) | | | | |
|   *Entamoeba histolyticum* | Q01957 | – | M64712 | M94162: complete gene (no introns) |
| Histolysain (CP2) | | | | |
|   *Entamoeba dispar* | Q06964 | – | M64721 | – |
|   *Entamoeba histolyticum* | Q01958 | – | – | M94163: 3′ UTR of mRNA missing |
|   *Entamoeba histolyticum* | P36185 | – | S58670 | |

| Species | SwissProt | PIR | EMBL (cDNA) | EMBL (genomic) |
|---|---|---|---|---|
| **CP3** | | | | |
| *Entamoeba dispar* | – | – | X87213 | – |
| *Entamoeba histolyticum* | P36184 | – | M27307 | – |
| | | | S58669 | |
| | | | X87214 | |
| **CP4** | | | | |
| *Entamoeba dispar* | – | – | X91643 | – |
| *Entamoeba histolyticum* | – | – | – | X91642: complete gene (no introns) |
| **CP5** | | | | |
| *Entamoeba histolyticum* | – | – | – | X91644: 3′ end |
| **CP6** | | | | |
| *Entamoeba histolyticum* | – | – | – | X91645: complete gene (no introns) |
| *Entamoeba invadens* | – | – | U28755 | – |

## Name and History

The name **histolysain** denotes the lysis of tissue ('**histolys**') and also includes the ending '-**ain**' that suggests a cysteine endopeptidase; the term is used for many related cysteine endopeptidases of *Entamoeba histolytica*. Thiol-dependent proteolytic activity in *E. histolytica* was first attributed to a **neutral sulfhydryl proteinase** (McLaughlin & Faubert, 1977) and later to a **cytotoxic proteinase** (Lushbaugh *et al.*, 1984). Other terms given to closely related or identical enzymes are **cathepsin B** (Lushbaugh *et al.*, 1985), **neutral proteinase** (Keene *et al.*, 1986), **histolysin** (Luaces & Barrett, 1988), later changed to **histolysain** (Luaces *et al.*, 1992), and **amoebapain** (Scholze *et al.*, 1992). A total of six different genes encoding cysteine endopeptidases in *E. histolytica* have been identified so far, designated **ehcp1** to **ehcp6** (Bruchhaus *et al.*, 1996). Amoebapain is encoded by *ehcp1* (or *acp3*) and histolysain by *ehcp2* (or *acp2*) (Tannich *et al.*, 1991; Reed *et al.*, 1993). Cysteine endopeptidases are also expressed in other *Entamoeba* species, such as *E. dispar* (human non-pathogenic: Bruchhaus *et al.*, 1996) and *E. invadens* (reptilian pathogenic: Avila *et al.*, 1985).

## Activity and Specificity

The substrate specificities of the known *Entamoeba* cysteine endopeptidases so far investigated are very similar or identical. The enzymes degrade a number of proteins, including those of the extracellular matrix. They also act on small molecule substrates, hydrolyzing a variety of blocked peptides analogs, and even have exopeptidase-like activities. Thus, amoebapain hydrolyzes Arg-Arg┼NHMec, and unblocked tetrapeptides that contain a basic residue at P2 (Scholze & Schulte, 1988). The enzyme also cleaves the C-terminal dipeptide, Phe-Phe, from Glu-Arg-Gly┼Phe-Phe (insulin B chain fragment 21–26) (Scholze, 1991). Thus, the enzyme exhibits both real peptidyl-dipeptidase and dipeptidyl-peptidase activities. The specificity of action on the small substrates markedly favours those with Arg in the P2 position, with little specificity in P1 (Keene *et al.*, 1986; Luaces & Barrett, 1988; Scholze & Schulte, 1988).

The activity is stimulated by free thiol groups and inhibited by thiol-blocking reagents and other compounds typically affecting cysteine proteinases. Proteinase assays are routinely performed with azocasein or azocoll as substrates, or the fluorogenic substrate Z-Arg-Arg┼NHMec (Luaces & Barrett, 1988; Scholze & Tannich, 1994). Maximal rates of hydrolysis of denatured proteins are at weakly acidic pH values (pH 5–6), but for the synthetic substrate, at a somewhat higher pH (Lushbaugh *et al.*, 1985; Luaces & Barrett, 1988).

## Structural Chemistry

Histolysain and its relatives are single-chain proteins of about 27 kDa and isoelectric points of about 5 (EhCP1, EhCP2, EhCP6) or 8 (EhCP3, EhCP5). Their primary structures as predicted from their nucleotide sequences exhibit no *N*-linked glycosylation sites (Reed *et al.*, 1993; Bruchhaus *et al.*, 1996). Due to the conservation of structurally important residues in the sequence of the enzymes, their general folding is expected to be similar to that of other cysteine peptidases of the papain family (C1), described by Baker & Drenth (1987).

## Preparation

*Entamoeba* cysteine endopeptidases are localized in lysosome-like vesicles of ameba trophozoites, which can be cultivated axenically in complex, serum-containing media (Scholze & Tannich, 1994). They have been purified to homogeneity in several laboratories (Lushbaugh *et al.*, 1985; Keene *et al.*, 1986; Luaces & Barrett, 1988; Scholze & Schulte, 1988; Montfort *et al.*, 1994) and the underlying gene organization has been characterized (Bruchhaus *et al.*, 1996). Fragments of the gene encoding amoebapain (*EhCP1*) have been overexpressed in *Escherichia coli* as a glutathione-transferase fusion protein (Stanley *et al.*, 1995).

## Biological Aspects

The localization of *Entamoeba* cysteine endopeptidases within lysosome-like vesicles indicates that the enzymes are involved in digestive processes. In addition, cysteine endopeptidases are found to be secreted by *Entamoeba* trophozoites (Leippe *et al.*, 1995) and mounting evidence exists that this class of enzymes is important for ameba-induced tissue destruction (Stanley *et al.*, 1995). However, only four of the six cysteine endopeptidase genes are expressed at the trophozoite stage (Bruchhaus *et al.*, 1996). Therefore, the remaining two genes (*EhCP4* and *EhCP6*) are suggested to be of functional importance during other stages of the *Entamoeba* life cycle (e.g. encystation or excystation).

## Distinguishing Features

The nature of the S2 subsite of histolysain and amoebapain differs from that of mammalian and plant cysteine proteinases of family C1. Instead of relatively hydrophobic amino acid side chains forming the S2 pocket, a Glu residue at position 160 flanks the specificity pocket and Asp205 is located at its base (Berti & Storer, 1995). Nevertheless, the preference for arginine at P2 does not mean that the amebic cysteine proteinases histolysain and amoebapain should be ranked with cathepsin B, as suggested by Lushbaugh *et al.* (1985). The *Entamoeba* enzymes differ from cathepsin B in several important respects, including their stability at neutral pH, higher pH optimum with Z-Arg-Arg-NHMec as substrate, and sensitivity to inhibition by cystatin and E-64 (Scholze & Tannich, 1994). According to their amino acid sequences, the amebic enzymes are closer to cathepsin L (Chapter 210) and the *Dictyostelium discoides* proteinases (Chapter 207) (Berti & Storer, 1995). The distinctive specificity of the *Entamoeba* enzymes may be useful in the design of specific inhibitors for use as antiamebic drugs.

## Further Reading

See Scholze & Tannich (1994) for a fuller review of these enzymes.

## References

Avila, E.E., Sánchez-Garza, M. & Calderón, J. (1985) *Entamoeba histolytica* and *E. invadens*: sulfhydryl-dependent proteolytic activity. *J. Protozool.* **32**, 163–166.

Baker, E.N. & Drenth, J. (1987) The thiol proteases: structure and mechanism. In: *Biological Macromolecules and Assemblies*, vol. 3: *Active Sites of Enzymes* (McPherson, J., ed.). New York: Wiley & Sons, pp. 312–368.

Berti, P.J. & Storer, A.C. (1995) Alignment/phylogeny of the papain superfamily of cysteine proteases. *J. Mol. Biol.* **246**, 273–283.

Bruchhaus, I., Jacobs, T., Leippe, M. & Tannich, E. (1996) *Entamoeba histolytica* and *Entamoeba dispar*: differences in numbers and expression of cysteine proteinase genes. *Mol. Microbiol.* **22**, 255–264.

Keene, W.E., Petitt, M.G., Allen, S. & McKerrow, J.H. (1986) The

major neutral proteinase of *Entamoeba histolytica*. *J. Exp. Med.* **163**, 536–549.

Leippe, M., Sievertsen, H.J., Tannich, E. & Horstmann, R.D. (1995) Spontanous release of cysteine proteinases but not of pore-forming peptides by viable *Entamoeba histolytica*. *Parasitology* **111**, 569–574.

Luaces, A.L. & Barrett, A.J. (1988) Affinity purification and biochemical characterization of histolysin, the major cysteine proteinase of *Entamoeba histolytica*. *Biochem. J.* **250**, 903–909.

Luaces, A.L., Picó, T. & Barrett, A.J. (1992) The ENZYMEBA test: detection of intestinal *Entamoeba histolytica* infection by immuno-enzymatic detection of histolysain. *Parasitology* **105**, 203–205.

Lushbaugh, W.B., Hofbauer, A.F. & Pittman, F.E. (1984) Proteinase activity of *Entamoeba histolytica* cytotoxin. *Gastroenterology* **87**, 17–27.

Lushbaugh, W.B., Hofbauer, A.F. & Pittman, F.E. (1985) *Entamoeba histolytica*: purification of cathepsin B. *Exp. Parasitol.* **59**, 328–336.

McLaughlin, J. & Faubert, G. (1977) Partial purification and some properties of a neutral sulfhydryl and an acid proteinase from *Entamoeba histolytica*. *Can. J. Microbiol.* **23**, 420–425.

Montfort, I., Pérez-Tamayo, R., Pérez-Montfort, R., Gonzáles-Canto, A. & Olivos, A. (1994) Purification and immunologic characterization of a 30-kDa cysteine proteinase of *Entamoeba histolytica*. *Parasitol. Res.* **80**, 607–613.

Reed, S., Bouvier, J., Pollack, A.S., Engel, J.C., Brown, M., Hirata, K., Que, X.C., Eakin, A., Hagblom, P., Gillin, F. & McKerrow, J.H. (1993) Cloning of a virulence factor of *Entamoeba-histolytica*-pathogenic strains possess a unique cysteine proteinase gene. *J. Clin. Invest.* **91**, 1532–1540.

Scholze, H. (1991) Amoebapain, the major proteinase of pathogenic *Entamoeba histolytica*. In: *Biochemical Protozoology* (Coombs, G. & North, M., eds) . London: Taylor & Francis, pp. 251–256.

Scholze, H., Löhden-Bendinger, U., Müller, G. & Bakker-Grunwald, T. (1992) Subcellular distribution of amoebapain, the major cysteine proteinase of *Entamoeba histolytica*. *Arch. Med. Res.* **23**, 105–108.

Scholze, H. & Schulte, W. (1988) On the specificity of a cysteine proteinase from *Entamoeba histolytica*. *Biomed. Biochim. Acta* **47**, 115–123.

Scholze, H. & Tannich, E. (1994) Cysteine endopeptidases of *Entamoeba histolytica*. *Methods Enzymol.* **244**, 512–523.

Stanley, S.L., Zhang, T., Rubin, D. & Li, E. (1995) Role of the *Entamoeba histolytica* cysteine proteinase in amebic liver abscess formation in severe combined immunodeficient mice. *Infect. Immun.* **63**, 1587–1590.

Tannich, E., Scholze, H., Nickel, R. & Horstmann, R.D. (1991) Homologous cysteine proteinases of pathogenic and non-pathogenic *Entamoeba histolytica*. Differences in structure and expression. *J. Biol. Chem.* **266**, 4798–4803.

*Henning Scholze*
*Universität Osnabrück,*
*Fachbereich Biologie/Chemie, Biochemie,*
*Barbarastrasse 11,*
*D-49069 Osnabrück, Germany*
*Email: scholtze@cipfb5.biologie.uni-osnabrueck.de*

*Egbert Tannich*
*Bernhard-Nocht-Institut für Tropenmedizin,*
*Molekularbiologie,*
*Bernhard-Nocht-Strasse 74,*
*20359 Hamburg, Germany*

# 203. Cruzipain

## Databanks

*Peptidase classification: clan CA, family C1, MEROPS ID: C01.075*
*NC-IUBMB enzyme classification: none*
*Databank codes:*

| Species | SwissProt | PIR | EMBL (cDNA) | EMBL (genomic) |
|---|---|---|---|---|
| *Trypanosoma brucei* | P14658 | S07051 | X54353 | – |
|  |  | S12099 | X16465 |  |
| *Trypanosoma brucei* | – | – | M27306 | – |
| *Trypanosoma congolense* | – | – | L25130 | – |
| *Trypanosoma congolense* | – | – | Z25813 | – |
| *Trypanosoma cruzi* | P25779 | A60667 | M27305 | – |
|  |  | S16162 | M69121 |  |
|  |  |  | M84342 |  |
|  |  |  | U41444 |  |
|  |  |  | U41454 |  |
|  |  |  | X54414 |  |
| *Trypanosoma cruzi* | – | – | M90067 | – |
| *Trypanosoma rangeli* | – | – | L38513 | – |
| *Trypanosoma rangeli* | – | – | L38514 | – |
| *Trypanosoma rangeli* | – | – | M99496 | – |

Brookhaven Protein Data Bank three-dimensional structures:

| Species | ID | Resolution | Notes |
|---|---|---|---|
| *Trypanosoma cruzi* | 1AIM | | Catalytic domain, Gly213 stop |
|  | 2AIM | | Catalytic domain, Gly213 stop |

## Name and History

The name **cruzipain** was proposed (Cazzulo *et al.*, 1990) to denote that the enzyme is the major cysteine proteinase of *Trypanosoma cruzi*, the causative agent of the American trypanosomiasis, Chagas disease, and that it belongs to the papain family. The name **cruzain** was suggested afterwards by Eakin *et al.* (1992). **Antigen GP57/51** (Murta *et al.*, 1990) is also synonymous. Cruzipain was discovered in cell-free extracts by Itow & Camargo (1977), and purified to homogeneity by Bontempi *et al.* (1984); the cysteine proteinase purified by Rangel *et al.* (1981) is likely to be the same enzyme, despite some discrepancies, as is the 50 kDa proteinase described by Greig & Ashall (1990).

## Activity and Specificity

Cruzipain is an endopeptidase able to digest proteins such as casein, bovine albumin and denatured hemoglobin (optimal pH 3–5), and synthetic, blocked, chromogenic and fluorogenic substrates. The optimal pH is 7–9. With the small-molecule substrates, it prefers Arg or Lys at the P1 position. When acting on the oxidized A and B chains of insulin, however, it acts better on peptide bonds having bulky hydrophobic residues at

P2 and P3 (Cazzulo & Frasch, 1992). Cruzipain activates its pro form, and is also able to liberate its C-terminal domain, by cutting at Val-Val-Gly+Ala-Pro-Ala or Val-Val-Gly+Gly-Pro-Gly, respectively (Cazzulo *et al.*, 1996). The enzyme is inhibited by organomercurial reagents, E-64, Tos-Lys-CH₂Cl, leupeptin, and a number of peptidyl chloromethane and peptidyl fluoromethane derivatives. Cystatins, stefins and kininogens (Stoka *et al.*, 1995; Serveau *et al.*, 1996; Turk *et al.*, 1996) are also strong inhibitors of the enzyme.

## Structural Chemistry

Cruzipain is encoded by numerous genes (up to 130 in the Tul 2 strain of the parasite), located in tandem arrays on two to four chromosomes, depending on the clone or strain of parasite (Campetella *et al.*, 1992). The genes, which contain no introns, encode a signal peptide, a propeptide, and a mature enzyme, consisting of a catalytic moiety, with high sequence identity to some cathepsins, particularly cathepsin S (Chapter 211), and a C-terminal extension (C-T) of 130 amino acids, which so far seems to be restricted to cysteine proteinases from trypanosomatids. The C-T consists of a 'core' of 76 amino acids stabilized by disulfide bridges, an N-terminal segment of 27 amino acid residues up to the only

Met residue, and a highly hydrophilic C-terminal 'tail' of 27 residues. The N-terminal segment contains seven modified Thr residues and seven Pro residues, and probably acts as a 'hinge' linking the 'core' to the catalytic moiety.

The Thr modifications are still unidentified, but seem to represent neither phosphorylation nor *O*-glycosylation (Cazzulo *et al.*, 1992). This 'normal' C-T is absent from the last gene of the tandem, where it is replaced by a highly hydrophobic extension of 49 amino acid residues (Tomas & Kelly, 1996). Natural cruzipain is a complex of isoforms, as judged from Mono-Q chromatography, reversed-phase HPLC and isoelectrofocusing (Cazzulo *et al.*, 1995). This heterogeneity is probably due to both the simultaneous expression of several genes with amino acid substitutions, and to the presence in different cruzipain molecules of either high mannose-type, hybrid monoantennary-type or complex biantennary-type oligosaccharide chains at the single *N*-glycosylation site in the C-T (Asn255) (Cazzulo *et al.*, 1995). The first potential *N*-glycosylation site in the catalytic moiety (Asn33) is also glycosylated *in vivo*, and bears only high mannose-type oligosaccharides; there is still no information about the *N*-glycosylation status of the second potential site (Asn169) (Metzner *et al.*, 1996).

The X-ray crystallographic structure of a recombinant truncated form of the enzyme (cruzain Dc) in complex with the synthetic inhibitor Z-Phe-Ala-CH$_2$Cl has been determined (McGrath *et al.*, 1995). The catalytic moiety of cruzipain consists of one polypeptide chain of 215 amino acid residues folded into two distinct domains which interact to create the active-site cleft. The overall folding pattern and the arrangement of the active-site residues are similar to those in papain.

The molecular mass of cruzipain can be estimated from sequence, allowing for two high-mannose oligosaccharide chains, as about 40 kDa. However, the enzyme shows an anomalous behavior in SDS-PAGE, yielding apparent molecular mass values of 35–60 kDa depending on the experimental conditions (Martínez & Cazzulo, 1992). In isoelectric focusing, purified preparations present up to 12 bands, with pI values ranging from 3.7 to 5.1 (Stoka *et al.*, 1995).

## Preparation

Cruzipain (3–4% of the total soluble protein of the cell in axenically cultured epimastigotes of *T. cruzi*, Tul 2 strain) can readily be purified by conventional methods (Cazzulo *et al.*, 1989) or affinity chromatography on con A-Sepharose and chicken egg cystatin-Sepharose (Labriola *et al.*, 1993). As noted above, an active truncated form of the enzyme has been expressed in *Escherichia coli* (Eakin *et al.*, 1992).

## Biological Aspects

Cruzipain is differentially expressed in the four main stages of the parasite life cycle; its activity in epimastigotes is 10–100-fold higher in epimastigotes than in other parasite forms (Franke de Cazzulo *et al.*, 1994). The enzyme is lysosomal, but it seems to be expressed also at the cell surface. It has also been reported to be present in an epimastigote-specific pre-lysosomal organelle called the 'reservosome', which contains protein that is digested during differentiation to metacyclic trypomastigotes (Soares *et al.*, 1992). Trypomastigotes are able to excrete cysteine proteinases, probably including cruzipain, into the medium (Yokoyama-Yasunaka *et al.*, 1994). In addition to its obvious role in parasite nutrition, cruzipain has been proposed to be involved in the penetration of the mammalian cell by the trypomastigote. It may also provide a mechanism by which the parasite can escape the immune response of the host, by digesting immunoglobulins at the 'hinge' region, and then destroying the Fc domain. The proteinase has also been proposed to play a part in the differentiation steps of the parasite's life cycle, which have been shown to be blocked by cell-permeant irreversible inhibitors of the enzyme (Franke de Cazzulo *et al.*, 1994; Cazzulo *et al.*, 1995).

Cruzipain is an immunodominant antigen, recognized by most sera from patients of Chagas disease (Murta *et al.*, 1990). Most antibodies in natural and experimental infections are directed against the C-T, and enzyme molecules with antibodies bound to this moiety are still active, at least against small peptide substrates (Martínez *et al.*, 1993).

## Distinguishing Features

The most important distinctive feature of cruzipain is the presence of the C-T (see above). As in the cysteine proteinases of *T. rangeli* and *Crithidia fasciculata*, but in contrast to the similar cysteine proteinases from *Leishmania mexicana* and *T. brucei*, the C-T is retained in the natural mature form of the enzyme (Cazzulo *et al.*, 1995).

## Further Reading

A further, recent review is that of Cazzulo *et al.* (1997).

## References

Bontempi, E., Franke de Cazzulo, B.M., Ruiz, A.M. & Cazzulo, J.J. (1984) Purification and some properties of an acidic protease from epimastigotes of *Trypanosoma cruzi*. *Comp. Biochem. Physiol.* **77B**, 599–604.

Campetella, O., Henriksson, J., Åslund, L., Frasch, A.C.C., Pettersson, U. & Cazzulo, J.J. (1992) The major cysteine proteinase (Cruzipain) from *Trypanosoma cruzi* is encoded by multiple polymorphic tandemly organized genes located on different chromosomes. *Mol. Biochem. Parasitol.* **50**, 225–234.

Cazzulo, J.J. & Frasch, A.C.C. (1992) SAPA/*trans*-sialidase and cruzipain: two antigens from *Trypanosoma cruzi* contain immunodominant but enzymatically inactive domains. *FASEB J.* **6**, 3259–3264.

Cazzulo, J.J., Couso, R., Raimondi, A., Wernstedt, C. & Hellman, U. (1989) Further characterization and partial amino acid sequence of a cysteine proteinase from *Trypanosoma cruzi*. *Mol. Biochem. Parasitol.* **33**, 33–41.

Cazzulo, J.J., Cazzulo Franke, M.C., Martinez, J. & Franke de Cazzulo, B.M. (1990) Some kinetic properties of a cysteine proteinase (cruzipain) from *Trypanosoma cruzi*. *Biochim. Biophys. Acta* **1037**, 186–191.

Cazzulo, J.J., Martínez, J., Parodi, A.J.A., Wernstedt, C. & Hellman, U. (1992) On the post-translational modifications at the C-terminal domain of the major cysteine proteinase (cruzipain) from *Trypanosoma cruzi*. *FEMS Microbiol. Lett.* **100**, 411–416.

Cazzulo, J.J., Labriola, C., Parussini, F., Duschak, V., Martinez, J. & Franke de Cazzulo, B.M. (1995) Cysteine proteinases in *Trypanosoma cruzi* and other Trypanosomatid parasites. *Acta Chim. Slovenica* **42**, 409–418.

Cazzulo, J.J., Bravo, M., Raimondi, A., Engström, U., Lindeberg, G. & Hellman, U. (1996) Hydrolysis of synthetic peptides by cruzipain, the major cysteine proteinase from *Trypanosoma cruzi*, provides evidence for self-processing and the possibility of more specific substrates for the enzyme. *Cell. Mol. Biol.* **42**, 691–696.

Cazzulo, J.J., Stoka, V. & Turk, V. (1997) Cruzipain, the major cysteine proteinase from the protozoan parasite *Trypanosoma cruzi*. *Biol. Chem. Hoppe-Seyler* **378**, 1–10.

Eakin, A.E., Mills, A.A., Harth, G., McKerrow, J.H., & Craik, C.S. (1992) The sequence, organization, and expression of the major cysteine proteinase (cruzain) from *Trypanosoma cruzi*. *J. Biol. Chem.* **267**, 7411–7420.

Franke de Cazzulo, B.M., Martínez, M., North, M.J., Coombs, G.H. & Cazzulo, J.J. (1994) Effect of proteinase inhibitors on growth and differentiation of *Trypanosoma cruzi*. *FEMS Microbiol. Lett.* **124**, 81–86.

Greig, S. & Ashall, F. (1990) Electrophoretic detection of *Trypanosoma cruzi* peptidases. *Mol. Biochem. Parasitol.* **39**, 31–38.

Itow, S. & Camargo, E. (1977) Proteolytic activities in cell extracts of *Trypanosoma cruzi*. *J. Protozool.* **24**, 591–595.

Labriola, C., Sousa, M. & Cazzulo, J.J. (1993) Purification of the major cysteine proteinase (cruzipain) from *Trypanosoma cruzi* by affinity chromatography. *Biological Res.* **26**, 101–107.

McGrath, M.E., Eakin, A.E., Engel, J.C., McKerrow, J.H., Craik, C.S. & Fletterick, R.J. (1995) The crystal structure of cruzain: a therapeutic target for Chagas' disease. *J. Mol. Biol.* **247**, 251–259.

Martínez, J. & Cazzulo, J.J. (1992) Anomalous electrophoretic behaviour of the major cysteine proteinase (cruzipain) from *Trypanosoma cruzi* in relation to its apparent molecular weight. *FEMS Microbiol. Lett.* **95**, 225–230.

Martínez, J., Campetella, O., Frasch, A.C.C. & Cazzulo, J.J. (1993) The reactivity of sera from chagasic patients against different fragments of cruzipain, the major cysteine proteinase from *Trypanosoma cruzi*, suggests the presence of defined antigenic and catalytic domains. *Immunol. Lett.* **35**, 191–196.

Metzner, S.I., Sousa, M.C., Hellman, U., Cazzulo, J.J. & Parodi, A.J. (1996) The use of the UDP-Glc: glycoprotein glucosyl transferase for radiolabeling protein-linked high mannose-type oligosaccharides. *Cell. Mol. Biol.* **42**, 631–635.

Murta, A.C.M., Persechini, P.M., de Souto Padron, T., de Souza, W., Guimaraes, J.A. & Scharfstein, J. (1990) Structural and functional identification of GP57/51 antigen of *Trypanosoma cruzi* as a cysteine proteinase. *Mol. Biochem. Parasitol.* **43**, 27–38.

Rangel, H.A., Araujo, P.M.F., Repka, D. & Costa, M.G. (1981) *Trypanosoma cruzi*: isolation and characterization of a proteinase. *Exp. Parasitol.* **52**, 199–209.

Serveau, C., Lalmanach, G., Juliano, M.A., Scharfstein, J., Juliano, L. & Gauthier, F. (1996) Investigation of the substrate specificity of cruzipain, the major cysteine proteinase of *Trypanosoma cruzi*, through the use of cystatin-derived substrates and inhibitors. *Biochem. J.* **313**, 951–956.

Soares, M.J., Souto-Padrón, T. & De Souza, W. (1992) Identification of a large pre-lysosomal compartment in the pathogenic protozoon *Trypanosoma cruzi. J. Cell Sci.* **102**, 157–167.

Stoka, V., Nycander, M., Lenarčič, B., Labriola, C., Cazzulo, J.J., Björk, I. & Turk, V. (1995) Inhibition of cruzipain, the major cysteine proteinase of the protozoan parasite, *Trypanosoma cruzi*, by proteinase inhibitors of the cystatin superfamily. *FEBS Lett.* **370**, 101–104.

Tomas, A.M. & Kelly, J.M. (1996) Stage regulated expression of cruzipain, the major cysteine protease of *Trypanosoma cruzi*, is independent of the level of RNA. *Mol. Biochem. Parasitol.* **76**, 91–103.

Turk, B., Stoka, V., Turk, V., Johansson, G., Cazzulo, J.J. & Björk, I. (1996) High-molecular-weight kininogen binds two molecules of cysteine proteinases with different rate constants. *FEBS Lett.* **391**, 109–112.

Yokoyama-Yasunaka, J.K.U., Pral, E.M.F., Oliveira, Jr, O.C., Alfieri, S.C. & Stolf, A.M.S. (1994) *Trypanosoma cruzi*: identification of proteinases in shed components of trypomastigote forms. *Acta Tropica* **57**, 307–315.

*Juan Jose Cazzulo*
*Laboratory of Parasite Biochemistry,*
*Instituto de Investigaciones Biotecnologicas,*
*Universidad Nacional de General San Martin,*
*Buenos Aires, Argentina*
*Email: jcazzulo@inti.gov.ar*

# 204. *Leishmania cysteine proteinases*

## Databanks

*Peptidase classification: clan CA, family C1, MEROPS ID: C01.076*
*NC-IUBMB enzyme classification: none*

*Databank codes:*

| Species | Gene | SwissProt | PIR | EMBL (cDNA) | EMBL (genomic) |
|---|---|---|---|---|---|
| **Type I (cathepsin L-like)** | | | | | |
| *Leishmania major* | *lmajl* | – | – | U43706 | – |
| *Leishmania mexicana* | *lmcpb* | P36400 | S29245 | Z14061 | – |
| | | | | Z49962 | – |
| | | | | Z49965 | – |
| *Leishmania pifanoi* | *lpcys2* | Q05094 | A48566 | L00718 | – |
| | | | | M97695 | – |
| **Type II (cathepsin L-like)** | | | | | |
| *Leishmania mexicana* | *lmcpa* | P25775 | S25267 | X62163 | – |
| *Leishmania pifanoi* | *lpcys1* | P35591 | B48566 | L29168 | – |
| **Type III (cathepsin B-like)** | | | | | |
| *Leishmania major* | *lmajc* | – | – | U43705 | – |
| *Leishmania mexicana* | *lmcpc* | – | – | Z48599 | – |

## Name and History

*Leishmania* is a trypanosomatid (a flagellated, parasitic protozoon) that causes a variety of diseases in humans and other mammals in the tropics and subtropics. The parasite alternates between sandfly and mammalian hosts, and has several developmental forms. The promastigote is a motile, flagellated form that multiplies in the gut of the fly, the metacyclic is a nondividing, infective form that resides in the mouthparts of the fly, and the amastigote is a nonmotile form that replicates and lives in the phagolysosomal compartment of mammalian macrophages.

The protozoa of the genus *Leishmania* contain an abundance of cysteine proteinases of the papain family (C1). One feature that spurred the interest of researchers in the early 1980s was that the cysteine proteinases show a high level of stage-regulated expression. In *L. mexicana*, for example, it was found that cysteine proteinase activity was considerably greater in the mammalian amastigote form than the promastigote forms that live in the sandfly vector (Coombs, 1982). This suggested that the cysteine proteinases might be important for survival of the parasite in the mammalian host. Further analysis of the *L. mexicana* enzymes showed that there were multiple isoenzymes that could be grouped according to their biochemical characteristics (Pupkis & Coombs, 1984; Robertson & Coombs, 1990, 1992, 1993; Bates *et al.*, 1994). Many of these are encoded by the *lmcpb* genes (Souza *et al.*, 1992; Robertson & Coombs, 1994), which are organized in a tandem array of 19 closely similar copies (Mottram *et al.*, 1996). Small sequence differences between the genes of the array are thought to account for some of the

biochemical peculiarities of the different isoenzymes (Robertson & Coombs, 1994). *lmcpb* has been classed as a type I cysteine proteinase gene (Coombs & Mottram, 1997), together with homologs from other leishmanias (Traub-Cseko *et al.*, 1993; Sakanari *et al.*, 1997) and other trypanosomatids such as *Trypanosoma cruzi* (cruzipain or cruzain) and *T. brucei* (Mottram *et al.*, 1989; Robertson *et al.*, 1996). The type I genes are cathepsin L-like and are characterized by the presence of an unusual C-terminal extension (Mottram *et al.*, 1989; Souza *et al.*, 1992; Traub-Cseko *et al.*, 1993). The type II cysteine proteinase genes (for example, *lmcpa*) are also predicted to encode cathepsin L-like cysteine proteinases, but the enzymes have not been characterized biochemically (Mottram *et al.*, 1992; Traub-Cseko *et al.*, 1993). Type III genes (for example, *lmcpc*) encode cathepsin B-like cysteine proteinases (Bart *et al.*, 1995; Robertson & Coombs, 1993).

## Activity and Specificity

Seven groups of LmCPb enzymes are recognized, some of which show differences in substrate specificity, apparently through selectivity in the P1 position (Table 204.1). All hydrolyze substrates containing a Pro-Phe-Arg+ moiety well, but group B enzymes hydrolyze Leu-Val-Tyr- or Leu-Tyr-containing substrates better than group C enzymes. Group C enzymes hydrolyze Phe-Val-Arg substrates better than the group B cysteine proteinases (Robertson & Coombs, 1990). *L. mexicana* cathepsin B-like cysteine proteinases (LmCPc) have relative molecular masses of 31 000 and 33 000, are amphiphilic, and do not hydrolyze gelatin in substrate SDS-PAGE, but are active towards synthetic

*Table 204.1* Features of the *L. mexicana* cysteine proteinases

| Group | Glycosylated | Features | Stage-specificity | Gene |
|---|---|---|---|---|
| A | Yes, binds con A | – | Amastigote | *lmcpb* |
| B | No | – | Amastigote | *lmcpb* |
| C | No | – | Amastigote | *lmcpb* |
| D | No | Hydrolyzes Bz-Phe-Val-Arg+NHMec, not gelatin. Amphiphilic | Both forms | *lmcpc* |
| E | Unknown | Particulate | Amastigote | *lmcpb* |
| F | Unknown | Higher mobility, metacyclic-specific | Promastigote | *lmcpb* |
| H | Unknown | Acid-activated. Precursor of F? | Promastigote | *lmcpb* |

substrates with a Phe-Val-Arg moiety (Robertson & Coombs, 1993). There is some evidence that *Leishmania* contain $Ca^{2+}$-dependent cysteine proteinases similar in properties to calpains, but the corresponding gene(s) have not yet been cloned (Bhattacharya *et al*., 1993).

## Structural Chemistry

Detailed structural analysis of the closely related type I cysteine proteinase of *T. cruzi*, cruzipain, has been carried out (see Chapter 203).

## Preparation

The *lmcpb* gene products were purified from *L. mexicana* amastigotes isolated from infected mice by use of con A-Sepharose CL-4B (group A bind, groups B and C remain in the flow through) followed by Superose 12 gel filtration and Mono-Q anion-exchange chromatography (Robertson & Coombs, 1990). The *lmcpb* gene products can also be efficiently purified from the cell-free supernatant of homogenized lesion tissue isolated from an infected mouse by chromatography involving (sequentially) DEAE-cellulose (DE52), octyl-Sepharose, Superose 12 gel filtration and Mono-Q FPLC (Ilg *et al*., 1994). Cathepsin B-like cysteine proteinases (products of the *lmcpc* gene) were isolated from *L. mexicana* amastigotes by Superose 12 gel filtration, alkyl Superose chromatography and Superdex 75 gel filtration (Robertson & Coombs, 1993).

## Biological Aspects

Electron microscope immunolocalization with antibodies to the type I and type II enzymes of *Leishmania* show the enzymes in extended lysosomes, termed 'megasomes' (Pupkis *et al*., 1986; Prina *et al*., 1990; Duboise *et al*., 1994). The cysteine proteinases are apparently not secreted by the parasite, although they are released from lysed amastigotes and thus accumulate in the parasitophorous vacuole of the host macrophage or extracellularly in the tissue (Ilg *et al*., 1994). The precise functions of the cysteine proteinases are not known. Genetic manipulation studies have revealed that the type I cysteine proteinases of *L. mexicana* are virulence factors (Mottram *et al*., 1996) because a null mutant for the *lmcpb* genes was found to be significantly less able to infect macrophages *in vitro* or produce infections in mice. The ability of a small proportion of *lmcpb* null mutants to invade host cells and infect animals shows that the genes, although important, are not essential for parasite survival (Mottram *et al*., 1996). In contrast, null mutants for the type II gene *lmcpa* showed no detectable phenotypic differences from wild-type parasites (Souza *et al*., 1994). It is likely that the different types of cysteine proteinases have some overlapping functions. Leishmanial cysteine proteinases are also implicated in modulation of the host immune system to favor parasite survival and proliferation (De Souza Leao *et al*., 1995; Wolfram *et al*., 1995; Coombs & Mottram, 1997).

The importance of leishmanial cysteine proteinases in survival of the parasite and its passage through the life cycle has made them targets for the development of cysteine proteinase inhibitors as antileishmanial drugs (see Robertson

*et al*., 1996; Mottram *et al*., 1996; Coombs & Mottram, 1997, for full discussion). Amino acid esters are antileishmanial prodrugs (Rabinovitch, 1989) that are activated by the LmCPb enzymes (Robertson *et al*., 1996).

## Distinguishing Features

Phylogenetic analysis of the type I and II cysteine proteinase genes of the leishmanias classes them with the *Dictyostelium discoidium dscp1* gene (Berti & Storer, 1995), which forms a subgroup of the papain subfamily. One distinguishing feature of the type I enzymes is the presence of a long (some 100 amino acids) C-terminal extension, the function of which is unknown. This comprises a hinge region, rich in Pro, Ser and/or Thr residues, followed by a domain containing conserved Cys residues that are thought to be important in the tertiary structure of the domain. The C-terminal extension is processed to give the mature cysteine proteinase (Duboise *et al*., 1994). The C-terminal extension may be involved in ensuring the correct folding or processing of the enzyme, in modulating its specificity (possibly as an inhibitor), or as an immune modulator (see Coombs & Mottram, 1997, for more discussion), but it is thought unlikely to be involved directly in trafficking of the endopeptidase to lysosomes (Duboise *et al*., 1994). The gene product of *lmcpa* (type II cysteine proteinase) has a tripeptide insertion, Gly-Val-Met, close to the N-terminus of the mature enzyme (Mottram *et al*., 1992).

## Further Reading

See Robertson *et al*. (1996) and Coombs & Mottram (1997) for fuller reviews of the proteinases of trypanosomes and *Leishmania*.

## References

Bart, G., Coombs, G.H. & Mottram, J.C. (1995) Isolation of *lmcpc*, a gene encoding a *Leishmania mexicana* cathepsin B-like cysteine proteinase. *Mol. Biochem. Parasitol.* **73**, 271–274.

Bates, P.A., Robertson, C.D. & Coombs, G.H. (1994) Expression of cysteine proteinases by metacyclic promastigotes of *Leishmania mexicana*. *J. Euk. Microbiol.* **41**, 199–203.

Berti, P.J. & Storer, A.C. (1995) Alignment/phylogeny of the papain superfamily of cysteine proteases. *J. Mol. Biol.* **246**, 273–283.

Bhattacharya, J., Dey, R. & Datta, S.C. (1993) Calcium dependent thiol protease caldonopain and its specific endogenous inhibitor in *Leishmania donovani. Mol. Cell. Biochem.* **126**, 9–16.

Coombs, G.H. (1982) Proteinases of *Leishmania mexicana* and other flagellate protozoa. *Parasitology* **84**, 149–155

Coombs, G.H. & Mottram, J.C. (1997) Proteinases of trypanosomes and *Leishmania*. In: *Trypanosomiasis and Leishmaniasis: Biology and Control* (Hide, G., Mottram, J.C., Coombs, G.H. & Holmes, P.H., eds). Oxford: CAB International, pp. 177–197.

De Souza Leao, S., Lang, T., Prina, E., Hellio, R. & Antoine, J.-C. (1995) Intracellular *Leishmania amazonensis* amastigotes internalize and degrade MHC class II molecules of their host cells. *J. Cell Sci.* **108**, 3219–3231.

Duboise, S.M., Vannier-Santos, M.A., Costa-Pinto, D., Rivas, L., Pan, A.A., Traub-Cseko, Y.M., de Souza, W. & McMahon-Pratt, D. (1994) The biosynthesis, processing, and immunolocalization of *Leishmania pifanoi* amastigote cysteine proteinases. *Mol. Biochem. Parasitol.* **68**, 119–132.

Ilg, T., Fuchs, M., Gnau, V., Wolfram, M., Harbecke, D. & Overath, P. (1994) Distribution of parasite cysteine proteinases in lesions of mice infected with *Leishmania mexicana* amastigotes. *Mol. Biochem. Parasitol.* **67**, 193–203.

Mottram, J.C., North, M.J., Barry, J.D. & Coombs, G.H. (1989) A cysteine proteinase cDNA from *Trypanosoma brucei* predicts an enzyme with an unusual C-terminal extension. *FEBS Lett.* **258**, 211–215.

Mottram, J.C., Robertson, C.D., Coombs, G.H. & Barry, J.D. (1992) A developmentally regulated cysteine proteinase gene of *Leishmania mexicana. Mol. Microbiol.* **6**, 1925–1932.

Mottram, J.C., Souza, A.E., Hutchison, J.E., Carter, R., Frame, M.J. & Coombs, G.H. (1996) Evidence from disruption of the *lmcpb* gene array of *Leishmania mexicana* that cysteine proteinases are virulence factors. *Proc. Natl Acad. Sci. USA* **93**, 6008–6013.

Prina, E., Antoine, J.C., Wiederanders, B. & Kirschke, H. (1990) Localization and activity of various lysosomal proteases in *Leishmania amazonensis*-infected macrophages. *Infect. Immun.* **58**, 1730–1737.

Pupkis, M.F. & Coombs, G.H. (1984) Purification and characterization of proteolytic enzymes of *Leishmania mexicana mexicana* amastigotes and promastigotes. *J. Gen. Microbiol.* **130**, 2375–2383.

Pupkis, M.F., Tetley, L. & Coombs, G.H. (1986) *Leishmania mexicana*: amastigote hydrolases in unusual lysosomes. *Exp. Parasitol.* **62**, 29–39.

Rabinovitch, M. (1989) Leishmanicidal activity of amino acid and peptide esters. *Parasitol. Today* **5**, 299–301.

Robertson, C.D. & Coombs, G.H. (1990) Characterization of three groups of cysteine proteinases in the amastigotes of *Leishmania mexicana mexicana. Mol. Biochem. Parasitol.* **42**, 269–276.

Robertson, C.D. & Coombs, G.H. (1992) Stage-specific proteinases of *Leishmania mexicana* promastigotes. *FEMS Microbiol. Lett.* **94**, 127–132.

Robertson, C.D. & Coombs, G.H. (1993) Cathepsin B-like cysteine proteases of *Leishmania mexicana. Mol. Biochem. Parasitol.* **62**, 271–279.

Robertson, C.D. & Coombs, G.H. (1994) Multiple high activity cysteine proteases of *Leishmania mexicana* are encoded by the *lmcpb* gene array. *Microbiology* **140**, 417–424.

Robertson, C.D., Martinez, J., Cazzulo, J.-J. & Coombs, G.H. (1994) Analysis of the cysteine proteinases of *Leishmania mexicana* and *Trypanosoma cruzi* using specific antisera. *FEMS Microbiol. Lett.* **124**, 191–194.

Robertson, C.D., Coombs, G.H., North, M.J. & Mottram, J.C. (1996) Parasite cysteine proteinases. *Perspect. Drug Discovery Design* **6**, 99–118.

Sakanari, J.A., Nadler, S.A., Chan, V.J., Engel, J.C., Leptak, C. & Bouvier, J. (1997) *Leishmania major*: comparison of the cathepsin L- and B-like cysteine protease genes with those of other trypanosomatids. *Exp. Parasitol.* **85**, 63–76.

Souza, A.E., Waugh, S., Coombs, G.H. & Mottram, J.C. (1992) Characterization of a multicopy gene for a major stage-specific cysteine proteinase of *Leishmania mexicana. FEBS Lett.* **311**, 124–127.

Souza, A.E., Bates, P.A., Coombs, G.H. & Mottram, J.C. (1994) Null mutants for the *lmcpa* cysteine proteinase gene in *Leishmania mexicana. Mol. Biochem. Parasitol.* **63**, 213–220.

Traub-Cseko, Y.M., Duboise, M., Boukai, L.K. & McMahon-Pratt, D. (1993) Identification of two distinct cysteine proteinase genes of *Leishmania pifanoi* axenic amastigotes using the polymerase chain reaction. *Mol. Biochem. Parasitol.* **57**, 101–116.

Wolfram, M., Ilg, T., Mottram, J.C. & Overath, P. (1995) Antigen presentation by *Leishmania mexicana*-infected macrophages: activation of helper T cells specific for amastigote cysteine proteinases requires intracellular killing of the parasites. *Eur. J. Immunol.* **25**, 1094–1100.

*Jeremy C. Mottram*
*Wellcome Unit of Molecular Parasitology,*
*The Anderson College,*
*University of Glasgow,*
*Glasgow G11 6NU, UK*
*Email: j.mottram@udcf.gla.ac.uk*

*Graham H. Coombs*
*Division of Infection & Immunity,*
*Institute of Biomedical and Life Sciences,*
*Joseph Black Building,*
*University of Glasgow,*
*Glasgow G12 8QQ, UK*
*Email: g.h.coombs@bio.gla.ac.uk*

# 205. Trichomonad and Giardia cysteine endopeptidases

## Databanks

*Peptidase classification: clan CA, family C1, MEROPS ID: C01.082, C01.094*
*NC-IUBMB enzyme classification: none*

*Databank codes:*

| Species | Gene | SwissProt | PIR | EMBL (cDNA) | EMBL (genomic) |
|---|---|---|---|---|---|
| *Trichomonas vaginalis* | *tvcp1* | – | S41427 | X77218 | – |
| *Trichomonas vaginalis* | *tvcp2* | – | S41428 | X77219 | – |
| *Trichomonas vaginalis* | *tvcp3* | – | S41425 | X77220 | – |
| *Trichomonas vaginalis* | *tvcp4* | – | S41426 | X77221 | – |
| *Trichomonas vaginalis* | – | P33404 | – | – | – |
| *Tritrichomonas foetus* | *tfcp1* | – | – | U13153 | – |
| *Tritrichomonas foetus* | *tfcp2* | – | – | U13154 | – |
| *Tritrichomonas foetus* | *tfcp3* | – | S57451 | X87776 | – |
| *Tritrichomonas foetus* | *tfcp4* | – | S57427 | X87777 | – |
| *Tritrichomonas foetus* | *tfcp5* | – | S57426 | X87778 | – |
| *Tritrichomonas foetus* | *tfcp6* | – | S57421 | X87779 | – |
| *Tritrichomonas foetus* | *tfcp7* | – | S57425 | X87780 | – |
| *Tritrichomonas foetus* | *tfcp8* | – | S57422 | X87781 | – |
| *Tritrichomonas foetus* | *tfcp9* | – | S57423 | X87782 | – |
| *Tritrichomonas foetus* | – | P33403 | – | – | – |

## Name and History

The genera *Trichomonas, Tritrichomonas* and *Giardia* contain protozoa that parasitize the urinogenital tract and gastrointestinal tract of humans and other mammals. Multiple cysteine endopeptidases are present in all species of trichomonad examined. These have not been given trivial names and are usually identified by their apparent molecular masses, as indicated by substrate SDS-PAGE. High endopeptidase activity was first detected in *Tritrichomonas foetus*, a urogenital parasite of cattle (McLaughlin & Müller, 1979) and similar activity was demonstrated in the human urogenital parasite *Trichomonas vaginalis* (Coombs & North, 1983). The proteolytic activity, largely due to cysteine endopeptidases, was among the highest recorded in any group of protozoa (Coombs, 1982). Subsequent analysis revealed the presence of multiple enzymes in both species (Lockwood *et al*., 1984, 1987) and showed that significant amounts of the endopeptidases are released by the parasites (Lockwood *et al*., 1987, 1988). Trichomonad species which inhabit the gastrointestinal tract also produce and secrete multiple forms of cysteine endopeptidase (see North, 1991).

In *T. vaginalis* there are at least four closely related cysteine endopeptidase genes (Mallinson *et al*., 1994) while nine different genes or gene fragments have been cloned from *T. foetus* (Mallinson *et al*., 1995; Thomford *et al*., 1996). All the predicted gene products are members of the papain family (C1).

Less extensive studies have been undertaken on *Giardia* species, but multiple cysteine endopeptidase activities have been detected (Hare *et al*., 1989; Williams & Coombs, 1995).

## Activity and Specificity

Trichomonad proteinases are able to hydrolyze a range of proteins including possible physiological substrates such as hemoglobin (Coombs & North, 1983), immunoglobulins (Talbot *et al*., 1991; Provenzano & Alderete, 1995) and extracellular matrix proteins (Bozner & Demes, 1991), although there are differences between the specificity of individual enzymes (Provenzano & Alderete, 1995). Based on differences in activity towards 11 fluorogenic peptide substrates, three groups of cysteine endopeptidases were defined in *T. vaginalis* (North *et al*., 1990). Group I comprises a single activity ($M_r$ 86 000) that hydrolyzed Z-Arg-Arg$+$NHMec only. The sole enzyme in group II ($M_r$ 54 000) had preferential activity towards Z-Phe-Arg$+$NHMec. All the group III enzymes ($M_r$ 20 000–110 000) preferred substrates with bulky residues at both the P2 and P3 positions and accommodated arginine, lysine or tyrosine at the P1 position, although subtle differences in specificity were apparent. The best substrate for all group III enzymes was Boc-Val-Leu-Lys$+$NHMec. Differences between groups were also apparent from their sensitivity to inhibitors, most notably the poor inhibition of the group I enzyme by E-64. Group I and III enzymes were also apparent among the *T. foetus* cysteine endopeptidases; some were found to be specific for Z-Arg-Arg-NHMec ($M_r$ of 25 000, 31 000 and 34 000), while others ($M_r$ 20 000 and 32 000) were similar to the group III enzymes, with a preference for substrates such as Boc-Val-Leu-Lys-NHMec with bulky residues at P2 and P3 (North *et al*., 1990).

*Giardia* cysteine endopeptidases also hydrolyze a range of proteins including hemoglobin, immunoglobulins (Parenti, 1989) and extracellular matrix proteins (Williams & Coombs, 1995). Some preference for arginine at the P2 position has been noted (Werries *et al*., 1991).

## Structural Chemistry

There has been little study of purified trichomonad or *Giardia* proteinases and structural details can only be predicted from the sequences of their corresponding genes (Mallinson *et al*., 1994, 1995; Thomford *et al*., 1996).

## Preparation

Cysteine endopeptidases of *T. vaginalis* have been purified from cell lysates ($M_r$ 18 000 and 64 000 by gel filtration: Lockwood *et al*., 1985) and from conditioned, serum-free culture medium ($M_r$ 60 000 and 30 000 by gel filtration and

SDS-PAGE: Garber & Lemchuk-Favel, 1989) using combinations of ion-exchange chromatography, gel filtration and, in the case of the intracellular enzymes, thiol-Sepharose affinity chromatography. An enzyme with an $M_r$ of 17 500 has been purified from *T. foetus* by ion-exchange chromatography, gel filtration and affinity chromatography on organomercurial agarose (McLaughlin & Müller, 1979). Affinity chromatography with bacitracin-Sepharose has been used in combination with PAGE to obtain purified enzymes from both *T. foetus* and *T. vaginalis* in quantities which have allowed peptide sequencing (Irvine *et al.*, 1993; Thomford *et al.*, 1996), The *T. vaginalis* peptide sequence confirmed that the protein was a cysteine endopeptidase, but did not match any of the mature domain sequences predicted from the genes (Mallinson *et al.*, 1994). A 15 amino acid sequence from a *T. foetus* 31 kDa extracellular enzyme TFECP (Thomford *et al.*, 1996) was found to be encoded by the *tfcp8* gene (Mallinson *et al.*, 1995).

Two endopeptidases ($M_r$ 35 000 and 95 000 by gel filtration) have been purified from *Giardia lamblia* trophozoites using gel filtration, organomercurial agarose affinity chromatography and ion-exchange chromatography on DEAE-cellulose (Werries *et al.*, 1991).

## Biological Aspects

Trichomonads are notable for the high levels of intracellular cysteine endopeptidases and the large quantities of the enzymes that are secreted. Within the cells, most of the activity is associated with the lysosomal fraction (Müller, 1973; Lockwood *et al.*, 1988), but some cysteine endopeptidase activity may be located elsewhere, as evidenced by fractionation and inhibitor studies (Scott *et al.*, 1995).

Trichomonad cysteine endopeptidases have been detected in vaginal samples taken from trichomoniasis patients (Alderete *et al.*, 1991; Garber & Lemchuk-Favel, 1994) and from artificially infected beef heifers (Yule *et al.*, 1989). They are also immunogenic (Alderete *et al.*, 1991). It is thought that the enzymes play an important role in the host–parasite interaction and may contribute to pathogenicity. There is evidence that a surface-located 43 kDa cysteine endopeptidase activity is essential for adherence of the parasite to host cells (Arroyo & Alderete, 1989; Neale & Alderete, 1990). Some of the enzymes are likely to be involved in nutrition, e.g. in the breakdown of host proteins such as low-density lipoprotein (Peterson & Alderete, 1984). They may also interfere with the acquired immunity of the host through the degradation of immunoglobulins (Talbot *et al.*, 1991; Provenzano & Alderete, 1995).

The importance of cysteine endopeptidases during trichomonad infections has been highlighted by the finding that leupeptin, a known inhibitor of the *T. vaginalis* cysteine endopeptidases, interferes with *in vivo* infection in a model system (Bremner *et al.*, 1986).

Many of the *Giardia* cysteine endopeptidases are located in membrane-bound organelles (Lindmark, 1988) and are likely to have functions typical of lysosomal endopeptidases of other organisms. However, specific functions in host–parasite interactions, as indicated for trichomonads, have not yet been demonstrated. A role for a parasite enzyme in the processing of the giardiavirus capsid polypeptide has been suggested by the finding that both leupeptin and E-64 inhibit the process (Yu *et al.*, 1995). The processing activity is induced or activated by the presence of the virus.

## Distinguishing Features

Based on the structures predicted from the gene sequences, many of the trichomonad cysteine endopeptidases show great similarities to mammalian cathepsins L and S (Chapters 210 and 211). They are apparently synthesized as typical pre-proenzymes.

All the cloned *T. vaginalis* genes encode products predicted to be as closely related to mammalian cathepsin L as they are to any other group of cysteine endopeptidases (Mallinson *et al.*, 1994). This is of especial interest in the context of molecular evolution, as trichomonads represent an early branch of the eukaryote tree (Gunderson *et al.*, 1995). The specificity of the majority of the enzymes resembles that of cathepsin L-like enzymes (North *et al.*, 1990). A major distinguishing feature of the *T. vaginalis* endopeptidases, however, is the range of their apparent molecular masses (20 kDa to more than 100 kDa) when analyzed by substrate SDS-PAGE. This size range is not suggested by the DNA sequences, which predict enzymes of the size typical for cysteine endopeptidases of the papain family. Purified endopeptidases do run according to this predicted size when analyzed by electrophoresis under denaturing conditions (Irvine *et al.*, 1993). The reason for the anomalous electrophoretic behavior under partially denaturing conditions is not known, but this property does distinguish the *T. vaginalis* enzymes from those of *T. foetus* (see Lockwood *et al.*, 1987; North *et al.*, 1990).

DNA sequences for some of the genes encoding *T. foetus* cysteine endopeptidases also predict that the products have properties analogous to mammalian cathepsin L. However, some of the gene products are more diverged and the possibility that these might have novel properties has been considered (Mallinson *et al.*, 1995). As yet these have not been linked to specific endopeptidases.

Too little is known about *Giardia* cysteine endopeptidases to predict distinguishing features, but as with those of *T. vaginalis* some have electrophoretic mobilities slower than anticipated for typical enzymes of this type.

## Further Reading

A review has been provided by North (1991). While the present chapter was in press, a paper appeared describing the identification of a *Giardia* cysteine endopeptidase as an essential factor in the excystation of the organism. The deduced amino acid sequence of the enzyme showed it to be a 'chimera between mammalian cathepsin B and papain or cathepsin L' that could well represent one of the most primitive examples from peptidase family C1 (Ward *et al.*, 1997).

## References

Alderete, J.F., Newton, E., Dennis, C. & Neale, K.A. (1991) The vagina of women infected with *Trichomonas vaginalis* has numerous proteinases and antibody to trichomonad proteinases. *Genitourin. Med.* **67**, 469–474.

Arroyo, R. & Alderete, J.F. (1989) *Trichomonas vaginalis* surface proteinase activity is necessary for parasite adherence to epithelial cells. *Infect. Immun.* **57**, 2991–2997.

Bozner, P. & Demes, P. (1991) Degradation of collagen types I, II, IV and V by extracellular proteinases of an oral flagellate *Trichomonas tenax. Arch. Oral Biol.* **36**, 765–770.

Bremner, A.F., Coombs, G.H. & North, M.J. (1986) Antitrichomonal activity of the proteinase inhibitor leupeptin. *IRCS Med. Sci.* **14**, 555–556.

Coombs, G.H. (1982) Proteinases of *Leishmania mexicana* and other flagellate protozoa. *Parasitology* **84**, 149–155.

Coombs, G.H. & North, M.J. (1983) An analysis of the proteinases of *Trichomonas vaginalis* by polyacrylamide gel electrophoresis. *Parasitology* **86**, 1–6.

Garber, G.E. & Lemchuk-Favel, L.T. (1989) Characterization and purification of extracellular proteases of *Trichomonas vaginalis. Can. J. Microbiol.* **35**, 903–909.

Garber, G.E. & Lemchuk-Favel, L.T. (1994) Analysis of the extracellular proteases of *Trichomonas vaginalis. Parasitol. Res.* **80**, 361–365.

Gunderson, J., Hinkle, G., Leipe, D., Morrison, H.G., Stickel, S.K., Odelson, D.A., Breznak, J.A., Nerad, T.A., Müller, M. & Sogin, M.L. (1995) Phylogeny of trichomonads inferred from small-subunit rRNA sequences. *J. Euk. Microbiol.* **42**, 411–415.

Hare, D.F., Jarroll, E.L. & Lindmark, D.G. (1989) *Giardia lamblia*: characterization of proteinase activity in trophozoites. *Exp. Parasitol.* **68**, 168–175.

Irvine, J.W., Coombs, G.H. & North, M.J. (1993) Purification of cysteine proteinases from trichomonads using bacitracin-Sepharose. *FEMS Microbiol. Lett.* **110**, 113–120.

Lindmark, D.G. (1988) *Giardia lamblia*: localization of hydrolase activities in lysosome-like organelles of trophozoites. *Exp. Parasitol.* **65**, 141–147.

Lockwood, B.C., North, M.J. & Coombs, G.H. (1984) *Trichomonas vaginalis, Tritrichomonas foetus* and *Trichomitus batrachorum*: comparative proteolytic activity. *Exp. Parasitol.* **58**, 245–253.

Lockwood, B.C., North, M.J. & Coombs, G.H. (1985) Purification and characterization of proteinases of the parasitic protozoan *Trichomonas vaginalis. Biochem. Soc. Trans.* **13**, 336.

Lockwood, B.C., North, M.J., Scott, K.I., Bremner, A.F. & Coombs, G.H. (1987) The use of a highly sensitive electrophoretic method to compare the proteinases of trichomonads. *Mol. Biochem. Parasitol.* **24**, 89–95.

Lockwood, B.C., North, M.J. & Coombs, G.H. (1988) The release of hydrolases from *Trichomonas vaginalis* and *Tritrichomonas foetus. Mol. Biochem. Parasitol.* **30**, 135–142.

McLaughlin, J. & Müller, M. (1979) Purification and characterization of a low molecular weight thiol proteinase from the flagellate protozoon *Tritrichomonas foetus. J. Biol. Chem.* **254**, 1526–1533.

Mallinson, D.J., Lockwood, B.C., Coombs, G.H. & North, M.J. (1994) Identification and molecular cloning of four cysteine proteinase genes from the pathogenic protozoon *Trichomonas vaginalis. Microbiology* **140**, 2725–2735.

Mallinson, D.J., Livingstone, J., Appleton, K.M., Lees, S.J., Coombs, G.H. & North, M.J. (1995) Multiple cysteine proteinases of the pathogenic protozoon *Tritrichomonas foetus*: identification of seven diverse and differentially expressed genes. *Microbiology* **141**, 3077–3085.

Müller, M. (1973) Biochemical cytology of trichomonad flagellates. I. Subcellular localization of hydrolases, dehydrogenases, and catalase in *Tritrichomonas foetus. J. Cell Biol.* **57**, 453–474.

Neale, K.A. & Alderete, J.F. (1990) Analysis of the proteinases of representative *Trichomonas vaginalis* isolates. *Infect. Immun.* **58**, 157–162.

North, M.J. (1991) Proteinases of trichomonads and *Giardia*. In: *Biochemical Protozoology* (Coombs, G.H. & North, M.J., eds). London: Taylor & Francis, pp. 234–244.

North, M.J., Robertson, C.D. & Coombs, G.H. (1990) The specificity of trichomonad cysteine proteinases analyzed using fluorogenic substrates and specific inhibitors. *Mol. Biochem. Parasitol.* **39**, 183–194.

Parenti, D.M. (1989) Characterization of a thiol proteinase in *Giardia lamblia. J. Infect. Dis.* **160**, 1076–1080.

Peterson, K.M. & Alderete, J.F. (1984) *Trichomonas vaginalis* is dependent on uptake and degradation of human low density lipoproteins. *J. Exp. Med.* **160**, 1261–1272.

Provenzano, D. & Alderete, J.F. (1995) Analysis of human immunoglobulin-degrading cysteine proteinases of *Trichomonas vaginalis. Infect. Immun.* **63**, 3388–3395.

Scott, D.A., North, M.J. & Coombs, G.H. (1995) The pathway of secretion of proteinases in *Trichomonas vaginalis. Int. J. Parasitol.* **25**, 657–666.

Talbot, J.A., Nielsen, K. & Corbeil, L.B. (1991) Cleavage of proteins of reproductive secretions by extracellular proteinases of *Tritrichomonas foetus. Can. J. Microbiol.* **37**, 384–390.

Thomford, J.W., Talbot, J.A., Ikeda, J.S. & Corbeil, L.B. (1996) Characterization of extracellular proteinases of *Tritrichomonas foetus. J. Parasitol.* **82**, 112–117.

Ward, W., Alvarado, L., Rawlings, N.D., Engel, J., Franklin, C. & McKerrow, J.H. (1997) A primitive enzyme for a primitive cell: the protease required for excystation of *Giardia. Cell* **89**, 437–444.

Werries, E., Franz, A., Hippe, H. & Acil, Y. (1991) Purification and substrate specificity of two cysteine proteinases of *Giardia lamblia. J. Protozool.* **38**, 378–383.

Williams, A.G. & Coombs, G.H. (1995) Multiple protease activities in *Giardia intestinalis* trophozoites. *Int. J. Parasitol.* **25**, 771–778.

Yu, D., Wang, C.C. & Wang, A.L. (1995) Maturation of giardiavirus capsid protein involves posttranslational proteolytic processing by a cysteine protease. *J. Virol.* **69**, 2825–2830.

Yule, A., Skirrow, S.Z. & BonDurant, R.H. (1989) Bovine trichomoniasis. *Parasitol. Today* **5**, 373–377.

***Jeremy C. Mottram***
*Wellcome Unit of Molecular Parasitology,*
*The Anderson College,*
*University of Glasgow,*
*Glasgow G11 6NU, UK*
*Email: j.mottram@udcf.gla.ac.uk*

***Michael J. North***
*University of Stirling,*
*Department of Biological and Molecular Sciences,*
*Stirling FK9 4LA, UK*
*Email: michael.north@msn.com*

# 206. V-cath proteinase

## Databanks

*Peptidase classification: clan CA, family C1, MEROPS ID: C01.083*
*NC-IUBMB enzyme classification: none*
*Databank codes:*

| Species | SwissProt | PIR | EMBL (cDNA) | EMBL (genomic) |
| --- | --- | --- | --- | --- |
| *Autographa californica* nuclear polyhedrosis virus | P25783 | S27044 | L22858 | M67451: complete genome |
| *Bombyx mori* nuclear polyhedrosis virus | P41721 | – | U12688 | – |
| *Choristoneura fumiferana* nuclear polyhedrosis virus | P41715 | – | M97906 | – |
| *Orgyia pseudotsugata* multicapsid polyherosis virus | O10364 | – | U75930 | – |

## Name and History

As defined by *Webster's Third International Dictionary* (Gove, 1971), a cathepsin is a protease that 'aids in autolysis in certain diseased conditions and after death'. The **V-cath proteinase** is a **viral cathepsin** expressed by arthropod viruses of the Baculoviridae family. The V-cath proteinase plays an active role in viral pathogenesis, promoting the autolysis of the host insect tissues after death.

The V-cath gene was originally sequenced from the *Autographa californica* nuclear polyhedrosis virus (AcNPV) and was submitted to GenBank (accession number M67451) in June of 1991 by K. Kuzio and P. Faulkner. The first publication of the V-cath gene sequence (described as 'papain-like') was by Rawlings *et al.* (1992). It was shown that the deduced V-cath protein amino acid sequence had a hydropathy profile and a predicted secondary structure similar to those of papain. V-cath was also later referred to as a **baculovirus endopeptidase** (Rawlings & Barrett, 1993).

V-cath homologs are found in several other baculovirus species including *Choristoneura fumiferana* nuclear polyhedrosis virus (CfNPV) (Hill *et al.*, 1995) and *Bombyx mori* nuclear polyhedrosis virus (BmNPV) (Ohkawa *et al.*, 1994). The BmNPV V-cath homolog is referred to as **BmNPV-cysteine proteinase (BmNPV-CP)**. A partial V-cath gene sequence is also identifiable in *Orgyia pseudotsugata* nuclear polyhedrosis virus (OpNPV) in a DNA sequence published by Blissard & Rohrmann (1989) (EMBL: M22446).

Where it is known, the genomic positioning of V-cath is conserved in different viral species. V-cath lies just downstream of the gp64 gene that encodes the major envelope fusion protein of baculoviruses (Blissard & Wenz, 1992). Immediately downstream of V-cath itself, there is a viral chitinase gene, *ChiA* (Hawtin *et al.*, 1995).

## Activity and Specificity

V-cath has activity characteristics of both cathepsin B (Chapter 209) and cathepsin L (Chapter 210). Proteolytic activity is maximal under acidic conditions in the range pH 5–6 (Brömme & Okamoto, 1995; Slack *et al.*, 1995). Assays with peptide substrates show cathepsin B-like characteristics, as V-cath has high specificity for the substrate Z-Arg-Arg+NHMec that is rather specific for cathepsin B (Brömme & Okamoto, 1995). V-cath is proteolytically active on azocoll and azocasein (Ohkawa *et al.*, 1994; Slack *et al.*, 1995) and the activity on azocasein is enhanced 4- to 5-fold in the presence of 3 M urea (Slack *et al.*, 1995). This is more characteristic of a cathepsin L-like cysteine proteinase (Barrett & Kirschke, 1981; Maciewicz & Etherington, 1988).

Cysteine proteinase inhibitors such as chicken egg white cystatin, E-64, leupeptin, antipain and iodoacetate effectively inhibit the proteolytic activity of V-cath (Ohkawa *et al.*, 1994; Brömme & Okamoto, 1995; Slack *et al.*, 1995).

## Structural Chemistry

The V-cath proteinase is processed in a complex manner analogous to other cathepsins (Slack *et al.*, 1995). The full-length, 323 amino acid protein is a 37 kDa preproprotein that is processed into a 27.5 kDa mature form with a pI of 5.0 The mature form is probably *N*-glycosylated at Asn158, and has a nonglycosylated molecular mass of 24 kDa. Based on alignment with papain, Pro113 had been proposed as the N-terminal cleavage site of the mature form, but N-terminal sequencing of the mature enzyme revealed a cleavage at Asn98 (Brömme & Okamoto, 1995).

Brömme & Okamoto (1995) identified a putative cathepsin B-like S2 specificity subsite at Glu315 in V-cath. Notably, this glutamate residue is conserved in BmNPV, CfNPV and OpNPV V-cath. However, V-cath lacks a two-histidine occluding loop region that confers on cathepsin B its peptidyl-dipeptidase activity (Musil *et al.*, 1991; Wiederanders *et al.*, 1992). V-cath also possesses an ERFNIN motif in its propeptide region (Ohkawa *et al.*, 1994), which is characteristic of cathepsin L and cathepsin H (Karrer *et al.*, 1993).

## Preparation

V-cath has not been expressed as an active enzyme in bacterial expression systems, but it can be overexpressed in indigenous baculovirus systems, and it accumulates as an inactive zymogen. In baculovirus-infected insect cell cultures, the cellular localization of V-cath varies with time post-infection. At 20–48 h postinfection, V-cath is localized to an insoluble fraction of the cell lysate, but as infection progresses beyond 48 h, V-cath is liberated into a soluble fraction (Slack *et al.*, 1995). Presumably, V-cath is released into the medium as insect cells are lysed. In contemporary baculovirus cell culture systems, inhibitors in the medium may prevent the detection of significant levels of V-cath activity (Ohkawa *et al.*, 1994), but there has been at least one report of cysteine proteinase activity in media from baculovirus-infected insect cells (Yamada *et al.*, 1990). This activity was inhibited by *p*-chloromercuribenzenesulfonic acid.

V-cath can be purified from lysates of baculovirus-infected *Spodoptera frugiperda* cells by retention on a Mono-S FPLC cation-exchange column and elution at pH 5.5 with 150 mM NaCl (Brömme & Okamoto, 1995).

## Biological Aspects

The CfNPV and AcNPV V-cath genes have late viral promoters, which is consistent with the late detection of V-cath in the viral infection cycle. The late promoters have TAAG motifs and transcription starts at −26 from the translational start point (Hill *et al.*, 1995). V-cath gene expression may be translationally regulated, as there exists an out of frame Met/stop sequence (ATGTAA) upstream of the CfNPV and AcNPV V-cath gene translational start point (Hill *et al.*, 1995).

There is evidence that V-cath is autocatalytically processed and activated. When the active mature form of V-cath is specifically degraded by a coexpressed adenovirus type 2 protease, the precursor form is not processed (Keyvani-Amineh *et al.*, 1995). Additionally, this precursor is not proteolytically active and is resistant to proteolysis by the adenovirus type 2 protease.

The natural substrates for V-cath during viral pathogenesis are probably a number of intracellular and extracellular proteins. V-cath has been found associated with budded virions of AcNPV (Lanier *et al.*, 1996) and may facilitate virion budding across the basal lamina between cells. V-cath is also linked to actin degradation in virally infected cells (Lanier *et al.*, 1996). V-cath may play a role analogous to that of the adenovirus L3 23 kDa proteinase (Chapter 251) which degrades cytokeratin during lytic infection of HeLa cells (Chen *et al.*, 1993). It has even been suggested that V-cath plays a role in breaking down protein substrates and providing amino acids for viral replication (Ohkawa *et al.*, 1994).

The gross pathological effects of the V-cath gene product were first described by Ohkawa *et al.* (1994) when characterizing BmNPV-CP. The most striking morphological effect of V-cath is that of liquefying or 'melting' host tissues (Ohkawa *et al.*, 1994; Slack *et al.*, 1995). Cadavers of insects killed by infection with wild-type AcNPV typically turn black and melt, whereas cadavers produced by infection with a V-cath deletion mutant are white, and do not melt. The V-cath proteinase acts synergistically with a viral chitinase

(ChiA) which is encoded by an adjacent viral gene (Hawtin *et al.*, 1995). V-cath and ChiA could facilitate the horizontal transmission of the virus by promoting the release of infectious viral polyhedral occlusions from cadavers. Blackening or melanization of cadavers is the result of V-cath directly or indirectly triggering the pro-phenoloxidase system of the insects. This would render the cadaver tissues a poor substrate for saprophytes such as fungi (St. Leger *et al.*, 1996) and would help preserve the virus in the environment.

## Distinguishing Features

Rabbit polyclonal antiserum against the AcMNPV V-cath proteinase has been produced from a bacterially expressed V-cath fragment (Slack *et al.*, 1995). This does not cross-react with papain, but has a low level of cross-reactivity with insect cell proteins. No commercially produced antisera are currently available.

## Related Peptidases

V-cath belongs to the papain (C1) family of cysteine peptidases, and it is possible that the viral V-cath gene was acquired by the virus from a host genome. There is particular similarity to protozoan cysteine proteinases such as that of *Trypanosoma brucei* (Chapter 205) (Mottram *et al.*, 1989). In addition, V-cath bears some similarity to lobster cysteine proteinases (Laycock *et al.*, 1991) and to a *Drosophila melanogaster* midgut cysteine proteinase (Matsumoto *et al.*, 1995).

## References

Barrett, A. & Kirschke, H. (1981) Cathepsin B, cathepsin H, and cathepsin L. *Methods Enzymol.* **80**, 535–561.

Blissard, G.W. & Rohrmann, G.F. (1989) Location, sequence, transcriptional mapping and temporal expression of the gp64 envelope glycoprotein gene of the *Orgyia pseudotsugata* multicapsid nuclear polyhedrosis virus. *Virology* **170**, 537–555.

Blissard, G.W. & Wenz, J.R. (1992) Baculovirus gp64 envelope glycoprotein is sufficient to mediate pH-dependent membrane fusion. *J. Virol.* **66**, 6829–6835.

Brömme, D. & Okamoto, K. (1995) The baculovirus cysteine protease has a cathepsin B-like S2-subsite specificity. *Biol. Chem. Hoppe-Seyler* **376**, 611–615.

Chen, P.H., Ornelles, D.A. & Shenk, T. (1993) The adenovirus L3 23-kilodalton proteinase cleaves the amino-terminal head domain from cytokeratin 18 and disrupts the cytokeratin network of HeLa cells. *J. Virol.* **67**, 3507–3514.

Gove, P.B. (ed.) (1971) *Webster's Third New International Dictionary of the English Language*, unabridged 3rd edn. Springfield, MA: Merriam Co.

Hawtin, R.E., Arnold, K., Ayres, M.D., Zanotto, P.M. d.A., Howard, S.C., Gooday, G.W., Chappell, L.H., Kitts, P.A., King, L.A. & Possee, R.D. (1995) Identification and preliminary characterization of a chitinase gene in the *Autographa californica* nuclear polyhedrosis virus genome. *Virology* **212**, 673–685.

Hill, J.E., Kuzio, J. & Faulkner, P. (1995) Identification and characterization of the v-cath gene of the baculovirus CfMNPV. *Biochim. Biophys. Acta* **1264**, 275–278.

Karrer, K.M., Peiffer, S.L. & DiTomas, M.E. (1993) Two distinct gene subfamilies within the family of cysteine protease genes. *Proc. Natl Acad. Sci. USA* **90**, 3063–3067.

Keyvani-Amineh, H., Labrecque, P., Cai, F., Carstens, E.B. & Weber, J.M. (1995) Adenovirus protease expressed in insect cells cleaves adenovirus proteins, ovalbumin and baculovirus protease in the absence of activating peptide. *Virus Res.* **37**, 87–97.

Lanier, L.M., Slack, J.M. & Volkman, L.E. (1996) Actin binding and proteolysis by the baculovirus AcMNPV: the role of virion-associated V-CATH. *Virology* **216**, 380–388.

Laycock, M.V., MacKay, R.M., Fruscio, M.D. & Gallant, J.W. (1991) Molecular cloning of three cDNAs that enocode cysteine proteinases in the digestive gland of the American lobster (*Homarus americanus*). *FEBS Lett.* **292**, 115–120.

Maciewicz, R.A. & Etherington, D.J. (1988) A comparison of four cathepsins (B, L, N and S) with collagenolytic activity from rabbit spleen. *Biochem. J.* **256**, 433–440.

Matsumoto, I., Watanabe, H., Arai, S. & Emori, Y. (1995) A putative digestive cysteine proteinase from *Drosophila melanogaster* is predominantly expressed in the embryonic and larval midgut. *Eur. J. Biochem.* **227**, 582–587.

Mottram, J.C., North, M.J., Barry, J.D. & Coombs, G.H. (1989) A cysteine proteinase complementary DNA from *Trypanosoma brucei* predicts an enzyme with an unusual carboxyl-terminal extension. *FEBS Lett.* **258**, 211–215.

Musil, D., Zucic, D., Turk, D., Engh, R.A., Mayr, I., Huber, R., Popovič, T., Turk, V., Towatari, T., Katunuma, N. & Bode, W. (1991) The refined 2.15 Å X-ray crystal structure of human liver cathepsin B: the structural basis for its specificity. *EMBO J.* **10**, 2321–2330.

Ohkawa, T., Majima, K. & Maeda, S. (1994) A cysteine protease encoded by the baculovirus *Bombyx mori* nuclear polyhedrosis virus. *J. Virol.* **68**, 6619–6625.

Rawlings, N.D. & Barrett, A.J. (1993) Evolutionary families of peptidases. *Biochem. J.* **290**, 205–218.

Rawlings, N.D., Pearl, L.H. & Buttle, D.J. (1992) The baculovirus *Autographa californica* nuclear polyhedrosis virus genome includes a papain-like sequence. *Biol. Chem. Hoppe-Seyler* **373**, 1211–1215.

Slack, J.M., Kuzio, J. & Faulkner, P. (1995) Characterization of V-cath, a cathepsin L-like proteinase expressed by the baculovirus *Autographa californica* multiple nuclear polyhedrosis virus. *J. Gen. Virol.* **76**, 1091–1098.

St. Leger, R.J., Joshi, L., Bidochka, M.J. & Roberts, D.W. (1996) Construction of an improved mycoinsecticide overexpressing a toxic protease. *Proc. Natl Acad. Sci. USA* **93**, 6349–6354.

Wiederanders, B., Brömme, D., Kirschke, H.V., von Figura, K., Schmidt, B. & Peters, C. (1992) Phylogenetic conservation of cysteine proteinases. Cloning and expression of a cDNA coding for human cathepsin S. *J. Biol. Chem.* **267**, 13708–13713.

Yamada, K., Nakajima, Y. & Natori, S. (1990) Production of recombinant sarcotoxin IA in *Bombyx mori* cells. *Biochem. J.* **272**, 633–636.

*Jeffrey M. Slack*
*Boyce Thompson Institute for Plant Research,*
*Cornell University,*
*Ithaca, NY 14853-1805, USA*
*Email: jms35@cornell.edu*

# 207. Cysteine proteinase-7 of Dictyostelium discoideum

## Databanks

*Peptidase classification: clan CA, family C1, MEROPS ID: C01.081*
*NC-IUBMB enzyme classification: none*
Databank codes:

| Species | Gene | SwissProt | PIR | EMBL (cDNA) | EMBL (genomic) |
|---|---|---|---|---|---|
| Proteinase 7 | | | | | |
| *Dictyostelium discoideum* | cprG/cp7 | – | – | U72746 | – |
| Other proteinases | | | | | |
| *Dictyostelium discoideum* | cprA/cp1 | P04988 | A22827 | X02407 | – |
| *Dictyostelium discoideum* | cprB/cp2 | P04989 | A24110 A25439 | M16039 X03344 | X04775: upstream |
| *Dictyostelium discoideum* | cprD/cp4 | P54639 | – | L36204 | – |

| Species | Gene | SwissProt | PIR | EMBL (cDNA) | EMBL (genomic) |
|---|---|---|---|---|---|
| Other proteinases (*continued*) | | | | | |
| *Dictyostelium discoideum* | *cprE/cp5* | P54640 | – | L36205 | – |
| *Dictyostelium discoideum* | *cprF/cp6* | – | – | U72745 | – |
| *Dictyostelium discoideum* | *cprC/cp3* | – | – | X03930 | – |

## Name and History

**Cysteine proteinase-7 (CP7)** was first identified by Gustafson & Thon (1979) from vegetatively grown *Dictyostelium discoideum* cells, and originally termed **proteinase-1**, being the first cysteine proteinase characterized from this organism. 'CP7' is the new designation, used here, based on the order of its cDNA cloning. Three other closely related cysteine proteinases have been identified in vegetative cells, but their biochemical characterization is less complete than that of CP7 (Souza *et al.*, 1995).

## Activity and Specificity

The enzyme prefers bulky residues at the P2 and P3 positions and has a substrate specificity similar to that of cathepsin L (Chapter 210). It has a slight preference for Lys over Arg at the P1 position (Mehta *et al.*, 1995). It has a narrow specificity as an esterase, and the relative activities with blocked amino acid *p*-nitrophenyl esters (Z-Xaa-OPhNO$_2$) are Xaa = Lys > Leu > Phe = Gly (Gustafson & Thon, 1979).

CP7 activity is routinely assayed with D-Val-Leu-Lys+NPhNO$_2$ at pH 5.0. It does not act on Arg-NHMec or Lys-NHMec. Other protein and peptide substrates are listed by Mehta *et al.* (1995). Zymography with Boc-Val-Leu-Lys+NHMec or gelatin can also be used to detect its activity (North *et al.*, 1988; Mehta *et al.*, 1995).

The enzyme is strongly inhibited by E-64, Tos-Lys-CH$_2$Cl, Tos-Phe-CH$_2$Cl, cystamine, iodoacetamide, antipain and leupeptin (Gustafson & Thon, 1979; Mehta *et al.*, 1995). CP7 is also inhibited by cystatin and a 14 kDa protein from *Dictyostelium* (Korth *et al.*, 1988). Thiols such as DTT or cysteine are necessary for full activity and stability.

## Structural Chemistry

The cDNA for CP7 has recently been cloned and contains a signal sequence, a prepropeptide and the catalytic domain. The sequence (Figure 207.1) is very similar to those of papain and mammalian cathepsins, but has a unique Ser-rich domain near the C-terminus between S4b and S5. This insert has repeated motifs of SGSG and SGSQ (Ord *et al.*, 1997). Carbohydrate analysis shows 30 residues of *N*-acetylglucosamine (GlcNAc), 8 of mannose, 2 of xylose and 7.5 residues of fucose per molecule of protein (Gustafson & Gander, 1984; Mehta *et al.*, 1996). Most of the mannose and xylose are probably present on a single *N*-linked oligosaccharide, while the GlcNAc is attached to main-chain serine residues via phosphodiester links (Gustafson & Milner, 1980). This modification is called phosphoglycosylation. In contrast to other lysosomal enzymes of *Dictyostelium*, CP7 does not contain any of the mannose-6-phosphate (Man-6-P) residues that are used as lysosomal enzyme targeting signals in mammalian cells (Kornfeld & Mellman, 1989). The predicted $M_r$ for the precursor is 47 136. The protein migrates as a 38 kDa band on SDS-PAGE, while on native PAGE it separates into two distinct bands having pI values of 2.5 and 2.6, respectively (Mehta *et al.*, 1995). Zymography of non-heat-denatured CP7 on SDS-PAGE with a fluorescent peptide or gelatin shows three bands of 46, 41 and 37 kDa (North *et al.*, 1988; Mehta *et al.*, 1995).

## Preparation

The enzyme has been purified 30–100-fold from vegetatively grown *Dictyostelium* cells using conventional column chromatography (Gustafson & Thon, 1979; Mehta *et al.*, 1995).

## Biological Aspects

CP7 activity and its mRNA levels increase during vegetative growth, reaching a maximum at the end of the growth phase, and then decreasing as development commences (Mehta *et al.*, 1995; Ord *et al.*, 1997). This suggests a role for CP7 during growth and in the transition of the amebae from growth to development.

Lysosomal enzymes from *Dictyostelium* are modified either by Man-6-P or GlcNAc-1-P and the two modifications are not found on the same protein; GlcNAc-1-P is found mainly on cysteine proteinases including CP7, whereas Man-6-P is present on the glycosidases (Mehta *et al.*, 1996). Confocal immunofluorescence microscopy revealed that proteins bearing these two sugar modifications were segregated into distinct vesicles which play important roles during the phagocytosis of bacteria (G.M. Souza, D.P. Mehta, M. Lammertz, J. Rodriguez-Paris, J. Cardelli & H. Freeze, unpublished results). Since CP7 is the major phosphoglycosylated cysteine proteinase of *Dictyostelium*, it may play a role in the initial degradation of ingested bacteria, on which the organism feeds.

## Distinguishing Features

At least three other related vegetative cysteine proteinases (CP4, CP5, CP6) have been identified and cloned (G.M. Souza, D.P. Mehta, M. Lammertz, J. Rodriguez-Paris, J. Cardelli & H. Freeze, unpublished results; Ord *et al.*, 1997). All three of them have the distinctive Ser-rich insert, and like CP7, their expression and activities are highly regulated during the life cycle of *Dictyostelium*. Among all the cysteine proteinases identified in *Dictyostelium*, CP7 is the most highly phosphoglycosylated and enzymatically active.

```
CP7 DICTYOSTELIUM  M------KV--I--SALCVLL---VSVAT--AK--QQISE VE-YRNAFIN WMIAHQRHYS S-EEFNGRIN IFKFNMDIVN EWN---TKG- SETVLGLNVF   77
CP6 DICTYOSTELIUM  M------KV--I--SALCVLL---VSVAT--AK--QQISE IQ-YRNAFIN WMIAHQRHYS S-EEFNGRFN IFKFNMDIIN EWN---TKG- SETVLGLNNF   77
CP5 DICTYOSTELIUM  M------KV--I--SFLCVLL---VSVAT--AK--QQFSE IQ-YRNAFID WMITHQKSYT S-EEFGARIN IFTFNMDIVQ QWN---SKG- SETVLGLNNF   77
CP4 DICTYOSTELIUM  M------RV--IL-SFLCLLL---VSVAS--AK--QQFSE IQ-YRNAFIN WMQAHQRTYS S-EEFNARIQ IFKSNMDIVH QWN---SKG- GETVLGLNNF   77
CP2 DICTYOSTELIUM  M------RL--I-VFLILLIF VNEISFAN---- SQ-YRTAFTE WTLKFNRQYS S-SEFSNRYS IFKSNMDIVD NWN---SKGD SQTVLGLNNF   84
CP1 DICTYOSTELIUM  M------KV--I--LLFVIAV---FTVFV- SS---RGIPP EE--QSQFIE H-EELERFE IFKSNIGHIE ELNLIAINHK ADTHFGVNKF   80
PAPAIN           MMIPSISKL IFVAICLFVY MGISFGDFSI VGYSQNDITS TERLIQLFES WMLKHNKIYK NIDEKIYRFE IFKFNIKID ETN---KKN- NSYWLGLNVF   96
Consensus        M-------KV---I---.LC.LL----.SL.A.--- .A.----SE ..-YR.AFT. WM..H...YS S-EEF..R.. IFK.NMD.I. .WN----KG- ..TWLGLN.F  100
```

```
CP7 DICTYOSTELIUM  ADINEEIRA TYLGTPFDAS SLEMTESDKI FD----A--S AQVDWRIQGA VTPIKNQGC GGCWSFSTTG ATEGAQYLAN G-KKNLSIS EQNLIDCSG-  169
CP6 DICTYOSTELIUM  ADINEEIRN TYLGTPFDAS SLEMTPSEKV FG----GVQA NSVDWRKGA VTPIKNQGC GGCWSFSATG ATEGAQYIAN G-DSELISWS EQQLIDCSG  171
CP5 DICTYOSTELIUM  ADINEEIRN TYLGTKFDAS SLIGTQEEKV HT----NSSA ASHDWRSEGA VTPVKNQGC GGCWSFSTTS STEGAHF--Q S-KGELISIS EQNLIDCST-  169
CP4 DICTYOSTELIUM  ADINQEIRT TYLGTPFDGS ALIGTEEEKI FS----T-PA PTVDWRRAQGA VTPIKNQGC GGCWSFSTTG STEGAHFIAS GTKKLLSIS EQNLIDCSK-  171
CP2 DICTYOSTELIUM  ADINEENRK TYLGTRVNAH SYNGYDGREV LNVEDLQTNP KSIDWRKNA VTPVKNQGC GSCWSFSTTG STEGAHAL-K T--KKLLSIS EQNLIDCCHE  180
CP1 DICTYOSTELIUM  ADLSDEEFN YLNNK-EAI FTDDLPVADY LDDEFINSIP TAFDWRIRGA VTPVKNQGC GSCWSFSTTG NVECQHFISQ G-NKLWSIS EQNLIDCCHE  176
PAPAIN           ADMSNDEFKE MYTGSIAGNY TTTELSYEEV LN-DGDVNIP EYVDWRQKGA VTPVKNQGC GSCWAFSAVV TIEGIIKRT G--NLNEMS EQELIDCDR-  191
Consensus        ADI.N..E.R. T.YLGT...A.. ......I... .DWR.....CA VTPIKNQGC G.CWSFSTTG .TEGA....I.. .L....SIS EQNLIDCS.-  200
```

```
CP7 DICTYOSTELIUM  --SYG ----NNGCEGGLMT LAFEYIINNK GIDTFSSYPY NPKNVAALS SYVNITSGSE SDLAAKVTQG PTSVAIDASN QSFQIYSGI  262
CP6 DICTYOSTELIUM  --SYG ----NNGCEGGLMT LAFEYIINNG GIDTFSSYPY NPSNIGAELS SYVNITSGSE SDLAAKVTQG PTSVAIDASQ PSFQIYSGI  263
CP5 DICTYOSTELIUM  ---E----- NSGCDGGLMT YAFEYIINNN GIDTFSSYPY KSENSGATLS SYKINTAGSE SSLESAVNVN PVSVAIDASH QSFQIYFGI  259
CP4 DICTYOSTELIUM  --SYG ----NNGCEGGLMT IGFEYIINNK GIDTFSSYPY KTSNIGAQIV SYQNNTSGSE ASLQSASNNA PVSVAIDASN ESFQIYESGI  264
CP2 DICTYOSTELIUM  --PEE ----NFGCDGGLMN NAFDYIIKNK GIDTFSSYPY NKSDIGAFIK GYVNIITAGSE ISLENGAQHG PVSVAIDASH NSFQLYFSGI  273
CP1 DICTYOSTELIUM  CMEYEGEEAC SMGCNGGLQP NAHNYIIKNG GIDTFSSYPY NSANIGAAIS NFTMIFKNET VMAGYIVSTG PIAIANDAVE -MQFTICGV  274
PAPAIN           ------R-- SMGCNGGYPW SAIQ-INAQY GIHYRNTYPY EGVQRYCRSR EKGPYAAKTD GVRQVPVYNE GALLYSIANQ KDFQIYRGI  281
Consensus        ......... S..N.GC.GGM. .AF..YIII.N. GIDTFSSYPY ..N.CA..S ..Y...VT.GSE .....L.....G P.SVAIDAS. ..SFQIY.SGI  300
```

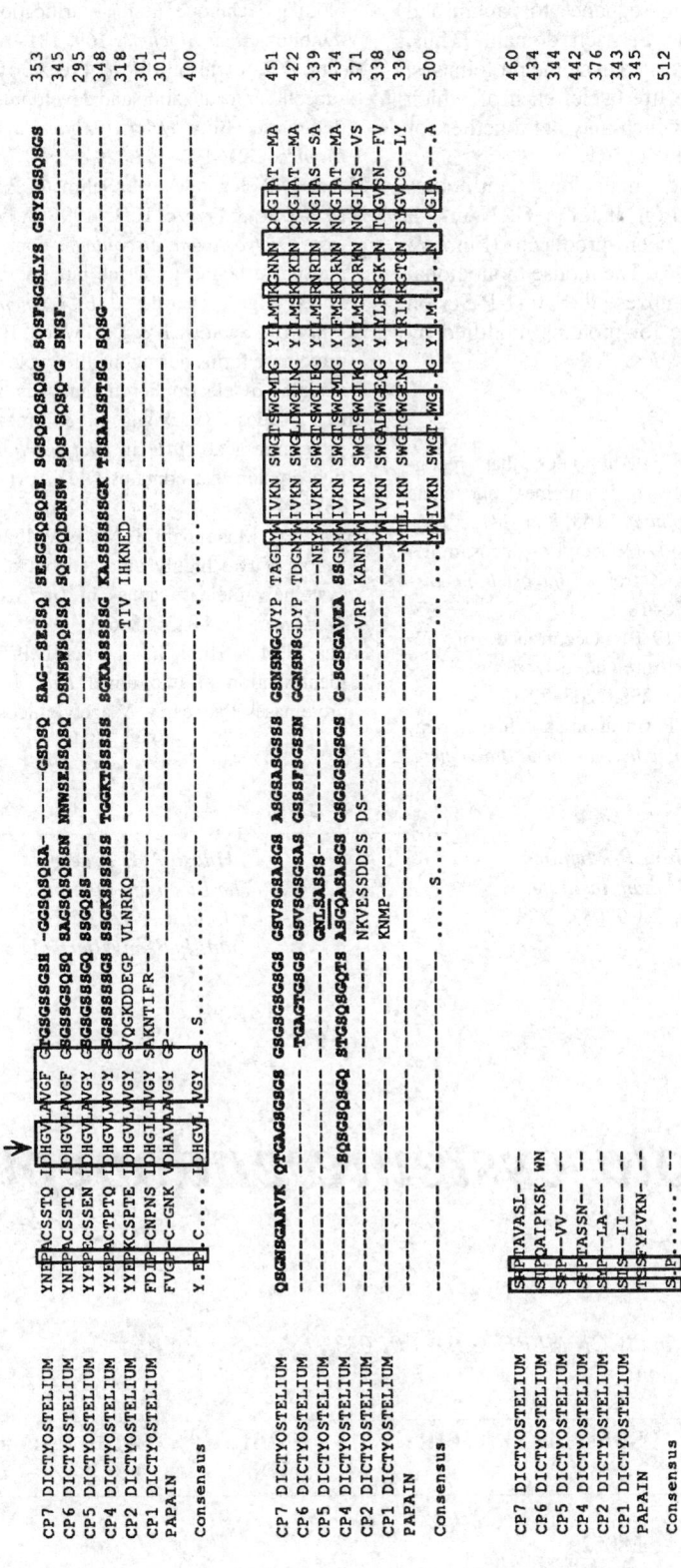

*Figure 207.1* Sequence alignment of *Dictyostelium* cysteine proteinases with papain. The consensus sequences are boxed and the arrow indicates the active-site residues. The distinctive Ser-rich domain is in bold.

Two developmental cysteine proteinases, CP1 and CP2, are expressed only at 10–14 h of development, and are absent from vegetative cells (Pears *et al*., 1985). These two cysteine proteinases are also very similar in sequence to papain and CPs 4–7, but do not contain the Ser-rich domain. Thus, *Dictyostelium* has a set of closely related cysteine proteinases, expressed at various stages of the life cycle, each of which may play a specific role, or all of which may act together for general proteolysis.

Rabbit polyclonal and mouse monoclonal antibodies recognizing both the protein portion and the GlcNAc-1-P carbohydrate modification have been produced (Finn & Gustafson, 1987; Mehta *et al*., 1996). The mouse monoclonal antibody, AD7.5, specifically recognizes GlcNAc-1-P-Ser on proteins and can be used to probe for proteins modified by GlcNAc-1-P in various cells or tissues.

## References

Finn, D.S. & Gustafson, G.L. (1987) Antibodies that recognize phosphodiester-linked $\alpha$-*N*-acetylglucosamine-1-phosphate residues. *Biochem. Biophys. Res. Commun.* **148**, 834–887.

Gustafson, G.L. & Gander, J.E. (1984) *O*-(*N*-acetyl-$\alpha$-glucosamine-1-phosphoryl)serine in proteinase 1 from *Dictyostelium discoideum. Methods Enzymol.* **107**, 172–183.

Gustafson, G.L. & Milner, L.A. (1980) Occurrence of *N*-acetylglucosamine-1-phosphate in proteinase-1 from *Dictyostelium discoideum. J. Biol. Chem.* **255**, 7208–7210.

Gustafson, G.L. & Thon, L.A. (1979) Purification and characterization of a proteinase from *Dictyostelium discoideum. J. Biol. Chem.* **254**, 12471–12478.

Kornfeld, S. & Mellman, I. (1989) The biogenesis of lysosomes. *Annu. Rev. Cell. Biol.* **5**, 483–525.

Korth, M.J., Finn, D.J. & Gustafson, G.L. (1988) Use of a Western blotting technique in the purification of a cysteine proteinase inhibitor. *Anal. Biochem.* **169**, 181–184.

Mehta, D.P., Etchison, J.R. & Freeze, H.H. (1995) Characterization, subcellular localization and developmental regulation of a cysteine proteinase from *Dictyostelium discoideum. Arch. Biochem. Biophys.* **321**, 191–198.

Mehta, D.P., Ichikawa, M., Salimath, P.V., Etchison, J.R., Haak, R., Manzi, A. & Freeze, H.H. (1996) A lysosomal cysteine proteinase from *Dictyostelium discoideum* contains *N*-acetylglucosamine-1-phosphate bound to serine but not mannose-6-phosphate on *N*-linked oligosaccharides. *J. Biol. Chem.* **271**, 10897–10903.

North, M.J., Scott, K.I. & Lockwood, B.C. (1988) Multiple cysteine proteinase forms during the life cycle of *Dictyostelium discoideum* revealed by electrophoretic analysis. *Biochem. J.* **254**, 262–268.

Ord, T., Adessi, C., Wang, L. & Freeze, H. (1997) The cysteine proteinase gene *cprG* in *Dictyostelium discoideum* has a serine-rich domain that contains GlcNAc-1-P. *Arch. Biochem. Biophys.* **339**, 64–72.

Pears, C.J., Mahbubani, H.M. & Williams, J.G. (1985) Characterization of two highly diverged but developmentally co-regulated cysteine proteinase genes in *Dictyostelium discoideum. Nucleic Acids Res.* **13**, 8853–8866.

Souza, G.M., Hirai, J., Mehta, D.P. & Freeze, H.H. (1995) Identification of two novel *Dictyostelium discoideum* cysteine proteinases that carry *N*-acetylglucosamine-1-P modification. *J. Biol. Chem.* **270**, 28938–28945.

*Darshini P. Mehta*
*The Burnham Institute,*
*La Jolla, CA 92037, USA*

*Hudson H. Freeze*
*The Burnham Institute,*
*La Jolla, CA 92037, USA*
*Email: hudson@ljcrf.edu*

# 208. *Fasciola cysteine endopeptidases*

## Databanks

*Peptidase classification: clan CA, family C1, MEROPS ID: C01.033*
*NC-IUBMB enzyme classification: none*
*Databank codes:*

| Species | SwissProt | PIR | EMBL (cDNA) | EMBL (genomic) |
|---|---|---|---|---|
| Cathepsin B-like endopeptidases | | | | |
| *Fasciola hepatica* | P80529 | | | |
| *Fasciola hepatica* | – | – | Z22768 | – |
| *Fasciola hepatica* | – | – | Z22770 | – |
| *Schistosoma japonicum* | P43157 | S31907 | X70968 | – |
| *Schistosoma mansoni* | P25792 | – | P25792 | – |

| Species | SwissProt | PIR | EMBL (cDNA) | EMBL (genomic) |
|---|---|---|---|---|
| Cathepsin L-like endopeptidases | | | | |
| *Fasciola hepatica* | P80342 | – | – | – |
| *Fasciola hepatica* | P80528 | – | – | – |
| *Fasciola hepatica* | P80532 | – | – | – |
| *Fasciola hepatica* | P80533 | – | – | – |
| *Fasciola hepatica* | Q09093 | – | – | – |
| *Fasciola hepatica* | – | S43991 | L33771 | – |
| *Fasciola hepatica* | – | – | L33772 | – |
| *Fasciola hepatica* | – | – | Z22763 | – |
| *Fasciola hepatica* | – | – | Z22764 | – |
| *Fasciola hepatica* | – | – | Z22765 | – |
| *Fasciola hepatica* | – | – | Z22766 | – |
| *Fasciola hepatica* | – | – | Z22767 | – |
| *Fasciola hepatica* | – | – | Z22769 | – |
| *Fasciola* sp. | – | – | S70380 | – |
| *Paragonimus westermani* | – | – | D21124 | – |
| *Schistosoma japonicum* | – | – | U38475 | – |
| *Schistosoma japonicum* | – | – | U38476 | – |
| *Schistosoma mansoni* | – | – | U07345 | – |
| *Schistosoma mansoni* | – | S44151 | Z32529 | – |
| *Spirometra mansoniodes* | – | – | U51913 | – |

## Name and History

Proteolytic enzymes are secreted by the liver fluke, *Fasciola hepatica*, a parasitic trematode, and assays with various protein substrates have identified cysteine proteinases as the major contributor to the activity. The *Fasciola* cysteine endopeptidases were originally described as **immunoglobulin cleaving enzyme** (Chapman & Mitchell, 1982) or **hemoglobinase** (Aoki, 1980; Coles & Rubano, 1988). Currently, they are referred to as the **Fasciola cathepsin L-like** or **cathepsin B-like proteinases**. There has been a recent study of the cysteine endopeptidases of the closely related trematode, *Fasciola gigantica* (Fagbemi & Hillyer, 1992).

## Activity and Specificity

The secreted cysteine endopeptidases of adult liver fluke exhibit cathepsin L- and B-like behavior (Chapters 209 and 210), hydrolyzing the fluorogenic substrates Z-Phe-Arg+NHMec and Z-Arg-Arg+NHMec, but not Z-Arg-NHMec (Rege *et al.*, 1989; Carmona *et al.*, 1993; McGinty *et al.*, 1993; Wijffels *et al.*, 1994a). The activity of the secreted cysteine endopeptidases in these assays is absolutely dependent on the presence of a reducing agent such as DTT or cysteine. Inhibitors include leupeptin, E-64, Tos-Lys-CH$_2$Cl, antipain, chymostatin and Z-Phe-Ala-CHN$_2$ (Dalton & Heffernan, 1989; Rege *et al.*, 1989; Yamasaki *et al.*, 1989; Carmona *et al.*, 1993; McGinty *et al.*, 1993; Wijffels *et al.*, 1994a). Z-Phe-TyrO(But)-CHN$_2$, which is a specific inhibitor of mammalian cathepsin L, does not inhibit the *Fasciola* enzymes (McGinty *et al.* 1993).

With the two fluorogenic substrates, maximal activity is seen in the range pH 6.0–8.0, and activity is detectable over a neutral to alkaline pH range (Rege *et al.*, 1989; McGinty *et al.*, 1993; Wijffels *et al.*, 1994a). However, a group of cysteine endopeptidases with an optimum at pH 4.5 has been reported (Dalton & Heffernan, 1989; Smith *et al.*, 1993). Fagbemi & Hillyer (1992) also reported an activity optimum at pH 4.5 for the putative *Fasciola gigantica* homolog of adult liver fluke cysteine endopeptidases.

Hemoglobin, immunoglobulin, globin, fibrinogen and collagen are degraded by adult *Fasciola* cysteine endopeptidases (Aoki, 1980; Simpkin *et al.*, 1980; Chapman & Mitchell, 1982; Coles & Rubano, 1988; Yamasaki *et al.*, 1989; Carmona *et al.*, 1993; Dowd *et al.*, 1995). Gelatin gels have also been used to study these secreted enzymes (Dalton & Heffernan, 1989; Carmona *et al.*, 1993; Smith *et al.*, 1993; Wijffels *et al.*, 1994b). The exact cleavage sites have not been identified for any protein substrate.

In contrast to the adult fluke, the major secreted proteinase of immature flukes is a cathepsin B-like endopeptidase (Wilson, 1994). The juvenile stage synthesizes cathepsin L-like enzymes, but these may represent tissue endopeptidases (Tkalcevic *et al.*, 1995). Maximal activity of the secreted cathepsin B-like enzymes on Z-Phe-Arg-NHMec is near pH 7.5, and cleavage of immunoglobulin and bovine albumin has been demonstrated (Wilson, 1994).

## Structural Chemistry

The purified adult *Fasciola* cysteine endopeptidases migrate as a complex of 2–4 species at 26–27 kDa in reducing conditions (Coles & Rubano, 1988; Yamasaki *et al.*, 1989; Smith *et al.*, 1993; Wijffels *et al.*, 1994a). The molecular masses and pI values calculated from the predicted sequences of the four full-length clones isolated thus far are 24.2–24.6 kDa and pI 4.3–7.1. Like the cathepsins, they are synthesized as preproproteins, with a 17 residue leader sequence and a 90 amino acid pro segment. The mature protein contains 219 amino acids (Heussler & Dobbelaere, 1994; Wijffels *et al.*,

1994a; Yamasaki & Aoki, 1993). The mature enzyme has clear similarity to human cathepsin L (Chapter 210) (44% identity), stem bromelain (Chapter 192) (39% identity), and the cathepsin L-like proteinase of the blood fluke, *Schistosoma mansoni* (Michel *et al.*, 1995) (43% identity). The adult *Fasciola* cysteine endopeptidases have the same structure as the thiol cathepsins: the cysteines are conserved in number and location, and the catalytic triad is represented by Cys132, Gly175 and His269 (as numbered in EMBL: L33771).

The secreted cathepsin B-like enzyme of the juvenile fluke migrates as a single species of 29 kDa (Wilson, 1994). Identical sequences have been derived from whole-worm preparations of juvenile fluke (Tkalcevic *et al.*, 1995). The predicted amino acid sequence of a clone isolated from an immature fluke cDNA library reveals 48–51% sequence identity with mature cathepsin B and *Schistosoma mansoni* cathepsin B-like enzyme (Wilson, 1994). Only 33% and 66% identity was obtained to the translated sequences of two PCR products derived from adult liver fluke cathepsin B-like endopeptidase transcripts (Heussler & Dobbelaere, 1994).

## Preparation

Adult *Fasciola* cysteine endopeptidases are isolated from the regurgitant of adult worms freshly excised from the host liver bile ducts. Purification procedures use a combination of molecular sieving and ion-exchange chromatographies (Rege *et al.*, 1989; Smith *et al.*, 1993; Wijffels *et al.*, 1994a), although other protocols have been used (Yamasaki *et al.*, 1989; Fagbemi & Hillyer, 1992). The purified product from regurgitant is heterogeneous as exemplified by two-dimensional gel analysis, and the number of different but clearly related N-termini that have been sequenced (Smith *et al.*, 1993; Yamasaki & Aoki, 1993; Wijffels *et al.*, 1994a). The secreted enzymes of the immature stage are obtained from the media of 4–5 day cultures of newly excysted juvenile flukes.

## Biological Aspects

The cathepsin L-like peptidases are the major secreted proteolytic enzymes of the adult stage liver fluke. They have been localized to secretory granules in the midgut epithelium and the luminal contents of the gut (Yamasaki *et al.*, 1992; Yamasaki & Aoki, 1993; Smith *et al.*, 1993), but not the lysosomes. Cathepsin B-like peptidases are synthesized by the adult worm (Heussler & Dobbelaere, 1994), but there is no evidence that enzymes of this group are secreted. The *Fasciola* cathepsin L-like endopeptidases arise from a multigenic family (Heussler & Dobbelaere, 1994). Like other parasites, the *Fasciola* species are thought to elaborate many cysteine endopeptidases of slightly differing specificities to permit migration through different tissue substrates during the life cycle and to adapt to different host species. The enzymes secreted by the adult worm can induce fibrin clots by degradation of the $\alpha$, $\beta$ and $\gamma$ chains of fibrinogen (Dowd *et al.*, 1995).

The secreted *Fasciola* cysteine endopeptidases may act to disrupt host immune function by degradation of immunoglobulin (Chapman & Mitchell, 1982; Simpkin *et al.*, 1980) and other immune effector molecules. *In vitro* assays have shown that the presence of these enzymes coincides with the inability of eosinophils to engage in or maintain antibody-mediated adherence to the newly excysted juvenile stage (Carmona *et al.*, 1993). Curiously, vaccination of sheep with the adult *Fasciola* cysteine endopeptidases causes a decrease in fecundity of worms on challenge infection (Wijffels *et al.*, 1994b). It is not known how this effect is mediated. It is possible that like *Schistosoma mansoni*, the cathepsin L-like enzymes are expressed in the female reproductive system (Michel *et al.*, 1995).

The juvenile fluke cysteine endopeptidases are likely to function in the migration of the immature worm from the host's intestine to the liver parenchyma and subsequently to the bile ducts. In early migration, the fluke gut is in a secretory phase, and nonabsorptive (Bennett & Threadgold, 1973).

## Distinguishing Features

One of the critical features of these cathepsin-like *Fasciola* cysteine endopeptidases is the secretion into the gut, and lack of localization to the lysosomes, the major site of their mammalian counterparts. Possibly reflecting their extracellular location, their pH optimum is in the neutral to alkaline range, and unsuited to the acidic environment of the lysosome. The other feature which distinguishes the adult *Fasciola* cysteine endopeptidases from most other endopeptidases is the occasional modification of prolines to the unusual 3-hydroxyproline form (Wijffels *et al.*, 1994a). This modification has been observed in other proteins isolated from *Fasciola* (Bozas & Spithill, 1996), but the biological significance of the modification is unclear.

## Further Reading

A review has been provided by Dalton & Brindley (1997).

## References

Aoki, T. (1980) Antigenic property of hemoglobin specific protease from *Schistosoma mansoni*, *S. japonicum* and *Fasciola* sp. for radioallergosorbent test. *Jpn J. Parasitol.* **29**, 325–332.

Bennett, C.E. & Threadgold, L.T. (1973) Electron microscope studies of *Fasciola hepatica* XIII. Fine structure of the newly excysted juvenile fluke. *Exp. Parasitol.* **34**, 85–99.

Bozas, E. & Spithill, T. (1996) Identification of 3-hydroxyproline residues in several proteins of *Fasciola hepatica*. *Exp. Parasitol.* **82**, 69–72.

Carmona, C., Dowd, A.J., Smith, A.M. & Dalton, J.P. (1993) Cathepsin L proteinase secreted by *Fasciola hepatica* in vitro prevents antibody mediated eosinophil attachment to newly excysted juveniles. *Mol. Biochem Parasitol.* **62**, 9–18.

Chapman, C.B. & Mitchell, G.F. (1982) Proteolytic cleavage of immunoglobulin by enzymes released by *Fasciola hepatica*. *Vet. Parasitol.* **11**, 165–178.

Coles, G.C. & Rubano, D. (1988) Antigenicity of a proteolytic enzyme of *Fasciola hepatica*. *J. Helminthol.* **62**, 257–260.

Dalton, J.P. & Brindley, P.J. (1997) Proteases of trematodes. In: *Advances in Trematode Biology* (Freed, B. & Graczyck, T., eds). Boca Raton, FL: CRC Press.

Dalton, J.P. & Heffernan, M. (1989) Thiol protease released in vitro by *Fasciola hepatica*. *Mol. Biochem. Parasitol.* **35**, 161–166.

Dowd, A.J., McGonigle, S. & Dalton, J.P. (1995) *Fasciola hepatica* cathepsin L proteinase cleaves fibrinogen and produces a novel type of fibrin clot. *Eur. J. Biochem.* **232**, 241–246.

Fagbemi, B.O. & Hillyer, G.V. (1992) The purification and characterization of a cysteine protease of *Fasciola gigantica* adult worms. *Vet. Parasitol.* **43**, 223–232.

Heussler, V.T. & Dobbelaere, D.A.E. (1994) Cloning of a protease gene family of *Fasciola hepatica* by the polymerase chain reaction. *Mol. Biochem. Parasitol.* **64**, 11–23.

McGinty, A., Moore, M., Halton, D.W. & Walker, B. (1993) Characterization of the cysteine proteinases of the common liver fluke *Fasciola hepatica* using novel, active site directed affinity labels. *Parasitology* **106**, 487–493.

Michel, A., Ghoneim, H., Resto, M., Klinkert, Q. & Kunz, W. (1995) Sequence, characterization and localization of a cysteine proteinase cathepsin L in *Schistosoma mansoni*. *Mol. Biochem. Parasitol.* **73**, 7–18.

Rege, A.A., Herrera, P.R., Lopez, M. & Dresden, M.H. (1989) Isolation and characterization of a cysteine proteinase from *Fasciola hepatica* adult worms. *Mol. Biochem. Parasitol.* **35**, 89–96.

Simpkin, K.G., Chapman, C.R. & Coles, G.C. (1980) *Fasciola hepatica*: a proteolytic digestive enzyme. *Exp. Parasitol.* **49**, 281–287.

Smith, A.M., Dowd, A.J., McGonigle, S., Keegan, P.S., Brennan, G., Trudgett, A. & Dalton, J.P. (1993) Purification of a cathepsin L-like proteinase secreted by adult *Fasciola hepatica*. *Mol. Biochem. Parasitol.* **62**, 1–8.

Tkalcevic, J., Ashman, K. & Meeusen, E. (1995) *Fasciola hepatica*: rapid identification of newly excysted juvenile proteins. *Biochem. Biophys. Res. Commun.* **213**, 169–174.

Wijffels, G.L., Panaccio, M., Salvatore, L., Wilson, L., Walker, I.D. & Spithill, T. (1994a) The secreted cathepsin L-like proteinases of the trematode, *Fasciola hepatica*, contain 3-hydoxyproline residues. *Biochem. J.* **299**, 781–790.

Wijffels, G.L., Salvatore, L., Dosen, M., Waddington, J., Wilson, L., Thompson, C., Campbell, N., Sexton, J., Wicker, J., Bowen, F., Friedel, T. & Spithill, T. (1994b) Vaccination of sheep with purified cysteine proteinases of *Fasciola hepatica* decreases worm fecundity. *Exp. Parasitol.* **78**, 132–148.

Wilson, L. (1994) Characterisation and cloning of a cathepsin B protease secreted from the newly excysted juvenile *Fasciola hepatica*. PhD thesis, La Trobe University, Melbourne, Australia.

Yamasaki, H. & Aoki, T. (1993) Cloning and sequence analysis of the major cysteine protease in the trematode parasite *Fasciola* sp. *Biochem. Mol. Biol. Int.* **3**, 537–542.

Yamasaki, H., Aoki, T. & Oya, H. (1989) A cysteine proteinase from the liver fluke *Fasciola* spp.: purification, characterisation, localisation and application to immunodiagnosis. *Jpn J. Parasitol.* **38**, 373–384.

Yamasaki, H., Kominami, E. & Aoki, T. (1992) Immunocytochemical localization of a cysteine protease in adult worms of the liver fluke *Fasciola* sp. *Parasitol. Res.* **78**, 574–580.

**C**

*Gene L. Wijffels*
*Long Pocket Laboratories,*
*CSIRO Division of Tropical Agriculture,*
*Indooroopilly 4068,*
*Queensland, Australia*
*Email: gene.wijffels@dance.tap.csiro.au*

# *209. Cathepsin B*

## *Databanks*

*Peptidase classification: clan CA, family C1, MEROPS ID: C01.060*
*NC-IUBMB enzyme classification: EC 3.4.22.1*
*Chemical Abstracts Service registry number: 9047-22-7*
*Databank codes:*

| Species | SwissProt | PIR | EMBL (cDNA) | EMBL (genomic) |
| --- | --- | --- | --- | --- |
| *Aedes aegypti* | – | – | L41940 | – |
| *Ancylostoma caninum* | – | – | U02611 | – |
| *Ancylostoma caninum* | – | – | U18911 | – |
| *Ancylostoma caninum* | – | – | U18912 | – |

*continued overleaf*

| Species | SwissProt | PIR | EMBL (cDNA) | EMBL (genomic) |
|---|---|---|---|---|
| *Bos taurus* | P07688 | A05143 | L06075 | U16336: exon 10 and complete CDS |
| | | A27013 | M64620 | U16337: exon 2 |
| | | A28103 | | U16338: exon 3 |
| | | A39475 | | U16339: exon 4 |
| | | I46007 | | U16340: exon 5 |
| | | S02674 | | U16341: exons 6 and 7 |
| | | S38328 | | U16342: exon 8 |
| | | | | U16343: exon 9 |
| *Caenorhabditis elegans* | P43507 | – | L39890 | L39925: complete gene |
| *Caenorhabditis elegans* | P43508 | – | L39895 | L39926: complete gene |
| *Caenorhabditis elegans* | P43509 | – | L39896 | L39927: complete gene |
| *Caenorhabditis elegans* | P43510 | – | L39894 | L39939: complete gene |
| *Caenorhabditis elegans* | – | – | M75822 | – |
| *Caenorhabditis elegans* | – | – | U11245 | – |
| *Fasciola hepatica* | – | – | Z22770 | – |
| *Gallus gallus* | P43233 | – | U18083 | – |
| *Gallus gallus* | – | – | X73074 | – |
| *Haemonchus contortus* | P19092 | A45524 | M31112 | – |
| *Haemonchus contortus* | P25793 | A44965 | – | M60212: exons 1–4 |
| | | | | M60213: exons 5–12 and complete CDS |
| *Haemonchus contortus* | – | – | M80385 | – |
| *Haemonchus contortus* | – | – | M80386 | – |
| *Haemonchus contortus* | – | – | – | M80393: exons 1–4 |
| *Haemonchus contortus* | – | – | Z69343 | – |
| *Haemonchus contortus* | – | – | Z69345 | – |
| *Haemonchus contortus* | – | – | Z69346 | – |
| *Haemonchus contortus* | – | – | Z81327 | – |
| *Homo sapiens* | P07858 | A25432 | L16510 | S62069: 5′ end |
| | | A26498 | L22569 | S62071: 5′end |
| | | A27139 | L38712 | |
| | | | M13230 | |
| | | | M14221 | |
| | | | U44029 | |
| *Leishmania major* | – | – | U43705 | – |
| *Leishmania mexicana* | – | – | Z48599 | – |
| *Mus musculus* | P10605 | A38458 | M14222 | M65262: exon 1 |
| | | A49826 | X54966 | M65263: exon 2 |
| | | B26498 | | M65264: exon 3 |
| | | PS0360 | | M65265: exon 4 |
| | | S12901 | | M65266: exon 5 |
| | | | | M65267: exons 6 and 7 |
| | | | | M65268: exon 8 |
| | | | | M65269: exon 9 |
| | | | | M65270: exon 10 and complete CDS |
| | | | | X76621: 5′ end |
| *Ostertagia ostertagi* | P25802 | – | – | M88503: 5′end |
| | | | | M88504: 3′end |
| *Ostertagia ostertagi* | Q06544 | B48454 | M88505 | – |
| *Rattus norvegicus* | P00787 | A00977 | M11305 | – |
| | | I59019 | | |
| | | S51041 | | |
| *Sarcophaga peregrina* | – | – | D16823 | – |
| *Strongyloides ratti* | – | – | U09818 | – |
| *Triticum aestivum* | – | – | X66012 | – |
| *Triticum aestivum* | – | – | X66014 | X66013: complete gene |
| *Triticum aestivum* | – | – | X66015 | |
| *Triticum aestivum* | – | – | – | X66116: promoter |

Brookhaven Protein Data Bank three-dimensional structures:

| Species | ID | Resolution | Notes |
|---|---|---|---|
| *Homo sapiens* | 1CSB | 2.1 | |
| | 1HUC | 2.1 | |
| *Rattus norvegicus* | 1CPJ | 2.2 | Ser115Ala mutant |
| | 1CTE | 2.1 | |
| | 1MIR | 2.8 | precursor |
| | 1THE | 1.9 | Ser115Ala mutant; complex with Z-Arg-Ser(OBzl) chloromethane |

## Name and History

Cathepsin B represents the first described member of what has become recognized as the large family of lysosomal cysteine peptidases. The history of the term 'cathepsin' and the lineage of cathepsin B have been reviewed in detail previously (Barrett, 1977). The first description of this enzyme was as an activity, denoted as *cathepsin II*, identified in cattle spleen by its ability to deamidate Bz-Arg-NH$_2$, a synthetic substrate originally developed for the assay of trypsin (Fruton *et al.*, 1941). This activity was later termed *cathepsin B* (Tallan *et al.*, 1952). Otto's group then showed that preparations of this enzyme contained two components of different size and assigned the terms 'cathepsin B' to the lower molecular weight protein and 'cathepsin B (new)' to the higher molecular weight component (Otto, 1967). Later the terms *cathepsin B1* and cathepsin B2 were substituted for these components.

While both of these enzymes were benzoylarginine amidases, studies with more sophisticated synthetic substrates demonstrated that cathepsin B2 acted solely as a carboxypeptidase while cathepsin B1 was able to cleave endopeptidase substrates such as Z-Arg-Arg-NHNap (McDonald & Ellis, 1975). Cathepsin B2 was therefore renamed as lysosomal carboxypeptidase B (Chapter 267) and the term cathepsin B retained for the lower molecular weight endopeptidase which was purified to homogeneity from human liver in 1973 (Barrett, 1973).

Due to considerable variations in its post-translational modification and the widespread interest in its role in numerous physiological and pathological processes, cathepsin B has masqueraded under other pseudonyms in the past (for example 'tumor cathepsin B'). Fortunately, the availability of specific antisera and selective inhibitors now makes the identification of this enzyme much more secure. The first complete cathepsin B protein sequence was published for the rat liver enzyme in 1983 (Takio *et al.*, 1983) and the first cDNA sequence, again for rat cathepsin B, was reported in 1985 (San Segundo *et al.*, 1985). Subsequently cDNAs coding for this enzyme have been cloned from many organisms.

## Activity and Specificity

Cathepsin B possesses both endopeptidase and exopeptidase activities, in the latter case acting as a peptidyl-dipeptidase, as first demonstrated by the C-terminal truncation of glucagon by the progressive removal of dipeptides (Aronson & Barrett, 1978). While it has been claimed that cathepsin B acts only as an exopeptidase (Takahashi *et al.*, 1986a), its ability to degrade various proteins has been demonstrated with recombinant enzyme (Fosang *et al.*, 1992; Illy *et al.*,

1997) produced in systems lacking other cysteine peptidases which could potentially contaminate cathepsin B preparations obtained from natural sources.

As with other members of the papain family, the principal determinant of substrate specificity appears to be the S2 subsite, where cathepsin B shows the usual preference for large hydrophobic residues. However, cathepsin B is unique in that it will also accept an arginine side chain at this position due to the location of a Glu residue (Glu245) in the S2 subsite (Hasnain *et al.*, 1993; Jia *et al.*, 1995).

As indicated above, the enzyme was originally assayed by following the cleavage of Bz-Arg-NH$_2$ through quantitation of the resulting ammonia. For many years peptide-NHNap substrates were used for the assay of cathepsin B (Barrett, 1972, 1976) but these have now been superseded due to the potentially carcinogenic nature of such compounds and the restricted availability of components used for these assays. However, peptidyl-4-methoxynaphthylamide substrates still represent a useful option for enzyme histochemistry (Dolbeare & Smith, 1977; Graf *et al.*, 1979). For colorimetric and fluorometric assay of cathepsin B peptidyl-NHPhNO$_2$ and -NHMec substrates are most commonly used. Where selectivity for cathepsin B over other lysosomal cysteine peptidases such as cathepsins H, L and S is required, Z-Arg-Arg+NHPhNO$_2$ or Z-Arg-Arg+NHMec should be used since the latter enzymes show very little activity with substrates containing a P2 Arg residue. However, it should be stressed that cathepsin B shows substantially higher $k_{cat}/K_m$ values for Z-Phe-Arg+NHPhNO$_2$ and Z-Phe-Arg+NHMec than for the corresponding Z-Arg-Arg-compounds. Although the maximal activity with the above synthetic substrates is found in the pH range of 7.5–8 (Khouri *et al.*, 1991), the enzyme is very unstable under these conditions and spontaneously denatures (Mort *et al.*, 1980; Turk *et al.*, 1994). An assay pH of 6.0 represents the optimal value for routine work. Cathepsin B preparations tend to undergo active site cysteine oxidation, so millimolar concentrations of DTT or cysteine together with EDTA should be included. Substrates have also been designed to assay the exopeptidase activity of cathepsin B. The quenched fluorescence substrate Dns-Phe-Arg+Phe(NO$_2$)-Leu shows high activity and gives a reasonable fluorescence yield on cleavage (Pohl *et al.*, 1987). While activity is observed over a wide pH range for the endopeptidase substrates, exopeptidase activity is restricted to the acidic range.

Due to the dual endopeptidase/exopeptidase attributes of cathepsin B studies on the specificity of this enzyme following the cleavage of protein substrates such as the insulin B chain should be viewed with caution. Under acidic pH conditions the exopeptidase activity is considerably greater than the endopeptidase activity. Thus the initial cleavage site may not be easily recognizable.

While the P2 specificity of cathepsin B has been well characterized, the selectivity at other subsites is less strict. Although some preferences have been determined, these cannot be considered to represent sources of primary specificity. A study using a series of five Z-Xaa-Arg-Arg-NHMec substrates demonstrated a preference for aromatic and hydrophobic residues in the P3 position (Taralp *et al.*, 1995). The S1 subsite has been explored using several series of inhibitors and demonstrates a preference for positively charged and straight-chained aliphatics with lower affinity for negatively charged and bulky hydrophobic residues (Shaw, 1990; Krantz *et al.*, 1991; Pliura *et al.*, 1992). Based on a study using quenched fluorescence substrates of the form Dns-Phe-Arg-Xaa-Trp-Ala, the S subsite was shown to prefer aromatic and large aliphatic side chains (Ménard *et al.*, 1993) and work with a limited series of *N*-peptidyl-*O*-carbamoyl amino acid hydroxamates showed little selectivity at the S subsite, except for a low tolerance for glycine (Brömme & Kirschke, 1993).

## Structural Chemistry

Cathepsin B (human) is synthesized as a 339 residue preproenzyme. Following removal of the 17 residue signal sequence, the inactive proenzyme undergoes several post-translational modifications which, depending on their extent, yield different forms of the enzyme with varied molecular characteristics. The proenzyme is substituted with two asparagine-linked oligosaccharide units, one each in the pro region and the mature protein. Forms of the proenzyme that are secreted from the cell rather than routed to the lysosome undergo additional oligosaccharide modification including sialylation rendering the protein heterogeneous on analysis by isoelectric focusing (Recklies & Mort, 1985). Activation of the proenzyme occurs following cleavage and dissociation of the 62 residue pro region. The propeptide functions by attaching to the surface of the enzyme and blocking the active-site cleft (Cygler *et al.*, 1996; Turk *et al.*, 1996) by running throughout its length in the opposite orientation to that required for substrate hydrolysis.

Although it shows a similar folding pattern to the propeptides of other papain family members (Coulombe *et al.*, 1996; Groves *et al.*, 1996) the cathepsin B pro region is only two-thirds as long. The isolated propeptide is a potent, selective, cathepsin B inhibitor (Fox *et al.*, 1992) but the dissociation constant increases dramatically under acidic conditions. Complexes between the propeptide and mature cathepsin B are stable at neutral pH (Mach *et al.*, 1994a) and can account for cathepsin B activity that has been observed in tissue culture medium in the past (Poole *et al.*, 1978). (Activity is measured under the normal assay conditions of pH 5.5–6.0.) *In vitro* studies demonstrate that procathepsin B undergoes autoprocessing at low pH, the environment of the lysosome, and kinetic analyses indicate contributions from both intra- and intermolecular mechanisms (Rowan *et al.*, 1992a; Mach *et al.*, 1994b). In addition, intermolecular processing can be mediated by a variety of exogenous peptidases such as pepsin (Mort *et al.*, 1981a).

Proteolytic cleavages of the propeptide described to date leave an N-terminal extension relative to the terminus found in the mature lysosomal form and it appears that trimming of this extension occurs by exogenous exopeptidase action that is limited by the presence of a conserved proline residue at position 2 in the mature sequence. An additional processing step is the removal of a six-residue C-terminal extension, the function of which is unclear. *In vitro* studies have demonstrated that this step can be mediated by the peptidyl-dipeptidase action of cathepsin B (Rowan *et al.*, 1993).

The final proteolytic event is cleavage between residues 47 and 50 to yield a two-chain form of the enzyme with the excision of a dipeptide. No evidence has been forthcoming to indicate that the latter step is the consequence of autoprocessing, but it has been suggested that legumain, the newly-discovered lysosomal peptidase (Chapter 255) may be involved (Chen *et al.*, 1997). Comparison of the various forms of cathepsin B suggests that proteolytic processing steps beyond the removal of the propeptide have very little effect on the catalytic properties of the enzyme.

In addition to proteolytic processing, the *N*-linked oligosaccharide moiety located on Asn113 of cathepsin B also undergoes degradation in the lysosome and is often reduced to a single *N*-acetylglucosamine unit (Takayuki *et al.*, 1984). Studies on variants containing different amounts of carbohydrate indicate that catalytic efficiency decreases as the length of the sugar chain increases (Takahashi *et al.*, 1986b; Hasnain *et al.*, 1992).

As expected from sequence similarity, the peptide backbone of cathepsin B is essentially superimposable on to that of papain and other papain family members. However, the cathepsin B structure is dramatically different in the primed subsite region. Here an extra 20 residue peptide segment, termed the occluding loop (Musil *et al.*, 1991), blocks off the C-terminal end of the active-site cleft, strategically positioning two histidine residues (His110 and His111) that provide an acceptor for the C-terminal carboxylate of the P2′ residue, and equips the enzyme to act as a peptidyl-dipeptidase. Deletion of the loop by site-directed mutagenesis abolishes this activity (Illy *et al.*, 1997). Clearly the occluding loop would be expected to compromise the ability of cathepsin B to act as an endopeptidase and to be inhibited by members of the cystatin family which bind at both the primed and unprimed subsites of the enzyme.

As demonstrated by the three-dimensional structures of rat and human procathepsin B (Cygler *et al.*, 1996; Turk *et al.*, 1996), the occluding loop can adopt an alternate conformation. In the proenzyme, the propeptide is observed to pass through the entire extent of the active-site cleft, displacing the occluding loop. In order to bind to the enzyme, inhibitors and extended substrates must occupy the same space. An energetic penalty must be paid to mediate the deflection of the loop in the mature enzyme and this accounts for the relatively low $k_{cat}/K_m$ values found for cathepsin B hydrolysis of endopeptidase substrates relative to those observed, for example, in the cases of cathepsins L and S.

Cathepsin B binds to $\alpha_2$-macroglobulin (Starkey & Barrett, 1973) and is inhibited by the cystatin family of inhibitors of the papain-like cysteine peptidases (Barrett, 1987) and by the newly described equistatin family (Lenarcic *et al.*, 1997). As indicated above, however, the presence of the occluding loop in the conformation normally adopted by the mature enzyme is incompatible with cystatin binding (Stubbs *et al.*, 1990). Most of the natural cysteine peptidase inhibitors are therefore much less effective against cathepsin B than other members of the papain family.

Cathepsin B is susceptible to the various classes of irreversible inhibitors that have been developed for the papain family. Thus peptidyl chloromethane, diazomethane, fluoromethane (Shaw, 1990) and acyloxymethane (Krantz, 1994) derivatives have been studied and, depending on the peptidyl moiety used, differing levels of potency have been observed relative to other family members. Selectivity has been achieved through the synthesis of E-64 derivatives that take advantage of the unique ability of cathepsin B to bind a C-terminal carboxylate through the strategically placed histidine residues on the occluding loop. Examples of the these compounds are CA-074 (*n*-propyl-epoxysuccinyl-Ile-Pro), CA-030 (ethyl-epoxysuccinyl-Ile-Pro) (Towatari *et al.*, 1991) and isobutyl-epoxysuccinyl-Leu-Pro (Gour-Salin *et al.*, 1993) which proved to be useful, selective cathepsin B inhibitors. Unlike the parent compound, E-64 (epoxysuccinyl-Leu-agmatine), in which the peptidyl moiety binds in the unprimed subsites (Varughese *et al.*, 1989) in an orientation similar to that observed for the propeptide (Cygler *et al.*, 1996), CA-030 was shown to bind to cathepsin B such that the Ile and Pro residues occupy the S and S' subsites respectively, adopting a substrate-like orientation (Turk *et al.*, 1995). A prodrug form of CA-074 has been prepared by esterification of the free prolyl carboxylate (Buttle *et al.*, 1992), allowing the compound to pass through cell membranes. Currently, several diazomethanes and fluoromethanes are available from Sigma, Calbiochem and Enzyme Systems Products (see Appendix 2 for full names and addresses of suppliers). CA-074 is available from the Peptide Institute.

## *Preparation*

Methods have been reported in detail for the purification of cathepsin B from liver using standard protein purification techniques including organomercurial-Sepharose chromatography (Barrett & Kirschke, 1981). Kidney, spleen and placenta have also been used as tissue sources. The development of immobilized Gly-Phe-Gly- and Phe-Gly-semicarbazones as affinity ligands represented a major advance in cathepsin B purification, effectively providing a single-step purification of the enzyme following elution with pyridyl disulfide (Rich *et al.*, 1986). On SDS-PAGE under reducing conditions, mammalian cathepsin B isolated from the above-mentioned tissues is usually observed as a 30 kDa band representing the single-chain enzyme and a 25 kDa (and 5 kDa) band representing the two-chain form. On isoelectric focusing, multiple components are observed with pI values in the range 4.5–5.5. Cathepsin B obtained from human liver and placenta and from cattle spleen is currently commercially available from Sigma.

In addition to natural sources, several recombinant expression systems have been used to prepare cathepsin B. These systems have also facilitated the investigation of the role of specific residues by site-directed mutagenesis. In all cases cDNAs coding for the proenzymes have been expressed and the active peptidase is produced later following processing.

The rat proenzyme is well expressed in the yeasts *Saccharomyces cerevisiae* (Hasnain *et al.*, 1992) and *Pichia pastoris* (Sivaraman *et al.*, 1996) using α-factor fusion constructs, while the human enzyme was only expressed efficiently in *P. pastoris* (Illy *et al.*, 1997). The *S. cerevisiae* system has also been used to express cathepsin B from *Schistosoma mansoni* (Lipps *et al.*, 1996). An advantage of the yeast α-factor system is that the fully functional proenzyme is secreted into the culture medium and the mature single-chain form of the enzyme is produced as the result of autoprocessing following dialysis into acidic conditions. Human procathepsin B has also been expressed in *E. coli* using a T7 promoter system (Kuhelj *et al.*, 1995). Here large amounts of protein can be produced as inclusion bodies which must however be solubilized, the proenzyme refolded and processed to produce the functional peptidase. Finally, procathepsin B has been overexpressed in mammalian cells using the cytomegalovirus promoter (Ren *et al.*, 1996).

## *Biological Aspects*

Human cathepsin B is coded for by a single-copy gene encompassing at least 27 kb of DNA. It consists of 13 exons and is located on chromosome 8p22 (Wang *et al.*, 1987), placing it in a region that frequently shows DNA loss in some human cancers. Investigation of the promoter is still incomplete. Although current information is compatible with constitutive expression it is clear that cathepsin B transcription varies with cell type and state of differentiation (Berquin & Sloane, 1996). In addition, alternate splicing of the mRNA precursor (Berquin *et al.*, 1995) and the use of both alternate transcription start (Berquin *et al.*, 1995) and stop sites (Tam *et al.*, 1994) has been observed. Most of the resulting variants do not affect the protein coding sequence but a transcript deficient in exon 3 has been demonstrated in human tumors (Gong *et al.*, 1993) and in rheumatoid synovium (Lemaire *et al.*, 1996). Although the polypeptide synthesized from this mRNA lacks the signal peptide and half of the pro region, preliminary data indicate that the protein product survives and, as would be expected, is not localized in the lysosome (Mehtani *et al.*, 1996). The properties and function of this novel cathepsin B variant remain unclear.

Procathepsin B synthesized from the majority of the mRNA is targeted to the *trans*-Golgi network and then on to the lysosome, through the mannose-6-phosphate receptor system. Processing to the mature enzyme form occurs in the acidic environment of the *trans*-Golgi and the lysosome. Further conversion to the two-chain form, in the lysosome, takes from 24 to 48 h (Mach *et al.*, 1992). In most systems, however, a variable fraction of the proenzyme is secreted from the cell and evidence has been presented for the presence of active extracellular cathepsin B (Reddy *et al.*, 1995) indicating either the release of the lysosomal form, or extracellular processing of the proenzyme.

The lysosomal localization of cathepsin B suggests that it functions principally as a component of the intracellular protein-degradation system. However, findings from many groups have indicated alternate locations including the nucleus, the cytoplasm and plasma membrane, in addition to extracellular secretion. Roles for cathepsin B in various physiological and pathological processes has been proposed.

However, recently cathepsin B null mice have been constructed and preliminary results indicate that they have no obvious phenotype, suggesting that the enzyme does not play a critical role in normal development (Deussing *et al.*, 1996).

Extensive studies have demonstrated increased cathepsin B levels and redistribution of the enzyme in human and animal tumors. A role for the enzyme in invasion and metastasis has been proposed (Berquin & Sloane, 1996). Aberrant colocalization of cathepsin B with pancreatic zymogens following obstruction of the pancreatic duct, and the concomitant activation of serine peptidases, has been proposed as a mechanism underlying acute pancreatitis (Steer, 1992). In addition, extracellular cathepsin B has been implicated in inflammatory airway disease (Burnett *et al.*, 1995) and bone and joint disorders (Buttle, 1994).

As indicated above, features unique to cathepsin B provide it with ways to survive in the extracellular environment under conditions that would lead to inactivation of other lysosomal cysteine peptidases. Weak binding by cystatins allows cathepsin B to escape inhibition and the stability of noncovalent complexes of cathepsin B with its propeptide at neutral pH allows the enzyme to remain available until an acidic environment is encountered, whereupon active enzyme is generated.

## Distinguishing Features

Two major features distinguish cathepsin B from other members of the papain family of cysteine peptidases. First, cathepsin B has the ability to cleave substrates containing a P2 arginine residue and secondly it is a very efficient peptidyl-dipeptidase. This latter characteristic can be exploited using the E-64 derivative CA-074 as a diagnostic inhibitor for cathepsin B. However this compound must be used at reasonable concentrations (in the micromolar range) since it is an irreversible inhibitor that will derivatize other cysteine peptidases if used at high concentration and for long incubation periods.

## Antibodies

Cathepsin B exhibits unusual properties when used as an immunogen. Experience from several laboratories indicates that the mature enzyme obtained from natural sources is only mildly antigenic in rabbits while sheep respond extremely strongly. In addition, antisera raised against the mature enzyme have been routinely found to recognize only the denatured protein (Barrett, 1973; Mort *et al.*, 1981b). This appears to be due to denaturation of the enzyme following introduction to the animal. Dominant epitopes are exposed following the unfolding of the protein that are not accessible in the native enzyme. Cathepsin B is unique in this regard, for although other lysosomal cysteine peptidases, for example cathepsins H and L, show instability similar to that of cathepsin B, antisera raised against these proteins are able to recognize the corresponding native enzymes. In contrast to mature cathepsin B, procathepsin B is stable under alkaline conditions and antibodies raised against recombinant rat proenzyme produced in *S. cerevisiae* were found to precipitate the active, mature enzyme (Rowan *et al.*, 1992b). In

this case strong antisera could be produced in rabbits. However, the recombinant protein was substituted with a highly immunogenic, yeast-derived *N*-linked oligosaccharide which could account for the enhanced immune response.

In addition to polyclonal antisera, mouse monoclonal antibodies have been produced, one of which appeared to recognize native cathepsin B (Wardale *et al.*, 1986). Currently, antisera to human cathepsin B and rat procathepsin B are commercially available from Calbiochem and Upstate Biotechnology, respectively.

## Related Peptidases

Most studies on cathepsin B to date have centered on mammalian systems, but characterization of the enzyme from other vertebrates has been reported. In addition, extensive studies have been carried out on cathepsin B homologs from various parasitic worms, in particular *Haemonchus contortus* (Cox *et al.*, 1990) and *Ostertagia ostertagi* (Pratt *et al.*, 1992) which are major pathogens of domestic animals, and *Schistosoma mansoni* (Klinkert *et al.*, 1989) which infects hundreds of millions of people in tropical countries. It is believed that cathepsin B plays a role in the penetration of the organisms into the host. Functional recombinant *S. mansoni* cathepsin B has been prepared and shows some differences to the mammalian enzyme (Lipps *et al.*, 1996).

Over the last few years a vast amount of sequence data has become available from molecular cloning. For example a four-member multigene family of cathepsin B homologs was discovered in *Caenorhabditis elegans* (Larminie & Johnstone, 1996). If the presence of a short propeptide (~60 residues) and the occluding loop segment containing the key histidine residues are taken as hallmarks for designation as cathepsin B, then this enzyme is present in most of the animal kingdom as well as in isolated parts of the plant kingdom. Recently, a cysteine peptidase has been described from the diplomonad protozoan *Giardia lamblia*, thought to be one of the earliest members in the eukaryotic cell lineage (Ward *et al.*, 1997). While phylogenetic analysis places this enzyme closest to the cathepsin B family, it lacks the occluding loop. It does, however, have a short propeptide and hydrolyzes P2 Arg-containing substrates due to the presence of a Glu245 equivalent and thus appears to represent the most primitive form of cathepsin B known to date.

## Further Reading

For reviews, see Barrett & Kirschke (1981), Berquin & Sloane (1996) and Mort & Buttle (1997).

## References

Aronson, N.N. & Barrett, A.J. (1978) The specificity of cathepsin B. Hydrolysis of glucagon at the C-terminus by a peptidyldipeptidase mechanism. *Biochem. J.* **171**, 759–765.

Barrett, A.J. (1972) A new assay for cathepsin B1 and other thiol proteinases. *Anal. Biochem.* **47**, 280–293.

Barrett, A.J. (1973) Human cathepsin B1. Purification and some properties of the enzyme. *Biochem. J.* **131**, 809–822.

Barrett, A.J. (1976) An improved color reagent for use in Barrett's assay of cathepsin B. *Anal. Biochem.* **76**, 374–376.

Barrett, A.J. (1977) Introduction to the history and classification of tissue proteinases. In: *Proteinases in Mammalian Cells and Tissues* (Barrett, A.J., ed.). Amsterdam: Elsevier/North Holland, pp. 1–55.

Barrett, A.J. (1980) Fluorimetric assays for cathepsin B and cathepsin H with methylcoumarylamide substrates. *Biochem. J.* **187**, 909–912.

Barrett, A.J. (1987) The cystatins: a new class of peptidase inhibitors. *Trends Biochem. Sci.* **12**, 193–196.

Barrett, A.J. & Kirschke, H. (1981) Cathepsin B, cathepsin H, and cathepsin L. *Methods Enzymol.* **80**, 535–561.

Berquin, I.M. & Sloane, B.F. (1996) Cathepsin B expression in human tumors. *Adv. Exp. Med. Biol.* **389**, 281–294.

Berquin, I.M., Cao, L., Fong, D. & Sloane, B.F. (1995) Identification of two new exons and multiple transcription start points in the 5′-untranslated region of the human cathepsin-B-encoding gene. *Gene* **159**, 143–149.

Brömme, D. & Kirschke, H. (1993) N-Peptidyl-O-carbonyl amino acid hydroxamates: irreversible inhibitors for the study of the $S_2'$ specificity of cysteine proteinases. *FEBS Lett.* **322**, 211–214.

Burnett, D., Abrahamson, M., Devalia, J.L., Sapsford, R.J., Davies, R.J. & Buttle, D.J. (1995) Synthesis and secretion of procathepsin B and cystatin C by human bronchial epithelial cells in vitro: modulation of cathepsin B activity by neutrophil elastase. *Arch. Biochem. Biophys.* **317**, 305–310.

Buttle, D.J. (1994) Lysosomal cysteine endopeptidases in the degradation of cartilage and bone. In: *Immunopharmacology of Joints and Connective Tissue* (Dingle, J.T. & Davies, M.E., eds). London: Academic Press, pp. 225–243.

Buttle, D.J., Murata, M., Knight, C.G. & Barrett, A.J. (1992) CA074 methyl ester: a proinhibitor for intracellular cathepsin B. *Arch. Biochem. Biophys.* **299**, 377–380.

Chen, J.-M., Dando, P.M., Rawlings, N.D., Brown, M.A., Young, N.E., Stevens, R.A., Hewitt, E., Watts, C. & Barrett, A.J. (1997) Cloning, isolation, and characterization of mammalian legumain, an asparaginyl endopeptidase. *J. Biol. Chem.* **272**, 8090–8098.

Coulombe, R., Grochulski, P., Sivaraman, J., Ménard, R., Mort, J.S. & Cygler, M. (1996) Structure of human procathepsin L reveals the molecular basis of inhibition by the prosegment. *EMBO J.* **15**, 5492–5503.

Cox, G.N., Pratt, D., Hageman, R. & Boisvenue, R.J. (1990) Molecular cloning and primary sequence of a cysteine protease expressed by *Haemonchus contortus* adult worms. *Mol. Biochem. Parasitol.* **41**, 25–34.

Cygler, M., Sivaraman, J., Grochulski, P., Coulombe, R., Storer, A.C. & Mort, J.S. (1996) Structure of rat procathepsin B. Model for inhibition of cysteine protease activity by the proregion. *Structure* **4**, 405–416.

Deussing, J., Roth, W., von Figura, K. & Peters, C. (1996) Generation of cathepsin B-deficient mice by gene targeting. *11th International Conference on Protein Turnover* 113 (abstr.).

Dolbeare, F.A. & Smith, R.E. (1977) Flow cytometric measurement of peptidases with use of 5-nitrosalicylaldehyde and 4-methoxy-beta-naphthylamine derivatives. *Clin. Chem.* **23**, 1485–1491.

Fosang, A.J., Neame, P.J., Last, K., Hardingham, T.E., Murphy, G. & Hamilton, J.A. (1992) The interglobular domain of cartilage aggrecan is cleaved by PUMP, gelatinases, and cathepsin B. *J. Biol. Chem.* **267**, 19470–19474.

Fox, T., de Miguel, E., Mort, J.S. & Storer, A.C. (1992) Potent slow-binding inhibition of cathepsin B by its propeptide. *Biochemistry* **31**, 12571–12576.

Fruton, J.S., Irving, G.W. & Bergmann, M. (1941) On the proteolytic enzymes of animal tissues. III. The proteolytic enzymes of beef spleen, beef kidney, and swine kidney. Classification of the cathepsins. *J. Biol. Chem.* **141**, 763–774.

Gong, Q., Chan, S.J., Bajkowski, A.S., Steiner, D.F. & Frankfater, A. (1993) Characterization of the cathepsin B gene and multiple mRNAs in human tissues: evidence for alternative splicing of cathepsin B pre-mRNA. *DNA Cell Biol.* **12**, 299–309.

Gour-Salin, B.J., Lachance, P., Plouffe, C., Storer, A.C. & Ménard, R. (1993) Epoxysuccinyl dipeptides as selective inhibitors of cathepsin B. *J. Med. Chem.* **36**, 720–725.

Graf, M., Leemann, U., Ruch, F. & Strauli, P. (1979) The fluorescence and bright field microscopic demonstration of cathepsin B in human fibroblasts. *Histochemistry* **64**, 319–322.

Groves, M.R., Taylor, A.M.R., Scott, M., Cummings, N.J., Pickersgill, R.W. & Jenkins, J.A. (1996) The prosequence of procaricain forms an $\alpha$-helical domain that prevents access to the substrate-binding cleft. *Structure* **4**, 1193–1203.

Hasnain, S., Hirama, T., Tam, A. & Mort, J.S. (1992) Characterization of recombinant rat cathepsin B and non-glycosylated mutants expressed in yeast. New insights into the pH-dependence of cathepsin B catalyzed hydrolyses. *J. Biol. Chem.* **267**, 4713–4721.

Hasnain, S., Hirama, T., Huber, C.P., Mason, P. & Mort, J.S. (1993) Characterization of cathepsin B specificity by site-directed mutagenesis. The importance of $Glu^{245}$ in the $S_2$-$P_2$ specificity for arginine and its role in transition state stabilization. *J. Biol. Chem.* **268**, 235–240.

Illy, C., Quraishi, O., Wang, J., Purisima, E., Vernet, T. & Mort, J.S. (1997) Role of the occluding loop in cathepsin B activity. *J. Biol. Chem.* **272**, 1187–1202.

Jia, Z., Hasnain, S., Hirama, T., Lee, X., Mort, J.S., To, R. & Huber, C.P. (1995) Crystal structure of recombinant rat cathepsin B and a cathepsin B-inhibitor complex. Implications for structure-based inhibitor design. *J. Biol. Chem.* **270**, 5527–5533.

Khouri, H.E., Plouffe, C., Hasnain, S., Hirama, T., Storer, A.C. & Ménard, R. (1991) A model to explain the pH-dependent specificity of cathepsin B-catalyzed hydrolyses. *Biochem. J.* **275**, 751–757.

Klinkert, M.-Q., Felleisen, R., Link, G., Ruppel, A. & Beck, E. (1989) Primary structures of Sm31/32 diagnostic proteins *of Schistosoma mansoni. Mol. Biochem. Parasitol.* **33**, 113–122.

Krantz, A. (1994) Peptidyl (acyloxy)methanes as quiescent affinity labels for cysteine proteinases. *Methods Enzymol.* **244**, 656–671.

Krantz, A., Copp, L.J., Coles, J., Smith, R.A. & Heard, S.B. (1991) Peptidyl(acyloxy)methyl ketones and the quiescent affinity label concept: the departing group as a variable structural element in the design of inactivators of cysteine proteinases. *Biochemistry* **30**, 4678–4687.

Kuhelj, R., Dolinar, M., Pungerčar, J. & Turk, V. (1995) The preparation of catalytically active human cathepsin B from its precursor expressed in *Escherichia coli* in the form of inclusion bodies. *Eur. J. Biochem.* **229**, 533–539.

Larminie, C.G. & Johnstone, I.L. (1996) Isolation and characterization of four developmentally regulated cathepsin B-like cysteine protease genes from the nematode *Caenorhabditis elegans. DNA Cell Biol.* **15**, 75–82.

Lemaire, R., Huet, G., Dacquembronne, E., Fontaine, C., Migaud, H. & Flipo, R.-M. (1996) Specific regulation of alternative splicing of the 5′ untranslated region of cathepsin B pre-mRNA in synovial tissue from patients with rheumatoid arthritis. *Arthritis Rheum.* **39**, S198 (abstr.).

Lenarčirč, B., Ritonja, A., Štrukelj, B., Turk, B. & Turk, V. (1997) Equistatin, a new inhibitor of cysteine proteinases from *Actinia equina*, is structurally related to thyroglobulin type-1 domain. *J. Biol. Chem.* **272**, 13899–13903.

Lipps, G., Füllkrug, R. & Beck, E. (1996) Cathepsin B of *Schistosoma mansoni*. Purification and activation of the recombinant proenzyme secreted by *Saccharomyces cerevisiae*. *J. Biol. Chem.* **271**, 1717–1725.

Mach, L., Stüwe, K., Hagen, A., Ballaun, C. & Glössl, J. (1992) Proteolytic processing and glycosylation of cathepsin B. The role of the primary structure of the latent precursor and of the carbohydrate moiety for cell-type-specific molecular forms of the enzyme. *Biochem. J.* **282**, 577–582.

Mach, L., Mort, J.S. & Glössl, J. (1994a) Non-covalent complexes between the lysosomal proteinase cathepsin B and its propeptide account for stable, extracellular, high molecular mass forms of the enzyme. *J. Biol. Chem.* **269**, 13036–13040.

Mach, L., Mort, J.S. & Glössl, J. (1994b) Maturation of human cathepsin B. Proenzyme activation and proteolytic processing of the precursor to the mature proteinase, *in vitro*, are primarily unimolecular processes. *J. Biol. Chem.* **269**, 13030–13035.

McDonald, J.K. & Ellis, S. (1975) On the substrate specificity of cathepsins B1 and B2 including a new fluorogenic substrate for cathepsin B1. *Life Sci.* **17**, 1269–1276.

Mehtani, S., Gong, Q. & Frankfater, A. (1996) Alternate splicing and the expression of cathepsin B in human tumors. *11th International Conference on Protein Turnover* 40 (abstr.).

Ménard, R., Carmona, E., Plouffe, C., Brömme, D., Konishi, Y., Lefebvre, J. & Storer, A.C. (1993) The specificity of the S1′ subsite of cysteine proteases. *FEBS Lett.* **328**, 107–110.

Mort, J.S. & Buttle, D.J. (1997) Molecules in focus. Cathepsin B. *Int. J. Biochem. Cell Biol.* **29**, 715–720.

Mort, J.S., Recklies, A.D. & Poole, A.R. (1980) Characterization of a thiol proteinase secreted by malignant human breast tumours. *Biochim. Biophys. Acta* **614**, 134–143.

Mort, J.S., Leduc, M. & Recklies, A.D. (1981a) A latent thiol proteinase from ascitic fluid of patients with neoplasia. *Biochim. Biophys. Acta* **662**, 173–180.

Mort, J.S., Poole, A.R. & Decker, R.S. (1981b) The immunofluorescent localization of cathepsins B and D in human fibroblasts. *J. Histochem. Cytochem.* **29**, 649–657.

Musil, D., Zucic, D., Engh, R.A., Mayr, I., Huber, R., Popovič, T., Turk, V., Towatari, T., Katunuma, N. & Bode, W. (1991) The refined 2.15 Å x-ray crystal structure of human liver cathepsin B: the structural basis for its specificity. *EMBO J.* **10**, 2321–2330.

Otto, K. (1967) Über ein neues Kathepsin. Reinigung aus Rindermilz, Eigenschaften, sowie Vergleich mit Kathepsin B. *Z. Physiol. Chem.* **348**, 1449–1460.

Pliura, D.H., Bonaventura, B.J., Smith, R.A., Coles, P.J. & Krantz, A. (1992) Comparative behaviour of calpain and cathepsin B toward peptidyl acyloxymethyl ketones, sulphonium methyl ketones and other potential inhibitors of cysteine proteinases. *Biochem. J.* **288**, 759–762.

Pohl, J., Davinic, S., Bláha, I., Štrop, P. & Kostka, V. (1987) Chromophoric and fluorophoric peptide substrates cleaved through the dipeptidyl carboxypeptidyl activity of cathepsin B. *Anal. Biochem.* **165**, 96–101.

Poole, A.R., Tiltman, K.J., Recklies, A.D. & Stoker, T.A.M. (1978) Differences in secretion of the proteinase cathepsin B at the edges of human breast carcinomas and fibroadenomas. *Nature* **273**, 545–547.

Pratt, D., Boisvenue, R.J. & Cox, G.N. (1992) Isolation of putative cysteine protease genes of *Ostertagia ostertagi*. *Mol. Biochem. Parasitol.* **56**, 39–48.

Recklies, A.D. & Mort, J.S. (1985) Characterization of a cysteine proteinase secreted by mouse mammary gland. *Cancer Res.* **45**, 2302–2307.

Reddy, V.Y., Zhang, Q.-Y. & Weiss, S.J. (1995) Pericellular mobilization of the tissue-destructive cysteine proteinases, cathepsins B, L, and S, by human monocyte-derived macrophages. *Proc. Natl Acad. Sci. USA* **92**, 3849–3853.

Ren, W.-P., Fridman, R., Zabrecky, J.R., Morris, L.D., Day, N.A. & Sloane, B.F. (1996) Expression of functional recombinant human procathepsin B in mammalian cells. *Biochem. J.* **319**, 793–800.

Rich, D.H., Brown, M.A. & Barrett, A.J. (1986) Purification of cathepsin B by a new form of affinity chromatography. *Biochem. J.* **235**, 731–734.

Rowan, A.D., Mason, P., Mach, L. & Mort, J.S. (1992a) Rat procathepsin B. Proteolytic processing to the mature form *in vitro*. *J. Biol. Chem.* **267**, 15993–15999.

Rowan, A.D., Mach, L. & Mort, J.S. (1992b) Antibodies to rat procathepsin B recognize the active mature enzyme. *Biol. Chem. Hoppe-Seyler* **373**, 427–432.

Rowan, A.D., Feng, R., Konishi, Y. & Mort, J.S. (1993) Demonstration by electrospray mass spectrometry that the peptidyldipeptidase activity of cathepsin B is capable of rat cathepsin B C-terminal processing. *Biochem. J.* **294**, 923–927.

San Segundo, B., Chan, S.J. & Steiner, D.F. (1985) Identification of cDNA clones encoding a precursor of rat liver cathepsin B. *Proc. Natl Acad. Sci. USA* **82**, 2320–2324.

Shaw, E. (1990) Cysteinyl proteinases and their selective inactivation. *Adv. Enzymol. Relat. Areas Mol. Biol.* **63**, 271–347.

Sivaraman, J., Coulombe, R., Magny, M.-C., Mason, P., Mort, J.S. & Cygler, M. (1996) Crystallization of rat procathepsin B. *Acta Crystallogr.* **D52**, 874–875.

Starkey, P.M. & Barrett, A.J. (1973) Human cathepsin B1. Inhibition by $\alpha_2$-macroglobulin and other serum proteins. *Biochem. J.* **131**, 823–831.

Steer, M.L. (1992) How and where does acute pancreatitis begin? *Arch. Surg.* **127**, 1350–1353.

Stubbs, M.T., Laber, B., Bode, W., Huber, R., Jerala, R., Lenarcic, B. & Turk, V. (1990) The refined 2.4 Å X-ray crystal structure of recombinant human stefin B in complex with the cysteine proteinase papain: a novel type of proteinase inhibitor interaction. *EMBO J.* **9**, 1939–1947.

Takahashi, T., Dehdarani, A.H., Yonezawa, S. & Tang, J. (1986a) Porcine spleen cathepsin B is an exopeptidase. *J. Biol. Chem.* **261**, 9375–9381.

Takahashi, T., Yonezawa, S., Dehdarani, A.H. & Tang, J. (1986b) Comparative studies of two cathepsin B isozymes from porcine spleen. Isolation, polypeptide chain arrangements, and enzyme specificity. *J. Biol. Chem.* **261**, 9368–9374.

Takayuki, T., Schmidt, P.G. & Tang, J. (1984) Novel carbohydrate structures of cathepsin B from porcine spleen. *J. Biol. Chem.* **259**, 6059–6062.

Takio, K., Towatari, T., Katunuma, N., Teller, D.C. & Titani, K. (1983) Homology of amino acid sequence of rat liver cathepsins B and H with that of papain. *Proc. Natl Acad. Sci. USA* **80**, 3666–3670.

Tallan, H.H., Jones, M.E. & Fruton, J.S. (1952) On the proteolytic enzymes of animal tissues. X. Beef spleen cathepsin C. *J. Biol. Chem.* **194**, 793–805.

Tam, S.W., Cote-Paulino, L.R., Peak, D.A., Sheahan, K. & Murnane, M.J. (1994) Human cathepsin B-encoding cDNAs: sequence variations in the 3′-untranslated region. *Gene* **139**, 171–176.

Taralp, A., Kaplan, H., Sytwu, I.-I., Vlattas, I., Bohacek, R., Knap, A.K., Hirama, T., Huber, C.P. & Hasnain, S. (1995) Characterization of the $S_3$ subsite specificity of cathepsin B. *J. Biol. Chem.* **270**, 18036–18043.

Towatari, T., Nikawa, T., Murata, M., Yokoo, C., Tamai, M., Hanada, K. & Katunuma, N. (1991) Novel epoxysuccinyl peptides. A selective inhibitor of cathepsin B, in vivo. *FEBS Lett.* **280**, 311–315.

Turk, B., Dolenc, I., Zerovnik, E., Turk, D., Gubensek, F. & Turk, V. (1994) Human cathepsin B is a metastable enzyme stabilized by specific ionic interactions associated with the active site. *Biochemistry* **33**, 14800–14806.

Turk, D., Podobnik, M., Popovič, T., Katunuma, N., Bode, W., Huber, R. & Turk, V. (1995) Crystal structure of cathepsin B inhibited with CA030 at 2.0-Å resolution: a basis for the design of specific epoxysuccinyl inhibitors. *Biochemistry* **34**, 4791–4797.

Turk, D., Podobnik, M., Kuhelj, R., Dolinar, M. & Turk, V. (1996) Crystal structures of human procathepsin B at 3.2 and 3.3 Å resolution reveal an interaction motif between a papain-like cysteine protease and its propeptide. *FEBS Lett.* **384**, 211–214.

Varughese, K.I., Ahmed, F.R., Carey, P.R., Hasnain, S., Huber, C.P. & Storer, A.C. (1989) Crystal structure of a papain-E-64 complex. *Biochemistry* **28**, 1330–1332.

Wang, X., Chan, S.J., Eddy, R.L., Byers, M.G., Fukushima, Y., Henry, W.M., Haley, L.L., Steiner, D.F. & Shows, T.B. (1987) Chromosome assignment of cathepsin B (CTSB) to 8p22 and cathepsin H (CTSH) to 15q24-q25. *Cytogenet. Cell Genet.* **46**, 710–711.

Ward, W., Alvarado, L., Rawlings, N.D., Engel, J.C., Franklin, C. & McKerrow, J.H. (1997) A primitive enzyme for a primitive cell: the protease required for excystation of *Giardia*. *Cell* **89**, 437–444.

Wardale, R.J., Maciewicz, R.A. & Etherington, D.J. (1986) Monoclonal antibodies to rabbit liver cathepsin B. *Biosci. Rep.* **6**, 639–646.

*John S. Mort*
*Joint Diseases Laboratory,*
*Shriners Hospital for Children,*
*1529 Cedar Avenue,*
*Montreal, Quebec H3G 1A6, Canada*
*Email: mc60@musica.mcgill.ca*

# 210. *Cathepsin L*

## Databanks

*Peptidase classification: clan CA, family C1, MEROPS ID: C01.032*
*NC-IUBMB enzyme classification: EC 3.4.22.15*
*Chemical Abstracts Service registry number: 60616-82-2*
*Databank codes:*

| Species | SwissProt | PIR | EMBL (cDNA) | EMBL (genomic) |
|---|---|---|---|---|
| *Bombyx mori* | – | – | S77508 | – |
| *Bos taurus* | – | – | X91755 | – |
| *Drosophila melanogaster* | – | – | D31970 | |
| *Drosophila melanogaster* | – | – | U75652 | |
| *Felis catus* | P25773 | A41404 | M31652 | – |
| *Gallus gallus* | – | A25654 B25654 S00081 | – | – |
| *Gallus gallus* | P09648 | A26818 | – | – |
| *Homo sapiens* | P07711 | A26069 A45043 B27011 B32333 | M20496 X05256 X12451 | L06426: exons 1 and 2 and intron 1 |

*continued overleaf*

| Species | SwissProt | PIR | EMBL (cDNA) | EMBL (genomic) |
|---|---|---|---|---|
| | | S00323 | | |
| | | S01002 | | |
| | | S09065 | | |
| *Homarus americanus* | P13277 | S06154 | X63567 | – |
| | | S19649 | | |
| | | S31654 | | |
| *Homarus americanus* | P25782 | S19650 | X63568 | – |
| | | S31655 | | |
| *Homarus americanus* | P25784 | S19651 | X63569 | – |
| | | S20838 | | |
| | | S31656 | | |
| *Mus musculus* | P06797 | A25999 | J02583 | |
| | | A32333 | M20495 | |
| | | A34972 | X06086 | |
| | | A45927 | | |
| | | S01177 | | |
| | | S13890 | | |
| | | S48734 | | |
| *Nephrops norvegicus* | – | S47432 | X80989 | – |
| *Nephrops norvegicus* | – | S47433 | X80990 | – |
| *Ovis aries* | Q10991 | – | – | – |
| *Penaeus vannamei* | – | – | X99730 | – |
| *Penaeus vannamei* | – | S53027 | X85127 | – |
| *Rattus norvegicus* | P07154 | A41550 | Y00697 | – |
| | | S00155 | | |
| | | S02445 | | |
| | | S02446 | | |
| | | S07098 | | |
| *Sarcophaga peregrina* | – | A53810 | D16533 | – |
| *Sus scrofa* | – | – | D37917 | – |
| *Toxocara canis* | – | – | U53172 | – |

### Name and History

A new enzyme from rat liver lysosomes that degrades proteins but not synthetic substrates (Bohley *et al*., 1971; Kirschke *et al*., 1972) was given the name **cathepsin L** (in which 'L' stands for lysosomes) (Bohley *et al*., 1974; Kirschke *et al*., 1974) and characterized in detail (Kirschke *et al*., 1976, 1977). At about the same time, Towatari *et al*. (1976, 1978) isolated a new cathepsin from rat liver lysosomes which also proved to be cathepsin L. The first indication that this enzyme might be involved in the process of malignant tumor growth came from work on the precursor of cathepsin L as **major excreted protein** (MEP) (Gottesman, 1978).

### Activity and Specificity

Cathepsin L has only endopeptidase activity, and preferentially cleaves peptide bonds with hydrophobic amino acid residues in P2 and P3 (Kärgel *et al*., 1980, 1981). Accordingly, Z-Phe-Arg+NHMec and Z-Phe-Phe-$CHN_2$ proved to be sensitive synthetic substrate and inhibitor, respectively, for cathepsin L (Barrett & Kirschke, 1981; Kirschke & Shaw, 1981). Cathepsin L is catalytically active at pH 3.0–6.5 in the presence of thiol compounds. In the physiological pH range, the activity and stability of cathepsin L are strikingly dependent on ionic strength (Dehrmann *et al*., 1995).

Assays of cathepsin L with the substrate Z-Phe-Arg-NHMec (5 µM) should include DTT (1–5 mM) and EDTA (1–5 mM) at pH 5.0–5.5 (Barrett & Kirschke, 1981; Kirschke *et al*., 1988; Kirschke & Shaw, 1981; Mason *et al*., 1985). Since Z-Phe-Arg-NHMec is also hydrolyzed by cathepsins B and S, negative controls are needed, containing a specific inhibitor of cathepsin L such as Z-Phe-Phe-$CHN_2$ or Z-Phe-Tyr(tBu)-$CHN_2$ (0.5 µM, with a 10 min preincubation time). These show the activities of cathepsins B and S, which are not inhibited by these reagents under the given conditions (Barrett & Kirschke, 1981; Kirschke *et al*., 1984; Shaw *et al*., 1993). CA-074 [*N*-(L-3-*trans*-propylcarbamoyloxirane-2-carbonyl)-L-isoleucyl-L-proline], a specific inhibitor of cathepsin B, was also used in a similar assay (Inubushi *et al*., 1994).

Cathepsin L is capable of degrading nearly all proteins, activating some precursors and inactivating several enzymes. Among the substrates are cytosolic proteins (Bohley *et al*., 1976; Bohley & Seglen, 1992), collagen (Kirschke *et al*., 1982), elastin (Mason *et al*., 1986), proplasminogen activator (Goretzki *et al*., 1992) and glucose-6-phosphate dehydrogenase (Towatari *et al*., 1978). Cathepsin L has been shown to hydrolyze several proteins with the same specific activity as cathepsin S (Kirschke *et al.*, 1989).

Naturally occurring inhibitors of cathepsin L are $\alpha_2$-macroglobulin, Ha-*ras* oncogene products (Hiwasa *et al.*, 1987), antigens from mouse cytotoxic lymphocytes CTLA-2$\beta$ (Delaria *et al.*, 1994), MHC class II-associated p41 invariant chain (Bevec *et al.*, 1996) and the cystatins. The latter have a very high affinity, displaying $K_i$ values with cathepsin L in the nanomolar to picomolar range (Abrahamson, 1994). Peptide aldehydes from microbial culture filtrates, such as leupeptin, antipain and chymostatin, and synthetic peptide aldehydes, are also reversible inhibitors of cathepsin L, but inhibit serine peptidases also. Irreversible inhibitors are the peptidyl-diazomethanes (Crawford *et al.*, 1988; Kirschke *et al.*, 1988), peptidyl-*O*-acyl-hydroxamates (Brömme *et al.*, 1993), peptidyl-(acyloxy)methanes (Krantz, 1994), and the epoxysuccinyl-peptides, which have been used as active-site titrants.

## Structural Chemistry

Cathepsin L is synthesized as a preproenzyme of 333 amino acid residues with a calculated $M_r$ of 37 564. The $M_r$ values of the different forms were determined by SDS-PAGE: procathepsin L 38 000–41 000, single-chain form 28 000, heavy chain 24 000 and light chain about 4000. The cleavage between the heavy and light chains is at Asn169E-Lys (mouse, rat), Asp169E-Asn (human) and Gly169A-Gly (chicken) (in papain numbering), and the bonds hydrolyzed are in an extended loop with respect to papain, as is shown in a simulated three-dimensional structure of cathepsin L (Kirschke *et al.*, 1995). The X-ray crystallographic structure of the mature enzyme has not yet been obtained, but that of procathepsin L was described by Coulombe *et al.* (1996).

Only one of the two potential glycosylation sites (Asn106) is glycosylated in cathepsin L of mouse and rat (Stearns *et al.*, 1990). The enzymes of all species have at least three disulfide bonds (Cys22/Cys63, Cys56/Cys95, Cys153/Cys200), and mouse and rat cathepsins L have an additional one (Cys12/Cys33). Cys25, His159 and Asn175 are directly implicated in catalytic activity. The pI for mature cathepsin L is in the range of 5.0–6.3.

## Preparation

Cathepsin L appears to be ubiquitous in eukaryotic cells. It has been purified from tissues of several species and tumor cells. Kidneys especially, contains high concentrations of the enzyme, e.g. rat kidney 606 ng mg$^{-1}$ of protein (Bando *et al.*, 1986). Several purification methods have been described in detail, starting from a lysosomal fraction (Kirschke *et al.*, 1972, 1977; Towatari *et al.*, 1976, 1978), or whole-tissue extracts after autolysis at acidic pH followed by fractionation with ammonium sulfate (Mason *et al.*, 1984, 1985). Three-phase partitioning in *t*-butanol–water–ammonium sulfate has also been used (Pike & Dennison, 1989).

Cathepsin L can be separated from the other lysosomal cysteine peptidases such as cathepsins B, H and S by most cation-exchange media, to which cathepsin L shows an anomalously high affinity. Procathepsin L has been isolated from the culture medium of normal and transformed fibroblasts (Gottesman & Cabral, 1981; Gal & Gottesman, 1986; Ishidoh *et al.*, 1993).

Recombinant human cathepsin L was expressed in *Escherichia coli* (Smith & Gottesman, 1989), and mouse cathepsin L as a fusion protein also in *E. coli* (Portnoy *et al.*, 1986) and in COS cells (Tao *et al.*, 1994).

## Biological Aspects

The human cathepsin L gene maps to chromosome 9q21–22 (Fan *et al.*, 1989; Chauhan *et al.*, 1993) and consists of eight exons and seven introns. The exons of the genes of different species have the same sizes except for exons 1 and 8 (human 1.575 kbp, mouse 1.4 kbp, rat 1.411 kbp) (Ishidoh *et al.*, 1989; Chauhan *et al.*, 1993). The synthesis and secretion of cathepsin L is stimulated by malignant transformation, tumor promoters, growth factors and other compounds.

Cathepsin L is synthesized as a preproenzyme by ribosomes bound to the endoplasmic reticulum. During the entry into the lumen of the endoplasmic reticulum the signal peptide is removed, and in the Golgi network procathepsin L is modified by glycosylation, mannose phosphorylation and assembly of the disulfide bonds. The propeptide is required for proper folding, stability and transport to the Golgi apparatus (Tao *et al.*, 1994). The transport to the lysosomes is mediated by mannose-6-phosphate receptor and probably by a pH-dependent membrane association (McIntyre *et al.*, 1994). In tumor cells, active cathepsin L and its precursor have been detected in small vesicles associated with the plasma membrane (Rozhin *et al.*, 1989).

The activation of procathepsin L by limited proteolysis can occur autocatalytically (Salminen & Gottesman, 1990), at pH 3.0–3.5 (Mason *et al.*, 1987; McDonald & Kadkhodayan, 1988), at negatively charged surfaces at pH 5.0–6.0 (Mason & Massay, 1992), or by the action of cathepsin D (Nishimura *et al.*, 1989; Wiederanders & Kirschke, 1989) or metalloendopeptidases (Hara *et al.*, 1988).

Cathepsin L was suggested to have major biological roles in normal lysosomal proteolysis, and in several diseases, and the secreted enzyme from normal and tumor cells was reported to be involved in such processes as initiation of a proteinase cascade, degradation of matrix proteins, interference with normal antigen processing and promotion of proliferation processes (Kirschke *et al.*, 1995). However, studies on cathepsin L-deficient mice have not yet confirmed an essential role of cathepsin L in intracellular protein catabolism or in the proposed extracellular functions. The mice lacking cathepsin L are viable, although they are partially devoid of hair and their mortality is enhanced (Roth *et al.*, 1996).

## Distinguishing Features

The activity of cathepsin L can be distinguished from those of the related cathepsins B, S and K only by the use of semi-specific inhibitors of cathepsin L in the correct concentration range (see Activity and Specificity), because no specific substrate for cathepsin L is known.

Differentiation between the related cathepsins is also possible by use of monospecific antibodies. Cathepsin L (human kidney or liver), rabbit and sheep antisera to cathepsin L (human), and a cathepsin L ELISA kit are available from Bio-Ass (see Appendix 2 for full names and addresses of suppliers). Cathepsin L propeptide, mouse monoclonal antibody (IgG1) to procathepsin L (human), and polyclonal antibody (IgG) from rabbit to procathepsin L are available from Oncogene Science.

## References

Abrahamson, M. (1994) Cystatins. *Methods Enzymol.* **244**, 685–700.

Bando, Y., Kominami, E. & Katunuma, N. (1986) Purification and tissue distribution of rat cathepsin L. *J. Biochem.* **100**, 35–42.

Barrett, A.J. & Kirschke, H. (1981) Cathepsin B, cathepsin H, and cathepsin L. *Methods Enzymol.* **80**, 535–561.

Bevec, T., Stoka, V., Pungerčič, G., Dolenc, I. & Turk, V. (1996) Major histocompatibility complex class II-associated p41 invariant chain fragment is a strong inhibitor of lysosomal cathepsin L. *J. Exp. Med.* **183**, 1331–1338.

Bohley, P. & Seglen, P.O. (1992) Proteases and proteolysis in the lysosome. *Experientia* **48**, 151–157.

Bohley, P., Kirschke, H., Langner, J., Ansorge, S., Wiederanders, B. & Hanson, H. (1971) Intracellular protein breakdown. In: *Tissue Proteinases* (Barrett, A.J. & Dingle, J.T., eds). Amsterdam and London: North-Holland Publishing, pp. 187–219.

Bohley, P., Kirschke, H., Langner, J., Ansorge, S., Wiederanders, B. & Hanson, H. (1974) Degradation of rat liver proteins. In: *Intracellular Protein Catabolism* (Hanson, H. & Bohley, P., eds). Leipzig: J.A. Barth, pp. 201–209.

Bohley, P., Kirschke, H., Langner, J., Wiederanders, B. & Ansorge, S. (1976) Intrazellulärer Proteinabbau. VIII. Einsatz doppeltmarkierter Substratproteine. *Acta Biol. Med. Germ.* **35**, 301–307.

Brömme, D., Neumann, U., Kirschke, H. & Demuth, H.U. (1993) Novel *N*-peptidyl-*O*-acyl hydroxamates: selective inhibitors of cysteine proteinases. *Biochim. Biophys. Acta* **1202**, 271–276.

Chauhan, S.S., Popescu, N.C., Ray, D., Fleischmann, R., Gottesman, M.M. & Troen, B.R. (1993) Cloning, genomic organization, and chromosomal localization of human cathepsin L. *J. Biol. Chem.* **268**, 1039–1045.

Coulombe, R., Grochulski, P., Sivaraman, J., Menard, R., Mort, J.S., and Cygler, M. (1996) Structure of human procathepsin L reveals the molecular basis of inhibition by the prosegment. *EMBO J.* **15**, 5492–5503.

Crawford, C., Mason, R.W., Wikstrom, P. & Shaw, E. (1988) The design of peptidyldiazomethane inhibitors to distinguish between the cysteine proteinases calpain II, cathepsin L and cathepsin B. *Biochem. J.* **253**, 751–758.

Dehrmann, F.M., Coetzer, T.H.T., Pike, R.N. & Dennison, C. (1995) Mature cathepsin L is substantially active in the ionic milieu of the extracellular medium. *Arch. Biochem. Biophys.* **324**, 93–98.

Delaria, K., Fiorentino, L., Wallace, L., Tamburini, P., Brownell, E. & Muller, D. (1994) Inhibition of cathepsin L-like cysteine proteases by cytotoxic T-lymphocyte antigen-2 beta. *J. Biol. Chem.* **269**, 25172–25177.

Fan, Y.S., Byers, M.G., Eddy, R.L., Joseph, L.J., Sukhatme, V.P., Chan, S.J. & Shows, T.B. (1989) Cathepsin L (CTSL) is located in the chromosome 9q21-q22 region; a related sequence is located on chromosome 10. *Cytogenet. Cell Genet.* **51**, 996.

Gal, S. & Gottesman, M.M. (1986) The major excreted protein (MEP) of transformed mouse cells and cathepsin L have similar protease specificity. *Biochem. Biophys. Res. Commun.* **139**, 156–162.

Goretzki, L., Schmitt, M., Mann, K., Calvete, J., Chucholowski, N., Kramer, M., Gunzler, W.A., Janicke, F. & Graeff, H. (1992) Effective activation of the proenzyme form of the urokinase-type plasminogen activator (pro-uPA) by the cysteine protease cathepsin L. *FEBS Lett.* **297**, 112–118.

Gottesman, M.M. (1978) Transformation-dependent secretion of a low molecular weight protein by murine fibroblasts. *Proc. Natl Acad. Sci. USA* **75**, 2767–2771.

Gottesman, M.M. & Cabral, F. (1981) Purification and characterization of a transformation-dependent protein secreted by cultured murine fibroblasts. *Biochemistry* **20**, 1659–1665.

Hara, K., Kominami, E. & Katunuma, N. (1988) Effect of proteinase inhibitors on intracellular processing of cathepsin B, H and L in rat macrophages. *FEBS Lett.* **231**, 229–231.

Hiwasa, T., Yokoyama, S., Ha, J.M., Noguchi, S. & Sakiyama, S. (1987) c-Ha-*ras* gene products are potent inhibitors of cathepsins B and L. *FEBS Lett.* **211**, 23–26.

Inubushi, T., Kakegawa, H., Kishino, Y. & Katunuma, N. (1994) Specific assay method for the activities of cathepsin L-type cysteine proteinases. *J. Biochem.* **116**, 282–284.

Ishidoh, K., Kominami, E., Suzuki, K. & Katunuma, N. (1989) Gene structure and 5′-upstream sequence of rat cathepsin L. *FEBS Lett.* **259**, 71–74.

Ishidoh, K., Takeda-Ezaki, M. & Kominami, E. (1993) Procathepsin L-specific antibodies that recognize procathepsin L but not cathepsin L. *FEBS Lett.* **322**, 79–82.

Kärgel, H.J., Dettmer, R., Etzold, G., Kirschke, H., Bohley, P. & Langner, J. (1980) Action of cathepsin L on the oxidized B-chain of bovine insulin. *FEBS Lett.* **114**, 257–260.

Kärgel, H.J., Dettmer, R., Etzold, G., Kirschke, H., Bohley, P. & Langner, J. (1981) Action of rat liver cathepsin L on glucagon. *Acta Biol. Med. Germ.* **40**, 1139–1143.

Kirschke, H. & Shaw, E. (1981) Rapid inactivation of cathepsin L by Z-Phe-PheCHN$_2$ and Z-Phe-AlaCHN$_2$. *Biochem. Biophys. Res. Commun.* **101**, 454–458.

Kirschke, H., Langner, J., Wiederanders, B., Ansorge, S. & Bohley, P. (1972) Intrazellulärer Proteinabbau. IV. Isolierung und Charakterisierung von Peptidasen aus Rattenleberlysosomen. *Acta Biol. Med. Germ.* **28**, 305–322.

Kirschke, H., Langner, J., Wiederanders, B., Ansorge, S., Bohley, P. & Hanson, H. (1974) Cathepsin L and proteinases with cathepsin B1-like activity from rat liver lysosomes. In: *Intracellular Protein Catabolism* (Hanson, H. & Bohley, P., eds). Leipzig: J.A. Barth, pp. 210–217.

Kirschke, H., Langner, J., Wiederanders, B., Ansorge, S., Bohley, P. & Broghammer, U. (1976) Intrazellulärer Proteinabbau. VII. Kathepsin L und H: Zwei neue Proteinasen aus Rattenleberlysosomen. *Acta Biol. Med. Germ.* **35**, 285–299.

Kirschke, H., Langner, J., Wiederanders, B., Ansorge, S. & Bohley, P. (1977) Cathepsin L. A new proteinase from rat-liver lysosomes. *Eur. J. Biochem.* **74**, 293–301.

Kirschke, H., Kembhavi, A.A., Bohley, P. & Barrett, A.J. (1982) Action of rat liver cathepsin L on collagen and other substrates. *Biochem. J.* **201**, 367–372.

Kirschke, H.,Ločnikar, P. & Turk, V. (1984) Species variations amongst lysosomal cysteine proteinases. *FEBS Lett.* **174**, 123–127.

Kirschke, H., Wikstrom, P. & Shaw, E. (1988) Active center differences between cathepsins L and B: the S$_1$ binding region. *FEBS Lett.* **228**, 128–130.

Kirschke, H., Wiederanders, B., Brömme, D. & Rinne, A. (1989) Cathepsin S from bovine spleen. Purification, distribution, intracellular localization and action on proteins. *Biochem. J.* **264**, 467–473.

Kirschke, H., Barrett, A.J. & Rawlings, N.D. (1995) Proteinases 1. Lysosomal cysteine proteinases. *Protein Profile* **2**, 1587–1644.

Krantz, A. (1994) Peptidyl (acyloxy)methanes as quiescent affinity labels for cysteine proteinases. *Methods Enzymol.* **244**, 656–671.

McDonald, J.K. & Kadkhodayan, S. (1988) Cathepsin L – a latent proteinase in guinea pig sperm. *Biochem. Biophys. Res. Commun.* **151**, 827–835.

McIntyre, G.F., Godbold, G.D. & Erickson, A.H. (1994) The pH-dependent membrane association of procathepsin L is mediated by a 9-residue sequence within the propeptide. *J. Biol. Chem.* **269**, 567–572.

Mason, R.W. & Massey, S.D. (1992) Surface activation of procathepsin L. *Biochem. Biophys. Res. Commun.* **189**, 1659–1666.

Mason, R.W., Taylor, M.A.J. & Etherington, D.J. (1984) The purification and properties of cathepsin L from rabbit liver. *Biochem. J.* **217**, 209–217.

Mason, R.W., Green, G.D.J. & Barrett, A.J. (1985) Human liver cathepsin L. *Biochem. J.* **226**, 233–241.

Mason, R.W., Johnson, D.A., Barrett, A.J. & Chapman, H.A. (1986) Elastinolytic activity of human cathepsin L. *Biochem. J.* **233**, 925–927.

Mason, R.W., Gal, S. & Gottesman, M.M. (1987) The identification of the major excreted protein (MEP) from a transformed mouse fibroblast cell line as a catalytically active precursor form of cathepsin L. *Biochem. J.* **248**, 449–454.

Nishimura, Y., Kawabata, T., Furuno, K. & Kato, K. (1989) Evidence that aspartic proteinase is involved in the proteolytic processing event of procathepsin L in lysosomes. *Arch. Biochem. Biophys.* **271**, 400–406.

Pike, R. & Dennison, C. (1989) A high yield method for the isolation of sheep's liver cathepsin L. *Prep. Biochem.* **19**, 231–245.

Portnoy, D.A., Erickson, A.H., Kochan, J., Ravetch, J.V. & Unkeless, J.C. (1986) Cloning and characterization of a mouse cysteine proteinase. *J. Biol. Chem.* **261**, 14697–14703.

Roth, W., Deussing, J., Hafner, A., Schmidt, P., Schmahl, W., von Figura, K. & Peters, C. (1996) Generation of cathepsin L-deficient mice gene targeting (abstract no. 114). *11th Conference on Proteolysis and Protein Turnover, Turku.*

Rozhin, J., Wade, R.L., Honn, K.V. & Sloane, B.F. (1989) Membrane-associated cathepsin L: a role in metastasis of melanomas. *Biochem. Biophys. Res. Commun.* **164**, 556–561.

Salminen, A. & Gottesman, M.M. (1990) Inhibitor studies indicate that active cathepsin L is probably essential to its own processing in cultured fibroblasts. *Biochem. J.* **272**, 39–44.

Shaw, E., Mohanty, S., Colic, A., Stoka, V. & Turk, V. (1993) The affinity-labelling of cathepsin S with peptidyl diazomethyl ketones. Comparison with the inhibition of cathepsin L and calpain. *FEBS Lett.* **334**, 340–342.

Smith, S.M. & Gottesman, M.M. (1989) Activity and deletion analysis of recombinant human cathepsin L expressed in *Escherichia coli. J. Biol. Chem.* **264**, 20487–20495.

Stearns, N.A., Dong, J., Pan, J.X., Brenner, D.A. & Sahagian, G.G. (1990) Comparison of cathepsin L synthesized by normal and transformed cells at the gene, message, protein, and oligosaccharide levels. *Arch. Biochem. Biophys.* **283**, 447–457.

Tao, K., Stearns, N.A., Dong, J., Wu, Q. & Sahagian, G.G. (1994) The proregion of cathepsin L is required for proper folding, stability, and ER exit. *Arch. Biochem. Biophys.* **311**, 19–27.

Towatari, T., Tanaka, K., Yoshikawa, D. & Katunuma, N. (1976) Separation of a new protease from cathepsin $B_1$ of rat liver lysosomes. *FEBS Lett.* **67**, 284–288.

Towatari, T., Tanaka, K., Yoshikawa, D. & Katunuma, N. (1978) Purification and properties of a new cathepsin from rat liver. *J. Biochem.* **84**, 659–671.

Wiederanders, B. & Kirschke, H. (1989) The processing of a cathepsin L precursor *in vitro. Arch. Biochem. Biophys.* **272**, 516–521.

***Heidrun Kirschke***
*Institute of Physiological Chemistry,*
*Martin-Luther University of Halle-Wittenberg,*
*D-06097 Halle (Saale), Germany*

# 211. Cathepsin S

## Databanks

*Peptidase classification: clan CA, family C1, MEROPS ID: C01.034*
*NC-IUBMB enzyme classification: EC 3.4.22.27*

*Databank codes:*

| Species | SwissProt | PIR | EMBL (cDNA) | EMBL (genomic) |
|---------|-----------|-----|-------------|----------------|
| *Bos taurus* | P25326 | S15844 | M95211 | – |
|  |  | S16972 | X62001 |  |
|  |  | S23680 |  |  |
|  |  | S23957 |  |  |
| *Cyprinus carpio* | – | – | L30111 | – |
| *Homo sapiens* | P25774 | A42482 | M86553 | U07369: exon 1 |
|  |  | A42896 | M90696 | U07370: exon 2 |
|  |  | A53625 | S39127 | U07371: exon 3 |
|  |  |  |  | U07372: exon 4 |
|  |  |  |  | U07373: exon 5 |
|  |  |  |  | U07374: exon 6 and partial CDS |
| *Rattus norvegicus* | Q02765 | A45087 | L03201 | – |

## Name and History

The name *cathepsin S* was given to a cysteine peptidase puri-fied from cattle lymph nodes (Turnšek *et al*., 1975) and later from spleen (Ločnikar *et al*., 1981). The enzyme showed sim-ilarities to cathepsin L, but differed in a number of respects, including pI, $M_r$, pH stability, activity against synthetic sub-strates and sensitivity to inhibitors (Kirschke *et al*., 1984, 1986, 1989; Brömme *et al*., 1989). Later, studies on the amino acid sequences (Ritonja *et al*., 1991; Wiederanders *et al*., 1991) confirmed that cathepsins S and L are different enzymes.

## Activity and Specificity

Cathepsin S is an endopeptidase with no detected exopep-tidase activities. The substrate specificity is similar to that of cathepsins B and L (Chapter 209 and 210) as has been shown by tests of a series of synthetic substrates, all of which were more or less sensitive to cathepsins S, B and L (Brömme *et al*., 1989, 1993, 1994; Xin *et al*., 1992; Ménard *et al*., 1993; Kirschke & Wiederanders, 1994). However, these detailed studies also revealed significant differences in the S2 subsite specificities of the three enzymes: cathepsin S favors branched hydrophobic residues in the P2 position, cathepsin L prefers aromatic and cathepsin B basic amino acid side chains. Substrates containing the -Arg-Arg- or -Lys-Lys-sequences, which are very sensitive to cathepsin B (McDon-ald & Ellis, 1975), are resistant to cathepsins L and S.

Although no specific substrate for cathepsin S is known, the activity of the enzyme can be determined in the presence of cathepsins B, H and L by changing the assay conditions to pH 7.5 and using the most susceptible substrates such as Bz-Phe-Val-Arg+NHMec or Z-Val-Val-Arg+NHMec. The stability of active cathepsin S above pH 7.0 is a distinctive property of this enzyme. The specific assay of cathepsin S requires a preincubation time of 60 min at pH 7.5 and 40°C to destroy the activities of cathepsins L, B and H. The preincubation can be followed by incubation with the substrate in continuous or stopped assays (Kirschke *et al*., 1989; Kirschke & Wiederanders, 1994). Control experiments with E-64 may be needed to confirm that the activity of a cysteine peptidase is being measured.

The pH optimum is at pH 6.5, and about 60–70% of activity is retained after 1 h at pH 7.5 (Kirschke *et al*., 1989). Assays are made with the inclusion of DTT and EDTA.

The action of cathepsin S on the oxidized B chain of insulin is very similar to that of cathepsin L with one exception: the Tyr26-Thr27 bond is resistant to cathepsin S, even after 10 h incubation (Brömme *et al*., 1989).

Cathepsin S is capable of hydrolyzing protein substrates as fast as cathepsin L (Kirschke *et al*., 1989; Kirschke & Wiederanders, 1994). It shows collagenolytic (Kirschke *et al*., 1989) and elastinolytic (Xin *et al*., 1992) activities. Cathepsin S has the unique property among the lysosomal cysteine peptidases of degrading proteins not only at acid but also at neutral pH values.

Naturally occurring inhibitors of cathepsin S are $\alpha_2$-macroglobulin and the cystatins, which react especially rapidly with cathepsin S (Brömme *et al*., 1991; Turk *et al*., 1994).

Several classes of irreversible inhibitors, such as chloromethanes, fluoromethanes, *O*-acylhydroxylamines and peptidyl vinylsulfones inhibit the lysosomal cysteine peptidases, cathepsins B, L and S, more or less as expected from the substrate specificity (Brömme & Demuth, 1994; Shaw, 1994; Brömme *et al*., 1996). Only Z-Val-Val-Nle-$CHN_2$ proved to be a semispecific inhibitor of cathepsin S, being over 300-fold more effective in inactivating cathepsin S than cathepsin L (Shaw *et al*., 1993). E-64 and other irreversible inhibitors can be used as active-site titrants.

## Structural Chemistry

Cathepsin S is synthesized as a preproenzyme of 331 amino acid residues with a calculated $M_r$ of 37 479. The N-terminal prepro extension is 114 and 112 amino acids long in the human (Shi *et al*., 1992; Wiederanders *et al*., 1992) and rat (Petanceska & Devi, 1992) enzymes, respectively. Mature active cathepsin S is a single-chain polypeptide with $M_r$ determined by SDS-PAGE of 24 000, whereas the $M_r$ of procathepsin S is 37 000. The only potential gly-cosylation sites are located in the propeptide at position Asn119 in the human enzyme (Shi *et al*., 1992; Wiederan-ders *et al*., 1992) and Asn117 and Asn131 in the rat enzyme (Petanceska & Devi, 1992), which indicates that maturation of procathepsin S occurs after delivery to the final lysosomal location.

Cathepsin S contains three disulfide bonds (Cys22/Cys63, Cys56/Cys95, Cys153/Cys200, numbered according to papain). Cys25, His159 and Asn175 are directly implicated in catalytic activity. The pI for mature cathepsin S is in the range 6.3–7.0.

## Preparation

Cathepsin S is less abundant in tissues than are cathepsins B, L and H. The highest levels have been found in lymph nodes, spleen, macrophages and other phagocytic cells.

Purification of cathepsin S from frozen or fresh cattle spleen (Ločnikar et al., 1981; Kirschke et al., 1989; Dolenc et al., 1992; Xin et al., 1992) has been summarized and described in detail (Kirschke & Wiederanders, 1994). Methods include acid autolysis of the tissue extract, ammonium sulfate fractionation, and chromatography on CM-Sephadex C-50 for separation of cathepsin S together with cathepsin B from cathepsins L and H. Cathepsins S and B have then been separated by chromatofocusing.

Recombinant human cathepsin S has been purified after expression in transfected baby hamster kidney cells (Wiederanders et al., 1992), yeast (Brömme et al., 1993) and as a fusion protein in COS cells (Shi et al., 1992) and bacterial cells (Shi et al., 1994). The rat enzyme was expressed in bacterial cells (Petanceska & Devi, 1992). High-level expression of human cathepsin S in Sf9 cells using a baculovirus system (Brömme & McGrath, 1996) and in Escherichia coli (Kopitar et al., 1996) has recently been described.

## Biological Aspects

The human cathepsin S gene maps to chromosome 1q21 (Shi et al., 1994) and consists of six exons and six introns. Sequencing of the 5′-flanking region revealed no classical TATA or CAAT box. The presence of AP1 sites and CA microsatellites suggests that the cathepsin S gene can be specifically regulated, but no signals are known yet (Shi et al., 1994).

The preproenzyme is synthesized by ribosomes bound to the endoplasmic reticulum. During entry into the lumen of the endoplasmic reticulum the signal peptide is removed, and procathepsin S is subjected to several modifications in the Golgi network, including glycosylation, phosphorylation and assembly of the disulfide bonds. The precursor is then targeted to the lysosomes via the mannose-6-phosphate receptor. Procathepsin S can be autocatalytically activated at pH 4.5 to form active cathepsin S with a six amino acid N-terminal extension (Brömme et al., 1993).

Physiological functions of cathepsin S are apparently to contribute to the overall degradation of protein in lysosomes (Rakoczy et al., 1994), and to MHC class II-associated invariant chain processing and peptide loading (Riese et al., 1996). Cathepsin S is suggested to be implicated in the pathogenesis of several diseases such as Alzheimer's disease (Munger et al., 1995; Lemere et al., 1995) and degenerative disorders associated with the cells of the mononuclear phagocytic system (Reddy et al., 1995; Petanceska et al., 1996).

## Distinguishing Features

The amino acid sequence identities of the human procathepsins S and L (49%) and S and K (55%) are high, but monospecific antibodies do not cross-react. Cathepsins S, L and K cannot be distinguished by their substrate specificities, cathepsins S and K, in particular, showing very similar S2/P2 subsite specificity. Cathepsin S is stable and active at pH 7.5, in contrast to cathepsin L, and this allows selective assays, as described above. However, it should be noted that cathepsin K also retains some activity at pH 7.5.

## References

Brömme, D. & Demuth, H.U. (1994) N,O-Diacyl hydroxamates as selective and irreversible inhibitors of cysteine proteinases. Methods Enzymol. 244, 671–685.

Brömme, D. & McGrath, M.E. (1996) High level expression and crystallization of recombinant human cathepsin S. Protein Sci. 5, 789–791.

Brömme, D., Steinert, A., Friebe, S., Fittkau, S., Wiederanders, B. & Kirschke, H. (1989) The specificity of bovine spleen cathepsin S. A comparison with rat liver cathepsins L and B. Biochem. J. 264, 475–481.

Brömme, D., Rinne, R. & Kirschke, H. (1991) Tight-binding inhibition of cathepsin S by cystatins. Biomed. Biochim. Acta 50, 631–635.

Brömme, D., Bonneau, P.R., Lachance, P., Wiederanders, B., Kirschke, H., Peters, C., Thomas, D.Y., Storer, A.C. & Vernet, T. (1993) Functional expression of human cathepsin S in Saccharomyces cerevisiae. Purification and characterization of the recombinant enzyme. J. Biol. Chem. 268, 4832–4838.

Brömme, D., Bonneau, P.R., Lachance, P. & Storer, A.C. (1994) Engineering the $S_2$ subsite specificity of human cathepsin S to a cathepsin L- and cathepsin B-like specificity. J. Biol. Chem. 269, 30238–30242.

Brömme, D., Klaus, J.L., Okamoto, K., Rasnick, D. & Palmer, J.T. (1996) Peptidyl vinyl sulphones: a new class of potent and selective cysteine protease inhibitors. $S_2P_2$ specificity of human cathepsin O2 in comparison with cathepsins S and L. Biochem. J. 315, 85–89.

Dolenc, I., Ritonja, A., Čolić, A., Podobnik, M., Ogrinc, T. & Turk, V. (1992) Bovine cathepsins S and L: isolation and amino acid sequences. Biol. Chem. Hoppe-Seyler 373, 407–412.

Kirschke, H. & Wiederanders, B. (1994) Cathepsin S and related lysosomal endopeptidases. Methods Enzymol. 244, 500–511.

Kirschke, H., Ločnikar, P. & Turk, V. (1984) Species variations amongst lysosomal cysteine proteinases. FEBS Lett. 174, 123–127.

Kirschke, H., Schmidt, I. & Wiederanders, B. (1986) Cathepsin S. The cysteine proteinase from bovine lymphoid tissue is distinct from cathepsin L (EC 3.4.22.15). Biochem. J. 240, 455–459.

Kirschke, H., Wiederanders, B., Brömme, D. & Rinne, A. (1989) Cathepsin S from bovine spleen. Purification, distribution, intracellular localization and action on proteins. Biochem. J. 264, 467–473.

Kopitar, G., Dolinar, M., Štrukelj, B., Pungerčar, J. & Turk, V. (1996) Folding and activation of human procathepsin S from inclusion bodies produced in Escherichia coli. Eur. J. Biochem. 236, 558–562.

Lemere, C.A., Munger, J.S., Shi, G.P., Natkin, L., Haass, C., Chapman, H.A. & Selkoe, D.J. (1995) The lysosomal cysteine protease, cathepsin S, is increased in Alzheimer's disease and Down syndrome brain. An immunocytochemical study. *Am. J. Pathol.* **146**, 848–860.

Ločnikar, P., Popović, T., Lah, T., Kregar, I., Babnik, J., Kopitar, M. & Turk, V. (1981) The bovine cysteine proteinases, cathepsin B, H and S. In: *Proteinases and Their Inhibitors. Structure, Function and Applied Aspects* (Turk, V. & Vitale, L., eds). Ljubljana and Oxford: Mladinska Knjiga and Pergamon Press, pp. 109–116.

McDonald, J.K. & Ellis, S. (1975) On the substrate specificity of cathepsins B1 and B2 including a new fluorogenic substrate for cathepsin B1. *Life Sci.* **17**, 1269–1276.

Ménard, R., Carmona, E., Plouffe, C., Brömme, D., Konishi, Y., Lefebvre, J. & Storer, A.C. (1993) The specificity of the $S_2'$ subsite of cysteine proteases. *FEBS Lett.* **328**, 107–110.

Munger, J.S., Haass, C., Lemere, C.A., Shi, G.P., Wong, W.S.F., Teplow, D.B., Selkoe, D.J. & Chapman, H.A. (1995) Lysosomal processing of amyloid precursor protein to Aβ peptides: a distinct role for cathepsin S. *Biochem. J.* **311**, 299–305.

Petanceska, S., Canoll, P. & Devi, L.A. (1996) Expression of rat cathepsin S in phagocytic cells. *J. Biol. Chem.* **271**, 4403–4409.

Petanceska, S. & Devi, L. (1992) Sequence analysis, tissue distribution, and expression of rat cathepsin S. *J. Biol. Chem.* **267**, 26038–26043.

Rakoczy, P.E., Mann, K., Cavaney, D.M., Robertson, T., Papadimitreou, J. & Constable, I.J. (1994) Detection and possible functions of a cysteine protease involved in digestion of rod outer segments by retinal pigment epithelial cells. *Invest. Ophthalmol. Vis. Sci.* **35**, 4100–4108.

Reddy, V.Y., Zhang, Q.Y. & Weiss, S.J. (1995) Pericellular mobilization of the tissue-destructive cysteine proteinases, cathepsins B, L, and S, by human monocyte-derived macrophages. *Proc. Natl Acad. Sci. USA* **92**, 3849–3853.

Riese, R.J., Wolf, P.R., Brömme, D., Natkin, L.R., Villadangos, J.A., Ploegh, H.L. & Chapman, H.A. (1996) Essential role for cathepsin S in MHC class II-associated invariant chain processing and peptide loading. *Immunity* **4**, 357–366.

Ritonja, A., Čolić, A., Dolenc, I., Ogrinc, T., Podobnik, M. & Turk, V. (1991) The complete amino acid sequence of bovine cathepsin S and a partial sequence of bovine cathepsin L. *FEBS Lett.* **283**, 329–331.

Shaw, E. (1994) Peptidyl diazomethanes as inhibitors of cysteine and serine proteinases. *Methods Enzymol.* **244**, 649–656.

Shaw, E., Mohanty, S., Čolić, A., Stoka, V. & Turk, V. (1993) The affinity-labelling of cathepsin S with peptidyl diazomethyl ketones. Comparison with the inhibition of cathepsin L and calpain. *FEBS Lett.* **334**, 340–342.

Shi, G.P., Munger, J.S., Meara, J.P., Rich, D.H. & Chapman, H.A. (1992) Molecular cloning and expression of human alveolar macrophage cathepsin S, an elastinolytic cysteine protease. *J. Biol. Chem.* **267**, 7258–7262.

Shi, G.P., Webb, A.C., Foster, K.E., Knoll, J.H.M., Lemere, C.A., Munger, J.S. & Chapman H.A. (1994) Human cathepsin S: chromosomal localization, gene structure, and tissue distribution. *J. Biol. Chem.* **269**, 11530–11536.

Turk, B., Čolić, A., Stoka, V. & Turk, V. (1994) Kinetics of inhibition of bovine cathepsin S by bovine stefin B. *FEBS Lett.* **339**, 155–159.

Turnšek, T., Kregar, I. & Lebez, D. (1975) Acid sulphydryl protease from calf lymph nodes. *Biochim. Biophys. Acta* **403**, 514–520.

Wiederanders, B., Brömme, D., Kirschke, H., Kalkkinen, N., Rinne, A., Paquette, T. & Toothman, P. (1991) Primary structure of bovine cathepsin S. Comparison to cathepsins L, H, B and papain. *FEBS Lett.* **286**, 189–192.

Wiederanders, B., Brömme, D., Kirschke, H., von Figura, K., Schmidt, B. & Peters, C. (1992) Phylogenetic conservation of cysteine proteinases. Cloning and expression of a cDNA coding for human cathepsin S. *J. Biol. Chem.* **267**, 13708–13713.

Xin, X.Q., Gunesekera, B. & Mason, R.W. (1992) The specificity and elastinolytic activities of bovine cathepsins S and H. *Arch. Biochem. Biophys.* **299**, 334–339.

*Heidrun Kirschke*
*Institute of Physiological Chemistry,*
*Martin-Luther University of Halle-Wittenberg,*
*D-06097 Halle (Saale), Germany*

# 212. Cathepsin K

## Databanks

*Peptidase classification: clan CA, family C1, MEROPS ID: C01.036*
*NC-IUBMB enzyme classification: EC 3.4.22.38*

*Databank codes:*

| Species | SwissProt | PIR | EMBL (cDNA) | EMBL (genomic) |
|---|---|---|---|---|
| *Gallus gallus* | – | – | U37691 | – |
| *Homo sapiens* | P43235 | JC2476 | S79895 | – |
| | | S48830 | U13665 | |
| | | S55763 | U20280 | |
| | | | X82153 | |
| *Mus musculus* | P55097 | – | X94444 | – |
| *Oryctolagus cuniculus* | P43236 | A49868 | D14036 | – |

## Name and History

Cathepsin K, a cysteine protease predominantly expressed in osteoclasts, was first isolated through differential screening of cDNA libraries from rabbit osteoclasts and spleen (Tezuka *et al.*, 1994). The cDNA sequence, which was originally designated as *OC-2*, displayed a high degree of homology with cathepsins L and S thereby allowing the classification of the putative gene product as a papain-like cysteine protease. The human homolog of OC-2 has been cloned independently by at least five groups and referred to as *cathepsin O* (Brömme & Okamoto, 1994; Shi *et al.*, 1995; Drake *et al.*, 1996; Hastings *et al.*, 1996), *cathepsin K* (Inaoka *et al.*, 1995), *cathepsin O2* (Brömme & Okamoto, 1995) and *cathepsin X* (Li *et al.*, 1995). The name cathepsin O ('O' as in osteoclast) was altered because it had been assigned to another cathepsin (Velasco *et al.*, 1994). We subsequently chose *cathepsin O2* since high levels of transcript were found in both osteoclasts and ovary. The Nomenclature Committee of IUBMB has recently recommended the name cathepsin K (EC 3.4.22.38). However, care should be taken that the osteoclast-related cathepsin K is not confused with a high molecular weight cysteine protease to which the same name was applied earlier (Liao & Lenney, 1984). [Subsequent work suggests that this was an activity of the proteasome (Chapter 172) – *eds.*] Besides the rabbit and human forms of cathepsin K, the mouse and chicken species variants have also been cloned (Hadman *et al.*, 1996; Rantakokko *et al.*, 1996).

## Activity and Specificity

Cathepsin K is inhibited by E-64, peptidyl diazomethanes, iodoacetic acid, peptide aldehydes and chicken cystatin, but not by PMSF, pepstatin, EDTA or *o*-phenanthroline. This pattern of inhibition is consistent with the classification of cathepsin K as a cysteine protease (Brömme *et al.*, 1996a; Bossard *et al.*, 1996). Peptidyl vinylsulfones, a novel class of irreversible cysteine protease inhibitors, are powerful inactivators of cathepsin K with second-order rate constants of inactivation in the range of $10^5$–$10^6$ $M^{-1}$ $s^{-1}$ (Palmer *et al.*, 1995; Brömme *et al.*, 1996b; D. Brömme, unpublished results).

The primary specificity pocket (S2) of cathepsin K displays a preference for hydrophobic, nonaromatic side chains such as leucine, norleucine and methionine as shown with peptidyl vinylsulfones (Brömme *et al.*, 1996b). The preference for leucine over phenylalanine has also been demonstrated with dipeptidyl fluorogenic substrates (Brömme *et al.*, 1996a;

Bossard *et al.*, 1996). Like cathepsins L (Chapter 210) and S (Chapter 211), and in contrast to cathepsin B (Chapter 209), an arginine residue in P2 makes a very poor substrate for cathepsin K (Figure 212.1).

Second-order rate constants for the hydrolysis of fluorogenic peptidyl substrates by cathepsin K are comparable with rates measured for cathepsin S, but are approximately one order of magnitude lower than those for cathepsin L (Brömme *et al.*, 1996a). The pH optimum of cathepsin K for synthetic substrates is between pH 6.0 and 6.5 (Brömme *et al.*, 1996a). Like cathepsin S, cathepsin K is characterized by a bell-shaped pH activity profile. The enzyme is significantly more stable at pH 6.5 than cathepsin L, but less stable than cathepsin S; its half-life of activity is approximately 60 min at pH 6.5 and 37°C.

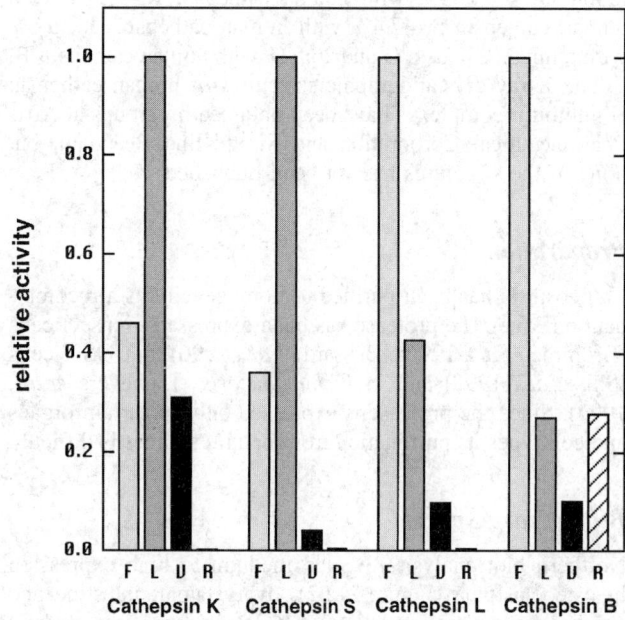

*Figure 212.1* Values of $k_{cat}/K_m$ for the hydrolysis of Z-Xaa-Arg-NHMec by cathepsins K, S, L and B. The identity of the Xaa residues is shown in the single-letter code, and data are normalized by setting the value of $k_{cat}/K_m$ for the best substrate to 1.0. These maximal values were: cathepsin K (Z-Leu-Arg-NHMec) $257\,900\,M^{-1}\,s^{-1}$; cathepsin S (Z-Leu-Arg-NHMec) $243\,000\,M^{-1}\,s^{-1}$; cathepsin L (Z-Phe-Arg-NHMec) $5\,111\,000\,M^{-1}\,s^{-1}$); cathepsin B (Z-Phe-Arg-NHMec) $460\,000\,M^{-1}\,s^{-1}$. (Reproduced from Brömme *et al.*, 1996a.).

## Structural Chemistry

Cathepsin K consists of a 15 amino acid signal sequence ($M_r$ of 1654), a 99 amino acid propeptide ($M_r$ of 11 833) and a 215 amino acid mature enzyme ($M_r$ of 23 495). Based on the amino acid sequence, the total molecular mass of the preproenzyme is 35 328. When measured in a 4–20% Tris–glycine SDS gel under reducing conditions, the apparent molecular masses of recombinant pro and mature cathepsin K are respectively, 43 000 and 29 000 (Brömme *et al.*, 1996a). These higher molecular masses are not due to *N*-glycosylation since extensive treatment with endoglycosidases H and F as well as with *N*-glycosidase F did not result in a shift in molecular weight. Human cathepsin K has two potential glycosylation sites in its mature part and one in its pro region. Both of the sites in the mature part have a proline residue adjacent either to the asparagine or to the threonine, so that they are unlikely to be glycosylated (Gavel & Heijne,1990). A crude preparation of procathepsin K expressed in *Sf* 9 cells using the baculovirus expression system can be efficiently processed to catalytically active, single-chain, mature enzyme by pepsin at pH 4.0. Processing by pepsin results in a typical N-terminal cleavage site for a papain family (Chapter 186) cysteine peptidase, with the N-terminal sequence Ala-Pro-Asp-Ser-Val-Asp containing a penultimate proline residue (Brömme *et al.*, 1996a). In contrast, autocatalytic processing at pH 4.0 generates two N-terminal sequences with one and two amino acid N-terminal extensions (Bossard *et al.*, 1996).

Mature human cathepsin K shares 95.9% identity with rabbit OC-2, 59.5% with human cathepsin L, 57.7% with human cathepsin S, 47.4% with human cathepsin H, 35.8% with human cathepsin O and 26.5 % with human cathepsin B.

The X-ray crystallographic structures of human cathepsin K–inhibitor complexes have been obtained by groups at Arris Pharmaceutical Corporation and SmithKline Beecham, but none of the structures has yet been published.

## Preparation

Cathepsin K has been purified to homogeneity as a recombinant enzyme. The protease has been expressed in insect cells (Brömme *et al.*, 1996a; Bossard *et al.*, 1996), in COS-7 cells (Shi *et al.*, 1995) and in *Pichia pastoris* (Linnevers *et al.*, 1997). Since the protease is expressed only in certain organs and cell types its purification from natural sources is difficult.

## Biological Aspects

Northern blot analysis revealed medium to high expression in osteoclastoma (Figure 212.2), ovary, small intestine and colon. Low expression levels were detected in heart, skeletal muscle, placenta, lung, prostate and testis, and very low levels in spleen, thymus, kidney, pancreas and liver. No message was detectable in brain, peripheral blood leukocytes or various leukemia cell lines (Brömme & Okamoto, 1995). Inaoka *et al.* (1995) observed similar message levels in osteoclastoma and in heart, brain, placenta, lung, liver, skeletal muscle, kidney and pancreas as described above. Rantakokko *et al.* (1996) reported very high expression levels in bone and articular cartilage of mouse, medium levels in auricular cartilage, pancreas and muscle and low levels in kidney, testis, intestine

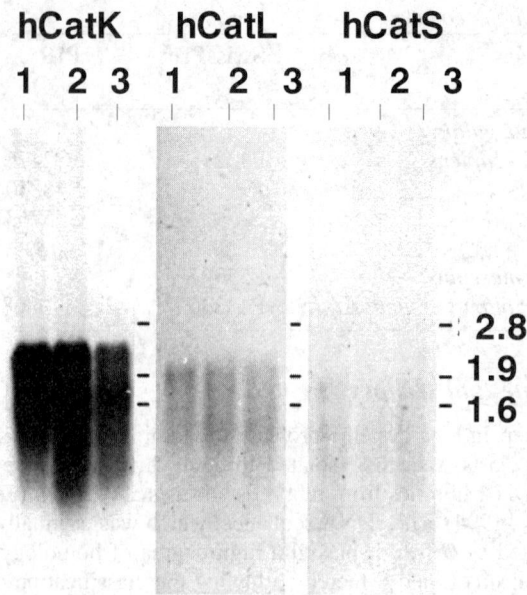

*Figure 212.2* Northern blot analysis of mRNAs for human cathepsins K, L and S in osteoclastoma preparations. Lane 1, patient 1 (fibrous and cellular tissue); lane 2, patient 2 (cellular tissue): lane 3, patient 2 (fibrous tissue). (Reproduced from Brömme & Okamoto, 1995.).

and skin. Immunohistochemical studies using a polyclonal antibody against human procathepsin K revealed strong staining in a variety of human cell types and tissues including giant multinucleated cells in osteoclastoma sections, glandular epithelial cells in colon, type II pneumocytes in lung, myocytes in heart and Kupffer cells in liver (D. Brömme, unpublished results). The immune-positive staining of these organs is in agreement with results obtained from northern blot analysis. In contrast, Drake *et al.* (1996) could detect neither expression levels nor immune-staining in human heart, lung, spleen, liver, skin or colon. The reason for this discrepancy remains unclear.

The prominent expression of cathepsin K in osteoclasts and osteoclast-related multinucleated cells suggests that it plays a key role in osteoclast-mediated bone resorption. This potential function of the protease is supported by its high collagenolytic activity. Cathepsin K efficiently degrades type I collagen at pH values between 4.5 and 6.0 (Brömme *et al.*, 1996a). Type I collagen constitutes approximately 90% of the organic bone matrix. In addition, the enzyme exhibits a high gelatinase activity (Brömme *et al.*, 1996a) and is capable of degrading osteonectin (Bosshard *et al.*, 1996).

The strongest support for a crucial role of cathepsin K in bone resorption is derived from a genetic analysis of patients suffering from the recessive autosomal disease pycnodysostosis. Gelb *et al.* (1995) and Polymeropoulos *et al.* (1995) linked this inherited trait to chromosome 1q21. In humans, pycnodysostosis is characterized by dwarfism, osteopetrosis and other severe bone abnormalities. Although very rare, the disorder gained public attention as the late French impressionist Henri de Toulouse Lautrec was thought to be afflicted by this disease (Maroteaux & Lamy, 1968; Sedano *et al.*, 1968).

Gelb *et al*. (1996) mapped cathepsin K to the same location on chromosome 1q21 and demonstrated that the human gene in pycnodysostosis patients is characterized by additional and altered stop codons. Everts *et al*. (1985) described the accumulation of collagenous fibrils in cytoplasmic vacuoles of osteoclasts in patients with this disorder. Ninety-two per cent of the osteoclasts from these patients contain intracellular collagen fibrils. It remains unclear where the primary site of action of cathepsin K is: does cathepsin K degrade collagen in the extracellular resorption lacuna, or within the lysosomes, or at both sites? Ishibashi *et al*. (1996) recently localized the protease in the vicinity of the ruffled border which may imply an extracellular localization.

Cathepsin K exhibits a very strong elastinolytic activity (Brömme *et al*., 1996a). In line with its expression in lung, the elastinolytic activity of the protease may be involved in pulmonary disorders.

Two regulators of cathepsin K gene expression at the transcriptional level have been identified. Saneshige *et al*. (1995) demonstrated the upregulation of cathepsin K in rabbit osteoclasts by all-*trans* retinoic acid. Retinoic acid is a potent stimulator of osteoclastic activity (Sanashige *et al*., 1995). On the other hand, Hadman *et al*. (1996) showed a 7- to 10-fold increase in expression of the cysteine protease after transformation of chicken embryo fibroblasts with v-*jun*, a highly transforming and tumorigenic viral oncogene.

## Distinguishing Features

Rabbit polyclonal antibodies against human cathepsin K have been generated in several laboratories (Brömme *et al*., 1996a; Drake *et al*., 1996; Hastings *et al*., 1996), but are presently not commercially available. Affinity purified anti-human cathepsin K shows no cross-reactivity with recombinant human cathepsins B, L, S, U and F on western blots (Brömme *et al*., 1996a). (Cathepsins U and F are novel, human papain-like proteases: D. Brömme, unpublished results.)

## References

Bossard, M.J., Tomaszek, T.A., Thompson, S.K., Amegadszie, B.Y., Hanning, C.R., Jones, C., Kurdyla, J.T., McNulty, D.E., Drake, F.H., Gowen, M. & Levy, M.A. (1996) Proteolytic activity of human osteoclast cathepsin K. Expression, purification, activation, and substrate specificity. *J. Biol. Chem.* **271**, 12517–12524.

Brömme, D. & Okamoto, K. (1994) Poster at the *10th International Conference on Intracellular Protein Catabolism*, Tokyo.

Brömme, D. & Okamoto, K. (1995) Human cathepsin O2, a novel cysteine protease highly expressed in osteoclastoma and ovary. Molecular cloning, sequencing and tissue distribution. *Biol. Chem. Hoppe-Seyler* **376**, 379–384.

Brömme, D., Okamoto, K., Wang, B.B. & Biroc, S. (1996a) Human cathepsin O2, a matrix protein-degrading cysteine protease expressed in osteoclasts. Functional expression of human cathepsin O2 in *Spodoptera frugiperda* and characterization of the enzyme. *J. Biol. Chem.* **271**, 2126–2132.

Brömme, D., Klaus, J.L., Okamoto, K., Rasnick, D. & Palmer, J.T. (1996b) Peptidyl vinyl sulphone: a new class of potent and selective cysteine protease inhibitors – $S_2P_2$ specificity of human cathepsin O2 in comparison with cathepsins S and L. *Biochem. J.* **315**, 85–89.

Drake, F.H., Dodds, R.A., James, I.E., Connor, J.R., Debouck, C., Richardson, S., Lee-Rykaczewski, E., Coleman, L., Rieman, D., Barthlow, R., Hastings, G. & Gowen, M. (1996) Cathepsin K, but not cathepsins B, L, or S, is abundantly expressed in human osteoclasts. *J. Biol. Chem.* **271**, 12511–12516.

Everts, V., Aronson, D.C. & Beertsen, W. (1985) Phagocytosis of bone collagen by osteoclasts in two cases of pycnodysostosis. *Calcif. Tissue Int.* **37**, 25–31.

Gavel, Y. & Heijne, G. (1990) Sequence differences between glycosylated and non-glycosylated Asn-X-Thr/Ser acceptor sites: implications for protein engineering. *Protein Eng.* **3**, 433–442.

Gelb, B.G., Edelson, J.G. & Desnick, R.J. (1995) Linkage of pycnodysostosis to chromosome 1q21 by homozygosity mapping. *Nature Genet.* **10**, 235–237.

Gelb, B.G., Shi, G.-P., Chapman, H.A. & Desnick, R.J. (1996) Pycnodysostosis, a lysosomal disease by cathepsin K deficiency. *Science* **273**, 1236–1238.

Hadman, M., Gabos, L., Loo, M., Sehgal, A. & Bos, T.J. (1996) Isolation and cloning of JTAP-1: a cathepsin like gene upregulated in response to V-Jun induced cell transformation. *Oncogene* **12**, 135–142.

Hastings, G.A., Adams, M.A., Fraser, C.M., Lee, N.H., Kirkness, E.F., Blake, J.A., Fitzgerald, L.M., Drake, F.H. & Gowan, M. (1996) Human osteoclast-derived cathepsin. *United States Patent*, Patent No. 5,501,969.

Inaoka, T., Bilbe, G., Ishibashi, O., Tezuka, K., Kumegawa, M. & Kokubo, T. (1995) Molecular cloning of cDNA for cathepsin K: novel cysteine proteinase predominantly expressed in bone. *Biochem. Biophys. Res. Commun.* **206**, 89–96.

Ishibashi, O., Inaoka, T., Togame, H., Bilbe, G., Nakamura, H., Mori, Y., Honda, Y., Hakeda, Y., Ozawa, H., Kumegawa, M. & Kokubo, T. (1996) A novel cysteine protease localized at the ruffled border – cathepsin K. *J. Bone Miner. Res.* **10**, S426.

Li, Y.P., Alexander, M.B., Wucherpfennig, A.L., Yelick, P., Chen, W. & Stashenko, P. (1995) Cloning and complete coding sequence of a novel human cathepsin expressed in giant cells of osteoclastomas. *J. Bone Miner. Res.* **10**, 1197–1202.

Liao, J.C. & Lenney, J.F. (1984) Cathepsins J and K: high molecular weight cysteine proteinases from human tissue. *Biochem. Biophys. Res. Commun.* **124**, 909–916.

Linnevers, C.J., McGrath, M.E., Armstrong, R., Mistry, F.R., Barnes, M.G., Klaus, J.L., Palmer, J.T., Katz, B.A. & Brömme, D. (1997) Expression of human cathepsin K in *Pichia pastoris* and preliminary crystallographic studies of an inhibitor complex. *Protein Sci.* **6**, 919–921.

Maroteaux, P. & Lamy, M. (1965) The malady of Toulouse-Lautrec. *JAMA* **191**, 111–113.

Maroteaux, P. & Lamy, M. (1968) [Diagnosis of chondrodystrophic dwarfism in the newborn.] *Arch. Fr. Pediatr.* **25**, 241–262.

Palmer, J.T., Rasnick, D., Klaus, J.L. & Brömme, D. (1995) Vinyl sulfones as mechanism-based cysteine protease inhibitors. *J. Med. Chem.* **38**, 3193–3196.

Polymeropoulos, M.H., Ortiz De Luna, R., Ide, S.E., Torres, R., Rubenstein, J. & Francomano, C.A. (1995) The gene for pycnodysostosis maps to human chromosome 1cen-q21. *Nature Genet.* **10**, 238–239.

Rantakokko, J., Aro, H.T., Savontaus, M. & Vuorio, E. (1996) Mouse cathepsin K: cDNA closing and predominant expression of the gene in osteoclasts, and in some hypertrophying chondrocytes during mouse development. *FEBS Lett.* **393**, 307–313.

Saneshige, S., Mano, H., Tezuka, K., Kakudo, S., Mori, Y., Honda, Y., Itabashi, A., Yamada, T., Miyata, K., Hakeda, Y., Ishii, J. & Kumegawa, M. (1995) Retinoic acid directly stimulates osteoclastic bone resorption and gene expression of cathepsin k/Oc-2. *Biochem. J.* **309**, 721–724.

Sedano, H.D., Gorlin, R.J. & Anderson, V.E. (1968) Pycnodysostosis. Clinical and genetic considerations. *Am. J. Dis. Child.* **116**, 70–77.

Shi, G.-P., Chapman, H.A., Bhairi, S., DeLeeuw, C., Reddy, V.Y. & Weiss, S.J. (1995) Molecular cloning of human cathepsin O, a novel endoproteinase and homologue of rabbit OC2. *FEBS Lett.*

**357**, 129–134.

Tezuka, K., Tezuka, Y., Maejima, A., Sato, T., Nemoto, K., Kamioka, H., Hakeda, Y. & Kumegawa, M. (1994) Molecular cloning of a possible cysteine protease predominantly expressed in osteoclasts. *J. Biol. Chem.* **269**, 1106–1109.

Velasco, G., Ferrando, A.A., Puente, X.S., Sanchez, L.M. & Lopez-Ótin, C. (1994) Human cathepsin O. Molecular cloning from a breast carcinoma, production of the active enzyme in *Escherichia coli*, and expression analysis in human tissues. *J. Biol. Chem.* **269**, 27136–27142.

*Dieter Brömme*
*Arris Pharmaceutical Corp.,*
*South San Francisco, CA 94080, USA*
*Email: bromme@arris.com*

# 213. Cathepsin H

## Databanks

*Peptidase classification: clan CA, family C1, MEROPS ID: C01.040*
*NC-IUBMB enzyme classification: EC 3.4.22.16*
*Chemical Abstracts Service registry number: 60748-73-4*
*Databank codes:*

| Species | SwissProt | PIR | EMBL (cDNA) | EMBL (genomic) |
|---|---|---|---|---|
| *Homo sapiens* | P09668 | A27011 | D28436 | – |
|  |  | C27011 | X07549 |  |
|  |  | S00322 | X16832 |  |
|  |  | S00635 |  |  |
|  |  | S00818 |  |  |
|  |  | S07634 |  |  |
|  |  | S12486 |  |  |
| *Mus musculus* | P49935 | – | U06119 | – |
| *Rattus norvegicus* | P00786 | A00976 | M36320 | – |
|  |  | A60371 | M38135 |  |
|  |  | S00211 | Y00708 |  |
|  |  | S05213 |  |  |

## Name and History

The name *cathepsin H* (Kirschke *et al.*, 1976) was given to a lysosomal peptidase first named *L20C21* (Kirschke *et al.*, 1972) and then *cathepsin B₃* (Kirschke *et al.*, 1974), because of similarities to cathepsin B (Chapter 209) (then known as cathepsin B₁). Histochemical demonstration of a thiol-dependent lysosomal enzyme active against Leu-NHNap (Sylvén, 1968) is now attributable to cathepsin H. Davidson & Poole (1975) partially purified an enzyme hydrolyzing

Bz-Arg-NHNap, but different from cathepsin B, from rat liver lysosomes. This enzyme obviously was cathepsin H. A *BANA hydrolase* from rat skin (Järvinen & Hopsu-Havu, 1975) was identified by immunological methods as cathepsin H (Rinne *et al.*, 1985). A *benzoylarginine-β-naphthylamide hydrolase* isolated from rabbit lung (Singh & Kalnitsky, 1978) was reported to degrade collagen (Singh & Kalnitsky, 1980) and was later named *cathepsin I* (Kalnitsky *et al.*, 1983). However, in later work, preparations of this enzyme from rabbit lung and rat lung did not show collagenolytic activity, but showed high cross-reactivity with an antibody to cathepsin H from rat liver (Kirschke *et al*., 1986). There was therefore no reason to retain the name cathepsin I for the enzyme from rabbit lung.

## Activity and Specificity

Cathepsin H acts as both an aminopeptidase and an endopeptidase (Kirschke *et al*., 1977; Kirschke & Barrett, 1987). Substrates for cathepsin H are unblocked amino acid derivatives such as Arg+NHMec, and among the 2-naphthylamides (-NHNap), the most sensitive to cathepsin H are those of phenylalanine, tryptophan, arginine, leucine, lysine and alanine (approximately in this order) (Takahashi *et al*., 1988). But it should be noted that the aminopeptidase substrates are susceptible to hydrolysis by noncysteine peptidases present mainly in the cytosol of all kinds of cells.

Cathepsin H hydrolyzes endopeptidase substrates such as Bz-Arg+NHNap (Singh & Kalnitsky, 1980), Bz-Arg+NHMec (Rothe & Dodt, 1992), Bz-Phe-Val-Arg+NHMec (Xin *et al*., 1992), and acts on Pro-Gly+Phe and Pro-Arg+NHNap much like a dipeptidyl-peptidase. The peptide bond specificity of the endopeptidase action of cathepsin H has not yet been determined. Cathepsin H was shown to cleave several proteins (Kirschke *et al*., 1995), but the endopeptidase activity of the enzyme is limited. Collagen and laminin, for instance, were not degraded by cathepsin H. Mature cathepsin H is irreversibly inactivated above pH 7.0, whereas the precursor is stable at such pH values. The activity can be determined with Arg-NHMec or Lys-NHMec at pH 6.5–6.8 in the presence of thiol compounds (1–5 mM), EDTA (1–5 mM) and puromycin (1 mM) or bestatin (1 μM) as described by Barrett & Kirschke (1981). Cathepsin H is unaffected by both puromycin and bestatin, whereas puromycin is a potent inhibitor of aminopeptidase PS (Chapter 343) and bestatin inhibits this and other cytosolic aminopeptidases.

Naturally occurring inhibitors of cathepsin H are the cystatins which have $K_i$ values with cathepsin H in the nanomolar to picomolar range (Barrett *et al*., 1986). Other natural inhibitors are $\alpha_2$-macroglobulin (Mason, 1989) and antigens from mouse cytotoxic lymphocytes CTLA-2$\beta$ (Delaria *et al*., 1994). Aminoacyl methanes (e.g. Leu-CH$_3$) are reversible, competitive inhibitors of cathepsin H displaying $K_i$ values in the micromolar range (Brömme *et al*., 1989). Leupeptin proved to be a poor inhibitor of this enzyme (Kirschke *et al*., 1976) in contrast to cathepsins B and L. E-64 and some related synthetic compounds are irreversible inhibitors and can be used as active-site titrants (Barrett *et al*., 1982).

## Structural Chemistry

Cathepsin H is synthesized as a preproenzyme of 335 amino acid residues with a calculated $M_r$ of 37 403. The $M_r$ values of the different forms were determined by SDS-PAGE: procathepsin H 41 000, single-chain form 28 000, heavy chain 23 000 and light chain about 5000. The pI for mature cathepsin H is in the range of 6.0–7.1. The cleavage between heavy and light chains is at Asn169A-Gly (numbering according

to papain), which is contained in an extended loop with respect to papain, as it is shown in a simulated three-dimensional structure of cathepsin H (Kirschke *et al*., 1995). It has been suggested that the newly-discovered lysosomal, peptidase, legumain (Chapter 255), may be responsible for the cleavage of the asparaginyl bond (Chen *et al*., 1997). The X-ray crystallographic structure has not yet been obtained. Procathepsin H has three potential glycosylation sites, but only one (Asn112) is used. There are the three disulfide bonds: Cys22/Cys63, Cys56/Cys95 and Cys153/Cys200, the last of which links the heavy and light chains. Cys25, His159 and Asn175 are directly involved in catalytic activity.

## Preparation

Cathepsin H appears to be ubiquitous in mammalian cells and shows a close relationship to peptidases of *Dictyostelium* and of plants. It has been purified from tissues of several species and tumor cells. Kidney especially contains high concentrations of the enzyme, e.g. rat kidney 1.4 pg mg$^{-1}$ of protein (Kominami *et al*., 1985). Several purification methods have been described in detail, starting from a lysosomal fraction (Kirschke *et al*., 1972, 1977), or whole-tissue extracts. The tissue extracts have been subjected to acidic autolysis prior to fractionation with ammonium sulfate (Singh & Kalnitsky, 1978) or acetone (Schwartz & Barrett, 1980). A one-step affinity chromatography procedure with an immobilized inhibitor from potato tubers has been described by Popovič *et al*. (1993). Cathepsin H can be separated from cathepsin L by most cation-exchange chromatography media, to which the latter shows a very high affinity. Separation from cathepsins B and S is dependent on species (Xin *et al*., 1992).

## Biological Aspects

The human cathepsin H gene maps to chromosome 15q24–25 (Wang *et al*., 1987). The gene structure of rat cathepsin H has been described as comprising at least 12 exons, consisting of 17.5 kb (Ishidoh *et al*., 1989). After synthesis of the preprocathepsin H by ribosomes bound to the endoplasmic reticulum the signal peptide is removed and the proenzyme is glycosylated, phosphorylated and the disulfide bonds are assembled in the Golgi apparatus. Procathepsin H is targeted to the lysosomes via the mannose-6-phosphate receptor. The activation of the precursor can be catalyzed by cathepsin D (Chapter 277) (Nishimura & Kato, 1988) and metalloendopeptidases (Hara *et al*., 1988).

Interferon $\gamma$ was shown to upregulate the expression of cathepsin H mRNA (Lafuse *et al*., 1995). Because this increase corresponds in time with the appearance of MHC class II and invariant chain mRNA in macrophages, cathepsin H may be involved in antigen processing. No other special functions of cathepsin H are known, although it is overexpressed in some diseases such as cancer and arthritis. It is suggested that this enzyme may act mainly in the overall degradation of proteins in the lysosomal–endosomal compartment as an aminopeptidase with broad substrate specificity.

## Distinguishing Features

Although the amino acid sequence identities of the mature human cathepsins H, L and S are about 40%, the enzymes can be distinguished by specific antibodies and differences in substrate specificity. Cathepsin H is the only lysosomal cysteine peptidase exhibiting strong aminopeptidase activity which in contrast to cytosolic aminopeptidases cannot be inhibited by puromycin or bestatin. In plants, the vacuolar enzyme aleurain (Chapter 200) has resemblances to cathepsin H.

Cathepsin H (human), sheep and rabbit antisera to human cathepsin H, and a cathepsin H ELISA kit are available from Bio-Ass (see Appendix 2 for full names and addresses of suppliers). Arg-NHMec and Lys-NHMec can be obtained from Bachem.

## References

Barrett, A.J. & Kirschke, H. (1981) Cathepsin B, cathepsin H, and cathepsin L. *Methods Enzymol.* **80**, 535–561.

Barrett, A.J., Rawlings, N.D., Davies, M.E., Machleidt, W., Salvesen, G. & Turk, V. (1986) Cysteine proteinase inhibitors of the cystatin superfamily. In: *Proteinase Inhibitors* (Barrett, A.J. & Salvesen, G., eds). Amsterdam, New York, Oxford: Elsevier, pp. 515–569.

Barrett, A.J., Kembhavi, A.A., Brown, M.A., Kirschke, H., Knight, C.G., Tamai, M. & Hanada, K. (1982) L-*trans*-Epoxysuccinyl-leucylamido(4-guanidino)butane (E-64) and its analogues as inhibitors of cysteine proteinases including cathepsins B, H and L. *Biochem. J.* **201**, 189–198.

Brömme, D., Bartels, B., Kirschke, H. & Fittkau, S. (1989) Peptide methyl ketones as reversible inhibitors of cysteine proteinases. *J. Enzym. Inhib.* **3**, 13–21.

Chen, J.-M., Dando, P.M., Rawlings, N.D., Brown, M.A., Young, N.E., Stevens, R.A., Hewitt, E., Watts, C. & Barrett, A.J. (1997) Cloning, isolation, and characterization of mammalian legumain, an asparaginyl endopeptidase. *J. Biol. Chem.* **272**, 8090–8098.

Davidson, E. & Poole, B. (1975) Fractionation of the rat liver enzymes that hydrolyze benzoyl-arginine-2-naphthylamide. *Biochim. Biophys. Acta* **397**, 437–442.

Delaria, K., Fiorentino, L., Wallace, L., Tamburini, P., Brownell, E. & Muller, D. (1994) Inhibition of cathepsin L-like cysteine proteases by cytotoxic T-lymphocyte antigen-28. *J. Biol. Chem.* **269**, 25172–25177.

Hara, K., Kominami, E. & Katunuma, N. (1988) Effect of proteinase inhibitors on intracellular processing of cathepsin B, H and L in rat macrophages. *FEBS Lett.* **231**, 229–231.

Ishidoh, K., Kominami, E., Katunuma, N. & Suzuki, K. (1989) Gene structure of rat cathepsin H. *FEBS Lett.* **253**, 103–107.

Järvinen, M. & Hopsu-Havu, V.K. (1975) α-N-Benzoylarginine-2-naphthylamide hydrolase (cathepsin B1?) from rat skin. II. Purification of the enzyme and demonstration of two inhibitors in the skin. *Acta Chem. Scand. B* **29**, 772–780.

Kalnitsky, G., Chatterjee, R., Singh, H., Lones, M. & Paszkowski, A. (1983) Bifunctional activities and possible modes of regulation of some lysosomal cysteinyl proteases. In: *Proteinase Inhibitors. Medical and Biological Aspects* (Katunuma, N., Umezawa, H. & Holzer, H., eds). Tokyo: Japan Scientific Societies Press; Berlin: Springer-Verlag, pp. 263–273.

Kirschke, H. & Barrett, A.J. (1987) Chemistry of lysosomal proteases. In: *Lysosomes: Their Role in Protein Breakdown* (Glaumann, H. & Ballard, F.J., eds). London: Academic Press, pp. 193–238.

Kirschke, H., Langner, J., Wiederanders, B., Ansorge, S. & Bohley, P. (1972) Intrazellulärer Proteinabbau. IV. Isolierung und Charakterisierung von Peptidasen aus Rattenleberlysosomen. *Acta Biol. Med. Germ.* **28**, 305–322.

Kirschke, H., Langner, J., Wiederanders, B., Ansorge, S., Bohley, P. & Hanson, H. (1974) Cathepsin L and proteinases with cathepsin B₁-like activity from rat liver lysosomes. In: *Intracellular Protein Catabolism* (Hanson, H. & Bohley, P., eds). Leipzig: J.A. Barth, pp. 210–217.

Kirschke, H., Langner, J., Wiederanders, B., Ansorge, S., Bohley, P. & Broghammer, U. (1976) Intrazellulärer Proteinabbau. VII. Kathepsin L und H: Zwei neue Proteinasen aus Rattenleberlysosomen. *Acta Biol. Med. Germ.* **35**, 285–299.

Kirschke, H., Langner, J., Wiederanders, B., Ansorge, S., Bohley, P. & Hanson, H. (1977) Cathepsin H: An endoaminopeptidase from rat liver lysosomes. *Acta Biol. Med. Germ.* **36**, 185–199.

Kirschke, H., Pepperle, M., Schmidt, I. & Wiederanders, B. (1986) Are there species differences amongst the lysosomal cysteine proteinases? *Biomed. Biochim. Acta* **45**, 1441–1446.

Kirschke, H., Barrett, A.J. & Rawlings, N.D. (1995) Proteinases 1. Lysosomal cysteine proteinases. *Protein Profile* **2**, 1587–1644.

Kominami, E., Tsukahara, T., Bando, Y. & Katunuma, N. (1985) Distribution of cathepsins B and H in rat tissues and peripheral blood cells. *J. Biochem.* **98**, 87–93.

Lafuse, W.P., Brown, D., Castle, L. & Zwilling, B.S. (1995) IFN-γ increases cathepsin H mRNA levels in mouse macrophages. *J. Leukoc. Biol.* **57**, 663–669.

Mason, R.W. (1989) Interaction of lysosomal cysteine proteinases with α₂-macroglobulin: conclusive evidence for the endopeptidase activities of cathepsins B and H. *Arch. Biochem. Biophys.* **273**, 367–374.

Nishimura, Y. & Kato, K. (1988) Identification of latent procathepsin H in microsomal lumen: characterization of proteolytic processing and enzyme activation. *Arch. Biochem. Biophys.* **260**, 712–718.

Popovič, T., Brzin, J., Ritonja, A., Svetič, B. & Turk, V. (1993) Rapid affinity chromatographic method for the isolation of human cathepsin H. *J. Chromatogr.* **615**, 243–249.

Rinne, A., Kirschke, H., Järvinen, M., Hopsu-Havu, V.K., Wiederanders, B. & Bohley, P. (1985) Localization of cathepsin H and its inhibitor in the skin and other stratified epithelia. *Arch. Dermatol. Res.* **277**, 190–194.

Rothe, M. & Dodt, J. (1992) Studies on the aminopeptidase activity of rat cathepsin H. *Eur. J. Biochem.* **210**, 759–764.

Schwartz, W.N. & Barrett, A.J. (1980) Human cathepsin H. *Biochem. J.* **191**, 487–497.

Singh, H. & Kalnitsky, G. (1978) Separation of a new α-N-benzoylarginine-β-naphthylamide hydrolase from cathepsin B1. Purification, characterization, and properties of both enzymes from rabbit lung. *J. Biol. Chem.* **253**, 4319–4326.

Singh, H. & Kalnitsky, G. (1980) α-N-benzoylarginine-β-naphthylamide hydrolase, an aminoendopeptidase from rabbit lung. *J. Biol. Chem.* **255**, 369–374.

Sylvén, B. (1968) Lysosomal enzyme activity in the interstitial fluid of solid mouse tumour transplants. *Eur. J. Cancer* **4**, 463–474.

Takahashi, T., Dehdarani, A.H. & Tang, J. (1988) Porcine spleen cathepsin H hydrolyzes oligopeptides solely by aminopeptidase activity. *J. Biol. Chem.* **263**, 10952–10957.

Wang, X., Chan, S.J., Eddy, R.L., Byers, M.G., Fukushima, Y., Henry, W.M., Haley, L.L., Steiner, D.F. & Shows, T.B. (1987)

Chromosome assignment cathepsin B (CTSB) to 8p22 and cathepsin H (CTSH) to 15q24-q25. *Cytogenet. Cell. Genet.* **46**, 710–711.

Xin, X.Q., Gunesekera, B. & Mason, R.W. (1992) The specificity and elastinolytic activities of bovine cathepsins S and H. *Arch. Biochem. Biophys.* **299**, 334–339.

*Heidrun Kirschke*
*Institute of Physiological Chemistry,*
*Martin-Luther University of Halle-Wittenberg,*
*D-06097 Halle (Saale), Germany*

# 214. *Dipeptidyl-peptidase I*

## Databanks

*Peptidase classification: clan CA, family C1, MEROPS ID: C01.070*
*NC-IUBMB enzyme classification: EC 3.4.14.1*
*Chemical Abstracts Service registry number: 9032-68-2*
*Databank codes:*

| Species | SwissProt | PIR | EMBL (cDNA) | EMBL (genomic) |
|---|---|---|---|---|
| *Homo sapiens* | P53634 | – | X87212 | – |
| *Rattus norvegicus* | P80067 | A41158 | D90404 | – |
| | | S23084 | | |
| | | S23953 | | |
| *Schistosoma mansoni* | – | – | Z32531 | – |

## Name and History

A peptidase first discovered in pig kidney by Gutman & Fruton (1948) and described as having chymotrypsin-like specificity was later termed *cathepsin C* (Tallan *et al.*, 1952; Fruton & Mycek, 1956; Izumiya & Fruton, 1956; De la Haba *et al.*, 1959). The enzyme was also known as *glucagon-degrading enzyme* (Kakiuchi & Tomizawa, 1964) and *dipeptidyl transferase* (Metrione *et al.*, 1966). McDonald *et al.* (1966b) renamed cathepsin C *dipeptidyl arylamidase I* and later (McDonald *et al.*, 1969a) *dipeptidyl aminopeptidase I*, in recognition of its exopeptidase activity. In semisystematic enzyme nomenclature, the term *dipeptidyl peptidase* describes the reaction in which an N-terminal dipeptide is hydrolyzed from a polypeptide. Consistent with this, the name *dipeptidyl-peptidase I* (here abbreviated *DPP I*) was recommended by the Nomenclature Committee of the IUBMB in 1981, together with the number EC 3.4.14.1. A high molecular mass, thiol-dependent enzyme that hydrolyzed Z-Phe-Arg-NHMec was termed *cathepsin J* by Liao & Lenney (1984), but was later shown to be identical to DPP I by Nikawa *et al.* (1992).

## Activity and Specificity

DPP I sequentially removes dipeptides from the unsubstituted N-termini of polypeptide substrates with broad specificity (McDonald *et al.*, 1971; McDonald & Schwabe, 1977). However, the enzyme does not cleave substrates containing a basic amino acid (Arg or Lys) in the N-terminal position or Pro on either side of the scissile bond (McDonald *et al.*, 1969a,b; McGuire *et al.*, 1992). The enzyme also possesses transferase activity (Planta *et al.*, 1964; Metrione *et al.*, 1966; Heinrich & Fruton, 1968; McDonald *et al.*, 1969a; McGuire *et al.*, 1992). The pH optimum for peptide hydrolysis is 5.0–6.0, whereas pH 6.8–7.7 is optimal for the transferase activity (McDonald & Schwabe, 1977). The enzyme requires halide ions and a thiol-reducing compound for activity (Fruton & Mycek, 1956; Kakiuchi & Tomizawa, 1964; McDonald *et al.*, 1966a,b). Most common assay substrates are derivatives of Gly-Phe- and Gly-Arg-dipeptides (McDonald & Schwabe, 1977; Kirschke *et al.*, 1995). Having unblocked N-termini, the substrates are also more or less susceptible to aminopeptidases. High concentrations of the substrate Gly-Phe+2-(4-methoxy)naphthylamide induced inhibition in the pH range 3.0–8.0, suggesting cooperative behavior of subunits (Dolenc *et al.*, 1995). Most of the substrates are available from Bachem (see Appendix 2 for full names and addresses of suppliers).

In general, DPP I shows an absolute requirement for a free N-terminus in its substrates, but exceptionally, it hydrolyzes Z-Phe-Arg+NHMec. This was the activity of 'cathepsin J' (Liao & Lenney, 1984) that was shown by Nikawa *et al.* (1992) and Kuribayashi *et al.* (1993) to be due to DPP

I. The exceptional nature of this activity was illustrated when Nikawa *et al.* (1992) tested 18 other blocked peptidyl-NHMec substrates with DPP I, and found that none was hydrolyzed at even 3% of the rate determined for Z-Phe-Arg-NHMec. Neither McDonald *et al.* (1969a) nor McGuire *et al.* (1992) detected any hydrolysis of a large range of N-terminally blocked peptidyl-naphthylamides, -nitroanilides or peptides. Hydrolysis of Z-Phe-Arg-NHMec is not in itself endopeptidase activity, because the bond being hydrolyzed is not a peptide bond, but Kuribayashi *et al.* (1993) do show evidence of true endopeptidase activity on a derivative of lysozyme. There has been no report of endopeptidase activity of DPP I on any natural protein, and the indications are that any such activity must be very slight. This is because it can be expected that endopeptidase activity would have been seen in the thorough study of the action of DPP I on polypeptide hormones by McDonald and coworkers (McDonald *et al.*, 1969a) and in the use of the enzyme for protein sequencing (Lindley, 1972; McDonald *et al.*, 1972; see Kirschke *et al.*, 1995) and for trimming polypeptide chains (see Kirschke *et al.*, 1995). DPP I was used by Kummer *et al.* (1996) to convert the propeptide of granzyme A to the active enzyme, by removal of the N-terminal dipeptide, and no further action on the granzyme was seen.

DPP I is specifically inhibited by Gly-Phe-diazomethane (Green & Shaw, 1981). It is also weakly inhibited by E-64 and leupeptin (Nikawa *et al.*, 1992; Kuribayashi *et al.*, 1993) and somewhat more strongly by two protein inhibitors from the cystatin superfamily, stefin A (Nikawa *et al.*, 1992) and chicken cystatin (Nicklin & Barrett, 1984; Dolenc *et al.*, 1996). E-64 is commercially available (Sigma, Peptide Research Institute).

## Structural Chemistry

DPP I is synthesized as a single-chain preproenzyme (Ishidoh *et al.*, 1991; Pariš *et al.*, 1995), which consists of a signal peptide, the propeptide and the precursor of the catalytic unit, which is cleaved into one heavy and one light chain (Muno *et al.*, 1993). Both human and rat variants of DPP I have four potential *N*-glycosylation sites (Ishidoh *et al.*, 1991; Pariš *et al.*, 1995), at least two of which are glycosylated in the monomeric form (Nikawa *et al.*, 1992). In addition, two potential tyrosine sulfation sites (residues 75, 340), four potential protein kinase C phosphorylation sites (residues 48, 138, 164, 209) and four potential myristoylation sites (residues 2, 44, 111, 256) are found in the human enzyme (preprocathepsin C numbering: Pariš *et al.*, 1995). DPP I oligomerizes just before entering the lysosomes (Muno *et al.*, 1993). The number of subunits and subunit composition are still a matter of debate, but the enzyme is most probably composed of four identical subunits, each composed of three different polypeptide chains: a heavy and a light chain, similar to other papain-like cysteine proteinases, and a propeptide part (Dolenc *et al.*, 1995). DPP I has an $M_r$ of 160 000–230 000 as determined by gel filtration (Nikawa *et al.*, 1992; Dolenc *et al.*, 1995) and sedimentation equilibrium (De la Haba *et al.*, 1959; Metrione *et al.*, 1966, 1970). The pI of the mature enzyme is 5.0–6.0 (McDonald & Schwabe, 1977; McGuire *et al.*, 1992; Lynn & Labow, 1984; Dolenc *et al.*, 1995).

## Preparation

DPP I is widely distributed in mammalian tissues, spleen and kidney being the richest sources (Bouma & Gruber, 1966; Vanha-Perttula & Kalliomäki, 1973; Kominami *et al.*, 1992). Most procedures have been based on that of Metrione *et al.* (1966), which includes several gel-filtration and ion-exchange chromatography steps (see McDonald & Schwabe, 1977; Kirschke *et al.*, 1995). More recently, human DPP I has been purified in high yield by another method, based on affinity chromatography on chicken cystatin-Sepharose (Dolenc *et al.*, 1996).

## Biological Aspects

DPP I is a lysosomal cysteine peptidase that has been found in a number of mammals (see McDonald & Schwabe, 1977; Kirschke *et al.*, 1995), as well as in lower animals (Butler *et al.*, 1995). Its main function is protein degradation in the lysosomes. In addition, DPP I has been found to participate in the activation of neuraminidase (D'Agrosa & Callahan, 1988) and factor XIII (Lynch & Pfueller, 1988). DPP I is also responsible for the activation of the proenzymes of serine proteinases (leukocyte elastase, cathepsin G, granzyme A) (Chapters 15, 16 and 22) in the lysosomal granules of cytotoxic lymphocytes, mast cells and myelocytic cells, where it is present in high concentrations (Thiele & Lipsky, 1990; McGuire *et al.*, 1992, 1993).

It has been suggested that DPP I is involved in several pathological events, such as Duchenne muscular dystrophy (Aoyagi *et al.*, 1983; Gelman *et al.*, 1980) and basal cell carcinomas (Schlagenauff *et al.*, 1992).

## Distinguishing Features

DPP I differs from most other relatives of papain in family C1 in its large molecular size, oligomeric structure, retention of a large part of the propeptide in the mature form, and requirement for halide ions for activity (McDonald & Schwabe, 1977; Kirschke *et al.*, 1995; Dolenc *et al.*, 1995). The enzyme is also exceptionally weakly inhibited by leupeptin and E-64, and is unusual amongst lysosomal cysteine endopeptidases in being stable above neutral pH.

The easiest way to distinguish DPP I from dipeptidyl-peptidase II (Chapter 138) is to use a substrate with a basic N-terminal residue, which is hydrolyzed only by dipeptidyl-peptidase II. In addition, the activity of DPP I can be potentiated by the use of sulfhydryl reagents and halide ions, which have no effect on the activity of dipeptidyl-peptidase II, a serine peptidase. Other dipeptidyl peptidases are not lysosomal enzymes (McDonald & Schwabe, 1977; McDonald & Barrett, 1986).

Polyclonal antibodies against DPP I have been prepared (Muno *et al.*, 1993), but no antibodies are commercially available at the time of writing.

## Further Reading

An historical overview of DPP I was provided by McDonald *et al.* (1971), and excellent overviews of early work on the enzyme are those of McDonald & Schwabe (1977) and

McDonald & Barrett (1986). A recent account is that of Kirschke *et al*. (1995).

## References

Aoyagi, T., Wada, T., Kojima, F., Nagai, M., Miyoshino, S. & Umezawa, H. (1983) Two different modes of enzymatic changes in serum with progression of Duchenne muscular dystrophy. *Clin. Chim. Acta* **129**, 165–173.

Bouma, J.M.W. & Gruber, M. (1964) The distribution of cathepsin B and C in rat tissues. *Biochim. Biophys. Acta* **89**, 545–547.

Bouma, J.M.W. & Gruber, M. (1966) Intracellular distribution of cathepsin B and cathepsin C in rat liver. *Biochim. Biophys. Acta* **113**, 350–358.

Butler, R., Michel, A., Kunz, W. & Klinkert, M.Q. (1995) Sequence of *Schistosoma mansoni* cathepsin C and its structural comparison with papain and cathepsins B and L of the parasite. *Protein Pept. Lett.* **2**, 313–320.

D'Agrosa, M.R. & Callahan, J.W. (1988) *In vitro* activation of neuraminidase in the β-galactosidase-neuraminidase-protective protein complex by cathepsin C. *Biochem. Biophys. Res. Commun.* **157**, 770–775.

De la Haba, G., Cammarata, P.S. & Timasheff, S.N. (1959) The partial purification and some physical properties of cathepsin C from beef spleen. *J. Biol. Chem.* **234**, 316–319.

Dolenc, I., Turk, B., Pungerčič, G., Ritonja, A. & Turk, V. (1995) Oligomeric structure and substrate-induced inactivation of cathepsin C. *J. Biol. Chem.* **270**, 21626–21631.

Dolenc, I., Turk, B., Kos, J. & Turk, V. (1996) Interaction of human cathepsin C with chicken cystatin. *FEBS Lett.* **392**, 277–280.

Fruton, J.S. & Mycek, M.J. (1956) Studies on beef spleen cathepsin C. *Arch. Biochem. Biophys.* **65**, 11–20.

Gelman, B.B., Papa, L., Davis, M.H. & Gruenstein, E. (1980) Decreased lysosomal dipeptidyl aminopeptidase I activity in cultured human skin fibroblasts in Duchenne's muscular dystrophy. *J. Clin. Invest.* **65**, 1398–1406.

Gutman, H.R. & Fruton, J.S. (1948) On the proteolytic enzymes of animal tissues VIII. An intracellular enzyme related to chymotrypsin. *J. Biol. Chem.* **174**, 851–858.

Green, G.D.J. & Shaw, E. (1981) Peptidyl diazomethyl ketones are specific inactivators of thiol proteinases. *J. Biol. Chem.* **256**, 1923–1928.

Heinrich, C.P. & Fruton, J.S. (1968) The action of dipeptidyl transferase as a polymerase. *Biochemistry* **7**, 3556–3565.

Ishidoh, K., Muno, D., Sato, N. & Kominami, E. (1991) Molecular cloning of cDNA for rat cathepsin C. Cathepsin C, a cysteine proteinase with an extremely long propeptide. *J. Biol. Chem.* **266**, 16312–16317.

Izumiya, N. & Fruton, J.S. (1956) Specificity of cathepsin C. *J. Biol. Chem.* **218**, 59–76.

Kakiuchi, S. & Tomizawa, H.H. (1964) Properties of a glucagon-degrading enzyme of beef liver. *J. Biol. Chem.* **239**, 2160–2164.

Kirschke, H., Barrett, A.J. & Rawlings, N.D. (1995) Proteinases 1. Lysosomal cysteine proteinases. *Protein Profile* **2**, 1587–1643.

Kominami, E., Ishidoh, K., Muno, D. & Sato, N. (1992) The primary structure and tissue distribution of cathepsin C. *Biol. Chem. Hoppe-Seyler* **373**, 367–373.

Kummer, J.A., Kamp, A.M., Citarella, F., Horrevoets, A.J.G. & Hack, C.E. (1996) Expression of human recombinant granzyme A zymogen and its activation by the cysteine proteinase cathepsin C. *J. Biol. Chem.* **271**, 9281–9286.

Kuribayashi, M., Yamada, H., Ohmori, T., Yanai, M. & Imoto, T.

(1993) Endopeptidase activity of cathepsin C, dipeptidyl aminopeptidase I, from bovine spleen. *J. Biochem.* **113**, 441–449.

Liao, J.C.R. & Lenney, J.F. (1984) Cathepsins J and K: high molecular weight cysteine proteinases from human tissues. *Biochem. Biophys. Res. Commun.* **124**, 909–916.

Lindley, H. (1972) The specificity of dipeptidyl aminopeptidase I (cathepsin C) and its use in peptide sequence studies. *Biochem. J.* **126**, 683–688.

Lynch, G.W. & Pfueller, S.L. (1988) Thrombin-independent activation of platelet factor XIII by endogeneous platelet acid protease. *Thromb. Haemost.* **59**, 372–377.

Lynn, K.R. & Labow, R.S. (1984) A comparison of four sulphydryl cathepsins (B, C, H, and L) from porcine spleen. *Can. J. Biochem. Cell. Biol.* **62**, 1301–1308.

McDonald, J.K. & Barrett, A.J. (1986) In: *Mammalian Proteases. A Glossary and Bibliography*, vol. 2. London: Academic Press, pp. 111–119.

McDonald, J.K. & Schwabe, C. (1977) Intracellular exopeptidases. In: *Proteinases in Mammalian Cells and Tissues* (Barrett, A.J., ed.). Amsterdam: North Holland Publishing, pp. 311–391.

McDonald, J.K., Reilly, T.J., Zeitman, B.B. & Ellis, S. (1966a) Cathepsin C: a chloride-requiring enzyme. *Biochem. Biophys. Res. Commun.* **24**, 771–775.

McDonald, J.K., Ellis, S. & Reilly, T.J. (1966b) Properties of dipeptidyl arylamidase I of the pituitary. Chloride and sulfhydryl activation of seryltyrosyl-β-naphthylamide hydrolysis. *J. Biol. Chem.* **241**, 1494–1501.

McDonald, J.K., Zeitman, B.B., Reilly, T.J. & Ellis, S. (1969a) New observations on the substrate specificity of cathepsin C (dipeptidyl aminopeptidase I) including the degradation of β-corticotropin and other peptide hormones. *J. Biol. Chem.* **244**, 2693–2709.

McDonald, J.K., Callahan, P.X., Zeitman, B.B. & Ellis, S. (1969b) Inactivation and degradation of glucagon by dipeptidyl aminopeptidase I (cathepsin C) of rat liver. Including a comparative study of secretin degradation. *J. Biol. Chem.* **244**, 6199–6208.

McDonald, J.K, Callahan, P.X., Ellis, S. & Smith, R.E. (1971) Polypeptide degradation by dipeptidyl aminopeptidase I (cathepsin C) and related peptidases. In: *Tissue Proteinases* (Barrett, A.J. & Dingle, J.T., eds). Amsterdam: North-Holland Publishing, pp. 69–107.

McDonald, J.K., Zeitman, B.B. & Ellis, S. (1972) Detection of a lysosomal carboxypeptidase and a lysosomal dipeptidase in highly-purified aminopeptidase I (cathepsin C) and the elimination of their activities from preparations used to sequence peptides. *Biochem. Biophys. Res. Commun.* **46**, 62–70.

McGuire, M.J., Lipsky, P.E. & Thiele, D.L. (1992) Purification and characterization of dipeptidyl-peptidase I from human spleen. *Arch. Biochem. Biophys.* **295**, 280–288.

McGuire, M.J., Lipsky, P.E. & Thiele, D.L. (1993) Generation of myeloid and lymphoid granule serine proteases requires processing by the granule thiol protease dipeptidyl-peptidase I. *J. Biol. Chem.* **268**, 2458–2467.

Metrione, M.R., Neves, A.G. & Fruton, J.S. (1966) Purification and properties of dipeptidyl transferase (cathepsin C). *Biochemistry* **5**, 1597–1604.

Metrione, M.R., Okuda, Y. & Fairclough, G.F., Jr. (1970) Subunit structure of dipeptidyl transferase. *Biochemistry* **9**, 2427–2432.

Muno, D., Ishidoh, K., Ueno, T. & Kominami, E. (1993) Processing and transport of the precursor of cathepsin C during its transfer into lysosomes. *Arch. Biochem. Biophys.* **306**, 2321–2330.

C

Nicklin, M.J.H. & Barrett, A.J. (1984) Inhibition of cysteine proteinases and dipeptidyl-peptidase I by egg-white cystatin. *Biochem. J.* **223**, 245–253.

Nikawa, T., Towatari, T. & Katunuma, N. (1992) Purification and characterization of cathepsin J from rat liver. *Eur. J. Biochem.* **204**, 381–393.

Pariš, A., Štrukelj, B., Pungerčar, J., Renko, M., Dolenc, I. & Turk, V. (1995) Molecular cloning and sequence analysis of human preprocathepsin C. *FEBS Lett.* **369**, 326–330.

Planta, R.J., Gorter, J. & Gruber, M. (1964) The catalytic properties of cathepsin C. *Biochim. Biophys. Acta* **89**, 511–519.

Schlagenauff, B., Klessen, C., Teichmann-Dörr, S., Breuninger, H. & Rassner, G. (1992) Demonstration of proteases in basal cell carcinomas. A histochemical study using amino acid-4-methoxy-2-naphthylamides as chromogenic substrates. *Cancer* **70**, 1133–1140.

Tallan, H.H., Jones, M.E. & Fruton, J.S. 1952. On the proteolytic enzymes of animal tissues. X. Beef spleen cathepsin C. *J. Biol. Chem.* **194**, 793–805.

Thiele, D.L. & Lipsky, P.E. (1990) Mechanism of L-leucyl-L-leucine methyl ester-mediated killing of cytotoxic lymphocytes: dependence on a lysosomal thiol protease, dipeptidyl-peptidase I, that is enriched in these cells. *Proc. Natl Acad. Sci. USA* **87**, 83–87.

Vanha-Perttula, T. & Kalliomäki, J.L. (1973) Comparison of dipeptide arylamidase I and II. Amino acid arylamidase and acid phosphatase activities in normal and pathological human sera. *Clin. Chim. Acta* **44**, 249–258.

*Boris Turk*
Department of Biochemistry and
Molecular Biology,
J. Stefan Institute, Jamova 39,
1000 Ljubljana, Slovenia

*Iztok Dolenc*
Department of Biochemistry and
Molecular Biology,
J. Stefan Institute, Jamova 39,
1000 Ljubljana, Slovenia

*Vito Turk*
Department of Biochemistry and
Molecular Biology,
J. Stefan Institute, Jamova 39,
1000 Ljubljana, Slovenia
Email: vito.turk@ijs.si

# 215. Bleomycin hydrolase

## Databanks

Peptidase classification: clan CA, family C1, MEROPS ID: C01.085
NC-IUBMB enzyme classification: none
Databank codes:

| Species | SwissProt | PIR | EMBL (cDNA) | EMBL (genomic) |
|---|---|---|---|---|
| *Homo sapiens* | – | – | X92106 | – |
| *Oryctolagus cuniculus* | P13019 | A32972 | J02866 | – |
| *Rattus* sp. | P70645 | – | D87336 | – |
| *Saccharomyces cerevisiae*[a] | Q01532 | S25606 | M97910 X68228 X69124 | – |

[a]The database records for the *Saccharomyces* sequence start from an incorrect start codon, upstream of the correct one, as is shown by the crystallographic structure, and discussed below.

Brookhaven Protein Data Bank three-dimensional structures:

| Species | ID | Resolution | Notes |
|---|---|---|---|
| *Saccharomyces cerevisiae* | 1GCB | 2.2 | |

## Name and History

The bleomycins are a family of glycometallopeptides produced by *Streptomyces verticillus* that are used as anticancer drugs because of their ability to catalytically cleave DNA and RNA, and so cause cell death (Stubbe & Kozarich, 1987; Kane & Hecht, 1994). A bleomycin-inactivating enzyme discovered in mice hydrolyzed the carboxyamide bond of the β-aminoalanine moiety of the drug (Umezawa *et al.*, 1972), and was therefore named **bleomycin hydrolase**. Homologs of this enzyme were cloned later from yeast and bacteria and were given various names. They were all shown to be intracellular homo-hexameric cysteine proteases with a papain-like catalytic triad. The enzyme was cloned from several lactic acid bacteria and characterized as a thiol aminopeptidase called **PepC** (Chapter 216) (Chapot-Chartier *et al.*, 1993, 1994; Wohlrab & Bockelmann, 1993; Fernandez *et al.*, 1994; Vesanto *et al.*, 1994).

In yeast, four groups cloned the gene by completely different means and gave it different names: Kambouris *et al.* (1992) named it **YCP1** (for **y**east **c**ysteine **p**roteinase) and Bandlow and coworkers (Magdolen *et al.*, 1993) called the gene **BLH1**. Both groups isolated the protein as a contaminant in their preparations of membrane-bound proteins (Magdolen *et al.*, 1993). Enenkel & Wolf (1993) discovered the gene in a screen for multicopy suppressors of a lesion in a proteosomal protein, Pre3p. Due to its sequence similarity to the mammalian protein they named it **S. cerevisiae bleomycin hydrolase** and the gene **BLH1**. The mutant *pre3-2* is deficient in cleavage of the substrate Z-Leu-Leu-Glu+NHNap used for monitoring the peptidyl glutamyl hydrolyzing activity. Multicopy *BLH1* restored activity against this substrate to a *pre3-2* strain. Interestingly, bleomycin hydrolase cannot cleave this substrate *in vitro*. Finally, Xu & Johnston (1994) discovered the yeast protein by its ability to bind to the DNA-recognition sites of the transcriptional activator Gal4p. Since they found that the expression of this protein was regulated by galactose and Gal4p in an expression pattern similar to that of other *GAL* regulatory proteins, the gene encoding the protein was named **GAL6**, and its product was **GAL6**. For the yeast gene, two ATG sites are possible, but the downstream site matches the consensus translation initiation sequence better than the upstream site. Expression from heterologous promoters using either of the two ATGs resulted in the same size protein, indicating that the second ATG is the correct translational start site, in contrast to what appears in some databases (H.E. Xu & S.A. Johnston, unpublished results). In addition, alignment of the derived amino acid sequences, 454 versus 483 residues, with the homologs from other species is more obvious using the shorter sequence.

Recently, a complete cDNA was obtained for the rat enzyme (Takeda *et al.*, 1996b) and two groups have cloned the human form of the enzyme (Brömme *et al.*, 1996; Ferrando *et al.*, 1996).

## Activity and Specificity

The pH optima for the human, rabbit and yeast enzymes are 7.2 (Brömme *et al.*, 1996), 7.0–7.5 (Sebti *et al.*, 1987), and 7.5 (Enenkel & Wolf, 1993; Xu & Johnston, 1994), respectively, whereas the value for the bacterial enzyme is 6.5–7.0 (Wohlrab & Bockelmann, 1993; Chapot-Chartier *et al.*, 1994). The substrate specificities of the bleomycin hydrolases are not fully understood yet, though some patterns are apparent. In general, these enzymes all cleave small peptides with an unblocked N-terminus. It is not clear what the maximum length of a peptide substrate is, though pentapeptides were shown to be cleaved by the yeast enzyme (W. Zhang & S.A. Johnston, unpublished results). The amino acid *p*-nitroanilides, methylcoumarylamides and β-naphthylamides are effective substrates for the eukaryotic enzymes, with the exception of the derivatives of Pro, Val, Asp and D-amino acids which are not hydrolyzed (Enenkel & Wolf, 1993; Brömme *et al.*, 1996; Takeda *et al.*, 1996a). The Met, Leu and Ala derivatives are better substrates for the human enzyme, whereas the Arg and Tyr derivatives are better for the yeast enzyme (Brömme *et al.*, 1996).

The closely related bacterial enzymes are forms of PepC, or aminopeptidase C (Chapter 216). The enzyme from *Lactobacillus delbrueckii* cleaves a broad range of di- and tripeptides, but is inactive towards *p*-nitroanilide or β-naphthylamide derivatives (Wohlrab & Bockelmann, 1993). The *Lactobacillus helveticus* enzyme is inactive towards Leu-NHNap (Vesanto *et al.*, 1994), whereas the lactococcal and the *Streptococcus thermophilus* enzymes are active towards *p*-nitroanilides and β-naphthylamides (Chapot-Chartier *et al.*, 1994).

Bleomycin B2 is more readily degraded by human bleomycin hydrolase than is bleomycin A2, but neither is degraded by cathepsins S, K (O2) (Brömme *et al.*, 1996) or B (Sebti *et al.*, 1989). The yeast bleomycin hydrolase can cleave bleomycin *in vitro* as well as *in vivo* (Kambouris *et al.*, 1992; Enenkel & Wolf, 1993; Xu & Johnston, 1994). In fact, deletion of the gene results in increased sensitivity of yeast towards the drug (Enenkel & Wolf, 1993), and overexpression induces resistance both in yeast (Kambouris *et al.*, 1992; H.E. Xu & S.A. Johnston, unpublished results) and in NIH-3T3 cells (Pei *et al.*, 1995).

The bleomycin hydrolases are all inhibited by cysteine protease inhibitors such as *N*-ethylmaleimide, iodoacetate, *p*-hydroxymercuribenzoate and E-64. Leupeptin is a weak inhibitor for the yeast and rat enzymes but has no effect on the human bleomycin hydrolase or the bacterial relatives. In general, the human enzyme is not as efficiently inhibited by the cysteine protease inhibitors tested as is the yeast bleomycin hydrolase, and E-64 for example is about 3 orders of magnitude less efficient than for papain-like cathepsins (Brömme *et al.*, 1996). Takeda *et al.* (1996a) have shown that the rat enzyme is not inhibited by cystatins A or C.

Apart from the peptidase activity, the yeast enzyme (Xu & Johnston, 1994) and more recently the rat enzyme (Takeda *et al.*, 1996b) were shown to bind nucleic acids, preferably single-stranded DNA or RNA. The yeast form binds single-stranded nucleic acid with nanomolar affinities. The lactococcal PepC also binds nucleic acids (H.E. Xu & S.A. Johnston, unpublished results).

## Structural Chemistry

The bleomycin hydrolases are intracellular, homohexameric proteins of approximately 300 kDa. The calculated pI values of the human, yeast and lactoccocal forms are 5.9, 7.1 and

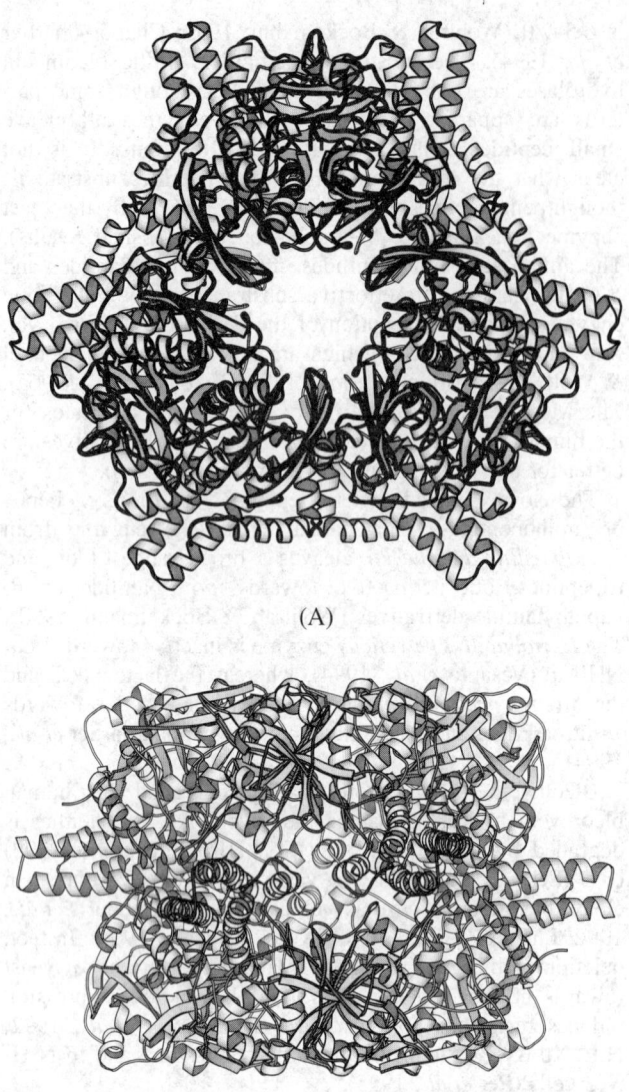

*Figure 215.1* The three-dimensional structure of the Gal6 hexamer. (A) Top view along the 3-fold axis. The central channel is a prominent feature. (B) A side view perpendicular to the top view.

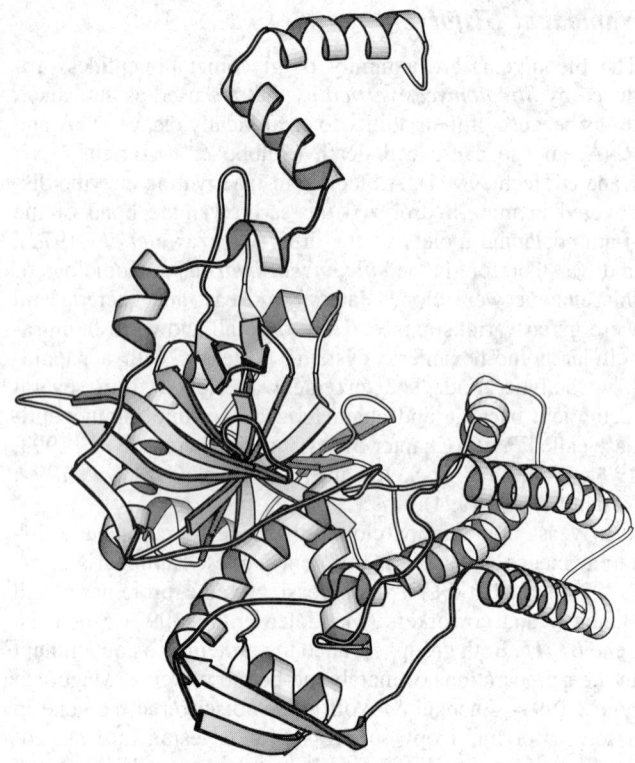

*Figure 215.2* Subunit structure of the yeast bleomycin hydrolase, Gal6.

4.8, respectively. The crystal structure of the yeast enzyme, Gal6, at 2.2 Å resolution (Joshua-Tor *et al*., 1995) reveals a hexameric structure with the six identical subunits arranged in a ring with 32 symmetry that creates a prominent central channel of ~22 Å diameter along the threefold axis (Figure 215.1). The monomers have a papain-like polypeptide fold as the core, with additional structural and functional modules inserted into loop regions. As a result, the Gal6 monomer has a three-domain structure composed of a catalytic (protease) domain, a hook-like oligomerization domain, and a helical domain (Figure 215.2). Like papain, the catalytic domain consists of two subdomains which are designated the L and the R subdomains by analogy to papain. The active-site triad of Gal6 formed by the side chains of Cys73, His369 and Asn392, as well as Gln67 positioned to stabilize the oxyanion hole, resembles the active-site triad of all other structurally characterized cysteine proteases of the papain family. Apart from the absence of any disulfide bonds, which are present in all the more papain-like peptidases of family C1, a striking feature of Gal6 is the projection of the C-terminus of the protein to occupy much of the active-site cleft (Figure 215.3). The structure of the C-terminus resembles that of inhibitors complexed with papain, most notably the complex with leupeptin. This suggests that the C-terminal arm may have a regulatory role in modulating the protease activity and substrate specificity of Gal6. It may also explain the *in vitro* activity as an aminopeptidase if the carboxylate of the C-terminus positions the unblocked N-terminus of the peptide substrate in the active site through ionic interactions. Although the protein sequence predicted from the DNA sequence should end with a lysine residue (Lys454), the lysine is absent in the mature protein, as shown from the crystal structure as well as C-terminal sequencing and mass spectrometry. It is probably removed by an autocleavage process. The sequence of the mature protein ends with Ala453, which is the last conserved residue in a highly conserved sequence from Trp446 that is located in the active-site cleft. All the sequences available to date show high sequence divergence of the residues beyond Ala453, and probably all of these are lost from the mature proteins.

The oligomerization domain forms many of the contacts that stabilize the oligomer, wrapping around the helical domain to form the dimer interactions, as well as forming a coiled-coil with a diagonally related monomer forming some of the trimer interactions. The helical domain is formed by two pairs of antiparallel helices that extend from the catalytic domain. Extensive intersubunit contacts stabilize the hexamer which is maintained in the crystal and in solution and appears to be the oligomeric form for the bleomycin hydrolases of all species

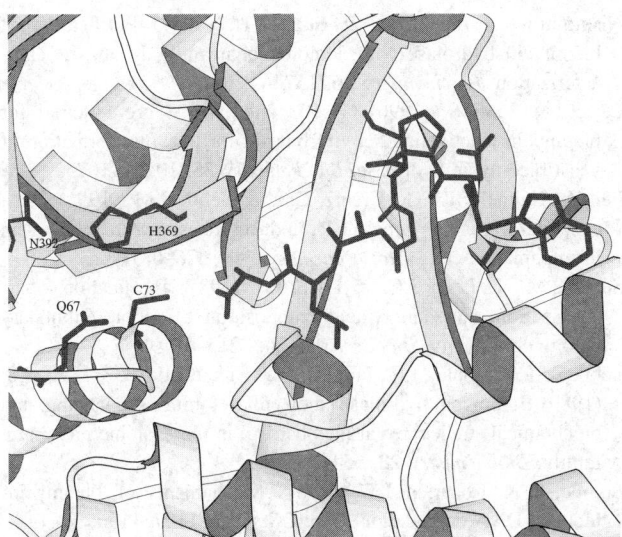

*Figure 215.3* The active site of yeast bleomycin hydrolase, Gal6, showing the conserved C-terminal arm residues, starting with Trp446 on the right, and the active-site residues in a stick representation.

studied so far. The central channel of the hexamer of Gal6 is lined with 60 lysine residues from the six subunits, providing a high positive electrostatic potential inside the channel. This prominent feature suggests that this region may be involved in nucleic acid binding. The papain-like active sites are situated within the central channel, in a manner resembling the organization of active sites in the proteasome (Löwe *et al.*, 1995), and are essentially inaccessible from the outer surface of the hexamer, except through the interior of the channel. The combination of this overall architecture of the hexamer and the presence of the C-terminal arm in the active-site cleft, making access to the active site quite restricted, may explain why the activity of bleomycin hydrolase is restricted to relatively small substrates, at least *in vitro*. If the enzymatic activities of bleomycin hydrolase are regulated by interactions with other factors as yet unknown, these may mediate changes in subunit positioning in the hexamer and open up the active sites to larger substrates. Crystals of the lactococcal enzyme, PepC, have been reported (Mistou *et al.*, 1994), and its structure should offer an interesting comparison.

## Preparation

Bleomycin hydrolase is ubiquitous in mammalian tissues. Purification from natural sources for biochemical studies has been reported from rabbit liver (Nishimura *et al.*, 1987) and lung (Sebti *et al.*, 1987), rat skin (Takeda *et al.*, 1996a) and yeast (Kambouris *et al.*, 1992; Xu & Johnston, 1994).

Recombinant, unmodified yeast bleomycin hydrolase has been purified from yeast (Xu & Johnston, 1994) and baculovirus overexpression systems (Brömme *et al.*, 1996). It has also been purified as His-tagged (H. Whelan, W. Zheng & S.A. Johnston, unpublished results) and GST-tagged (Pei & Sebti, 1996) proteins in *Escherichia coli*. Human bleomycin hydrolase has also been produced in a baculovirus/insect cell system (Brömme *et al.*, 1996). The yeast form is amenable to

large-scale production in both bacteria (H. Whelan, W. Zheng & S.A. Johnston, unpublished results) and yeast (Joshua-Tor *et al.*, 1995).

## Biological Aspects

Bleomycin hydrolase is a highly conserved protein from bacteria to humans. Pairwise identities are ~40% between the human and the yeast proteins and between 35 and 38% between the human and the bacterial enzymes. It is ubiquitous in mammals, birds and reptiles (Umezawa *et al.*, 1974; Sebti *et al.*, 1989). In mammals it is expressed in every tissue and expressed at higher levels in most tumor cell lines than in normal tissues (Brömme *et al.*, 1996; Ferrando *et al.*, 1996). However, there is clear variability in bleomycin hydrolase levels in different tumor specimens, and it seems that where especially high levels of the enzyme are observed, for example in leukemia cell lines or in Burkitt's lymphoma cells, these do not respond to bleomycin treatment, whereas Hodgkin's lymphomas, for example, that do not express the enzyme in significant amounts, are sensitive to bleomycin treatment (Brömme *et al.*, 1996; Ferrando *et al.*, 1996). These observations are consistent with metabolic degradation of the drug by bleomycin hydrolase as a reason for tumor resistance.

Bleomycin, although a useful anticancer drug, is the only natural substrate for bleomycin hydrolase known to date, even though this compound is not normally seen by any of the organisms mentioned. Thus, the hydrolysis of bleomycin is probably not the normal function of the enzyme. It may not be accidental that the enzyme which hydrolyzes and detoxifies bleomycin binds to a product of bleomycin damage, namely single-stranded or nicked nucleic acids. Another curious feature of the yeast bleomycin hydrolase, Gal6, apart from its nucleic acid binding capacity, is the effect it has on mRNA levels of genes expressed from *GAL* promoters (W. Zheng, H.E. Xu & S.A. Johnston, unpublished results). It appears to behave as a repressor in this system, its presence lowering levels of RNA. To date this is the only indication of a biological activity of bleomycin hydrolase besides detoxifying bleomycin.

## Distinguishing Features

Bleomycin hydrolase is a nucleic acid-binding cysteine peptidase of family C1, with an unusual hexameric ring structure, and distinctive specificity.

## Acknowledgements

The authors are grateful to the National Institutes of Health (grants CA71746 and CA67982 to S.A.J.) for their finanacial support.

## References

Brömme, D., Rossi, A.B., Smeekens, S.P., Anderson, D.C. & Payan, D.G. (1996) Human bleomycin hydrolase: molecular cloning, sequencing, functional expression, and enzymatic characterization. *Biochemistry* **35**, 6706–6714.

Chapot-Chartier, M.-P., Nardi, M., Chopin, M.-C., Chopin, A. & Gripon, J.-C. (1993) Cloning and sequencing of pepC, a cysteine aminopeptidase gene from *Lactococcus lactis subsp. cremoris AM2. Appl. Environ. Microbiol.* **59**, 330–333.

Chapot-Chartier, M.-P., Rul, F., Nardi, M. & Gripon, J.-C. (1994) Gene cloning and characterization of PepC, a cysteine aminopeptidase from *Streptococcus thermophilus*, with sequence similarity to the eucaryotic bleomycin hydrolase. *Eur. J. Biochem.* **224**, 497–506.

Enenkel, C. & Wolf, D.H. (1993) BLH1 codes for a yeast thiol aminopeptidase, the equivalent of mammalian bleomycin hydrolase. *J. Biol. Chem.* **268**, 7036–7043.

Fernandez, L., Bhowmik, T. & Steele, J.L. (1994) Characterization of the *Lactobacillus helveticus CNRZ32* pepC gene. *Appl. Environ. Microbiol.* **60**, 333–336.

Ferrando, A.A., Velasco, G., Campo, E. & López-Otín, C. (1996) Cloning and expression analysis of human bleomycin hydrolase, a cysteine proteinase involved in chemotherapy resistance. *Cancer Res.* **56**, 1746–1750.

Joshua-Tor, L., Xu, E.H., Johnston, S.A. & Rees, D.C. (1995) Crystal structure of a conserved protease that binds DNA: the bleomycin hydrolase, Gal6. *Science* **269**, 945–950.

Kambouris, N.G., Burke, D.J. & Creutz, C.E. (1992) Cloning and characterization of a cysteine proteinase from *Saccharomyces cerevisiae. J. Biol. Chem.* **267**, 21570–21576.

Kane, S.A. & Hecht, S.M. (1994) Polynucleotide recognition and degradation by bleomycin. *Prog. Nucleic Acid Res. Mol. Biol.* **49**, 313–352.

Löwe, J., Stock, D., Jap, R., Zwickl, P., Baumeister, W. & Huber, R. (1995) Crystal structure of the 20S proteasome from the archaeon *T. acidophilum* at 3.4 Å resolution. *Science* **268**, 533–539.

Magdolen, U., Müller, G., Magdolen, V. & Bandlow, W. (1993) A yeast gene (BLH1) encodes a polypeptide with high homology to vertebrate bleomycin hydrolase, a family member of thiol proteinases. *Biochim. Biophys. Acta* **1171**, 299–303.

Mistou, M.Y., Rigolet, P., Chapot-Chartier, M.P., Nardi, M., Gripon, J.C. & Brunie, S. (1994) Crystallization and preliminary X-ray analysis of PepC, a thiol aminopeptidase from *Lactoccocuslactis* homologous to bleomycin hydrolase. *J. Mol. Biol.* **237**, 160–162.

Nishimura, C., Tanaka, N. & Suzuki, H. (1987) Purification of bleomycin hydrolase with a monoclonal antibody and its characterization. *Biochemistry* **26**, 1574–1578.

Pei, Z. & Sebti, S. (1996) Cys102 and His398 are required for bleomycin-inactivating activity but not for hexamer formation of yeast bleomycin hydrolase. *Biochemistry* **35**, 10751–10756.

Pei, Z., Calmels, T.P.G., Creutz, C.E. & Sebti, S.M. (1995) Yeast cysteine proteinase gene YCP1 induces resistance to bleomycin in mammalian cells. *Mol. Pharmacol.* **48**, 676–681.

Sebti, S.M., DeLeon, J.C. & Lazo, J.S. (1987) Purification, characterization, and amino acid composition of rabbit pulmonary bleomycin toxicity. *Biochemistry* **26**, 4213–4219.

Sebti, S.M., Mignano, J.E., Jani, J.P., Srimatkandada, S. & Lazo, J.S. (1989) Bleomycin hydrolase: molecular cloning, sequencing, and biochemical studies reveal membership in the cysteine proteinase family. *Biochemistry* **28**, 6544–6548.

Stubbe, J. & Kozarich, J.W. (1987) Mechanisms of bleomycin-induced DNA degradation. *Chem. Rev.* **87**, 1107–1136.

Takeda, A., Higuchi, D., Yamamoto, T., Nakamura, Y., Masuda, Y., Hirabayashi, T. & Nakaya, K. (1996a) Purification and characterization of bleomycin hydrolase, which represents a new family of cysteine proteases, from rat skin. *J. Biochem.* **119**, 29–36.

Takeda, A., Masuda, Y., Yamamoto, T., Hirabayashi, T., Nakamura, Y. & Nakaya, K. (1996b) Cloning and analysis of cDNA encoding rat bleomycin hydrolase, a DNA-binding cysteine protease. *J. Biochem.* **120**, 353–359.

Umezawa, H., Takeuchi, T., Hori, S., Sawa, T. & Ishizuka, M. (1972) Studies on the mechanism of antitumor effect of bleomycin on squamous cell carcinoma. *J. Antibiot.* **25**, 409–420.

Umezawa, H., Hori, S., Sawa, T., Yoshioka, T. & Takeuchi, T. (1974) A bleomycin-inactivating enzyme in mouse liver. *J. Antibiot.* **27**, 419–424.

Vesanto, E., Varmanen, P., Steele, J.L. & Palva, A. (1994) Characterization and expression of the *Lactobacillus helveticus* PepC gene encoding a general aminopeptidase. *Eur. J. Biochem.* **224**, 991–997.

Wohlrab, Y. & Bockelmann, W. (1993) Purification and characterization of a second aminopeptidase (PepC-like) from *Lactobacillus delbrueckii subsp. bulgaricus B14. Int. Dairy J.* **3**, 685–701.

Xu, H.E. & Johnston, S.A. (1994) Yeast bleomycin hydrolase is a DNA-binding cysteine protease. *J. Biol. Chem.* **269**, 21177–21183.

***Leemor Joshua-Tor***
*W. M. Keck Structural Biology,*
*Cold Spring Harbor Laboratory,*
*Cold Spring Harbor, NY 11724, USA*
*Email: leemor@cshl.org*

***Stephen Albert Johnston***
*Internal Medicine,*
*University of Texas – Southwestern Medical Center,*
*5323 Harry Hines Blvd.,*
*Dallas, TX 75235-8573, USA*
*Email: johnston@ryburn.swmed.edu*

# 216. *Aminopeptidase C (PepC) of lactic acid bacteria*

## Databanks

*Peptidase classification: clan CA, family C1, MEROPS ID: C01.086*
*NC-IUBMB enzyme classification: none*
*Databank codes:*

| Species | SwissProt | PIR | EMBL (cDNA) | EMBL (genomic) |
|---|---|---|---|---|
| *Lactobacillus delbrueckii* | – | S52865 | X80643 | – |
| *Lactobacillus helveticus* | – | S47275 | L26223 | – |
| | | S48200 | Z30340 | |
| *Lactococcus lactis* | Q04723 | B48957 | M86245 | – |
| *Streptococcus thermophilus* | – | S47284 | Z30315 | – |
| | | S48143 | | |

## Name and History

A peptidase termed **PepC** was initially purified from the lactic acid bacterium *Lactococcus lactis* and was characterized as a general aminopeptidase inhibited by thiol-reacting compounds (Neviani *et al.*, 1989). The enzyme was later classified as a cysteine peptidase on the basis of sequence homology with papain around the catalytic site residues (Chapot-Chartier *et al.*, 1993). The 'C' in the gene name *pepC* reminds us of the catalytic cysteine of the catalytic site, and the term PepC has been found convenient for the enzyme, too (Tan *et al.*, 1993). PepC is found also in other lactic acid bacteria (Wohlrab & Bockelmann, 1993; Chapot-Chartier *et al.*, 1994; Fernández *et al.*, 1994; Klein *et al.*, 1994; Vesanto *et al.*, 1994), but has not been described from other bacteria such as *Escherichia coli* or *Salmonella typhimurium*. We here propose the semisystematic name **aminopeptidase C** for the enzyme, but will also use the short form PepC for convenience.

Bleomycin hydrolase (Chapter 215) is a peptidase of eukaryotic organisms that is closely related in structure and activity to aminopeptidase C. Bleomycin hydrolase is known from yeast (*BLH1, YCP1* or *GAL6*) (Kambouris *et al.*, 1992; Enenkel & Wolfe, 1993; Magdolen *et al.*, 1993) and mammals (bleomycin hydrolase) (Sebti *et al.*, 1989; Brömme *et al.*, 1996; Ferrando *et al.*, 1996).

## Activity and Specificity

The substrate specificity of aminopeptidase C from various species of bacteria has been studied with amino acid *p*-nitroanilides and *β*-naphthylamides (Chapot-Chartier *et al.*, 1994; Vesanto *et al.*, 1994; Neviani *et al.*, 1989) as well as di- and tripeptides (Wohlrab & Bockelmann, 1993). It is a strict aminopeptidase, i.e. the presence of the free N-terminal group on the substrate is an absolute requirement for hydrolysis to occur. The specificity for the N-terminal amino acid is rather broad, except that bonds in which a proline is involved (in the P1 or P1′ position) are not hydrolyzed. Lactococcal PepC can act on peptides from two to at least ten residues in length, with maximal activity on peptides of four residues (M.-Y. Mistou, unpublished results).

*In vitro*, aminopeptidase C requires activation by a thiol compound such as DTT, and maximal activity is near to pH 7.2. The thiol-blocking reagents iodoacetamide, *p*-chloromercuric benzoic acid and *N*-ethylmaleimide totally inhibit the enzyme. E-64, an inhibitor of many cysteine peptidases, has low inhibitory potency towards the lactococcal PepC, with $k_{inact}$ of $99\,M^{-1}\,s^{-1}$, as compared to $6.7 \times 10^6\,M^{-1}\,s^{-1}$ for papain.

## Structural Chemistry

The lactococcal *pepC* gene encodes a 436 residue (49 826 Da; pI 4.6) polypeptide. The enzyme is cytoplasmic and does not contain disulfide bonds.

The crystallographic structure of lactococcal PepC has recently been solved (M.-Y. Mistou & D. Housset, unpublished results). The structure was solved by using the model of the closely related yeast bleomycin hydrolase (Chapter 215), Gal6 (Joshua-Tor *et al.*, 1995). In solution and in the crystalline state, PepC is organized as a hexamer composed of two superimposed trimers. A central channel goes through the hexamer (diameter approximately 25 Å) and gives access to the active sites. The crystal structure has confirmed the papain-like fold of the catalytic domain of the monomer. The location and orientation of the catalytic dyad, Cys73/His355, are reminiscent of that of papain, but the extreme C-terminal residues of PepC (Gly433-Ala-Leu-Ala436) fit into the active-site cleft as has been described for bleomycin hydrolase (Chapter 215). The location of the C-terminal tail leaves sufficient space for only one residue, the P1 residue, on the N-terminal side of the scissile peptide bond. The C-terminal carboxylic group of the enzyme could thus theoretically interact with the positively charged N-terminus of the substrate to provide the structural basis for the aminopeptidase specificity.

## Preparation

PepC was initially purified to homogeneity from the *L. lactis* subsp. *cremoris* AM2 strain with a purification factor of 136 (Neviani *et al*., 1989). The enzyme can now be prepared from a recombinant *L. lactis* strain which allows the production of 10 mg of pure crystallizable enzyme per liter of culture (Mistou *et al*., 1994). PepC from *Streptococcus thermophilus* (Chapot-Chartier *et al*., 1994) and from *Lactobacillus delbrueckii* (Klein *et al*., 1996) has been produced at very high level in *E. coli* (40–50% of the total soluble proteins).

## Biological Aspects

PepC is a cytoplasmic enzyme (Tan *et al*., 1992) that appears to be involved in the nitrogen metabolism of *L. lactis* grown in milk, releasing free amino acids from peptides taken up by the cell (Mierau *et al*., 1996).

It is remarkable that PepC is highly similar in structure and in its biochemical properties to the mammalian bleomycin hydrolase (38% sequence identity) (Sebti *et al*., 1989; Brömme *et al*., 1996; Ferrando *et al*.; 1996) and the yeast Gal6 enzyme (31% sequence identity) (Kambouris *et al*., 1992; Enenkel & Wolf, 1993). The conservation of this enzyme from bacteria to mammals suggests that it plays an important physiological role. The yeast enzyme has been shown to bind DNA (Xu & Johnson, 1994), but its function is not known.

## Distinguishing Features

Rabbit polyclonal antibodies have been raised against the lactococcal PepC. They cross-react with the streptococcal enzyme (Chapot-Chartier *et al*., 1993, 1994).

## Related Peptidases

Two genes encoding proteins about 55% identical in amino acid sequence to the PepC variants have recently been identified in *Lactobacillus delbruekii* subsp. *lactis* (Klein *et al*., 1996). The first protein, which was named PepG, is also an aminopeptidase, but has different specificity for synthetic substrates to that of PepC. No aminopeptidase activity could be assigned to the product of the second gene (named *OrfW*).

A gene encoding an oligopeptidase showing about 40% similarity with the PepCs has been cloned from *Lactobacillus helveticus* CNRZ32 (Fenster *et al*., 1996). The protein, from the *pepE2* gene, was purified as a fusion with maltose-binding protein, and found to hydrolyze Bz-Phe-Val-Arg+NHPhNO$_2$ and Tyr-Gly-Gly+Phe-Met. This peptidase has recently been named oligopeptidase E (Chapter 217).

## References

Brömme, D., Rossi, A.B., Smeekens, S.P., Anderson, D.C. & Payan, D.G. (1996) Human bleomycin hydrolase: molecular cloning, sequencing, functional expression, and enzymatic characterization. *Biochemistry* **35**, 6706–6714.

Chapot-Chartier, M.-P., Nardi, M., Chopin, M.-C., Chopin, A. & Gripon, J.-C. (1993) Cloning and sequencing of *pepC*, a cysteine aminopeptidase gene from *Lactococcus lactis* subsp. *cremoris* AM2. *Appl. Environ. Microbiol.* **59**, 330–333.

Chapot-Chartier, M.-P., Rul, F., Nardi, M. & Gripon, J.-C. (1994) Gene cloning and characterization of PepC, a cysteine aminopeptidase from *Streptococcus thermophilus*, with sequence similarity to the eucaryotic bleomycin hydrolase. *Eur. J. Biochem.* **224**, 497–506.

Enenkel, C. & Wolf, D.H. (1993) *BLH1* codes for a yeast thiol aminopeptidase, the equivalent of mammalian bleomycin hydrolase. *J. Biol. Chem.* **268**, 7036–7043.

Fenster, K.M., Chen, Y.S. & Steele, J.L. (1996) Endopeptidase from *Lactobacillus helveticus* CNRZ32. In: *Abstracts K22. Fifth Symposium on Lactic Acid Bacteria. Genetics, Metabolism and Applications.* Veldhoven, The Netherlands, 1996 September 8/12.

Fernández, L., Bhowmik, T. & Steele, J.L. (1994) Characterization of the *Lactobacillus helveticus* CNRZ32 *pepC* gene. *Appl. Environ. Microbiol.* **60**, 333–336.

Ferrando, A.A., Velasco, G., Campo, E. & López-Otín, C. (1996) Cloning and expression analysis of human bleomycin hydrolase, a cysteine proteinase involved in chemotherapy resistance. *Cancer Res.* **56**, 1746–1750.

Joshua-Tor, L., Xu, H.E., Johnston, S.A. & Rees, D.C. (1995) Crystal structure of a conserved protease that binds DNA: the bleomycin hydrolase, Gal6. *Science* **269**, 945–950.

Kambouris, N.G., Burke, D.J. & Creutz, C.E. (1992) Cloning and characterization of a cysteine proteinase from *Saccharomyces cerevisiae. J. Biol. Chem.*, **267**, 21570–21576.

Klein, J.R., Henrich, B. & Plapp, R. (1994) Cloning and nucleotide sequence analysis of the *Lactobacillus delbrueckii* subsp. *lactis* DSM7290 cysteine aminopeptidase gene *pepC. FEMS Microbiol. Lett.* **124**, 291–299.

Klein, J.R., Schick, J., Henrich, B. & Plapp, R. (1996) *Lactobacillus delbrueckii* ssp. *lactis* DSM7290 *pepG* encodes a novel cysteine aminopeptidase. *Microbiology* **143**, 527–537.

Magdolen, U., Müller, G., Magdolen, V. & Bandlow, W. (1993) A yeast gene (BLH1) encodes a polypeptide with high homology to vertebrate bleomycin hydrolase, a family member of thiol proteinases. *Biochem. Biophys. Acta* **1171**, 299–303.

Mierau, I., Kunji, E.R., Leenhouts, K.J., Hellendoorn, M.A., Haandrikman, A.J., Poolman, B., Konings, W.N., Venema, G. & Kok, J. (1996) Multiple-peptidase mutants of *Lactococcus lactis* are severely impaired in their ability to grow in milk. *J Bacteriol.* **178**, 2794–2803.

Mistou M.-Y., Rigolet P., Chapot-Chartier M.-P., Nardi M., Gripon J.-C. & Brunie S. (1994) Crystallization and preliminary X-ray analysis of PepC, a thiol aminopeptidase from *Lactococcus lactis* homologous to bleomycin hydrolase. *J. Mol. Biol.* **237**, 160–162.

Neviani, E., Boquien, C.-Y., Monnet, V., Phan Thanh, L. & Gripon, J.-C. (1989) Purification and characterization of an aminopeptidase from *Lactococcus lactis* subsp. *cremoris* AM2. *Appl. Environ. Microbiol.* **55**, 2308–2314.

Sebti, S.M., Mignano, J.E., Jani, J.P., Srimatkandada, S. & Lazo, J.S. (1989) Bleomycin hydrolase: molecular cloning, sequencing, and biochemical studies reveal membership in the cysteine proteinase family. *Biochemistry* **28**, 6544–6548.

Tan, P.S.T., Chapot-Chartier, M.-P., Pos, K.M., Rousseau, M., Boquien, C.-Y. Gripon, J.-C. & Konings, W.N. (1992) Localization of peptidases in lactococci. *Appl. Environ. Microbiol.* **58**, 285–290.

Tan, P.S.T., Poolman, B. & Konings, W.N. (1993) Proteolytic enzymes of *Lactococcus lactis. J. Dairy Res.* **60**, 269–286.

Vesanto, E., Varmanen, P., Steele, J.L. & Palva, A. (1994) Characterization and expression of the *Lactobacillus helveticus pepC* gene encoding a general aminopeptidase. *Eur. J. Biochem.* **224**, 991–997.

Wohlrab, Y. & Bockelmann, W. (1993) Purification and characterization of a second aminopeptidase (PepC-like) from *Lactobacillus delbrueckii* subsp. *bulgaricus* B14. *Int. Dairy J.* **3**, 685–701.

Xu, U.E. & Johnston, S.A. (1994) Yeast bleomycin hydrolase is a DNA-binding cysteine protease. *J. Biol. Chem.* **269**, 21177–21183.

**Marie-Pierre Chapot-Chartier**
*I.N.R.A., Unité de Biochimie et Structure des Protéines,*
*78352 Jouy-en-Josas cedex, France*
Email: *chapot@biotec.jouy.inra.fr*

**Michel-Yves Mistou**
*I.N.R.A., Unité de Biochimie et Structure des Protéines,*
*78352 Jouy-en-Josas cedex, France*
Email: *mistou@jouy.inra.fr*

# 217. *Oligopeptidase E*

## Databanks

*Peptidase classification: clan CA, family C1, MEROPS ID: C01.088*
*NC-IUBMB enzyme classification: none*
*Databank codes:*

| Species | SwissProt | PIR | EMBL (cDNA) | EMBL (genomic) |
|---|---|---|---|---|
| *Lactobacillus helveticus* | – | – | U77050 | – |

## Name and History

A genomic library of *Lactobacillus helveticus* CNRZ32 constructed in *Escherichia coli* DH5α (Nowakowski *et al.*, 1993) was screened for endopeptidase activities using the substrates Bz-Phe-Val-Arg-NHPhNO₂, Bz-Pro-Phe-Arg-NHPhNO₂ and Bz-Val-Gly-Arg-NHPhNO₂. Two isolates, which had qualitatively different endopeptidase activities, were identified from this screening. One clone hydrolyzed Bz-Phe-Val-Arg+NHPhNO₂ and Bz-Pro-Phe-Arg+NHPhNO₂, but did not hydrolyze Bz-Val-Gly-Arg-NHPhNO₂. The gene encoding this peptidase was sequenced and found to have similarities to thiol-dependent general aminopeptidases (PepC and PepG) from a variety of lactic acid bacteria (Chapter 216). The endopeptidase encoded by this gene was designated ***oligopeptidase E***, but for convenience, it is referred to by the gene name, ***PepE***, in the following text.

## Activity and Specificity

Purified PepE was assayed at 0.1 mg protein ml$^{-1}$ in a mixture of 10 mM HEPES (pH 7.0), and 0.22 mM Bz-Phe-Val-Arg-NHPhNO₂, that had been pre-equilibrated for 15 min at 35°C. Reactions were initiated by the addition of PepE and stopped after 5 min by the addition of 30% glacial acetic acid. Reaction rates were verified to be linear under these conditions and were quantified based on the release of *p*-nitroaniline (extinction coefficient 8.8 mM$^{-1}$ cm$^{-1}$ at 410 nm) (Erlanger

*et al.*, 1961). The optimum temperature for PepE was in the range 32–37°C. The activation energy of PepE over the range of 0–30°C was calculated using an Arrhenius plot to be 15 kcal mol$^{-1}$. Similarly, the $E_a$ for deactivation of PepE over the range of 40–55°C was determined to be 59 kcal mol$^{-1}$. The optimum conditions for PepE at 35°C included 0.5% NaCl. The pH dependence of PepE revealed an optimum pH of 4.5, and enzyme instability and loss of activity below pH 3.5 and above pH 7.5.

Inhibitor studies employed the standard assay mixture supplemented with 1 mM inhibitor. The potential inhibitors tested were EDTA, 1,10-phenanthroline, PMSF, DFP, pepstatin A, iodoacetic acid (IAA), PCMB, DTT, and 2-ME. Of these inhibitors, only IAA and PCMB were found to inhibit PepE activity (96% and 98%, respectively), implying that a cysteine residue is essential for activity. The redox state of the residue may also be important as DTT and 2-ME enhanced activity by 64%. The inhibitors of aspartic, serine and metallopeptidases had no effect, except EDTA, which stimulated activity by 70%.

Met-enkephalin, bradykinin and β-casomorphin (1 mg ml$^{-1}$) were incubated with purified PepE (0.1 mg protein ml$^{-1}$) in 50 mM Na-phosphate buffer (pH 5.0) for 10, 15, 20 and 30 min at 35°C prior to quenching reactivity by adding water/trifluoroacetic acid (1000:1, by volume). Met-enkephalin was determined to be hydrolyzed by PepE primarily at position Gly3+Phe4 and to a lesser extent at Gly2+Gly3. Bradykinin was hydrolyzed by PepE only at position Gly4+Phe5. The products of hydrolysis were confirmed by HPLC, amino acid analysis and mass spectroscopy. PepE did not hydrolyze $\beta$-casomorphin.

$\alpha$-, $\beta$- and $\kappa$-casein were incubated (at 0.9 mg protein ml$^{-1}$) in 50 mM Na-phosphate buffer (pH 5.0) with purified PepE (0.1 mg protein ml$^{-1}$) for 2 h at 35°C. Hydrolysis was qualitatively monitored by running a 15% SDS-PAGE. There was no evidence of digestion of intact $\alpha$-, $\beta$- or $\kappa$-casein, as determined by 15% SDS-PAGE stained with Coomassie brilliant blue.

## Structural Chemistry

The monomeric molecular mass of PepE was estimated to be 50 kDa from an 8% SDS-PAGE stained with Coomassie brilliant blue.

## Preparation

PepE was purified to electrophoretic homogeneity using the QIA*expressionist* Protein Purification System (Qiagen; see Appendix 2 for full names and addresses of suppliers) following the manufacturer's instructions.

## Biological Aspects

Nucleotide sequencing of *pepE* revealed a 1314 bp open reading frame, which could encode a protein of 52.1 kDa. Putative −10 and −35 transcriptional promoters were identified, which indicates that *pepE* may be transcribed from its own promoter. Also, a putative *rho*-independent transcriptional terminator ($\Delta G = -25.4$ kcal mol$^{-1}$) was observed in the 3′ noncoding region (Tinoco *et al.*, 1973). The presence of these putative transcriptional promoter and terminator sequences suggests that *pepE* is transcribed monocistronically. PepE is probably located intracellularly, because no signal sequence was detected at the N-terminus of the amino acid sequence deduced from the *pepE* gene.

Protein sequence homology searches using BLAST (Altschul *et al.*, 1990) revealed that PepE had a high amino acid sequence identity with PepC from *Lactobacillus delbrueckii* DSM7290 (Klein *et al.*, 1994), *Lactobacillus helveticus* CNRZ32 (Fernández *et al.*, 1994; Vesanto *et al.*, 1994), *Streptococcus thermophilus* CNRZ302 (Chapot-Chartier *et al.*, 1994), and *Lactococcus lactis* AM2 (Chapot-Chartier *et al.*, 1993) as well as PepG from *L. delbrueckii* DSM7290 (Klein *et al.*, 1997). The amino acid sequence identities of PepE with the PepC proteins from these bacteria were 41.7%, 40.8%, 39.1% and 37.4%, respectively. The amino acid sequence identity between PepE and PepG was 72.3%. A search of the PROSITE Dictionary of Protein Sites and Patterns (Bairock, 1993) with the deduced *pepE* amino acid sequence identified two highly conserved domains involved in substrate binding and catalysis that are characteristic of proteinases from the papain (C1) family. The amino acid residues instrumental in substrate binding and catalysis by papain-family cysteine proteinases of prokaryotic and eukaryotic origin were found to be conserved in PepE (Gln64, Cys70, His362, Asn383 and Trp385, in PepE numbering).

## Distinguishing Features

PepE from *L. helveticus* CNRZ32 is distinguished from PepC of *L. delbrueckii* DSM7290, *L. helveticus* CNRZ32, *S. thermophilus* CNRZ302, and *Lactococcus lactis* AM2 as well as PepG from *L. delbrueckii* DSM7290 by its ability to hydrolyze internal bonds of Met-enkephalin and bradykinin, and its inability to hydrolyze traditional aminopeptidase substrates, such as Phe-NHPhNO$_2$, Val-NHPhNO$_2$ and Arg-NHPhNO$_2$. Thus, PepE has endopeptidase activity, but not the aminopeptidase activity characteristic of PepC and PepG. The lack of action on casein implies that PepE is an oligopeptidase, but the catalytic type of PepE, a cysteine peptidase, differentiates it from the PepO-like endopeptidase of *Lactobacillus delbrueckii* B14 (Bockelmann *et al.*, 1996) and the lactococcal PepO (Chapter 365) (Tan *et al.*, 1991; Mierau *et al.*, 1993), PepF (Chapter 378) (Monnet *et al.*, 1994), LEPI (Chapter 378) (Yan *et al.*, 1987) and MEP (Baankreis, 1992; Baankreis *et al.*, 1995), which are all metallopeptidases.

## Further Reading

See Kunji *et al.* (1996) for a general account of the proteolytic systems of lactic acid bacteria, and Fenster *et al.* (1997) for the full characterization of oligopeptidase E.

## References

Altschul, S.F., Gish, W., Miller, W., Meyers, E.W. & Lipman, D.J. (1990) Basic local alignment search tool. *J. Mol. Biol.* **215**, 403–410.

Baankreis, R. (1992) The role of lactococcal peptidases in cheese ripening. PhD thesis. University of Amsterdam, Amsterdam, The Netherlands.

Baankreis, R., Van Schalkwijk, S.; Alting, A.C. & Exterkate, F.A. (1995) The occurrence of two intracellular oligoendopeptidases in *Lactococcus lactis* and their significance for peptide conversion in cheese. *Appl. Microbiol. Biotechnol.* **44**, 386–392.

Bairock, A. (1993) The PROSITE dictionary of sites and patterns in proteins, its current status. *Nucleic Acids Res.* **21**, 3097–3103.

Bockelmann, W., Hoppe-Seyler, T. & Heller, K.J. (1996) Purification and characterization of an endopeptidase from *Lactobacillus delbrueckii* subsp. *bulgaricus* B14. *Int. Dairy J.* **6**, 1167–1180.

Chapot-Chartier, M.-P., Nardi, M., Chopin, M.-C., Chopin, A. & Gripon, J.-C. (1993) Cloning and sequencing of *pep*C, a cysteine aminopeptidase gene from *Lactococcus lactis* subsp. *cremoris* AM2. *Appl. Environ. Microbiol.* **59**, 330–333.

Chapot-Chartier, M.-P., Rul, F., Nardi, M. & Gripon, J.-C. (1994) Gene cloning and characterization of PepC, a cysteine aminopeptidase from *Streptococcus thermophilus*, with sequence similarity to the eucaryotic bleomycin hydrolase. *Eur. J. Biochem.* **224**, 497–506.

Erlanger, B.F., Kowkowsky, N. & Cohen, W. (1961) The preparation and properties of two new chromogenic substrates of trypsin. *Arch. Biochem. Biophys.* **95**, 271–278.

Fenster, K.M., Parkin, K.L. & Steele, J.L. (1997) Characterization of a thiol-dependent endopeptidase from *Lactobacillus helveticus* CNRZ32. *J. Bacteriol.* **179**, 2529–2533.

Fernández, L., Bhowmik, T. & Steele, J.L. (1994) Characterization of the *Lactobacillus helveticus* CNRZ32 *pep*C gene. *Appl. Environ. Microbiol.* **60**, 333–336.

Klein, J.R., Henrich, B. & Plapp, R. (1994) Cloning and nucleotide sequence analysis of the *Lactobacillus delbrueckii* ssp. *lactis* DSM7290 cysteine aminopeptidase gene *pep*C. *FEMS Microbiol. Lett.* **124**, 291–300.

Klein, J.R., Schick, J., Henrich, B. & Plapp, R. (1997) *Lactobacillus delbrueckii* subsp. *lactis* DSM7290 *pep*G gene encodes a novel cysteine aminopeptidase. *Microbiology* **143**, 527–537.

Kunji, E.R.S., Mierau, I., Hagting, A., Poolman, B. & Konings, W.N. (1996) The proteolytic systems of lactic acid bacteria. *Antonie van Leeuwenhoek* **70**, 187–221.

Mierau, I., Tan, P.S.T., Haandrikman, A.J., Kok, J., Leenhouts, K.J., Konings, W.N. & Venema, G. (1993) Cloning and sequencing of the gene for a lactococcal endopeptidase, an enzyme with sequence similarity to mammalian enkephalinase. *J. Bacteriol.* **175**, 2087–2096.

Monnet, V., Nardi, M., Chopin, A., Chopin, M.-C. & Gripon, J.-C. (1994) Biochemical and genetic characterization of PepF, an oligopeptidase from *Lactococcus lactis. J. Biol. Chem.* **269**, 32070–32076.

Nowakowski, C.M., Bhowmik, T.K. & Steele, J.L. (1993) Cloning of peptidase genes from *Lactobacillus helveticus* CNRZ32. *Appl. Microbiol. Biotechnol.* **39**, 204–210.

Tan, P.S.T., Pos, K.M. & Konings, W.N. (1991) Purification and characterization of an endopeptidase from *Lactococcus lactis* subsp. *cremoris* WG2. *Appl. Environ. Microbiol.* **57**, 3593–3599.

Tinoco, I.J., Borer, P.N., Dengler, B., Levine, M.D., Uhlenbeck, O.C., Crothers, D.M. & Gralla, J. (1973) Improved estimation of secondary structure in ribonucleic acids. *Nature (Lond.) New Biol.* **246**, 40–41.

Vesanto, E., Varmanen, P., Steele, J.L. & Palva, A. (1994) Characterization and expression of the *Lactobacillus helveticus pep*C gene encoding a general aminopeptidase. *Eur. J. Biochem.* **224**, 991–997.

Yan, T.-R., Azuma, N., Kaminogawa, S., & Yamauchi, K. (1987) Purification and characterization of a substrate-size-recognizing metalloendopeptidase from *Streptococcus cremoris* H61. *Appl. Environ. Microbiol.* **53**, 2296–2302.

*Kurt M. Fenster*
*Department of Food Science,*
*University of Wisconsin-Madison,*
*Madison, WI 53706, USA*

*Kirk L. Parkin*
*Department of Food Science,*
*University of Wisconsin-Madison,*
*Madison, WI 53706, USA*

*James L. Steele*
*Department of Food Science,*
*University of Wisconsin-Madison,*
*Madison, WI 53706, USA*
*Email: jlsteele@facstaff.wisc.edu*

# *218. μ-Calpain*

## *Databanks*

*Peptidase classification: clan CA, family C2, MEROPS ID: C02.001*
*NC-IUBMB enzyme classification: EC 3.4.22.17*
*Chemical Abstracts Service registry number: 78990-62-2*
*Databank codes:*

| Species | SwissProt | PIR | EMBL (cDNA) | EMBL (genomic) |
| --- | --- | --- | --- | --- |
| μ-Calpain large subunit | | | | |
| *Bos taurus* | –<br>S16181 | A40432 | U07849 | – |
| *Gallus gallus* | – | S57195 | D38027 | – |
| *Homo sapiens* | P07384 | A26213 | X04366 | – |
| *Oryctolagus cuniculus* | P06815 | A24815 | M13363 | – |
| *Sus scrofa* | P35750 | – | U01180 | – |
| μ/m-Calpain large subunit | | | | |
| *Gallus gallus* | P00789 | A00979 | X01415 | – |
| Small subunit | | | | |
| *Bos taurus* | P13135 | A34466 | J05065 | – |
| *Homo sapiens* | P04632 | A26107 | M32886 | M31501: exon 1<br>M31502: exon 2<br>M31503: exon 3 |

*continued overleaf*

| Species | SwissProt | PIR | EMBL (cDNA) | EMBL (genomic) |
|---|---|---|---|---|
| **Small subunit** (*continued*) | | | | |
| | | | | M31504: exon 4 |
| | | | | M31505: exon 5 |
| | | | | M31506: exon 6 |
| | | | | M31507: exon 7 |
| | | | | M31508: exon 8 |
| | | | | M31509: exon 9 |
| | | | | M31510: exon 10 |
| | | | | M31511: exon 11 and complete CDS |
| *Mus musculus* | – | – | Z36352 | – |
| *Oryctolagus cuniculus* | P06813 | A24816 | M13364 | – |
| *Rattus norvegicus* | – | A55143 | U10861 | – |
| *Sus scrofa* | P04574 | A25166 | M11778 | – |
| | S39392 | | M11779 | |

## Name and History

*Calpain* is an intracellular, $Ca^{2+}$-dependent cysteine protease that is ubiquitously distributed, and shows regulated activity at neutral pH. The occurrence of $Ca^{2+}$-dependent neutral protease activity in rat brain was first described by Guroff (1964), and then Meyer *et al.* (1964) described a *kinase-activating factor* (*KAF*) in skeletal muscle that was identified as a $Ca^{2+}$-dependent protease by Huston & Krebs (1968). Furthermore, in 1972, calpain was reidentified as a *calcium-activated sarcoplasmic factor* (*CaSF*) hydrolyzing Z-lines by the Goll group (Busch *et al.*, 1972), and in 1977 as a *protein kinase C-activating factor* by Nishizuka's group (Takai *et al.*, 1977). Finally, calpain, which was called *CANP* (**c**alcium-**a**ctivated **n**eutral **p**rotease) at that time, was purified to homogeneity by Ishiura *et al.* (1978). Both names, calpain and CANP, had been used for several years, but they were unified as calpain by Suzuki in 1991 (Suzuki, 1991; Sorimachi *et al.*, 1996b). The name calpain originated from **cal**cium ion-dependent pa**pain**-like cysteine protease (most of the cysteine proteases have names ending in 'ain'), and led to convenient, shorter terms for calcium ion-dependent protease and calcium ion-dependent protease inhibitor (Murachi, 1989).

In 1984, cDNA for the large subunit of one of the chicken calpains, which has now been found to be of intermediate calcium sensitivity (μ/m-type) (Sorimachi *et al.*, 1995a), was cloned (Ohno *et al.*, 1984). As a result, the whole structure of the large subunit was determined, revealing that the calpain large subunit is a chimeric molecule, containing cysteine protease and calmodulin-like $Ca^{2+}$-binding modules. During the decade following this first cloning, various types of calpain subunits of several animals were identified and the structures were determined by cDNA cloning (Ohno *et al.*, 1984, 1986; Sakihama *et al.*, 1985; Aoki *et al.*, 1986; Emori *et al.*, 1986a,b, 1987; Takano *et al.*, 1986; Imajoh *et al.*, 1988; Asada *et al.*, 1989; McClelland *et al.*, 1989; Sorimachi *et al.*, 1989, 1993a, 1995b, 1996a; Delaney *et al.*, 1991; Ishida *et al.*, 1991; Karcz *et al.*, 1991; Lee *et al.*, 1992; Blumenthal *et al.*, 1993; DeLuca *et al.*, 1993; Sun *et al.*, 1993; Emori & Saigo, 1994; Graham-Siegenthaler *et al.*, 1994; Killefer & Koohmaraie, 1994; Denison *et al.*, 1995; Richard *et al.*, 1995; Theopold *et al.*, 1995; Barnes & Hodgkin, 1996).

Table 218.1 summarizes the members of the calpain family (peptidase family C2).

In mammals, two major, ubiquitous calpains exist, μ- and m-calpains. The old names, *calpain I* (equivalent to *μ-calpain*) and calpain II (equivalent to m-calpain), are no longer recommended (Suzuki, 1991). As the names suggest, μ- and m-calpains are activated *in vitro* by micromolar and millimolar $Ca^{2+}$ concentrations, respectively. Each of these calpains consists of two subunits, the different, large (ca. 80 kDa) and the common, small (ca. 30 kDa) subunits. As is shown in Figure 218.1, the large subunit can be divided into four domains. The second and the fourth domains are the cysteine protease and $Ca^{2+}$-binding domains, respectively. Thus, the protease activity of calpain is due to the large subunit. The functions of the first and third domains are unclear at present. The small subunit is composed of two domains, the N-terminal Gly-clustering, hydrophobic domain, and a C-terminal $Ca^{2+}$-binding region similar to that of the large subunit. The role of the small subunit is to regulate the activity of calpains. The two subunits are associated through the $Ca^{2+}$-binding domains in the absence of $Ca^{2+}$ (Yoshizawa *et al.*, 1995). The small subunits of μ- and m-calpain (Chapter 219) are identical.

## Activity and Specificity

The substrate specificity of calpain is relatively restricted, and most oligopeptides are not hydrolyzed. Casein is the most popular substrate for *in vitro* assays, and it is used either as the natural protein or modified with various chromophores, fluorescent reagents or isotopes. Calpain purified from skeletal muscle by standard methods as described later has specific activity of several hundred units per mg protein, where 1 unit corresponds to an increase of 1.0 absorbance unit at 280 nm per hour under the standard assay conditions. These are 3 mg ml$^{-1}$ casein, 0.1 M Tris–HCl (pH 7.5), 0.1–1 mM $CaCl_2$, and 5 mM 2-ME; 30°C, 20 min. The pH optimum is about 7.5. Activity is dependent on the $Ca^{2+}$ concentration, and that giving half maximal activity for μ-calpain is around 50 μM.

The rules governing the specificity of the calpains remain unclear. It seems that calpain recognizes the overall

*Table 218.1*   Members of the calpain family

| Type | | | Distribution | No. of amino acid residues in the large subunit | $M_r$ (kDa) | Domains | Ca²⁺ requirement |
|---|---|---|---|---|---|---|---|
| Vertebrate | μ-calpain | X04366 | ubiquitous | 714 (human) | 82 | I–IV | μM range |
| | μ/m-calpain | X01415 | ubiquitous | 705 (chicken) | 80 | I–IV | 0.1 mM range |
| | m-calpain | M23254 | ubiquitous | 700 (human) | 80 | I–IV | mM range |
| | p94 (nCL-1) | J05121 | skeletal muscle | 821 (rat) | 94 | I–IV, NS, IS1, IS2 | none |
| | nCL-2 | D14479 | stomach | 703 (rat) | 80 | I–IV | ? |
| | nCL-2′ | D14480 | stomach | 381 (rat) | 43 | I, II | – |
| | nCL-3 | Y10656 | digestive tissue | 753 (mouse) | 86 | I–IV, N3S | ? |
| *Drosophila* | dm-calpain (*calpA*) | Z46892 | ubiquitous | 828 | 94 | I–IV | ? |
| | *calpA′* | Z46891 | ubiquitous | 558 | 64 | I–III | ? |
| | *sol* | M64084 | ? | 1597 | 175 | II, solH | – |
| Schistosome | p87 (Sm-CL) | M67499 | ? | 758 | 87 | I–IV | ? |
| Nematode | *tra-3* | U12921 | ? | 648 | 74 | I–III | ? |
| | p92 (Ce-CL3) | Z35663 | ? | 805 | 92 | II, III | ? |
| | p71 (Ce-CL2) | L25598 | ? | 653 | 71 | I–III, V, N3S | ? |
| | p70 (Ce-CL4) | Z72515 | ? | 616 | 70 | I–III, solH | ? |
| | Ce-CL1 | M89338 | ? | ? | ? | II, ? | ? |
| Fungus | *palB* | Z54244 | ? | 842 | 94 | II, III | ? |
| Yeast | p83 (YMR154C) | Z49705 | ? | 727 | 83 | II | ? |

three-dimensional structure of its substrates rather than the primary structure. Even so, hydrophobic (Tyr, Met, Leu, Val) and Arg residues tend to be preferred in position P2 (Mellgren & Murachi, 1990). Protein kinases, phosphatases, phospholipases, cytoskeletal proteins, membrane proteins, cytokines, transcription factors, lens proteins, calmodulin-binding proteins and others have been suggested to be *in vivo* substrates, but clear evidence has not been obtained. Calpain proteolyzes these proteins in a limited manner rather than digesting them to small peptides, suggesting that it may modulate the functions of the substrate proteins by cutting their interdomain regions (Saido *et al.*, 1994). There is no evidence of any difference in substrate specificity between μ- and m-calpains.

Calpain has a very specific *in vivo* protein inhibitor, named calpastatin. Calpastatin contains four repeats of the inhibitory unit, each of which can inhibit calpain independently. Both μ- and m-calpains have similar susceptibility to calpastatin.

The Ca²⁺ concentration required for activation of μ-calpain ($pK_a$) is lowered by the addition of phosphoinositides such as PIP, PIP₂ and PIP₃ (Saido *et al.*, 1992a). In the presence of phosphoinositides, the Ca²⁺ concentration required for the activity is reduced to the range 100 nM–1 μM.

During the activation of μ-calpain, autolysis of the N-terminal few residues occurs. This autolysis precedes the appearance of proteolytic activity under the normal conditions *in vitro*, and thus has been considered to be an essential part of the activation mechanism (Saido *et al.*, 1992b).

## Structural Chemistry

As was mentioned above, mammalian μ-calpain is a heterodimer of about 110 kDa. At least in mammals, the μ-calpain large subunit is similar to but distinct from that of m-calpain, whereas the small subunits of both calpains are identical. The different Ca²⁺ dependencies of μ- and m-calpains must therefore be due to the large subunit.

The N-terminus of domain I of the large subunit is autolyzed upon activation by Ca²⁺, and the autolyzed calpain possesses higher sensitivity to Ca²⁺. It has therefore been considered that the processing of the N-terminus of the calpain large subunit is involved in the activation mechanism. A recent study, however, indicates that the change in Ca²⁺ sensitivity is caused by the dissociation of the two subunits, not by the autolysis, which is the result of the activation.

Domain II is a cysteine protease domain similar to those of papain and other members of peptidase family C1. Active-site Cys, His and Asn residues can be identified by similarity to papain, and have been confirmed by site-directed mutagenesis (Arthur *et al.*, 1995; Sorimachi *et al.*, 1993b). This domain is the most conserved among calpain family members, suggesting that it has indispensable functions.

The function of domain III has not yet been elucidated, and the domain has no significant similarity to other proteins in the databases. It is speculated that this domain plays a role in the recognition of substrates.

Domain IV contains four E-F hand structures, each of which can potentially bind one Ca²⁺ ion. However, *in vitro* experiments have revealed that only the first and the fourth units actually bind Ca²⁺, with $K_d$ values of about 35 μM (Minami *et al.*, 1987, 1988).

Domain V of the calpain small subunit contains clusters of Gly residues, making it hydrophobic in nature. This domain is considered to interact with membrane and/or membrane-associated proteins by hydrophobic interaction. Most of this domain is cut off following activation by Ca²⁺, indicating no involvement of this region in the protease activity itself.

Domain IV′ of the small subunit is similar to domain IV of the large subunit, also containing four E-F hand structures. This domain also can potentially bind four Ca²⁺ ions, but, as in the case of the large subunit, actually only the first and

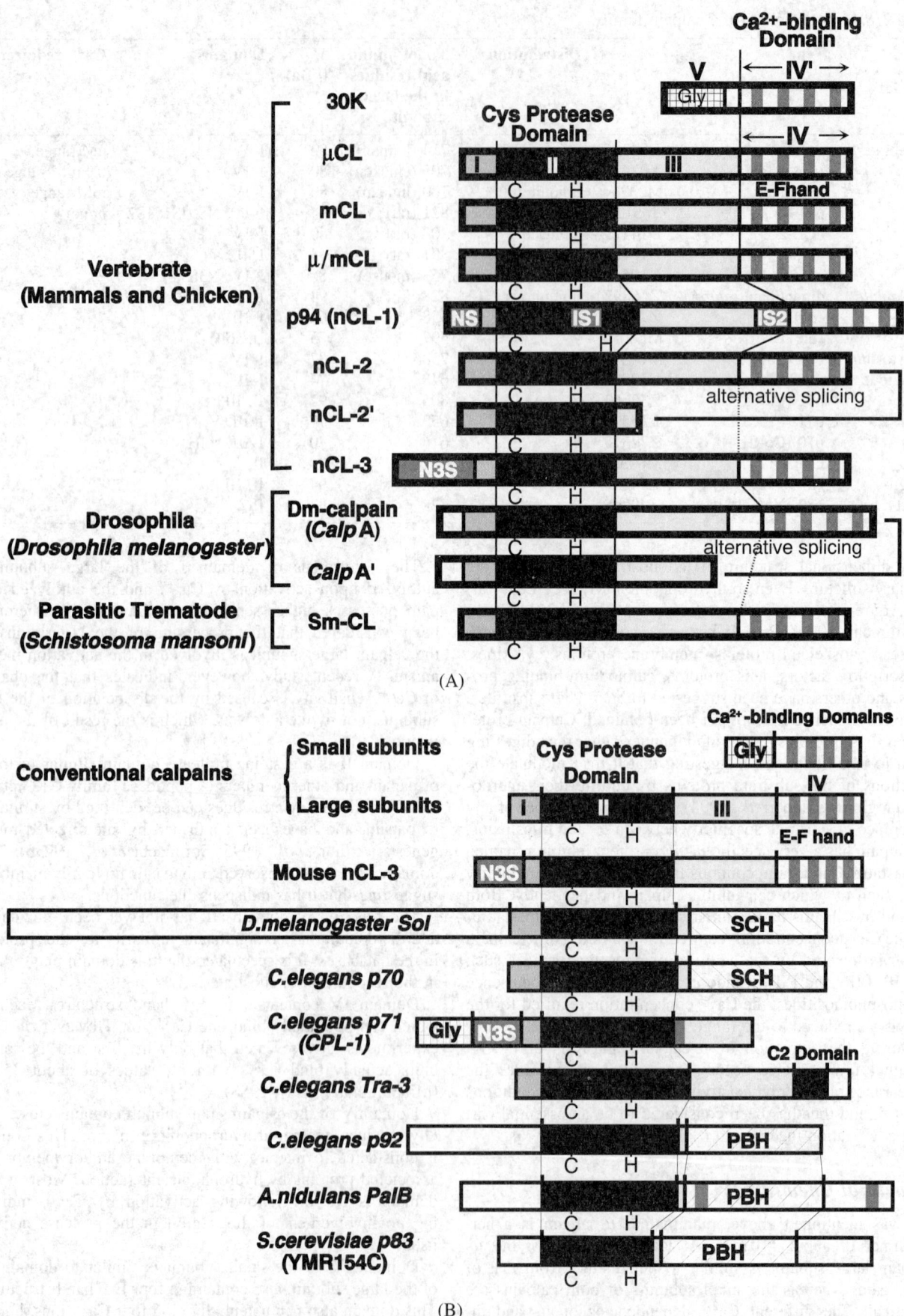

*Figure 218.1*    (A) Schematic structure of typical calpain subunits. (B) Schematic structure of atypical calpain homologs.

the fourth units bind $Ca^{2+}$, with $K_d$ values of about $150\,\mu M$ (Minami *et al.*, 1987, 1988).

Chicken also has $\mu$- and m-calpains, about 80% identical in sequence to those of mammals. In addition to these two isozymes, chicken has an intermediate type, $\mu$/m-calpain, whose structure and $Ca^{2+}$ dependence are intermediate between $\mu$- and m-calpains. $\mu$/m-Calpain has not yet been identified in mammals. Lower vertebrates than chicken also possess multiple forms of calpain, but the distinction between $\mu$- and m-calpains is unclear because the proteins have not yet been cloned.

Numerous calpain homologs have been reported from invertebrates. *Drosophila* has at least one typical calpain form, dm-calpain (identical to *calpA*) that is equally similar to both $\mu$- and m-calpains (ca. 45%). Thus, the structure does not show to which type this molecule belongs. In addition, *Drosophila* possesses another molecule, the *sol* (small optic lobe) gene product, that contains a cysteine protease domain significantly similar to that of calpain (Delaney *et al.*, 1991). However, *sol* does not contain a $Ca^{2+}$-binding domain, and, instead, includes zinc-finger-like repeats (TrpXCysXXCysX$_{10-11}$CysXXCys; Figure 218.1B). We propose to call such molecules 'atypical' calpain homologs. Mutations in the *sol* gene cause specific cells to degenerate in the developing optic lobes, resulting in the absence of certain classes of columnar neurons. The protease activity has yet to be investigated.

*Caenorhabditis elegans* has at least five distinct molecules that contains cysteine protease domains similar to that of calpain. However, none of these molecules has a recognizable $Ca^{2+}$-binding region, so they are all 'atypical' calpain homologs. One of them is a product of the *tra-3* gene, which is one of the sex-determination genes of *C. elegans*. *tra-3* is required for the correct sexual development of the soma and germ line in hermaphrodites, while being fully dispensable in males (Barnes & Hodgkin, 1996). One 'typical' calpain homolog, which possesses both cysteine protease and $Ca^{2+}$-binding domains, has been identified in a parasitic trematode, *Schistosoma mansoni* (Karcz *et al.*, 1991).

Apart from animals, one 'atypical' calpain homolog, the *pal*B gene product, has been identified in a fungus, *Aspergillus nidulans*. The *pal*B gene is involved in a signal transduction pathway involved in adaptation to alkaline ambient pH, and mutations in the *pal*B gene alter the response to pH (Denison *et al.*, 1995). In addition, one of the open reading frames (YMR154C) of the *Saccharomyces cerevisiae* genome encodes a cysteine protease (we tentatively call it 'p83') that is most similar to calpain amongst the known cysteine proteases. The function of this potential gene product is totally unknown.

## Preparation

$\mu$-Calpain is ubiquitously expressed in vertebrate tissues, so the largest organ, skeletal muscle, of rat, rabbit, cattle, pig, etc. is often used for the preparation of $\mu$- and m-calpains. The spleen of large animals such as pig and cattle can also be used. Usually, successive steps of DEAE-cellulose anion-exchange column chromatography, gel filtration, phenyl-Sepharose chromatography, and Mono-Q FPLC yield more than 1 mg of $\mu$-calpain from 1 kg of rabbit muscle, at a specific activity of 300–500 units mg$^{-1}$ and more than 10 000-fold purification. Chicken muscle does not contain distinct activity of $\mu$-calpain. Recently, some new purification methods involving affinity column chromatography were proposed (Anagli *et al.*, 1996; Molinari *et al.*, 1995). Recombinant $\mu$-calpain large and small subunits have been expressed in the baculovirus/insect cell system, and purified on a scale of tens of milligrams with specific activity comparable to native $\mu$-calpain (Meyer *et al.*, 1996).

## Biological Aspects

$\mu$-Calpain is ubiquitously distributed in mammalian and avian cells, so the enzyme is thought to have a fundamental and essential function, although none of the functions proposed to date has been proved (Saido *et al.*, 1994; Suzuki *et al.*, 1995). Maturation of megakaryocytes is accompanied by an increase in $\mu$-calpain level. The induction of vasospasm resulted in continuous autolytic activation of $\mu$-calpain as well as reduction in calpastatin activity. The interleukin $1\alpha$ precursor, which does not have a signal peptide, is believed to be processed through a novel secretion pathway that involves calpain. Protein kinase C is activated by limited proteolysis by calpain to become $Ca^{2+}$ independent. *Jun* is proteolyzed by $\mu$-calpain in a limited manner. When calpastatin mRNA or its antisense mRNA are expressed in NIH-3T3 or F9 cells, c-*jun* activity is promoted or suppressed, respectively, suggesting that c-*jun* is an *in vivo* substrate for $\mu$-calpain.

Various pathological states that may involve calpain and/or calpastatin have been reported, including cataract formation, Alzheimer's disease, ischemia, inflammation and muscular dystrophy (Saido *et al.*, 1994; Suzuki *et al.*, 1995). In Duchenne-type muscular dystrophy, the primary cause is a defect in dystrophin, which leads to membrane permeability, and an influx of $Ca^{2+}$ into the skeletal muscle cells; this activates calpain, resulting in degradation of muscle structural proteins. However, this mechanism is totally different from that of limb-girdle muscular dystrophy type 2A, which is caused by loss of function of the skeletal muscle-specific calpain, p94 (Chapter 220) (Richard *et al.*, 1995; Sorimachi *et al.*, 1995a).

## Distinguishing Features

Pig $\mu$-calpain, and monoclonal antibodies against human $\mu$-calpain large subunit and cattle small subunit, are commercially available from Chemicon International (see Appendix 2 for full names and addresses of suppliers).

## Further Reading

For detailed reviews, see Murachi (1989), Mellgren & Murachi (1990), Saido *et al.* (1994), and Sorimachi *et al.* (1996b), as well as other articles cited above.

## References

Anagli, J., Vilei, E.M.., Molinari, M., Calderara, S. & Carafoli, E. (1996) Purification of active calpain by affinity chromatography on an immobilized peptide inhibitor. *Eur. J. Biochem.* **241**, 948–954.

Aoki, K., Imajoh, S., Ohno, S., Emori, Y., Koike, M., Kosaki, G. & Suzuki, K. (1986) Complete amino acid sequence of the large subunit of the low-Ca$^{2+}$-requiring form of human Ca$^{2+}$-activated neutral protease (muCANP) deduced from its cDNA sequence. *FEBS Lett.* **205**, 313–317.

Arthur, J.S.C., Gauthier, S. & Elce, J.S. (1995) Active site residues in m-calpain: Identification by site-directed mutagenesis. *FEBS Lett.* **368**, 397–400.

Asada, K., Ishino, Y., Shimada, M., Shimojo, T., Endo, M., Kimizuka, F., Kato, I., Maki, M., Hatanaka, M. & Murachi, T. (1989) cDNA cloning of human calpastatin: sequence homology among human, pig, and rabbit calpastatins. *J. Enzym. Inhib.* **3**, 49–56.

Barnes, T.M. & Hodgkin, J. (1996) The tra-3 sex determination gene of *Caenorhabditis elegans* encodes a member of the calpain regulatory protease family. *EMBO J.* **15**, 4477–4484.

Blumenthal, E.J., Miller, A.C., Stein, G.H. & Malkinson, A.M. (1993) Serine/threonine protein kinases and calcium-dependent protease in senescent IMR-90 fibroblasts. *Mech. Ageing Dev.* **72**, 13–24.

Busch, W.A., Stromer, M.H., Goll, D.E. & Suzuki, A. (1972) Ca$^{2+}$-specific removal of Z lines from rabbit skeletal muscle. *J. Cell Biol.* **52**, 367–381.

Delaney, S.J., Hayward, D.C., Barleben, F., Fischbach, K.F. & Miklos, G.L. (1991) Molecular cloning and analysis of small optic lobes, a structural brain gene of *Drosophila melanogaster*. *Proc. Natl Acad. Sci. USA* **88**, 7214–7218.

DeLuca, C.I., Davies, P.L., Samis, J.A. & Elce, J.S. (1993) Molecular cloning and bacterial expression of cDNA for rat calpain II 80 kDa subunit. *Biochim. Biophys. Acta Gene Struct. Expression* **1216**, 81–93.

Denison, S.H., Orejas, M. & Arst, H.N. (1995) Signaling of ambient pH in *Aspergillus* involves a cysteine protease. *J. Biol. Chem.* **270**, 28519–28522.

Emori, Y. & Saigo, K. (1994) Calpain localization changes in coordination with actin-related cytoskeletal changes during early embryonic development of *Drosophila*. *J. Biol. Chem.* **269**, 25137–25142.

Emori, Y., Kawasaki, H., Imajoh, S., Kawashima, S. & Suzuki, K. (1986a) Isolation and sequence analysis of cDNA clones for the small subunit of rabbit calcium-dependent protease. *J. Biol. Chem.* **261**, 9472–9476.

Emori, Y., Kawasaki, H., Sugihara, S., Imajoh, S., Kawashima & Suzuki, K. (1986b) Isolation and sequence analyses of cDNA clones for the large subunits of two isozymes of rabbit calcium-dependent protease. *J. Biol. Chem.* **261**, 9465–9471.

Emori, Y., Kawasaki, H., Imajoh, S., Imahori, K. & Suzuki, K. (1987) Endogenous inhibitor for calcium-dependent cysteine protease contains four internal repeats that could be responsible for its multiple reactive sites. *Proc. Natl Acad. Sci. USA* **84**, 3590–3594.

Graham-Siegenthaler, K., Gauthier, S., Davies, P.L. & Elce, J.S. (1994) Active recombinant rat calpain II. Bacterially produced large and small subunits associate both *in vivo* and *in vitro*. *J. Biol. Chem.* **269**, 30457–30460.

Guroff, G. (1964) A neutral calcium-activated proteinase from the soluble fraction of rat brain. *J. Biol. Chem.* **239**, 149–155.

Huston, R.B. & Krebs, E.G. (1968) Activation of skeletal muscle phosphorylase kinase by Ca$^{2+}$. II. Identification of the kinase activating factor as a proteolytic enzyme. *Biochemistry* **7**, 2116–2122.

Imajoh, S., Aoki, K., Ohno, S., Emori, Y., Kawasaki, H., Sugihara, H. & Suzuki, K. (1988) Molecular cloning of the cDNA for the large subunit of the high-Ca$^{2+}$-requiring form of human Ca$^{2+}$-activated neutral protease. *Biochemistry* **27**, 8122–8128.

Ishida, S., Emori, Y. & Suzuki, K. (1991) Rat calpastatin has diverged primary sequence from other mammalian calpastatins but retains functionally important sequences. *Biochim. Biophys. Acta* **1088**, 436–438.

Ishiura, S., Murofushi, H., Suzuki, K. & Imahori, K. (1978) Studies of a calcium-activated neutral protease from chicken skeletal muscle. I. Purification and characterization. *J. Biochem.* **84**, 225–230.

Karcz, S.R., Podesta, R.B., Siddiqui, A.A., Dekaban, G.A., Strejan, G.H. & Clarke, M.W. (1991) Molecular cloning and sequence analysis of a calcium-activated neutral protease (calpain) from *Schistosoma mansoni*. *Mol. Biochem. Parasitol.* **49**, 333–336.

Killefer, J. & Koohmaraie, M. (1994) Bovine skeletal muscle calpastatin: cloning, sequence analysis & steady-state mRNA expression. *J. Anim. Sci.* **72**, 606–614.

Lee, W.J., Ma, H., Takano, E., Yang, H.Q., Hatanaka, M. & Maki, M. (1992) Molecular diversity in amino-terminal domains of human calpastatin by exon skipping. *J. Biol. Chem.* **267**, 8437–8442.

McClelland, P., Lash, J.A. & Hathaway, D.R. (1989) Identification of major autolytic cleavage sites in the regulatory subunit of vascular calpain II. A comparison of partial amino-terminal sequences to deduced sequence from complementary DNA. *J. Biol. Chem.* **264**, 17428–17431.

Mellgren, R.L. & Murachi, T. (1990) *Intracellular Calcium-dependent Proteolysis*. Boston: CRC Press.

Meyer, S.L., Bozyczko-Coyne, D., Mallya, S.K., Spais, C.M., Bihovsky, R., Kawooya, J.K., Lang, D.M., Scott, R.W. & Siman, R. (1996) Biologically active monomeric and heterodimeric recombinant human calpain I produced using the baculovirus expression system. *Biochem. J.* **314**, 511–519.

Meyer, W.L., Fischer, E.H. & Krebs, E.G. (1964) Activation of skeletal muscle phosphorylase b kinase by Ca$^{2+}$. *Biochemistry* **3**, 1033–1039.

Minami, Y., Emori, Y., Kawasaki, H. & Suzuki, K. (1987) E-F hand structure-domain of calcium-activated neutral protease (CANP) can bind Ca$^{2+}$ ions. *J. Biochem.* **101**, 889–895.

Minami, Y., Emori, Y., Imajoh-Ohmi, S., Kawasaki, H. & Suzuki, K. (1988) Carboxyl-terminal truncation and site-directed mutagenesis of the E-F hand structure-domain of the small subunit of rabbit calcium-dependent protease. *J. Biochem.* **104**, 927–933.

Molinari, M., Maki, M. & Carafoli, E. (1995) Purification of m-calpain by a novel affinity chromatography approach. New insights into the mechanism of the interaction of the protease with targets. *J. Biol. Chem.* **270**, 14576–14581.

Murachi, T. (1989) Intracellular regulatory system involving calpain and calpastatin. *Biochem. Int.* **18**, 263–294.

Ohno, S., Emori, Y., Imajoh, S., Kawasaki, H., Kisaragi, M. & Suzuki, K. (1984) Evolutionary origin of a calcium-dependent protease by fusion of genes for a thiol protease and a calcium-binding protein? *Nature* **312**, 566–570.

Ohno, S., Emori, Y. & Suzuki, K. (1986) Nucleotide sequence of a cDNA coding for the small subunit of human calcium-dependent protease. *Nucleic Acids Res.* **14**, 5559.

Richard, I., Broux, O., Allamand, V. *et al.* (1995) Mutations in the proteolytic enzyme calpain 3 cause limb-girdle muscular dystrophy type 2A. *Cell* **81**, 27–40.

Saido, T.C., Nagao, S., Shiramine, M., Tsukaguchi, M., Sorimachi, H., Murofushi, H., Tsuchiya, T., Ito, H. & Suzuki, K. (1992a) Autolytic transition of μ-calpain upon activation as resolved by antibodies distinguishing between the pre- and post-autolysis forms. *J. Biochem.* **111**, 81–86.

Saido, T.C., Shibata, M., Takenawa, T., Murofushi, H. & Suzuki, K. (1992b) Positive resultation of mu-calpain action by polyphospho-inositides. *J. Biol. Chem.* **267**, 24585–24590.

Saido, T.C., Sorimachi, H. & Suzuki, K. (1994) Calpain: New perspectives in molecular diversity and physiological-pathological involvement. *FASEB J.* **8**, 814–822.

Sakihama, T., Kakidani, H., Zenita, K., Yumoto, N., Kikuchi, T., Sasaki, T., Kannagi, R., Nakanishi, S., Ohmori, M., Takio, K. & Murachi, T. (1985) A putative Ca$^{2+}$-binding protein: structure of the light subunit of porcine calpain elucidated by molecular cloning and protein sequence analysis. *Proc. Natl Acad. Sci. USA* **82**, 6075–6079.

Sorimachi, H., Imajoh-Ohmi, S., Emori, Y., Kawasaki, H., Ohno, S., Minami, Y. & Suzuki, K. (1989) Molecular cloning of a novel mammalian calcium-dependent protease distinct from both $\mu$- and m-types. Specific expression of the mRNA in skeletal muscle. *J. Biol. Chem.* **264**, 20106–20111.

Sorimachi, H., Ishiura, S. & Suzuki, K. (1993a) A novel tissue-specific calpain species expressed predominantly in the stomach comprises two alternative splicing products with and without Ca$^{2+}$-binding domain. *J. Biol. Chem.* **268**, 19476–19482.

Sorimachi, H., Toyama-Sorimachi, N., Saido, T.C., Kawasaki, H., Sugita, H., Miyasaka, M., Arahata, K., Ishiura, S. & Suzuki, K. (1993b) Muscle-specific calpain, p94, is degraded by autolysis immediately after translation, resulting in disappearance from muscle. *J. Biol. Chem.* **268**, 10593–10605.

Sorimachi, H., Kimura, S., Kinbara, K., Kazama, J., Takahashi, M., Yajima, H., Ishiura, S., Sasagawa, N., Nonaka, I., Sugita, H., Maruyama, K. & Suzuki, K. (1995a) Muscle-specific calpain, p94, responsible for limb girdle muscular dystrophy type 2A, associates with connectin through IS2, a p94-specific sequence. *J. Biol. Chem.* **270**, 31158–31162.

Sorimachi, H., Tsukahara, T., Okada-Ban, M., Sugita, H., Ishiura, S. & Suzuki, K. (1995b) Identification of a third ubiquitous calpain species – chicken muscle expresses four distinct calpains. *Biochim. Biophys. Acta Gene Struct. Expression* **1261**, 381–393.

Sorimachi, H., Amano, S., Ishiura, S. & Suzuki, K. (1996a) Primary sequences of rat $\mu$-calpain large and small subunits are moderately and highly similar to those of human, respectively. *Biochim. Biophys. Acta Gene Struct. Expression* **1309**, 37–41.

Sorimachi, H., Kimura, S., Kinbara, K., Kazama, J., Takahashi, M., Yajima, H., Ishiura, S., Sasagawa, N., Nonaka, I., Sugita, H., Maruyama, K. & Suzuki, K. (1996b) Structure and physiological functions of ubiquitous and tissue-specific calpain species: muscle-specific calpain, p94 interacts with connectin/titin. *Adv. Biophys.* **33**, 101–122.

Sun, W., Ji, S.Q., Ebert, P.J., Bidwell, C.A. & Hancock, D.L. (1993) Cloning the partial cDNAs of $\mu$-calpain and m-calpain from porcine skeletal muscle. *Biochimie* **75**, 931–936.

Suzuki, K. (1991) Nomenclature of calcium dependent proteinase. *Biomed. Biochim. Acta* **50**, 483–484.

Suzuki, K., Sorimachi, H., Hata, A., Ohno, S., Emori, Y., Kawasaki, H., Saido, T. Ohmi-Imajoh, S. & Akita, Y. (1990) Calcium dependent protease: a novel molecular species, regulation of gene expression, and activation at the cell membrane. In: *Neurotoxicity of Excitatory Amino Acids* (Guidotti, A., ed.). New York: Raven Press, pp. 79–93.

Suzuki, K., Sorimachi, H., Yoshizawa, T., Kinbara, K. & Ishiura, S. (1995) Calpain: novel family members, activation, and physiological function. *Biol. Chem. Hoppe-Seyler* **376**, 523–529.

Takai, Y., Yamamoto, M., Inoue, M., Kishimoto, A. & Nishizuka, Y. (1977) A proenzyme of cyclic nucleotide-independent protein kinase and its activation by calcium-dependent neutral protease from rat liver. *Biochem. Biophys. Res. Commun.* **77**, 542–550.

Takano, E., Maki, M., Hatanaka, M., Mori, H., Zenita, K., Sakihama, T., Kannagi, R., Marti, T., Titani, K. & Murachi, T. (1986) Evidence for the repetitive domain structure of pig calpastatin as demonstrated by cloning of complementary DNA. *FEBS Lett.* **208**, 199–202.

Theopold, U., Pinter, M., Daffre, S., Tryselius, Y., Friedrich, P., Naessel, D.R. & Hultmark, D. (1995) CalpA, a *Drosophila* calpain homologue specifically expressed in a small set of nerve, midgut and blood cells. *Mol. Cell Biol.* **15**, 824–834.

Yoshizawa, T., Sorimachi, H., Tomioka, S., Ishiura, S. & Suzuki, K. (1995) Calpain dissociates into subunits in the presence of calcium ions. *Biochem. Biophys. Res. Commun.* **208**, 376–383.

*Hiroyuki Sorimachi*
Laboratory of Molecular Structure and Function,
Department of Molecular Biology,
Institute of Molecular and Cellular Biosciences,
University of Tokyo, Bunkyo-ku, Tokyo 113, Japan
Email: sorimach@imcbns.iam.u-tokyo.ac.jp

*Koichi Suzuki*
Laboratory of Molecular Structure and Function,
Department of Molecular Biology,
Institute of Molecular and Cellular Biosciences,
University of Tokyo, Bunkyo-ku, Tokyo 113, Japan
Email: kosuzuki@imcbns.iam.u-tokyo.ac.jp

# 219. m-Calpain

## Databanks

*Peptidase classification: clan CA, family C2, MEROPS ID: C02.002*
*NC-IUBMB enzyme classification: EC 3.4.22.17*

*Chemical Abstracts Service registry number: 78990-62-2*
*Databank codes:*

| Species | SwissProt | PIR | EMBL (cDNA) | EMBL (genomic) |
|---|---|---|---|---|
| Large subunit | | | | |
| *Bos taurus* | – | B40432 | U07850 | – |
| *Gallus gallus* | Q92178 | S57194 | D38026 | – |
| *Homo sapiens* | P17655 A33529 | A31218 | M23254 | J04700: exon 1 |
| *Oryctolagus cuniculus* | S10590 | B24815 | M13797 | – |
| *Rattus norvegicus* | Q07009 | S38361 | L09120 | X51772: exon 10 X51773: exon 11 |
| *Rattus norvegicus* | – | – | D14479 D14480 | D14478: pre-mature RNA, alternative splicing |
| Small subunit | | | | |
| *Bos taurus* | P13135 | A34466 | J05065 | – |
| *Homo sapiens* | P04632 | A26107 | M32886 | M31501: exon 1 |
| | | | | M31502: exon 2 |
| | | | | M31503: exon 3 |
| | | | | M31504: exon 4 |
| | | | | M31505: exon 5 |
| | | | | M31506: exon 6 |
| | | | | M31507: exon 7 |
| | | | | M31508: exon 8 |
| | | | | M31509: exon 9 |
| | | | | M31510: exon 10 |
| | | | | M31511: exon 11 and complete CDS |
| *Mus musculus* | – | – | Z36352 | – |
| *Oryctolagus cuniculus* | P06813 | A24816 | M13364 | – |
| *Rattus norvegicus* | – | A55143 | U10861 | – |
| *Sus scrofa* | P04574 S39392 | A25166 | M11778 M11779 | – |

## Name and History

*Calpain* is an intracellular, $Ca^{2+}$-dependent cysteine protease that is ubiquitously distributed, and shows regulated activity at neutral pH. The occurrence of $Ca^{2+}$-dependent neutral protease activity in rat brain was first described by Guroff (1964), and then Meyer *et al*. (1964) described a *kinase-activating factor* (KAF) in skeletal muscle that was identified as a $Ca^{2+}$-dependent protease by Huston & Krebs (1968). Furthermore, in 1972, calpain was reidentified as a *calcium-activated sarcoplasmic factor* (CaSF) hydrolyzing Z-lines by the Goll group (Busch *et al*., 1972), and in 1977 as a *protein kinase C-activating factor* by Nishizuka's group (Takai *et al*., 1977). Finally, calpain, which was called *CANP* (**c**alcium-**a**ctivated **n**eutral **p**rotease) at that time, was purified to homogeneity by Ishiura *et al*. (1978). Both names, calpain and CANP, had been used for several years, but they were unified as calpain by Suzuki in 1991 (Suzuki, 1991; Sorimachi *et al*., 1996b). The name calpain originated from **cal**cium ion-dependent pa**pain**-like cysteine protease (most cysteine proteases have names ending in 'ain'), and led to convenient, shorter terms for calcium ion-dependent protease and calcium ion-dependent protease inhibitor (Murachi, 1989).

In 1984, cDNA for the large subunit of one of the chicken calpains, which has now been found to be of intermediate calcium sensitivity ($\mu$/m-type) (Sorimachi *et al*., 1995a), was cloned (Ohno *et al*., 1984). As a result, the whole structure of the large subunit was determined, revealing that the calpain large subunit is a chimeric molecule, containing cysteine protease and calmodulin-like $Ca^{2+}$-binding modules. During the decade following this first cloning, various types of calpain subunits of several animals were identified and the structures were determined by cDNA cloning (Ohno *et al*., 1984, 1986; Sakihama *et al*., 1985; Aoki *et al*., 1986; Emori *et al*., 1986a,b, 1987; Takano *et al*., 1986; Imajoh *et al*., 1988; Asada *et al*., 1989; McClelland *et al*., 1989; Sorimachi *et al*., 1989, 1993a, 1995b, 1996a; Delaney *et al*., 1991; Ishida *et al*., 1991; Karcz *et al*., 1991; Lee *et al*., 1992; Blumenthal *et al*., 1993; DeLuca *et al*., 1993; Sun *et al*., 1993; Emori & Saigo, 1994; Graham-Siegenthaler *et al*., 1994; Killefer & Koohmaraie, 1994; Denison *et al*., 1995; Richard *et al*., 1995; Theopold *et al*., 1995; Barnes & Hodgkin, 1996). Table 218.1 lists the members of the calpain family (peptidase family C2).

In mammals, two major, ubiquitous calpains exist, $\mu$- and m-calpains. The old names, *calpain I* (equivalent to $\mu$-calpain) and *calpain II* (equivalent to *m-calpain*), are no longer recommended (Suzuki, 1991). As the names suggest, $\mu$- and m-calpains are activated *in vitro* by micromolar and millimolar $Ca^{2+}$ concentrations, respectively. Each of these calpains consists of two subunits, the different, large (ca. 80 kDa) and the common, small (ca. 30 kDa) subunits. As is shown in Figure 218.1, the large subunit can be divided into four domains. The second and the fourth domains are the cysteine protease and $Ca^{2+}$-binding domains, respectively. Thus, the protease activity of calpain is due to the large subunit. The functions of the first and third domains are unclear at present. The small subunit is composed of two domains, the N-terminal Gly-clustering, hydrophobic domain, and a C-terminal $Ca^{2+}$-binding region similar to that of the large subunit. The role of the small subunit is to regulate the activity of calpains. The two subunits are associated through the $Ca^{2+}$-binding domains in the absence of $Ca^{2+}$ (Yoshizawa et al., 1995). The small subunits of m- and $\mu$-calpain (Chapter 218) are identical.

## Activity and Specificity

The substrate specificity of calpain is relatively restricted, and most oligopeptides are not hydrolyzed. Casein is the most popular substrate for in vitro assays, and it is used either as the natural protein or modified with various chromophores, fluorescent reagents or isotopes. Calpain purified from skeletal muscle by standard methods as described later has a specific activity of several hundred units per mg protein, where 1 unit corresponds to an increase of 1.0 absorbance unit at 280 nm per hour under the standard assay conditions. These are 3 mg ml$^{-1}$ casein, 0.1 M Tris–HCl (pH 7.5), 5–10 mM $CaCl_2$, and 5 mM 2-ME; 30°C, 20 min. The pH optimum is about 7.5. Activity is dependent on the $Ca^{2+}$ concentration, and that giving half maximal activity for m-calpain is around 0.7 mM.

The rules governing the specificity of the calpains remain unclear. It seems that calpain recognizes the overall three-dimensional structure of its substrates, more than the primary structure. Even so, hydrophobic (Tyr, Met, Leu, Val) and Arg residues tend to be preferred in position P2 (Mellgren & Murachi, 1990). Protein kinases, phosphatases, phospholipases, cytoskeletal proteins, membrane proteins, cytokines, transcription factors, lens proteins, calmodulin-binding proteins and others have been suggested to be in vivo substrates, but clear evidence has not yet been obtained. Calpain proteolyzes these proteins in a limited manner rather than digesting them to small peptides, suggesting that it may modulate the functions of the substrate proteins by cutting their interdomain regions (Saido et al., 1994). There is no evidence of any difference in substrate specificity between $\mu$- and m-calpains.

Calpain has a very specific in vivo protein inhibitor, named calpastatin. Calpastatin contains four repeats of the inhibitory unit, each of which can inhibit calpain independently. Both $\mu$- and m-calpains have similar susceptibility to calpastatin.

During the activation of m-calpain, autolysis of the N-terminal few residues occurs. This autolysis precedes the appearance of proteolytic activity in the normal conditions in vitro, and thus has been considered to be an essential part of the activation mechanism. However, there are differences between the activation processes of m-calpain and $\mu$ calpain (Saido et al., 1994).

## Structural Chemistry

As was mentioned above, mammalian m-calpain is a heterodimer of about 110 kDa. At least in mammals, the m-calpain large subunit is similar to but distinct from that of $\mu$-calpain, whereas the small subunits of both calpains are identical. The different $Ca^{2+}$ dependencies of $\mu$- and m-calpains must therefore be due to the large subunits.

The N-terminus of domain I of the large subunit is autolyzed upon activation by $Ca^{2+}$, and the autolyzed calpain possesses higher sensitivity to $Ca^{2+}$. It has therefore been considered that the processing of the N-terminus of the calpain large subunit is involved in the activation mechanism. A recent study, however, indicates that the change in $Ca^{2+}$ sensitivity is caused by the dissociation of the two subunits, not by the autolysis, which is the result of the activation.

Domain II is a cysteine protease domain similar to those of papain and other members of peptidase family C1. Active-site Cys, His and Asn residues can be identified by similarity to papain, and have been confirmed by site-directed mutagenesis (Sorimachi et al., 1993c; Arthur et al., 1995). This domain is the most conserved among calpain family members, suggesting that it has indispensable functions.

The function of domain III has not yet been elucidated, and the domain has no significant similarity to other proteins on the databases. It is speculated that this domain plays a role in the recognition of substrates.

Domain IV contains four EF hand structures, each of which can potentially bind one $Ca^{2+}$ ion. However, in vitro experiment have revealed that only the first and the fourth units actually bind $Ca^{2+}$, with $K_d$ values of about 130 $\mu$M (Minami et al., 1987, 1988).

Domain V of the calpain small subunit contains clusters of Gly residues, making it hydrophobic in nature. This domain is considered to interact with membrane and/or membrane-associated proteins by hydrophobic interaction. Most of this domain is cut off following activation by $Ca^{2+}$, indicating no involvement of this region in the protease activity itself.

Domain IV' of the small subunit is similar to domain IV of the large subunit, also containing four EF hand structures. This domain also can potentially bind four $Ca^{2+}$ ions, but, as in the case of the large subunit, actually only the first and the fourth units bind $Ca^{2+}$, with $K_d$ values of about 150 $\mu$M (Minami et al., 1987, 1988).

Chicken also has $\mu$- and m-calpains, about 80% identical in sequence to those of mammals. In addition to these two isozymes, chicken has an intermediate type, $\mu$/m-calpain, whose structure and $Ca^{2+}$ dependence are intermediate between $\mu$- and m-calpains. $\mu$/m-Calpain has not yet been identified in mammals. Vertebrates lower than chicken also possess multiple forms of calpain, but the distinction between $\mu$- and m-calpains is unclear because the proteins have not yet been cloned. Calpains and their homologs are listed in Table 218.1.

## Preparation

m-Calpain is ubiquitously expressed in vertebrate tissues, so the largest organ, skeletal muscle, of rat, rabbit, cattle, pig, etc. is often used as source. The spleen of large animals such as pig and cattle can also be used. Usually, successive steps of DEAE-cellulose anion-exchange column chromatography, gel filtration, phenyl-Sepharose chromatography, and Mono-Q FPLC yield more than 1 mg of calpain from 1 kg of rabbit muscle, at a specific activity of 300–500 units mg$^{-1}$ and more than 10 000-fold purification. Recently, some new purification methods involving affinity column chromatography were proposed (Molinari *et al.*, 1995; Anagli *et al.*, 1996). Recombinant m-calpain large and small subunits have been expressed in *Escherichia coli* and purified on a scale of tens of milligrams with specific activity comparable to native m-calpain (Meyer *et al.*, 1996).

## Biological Aspects

m-Calpain is ubiquitously distributed in mammalian and avian cells, apart from mammalian erythrocytes, so the enzyme is thought to have a fundamental and essential function, although none of the functions proposed to date has been proved (Saido *et al.*, 1994; Suzuki *et al.*, 1995). The interleukin 1$\alpha$ precursor, which does not have a signal peptide, is believed to be processed through a novel secretion pathway that involves calpain. Protein kinase C is activated by limited proteolysis by calpain to become $Ca^{2+}$ independent.

Various pathological states that may involve calpain and/or calpastatin have been described, including cataract formation, Alzheimer's disease, ischemia, inflammation and muscular dystrophy (Saido *et al.*, 1994; Suzuki *et al.*, 1995). In Duchenne-type muscular dystrophy, the primary cause is a defect in dystrophin, which leads to membrane permeability, and an influx of $Ca^{2+}$ into the skeletal muscle cells; this activates calpain, resulting in degradation of muscle structural proteins. However, this mechanism is totally different from that of limb-girdle muscular dystrophy type 2A, which is caused by loss of function of the skeletal muscle-specific calpain, p94 (Chapter 220) (Richard *et al.*, 1995; Sorimachi *et al.*, 1995a).

## Distinguishing Features

Pig m-calpain and polyclonal antiserum against human m-calpain large subunit are commercially available from Chemicon International (see Appendix 2 for full names and addresses of suppliers).

## Further Reading

For detailed reviews, see Mellgren & Murachi (1990), Murachi (1989), Saido *et al.* (1994), Sorimachi & Suzuki (1993), Sorimachi *et al.* (1993b, 1994, 1995c, 1996b), and Suzuki *et al.* (1990, 1995).

## References

Anagli, J., Vilei, E.M., Molinari, M., Calderara, S. & Carafoli, E. (1996) Purification of active calpain by affinity chromatography on an immobilized peptide inhibitor. *Eur. J. Biochem.* **241**, 948–954.

Aoki, K., Imajoh, S., Ohno, S., Emori, Y., Koike, M., Kosaki, G. & Suzuki, K. (1986) Complete amino acid sequence of the large subunit of the low-$Ca^{2+}$-requiring form of human $Ca^{2+}$-activated neutral protease ($\mu$CANP) deduced from its cDNA sequence. *FEBS Lett.* **205**, 313–317.

Arthur, J.S.C., Gauthier, S. & Elce, J.S. (1995) Active site residues in m-calpain: identification by site-directed mutagenesis. *FEBS Lett.* **368**, 397–400.

Asada, K., Ishino, Y., Shimada, M., Shimojo, T., Endo, M., Kimizuka, F., Kato, I., Maki, M., Hatanaka, M. & Murachi, T.

(1989) cDNA cloning of human calpastatin: sequence homology among human, pig, and rabbit calpastatins. *J. Enzym. Inhib.* **3**, 49–56.

Barnes, T.M. & Hodgkin, J. (1996) The tra-3 sex determination gene of *Caenorhabditis elegans* encodes a member of the calpain regulatory protease family. *EMBO J.* **15**, 4477–4484.

Blumenthal, E.J., Miller, A.C., Stein, G.H. & Malkinson, A.M. (1993) Serine/threonine protein kinases and calcium-dependent protease in senescent IMR-90 fibroblasts. *Mech. Ageing Dev.* **72**, 13–24.

Busch, W.A., Stromer, M.H., Goll, D.E. & Suzuki, A. (1972) $Ca^{2+}$-specific removal of Z lines from rabbit skeletal muscle. *J. Cell Biol.* **52**, 367–381.

Delaney, S.J., Hayward, D.C., Barleben, F., Fischbach, K.F. & Miklos, G.L. (1991) Molecular cloning and analysis of small optic lobes, a structural brain gene of *Drosophila melanogaster. Proc. Natl Acad. Sci. USA* **88**, 7214–7218.

DeLuca, C.I., Davies, P.L., Samis, J.A. & Elce, J.S. (1993) Molecular cloning and bacterial expression of cDNA for rat calpain II 80 kDa subunit. *Biochim. Biophys. Acta Gene Struct. Expression* **1216**, 81–93.

Denison, S.H., Orejas, M. & Arst, H.N. (1995) Signaling of ambient pH in *Aspergillus* involves a cysteine protease. *J. Biol. Chem.* **270**, 28519–28522.

Emori, Y. & Saigo, K. (1994) Calpain localization changes in coordination with actin-related cytoskeletal changes during early embryonic development of *Drosophila. J. Biol. Chem.* **269**, 25137–25142.

Emori, Y., Kawasaki, H., Imajoh, S., Kawashima, S. & Suzuki, K. (1986a) Isolation and sequence analysis of cDNA clones for the small subunit of rabbit calcium-dependent protease. *J. Biol. Chem.* **261**, 9472–9476.

Emori, Y., Kawasaki, H., Sugihara, S., Imajoh, S., Kawashima & Suzuki, K. (1986b) Isolation and sequence analyses of cDNA clones for the large subunits of two isozymes of rabbit calcium-dependent protease. *J. Biol. Chem.* **261**, 9465–9471.

Emori, Y., Kawasaki, H., Imajoh, S., Imahori, K. & Suzuki, K. (1987) Endogenous inhibitor for calcium-dependent cysteine protease contains four internal repeats that could be responsible for its multiple reactive sites. *Proc. Natl Acad. Sci. USA* **84**, 3590–3594.

Graham-Siegenthaler, K., Gauthier, S., Davies, P.L. & Elce, J.S. (1994) Active recombinant rat calpain II. Bacterially produced large and small subunits associate both in vivo and in vitro. *J. Biol. Chem.* **269**, 30457–30460.

Guroff, G. (1964) A neutral calcium-activated proteinase from the soluble fraction of rat brain. *J. Biol. Chem.* **239**, 149–155.

Huston, R.B. & Krebs, E.G. (1968) Activation of skeletal muscle phosphorylase kinase by $Ca^{2+}$. II. Identification of the kinase activating factor as a proteolytic enzyme. *Biochemistry* **7**, 2116–2122.

Imajoh, S., Aoki, K., Ohno, S., Emori, Y., Kawasaki, H., Sugihara, H. & Suzuki, K. (1988) Molecular cloning of the cDNA for the large subunit of the high-$Ca^{2+}$-requiring form of human $Ca^{2+}$-activated neutral protease. *Biochemistry* **27**, 8122–8128.

Ishida, S., Emori, Y. & Suzuki, K. (1991) Rat calpastatin has diverged primary sequence from other mammalian calpastatins but retains functionally important sequences. *Biochim. Biophys. Acta* **1088**, 436–438.

Ishiura, S., Murofushi, H., Suzuki, K. & Imahori, K. (1978) Studies of a calcium-activated neutral protease from chicken skeletal muscle. I. Purification and characterization. *J. Biochem.* **84**, 225–230.

Karcz, S.R., Podesta, R.B., Siddiqui, A.A., Dekaban, G.A., Strejan, G.H. & Clarke, M.W. (1991) Molecular cloning and sequence analysis of a calcium-activated neutral protease (calpain) from *Schistosoma mansoni*. *Mol. Biochem. Parasitol.* **49**, 333–336.

Killefer, J. & Koohmaraie, M. (1994) Bovine skeletal muscle calpastatin: cloning, sequence analysis and steady-state mRNA expression. *J. Anim. Sci.* **72**, 606–614.

Lee, W.J., Ma, H., Takano, E., Yang, H.Q., Hatanaka, M. & Maki, M. (1992) Molecular diversity in amino-terminal domains of human calpastatin by exon skipping. *J. Biol. Chem.* **267**, 8437–8442.

McClelland, P., Lash, J.A. & Hathaway, D.R. (1989) Identification of major autolytic cleavage sites in the regulatory subunit of vascular calpain II. A comparison of partial amino-terminal sequences to deduced sequence from complementary DNA. *J. Biol. Chem.* **264**, 17428–17431.

Mellgren, R.L. & Murachi, T. (1990) *Intracellular Calcium-dependent Proteolysis.* Boston: CRC Press.

Meyer, W.L., Fischer, E.H. & Krebs, E.G. (1964) Activation of skeletal muscle phosphorylase b kinase by $Ca^{2+}$. *Biochemistry* **3**, 1033–1039.

Meyer, S.L., Bozyczko-Coyne, D., Mallya, S.K., Spais, C.M., Bihovsky, R., Kawooya, J.K., Lang, D.M., Scott, R.W. & Siman, R. (1996) Biologically active monomeric and heterodimeric recombinant human calpain I produced using the baculovirus expression system. *Biochem. J.* **314**, 511–519.

Minami, Y., Emori, Y., Kawasaki, H. & Suzuki, K. (1987) E-F hand structure-domain of calcium-activated neutral protease (CANP) can bind $Ca^{2+}$ ions. *J. Biochem.* **101**, 889–895.

Minami, Y., Emori, Y., Imajoh-Ohmi, S., Kawasaki, H. & Suzuki, K. (1988) Carboxyl-terminal truncation and site-directed mutagenesis of the E-F hand structure-domain of the small subunit of rabbit calcium-dependent protease. *J. Biochem.*, **104**, 927–933.

Molinari, M., Maki, M. & Carafoli, E. (1995) Purification of $\mu$-calpain by a novel affinity chromatography approach. New insights into the mechanism of the interaction of the protease with targets. *J. Biol. Chem.* **270**, 14576–14581.

Murachi, T. (1989) Intracellular regulatory system involving calpain and calpastatin. *Biochem. Int.* **18**, 263–294.

Ohno, S., Emori, Y., Imajoh, S., Kawasaki, H., Kisaragi, M. & Suzuki, K. (1984) Evolutionary origin of a calcium-dependent protease by fusion of genes for a thiol protease and a calcium-binding protein? *Nature* **312**, 566–570.

Ohno, S., Emori, Y. & Suzuki, K. (1986) Nucleotide sequence of a cDNA coding for the small subunit of human calcium-dependent protease. *Nucleic Acids Res.* **14**, 5559.

Richard, I., Broux, O., Allamand, V. *et al.* (1995) Mutations in the proteolytic enzyme calpain 3 cause limb-girdle muscular dystrophy type 2A. *Cell* **81**, 27–40.

Saido, T.C., Nagao, S., Shiramine, M., Tsukaguchi, M., Yoshizawa, T., Sorimachi, H., Ito, H., Tsuchiya, T., Kawashima, S. & Suzuki, K. (1994) Distinct kinetics of subunit autolysis in mammalian m-calpain activation. *FEBS Lett.* **346**, 263–267.

Sakihama, T., Kakidani, H., Zenita, K., Yumoto, N., Kikuchi, T., Sasaki, T., Kannagi, R., Nakanishi, S., Ohmori, M., Takio, K. & Murachi, T. (1985) A putative $Ca^{2+}$-binding protein: structure of the light subunit of porcine calpain elucidated by molecular cloning and protein sequence analysis. *Proc. Natl Acad. Sci. USA* **82**, 6075–6079.

Sorimachi, H. & Suzuki, K. (1993) A novel calpain species, n-calpain, active at nM levels of calcium? In: *Biological Functions of Proteases and Inhibitors* (Katunuma, N., Suzuki, K., Travis, J. & Fritz, H., eds). Tokyo: Japan Scientific Societies Press, pp. 35–46.

Sorimachi, H., Imajoh-Ohmi, S., Emori, Y., Kawasaki, H., Ohno, S., Minami, Y. & Suzuki, K. (1989) Molecular cloning of a novel mammalian calcium-dependent protease distinct from both $\mu$- and m-types. Specific expression of the mRNA in skeletal muscle. *J. Biol. Chem.* **264**, 20106–20111.

Sorimachi, H., Ishiura, S. & Suzuki, K. (1993a) A novel tissue-specific calpain species expressed predominantly in the stomach comprises two alternative splicing products with and without $Ca^{2+}$-binding domain. *J. Biol. Chem.* **268**, 19476–19482.

Sorimachi, H., Sorimachi, N., Ishiura, S. & Suzuki, K. (1993b) Identification and localization of a novel muscle-specific calpain, p94. In: *Proteolysis and Protein Turnover* (Bond, J.S. & Barrett, A.J., eds). London: Portland Press, pp. 45–49.

Sorimachi, H., Toyama-Sorimachi, N., Saido, T.C., Kawasaki, H., Sugita, H., Miyasaka, M., Arahata, K., Ishiura, S. & Suzuki, K. (1993c) Muscle-specific calpain, p94, is degraded by autolysis immediately after translation, resulting in disappearance from muscle. *J. Biol. Chem.* **268**, 10593–10605.

Sorimachi, H., Saido, T.C. & Suzuki, K. (1994) New era of calpain research: discovery of tissue-specific calpains. *FEBS Lett.* **343**, 35069.

Sorimachi, H., Kimura, S., Kinbara, K., Kazama, J., Takahashi, M., Yajima, H., Ishiura, S., Sasagawa, N., Nonaka, I., Sugita, H., Maruyama, K. & Suzuki, K. (1995a) Muscle-specific calpain, p94, responsible for limb girdle muscular dystrophy type 2A, associates with connectin through IS2, a p94-specific sequence. *J. Biol. Chem.* **270**, 31158–31162.

Sorimachi, H., Tsukahara, T., Okada-Ban, M., Sugita, H., Ishiura, S. & Suzuki, K. (1995b) Identification of a third ubiquitous calpain species – chicken muscle expresses four distinct calpains. *Biochim. Biophys. Acta Gene Struct. Expression* **1261**, 381–393.

Sorimachi, H., Yoshizawa, T. & Suzuki, K. (1995c) Molecular diversity and structure–function relationship of calpain. In: *Advances in Biochemistry and Molecular Biology*, Proceedings of the 11th FAOBMB Symposium on Biopolymers and Bioproducts.

Sorimachi, H., Amano, S., Ishiura, S. & Suzuki, K. (1996a) Primary sequences of rat $\mu$-calpain large and small subunits are moderately and highly similar to those of human, respectively. *Biochim. Biophys. Acta Gene Struct. Expression* **1309**, 37–41.

Sorimachi, H., Kimura, S., Kinbara, K., Kazama, J., Takahashi, M., Yajima, H., Ishiura, S., Sasagawa, N., Nonaka, I., Sugita, H., Maruyama, K. & Suzuki, K. (1996b) Structure and physiological functions of ubiquitous and tissue-specific calpain species: muscle-specific calpain, p94 interacts with connectin/titin. *Adv. Biophys.* **33**, 101–122.

Sun, W., Ji, S.Q., Ebert, P.J., Bidwell, C.A. & Hancock, D.L. (1993) Cloning the partial cDNAs of $\mu$-calpain and m-calpain from porcine skeletal muscle. *Biochimie* **75**, 931–936.

Suzuki, K. (1991) Nomenclature of calcium dependent proteinase *Biomed. Biochim. Acta* **50**, 483–484.

Suzuki, K., Sorimachi, H., Hata, A., Ohno, S., Emori, Y., Kawasaki, H., Saido, T. Ohmi-Imajoh, S. & Akita, Y. (1990) Calcium dependent protease: a novel molecular species, regulation of gene expression, and activation at the cell membrane. In: *Neurotoxicity of Excitatory Amino Acids* (Guidotti, A., ed.). New York: Raven Press, pp. 79–93.

Suzuki, K., Sorimachi, H., Yoshizawa, T., Kinbara, K. & Ishiura, S. (1995) Calpain: novel family members, activation, and physiological function. *Biol. Chem. Hoppe-Seyler* **376**, 523–529.

Takai, Y., Yamamoto, M., Inoue, M., Kishimoto, A. & Nishizuka, Y. (1977) A proenzyme of cyclic nucleotide-independent protein kinase and its activation by calcium-dependent neutral protease from rat liver. *Biochem. Biophys. Res. Commun.* **77**, 542–550.

Takano, E., Maki, M., Hatanaka, M., Mori, H., Zenita, K., Saki-hama, T., Kannagi, R., Marti, T., Titani, K. & Murachi, T. (1986) Evidence for the repetitive domain structure of pig calpastatin as demonstrated by cloning of complementary DNA. *FEBS Lett.*

**208**, 199–202.

Theopold, U., Pinter, M., Daffre, S., Tryselius, Y., Friedrich, P., Naessel, D.R. & Hultmark, D. (1995) CalpA, a *Drosophila* calpain homologue specifically expressed in a small set of nerve, midgut, and blood cells. *Mol. Cell Biol.* **15**, 824–834.

Yoshizawa, T., Sorimachi, H., Tomioka, S., Ishiura, S. & Suzuki, K. (1995) Calpain dissociates into subunits in the presence of calcium ions. *Biochem. Biophys. Res. Commun.* **208**, 376–383.

**Hiroyuki Sorimachi**
*Laboratory of Molecular Structure and Function,*
*Department of Molecular Biology,*
*Institute of Molecular and Cellular Biosciences,*
*University of Tokyo,*
*Bunkyo-ku, Tokyo 113, Japan*
*Email: sorimach@imcbns.iam.u-tokyo.ac.jp*

**Koichi Suzuki**
*Laboratory of Molecular Structure and Function,*
*Department of Molecular Biology,*
*Institute of Molecular and Cellular Biosciences,*
*University of Tokyo,*
*Bunkyo-ku, Tokyo 113, Japan*
*Email: kosuzuki@imcbns.iam.u-tokyo.ac.jp*

# 220. *Muscle calpain p94*

## Databanks

*Peptidase classification: clan CA, family C2, MEROPS ID: C02.004*
*NC-IUBMB enzyme classification: none*
*Databank codes:*

| Species | SwissProt | PIR | EMBL (cDNA) | EMBL (genomic) |
|---|---|---|---|---|
| *Bos taurus* | P51186 | – | U07858 | – |
| *Gallus gallus* | – | S57196 | D38028 | – |
| *Homo sapiens* | P20807 | A56218 | X85030 | – |
| *Mus musculus* | – | – | X92523 | – |
| *Rattus norvegicus* | P16259 | B34488 | J05121 | – |
| *Sus scrofa* | P43368 | – | U05678 | U23954: exons A, B and C |

## Name and History

As is described in the chapters on $\mu$-calpain and m-calpain (Chapters 218 and 219), until relatively recently, the study of calpain concerned mainly those two enzymes, but in 1989, in the course of the cDNA cloning of human $\mu$- and m-calpain large subunits, a cDNA encoding a novel molecule that is similar to but distinct from both $\mu$- and m-calpains was discovered (Sorimachi *et al.*, 1989). The novel large subunit was about 60% identical in amino acid sequence to those of $\mu$- and m-calpains. The novel sequence had the same basic domain structure as the calpains already known, in that it could be divided into four domains, an N-terminal domain (I), a protease domain (II), an intermediate domain (III), and a $Ca^{2+}$-binding domain (IV) (Sorimachi & Suzuki, 1992). Initially, this molecule was called **nCANP**, standing for **n**ovel $Ca^{2+}$-**a**ctivated **n**eutral **p**rotease, but soon

after, we decided to use a rather neutral name, **p94**, from its putative molecular mass (Sorimachi *et al.*, 1989). Four years later, another novel calpain homolog was identified, and we felt we had to systematize the names of calpain-related molecules (Sorimachi *et al.*, 1993a). We thought in terms of the existence of hypothetical enzymes, n-calpain-1, n-calpain-2, etc., where 'n' stands for novel (or hopefully nano), and took p94 to be the large subunit of n-calpain-1. Accordingly, we proposed to call the molecule we had discovered **nCL-1** (**n**-**c**alpain-**1** **l**arge subunit). By this time, however, the name p94 had become established, especially in the field of meat tenderization (Balcerzak *et al.*, 1995), and hoping to minimize confusion, we now try to use both names, p94 and nCL-1, in each publication. Recently, as is described below, Beckmann's group discovered that the gene for p94 is responsible for limb-girdle muscular dystrophy (LGMD) type 2A (LGMD2A), and they termed p94 *calpain 3* (Richard *et al.*, 1995). Calpain 3 could be considered a derivative of the older names calpain I (equivalent to $\mu$-calpain), and calpain II (m-calpain). Our recommendation is that the names p94 (for the protein corresponding to the known cDNA sequence), and nCL-1 (for the putative heterodimer) are used. For reasons that are described below, it has not yet been possible to prove that the heterodimer exists, so n-calpain-1 is still hypothetical.

## Activity and Specificity

For 4 years after the discovery of the p94 cDNA, the p94 protein had still not been detected. The mRNA for p94 is expressed predominantly in skeletal muscle, and it is about ten times more abundant than the mRNAs for the $\mu$- and m-calpain large subunits. Nevertheless, we could not detect either activity or protein of p94 in a skeletal muscle extract by use of the usual procedures for $\mu$- and m-calpains. In 1993, expression experiments with mutated p94 began to reveal the unique properties of p94 (Sorimachi *et al.*, 1993b), and further work has clarified further aspects (Richard *et al.*, 1995; Sorimachi *et al.*, 1995). The conclusions can be summarized as follows:

- The mRNA shows skeletal muscle-specific expression.
- The p94 protein undergoes extremely rapid autolysis (half-life *in vitro* being less than 10 min).
- Specific inhibitors of $\mu$- and m-calpains, including calpastatin, E-64 and leupeptin, have no effect on p94 activity.
- p94 possesses a putative nuclear localization signal, and is located in the nucleus as well as in the cytosol.
- p94 binds to the gigantic muscle protein, connectin/titin, specifically through one of the unique regions of p94, termed IS2 (see below).
- The gene for p94 is responsible for limb-girdle muscular dystrophy type 2A.

Because of its extremely rapid autolysis, studies of p94 at the protein level are very difficult, but what we have learned so far is that:

- p94 certainly possesses protease activity, as was expected from the highly conserved structure of its protease domain

relative to those of the $\mu$- and m-calpain large subunits.

- Substrates of p94 so far identified are p94 itself, myotonin protein kinase, fodrin, and a 65 kDa protein of COS cells. On the other hand, p94 does not cleave $\mu$- and m-calpain subunits, connectin/titin, or most of the endogenous proteins of COS cells.
- Despite the fact that p94 contains significantly conserved EF hand structures at its C-terminus (domain IV), the protease activity does not require $Ca^{2+}$, and the addition of $Ca^{2+}$ does not affect the activity detectably.
- During autolysis, there is a transient appearance and subsequent disappearance of a 55 kDa fragment of p94.

The proteolytic activity of p94 must presumably be regulated *in vivo*, but the mechanism remains unclear; one possibility is that the activity is suppressed by the interaction of the enzyme with connectin/titin.

## Structural Chemistry

The basic domain structure of p94 is identical to those of the $\mu$- and m-calpain large subunits, as shown in Figure 220.1. However, p94 contains three unique regions, NS, IS1 and IS2, that are not found in the other calpains, and are not related to other peptide sequences in the databases.

The 'N-terminus specific region' (NS) is a proline-rich structure of unknown function within domain I.

The protease domain (domain II) is the most highly conserved domain amongst the calpain large subunit homologs, and there is particularly high conservation of sequence around the putative active-site residues, Cys129, His334 and Asn358. Site-directed mutagenesis has confirmed that Cys129 and His334 are essential for activity. The IS1 region is an insertion into domain II just ahead of His334. The presence or absence of the IS1 region does not affect proteolytic activity, however.

The IS2 sequence, in the C-terminal part of domain III, contains a PXKKKKXKP motif that resembles a nuclear localization signal. Deletion of this region suppresses autolysis, and thus stabilizes p94 expression. Also, the IS2 region is necessary and sufficient for the binding to the $N_2$-line portion of connectin/titin.

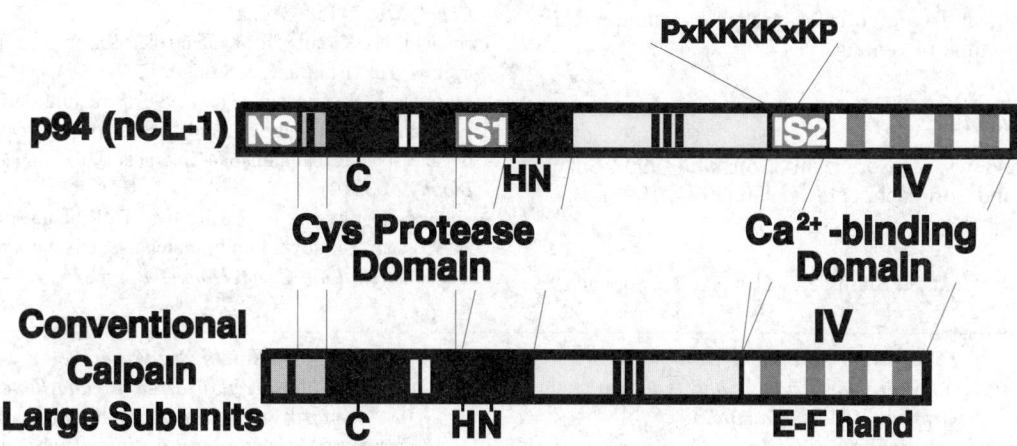

*Figure 220.1*   Schematic structure of p94 and conventional calpain large subunits.

The potential $Ca^{2+}$-binding domain (domain IV) of p94 has yet to be examined for its $Ca^{2+}$-binding activity, but the similarity to the domains IV of other calpains strongly suggests that it must bind $Ca^{2+}$. However, the proteolytic activity of p94 is apparently $Ca^{2+}$ independent, and excess of EGTA does not prevent the autolysis of the protein.

## Preparation

So far, p94 has never been purified from any kind of tissue, including skeletal muscle, of any organism. Even the unambiguous detection of p94 in skeletal muscle by a specific antibody has not yet been reported.

## Biological Aspects

In 1995, Beckmann's group discovered by a positional cloning method that a gene for p94 is responsible for limb-girdle muscular dystrophy type 2A (LGMD2A), and that LGMD2A patients have various mutations in the p94 gene (Richard et al., 1995). The mutations are widely distributed in the p94 gene, and include missense point mutations, nonsense mutations, frame-shift mutations, and splice site mutations. The absence of any 'hot-spot' in the p94 gene makes diagnosis very difficult.

All other muscular dystrophies such as Duchenne-type muscular dystrophy (DMD), congenital muscular dystrophy (CMD), and other LGMDs originate from deficiencies of structural proteins associated with the muscle cell membrane (sarcolemma), dystrophin, merosin and sarcoglycans, respectively (Campbell, 1995). But in LGMD2A, a defect in a cytosolic enzyme, p94, causes symptoms similar to those of other muscular dystrophies. In LGMD2A, dystrophin and sarcoglycans are normally expressed, whereas both are in very low level in DMD and all of the sarcoglycans are downregulated in four different types of LGMD (2C, 2D, 2E and 2F) (Campbell, 1995; Fardeau et al., 1996). It seems that some signal transduction pathway from sarcolemma to cytosol and/or nucleus must play an important role in the muscular dystrophies.

## Distinguishing Features

Various specific antisera have been described in the literature, but no specific antiserum or p94 protein is commercially available at the time of writing.

## Further Reading

For detailed reviews, see Sorimachi & Suzuki (1992), Suzuki et al. (1995) and Sorimachi et al. (1990, 1994, 1996).

## References

Balcerzak, D., Poussard, S., Brustis, J.J., Elamrani, N., Soriano, M.,
Cottin, P. & Ducastaing, A. (1995) An antisense oligodeoxyribonucleotide to m-calpain mRNA inhibits myoblast fusion. J. Cell Sci. 108, 2077–2082.

Campbell, K.P. (1995) Three muscular dystrophies: Loss of cytoskeleton-extracellular matrix linkage. Cell 80, 675–679.

Fardeau, M., Hillaire, D., Mignard, C., Feingold, N., Feingold, J., Mignard, D., de Ubeda, B., Collin, H., Tomé, F.M.S., Richard, I. & Beckmann, J. (1996) Juvenile limb-girdle muscular dystrophy: Clinical, histopathological and genetic data from a small community living in the Réunion Island. Brain 119, 295–308.

Murachi, T. (1989) Intracellular regulatory system involving calpain and calpastatin. Biochem. Int. 18, 263–294.

Richard, I., Broux, O., Allamand, V. et al. (1995) Mutations in the proteolytic enzyme calpain 3 cause limb-girdle muscular dystrophy type 2A. Cell 81, 27–40.

Sorimachi, H. & Suzuki, K. (1992) Sequence comparison among muscle-specific calpain, p94, and calpain subunits. Biochim. Biophys. Acta 1160, 55–62.

Sorimachi, H., Imajoh-Ohmi, S., Emori, Y., Kawasaki, H., Ohno, S., Minami, Y. & Suzuki, K. (1989) Molecular cloning of a novel mammalian calcium-dependent protease distinct from both $\mu$- and m-types. Specific expression of the mRNA in skeletal muscle. J. Biol. Chem. 264, 20106–20111.

Sorimachi, H., Ohmi, S., Emori, Y., Kawasaki, H., Saido, T.C., Ohno, S., Minami, Y. & Suzuki, K. (1990) A novel member of the calcium-dependent cysteine protease family. Biol. Chem. Hoppe-Seyler 371, 171–176.

Sorimachi, H., Ishiura, S. & Suzuki, K. (1993a) A novel tissue-specific calpain species expressed predominantly in the stomach comprises two alternative splicing products with and without $Ca^{2+}$-binding domain. J. Biol. Chem. 268, 19476–19482.

Sorimachi, H., Toyama-Sorimachi, N., Saido, T.C., Kawasaki, H., Sugita, H., Miyasaka, M., Arahata, K., Ishiura, S. & Suzuki, K. (1993b) Muscle-specific calpain, p94, is degraded by autolysis immediately after translation, resulting in disappearance from muscle. J. Biol. Chem. 268, 10593–10605.

Sorimachi, H., Saido, T.C. & Suzuki, K. (1994) New era of calpain research: Discovery of tissue-specific calpains. FEBS Lett. 343, 35069.

Sorimachi, H., Kinbara, K., Kimura, S., Takahashi, M., Ishiura, S., Sasagawa, N., Sorimachi, N., Shimada, H., Tagawa, K., Maruyama, K. & Suzuki, K. (1995) Muscle-specific calpain, p94, responsible for limb girdle muscular dystrophy type 2A, associates with connectin through IS2, a p94-specific sequence. J. Biol. Chem. 270, 31158–31162.

Sorimachi, H., Kimura, S., Kinbara, K., Kazama, J., Takahashi, M., Yajima, H., Ishiura, S., Sasagawa, N., Nonaka, I., Sugita, H., Maruyama, K. & Suzuki, K. (1996) Structure and physiological functions of ubiquitous and tissue-specific calpain species: muscle-specific calpain, p94 interacts with connectin/titin. Adv. Biophys. 33, 101–122.

Suzuki, K., Sorimachi, H., Yoshizawa, T., Kinbara, K. & Ishiura, S. (1995) Calpain: novel family members, activation, and physiological function. Biol. Chem. Hoppe-Seyler 376, 523–529.

**Hiroyuki Sorimachi**
Laboratory of Molecular Structure and Function,
Department of Molecular Biology,
Institute of Molecular and Cellular Biosciences,
University of Tokyo, Bunkyo-ku, Tokyo 113, Japan
Email: sorimach@imcbns.iam.u-tokyo.ac.jp

**Koichi Suzuki**
Laboratory of Molecular Structure and Function,
Department of Molecular Biology,
Institute of Molecular and Cellular Biosciences,
University of Tokyo, Bunkyo-ku, Tokyo 113, Japan
Email: kosuzuki@imcbns.iam.u-tokyo.ac.jp

# 221. *Streptopain*

## Databanks

*Peptidase classification: clan CA, family C10, MEROPS ID: C10.001*
*NC-IUBMB enzyme classification: EC 3.4.22.10*
*Chemical Abstracts Service registry number: 9025-51-8*
*Databank codes:*

| Species | SwissProt | PIR | EMBL (cDNA) | EMBL (genomic) |
|---|---|---|---|---|
| *Porphyromonas gingivalis* | P43158 | – | S75942 | – |
| *Porphyromonas gingivalis* | – | – | M83096 | – |
| *Streptococcus pyogenes* | P00788 | A00978 | – | – |
| *Streptococcus pyogenes* | P26296 | – | M86905 | – |

## Name and History

The proteolytic behavior of cultures of hemolytic streptococci was first described by Frobisher (1926) who noticed the effect that they had on certain animal tissues. The active component was named *histase*, but it remained uncharacterized, as did the strain of *Streptococcus* producing it, although the author did state that it resembled *S. pyogenes*. The first definite recognition of a proteolytic enzyme from group A streptococci was by Elliott (1945) who observed its ability to destroy the serological reactivity of the type-specific M protein of the organism. Elliott observed also that many types of group A streptococci, regardless of their serological type, produced an extracellular enzyme.

The name *streptopain* is a shorter form of the previously used *streptococcal proteinase*, which defined the source and nature of the enzyme. Streptopain has been referred to previously also as *streptococcal cysteine proteinase* and *streptococcus peptidase A*. The enzyme that Kapur *et al.* (1994) refer to as *interleukin-1β convertase* appears to be streptopain, too. It has further been suggested that *pyrogenic exotoxin B* is identical to the streptopain precursor (Gerlach *et al.*, 1983; Hauser & Schlievert, 1990). However, the nucleotide sequence of pyrogenic exotoxin B shows an insert of 34 amino acid residues between residues 56 and 57 of the chemically derived sequence of streptopain. In addition, there are several residues that are placed differently in the nucleotide sequence as compared to the chemically derived sequence (Tai *et al.*, 1976; Yonaha *et al.*, 1982). These differences were attributed to incorrect sequence alignments and errors in sequence analysis, but it is not known currently whether these two proteins are actually identical. The report of Ohara-Nemoto *et al.* (1994), which states that streptococcal pyrogenic exotoxin B is proteolytically active, is misleading. Those authors observed proteolytic activity in what they termed a '29 kDa mature form'. This mature form appears to be streptopain itself and not the exotoxin. The nucleotide sequence of streptococcal pyrogenic exotoxin B has a His residue located at residue 42 as compared to residue 46 in the chemically derived sequence (Yonaha *et al.*, 1982; Hauser & Schlievert, 1990). The reaction of streptopain with

2,2'-dipyridyl disulfide (French *et al.*, 1990) suggests that a cationic residue is in close proximity to Cys47, the catalytically essential Cys. This might be taken to support the chemically derived sequence, although in the absence of the three-dimensional structures the uncertainty remains.

## Activity and Specificity

The ability of streptopain to act as a proteolytic enzyme was first demonstrated by its capacity to destroy the serological activity of the type-specific M protein of group A streptococci (Elliott, 1945). Elliott showed that human and rabbit fibrin, casein, milk and gelatin were susceptible to attack by the enzyme. Subsequent to this study it was found that the enzyme catalyzed the hydrolysis of typical trypsin substrates. The rates of hydrolysis catalyzed by streptopain, however, were only about 1% of those catalyzed by trypsin (Elliott & Liu, 1970). The proteinase has also been observed to cleave fibronectin and vitronectin, two extracellular matrix proteins involved in the maintenance of host tissue integrity, and also to cleave human interleukin 1β precursor to generate mature interleukin 1β with full biological activity, suggesting a role in inflammation and shock (Kapur *et al.*, 1993; Musser *et al.*, 1996).

The preference of streptopain for hydrophobic residues was demonstrated by Gerwin *et al.* (1966) who used the reduced and carboxymethylated B chain of insulin as substrate. They found that the most rapidly hydrolyzed peptide bond was the Phe+Tyr linkage in the Phe-Phe-Tyr sequence (residues 24–26). The chief requirement for hydrolysis is a bulky side chain in the amino acid residue adjacent on the N-terminal side of the residue contributing the carbonyl group to the susceptible bond (the P2 position; Schechter & Berger, 1967, 1968; Berger & Schechter, 1970). The nature of the amino blocking group is important in determining the rate of hydrolysis of dipeptide substrates. Specificity studies carried out by Liu *et al.* (1969) on dipeptide substrates of the form Z-Xaa-Phe showed that the specificity was reflected in the value of $K_m$ and not $k_{cat}$. In this situation, they found that the most susceptible peptide bond was that where Xaa was norleucine. This substrate had the lowest value of $K_m$, although where side chains were of similar dimensions the enzyme appeared to distinguish only poorly between polar and nonpolar groups.

An extensive data set for the hydrolysis of $N$-acylamino acid esters by streptopain was presented by Kortt & Liu (1973). Streptopain exhibits a high ratio of esterase to peptidase activity (Liu *et al.*, 1969). With peptide ester substrates of the form Z-Xaa-OPh (the phenyl ester) or Z-Xaa-OPhNO$_2$ the enzyme exhibited a preference for Xaa = Lys, whereas Lys in this position provided the worst substrates for peptidase activity.

Changing Xaa to Gly in the ester substrates produced marked resistance to hydrolysis. Similarly, when Xaa = Lys, removal of Z from Xaa and derivatization of the $\varepsilon$-amino group of Lys with Z produced the very poor substrate $N^\varepsilon$,Z-Lys-OPhNO$_2$. These results agree with those of Gerwin *et al.* (1966) who suggested that for peptidase activity there is a requirement for interaction of the N-terminal blocking group of a substrate in a hydrophobic binding pocket. Thus, the major specificity determinant of substrates for streptopain appears to be a hydrophobic residue in the P2 position. The specificity requirement for esterase activity is similar to that for peptidase activity (Liu & Elliott, 1971).

Streptopain also exhibits transferase activity. Its ability to catalyze the synthesis of anilides has been demonstrated (Liu & Elliott, 1971). The fact that it catalyzes the synthesis of hydroxamic acids only poorly has been attributed to the rapid hydrolysis of the product at the pH at which the experiment was performed (Liu & Elliott, 1971).

Assays of catalytic activity of streptopain (proteolytic, peptidase, esterase and serological) have been extensively reported (Liu & Elliott, 1971; Elliott & Liu, 1970; French, 1992).

French *et al.* (1990) demonstrated that 2,2'-dipyridyl disulfide can be used to determine the amount of free thiol present in the enzyme and thus serves as an active center titrant.

Streptopain is inhibited by common cysteine proteinase inhibitors, such as iodoacetate, iodoacetamide, $N$-ethylmaleimide, $p$-chloromercuribenzoate, metallic ions (Cu$^{2+}$, Hg$^{2+}$ and Ag$^+$) and atmospheric oxygen (Liu & Elliott, 1965a).

## Structural Chemistry

The primary structures of streptopain and its proenzyme have been chemically sequenced (Tai *et al.*, 1976; Yonaha *et al.*, 1982). The enzyme is first expressed as an inactive proenzyme ($M_r$ 34 000, 337 residues), in which the catalytic site cysteine side chain is derivatized as a mixed disulfide with the volatile mercaptan methanthiol (Lo *et al.*, 1984). This undergoes reduction and subsequent autocatalytic proteolysis to form the active enzyme ($M_r$ 28 000, 253 residues). It has been suggested that the reducing activity of the streptococcal cell walls might initiate the activation (Liu & Elliott, 1965b). The proenzyme and the active proteinase each contain only a single cysteine residue (Cys47, enzyme numbering) and both proteins are devoid of cystine disulfide bonds. It has been suggested on the basis of pH-dependent reactivity probe studies that the Cys side chain forms an ion pair with that of a catalytically essential His residue within the catalytic site (French *et al.*, 1990). The presence of a His residue at the catalytic site was suggested by chemical modification studies (Liu, 1967), although it was not possible to determine whether His195 or His46 was the residue modified. The catalytic residue is probably His195 because it is unlikely that His46, the residue adjacent to the catalytically essential Cys, would be able to adopt the required geometry for ion pair formation.

Potentially important virulence factors from the bacterium *Porphyromonas gingivalis* are involved in proteolysis and agglutination of erythrocytes (hemagglutination) (Madden *et al.*, 1995). The 96 to 99 kDa protein derived from the *PrtT* gene of *P. gingivalis* has significant similarity in sequence to streptopain and its proenzyme. The similarity spans the entire amino acid sequence, being greatest around the putative catalytic-site residues. The *P. gingivalis* protein contains 11 cysteine residues, compared to one in streptopain, which makes possible the formation of disulfide bonds. Assuming that the mechanism of *PrtT* catalysis is similar to that of other cysteine proteinases, it would be expected that Cys202 and His345 would be the catalytically essential residues. This is based on the similarity of two regions of *PrtT* to the identified putative catalytic sites of papain, actinidain, cathepsins, streptopain and other cysteine proteinases (Tai *et al.*, 1976; Brocklehurst *et al.*, 1987). It is suggested that the *Streptococcus* and *Porphyromonas* species have inherited the homologous genes from a common ancestor (Madden *et al.*, 1995).

## Preparation

Streptopain precursor can readily be purified from most strains of group A streptococci. Active streptopain can be obtained from the proenzyme by the action of subtilisin, trypsin, preformed streptopain, the bacterial cell walls, or by the autocatalytic activity of the proenzyme itself (Liu & Elliott, 1971). It has been reported that the best method for the preparation of a homogeneous sample of streptopain with high yield (88–98%) is by the action of trypsin on the proenzyme (Liu & Elliott, 1965b). The recombinant pyrogenic exotoxin type B has been expressed in *Escherichia coli* (Hauser & Schlievert, 1990).

## Biological Aspects

Streptopain is found in most strains of *Streptococcus pyogenes*. It occurs as an inactive precursor in which the catalytically essential Cys is derivatized as a disulfide with methanethiol. The proenzyme is activated by interaction with the bacterial cell walls, after which it undergoes autocatalytic cleavage to form the active enzyme (Liu & Elliott, 1965a). The first stage of activation forms an intermediate protein, termed the modified proenzyme, which can be separated from the enzyme and proenzyme by chromatography on carboxymethyl cellulose (Liu & Elliott, 1965b, 1971). The intermediate results from cleavage of a 17 amino acid residue peptide from the N-terminus. It exhibits the immunological properties of both the enzyme and proenzyme and about 90% of the proteolytic activity of the fully active enzyme (Liu & Elliott, 1965b). Thus, removal of the 17 residue peptide exposes the catalytic site of the enzyme. Subsequent removal of a peptide of approximately 100 amino acid residues from the N-terminus produces the fully active protein (Liu & Elliott, 1965b, 1971). It has been suggested that the gene that encodes pyrogenic exotoxin type B is the same gene that encodes a streptopain precursor (Gerlach *et al*., 1983; Hauser & Schlievert, 1990).

## Further Reading

Full reviews have been published by Liu & Elliott (1970, 1971) and Brocklehurst *et al*. (1987), the latter being an extensive review of the major types of cysteine proteinase including streptopain.

## References

Berger, A. & Schechter, I. (1970) Mapping the active site of papain with the aid of peptide substrates and inhibitors. *Philos. Trans. R. Soc. Lond. [Biol.]* **257**, 249–264.

Brocklehurst, K., Willenbrock, F. & Salih, E. (1987) Cysteine proteinases. In: *Hydrolytic Enzymes* (Neuberger, A. & Brocklehurst, K., eds). Amsterdam: Elsevier, pp. 39–158.

Elliott, S.D. (1945) A proteolytic enzyme produced by group A streptococci with special reference to its effect on the type-specific M antigen. *J. Exp. Med.* **81**, 573–592.

Elliott, S.D. & Liu, T.-Y. (1970) Streptococcal proteinase. *Methods Enzymol.* **19**, 252–261.

French, H.P. (1992) Studies on streptococcal proteinase. PhD thesis, University of London.

French, H., Williams, R., Salih, E., Kowlessur, D. & Brocklehurst, K. (1990) Studies on streptococcal proteinase. *Biochem. Soc. Trans.* **18**, 593–594.

Frobisher, M. (1926) Tissue-digesting enzyme (histase) of streptococci. *J. Exp. Med.* **44**, 777–786.

Gerlach, D., Knöll, H., Köhler, W., Ozegowski, J.-H. & Hríbalova, V. (1983) Isolation and characterization of erythrogenic toxins V. Communication: identity of erythrogenic toxin type B and streptococcal proteinase precursor. *Zbl. Bakt. Hyg., I. Abt. Orig. A.* **255**, 221–233.

Gerwin, B.I., Stein, W.H. & Moore, S. (1966) On the specificity of streptococcal proteinase. *J. Biol. Chem.* **241**, 3331–3339.

Hauser, A.R. & Schlievert, P.M. (1990) Nucleotide sequence of the streptococcal pyrogenic exotoxin type B gene and relationship between the toxin and streptococcal precursor. *J. Bacteriol.* **172**, 4536–4542.

Kapur, V., Majesky, M.W., Li, L.-L., Black, R.A. and Musser, J.M. (1993) Cleavage of interleukin 1$\beta$ (IL-1$\beta$) precursor to produce active IL-1$\beta$ by a conserved extracellular cysteine protease from *Streptococcus pyogenes*. *Proc. Natl Acad. Sci. USA* **90**, 7676–7680.

Kapur, V., Maffei, J.T., Greer, R.S., Li, L.-L., Adams, G.J. and Musser, J.M. (1994) Vaccination with streptococcal extracellular cysteine protease (interleukin-1$\beta$ convertase) protects mice against challenge with heterologous group A streptococci. *Microb. Pathog.* **16**, 443–450.

Kortt, A.A. & Liu, T.-Y. (1973) On the mechanism of action of streptococcal proteinase II. Comparison of the kinetics of proteinase- and papain-catalyzed hydrolysis of N-acyl amino acid esters. *Biochemistry* **12**, 328–337.

Liu, T.-Y. (1967) Demonstration of the presence of a histidine residue at the active site of streptococcal proteinase. *J. Biol. Chem.*, **242**, 4029–4032.

Liu, T.-Y. & Elliott, S.D. (1965a) Activation of streptococcal proteinase and its zymogen by bacterial cell walls. *Nature* **206**, 33–34.

Liu, T.-Y. & Elliott, S.D. (1965b) Streptococcal proteinase: the zymogen to enzyme transformation. *J. Biol. Chem.* **240**, 1138–1142.

Liu, T.-Y. & Elliott, S.D. (1970) Streptococcal proteinase. *Methods Enzymol.* **19**, 252–261.

Liu, T.-Y. & Elliott, S.D. (1971) Streptococcal proteinase. *The Enzymes* **3**, 609–647.

Liu, T.-Y., Nomura, N., Jonsson, E.H. & Wallace, B.G. (1969) Streptococcal proteinase-catalyzed hydrolysis of some ester and amide substrates. *J. Biol. Chem.* **244**, 5745–5756.

Lo, S.-S., Fraser, B.A. & Liu, T.-Y. (1984) The mixed disulfide in the zymogen of streptococcal proteinase. *J. Biol. Chem.* **259**, 11041–11045.

Madden, T.E., Clark, V.L. & Kuramitsu, H.K. (1995) Revised sequence of the *Porphyromonas gingivalis* PrtT cysteine protease/hemagglutinin gene: homology with streptococcal pyrogenic exotoxin B/streptococcal proteinase. *Infect. Immun.* **63**, 238–247.

Musser, J.M., Stockbauer, K., Kapur, V. & Rudgers, G.W. (1996) Substitution of cysteine 192 in a highly conserved *Streptococcus pyogenes* extracellular cysteine proteinase (interleukin 1$\beta$ convertase) alters proteolytic activity and ablates zymogen processing. *Infect. Immun.* **64**, 1913–1917.

Ohara-Nemoto, T., Sasaki, M., Kaneko, M., Nemoto, T. & Ota, M. (1994) Cysteine proteinase activity of streptococcal pyrogenic exotoxin B. *Can. J. Microbiol.* **40**, 930–936.

Schechter, I. & Berger, A. (1967) On the size of the active site in proteases. I. Papain. *Biochem. Biophys. Res. Commun.* **27**, 157–162.

Schechter, I. & Berger, A. (1968) On the active site of proteases. III. Mapping the active site of papain; specific peptide inhibitors of papain. *Biochem. Biophys. Res. Commun.* **32**, 898–912.

Tai, J.Y., Kortt, A.A., Liu, T.-Y. & Elliott, S.D. (1976) Primary structure of streptococcal proteinase. III. Isolation of cyanogen bromide peptides: complete covalent structure of the polypeptide chain. *J. Biol. Chem.* **251**, 1955–1959.

Yonaha, K., Elliott, S.D. & Liu, T.-Y. (1982) Primary structure of zymogen of streptococcal proteinase. *J. Protein Chem.* **1**, 317–334.

*Aaron B. Watts*
*Laboratory of Structural and Mechanistic Enzymology,*
*Department of Biochemistry,*
*Queen Mary and Westfield College,*
*University of London,*
*Mile End Road, London E1 4NS, UK*
*Email: ha5345@qmw.ac.uk*

*Keith Brocklehurst*
*Laboratory of Structural and Mechanistic Enzymology,*
*Department of Biochemistry,*
*Queen Mary and Westfield College,*
*University of London,*
*Mile End Road, London E1 4NS, UK*
*Email: kb1@qmw.ac.uk*

# 222. *Ubiquitin C-terminal hydrolase*

## Databanks

Peptidase classification: clan CA, family C12, MEROPS ID: C12.001
NC-IUBMB enzyme classification: EC 3.1.2.15
Databank codes:

| Species | Gene/form | SwissProt | PIR | EMBL (cDNA) | EMBL (genomic) |
|---|---|---|---|---|---|
| *Bos taurus* | *uchL1* | P23356 | S17561 | – | – |
| *Bos taurus* | pgp9.5 | – | B40085 | – | – |
| *Caenorhabditis elegans* | – | Q09444 | – | – | Z46676: cosmid C08B11 |
| *Drosophila melanogaster* | uch | P35122 | S33955 | X69678 | X69679: complete gene |
| *Homo sapiens* | *uchL1* | P09936 | A25856 | X04741 S10891 | X17377: 5′ end |
| *Homo sapiens* | *uchL3* | P15374 | A40085 | M30496 | – |
| *Homo sapiens* | pgp9.5 | – | S14307 | – | – |
| *Homo sapiens* | EST | – | – | T06263 | – |
| *Homo sapiens* | EST | – | – | T11220 | – |
| *Rattus norvegicus* | *uchL1* | Q00981 | – | D10699 | – |
| *Saccharomyces cerevisiae* | *yuH1* | P35127 | S51332 | – | Z49599: chromosome X ORF |
| *Schizosaccharomyces pombe* | – | Q10171 | – | Z69368 | – |

## Name and History

The enzymes to be considered in the present chapter are members of the **deub**iquitinating enzyme (DUB) class of hydrolases (Mayer & Wilkinson, 1989; Tobias & Varshavsky, 1991; Baker *et al*., 1992; Papa & Hochstrasser, 1993; Wilkinson *et al*., 1995). Ubiquitin is a 76 amino acid protein that is covalently attached to a variety of cellular proteins to target them for proteolysis by the 26S proteasome (Wilkinson, 1995) (see also Chapter 173).

*Ubiquitin thiolesterase* activity was first discovered as an activity that hydrolyzed thioesters and oxygen esters of ubiquitin (Rose & Warms, 1983). It was known at the time that the C-terminus of ubiquitin was activated by formation of an adenylate, and that this adenylate could further react with thiols or amines. The authors postulated that the thioesterase played a role in reversing the adventitious trapping of the ubiquitin adenylate by endogenous thiols and small amines. Upon purification and further characterization, it was found that this esterase was also an amidase and it was subsequently referred to as **ubiquitin carboxyl-terminal hydrolase (UCH)** (Pickart & Rose, 1985). By use of the C-terminal ethyl ester of ubiquitin as a substrate, at least four similar activities were demonstrated in cattle thymus (Mayer & Wilkinson, 1989). The major enzyme (now known as **UCH-L3**) was cloned (Wilkinson *et al*., 1989) and found to be similar to a neuron-specific protein of unknown function (**PGP 9.5**, now called **UCH-L1**) (Doran *et al*., 1983) and a yeast enzyme (Miller *et al*., 1989; Liu *et al*., 1989) with similar sequence and enzymatic activity (**YUH1p**). Similar enzymes exist in *Drosophila* (Zhang *et al*., 1993) and *Caenorhabditis elegans* and are discussed below. These enzymes are commonly known as ubiquitin C-terminal

hydrolases (family 1) or ubiquitin thiolesterases. NC-IUBMB (1992) classify ubiquitin C-terminal hydrolase only with regard to its thioester hydrolase activity, as EC 3.1.2.15. Here, we shall use the term UCH.

## Activity and Specificity

UCH catalyzes the hydrolysis of an ester or amide bond involving the C-terminal Gly76 of ubiquitin (Rose & Warms, 1983; Pickart & Rose, 1985; Liu *et al*., 1989; Mayer & Wilkinson, 1989; Wilkinson *et al*., 1989; Zhang *et al*., 1993). No other cleavages have been demonstrated. The range of substrates varies somewhat between different forms of the enzyme, but most of them strongly prefer small leaving groups of less than about 60 amino acids. The human enzymes do not discriminate on the basis of charge in the P1′ position (see Figure 222.1). They show diminished rates with larger substrates. UCH-L1 is more selective than UCH-L3, although neither will hydrolyze peptides with Pro in the P1′ position.

UCH shows maximal activity at about pH 7.5 (Wilkinson *et al*., 1986). The active-site thiol is very sensitive to oxidation, and reductants such as DTT or 2-ME are required for optimal activity. The oxidized enzyme can be reactivated by reduction with high concentrations of DTT. Metal ion inhibition (Wilkinson *et al*., 1986) is probably due to oxidation of the thiol. Inhibition by high salt is due to interference with ubiquitin binding (Larsen *et al*., 1996). The C-terminal aldehyde and the hydroxamate of ubiquitin are very effective inhibitors with inhibition constants in the nanomolar range (Pickart & Rose, 1985; Wilkinson *et al*., 1986).

The original assay for UCH was a coupled assay in which the ubiquitin was enzymatically adenylated and trapped with

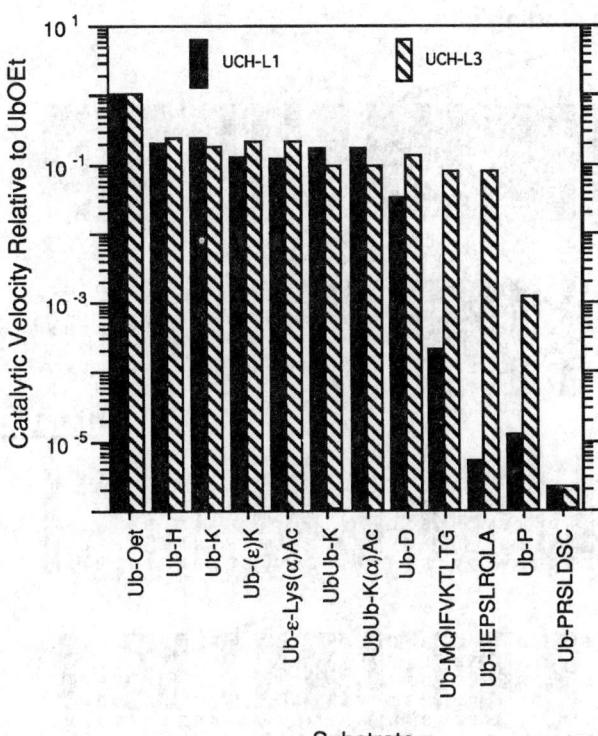

Substrate

*Figure 222.1* Substrate specificity of human UCH isozymes. The rate of hydrolysis by UCH-L1 and UCH-L3 was measured for each of the indicated substrates at 15 µM substrate concentration. Data are normalized to the rate observed for hydrolysis of the generic substrate ubiquitin ethyl ester. Note the logarithmic scale. *Abbreviations:* Ub, ubiquitin; Ub-xxx, a fusion protein with the indicated amino acid or peptide fused to the C-terminal glycine of ubiquitin, Ub-(ε)-K, ubiquitin linked to lysine through the ε-amino group.

high thiol concentrations (Rose & Warms, 1983). Upon cleavage of the ubiquitin thioester, the liberated ubiquitin was again adenylated and the consumption of ATP was measured. Three more direct forms of assay have been used subsequently: (a) hydrolysis of the ubiquitin C-terminal amide of [$^3$H]butanol-4-amine can be quantitated by counting released radioactivity (Pickart & Rose, 1985), although this substrate has been difficult to synthesize reproducibly; (b) the hydrolysis of the C-terminal ethyl ester or amide of ubiquitin can be quantitated by HPLC (Wilkinson *et al.*, 1986), and (c) the cleavage of ubiquitin fusion peptides can be quantitated by SDS-PAGE (Miller *et al.*, 1989; Liu *et al.*, 1989).

## Structural Chemistry

The amino acid sequences of several forms of UCH are shown in Figure 222.2. The enzymes vary from 24.8 to 26.4 kDa, with pI values of 4.3–5.3. The crystal structure of human UCH-L3 has been solved (Johnston *et al.*, 1997) (Brookhaven PDB 1UCH). It contains a core of six antiparallel sheets and nine helices. The closest structural homolog is cathepsin B (Chapter 209), a member of the papain family (C1) (Chapter 186).

Using site-directed mutagenesis, the catalytic residues have been identified as Cys95, His169 and Asp184 in the UCH-L3

numbering system. These residues are arranged similarly to those of papain, and the secondary structure is also similar in the six strands and in the helix containing the catalytic cysteine.

Binding of ubiquitin to the enzyme is primarily electrostatic, being prevented by high salt concentrations. Three absolutely conserved acidic residues are implicated in binding of ubiquitin: Glu10, Glu14 and Asp33.

Most of the UCH family 1 enzymes are extremely soluble, but the brain enzyme (UCH-L1, PGP 9.5) precipitates and forms fibrils at concentrations above 1 g liter$^{-1}$. In the presence of ubiquitin, however, the protein is soluble to over 20 g liter$^{-1}$.

## Preparation

The forms of UCH can be purified from a number of sources, including yeast (Liu *et al.*, 1989), erythrocytes (Pickart & Rose, 1985), brain (Doran *et al.*, 1983) and thymus (Mayer & Wilkinson, 1989). Both human isozymes have been expressed at high levels in *E. coli* (Larsen *et al.*, 1996), as has the yeast protein (Miller *et al.*, 1989). The protein is conveniently purified from brain using conventional methods (Doran *et al.*, 1983), and can be purified from several tissues using ubiquitin affinity columns (see above). The purification of recombinant protein is particularly easy due to its high level of expression.

## Biological Aspects

The expression of UCH isozymes is tissue specific and developmentally regulated. UCH-L1 is an abundant neuronal protein accounting for approximately 2% of the soluble protein in brain (Doran *et al.*, 1983). It is transported down the axon by the slow component b pathway known to carry cytoskeletal and cytoplasmic proteins (Bizzi *et al.*, 1991). The protein is first expressed early in neural development (Smith-Thomas *et al.*, 1994; Schofield *et al.*, 1995). It is a prominent message in *Drosophila* nurse cells, ovary and testis (Zhang *et al.*, 1993). Upon fertilization, the message is lost over the first 4–6 h of development, consistent with it being a maternal message. Protein is selectively accumulated in a variety of neural inclusion bodies (Lowe *et al.*, 1990) and the level of message is strongly downregulated upon SV40-induced transformation of lung fibroblasts (Honore *et al.*, 1991). Antibodies to UCH-L1 have been widely used to examine neural anatomy and innervation (Day, 1992).

The human gene structure has been reported, including the putative promoter (Day *et al.*, 1990). The 5′ sequence resembles that found in neuron-specific enolase and Thy-1 antigen genes and contains both positive and negative *cis*-acting elements (Mann *et al.*, 1996). The protein is not induced by stress, nor by any of a variety of experimental manipulations (Wilkinson *et al.*, 1992).

The normal substrates of UCH are thought to be small esters and amides of ubiquitin (Rose & Warms, 1983; Pickart & Rose, 1985; Wilkinson *et al.*, 1992). These may include esters with cellular thiols, and amides between ubiquitin and peptides remaining from the degradation of ubiquitinated proteins. It has also been postulated that these enzymes participate in the cotranslational processing of ubiquitin fusion protein gene products (Wilkinson, 1995).

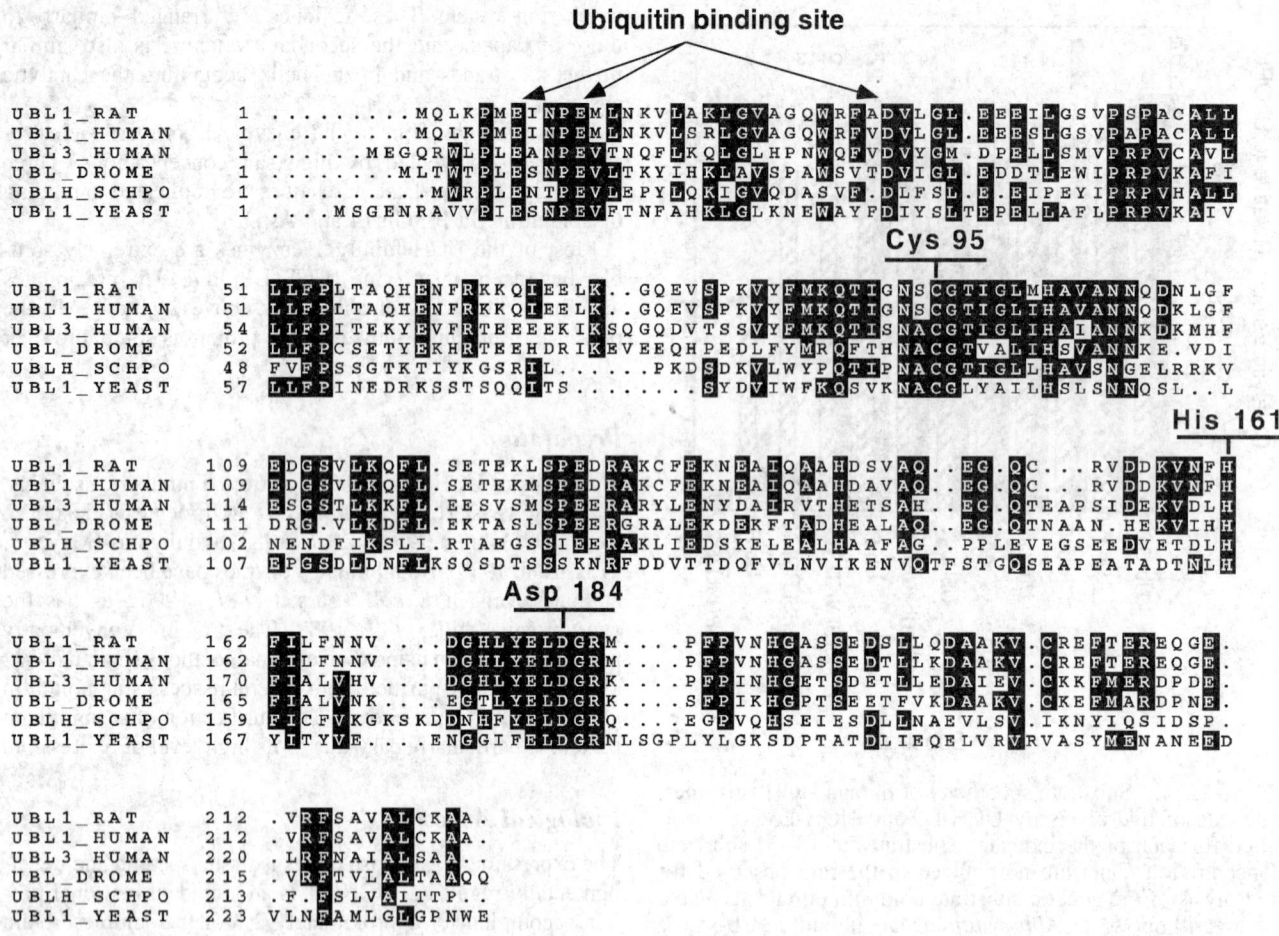

**Figure 222.2** Alignment of amino acid sequences of six UCH isozymes. Residues identical in at least four sequences are boxed in black, while conservative substitutions are boxed in gray. The putative residues contacting ubiquitin are indicated, as are the residues comprising the catalytic triad (Cys95, His161 and Asp184).

## Related Peptidases

A similar, although larger enzyme has been predicted from the sequence of the *C. elegans* genome. This 37.7 kDa protein contains an additional 100 amino acid residues C-terminal to a recognizable UCH domain. Polyclonal antibodies directed against the human UCH forms have been raised. Antibodies to UCH-L1 only recognize that isozyme, while those to UCH-L3 also recognize a slightly smaller form (designated UCH-L2). Both antibodies cross-react with the enzymes from other mammals, although not with the yeast protein.

A second family (C19) of deubiquitinating enzymes, the UBP family (Hochstrasser, 1996), are thiol peptidases that will remove ubiquitin from larger leaving groups. While mechanistically similar, they are of much larger molecular size than the UCH enzymes of family C12, and appear to be genetically unrelated (see Chapter 223).

## Further Reading

See Wilkinson (1995) for a comprehensive review, and Hochstrasser (1996) for an account of the most recent developments.

## References

Baker, R.T., Tobias, J.W. & Varshavsky, A. (1992) Ubiquitin-specific proteases of *Saccharomyces cerevisiae*. Cloning of UBP2 and UBP3, and functional analysis of the UBP gene family. *J. Biol. Chem.* **267**, 23364–23375.

Bizzi, A., Schaetzle, B., Patton, A., Gambetti, P. & Autilio Gambetti, L. (1991) Axonal transport of two major components of the ubiquitin system: free ubiquitin and ubiquitin carboxyl-terminal hydrolase PGP 9.5. *Brain Res.* **548**, 292–299.

Day, I.N. (1992) Enolases and PGP9.5 as tissue-specific markers. *Biochem. Soc. Trans.* **20**, 637–642.

Day, I.N., Hinks, L.J. & Thompson, R.J. (1990) The structure of the human gene encoding protein gene product 9.5 (PGP9.5), a neuron-specific ubiquitin C-terminal hydrolase. *Biochem. J.* **268**, 521–524.

Doran, J.F., Jackson, P., Kynoch, P.A.M. & Thompson, R.J. (1983) Isolation of PGP 9.5, a new human neurone-specific protein detected by high-resolution two-dimensional electrophoresis. *J. Neurochem.* **40**, 1542–1547.

Hochstrasser, M. (1996) Protein degradation or regulation: Ub the judge. [Review]. *Cell* **84**, 813–815.

Honore, B., Rasmussen, H.H., Vandekerckhove, J. & Celis, J.E. (1991) Neuronal protein gene product 9.5 (IEF SSP 6104) is expressed in cultured human MRC-5 fibroblasts of normal origin and is strongly down-regulated in their SV40 transformed counterparts. *FEBS Lett.* **280**, 235–240.

Johnston, S.C., Larsen, C.N., Cook, W.J., Wilkinson, K.D. & Hill, C.P. (1997) Crystal structure of a deubiquitinating enzyme (human UCH-L3) at 1.8 Å resolution. *EMBO J.* **16**, 3787–3796.

Larsen, C.N., Price, J.S. & Wilkinson, K.D. (1996) Substrate binding and catalysis by ubiquitin C-terminal hydrolases: identification of two active site residues. *Biochemistry* **35**, 6735–6744.

Liu, C.C., Miller, H.I., Kohr, W.J. & Silber, J.I. (1989) Purification of a ubiquitin protein peptidase from yeast with efficient in vitro assays. *J. Biol. Chem.* **264**, 20331–20338.

Lowe, J., McDermott, H., Landon, M., Mayer, R.J. & Wilkinson, K.D. (1990) Ubiquitin carboxyl-terminal hydrolase (PGP 9.5) is selectively present in ubiquitinated inclusion bodies characteristic of human neurodegenerative diseases. *J. Pathol.* **161**, 153–160.

Mann, D.A., Trowern, A.R., Lavender, F.L., Whittaker, P.A. & Thompson, R.J. (1996) Identification of evolutionary conserved regulatory sequences in the 5' untranscribed region of the neural-specific ubiquitin C-terminal hydrolase (PGP9.5) gene. *J. Neurochem.* **66**, 35–46.

Mayer, A.N. & Wilkinson, K.D. (1989) Detection, resolution, and nomenclature of multiple ubiquitin carboxyl-terminal esterases from bovine calf thymus. *Biochemistry* **28**, 166–172.

Miller, H.I., Henzel, W.J., Ridgeway, J.B., Kuang, W., Chisholm, V. & Liu, C. (1989) Cloning and expression of a yeast ubiquitin-protein cleaving activity in *Escherichia coli*. *Biotechnology* **7**, 698–704.

NC-IUBMB (Nomenclature Committee of the International Union of Biochemistry and Molecular Biology), (1992) *Enzyme Nomenclature 1992*. Orlando, FL: Academic Press.

Papa, F.R. & Hochstrasser, M. (1993) The yeast DOA4 gene encodes a deubiquitinating enzyme related to a product of the human tre-2 oncogene. *Nature* **366**, 313–319.

Pickart, C.M. & Rose, I.A. (1985) Ubiquitin carboxyl-terminal hydrolase acts on ubiquitin carboxyl-terminal amides. *J. Biol. Chem.* **260**, 7903–7910.

Rose, I.A. & Warms, J.V. (1983) An enzyme with ubiquitin carboxy-terminal esterase activity from reticulocytes. *Biochemistry* **22**, 4234–4237.

Schofield, J.N., Day, I.N., Thompson, R.J. & Edwards, Y.H. (1995) PGP9.5, a ubiquitin C-terminal hydrolase; pattern of mRNA and protein expression during neural development in the mouse. *Brain Res. Dev. Brain Res.* **85**, 229–238.

Smith-Thomas, L.C., Kent, C., Mayer, R.J. & Scotting, P.J. (1994) Protein ubiquitination and neuronal differentiation in chick embryos. *Brain Res. Dev. Brain Res.* **81**, 171–177.

Tobias, J.W. & Varshavsky, A. (1991) Cloning and functional analysis of the ubiquitin-specific protease gene UBP1 of *Saccharomyces cerevisiae*. *J. Biol. Chem.* **266**, 12021–12028.

Wilkinson, K.D. (1995) Roles of ubiquitinylation in proteolysis and cellular regulation. [Review]. *Annu. Rev. Nutr.* **15**, 161–189.

Wilkinson, K.D., Cox, M.J., Mayer, A.N. & Frey, T. (1986) Synthesis and characterization of ubiquitin ethyl ester, a new substrate for ubiquitin carboxyl-terminal hydrolase. *Biochemistry* **25**, 6644–6649.

Wilkinson, K.D., Lee, K.M., Deshpande, S., Duerksen-Hughes, P.J., Boss, J.M. & Pohl, J. (1989) The neuron-specific protein PGP 9.5 is a ubiquitin carboxyl-terminal hydrolase. *Science* **246**, 670–673.

Wilkinson, K.D., Deshpande, S. & Larsen, C.N. (1992) Comparisons of neuronal (PGP 9.5) and non-neuronal ubiquitin C-terminal hydrolases. *Biochem. Soc. Trans.* **20**, 631–637.

Wilkinson, K.D., Tashayev, V.L., O'Connor, L.B., Larsen, C.N., Kasperek, E. & Pickart, C.M. (1995) Metabolism of the polyubiquitin degradation signal: structure, mechanism, and role of isopeptidase T. *Biochemistry* **34**, 14535–14546.

Zhang, N., Wilkinson, K.D. & Bownes, M. (1993) Cloning and analysis of expression of a ubiquitin carboxyl terminal hydrolase expressed during oogenesis in *Drosophila melanogaster*. *Dev. Biol.* **157**, 214–223.

***Keith D. Wilkinson***
*Department of Biochemistry,*
*Emory University School of Medicine,*
*Atlanta, GA 30322, USA*
*Email: genekdw@emory.edu*

# *223. Ubiquitin isopeptidase T*

## *Databanks*

*Peptidase classification: clan CA, family C19, MEROPS ID: C19.001*
*NC-IUBMB enzyme classification: EC 3.1.2.15*

*Databank codes:*

| Species | Gene | SwissProt | PIR | EMBL (cDNA) | EMBL (genomic) |
|---|---|---|---|---|---|
| **Isopeptidase T** | | | | | – |
| *Dictyostelium discoideum* | *ubpA* | P54201 | – | U48271 | – |
| *Homo sapiens* | *isoT* | P45974 | – | U35116 U47927 X91349 | U47924: chromosome 12p13 |
| *Saccharomyces cerevisiae* | *UBP14* | P38237 | S45916 | – | Z35927: chromosome II ORF Z46260: complete chromosome II |
| **Other related sequences** | | | | | |
| *Caenorhabditis elegans* | *tgt* | – | – | U32223 | – |
| *Caenorhabditis elegans* | – | P34547 | S40715 | Z29095 | – |
| *Caenorhabditis elegans* | – | Q09931 | – | Z47811 | – |
| *Caenorhabditis elegans* | – | – | – | – | Z47812: complete gene |
| *Caenorhabditis elegans* | – | – | – | – | Z54218: complete gene |
| *Caenorhabditis elegans* | – | – | – | – | Z66511: complete gene |
| *Caenorhabditis elegans* | – | – | – | – | Z74033: complete gene |
| *Caenorhabditis elegans* | – | – | – | – | Z79601: complete gene |
| *Drosophila melanogaster* | *D-ubp-64E* | – | – | X99211 | – |
| *Drosophila melanogaster* | *faf* | – | – | L04958 | – |
| *Homo sapiens* | *hausp* | – | – | Z72499 | – |
| *Homo sapiens* | *isoT* | Q92995 | – | U75362 | – |
| *Homo sapiens* | *tgt* | P54578 | – | U30888 | – |
| *Homo sapiens* | *tre2* | P35125 | S22158 | X63547 | – |
| *Homo sapiens* | *uhx1* | P51784 | – | U4483 | – |
| *Homo sapiens* | *unpH* | Q13107 | – | U20657 | – |
| *Homo sapiens* | – | P40818 | – | D29956 | – |
| *Homo sapiens* | – | – | – | – | D38378: unannotated |
| *Mus musculus* | *dub-1* | – | – | U41636 | – |
| *Mus musculus* | *fam* | – | – | U67874 | – |
| *Mus musculus* | *ode-1* | P52479 | – | D84096 | – |
| *Mus musculus* | *unp* | P35123 | – | L00681 | – |
| *Oryctolagus cuniculus* | *tgt* | P40826 | – | L37420 | – |
| *Saccharomyces cerevisiae* | *UBP1* | P25037 | A404566 | M63484 | – |
| *Saccharomyces cerevisiae* | *UBP2* | Q01476 | S60999 | M94916 | X90518: chromosome XV fragment X94335: chromosome XV fragment Z75032: chromosome XV ORF |
| *Saccharomyces cerevisiae* | *UBP3* | Q01477 | B44450 | M94917 | U18917: chromosome V cosmids |
| *Saccharomyces cerevisiae* | *DOA4* | P32571 | S39344 | L08070 U02518 | X84162: chromosome IV segment Z46796: chromosome IV cosmid Z49209: chromosome IV cosmid Z74365: chromosome IV ORF |
| *Saccharomyces cerevisiae* | *UBP5* | P39944 | – | U10082 | U18917: chromosome V cosmids |
| *Saccharomyces cerevisiae* | *UBP7* | P40453 | S48378 | – | Z38059: chromosome IX lambda clones Z47047: complete chromosome IX |
| *Saccharomyces cerevisiae* | *UBP9* | P39967 | – | – | U18839: chromosome V cosmids |
| *Saccharomyces cerevisiae* | *UBP11* | P36026 | S38187 | Z35828 | – |
| *Saccharomyces cerevisiae* | *UBP12* | P39538 | S46636 | X77688 Z35927 | – |
| *Saccharomyces cerevisiae* | *UBP13* | P38187 | S45803 | – | Z35828: chromosome II ORF |
| *Saccharomyces cerevisiae* | *UBP6* | P43593 | – | – | D50617: complete chromosome VI |
| *Saccharomyces cerevisiae* | *UBP10* | P50101 | – | Z49212 | – |
| *Saccharomyces cerevisiae* | *UBP8* | P50102 | – | Z49939 | – |
| *Saccharomyces cerevisiae* | – | P53874 | – | – | Z71462: chromosome XIV ORF |
| *Schizosaccharomyces pombe* | – | Q09738 | – | – | Z54096: chromosome I cosmid |
| *Schizosaccharomyces pombe* | – | Q09879 | – | Z66569 | Z70720: chromosome I cosmid |

## Name and History

Ubiquitin isopeptidase T is one of the **deub**iquitinating enzyme (DUB) group of cysteine peptidases (Mayer & Wilkinson, 1989; Tobias & Varshavsky, 1991; Baker *et al.*, 1992; Papa & Hochstrasser, 1993; Wilkinson *et al.*, 1995). Ubiquitin is a 76 amino acid protein that is covalently attached to a variety of cellular proteins to target them for proteolysis by the 26S proteasome (Wilkinson, 1995) (see also Chapter 173). Several polymeric forms exist: a linear polyprotein gene product consisting of several repeats of the ubiquitin sequence, which is processed to free ubiquitin (Finley *et al.*, 1987; Ozkaynak *et al.*, 1987); two different ubiquitin fusion protein gene products consisting of ubiquitin followed by ribosomal proteins, which must be processed (Ozkaynak *et al.*, 1987; Finley *et al.*, 1989), and polyubiquitin, in which the ubiquitin subunits are linked by isopeptide bonds between the carboxyl group of Gly76 of ubiquitin and a side chain amino group of ubiquitin or other proteins (Finley & Chau, 1991; Finley *et al.*, 1994; Gregori *et al.*, 1990). It is the latter polymer that is attached to a variety of cellular proteins as a signal to target them for proteolysis by the 26S proteasome, and this must also be processed

to free monomers. The term **ubiquitin isopeptidase** was first used to describe the activity which removed ubiquitin from ubiquitinated histone H2a (Matsui *et al.*, 1982). *Isopeptidase T* (Hadari *et al.*, 1992; Falquet *et al.*, 1995a; Stein *et al.*, 1995; Wilkinson *et al.*, 1995) is one such deubiquitinating enzyme, but with specificity for free polyubiquitin chains (see Figure 223.1).

Isopeptidase T was first described (Pickart & Rose, 1985) as a 98 kDa protein of unknown function that bound to the ubiquitin affinity column used in the purification of ubiquitin C-terminal hydrolase (see Chapter 222). Subsequently, it was observed that partially purified preparations were able to 'trim' polyubiquitin chains (Hadari *et al.*, 1992), and disassemble polyubiquitin chains (Chen & Pickart, 1990) with a free C-terminus on the proximal copy of ubiquitin (Wilkinson *et al.*, 1995). (The term 'proximal' applies to the copy of ubiquitin at the end with the free C-terminus.) This enzyme has also been called *de-ubiquitinase* and *human UBP1* (Falquet *et al.*, 1995a,b). In addition, forms of the enzyme are commonly known as family 2 ubiquitin C-terminal hydrolases or ubiquitin thiolesterases. In the nomenclature of NC-IUBMB (1992), ubiquitin isopeptidase T appears only as a thiolesterase, EC 3.1.2.15.

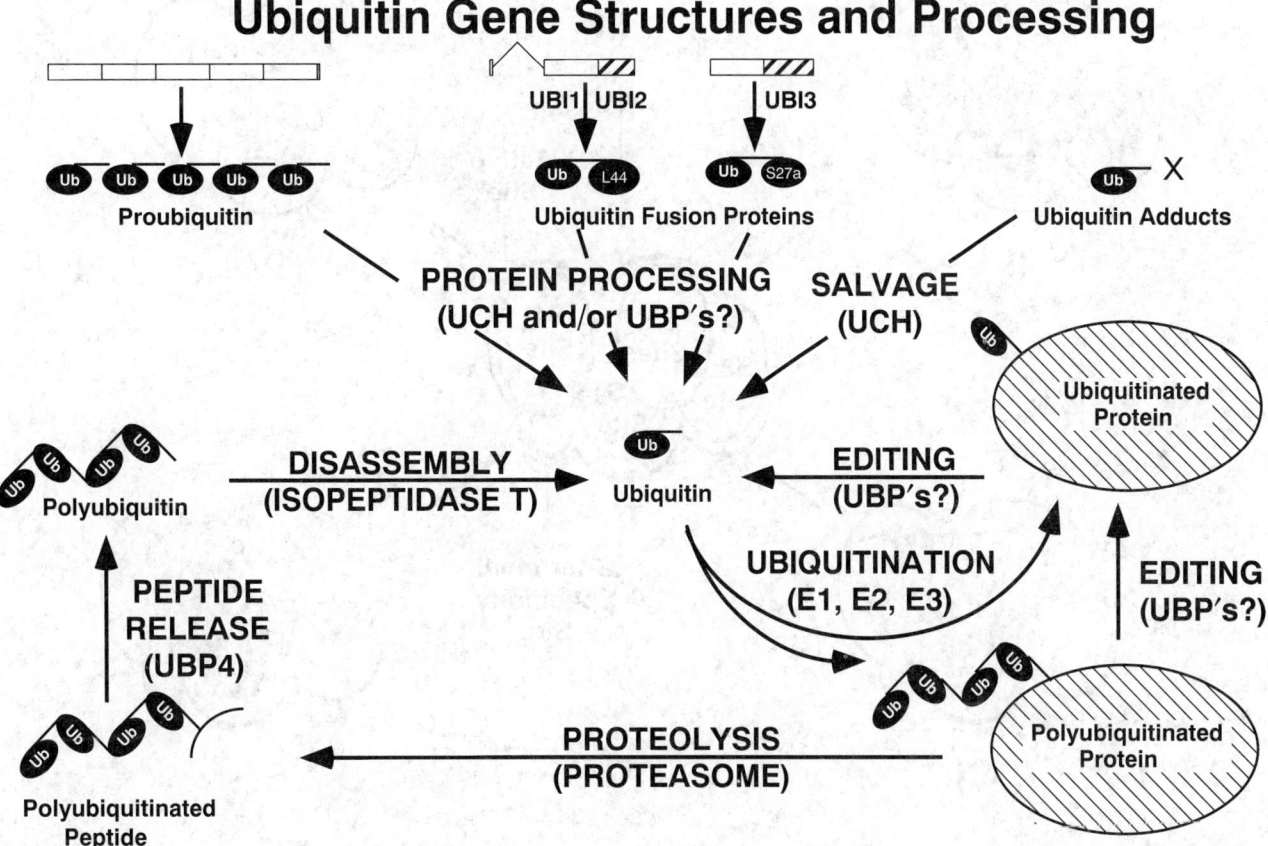

**Ubiquitin Gene Structures and Processing**

*Figure 223.1*   Ubiquitin gene structures and processing. The processing reactions which polymeric forms of ubiquitin must undergo are shown. The UBI1–4 gene products in yeast are fusion proteins which must be processed to monomeric ubiquitin. Ub-X is any of a variety of derivatives obtained by nucleophilic attack on the activated C-terminus of ubiquitin. The polyubiquitin shown here is linked by an isopeptide bond between Gly76 and Lys48 of another ubiquitin or a lysine of the target protein.

## Activity and Specificity

Highly purified preparations of ubiquitin isopeptidase T will hydrolyze the Gly76+Lys48 isopeptide bond of polyubiquitin ($k_{cat}/K_m = 3 \times 10^5\,\mathrm{M^{-1}\,s^{-1}}$), as well as the Gly76+Met1 peptide bond of proubiquitin ($k_{cat}/K_m = 3 \times 10^3\,\mathrm{M^{-1}\,s^{-1}}$), the polyprotein precursor of ubiquitin (Falquet *et al.*, 1995a; Wilkinson *et al.*, 1995). A synthetic substrate, Z-Arg-Leu-Arg-Gly-Gly+NHMec (Stein *et al.*, 1995), is also slowly hydrolyzed by this enzyme ($k_{cat}/K_m = 95\,\mathrm{M^{-1}\,s^{-1}}$). Ubiquitin ethyl ester is a useful generic substrate for deubiquitinating enzymes (Wilkinson *et al.*, 1986) and isopeptidase T catalyzes the hydrolysis of this substrate at a rate of about $10\,\mathrm{min^{-1}}$ (Wilkinson *et al.*, 1995). There are at least four ubiquitin-binding sites detected kinetically (see Figure 223.2).

Binding of ubiquitin to the S1′ site greatly enhances the rate of hydrolysis of the peptidyl-NHMec substrate, while ubiquitin bound at the S1 site is a partial inhibitor (Stein *et al.*, 1995).

Isopeptidase T is very specific, strongly preferring to hydrolyze substrates with a normal, intact C-terminus on the ubiquitin occupying the S1′ site (Wilkinson *et al.*, 1995). With Gly76-Lys48 linked diubiquitin as substrate, the Gly76Ala substitution on the free C-terminus results in at least a hundred-fold slower rate of hydrolysis. Hydrolysis of the polyubiquitin chain proceeds by release of one monomer at a time from the proximal end (i.e. the end with the free C-terminus). Thus, isopeptidase T can be characterized as a 'p-exo-deubiquitinating enzyme'; p- specifying that it is the proximal subunit that is released. By comparison, the

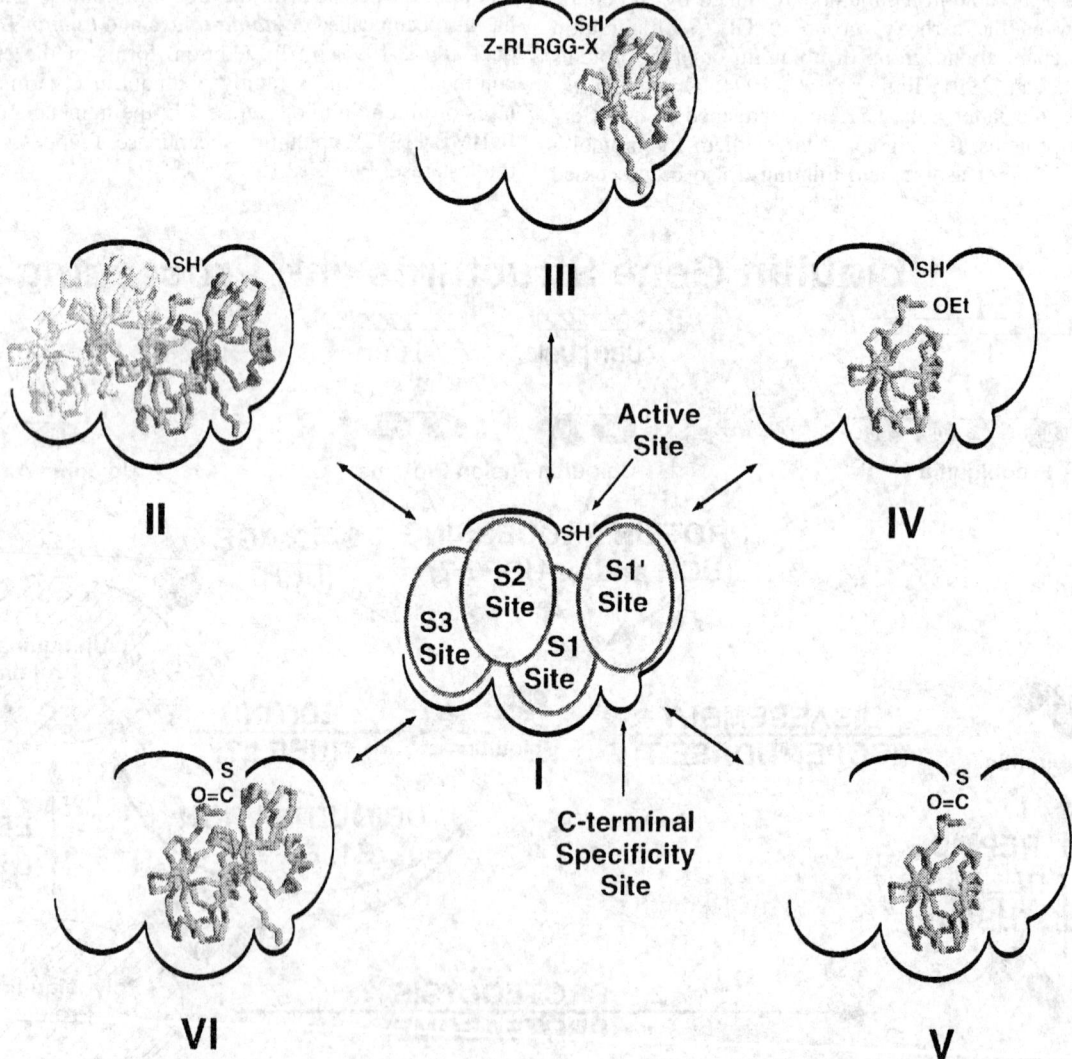

*Figure 223.2* Model of isopeptidase T mechanism. The subsite nomenclature was adapted from that used for specifying recognition sites in peptidases. The sites are drawn to correspond to the shape of the tetra-ubiquitin chain. The S1′ site is the position of free ubiquitin binding (III) and the proximal subunit of the polyubiquitin chain (II). Ubiquitin ethyl ester apparently binds in the S1 site as shown in IV with its scissile bond positioned near the active-site thiol. Structure III depicts the simultaneous binding of ubiquitin to the S1′ site and a peptidyl-NHMec to the S1 site, resulting in enhanced hydrolysis of the peptide-NHMec. Structure V is the presumed structure of the ubiquitin aldehyde-inhibited protein, probably present as the thiol-hemiacetal. This complex may be stabilized by simultaneous binding of ubiquitin at the S1′ site (VI).

*Figure 223.3*  Sequence alignment of four isopeptidase T species. Identities are boxed in black and similarities are boxed in gray. The conserved regions present in other UBPs or ubiquitin system enzymes are shown. The position of a brain specific exon is also indicated and discussed in the text.

recently reported proteasome-associated isopeptidase is a 'd-exo-deubiquitinating enzyme' where d- indicates hydrolysis of the distal subunit.

Like most deubiquitinating enzymes, isopeptidase T is inhibited by thiol-blocking agents, by oxidation, and by ubiquitin aldehyde (Hadari *et al*., 1992; Falquet *et al*., 1995a; Stein *et al*., 1995; Wilkinson *et al*., 1995). The optimal assay pH is around 7.5, and the activity is stabilized by thiols such as DTT. It has been suggested that the enzyme is inhibited by 1,10-phenanthroline, but reconstitution of activity with added zinc has not been demonstrated (L. Falquet & J.-C. Jaton, unpublished results).

## Structural Chemistry

Two groups cloned human cDNAs on the basis of peptide sequences (Falquet *et al*., 1995b; Wilkinson *et al*., 1995) and the genomic structure has now been reported. Two sequences, from yeast and a slime mold, probably represent additional species variants of isopeptidase T. Figure 223.3 shows the amino acid alignment of the four known isopeptidase T sequences. The two human sequences differ only by a brain-specific splice that results in an additional 23 amino acid exon in neural tissues (Falquet *et al*., 1995b; Wilkinson *et al*., 1995; V.L. Tashayev & K.D. Wilkinson, unpublished work). These enzymes have N-terminal sequences of approximately 250 amino acids that are unique to the UBP family, peptidase family C19. The N-terminal 600 amino acids comprise the catalytic core and have several blocks of conserved sequence in common with all other members of family C19 (Falquet *et al*., 1995b; Papa & Hochstrasser, 1993; Wilkinson *et al*., 1995; Zhu *et al*., 1996).

Examination of the UBP structures shows several highly conserved blocks of sequence in the catalytic domain. These include the Cys box which is at the N-terminal end of the catalytic core and the His box which is at the C-terminal end (Papa & Hochstrasser, 1993; Wilkinson *et al*., 1995; Zhu *et al*., 1996). In addition, isopeptidase T has two copies of the UBA domain (Hofmann & Bucher, 1996) which is a weak consensus motif found in a number of the enzymes of the ubiquitin pathway. Most of the UBP sequences have additional N- and/or C-terminal sequences that may confer specificity or localization to these forms of the enzyme.

## Preparation

Isopeptidase T can conveniently be purified from erythrocytes by affinity chromatography on a column of immobilized ubiquitin (Hadari *et al*., 1992; Wilkinson *et al*., 1995; Falquet *et al*., 1995b). The protein binds tightly in high salt and is eluted by high pH. The human protein can be expressed at high levels in *Escherichia coli*, although a significant fraction is insoluble and must be refolded (B. Krantz & K.D. Wilkinson, unpublished results).

## Biological Aspects

The biological role of the UBP enzymes is not precisely defined, but must relate to the processing of the various forms of polymeric ubiquitin, and/or the numerous ubiquitin-like sequences (Wilkinson, 1995; Hochstrasser, 1996).

Because of its requirement for an intact, free C-terminus, isopeptidase T cannot deubiquinate proteins or polyubiquitin chains that are still attached to substrates or to degradation fragments. Isopeptidase T acts only after the chains have been released from the proteasome by hydrolysis of the peptide remnant from the C-terminus of the polyubiquitin chain. This has been confirmed by disruption of the *UBP14* gene in yeast (M. Hochstrasser, unpublished results). The mutated cells accumulate polyubiquitin chains and show defects in proteolysis. The phenotype can be complemented by either the human or the *Dictyostelium* homolog (M. Hochstrasser & K. Wilkinson, unpublished results).

The human isopeptidase T gene is located on chromosome 12p13 in a gene cluster containing CD4, a G protein subunit, and triose phosphate isomerase. The gene spans a 15.5 kb region, with 21 exons. In most tissues, a protein of 835 amino acids is specified, but in brain an alternate splice donor 69 bp to the 3′ of the exon 15 donor is used, so that in brain a protein of 858 amino acids is found (V. Tashayev & K.D. Wilkinson, unpublished results).

## Distinguishing Features and Related Peptidases

There are numerous other members of peptidase family C19, commonly termed ubiquitin-specific processing proteases (UBP) with similarity throughout the catalytic domain. *Saccharomyces* has 16 isozymes, and there are ten related sequences known already in *C. elegans*. At least 15 mammalian coding sequences have been identified, including at least ten in humans (Woo *et al*., 1995, and the Databanks table).

## Further Reading

For a general review, see Wilkinson (1995). A good account of recent work on nonproteolytic functions is that of Hochstrasser (1996).

## References

Baker, R.T., Tobias, J.W. & Varshavsky, A. (1992) Ubiquitin-specific proteases of *Saccharomyces cerevisiae*. Cloning of UBP2 and UBP3, and functional analysis of the UBP gene family. *J. Biol. Chem.* **267**, 23364–23375.

Chen, Z. & Pickart, C.M. (1990) A 25-kilodalton ubiquitin carrier protein (E2) catalyzes multi-ubiquitin chain synthesis via lysine 48 of ubiquitin. *J. Biol. Chem.* **265**, 21835–21842.

Falquet, L., Paquet, N., Frutiger, S., Hughes, G.J., Hoang-Van, K. & Jaton, J.C. (1995a) A human de-ubiquitinating enzyme with both isopeptidase and peptidase activities in vitro. *FEBS Lett.* **359**, 73–77.

Falquet, L., Paquet, N., Frutiger, S., Hughes, G.J., Hoang-Van, K. & Jaton, J.C. (1995b) cDNA cloning of a human 100 kDa de-ubiquitinating enzyme: the 100 kDa human de-ubiquitinase belongs to the ubiquitin C-terminal hydrolase family 2 (UCH2). *FEBS Lett.* **376**, 233–237.

Finley, D. & Chau, V. (1991) Ubiquitination. *Annu. Rev. Cell Biol.* **7**, 25–69.

Finley, D., Ozkaynak, E. & Varshavsky, A. (1987) The yeast polyubiquitin gene is essential for resistance to high temperatures, starvation, and other stresses. *Cell* **48**, 1035–1046.

Finley, D., Bartel, B. & Varshavsky, A. (1989) The tails of ubiquitin precursors are ribosomal proteins whose fusion to ubiquitin facilitates ribosome biogenesis. *Nature* **338**, 394–401.

Finley, D., Sadis, S., Monia, B.P., Boucher, P., Ecker, D.J., Crooke, S.T. & Chau, V. (1994) Inhibition of proteolysis and cell cycle progression in a multiubiquitination-deficient yeast mutant. *Mol. Cell Biol.* **14**, 5501–5509.

Gregori, L., Poosch, M.S., Cousins, G. & Chau, V. (1990) A uniform isopeptide-linked multiubiquitin chain is sufficient to target substrate for degradation in ubiquitin-mediated proteolysis. *J. Biol. Chem.* **265**, 8354–8357.

Hadari, T., Warms, J.V., Rose, I.A. & Hershko, A. (1992) A ubiquitin C-terminal isopeptidase that acts on polyubiquitin chains. Role in protein degradation. *J. Biol. Chem.* **267**, 719–727.

Hochstrasser, M. (1996) Protein degradation or regulation: Ub the judge. [Review]. *Cell* **84**, 813–815.

Hofmann, K. & Bucher, P. (1996) The UBA domain: a sequence motif present in multiple enzyme classes of the ubiquitination pathway. *Trends Biochem. Sci.* **21**, 172–173.

Matsui, S., Sandberg, A.A., Negoro, S., Seon, B.K. & Goldstein, G. (1982) Isopeptidase: a novel eukaryotic enzyme that cleaves isopeptide bonds. *Proc. Natl Acad. Sci. USA* **79**, 1535–1539.

Mayer, A.N. & Wilkinson, K.D. (1989) Detection, resolution, and nomenclature of multiple ubiquitin carboxyl-terminal esterases from bovine calf thymus. *Biochemistry* **28**, 166–172.

NC-IUBMB (Nomenclature Committee of the International Union of Biochemistry and Molecular Biology) (1992) *Enzyme Nomenclature 1992.* Orlando, FL: Academic Press.

Ozkaynak, E., Finley, D., Solomon, M.J. & Varshavsky, A. (1987) The yeast ubiquitin genes: a family of natural gene fusions. *EMBO J.* **6**, 1429–1439.

Papa, F.R. & Hochstrasser, M. (1993) The yeast DOA4 gene encodes a deubiquitinating enzyme related to a product of the human tre-2 oncogene. *Nature* **366**, 313–319.

Pickart, C.M. & Rose, I.A. (1985) Ubiquitin carboxyl-terminal hydrolase acts on ubiquitin carboxyl-terminal amides. *J. Biol. Chem.* **260**, 7903–7910.

Stein, R.L., Chen, Z. & Melandri, F. (1995) Kinetic studies of isopeptidase T: modulation of peptidase activity by ubiquitin. *Biochemistry* **34**, 12616–12623.

Tobias, J.W. & Varshavsky, A. (1991) Cloning and functional analysis of the ubiquitin-specific protease gene UBP1 of *Saccharomyces cerevisiae. J. Biol. Chem.* **266**, 12021–12028.

Wilkinson, K.D. (1995) Roles of ubiquitinylation in proteolysis and cellular regulation. [Review]. *Annu. Rev. Nutr.* **15**, 161–189.

Wilkinson, K.D., Cox, M.J., Mayer, A.N. & Frey, T. (1986) Synthesis and characterization of ubiquitin ethyl ester, a new substrate for ubiquitin carboxyl-terminal hydrolase. *Biochemistry* **25**, 6644–6649.

Wilkinson, K.D., Tashayev, V.L., O'Connor, L.B., Larsen, C.N., Kasperek, E. & Pickart, C.M. (1995) Metabolism of the polyubiquitin degradation signal: structure, mechanism, and role of isopeptidase T. *Biochemistry* **34**, 14535–14546.

Woo, S.K., Lee, J.I., Park, I.K., Yoo, Y.J., Cho, C.M., Kang, M.S., Ha, D.B., Tanaka, K. & Chung, C.H. (1995) Multiple ubiquitin C-terminal hydrolases from chick skeletal muscle. *J. Biol. Chem.* **270**, 18766–18773.

Zhu, Y., Carroll, M., Papa, F.R., Hochstrasser, M. & D'Andrea, A.D. (1996) DUB-1, a deubiquitinating enzyme with growth-suppressing activity. *Proc. Natl Acad. Sci. USA* **93**, 3275–3279.

*Keith D. Wilkinson*
*Department of Biochemistry,*
*Emory University School of Medicine,*
*Atlanta, GA 30322, USA*
*Email: genekdw@emory.edu*

# 224. Staphylopain

## Databanks

*Peptidase classification: not assigned, but properties support tentative placing in clan CA. MEROPS ID: C9G.018*
*NC-IUBMB enzyme classification: none*
*Databank codes: no sequence data available*

## Name and History

Proteolytic activity in the culture medium of *Staphylococcus aureus* that is dependent on reducing agents and totally inhibited by $Hg^{2+}$, $Ag^{2+}$ and $Zn^{2+}$ was first demonstrated by Arvidson (1973). The enzyme, later purified from the V8 strain of *S. aureus*, was classified as a typical papain-like cysteine proteinase (Arvidson *et al.*, 1973; Potempa *et al.*, 1988). Based on this similarity we refer to it as *staphylopain*.

## Activity and Specificity

Staphylopain specificity was tested on carboxymethylated horse liver alcohol dehydrogenase. It was found to have a very broad specificity with no preference for any particular amino acid at several positions on either side of the scissile peptide bond (Björklind & Jörnvall, 1974). Contrary to this apparent lack of specificity, the enzyme has a limited ability to cleave synthetic peptide *p*-nitroanilide substrates. Among several substrates typical for trypsin (Bz-Arg-NHPhNO$_2$, Tos-Gly-Pro-Arg-NHPhNO$_2$, Tos-Gly-Pro-Lys-NHPhNO$_2$, D-Val-Leu-Arg-NHPhNO$_2$, D-Val-Leu-Lys-NHPhNO$_2$) or chymotrypsin (Bz-Tyr-NHPhNO$_2$, Suc-Phe-NHPhNO$_2$, Ac-Phe-NHPhNO$_2$, Suc-Ala-Ala-Pro-Phe-NHPhNO$_2$), none was cleaved. An exception was Z-Phe-Leu-Glu+NHPhNO$_2$, a substrate of glutamyl endopeptidases, which was hydrolyzed ($K_m = 0.5$ mM, $k_{cat} = 0.16$ s$^{-1}$, $k_{cat}/K_m = 320$ M$^{-1}$ s$^{-1}$) (Potempa *et al.*, 1988). Staphylopain also cleaves Z-Phe-Arg+NHMec and Bz-Tyr+OEt, but no kinetic data are available for these (Arvidson *et al.*, 1973; Takahashi *et al.*, 1994).

The pH optimum of staphylopain is in the range 8.0–8.8 with casein and hemoglobin, about 6.5 with insoluble elastin, and 7.5 with Bz-Tyr-OEt as substrate. Thiol compounds are necessary for activity. In contrast to general proteolytic activity which is sensitive to ionic strength, the elastinolytic activity of staphylopain is not affected by changes in NaCl concentration (from 0.1 to 1.0 M) (Arvidson *et al.*, 1973; Potempa *et al.*, 1988).

Assays are most conveniently performed with the synthetic substrates (Z-Phe-Leu-Glu-NHPhNO$_2$ or Z-Phe-Arg-NHMec). Unfortunately, only the latter is commercially available (Sigma) (see Appendix 2 for full names and addresses of suppliers). As an alternative, staphylopain activity can be measured with hemoglobin or casein.

Beside heavy metals and oxidizing agents, staphylopain is stoichiometrically and irreversibly inhibited by E-64, which is used for the active-site titration of the enzyme. Proteolytic activity is also inhibited by rat T-kininogen ($K_i = 5.2 \times 10^{-7}$ M) and human $\alpha_2$-macroglobulin, but not by cystatin C or human kininogens. Nevertheless, human plasma contains some unidentified inhibitors of staphylopain (Potempa *et al.*, 1988). These are likely to belong to the cystatin family, since it was reported that rat skin phosphorylated cystatin $\alpha$ (P-cystatin $\alpha$) inhibits *S. aureus* cysteine proteinase activity (Takahashi *et al.*, 1994).

### Structural Chemistry

Staphylopain is a basic protein of pI 9.4, and molecular mass about 13 kDa as estimated by gel filtration and SDS-PAGE. The primary structure of the enzyme is unknown and its exceptionally low apparent molecular mass has yet to be confirmed (Arvidson *et al.*, 1973; Potempa *et al.*, 1988). If this is confirmed, then staphylopain is one of the smallest proteinases described to date.

### Preparation

So far, staphylopain has been identified only in the V8 strain of *S. aureus*, and purified from culture medium by consecutive ammonium sulfate and acetone precipitations, negative adsorption on DEAE-cellulose, ion exchange on

CM-Sephadex and gel filtration (Arvidson *et al.*, 1973; Potempa *et al.*, 1988).

### Biological Aspects

To date, no correlation has been reported between the pathogenicity of *S. aureus* and its proteolytic activity. However, the elastinolytic activity of staphylopain, which is comparable to that of human neutrophil elastase (Chapter 15), allows a speculation that the enzyme may participate in tissue invasion and destruction associated with staphylococcal ulceration. Although the expression of staphylopain seems to be limited to the V8 strain, it cannot be excluded that under physiological conditions synthesis of this enzyme might be induced by elastin, as has been found to occur for *Staphylococcus epidermidis* elastase (Hartman & Murphy, 1977). The latter enzyme, which has been suggested to play a major role in the development of perifollicular macular atrophy, was at one time thought to be a cysteine proteinase (Varadi & Saqueton, 1968), but was recently shown to be a metalloproteinase (Teufel & Götz, 1993).

Together with two other *S. aureus* proteinases, aureolysin (Chapter 538) and glutamyl endopeptidase (V8 protease) (Chapter 79), staphylopain is coordinately regulated at the transcriptional level by the accessory gene regulator (*agr*) (Björklind & Arvidson, 1980; Janzon *et al.*, 1986). Staphylopain may play some function in *S. aureus* growth, since P-cystatin $\alpha$ markedly suppresses the growth of the bacteria (Takahashi *et al.*, 1994).

### References

Arvidson, S.O. (1973) Hydrolysis of casein by three extracellular proteolytic enzymes from *S. aureus* strain V8. *Acta Pathol. Microbiol. Scand. Sect. B* **80**, 835–844.

Arvidson, S., Holme, T. & Lindholm, B. (1973) Studies on extracellular proteolytic enzymes from *Staphylococcus aureus*. I. Purification and characterization of one neutral and one alkaline protease. *Biochim. Biophys. Acta* **302**, 135–148.

Björklind, A. & Arvidson, S. (1980) Mutants of *Staphylococcus aureus* affected in the regulation of exoprotein synthesis. *FEMS Microbiol. Lett.* **33** 193–198.

Björklind, A. & Jörnvall, H. (1974) Substrate specificity of three different extracellular proteolytic enzymes from *Staphylococcus aureus. Biochim. Biophys. Acta* **370**, 542–529.

Hartman, D.P. & Murphy, R.A. (1977) Production and detection of staphylococcal elastase. *Infect. Immun.* **15**, 59–65.

Janzon, L., Lofdahl, S. & Arvidson, S. (1986) Identification and nucleotide sequence of the delta lysin gene *hld* adjacent to the accessory gene regulator (*agr*) of *Staphylococcus aureus. Mol. Gen. Genet.* **219**, 480–485.

Potempa, J., Dubin, A., Korzus, G. & Travis, J. (1988) Degradation of elastase by a cysteine proteinase from *Staphylococcus aureus. J. Biol. Chem.* **263**, 2664–2667.

Takahashi, M., Tezuka, T. & Katunuma, N. (1994) Inhibition of growth and cysteine proteinase activity of *Staphylococcus aureus* V8 by phosphorylated cystatin $\alpha$ in skin cornified envelope. *FEBS Lett.* **355**, 275–278.

Teufel, P. & Götz, F. (1993) Characterization of an extracellular metalloproteinase with elastase activity from *Staphylococcus epidermidis. J. Bacteriol.* **175**, 4218–4224.

Varadi, D.P. & Saqueton, A.C. (1968) Elastase from *Staphylococcus epidermidis. Nature* **218**, 468–470.

***Jan Potempa***
*Jagiellonian University,*
*Institute of Molecular Biology,*
*Al. Mickiewicza 3, 31-120 Krakow, Poland*
*Email: potempa@mol.uj.edu.pl*

***Adam Dubin***
*Jagiellonian University,*
*Institute of Molecular Biology,*
*Al. Mickiewicza 3, 31-120 Krakow, Poland*
*Email: dubin@mol.uj.edu.pl*

***James Travis***
*University of Georgia,*
*Department of Biochemistry,*
*Athens, GA 30602, USA*
*Email: jtravis@uga.cc.uga.edu*

C

# 225. Introduction: clan CC containing viral 'papain-like' endopeptidases

## Databanks

*MEROPS ID: CC*

| Species | SwissProt | PIR | EMBL (genomic) |
|---|---|---|---|
| **Family C6** | | | |
| Helper component proteinase (Chapter 227) | | | |
| **Family C7** | | | |
| Chestnut blight fungus virus p29 proteinase (Chapter 228) | | | |
| **Family C8** | | | |
| Chestnut blight fungus virus p48 proteinase (Chapter 228) | | | |
| **Family C9** | | | |
| Sindbis virus nsP2 proteinase (Chapter 230) | | | |
| **Family C16** | | | |
| Mouse hepatitis coronavirus papain-like endopeptidase 1 (Chapter 229) | | | |
| **Family C21** | | | |
| Tymovirus endopeptidase (Chapter 232) | | | |
| **Family C23** | | | |
| Blueberry scorch carlavirus endopeptidase (Chapter 233) | | | |
| **Family C27** | | | |
| Rubella rubivirus endopeptidase (Chapter 231) | | | |
| **Family C28** | | | |
| Leader (L) proteinase, foot-and-mouth disease virus FMDV (Chapter 226) | | | |
| **Family C29** | | | |
| Murine hepatitis coronavirus papain-like endopeptidase 2 (Chapter 229) | | | |
| **Family C31** | | | |
| Porcine respiratory and reproductive syndrome arterivirus (Chapter 234) | | | |
| **Family C32** | | | |
| Equine arteritis virus PCP$\beta$ endopeptidase (Chapter 235) | | | |
| **Family C33** | | | |
| Equine arterivirus Nsp2 cysteine endopeptidase (Chapter 236) | | | |
| **Family C34** | | | |
| Apple chlorotic leaf spot closterovirus putative papain-like proteinase (Chapter 232) | | | |
| **Family C35** | | | |
| Apple stem grooving capillovirus putative papain-like endopeptidase (Chapter 232) | | | |
| **Family C36** | | | |
| Beet necrotic yellow vein furovirus putative papain-like endopeptidase (Chapter 232) | | | |
| **Family C41** | | | |
| Hepatitis E virus cysteine proteinase | | | |
| Hepatitis E virus (strain Burma) | P29324 | A40778 | M73218: complete genome |
| Hepatitis E virus (strain Indian) | – | – | X98292: complete genome |
| Hepatitis E virus (strain Mexico) | Q03495 | A44212 | M74506: complete genome |

| Species | SwissProt | PIR | EMBL (genomic) |
|---|---|---|---|
| **Family C41** (*continued*) | | | |
| Hepatitis E virus (strain Myanmar) | Q04610 | – | D10330: complete genome |
| Hepatitis E virus (strain Pakistan) | P33424 | – | M80581: complete genome |

**Family C42**

Sugar beet yellow virus papain-like endopeptidase (Chapter 237)

**Family C43**

Citrus tatter leaf capillovirus putative endopeptidase (Chapter 232)

As was described in Chapter 185, the polyprotein-processing endopeptidases encoded by RNA viruses are commonly described either as 'chymotrypsin-like' or 'papain-like', and accordingly are placed in clan CB or clan CC, respectively. The papain-like viral endopeptidases of clan CC have a catalytic dyad in the order Cys, His in the sequence, which is the same as that in papain and all members of clan CA. For some of the enzymes, there is limited further similarity of sequence. For example, the sequence of the L-peptidase of foot-and-mouth disease virus contains the residues of the catalytic dyad in the sequences **Cys**-Trp and **His**-Ala-Val, exactly as in papain; this seems unlikely to be the result either of coincidence or of convergence. However, residues equivalent to Gln19 and Asn175 of papain have not been identified in endopeptidases of clan CC, and there is no sequence conservation to indicate that such residues exist. No tertiary structure has yet been determined for any of these viral endopeptidases, so there are not yet sufficient grounds to place the viral papain-like endopeptidases in clan CA, and they are assigned to a separate clan, CC, for the time being at least. Alignment 225.1 (see CD-ROM) shows the slight conservation of identity of amino acids around the catalytic dyad residues, and a greater degree of similarity of amino acids in many members of clan CC. Most notably, the catalytic cysteine is generally followed by a large hydrophobic residue. The cleavages mediated by the clan CC viral endopeptidases usually show Gly in P1 (contrasting with Gln in P1 for many cleavages by clan CB endopeptidases).

*Family C28* includes the L-peptidase from aphthoviruses such as the foot-and-mouth disease virus. Although aphthoviruses possess picornain 3C (Chapter 239) as a general polyprotein-processing enzyme, they lack a picornain 2A, and the functions of that endopeptidase are performed by the L-peptidase or an autolytic activity (see Chapter 566). The L-peptidase is sited at the N-terminus of the polyprotein, from which it autocatalytically releases itself. The L-peptidase also cleaves the eukaryotic initiation factor eIF-4G (formerly known as p220 or eIF-4$\gamma$), thus preventing cap-dependent synthesis of host-cell proteins. The catalytic residues have been identified by site-directed mutagenesis (Roberts & Belsham, 1995), and a promising preliminary crystallographic study (Guarné *et al.*, 1996) suggests that this endopeptidase may be the first member of clan CC for which a tertiary structure is solved.

*Family C6* contains one of the two cysteine endopeptidases found in potyviruses, the helper-component endopeptidase.

(The second is the NIa proteinase, Chapter 224.) The only cleavage performed by the helper-component endopeptidase is its own release from the polyprotein by cleavage of a Gly+Gly bond; further processing is performed by the serine-type P1 endopeptidase (Chapter 90). The helper-component endopeptidase is a two-part molecule with the peptidase unit at the C-terminus, and the helper component, required for virus transmission from plant to plant by aphids, at the N-terminus. The catalytic residues have been identified by site-directed mutagenesis in tobacco etch virus (Oh & Carrington, 1989).

*Families C7 and C8* contain the chestnut blight hypovirus endopeptidases p29 and p48. Chestnut blight is caused by the fungus *Cryphonectria parasitica*, and the symptoms are reduced if the hypovirus, which is a double-stranded RNA virus, is present. The virus encodes two polyproteins, and the enzymes responsible for their processing constitute the smaller polyprotein. The p29 endopeptidase cleaves a single Gly+Gly bond in the polyprotein to release itself and the p48 endopeptidase (Choi *et al.*, 1991b). The p48 endopeptidase cleaves a single Gly+Ala bond in the larger polyprotein (Shapira & Nuss, 1991). Catalytic residues have been identified by site-specific mutagenesis for both endopeptidases (Choi *et al.*, 1991a; Shapira & Nuss, 1991). The endopeptidases show no significant sequence similarity to any other peptidase, and each is considered a representative of a distinct family, although, as can be seen in Alignment 225.1, there is conservation of a Gly-Tyr-**Cys**-Tyr motif containing the catalytic Cys between the p29 endopeptidase (family C7) and family C6.

*Family C9* includes the nonstructural polyprotein-processing endopeptidase from togaviruses, the nsP2 endopeptidase. The catalytic dyad has been identified by site-specific mutagenesis (Strauss *et al.*, 1992), and the nsP2 endopeptidase is a bifunctional molecule with the peptidase unit at the C-terminus and an N-terminal unit that is probably involved in RNA-binding during replication. The N-terminal domain is homologous to a similarly located domain in the cucumber green mottle mosaic tobamovirus.

*Family C16* contains the mouse hepatitis virus p28 endopeptidase, which releases itself from the N-terminus of the polyprotein encoded by gene A. The catalytic dyad has been identified by site-specific mutagenesis (Baker *et al.*, 1993). The polyprotein encoded by gene A includes a picornain-like endopeptidase, and a putative second papain-like endopeptidase, included in *family C29*. The compound E-64c inhibits polyprotein processing in the mouse hepatitis virus (Kim *et al.*, 1995), presumably by inhibiting the action of one or both of the two papain-like endopeptidases. Most of the peptidases known to be inhibited by E-64 and its analogs are in clan CA, and this result is a further example of similarities between some clan CC peptidases and those of clan CA.

*Family C21* contains the polyprotein-processing endopeptidase from tymoviruses. The 206 kDa polyprotein is cleaved at an Ala+Thr bond (Bransom *et al.*, 1996) to release the 70 kDa polymerase from the C-terminus and a 150 kDa protein that includes the endopeptidase and a helicase from the N-terminus. The catalytic residues have been identified by site-specific mutagenesis (Bransom & Dreher, 1994).

*Family C23* includes the p223 polyprotein processing endopeptidase from the blueberry scorch virus. Cleavage is at a single Gly+Ala bond, and site-directed mutagenesis has been used to identify the potential catalytic residues (Lawrence *et al.*, 1995).

*Family C27* includes the nonstructural polyprotein-processing endopeptidase from rubella virus. Processing occurs at a single Gly+Gly bond, and the Cys, His catalytic dyad has been identified (Chen *et al.*, 1996).

*Families C31, C32 and C33* include polyprotein-processing endopeptidases from arteriviruses. In the Lelystad (porcine reproductive and respiratory syndrome) virus, the three papain-like endopeptidases are consecutive in the pol polyprotein, with $PCP\alpha$ at the N-terminus, followed by $PCP\beta$ and then Nsp2, which are so different in sequence that they are assigned to families C31, C32 and C33, respectively. $PCP\alpha$ and $PCP\beta$ together constitute the Nsp1 protein. The catalytic dyads have been determined by site-directed mutagenesis for all three endopeptidases (den Boon *et al.*, 1995; Snijder *et al.*, 1995). Besides the three papain-like cysteine endopeptidases, the arteriviruses possess a fourth, serine-type, endopeptidase (see Chapter 92) in the pol polyprotein. Not all arteriviruses require all four endopeptidases, because in the equine arteritis virus, $PCP\alpha$ is inactive, the catalytic Cys being replaced by Lys. The $PCP\alpha$ endopeptidase releases itself from the N-terminus of the Nsp1 protein (den Boon *et al.*, 1995). The $PCP\beta$ endopeptidase cleaves between the Nsp1 and Nsp2 proteins in the pol polyprotein (den Boon *et al.*, 1995). The Nsp2 endopeptidase cleaves between the Nsp2 and Nsp3 proteins, and, unusually, the catalytic Cys is followed by Gly. Cys-Gly is generally regarded as diagnostic for the clan CB viral endopeptidases, but it has also been seen in clan CA in some calpain homologs (Snijder *et al.*, 1995) and in family C12 (see Chapter 186).

*Family C42* includes a polyprotein-processing endopeptidase from the beet yellows closterovirus. Processing occurs at a single Gly+Gly bond, and the catalytic dyad has been identified by site-directed mutagenesis (Agranovsky *et al.*, 1994).

*Families C34, C35, C36 and C43* are represented by putative polyprotein processing endopeptidases from a variety of RNA viruses, identified mainly from conservation around predicted catalytic residues and relatedness to turnip yellow mosaic virus (Rozanov *et al.*, 1995). *Family C41* includes a putative endopeptidase from hepatitis E virus (Reyes *et al.*, 1993), but the His predicted to be part of the catalytic dyad is not conserved in the Myanmar and Mexico strains, and has probably been misidentified. One conserved His is N-terminal to the putative catalytic cysteine, which would imply a picornain-like endopeptidase, and the only other conserved His is C-terminal, but close, to the potential active-site Cys.

## References

Agranovsky, A.A., Koonin, E.V., Boyko, V.P., Maiss, E., Frotschl, R., Lunina, N.A. & Atabekov, J.G. (1994) Beet yellows closterovirus: complete genome structure and identification of a leader papain-like thiol protease. *Virology* **198**, 311–324.

Baker, S.C., Yokomori, K., Dong, S., Carlisle, R., Gorbalenya, A.E., Koonin, E.V. & Lai, M.M.C. (1993) Identification of the catalytic sites of a papain-like cysteine proteinase of murine coronavirus. *J. Virol.* **67**, 6056–6063.

Bransom, K.L. & Dreher, T.W. (1994) Identification of the essential cysteine and histidine residues of the turnip yellow mosaic virus protease. *Virology* **198**, 148–154.

Bransom, K.L., Wallace, S.E. & Dreher, T.W. (1996) Identification of the cleavage site recognized by the turnip yellow mosaic virus protease. *Virology* **217**, 404–406.

Chen, J.P., Strauss, J.H., Strauss, E.G. & Frey, T.K. (1996) Characterization of the rubella virus nonstructural protease domain and its cleavage site. *J. Virol.* **70**, 4707–4713.

Choi, G.H., Pawlyk, D.M. & Nuss, D.L. (1991a) The autocatalytic protease p29 encoded by a hypovirulence-associated virus of the chestnut blight fungus resembles the potyvirus-encoded protease HC-Pro. *Virology* **183**, 747–752.

Choi, G.H., Shapira, R. & Nuss, D.L. (1991b) Cotranslational autoproteolysis involved in gene expression from a double-stranded RNA genetic element associated with hypovirulence of the chestnut blight fungus. *Proc. Natl Acad. Sci. USA* **88**, 1167–1171.

den Boon, J.A., Faaberg, K.S., Meulenberg, J.J.M., Wassenaar, A.L.M., Plagemann, P.G.W., Gorbalenya, A.E. & Snijder, E.J. (1995) Processing and evolution of the N-terminal region of the arterivirus replicase ORF1a protein: Identification of two papainlike cysteine proteases. *J. Virol.* **69**, 4500–4505.

Guarné, A., Kirchweger, R., Verdaguer, N., Liebig, H.D., Blaas, D., Skern, T. & Fita, I. (1996) Crystallization and preliminary X-ray diffraction studies of the Lb proteinase from foot-and-mouth disease virus. *Protein Sci.* **5**, 1931–1933.

Kim, J.C., Spence, R.A., Currier, P.F., Lu, X. & Denison, M.R. (1995) Coronavirus protein processing and RNA synthesis is inhibited by the cysteine proteinase inhibitor E64d. *Virology* **208**, 1–8.

Lawrence, D.M., Rozanov, M.N. & Hillman, B.I. (1995) Autocatalytic processing of the 223-kDa protein of blueberry scorch carlavirus by a papain-like proteinase. *Virology* **207**, 127–135.

Oh, C.-S. & Carrington, J.C. (1989) Identification of essential residues in potyvirus proteinase HC-Pro by site-directed mutagenesis. *Virology* **173**, 692–699.

Reyes, G.R., Huang, C.C., Tam, A.W. & Purdy, M.A. (1993) Molecular organization and replication of hepatitis E virus (HEV). *Arch. Virol. Suppl.* **7**, 15–25.

Roberts, P.J. & Belsham, G.J. (1995) Identification of critical amino acids within the foot-and-mouth disease virus leader protein, a cysteine protease. *Virology* **213**, 140–146.

Rozanov, M.N., Drugeon, G. & Haenni, A.-L. (1995) Papain-like proteinase of turnip yellow mosaic virus: a prototype of a new viral proteinase group. *Arch. Virol.* **140**, 273–288.

Shapira, R. & Nuss, D.L. (1991) Gene expression by a hypovirulence-associated virus of the chestnut blight fungus involves two papain-like protease activities. Essential residues and cleavage site requirements for p48 autoproteolysis. *J. Biol. Chem.* **266**, 19419–19425.

Snijder, E.J., Wassenaar, A.L.M., Spaan, W.J.M. & Gorbalenya, A.E. (1995) The arterivirus Nsp2 protease. An unusual cysteine protease with primary structure similarities to both papain-like and chymotrypsin-like proteases. *J. Biol. Chem.* **270**, 16671–16676.

Strauss, E.G., De Groot, R.J., Levinson, R. & Strauss, J.H. (1992) Identification of the active site residues in the nsP2 proteinase of Sindbis virus. *Virology* **191**, 932–940.

# 226. *Foot-and-mouth disease virus L-peptidase*

## Databanks

*Peptidase classification: clan CC, family C28, MEROPS ID: C28.001*
*NC-IUBMB enzyme classification: none*
*Databank codes:*

| Species | Strain/type | SwissProt | PIR | EMBL (genomic) |
| --- | --- | --- | --- | --- |
| Equine rhinovirus | 1 | – | – | X96870: complete genome |
| Foot-and-mouth disease virus | A10-61 | P03306 | A93508 | X00429: complete genome |
| | A22/50 | P49303 | – | X74812: complete genome |

## Name and History

Foot-and-mouth disease virus (FMDV) comprises the genus *Aphthovirus* within the family Picornaviridae. Two genera within this family, the aphthoviruses and the cardioviruses, encode a leader (L) protein. Recently, nucleotide sequence analysis of equine rhinovirus (ERV), an unclassified picornavirus, indicated that it also codes for an L-protein (Li *et al.*, 1996; Wutz *et al.*, 1996). The FMDV L-protein has been identified as a peptidase by sequence homology, and it has been suggested that ERV L also has proteolytic activity, whereas the cardiovirus L-protein is proteolytically inactive. The current nomenclature for picornavirus proteins was adopted at the third meeting of the European Study Group on the Molecular Biology of Picornaviruses and formalized by Rueckert & Wimmer (1984). A uniform nomenclature was devised at that time because of the complex cleavage pathways of these proteins and the additional variability of the numerous virus species within the family. The L-protein is defined as that region of the polyprotein preceding the capsid precursor protein, and the name was chosen because of its location in the polyprotein and to minimize possible confusion with other viral proteins. For FMDV there are two in-frame AUG codons at the beginning of the viral genome open reading frame, resulting in the synthesis of two L-proteins, Lab and Lb. Lb, the smaller of the two proteins, is the major species synthesized, and its sequence is highly conserved among the FMDV subtypes and serotypes that have been examined.

Strebel & Beck (1986) were the first to demonstrate that the L-protein of FMDV has proteolytic activity, and autocatalytically cleaves itself from the growing viral polypeptide chain. Subsequently it was demonstrated that the L-protein is also involved in the cleavage of a host protein that is required for the initiation of cap-dependent protein synthesis (Devaney *et al.*, 1988; Medina *et al.*, 1993). By a number of criteria, including limited sequence alignment (Gorbalenya *et al.*, 1991), inhibitor studies (Kleina & Grubman, 1992), and genetic analysis (Piccone *et al.*, 1995c; Roberts & Belsham, 1995), the L-protein has been defined as one of the 'papain-like' viral proteases (see Chapter 225).

## Activity and Specificity

Kirchweger *et al.* (1994) purified the Lb form of the L-peptidase and demonstrated that it cleaves the host initiation factor eIF-4G (formerly known as p220 or eIF-4γ) first at the Gly479╀Arg480 bond and then at Lys318╀Arg319. These sites differ from the cleavage site of the L-peptidase from its precursor polyprotein P1. Cleavage between L and viral protein VP4, which is at the N-terminus of P1, occurs at a Lys╀Gly bond. There is little sequence similarity between the cleavage sites of these two natural substrates of L-peptidase:

L-VP4

      Lys-Val-Gln-Arg-Lys-Leu-Lys╀Gly-Ala-Gly-Asn-Ser-Ser

eIF-4G

  Thr-Pro-Ser-Phe-Ala-Asn-Leu-Gly╀Arg-Pro-Ala-Leu-Ser-Ser-Arg

eIF-4G

      Gln-Val-Ala-Val-Ser-Val-Pro-Lys╀Arg-Arg-Arg-Lys-Ile-Lys-Glu-Leu

Ziegler *et al.* (1995) found that purified L-peptidase cleaved the translation products of a number of reporter genes, but no significant pattern of sequence conservation was apparent when the different cleavage sites were compared.

L-peptidase autocatalytic activity at the L-VP4 scissile bond is blocked by E-64 in a cell-free system and by E-64d, a membrane-permeable analog, in infected cells (Kleina & Grubman, 1992). Likewise, cleavage of eIF-4G is inhibited by E-64d in infected cells.

## Structural Chemistry

The Lb-peptidase is a 173 amino acid protein of approximately 21 kDa and pI 4.8. Preliminary crystallization and X-ray diffraction studies are in progress (Guarné *et al.*, 1996).

## Preparation

The Lb gene has been expressed in *Escherichia coli* either under the control of the inducible λPL promoter (Strebel *et al.*, 1986) or a T7 RNA polymerase promoter (Kirchweger *et al.*, 1994; Piccone *et al.*, 1995b). In the latter two cases, because of the toxicity of the Lb-peptidase in *E. coli*, the protein can be reliably expressed only in a plasmid in which the basal level of T7 RNA polymerase is tightly controlled.

The Lb-peptidase was purified to homogeneity from an *E. coli* expression system (Kirchweger *et al.*, 1994) and this material is being utilized to determine its three-dimensional structure (Guarné *et al.*, 1996).

## Biological Aspects

The major function of the Lb-peptidase in infected cells appears to be the cleavage of eIF-4G which is a subunit of the cap-binding protein complex required for the translation of cap-dependent mRNAs. Cleavage of this component correlates with the shut-off of most host cell protein synthesis. However, translation of FMDV mRNA occurs by a cap-independent mechanism and is not impaired. Recently, a genetically altered variant of FMDV which lacks the coding region for the Lb-peptidase was developed (Piccone *et al.*, 1995a). This virus is significantly attenuated in cattle as compared to wild-type virus, suggesting that the Lb-peptidase may be required for rapid viral growth and for the dramatic cytopathic effects associated with FMDV infection (Brown *et al.*, 1996).

## Distinguishing Features

The Lb-peptidase may be blocked at its N-terminus, since microsequencing of this protein was only possible after acetylation was inhibited (Robertson *et al.*, 1985). Rabbit polyclonal antisera have been prepared against the Lb-peptidase from types C1 and A12 expressed in *E. coli* (Strebel *et al.*, 1986; Piccone *et al.*, 1995b).

## References

Brown, C.C., Piccone, M.E., Mason, P.W., McKenna, T.S.-C. & Grubman, M.J. (1996) Pathogenesis of wild-type and leaderless foot-and-mouth disease virus in cattle. *J. Virol.* **70**, 5638–5641.

Devaney, M.A., Vakharia, V.N., Lloyd, R.E., Ehrenfeld, E. & Grubman, M.J. (1988) Leader protein of foot-and-mouth disease virus is required for cleavage of the p220 component of the cap-binding protein complex. *J. Virol.* **62**, 4407–4409.

Gorbalenya, A.E., Koonin, E.V. & Lai, M.M.-C. (1991) Putative papain-related thiol proteases of positive-strand RNA viruses: identification of rubi- and aphthovirus proteases and delineation of a novel conserved domain associated with proteases of rubi, α- and coronaviruses. *FEBS Lett.* **288**, 201–205.

Guarné, A., Kirchweger, R., Verdaguer, N., Liebig, H.-D., Blaas, D., Skern, T. & Fita, I. (1996) Crystallization and preliminary X-ray diffraction studies of the Lb proteinase from foot-and-mouth disease virus. *Protein Sci.* **5**, 1–3.

Kirchweger, R., Ziegler, E., Lamphear, B.J., Waters, D., Liebig, H.-D., Sommergruber, W., Sobrino, F., Hohenadl, C., Blaas, D., Rhoads, R.E. & Skern, T. (1994) Foot-and-mouth disease virus leader proteinase: purification of the Lb form and determination of its cleavage site on eIF-4γ. *J. Virol.* **68**, 5677–5684.

Kleina, L.G. & Grubman, M.J. (1992) Antiviral effects of a thiol protease inhibitor on foot-and-mouth disease virus. *J. Virol.* **66**, 7168–7175.

Li, F., Browning, G.F., Studdert, M.J. & Crabb, B.S. (1996) Equine rhinovirus 1 is more closely related to foot-and-mouth disease virus than to other picornaviruses. *Proc. Natl Acad. Sci. USA* **93**, 990–995.

Medina, M., Domingo, E., Brangwyn, J.K. & Belsham, G.J. (1993) The two species of the foot-and-mouth disease virus leader protein expressed individually, exhibit the same activities. *Virology* **194**, 355–359.

Piccone, M.E., Rieder, E., Mason, P.W. & Grubman, M.J. (1995) The foot-and-mouth disease virus leader proteinase gene is not required for viral replication. *J. Virol.* **69**, 5376–5382.

Piccone, M.E., Sira, S., Zellner, M. & Grubman, M.J. (1995) Expression in *Escherichia coli* and purification of biologically active L proteinase of foot-and-mouth disease virus. *Virus Res.* **35**, 263–275.

Piccone, M.E., Zellner, M., Kumosinski, T.F., Mason, P.W. & Grubman, M.J. (1995) Identification of the active-site residues of the L proteinase of foot-and-mouth disease virus. *J. Virol.* **69**, 4950–4956.

Roberts, P.J. & Belsham, G.J. (1995) Identification of critical amino acids within the foot-and mouth disease virus leader protein, a cysteine protease. *Virology* **213**, 140–146.

Robertson, B.H., Grubman, M.J., Weddel, G.N., Moore, D.M., Welsh, J.D., Fischer, T., Dowbenko, D.J., Yansura, D.G., Small, B. & Kleid, D.G. (1985) Nucleotide and amino acid sequence coding for polypeptides of foot-and-mouth disease virus type A12. *J. Virol.* **54**, 651–660.

Rueckert, R.R. & Wimmer, E. (1984) Systematic nomenclature of picornavirus proteins. *J. Virol.* **50**, 957–959.

Strebel, K. & Beck, E. (1986) A second protease of foot-and-mouth disease virus. *J. Virol.* **58**, 893–899.

Strebel, K., Beck, E., Strohmaier, K. & Schaller, H. (1986) Characterization of foot-and-mouth disease virus gene products with antisera against bacterially synthesized fusion proteins. *J. Virol.* **57**, 983–991.

Wutz, G., Auer, H., Nowotny, N., Grosse, B., Skern, T. & Kuechler, E. (1996) Equine rhinovirus serotypes 1 and 2: relationship to each other and to aphthoviruses and cardioviruses. *J. Gen. Virol.* **77**, 1719–1730.

Ziegler, E., Borman, A.M., Kirchweger, R., Skern, T. & Kean, K.-M. (1995) Foot-and-mouth disease virus Lb proteinase can stimulate rhinovirus and enterovirus IRES-driven translation and cleave several proteins of cellular and viral origin. *J. Virol.* **69**, 3465–3474.

*Marvin J. Grubman*
*USDA, ARS, NAA,*
*Plum Island Animal Disease Center,*
*PO Box 848, Greenport, NY 11944, USA*
*Email: mgrubman@asrr.arsusda.gov*

# 227. *Potyvirus helper component proteinase*

## Databanks

*Peptidase classification: clan CC, family C6, MEROPS ID: C06.001*
*NC-IUBMB enzyme classification: none*
*ATCC entries: PVAS-69 (TEV inoculum); 45035 (molecularly cloned)*
*Databank codes:*

| Species | Strain/isolate | SwissProt | PIR | EMBL (genomic) |
|---|---|---|---|---|
| Barley yellow mosaic virus | II-1 | Q01207 | – | D01092 |
| Barley yellow mosaic virus | German | Q01365 | JQ1947 | D01099 |
| Johnson grass mosaic virus | – | – | – | Z26920: complete genome |
| Papaya ringspot virus | – | Q01901 | JQ1899 | S46722: complete genome |
| | | | | X67673: complete genome |
| Pea seedborne mosaic virus | – | P29152 | JQ1331 | D10930: complete genome |
| Peanut stripe virus | – | – | – | U05771: complete genome |
| Pepper mottle virus | – | Q01500 | A44062 | M96425: complete genome |
| Plum pox potyvirus | D | P13529 | JA0078 | D00298: 3' end of genome |
| | | | S06929 | X16415: complete genome |
| Plum pox potyvirus | NAT | P17766 | JQ0003 | D00424: complete genome |
| | | | | D13751: complete genome |
| Potato virus C | C | P22601 | – | M38377: 5' end of genome |
| Potato virus C | O | P22602 | – | M37180: 5' end of genome |
| Potato virus Y | N | P18247 | JS0166 | D00441: complete genome |
| | | | | U33454 |
| | | | | X12456: complete genome |
| Potato virus Y | Hungarian | Q02963 | JN0545 | M95491: complete genome |
| Tobacco etch virus | – | P04517 | A04207 | L38714 |
| | | | | M11216 |
| | | | | M11458: complete genome |
| | | | | M15239: complete genome |
| Tobacco vein mottling virus | – | P09814 | A23647 | U38621: complete genome |
| | | | | X04083 |
| Turnip mosaic virus | – | Q02597 | JQ1168 | D10601: 3' end of genome |
| | | | JQ1895 | D10927: capsid protein |
| | | | | U18654: proteinase region |
| Zucchini yellow mosaic virus | – | – | – | L35590: 5' end of genome |

## Name and History

The *potyvirus helper component proteinase (HC-Pro)* was initially characterized as the helper component required for aphid transmission of the virus (Pirone & Thornbury, 1983), but work on the proteolytic activity of HC-Pro proteinase activity was done by *in vitro* translation of transcripts derived from tobacco etch virus (TEV) (Oh & Carrington, 1989). More recent studies have revealed that HC-Pro is multifunctional, having three domains, each of which serves a separate function in viral infectivity: the N-terminal third contributes to viral replication and aphid transmission, the central domain is required for viral vascular transport, and the C-terminal third contains the proteolytic domain (Carrington *et al.*, 1989a; Dolja *et al.*, 1993; Cronin *et al.*, 1995).

## Activity and Specificity

Proteolytic processing by HC-Pro occurs at a Gly+Gly bond located at its C-terminus. Amino acid substitution analysis of the tobacco etch virus HC-Pro revealed that amino acid residues in the P4, P2, P1 and P1' positions are essential for cleavage. Among potyviruses, a Tyr-Xaa-Val-Gly-Gly sequence surrounding the cleavage site is highly conserved, and is proposed to serve as the cleavage recognition sequence interacting with the substrate-binding pocket (Carrington & Herndon, 1992). HC-Pro proteinase activity has been observed during *in vitro* translation of potyviral transcripts in the presence of rabbit reticulocyte lysate, and also in *Escherichia coli* expressing similar polypeptides from transformed plasmids, suggesting that plant host factors are not required for proteolysis (Carrington *et al.*, 1989a,b; Oh & Carrington, 1989).

## Structural Chemistry

Amino acid substitution analysis revealed that the residues Cys649 and His722 within the proteolytic domain are essential for proteolysis. It is notable that His722 lies within a conserved Lys-(Xaa)$_2$-His-Ala-Val-(Xaa)$_3$-Gly sequence which resembles an active-site motif of peptidases in the papain family (Rawlings & Barrett, 1994) (Chapter 225), lending further support to the idea that HC-Pro is a cysteine-type proteinase (Carrington & Herndon, 1992), and belongs to clan CC.

Since the role of HC-Pro in aphid transmission was discovered before its proteolytic activity was recognized, most of the physical properties of the protein that have been described pertain to the N-terminal half and relate to its role in insect transmission. HC-Pro purified from tobacco leaves infected with tobacco vein mottling potyvirus exists as a dimer (Thornbury *et al.*, 1985), and this may function in aphid transmission, but the *cis*-preferential cleavage mechanism argues that the potyviral polyprotein processing cannot be assayed using this fraction. Conserved cysteine residues near the N-terminus may serve as a metal-binding site, which correlates with the requirement of a $Mg^{2+}$ ion for the activity of HC-Pro in aphid-mediated transmission (Atreya *et al.*, 1992).

## Preparation

While HC-Pro can be purified from plant extracts by sucrose gradient fractionation followed by affinity chromatography on oligo d(T)-cellulose (Thornbury *et al.*, 1985), proteinase activity has only been measured following expression in bacteria, or by *in vitro* translation of transcripts in the presence of rabbit reticulocyte lysate or wheatgerm extract (Carrington *et al*, 1989a; Oh & Carrington, 1989; Verchot *et al.*, 1991).

## Biological Aspects

The polyprotein expression strategy employed by a number of picorna-like viruses provides temporal regulation of available proteins by differential proteolysis (Krausslich & Wimmer, 1988). The three proteinases encoded within the potyviral genome may provide a cascade of proteolytic events to produce a range of functional intermediate and mature polypeptides regulating events during the viral life cycle. Unlike that of the serine-type, P1 proteinase (see Chapter 90), HC-Pro proteolysis is essential for viral infectivity in plants (Kasschau & Carrington, 1995; Verchot & Carrington, 1995). Viruses encoding defects within the HC-Pro proteinase active site were debilitated in viral replication. When a second cleavage site, recognized in *trans* by the NIa proteinase (Chapter 244), was inserted at the C-terminus, processing was restored, but not viral replication or transport. One explanation for these findings is that HC-Pro functions in *cis* for viral replication and/or transport. Alternatively, insertion of the second cleavage site restores proteolytic separation of HC-Pro from the adjacent P3 protein, but does not restore any temporal requirement for processing that may exist (Kasschau & Carrington, 1995). Since HC-Pro lies adjacent to a polyprotein module containing essential, replication-associated proteins, it is reasonable to suppose that cotranslational processing may be needed for early events in replication contributed either by a mature HC-Pro or by mature forms of neighboring proteins.

## Related Peptidases

The potyviral HC-Pro proteinase is suggested to belong to a class of viral cysteine-type enzymes that includes the p29 and p48 proteinases encoded by the hypovirulence-associated virus (HAV) of the chestnut blight fungus (see Chapter 228) and the 28K protein of barley yellow mosaic bymovirus (BaYMV) (Choi *et al.*, 1991; Shapira & Nuss, 1991). Each of these related enzymes functions to autoproteolytically process a site located at its C-terminus. In particular, sequences near the N-terminus of both the HAV p29 and the potyviral HC-Pro proteinases show some similarity, in addition to that around the proteinase active site, suggesting that these enzymes are closely related and arose from a common ancestor (Koonin *et al.*, 1991).

## Further Reading

More comprehensive reviews have been provided by Reichmann *et al.* (1992) and Dougherty & Semler (1993).

## References

Atreya, C.D., Atreya, P.L., Thornbury, D.W. & Pirone, T.P. (1992) Site-directed mutations in the potyvirus HC-Pro gene affect helper component activity, virus accumulation, and symptom expression in infected tobacco plants. *Virology* **191**, 106–111.

Carrington, J.C. & Herndon, K.L. (1992) Characterization of the poty-viral HC-Pro autoproteolytic cleavage site. *Virology* **187**, 308–315.

Carrington, J.C., Cary, S.M., Parks, T.D. & Dougherty, W.G. (1989a) A second proteinase encoded by a plant potyvirus genome. *EMBO* **8**, 365–370.

Carrington, J.C., Freed, D.D. & Sanders, T.C. (1989b) Autocat-alytic processing of the potyvirus helper component proteinase in *Escherichia coli* and *in vitro. J. Virol.* **63**, 4459–4463.

Choi, G.H., Pawlyk, D.M. & Nuss, D.L. (1991) The autocatalytic protease p29 encoded by a hypovirulence-associated virus of the chestnut blight fungus resembles the potyvirus-encoded protease HC-Pro. *Virology* **183**, 747–752.

Cronin, S., Verchot, J., Haldeman-Cahill, R., Schaad, M.C. & Car-rington, J.C. (1995) Long-distance movement factor: a transport function of the potyvirus helper component proteinase. *Plant Cell* **7**, 549–559.

Dolja, V.V., Herndon, K.L., Pirone, T.P. & Carrington, J.C. (1993) Spontaneous mutagenesis of a plant potyvirus genome after inser-tion of a foreign gene. *J. Virol.* **67**, 5968–5975.

Dougherty, W.G. & Semler, B.L. (1993) Expression of virus-encoded proteinases: functional and structural similarities with cellular enzymes. *Microbiol. Rev.* **57**, 781–822.

Kasschau, K.D. & Carrington, J.C. (1995) Requirement for HC-Pro processing during genome amplification of tobacco etch potyvirus. *Virology* **209**, 268–273.

Koonin, E.V., Choi, G.H., Nuss, D.L., Shapira, R. & Carrington, J.C. (1991) Evidence for common ancestry of a chestnut blight hypovirulence-associated double-stranded RNA and a group of positive-strand RNA plant viruses. *Proc. Natl Acad. Sci USA* **88**, 10647–10651.

Kräusslich, H.-G. & Wimmer, E. (1988) Viral proteinases. *Annu. Rev. Biochem.* **57**, 701–754.

Oh, C.-S. & Carrington, J.C. (1989) Identification of essential residues in potyvirus proteinase HC-Pro by site-directed mutagenesis. *Virology* **173**, 692–699.

Pirone, T.P. & Thornbury, D.W. (1983) Role of virion and helper component in regulating aphid transmission of tobacco etch virus. *Phytopathology* **73**, 872–875.

Rawlings, N.D & Barrett, A.J. (1994) Families of cysteine pepti-dases. *Methods Enzymol.* **244**, 461–486.

Reichmann, J.L., Lain, S. & Garcia, J.A. (1992) Highlights and prospects of potyvirus molecular biology. *J. Gen. Virol.* **73**, 1–16.

Shapira, R. & Nuss, D.L. (1991) Gene expression by a hypovirulence-associated virus of the chestnut blight fungus involves two papain-like protease activities. *J. Biol. Chem.* **266**, 19419–19425.

Thornbury, D.W., Hellman, G.M., Rhoads, R.E. & Pirone, T.P. (1985) Purification and characterization of potyvirus helper component. *Virology* **144**, 260–267.

Verchot, J. & Carrington, J.C. (1995) Debilitation of plant potyvirus infectivity by P1 proteinase-inactivating mutations and restoration by second-site modifications. *J. Virol.* **69**, 1582–1590.

Verchot, J., Koonin, E.V. & Carrington, J.C. (1991) The 35-kDa protein from the N-terminus of the potyviral polyprotein functions as a third virus-encoded proteinase. *Virology* **185**, 527–535.

***Jeanmarie Verchot***
*The Sainsbury Laboratory,*
*Norwich Research Park,*
*Colney, Norwich NR47UH, UK*
*Email: verchot@bbsrc.ac.uk*

# 228. *Hypovirus cysteine proteases p29 and p48*

## Databanks

*Peptidase classification: clan CC, families C7 and C8, MEROPS ID: C07.001, C08.001*
*NC-IUBMB enzyme classification: none*
*Databank codes:*

| Species | SwissProt | PIR | EMBL (genomic) |
|---|---|---|---|
| Family C7, p29 protease | | | |
| *Cryphonectria* hypovirus | P10941 | S03833 | L29010: complete genome |
| | | | M57938: complete genome |
| | | | X14524: 5′ end of genome |

(The p29 proteinase represents residues 1–248 of the 5′ proximal ORF, ORF A.)

*continued overleaf*

| Species | SwissProt | PIR | EMBL (genomic) |
|---|---|---|---|
| Family C8, p48 protease | | | |
| *Cryphonectria* hypovirus | P10942 | S03834 | L29010: complete genome |
| | | | M57938: complete genome |
| | | | X14524: 5′ end of genome |

(The p48 proteinase represents residues 1–418 of the 3′ proximal ORF, ORF B.)

## Name and History

Members of the RNA virus family Hypoviridae infect and attenuate virulence of the chestnut blight fungus *Cryphonectria parasitica* (Hillman *et al.*, 1995). **Hypovirus cysteine protease p29** and **hypovirus cysteine protease p48** were first identified during the cloning and characterization of the RNA genome of the prototypic hypovirus CHV1-713. The CHV1-713 RNA coding strand contains two contiguous, large open reading frames designated ORF A and ORF B, and p29 and p48 comprise the N-terminal portions of the polyprotein products of ORF A and ORF B, respectively (Shapira *et al.*, 1991).

## Activity and Specificity

Protease p29 is autocatalytically released from the 69 kDa polyprotein product of CHV1-713 RNA ORF A (Choi *et al.*, 1991a). Cleavage occurs at Gly248┼Gly249 during translation and is dependent on residues Cys162 and His215 (Choi *et al.*, 1991a,b). Similarly, p48 is autocatalytically released from the N-terminus of the ORF B-encoded polyprotein (Shapira *et al.*, 1991). In this case, cleavage occurs at Gly418┼Ala419 and is dependent on residues Cys341 and His388 (Shapira & Nuss, 1991). A detailed amino acid substitution analysis of the p48 cleavage site has been reported (Shapira & Nuss, 1991). Both proteases have been shown to cleave in *cis*, but *trans* cleavage has not been demonstrated.

## Preparation

Recombinant p48 containing a His$_6$ tag has been produced in and purified from *Escherichia coli* (Shapira & Nuss, 1991).

## Biological Aspects

Craven *et al.* (1993) showed that p29 contributes to hypovirus-mediated reduction in fungal pigment production, sporulation and laccase production. Significantly, p29-mediated symptom expression was found to be dependent upon release from the polyprotein precursor, but to be independent of intrinsic proteolytic activity. The same authors also reported that p29 is nonessential for viral RNA replication and virus-mediated attenuation of fungal virulence.

Computer-assisted analysis of the p29 and p48 amino acid sequences confirmed similarities between these cysteine proteases and the HC-Pro proteases encoded by potyviruses and by a related fungus-vectored plant virus, barley yellow mosaic virus (BaYMV) (Koonin *et al.*, 1991) (see Chapter 227). These results are consistent with the results of a larger sequence comparison that predicted a common ancestry for hypoviruses and potyviruses. It has also been suggested that p29 and p48 could have evolved by intragenomic duplication followed by rapid divergence that could have been due to functional diversification of the two proteases (Koonin *et al.*, 1991).

## References

Choi, G.H., Shapira, R. & Nuss, D.L. (1991a) Co-translational autoproteolysis involved in gene expression from a double-stranded RNA genetic element associated with hypovirulence of the chestnut blight fungus. *Proc. Natl Acad. Sci. USA* **88**, 1167–1171.

Choi, G.H., Pawlyk, D.M. & Nuss, D.L. (1991b) The autocatalytic protease p29 encoded by a hypovirulence-associated virus of the chestnut blight fungus resembles the potyvirus-encoded protease HC-Pro. *Virology* **183**, 747–752.

Craven, M.G., Pawlyk, D.M., Choi, G.H. & Nuss, D.L. (1993) Papain-like protease p29 as a symptom determinant encoded by a hypovirulence-associated virus of the chestnut blight fungus. *J. Virol.* **67**, 6513–6521.

Hillman, B.I., Fulbright, D.W., Nuss, D.L. & Van Alfen, N.K. (1995) Hypoviridae. In: *Virus Taxonomy* (Murphy, F.A., Fauquet, C.M., Bishop, D.H.L., Ghabrial, S.A., Jarvis, A.W., Martelli, G.P., Mayo, M.A. & Summers, M.D., eds). New York: Springer-Verlag, pp. 261–264.

Koonin, E.V., Choi, G.H., Nuss, D.L., Shapira, R. & Carrington, J.C. (1991) Evidence for common ancestry of a chestnut blight hypovirulence-associated double-stranded RNA and a group of positive-strand RNA plant viruses. *Proc. Natl Acad. Sci. USA* **88**, 10647–10651.

Shapira, R. & Nuss, D.L. (1991) Gene expression by a hypovirus-associated virus of the chestnut blight fungus involves two papain-like protease activities. *J. Biol. Chem.* **266**, 19419–19425.

Shapira, R., Choi, G.H. & Nuss, D.L. (1991) Virus-like genetic organization and expression strategy for a double-stranded RNA genetic element associated with biological control of chestnut blight. *EMBO J.* **10**, 731–739.

*Donald L. Nuss*
*Center for Agricultural Biotechnology,*
*University of Maryland Biotechnology Institute,*
*5115 Plant Sciences Bldg.,*
*College Park, MD 20742, USA*
*Email: nuss@umbi.umd.edu*

# 229. *Coronavirus papain-like endopeptidases*

## Databanks

*Peptidase classification: clan CC, families C16 and C29, MEROPS ID: C16.001, C29.001*
*NC-IUBMB enzyme classification: none*
Databank codes:

| Species | Strain/isolate | SwissProt | PIR | EMBL (genomic) |
|---|---|---|---|---|
| **Family C16** | | | | |
| Mouse hepatitis coronavirus papain-like endopeptidase 1, PLP-1 and C29 | | | | |
| Avian infectious bronchitis virus | – | P27920 | A33094 | M94356: F1 and F2 genes |
| | | | | M95169: complete genome |
| Human coronavirus | 229E | Q05002 | S28600 | X69721: ORF1a |
| Porcine transmissible gastroenteritis virus | – | – | – | Z34093: ORF1a |
| Murine hepatitis virus | A59 | | | X73559: gene 1 and ORF1a |
| | JHM | P19751 | A36815 | M18040: gene 1, 5′ end |
| | | | | M55148: gene 1 and ORF1b |
| | | | | S51684: gene 1 |
| | defective A59 | | | X57302: defective interfering mRNA |
| | defective JHM | P26627 | A36388 | M61144: defective interfering mRNA |
| | not known | | | M23258: defective interfering mRNA |
| | | | | M28352: autoproteolytic cleavage domain |
| **Family C29** | | | | |
| Murine hepatitis coronavirus papain-like endopeptidase 2 | | | | |
|   Murine hepatitis virus | | P19751 | A36815 | M18040: gene 1, 5′ end |
| | | | | M23258: defective interfering RNA |
| | | | | M28352: cleavage domain |
| | | | | M55148: gene 1 and ORF 1b |
| | | | | M61144: defective interfering RNA |
| | | | | S51684: gene 1 |
| | | | | X57302: defective interfering RNA |
| | | | | X73559: gene 1 and ORF 1a |
|   Avian infectious bronchitis virus | | P27920 | A33094 | M94356: F1 and F2 genes |
| | | | | M95169: complete genome |
|   Porcine transmissible gastroenteritis virus | – | – | Z34093: ORF 1a |

## Name and History

Proteolytic processing of a polyprotein precursor is an event common to the replication cycle of many RNA viruses. For coronaviruses, a family of positive-strand RNA viruses with large genomes (28–32 kb), the gene encoding the viral, RNA-dependent RNA polymerase is translated into a large precursor polyprotein which must be proteolytically processed to mediate viral transcription and replication. For the prototype coronavirus, mouse hepatitis virus (MHV), the gene encoding the RNA-dependent RNA polymerase is approximately 22 kb in length, and contains two overlapping open reading frames (ORFs), which are translated by a ribosomal frameshifting mechanism to generate a polyprotein precursor of approximately 750 kDa (Brierly *et al*., 1987, 1989; Bredenbeek *et al*., 1990). Sequencing of this gene in MHV strains A59 and JHM (Pachuk *et al*., 1989; Lee *et al*., 1991; Bonilla *et al*., 1994) led to the prediction of various putative functional domains within the polymerase polyprotein (Gorbalenya *et al*., 1989; Lee *et al*., 1991) (Figure 229.1).

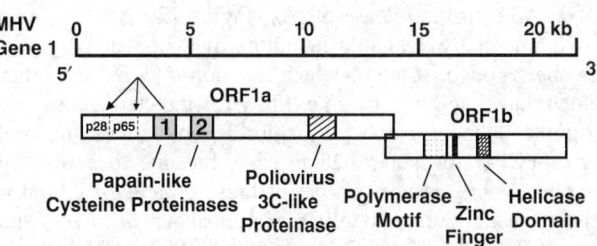

*Figure 229.1* Schematic diagram of the mouse hepatitis virus polymerase polyprotein. Arrows indicate cleavages mediated by PCP-1 in the release of p28 (Gly247↓Val248) and potentially p65 (Ala832↓Gly833) from the N-terminus of the polyprotein.

Among these putative domains there are three proteinase domains, two of which are papain-like, as described in this chapter, and one which is picornain-like (Chapter 246). The first **papain-like cysteine proteinase (PCP-1)** has limited sequence similarity to papain, mainly in the conserved Cys-Trp dipeptide containing the predicted catalytic Cys residue, and the spacing between the predicted catalytic Cys and His residues, which is comparable to the longest spacing found in cellular papain-like proteases (Gorbalenya *et al.*, 1991). The second 'papain-like' domain is, in fact, quite distantly papain-like, with only conserved Cys and His residues and spacing. Endopeptidase activity of this domain has not yet been experimentally demonstrated. These domains have also been referred to in the literature as **PLP-1** and **PLP-2**.

### Activity and Specificity

Initial studies involving translation of coronavirus MHV-A59 genomic RNA *in vitro* led to the detection of the release of a 28 kDa protein (p28) from the N-terminus of the polymerase polyprotein precursor. These studies suggested that a viral endopeptidase was essential for the processing event (Denison & Perlman, 1986). Expression of cDNA clones representing the 5′ end of the viral genome showed that the PCP-1 domain mediates the cleavage of p28 (Baker *et al.*, 1989). Cleavage of p28 is rapid and essentially cotranslational *in vitro*, suggesting a *cis*-cleavage event. PCP-1 activity has also been implicated in the release of the protein adjacent to p28, p65 (Bonilla *et al.*, 1995). Interestingly, cleavage of p65 is not detected during *in vitro* translation of contiguous 5.3 kb cDNA clones, but may be observed using cDNA clones with internal deletions (Bonilla *et al.*, 1995). These results imply that the processing of p28 and p65 *in vitro* are distinct events, perhaps requiring specific folding of the substrate or of the PCP-1 endopeptidase domain to allow recognition and processing.

The minimum functional PCP-1 endopeptidase domain was defined by deletion analysis to reside between 3.46 and 4.15 kb from the 5′ end of the MHV-A59 genome (Bonilla *et al.*, 1995). The catalytic residues of PCP-1, Cys1137 and His1288, were identified by assessing the effect of site-directed mutations of the putative catalytic sites on the efficiency of cleavage of p28 (Baker *et al.*, 1993). The specificity of PCP-1 was characterized by extensive site-directed mutagenesis of both the MHV-JHM and MHV-A59 p28 cleavage sites (Dong & Baker, 1994; Hughes *et al.*, 1995). These studies demonstrated that the Gly in the P1 position and the Arg in the P2 position are critical in both MHV-JHM and MHV-A59 for the release of p28 (Figure 229.2).

Commercially available inhibitors of PCP-1 activity have also been found, some of which function *in vitro*, and others which have activity *in vivo* (in virus-infected cells). For example, leupeptin and zinc chloride both block the ability of PCP-1 to cleave p28 *in vitro*, but due to permeability and toxicity issues, neither of these proteinase inhibitors has yet been shown to block p28 cleavage *in vivo* (Denison & Perlman, 1986; Denison *et al.*, 1992). Interestingly, a membrane-permeable derivative of the irreversible cysteine proteinase inhibitor E-64, E-64d, has been shown to block the cleavage of p65 *in vivo*, but has no corresponding effect on p28 cleavage (Kim *et al.*, 1995). Like the *in vitro* translation studies described above, these results suggest that

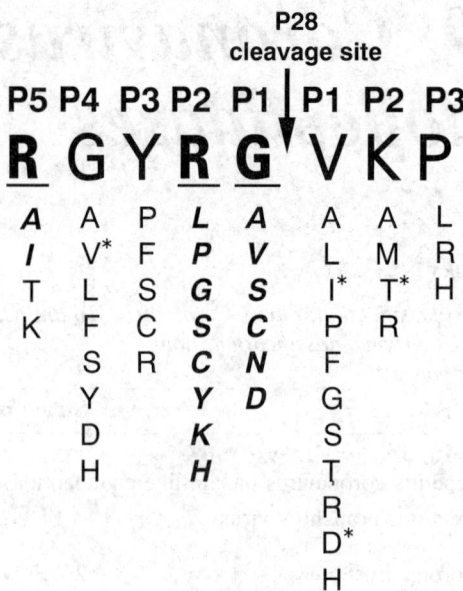

*Figure 229.2*  Schematic diagram of the mouse hepatitis virus p28 cleavage site. Site-directed mutagenesis of the amino acid residues flanking the cleavage site was performed, and *in vitro* cleavage efficiency by PCP-1 was assessed for each amino acid change. Underlined amino acids in the wild-type sequence are required for efficient cleavage; italicized amino acid substitutions inactivate cleavage >80%; unitalicized substitutions have little or no effect on cleavage efficiency; and amino acids marked with an asterisk (*) are those for which substitutions reduce cleavage >60% in MHV-A59.

the processing of p28 and p65 *in vivo* may require distinct conformations or concentrations of the substrate or of the endopeptidase domain for efficient processing. Furthermore, E-64d was shown to inhibit coronavirus RNA synthesis in infected cells, indicating that processing of the polymerase polyprotein is required throughout infection to sustain viral RNA synthesis (Kim *et al.*, 1995).

### Preparation

To date, coronavirus PCP-1 has not been purified from coronavirus-infected cells or expressed and purified from *Escherichia coli*. However, active PCP-1 has been generated *in vitro* by translation of viral genomic RNA preparations (Denison & Perlman, 1986), as well as by transcription and translation of cloned cDNAs (Baker *et al.*, 1989).

### Biological Aspects

The PCP-1 domain is part of a large polyprotein precursor of the viral RNA-dependent RNA polymerase. The PCP-1 endopeptidase is believed to play a critical role in the processing of the precursor polyprotein, perhaps both in *cis* and in *trans*, to generate the nonstructural proteins associated with viral replication. The critical role of a cysteine proteinase domain has been elegantly demonstrated for Sindbis virus replication (Shirako & Strauss, 1994). The Sindbis virus nsP2 proteinase (Chapter 230) regulates the switch between negative- and positive-strand RNA synthesis via its controlled cleavage of polymerase subunits. It is possible that the PCP-1 domain may have a similar role in the coronavirus life cycle.

## Related Peptidases

PCP domains have been predicted, but not yet functionally demonstrated in three other coronaviruses for which the polymerase gene has been sequenced: avian infectious bronchitis virus (Boursnell *et al.*, 1987), human coronavirus 229E (Herold *et al.*, 1993), and porcine transmissible gastroenteritis virus (Eleouet *et al.*, 1995). Arteriviruses, coronavirus superfamily members, have also been shown to encode functional PCP domains (Snijder *et al.*, 1992, 1994; den Boon *et al.*, 1995) (see Chapter 234).

## Further Reading

Recommended papers are those of Baker *et al.* (1993), Dong & Baker (1994), Hughes *et al.*, (1995) and Bonilla *et al.* (1995).

## References

Baker, S.C., Shieh, C.K., Soe, L.H., Chang, M.F., Vannier, D.M. & Lai, M.M.C. (1989) Identification of a domain required for autoproteolytic cleavage of murine coronavirus gene A polyprotein. *J. Virol.* **63**, 3693–3699.

Baker, S.C., Yokomori, K., Dong, S., Carlisle, R., Gorbalenya, A.E., Koonin, E.V. & Lai, M.M.C. (1993) Identification of the catalytic sites of a papain-like cysteine proteinase of murine coronavirus. *J. Virol.* **67**, 6056–6063.

Bonilla, P.J., Gorbalenya, A.E. & Weiss, S.R. (1994) Mouse hepatitis virus A59 RNA polymerase gene ORF 1a: heterogeneity among MHV strains. *Virology* **198**, 736–740.

Bonilla, P.J., Hughes, S.A., Piñón, J.D. & Weiss, S.R. (1995) Characterization of the leader papain-like proteinase of MHV-A59: identification of a new *in vitro* cleavage site. *Virology* **209**, 489–497.

Boursnell, M.E.G., Brown, T.D.K., Foulds, I.J., Green, P.F., Tomley, F.M. & Binns, M.M. (1987) Completion of the sequence of the genome of the coronavirus avian infectious bronchitis virus. *J. Gen. Virol.* **68**, 57–77.

Bredenbeek, P.J., Pachuk, C.J., Noten, A.F.H., Charité, J., Luytjes, W., Weiss, S.R. & Spaan, W.J.M. (1990) The primary structure and expression of the second open reading frame of the polymerase gene of the coronavirus MHV-A59; a highly conserved polymerase is expressed by an efficient ribosomal frameshifting mechanism. *Nucleic Acids Res.* **18**, 1825–1832.

Brierly, I., Boursnell, M.E.G., Binns, M.M., Bilimoria, B., Blok, V.C., Brown, T.D.K. & Inglis, S.C. (1987) An efficient ribosomal frame-shifting signal in the polymerase-encoding region of the coronavirus IBV. *EMBO J.* **6**, 3779–3785.

Brierly, I., Digard, P. & Inglis, S.C. (1989) Characterization of an efficient coronavirus ribosomal frameshifting signal: requirement for an RNA pseudoknot. *Cell* **57**, 537–547.

den Boon, J.A., Faaberg, K.S., Meulenberg, J.J.M., Wassenaar, A.L.M., Plagemann, P.G.W., Gorbalenya, A.E. & Snijder, E.J. (1995) Processing and evolution of the N-terminal region of the arterivirus replicase ORF1a protein: identification of two papainlike cysteine proteases. *J. Virol.* **69**, 4500–4505.

Denison, M.R. & Perlman, S. (1986) Translation and processing of mouse hepatitis virus virion RNA in a cell-free system. *J. Virol.* **60**, 12–18.

Denison, M.R., Zoltick, P.W., Hughes, S.A., Giangreco, B., Olson, A.L., Perlman, S., Leibowitz, J.L. & Weiss, S.R. (1992) Intracellular processing of the N-terminal ORF 1a proteins of the coronavirus MHV-A59 requires multiple proteolytic events. *Virology* **189**, 274–284.

Dong, S. & Baker, S.C. (1994) Determinants of the p28 cleavage site recognized by the first papain-like cysteine proteinase of murine coronavirus. *Virology* **204**, 541–549.

Eleouet, J.-F., Rasschaert, D., Lambert, P., Levy, L., Vende, P. & Laude, H. (1995) Complete sequence (20 kilobases) of the polyprotein-encoding gene 1 of transmissible gastroenteritis virus. *Virology* **206**, 817–822.

Gorbalenya, A.E., Koonin, E.V., Donchenko, A.P. & Blinov, V.M. (1989) Coronavirus genome: prediction of putative functional domains in the non-structural polyprotein by comparative amino acid sequence analysis. *Nucleic Acids Res.* **17**, 4847–4861.

Gorbalenya, A.E., Koonin, E.V. & Lai, M.M.C. (1991) Putative papain-related thiol proteases of positive-strand RNA viruses. *FEBS Lett.* **288**, 201–205.

Herold, J., Raabe, T., Schelle-Prinz B., & Siddell, S.G. (1993) Nucleotide sequence of the human coronavirus 229E RNA polymerase locus. *Virology* **195**, 680–691.

Hughes, S.A., Bonilla, P. & Weiss, S.R. (1995) Identification of the murine coronavirus p28 cleavage site. *J. Virol.* **69**, 809–813.

Kim, J.C., Spence, R.A., Currier, P.F., Lu, X. & Denison, M.R. (1995) Coronavirus protein processing and RNA synthesis is inhibited by the cysteine proteinase inhibitor E64d. *Virology* **208**, 1–8.

Lee, H.-J., Shieh, C.-K., Gorbalenya, A.E., Koonin, E.V., La Monica, N., Tuler, J., Bagdzhadzhyan, A. & Lai, M.M.C. (1991) The complete sequence (22 kilobases) of murine coronavirus gene 1 encoding the putative proteases and RNA polymerase. *Virology* **180**, 567–582.

Pachuk, C.J., Bredenbeek, P.J., Zoltick, P.W., Spaan, W.J.M. & Weiss, S.R. (1989) Molecular cloning of the gene encoding the putative polymerase of mouse hepatitis coronavirus, strain A59. *Virology* **171**, 141–148.

Shirako, Y. & Strauss, J.H. (1994) Regulation of sindbis virus RNA replication: uncleaved P123 and nsP4 function in minus-strand RNA synthesis, whereas cleaved products from P123 are required for efficient plus-strand RNA synthesis. *J. Virol.* **68**, 1874–1885.

Snijder, E.J., Wassenaar, A.L.M. & Spaan, W.J.M. (1992) The 5′ end of the equine arteritis virus replicase gene encodes a papainlike cysteine protease. *J. Virol.* **66**, 7040–7048.

Snijder, E.J., Wassenaar, A.L.M. & Spaan, W.J.M. (1994) Proteolytic processing of the replicase ORF1a protein of equine arteritis virus. *J. Virol.* **68**, 5755–5764.

*Jennifer J. Schiller*
*Department of Microbiology and Immunology,*
*Loyola University Medical Center,*
*2160 South 1st Avenue, Bldg. 105,*
*Maywood, IL 60153, USA*

*Susan C. Baker*
*Department of Microbiology and Immunology,*
*Loyola University Medical Center,*
*2160 South 1st Avenue, Bldg. 105,*
*Maywood, IL 60153, USA*
*Email: sbaker@luc.edu*

# 230. Sindbis virus nsP2 endopeptidase

## Databanks

*Peptidase classification: clan CC, family C9, MEROPS ID: C09.001*
*NC-IUBMB enzyme classification: none*
*Databank codes:*

| Species | Strain/isolate | SwissProt | PIR | EMBL (genomic) |
|---|---|---|---|---|
| Aura virus | – | – | – | S78478: complete genome |
| Barmah Forest virus | – | – | – | U73745: complete genome |
| Eastern equine encephalomyelitis virus | | – | – | U01034: complete genome |
| | | | | X63135: complete genome |
| Middleburg virus | – | – | – | J02246: 3′ end of genome |
| O'Nyong-nyong virus | – | P13886 | A34680 | M20303: complete genome |
| Ross River virus | – | P13887 | A28605 | M20162: complete genome |
| Sindbis virus | HRSP | P03317 | A03917 | J02363: complete genome |
| | Ockelbo | P27283 | A39991 | M69205: complete genome |
| Sindbis-like virus | Girdwood | – | – | U38304: complete genome |
| | S.A.AR86 | – | – | U38305: complete genome |
| Semliki Forest virus | – | P08411 | A23592 | A18788: artificial sequence |
| | | | | X04129: complete genome |
| Venezuelan equine encephalitis virus | | | | |
| | Trinidad | P27282 | A31467 | J04332: complete polyprotein gene |
| | | | | L01443: complete polyprotein gene |
| | | | | U34999: complete genome |
| | 3880 | P36327 | C44213 | L00930: complete polyprotein gene |
| | p676 | P36328 | A44213 | L01442: complete genome |
| | | | | L04653: complete genome |
| Western equine encephalomyelitis virus | | – | | S57222 | X74892: 3′ end of polyprotein gene |
| | | | | U01065: mid-section of polyprotein |

## Name and History

Sindbis virus (family Togaviridae, genus *Alphavirus*) is an animal virus with a single molecule of positive-strand RNA 11 703 nucleotides in length as its genome. The genomic RNA is a message for production of four nonstructural proteins which are translated in the cytoplasm of the infected cell as two polyproteins, P123 (1896 amino acids) and P1234 (2513 amino acids, produced by read-through of an opal termination codon at position 1897). The P1234 polyprotein is processed by the ***nsP2 proteinase*** through a series of cleavage intermediates into the four final products nsP1, nsP2, nsP3 and nsP4, named in order 5′ to 3′ along the genome. Since cytoplasmic, nonorganelle-bound proteinases are rare in animal cells, a virally encoded activity was predicted as early as 1981 (Rice & Strauss, 1981). The proteinase was localized to the C-terminal half of nsP2 by translation *in vitro* of RNA transcribed from deleted or truncated full-length cDNA clones of Sindbis virus. Deletions within nsP1, nsP3 and nsP4 did not affect proteolysis, but deletions in the C-terminal half of nsP2 abolished proteinase activity, and deletions in the N-terminal half of nsP2 led to aberrant processing (Hardy & Strauss, 1989; Ding & Schlesinger, 1989). At the same time, several mutations that rendered the proteinase temperature sensitive

were found to lie in the C-terminal half of nsP2 (Hahn *et al.*, 1989). Site-specific mutagenesis of Cys and His residues in this domain that are conserved among alphaviruses later showed that Cys481 and His558 are essential for proteolytic activity, and it was proposed that these constitute the catalytic dyad of a cysteine proteinase (Strauss *et al.*, 1992).

## Activity and Specificity

The only known substrates for the Sindbis nsP2 proteinase are the Sindbis nonstructural polyproteins (P123 and P1234) and their cleavage intermediates (which include P12, P23 and P34). The consensus cleavage site is (Ala/Ile)-Gly-(Ala/Cys/Gly)⫩(Ala/Tyr). Mutational studies have demonstrated that the invariant Gly in the P2 position is essential for efficient cleavage; changing this Gly to Ala led to substantially slower cleavage of the mutated site, and substitution with Val or Glu rendered the site completely uncleavable (Shirako & Strauss, 1990). Similarly, an Ala or Gly was found to be essential in the P1 position for efficient cleavage in Sindbis virus (Cys was not tested) (de Groot *et al.*, 1990). The enzyme is relatively insensitive to the identity of the residue in the P1′ position, however (de Groot *et al.*, 1990). It is thought that additional determinants of specificity must exist, since these are the only sites cleaved by the enzyme. In this regard it has been shown that different polyproteins containing the nsP2 proteinase differ in their preferences for the three cleavage sites, suggesting that cleavage-site recognition involves more extensive interactions than the simple recognition of a short linear sequence by the enzyme (de Groot *et al.*, 1990).

## Structural Chemistry

Sindbis nsP2 is a polypeptide of 807 amino acids encoded by nucleotides 1680–4100 of the genome. The N-terminal half of the protein is believed to be a helicase required for RNA replication. The C-terminal domain between residues 460 and 807 constitutes the proteinase (Hardy & Strauss, 1989). Both Cys481 and His558, the putative catalytic dyad, are followed by tryptophan residues, and Trp559 has been shown to be essential for proteolysis. These two aromatic residues may interact (stack) to maintain proper conformation of the catalytic pocket (Strauss *et al.*, 1992). Substitution of Asn614 by Asp enhances cleavage; an Asn residue has been shown previously to be associated with the catalytic pocket in papain (Garavito *et al.*, 1977). Mutations Phe509Leu, Ala517Thr, Asp522Asn and Gly748Ser have been shown to render the proteinase temperature sensitive *in vivo*; these mutations are believed to alter the structure of the protein (Hardy *et al.*, 1990). The proteinase has been hypothesized to be related to papain on the basis of alignment studies (Gorbalenya *et al.*, 1991; Koonin & Dolja, 1993), but there is very limited sequence identity between the nsP2 proteinase and papain to support this hypothesis (Strauss *et al.*, 1992). No structural studies of the enzyme have been produced to date.

## Preparation

Sindbis virus has a very wide host range and tissue distribution, and infects many types of cells in culture, including avian, mammalian and mosquito cells. The genome of the infecting virus is a messenger RNA from which the nonstructural proteins are translated in the cytoplasm of infected cells. Both nsP2 and polyproteins containing nsP2 have also been expressed by translation in reticulocyte lysates of virion RNA or of *in vitro* transcribed RNA (Ding & Schlesinger, 1989; Hardy & Strauss, 1989; de Groot *et al.*, 1990, 1991; Strauss *et al.*, 1992), and by translation in bacteria or in mammalian cells of appropriate constructs (Hardy & Strauss, 1988; Lemm *et al.*, 1994).

## Biological Aspects

Both nsP2 and polyproteins containing nsP2 are active proteinases, but they differ in their site preferences. The

activities of the enzyme have been studied both *in vitro* (referenced above) and in infected or transfected cells (Hardy & Strauss, 1988; de Groot *et al.*, 1990; Hardy *et al.*, 1990; Shirako & Strauss, 1990, 1994; Lemm *et al.*, 1994). In many of these studies, site-directed mutagenesis was used to destroy cleavage sites or to inactivate the proteinase. These studies have shown that P1234 can cleave itself autoproteolytically in *cis* to give P123 and nsP4, whereas P123 can be cleaved only in *trans*. Cleavage of P123 by another P123 molecule or by P12 produces nsP1 and P23, whereas cleavage by P23 produces predominantly P12 and nsP3. Finally, enzymes that lack nsP3 (that is, nsP2 or P12) cannot cleave the nsP3/nsP4 site. These preferences lead to a temporal regulation of RNA synthesis during the infection cycle. Early in infection P123 and nsP4 are produced, and together with host factors these form a replicase complex that synthesizes negative-strand RNA. Cleavage of P123 in the replicase complex produces a replicase that synthesizes positive-strand RNAs very efficiently, but cannot make negative strands (Lemm *et al.*, 1994; Shirako & Strauss, 1994). As the concentration of *trans*-acting proteinases builds up, the nonstructural polyprotein is cleaved while nascent at the nsP2/nsP3 site, and the products produced are nsP1, nsP2, nsP3 and P34. Negative-strand RNA synthesis is then shut off, because the cleaved products cannot form a negative-strand replicase. The use of the nonstructural proteinase to control the RNA developmental pathway represents an interesting evolutionary adaptation of the need of viruses to encode proteinases to process polyproteins in the cell cytoplasm.

## Related Peptidases

All alphaviruses encode nsP2 proteinases that share more than 50% amino acid sequence identity. Sequences of nsP2 proteinases that have been deposited in the databases include those for Sindbis virus, eastern equine encephalitis virus, O'nyong-nyong virus, Ross River virus, Semliki Forest virus, Venezuelan equine encephalitis virus, western equine encephalitis virus and Aura virus.

All alphavirus nonstructural cleavage sites are of similar form (Table 230.1) (reviewed in Strauss & Strauss, 1994).

The extensive sequence similarity between the nsP2 proteinases and the similarity in the cleavage sites suggest that the proteinases must be very similar to one another, but no tests of cross-cleavage have been done to determine how specific each enzyme is for its own substrate. Similar papain-like endopeptidases, which may have a common evolutionary origin, have been reported for a number of plant and animal viruses, including rubella virus, a number of coronaviruses, a number of potyviruses, a bymovirus and hypovirulence-associated virus of fungi (Strauss *et al.*, 1992; Koonin & Dolja, 1993) (Chapter 225).

*Table 230.1*  Substrate sequences of alphavirus nonstructural proteinase cleavage sites

| | P4 | P3 | P2 | P1 | | P1′ | P2′ | P3′ | P4′ |
|---|---|---|---|---|---|---|---|---|---|
| nsP1/nsP2 | Xaa | Ala/Ile | Gly | Ala | + | Ala/Gly | Xaa | Val | Glu |
| nsP2/nsP3 | Xaa | Ala/Val/Ser | Gly | Ala/Cys/Arg | + | Ala | Pro | Ser/Ala | Tyr |
| nsP3/nsP4 | Xaa | Ala/Val | Gly | Ala/Gly | + | Tyr | Ile | Phe | Ser |

## References

de Groot, R.J., Hardy, W.R., Shirako, Y. & Strauss, J.H. (1990) Cleavage-site preferences of Sindbis virus polyproteins containing the nonstructural proteinase: evidence for temporal regulation of polyprotein processing in vivo. *EMBO J.* **9**, 2631–2638.

de Groot, R.J., Rümenapf, T., Kuhn, R.J., Strauss, E.G. & Strauss, J.H. (1991) Sindbis virus RNA polymerase is degraded by the N-end rule pathway. *Proc. Natl Acad. Sci. USA* **88**, 8967–8971.

Ding, M. & Schlesinger, M.J. (1989) Evidence that Sindbis virus nsP2 is an autoprotease which processes the virus nonstructural polyprotein. *Virology* **171**, 280–284.

Garavito, R.M., Rossmann, M.G., Argos, P. & Eventoff, W. (1977) Convergence of active site geometries. *Biochemistry* **16**, 5065–5071.

Gorbalenya, A.E., Koonin, E.V. & Lai, M.M.C. (1991) Putative papain-related thiol proteases of positive-strand RNA viruses – identification of rubivirus and aphthovirus proteases and delineation of a novel conserved domain associated with proteases of rubivirus, alpha-, and coronavirus. *FEBS Lett.* **288**, 201–205.

Hahn, Y.S., Strauss, E.G., & Strauss, J.H. (1989) Mapping of RNA-temperature-sensitive mutants of Sindbis virus: Assignment of complementation groups A, B, and G to nonstructural proteins. *J. Virol.* **63**, 3142–3150.

Hardy, W.R., Hahn, Y.S., de Groot, R.J., Strauss, E.G. & Strauss, J.H. (1990) Synthesis and processing of the nonstructural polyproteins of several temperature-sensitive mutants of Sindbis virus. *Virology* **177**, 199–208.

Hardy, W.R. & Strauss, J.H. (1988) Processing of the nonstructural polyproteins of Sindbis virus: study of the kinetics *in vivo* using monospecific antibodies. *J. Virol.* **62**, 998–1007.

Hardy, W.R. & Strauss, J.H. (1989) Processing the nonstructural polyproteins of Sindbis virus: nonstructural proteinase is in the C-terminal half of nsP2 and functions both in *cis* and in *trans*. *J. Virol.* **63**, 4653–4664.

Koonin, E.V. & Dolja, V.V. (1993) Evolution and taxonomy of positive-strand RNA viruses: implications of comparative analysis of amino acid sequences. *Crit. Rev. Biochem. Mol. Biol.* **28**, 375–430.

Lemm, J.A., Rümenapf, T., Strauss, E.G., Strauss, J.H. & Rice, C.M. (1994) Polypeptide requirements for assembly of functional Sindbis virus replication complexes: a model for the temporal regulation of minus-strand and plus-strand RNA-synthesis. *EMBO J.* **13**, 2925–2934.

Rice, C.M. & Strauss, J.H. (1981) Nucleotide sequence of the 26S mRNA of Sindbis virus and deduced sequence of the encoded virus structural proteins. *Proc. Natl Acad. Sci. USA* **78**, 2062–2066.

Shirako, Y. & Strauss, J.H. (1990) Cleavage between nsP1 and nsP2 initiates the processing pathway of Sindbis virus nonstructural polyprotein P123. *Virology* **177**, 54–64.

Shirako, Y. & Strauss, J.H. (1994) Regulation of Sindbis virus RNA replication: uncleaved P123 and nsP4 function in minus strand RNA synthesis whereas cleaved products from P123 are required for efficient plus strand RNA synthesis. *J. Virol.* **68**, 1874–1885.

Strauss, E.G., de Groot, R.J., Levinson, R. & Strauss, J.H. (1992) Identification of the active site residues in the nsP2 proteinase of Sindbis virus. *Virology* **191**, 932–940.

Strauss, J.H. & Strauss, E.G. (1994) The alphaviruses: gene expression, replication, and evolution. *Microbiol. Rev.* **58**, 491–562.

*Ellen G. Strauss*
*Division of Biology, 156-29,*
*California Institute of Technology,*
*Pasadena, CA 91125, USA*

*James H. Strauss*
*Division of Biology, 156-29,*
*California Institute of Technology,*
*Pasadena, CA 91125, USA*
*Email: straussj@starbase1.caltech.edu*

# 231. Rubella nonstructural protease

## Databanks

*Peptidase classification: clan CC, family C27, MEROPS ID: C27.001*
*NC-IUBMB enzyme classification: none*
*Databank codes:*

| Species | SwissProt | PIR | EMBL (genomic) |
|---|---|---|---|
| Rubella virus | P13889 | A35320 | M15240: complete genome |

## Name and History

Rubella virus, a positive-strand RNA virus belonging to the *Rubivirus* genus of the Togaviridae family (reviewed by Frey, 1994), contains two open reading frames within its genome of roughly 10 000 nucleotides: a 5′-proximal ORF encoding nonstructural proteins involved in RNA replication (NSP-ORF) and a 3′-proximal ORF encoding the virion proteins, a capsid protein and two envelope glycoproteins. Following publication of the genomic sequence (Dominguez *et al*., 1990), computer alignment predicted the existence within the NSP-ORF of a cysteine protease belonging to the papain-like family of viral proteases (Gorbalenya *et al*., 1991). Subsequent expression of the NSP-ORF confirmed that proteolytic processing occurred and that processing was mediated by the computer-predicted protease (Marr *et al*., 1994). The name **rubella nonstructural protease** is applied to this enzyme, being a virally encoded endopeptidase functioning in the processing of the nonstructural proteins. (Processing of the SP-ORF is mediated by the cellular signal peptidase; Chapter 153.)

## Activity and Specificity

The product of the rubella virus NSP-ORF is 2115 amino acids in length (roughly 240 kDa). The protease domain resides between residues 1000 and 1300 of the NSP-ORF. The catalytic dyad consists of Cys1151 and His1272 (Marr *et al*., 1994; Chen *et al*., 1996). The protease catalyzes a single cleavage at Gly1300↓Gly1301 (28 residues from the catalytic histidine) (Chen *et al*., 1996) producing two products: an N-terminal 150 kDa protein and a C-terminal 90 kDa protein (Forng & Frey, 1995). The protease appears to function only in *cis* (Chen *et al*., 1996). No other cleavages, either within the viral polyprotein precursors or of cellular proteins, have been detected. The protease has not been purified and thus little is known about its requirements for activity or its structure. Interestingly, the protease is not active following *in vitro* translation of the NSP-ORF and thus requires some moiety present in cells for acquisition of activity.

## Biological Aspects

The rubella nonstructural protease is related to a number of other viral papain-like proteases, including proteases expressed by the alphaviruses, the arteriviruses, the coronaviruses and foot-and-mouth disease virus of animals, and the potyviruses and hypovirulence-associated viruses of plants and fungi (reviewed in Chen *et al*., 1996). The existence of these proteases in these diverse viruses (which belong to different families as well as infecting animals and plants) and

their relatedness to cellular papain-like proteases indicates both that these viral proteases were 'captured' from cells and that this capture event occurred several times independently during evolution of these viruses. Considering the apparent independent acquisition of these proteases, it is not surprising that the function of the proteolytic cleavage mediated by these proteases differs among these diverse viruses (Gorbalenya *et al*., 1991). Interestingly, within a group of related viruses of animals and plants, the alphavirus superfamily, which includes rubella virus, the alphaviruses and several plant virus families such as tobacco mosaic virus and brome mosaic virus, proteolytic processing of the nonstructural replicase proteins is a characteristic of the animal viruses but not the plant viruses. Why processing is necessary for the animal viruses but not for the plant viruses is unknown, and is a question of great interest in the biology of these viruses.

## Further Reading

For a general review of virus-encoded proteases, see Strauss (1990).

## References
Chen, J.-P., Strauss, J.H., Strauss, E.G. & Frey, T.K. (1996) Characterization of the rubella virus nonstructural protease domain and its cleavage site. *J. Virol.* **70**, 4707–4713.

Dominguez, G., Wang, C.-Y. & Frey, T.K. (1990) Sequence of the genome RNA of rubella virus: evidence for genetic rearrangement during Togavirus evolution. *Virology* **177**, 225–238.

Forng, R.-Y. & Frey, T.K. (1995) Identification of the rubella virus nonstructural proteins. *Virology* **206**, 843–853.

Frey, T.K. (1994) Molecular biology of rubella virus. *Adv. Virus Res.* **44**, 69–160.

Gorbalenya, A.E., Koonin, E.V. & Lai, M.M.-C. (1991) Putative papain-related thiol proteases of positive-strand RNA viruses. *FEBS Lett.* **288**, 201–205.

Marr, L.D., Wang, C.-Y. & Frey, T.K. (1994) Expression of the rubella virus nonstructural protein ORF and demonstration of proteolytic processing. *Virology* **198**, 586–592.

Strauss, J.H. (1990) Viral proteinases. *Semin. Virol.* **1**, 307–384.

*Teryl K. Frey*
*Department of Biology,*
*Georgia State University, Atlanta, GA.30303, USA*
*Email: tfrey@gsu.edu*

# 232. *Tymovirus endopeptidase*

## Databanks

*Peptidase classification: clan CC, family C21, MEROPS ID: C21.001*
*NC-IUBMB enzyme classification: none*

*Databank codes:*

| Species | Abbreviation | SwissProt | PIR | EMBL (genomic) |
|---|---|---|---|---|
| **Family C21** | | | | |
| Eggplant mosaic virus | EPMV | P20126 | JQ0102 | J04374: complete genome |
| *Erysimum* latent virus | ELV | P35928 | JQ1555 | – |
| *Kennedya* yellow mosaic virus | KYMV | P36304 | JQ0533 | D00637: complete genome |
| *Ononis* yellow mosaic virus | OYMV | P20127 | JQ0106 | J04375: complete genome |
| Turnip yellow mosaic virus | TYMV | P10358 | S01956 | X07441: complete genome |
| **Family C23** | | | | |
| Apple stem pitting virus | ASPV | – | – | D21829: complete genome |
| Blueberry scorch virus | BBSCV | – | – | L25658: complete genome |
| Garlic latent virus | GLV | – | – | Z68502: complete genome |
| Potato virus M | PVM | P17965 | S21601 | X53062: complete genome |
| **Family C34** | | | | |
| Apple chlorotic leaf spot virus | ACLSP | P27738 | A45353 | M58152: complete genome |
| **Family C35** | | | | |
| Apple stem grooving virus | ASGV | P36309 | A44059 | D14995: complete genome |
| Cherry capillovirus A | ChVA | – | – | X82547: complete genome |
| **Family C36** | | | | |
| Beet necrotic yellow vein virus | BNYVV | – | – | X05147: complete genome |
| **Family C43** | | | | |
| Citrus tatter leaf virus | CTLV | – | – | D16681: polymerase ORF |

## Name and History

The genomes of animal positive-strand RNA viruses typically encode cysteine proteinases, most of which have been classified as either 'chymotrypsin-like' (see Chapter 238) or 'papain-like' (see Chapter 225) (Dougherty & Semler, 1993; Gorbalenya & Snijder, 1996). The plant 'picorna-like' viruses (Goldbach *et al*., 1991), e.g. comoviruses (see Chapter 242), nepoviruses (see Chapter 243) and potyviruses (see Chapters 227 and 244), as well as sobemo-like viruses, share this feature with their animal counterparts. The same is true for at least some of plant viruses belonging to another large virus supergroup, the 'alpha-like', or 'Sindbis-like', viruses (see Chapters 229 and 230).

The first and by far the best-studied example of a plant alpha-like virus proteolytic enzyme is the proteinase of turnip yellow mosaic virus (TYMV), the type member of the tymovirus family. The 6.3 kb genomic RNA of TYMV codes for a large nonstructural protein of 206 kDa, involved in RNA replication, and two other proteins (Zaccomer *et al*., 1995). It has been demonstrated by *in vitro* translation experiments that the replicase protein undergoes autocleavage yielding two products of ~150 kDa and ~70–78 kDa (Morch *et al*., 1989), and the responsible proteinase activity has been mapped to the central part of the polyprotein (Bransom *et al*., 1991). Sequence comparisons had revealed in this region a putative proteinase domain with papain-like catalytic dyad, Cys783 and His869 (Rozanov *et al*., 1995). Deletion/mutagenesis experiments have confirmed that these, but not other conserved residues, are indeed indispensable for both *in vitro* autoproteolysis and virus viability, and have determined the limits of the *tymovirus endopeptidase* domain between amino acid residues 731 and 885 (Bransom & Dreher, 1994; Rozanov *et al*., 1995). Thus, the endopeptidase of TYMV appears to utilize the papain-like dyad for catalysis, and is considered an unusual papain-like proteinase.

The tymovirus endopeptidase domain is conserved through all known members of the tymovirus family, and the replicase polyproteins of five of them (TYMV, EPMV, OYMV, belladonna mottle virus and physalis mottle virus) have been shown to undergo autoproteolysis (Kadaré *et al*., 1992). The putative proteinases of carlaviruses, capilloviruses, ASPV, ACLSP and furovirus BNYVV have been found by sequence similarity with those of tymoviruses. Therefore, all these (putative) viral enzymes have been termed '**tymo**-like' **p**apain-**l**ike **p**roteinases (Rozanov *et al*., 1995), or *TPLPs*.

## Activity and Specificity

The proteinase cleaves the TYMV polyprotein in *cis* at Ala1259┼Thr1260 (Kadaré *et al*., 1995; Bransom *et al*., 1996). Certain amino acid residues located at or immediately upstream from the cleavage site are important (Bransom *et al*., 1991): substitutions of Ala-Ala-Ala for Gly-Pro-Lys1255 (P7, P6, P5) and for Leu-Asn-Gly1258 (P4, P3, P2) result in poor cleavage and undetectable cleavage, respectively. Replacement of Ala┼Thr-Pro (P1┼P1'P2') with Gly-Ala-Ala leads to only partial loss of cleavage; deletions of downstream sequences are without effect.

The proteinase is active when expressed in a reticulocyte lysate, a wheatgerm extract (Morch *et al*., 1989) or *Escherichia coli* cells (Kadaré *et al*., 1995) at 25°C or 30°C, but not at 37°C. *N*-Ethylmaleimide (10 mM), 5 mM dithio-*bis*-nitrobenzoic acid, 4 mM $ZnCl_2$, 4 mM Tos-Lys-$CH_2Cl$ and 4 mM Tos-Phe-$CH_2Cl$ are inhibitory in the reticulocyte lysate, but cystatin and E-64 are not. However, serine proteinase inhibitors (DFP and aprotinin) are reportedly effective too, albeit in concentrations 30- to 50-fold greater than are required for trypsin inhibition (Morch *et al*., 1989).

## Structural Chemistry, Related Proteinases and Distinguishing Features

The replicase polyproteins of tymoviruses display high sequence similarity to those of potexviruses, carlaviruses, capilloviruses, ACLSV and ASPV around the three domains that are conserved in the alpha-like viruses and believed to be functional, i.e. RNA methyltransferase/guanylyltransferase, NTPase/helicase and RNA polymerase (Rozanov *et al.*, 1992; Koonin & Dolja, 1993). The proteinase domain is also conserved (Figure 232.1A), though at a lower level, in all the viruses listed above and BNYVV, with the exception of potexviruses (Koonin & Dolja, 1993; Lawrence *et al.*, 1995; Rozanov *et al.*, 1995). Its location in the polyprotein is usually conserved too: just upstream from the NTPase domain (downstream in the BNYVV protein). The level of conservation of the cleavage site is so low that it is not possible to predict its exact location by similarity with TYMV with any confidence, even for other tymoviruses (Lawrence *et al.*, 1995; Bransom *et al.*, 1996). Nevertheless, it is likely that, as in the case of TYMV, the polyproteins of all tymo- and other related viruses with the domain array 'methyltransferase-NTPase-polymerase' are cleaved between the NTPase and polymerase domains (Kadaré *et al.*, 1992; Lawrence *et al.*, 1995).

In all (putative) tymo-like proteinases, essential (invariant) Cys and His residues (Figure 232.2) are preceded by structure breakers; the His is followed by at least one hydrophobic amino acid. This is similar to other papain-like proteinases in clan CA (see Chapter 186) and clan CC (see Chapter 225). The Cys is followed by at least two hydrophobic residues, most commonly Leu or Val, in one case Met, and in two new cases (ChVA and GLV) Phe. This distinguishes the sequences of tymo-like proteinases from those of most other papain-like cysteine proteinases which, with rare exceptions, have Trp or Tyr following the active Cys (Rawlings & Barrett, 1994; Gorbalenya & Snijder, 1996).

The tymo-like proteinase of carlavirus BBSCV (see Chapter 233) has been characterized by mutagenesis, with the finding that the predicted catalytic residues, Cys994 and His1075, are indeed essential for both *in vitro* autocleavage and virus infectivity, and that Gly1472 is essential for cleavage as well, probably being located close to the cleavage site (Lawrence *et al.*, 1995).

Thus, tymo-like proteinases constitute a poorly conserved, but distinct, group within the supergroup of viral proteolytic enzymes with the papain-like dyad of essential amino acid residues.

## Biological Aspects

No direct data on *in vivo* activities of the tymo-like proteinases are available, but it is tempting to speculate that the biological role of these enzymes is to produce a functional 'methyltransferase-NTPase' protein and an RNA polymerase from the polyprotein precursor (Figure 232.1A).

Tymo-like proteinase genes were not found in the genomes of tobamo-, tobra-, furo-(SBWMV), tricorna-, hordei-, clostero- or potexviruses, which also belong to the alpha-like supergroup. The expression strategies used by these viruses apparently do not include autocleavage of the replicative protein between the NTPase and polymerase domains. However, all of them, except potexviruses, do produce the 'methyltransferase-NTPase' block as a separate protein by different mechanisms (Figure 232.1B,C) (Goldbach *et al.*,

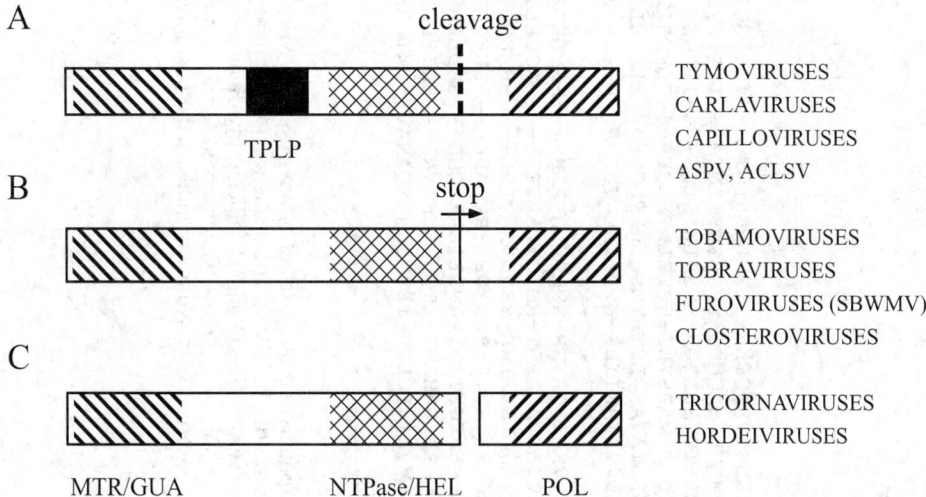

*Figure 232.1* Array of replicase domains conserved by alpha-like viruses. The scheme shows production of (putative) replicative proteins by cleavage (thick dashed bar) with a viral tymo-like proteinase domain (panel A), by translational frameshift or read-through (horizontal arrow, panel B), or by synthesis from different RNAs (panel C). Conserved domains are hatched. MTR/GUA, putative virus RNA capping enzyme (methytransferase/guanylyltransferase) (Ahola *et al.*, 1997); NTPase/HEL, NTPase/putative helicase (Kadaré *et al.*, 1996); POL, polymerase. Virus families are listed on the right; animal viruses and potexviruses are not presented. The BNYVV putative proteinase domain is located downstream from NTPase/HEL (not shown). The closterovirus replicase protein is synthesized as a precursor with one or two additional N-terminus proximal domain(s) containing so-called leader papain-like proteinase(s) similar to the potyvirus helper component proteinase (see Chapter 227); this closterovirus proteinase(s) mediates cleavage(s) upstream from the MTR/GUA (Agranovsky *et al.*, 1994; Karasev *et al.*, 1995) thereby moving this domain to the N-terminus.

```
TYMV   774_HLPQPTLNCLLSAVSDQTKV-SEEHLWESLQT----ILPDSQLSNEETNTLGLSTEHLTALAHLYNFQATVYSDRGPILFGPSDTIKRIDITHTTG-PPSHFSPGKR
ELV    672_EAPYPPLDCMLVALSAQMPQ-SPQELWSALNT----LMPLSALTSPSLRVLGLGTEELTALSYYHFQAEIHSDNEIYRFGIQTASTKLCLIRDSG-PPAHFTAPDP
EPMV   768_SLSHPKLNCLLTCFSELSGH-SESDIWLSIQS----ILPDSQLQNPEVSTLGLSTDIITALCFIYHSSVTLHAPSGVYHYGIASSTVYVIHYQPG-PPPHFSLSPR
KYMV   802_PLPMPKNNCLLTAVAPSLHI-NPHRLWTSLQE----VLPDSLLSNSEIDSVGMSTDLLTALSHLFNFQAVVHSERGDILFGLQSAKTVIHIYHTNG-PPAHYSPPPK
OYMV   703_SAPFPQKHCLLTAVASQLSY-TEHQLWEFLCD----MLPDSLLTNSEVENFGLSTDHLTCLSYRLHFECIIHTSHSTIPYGIKKASTVIQISYIDG-PPKHFKAFIK
ACLSV  845_RFIKGKFDCLFVSIAEIIHK-KPEEVMFIPH-------IVDRCVSNRGCSLDDARAICEKYEIKIECEGD--CGLVECGTIGLSVGRMLL--RGNHFTVASV

BBScV   986_EPLEPKEMCVVKAIAQAVKR-SPMDVLRVALK-----KMGEDFKEQICRGKGVMLDVFMVLAKIFDVSACVLQG--TEQIMINPKGRIKGLFRM--TTDHLSYDGV
PVM     986_EPQVLRNGCVIESVAQALGT-RNADILAVVEE-----RCCEEVESVQAGLGLNLHHVEIVLQCFDIVGHCNLG--DKEITLNAGGKMPFCFDI--SDEHMSFCGR
ASPV   1194_KPCMPVNGCVIRAISSALNR-REVDVLAVIGK-----PAHEDLFEEVAEGRGFSIFDLTRLFEIFSICGSVDTG--GELIMVNENGRIPAEFSL--EKEHLAHIPT
GLV     929_SLLFYKNDCFVSAVAQTFSR-DTNEMYRVLAD-----KKFDELTELIRLGCGLTLEDLEIGFRLLNIKAHINKD-GEYLL-INETGEINGFYAL--TEEHLVACPP
ChVA    635_KKETRKNDCFFKAVGETIGIPANSLIERILCSD-----SEDLKPVIEQLNLDHPISSKLLEVCCKFLGYRVHIYYG--DSIIKLNDDINMHAIHIGG--KPGHLFCINQ
CTLV    598_PRRKRKNDCVFRAISAHLGI-ETQDLLNFLVN-----EDISEELLDCIDEDKGLSHEMIEEVLITKGLSMVYTSDFKEMAV-LNRKYGVNGKMYCTI-KGNHCELSSK
BNYVV  1233_NIVSRPNNCLVVAISECLGV-TLEKLDNLMQANAVTLDKYHAWLSKKSPSTWQDCRMFADALKVSMYVKVLSDKPYDLTYEVDGAGSSVTLYLTGKESDGHFIAAPL

consensus      C**  æ*æ   *       **  **          *     *   *  *  *  *  *  *      g*     *  **    *        H*
```

*Figure 232.2*  Alignment of (putative) proteinases of tymoviruses (TYMV, ELV, EPMV, KYMV, OYMV), ACLSV, carlaviruses (BBSCV, PVM, GLV, ChVA), capilloviruses (CTLV) and furovirus BNYVV. The sequences were aligned following previously published alignments (Rozanov *et al.*, 1995; Lawrence *et al.*, 1995; Ohira *et al.*, 1995). In the consensus line C and H indicate invariant (essential) Cys and His, respectively, 'g' indicates highly conserved Gly, an asterisk denotes a bulky hydrophobic residue (I, L, V, M, F, Y or W) conserved for at least nine sequences and 'æ' denotes one of the small residue (A, S, G or C) found in positions 4 and 6 downstream from the conserved Cys. Numbers indicate the positions in the viral polyproteins.

1991; Zaccomer *et al*., 1995). Thus, the strategy of processing of the viral polyprotein by the tymo-like proteinase domain appears to be just one of several possible solutions to the problem of organizing and regulating production of the viral replicase complex. It cannot be excluded, however, that tymo-like proteinases have other functions as well.

## Further Reading

The reader is referred to Rozanov *et al*. (1995).

## References

Agranovsky, A.A., Koonin, E.V., Boyko, V.P., Maiss, E., Frotschl, R., Lunina, N.A. & Atabekov, J.G. (1994) Beet yellows closterovirus: complete genome structure and identification of a leader papain-like thiol protease. *Virology* **198**, 311–324.

Ahola, T., Laakkonen, P., Vihinen, H. & Kääriäinen, L. (1997) Critical residues of semliki forest virus RNA capping enzyme involved in methyltransferase and guanylyltransferase-like activities. *J. Virol.* **71**, 392–397.

Bransom, K.L. & Dreher, T.W. (1994) Identification of the essential cysteine and histidine residues of the turnip yellow mosaic virus protease. *Virology* **198**, 148–154.

Bransom, K.L., Weiland, J.J. & Dreher, T.W. (1991) Proteolytic maturation of the 206-kDa nonstructural protein encoded by turnip yellow mosaic virus RNA. *Virology* **184**, 351–358.

Bransom, K.L., Wallace, S.E. & Dreher, T.W. (1996) Identification of the cleavage site recognized by the turnip yellow mosaic protease. *Virology* **217**, 404–406.

Dougherty, W.G. & Semler, B.L. (1993) Expression of virus-encoded proteinases: functional and structural similarities with cellular enzymes. *Microbiol. Rev.* **57**, 781–822.

Goldbach, R., Le Gall, O. & Wellink, J. (1991) Alpha-like viruses in plants. *Semin. Virol.* **2**, 19–25.

Gorbalenya, A.E. & Snijder, E.J. (1996) Viral cysteine proteinases. *Perspect. Drug Discovery Design* **6**, 64–86.

Kadaré, G., Drugeon, G., Savithri, H.S. & Haenni, A.-L. (1992) Comparison of the strategies of expression of five tymovirus RNAs by in vitro translation studies. *J. Gen. Virol.* **73**, 493–498.

Kadaré, G., Rozanov, M.N. & Haenni, A.-L. (1995) Expression of the turnip yellow mosaic virus proteinase in *Escherichia coli*: identification of essential residues and of the cleavage site. *J. Gen. Virol.* **76**, 2853–2857.

Kadaré, G., David, C. & Haenni, A.-L. (1996) ATPase, GTPase, and RNA binding activities associated with the 206-kilodalton protein of turnip yellow mosaic virus. *J. Virol.* **70**, 8169–8174.

Karasev, A.V., Boyko, V.P., Gowda, S., Nikolaeva, O.V., Hilf, M.E., Koonin, E.V., Niblett, C.L., Cline, K., Gumpf, D.J. & Lee, R.F. (1995) Complete sequence of the citrus tristeza virus RNA genome. *Virology* **208**, 511–520.

Koonin, E.V. & Dolja, V.V. (1993) Evolution and taxonomy of positive-strand RNA viruses: implications of comparative analysis of amino acid sequences. *Crit. Rev. Biochem. Mol. Biol.* **28**, 375–430.

Lawrence, D.M., Rozanov, M.N & Hillman, B.I. (1995) Autocatalytic processing of the 223-kDa protein of blueberry scorch carlavirus by a papain-like proteinase. *Virology* **207**, 127–135.

Morch, M.-D., Drugeon, G., Szafranski, P. & Haenni, A.-L. (1989) Proteolytic origin of the 150-kilodalton protein encoded by turnip yellow mosaic virus genomic RNA. *J. Virol.* **63**, 5153–5158.

Ohira, K., Namba, S., Rozanov, M., Kusumi, T. & Tsuchizaki, T. (1995) Complete sequence of an infectious full-length cDNA clone of citrus tatter leaf capillovirus: comparative sequence analysis of capillovirus genomes. *J. Gen. Virol.* **76**, 2305–2309.

Rawlings, N.D. & Barrett, A.J. (1994) Families of cysteine peptidases. *Methods Enzymol.* **244**, 461–486.

Rozanov, M.N., Koonin, E.V. & Gorbalenya, A.E. (1992) Conservation of the putative methyltransferase domain: a hallmark of the 'Sindbis-like' supergroup of positive-strand RNA viruses. *J. Gen. Virol.* **73**, 2129–2134.

Rozanov, M.N., Drugeon, G. & Haenni, A.-L. (1995) Papain-like proteinase of turnip yellow mosaic virus: a prototype of a new viral proteinase group. *Arch. Virol.* **140**, 273–288.

Zaccomer, B., Haenni, A.L. & Macaya, G. (1995) The remarkable variety of plant RNA virus genomes. *J. Gen. Virol.* **76**, 231–247.

**C**

*Mikhail Rozanov*
*University of Pennsylvania Medical School,*
*Department of Pathology and Laboratory Medicine, Philadelphia, PA 19104, USA*
*Email: rozanov@mail.med.upenn.edu; rozanov@ncbi.nlm.nih.gov; mrozanov@aol.com*

# 233. *Blueberry scorch carlavirus endopeptidase*

## Databanks

*Peptidase classification: clan CC, family C23, MEROPS ID: C23.001*
*NC-IUBMB enzyme classification: none*

*Databank codes:*

| Species | SwissProt | PIR | EMBL (genomic) |
|---|---|---|---|
| Apple stem pitting virus | – | – | D21829: complete genome |
| Blueberry scorch virus | – | – | L25658: complete genome |
| Garlic latent virus | – | – | Z68502: complete genome |
| Potato virus M | P17965 | S21601 | X53062: complete genome |

## Name and History

Blueberry scorch carlavirus (BBSCV) contains a positive-strand RNA genome of 8514 nucleotide residues (Cavileer *et al.*, 1994; reviewed in Hillman & Lawrence, 1995). The first open reading frame was predicted to encode a protein of 223 kDa. Based on the sequence similarity of carlaviruses to the RNA-containing plant tymoviruses (Morozov *et al.*, 1990; Koonin & Dolja, 1993), it was predicted that p223 contained a papain-like proteinase capable of *cis*-cleavage. Proteinase activity of p223 was confirmed (Lawrence *et al.*, 1995), and is summarized in Figure 233.1. Similar proteinases have been identified in related RNA viruses, and their importance in viral replication has been documented (Dougherty & Semler, 1993).

## Activity and Specificity

Like other similar viral papain-like cysteine proteinases, the BBSCV proteinase cleaves in *cis* but not in *trans*. The cleavage site has not been determined unambiguously, but the approximate position, and the importance of Gly1472 in cleavage, was demonstrated by deletion analysis and mutagenesis (Lawrence *et al.*, 1995). Based on alignments

with similar tymovirus proteinase cleavage sites (Kaderé *et al.*, 1995), it is likely that the cleavage site is Gly1472┼Ala1473.

## Structural Chemistry

No structural studies have been done on the BBSCV proteinase. Proteolytic activity of protein expressed from deletion constructs containing the putative catalytic domain and the cleavage site was demonstrated (Lawrence *et al.*, 1995).

## Preparation

Highest proteinase activity of BBSCV p223 was demonstrated using a coupled bacterial transcription/translation system (Lawrence *et al.*, 1995).

## Biological Aspects

Proteolytic activity appears to be required for viral infectivity. Cys994 and His1075 were predicted to be catalytically active on the basis of sequence alignments, and were shown to be so by site-directed mutagenesis and *in vitro* translation

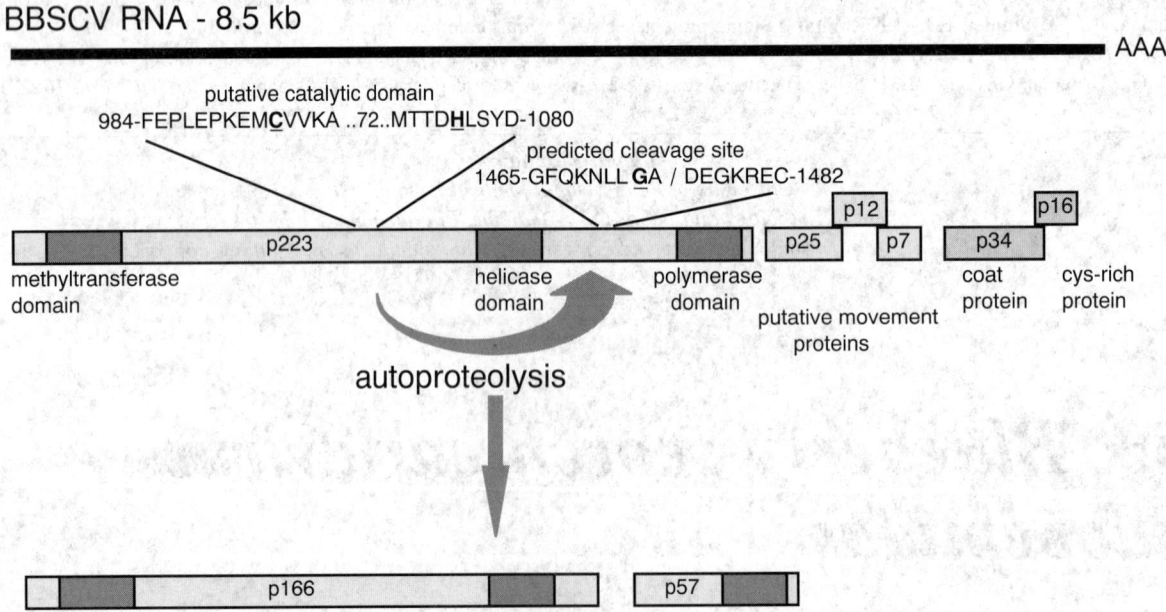

*Figure 233.1*   Genome organization of BBSCV RNA and autocatalytic processing of p223, the putative replicase. Underlined residues are Cys and His, shown to be essential components of the putative catalytic domain (either side of a 72-amino acid spacing sequence), and Gly, predicted by alignment to occupy the P2 position of the cleavage site and demonstrated by mutagenesis to be important for cleavage. The point of the predicted cleavage between Ala in the P1 position and Asp in P1' position is indicated by the oblique stroke between them.

of synthetic transcripts. Full-length viral cDNA clones that would normally have been infective became noninfective when the putative catalytic Cys or His residues were altered by site-directed mutation (Lawrence *et al*., 1995).

## Distinguishing Features

An upstream sequence similar to the putative catalytic site that appears to represent a duplication of the catalytic site was subjected to mutagenesis experiments similar to those described above, but was found not to be required for *in vitro* proteolysis, nor for viral infectivity (Lawrence *et al*., 1995).

## Further Reading

For a review, see Dougherty & Semler (1993).

## References

Cavileer, T.D., Halpern, B.T., Lawrence, D.M., Podleckis, E.V., Martin, R.R. & Hillman, B.I. (1994) Nucleotide sequence of the carlavirus associated with blueberry scorch and similar diseases. *J. Gen. Virol.* **75**, 711–720.

Dougherty, W.G. & Semler, B.L. (1993) Expression of virus-encoded proteinases: functional and structural similarities with cellular enzymes. *Microbiol. Rev.* **57**, 781–822.

Hillman, B.I. & Lawrence, D.L. (1995) Carlaviruses. In: *Pathogenesis and Host-Parasite Specificity in Plant Disease*, vol. III: *Viruses and Viroids* (Kohmoto, K., Singh, R.P., Singh, U.S. & Zeigler, R., eds). London: Pergamon Press, pp. 35–50.

Kaderé, G., Rozanov, M. & Haenni, A.-L. (1995) Expression of the turnip yellow mosaic virus proteinase in *Echerichia coli* and determination of the cleavage site within the 206 kDa protein. *J. Gen. Virol.* **76**, 2853–2857.

Koonin, E.V. & Dolja, V.V. (1993) Evolution and taxonomy of positive-strand RNA viruses: implications of comparative analysis of amino acid sequences. *Crit. Rev. Biochem. Mol. Biol.* **28**, 375–430.

Lawrence, D.L., Rozanov, M.N. & Hillman, B.I. (1995) Autocatalytic processing of the 223 kDa protein of blueberry scorch carlavirus by a papain-like proteinase. *Virology* **207**, 127–135.

Morozov, S.Y., Kanyuka, K.V., Levay, K.E. & Zavriev, S.K. (1990) The putative RNA replicase of potato virus M: obvious sequence similarity with those of potex- and tymoviruses. *Virology* **179**, 911–914.

*Bradley I. Hillman*
*Department of Plant Pathology,*
*Rutgers University,*
*New Brunswick, NJ 08903, USA*
*Email: hillman@aesop.rutgers.edu*

# 234. *Arterivirus papain-like cysteine endopeptidase* α

## Databanks

*Peptidase classification: clan CC, family C31, MEROPS ID: C31.001*
*NC-IUBMB enzyme classification: none*
*Databank codes:*

| Species | SwissProt | PIR | EMBL (genomic) |
| --- | --- | --- | --- |
| Peptidases | | | |
| Lactate dehydrogenase-elevating virus | – | – | U15146: complete genome |
| Lelystad virus | Q04561 | A36861 | L04493: 3′ end of genome |
| | | A45392 | M96262: complete genome |
| Nonpeptidase homologs | | | |
| Equine arteritis virus | P19811 | A39925 | X52277: 5′ end of genome |
| | | S10158 | X53549: complete genome |

## Name and History

Arteriviruses are enveloped, positive-strand RNA viruses that contain a polycistronic genome (12–15 kb) (den Boon *et al*., 1991; Godeny *et al*., 1993; Meulenberg *et al*., 1993; Snijder & Spaan, 1995). Their replicase proteins are expressed from the open reading frames (ORFs) 1a and 1b that encode two large precursor proteins: the ORF1a protein (187–260 kDa) and, due to ribosomal frameshifting, the ORF1ab protein (345–422 kDa). Both precursors are processed extensively by three or four ORF1a-encoded endopeptidases (Snijder *et al*., 1992, 1994, 1995, 1996; den Boon *et al*., 1995; van Dinten *et al*., 1996).

The *papain-like cysteine proteinase α (PCPα)* is the most N-terminally located member of an array of three cysteine proteinase domains that has been identified in the N-terminal 500 residues of the arterivirus ORF1a protein (see also Chapters 235 and 236). The PCPα domain is functional in the replicases of the arteriviruses porcine reproductive and respiratory syndrome virus (PRRSV) and lactate dehydrogenase-elevating virus (LDV), where it catalyzes the autoproteolytic generation of a 20–22 kDa N-terminal cleavage product named nonstructural protein 1α (Nsp1α). However, the PCPα domain of the arterivirus prototype equine arteritis virus (EAV) is not an active proteinase and therefore the EAV Nsp1α equivalent remains attached to the next cleavage product (see Chapter 235) (Snijder *et al*., 1992; den Boon *et al*., 1995).

## Activity and Specificity

The activity of the PRRSV and LDV PCPα domains (and the inactivated nature of the corresponding EAV domain) has been demonstrated by *in vitro* translation of RNA transcripts encoding the N-terminal region of the respective ORF1a proteins (Snijder *et al*., 1992; den Boon *et al*., 1995). The exact site of cleavage by the PCPα of PRRSV and LDV is unknown. On the basis of the size of Nsp1α, it has been estimated to be located about 20 residues downstream of catalytic His146/147. For PRRSV, there is evidence that at least 30 residues downstream of the scissile bond (up to residue 194) are required for cleavage (den Boon *et al*., 1995). This region appears to be a 'spacer sequence' between PCPα and the next cysteine proteinase domain, PCPβ (Chapter 235), of which the active-site Cys is located at position 276. Consistent with this idea, the spacer region is no longer present between the inactivated PCPα and active PCPβ domains in the EAV sequence. In a reticulocyte lysate, PCPα was found to cleave rapidly, suggesting cleavage in *cis*. Attempts to demonstrate *trans*-cleaving activity have been unsuccessful so far.

## Structural Chemistry

Comparative sequence analysis (see Figure 234.1) suggested that the PRRSV and LDV PCPα domains contain the typical Cys/His catalytic dyad found in viral papain-like proteinases (Chapter 225) (Chen *et al*., 1993; Godeny *et al*., 1993; Meulenberg *et al*., 1993; den Boon *et al*., 1995). The putative active-site Cys is followed, as usual, by a bulky, hydrophobic residue (Trp). The putative PCPα catalytic nucleophile of both PRRSV and LDV, Cys76, was probed by site-directed mutagenesis. Replacement of this residue by Gly or Ser completely inactivated PCPα. Four downstream conserved His residues (at positions 92, 115, 146 and 157) were tested to identify the second active-site residue of PRRSV PCPα. Although some substitutions of other His residues affected PCPα activity, only the replacement of His146 (by either Asp, Phe, Ile, Asn or Tyr) completely abolished cleavage of the Nsp1α/Nsp1β site. Thus, the available data strongly suggest that Cys76 and His146 form the PRRSV PCPα catalytic dyad, and that Cys76 and His147 fulfill the same role in the LDV PCPα (den Boon *et al*., 1995). The exact boundaries of PCPα remain to be determined, but the domain may include approximately 110 amino acid residues.

## Biological Aspects

The proteolytic inactivity of the PCPα domain in the arterivirus EAV is (as far as we know) unparalleled in virology. It seems to be due to replacement of the catalytic Cys residue, and possibly other substitutions. However, despite this loss of proteolytic function and the low overall sequence similarity between EAV and PRRSV/LDV in this region, a number of residues has been conserved in the EAV Nsp1α equivalent, including a motif that contains the former catalytic His (His122 in EAV). This observation suggests the conservation of an additional, nonproteolytic function that is

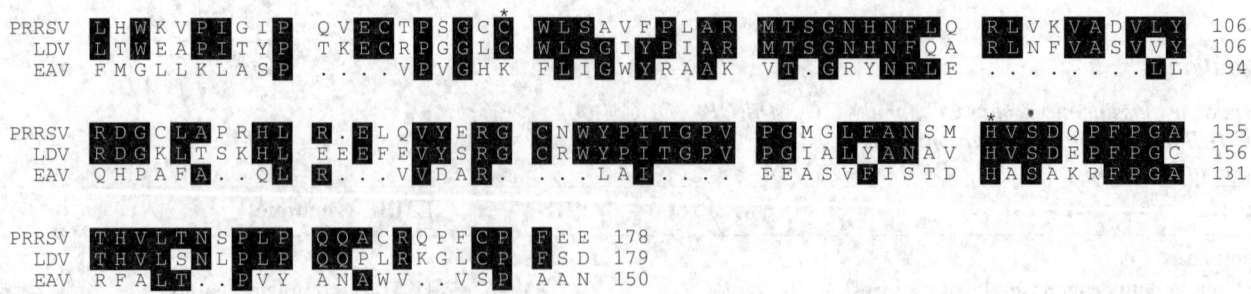

*Figure 234.1* Alignment of the PCP domains of the arteriviruses porcine reproductive and respiratory syndrome virus (PRRSV), lactate dehydrogenase-elevating virus (LDV) and equine arteritis virus (EAV). The alignment was produced with the ClustalW program and presented in the PrettyBox format derived from the GCG package. The putative catalytic Cys and His residues are marked with asterisks. Black indicates identical residues. The percentages of identical residues in the various protein pairs are: PRRSV versus LDV, 73%; LDV versus EAV, 25%; EAV versus PRRSV, 31%. The N- and C-terminal boundaries of the protease domain were chosen arbitrarily.

associated with the arterivirus PCPα domain (den Boon *et al.*, 1995).

## Distinguishing Features

PCPα is a small papain-like cysteine protease domain, which cleaves the polyprotein in *cis* just downstream of its active-site His residue. In contrast to the cleavage carried out by the arterivirus PCPβ domain, cleavage by PCPα requires the presence of a substantial part of the polyprotein downstream of the cleavage site. A polyclonal rabbit antiserum against EAV Nsp1 was raised by immunization with a peptide representing the first 23 residues of the ORF1a protein (Snijder *et al.*, 1994). This serum is available from the authors for research purposes on request. Antisera directed against the LDV or PRRSV Nsp1α or Nsp1β proteins have not been described.

## Further Reading

The paper of den Boon *et al.* (1995) is recommended.

## References

Chen, Z., Kuo, L., Rowland, R.R.R., Even, C., Faaberg, K.S. & Plagemann, P.G.W. (1993) Sequence of 3′-end of genome and of 5′-end of ORF 1a of lactate dehydrogenase-elevating virus (LDV) and common junction motifs between 5′-leader and bodies of seven subgenomic mRNAs. *J. Gen. Virol.* **74**, 643–660.

den Boon, J.A., Snijder, E.J., Chirnside, E.D., de Vries, A.A.F., Horzinek, M.C. & Spaan, W.J.M. (1991). Equine arteritis virus is not a togavirus but belongs to the coronaviruslike superfamily. *J. Virol.* **65**, 2910–2920.

den Boon, J.A., Faaberg, K.S., Meulenberg, J.J.M., Wassenaar, A.L.M., Plagemann, P.G.W., Gorbalenya, A.E. & Snijder, E.J.

(1995) Processing and evolution of the N-terminal region of the arterivirus replicase ORF1a protein: identification of two papainlike cysteine proteases. *J. Virol.* **69**, 4500–4505.

Godeny, E.K., Chen, L., Kumar, S.N., Methven, S.L., Koonin, E.V. & Brinton, M.A. (1993) Complete genomic sequence and phylogenetic analysis of the lactate dehydrogenase-elevating virus. *Virology* **194**, 585–596.

Meulenberg, J.J.M., Hulst, M.M., de Meijer, E.J., Moonen, P.L.J.M., den Besten, A., de Kluyver, E.P., Wensvoort, G. & Moormann, R.J.M. (1993) Lelystad virus, the causative agent of porcine epidemic abortion and respiratory syndrome (PEARS), is related to LDV and EAV. *Virology* **192**, 62–72.

Snijder, E.J. & Spaan, W.J.M. (1995) The coronaviruslike superfamily. In: *The Coronaviridae* (Siddell, S.G., ed.). New York: Plenum Press, pp. 239–255.

Snijder, E.J., Wassenaar, A.L.M. & Spaan, W.J.M. (1992) The 5′ end of the equine arteritis virus replicase gene encodes a papainlike cysteine protease. *J. Virol.* **66**, 7040–7048.

Snijder, E.J., Wassenaar, A.L.M. & Spaan, W.J.M. (1994) Proteolytic processing of the replicase ORF1a protein of equine arteritis virus. *J. Virol.* **68**, 5755–5764.

Snijder, E.J., Wassenaar, A.L.M., Spaan, W.J.M. & Gorbalenya, A.E. (1995) The arterivirus Nsp2 protease. An unusual cysteine protease with primary structure similarities to both papain-like and chymotrypsin-like proteases. *J. Biol. Chem.* **270**, 16671–16676.

Snijder, E.J., Wassenaar, A.L.M., van Dinten, L.C., Spaan, W.J.M. & Gorbalenya, A.E. (1996) The arterivirus nsp4 protease is the prototype of a novel group of chymotrypsin-like enzymes, the 3C-like serine proteases. *J. Biol. Chem.* **271**, 4864–4871.

van Dinten, L.C., Wassenaar, A.L.M., Gorbalenya, A.E., Spaan, W.J.M. & Snijder, E.J. (1996) Processing of the equine arteritis virus replicase ORF1b protein: identification of cleavage products containing the putative viral polymerase and helicase domains. *J. Virol.* **70**, 6625–6633.

*Alexander E. Gorbalenya*
*M.P. Chumakov Institute of Poliomyelitis and Viral Encephalitides, Russian Academy of Medical Sciences, 142782 Moscow Region, and A.N. Belozersky Institute of Physico-Chemical Biology, Moscow State University, 119899 Moscow, Russia*

*Johan A. den Boon*
*Department of Virology, Institute of Medical Microbiology, Leiden University, PO Box 9600, 2300 RC Leiden, The Netherlands*

*Eric J. Snijder*
*Department of Virology, Institute of Medical Microbiology, Leiden University, PO Box 9600, 2300 RC Leiden, The Netherlands*
*Email: snijder@virology.azl.nl*

# 235. *Arterivirus papain-like cysteine endopeptidase β*

## Databanks

*Peptidase classification: clan CC, family C32, MEROPS ID: C32.001*
*NC-IUBMB enzyme classification: none*

*Databank codes:*

| Species | SwissProt | PIR | EMBL (genomic) |
|---|---|---|---|
| Equine arteritis virus | P19811 | A39925 | X52277: 5′ end of genome |
|  |  | S10158 | X53459: complete genome |
| Lactate dehydrogenase-elevating virus | – | – | U15146: complete genome |
| Lelystad virus | Q04561 | A36861 | L04493: 3′ end of genome |
|  |  | A45392 | M96262: complete genome |

## Name and History

Arteriviruses are enveloped, positive-strand RNA viruses that contain a polycistronic genome (12–15 kb) (Chen *et al.*, 1993; Godeny *et al.*, 1993; Meulenberg *et al.*, 1993; Snijder & Spaan, 1995). Their replicase proteins are expressed from the open reading frames (ORFs) 1a and 1b that encode two large precursor proteins: the ORF1a protein (187–260 kDa) and, due to ribosomal frameshifting, the ORF1ab protein (345–422 kDa). Both precursors are processed extensively by three or four ORF1a-encoded endopeptidases (den Boon *et al.*, 1995; Snijder *et al.*, 1992, 1994, 1995, 1996; van Dinten *et al.*, 1996). The *papain-like cysteine proteinase β (PCPβ)* is the central member of an array of three cysteine proteinase domains that has been identified in the N-terminal 500 residues of the arterivirus ORF1a protein. In contrast to the upstream PCPα domain (Chapter 234), PCPβ is functional in all arteriviruses studied so far: equine arteritis virus (EAV), porcine reproductive and respiratory syndrome virus (PRRSV), and lactate dehydrogenase-elevating virus (LDV).

## Activity and Specificity

The activity of the EAV PCPβ domain was first detected *in vitro*, upon translation of RNA transcripts encoding the N-terminal region of the arterivirus ORF1a protein (Snijder *et al.*, 1992; den Boon *et al.*, 1995). After the production of an EAV Nsp1-specific antiserum, the proteolytic activity of PCPβ was also studied *in vivo* in infected cells and in transient expression systems (Snijder *et al.*, 1994). Furthermore, the EAV PCPβ was found to be active in *Escherichia coli* when it was expressed as part of a bacterial fusion protein (Snijder *et al.*, 1992). Using the latter system, the site cleaved by this proteinase was determined by sequence analysis of the Nsp2 N-terminus. Like PCPα, PCPβ was found to cleave just downstream of the active-site His residue (His230 in EAV; here and hereafter polyprotein numbering is used), at Gly260⁺Gly261 of the EAV ORF1a polyprotein (Snijder *et al.*, 1992). A limited mutagenesis study of this site revealed that its P1 position is more sensitive to replacements than its P1′ position. Furthermore, it was shown that sequences downstream of P2′ are not required for cleavage of the EAV Nsp1/Nsp2 site. Sequence comparison suggests that the PCPβ domains of both PRRSV and LDV cleave between Tyr and Gly (Tyr384⁺Gly385 and Tyr380⁺Gly381, respectively) (den Boon *et al.*, 1995). Like PCPα, PCPβ was found to cleave rapidly and probably exclusively in *cis*. *Trans*-cleaving activity could not be detected (Snijder *et al.*, 1992).

## Structural Chemistry

Deletion mutagenesis has revealed that the EAV PCPβ domain consists of 140 residues at most (sequences upstream

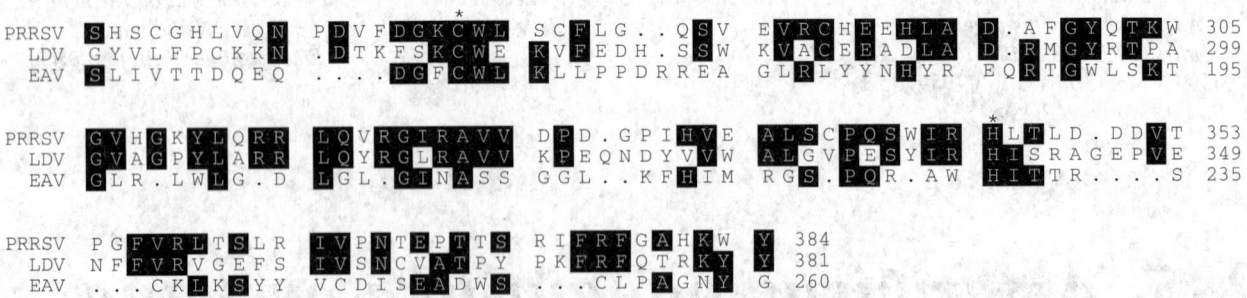

*Figure 235.1* Alignment of the PCPβ domains of the arteriviruses porcine reproductive and respiratory syndrome virus (PRRSV), lactate dehydrogenase-elevating virus (LDV) and equine arteritis virus (EAV). The alignment was produced with the ClustalW program and presented in the PrettyBox format derived from the GCG package. The putative catalytic Cys and His residues are marked with asterisks. Black indicates identical residues. The percentages of identical residues in the various protein pairs are: PRRSV versus LDV, 42%; LDV versus EAV, 14%; EAV versus PRRSV, 25%. The N-terminal boundary of the protease domains was chosen arbitrarily, the C-terminal boundary is the Nsp1/Nsp2 cleavage site (EAV) or putative Nsp1β/Nsp2 cleavage site (PRRSV and LDV).

of residue 123 and downstream of residue 263 are not required for its activity). The probable PCPβ active-site Cys and His residues were identified by comparative sequence analysis (Chen *et al.*, 1993; den Boon *et al.*, 1991; Godeny *et al.*, 1993; Meulenberg *et al.*, 1993) (Figure 235.1) and site-directed mutagenesis. Replacement of EAV Cys164 (by Ser or Gly) and His230 (by Val, Ala or Gly) inactivated the proteinase (Snijder *et al.*, 1992). Similar results were later obtained for the PCPβ domain of PRRSV, for which the replacement of Cys276 (by Ile, Leu, Arg or Ser) or His345 (by Asp or Tyr) was found to be lethal (den Boon *et al.*, 1995). For the LDV PCPβ, the available evidence suggests that Cys269 and His340 form the catalytic dyad (den Boon *et al.*, 1995).

## Biological Aspects

The extensive proteolytic processing of the replicase polyprotein is assumed to play a key role in the replication of arteriviruses. In EAV, PCPβ catalyzes the autoproteolytic generation of nonstructural protein 1 (Nsp1), the 29 kDa N-terminal replicase cleavage product (Snijder *et al.*, 1992). The Nsp1 equivalent of PRRSV and LDV is larger (about 47 kDa) and is cleaved into Nsp1α (20–22 kDa) and Nsp1β (26–27 kDa) by the upstream located PCPα (Chapter 234) (den Boon *et al.*, 1995). In EAV-infected cells, Nsp1 is (partially) transported to the nucleus, but information on the biological significance of this is not yet available.

## Distinguishing Features

PCPβ is a small papain-like cysteine protease domain, which cleaves the polyprotein in *cis* just downstream of its active site His residue. In contrast to the cleavage carried out by the PCPα domain of PRRSV and LDV, only two residues downstream of the cleavage site are required for processing.

A polyclonal rabbit antiserum against EAV Nsp1 was raised using a peptide representing the first 23 residues of the ORF1a protein (Snijder *et al.*, 1994). This serum is available from the authors for research purposes on request.

## Further Reading

The papers of Snijder *et al.* (1992) and Den Boon *et al.* (1995) are recommended.

## References

Chen, Z., Kuo, L., Rowland, R.R.R., Even, C., Faaberg, K.S. & Plagemann, P.G.W. (1993) Sequence of 3′-end of genome and of 5′-end of ORF 1a of lactate dehydrogenase-elevating virus (LDV) and common junction motifs between 5′-leader and bodies of seven subgenomic mRNAs. *J. Gen. Virol.* **74**, 643–660.

den Boon, J.A., Snijder, E.J., Chirnside, E.D., de Vries, A.A.F., Horzinek, M.C. & Spaan, W.J.M. (1991). Equine arteritis virus is not a togavirus but belongs to the coronaviruslike superfamily. *J. Virol.* **65**, 2910–2920.

den Boon, J.A., Faaberg, K.S., Meulenberg, J.J.M., Wassenaar, A.L.M., Plagemann, P.G.W., Gorbalenya, A.E. & Snijder, E.J. (1995) Processing and evolution of the N-terminal region of the arterivirus replicase ORF1a protein: identification of two papainlike cysteine proteases. *J. Virol.* **69**, 4500–4505.

Godeny, E.K., Chen, L., Kumar, S.N., Methven, S.L., Koonin, E.V. & Brinton, M.A. (1993) Complete genomic sequence and phylogenetic analysis of the lactate dehydrogenase-elevating virus. *Virology* **194**, 585–596.

Meulenberg, J.J.M., Hulst, M.M., de Meijer, E.J., Moonen, P.L.J.M., den Besten, A., de Kluyver, E.P., Wensvoort, G. & Moormann, R.J.M. (1993) Lelystad virus, the causative agent of porcine epidemic abortion and respiratory syndrome (PEARS), is related to LDV and EAV. *Virology* **192**, 62–72.

Snijder, E.J. & Spaan, W.J.M. (1995) The coronaviruslike superfamily. In: *The Coronaviridae* (Siddell, S.G., ed.). New York: Plenum Press, pp. 239–255.

Snijder, E.J., Wassenaar, A.L.M. & Spaan, W.J.M. (1992) The 5′ end of the equine arteritis virus replicase gene encodes a papainlike cysteine protease. *J. Virol.* **66**, 7040–7048.

Snijder, E.J., Wassenaar, A.L.M. & Spaan, W.J.M. (1994) Proteolytic processing of the replicase ORF1a protein of equine arteritis virus. *J. Virol.* **68**, 5755–5764.

Snijder, E.J., Wassenaar, A.L.M., Spaan, W.J.M. & Gorbalenya, A.E. (1995) The arterivirus Nsp2 protease. An unusual cysteine protease with primary structure similarities to both papain-like and chymotrypsin-like proteases. *J. Biol. Chem.* **270**, 16671–16676.

Snijder, E.J., Wassenaar, A.L.M., van Dinten, L.C., Spaan, W.J.M. & Gorbalenya, A.E. (1996) The arterivirus nsp4 protease is the prototype of a novel group of chymotrypsin-like enzymes, the 3C-like serine proteases. *J. Biol. Chem.* **271**, 4864–4871.

van Dinten, L.C., Wassenaar, A.L.M., Gorbalenya, A.E., Spaan, W.J.M. & Snijder, E.J. (1996) Processing of the equine arteritis virus replicase ORF1b protein: identification of cleavage products containing the putative viral polymerase and helicase domains. *J. Virol.* **70**, 6625–6633.

**Eric J. Snijder**
*Department of Virology, Institute of Medical Microbiology, Leiden University, PO Box 9600, 2300 RC Leiden, The Netherlands*
*Email: snijder@virology.azl.nl*

**Alfred L.M. Wassenaar**
*Department of Virology, Institute of Medical Microbiology, Leiden University, PO Box 9600, 2300 RC Leiden, The Netherlands*

**Alexander E. Gorbalenya**
*M.P. Chumakov Institute of Poliomyelitis and Viral Encephalitides, Russian Academy of Medical Sciences, 142782 Moscow Region, and A.N. Belozersky Institute of Physico-Chemical Biology, Moscow State University, 119899 Moscow, Russia*

# 236. Arterivirus Nsp2 cysteine endopeptidase

## Databanks

*Peptidase classification: clan CC, family C33, MEROPS ID: C33.001*
*NC-IUBMB enzyme classification: none*
*Databank codes:*

| Species | SwissProt | PIR | EMBL (genomic) |
|---|---|---|---|
| Equine arteritis virus | P19811 | A39925 | X52277: 5′ end of genome |
| | | S10158 | X53459: complete genome |
| Lactate dehydrogenase-elevating virus | – | – | U15146: complete genome |
| Lelystad virus | Q04561 | A36861 | L04493: 3′ end of genome |
| | | A45392 | M96262: complete genome |

## Name and History

Arteriviruses are enveloped, positive-strand RNA viruses that contain a polycistronic genome (12–15 kb) (den Boon *et al.*, 1991; Godeny *et al.*, 1993; Meulenberg *et al.*, 1993; Snijder & Spaan, 1995). Their replicase proteins are expressed from the open reading frames (ORFs) 1a and 1b that encode two large precursor proteins: the ORF1a protein (187–260 kDa) and, due to ribosomal frameshifting, the ORF1ab protein (345–422 kDa). Both precursors are processed extensively by three or four ORF1a-encoded endopeptidases (Snijder *et al.*, 1992, 1994, 1995, 1996; den Boon *et al.*, 1995; van Dinten *et al.*, 1996). The *Nsp2 cysteine proteinase (Nsp2 CP)* is the most C-terminally located member of an array of three cysteine proteinase domains that has been identified in the N-terminal 500 residues of the arterivirus ORF1a protein. It was mapped to the N-terminal region of Nsp2 and is highly conserved among arteriviruses (Figure 236.1), although it has

only been studied experimentally for equine arteritis virus (EAV) (Snijder *et al.*, 1995). The Nsp2 N-terminus (starting with Gly261; here and hereafter polyprotein numbering is used), is liberated by the rapid action of the upstream papain-like cysteine proteinase PCPβ (Chapter 235) (Snijder *et al.*, 1992). Subsequently, the Nsp2 CP mediates the cleavage at the C-terminus of Nsp2 (the Nsp2/Nsp3 site) to generate a 61 kDa cleavage product (Snijder *et al.*, 1994, 1995).

## Activity and Specificity

The activity of the EAV Nsp2 CP has been analyzed *in vivo* in infected cells and eukaryotic expression systems (Snijder *et al.*, 1994, 1995). The Nsp2 CP cleaves rapidly and probably in *cis*. However, *trans*-cleaving activity could be demonstrated in a eukaryotic expression system, albeit with relatively low efficiency (Snijder *et al.*, 1995). Because

*Figure 236.1* Alignment of the Nsp2 CP domains of the arteriviruses porcine reproductive and respiratory syndrome virus (PRRSV), lactate dehydrogenase-elevating virus (LDV), and equine arteritis virus (EAV). The alignment was produced with the ClustalW program and presented in the PrettyBox format derived from the GCG package. The putative catalytic Cys and His residues are marked with asterisks. Black indicates identical residues. The percentages of identical residues in the various protein pairs are: PRRSV versus LDV, 42%; LDV versus EAV, 40%; EAV versus PRRSV, 34%. The N-terminal boundary of the protease domains is the Nsp1/Nsp2 cleavage site (EAV) or putative Nsp1β/Nsp2 cleavage site (PRRSV and LDV), the C-terminal boundary was chosen arbitrarily.

conditions under which the Nsp2 CP is active *in vitro* or in *Escherichia coli* have not been found, there is only indirect evidence to support the idea that CP cleaves between two Gly residues (Gly831↓Gly832), a site that is conserved among arteriviruses. Mutagenesis of the putative P1 residue (Gly831 to Pro) abolished processing of the Nsp2/Nsp3 site (Snijder *et al.*, 1996).

## Structural Chemistry

Comparative sequence analysis (Figure 236.1) and site-directed mutagenesis have identified the Nsp2 CP as an unusual cysteine endopeptidase of about 100 residues. It most clearly resembles viral papain-like cysteine proteases, like the proteinases PCPα and PCPβ in the Nsp1 region of the arteriviruses (see Chapters 234 and 235). Residues Cys270 and His332 are assumed to form the cysteine proteinase catalytic dyad: their replacement (Cys270 by Ala, His, Arg or Ser; His332 by Cys, Ile, Asn or Tyr) completely eliminated proteolytic activity (Snijder *et al.*, 1995). Although the distance between these active-site residues resembles that found in viral papain-like proteinases, there are also important differences. First, the putative catalytic Cys270 is flanked by Gly271, and not by a bulky, hydrophobic residue which is usually found at this position in peptidases of clan CC (see Chapter 225). This resembles the situation in chymotrypsin-like cysteine proteinases (clan CB, see Chapter 238), in which the catalytic nucleophile is flanked by a Gly residue. The replacement of Gly271 by Trp completely abolished Nsp2 CP activity. Second, unlike most viral papain-like proteinases, the Nsp2 CP does not cleave immediately downstream of the active-site His: about 500 residues separate the proteinase domain from the cleavage site. Third, the entire Nsp2 CP domain is highly conserved among arteriviruses, which is a remarkable difference from the papain-like proteinases α and β in the Nsp1 regions of these viruses. Among the conserved residues are a number of acidic residues and cysteines. Although the replacement of the acidic residues (Asp291, Asp295, Asp296 and Glu297) did not influence CP activity, the substitution of the conserved Cys residues abolished (Cys319, Cys349 and Cys354) or reduced (Cys344 and Cys356) processing at the Nsp2/Nsp3 site (Snijder *et al.*, 1995). Together, the characteristics described above distinguish the Nsp2 CP from previously described viral papain-like and chymotrypsin-like cysteine proteinases.

## Biological Aspects

Nsp2 appears to be a multidomain protein: it contains highly conserved regions, but there are also sequences that cannot be aligned among arteriviruses (Snijder *et al.*, 1994). Functions associated with these domains remain to be elucidated. There are several indications that EAV Nsp2 is part of a membrane-associated replication complex. This may also have consequences for the proteolytic activity of the Nsp2 CP. Both large and small deletions in the Nsp2 region which separates the cysteine proteinase domain and its cleavage site were found to interfere with proteolytic processing. Furthermore,

cleavage of the Nsp2/Nsp3 junction modulates processing of the ORF1a protein region downstream of Nsp2 by the arterivirus serine proteinase (see Chapter 92) (Snijder *et al.*, 1996).

## Distinguishing Features

The arterivirus Nsp2 CP is a small and unique cysteine proteinase domain that shares primary structure similarities with both papain-like (clan CC) and chymotrypsin-like (clan CB) viral proteinases. Its activity is essential for proper proteolytic processing of the arterivirus replicase polyproteins.

A rabbit antiserum has been raised using an N-terminal peptide of EAV Nsp2 and is available from the authors for research purposes on request.

## Further Reading

The reader is referred to the paper of Snijder *et al.* (1995).

## References

den Boon, J.A., Snijder, E.J., Chirnside, E.D., de Vries, A.A.F., Horzinek, M.C. & Spaan, W.J.M. (1991) Equine arteritis virus is not a togavirus but belongs to the coronaviruslike superfamily. *J. Virol.* **65**, 2910–2920.

den Boon, J.A., Faaberg, K.S., Meulenberg, J.J.M., Wassenaar, A.L.M., Plagemann, P.G.W., Gorbalenya, A.E. & Snijder, E.J. (1995) Processing and evolution of the N-terminal region of the arterivirus replicase ORF1a protein: identification of two papainlike cysteine proteases. *J. Virol.* **69**, 4500–4505.

Godeny, E.K., Chen, L., Kumar, S.N., Methven, S.L., Koonin, E.V. & Brinton, M.A. (1993) Complete genomic sequence and phylogenetic analysis of the lactate dehydrogenase-elevating virus. *Virology* **194**, 585–596.

Meulenberg, J.J.M., Hulst, M.M., de Meijer, E.J., Moonen, P.L.J.M., den Besten, A., de Kluyver, E.P., Wensvoort, G. & Moormann, R.J.M. (1993) Lelystad virus, the causative agent of porcine epidemic abortion and respiratory syndrome (PEARS), is related to LDV and EAV. *Virology* **192**, 62–72.

Snijder, E.J. & Spaan, W.J.M. (1995) The coronaviruslike superfamily. In: *The Coronaviridae* (Siddell, S.G., ed.). New York: Plenum Press, pp. 239–255.

Snijder, E.J., Wassenaar, A.L.M. & Spaan, W.J.M. (1992) The 5′ end of the equine arteritis virus replicase gene encodes a papainlike cysteine protease. *J. Virol.* **66**, 7040–7048.

Snijder, E.J., Wassenaar, A.L.M. & Spaan, W.J.M. (1994) Proteolytic processing of the replicase ORF1a protein of equine arteritis virus. *J. Virol.* **68**, 5755–5764.

Snijder, E.J., Wassenaar, A.L.M., Spaan, W.J.M. & Gorbalenya, A.E. (1995) The arterivirus Nsp2 protease. An unusual cysteine protease with primary structure similarities to both papain-like and chymotrypsin-like proteases. *J. Biol. Chem.* **270**, 16671–16676.

Snijder, E.J., Wassenaar, A.L.M., van Dinten, L.C., Spaan, W.J.M. & Gorbalenya, A.E. (1996) The arterivirus nsp4 protease is the prototype of a novel group of chymotrypsin-like enzymes, the 3C-like serine proteases. *J. Biol. Chem.* **271**, 4864–4871.

van Dinten, L.C., Wassenaar, A.L.M., Gorbalenya, A.E., Spaan, W.J.M. & Snijder, E.J. (1996) Processing of the equine arteritis virus replicase ORF1b protein: identification of cleavage products containing the putative viral polymerase and helicase domains. *J. Virol.* **70**, 6625/6633.

*Alexander E. Gorbalenya*
*M.P. Chumakov Institute of Poliomyelitis and Viral Encephalitides, Russian Academy of Medical Sciences, 142782 Moscow Region, and A.N. Belozersky Institute of Physico-Chemical Biology, Moscow State University, 119899 Moscow, Russia*

*Alfred L.M. Wassenaar*
*Department of Virology, Institute of Medical Microbiology, Leiden University, PO Box 9600, 2300 RC Leiden, The Netherlands*

*Eric J. Snijder*
*Department of Virology, Institute of Medical Microbiology, Leiden University, PO Box 9600, 2300 RC Leiden, The Netherlands*
*Email: snijder@virology.azl.nl*

# 237. *Closterovirus papain-like cysteine proteinases*

## Databanks

*Peptidase classification: clan CC, family C42, MEROPS ID: C42.001*
*NC-IUBMB enzyme classification: none*
*Databank codes:*

| Species | SwissProt | PIR | EMBL (cDNA) | EMBL (genomic) |
|---|---|---|---|---|
| Citrus tristeza virus* | – | – | – | U16304: complete genome |
| Sugar beet yellow virus | – | – | – | X73476: complete genome |

*PCP1 is residues 396–484; PCP2 is residues 889–976.

## Name and History

The family Closteroviridae contains several elongated plant viruses with large positive-stranded RNA genomes (Bar-Joseph *et al.*, 1979; Dolja *et al.*, 1994; Agranovsky, 1996). Being closest to tobacco mosaic virus and the related plant viruses in evolutionary terms (Koonin & Dolja, 1993), closteroviruses resemble animal corona-like viruses in their genome size (15–19 kb), layout of open reading frames (ORFs), and expression strategy. The 5'-terminal replication-associated ORFs 1a and 1b of beet yellows closterovirus (BYV) are expressed by autoproteolysis and ribosomal frameshifting (Agranovsky, 1996; Agranovsky *et al.*, 1994). Based on the limited sequence similarity with the helper component proteinase of potyviruses (Chapter 227), a *closterovirus papain-like cysteine proteinase (PCP)* domain has been identified in the BYV ORF1a-encoded 295 kDa polyprotein and proved to be active in an *in vitro* assay (Agranovsky *et al.*, 1994). The related PCP domains have been predicted in the ORF1a products of citrus tristeza virus (CTV; Karasev *et al.*, 1995), lettuce infectious yellows virus (LIYV; Klaassen *et al.*, 1995), and some other closteroviruses (Agranovsky, 1996) (Figure 237.1).

## Activity and Specificity

By using *in vitro* translation of the T7 transcripts of BYV ORF1a, it was found that the N-terminal 66 kDa protein is released from the larger molecular-mass translation product. Point substitutions of the catalytic residues Cys509 (to Thr) and His569 (to Glu), or the Gly588 at the cleavage site (to Asp), completely blocked the cleavage. Replacement of the nonconserved Cys and His residues near the active center had different effects on proteolysis; thus, the substitution of His556 to Glu was tolerated, whereas substitutions of Cys517 or Cys518 to Thr strongly inhibited activity (Agranovsky *et al.*, 1994). By analogy with the related enzymes, it has been postulated that the BYV PCP cleaves in *cis*, but the experimental evidence for this has not been published.

## Structural Chemistry

Sequence comparisons of the closterovirus and potyvirus PCPs suggest that the BYV PCP domain consists of 140 amino acid residues at the extreme C-terminus of the 66 kDa protein (Agranovsky *et al.*, 1994; Karasev *et al.*, 1995; Klaassen *et al.*, 1995). Deletion of at least 245 amino acids

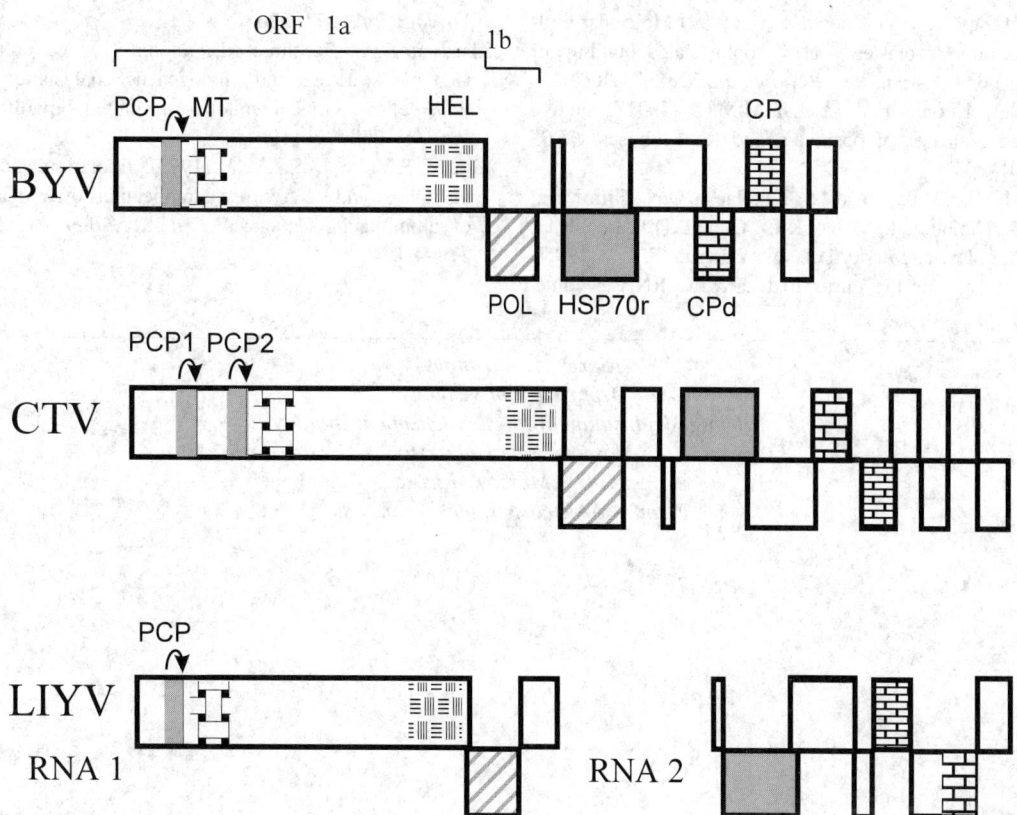

*Figure 237.1*  Genome maps of the closteroviruses BYV, CTV and LIYV. ORFs are shown as boxes, with the related domains indicated in the same fill-in. PCP, papain-like proteinase (with the cleavage site denoted by arrow); MT, HEL and POL, replication-associated domains of methyltransferase, RNA helicase and RNA-dependent RNA polymerase, respectively; HSP70r, heat-shock protein 70-related protein; CP and CPd, capsid protein and its diverged duplicate. Drawn approximately to scale (horizontally).

from the N-terminus of the BYV ORF1a product did not impair the cleavage. No data are available on the requirement for sequences downstream of Gly589 for the endopeptidase activity. Cys509 and His569 are likely to form the catalytic dyad. In the CTV polyprotein, the PCP domain is duplicated, and the PCP1 and PCP2 cleavage sites have been predicted at the Gly-Gly doublets, positions 484–485 and 976–977 from the N-terminus, respectively (Karasev *et al*., 1995). The tentative PCP cleavage site in the polyprotein of LIYV is Gly-Ala (positions 412–413) (Klaassen *et al*., 1995). In all the closterovirus PCPs, position P2 is occupied by a bulky aliphatic residue.

## Biological Aspects

The leader PCP is encoded in the RNA genomes of viruses belonging to quite disparate groups (Gorbalenya *et al*., 1991; Koonin & Dolja, 1993), where it is instrumental in the release of proteins with unrelated sequences and biological functions. The PCP domain is believed to mediate *in vivo* the cleavage of the leader proteins from the major portion of closterovirus polyprotein containing replicative domains. Apart from their C-terminal PCP domains, the cleaved leader proteins (BYV 66 kDa, CTV 54 kDa, CTV 55 kDa and LIYV 45 kDa) are not related to each other, nor to any other proteins in the database. They may have additional nonproteolytic functions in virus infection (Agranovsky, 1996).

## Distinguishing Features

Closterovirus PCPs are compact papain-like endopeptidase domains. The distance from the catalytic His to the cleavage site is 18–19 residues, which is much smaller than that in the related potyvirus PCPs (39 residues).

## Further Reading

A full, recent review is that of Agranovsky (1996).

## References

Agranovsky, A.A. (1996) The principles of molecular organization, expression and evolution of closteroviruses: over the barriers. *Adv. Virus Res.* **47**, 119–158.

Agranovsky, A.A., Koonin, E.V., Boyko, V.P., Maiss, E., Froet-schl, R., Lunina, N.A. & Atabekov, J.G. (1994) Beet yellows closterovirus: complete genome structure and identification of a leader papain-like thiol protease. *Virology* **198**, 311–324.

Bar-Joseph, M., Garnsey, S.M. & Gonsalves, D. (1979) The closteroviruses: a distinct group of elongated plant viruses. *Adv. Virus Res.* **25**, 93–168.

Dolja, V.V., Karasev, A.V. & Koonin, E.V. (1994) Molecular biology and evolution of closteroviruses: sophisticated build-up of large RNA genomes. *Annu. Rev. Phytopathol.* **32**, 261–285.

Gorbalenya, A.E., Koonin, E.V. & Lai, M.M.C. (1991) Putative papain-related proteases of positive-strand RNA viruses. *FEBS Lett.* **288**, 201–205.

Karasev, A.V., Boyko, V.P., Gowda, S., Nikolaeva, O.N., Hilf, M.E., Koonin, E.V., Niblett, C.L., Cline, K.C., Gumpf, D.J., Lee, R.F., Garnsey, S.M., Lewandowski, D.J. & Dawson, W.O. (1995). Complete sequence of the citrus tristeza virus RNA genome. *Virology* **208**, 511–520.

Klaassen, V.A., Boeshore, M., Koonin, E.V. & Falk, B.W. (1995) Genome structure and phylogenetic analysis of lettuce infectious yellows virus, a whitefly-transmitted, bipartite closterovirus. *Virology* **208**, 99–110.

Koonin, E.V. & Dolja, V.V. (1993). Evolution and taxonomy of positive-strand RNA viruses: implications of comparative analysis of amino acid sequences. *Crit. Rev. Biochem. Mol. Biol.* **28**, 375–430.

*Alexey A. Agranovsky*
*Department of Virology,*
*Belozersky Institute of Physico-Chemical Biology,*
*Moscow State University,*
*119899 Moscow, Russia*
*Email: AAA@closter.genebee.msu.su*

# 238. Introduction: clan CB containing viral 'chymotrypsin-like' cysteine peptidases

## Databanks

*MEROPS ID: CB*

| Species | SwissProt | PIR | EMBL (cDNA) | EMBL (genomic) |
|---|---|---|---|---|
| **Family C3** | | | | |
| Cowpea mosaic comovirus 24 kDa proteinase (Chapter 242) | | | | |
| Grapevine fanleaf nepovirus cysteine proteinase (Chapter 243) | | | | |
| Picornain 2A (Chapter 241) | | | | |
| Picornain 3C (Chapter 239) | | | | |
| Picornain 3C, hepatitis A virus (Chapter 240) | | | | |
| **Family C4** | | | | |
| Potyvirus NIa proteinase (Chapter 244) | | | | |
| **Family C24** | | | | |
| Feline calicivirus 2C endopeptidase (Chapter 245) | | | | |
| **Family C30** | | | | |
| Murine hepatitis coronavirus picornain 3C-like (3CL) polyprotein (Chapter 246) | | | | |
| **Family C37** | | | | |
| Norwalk virus | – | – | – | M87661: complete genome |
| Southampton virus | Q04544 | A37491 | – | L07418: complete genome |
| **Family C38** | | | | |
| Parsnip yellow fleck virus | Q05057 | JQ1917 | D14066 | – |

Clan CB contains only viral endopeptidases that process the viral polyproteins. The great majority are from single, positive-stranded RNA viruses. Endopeptidase activity depends on a pair of catalytic residues, which occur in the order His, Cys in the sequence (the reverse of that in clans CA and CC), but it has been suggested that a catalytic triad exists, with a Glu (not infrequently replaced by Asp) being the third member (Matthews *et al.*, 1994b). Alignment 238.1 (see CD-ROM) shows an alignment around the catalytic and proposed catalytic residues (for many of the endopeptidases in this clan, the active-site residues have been identified only by site-directed mutagenesis). It can be seen that the catalytic Cys is invariably followed by Gly, like Ser195 in chymotrypsin (Chapter 2).

The tertiary structures have been determined for picornains 3C from human hepatitis A virus (Chapter 240) (Figure 238.1) and human rhinovirus type 14 (Matthews *et al.*, 1994a,b). Both structures show a fold very similar to that of chymotrypsin, with the active site situated between two β-barrel domains. The catalytic histidines are in equivalent positions for the viral endopeptidases and chymotrypsin, the proposed catalytic glutamate of the rhinovirus picornain 3C and Asp102 of chymotrypsin are similarly placed, and the catalytic cysteine of the picornains occupies a position similar to that of Ser195 in chymotrypsin. Thus, clans SA and CB have a common fold and a common evolutionary origin; they are treated separately simply because of the importance currently given to the identity of the nucleophile in the catalytic site.

Endopeptidases of clan CB are frequently located between the helicase and the polymerase in the viral polyprotein. In contrast, those of clan CC are sited upstream of the helicase and polymerase. Taken together with other indications that the positive-stranded RNA viruses are evolutionarily related, the similar locations of the clan CB endopeptidases in the genome is entirely consistent with a common evolutionary origin of all of the peptidases in the clan. The high rate of mutation of RNA viruses, lacking proofreading in the replication of the genome, could rapidly have obscured any obvious relationships between the sequences.

*Family C3* includes the processing endopeptidases from picornaviruses, aphthoviruses, nepoviruses and comoviruses. The first group includes animal pathogens such as the causative agents of polio, hepatitis A, foot-and-mouth disease and encephalomyocarditis. Picornaviruses encode one polyprotein that contains coat and core proteins, an RNA polymerase and two endopeptidases that are homologous to each other and responsible for excising the individual

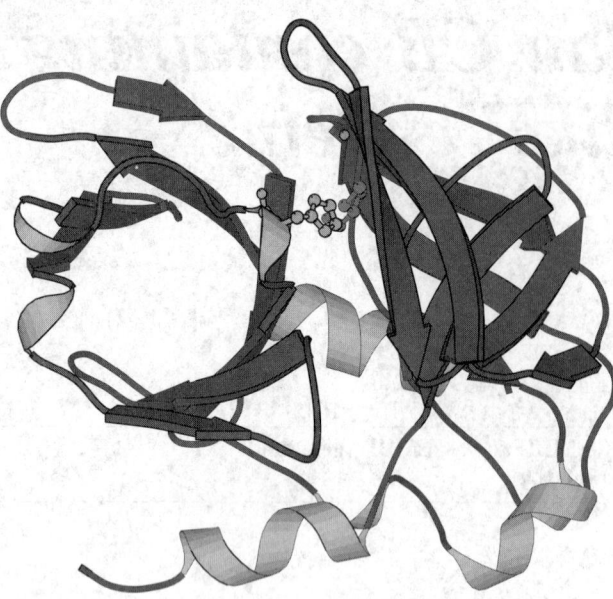

*Figure 238.1*   Richardson diagram of hepatitis A virus picornain 3C. The image was prepared from the Brookhaven Protein Data Bank entry (1HAV) as described in the Introduction (p. xxv). Catalytic residues are shown in ball-and-stick representation: His40 and Cys147 (engineered to be serine) (numbering as in Alignment 238.1).

proteins. The larger of the endopeptidases, picornain 3C, is responsible for most of the cleavages, mainly at Gln⧾Gly bonds, whereas the smaller endopeptidase, picornain 2A, releases itself by cleavage of a Tyr⧾Gly bond. Picornain 2A has a further function: cleavage of eukaryote initiation factor 4G (Lamphear & Rhoads, 1996). Picornaviruses are unusual in possessing two homologous endopeptidases; in aphthoviruses, for example, cleavage of the eukaryote initiation factor 4G is performed by an endopeptidase unrelated to the picornains, the L-peptidase (Chapter 226).

*Family C4* contains one of the three polyprotein processing endopeptidases from potyviruses, which are plant pathogens. The NIa endopeptidase has been shown by site-directed mutagenesis to have at least a catalytic dyad that is in the order His, Cys (Dougherty *et al.*, 1989). Another similarity with family C3 is the preference for Gln⧾Gly bonds. The endopeptidase is a multifunctional molecule, acting also as the Vpg protein, which is attached covalently to the viral RNA (Rorrer *et al.*, 1992).

*Family C24* includes the processing endopeptidase from caliciviruses. Catalytic residues have been identified by site-specific mutagenesis in the rabbit hemorrhagic disease virus, and a His, Cys dyad is the most likely order in the sequence, but a histidine C-terminal to the cysteine is also important, probably for substrate binding (Boniotti *et al.*, 1994). The endopeptidase cleaves Glu⧾Gly bonds in the polyprotein (Wirblich *et al.*, 1995).

*Family C30* includes the processing endopeptidase from coronaviruses. Again, there is a His, Cys catalytic dyad, and cleavage is preferentially at Gln⧾Gly bonds (Liu & Brown, 1995).

*Family C37* includes the processing endopeptidase from Southampton and Norwalk caliciviruses. The catalytic cysteine has been identified in Southampton virus by site-directed mutagenesis, and there is a preference for cleavage at Gln⧾Gly bonds (Liu *et al.*, 1996).

*Family C38* comprises a putative endopeptidase from parsnip yellow fleck virus identified from comparisons with the picornavirus polyprotein (Turnbullross *et al.*, 1993).

### References

Boniotti, B., Wirblich, C., Sibilia, M., Meyers, G., Thiel, H.J. & Rossi, C. (1994) Identification and characterization of a 3C-like protease from rabbit hemorrhagic disease virus, a calicivirus. *J. Virol.* **68**, 6487–6495.

Dougherty W.G., Parks T.D., Cary S.M., Bazan J.F. & Fletterick R.J. (1989) Characterization of the catalytic residues of the tobacco etch virus 49-kDa proteinase. *Virology* **172**, 302–310.

Lamphear, B.J. & Rhoads, R.E. (1996) A single amino acid change in protein synthesis initiation factor 4G renders cap-dependent translation resistant to picomaviral 2A proteases. *Biochemistry* **35**, 15726–15733.

Liu, D.X. & Brown, T.D.K. (1995) Characterization and mutational analysis of an ORF 1a-encoding proteinase domain responsible for proteolytic processing of the infectious-bronchitis virus 1a/1b polyprotein. *Virology* **209**, 420–427.

Liu, B.L., Clarke, I.N. & Lambden, P.R. (1996) Polyprotein processing in Southampton virus: identification of 3C-like protease cleavage sites by in vitro mutagenesis. *J. Virol.* **70**, 2605–2610.

Matthews, D., Smith, W.W., Ferre, R.A., Condon, B., Budahazi, G., Sisson, W., Villafranca, J.E., Janson, C., Mcelroy, H., Gribskov, C. & Worland, S. (1994a) Crystal-structure of human rhinovirus type 14 3C protease. *J. Cell. Biochem.* **77**, 761–771.

Matthews, D.A., Smith, W.W., Ferre, R.A., Condon, B., Budahazi, G., Sisson, W., Villafranca, J.E., Janson, C.A., McElroy, H.E., Gribskov, C.L. & Worland, S. (1994b) Structure of human rhinovirus 3C protease reveals a trypsin-like polypeptide fold, RNA-binding site, and means for cleaving precursor polyprotein. *Cell* **77**, 761–771.

Rorrer, K., Parks, T.D., Scheffler, B., Bevan, M. & Dougherty, W.G. (1992) Autocatalytic activity of the tobacco etch virus NIa proteinase in viral and foreign protein sequences. *J. Gen. Virol.* **73**, 775–783.

Turnbullross, A.D., Mayo, M.A., Reavy, B. & Murant, A.F. (1993) Sequence-analysis of the parsnip yellow fleck virus polyprotein – evidence of affinities with picornaviruses. *J. Gen. Virol.* **74**, 555–561.

Wirblich, C., Sibilia, M., Boniotti, M.B., Rossi, C., Thiel, H.J. & Meyers, G. (1995) 3C-like protease of rabbit hemorrhagic disease virus: identification of cleavage sites in the ORF1 polyprotein and analysis of cleavage specificity. *J. Virol.* **69**, 7159–7168.

# 239. Picornain 3C

## Databanks

*Peptidase classification: clan CB, family C3, MEROPS ID: S03.001*
*NC-IUBMB enzyme classification: EC 3.4.22.28*
*Databank codes:*

| Species | SwissProt | PIR | EMBL (cDNA) | EMBL (genomic) |
|---|---|---|---|---|
| Cattle enterovirus | P12915 | – | – | D00214: complete genome |
| Coxsackievirus a9 | P21404 | JQ0523 | – | D00627: complete genome |
| Coxsackievirus a21 | P22055 | A33373 | – | D00538: complete genome |
| Coxsackievirus a23 | P08490 | – | X67706 | M12167: proteinase region |
| Coxsackievirus a24 | P36290 | A45848 | D10267 | D90457: complete genome |
| | | | D10268 | |
| | | | D10269 | |
| | | | D10270 | |
| | | | D10271 | |
| | | | D10273 | |
| | | | D10274 | |
| | | | D10276 | |
| | | | D10277 | |
| | | | D10278 | |
| | | | D10302 | |
| | | | D10312 | |
| | | | D10322 | |
| | | | D10325 | |
| | | | D13277 | |
| | | | D13285 | |
| Coxsackievirus b1 | P08291 | A26353 | – | M16560: complete genome |
| Coxsackievirus b3 | P03313 | A26354 | – | M16572: complete genome |
| | | A34664 | | M33854: complete genome |
| Coxsackievirus b4 | P08292 | A27120 | – | X05690: complete genome |
| Coxsackievirus b5 | Q03053 | JQ2021 | – | X67706: complete genome |
| Echovirus 9 | – | – | – | X84981: complete genome |
| Echovirus 11 | P29813 | A36642 | D10582 | – |
| Encephalomyocarditis virus | P03304 | A03906 | X00463 | X00463: polyprotein gene |
| Encephalomyocarditis virus (strain EMC-B) | P17593 | B31473 | – | M22457: complete genome |
| Encephalomyocarditis virus (strain EMC-D) | P17594 | A31473 | – | M22458: complete genome |
| Foot-and-mouth disease virus (strains O1K and O1BFS) | P03305 | A03907 | – | X00817: complete genome |
| Foot-and-mouth disease virus (strain A10-61) | P03306 | A93508 | – | X00429: complete genome |
| Foot-and-mouth disease virus (strain A12) | P03308 | A25794 | – | M10975: complete genome |
| Foot and mouth disease virus (strain C1-Santa Pau) | P03311 | A03913 | – | M11027: 3' end of genome |
| Foot-and-mouth disease virus (strain A22/550 Azerbaijan 65) | P49303 | – | X74812 | X74812: complete genome |
| Human enterovirus 70 | P32537 | A36253 | D00820 | – |
| Human rhinovirus 1b | P12916 | A28699 | D00239 | D00239: complete genome |
| Human rhinovirus 2 | P04936 | A03902 | X02316 | X02316: complete genome |

*continued overleaf*

| Species | SwissProt | PIR | EMBL (cDNA) | EMBL (genomic) |
|---|---|---|---|---|
| Human rhinovirus 14 | P03303 | A03901 | M12168 M73790 | K02121: complete genome L05355: polyprotein gene X01087: complete genome |
| Human rhinovirus 89 | P07210 | A29862 | – | A10937: complete genome M16248: complete genome |
| Pig vesicular disease virus | P13900 | – | – | X54521: complete genome |
| Pig vesicular disease virus | P16604 S11670 | A31331 | D00435 | – |
| Poliovirus type 1 | P03299 | A93258 | – | V01148: complete genome |
| Poliovirus type 1 | P03300 | A03898 | – | J02281: complete genome |
| Poliovirus type 1 | P03301 | A03899 | – | V01150: complete genome |
| Poliovirus type 2 | P06210 | A29507 | – | M12197: complete genome |
| Poliovirus type 2 | P23069 | A34032 | – | D00625: complete genome |
| Poliovirus type 3 | P03302 | A93987 | – | K01392: complete genome |
| Poliovirus type 3 | P06209 | A27245 | – | X04468: complete genome |
| Theiler's murine encephalomyelitis virus (strain BEAN 8386) | P08544 | A29535 | M16020 | |
| Theiler's murine encephalomyelitis virus (strain GDVII) | P08545 | – | – | M14703: polymerase and proteinase genes 3′ end  M20562: complete genome |
| Theiler's murine encephalomyelitis virus (strain DA) | P13899 | A31228 | M20301 | – |

## Name and History

Cysteine proteinases encoded by picornaviruses were first detected in *in vitro* experiments designed to investigate the processing of primary products of protein synthesis from picornaviral mRNA. Processing was shown to be inhibited by characteristic inhibitors of cysteine proteinases such as *N*-ethylmaleimide and iodoacetamide (Pelham, 1978). Picornaviral proteins are named according to Rueckert & Wimmer (1984) after their position in the viral genome (see Figure 239.1). The first viral protein shown to have proteolytic activity was the poliovirus ***protein 3C*** (Hanecak *et al.*, 1982, 1984). Subsequently, all genera of the picornavirus family were found to encode this proteinase, now called ***picornain 3C***. These viruses comprise the enteroviruses (polioviruses and coxsackieviruses), the human rhinoviruses, the cardioviruses, the aphthoviruses (foot-and-mouth disease virus) and the hepatoviruses (hepatitis A virus, HAV). However, there are differences between the picornains 3C from genus to genus in substrate recognition and the number of junctions cleaved on the viral polyprotein.

In the entero- and rhinovirus genera, picornain 3C, or its precursor 3CD, cleaves between all viral proteins except VP4/VP2 and VP1 and 2A (Hanecak *et al.*, 1982). Subsequently, the 2A protein (Toyoda *et al.*, 1986) was identified as the proteinase cleaving the VP1/2A junction in rhino- and enteroviruses (Chapter 241). In cardioviruses (encephalomyocarditis virus (EMCV) and the closely related Mengo virus), the 3C proteinase processes all sites with the exception of the VP4/VP2 and 2A/2B junctions (Palmenberg, 1990). The protein 2A cleaves the 2A/2B junction (Palmenberg, 1990). In

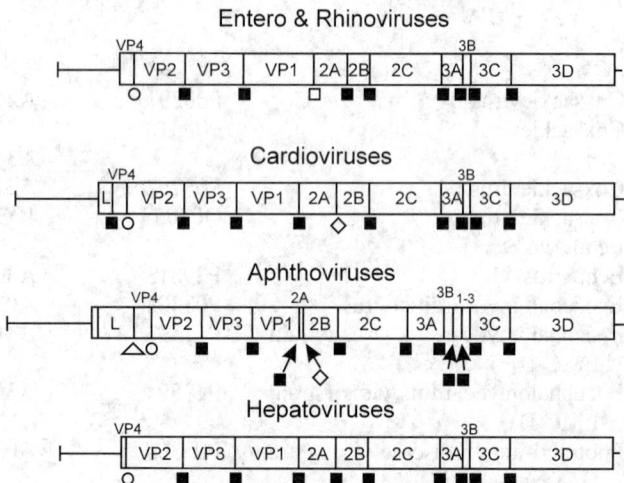

*Figure 239.1* Schematic representation of the RNA genomes for the five picornavirus genera. Noncoding regions are shown by lines, the large open reading frame by an open box. The virally encoded protein 3B, which is linked covalently to the 5′ end of the viral genome is indicated by a vertical stroke. The positions of viral proteins are given. Sites of cleavage for picornaviral proteinases are indicated as follows: ■, picornain 3C; □, picornain 2A; ◇, protein 2A of cardioviruses and aphthoviruses; △, leader (L) proteinase of FMDV. The proteolytic agent cleaving at the VP4/VP2 junction (○) has not yet been identified; this cleavage has not yet been proven in HAV. The L-protein of FMDV exists in two forms as a result of the presence of two initiation sites for protein synthesis on the FMDV RNA.

aphthoviruses, the situation is similar except that 3C does not process the L/VP4 site; this is done by the L-protein (Chapter 226) itself (Strebel & Beck, 1986). The HAV picornain 3C (Chapter 240) processes all presently known cleavage sites on the polyprotein (Schultheiss *et al.*, 1994); the VP4/VP2 cleavage has not yet been demonstrated during processing in this virus. The proteolytic agent responsible for cleavage at the VP4/VP2 junction has not been identified in any of the picornaviruses. The cleavages carried out by the various picornaviral proteinases are summarized in Figure 239.1.

## Activity and Specificity

Activity of the picornains 3C on polyprotein substrates has been assayed by producing *in vitro* translated radiolabeled proteins corresponding to various parts of the polyprotein; normally, the regions containing the picornain genes are excluded or partially lacking. The translation extracts are incubated with picornain 3C extracts or purified enzyme and the cleavage products separated by SDS-PAGE and detected by autoradiography. Although quantitation in such assays is difficult, this method has been useful for detecting cleavage activity in infected cell extracts (Nicklin *et al.*, 1987). Additionally, use of this method also revealed that the protein responsible for the cleavage at the VP2/VP3 and VP3/VP1 junctions is the polypeptide 3CD, the precursor of 3CD (Ypma-Wong *et al.*, 1988a).

The more favored assay for picornains 3C has been the hydrolysis of oligopeptides corresponding to cleavage sites on the respective viral polyprotein (Long *et al.*, 1989; Pallai *et al.*, 1989; Cordingley *et al.*, 1990). Picornain 3C cleavage sites usually show a P1 Gln and P1′ Gly. However, these residues and the surrounding sequence vary among the various genera. Thus, poliovirus picornain 3C cleaves exclusively at Gln╪Gly sites, whereas the enzyme from EMCV can also accept Glu╪Ser and Glu╪Ala bonds and that of foot-and-mouth disease virus (FMDV) can process Gln╪Leu and Gln╪Ile. Further conservation is seen in the presence of a small aliphatic residue at P4, and P2 Pro in rhinoviruses and enteroviruses; P2 Pro and P2′ Pro are often present in cardioviruses (Kräusslich & Wimmer, 1988). However, the presence of such sequences is not sufficient for picornain 3C cleavage. Both the amino acid sequence flanking the amino acid pair and the nature of the secondary structure in which it is embedded determine whether an amino acid pair is processed. Thus, in poliovirus, 13 Gln-Gly pairs are found in the polyprotein, of which only eight are processed. Pairs which lie between $\beta$ barrels are cleaved efficiently whereas those inside such a structure are not (Ypma-Wong *et al.*, 1988b).

Use of oligopeptides corresponding to 3C cleavage site sequences has shown that rhino- and enterovirus picornains 3C require a minimum of five nonprime residues together with two prime-side residues (Long *et al.*, 1989; Pallai *et al.*, 1989). Substitutions at P5 and P3 have little effect on cleavage, whereas positions P4, P1, P1′ and P2′ are much more sensitive to change. Optimal occupancy for poliovirus picornain 3C is P4 Ala, P1 Gln, P1′ Gly and P2′ Pro; the situation is similar in human rhinovirus 14 (HRV14) picornain 3C except that Thr and Val can also be accepted (Long *et al.*, 1989; Pallai *et al.*, 1989; Cordingley *et al.*, 1990). The cleavage sites for EMCV picornain 3C have been investigated by analyzing cleavage of polyproteins containing mutated cleavage sites; cleavage occurred at Gln╪Ser, Gln╪Cys, Gln╪Gly and Gln╪Ala bonds, but not at those containing Gln-Thr, Gln-Ile or Gln-Tyr (Parks *et al.*, 1989).

As cysteine proteinases, picornains 3C are inhibited by *N*-ethylmaleimide, iodoacetamide and $Hg^{2+}$ ions (Pelham, 1978), but are distinctively not inhibited by E-64 (Orr *et al.*, 1989). Chymostatin has been shown to inhibit HRV14 picornain 3C (Orr *et al.*, 1989).

## Structural Chemistry

Picornains 3C are single-chain proteins containing between 180 (human rhinoviruses) and 220 (hepatitis A) amino acids with pI values between 8 and 10. Several studies have been performed in attempts to predict the mechanism and structure of picornains 3C. Argos *et al.* (1984) concluded that the only possible type of active site comprised a cysteine residue and a histidine residue. The realization that picornains 3C appeared related to both cysteine and serine proteinases led to suggestions that the picornains 3C might represent an evolutionary link between the two groups of proteinases (Gorbalenya *et al.*, 1986; Brenner, 1988). Subsequently, both Bazan & Fletterick (1988) and Gorbalenya *et al.* (1989) independently predicted homology with the chymotrypsin-like serine proteinases (family C1); however, some differences in the predictions were evident. For instance, both groups predicted a catalytic triad containing residues His40 and Cys146 (numbering is for HRV14 picornain 3C; the equivalent residues in HAV picornain 3C are His44 and Cys172), but whereas Bazan & Fletterick (1998) proposed Asp85 (Asp98 in HAV picornain 3C), Gorbalenya *et al.* (1989) predicted the involvement of Glu71 (equivalent to Asp84 in HAV picornain 3C) as the third member of the triad. The results of site-directed mutagenesis experiments with poliovirus (Kean *et al.*, 1991) supported the hypothesis of Gorbalenya *et al.* (1989).

With the availability of the X-ray crystallographic structures for the picornains 3C from HRV14 (Matthews *et al.*, 1994) and HAV (Chapter 240) (Allaire *et al.*, 1994), the accuracy of the models could be investigated. As predicted by sequence alignment and modeling, the HRV14 structure is similar to serine proteinases of the trypsin family in that it contains two topologically equivalent six-stranded $\beta$ barrels. A long, shallow groove for substrate binding is present between the two domains. The S1 pocket contains the side chains of Thr141 and His160 which can form hydrogen bonds with the preferred P1 Gln residue. The S4 pocket is shallow and hydrophobic, explaining the requirement for small, hydrophobic residues at P4. In contrast, the requirement for P1′ Gly and P2′ Pro is imposed by the need to turn the main chain of the substrate into a second canyon that has no counterpart in serine proteinases. Catalysis in HRV14 picornain 3C is carried out by a triad of Cys146, His40 and Glu71, bearing out the proposition of Gorbalenya *et al.* (1989).

A putative RNA-binding site lies on the opposite side of the protein to the active site and appears to be made up of a large number of basic residues. Furthermore, two Phe residues may also play a role in RNA binding; they are unusually exposed and either conserved or present as Tyr or His residues in most picornaviral picornain 3C sequences (Matthews *et al.*, 1994).

## Preparation

There are no reports of the purification of picornain 3C from infected cell extracts. Several recombinant picornains 3C have been expressed in *Escherichia coli* on a milligram scale and purified to homogeneity. These include the enzymes from poliovirus (Nicklin *et al*., 1988), HRV14 (Libby *et al*., 1988; Cordingley *et al*., 1989; Matthews *et al*., 1994) and Mengo virus (Hall & Palmenberg, 1996). The elimination of autolytic processing at the 3CD cleavage site by the substitution of the P4 Thr residue with Lys prevented self-cleavage and allowed poliovirus 3CD to be expressed in *E. coli* and purified to homogeneity (Harris *et al*., 1992).

## Biological Aspects

Picornain 3C is the major processing enzyme in picornaviruses, which cleaves between eight and ten times in the polyprotein, depending on the picornavirus. However, the cleavage sites are not all processed at the same rate. Cleavage at the 2BC and 3CD sites is generally slow, accounting for the relatively large amounts of the 2BC and 3CD precursors present in the infected cell. Much faster rates of cleavage are found at the 2C3A and 3BC junctions. Experiments with oligopeptides derived from these sites generally bear out these observations (Dunn & Kay, 1990).

A number of cellular substrates have been identified for picornains 3C also. The poliovirus enzyme has been shown to adversely affect host-cell transcription by specifically cleaving the transcription factors TFIIIC (Clark *et al*., 1991) and TFIID (Clark *et al*., 1993). The microtubule-associated protein 4 is also a substrate for poliovirus picornain 3C (Joachims *et al*., 1995). FMDV picornain 3C has been shown to cleave the histone protein H3 in baby hamster kidney cells (Tesar & Marquardt, 1990).

## RNA Binding

Both poliovirus and HRV14 picornains 3C have been shown to bind specifically to a hairpin loop structure at the 5′ end of their cognate genomic RNAs (Leong *et al*., 1993; Andino *et al*., 1993). For poliovirus, the precursor 3CD was shown to be at least 10-fold more efficient in binding to the RNA structure than the mature 3C protein (Andino *et al*., 1993). It has been proposed that this RNA recognition by the 3CD precursor locates and spatially orientates the protein 3D polymerase molecule correctly for the initiation of the synthesis of viral RNA. Subsequent autocleavage of 3CD frees the 3D polymerase activity at the required position.

## Distinguishing Features

Picornains 3C utilize a cysteine residue as the active-site nucleophile and yet possess a mechanistic and structural relationship to the chymotrypsin-like serine proteinases. In addition, they are unusual in being proteinases which can bind specifically to a particular RNA stem-loop structure. In the form of the immediate precursor 3CD, both the specificity of substrate recognition and the efficiency of RNA binding are modulated.

## Further Reading

For a detailed discussion of substrate specificity in rhinoviral and enteroviral picornains 3C see Dunn & Kay (1990). Valuable discussions of the implications of the two picornain 3C crystallographic structures for structure/function relationships in these enzymes, and for the evolution of the picornains 3C have been provided by Carrell & Lesk (1994) and Allaire & James (1994).

## References

Allaire, M. & James, M.N.G. (1994) Deduction of the 3C proteinases' fold. *Nature Struct. Biol.* **1**, 505–506.

Allaire, M., Chrenaia, M.M, Malcolm, B.A. & James, M.N. (1994) Picornaviral 3C cysteine proteinases have a fold similar to chymotrypsin-like serine proteinases. *Nature* **369**, 72–76.

Andino, R., Rieckhof, G.E., Achacoso, P.L. & Baltimore, D. (1993) Poliovirus RNA synthesis utilizes an RNP complex formed around the 5′-end of viral RNA. *EMBO J.* **12**, 3587–3598.

Argos, P., Kamer, G., Nicklin, M.J.H. & Wimmer, E. (1984) Similarity in gene organisation and homology between proteins of animal picornaviruses and a plant comovirus suggest common ancestry of these virus families. *Nucleic Acids Res.* **12**, 7251–7267.

Bazan, J.F. & Fletterick, R.J. (1988) Viral cysteine proteinases are homologous to the trypsin-like family of serine proteinases: structural and functional implications. *Proc. Natl Acad. Sci. USA* **85**, 7872–7876.

Brenner, S. (1988) The molecular evolution of genes and proteins: a tale of two serines. *Nature* **334**, 528–530.

Carrell, R.W. & Lesk, A.M. (1994) A tale of two proteinases. *Nature Struct. Biol.* **1**, 492–494.

Clark, M.E., Hammerle, T., Wimmer, E. & Dasgupta, A. (1991) Poliovirus proteinase 3C converts an active form of transcription factor IIIC to an inactive form: a mechanism for inhibition of host cell polymerase III transcription by poliovirus. *EMBO J.* **10**, 2941–2947.

Clark, M.E., Lieberman, P.M., Berk, A. & Dasgupta, A. (1993) Direct cleavage of human TATA-binding protein by poliovirus proteinase 3C in vivo and in vivo. *Mol. Cell. Biol.* **13**, 1232–1237.

Cordingley, M.G., Register, R.B., Callahan, P.J., Garsky, V.M. & Colonno, R.J. (1989) Cleavage of small peptides *in vitro* by human rhinovirus 3C protease expressed in *Escherichia coli*. *J. Virol.* **63**, 5037–5045.

Cordingley, M.G., Callahan, P.L., Sardana, V.V., Garsky, V.M. & Colonno, R.J. (1990) Substrate requirements of human rhinovirus 3C protease for peptide cleavage *in vivo*. *J. Biol. Chem.* **265**, 9062–9065.

Dunn, B. & Kay, J. (1990) Viral proteinases: weakness in strength. *Biochim. Biophys. Acta* **1048**, 1–18.

Gorbalenya, A.E., Blinov, V.M. & Donchenko, A.P. (1986) Poliovirus-encoded proteinase 3C: a possible evolutionary link between cellular serine and cysteine proteinase families. *FEBS Lett.* **194**, 253–257.

Gorbalenya, A.E., Donchenko, A.P, Blinov, V.M. & Koonin, E.V. (1989) Cysteine proteinases of positive strand RNA viruses and chymotrypsin-like serine proteinases. *FEBS Lett.* **243**, 103–114.

Hall, D.J. & Palmenberg, A.C. (1996) Mengo virus 3C proteinase: recombinant expression, intergenus substrate cleavage and localization *in vivo*. *Virus Genes* **13**, 99–110.

Hanecak, R., Semler, B.L., Anderson, C.W. & Wimmer, E. (1982) Proteolytic processing of poliovirus polypeptides: antibodies to polypeptide P3–7c inhibit cleavage at glutamine-glycine pairs. *Proc. Natl Acad. Sci. USA* **79**, 3973–3977.

Hanecak, R., Semler, B.L., Ariga, H., Anderson, C.W. & Wimmer, E. (1984) Expression of a cloned gene segment of poliovirus in *E. coli*: evidence for autocatalytic production of the viral proteinase. *Cell* **37**, 1063–1073.

Harris, K.S., Reddigari, S.R., Nicklin, M.J., Hammerle, T. & Wimmer, E. (1992) Purification and characterisation of poliovirus polypeptide 3CD, a proteinase and a precursor for RNA polymerase. *J. Virol.* **66**, 7481–7489

Joachims, M., Harris, K.S. & Etchison, D. (1995) Poliovirus protease 3C mediates cleavage of microtubule-associated protein 4. *Virology* **211**, 451–461.

Kean, K.M., Teterina, N.L., Marc, D. & Girard, M. (1991) Analysis of putative active site residues of the poliovirus 3C protease. *Virology* **181**, 609–619.

Kräusslich, H.-G. & Wimmer, E. (1988) Viral proteinases. *Annu. Rev. Biochem.* **57**, 701–754.

Leong, L.E., Walker, P.A. & Porter, A.G. (1993) Human rhinovirus-14 protease (3C$^{pro}$) binds specifically to the 5′-noncoding region of the viral RNA. *J. Biol. Chem.* **268**, 25735–25739.

Libby, R.T., Cosman, D., Cooney, M.K., Merriam, J.E., March, C.J. & Hopp, T.P. (1988) Human rhinovirus 3C proteinase: cloning and expression of an active form in *Escherichia coli*. *Biochemistry* **27**, 6262–6268.

Long, A., Orr, D.C., Cameron, J.M., Dunn, B.M. & Kay, J. (1989) A consensus sequence for substrate hydrolysis by rhinovirus 3C proteinase. *FEBS Lett.* **258**, 75–78.

Matthews, D.A., Smith, W.W., Ferre, R.A., Condon, B., Budahazi, G., Sisson, W., Villafranca, J.E., Janson, C.A., McElroy, H.E., Gribskov, C.L. & Worland, S. (1994) Structure of human rhinovirus 3C protease reveals a trypsin-like polypeptide fold, RNA-binding site and means for cleaving precursor polyprotein. *Cell* **77**, 761–771.

Nicklin, M.J.H., Kräusslich, H.G., Toyoda, H., Dunn, J.J. & Wimmer, E. (1987) Poliovirus polypeptide precursors. Expression in vitro and processing by 3C and 2A proteinases. *Proc. Natl Acad. Sci. USA* **84**, 4002–4006.

Nicklin, M.J.H., Harris, K.S., Pallai, P.V. & Wimmer, E. (1988) Poliovirus proteinase 3C: large-scale expression, purification, and

specific cleavage activity on natural and synthetic substrates in vitro. *J. Virol.* **62**, 4586–4593.

Orr, D.C., Long, A.C., Kay, J., Dunn, B.M. & Cameron, J.M. (1989) Hydrolysis of a series of synthetic peptide substrates by the human rhinovirus 14 3C protease, cloned and expressed in *Escherichia coli*. *J. Gen. Virol.* **70**, 2931–2942.

Palmenberg, A.C. (1990) Proteolytic processing of picornaviral polyprotein. *Annu. Rev. Microbiol.* **44**, 603–623.

Pallai, P.V., Burkhardt, F., Skoog, M., Schreiner, K., Bax, P., Cohen, K.A., Hansen, G., Palladino, D.E.H., Harris, K.S., Nicklin, M.J.H & Wimmer, E. (1989) Cleavage of synthetic peptides by purified poliovirus 3C proteinase. *J. Biol. Chem.* **264**, 9738–9741.

Parks, G.D., Baker, J.C., & Palmenberg, A.C. (1989) Proteolytic cleavage of encephalomyocarditis virus capsid region substrates by precursors to the 3C enzyme. *J. Virol.* **63**, 1054–1058.

Pelham, H.R.B. (1978) Translation of encephalomyocarditis virus RNA *in vitro* yields an active proteolytic processing enzyme. *Eur. J. Biochem.* **85**, 457–462.

Rueckert, R.R. & Wimmer, E. (1984) Systematic nomenclature of picornavirus proteins. *J. Virol.* **50**, 957–959.

Schultheiss, T., Kusov, Y.Y. & Gauss-Müller, V. (1994) Proteinase of hepatitis A virus (HAV) cleaves the HAV polyprotein P2-P3 at all sites including VP1/2A and 2A/2B. *Virology* **198**, 275–281.

Strebel, K. & Beck, E (1986) A second protease of foot-and-mouth disease virus. *J. Virol.* **57**, 983–991.

Tesar, M. & Marquardt, O. (1990) Foot-and-mouth disease virus protease 3C inhibits cellular transcription and mediates cleavage of histone H3. *Virology* **174**, 364–374.

Toyoda, H., Nicklin, M.J.H., Murray, M.G., Anderson, C.W., Dunn, J.J., Studier, F.W. & Wimmer, E. (1986) A second virus-encoded proteinase involved in proteolytic processing of poliovirus polyprotein. *Cell* **45**, 761–770.

Ypma-Wong, M.F., Dewalt, P.G., Johnson, V.H., Lamb, J.G. & Semler, B.L. (1988a) Protein 3CD is the major poliovirus proteinase responsible for cleavage of the P1 capsid precursor. *Virology* **166**, 265–270.

Ypma-Wong, M.F., Filman, J., Hogle, J.M. & Semler, B.L (1988b) Structural determinants of the poliovirus protein are major determinants for proteolytic cleavage at Gln-Gly pairs. *J. Biol. Chem.* **263**, 17846–17856.

***Tim Skern***
*Department of Biochemistry, Medical Faculty,*
*University of Vienna,*
*Dr. Bohr-Gasse 9/3, A-1090 Vienna, Austria*
*Email: timothy.skern@univie.ac.at*

# 240. *Hepatitis A virus picornain 3C*

## Databanks

*Peptidase classification: clan CB, family C3, MEROPS ID: C03.005*
*NC-IUBMB enzyme classification: none*

*Databank codes:*

| Species | Strain | SwissProt | PIR | EMBL (cDNA) | EMBL (genomic) |
|---|---|---|---|---|---|
| Hepatitis A virus | LA | P06441 | A03903 | – | K02990: complete genome |
| Hepatitis A virus | HM-175 | P08617 | – | – | M14114: 5′ end of genome |
| | | | | | M14707: complete genome |
| | | | | | M16632: complete genome |
| Hepatitis A virus | MBB | P13901 | JS0303 | – | M20273: complete genome |
| Hepatitis A virus | AGM-27 | P14553 | A30470 | – | D00924: complete genome |
| Hepatitis A virus | 24A | P26580 | – | – | M59810: complete genome |
| Hepatitis A virus | 43C | P26581 | – | – | M59809: complete genome |
| Hepatitis A virus | 18F | P26582 | – | – | M59808: complete genome |

Brookhaven Protein Data Bank three-dimensional structures:

| Species | ID | Resolution | Notes |
|---|---|---|---|
| Hepatitis A virus | 1HAV | 2.0 | |

## Name and History

The autoproteolytic activity of the hepatitis A virus (HAV) polyprotein and the proteolytic activity of the HAV 3C gene product were initially inferred from the similarity of the sequence of the HAV RNA genome to those of other picornaviruses (Cohen *et al.*, 1987). The 3C gene product of picornaviruses is commonly referred to as the *3C proteinase* (Palmenberg, 1990). More recently the name *picornain 3C* has been recommended by IUBMB.

## Activity and Specificity

Three different peptide-based assays of the proteolytic activity of HAV 3C have been reported (Jewell *et al.*, 1992; Malcolm *et al.*, 1992; Schultheiss *et al.*, 1995). Malcolm *et al.* (1992) used a discontinuous trinitrobenzene sulfonate assay of free amino groups to measure the activity of HAV 3C on a peptide substrate Ac-Glu-Leu-Arg-Thr-Gln+Ser-Phe-Ser-NH₂. This peptide sequence was derived from the sequence of the 2B+2C cleavage site of the HAV polyprotein. Jewell *et al.* (1992) developed a continuous quenched fluorescence assay with a dansylated peptide as substrate. Optimal peptide substrates of HAV 3C require residues from P4 to P2′. The

pH optimum of the proteolytic activity lies between pH 7.0 and 8.5.

The substrate sequences of the 3C cleavage sites in the HAV polyprotein were initially predicted from sequence alignments of the HAV genome (Cohen *et al.*, 1987). The location of several cleavage sites has subsequently been confirmed or corrected experimentally (Gauss-Müller *et al.*, 1991; Jia *et al.*, 1991, 1993; Harmon *et al.*, 1992; Tesar *et al.*, 1994; Martin *et al.*, 1995; Schultheiss *et al.*, 1995). Table 240.1 shows the amino acid sequence of seven HAV 3C cleavage sites in the polyprotein which have been experimentally confirmed.

Jewell *et al.* (1992) and Petithory *et al.* (1991) determined the sequence preferences of HAV 3C cleavages on peptide substrates. HAV 3C requires a Gln in the P1 position of a substrate and has a strong preference for a hydrophobic residue in P4. In addition, all the natural cleavage sites in the viral polyprotein have Ser or Thr in P2 (Table 240.1). HAV 3C is inhibited by typical cysteine proteinase inhibitors such as *N*-methylmaleimide and iodoacetamide. Malcolm *et al.* (1995) determined quantitatively the inhibition of HAV 3C with peptide substrate-derived aldehyde inhibitors. Some inhibitors of chymotrypsin-like serine proteinases such as Tos-Phe-CH₂Cl are also effective against the enzyme (Malcolm *et al.*, 1992).

*Table 240.1*  Experimentally confirmed substrate sequences of 3C cleavage sites in the HAV polyprotein

| Cleavage site (viral gene products) | Polyprotein sequence | | | | | + | | | |
|---|---|---|---|---|---|---|---|---|---|
| | P5 | P4 | P3 | P2 | P1 | | P1′ | P2′ | P3′ |
| 1B+1C (VP2+VP3) | Pro | Leu | Ser | Thr | Gln | | Met | Met | Arg |
| 1C+1D (VP3+VP1) | Asp | Val | Thr | Thr | Gln | | Val | Gly | Asp |
| 2A+2B | Gly | Leu | Phe | Ser | Gln | | Ala | Lys | Ile |
| 2B+2C | Glu | Leu | Arg | Thr | Gln | | Ser | Phe | Ser |
| P2+P3 (2C+3A) | Glu | Leu | Trp | Ser | Gln | | Gly | Ile | Ser |
| 3B+3C | Pro | Val | Glu | Ser | Gln | | Ser | Thr | Leu |
| 3C+3D | Lys | Ile | Glu | Ser | Gln | | Arg | Ile | Met |

## Structural Chemistry

HAV 3C is a single subunit of 219 residues with an $M_r$ of 24 093. The purified enzyme from a bacterial expression system contains only the first 217 residues (Malcolm *et al.*, 1992). The first crystal structure obtained for a picornain was the structure of an inactive mutant of the HAV 3C enzyme (Allaire *et al.*, 1994). The refined crystal structure of a proteolytically competent HAV 3C was recently completed (Bergmann *et al.*, 1997). The overall fold and domain structure of the HAV 3C picornain resembles that of the chymotrypsin-like serine proteinases: two antiparallel $\beta$-barrel domains with the active site in between (Figure 240.1).

Unique features of the HAV 3C picornain are the N- and C-terminal helices which pack against the opposite domain, and a long antiparallel $\beta$ ribbon which extends from the $\beta$ barrel of the C-terminal domain and forms part of the active site. Cys172 and His44 form a catalytic dyad in the active site. In the refined crystal structure of the active enzyme, an ordered water molecule takes up the place of the carboxylate of a third member of a typical catalytic triad. It has been suggested that a charged form of the side chain of Tyr143 stabilizes this arrangement and is important in catalysis (Bergmann *et al.*, 1997). His191 lies at the bottom of the S1 specificity pocket and is in a position to interact with the oxygen of the carbonylamide of a glutamine residue in the P1 position of a substrate. There is no partner for an interaction with the side chain -NH$_2$ of the P1 glutamine substrate in the crystal structure. This is in agreement with the finding that a peptide substrate with $N,N$-dimethylglutamine in the P1 position is equally well hydrolyzed (Malcolm, 1995). Bergmann *et al.* (1997) suggest that the interaction of His191 with a buried, and presumably uncharged, Glu132 via buried solvent is

responsible for the specific distinction between glutamine and glutamate in the P1 position of a substrate (Figure 240.1).

## Preparation

HAV 3C picornain has been expressed in bacteria (Malcolm *et al.*, 1992; Harmon *et al.*, 1992; Kusov *et al.*, 1992), cell-free transcription-translation systems (Jia *et al.*, 1991; Schultheiss *et al.*, 1994) and eukaryotic cells (Tesar *et al.*, 1994; Martin *et al.*, 1995). For kinetic and structural studies the enzyme has been purified from a bacterial overexpression system as described by Malcolm *et al.* (1992).

## Biological Aspects

Due to the protracted replication cycle of HAV and the low yield of viral gene products in cell culture, the study of all aspects of viral replication, including polyprotein processing, is difficult in HAV (Lemon & Robertson, 1993). Furthermore, HAV infection does not result in inhibition of host cell protein synthesis. Therefore, investigations of HAV polyprotein processing had to rely on expression of cDNA constructs of the HAV genome in *in vitro* and *in vivo* systems. Interpretation of these results is further complicated by the appearance of aberrant initiation products (Jia *et al.*, 1993). Schultheiss *et al.* (1995) found a differential dependence on enzyme concentration for the 3C-mediated cleavages of the HAV polyprotein. How well this reflects the sequence of events during viral polyprotein processing *in vivo* is not clear. Nevertheless, it is becoming increasingly obvious that HAV is distinct from the other families of picornaviruses in its polyprotein processing (Tesar *et al.*, 1994; Martin *et al.*,

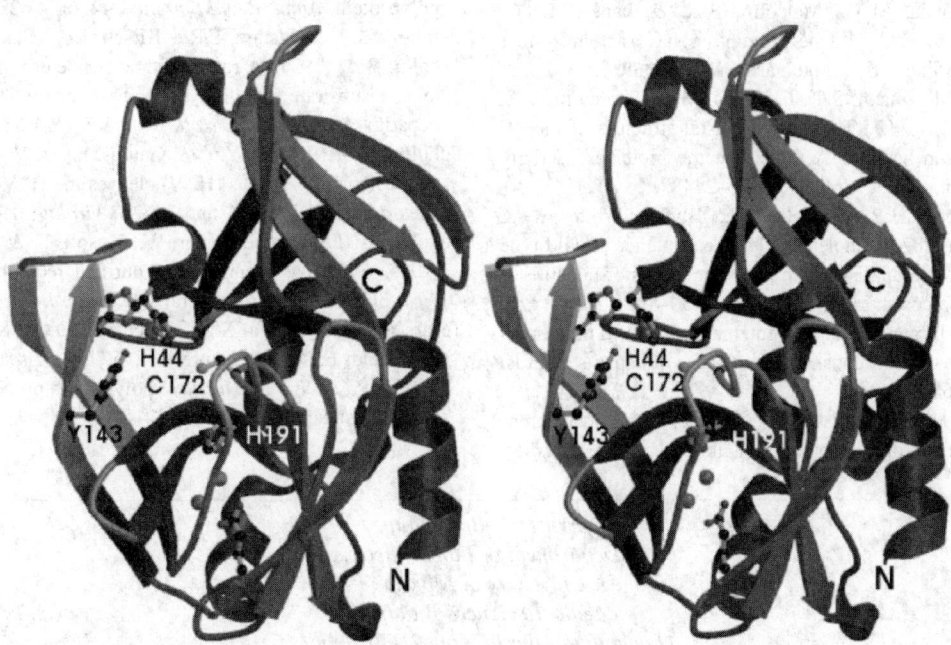

*Figure 240.1*   Stereo view of a ribbon secondary structure representation of the crystal structure of the picornain 3C from hepatitis A virus. Helices are represented by ribbons and $\beta$ strands by arrows. The antiparallel $\beta$ ribbon that extends from the $\beta$ barrel of the C-terminal domain is shown in light gray. Also included in a ball-and-stick representation are the side chains of Cys172, His44, Tyr143 and His191. Included but not labeled are the side chains of Glu132 (bottom) and His145 (top). Water molecules that are hydrogen bonded to His44, His191 and Glu132 are represented by spheres.

1995; Schultheiss *et al*., 1995). Most importantly, 3C appears to be the only proteolytic activity in HAV (Schultheiss *et al*., 1995). Whether any of the precursors of 3C (such as 3ABC or 3CD) have a function as a distinct proteolytic activity in the HAV polyprotein processing is presently unknown. At the level of the crystal structure, HAV 3C is clearly distinct from all other proteinases with a chymotrypsin-like fold, even the picornain 3Cs of other picornaviruses. As is common for the gene products of small RNA viruses, 3C has functions besides its proteolytic activity. Residues that are conserved throughout the 3Cs of all picornaviruses form a defined RNA-binding site on the surface of the molecule opposite from the proteolytic active site (Bergmann *et al*., 1997). In other picornaviruses (see Chapter 239) the RNA recognition activity of 3C has been implicated in the initiation, localization and regulation of viral RNA replication, but this has not been investigated in HAV.

## Distinguishing Features

The picornain 3C from HAV is larger than the 3C gene products from other picornaviruses (219 residues versus 183 in rhino- or poliovirus) (Palmenberg, 1990). Antisera specific for HAV 3C have been described (Gauss-Müller *et al*., 1991; Schultheiss *et al*., 1994; Martin *et al*., 1995).

## Acknowledgements

Critical comments and advice from Michael James, Bruce Malcolm, Doug Scraba and Katherine Bateman are gratefully acknowledged.

## References

Allaire, M., Chernaia, M.M., Malcolm, B.A. & James, M.N.G. (1994) Picornaviral 3C cysteine proteinases have a fold similar to chymotrypsin-like serine proteinases. *Nature* **369**, 72–76.

Bergmann, E.M., Mosimann, S.C., Chernaia, M.M., Malcolm, B.A. & James, M.N.G. (1997) The refined crystal structure of the 3C gene product from hepatitis A virus: specific proteinase activity and RNA recognition. *J. Virol.* **71**, 2436–2448.

Cohen, J.I., Ticehurst, J.R., Purcell, R.H., Buckler-White, A. & Baroudy, B.M. (1987) Complete nucleotide sequence of wild-type hepatitis A virus: comparison with different strains of hepatitis A virus and other picornaviruses. *J. Virol.* **61**, 50–59.

Gauss-Müller, V., Jürgensen, D. & Deutzmann, R. (1991) Autoproteolytic cleavage of recombinant 3C proteinase of hepatitis A virus. *Virology* **182**, 861–864.

Harmon, S.A., Updike, W., Xi-Ju, J., Summers, D.F. & Ehrenfeld, E. (1992) Polyprotein processing in *cis* and in *trans* by

hepatitis A virus 3C protease cloned and expressed in *E. coli*. *J. Virol.* **66**, 5242–5247.

Jewell, D.A., Swietnicki, W., Dunn, B.M. & Malcolm, B.A. (1992) Hepatitis A virus 3C proteinase substrate specificity. *Biochemistry* **31**, 7862–7869.

Jia, X.-Y., Ehrenfeld, E. & Summers, D.F. (1991) Proteolytic activity of hepatitis A virus 3C protein. *J. Virol.* **65**, 2595–2600.

Jia, X.-Y., Summers, D.F. & Ehrenfeld, E. (1993) Primary cleavage of the HAV capsid precursor in the middle of the proposed 2A coding region. *Virology* **193**, 515–519.

Kusov, Y.Y., Sommergruber, W., Schreiber, M. & Gauss-Müller, V. (1992) Intermolecular cleavage of hepatitis A virus precursor protein P1–P2 by recombinant HAV proteinase 3C. *J. Virol.* **66**, 6794–6796.

Lemon, S.M. & Robertson, G.H. (1993) Current perspectives in the virology and molecular biology of hepatitis A virus. *Semin. Virol.* **4**, 285–295.

Malcolm, B.A. (1995) The picornaviral 3C proteinases: cysteine nucleophiles in serine proteinase folds. *Protein Sci.* **4**, 1439–1445.

Malcolm, B.A., Chin, S.M., Jewell, D.A., Stratton-Thomas, J.R., Thudium, K.B. Ralston, R. & Rosenberg, S. (1992) Expression and characterization of recombinant hepatitis A virus 3C proteinase. *Biochemistry* **31**, 3358–3363.

Malcolm, B.A., Lowe, C., Shechosky, S., McKay, R.T., Yang, C.C., Shah, V.S., Simon, R.J., Vederas, J.C. & Santi, D.V. (1995) Peptide aldehyde inhibitors of hepatitis A virus 3C proteinase. *Biochemistry* **34**, 8172–8178.

Martin, A., Escriou, N., Chao, S.-F., Girard, M., Lemon, S.M. & Wychoswski, C. (1995) Identification and site-directed mutagenesis of the primary (2A/2B) cleavage site of the hepatitis A virus polyprotein: functional impact on the infectivity of HAV RNA transcripts. *Virology* **213**, 213–222.

Palmenberg, A.C. (1990) Proteolytic processing of picornaviral polyprotein. *Annu. Rev. Microbiol.* **44**, 603–623.

Petithory, J.R., Masiarz, F.R., Kirsch, J.F., Santi, D.V. & Malcolm, B.A. (1991) A rapid method for determination of endoproteinase substrate specificity: specificity of the 3C proteinase from hepatitis A virus. *Proc. Natl Acad. Sci. USA* **88**, 11510–11514.

Schultheiss, T., Kusov, Y.Y. & Gauss-Müller, V. (1994) Proteinase 3C of hepatitis A virus (HAV) cleaves the HAV polyprotein at all sites including VP1/2A and 2A/2B. *Virology* **198**, 275–281.

Schultheiss, T., Sommergruber, W., Kusov, Y. & Gauss-Müller, V. (1995) Cleavage specificity of purified recombinant hepatitis A virus 3C proteinase on natural substrates. *J. Virol.* **69**, 1727–1733.

Tesar, M., Pak, I., Jia, X.-Y., Richards, O.C., Summers, D.F. & Ehrenfeld, E. (1994) Expression of hepatitis A virus precursor protein P3 *in vivo* and *in vitro*: polyprotein processing of the 3CD cleavage site. *Virology* **198**, 524–533.

*Ernst M. Bergmann*
*Department of Biochemistry,*
*University of Alberta,*
*Medical Sciences Building,*
*Edmonton, Alberta, T6G 2H7 Canada*
*Email: berg@prometheus.biochem.ualberta.ca*

# 241. Picornain 2A

## Databanks

*Peptidase classification: clan CB, family C3, MEROPS ID: C03.020*
*NC-IUBMB enzyme classification: EC 3.4.22.29*
*Databank codes:*

| Species | SwissProt | PIR | EMBL (cDNA) | EMBL (genomic) |
|---|---|---|---|---|
| Cattle enterovirus | P12915 | – | – | D00214: complete genome |
| Coxsackievirus a9 | P21404 | JQ0523 | – | D00627: complete genome |
| Coxsackievirus a21 | P22055 | A33373 | – | D00538: complete genome |
| Coxsackievirus a23 | P08490 | – | X67706 | M12167: proteinase region |
| Coxsackievirus a24 | P36290 | A45848 | – | D90457: complete genome |
| Coxsackievirus b1 | P08291 | A26353 | – | M16560: complete genome |
| Coxsackievirus b3 | P03313 | A26354 | – | M16572: complete genome |
| | | A34664 | | M33854: complete genome |
| Coxsackievirus b4 | P08292 | A27120 | – | X05690: complete genome |
| Coxsackievirus b5 | Q03053 | JQ2021 | – | X67706: complete genome |
| Echovirus 11 | P29813 | A36642 | D10582 | – |
| Human enterovirus 70 | P32537 | A36253 | D00820 | – |
| Human rhinovirus 1b | P12916 | A28699 | D00239 | D00239: complete genome |
| Human rhinovirus 2 | P04936 | A03902 | X02316 | X02316: complete genome |
| Human rhinovirus 14 | P03303 | A03901 | M12168 | K02121: complete genome |
| | | | M73790 | L05355: polyprotein gene |
| | | | | X01087: complete genome |
| Human rhinovirus 89 | P07210 | A29862 | – | A10937: complete genome |
| | | | | M16248: complete genome |
| Pig vesicular disease virus | P13900 | – | – | X54521: complete genome |
| Pig vesicular disease virus | P16604 | A31331 | D00435 | – |
| | | S11670 | | |
| Poliovirus type 1 | P03299 | A93258 | – | V01148: complete genome |
| Poliovirus type 1 | P03300 | A03898 | – | J02281: complete genome |
| Poliovirus type 1 | P03301 | A03899 | – | V01150: complete genome |
| Poliovirus type 2 | P06210 | A29507 | – | M12197: complete genome |
| Poliovirus type 2 | P23069 | A34032 | – | D00625: complete genome |
| Poliovirus type 3 | P03302 | A93987 | – | K01392: complete genome |
| Poliovirus type 3 | P06209 | A27245 | – | X04468: complete genome |

Note: Positions of picornain 2A in the viral genomes are 850–1000 (coxsackievirus b4), 851–992 (human rhinovirus 2) and 881–1029 (poliovirus type 1).

## Name and History

Proteolytic activity for the 2A protein of polio virus 1 (PV1) was first postulated after identification of sequence motifs also found in the 3C protein (Blinov *et al.*, 1985). Toyoda *et al.* (1986) used a T7 bacterial expression system in *Escherichia coli* to show proteolytic processing at the VP1/2A junction; processing was inhibited by the disruption of the 2A coding sequence. Faithful processing was also seen on *in vitro* translation of a subgenomic mRNA encoding the VP1/2A/2B region and could be inhibited with an antibody against the 2A polypeptide. This processing reaction is thought to occur intramolecularly as cleavage of the viral polyprotein occurs before translation of all proteins on the polyprotein has been completed. Nicklin *et al.* (1987) showed that PV1 picornain 2A was also capable of cleaving intermolecularly by translating subgenomic mRNAs encoding the VP1/2A cleavage site and adding exogenous picornain 2A. Subsequently, the proteinase was shown to be responsible for inducing the cleavage of the eukaryotic initiation factor eIF-4G (formerly known as p220 or eIF-4$\gamma$) (Kräusslich *et al.*, 1987), a protein known to be cleaved during the replication of entero- and rhinoviruses (Etchison *et al.*, 1982).

*Picornain 2A* of human rhinovirus 2 (HRV2) was shown to possess proteolytic activity by examining processing of the VP1/2A/2B region expressed *E. coli* (Sommergruber *et al.*, 1989).

## Activity and Specificity

Picornains 2A are thought to act in both an intramolecular (*cis*) and an intermolecular (*trans*) fashion. To measure the intramolecular reaction, truncated forms of the picornaviral cDNAs encoding the VP1/2A/2B region have been expressed in bacteria or *in vitro* in rabbit reticulocytes using *in vitro* transcribed mRNAs (Toyoda *et al.*, 1986; Sommergruber *et al.*, 1989; Hellen *et al.*, 1992). Intermolecular proteolysis has been examined by producing polypeptides encoding the VP1/2A cleavage site either in bacteria or in rabbit reticulocyte lysates. Cleavage is detected by the appearance of processed products on SDS-PAGE (Nicklin *et al.*, 1987). Intermolecular activity has also been assayed by using the polypeptide substrate eIF-4G present in HeLa cell cytoplasmic extracts (Hellen *et al.*, 1991; Yu & Lloyd, 1991). The conversion of material migrating at 220 kDa into species migrating between 130 kDa and 100 kDa is monitored by immunoblotting with an anti-eIF-4G antiserum following separation of the HeLa cell proteins by SDS-PAGE.

Synthetic oligopeptides corresponding to the amino acid sequence of the cognate VP1/2A junction (usually P8–P8′ or P8–P7′) have also been used extensively to measure intermolecular cleavage by picornains 2A. Substrate is separated from cleaved product by HPLC (Sommergruber *et al.*, 1992). Alternatively, the oligopeptide substrate can be labeled with fluorescein; substrate and product can be separated by SDS-PAGE and visualized by UV light (Alvey *et al.*, 1991). PV1 picornain 2A has also been assayed using its esterase activity on the peptide Gly-Leu-Gly-Gln-Met┼OCH₃ to generate Gly-Leu-Gly-Gln-Met-COOH. Separation of processed from unprocessed material was by thin-layer chromatography on silica gel 60 $F_{254}$ plates with detection by ninhydrin (König & Rosenwirth, 1988).

Like picornains 3C (Chapter 239), picornains 2A are inhibited by *N*-ethylmaleimide, iodoacetamide and $Hg^{2+}$ ions, but are distinctively not inhibited by E-64. Certain serine proteinase inhibitors such as chymostatin and elastatinal and derivatives thereof (König & Rosenwirth, 1988; Sommergruber *et al.*, 1992; Molla *et al.*, 1993) are also effective against picornains 2A.

The cleavage sites recognized by the rhinovirus and enterovirus picornains 2A are characterized by a P1′ Gly and a preference for P4 Ile, Leu or Val, P2 Thr or Asn, and P2′ Pro or Phe. P1 is variable: Ala, Thr, Phe, Tyr and Val have been noted. The stringent requirements at P4, P2, P1′ and P2′ in determining cleavage efficiency have been demonstrated experimentally (Skern *et al.*, 1991; Hellen *et al.*, 1992; Sommergruber *et al.*, 1992, 1994a).

The cleavage site for human rhinovirus 2 (HRV2) and coxsackievirus B4 (CVB4) picornains 2A on eIF-4G has been determined to be Gly-Arg-Thr-Thr-Leu-Ser-Thr-Arg486┼Gly-Pro-Pro-Arg-Gly-Gly-Pro-Gly (Lamphear *et al.*, 1993). Occupancy of the vital P4, P2, P1′ and P2′ sites is optimal.

## Structural Chemistry

Picornains 2A are single-chain proteins that use a cysteine residue as the active-site nucleophile. However, they show some sequence identity to serine proteinases and have been suggested to possess a similar fold to that of small bacterial proteinases such as α-lytic proteinase (Bazan & Fletterick, 1988) (Chapter 76).

The picornains of PV1, HRV2 and CVB4 contain 149, 142 and 150 amino acids, respectively. The pI values are between 5 and 6. In HRV2 picornain 2A, the mutation Phe130Tyr gives a thermolabile phenotype (Luderer-Gmach *et al.*, 1996). HRV2 picornain 2A has been shown to contain one $Zn^{2+}$ ion per molecule of enzyme (Sommergruber *et al.*, 1994b), probably coordinated by Cys52, Cys54, Cys112 and His114. The metal ion appears to be required for maintenance of structure, rather than enzymatic activity, as the activity is insensitive to the presence of chelating agents such as EDTA (Sommergruber *et al.*, 1992).

## Preparation

PV1 picornain 2A has been purified from infected HeLa cells (König & Rosenwirth, 1988), but yields were low. The recombinant enzyme has been expressed in an insoluble form in *E. coli* and purified after solubilization (Martinez-Abarca *et al.*, 1993).

Recombinant picornains 2A from HRV2 and CVB4 have been expressed in soluble form in *E. coli* on a milligram scale and purified to homogeneity (Liebig *et al.*, 1993).

## Biological Aspects

Picornain 2A plays at least two roles in the replicative cycle of enteroviruses and rhinoviruses. It carries out the first processing event on the viral polyprotein, cleaving between the C-terminus of VP1 and its N-terminus. In PV1, a second cleavage by picornain 2A on the protein 3CD, generating the products 3C′ and 3D′, has also been documented. However, mutant PVs in which this cleavage site has been changed by mutation appear to replicate normally, indicating that the cleavage is not essential (Lee & Wimmer, 1988).

Enterovirus and rhinovirus picornains 2A are also responsible for the cleavage of eIF-4G, a cellular protein involved in the initiation of protein synthesis. Both HRV2 and CVB4 picornain 2As have been shown to cleave eIF-4G directly (Liebig *et al.*, 1993; Lamphear *et al.*, 1993). The cleavage site for PV1 picornain 2A on eIF-4G has not yet been determined. Furthermore, it has been reported that this cleavage is mediated by a host proteinase which is activated by PV1 picornain 2A cleavage (Wyckoff *et al.*, 1992), but this proteinase has not yet been identified.

## Distinguishing Features

Enterovirus and rhinovirus picornains 2A can be distinguished from picornains 3C by their smaller size and different cleavage specificities. No identity is found between enterovirus and rhinovirus picornain 2A and foot-and-mouth disease virus or encephalomyocarditis virus 2A proteins.

Recombinant PV1 in which the coding sequence of the picornain 2A has been replaced by that of HRV2 is not viable. Replacement by the coxsackievirus B4 coding sequence leads to small plaque mutants (Lu *et al.*, 1995). In both cases, *in vitro* translation and processing was not affected, suggesting a third, as yet undefined, role for picornain 2A in viral replication.

# References

Alvey, J.C., Wyckoff, E.E., Yu, S.F., Lloyd, R.E & Ehrenfeld, E. (1991) *Cis*- and *trans*-cleavage activities of poliovirus 2A protease expressed in *Escherichia coli*. *J. Virol.* **65**, 6077–6083.

Bazan, J.F. & Fletterick, R.J. (1988) Viral cysteine proteinases are homologous to the trypsin-like family of serine proteinases: structural and functional implications. *Proc. Natl Acad. Sci. USA* **85**, 7872–7876.

Blinov, V.M., Donchenke, A.P. & Gorbalenya, A.E. (1985) Vnutrenaja gamalogia v pervitschnoi strukture polyproteina poliovirusa wosmoshnostj cuschtschestvovania dwuch virusnich proteinas. *Proc. Acad. Sci. USSR* **281**, 984–987.

Etchison, D., Milburn, S.C., Edery, I., Sonenberg, N. & Hershey, J.W. (1982) Inhibition of HeLa cell protein synthesis following poliovirus infection correlates with the proteolysis of a 220,000-dalton polypeptide associated with eucaryotic initiation factor 3 and a cap binding protein complex. *J. Biol. Chem.* **257**, 14806–14810.

Hellen, C.U.T., Fäcke, M., Kräusslich, H., Lee, C. & Wimmer, E. (1991) Characterization of poliovirus 2A proteinase by mutational analysis: residues required for autocatalytic activity are essential for induction of cleavage of eukaryotic initiation factor 4F polypeptide p220. *J. Virol.* **65**, 4226–4231.

Hellen, C.U.T., Lee, C.K. & Wimmer, E. (1992) Determinants of substrate recognition by poliovirus 2A proteinase. *J. Virol.* **66**, 3330–3338.

König, H. & Rosenwirth, B. (1988) Purification and partial characterisation of poliovirus protease 2A by means of a functional assay. *J. Virol.* **62**, 1243–1250.

Kräusslich, H.G., Nicklin, M.J., Toyoda, H., Etchison, D. & Wimmer, E. (1987) Poliovirus proteinase 2A induces cleavage of eucaryotic initiation factor 4F polypeptide p220. *J. Virol.* **61**, 2711–2718.

Lamphear, B., Yang, F., Waters, D., Yan, R., Liebig, H.-D., Klump, H., Kuechler, E., Skern, T. & Rhoads, R. (1993) Mapping the cleavage site in protein synthesis initiation factor eIF-4 of the 2A proteases from human coxsackie- and rhinoviruses. *J. Biol. Chem.* **268**, 19200–19203.

Lee, C.K. & Wimmer, E. (1988) Proteolytic processing of poliovirus polyprotein: elimination of 2Apro-mediated, alternative cleavage of polypeptide 3CD by in vitro mutagenesis. *Virology* **166**, 405–414.

Liebig, H.-D., Ziegler, E., Yan, R., Hartmuth, K., Klump, H., Kowalski, H., Blaas, D., Sommergruber, W., Lamphear, B., Rhoads, R., Kuechler, E. & Skern, T. (1993) Purification of two picornaviral proteinases: interaction with eIF-4γ and influence on *in vitro* translation. *Biochemistry* **32**, 7581–7588.

Lu, H.-H., Cuconati, A. & Wimmer, E. (1995) Analysis of picornavirus 2Apro proteins: separation of proteinase from translation and replication functions. *J. Virol.* **69**, 7445–7452.

Luderer-Gmach, M., Liebig, H.-D., Sommergruber, W., Voss, T., Fessl, F., Skern, T. & Kuechler, E. (1996) Human rhinovirus 2A proteinase mutant and its second-site revertants. *Biochem. J.* **318**, 213–218.

Martinez-Abarca, F., Alonso, M.A. & Carrasco, L. (1993) High level expression in *Escherichia coli* and purification of poliovirus protein 2Apro. *J. Gen. Virol.* **74**, 2645–2652.

Molla, A., Hellen, C.U.T. & Wimmer, E. (1993) Inhibition of proteolytic activity of poliovirus and rhinovirus 2A proteinases by elastase-specific inhibitors. *J. Virol.* **67**, 4688–4695.

Nicklin, M.J.H., Kräusslich, H.G., Toyoda, H., Dunn, J.J. & Wimmer, E. (1987) Poliovirus polypeptide precursors. Expression in vitro and processing by 3C and 2A proteinases. *Proc. Natl Acad. Sci. USA* **84**, 4002–4006.

Skern, T., Sommergruber, W., Auer, H., Volkmann, P., Zorn, M., Liebig, H.-D., Fessl, F., Blaas, D. & Kuechler, E. (1991) Substrate requirements of a human rhinoviral 2A proteinase. *Virology* **181**, 46–54.

Sommergruber, W., Zorn, M., Blaas, D., Fessl, F., Volkmann, P., Maurer-Fogy, I., Pallai, P., Merluzzi, V., Matteo, M., Skern, T. & Kuechler, E. (1989) Polypeptide 2A of human rhinovirus type 2: identification as a proteinase and characterisation by mutational analysis. *Virology* **169**, 68–77.

Sommergruber, W., Ahorn, H., Zöphel, A., Maurer-Fogy, I., Fessl, F., Schnorrenberg, G., Liebig, H.-D., Blaas, D., Kuechler, E. & Skern, T. (1992). Cleavage specificity on synthetic peptide substrates of human rhinovirus 2 proteinase 2A. *J. Biol. Chem.* **267**, 22639–22644.

Sommergruber, W., Ahorn, H., Klump, H., Seipelt, J., Zoephel, A., Fessl, F., Krystek, C., Blaas, D., Kuechler, E., Liebig, H.-D. & Skern, T. (1994a) 2A proteinases of coxsackie- and rhinovirus cleave peptides derived from eIF-4γ via a common recognition motif. *Virology* **198**, 741–745.

Sommergruber, W., Casari, G., Fessl, F., Seipelt, J. & Skern, T. (1994b) The 2A proteinase of human rhinovirus is a zinc containing enzyme. *Virology* **204**, 815–818.

Toyoda, H., Nicklin, M.J.H., Murray, M.G., Anderson, C.W., Dunn, J.J., Studier, F.W. & Wimmer, E. (1986) A second virus-encoded proteinase involved in proteolytic processing of poliovirus polyprotein. *Cell* **45**, 761–770.

Wyckoff, E., Lloyd, R.E. & Ehrenfeld, E. (1992) Relationship of eukaryotic initiation factor 3 to poliovirus-induced p220 cleavage activity. *J. Virol.* **66**, 2943–2951.

Yu, S. & Lloyd, R.E. (1991) Identification of essential amino acid residues in the functional activity of poliovirus 2A proteinase. *Virology* **182**, 615–625.

*Tim Skern*
*Institute of Biochemistry, Medical Faculty,*
*University of Vienna,*
*Dr. Bohr-Gasse 9/3,*
*A-1030 Vienna, Austria*
*Email: timothy.skern@univie.ac.at*

# 242. Cowpea mosaic virus 24 kDa proteinase

## Databanks

*Peptidase classification: clan CB, family C3, MEROPS ID: C03.003*
*NC-IUBMB enzyme classification: none*
*Databank codes:*

| Species | SwissProt | PIR | EMBL (cDNA) | EMBL (genomic) |
|---|---|---|---|---|
| Cowpea mosaic virus | P03600 | A04211 | – | X00206: polyprotein gene |
| Cowpea severe mosaic virus | P36312 | A44214 | – | M83830: polyprotein gene |
| Red clover mottle virus | P35930 | JQ0572 | D00657 | – |
|  |  | JQ1657 |  |  |

## Name and History

Cowpea mosaic virus (CPMV) is the type member of the comovirus group of bipartite, positive-stranded RNA plant viruses. The two genome segments (RNA 1 and RNA 2) are separately encapsidated in capsids termed bottom (B) and middle (M) component, respectively. In earlier literature, RNA 1 is often referred to as bottom component RNA (B-RNA) and RNA 2 as middle component RNA (M-RNA). Both genome segments are expressed through the synthesis and subsequent cleavage of large precursor polyproteins (Figure 242.1). The activity of the **cowpea mosaic virus 24 kDa proteinase** was first reported as a result of the *in vitro* translation studies of Pelham (1979) which showed that the enzyme is part of the RNA 1-encoded polyprotein. The term 24 kDa proteinase refers to the size of the mature enzyme, and the existence of a protein of this size was first predicted by Rezelman *et al*. (1980). The first demonstration that the 24 kDa proteinase is the enzyme responsible for cleaving the CPMV-encoded polyproteins was reported by Franssen *et al*. (1984), though in this publication the protein is referred to as the **28K polypeptide**. Subsequent studies (Wellink *et al*., 1986; Verver *et al*., 1987) defined the precise sequence of the enzyme. It consists of 208 amino acids and is encoded by nucleotides 3038–3671 of RNA 1 (Figure 242.1).

## Activity and Specificity

The 24 kDa protein is active both in *cis* and in *trans*, though at least one of the *trans* cleavages also requires the presence of the RNA 1-encoded 32 kDa protein (Vos *et al*., 1988). The enzyme cleaves the CPMV polyproteins at six sites, four in the polyprotein from RNA 1 and two in that encoded by RNA 2. All sites have a Gln residue at the P1 position and either a Gly, Ser or Met at P1′. A Pro or Ala is present at the P2 position and an Ala at the P4 position in five out of the six cleavage sites. The 24 kDa proteinase from CPMV is unable to cleave the polyproteins of other comoviruses either in *trans* (Gabriel *et al*., 1982; Goldbach & Krijt, 1982) or in *cis* (Shanks *et al*., 1996). In addition, the rate of cleavage of the various sites in the CPMV polyproteins is controlled by sequences upstream of the proteinase sequence (Dessens & Lomonossoff, 1992).

The enzyme is inhibited *in vitro* by $Zn^{2+}$ ions and *N*-ethylmaleimide, but unaffected by the typical inhibitors of serine peptidases (e.g. DFP, PMSF) (Pelham, 1979). The inhibition by $Zn^{2+}$ ions has also been observed *in vivo* (Wellink *et al*., 1987a). The pattern of inhibition led Pelham (1979) to conclude that the enzyme has an essential thiol group. The activity of the 24 kDa proteinase in reticulocyte lysates is stimulated by DTT (Pelham, 1979; Peng & Shih, 1984), but is inhibited by hemin (Bu & Shih, 1989). Both wheatgerm

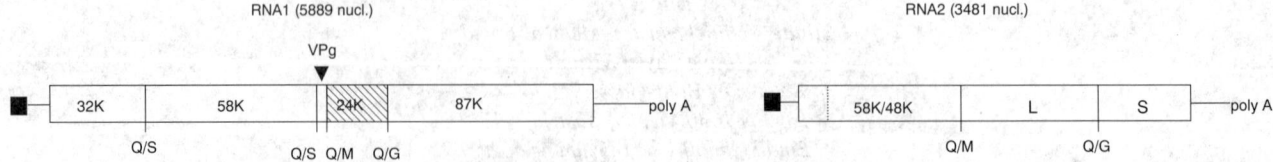

*Figure 242.1*   Genetic organization of the CPMV genome. The open reading frames on the viral RNAs are indicated by the open boxes, with the region that encodes the 24 kDa protein being shaded. The cleavage sites on the polyproteins are indicated: Q/S is Gln+Ser, Q/M is Gln+Met and Q/G is Gln+Gly. The positions of the various virus products are indicated by their apparent molecular masses (32K, 58K etc.) except for the two viral coat proteins which are indicated by L (large) and S (small). The overlapping 58K/48K proteins derived from RNA 2 are formed as a result of two initiation events. The black box at the end of each viral RNA represents the small genome-linked protein (VPg).

and cowpea embryo extracts contain a high molecular mass inhibitor of the enzyme (Shih *et al.*, 1987).

## Structural Chemistry

The 24 kDa proteinase domain is embedded in the RNA 1-encoded polyprotein and is present in a number of cleavage products formed during the maturation of this protein (Lomonossoff, 1994). Although the free 24 kDa proteinase can be observed both *in vitro* (Franssen *et al.*, 1984) and *in vivo* (Wellink *et al.*, 1987b), it is unclear whether the free polypeptide has a role in the processing pathway. Sequences upstream of the protease domain have been shown to modulate its activity (Dessens & Lomonossoff, 1992; Peters *et al.*, 1992a,b, 1995) and it is possible that the enzyme functions entirely as part of a larger protein.

Although inhibitor studies indicated that the CPMV 24 kDa proteinase is a cysteine proteinase, two groups (Bazan & Fletterick, 1988; Gorbalenya *et al.*, 1989) independently developed structure profiles indicating that the protein is related to the chymotrypsin (S1) family of serine proteinases. The structure is predicted to be composed of two six-stranded β-barrel domains, although no crystallographic data are yet available to confirm this. While Bazan & Fletterick (1988) and Gorbalenya *et al.* (1989) both assigned Cys166 as the active-site cysteine, they did not agree on the identity of the other two residues of the proposed catalytic triad. Amino acid sequence comparisons (Shanks & Lomonossoff, 1990) and site-directed mutagenesis experiments (Dessens & Lomonossoff, 1991) indicated that His40 and Glu76 (with Cys166) comprise the catalytic triad, in line with the prediction of Gorbalenya *et al.* (1989). When Cys166 is replaced by Ser, the proteinase retains some activity (Dessens & Lomonossoff, 1991).

The origin of the specificity of the CPMV 24 kDa protease is not understood. Bazan & Fletterick (1988) identified seven residues as potentially being part of the substrate-binding pocket. However, all seven residues are conserved in the sequence of the homologous proteinase encoded by the related comovirus, red clover mottle virus (RCMV) (Shanks & Lomonossoff, 1990), and yet each enzyme is strictly specific for its homologous polyprotein (Goldbach & Krijt, 1982; Shanks *et al.*, 1996). The viral specificity does not lie in the sequence of the dipeptide at the cleavage site since identical dipeptides are present at the cleavage sites in both the CPMV and RCMV polyproteins. It is, therefore, probable that the specificity is determined by the three-dimensional structures of the substrates and enzymes.

## Preparation

The enzyme has not been purified, but a source of proteinase activity is readily available from CPMV-infected cowpea (*Vigna unguiculata*) plants or protoplasts (Franssen *et al.*, 1982). CPMV RNA 1, which directs the synthesis of the 24 kDa proteinase, can be isolated from purified bottom components from a virus preparation. This RNA is capable of independent replication in protoplasts (Goldbach *et al.*, 1980) and can be translated *in vitro*. In both cases, proteolytic activity of the 24 kDa proteinase can be detected.

Plasmids containing cDNA fragments from CPMV RNA 1 including the portion encoding the 24 kDa proteinase have been constucted in expression vectors. Activity of the enzyme has been observed in *Escherichia coli* (Garcia *et al.*, 1987; Richards *et al.*, 1989) and in animal cells (*Spodoptera frugiperda*) (van Bokhovan *et al.*, 1990), but no attempt has been made to purify the enzyme from these heterologous systems.

## Biological Aspects

The 24 kDa proteinase plays an essential role in the replication of CPMV; mutations which affect its activity are inevitably lethal. To date, the number of functions carried out by the 24 kDa proteinase is unclear, but it is interesting to note that the domain is present in the sole virus-encoded protein (the 110K protein) associated with replication complexes in infected cells (Dorssers *et al.*, 1984).

## Distinguishing Features

Since the enzyme is so specific and is produced only on infection of susceptible plants with CPMV it is unlikely to be confused with other proteinases. A specific antipeptide serum has been produced (Wellink *et al.*, 1987a).

## Related Peptidases

The CPMV 24 kDa proteinase is closely related to the equivalent enzymes from other comoviruses such red clover mottle virus and cowpea severe mosaic virus. More distant relationships are found to the picornains 3C of other members of the picorna-like supergroup of viruses.

## Further Reading

A comprehensive review of the properties of virus-encoded proteinases has been provided by Dougherty & Semler (1993), and the article of Goldbach & Wellink (1996) is a detailed account of the molecular biology of the comovirus group that describes the role of the 24 kDa proteinase in the viral replication cycle.

## References

Bazan, J.F. & Fletterick, R.J. (1988) Viral cysteine proteases are homologous to the trypsin-like family of serine proteases: structural and functional implications. *Proc. Natl Acad. Sci. USA* **85**, 7872–7876.

Bu, M. & Shih, D.S. (1989) Inhibition of proteolytic processing of the polyproteins of cowpea mosaic virus by hemin. *Virology* **173**, 348–351.

Dessens, J.T. & Lomonossoff, G.P. (1991) Mutational analysis of the putative catalytic triad of cowpea mosaic virus 24K protease. *Virology* **184**, 738–746.

Dessens, J.T. & Lomonossoff, G.P. (1992) Sequence upstream of the 24K protease enhances cleavage of the cowpea mosaic virus B RNA-encoded polyprotein at the junction betwen the 24K and 87K proteins. *Virology* **189**, 225–232.

Dorssers, L., van der Krol, S., van der Meer, J., van Kammen, A. & Zabel, P. (1984) Purification of cowpea mosaic virus RNA replication complex: identification of a 110,000 dalton polypeptide responsible for RNA chain elongation. *Proc. Natl Acad. Sci. USA* **81**, 1951–1955.

Dougherty, W.G. & Semler, B.L. (1993) Expression of virus-encoded proteinases: functional and structural similarities with cellular enzymes. *Microbiol. Rev.* **57**, 781–822.

Franssen, H., Goldbach, R., Broekhuijsen, M., Moerman, M. & van Kammen, A. (1982) Expression of middle component RNA of cowpea mosaic virus: *in vitro* generation of a precursor to both capsid proteins by a bottom component RNA-encoded protease from infected cells. J. Virol. **41**, 8–17.

Franssen, H., Goldbach, R. & van Kammen, A. (1984) Translation of bottom component RNA of cowpea mosaic virus in reticulocyte lysate: faithful proteolytic processing of the primary translation product. *Virus Res.* **1**, 34–49.

Gabriel, C.J., Derrick, K.S. & Shih, D.S. (1982) The synthesis and processing of the proteins of bean pod mottle virus in rabbit reticulocyte lysates. *Virology* **122**, 476–480.

Garcia, J.A., Schrijvers, L., Agnes T., Vos, P., Wellink, J. & Goldbach, R. (1987) Proteolytic activity of the cowpea mosaic virus encoded 24K protein synthesized in *Escherichia coli. Virology* **159**, 67–75.

Goldbach, R. & Krijt, J. (1982) Cowpea mosaic virus-encoded protease does not recognize primary translation products of M RNAs from other comoviruses. *J. Gen. Virol.* **43**, 1151–1154.

Goldbach, R., Rezelman, G. & van Kammen, A. (1980) Independent replication and expression of B-component RNA of cowpea mosaic virus. *Nature* **286**, 297–300.

Goldbach, R.W. & Wellink, J. (1996) Comoviruses: molecular biology and replication. In: *The Plant Viruses* (Harrison, B.D. & Murant, A.F., eds). New York: Plenum Press, pp. 35–76.

Gorbalenya, A.E., Donchenko, A.P., Blinov, V.N. & Koonin, E.V. (1989) Cysteine proteases of positive-stranded RNA viruses and chymotrypsin-like serine proteases: a distinct protein superfamily with a common structural fold. *FEBS Lett.* **243**, 103–114.

Lomonossoff, G.P. (1994) Comoviruses. In: *Encyclopedia of Virology* (Webster, R.G. & Granoff, A., eds). London: Academic Press, pp. 249–254.

Pelham, H.R.B. (1979) Synthesis and proteolytic processing of cowpea mosaic virus proteins in reticulocyte lysates. *Virology* **96**, 463–477.

Peng, X.X. & Shih, D.S. (1984) Proteoloytic processing of the proteins translated from bottom component RNA of cowpea mosaic virus. *J. Biol. Chem.* **259**, 3197–3201.

Peters, S.A., Voorhorst, W.G.B., Wery, J., Wellink, J. & van Kammen, A. (1992a) A regulatory role for the 32K protein in the proteolytic processing of cowpea mosaic virus polyproteins. *Virology* **191**, 81–89.

Peters, S.A., Voorhorst, W.G.B., Wellink, J. & van Kammen, A. (1992b) Processing of VPg-containing polyproteins encoded by the B RNA of cowpea mosaic virus. *Virology* **191**, 90–97.

Peters, S.A., Mesnard, J., Kooter, I., Verver, J., Wellink, J. & van Kammen, A. (1995) The cowpea mosaic virus RNA 1-encoded 112kDa protein may function as a VPg precursor *in vivo. J. Gen. Virol.* **76**, 1807–1813.

Rezelman, G., Goldbach, R. & van Kammen, A. (1980) Expression of bottom component RNA of cowpea mosaic virus in cowpea protoplasts. *J. Virol.* **36**, 366–373.

Richards, O.C., Eggen, R., Goldbach, R. & van Kammen, A. (1989) High level synthesis of cowpea mosaic virus RNA polymerase and protease in *Escherichia coli. Gene* **78**, 135–146.

Shanks, M. & Lomonossoff, G.P. (1990) The primary structure of the 24K proteases from red clover motttle virus: implications for the mode of action of comovirus proteases. *J. Gen. Virol.* **71**, 735–738.

Shanks, M., Dessens, J.T. & Lomonossoff, G.P. (1996) The 24 kDa proteinases of comoviruses are virus-specific in *cis* as well as in *trans. J. Gen. Virol.* **77**, 2365–2369.

Shih, D.S., Bu, M., Price, M.A. & Shih, C.T. (1987) Inhibition of cleavage of a plant polyprotein by an inhibitor activity present in wheat germ and cowpea embryos. *J. Virol.* **61**, 912–915.

van Bokhovan, H., Wellink, J., Usmany, M., Vlak, J.M., Goldbach, R. & van Kammen, A. (1990) Expression of plant virus genes in animal cells: high level synthesis of cowpea mosaic virus B-RNA-encoded proteins in baculovirus expression vectors. *J. Gen. Virol.* **71**, 2509–2517.

Verver, J., Goldbach, R., Garcia, J.A. & Vos, P. (1987) *In vitro* expression of a full-length DNA copy of cowpea mosaic virus B RNA: identification of the B RNA encoded 24 kD protein as a viral protease. *EMBO J.* **6**, 549–554.

Vos, P., Verver, J., Jaegle, M., Wellink, J. & van Kammen, A. (1988) Two viral proteins involved in the proteolytic processing of the cowpea mosaic polyproteins. *Nucleic Acids Res.* **16**, 1967–1985.

Wellink, J., Rezelman, G., Goldbach, R. & Beyreuther, K. (1986) Determination of the proteolytic processing sites in the polyprotein encoded by the bottom component RNA of cowpea mosaic virus. *J. Virol.* **59**, 50–58.

Wellink, J., Jaegle, M. & Goldbach, R. (1987a) Detection of a novel protein encoded by the bottom component RNA of cowpea mosaic virus using antibodies raised against a synthetic peptide. *J. Virol.* **61**, 236–238.

Wellink, J., Jaegle, M., Prinz, H., van Kammen, A. & Goldbach, R. (1987b) Expression of middle component RNA of cowpea mosaic virus *in vivo. J. Gen. Virol.* **68**, 2577–2585.

*G.P. Lomonossoff*
*Department of Virus Research,*
*John Innes Centre, Colney,*
*Norwich NR4 7UH, UK*
*Email: George.Lomonossoff@bbsrc.ac.uk*

*M. Shanks*
*Department of Virus Research,*
*John Innes Centre, Colney,*
*Norwich NR4 7UH, UK*

# 243. *Grapevine fanleaf nepovirus cysteine proteinase*

## Databanks

*Peptidase classification: clan CB, family C3, MEROPS ID: C03.004*
*NC-IUBMB enzyme classification: none*
*Databank codes:*

| Species | SwissProt | PIR | EMBL (cDNA) | EMBL (genomic) |
|---|---|---|---|---|
| Andean potato mottle virus | Q02941 | JQ1898 | M84806 | – |
| Grapevine fanleaf virus (GFLV) | P29149 | JQ1373 | D00915 | – |
| Hungarian grapevine chrome mosaic virus | P13025 | S06188 | X15346 | – |
| Tomato black ring virus | P18522 | JQ0009 | – | D00322: complete genome |
| Tomato black ring virus | – | – | X96485 | – |

## Name and History

Grapevine fanleaf virus (GFLV) belongs to the plant *Nepovirus* genus (i.e. nematode-transmitted) within the Comoviridae family which is part of the supergroup of picorna-like viruses (Ward, 1993). The genome of GFLV consists of two positive RNA strands, polyadenylated at their 3′ ends and covalently attached to a small viral protein (VPg, **v**iral **p**rotein **g**enome-linked). Each RNA contains a single open reading frame coding for a polyprotein. The polyproteins, P1 of 253 kDa (RNA 1) (Ritzenthaler *et al.*, 1991) and P2 of 122 kDa (RNA 2) (Serghini *et al.*, 1990) are processed into mature structural and nonstructural proteins by a virus-encoded proteinase. For GFLV, the first evidence of a proteolytic activity associated with RNA 1 was provided by *in vitro* translation of viral RNAs in rabbit reticulocyte lysate (Morris-Krsinich *et al.*, 1983). Sequence comparison of GFLV with animal picornaviruses and plant como-, nepo- and potyviruses indicated similar genomic organization (Gorbalenya *et al.*, 1989). In polyprotein P1 the order of nonstructural proteins is: putative helicase, VPg, proteinase, polymerase (Ritzenthaler *et al.*, 1991). The precise location of the VPg protein-coding sequence in RNA 1 (Pinck *et al.*, 1991) allowed the cloning of a region including the VPg, the putative proteinase cistron and several residues from the N-terminus of the polymerase. A start codon was created by polymerase chain reaction (PCR) mutagenesis and this proteinase domain was cloned in a transcription vector (pVP7). Transcripts of pVP7 produced a 37.8 kDa protein which was autocleaved to a stable 24 kDa protein upon translation in a rabbit reticulocyte lysate (Margis *et al.*, 1991).

## Activity and Specificity

Autocleavage of the 37.8 kDa recombinant protein in nuclease-treated rabbit reticulocyte lysate is maximal at pH 7.0–8.5 and at 30°C. This activity was reduced by 80% when translation was in a wheatgerm system. Inhibition of translation was also observed in rabbit reticulocyte lysate with various proteinase inhibitors: PMSF, EDTA, E-64, $Ca^{2+}$, $Zn^{2+}$ and $Co^{2+}$ (Margis *et al.*, 1991). In the proteinase domain of GFLV, a triad of highly conserved His, Asp and Cys residues is equivalent to the catalytic triad His, Asp, Ser of the chymotrypsin (S1) proteinase family. Site-directed mutagenesis of residues His43, Glu87 and Leu197 abolished proteinase activity. A similar effect is obtained by replacement of catalytic residue Cys179 by Ile while activity is conserved if substitution is by Ser (Margis & Pinck, 1992). The 37.8 kDa protein acts in *cis* for its autocleavage and in *trans* for the processing of polyprotein P2 to produce the C-terminal coat protein and a 66 kDa protein which is further processed to a 38 kDa movement protein and a 28 kDa N-terminal protein.

## Structural Chemistry

The GFLV proteinase (Pro) of 219 residues, 24 kDa, is encoded between nucleotides 3966 and 4622 of GFLV RNA 1, with cleavage at Gly╪Glu1242 and Arg╪Gly1461 (or Gly╪Glu1462 one residue downstream; numbering corresponds to P1) at the N- and C-terminus respectively (Margis & Pinck, 1992). The proteinase precursor form, constituted by the VPg-Pro protein produced upon *in vitro* translation of pVP7 transcribed RNA, is a stable protein equally active in polyprotein processing in *cis* and in *trans*. Its maturation into Pro occurs at a very low rate (Margis *et al.*, 1993).

## Preparation

The cistrons encoding the GFLV proteinase (Pro) and VPg-Pro were cloned in pBS⁺ expression vector (Stratagene) as pPro and pVPg-Pro plasmids respectively, and translated in reticulocyte lysate from transcripts produced by T7 RNA polymerase transcription (Margis *et al.*, 1994).

## Biological Aspects

The active proteinase used for polyprotein-processing studies was produced by *in vitro* translation. Direct evidence that the processing of the viral polyproteins proceeds through the same intermediate forms *in vitro* and *in vivo* has been obtained for maturation of polyprotein P2. Analysis by western immunoblot of total proteins from protoplasts transfected with GFLV RNAs revealed, after chemiluminescent detection with purified antibodies raised against the movement protein (38 kDa protein), that polyprotein P2 and the maturation intermediate forms of 94 kDa (38 kDa plus coat protein) and 66 kDa (N-terminal 28 kDa protein plus 38 kDa) detected *in vitro* were present *in vivo* (Ritzenthaler *et al.*, 1995). Similarly, analysis with antibodies raised against the proteinase showed that, during processing of polyprotein P1, the proteinase–polymerase intermediate, but not VPg–proteinase–polymerase, constitute the proteinase precursor *in vivo*. This indicates that the 24 kDa viral proteinase cleaves itself cotranslationally from polyprotein P1 and the VPg protein during translational expression of RNA 1, a conclusion supported by the lack of detection of the entire 253 kDa polyprotein P1.

## Distinguishing Features

Cleavage specificity is not restricted to the Arg↓Gly bonds that yield coat protein. Cleavage of Cys↓Ala, Cys↓Ser and Gly↓Glu dipeptides has also been identified by microsequencing of the maturation products (Margis *et al.*, 1993). The cleavage sites in GFLV polyproteins have in common Ser residues in the P4 or P5 positions. GFLV proteinase also processes the polyprotein P2 of the closely related arabis mosaic nepovirus at the Cys↓Ala site between the N-terminal protein and the movement protein, where the same residues are present, but not at the Arg↓Gly site that yields the coat protein and for which the residue in position P4 differs (Loudes *et al.*, 1995).

## Further Reading

Plant viral proteinases have been reviewed by Goldbach (1990), and the functional characterization of the proteolytic activity of the tomato black ring nepovirus RNA 1-encoded polyprotein is described by Hemmer *et al.* (1995).

## References

Goldbach, R. (1990) Plant viral proteinases. *Semin. Virol.* **1**, 335–346.

Gorbalenya, A.E., Donchenko, A.P., Blinov, V.M. & Koonin, E.V. (1989) Cysteine proteases of positive strand RNA viruses and chymotrypsin-like serine proteases. A distinct protein superfamily with a common structural fold. *FEBS Lett.* **243**, 103–114.

Hemmer, O., Greif, C., Dufourcq, P., Reinbolt, J. & Fritsch, C. (1995) Functional characterization of the proteolytic activity of the tomato black ring nepovirus RNA-1-encoded polyprotein. *Virology* **206**, 362–371.

Loudes, A.M., Ritzenthaler, C., Pinck, M., Serghini, M.A. & Pinck, L. (1995) The 119 kDa and 124 kDa polyproteins of arabis mosaic nepovirus (isolate S) are encoded by two distinct RNA2 species. *J. Gen. Virol.* **76**, 899–906.

Margis, R. & Pinck, L. (1992). Effects of site directed mutagenesis on the presumed catalytic triad and substrate binding pocket of grapevine fanleaf nepovirus 24-kDa proteinase. *Virology* **190**, 884–888.

Margis, R., Viry, M., Pinck, M. & Pinck, L. (1991) Cloning and *in vitro* characterization of the grapevine fanleaf virus proteinase cistron. *Virology* **185**, 779–787.

Margis, R., Ritzenthaler, C., Reinbolt, J., Pinck, M. & Pinck, L. (1993) Genome organization of grapevine fanleaf nepovirus RNA2 deduced from the 122 K polyprotein P2 *in vitro* cleavage products. *J. Gen. Virol.* **74**, 1919–1926.

Margis, R., Viry, M., Pinck, M., Bardonnet, N. & Pinck, L. (1994) Differential proteolytic activities of precursor and mature forms of the 24 K proteinase of grapevine fanleaf nepovirus. *Virology* **200**, 79–86.

Morris-Krsinich, B.A.M., Forster, R.L.S. & Mossop, D.W. (1983) The synthesis and processing of the nepovirus grapevine fanleaf virus proteins in rabbit reticulocyte lysate. *Virology* **130**, 523–526.

Pinck, M., Reinbolt, J., Loudes, A.M., Le Ret, M. & Pinck, L. (1991) Primary structure and location of the genome-linked protein (VPg) of grapevine fanleaf nepovirus. *FEBS Lett.* **284**, 117–119.

Ritzenthaler, C., Viry, M., Pinck, M., Margis, R., Fuchs, M. & Pinck, L. (1991) Complete nucleotide sequence and genetic organization of grapevine fanleaf nepovirus RNA1. *J. Gen. Virol.* **72**, 2357–2365.

Ritzenthaler, C., Pinck, M. & Pinck, L. (1995). Grapevine fanleaf nepovirus P38 putative movement protein is not transiently expressed and is a stable final maturation product *in vivo*. *J. Gen. Virol.* **76**, 907–915.

Serghini, M.A., Fuchs, M., Pinck, M., Reinbolt, J., Walter, B. & Pinck, L. (1990) RNA2 of grapevine fanleaf virus: sequence analysis and coat protein cistron location. *J. Gen. Virol.* **71**, 1433–1441.

Ward, C.W. (1993) Progress towards a higher taxonomy of viruses. *Res. Virol.* **144**, 419–453.

*Lothaire Pinck*
*Institut de Biologie Moléculaire des Plantes du CNRS,*
*Département de Virologie,*
*12 rue du Général Zimmer,*
*67084 Strasbourg Cedex, France*
*Email: pinck@piloth.U-strasbg.fr*

# 244. Potyvirus NIa protease

## Databanks

*Peptidase classification: clan CB, family C4, MEROPS ID: C04.001*
*NC-IUBMB enzyme classification: none*
*Databank codes:*

| Species | Abbreviation | SwissProt | PIR | EMBL (cDNA) | EMBL (genomic) |
|---|---|---|---|---|---|
| Bean yellow mosaic virus | BYMV | – | S18921 | – | X63358: 3′ end |
| Brome streak mosaic rymovirus | | – | – | Z48506 | – |
| Garlic yellow streak virus | | – | – | D11118 | – |
| Johnson grass mosaic virus | JGMV | – | – | Z26920 | – |
| *Ornithogalum* mosaic virus | | P20234 | – | – | – |
| Papaya ringspot virus | PRSV | Q01901 | JQ1899 | – | S46722: complete genome |
| | | | | | X67673: complete genome |
| Pea seed-borne mosaic virus | PSbMV | P29152 | JQ1331 | D10930 | – |
| Peanut stripe virus | PStV | – | – | – | U05771: complete genome |
| Pepper mottle virus | PepMoV | Q01500 | A44062 | M96425 | – |
| Plum pox potyvirus | PPV | P13529 | JA0078 | D00298 | X16415: complete genome |
| | | | S06929 | | |
| Plum pox potyvirus | | P17766 | JQ0003 | – | D00424: complete genome |
| | | | | | D13751: complete genome |
| Plum pox potyvirus | | P17767 | A60009 | – | M26965: NIa endopeptidase region |
| Plum pox potyvirus | | Q01681 | PQ0221 | X56258 | – |
| Potato virus Y | PVY | P18247 | JS0166 | U33454 | D00441: complete genome |
| | | | | | X12456: complete genome |
| Potato virus Y | | Q02963 | JN0545 | M95491 | – |
| Soya bean mosaic virus | SbMV | – | – | D00717 | – |
| Sweet potato feathery mottle virus | | – | – | D38543 | – |
| Tobacco etch virus | TEV | P04517 | A04207 | L38714 | M11458: complete genome |
| | | | | M11216 | M15239: complete genome |
| Tobacco vein mottling virus | TVMV | P09814 | A23647 | X04083 | U38621: complete genome |
| Turnip mosaic virus | TuMV | Q02597 | JQ1168 | – | D10601: 3′ end |
| | | | JQ1895 | | D10927: capsid protein |
| | | | | | U18654: proteinase region |
| Watermelon mosaic virus II | | P18478 | JQ0498 | D00592 | – |
| Zucchini yellow mosaic virus | | – | – | X68509 | – |

## Name and History

A protein of 48 kDa (in PStV) (see the Database table for abbreviations) or 49 kDa (in TEV, TVMV, PPV, TuMV and PVY) was found as inclusion bodies in nuclei of plant cells infected by some potyviruses (Knuhtsen *et al.*, 1974; Dougherty & Hiebert, 1980). The inclusion protein has been found to have proteolytic activity responsible for the processing of the viral polyprotein (Carrington & Dougherty, 1987). The protein was termed the ***potyvirus nuclear inclusion protein a (NIa)*** to distinguish it from another nuclear inclusion protein (NIb) containing the polymerase activity. The proteolytic activity of the NIa protein lies within the C-terminal half of the protein, whereas the N-terminal region is the VPg (**v**iral **p**rotein **g**enome-linked) domain (Carrington & Dougherty, 1987; Dougherty & Parks, 1991).

## Activity and Specificity

At least seven cleavage sites in the viral polyprotein are processed by the NIa protease (Carrington & Dougherty, 1987). Amino acids at positions P6–P1′ of the cleavage site have been determined to be important for efficient cleavage (Dougherty *et al.*, 1989a; Dougherty & Parks, 1989). Consensus sequence motifs were Glu-Xaa-Xaa-Tyr-Xaa-Gln+(Ser or Gly) for TEV, Val-(Arg or Lys)-Phe-Gln+(Ser or Gly) for TVMV, Val-Xaa-His-Gln+Yaa for PPV, Val-Xaa-His-Gln+Yaa for PVY, Val-Xaa-Xaa-Gln+Yaa for PSbMV and Val-Xaa-His-Gln+Yaa for TuMV, where Xaa is any amino acid and Yaa is a small residue (Gly, Ala, Ser or Thr) (Dougherty & Parks, 1989; Domier *et al.*, 1986; García *et al.*, 1992; Kim *et al.*, 1995; Shukla *et al.*, 1994). Quantitative assays have been made with synthetic peptide substrates

(NH$_2$-Glu-Pro-Thr-Val-Tyr-His-Gln+Thr-Leu-Asn-Glu, Ac-Glu-Pro-Thr-Val-Tyr-His-Gln+Thr-Leu-NH$_2$ and Ac-Ala-Ala-Val-Tyr-His-Gln+Ala-Ala-NH$_2$), in which the products have been quantified by HPLC analysis or fluorescamine-detection of newly generated primary amino groups after reaction. Continuous assays have been made with an intramolecularly quenched fluorometric substrate, *o*-aminobenzoyl-Ala-Ala-Val-Tyr-His-Gln+Ala-Ala-NHPhNO$_2$ (Kim *et al.*, 1995; Parks *et al.*, 1995; Ménard *et al.*, 1995). Optimum pH and temperature are about 8.5 and 15°C, respectively, for TuMV NIa protease (Ménard *et al.*, 1995; Kim *et al.*, 1996a). The higher activity at lower temperatures was found to be due to the lower value of $K_m$ (Kim *et al.*, 1996a). TuMV NIa 27 kDa protease exhibited a $K_m$ of 1.15 ± 0.16 mM and a $V$ of 0.74 ± 0.091 μmol mg$^{-1}$ min$^{-1}$ with NH$_2$-Glu-Pro-Thr-Val-Tyr-His-Gln+Thr-Leu-Asn-Glu as substrate (Kim *et al.*, 1995). TEV NIa protease exhibited a $K_m$ of 0.069 ± 0.024 mM and a $V$ of 0.38 μmol mg$^{-1}$ min$^{-1}$ with Pro-Thr-Thr-Glu-Asn-Leu-Tyr-Phe-Gln+Ser-Gly-Thr-Val-Asp-Arg-Arg as substrate (Parks *et al.*, 1995). Tos-Phe-CH$_2$Cl and Tos-Lys-CH$_2$Cl exhibited significant inhibitory effects on the catalytic activity of TuMV NIa protease with IC$_{50}$ values of 50 μM and 200 μM, respectively (Kim *et al.*, 1996c).

## Structural Chemistry

The NIa protease domain is composed of 242–244 amino acids with $M_r$ 27 300–27 600. The deduced amino acid sequences of soyabean mosaic virus, Johnson grass mosaic virus, and barley yellow mosaic virus (BaYMV) NIa proteases as well as those of TEV, TuMV, TVMV, PVY, PStV, PepMoV, PSbMV, BYMV, PRSV and PPV are known (Shukla *et al.*, 1994). The primary structure shows a similarity with that of picornavirus 3C protease with His46, Asp81 and Cys151 as the putative catalytic triad (Bazan & Fletterick, 1988; Dougherty *et al.*, 1989b). A structural modeling of TEV NIa protease proposed specific properties in the vicinity of the active-site triad and substrate-binding pocket distinct from those of the chymotrypsin family (S1) (Dougherty *et al.*, 1989b). pI values were calculated to be 8.76, 8.60, 7.83, 7.85 and 8.47 for TuMV, TEV, TVMV, PVY and PPV, respectively.

## Preparation

TEV NIa protease has been purified from infected leaf tissue (Knuhtsen *et al.*, 1974; Dougherty & Hiebert, 1980). The recombinant NIa protease of TuMV or TEV was expressed in *Escherichia coli* and purified to homogeneity (Kim *et al.*, 1995; Ménard *et al.*, 1995; Parks *et al.*, 1995). In the case of TuMV NIa protease, 2 mg of the NIa protease was obtained from a 1 liter culture of *E. coli* (Kim *et al.*, 1995).

## Biological Aspects

Some potyviruses (TEV or PPV) induce the formation of the NIa protein in nuclei while TVMV, TuMV or PStV do not (Shukla *et al.*, 1994; Hajimorad *et al.*, 1996). Among the

cleavage sites recognized by the NIa protease, the junctions between P3 and 6K$_1$, between 6K$_1$ and CI, and between NIb and CP have been known to be cleaved in *trans*, whereas the junctions between CI and 6K$_2$, between 6K$_2$ and NIa, and between NIa and NIb were observed to be cleaved preferentially in *cis* (Carrington *et al.*, 1988; García *et al.*, 1992; Hellmann *et al.*, 1988). TuMV 27 kDa NIa protease cleaves itself at Ser223+Gly224 and at Thr207+Ser208 to generate a 25 kDa protein (lacking the C-terminal 20 residues) or a 24 kDa protein (lacking the C-terminal 36 residues), respectively (Kim *et al.*, 1995, 1996b). TEV NIa protease cleaves itself at Met218+Ser219 (Parks *et al.*, 1995). The truncated 25 kDa protein of TuMV retains specific activity comparable to that of the wild-type protein, whereas the 24 kDa protein is inactive (Kim *et al.*, 1996b). The truncated form of TEV NIa protease was 20-fold less efficient than the full-length protease (Parks *et al.*, 1995). The self-cleavage of TuMV NIa 27 kDa protease to generate the 25 kDa protease occurred more rapidly at 25°C than at 12°C (Kim *et al.*, 1996a). Trp211, Gly212 and Ile217 in the C-terminal region of TuMV NIa protease were found to be important for the cleavage of the nonapeptide representing the 6K$_1$-CI and NIb-CP junction sequences (Kim *et al.*, 1996b). Deletion analyses of the C-terminal region of TuMV NIa protease revealed that at least 217 amino acids from the N-terminus are required for the catalytic activity of the NIa protease. The intrinsic fluorescence of the TuMV NIa protease as a function of temperature shows that it undergoes a large conformational change between 2 and 42°C and a drastic transition near 45°C (Kim *et al.*, 1996a).

## Distinguishing Features

TEV NIa protease is commercially available from Life Technologies (see Appendix 2 for full names and addresses of suppliers). Polyclonal antisera against TEV and TuMV NIa 49 kDa proteins have been described (Carrington & Dougherty, 1987; Laliberté *et al.*, 1992).

## Further Reading

Comprehensive reviews of the NIa proteinase have been provided by Lindbo *et al.* (1994) and Shukla *et al.* (1994).

## References

Bazan, J.F. & Fletterick, R.J. (1988) Viral cysteine proteases are homologous to the trypsin-like family of serine proteases: structural and functional implications. *Proc. Natl Acad. Sci. USA* **85**, 7872–7876.

Carrington, J.C. & Dougherty, W.G. (1987) Small nuclear inclusion protein encoded by a plant potyvirus genome is a protease. *J. Virol.* **61**, 2540–2548.

Carrington, J.C., Cary, S.M. & Dougherty, W.G. (1988) Mutational analysis of tobacco etch virus polyprotein processing: *cis* and *trans* proteolytic activities of polyproteins containing the 49 kilodalton proteinase. *J. Virol.* **62**, 2313–2320.

Domier, L.L., Franklin, K.M., Shahabuddin, M., Hellman, G.M., Overmeyer, J.H., Hiremath, S.T., Siaw, M.E.E., Lomonossoff, G.P., Shaw, J.G. & Rhoads, E. (1986) The nucleotide sequence of tobacco vein mottling virus. *Nucleic Acids Res.* **14**, 5417–5430.

Dougherty, W.G. & Hiebert, E. (1980) Translation of potyviral RNA in a rabbit reticulocyte lysate: identification of nuclear inclusion proteins as products of the in vitro translation of tobacco etch virus RNA and cylindrical protein as a product of the potyvirus genome. *Virology* **104**, 174–182.

Dougherty, W.G. & Parks, T.D. (1989) Molecular genetic and biochemical evidence for the involvement of the heptapeptide cleavage sequence in determining the reaction profile at two tobacco etch virus cleavage sites in cell-free assays. *Virology* **172**, 145–155.

Dougherty, W.G. & Parks, T.D. (1991) Post-translational processing of the tobacco etch virus 49-kDa small nuclear inclusion polyprotein: identification of an internal cleavage site and delimitation of VPg and proteinase domains. *Virology* **183**, 449–456.

Dougherty, W.G., Cary, S.M. & Parks, T.D. (1989a) Molecular genetic analysis of a plant virus polyprotein cleavage site: a model. *Virology* **171**, 356–364.

Dougherty, W.G., Parks, T.D., Cary, S.M., Bazan, J.F. & Fletterick, R.J. (1989b) Characterization of the catalytic residues of the tobacco etch virus 49-kDa proteinase. *Virology* **172**, 302–310.

García, J.A., Martín, M.T., Cervera, M.T. & Riechmann, J.L. (1992) Proteolytic processing of the plum pox potyvirus polyprotein by the NIa protease at a novel cleavage site. *Virology* **188**, 697–703.

Hajimorad, M.R., Ding, X.S., Flasinski, S., Mahajan, S., Graff, E., Haldeman-Cahill, R., Carrington, J.C. & Cassidy, B.G. (1996) NIa and NIb of peanut strip potyvirus are present in the nucleus of infected cells, but do not form inclusions. *Virology* **224**, 368–379.

Hellmann, G.M., Shaw, J.G. & Rhoads, R.E. (1988) *In vitro* analysis of tobacco vein mottling virus NIa cistron: evidence for a virus encoded protease. *Virology* **163**, 554–562.

Kim, D.-H., Park, Y.S., Kim, S.S., Lew, J., Nam, H.G. & Choi, K.Y. (1995) Expression, purification, and identification of a novel self-cleavage site of the NIa C-terminal 27-kDa protease of turnip mosaic potyvirus C5. *Virology* **213**, 517–525.

Kim, D.-H., Hwang, D.C., Kang, B.H., Lew, J. & Choi, K.Y. (1996a) Characterization of the NIa protease from turnip mosaic potyvirus exhibiting a low-temperature optimum catalytic activity. *Virology* **221**, 245–249.

Kim, D.-H., Hwang, D.C., Kang, B.H., Lew, J., Han, J., Song, B.D. & Choi, K.Y. (1996b) Effects of internal cleavages and mutations in the C-terminal region of NIa protease of turnip mosaic potyvirus on the catalytic activity. *Virology* **226**, 183–190.

Kim, D.-H., Kang, B.H., Hwang, D.C., Kim, S.S., Kwon, T.I. & Choi, K.Y. (1996c) Inhibition studies on the NIa protease from turnip mosaic potyvirus C5. *Mol. Cells* **6**, 653–658.

Knuhtsen, H., Hiebert, E. & Purcifull, D.E. (1974) Partial purification and some properties of tobacco etch virus induced intranuclear inclusions. *Virology* **61**, 200–209.

Laliberté, J.-F., Nicolas, O., Chatel, H., Lazure, C. & Morosoli, R. (1992) Release of a 22-kDa protein derived from the amino-terminal domain of the 49-kDa NIa of turnip mosaic potyvirus in *Escherichia coli*. *Virology* **190**, 510–514.

Lindbo, J.A. & Dougherty, W.G. (1994) Potyviruses. In: *Encyclopedia of Virology*, vol. 3 (Webster, R.G. & Granoff, A., eds). San Diego: Academic Press, pp. 1148–1153.

Ménard, R., Chatel, H., Dupras, R., Plouffe, C. & Laliberté, J.-F. (1995) Purification of turnip mosaic potyvirus viral protein genome-linked proteinase expressed in *Escherichia coli* and development of a quantitative assay for proteolytic activity. *Eur. J. Biochem.* **229**, 107–112.

Parks, T.D., Howard, E.D., Wolpert, T.J., Arp, D.J. & Dougherty, W.G. (1995) Expression and purification of a recombinant tobacco etch virus NIa proteinase: biochemical analyses of the full-length and a naturally occurring truncated proteinase form. *Virology* **210**, 194–201.

Shukla, D.D., Ward, C.W. & Brunt, A.A. (1994) *The Potyviridae*. Wallingford, Oxon: CAB International, pp. 80–96.

*Do-Hyung Kim*
*Department of Life Sciences and Center for Biofunctional Molecules,*
*Pohang University of Science and Technology,*
*Pohang, 790-784, Korea*

*Kwan Yong Choi*
*Department of Life Sciences and Center for Biofunctional Molecules,*
*Pohang University of Science and Technology,*
*Pohang, 790-784, Korea*
*Email: kchoi@vision.postech.ac.kr*

# 245. *Calicivirus endopeptidases*

## Databanks

*Peptidase classification: clan CB, families C24 and C37, MEROPS ID: C24.001, C37.001, C37.002*
*NC-IUBMB enzyme classification: none*

*Databank codes:*

| Species | Strain | SwissProt | PIR | EMBL (cDNA) | EMBL (genomic) |
|---|---|---|---|---|---|
| **Family C24** | | | | | |
| European brown hare syndrome virus | – | – | – | – | Z69620: complete genome |
| Feline calicivirus | CFI/68 FIV | P27407 | A43488 | M32296 | – |
| Feline calicivirus | Japanese F4 | P27408 | A40481 | D90357 | – |
| Feline calicivirus | F9 | P27409 | A43382 | – | M86379: complete genome Z11536: 3′ end |
| Feline calicivirus | Urbana | – | – | – | L40021: complete genome |
| Pig vesicular exanthema virus | – | – | – | U14678 | – |
| Rabbit hemorrhagic disease virus | – | P27410 | A41039 | M67473 | X87607: complete genome Z29514: complete genome |
| San Miguel sea lion virus | – | P36286 | – | M87481 U14675 U15301 | |
| Skunk calicivirus | – | – | – | U14667 | – |
| **Family C37** | | | | | |
| Norwalk virus | – | – | – | – | M87661: complete genome |
| Southampton virus | – | Q04544 | A37491 | – | L07418: complete genome |

## Name and History

Caliciviruses are nonenveloped positive-strand RNA viruses that belong to the picornavirus superfamily (Goldbach & Wellink, 1988; Cubbitt *et al*., 1995). Caliciviruses encode most of their proteins in one long continuous open reading frame (ORF1) (Carter *et al*., 1992a,b; Jiang *et al*., 1993; Lambden *et al*., 1993; Meyers *et al*., 1991). Translation results in a hypothetical polyprotein which is proteolytically processed into the mature viral proteins. First indications for a viral protease encoded in ORF1 were obtained by sequence comparison with the 3C-proteases from picornaviruses; these studies led to the identification of conserved residues known to be present in the active site of the 'trypsin-like' cysteine proteases (TCPs) (Bazan & Fletterick, 1988; Gorbalenya *et al*., 1989; Neill, 1990) (see Chapter 238). It was shown experimentally that there is a protease in the ORF1 polyprotein of several caliciviruses (Boniotti *et al*., 1994; Wirblich *et al*., 1995; Liu *et al*., 1996; Alonso *et al*., 1996). Based on the protease family the enzyme has been named *calicivirus trypsin-like cysteine protease* or *calicivirus TCP*. In addition, the name *calicivirus 3C-like protease* has been used which reflects the similarities to the picornavirus protease 3C.

## Activity and Specificity

The calicivirus TCP is an endopeptidase that specifically cleaves sites in the viral polyprotein. Data on cleavage specificity of the protease have mostly been obtained after *in vitro* translation or expression in bacteria. Experiments conducted for the TCP of rabbit hemorrhagic disease virus (RHDV) showed that the protease is able to act at several sites in *trans* (Boniotti *et al*., 1994; Wirblich *et al*., 1995). Cleavage in *cis* is very likely but has not yet been experimentally proven. Cleavage sites are apparently defined by features based on both sequence and structure since several sites in the polyprotein fulfilling the identified sequence requirements

are not cleaved. For RHDV four out of seven cleavage sites have been determined by N-terminal sequencing of processing products and/or analyzed by site-directed mutagenesis (Wirblich *et al*., 1995; Alonso *et al*., 1996). In all four cases the P1 position was occupied by Glu. Apart from this residue no strictly conserved amino acids were found. At three sites (numbers 3, 5 and 7 in the viral polyprotein), Gly represents the P1′ residue whereas at another site (number 6) a Thr is found in P1′. Except for site 7 which displays a Met at position P2, aromatic residues (Tyr or Phe) precede the conserved glutamic acid. Mutagenesis studies conducted for the cleavage site at the N-terminus of the protease (site 5) revealed that the RHDV TCP tolerates only Glu, Gln and Asp at the P1 position. Changes at the P1′ position resulted in the relative cleavage efficiencies Gly = Ser = Ala > Asn = His > Asp = Arg = Tyr > Leu = Thr > Pro = Val with the last two residues inhibiting cleavage almost completely (Wirblich *et al*., 1995). Since the presence of acidic (Asp), basic (Arg), aromatic (Tyr) or aliphatic (Leu) amino acids at the P1′ position results in only minor differences in cleavage efficiency, the specificity of the TCP with respect to the P1′ residue can hardly be due to specific protease/substrate interactions but is probably dictated mainly by steric and conformational constraints. In contrast to a variety of other viral proteases, the P2 position of the cleavage site represents only a minor determinant for the cleavage specificity of the RHDV TCP. Changes at this position resulted in almost unaltered cleavage efficiency (Wirblich *et al*., 1995).

For the human calicivirus Southampton virus two sites were mapped which correspond to RHDV sites 2 and 3 (Liu *et al*., 1996). In both cases Gln and Gly occupy the P1 and P1′ positions, respectively. In addition, the residues at positions P2 (Leu) and P4 (Phe) are conserved and P5 is occupied by acidic residues.

For the feline calicivirus, cleavage at the N-terminus of the capsid protein was found to occur after a Glu residue with the P1′ position occupied by Ala (Carter *et al*., 1992b).

## Structural Chemistry

For RHDV both N- and C-terminus of the TCP have been mapped (Wirblich *et al.*, 1995). The protease has a length of 143 amino acids which represent residues 1109–1251 of the ORF1 polyprotein. Because of inefficient cleavage at the C-terminal Glu+Thr site (site number 6 in the polyprotein) a considerable part of the TCP is expected to be present in infected cells in the form of a fusion protein with the viral RNA polymerase. The size of the TCP from other caliciviruses is not definitely known.

Functionally important residues of the enzyme from RHDV have been identified by sequence comparison and mutagenesis studies. The results of the experiments support the prediction that His27, Asp44 and Cys104 represent the constituents of the catalytic triad (positions 1135, 1152 and 1212 in the polyprotein, respectively) (Boniotti *et al.*, 1994). Replacement of Cys by Ser and Asp by Glu only reduces protease activity, while other substitutions at these sites abolish the enzymatic function. A His residue at position 119 of the RHDV TCP is apparently also needed for enzymatic function since replacement of this amino acid by Leu completely abolished protease activity (Boniotti *et al.*, 1994). The corresponding residue in picornains 3C from picornaviruses is thought to contribute to substrate binding (Bazan & Fletterick, 1990). The residues belonging to the catalytic triad and a histidine corresponding to residue 119 of the RHDV TCP were found at similar distances in the TCPs from other caliciviruses. Sequence alignments predict Cys residues at positions 1187 and 1193 of the ORF1 polyproteins as active-site residues of the Norwalk virus and feline calicivirus (FCV) TCPs, respectively. Replacement of the putative active-site cysteine at position 1238 of the Southampton virus polyprotein by glycine abolished proteolytic activity of the TCP (Liu *et al.*, 1996).

## Preparation

The calicivirus TCP was found to be active after *in vitro* translation of viral or *in vitro* transcribed RNA. Enzymatically active protease was purified after expression in bacteria using various expression systems (Boniotti *et al.*, 1994; Wirblich *et al.*, 1995, 1996; Liu *et al.*, 1996; Alonso *et al.*, 1996). TCP expressed in eukaryotic cells via vaccinia virus or baculovirus was shown to be active (G. Meyers, unpublished results; Sibilia *et al.*, 1995).

## Biological Aspects

According to data obtained for RHDV, the TCP represents the only protease encoded in the ORF1 polyprotein (Wirblich *et al.*, 1996). It is therefore very likely that this enzyme is responsible for all processing reactions necessary to generate the eight final products encoded in ORF1. The calicivirus TCPs resemble the 3C proteases of picornaviruses (Chapter 239) with regard to function and location in the polyprotein. Moreover, these proteases belong to the same peptidase clan, and preferred cleavage after glutamic acid and glutamine represents a characteristic feature of both the picornaviral and caliciviral enzymes.

*In vitro* studies and bacterial expression indicate that the cleavage efficiency of the calicivirus TCP differs considerably at individual cleavage sites. Some sites, namely site 6 and, in the case of Southampton virus, apparently also sites corresponding to 1, 4 and 5 of the RHDV polyprotein, are not cleaved *in vitro* at all; processing at site 6 is only poor upon expression in bacteria (Wirblich *et al.*, 1995, 1996; Liu *et al.*, 1996). Accordingly, several stable precursors composed of two or even more final products were observed, some of which probably also exist within infected cells. Inefficient cleavage at site 6 results in a precursor composed of TCP and RNA polymerase, a finding that is reminiscent of the 3CD protein of picornaviruses. For the latter viruses the proteases 3C and 3CD differ in function (Harris *et al.*, 1990). It is therefore likely that the calicivirus TCP also exhibits different activities depending upon whether it is present as fully processed polypeptide or fused to the viral RNA polymerase. Variation of cleavage efficiencies at individual sites and generation of two functionally different forms of the protease represent means for regulation of viral replication. Thus, the calicivirus protease probably is not only responsible for cutting the viral polyprotein into defined pieces, but plays an important role in regulation of the viral life cycle. Whether the calicivirus TCP cleaves cellular substrates and is involved in modulation of host cellular functions, as discussed for the picornains 3C from picornaviruses, is not known.

## Distinguishing Features

While the calicivirus TCPs are similar to the picornavirus 3C-proteases with respect to peptidase clan, genomic location, cleavage specificity and function, they are considerably smaller than the picornavirus enzymes. The latter are composed of more than 180 residues but the RHDV enzyme comprises only 143 amino acids and thus is even smaller than the 2A proteases of entero- and rhinoviruses which exhibit sizes of 146–149 residues (Harris *et al.*, 1990; Wirblich *et al.*, 1995). Since the TCPs from feline calicivirus and Norwalk virus are apparently similar in size to the RHDV enzyme, the calicivirus TCPs are amongst the smallest trypsin-like proteases so far identified. Antibodies directed against a bacterial fusion protein encompassing the RHDV TCP together with parts from two other viral proteins have been prepared (Wirblich *et al.*, 1995).

## Further Reading

The characterization and cleavage specificity of the rabbit hemorrhagic disease virus cysteine proteinase have been discussed by Boniotti *et al.* (1994) and Wirblich *et al.* (1995).

## References

Alonso, M.J.M., Casais, R., Boga, J.A. & Parra, F. (1996) Processing of rabbit hemorrhagic disease virus polyprotein. *J. Virol.* **70**, 1261–1265.

Bazan, J.F. & Fletterick, R.J. (1988) Viral cysteine proteases are homologous to the trypsin-family of serine protease: structural and functional implications. *Proc. Natl Acad. Sci. USA* **85**, 7872–7876.

Bazan, J.F. & Fletterick, R.J. (1990) Structural and catalytic models of trypsin-like viral proteases. *Semin. Virol.* **1**, 311–322.

Boniotti, B., Wirblich, C., Sibilia, M., Meyers, G., Thiel, H.-J. & Rossi, C. (1994) Identification and characterization of a 3C-like protease form rabbit hemorrhagic disease virus, a calicivirus. *J. Virol.* **68**, 6487–6495.

Carter, M.J., Milton, I.D., Meander, J., Bennett, M., Gaskell, R.M. & Turner, P.C. (1992a) The complete nucleotide sequence of a feline calicivirus. *Virology* **190**, 443–448.

Carter, M.J., Milton, I.D., Turner, P.C., Meanger, J., Bennett, M. & Gaskell, R.M. (1992b) Identification and sequence determination of the capsid protein gene of feline calicivirus. *Arch. Virol.* **122**, 223–235.

Cubbitt, D., Bradley, D.W., Carter, M.J., Chiba, S., Estes, M.K., Saif, L.J., Schaffer, F.L., Smith, A.W., Studdert, M.J. & Thiel, H.-J. (1995) Family Caliciviridae. *Arch. Virol.* **10** (suppl), 359–363.

Goldbach, R. & Wellink, J. (1988) Evolution of plus-strand RNA viruses. *Intervirology* **29**, 260–267.

Gorbalenya, A.A., Donchenko, A.P., Blinov, V.M. & Koonin, E.V. (1989) Cysteine proteases of positive strand RNA viruses and chymotrypsin-like serine proteases. *FEBS Lett.* **243**, 103–114.

Harris, K.S., Hellen, C.U.T. & Wimmer, E. (1990) Proteolytic processing in the replication of picornaviruses. *Semin. Virol.* **1**, 103–114.

Jiang, X., Wang, M., Wang, K. & Estes, M.K. (1993) Sequence and genomic organization of Norwalk virus. *Virology* **195**, 51–61.

Lambden, P.R., Caul, E.O., Ashley, C.R. & Clarke, I.N. (1993) Sequence and genome organization of a human small round-structured (Norwalk-like) virus. *Science* **259**, 516–519.

Liu, B., Clarke, I.N. & Lambden, P.R. (1996) Polyprotein processing in Southampton virus: identification of 3C-like protease cleavage sites by *in vitro* mutagenesis. *J. Virol.* **70**, 2605–2610.

Meyers, G., Wirblich, C. & Thiel, H.-J. (1991) Rabbit hemorrhagic disease virus: molecular cloning and nucleotide sequencing of a calicivirus genome. *Virology* **184**, 664–676.

Neill, J.D. (1990) Nucleotide sequence of a region of the feline calicivirus genome that encodes picornavirus-like RNA-dependent RNA polymerase, cysteine protease and 2C polypeptides. *Virus Res.* **17**, 145–160.

Sibilia, M., Boniotti, M.B., Angoscini, P., Capucci, L. & Rossi, C. (1995) Two independent pathways of expression lead to self-assembly of the rabbit hemorrhagic disease virus capsid protein. *J. Virol.* **69**, 5812–5815.

Wirblich, C., Sibilia, M., Boniotti, M.B., Rossi, C., Thiel, H.-J. & Meyers, G. (1995) 3C-like protease of rabbit hemorrhagic disease virus: identification of cleavage sites in the ORF1 polyprotein and analysis of cleavage specificity. *J. Virol.* **69**, 7159–7168.

Wirblich, C., Thiel, H.-J. & Meyers, G. (1996) Genetic map of the calicivirus rabbit hemorrhagic disease virus as deduced from *in vitro* translation studies. *J. Virol.* **70**, 7974–7983.

***Gregor Meyers***
*Department of Clinical Virology,*
*Federal Research Centre for Virus*
*Diseases of Animals,*
*PO Box 1149,*
*D-72001 Tübingen, Germany*
*Email: gregor.meyers@tue.bfav.de*

***Cesare Rossi***
*Istituto Zooprofilattico Sperimentale*
*della Lombardia e dell' Emilia,*
*Via Bianchi 7, 25124 Brescia, Italy*
*Email: izsviro@master.cci.unibs.it*

***Heinz-Jürgen Thiel***
*Institute of Virology*
*(FB Veterinärmedizin),*
*University of Giessen,*
*Frankfurter Strasse 107,*
*D-35392 Giessen, Germany*
*Email: Heinz-*
*Juergen.Thiel@vetmed.uni-giessen.de*

# 246. *Coronavirus picornain-like cysteine proteinase*

*Peptidase classification: clan CB, family C30, MEROPS ID: C30.001*
*NC-IUBMB enzyme classification: none*
*Databank codes:*

| Species | SwissProt | PIR | EMBL (cDNA) | EMBL (genomic) |
|---|---|---|---|---|
| Avian infectious bronchitis virus | P27920 | A33094 | M94356 M95169 | – |
| Human coronavirus | Q05002 | S28600 | X69721 | – |

| Species | SwissProt | PIR | EMBL (cDNA) | EMBL (genomic) |
|---|---|---|---|---|
| Murine hepatitis virus | P19751 | A36815 | M18040 M23258 M28352 M55148 M61144 S51684 X57302 X73559 | X73559: gene 1 and ORF1a |
| Porcine transmissible gastroenteritis virus | – | – | – | Z34093: ORF1a |

## Name and History

The existence of the coronavirus picornain-like cysteine proteinase was initially predicted by amino acid sequence comparison with chymotrypsin (Chapter 8) and with the picornavirus 3C proteinases (Chapter 239) (Gorbalenya *et al.*, 1989; Lee *et al.*, 1991), so the proteinase was named *3C-like proteinase (3CLpro* or *3CLP)*. The 3CLpro has been identified for mouse hepatitis virus (MHV), avian infectious bronchitis virus (IBV), and the human coronavirus 229E (HCV-229E). The 3CLpro are also referred to as *p27* or *p29* (MHV), *p34* (229E) and *p35* (IBV) according to the apparent molecular mass in kilodaltons in SDS-PAGE. The actual sizes of the 3CL proteinases have not been precisely determined by gel filtration, but have been calculated to be 33 kDa from the sequences. Because of the variation in sizes and evidence for differences in calculated and apparent masses, the 3CLpro or 3CLP nomenclature is preferred.

## Activity and Specificity

The precise determinants of cleavage specificity of 3CLpro have not been defined, but the experimentally confirmed coronavirus 3CLpro cleavage sites have a Gln at P1, Leu at P2, and Ser, Ala or Gly at P1'. Other predicted P2 residues include Ile, Val and Met. Substitution of the Gln at position P1 by mutagenesis of the IBV 3CLpro abolished proteolytic activity in vitro (Tibbles *et al.*, 1996). Autoproteolytic cleavage of 3CLpro from the polyprotein occurs at an N-terminal Leu-Gln+Ser in MHV (Y. Lu *et al.*, 1995) and IBV (Tibbles *et al.*, 1996), and at a Leu-Gln+Ala motif in HCV-229E (Ziebuhr *et al.*, 1995). Comparison of confirmed and predicted 3CLpro cleavage sites from several coronaviruses

indicates that there may be a strong preference at P4 for Val, Ala, Ser or Thr.

The coronavirus 3CL proteinases have been expressed in a variety of systems, and have been demonstrated to cleave substrate in *trans* (Table 246.1) (Liu & Brown, 1995; Y. Lu *et al.*, 1995; Ziebuhr *et al.*, 1995; X. Lu *et al.*, 1996; Tibbles *et al.*, 1996). Only MHV has been assessed for *cis* cleavage activity, and it was found that MHV 3CLpro was not able to cleave itself in *cis* from a larger precursor *in vitro* (X. Lu *et al.*, 1996). The pH optimum for MHV 3CLpro activity is 7.0, although the enzyme remains active between pH 5.5 and 8.0. Reducing agents are not required for cleavage by *Escherichia coli*-expressed recombinant 3CLpro (r3CLpro), but also do not interfere with proteolytic activity; activity is maintained in the presence of 0.2–25 mM 2-ME and 0.1–100 mM DTT (Sims *et al.*, 1998). The MHV 3CLpro is sensitive to inhibition by typical inhibitors of both serine and cysteine proteinases, but is not affected by aspartic proteinase inhibitors (Y. Lu & Denison, 1997). MHV 3CLpro activity is inhibited in MHV-infected cells by leupeptin, PMSF, *N*-ethylmaleimide and E-64d (X. Lu *et al.*, 1996). MHV 3CLpro is also inhibited by high concentrations of $ZnCl_2$ (2 mM) and 1,10-phenanthroline (10 mM), but is unaffected by EDTA even at high concentrations (10 mM) (Y. Lu & Denison, 1997). Thus any role of metal ions in 3CLpro activity remains to be established.

## Structural Chemistry

MHV 3CLpro is synthesized as part of an 803 kDa polyprotein that is autoproteolytically cleaved to generate the mature 27 kDa proteinase. Mutagenesis of the 3CLpro of MHV,

*Table 246.1*   Molecular properties of coronavirus 3CLpro

| Strain | Mass (kDa) | | pI[a] | Activity[b] | Expression systems | | | |
|---|---|---|---|---|---|---|---|---|
| | Expected | Observed | | | RRL[c] | Vaccinia | Cells[d] | E. coli |
| MHV-A59 | 33 | 27–29 | 5.81 | + | + | | + | + |
| HCV-229E | 33 | 34 | 6.39 | + | | | + | + |
| IBV | 33 | 35 | 6.08 | + | + | + | + | |

[a]pI calculated – not experimentally confirmed.
[b]Activity has been demonstrated during *trans* cleavage assays.
[c]RRL, rabbit reticulocyte lysate *in vitro* translation systems.
[d]Virus infection of permissive cells in culture.

IBV and HCV-229E has confirmed that His is an essential active-site residue (Liu & Brown, 1995; Y. Lu *et al.*, 1995; Ziebuhr *et al.*, 1995). Similar studies have confirmed that the predicted catalytic Cys residues are required for MHV and IBV 3CLpro activity *in vitro* (Liu & Brown, 1995; Y. Lu *et al.*, 1995). Substitutions at a putative third catalytic residue (Asp3386 or Asp3398 in MHV and Glu2843 in IBV) do not alter 3CLpro-mediated cleavage (Y. Lu *et al.*, 1995). There is no evidence for 3CLpro oligomerization. The predicted substrate-binding domains for the MHV and IBV 3CLpro are similar to those of the picornavirus picornains 3C, both in proximity to the catalytic residues and in the conservation of amino acids contributing the structure of the sites. Mutagenesis of predicted substrate-binding sites of 229E 3CLpro altered or abolished proteinase activity (Ziebuhr *et al.*, 1997). The calculated pI values for several coronavirus 3CLpro are shown in Table 246.1. The crystal structure for 3CLpro has not been solved.

## Preparation

MHV 3CLpro has been isolated in an active form from virus-infected cells (X. Lu *et al.*, 1996). Active 3CLpro from MHV and IBV have also been expressed from multiple vectors in rabbit reticulocyte lysate, in *in vitro* translation systems. An MHV 3CLpro/maltose-binding protein (MBP/3CLpro) fusion product has been expressed in *E. coli* (Seybert *et al.*, 1997; Sims *et al.*, 1998). The recombinant proteinase is active in *trans* after separation of the MBP from the 3CLpro at a factor Xa cleavage site (X. Lu *et al.*, 1998). Similarly, an HCV-229E 3CLpro/β-galactosidase fusion protein expressed in *E. coli* has proteolytic activity in an *in vitro*, *trans*-cleavage assay (Ziebuhr *et al.*, 1995). The IBV 3CLpro has been expressed in a vaccinia virus T7 expression system in Vero cells (Liu *et al.*, 1994).

## Biological Aspects

Coronaviruses contain a 32 kb, positive-strand RNA genome. During viral replication, the viral proteinases, polymerase and helicase are expressed as a polyprotein that contains multiple putative 3CLpro cleavage sites. The polyprotein undergoes post-translational processing to liberate the active forms of the viral replicative proteins. Although at least one proteinase other than 3CLpro has been demonstrated to be expressed as part of the coronavirus polyprotein (Baker *et al.*, 1993; Denison *et al.*, 1992; Weiss *et al.*, 1994) (see Chapter 229), 3CLpro is thought to mediate the majority of the maturation cleavages within the polyprotein (Figure 246.1). Processing of the coronavirus polyprotein is required throughout the infection cycle for productive infection, and thus the proteinase is critical for viral replication (Kim *et al.*, 1995).

Within the polyprotein, 3CLpro is flanked by 200–300 residue domains of predominantly hydrophobic residues, known as MP1 and MP2, that are hypothesized to be involved in presentation of the proteinase or flanking cleavage sites. Microsomal membranes are not required for activity of the mature 3CLpro from cells or *in vitro* translates from *E. coli*, or for activity of 3CLpro expressed with only small portions of MP1 and MP2 (Y. Lu *et al.*, 1995; X. Lu *et al.*, 1996). In contrast, expression of one or both of the MP domains

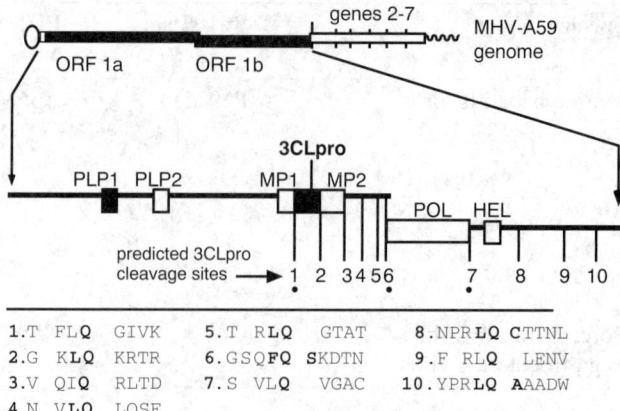

| | | | | | | | |
|---|---|---|---|---|---|---|---|
| 1.T | FLQ | GIVK | 5.T | RLQ | GTAT | 8.NPRLQ | CTTNL |
| 2.G | KLQ | KRTR | 6.GSQFQ | | SKDTN | 9.F | RLQ | LENV |
| 3.V | QIQ | RLTD | 7.S | VLQ | VGAC | 10.YPRLQ | AAADW |
| 4.N | VLQ | LQSE | | | | | |

*Figure 246.1* Structure of the MHV-A59 genome RNA, functional domains and predicted 3CLpro cleavage sites. The MHV genome is a 32 kb single-stranded RNA. Gene 1 is 22 kb in length and is composed of two overlapping open reading frames, ORFs 1a and 1b, which are translated together as an 800 kDa fusion polyprotein. The polyprotein has been shown to be processed by at least two virus-encoded proteinases (PLP-1 and 3CLpro). The location of predicted (white box) and confirmed (black box) functional domains within the polyprotein are shown. *Abbreviations*: PLP1, PLP2, papain-like proteinases; MP1, MP2, putative membrane spanning domains; 3CLpro, 3C-like proteinase; POL, RNA-dependent RNA polymerase; HEL, helicase. Numbers 1–10 show predicted 3CLpro cleavage sites. The sequences of the numbered sites are shown below the figure. Sites that have been experimentally shown to be cleaved by a coronavirus 3CLpro are indicated by dots.

results in a membrane requirement for autoproteolytic cleavage of 3CLpro from the flanking domains (Pinon *et al.*, 1997; Tibbles *et al.*, 1996). The significance of these findings for proteinase activity and polyprotein processing is not known.

## Distinguishing Features

The coronavirus 3CL proteinases contain a significantly larger number of amino acids C-terminal to the putative substrate-binding residues than other 3C or 3C-like proteinases (Gorbalenya & Koonin, 1993). The extended region contains several predicted β-sheet and α-helical domains. In addition, MHV 3CLpro does not appear to use an Asp/Glu residue in a catalytic triad with His and Cys (Y. Lu & Denison, 1997). Whether the 3C endopeptidases of family C3 contain a catalytic triad has been controversial (see Chapter 238), but if so, then the coronavirus 3CLpro may have a different structure and catalytic mechanism from other 3C-like proteinases.

Rabbit polyclonal antibodies have been produced against 3CLpro from MHV, IBV and HCV-229E (Liu & Brown, 1995; Ziebuhr *et al.*, 1995; X. Lu *et al.*, 1996), and have been used for immunoprecipitation, western blotting, and purification of 3CLpro. There are no commercially available antibodies to these proteinases.

## References

Baker, S.C., Yokomori, K., Dong, S., Carlisle, R., Gorbalenya, A.E., Koonin, E.V. & Lai, M.M. (1993) Identification of the catalytic sites of a papain-like cysteine proteinase of murine coronavirus. *J. Virol.* **67**, 6056–6063.

Denison, M.R., Zoltick, P.W., Hughes, S.A., Giangreco, B., Olson, A.L., Perlman, S., Leibowitz, J.L. & Weiss, S.R. (1992) Intracellular processing of the N-terminal ORF 1a proteins of the coronavirus MHV-A59 requires multiple proteolytic events. *Virology* **189**, 274–284.

Gorbalenya, A. & Koonin, E. (1993) Comparative analysis of amino-acid sequences of key enzymes of replication and expression of positive-strand RNA viruses: validity of approach and functional and evolutionary implications. *Sov. Sci. Rev. D. Physiochem. Biol.* **11**, 1–81.

Gorbalenya, A.E., Koonin, E.V., Donchenko, A.P. & Blinov, V.M. (1989) Coronavirus genome: prediction of putative functional domains in the nonstructural polyprotein K by comparative amino acid sequence analysis. *Nucleic Acids Res.* **17**, 4847–4861.

Kim, J.C., Spence, R.A., Currier, P.F., Lu, X. & Denison, M.R. (1995) Coronavirus protein processing and RNA synthesis is inhibited by the cysteine proteinase inhibitor, E64d. *Virology* **208**, 1–8.

Lee, H.-J., Shieh, C.-K., Gorbalenya, A.E., Koonin, E.V., LaMonica, N., Tuler, J., Bagdzhadhzyan, A. & Lai, M.M.C. (1991) The complete sequence (22 kilobases) of murine coronavirus gene 1 encoding the putative proteases and RNA polymerase. *Virology* **180**, 567–582.

Liu, D.X. & Brown, T.D.K. (1995) Characterisation and mutational analysis of an ORF1a-encoding proteinase domain responsible for proteolytic processing of the infectious bronchitis virus 1a/1b polyprotein. *Virology* **209**, 420–427.

Liu, D.X., Tibbles, K.W. & Brown, T.D.K. (1994) A 100 kilodalton polypeptide encoded by open reading frame (ORF) 1b of the coronavirus infectious bronchitis virus is processed by ORF 1a products. *J. Virol.* **68**, 5772–5780.

Lu, X., Lu, Y. & Denison, M.R. (1996) Intracellular and *in vitro* translated 27-kDa proteins contain the 3C-like proteinase activity of the coronavirus MHV-A59. *Virology* **222**, 375–382.

Lu, X., Sims, A. & Denison, M.R. (1998) Mouse hepatitis virus 3C-like protease cleaves a 22-kilodalton protein from the open reading frame 1a polyprotein in virus-infected cells and *in vitro*. *J. Virol.* **72**, 2265–2271.

Lu, Y.Q. & Denison, M.R. (1997) Determinants of mouse hepatitis virus 3C-like proteinase activity. *Virology* **230**, 335–342.

Lu, Y., Lu, X. & Denison, M.R. (1995) Identification and characterization of a serine-like proteinase of the murine coronavirus MHV-A59. *J. Virol.* **69**, 3554–3559.

Piñón, J.D., Turner, J.D., Khan, F.S., Bonilla, P.J. & Weiss, S.R. (1997) Efficient autoproteolytic processing of the MHV-A59 3C-like proteinase from the flanking hydrophobic domains requires membranes. *Virology* **230**, 309–322.

Seybert, A., Ziebuhr, J. & Siddell, S.G. (1997) Expression and characterization of a recombinant murine coronavirus 3C-like proteinase. *J. Gen. Virol.* **78**, 71–75.

Sims, A.C., Lu, X.T. & Denison, M.R. (1998) Expression, purification and activity of recombinant MHV-A59 3CLpro. *Adv. Exp. Med. Biol.* (in press).

Tibbles, K.W., Brierley, I., Cavanaugh, D. & Brown, T.D.K. (1996) Characterization in vitro of an autocatalytic processing activity associated with the predicted 3C-like proteinase domain of the coronavirus avian infectious bronchitis virus. *J Virol.* **70**, 1923–1930.

Weiss, S.R., Hughes, S.A., Bonilla, P.J., Turner, J.D., Leibowitz, J.L. & Denison, M.R. (1994) Coronavirus polyprotein processing. *Arch. Virol.* **9** (suppl), 349–358.

Ziebuhr, J., Herold, J. & Siddell, S.G. (1995) Characterization of a human coronavirus (strain 229E) 3C-like proteinase activity. *J Virol.* **69**, 4331–8.

Ziebuhr, J., Heusipp, G. & Siddell, S.G. (1997) Biosynthesis, purification, and characterization of the human coronavirus 229E 3C-like proteinase. *J. Virol.* **71**, 3992–3997.

*C. Anne Gibson*
*Department of Pediatrics and Microbiology and Immunology,*
*Elizabeth B. Lamb Center for Pediatric Research,*
*Vanderbilt University Medical Center,*
*D7235 MCN, Nashville, TN 37232-2581, USA*

*Mark R. Denison*
*Department of Pediatrics and Microbiology and Immunology,*
*Elizabeth B. Lamb Center for Pediatric Research,*
*Vanderbilt University Medical Center,*
*D7235 MCN, Nashville, TN 37232-2581, USA*
*Email: mark.denison@mcmail.vanderbilt.edu*

# 247. Introduction: clan CD containing the caspases

## Databanks

MEROPS ID: CD

| Species | SwissProt | PIR | EMBL (cDNA) | EMBL (genomic) |
|---|---|---|---|---|
| **Family C14** | | | | |
| Caspase-1 (Chapter 248) | | | | |
| Caspase-2 | | | | |
| *Gallus gallus* | – | – | U64963 | – |
| *Homo sapiens* | P42575 | – | U13021 | – |
| *Homo sapiens* | P42576 | – | U13022 | – |
| *Mus musculus* | P29594 | – | D10713 | – |
| | | | D28492 | |
| Caspase-3 (Chapter 249) | | | | |
| Caspase-4 | | | | |
| *Homo sapiens* | P49662 | A57511 | S78281 | – |
| | | | U25804 | |
| | | | U28014 | |
| | | | Z48810 | |
| Caspase-5 | | | | |
| *Homo sapiens* | P51878 | B57511 | U28015 | – |
| | | X94993 | | |
| Caspase-6 | | | | |
| *Homo sapiens* | P55212 | – | U20536 | – |
| | | | U20537 | |
| Caspase-7 | | | | |
| *Homo sapiens* | P55210 | – | U37448 | – |
| | | | U39613 | |
| | | | U40281 | |
| Caspase-8 | | | | |
| *Homo sapiens* | – | – | U58143 | – |
| | | | U60520 | |
| | | | X98172 | |
| *Homo sapiens* | – | – | X98173 | – |
| Caspase-9 | | | | |
| *Homo sapiens* | – | – | U60521 | – |
| Caspase-10 | | | | |
| *Homo sapiens* | – | – | U60519 | – |
| FLIP$_L$ | | | | |
| *Homo sapiens* | – | – | U97074 | |
| *Mus musculus* | – | – | U97076 | |
| Other homologs | | | | |
| *Caenorhabditis elegans* | P42573 | A49429 | L29052 | – |
| *Caenorhabditis vulgaris* | P45436 | – | – | – |
| *Homo sapiens* | P42574 | – | U13737 | – |
| | | | U26943 | |
| *Homo sapiens* | P55211 | – | U56390 | – |
| *Homo sapiens* | – | – | X98174 | – |
| *Mesocricetus auratus* | P55214 | – | U47332 | – |
| *Rattus norvegicus* | | I67436 | U34684 | |

Clan CD contains only one family, C14, comprising a number of cytosolic endopeptidases that have strict specificity for hydrolysis of aspartyl bonds, the best known of which are caspase-1 (Chapter 248) and caspase-3 (Chapter 249). As can be seen from Alignment 247.1 (see CD-ROM), there is a catalytic dyad in the order His, Cys in the sequence, which is the same as in clan CB.

Three-dimensional structures have been solved for both caspase-1 and caspase-3, and show an $\alpha/\beta$ protein fold, quite unlike the $\beta$ fold of picornain. Figure 247.1 shows the structure of caspase-1.

Caspase-1 is synthesized as a single-chain precursor, and the activation by cleavage of four aspartyl bonds is presumably autocatalytic. The mature endopeptidase is a heterodimer of a 22 kDa heavy chain and a 10 kDa light chain (Yamin *et al.*, 1996). Caspase-1 also mediates the processing of interleukin 1$\beta$ at aspartyl bonds (Thornberry & Molineaux,

1995), and was formerly known as interleukin 1$\beta$-converting enzyme (ICE).

CrmA, a serpin from cowpox virus, is an inhibitor of the caspases (Ray *et al.*, 1992) that also inhibits the serine endopeptidase granzyme B (Chapter 23) (Quan *et al.*, 1995). Rather than reflecting any structural relationship, however, this is undoubtedly the result of a similar specificity of the enzymes for aspartyl bonds.

The discovery that the product of the *ced3* gene in *Caenorhabditis elegans*, which when defective increases the nematode lifespan (Xue *et al.*, 1996), is a homolog of caspase-1 led to the discovery of a proteolytic cascade that leads to apoptosis in mammalian cells. Apoptosis is receptor mediated, and once the system is activated, proteins associate with one another by means of 'death domains' and 'death-effector domains'. Death-effector domains occur in a number of intracellular proteins and in caspases-8 and -10 (Wallach, 1997). FLIP, a newly reported inhibitor of caspase-8, occurs in two alternatively spliced forms, the longer of which, $FLIP_L$, contains two death-effector domains, and a caspase-like domain in which the catalytic Cys is replaced by Tyr, and the His by Arg (human) or Leu (mouse) (Irmler *et al.*, 1997) (Alignment 247.1).

The evolutionary tree (Figure 247.2) (see overleaf) shows the relationships between the endopeptidases of family C14.

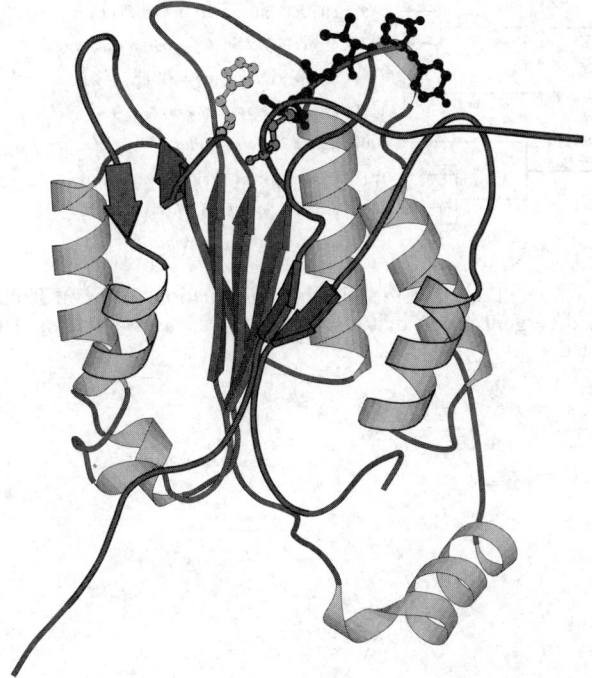

*Figure 247.1* Richardson diagram of the human caspase-1/ Ac-Tyr-Val-Ala-Asp complex. The image was prepared from the Brookhaven Protein Data Bank entry (1ICE) as described in the Introduction (p. xxv). Catalytic residues are shown in ball-and-stick representation: His237 and Cys285 (numbering as in Alignment 247.1). Ac-Tyr-Val-Asp is shown in black in ball-and-stick representation.

### References

Irmler, M., Thome, M., Hahne, M., Schneider, P., Hoffman, K., Steiner, V., Bodmer, J.-L., Schröter, M., Burns, K., Mattmann, C., Rimoldi, D., French, L.E. & Tschopp, J. (1997) Inhibition of death receptor signals by cellular FLIP. *Nature* **388**, 190–195.

Quan, L.T., Caputo, A., Bleackley, R.C., Pickup, D.J. & Salvesen, G.S. (1995) Granzyme B is inhibited by the cowpox virus serpin cytokine response modifier A. *J. Biol. Chem.* **270**, 10377–10379.

Ray, C.A., Black, R.A., Kronheim, S.R., Greenstreet, T.A., Sleath, P.R., Salvesen, G.S. & Pickup, D.J. (1992) Viral inhibition of inflammation: cowpox virus encodes an inhibitor of the interleukin-1$\beta$ converting enzyme. *Cell* **69**, 597–604.

Thornberry, N.A. & Molineaux, S.M. (1995) Interleukin-1$\beta$ converting enzyme: a novel cysteine protease required for IL-1$\beta$ production and implicated in programmed cell death. *Protein Sci.* **4**, 3–12.

Wallach, D. (1997) Placing death under control. *Nature* **388**, 123–126.

Xue, D., Shaham, S. & Horvitz, H.R. (1996) The *Caenorhabditis elegans* cell-death protein CED-3 is a cysteine protease with substrate specificities similar to those of the human CPP32 protease. *Genes Dev.* **10**, 1073–1083.

Yamin, T.T., Ayala, J.M. & Miller, D.K. (1996) Activation of the native 45-kDa precursor form of interleukin-1-converting enzyme. *J. Biol. Chem.* **271**, 13273–13282.

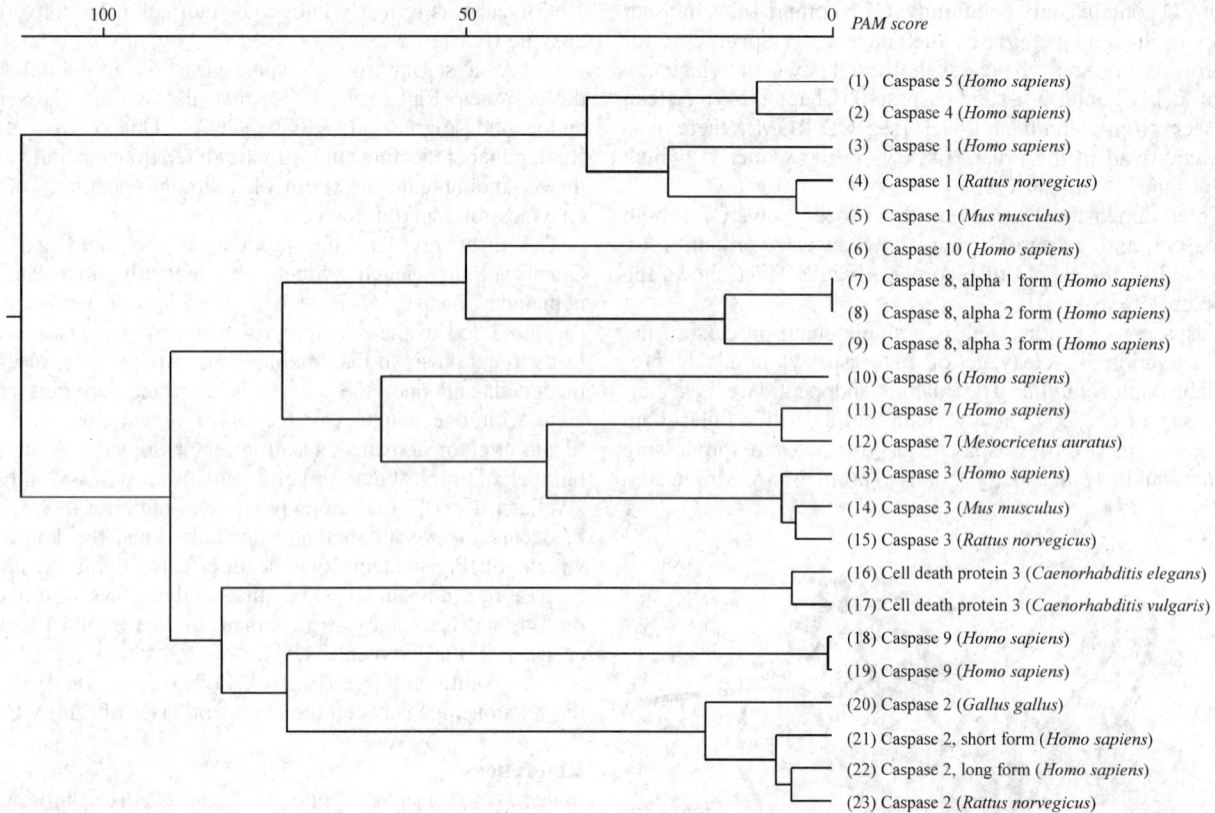

*Figure 247.2*  Evolutionary tree for family C14. The tree was constructed as described in the Introduction (p. xxv). It can be seen that caspase-1 is representative of a small and particularly divergent group that also contains caspases-4 and -5. The remaining enzymes fall into two groups that are more closely related.

# 248. Caspase-1

### Databanks

*Peptidase classification: clan CD, family C14, MEROPS ID: C14.001*
*NC-IUBMB enzyme classification: EC 3.4.22.36*
*Databank codes:*

| Species | SwissProt | PIR | EMBL (cDNA) | EMBL (genomic) |
|---|---|---|---|---|
| *Homo sapiens* | P29466 | A42677 | M87507 | – |
| | | A56084 | U13697 | |
| | | B56084 | U13698 | |
| | | C56084 | U13699 | |
| | | D56084 | U13700 | |
| | | | X65019 | |
| *Mus musculus* | P29452 | A46495 | L03799 | U04269: complete gene |
| *Rattus norvegicus* | P43527 | I53300 | U14647 | – |

Brookhaven Protein Data Bank three-dimensional structures

| Species | ID | Resolution | Notes |
|---|---|---|---|
| *Homo sapiens* | 1ICE | 2.6 | |

## Name and History

**Caspase-1** was first described in 1989 as the cysteine proteinase responsible for the production of interleukin 1$\beta$ (IL-1$\beta$) in monocytes (Black *et al*., 1989; Kostura *et al*., 1989), hence its original name, *interleukin 1$\beta$-converting enzyme (ICE)*. When it was first purified, cloned, and sequenced in 1992, it was found to be unrelated to any known protein (Cerretti *et al*., 1992; Thornberry *et al*., 1992). Late in 1993, it was shown to be homologous to CED-3, the product of a gene required for programmed cell death in the nematode, *Caenorhabditis elegans* (Yuan *et al*., 1993). The crystallographic structure of caspase-1 reported in 1994 identified the residues important for binding and catalysis, and confirmed its relationship to CED-3 (Walker *et al*., 1994; Wilson *et al*., 1994). To date, ten related proteinases of human origin have been described (Figure 248.1). Recently, a unified nomenclature was introduced for these enzymes, which uses the trivial name 'caspase' as a root for serial names (Alnemri *et al*., 1996). The '**c**' reflects a cysteine proteinase mechanism, and '**aspase**' relates to their ability to cleave after Asp, the most distinguishing catalytic property of this family.

## Activity and Specificity

Caspase-1 catalyzes the cleavage of the 31 kDa human proIL-1$\beta$ at Tyr-Val-His-Asp116+Ala117 to generate the 17.5 kDa mature, biologically active cytokine (Black *et al*., 1989; Kostura *et al*., 1989). It has a near absolute specificity for Asp in the P1 position of both peptide and macromolecular substrates; any substitution of this residue results in more than 100-fold decrease in $k_{cat}/K_m$ (Sleath *et al*., 1990; Howard *et al*., 1991; Thornberry *et al*., 1992). The enzyme has an equally stringent requirement for four amino acids to the left of the scissile bond (Thornberry *et al*., 1992), with hydrophobic amino acids preferred in P4 (Thornberry & Molineaux, 1995).

Compounds with the general structure Ac-Tyr-Val-Ala-Asp+X, where X is amino-4-methylcoumarin or *p*-nitroanilide, have been used to develop continuous fluorometric (Thornberry *et al*., 1992) and spectrophotometric assays (Reiter, 1994; Thornberry, 1994), respectively. These and similar substrates are available from Bachem, Calbiochem, Peptide Institute, and Enzyme Systems (see Appendix 2 for full names and addresses of suppliers). An assay employing a quenched fluorescence substrate, DABCYL-Tyr-Val-Ala-Asp+Ala-Pro-Val-EDANS (available from Bachem and Calbiochem), has also been described (Pennington & Thornberry, 1994). Assays typically contain at least 1 mM DTT to prevent oxidation of the catalytic Cys, and 10% sucrose and 0.1% 3-[(3-cholamidopropyl) dimethylammonio]-1-propane sulfonate (CHAPS), to stabilize the protein (Thornberry, 1994). The pH optimum of the enzyme is about 7.0.

Caspase-1 is inactivated by thiol alkylating agents such as iodoacetate and *N*-ethylmaleimide. It is also inhibited by 1,10-phenanthroline via phenanthroline-metal-catalyzed oxidation of the catalytic Cys (Thornberry, 1994). Several classes of potent reversible ($K_i < 60$ nM) and irreversible ($k_{inact}/K_i > 1 \times 10^4$ M$^{-1}$ s$^{-1}$) inhibitors, containing the Ac-Tyr-Val-Ala-Asp recognition sequence, have been described. Reversible inhibitors include aldehydes (Chapman, 1992; Thornberry *et al*., 1992; Mullican *et al*., 1994), ketones (Robinson *et al*., 1992; Mjalli *et al*., 1993a,b, 1994) and nitriles (Thornberry & Molineaux, 1995). Irreversible inhibitors include halomethanes (Walker *et al*., 1994), diazomethanes (Thornberry *et al*., 1992) and acyloxymethanes (Dolle *et al*., 1994; Revesz *et al*., 1994; Thornberry *et al*., 1994; Prasad *et al*., 1995). Many of these inhibitors are available from Bachem, Calbiochem and Peptide Institute. The only protein known to be an inhibitor is the cowpox serpin, CrmA (Ray *et al*., 1992; Komiyama *et al*., 1994).

## Structural Chemistry

Human caspase-1 is a heterodimer composed of two subunits with molecular masses of 10 248 Da (p10) and 19 866 Da (p20), both of which are derived from a common 45 kDa proenzyme (Thornberry *et al*., 1992). The proenzyme also encodes an 11.5 kDa propeptide and a 2 kDa linker peptide between the two subunits. Both p10 and p20 are flanked by Asp+X sites in the proenzyme, suggesting that activation is either autoproteolytic, or mediated by an enzyme with similar specificity.

The X-ray crystal structure shows the two subunits intimately associated to form a single catalytic domain in which both contribute residues to the active site (Walker *et al*., 1994; Wilson *et al*., 1994). The catalytic Cys and His reside in p20, whereas the S1 subsite is composed of residues from both subunits. Amino acids that form the S2, S3 and S4 specificity subsites are predominantly contributed by p10 (see Fig. 247.2). In the crystal, two heterodimers associate to form a tetramer with 2-fold rotational symmetry.

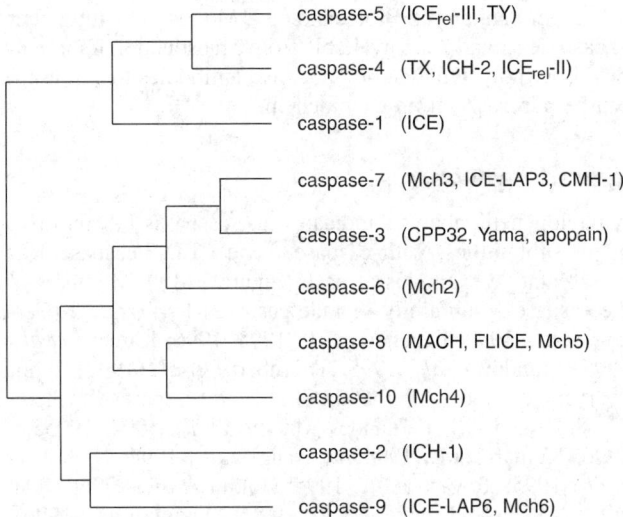

caspase-5 (ICE$_{rel}$-III, TY)

caspase-4 (TX, ICH-2, ICE$_{rel}$-II)

caspase-1 (ICE)

caspase-7 (Mch3, ICE-LAP3, CMH-1)

caspase-3 (CPP32, Yama, apopain)

caspase-6 (Mch2)

caspase-8 (MACH, FLICE, Mch5)

caspase-10 (Mch4)

caspase-2 (ICH-1)

caspase-9 (ICE-LAP6, Mch6)

*Figure 248.1* The caspase family of cysteine proteases. Aliases for the proteases are shown in parenthesis. The phylogenic relationships were determined using the PILEUP algorithm of the Genetics Computer Group (version 8.1; gap weight = 3.0; gap length weight = 0.1). (Program Manual for the Wisconsin Package, Version 8, September 1994, Genetics Computer Group, 575 Science Drive, Madison, WI 53711, USA.).

## Preparation

The enzyme has been purified to homogeneity from THP-1 cells, a human monocytic cell line, by affinity chromatography employing a peptide aldehyde ligand (Miller *et al*., 1993; Thornberry, 1994; Thornberry *et al*., 1992). However, because of the relative low abundance of this protein in natural sources (Ayala *et al*., 1994), investigators have instead turned to *Escherichia coli* (Kamens *et al*., 1995; Malinowski *et al*., 1995; Ramage *et al*., 1995; Walker *et al*., 1994) or baculovirus/insect cell (X.-M. Wang *et al*., 1994; Howard *et al*., 1995) expression systems. The latter has not proven useful for producing large amounts of protein, most likely because caspase-1, when overexpressed in insect cells, induces apoptosis (Alnemri *et al*., 1995). In contrast, yields of more than 1 mg liter$^{-1}$ culture have been obtained via expression of the proenzyme in *E. coli* (Malinowski *et al*., 1995). Increasingly, investigators are using epitope tags to facilitate purification of the expressed protein (Kamens *et al*., 1995). Large quantities of caspase-1 have also been prepared through folding of 20 and 10 kDa subunits individually expressed in *E. coli* (Walker *et al*., 1994).

## Biological Aspects

The human enzyme is constitutively expressed in monocytes and THP-1 cells. Caspase-1 mRNA has also been detected in T cells, B cells and neutrophils (Cerretti *et al*., 1992; Thornberry *et al*., 1992). There are recent reports that the mRNA is upregulated in malignant glioma cells and human myeloid leukemia cells during apoptosis (Kondo *et al*., 1995; Shibata *et al*., 1996). The mRNA is constitutively expressed in several murine cell lines including IC21 and J774 cells (macrophage), WeHI-3 (myelomonocyte), RAW8.1 (lymphoma) and RLM-11 (mature, T cell) (Molineaux *et al*., 1993; Nett *et al*., 1992), and in several tissues, including murine spleen, heart, brain, adrenal glands, liver, aortic endothelial cells and rat brain vasculature (Nett *et al*., 1992; Wong *et al*., 1995; Kondo *et al*., 1996).

Using standard biochemical methods, the proenzyme is the only detectable form of the enzyme in monocytes or THP-1 cells, where it is localized primarily in the cytoplasm (Ayala *et al*., 1994; Singer *et al*., 1995). The mechanism of activation of the enzyme *in vivo* is unknown; in heterologous expression systems activation is, at least in part, autocatalytic (Howard *et al*., 1995; X.-M. Wang *et al*., 1994; Wilson *et al*., 1994). Mature, active enzyme has been detected only by electron microscopy, which shows it predominantly associated with the plasma membrane (Singer *et al*., 1995), yet the enzyme has no obvious membrane-binding domains. In this regard, it is important to note that the enzyme upon which the crystal structure is based was derived from the proenzyme *in vitro*, following cell lysis. This leaves open the possibility that the mature enzyme *in vivo* is post-translationally modified.

Mice that are deficient in caspase-1 do not produce mature IL-1$\beta$, confirming its essential role in proIL-1$\beta$ processing (Kuida *et al*., 1995; Li *et al*., 1995). IL-1$\beta$ has been implicated in the pathogenesis of several diseases including rheumatoid arthritis, endotoxic shock, inflammatory bowel disease, insulin-dependent diabetes mellitus and stroke (Dinarello & Thompson, 1991). Caspase-1-deficient mice are resistant to lipopolysaccharide-induced endotoxic shock, legitimizing the enzyme as a potential therapeutic target in inflammation (Kuida *et al*., 1995; Li *et al*., 1995). In addition, an acyl-oxymethane inhibitor of caspase-1 has recently been shown to block progression of type II collagen-induced arthritis in mice (Ku *et al*., 1996). Surprisingly, mice deficient in caspase-1 are not only defective in the production of IL-1$\beta$, their ability to make IL-1$\alpha$ is also impaired. The biological basis for this phenotype is presently unknown, but these results do imply that the enzyme has multiple biological functions.

Regarding a potential role for this enzyme in cell death, caspase-1-deficient mice develop normally, and do not have any profound defect in apoptosis, suggesting that it does not play an essential, nonredundant role. Instead, there is increasing evidence that members of the caspase family more closely related to CED-3 (such as caspase-3, Chapter 249) are key mediators in this process.

## Distinguishing Features

All members of the caspase family (C14) are distinguished from other known peptidases by their stringent specificity for Asp in P1. (The serine endopeptidase, granzyme B, also has a preference for hydrolysis of aspartyl bonds, but this is less strict; see Chapter 23.) Caspase-1 can further be distinguished from caspase-3 (the only other caspase that is well characterized) by its efficient cleavage of proIL-1$\beta$, inhibition by CrmA ($K_i$ for caspase-1 = 4 pM; $K_i$ for caspase-3 > 100 000 pM), and inhibition by compounds containing the sequence Ac-Tyr-Val-Ala-Asp (e.g. Ac-Tyr-Val-Ala-Asp-CHO: $K_i$ for caspase-1 = 0.76 nM; $K_i$ for caspase-3 = 10 000 nM). Identification of properties that will distinguish caspase-1 from other homologs must await further characterization of these enzymes.

Polyclonal antibodies to caspases-1, -2 and -6 are available from Santa Cruz Biotechnology. Monoclonal antibodies to caspases 2 and 3 are available from Transduction Laboratories. Both polyclonal and monoclonal antibodies to caspase-1 can be purchased from Calbiochem.

## Related Peptidases

A phylogenetic analysis indicates that caspases fall into two major subfamilies, with caspase-1 and CED-3/caspase-3 as representative examples of each (Figure 248.1). Members of the caspase-1 subfamily include caspases-1, -4 and -5 (Cerretti *et al*., 1992; Faucheu *et al*., 1995, 1996; Kamens *et al*., 1995; Munday *et al*., 1995; Thornberry *et al*., 1992). Family members more similar to CED-3 are caspases-2, -3, -6, -7, -8, -9 and -10 (Fernandes-Alnemri *et al*., 1994, 1995a,b, 1996; Kumar *et al*., 1994; L. Wang *et al*., 1994; Nicholson *et al*., 1995; Tewari *et al*., 1995; Boldin *et al*., 1996; Duan *et al*., 1996a,b; Lippke *et al*., 1996; Muzio *et al*., 1996). A sequence comparison indicates that all of these enzymes are heterodimeric cysteine proteinases with a near absolute specificity for Asp in S1. Beyond these similarities, there are significant differences between their proenzyme structures and primary sequences (particularly among those residues that dictate the specificity of S4), suggesting that these enzymes have diverse biological functions. This is clearly true for caspase-1 and the homolog most closely related to CED-3, caspase-3 (described in Chapter 249). Of the other, less-well characterized caspases, caspase-8 is particularly noteworthy as it appears to be associated with the Fas receptor signaling complex during Fas-mediated apoptosis (Boldin *et al*., 1996; Muzio *et al*., 1996).

## Further Reading

For a full review of caspase-1 see Thornberry & Molineaux (1995).

## References

Alnemri, E.S., Fernandes-Alnemri, T. & Litwack, G. (1995) Cloning and expression of four novel isoforms of human interleukin-1β converting enzyme with different apoptotic activities. *J. Biol. Chem.* **270**, 4312–4317.

Alnemri, E.S., Livingston, D.J., Nicholson, D.W., Salvesen, G., Thornberry, N.A., Wong, W.W. & Yuan, J. (1996) Human ICE/CED-3 protease nomenclature. *Cell* **87**, 171.

Ayala, J.M., Yamin, T.-T., Egger, L.A., Chin, J., Kostura, M.J. & Miller, D.K. (1994) Interleukin-1β converting enzyme is present in monocytic cells as an inactive 45 kDa precursor. *J. Immunol.* **153**, 2592–2599.

Black, R.A., Kronheim, S.R. & Sleath, P.R. (1989) Activation of interleukin-1β by a co-induced protease. *FEBS Lett.* **247**, 386–390.

Boldin, M.P., Goncharov, T.M., Goltsev, Y.V. & Wallach, D. (1996) Involvement of MACH, a novel MORT1/FADD-interacting protease, in Fas/APO-1- and TNF receptor-induced cell death. *Cell* **85**, 803–815.

Cerretti, D.P., Kozlosky, C.J., Mosley, B., Nelson, N., Van Ness, K., Greenstreet, T.A., March, C.J., Kronheim, S.R., Druck, T., Cannizzaro, L.A., Huebner, K. & Black, R.A. (1992) Molecular cloning of the interleukin-1β converting enzyme. *Science* **256**, 97–100.

Chapman, K.T. (1992) Synthesis of a potent, reversible inhibitor of interleukin-1β converting enzyme. *Bioorg. Med. Chem. Lett.* **2**, 613–618.

Dinarello, C.A. & Thompson, R.C. (1991) Blocking IL-1: interleukin-1 receptor antagonist *in vivo* and *in vitro*. *Immunol. Today* **12**, 404–410.

Dolle, R.E., Hoyer, D., Prasad, C.V.C., Schmidt, S.J., Helaszek, C.T., Miller, R.E. & Ator, M.A. (1994) P$_1$ aspartate-based peptide α-((2,6-dichlorobenzoyl)oxy)methyl ketones as potent time-dependent inhibitors of interleukin-1β converting enzyme. *J. Med. Chem.* **37**, 563–564.

Duan, H., Chinnaiyan, A.M., Hudson, P.L., Wing, J.P., Wei-Wu, H. & Dixit, V.M. (1996a) ICE-LAP3, a novel mammalian homologue of the *Caenorhabditis elegans* cell death protein Ced-3 is activated during Fas- and tumor necrosis factor-induced apoptosis. *J. Biol. Chem.* **271**, 1621–1625.

Duan, H., Orth, K., Chinnaiyan, A.M., Poirier, G.G., Froelich, C.J., He, W.-W. & Dixit, V.M. (1996b) ICE-LAP6, a novel member of the ICE/Ced-3 gene family, is activated by the cytotoxic T cell protease granzyme B. *J. Biol. Chem.* **271**, 16720–16724.

Faucheu, C., Diu, A., Chan, A.W.E., Blanchet, A.-M., Miossec, C., Hervi, F., Collard-Dutilleul, V., Gu, Y., Aldape, R.A., Lippke, J.A., Rocher, C., Su, M.S.-S., Livingston, D.J., Hercend, T. & Lalanne, J.-L. (1995) A novel human protease similar to the interleukin-1β converting enzyme induces apoptosis in transfected cells. *EMBO J.* **14**, 1914–1922.

Faucheu, C., Blanchet, A.-M., Collard-Dutilleul, V., Lalanne, J.-L.

& Diu-Hercend, A. (1996) Identification of a cysteine protease closely related to interleukin-1β-converting enzyme. *Eur. J. Biochem.* **236**, 207–213.

Fernandes-Alnemri, T., Litwack, G. & Alnemri, E.S. (1994) CPP32, a novel human apoptotic protein with homology to *Caenorhabditis elegans* cell death proteins Ced-3 and mammalian interleukin-1β-converting enzyme. *J. Biol. Chem.* **269**, 30761–30764.

Fernandes-Alnemri, T., Litwack, G. & Alnemri, E.S. (1995a) Mch2, a new member of the apoptotic Ced-3/Ice cysteine protease gene family. *Cancer Res.* **55**, 2737–2742.

Fernandes-Alnemri, T., Takahashi, A., Armstrong, R., Krebs, J., Fritz, L., Tomaselli, K.J., Wang, L., Yu, Z., Croce, C.M., Salveson, G., Earnshaw, W.C., Litwack, G. & Alnemri, E.S. (1995b) Mch3, a novel human apoptotic cysteine protease highly related to CPP32. *Cancer Res.* **55**, 6045–6052.

Fernandes-Alnermi, T., Armstrong, R.C., Krebs, J., Srinivasula, S.M., Wang, L., Bullrich, F., Fritz, L.C., Trapani, J.A., Tomaselli, K.J., Litwack, G. & Alnemri, E.S. (1996) *In vitro* activation of CPP32 and Mch3 by Mch4, a novel human apoptotic cysteine protease containing two FADD-like domains. *Proc. Natl Acad. Sci. USA* **93**, 7464–7469.

Howard, A.D., Kostura, M.J., Thornberry, N., Ding, G.J.F., Limjuco, G., Weidner, J., Salley, J.P., Hogquist, K.A., Chaplin, D.D., Mumford, R.A., Schmidt, J.A. & Tocci, M.J. (1991) IL-1-converting enzyme requires aspartic acid residues for processing of the IL-1β precursor at two distinct sites and does not cleave 31 kDa IL-1α. *J. Immunol.* **147**, 2964–2969.

Howard, A.D., Palyha, O.C., Griffin, P.R., Peterson, E.P., Lenny, A.B., Ding, G.J.-F., Pickup, D.J., Thornberry, N.A., Schmidt, J.A. & Tocci, M.J. (1995) Human IL-1β processing and secretion in recombinant baculovirus-infected *S f* 9 cells is blocked by the cowpox virus serpin *crmA*. *J. Immunol.* **154**, 2321–2332.

Kamens, J., Paskind, M., Hugunin, M., Talanian, R.V., Allen, H., Banach, D., Bump, N., Hackett, M., Johnston, C.G., Li, P., Mankovich, J.A., Terranova, M. & Ghayur, T. (1995) Identification and characterization of ICH-2, a novel member of the interleukin-1β-converting enzyme family of cysteine proteases. *J. Biol. Chem.* **270**, 15250–15256.

Komiyama, T., Ray, C.A., Pickup, D.J., Howard, A.D., Thornberry, N.A., Peterson, E.P. & Salvesen, G. (1994) Inhibition of interleukin-1β converting enzyme by the cowpox virus serpin crmA. *J. Biol. Chem.* **269**, 19331–19337.

Kondo, S., Barna, B.P., Morimura, T., Takeuchi, J., Yuan, J., Akbasak, A. & Barnett, G.H. (1995) Interleukin-1β-converting enzyme mediates cisplatin-induced apoptosis in malignant glioma cells. *Cancer Res.* **55**, 6166–6171.

Kondo, S., Kondo, Y., Yin, D., Barnett, G.H., Kaakaji, R., Peterson, J.W., Morimura, T., Kubo, H., Takeuchi, J. & Barna, B.P. (1996) Involvement of interleukin-1β-converting enzyme in apoptosis of bFGF-deprived murine aortic endothelial cells. *FASEB J.* **10**, 1192–1197.

Kostura, M.J., Tocci, M.J., Limjuco, G., Chin, J., Cameron, P., Hillman, A.G., Chartrain, N.A. & Schmidt, J.A. (1989) Identification of a monocyte specific pre-interleukin 1β convertase activity. *Proc. Natl Acad. Sci. USA* **86**, 5227–5231.

Ku, G., Faust, T., Lauffer, L.L., Livingston, D.J. & Harding, M.W. (1996) Interleukin-1β converting enzyme inhibition blocks progression of type II collagen-induced arthritis in mice. *Cytokine* **8**, 377–386.

Kuida, K., Lippke, J.A., Ku, F., Harding, M.W., Livingston, D.J., Su, M.S.-S. & Flavell, R.A. (1995) Altered cytokine export and apoptosis in mice deficient in interleukin-1β converting enzyme. *Science* **267**, 2000–2003.

Kumar, S., Kinoshita, M., Noda, M., Copeland, N.G. & Jenkins, N.A. (1994) Induction of apoptosis by the mouse *Nedd2* gene, which encodes a protein similar to the product of *Caenorhabditis elegans* cell death gene *ced-3* and the mammalian IL-1β-converting enzyme. *Genes Dev.* **8**, 1613–1626.

Li, P., Allen, H., Banerjee, S., Franklin, S., Herzog, L., Johnston, C., McDowell, J., Paskind, M., Rodman, L., Salfeld, J., Towne, E., Tracey, D., Wardwell, S., Wei, F.-Y., Wong, W., Kamen, R. & Seshadri, T. (1995) Mice deficient in IL-1β-converting enzyme are defective in production of mature IL-1β and resistant to endotoxic shock. *Cell* **80**, 401–411.

Lippke, J.A., Gu, Y., Sarnecki, C., Caron, P.R. & Su, M.S.-S. (1996) Identification and characterization of CPP32/Mch2 homolog 1, a novel cysteine protease similar to CPP32. *J. Biol. Chem.* **271**, 1825–1828.

Malinowski, J.J., Grasberger, B.L., Trakshel, G. *et al.* (1995) Production, purification, and crystallization of human interleukin-1β converting enzyme derived from an *Escherichia coli* expression systems. *Protein Sci.* **4**, 2149–2155.

Miller, D.K., Ayala, J.M., Egger, L.A. *et al.* (1993) Purification and characterization of active human interleukin-1β-converting enzyme from THP.1 monocytic cells. *J. Biol. Chem.* **268**, 18062–18069.

Mjalli, A.M.M., Chapman, K.T. & MacCoss, M. (1993a) Synthesis of a peptidyl 2,2-difluoro-4-phenylbutyl ketone and its evaluation as an inhibitor of interleukin-1β converting enzyme. *Bioorg. Med. Chem. Lett.* **3**, 2693–2698.

Mjalli, A.M.M., Chapman, K.T., MacCoss, M. & Thornberry, N.A. (1993b) Phenylalkyl ketones as potent reversible inhibitors of interleukin-1β converting enzyme. *Bioorg. Med. Chem. Lett.* **3**, 2689–2692.

Mjalli, A.M.M., Chapman, K.T., MacCoss, M., Thornberry, N.A. & Peterson, E.P. (1994) Activated ketones as potent reversible inhibitors of interleukin-1β converting enzyme. *Bioorg. Med. Chem. Lett.* **4**, 1965–1968.

Molineaux, S.M., Casano, F.J., Rolando, A.M., Peterson, E.P., Limjuco, G., Chin, J., Griffin, P.R., Calaycay, J.R., Ding, G.J.-F., Yamin, T.-T., Palyha, O.C., Luell, S., Fletcher, D., Miller, D.K., Howard, A.D., Thornberry, N.A. & Kostura, M.J. (1993) Interleukin 1β (IL-1β) processing in murine macrophages requires a structurally conserved homologue of human IL-1β converting enzyme. *Proc. Natl Acad. Sci. USA* **90**, 1809–1813.

Mullican, M.D., Lauffer, D.J., Gillespie, R.J., Matharu, S.S., Kay, D., Porritt, G.M., Evans, P.L., Golec, J.M.C., Murcko, M.A., Luong, Y.-P., Raybuck, S.A. & Livingston, D.J. (1994) The synthesis and evaluation of peptidyl aspartyl aldehydes as inhibitors of ICE. *Bioorg. Med. Chem. Lett.* **4**, 2359–2364.

Munday, N.A., Vaillancourt, J.P., Ali, A., Casano, F.J., Miller, D.K., Molineaux, S.M., Yamin, T.-T., Yu, V.L. & Nicholson, D.W. (1995) Molecular cloning and pro-apoptotic activity of ICE$_{rel}$II and ICE$_{rel}$III, members of the ICE/CED-3 family of cysteine proteases. *J. Biol. Chem.* **270**, 15870–15876.

Muzio, M., Chinnaiyan, A.M., Kischkel, F.C., O'Rourke, K., Shevchenko, A., Ni, J., Scaffidi, C., Bretz, J.D., Zhang, M., Gentz, R., Mann, M., Krammer, P.J., Peter, M.E. & Dixit, V.M. (1996) FLICE, a novel FADD-homologous ICE/CED-3-like protease, is recruited to the CD95 (Fas/APO-1) death-inducing signaling complex. *Cell* **85**, 817–827.

Nett, M.A., Cerretti, D.P., Berson, D.R., Seavitt, J., Gilbert, D.J.,

Jenkins, N.A., Copeland, N.G., Black, R.A. & Chaplin, D.D. (1992) Molecular cloning of the murine IL-1β converting enzyme cDNA. *J. Immunol.* **149**, 3254–3259.

Nicholson, D.W., Ali, A., Thornberry, N.A., Vaillancourt, J.P., Ding, C.K., Gallant, M., Gareau, Y., Griffin, P.R., Labelle, M., Lazebnik, Y.A., Munday, N.A., Raju, A.M., Smulson, M.E., Yamin, T.-T., Yu, V.L. & Miller, D.K. (1995) Identification and inhibition of the ICE/CED-3 protease necessary for mammalian apoptosis. *Nature* **376**, 37–43.

Pennington, M.W. & Thornberry, N.A. (1994) Synthesis of a fluorogenic interleukin-1β converting-enzyme substrate based on resonance energy-transfer. *Peptide Res.* **7**, 72–76.

Prasad, C.V.C., Prouty, C.P., Hoyer, D., Ross, T.M., Salvino, J.M., Awad, M., Graybill, T.L., Schmidt, S.J., Osifo, I.K., Dolle, R.E., Helaszek, C.T., Miller, R.E. & Ator, M.A. (1995) Structural and stereochemical requirements of time-dependent inactivators of the interleukin-1β converting enzyme. *Bioorg. Med. Chem. Lett.* **5**, 315–318.

Ramage, P., Cheneval, D., Chvei, M., Graff, P., Hemmig, R., Heng, R., Kocher, H.P., Mackenzie, A., Memmert, K., Revesz, L. & Wishart, W. (1995) Expression, refolding, and autocatalytic proteolytic processing of the interleukin-1β-converting enzyme precursor. *J. Biol. Chem.* **270**, 9378–9383.

Ray, C.A., Black, R.A., Kronheim, S.R., Greenstreet, T.A., Sleath, P.R., Salvesen, G.S. & Pickup, D.J. (1992) Viral inhibition of inflammation: Cowpox virus encodes an inhibitor of the interleukin-1β converting enzyme. *Cell* **69**, 597–604.

Reiter, L.A. (1994) Peptidic *p*-nitroanilide substrates of interleukin-1β-converting enzyme. *Int. J. Peptide Protein Res.* **43**, 87–96.

Revesz, L., Briswaiter, C., Heng, R., Leutiwiler, A., Mueller, R. & Wuethrich, H.-J. (1994) Synthesis of P$_1$ aspartate-based peptide acyloxymethyl and fluoromethyl ketones and inhibitors of interleukin-1β-converting enzyme. *Tetrahedron Lett.* **35**, 9693–9696.

Robinson, R.P. & Donahue, K.M. (1992) Synthesis of a peptidyl difluoro ketone bearing the aspartic acid side chain: an inhibitor of interleukin-1β converting enzyme. *J. Org. Chem.* **57**, 7309–7314.

Shibata, Y., Takiguchi, H., Tamura, K., Yamanaka, K., Tezuka, M. & Abiko, Y. (1996) Stimulation of interleukin-1β-converting enzyme activity during growth inhibition by CPT-11 in the human myeloid leukemia cell line K562. *Biochem. Mol. Med.* **57**, 25–30.

Singer, I.I., Scott, S., Chin, J., Bayne, E.K., Limjuco, G., Weidner, J., Miller, D.K., Chapman, K. & Kostura, M.J. (1995) Interleukin-1β converting enzyme (ICE) is localized on the external cell surface membranes and in the cytoplasmic ground substance of human monocytes by immuno-electron microscopy. *J. Exp. Med.* **182**, 1447–1459.

Sleath, P.R., Hendrickson, R.C., Kronheim, S.R., March, C.J. & Black, R.A. (1990) Substrate specificity of the protease that processes human interleukin-1β. *J. Biol. Chem.* **265**, 14526–14528.

Tewari, M., Quan, L.T., O'Rourke, K., Desnoyers, S., Zeng, Z., Beidler, D.R., Poirier, G.G., Salvesen, G.S. & Dixit, V.M. (1995) Yama/CPP32b, a mammalian homolog of CED-3, is a CrmA-inhibitable protease that cleaves the death substrate poly(ADP-ribose) polymerase. *Cell* **81**, 801–809.

Thornberry, N.A. (1994) Interleukin-1β converting enzyme. *Methods Enzymol.* **244**, 615–631.

Thornberry, N.A. & Molineaux, S.M. (1995) Interleukin-1β converting enzyme: a novel cysteine protease required for IL-1β production and implicated in programmed cell death. *Protein Sci.* **4**, 3–12.

Thornberry, N.A., Bull, H.G., Calaycay, J.R. *et al.* (1992) A novel heterodimeric cysteine protease is required for interleukin-1β processing in monocytes. *Nature* **356**, 768–774.

Thornberry, N.A., Peterson, E.P., Zhao, J.J., Howard, A.D., Griffin, P.R. & Chapman, K.T. (1994) Inactivation of interleukin-1β converting enzyme by peptide (acyloxy)methyl ketones. *Biochemistry* **33**, 3934–3940.

Walker, N.P.C., Talanian, R.V., Brady, K.D. *et al.* (1994) Crystal structure of the cysteine protease interleukin-1β-converting enzyme: a (p20/p10)₂ homodimer. *Cell* **78**, 343–352.

Wang, L., Miura, M., Bergeron, L., Zhu, H. & Yuan, J. (1994) Ich-1, an *Ice/ced-3*-related gene, encodes both positive and negative regulators of programmed cell death. *Cell* **78**, 739–750.

Wang, X.-M., Helaszek, C.T., Winter, L.A., Lirette, R.P., Dixon, D.C., Ciccarelli, R.B., Kelley, M.M., Malinowski, J.J., Simmons, S.J., Huston, E.E., Koehn, J.A., Kratz, D., Bruckner, R.C.,

Graybill, T., Ator, M.A., Lehr, R.V. & Stevis, P.E. (1994) Production of active human interleukin-1β-converting enzyme in a baculovirus expression system. *Gene* **145**, 273–277.

Wilson, K.P., Black, J.F., Thomson, J.A., Kim, E.E., Griffith, J.P., Navia, M.A., Murcko, M.A., Chambers, S.P., Aldape, R.A., Raybuck, S.A. & Livingston, D.J. (1994) Structure and mechanism of interleukin-1β converting enzyme. *Nature* **370**, 270–275.

Wong, M.-L., Bongiorno, P.B., Gold, P.W. & Licinio, J. (1995) Localization of interleukin-1β-converting enzyme mRNA in rat brain vasculature: evidence that the genes encoding the interleukin-1 system are constitutively expressed in brain blood vessels. *Neuroimmunomodulation* **2**, 141–148.

Yuan, J., Shaham, S., Ledoux, S., Ellis, H.M. & Horvitz, H.R. (1993) The *C. elegans* cell death gene *ced-3* encodes a protein similar to mammalian interleukin-1β-converting enzyme. *Cell* **75**, 641–652.

*Nancy A. Thornberry*
*Department of Enzymology,*
*Merck Research Laboratories,*
*Rahway, NJ 07065, USA*
*Email: nancy@merck.com*

# 249. Caspase-3

## Databanks

*Peptidase classification: clan CD, family C14, MEROPS ID: C14.003*
*NC-IUBMB enzyme classification: none*
*Databank codes:*

| Species | SwissProt | PIR | EMBL (cDNA) | EMBL (genomic) |
|---|---|---|---|---|
| *Homo sapiens* | P42574 | – | U13738 | – |
| *Mus musculus* | – | – | U49929 | – |

Brookhaven Protein Data Bank three-dimensional structures

| Species | ID | Resolution | Notes |
|---|---|---|---|
| *Homo sapiens* | ICP3 | 2.3 | complex with tetrapeptide inhibitor acteyl-Asp-Val-Ala-Asp-fluoromethane |
| | 1PAU | 2.5 | complex with acetyl-Asp-Glu-Val-Asp-CHO |

## Name and History

Caspase-3 was first identified in 1994 as a protein homologous to caspase-1 (interleukin 1β-converting enzyme, ICE, Chapter 248) and CED-3, the product of a gene required for programmed cell death in the nematode, *Caenorhabditis elegans* (Fernandes-Alnemri *et al.*, 1994). This putative cysteine proteinase was named CPP32 (for a **c**ysteine **p**roteinase that was a **p**rotein of **32** kDa). Several months later, *in vitro* translated CPP32 was reported to be capable of cleaving poly(ADP-ribose) polymerase (PARP), a protein that is proteolytically inactivated during apoptosis, and was renamed *yama*, after the

Hindu god of death (Tewari *et al.*, 1995). At about the same time, the proteinase responsible for specific PARP cleavage from apoptotic cell cytosols was isolated, found to be identical to CPP32, and named *apopain*, to designate its function in apoptosis ('**apop**'), and its cysteine proteinase mechanism ('**ain**') (Nicholson *et al.*, 1995). Recently, a unified nomenclature has been introduced for this proteinase family that employs the trivial name 'caspase' as a root for serial names for all proteins related to ICE and CED-3 (Alnemri *et al.*, 1996). As the third published member of the family, CPP32 (yama, apopain) has now been assigned the name *caspase-3*.

## Activity and Specificity

Caspase-1 (Chapter 248) and caspase-3 are the best characterized members of peptidase family C14 (as defined by Rawlings & Barrett, 1994). Both enzymes are heterodimeric cysteine proteinases, and have a near absolute specificity for Asp in S1, but there are profound differences between their S4 subsites. Caspase-3 cleaves poly(ADP-ribose) polymerase at the sequence Asp-Glu-Val-Asp216$\dashv$Gly217 with a $k_{cat}/K_m$ of $5 \times 10^6$ M$^{-1}$ s$^{-1}$ (Lazebnik *et al.*, 1994; Nicholson *et al.*, 1995; Casciola-Rosen *et al.*, 1996). Other putative endogenous substrates for caspase-3 include 70 kDa subunits of the U1 small ribonucleoprotein (U1-70 kDa), the catalytic subunit of DNA-dependent protein kinase (DNA-$_{PACS}$), and sterol regulatory element-binding proteins (SREBP) (Wang *et al.*, 1995; Casciola-Rosen *et al.*, 1996). All of these are cleaved by caspase-3 at an Asp-Xaa-Xaa-Asp$\dashv$-Xaa motif, suggesting that the enzyme has a stringent requirement for Asp in P4. The X-ray crystal structure supports this prediction (Rotonda *et al.*, 1996). In contrast, caspase-1 prefers hydrophobic amino acids in P4.

The catalytic machinery of caspases-1 and -3 is highly conserved, and many of the strategies used to develop substrates and inhibitors for caspase-1 have been employed successfully with caspase-3. A continuous fluorometric assay utilizing the substrate Ac-Asp-Glu-Val-Asp$\dashv$NHMec has been described (Nicholson *et al.*, 1995). This or a related substrate is available from Peptide Institute, Enzyme Systems, RBI and Bachem (see Appendix 2 for full names and addresses of suppliers). The pH optimum of the enzyme is about 7.0. Assays typically contain at least 1 mM DTT to prevent oxidation of the catalytic Cys, and 10% sucrose and 0.1% CHAPS, to stabilize the protein.

The enzyme is inhibited by thiol alkylating agents such as iodoacetate and *N*-ethylmaleimide (Nicholson *et al.*, 1995). The tetrapeptide aldehyde, Ac-Asp-Glu-Val-Asp-CHO (available from Peptide Institute) is a potent ($K_i < 1$ nM), competitive, reversible inhibitor (Nicholson *et al.*, 1995). Because of the stringent specificity that this enzyme has for Asp in P4, potent inhibitors of caspase-1 containing the sequence Tyr-Val-Ala-Asp are not effective. A fluoromethane, available from Enzyme Systems, Z-Val-Ala-Asp-fluoromethane, inactivates the enzyme with a second-order rate constant of $4.1 \times 10^3$ M$^{-1}$ s$^{-1}$) (N. Thornberry, unpublished results).

## Structural Chemistry

Human caspase-3 is a heterodimer composed of two subunits with molecular masses of 11 896 Da (p12) and 16 617 Da (p17), both of which are derived from a common 32 kDa proenzyme (Nicholson *et al.*, 1995). The proenzyme also encodes a short propeptide of unknown function. Both p12 and p17 are flanked by Asp$\dashv$X sites in the proenzyme, suggesting that activation is either autoproteolytic, or mediated by an enzyme with similar specificity.

The X-ray crystal structure (Rotonda *et al.*, 1996) shows that the enzyme is generally similar in tertiary and quaternary structure to caspase-1 (Walker *et al.*, 1994; Wilson *et al.*, 1994). The two subunits form a heterodimer with a single catalytic domain, with both p12 and p17 contributing residues important for binding and catalysis. The most profound differences between the structures of caspases-1 and -3

are associated with the S4 subsite, and these appear to account entirely for the distinct peptide and macromolecular specificities of the two enzymes. In the crystal, two heterodimers associate to form a tetramer with 2-fold rotational symmetry.

## Preparation

The enzyme was originally isolated from THP-1 cells, a human monocytic cell line, by a combination of ion-exchange and affinity chromatography employing an immobilized peptide aldehyde (Nicholson *et al.*, 1995). More recently, a method for high-level production of the enzyme has been described that involves folding of active enzyme from its constituent subunits which are individually expressed in *Escherichia coli* (Rotonda *et al.*, 1996).

## Biological Aspects

The distribution of caspase-3 has been examined in several human tumor cell lines. High expression of enzyme mRNA was observed in cell lines of the immune system (e.g. lymphocytes, promyelocytes), and in cell lines of brain and embryonic origins (Fernandes-Alnemri *et al.*, 1994). The enzyme was originally isolated from THP-1 cells (Nicholson *et al.*, 1995). The precise subcellular localization of the cytosolic fraction of enzyme is not known.

Caspase-3 appears to function to proteolytically inactivate proteins involved in cellular repair and homeostasis during the effector phase of apoptosis. Proteins that are likely substrates for caspase-3 during apoptosis include PARP, U1-70 kDa, and DNA$_{PACS}$ (Casciola-Rosen *et al.*, 1996). Other potential endogenous substrates are SREBP (Wang *et al.*, 1995) and protein kinase C$\delta$ (PKC$\delta$) (Emoto *et al.*, 1995). Interest in this enzyme as a potential therapeutic target stems from the theory that many diseases (e.g. Alzheimer's disease, cancer) are caused by excessive or insufficient apoptosis (Thompson, 1995). Inhibitors of caspases, including those containing the Ac-Asp-Glu-Val-Asp motif, are effective in models of apoptosis, suggesting that at least some of these enzymes are suitable targets.

The mechanism of proenzyme activation during apoptosis has not been established, but there is increasing evidence to suggest that a protease cascade is involved. For example granzyme B (Chapter 23), a serine protease involved in cytotoxic lymphocyte-mediated cell death, has been shown to activate several caspases, including caspase-3, *in vitro* (Darmon *et al.*, 1995; Chinnaiyan *et al.*, 1996; Duan *et al.*, 1996a,b; Fernandes-Alnermi *et al.*, 1996; Martin *et al.*, 1996; Orth *et al.*, 1996; Quan *et al.*, 1996). In addition, compelling evidence has emerged recently to suggest that caspase-8 participates in the signaling activity of the receptor complex during Fas-mediated cell death, and functions to amplify the death signal via activation of other caspases (Boldin *et al.*, 1996; Muzio *et al.*, 1996).

## Distinguishing Features

See section on Distinguishing Features in Chapter 248.

Monoclonal and polyclonal antibodies are available from Transduction Laboratories, Santa Cruz Technology and Calbiochem-Novabiochem.

## Related Peptidases

Other members of the caspase family are described in Chapter 248.

## Further Reading

Recommended articles are those of Fernandes-Alnemri *et al*. (1994), Nicholson *et al*. (1995) and Rotonda *et al*. (1996).

## References

Alnemri, E.S., Livingston, D.J., Nicholson, D.W., Salvesen, G., Thornberry, N.A., Wong, W.W. & Yuan, J. (1996) Human ICE/CED-3 protease nomenclature. *Cell* **87**, 171.

Boldin, M.P., Goncharov, T.M., Goltsev, Y.V. & Wallach, D. (1996) Involvement of MACH, a novel MORT1/FADD-interacting protease, in Fas/APO-1- and TNF receptor-induced cell death. *Cell* **85**, 803–815.

Casciola-Rosen, L.A., Nicholson, D.W., Chong, T., Rowan, K.R., Thornberry, N.A., Miller, D.K. & Rosen, A. (1996) Apopain/CPP32 cleaves proteins that are essential for cellular repair: a fundamental principle of apoptotic death. *J. Exp. Med.* **183**, 1957–1964.

Chinnaiyan, A.M., Hanna, W.L., Orth, K., Duan, H., Poirier, G.G., Froelich, C.J. & Dixit, V.M. (1996) Cytotoxic T-cell-derived granzyme B activates the apoptotic protease ICE-LAP3. *Curr. Biol.* **6**, 897–899.

Darmon, A.J., Nicholson, D.W. & Bleackley, R.C. (1995) Activation of the apoptotic protease CPP32 by cytotoxic T-cell-derived granzyme B. *Nature* **377**, 446–448.

Duan, H., Chinnaiyan, A.M., Hudson, P.L., Wing, J.P., Wei-Wu, H. & Dixit, V.M. (1996a) ICE-LAP3, a novel mammalian homologue of the *Caenorhabditis elegans* cell death protein Ced-3 is activated during Fas- and tumor necrosis factor-induced apoptosis. *J. Biol. Chem.* **271**, 1621–1625.

Duan, H., Orth, K., Chinnaiyan, A.M., Poirier, G.G., Froelich, C.J., He, W.-W. & Dixit, V.M. (1996b) ICE-LAP6, a novel member of the ICE/Ced-3 gene family, is activated by the cytotoxic T cell protease granzyme B. *J. Biol. Chem.* **271**, 16720–16724.

Emoto, Y., Manome, Y., Meinhardt, G., Kisaki, H., Kharbanda, S., Robertson, M., Ghayur, T., Wong, W.W., Kamen, R., Weichselbaum, R. & Kufe, D. (1995) Proteolytic activation of protein kinase Cδ by an ICE-like protease in apoptotic cells. *EMBO J.* **14**, 6148–6156.

Fernandes-Alnemri, T., Litwack, G. & Alnemri, E.S. (1994) CPP32, a novel human apoptotic protein with homology to *Caenorhabditis elegans* cell death proteins Ced-3 and mammalian interleukin-1β-converting enzyme. *J. Biol. Chem.* **269**, 30761–30764.

Fernandes-Alnermi, T., Armstrong, R.C., Krebs, J., Srinivasula, S.M., Wang, L., Bullrich, F., Fritz, L.C., Trapani, J.A., Tomaselli, K.J., Litwack, G. & Alnemri, E.S. (1996) *In vitro* activation of CPP32 and Mch3 by Mch4, a novel human apoptotic cysteine protease containing two FADD-like domains. *Proc. Natl Acad. Sci. USA* **93**, 7464–7469.

Lazebnik, Y.A., Kaufmann, S.H., Desnoyers, S., Poirer, G.G. & Earnshaw, W.C. (1994) Reconstitution of the apoptotic cascade *in vitro*: pivotal role for prICE, a protease resembling interleukin-1β converting enzyme and demonstration that poly(ADP ribose) polymerase is a substrate for this enzyme during apoptosis. *Nature* **371**, 346–347.

Martin, S.J., Amarante-Mendes, G.P., Shi, L., Chuang, T.-H., Casiano, C.A., O'Brien, G.A., Fitzgerald, P., Tan, E.M., Bokoch, G.M., Greenberg, A.H. & Green, D.R. (1996) The cytotoxic cell protease granzyme B initiates apoptosis in a cell-free system by proteolytic processing and activation of the ICE/CED-3-family protease, CPP32, via a novel two-step mechanism. *EMBO J.* **15**, 2407–2416.

Muzio, M., Chinnaiyan, A.M., Kischkel, F.C., O'Rourke, K., Shevchenko, A., Ni, J., Scaffidi, C., Bretz, J.D., Zhang, M., Gentz, R., Mann, M., Krammer, P.J., Peter, M.E. & Dixit, V.M. (1996) FLICE, a novel FADD-homologous ICE/CED-3-like protease, is recruited to the CD95 (Fas/APO-1) death-inducing signaling complex. *Cell* **85**, 817–827.

Nicholson, D.W., Ali, A., Thornberry, N.A., Vaillancourt, J.P., Ding, C.K., Gallant, M., Gareau, Y., Griffin, P.R., Labelle, M., Lazebnik, Y.A., Munday, N.A., Raju, A.M., Smulson, M.E., Yamin, T.-T., Yu, V.L. & Miller, D.K. (1995) Identification and inhibition of the ICE/CED-3 protease necessary for mammalian apoptosis. *Nature* **376**, 37–43.

Orth, K., Cninnaiyan, A.M., Garg, M., Froelich, C.J. & Dixit, V.M. (1996) The CED-3/ICE-like protease Mch2 is activated during apoptosis and cleaves the death substrate lamin A. *J. Biol. Chem.* **271**, 16443–16446.

Quan, L.T., Tewari, M., O'Rourke, K., Dixit, V., Snipas, S.J., Poirier, G.G., Ray, C., Pickup, D.J. & Salvesen, G.S. (1996) Proteolytic activation of the cell-death protease yama/CPP32 by granzyme B. *Proc. Natl Acad. Sci. USA* **93**, 1972–1976.

Rawlings, N.D. & Barrett, A.J. (1994) Families of cysteine peptidases. *Methods Enzymol.* **244**, 461–486.

Rotonda, J., Nicholson, D.W., Fazil, K.M., Gallant, M., Gareau, Y., Labelle, M., Peterson, E.P., Rasper, D.M., Ruel, R., Vaillancourt, J.P., Thornberry, N.A. & Becker, J.W. (1996) The three-dimensional structure of apopain/CPP32, a key mediator of apoptosis. *Nature Struct. Biol.* **3**, 619–625.

Tewari, M., Quan, L.T., O'Rourke, K., Desnoyers, S., Zeng, Z., Beidler, D.R., Poirier, G.G., Salvesen, G.S. & Dixit, V.M. (1995) Yama/CPP32b, a mammalian homolog of CED-3, is a CrmA-inhibitable protease that cleaves the death substrate poly(ADP-ribose) polymerase. *Cell* **81**, 801–809.

Thompson, C.B. (1995) Apoptosis in the pathogenesis and treatment of disease. *Science* **267**, 1456–1462.

Walker, N.P.C., Talanian, R.V., Brady, K.D. *et al*. (1994) Crystal structure of the cysteine protease interleukin-1β-converting enzyme: a (p20/p10)₂ homodimer. *Cell* **78**, 343–352.

Wang, X., Pai, J., Wiedenfeld, E.A., Medina, J.C., Slaughter, C.A., Goldstein, J.L. & Brown, M.S. (1995) Purification of an interleukin-1β converting enzyme-related cysteine protease that cleaves sterol regulatory element-binding proteins between the leucine zipper and transmembrane domains. *J. Biol. Chem.* **270**, 18044–18050.

Wilson, K.P., Black, J.F., Thomson, J.A., Kim, E.E., Griffith, J.P., Navia, M.A., Murcko, M.A., Chambers, S.P., Aldape, R.A., Raybuck, S.A. & Livingston, D.J. (1994) Structure and mechanism of interleukin-1β converting enzyme. *Nature* **370**, 270–275.

*Nancy A. Thornberry*
*Department of Enzymology,*
*Merck Research Laboratories,*
*Rahway, NJ 07065, USA*
*Email: nancy@merck.com*

# 250. Introduction: clan CE containing the adenovirus endopeptidase

## Databanks

*MEROPS ID: CE*

### Family C5
Adenovirus endopeptidase (Chapter 251)

Clan CE contains only one family, C5, that of the adenovirus endopeptidase. Adenoviruses are double-stranded DNA viruses. They do not encode polyproteins, but instead each protein is encoded by a separate gene. Activation of the proteins involves processing at the N-terminus to remove a propeptide, and it is the adenovirus endopeptidase that is

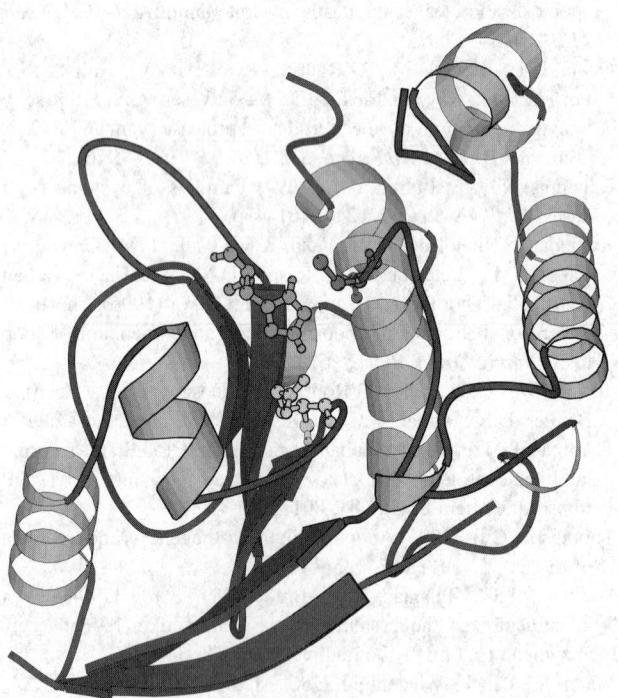

*Figure 250.1* Richardson diagram of the endopeptidase from human adenovirus type 2. The image was prepared from the Brookhaven Protein Data Bank entry (1AVP) as described in the Introduction (p. xxv). Catalytic residues are shown in ball-and-stick representation: His54, Glu71 and Cys122 (numbering as in Alignment 250.1). The eleven-residue cofactor is shown at the bottom of the image.

responsible for this (Weber & Tihanyi, 1994). The catalytic residues have been identified, and occur in the order His, Cys in the sequence (see Alignment 250.1 on CD-ROM). This is in the reverse order to papain, but the same as in the picornains and caspases.

The tertiary structure has been determined for the endopeptidase from human adenovirus type 2 (Figure 250.1). The structure shows a unique $\alpha/\beta$ fold unlike that of any other protein, but with some remarkable similarities to papain. The catalytic dyad is complemented by two other residues: Glu71 (most commonly Asp, in the family as a whole), which has the same role of orientating the imidazolium ring of the catalytic His as Asn175 of papain, and Gln115, which helps form the oxyanion hole, similarly to Gln19 of papain. The catalytic Cys is also positioned at the end of a long helix, as in papain (Ding *et al.*, 1996). However, because the structures are different, these similarities must be the consequence of convergent evolution. It can be seen that the catalytic Cys is followed by Gly, as in the viral endopeptidases of clan CB.

The adenovirus endopeptidase is synthesized as an active enzyme without a propeptide. Further processing is unnecessary because the endopeptidase is highly selective for Xaa-Xbb-Gly-Xbb+Xbb bonds, in which Xaa is either Met, Leu or Ile, and Xbb is any amino acid. A virally encoded, 11 residue cofactor is also required for activity (Webster *et al.*, 1994), and the enzyme is apparently further activated by interaction with DNA.

## References

Ding, J.Z., McGrath, W.J., Sweet, R.M. & Mangel, W.F. (1996) Crystal structure of the human adenovirus proteinase with its 11 amino acid cofactor. *EMBO J.* **15**, 1778–1783.

Weber, J.M. & Tihanyi, K. (1994) Adenovirus endopeptidases. *Methods Enzymol.* **244**, 595–604.

Webster, A., Leith, I.R. & Hay, R.T. (1994) Activation of adenovirus-coded protease and processing of preterminal protein. *J. Virol.* **68**, 7292–7300.

# 251. *Adenovirus protease*

## Databanks

*Peptidase classification: clan CE, family C5, MEROPS ID: C05.001*
*NC-IUBMB enzyme classification: none*
*Databank codes:*

| Species | SwissProt | PIR | EMBL (cDNA) | EMBL (genomic) |
|---------|-----------|-----|-------------|----------------|
| Avian adenovirus 1 | P42672 | – | L13161 | U46933: complete genome |
| Cattle adenovirus type 3 | P19119 | S11460 | X53990 | – |
| Cattle adenovirus 7 | P19151 | S11434 | X53989 | – |
| Dog mastadenovirus c1 | P35990 | B48550 | M72715 | – |
| Human adenovirus type 2 | P03252 | A03823 B03823 | – | J01917: complete genome |
| Human adenovirus type 3 | P10381 | S01988 | X13271 | – |
| Human adenovirus type 4 | P07885 | A27473 | M16692 | – |
| Human adenovirus type 5 | P03253 | C03823 | V00031 | M73260: complete genome |
| Human adenovirus type 12 | P09569 | S01731 S33943 | X07655 | X73487: complete genome |
| Human adenovirus type 40 | P11825 | B28645 | M19540 | L19443: complete genome |
| Human adenovirus type 41 | P11826 | E28645 S08658 | M19539 M19540 M21163 X51783 | – |
| Mastadenovirus mus1 | P30202 | S20314 | M81056 | – |
| Ovine adenovirus | – | – | U40837 | – |
| Pig adenovirus type 3 | – | – | U33016 | – |

Brookhaven Protein Data Bank three-dimensional structures:

| Species | ID | Resolution | Notes |
|---------|-----|------------|-------|
| Human adenovirus type 2 | 1AVP | 2.6 | Complex with peptide cofactor |

## Name and History

A proteolytic activity associated with adenovirus infection was first noticed in pulse-chase experiments and later shown to be defective in the temperature-sensitive mutant ts1 of human adenovirus type 2 (Anderson *et al.*, 1973; Weber, 1976). The ts1 mutation was mapped and sequenced, identifying the location of the protease gene on the viral genome between nucleotides 21 778 and 22 390 (Yeh-Kai *et al.*, 1983). The gene was subsequently sequenced in a wide variety of adenoviruses. It has been cloned into expression vectors, purified, and the crystal structure solved (Anderson, 1990; Tihanyi *et al.*, 1993; Webster *et al.*, 1993; Ding *et al.*, 1996). The enzyme has been referred to as *late 23K* protein or *L3-23K*, being a product of the L3 transcription region (Kräusslich & Wimmer, 1988; Weber, 1990). More recently, it has been abbreviated to *EP* and *AVP*. AVP is the Brookhaven Protein Data Bank designation for the adenovirus protease. Aside from sequencing, all genetic and biochemical work has been carried out on the enzyme of human adenovirus type 2, and consequently, unless indicated otherwise, it is to this *adenovirus protease (AVP)* that the present review refers. Sequence conservation suggests that all adenovirus proteases have similar properties (Weber, 1995).

## Activity and Specificity

Internal bonds in the viral proteins are cleaved at two consensus sites, (Met,Ile,Leu)-Xaa-Gly-Gly+Xaa or (Met,Ile,Leu)-Xaa-Gly-Xaa+Gly, where Xaa is apparently any amino acid (Webster *et al.*, 1989; Anderson, 1990; Weber, 1995). The rate of hydrolysis appears to depend on the nature of the variable residues at the Xaa sites (Webster *et al.*, 1989; Diouri *et al.*, 1995). Recent experiments suggest that Gly-Xaa+Gly sites may be cleaved 3- to 4-fold faster than Gly-Gly+Xaa sites (unpublished results of the author). pH and temperature optima are about 8 and 40–45°C, respectively (Bhatti & Weber, 1979a,b; Webster *et al.*, 1993; Tihanyi *et al.*, 1993). Activity is greatly stimulated by an 11 residue cleavage fragment of viral capsid protein pre-VI (Gly-Val-Gln-Ser-Leu-Lys-Arg-Arg-Arg-Cys-Phe in human adenovirus 2) which forms a disulfide bond with Cys104 of the enzyme (Mangel *et al.*, 1993; Webster *et al.*, 1993; Ding *et al.*, 1996; Jones *et al.*, 1996). Activity is further stimulated by low concentrations of thiol compounds (1 mM DTT) and NaCl (25 mM). There have also been reports of stimulation by negatively charged polymers, particularly DNA (Mangel *et al.*, 1993, 1996).

Assays are most conveniently made with the R110 substrate (available from Molecular Probes; see Appendix 2 for full names and addresses of suppliers), which emits rhodamine fluorescence upon cleavage (McGrath *et al.*, 1996; Diouri *et al.*, 1995). AVP is inhibited by general inhibitors of cysteine proteases (Sircar *et al.*, 1996). Attempts are being made to design specific inhibitors.

## Structural Chemistry

AVP is a single-chain protein of 201–214 amino acids, depending on the virus serotype. It has a pI of 10.2 and it does not contain disulfide bonds nor any known post-translational modifications (Tihanyi *et al.*, 1993; Weber & Tihanyi, 1994). AVP has no sequence relationship with any other protein detected to date. The active site is His54, Glu71 (or Asp71), Cys122, in an arrangement similar to that in papain. The atomic structure of AVP complexed with its 11 amino acid cofactor has been determined at 2.6 Å resolution (Ding *et al.*, 1996), and those authors concluded that despite some striking similarities in the geometry of the catalytic site to that of papain, the fold of the protein is distinct.

## Preparation

AVP is encapsidated and can be isolated by disrupting purified virions (Tremblay *et al.*, 1983; Webster *et al.*, 1989). The protease has also been cloned into various vectors and expressed in *Escherichia coli* or insect cells (Houde & Weber, 1990; Anderson, 1990; Tihanyi *et al.*, 1993; Webster *et al.*, 1993; Weber & Tihanyi, 1994; Keyvani-Amineh *et al.*, 1995).

## Biological Aspects

AVP is encoded by adenoviruses near the middle of the genome in the rightward transcribed L3 region. Upon synthesis during the late phase of infection, AVP is rapidly transported to the site of virus assembly in the nucleus, and becomes associated with assembly intermediates, eventually packaged along with the viral genome (Rancourt *et al.*, 1995). The role of the enzyme in viral assembly is not understood in detail, but it is known to cleave a presumed scaffolding protein, L1-52K to make space for the genome (Hasson *et al.*, 1992). Other viral proteins that are cleaved include capsid proteins pre-VI, pre-VIII, pre-IIIa, and core proteins pTP, pre-VII and pre-X (also called $\mu$). Virus infectivity is dependent on these cleavages. Two cellular proteins, cytokeratins K7 and K18 are also cleaved, with consequent depolymerization and alterations in cell shape (Chen *et al.*, 1993). Mature virus contains encapsidated active enzyme which may have a role in the very early events of infection, possibly in decapsidation and release from endosomes (Cotten & Weber, 1995; Greber *et al.*, 1996). Mutations affecting the active site and

other parts of the AVP molecule have been prepared, and the phenotypes studied (Rancourt *et al.*, 1994; Grierson *et al.*, 1994; Jones *et al.*, 1996).

## Distinguishing Features

AVP does not cleave all viral proteins that contain a potential cleavage site, suggesting that proper exposure is important. Nonviral proteins that contain sites have also been cleaved, including fibrinogen and baculovirus protease, whereas actin and ovalbumin were cleaved only after denaturation (Houde & Weber, 1990; Tihanyi *et al.*, 1993; Keyvani-Amineh *et al.*, 1995).

## Further Reading

A full review has been written by Weber (1995).

## References

Anderson, C.W. (1990) The proteinase polypeptide of adenovirus serotype 2 virions. *Virology* **177**, 259–272.

Anderson, C.W., Baum, P.R. & Gesteland, R.F. (1973) Processing of adenovirus 2-induced proteins. *J. Virol.* **12**, 241–252.

Bhatti, A.R. & Weber, J. (1979a) Protease of adenovirus type 2: partial characterization. *Virology* **96**, 478–485.

Bhatti, A.R. & Weber, J.M. (1979b) Protease of adenovirus type 2: subcellular localization. *J. Biol. Chem.* **254**, 12265–12268.

Chen, P.H., Ornelles, D.A. & Shenk, T. (1993) The adenovirus L3 23-kilodalton proteinase cleaves the amino-terminal head domain from cytokeratin system 18 and disrupts the cytokeratin network of HeLa cells. *J. Virol.* **67**, 3507–3514.

Cotten, M. & Weber, J.M. (1995) The adenovirus protease is required for virus entry into host cells. *Virology* **213**, 494–502.

Ding, J., McGrath, W.J., Sweet, R.M. & Mangel, W.F. (1996) Crystal structure of the human adenovirus proteinase with its 11 amino acid cofactor. *EMBO J.* **15**, 1778–1783.

Diouri, M., Geoghegan, K.F. & Weber, J.M. (1995) Functional characterization of the adenovirus proteinase using fluorogenic substrates. *Protein Peptide Lett.* **6**, 363–370.

Grierson, A.W., Nicholson, R., Talbot, P., Webster, A. & Kemp, G. (1994) The protease of adenovirus serotype 2 requires cysteine residues for both activation and catalysis. *J. Gen. Virol.* **75**, 2761–2764.

Greber, U.F., Webster, P., Weber, J. & Helenius, A. (1996) The role of the adenovirus protease in virus entry into cells. *EMBO J.* **15**, 1766–1777.

Hasson, T.B., Ornelles, D.A. & Shenk, T. (1992) Adenovirus L1 52- and 55-kilodalton proteins are present within assembling virions and colocalize with nuclear structures distinct from replication centers. *J. Virol.* **66**, 6133–6142.

Houde, A. & Weber, J.M. (1990) Adenovirus proteinases: comparaison of amino acid sequences and expression of the cloned cDNA in *Escherichia coli*. *Gene* **88**, 269–273.

Jones, S.J., Iqbal, M., Grierson, A.W. & Kemp., G. (1996) Activation of the protease from human adenovirus type 2 is accompanied by a conformational change which is dependent on cysteine-104. *J. Gen. Virol.* **77**, 1821–1824.

Keyvani-Amineh, H., Labrecque, P., Cai, F., Carstens, E.B. & Weber, J.M. (1995) Adenovirus protease expressed in insect cells cleaves adenovirus proteins, ovalbumin and baculovirus protease in the absence of activating peptide. *Virus Res.* **37**, 87–97.

Kräusslich, H.-G. & Wimmer, E. (1988) Viral proteinases. *Annu. Rev. Biochem.* **57**, 701–754.

Mangel, W.F., McGrath, W.J., Toledo, D.L. & Anderson, C.W. (1993) Viral DNA and a viral peptide can act as cofactors of adenovirus virion proteinase activity. *Nature* **361**, 274–275.

Mangel, W.F., Toledo, D.L., Brown, M.T., Martin, J.H. & McGrath, W.J. (1996) Characterization of three components of human adenovirus proteinase activity *in vitro*. *J. Biol. Chem.* **271**, 536–543.

McGrath, W.J., Abola, A.P., Toledo, D.L., Brown, M.T. & Mangel, W.F. (1996) Characterization of human adenovirus proteinase activity in disrupted virus particles. *Virology* **217**, 131–138.

Rancourt, C., Tihanyi, K., Bourbonnière, M. & Weber, J.M. (1994) Identification of active-site residues of the adenovirus endopeptidase. *Proc. Natl Acad. Sci. USA* **91**, 844–847.

Rancourt, C., Keyvani-Amineh, H., Sircar, S., Labrecque, P. & Weber, J.M. (1995) Proline 137 is critical for adenovirus protease encapsidation and activation but not enzyme activity. *Virology* **209**, 167–173.

Sircar, S., Keyvani-Amineh, H. & Weber, J.M. (1996) Inhibition of adenovirus infection with protease inhibitors. *Antiviral Res.* **30**, 147–153.

Tihanyi, K., Bourbonniere, M., Houde, A., Rancourt, C. & Weber, J.M. (1993) Isolation and properties of the adenovirus type 2 proteinase. *J. Biol. Chem.* **268**, 1780–1785.

Tremblay, M., Dery, C., Talbot, B. & Weber, J.M. (1983) In vitro cleavage specificity of the adenovirus type 2 proteinase. *Biochim. Biophys. Acta* **743**, 239–245.

Weber, J. (1976) Genetic analysis of adenovirus type 2 III. Temperature sensitivity of processing of viral proteins. *J. Virol.* **17**, 462–471.

Weber, J.M. (1990) The adenovirus proteinase. *Semin. Virol.* **1**, 379–384.

Weber, J.M. (1995) The adenovirus endopeptidase and its role in virus infection. In: *Molecular Repertoire of Adenoviruses* (Doerfler, W. & Bohm, P., eds). *Curr. Top. Microbiol. Immunol.* **199/I**, 227–235.

Weber, J.M. & Tihanyi, K. (1994) Adenovirus endopeptidases. *Methods Enzymol.* **244**, 595–604.

Webster, A., Russell, W.C. & Kemp, G.D. (1989) Characterization of the adenovirus proteinase; substrate specificity. *J. Gen. Virol.* **70**, 3215–3223.

Webster, A., Hay, R.T. & Kemp, G. (1993) The adenovirus protease is activated by a virus-coded disulphide-linked peptide. *Cell* **72**, 97–104

Yeh-Kai, L., Akusjärvi, G., Aleström, P., Pettersson, U., Tremblay, M. & Weber, J. (1983) Genetic identification of an endoproteinase encoded by the adenovirus genome. *J. Mol. Biol.* **167**, 217–222.

*Joseph M. Weber*
*Department of Microbiology,*
*Faculty of Medicine,*
*University of Sherbrooke,*
*Sherbrooke, Quebec, Canada, J1H 5N4*
*Email: j.weber@courrier.usherb.ca*

# 252. Introduction: other cysteine peptidases

## Databanks

MEROPS ID: C11, C13, C15, C22, C25, C26, C39, C40

| Species | SwissProt | PIR | EMBL (cDNA) | EMBL (genomic) |
|---|---|---|---|---|
| **Family C11** | | | | |
| Clostripain (Chapter 257) | | | | |
| **Family C13** | | | | |
| Legumain 1, mammal (Chapter 255) | | | | |
| Glycosylphosphatidylinositol:protein transamidase (Chapter 256) | | | | |
| Legumain, plant (Chapter 253) | | | | |
| Legumain, trematode (*Schistosoma, Fasciola*) (Chapter 254) | | | | |
| **Family C15** | | | | |
| Pyroglutamyl-peptidase I (bacterial) (Chapter 268) | | | | |
| **Family C22** | | | | |
| *Trichomonas* putative cysteine proteinase | | | | |
|   *Trichomonas vaginalis* | – | – | X70823 | – |
| **Family C25** | | | | |
| Gingipain K (Chapter 258) | | | | |
| Gingipain R (Chapter 258) | | | | |
| **Family C26** | | | | |
| γ-Glu-X carboxypeptidase (Chapter 264) | | | | |
| **Family C39** | | | | |
| ABC-Protease (*Lactococcus*) (Chapter 259) | | | | |
| **Family C40** | | | | |
| Dipeptidyl-peptidase VI (Chapter 265) | | | | |

There are a number of families of cysteine peptidases for which tertiary structures are unknown, and which cannot be assigned to clans, but are described as being in "clan CX" when a nominal assignment is required. Although there are some poorly characterized peptidases in this group, there are also some extremely well-known enzymes. We shall consider the endopeptidases first, and then the exopeptidases.

*Family C13* includes a number of endopeptidases that specifically cleave asparaginyl bonds. An asparaginyl endopeptidase was first identified in seeds of leguminous plants and was named 'legumain' (Kembhavi *et al.*, 1993). The legumain of plant seeds has now been found in a wide variety of dicotyledonous plants and is proposed to be responsible for the post-translational processing of seed proteins prior to storage (Hara-Nishimura *et al.*, 1995). The processing of a few of these proteins, such as concanavalin A, also involves protein splicing at asparaginyl bonds, and this transamidase reaction also can be mediated by legumain (Min & Jones, 1994). Homologous asparaginyl endopeptidases have also been characterized from the blood flukes *Schistosoma mansoni* and *S. japonicum*; they were initially thought to be responsible for hemoglobin breakdown, and

were termed 'hemoglobinases', but this name no longer seems appropriate (see Chapter 254). More recently, a mammalian legumain has been characterized, and shown to be a lysosomal enzyme (Chen *et al.*, 1997). Like the endopeptidases of the papain family, legumain is inhibited by cystatins. However, it differs in that it is not inhibited by E-64 and, like clostripain, it reacts more rapidly with iodoacetamide than with iodoacetate.

Not a conventional endopeptidase, but also in the legumain family, is glycosylphosphatidylinositol:protein transamidase. This has been identified in *Saccharomyces* and humans, and is responsible for attaching glycosylphosphatidylinositol (GPI) anchors to the C-termini of newly synthesized proteins in the endoplasmic reticulum. A C-terminal peptide is removed before the preformed GPI anchor is attached, apparently in a single reaction. The transamidase differs considerably in sequence from the legumains, and the Cys residue that had been predicted to be catalytic in *Schistosoma* legumain (Götz & Klinkert, 1993) is replaced by Leu. Figure 252.1 shows an evolutionary tree for the family, and the marked difference between the forms of legumain and the GPI8 transamidase (Chapter 256) is clear.

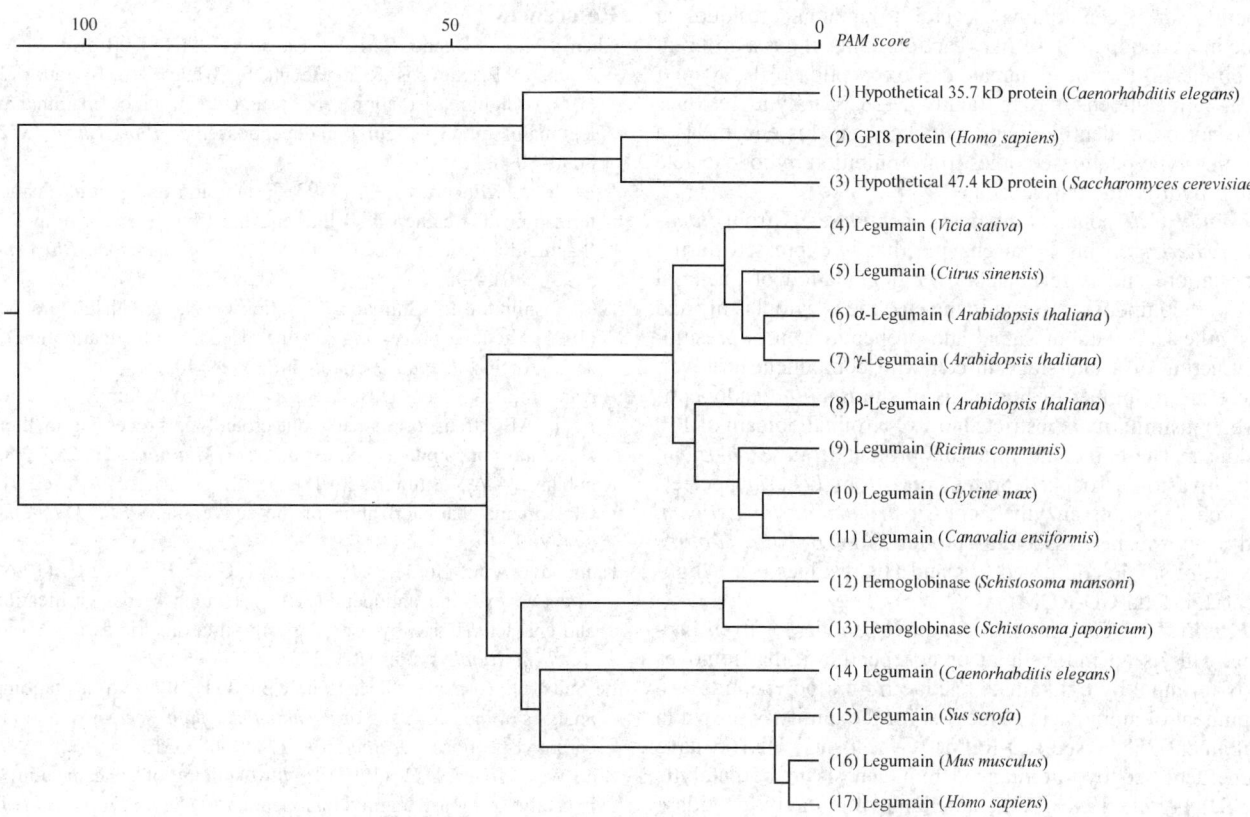

*Figure 252.1*  Evolutionary tree for the legumain family, C13. The tree was constructed as described in the Introduction (p. xxv).

**Family C11** contains only clostripain. This is an endopeptidase from *Clostridium histolyticum* that has a preference for hydrolysis of arginyl bonds. Clostripain is synthesized as a proprotein, and activation requires loss of a 23 residue propeptide and autolytic removal of an internal nonapeptide to produce a heterodimer consisting of a 131 residue light chain and a 336 residue heavy chain. The catalytic Cys has been identified in the heavy chain. Activity depends also on a His, as has been demonstrated by the use of diethyl pyrocarbonate (Kembhavi *et al.*, 1991), but which of the seven His residues in the heavy chain and two in the light chain complete the catalytic dyad is unknown. Clostripain differs from enzymes of the papain family in a number of respects: it is calcium dependent, it is more rapidly inactivated by iodoacetamide than by iodoacetate, and E-64 gives only reversible inhibition.

**Family C25** is represented by gingipains R and K, secreted endopeptidases from the pathogenic bacterium *Porphyromonas gingivalis* that may contribute to the disease process in a number of ways. Gingipain is a multidomain protein, containing not only a peptidase unit but also a C-terminal hemagglutinin domain. It is synthesized with an N-terminal propeptide, and the catalytic cysteine has been identified by labeling with a tritiated chloromethane (Nishikata & Yoshimura, 1995).

**Family C39**. There are a number of specialized secretory pathways in bacteria that have equally specialized endopeptidases associated with them to remove targeting signals from the secreted proteins. Besides signal peptidase I (Chapter 153), which is a general-purpose processing enzyme for proteins targeted to the periplasm, there are signal peptidase II (Chapter 333), which removes the targeting signal from

various lipoproteins; type IV prepilin peptidase (Chapter 560), which removes targeting peptides from type IV prepilin; and at least two peptidases involved in the processing of bacteriocins and lantibiotics, which are exported to the surrounding medium (Chapter 109). Some lantibiotics are processed by lantibiotic leader peptidase, which is a subtilisin homolog, whereas a cysteine-type endopeptidase is responsible for the processing of bacteriocins exported by the ATP-binding cassette (ABC) system. The translocator-associated bacteriocin leader peptidase is a member of family C39 and is a multidomain protein that has the peptidase unit at the N-terminus, followed by an ATP-binding domain and a membrane-spanning domain. The nonpeptidase domains are homologous to a variety of other ABC-transport proteins. Cleavage occurs at a Gly-Gly+Xaa bond. There are conserved Cys and His residues in the peptidase unit, occurring in the order Cys, His, and mutation of Cys to Ala abolishes processing activity (Håvarstein *et al.*, 1995).

Among the endopeptidases that have been identified as cysteine peptidases by inhibition characteristics, but for which amino acid sequences are not available (leading to assignment to family 'C99'), are isoprenylated protein peptidase (Chapter 260), cathepsin T (Chapter 261), cancer procoagulant (Chapter 262) and prohormone thiol protease (Chapter 263).

**Family C26** contains a $\gamma$-glutamyl hydrolase from rat and human, also detected as an EST in *Arabidopsis*. The enzyme is lysosomal, and is responsible for the release of $\gamma$-glutamate residues from pteroylpoly-$\gamma$-glutamate to yield pteroyl-$\alpha$-glutamate (folic acid) and free glutamate. The enzyme can remove $\gamma$-glutamate residues progressively by cleaving between the $\gamma$-glutamyl bonds in a carboxypeptidase-like reaction, or it can remove several $\gamma$-glutamate residues at once in an endopeptidase-like reaction. Thus, it has a different specificity to that of glutamate carboxypeptidase II (Chapter 492), a metalloenzyme in family M28. Catalytic residues have not been identified, and evidence that this enzyme is a cysteine-type peptidase comes from inhibition by iodoacetate and *p*-hydroxymercuribenzoate.

**Family C40** contains dipeptidyl-peptidase VI from *Bacillus sphaericus*. This is an enzyme that is expressed during sporulation, and is responsible for degradation of bacterial cell wall components. Because the enzyme is cytoplasmic and is synthesized without signal and propeptides, it is presumably acting at a late stage in cell wall component turnover. There are a number of homologs of the *Bacillus* endopeptidase, but similarity is restricted to a C-terminal domain of 110 residues. These include nlpC lipoprotein from *Escherichia coli*, invasion-associated protein p60 from *Listeria* species, a starch-degrading enzyme from *Clostridium acetobutylicum*, and a phosphatase-associated protein from *Bacillus subtilis*. There are single conserved Cys and His residues (see Alignment 252.1 on CD-ROM).

**Family C15** contains pyroglutamyl-peptidase I from bacteria. This is an intracellular omega peptidase that removes an N-terminal pyroglutamate residue from a polypeptide. An alignment of amino acid sequences for the family is shown in Alignment 252.2 (see CD-ROM). Cys144 and His166 have been identified by site-directed mutagenesis to be catalytic (Le Saux *et al*., 1996). A mammalian pyroglutamyl-peptidase exists, and an homologous mouse EST is known.

**Family C22** contains a fragment of a putative peptidase from *Trichomonas vaginalis*. The sequence is rich in charged residues, particularly Lys and Glu.

**References**

Chen, J.M., Dando, P.M., Rawlings, N.D., Brown, M.A., Young, N.E., Stevens, R.A., Hewitt, E., Watts, C. & Barrett, A.J. (1997) Cloning, isolation, and characterization of mammalian legumain, an asparaginyl endopeptidase. *J. Biol. Chem.* **272**, 8090–8098.

Götz, B. & Klinkert, M.-Q. (1993) Expression and partial characterization of a cathepsin B-like enzyme (Sm31) and a proposed 'haemoglobinase' (Sm32) from *Schistosoma mansoni. Biochem. J.* **290**, 801–806.

Hara-Nishimura, I., Shimada, T., Hiraiwa, N. & Nishimura, M. (1995) Vacuolar processing enzyme responsible for maturation of seed proteins. *J. Plant Physiol.* **145**, 632–640.

Håvarstein, L.S., Diep, D.B. & Nes, I.F. (1995) A family of bacteriocin ABC transporters carry out proteolytic processing of their substrates concomitant with export. *Mol. Microbiol.* **16**, 229–240.

Kembhavi, A.A., Buttle, D.J., Rauber, P. & Barrett, A.J. (1991) Clostripain: characterization of the active site. *FEBS Lett.* **283**, 277–280.

Kembhavi, A.A., Buttle, D.J., Knight, C.G. & Barrett, A.J. (1993) The two cysteine endopeptidases of legume seeds: purification and characterization by use of specific fluorometric assays. *Arch. Biochem. Biophys.* **303**, 208–213

Le Saux, O., Gonzales, T. & Robert-Baudouy, J. (1996) Mutational analysis of the active site of *Pseudomonas fluorescens* pyrrolidone carboxyl peptidase. *J. Bacteriol.* **178**, 3308–3313.

Min, W. & Jones, D.H. (1994) In vitro splicing of concanavalin A is catalyzed by asparaginyl endopeptidase. *Nature Struct. Biol.* **1**, 502–504.

Nishikata, M. & Yoshimura, F. (1995) Active site structure of a hemagglutinating protease from *Porphyromonas gingivalis*: similarity to clostripain. *Biochem. Mol. Biol. Int.* **37**, 547–553.

# 253. *Asparaginyl endopeptidase*

## *Databanks*

*Peptidase classification: clan CX, family C13, MEROPS ID: C13.001*
*NC-IUBMB enzyme classification: EC 3.4.22.34*
*Chemical Abstract Service registry number: 149371-18-6*
*Databank codes:*

| Species | SwissProt | PIR | EMBL (cDNA) | EMBL (genomic) |
|---|---|---|---|---|
| **Legumain α** | | | | |
| *Arabidopsis thaliana* | – | – | – | D61394: complete gene |
| *Vicia sativa* | P49044 | S49175 | Z34899 | – |
| **Legumain β** | | | | |
| *Arabidopsis thaliana* | P49047 | – | – | D61393: complete gene |
| *Canavalia ensiformis* | P49046 | JX0344 | D31787 | – |
| *Citrus sinensis* | P49043 | S51117 | Z47793 | – |

| Species | SwissProt | PIR | EMBL (cDNA) | EMBL (genomic) |
|---|---|---|---|---|
| Legumain β (*continued*) | | | | |
| Glycine max | P49045 | – | D28876 | – |
| Oryza sativa | – | – | D22244 | – |
| Ricinus communis | P49042 | JQ2387 | D17401 | – |
| Legumain γ | | | | |
| Arabidopsis thaliana | – | – | – | D61395: complete gene |

Notes: The designation of sequences other than those of *Arabidopsis* as legumain β is based on groupings from the KITSCH evolutionary tree (Fig. 252.1).

## Name and History

A cysteine endopeptidase with specificity towards asparaginyl bonds was found in germinating seeds of kidney bean (Csoma & Polgár, 1984). The enzyme was designated **proteinase B** in the work of Shutov & Vaintraub (1987) as distinct from proteinase A (or vignain) (Chapter 199), which is a cysteine endopeptidase of the papain family (C1) that has a broad substrate specificity in germinating legume seeds. Proteinase B was thought to be involved in the degradation of storage proteins during germination of legume seeds, and the name **vicillin peptidohydrolase** has been used (see below). The term **asparaginyl endopeptidase** was applied to the enzyme from jackbean (*Canavalia ensiformis*) by Ishii *et al.* (1990). In 1991, however, an asparaginyl endopeptidase responsible for maturation of seed storage proteins, but not for degradation of storage proteins, was discovered in protein-storage vacuoles of castor bean (Hara-Nishimura *et al.*, 1991). This enzyme was designated **vacuolar processing enzyme (VPE)**, because of its role in post-translational processing of vacuolar proteins in various plant organs (see below). In 1992, the name **legumain** (EC 3.4.22.34) was recommended by NC-IUBMB for the asparaginyl endopeptidase that had been found in legume seeds. Kembhavi *et al.* (1993) used the name legumain in describing the purification of the enzyme from seeds of the moth bean (*Vigna aconitifolia*). We shall here use the terms asparaginyl endopeptidase and VPE for the vacuolar processing enzyme of plants.

The primary structure of VPE was first deduced from a cDNA from castor bean (Hara-Nishimura *et al.*, 1993b). The sequence showed that the enzyme is homologous to hemoglobinase or Sm32, a putative cysteine proteinase of the human parasite, *Schistosoma mansoni* (Klinkert *et al.*, 1989) (see Chapter 254), and these were the first peptidases of the family termed C13 by Rawlings & Barrett (1994).

## Activity and Specificity

Asparaginyl endopeptidases have a restricted specificity toward Asn in small peptides and proteins. The jackbean enzyme hydrolyzes peptide bonds with Asn in P1 and all 20 amino acids in P1′ including *S*-alkylated Cys, with a few exceptions (Abe *et al.*, 1993). Asparaginyl endopeptidase did not act on Asn when it was in the first or second positions from the N-terminus, or when the residue was *N*-glycosylated, but it could act when Asn was at the penultimate position to the C-terminus (Abe *et al.*, 1993). Recently, the vetch enzyme was reported to act toward Asp

as well as Asn (Becker *et al.*, 1995), but this may require confirmation.

Assay of the enzyme activity on small substrates has been done with Boc-Asn⊥OPhNO$_2$ (Csoma & Polgár, 1984), Z-Ala-Ala-Asn⊥NHMec (Kembhavi *et al.*, 1993), and Dnp-Pro-Glu-Ala-Asn⊥NH$_2$ (Abe *et al.*, 1993). Synthetic peptides can also be used as substrates: the sequences Ser-Glu-Ser-Glu-Asn⊥Gly-Leu-Glu-Glu-Thr (Hara-Nishimura *et al.*, 1991, 1993b; Hiraiwa *et al.*, 1993) and Glu-Thr-Arg-Asn⊥Gly-Val-Glu-Glu (Becker *et al.*, 1995) were derived from those around the processing sites of 11S globulin precursors of pumpkin and soybean, respectively. For the assay of the proteolytic processing by VPE, authentic proprotein precursors to seed proteins, including 11S globulin, 2S albumin and 51 kDa protein, have been prepared from developing pumpkin seeds and used as substrates (Hara-Nishimura & Nishimura, 1987; Hara-Nishimura *et al.*, 1991, 1993a).

The pH optimum is about 5.0–5.5 (Hara-Nishimura *et al.*, 1991; Kembhavi *et al.*, 1993; Cornel & Plaxton, 1994). Activity is stimulated by EDTA (0.1–1 mM) and DTT (1–100 mM). Abe *et al.* (1993) tabulated studies on various inhibitors of jackbean enzyme. Both proteolytic processing and activities on small peptides are effectively inhibited by *N*-ethylmaleimide and *p*-chloromercuribenzene sulfonic acid, but not by E-64 (Hara-Nishimura & Nishimura, 1987; Hara-Nishimura *et al.*, 1991; Abe *et al.*, 1993; Hiraiwa *et al.*, 1993).

## Structural Chemistry

Asparaginyl endopeptidases are soluble single-chain proteins with molecular masses generally in the range 37–39 kDa (Hara-Nishimura *et al.*, 1991; Abe *et al.*, 1993; Shimada *et al.*, 1994; Becker *et al.*, 1995). Exceptions to the above molecular mass range are mothbean enzyme (33 kDa: Kembhavi *et al.*, 1993), kidney bean enzyme (23.4 kDa: Csoma & Polgár, 1984) and soybean enzymes (33.0–33.8 kDa: Muramatsu & Fukazawa, 1993; 80 kDa: Scott *et al.*, 1992). Mothbean legumain might be *N*-glycosylated (Kembhavi *et al.*, 1993), but castor bean VPE is not (Hara-Nishimura *et al.*, 1991).

Primary structures of the precursors of the enzymes were deduced from cDNAs in castor bean (Hara-Nishimura *et al.*, 1993b), soybean (Shimada *et al.*, 1994), *Citrus* (Alonso & Granell, 1995), jackbean (Takeda *et al.*, 1994) and genomic DNAs of *Arabidopsis* (Kinoshita *et al.*, 1995a,b). Castor bean VPE is synthesized as a 51 kDa inactive precursor (preproVPE) that is converted into mature VPE by cotranslational cleavage of an N-terminal signal peptide and subsequent cleavage of propeptides (Hara-Nishimura *et al.*, 1993b). Removal of the propeptides is required for activation of the enzyme (Hara-Nishimura *et al.*, 1993b).

### Preparation

Asparaginyl endopeptidase of jackbean is commercially available (Takara) (see Appendix 2 for full names and addresses of suppliers). Homogeneous preparations of asparaginyl endopeptidases have been obtained from dry seeds of castor bean (Hara-Nishimura *et al.*, 1991), soybean (Muramatsu & Fukazawa, 1993) and jackbean (Abe *et al.*, 1993) and germinating seeds of kidney bean (Csoma & Polgár, 1984), mothbean (Kembhavi *et al.*, 1993) and vetch (Becker *et al.*, 1995). VPE activity is detected in different tissues of castor bean, and the highest VPE activity is found in the endosperms at the late stage of seed maturation (Hiraiwa *et al.*, 1993).

### Biological Aspects

Asparaginyl endopeptidases were initially thought to be involved in the degradation of seed storage proteins. Baumgartner & Chrispeels (1977) purified a cysteine endopeptidase from mung bean seedlings that had a specificity toward Asn and Gln and that could hydrolyze a seed protein, vicilin. They called the enzyme vicilin peptidohydrolase. In contrast, the asparaginyl endopeptidases from seeds of castor bean and vetch are inactive towards seed proteins (Hara-Nishimura *et al.*, 1991; Shutov & Vaintraub, 1987).

Castor bean enzyme (VPE) can cleave certain asparaginyl bonds of proprotein precursors to various seed proteins to generate their respective mature forms; the enzyme cleaves on the C-terminal side of Asn residues exposed on the molecular surface of the proprotein precursors (Hara-Nishimura & Nishimura, 1987; Hara-Nishimura *et al.*, 1991, 1993a). Similar proteolytic processing at Asn residues of the bean α-amylase inhibitor proprotein is required for its activation (Pueyo *et al.*, 1993). Proprotein precursor of concanavalin A, a major seed lectin in jackbean, is converted to the mature form by proteolytic processing at exposed Asn residues on the molecular surface and transpeptidation (Bowles *et al.*, 1986). It is possible that the asparaginyl endopeptidase might be involved in the transpeptidation reaction.

Most asparaginyl endopeptidases were purified from seeds and seedlings (described above). However, VPEs are distributed throughout various plant tissues and organs, including endosperms, cotyledons, hypocotyls, leaves and roots (Hiraiwa *et al.*, 1993). A variety of vacuolar proteins not only in storage organs but also in vegetative organs are reported to be proteolytically processed on the C-terminal side of Asn residues to generate mature forms, as listed in the literature (Hara-Nishimura *et al.*, 1995). The proprotein precursors to these proteins might be processed in the vacuoles of different organs by the action of VPEs. *Arabidopsis thaliana* has three VPE homolog genes with different organ distributions (Kinoshita *et al.*, 1995a,b). VPEs and the homologs are separated into two types; a seed type and a vegetative type (Hara-Nishimura *et al.*, 1995). The transcript of a putative VPE homolog of *Citrus* increases in level during fruit ripening and during flower development, and is induced by treatment with ethylene (Alonso & Granell, 1995).

Immunogold electron microscopy of developing castor bean endosperm has shown that VPE is localized in protein-storage vacuoles (Hara-Nishimura *et al.*, 1993b). Molecular characterization of VPE revealed that inactive proVPE (proprotein precursor to VPE) is synthesized on rough endoplasmic reticulum and then transported to protein-storage vacuoles via dense vesicles (Hiraiwa *et al.*, 1993). After arrival at the vacuoles, proVPE is converted into a mature active enzyme (Hara-Nishimura *et al.*, 1993b, 1995). This conversion is accompanied by the removal of N-terminal and C-terminal propeptides from proVPE (Hara-Nishimura *et al.*, 1993b).

### Distinguishing Features

The susceptibility of asparaginyl endopeptidase to cystatins and high molecular weight kininogen is lower than that of the cysteine proteinases of family C1 that were tested (Abe *et al.*, 1993). Until recently, the hemoglobinase, Sm32, of *Schistosoma mansoni* was the only VPE homolog that had been purified and characterized from animals (Hara-Nishimura *et al.*, 1993b). Sm32 was a putative cysteine endopeptidase, but the substrate specificity of the enzyme was not clear until recently, because of the difficulty of separating it from another cysteine endopeptidase, the cathepsin L-like Sm31 (Klinkert *et al.*, 1989) (see Chapter 254). Recently, a mammalian legumain has been discovered (Chapter 255).

Specific antisera against asparaginyl endopeptidase are not commercially available at the time of writing. Polyclonal antisera against the castor bean VPE have been described (Hara-Nishimura *et al.*, 1993b).

### Further Reading

The molecular characterization of VPE has been reviewed by Hara-Nishimura *et al.* (1993b), and the physiological role of VPE in maturation of seed proteins by Hara-Nishimura *et al.* (1995). A recent article on the enzymatic properties of the enzyme is that of Ishii (1994).

### References

Abe, Y., Shirane, K., Yokosawa, H., Matsushita, H., Mitta, M., Kato, I. & Ishii, S.-I. (1993) Asparaginyl endopeptidase of jack bean seed. *J. Biol. Chem.* **268**, 3525–3529.

Alonso, J.M. & Granell, A. (1995) A putative vacuolar processing protease is regulated by ethylene and also during fruit ripening in *Citrus* fruit. *Plant Physiol.* **109**, 541–547.

Baumgartner, B. & Chrispeels, M.J. (1977) Purification and characterization of vicilin peptidohydrolase, the major endopeptidase in the cotyledons of mung-bean seedlings. *Eur. J. Biochem.* **77**, 223–233.

Becker, C., Shutov, A.D., Nong, V.H., Senyuk, V.I., Jung, R., Horstmann, C., Fischer, J., Nielsen, N.C. & Müntz, K. (1995) Purification, cDNA cloning and characterization of proteinase B, an asparagine-specific endopeptidase from germinating vetch (*Vicia sativa* L.) seeds. *Eur. J. Biochem.* **228**, 456–462.

Bowles, D.J., Marcus, S.E., Pappin, D.J.C., Findlay, J.B.C., Eliopoulos, E., Maycox, P.R. & Burgess, J. (1986) Posttranslational processing of concanavalin A precursors in jackbean cotyledons. *J. Cell Biol.* **102**, 1284–1297.

Cornel, F.A. & Plaxton, W.C. (1994) Characterization of asparaginyl endopeptidase activity in endosperm of developing and germinating castor oil seeds. *Physiol. Plant.* **91**, 599–604.

Csoma, C. & Polgár, L. (1984) Proteinase from germinating cotyledons. Evidence for involvement of a thiol group in catalysis. *Biochem. J.* **222**, 769–776.

Hara-Nishimura, I. & Nishimura, M. (1987) Proglobulin processing enzyme in vacuoles isolated from developing pumpkin cotyledons. *Plant Physiol.* **85**, 440–445.

Hara-Nishimura, I., Inoue, K. & Nishimura, M. (1991) A unique vacuolar processing enzyme responsible for conversion of several proprotein precursors into the mature forms. *FEBS Lett.* **294**, 89–93.

Hara-Nishimura, I., Takeuchi, Y., Inoue, K. & Nishimura, M. (1993a) Vesicle transport and processing of the precursor to 2S albumin in pumpkin. *Plant J.* **4**, 793–800.

Hara-Nishimura, I., Takeuchi, Y. & Nishimura, M. (1993b) Molecular characterization of a vacuolar processing enzyme related to a putative cysteine proteinase of *Schistosoma mansoni*. *Plant Cell* **5**, 1651–1659.

Hara-Nishimura, I., Shimada, T., Hiraiwa, N. & Nishimura, M. (1995) Vacuolar processing enzyme responsible for maturation of seed proteins. *J. Plant Physiol.* **145**, 632–640.

Hiraiwa, N., Takeuchi, Y., Nishimura, M. & Hara-Nishimura, I. (1993) A vacuolar processing enzyme in maturing and germinating seeds: its distribution and associated changes during development. *Plant Cell Physiol.* **34**, 1197–1204.

Ishii, S.-I. (1994) Legumain: asparaginyl endopeptidase. *Methods Enzymol.* **244**, 604–615.

Ishii, S., Abe, Y., Matsushita, H. & Kato, I. (1990) An asparaginyl endopeptidase purified from jackbean seeds. *J. Protein Chem.* **9**, 294–295.

Kembhavi, A.A., Buttle, D.J., Knight, C.G. & Barrett, A.J. (1993) The two cysteine endopeptidases of legume seeds: purification and characterization by use of specific fluorometric assays. *Arch. Biochem. Biophys.* **303**, 208–213.

Kinoshita, T., Nishimura, M. & Hara-Nishimura, I. (1995a) Homologues of a vacuolar processing enzyme that are expressed in different organs in *Arabidopsis thaliana*. *Plant Mol. Biol.* **29**, 81–89.

Kinoshita, T., Nishimura, M. & Hara-Nishimura, I. (1995b) The sequence and expression of the $\gamma$-VPE gene, one member of a family of three genes for vacuolar processing enzymes in *Arabidopsis thaliana*. *Plant Cell Physiol.* **36**, 1555–1562.

Klinkert, M.-Q., Felleisen, R., Link, G., Ruppel, A. & Beck, E. (1989) Primary structure of Sm31/32 diagnostic proteins of *Schistosoma mansoni* and their identification as proteases. *Mol. Biochem. Parasitol.* **33**, 113–122.

Muramatsu, M. & Fukazawa, C. (1993) A high-order structure of plant storage proprotein allows its second conversion by an asparagine-specific cysteine protease, a novel proteolytic enzyme. *Eur. J. Biochem.* **215**, 123–132.

Pueyo, J.J., Hunt, D.C. & Chrispeels, M.J. (1993) Activation of bean (*Phaseolus vulgaris*) $\alpha$-amylase inhibitor requires proteolytic processing of the proprotein. *Plant Physiol.* **101**, 1341–1348.

Rawlings, N.D. & Barrett, A.J. (1994) Families of cysteine peptidases. *Methods Enzymol.* **244**, 461–486.

Scott, M.P., Jung, R., Müntz, K. & Nielsen, N.C. (1992) A protease responsible for post-translational cleavage of a conserved Asn-Gly linkage in glycinin, the major seed storage protein of soybean. *Proc. Natl Acad. Sci. USA* **89**, 658–662.

Shimada, T., Hiraiwa, N., Nishimura, M. & Hara-Nishimura, I. (1994) Vacuolar processing enzyme of soybean that converts proprotein to the corresponding mature forms. *Plant Cell Physiol.* **35**, 713–718.

Shutov, A.D. & Vaintraub, I.A. (1987) Degradation of storage proteins in germinating seeds. *Phytochemistry* **26**, 1557–1566.

Takeda, O., Miura, Y., Mitta, M., Matsushita, H., Kato, I., Abe, Y., Yokosawa, H. & Ishii, S.-I. (1994) Isolation and analysis of cDNA encoding a precursor of *Canavalia ensiformis* asparaginyl endopeptidase (legumain). *J. Biochem.* **116**, 541–546.

*Ikuko Hara-Nishimura*
*Division of Cell Mechanism, Department of Cell Biology,*
*National Institute for Basic Biology,*
*Okazaki 444, Japan*
*Email: ihnishi@nibb.ac.jp*

# 254. *Schistosome legumain*

## Databanks

*Peptidase classification: clan CX, family C13, MEROPS ID: C13.003*
*NC-IUBMB enzyme classification: none*

*Databank codes:*

| Species | SwissProt | PIR | EMBL (cDNA) | EMBL (genomic) |
| --- | --- | --- | --- | --- |
| *Caenorhabditis elegans* | – | – | Z75551 | – |
| *Caenorhabditis elegans* | – | – | Z77653 | – |
| *Fasciola hepatica* | P80527 | – | – | – |
| *Fasciola hepatica* | P80530 | – | – | – |
| *Schistosoma japonicum* | P42665 | – | X70967 | – |
| *Schistosoma mansoni* | P09841 | A60145 | M17423 | – |
| | | M21308 | | |

## Name and History

Schistosomiasis is an infectious tropical disease estimated to afflict more than 200 million people in over 70 countries. The disease is caused by several species of blood flukes of the genus *Schistosoma*, which as adults live and lay eggs in the veins of the intestines and bladder. Infection is acquired in contaminated waters when people come in contact with the aquatic larvae (cercariae) of schistosomes. The cercariae penetrate the skin and migrate into the blood vessels of the infected person.

A schistosome antigen, later to be identified as *schistosome legumain*, was reported to be a minor component of soluble extracts of adult *Schistosoma mansoni* and was shown by immunoblotting to be highly immunogenic in *S. mansoni*-infected mice (Ruppel *et al.*, 1985a) and humans (Ruppel *et al.*, 1985b, 1987b). Accordingly, initial interest in the molecule centered on its potential in the serodiagnosis of human schistosomiasis (Ruppel *et al.* 1985b, 1987a; Idris & Ruppel, 1988). The molecule was termed *Sm32* because of its apparent molecular mass of 32 kDa as determined by migration in SDS-PAGE (Ruppel *et al.*, 1985a,b). The isolation of a cDNA encoding Sm32 was reported soon after by Davis *et al.* (1987) and independently by Klinkert *et al.* (1989). Davis *et al.* (1987) described Sm32 as the schistosome *'hemoglobinase'* because a β-galactosidase–Sm32 fusion protein expressed in *Escherichia coli* seemingly possessed hemoglobin-degrading activity, and because it had long been considered that the schistosome employed an acid protease to metabolize host hemoglobin (Timms & Bueding, 1959). By contrast, Sm32 expressed in insect cells did not exhibit proteolytic activity (Felleisen *et al.*, 1990; Gotz & Klinkert, 1993). While the exact biological role of Sm32 remained unresolved at that time, immunolocalization studies with monoclonal antibodies had localized Sm32 in the digestive tract of larval and adult *S. mansoni* (El Meanawy *et al.*, 1990).

The identity of Sm32 was revealed a decade after its discovery when Takeda *et al.* (1994) reported that the deduced amino acid sequence of an asparaginyl endopeptidase (Chapter 253) from seeds of the jackbean *Canavalia ensiformis* was similar to that of Sm32. Dalton *et al.* (1995a) subsequently showed that the sequence of Sm32 was also similar to that of an asparaginyl endopeptidase from peas of the vetch, *Vicia sativa*, and, in a subsequent study, demonstrated the presence of legumain-like, asparaginyl endopeptidase activity in soluble extracts of adult *S. mansoni*, the first report of such an activity in animal tissues (Dalton *et al.*, 1995b). Because of the low specific activity of the enzyme in

schistosome extracts, Dalton *et al.* (1995b) refuted the suggestion of a direct role of the Sm32 enzyme in hemoglobin digestion. Rather, they proposed that Sm32 may be involved in the post-translational modification of proteins, in similar fashion to the role of legumains from leguminous plants (see Ishii, 1994). Sm32 is now termed schistosome legumain because of its similarities to the plant legumains.

Nearly 40 years ago Timms & Bueding (1959) suggested that several proteases in the alimentary canal of adult schistosomes may perform the function of degrading hemoglobin to amino acids that are metabolized by the parasite for protein synthesis. Since then, several proteases besides schistosome legumain have been identified and ascribed roles as the schistosome *'hemoglobinase'*; these other enzymes include cathepsin B (Dresden & Deelder, 1979; Chappell & Dresden, 1986; Götz & Klinkert, 1993) and cathepsin D (Bogitsh *et al.*, 1992; Becker *et al.*, 1995). In addition, Smith *et al.* (1994) and Day *et al.* (1995) characterized a cathepsin L proteinase secreted by *S. mansoni* and *S. japonicum* which may also be involved in hemoglobin digestion (Chapter 208). Since it is likely that a series of proteolytic enzymes is involved in the progressive degradation of hemoglobin to absorbable peptides, Dalton *et al.* (1995a) advised against the continued use of the term 'hemoglobinase' for any individual protease, particularly in respect to keywords or descriptions associated with sequence entries into the public databases.

## Activity and Specificity

Dalton *et al.* (1995b) have characterized activity ascribable to the schistosome legumain in soluble extracts of adult *S. mansoni*, using peptide substrates synthesized as described by Kembhavi *et al.* (1993). The specific activity (nmoles NH$_2$Mec released per mg protein per min) of the schistosome legumain against the substrate Z-Ala-Ala-Asn+NHMec in schistosome extracts was $4.5 \times 10^{-3}$, which is comparable to that present in extracts of jack bean ($13 \times 10^{-3}$). Schistosome legumain cleaves the substrate Z-Ala-Pro-Asn+NHMec with approximately half the efficiency at which it cleaves Z-Ala-Ala-Asn-NHMec, but it has little activity against Asn-NHMec.

The pH optimum for activity against Z-Ala-Ala-Asn-NHMec is 6.8, and the optimum temperature for activity is in the range 37–45°C. The activity is inhibited by *N*-ethylmaleimide and iodoacetamide, whereas the cysteine proteinase inhibitors E-64 and leupeptin have little inhibitory activity at concentrations as high as 1 mM. Neither the cathepsin cysteine proteinase inhibitor Z-Phe-Ala-CHN$_2$ nor the metal chelator 1,10-phenanthroline has any inhibitory effect on the activity of schistosome legumain. Reducing reagents (DTT, 1 mM) do not appear to enhance the activity (Dalton *et al.*, 1995a).

## Structural Chemistry

The mRNA of *S. mansoni* legumain encodes a 50 kDa, single-chain protein, containing 429 amino acids (Davis *et al.*, 1987; Klinkert *et al.*, 1989; El Meanawy *et al.*, 1990) (Figure 254.1). A cDNA clone encoding the *S. japonicum* homolog of the *S. mansoni* legumain has been isolated and shows 78% identity at the nucleotide level. The deduced amino acid sequences of the two schistosome legumains are 73% identical (Merckelbach *et al.*, 1994).

Based on the deduced amino acid sequence, the predicted molecular mass of *S. mansoni* legumain is 31 kDa, rather than 32 kDa (El Meanawy *et al.*, 1990). The 50 kDa proenzyme appears to be post-translationally processed at both its N- and C-termini. Sequencing of the 31 kDa mature enzyme at the N- and C-termini determined that it is comprised of 260 amino acids beginning at residue 32 and terminating at residue 292 (El Meanawy *et al.*, 1990). The structures of the deduced amino acid sequence of the prepro-legumains of *S. mansoni* and *S. japonicum* include typical signal sequences of 19 and 18 amino acids, respectively (Klinkert *et al.*, 1989; El Meanawy *et al.*, 1990; Merckelbach *et al.*, 1994). Therefore N-terminal processing of *S. mansoni* legumain involves the removal of a short region of 12 amino acid residues. This processing event may not need to be precise, since the N-termini of legumains from *S. mansoni*, *Fasciola hepatica* and jackbean give a 'ragged' alignment (Figure 254.2). In addition, five amino acid residues are deleted from this processed N-terminal region of the *S. japonicum* legumain (Merckelbach *et al.*, 1994). However, a highly conserved block of amino acids, -Trp-Ala-(Val/Ile)-Leu-(Val/Ile)-Ala-Gly-Ser-Asn-Gly-, begins within a short distance from the N-terminus of all legumains so far examined.

The block of amino acids Asp56-Val-**Cys**-His-Ala-Tyr61-, which are conserved between the plant (jack bean and vetch pea) and schistosome legumains, may contain the active-site Cys residue. Additionally, the conserved block represented in Sm32 by Val144-Phe-Ile-Tyr-Phe-Thr-Asp-**His**-Gly-Ala-Pro-Gly155 may contain the active-site His. The conserved sequence Asp-**His**-Gly is also observed in many papain-like, cysteine proteinases around the active-site His159 (Rawlings & Barrett, 1994). A conserved Glu within a third block of conserved residues Tyr188-Ile-**Glu**-Ala-Asn-**Glu**-Ser-Gly-Ser196 (in Sm32) may complete the catalytic triad of legumain active-site residues.

The sequence of the legumain from *S. mansoni* contains three potential glycosylation sites that are absent from *S. japonicum* legumain and from the plant legumains. Conversely, a sequence of four proline residues present in the plant enzymes is absent from schistosome legumain. Since this motif is flanked on both sides by conserved cysteine residues, which may form a disulfide bridge, this feature may introduce a loop into the structure of plant legumains

that is apparently absent from the schistosome enzyme (Figure 254.1).

The similarities between the primary sequence of the plant and schistosome legumains are highest in the region of the fully processed molecule. The C-terminus, however, contains highly conserved residues with regular interspersion distances that may indicate a regular structural motif such as an $\alpha$-helix. In addition, this region contains four conserved Cys residues that may form two disulfide bridges. Therefore, the molecule could be considered to consist of two domains, an enzymatic domain and a C-terminal domain. The mature 31 kDa form of schistosome legumain can be detected in soluble extracts of adult *S. mansoni*, whereas the 50 kDa form is detected only in detergent extracts, suggesting that the 50 kDa form is membrane associated and is clipped by some unknown mechanism to produce the 31 kDa form (El Meanawy *et al.*, 1990). The C-terminal region may therefore represent the membrane-associating domain.

## Preparation

Schistosome legumain has not been isolated to homogeneity. A soluble schistosome extract containing the activity can be prepared in 0.1 M sodium acetate, pH 5.5 by three freeze-thaw and sonication cycles, followed by centrifugation at $14\,000 \times g$ to remove insoluble material. Because soluble extracts of schistosomes also contain cathepsin L and cathepsin B activities that are capable of cleaving Z-Ala-Ala-Asn-NHMec, the cathepsin inhibitor Z-Phe-Ala-CHN$_2$ (10 μM) should be included in the extraction buffer (Dalton *et al.*, 1995b).

Schistosome legumain has been expressed in *E. coli* as a $\beta$-galactosidase fusion protein (Davis *et al.*, 1987), as a fusion protein with the N-terminal region of the RNA replicase of the phage MS2, and as an unfused form (Klinkert *et al.*, 1988). All *E. coli*-produced proteins were insoluble. The enzyme was also expressed, in low yield, in cells of the insect *Spodoptera frugiperda* (Felleisen *et al.*, 1990; Gotz & Klinkert, 1993).

## Biological Aspects

Messenger RNAs encoding schistosome legumain have been detected by northern blots in schistosome cercariae, in newly transformed schistosomula, in adult schistosomes, and in eggs (El Meanawy *et al.*, 1990). Therefore, it would appear that the molecule is constitutively expressed during all stages of the life cycle that parasitize the mammalian host. The adult worm legumain mRNA (1500 nucleotides) is larger than that observed in cercariae (1450 nucleotides) due to a longer poly(A$^+$) tract (El Meanawy *et al.*, 1990).

Immunocalization studies showed that schistosome legumain is associated with the gut epithelium cells of schistosome larvae and adults. In addition, the legumain was observed in the gut lumen, and associated with red blood cell debris (El Meanawy *et al.*, 1990). These observations suggest that the enzyme plays some role in the degradation of host hemoglobin. Other proteases that have implied roles in hemoglobin degradation, including cathepsin B (Ruppel *et al.*, 1987b) and cathepsin D (Bogitsh & Kirschner, 1987), have also been localized to the schistosome digestive tract.

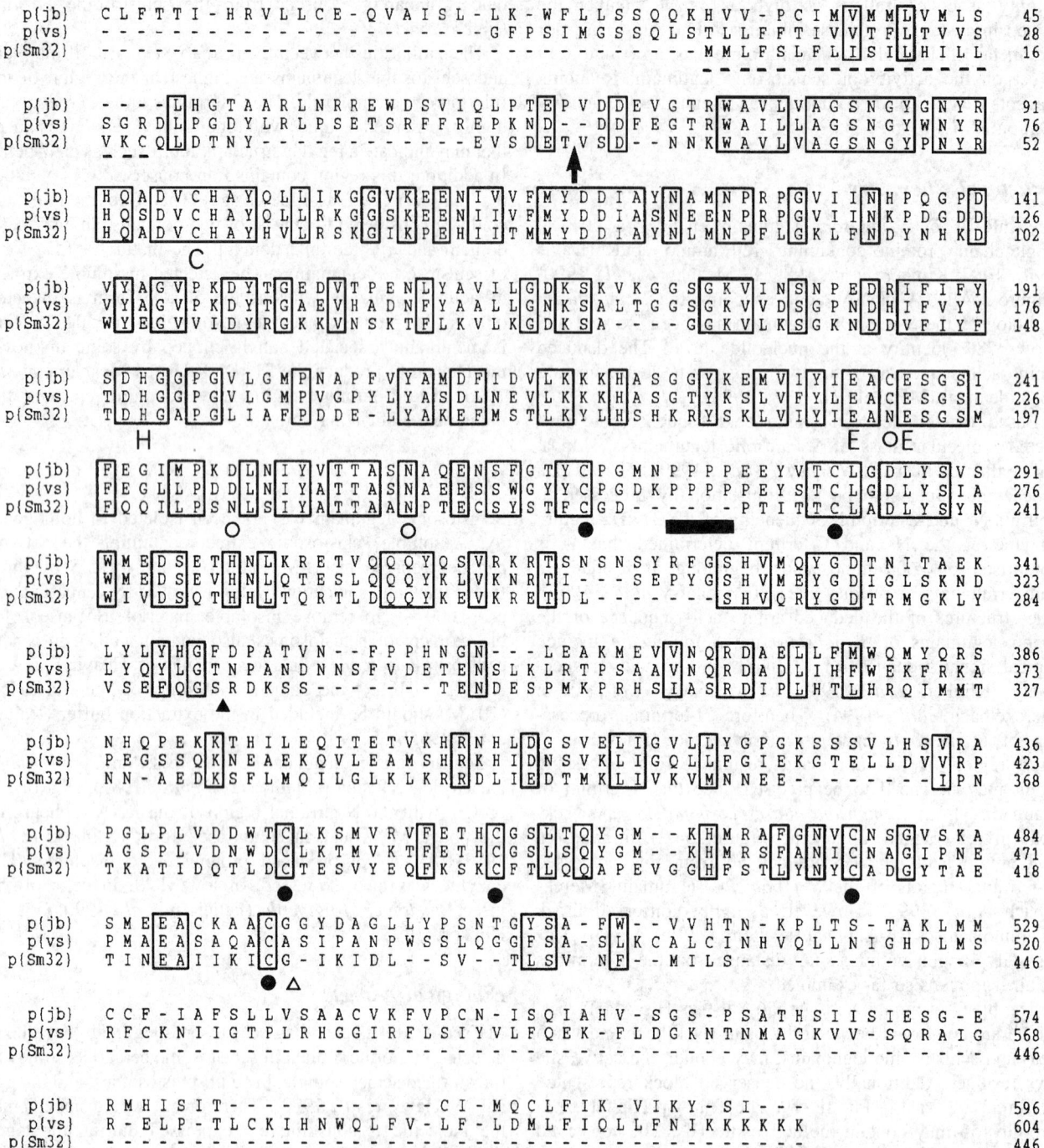

*Figure 254.1* Alignment of amino acids sequences of the legumains of the jackbean, *Canavalia ensiformis* (jb), and of the vetch pea, *Vicia sativa* (vs), with the *Schistosoma mansoni* legumain (Sm32). Boxes denote the positions of similarity between the genes; gaps ( – ) have been introduced to maximize alignment. Symbols have been used to indicate structural features on the schistosome legumain as follows: dashed underline, the putative signal peptide; arrow, beginning of the mature protein; open circles, potential glycosylation sites; solid circles, conserved cysteine residues; solid triangle, processing site in the C-terminus; open triangle, last amino acid in the protein; solid bar, putative proline loop absent from schistosome legumain; C, putative active-site Cys; H, putative active-site His; E, putative active-site Glu residues.

Schistosome legumain activity has been characterized only in adult worms. Because the specific activity of the enzyme in soluble extracts was very low, Dalton *et al.* (1995b) suggested that schistosome legumain is probably not primarily involved in hemoglobin digestion, particularly given the large numbers of red blood cells that schistosomes ingest (Lawrence, 1973). Rather, they proposed that schistosome legumain post-translationally modifies other proteases leading to their activation. Support for this hypothesis was provided by the observation that an asparagine residue (the

```
S. mansoni      V S D N N K W A V L V A G S N G Y P N Y R H
F. hepatica   L E D N G R T H W A V L V A
F. hepatica       K N W A V L V A G S N G W P N Y R H H A
F. hepatica       K N W A V L V A G S D G L P N Y R H H A
C. ensiformis     E V G T R W A V L V A G S N G Y G N Y R H Q A
```

*Figure 254.2* Alignment of N-terminal sequences obtained for legumains of *Schistosoma mansoni* (El Meanawy *et al.*, 1990), *Fasciola hepatica* (Tkalcevic *et al.*, 1995), and *Canavalia ensiformis* (jackbean) (Abe *et al.*, 1993).

characteristic P1 substrate residue for legumain) occurs in the vicinity of the cleavage point between the pro region and mature protein of a panel of other proteases reputed to be involved in the degradation of host hemoglobin, including cathepsin L, cathepsin B (Sm31), cathepsin D (aspartic protease) and dipeptidyl-peptidase I. By contrast, examination of the mammalian homologs of these proteases showed that asparagine residues are absent from the vicinity of their cleavage points (Dalton & Brindley, 1996). Schistosome legumain may therefore play a pivotal, although indirect, role in the digestion of host hemoglobin. Furthermore, since the enzyme has been localized to tissues other than the digestive tract of adult worms, including the ventral surface of the male schistosome (Zhong *et al.*, 1995), schistosome legumain may be involved in the activation of proteases at sites other than the digestive tract.

N-terminal sequencing of components of electrophoretically separated protein extracts of the infective stages of the related trematode *Fasciola hepatica* identified three proteins with similar sequences to the schistosome and plant legumains (Figure 254.2). The three *F. hepatica* sequences were not ientical, suggesting that more than one *F. hepatica* legumain may exist (Tkalcevic *et al.*, 1995).

## Distinguishing Features

The pH optimum for activity of the schistosome legumain (pH 6.8) markedly differs from those of jack bean legumain (pH 5.4) (Abe *et al.*, 1993; Dalton *et al.*, 1995b) and moth bean legumain (pH 5.0) (Kembhavi *et al.*, 1993). The temperature optimum for activity of the schistosome legumain is 37–45°C, whereas that for the jackbean enzyme is 25–30°C. Unlike legumains from legumes, the activity of schistosome legumain is not enhanced by DTT (Dalton *et al.*, 1995b).

## Further Reading

Reviews include those of Dalton *et al.* (1995a,b) and Dalton & Brindley (1996).

## References

Abe, Y., Shirane, K., Yokasawa, H., Matsushita, H., Mitta, M., Kato, I. & Ishii, S.-I. (1993) Asparaginyl endopeptidase of jack bean seeds: purification, characterization, and high utility in protein sequence analysis. *J. Biol. Chem.* **268**, 3525–3529.

Becker, M.M., Harrop, S.A., Dalton, J.P., Kalinna, B.H., McManus, D.P. & Brindley, P.J. (1995) Cloning and characterization of the *Schistosoma japonicum* aspartic proteinase involved in hemoglobin

degradation. *J. Biol. Chem.* **270**, 24496–24501.

Bogitsh, B.J. & Kirschner, K.F. (1987) *Schistosoma japonicum*: immunocytochemistry of adults using heterologous antiserum to bovine cathepsin D. *Exp. Parasitol.* **64**, 213–218.

Bogitsh, B.J., Kirschner, K.F. & Rotmans, J.P. (1992) *Schistosoma japonicum*: immunoinhibitory studies on hemoglobin digestion using heterologous antiserum to bovine cathepsin D. *J. Parasitol.* **78**, 454–459.

Chappel, C.L. & Dresden, M.H. (1986) *Schistosoma mansoni*: proteinase activity of 'hemoglobinase' from the digestive tract of adult worms. *Exp. Parasitol.* **61**, 160–167.

Dalton, J.P. & Brindley, P.J. (1996) Schistosome asparaginyl endopeptidase Sm32 in hemoglobin digestion. *Parasitol. Today* **12**, 125.

Dalton, J.P., Smith, A.M., Clough, K.A. & Brindley, P.J. (1995a) Digestion of haemoglobin by schistosomes: 35 years on. *Parasitol. Today* **11**, 299–302.

Dalton, J.P., Hola-Jamriska, L. & Brindley, P.J. (1995b) Asparaginyl endopeptidase activity in adult *Schistosoma mansoni. Parasitology* **111**, 575–580.

Davis, A.H., Nanduri, J. & Watson, D.C. (1987) Cloning and gene expression of *Schistosoma mansoni* protease. *J. Biol. Chem.* **262**, 12851–12855.

Day, S.A., Dalton, J.P., Clough, K.A., Leonardo, L., Tiu, W.U. & Brindley, P.J. (1995) Characterization and cloning of the cathepsin L proteinases of *Schistosoma japonicum. Biochem. Biophys. Res. Commun.* **217**, 1–9.

Dresden, M.H. & Deelder, A.M. (1979) *Schistosoma mansoni*: thiol proteinase properties of adult worm 'hemoglobinase'. *Exp. Parasitol.* **48**, 190–197.

El Meanawy, M.A., Aji, T., Phillips, N.F.B., Davis, R.E., Salata, R.A., Malhotra, I., McClain, D., Aikawa, M. & Davis, A.H. (1990) Definition of the complete *Schistosoma mansoni* hemoglobinase mRNA sequence and gene expression in developing parasites. *Am. J. Trop. Med. Hyg.* **43**, 67–78.

Felleisen, R., Beck, E., Usmany, M, Vlak, J. & Klinkert, M.-Q. (1990) Cloning and expression of *Schistosoma mansoni* protein Sm32 in a baculovirus vector. *Mol. Biochem. Parasitol.* **43**, 289–292.

Götz, B. & Klinkert, M.-Q. (1993) Expression and partial characterization of a cathepsin B-like enzyme (Sm31) and a proposed 'haemoglobinase' (Sm32) from *Schistosoma mansoni. Biochem. J.* **290**, 801–806.

Idris, M.A. & Ruppel, A. (1988) Diagnostic $M_r$ 31/32000 *Schistosoma mansoni* proteins (Sm31/32): reaction with sera from Sudanese patients infected with *S. mansoni* or *S. haematobium. J. Helminthol.* **62**, 95–101.

Ishii, S.-I. (1994) Legumain: asparaginyl endopeptidase. *Methods Enzymol.* **244**, 604–615.

Kembhavi, A.A., Buttle, D.J., Knight, C.G. & Barrett, A.J. (1993) The two cysteine endopeptidases of legume seeds: purification and characterization by use of specific fluorometric assays. *Arch. Biochem. Biophys.* **303**, 208–213.

Klinkert, M.-Q., Ruppel, A., Felleisen, R., Link, G. & Beck, E. (1988) Expression of diagnostic 31/32 kilodalton proteins of *Schistosoma mansoni* as fusions with bacteriophage MS2 polymerase. *Mol. Biochem. Parasitol.* **27**, 233–240.

Klinkert, M.-Q., Felleisen, R., Link, G., Ruppel, A. & Beck, E. (1989) Primary structures of Sm31/32 diagnostic proteins of *Schistosoma mansoni* and their identification as proteases. *Mol. Biochem. Parasitol.* **33**, 113–122.

Lawrence, J.D. (1973) The ingestion of red blood cells by *Schistosoma mansoni*. *J. Parasitol.* **59**, 60–63.

Merckelbach, A., Hasse, S., Dell, R., Eschlbeck, A. & Ruppel, A. (1994) cDNA sequences of *Schistosoma japonicum* coding for two cathepsin B-like proteins and Sj32. *Trop. Med. Parasitol.* **45**, 193–198.

Rawlings, N.D. & Barrett, A.J. (1994) Families of cysteine peptidases. *Methods Enzymol.* **244**, 461–486.

Ruppel, A., Rother, U., Vongerichten, H., Lucius, R. & Diesfeld, H.J. (1985a) *Schistosoma mansoni*: immunoblot analysis of adult worm proteins. *Exp. Parasitol.* **60**, 195–206.

Ruppel, A., Diesfeld, H.J. & Rother, U. (1985b) Immunoblot analysis of *Schistosoma mansoni* antigens with sera of schistosomiasis patients: diagnostic potential of an adult schistosome polypeptide. *Clin. Exp. Immunol.* **62**, 499–506.

Ruppel, A., Breternitz, U. & Burger, R. (1987a) Diagnostic $M_r$ 31,000 *Schistosoma mansoni* proteins: requirement of infection, but not immunization, and use of the 'miniblot' technique for the production of monoclonal antibodies. *J. Helminthol.* **61**, 95–101.

Ruppel, A., Shi, Y.E., Wei, D.X. & Diesfeld, H.J. (1987b) Sera of *Schistosoma japonicum*-infected patients crossreact with diagnostic 31/32 kDa proteins of *S. mansoni*. *Clin. Exp. Immunol.* **69**, 291–298.

Smith, A.M., Dalton, J.P., Clough, K.A., Killbane, C.L., Harrop, S.A., Hole, N. & Brindley, P.J. (1994) Adult *Schistosoma mansoni* express cathepsin L proteinase activity. *Mol. Biochem. Parasitol.* **67**, 11–19.

Takeda, O., Miura, Y., Mitta, M., Matsushita, H., Kato, I., Abe, Y., Yokosawa, H. & Ishii, S.-I. (1994) Isolation and analysis of cDNA encoding a precursor of *Canavalia ensiformis* asparaginyl endopeptidase (legumain). *J. Biochem.* **116**, 541–546.

Timms, A.R. & Bueding, E. (1959) Studies of a proteolytic enzyme from *Schistosoma mansoni*. *Br. J. Pharmacol.* **14**, 68–73.

Tkalcevic, J., Ashman, K. & Meeusen, E. (1995) *Fasciola hepatica*: rapid identification of newly excysted juvenile proteins. *Biochem. Biophys. Res. Commun.* **213**, 169–174.

Zhong, C., Skelly, P.J., Leaffer, D., Cohn, R.G., Caulfield, J.P. & Shoemaker, C.B. (1995) Immunolocalization of a *Schistosoma mansoni* facilitated diffusion glucose transporter to the basal, but not the apical, membranes of the surface syncytium. *Parasitology* **110**, 383–394.

*John P. Dalton*
*School of Biological Sciences,*
*Dublin City University,*
*Dublin 9, Republic of Ireland*
*Email: johnD@raven.dcu.ie*

*Paul J. Brindley*
*Molecular Parasitology Unit,*
*Queensland Institute of Medical Research,*
*and Australian Centre for International & Tropical Health & Nutrition,*
*Post Office, Royal Brisbane Hospital,*
*Brisbane, Queensland 4029, Australia*
*Email: paulB@qimr.edu.au*

# 255. *Mammalian legumain*

## Databanks

*Peptidase classification: clan CX, family C13, MEROPS ID: C13.004*
*NC-IUBMB enzyme classification: none*
*Databank codes:*

| Species | SwissProt | PIR | EMBL (cDNA) | EMBL (genomic) |
|---|---|---|---|---|
| *Homo sapiens* | – | – | Y09862 | |

## Name and History

The name **legumain** was recommended by NC-IUBMB (1992) for an asparaginyl endopeptidase that was discovered by Csoma & Polgár (1984) in germinating bean cotyledons, and was subsequently found in various other plant seeds and tissues (see Chapter 253). Kembhavi *et al.* (1993) isolated the enzyme from mothbean (*Vigna aconitifolia*), and described a convenient fluorometric assay. The cloning and sequencing of legumain from castor bean (Hara-Nishimura *et al.*, 1993) showed it to be homologous with what was at that time a putative cysteine endopeptidase from the blood fluke, *Schistosoma mansoni*. The parasite enzyme has now been shown also to be an asparaginyl endopeptidase (Dalton *et al.*, 1995) and is known as schistosome legumain (Chapter 254).

In their classification of the cysteine peptidases, Rawlings & Barrett (1994) assigned the number C13 to the peptidase family containing the legume asparaginyl endopeptidase and the *Schistosoma* endopeptidase. During 1995, the publication of human EST sequences made it evident that the human genome contains a form of legumain. This prompted work in the laboratory of the present author to clone and sequence the cDNA for human legumain, and to isolate and characterize the enzyme from pig kidney (Chen *et al.*, 1997). The properties of the mammalian enzyme were so similar to those of the plant and *Schistosoma* forms that it seems appropriate to use the same name, **legumain**, for the mammalian enzyme. Quite independently of Chen and coworkers, Tanaka *et al.* (1996) cloned and sequenced a human cDNA differing in only one deduced amino acid residue, and determined the chromosomal location of the gene for the putative peptidase, *PRSC1*.

## Activity and Specificity

The assay substrate used by Chen *et al.* (1997) was Z-Ala-Ala-Asn+NHMec, originally described for the legume enzyme by Kembhavi *et al.* (1993), but Bz-Asn+OPhNO$_2$ was also hydrolyzed. The purified enzyme showed a requirement for activation by a thiol compound, and was maximally active at pH 5.8, in a sodium citrate buffer.

Pig kidney legumain acts on polypeptides as a strict asparaginyl endopeptidase. The asparaginyl bonds of neurotensin and vasoactive intestinal peptide are hydrolyzed selectively. The three bonds cleaved in a model protein, the 500 amino acid recombinant C-fragment of tetanus toxoid, were asparaginyl bonds, but 44 other asparaginyl bonds in the protein were completely unaffected, showing that there are additional determinants of specificity. For plant seed legumain, Hara-Nishimura *et al.* (1995) have pointed out that sites of cleavage are typically in markedly hydrophilic parts of the substrate protein, and this could account for the selection of bonds cleaved in the tetanus toxoid fragment, also.

Pig legumain is stable in the pH range 4–6, but is irreversibly denatured at higher pH values. This is strongly reminiscent of the behavior of most of the lysosomal cysteine peptidases of family C1 (e.g. cathepsins B, H and L).

Typically for a cysteine peptidase, legumain is inhibited by general thiol-blocking agents such as iodoacetate and iodoacetamide (reacting only slowly), as well as by *N*-ethylmaleimide. *N*-Phenylmaleimide was the most potent irreversible inhibitor identified by Chen *et al.* (1997). E-64 and leupeptin do not inhibit pig kidney legumain, sharply distinguishing the enzyme from most members of the papain family, C1.

Human cystatin C and cystatin from chicken egg white inhibit pig legumain with low nanomolar $K_i$ values, and it is possible to determine the molar concentration of legumain active sites by titration of the enzyme with cystatin that has itself been standardized with papain and E-64 (Chen *et al.*, 1997).

## Structural Chemistry

The cDNA of human legumain seems to encode a preproprotein of 433 amino acid residues, and 49 kDa. The prepro sequence is estimated to be quite short, about 25 residues, and yet the mature (and deglycosylated) enzyme runs as a protein of 31 kDa in SDS-PAGE. This clearly indicates that the mammalian legumain is C-terminally processed, as are the plant and *Schistosoma* forms. Pig legumain is *N*-glycosylated, and can be deglycosylated by treatment with *N*-glycosidase F (available from Boehringer-Mannheim) (see Appendix 2 for full names and addresses of suppliers).

## Preparation

The purification of legumain from pig kidney by Chen *et al.* (1997) required a somewhat unconventional procedure, because the enzyme is stable only in the range pH 3–6, and at low salt concentrations it tends to be adsorbed to any solid material present. The tissue was homogenized and centrifuged, and the supernatant was fractionated with ammonium sulfate. When the active fraction was dialysed at pH 5, a heavy precipitate formed, to which the legumain was adsorbed. The enzyme was eluted by raising the salt concentration at pH 6.0, and was then run on SP-Sepharose at pH 5.5, requiring 0.4 M NaCl for elution. The enzyme was then run on Mono-S in FPLC, and bound to thiopropyl-Sepharose activated with 2-pyridyl disulfide. The purified enzyme was stable to storage at pH 5.8, 4°C, showing little loss of activity over several months.

## Biological Aspects

Legumain is lysosomal in rat kidney (P.M. Dando, unpublished results). This was expected from the findings of Chen *et al.* (1997) that the pig enzyme is a glycoprotein that requires an acidic pH for stability and activity, and is a homolog of a plant vacuolar acid hydrolase. Chen *et al.* pointed out that a number of lysosomal proteins are known to be processed by cleavage of asparaginyl bonds, and it is likely that this is an action of legumain.

The conservation of the strict asparaginyl endopeptidase specificity of legumain, at least since the divergence of plants and animals perhaps 1000 million years ago, suggests that there is a biological need in the eukaryotic cell for an enzyme with this specificity, but the natural substrates have yet to be identified. Possible protein-processing roles have been discussed elsewhere in the present volume (see Chapters 253 and 254). As a lysosomal enzyme, legumain is likely to be secreted from cells under some conditions, and like cathepsin L (Dehrmann *et al.*, 1996), it may be active in the pericellular environment.

The gene for human legumain is located at 14q32.1 (Tanaka *et al.*, 1996), but nothing is known of the structure of the gene or possible regulatory elements.

## Distinguishing Features

Legumain differs from the majority of mammalian cysteine endopeptidases, which belong to families C1 and C2, in being unaffected by E-64 but inhibited by cystatins, as well as in its selectivity for the hydrolysis of asparaginyl bonds. Interestingly, the cysteine endopeptidases of the caspase family (C14) show an equally strict specificity, but for hydrolysis of aspartyl bonds (see Chapters 248 and 249). Pig kidney legumain is much less stable to neutral pH than the otherwise similar enzyme of *Schistosoma*, which shows a pH optimum at 6.8 (Dalton *et al.*, 1995).

## Related Peptidases

There appear to be at least two variants of legumain in animals, which we tentatively refer to as type 1 and type 2 legumains. The human sequence of Tanaka *et al.* (1996) and Chen *et al.* (1997) is representative of the type 1 protein, and species variants are known from the mouse (*Mus musculus*) and pig (*Sus scropha*). The type 2 protein is quite divergent in sequence and lacks the generally conserved Cys residue (in -Asn-Val-**Cys**-His-Ala-Tyr-) near the N-terminus that has been suggested to be the catalytic residue (see Chapter 254). The human type 2 legumain, which is known as GPI8 (EMBL: Y07596), is only 26% identical in amino acid sequence to human type 1 legumain, and is believed to be a glycosylphosphatidylinositol:protein transamidase (see Chapter 256). Type 2 legumains are known from *Caenorhabditis elegans*, mouse, rat (*Rattus norvegicus*) and human. A small family of legumains is known to exist in the plant *Arabidopsis thaliana* (Kinoshita *et al.*, 1995a,b), and it seems that a similar situation may well exist in animals.

## Further Reading

The discovery of mammalian legumain as an enzyme was described by Chen *et al.* (1997), and the location of the human gene by Tanaka *et al.* (1996).

## References

Chen, J.-M., Dando, P.M., Rawlings, N.D., Brown, M.A., Young, N.E., Stevens, R.A., Hewitt, E., Watts, C. & Barrett, A.J. (1997) Cloning, isolation, and characterization of mammalian legumain, an asparaginyl endopeptidase. *J. Biol. Chem.* **272**, 8090–8098.

Csoma, C. & Polgár, L. (1984) Proteinase from germinating bean cotyledons. Evidence for involvement of a thiol group in catalysis. *Biochem. J.* **222**, 769–776.

Dalton, J.P., Hola-Jamriska, L. & Brindley, P.J. (1995) Asparaginyl endopeptidase activity in adult *Schistosoma mansoni*. *Parasitology* **111**, 575–580.

Dehrmann, F.M., Elliott, E. & Dennison, C. (1996) Reductive activation markedly increases the stability of cathepsins B and L to extracellular ionic conditions. *Biol. Chem. Hoppe-Seyler* **377**, 391–394.

Hara-Nishimura, I., Takeuchi, Y. & Nishimura, M. (1993) Molecular characterization of a vacuolar processing enzyme related to a putative cysteine proteinase of *Schistosoma mansoni*. *Plant Cell* **5**, 1651–1659.

Hara-Nishimura, I., Shimada, T., Hiraiwa, N. & Nishimura, M. (1995) Vacuolar processing enzyme responsible for maturation of seed proteins. *J. Plant Physiol.* **145**, 632–640.

Kembhavi, A.A., Buttle, D.J., Knight, C.G. & Barrett, A.J. (1993) The two cysteine endopeptidases of legume seeds: purification and characterization by use of specific fluorometric assays. *Arch. Biochem. Biophys.* **303**, 208–213.

Kinoshita, T., Nishimura, M. & Hara-Nishimura, I. (1995a) Homologues of a vacuolar processing enzyme that are expressed in different organs in *Arabidopsis thaliana*. *Plant Mol. Biol.* **29**, 81–89.

Kinoshita, T., Nishimura, M. & Haranishimura, I. (1995b) The sequence and expression of the gamma-VPE gene, one member of a family of three genes for vacuolar processing enzymes in *Arabidopsis thaliana*. *Plant Cell Physiol.* **36**, 1555–1562.

NC-IUBMB (Nomenclature Committee of the International Union of Biochemistry and Molecular Biology) (1992) *Enzyme Nomenclature 1992*. Orlando, FL: Academic Press.

Rawlings, N.D. & Barrett, A.J. (1994) Families of cysteine peptidases. *Methods Enzymol.* **244**, 461–486.

Tanaka, T., Inazawa, J. & Nakamura, Y. (1996) Molecular cloning of a human cDNA encoding putative cysteine protease (PRSC1) and its chromosome assignment to 14q32.1. *Cytogenet. Cell Genet.* **74**, 120–123.

*Alan J. Barrett*
*MRC Peptidase Laboratory,*
*Department of Immunology,*
*The Babraham Institute,*
*Cambridgeshire CB2 4AT, UK*
*Email: alan.barrett@bbsrc.ac.uk*

# 256. Glycosylphosphatidylinositol: protein transamidase

## Databanks

*Peptidase classification: clan CX, family C13, MEROPS ID: C13.005*
*NC-IUBMB enzyme classification: none*

*Databank codes:*

| Species | SwissProt | PIR | EMBL (cDNA) | EMBL (genomic) |
|---|---|---|---|---|
| Legumain homologs | | | | |
| *Caenorhabditis elegans* | P49048 | – | – | Z68751: complete cosmid T05E11 |
| *Homo sapiens* | – | – | R18975 Y07596 | – |
| *Mus musculus* | – | – | AA212707 | – |
| *Rattus norvegicus* | – | – | H34389 | – |
| *Saccharomyces cerevisiae* | P49018 | – | U32517 | – |
| Other proteins | | | | |
| *Saccharomyces cerevisiae* | P39012 | S45053 | X79409 | U53880: chromosome XII cosmid Z73260: chromosome XII ORF |

## Name and History

Many eukaryotic proteins are attached to the membrane by a glycosylphosphatidylinositol (GPI)-containing anchor. The carbohydrate core of the GPI anchor is conserved among all eukaryotes examined thus far, but glycosidic side chains and the lipid moieties show distinct variations between and within species (Figure 256.1A) (Fankhauser *et al.*, 1993; McConville & Ferguson, 1993). GPI-anchored proteins are synthesized with N- and C-terminal hydrophobic regions that are subsequently removed (Figure 256.1B). The N-terminal hydrophobic signal sequence targets the protein to the lumen of the endoplasmic reticulum, and is there removed by signal peptidase (Chapter 156). The C-terminal hydrophobic region is part of a signal that directs its own removal and replacement with a GPI anchor (Caras *et al.*, 1987). The GPI anchor is attached to the protein by an amide linkage between the amino group of the terminal ethanolamine of the GPI anchor structure and the C-terminal carboxyl group of the protein (Ferguson *et al.*, 1988; Homans *et al.*, 1988). The very structure of the anchor and the rapid kinetics of its addition (Bangs *et al.*, 1985; Conzelmann *et al.*, 1987) have suggested that the GPI is preformed and transferred *en bloc*. Indeed, complete GPI lipids of the expected structure have been found (Masterson *et al.*, 1989; Menon *et al.*, 1990). In living cells the removal of the C-terminal signal without the concomitant attachment of a GPI has never been observed (Conzelmann *et al.*, 1986; Kodukula *et al.*, 1992), so it is believed that a single enzyme cleaves off the C-terminal signal and replaces it with the GPI anchor in one reaction; this enzyme has been given the name ***glycosylphosphatidylinositol:protein transamidase (GPI transamidase)***. In a microsomal preparation *in vitro*, the cleavage of a protein substrate can occur in the absence of GPI if a suitably strong alternative nucleophile is present, consistent with the transamidation mechanism (Maxwell *et al.*, 1995b).

## Activity and Specificity

Transamidation requires two substrates: a GPI unit and a protein precursor. GPI lipids are similar to the mature GPI anchor, but often their side chains are not yet completed and they do not contain the same lipid moieties as the mature

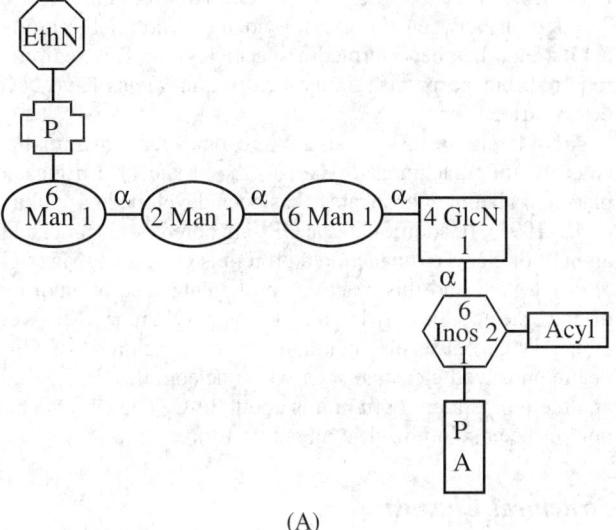

(A)

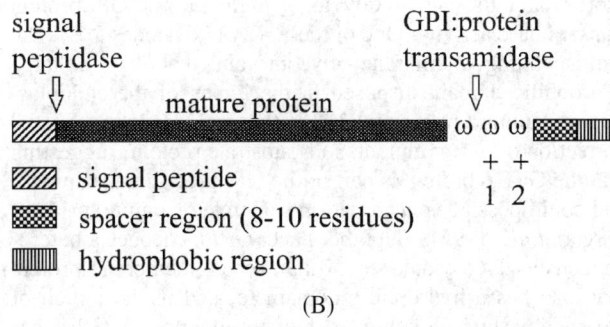

(B)

*Figure 256.1* (A) The GPI precursor core structure. Carbohydrate or other side chains can be attached at either Man1 or Man3, depending upon the species. Some side chains can be attached before transfer to protein. The acyl group on the 2 position of inositol is removed after GPI attachment. (B) The GPI protein precursor. The N-terminal signal peptide targets the protein for translocation into the endoplasmic reticulum where cleavage and anchor attachment occur at the ω residue, mediated by the GPI:protein transamidase. The spacer and hydrophobic regions are thereby replaced by the GPI anchor. As ω residues, only Ala, Asn, Asp, Cys, Gly or Ser are allowed, and efficient anchoring requires a small side chain at ω + 2.

anchors. In trypanosomes, the lipid moiety is remodeled shortly before transfer of the GPI to protein (Masterson *et al.*, 1990), whereas in yeast, lipid remodeling occurs after transfer (Sipos *et al.*, 1994).

The second substrate is the protein to be anchored. The anchor attachment site has been termed the $\omega$ site (Gerber *et al.*, 1992) and is located approximately 8–10 amino acids from the C-terminal hydrophobic stretch of amino acids (Moran & Caras, 1991; Nuoffer *et al.*, 1993). The length of the hydrophobic stretch is variable. Anchoring requires one of six amino acids with small side chains at the $\omega$ site, with some residues being more efficient than others (Micanovic *et al.*, 1990; Moran *et al.*, 1991; Nuoffer *et al.*, 1993). Anchoring is improved by the presence of an amino acid with a small side chain at the $\omega + 2$ site. The amino acid at the $\omega + 1$ site also contributes to the efficiency of GPI anchor attachment, but is less critical (Gerber *et al.*, 1992; Nuoffer *et al.*, 1993). The region between the $\omega$ site and the hydrophobic amino acid stretch has been termed a spacer region. This region is required, but no precise amino acid requirements have been demonstrated for it.

A crude *in vitro* assay using microsomes from mammalian cells for the attachment of GPI to the $\omega$ site of a truncated placental alkaline phosphatase has been developed (Kodukula *et al.*, 1992). In addition to the GPI-anchored product, a small amount of cleaved, unanchored protein is generated (Maxwell *et al.*, 1995a) and this reaction can be enhanced by addition of the nucleophiles hydrazine or hydroxylamine (Maxwell *et al.*, 1995b). The pH optimum for the attachment of GPI, or the enhanced cleavage seen with nucleophiles, is 7.5–8.0, and the temperature optimum is about 30°C. The enzyme has not yet been solubilized in an active form.

## Structural Chemistry

Two distinct genes have been identified in yeast that are required for the attachment of a complete GPI precursor to protein and thus are likely to be required for GPI:protein transamidase activity. One of these, *GAA1*, encodes a 72 kDa, multispanning membrane glycoprotein that has a large hydrophilic domain disposed in the lumen of the endoplasmic reticulum. Overexpression of this gene leads to a partial correction of $\omega$ site mutants in a substrate protein, suggesting that the Gaa1p protein is part of the GPI:protein transamidase and could possibly play a part in substrate recognition (Hamburger *et al.*, 1995). The other gene, *GPI8*, encodes a heterogeneously glycosylated 46–50 kDa type I membrane protein that also has a hydrophilic domain located in the lumen of the endoplasmic reticulum. A human homolog of *GPI8* has been identified. Both yeast and human Gpi8p proteins have homology to a novel family of cysteine proteases found in plants and invertebrates (Benghezal *et al.*, 1996), the legumain family, peptidase family C13. Interestingly, the jackbean asparaginyl endopeptidase (Chapter 253) is involved in maturation of concanavalin A and carries out a reaction that may be similar to a transamidation reaction, consistent with the possibility that the Gpi8p protein may be an important part of the catalytic center of the GPI:protein transamidase. The synthetic lethality between *GPI8* and *GAA1* mutants is consistent with an interaction between the two proteins. A mutation in human cells also affects GPI-protein transamidase activity

(Chen *et al.*, 1996), but cloning of the corresponding gene or cDNA has not been reported.

## Biological Aspects

The transamidase is most likely ubiquitous among eukaryotes. In lower eukaryotes such as protozoa and fungi, GPI anchoring is obviously a very important physiological function. Protozoan cell surfaces are covered with GPI-anchored proteins which most likely play a protective role for the organism (Ferguson, 1994). Both *GAA1* and *GPI8* are essential genes in *S. cerevisiae*, attesting to the importance of GPI anchor attachment in this organism. On the other hand, certain mammalian cell lines can survive in the absence of GPI anchoring (Hyman, 1988). Therefore, interference with GPI anchoring may be a possible new target for chemotherapy against protozoan and fungal parasites. In addition, despite apparent similarities between species in the definitions of GPI anchor attachment signals (Nuoffer *et al.*, 1993), a protozoan signal was not efficiently processed in mammalian cells, suggesting subtle differences in the specificity of the GPI:protein transamidase between distant species (Moran & Caras, 1994).

## Further Reading

For reviews, see Takeda & Kinoshita (1995), Englund (1993) and McConville & Ferguson (1993).

## References

Bangs, J.D., Hereld, D., Krakow, J.L., Hart, G.W. & Englund, P.T. (1985) Rapid processing of the carboxyl terminus of a trypanosome variant surface glycoprotein. *Proc. Natl Acad. Sci. USA* **82**, 3207–3211.

Benghezal, M., Benachour, A., Rusconi, S., Aebi, M. & Conzelmann, A. (1996) Yeast Gpi8p is essential for GPI anchor attachment onto proteins. *EMBO J.* **15**, 6575–6583.

Caras, I.W., Weddell, G.N., Davitz, M.A., Nussenzweig, V. & Martin, D.W., Jr. (1987) Signal for attachment of a phospholipid membrane anchor in decay-accelerating factor. *Science* **238**, 1280–1283.

Chen, R., Udenfriend, S., Prince, G.M., Maxwell, S.E., Ramalingam, S., Gerber, L.D., Knez, J. & Medof, M.E. (1996) A defect in glycosylphosphatidylinositol (GPI) transamidase activity in mutant K cells is responsible for their inability to display GPI surface proteins. *Proc. Natl Acad. Sci. USA* **93**, 2280–2284.

Conzelmann, A., Spiazzi, A., Hyman, R. & Bron, C. (1986) Anchoring of membrane proteins via phosphatidylinositol is deficient in two classes of Thy-1 negative mutant lymphoma cells. *EMBO J.* **5**, 3291–3296.

Conzelmann, A., Spiazzi, A. & Bron, C. (1987) Glycolipid anchors are attached to Thy-1 glycoprotein rapidly after translation. *Biochem. J.* **246**, 605–610.

Englund, P.T. (1993) The structure and biosynthesis of glycosylphosphatidylinositol protein anchors. *Annu. Rev. Biochem.* **62**, 121–138.

Fankhauser, C., Homans, S.W., Thomas-Oates, J.E., McConville, M.J., Desponds, C., Conzelmann, A. & Ferguson, M.A.J. (1993) Structures of glycosylphosphatidylinositol membrane anchors from *Saccharomyces cerevisiae*. *J. Biol. Chem.* **268**, 26365–26374.

Ferguson, M.A.J. (1994) What can GPI do for you? *Parasitol. Today* **10**, 48–52.

Ferguson, M.A.J., Homans, S.W., Dwek, R.A. & Rademacher, T.W. (1988) Glycosyl-phosphatidylinositol moiety that anchors *Trypanosoma brucei* variant surface glycoprotein to the membrane. *Science* **239**, 753–759.

Gerber, L.D., Kodukula, K. & Udenfriend, S. (1992) Phosphatidyl-inositol glycan (PI-G) anchored membrane proteins. Amino acid requirements adjacent to the site of cleavage and PI-G attachment in the COOH-terminal signal peptide. *J. Biol. Chem.* **267**, 12168–12173.

Hamburger, D., Egerton, M. & Riezman, H. (1995) Yeast Gaa1p is required for attachment of a completed GPI anchor onto proteins. *J. Cell Biol.* **129**, 629–639.

Homans, S.W., Ferguson, M.A.J., Dwek, R.A., Rademacher, T.W., Anand, R. & Williams, A.F. (1988) Complete structure of the glcosyl phosphatidylinositol membrane anchor of rat brain Thy-1 glycoprotein. *Nature (Lond.)* **333**, 269–272.

Hyman, R. (1988) Somatic genetic analysis of the expression of cell surface molecules. *Trends Genet.* **4**, 5–8.

Kodukula, K., Amthauer, R., Cines, D., Yeh, E.T., Brink, L., Thomas, L.J. & Udenfriend, S. (1992) Biosynthesis of phospha-tidylinositol-glycan (PI-G)-anchored membrane proteins in cell-free systems: PI-G is an obligatory cosubstrate for COOH-terminal processing of nascent proteins. *Proc. Natl Acad. Sci. USA* **89**, 4982–4985.

McConville, M.J. & Ferguson, M.A. (1993) The structure, biosynthesis and function of glycosylated phosphatidylinositols in the parasitic protozoa and higher eucaryotes. *Biochem. J.* **294**, 305–324.

Masterson, W.J., Doering, T.L., Hart, G.W. & Englund, P.T. (1989) A novel pathway for glycan assembly: biosynthesis of the glycosyl-phosphatidylinositol anchor of the trypanosome variant surface glycoprotein. *Cell* **56**, 793–800.

Masterson, W.J., Raper, J., Doering, T.L., Hart, G.W. & Englund, P.T. (1990) Fatty acid remodeling: a novel reaction sequence in the biosynthesis of trypanosome glycosyl phos-phatidylinositol membrane anchors. *Cell* **62**, 73–80.

Maxwell, S.E., Ramalingam, S., Gerber, L.D. & Udenfriend, S. (1995a) Cleavage without anchor addition accompanies the processing of a nascent protein to its glycosylphosphatidylinositol-anchored form. *Proc. Natl Acad. Sci. USA* **92**, 1550–1554.

Maxwell, S.E., Ramalingam, S., Gerber, L.D., Brink, L. & Uden-friend, S. (1995b) An active carbonyl formed during glycosyl-phosphatidylinositol addition to a protein is evidence of catalysis by a transamidase. *J. Biol. Chem.* **270**, 19576–19582.

Menon, A.K., Schwarz, R.T., Mayor, S. & Cross, G.A. (1990) Cell-free synthesis of glycosyl-phosphatidylinositol precursors for the glycolipid membrane anchor of *Trypanosoma brucei* variant surface glycoproteins. Structural characterization of putative biosynthetic intermediates. *J. Biol. Chem.* **265**, 9033–9042.

Micanovic, R., Gerber, L.D., Berger, J., Kodukula, K. & Uden-friend, S. (1990) Selectivity of the cleavage/attachment site of phosphatidylinositol-glycan-anchored membrane proteins determined by site-specific mutagenesis at Asp-484 of placental alkaline phosphatase. *Proc. Natl Acad. Sci. USA* **87**, 157–161.

Moran, P. & Caras, I.W. (1991) Fusion of sequence elements from non-anchored proteins to generate a fully functional signal for glycophosphatidylinositol membrane anchor attachment. *J. Cell Biol.* **115**, 1595–1600.

Moran, P. & Caras, I.W. (1994) Requirements for glycosylphos-phatidylinositol attachment are similar but not identical in mammalian cells and parasitic protozoa. *J. Cell Biol.* **125**, 333–343.

Moran, P., Raab, H., Kohr, W.J. & Caras, I.W. (1991) Glycophos-pholipid membrane anchor attachment. Molecular analysis of the cleavage/attachment site. *J. Biol. Chem.* **266**, 1250–1257.

Nuoffer, C., Horvath, A. & Riezman, H. (1993) Analysis of the sequence requirements for glycosylphosphatidylinositol anchoring of *Saccharomyces cerevisiae* Gas1 protein. *J. Biol. Chem.* **268**, 10558–10563.

Sipos, G., Puoti, A. & Conzelmann, A. (1994) Glycosylphospha-tidylinositol membrane anchors in *Saccharomyces cerevisiae*: absence of ceramides from complete precursor glycolipids. *EMBO J.* **13**, 2789–2796.

Takeda, J. & Kinoshita, T. (1995) GPI-anchor biosynthesis. *Trends Biochem. Sci.* **20**, 367–371.

*Howard Riezman*
*Biozentrum of the University of Basel,*
*Klingelbergstrasse 70,*
*CH-4056 Basel, Switzerland*
*Email: riezman@ubaclu.unibas.ch*

*Andreas Conzelmann*
*Institut de Biochimie, Université de Fribourg,*
*Rue de Musée 5,*
*CH-1700 Fribourg, Switzerland*
*Email: andreas.conzelmann@unifr.ch*

# 257. Clostripain

## Databanks

*Peptidase classification: clan CX, family C11, MEROPS ID: C11.001*
*NC-IUBMB enzyme classification: EC 3.4.22.8*
*Chemical Abstract Service registry number: 9028-00-6*

*Databank codes:*

| Species | SwissProt | PIR | EMBL (cDNA) | EMBL (genomic) |
|---|---|---|---|---|
| *Clostridium histolyticum* | P09870 | A29174 | X63673 | – |
| | | A29175 | | |
| | | B29175 | | |
| | | S35190 | | |

## Name and History

It has been known for many years that culture filtrates of the anaerobic bacterium *Clostridium histolyticum* contain, in addition to the forms of clostridial collagenase (Chapter 368) and clostridial aminopeptidase (Chapter 521), a cysteine-activated proteinase. In the early work, the cysteine proteinase was distinguished by its property of hydrolyzing clupein (protamine) but not gelatin. Kocholaty *et al.* (1938) were the first to isolate the enzyme from the culture filtrate in relatively pure form, by a procedure consisting of $K_2SO_4$ precipitation and electrophoretic separation. Ogle & Tytell (1953) purified the proteinase by precipitation from cell-free lysates with cold methanol, and were the first to recognize the specificity of the enzyme for the hydrolysis of arginyl bonds. Labouesse & Gros (1960) chromatographically purified the enzyme, which they designated **clostripain**, and observed that the proteinase hydrolyzes arginine-containing synthetic substrates much more rapidly than analogous compounds containing lysine. Other names that have been used for clostripain include **clostridiopeptidase B, endoproteinase Arg-C** and **γ protease**.

## Activity and Specificity

Clostripain is well known for its selective hydrolysis of arginyl bonds, although lysyl bonds are cleaved at a lower rate. The preference for argininine over lysine is much less pronounced in inhibitors than in substrates.

Clostripain has generally been assayed by determining the initial rate of hydrolysis of Bz-Arg┼OEt (BAEE) followed at 253 nm (Mitchell & Harrington, 1970). Continuous, fluorimetric assays can be made with Z-Phe-Arg┼NHMec (Kembhavi *et al.*, 1991). The activity of clostripain depends upon a cysteine thiol group, so a reducing agent such as DTT is included in the assay buffer, and there is also an absolute requirement for $Ca^{2+}$ ions (conveniently used at 10 mM). No other cations have been found to substitute for $Ca^{2+}$ (Mitchell & Harrington, 1968; Kembhavi *et al.*, 1991). The pH optimum for the hydrolytic reactions is 7.2–7.8 (Ogle & Tytell, 1953; Mitchell & Harrington, 1968). The selectivity of hydrolysis of peptide bonds by clostripain has made it a useful tool in protein sequence analysis.

Clostripain can also act as a transpeptidase, showing maximal activity in the pH range 7.6–9.0 (Andersen, 1985; Fortier & MacKenzie, 1986; Fortier & Gagnon, 1990; Ullmann & Jakubke, 1994b). Recent studies by the authors demonstrate that lysine-containing substrates can act as effective acyl donors in peptide synthesis. The specificity requirements for the nucleophilic acceptor, due to the S' subsites of the enzyme, have been investigated by Fortier & Gagnon (1990)

and Ullmann & Jakubke (1994b). It has been found that the nucleophilic effectiveness varies over more than three orders of magnitude for a range of P1'–P3' structures. Unusually amongst peptidases, clostripain accepts proline in the S1' subsite (Mitchell, 1968), and this property has attracted interest for applications in protein sequencing and peptide synthesis. The interaction of D-amino acid residues with the S' subsite of clostripain underlines the fact that the nucleophile stereospecificity is not restricted to L-amino acids.

Potent inhibitors of clostripain include oxidizing agents, thiol-blocking agents, $Co^{2+}$, $Cu^{2+}$, $Cd^{2+}$ and heavy metal ions. Citrate, borate and Tris partially inhibit. Tos-Lys-$CH_2Cl$ reacts covalently with the active site, and is an effective active-site titrant (Porter *et al.*, 1971; Gilles & Keil, 1984; Ullmann & Jakubke, 1994a). Higher rates of covalent reaction can be obtained with Phe-Ala-Lys-Arg-chloroethane and some methylsulfonium salts (Wikstrom *et al.*, 1989; Kembhavi *et al.*, 1991), but arginyl compounds are not very much more potent than their lysyl analogs. The inhibition by diethylpyrocarbonate, reversed by hydroxylamine, indicates that histidine is essential for catalytic activity (Kembhavi *et al.*, 1991).

Leupeptin is the most potent reversible inhibitor yet reported for the clostripain (Kembhavi *et al.*, 1991). A strong affinity of alkylguanidines for the S1 site of clostripain was described by Cole *et al.* (1971), and further work by the present authors has extended that finding, showing that structures containing a guanidine function are powerful competitive inhibitors.

Clostripain is inhibited weakly by some protein inhibitors of serine endopeptidases, with either arginine (soybean Bowman–Birk inhibitor) or lysine (soybean Kunitz inhibitor, aprotinin, limabean trypsin inhibitor) in P1. In contrast to its behavior towards substrates, clostripain does not appear to discriminate in favor of arginine over lysine in the protein inhibitors. Also, it is notable that the affinity of clostripain for the various trypsin inhibitors does not parallel that of trypsin (Siffert *et al.*, 1976).

## Structural Chemistry

Mature clostripain is a heterodimeric protein of 526 amino acids with $M_r$ values of 43 000 and 15 398 for the heavy and light chains, respectively (Gilles *et al.*, 1984). The chains are held together by strong noncovalent forces rather than by disulfide bridges (Gilles *et al.*, 1979). Amino acid sequences surrounding the eight Cys residues have been determined (Gilles *et al.*, 1983). By application of radiolabeled inhibitors, Cys41 of the heavy chain was identified as the catalytic sulfhydryl residue of the active site. Sequencing of the light chain revealed an arginine residue at the C-terminus (Gilles *et al.*, 1984). The precursor protein consists of a putative signal peptide (27 amino acids), propeptide (23 amino acids), light chain subunit (131 amino acids), linker peptide (9 amino acids) and heavy chain subunit (336 amino acids).

Cloning and sequencing of two overlapping genomic DNA fragments encoding clostripain has revealed that both polypeptide chains are encoded by a single gene, in a single open reading frame (ORF) of 1581 nucleotides. The ORF is preceded by canonical transcription signals and both chains of the clostripain heterodimer are completely represented by the deduced coding sequence. Most interestingly, the sequences coding for the light and the heavy chains are connected by a DNA stretch coding for a linker nonapeptide that is preceded by the C-terminal Arg residue of the light chain and also ends with an Arg (Dargatz *et al.*, 1993). This could provide an ideal basis for autocatalytic processing of the native clostripain precursor by removal of the internal linker nonapeptide. However, specific activity remains fairly constant even during prolonged incubation of the enzyme, so the removal of the nonapeptide linker does not seem to be a prerequisite for attaining the active conformation.

## Preparation

Usually, clostripain can be obtained from the culture filtrate of the anaerobic gram-positive bacterium *Clostridium histolyticum*. The latest purification protocols are based on affinity chromatography on Reactive Red 120 (Sigma; see Appendix 2 for full names and addresses of suppliers), but simple methanol precipitation gives satisfactory results for most applications (Ullmann & Jakubke, 1994a).

Heterologous expression of the clostripain gene in *Escherichia coli* yielded an enzyme capable of hydrolyzing the standard substrates of clostripain (Dargatz *et al.*, 1993). Expression was investigated using the constructs pHM7-10 and pHM3-23 in XL1-Blue. However, the clostripain activity was produced at low levels in *E. coli*, and the enzyme was not exported to the periplasm. *Bacillus subtilis* was chosen as an alternative host for the expression of the preproenzyme and the core protein (Witte *et al.*, 1994). BR151 cells harboring pHM7-10B secreted the clostripain precursor to the growth medium, where it matured to the active enzyme.

## Distinguishing Features

Clostripain belongs to a peptidase family (C11) of its own, and has a number of distinctive characteristics, including the specificity for arginyl bonds, and dependence on thiol and calcium ions for activity. The structures of the S' subsites are probably fundamentally different from those of serine proteinases in family S1 (Schellenberger *et al.*, 1993). It has been suggested that the amino acid sequence around the active-site cysteine residue of the hemagglutinating arginine-specific proteinase of *Porphyromonas gingivalis* shows similarity to that of clostripain (Nishikata & Yoshimura, 1995), but the significance of this is uncertain.

## Further Reading

An excellent review of the earlier work on clostripain was provided by Mitchell & Harrington (1970).

## References

Andersen, A.J. (1985) Enzymatic synthesis of arginine proline peptide bonds using clostripain as a catalyst. In: *Peptides: Structure and Function* (Deber, C.M., Hruby, V.J. & Kopple, K.D., eds). Rockford, IL: Pierce Chemical Co., pp. 355–358.

Cole, P., Murakami, K. & Inagami, T. (1971) Specificity and mechanism of clostripain catalysis. *Biochemistry* **10**, 4246–4252.

Dargatz, H., Diefenthal, T., Witte, V., Reipen, G. & von Wettstein, D. (1993) The heterodimeric protease clostripain from *Clostridium histolyticum* is encoded by a single gene. *Mol. Gen. Genet.* **240**, 140–145.

Fortier, G. & MacKenzie, S.L. (1986) Peptide bond synthesis by clostridiopeptidase B. *Biotechnol. Lett.* **8**, 777–782.

Fortier, G. & Gagnon, J. (1990) Kinetic study of nucleophile specificity in dipeptide synthesis catalyzed by clostridiopeptidase B. *Arch. Biochem. Biophys.* **276**, 317–321.

Gilles, A.-M., Imhoff, J.-M. & Keil, B. (1979) α-Clostripain: chemical characterization, activity, and thiol content of the highly active form of clostripain. *J. Biol. Chem.* **254**, 1462–1468.

Gilles, A.-M., De Wolf, A. & Keil, B. (1983) Amino-acid sequences of the active-site sulfhydryl peptide and other thiol peptides from the cysteine proteinase α-clostripain. *Eur. J. Biochem.* **130**, 473–479.

Gilles, A.-M. & Keil, B. (1984) Evidence for an active-center cysteine in the SH-proteinase α-clostripain through use of *N*-tosyl-L-lysine chloromethyl ketone. *FEBS Lett.* **173**, 58–62.

Gilles, A.-M., Lecroisey, A. & Keil, B. (1984) The primary structure of α-clostripain light chain. *Eur. J. Biochem.* **145**, 469–476.

Kembhavi, A.A., Buttle, D.J., Rauber, P. & Barrett, A.J. (1991) Clostripain: characterization of the active site. *FEBS Lett.* **283**, 277–280.

Kocholaty, W., Weil, L. & Smith, L. (1938) Proteinase secretion and growth of *Clostridium histolyticum*. *Biochem. J.* **32**, 1685–1690.

Labouesse, B. & Gros, P. (1960) La clostripaïne, protéase de *Clostridium histolyticum*. I. – Purification et activation par les thiols. *Bull. Soc. Chim. Biol.* **42**, 543–568.

Mitchell, W.M. (1968) Hydrolysis at arginylproline in polypeptides by clostridiopeptidase B. *Science* **162**, 374–375.

Mitchell, W.M. & Harrington, W.F. (1968) Purification and properties of clostridiopeptidase B (clostripain). *J. Biol. Chem.* **243**, 4683–4692.

Mitchell, W.M. & Harrington, W.F. (1970) Clostripain. *Methods Enzymol.* **19**, 635–642.

Nishikata, M. & Yoshimura, F. (1995) Active site structure of a hemagglutinating protease from *Porphyromonas gingivalis*: similarity to clostripain. *Biochem. Mol. Biol. Int.* **37**, 547–553.

Ogle, J.D. & Tytell, A.A. (1953) The activity of *Clostridium histolyticum* proteinase on synthetic substrates. *Arch. Biochem. Biophys.* **42**, 327–336.

Porter, W.H., Cunningham, L.W. & Mitchell, W.M. (1971) Studies on the active site of clostripain. *J. Biol. Chem.* **246**, 7675–7682.

Schellenberger, V., Turck, C.W., Hedstrom, L. & Rutter, W.J. (1993) Mapping the S' subsites of serine proteases using acyl transfer to mixtures of peptide nucleophiles. *Biochemistry* **32**, 4349–4353.

Siffert, O., Emöd, I. & Keil, B. (1976) Interaction of clostripain with natural trypsin inhibitors and its affinity labeling by $N^\alpha$-*p*-nitrobenzyloxycarbonyl arginine chloromethyl ketone. *FEBS Lett.* **66**, 114–119.

Ullmann, D. & Jakubke, H.-D. (1994a) Kinetic characterization of affinity chromatography purified clostripain. *Biol. Chem. Hoppe-Seyler* **375**, 89–92.

Ullmann, D. & Jakubke, H.-D. (1994b) The specificity of clostripain from *Clostridium histolyticum*. Mapping the S′ subsites via acyl transfer to amino acid amides and peptides. *Eur. J. Biochem.* **223**, 865–872.

Wikstrom, P., Kirschke, H., Stone, S. & Shaw, E. (1989) The properties of peptidyl diazoethanes and chloroethanes as protease inactivators. *Arch. Biochem. Biophys.* **270**, 286–293.

Witte, V., Wolf, N., Diefenthal, T., Reipen, G. & Dargatz, H. (1994) Heterologous expression of the clostripain gene from *Clostridium histolyticum* in *Escherichia coli* and *Bacillus subtilis*: maturation of the clostripain precursor is coupled with self-activation. *Microbiology* **140**, 1175–1182.

**Dirk Ullmann**
University of Leipzig,
Faculty of Biosciences, Pharmacy and Psychology,
Institute of Biochemistry,
Talstrasse 33, D-04 103 Leipzig, Germany
Email: biomidu@rzaix340.rz.uni-leipzig.de

**Frank Bordusa**
University of Leipzig,
Faculty of Biosciences, Pharmacy and Psychology,
Institute of Biochemistry,
Talstrasse 33, D-04 103 Leipzig, Germany

# 258. *Gingipain R and gingipain K*

## Databanks

*Peptidase classification: clan CX, family C25, MEROPS ID: C25.001, C25.002*
*NC-IUBMB enzyme classification: EC 3.4.22.37*
*Databank codes:*

| Species | Strain | SwissProt | PIR | EMBL (cDNA) | EMBL (genomic) |
|---|---|---|---|---|---|
| Gingipain K | | | | | |
| *Porphyromonas gingivalis* | 381 | – | – | D83258 | – |
| | HG66 | – | – | U54691 | – |
| | W12 | – | – | U42210 | – |
| Gingipain R | | | | | |
| *Porphyromonas gingivalis* | 381 | – | – | D26470 | – |
| | ATCC33277 | – | – | D64081 | – |
| | ATCC33277 | – | – | X85186 | – |
| | H66 | P28784 | – | – | – |
| | HG66 | – | A55426 | U15282 | – |
| | W50 | – | – | L26341 | – |
| | W50 | – | – | X82680 | – |
| | W83 | P46071 | – | L27483 | – |

## Name and History

*Porphyromonas gingivalis*, a well-established oral pathogen, produces substantial quantities of thiol-dependent enzymes cleaving synthetic substrates with arginine and/or lysine residues at the P1 position (Mayrand & Holt, 1988). Since 1984 these proteinases have become targets for purification and several seemingly related proteins have been separated and referred to as '*trypsin-like proteinases*'. More recently, acronyms were given to the rediscovered enzymes. Shah *et al.* (1991) were first to propose the name **gingivain**. Alternatively, the term **gingipain** (*P. gingivalis* + clostri**pain**) was suggested (Chen *et al.*, 1992), since the purified enzyme shared some properties with clostripain including a similar molecular mass, requirements of calcium for stability and narrow specificity limited to Arg+Xaa peptide bonds. Both acronyms with prefix lysine- or arginine- were later adopted to designate Lys- and Arg-specific enzymes (Scott *et al.*, 1993; Bedi, 1994; Pike *et al.*, 1994). In addition to gingivain and gingipain, the term **argingipain** has also been introduced to describe an Arg-specific enzyme (Kadowaki *et al.*, 1994).

Fortunately, the recent advances in cloning, sequencing and Southern blot analysis of *P. gingivalis* genes encoding 'trypsin-like' proteinases revealed that various molecular mass forms of arginine-specific proteinases are derived from initial translation products of two very closely related genes (*rgp-1* and *rgp-2*) (Pavloff *et al.*, 1995; Aduse-Opoku *et al.*, 1995; Okamoto *et al.*, 1995; Nakayama *et al.*, 1995; Slakeski *et al.*, 1996), whereas the lysine-specific enzyme is encoded by a single gene (Barkocy-Gallagher *et al.*, 1996; Okamoto *et al.*, 1996). Therefore, to avoid redundancies in nomenclature, in line with a recommendation by the IUBMB in 1995, the names **gingipain R** (EC 3.4.22.37) and **gingipain K**, abbreviated **RGP** and **KGP**, are suggested to account for the unique specificity of two major cysteine proteinases of *P. gingivalis* (Potempa *et al.*, 1995a).

In addition to these strictly specific gingipains, two single-chain, membrane-associated high molecular mass forms (120 kDa and 150 kDa) of an arginine/lysine-specific proteinase have been purified and referred to as **porphypain-1** and **porphypain-2** (Ciborowski *et al.*, 1994). The reported specificity of porphypain for both basic residues is not compatible with the fact that the primary structure of the enzyme catalytic domain (Barkocy-Gallagher *et al.*, 1996) is essentially identical with a corresponding part of gingipain K (Okamoto *et al.*, 1996; Pavloff *et al.*, 1997), and it is likely that **porphypain** represents a single-chain form of gingipain K contaminated with gingipain R.

## Activity and Specificity

The occurrence of several molecular forms of gingipain R and gingipain K and possible cross-contamination of poorly investigated proteinases makes it difficult to summarize their properties. Therefore, only the most unambiguous enzyme features are presented below.

The activity of all forms of gingipain R on synthetic peptide substrates is limited to substrates with the P1 Arg residue. This narrow specificity was confirmed by examination of the cleavage products after incubation with the insulin B chain or mellitin, where it was found that cleavage occurred specifically after all arginine residues but not after lysine or any other amino acid (Chen *et al.*, 1992). In contrast, gingipain K cleaves exclusively on the C-terminal side of lysine residues in various peptides and *p*-nitroanilide substrates (Fujimura *et al.*, 1993; Pike *et al.*, 1994). Substrate turnover is affected by amino acids at the P2 position; this is most noticeable in the lack of either Arg-Lys-Xaa or Lys-Lys-Xaa peptide bond cleavage. On the other hand, gingipain K effectively cleaves Lys+Pro peptide bonds (Pike *et al.*, 1994), a peptide bond resistant to hydrolysis by most mammalian serine and cysteine proteinases.

Assay of gingipain activity can be done with azocoll, hide powder blue or azocasein, but it is more conveniently performed with synthetic chromogenic or fluorogenic substrates with a basic residue at the P1 site. The less expensive, commonly available substrates are Z-Lys+NHPhNO$_2$ (Calbiochem-Novabiochem) or Bz-Arg+NHPhNO$_2$; however, longer peptide substrates, such as Tos-Gly-Pro-Lys+NHPhNO$_2$ (Chromogenix) are far more sensitive (see Appendix 2 for full names and addresses of suppliers). Thus comparable kinetic parameters were: Tos-Gly-Pro-Lys-NHPhNO$_2$ ($K_m = 0.05$ mM, $V = 215$ mmol min$^{-1}$, $V/K_m = 4280$), Z-Lys-NHPhNO$_2$ ($K_m = 0.18$, $V = 40$ mmol min$^{-1}$, $V/K_m = 220$) for gingipain K (Fujimura *et al.*, 1993; Pike *et al.*, 1994). The unusual feature of gingipains R is the fact that amidase activity is considerably enhanced in the presence of glycine-containing compounds (Chen *et al.*, 1991). This allowed the design of a specific assay for gingipain R in which activity is measured using Bz-Arg-NHPhNO$_2$ as a substrate in a buffer (0.1 M Tris, 5 mM CaCl$_2$, 10 mM cysteine, pH 7.6) with and without 200 mM glycylglycine. Increased activity in the presence of glycylglycine allows detection of gingipain R, even in the complex pathological fluids (Wikström *et al.*, 1994), and could readily be utilized to detect periodontitis at an early stage of the disease.

In addition to such commonly used protein substrates as azocasein, both gingipains are reported to degrade many protein components of human connective tissue and plasma, including immunoglobulins, proteinase inhibitors and collagens. Unfortunately, however, such activities have to be confirmed, because several of the initial reports depended upon the use of SDS-PAGE to show degradation of the proteins, without appropriate controls (Hinode *et al.*, 1991; Fujimura *et al.*, 1993; Bedi & Williams, 1994; Kadowaki *et al.*, 1994). For example, it was reported that collagen type I and α$_2$-macroglobulin are degraded by gingipain R, but these findings were artefactual, arising from the fact that the potential substrate proteins are far more sensitive to denaturation by SDS than gingipain, and it is the denatured proteins that were degraded (Bleeg, 1990). In fact, gingipain R is inhibited by α$_2$-macroglobulin (Grøn *et al.*, 1997), and SDS enhances collagen degradation by *P. gingivalis* proteinase (Bleeg & Polenik, 1991).

The pH optimum of gingipain K is at pH 8.0 for hydrolysis of small synthetic substrates, whereas with protein substrates, such as azocasein, it is near pH 8.5. Gingipain R has a slightly lower pH optimum of about 7.5, which is the same for degradation of synthetic and protein substrates.

The activity of gingipains depends on the presence of reducing agents and is irreversibly inhibited by thiol-blocking reagents, including iodoacetamide, iodoacetic acid and *N*-ethylmaleimide, but only at rather high concentration (10 mM), and after relatively long times of incubation (20–30 min). E-64 inhibits exclusively gingipain R but only at a several thousand-fold molar excess over the enzyme (Pike *et al.*, 1994). Gingipain R, but not gingipain K, is also sensitive to inhibition by Zn$^{2+}$, EDTA and the bacterial-derived peptidyl aldehyde inhibitors leupeptin and antipain. In contrast, peptidyl chloromethanes are equally good inhibitors of both gingipains.

Among natural inhibitors, only α$_2$-macroglobulin is able to inhibit gingipain R, but it does not inhibit gingipain K (Grøn *et al.*, 1997); human α$_2$-macroglobulin does not contain any Lys residue in its bait region, and has previously been found not to trap a lysine-specific endopeptidase. Other inhibitors, including serpins, cystatins and H-kininogen are ineffective against both proteinases and are degraded by them.

## Structural Chemistry

In most reports the Arg-specific proteinases are described as single-chain proteins with molecular masses of between 44 kDa and 50 kDa; however, enzyme species of molecular mass in the range 70–90 kDa have also been reported (for review see Potempa *et al.*, 1995b). In addition, the high molecular mass gingipain R, being a noncovalent but tight complex of the 50 kDa catalytic domain with a noncatalytic polypeptide chain (Aduse-Opoku *et al.*, 1995) or chains (Pike *et al.*, 1994), has been purified. The presence of various forms of gingipain R in different strains and cellular fractions of *P. gingivalis* was confirmed by the results of western blotting combined with zymography, specific labeling of the active-site cysteine residue and inhibition studies (Potempa *et al.*, 1995c). These results, together with amino acid sequence data, make it clear that gingipain R occurs as 110, 95, 70–90 and 50 kDa (or 44 kDa) proteins regardless of strain and method of cultivation; however, other molecular mass species may occur as a result of alternative proteolytic processing of the initial transcript of *rgp-1* or *rgp-2* genes. The first two species are complexes of the 50 kDa catalytic domain with polypeptide chains possessing hemagglutinin/adhesin activity (Pike *et al.*, 1996; Curtis *et al.*, 1996), and it is apparent that they are derived by the post-translational proteolytic processing of a proenzyme consisting of the N-terminal prepro fragment (227 residues), catalytic domain (492 residues) and hemagglutinin/adhesin domain (985 residues). The 50 kDa (44 kDa) form of gingipain R may be a product of either the *rgp-1* or *rgp-2* gene. The second gene encodes a 50 kDa gingipain R of primary structure very similar to the catalytic domain encoded by *rgp-1*, flanked by a large N-terminal extension (prepro fragment of 229 amino acid residues) but missing a 3′ fragment encoding a C-terminal polypeptide with hemagglutinin/adhesin activity. The primary structure of the mature RGP-2 (507 amino acid residues) is virtually identical to the equivalent sequence of RGP-1 catalytic domain, except for the C-terminal fragment of 147 amino acid residues which share only 28% identity. The gene origin of a single-chain 70–90 kDa gingipain R is uncertain, but it probably represents the 50 kDa proteinase covalently modified by lipopolysaccharide (Rangarajan *et al.*, 1995). Significantly, all form of gingipains R are stabilized by calcium ions, although calcium is not necessary for enzymic activity.

The lysine-specific, gingipain K has been purified from *P. gingivalis* less frequently than gingipain R, and the enzyme has been characterized as a single-chain protein of 48 kDa and pI 7.3 (Fujimura *et al.*, 1993), a 67–70 kDa protein (Scott *et al.*, 1993), or a 105 kDa complex of the 60 kDa catalytic domain noncovalently associated with hemagglutinin/adhesin domains (Pike *et al.*, 1994). Other forms of gingipain K have also been detected by western blot analysis of the active-site biotinylated proteinase in various *P. gingivalis* fractions (Potempa *et al.*, 1995c). All forms are apparently derived from the initial transcript of a single *kgp* gene.

Domain organization of the gingipain K polyprotein closely resembles that of gingipain R1. The catalytic domain of 519 residues is preceded by the N-terminal extension of 228 residues and followed by the large hemagglutinin/adhesin domain forming part of the C-terminus (986 residues or 995 residues for proenzyme from HG66 or 381 and W12 strain, respectively). While there is no primary structure similarity within the prepro fragment, the catalytic domains of RGP-1 and KGP share 27% identity. In a manner similar to the RGP-1 polyprotein, the KGP hemagglutinin/adhesin domain from HG66 can conveniently be divided into several subdomains of 44 kDa, 15 kDa, 17 kDa and 27 kDa, according to putative proteolytic processing after three arginines and one lysine residue. The gingipain K gene (*kgp*) has been cloned and sequenced independently from three strains of *P. gingivalis*, including HG66, W50 and W12, the gene from the last strain being referred to as *prtP* (Barkocy-Gallagher *et al.*, 1996; Okamoto *et al.*, 1996; Pavloff *et al.*, 1997). The initial transcript of the gene from the different strains is essentially identical, except for a difference in the sequence of the 17/27 kDa subdomain junction (residues 1405–1597). The active-site cysteine residue of gingipain R (Cys244) and gingipain K (Cys249) have been determined by active-site labeling with biotinylated peptidyl chloromethane. On the other hand, it was the amino acid sequence alignment of gingipain R1, gingipain R2 and gingipain K that led to the prediction that His212 and His217 form the second residues of the catalytic dyad of gingipain R and gingipain K, respectively (Pavloff *et al.*, 1997).

There have been considerable efforts to link the structures of the gingipains with those of clostripain and papain, but with the exception of a few residues encompassing the reactive cysteine of gingipain R and clostripain (Nishikata & Yoshimura, 1995), there are no significant similarities. For this reason the gingipains are classified in a peptidase family, C25, separate from those of papain (C1) and clostripain (C11) (see Chapter 252). However, additional genes encoding putative papain-like (*tpr* gene, databank accession no. M84471) and streptopain-like proteinases (*prtT* gene, databank accession no. M83096) (Chapter 221) have been found to be present within the *P. gingivalis* genome (Bourgeau *et al.*, 1992; Madden *et al.*, 1995). Whether these are ever expressed has yet to be determined.

## Preparation

Gingipains are produced by all strains of *P. gingivalis*, and substantial amounts of these enzymes have been found associated with the bacterial outer membrane and vesicles, with lesser quantities being present in soluble form in culture medium. The following strains have been used for proteinase purification: HG66 (Chen *et al.*, 1992; Pike *et al.*, 1994), W12 (Ciborowski *et al.*, 1994), W83 (Curtis *et al.*, 1993), 381 (Hinode *et al.*, 1991; Kadowaki *et al.*, 1994; Bedi & Williams, 1994), ATCC 33277 (Fujimura & Nakamura, 1987; Scott *et al.*, 1993; Fujimura *et al.*, 1993) and ATCC 53978 (W50) (Aduse-Opoku *et al.*, 1995). It is very likely that the ratio between soluble and membrane-bound enzyme depends on the strain and/or its maintenance, and at least in the HG66 strain, most of the activity is released in soluble form (Chen *et al.*, 1992; Pike *et al.*, 1994). The culture medium is the most common source of gingipains and initial purification steps involve protein precipitation with ammonium sulfate or acetone followed by ion-exchange chromatography, gel filtration, isoelectric or chromatofocusing and affinity chromatography on Arg-Sepharose, Lys-Sepharose or benzamidine-Sepharose (for references see Potempa *et al.*, 1995b). Gingipain purification from intact cells (Ciborowski *et al.*, 1994), cell extracts (Scott *et al.*, 1993), whole-cell envelope fractions (Nishikata & Yoshimura, 1991), outer membrane fractions (Grenier, 1992) and vesicles (Fujimura & Nakamura, 1987) requires protein solubilization with detergent, and often preparative electrophoresis in SDS is used to separate the enzyme. Unfortunately, in many cases the absence of convincing proof of homogeneity of the separated gingipains makes it impossible to compare the purification protocols in any reliable way.

## Biological Aspects

In most *P. gingivalis* strains the majority of gingipains are associated with either cell surface outer membrane or the vesicles that are the blebs of this membrane (Grenier *et al.*, 1989; Smalley *et al.*, 1989; Smalley & Birss, 1991). This indicates that, as a part of the secretory pathway, gingipains translocate through the cytoplasmic (inner) membrane and become inserted, at least transiently, into the outer membrane. This process resembles the secretory pathway followed by the IgA1-specific serine proteinases of *Neisseria gonorrhoeae*, *Neisseria meningitidis* and *Haemophilus influenzae* (Chapter 87) and the serine proteinase of *Serratia marcescens*, all of which are secreted via an outer-membrane-anchored intermediate (Klauser *et al.*, 1993). In addition, the initial transcript configuration of gingipains parallels the *Neisseria* IgA proteinase precursor structure which consists of an N-terminal signal peptide followed by a domain corresponding to the mature secreted proteinase and a C-terminal extension. The C-terminal peptide of the precursors contains the determinants necessary to direct the peptide across the outer membrane where it remains anchored via part of the C-terminal peptide. The same scheme is assumed to apply to the gingipains, and the following secretory events are suggested: (1) the crossing of the cytoplasmic membrane into the periplasmic space is directed by the cleavable N-terminal signal peptide, (2) the C-terminal part of the gingipain precursor integrates with the outer membrane and translocates the remaining part, (3) at the cell surface the enzyme acquires its active conformation after proteolytic removal of a profragment, where (4) it can undergo further proteolytic processing. In some strains of *P. gingivalis* the enzyme stays attached to the outer membrane, while in others it is shed into the medium. Except for signal peptide removal, the major post-translational proteolytic processing of gingipain precursors occurs by cleavage at Arg+Xaa peptide bonds. From studies using *rgp* mutants it is apparent that gingipains R are involved in the processing and trafficking of gingipain K precursors out of the cell (Okamoto *et al.*, 1996). Indirectly, it can be assumed that maturation of gingipain R is achieved by way of autoproteolysis.

In addition to autoprocessing and activation of gingipain K, gingipains R are involved in the maturation process of at least two other *P. gingivalis* proteins, including fimbrilin and the 75 kDa major outer membrane protein (Onoe *et al.*, 1995; Nakayama *et al.*, 1996). In both cases, proforms are processed by cleavage of a long, N-terminal leader peptide.

The availability of several isogenic mutant strains deficient in defined proteinases (Park & McBride, 1993; Nakayama *et al.*, 1995; Fletcher *et al.*, 1995; Yoneda & Kuramitsu, 1996) together with studies using individual enzymes have made it possible to indicate a role for gingipains, especially gingipains R, as the major virulence factors of *P. gingivalis*, which play a pivotal role in the pathogenesis of human periodontal diseases. Gingipains R are potent vascular permeability enhancement (VPE) factors (Hinode *et al.*, 1992) inducing this activity through plasma prekallikrein activation and consequent bradykinin release (Kaminishi *et al.*, 1993; Imamura *et al.*, 1994). Gingipain K by itself is not able to induce VPE in human plasma, but working in concert, gingipains R and K can cleave bradykinin directly from H-kininogen (Imamura *et al.*, 1995a). Gingipain R is also a very efficient enzyme in terms of the generation of a potent chemotactic factor, C5a, through direct cleavage of C5 (Wingrove *et al.*, 1992). Gingipain K, although unable to release chemotactic factors from native C5, can perform this function with oxidized C5 (DiScipio *et al.*, 1996). This process may contribute to excessive neutrophil accumulation at the periodontitis sites infected with *P. gingivalis*. At the same time, gingipains R degrade C3 and in this manner eliminate the formation of C3-derived opsonins. In addition, gingipain K is capable of cleaving the C5a receptor (C5aR, CD88) on neutrophil surfaces (Jagels *et al.*, 1996), and thus provides a means for *P. gingivalis* to avoid phagocytosis.

Among the plasma proteins, fibrinogen appears to be the prominent target for gingipain K. *In vitro*, and at nanomolar concentrations, the enzyme degrades the fibrinogen A$\alpha$ chain within minutes, rendering it nonclottable (Pike *et al.*, 1996). In normal plasma, the degradation of fibrinogen causes a prolongation of plasma thrombin time, exercising a strong anticoagulant activity at low nanomolar concentration (Imamura *et al.*, 1995b). This effect is further potentiated by the fact that gingipain K also degrades the procoagulant portion of high molecular mass kininogen (Scott *et al.*, 1983).

It is also very likely that gingipains contribute to the pathology of periodontitis through (a) inactivation of the bactericidal activity of human serum (Grenier, 1992), (b) activation of matrix metalloproteinases (Sorsa *et al.*, 1992), (c) degradation of immunoglobulins, bactericidal proteins and peptides (Fujimura *et al.*, 1993; Otsuka *et al.*, 1987), (d) degradation of iron-transporting proteins (Carlsson *et al.*, 1984b), (e) inactivation of proteinase inhibitors (Carlsson *et al.*, 1984a), and (f) digestion of extracellular matrix proteins which may lead to exposure of cryptotopes for bacterial adhesion to host tissue (Kontani *et al.*, 1996).

## Distinguishing Features

High molecular mass forms of gingipains are unique multifunctional proteins which, acting as proteinase-adhesins, may progressively attach to, degrade, and detach from target extracellular matrix proteins, modifying the interaction between host cells and *P. gingivalis*. In contrast to many other cysteine proteinases, they may be distinguished by their resistance to inhibition by E-64 and narrow specificity for synthetic substrates. In addition, activation of amidolytic activity of gingipain R by glycine-containing compounds, a unique feature of this enzyme, allows its detection in gingival crevicular fluid from periodontitis sites infected with *P. gingivalis* (Wikström *et al.*, 1994).

Polyclonal antisera and monoclonal antibodies against various forms of gingipains have been described, but none are commercially available.

## Further Reading

Articles for further reading include Potempa & Travis (1996) and Travis *et al.* (1995).

## References

Aduse-Opoku, J., Muir, J., Slaney, J.M., Rangarajan, M. & Curtis, M.A. (1995) Characterization, genetic analysis, and expression of a protease antigen (PrpRI) of *Porphyromonas gingivalis* W50. *Infect. Immun.* **63**, 4744–4754.

Barkocy-Gallagher, G.A., Han, N., Patti, J.M., Whitlock, J., Progulske-Fox, A. & Lantz, M.S. (1996) Analysis of the *prtP* gene encoding porphypain, a cysteine proteinase of *Porphyromonas gingivalis*. *J. Bacteriol.* **178**, 2734–2741.

Bedi, G.S. (1994) Purification and characterization of lysine- and arginine-specific gingivain proteases from *Porphyromonas gingivalis Prep. Biochem.* **24**, 251–261.

Bedi, G.S. & Williams, T. (1994) Purification and characterization of a collagen-degrading protease from *Porphyromonas gingivalis*. *J. Biol. Chem.* **269**, 599–606.

Bleeg, H.S. (1990) Non-specific cleavage of collagen by proteinases in the presence of sodium dodecyl sulfate. *Scand. J. Dent. Res.* **98**, 235–241.

Bleeg, H.S. & Polenik, P. (1991) Sodium dodecyl sulfate potentiates collagen degradation by proteases from *Porphyromonas (Bacteroides) gingivalis*. *FEMS Microbiol. Ecol.* **85**, 125–132.

Bourgeau, G., Lapointe, H., Peloquin, P. & Mayrand, D. (1992) Cloning, expressing, and sequencing of a protease gene (*tpr*) from *Porphyromonas gingivalis* W83 in *Escherichia coli*. *Infect. Immun.* **60**, 3186–3192.

Carlsson, J., Herrmann, B.F., Hofling, J.F. & Sundqvist, G.K. (1984a) Degradation of the human proteinase inhibitors alpha-1-antitrypsin and alpha-2-macroglobulin by *Bacteroides gingivalis*. *Infect. Immun.* **43**, 644–648.

Carlsson, J., Hofling, J.F. & Sundqvist, G.K. (1984b) Degradation of albumin, haemopexin, haptoglobin and transferrin, by black-pigmented *Bacteroides* species. *J. Med. Microbiol.* **18**, 39–46.

Chen, Z., Potempa, J., Polanowski, A., Renvert, S., Wikström, M. & Travis, J. (1991) Stimulation of proteinase and amidase activities in *Porphyromonas gingivalis* by amino acids and dipeptides. *Infect. Immun.* **59**, 2846–2850.

Chen, Z., Potempa, J., Polanowski, A., Wikström, M. & Travis, J. (1992) Purification and characterization of a 50-kDa cysteine proteinase (gingipain) from *Porphyromonas gingivalis*. *J. Biol. Chem.* **267**, 18896–18901.

Ciborowski, P., Nishikata, M., Allen, R.D. & Lantz, M.S. (1994) Purification and characterization of two forms of a high-molecular-weight cysteine proteinase (porphypain) from *Porphyromonas gingivalis*. *J. Bacteriol.* **176**, 4549–4557.

Curtis, M.A., Ramakrishnan, M. & Slaney, J.M. (1993) Characterization of the trypsin-like enzymes of *Porphyromonas gingivalis* W83 using a radiolabelled active-site-directed inhibitor. *J. Gen. Microbiol.* **139**, 949–955.

Curtis, M.A., Aduse-Opoku, J., Slaney, M., Rangarajan, M., Booth, V., Cridland, J. & Shepherd, P. (1996) Characterization of an adherence and antigenic determinant of the ArgI protease of *Porphyromonas gingivalis* which is present on multiple gene products. *Infect Immun.* **64**, 2532–2539.

DiScipio, R.G., Daffern, P.J., Kawahara, M., Pike, R., Travis, J. & Hugli, T.E. (1996) Cleavage of human complement C5 by cysteine proteinases from *Porphyromonas (Bacteroides) gingivalis*. Prior oxidation of C5 augments proteinase digestion of C5. *Immunology* **87**, 660–667.

Fletcher, H.M., Schenkein, H.A., Morgan, R.M., Bailey, K.A., Barry, C.R. & Macrina, F.L. (1995) Virulence of a *Porphyromonas gingivalis* W83 mutant defective in the *prtH* gene. *Infect. Immun.* **63**, 1521–1528.

Fujimura, S. & Nakamura, T. (1987) Isolation and characterization of a protease from *Bacteroides gingivalis*. *Infect. Immun.* **55**, 949–955.

Fujimura, S., Shibata, Y. & Nakamura, T. (1993) Purification and partial characterization of a lysine-specific protease of *Porphyromonas gingivalis*. *FEMS Microbiol. Lett.* **113**, 133–138.

Grenier, D. (1992) Inactivation of human serum bactericidal activity by a trypsinlike protease isolated from *Porphyromonas gingivalis*. *Infect. Immun.* **60**, 1854–1857.

Grenier, D., Chao, G., McBride, B.C. (1989) Characterization of sodium dodecyl sulfate-stable *Bacteroides gingivalis* proteases by polyacrylamide gel electrophoresis. *Infect. Immun.* **57**, 95–99.

Grøn, H., Pike, R., Potempa, J., Travis, J., Thøgersen, I.B., Enghild, J.J. & Pizzo, S.V. (1997) The potential role of $\alpha_2$-macroglobulin in the control of trypsin-like cysteine proteinases (gingipains) from *Porphyromonas gingivalis*. *J. Periodontol. Res.* **32**, 61–68.

Hinode, D., Hayashi, H. & Nakamura, R. (1991) Purification and characterization of three types of proteinases from culture supernatants of *Porphyromonas gingivalis*. *Infect. Immun.* **59**, 3060–3068.

Hinode, D., Nagata, A., Ichimiya, S., Hayashi, H., Morioka, M. & Nakamura, R. (1992) Generation of plasma kinin by three types of protease isolated from *Porphyromonas gingivalis*. *Arch. Oral Biol.* **37**, 859–861.

Imamura, T., Pike, R.N., Potempa, J. & Travis J. (1994) Pathogenesis of periodontitis: a major arginine-specific cysteine proteinase from *Porphyromonas gingivalis* induces vascular permeability enhancement through activation of the kallikrein/kinin pathway. *J. Clin. Invest.* **94**, 361–369.

Imamura, T., Potempa, J., Pike, R.N. & Travis, J. (1995a) Dependence of vascular permeability enhancement of cysteine proteinases in vesicles of *Porphyromonas gingivalis*. *Infect. Immun.* **63**, 1999–2003.

Imamura, T., Potempa, J., Pike, R.N., Moor, J.N., Barton, M.H. & Travis, J. (1995b) Effect of free and vesicle-bound cysteine proteinases of *Porphyromonas gingivalis* on plasma clot formation: implication for bleeding tendency at periodontitis sites. *Infect. Immun.* **63**, 4877–4882.

Jagels, M. A., Travis, J., Potempa, J., Pike, R. & Hugli, T. E. (1996) Proteolytic inactivation of the leukocyte C5a receptor by proteinases derived from *Porphyromonas gingivalis*. *Infect. Immun.* **64**, 1984–1991.

Kadowaki, T., Yoneda, M., Okamoto, K., Maeda, K. & Yamamoto, K. (1994) Purification and characterization of a novel arginine-specific cysteine proteinase (argingipain) involved in the pathogenesis of periodontal disease from the culture supernatant of *Porphyromonas gingivalis. J. Biol. Chem.* **269**, 21371–21378.

Kaminishi, H., Cho, T., Itoh, T., Iwata, A., Kawasaki, K., Hagihara, Y. & Maeda, H. (1993) Vascular permeability enhancing activity of *Porphyromonas gingivalis* protease in guinea pigs. *FEMS Microbiol. Lett.* **114**, 109–114.

Klauser, T., Pohlner, J. & Meyer, T.F. (1993) The secretion pathway of IgA protease-type proteins in gram-negative bacteria. *BioEssays* **15**, 799–805.

Kontani, M., Ono, H., Shibata, H., Okamura, Y., Tanaka, T., Fujiwara, T., Kimura, S. & Hamada, S. (1996) Cysteine proteinase of *Porphyromonas gingivalis* 381 enhances binding of fimbriae to cultured human fibroblasts and matrix proteins. *Infect. Immun.* **64**, 756–762.

Madden, T.E., Clark, V.L. & Kuramitsu, H.K. (1995) Revised sequence of the *Porphyromonas gingivalis* prtT cysteine protease/hemagglutinin gene: homology with streptococcal pyrogenic exotoxin B/streptococcal proteinase. *Infect. Immun.* **63**, 238–247.

Mayrand, D. & Holt, S.C. (1988) Biology of asaccharolytic black-pigmented *Bacteroides* species. *Microbiol. Rev.* **52**, 134–152.

Nakayama, K., Kadowaki, T., Okamoto, K. & Yamamoto, K. (1995) Construction and characterization of arginine-specific cysteine proteinase (Arg-gingipain)-deficient mutant of *Porphyromonas gingivalis. J. Biol. Chem.* **270**, 23619–23626.

Nakayama, K., Yoshimura, F., Kadowaki, T. & Yamamoto, K. (1996) Involvement of arginine-specific cysteine proteinase (Arg-gingipain) in fimbrination of *Porphyromonas gingivalis. J. Bacteriol.* **178**, 2818–2824.

Nishikata, M. & Yoshimura, F. (1991) Characterization of *Porphyromonas (Bacteroides) gingivalis* hemagglutinin as a protease. *Biochem. Biophys. Res. Commun.* **178**, 336–342.

Nishikata, M. & Yoshimura, F. (1995) Active site structure of a hemagglutinating protease from *Porphyromonas gingivalis*: similarity to clostripain. *Biochem. Mol. Biol. Int.* **37**, 547–553.

Okamoto, K., Misumi, Y., Kadowaki, T., Yoneda, M., Yamamoto, K. & Ikehara, Y. (1995) Structural characterization of argingipain, a novel arginine-specific proteinase as a major periodontal pathogenic factor from *Porphyromonas gingivalis: Arch. Biochem. Biophys.* **316**, 917–925.

Okamoto, K., Kadowaki, T., Nakayma, K. & Yamamoto, K. (1996) Cloning and sequencing of the gene encoding a novel lysine-specific cysteine proteinase (Lys-gingipain) in *Porphyromonas gingivalis*: structural relationship with the arginine-specific cysteine proteinase (Arg-gingipain). *J. Biochem.* **120**, 398–406.

Onoe, T., Hoover, C.I., Nakayama, K., Ideka, T., Nakamura, H. Yoshimura, F. (1995) Identification of *Porphyromonas gingivalis* prefimbrilin possessing a long peptide: possible involvement of trypsin-like protease in fimbrilin maturation. *Microb. Pathogen.* **19**, 351–354.

Otsuka, M., Endo, J., Hinode, D., Nagata, A., Maehara, R., Sato, M. & Nakamura, R. (1987) Isolation and characterization of protease from culture supernatant of *Bacteroides gingivalis. J. Periodont. Res.* **22**, 491–498.

Park, Y. & McBride, B.C. (1993) Characterization of the *tpr* gene product and isolation of a specific protease-deficient

mutant of *Porphyromonas gingivalis* W83. *Infect. Immun.* **61**, 4139–4146.

Pavloff, N., Potempa, J., Pike, N.R., Prochazka, V., Kiefer, M.C., Travis, J. & Barr, P. (1995) Molecular cloning and structural characterization of the Arg-gingipain proteinase of *Porphyromonas gingivalis. J. Biol. Chem.* **270**, 1007–1010.

Pavloff, N., Pemberton, P.A., Potempa, J., Chen, W.C.A., J., Pike, N.R., Prochazka, V., Kiefer, M.C., Travis, J. & Barr, P. (1997) Molecular cloning and characterization of *Porphyromonas gingivalis* Lys-gingipain. A new member of an emerging family of pathogenic bacterial cysteine proteinases. *J. Biol. Chem.* **272**, 1595–1600.

Pike, R., McGraw, W., Potempa, J. & Travis, J. (1994) Lysine- and arginine-specific proteinases from *Porphyromonas gingivalis*. Isolation, characterization, and evidence for the existence of complexes with hemagglutinins. *J. Biol. Chem.* **269**, 406–411.

Pike, R., Potempa, J., McGraw, W., Coetzer, T.H.T. & Travis, J. (1996) Characterization of the binding activities of proteinase-adhesin complex of *Porphyromonas gingivalis. J. Bacteriol.* **178**, 2876–2882.

Potempa, J. & Travis, J. (1996) *Porphyromonas gingivalis* proteinases in periodontitis, a review. *Acta Biochem. Pol.* **43**, 455–466.

Potempa, J., Pavloff, N. & Travis, J. (1995a) *Porphyromonas gingivalis*: a proteinase/gene accounting audit. *Trends Microbiol.* **3**, 430–434.

Potempa, J., Pike, R. & Travis, J. (1995b) Host and *Porphyromonas gingivalis* proteinases (gingipains) in periodontitis: a biochemical model of infection and tissue destruction. *Prospect. Drug Discovery Design* **2**, 445–458.

Potempa, J., Pike, R. & Travis, J. (1995c) The multiple forms of trypsin-like activity present in various strains of *Porphyromonas gingivalis* are due to the presence of either Arg-gingipain or Lys-gingipain. *Infect. Immun.* **63**, 1176–1182.

Rangarajan, M., Smith, S. & Curtis, M.A. (1995) Three forms of protease specific for Arg-x bonds from *Porphyromonas gingivalis* are W50. *J. Dent. Res.* **74**, 848 (Abstract).

Scott, C.F., Whitaker, E.J., Hammond, B.F. & Colman, R.W. (1993) Purification and characterization of a potent 70-kDa thiol lysylproteinase (Lys-gingipain) from *Porphyromonas gingivalis* that cleaves kininogen and fibrinogen. *J. Biol. Chem.* **268**, 7935–7942.

Shah, N.H., Gharbia, S.E., Kowlessur, D., Wilkie, E. & Brocklehurst, K. (1991) Gingivain: a cysteine proteinase isolated from *Porphyromonas gingivalis. Microb. Ecol. Health Dis.* **4**, 319–328.

Slakeski, N., Clean, S.M. & Reynolds, E.C. (1996) Characterization of a *Porphyromonas gingivalis* gene prtR that encodes an arginine-specific thiol proteinase and multiple adhesins. *Biochem. Biophys. Res. Commun.* **224**, 605–610.

Smalley, J.W. & Birss, A.J. (1991) Extracellular vesicle associated and soluble trypsin-like enzyme fractions of *Porphyromonas gingivalis* W50. *Oral Microbiol. Immunol.* **6**, 202–208.

Smalley, J.W., Birss, A.J., Kay, H.M., McKee, A.S. & Marsh, P.D. (1989) The distribution of trypsin-like enzyme activity in cultures of a virulent and an avirulent strain of *Bacteroides gingivalis* W50. *Oral Microbiol. Immunol.* **4**, 178–181.

Sorsa, T., Ingman, T., Soumalainen, K., Haapasalo, M., Konttinen, Y.T., Lindy, O., Saari, H. & Uitto, V.-J. (1992) Identification of proteases from periodontopathogenic bacteria as activators of latent human neutrophil and fibroblast-type interstitial collagenases. *Infect. Immun.* **60**, 4491–4495.

Travis, J., Potempa, J. & Maeda, H. (1995) Are bacterial proteinases pathogenic factors? *Trends Microbiol.* **3**, 405–407.

Wikström, M., Potempa, J., Polanowski, A., Travis, J. & Renvert, S. (1994) Detection of *Porphyromonas gingivalis* in gingival exudate by dipeptide enhanced trypsin-like activity. *J. Periodontol.* **65**, 47–55.

Wingrove, J.A., DiScipio, R.G., Hugli, T.E., Chen, Z., Potempa, J.

& Travis, J. (1992) Activation of complement components C3 and C5 by a cysteine proteinase (gingipain) from *Porphyromonas (Bacteroides) gingivalis.* *J. Biol. Chem.* **267**, 18902–18907.

Yoneda, M. & Kuramitsu, H.K. (1966) Genetic evidence for the relationship of *Porphyromonas gingivalis* cysteine protease and hemagglutinin activities. *Oral Microbiol. Immunol.* **11**, 129–134.

***Jan Potempa***
*Department of Microbiology and Immunology,*
*Institute of Molecular Biology,*
*Jagiellonian University,*
*Al. Mickiewicza 3,*
*31-120 Krakow, Poland*
*Email: potempa@mol.uj.edu.pl*

***James Travis***
*Department of Biochemistry,*
*University of Georgia,*
*Athens, GA 30602, USA*
*Email: jtravis@uga.cc.uga.edu*

# 259. Translocator-associated bacteriocin leader peptidase

## Databanks

*Peptidase classification: clan CX, family C39, MEROPS ID: C39.001*
*NC-IUBMB enzyme classification: none*
*Databank codes:*

| Species | Gene | SwissProt | PIR | EMBL (cDNA) | EMBL (genomic) |
|---|---|---|---|---|---|
| *Carnobacterium divergens* | dvnT | – | – | Z54201 | – |
| *Carnobacterium piscicola* | cbnT | – | – | L47121 | – |
| *Enterococcus faecalis* | cylB | – | – | M38052 | – |
| *Escherichia coli* | cvaB | P22520 | S12272 | X57524 | – |
| *Escherichia coli* | mtfB | – | – | U47048 | – |
| *Lactobacillus sake* | sapT | – | I56273 | Z46867 | – |
| *Lactobacillus sake* | sppT | – | S57913 | Z48542 | – |
| *Lactococcus lactis* | lcnDR3 | P37608 | – | U04057 | – |
| *Lactococcus lactis* | lcnC | Q00564 | B43943 | M90969 | – |
| *Leuconostoc gelidum* | lcaC | – | – | L40491 | – |
| *Leuconostoc mesenteroides* | mesD | Q10418 | S52205 | X81803 | – |
| *Methanobacterium thermoautotrophicum* | – | – | – | U19362 | – |
| *Methanobacterium thermoautotrophicum* | – | – | A55712 | L37108 | – |
| *Pediococcus acidilactici* | pedD | P36497 | D48941 | M83924 | – |
| *Pediococcus acidilactici* | papD | – | – | U02482 | – |
| *Streptococcus pneumoniae* | comA | Q03727 | A39203 | M36180 | – |

## Name and History

Bacteriocins of gram-positive bacteria (Jack *et al.*, 1995) are produced as precursor peptides that require processing of a leader segment, and are exported to the surrounding medium. Some bacteriocin-synthesizing operons encode specific serine peptidases able to carry out this processing (either before or after export) as seen, for example, for several lantibiotics (Sahl *et al.*, 1995) (Chapter 109); other operons do not encode such proteases. Furthermore, it appears that all bacteriocins are exported from the cell by translocators of the ATP-binding cassette (ABC) superfamily (Fath & Kolter, 1993) and not by *sec*-dependent pathways. ABC transport proteins typically consist of two domains: an ATP-binding domain and a membrane-spanning domain. Early observations (Bukhtiyarova *et al.*, 1994) showed that expression of only the structural gene and the translocator protein were sufficient for homologous production of active pediocin AcH in *Escherichia coli*. Recently, a number of bacteriocin-associated ABC transporters have been found to contain an additional N-terminal extension that encodes a cysteine protease able to cleave the leader peptide during export (Håvarstein *et al.*, 1995; Venema *et al.*, 1995), and this enzyme is the ***translocator-associated bacteriocin leader peptidase***.

## Activity and Specificity

Bacteriocins that are transported by these novel, protease-transporter hybrid proteins include pediocin PA-1 (Marugg *et al.*, 1992), lactococcin A (Stoddard *et al.*, 1992), lactococcin G (Håvarstein *et al.*, 1995) and plantaracin A (Diep *et al.*, 1996), as well as the lantibiotics lactococcin DR (Rince *et al.*, 1994) and cytolysin (Gilmore *et al.*, 1994). These bacteriocins have similarities in the sequences of their leader peptides (Siezen *et al.*, 1995). Most notably, they all contain Gly in position P2 and either Gly, Ala or Ser at P1, relative to the site of cleavage, suggesting that the peptidase-containing transport proteins have specificity for these residues. A number of other bacteriocins also contain this conserved sequence at their processing sites, and may be processed by a similar mechanism. No data are currently available concerning further definition of the stability, specificity or kinetics of these proteolytic domains.

## Structural Chemistry

The proteolytic domain of the lactococcin G transporter, LagD, has been cloned and overexpressed (Håvarstein *et al.*, 1995). The domain consists of 150 amino acids and was shown, *in vitro*, to specifically remove the leader peptide of recombinant lactococcin G immediately after the Gly-Gly-sequence preceding the cleavage site; i.e. at the same position as occurs *in vivo*. Activity of the LagD proteolytic domain is enhanced under reducing conditions and is not effected by chelators such as EDTA. Similarly, the N-terminal domain of the pediocin PA-1 transport protein PedD has been expressed in *E. coli* cells which also expressed the precursor of pediocin PA-1, PedA (Venema *et al.*, 1995). In this case, it could be shown that this domain alone was sufficient for the processing of the leader peptide at the same site as occurs *in vivo*.

Sequence comparison of seven transport proteins thought to include a cleavage peptidase shows two conserved motifs: Gln-(Xaa)$_4$-(Asp/Glu)-**Cys**-(Xaa)$_2$-Ala-(Xaa)$_3$-Met-(Xaa)$_4$-(Tyr/Phe)-Gly-(Xaa)$_4$-(Ile/Leu) and **His**-(Tyr/Phe)-(Tyr/Val)-Val-(Xaa)$_{10}$-(Ile/Leu)-Xaa-Asp-Pro (Håvarstein *et al.*, 1995). However, comparison of these motifs with those of known proteases reveals no significant similarities. Mutation of Cys to Ala or Val to Met in the LagD proteolytic domain abolished activity. From these results, as well as because the proteolytic domain does not contain a conserved Ser, and was not inhibited by metal ion chelators, it has been proposed that these proteolytic domains belong to the cysteine protease group.

## Preparation

These proteolytic fragments are not currently available, but bacterial strains expressing the proteolytic domains of some bacteriocin translocators should be available for research purposes from several laboratories.

## References

Bukhtiyarova, M., Yang, R. & Ray, B. (1994) Analysis of the pediocin AcH cluster from plasmid pSMB74 and its expression in a pediocin-negative *Pediococcus acidilactici* strain. *Appl. Environ. Microbiol.* **60**, 3405–3408.

Diep, D.B., Håvarstein, L.S. & Nes, I.F. (1996) Characterization of the locus responsible for the bacteriocin production in *Lactobacillus plantarum* C11. *J. Bacteriol.* **178**, 4472–4483.

Fath, M.J. & Kolter, R. (1993) ABC transporters: bacterial exporters. *Microbiol. Rev.* **57**, 995–1017.

Gilmore, M.S., Segarra, R.A., Booth, M.C., Bogie, C.P., Hall, L.R. & Clewell, D.B. (1994) Genetic structure of the *Enterococcus faecalis* plasmid pAD1-encoded cytolytic toxin system and its relationship to lantibiotic determinants. *J. Bacteriol.* **176**, 7335–7344.

Håvarstein, L.S., Diep, D.B. & Nes, I.F (1995) A family of bacteriocin ABC transporters carry out proteolytic processing of their substrates concomitant with transport. *Mol. Microbiol.* **16**, 229–240.

Jack, R.W., Tagg, J.R. & Ray, B. (1995) Bacteriocins of Gram-positive bacteria. *Microbiol. Rev.* **59**, 171–200.

Marugg, J.D., Gonzales, C.F., Kunka, B.S., Ledeboer, A.M., Pucci, M.J., Toonen, M.Y, Walker, S.A., Zoetmulder, L.C.M. & Vandenbergh, P.A. (1992) Cloning, expression and nucleotide sequence of genes involved in production of pediocin PA-1, a bacteriocin from *Pediococcus acidilactici* PAC1.0. *Appl. Environ. Microbiol.* **58**, 2360–2367.

Rince, A., Dufour, A., Le Pogram, S., Thuault, D., Bourgeois, C.M. & Le Pennec, J.P. (1994) Cloning, expression and nucleotide sequence of genes involved in production of lactococcin DR, a bacteriocin from *Lactococcus lactis* subsp. *lactis*. *Appl. Environ. Microbiol,* **60**, 1652–1657.

Sahl, H.-G., Jack, R.W. & Bierbaum, G. (1995) Biosynthesis and biological activities of lantibiotics with unique posttranslational modifications. *Eur. J. Biochem.* **230**, 827–853.

Siezen, R.J., Kuipers, O.P. & de Vos, W.M. (1995) Comparison of lantibiotic gene clusters and encoded proteins. *Antonie van Leeuwenhoek* **69**, 171–184.

Stoddard, G.W., Petzel, J.P., van Belkum, M.J., Kok, J. & McKay, L.L. (1992) Molecular analyses of the lactococcin A gene cluster from *Lactococcus lactis* subsp. *lactis* biovar *diacetylactis* WM4. *Appl. Environ. Microbiol.* **58**, 1952–1961.

Venema, K., Kok, J., Marugg, J.D., Toonen, M.Y., Ledeboer, A.M., Venema, G. & Chikindas, M.L. (1995) Functional analysis of the

pediocin operon of *Pediococcus acidilactici* PAC1.0: PedB is the immunity protein and PedD is the precursor processing enzyme. *Mol. Microbiol.* **17**, 512–522.

**Ralph W. Jack**
*Institut für Medizinische
Mikrobiologie der Universität Bonn,
Sigmund-Freud-Strasse 25, D-53105
Bonn, Germany
Email: ralph@mibi03.meb.uni-bonn.de*

**Gabriele Bierbaum**
*Institut für Medizinische
Mikrobiologie der Universität Bonn,
Sigmund-Freud-Strasse 25, D-53105
Bonn, Germany*

**Hans-Georg Sahl**
*Institut für Medizinische
Mikrobiologie der Universität Bonn,
Sigmund-Freud-Strasse 25, D-53105
Bonn, Germany*

# 260. *Isoprenylated protein peptidase*

## Databanks

*Peptidase classification: clan CX, family C99, MEROPS ID: U9G.030*
*NC-IUBMB enzyme classification: none*
*Databank codes: no sequence data available*

## Name and History

Proteins can undergo hydrophobic post-translational modifications by a number of routes. One important group of hydrophobic post-translational modifications involves isoprenylation/methylation (Figure 260.1). As shown in Figure 260.1, proteins can be singly or doubly isoprenylated. A proteolytic step is confined to the enzymatic processing of those proteins that are singly isoprenylated by either farnesylation or geranylgeranylation (Figure 260.2). The proteins to be modified by single isoprenylation all have a C-terminal -Cys-Yaa-Yaa-Xaa motif, in which Yaa is an aliphatic amino acid, and Xaa is any amino acid. We use the abbreviation $CY_1Y_2X$ for this motif here. The isoprenylation pathway for $CY_1Y_2X$ has been well described (Casey, 1995). In this pathway (Figure 260.2), soluble farnesyl or geranylgeranyl transferases condense either farnesyl pyrophosphate or geranylgeranyl pyrophosphate with a cysteine residue of a protein containing a $CY_1Y_2X$ motif at its C-terminus. After isoprenylation has occurred, a peptidase, which is the subject of the present chapter, cleaves between the modified cysteine moiety and the adjacent amino acid to produce the trimmed isoprenylated protein and the intact $Y_1Y_2X$ tripeptide (Ma & Rando, 1992; Ashby *et al.*, 1992). Finally, an AdoMet-linked isoprenylated protein methyl transferase methylates the isoprenylated cysteine residue to produce the fully matured isoprenylated/methylated protein (Clarke, 1992). The effect of the modification is to render soluble proteins membrane bound (Casey, 1995; Parish & Rando, 1996). As all signal-transducing G proteins are modified by isoprenylation/methylation, these modifications are of exceptional significance in the general processing of the signals in cells.

## Activity and Specificity

The isoprenylated protein peptidase carries out the proteolytic cleavage of the C-terminal tripeptide from a protein containing an isoprenylated cysteine residue in the fourth position from the C-terminus. The enzyme is assayed with tetrapeptide substrates containing an isoprenylated cysteine residue at the N-terminus (Ashby *et al.*, 1992; Ma & Rando, 1992). A simple assay involves the use of tritiated $N$-acetyl-$S$-farnesyl-Cys+Val-Ile-Met as substrate (Ma & Rando, 1992). The enzyme cleaves the substrate to form radioactive $N$-acetyl-$S$-farnesyl-L-cysteine plus the tripeptide. Since the acetyl farnesyl cysteine and radiolabeled tetrapeptide substrate can be readily separated by HPLC, one can easily assay the activity of the enzyme (Ma & Rando, 1992).

With respect to tetrapeptide substrates, the enzyme is stereospecific at the Cys residue and both Yaa sites, but is only stereoselective at the C-terminal Xaa residue (Ma *et al.*, 1992). The enzyme will hydrolyze tripeptides, though with a much lower efficiency than tetrapeptides, but it will not appreciably hydrolyze dipeptide substrates (Ma *et al.*, 1992). A thorough investigation of specificities with respect to various amino acids at the $Y_1Y_2$ and X sites has not been performed. However, the results of molecular biological experiments using a variety of amino acids in these sites suggest that there is little specificity at $Y_1$ and more specificity at $Y_2$ and X (Kato *et al.*, 1992). For example, acidic and basic amino acid substitutions at either $Y_2$ or X decreased the apparent proteolytic processing of the constructs studied (Kato *et al.*, 1992). Gly and Tyr at $Y_2$ also appear to reduce processing (Kato *et al.*, 1992). These experiments, however, are quite indirect and refer only to processing in NIH-3T3 cells (Kato

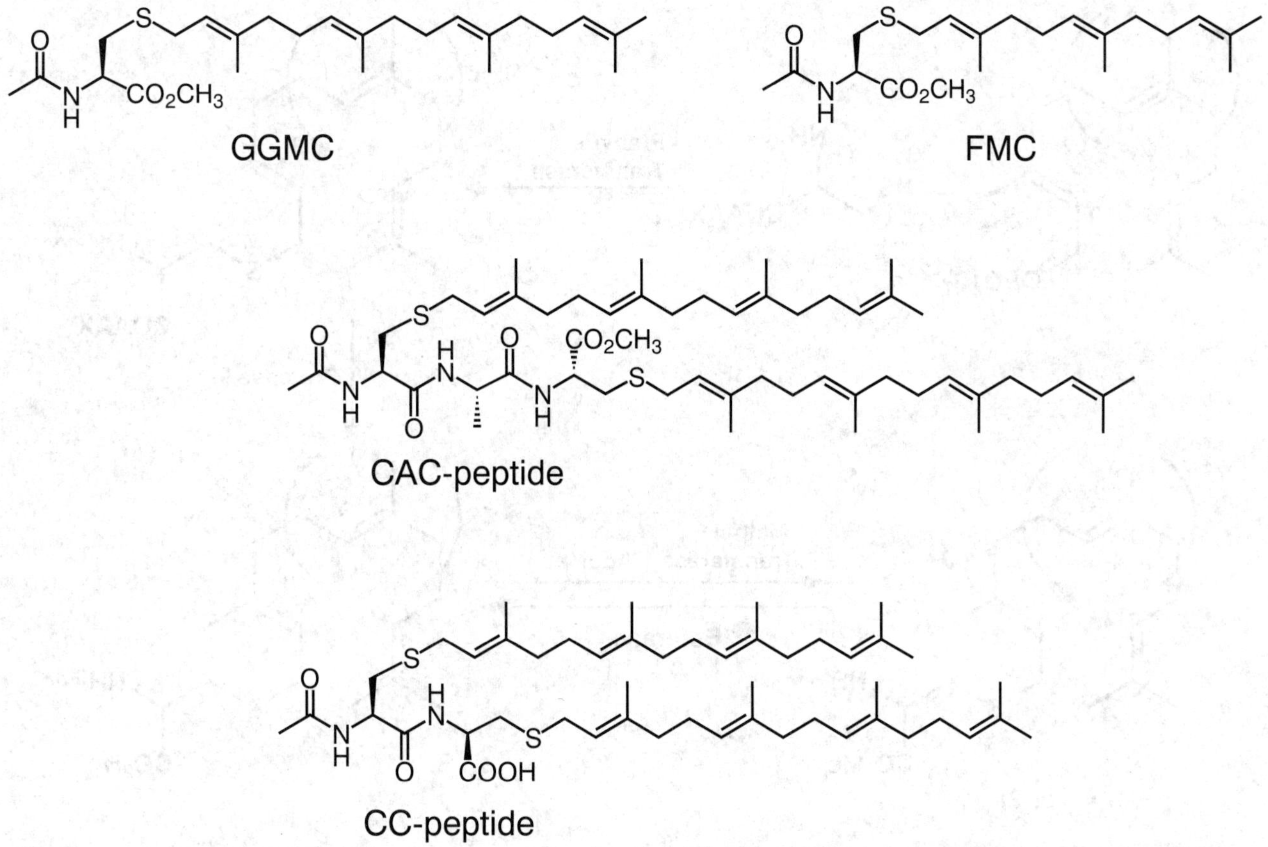

*Figure 260.1* Biochemical scope of isoprenylation. Proteins with C-terminal Cys-Yaa-Yaa-Xaa, Cys-Xaa-Cys (CAC) or Cys-Cys (CC) moieties can be enzymatically processed by isoprenylation/methylation. The isoprenyl moiety can be either all-*trans*-geranylgeranyl ($C_{20}$) as seen in GGMC or all-*trans*-farnesyl ($C_{15}$) as in FMC. The modification is of interest in signal transduction because all known small and heterotrimeric G proteins are isoprenylated/methylated. The $C_{20}$ geranylgeranyl modification predominates.

*et al.*, 1992). Given that the enzyme has yet to be purified, and there is some indication that isoforms of the enzyme may exist (Ma & Rando, 1993), the relevance of these studies to the general specificity of the enzyme(s) is not yet clear.

Limited studies on the role of the isoprenoid have also been carried out (Ashby *et al.*, 1992; Ma *et al.*, 1992). In the absence of an isoprenyl group, or in the presence of a *t*-butyl thio-group in place of the isoprene, no activity was observed (Ma & Rando, 1992; Ma *et al.*, 1992), showing the importance of an isoprene-like moiety at the cysteine. However, in the isoprenyl series, substitution of geranylgeranyl for farnesyl was quite acceptable, with the geranylgeranylated compound having about half the $K_m$ and half the $V$ of its farnesylated counterpart (Ma & Rando, 1992; Ma *et al.*, 1992). A decrease in substrate susceptibility was found, however, with the geranylated substrate, which had a 2-fold higher $K_m$ than the farnesylated substrate, and only about one-third the $V$ (Ma *et al.*, 1992).

Many of the early studies were performed with calf liver microsomes as the source of enzymatic activity. With *N*-acetyl-*S*-farnesyl-Cys┼Val-Ile-Ser as substrate, $K_m$ was 5.7 μM and $V$ was 251 pmol min$^{-1}$ mg$^{-1}$ of protein (Ma & Rando, 1992). Again, substituting a D-Cys for an L-Cys residue leads to complete abolition of activity. The pH

dependence of activity showed a broad pH maximum centered at about pH 7. Most studies of the enzyme are done in buffers such as Tris–HCl (20 mM) at pH 7.0 (Ma & Rando, 1992). DTT has no significant effect on enzyme activity, and no other redox influences on activity have been found. EDTA has no effect (Ma & Rando, 1992). There are no known activators of the enzyme. The membrane-bound enzyme has been solubilized and partially purified (see below), and with these preparations the assays are performed in 0.1% CHAPSO (Chen *et al.*, 1996).

Given the likely importance of this enzyme in signal transduction, it is of some interest to develop inhibitors of it. Very few of the well-known peptidase inhibitors affect activity very much; for example, chymostatin, DTT, E-64, 1,10-phenanthroline and PMSF do not affect activity (Ma & Rando, 1992; Chen *et al.*, 1996). However, the thiol-reactive reagent *p*-chloromercuribenzoate does inactivate the partially purified enzyme, as little as 5 μM giving substantial inhibition (Ma & Rando, 1992; Chen *et al.*, 1996). These findings do not clearly reveal the catalytic type of the enzyme, but the inhibition by thiol reagents suggests the possibility that the enzyme may be a cysteine protease.

The possibility that the enzyme might be a cysteine protease is strengthened by the fact that the aldehyde analog,

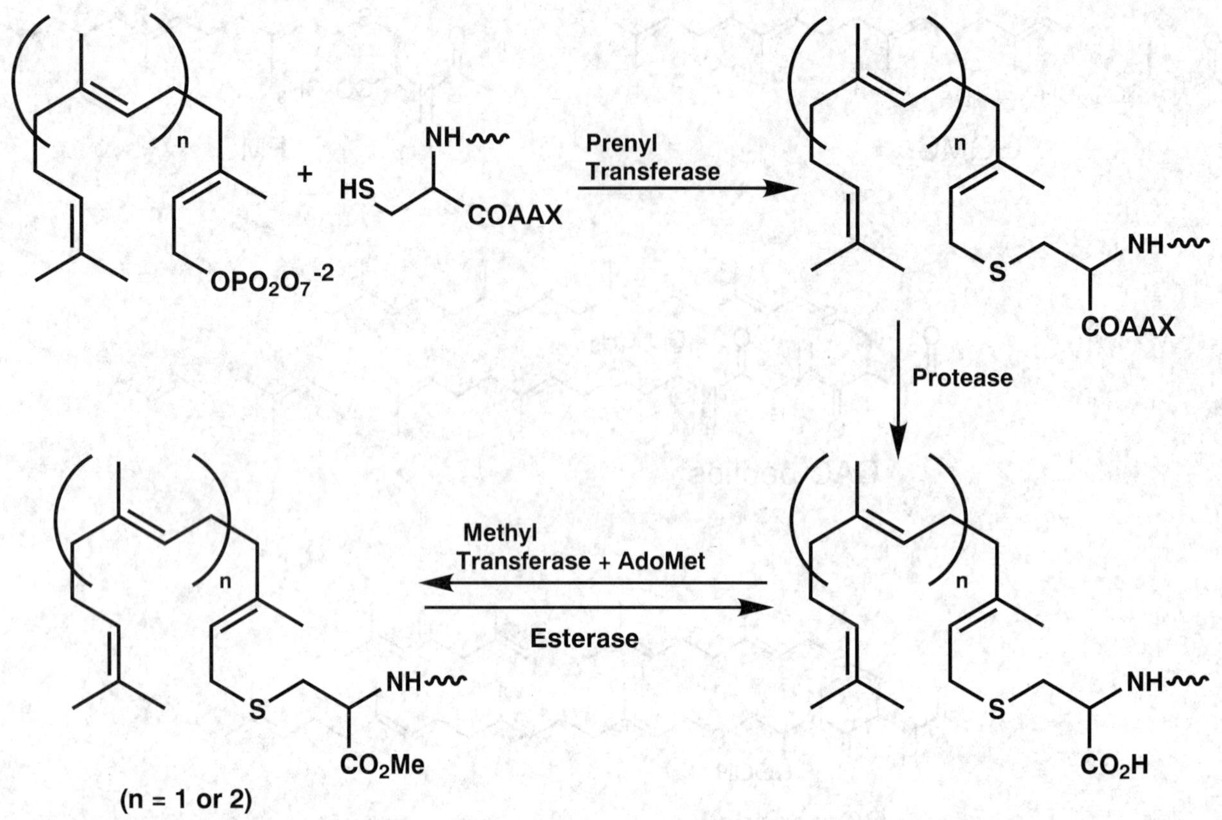

*Figure 260.2* Post-translational isoprenylation/methylation of proteins containing the Cys-Yaa-Yaa-Xaa motif. (The Yaa-Yaa-Xaa C-terminal tripeptide is 'aliphatic amino acid–aliphatic amino acid–any amino acid', and is abbreviated AAX in the figure.)

$N$-Boc-$S$-farnesyl-L-cysteine aldehyde (Figure 260.3) is a potent competitive inhibitor, with $K_i$ approximately 2 μM (Ma *et al.*, 1993). Aldehyde inhibitors are typically thought to be transition state inhibitors of cysteine and serine proteases. Tos-Phe-CH$_2$Cl is an affinity labeling agent of the enzyme with a $K_i$ of approximately 1 mM and first-order rate of inhibition of $1 \times 10^{-3}$ s$^{-1}$, but Tos-Lys-CH$_2$Cl is inert (Chen *et al.*, 1996). Other hydrophobic chloromethanes, including $\beta$-(2-naphthyl)-Ala-CH$_2$Cl, are also affinity labeling agents of the enzyme (Chen *et al.*, 1996), but while these compounds do inactivate the enzyme, they are weak inhibitors, presumably because they do not much resemble the farnesyl cysteine residue in the substrates. As predicted, a farnesyl cysteinyl chloromethane derivative, $N$-Boc-$S$-farnesyl-L-Cys-CH$_2$Cl, inactivated the protease with a $K_i$ of 30 μM and first-order rate constant of inhibition of $6 \times 10^{-3}$ s$^{-1}$ (Chen *et al.*, 1996).

In addition to the affinity labeling agents described above, a series of potent competitive inhibitors of the enzyme were also prepared (Ma *et al.*, 1993). One of the best of these is the pseudo peptide R (CH$_2$-NH) analog of the substrate $N$-Boc-$S$-farnesyl-L-Cys-Val-Ile-Met, which inhibits the enzyme with a $K_i$ of 85 nM (Chen *et al.*, 1996) (Figure 260.3). This type of inhibitor is quite useful for exploring the physiologic role of the protease.

Of the inhibitors described above, only Tos-Phe-CH$_2$Cl and $\beta$-(2-naphthyl)-Ala-CH$_2$Cl are commercially available.

## Structural Chemistry

At the time of writing, the enzyme has not been purified, and hence there is no information on its structure.

## Preparation

As mentioned above, isoprenylated protein peptidase is an integral membrane protein (Ma & Rando, 1992), and our attempts at purification (Chen *et al.*, 1996) can be summarized as follows. The only method that has been found to solubilize the activity is the use of detergents. A variety of detergents such as $n$-octylglucoside, CHAPS, CHAPSO, sodium cholate and cetyl trimethylammonium bromide all solubilize the enzyme. CHAPSO is the most effective, in that it solubilized the highest percentage of activity, has a high critical micelle concentration, and does not interfere with monitoring of chromatography columns at 280 nm. Various criteria were used to establish solubilization, including the failure of the activity to sediment during prolonged centrifugation, ability of the solubilized material to enter a Sephadex G-200 gel filtration column, and the ability of the enzyme to pass a Millex GV 0.22 μm filter with 100% recovery. The detergent-solubilized enzyme is quite stable, and can be partially purified by column chromatography. The two columns that are generally used are a Mono-Q anion-exchange column and a Superose 12

| Entry | Compound | Ki (µM) |
|---|---|---|
| 1 | 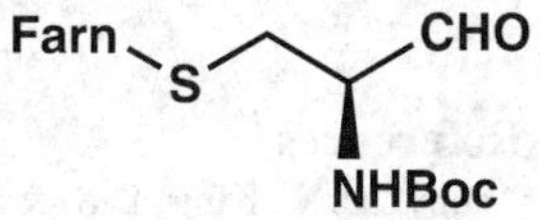 Farn–S–CH(NHBoc)–CHO | 1.95±0.15 |
| 2 | 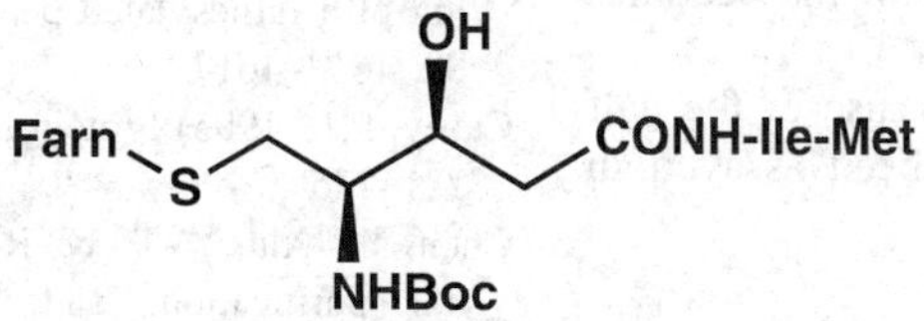 Farn–S–CH(NHBoc)–CH(OH)–CONH-Ile-Met | 28.7±3.00 |
| 3 | 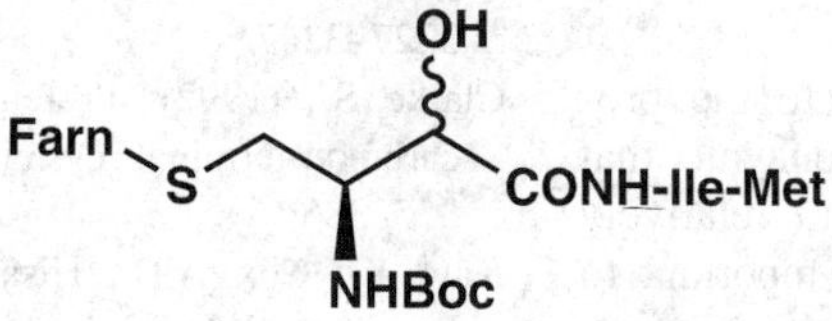 Farn–S–CH(NHBoc)–C(OH)~–CONH-Ile-Met | 36.4±6.8 |
| 4 | 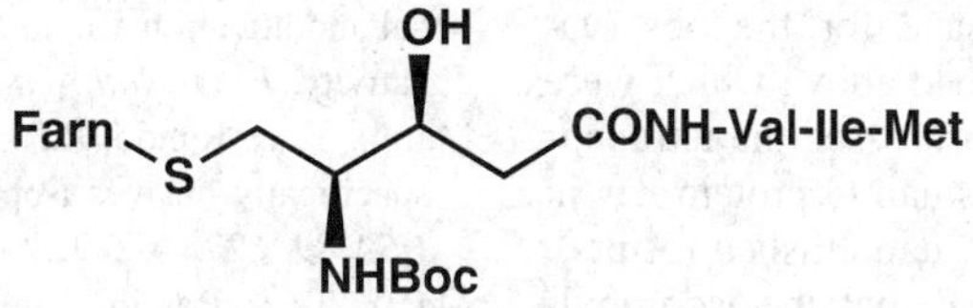 Farn–S–CH(NHBoc)–CH(OH)–CONH-Val-Ile-Met | 0.064±0.011 |
| 5 | 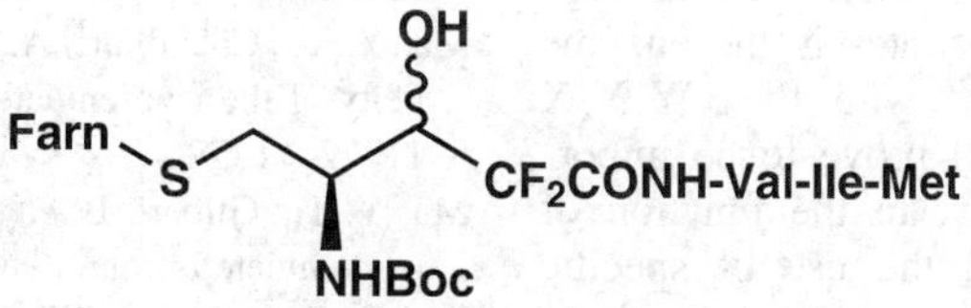 Farn–S–CH(NHBoc)–C(OH)~–CF$_2$CONH-Val-Ile-Met | 3.60±0.70 |
| 6 | 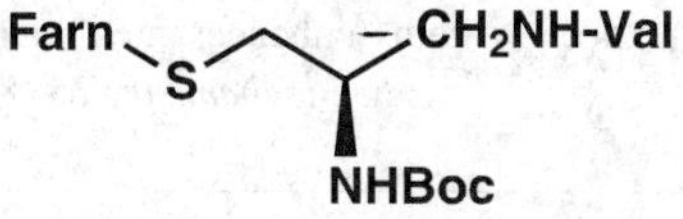 Farn–S–CH(NHBoc)–CH$_2$NH-Val | 0.0 |
| 7 | 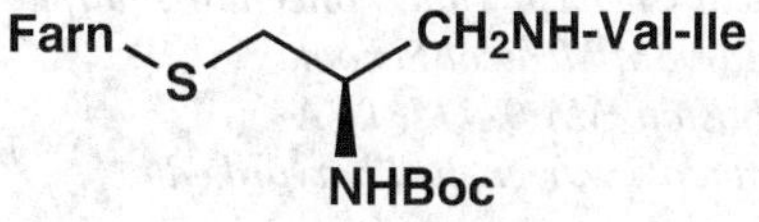 Farn–S–CH(NHBoc)–CH$_2$NH-Val-Ile | 20.95±2.62 |
| 8 | 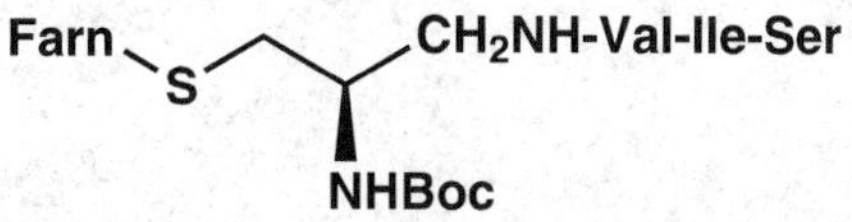 Farn–S–CH(NHBoc)–CH$_2$NH-Val-Ile-Ser | 0.31±0.07 |
| 9 | 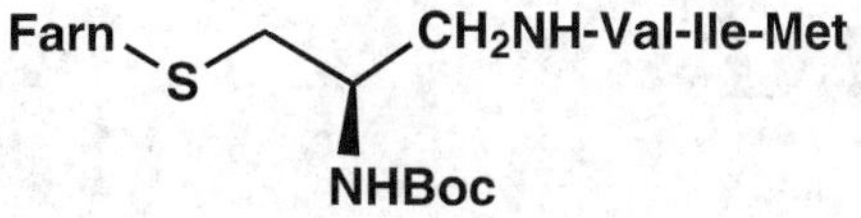 Farn–S–CH(NHBoc)–CH$_2$NH-Val-Ile-Met | 0.086±0.017 |

*Figure 260.3*  Inhibitors of isoprenylated protein peptidase.

gel-filtration column. Together, these columns provided a purification of roughly 10-fold. Tests with a variety of other columns, including thiopropyl-Sepharose 4B, zinc-charged chelating columns, DEAE-cellulose and hydrophobic interaction columns achieved little. We have tried immobilized inhibitors, farnesyl cysteinyl aldehyde and $\psi$ peptide, as affinity columns, but were not able to elute the enzyme from the $\psi$ peptide columns. It is probable that the enzyme becomes labile as it is purified.

While much of our work has centered on use of the calf liver enzyme, a similar protease has also been assayed in yeast cell extracts (Ashby *et al.*, 1992).

## Biological Aspects

The isoprenylated protein peptidase is central to the processing of $CY_1Y_2X$-containing proteins. The enzyme that has been studied so far has been assayed with relatively simple tetrapeptide substrates. It is, of course, important to know that this same enzyme processes fully mature proteins. There is little information on this, save for the observations that some of the inhibitors described above which were designed to block the enzyme which processes the tetrapeptides, also block the proteolysis of the small G protein *ras* in cells (A. Neri, L. Foley & R.R. Rando, unpublished results). By this criterion, it would be expected that the isoprenylated protein peptidase that is initially described through the use of simplified tetrapeptide substrates is the enzyme that processes proteins that terminate with the $CY_1Y_2X$ motif. It is likely that proteolysis will prove to be important quantitatively, if not qualitatively, in the function of $CY_1Y_2X$ modified proteins. Through the use of specific inhibitors of the isoprenylated protein peptidase, the functional importance of the enzyme will be clarified in the near future.

## Acknowledgements

Studies quoted here from the author's laboratory were supported by NIH grant EY-03624.

## References

Ashby, M.N., King, D.S. & Rine, J. (1992) Endoproteolytic processing of a farnesylated peptide *in vitro*. *Proc. Natl Acad. Sci. USA* **89**, 4613–4617.
Casey, P.J. (1995) Protein lipidation in cell signalling. *Science* **268**, 221–225.
Chen, Y., Ma, Y.-T. & Rando, R.R. (1996) Solubilization, partial purification, and affinity labeling of the membrane-bound isoprenylated protein endoprotease. *Biochemistry* **35**, 3227–3247.
Clarke, S. (1992) Protein isoprenylation and methylation at carboxyl-terminal cysteine residues. *Annu. Rev. Biochem.* **61**, 355–386.
Kato, K., Cox, A.D., Hisaka, M.M., Graham, S.M., Buss, J.E. & Der, C.J. (1992) Isoprenoid addition to Ras protein is the critical modification for its membrane association and transforming activity. *Proc. Natl Acad. Sci. USA* **89**, 6403–6407.
Ma, Y.-T. & Rando, R.R. (1992) A microsomal endoprotease that specifically cleaves isoprenylated peptides. *Proc. Natl Acad. Sci. USA* **89**, 6275–6279.
Ma, Y.-T. & Rando, R.R. (1993) Endoproteolysis of non-CAAX-containing isoprenylated peptides. *FEBS* **332**, 105–110.
Ma, Y.-T., Chaudhuri, A. & Rando, R.R. (1992) Substrate specificity of the isoprenylated protein endoprotease. *Biochemistry* **31**, 11772–11777.
Ma, Y.-T., Gilbert, B.A. & Rando, R.R. (1993) Inhibitors of the isoprenylated protein endoprotease. *Biochemistry* **32**, 2386–2393.
Parish, C. & Rando, R.R. (1996) Isoprenylation/methylation of proteins enhances membrane association by a hydrophobic mechanism. *Biochemistry* **35**, 8473–8477.

***Robert R. Rando***
*Department of Biological Chemistry and Molecular Pharmacology,*
*Harvard Medical School,*
*Boston, MA 02115, USA*
*Email: rando@warren.med.harvard.edu*

# 261. Cathepsin T

## Databanks

*Peptidase classification: clan CX, family C99, MEROPS ID: C9G.005*
*Enzyme classification: EC 3.4.22.24*
*Databank codes: no sequence data available*

## Name and History

Multiple forms of rat liver tyrosine aminotransferase have been known for more than 30 years (Pitot & Gohda, 1987). The presence of a factor in the lysosomal fraction of rat liver that interconverts three of these forms was also demonstrated (Pitot & Gohda, 1987). Hargrove *et al*. (1980) purified the three forms, designated I–III in their order of elution from hydroxyapatite, and showed that they differ in molecular mass; form I is composed of two 53 kDa subunits, form II has one 53 kDa subunit and one 49 kDa subunit, and form III has two 49 kDa subunits. Gohda & Pitot (1980, 1981b) subsequently isolated the converting factor from rat liver and rat kidney and demonstrated the *in vitro* conversion of purified form I of tyrosine aminotransferase to form II and then form III with concomitant decrease in the molecular masses of their subunits. Since this proteinase was clearly different from other known lysosomal cathepsins on the basis of substrate specificity, molecular mass, and the effects of inhibitors, and since there is little if any evidence that the enzyme acts on native proteins other than tyrosine aminotransferase, the name *cathepsin T* was suggested (Gohda & Pitot, 1981a).

## Activity and Specificity

The only native protein known to serve as a substrate for cathepsin T is tyrosine aminotransferase (Gohda & Pitot, 1981a). The proteinase acts on the 53 kDa monomer of rat tyrosine aminotransferase to generate the 49 kDa monomer and a 4.5 kDa peptide (Gohda & Pitot, 1980; Hargrove *et al*., 1982). This cleavage occurs at the N-terminus of the protein, and the new free N-terminus of the 49 kDa monomer was identified as Ser29 (Dietrich *et al*., 1988) or Lys35 (Hargrove *et al*., 1989). Cathepsin T degrades azocasein (Gohda & Pitot, 1980) and denatured hemoglobin (Gohda & Pitot, 1981a), but no synthetic substrates are known. The activity can be assayed with form I of tyrosine aminotransferase from rat liver as described by Pitot & Gohda (1987). The optimal pH is approximately 6.9 at 0°C and the assay system typically includes 2 mM DTT (Gohda & Pitot, 1980). Activity is stimulated by thiol compounds and EDTA and inhibited by sulfhydryl-reactive agents, Tos-Lys-CH$_2$Cl and leupeptin (Gohda & Pitot, 1981a).

## Structural Chemistry

Cathepsin T is a single-chain protein of about 35 kDa (Gohda & Pitot, 1980, 1981b).

## Preparation

Cathepsin T has been purified to apparent homogeneity from rat liver (Gohda & Pitot, 1980) and rat kidney (Gohda & Pitot, 1981b).

## Biological Aspects

Cathepsin T activity occurs in liver, kidney, spleen, lung and small intestine of the rat, with the specific activity being highest in the kidney (Gohda & Pitot, 1981b). The activity of tyrosine aminotransferase does not change during the *in vitro* conversion of its forms catalyzed by the proteinase (Gohda & Pitot, 1980). There has been no evidence supporting the notion that cathepsin T-catalyzed limited proteolysis (conversion of form I to forms II and III) of tyrosine aminotransferase is related to its short intracellular half-life (Hargrove *et al*., 1980; Stellwagen, 1992). Treatment of rats with carbon tetrachloride causes a several-fold increase in the activity of cathepsin T in liver (Gohda *et al*., 1984).

## References

Dietrich, J.B., Genot, G. & Beck, G. (1988) Structural and immunochemical properties of rat liver tyrosine aminotransferase. *Biochimie* **70**, 673–679.

Gohda, E. & Pitot, H.C. (1980) Purification and characterization of a factor catalyzing the conversion of the multiple forms of tyrosine aminotransferase from rat liver. *J. Biol. Chem.* **255**, 7371–7379.

Gohda, E. & Pitot, H.C. (1981a) A new thiol proteinase from rat liver. *J. Biol. Chem.* **256**, 2567–2572.

Gohda, E. & Pitot, H.C. (1981b) Purification and characterization of a new thiol proteinase from rat kidney. *Biochim. Biophys. Acta* **659**, 114–122.

Gohda, E., Nagahama, J., Nakamura, O., Tsubouchi, H., Daikuhara, Y. & Pitot, H.C. (1984) Increased activities of liver cathepsins T and D in carbon tetrachloride-treated rats. *Biochim. Biophys. Acta* **802**, 362–371.

Hargrove, J.L., Diesterhaft, M., Noguchi, T. & Granner, D.K. (1980) Identification of native tyrosine aminotransferase and an explanation for the multiple forms. *J. Biol. Chem.* **255**, 71–78.

Hargrove, J.L., Gohda, E., Pitot, H.C. & Granner, D.K. (1982) Cathepsin T (convertase) generates the multiple forms of tyrosine aminotransferase by limited proteolysis. *Biochemistry* **21**, 283–289.

Hargrove, J.L., Scoble, H.A., Mathews, W.R., Baumstark, B.R. & Biemann, K. (1989) The structure of tyrosine aminotransferase. Evidence for domains involved in catalysis and enzyme turnover. *J. Biol. Chem.* **264**, 45–53.

Pitot, H.C. & Gohda, E. (1987) Cathepsin T. *Methods Enzymol.* **142**, 279–289.

Stellwagen, R.H. (1992) Involvement of sequences near both amino and carboxyl termini in the rapid intracellular degradation of tyrosine aminotransferase. *J. Biol. Chem.* **267**, 23713–23721.

**Eiichi Gohda**
*Department of Immunochemistry,
Faculty of Pharmaceutical Sciences,
Okayama University,
Tsushima-naka, Okayama 700, Japan
Email: gohda@pheasant.pharm.okayama-u.ac.jp*

**Henry C. Pitot**
*McArdle Laboratory for Cancer Research,
University of Wisconsin-Madison,
Madison, WI 53706-1599, USA
Email: pitot@oncology.wisc.edu*

# 262. Cancer procoagulant

## Databanks

*Peptidase classification: clan CX, family C99, MEROPS ID: C9G.012*
*Enzyme classification: EC 3.4.22.26*
*Databank codes: no sequence data available*

## Name and History

Extensive studies on factors responsible for the hyper-coagulation frequently observed in cancer patients led to the discovery of a unique proteolytic activator of coagulation factor X (Chapter 54) (Gordon *et al.*, 1975). The existence of this direct, proteolytic activator of factor X in extracts from various malignant tissues has subsequently been confirmed by several other laboratories (Curatolo *et al.*, 1979; Hilgard & Whur, 1980; Delaini *et al.*, 1981; Falanga *et al.*, 1987; Mielicki & Wierzbicki, 1990; Moore, 1992). Since the protein was obtained from cancerous tissue, and it initiated coagulation (procoagulative properties), it was named cancer procoagulant. Originally, the name *cancer procoagulant A* was chosen because it was believed that there might be several isoforms of the enzyme in different tissues and species, but no such isoforms have been identified, and the name has accordingly been shortened to *cancer procoagulant (CP)*. The CP activity characterized in early studies was inhibited by commercial preparations of DFP and classified as a serine proteinase (Gordon *et al.*, 1975, 1979), but later it became evident from the inability of [$^{14}$C]DFP to radiolabel CP, the pH optimum, and inhibition by alkylating reagents, that CP was a cysteine proteinase, and the DFP inhibition was due to an impurity in the DFP preparations (Falanga & Gordon, 1985b; Gordon & Cross, 1981).

## Activity and Specificity

The only known physiological substrate of CP is coagulation factor X (Gordon & Cross, 1981). The cleavage site of the factor X heavy chain by CP is the internal peptide bond Pro-Tyr$\downarrow$Asp22 (Gordon & Mourad, 1991). The other known activators of factor X, including Russell's viper venom (Chapter 433), tissue factor/factor VIIa complex (Chapter 53), and the factor IXa/factor VIIIa complex (Chapter 52), cleave the heavy chain at Arg$\downarrow$Ile52 (Gordon & Mourad, 1991).

Originally, the factor X-activating activity of CP was determined by measuring the recalcified clotting time of factor VII-deficient plasma (Gordon *et al.*, 1975). Another method of analysis is the two-stage chromogenic assay in which CP activates factor X to factor Xa and the amount (activity) of factor Xa is determined with the chromogenic substrate, Bz-Ile-Glu-Gly-Arg-NHPhNO$_2$ (Colucci *et al.*, 1980). The most sensitive and reproducible method is a three-stage chromogenic assay: CP activates factor X to factor Xa, factor Xa activates prothrombin to thrombin, and the amount of thrombin is determined with the chromogenic substrate, Sar-Pro-Arg-NHPhNO$_2$ (Mielicki

& Gordon, 1993). This 'coupled' assay amplifies the signal from the factor Xa generated by CP.

The enzymatic activity of CP is calcium dependent, the optimum calcium concentration being 7 mM (Mielicki & Gordon, 1993). Calcium ions can be replaced by magnesium with the same optimum concentration (Mielicki *et al.*, 1994a). Manganese (10–100 mM) potentiates the activity of CP, whereas zinc and ferrous ions inhibit (Mielicki *et al.*, 1994a). The pH optimum for the activation of factor X by CP is 6.9–7.2 (Mielicki & Gordon, 1993). CP works best in a slightly reducing environment, and its active site oxidizes easily. Therefore, 10 mM KCN is used to reduce any oxidized active-site thiol, and addition of 1–5 µM cysteine, glutathione or 2-ME maintains the reduced state of the active site.

Cancer procoagulant activity is inhibited by iodoacetamide, mercuric chloride, E-64, peptidyl diazomethanes, peptidyl chloromethanes, peptidyl dimethyl sulfonium salts, peptidyl trifluoromethanes, leupeptin, antipain, PMSF and 2-vinylpyridine, but is not inhibited by cystamine, cystatin, Z-Val-Phe-H, $\alpha_1$-antichymotrypsin, $\alpha_2$-macroglobulin, $\alpha_1$-proteinase inhibitor or antithrombin III/heparin (Gordon & Cross, 1981; Falanga *et al.*, 1989; Moore, 1992; Mielicki *et al.*, 1994b). Addition of 4 mM KCN and 5 µM cysteine to the CP sample increased the inhibition of CP by iodoacetamide, E-64, leupeptin and antipain from 3.5- to 5.6-fold, but there was no effect on inhibition by cystatin, Z-Val-Phe-H, $\alpha_1$-antichymotrypsin, $\alpha_2$-macroglobulin, $\alpha_1$-proteinase inhibitor or antithrombin III/heparin. Inhibition by mercuric chloride was decreased almost 10-fold.

## Structural Chemistry

Cancer procoagulant is a single-chain, 68 kDa, pI 4.8 endopeptidase, which contains no detectable carbohydrate (Falanga & Gordon, 1985b). Prevalent amino acids in the CP molecule are Ser (19.1%), Gly (18.7%), Glu (12.5%), Lys (8.1%) and Asp (7.1%) (Falanga & Gordon, 1985b). There is indirect evidence that CP is a vitamin K-dependent protein, with $\gamma$-carboxyglutamic acid residues in the proteinase to bind calcium (Delaini *et al.*, 1981; Colucci *et al.*, 1983; Roncaglioni *et al.*, 1989). There are at least two different metal-binding sites in the CP molecule: one for calcium and magnesium, and the other for manganese (Mielicki *et al.*, 1994a). The amino acid sequence of CP is still unknown.

## Preparation

Cancer procoagulant has been purified to apparent homogeneity from rabbit V2 carcinoma (Falanga & Gordon, 1985b), mouse Lewis lung carcinoma (Moore, 1992) and human amnion-chorion (Gordon *et al*., 1985). CP has been identified in extracts of various human and animal neoplastic tissues, both solid and leukemic (Curatolo *et al*., 1979; Hilgard & Whur, 1980; Donati *et al*., 1986; Grignani *et al*., 1988; Alessio *et al*., 1990; Mielicki *et al*., 1990; Mielicki & Wierzbicki, 1990), but not in normal tissue (Gordon *et al*., 1979) other than human amniochorion, a fetal tissue (Falanga & Gordon, 1985a; Gordon *et al*., 1985). Malignant cells in culture do not express, or express only very low levels, of CP (Curatolo *et al*., 1985, 1988; Zucchella *et al*., 1993; Falanga *et al*., 1994). There is no recombinant CP yet.

## Biological Aspects

The only well-documented biological function of CP is the acceleration of blood coagulation in a cancer host. It is believed that the increased generation of fibrin, and its deposition on the tumor, protects it from the host's immune defense system. Fibrin forms a layer protecting tumor cells from natural killer or other cytotoxic cells (Gunji & Gorelik, 1988). Tissue factor and CP are believed to be the major factors responsible for fibrin formation in cancer.

There is a second biological activity of CP that is less well characterized. CP, like many other proteolytic enzymes, appears to have an effect on the growth and viability of malignant cells. After exposure to anti-CP IgG cultured H345 human small cell lung carcinoma cells lost about 50% of their viability, in a dose-dependent fashion, and showed a decrease in DNA synthesis (Gordon & Tagawa, 1993). There was no such effect of other (non-anti-CP) control antibodies on the H345 cells, nor was there any effect of anti-CP IgG on normal human fibroblasts. Currently, the mechanism of this effect is unknown.

The enzymatic and immunological characteristics of CP from different species and tissues are similar. Antibodies against CP from rabbit V2 carcinoma recognize CP from human melanoma (Donati *et al*., 1986), leukemic cells (Falanga *et al*., 1988), amniochorion (Falanga & Gordon, 1985a), mouse B16 melanoma, JW sarcoma and Lewis lung carcinoma (Falanga *et al*., 1987). CP from all those sources had the same enzymatic characteristics. Thus, CP seems to be a protein highly conserved across species and tissues, having the characteristics of an oncofetal protein.

There is evidence that the expression of CP precedes the appearance of the malignant phenotype. The activity of CP in bone marrow cells aspirated from patients with acute nonlymphoid leukemia was high during the active disease, disappeared during remission, and reappeared 2–5 months before recurrence of the disease (the reappearance of malignant cells in the bone marrow) (Donati *et al*., 1990).

The production of CP only by a malignant tissue, the CP expression before the appearance of the malignant phenotype, and the presence of CP in the blood of cancer patients, suggested that CP could potentially be a good tumor marker. Preliminary studies involving 561 cancer patients, 102 noncancer patients and 139 healthy controls have been performed. The overall mean sensitivity of CP as a tumor marker was 80% while the specificity of the assay was 82% for the

controls and 84% for the noncancer individuals. However, the sensitivity of the assay for the early stages of cancers was 88% (Kozwich *et al*., 1994).

There is no information about the localization and structure of the gene encoding CP. The early expression of the protein suggests that its gene is 'activated' during early steps of the tumorigenesis. At such an early stage of the malignant transformation, it is not apparent what role the procoagulative properties of CP might play in the malignant process. It seems more likely that CP may elicit a growth stimulatory role in the dedifferentiating cell undergoing malignant transformation. There is preliminary evidence that CP can interact with a cancer cell other than via the blood-clotting system.

The post-translational $\gamma$-carboxylation seems to be the only step at which biological control of CP activity may be exerted. There are no data on natural, cellular or plasma inhibitors of CP, and there is no evidence for the existence of a proenzyme.

## Distinguishing Features

Cancer procoagulant cleaves the heavy chain of factor X at a site different from that of other known activators (Gordon & Mourad, 1991). The molecular mass, substrate requirement and pH optimum for CP are different from those for cathepsins B, H and L (see Chapters 209, 213 and 210). CP differs from papain in the effects of peptidyl diazomethanes and peptidyl sulfonium inhibitors (Falanga *et al*., 1989). Neither in its divalent ion requirements nor its sensitivity to oxidation does CP resemble other known cysteine proteinases (Mielicki *et al*., 1994a). Monoclonal and polyclonal antibodies against CP were used for the development of an immunoassay (ELISA) diagnostic test for the measurement of CP antigen in serum of cancer patients (Gordon & Cross, 1990; Kozwich *et al*., 1994). The antibodies or test kits are not commercially available yet.

## Further Reading

Further reviews are those of Gordon (1994) and Mielicki *et al*. (1997).

## References

Alessio, M.G., Falanga, A., Consonni, R., Bassan, R., Minetti, B., Donati, M.B. & Barbui, T. (1990) Cancer procoagulant in acute lymphoblastic leukemia. *Eur. J. Haematol.* **45**, 78–81.

Colucci, M., Curatolo, L., Donati, M.B. & Semeraro, N. (1980) Cancer cell procoagulant activity: evaluation by an amidolytic assay. *Thromb. Res.* **18**, 589–595.

Colucci, M., Delaini, F., De Bellis Vitti, G., Locati, D., Poggi, A., Semeraro, N. & Donati, M.B. (1983) Warfarin inhibits both procoagulant activity and metastatic capacity of Lewis lung carcinoma cells. *Biochem. Pharmacol.* **32**, 1689–1691.

Curatolo, L., Colucci, M., Cambini, A.L., Poggi, A., Morasca, L., Donati, M.B. & Semeraro, N. (1979) Evidence that cells from experimental tumours can activate coagulation factor X. *Br. J. Cancer* **40**, 228–233.

Curatolo, L., Alessio, G., Gambacorti Passerini, C., Casali, B., Morasca, L., Semeraro, N. & Donati, M.B. (1985) Procoagulant activity of mouse and human cultured cells following various types of transformation. *Int. J. Cancer* **35**, 411–414.

Curatolo, L., Alessio, M.G., Casali, B., Falanga, A., Donati, M.B. & Semeraro, N. (1988) Procoagulant activity of mouse transformed cells: different expression in freshly isolated or cultured cells. *In Vitro Cell. Dev. Biol.* **24**, 1154–1158.

Delaini, F., Colucci, M., De Bellis Vitti, G., Locati, D., Poggi, A., Semeraro, N. & Donati, M.B. (1981) Cancer cell procoagulant: a novel vitamin K-dependent activity. *Thromb. Res.* **24**, 263–266.

Donati, M.B., Gambacorti-Passerini, C., Casali, B., Falanga, A., Vannotti, P., Fossati, G., Semeraro, N. & Gordon, S.G. (1986) Cancer procoagulant in human tumor cells: evidence from melanoma patients. *Cancer Res.* **46**, 6471–6474.

Donati, M.B., Falanga, A., Consonni, R., Alessio, M.G., Bassan, R., Buelli, M., Borin, L., Catani, L., Pogliani, E., Gugliotta, L., Masera, G. & Barbui, T. (1990) Cancer procoagulant in acute nonlymphoid leukemia: relationship of enzyme detection to disease activity. *Thromb. Haemost.* **64**, 11–16.

Falanga, A. & Gordon, S.G. (1985a) Comparison of properties of cancer procoagulant and human amnion-chorion procoagulant. *Biochim. Biophys. Acta* **831**, 161–165.

Falanga, A. & Gordon, S.G. (1985b) Isolation and characterization of cancer procoagulant: a cysteine proteinase from malignant tissue. *Biochemistry* **24**, 5558–5567.

Falanga, A., Bolognese D'Alessandro, A.P., Casali, B., Roncaglioni, M.C. & Donati, M.B. (1987) Several murine metastasizing tumors possess a cysteine proteinase with cancer procoagulant characteristics. *Int. J. Cancer* **39**, 774–777.

Falanga, A., Alessio, M.G., Donati, M.B. & Barbui, T. (1988) A new procoagulant in acute leukemia. *Blood* **71**, 870–875.

Falanga, A., Shaw, E., Donati, M.B., Consonni, R., Barbui, T. & Gordon, S.G. (1989) Inhibition of cancer procoagulant by peptidyl diazomethyl ketones and peptidyl sulfonium salts. *Thromb. Res.* **54**, 389–398.

Falanga, A., Consonni, R., Marchetti, M., Mielicki, W.P., Rambaldi, A., Lanotte, M., Gordon, S.G. & Barbui, T. (1994) Cancer procoagulant in the human promyelocytic cell line NB4 and its modulation by all-trans retinoic acid. *Leukemia* **8**, 156–159.

Gordon, S.G. (1994) Cancer procoagulant. *Methods Enzymol.* **244**, 568–583.

Gordon, S.G. & Cross, B.A. (1981) A factor X-activating cysteine protease from malignant tissue. *J. Clin. Invest.* **67**, 1665–1671.

Gordon, S.G. & Cross, B.A. (1990) An enzyme-linked immunosorbent assay for cancer procoagulant and its potential as new tumor marker. *Cancer Res.* **50**, 6229–6234.

Gordon, S.G. & Mourad, A.M. (1991) The site of activation of factor X by cancer procoagulant. *Blood Coag. Fibrinol.* **2**, 735–739.

Gordon, S.G. & Tagawa, M. (1993) The cytotoxic effect of cancer procoagulant (CP) antibodies on H345 small cell lung carcinoma cells. *Cancer Res.* **53**, 453 (abstr.).

Gordon, S.G., Franks, J.J. & Lewis, B. (1975) A factor X activating procoagulant from malignant tissue. *Thromb. Res.* **6**, 127–137.

Gordon, S.G., Franks, J.J. & Lewis, B. (1979) Comparison of procoagulant activities in extracts of normal and malignant human tissue. *J. Natl Cancer Inst.* **62**, 773–776.

Gordon, S.G., Hasiba, U., Cross, B.A., Poole, M.A. & Falanga, A. (1985) Cysteine proteinase procoagulant from amnion-chorion. *Blood* **66**, 1261–1265.

Grignani, G., Falanga, A., Pacchiarini, L., Alessio, M.G., Zucchella, M., Fratino, P. & Donati, M.B. (1988) Human breast and colon carcinomas express cysteine proteinase activities with pro-aggregating and pro-coagulant properties. *Int. J. Cancer* **42**, 554–557.

Gunji, Y. & Gorelik, E. (1988) Role of fibrin coagulation in protection of murine tumor cell from destruction by cytotoxic cells. *Cancer Res.* **48**, 5216–5221.

Hilgard, P. & Whur, P. (1980) Factor X-activating activity from Lewis lung carcinoma. *Br. J. Cancer* **41**, 642–643.

Kozwich, D.L., Kramer, L.C., Mielicki, W.P., Fotopoulos, S.S. & Gordon, S.G. (1994) Application of cancer procoagulant as an early detection tumor marker. *Cancer* **74**, 1367–1376.

Mielicki, W.P. & Gordon, S.G. (1993) Three-stage chromogenic assay for the analysis of activation properties of factor X by cancer procoagulant. *Blood Coag. Fibrinol.* **4**, 441–446.

Mielicki, W.P. & Wierzbicki, R. (1990) Cancer procoagulant in serum of rats during development of experimental epithelioma. *Int. J. Cancer* **45**, 125–126.

Mielicki, W.P., Serwa, J., Kurzawinski, T. & Wierzbicki, R. (1990) Procoagulant activities in human stomach and colon cancers. *Oncology* **47**, 299–302.

Mielicki, W.P., Kozwich, D.L, Kramer, L.C. & Gordon, S.G. (1994a) Effect of divalent ions on the activity of cancer procoagulant. *Arch. Biochem. Biophys.* **314**, 165–170.

Mielicki, W.P., Tagawa, M. & Gordon, S.G. (1994b) New immunocapture enzyme (ICE) assay for quantification of cancer procoagulant activity: studies of inhibitors. *Thromb. Haemost.* **71**, 456–460.

Mielicki, W.P., Mielicka, E. & Gordon, S.G. (1997) Cancer procoagulant activity studies using synthetic peptidyl substrates. *Thromb. Res.* **87**, 251–256.

Moore, W.R. (1992) The purification and properties of cancer procoagulant from murine tumors. *Biochem. Biophys. Res. Commun.* **184**, 819–824.

Roncaglioni, M.C., Falanga, A., Bolognese D'Alessandro, A.P., Alessio, M.G., Casali, B. & Donati, M.B. (1989) Evidence of warfarin-sensitive cancer procoagulant in V2 carcinoma. *Haematologica* **74**, 143–147.

Zucchella, M., Pacchiarini, L., Tacconi, F., Saporiti, A. & Grignani, G. (1993) Different expression of procoagulant activity in human cancer cells cultured 'in vitro' or in cells isolated from human tumor tissues. *Thromb. Haemost.* **69**, 335–338.

**Stuart G. Gordon**
*Department of Pathology and Colorado Cancer Center,*
*University of Colorado Health Sciences Center,*
*4200 East 9th Avenue, Campus Box B169,*
*Denver, CO 80262, USA*

**Wojciech P. Mielicki**
*Medical University of Lodz,*
*Department of Biochemistry IBSiB,*
*Ul, Muszynskiego 1,*
*90151 Lodz, Poland*

# 263. *Prohormone thiol protease*

## Databanks

*Peptidase classification: clan CX, family C99, MEROPS ID: C9G.013*
*NC-IUBMB enzyme classification: none*
*Databank codes: no sequence data available*

## Name and History

Proteolytic processing of prohormones is required to generate small, bioactive peptide hormones and peptide neurotransmitters. Many of these processing steps occur within maturing secretory vesicles. Therefore, neurosecretory vesicles of the adrenal medulla (also known as chromaffin granules) have been utilized to identify the major proteolytic activity for processing proenkephalin into enkephalin neuropeptides (Krieger & Hook, 1991; Krieger *et al.*, 1992; Schiller *et al.*, 1995; Hook *et al.*, 1996a,b). In addition, full-length recombinant $^{35}$S-labeled enkephalin precursor was utilized as substrate, synthesized from the corresponding cDNA by *in vitro* transcription and translation.

Identification and characterization of the major enkephalin precursor-cleaving activity led to the discovery of the novel cysteine protease termed **prohormone thiol protease (PTP)** (Krieger & Hook, 1991). PTP represents the majority (60–70%) of proenkephalin-processing activity in chromaffin granules. The subtilisin-like proprotein convertase 1 (Chapter 118) and proprotein convertase 2 (Chapter 119) (Azaryan *et al.*, 1995a) and a 70 kDa aspartic protease (Azaryan *et al.*, 1995b) similar to the pituitary pro-opiomelanocortin-converting enzyme or yapsin 3 (Chapter 306) (Loh *et al.*, 1993), account for only about 20% and 10%, respectively, of total vesicular enkephalin precursor-cleaving activity. Importantly, PTP fulfills the criteria expected of an authentic processing enzyme (Docherty & Steiner, 1982): (a) PTP is colocalized with proenkephalin and enkephalin peptide products within secretory vesicles, (b) PTP cleaves proenkephalin and other prohormones at the expected paired basic and monobasic residue cleavage sites, (c) PTP is active at the intragranular pH of secretory vesicles, (d) characterization of purified PTP with protease inhibitors and activators indicates that it is a member of the class of cysteine proteases, and (e) inhibition of PTP in chromaffin cells blocks cAMP-stimulation of enkephalin peptide production. Clearly, PTP plays an important role in prohormone processing.

## Activity and Specificity

PTP cleaves proenkephalin at paired basic residues to generate high molecular mass intermediates and [Met]enkephalin (Krieger & Hook, 1991; Krieger *et al.*, 1992; Schiller *et al.*, 1995) (Figure 263.1). Processing of 35 kDa proenkephalin generates 22.5, 21.7, 12.5 and 11.0 kDa intermediate products that represent N-terminal fragments of proenkephalin, as assessed by peptide microsequencing (Schiller *et al.*, 1995).

Differences in molecular masses of the 22.5, 21.7, 12.5 and 11.0 kDa products reflect processing of proenkephalin within the C-terminal region of the precursor, which resembles proenkephalin processing *in vivo*. Products of 12.5, 11.0, and 8.5 kDa are generated by PTP-mediated Lys┼Arg cleavages at the C-terminus of [Met]$^5$enkephalin-Arg-Gly-Leu. The 8.5 kDa product may represent peptide I, which is present in adrenal medulla. The 12.5 and 11.0 kDa fragments most likely contain peptide I and peptide E. Additional cleavage studies (Krieger *et al.*, 1992) with the enkephalin-containing peptides known as peptide F, peptide E and BAM-22P indicate that PTP cleaves between the two basic residues, and at the N-terminal side of the pair. PTP conversion of proenkephalin to intermediates and final enkephalin peptides that are generated *in vivo* indicate PTP as an appropriate proenkephalin-processing enzyme.

PTP enzyme activity is stabilized by 10 mM CHAPS, 1 mM EDTA and 1 mM DTT, which are routinely included in enzyme assays (Krieger & Hook, 1991). It shows a pH optimum of 5.5, which is consistent with the intragranular pH of secretory vesicles (Pollard *et al.*, 1978; Orci *et al.*, 1987). PTP demonstrates a pI of 6.0 in chromatofocusing. Activity is inhibited by the cysteine protease inhibitors iodoacetate, *p*-hydroxymercuribenzoate, Hg$^{2+}$, cystatin and E-64c (Krieger & Hook, 1991; Azaryan & Hook, 1994a). Stimulation by DTT shows the dependence of PTP on reduced thiol groups. PTP is not inhibited by the aspartic protease inhibitor pepstatin A, nor by the serine protease inhibitors PMSF and soybean trypsin inhibitor, although it is potently inhibited by $\alpha_1$-antichymotrypsin (see below). Importantly, the differential sensitivity of PTP to active-site directed peptidyl diazomethane inhibitors and E-64c distinguishes PTP from the lysosomal cysteine proteases including cathepsins B, L and H (Azaryan & Hook, 1994a). E-64c is most potent in inhibiting PTP compared to the cathepsins. These properties of PTP distinguish it from previously known processing proteases (such as a metallopeptidase involved in prodynorphin processing) (Berman *et al.*, 1995). These comparative studies indicate that PTP is a novel cysteine protease.

Further studies (Azaryan & Hook, 1994b) demonstrate the unique cleavage site specificity of PTP, which differs from those of proprotein convertases 1 and 2 (Benjannet *et al.*, 1991; Seidah *et al.*, 1991; Thomas *et al.*, 1991; Steiner *et al.*, 1992; Azaryan *et al.*, 1995a) and the pro-opiomelanocortin-converting aspartic protease yapsin 3 (Loh *et al.*, 1993). PTP cleaves monobasic peptide-NHMec substrates, including Z-Phe-Arg-NHMec and Bz-Val-Leu-Lys-NHMec, at the C- and N-terminal sides of the single basic residue. PTP

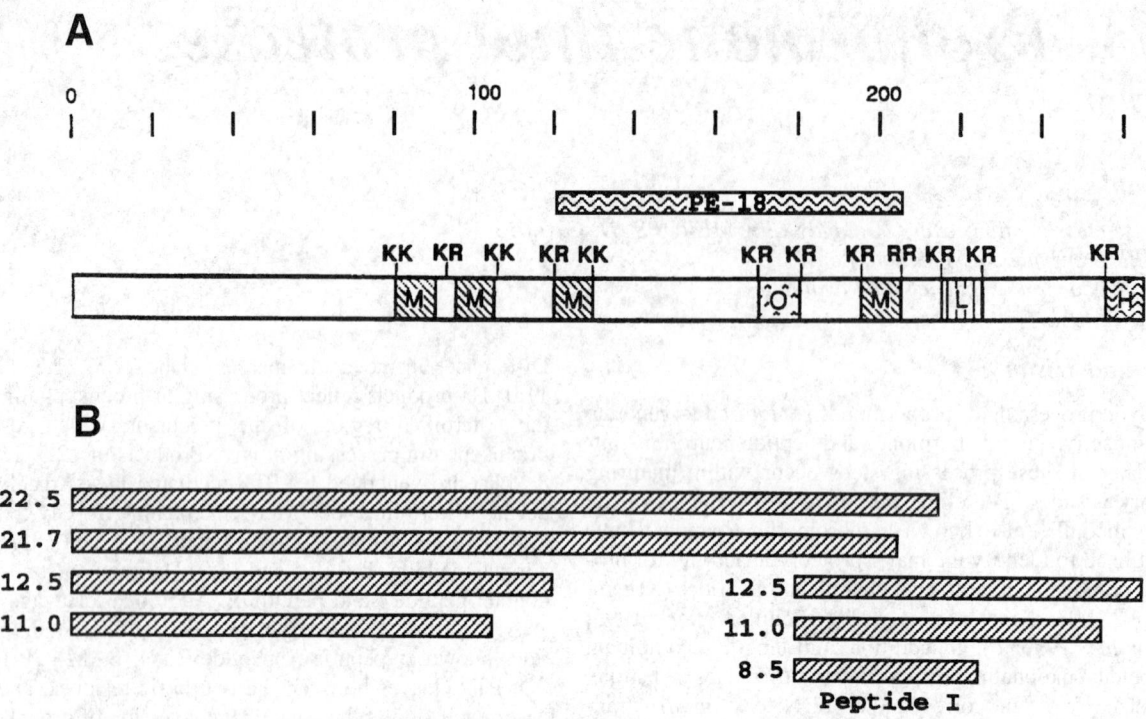

*Figure 263.1* Proenkephalin processing by the prohormone thiol protease (PTP). (A) Structure of proenkephalin. *Key*: M, [Met$^5$]-enkephalin (Tyr-Gly-Gly-Phe-Met); O, [Met$^5$]-enkephalin-Arg-Gly-Leu; L, [Leu$^5$]-enkephalin (Tyr-Gly-Gly-Phe-Leu); H, [Met$^5$]-enkephalin-Arg-Phe; K, Lys; R, Arg. The PE-18 monoclonal antibody that recognizes the mid-region of proenkephalin was used with immunoblots to identify proenkephalin-derived products (Schiller *et al.*, 1995). (b) Proenkephalin-derived products generated by PTP. Bars with apparent molecular masses of 22.5, 21.7, 12.5, 11.0 and 8.5 kDa indicate products generated by PTP. These products were identified based on relative $M_r$ on SDS-PAGE, reactivity with PE-18 monoclonal antibody and peptide microsequencing (Schiller *et al.*, 1995).

cleaves dibasic peptides, Boc-Gly-Lys-Arg-NHMec, Z-Arg-Arg-NHMec, Boc-Gln-Arg-Arg-NHMec and Z-Arg-Val-Arg-NHMec, at the N-terminal side of the pair or between the two basic residues, with lower cleavage at the C-terminal side of the pair. In contrast, proprotein convertases 1 and 2 and pro-opiomelanocortin-converting enzyme cleave between the basic residues of the pair and at the C-terminal side of the pair. The novel cleavage specificity of PTP implies a requirement for exopeptidase processing of basic residue extensions of the products by aminopeptidase(s), as well as by carboxypeptidase H (also known as carboxypeptidase E) (Chapter 458) (Fricker & Snyder, 1982; Hook *et al.*, 1982; Hook & Loh, 1984; Fricker, 1991).

The unique primary structures of different prohormones raises the question of whether a processing enzyme possesses selectivity for different precursor substrates. Therefore, PTP processing of different recombinant prohormones was compared (Schiller *et al.*, 1996; Hook *et al.*, 1996a,b). Rates for PTP processing of proenkephalin and proneuropeptide Y (proNPY) were 30–40 times greater than that observed for pro-opiomelanocortin. Thus, PTP prefers proenkephalin and proNPY to pro-opiomelanocortin. Proprotein convertases 1 and 2 and pro-opiomelanocortin-converting enzyme also show preferences for prohormone substrates. The selectivity of PTP for proenkephalin is consistent with evidence that PTP is the major activity for proenkephalin processing in chromaffin cells.

### Structural Chemistry

PTP is a single-chain glycoprotein of 33 kDa with a pI of 6.0 (Krieger & Hook, 1991). Preliminary peptide microsequencing indicates that PTP is a novel cysteine protease (Tezapsidis *et al.*, 1995).

### Preparation

PTP has been purified from neurosecretory vesicles of cattle adrenal medulla (also known as chromaffin granules) by use of concanavalin A-Sepharose, Sephacryl S-200 gel filtration, chromatofocusing, and thiopropyl-Sepharose (Krieger & Hook, 1991). This represents a 88 000-fold purification. PTP is detected in pituitary, hypothalamus and other neuroendocrine tissues by antibodies generated to the N-terminal peptide sequence of PTP (Tezapsidis *et al.*, 1995).

### Biological Aspects

PTP is regulated by cAMP during stimulation of proenkephalin processing by forskolin (a direct activator of adenylate cyclase and cAMP) and [Met]enkephalin production in cattle chromaffin cells (Tezapsidis *et al.*, 1995). Forskolin stimulates cellular [Met]enkephalin levels 2–3-fold, suggesting a role of processing enzymes during enhanced synthesis of [Met]enkephalin. PTP activity is increased 4-fold in the membrane component of the secretory vesicles. The increased PTP activity involves stimulation of PTP synthesis, indicated by a 10-fold increase in $^{35}$S-PTP pulse labeling and higher levels of PTP detected in western blots. Importantly, the forskolin-mediated rise in cellular [Met]enkephalin is completely blocked by the cysteine protease inhibitor Ep453, which is converted by intracellular esterases to E-64c (Buttle *et al.*, 1992), a potent inhibitor of PTP. Both E-64c and Ep453 inhibit PTP, with E-64c being the more potent (Azaryan & Hook, 1994a,b). Results indicate a role for PTP in proenkephalin processing in chromaffin cells, and indicate cAMP regulation of PTP and [Met]enkephalin formation.

Cellular localization of PTP by immunofluorescence and immunoelectron microscopy indicates secretory vesicle localization of PTP. Immunohistochemistry with anti-PTP serum (generated against the N-terminal peptide sequence of PTP) showed punctate immunostaining of PTP in chromaffin cells, and absence of PTP staining in the nucleus (V.Y.H. Hook, unpublished results). This pattern of immunostaining is compatible with PTP's presence within secretory vesicles. Furthermore, immunoelectron microscopy indicated that PTP is colocalized with [Met]enkephalin within these vesicles. These results indicate localization of PTP with neurosecretory vesicles.

Protease inhibitors are important for controlling key proteases in many biological systems (Davie *et al.*, 1975). Proenkephalin in adrenal medulla is incompletely processed, as shown by the presence of 90% of total tissue [Met]enkephalin as high molecular mass intermediates (Fleminger *et al.*, 1983; Spruce *et al.*, 1988). Furthermore, the extent of proenkephalin processing varies in different tissue regions (Liston *et al.*, 1984). Regulatory studies indicate control of PTP by an endogenous $\alpha_1$-antichymotrypsin-like protease inhibitor (Hook *et al.*, 1993). $\alpha_1$-Antichymotrypsin (ACT from human liver, from Calbiochem; see Appendix 2 for full names and addresses of suppliers) is a potent inhibitor of PTP, with inhibition occurring at nanomolar concentrations of ACT. ACT is colocalized with PTP in neurosecretory vesicles of adrenal medulla and pituitary, as indicated by ACT immunoblots. The purified endogenous cattle pituitary ACT-like protein effectively inhibits PTP with a $k_{i,app}$ value of 2.2 nM. PTP forms SDS-stable complexes with the pituitary ACT-like protein, as well as with ACT, suggesting similarities in PTP cleavage site specificity and the reactive center of ACT. PTP cleavage of enkephalin-containing peptides at the N-terminal side of paired basic residues (Lys-Arg, Arg-Arg, Lys-Lys) flanking the C-terminus of [Met]enkephalin (Tyr-Gly-Gly-Phe-Met) indicates methionine at the P1 position. This cleavage specificity resembles the reactive site of ACT that is recognized and cleaved by the target protease (Rubin *et al.*, 1990). The developmental regulation of ACT in brain and significant amounts of ACT in amyloid plaques of Alzheimer's disease (Abraham *et al.*, 1988) suggest a possible role for PTP in the maturation of peptidergic neurons.

The PTP gene has been detected in genomic blots with an oligonucleotide probe complementary to the determined PTP peptide sequence (S.R. Hwang & V.H.Y. Hook, unpublished results). Additionally, PTP mRNA is also detected with the same PTP oligonucleotide as probe. Molecular cloning of PTP will be important to establish its novel primary structure.

## Distinguishing Features

These studies indicate that PTP is a unique cysteine protease that participates in prohormone processing. This conclusion is based on the distinctive biochemical properties of the enzyme compared to other proteases, and the unique N-terminal peptide sequence of PTP. Furthermore, anti-PTP sera show no cross-reactivity with other cathepsin cysteine proteases. PTP shows unique cleavage site specificity compared to other known prohormone processing enzymes, as was described fully under Activity and Specificity. PTP cleavage at the N-terminal side of the paired basic residue site, as well as between the two basic residues, indicates the requirement for subsequent aminopeptidase and carboxypeptidase processing to generate mature peptides.

Importantly, cAMP regulation of PTP indicates that PTP may be important for receptor-mediated second messenger control of peptide hormone and neurotransmitter biosynthesis. It will be important in future studies to elucidate how PTP and prohormone genes may be coordinately regulated.

## Further Reading

Fuller reviews are to be found in Hook *et al.* (1994, 1996b).

## References

Abraham, C.R., Selkoe, D.J. & Potter, H. (1988) Immunochemical identification of the serine protease inhibitor $\alpha_1$-antichymotrypsin in the brain amyloid deposits of Alzheimer's disease. *Cell* **52**, 487–501.

Azaryan, A.V. & Hook, V.Y.H. (1994a) Distinct properties of prohormone thiol protease (PTP) compared to cathepsins B, L & H: evidence for PTP as a novel cysteine protease. *Arch. Biochem. Biophys.* **314**, 171–177.

Azaryan, A.V. & Hook, V.Y.H. (1994b) Unique cleavage specificity of 'prohormone thiol protease' related to proenkephalin processing. *FEBS Lett.* **341**, 197–202.

Azaryan, A.V., Krieger, T.J. & Hook, V.Y.H. (1995a) Purification and characteristics of the candidate prohormone processing proteases PC2 and PC1/3 from bovine adrenal medulla chromaffin granules. *J. Biol. Chem.* **270**, 8201–8208.

Azaryan, A.V., Schiller, M., Mende-Mueller, L. & Hook, V.Y.H. (1995b) Characteristics of the chromaffin granule aspartic proteinase involved in proenkephalin processing. *J. Neurochem.* **65**, 1771–1779.

Benjannet, S., Rondeau, N., Day, R., Chrétien, M. & Seidah, N.G. (1991) PC1 and PC2 are proprotein convertases capable of cleaving proopiomelanocortin at distinct pairs of basic residues. *Proc. Natl Acad. Sci. USA* **88**, 3564–3568.

Berman, Y.L., Juliano, L. & Devi, L.A. (1995) Purification and characterization of a dynorphin-processing endopeptidase. *J. Biol. Chem.* **270**, 23845–23850.

Buttle, D.J., Saklatvala, J., Tamai, M. & Barrett, A.J. (1992) Inhibition of interleukin 1-stimulated cartilage proteoglycan degradation by a lipophilic inactivator of cysteine endopeptidases. *Biochem. J.* **281**, 175–177.

Davie, E.W., Fujikawa, K., Legaz, M.E. & Kato, H. (1975) Role of proteases in blood coagulation. In: *Proteases and Biological Control* (Reich, E., Rifkin, D.B. & Shaw, E., eds). Cold Spring Harbor, NY: Cold Spring Harbor Laboratory, pp. 65–78.

Docherty, K. & Steiner, D.F. (1982) Post-translational proteolysis in polypeptide hormone biosynthesis. *Annu. Rev. Physiol.* **44**, 625–638.

Fleminger, G., Kilpatrick, D. & Udenfriend, S. (1983) Processing of enkephalin-containing peptides in isolated bovine adrenal chromaffin granules. *Proc. Natl Acad. Sci. USA* **80**, 6418–6421.

Fricker, L.D. (1991) Peptide processing exopeptidases: amino and carboxypeptidases involved with peptide biosynthesis. In: *Peptide Biosynthesis and Processing* (Fricker, L.D., ed.). Boca Raton, FL: CRC Press, pp. 199–229.

Fricker, L.E. & Snyder, S.H. (1982) Enkephalin convertase: purification and characterization of a specific enkephalin-synthesizing carboxypeptidase localized to adrenal chromaffin granules. *Proc. Natl Acad. Sci. USA* **79**, 3886–3891.

Hook, V.Y.H. & Loh, Y.P. (1984) Carboxypeptidase B-like converting enzyme activity in secretory granules of rat pituitary. *Proc. Natl Acad. Sci. USA* **81**, 2776–2780.

Hook, V.Y.H., Eiden, L.W. & Brownstein, M.G. (1982) A carboxypeptidase processing enzyme for enkephalin precursors. *Nature* **295**, 4341–4342.

Hook, V.Y.H., Purviance, R.T., Azaryan, A.V., Hubbard, G. & Krieger, T.J. (1993) Purification and characterization of $\alpha_1$-antichymotrypsin-like protease inhibitor that regulates prohormone thiol protease involved in enkephalin precursor processing. *J. Biol. Chem.* **268**, 20570–20577.

Hook, V.Y.H., Azaryan, A.V., Hwang, S.R. & Tezapsidis, N. (1994) Proteases and the emerging role of protease inhibitors in prohormone processing. *FASEB J.* **8**, 1269–1278.

Hook, V.Y.H., Schiller, M.R. & Azaryan, A.V. (1996a) The processing proteases prohormone thiol protease, PC1/3 and PC2, and 70 kDa aspartic proteinase show preferences among proenkephalin, proneuropeptide Y, and proopiomelanocortin substrates. *Arch. Biochem. Biophys.* **328**, 107–114.

Hook, V.Y.H., Schiller, M.R., Azaryan, A.V. & Tezapsidis, N. (1996b) Proenkephalin processing enzymes in chromaffin granules. *Ann. N.Y. Acad.* **780**, 121–133.

Krieger, T.J. & Hook, V.Y.H. (1991) Purification and characterization of a novel thiol protease involved in processing the enkephalin precursor. *J. Biol. Chem.* **266**, 8376–8383.

Krieger, T.J., Mende-Mueller, L. & Hook, V.Y.H. (1992) Prohormone thiol protease and enkephalin precursor processing: cleavage at dibasic and monobasic sites. *J. Neurochem.* **59**, 26–31.

Liston, D., Patey, G., Rossier, J., Verbanck, P. & Vanderhaeghen, J.J. (1984) Processing of proenkephalin is tissue-specific. *Science* **225**, 734–737.

Loh, Y.P., Beinfeld, M.C. & Birch, N.P. (1993) Proteolytic processing of prohormones and pro-neuropeptides. In: *Mechanisms of Intracellular Trafficking and Processing of Proproteins* (Loh, Y.P., ed.). Boca Raton, FL: CRC Press, pp. 179–223.

Orci, L., Ravazzola, M., Storch, M.J., Anderson, R.G.W., Vassalli, J.D. & Perrelet, A. (1987) Proteolytic maturation of insulin is a post-Golgi event which occurs in acidifying clathrin-coated secretory vesicles. *Cell* **49**, 865–868.

Pollard, H.B., Shindo, H., Creutz, C.E., Pazoles, C.J. & Cohen, J.S. (1978) Internal pH and state of ATP in adrenergic chromaffin granules determined by $^{31}$P nuclear magnetic resonance spectroscopy. *J. Biol. Chem.* **254**, 1170–1177.

Rubin, H., Wang, Z., Nickbarg, E.B., McLarney, S., Naidoo, N., Schoenberger, O.L., Johnson, J.L. & Cooperman, B.S. (1990) Cloning, expression, purification and biological activity of recombinant native and variant human $\alpha_1$-antichymotrypsins. *J. Biol. Chem.* **265**, 1199–1207.

Schiller, M.R., Mende-Mueller, L., Moran, K., Meng, M., Miller, D.W. & Hook, V.Y.H. (1995) 'Prohormone thiol protease' (PTP) processing of recombinant proenkephalin. *Biochemistry* **34**, 7988–7995.

Schiller, M.R., Kohn, A.B., Mende-Mueller, L.M., Miller, K. & Hook, V.Y.H. (1996) Expression of recombinant proneuropeptide Y, proopiomelanocortin and proenkephalin: relative processing by 'prohormone thiol protease' (PTP). *FEBS Lett.* **382**, 6–10.

Seidah, N.G., Day, R., Marcinkiewicz, M., Benjannet, S. & Chrétien, M. (1991) Mammalian neural and endocrine pro-protein and prohormone convertases belonging to the subtilisin family of serine proteinases. *Enzyme* **45**, 271–284.

Spruce, B.A., Jackson, S., Lowry, P.J., Lane, D.P. & Glover, D.M. (1988) Monoclonal antibodies to a proenkephalin A fusion peptide synthesized in *Escherichia coli* recognize novel proenkephalin A precursor forms. *J. Biol. Chem.* **263**, 19788–19795.

Steiner, D.F., Smeekens, S.P., Ohagi, S. & Chan, S.J. (1992) The new enzymology of precursor processing endoproteases. *J. Biol. Chem.* **267**, 23435–23438.

Tezapsidis, N., Noctor, S., Kannan, R., Krieger, T.J., Mende-Mueller, L. & Hook, V.Y.H. (1995) Stimulation of 'prohormone thiol protease' (PTP) and (Met)enkephalin by forskolin. *J. Biol. Chem.* **270**, 13285–13290.

Thomas, L., Leduc, R., Thorne, B.A., Smeekens, S.P., Steiner, D.F. & Thomas, G. (1991) Kex2-like endoproteases PC2 and PC3 accurately cleave a model prohormone in mammalian cells: evidence for a common core of neuroendocrine processing enzymes. *Proc. Natl Acad. Sci. USA* **88**, 5297–5301.

*Vivian Y.H. Hook*
*Department of Medicine and Center for Molecular Genetics,*
*University of California, San Diego,*
*9500 Gilman Drive no. 0822,*
*La Jolla, CA 92093-0822, USA*
*Email: vhook@ucsd.edu*

# 264. *γ-Glutamyl hydrolase*

## Databanks

*Peptidase classification: clan CX, family C26, MEROPS ID: C26.001*
*NC-IUBMB enzyme classification: EC 3.4.19.9*
*Databank codes:*

| Species | SwissProt | PIR | EMBL (cDNA) | EMBL (genomic) |
|---------|-----------|-----|-------------|----------------|
| *Arabidopsis thaliana* | – | – | T22220 | – |
| *Arabidopsis thaliana* | – | – | T46595 | – |
| *Arabidopsis thaliana* | – | – | T76130 | – |
| *Homo sapiens* | – | – | U55206 | – |
| *Rattus norvegicus* | – | – | U38379 | – |

## Name and History

The enzyme designated **γ-glutamyl hydrolase** was first detected by microbiological assay. A source of pteroyl poly-γ-glutamates such as yeast was assessed for its ability to serve as a folate source for bacterial growth following exposure to enzyme preparations that have the capacity to cleave γ-glutamyl linkages (McGuire & Coward, 1984). The enzyme removes poly-γ-glutamate or γ-glutamate from pteroyl poly-γ-glutamate to yield pteroyl glutamate (folic acid) which can support bacterial growth. This activity has been referred to in the past as **conjugase, folate conjugase, pteroyl poly-γ-glutamate hydrolase, γ-Glu-X carboxypeptidase** and **lysosomal γ-glutamyl hydrolase**. It is known to cleave pteroyl poly-γ-glutamate (folyl poly-γ-glutamate), 4-amino-10-methylpteroyl poly-γ-glutamate (methotrexate poly-γ-glutamate) and *p*-aminobenzoylpoly-γ-glutamate (Baugh *et al.*, 1970; Silink *et al.*, 1975; Rao & Noronha, 1977b; Elsenhans *et al.*, 1984; McGuire & Coward, 1984; Wang *et al.*, 1986, 1993; Chandler *et al.*, 1986; Lin *et al.*, 1993; Wang *et al.*, 1993). It is likely that it cleaves a number of other antifolyl poly-γ-glutamates also (Rhee *et al.*, 1993). The N-terminal moiety in the substrate is not required for activity since the enzyme is active on poly-γ-glutamate (Silink *et al.*, 1975; McGuire & Coward, 1984; Wang *et al.*, 1993).

Earlier studies on this enzyme and the cellular status of folyl poly-γ-glutamates were reviewed in an excellent chapter by McGuire & Coward (1984). Folates and antifolates enter cells via specific transport systems. Once inside the cells they are converted to poly-γ-glutamates by the repetitive addition of glutamate to a total chain length of 3–8 γ-linked glutamates. The enzyme catalyzing this reaction is folylpolyglutamate synthetase (EC 6.3.2.17). The folate poly-γ-glutamate derivatives differ from the monoglutamates in that they are retained within the cell because they cannot cross the cell membrane, and they generally bind better to the folate utilizing enzymes. Cellular folyl and antifolyl poly-γ-glutamates can be hydrolyzed by γ-glutamyl hydrolase by sequential removal of single γ-glutamates or by the preferential removal of the entire poly-γ-glutamate

chain by an endopeptidase-type reaction. Since the catalytic mechanism of the hydrolase has not been established, the terminology used to describe enzymes acting on α-linked polypeptides has been adapted. Thus, with γ-glutamyl hydrolase the use of the term endopeptidase in the literature has a meaning specific to poly-γ-glutamate chains and refers to hydrolysis that occurs at an internal glutamate site. In keeping with this, 'exopeptidase' and the less frequently used term 'carboxypeptidase', refer to removal of the C-terminal γ-Glu-linked residue.

Early studies showed that the enzyme from several sources could cleave pteroyl or 4-amino,10-methylpteroyl poly-γ-glutamates at internal γ linkages (Elsenhans *et al.*, 1984; McGuire & Coward, 1984; O'Connor *et al.*, 1991; Lin *et al.*, 1993). More recently, the endopeptidase activity has been documented with the purified enzyme secreted from rat hepatoma cells (Wang *et al.*, 1993) and with the rat enzyme expressed in *Escherichia coli* (Yao *et al.*, 1996a). With pteroyl or 4-amino,10-methyl pteroyl penta-γ-glutamate as substrates, the primary reaction products of the rat enzyme are tetra-γ-glutamate and folate or methotrexate, respectively. Ultimately, the released tetra-γ-glutamate is hydrolyzed completely to glutamate. The N-linked glutamate, which is an inherent part of the folate or methotrexate molecule and to which γ-glutamates are sequentially added, is not cleaved by this enzyme. γ-Glutamyl hydrolase should not be confused with the bacterial enzyme carboxypeptidase G (Kalghatgi & Bertino, 1981), which removes the N-linked glutamate from folate to yield pteroate (4-[((2-amino-4(3H)-oxopteridin-6-yl)methyl)amino] benzoic acid) and glutamate (see Chapter 483). γ-Glutamyl hydrolase and carboxypeptidase G are clearly different proteins and with different activities (Kalghatgi & Bertino, 1981; McGuire & Coward, 1984; Yao *et al.*, 1996a).

## Activity and Specificity

γ-Glutamyl hydrolase has an absolute requirement for γ-linked peptide bonds; α-linked polyglutamates are not substrates (Baugh *et al.*, 1970; Silink *et al.*, 1975; McGuire & Coward, 1984; Wang *et al.*, 1993). For the human enzyme, the

C-terminal amino acid of the peptide chain does not have to be Glu but it must be γ-linked (Baugh *et al.*, 1970). Despite high sequence homology (67%), the enzyme specificities of the human and rat enzymes differ. Human γ-glutamyl hydrolase isolated from jejunal brush border (Chandler *et al.*, 1986) or expressed in *E. coli* (Yao *et al.*, 1996b) acts as an exopeptidase, removing C-terminal glutamic acid sequentially from the γ-Glu chain. An exception to this is the intracellular enzyme from human jejunum which appears to have both endo- and exopeptidase activities (Wang *et al.*, 1986). In contrast, the isolated or expressed rat enzyme acts as an endopeptidase with folyl penta-γ-glutamate or *p*-aminobenzoyl penta-γ-glutamate as substrate, cleaving the innermost linkage of the polygluta-mate chain to release the remainder of the side chain as a single unit (Wang *et al.*, 1993; Yao *et al.*, 1996a). However, when poly-γ-glutamate is used as substrate the site of cleav-age appears to be more random. This suggests that the *N*-linked pteroate or *p*-aminobenzoate may be a docking site for the sub-strate with the rodent enzyme and cause hydrolysis to occur at a specific endo-γ-glutamate linkage. Other sources also express endopeptidase activity with folyl poly-γ-glutamate substrates (Elsenhans *et al.*, 1984; McGuire & Coward, 1984; Lin *et al.*, 1993). The human enzyme shows a preference for extended substrates containing three or more γ-linked glutamates while the rat enzyme is equally active on di- and penta-γ-glutamates of methotrexate (McGuire & Coward, 1984; Chandler *et al.*, 1986; Rhee *et al.*, 1995).

The pH optimum of γ-glutamyl hydrolase from a variety of plant and animal sources is in the range 4.0–6.0 (Rao & Noronha, 1977a,b; McGuire & Coward, 1984; Bhandari *et al.*, 1990; Lin *et al.*, 1993; Wang *et al.*, 1993) with the human enzyme displaying maximal activity at pH 4.5 (Wang *et al.*, 1986).

Enzyme activity has been measured by a number of meth-ods using methotrexate poly-γ-glutamates and other poly-γ-glutamates as substrates. The activity of the enzyme is readily measured using 4-amino-10-methylpteroylpenta-γ-glutamate as substrate, and separating the products by HPLC (Wang *et al.*, 1993; Rhee *et al.*, 1995) or capillary electrophore-sis (Takemura *et al.*, 1996). A number of pteroyl poly-γ-glutamates and the substrate 4-amino-10-methylpteroylpenta-γ-L-glutamic acid are available from Dr B. Schircks (see Appendix 2 for full names and addresses of suppliers).

## Structural Chemistry

On the basis of gel filtration, a range of molecular masses from 50 to 150 kDa has been reported for γ-glutamyl hydrolase from different sources. The cDNAs for both human and rat γ-glutamyl hydrolase have been cloned and expressed (Yao *et al.*, 1996a,b). The human enzyme is a single-chain protein of molecular mass 36 kDa (Yao *et al.*, 1996b). The rat enzyme is a single-chain 33 kDa molecule with 67% amino acid sequence identity to the human enzyme (Yao *et al.*, 1996a). The rat enzyme as isolated from rat H35 hepatoma cells is highly glycosylated and has an apparent molecular mass of 55 kDa in SDS-PAGE. The sequence of the rat enzyme shows seven potential *N*-glycosylation sites, while the deduced amino acid sequence of the human enzyme contains four (Yao *et al.*, 1996b). Enzymatic removal of the *N*-linked carbohydrate from the rat H35 cell enzyme generates a protein of molecular

mass 33 kDa (Yao *et al.*, 1996a). This preparation is active, as are rat and human enzymes expressed in *E. coli* (Yao *et al.*, 1996a,b), demonstrating that glycosylation is not required for catalytic activity. The observations that the activity of γ-glutamyl hydrolase is enhanced by sulfhydryl-containing compounds (McGuire & Coward, 1984; Wang *et al.*, 1986; Wang *et al.*, 1993; Takemura *et al.*, 1996) and inhibited by *p*-hydroxymercuribenzoate (Lin *et al.*, 1993) or iodoacetate (Yao *et al.*, 1996a) suggest that the enzyme contains a catalytically essential cysteine.

## Preparation

γ-Glutamyl hydrolase has been identified in a wide variety of mammalian and other sources. Preparations of varying purity have been studied from human liver (Baugh *et al.*, 1970), rat, pig and human intestine (Reisenauer *et al.*, 1977; Chandler *et al.*, 1986, 1991; Wang *et al.*, 1986), chicken liver (Rao & Noronha, 1977a,b), rat intestinal mucosa (Elsenhans *et al.*, 1984), pig pancreatic juice (Bhandari *et al.*, 1990) and pea cotyledons (Lin *et al.*, 1993). Two forms have been purified from the jejunum, one of which is intracellular and the other located in the brush border (Reisenauer *et al.*, 1977; Elsenhans *et al.*, 1984; Chandler *et al.*, 1986, 1991; Wang *et al.*, 1986). Pure preparations of the enzyme have been produced from cattle liver (Silink *et al.*, 1975), and from rat H35 hepatoma cell line-conditioned medium (Wang *et al.*, 1993; Yao *et al.*, 1996a). Recently recombinant rat and human enzyme have been expressed in *E. coli* and shown to have properties consistent with partially purified enzyme from rat and human cells, respectively (Yao *et al.*, 1996b).

## Biological Aspects

γ-Glutamyl hydrolase, which functions physiologically to cleave the poly-γ-glutamyl forms of folic acid, is widely distributed in nature (McGuire & Coward, 1984). Similarly, folate poly-γ-glutamates, the intracellular forms of folate, are also found throughout nature. The enzyme has been found in most mammalian tissues, including serum, bile, pancre-atic juice and the brush border of the intestine (Reisenauer *et al.*, 1977; Elsenhans *et al.*, 1984; McGuire & Coward, 1984; Chandler *et al.*, 1986, 1991; Wang *et al.*, 1986; Bhan-dari *et al.*, 1990). It is in these locations that it probably serves to hydrolyze folyl poly-γ-glutamates in food so that they can be absorbed from the intestine as folic acid. γ-Glutamyl hydrolase has also been identified in plants and microorganisms (McGuire & Coward, 1984; Lin *et al.*, 1993; Huangpu *et al.*, 1996).

The gene for the enzyme has not been located, but a cDNA for the enzyme from rat and human sources has been identified. Each protein contains a leader sequence which directs the entry into the endoplasmic reticulum and several consensus *N*-linked glycosylation sequences that aid in determining its cellular destination (Yao *et al.*, 1996a,b).

It has been known for some time that cellular γ-glutamyl hydrolase is primarily lysosomal, and generally has an acid pH optimum (Silink *et al.*, 1975; Rao & Noronha, 1977a,b; Kalghatgi & Bertino, 1981; McGuire & Coward, 1984; Wang *et al.*, 1986; Bhandari *et al.*, 1990; O'Connor *et al.*, 1991; Wang *et al.*, 1993). These data are consistent with its leader sequence and glycosylation (Yao *et al.*, 1996a,b). Also consistent with these structural motifs is the recent finding that the majority of enzyme activity is secreted when a wide variety of cell lines in culture was examined (O'Connor *et al.*, 1991; Rhee *et al.*, 1993; Wang *et al.*, 1993; Yao *et al.*, 1995). The lysosomal enzyme is thought to be involved in the turnover and intracellular homeostasis of the folylpoly-γ-glutamates, whereas the function of the secreted enzyme is less certain. The secreted enzyme that appears in serum, pancreatic juice and bile would almost certainly be involved in folate digestion, but its unanticipated apparently universal secretion *in vitro* remains unexplained. The possibility of a distinct cytosolic form of the enzyme has been discussed (McGuire & Coward, 1984) but no clearcut evidence for this exists. It is now well established that this enzyme can have either endo- or exopeptidase activity and that the type of activity is tissue- and species-specific (Baugh, 1970; Silink *et al.*, 1975; Rao & Noronha, 1977a,b; Elsenhans *et al.*, 1984; McGuire & Coward, 1984; Wang *et al.*, 1986; Yao *et al.*, 1996b). The possible functional reason for having an endo-, as opposed to an exopeptidase in specific tissues or species is not understood.

The regulation of γ-glutamyl hydrolase activity is not well studied. It has been shown to be reduced by insulin, which results in an increase in cellular folate and methotrexate poly-γ-glutamates (Galivan & Rhee, 1995). Conversely, antifolate-resistant lines have been developed that have elevated γ-glutamyl hydrolase activity (Rhee *et al.*, 1993; Yao *et al.*, 1995). These cells have reduced levels of both folate and methotrexate poly-γ-glutamates (Yao *et al.*, 1995).

Recent interesting studies that may ultimately bear on the biological role of this enzyme indicate that it is also secreted by the young tissues of the soybean plant (Huangpu *et al.*, 1996). Further studies are required to fully understand the enzymatic mechanism and the biological, pharmacological and clinical roles of γ-glutamyl hydrolase.

## Distinguishing Features

The enzyme can be identified by its specificity for poly-γ-glutamates as substrates. Poly α-linked glutamates are not substrates (Wang *et al.*, 1993). The enzyme from rat, which is an endopeptidase, also does not hydrolyze protein-bound γ-glutamates (Wang *et al.*, 1993). γ-Glutamyl hydrolase can be distinguished from carboxypeptidase G by its lack of activity on folic acid and methotrexate. Polyclonal antibodies to the enzyme isolated from rat H35 cell culture media have been described (Yao *et al.*, 1996a) but are not commercially available. Interestingly these antibodies against the rat enzyme are only weakly reactive with the human enzyme expressed in *E. coli* despite close amino acid sequence similarity between the two proteins (Yao *et al.*, 1996b).

It has recently been found that prostatic-specific membrane antigen (PSMA) catalyzes the hydrolysis of methotrexate tri-γ-glutamate and pteroyl penta-γ-glutamates (Pinto *et al.*, 1996) (see Chapter 492). Studies of the rodent and human γ-glutamyl hydrolase indicate that a sulfhydryl group is important, and possibly at the active site (O'Connor *et al.*, 1991; Wang *et al.*, 1993; Yao *et al.*, 1996a,b). This feature distinguishes γ-glutamyl hydrolase from the enzyme found in intestinal brush border (Chandler *et al.*, 1986,

1991) and the prostate membrane (Pinto *et al.*, 1996) which are not inhibited by sulfhydryl reagents, and are almost certainly metallopeptidases. Moreover the hydrolase from the prostatic membrane has no sequence relationship to γ-glutamyl hydrolase and has a molecular mass of 94 kDa (Pinto *et al.*, 1996). The hydrolase from human intestinal brush border has a reported mass of 120 kDa (Chandler *et al.*, 1991).

## Further Reading

For a review, see McGuire & Coward (1984).

## References

Baugh, C.M., Stevens, J.C. & Krumdieck, C.L. (1970) Studies on γ-glutamyl carboxypeptidase. I. The solid phase synthesis of analogs of polyglutamates of folic acid and their effects on human liver γ-glutamyl carboxypeptidase. *Biochim. Biophys. Acta* **212**, 116–125.

Bhandari, S.D., Gregory III, J.F., Renuart, D.R. & Merritt, A.M. (1990) Properties of pteroylpolyglutamate hydrolase in pancreatic juice of the pig. *J. Nutr.* **120**, 467–475.

Chandler, C.J., Wang, T.T.Y. & Halsted, C.H. (1986) Pteroylpolyglutamate hydrolase from human jejunal brush borders. *J. Biol. Chem.* **261**, 928–933.

Chandler, C.J., Harrison, D.A., Buffington, C.A., Santiago, N.A. & Halsted, C.H. (1991) Functional specificity of jejunal brush-border pteroylpolyglutamate hydrolase in pig. *Am. J. Physiol.* **260**, G865–G872.

Elsenhans, B., Ahmad, O. & Rosenberg, I.H. (1984) Isolation and characterization of pteroylpolyglutamate hydrolase from rat intestinal mucosa. *J. Biol. Chem.* **259**, 6364–6368.

Galivan, J. & Rhee, M.S. (1995) Insulin-dependent suppression in glutamyl hydrolase activity and elevated cellular methotrexate polyglutamates. *Biochem. Pharmacol.* **50**, 1659–1663.

Huangpu, J., Pak, J.H., Graham, M.C., Rickle, S.A. & Graham, J.S. (1996) Purification and molecular analysis of an extracellular γ-glutamyl hydrolase present in young tissues of the soybean plant. *Biochem. Biophys. Res. Commun.* **228**, 1–6.

Kalghatgi, K.K. & Bertino, J.R. (1981) Folate-degrading enzymes: a review with special emphasis on carboxypeptidase G. In: *Enzymes as Drugs* (Molcenberg, J.S. & Roberts, J., eds). New York: John Wiley & Sons, pp. 77–102.

Lin, S., Rogiers, S. & Cossins, E.A. (1993) γ-Glutamyl hydrolase from pea cotyledons. *Phytochemistry* **32**, 1109–1117.

McGuire, J.J. & Coward, J.K. (1984) Pteroylpolyglutamates: biosynthesis, degradation and function. In: *Folates and Pterins* (Blakley, R.L. & Benkovic, S.J., eds). New York: John Wiley & Sons, pp. 135–191.

O'Connor, B.M., Rotundo, R.F., Nimec, Z., McGuire, J.J. & Galivan, J. (1991) Secretion of γ-glutamyl hydrolase *in vitro*. *Cancer Res.* **51**, 3874–3881.

Pinto, J.T., Suffoletto, B.P., Berzin, T.M., Qiao, C.H., Lin, S., Tong, W.P., May, F., Mukherjee, B. & Heston, W.D.W. (1996) Prostate-specific membrane antigen: a novel folate hydrolase in human prostatic carcinoma cells. *Clin. Cancer Res.* **2**, 1445–1451.

Rao, K.N. & Noronha, J.M. (1977a) Studies on the enzymatic hydrolysis of polyglutamyl folates by chicken liver folyl poly-γ-glutamyl carboxypeptidase. 1. Intracellular localization, purification and partial characterization of the enzyme. *Biochim. Biophys. Acta* **481**, 594–607.

Rao, K.N. & Noronha, J.M. (1977b) Studies on the enzymatic hydrolysis of polyglutamyl folates by chicken liver folyl poly-γ-glutamyl carboxypeptidase. II. Structural studies. *Biochim. Biophys. Acta* **481**, 608–616.

Reisenauer, A.M., Krumdieck, C.L. & Halsted, C.H. (1977) Folate conjugase: two separate activities in human jejunum. *Science* **198**, 196–197.

Rhee, M.S., Wang, Y., Nair, M.G. & Galivan, J. (1993) Acquisition of resistance to antifolates caused by enhanced γ-glutamyl hydrolase activity. *Cancer Res.* **53**, 2227–2230.

Rhee, M.S., Ryan, T.J. & Galivan, J.H. (1995) γ-Glutamyl hydrolase secreted from human tumour cell lines. *Cell Pharmacol.* **2**, 289–292.

Silink, M., Reddel, R., Bethel, M. & Rowe, P.B. (1975) γ-Glutamyl hydrolase (conjugase). *J. Biol. Chem.* **250**, 5982–5994.

Takemura, Y., Kobayashi, H. & Sekiguchi, S. (1996) Separation of methotrexate-poly-γ-glutamates by capillary electrophoresis and its application to the measurement of gamma-glutamyl hydrolase

activity in human leukemia cells in culture. *Rinsho Byori – Jpn J. Clin. Pathol.* **44**, 51–56.

Wang, T.T.Y., Chandler, C.J. & Halsted, C.H. (1986) Intracellular pteroylpolyglutamate hydrolase from human jejunal mucosa. *J. Biol. Chem.* **261**, 13551–13555.

Wang, Y., Nimec, Z., Ryan, T.J., Dias, J.A. & Galivan, J. (1993) The properties of the secreted γ-glutamyl hydrolases from H35 hepatoma cells. *Biochim. Biophys. Acta* **1164**, 227–235.

Yao, R., Rhee, M.S. & Galivan, J. (1995) Effects of γ-glutamyl hydrolase on folyl and antifolylpolyglutamates in cultured H35 hepatoma cells. *Mol. Pharmacol.* **48**, 505–511.

Yao, R., Nimec, Z., Ryan, T.J. & Galivan, J. (1996a) Identification, cloning and sequencing of a cDNA coding for rat γ-glutamyl hydrolase. *J. Biol. Chem.* **271**, 8525–8528.

Yao, R., Schneider, E., Ryan, T.J. & Galivan, J. (1996b) Human γ-glutamyl hydrolase: cloning and characterization of the enzyme expressed *in vitro*. *Proc. Natl Acad. Sci. USA* **93**, 10134–10138.

*John Galivan*
*Division of Molecular Medicine,*
*Wadsworth Center PO Box 509,*
*New York State Department of Health,*
*Albany, NY 12201-0509, USA*
*Email: jhg01@health.state.ny.us*

*Thomas J. Ryan*
*Division of Molecular Medicine,*
*Wadsworth Center PO Box 509,*
*New York State Department of Health,*
*Albany, NY 12201-0509, USA*

# 265. *Dipeptidyl-peptidase VI*

## Databanks

*Peptidase classification: clan CX, family C40, MEROPS ID: C40.001*
*NC-IUBMB enzyme classification: none*
*Databank codes:*

| Species | Gene/type | SwissProt | PIR | EMBL (cDNA) | EMBL (genomic) |
|---|---|---|---|---|---|
| **Dipeptidyl-peptidase VI** | | | | | |
| *Bacillus sphaericus* | – | P39043 | S26056 | X64809 X83680 | – |
| **Other related sequences** | | | | | |
| *Bacillus subtilis* | papQ | – | – | U38819 | – |
| *Clostridium acetobutylicum* | – | – | – | X70334 | – |
| *Enterococcus faecium* | p54 | P13692 | S05542 | X16421 | – |
| *Escherichia coli* | nlpC | P23898 | – | M14031 | – |
| *Escherichia coli* | yafL | – | – | D38582 | – |
| *Haemophilus influenzae* | nlpC | P45296 | – | – | U32838: complete genome section 153 of 163 |
| *Listeria grayi* | iap | Q01835 | – | M80352 | – |
| *Listeria ivanovii* | iap | Q01837 | – | M80350 | – |
| *Listeria monocytogenes* | iap | P21171 | A41487 | X52268 | – |
| *Listeria seeligeri* | iap | Q01838 | – | M80353 | – |
| *Porphyromonas gingivalis* | – | – | – | X95938 | – |

## Name and History

*Bacillus sphaericus* NCTC 9602 produces two sporulation-related peptidases: γ-D-glutamyl-(L)-diamino acid-hydrolyzing peptidase I and *γ-D-glutamyl-L-diamino acid endopeptidase II* (Guinand *et al.*, 1974; Vacheron *et al.*, 1979). These enzymes selectively cleave the peptide sequence of the peptidoglycans of chemotype $A_{1\gamma}$ (Schleifer & Kandler, 1972). They differ with respect to cellular localization (Guinand *et al.*, 1979; Vacheron *et al.*, 1979), molecular mass and catalytic mechanism (Garnier *et al.*, 1985; Bourgogne *et al.*, 1992; Hourdou *et al.*, 1992, 1993). In the light of work on their modes of action (Arminjon *et al.*, 1977; Valentin *et al.*, 1983), they are now recognized as a carboxypeptidase/peptidyl-dipeptidase (see Chapter 457) and a dipeptidyl-peptidase (the present enzyme), respectively. We here propose the name *dipeptidyl-peptidase VI (DPP VI)* for this novel exopeptidase.

## Activity and Specificity

The substrates of the general structure L-Ala-γ-D-Glu+L-Zaa-Y are the peptide moieties of some bacterial peptidoglycans. The amino acid written Zaa here (sometimes also abbreviated $A_2Ac$) is a diamino acid that may be L-lysine, *meso*-diaminopimelic acid (*ms*-$A_2$pm), or ω-amidated-*ms*-$A_2$pm. The C-terminal moiety Y is either D-Ala or D-Ala-D-Ala. DPP VI removes the dipeptide L-Ala-D-Glu by hydrolysis of the γ-D-glutamyl-(L)-diamino acid bond. The substrates are prepared as described by Guinand *et al.* (1974) and Arminjon *et al.* (1976). The peptidase has no strict specificity towards the diamino acid nor towards the C-terminal moiety Y. The only strict requirement for activity is the free N-terminal L-Ala (Vacheron *et al.*, 1979). Polypeptides such as the cross-linked peptide side chains of the peptidoglycans are also substrates, the dipeptide L-Ala-D-Glu being released (Valentin *et al.*, 1983). Thus, the reaction is that of a dipeptidyl-peptidase.

Activity is stimulated by 5 mM EDTA or 1 mM DTT, and the pH optimum is about 8. The rate of reaction increases 2-fold with a tetrapeptide instead of a pentapeptide as substrate. The nature of the L-diamino acid also modulates the activity, which decreases 2-fold through the series *ms*-$A_2$pm, ω-amidated-*ms*-$A_2$pm, Lys (Vacheron *et al.*, 1979). A $K_m$ value of 0.24 mM and a specific activity of 8.3 $\mu$mol min$^{-1}$ mg$^{-1}$ were derived from kinetic studies with L-Ala-γ-D-Glu+L-Lys-D-Ala-D-Ala as substrate (Bourgogne *et al.*, 1992).

DPP VI is thiol-dependent, being inhibited by *N*-ethylmaleimide, iodoacetamide and *p*-hydroxymercuribenzoate, and may be considered a cysteine peptidase.

## Structural Chemistry

The gene for DPP VI encodes a 271 amino acid protein ($M_r$ 30 604; pI 4.1) with no signal peptide, as expected for a cytoplasmic enzyme (Hourdou *et al.*, 1992). There are three Cys residues, at positions 55, 178 and 182, which is in agreement with the other evidence that DPP VI is a cysteine peptidase (Hourdou *et al.*, 1992). DPP VI shows homology with a C-terminal domain of p54 protein of unknown function from *Enterococcus faecium* (Hourdou *et al.*, 1992). The single cysteine of the *E. faecium* protein, Cys420, aligns with Cys178 of the *Bacillus sphaericus* enzyme.

## Preparation

DPP VI has been found in the cytoplasm of sporulating *B. sphaericus* cells, and has been purified (172-fold) to homogeneity from the sporulation medium (Bourgogne *et al.*, 1992). The overexpression of the gene has not been attempted.

## Biological Aspects

DPP VI appears at the onset of sporulation and remains at a rather constant level throughout the process (Vacheron *et al.*, 1979). It is probably involved in peptidoglycan turnover and, given that it hydrolyzes γ-D-Glu-(L)-diamino acid linkages only in peptide units having a free N-terminal L-alanine, it must act after an *N*-acetylmuramoyl-L-alanine amidase. Such amidase activity has been reported to increase during sporulation of *B. subtilis* (Guinand *et al.*, 1976), and various *N*-acetylmuramoyl-L-alanine amidases have been isolated from bacilli and are being studied (Kuroda & Sekiguchi, 1991; Moriyama *et al.*, 1996).

## References

Arminjon, F., Guinand, M., Michel, G., Coyette, J. & Ghuysen, J.M. (1976) Préparation enzymatique de peptides du type L-Ala-γ-D-Glu-(L)-*ms*-$A_2$pm(L)-D-Ala[$^{14}$C] par réaction d'échange entre les peptides correspondants non radioactifs et la D-alanine [$^{14}$C]. *Biochimie* **58**, 1167–1172.

Arminjon, F., Guinand, M., Vacheron, M.J. & Michel, G. (1977) Specificity profiles of the membrane bound γ-D-glutamyl-(L)*meso*-diaminopimelate endopeptidase and LD-carboxypeptidase from *Bacillus sphaericus* 9602. *Eur. J. Biochem.* **73**, 557–565.

Bourgogne, T., Vacheron, M.J., Guinand, M. & Michel, G. (1992) Purification and partial characterization of the γ-D-glutamyl-L-di-amino acid endopeptidase II from *Bacillus sphaericus. Int. J. Biochem.* **24**, 471–476.

Garnier, M., Vacheron, M.J., Guinand, M. & Michel, G. (1985) Purification and partial characterization of the extracellular γ-D-glutamyl-(L)*meso*-diaminopimelate endopeptidase I, from *Bacillus sphaericus* NCTC 9602. *Eur. J. Biochem.* **148**, 539–543.

Guinand, M., Michel, G. & Tipper, D.J. (1974) Appearance of a γ-D-glutamyl-(L)*meso*-diaminopimelate peptidoglycan hydrolase during sporulation in *Bacillus sphaericus. J. Bacteriol.* **120**, 173–184.

Guinand, M., Michel, G. & Balassa, G. (1976) Lytic enzymes in sporulating *Bacillus subtilis. Biochem. Biophys. Res. Commun.* **68**, 1287–1293.

Guinand, M., Vacheron, M.J., Michel, G. & Tipper, D.J. (1979) Location of peptidoglycan lytic enzymes in *Bacillus sphaericus. J. Bacteriol.* **138**, 126–132.

Hourdou, M.L., Duez, C., Joris, B., Vacheron, M.J., Guinand, M., Michel, G. & Ghuysen, J.-M. (1992) Cloning and nucleotide sequence of the gene encoding the γ-D-glutamyl-L-diaminoacid endopeptidase II of *Bacillus sphaericus. FEMS Microbiol. Lett.* **91**, 165–170.

Hourdou, M.L., Guinand, M., Vacheron, M.J., Michel, G., Denoroy, L., Duez, C., Englebert, S., Joris, B., Weber, G. & Ghuysen, J.M.

(1993) Characterization of the sporulation-related γ-D-glutamyl-(L)*meso*-diaminopimelic-acid-hydrolysing peptidase I of *Bacillus sphaericus* NCTC 9602 as a member of the metallo (zinc) carboxypeptidase A family. Modular design of the protein. *Biochem. J.* **292**, 563–570.

Kuroda, A. & Sekiguchi, J. (1991) Molecular cloning and sequencing of a major *Bacillus subtilis* autolysin gene. *J. Bacteriol.* **173**, 7304–7312.

Moriyama, R., Kudoh, S., Miyata, S., Nonobe, S., Hattori, A. & Makino, S. (1996) A germination-specific spore cortex-lytic enzyme from *Bacillus cereus* spores: cloning and sequencing of the gene and molecular characterization of the enzyme. *J. Bacteriol.* **178**, 5330–5332.

Schleifer, K.H. & Kandler, O. (1972) Peptidoglycan types of bacterial cell walls and their taxonomic implications. *Bacteriol. Rev.* **36**, 407–477.

Vacheron, M.J., Guinand, M., Françon, A. & Michel, G. (1979) Caractérisation d'une nouvelle endopeptidase spécifique des liaisons γ-D-glutamyl-L-lysine et γ-D-glutamyl-(L)*meso*-diaminopimélate de substrats peptidoglycaniques chez *Bacillus sphaericus* 9602 au cours de la sporulation. *Eur. J. Biochem.* **100**, 189–196.

Valentin, C., Vacheron, M.J., Martinez, C., Guinand, M. & Michel, G. (1983) Action d'endopeptidases de *Bacillus sphaericus* sur les peptidoglycanes bactériens et sur des fragments peptidoglycaniques. *Biochimie* **65**, 239–245.

*Micheline Guinand*
*Laboratoire de Biochimie Analytique,*
*Université Claude Bernard Lyon 1,*
*43 Boulevard du 11 Novembre 1918,*
*F-69622 Villeurbanne Cedex, France*

# 266. Dipeptidyl-dipeptidase

## Databanks

*Peptidase classification: clan CX, family C99, MEROPS ID: C9C.001*
*Enzyme classification: EC 3.4.14.6*
*Databank codes: no sequence data available*

## Name and History

The activity of this enzyme was discovered in extracts of cabbage (*Brassica oleracea*). The partially purified enzyme, which splits certain tetrapeptides into two dipeptides, was called both a **dipeptidyl tetrapeptide hydrolase** and a **tetrapeptide dipeptidase**. It was also referred to as a **dipeptidyl ligase** because of its tetrapeptide synthetase activity (Eng, 1984). In 1992, IUBMB recommended the name **dipeptidyl-dipeptidase**, based on its ability to separate a dipeptidyl moiety from a dipeptide (leaving group).

## Activity and Specificity

The enzyme cleaves certain tetrapeptides, under weakly alkaline conditions, yielding only dipeptide products. The reaction does not proceed to completion, but rather to a state of equilibrium that attests to its reversibility. In the reverse reaction, the enzyme catalyzes the direct condensation of two dipeptides without prior activation of the carboxyl group of one molecule of substrate coupled to an energy-yielding reaction. At pH 8.0, Ala-Gly+Ala-Gly is hydrolyzed to Ala-Gly dipeptides at more than twice the rate seen for tetraglycine. Triglycine and pentaglycine are not attacked.

Substrate binding to the enzyme is favored by the methyl side group of alanine. In terms of its ligase activity, the enzyme displays its greatest specificity for dipeptides containing two Ala residues or one Ala and one Gly residue. The highest rate of condensation occurs with Ala-Ala, with rates on Ala-Gly or Gly-Ala being about one-third of the rate on Ala-Ala. The Ala residue must be the L-isomer with the amino group in the $\alpha$ position. Deletion of the methyl group, as in Gly-Gly, or substitution by an hydroxyl, as in Ser-Gly, or a methionine side chain, as in Met-Gly, results in little or no ligase activity. The tetrapeptides generated by these condensation reactions (Ala-Ala-Ala-Ala, Ala-Gly-Ala-Gly, or Gly-Ala-Gly-Ala) serve as substrates for the hydrolytic reaction. Tripeptides (Ala-Ala-Gly, Gly-Gly-Ala, Gly-Gly-Gly, Gly-Gly-Leu) yield no condensation products (Eng, 1984).

The hydrolytic activity of dipeptidyl-dipeptidase can be assayed using Ala-Gly-Ala-Gly as a substrate. Samples are removed at timed intervals and the reaction is stopped with glacial acetic acid. Substrate and products are resolved, along with peptide and amino acid standards, by means of high-voltage electrophoresis on paper at pH 2.1, and located by staining with ninhydrin. When necessary, these ninhydrin spots can be eluted for quantitation based on their absorbance at 505 nm. Unstained products can be eluted for amino acid analysis and end group analysis by dansylation (Gray & Hartley, 1963).

There is a requirement for a free sulfhydryl group; activity is stabilized by DTT, and inhibition by *p*-hydroxymercuribenzoate is completely reversed by DTT.

### Structural Chemistry

The substrate-binding site is estimated to be about 15 Å in length since this would accommodate four consecutive amino acid residues in a fully extended polypeptide (Corey & Pauling, 1953). The binding site should contain four subsites, each recognizing alanine. The catalytic groups, which probably include a sulfhydryl (p*K* 7.9) and an imidazole (p*K* 6.3), must be located in the middle of the active site since tetrapeptide substrates are split only at the midpoint to yield dipeptides (Eng, 1984).

The molecular mass and other structural characteristics have not yet been determined.

### Preparation

Shredded cabbage is subjected to a freeze–thaw cycle and crushed in a hydraulic press to obtain an extract that serves as starting material in a process of partial purification that involves fractionation by ammonium sulfate precipitation, gel filtration and ion-exchange chromatography (Eng, 1984).

### Biological Aspects

The function of the enzyme is unknown.

### Distinguishing Features

Hydrolytic activity is restricted to the cleavage of certain tetrapeptides yielding only dipeptide products, a reaction that is readily reversed by the dipeptidyl ligase activity of the enzyme.

### References

Corey, R.B. & Pauling, L. (1953) Fundamental dimensions of polypeptide chains. *Proc. R. Soc. Lond. [Biol.]* **141**, 10–20.

Eng, F.W.H.T. (1984) Dipeptidyl tetrapeptide hydrolase, a new enzyme with dipeptidyl ligase activity. *Can. J. Biochem. Cell Biol.* **62**, 516–528.

Gray, W.R. & Hartley, B.S. (1963) A fluorescent end-group reagent for proteins and peptides. *Biochem. J.* **89**, 59P.

*J. Ken McDonald*
*Department of Biochemistry and Molecular Biology,*
*Medical University of South Carolina,*
*171 Ashley Avenue,*
*Charleston, SC 29425, USA*

# 267. Cysteine-type carboxypeptidase, carboxypeptidase LB

### Databanks

*Peptidase classification: clan CX, family C99, MEROPS ID: C9E.001*
*NC-IUBMB enzyme classification: EC 3.4.18.1*
*Databank codes: no sequence data available*

### Name and History

In their study of carboxypeptidase activity in cattle spleen, Fruton & Bergmann (1939) detected two enzymes that hydrolyzed Z-Glu↓Tyr. One of these was what we now know as the serine-type carboxypeptidase lysosomal carboxypeptidase A (Chapter 133), but the second was thiol dependent. This latter enzyme was termed **cathepsin IV** by Fruton *et al*. (1941), but when the cathepsin nomenclature was revised later, it was renamed simply **carboxypeptidase** (Tallan *et al*., 1952) or **catheptic carboxypeptidase** (Greenbaum & Sherman, 1962). Reporting on the catheptic carboxypeptidase activity of an 'Hg-ethanol' fraction from cattle spleen, Greenbaum & Sherman (1962) showed that only the thiol-dependent component of the carboxypeptidase activity (and not the lysosomal carboxypeptidase A) hydrolyzed Bz-Gly-Arg, a substrate of pancreatic carboxypeptidase B (Chapter 455). At this time, Bz-Arg-NH$_2$ was the standard substrate of cathepsin B, (Chapter 209), but Otto (1967) discovered that Bz-Arg-NH$_2$ is hydrolyzed by two separate enzymes in a preparation of cathepsin B prepared as described by Greenbaum & Fruton (1957). Thus, gel filtration on Sephadex G-100 separated components of 52 kDa and 25 kDa, both of which hydrolyzed Bz-Arg-NH$_2$. The 25 kDa enzyme was an endopeptidase, cathepsin B, but the larger enzyme was a carboxypeptidase. During the period 1967–1970, the 52 kDa enzyme (carboxypeptidase) was misleadingly termed *cathepsin B* in some reports, and the 25 kDa endopeptidase was known as cathepsin B'. The term *cathepsin B2* was introduced by Otto (1971) for the carboxypeptidase, and in reviewing the earlier work on this enzyme, McDonald & Schwabe (1977) proposed the name *lysosomal carboxypeptidase B*. Other names that have been applied to activities that can now be attributed to this enzyme are *catheptic carboxypeptidase B, catheptic carboxypeptidase G* (Taylor & Tappel, 1974), and *histone hydrolase* (De Lumen & Tappel, 1973). The Nomenclature Committee of IUBMB (NC-IUBMB, 1992) recommended the name *lysosomal cysteine-type carboxypeptidase*, but we shall here use the more specific term *carboxypeptidase LB (CLB)*.

## Activity and Specificity

Carboxypeptidase LB releases C-terminal amino acids from polypeptides and *N*-acylated dipeptides that have a free α-carboxyl group. Dipeptides are not hydrolyzed, but many tripeptides are. Trilysine is not hydrolyzed (Ninjoor *et al.*, 1974), but tetralysine is (McDonald & Ellis, 1975). CLB resembles cathepsin B in deamidating Bz-Arg-NH$_2$, but differs in not being an endopeptidase, and having no action on arylamides (e.g. Bz-Arg-NHPhNO$_2$, Bz-Arg-NHNap, Bz-Arg-Arg-NHNap) (Otto, 1967; McDonald *et al.*, 1970).

The most commonly used assay substrate has been Bz-Arg+NH$_2$ (Taylor *et al.*, 1974), but this is also hydrolyzed by cathepsin B. A more selective substrate is Bz-Gly+Arg (McDonald & Barrett, 1986), with which the liberated Arg can be detected colorimetrically with ninhydrin or fluorometrically with phthalaldehyde. Assays are best made at pH 5.0, 37°C, in the presence of EDTA and DTT.

Carboxypeptidase LB has broad specificity for the C-terminal residue, releasing amino acids of all classes except Pro. Approximate relative rates for the cattle spleen enzyme at pH 5.0 are Z-Gly-Met (100), Z-Gly-Phe (93), Z-Gly-Ser (85), Z-Gly-Asp (70), Z-Gly-Arg (58), Z-Gly-Leu (47) and Z-Gly-Gly (22) (McDonald & Ellis, 1975).

Polypeptides also serve as substrates for CLB, which acts on Leu-Trp-Met-Arg-Phe-Ala (releasing four residues), Val-Leu-Ser-Glu-Gly (releasing three residues) and glucagon (Afroz *et al.*, 1976; McDonald & Ellis, 1975; Ninjoor *et al.*, 1974). Several residues are also released from each chain of the two-chain relaxin (Schwabe *et al.*, 1977, 1978). Peptides with an amidated C-terminus, such as substance P, are not attacked (Lones *et al.*, 1983).

CLB is inhibited by such nonspecific thiol-blocking reagents as *p*-chloromercuribenzoate, iodoacetate and heavy metal ions, but reportedly not by *N*-ethylmaleimide. Leupeptin and antipain (1 μM) are the most potent inhibitors described to date (Lones *et al.*, 1983).

## Structural Chemistry

CLB is a dimeric protein of about 52 kDa, the two subunits being of the same size, and perhaps identical. The dimer contains two thiol groups (Otto & Riesenkönig, 1975) and about 10% carbohydrate (Lones *et al.*, 1983). The pI is approximately 5.0 (McDonald & Schwabe, 1977). The cDNA has not yet been cloned, and nothing is known of the amino acid sequence of carboxypeptidase LB.

The enzyme is most stable in the range pH 5–6, and is very unstable above pH 7. Stability in storage is greater with the active-site thiol blocked than free (Otto & Riesenkönig, 1975).

## Preparation

CLB has been prepared from rat liver lysosomes (Ninjoor *et al.*, 1971), cattle spleen (McDonald & Ellis, 1975; Otto & Reisenkönig, 1975; McDonald & Schwabe, 1977), and rabbit lung (Lones *et al.*, 1983), by conventional methods.

## Biological Aspects

Carboxypeptidase LB is a lysosomal enzyme, and probably contributes to the late stages of degradation of proteins in the lysosomal system, working cooperatively with the serine-type lysosomal carboxypeptidase A (Chapter 133). It has also been reported that the enzyme is present in preparations of T cell-activating factor, and can account almost quantitatively for the proliferative stimulation (Dessaint *et al.*, 1979).

## Further Reading

There has been very little recent work on carboxypeptidase LB, so the reviews of McDonald & Schwabe (1977) and McDonald & Barrett (1986) remain current.

## References

Afroz, H., Otto, K., Müller, R. & Fuhge, P. (1976) On the specificity of bovine spleen cathepsin B2. *Biochim. Biophys. Acta* **452**, 503–509.

De Lumen, B.O. & Tappel, A.L. (1973) Histone hydrolase activity of rat liver lysosomal cathepsin B$_2$. *Biochim. Biophys. Acta* **293**, 217–225.

Dessaint, J.-P., Katz, S.P. & Waksman, B.H. (1979) Catheptic carboxypeptidase B as a major component in 'T-cell activating factor' of macrophages. *J. Immunopharmacol.* **1**, 399–414.

Fruton, J.S. & Bergmann, M. (1939) On the proteolytic enzymes of animal tissues. I. Beef spleen. *J. Biol. Chem.* **130**, 19–27.

Fruton, J.S., Irving, G.W., Jr. & Bergmann, M. (1941) On the proteolytic enzymes of animal tissues. III. The proteolytic enzymes of beef spleen, beef kidney and swine kidney. Classification of the cathepsins. *J. Biol. Chem.* **141**, 763–774.

Greenbaum, L.M. & Fruton, J.S. (1957) Purification and properties of beef spleen cathepsin B. *J. Biol. Chem.* **226**, 173–180.

Greenbaum, L.M. & Sherman, R. (1962) Studies on catheptic carboxypeptidase. *J. Biol. Chem.* **237**, 1082–1085.

Lones, M., Chatterjee, R., Singh, H. & Kalnitsky, G. (1983) Lysosomal carboxypeptidase B from rabbit lung. Purification and characterization. *Arch. Biochem. Biophys.* **221**, 64–78.

McDonald, J.K. & Barrett, A.J. (1986) *Mammalian Proteases: a Glossary and Bibliography*, vol. 2: *Exopeptidases*. London: Academic Press.

McDonald, J.K. & Ellis, S. (1975) On the substrate specificity of cathepsin B1 and B2 including a new fluorogenic substrate for cathepsin B1. *Life Sci.* **17**, 1269–1276.

McDonald, J.K. & Schwabe, C. (1977) Intracellular exopeptidases. In: *Proteinases in Mammalian Cells and Tissues* (Barrett, A.J., ed.). Amsterdam: North-Holland Publishing, pp. 311–391.

McDonald, J.K., Zeitman, B.B. & Ellis, S. (1970) Leucine naphthylamide: an inappropriate substrate for the histochemical detection of cathepsins B and B'. *Nature* **225**, 1048–1049.

NC-IUBMB (Nomenclature Committee of the International Union of Biochemistry and Molecular Biology) (1992) *Enzyme Nomenclature 1992*. Orlando, FL: Academic Press.

Ninjoor, V., Taylor, S.L. & Tappel, A.L. (1971) Purification and characterization of rat liver lysosomal cathepsin B2. *Biochim. Biophys. Acta* **370**, 308–321.

Otto, K. (1967) Über ein neues Kathepsin. Reinigung aus Rindermilz, Eigenschaften, sowie Vergleich mit Kathepsin B. *Z. Physiol. Chem.* **348**, 1449–1460.

Otto, K. (1971) Cathepsins B1 and B2. In: *Tissue Proteinases* (Barrett, A.J. & Dingle, J.T., eds). Amsterdam: North-Holland Publishing, pp. 1–28.

Otto, K. & Riesenkönig, H. (1975) Improved purification of cathepsin B1 and cathepsin B2. *Biochim. Biophys. Acta* **379**, 462–475.

Schwabe, C., McDonald, J.K. & Steinetz, B.G. (1977) Primary structure of the B-chain of porcine relaxin. *Biochem. Biophys. Res. Commun.* **75**, 503–510.

Schwabe, C., Steinetz, B., Weiss, G., Segaloff, A., McDonald, J.K., O'Byrne, E., Hochman, J., Carriere, B. & Goldsmith, L. (1978) Relaxin. *Recent Prog. Horm. Res.* **34**, 123–211.

Tallan, H.H., Jones, M.E. & Fruton, J.S. (1952) On the proteolytic enzymes of animal tissues. X. Beef spleen cathepsin C. *J. Biol. Chem.* **194**, 793–805.

Taylor, S.L. & Tappel, A.L. (1974) Identification and separation of lysosomal carboxypeptidases. *Biochim. Biophys. Acta* **341**, 99–111.

Taylor, S., Ninjoor, V., Dowd, D.M. & Tappel, A.L. (1974) Cathepsin B2 measurement by sensitive fluorometric ammonia analysis. *Anal. Biochem.* **60**, 153–162.

*Alan J. Barrett*
*MRC Peptidase Laboratory,*
*Department of Immunology,*
*The Babraham Institute,*
*Cambridgeshire CB2 4AT, UK*
*Email: alan.barrett@bbsrc.ac.uk*

# 268. *Bacterial pyroglutamyl-peptidase*

## Databanks

Peptidase classification: clan CX, family C15, MEROPS ID: C15.001
NC-IUBMB enzyme classification: EC 3.4.19.3
Databank codes:

| Species | SwissProt | PIR | EMBL (cDNA) | EMBL (genomic) |
|---|---|---|---|---|
| *Bacillus amyloliquefaciens* | P46107 | – | D11035 | – |
| *Bacillus subtilis* | P28618 | S23432 | A25847 X66034 | D30808: multiple entry |
| *Pseudomonas fluorescens* | P42673 | – | X75919 | – |
| *Staphylococcus aureus* | – | – | U19770 | – |
| *Streptococcus pyogenes* | Q01328 | S24717 | X65717 | – |
| Eukaryotic homolog | | | | |
| *Mus musculus* | – | – | W90999 W97651 | – |

## Name and History

***Pyroglutamyl-peptidase I*** hydrolyzes an L-pyroglutamyl (Glp) residue from the N-terminus of a polypeptide. The enzyme was formerly classified as an aminopeptidase (EC 3.4.11.8), but is more accurately regarded as an omega peptidase, because the substrate contains no free N-terminal amino group. Other names that have been used include ***pyrrolidonyl peptidase*** (Doolittle & Armentrout, 1968), ***pyrrolidone carboxyl peptidase, pyrrolidonecarboxylate peptidase*** (Kwiatkowska *et al.*, 1974; Sullivan *et al.*, 1977), ***pyrrolidonecarboxylyl peptidase*** (Doolittle, 1970; Sullivan & Jago, 1970), ***L-pyroglutamyl peptide hydrolase, 5-oxoprolyl-peptidase, pyroglutamate aminopeptidase*** (Taylor & Dixon, 1978; O'Connor & O'Cuinn, 1985) and ***pyroglutamyl aminopeptidase*** (Mantle *et al.*, 1991).

Since the discovery of pyroglutamyl-peptidase activity nearly three decades ago (Doolittle & Armentrout, 1968), the activity has been demonstrated in bacteria, and in plant, animal and human tissues (Szewczuk & Kwiatkowska, 1970) (for a review see Awadé *et al.*, 1994). In the animal systems, two types of pyroglutamyl-peptidase are found: a soluble enzyme whose subunit molecular mass of about 22 000–25 000 is similar to that of bacterial pyroglutamyl-peptidase, and a much larger pyroglutamyl-peptidase of about 230 000–280 000 molecular mass, which is membrane associated (Bauer, 1994). The smaller, soluble enzyme is apparently a cysteine peptidase, like the bacterial pyroglutamyl-peptidase, and is called pyroglutamyl-peptidase I (Chapter 269). No amino acid sequence is yet available for a eukaryotic pyroglutamyl-peptidase I, but it may very well be homologous to the bacterial enzyme. The larger, membrane-bound enzyme is pyroglutamyl-peptidase II (Chapter 340), a metallopeptidase belonging to family M1. Because the bacterial pyroglutamyl-peptidase is similar in many properties to the eukaryotic pyroglutamyl-peptidase I, the present chapter and that on mammalian pyroglutamyl-peptidase I (Chapter 269) will compare and contrast the enzymes from the two types of organism. For convenience, we shall term the bacterial pyroglutamyl-peptidase ***pyrase*** in the following text.

## Activity and Specificity

Pyrase hydrolyzes the L-pyroglutamyl (Glp) residue from the N-terminus of polypeptides and small synthetic substrates. A useful test substrate is the chromogenic and fluorometric Glp┼NHNap (Patterson *et al.*, 1963; Mulczyk & Szewczyk, 1970). An *in situ* detection method based on this assay was used to select bacterial clones expressing the pyrase gene (*pcp*) (Cleuziat *et al.*, 1992; Yoshimoto *et al.*, 1993). Other synthetic substrates include Glp┼NHPhNO$_2$ and Glp┼NHMec (Fujiwara & Tsuru, 1978).

Pyrases of type I from bacteria and animals appear to have a broad substrate specificity, most polypeptides with an N-terminal Glp being recognized. However, the rate of hydrolysis is affected by the amino acid adjacent to the Glp residue in a way that depends upon the source of the enzyme. For instance, it has been shown that the enzyme from *Klebsiella cloacae* is able to split Glp-Pro, but that this compound is not hydrolyzed by the pyrases of *Pseudomonas fluorescens*

(Uliana & Doolittle, 1969), *Bacillus amyloliquefaciens* (Fujiwara *et al.*, 1979) or cattle pituitary (Mudge & Fellows, 1973). Pyrase appears to have a strict specificity for L-Glp; no activity is seen with D-Glp.

Bacterial pyrase generally shows optimal activity at pH 7–9 (Table II of Awadé *et al.*, 1994), but this may be influenced by the way in which the enzyme is prepared, since, for example, the optimal pH of *Bacillus subtilis* pyrase is 7 for the purified expressed enzyme and 8–9 for crude extracts. The optimal pH for the animal pyrases has been reported to be pH 7–8.5.

It is difficult to compare directly the kinetic parameters of the different pyrases, because the Michaelis–Menten constants have been determined for different substrates. Nevertheless, it appears that the pyrases exhibit Michaelis-Menten-type kinetics, rather than the allosteric kinetics that might have been expected for the apparently oligomeric bacterial enzyme (see below). The affinity of pyrase for its substrate is relatively low, with $K_m$ values commonly in the range 0.2–2 mM.

Bacterial pyrase is heat sensitive, being rapidly inactivated above 50°C (Awadé *et al.*, 1994). It is a sulfhydryl-dependent enzyme that can be inactivated by iodoacetamide or other sulfhydryl-blocking reagents. Thus it is necessary to include a reducing agent such as 2-ME during purification and storage. The enzyme can also be stabilized by a noncompetitive substrate analog, such as 2-pyrrolidone: activity is recovered by subsequent removal of the analog by dialysis. Maximal activity of type I pyrase is generally obtained in the absence of divalent ions, and even trace amounts of Hg$^{2+}$ are inhibitory (Awadé *et al.*, 1994). The effect of other divalent ions is variable and depends on the enzyme. Generally they have no effect, or are weakly inhibitory, but for the *K. cloacae* peptidase, Ca$^{2+}$ ions are reported to increase activity (Kwiatkowska *et al.*, 1974).

Other compounds that have been found to inhibit pyrase activity include 1,10-phenanthroline and antipain (Mantle *et al.*, 1991), *N*-ethylmaleimide, puromycin and bestatin (Mantle *et al.*, 1990), sodium tetrathionate, *p*-chloromercuribenzoate and *N*-bromosuccinimide (Tsuru *et al.*, 1978), bacitracin, L-pyroglutamyl chloromethane, (*Z*)-pyroglutamyl diazomethane (Fujiwara *et al.*, 1981; Wilk *et al.*, 1985), and 5-oxo-prolinal (Friedman *et al.*, 1985). Benarthin and pyrizinostatin have also been reported as inhibitors of pyrases (Aoyagi *et al.*, 1992a,b).

## Structural Chemistry

Five bacterial *pcp* genes encoding pyrase show a common structure for the protein. The sizes of the open reading frames (ORFs) are similar and relatively small. The *pcp* genes from *Streptococcus pyogenes, Bacillus subtilis, B. amyloliquefaciens* and *Staphylococcus aureus* are all 645 nucleotides, and the gene from *P. fluorescens* is 639 nucleotides long. These genes encode polypeptides of 215 or 213 amino acids, with deduced $M_r$ values of 23 000–24 000. There are putative ribosome-binding sites that are typical for bacteria, and at the −35 and −10 positions there are putative promoters that are nearly identical to consensus sequences in *Escherichia coli*; moreover, they have the 17 nucleotide spacing that is typical of strong transcriptional signals (Harley & Reynolds, 1987). The presence of multiple putative RNA polymerase-binding sites was noted in the *pcp* gene from *S. pyogenes* and characterized as being functional in *E. coli* by primer extension experiments (Cleuziat *et al.*, 1992). For the complete analysis of these genes and the very conserved protein sequences, see Awadé *et al.* (1994) and Patti *et al.* (1995), and for the localization of the gene on the chromosomal map of *S. pyogenes* see Suvorov & Ferretti (1996).

Some of the bacterial pcp genes have been cloned and expressed in *E. coli* using the pT7 system (Tabor & Richardson, 1985). Analysis of the sequences by the criteria of Kyte & Doolittle (1982) indicates that the charges are uniformly distributed along the polypeptide chain, which is consistent with the fact that the enzymes are soluble (Awadé *et al.*, 1994).

No other nucleotide sequences currently in GenBank show significant similarity to the *pcp* genes. The GC content of *pcp* genes from gram-positive bacteria is significantly higher than the average for the respective genomes. The alignment of the deduced amino acid sequences for pyrase shows that the primary structure is highly conserved; the percentage identity ranges from 31 to 72%. Two particularly highly conserved sequences of 20 amino acids that are located in the central part of the polypeptides are likely to be involved in their biological activity. The second domain, containing a cysteine residue, appears to be functionally important and may form part of the catalytic site of these enzymes (Awadé *et al.*, 1994). This has been confirmed by site-directed mutagenesis of the enzymes from *B. amyloliquefaciens* (Yoshimoto *et al.*, 1993) and *P. fluorescens* (LeSaux *et al.*, 1996). Substitution of Glu10 and Glu22 by Gln led to enzymes which displayed catalytic properties and sensitivities to 1-ethyl-3-(3-dimethylaminopropyl) carbodiimide similar to those of the wild-type pyrase, so that it could be concluded that these residues are not essential for the catalytic activity. Replacement of Asp89 by Asn or Ala resulted in enzymes which retained nearly 25% and 0% of activity, respectively. Substitution of Cys144 and His166 by Ala and Ser, respectively, resulted in inactive enzymes. Proteins with changes of Glu81 to Gln and Asp94 to Asn were not detectable in the crude extracts and were probably unstable in the bacteria. These results are consistent with the view that Cys144 and His166 constitute a catalytic dyad, while residues Glu81, Asp89 or Asp94 might also contribute to the catalytic mechanism. These results led to the proposal that pyrase represents a new class of cysteine peptidase (Le Saux *et al.*, 1996). Now that the enzymes from *B. amyloliquefaciens* (Yoshimoto *et al.*, 1993) and *S. pyogenes* (Gonzales, 1995) have been overexpressed, and that from *B. amyloliquefaciens* has been crystallized, more detailed knowledge of pyrase structure may be available before long. No amino acid sequence of a mammalian pyrase I has yet been published.

Molecular mass determinations for pyrases from various bacterial species by SDS-PAGE under native and denaturing conditions again shows them to be similar. Determinations under denaturing conditions indicate a mean $M_r$ for the subunit of the bacterial enzyme of about 25 000. Native gel conditions, on the other hand, indicate more variability in size, showing $M_r$ values in the range 50 000–91 000. Tsuru *et al.* (1978) suggested that the *B. amyloliquefaciens* enzyme may be a trimer, but more recent studies indicate that the *S. pyogenes* and *B. subtilis* enzymes are probably tetramers (Awadé *et al.*, 1992; Gonzales *et al.*, 1992). On the other hand, it has been suggested that the overexpressed *B. amyloliquefaciens* enzyme was probably a dimer, although the natural enzyme is more likely a trimer or a tetramer. The subunit $M_r$ of the *Streptococcus faecium* pyrase has been estimated at 42 000 (Sullivan *et al.*, 1977), i.e. almost twice that of the other bacterial enzymes. Interestingly, antibodies against the pyrase of *S. pyogenes* reacted with two protein bands in a western blot of crude cell extract (Gonzales *et al.*, 1992). As this had never before been reported in bacteria, it was suggested that *S. pyogenes* possessed at least two different pyrases, the larger of which, having an $M_r$ of about 40 000, approximated the size of the *S. faecium* enzyme. It would also be of interest to determine the $M_r$ of the *S. faecium* enzyme since it has been reported to be a dimer of 24 000 subunits, as suggested by Sullivan *et al.* (1977), rather than a tetramer like most other bacterial enzymes.

## Preparation

Bacterial pyrases that have been purified to homogeneity after overexpression in host cells are those of *S. pyogenes* (Awadé *et al.*, 1992), *B. subtilis* (Gonzales *et al.*, 1992), *B. amyloliquefaciens* (Yoshimoto *et al.*, 1993), *P. fluorescens* (Gonzales & Robert-Baudouy, 1994) and *Staphylococcus aureus* (Patti *et al.*, 1995). The pyrase of *Klebsiella cloacae* has also been purified to homogeneity (Kwiatkowska *et al.*, 1974). Partial purification of mammalian pyrases has been claimed for the following tissues: human skeletal muscle (Mantle *et al.*, 1991), cerebral cortex (Lauffart *et al.*, 1988), kidney (Mantle *et al.*, 1990), rat brain (Busby *et al.*, 1982) and guinea pig brain (Browne & O'Cuinn, 1983).

## Biological and Biotechnological Aspects

Pyrase is widely distributed in living organisms and appears to play an important role in the activation and inactivation of many Glp-terminated peptides. The isolation of mammalian type I pyrase (Chapter 269) from the soluble fraction of many tissues suggests that it is a cytosolic enzyme (Mudge & Fellows, 1973; Emerson & Wu, 1987; Mantle *et al.*, 1991). The enzyme appears to have a wide tissue distribution, being found at much the same specific activity in brain, hypothalamus, pancreas, liver, skeletal muscle and kidney (Mantle, 1992; Scharfmann *et al.*, 1989). Pyrase I has been shown to act on many substrates having an N-terminal Glp, especially neuropeptides including thyrotropin-releasing hormone (TRH). This pyrase, however, does not appear to be involved in the control of TRH levels, as is type II pyrase (Torres *et al.*, 1986; Mantle *et al.*, 1990; Mendes *et al.*, 1990; Mantle, 1992). The presence in mammals of two enzymes with similar activity suggests that they may be involved in different physiological pathways. If the role of the more specific type II pyrase has been identified (see Chapter 340), the role of type I pyrase remains unclear. It has been proposed, by analogy with other soluble aminopeptidases, that type I pyrase may contribute to the final stages of the intracellular catabolism of peptides to free amino acids, which are then released to the cellular pool (Mantle *et al.*, 1990; Mantle, 1992). Thus, this enzyme may, at least in part, be involved in the regulation of the cellular pool of free Glp. It is noteworthy that free Glp is known to have pharmacological properties, and a specific pathway for its production by the activity of pyrase I may exist. The specific source of the free Glp that is associated with some diseases remains unknown, but the involvement of pyrase remains a possibility. In bacteria, the enzyme may have a nutritional function. Further investigations are clearly required to improve our understanding of the role of pyrase in the metabolism of cells and tissues.

Pyrase may have several biotechnological applications. It was initially discovered as an enzyme that could remove the N-terminal Glp block that prevented N-terminal amino acid sequencing of proteins and peptides by the Edman (1950) method (Fellows & Mudge, 1971). Pyrase is still used in protein sequencing work to confirm the presence of this residue (Bieber *et al.*, 1990; Lu *et al.*, 1991), and is marketed for this application. A new application has been found in the diagnosis of bacterial infections, particularly in the identification of group A streptococci and enterococci, facilitated by convenient colorimetric procedures (Awadé *et al.*, 1994). It has also been proposed that the pyrase gene could be used as a reporter gene (Cleuziat *et al.*, 1992).

Activators and inhibitors of bacterial and mammalian pyrases might form valuable therapeutic agents. For example, the protection of the pyroglutamyl residue of peptides from pyrase attack may improve the delivery of these peptides in therapeutics (Bundgaard & Moss, 1989). Pyroglutamyl derivatives of various anticancer drugs have been proposed as potential 'prodrugs' that are designed to be cleaved to the active cytotoxic agents by pyrase at the tumor site. For targeting the tumor site, a pyrase chemically linked to a monoclonal antibody may be useful (Cheung *et al.*, 1993).

## Distinguishing Features

A phylogenetic analysis of the bacterial pyrases sequenced so far shows that the *B. subtilis* and *B. amyloliquefaciens* enzymes are particularly similar (72% identity), which is consistent with the fact that the species also are closely related. Surprisingly, the enzyme form *S. pyogenes* (gram-positive) appears to be closer to the *P. fluorescens* (gram-negative) enzyme than to the *Bacillus* (gram-positive) enzymes. The location of the *P. fluorescens* branch point in a dendrogram suggests that the common ancestor of the pyrase genes may have been in a gram-negative bacterium (Awadé *et al.*, 1994). The features distinguishing the bacterial pyrase from pyroglutamyl-peptidases I and II of mammals were described above.

## Further Reading

An excellent recent review is that of Awadé *et al.* (1994).

## References

Aoyagi, T., Hatsu, M., Imada, C., Naganawa, H., Okami, Y. & Takeuchi, T. (1992a) Pyrizinostatin: a new inhibitor of pyroglutamyl peptidase. *J. Antibiot.* **45**, 1795–1796.
Aoyagi, T., Hatsu, M., Kojima, F., Hayashi, C., Hamada, M. & Takeuchi, T. (1992b) Benarthin: a new inhibitor of pyroglutamyl peptidase. I. Taxonomy, fermentation, isolation and biological activities. *J. Antibiot.* **45**, 1079–1083.

Awadé, A., Gonzalès, T., Cleuziat, P. & Robert-Baudouy, J. (1992) One step purification and characterization of the pyrrolidone carboxylyl peptidase from *Streptococcus pyogenes* overexpressed in *Escherichia coli. FEBS Lett.* **308**, 70–74.
Awadé, A., Cleuziat, P., Gonzalès, T. & Robert-Baudouy, J. (1994) Pyrrolidone carboxyl peptidase (Pcp): an enzyme that removes pyroglutamic acid (pGlu) from pGlu-peptides and pGlu-proteins. *Protein Struct. Funct. Genet.* **20**, 34–51.
Bauer, K. (1994) Purification and characterization of the thyrotropin-releasing-hormone-degrading ectoenzyme. *Eur. J. Biochem.* **224**, 387–396.
Bieber, A.L., Becker, R.R., McParland, R., Hunt, D.F., Shabanowitz, J., Yates III J.R., Martino, P.A. & Johnson, G.R. (1990) The complete sequence of the acidic subunit from Mojave toxin determined by Edman degradation and mass spectrometry. *Biochim. Biophys. Acta* **1037**, 413–421.
Browne, P. & O'Cuinn, G. (1983) An evaluation of the role of a pyroglutamyl peptidase, a post-proline cleaving enzyme and a post-proline dipeptidyl amino peptidase, each purified from the soluble fraction of guinea-pig brain, in the degradation of thyroliberin *in vitro. Eur. J. Biochem.* **137**, 75–87.
Bundgaard, H. & Moss, J. (1989) Prodrugs of peptides. IV. Bioreversible derivatization of the pyroglutamyl group by *N*-acylation and *N*-aminoethylation to effect protection against pyroglutamyl aminopeptidase. *J. Pharmaceut. Sci.* **78**, 122–126.
Busby, W.H., Youngblood, W.W. & Kizer, J.S. (1982) Studies of substrate requirements, kinetic properties and competitive inhibitors of the enzymes catabolizing TRH in rat brain. *Brain Res.* **242**, 261–270.
Cheung, H.T.A., Dong, Z., Escoffer, L. Smal, M.A. & Tattersall, M.H.N. (1993) Activation by peptidases and cytotoxicity of 2-(L-α-aminoacyl) prodrugs of methotrexate. In: *Chemistry and Biology of Pteridines and Folates* (Ayling, J.E., Nair, M.G. & Baugh, C.M., eds). New York: Plenum Press, pp. 457–460.
Cleuziat, P., Awadé, A. & Robert-Baudouy, J. (1992) Molecular characterization of *pcp*, the structural gene encoding the pyrrolidone carboxylyl peptidase from *Streptococcus pyogenes. Mol. Microbiol.* **6**, 2051–2063.
Doolittle, R.F. (1970) Pyrrolidonecarboxylyl peptidase. *Methods Enzymol.* **19**, 555–569.
Doolittle, R.F. & Armentrout, R.W. (1968) Pyrrolidone peptidase. An enzyme for selective removal of pyrrolidonecarboxylic acid residues from polypeptides. *Biochemistry* **7**, 516–521.
Edman, P. (1950) Method for determination of the amino sequence in peptides. *Acta Chem. Scand.* **4**, 283–293.
Emerson, C.H. & Wu, C.F. (1987) Thyroid status influences rat serum but not brain TRH pyroglutamyl aminopeptidase activities. *Endocrinology* **120**, 1215–1217.
Fellows, R.E. & Mudge, A. (1971) Isolation and characterization of *B. subtilis* pyrrolidonecarboxylyl peptidase as an adjunct for the investigation of peptide structure. *Fedn Proc. Fedn Am. Soc. Exp. Biol.* **30** (abstr. 151), 1078.
Friedman, T.C., Kline, T.B. & Wilk, S. (1985) 5-Oxoprolinal: transition-state aldehyde inhibitor of pyroglutamyl-peptide hydrolase. *Biochemistry* **24**, 3907–3913.
Fujiwara, K. & Tsuru, D. (1978) New chromogenic and fluorogenic substrates for pyrolidonyl peptidase. *J. Biochem. (Tokyo)* **83**, 1145–1149.
Fujiwara, K., Kobayashi, R. & Tsuru, D. (1979) The substrate specificity of pyrrolidone carboxylyl peptidase from *Bacillus amyloliquefaciens. Biochim. Biophys. Acta* **570**, 140–148.

Fujiwara, K., Kitagawa, T. & Tsuru, D. (1981) Inactivation of pyroglutamyl aminopeptidase by L-pyroglutamyl chloromethyl ketone. *Biochim. Biophys. Acta* **655**, 10–16.

Gonzalès, T. (1995) Contribution à l'étude moléculaire, physiologique et biochimique de pyrrolidone carboxyle peptidases bactériennes. Thèse de Doctorat, Biochimie, INSA, 29 May 1995.

Gonzalès, T. & Robert-Baudouy, J. (1994) Characterisation of the *pcp* gene of *Pseudomonas fluorescens* and of its products, pyrrolidone carboxyl peptidase (Pcp). *J. Bacteriol.* **176**, 2569–2576.

Gonzalès, T., Awadé, A., Besson, C. & Robert-Baudouy, J. (1992) Purification and characterization of recombinant pyrrolidone carboxyl peptidase of *Bacillus subtilis*. *J. Chromatogr.* **584**, 101–107.

Harley, C.B. & Reynolds, R.P. (1987) Analysis of *E. coli* promoter sequences. *Nucleic Acids Res.* **15**, 2343–2361.

Kwiatkowska, J., Torain, B. & Glenner, G.G. (1974) A pyrrolidonecarboxylate peptidase from the particulate fraction of *Klebsiella cloacae*: purification of the stable enzyme and its use in releasing the NH$_2$ terminus from pyrrolidonecarboxylyl peptides and proteins. *J. Biol. Chem.* **249**, 7729–7736.

Kyte, J. & Doolittle, R.F. (1982) A simple method for displaying the hydropathic character of a protein. *J. Mol. Biol.* **157**, 105–132.

Lauffart, B., McDermott, J.R., Biggins, J.A., Gibson, A.M. & Mantle, D. (1988) Purification and characterization of pyroglutamyl aminopeptidase from human cerebral cortex. *Biochem. Soc. Trans.* **17**, 207–208.

Le Saux, O., Gonzalès, T. & Robert-Baudouy, J. (1996) Mutational analysis of the active site of *Pseudomonas fluorescens* pyrrolidone carboxyl peptidase. *J. Bacteriol.* **178**, 3308–3313.

Lu, H.S., Clogston, C.L., Wypych, J., Fausset, P.R., Lauren, S., Mendiaz, E.A., Zsebo, K.M. & Langley, K.E. (1991) Amino acid sequence and post translational modification of stem cell factor isolated from buffalo rat liver cell-conditioned medium. *J. Biol. Chem.* **266**, 8102–8107.

Mantle, D. (1992) Comparison of soluble aminopeptidases in human cerebral cortex, skeletal muscle and kidney tissues. *Clin. Chim. Acta* **207**, 107–118.

Mantle, D., Lauffart, B., McDermott, J.R. & Gibson, A. (1990) Characterization of aminopeptidases in human kidney soluble fraction. *Clin. Chim. Acta* **187**, 105–114.

Mantle, D., Lauffart, B. & Gibson, A. (1991) Purification and characterization of leucyl aminopeptidase and pyroglutamyl aminopeptidase from human skeletal muscle. *Clin. Chim. Acta* **197**, 35–46.

Mendez, M., Cruz, C., Joseph-Bravo, P., Wilk, S. & Charli, J.L. (1990) Evaluation of the role of prolyl endopeptidase and pyroglutamyl peptidase I in metabolism of LHRH and TRH in brain. *Neuropeptides* **17**, 55–62.

Mudge, A.W. & Fellows, R.F. (1973) Bovine pituitary pyrrolidonecarboxylyl peptidase. *Endocrinology* **93**, 1428–1434.

Mulczyk, M. & Szewczyk, A. (1970) Pyrrolidonyl peptidase in bacteria: a new colorimetric test for differentiation of enterobacteriaceae. *J. Gen. Microbiol.* **61**, 9–13.

O'Connor, B. & O'Cuinn, G. (1985) Purification and kinetic studies on a narrow specificity synaptosomal membrane pyroglutamate aminopeptidase from guinea-pig brain. *Eur. J. Biochem.* **150**, 47–52.

Patterson, K.E., Hsiao, S.H. & Keppel, A. (1963) Studies on dipeptidase and aminopeptidases. *J. Biol. Chem.* **238**, 3611–3620.

Patti, J.M., Schneider, A., Garza, N. & Boles, J.O. (1995) Isolation and characterization of *pcp*, a gene encoding a pyrrolidone carboxyl peptidase in *Staphylococcus aureus*. *Gene* **166**, 95–99.

Scharfmann, R., Morgat, J.-L. & Aratan-Spire, S. (1989) Presence of particulate thyrotropin-releasing hormone-degrading pyroglutamate amino peptidase activity in rat liver. *Neuroendocrinology* **49**, 442–448.

Sullivan, J.J. & Jago, G.R. (1970) Pyrrolidonecarboxylyl peptidase activity in *Streptococcus cremoris* ML1. *Aust. J. Dairy Technol.* **25**, 141.

Sullivan, J.J., Muchnicky, E.E., Davison, B.E. & Jago, G.R. (1977) Purification and properties of the pyrrolidonecarboxylate peptidase of *Streptococcus faecium*. *Aust. J. Biol. Sci.* **30**, 543–552.

Suvorov, A.N. & Ferretti, J.J. (1996) Physical and genetic chromosomal map of an M type 1 strain of *Streptococcus pyogenes*. *J. Bacteriol.* **178**, 5546–5549.

Szewczuk, A. & Kwiatkowska, J. (1970) Pyrrolidonyl peptidase in animal, plant and human tissues: occurrence and some properties of the enzyme. *Eur. J. Biochem.* **15**, 92–96.

Tabor, S. & Richardson, C.C. (1985) A bacteriophage T7 RNA polymerase/promoter system for controlled exclusive expression of specific genes. *Proc. Natl Acad. Sci. USA* **82**, 1074–1078.

Taylor, W.L. & Dixon, J.E. (1978) Characterization of a pyroglutamate aminopeptidase from rat serum that degrades thyrotropin-releasing hormone. *J. Biol. Chem.* **253**, 6934–6940.

Torres, H., Charli, J.L., Gonzàlez-Noriega, A., Vargas, M.A. & Joseph-Bravo, P. (1986) Subcellular distribution of the enzymes degrading thyrotropin releasing hormone and metabolites in rat brain. *Neurochem. Int.* **9**, 103–110.

Tsuru, D., Fujiwara, K. & Kado, K. (1978) Purification and characterization of L-pyrrolidone-carboxylate peptidase from *Bacillus amyloliquefaciens*. *J. Biochem.* **84**, 467–476.

Uliana, J.A. & Doolittle, R.F. (1969) Pyrrolidonecarboxylyl peptidase: studies on the specificity of the enzyme. *Arch. Biochem. Biophys.* **131**, 561–565.

Wilk, S., Friedman, T.C. & Kline, T.B. (1985) Pyroglutamyl diazomethyl ketone: potent inhibitor of mammalian pyroglutamyl peptide hydrolase. *Biochem. Biophys. Res. Commun.* **130**, 662–668.

Yoshimoto, T., Shimoda, T., Kitazono, A., Kabashima, T., Ito, K. & Tsuru, D. (1993) Pyroglutamyl peptidase gene from *B. amyloliquefaciens*: cloning, sequencing, expression, and crystallization of the expressed enzyme. *J. Biochem. (Tokyo)* **113**, 67–73.

*This chapter is dedicated to the memory of Dr A.C. Awadé, who contributed so much to the study of pyrase.*

***Janine Robert-Baudouy***
*Laboratoire de Génétique Moléculaire des Microorganismes et des Interactions Cellulaires, CNRS UMR 5577, INSA, Bât 406, 20 avenue Albert Einstein, 69621 Villeurbanne Cedex, France Email: lgmm@cismibm.univ-lyon1.fr*

***Philippe Clauziat***
*Biomerieu Company, 16 Rue de l'Espérance, 69003 Lyon, France*

***Gonzales Thierry***
*Laboratoire de Génétique Moléculaire des Microorganismes et des Interactions Cellulaires, CNRS UMR 5577, INSA, Bât 406, 20 avenue Albert Einstein, 69621 Villeurbanne Cedex, France*

# 269. *Mammalian pyroglutamyl-peptidase I*

## Databanks

*Peptidase classification: clan CX, family C15, MEROPS ID: C15.010*
*Enzyme classification: EC 3.4.19.3*
*Databank codes: no sequence data available*

## Name and History

The enzyme that releases pyroglutamate from the N-terminus of a peptide was first described from *Pseudomonas fluorescens* by Doolittle & Armentrout (1968), who found that pyroglutamic acid (Glp) could be released from Glp$+$Ala. Rat liver was the first mammalian source of the enzyme described (Armentrout, 1969). The enzyme has been referred to as **pyroglutamyl aminopeptidase, 5-oxoprolyl-peptidase, pyroglutamate aminopeptidase, pyrrolidonecarboxylate peptidase** and **pyrrolidonecarboxylyl peptidase** (Awadé *et al.*, 1994). The name recommended by IUBMB is **pyroglutamyl-peptidase I**, and we shall here use the abbreviation **PAP-I**.

## Activity and Specificity

PAP-I can be classified as an exopeptidase, or more correctly, an omega peptidase (McDonald & Barrett, 1986), which specifically removes the Glp residue from the N-terminus of Glp-peptides and Glp-proteins. Mammalian PAP-I has a broad pyroglutamyl substrate specificity, with only Glp-Pro bonds not normally hydrolyzed (Mudge & Fellows, 1973; Browne & O'Cuinn, 1983). Synthetic substrates such as Glp$+$NHMec, Glp$+$NHPhNO$_2$ and Glp$+$NHNap (all available from Sigma; see Appendix 2 for full names and addresses of suppliers) are readily hydrolyzed by PAP-I, as are dipeptides such as Glp$+$Ala and Glp$+$Val.

The optimal pH for mammalian PAP-I activity is pH 6.5–8.5. PAP-I is a sulfhydryl-dependent enzyme, displaying a strict requirement for a thiol-reducing agent such as DTT or 2-ME, one of which is generally included in the assay system and during purification. The enzyme can be inhibited by iodoacetamide and other sulfhydryl-blocking agents (Browne & O'Cuinn, 1983; Awadé *et al.*, 1994), and is also sensitive to traces of heavy metals (Wilk, 1995). Other inhibitors include the active-site directed compounds Glp-chloromethane, Z-Glp-diazomethane, 5-oxoprolinal and the reversible noncompetitive substrate analog, 2-pyrrolidone. The latter compound has been used to stabilize PAP-I during purification and storage (Armentrout, 1969; Mudge & Fellows, 1973).

## Structural Chemistry

Mammalian PAP-I, which generally displays a soluble or cytosolic location, is a monomeric enzyme with $M_r$ approximately 24 000 (Mudge & Fellows, 1973; Browne & O'Cuinn, 1983; Cummins & O'Connor, 1996), although an $M_r$ of 60 000 has been reported for the rat brain enzyme (Busby *et al.*, 1982). A pI of 5.5 has been described for the chicken liver enzyme (Tsuru *et al.*, 1982). The overexpressed *Bacillus amyloliquefaciens* enzyme has been crystallized, so a knowledge of the three-dimensional structure of bacterial PAP-I may be available before long (Yoshimoto *et al.*, 1993). This enzyme has two cysteine residues, one of which is involved in its catalytic activity.

## Preparation

PAP-I has been purified from numerous mammalian sources including guinea pig brain (Browne & O'Cuinn, 1983), cattle pituitary (Mudge & Fellows, 1973), rat brain (Busby *et al.*, 1982) and more recently, cattle brain (Cummins & O'Connor, 1996). Commercial suppliers typically provide the enzyme from calf liver or bacterial sources. PAP-I genes from several bacteria including *B. amyloliquefaciens* (Yoshimoto *et al.*, 1993) and *Staphylococcus aureus* (Patti *et al.*, 1995) have been cloned and characterized, allowing the study of the primary structure of these enzymes, and their overexpression in *Escherichia coli* (see Chapter 268). Little is known about the PAP-I genes from mammalian sources to date.

## Biological Aspects

*In vitro*, PAP-I is capable of liberating the N-terminal Glp residue from a range of biologically active peptides including thyrotropin-releasing hormone (TRH), luteinizing hormone-releasing hormone (LHRH), neurotensin and bombesin (Browne & O'Cuinn, 1983). No definitive evidence linking neuropeptide inactivation *in vivo* with PAP-I has yet been presented. PAP-I is ubiquitous, having been found in a variety of animal and human tissues, in particular the brain and digestive system (Szewczuk & Kwiatkowska, 1970). The role of the enzyme remains unclear. It has been proposed, by analogy with other soluble aminopeptidases, that PAP-I may contribute to the final stages of the intracellular catabolism of peptides to free amino acids, which are then released to the cellular pool (Awadé *et al.*, 1994). Thus, this enzyme may, at least in part, be involved in the regulation of the cellular pool of free Glp. It is noteworthy that free Glp is known to have pharmacological properties (for a full review see Awadé *et al.*, 1994). Thus a specific pathway for Glp production, e.g. through PAP-I activity, may exist to generate this molecule. The source of free Glp that is produced in certain disease states remains unknown, but the involvement of PAP-I is a possibility. It has also been suggested by Albert & Szewczuk (1972) that PAP-I may participate in the absorption of peptides and proteins from the alimentary tract.

The initial practical application of PAP-I was in protein and peptide sequencing. Nowadays, even though enzymatic and chemical methods are available to open Glp rings, and physical methods such as mass spectrometry have partially circumvented sequencing difficulties due to N-terminal Glp block, PAP-I is still used by sequencers to confirm the presence of this residue (Awadé *et al.*, 1994).

## Distinguishing Features and Related Peptidases

To date, three distinct forms of pyroglutamyl-peptidase have been identified in mammalian tissues: PAP-I, PAP-II and serum PAP. PAP-II is a membrane-bound metalloenzyme of high relative molecular mass with a narrow substrate specificity centering around TRH and closely related peptides (see Chapter 340). Serum PAP displays very similar biochemical characteristics to PAP-II (Bauer, 1995), and it may be a soluble form of the membrane enzyme, since it is known for some other enzymes that membrane-bound and serum forms can be derived from the same gene (O'Connor & O'Cuinn, 1984).

Mammalian PAP-I displays many of the biochemical characteristics of the bacterial pyroglutamyl-peptidase: sulfhydryl dependence, cytosolic location and broad substrate specificity. Moreover, the monomer molecular masses of both enzymes are virtually the same (23 000–24 000). The mammalian enzyme differs in being monomeric rather than oligomeric, however. Bacterial PAP is present in some bacterial species, but absent from others (Doolittle & Armentrout, 1968), whereas the mammalian enzyme PAP-I seems to have an almost ubiquitous tissue distribution.

## References

Albert, Z. & Szewczuk, A. (1972) Pyrrolidonyl peptidase in some avian and rodent tissues: histochemical localization and biochemical studies. *Acta Histochem.* **44**, 98–105.

Armentrout, R.W. (1969) Pyrrolidonecarboxylyl peptidase from rat liver. *Biochim. Biophys. Acta* **191**, 756–759.

Awadé, A.C., Cleuziat, P., Gonzalès, T. & Robert-Baudouy, J. (1994) Pyrrolidone carboxyl peptidase (Pcp): an enzyme that removes pyroglutamic acid (pGlu) from pGlu-peptides and pGlu-proteins. *Proteins: Struct. Funct. Genet.* **20**, 34–51.

Bauer, K. (1995) Inactivation of thyrotropin releasing hormone (TRH). The TRH-degrading enzyme as a regulator and/or terminator of TRH signals? In: *Metabolism of Brain Peptides* (O'Cuinn, G., ed.). Boca Raton, FL: CRC Press, pp. 201–213.

Browne, P. & O'Cuinn, G. (1983) An evaluation of the role of a pyroglutamyl peptidase, a post-proline cleaving enzyme and a post-proline dipeptidyl amino peptidase, each purified from the soluble fraction of guinea-pig brain, in the degradation of thyroliberin *in vitro*. *Eur. J. Biochem.* **137**, 75–87.

Busby, W.H., Youngblood, W.W. & Kizer, J.S. (1982) Studies of substrate requirements, kinetic properties and competitive inhibitors of the enzymes catabolizing TRH in rat brain. *Brain Res.* **242**, 261–270.

Cummins, P.M. & O'Connor, B. (1996) Bovine brain pyroglutamyl aminopeptidase (type-1): purification and characterisation of a neuropeptide-inactivating peptidase. *Int. J. Biochem. Cell. Biol.* **28**, 883–893.

Doolittle, D.F. & Armentrout, R.W. (1968) Pyrrolidonyl peptidase. An enzyme for selective removal of pyrrolidonecarboxylic acid residues from polypeptides. *Biochemistry* **7**, 516–521.

McDonald, J.K. & Barrett, A.J. (1986) *Mammalian Proteases: A Glossary and Bibliography*, vol 2: *Exopeptidases*. London: Academic Press.

Mudge, A.W. & Fellows, R.E. (1973) Bovine pituitary pyrrolidonecarboxylyl peptidase. *Endocrinology* **93**, 1428–1434.

O'Connor, B. & O'Cuinn, G. (1984) Localization of a narrow specificity thyroliberin-hydrolysing pyroglutamate aminopeptidase in synaptosomal membranes of guinea-pig brain. *Eur. J. Biochem.* **144**, 271–278.

Patti, J.M., Schneider, A., Garza, N. & Boles, J.O. (1995) Isolation and characterization of *pcp*, a gene encoding a pyrrolinone carboxyl peptidase in *Staphylococcus aureus*. *Gene* **166**, 95–99.

Szewczuk, A. & Kwiatkowska, J. (1970) Pyrrolidonyl peptidase in animal, plant and human tissues: occurrence and some properties of the enzyme. *Eur. J. Biochem.* **15**, 92–96.

Tsuru, D., Sakabe, K., Yoshimoto, T. & Fujiwara, K. (1982) Pyroglutamyl peptidase from chicken liver: purification and some properties. *J. Pharmacobiodyn.* **5**, 859–868.

Wilk, S. (1995) Intracellular peptide turnover: properties and physiological significance of the major peptide hydrolases of brain cytosol. In: *Metabolism of Brain Peptides* (O'Cuinn, G., ed.). Boca Raton, FL: CRC Press, pp. 215–250.

Yoshimoto, T., Shimoda, T., Kitazono, A., Kabashima, T., Ito, K. & Tsuru, D. (1993) Pyroglutamyl peptidase gene from *Bacillus amyloliquefaciens*: cloning, sequencing, expression, and crystallization of the expressed enzyme. *J. Biochem.* **113**, 67–73.

*Ultan G. McKeon*
*Biological Sciences,*
*Dublin City University,*
*Dublin 9, Ireland*

*Brendan O'Connor*
*Biological Sciences,*
*Dublin City University,*
*Dublin 9, Ireland*
*Email: oconnorb@ccmail.dcu.ie*

# ASPARTIC

# 270. Introduction: aspartic peptidases and their clans

*MEROPS ID: AA, AB*

Aspartic and metallopeptidases differ importantly from serine and cysteine peptidases in that the nucleophile that attacks the scissile peptide bond is an activated water molecule rather than the nucleophilic side chain of an amino acid. Residues that are involved in catalysis include amino acids that act as ligands, either directly for the activated water molecule as in aspartic peptidases, or by binding one or two metal ions that in turn bind the activated water molecule in the metallopeptidases.

Aspartic peptidases are so named because Asp residues are ligands of the activated water molecule in all examples for which the catalytic residues have been identified. The primary example is pepsin (Chapter 272), in which the side chains of two Asp residues hold the water in place. However, an endopeptidase from nodaviruses (Chapter 324) is believed to have an Asp and an Asn as its catalytic dyad. In the related tetravirus endopeptidase, the Asp residue is replaced by Glu, and it has been suggested that in scytalidopepsin B (Chapter 326) a Glu replaces Asp, although this is further replaced by Gln in other members of the family.

One distinguishing characteristic of aspartic peptidases is that all the enzymes so far described are endopeptidases, although there is no obvious reason why this type of catalytic mechanism could not be used in an exopeptidase.

Tertiary structures are known for only four families of aspartic peptidases. In family A1, the endopeptidases are bilobed, with the active site between the lobes. This is a feature common to many peptidases, but unusually in the aspartic peptidases of family A1, one lobe has been derived from the other by a gene duplication event. Each lobe bears one Asp residue of the catalytic dyad, and the two residues are homologous. In the modern-day enzymes, only very limited similarities in the amino acid sequences of the two lobes remain, but the three-dimensional structures are very similar. The discovery of this situation raised the question: how could the product of the original gene have functioned as a peptidase if it consisted of only one lobe, with only one Asp? One possible answer was supplied when the crystal structure of a retropepsin (see Figure 309.1) from family A2 was solved, showing a single lobe carrying one catalytic Asp, very similar in structure to one lobe of pepsin (see Figure 271.2). The retropepsin is active only as a homodimer, and the catalytic site forms between the molecules. The similar protein folds of the endopeptidases in families A1 and A2 fully support the view that they have evolved from a common ancestor, and they are grouped together in *clan AA* (Chapters 271 and 309).

There are a number of other families also included in clan AA on the grounds that the enzymes possess similar amino acid sequence motifs around the catalytic Asp residues. Most of these families comprise monomeric enzymes from RNA viruses or retrotransposons, and the enzymes are assumed to require dimerization for activity, like the retropepsins.

Tertiary structures have also been solved for families A6 and A21, those of the endopeptidases that are the coat proteins of nodaviruses and tetraviruses, respectively. The only peptide bond cleavage known to be performed by either enzyme is an autolytic cleavage near the C-terminus of the coat protein itself, to generate a two-chain molecule. This cleavage occurs during the late stages of virion assembly. The tertiary structures (see Figure 323.1) are different to that known for any other protein, and the family is assigned to *clan AB* (Chapter 323).

There are a number of other families of aspartic endopeptidases that cannot yet be assigned to any clan (Chapter 325).

# 271. Introduction: family A1 of pepsin (clan AA)

## Databanks

*MEROPS ID: A1*

| Species | SwissProt | PIR | EMBL (cDNA) | EMBL (genomic) |
|---|---|---|---|---|
| **Family A1** | | | | |
| Aspergillopepsin I (Chapter 294) | | | | |
| Barrierpepsin (Chapter 303) | | | | |
| *Candida parapsilosis* aspartic endopeptidase (Chapter 301) | | | | |
| Candidapepsin (Chapter 300) | | | | |
| Canditropsin (Chapter 302) | | | | |
| Cathepsin D (Chapter 277) | | | | |
| Cathepsin E (Chapter 275) | | | | |
| Chymosin (Chapter 274) | | | | |
| Cyprosin (Chapter 279) | | | | |
| Embryonic pepsinogen (chicken) | | | | |
|    *Gallus gallus* | P16476 | A4144 | D00215 | M19660: 5′ UTR |
| Endothiapepsin (Chapter 296) | | | | |
| Gastricsin (Chapter 276) | | | | |
| Mucorpepsin (Chapter 298) | | | | |
| Neurosporapepsin (Chapter 289) | | | | |
| Penicillopepsin (Chapter 295) | | | | |
| Pepsin A (Chapter 272) | | | | |
| Pepsin B (Chapter 273) | | | | |
| Peptidase E | | | | |
|    *Aspergillus fumigatus* | P41748 | – | L31490 | – |
|    *Aspergillus niger* | – | – | – | U03278: complete gene |
| Phytepsin (Chapter 278) | | | | |
| Plasmepsin I (Chapter 286) | | | | |
| Plasmepsin II (Chapter 287) | | | | |
| Polyporopepsin (Chapter 299) | | | | |
| Pregnancy-associated glycoprotein 1 (Chapter 307) | | | | |
| Renin (Chapter 284) | | | | |
| Rhizopuspepsin (Chapter 297) | | | | |
| Saccharopepsin (Chapter 283) | | | | |
| Submandibular renin (mouse) (Chapter 285) | | | | |
| Yapsin 1 (Chapter 304) | | | | |
| Yapsin 2 (Chapter 305) | | | | |
| Others | | | | |
|    *Bos taurus* | – | JT0399 | – | – |
|    *Bos taurus* | – | S03266 | – | – |
|    *Glomerella cingulata* | – | – | – | U43775: complete gene |
|    *Saccharomyces cerevisiae* | P40583 | – | – | Z47047: chromosome IX complete sequence |
|    *Saccharomycopsis fibuligera* | P22929 | JT0334 | D00313 | – |

All the peptidases in *family A1* are endopeptidases from eukaryotes. The great majority are most active at acidic pH, but a few show activity under neutral conditions. Examples of the enzymes from mammals include the digestive enzymes, pepsin and chymosin, as well as the intracellular cathepsins D and E. Renin, which originates from the kidney and processes angiotensinogen in the plasma, is one of the minority of endopeptidases in the family that act at neutral pH. Pepsin homologs are known from plants, for example phytepsin, from fungi, for example penicillopepsin, and from protozoa, for example plasmepsin 1.

The tertiary structures of several members of family A1 have been determined. Each shows a bilobed molecule with the active site between the lobes. The lobes are similar in structure, and are believed to have arisen by duplication of an ancestral gene, a mechanism that is thought to be responsible

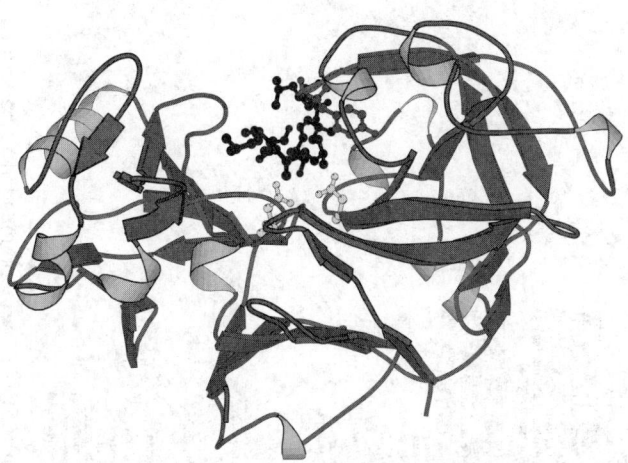

*Figure 271.1* Richardson diagram of the human pepsin A/pepstatin complex. The image was prepared from the Brookhaven Protein Data Bank entry (1PSO) as described in the Introduction (p. xxv). Catalytic residues are shown in ball-and-stick representation: Asp32, Asp215 and Tyr75 (numbering as in Alignment 271.1). Pepstatin is shown in black in ball-and-stick representation.

also for the bilobed structures of members of clans SA and MG. A representation of the molecule of human pepsin A is shown in Figure 271.1.

Each of the two lobes of pepsin and other members of the family contains one of the pair of catalytic Asp residues. They occur in very similar sequence motifs, reflecting the fact that the two Asp residues have originated from a single residue in the ancestral, monomeric protein. The core motif is commonly summarized as Asp-Thr-Gly (the 'DTG' motif), but a fuller description of the environment of the catalytic Asp is: Xaa-Xaa-**Asp**-Xbb-Gly-Xbb, in which Xaa is a hydrophobic residue and Xbb is either Ser or Thr. As can be seen in Alignment 271.1 (see CD-ROM), there are a few exceptions. Human renin, which is active at neutral pH, has Ala218, and this also occurs in almost all of the retropepsins from family A2. There is evidence that this residue affects the pH optimum of the enzymes, because lack of a hydrogen bond to the usual Thr218 alters the acidity of Asp215 (Ido *et al.*, 1991). The mammalian pregnancy-associated glycoprotein is an inactive member of the family in which Asp215 is replaced by Gly. A similar DTG motif occurs within the subtilisin family, which can be described as Xaa-Xaa-**Asp**-Xbb-Gly-Xaa (in which the Asp is a member of the catalytic triad), the principal difference being in the final residue of the motif. Care must therefore be exercised in using the DTG motif to assign a peptidase sequence to the pepsin family!

There is a third residue besides the two aspartates that is important for catalysis in family A1. Although the two lobes are structurally similar (Figure 271.1), it will be noted that the lobe shown on the right-hand side has an extra β hairpin loop capping the active site. This loop has been termed the 'flap', and carries residues important for specificity, notably Tyr75 and Thr77, which can interact with the substrate and are part of the S1 subsite. Many fungal enzymes, including penicillopepsin, are able to activate trypsinogen by cleavage

of a Lys╂Ile bond. As can be seen from the alignment (Alignment 271.1), in penicillopepsin and other fungal endopeptidases, Thr77 is replaced by Asp, which has been shown to be the anionic binding site (James *et al.*, 1985).

Most enzymes of family A1 are synthesized with signal and propeptides, and enter the secretory pathway. Human pepsin A has a 15 residue signal peptide and a 47 residue

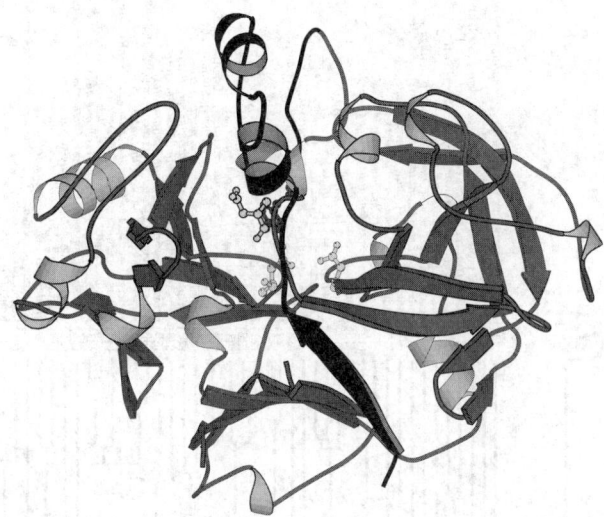

*Figure 271.2* Richardson diagram of pig pepsinogen A. The image was prepared from the Brookhaven Protein Data Bank entry (3PSG) as described in the Introduction (p. xxv). The propeptide is shown in gray. Residues shown in ball-and-stick representation are: the catalytic residues Asp32 and Asp215 (numbering as in Alignment 271.1); Asp11, which interacts with the propeptide; and Tyr37P (numbering the propeptide 1P to 44P), which blocks the substrate-binding site.

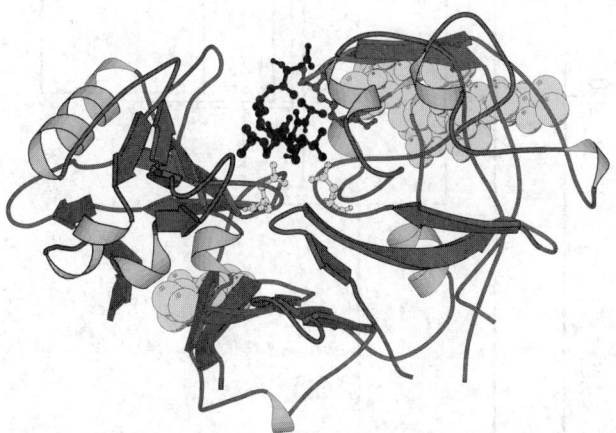

*Figure 271.3* Richardson diagram of the human cathepsin D/pepstatin complex. The image was prepared from the Brookhaven Protein Data Bank entry (1LYB) as described in the Introduction (p. xxv). Catalytic residues are shown in ball-and-stick representation: Asp32, Tyr75 and Asp215 (numbering as in Alignment 271.1). Carbohydrates are shown as CPK spheres; the carbohydrate containing mannose is at the top right of the molecule. Pepstatin is shown in black in ball-and-stick representation.

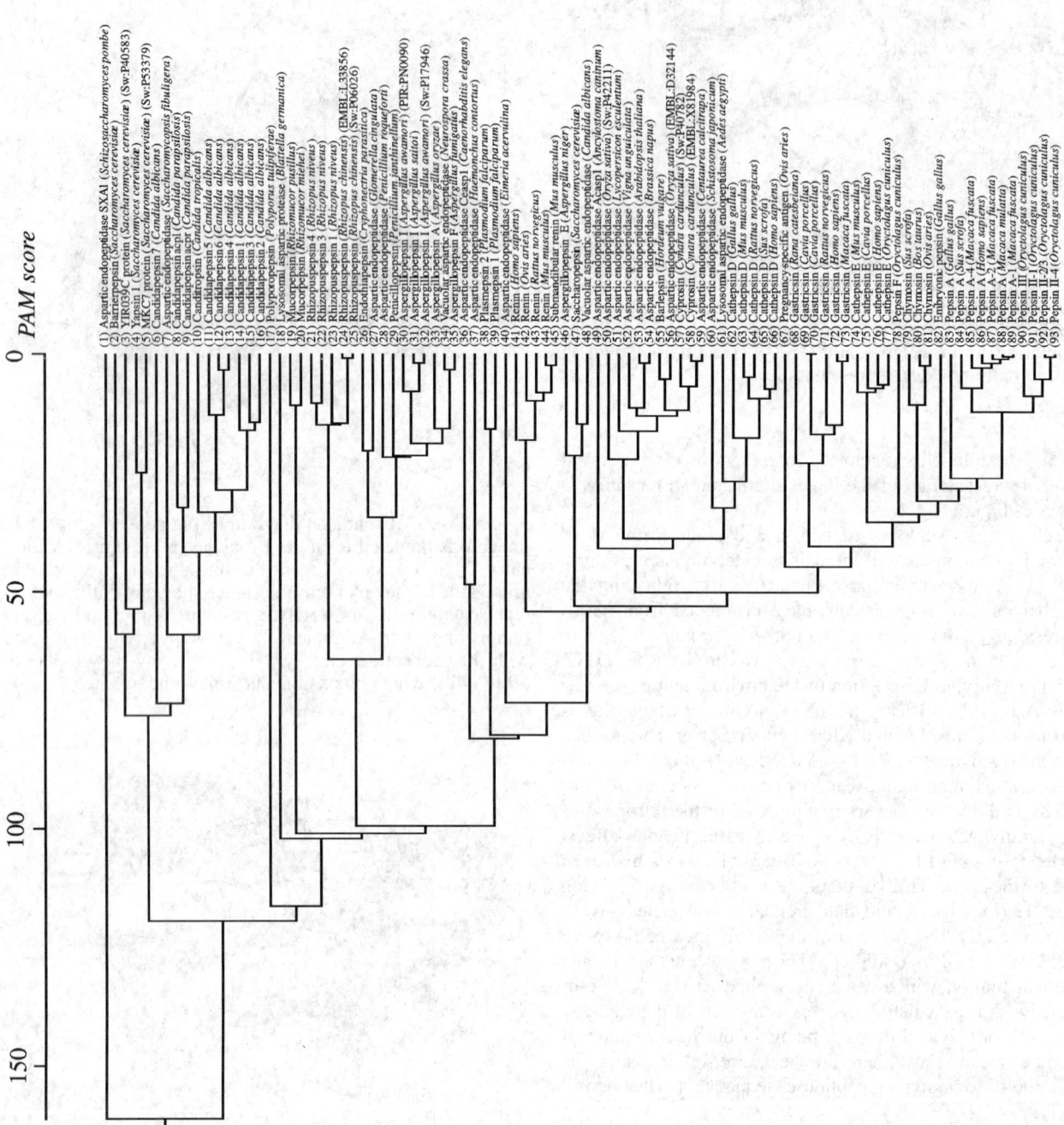

*Figure 271.4* Evolutionary tree for family A1. The tree was constructed as described in the Introduction (p. xxv). Barlepsin (55) is now recommended to be termed phytepsin (Chapter 278).

propeptide. The other animal members of the family have propeptides homologous to that of pepsinogen A, but the propeptides of plant and fungal members, though similar in size, show little relationship to each other or to the pepsinogen A propeptide. In pepsinogen A, the first 11 residues of the mature pepsin sequence are displaced by residues of the propeptide, in a six-stranded β sheet. The propeptide contains two helices that block the active-site cleft, and the conserved Asp11 hydrogen bonds to a conserved Arg in the propeptide. This stabilizes the propeptide conformation and is probably responsible for the triggering of the conversion of pepsinogen to pepsin by acidic conditions. Two tyrosine residues block the S1 and S1′ binding sites. Figure 271.2 shows the structure of pepsinogen.

Cathepsin D is a lysosomal aspartic endopeptidase that commonly matures to a two-chain form. It is also a glycoprotein, and the carbohydrate is involved in the lysosomal targeting. The tertiary structure has been solved, and a diagrammatic representation is shown in Figure 271.3.

Pepsin homologs from fungi are diverse in structure. In the phylogenetic tree (Figure 271.4), most of the fungal enzymes are derived from the most divergent branches, with the exception of saccharolysin, candidapepsin 1, neurosporapepsin and aspergillopepsin F, which form a branch between the plant and animal homologs. The divergent fungal enzymes are almost as different to one another as they are to the animal and plant endopeptidases, implying either that the divergences between the fungal enzymes predate the divergence of the fungi, or that the mutation rate among these fungal genes is much higher than for others in the family. These differences are most apparent around the active-site cleft, as can be seen from the structure of mucorpepsin shown in Figure 271.5.

There is a striking conservation of disulfide bonds among the mammalian members of family A1. The mammalian enzymes have three bonds, the first and third of which are also present in many of the fungal homologs. The disulfide bonds are different in the different lobes, implying that the bonds were introduced after the gene duplication that gave rise to the two domains. Cathepsin D has an additional disulfide bond. Cathepsin E is unusual in that the proenzyme is a disulfide-linked dimer, and the active enzyme can be either a dimer or a monomer.

The plant aspartic endopeptidase phytepsin remains the only mosaic member of family A1 so far described, containing

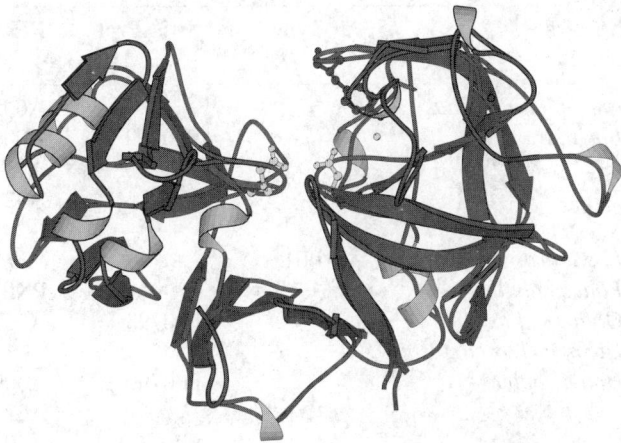

*Figure 271.5*  Richardson diagram of *Rhizomucor pusillus* mucorpepsin. The image was prepared from the Brookhaven Protein Data Bank entry (1MPP) as described in the Introduction (p. xxv). Catalytic residues are shown in ball-and-stick representation: Asp32, Tyr75 and Asp215 (numbering as in Alignment 271.1).

a 104 residue insert in the C-terminal lobe. This is homologous to part of the mammalian saposin precursor, which contains four saposin peptides. Once proteolytically excised, saposins are lysosomal proteins that act as activators for β-galactosylceramidase and β-galactosidase. The phytepsin insert does not correspond to a single saposin domain, but rather overlaps the C-terminal third of one domain and the N-terminal two-thirds of the next. Not only Cys residues but also potential *N*-linked glycosylation sites are conserved in this novel part of the sequence, however.

### References

Ido, E., Han, H., Kezdy, F.J. & Tang, J. (1991) Kinetic studies of human immunodeficiency virus type 1 protease and its active-site hydrogen bond mutant A28S. *J. Biol. Chem.* **266**, 24359–24366.

James, M.N.G., Sielecki, A.R. & Hofmann, T. (1985) X-ray diffraction studies on penicillopepsin and its complexes: the hydrolytic mechanism. *In: Aspartic Proteinases and their Inhibitors* (Kostka, V., ed.). Berlin: Walter de Gruyter, pp. 165–177.

# 272. *Pepsin A*

### *Databanks*

*Peptidase classification: clan AA, family A1, MEROPS ID: A01.001*
*NC-IUBMB enzyme classification: EC 3.4.23.1*
*Chemical Abstracts Service registry number: 9001-75-6*

| Species | Type | SwissProt | PIR | EMBL (cDNA) | EMBL (genomic) |
|---------|------|-----------|-----|-------------|----------------|
| *Anas platyrhynchos* | – | | A60516 | – | – |
| *Bos taurus* | | P00792 | A00983 | – | M19698: exons 6–8 |
| | | | A91424 | | M23163: exon 6 |
| | | | A92157 | | M23165: exon 7 |
| | | | | | M23167: exon 8 |
| *Bos taurus* | PIII | – | JT0398 | – | |
| *Equus caballus* | | – | PN0135 | – | – |
| *Gallus gallus* | | P00793 | A00984 | – | – |
| *Haemonchus contortus* | | – | – | Z72490 | – |
| *Homo sapiens* | | P00790 | A00980 | – | J00279: exon 1 |
| | | | A22434 | | J00280: exon 2 |
| | | | A30142 | | J00281: exon 3 |
| | | | PX0023 | | J00282: exon 4 |
| | | | PX0024 | | J00283: exon 5 |
| | | | PX0025 | | J00284: exon 6 |
| | | | PX0026 | | J00285: exon 7 |
| | | | PX0027 | | J00286: exon 8 |
| | | | S02542 | | J00287: exon 9 and complete CDS |
| | | | S02663 | | M26025: exon 1 |
| | | | S02664 | | M26026: exon 2 |
| | | | | | M26027: exon 3 |
| | | | | | M26028: exon 4 |
| | | | | | M26029: exon 5 |
| | | | | | M26030: exon 6 |
| | | | | | M26031: exons 7 and 8 |
| | | | | | M26032: exon 9 and complete CDS |
| | | | | | M27594: exon 1 |
| | | | | | Z14129: 5′ end and exon 1 |
| *Macaca fuscata* | A-1 | P03954 | A00981 | X59753 | X59752: complete gene |
| | | | A91960 | | |
| | | | A92579 | | |
| | | | S19681 | | |
| *Macaca fuscata* | A-2 | P27677 | S16064 | X59755 | – |
| | | | S19684 | | |
| *Macaca fuscata* | A-4 | P27678 | S16065 | X59735 | – |
| | | | S19682 | | |
| *Macaca mulatta* | | P11489 | JT0309 | M20788 | – |
| *Oryctolagus cuniculus* | II-1 | P28712 | B38302 | – | – |
| *Oryctolagus cuniculus* | II-2/3 | P27821 | C38302 | M59235 | – |
| *Oryctolagus cuniculus* | II-4 | P28713 | D38302 | – | – |
| *Oryctolagus cuniculus* | III | P27822 | E38302 | M59237 | – |
| | | | JT0397 | | |
| *Oryctolagus cuniculus* | F | P27823 | A38302 | M59238 | – |
| *Sus scrofa* | | P00791 | A00982 | J04601 | Z14130: 5′ end and exon 1 |
| | | | A32455 | M20920 | |
| | | | A91410 | | |
| | | | A92039 | | |
| | | | A92179 | | |
| | | | B22434 | | |
| | | | JT0307 | | |
| *Thunnus thynnus* | 1 | P20139 | S01798 | – | – |
| *Thunnus thynnus* | 2 | P20140 | S01799 | – | – |
| *Thunnus thynnus* | 3 | P20141 | S01800 | – | – |
| *Ursus americanus* | | P13636 | A28859 | – | – |

Brookhaven Protein Data Bank three-dimensional structures:

| Species | ID | Resolution | Notes |
|---|---|---|---|
| *Homo sapiens* | 1PSN | 2.2 | pepsin 3A |
| | 1PSO | 2 | pepsin 3A; complex with pepstatin |
| *Sus scrofa* | 1PSA | 2.9 | complex with A62905 |
| | 2PSG | 1.8 | precursor |
| | 3PEP | 2.3 | |
| | 3PSG | 1.65 | precursor |
| | 4PEP | 1.8 | |
| | 5PEP | 2.34 | |

## Name and History

Pepsin is the principal acid protease of the stomach. It is generally recognized as the first enzyme to be discovered (in the eighteenth century) and was named by T. Schwann in 1825. Pig pepsin was the second enzyme, after urease, to be crystallized (Northrop, 1930). The crystallization of these enzymes established for the first time the protein nature of the enzymes. To distinguish them from other pepsin-like minor gastric proteolytic enzymes, the major pig and human gastric proteases were named ***pepsin A***. Pepsin B and C, isolated by Ryle & Porter (1959) from pig stomach, were first named parapepsin I and II. Pig pepsin B (Chapter 273), a minor gastric protease, is a different gene product from pepsin A (Nielsen & Foltmann, 1995). Pepsin C is gastricsin (Chapter 276) also isolated from human gastric juice (Richmond *et al.*, 1958). In some clinical literature, human pepsin A was referred to as ***pepsin I*** (Samloff, 1971a).

## Activity and Specificity

Pepsin is an endopeptidase with a broad specificity. The pH range of peptide hydrolysis is from below 1 to about 6 with an optimum $k_{cat}/K_m$ near pH 3.5 and active-site $pK_a$ values of 1.57 and 5.02 (Lin *et al.*, 1992a). The pH optima for some protein substrates, such as cattle hemoglobin, are near pH 2 and are attributable to acid denaturation and solubility of substrates.

A pepsin assay is most commonly carried out at pH near 2 using cattle hemoglobin as substrate. After digestion, the resulting peptide fragments soluble in trichloroacetic acid are quantified, most conveniently by $A_{280}$ (Anson, 1938). Hemoglobin substrate can be radiolabeled (Lin *et al.*, 1989) to assay pepsin at the femtomolar level. For clinical samples, kits for radioimmunoassay of serum pepsinogen A are commercially available. For kinetic measurements, synthetic peptide substrates have been developed (Fruton, 1976; Dunn *et al.*, 1995). Pepsin is inhibited by pepstatin (Umezawa *et al.*, 1970), a transition-state analog inhibitor (Marciniszyn *et al.*, 1976a). DAN (Rajagopolan *et al.*, 1966) and 1,2-epoxy-3-(*p*-nitrophenoxy)propane (EPNP; Tang, 1971) irreversibly inactivate pepsin by covalent esterification. These three inhibitors, which are commercially available, are specific for aspartic proteinases. A protein inhibitor from *Ascaris lumbricoides* has been reported (Martzen *et al.*, 1991).

Pepsin hydrolyzes N- and C-terminally blocked synthetic dipeptides like Glu┼Tyr or Phe┼Phe but with poor efficiency. For example, the $k_{cat}/K_m$ of Z-Phe┼Phe-OP4P (OP4P:

4-pyridinium alkoxy) is only $0.7\,s^{-1}\,mM^{-1}$. Placing additional Gly or Ala residues on either side of the Phe-Phe sequence significantly enhances $k_{cat}$ values with little change of $K_m$ values (Fruton, 1976). For example, the $k_{cat}$ and $K_m$ of Z-Gly-Gly-Phe┼Phe-OP4P are $56.5\,s^{-1}$ and 0.8 mM respectively at pH 2. The corresponding values for the above dipeptide are $0.49\,s^{-1}$ and 0.7 mM (Sachdev & Fruton, 1969). Although the preference of pepsin for long substrates can be expected since it recognizes eight substrate residues, the reasons for this dramatic improvement of catalytic efficiency from substrate elongation have only been speculated upon (Pearl, 1985; Tang & Koelsch, 1995). Some longer chromogenic substrates have good catalytic efficiency and are suitable for kinetic and mechanistic studies. For example, Lys-Pro-Ala-Glu-Phe┼PNP-Arg-Leu (PNP: *p*-nitrophenylalanine) has a $k_{cat}/K_m$ value of $2770\,s^{-1}\,mM^{-1}$ for pig pepsin (Dunn *et al.*, 1986).

As a general purpose protease, the broad specificity of pepsin is physiologically appropriate. The specificity determinants are the steric structures of the eight substrate side-chain binding sites in the active-site cleft of the enzyme

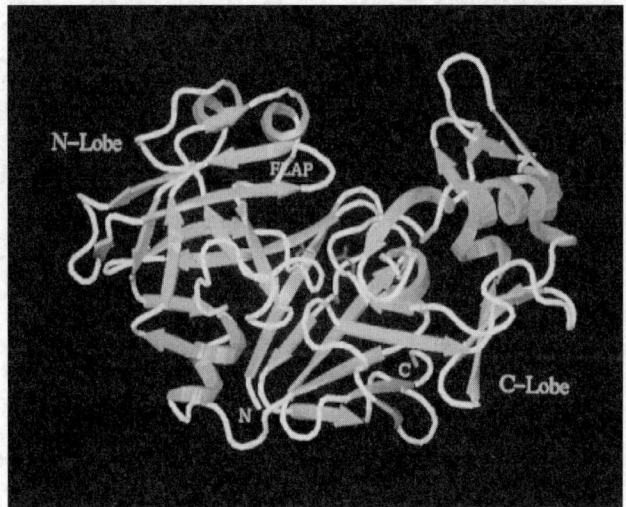

*Figure 272.1* Three-dimensional structure of pig pepsin. The view is looking into the active-site cleft from an angle. The N-terminal lobe, the C-terminal lobe, the flap, and the N- and C-termini are marked. The substrate cleft is visible between the lobes with the active-site aspartyl groups in ball-and-stick. Random coils are shown as narrow lines.

(Figure 272.1). The major specificity recognition sites lie at P1 and P1′ subsites, which prefer large hydrophobic residues, such as Phe and Leu (Tang, 1963; Fruton, 1976). The specificity requirement of these two central subsites usually takes precedence over the outside sites. Based on the crystal structure of pepsin–inhibitor complexes, the P1 interacting residues in the S1′ pocket are Ile30, Asp32, Gly34, Tyr75, Gly76 and Thr77. P1′ interacts with Tyr189, Ile213, Asp215, Gly217, Thr218 and Ile300, which form the S1′ pocket (Abad-Zapatero *et al.*, 1991). The specificity of subsites outside of P1 and P1′ has been partially explored using synthetic peptides (Fruton, 1976; Dunn *et al.*, 1986; Rao & Dunn, 1995). These subsites are in general quite nonspecific and can accommodate different types of residues. For example, the S2 subsite can accommodate Leu, Ala, norleucine, Glu, Ser, Asp, Arg and Ile in a peptide substrate with variation of $k_{cat}/K_m$ values less than 2-fold (Rao & Dunn, 1995). Pepsin residues involved in the binding of the outside subsite side-chains have also been described (Abad-Zapatero *et al.*, 1991).

Since the three-dimensional structures of the catalytic aspartic residues are essentially identical for all aspartic proteases in clan AA, it is generally assumed that this group of enzymes shares a common catalytic mechanism. The mechanism proposed by Davies (1990), which is most representative, relies heavily on the crystal structures of complexes between aspartic proteases and their transition-state inhibitors. In this mechanism, the oxygen atom in the water molecule hydrogen bonded between Asp32 and Asp215 is thought to be engaged in a nucleophilic attack on the carbonyl carbon of the substrate peptide bond (Figure 272.2). Proton donation from the hydrogen atom on Asp215 to the nitrogen of the peptide bond results in the hydrolysis. No covalent intermediate would be expected in this mechanism; indeed, none has been found. The transpeptidation reaction catalyzed by pepsin (for review, see Fruton, 1971) is apparently a consequence of different release rates of amino and carboxyl products from the enzyme rather than mediation by covalent intermediates. In spite of detailed crystallographic information, which establishes the location of specificity subsites and the steric relationship of catalytic Asp residues with transition-state analogs, the catalytic mechanism is poorly understood in many ways. Among these is the function of the 'flap'. Because the opening in the crystal structure of pepsin does not allow sufficient access for substrates to enter freely, it is generally assumed that the conformational change involving the flap must occur during the catalytic cycle. However, the evidence for the opening of the flap has not been obtained and the contribution of the flap to hydrolytic mechanism has only been speculated (Tang & Koelsch, 1995).

## Structural Chemistry

The amino acid sequence of pig pepsin, determined by chemical methods (Tang *et al.*, 1973; Moravek & Kostka, 1974) and by cDNA structure (Tsukagoshi *et al.*, 1988; Lin *et al.*, 1989), contains a single chain of 326 residues (34.6 kDa). Pig pepsin has 42 acidic residues and only 4 basic residues. The active-site residues identified by chemical modifications are Asp32, which was esterified by EPNP (Chen & Tang, 1972), and Asp215, which was esterified by DAN (Lundblad & Stein, 1969).

Pig pepsin is *O*-phosphorylated at Ser68 (Tang *et al.*, 1973). The presence of Glu at position 70 suggests that the phosphorylation is mediated by a cyclic AMP-independent kinase. The function of this phosphoserine is not known. Pepsins from some vertebrates, including avian species (Kostka *et al.*, 1985) and turtle (Takahashi *et al.*, 1995) are *N*-glycosylated at multiple sites.

The crystallographic structure of pig pepsin (Abad-Zapatero *et al.*, 1990; Cooper *et al.*, 1990; Sielecki *et al.*, 1990) shows that the main secondary structure consists of β strands with only four short α helices (Figure 272.1). Important structural features of pepsin are shared by other aspartic proteases, such as fungal enzymes (Chapters 295 and 297), renin (Chapter 284) and cathepsin D (Chapter 277). Prominent in the pepsin structure is a substrate-binding cleft, which accommodates eight amino acid residues of the substrates and contains the catalytic Asp32 and Asp215. The carboxyl groups from these two Asp residues are hydrogen-bonded to each other and to a shared water (Figure 272.2), which is the putative nucleophile in the hydrolytic mechanism discussed above. The binding locations of substrate side chains are known (Abad-Zapatero *et al.*, 1990; Sielecki *et al.*, 1990). The substrate cleft is covered by a 'flap' which consists

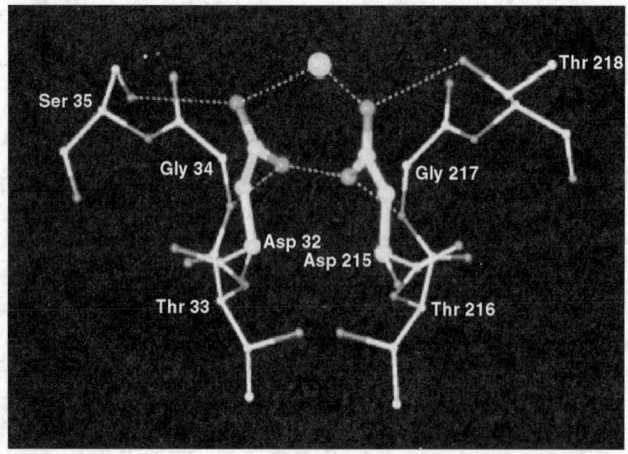

*Figure 272.2* (A) The conformation of catalytic aspartic residues of pig pepsin. The active-site Asp32, Asp215 (thick lines) and their surrounding residues (thin lines) are shown. Their oxygen atoms appear as larger balls with dotted-line putative hydrogen bonds. The water molecule which is thought to be engaged in nucleophilic attack during the hydrolytic mechanism (see B) is shown as the largest ball. The symmetry in the conformation of these two active-site peptides can be clearly seen. (B) A proposed catalytic mechanism of pepsin and other aspartic proteases. The substrate (upper molecule) shows a Phe-Phe dipeptide at P1 and P1′ subsites with the nitrogen (lowest atom) and oxygen (with H attached) of the peptide bond. The oxygens of the pepsin active-site aspartyls (lower molecule) and the jointly held water are shown as dark atoms. The hydrogen bonds are shown in dotted lines. (a) One of the active-site aspartyls is protonated, while the other is ionized. The water oxygen is engaged in nucleophilic attack to the carbonyl carbon of the peptide bond. (b) Intermediate with a tetrahedral carbon at the substrate carbonyl position. (c) Hydrogen donation from Asp215 to substrate nitrogen resulted in (d) the breaking of the peptide bond. This mechanism is based on Davies (1990).

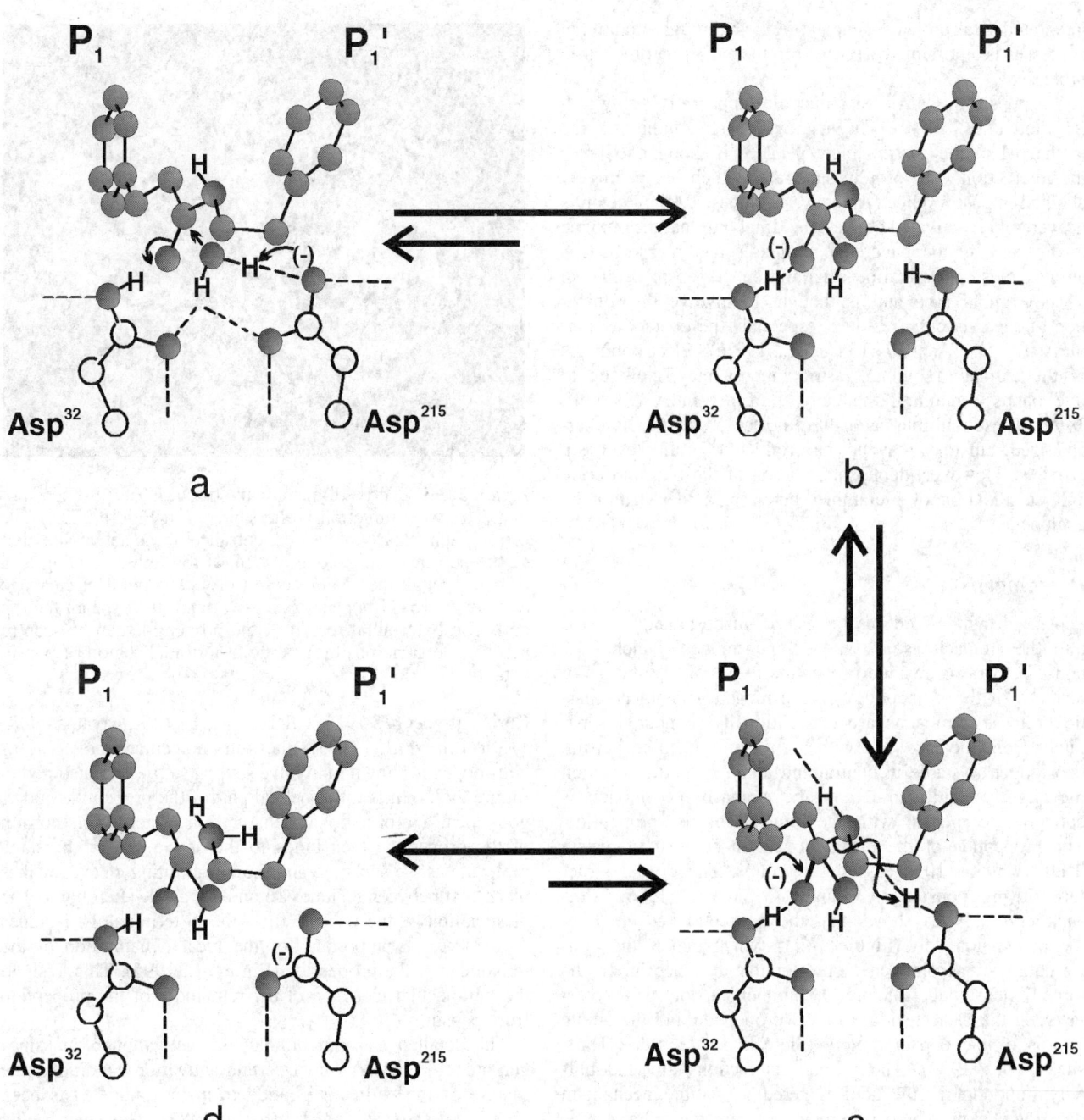

*Figure 272.2  (continued).*

of a 20 residue β-hairpin structure approximately between residues 66 and 85.

The pepsin structure contains two distinct lobes which are homologous in polypeptide chain folding (Tang *et al.*, 1978). This structure is thought to come from gene duplication and fusion during the evolution of aspartic proteases (Tang *et al.*, 1978). This evolutionary mechanism predicts the primordial enzyme of pepsin to be a homodimer, which is in fact the form of the retroviral aspartic proteases. The evolutionary relationship of pepsin with retroviral aspartic proteases is further supported by (a) the similarity in chain foldings of retroviral protease monomers to individual lobes

(Rao *et al.*, 1991), (b) the ability of recombinant lobes of pepsin to fold independently (Lin *et al.*, 1992a), and (c) the ability of the refolded N-terminal lobe of pepsin to generate proteolytic activity, apparently from its homodimer (Lin *et al.*, 1992b). There is some structural evidence that each lobe contains an internal 2-fold structural similarity suggesting that a single lobe may have been evolved from yet another preceding gene duplication (Andreeva & Gustchina, 1979). Yeast transposon *Ty* contains a homodimeric aspartic protease (Chapter 321) and is indeed retrovirus-like (Toh *et al.*, 1985). This observation suggest that primitive eukaryotes can serve as hosts for retroviruses. Thus, the retention of the integrated viral protease gene by the host through mutational processes is a possible early event in the evolution of aspartic proteases.

Pig pepsin has an isoelectric point apparently below 1.0 (Herriott *et al.*, 1940). The negative charge of the enzyme in the pH of the gastric juice (pH 2–3) is thought to favor the interaction with protein substrates which are positively charged in gastric juice (Andreeva & James, 1991). Pepsin is irreversibly inactivated above pH 7, implying that the enzyme is digested by the pancreatic proteases and reabsorbed as amino acids. The alkaline denaturation of pepsin occurs in its N-terminal lobe and is thought to involve the ionization of several acidic residues, especially the interacting pair of Asp11 and Asp159 (Lin *et al.*, 1993a). Electrophoresis of the extract of human gastric mucosa may show up to five bands containing the activity of pepsin A (Samloff, 1969). These multiple bands have not been structurally characterized, but they may be the products of multiple pepsin A genes. However, dephosphorylation of the phosphoserine and deamidation of glutamines and asparagines cannot be excluded.

## Pepsinogen

Pepsin is synthesized in the gastric mucosa and secreted into the stomach as a zymogen, pepsinogen, which contains an extra 44 residue propeptide at the N-terminus. The residues of the propeptides of pepsinogen and other aspartic protease zymogens are conventionally numbered separately from the enzymes with a 'p' suffix. Unlike pepsin, pepsinogen is stable in neutral and alkaline solutions. Upon meeting the acidic milieu of the stomach, pepsinogen is converted to pepsin with the removal of the propeptide. The propeptide of the pig zymogen, in contrast to pepsin itself, contains 10 excess basic groups. The crystal structure of pig pepsinogen (James & Sielecki, 1986; Hartsuck *et al.*, 1992) shows that the propeptide covers over the active-site cleft (Figure 272.3) which makes the conformation of pepsinogen more globular than pepsin. In the three-dimensional structure, a number of ion pairs exist between the basic residues of the propeptide and the acidic groups of the pepsin moiety (Sielecki *et al.*, 1991; Hartsuck *et al.*, 1992). These ionic interactions are undoubtedly important in the acid-triggered activation mechanism described below. The pepsin moiety in pepsinogen has essentially the same conformation as pepsin except the N-terminal region of pepsin. The N-terminal six residues of pepsin are involved in a six-strand $\beta$ structure. The positions of these six residues in pepsinogen are occupied by the N-terminal six residues of the propeptide. This region of pepsin apparently changes its conformation after activation cleavage of the propeptide.

Purified pig pepsinogen spontaneously converts to pepsin at pH values below 5 (Herriott, 1938). The conversion is predominantly an intramolecular mechanism (Bustin & Conway-Jacobs, 1971; Al-Janabi *et al.*, 1972; McPhie, 1972) under the conditions (pH 1–2) expected for the stomach content. However, in the pH range of 3–5 and high pepsinogen concentration, the activation is predominantly by a bimolecular mechanism (Al-Janabi *et al.*, 1972). From a number of studies (Dyke & Kay, 1976; Marciniszyn *et al.*, 1976b;

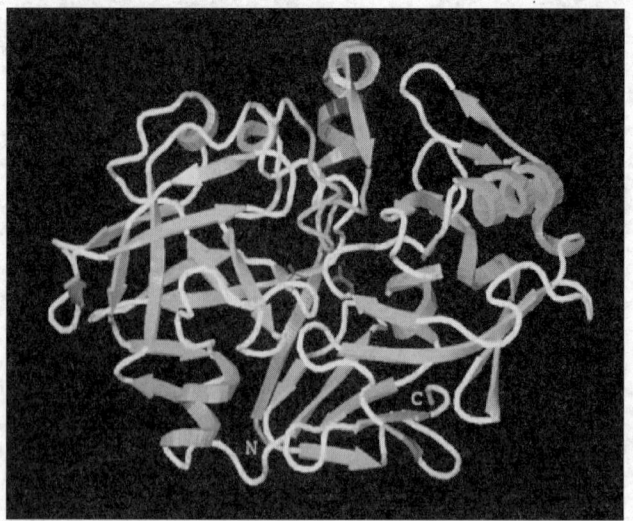

*Figure 272.3* Three-dimensional structure of pig pepsinogen. The view direction is the same as in Figure 272.1. The pro segment clearly blocks the entrance of the active-site cleft of the pepsin moiety (loop at central top and extending as a ribbon through the center down to N). The point of cleavage at residues 16–17 appears as a grey band just beneath the top coil. The N-terminal region of the propeptide can be seen to assume the same position as the N-terminal region of pepsin (Figure 272.1).

Christensen *et al.*, 1977; Glick *et al.*, 1986; Kagayama *et al.*, 1989; Lin *et al.*, 1995), the intramolecular activation of pepsinogen is known to involve several steps. As illustrated in Figure 272.4, the central helical part of the propeptide undergoes local conformational denaturation in acid. This portion of the propeptide then binds to the active site of the same molecule as the substrate and the rate-limiting cleavage takes place first between residues 16 and 17, an Ile⌐Leu bond. The dissociation of the peptide from the N-terminal 16 residues of the 'pro' is important for the local denaturation of the remainder of the propeptide (Lin *et al.*, 1995). This leads to the bimolecular cleavage of the remainder of the propeptide from pepsin.

The detailed steps in bimolecular activation of pepsinogen are still unclear. Under certain activation conditions, the release of the entire propeptide from pepsinogen has been observed (Kageyama & Takahashi, 1983). This may be the result of a bimolecular activation mechanism. The first cleavage sites in the activation of human and pig pepsinogens are apparently different (Foltmann, 1988; Kageyama *et al.*, 1989). This is not surprising since the sequences of the pro regions of aspartic protease zymogens diverge much more than the enzymes. Both intramolecular and bimolecular activations are found in other zymogens of the aspartic protease family. For example, prochymosin (Chapter 274), progastricsin (Chapter 276), procathepsin D (Chapter 277) and rhizopuspepsinogen (Chapter 297) are activated by intramolecular mechanisms. The physiological advantages of the intramolecular activation are the independent and fast activation. Prorenin (Chapter 284) and an aspartic protease zymogen of *Candida tropicalis* (Chapter 300) (Lin *et al.*, 1993b) are apparently activated by other proteases since they contain basic residues at the cleavage sites.

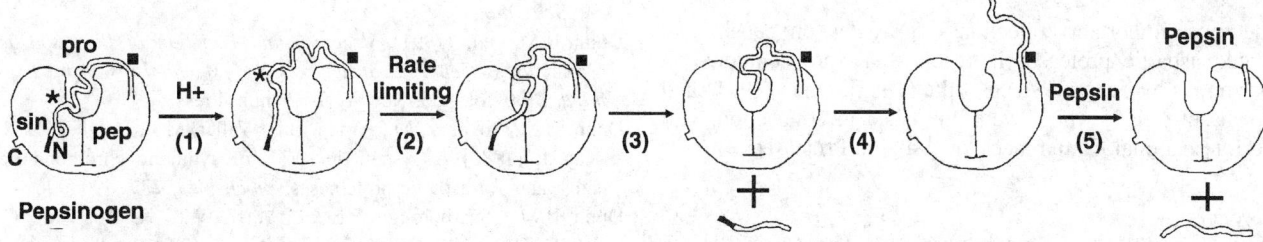

*Figure 272.4* Schematic drawing of the intramolecular activation mechanism of pepsinogen. Pepsinogen on the left is marked for the N- and C-terminal positions and three domains: pep, the N-terminal lobe; sin, the C-terminal lobe; and pro, the pro segment. The positions of the first cleavage site at residues 16–17 of the pro region and the N-terminus of mature pepsin are marked with ∗ and ■ respectively. Step 1: Pepsinogen in acid solution proceeds with a local conformational change which involves the central part of the pro region. Step 2: The denatured pro region is bound in the active site of pepsin moiety and the rate-limiting first cleavage takes place. Step 3: The cleavage product, the N-terminal 16 residues of the pro region, is dissociated from the zymogen. Step 4: The 28 residue pro segment, which is still attached, is conformationally detached from the pepsin moiety. Step 5: The remaining propeptide is cleaved off by pepsin.

## Preparation

Both pig pepsin and pepsinogen are commercially available. Commercial pepsin is prepared from acidified pig stomach homogenate, which is allowed to digest away impurity proteins before the recovery of pepsin. The enzyme prepared in this manner may contain internal nicks from autolysis and extra residues at the N-terminus (Tang *et al.*, 1973). Pig pepsinogen was first purified by Herriott (1938) but a later purification procedure (Rajagopalan *et al.*, 1966) yields much better homogeneity of both pepsinogen and pepsin. Pepsinogen and pepsin from many different vertebrate species have been purified, as described in *Methods in Enzymology* volumes 19 and 45 and in a chapter by Takahashi *et al.* (1995).

Recombinant pig pepsinogen has been expressed in *E. coli* (Tsukagoshi *et al.*, 1988; Lin *et al.*, 1989). A later *E. coli* expression vector driven by T7 promoter has a much higher yield (500 mg liter$^{-1}$ of shaking culture; Lin *et al.*, 1992b, 1994). The purification of recombinant pepsinogen is also relatively simple. The recovered pepsinogen 'inclusion bodies' are washed with a Triton buffer to yield almost chemically pure protein. The zymogen is then refolded from an 8 M urea solution. Although the yield of refolding is only 10–15%, about 50 mg of active pepsinogen can be obtained from each liter of culture. The recombinant pepsinogen, which lacks the phosphate group on Ser68, is identical to the native pepsinogen in activation properties and in the enzymic properties of the activated pepsin. This expression system is highly effective in facilitating mutagenesis and protein engineering studies (Lin *et al.*, 1992a,b, 1993a, 1995).

## Biological Aspects

The human pepsinogen gene contains nine exons in 9.4 kb of DNA (Sogawa *et al.*, 1983). Evidence supports the existence of a multiple gene family for human pepsinogen on chromosome 11 (Taggart *et al.*, 1985; Zelle *et al.*, 1988; Bebelman *et al.*, 1989). Human pepsinogen A isozymes, which are distinguishable using monoclonal antibodies (Taggart & Samloff, 1984), have different sequences (Evers *et al.*, 1989; Bank *et al.*, 1988). The synthesis of human pepsinogen is apparently transcriptionally controlled (Sogawa *et al.*, 1983).

The upstream regulation elements for the human pepsinogen gene have been studied (Meijerink *et al.*, 1993).

Pepsinogen is synthesized in the chief and mucous neck cells of the fundic region of the gastric mucosa and stored as secretory granules (Samloff, 1971b; Helander, 1981). Stomach secretes stored pepsinogen in response to hormonal and neural stimuli (Tsunoda *et al.*, 1988). The secretion of pepsinogen synthesized from the transfected gene in AtT20 cells responds to cyclic AMP (Lin *et al.*, 1990), indicating that pepsinogen is secreted via the regulated secretory pathway. A small amount of pepsinogen is also found in the bloodstream and secreted in the urine; the latter is sometimes referred to as uropepsinogen. The origin of these pepsinogens is apparently stomach since they are absent in gastrectomy patients. No physiological function has been demonstrated for plasma and urine pepsinogen.

The serum pepsinogen and progastricsin levels have been linked to several stomach diseases (for early works, see Samloff, 1989), in particular to peptic ulcerogenesis (Samloff & Taggart, 1987; Busson & Modlin, 1988) and the gastric infection of *Helicobacter pylori* (Asaka *et al.*, 1992; Gisbert *et al.*, 1996 and references therein). The serum pepsinogen and progastricsin levels have also been linked to atrophic gastritis and stomach cancer (Defize *et al.*, 1987; Miki *et al.*, 1993). A mass screening program in Japan, using the combination of serum pepsinogen and progastricsin levels and indirect X-ray examination, has dramatically increased the early detection of stomach cancer since 1993 (Miki *et al.*, 1995).

Zymogens and enzymes very similar to pepsinogen and pepsin are not limited to the stomachs of vertebrates. For example, fungi secrete pepsin-like enzymes apparently for digesting protein in the media. Among these enzymes rhizopuspepsin (Chapter 297), penicillopepsin (Chapter 295) and endothiapepsin (Chapter 296) are well-studied. Although native fungal aspartic protease zymogens have not been observed, rhizopuspepsinogen has been cloned and recombinant rhizopuspepsinogen spontaneously activates in acid solutions (Chen *et al.*, 1991). The absence of the zymogens of aspartic proteases in the extracts of fungi suggests that the fungal zymogens are activated in the secretory granules.

## Further Reading

The most important collections of articles on pepsin and related aspartic proteases in recent years are contained in four books by Tang (1976), Kostka (1985), Dunn (1991) and Takahashi (1995) which were published in connection with the International Conferences on Aspartic Proteases.

## References

Abad-Zapatero, C., Rydel, T.J. & Erickson, J. (1990) Revised 2.3 Å structure of porcine pepsin: evidence for a flexible subdomain. *Proteins* **8**, 62–71.

Abad-Zapatero, C., Rydel, T.J., Neidhart, D.J., Luly, J. & Erickson, J.W. (1991) Inhibitor binding induces structural changes in porcine pepsin. *Adv. Exp. Med. Biol.* **306**, 9–21.

Al-Janabi, J., Hartsuck, J.A. & Tang, J. (1972) Kinetics and mechanism of pepsinogen activation. *J. Biol. Chem.* **247**, 4628–4632.

Andreeva, N.S. & Gustchina, A.E. (1979) On the supersecondary structure of acid proteases. *Biochem. Biophys. Res. Commun.* **87**, 32–42.

Andreeva, N.S. & James, M.N.G. (1991) Why does pepsin have a negative charge at very low pH? An analysis of conserved charged residues in aspartic proteinases. *Adv. Exp. Med. Biol.* **306**, 39–45.

Anson, M.L. (1938) The estimation of pepsin, trypsin, papain, and cathepsin with hemoglobin. *J. Gen. Physiol.* **22**, 79–89.

Asaka, M., Kimura, T., Kudo, M., Takeda, H., Mitani, S., Miyazaki, T., Miki, K. & Graham, D. (1992) Relationship of *Helicobacter pylori* to serum pepsinogens in an asymptomatic Japanese population. *Gastroenterology* **102**, 760–766.

Bank, R.A., Crusius, B.C., Zwiers, T., Meuwissen, S.G.M., Arwert, F. & Pronk, J.C. (1988) Identification of a Glu greater than Lys substitution in the activation segment of human pepsinogen A-3 and -5 isozymogens by peptide mapping using endoproteinase Lys-C. *FEBS Lett.* **238**, 105–108.

Bebelman, J.P., Evers, M.P.J., Zelle, B., Bank, R., Ponk, J.C., Mauwissen, S.G.M., Mager, W.H., Planta, R.J., Eriksson, A.W. & Frants, R.R. (1989) Family and population studies on the human pepsinogen A multigene family. *Hum. Genet.* **82**, 142–146.

Busson, M.D. & Modlin, I.M. (1988) Pepsinogen: biological and pathophysiologic significance. *J. Surg. Res.* **44**, 82–97.

Bustin, M. & Conway-Jacobs, A. (1971) Intramolecular activation of porcine pepsinogen. *J. Biol. Chem.* **246**, 615–620.

Chen, K.C.S. & Tang, J. (1972) Amino acid sequence around the epoxide-reactive residues in pepsin. *J. Biol. Chem.* **247**, 2566–2574.

Chen, Z., Koelsch, G., Han, H., Wang, X., Lin, X., Hartsuck, J.A. & Tang, J. (1991) Recombinant rhizopuspepsinogen. Expression, purification, and activation properties of recombinant rhizopuspepsinogens. *J. Biol. Chem.* **266**, 11718–11725.

Christensen, K.A., Pedersen, V.B. & Foltmann, B. (1977) Identification of an enzymatically active intermediate in the activation of porcine pepsinogen. *FEBS Lett.* **76**, 214–218.

Cooper, J.B., Khan, G., Taylor, G., Tickle, I.J. & Blundell, T.L. (1990) X-ray analyses of aspartic proteinases. II. Three-dimensional structure of the hexagonal crystal form of porcine pepsin at 2.3 Å resolution. *J. Mol. Biol.* **214**, 199–222.

Davies, D.R. (1990) The structure and function of aspartic proteinases. *Annu. Rev. Biophys. Biophys. Chem.* **19**, 189–215.

Defize, J., Pronk, J.C., Frants, R.R., Ooma, E.C.M., Kreuning, J., Kostense, P., Eriksson, A.W. & Meuwissen, S.G.M. (1987) Clinical significance of pepsinogen A isozymogens, serum pepsinogens A and C levels and serum gastrin levels. *Cancer* **59**, 952–958.

Dunn, B.M., ed. (1991) *Structure and Function of the Aspartic Proteinases, Genetics, Structures, and Mechanisms. Adv. Exp. Med. Biol.* vol. 306. New York: Plenum Press.

Dunn, B.M., Jimenez, M., Parten, B.F., Valler, M.J., Rolph, C.E. & Kay, J. (1986) A systematic series of synthetic chromophoric substrates for aspartic proteinases. *Biochem. J.* **237**, 899–906.

Dunn, B.M., Scarborough, P.E., Lowther, W.T. & Roa-Naik, C. (1995) Comparison of the active site specificity of the aspartic proteinases based on a systematic series of peptide substrates. In: *Aspartic Proteinases, Structure, Function, Biology and Biomedical Implications* (Takahashi, K., ed.). New York: Plenum Press, pp 1–9.

Dyke, C.W. & Kay, J. (1976) Conversion of pepsinogen into pepsin is not a one-step process. *Biochem. J.* **153**, 141–144.

Evers, M.P.J., Zelle, B., Bebelman, J.P., van Beusechem, V., Kraakman, L., Hoffer, M.J.V., Pronk, J.C., Mager, W.H., Planta, R.J., Eriksson, A.W. & Frants, R.R. (1989) Nucleotide sequence comparison of five human pepsinogen A (PGA) genes: evolution of the PGA multigene family. *Genomics* **4**, 232–239.

Foltmann, B. (1988) Activation of human pepsinogens. *FEBS Lett.* **241**, 69–72.

Fruton, J.S. (1971) Pepsin. In: *The Enzymes*, 3rd edn (Boyer, P.D., ed.), New York: Academic Press, pp. 119–164.

Fruton, J.S. (1976) The mechanism of the catalytic action of pepsin and related acid proteinases. *Adv. Enzymol.* **44**, 1–36.

Gisbert, J.P., Boixeda, D., Vila, T., de Rafael, L., Redondo, C., Canton, R. and Martin de Argila, C. (1996) Verification of decreased basal and stimulated serum pepsinogen-I levels is a useful non-invasive method for determining the success of eradication therapy for *Helicobacter pylori*. *Scand. J. Gastroenterol.* **31**, 103–110.

Glick, D.M., Auer, H.E., Rich, D.H., Kawai, M. & Kamath, A. (1986) Pepsinogen activation: genesis of the binding site. *Biochemistry* **25**, 1858–1864.

Hartsuck, J.A., Koelsch, G. & Remington, S.J. (1992) The high-resolution crystal structure of porcine pepsinogen. *Protein Struct. Funct. Genet.* **13**, 1–25.

Helander, H.F. (1981) The cells of the gastric mucosa. *Int. Rev. Cytol.* **70**, 217–289.

Herriott, R.M. (1938) Isolation, crystallization and properties of swine pepsinogen. *J. Gen. Physiol.* **21**, 501–540.

Herriott, R.M., Desreux, V. & Northrop, J.H. (1940) Electrophoresis of pepsin. *J. Gen. Physiol.* **23**, 439–447.

James, M.N.G. & Sielecki, A.R. (1986) Molecular structure of an aspartic proteinase zymogen, porcine pepsinogen at 1.8 Å resolution. *Nature* **319**, 33–38.

Kageyama, T. & Takahashi, K. (1983) Occurrence of two different pathways in the activation of porcine pepsinogen to pepsin. *J. Biochem. (Tokyo)* **93**, 743–754.

Kageyama, T., Ichinose, M., Miki, K., Athauda, S.B., Tanji, M. & Takahashi, K. (1989) Difference of activation processes and structure of activation peptides in human pepsinogen A and progastricsin. *J. Biochem. (Tokyo)* **105**, 15–22.

Kostka, V., ed. (1985) *Aspartic Proteinases and Their Inhibitors.* Berlin: de Gruyter.

Kostka, V., Pichova, I. & Baudys, M. (1985) Isolation and molecular characterization of avian pepsins. In: *Aspartic Proteinases and Their Inhibitors* (Kosta, V., ed.). Berlin: de Gruyter, pp. 53–72.

Lundblad, R.L. & Stein, W.H. (1969) On the reaction of diazoacetyl compounds with pepsin. *J. Biol. Chem.* **244**, 154–160.

Lin, X., Wong, R.N.S. & Tang, J. (1989) Synthesis, purification, and active site mutagenesis of recombinant porcine pepsinogen. *J. Biol. Chem.* **264**, 4482–4489.

Lin, X.L., Dashti, N. & Tang, J. (1990) Internal pH of the secretory granule in mouse pituitary AtT20 cells is above 5.0. *FASEB J.* **4**, 3566.

Lin, Y., Fusek, M., Lin, X., Hartsuck, J.A., Kezdy, F.J. & Tang, J. (1992a) pH dependence of kinetic parameters of pepsin, rhizo-puspepsin, and their active-site hydrogen bond mutants. *J. Biol. Chem.* **267**, 18413–18418.

Lin, X.L., Lin, Y., Koelsch, G., Gustchina, A., Wlodawer, A. & Tang, J. (1992b) Enzymic activities of two-chain pepsinogen, two-chain pepsin, and the amino-terminal lobe of pepsinogen. *J. Biol. Chem.* **267**, 17257–17263.

Lin, X.L., Loy, J.A., Sussman, F. & Tang, J. (1993a) Conformational instability of the N- and C-terminal lobes of porcine pepsin in neutral and alkaline solutions. *Protein Sci.* **2**, 1383–1390.

Lin, X.L., Tang, J., Koelsch, G., Monod, M. & Foundling, S. (1993b) Recombinant canditropsin, an extracellular aspartic protease from yeast *Candida tropicalis. J. Biol. Chem.* **268**, 20143–20147.

Lin, X.L., Lin, Y.-Z. & Tang, J. (1994) Relationships of human immunodeficiency virus protease with eukaryotic aspartic proteases. *Methods Enzymol.* **241**, 195–224.

Lin, X.L., Koelsch, G., Loy, J.A. & Tang, J. (1995) Rearranging the domains of pepsinogen. *Protein Sci.* **4**, 159–166.

Marciniszyn, J., Jr, Hartsuck, J.A. & Tang, J. (1976a) Mode of inhibition of acid proteases by pepstatin. *J. Biol. Chem.* **251**, 7088–7094.

Marciniszyn, J., Jr, Huang, J.S., Hartsuck, J.A. & Tang, J. (1976b) Mechanism of intramolecular activation of pepsinogen. *J. Biol. Chem.* **251**, 7095–7102.

Martzen, M.R., McMullen, B.A., Fujikawa, K. & Peanasky, R.J. (1991) Aspartic protease inhibitors from the parasitic nematode *Ascaris. Adv. Exp. Med. Biol.* **306**, 63–73.

McPhie, P. (1972) Pepsinogen: activation by a unimolecular mechanism. *Biochem. Biophys. Res. Commun.* **56**, 789–792.

Meijerink, P.H., Bebelman, J.P., Oldenburg, A.M., Defize, J., Planta, R.J., Eriksson, A.W., Pals, G. & Meger, W.H. (1993) Gastric chief cell-specific transcription of the pepsinogen A gene. *Eur. J. Biochem.* **213**, 1283–1296.

Miki, K., Ichinose, M., Ishikawa, K.B., Yahagi, N., Matsushima, M., Kakei, N., Tsukada, S., Kido, M., Ishihama, S., Shimizu, Y., Suzuki, T. & Kurokawa, K. (1993) Clinical application of serum pepsinogen I and II levels for mass screening to detect gastric cancer. *Jpn J. Cancer Res.* **84**, 1086–1090.

Miki, K., Ichinose, M., Kakei, N., Yahagi, N., Matsushima, M., Tsukada, S., Ishihama, S., Shimizu, Y., Suzuki, T., Kurokawa, K. & Takahashi, K. (1995) The clinical application of the serum pepsinogen I and II levels as a mass screening method for gastric cancer. *Adv. Exp. Med. Biol.* **362**, 139–143.

Moravek, L. & Kostka, V. (1974) Complete amino acid sequence of hog pepsin. *FEBS Lett.* **43**, 207–211.

Nielsen, P.K. & Foltmann, B. (1995) Purification and characterization of porcine pepsinogen B and pepsin B. *Arch. Biochem. Biophys.* **322**, 417–422.

Northrop, J. H. (1930) Crystalline pepsin. I. Isolation and tests of purity. *J. Gen. Physiol.* **13**, 739.

Pearl, L. (1985) The extended binding cleft of aspartic proteinases and its role in peptide hydrolysis. In: *Aspartic Proteinases and Their Inhibitors* (Kostka, V., ed.). Berlin: de Gruyter, pp. 189–196.

Rajagopalan, T.G., Moore, S. & Stein, W.H. (1966) Pepsin from pepsinogen, preparation and properties. *J. Biol. Chem.* **241**, 4940–4950.

Rao, C. & Dunn, B.M. (1995) Evidence for electrostatic interactions in the S$_2$ subsite of porcine pepsin. *Adv. Exp. Med. Biol.* **362**, 91–94.

Rao, J.K.M., Erickson, J.W. & Wlodawer, A. (1991) Structural and evolutionary relationships between retroviral and eukaryotic aspartic proteinases. *Biochemistry* **30**, 4663–4671.

Richmond, V., Tang, J., Wolf, S., Trucco, R.E. & Caputto, R. (1958) chromatographic isolation of gastricsin, the proteolytic enzyme from gastric juice with pH optimum 3.2. *Biochim. Biophys. Acta* **29**, 453–454.

Ryle, A.P. & Porter, R.R. (1959) Parapepsins: two proteolytic enzymes associated with porcine pepsin. *Biochem. J.* **73**, 75–86.

Sachdev, G.P. & Fruton, J.S. (1969) Pyridyl esters of peptides as synthetic substrates of pepsin. *Biochemistry* **8**, 4231–4238.

Samloff, I.M. (1969) Slow moving protease and the seven pepsinogens – electrophoretic demonstration of the existence of eight proteolytic fractions in human gastric mucosa. *Gastroenterology* **57**, 659–669.

Samloff, I.M. (1971a) Pepsinogens, pepsins, and pepsin inhibitors. *Gastroenterology* **60**, 586–604.

Samloff, I.M. (1971b) Cellular localization of group I pepsinogens in human gastric mucosa by immunofluorescence. *Gastroenterology* **61**, 185–188.

Samloff, I.M. (1989) Peptic ulcer: the many proteinases of aggression. *Gastroenterology* **96**, 586–595.

Samloff, I.M. and Taggart, R.T. (1987) Pepsinogens, pepsins and peptic ulcer. *Clin. Invest. Med.* **10**, 215–220.

Sielecki, A.R., Fedorov, A.A., Boodhoo, A., Andreeva, N.S. & James, M.N.G. (1990) Molecular and crystal structures of monoclinic porcine pepsin refined at 1.8 Å resolution. *J. Mol. Biol.* **214**, 143–170.

Sielecki, A.R., Fujinaga, M., Read, R.J. & James, M.N.G. (1991) Refined structure of porcine pepsinogen at 1.8 Å resolution. *J. Mol. Biol.* **219**, 671–692.

Sogawa, K., Fujii-Kuriyama, Y., Mizukama, Y., Ichihara, Y. & Takahashi, K. (1983) Primary structure of human pepsinogen gene. *J. Biol. Chem.* **258**, 5306–5311.

Taggart, R.T. & Samloff, I.M. (1984) Multiple genes determine the immunochemical heterogeneity of human pepsinogen-1 (PG1) isozymogens. *Gastroenterology* **86**, 1272 (abstr.).

Taggart, R.T., Mohandas, T.K., Shows, T.B. & Bell, G.I. (1985) Variable numbers of pepsinogen genes are located in the centromeric region of human chromosome 11 and determine the high-frequency electrophoretic polymorphism. *Proc. Natl Acad. Sci. USA* **82**, 6240–6244.

Takahashi, K., ed. (1995) *Aspartic Proteinases, Structure, Function, Biology, and Biomedical Implications. Adv. Exp. Med. Biol.* vol. 362. New York: Plenum Press.

Takahashi, K., Tanji, M., Yakabe, E., Hirasawa, A., Athauda, S.B.P. & Kageyama, T. (1995) Non-mammalian vertebrate pepsinogens and pepsins: isolation and characterization. *Adv. Exp. Med. Biol.* **362**, 53–65.

Tang, J. (1963) Specificity of pepsin and its dependence on a possible hydrophobic binding site. *Nature* **199**, 1094–1095.

Tang, J. (1971) Specific and irreversible inactivation of pepsin by substrate-like epoxides. *J. Biol. Chem.* **246**, 4510–4517.

Tang, J., ed. (1976) *Acid Proteases, Structure, Function, and Biology. Adv. Exp. Med. Biol.* vol. 95. New York: Plenum Press.

Tang, J. & Koelsch, G. (1995) A possible function of the flaps of aspartic proteases: the capture of substrate side chains determines the specificity of cleavage positions. *Protein Peptide Lett.* **2**, 257–266.

Tang, J., Sepulveda, P., Marciniszyn, J., Jr, Chen, K.C.S., Huang, W.-Y., Liu, D. & Lanier, J.P. (1973) Amino-acid sequence of porcine pepsin. *Proc. Natl Acad. Sci. USA* **70**, 3437–3439.

Tang, J., James, M.N.G., Hsu, I.N., Jenkins, J.A. & Blundell, T.L. (1978) Structural evidence for gene duplication in the evolution of the acid proteases. *Nature* **271**, 618–621.

Toh, H., Ono, M., Saigo, K. & Miyata, T. (1985) Retroviral protease-like sequence in the yeast transposon TY 1. *Nature* **315**, 691–692.

Tsukagoshi, N., Ando, Y., Tomita, Y., Uchida, R., Takemura, T., Sasaki, T., Yamagata, H., Udaka, S., Ichihara, Y. & Takahashi, K. (1988) Nucleotide sequence and expression in *Escherichia coli* of cDNA of swine pepsinogen: involvement of the amino-terminal portion of the activation peptide segment in restoration of the functional protein. *Gene* **65**, 285–292.

Tsunoda, Y., Takeda, H., Otaki, T., Asaka, M., Nakagaki, I. & Sasaki, S. (1988) A role for Ca$^{2+}$ in mediating hormone-induced biphasic pepsinogen secretion from the chief cell determined by luminescent and fluorescent probes and X-ray microprobe. *Biochim. Biophys. Acta* **941**, 83–101.

Umezawa, H., Aoyagi, T., Morishima, H., Matsuzaki, H., Hamada, M. & Takeuchi, T. (1970) Pepstatin, a new pepsin inhibitor produced by Actinomycetes. *J. Antibiot.* **23**, 259–262.

Zelle, B., Evers, M.P.J., Groot, P.C., Bebelman, J.P., Mager, W.H., Planta, R.J., Pronk, J.C., Meuwissen, S.G.M., Hofker, M.H., Eriksson, A.W. & Frants, R.R. (1988) Genomic structure and evolution of the human pepsinogen A multigene family. *Hum. Genet.* **78**, 79–82.

*Jordan Tang*
*Protein Studies Program,*
*Oklahoma Medical Research Foundation,*
*825 NE 13 Street, Oklahoma City, OK 73104, USA*
*Email: jordan-tang@omrf.uokhsc.edu*

# 273. Pepsin B

## Databanks

*Peptidase classification: clan AA, family A1, MEROPS ID: A01.002*
*NC-IUBMB enzyme classification: EC 3.4.23.2*
*Chemical Abstracts Service registry number: 9025-48-3*
*Databank codes:*

| Species | SwissProt | PIR | EMBL (cDNA) | EMBL (genomic) |
|---|---|---|---|---|
| *Sus scrofa* | – | – | – | – |

*Note*: There are two pig fragments deposited in GenPept, but only accessible via the Entrez server. The accession number is 1168175.

## Name and History

The name *pepsin B* was introduced in 1961 by the Commission on Enzymes of the International Union of Biochemistry instead of *parapepsin I*, previously used by Ryle & Porter (1959). The enzyme is presumably the same as gelatinase (Northrop, 1932). By error, the term was also used for a cattle gastric proteinase which later turned out to be gastricsin (Chapter 276) (Klemm *et al.*, 1976).

## Activity and Specificity

Pepsin B has a very weak general proteolytic activity; at the optimum of pH 3, the activity against acid-denatured hemoglobin is about 4% of that shown by pepsin A (Chapter 272) at pH 2 (Nielsen & Foltmann, 1995). After chromatography, pepsin B was originally detected through its activity against Ac-Phe+Tyr(I$_2$), which shows an optimum at pH 2. Relative to pepsin A, pepsin B shows a restricted specificity in hydrolysis of the B chain of oxidized insulin–

FVNQHLCGSHL+VEA+L+Y+LVCGERGF+F+YTPKA

(where * indicates a major site), but it liquefies gelatin much more readily than does pepsin A (Ryle & Porter, 1959). In a recent preparation, pepsin B was detected through its milk-clotting activity (Nielsen & Foltmann, 1995).

## Structural Chemistry

The sequence of the first 67 amino acid residues of pepsinogen B has been determined; this fragment reveals 40, 55 and 51% of identity with pig pepsinogen A (Chapter 272), progastricsin (Chapter 276) and prochymosin (Chapter 274), respectively. A combination of results from SDS-PAGE and amino acid composition indicates an $M_r$ of 41 000 for pepsinogen B and 36 000 for pepsin B (Nielsen & Foltmann, 1995).

## Preparation

The present preparation of pepsinogen B is, in principle, as described by Ryle (1970); it has been improved by introducing gel filtration before a final purification step by FPLC on a column of Mono-Q (Nielsen & Foltmann, 1995).

## Biological Aspects

The first experiments on the activation of pepsinogen B were monitored by the development of gelatin-liquefying activity (Ryle, 1965). Subsequent experiments (Nielsen & Foltmann, 1995) were monitored by SDS-PAGE in parallel with non-denaturing gel electrophoresis. These results show that both covalent and noncovalent intermediates occur during the activation. At pH 2, the first cleavage takes place at the bond Met16p⏐Glu17p (numbering relates to pig pepsinogen A; James & Sielecki, 1986; the suffix 'p' indicates propart). The resultant pseudopepsin B is stable at pH 2 and is converted to mature pepsin B at pH 5.5. Pepsin B is stable from pH 4 to pH 6.9 at room temperature; higher values of pH were not tested (Ryle & Porter, 1959).

Pepsinogen B is produced in the gastric mucosa of pigs. The enzyme is present in trace amounts at birth; its production reaches a maximum after 4 weeks and then levels off to a concentration which is about half of the maximum (Foltmann *et al.*, 1995). Nothing is known about function; pepsinogen B has never been found in species other than pig, in spite of intensive search in gastric extracts from cow and humans.

## References

Foltmann, B., Harlow, K., Houen, G., Nielsen, P.K. & Sangild, P. (1995) Comparative investigations on pig gastric proteases and their zymogens. *Adv. Exp. Med. Biol.* **362**, 41–51.

James, M.N.G. & Sielecki, A.R. (1986) Molecular structure of an aspartic proteinase zymogen, porcine pepsinogen at 1.8 Å resolution. *Nature* **319**, 33–38.

Klemm, P., Poulsen, F., Harboe, M.K. & Foltmann, B. (1976) N-terminal amino acid sequences of pepsinogens from dogfish and seal and of bovine pepsinogen B. *Acta Chem. Scand.* **B30**, 979–984.

Nielsen, P.K. & Foltmann, B. (1995) Purification and characterization of porcine pepsinogen B and pepsin B. *Arch. Biochem. Biophys.* **322**, 417–422.

Northrop, J.H. (1932) The presence of a gelatin-liquefying enzyme in crude pepsin preparations. *J. Gen. Physiol.* **15**, 29–43

Ryle, A.P. (1965) Pepsinogen B: the zymogen of pepsin B. *Biochem. J.* **96**, 6–16.

Ryle, A.P. (1970) The porcine pepsins and pepsinogens. *Methods Enzymol.* **19**, 316–336.

Ryle, A.P. & Porter, R.R. (1959) Parapepsins: two proteolytic enzymes associated with porcine pepsin. *Biochem. J.* **73**, 75–86.

*Bent Foltmann*
*Institute of Molecular Biology,*
*University of Copenhagen,*
*2A Øster Farimagsgade, DK-1353 Copenhagen K,*
*Denmark*

*Pal B. Szecsi*
*Department of Clinical Chemistry,*
*Roskilde County Hospital,*
*Køgevej 13-16, DK-4000 Roskilde, Denmark*
*Email: biotrek@symbion.dk*

# 274. Chymosin

## Databanks

*Peptidase classification: clan AA, family A1, MEROPS ID: A01.006*
*NC-IUBMB enzyme classification: EC 3.4.23.4*
*ATCC entries: 20661; 20662; 20623; 31929; 39543; 39544*
*Chemical Abstracts Service registry number: 9001-98-3*

*Databank codes:*

| Species | SwissProt | PIR | EMBL (cDNA) | EMBL (genomic) |
|---|---|---|---|---|
| Chymosin | | | | |
| *Bos taurus* | P00794 | A25631 | J00002 | M14069: exon 1 |
| | | A44608 | J00003 | M14070: exon 2 |
| | | A44620 | J00004 | M14071: exon 3 |
| | | A91495 | U19786 | M14072: exon 4 |
| | | A91935 | | M14073: exon 5 |
| | | A92259 | | M14074: exon 6 |
| | | A93419 | | M14075: exon 7 |
| | | | | M14076: exon 8 |
| | | | | M14077: exon 9 and complete CDS |
| | | | | M19699: exon 2 |
| *Felis catus* | P09873 | A20146 | – | – |
| *Ovis aries* | P18276 | S10996 | X53037 | – |
| *Sus scrofa* | – | – | U14406 | – |
| Chymosin pseudogene | | | | |
| *Homo sapiens* | – | – | – | M57258: exon 1 |
| | | | | M57259: exon 2 |
| | | | | M57260: exon 3 |
| | | | | M57262: exon 4 |
| | | | | M57264: exon 5 |
| | | | | M57265: exon 6 |
| | | | | M57266: exon 7 |
| | | | | M57267: exon 8 |
| | | | | M57268: exon 9 |

Brookhaven Protein Data Bank three-dimensional structures:

| Species | ID | Resolution | Notes |
|---|---|---|---|
| *Bos taurus* | 1CMS | 2.3 | recombinant |
| | 3CMS | 2 | Val111Phe mutant |
| | 4CMS | 2.2 | |

## Name and History

Since ancient days, extracts of calf stomach have been used for clotting milk in cheese making. The first attempts at isolation of the enzyme were made by Deschamps (1840), who suggested the name **chymosine** (Gr. *chyme*, '(gastric) juice'). Lea & Dickinson (1890) suggested the name **rennin** (derived from rennet). This name was dominant in the English literature for many years but was almost indistinguishable from renin from the kidneys, so Foltmann (1970) suggested a return to the original designation, **chymosin**, as subsequently adopted by the IUBMB. Other designations and reviews of historical interest include: **pexin** (Warren, 1897) and **chymase** (Zunz, 1911; Oppenheimer; 1926; Holter, 1941).

Hammarsten (1872) discovered that the enzyme was secreted as an inactive component, which was converted into active enzyme through treatment with acids. He was the first to discover a proenzyme. For many years it was generally assumed that chymosin was restricted to young ruminants, but Foltmann & Axelsen (1980) showed that chymosin-like enzymes are widely distributed among young mammals. Thus, the chymosins may be characterized as mammalian fetal or neonatal gastric proteinases.

## Activity and Specificity

Table 274.1 shows a comparison of the enzymatic properties of chymosins from six different species of mammal. Cow's milk is generally used for investigating milk-clotting activity, but mutual adaptations have occurred between chymosin and casein for any given species. Pig chymosin is 6–8 times more active against sow's milk than against cow's milk; conversely, calf chymosin clots sow's milk with only half of its activity against cow's milk (Foltmann *et al.*, 1981). Calf chymosin clots cow's milk through cleavage of the peptide bond Phe105+Met106 in $\kappa$-casein. Visser *et al.* (1980, 1987) have investigated the cleavage of synthetic peptides and fragments of $\kappa$-casein containing the Phe+Met bond. The peptide His98-Pro-His-Pro-His-Leu-Ser-Phe+Met-Ala-Ile-Pro-Pro-Lys-Lys111 was cleaved with a $k_{cat}/K_m$ value about 30 times larger than the value for the peptide Leu103-Lys111. This indicates interactions of residues outside the substrate-binding cleft, which normally will accommodate four residues before and after the scissile peptide bond. Gustchina *et al.* (1996) have investigated the binding of the peptide segment His98-Pro-His-Pro-His102 which may act as an activator. The action of calf chymosin on cattle $\alpha$-

*Table 274.1* Comparison of the enzymatic properties of chymosins from different species

| Enzyme | Milk-clotting activity[a] | Proteolytic activity[b] |
|---|---|---|
| Calf chymosin | 100 | 25 (pH 3.5) |
| Lamb chymosin | 95 | 30 (pH 3.5) |
| Pig chymosin | 25 | 3 (pH 3.5) |
| Cat chymosin | 15 | 50 (pH 2.8) |
| Seal chymosin | 130 | 40 (pH 2.9) |
| Pig pepsin A | 25 | 100 (pH 2.0) |

[a]Expressed as per cent of calf chymosin using reconstituted cattle skim-milk at pH 6.3 as substrate.
[b]Degradation of acid-denatured hemoglobin at optimum pH (in parentheses), expressed relative to the degradation by pig pepsin A at pH 2. See also Foltmann (1992).

and β-casein was analyzed by Carles & Ribadeau-Dumas (1984, 1985).

## Structural Chemistry

The primary structure of calf prochymosin has been determined both at the amino acid level (Pedersen & Foltmann, 1975; Foltmann *et al.*, 1979) and at the nucleotide level (Harris *et al.*, 1982; Moir *et al.*, 1982; Hidaka *et al.*, 1986). Calf prochymosin occurs with at least two genetic variants, A: Asp243 and B: Gly243 (numbers relative to the homologous positions in pig pepsinogen and pepsin A) (James & Sielecki, 1986; Chapter 272). A degradation product of the A variant was designated chymosin C2 (Foltmann, 1966); relative to chymosin A, its general proteolytic activity is 50% and its milk-clotting activity is only 25%. Danley & Geoghegan (1988) showed that the degradation consists of an autolytic excision of Asp243-Glu244-Phe245 located near subsite S4.

Pig prochymosin also occurs with at least two genetic variants (Houen *et al.*, 1996); one has been sequenced at the nucleotide and partly at the amino acid level. Lamb prochymosin has been completely sequenced at the nucleotide level only (Pungercar *et al.*, 1990). The results show a Glu residue at position 36p (where 'p' indicates propart numbering). This is remarkable, since at this position all other aspartic proteinase zymogens have a basic residue which forms salt bridges to Asp32 and Asp215 in the active site.

The tertiary structure of calf chymosin has been determined by Gilliland *et al.* (1990) and Newman *et al.* (1991). As expected, the folding of the peptide chain follows the same pattern as found in all other aspartic proteinases. All prochymosins consist of about 365 amino acid residues and have $M_r$ values near 40 000. An N-terminal propart is removed during the conversion into mature enzymes that have $M_r$ values of about 35 000. Calf prochymosin has a pI about 5.0 and calf chymosin has a pI about 4.6 (Foltmann, 1966).

## Preparation

Preparation of prochymosin begins with the extraction of gastric mucosa from young animals at neutral or weakly alkaline pH. Pure zymogens have been obtained after repeated steps of anion-exchange chromatography and gel filtration (Foltmann, 1966; Jensen *et al.*, 1982; Baudys *et al.*, 1988). Calf

chymosin may be prepared from commercial cheese rennet (Foltmann, 1970). Other chymosins are in general most easily prepared after activation of prochymosin in the crude extracts (Houen *et al.*, 1996). Due to its commercial importance, calf prochymosin has been cloned, expressed and subjected to site-directed mutagenesis in numerous laboratories using *Escherichia coli*, yeast and different filamentous fungi. Selected references for production of recombinant chymosin are found in Foltmann (1993).

## Biological Aspects

Like other gastric zymogens, prochymosin is converted into active enzyme by a limited proteolysis at pH below 5, and the rate of this process is increased considerably by lowering the pH from 5 to 2. Investigations on the activation of calf prochymosin were the first that suggested that the initial step in the conversion of gastric zymogens into active enzymes consists of a pH-dependent conformational change which uncovers the active-site cleft (Foltmann, 1966). At pH 2, calf prochymosin is cleaved at Phe25p+Leu26p; the resultant intermediate (pseudochymosin) is stable at pH 2. Mature chymosin is formed by activation at pH 4–5; pseudochymosin is converted to chymosin at pH 5.5 (Pedersen *et al.*, 1979). The activation of prochymosins from other species has not been investigated to the same extent; noncovalent intermediates may occur during the activation of pig prochymosin. Like pig pepsin A, pig chymosin has Glu at position 4 whereas calf chymosin has Val. This substitution is probably important for the conformational change of Val2p to Leu6p in the zymogen with Ala2 to Leu6 in the mature enzyme (Nielsen & Foltmann, 1993).

Calf prochymosin is stable at room temperature up to pH 8.5, whereas chymosin is rapidly denatured at pH above 6.5. Chymosin undergoes autolysis at pH near 3.5, but it is rather stable at pH 2 (Foltmann, 1966).

Like other gastric zymogens, prochymosin is mainly produced in chief cells of the gastric mucosa, but other cells in the fundic glands may also produce such zymogens. The histochemistry and ontogeny of calf prochymosin has been reviewed by Andrén (1992). The production of prochymosin starts at the 10th week of gestation, and it reaches a maximum a few weeks before birth. A high production of prochymosin is maintained with milk-feeding, at least to the age of 6 months. With a normal mixed diet, a decline in the production takes place after 3 months, but small amounts of prochymosin are produced even at the adult stage. In pigs the production of prochymosin starts at about 10 weeks of gestation and reaches a maximum at birth. A rapid decline occurs after the first week of life, and no prochymosin has been observed after the age of 50 days (Foltmann *et al.*, 1995). It has been claimed that gastric juice from human infants is able to precipitate anti-calf chymosin (Henschel *et al.*, 1987), but we have never found such reactions. Furthermore, only a single chymosin-like pseudogene is found in human genomic DNA (Örd *et al.*, 1990). Thus, functional chymosins apparently do not occur among primates (Foltmann, 1992).

A very large literature exists about the use of chymosin in cheese making, but there is almost nothing about the physiological function of chymosin. Chymosin is found in all animals with postnatal uptake of immunoglobulin G, but not in primates with placental transfer. It has therefore been suggested that the chymosins have evolved as neonatal proteinases which have sufficient activity to clot the milk, but have limited ability to degrade the immunoglobulins (Foltmann *et al.*, 1981).

## Distinguishing Features

Although considerable differences are found among chymosins from different species, these enzymes may be characterized as neonatal gastric proteinases with a weak general proteolytic and a high milk-clotting activity. By gel electrophoreses at pH 5–8.8 of extracts from gastric mucosa, prochymosin and chymosin generally have lower mobilities than the other zymogens or proteinases. Detection may take place by clotting of casein in a milk-containing gel placed on top of the separation gel after electrophoresis (Foltmann *et al.*, 1985). Polyclonal antisera against calf chymosin are available from Chr. Hansens Laboratory (see Appendix 2 for full names and addresses of suppliers).

## Related Proteinases

Comparison between the primary structure of chicken fetal pepsinogen (Hayashi *et al.*, 1988) and the known primary structures of prochymosins suggests that the former enzyme is related to the chymosins. Chicken fetal pepsinogen has been prepared from 15-day-old embryos (Yasugi & Mizuno, 1981); after activation the enzyme shows optimum activity at pH 3. Transcription of the gene for chicken pepsinogen has been investigated and discussed by Fukuda *et al.* (1995).

## Further Reading

See Foltmann (1966) for a review on the history of calf chymosin and the fundamental properties of the enzyme; Foltmann (1992) for a review on chymosins from different species; Foltmann (1993) for a general review on milk-clotting enzymes, and Suzuki *et al.* (1989) for information on site-directed mutagenesis.

## References

Andrén, A. (1992) Production of prochymosin, pepsinogen and progastricsin, and their cellular and intracellular localization in bovine abomasal mucosa. *Scand. J. Clin. Lab. Invest.* **52**(suppl. 210), 59–64.

Baudys, M., Erdene, T.G., Kostka, V., Pavlik, M. & Foltmann, B. (1988) Comparison between prochymosin and pepsinogen from lamb and calf. *Comp. Biochem. Physiol.* **89B**, 385–391.

Carles, C. & Ribadeau-Dumas, B. (1984) Kinetics of the action of chymosin (rennin) on some peptide bond of bovine $\beta$-casein. *Biochemistry* **23**, 6839–6843.

Carles, C. & Ribadeau-Dumas, B. (1985) Kinetics of the action of chymosin (rennin) on a peptide bond of bovine $\alpha_{s1}$-casein. *FEBS Lett.* **185**, 282–286.

Danley, D.E. & Geoghegan, K.F. (1988) Structure and mechanism of formation of recombinant-derived chymosin C. *J. Biol. Chem.* **263**, 9785–9789.

Deschamps, J.B. (1840) De la présure. *J. Pharmacol.* **26**, 412–420.

Foltmann, B. (1966) A review on prorennin and rennin. *C.R. Trav. Lab. Carlsberg* **35**, 143–231.

Foltmann, B. (1970) Prochymosin and chymosin (prorennin and rennin). *Methods Enzymol.* **19**, 421–436.

Foltmann, B. (1992) Chymosin. A short review on foetal and neonatal gastric proteases. *Scand. J. Clin. Lab. Invest.* **52** (suppl. 210), 65–79.

Foltmann, B. (1993) General and molecular aspects of rennets. In: *Cheese: Chemistry, Physics and Microbiology*, vol. 1 (Fox, P.F., ed.). London: Chapman & Hall, pp. 37–68.

Foltmann, B. & Axelsen, N.H. (1980) Gastric proteinases and their zymogens. Phylogenetic and developmental aspects. In: *Enzyme Regulation and Mechanism of Action* (Mildner, P. & Ries, B., eds). Oxford: Pergamon, pp. 271–280.

Foltmann, B., Pedersen, V.B., Kauffman, D. & Wybrandt, G. (1979) The primary structure of calf chymosin. *J. Biochem.* **254**, 8447–8456.

Foltmann, B., Jensen, A.L., Lønblad, P., Smidt, E. & Axelsen, N.H. (1981) A developmental analysis of the production of chymosin and pepsin in pigs. *Comp. Biochem. Physiol.* **68B**, 9–13.

Foltmann, B., Szecsi, P.B. & Tarasova, N.I. (1985) Detection of proteases by clotting of casein after gel electrophoresis. *Anal. Biochem.* **146**, 353–360.

Foltmann, B., Harlow, K., Houen, G., Nielsen, P.K. & Sangild, P. (1995) Comparative investigations on pig gastric proteases and their zymogens. *Adv. Exp. Med. Biol.* **362**, 41–51.

Fukuda, K., Saiga, H. & Yasugi, S. (1995) Transcription of embryonic chick pepsinogen gene is affected by mesenchymal signals through its 5′-flanking region. *Adv. Exp. Med. Biol.* **362**, 125–129.

Gilliland, G.L., Winborne, E.L., Nachman, J. & Wlodawer, A. (1990) The three-dimensional structure of recombinant bovine chymosin at 2.3 Å resolution. *Proteins* **8**, 82–101.

Gustchina, E., Rumsch, L., Ginodman, L., Majer, P. & Andreeva, N. (1996) Post X-ray crystallographic studies of chymosin: the existence of two structural forms and the regulation of activation with the histidine-proline cluster of κ-casein. *FEBS Lett.* **399**, 60–62.

Hammarsten, O. (1872) Om mjölk-ystningen och de dervid verksamma fermenterna i magslemhinnan. *Ups. Läkareförn. Förhandl.* **8**, 63–86.

Harris, T.J.R., Lowe, P.A., Lyons, A., Thomas, P.G., Eaton, M.A.W., Millican, T.A., Patel, T.P., Bose, C.C., Carey, N.H. & Doel, M.T. (1982) Molecular cloning and nucleotide sequence of cDNA coding for calf preprochymosin. *Nucleic Acids Res.* **10**, 2177–2187.

Hayashi, K., Agata, K., Mochii, M., Yasugi, S., Eguchi, G. & Mizuno, T. (1988) Molecular cloning and the nucleotide sequence of cDNA for embryonic chicken pepsinogen: phylogenetic relationship with prochymosin. *J. Biochem. (Tokyo)* **103**, 290–296.

Henschel, M.J., Newport, M.J. & Parmar, V. (1987) Gastric proteases in the human infant. *Biol. Neonate* **52**, 268–272.

Hidaka, M., Sasaki, K., Uozumi, T. & Beppu, T. (1986) Cloning and structural analysis of the calf prochymosin gene. *Gene* **43**, 197–203.

Holter, H. (1941) Chymase. In: *Die Methoden der Fermentforschung*, vol. 2 (Bamann, E. & Myrbäck, K., eds). Leipzig: G. Thieme Verlag, pp. 2081–2090.

Houen, G., Madsen, M.T., Harlow, K.W., Lønblad, P. & Foltmann, B. (1996) The primary structure and enzymic properties of porcine prochymosin and chymosin. *Int. J. Biochem. Cell Biol.* **28**, 667–675.

James, M.N.G. & Sielecki, A.R. (1986) Molecular structure of an aspartic proteinase zymogen, porcine pepsinogen, at 1.8 Å resolution. *Nature* **319**, 33–38.

Jensen, T., Axelsen, N.H. & Foltmann, B. (1982) Isolation and partial characterization of prochymosin and chymosin from cat. *Biochim. Biophys. Acta* **705**, 249–256.

Lea, A.S. & Dickinson, W.L. (1890) Notes on the mode of action of rennin and fibrin-ferment. *J. Physiol.* **11**, 307–311.

Moir, D., Mao, J.-I., Schumm, J.W., Vovis, G.F., Alford, B.L. & Taunton-Rigby, A. (1982) Molecular cloning and characterization of double-stranded cDNA coding for bovine chymosin. *Gene* **19**, 127–138.

Newman, M., Safro, M., Frazao, C., Kahn, G., Zdanov, A., Tickle, I.J., Blundell, T.J. & Andreeva, N. (1991) X-ray analysis of aspartic proteinases. IV. Structure and refinement at 2.2 Å resolution of bovine chymosin. *J. Mol. Biol.* **221**, 1295–1309.

Nielsen, F.S. & Foltmann, B. (1993) Activation of porcine pepsinogen A. *Eur. J. Biochem.* **217**, 137–142.

Örd, T., Kolmer, M., Villems, R. & Saarma, M. (1990) Structure of the human genomic region homologous to the bovine prochymosin-encoding gene. *Gene* **91**, 241–246.

Oppenheimer, C. (1926) *Die Fermente und Ihre Wirkungen*, vol. 2. Leipzig: G. Thieme Verlag, pp. 977–1024.

Pedersen, V.B. & Foltmann, B. (1975) Amino-acid sequence of the peptide segment liberated during activation of prochymosin (prorennin). *Eur. J. Biochem.* **55**, 95–103.

Pedersen, V.B., Christensen, K.A. & Foltmann, B. (1979) Investigations on the activation of bovine prochymosin. *Eur. J. Biochem.* **94**, 573–580.

Pungercar, J., Strukelj, B., Gubensek, F., Turk, V. & Kregar, I. (1990) Complete primary structure of lamb preprochymosin deduced from cDNA. *Nucleic Acids Res.* **18**, 4602.

Suzuki, J., Sasaki, K., Sasao, Y., Hamu, A., Kawasaki, H., Nishiyama, M., Horinouchi, S. & Beppu, T. (1989) Alteration of catalytic properties of chymosin by site-directed mutagenesis. *Protein Eng.* **2**, 563–569.

Visser, S., Van Rooijen, P.J. & Slangen, C.J. (1980) Peptide substrates for chymosin (rennin). *Eur. J. Biochem.* **108**, 415–421.

Visser, S., Slangen, C.J. & Van Rooijen, P.J. (1987) Peptide substrates for chymosin (rennin). *Biochem. J.* **244**, 553–558.

Warren, J.W. (1897) On the presence of a milk-curdling ferment (pexin) in the gastric mucous membrane of vertebrates. *J. Exp. Med.* **2**, 475–492.

Yasugi, S. & Mizuno, T. (1981) Purification and characterization of embryonic chicken pepsinogen, a unique pepsinogen with large molecular weight. *J. Biochem. (Tokyo)* **89**, 311–315.

Zunz, E. (1911) Chymase. In: *Biochem. Handlexicon*, vol. 5 (Abderhalden, E., ed.). Berlin: J. Springer Verlag, pp. 618–625.

*Bent Foltmann*
*Institute of Molecular Biology,*
*University of Copenhagen,*
*2A Øster Farimagsgade, DK-1353 Copenhagen K,*
*Denmark*

*Pal B. Szecsi*
*Department of Clinical Chemistry,*
*Roskilde County Hospital,*
*Køgevej 13-16, DK-4000 Roskilde, Denmark*
*Email:biotrek@symbion.dk*

# 275. Cathepsin E

## Databanks

*Peptidase classification: clan AA, family A1, MEROPS ID: A01.010*
*NC-IUBMB enzyme classification: EC 3.4.23.34*
*Chemical Abstracts Service registry number: 110910-42-4*
*Databank codes:*

| Species | SwissProt | PIR | EMBL (cDNA) | EMBL (genomic) |
|---|---|---|---|---|
| *Cavia porcellus* | P25796 | A43356 | M88653 | – |
| *Homo sapiens* | P14091 | A34401 | J05036 | M84413: exon 1 |
| | | A34643 | | M84417: exon 2 |
| | | A42038 | | M84418: exon 3 |
| | | S34467 | | M84419: exon 4 |
| | | S35663 | | M84420: exon 5 |
| | | | | M84421: exon 6 |
| | | | | M84422: exon 7 |
| | | | | M84423: exon 8 |
| | | | | M84424: exon 9 and complete CDS |

*continued overleaf*

| Species | SwissProt | PIR | EMBL (cDNA) | EMBL (genomic) |
|---|---|---|---|---|
| *Mus musculus* | – | JH0240 | X97399 | – |
| *Oryctolagus cuniculus* | P43159 | – | L08418 | – |
| *Rattus norvegicus* | P16228 | A34401 | D38104 | – |
| | | A34657 | D45187 | |
| | | B34643 | | |
| | | C34643 | | |

## Name and History

Early reports indicated that a second acid (aspartic) proteinase was present in vertebrate cells and tissues in addition to cathepsin D (Chapter 277) (Lapresle *et al.*, 1986). A variety of names were given to this enzyme, including: *cathepsin-D like proteinase* (Kageyama & Takahashi, 1980; Muto *et al.*, 1983), *gastric mucosa non-pepsin acid proteinase* (Yonezawa *et al.*, 1987a), *slow-moving proteinase* (Samloff *et al.*, 1987) and *erythrocyte membrane acid (or aspartic) proteinase* (Yamamoto & Marchesi, 1984). It was shown subsequently by immunochemical and biochemical means (Tarasova *et al.*, 1986; Samloff *et al.*, 1987; Yamamoto *et al.*, 1987; Yonezawa *et al.*, 1987b; Jupp *et al.*, 1988) that all of these activities were derived from the same enzyme for which the name *cathepsin E* has since been used universally. In contrast to the relatively ubiquitous cathepsin D (Chapter 277), cathepsin E has a limited cell and tissue distribution; a further complication arises in that this cell-specific expression is *not* consistent from one vertebrate species to another.

## Activity and Specificity

Hemoglobin has been, and continues to be, the most commonly used substrate; it can also be used in zymogram format (Jupp *et al.*, 1988). Towards this protein, cathepsin E has a pH optimum of about 3.1 (Yamamoto *et al.*, 1978; Samloff *et al.*, 1987), somewhat lower than that of cathepsin D (Chapter 277). In common with most other aspartic proteinases, cathepsin E has an ability to accommodate a relatively broad range of residues in the P1–P1′ positions, but strongly prefers both to be hydrophobic (Rao-Naik *et al.*, 1995); $\beta$-branched residues, e.g. Ile, Val, are readily accepted in P1′ but not in the P1 position. A basic residue, e.g. Lys, is acceptable in the P2 position, in contrast to the situation for cathepsin D (Scarborough & Dunn, 1994). The presence of a Pro residue in P4 may facilitate the presentation of the scissile peptide bond in an extended $\beta$ strand for cleavage by the enzyme (Kageyama *et al.*, 1995). The synthetic peptide substrates Lys-Pro-Ile-Glu-Phe┼Nph-Arg-Leu and Pro-Pro-Thr-Ile-Phe┼Nph-Arg-Leu both contain the hydrophobic *p*-nitrophenylalanine residue in P1′, where it serves as a chromogenic reporter group (Dunn *et al.*, 1994). This enables spectrophotometric assays to be carried out; however, this assay is not without its limitations: since the absorbance change (a) is only detectable at pH values below about 6; and (b) is measured at a wavelength of 300 nm (Dunn *et al.*, 1994) so that interference by high concentrations of unrelated macromolecules can cause difficulties. Thus, for example, assays of fractions during purification procedures are better performed using hemoglobin as substrate. Nevertheless, characterization including subsite mapping of purified cathepsin E has been made feasible by using systematic series of chromogenic substrates, based on these peptide sequences (Rao-Naik *et al.*, 1995). Against such substrates, $k_{cat}$ and $K_m$ are constant between pH 3 and 5 (Jupp *et al.*, 1988); activity is thus routinely measured at a pH of 3.0–3.5, which assists the distinction from other proteinase activities. The cleavage specificity of cathepsin E on the B chain of oxidized insulin at neutral pH has been described (Athauda *et al.*, 1991):

$$FVNQHLCGSHLVE \dotplus A \dotplus LYLVCGE \dotplus RGF \dotplus F \dotplus YTPKA$$

Cathepsin E is not affected by ATP at low pH values but is stabilized at higher pH values by physiological concentrations of ATP such that the activity is still manifested at pH 7 in the presence of this ligand (Thomas *et al.*, 1989a). Cathepsin E is inhibited in stoichiometric fashion by isovaleryl- and acetyl-pepstatins and a wide variety of synthetic peptidomimetic inhibitors has been described that inhibit this enzyme potently (e.g. Jupp *et al.*, 1990; Bird *et al.*, 1992; Rao *et al.*, 1993). However, none of these low molecular mass compounds inhibit cathepsin E specifically, e.g. in relation to cathepsin D (Chapter 277). They are thus of little value in discriminating the activity of cathepsin E from that of cathepsin D in dissecting their respective cellular functions. In this regard, the most useful diagnostic inhibitor is a protein ($M_r \sim 17\,000$) from *Ascaris lumbricoides* (Keilová & Tomášek, 1972); this does not inhibit cathepsin D but has a $K_i$ value of ~5 nM towards cathepsin E (Samloff *et al.*, 1987; Jupp *et al.*, 1988; Hill *et al.*, 1993) and is even more potent as a pepsin inhibitor. The protein sequence of an isoform of this inhibitor has been reported (Martzen *et al.*, 1990). Other natural protein inhibitors of cathepsin E include $\alpha_2$-macroglobulin (Thomas *et al.*, 1989b).

## Structural Chemistry

The gene for procathepsin E, like those of most vertebrate and fungal aspartic proteinase precursors, encodes a signal peptide, a propeptide of 35–40 residues and the mature enzyme. However, cathepsin E is totally distinct from all other known vertebrate aspartic proteinases which, in their mature forms, have $M_r \sim 40\,000$ and are (commonly) single polypeptide chains. Cathepsin E exists as a dimer and has a molecular mass of ~84 kDa. Located close to the N-terminus of the mature enzyme, cathepsin E has a unique cysteine residue which, through formation of an intermolecular disulfide bond, brings about dimerization of two identical subunits to give rise to the ~84 kDa form of the mature enzyme.

| Human: | ~L | D | M╫I | Q | F╫T | E | S | **C** | S | M | D | Q | S~ |
|--------|----|----|-----|----|-----|----|----|----|----|----|----|----|-----|
| Guinea pig: | ~L╫N | M | D | Q | – | – | – | – | **C** | S | T | I | Q | S~ |
| Rabbit: | ~V | D | M╫V | Q | Y | T | E | T | **C** | T | M | E | Q | S~ |
| Rat: | ~L | D | M | I | E | F╫S | E | S | **C** | N | V | D | K | G~ |
| Mouse: | ~L | D | M╫T | R | L╫S | E | S | **C** | N | V | Y | S | S~ |

*Figure 275.1* Location of the unique cysteine residue conserved in cathepsin E from the species indicated. This residue is not in the propart but is positioned just downstream from the N-termini (indicated by ╫) of the mature enzymes.

Procathepsin E ($M_r \sim 90\,000$) is also a homodimer, since it too contains this intermolecular disulfide bond. Exposure of the precursor to acidic pH results in autoactivation to generate the mature enzyme form of cathepsin E. In the human protein, activation occurs at two closely related bonds (indicated by ╫ in Figure 275.1), so that microheterogeneity is commonly observed at the N-terminus of mature human cathepsin E. The enzyme is a glycoprotein (Athauda *et al.*, 1990; Takeda-Esaki & Yamamoto, 1993); an *N*-glycosylation motif (~Asn-X-Thr~) is located immediately prior to the first active-site aspartic acid residue in the sequence ~Asn-Phe-Thr-Val-Ile-Phe-Asp-Thr-Gly~ in the enzyme from all of the species so far studied. Rat and guinea pig cathepsin E have a second motif (at different positions) towards the C-terminus of the molecule; mouse cathepsin E has both of these, i.e. three sites in total. The pI is approximately 4.1, depending on species. Although the enzyme has been crystallized, a structure for cathepsin E has not yet been solved by X-ray crystallography. However, since each ~42 kDa monomer is closely similar to other aspartic proteinases for which X-ray structures are available, e.g. pepsin (Chapter 272), renin (Chapter 284) and cathepsin D (Chapter 277), a homology-based model has been described for human cathepsin E (Rao-Naik *et al.*, 1995).

## Preparation

Cathepsin E has been purified from a variety of vertebrate cells and tissues, but it should be emphasized that the cell-specific expression of the procathepsin E gene does not appear to be maintained consistently between species. The mature enzyme has been purified to homogeneity from such natural sources as human gastric mucosa (Samloff *et al.*, 1987), human erythrocytes (Yamamoto & Marchesi, 1984), rat spleen (Yamamoto *et al.*, 1978), rat neutrophils (Yonezawa *et al.*, 1988) and rat epidermis (Hara *et al.*, 1993). Purification usually involves an initial step on DEAE-cellulose, followed by chromatography on ConA-Sepharose or by FPLC on a Mono-Q column and affinity chromatography on pepstatin-agarose or on an immobilized antibody. Protein samples are stored at $-20°C$ in neutral pH buffers containing 50% glycerol. Isolation of the precursor form, procathepsin E, is considerably more difficult but has been accomplished (Takeda-Ezaki & Yamamoto, 1993; Kageyama, 1995). Recombinant procathepsin E has been produced in *E. coli* (Hill *et al.*, 1993) and active, mature cathepsin E, indistinguishable from its naturally occurring counterpart, has been generated by autoactivation of the precursor at acidic pH. Recombinant human cathepsin E has also been produced in and purified from Chinese hamster ovary cells (Tsukuba *et al.*, 1993) and *Pichia pastoris* (Yamada *et al.*, 1994).

## Biological Aspects

The enzyme is present in mammals and almost certainly in lower animals including fish and bullfrog (Inokuchi *et al.*, 1994). It is present in epithelial linings of the gastrointestinal tract (particularly in the stomach of humans and especially guinea pigs); in lung, kidney, bladder, spleen (rat, mouse, but little in humans), thymus and other lymphoid-associated tissues (Arbustini *et al.*, 1994). It has been detected in red blood cells from rat and humans but not from guinea pig, cattle, goat or pig (Yonezawa & Nakamura, 1991); in contrast, neutrophils/monocytes from guinea pig and rat but not humans appear to contain cathepsin E (Sakai *et al.*, 1989; Yonezawa & Nakamura, 1991; Kageyama, 1995). Levels are substantially higher in lymphocytes originating from mouse compared to humans.

Cathepsin E is not a lysosomal enzyme; it appears to be contained within vesicular structures associated with the endoplasmic reticulum/endosomal compartments in a number of cells (Bennett *et al.*, 1992) except for mature erythrocytes where it is associated with the cytoplasmic face of the plasma membrane (Ueno *et al.*, 1989). The major features which determine that it is retained in these compartments and not secreted like pepsin (Chapter 272), renin (Chapter 284) or trafficked into lysosomes like cathepsin D (Chapter 277) appear to be located within the N-terminal 48 residues of the mature enzyme sequence. This region contains both the intermolecular disulfide bond and the one conserved glycosylation site (Finley & Kornfeld, 1994). The unique disulfide bond is also responsible for the stability of the mature enzyme at near neutral pH values. Recombinant single-chain forms of human cathepsin E have been engineered by mutating the unique Cys residue to Ala or Ser. The resultant mutant enzymes are less stable than wild-type but were not noticeably altered in activity or in their sensitivity to inhibition (Fowler *et al.*, 1995; Tsukuba *et al.*, 1996). A monomeric form that is unstable but still capable of dimerization has been reported to occur in human gastric cancer cells (Aoki *et al.*, 1995).

The single human procathepsin E locus is at chromosome 1q31–q32 (Azuma *et al.*, 1992). The gene structure mapped for human procathepsin E shows it to consist of nine exons, consistent with other aspartic proteinase genes. Control elements remain to be elucidated (Tsukada *et al.*, 1992) but transcription of the human and rat genes results in the generation of alternatively spliced products (Finley & Kornfeld, 1994; Okamoto *et al.*, 1995) as well as message for authentic procathepsin E. The enzyme has been shown to degrade neuropeptides, act on serum albumin and casein and to clot milk. Postulated physiological roles for cathepsin E include biogenesis of the vasoconstrictor peptide endothelin (Lees *et al.*, 1990; Bird *et al.*, 1992), antigen/invariant chain processing (Bennett *et al.*, 1992; Kageyama *et al.*, 1996) and in neurodegeneration associated with brain ischemia and aging (Nakanishi *et al.*, 1994).

## Distinguishing Features

SDS-PAGE under nonreducing/reducing conditions readily reveals the hallmark of cathepsin E – its dimeric/monomeric interconversion. Distinction from cathepsin D (Chapter 277) is made on the basis of zymography and susceptibility to inhibition by the *Ascaris* protein inhibitor.

## Further Reading

Cathepsin E is reviewed in Fusek & Vetvicka (1995). Details of enzyme purification and assay are found in Kageyama (1995).

## References

Aoki, T., Takasaki, T., Furukawa, T., Morikawa, J., Yano, T. & Watabe, H. (1995) Conversion of cathepsin E to enzymatic unstable form in gastric cancer cells. *Biol. Pharm. Bull.* **18**, 1522–1525.

Arbustini, E., Morbin, P., Diegoli, M., Grasso, M., Fasani, R., Vitulo, P., Fiocca, R., Cremaschi, P., Volpato, G., Martinelli, L., Vigano, M., Samloff, I.M. & Solcia, E. (1994) Coexpression of aspartic proteinases and human leukocyte antigen–DR in human transplanted lung. *Am. J. Pathol.* **145**, 310–321.

Athauda, S.B.P., Matsuzaki, O., Kageyama, T. & Takahashi, K. (1990) Structural evidence for two isozymic forms and the carbohydrate attachment site of human gastric cathepsin E. *Biochem. Biophys. Res. Commun.* **168**, 878–885.

Athauda, S.B.P., Takahashi, T., Inoue, H., Ichinose, M. & Takahashi, K. (1991) Proteolytic activity and cleavage specificity of cathepsin E at the physiological pH as examined towards the B chain of oxidized insulin. *FEBS Lett.* **292**, 53–56.

Azuma, T., Liu, W., Vander Laan, D.J., Bowcock, A.M. & Taggart, R.T. (1992) Human gastric cathepsin E gene. *J. Biol. Chem.* **267**, 1609–1614.

Bennett, K., Levine, T., Ellis, J.S., Peanasky, R.J., Samloff, I.M., Kay, J. & Chain, B.M. (1992) Antigen processing for presentation by class II major histocompatibility complex requires cleavage by cathepsin E. *Eur. J. Immunol.* **22**, 1519–1524.

Bird, J.E., Waldron, T.L., Little, D.K., Asaad, M.M., Dorso, C.R., DiDonato, G. & Norman, J.A. (1992) The effects of novel cathepsin E inhibitors on the big endothelin pressor response in conscious rats. *Biochem. Biophys. Res. Commun.* **182**, 224–231.

Dunn, B.M., Scarborough, P.E., Davenport, R. & Swietnicki, W. (1994) Analysis of proteinase specificity by studies of peptide substrates. In: *Methods in Molecular Biology*, vol. 36 (Dunn, B.M. & Pennington, M.W., eds). Totowa, NJ: Humana Press, pp. 225–243.

Finley, E.M. & Kornfeld, S. (1994) Subcellular localization and targeting of cathepsin E. *J. Biol. Chem.* **49**, 31259–31266.

Fowler, S.D., Kay, J., Dunn, B.M. & Tatnell, P.J. (1995) Monomeric human cathepsin E. *FEBS Lett.* **366**, 72–74.

Fusek, M. & Vetvicka, V. (1995) *Aspartic Proteinases: Physiology and Pathology.* Boca Raton, FL: CRC Press, pp. 207–219.

Hara, K., Fukuyama, K., Sakai, H., Yamamoto, K. & Epstein, W.L. (1993) Purification and immunohistochemical localization of aspartic proteinases in rat epidermis. *J. Invest. Dermatol.* **100**, 394–399.

Hill, J., Montgomery, D.S. & Kay, J. (1993) Human cathepsin E produced in *E. coli. FEBS Lett.* **326**, 101–104.

Inokuchi, T., Kobayashi, K. & Horiuchi, S. (1994) Purification and characterization of cathepsin E-type acid proteinase from gastric mucosa of bullfrog, *Rana catesbeiana. J. Biochem.* **115**, 76–81.

Jupp, R.A., Richards, A.D., Kay, J., Dunn, B.M., Wyckoff, J.B., Samloff, I.M. & Yamamoto, K. (1988) Identification of the aspartic proteinases from human erythrocyte membranes and gastric

mucosa (slow-moving proteinase) as catalytically equivalent to cathepsin E. *Biochem. J.* **254**, 895–898.

Jupp, R.A., Dunn, B.M., Jacobs, J.W., Vlasuk, G., Arcuri, K.E., Veber, D.F., Perlow, D.S., Payne, L.S., Boger, J., DeLaszlo, S., Chakravarty, P.K., TenBroeke, J., Hangauer, D.G., Ondeyka, D., Greenlees, W.J. & Kay, J. (1990) The selectivity of statine-based inhibitors against various human aspartic proteinases. *Biochem. J.* **265**, 871–878.

Kageyama, T. (1995) Procathepsin E and cathepsin E. *Methods Enzymol.* **248**, 120–136.

Kageyama, T. & Takahashi, K. (1980) A cathepsin D-like acid proteinase from human gastric mucosa *J. Biochem.* **87**, 725–735.

Kageyama, T., Ichinose, M. & Yonezawa, S. (1995) Processing of the precursors to neurotensin and other bioactive peptides by cathepsin E. *J. Biol. Chem.* **270**, 19135–19140.

Kageyama, T., Yonezawa, S., Ichinose, M., Miki, K. & Moriyama, A. (1996) Potential sites for processing of the human invariant chain by cathepsins D and E. *Biochem. Biophys. Res. Commun.* **223**, 549–553.

Keilová, H. & Tomášek, V. (1972) Effect of pepsin inhibitor from *Ascaris lumbricoides* on cathepsin D and E. *Biochim. Biophys. Acta* **284**, 461–464.

Lapresle, C., Puizdar, V., Porchon-Bertolotto, C., Joukoff, E. & Turk, V. (1986) Structural differences between rabbit cathepsin E and cathepsin D. *Biol. Chem. Hoppe-Seyler* **367**, 523–526.

Lees, W., Kalinka, S., Meech, J., Capper, S.J., Cook, N.D. & Kay, J. (1990) Generation of human endothelin by cathepsin E. *FEBS Lett.* **273**, 99–102.

Martzen, M.R., McMullen, B.A., Smith, N.E., Fujikawa, K. & Peanasky, R.J. (1990) Primary structure of the major pepsin inhibitor from *Ascaris suum. Biochemistry* **29**, 7366–7372.

Muto, N., Arai, K.M. & Tani, S. (1983) Purification and properties of a cathepsin D-like acid proteinase from rat gastric mucosa. *Biochim. Biophys. Acta* **745**, 61–69.

Nakanishi, H., Tominaga, K., Amano, T., Hirotsu, I., Inoue, T. & Yamamoto, K. (1994) Age-related changes in activities and localizations of cathepsins D, E, B, and L in the rat brain tissues. *Exp. Neurol.* **126**, 119–128.

Okamoto, K., Hu, H., Misumi, Y., Ikehara, Y. & Yamamoto, K. (1995) Isolation and sequencing of two cDNA clones encoding rat spleen cathepsin E and analysis of the activation of purified procathepsin E. *Arch. Biochem. Biophys.* **322**, 103–111.

Rao, C.M., Scarborough, P.E., Kay, J., Batley, B., Rapundalo, S., Klutchko, S., Taylor, M.D., Lunney, E.A., Humblet, C.C. & Dunn, B.M. (1993) Specificity in the binding of inhibitors to the active site of human/primate aspartic proteinases: analysis of $P_2$-$P_1$-$P_1'$-$P_2'$ variation. *J. Med. Chem.* **36**, 2614–2620.

Rao-Naik, C., Guruprasad, K., Batley, B., Rapundalo, S., Hill, J., Blundell, T.L., Kay, J. & Dunn, B.M. (1995) Exploring the binding preferences/specificity in the active site of human cathepsin E. *Proteins: Struct. Funct. Genet.* **22**, 168–181.

Sakai, H., Saku, T., Kato, Y. & Yamamoto, K. (1989) Quantitation and immunohistochemical localization of cathepsins E and D in rat tissues and blood cells. *Biochim. Biophys. Acta* **991**, 367–375.

Samloff, I.M., Taggart, R.T., Shiraishi, T., Branch, T., Reid, W.A., Heath, R., Lewis, R.W., Valler, M.J. & Kay, J. (1987) Slow moving proteinase. Isolation, characterization and immunohistochemical localization in gastric mucosa. *Gastroenterology* **93**, 77–84.

Scarborough, P.E. & Dunn, B.M. (1994) Redesign of the substrate specificity of human cathepsin D. *Protein Eng.* **7**, 495–502.

Takeda-Ezaki, M. & Yamamoto, K. (1993) Isolation and biochemical characterization of procathepsin E from human erythrocyte membranes. *Arch. Biochem. Biophys.* **304**, 352–358.

Tarasova, N.I., Szecsi, P.B. & Foltmann, B. (1986) An aspartic proteinase from human erythrocytes is immunochemically indistinguishable from a non-pepsin, electrophoretically slow moving proteinase from gastric mucosa. *Biochim. Biophys. Acta* **880**, 96–100.

Thomas, D.J., Richards, A.D., Jupp, R.A., Ueno, E., Yamamoto, K., Samloff, I.M., Dunn, B.M. & Kay, J. (1989a) Stabilisation of cathepsin E by ATP. *FEBS Lett.* **243**, 145–148.

Thomas, D.J., Richards, A.D. & Kay, J. (1989b) Inhibition of aspartic proteinases by $\alpha_2$-macroglobulin. *Biochem. J.* **259**, 905–907.

Tsukada, S., Ichinose, M., Miki, K., Tatematsu, M., Yonezawa, S., Matsushima, M., Kakei, N., Fukamachi, H., Yasugi, S., Kurokawa, K., Kageyama, T. & Takahashi, K. (1992) Tissue- and cell-specific control of guinea pig cathepsin E gene expression. *Biochem. Biophys. Res. Commun.* **187**, 1401–1408.

Tsukuba, T., Hori, H., Azuma, T., Takahashi, T., Taggart, R.T., Akamine, A., Ezaki, M., Nakanishi, H., Sakai, H. & Yamamoto, K. (1993) Isolation and characterization of recombinant human cathepsin E expressed in Chinese hamster ovary cells. *J. Biol. Chem.* **268**, 7276–7282.

Tsukuba, T., Sakai, H., Yamada, M., Maeda, H., Hori, H., Azuma, T., Akamine, A. & Yamamoto, K. (1996) Biochemical properties of the monomeric mutant of human cathepsin E expressed in Chinese hamster ovary cells: comparison with dimeric forms of the natural and recombinant cathepsin E. *J. Biochem.* **119**, 126–134.

Ueno, E., Sakai, H., Kato, Y. & Yamamoto, K. (1989) Activation mechanism of erythrocyte cathepsin E. Evidence for the occurrence of the membrane-associated active enzyme. *J. Biochem.* **105**, 878–882.

Yamada, M., Azuma, T., Matsuba, T., Iida, H., Suzuki, H., Yamamoto, K., Kohli, Y. & Hori, H. (1994) Secretion of human intracellular aspartic proteinase cathepsin E expressed in the methylotrophic yeast, *Pichia pastoris* and characterization of produced recombinant cathepsin E. *Biochim. Biophys. Acta* **1206**, 279–285.

Yamamoto, K. & Marchesi, V.T. (1984) Purification and characterization of acid proteinase from human erythrocyte membranes. *Biochim. Biophys. Acta* **790**, 208–218.

Yamamoto, K., Katsuda, N. & Kato, K. (1978) Affinity purification and properties of cathepsin E-like acid proteinase from rat spleen. *Eur. J. Biochem.* **92**, 499–508.

Yamamoto, K., Ueno, E., Uemura, H. & Kato, Y. (1987) Biochemical and immunochemical similarity between erythrocyte membrane aspartic proteinase and cathepsin E. *Biochem. Biophys. Res. Commun.* **148**, 267–272.

Yonezawa, S. & Nakamura, K. (1991) Species-specific distribution of cathepsin E in mammalian blood cells. *Biochim. Biophys. Acta* **1073**, 155–160.

Yonezawa, S., Tanaka, T., Muto, N. & Tani, S. (1987a) Immunochemical similarity between a gastric mucosa non-pepsin acid proteinase and neutrophil cathepsin E of the rat. *Biochem. Biophys. Res. Commun.* **144**, 1251–1256.

Yonezawa, S., Tanaka, T. & Miyauchi, T. (1987b) Cathepsin E from rat neutrophils: its properties and possible relations to cathepsin D-like and cathepsin E-like acid proteinases. *Arch. Biochem. Biophys.* **256**, 499–508.

Yonezawa, S., Fujii, K., Maejima, Y., Tamoto, K., Mori, Y. & Muto, N. (1988) Further studies on rat cathepsin E: subcellular localization and existence of the active subunit form. *Arch. Biochem. Biophys.* **267**, 176–183.

*John Kay*
*Protein Processing and Design Unit,*
*School of Molecular & Medical Biosciences,*
*University of Wales, College of Cardiff, PO Box 911,*
*Cardiff CF1 3US, UK*
*Email: KAYJ@Cardiff.ac.uk*

*Peter J. Tatnell*
*Protein Processing and Design Unit,*
*University of Wales, College of Cardiff, PO Box 911,*
*School of Molecular & Medical Biosciences,*
*PO Box 911, Cardiff CF1 3US, UK*

# 276. Gastricsin

## Databanks

*Peptidase classification: clan AA, family A1, MEROPS ID: A01.003*
*NC-IUBMB enzyme classification: EC 3.4.23.3*
*Chemical Abstracts Service registry number: 9012-71-9*
*Databank codes:*

| Species | SwissProt | PIR | EMBL (cDNA) | EMBL (genomic) |
|---|---|---|---|---|
| *Cavia porcellus* | – | – | M88652 | – |
| *Cavia porcellus* | – | B43356 | – | – |

*continued overleaf*

| Species | SwissProt | PIR | EMBL (cDNA) | EMBL (genomic) |
|---|---|---|---|---|
| *Homo sapiens* | P20142 | A23458 | J04443 | M18659: exon 1 |
|  |  | A29937 | U75272 | M18660: exon 2 |
|  |  | A31811 |  | M18661: exon 3 |
|  |  | A91125 |  | M18662: exon 4 |
|  |  | I54213 |  | M18663: exon 5 |
|  |  | PX0028 |  | M18664: exon 6 |
|  |  |  |  | M18665: exon 7 |
|  |  |  |  | M18666: exon 8 |
|  |  |  |  | M18667: exon 9 and complete CDS |
|  |  |  |  | M23069: exon 1 |
|  |  |  |  | M23070: exon 2 |
|  |  |  |  | M23071: exon 3 |
|  |  |  |  | M23072: exon 4 |
|  |  |  |  | M23073: exon 5 |
|  |  |  |  | M23074: exon 6 |
|  |  |  |  | M23075: exon 7 |
|  |  |  |  | M23076: exon 8 |
|  |  |  |  | M23077: exon 9 and complete CDS |
|  |  |  |  | M96835: unannotated |
| *Macaca fuscata* | P03955 | A00986 | X59754 | – |
|  |  | A22402 |  |  |
|  |  | S16066 |  |  |
|  |  | S19683 |  |  |
| *Rana catesbeiana* | – | A39314 | M73750 | – |
| *Rattus norvegicus* | P04073 | A05145 | X04644 | M25985: exon 1 |
|  |  | A24608 |  | M25986: exon 2 |
|  |  | A33510 |  | M25987: exon 3 |
|  |  | A61298 |  | M25988: exon 4 |
|  |  | C22434 |  | M25989: exon 5 |
|  |  |  |  | M25990: exon 6 |
|  |  |  |  | M25991: exon 7 |
|  |  |  |  | M25992: exon 8 |
|  |  |  |  | M25993: exon 9 and complete CDS |
| *Sus scrofa* | P30879 | S21754 | – | – |

Brookhaven Protein Data Bank three-dimensional structures:

| Species | ID | Resolution | Notes |
|---|---|---|---|
| *Homo sapiens* | 1HTR | 1.62 | precursor |

## Name and History

*Gastricsin* is an acid endopeptidase present in the gastric juice of many vertebrate species. The presence of pepsin A (Chapter 272) in human gastric juice has been known since the nineteenth century and a second protease was postulated for a long time because the pH activity curve of human gastric juice often revealed two maxima (Buchs, 1949). When it was isolated from human gastric juice it was named gastricsin (Richmond *et al.*, 1958; Tang *et al.*, 1959). Independently, Ryle & Porter (1959) isolated from pig stomach *parapepsin II*, which was later found to be pig gastricsin. Since the enzymic activities of pepsin and gastricsin overlap, the name *pepsin C* was also used for gastricsin. The amino acid sequences of three gastric enzymes, pepsin A, gastricsin and chymosin (Chapter 274), eventually showed that they diverge in evolution to a similar degree from one another and are indeed separate enzymes. The name gastricsin has been recommended by the IUBMB. In some clinical literature, the name *pepsin II* was also used for human gastricsin (Samloff, 1971). Gastricsin is also found in normal human seminal plasma (*seminal pepsin*) and breast tumor tissue.

## Activity and Specificity

The specific activity for hemoglobin digestion by gastricsin is about 20% higher than that by pepsin (Tang *et al.*, 1959). Only partial specificity information is available for gastricsin since systematic study of subsite residue preferences has not been done. Gastricsin and pepsin share most of the cleavage sites in the hydrolysis of glucagon and oxidized ribonuclease A. These include hydrophobic as well as other types of residues at both P1 and P1'. Gastricsin may be distinguished from pepsin by virtue of its ability to hydrolyze certain substrates which have Tyr at P1. Synthetic *N*-carbobenzoxy dipeptides Tyr$+$Ala, Tyr$+$Thr, Tyr$+$Leu, Tyr$+$Ser and Trp$+$Ala are hydrolyzed by gastricsin but not by pepsin (Huang & Tang, 1969). Taken together, these results suggest that gastricsin has a broad specificity similar to that of pepsin but with some differences. The fact that Ac-Phe$+$Leu-Val-His ($k_{cat}/K_m$ about $3\,s^{-1}\,mM^{-1}$) is a better substrate for pig gastricsin than Ac-Tyr$+$Leu-Val-His ($k_{cat}/K_m$ about $1\,s^{-1}\,mM^{-1}$) suggests that Phe is preferable to Tyr as a P1 residue for gastricsin (Auffret & Ryle, 1979). The difference in the hydrolysis of Tyr and Trp dipeptides by the two enzymes is probably because Tyr does not bind well to pepsin in P1. Peptide Ac-Phe-diiodotyrosine, on the other hand, is hydrolyzed by human pepsin but not by gastricsin (Chiang *et al.*, 1966). This suggests that diiodotyrosine is poorly accommodated at the P1' site of gastricsin. Dunn *et al.* (1986) compared the P3 specificity of several aspartic proteases including gastricsin using a series of substrates differing only at P3 residues. The peptides of this series, derived from Lys-Pro-Xaa-Glu-Phe$+$Nph-Arg-Leu, were hydrolyzed well by pepsin and most other aspartic proteases but were only minimally cleaved by gastricsin. These results confirm the existence of specificity differences between gastricsin and other aspartic proteases.

Human gastricsin hydrolyzes hemoglobin over a pH range from below 1 to 5 with an optimum near pH 3, which differs from the maximum of pepsin near pH 2 (Richmond *et al.*, 1958; Tang *et al.*, 1959). Auffret & Ryle (1979), using a peptide substrate, Ac-Phe$+$Leu-Val-His-NH$_2$, observed the maximal $k_{cat}/K_m$ value for pig gastricsin to be at pH 3. Two active-site p$K_a$ values of 1.42 and 4.88 were also determined. The pepsin assays with hemoglobin substrates (Anson, 1938; Lin *et al.*, 1989) are applicable to purified gastricsin. However, in the presence of pepsin, such as in gastric juice, hemoglobin would determine the combined activities of both enzymes. Several methods have been devised to separately quantitate pepsin and gastricsin in a mixture. Chiang *et al.* (1966) measured the amount of pepsin and gastricsin by using the combined activities on hemoglobin substrate and specific pepsin hydrolysis of Ac-diiodotyrosyl-phenylalanine. A radioimmunoassay was devised by Ichinose *et al.* (1982) to measure pepsin (pepsin I) and gastricsin (pepsin II) in clinical samples. This procedure is available in commercial kits and is widely used in Japan. A 'rocket' immunochemical method for separately determining pepsin and gastricsin was devised by Axelsson *et al.* (1983). This method has an accuracy of better than 9%. Jones *et al.* (1993) reported a method using high-performance ion-exchange chromatography to quantify gastricsin and pepsin A isozymes.

Like pepsin, gastricsin is inhibited by pepstatin although the inhibition potency toward gastricsin is somewhat weaker (Marciniszyn *et al.*, 1976; Reid *et al.*, 1984). The change of the valine residues in lactoyl-Val-Sta-Ala-Sta-NH$_2$ (Sta: statine), a pepstatin analog, to a lysine or glutamic acid improves the IC$_{50}$ toward human gastricsin about 21- and 58-fold respectively (Baxter *et al.*, 1990). Since statine is a transition-state analog (Marciniszyn *et al.*, 1976), these results suggest that gastricsin favors lysine and glutamic acid at the P2 position. Gastricsin is inactivated by specific diazo

inactivators of aspartic proteases, such as DAN (Hunkapiller *et al.*, 1970), and by 1,2-epoxy-3-(*p*-nitrophenoxy)propane (Martin *et al.*, 1982). These inhibitors, which are commercially available, are expected to esterify the active-site aspartyl residues, as in pepsin.

## Structural Chemistry

The first complete chemical structure of gastricsin was determined for the Japanese macaque enzyme using chemical methods (Kageyama & Takahashi, 1986). The first complete human gastricsin sequence was deduced from its gene sequence (Hayano *et al.*, 1988) and confirmed by its cDNA sequence (Taggart *et al.*, 1989). Gastricsin sequences from rat (Ichihara *et al.*, 1986), frog (Yakabe *et al.*, 1991) and guinea pig (Kageyama *et al.*, 1992) are also known. Human gastricsin is a 329 residue single-chain protein. Over half of the residues in gastricsin sequence are identical to the homologous pepsin sequence. Especially conserved are regions near the active-site Asp32 and Asp215 (pepsin numbers). A phosphate-containing peptide has been observed in the peptide mapping of human gastricsin (Tang *et al.*, 1967), but the location of the phosphate group is unknown. The three-dimensional structure of gastricsin has not been determined. The crystal structure of its precursor, progastricsin, has been solved (Moore *et al.*, 1995). Considering that the pepsin moiety in pepsinogen crystal structure is very close to that of pepsin itself (with the exception of the first $\beta$ strand in pepsin; James & Sielecki, 1986; Hartsuck *et al.*, 1992), the same close structural relationship between the gastricsin moiety in progastricsin and gastricsin itself can be expected. Therefore, the important structural features, including the active-site cleft, the flap, the catalytic aspartyl residue, the overall folding and the internal 2-fold symmetry in conformation, found in pepsin and fungal aspartic proteases, should also be present in the gastricsin three-dimensional structure.

## Progastricsin

Like pepsin, gastricsin is synthesized as a zymogen, progastricsin, which contains an extra 43 residue propeptide at the N-terminus. The crystal structure of progastricsin (Moore *et al.*, 1995) shows that the propeptide is largely helical in conformation in contrast to the predominantly $\beta$ structure in the gastricsin moiety. As in pepsinogen (Chapter 272), the center part of the propeptide covers over the active-site cleft of the enzyme and renders the cleft inaccessible to the substrates. Progastricsin is converted to gastricsin in acidic solutions of pH below 5. This acid-triggered activation is likely caused by an intramolecularly catalyzed mechanism as in pepsinogen A, although this has not been clearly demonstrated. The strong structural homology of the two zymogens and the observation of the first cleavage at a Phe$+$Leu bond between residues 27 and 28 of the pro region of progastricsin (Foltmann & Jensen, 1982), however, are supportive of an intramolecular mechanism similar to that of pepsinogen, in which the initial event is a selective local 'denaturation' involving the central part of the pro region followed by the cleavage of the Phe-Leu bond by the active site of the same zymogen molecule (see Chapter 272). The complete removal of the remainder of the propeptide may proceed by a bimolecular hydrolysis.

## Preparation

Human gastricsin was first purified from dialyzed human gastric juice by anion-exchange chromatography on a column of Amberlite IRC-50 (CG-50; Richmond *et al.*, 1958), which was followed by crystallization (Tang *et al.*, 1959). In a large-scale preparation, 100–200 mg of gastricsin can be obtained from a liter of human gastric juice (Tang, 1970). Human progastricsin has been purified from stomach tissue (Foltmann & Jensen, 1982; Ivanov *et al.*, 1990), seminal plasma (Samloff & Liebmann, 1972; Ruenwongsa & Chulavatnatol, 1975), prostate (Chiang *et al.*, 1981) and breast cyst fluid (Sanchez *et al.*, 1992). Procedures have also been reported for the purification of progastricsin and/or gastricsin from pig (Ryle & Porter, 1959; Chiang *et al.*, 1967; Ryle, 1970; Foltmann *et al.*, 1992), guinea pig (Kageyama *et al.*, 1992), Japanese macaque (Kageyama & Takahashi, 1986) and tuna fish (Takahashi, 1996). Gastricsin is not commercially available.

## Biological Aspects

Both gastricsin and pepsin are found in the stomach of many vertebrates including fish (Takahashi, 1996). In human gastric juice, the ratio of pepsin to gastricsin ranges from 1 to 5 (Chiang *et al.*, 1966; Foltmann & Jensen, 1982). Thus, both proteases are significant in digesting proteins in human stomach. Because of the similar yet somewhat distinct enzyme properties, the physiological functions of these two enzymes in the stomach are partially redundant and partially complementary. In rodent stomachs, gastricsin is the major protease (Furihata *et al.*, 1980; Ichihara *et al.*, 1986). Progastricsin is synthesized by the chief cells in the gastric mucosa of all parts of the stomach, in contrast to pepsinogen A which is mainly synthesized in the fundus (the central part) of the stomach. When both zymogens are synthesized, they are localized in the same cells (Cornaggia *et al.*, 1986; Waalewijen *et al.*, 1991). Like pepsinogen A, progastricsin is also secreted into human serum. But unlike pepsinogen A, serum progastricsin is reabsorbed in the kidney and not secreted in the urine (ten Kate *et al.*, 1989). Serum progastricsin level and its ratio to serum pepsinogen A are apparently related also to stomach ulcer (Samloff *et al.*, 1986; Matsushima *et al.*, 1995). The amount of serum progastricsin and its ratio to pepsinogen A have been developed into a clinical diagnostic tool for stomach diseases including gastric cancer (Defize *et al.*, 1987; Miki *et al.*, 1993). A mass screening method including the analysis of serum pepsinogen A and progastricsin levels has significantly increased the early detection of gastric cancer (Miki *et al.*, 1995).

The presence of an acid protease zymogen in human seminal fluid was reported by Lundquist and Seedorff (1952). It was later identified to be progastricsin (Seiffers *et al.*, 1965; Hirsch-Marie & Conte, 1967; Samloff & Liebmann, 1972), which is present at an average concentration of $42\,mg\,ml^{-1}$ (Szecsi *et al.*, 1989). Seminal fluid progastricsin appears unrelated to the fertilization process. It is activated in the acidic environment of vagina (pH < 4.5) and likely has the main function of degrading seminal proteins to prevent immunogenic responses (Szecsi *et al.*, 1989). Seminal progastricsin is synthesized in the prostate (Chiang *et al.*, 1981; Moriyama *et al.*, 1985). The sequence of human prostate progastricsin cDNA suggests that the progastricsin is produced in human seminal vesicles and the zymogens from prostate and stomach are the products of the same gene (Szecsi *et al.*, 1995). Moriyama *et al.* (1983) have described an enzyme from macaque lung, called procathepsin D-II, which has an identical N-terminal sequence to macaque progastricsin. However, this enzyme does not immunochemically cross-react with progastricsin antiserum, suggesting that it may be a close but different enzyme. The biological function of this enzyme has not been identified.

The human progastricsin gene (Hayano *et al.*, 1988), which is located in chromosome 6 (Taggart *et al.*, 1989; Pals *et al.*, 1989), has an intron–exon organization similar to those of other human aspartic protease zymogens such as pepsinogen, prorenin (Chapter 284) and cathepsins D (Chapter 277) and E (Chapter 275). Progastricsin is normally not synthesized in human breast tissue but is found in breast tumors and breast cysts (Sanchez *et al.*, 1992; Diez-Itza *et al.*, 1993), apparently as a result of an upregulation of progastricsin gene expression by hormones, including androgens, glucocorticoids and progesterone (Balbin & López-Otin, 1996). However, progastricsin expression appears to be associated with the breast tumor type which leads to more favorable clinical outcomes (Vizoso *et al.*, 1995).

## References

Anson, M.L. (1938) The estimation of pepsin, trypsin, papain, and cathepsin with hemoglobin. *J. Gen. Physiol.* **22**, 79–89.

Auffret, C.A. & Ryle, A.P. (1979) The catalytic activity of pig pepsin C towards small synthetic substrates. *Biochem. J.* **179**, 239–246.

Axelsson, C.K., Axelsen, N.H., Szecsi, P.B. & Foltmann, B. (1983) Determination of pepsin (EC 3.4.23.1) and gastricsin (EC 3.4.23.3) in gastric juice by rocket immunoelectrophoresis. *Clin. Chim. Acta* **129**, 323–331.

Balbin, M. & López-Otin, C. (1996) Hormonal regulation of human pepsinogen C gene in breast cancer cells. *J. Biol. Chem.* **271**, 15175–15181.

Baxter, A., Campbell, C.J., Grinham, C.J., Keane, R.M., Lawton, B.C. & Pendlebury, J.E. (1990) Substrate and inhibitor studies with human gastric aspartic proteinases. *Biochem. J.* **267**, 665–669.

Buchs, S. (1949) Pepsin, cathepsin and parachymosin as equivalent and integrant constituents of stomach protease. *Enzymologia* **13**, 208–222.

Chiang, L., Sanchez-Chiang, L., Wolf, S. & Tang, J. (1966) The separate determination of human pepsin and gastricsin. *Proc. Exp. Biol. Med.* **122**, 700–704.

Chiang, L., Sanchez-Chiang, L., Mills, J.N. & Tang, J. (1967) Purification and properties of porcine gastricsin. *J. Biol. Chem.* **242**, 3098–3102.

Chiang, L., Contreras, L., Chiang, J. & Ward, P.H. (1981) Human prostatic gastricsinogen: the precursor of seminal fluid acid proteinase. *Arch. Biochem. Biophys.* **210**, 14–20.

Cornaggia, M., Capella, C., Riva, C., Finzi, G. & Solcia, E. (1986) Electron immuno-cytochemical localization of pepsinogen I (PGI) in chief cells, mucous-neck cells and transitional mucous-neck/chief cells of the human fundic mucosa. *Histochemistry* **85**, 5–11.

Defize, J., Pronk, J.C., Frants, R.R., Ooma, E.C.M., Kreuning, J., Kostense, P., Eriksson, A.W. & Meuwissen, S.G.M. (1987) Clinical significance of pepsinogen A isozymogens, serum pepsinogens A and C levels and serum gastrin levels. *Cancer* **59**, 952–958.

Diez-Itza, I., Merino, A.M., Tolivia, J., Sanchez, L.M. & López-Otin, C. (1993) Expression of pepsinogen C in human breast tumors and correlation with clinicopathologic parameters. *Br. J. Cancer* **68**, 637–640.

Dunn, B.M., Jimenez, M., Parten, B.F., Valler, M.J., Rolph, C.E. & Kay, J. (1986) A systematic series of synthetic chromophobic substrates for aspartic proteinases. *Biochem. J.* **237**, 899–906.

Foltmann, B. & Jensen, A.L. (1982) Human progastricsin: analysis of intermediates during activation into gastricsin and determination of the amino acid sequence in the propart. *Eur. J. Biochem.* **128**, 63–70.

Foltmann, B., Drohse, H.B., Nielsen, P.K. & James, M.N. (1992) Separation of porcine pepsinogen A and progastricsin: sequencing of the first 73 amino acid residues in progastricsin. *Biochim. Biophys. Acta* **1121**, 75–82.

Furihata, C., Saito, D., Fujiki, H., Kanai, Y., Matsushima, T. & Sugimura, T. (1980) Purification and characterization of pepsinogens and a unique pepsin from rat stomach. *Eur. J. Biochem.* **105**, 43–50.

Hartsuck, J.A., Koelsch, G. & Remington, S.J. (1992) The high-resolution crystal structure of porcine pepsinogen. *Protein Struct. Funct. Genet.* **13**, 1–25.

Hayano, T., Sogawa, K., Ichihara, Y., Fujii-Kuriyama, Y. & Takahashi, K. (1988) Primary structure of human pepsinogen C gene. *J. Biol. Chem.* **263**, 1382–1385.

Hirsch-Marie, H. & Conte, M. (1967) Étude de la protéase acide du liquide séminal humain [Study of the acid protease of human seminal plasma]. *Bull. Chim. Biol.* **49**, 147–155.

Huang, W.-Y. & Tang, J. (1969) On the specificity of human gastricsin and pepsin. *J. Biol. Chem.* **244**, 1085–1091.

Hunkapiller, M., Heinze, J.E. & Mills, J.N. (1970) Comparative studies on the effect of specific inactivators on human gastricsin and pepsin. *Biochemistry* **9**, 2897–2902.

Ichihara, Y., Sogawa, K., Morohashi, K., Fujii-Kuriyama, Y. & Takahashi, K. (1986) Nucleotide sequence of a near full-length cDNA coding for pepsinogen of rat gastric mucosa. *Eur. J. Biochem.* **161**, 7–12.

Ichinose, M., Miki, K., Furihata, C., Kageyama, T., Hayashi, R., Niwa, H., Oka, H., Matsushima, T. & Takahashi, K. (1982) Radioimmunoassay of serum group I and group II pepsinogens in normal controls and patients with various disorders. *Clin. Chim. Acta* **126**, 183–191.

Ivanov, P.K., Chernaya, M.M., Gustchina, A.E., Pechik, I.V., Nikonov, S.V. & Tarasova, N.I. (1990) Isolation, crystallization and preliminary X-ray diffraction data of human progastricsin. *Biochim. Biophys. Acta* **1040**, 308–310.

James, M.N.G. & Sielecki, A.R. (1986) Molecular structure of an aspartic proteinase zymogen, porcine pepsinogen at 1.8 Å resolution. *Nature* **319**, 33–38.

Jones, A.T., Balan, K.K., Jenkins, S.A., Sutton, R., Critchley, M. & Roberts, N.B. (1993) Assay of gastricsin and individual pepsins in human gastric juice. *J. Clin. Pathol.* **46**, 254–258.

Kageyama, T. & Takahashi, K. (1986) The complete amino acid sequence of monkey progastricsin. *J. Biol. Chem.* **261**, 4406–4419.

Kageyama, T., Ichinose, M., Tsukada, S., Miki, K., Kurokawa, K., Koiwai, O., Tanji, M., Yakabe, E., Athauda, S.B. & Takahashi, K. (1992) Gastric procathepsin E and progastricsin from guinea pig,

purification, molecular cloning of cDNAs, and characterization of enzymic properties, with special reference to procathepsin E. *J. Biol. Chem.* **267**, 16450–16459.

Lin, X., Wong, R.N.S. & Tang, J. (1989) Synthesis, purification, and active site mutagenesis of recombinant porcine pepsinogen. *J. Biol. Chem.* **264**, 4482–4489.

Lundquist, F. & Seedorff, H.H. (1952) Pepsinogen in human seminal fluid. *Nature* **170**, 1115–1116.

Marciniszyn, J., Jr, Hartsuck, J.A. & Tang, J. (1976) Mode of inhibition of acid proteases by pepstatin. *J. Biol. Chem.* **251**, 7088–7094.

Martin, P., Trieu-Cuot, P., Collin, J.C. & Ribadeau-Dumas, B. (1982) Purification and characterization of bovine gastricsin. *Eur. J. Biochem.* **122**, 31–39.

Matsushima, M., Miki, K., Ichinose, M., Kakei, N., Yahagi, N., Kido, M., Shimizu, Y., Ishihama, S., Tsukada, S., Kurokawa, K. & Takahashi, K. (1995) Serum pepsinogen values as possible markers for evaluating the possibility of peptic ulcer recurrence under $H_2$-blocker half-dose maintenance therapy. *Adv. Exp. Biol. Med.* **362**, 131–137.

Miki, K., Ichinose, M., Ishikawa, K.B., Yahagi, N., Matsushima, M., Kakei, N., Tsukada, S., Kido, M., Ishihama, S., Shimizu, Y., Suzuki, T. & Kurokawa, K. (1993) Clinical application of serum pepsinogen I and II levels for mass screening to detect gastric cancer. *Jpn J. Cancer Res.* **84**, 1086–1090.

Miki, K., Ichinose, M., Kakei, N., Yahagi, N., Matsushima, M., Tsukada, S., Ishihama, S., Shimizu, Y., Suzuki, T., Kurokawa, K. & Takahashi, K. (1995) The clinical application of the serum pepsinogen I and II levels as a mass screening method for gastric cancer. *Adv. Exp. Med. Biol.* **362**, 139–143.

Moore, S.A., Sielecki, A.R., Chernaia, M.M., Tarasova, N.I. & James, M.N.G. (1995) Crystal and molecular structures of human progastricsin at 1.62 Å resolution. *J. Mol. Biol.* **247**, 466–485.

Moriyama, A., Kageyama, T. & Takahashi, K. (1983) Identification of monkey lung procathepsin D-II as a pepsinogen-C-like acid protease zymogen. *Eur. J. Biochem.* **16**, 687–692.

Moriyama, A., Kageyama, T., Takahashi, K. & Sasaki, M. (1985) Purification of Japanese monkey prostate acid protease zymogen and its identification as a pepsinogen C-like zymogen. *J. Biochem.* **98**, 1255–1261.

Pals, G., Azuma, T., Mohandas, T.K., Bell, G.I., Bacon, J., Samloff, I.M., Walz, D.A., Barr, P.J. & Taggart, R.T. (1989) Human pepsinogen C (progastricsin) polymorphism: evidence for a single locus located at 6p21.1-pter. *Genomics* **4**, 137–145.

Reid, W.A., Vongsorasak, L., Svasti, J., Valler, M.J. & Key, J. (1984) Identification of the acid proteinase in human seminal fluid as a gastricsin originating in the prostate. *Cell Tissue Res.* **236**, 597–600.

Richmond, V., Tang, J., Wolf, S., Trucco, R.E. & Caputto, R. (1958) Chromatographic isolation of gastricsin, the proteolytic enzyme from gastric juice with pH optimum 3.2. *Biochim. Biophys. Acta* **29**, 453–454.

Ruenwongsa, P. & Chulavatnatol, M. (1975) Acidic protease from human seminal plasma. Purification and some properties of active enzyme and of proenzyme. *J. Biol. Chem.* **250**, 7574–7578.

Ryle, A. (1970) The porcine pepsins and pepsinogens. *Methods Enzymol.* **19**, 316–336.

Ryle, A.P. & Porter, R.R. (1959) Parapepsins: two proteolytic enzymes associated with porcine pepsin. *Biochem. J.* **73**, 75–86.

Sanchez, L.M., Freije, J.P., Merino, A.M., Vizoso, F., Foltmann, B. & López-Otín, C. (1992) Isolation and characterization of a pepsin

C zymogen produced by human breast tissue. *J. Biol. Chem.* **267**, 24725–24731.

Samloff, I.M. (1971) Pepsinogens, pepsins, and pepsin inhibitors. *Gastroenterology* **60**, 586–604.

Samloff, I.M. & Liebmann, W.M. (1972) Purification and immunochemical characterization of group II pepsinogens in seminal fluid. *Clin. Exp. Immunol.* **11**, 405–414.

Samloff, I.M., Stemmermann, G.N., Geilbrun, L.K. & Nomura, A. (1986) Elevated serum pepsinogen I and II levels differ as risk factors for duodenal ulcer and gastric ulcer. *Gastroenterology* **90**, 570–576.

Seiffers, M.J., Segal, H.L. & Miller, L.L. (1965) Preliminary appraisal of the role of seminal pepsinogen I in human sterility. *Fertil. Steril.* **16**, 202–207.

Szecsi, P.B., Dalgaard, D., Stakemann, G., Wagner, G. & Foltmann, B. (1989) The concentration of pepsinogen C in human semen and the physiological activation of zymogens in the vagina. *Biol. Reprod.* **40**, 653–659.

Szecsi, P.B., Halgreen, H., Wong, R.N.S., Kjaer, T. & Tang, J. (1995) Cellular origin, complementary deoxyribonucleic acid and N-terminal amino acid sequences of human seminal progastricsin. *Biol. Reprod.* **53**, 227–233.

Taggart, R.T., Cass, L.G., Mohandas, T.K., Derby, P., Barr, P.J., Pals, G. & Bell, G. (1989) Human pepsinogen C (progastricsin). *J. Biol. Chem.* **264**, 375–379.

Takahashi, K. (1996) The primary structure of the major pepsinogen from the gastric mucosa of tuna stomach. *J. Biochem.* **120**, 647–656.

Tang, J. (1970) Gastricsin and pepsin. *Methods Enzymol.* **19**, 406–421.

Tang, J., Wolf, S., Caputto, R. & Trucco, R.E. (1959) Isolation and crystallization of gastricsin from human gastric juice. *J. Biol. Chem.* **234**, 1174–1178.

Tang, J., Mills, J., Chiang, L. & de Chiang, L. (1967) Comparative studies of the structure and specificity of human gastricsin, pepsin and zymogens. *Ann. N.Y. Acad. Sci.* **140**, 688–696.

ten Kate, R.W., Pals, G., Res, J.C., van Kamp, G.J., Verheugt, F.W., Pronk, J.C., Donker, A.J., Ericksson, A.W. & Meuwissen, S.G. (1989) The glomerular sieving of pepsinogens A and C in man. *Eur. J. Clin. Invest.* **19**, 306–310.

Visozo, F., Sanchez, L.M., Diez-Itza, I., Merino, A.M. & López-Otín, C. (1995) Pepsinogen C is a new prognostic marker in primary breast cancer. *J. Clin. Oncol.* **13**, 54–61.

Waalewijen, R.A., Meuwissen, S.G.M., Pals, G. & Hoetsmit, E.C.M. (1991) Location of pepsinogens (A and C) and cellular differentiation of pepsinogen-synthesizing cells in human gastric mucosa. *Eur. J. Cell Biol.* **54**, 55–66.

Yakabe, E., Tanji, M., Ichinose, M., Goto, S., Miki, K., Kurokawa, K., Ito, H., Kageyama, T. & Takahashi, K. (1991) Purification, characterization, and amino acid sequences of pepsinogens and pepsins from the esophageal mucosa of bullfrog (*Rana catesbeiana*). *J. Biol. Chem.* **266**, 22436–22443.

*Jordan Tang*
*Protein Studies Program,*
*Oklahoma Medical Research Foundation,*
*825 NE 13 Street, Oklahoma City, OK 73104, USA*
*Email: jordan-tang@omrf.uokhsc.edu*

# 277. Cathepsin D

## Databanks

*Peptidase classification: clan AA, family A1, MEROPS ID: A01.009*
*NC-IUBMB enzyme classification: EC 3.4.23.5*
*Chemical Abstracts Service registry number: 9025-26-7*
*Databank codes:*

| Species | SwissProt | PIR | EMBL (cDNA) | EMBL (genomic) |
|---|---|---|---|---|
| *Aedes aegypti* | Q03168 | A45117 | M95187 | – |
| *Ancyclostoma caninum* | – | – | U34888 | – |
| *Blattella germanica* | P54958 | – | U28863 | – |
| *Bos taurus* | P80209 | A31918 S32383 S37419 | – | – |
| *Caenorhabditis elegans* | – | – | U34889 | – |

| Species | SwissProt | PIR | EMBL (cDNA) | EMBL (genomic) |
|---|---|---|---|---|
| *Gallus gallus* | Q05744 | – | S49650 | – |
| *Homo sapiens* | P07339 | A25771 | M11233 | L12980: exon 1 |
| | | I57716 | X05344 | M63134: exons 1 and 2 |
| | | PC2066 | | M63135: exons 3 and 4 |
| | | S30749 | | M63136: exon 5 |
| | | | | M63137: exon 6 |
| | | | | M63138: exons 7–9 and complete CDS |
| | | | | S52557: promoter |
| | | | | S74689: promoter |
| *Mus musculus* | P18242 | S12587 | X52886 | X68378: exon 1 |
| | | S14704 | X53337 | X68379: exon 2 |
| | | | | X68380: exon 3 |
| | | | | X68381: exon 4 |
| | | | | X68382: exon 5 |
| | | | | X68383: exons 6–9 |
| *Rattus norvegicus* | P24268 | C31918 | X54467 | – |
| | | JQ1177 | | |
| | | PQ0222 | | |
| | | S13111 | | |
| *Schistosoma japonicum* | – | – | L41346 | – |
| *Sus scrofa* | P00795 | A00987 | – | – |
| | | A92425 | | |
| | | A93990 | | |
| | | B31918 | | |

Brookhaven Protein Data Bank three-dimensional structures:

| Species | ID | Resolution | Notes |
|---|---|---|---|
| *Homo sapiens* | 1LYA | 2.5 | |
| | 1LYB | 2.5 | complex with pepstatin |

## Name and History

Proteinases having an acid pH optimum were reported in extracts of tissues over 90 years ago. They were subsequently named cathepsins (Gr. *kathepsein*, 'to digest'). (See Barrett, 1977, for a brief review of the early history.) The term ***cathepsin D*** (Press *et al.*, 1960) was applied to distinguish this enzyme from other cellular acidic proteinases and the reader is cautioned that a large number of unrelated proteinases are designated 'cathepsin' and are differentiated by appended letters. Early isolations of cathepsin D showed as many as 12 different forms of the enzyme which differed in size, number of associated polypeptides and isoelectric point but had similar activities, e.g. Sapolsky & Woessner (1972). Keilová (1970) and Cunningham & Tang (1976) used site-directed covalent modification to demonstrate that cathepsin D had a similar active center to pepsin (Chapter 271) confirming its identity as an aspartic proteinase. It was proposed that the two-chain forms were derived from the single polypeptide forms by proteolysis (Woessner, 1977). Determination of the N-terminal sequence of different forms of pig cathepsin D confirmed this and showed that cathepsin D was related to other aspartic proteinases by amino acid sequence (Huang *et al.*, 1979).

Studies of cathepsin D biosynthesis showed the existence of a precursor (Erickson & Blobel, 1979; Hasilik & Neufeld, 1980) and N-terminal sequencing of the different biosynthetic forms demonstrated the precursor–product relationships and predicted that the precursor was inactive (Erickson *et al.*, 1981). Isolation of a cDNA clone (Faust *et al.*, 1985) and of the gene (Redecker *et al.*, 1991), detailed studies of its biosynthesis (Horst & Hasilik, 1991), its activation (Richo & Conner, 1994), its substrate specificity (Scarborough *et al.*, 1993), and determination of the crystal structure (Baldwin *et al.*, 1993; Metcalf & Fusek, 1993) now make cathepsin D one of the best studied of the lysosomal proteinases.

## Activity and Specificity

### Assay

The most common assay uses denatured hemoglobin as a substrate and measures activity by the release of peptides soluble in trichloroacetic acid. For a full discussion of denatured hemoglobin assays of cathepsin D, and general comments on proteinase assays, see Barrett (1977). Cathepsin D is an endopeptidase with a pH optima between pH 3.5 and 5.0, dependent on the substrate and assay conditions. The pH optimum of the enzyme may reflect its stability to assay conditions (Barrett, 1977). Thus, cathepsin D displays a lower pH optimum with acid-denatured hemoglobin than with urea-denatured hemoglobin (Woessner & Shamberger, 1971) and with certain protein substrates, optima as high as pH 5.0 have been reported (e.g. van Noort & Van der Drift, 1989).

## Specificity

Using the oxidized B chain of insulin as a substrate, a number of investigators found that the cleavage specificity of cathepsin D was similar to that of pepsin in that it preferred hydrophobic amino acids at the scissile bond (for review see Barrett, 1977). Numerous isoforms of cathepsin D have been isolated and have indistinguishable kinetic constants (Sapolsky & Woessner, 1972) although Tanji et al. (1991) have shown two different isoform-dependent specificities using oxidized insulin B chain as substrate. Early attempts to use peptides as substrates were unsuccessful because of the short length of the tested peptides. Later, cathepsin D was shown to efficiently cleave longer peptides containing naphthylamide and p-nitrophenylalanine moieties (e.g. Ferguson et al., 1973).

As with other aspartic proteinases (Chapter 271), cathepsin D's substrate-binding cleft can accommodate up to seven amino acids. The secondary specificities of substrate binding have been explored by mild cleavage of proteins (van Noort & Van der Drift, 1989), by a series of synthetic substrates (Scarborough et al., 1991, 1993), and most recently by mutagenesis (Scarborough & Dunn, 1994). Mild cleavage of proteins confirmed early studies with insulin that showed that cathepsin D prefers to cleave between two hydrophobic residues (Offermann et al., 1983; Imoto et al., 1987; van Noort & Van der Drift, 1989) and showed evidence of secondary specificity by noting the preference for hydrophobic amino acids in the P2 position (van Noort & Van der Drift, 1989), as well as a tolerance for glutamic acid and the absence of positively charged amino acids at the P2 position (van Noort & Van der Drift, 1989). Dunn and colleagues (Scarborough et al., 1991, 1993) provided quantitative assessment and refinement of P2 and P3 specificity using synthetic peptides. Mutagenesis, directed by a rule-based structural model (Scarborough et al., 1993), showed that changing Met287 (following the pepsin numbering; Chapter 272) in the S3 subsite to a glutamic acid or glutamine, allowed cleavage of substrates with a positive charge at P2 with a specificity equivalent to the best substrates of the wild-type enzyme (Scarborough & Dunn, 1994). Secondary specificity at P2' and P5', suggested by cleavage specificity on protein substrates, remains untested by studies with synthetic peptides.

## Inhibitors

Cathepsin D is inhibited by pepstatin (Aoyagi et al., 1972) similarly to other members of the family (Chapter 271). Pepstatin binding ($K_i = 1-4 \times 10^{-12}$ M; Baldwin et al., 1993) can be exploited to purify cathepsin D (e.g. Huang et al., 1979) and to titrate active enzyme (Knight & Barrett, 1976). As in the case of pepsinogen, pepstatin binds to procathepsin D and blocks autoproteolysis in vitro (Hasilik et al., 1982). pH-dependent differences in procathepsin D and cathepsin D affinity for pepstatin allow their separation on pepstatinyl

agarose (Conner, 1989). Synthetic peptides containing statine derivatives inhibit cathepsin D to varying degrees but none with $K_i$ values as low as that of pepstatin (Agarwal & Rich, 1986; Jupp et al., 1990). A number of compounds screened for inhibition of the HIV-1 protease (Chapter 310) also inhibit cathepsin D with $K_i$ values in the nanomolar range (e.g. Alteri et al., 1993) with varying degrees of specificity.

$\alpha_2$-Macroglobulin inhibits residual cathepsin D activity at pH 6.2 but probably is not effective at lower pH due to structural instability of the inhibitor (Thomas et al., 1989). An inhibitory protein from potato has been identified (Keilová & Tomášek, 1975). This protein has sequence similarity to soybean trypsin inhibitor (Mares et al., 1989) and also inhibits serine proteinases (Mares et al., 1989; Keilová & Tomásek, 1977). A protein inhibitor of cathepsin D has been identified from tomato which is 80% identical to the potato inhibitor (Werner et al., 1993).

## Structural Chemistry

### Primary Structure and Proteolytic Processing

Cathepsin D is derived from a precursor which is highly related in amino acid sequence to other aspartic proteinases (Chapter 271). Depending on the species, the precursor undergoes at least two and as many as five distinct proteolytic cleavages to generate the forms isolated from tissue and cells. The human cathepsin D precursor comprises 412 amino acids (Faust et al., 1985) and has been termed 'preprocathepsin D'. Preprocathepsin D has 410 residues in mouse (Diedrich et al., 1990), 407 in rat (Birch & Loh, 1990) and 398 in chicken (Retzek et al., 1992). The precursor contains an N-terminal secretion signal peptide which is 20 residues in human (Erickson et al., 1981) and which is cleaved during translocation across the membrane to generate an inactive procathepsin D (392 residues in human). As predicted by the propeptide's sequence similarity with pepsinogen (Erickson et al., 1981), procathepsin D can undergo autocatalytic cleavage (Hasilik et al., 1982), presumably by intramolecular cleavage (Conner, 1989). Sequence analysis of human placenta cathepsin D shows a heterogeneous N-terminus with greater than 50% of the molecules beginning at Thr43p (Richo & Conner, 1994). In contrast, the crystal-derived models (see below) indicates that the N-terminus is Gly45p (Baldwin et al., 1993; Metcalf & Fusek, 1993) similar to that detected in purified pig and cattle cathepsin D (Takahashi & Tang, 1983; Huang et al., 1979). In vitro, autoproteolysis generates a pseudo-cathepsin D by cleavage between Leu26p and Ile27p of the propeptide (Conner & Richo, 1992). Thus, autoproteolysis under a variety of conditions does not generate the mature N-terminus as isolated from cells and tissue. The differences in the N-terminal sequence of isolated cathepsin D probably reflect aminopeptidase activity either in the lysosome or during purification. Addition of proteinase inhibitors to cell cultures during studies of biosynthesis (Samarel et al., 1989) suggests that an unknown intracellular cysteinyl proteinase is responsible for formation of the mature N-terminus.

Cathepsin D can undergo further cleavage to generate a noncovalently associated two-chain form with an N-terminal light chain of about 15 kDa and a C-terminal 30 kDa heavy

| Proteinase | | Processing to Two-Chain Form |
|---|---|---|
| Human Pepsin | G Y D T V Q V . . . . . . . . . . . G G I S D T N Q | None |
| Human Renin | S Q D S V T V . . . . . . . . . . . G G I T V T . Q | None |
| Rat Cathepsin D | S Q D T V S V P C K S D . . . . . L G G I K V E K Q | Minimal |
| Mouse Cathepsin D | S Q D T V S V P C K S D Q S K . . A R G I K V E K Q | Minimal |
| Bovine Cathepsin D | S Q D T V S V P C N P S\|S S\|S . . P G G V T V Q R Q | Partial |
| Porcine Cathepsin D | S Q D T V S V P C N\|S A L S G . . V G G I K V E R Q | Almost complete |
| Human Cathepsin D | S Q D T V S V P C Q S\|A S S A S A\|L\|G\|G V K V E R Q | Almost complete |

*Figure 277.1*  Alignment of cathepsin D sequences in the region surrounding the cleavage site between chains. Cathepsin D amino acid sequences are aligned with pepsin and renin as representatives of the aspartic proteinase family to show the additional residues found in cathepsin D. Vertical bars indicate cleavage sites derived from amino acid sequencing of light and heavy chains (Yonezawa *et al.*, 1988; Horst & Hasilik, 1991; Baldwin *et al.*, 1993). Heavy vertical bars are most frequently detected termini and thin vertical bars indicate termini less frequently detected.

chain. The majority of human and pig cathepsin exists as this two-chain molecule while purified cattle cathepsin D is almost equally distributed between the single- and two-chain forms. Mouse, rat and Chinese hamster cathepsin D exist almost entirely as single-chain enzymes. Comparison of the primary structure of cathepsin D from these species with other members of the aspartic proteinase family (Chapter 271) demonstrates an insertion in the region that is cleaved to generate the two-chain form of the enzyme (Figure 277.1). Two-chain enzymes lack a short segment of the insertion that is excised following cleavage. It is not known whether the excision occurs by one or two cleavages and whether exopeptidase activity is involved. The initial cleavage appears to require the action of a cysteinyl proteinase (Hentze *et al.*, 1984; Samarel *et al.*, 1989). Molecular modeling of this region suggests that it forms a loop on the surface and thus would be disposed to proteolysis in the lysosome (Yonezawa *et al.*, 1988) where the cleavage has been shown to occur. The electron density map obtained from cattle cathepsin D crystals, which contained 50% single-chain enzyme, did not show evidence of the loop, suggesting that it was mobile or disordered (Metcalf & Fusek, 1993).

Comparison of pig cathepsin D amino acid sequence (Shewale & Tang, 1984) with the deduced amino acid sequence of human preprocathepsin (Faust *et al.*, 1985) and the determined sequence of human liver cathepsin D (Baldwin *et al.*, 1993) shows that two amino acids are removed from the C-terminus of the heavy chain by proteolysis as predicted by carboxypeptidase analysis of newly synthesized pig cathepsin D (Erickson & Blobel, 1983). Crystal-derived structures confirm the absence of both residues in the human enzyme and the C-terminal leucine in the cattle enzyme.

Cathepsin D is a glycoprotein with two *N*-linked oligosaccharides, one each in the light and heavy chains. These oligosaccharides are normally modified by phosphorylation of the six position of mannoses during biosynthesis (reviewed by von Figura & Hasilik, 1986; Kornfeld & Mellman, 1989). The mannose-6-phosphate modification serves to target the enzyme to lysosomes through association with specific mannose-6-phosphate receptors. Glycosylation is not necessary for folding (Fortenberry *et al.*, 1995) or enzyme activity (Conner & Udey, 1990) but is required for targeting

of the enzyme (Fortenberry *et al.*, 1995). The multiplicity of cathepsin D isoforms purified from tissue extracts is, in large measure, explained by proteolytic processing and carbohydrate modifications which produce molecules of differing isoelectric points.

### Crystal Structure

A 2.5 Å resolution structural model of human cathepsin D has been derived from crystals of human liver cathepsin D (Baldwin *et al.*, 1993) and a 3.0 Å resolution model from human spleen (Metcalf & Fusek, 1993). Structures of human liver and cattle liver cathepsin D complexed with pepstatin were also reported. These structures are in good general agreement with each other and with previous computationally derived models of cathepsin D based on the structure of other aspartic proteinases (Yonezawa *et al.*, 1988; Baranski *et al.*, 1991; Schorey & Chirgwin, 1991; Scarborough *et al.*, 1993). The overall fold of the enzyme is similar to pepsin, renin and chymosin (overview in Chapter 271); structural agreement of the main chain with these proteinases was 0.9–1.0 Å r.m.s. (Baldwin *et al.*, 1993). Superposition of individual cathepsin D subdomains with those of other aspartic proteinases indicated different relative subdomain displacements (Baldwin *et al.*, 1993) which correlated with substrate specificities. The active-site cleft resembles those of other aspartic proteinases. Although narrower than that of rhizopuspepsin (Chapter 297), the cleft is wider than that of renin (Chapter 284) and has relatively larger S2 and S4 subsite volumes. Cathepsin D contains an additional disulfide bridge (Cys27-Cys96) not found in other aspartic proteinases. Cys96 is located near the C-terminus of the light chain and may play a role in restricting proteolysis of the single-chain enzyme. The two *N*-linked oligosaccharides are found on the same face of the molecule away from the active site and the majority of the lysosomal enzyme phosphotransferase recognition site (Baranski *et al.*, 1990, 1991) (Figure 277.2).

### *Preparation*

Human and cattle cathepsin D are commercially available from several suppliers, e.g. Sigma or Calbiochem (see

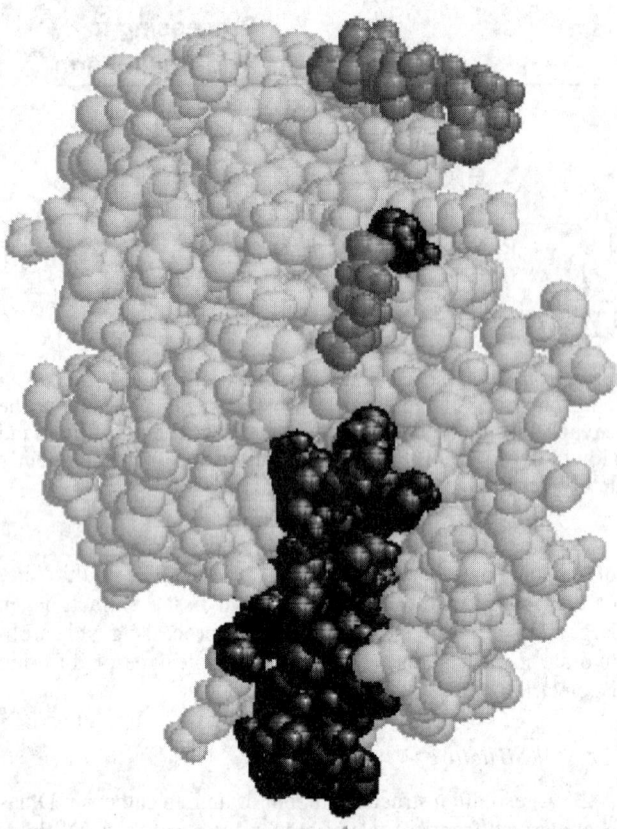

*Figure 277.2* Lysosomal targeting signal of cathepsin D. A spacefilling model of human cathepsin D, using the coordinates of Baldwin *et al.* (1993), shows the structural relationship of amino acid residues important for recognition of cathepsin D by phosphotransferase (residues 203, and 265–292 shown in black; Baranski *et al.*, 1991) and the target glycosylation sites (shown in dark gray). The active-site cleft is on the opposite side of the model and is not visible in this view.

Appendix 2 for full names and addresses of suppliers). Rabbit antiserum and mouse monoclonal antibodies to human cathepsin D are also commercially available (Oncogene Research Products and Novocastra Laboratories Ltd). Cathepsin D can be prepared from virtually any tissue homogenate although spleen and liver are commonly used, and placenta is a readily available source in humans. Preparation typically involves an acid precipitation and chromatography on pepstatinyl-agarose affinity columns (e.g. Huang *et al.*, 1979). DEAE chromatography removes contaminating cathepsin E (Chapter 275) (Lapresle & Webb, 1962; Yamamoto *et al.*, 1978).

The extensive proteolytic processing during maturation of the enzyme makes production of recombinant cathepsin D difficult because the processing proteinases have yet to be identified. Recombinant human procathepsin D has been expressed in bacteria (Conner & Udey, 1990), yeast (Nishimura *et al.*, 1995), baculovirus (Beyer & Dunn, 1996) and numerous mammalian cell culture systems. The bacterial procathepsin D is found in inclusion bodies and requires refolding and autocatalytic activation to give pseudocathepsin D (Conner & Richo, 1992). The procathepsin D expressed in baculovirus is

secreted and the proenzyme can be purified and autoactivated (Conner & Richo, 1992).

Pseudocathepsin D has slightly different kinetic constants when compared to cathepsin D (Scarborough *et al.*, 1991). A shorter mutant procathepsin D molecule autoactivates to give a molecule with only six additional residues compared to single-chain cathepsin D and is kinetically indistinguishable from the glycosylated forms (Beyer & Dunn, 1996). Expression in yeast and mammals has primarily been used to study biosynthesis and intracellular targeting rather than for preparation of the enzyme. Procathepsin D has been prepared from conditioned media of breast cancer cell cultures (e.g. Briozzo *et al.*, 1988; Vetvicka & Fusek, 1994; Stewart *et al.*, 1994).

Conflicting data have been reported on the role of the propeptide in folding. A large portion of the protein expressed without the propeptide is degraded in mammalian (Conner, 1992) and yeast cells (Nishimura *et al.*, 1995), although folded and active enzyme can be recovered (Fortenberry & Chirgwin, 1995).

## Biological Aspects

Cathepsin D is a lysosomal proteinase that can constitute as much as 10% of the soluble lysosomal protein in rat liver; the concentration of cathepsin D inside liver lysosomes could be as high as 0.7 mM (Dean & Barrett, 1976). As a normal and frequently major component of the lysosome, it is found in nearly all cells and tissues of mammals (Sakai *et al.*, 1989). Immunocytochemical reports of the absence of cathepsin D in certain tissues is probably due to adjustment of the technique's sensitivity in order to display difference in levels of expression or number of lysosomes (Sakai *et al.*, 1989). Lysosomal cathepsin D is generally thought to be a soluble enzyme although several authors have reported membrane-associated cathepsin D (e.g. Diment & Stahl, 1985). Cathepsin D is encoded by a single gene in humans (Redecker *et al.*, 1991) and in mice (Hetman *et al.*, 1994), although alleles have been identified (Touitou *et al.*, 1994). Despite being constitutively expressed in most cells, in some cells it appears to be upregulated by calcitriol (Redecker *et al.*, 1989), estrogen (Cavailles *et al.*, 1991) or retinoic acid (Atkins & Troen, 1995).

As expected for a lysosomal enzyme, procathepsin D has been identified in various parts of the secretory pathway including the rough endoplasmic reticulum, Golgi apparatus, transport vesicles, as well as endosomes but it is apparently processed rapidly after reaching the prelysosome or lysosome. Some procathepsin D escapes targeting to the lysosome and is secreted from cells and has been detected in cow's (Larsen *et al.*, 1993) and human milk (Vetvicka *et al.*, 1993). Enzymatically active cathepsin D only appears in the lysosome and endosome and occasionally in extracellular spaces following cellular damage such as that caused by ischemic episodes.

Evidence to suggest that cathepsin D functions in general degradation of proteins inside lysosomes comes primarily from three types of experiments: addition of pepstatin to cells or perfused organs, *in vitro* measurement of cathepsin D proteolysis of proteins, and measurement of cathepsin D levels in conditions known to alter protein turnover in cells. Estimates using pepstatin to evaluate cathepsin D contribution to protein degradation vary from 10 to 50% of total lysosomal degradation, depending on the system and protein(s) studied.

Although disruption of the cathepsin D gene in mice leads to postnatal lethality, general lysosomal protein degradation appears to be largely unimpaired in fibroblasts from these animals (Saftig *et al.*, 1995), indicating that cathepsin D may not be required for general intracellular protein turnover in lysosomes. Mice with a nonfunctional cathepsin D gene develop normally but stop thriving in the third week after birth and die in a state of anorexia after 4 weeks (Saftig *et al.*, 1995). Atrophy of the ileum and destruction of lymphocytes in the spleen and thymus are the major effects, suggesting an important role for cathepsin D in regulating cell turnover in these organs.

Cathepsin D is also involved in the presentation of antigenic peptides. It is colocalized with components of the immune system in the endosomes (Guagliardi *et al.*, 1990), and its specificity and activity (e.g. van Noort *et al.*, 1991) suggest that cathepsin D is used to process protein antigens for presentation to the immune system. A role for cathepsin D in antigen presentation is suggested by pepstatin inhibition of invariant chain processing (Maric *et al.*, 1994; Mizuochi *et al.*, 1994), a required step in antigen presentation.

Cathepsin D is overexpressed in some breast tumors and breast tumor cell lines (for review see Rochefort, 1990). Estradiol-induced overexpression in cell lines results in a high level of procathepsin D secretion and a high level of intracellular cathepsin D and procathepsin D (Capony *et al.*, 1989). Cathepsin D expression levels are useful as a tumor prognostic marker but the role of the enzyme in the etiology of the disease is controversial. Rocheforte and colleagues proposed that extracellular cathepsin D activity potentiates metastasis (Briozzo *et al.*, 1988). Although cathepsin D cleaves extracellular matrix protein between pH 4 and 5 (e.g. Woessner, 1973; Briozzo *et al.*, 1988), the primary secretion product is procathepsin D, which is expected to be inactive. *In vitro* assessment of cathepsin D's role in invasion indicates that there is no correlation between activity and invasion (Johnson *et al.*, 1993). Since little is known of procathepsin D activation by extracellular proteinases, and the pH in microenvironments in the tumors, further *in vivo* studies will be needed. High levels of intracellular cathepsin D activity could also regulate cell growth through degradation of inhibitory molecules or even of receptors. A potential role for cathepsin D in regulation of apoptosis, suggested by knockout of the gene, may be relevant to its overexpression in tumors (Saftig *et al.*, 1995).

Cathepsin D expression is constitutive in almost all cells but it can also be regulated by estradiol, calcitriol (Redecker *et al.*, 1989) and retinoic acid (Atkins & Troen, 1995). The exact structure of the promoter remains confusing as no classical estrogen response element has been identified (May *et al.*, 1993; Augereau *et al.*, 1994). No TATA or CCAAT boxes are found but SP1- and AP2-binding sites are present (Redecker *et al.*, 1991). The human gene is transcribed from different start points when upregulated by estrogen (May *et al.*, 1993; Augereau *et al.*, 1994).

Cathepsin D proteolytic activity has been implicated in normal and abnormal processing of a number of proteins. Cathepsin D may play a role in processing precursors to form bioactive polypeptides or in downregulating polypeptides including hormones, carrier proteins or cell surface receptors. Cathepsin D can cleave endothelin and its precursors, tachykinin and enkephalin precursors, but its function as the normal

processing enzyme for these precursors must be evaluated in light of other proteinases that have been purified and which appear to function in this regard (Chapters 306, 363 and 364). Cathepsin D has been isolated from chromaffin granules (Krieger & Hook, 1992). Cathepsin D also appears to degrade lipoproteins and to cleave parathyroid hormone, thyroglobulin and glucagon. Cathepsin D activity is necessary for toxicity of the ricin A chain. Cathepsin D has been shown to cleave β-amyloid precursor protein (APP) in Alzheimer's disease, although transfection studies with lysosomal proteinases suggests that cathepsin S (Chapter 211) is also capable of increasing amyloid β peptides. Reports on differences in cathepsin D content in normal and Alzheimer's-diseased brains are conflicting (Dreyer *et al.*, 1994; Kohnken *et al.*, 1995). Cathepsin D activity is probably involved in processing other lysosomal protein precursors such as other cathepsins (Nishimura *et al.*, 1989) and prosaposin (Nishimura *et al.*, 1989). A portion of newly synthesized procathepsin D associates with prosaposin immediately after synthesis and both proteins are transported to the lysosome together (Zhu & Conner, 1994).

## Distinguishing Features

Cathepsin D is distinguished from all other intracellular proteinases except cathepsin E by its pepstatin sensitivity. Cathepsin D and E can be distinguished immunologically and by apparent size on nonreducing SDS gel electrophoresis, by the more restricted peptide substrate specificity of cathepsin D and by the subcellular localization of cathepsin D to lysosomes.

## Related Enzymes

Digestive enzymes of mosquitoes (Cho & Raikhel, 1992) and cockroaches (Arruda *et al.*, 1995) are slightly more related to cathepsin D than to other members of the family; each is about 50% identical to cathepsin D. Proteinases with similar relatedness have also been identified in *Caenorhabditis* (Jacobson *et al.*, 1988) and schistosomes (Becker *et al.*, 1995).

## Further Reading

For a review see Fusek & Vetvicka (1995).

## References

Agarwal, N.S. & Rich, D.H. (1986) Inhibition of cathepsin D by substrate analogues containing statine and by analogues of pepstatin. *J. Med. Chem.* **29**, 2519–2524.

Alteri, E., Bold, G., Cozens, R. *et al.* (1993) CGP 53437, an orally bioavailable inhibitor of human immunodeficiency virus type 1 protease with potent antiviral activity. *Antimicrob. Agents Chemother.* **37**, 2087–2092.

Aoyagi, T., Morishima, H., Nishizawa, R., Kunimoto, S. & Takeuchi, T. (1972) Biological activity of pepstatins, pepstanone A and partial peptides on pepsin, cathepsin D and renin. *J. Antibiotics* **25**, 689–694.

Arruda, L.K., Vailes, L.D., Mann, B.J., Shannon, J., Fox, J.W., Vedvick, T.S., Hayden, M.L. & Chapman, M.D. (1995) Molecular cloning of a major cockroach (*Blattella germanica*) allergen,

Bla g 2. Sequence homology to the aspartic proteases. *J. Biol. Chem.* **270**, 19563–19568.

Atkins, K.B. & Troen, B.R. (1995) Regulation of cathepsin D gene expression in HL-60 cells by retinoic acid and calcitriol. *Cell Growth Differ.* **6**, 871–877.

Augereau, P., Miralles, F., Cavailles, V., Gaudelet, C., Parker, M. & Rochefort, H. (1994) Characterization of the proximal estrogen-responsive element of human cathepsin D gene. *Mol. Endocrinol.* **8**, 693–703.

Baldwin, E.T., Bhat, T.N., Gulnik, S., Hosur, M.V., Sowder, R.C., Cachau, R.E., Collins, J., Silva, A.M. & Erickson, J.W. (1993) Crystal structures of native and inhibited forms of human cathepsin D: implications for lysosomal targeting and drug design. *Proc. Natl Acad. Sci. USA* **90**, 6796–6800.

Baranski, T.J., Faust, P.L. & Kornfeld, S. (1990) Generation of a lysosomal enzyme targeting signal in the secretory protein pepsinogen. *Cell* **63**, 281–291.

Baranski, T.J., Koelsch, G., Hartsuck, J.A. & Kornfeld, S. (1991) Mapping and molecular modeling of a recognition domain for lysosomal enzyme targeting. *J. Biol. Chem.* **266**, 23365–23372.

Barrett, A.J. (1977) Cathepsin D and other carboxyl proteinases. In: *Proteinases in Mammalian Cells and Tissues* (Barrett, A.J., ed.). Amsterdam: North-Holland Biomed. Press, pp. 209–248.

Becker, M.M., Harrop, S.A., Dalton, J.P., Kalinna, B.H., McManus, D.P. & Brindley, P.J. (1995) Cloning and characterization of the *Schistosoma japonicum* aspartic proteinase involved in hemoglobin degradation. *J Biol. Chem.* **270**, 24496–24501.

Beyer, B.M. & Dunn, B.M. (1996) Self-activation of recombinant human lysosomal procathepsin D at a newly engineered cleavage junction, 'short' pseudocathepsin D. *J. Biol. Chem.* **271**, 15590–15596.

Birch, N.P. & Loh, Y.P. (1990) Cloning, sequence and expression of rat cathepsin D. *Nucleic Acids Res.* **18**, 6445–6446.

Briozzo, P., Morisset, M., Capony, F., Rougeot, C. & Rochefort, H. (1988) In vitro degradation of extracellular matrix with $M_r$ 52,000 cathepsin D secreted by breast cancer cells. *Cancer Res.* **48**, 3688–3692.

Capony, F., Rougeot, C., Montcourrier, P., Cavailles, V. & Salazar, G. (1989) Increased secretion, altered processing, and glycosylation of pro-cathepsin D in human mammary cancer cells. *Cancer Res.* **49**, 3904–3909.

Cavailles, V., Augereau, P. & Rochefort, H. (1991) Cathepsin D gene of human MCF7 cells contains estrogen-responsive sequences in its 5′ proximal flanking region. *Biochem. Biophys. Res. Commun.* **174**, 816–824.

Cho, W.L. & Raikhel, A.S. (1992) Cloning of cDNA for mosquito lysosomal aspartic protease. Sequence analysis of an insect lysosomal enzyme similar to cathepsins D and E. *J. Biol. Chem.* **267**, 21823–21829.

Conner, G.E. (1989) Isolation of procathepsin D from mature cathepsin D by pepstatin affinity chromatography. Autocatalytic proteolysis of the zymogen form of the enzyme. *Biochem. J.* **263**, 601–604.

Conner, G.E. (1992) The role of the cathepsin D propeptide in sorting to the lysosome. *J. Biol. Chem.* **267**, 21738–21745.

Conner, G.E. & Richo, G. (1992) Isolation and characterization of a stable activation intermediate of the lysosomal aspartyl protease cathepsin D. *Biochemistry* **31**, 1142–1147.

Conner, G.E. & Udey, J.A. (1990) Expression and refolding of recombinant human fibroblast procathepsin D. *DNA Cell Biol.* **9**, 1–9.

Cunningham, M. & Tang, J. (1976) Purification and properties of cathepsin D from porcine spleen. *J. Biol. Chem.* **251**, 4528–4536.

Dean, R.T. & Barrett, A.J. (1976) Lysosomes. *Essays Biochem.* **12**, 1–40.

Diedrich, J.F., Staskus, K.A., Retzel, E.F. & Haase, A.T. (1990) Nucleotide sequence of a cDNA encoding mouse cathepsin D. *Nucleic Acids Res.* **18**, 7184.

Diment, S. & Stahl, P. (1985) Macrophage endosomes contain proteases which degrade endocytosed protein ligands. *J. Biol. Chem.* **260**, 15311–15317.

Dreyer, R.N., Bausch, K.M., Fracasso, P., Hammond, L.J., Wunderlich, D., Wirak, D.O.D., Brini, C.M., Buckholz, T.M., König, G., Kamarck, M.E. & Tamburini, P.P. (1994) Processing of the pre-β-amyloid protein by cathepsin D is enhanced by a familial Alzheimer's disease mutation. *Eur. J. Biochem.* **224**, 265–271.

Erickson, A.H. & Blobel, G. (1979) Early events in the biosynthesis of the lysosomal enzyme cathepsin D. *J. Biol. Chem.* **254**, 11771–11774.

Erickson, A.H. & Blobel, G. (1983) Carboxyl-terminal proteolytic processing during biosynthesis of the lysosomal enzymes beta-glucuronidase and cathepsin D. *Biochemistry* **22**, 5201–5205.

Erickson, A.H., Conner, G.E. & Blobel, G. (1981) Biosynthesis of a lysosomal enzyme. Partial structure of two transient and functionally distinct $NH_2$-terminal sequences in cathepsin D. *J. Biol. Chem.* **256**, 11224–11231.

Faust, P.L., Kornfeld, S. & Chirgwin, J.M. (1985) Cloning and sequence analysis of cDNA for human cathepsin D. *Proc. Natl Acad. Sci. USA* **82**, 4910–4914.

Ferguson, J.B., Andrews, J.R., Voynick, I.M. & Fruton, J.S. (1973) The specificity of cathepsin D. *J. Biol. Chem.* **248**, 6701–6708.

Fortenberry, S.C. & Chirgwin, J.M. (1995) The propeptide is nonessential for the expression of human cathepsin D. *J. Biol. Chem.* **270**, 9778–9782.

Fortenberry, S.C., Schorey, J.S. & Chirgwin, J.M. (1995) Role of glycosylation in the expression of human procathepsin D. *J. Cell Sci.* **108**, 2001–2006.

Fusek, M. & Vetvicka, V. (1995) *Aspartic Proteinases: Physiology and Pathology*. Boca Raton: CRC Press.

Guagliardi, L.E., Koppelman, B., Blum, J.S., Marks, M.S., Cresswell, P. & Brodsky, F.M. (1990) Co-localization of molecules involved in antigen processing and presentation in an early endocytic compartment [Review]. *Nature* **343**, 133–139.

Hasilik, A. & Neufeld, E.F. (1980) Biosynthesis of lysosomal enzymes in fibroblasts. Synthesis as precursors of higher molecular weight. *J. Biol. Chem.* **255**, 4937–4945.

Hasilik, A., von Figura, K., Conzelmann, E., Nehrkorn, H. & Sandhoff, K. (1982) Lysosomal enzyme precursors in human fibroblasts. Activation of cathepsin D precursor in vitro and activity of beta-hexosaminidase A precursor towards ganglioside GM2. *Eur. J. Biochem.* **125**, 317–321.

Hentze, M., Hasilik, A. & von Figura, K. (1984) Enhanced degradation of cathepsin D synthesized in the presence of the threonine analog beta-hydroxynorvaline. *Arch. Biochem. Biophys.* **230**, 375–382.

Hetman, M., Perschl, A., Saftig, P., von Figura, K. & Peters, C. (1994) Mouse cathepsin D gene: molecular organization, characterization of the promoter and chromosomal localization. *DNA Cell Biol.* **13**, 419–427.

Horst, M. & Hasilik, A. (1991) Expression and maturation of human cathepsin D in baby-hamster kidney cells. *Biochem. J.* **273**, 355–361.

Huang, J.S., Huang, S.S. & Tang, J. (1979) Cathepsin D isozymes from porcine spleens. Large scale purification and polypeptide chain arrangements. *J. Biol. Chem.* **254**, 11405–11417.

Imoto, T., Okazaki, K., Koga, H. & Yamada, H. (1987) Specificity of rat liver cathepsin D. *J. Biochem. (Tokyo)* **101**, 575–580.

Jacobson, L.A., Jen-Jacobson, L., Hawdon, J.M., Owens, G.P., Bolanowski, M.A., Shah, M.V., Pollock, R.A. & Conklin, D.S. (1988) Identification of a putative structural gene for cathepsin D in *Caenorhabditis elegans*. *Genetics* **119**, 355–363.

Johnson, M.D., Torri, J.A., Lippman, M.E. & Dickson, R.B. (1993) The role of cathepsin D in the invasiveness of human breast cancer cells. *Cancer Res.* **53**, 873–877.

Jupp, R.A., Dunn, B.M., Jacobs, J.W., Vlasuk, G., Arcuri, K.E., Veber, D.F., Perlow, D.S., Payne, L.S., Boger, J., De Laszlo, S., Chakravarty, P.K., TenBroeke, J., Hangauer, D.G., Ondeyka, D., Greenlee, W.J. & Kay, J. (1990) The selectivity of statine-based inhibitors against various human aspartic proteinases. *Biochem. J.* **265**, 871–878.

Keilová, H. (1970) Inhibition of cathepsin D by diazoacetylnorleucine methyl ester. *FEBS Lett.* **6**, 312–314.

Keilová, H. & Tomášek, V. (1975) Isolation and some properties of cathepsin D inhibitor from potatoes. *Collect. Czech. Chem. Commun.* **41**, 489.

Keilová, H. & Tomášek, V. (1977) Naturally occurring inhibitors of intracellular proteinases. *Acta Biol. Med. Germ.* **36**, 1873–1881.

Knight, C.G. & Barrett, A.J. (1976) Interaction of human cathepsin D with the inhibitor pepstatin. *Biochem. J.* **155**, 117–125.

Kohnken, R.E., Ladror, U.S., Wang, G.T., Holzman, T.F., Miller, B.E. & Krafft, G.A. (1995) Cathepsin D from Alzheimer's-diseased and normal brains. *Exp. Neurol.* **133**, 105–112.

Kornfeld, S. & Mellman, I. (1989) The biogenesis of lysosomes. *Annu. Rev. Cell Biol.* **5**, 483–525.

Krieger, T.J. & Hook, V.Y. (1992) Purification and characterization of a cathepsin D protease from bovine chromaffin granules. *Biochemistry* **31**, 4223–4231.

Lapresle, C. & Webb, T. (1962) The purification and properties of a proteolytic enzyme, rabbit cathepsin E, and further studies on rabbit cathepsin D. *Biochem. J.* **84**, 455–462.

Larsen, L.B., Boisen, A. & Petersen, T.E. (1993) Procathepsin D cannot autoactivate to cathepsin D at acid pH. *FEBS Lett.* **319**, 54–58.

Mareš, M., Meloun, B., Pavlík, M., Kostka, V. & Baudyš, M. (1989) Primary structure of cathepsin D inhibitor from potatoes and its structure relationship to soybean trypsin inhibitor family. *FEBS Lett.* **251**, 94–98.

Maric, M.A., Taylor, M.D. & Blum, J.S. (1994) Endosomal aspartic proteinases are required for invariant-chain processing. *Proc. Natl Acad. Sci. USA* **91**, 2171–2175.

May, F.E., Smith, D.J. & Westley, B.R. (1993) The human cathepsin D-encoding gene is transcribed from an estrogen-regulated and a constitutive start point. *Gene* **134**, 277–282.

Metcalf, P. & Fusek, M. (1993) Two crystal structures for cathepsin D: the lysosomal targeting signal and active site. *EMBO J.* **12**, 1293–1302.

Mizuochi, T., Yee, S.T., Kasai, M., Kakiuchi, T., Muno, D. & Kominami, E. (1994) Both cathepsin B and cathepsin D are necessary for processing of ovalbumin as well as for degradation of class II MHC invariant chain. *Immunol. Lett.* **43**, 189–193.

Nishimura, Y., Kawabata, T., Furuno, K. & Kato, K. (1989) Evidence that aspartic proteinase is involved in the proteolytic processing event of procathepsin L in lysosomes. *Arch. Biochem.*

*Biophys.* **271**, 400–406.

Nishimura, Y., Takeshima, H., Sakaguchi, M., Mihara, K., Omura, T. & Kato, K. (1995) Expression of rat cathepsin D cDNA in *Saccharomyces cerevisiae*: implications for intracellular targeting of cathepsin D to vacuoles. *J. Biochem. (Tokyo)* **118**, 168–177.

Offermann, M.K., Chlebowski, J.F. & Bond, J.S. (1983) Action of cathepsin D on fructose-1,6-bisphosphate aldolase. *Biochem. J.* **211**, 529–534.

Press, E.M., Porter, R.R. & Cebra, J. (1960) The isolation and properties of a proteolytic enzyme, cathepsin D, from bovine spleen. *Biochem. J.* **74**, 501–551.

Redecker, B., Horst, M. & Hasilik, A. (1989) Calcitriol enhances transcriptional activity of lysozyme and cathepsin D genes in U937 promonocytes. *Biochem. J.* **262**, 843–847.

Redecker, B., Heckendorf, B., Grosch, H.W., Mersmann, G. & Hasilik, A. (1991) Molecular organization of the human cathepsin D gene. *DNA Cell Biol.* **10**, 423–431.

Retzek, H., Steyrer, E., Sanders, E.J., Nimpf, J. & Schneider, W.J. (1992) Molecular cloning and functional characterization of chicken cathepsin D, a key enzyme for yolk formation. *DNA Cell Biol.* **11**, 661–672.

Richo, G.R. & Conner, G.E. (1994) Structural requirements of procathepsin D activation and maturation. *J. Biol. Chem.* **269**, 14806–14812.

Rochefort, H. (1990) Cathepsin D in breast cancer. *Breast Cancer Res. Treat.* **16**, 3–13.

Saftig, P., Hetman, M., Schmahl, W., Weber, K., Heine, L., Mossmann, H., Koster, A., Hess, B., Evers, M., von Figura, K. & Peters, C. (1995) Mice deficient for the lysosomal proteinase cathepsin D exhibit progressive atrophy of the intestinal mucosa and profound destruction of lymphoid cells. *EMBO J.* **14**, 3599–3608.

Sakai, H., Saku, T., Kato, Y. & Yamamoto, K. (1989) Quantitation and immunohistochemical localization of cathepsins E and D in rat tissues and blood cells. *Biochim. Biophys. Acta* **991**, 367–375.

Samarel, A.M., Ferguson, A.G., Decker, R.S. & Lesch, M. (1989) Effects of cysteine protease inhibitors on rabbit cathepsin D maturation. *Am. J. Physiol.* **257**, C1069–C1079.

Sapolsky, A.I. & Woessner, J.F., Jr (1972) Multiple forms of cathepsin D from bovine uterus. *J. Biol. Chem.* **247**, 2069–2076.

Scarborough, P.E. & Dunn, B.M. (1994) Redesign of the substrate specificity of human cathepsin D: the dominant role of position 287 in the S2 subsite. *Protein Eng.* **7**, 495–502.

Scarborough, P.E., Richo, G.R., Kay, J., Conner, G.E. & Dunn, B.M. (1991) Comparison of kinetic properties of native and recombinant human cathepsin D. *Adv. Exp. Med. Biol.* **306**, 343–347.

Scarborough, P.E., Guruprasad, K., Topham, C., Richo, G.R., Conner, G.E., Blundell, T.L. & Dunn, B.M. (1993) Exploration of subsite binding specificity of human cathepsin D through kinetics and rule-based molecular modeling. *Protein Sci.* **2**, 264–276.

Schorey, J. & Chirgwin, J. (1991) Mapping of lysosomal targeting determinants of cathepsin D. *Adv. Exp. Med. Biol.* **306**, 339–342.

Shewale, J.G. & Tang, J. (1984) Amino acid sequence of porcine spleen cathepsin D. *Proc. Natl Acad. Sci. USA* **81**, 3703–3707.

Stewart, A.J., Piggott, N.H., May, F.E. & Westley, B.R. (1994) Mitogenic activity of procathepsin D purified from conditioned medium of breast-cancer cells by affinity chromatography on pepstatinyl agarose [see comments]. *Int. J. Cancer* **57**, 715–718.

Takahashi, T. & Tang, J. (1983) Amino acid sequence of porcine spleen cathepsin D light chain. *J. Biol. Chem.* **258**, 6435–6443.

Tanji, M., Kageyama, T. & Takahashi, K. (1991) Occurrence of cathepsin D isozymes with different specificities in monkey skeletal muscle. *Biochem. Biophys. Res. Commun.* **176**, 798–804.

Touitou, I., Capony, F., Brouillet, J.P. & Rochefort, H. (1994) Missense polymorphism (C/T224) in the human cathepsin D profragment determined by polymerase chain reaction: single strand conformational polymorphism analysis and possible consequences in cancer cells. *Eur. J. Cancer* **30A**, 390–394.

van Noort, J.M. & Van der Drift, A.C. (1989) The selectivity of cathepsin D suggests an involvement of the enzyme in the generation of T-cell epitopes. *J. Biol. Chem.* **264**, 14159–14164.

van Noort, J.M., Boon, J., Van der Drift, A.C., Wagenaar, J.P., Boots, A.M. & Boog, C.J. (1991) Antigen processing by endosomal proteases determines which sites of sperm-whale myoglobin are eventually recognized by T cells. *Eur. J. Immunol.* **21**, 1989–1996.

Vetvicka, V. & Fusek, M. (1994) Activation of peripheral blood neutrophils and lymphocytes by human procathepsin D and insulin-like growth factor II. *Cell. Immunol.* **156**, 332–341.

Vetvicka, V., Vagner, J., Baudys, M., Tang, J., Foundling, S.I. & Fusek, M. (1993) Human breast milk contains procathepsin D: detection by specific antibodies. *Biochem. Mol. Biol. Int.* **30**, 921–928.

von Figura, K. & Hasilik, A. (1986) Lysosomal enzymes and their receptors. *Annu. Rev. Biochem.* **55**, 167–193.

Werner, R., Guitton, M.C. & Mühlbach, H.-P. (1993) Nucleotide sequence of a cathepsin D inhibitor protein from tomato. *Plant Physiol.* **103**, 1473.

Woessner, J.F., Jr. (1973) Purification of cathepsin D from cartilage and uterus and its action on the protein-polysaccharide complex of cartilage. *J. Biol. Chem.* **248**, 1634–1642.

Woessner, J.F., Jr. (1977) Specificity and biological role of cathepsin D. *Adv. Exp. Med. Biol.* **95**, 313–327.

Woessner, J.F., Jr & Shamberger, R.J., Jr (1971) Purification and properties of cathepsin D from bovine uterus. *J. Biol. Chem.* **246**, 1951–1960.

Yamamoto, K., Katsuda, N. & Kato, K. (1978) Affinity purification and properties of cathepsin-E-like acid proteinase from rat spleen. *Eur. J. Biochem.* **92**, 499–508.

Yonezawa, S., Takahashi, T., Wang, X.J., Wong, R.N., Hartsuck, J.A. & Tang, J. (1988) Structures at the proteolytic processing region of cathepsin D. *J. Biol. Chem.* **263**, 16504–16511.

Zhu, Y. & Conner, G.E. (1994) Intermolecular association of lysosomal protein precursors during biosynthesis. *J. Biol. Chem.* **269**, 3846–3851.

*Gregory E. Conner*
*Department of Cell Biology and Anatomy,*
*University of Miami School of Medicine,*
*PO Box 016960, Miami, FL 33101, USA*
*Email: gconner@mednet.med.miami.edu*

# 278. *Phytepsin*

## *Databanks*

*Peptidase classification: clan AA, family A1, MEROPS ID: A01.020*
*NC-IUBMB enzyme classification: EC 3.4.24.40*
*Databank codes:*

| Species | SwissProt | PIR | EMBL (cDNA) | EMBL (genomic) |
|---|---|---|---|---|
| *Arabidopsis thaliana* | – | – | U51036 | – |
| *Brassica napus* | – | – | – | U55032: complete gene |
| *Brassica napus* | – | – | – | U55033: mid-section |
| *Brassica oleracea* | – | – | X80067 | – |
| *Brassica oleracea* | – | – | X88774 | – |
| *Brassica oleracea* | – | S41400 | X77260 | – |
| *Centaurea calcitrapa* | – | – | Y09123 | – |
| *Cynara cardunculus* | P40782 | S47096 | X69193 | – |
| *Cynara cardunculus* | – | S49349 | X81984 | – |
| *Hordeum vulgare* | P42210 | S19697 | X56136 | – |
| *Lycopersicon esculentum* | – | – | L46681 | – |
| *Oryza sativa* | – | – | D32144 | D32165: complete gene |

| Species | SwissProt | PIR | EMBL (cDNA) | EMBL (genomic) |
|---|---|---|---|---|
| *Oryza sativa* | P42211 | JS0732 PC4080 | D12777 | – |
| *Vigna unguiculata* | – | – | U61396 | – |

*Note: Cynara* is also treated in Chapter 279.

## Name and History

Barley grains contain a prominent aspartic peptidase activity (Morris *et al.*, 1985; Sarkkinen *et al.*, 1992; Wrobel & Jones, 1992), which has been suggested to initiate the hydrolysis of grain storage proteins at the onset of germination (Mikola, 1987). **Barley aspartic proteinase** (*Hordeum vulgare* AP, **HvAP**) has been purified and characterized (Runeberg-Roos *et al.*, 1991, 1994; Sarkkinen *et al.*, 1992; Kervinen *et al.*, 1993; Törmäkangas *et al.*, 1994). It is becoming apparent that the aspartic proteinases from many different plants are closely related, so an all-embracing name **phytepsin** is introduced here as a new name for HvAP and related plant enzymes and has been adopted by NC-IUBMB.

## Activity and Specificity

Phytepsin activity from barley (HvAP) has been measured both as a pepstatin-sensitive proteinase activity at pH 3.7 using hemoglobin as a substrate (Sarkkinen *et al.*, 1992; Törmäkangas *et al.*, 1994) and by an electrophoretic method using a native gel with immobilized edestin (Wrobel & Jones, 1992; Zhang & Jones, 1995). Purified HvAP hydrolyzes hemoglobin and a chromophoric substrate, Pro-Thr-Glu-Phe+Nph-Arg-Leu (NovaBiochem; see Appendix 2 for full names and addresses of suppliers), optimally at pH 3.5–4.1 (Sarkkinen *et al.*, 1992; Kervinen *et al.*, 1993). Insulin B chain, glucagon and melittin have been used for the characterization of hydrolytic specificity of HvAP. The cleavage of insulin B chain is shown:

FVNQHLCGSHL+VEA+L+YLVCGERGF+F+YTPKA

The cleavage typically occurred between two residues with large hydrophobic side chains (Leu, Ile, Val, Phe) or next to one hydrophobic residue. In glucagon, the Asp+Tyr bond was also readily cleaved (Kervinen *et al.*, 1993):

HSQGTFTSD+YSKY+L+DS+RR+AQDF+VQW+L+MNT

Pepstatin, as well as several substrate-analog inhibitors effective for cathepsin D, inhibits the barley enzyme (Sarkkinen *et al.*, 1992). In plants, endogenous aspartic proteinase inhibitors have been detected in potato (Mareš *et al.*, 1989; Maganja *et al.*, 1992) and tomato (Hansen & Hannapel, 1992), but no published data on the sensitivity of plant aspartic proteinases to these inhibitors are available.

## Preparation and Structural Chemistry

HvAP has been purified from dry grains by affinity chromatography using immobilized pepstatin as an affinity matrix (Sarkkinen *et al.*, 1992). On the basis of the direct amino acid sequencing of purified proteins (Sarkkinen *et al.*, 1992) and the cDNA-derived primary structure (Runeberg-Roos *et al.*, 1991), the primary translation product of 508 amino acid residues contains a signal sequence (prepart), a propart, and two active enzyme forms. These 48 and 40 kDa enzymes apparently represent sequential processing products of the proprotein and both are two-chain enzymes; the 48 kDa enzyme contains 32 and 16 kDa polypeptides, while the 40 kDa enzyme comprises 29 and 11 kDa polypeptides. The primary structure of HvAP resembles other aspartic proteinases, especially mammalian lysosomal cathepsin D (Chapter 277) (~50% similarity) and yeast saccharopepsin (formerly known as proteinase A) (Chapter 283). The similarity is dispersed over two regions in the 48 kDa enzyme, separated by a unique region of 104 residues typical only for plant aspartic proteinases. Part of the unique region is removed during the processing of HvAP. The 48 kDa form contains one N-glycosylation site located in the 16 kDa polypeptide, and the attached carbohydrates are mostly of the plant-modified type (Costa *et al.*, 1997). A three-dimensional model, based on the primary structure of HvAP (excluding the unique region), has been created (Guruprasad *et al.*, 1994).

## Biological Aspects

The barley protease is expressed in several grain tissues including embryo, scutellum and aleurone layer, and in stem, leaves, flowers and roots (Törmäkangas *et al.*, 1994). HvAP has been localized to the scutellar and aleuronal protein storage vacuoles in grains by immunocytochemistry (Bethke *et al.*, 1996; Marttila *et al.*, 1995) as well as to leaf and root cell vacuoles (Runeberg-Roos *et al.*, 1994; Paris *et al.*, 1996). Barley may contain more than one aspartic proteinase. Southern analysis suggests the presence of one or two HvAP-encoding genes in the genome (Törmäkangas *et al.*, 1994). Electrophoretically, four aspartic proteinase activities with different mobilities have been detected from the grain extracts (Zhang & Jones, 1995). A putative aspartic proteinase is also encoded by part of the *BARE*-1 retroelement in the barley genome (Manninen & Schulman, 1993).

The function of phytepsin is obscure; it is present in developing, resting and germinating grains, but its possible role in the hydrolysis of storage proteins remains to be elucidated. The structural and catalytic features of phytepsin are similar to those of animal lysosomal cathepsin D and yeast vacuolar saccharopepsin. Its wide tissue distribution, as well as its intracellular location in a hydrolytic organelle (vacuole), also suggests that phytepsin has several functions in protein processing and turnover in plant cells, similar to the roles proposed for cathepsin D (Saftig *et al.*, 1995) and saccharopepsin (Teichert *et al.*, 1989). This hypothesis is supported by a study showing that HvAP cleaves the

C-terminal vacuolar targeting signal of prolectin *in vitro* (Runeberg-Roos *et al.*, 1994).

## Distinguishing Features

HvAP, as well as other sequenced plant phytepsins, contains an internal region of about 100 residues not present in animal or microbial aspartic proteinases (Chapter 271). The region, however, is highly similar to saposins, sphingolipid-activating proteins in mammalian cells. The similarity includes six conserved cysteines, a glycosylation site, a hydrophobicity pattern, and an invariant Tyr residue (Tyr55 in saposins) (Guruprasad *et al.*, 1994; Munford *et al.*, 1995; Ponting & Russell, 1995). Saposins have been shown to be transiently associated with procathepsin D during a mannose-6-phosphate-independent route to the lysosomes (Zhu & Conner, 1994), and a saposin-like domain influences the intracellular localization and stability of human acyloxyacyl hydrolase (Staab *et al.*, 1994). Therefore, it has been proposed that the saposin-like region also plays a role in the vacuolar (lysosomal) targeting of phytepsins (Guruprasad *et al.*, 1994; Paris *et al.*, 1996). However, no experimental evidence exists to support this hypothesis.

## Related Peptidases

Phytepsins are widely dispersed in the plant kingdom (Kervinen *et al.*, 1995). In addition to HvAP, aspartic proteinases from the seeds of wheat (Belozersky *et al.*, 1989; Galleschi & Felicioli, 1994), rice (Doi *et al.*, 1980; Asakura *et al.*, 1995), sorghum (Garg & Virupaksha, 1970), hemp (St. Angelo & Ory, 1970), buckwheat (Elpidina *et al.*, 1990), lotus (Shinano & Fukushima, 1971), cucumber and squash (Polanowski *et al.*, 1985), rape (D'Hondt *et al.*, 1993), cauliflower (Fujikura & Karssen, 1995), cocoa (Voigt *et al.*, 1995), potato (Brierley *et al.*, 1996), and pine (Salmia, 1981; Bourgeois & Malek, 1991) have been characterized. Aspartic proteinases isolated from cardoon flowers and the digestive fluid of some insectivorous species are reviewed in the cyprosin/cardosin (Chapters 279, 280 and 281) and nepenthesin (Chapter 282) chapters, respectively. Based on the sequenced plant aspartic proteinases, the predicted primary structures show about 80% similarity and the size of an unprocessed active enzyme is about 48 kDa.

## Acknowledgements

Jukka Kervinen is supported by the National Cancer Institute, DHHS, under contract with ABL, and by the Academy of Finland.

## Further Reading

A detailed review of plant aspartic proteinases is found in Kervinen *et al.* (1995).

## References

Asakura, T., Watanabe, H., Abe, K. & Arai, S. (1995) Rice aspartic proteinase, oryzasin, expressed during seed ripening and germination, has a gene organization distinct from those of animal and microbial aspartic proteinases. *Eur. J. Biochem.* **232**, 77–83.

Belozersky, M.A., Sarbakanova, S.T. & Dunaevsky, Y.E. (1989) Aspartic proteinase from wheat seeds: isolation, properties and action on gliadin. *Planta* **177**, 321–326.

Bethke, P.C., Hillmer, S. & Jones, R.L. (1996) Isolation of intact protein storage vacuoles from barley aleurone. *Plant Physiol.* **110**, 521–529.

Bourgeois, J. & Malek, L. (1991) Purification and characterization of an aspartyl proteinase from dry jack pine seeds. *Seed Sci. Res.* **1**, 139–147.

Brierley, E.R., Bonner, P.L.R. & Cobb, A.H. (1996) Factors influencing the free amino acid content of potato (*Solanum tuberosum* L) tubers during prolonged storage. *J. Sci. Food Agric.* **70**, 515–525.

Costa, J., Ashford, D.A., Nimtz, M., Bento, I., Frazao, C., Esteves, C.L., Faro, C.J., Kervinen, J., Pires, E., Veríssimo, P., Wlodawer, A. & Carrondo, M.A. (1997) The glycosylation of aspartic proteinases from barley (*Hordeum vulgare* L.) and cardoon (*Cynara cardunculus* L.). *Eur. J. Biochem.* **243**, 695–700.

D'Hondt, K., Bosch, D., Van Damme, J., Goethals, M., Vandekerckhove, J. & Krebbers, E. (1993) An aspartic proteinase present in seeds cleaves *Arabidopsis* 2 S albumin precursors *in vitro*. *J. Biol. Chem.* **268**, 20884–20891.

Doi, E., Shibata, D., Matoba, T. & Yonezawa, D. (1980) Characterization of pepstatin-sensitive acid protease in resting rice seeds. *Agric. Biol. Chem.* **44**, 741–747.

Elpidina, E.N., Dunaevsky, Y.E. & Belozersky, M.A. (1990) Protein bodies from buckwheat seed cotyledons: isolation and characteristics. *J. Exp. Bot.* **41**, 969–977.

Fujikura, Y. & Karssen, C.M. (1995) Molecular studies on osmoprimed seeds of cauliflower: a partial amino acid sequence of a vigour-related protein and osmopriming-enhanced expression of putative aspartic protease. *Seed Sci. Res.* **5**, 177–181.

Galleschi, L. & Felicioli, F. (1994) Purification, characterization and activation by anions of an aspartic proteinase isolated from bran of soft wheat. *Plant Sci.* **98**, 15–24.

Garg, G.K. & Virupaksha, T.K. (1970) Acid protease from germinated sorghum. 2. Substrate specificity with synthetic peptides and ribonuclease A. *Eur. J. Biochem.* **17**, 13–18.

Guruprasad, K., Törmäkangas, K., Kervinen, J. & Blundell, T.L. (1994) Comparative modelling of barley-grain aspartic proteinase: a structural rationale for observed hydrolytic specificity. *FEBS Lett.* **352**, 131–136.

Hansen, J.D. & Hannapel, D.J. (1992) A wound-inducible potato proteinase inhibitor gene expressed in non-tuber-bearing species is not sucrose inducible. *Plant Physiol.* **100**, 164–169.

Kervinen, J., Sarkkinen, P., Kalkkinen, N., Mikola, L. & Saarma, M. (1993) Hydrolytic specificity of the barley grain aspartic proteinase. *Phytochemistry* **32**, 799–803.

Kervinen, J., Törmäkangas, K., Runeberg-Roos, P., Guruprasad, K., Blundell, T. & Teeri, T.H. (1995) Structure and possible function of aspartic proteinases in barley and other plants. In: *Aspartic Proteinases: Structure, Function, Biology, and Biomedical Implications* (Takahashi, K., ed.). New York: Plenum Press, pp. 241–254.

Maganja, D.B., Strukelj, B., Pungercar, J., Gubensek, F., Turk, V. & Kregar, I. (1992) Isolation and sequence analysis of the genomic DNA fragment encoding an aspartic proteinase inhibitor homologue from potato (*Solanum tuberosum* L.). *Plant Mol. Biol.* **20**, 311–313.

Manninen, I. & Schulman, A.H. (1993) *BARE*-1, a *copia*-like retroelement in barley (*Hordeum vulgare* L.). *Plant Mol. Biol.* **22**, 829–846.

Mareš, M., Meloun, B., Pavlík, M., Kostka, V. & Baudyš, M. (1989) Primary structure of cathepsin D inhibitor from potatoes and its structure relationship to soybean trypsin inhibitor family. *FEBS Lett.* **251**, 94–98.

Marttila, S., Jones, B.L. & Mikkonen, A. (1995) Differential localization of two acid proteinases in germinating barley (*Hordeum vulgare*) seed. *Physiol. Plant.* **93**, 317–327.

Mikola, J. (1987) Proteinases and peptidases in germinating cereal grains. In: *Fourth International Symposium on Pre-harvest Sprouting in Cereals* (Mares, D.J., ed.). Boulder, CO: Westview Press, pp. 463–473.

Morris, P.C., Miller, R.C. & Bowles, D.J. (1985) Endopeptidase activity in dry harvest-ripe wheat and barley grains. *Plant Sci.* **39**, 121–124.

Munford, R.S., Sheppard, P.O. & O'Hara, P.J. (1995) Saposin-like proteins (SAPLIP) carry out diverse functions on a common backbone structure. *J. Lipid Res.* **36**, 1653–1663.

Paris, N., Stanley, C.M., Jones, R.L. & Rogers, J.C. (1996) Plant cells contain two functionally distinct vacuolar compartments. *Cell* **85**, 563–572.

Polanowski, A., Wilusz, T., Kolaczkowska, M.K., Wieczorek, M. & Wilimowska-Pelc, A. (1985) Purification and characterization of aspartic proteinases from *Cucumis sativus* and *Cucurbita maxima* seeds. In: *Aspartic Proteinases and Their Inhibitors* (Kostka, V., ed.). New York: Walter de Gruyter, pp. 49–52.

Ponting, C.P. & Russell, R.B. (1995) Swaposins: circular permutations within genes encoding saposin homologues. *Trends Biochem. Sci.* **20**, 179–180.

Runeberg-Roos, P., Törmäkangas, K. & Östman, A. (1991) Primary structure of a barley-grain aspartic proteinase. A plant aspartic proteinase resembling mammalian cathepsin D. *Eur. J. Biochem.* **202**, 1021–1027.

Runeberg-Roos, P., Kervinen, J., Kovaleva, V., Raikhel, N.V. & Gal, S. (1994) The aspartic proteinase of barley is a vacuolar enzyme that processes probarley lectin *in vitro. Plant Physiol.* **105**, 321–329.

Saftig, P., Hetman, M., Schmahl, W., Weber, K., Heine, L., Mossmann, H., Köster, A., Hess, B., Evers, M., von Figura, K. & Peters, C. (1995) Mice deficient for the lysosomal proteinase cathepsin D exhibit progressive atrophy of the intestinal mucosa and profound destruction of lymphoid cells. *EMBO J.* **14**, 3599–3608.

St. Angelo, A.J. & Ory, R.L. (1970) Properties of a purified proteinase from hempseed. *Phytochemistry* **9**, 1933–1938.

Salmia, M.A. (1981) Proteinase activities in resting and germinating seeds of Scots pine, *Pinus sylvestris. Physiol. Plant.* **53**, 39–47.

Sarkkinen, P., Kalkkinen, N., Tilgmann, C., Siuro, J., Kervinen, J. & Mikola, L. (1992) Aspartic proteinase from barley grains is related to mammalian lysosomal cathepsin D. *Planta* **186**, 317–323.

Shinano, S. & Fukushima, K. (1971) Studies on lotus protease. III. Some physicochemical and enzymic properties. *Agric. Biol. Chem.* **35**, 1488–1494.

Staab, J.F., Ginkel, D.L., Rosenberg, G.B. & Munford, R.S. (1994) A saposin-like domain influences the intracellular localization, stability, and catalytic activity of human acyloxyacyl hydrolase. *J. Biol. Chem.* **269**, 23736–23742.

Teichert, U., Mechler, B., Müller, H. & Wolf, D.H. (1989) Lysosomal (vacuolar) proteinases of yeast are essential catalysts for protein degradation, differentiation, and cell survival. *J. Biol. Chem.* **264**, 16037–16045.

Thomas, D.J., Richards, A.D. & Kay, J. (1989) Inhibition of aspartic proteinases by $\alpha_2$-macroglobulin. *Biochem. J.* **259**, 905–907.

Törmäkangas, K., Kervinen, J., Östman, A. & Teeri, T.H. (1994) Tissue-specific localization of aspartic proteinase in developing and germinating barley grains. *Planta* **195**, 116–125.

Voigt, J., Kamaruddin, S., Heinrichs, H., Wrann, D., Senyuk, V. & Biehl, B. (1995) Developmental stage-dependent variation of the levels of globular storage protein and aspartic endoprotease during ripening and germination of *Theobroma cacao* L. seeds. *J. Plant Physiol.* **145**, 299–307.

Wrobel, R. & Jones, B.L. (1992) Appearance of endoproteolytic enzymes during the germination of barley. *Plant Physiol.* **100**, 1508–1516.

Zhang, N. & Jones, B.L. (1995) Development of proteolytic activities during barley malting and their localization in the green malt kernel. *J. Cereal Sci.* **22**, 147–155.

Zhu, Y. & Conner, G.E. (1994) Intermolecular association of lysosomal protein precursors during biosynthesis. *J. Biol. Chem.* **269**, 3846–3851.

*Jukka Kervinen*
*Macromolecular Structure Laboratory,*
*ABL-Basic Research Program,*
*NCI-Frederick Cancer Research and Development Center,*
*Frederick, MD 21702, USA*
*Email: kervinen@crysv1.ncifcrf.gov*

# *279. Cyprosin*

## *Databanks*

*Peptidase classification: clan AA, family A1, MEROPS ID: A01.024*
*NC-IUBMB enzyme classification: none*

*Databank codes:*

| Species | SwissProt | PIR | EMBL (cDNA) | EMBL (genomic) |
|---|---|---|---|---|
| *Centaurea calcitrapa* | – | – | Y09123 | – |
| *Cynara cardunculus* | P40782 | S47096 | X69193 | – |
| *Cynara cardunculus* | – | S49349 | X81984 | – |

## Name and History

Flowers of the plant *Cynara cardunculus flavescens* (cv. cardoon) have been traditionally used for many centuries to produce ewe's cheese in Portugal and Spain. One of the first reports on this use by farmers was made by Brotero (1804) in *Flora Lusitanica*. Today, the Serpa and Serra cheeses are some examples of typical and highly appreciated products in Portugal. These cheeses are made with water extracts of dried flowers.

The *Cynara* proteinases were first purified and partially characterized by Heimgartner *et al*. (1990) and were named **cynarases** (**cynar**a prote**ase**). This name was used in three papers (Heimgartner *et al*., 1990; Cordeiro *et al*., 1992, 1993) and one Doctoral dissertation (Cordeiro, 1993). Later on, according to the catalytic properties of the enzymes, and due to conventional rules on naming aspartic proteinases, this name was changed to **cyprosin**, derived from **cyn**ara **pro**tease with the ending **-sin**.

The three proteinases purified from the flower styles are named cyprosin 1, 2 and 3. So far, two cDNA clones (*cyprols* and *cyproll*) encoding cyprosins have been isolated and sequenced. The deduced amino acid sequences of these clones are named cyprosin A and B, respectively. Correlation between cyprosins 1, 2 and 3 and the cyprosins A and B has, however, not yet been fully achieved.

## Activity and Specificity

Initially, the proteolytic activity of various enzyme preparations was determined with a standard fluorometric proteolytic assay based on the release of trichloroacetic acid-soluble FITC-labeled peptides from casein (Heimgartner *et al*., 1990). Using this assay we could establish that cyprosin 3 showed the highest specific activity and cyprosin 1 the lowest. With casein as substrate the pH optimum for the three cyprosins is 5.1.

Using a library of 46 synthetic chromophoric octapeptides with systematic variation in the amino acid residues as substrates, it was concluded that the cyprosins preferentially, as do other aspartic proteinases, cleaved peptide bonds between two hydrophobic amino acids (Cordeiro *et al*., 1996). Four peptides were selected for detailed kinetic measurements. The lowest $K_m$ values (15–25 mM) were obtained with Lys-Pro-Ile-Val-Phe$\dagger$Nph-Arg-Leu while the highest $K_{cat}$ values (34–85 s$^{-1}$) were obtained with Lys-Pro-Ile-Leu-Phe$\dagger$Nph-Arg-Leu. The $K_i$ values for pepstatin A were below 0.1 nM for the three isozymes. The pH optimum for cyprosin 3 is around 4.1 using a synthetic peptide (Lys-Pro-Leu-Gln-Leu$\dagger$Nph-Arg-Leu) as substrate. Hydrolysis of milk proteins by cyprosin 3 was investigated by determining the release

of nonprotein nitrogen (NPN) from milk. Clotting activity was followed with a formagraph using cow's or ewe's milk. The pattern of breakdown products after hydrolysis of milk proteins was determined by ultrathin isoelectric focusing (UTIEF). All of these studies were performed in comparison to cattle chymosin (Chapter 274). Purified cyprosin 3 and chymosin revealed similar clotting activities, in particular in the initial phase of the hydrolysis when curd is formed (Cordeiro *et al*., 1992). NPN and UTIEF studies with cow's milk indicated that cyprosin preferentially cleaves the Phe105$\dagger$Met106 bond of $\kappa$-casein. However, in addition to this protein, cyprosin 3 also slowly hydrolyzes some of the $\gamma$-caseins, $\beta$-casein A$^2$ and $\alpha_{s1}$-casein. This enzyme has shown a considerably higher clotting activity in ewe's milk than chymosin, and also a more specific hydrolysis of proteins in this type of milk (Cordeiro *et al*., 1992).

The clotting activities of cyprosins 1, 2 and 3 are different. At an enzyme concentration of 6 µg protein per ml milk, the clotting time is less than 10 min for cyprosin 3, more than 60 min for cyprosin 2 and more than 120 min for cyprosin 1 at 35°C.

## Structural Chemistry

The enzymes are heterodimeric proteins and the subunit sizes, as estimated by SDS-PAGE, are 32.5 + 16.5 kDa, 33.5 + 16.5 kDa and 35.5 + 13.5 kDa for cyprosin 1, 2 and 3, respectively. A native $M_r$ of around 49 000 was determined for the three cyprosins by native PAGE. Gel-filtration chromatography indicated somewhat lower native $M_r$ values, i.e. 41 000, 42 000 and 45 000 for cyprosin 1, 2 and 3, respectively. The cyprosins are glycoproteins of the high mannose type. The pIs for the cyprosin isoforms are 3.85, 4.0 and 4.15.

## Primary Structure

The nucleotide sequence of clone *cyprols* (X69193) contained a 1422 bp open reading frame coding for 474 amino acids including a putative full-length mature protein (440 amino acids) and a partial prosequence (34 amino acids). The missing N-terminal sequence of this clone was later obtained by RT-PCR and the full-length gene product contains 505 amino acids. The nucleotide sequence of clone *cyproll* (X81984) contains a 1527 bp open reading frame coding for 509 amino acids including a putative full-length mature protein (441 amino acids) and a prosequence (68 amino acids). Each of the deduced amino acid sequences contain two putative *N*-glycosylation sites. The identity and similarity of the putative mature cyprosins are 85 and 92%, respectively.

## Modeling the Structure of Cyprosin

A three-dimensional model of cyprosin has been proposed which was constructed using the deduced amino acid sequence of *cyprols* cDNA omitting the large insert (Cordeiro *et al*., 1994a). Six highly homologous aspartic proteinases were used for modeling, i.e. human cathepsin D (Chapter 277), yeast proteinase A (Chapter 283), cattle chymosin (), human renin (Chapter 284), mouse renin (Chapter 285) and pig pepsin (Chapter 272). The overall fold of the model is very similar to that of other aspartic proteinases.

Models of cyprosin complexed with ligands in the active site were constructed to mimic the transition-state analog. Most of the pockets in cyprosin are very similar to those in the barley-grain aspartic proteinase (Chapter 278) (Guruprasad *et al*., 1994); the major difference is near the S2′ and S3′ pockets. This difference is due to His295 in cyprosin compared with Arg295 in the barley enzyme.

## Preparation

Cyprosins are present in the flowers of *Cynara cardunculus*, mostly in the upper violet parts of mature styles. Therefore, the best source for cyprosin is fresh or dried mature styles of *C. cardunculus*. The purification procedure (Heimgartner *et al*., 1990) includes extraction with Tris–HCl buffer, pH 8.3; fractionation with ammonium sulfate precipitation (30–80% saturation); anion-exchange chromatography on DEAE-Sepharose and Mono-Q FPLC. Elution from the Mono-Q column is achieved with a linear gradient of sodium chloride. Three peaks with proteolytic and clotting activity are eluted and rechromatographed separately.

The enzymes can also be purified by preparative isoelectric focusing on an IEF Ultrodex gel with a pH gradient of 3.5–5 (Cordeiro, 1993). The protein bands are visualized under UV light and three bands with high proteolytic and clotting activities are collected into separate tubes and eluted from the gel by centrifugation at 5000 rpm. These cyprosins may be further purified by anion-exchange chromatography.

## Biological Aspects

The cyprosins are expressed as preproenzymes which are post-translationally modified in various ways. A signal sequence is removed after appropriate translocation of the enzyme molecule. The enzymes are *N*-glycosylated with high-mannose type carbohydrates. From N-terminal protein sequencing and the deduced amino acid sequence, we can conclude that the cyprosins are most likely not autoactivated (cleavage of an Arg-Asp bond). The characterization of an activating proteinase has not yet been carried out. Furthermore, at least part of the large insert is removed, resulting in heterodimeric enzymes. This processing may be autocatalyzed.

The genes encoding the cyprosins are expressed during the early stages of flower development (up to 40 mm long). The expression of the cyprosin genes seems to be specific for organs present in flowers (i.e. styles and bracts). Genomic DNA from young flower tissue of *C. cardunculus* was isolated, digested with restriction endonucleases (*Eco*RI and *Hin*dIII) and analyzed by Southern hybridization (Cordeiro *et al*., 1994c). The *Eco*RI and *Hin*dIII digests each showed 4–5 strong hybridizing bands and several minor bands when the entire 1.7 kb insert of *cyprols* was used as probe. From these hybridization patterns it may be suggested that the cyprosin genes are organized as a multigene family. Preliminary sequencing of genomic PCR fragments (corresponding to *cyprols*) has shown that the open reading frame of the cyprosin gene contains 13 exons and 12 introns.

The accumulation of cyprosins in flowers at several stages of development (2–50 mm in length), styles and corollas from mature flowers, as well as seeds and leaves, was studied (Cordeiro *et al*., 1994b). The presence of cyprosin 3 can be detected in very early stages (flowers of 2 mm length) showing that the enzyme is already formed at the beginning of flower development. Cyprosins 1 and 2 are present at a later stage (flowers of 7 mm length). The localization of cyprosins in flower tissues was performed by immunogold labeling using cyprosin antibodies (Cordeiro *et al*., 1994b). Studies were carried out with flowers fixed before anthesis at several stages of development. Silver enhancement of immunogold-labeled flower cuts (1–2 mm) allowed the detection of specific labeling of the cyprosins by light microscopy. The cyprosins appeared to be specifically located in the epidermal cell layer of styles. No other tissues from styles or corollas showed any labeling. The cortex, the transmitting tissue and vascular bundles of styles, as well as the epidermal cell layer and cortex of corollas, did not show any labeling. Cuts obtained from the basal region of flowers showed a less intense labeling when compared to cuts from the intermediate region of the same flowers. These data confirm a differential distribution of cyprosins along the flower, and a higher accumulation of the enzymes in the upper violet part of the flower.

Also in electron microscopy, specific labeling of cyprosin antibodies was detected in the epidermal cell layer of styles using a gold-labeled secondary antibody. Gold particles (10 nm), concentrated in electron-dense agglomerates, are visualized in the hyaloplasm dispersed all over the cell. They are not concentrated in plastids, mitochondria or the nucleus.

Other clotting enzymes that cross-react with cyprosin antibodies have been found in flowers of *Cynara humilis*, *Centaurea calcitrapa* and the former *Cynara scolymus* (artichoke). It is likely that these enzymes belong to a family of closely related aspartic proteinases with similar structure and biological function in the plants (Chapter 278).

## Distinguishing Features

The cyprosins differ, like other plant aspartic proteinases, from aspartic proteinases from other phyla in that they are synthesized with a large insert of around 100 amino acids in the C-terminal part of the protein molecule. This insert is at least partly removed during a post translational proteolytic modification of the protein resulting in heterodimeric enzymes. Furthermore, the normal Asp-Thr-Gly sequence of catalytic aspartic acid residues has been substituted for Asp-Ser-Gly for the second catalytic aspartic acid in the cyprosins. This substitution is seen in all plant enzymes.

The organization of the cyprosin gene is different from that of mammalian aspartic proteinase genes. The cyprosin gene contains 13 exons and 12 introns while the mammalian genes (e.g. human cathepsin D (Chapter 277); Redecker *et al.*, 1991) contain 9 exons and 8 introns. The localizations of the introns are also different in the plant and mammalian genes.

Rabbit polyclonal antibodies have been produced against the large subunit of cyprosin 3. These antibodies cross-react with cyprosin 1 and 2 and with aspartic proteinases from other *Cynara* species as well as with an aspartic proteinase from *C. calcitrapa*.

## Related Peptidases

The differences between cyprosins and cardosins (Chapters 280 and 281) have not yet been fully established. However, we expect these enzymes to be quite similar as they have been isolated from very closely related plants.

We have isolated and partly characterized a heterodimeric aspartic proteinase from flowers of *Centaurea calcitrapa* (Domingos *et al.*, 1996). These flowers are also used for cheese-making. A cDNA clone encoding this enzyme has been isolated and sequenced (open reading frame = 509 amino acids). The enzyme shows a high homology to cyprosin B (identity: 96.2%; similarity: 98.4%).

## Further Reading

For reviews see Brodelius *et al.* (1995) and Cordeiro *et al.* (1995, 1977).

## References

Brodelius, P.E., Cordeiro, M. & Pais, M.S. (1995) Aspartic proteinases from *Cynara cardunculus*. Purification, characterization and tissue-specific expression. In: *Aspartic Proteinases: Structure, Function, Biology, and Biomedical Implications* (Takahashi, K., ed.). New York: Plenum Press, pp. 255–266.

Brotero, F. (1804) *Flora lusitanica*. Olissipone.

Cordeiro, M. (1993) Milk clotting enzymes from *Cynara cardunculus* spp *flavescens* cv. Cardoon. Characterization and molecular cloning of the enzymes and studies on their expression in flower tissue. PhD thesis, FCUL, Lisbon.

Cordeiro, M., Jacob, E., Puhan, Z., Pais, M.S. & Brodelius, P. (1992) Milk clotting and proteolytic activities of purified cynarases from *Cynara cardunculus*: a comparison to chymosin.

*Milchwissenschaft* **47**, 683–687.

Cordeiro, M., Zue, Z.-T., Pais, M.S. & Brodelius, P.E. (1993) Proteases from cell suspension cultures of *Cynara cardunculus*. *Phytochemistry* **33**, 1323–1326.

Cordeiro, M., Guruprasad, K., Blundell, T., Pais, M.S. & Brodelius, P.E. (1994a) Rule-based comparative modelling of cyprosin from its amino acid sequence deduced from a cDNA clone. *4th International Congress of Plant Molecular Biology*, 19–24 June 1994, Amsterdam. Abstract 1199.

Cordeiro, M., Pais, M.S. & Brodelius, P.E. (1994b) Tissue-specific expression of multiple forms of cyprosin (aspartic proteinase) in flowers of *Cynara cardunculus*. *Physiol. Plant.* **92**, 645–653.

Cordeiro, M., Xue, Z.-T., Pietrzak, M., Pais, M.S. & Brodelius, P.E. (1994c) Isolation and characterization of a cDNA from flowers of *Cynara cardunculus* encoding cyprosin (an aspartic proteinase) and its use to study organ specific expression of cyprosin. *Plant Mol. Biol.* **24**, 733–741.

Cordeiro, M., Xue, Z.-T., Pietrzak, M., Pais, M.S. & Brodelius, P.E. (1995) Plant aspartic proteinase from *Cynara cardunculus* (cardoon). Nucleotide sequence of a cDNA encoding cyprosin and its tissue specific expression. In: *Aspartic Proteinases: Structure, Function, Biology, and Biomedical Implications* (Takahashi, K., ed.). New York: Plenum Press, pp. 367–371.

Cordeiro, M.C., Lowther, T., Dunn, B.M., Guruprasad, K., Blundell, T., Pais, M.S. & Brodelius, P.E. (1996) Substrate specificity and molecular modelling of aspartic proteinases (cyprosins) from flowers of *Cynara cardunculus* subsp. *flavescens* cv. cardoon. *VIIth International Aspartic Proteinase Conference*, 22–27 October 1996, Banff, Alberta, Canada. Abstract P7–7.

Cordeiro, M., Pais, M.S. & Brodelius, P.E. (1997) Milk clotting enzymes from *Cynara cardunculus* subsp. *flavescens* (cardoon) In: *Biotechnology in Agriculture and Forestry* (Bajaj, Y.P.S., ed.). Berlin: Springer-Verlag.

Domingos, A., Xue, Z.-T., Guruprasad, K., Clemente, A., Blundell, T., Pais, M.S. & Brodelius, P.E. (1996) An aspartic proteinase from flowers of *Centaurea calcitrapa*; purification, characterization, molecular cloning and modelling of its three-dimensional structure. *VIIth International Aspartic Proteinase Conference*, 22–27 October 1996, Banff, Alberta, Canada. Abstract P7–6.

Guruprasad, K., Törmäkangas, K., Kervinen, J. & Blundell, T.L. (1994) Comparative modelling of barley-grain aspartic proteinase: a structural rationale for observed hydrolytic specificity. *FEBS Lett.* **352**, 131–136.

Heimgartner, U., Pietrzak, M., Geertsen, R., Brodelius, P., da Silva Figueiredo, A.C. & Pais, M.S.S. (1990) Purification and partial characterization of milk clotting proteases from flowers of *Cynara cardunculus*. *Phytochemistry* **29**, 1405–1410.

Redecker, B., Heckendorf, B., Grosch, H.-W., Mersmann, G. & Hasilik, A. (1991) Molecular organization of the human cathepsin D gene. *DNA Cell Biol.* **10**, 423–431.

*Maria C. Cordeiro*
*Department of Plant Biochemistry,*
*Lund University,*
*PO Box 117, S-22100 Lund, Sweden*

*Peter E. Brodelius*
*Department of Plant Biochemistry,*
*Lund University,*
*PO Box 117, S-22100 Lund, Sweden*
*Email: peter.brodelius@plantbio.lu.su*

*Salomé M. Pais*
*Centro de Biotecnologia Vegetal,*
*Faculdade de Ciências de Lisboa,*
*Bloco C2, Campo Grande, P-1700*
*Lisboa, Portugal*
*Email: bmspais@bio.fc.ul.pt*

# 280. Cardosin A

## Databanks

*Peptidase classification: clan AA, family A1, MEROPS ID: A01.033*
*NC-IUBMB enzyme classification: none*
*Databank codes: no sequence data available*

## Name and History

The flowers of cardoon (*Cynara cardunculus*) are tradition-ally used in Portugal, and have been since at least the Roman era, for making ewe's cheese. The milk-clotting activity was initially thought to be due to a single proteinase (Faro *et al.*, 1987), but it was recently shown that the flowers of *C. cardunculus* contain two different proteinases (Veríssimo *et al.*, 1995, 1996).

The name **Cynara cardunculus protease** was previously used in several papers (Faro *et al.*, 1987, 1992; Barros *et al.*, 1992, 1993; Macedo *et al.*, 1993) during a period when only one proteinase was identified in the flowers. At that stage of the work, commercial dried flowers were the source of enzyme and probably the material contained mainly flowers from other species of cardoon in which the second pro-teinase was found to be absent (Veríssimo *et al.*, 1995). When flowers from plants grown from selected seeds of *C. cardunculus* were used as starting material, two aspartic pro-teinases were identified and isolated (Veríssimo *et al.*, 1996). To follow the descriptive nomenclature for other aspartic proteinases, the more abundant proteinase has been named **cardosin A** whereas the minor one has been named cardosin B (Chapter 281), both designations being derived from *cardo*, the Portuguese name of the plant.

## Activity and Specificity

Cardosin A preferentially cleaves peptide bonds between residues with hydrophobic side-chains. The first study on the specificity of the enzyme was carried out using oxidized insulin B chain as substrate (Veríssimo *et al.*, 1995). Cleav-age of this polypeptide by cardosin A occurs at the bonds Leu15╪Tyr16, Leu17╪Val18 and Phe24╪Phe25:

FVNQHLCGSHLVEAL╪YL╪VCGERGF╪FYTPKA

Since the enzyme has milk-clotting activity, the specificity towards isolated cow caseins has also been studied in detail. Like the majority of other milk-clotting enzymes, cardosin A cleaves $\kappa$-casein at the bond Phe105╪Met106. With respect to the other caseins, cardosin A cleaves $\alpha_{s1}$-casein in a sequential manner starting from the bond Phe23╪Phe24, the most susceptible bond, and then proceeding to the C-terminal part where the enzyme cleaves the bonds Trp164╪Tyr165, Tyr165╪Tyr166 and Phe153╪Tyr154 (Ramalho-Santos *et al.*, 1996). Hydrolysis of $\beta$-casein by cardosin A was found to occur at the peptide bonds Leu165╪Ser166, Leu192╪Tyr193 and Leu127╪Thr128 (I. Simões, P. Veríssimo, C. Faro & E. Pires, unpublished results).

The pH optimum for hydrolysis of the synthetic peptide Lys-Pro-Ala-Glu-Phe╪Nph-Ala-Leu is 4.5 and the values determined for the apparent active-site ionization constants $pK_{e1}$ and $pK_{e2}$ of the free enzyme are $2.5 \pm 0.2$ and $5.3 \pm 0.2$ respectively. The proteolytic activity is inhibited by pepstatin A ($K_i = 3$ nM) and DAN (Veríssimo *et al.*, 1996).

## Structural Chemistry

Cardosin A is a two-chain enzyme formed by glycosyl-ated polypeptides with apparent molecular masses of 31 and 15 kDa. Cardosin A is synthesized in its preform and converted into two chains by removal of an internal segment of about 104 amino acids known as the plant-specific inserted sequence. Recently, a 64 kDa precursor form of cardosin A containing this segment was identified and characterized (Ramalho-Santos *et al.*, 1997).

The structures of the major oligosaccharides in each chain of cardosin A were determined by exoglycosidase sequenc-ing combined with matrix-assisted laser desorption/ionization time-of-flight mass spectrometry and they were found to be of the plant modified N-type (Costa *et al.*, 1997). The X-ray crystallographic structure of cardosin A has been obtained for the active enzyme in its uncomplexed form (by Frazão and coworkers, unpublished results).

## Preparation

Cardosin A is isolated from fresh stigmas of *Cynara cardun-culus* by a simple and quick two-step purification procedure including extraction at low pH followed by gel filtration and ion-exchange chromatography on Mono-Q (Veríssimo *et al.*, 1996). Cardosin A-like enzymes have also been isolated from *C. humilis*, *C. algarbiensis* and *C. scolymus*. The enzyme must be isolated from fresh flowers or flowers frozen in liquid nitrogen rather than dried flowers, since chemical modifica-tions that occur during the drying process affect both the purification and the activity of the enzyme. Cardosin A iso-lated from fresh flowers proved to be suitable for amino acid sequencing and crystallization.

## Biological Aspects

Expression of cardosin A occurs essentially in the pistils of *Cynara cardunculus* and begins at early stages of floral development; the proteolytic activity remains in the flower until the senescent stage and can be preserved in dried flowers for many years. Western tissue prints and immuno-gold electron microscopy studies reveal that cardosin A accumulates in protein storage vacuoles of the stigmatic papillae, being also present in the large central vacuole of the epidermal cells of the style (Ramalho-Santos *et al.*, 1997). Based on these localization studies, it was proposed that cardosin A may have a primary function in the interaction with pathogens and/or in pollination, and a secondary function in the senescence of the aerial parts of the plant (Ramalho-Santos *et al.*, 1998).

## Distinguishing Features

Cardosin A can be distinguished from cardosin B (Chapter 281) by the mobility of the polypeptide chains in SDS-polyacrylamide gels. The cardosin A chains migrate as two bands of 31 and 15 kDa whereas those of cardosin B correspond to polypeptides of 34 and 14 kDa. Furthermore, both chains of cardosin A are glycosylated whereas in cardosin B only the 34 kDa chain is glycosylated.

Rabbit polyclonal antibodies have been produced against cardosin A (Verissimo *et al.*, 1996) and a monospecific antibody against a region of cardosin A with low homology to cardosin B has been used for immunocytochemistry (Ramalho-Santos *et al.*, 1997).

## Further Reading

See Faro *et al.* (1995) for a review.

## References

Barros, M.T., Carvalho, M.G.V., Garcia, F.A.P. & Pires, E. (1992) Stability performance of *Cynara cardunculus* L. acid protease in aqueous-organic biphasic systems. *Biotechnol. Lett.* **14**, 179–118.

Barros, M., Faro, C.J. & Pires, E. (1993) Peptide synthesis catalyzed by *Cynara cardunculus* L. acid protease in aqueous-organic biphasic systems. *Biotechnol. Lett.* **15**, 653–656.

Costa, J., Ashford, A., Nimetz, M., Bento, I., Frazão, C., Esteves, C., Faro, C., Kervinen, J., Pires, E., Veríssimo, P., Wlodawer, A. & Carrondo, M.A. (1997) The glycosylation of barley (*Hordeum*

*vulgare* L.) and cardoon (*Cynara cardunculus* L.) aspartic proteinases. *Eur. J. Biochem.* **243**, 695–700.

Faro, C.J., Alface, J.S. & Pires, E.V. (1987) Purification of a protease from the flowers of *Cynara cardunculus* L. *Ciênc. Biol.* **12**(5A), 201.

Faro, C.J., Moir, A.G.J. & Pires, E.M.V. (1992) Specificity of a milk clotting enzyme extracted from the thistle *Cynara cardunculus* L.: action on oxidized insulin and κ-casein. *Biotechnol. Lett.* **14**, 841–847.

Faro, C., Verissimo, P., Lin, Y., Tang, J. & Pires, E. (1995) Cardosins A and B, aspartic proteinases from the flowers of cardoon. In: *Aspartic Proteinases: Structure, Function, Biology and Biomedical Implications* (Takahashi, K., ed.). New York: Plenum Press, pp. 374–377.

Macedo, I.Q., Faro, C.J. & Pires, E.M. (1993) Specificity and kinetics of the milk-clotting enzyme from cardoon (*Cynara cardunculus* L) toward bovine κ-casein. *J. Agric. Food Chem.* **41**, 1537–1540.

Macedo, I., Faro, C. & Pires, E. (1996) Caseinolytic specificity of cardosin, an aspartic protease from the cardoon *C. cardunculus* L. Action on bovine and $\alpha_{s1}$ and β-casein: comparison with chymosin. *J. Agric. Food Chem.* **44**, 42–47.

Ramalho-Santos, M., Veríssimo, P., Faro, C.J. & Pires, E.V. (1996) Action on bovine $\alpha_{s1}$-casein of cardosins A and B, aspartic proteinases from the flowers of the cardoon *Cynara cardunculus* L. *Biochim. Biophys. Acta* **1297**, 83–89.

Ramalho-Santos, M., Pissara, J., Verissimo, P., Pereira, S., Salema, R., Pires, E. & Faro, C. (1997) Cardosin A accumulates in protein storage vacuoles of the stigmatic papillae of *Cynara cardunculus* L. *Planta* **203**, 204–212.

Veríssimo, P.C., Esteves, C., Faro, C.J. & Pires, E. (1995) The vegetable rennet from *Cynara cardunculus* L. contains two proteinases with chymosin-like and pepsin-like specificities. *Biotechnol. Lett.* **17**, 614–645.

Veríssimo, P., Faro, C., Moir, A., Lin, Y., Tang, J. & Pires, E. (1996) Purification, characterization and partial amino acid sequencing of two novel aspartic proteinases from the flowers of *Cynara cardunculus* L. *Eur. J. Biochem.* **235**, 762–768.

*Euclides M.V. Pires*
*Departamento de Bioquímica,*
*Faculdade de Ciências e Tecnologia,*
*Universidade de Coimbra,*
*Apartado 3126, 3000 Coimbra, Portugal*
*Email: epires@imagem.ibili.uc.pt*

# 281. Cardosin B

## Databanks

*Peptidase classification: clan AA, family A1, MEROPS ID: A01.034*
*NC-IUBMB enzyme classification: none*
*Databank codes: no sequence data available*

## Name and History

The milk-clotting activity of cardoon (*Cynara cardunculus*) flowers was initially thought to be due to a single acid proteinase then named **Cynara cardunculus proteinase** (Faro *et al.*, 1987). At this stage, commercial dried flowers were used as the enzyme source. Recently, by using flowers from plants grown from selected seeds of *C. cardunculus*, a second aspartic proteinase was identified (Veríssimo *et al.*, 1995, 1996). The enzyme, which is less abundant than cardosin A (Chapter 280) was named **cardosin B**. Re-evaluation and quality control of the commercial dried flowers revealed that the lots can be heterogeneous, containing variable proportions of flowers of at least two different cardoon species, *Cynara cardunculus* and *Cynara humilis*. Studies on the proteinase content of the flowers of these species have shown that *C. humilis* contains only a proteinase related to cardosin A while *C. cardunculus* consistently contains cardosins A and B (Veríssimo *et al.*, 1996).

## Activity and Specificity

Cardosin B clearly has a broader specificity than cardosin A, although they both preferentially cleave peptide bonds between residues with hydrophobic side chains. Hydrolysis of oxidized insulin B chain by cardosin B occurs at six peptide bonds (Glu13+Ala14, Ala14+Leu15, Leu15+Tyr16, Leu17+Val18, Phe24+Phe25 and Phe25+Tyr26):

FVNQHLCGSHLVE+A+L+YL+VCGERGF+F+YTPKA

Studies on the action of cardosin B towards cattle casein have shown that this enzyme cleaves essentially the same bonds as cardosin A, but the cleavage rate is undoubtedly higher. Hydrolysis of $\kappa$-casein by cardosin B occurs at the bond Phe105+Met106 whereas $\beta$-casein is cleaved at the peptide bonds Leu192+Tyr193 and Leu165+Ser166 (I. Simões, P. Veríssimo, C. Faro & E. Pires, unpublished results). Cleavage of $\alpha_{s1}$-casein by cardosin B occurs in a sequential manner starting from the bond Phe23+Phe24, the most susceptible bond, and then proceeding to the C-terminal part where cardosin B cleaves the bonds Trp164+Tyr165, Phe153+Tyr154 and Phe150+Arg151 (Ramalho-Santos *et al.*, 1996).

The pH optimum for hydrolysis of the synthetic peptide Leu-Ser-Nph+Ahx-Ala-Leu-OMe (Ahx: 2-aminohexanoic acid) is about 5.0 and the values determined for the apparent active-site ionization constants $pK_{e1}$ and $pK_{e2}$ of the free enzyme are $3.73 \pm 0.09$ and $6.7 \pm 0.1$ respectively. The activity is inhibited by pepstatin A ($K_i < 1\,nM$) and DAN (Veríssimo *et al.*, 1996).

## Structural Chemistry

Cardosin B is a two-chain enzyme formed by a glycosylated polypeptide with apparent molecular mass of 34 kDa and a nonglycosylated polypeptide chain of 14 kDa. Upon deglycosylation with trifluoromethanesulfonic acid from Sigma (see Appendix 2 for full names and addresses of suppliers) the 34 kDa chain migrated in SDS gels as a band of about 30 kDa (Faro *et al.*, 1995). Like other plant aspartic proteinases,

cardosin B is probably synthesized as a single chain and then converted into two chains by removal of an internal segment, the plant-specific inserted sequence.

## Preparation

Cardosin B is isolated from fresh stigmas of *Cynara cardunculus* by a simple and quick two-step purification procedure including extraction at low pH followed by gel filtration and ion-exchange chromatography on Mono-Q (Veríssimo *et al.*, 1996). Cardosin B-like enzymes were also isolated from *C. scolymus* and were found to be absent in *C. humilis* and *C. algarbiensis*. It is recommended to isolate cardosin B from fresh flowers or flowers frozen in liquid nitrogen rather than from dried flowers, since chemical modifications that occur during the drying process affect both the purification and the specificity of the enzyme (Veríssimo *et al.*, 1996). The cardosin B isolated from fresh flowers of *C. cardunculus* proved to be suitable for amino acid sequencing and crystallization.

## Biological Aspects

Expression of cardosin B occurs essentially in the pistils of *C. cardunculus* and begins at early stages of floral development. Immunogold electron microscopy and western tissue printing studies revealed that the cardosin B is an extracellular enzyme expressed specifically in the transmitting tissue. This tissue supports pollen tube growth and therefore it is probable that cardosin B is involved in this process.

## Distinguishing Features

Cardosin B can be distinguished from cardosin A (Chapter 280) by the mobility of the polypeptide chains in SDS-polyacrylamide gels. Cardosin B chains migrate as two bands of 34 and 14 kDa whereas those of cardosin A correspond to polypeptides of 31 and 15 kDa. Furthermore, only the 34 kDa chain of cardosin B is glycosylated whereas in cardosin A both chains are glycosylated.

A rabbit polyclonal antibody has been produced against isolated cardosin B and upon purification by affinity chromatography, has been used in immunocytochemistry studies.

## References

Faro, C.J., Alface, J.S. & Pires, E.V. (1987) Purification of a protease from the flowers of *Cynara cardunculus* L. *Ciênc. Biol.* **12**(5A), 201.

Faro, C., Veríssimo, P., Lin, Y., Tang, J. & Pires, E. (1995) Cardosins A and B, aspartic proteinases from the flowers of cardoon. In: *Aspartic Proteinases: Structure, Function, Biology and Biomedical Implications* (Takahashi, K., ed.). New York: Plenum Press, pp. 374–377.

Ramalho-Santos, M., Veríssimo, P., Faro, C.J. & Pires, E.V. (1996) Action on bovine $\alpha_{s1}$-casein of cardosins A and B, aspartic proteinases from the flowers of the cardoon *Cynara cardunculus* L. *Biochim. Biophys. Acta* **1297**, 83–89.

Veríssimo, P.C., Esteves, C., Faro, C.J. & Pires, E. (1995) The vegetable rennet from *Cynara cardunculus* L. contains two proteinases with chymosin-like and pepsin-like specificities. *Biotechnol. Lett.* **17**, 614–645.

Veríssimo, P., Faro, C., Moir, A., Lin, Y., Tang, J. & Pires, E. (1996) Purification, characterization and partial amino acid sequencing of two novel aspartic proteinases from the flowers of *Cynara cardunculus* L. *Eur. J. Biochem.* **235**, 762–768.

*Euclides M.V. Pires*
*Departamento de Bioquímica,*
*Faculdade de Ciências e Tecnologia,*
*Universidade de Coimbra,*
*Apartado 3126, 3000 Coimbra, Portugal*
*Email: epires@imagem.ibili.uc.pt*

# 282. Nepenthesin

## Databanks

*Peptidase classification: clan AX, family A99, MEROPS ID: A9G.006*
*NC-IUBMB enzyme classification: EC 3.4.23.12*
*Chemical Abstracts Service registry number: 9073-80-7*

## Name and History

The name **nepenthesin** was taken from the genus name *Nepenthes* by Nakayama & Amagase (1969). The plant was named by Linnaeus from the Greek word for a potion 'removing all sorrow' that was prepared by Helen of Troy and presumably served in a Greek cup (*rhyton*) resembling the pitcher of this plant. *Nepenthes* is one of a dozen different genera of plants (from as many different families) that are carnivorous, digesting insects to obtain nitrogen. The ability of plants to capture and digest insect prey was first recognized towards the end of the eighteenth century. The first detailed study was, interestingly, a long book by Charles Darwin (1875) devoted entirely to such plants and containing 50 pages of detailed study of the protein digestive capacity of *Drosera rotundifolia*. There was an immediate burst of seven or eight papers from various parts of Europe in the next few years which reached the conclusion that protease was secreted by these plants, that bacterial growth was not responsible for the activity, that the secretions were acidic on litmus and that the 'ferment' responsible was a type of pepsin (trypsin or erepsin being the only remaining choices at the time). The early history up to 1935 is thoroughly reviewed by Lloyd (1942) and includes many illustrious names including Hooker, Dakin, Abderhalden and Linderstrøm-Lang. A series of three papers by Amagase's group in Osaka established most of the known properties of this enzyme.

## Activity and Specificity

The enzyme digests a variety of proteins such as casein, fibrin and egg albumin. The pH optimum lies between 2 and 3 for incubations of 10–30 min. Specificity has been studied using a series of eight peptides of known sequence prepared by digesting lysozyme, phage T4 and cytochrome $b_5$. Cleavage was found to involve the carboxy and amino side of Asp, in particular. However, Ala, Leu, Ser, Thr and Tyr were also cleaved on the carboxy side and Ala, Phe, Thr, Tyr and Lys on the amino side. A total of 10 bonds were cleaved in seven peptides. Pepsin cleaved 14 bonds; in only five cases were these the same as those cleaved by nepenthesin (Amagase *et al.*, 1969). It is possible that the purification methods used here did not completely purify the protease; however, a zymographic technique (Amagase, 1972) showed that three contaminating activities had been removed and only one spot could be visualized in the purified preparation.

Another preparation from a different species of *Nepenthes* (Tökés *et al.*, 1974) was applied to ribonuclease as a protein substrate: 11 peptide bonds were cleaved. In this case no bonds of type Asp-Xaa were cleaved, but Ala, Leu and Thr were found in this position as well as Glu, Lys and Gly. Residues contributing the amino moiety of the peptide bond included the same residues seen by Amagase plus Val, Gly and Glu. It was concluded that the enzyme has similarity to pepsin.

Most recently, the enzyme has been purified from *Nepenthes distillatoria* (Athauda *et al.*, 1998) and used to digest the B chain of oxidized insulin:

FVNQHL+CGSHLVE+AL+Y+LVCGERGF+FYTPKA

The similarity to pepsin includes the pH optimum and inhibition by pepstatin and DAN (Takahashi *et al.*, 1974) and *N*-(diazoacetyl)-*N*-(2,4-dinitrophenyl)ethylenediamine (Lobareva *et al.*, 1973).

## Structural Chemistry

The first 24 amino acids of the N-terminus of the enzyme from *N. distillatoria* have been sequenced (Athauda *et al.*, 1998); this shows similarities to the phytepsins (Chapter 278). The enzyme appears to be a monomer with an $M_r$ of 45 000.

## Preparation

The preparations of significance have all started with secretions recovered from unopened pitchers (to avoid bacterial contamination or enzymes that might normally remain within the plant tissues). The methods employed have been directed at isolating a single enzyme activity, but not at establishing complete purity. The published methods employ molecular sieve, DEAE-Sephadex and ECTEOLA-cellulose chromatography as the only steps (Nakayama and Amagase, 1968; Jentsch, 1972; Lobareva *et al.*, 1973; Tökés *et al.*, 1974). Most recently, a further step of pepstatin affinity chromatography has been used to obtain a homogeneous enzyme (Athauda *et al.*, 1998). A serious obstacle to purification is the large number of pitchers that must be gathered and the difficulty of cultivation of the plant. Cloning and expression will probably be required to complete the studies of this protease.

## Biological Aspects

The enzyme is secreted into the pitcher by glands within the pitcher. Opening of the pitcher cover and trapping of insects stimulates enzyme secretion and acidification of the fluid contents. In this situation, further contributions to digestion may be made by microbial enzymes that grow in the fluid. Carnivorous plants typically grow in environments such as bogs that are deficient in soil nitrogen, so the trapped insects help to supply this nutrient.

## Related Peptidases

The only other carnivorous plant enzyme studied in detail is that from *Drosera* species. Such studies have continued in parallel with those on *Nepenthes* since the days of Darwin. In these species droplets of glandular secretion (which give rise to the common name 'sundew') can be collected on filter paper. However, this work is so cumbersome that most studies employ the entire modified leaf with its glands to prepare extracts. This introduces the possibility that secreted protease is mixed with proteases that may normally remain within the leaf. Amagase (1972) prepared enzyme from *Drosera* and compared it to nepenthesin. For some reason, he did not use exactly the same peptides for this comparison, but the general pattern of digestion was similar

for the two enzymes. Takahashi *et al.* (1974) also studied the two enzymes in parallel and found similar inhibition by pepstatin and DAN for both. Of course, the two enzymes come from completely different plant families (within the same order Nepenthales) and so might be expected to have somewhat different specificities.

The current NC-IUBMB entry for nepenthesin includes mention of aspartic proteases from sorghum seeds (Garg & Virupaksha, 1970) and lotus seeds (Shinano & Fukushima, 1971). When the entry was prepared there was no other locus to which plant aspartic proteases could be assigned. At the present time, it seems likely that these seed enzymes may be conveniently grouped under phytepsin (Chapter 278). Why not move nepenthesin there as well? There is still the possibility that the specialized secretory glands have developed a modified plant aspartic protease for the specific purpose of digesting insects and that this enzyme might differ from intracellular aspartic proteases. In fact, Holter & Linderstrøm-Lang (1933) noted that *Drosera* leaves seemed to have a second activity in addition to that found in the secretions and that this second activity had a distinctly higher pH curve that was close to that of malt (= barley = phytepsin). It should also be noted that the barley protease and nepenthesin cleave the B chain of insulin at five points, but only two of these cleavages are in common. It would be premature to subsume nepenthesin under the phytepsin family until we have full sequence data and a more detailed specificity study of the highly purified enzyme.

## References

Amagase, S., Nakayama, S. & Tsugita, A. (1969) Acid protease in *Nepenthes*. II. Study on the specificity of nepenthesin. *J. Biochem. (Tokyo)* **66**, 431–439.

Amagase, S. (1972) Digestive enzymes in insectivorous plants. II. Acid proteases in the genus *Nepenthes* and *Drosera peltata*. *J. Biochem. (Tokyo)* **72**, 73–81.

Athauda, S.B.P., Inoue, H., Iwamatsu, A. & Takahashi, K. (1998) Acid proteinase from *Nepenthes distillatoria* (Badura). In: *Structure and Function of Aspartic Proteinases: Retroviral and Cellular Enzymes* (James, M.N.G., ed.) New York: Plenum Press, p. 453–458.

Darwin, C. (1875) *Insectivorous Plants*. London: John Murray.

Garg, G.K. & Virupaksha, T.K. (1970) Acid protease from germinated sorghum. 2. Substrate specificity with synthetic peptides and ribonuclease A. *Eur. J. Biochem.* **17**, 13–18.

Holter, H. & Linderstrøm-Lang, K. (1933) Beiträge zur enzymatischen Histochemie. III. Über die Proteinasen von *Drosera rotundifolia* [Studies on enzyme histochemistry. III. On the proteinases of *Drosera rotundifolia*]. *Z. Physiol. Chem.* **214**, 223–240.

Jentsch, J. (1972) Enzymes from carnivorous plants (*Nepenthes*). Isolation of the protease nepenthacin. *FEBS Lett.* **21**, 273–276.

Lloyd, F.E. (1942) *The Carnivorous Plants*. New York: Ronald Press.

Lobareva, L.S., Rudenskaya, G.N. & Stepanov, V.M. (1973) [Pepsin-like proteinase from the insectivorous plant *Nepenthes*]. *Biokhimiya* **38**, 640–642.

Nakayama, S. & Amagase, S. (1968) Acid proteases in *Nepenthes*. Partial purification and properties of the enzyme. *Proc. Jap. Acad.* **44**, 358–362.

Shinano, S. & Fukushima, K. (1971) Studies on lotus seed protease. III. Some physicochemical and enzymic properties. *Agric. Biol.*

*Chem.* **35**, 1488–1494.

Takahashi, K., Chang, W.-J. & Ko, J.-S. (1974) Specific inhibition of acid proteases from brain, kidney, skeletal muscle, and insectivorous plants by diazoacetyl-DL-norleucine methyl ester and by

pepstatin. *J. Biochem. (Tokyo)* **76**, 897–899.

Tökés, Z.A., Woon, W.C. & Chambers, S.M. (1974) Digestive enzymes secreted by the carnivorous plant *Nepenthes macferlanei* L. *Planta* **119**, 39–46.

*J. Fred Woessner*
*Department of Biochemistry and Molecular Biology,*
*University of Miami School of Medicine,*
*Miami, FL 33101, USA*
*Email: fwoessne@mednet.med.miami.edu*

# 283. Saccharopepsin

## Databanks

*Peptidase classification: clan AA, family A1, MEROPS ID: A01.018*
*NC-IUBMB enzyme classification: EC 3.4.23.25*
*Chemical Abstracts Service registry number: 37228-80-1*
*Databank codes:*

| Species | SwissProt | PIR | EMBL (cDNA) | EMBL (genomic) |
|---|---|---|---|---|
| *Saccharomyces cerevisiae* | P07267 | A25379 | M13358 Z11963 | – |

## Name and History

Saccharopepsin activity in yeast was described as early as 1917 under the name 'Hefepepsin' (Dernby, 1917). The protein was further characterized and partially purified by Hata *et al.* (1967) as a major proteolytic activity in *Saccharomyces cerevisiae*. Since then the enzyme has almost exclusively been termed **proteinase A** or **proteinase yscA**. The name **saccharopepsin** used in this entry is the one recommended by the IUBMB and is desirable because it uniquely identifies this protease by its origin (**Saccharomyces**) and type (**pepsin**-like). Consistent with this nomenclature we have termed the proenzyme saccharopepsinogen. The gene for saccharopepsin (*PEP4*) was known long before it was known to encode saccharopepsin. Mutants in *PEP4* were characterized by lack of multiple vacuolar hydrolytic activities and the broad phenotype of the *PEP4* mutation was for some time a source of confusion. However, it was later found to be consistent with the vacuolar localization of saccharopepsin and its role in activation of vacuolar zymogens. Eventually, the biological aspects of saccharopepsin's role in activation of the yeast vacuolar (lysosomal) activation cascade have become one of the best characterized of any aspartic protease.

## Activity and Specificity

Saccharopepsin has a specificity which resembles that of pepsin (Chapter 272) and cathepsin D (Chapter 277); Phe, Leu and Glu are favored in the P1 subsite; Phe, Ile, Leu and Ala in P1' (Dreyer, 1989). Saccharopepsin activity can be measured using acid-denatured hemoglobin (Jones, 1991). More recently, saccharopepsin activity has been monitored using the internally quenched fluorescent peptide 2-aminobenzoyl-Leu-Phe-Ala-Leu-Glu-Val-Ala-Tyr(3-$NO_2$)-Asp (Meldal & Breddam, 1991; van den Hazel *et al.*, 1995). *In vivo* (on agar plates), saccharopepsin activity is detected indirectly using an overlay assay for carboxypeptidase Y (Chapter 132) activity, using *N*-acetyl-DL-phenylalanyl-$\beta$-naphthyl ester as substrate (Jones, 1991). Because activation of procarboxypeptidase Y is dependent on saccharopepsin activity, this gives a very sensitive, albeit not very quantitative assay. The activity of saccharopepsin secreted from yeast can be monitored as cleared zones on agar plates containing 1% nonfat dry milk (Winther *et al.*, 1994).

Like most other aspartic proteases, saccharopepsin is efficiently inhibited by pepstatin A. In addition, yeast itself produces a potent macromolecular saccharopepsin inhibitor which, in the absence of a hydrophobic signal sequence, is believed to reside in the cytoplasm. This inhibitor, termed $I_3^A$, is a polypeptide of 76 amino acid residues (Biedermann *et al.*, 1980; Schu & Wolf, 1991). Preliminary $^1$H NMR data suggest that the unbound state of the inhibitor contains little well-defined tertiary structure (J.R. Winther & F.M. Poulsen, unpublished results).

## Structural Chemistry

Saccharopepsin is similar in sequence to other members of the family of aspartic proteases. The sequence identities to human pepsin and cathepsin D are 40% and 46%, respectively (Ammerer *et al.*, 1986; Dreyer *et al.*, 1986; Woolford *et al.*, 1986). The crystal structure of saccharopepsin has been solved to a resolution of 3.5 Å for the native structure and 2.5 Å for an enzyme–inhibitor complex (Badasso *et al.*, 1993; N. Cronin, personal communication). This shows that saccharopepsin conforms to the two-domain structure found in other aspartic proteases. In this structure each domain contains an active aspartic acid residue within a highly conserved region (see chapter on pepsin: Chapter 272).

The mature active saccharopepsin found in the vacuole has an apparent molecular mass of 42 kDa. Most of the saccharopepsin produced in yeast is *N*-glycosylated at two Asn-Xaa-Thr positions (residues 68 and 269 in the enzyme sequence). However, the latter site is frequently not utilized, giving a subpopulation of saccharopepsin molecules having only one *N*-linked glycosyl structure (10–30% depending on strain and growth conditions). The molecular mass (as determined by mass spectroscopy of the relevant peptides; Pedersen & Biedermann, 1993) suggests that both the glycosyl structures are of the high-mannose type ($Man_{16-18}GlcNAc_2$), possibly containing diesterified mannose phosphates. The glycosylation at position 68 is conserved in the sequence among a number of aspartic proteases, most notably including human cathepsin D.

Saccharopepsin is initially synthesized as a zymogen (saccharopepsinogen) with a 54 amino acid residue N-terminal propeptide (Ammerer *et al.*, 1986; Woolford *et al.*, 1986; Klionsky *et al.*, 1988). The propeptide, however, is not homologous to those of even closely related aspartic proteases. The mechanism and the proteases involved in processing of saccharopepsinogen have been extensively characterized and a number of processing intermediates have been identified. Thus, studies of saccharopepsinogen activation *in vitro* have shown that autoactivation is initiated by low pH and/or high ionic strength. In contrast to pepsinogen activation (Chapter 272) at low pH, saccharopepsinogen autoactivation occurs primarily via a bimolecular, product-catalyzed mechanism. The initial activation product contains 22 residues of the propeptide. No activity could be detected from preparations containing this intermediate, however. It is not clear whether it is inherently inactive or inhibition is conferred by more N-proximal parts of the propeptide still bound noncovalently (van den Hazel *et al.*, 1997). Further autoproteolysis of saccharopepsinogen results in accumulation of a stable product which is fully active but which still contains nine amino acid residues of the propeptide as compared to the normally matured cellular saccharopepsin. This form of saccharopepsin has been termed the 'pseudo' form and has

also been found in cells lacking another vacuolar protease, proteinase B (cerevisin: Chapter 104), which is a subtilisin-type protease (van den Hazel *et al.*, 1992; Wolff *et al.*, 1996). Cerevisin is responsible for processing of saccharopepsin to its mature N-terminus.

Deletion of the saccharopepsinogen propeptide sequence results in accumulation and retention of the glycosylated saccharopepsin polypeptide (saccharopepsinogen△pro) in the endoplasmic reticulum (van den Hazel *et al.*, 1993). It is known from other eukaryotic systems that a 'quality control' ensures that only correctly folded proteins are allowed to exit the endoplasmic reticulum. Furthermore, while saccharopepsin is normally a very stable enzyme, saccharopepsinogen△pro is degraded and no activity can be detected in cells producing saccharopepsinogen△pro. Based on these observations it was concluded that *in vivo* folding was compromised by deletion of the propeptide. Interestingly, when a plasmid producing only the propeptide was introduced into cells producing saccharopepsinogen△pro, a significant fraction of the enzyme was able to fold into an active structure. This indicated that the propeptide did not require covalent linkage to the enzyme region for function (van den Hazel *et al.*, 1993).

The requirement for sequence conservation of the propeptide has been investigated by a genetic approach in which either of the two halves of the propeptide was substituted with randomly generated sequences (van den Hazel *et al.*, 1995). The resulting mutants was then screened for saccharopepsin activity. A surprisingly high number of active mutant proteins (~1%) was identified, indicating that the propeptide had a very high tolerance for mutations. Moreover, the mutational analysis showed that no particular element in the propeptide is strictly required for formation of active enzyme. Rather, various parts of the propeptide contribute to this function in a somewhat additive manner. However, in the screen the importance of a particular lysine residue (Lys53) was highlighted. Of the random inserts sequenced 18 out of 20 contained lysine or arginine residues at this position. Directed mutational analysis showed that this residue was important, albeit not essential, for promoting efficient folding of saccharopepsinogen *in vivo*. Pig pepsinogen has a lysine residue at position 36 (Lys36p) and alignment of the vertebrate aspartic protease propeptides show a clear conservation of this residue. In the three-dimensional structure of pepsinogen this residue hydrogen bonds to the enzyme active-site aspartates. Like Lys53 in saccharopepsinogen, Lys36p in pepsinogen is followed by a tyrosine residue which also participates in hydrogen bonds in the enzyme active site. Based on this it was hypothesized that the Lys56 in saccharopepsinogen is functionally equivalent to Lys36 in pepsinogen (van den Hazel *et al.*, 1995).

## Preparation

Saccharopepsin can be prepared from homogenized yeast cells (Meussdoerffer *et al.*, 1980). Saccharopepsin is more readily produced in large quantities by overexpression of *PEP4* from high-copy plasmids. Due to saturation of the vacuolar sorting receptor, overexpression of *PEP4* leads to secretion of large amounts (10–100 mg liter$^{-1}$) of saccharopepsinogen into the culture supernatant (Rothman *et al.*, 1986; Sørensen *et al.*, 1994). Under normal growth conditions saccharopepsinogen activation takes place extracellularly, yielding correctly processed active saccharopepsin, which can be easily purified by ion-exchange chromatography (Sørensen *et al.*, 1994). Using a similar overexpression system, saccharopepsinogen can be purified from the culture supernatant of a strain which is deleted for cerevisin (*PRB1*). In this case pepstatin must be added to the growth medium to prevent autoactivation (H.B. van den Hazel, A.M. Wolff, M.C. Kielland-Brandt & J.R. Winther, unpublished results).

## Biological Aspects

The biogenesis of saccharopepsin is accomplished through a number of intracellular transport and maturation steps. These processing steps have made saccharopepsin a valuable tool for the study of intracellular transport and sorting of proteins in yeast. The primary translation product of saccharopepsin is a precursor, termed presaccharopepsinogen, which contains a cleavable 22 amino acid N-terminal pre- or signal peptide which directs its translocation into the endoplasmic reticulum. In the endoplasmic reticulum the majority of the protein is glycosylated at two positions (see above) to attain a molecular mass of 47 kDa. During transit of the Golgi apparatus further modification of the carbohydrate takes place to give a molecular mass of 48 kDa (Klionsky *et al.*, 1988). In a late Golgi compartment saccharopepsinogen is recognized by the vacuolar sorting receptor Vps10p which ensures that the precursor is localized to the vacuole (Westphal *et al.*, 1996). Contrary to cathepsin D, the carbohydrate does not appear to play any significant role in the intracellular targeting of saccharopepsin. There is, however, evidence for sorting pathways that are not dependent on Vps10p (Westphal *et al.*, 1996).

While the propeptide prevents proteolytic activity from being unleashed during intracellular transport, activation of saccharopepsinogen occurs rapidly in the vacuole. Active-site mutants of saccharopepsin are found in the pro-form and are unable to initiate vacuolar activation (van den Hazel *et al.*, 1992). Thus, saccharopepsin plays an important physiological role since its autoactivation is ultimately responsible for the activation of all known vacuolar hydrolase precursors. It is furthermore essential for sporulation of diploids and important for viability during nitrogen starvation (van den Hazel *et al.*, 1996, for review).

## Distinguishing Features

Saccharopepsin is by far the most abundant aspartic protease in yeast. The two other aspartate proteases known, yapsin (Chapter 304) and barrierpepsin (Chapter 303), have more restricted substrate specificities and much higher molecular mass (MacKay *et al.*, 1988; Azaryan *et al.*, 1993).

## References

Ammerer, G., Hunter, C.P., Rothman, J.H., Saari, G.C., Valls, L.A. & Stevens, T.H. (1986) PEP4 gene of *Saccharomyces cerevisiae* encodes proteinase A, a vacuolar enzyme required for processing of vacuolar precursors. *Mol. Cell. Biol.* **6**, 2490–2499.

Azaryan, A.V., Wong, M., Friedman, T.C., Cawley, N.X., Estivariz, F.E., Chen, H.C. & Loh, Y.P. (1993) Purification and characterization of a paired basic residue-specific yeast aspartic protease encoded by the *YAP3* gene. Similarity to the mammalian pro-opiomelanocortin-converting enzyme. *J. Biol. Chem.* **268**, 11968–11975.

Badasso, M., Wood, S.P., Aguilar, C., Cooper, J.B. & Blundell, T.L. (1993) Crystallization and preliminary crystallographic characterization of aspartic proteinase-A from baker's yeast and its complexes with inhibitors. *J. Mol. Biol.* **232**, 701–703.

Biedermann, K., Montali, U., Martin, B., Svendsen, I. & Ottesen, M. (1980) The amino acid sequence of proteinase A inhibitor 3 from baker's yeast. *Carlsberg Res. Commun.* **45**, 225–235.

Dernby, K.G. (1917) Studien über die proteolytischen Enzyme der Hefe und ihre Beziehung zu der Autolyse [Studies on the proteolytic enzymes of yeast and their relationship to autolysis]. *Biochem. Z.* **81**, 107–208.

Dreyer, T. (1989) Substrate specificity of proteinase yscA from saccharomyces cerevisiae. *Carlsberg Res. Commun.* **54**, 85–97.

Dreyer, T., Halkier, B., Svendsen, I. & Ottesen, M. (1986) Primary structure of the aspartic proteinase A from *Saccharomyces cerevisiae. Carlsberg Res. Commun.* **51**, 27–41.

Hata, T., Hayashi, R. & Doi, E. (1967) Purification of yeast proteinases. I. Fractionation and some properties of the proteinases. *Agric. Biol. Chem.* **31**, 150–159.

Jones, E.W. (1991) Tackling the protease problem in *Sacchromyces cerevisiae. Methods Enzymol.* **194**, 428–453.

Klionsky, D.J., Banta, L.M. & Emr, S.D. (1988) Intracellular sorting and processing of a yeast vacuolar hydrolase: proteinase A propeptide contains vacuolar targeting information. *Mol. Cell. Biol.* **8**, 2105–2116.

MacKay, V.L., Welch, S.K., Insley, M.Y., Manney, T.R., Holly, J., Saari, G.C. & Parker, M.L. (1988) The *Saccharomyces cerevisiae BAR1* gene encodes an exported protein with homology to pepsin. *Proc. Natl Acad. Sci. USA* **85**, 55–59.

Meldal, M. & Breddam, K. (1991) Anthranilamide and nitrotyrosine as donor-acceptor pair in internally quenched fluorescent substrates for endopeptidases: multicolumn peptide synthesis of enzyme substrates for subtilisin Carlsberg and pepsin. *Anal. Biochem.* **195**, 141–147.

Meussdoerffer, F., Tortora, P. & Holzer, H. (1980) Purification and properties of proteinase A from yeast. *J. Biol. Chem.* 12087–12093.

Pedersen, J. & Biedermann, K. (1993) Characterization of proteinase A glycoforms from recombinant *Saccharomyces cerevisiae. Biotechnol. Appl. Biochem.* **18**, 377–388.

Rothman, J.H., Hunter, C.P., Valls, L.A. & Stevens, T.H. (1986) Overproduction-induced mislocalization of a yeast vacuolar protein allows isolation of its structural gene. *Proc. Natl Acad. Sci. USA* **83**, 3248–3252.

Schu, P. & Wolf, D.H. (1991) The proteinase yscA-inhibitor, I$^A_3$, gene. Studies of cytoplasmic proteinase inhibitor deficiency on yeast physiology. *FEBS Lett.* **283**, 78–84.

Sørensen, S.O., van den Hazel, H.B., Kielland-Brandt, M.C. & Winther, J.R. (1994) pH-dependent processing of yeast carboxypeptidase Y by proteinase A *in vivo* and *in vitro. Eur. J. Biochem.* **220**, 19–27.

van den Hazel, H.B., Kielland-Brandt, M.C. & Winther, J.R. (1992) Autoactivation of proteinase A initiates activation of yeast vacuolar zymogens. *Eur. J. Biochem.* **207**, 277–283.

van den Hazel, H.B., Kielland-Brandt, M.C. & Winther, J.R. (1993) The propeptide is required for *in vivo* formation of stable active yeast proteinase A and can function even when not covalently linked to the mature region. *J. Biol. Chem.* **268**, 18002–18007.

van den Hazel, H.B., Kielland-Brandt, M.C. & Winther, J.R. (1995) Random substitution of large parts of the propeptide of yeast proteinase A. *J. Biol. Chem.* **270**, 8602–8609.

van den Hazel, H.B., Winther, J.R. & Kielland-Brandt, M.C. (1996) Review: biosynthesis and function of yeast vacuolar proteases. *Yeast* **12**, 1–16.

van den Hazel, H.B., Wolff, A.M., Kielland-Brandt, M.C. & Winther, J.R. (1997) Mechanism and ion dependence of *in vitro* autoactivation of yeast proteinase A – possible implications for compartmentalized activation *in vivo*. *Biochem. J.* **326**, 339–344.

Westphal, V., Marcusson, E.G., Winther, J.R., Emr, S.D. & van den Hazel, H.B. (1996) Multiple pathways for vacuolar sorting of

yeast proteinase A. *J. Biol. Chem.* **271**, 11865–11870.

Winther, J.R., Raymond, C.K. & Stevens, T.H. (1994) *PEP4*. In: *Guidebook to the Secretory Pathway* (Rothblatt, J., Novick, P. & Stevens, T.H., eds). Oxford: Oxford University Press, pp. 247–249.

Wolff, A.M., Din, N. & Petersen, J.G.L. (1996) Vacuolar and extracellular maturation of *Saccharomyces cerevisiae* proteinase A. *Yeast* **12**, 823–832.

Woolford, C.A., Daniels, L.B., Park, F.J., Jones, E.W., Van Arsdell, J.N. & Innis, M.A. (1986) The *PEP4* gene encodes an aspartyl protease implicated in the posttranslational regulation of *Saccharomyces cerevisiae* vacuolar hydrolases. *Mol. Cell. Biol.* **6**, 2500–2510.

*Jakob R. Winther*
*Carlsberg Laboratory,*
*Department of Yeast Genetics,*
*Gamle Carlsberg Vej 10, DK-2500 Copenhagen Valby, Denmark*
*Email: carljw@unidhp.uni-c.dk*

# 284. Renin

## Databanks

*Peptidase classification: clan AA, family A1, MEROPS ID: A01.007*
*NC-IUBMB enzyme classification: EC 3.4.23.15*
*ATCC entries: 1653 (anti-pig); 10285 (anti-human)*
*Chemical Abstracts Service registry number: 9015-94-5*
*Databank codes:*

| Species | SwissProt | PIR | EMBL (cDNA) | EMBL (genomic) |
|---|---|---|---|---|
| *Homo sapiens* | P00797 | A00990 | X00063 | L00064: exon 1 |
| | | A21190 | | L00065: exon 2 |
| | | A21454 | | L00066: exon 3 |
| | | A21673 | | L00067: exon 4 |
| | | A26531 | | L00068: exon 5 |
| | | I52884 | | L00069: exon 6 |
| | | I55306 | | L00070: exon 7 |
| | | | | L00071: exon 8 |
| | | | | L00072: exon 9 |
| | | | | L00073: exon 10 and complete CDS |
| | | | | M10030: exon 1 |
| | | | | M10128: exon 2 |
| | | | | M10150: exons 3–5 |
| | | | | M10151: exons 6–9 |
| | | | | M10152: exon 10 |
| | | | | M15410: exon 1 |
| | | | | M26899: exon 1 |
| | | | | M26900: exons 2–8 |

*continued overleaf*

| Species | SwissProt | PIR | EMBL (cDNA) | EMBL (genomic) |
|---|---|---|---|---|
| | | | | M26901: exon 9 |
| | | | | X01391: 5′ end and exon 1 |
| | | | | X01695: exon 2 |
| | | | | X01696: exons 3–5 |
| | | | | X01732: exons 5a–8 |
| | | | | X01733: exon 9 |
| *Mus musculus* | P06281 | A00989 | X16642 | K02596: 5′ end |
| | | A22058 | | X00810: exon 1 and complete CDS |
| | | S07636 | | X00811: exon 2 |
| | | | | X00812: exon 3 |
| | | | | X00813: exon 4 |
| | | | | X00814: exon 5 |
| | | | | X00815: exon 6 |
| | | | | X00816: exon 7 |
| | | | | X00850: exon 8 |
| | | | | X00851: exon 9 |
| *Oryctolagus cuniculus* | – | A61388 | – | – |
| *Ovis aries* | P52115 | – | L43524 | – |
| *Rattus norvegicus* | P08424 | A29991 | J02941 | M37278: exons 1–9 |
| | | A32702 | S60054 | |
| | | A60837 | X07033 | |
| | | S00923 | | |
| | | S02090 | | |

Brookhaven Protein Data Bank three-dimensional structures:

| Species | ID | Resolution | Notes |
|---|---|---|---|
| *Homo sapiens* | 1BBS | 2.8 | recombinant glycosylated form |
| | 1BIL | 2.4 | complex with butanediamide inhibitor BILA 1908 |
| | 1BIM | 2.8 | complex with butanediamide inhibitor BILA 2151 |
| | 1HRN | 1.8 | complex with polyhydroxymonoamide inhibitor BILA 980 |
| | 1RNE | 2.4 | complex with CGP 38,560 |
| | 2REN | 2.5 | recombinant |

## Name and History

More than 100 years ago, a pressor activity was discovered in kidney extract and the name *renin* was coined in 1898. Renin remained a mysterious entity until it was purified in stable form in the 1970s. During this 70 year period, the pathophysiological importance of renin in renovascular hypertension was established. Identification of renin as an enzyme rather than a pressor hormone and production of a pressor angiotensin peptide by renin reaction, laid a foundation for the future development of research on the renin–angiotensin system. This led to extensive studies on angiotensins and their formation from angiotensinogen by renin and angiotensin-converting enzymes (Chapter 359) (Soffer, 1981). Other names that have been used for renin include *angiotensin-forming enzyme* and *angiotensinogenase*.

## Activity and Specificity

Renin has a highly restricted substrate specificity. It has no known physiological effect other than the proteolysis of renin substrate, angiotensinogen. It belong to the class of aspartyl proteinases. Renin differs from the other members of this class by having a pH optimum of 5.5–7.5 instead of 2.0–3.4 (Pickens *et al.*, 1965; Yokosawa *et al.*, 1978). This neutral pH optimum is essential for it to be functional in the plasma. Renin also differs from the other aspartyl proteinases in having a very high degree of selectivity for the amino acid sequence on either side of the unique scissile peptide bond of angiotensinogen.

The definition of renin's substrate specificity began with isolation of a tetradecapeptide from the N-terminus of plasma renin substrate angiotensinogen following tryptic digestion:

Asp-Arg-Val-Tyr-Ile-His-Pro-Phe-His-Leu┼Leu-Val-Tyr-Ser

Renin cleaves the peptide bond between the two Leu residues in this sequence. The minimal size of a substrate cleaved by renin is the octapeptide Pro-Phe-His-Leu┼Leu-Val-Tyr-Ser; this reflects the size of the active site and gives clues to the nature of the residues involved in substrate binding. Additionally, renin from almost all species including human cleaves the Leu10∔Leu11 bond in the case of pig, horse and sheep angiotensinogen; human renin cleaves the Leu10∔Val11 bond in human angiotensinogen. But human renin does not cleave rat angiotensinogen nor does rat renin cleave human angiotensinogen.

It is well known that nonprimate renins react very poorly with primate angiotensinogen, whereas human renin reacts well with nonprimate angiotensinogen. The presence of an asparaginyl $N$-glycosylation site near the scissile bond may very well account for the selectivity. The structural basis for the hierarchical relationship is not yet clear. However, a conceptually interesting breakthrough was made by synthesis of a renin inhibitor with a strong preference for human renin and little or no effect on rat, dog, rabbit, mouse or guinea-pig renin. This inhibitor, Pro-His-Pro-Phe-His-Phe-Phe-Val-Tyr, has a structural analogy to horse angiotensinogen. No matter what the structural feature of this peptide may be, it demonstrated that the structure of human renin is considerably different from renin of other species and that a specific renin inhibitor can be constructed. This concept has been supported by many investigators working on the design of specific human renin inhibitors.

All renins are also inhibited by pepstatin, which acts as a competitive inhibitor. The pepstatin concentration ($IC_{50}$) required to inhibit the renin activity by 50% depends on the renin species. Pepstatin inhibits pig renin more strongly than human renin ($IC_{50} = 10^{-7}$ M vs. μM). Pepstatin is much more efficient in inhibiting other aspartyl proteinases, such as pepsin (Chapter 272) and cathepsin D (Chapter 277) ($IC_{50} = $ nM). Pepstatin used as an affinity ligand has enabled investigators to purify renin, which had not previously been possible due to its low concentration and instability (Inagami & Murakami, 1977). Human and pig renin are also completely inhibited by DAN in the presence of the cupric ion (Inagami et al., 1974). DAN is a specific inhibitor for the aspartyl proteinases. These results provided conclusive evidence that renin is an aspartyl proteinase.

Renin action on angiotensinogen and synthetic substrates is optimal in the pH range 5.5–7.0, depending on renin species and the substrate used. The affinity constant of renin for angiotensinogen is approximately the same as for the synthetic tetradecapeptide substrates ($K_m$ around 1–7 mM). Interaction of renin with angiotensinogen does not seem to be limited to the six or seven subsites immediately adjacent to the scissile bond. For example, while human renin reacting with human angiotensinogen shows a near neutral pH profile, in its reaction with rat angiotensinogen its pH optimum is shifted to pH 5.0–5.5. With sheep or pig angiotensinogen the pH optima are markedly shifted to higher pH.

The enzyme activity of renin is determined by measuring the rate of production of angiotensin I from angiotensinogen or tetradecapeptide substrate by radioimmunoassay (Haber et al., 1969) or enzyme immunoassay (Scharpé et al., 1987). The activity is also measured using fluorogenic renin substrates, Suc-Arg-Pro-Phe-His-Leu∔Leu-Val-Tyr-NHMec (Murakami et al., 1981) and Z-Pro-Phe-His-Leu∔Leu-Val-Tyr-Ser-β-NHNap (Reinharz & Roth, 1969). The fluorogenic substrates, tetradecapeptides and their derivatives are now commercially available from Peptide Research Institute or Sigma (see Appendix 2 for full names and addresses of suppliers).

## Structural Chemistry

The active form of renin is produced from prorenin by a proteolytic cleavage of the N-terminal propeptide. Renin consists of two similar domains each containing a catalytically essential Asp. Mature native renin is a two-chain protein produced by proteolysis of the native single-chain renin. It is a protein of about 36–42 kDa, pI 5.0–5.2, which contains two disulfide bonds and two potential $N$-glycosylation sites. Only one renin gene (ren-1) exists in all species examined, except for the mouse which contains two genes (ren-1 and ren-2) which are contiguous. The structures of renal renins (derived from the ren-1 gene) were deduced from the nucleotide sequence of cDNA (human and rat) which revealed the presence of preprorenin and prorenin consisting of about 400 amino acid residues. A single-chain form of renin of 42 kDa was produced by a recombinant DNA method.

The three-dimensional structure of renin has been determined by X-ray crystallography using a partially deglycosylated form of recombinant human renin at 2.5 Å (Sielecki et al., 1989) and a complex of recombinant human renin and its peptide inhibitor at resolution 2.8 Å (Dhanaraj et al., 1992). Candidates for subsites from S3′ to S4 and S1′ to S4 are indicated in their structure.

The reaction mechanism of renin has been investigated by determining the essential structure for substrate and enzyme. The His residue at the P2 site in angiotensinogen has been proposed to participate in the catalytic step (Green et al., 1990; Thaisrivongs et al., 1990). Tyr83 (S1 and S1′) in renin has been reported to stabilize the transition state in the catalytic step (Suzuki et al., 1996).

## Preparation

Renin has been purified to homogeneity from various sources such as kidney of pig (Murakami & Inagami, 1975), human, dog and rat, and pituitary of pig (for review see Inagami, 1989). These preparations gave 70 000–133 000-fold, 500 000-fold, 600 000-fold, 3000-fold and 1 700 000-fold purification, respectively. Renin cDNA has been cloned from kidneys of various species and expressed in mammalian cells to yield recombinant human (Poorman et al., 1986; Ishizuka et al., 1991) and rat (Yamauchi et al., 1992) renins in milligram quantities. Affinity ligands such as pepstatin as described above, a renin-specific inhibitor peptide (H77) and monoclonal antibodies have been used as powerful tools in the purification of renin (Murakami & Inagami, 1975; Inagami & Murakami, 1977; McIntyre et al., 1983).

## Biological Aspects

The enzyme is present in mammals, birds, amphibians and teleosts (Nishimura, 1980). It is expressed mainly in the juxtaglomerular cells in the kidney, secreted into the blood circulation by exocytosis, and selectively reacts with angiotensinogen to produce angiotensin I. Plasma angiotensinogen mainly derives from hepatocytes, but many other cells produce the renin substrate. This is the initial and rate-limiting step in the plasma angiotensin-producing system which controls blood pressure and electrolyte balance.

The renin gene is also expressed in many other tissues such as adrenal gland, gonads, placenta, pituitary, brain and hypothalamus (Deschepper *et al.*, 1986; Dzau *et al.*, 1987). These extrarenal renins have been proposed to play a role in the tissue renin–angiotensin system by several investigators (Ganten *et al.*, 1976; Hirose *et al.*, 1978; Dzau *et al.*, 1987).

Renin is produced as prorenin with a 43 residue pro-segment attached at the N-terminus of mature renin (Panthier *et al.*, 1982; Corvol *et al.*, 1983; Imai *et al.*, 1983; Burnham *et al.*, 1987; Higashimori *et al.*, 1989). Prorenin is an inactive proenzyme that is activated by trypsin, cathepsin B and other proteinases (Inagami & Murakami, 1980). Prorenin sorted into the renin granule in the juxtaglomerular cells should be processed into mature renin by cathepsin B (Chapter 209) (Wang *et al.*, 1991). Prorenin, as well as renin, may be taken up from the plasma into peripheral tissues where prorenin may be activated (Skinner *et al.*, 1986; Lever, 1989).

The prosegment of prorenin is responsible for the inhibition of renin. Prorenin has less than 10% of the full activity of renin, but its activity is nonproteolytically increased by lowering the pH or temperature, or by lipids (Sealey & Laragh, 1975; Derkx *et al.*, 1976; Leckie & McGhee, 1980; Hseuh & Carison, 1981). The first two phenomena have been called acid-activation and cryoactivation, respectively. Recombinant prorenin has also been reversibly activated by exposure to low temperature or acidic pH (Pitarresi *et al.*, 1992) and a conformational change can be induced in the active cleft by adding nonpeptide renin inhibitors.

Many tissues produce prorenin but not renin. In the early phase of diabetes mellitus, plasma prorenin is elevated but renin levels are not. In some cases there is hyporeninemic hypoaldosteronism with normal to elevated prorenin levels in the circulation (Luetscher *et al.*, 1985). The possibility that prorenin itself is active *in vivo* by virtue of reversible opening and closing of the active site is, therefore, an intriguing possibility (Leckie & McGhee, 1980; Hseuh & Carison, 1981; Higashimori, *et al.*, 1989).

The gene locus of human renin is 1q42. The gene structure of human renin was revealed to span over 12 kb of DNA and to contain 10 exons separated by nine introns (reviewed by Inagami, 1989). The structures of renin from rat and some but not all mouse strains were also shown to contain nine exons separated by eight introns (Mullins *et al.*, 1982; Holm *et al.*, 1984). Based on the gene structures of renin and other aspartyl proteinases, it is likely that the two exons containing the two Asp residues of the active site were derived from a duplication of an ancestral gene that led to the bilobal aspartyl proteinases (Sogawa *et al.*, 1983; Holm *et al.*, 1984). The gene structure suggests that there is a second TATA box upstream from the major functional TATA box (Soubrier *et al.*, 1986; Duncan *et al.*, 1990). There are several regions in the 5'-flanking sequence homologous to consensus steroid hormone response elements.

Cyclic AMP is a candidate to activate the renin promoter, since mRNA levels for placental renin are regulated by cAMP and since $\beta$-adrenergic stimuli that activate adenylate cyclase stimulate renal renin release. Actually, $10^{-5}$ M forskolin increases the activity of the renin promoter in the choriodecidual cells by 10-fold. More than one cAMP-responsive element exists in the renin promoter region (Burt *et al.*, 1989).

Renin release from the juxtaglomerular cells is the major factor in determining plasma angiotensinogen. It is stimulated by the renal adrenergic system, hypotension and reduced salt intake, and low plasma angiotensin II concentration. The so-called tissue renin may be derived from the uptake of circulating renin (Derkx *et al.*, 1976) or endogenous renin production. The brain and adrenal gland can produce a significant amount of renin. Transgenic mice harboring the rat or human renin gene have been established as hypertensive mice (for reviews see Inagami, 1989; Fukamizu, 1993; Fukamizu & Murakami, 1995). Such transgenic animals will serve as hypertension models for a number of hypertension and renin researchers. Since angiotensin II plays the major roles not only in blood pressure regulation and electrolyte balance, but also in degenerative diseases such as atherosclerosis, glomerular nephritis and left ventricular hypertrophy, the angiotensin-producing enzyme renin plays an important pathophysiological role in the cardiovascular system.

## Further Reading

An extensive review is provided by Inagami (1989). New material is covered by Fukamizu & Murakami (1995) and transgenic mouse models are reviewed by Fukamizu (1993).

## References

Burnham, C.E., Howelu-Johnson, C.L., Frank, B.M. & Lynch, K.R. (1987) Molecular cloning of rat renin cDNA and its gene. *Proc. Natl Acad. Sci. USA* **84**, 5605–5609.

Burt, D.W., Nakamura, N., Kelly, P. & Dzau, V.J. (1989) Identification of negative and positive regulatory elements in the human renin gene. *J. Biol. Chem.* **264**, 7357–7362.

Corvol, P., Panthier, J.-J., Foote, S. & Rougeon, F. (1983) Structure of the mouse submaxillary gland renin precursor and a model for renin processing. *Hypertension* **5** (suppl. 1), 13–19.

Derkx, F.H., von Gool, J.M.G., Wenting, G.J., Verhoeven, R.P., Man i'nt Veld, A.J. & Schalekamp, M.A.D.H. (1976) Inactive renin in human plasma. *Lancet* **2**, 496–498.

Deschepper, C.F., Mellon, S.H., Cumin, F., Baxter, J.D. & Ganong, W.F. (1986) Analysis by immunocytochemistry and *in situ* hybridization of renin and its mRNA in kidney, testis, adrenal, and pituitary of the rat. *Proc. Natl Acad. Sci. USA* **83**, 7552–7556.

Dhanaraj, V., Dealwis, C.G., Frazao, C., Badasso, M., Sibanda, B.L., Tickle, I.J., Cooper, J.B., Driessen, H.P.C., Newman, M., Aguilar, C., Wood, S.P., Blundell, T.L., Hobart, P.M., Geoghegan, K.F., Ammirati, M.J., Danley, D.E., O'Connor, B.A. & Hoover, D.J. (1992) X-ray analyses of peptide-inhibitor complexes define the structural basis of specificity for human and mouse renins. *Nature* **357**, 466–472.

Duncan, K.G., Haidar, M.A., Baxter, J.D. & Reudelhuber, T.A. (1990) Regulation of human renin expression in chorion cell primary cultures. *Proc. Natl Acad. Sci. USA* **87**, 7588–7592.

Dzau, V.J. (1987) Implications of local angiotensin production in cardiovascular physiology and pharmacology. *Am. J. Cardiol.* **59**, 59A–65A.

Dzau, V.J., Ellison, K.E., Brody, T., Ingelfinger, J. & Pratt, R.E. (1987) A comparative study of the distributions of renin and angiotensinogen messenger ribonucleic acids in rat and mouse tissues. *Endocrinology* **120**, 2334–2338.

Fukamizu, A. (1993) Transgenic animals in endocrinological investigation. *J. Endocrinol. Invest.* **16**, 461–473.

Fukamizu, A. & Murakami, K. (1995) New aspects of the renin-angiotensin system in blood pressure regulation. *TEM* **6**, 279–284.

Ganten, D., Schelling, P., Vecsei, P. & Ganten, U. (1976) Iso-renin of extrarenal origin. 'The tissue angiotensinogenase systems'. *Am. J. Med.* **60**, 760–772.

Green, D.W., Aykent, S., Gierse, J.K. & Zupec, M.E. (1990) Substrate specificity of recombinant human renal renin: effect of histidine in the $P_2$ subsite on pH dependence. *Biochemistry* **29**, 3126–3133.

Haber, E., Koerner, T., Page, L.B., Kliman, B. & Purnode, A. (1969) Application of a radioimmunoassay for angiotensin I to the physiologic measurements of plasma renin activity in normal human subjects. *J. Clin. Endocrinol. Metab.* **29**, 1349–1355.

Higashimori, K., Mizuno, K., Nakajo, S., Boehm, F.H., Marcotte, P.A., Egan, D.A., Holleman, W.H., Heusser, C.H., Poisner, A.M. & Inagami, T. (1989) Pure human inactive renin: evidence that native inactive renin is prorenin. *J. Biol. Chem.* **264**, 14662–14667.

Hirose, S., Yokosawa, H. & Inagami, T., (1978) Immunological identification of renin in rat brain and distinction from acid protease. *Nature* **274**, 392–393.

Holm, I., Ollo, R., Panthier, J.J. & Rougeon, F. (1984) Evolution of aspartyl proteases by gene duplication: the mouse renin gene is organized in two homologous clusters of four exons. *EMBO J.* **3**, 557–562.

Hseuh, W.A. & Carison, E.J. (1981) Reversible activation of plasma inactive renin. In: *Heterogeneity of Renin and Renin Substrate* (Sambi, M.P., ed.). New York: Elsevier/North-Holland, pp. 175–182.

Imai, T., Miyazaki, H., Hirose, S., Hori, H., Hayashi, T., Kageyama, R., Ohkubo, H., Nakanishi, S. & Murakami, K. (1983) Cloning and sequence analysis of cDNA for human renin precursor. *Proc. Natl Acad. Sci. USA* **80**, 7405–7409.

Inagami, T. (1989) Structure and function of renin. *J. Hypertens.* **7** (suppl. 2), S3–S8.

Inagami, T. & Murakami, K. (1977) Pure renin: isolation from hog kidney and characterization. *J. Biol. Chem.* **252**, 2978–2983.

Inagami, T. & Murakami, K. (1980) Prorenin. *Biomed. Res.* **1**, 456–475.

Inagami, T., Misono, K.S. & Michelakis, A.M. (1974) Definitive evidence for similarity in the active site of renin and acid proteases. *Biochem. Biophys. Res. Commun.* **56**, 503–509.

Ishizuka, Y., Shoda, A., Yoshida, S., Kawamura, Y., Haraguchi, K. & Murakami, K. (1991) Isolation and characterization of recombinant human prorenin in Chinese hamster ovary cells. *J. Biochem.* **109**, 30–35.

Leckie, B.J. & McGhee, N.K. (1980) Reversible activation inactivation of renin in human plasma. *Nature* **288**, 702–705.

Lever, A.K. (1989) Renin: endocrine, paracrine, or part-paracrine control of blood pressure? *Am. J. Hypertens.* **2**, 276–285.

Luetscher, J.A., Kraemer, F.B., Wilson, D.M., Schartz, H.C. & Bryer-Ash, M. (1985) Increased plasma inactive renin in diabetes mellitus. A marker of microvascular complications. *N. Engl. J. Med.* **312**, 1412–1417.

McIntyre, G.D., Leckie, B., Hallet, A. & Szelke, M. (1983) Purification of human renin by affinity chromatography using a new peptide inhibitor of renin: H77 (D-His-Pro-Phe-His-Leu$^R$Leu-Val-Tyr).

*Biochem. J.* **211**, 519–522.

Mullins, J.J., Burt, D.W., Windass, J.D., McTurk, P., George, H. & Brummar, W.J. (1982) Molecular cloning of two distinct renin genes from the DBA/2 mouse. *EMBO J.* **1**, 1461–1466.

Murakami, K. & Inagami, T. (1975) Isolation of pure and stable renin from hog kidney. *Biochem. Biophys. Res. Commun.* **62**, 757–763.

Murakami, K., Ohsawa, T., Hirose, S., Takada, K. & Sakakibara, S. (1981) New fluorogenic substrates for renin. *Anal. Biochem.* **110**, 232–236.

Nishimura, H. (1980) Comparative endocrinology of renin and angiotensin. In: *The Renin Angiotensin System* (Alan, J. & Anderson, R.R., eds). New York: Plenum Publishing, pp. 29–77.

Panthier, J.-J., Foote, S., Chamnraud, B., Strosberg, A.D., Corvol, P. & Rougeon, F. (1982) Complete amino acid sequence and maturation of the mouse submaxillary gland renin precursor. *Nature* **298**, 90–92.

Pickens, P.T., Bumpus, F.M., Lloyd, A.M., Smeby, R.R. & Page, I.H. (1965) Measurements of renin activity in human plasma. *Circ. Res.* **17**, 438–448.

Pitarresi, T.M., Rubattu, S., Heinrikson, R. & Sealey, J. (1992) Reversible cryoactivation of recombinant human prorenin. *J. Biol. Chem.* **267**, 11753–11759.

Poorman, R.A., Palermo, D.P., Post, L.E., Murakami, K., Kinner, J.H., Smith, C.W., Reardon, I. & Heinrikson, R.L. (1986) Isolation and characterization of native human renin derived from Chinese hamster ovary cells. *Protein: Struct. Func. Genet.* **1**, 139–145.

Reinharz, A. & Roth M. (1969) Studies on renin with synthetic substrate. *Eur. J. Biochem.* **7**, 334–339.

Scharpé, S., Verkerk, R., Sasmito, E. & Theeuws, M. (1987) Enzyme immunoassay of angiotensin I and renin. *Clin. Chem.* **33**, 1774–1777.

Sealey, J.E. & Laragh, J.H. (1975) 'Prorenin' in human plasma? Methodological and physiological implications. *Circ. Res.* **36–37**(suppl. I), I10–I16.

Sielecki, A.R., Hayakawa, K., Fujinaga, M., Murphy, M.E.P., Fraser, M., Muir, A.K., Carilli, C.T., Lewicki, J.A., Baxter, J.D. & James, M.N.G. (1989) Structure of recombinant human renin, a target for cardiovascular-active drug, at 2.5 Å resolution. *Science* **243**, 1346–1351.

Skinner, S.L., Thatcher, R.L., Whitworth, J.A. & Horowitz, J.D. (1986) Extraction of plasma prorenin by human heart. *Lancet* **1**, 995–997.

Soffer, R. (1981) *Biochemical Regulation of Blood Pressure.* New York: Wiley Interscience.

Sogawa, K., Fuji-Kuriwama, Y., Mizukami, Y., Ichihara, Y. & Takahashi, K. (1983) Primary structure of human pepsinogen gene. *J. Biol. Chem.* **258**, 5306–5311.

Soubrier, F., Panthier, J.-J., Houot, A.-M., Rougeon, F. & Corvol, P. (1986) Segmental homology between the promoter region of the human renin gene and the mouse *ren1* and *ren2* promoter regions. *Gene* **41**, 85–92.

Suzuki, F., Goto, K., Shiratori, Y., Inagami, T., Murakami, K. & Nakamura, Y. (1996) Tyrosine-83 of renin has an important role in renin-angiotensinogen reaction. *Protein Pept. Lett.* **3**, 45–49.

Thaisrivongs, S., Mao, B., Pals, D.T., Turner, S.R. & Kroll, L.T. (1990) Renin inhibitory peptide. A β-aspartyl residue as replacment for the histidyl residue in the P-2 site. *J. Med. Chem.* **33**, 1337–1343.

Wang, P.H., Do, Y.S., Macaulay, L., Shinagawa, T., Anderson, P.W., Baxter, J.D. & Hseuh, W.A. (1991) Identification of renal cathepsin B as a human prorenin-processing enzyme. *J. Biol. Chem.* **266**, 12633–12638.

Yamauchi, T., Suzuki, F., Takahashi, A., Tsutsumi, I., Hori, H., Watanabe, T., Ishizuka, Y., Nakamura, Y. & Murakami, K (1992)

Expression of rat renin in mammalian cells and its purification. *Clin. Exp. Hypertens.* **A14**, 377–392.

Yokosawa, H., Inagami, T. & Haas, E. (1978) Purification of human renin. *Biochem. Biophys. Res. Commun.* **83**, 306–312.

*Fumiaki Suzuki*
Molecular Genetics Research Centre,
Gifu University,
Gifu 501-11, Japan
Email: aob3073@cc.gifu-u.ac.jp

*Kazuo Murakami*
Institute of Applied Biochemistry,
University of Tsukuba,
Ibaraki 305, Japan

*Yukio Nakamura*
Protein Engineering Laboratory,
Department of Biotechnology,
Gifu University,
Gifu 501-11, Japan

*Tadashi Inagami*
Department of Biochemistry, School of Medicine,
Vanderbilt University,
Nashville, TN 37232-0146, USA
Email: inagamit@ctrvax.vanderbilt.edu

# 285. *Mouse submandibular renin*

## Databanks

*Peptidase classification: clan AA, family A1, MEROPS ID: A01.008*
*NC-IUBMB enzyme classification: none*
*Databank codes:*

| Species | SwissProt | PIR | EMBL (cDNA) | EMBL (genomic) |
|---|---|---|---|---|
| *Mus musculus* | P00796 | A00988 A93285 A93923 B22058 B93285 | J00621 | K02597: 5′ end |

Brookhaven Protein Data Bank three-dimensional structures:

| Species | ID | Resolution | Notes |
|---|---|---|---|
| *Mus musculus* | 1SMR | 2 | complex with inhibitor CH-66 |

## Name and History

Since Werle and his associates reported the presence of renin-like enzyme activity in the submandibular gland of mice (Werle *et al.*, 1963), the renin-like enzyme has attracted hypertension researchers (Bing & Faarup, 1965; Oliver & Gross, 1967). The enzyme was purified from the gland and demonstrated to elicit prolonged pressor responses typical of renin (Cohen *et al.*, 1972; Suzuki *et al.*, 1981a) and to possess many other properties similar to those of renal renin (Chapter 284) (Bing *et al.*, 1980). It is also expressed in the kidney. *Mouse submandibular renin* has therefore been used as a useful model enzyme for renal renin.

## Activity and Specificity

Submandibular renin has no known physiological function other than the generation of angiotensin I from angiotensinogen. The enzyme has a pH optimum of 6.5–8.3 (Figueiredo *et al.*, 1985) and has a very high degree of selectivity for the amino acid sequence on either side of the scissile peptide bond specifically cleaved by renin. All renins are also inhibited by pepstatin which acts as a competitive inhibitor. Pepstatin has been used as a ligand of an affinity column for purification of renin (Inagami & Murakami, 1977). Submandibular renin was purified to homogeneity by a single step using the affinity column (Suzuki *et al.*, 1981a).

Submandibular renin is completely inhibited by DAN in the presence of cupric ion (Misono & Inagami, 1980). The inhibition of the enzyme was linearly related to the blocking of one aspartyl $\beta$-carboxyl group on the active site (Misono & Inagami, 1980) (Chapter 272). Another pepsin-specific inhibitor, 1,2-epoxy-3-*p*-nitrophenoxypropane (EPNP), produced similar specific inhibition of renin (Misono & Inagami, 1980). The incorporation of radiolabeled DAN and EPNP proceeded independent of each other, indicating the presence of two Asp residues in the active site of renin, as in the active site of pepsin (Misono & Inagami, 1980) (Chapter 272). Furthermore, by chemical modification of the tyrosyl phenolic group with tetranitromethane or acetyl imidazole and blocking of the guanidine group of arginyl residue with phenylglyoxal, two Tyr residues and one Arg residue were shown to be in the active site of the enzyme (Misono & Inagami, 1980). These properties are analogous to pepsin except that the optimum pH of pepsin is close to 1.0. It is, therefore, definitive that the submandibular renin belongs to the class of aspartyl proteinases. Renin differs from the other members of this class by having a pH optimum of 5.5–7.5 instead of 2.0–3.4 (Figueiredo *et al.*, 1985). This neutral or near neutral pH optimum is compatible with being functional in the plasma. Renin also differs from other enzymes of this class by having a very high degree of selectivity for the amino acid sequence on either side of the scissile peptide bond.

Enzyme activity is determined by the radioimmunoassay or enzyme immunoassay of angiotensin I generated from angiotensinogen (see renin: Chapter 284). Pure renin and antibodies permit direct radioimmunoassay of renin (Michelakis *et al.*, 1974) and prove that it is not secreted into the plasma in large measure under normal conditions in spite of its enormously high concentration in male mouse submandibular gland.

## Structural Chemistry

Submandibular renin is composed of two polypeptide chains, formed by proteolysis of active single-chain renin, and linked by a disulfide bridge. It has a molecular mass of 36–38 kDa and a pI of 5.4–5.9 (Cohen *et al.*, 1972; Misono & Inagami, 1980). The primary structure of submandibular renin has been deduced from the cDNA sequence (Panthier *et al.*, 1982) and from purified enzyme (Misono *et al.*, 1982). No consensus sequence for an *N*-glycosylation site was found. The similarity of the active site of the renin to that of other aspartyl proteinases such as pepsin, cathepsin D (Chapter 277) or several fungal aspartyl proteinases was demonstrated by sequence homology in the vicinity of the catalytically essential Asp (labeled with EPNP and DAN) and Tyr in the submandibular renin (Misono *et al.*, 1982).

## Preparation

Submandibular renin was purified from albino male mice by conventional methods (Cohen *et al.*, 1972; Misono *et al.*, 1982). The purification factor was 40-fold. The renin was also purified by one-step purification using an affinity column (Suzuki *et al.*, 1981a).

## Biological Aspects

Submandibular renin is the product of the *ren-2* renin gene. Mice harboring *ren-2* renin gene express this gene as well as the *ren-1* renin gene in several other tissues (Wagner *et al.*, 1990; Suzuki *et al.*, 1990). The expression of the *ren-2* renin gene is regulated by androgen (Wagner *et al.*, 1990). Its expression level is remarkably high in the submandibular gland of the male mouse (Bing *et al.*, 1980). The renin content in the gland is approximately 100-fold that in the kidney. The administration of testosterone to a female mouse increases the renin content of the submandibular gland, while castration of the male mouse decreases its content (Bing *et al.*, 1980). The submandibular renin is released in significant amounts into the bloodstream in the course of aggression or previous contact with other mice (Bing & Poulsen, 1979). The content increases with aging up to the adult stage in the albino male mice (Bhoola *et al.*, 1973; Suzuki *et al.*, 1981b).

Transgenic mice and rats with the *ren-2* renin gene have been established as animal models of hypertension (Tronik *et al.*, 1987; Mullins *et al.*, 1989).

## Distinguishing Features

Submandibular renin, *ren-2* renin, has fundamentally the same enzymological properties as human renin, *ren-1* renin (Bing *et al.*, 1980). Submandibular renin has no sugar chain but *ren-1* renin such as human renin is a glycoconjugate protein (Misono *et al.*, 1982). Polyclonal antibodies against submandibular renin have been described in this text, but are not commercially available at the time of writing.

## References

Bhoola, K.D., Dorey, G. & Jones, C.W. (1973) The influence of androgens on enzymes (chymotrypsin- and trypsin-like proteases, renin, kallikrein and amylase) and on cellular structure of the mouse submaxillary gland. *J. Physiol.* **235**, 503–522.

Bing, J. & Faarup, P. (1965) Location of renin (or renin-like substance) in the submaxillary glands of albino mice. *Acta Pathol. Microbiol. Scand.* **64**, 203–212.

Bing, J. & Poulsen, K. (1979) In mice aggressive behaviour provokes vast increase in plasma renin concentration, causing only slight, if any, increase in blood pressure. *Acta Physiol. Scand.* **105**, 64–72.

Bing, J., Poulsen, K., Hackenthal, E., Rix, E. & Taugner, R. (1980) Renin in the submaxillary gland: a review. *J. Histochem. Cytochem.* **28**, 874–880.

Cohen, S., Taylor, J.M., Murakami, K., Michelakis, A.M. & Inagami, T. (1972) Isolation and characterization of renin-like enzymes from mouse submaxillary glands. *Biochemistry* **11**, 4286–4293.

Figueiredo, A.F.S., Takii, Y., Tsuji, H., Kato, K. & Inagami, T. (1985) Rat kidney renin and cathepsin D: purification and comparison of properties. *Biochemistry* **22**, 5476–5481.

Inagami, T. & Murakami, K. (1977) Pure renin: isolation from hog kidney and characterization. *J. Biol. Chem.* **252**, 2978–2983.

Michelakis, A.M., Yoshida, H., Menzie, J., Murakami, K. & Inagami, T. (1974) A radioimmunoassay or the direct measurement of renin in mice and its application to submaxillary gland and kidney studies. *Endocrinology* **94**, 1101–1105.

Misono, K.S. & Inagami, T. (1980) Characterization of the active site of mouse submaxillary gland renin. *Biochemistry* **19**, 2615–2622.

Misono, K.S., Chang, J.J. & Inagami, T. (1982) Amino acid sequence of mouse submaxillary gland renin. *Proc. Natl Acad. Sci. USA* **79**, 4858–4862.

Mullins, J.J., Sigmund, C.D., Kane-Haas, C. & Gross, K.W. (1989) Expression of the DBA/2J Ren-2 gene in the adrenal gland of transgenic mice. *EMBO J.* **8**, 4065–4072.

Oliver, W.J. & Gross, F. (1967) Effect of testosterone and duct ligation on submaxillary renin-like principle. *Am. J. Physiol.* **213**, 341–346.

Panthier, J.-J., Foote, S., Chamraud, B., Strosberg, A.D., Corvol, P. & Rougeon, F. (1982) Complete amino acid sequence and maturation of the mouse submaxillary gland renin precursor. *Nature* **298**, 90–92.

Suzuki, F., Nakamura, Y., Nagata, Y., Ohsawa, T. & Murakami, K. (1981a) A rapid and large-scale isolation of renin from mouse submaxillary gland by pepstatin-aminohexyl-agarose affinity chromatography. *J. Biochem.* **89**, 1107–1112.

Suzuki, F., Takahashi, M., Nakamura, Y., Nagata, Y. & Murakami, K. (1981b) Renins in mouse submaxillary gland: multiple molecular forms and their renin activity in 1- to 8-week-old mice. *Biomed. Res.* **2**, 225–228.

Suzuki, F., Yamashita, S., Ito, M., Nagata, Y. & Nakamura, Y. (1990) Ren-2 as well as Ren-1 renin exists in the kidney and plasma of a two-renin gene mouse. *Agric. Biol. Chem.* **54**, 2407–2412.

Tronik, D., Dreyfus, M., Babinet, C. & Rougeon, F. (1987) Regulated expression of *Ren-2* gene in transgenic mice derived from parental strains carrying only the *Ren-1* gene. *EMBO J.* **6**, 983–987.

Wagner, D., Metzger, R., Paul, M., Ludwig, G., Suzuki, F., Takahashi, S., Murakami, K. & Ganten, D. (1990) Androgen dependence and tissue specificity of renin messenger RNA expression in mice. *J. Hypertens.* **8**, 45–52.

Werle, E., Trautschold, J. & Schmal, A. (1963) Uber ein Iso-Enzyme des Renin und uber die Isolierung eines biologisch aktiven Spaltproduktes seines Substrates [An isoenzyme of renin and the isolation of a biologically-active cleavage product of its substrate]. *Z. Physiol. Chem.* **332**, 79–87.

*Fumiaki Suzuki*
*Molecular Genetics Research Centre,*
*Gifu University,*
*Gifu 501-11, Japan*
*Email: aob3073@cc.gifu-u.ac.jp*

*Kazuo Murakami*
*Institute of Applied Biochemistry,*
*University of Tsukuba,*
*Ibaraki 305, Japan*

*Yukio Nakamura*
*Protein Engineering Laboratory,*
*Department of Biotechnology,*
*Gifu University,*
*Gifu 501-11, Japan*

*Tadashi Inagami*
*Department of Biochemistry, School of Medicine,*
*Vanderbilt University,*
*Nashville, TN 37232-0146, USA*
*Email: inagamit@ctrvax.vanderbilt.edu*

# 286. Plasmepsin I

## Databanks

*Peptidase classification: clan AA, family A1, MEROPS ID: A01.022*
*NC-IUBMB enzyme classification: EC 3.4.23.38*
*Databank codes:*

| Species | SwissProt | PIR | EMBL (cDNA) | EMBL (genomic) |
|---|---|---|---|---|
| *Eimeria acervulina* | – | – | Z24676 | – |
| *Plasmodium falciparum* | P39898 | PT0434 | X75787 | – |

## Name and History

Aspartic protease activities from the human malaria pathogen *Plasmodium falciparum* have been described in a number of publications (reviewed by Vander Jagt *et al.*, 1989). Partial purification and characterization of aspartic protease activities from *P. falciparum* was achieved by Vander Jagt and colleagues (1987). Purification of **plasmepsin I** (previously called **aspartic hemoglobinase I** or **PFAPG**), N-terminal sequence and analysis of specificity was reported by Goldberg *et al.* (1991). The cDNA sequence was obtained in 1994 (Francis *et al.*, 1994).

## Activity and Specificity

The enzyme is capable of cleaving native human hemoglobin as well as acid-denatured globin, with a pH optimum around 5 (Goldberg *et al.*, 1991). Initial proteolysis of intact hemoglobin occurs at the α-chain Phe33┼Leu34 peptide bond. This cleavage renders 4–5 other sites in the substrate susceptible to cleavage; there is a preference for P1 Phe (Goldberg *et al.*, 1991; Gluzman *et al.*, 1994).

Assay of activity can be performed with $^{14}$C-methylated globin in a TCA-precipitation assay (Goldberg *et al.*, 1990), or spectrophotometrically using a fluorescence-quenched substrate Dabcyl-GABA-Glu-Arg-Met-Phe┼Leu-Ser-Phe-Pro-GABA-EDANS (Dabcyl: 4-(4-dimethylamino-phenylazo)benzoyl; GABA: γ-amino-*n*-butyric acid; EDANS: 5-(2-aminoethylaminio)-1-naphthalene-sulfonic acid) (Luker *et al.*, 1996). This peptide is based on the α-chain cleavage site.

## Structural Chemistry

The enzyme is a single-chain protein, made as a 51 kDa precursor that is cleaved to a 37 kDa mature form (Francis *et al.*, 1994; Francis & Goldberg, 1996). There are two disulfide bonds predicted. The second domain contains an active site Asp-Ser-Gly instead of the usual aspartic protease Asp-Thr-Gly. A molecular model has been constructed, based on the crystal structure of the highly homologous plasmepsin II (Silva *et al.*, 1996) (Chapter 287). The modeled structure bears general features of mammalian and fungal aspartic proteases. Pepstatin is bound in the active site with enough conformational difference from mammalian homologs to give hope for development of a selective inhibitor.

## Preparation

Native plasmepsin I can be prepared in nanogram to low microgram amounts from cultured intraerythrocytic organisms (Goldberg *et al.*, 1991). Production of significant quantities of recombinant enzyme has been difficult (Luker *et al.*, 1996), but functional enzyme has recently been obtained using a proregion mutation that allows activation after refolding in an *E. coli* expression system (R. Ridley, personal communication).

## Biological Aspects

Plasmepsin I is found in the digestive vacuole of intraerythrocytic *P. falciparum*, where it appears to play an important role in hemoglobin degradation (Goldberg *et al.*, 1990, 1991; Vander Jagt *et al.*, 1992; Francis *et al.*, 1994). In culture, the plasmepsin I-selective inhibitor SC-50083 blocks proteolysis and leads to cell death at low micromolar concentrations (Francis *et al.*, 1994).

Plasmepsin I is synthesized as a type II integral membrane protein, appears to be exported to the parasite surface and then is internalized with its substrate hemoglobin into the digestive vacuole (Francis *et al.*, 1994; Francis & Goldberg, 1996).

An unusual acidic processing enzyme that is not susceptible to inhibitors of aspartic, cysteine, serine or metalloproteases but is blocked by certain tripeptide aldehydes, cleaves the plasmepsin to its mature form.

## Distinguishing Features

The enzyme is 73% identical to plasmepsin II (Chapter 287) at an amino acid level (Francis *et al.*, 1994; Dame *et al.*, 1994), but has a different cleavage specificity. Plasmepsin I is susceptible to the peptidomimetic inhibitors SC-50083 and P1, whereas II is relatively insensitive to the former and not inhibited detectably by the latter (Francis *et al.*, 1994; Luker *et al.*, 1996). Specific polyclonal antisera to each plasmepsin have been made but are not commercially available.

## Related Peptidases

A *P. falciparum* gene with substantial sequence homology to plasmepsin I, but different from either of the known plasmepsins, has been identified (C. Berry, personal communication). A fragment of this gene derived from an expressed sequence tag library has been deposited in the database (T18125). A 55 kDa acidic, pepstatin-inhibitable protease in *P. falciparum* has been reported (Bailly *et al.*, 1991). Its function is unknown. A cathepsin D-like protease was isolated from *P. lophurae* and postulated to have a role in host erythrocyte cytoskeleton degradation (Sherman & Tanigoshi, 1983). An aspartic protease in the related organism *Eimeria acervulina* has been reported (Laurent *et al.*, 1993).

## Further Reading

For a discussion of the biology of hemoglobin degradation in the organelle where catabolism occurs, see Olliaro & Goldberg (1995).

## References

Bailly, E., Savel, J., Mahouy, G. & Jaureguiberry, G. (1991) *Plasmodium falciparum*: isolation and characterization of a 55-kDa protease with a cathepsin D-like activity from *P. falciparum*. *Exp. Parasitol.* **72**, 278–284.

Dame, J.B., Reddy, G.R., Yowell, C.A., Dunn, B.M., Kay, J. & Berry, C. (1994) Sequence, expression and modeled structure of an aspartic proteinase from the human malaria parasite *Plasmodium falciparum*. *Mol. Biochem. Parasitol.* **64**, 177–190.

Francis, S.E. & Goldberg, D.E. (1996) Biosynthesis of plasmepsins I and II, proteases in the *Plasmodium falciparum* hemoglobin degradation pathway. Molecular Parasitology Meeting, Woods Hole, MA, USA 7, 56 (abstr.).

Francis, S.E., Gluzman, I.Y., Oksman, A., Knickerbocker, A., Mueller, A., Bryant, M.L., Sherman, D.R., Russell, D.G. & Goldberg, D.E. (1994) Molecular characterization and inhibition of a *Plasmodium falciparum* aspartic hemoglobinase. *EMBO J.* **13**, 306–317.

Gluzman, I.Y., Francis, S.E., Oksman, A., Smith, C., Duffin, K. & Goldberg, D.E. (1994) Order and specificity of the *Plasmodium falciparum* hemoglobin degradation pathway. *J. Clin. Invest.* **93**, 1602–1608.

Goldberg, D.E., Slater, A.F.G., Cerami, A. & Henderson, G.B. (1990) Hemoglobin degradation in the malaria parasite

*Plasmodium falciparum*: an ordered process in a unique organelle. *Proc. Natl Acad. Sci. USA* **87**, 2731–2735.

Goldberg, D.E., Slater, A.F.G., Beavis, R., Chait, B., Cerami, A. & Henderson, G.B. (1991) Hemoglobin degradation in the human malaria pathogen *Plasmodium falciparum*: a catabolic pathway initiated by a specific aspartic protease. *J. Exp. Med.* **173**, 961–969.

Laurent, F., Bourdieu, C., Kaga, M., Chilmonczyk, S., Zgrzebski, G., Yvore, P. & Pery, P. (1993) Cloning and characterization of an *Eimeria acervulina* sporozoite gene homologous to aspartyl proteases. *Mol. Biochem. Parasitol.* **62**, 303–312.

Luker, K.E., Francis, S.E., Gluzman, I.Y. & Goldberg, D.E. (1996) Kinetic analysis of plasmepsins I and II, aspartic proteases of the *Plasmodium falciparum* digestive vacuole. *Mol. Biochem. Parasitol.* **79**, 71–78.

Olliaro, P.L. & Goldberg, D.E. (1995) The *Plasmodium* digestive vacuole: metabolic headquarters and choice drug target. *Parasitol. Today* **11**, 294–297.

Sherman, I.W. & Tanigoshi, L. (1983) Purification of *Plasmodium lophurae* cathepsin D and its effects on erythrocyte membrane proteins. *Mol. Biochem. Parasitol.* **8**, 207–226.

Silva, A.M., Lee, A.Y., Gulnik, S.V., Majer, P., Collins, J., Bhat, T.N., Collins, P.J., Cachau, R.E., Luker, K.E., Gluzman, I.Y., Francis, S.E., Oksman, A., Goldberg, D.E. & Erickson, J.W. (1996) Structure and inhibition of plasmepsin II, a hemoglobin-degrading enzyme from *Plasmodium falciparum. Proc. Natl Acad. Sci. USA* **93**, 10034–10039.

Vander Jagt, D.L., Hunsaker, L.A. & Campos, N.M. (1987) Comparison of proteases from chloroquine-sensitive and chloroquine-resistant strains of *Plasmodium falciparum. Biochem. Pharmacol.* **36**, 3285–3291.

Vander Jagt, D.L., Caughey, W.S., Campos, N.M., Hunsaker, L.A. & Zanner, M.A. (1989) Parasite proteases and antimalarial activities of protease inhibitors. *Prog. Clin. Biol. Res.* **313**, 105–118.

Vander Jagt, D.L., Hunsaker, L.A., Campos, N.M. & Scaletti, J.V. (1992) Localization and characterization of hemoglobin-degrading aspartic proteinases from the malarial parasite *Plasmodium falciparum. Biochim. Biophys. Acta* **1122**, 256–264.

*Daniel E. Goldberg*
*Howard Hughes Medical Institute,*
*Washington University,*
*Departments of Molecular Microbiology and Medicine,*
*Box 8230, 660 S. Euclid Avenue, St. Louis, MO 63110, USA*
*Email: goldberg@borcim.wustl.edu*

# 287. Plasmepsin II

## Databanks

*Peptidase classification: clan AA, family A1, MEROPS ID: A01.023*
*NC-IUBMB enzyme classification: EC 3.4.23.39*
*Databank codes:*

| Species | SwissProt | PIR | EMBL (cDNA) | EMBL (genomic) |
|---|---|---|---|---|
| *Plasmodium falciparum* | P46925 | – | L10740 | – |

Brookhaven Protein Data Bank three-dimensional structures:

| Species | ID | Resolution | Notes |
|---|---|---|---|
| *Plasmodium falciparum* | 1SME | 2.7 | complex with pepstatin A |

## Name and History

Aspartic protease activities from the human malaria pathogen *Plasmodium falciparum* have been described in a number of publications (reviewed by Vander Jagt *et al.*, 1989). Partial purification and characterization of aspartic protease activities from *P. falciparum* was achieved by Vander Jagt and colleagues (1987). The genomic sequence of **plasmepsin II** (previously called **aspartic hemoglobinase II** or alternatively **PFAPD**) was reported by Dame *et al.* (1994). Purification, N-terminal sequence and analysis of specificity was reported by Gluzman *et al.* (1994).

## Activity and Specificity

The enzyme is capable of cleaving native human hemoglobin as well as acid-denatured globin, with a pH optimum around 5 (Gluzman *et al.*, 1994). Incubation with hemoglobin results in proteolysis at 3–4 sites. There is a preference for hydrophobic P2, P1 and P1′ residues (Gluzman *et al.*, 1994). The same peptide bond recognized initially by plasmepsin I at hemoglobin Phe33↓Leu34 is cleaved by plasmepsin II.

Assay of activity can be performed with $^{14}$C-methylated globin in a TCA-precipitation assay (Goldberg *et al.*, 1990), or spectrophotometrically using either a chromogenic peptide Lys-Pro-Ile-Val-Phe↓Nph-Arg-Leu (Hill *et al.*, 1994), or a fluorescence quenched substrate Dabcyl-GABA-Glu-Arg-Met-Phe↓Leu-Ser-Phe-Pro-GABA-EDANS (Dabcyl: 4-(4-dimethylaminophenylazo)benzoyl; GABA: $\gamma$-amino-*n*-butyric acid; EDANS: 5-(2-aminoethylaminio)-1-naphthalene-sulfonic acid) (Luker *et al.*, 1996). This peptide is based on the $\alpha$-chain cleavage site.

## Structural Chemistry

The enzyme is a single-chain protein, made as a 51 kDa precursor that is cleaved to a 37 kDa mature form (Francis *et al.*, 1994; Francis & Goldberg, 1996). There are two disulfide bonds predicted. The second domain contains an active site Asp-Ser-Gly instead of the usual aspartic protease Asp-Thr-Gly. A crystal structure of the recombinant enzyme has been solved at 2.7 Å resolution (Silva *et al.*, 1996). The structure bears general features of mammalian and fungal aspartic proteases (Chapter 271). Val replaces the mammalian Gly at the tip of the flap. Pepstatin is bound in the active site with enough conformational difference from mammalian homologs to give hope for development of a selective inhibitor. The crystals contain molecules in two different conformations, demonstrating substantial interdomain flexibility.

## Preparation

Native plasmepsin II can be prepared in nanogram to low microgram amounts from cultured intraerythrocytic organisms (Gluzman *et al.*, 1994). High-level expression of active recombinant enzyme has been achieved in *E. coli* (Hill *et al.*, 1994) and the recombinant protease has kinetic properties similar to those of the native enzyme (Luker *et al.*, 1996).

## Biological Aspects

Plasmepsin II is found in the digestive vacuole of intraerythrocytic *P. falciparum*, where it appears to play an important role in hemoglobin degradation (Goldberg *et al.*, 1990; Vander Jagt *et al.*, 1992; Francis *et al.*, 1994; Gluzman *et al.*, 1994). Plasmepsin II is synthesized as a type II integral membrane protein, appears to be exported to the parasite surface and then is internalized with its substrate hemoglobin into the digestive vacuole (Francis *et al.*, 1994; Francis & Goldberg, 1996). An unusual acidic processing enzyme that is not susceptible to inhibitors of aspartic, cysteine, serine or metalloproteases but is blocked by certain tripeptide aldehydes, cleaves the plasmepsin to its mature form.

## Distinguishing Features

The enzyme is 73% identical to plasmepsin I (Chapter 286) at the amino acid level (Francis *et al.*, 1994; Dame *et al.*, 1994), but has a different cleavage specificity. Plasmepsin I is susceptible to the peptidomimetic inhibitors SC-50083 and P1, whereas II is relatively insensitive to the former and not inhibited detectably by the latter (Francis *et al.*, 1994; Luker *et al.*, 1996). Specific polyclonal antisera to each plasmepsin have been made but are not commercially available.

## Related Peptidases

Other enzymes related to both plasmepsins I and II are described in the preceding chapter (Chapter 286) and in Chapter 288.

## Further Reading

For a discussion of the biology of hemoglobin degradation in the organelle where catabolism occurs, see Olliaro & Goldberg (1995).

## References

Dame, J.B., Reddy, G.R., Yowell, C.A., Dunn, B.M., Kay, J. & Berry, C. (1994) Sequence, expression and modeled structure of an aspartic proteinase from the human malaria parasite *Plasmodium falciparum*. *Mol. Biochem. Parasitol.* **64**, 177–190.

Francis, S.E. & Goldberg, D.E. (1996) Biosynthesis of plasmepsins I and II, proteases in the *Plasmodium falciparum* hemoglobin degradation pathway. Molecular Parasitology Meeting, Woods Hole, MA, USA **7**, 56 (abstr.).

Francis, S.E., Gluzman, I.Y. Oksman, A., Knickerbocker, A., Mueller, A., Bryant, M.L., Sherman, D.R., Russell, D.G. & Goldberg, D.E. (1994) Molecular characterization and inhibition of a *Plasmodium falciparum* aspartic hemoglobinase. *EMBO J.* **13**, 306–317.

Gluzman, I.Y., Francis, S.E., Oksman, A., Smith, C., Duffin, K. & Goldberg, D.E. (1994) Order and specificity of the *Plasmodium falciparum* hemoglobin degradation pathway. *J. Clin. Invest.* **93**, 1602–1608.

Goldberg, D.E., Slater, A.F.G., Cerami, A. & Henderson, G.B. (1990) Hemoglobin degradation in the malaria parasite *Plasmodium falciparum*: an ordered process in a unique organelle. *Proc. Natl Acad. Sci. USA* **87**, 2731–2735.

Hill, J., Tyas, L., Phylip, L.H., Kay, J., Dunn, B.M. & Berry, C. (1994) High level expression and characterization of plasmepsin II, an aspartic proteinase from *Plasmodium falciparum*. *FEBS Lett.* **352**, 155–158.

Luker, K.E., Francis, S.E., Gluzman, I.Y. & Goldberg, D.E. (1996) Kinetic analysis of plasmepsins I and II, aspartic proteases of the *Plasmodium falciparum* digestive vacuole. *Mol. Biochem. Parasitol.* **79**, 71–78.

Olliaro, P.L. & Goldberg, D.E. (1995) The *Plasmodium* digestive vacuole: metabolic headquarters and choice drug target. *Parasitol. Today* **11**, 294–297.

Silva, A.M., Lee, A.Y., Gulnik, S.V., Majer, P., Collins, J., Bhat, T.N., Collins, P.J., Cachau, R.E., Luker, K.E., Gluzman, I.Y., Francis, S.E., Oksman, A., Goldberg, D.E. & Erickson, J.W. (1996) Structure and inhibition of plasmepsin II, a hemoglobin-degrading enzyme from *Plasmodium falciparum*. *Proc. Natl Acad. Sci. USA* **93**, 10034–10039.

Vander Jagt, D.L., Hunsaker, L.A. & Campos, N.M. (1987) Comparison of proteases from chloroquine-sensitive and chloroquine-resistant strains of *Plasmodium falciparum*. *Biochem. Pharmacol.* **36**, 3285–3291.

Vander Jagt, D.L., Caughey, W.S., Campos, N.M., Hunsaker, L.A.

& Zanner, M.A. (1989) Parasite proteases and antimalarial activities of protease inhibitors. *Prog. Clin. Biol. Res.* **313**, 105–118.

Vander Jagt, D.L., Hunsaker, L.A., Campos, N.M. & Scaletti, J.V. (1992) Localization and characterization of hemoglobin-degrading aspartic proteinases from the malarial parasite *Plasmodium falciparum*. *Biochim. Biophys. Acta* **1122**, 256–264.

*Daniel E. Goldberg*
*Howard Hughes Medical Institute,*
*Washington University,*
*Departments of Molecular Microbiology and Medicine,*
*Box 8230, 660 S. Euclid Avenue, St. Louis, MO 63110, USA*
*Email: goldberg@borcim.wustl.edu*

# 288. *Additional plasmepsins*

## Databanks

*Peptidase classification: clan AA, family A1, MEROPS ID: A01.039*
*NC-IUBMB enzyme classification: none*
*Databank codes:*

| Species | SwissProt | PIR | EMBL (cDNA) | EMBL (genomic) |
|---|---|---|---|---|
| Plasmodium vivax | – | – | – | AF001208 |
| Plasmodium ovale | – | – | – | AF001209 |
| Plasmodium malariae | – | – | – | AF001210 |

## Name and History

**Plasmepsins** are aspartic proteinases found in malaria parasites of the genus *Plasmodium*. Aspartic proteinases of the human malaria parasite, *Plasmodium falciparum*, are the best studied (Vander Jagt *et al.*, 1987, 1989; Goldberg *et al.*, 1991; Gluzman *et al.*, 1994; Francis *et al.*, 1994; Dame *et al.*, 1994) and multiple forms of this enzyme have been identified (Vander Jagt *et al.*, 1992; Gluzman *et al.*, 1994). Two forms have been characterized in the food vacuole of the asexual, bloodstage parasite (Vander Jagt *et al.*, 1992). These enzymes from *P. falciparum* have been termed plasmepsin I (Chapter 286) and plasmepsin II (Chapter 287), although other names have been used including aspartic hemoglobinase I (or PFAPG) and aspartic hemoglobinase II (or PFAPD) (Francis *et al.*, 1994; Dame *et al.*, 1994). The presence of two distinct enzymes has been established by N-terminal peptide sequence analysis (Gluzman *et al.*, 1994) and by cloning and sequencing the complete coding regions of both genes (Francis *et al.*, 1994; Dame *et al.*, 1994). Whether other *Plasmodium* spp. express multiple, distinct plasmepsins is unknown. Genes encoding a single plasmepsin from each of the three other species of malaria parasite infecting humans have been

cloned and sequenced, and active recombinant enzyme has been expressed from each (J.B. Dame *et al.*, unpublished results).

## Activity and Specificity

The activity and specificity of the plasmepsins from *P. falciparum* are described in Chapters 286 and 287. These proteases are recognized as having a primary function in the digestion of hemoglobin in the food vacuole of the parasite. The plasmepsins from the other three species presumably have a similar role. These enzymes have been studied initially using a gene-first approach via cloning, sequencing and expression as recombinant enzymes in *Escherichia coli* (J.B. Dame *et al.*, unpublished results). The functional correspondence between the plasmepsins from *P. vivax*, *P. ovale* and *P. malariae*, and either of the two forms of plasmepsin from *P. falciparum*, has not been established. Substrate-specificity studies utilizing systematically varied panels of chromogenic peptide substrates have indicated that, like plasmepsin II, the *P. vivax* and *P. malariae* enzymes have pH optima in the range of pH 4–5 and prefer hydrophobic residues in positions P2 and P3 (J.B. Dame *et al.*, unpublished results).

## Structural Chemistry

The five plasmepsins examined from the four *Plasmodium* species infecting humans (Francis *et al.*, 1994; Dame *et al.*, 1994; J.B. Dame *et al.*, unpublished results) are single polypeptide chains of 450–453 amino acids initially translated in a proenzyme form. The active, mature plasmepsins are 327–329 amino acids in length with a pI of 4.25–4.67, as determined from the sequence. They are well conserved with >60% sequence identity shared by all plasmepsins over the region encoding the active enzyme. The *P. vivax* and *P. malariae* plasmepsins share the highest sequence identity (87.5%), which is 14% greater than that shared by plasmepsins I and II from *P. falciparum*. The X-ray crystal structure of plasmepsin II from *P. falciparum* (Silva *et al.*, 1996) has been used as a template to model the structures of the other enzymes (J.B. Dame *et al.*, unpublished results).

## Preparation

Native plasmepsins have been prepared only from *P. falciparum* (Vander Jagt *et al.*, 1992; Gluzman *et al.*, 1994) as described in Chapters 286 and 287. Plasmepsins from the other species described have been characterized from gene sequences and recombinant proteins expressed in *E. coli* (J.B. Dame *et al.*, unpublished results). The genes encoding the mature plasmepsin, plus 48 amino acids of the proregion, have been engineered for cloning into the *Bam*HI site of the pET3a cloning vector (Hill *et al.*, 1994). Recombinant proplasmepsins are recovered from inclusion bodies, solubilized in urea, and refolded, as described (Hill *et al.*, 1994). Purification is accomplished via ammonium sulfate precipitation and/or ion-exchange chromatography, and self-activation is accomplished by brief preincubation under optimal assay conditions in the absence of substrate (Hill *et al.*, 1994; J.B. Dame *et al.*, unpublished results).

## Biological Aspects

Plasmepsins from all species of the malaria parasite are believed to function as hemoglobinases, digesting the contents of the erythrocyte in which they develop. Although this has only been demonstrated directly for the plasmepsins of *P. falciparum*, the similarity of the gene sequence and the common life histories of the malaria parasites suggests that this function is shared by the plasmepsins of the other species. *P. falciparum* is not closely related evolutionarily to the other three species infectious for humans (Waters *et al.*, 1991; Escalante *et al.*, 1995), and this more distant relationship is reflected in the primary sequences of the plasmepsins. The plasmepsins from these three species share a greater sequence identity with each other than with either plasmepsin from *P. falciparum*, or than the two *P. falciparum* plasmepsins share with each other (J.B. Dame *et al.*, unpublished results).

## Distinguishing Features

All of the plasmepsins share a long proregion (123–124 amino acids) as compared to most other aspartic proteinases (ca. 50 amino acids; Francis *et al.*, 1994; Dame *et al.*, 1994; J.B. Dame *et al.*, unpublished results). A novel proteolytic enzyme activity has been identified that activates the proenzyme by removing this proregion of the *P. falciparum* plasmepsins (Francis & Goldberg, 1996). The common structure of this proregion among all of the plasmepsins suggests a similar mode of activation. Self-activated, recombinant plasmepsins have from 2 to 12 additional amino acids at the N-terminus (Hill *et al.*, 1994; J.B. Dame *et al.*, unpublished results).

## Related Peptidases

Human cathepsin D (Chapter 277) is the peptidase of the human host for the malaria parasite that is most closely similar to the plasmepsins in structure and function. Inhibitors which may eventually be used as antimalarial drugs must differentially inhibit plasmepsins versus this important host enzyme.

## References

Dame, J.B., Reddy, G.R., Yowell, C.A., Dunn, B.M., Kay, J. & Berry, C. (1994) Sequence, expression and modeled structure of an aspartic proteinase from the human malaria parasite *Plasmodium falciparum*. *Mol. Biochem. Parasitol.* **64**, 177–190.

Escalante, A.A., Barrio, E. & Ayala, F.J. (1995) Evolutionary origin of human and primate malarias: evidence from the circumsporozoite protein gene. *Mol. Biol. Evol.* **12**, 616–626.

Francis, S.E. & Goldberg, D.E. (1996) Biosynthesis of plasmepsins I and II, proteases in the *Plasmodium falciparum* hemoglobin degradation pathway. Molecular Parasitology Meeting, Woods Hole, MA, USA **7**, 56 (abstr.).

Francis, S.E., Gluzman, I.Y., Oksman, A., Knickerbocker, A., Mueller, R., Bryant, M.L., Sherman, D.R., Russell, D.G. & Goldberg, D.E. (1994) Molecular characterization and inhibition of the *Plasmodium falciparum* aspartic hemoglobinase. *EMBO J.* **13**, 306–317.

Gluzman, I.Y., Francis, S.E., Oksman, A., Smith, C., Duffin, K. & Goldberg, D.E. (1994) Order and specificity of the *Plasmodium falciparum* hemoglobin degradation pathway. *J. Clin. Invest.* **93**, 1602–1608.

Goldberg, D.E., Slater, A.F.G., Beavis, R., Chait, B., Cerami, A. & Henderson, G.B. (1991) Hemoglobin degradation in the human malaria pathogen *Plasmodium falciparum*: a catabolic pathway initiated by a specific aspartic protease. *J. Exp. Med.* **173**, 961–969.

Hill, J., Tyas, L., Phylip, L.H., Kay, J., Dunn, B.M. & Berry, C. (1994) High level expression and characterisation of plasmepsin II, an aspartic proteinase from *Plasmodium falciparum*. *FEBS Lett.* **352**, 155–158.

Silva, A.M., Lee, A.Y., Gulnik, S.V., Majer, P., Collins, J., Bhat, T.N., Collins, P.J., Cachau, R.E., Luker, K.E., Gluzman, I.Y., Francis, S.E., Oksman, A., Goldberg, D.E. & Erickson, J.W. (1996) Structure and inhibition of plasmepsin II, a hemoglobin-degrading enzyme from *Plasmodium falciparum*. *Proc. Natl Acad. Sci. USA* **93**, 10034–10039.

Vander Jagt, D.L., Hunsaker, L.A. & Campos, N.M. (1987) Comparison of proteases from chloroquine-sensitive and chloroquine-resistant strains of *Plasmodium falciparum. Biochem. Pharmacol.* **36**, 3285–3291.

Vander Jagt, D.L., Caughey, W.S., Campos, N.M., Hunsaker, L.A. & Zanner, M.A. (1989) Parasite proteases and antimalarial activities of protease inhibitors. *Prog. Clin. Biol. Res.* **313**, 105–118.

Vander Jagt, D.L., Hunsaker, L.A., Campos, N.M. & Scaletti, J.V. (1992) Localization and characterization of hemoglobin-degrading aspartic proteinases from the malarial parasite *Plasmodium falciparum. Biochim. Biophys. Acta* **1122**, 256–264.

Waters, A.P., Higgins, D.G. & McCutchan, T.F. (1991) *Plasmodium falciparum* appears to have arisen as a result of lateral transfer between avian and human hosts. *Proc. Natl Acad. Sci. USA* **88**, 3140–3144.

*John B. Dame*
*Department of Pathobiology,*
*University of Florida,*
*Gainesville, FL 21611-0880, USA*
*Email: jbd@vetmed3.vetmed.ufl.edu*

# *289. Neurosporapepsin*

## Databanks

*Peptidase classification: clan AA, family A1, MEROPS ID: A01.029*
*NC-IUBMB enzyme classification: none*
*Databank codes:*

| Species | SwissProt | PIR | EMBL (cDNA) | EMBL (genomic) |
|---------|-----------|-----|-------------|----------------|
| *Neurospora crassa* | – | – | U36471 | – |

## Name and History

More than 20 years ago intracellular proteinases were purified from whole-cell extracts of *Neurospora crassa* and one major acid proteinase was reported (Siepen *et al.*, 1975). The substrate specificity and resistance of the enzyme to sulfhydryl-blocking reagents indicated that it resembled yeast saccharopepsin (Chapter 283). No further information was available until the recent isolation of the *pep-4* gene, which appears to encode this major intracellular acid proteinase. Analysis of the amino acid sequence indicates that it is clearly an aspartyl proteinase, most closely related to saccharopepsin, and more distantly related to the cathepsins and pepsins of mammalian cells (Vázquez-Laslop *et al.*, 1996). The name *neurosporapepsin* is used here.

## Activity and Specificity

The enzyme can act as an endopeptidase on both small and large polypeptides, but the specificity of cleavage has not been established. For example, the B chain of insulin is cleaved at four points–

FVNQHLCGSHLVEAL↓Y↓LVCGERGF↓F↓YTPKA

and glucagon is also cleaved at four points–

HSQGTFTSAYSK↓YLDSRRAQDFVQ↓W↓L↓MNT

(Siepen & Kula, 1976). The enzyme also hydrolyzes the octapeptide Suc-Arg-Suc-Arg-Pro-Phe-His-Leu-Leu↓Val-Tyr-NHMec (Vázquez-Laslop *et al.*, 1996). The activity of the enzyme *in vitro* has routinely been determined by measuring the hydrolysis of acid-denatured hemoglobin at pH 3.0. The activity of whole-cell extracts is approximately 11 mg of tyrosine $min^{-1} mg^{-1}$ protein (Vázquez-Laslop *et al.*, 1996). At a concentration of $2 mg ml^{-1}$, pepstatin completely inhibits the enzyme (Kaehn *et al.*, 1979; Váquez-Laslop *et al.*, 1996).

## Structural Chemistry

From analysis of the *pep-4* gene the mature enzyme is predicted to contain 326 residues with a molecular mass of 35 426 Da, pI 4.5 (Vázquez-Laslop *et al.*, 1996). The polypeptide appears to have two catalytic Asp residues, each in a region of 11 residues that are highly conserved in aspartyl proteases (Woolford *et al.*, 1986; Vázquez-Laslop *et al.*, 1996). Comparison with the homologous enzyme from

*S. cerevisiae* suggests two sites of disulfide bond formation and two putative glycosylation sites. The extent to which the enzyme is glycosylated has not been reported, but the molecular mass of the polypeptide when analyzed by PAGE is approximately 40 kDa, significantly larger than predicted from the amino acid sequence (Vázquez-Laslop *et al*., 1996).

## Preparation

Extracts were prepared from 60 liter batches of *N. crassa* grown for 24 h at 25°C with vigorous aeration in a fermenter. The enzyme was purified to apparent homogeneity with a 630-fold increase in specific activity and a yield of 6% (Siepen *et al*., 1975).

## Biological Aspects

Neurosporopepsin is one of the most abundant polypeptides in the vacuole of *N. crassa*. Most of the protein is found in the water-soluble contents of the vacuole, but it also binds tightly to vacuolar membranes. In *N. crassa* the size, composition and function of vacuoles is similar to that of mammalian lysosomes; it is likely that neurosporapepsin has the same function as lysosomal cathepsins (Vázquez-Laslop *et al*., 1996).

Although the role of the enzyme *in vivo* is not known, it may be responsible for the proteolytic processing and maturation of vacuolar enzymes. Mutant strains that completely lack neurosporapepsin grow at normal rates and are able to proceed through all stages of the life cycle. Other proteinases within the vacuole may be able to compensate for the loss of this enzyme. Comparison of *pep-4* mutant strains in *N. crassa* versus *S. cerevisiae* has revealed an interesting difference. In *S. cerevisiae*, *PEP4* mutants are deficient in several vacuolar proteinases, including cerevisin (Chapter 104) and carboxypeptidase Y (Chapter 132) (Ammerer *et al*., 1986; Woolford *et al*., 1986). By contrast, the activity of these other peptidases is elevated 1.5- to 2.0-fold in *pep-4* mutant strains of *N. crassa* (Vázquez-Laslop *et al*., 1996).

Like saccharopepsin, neurosporapepsin appears to be synthesized as an inactive precursor with signal sequences recognized in the endoplasmic reticulum and in the vacuole. The initial gene product contains 396 residues, with a molecular mass of 42 900 Da. The first 70 amino acids are removed to yield the mature protein. The 20 amino acids preceding the cleavage site are completely dissimilar in *N. crassa* and *S. cerevisiae*, but the mature N-terminus region is conserved. (Ten of the first 15 amino acids are identical.) The specificity of cleavage may be dictated by the structure of the mature protein.

The *pep-4* gene which encodes neurosporapepsin contains two small introns. It has been mapped to linkage group VII in the middle of a 25 kb region that separates the *frq* and *for* genes (Vázquez-Laslop *et al*., 1996).

## Distinguishing Features

Neurosporopepsin is the major enzyme in *N. crassa* cell extracts that can hydrolyze denatured protein at pH 3.0. Under these assay conditions it is the only protease inhibited by pepstatin. The polypeptide is specifically recognized by polyclonal antibodies raised to saccharopepsin from *S. cerevisiae*, however (Vázquez-Laslop *et al*., 1996).

## Further Reading

Details of enzyme purification may be found in Siepen *et al*. (1975); the primary structure and a comparison to yeast saccharopepsin are detailed by Vázquez-Laslop *et al*. (1996).

## References

Ammerer, G., Hunter, C.P., Rothman, J.H., Saari, G.C., Valls, L.A. & Stevens, T.H. (1986) *PEP4* gene of *Saccharomyces cerevisiae* encodes proteinase A, a vacuolar enzyme required for processing of vacuolar precursors. *Mol. Cell. Biol.* **6**, 2490–2499.

Kaehn, K.K., Morr, M. & Kula, M.-R. (1979) Inhibition of the acid proteinase from *Neurospora crassa* by diazoacetyl-DL-norleucine methyl ester, 1,2-epoxy-3-(4-nitrophenoxy)propane and pepstatin. *Z. Physiol. Chem.* **360**, 791–794.

Siepen, D. & Kula, M.-R. (1976) Specificity of five intracellular proteinases of *Neurospora crassa*. *FEBS Lett.* **66**, 31–34.

Siepen, D., Yu, P.-H. & Kula, M.-R. (1975) Proteolytic enzymes of *Neurospora crassa*. Purification and some properties of five intracellular proteinases. *Eur. J. Biochem.* **56**, 271–281.

Vázquez-Laslop, N., Tenney, K. & Bowman, B.J. (1996) Characterization of a vacuolar protease in *Neurospora crassa* and the use of gene RIPing to generate protease-deficient strains. *J. Biol. Chem.* **271**, 21944–21949.

Woolford, C.A., Daniels, L.B., Park, F.J., Jones, E.W., Van Arsdell, J.N. & Innis, M.A. (1986) The *PEP4* gene encodes an aspartyl protease implicated in the posttranslational regulation of *Saccharomyces cerevisiae* vacuolar hydrolases. *Mol. Cell. Biol.* **6**, 2500–2510.

*Barry J. Bowman*
*Sinsheimer Laboratories,*
*Department of Biology,*
*University of California,*
*Santa Cruz, CA 95060, USA*
*Email: bowman@biology.ucsc.edu*

# 290. *Rhodotorulapepsin*

## Databanks

*Peptidase classification: clan AX, family A99, MEROPS ID: A9G.008*
*NC-IUBMB enzyme classification: EC 3.4.23.26*
*Chemical Abstracts Service registry number: 37259-59-9*
*Databank codes: no sequence data available*

## Name and History

Although many kinds of molds were known to produce a variety of acid proteinases, little was known about the extracellular proteinases of yeast in the 1970s, with the exception of candidapepsin (Chapter 300). During our screening tests for extracellular proteinases of yeast, a potent extracellular acid proteinase-producing yeast (strain K-24) was isolated from the soil of a grape farm collected in Ikoma, Osaka Prefecture, Japan. This yeast was identified as *Rhodotorula glutinis* (Fres.) Harrison (Kamada *et al.*, 1972). The name **rhodotorulapepsin** was recommended by IUBMB in 1992.

## Activity and Specificity

Rhodotorulapepsin is most active toward casein at pH values between 2.0 and 2.5. Compared to several acid proteinases from various sources, the specific activity of this protease toward casein is high and is 1.7 times that of pig pepsin. Rhodotorulapepsin is inhibited by DAN, diazoacetyl glycine ethylester, *p*-bromo-phenacylbromide, and pepstatin Ac. But the inhibition rate by pepstatin Ac is different from that for pig pepsin: 20 μg of pepstatin Ac ($3.1 \times 10^{-8}$ mole) inhibited 60% of the activity of 1 mg of rhodotorulapepsin ($3.3 \times 10^{-8}$ mole). At this dose of pepstatin Ac, 1 mg of pig pepsin ($2.9 \times 10^{-8}$ mole) was completely inhibited (Kamada *et al.*, 1972).

The substrate specificity of rhodotorulapepsin was evaluated: it is able to activate trypsinogen and cleaves various oligopeptides containing Lys; Z-Lys-Ala-Ala (no hydrolysis), Z-Lys+Ala-Ala-Ala ($k_{cat}/K_m = 7.7\,\mathrm{M^{-1}\,s^{-1}}$), and Z-Lys+Leu-Ala-Ala ($k_{cat}/K_m = 20.8\,\mathrm{M^{-1}\,s^{-1}}$). The rate of hydrolysis is markedly accelerated by elongation of the peptide chain with Ala on the N-terminal side of the scissile bond (Morihara & Oka, 1973). The tetradecapeptide renin substrate (DRVY+IHPFHLL+VYS) was cleaved by rhodotorulapepsin at Tyr4+Ile5 and Leu11+Val12 bonds (additional cleavage was seen at Val3+Tyr4 and Leu10+Leu11 at a slower rate). This cleavage pattern differs from that of several fungal acid proteinases (common cleavage sites are Tyr4-Ile5, His6-Pro7 and Leu11-Val12); pepsin (Chapter 272) primarily cleaves Val3-Tyr4 and Leu10-Leu11 (Majima *et al.*, 1988).

## Structural Chemistry

The molecular mass of rhodotorulapepsin is estimated to be about 30 000 by the sedimentation equilibrium method and about 29 000 by gel filtration on Sephadex G-75. Its pI is 4.5 by isoelectric focusing. The enzyme contains 285 amino acid residues calculated from the amino acid analysis; its N-terminus is Ala and the C-terminus is Ser (Oda *et al.*, 1972). The complete amino acid sequence is not yet known.

## Preparation

Rhodotorulapepsin, produced from *Rhodotorula glutinis* K-24, was purified by precipitation with ammonium sulfate, fractional precipitation with acetone, gel filtration with Sephadex G-100, and finally by pH adjustment to yield needle-like crystals. The recovery of enzyme activity was about 40%, and about 80–100 mg of three-times crystallized rhodotorulapepsin were obtained from 1 liter of culture broth. Expression systems have not been utilized.

## Biological Aspects

The physiological function of rhodotorulapepsin is not clear. The addition of the inhibitor pepstatin Ac to the culture medium enhanced the growth and changed the optimum growth temperature of *Rhodotorula glutinis*, indicating the unique physiological functions played by this enzyme in the microorganism (Arai *et al.*, 1978).

## References

Arai, M., Tsuchiya, E. & Murao, S. (1978) Physiological changes of *Rhodotorula glutinis* induced by protease inhibitor (S-PI). *Agric. Biol. Chem.* **42**, 1429–1432.

Kamada, M., Oda, K. & Murao, S. (1972) The purification of the extracellular acid protease of *Rhodotorula glutinis* K-24 and its general properties. *Agric. Biol. Chem.* **36**, 1095–1101.

Oda, K., Kamada, M. & Murao, S. (1972) Some physicochemical properties and substrate specificity of acid protease of *Rhodotorula glutinis* K-24. *Agric. Biol. Chem.* **36**, 1103–1108.

Majima, E., Oda, K., Murao, S. & Ichishima, E. (1988) Comparative studies on the specificities of several fungal aspartic and acidic proteinases towards the tetradecapeptide of a renin substrate. *Agric. Biol. Chem.* **52**, 787–793.

Morihara, K. & Oka, T. (1973) Comparative specificity of microbial

acid proteinases for synthetic peptides. III. Relationship with their trypsinogen activating ability. *Arch. Biochem. Biophys.* **157**, 561–572.

***Sawao Murao***
*Department of Applied Microbial Technology,*
*Kumamoto Institute of Technology,*
*Ikeda 4-22-1, Kumamoto 860, Japan*
*Email: murao@bio.kumamoto-it.ac.jp*

# 291. Acrocylindropepsin

## Databanks

*Peptidase classification: clan AA, family A1, MEROPS ID: A01.039*
*NC-IUBMB enzyme classification: EC 3.4.23.28*
*Chemical Abstracts Service registry number: 37288-84-9*
*Databank codes: sequence data are unpublished (see the text)*

## Name and History

This enzyme was isolated and purified from a strain of *Acrocylindrium*, a plant pathogenic fungus (Uchino *et al.*, 1967).

## Activity and Specificity

The specificity of the enzyme was examined with oxidized insulin A and B chains and glucagon as substrates (Ichihara & Uchino, 1975). Major cleavages of the B chain are shown:

FVNQHL┼CGSHL┼VEAL┼Y┼LVCGERGF┼F┼YTPKA

These studies indicated that the enzyme prefers Tyr, Phe or Leu at the P1′ position. Assay is routinely performed at pH 2.0 and 37°C with casein as substrate by the Kunitz method with a slight modification (Uchino *et al.*, 1967). The optimum pH is 2.0 toward casein, and the specific activity is similar to that of pig pepsin (Chapter 272). The enzyme is strongly inhibited by pepstatin, by reaction with DAN in the presence of cupric ions, or by EPNP (1,2-epoxy-3-(*p*-nitrophenoxy)propane), like other pepsin-type aspartic proteinases (Takahashi & Chang, 1976).

## Structural Chemistry

The N-terminal amino acid sequenced has been determined (K. Takahashi, unpublished results); this shows a close

homology with other pepsin-type aspartic proteinases, especially rhizopuspepsin (Takahashi, 1987).

## Preparation

The enzyme is purified from the culture medium by successive steps of acetone precipitation, ammonum sulfate precipitation, chromatography on a Duolite CS-101 column and repeated crystallization (Uchino *et al.*, 1967).

## Biological Aspects

The enzyme is stable below 50°C and is inactivated above 70°C. It is most stable in the pH range of 2.5–5.0, and is unstable below pH 0.7 or above pH 5.6 (Uchino *et al.*, 1967).

## Distinguishing Features

The proteinase is inhibited by the specific inhibitors of ordinary pepsin-type aspartic proteinases such as pepstatin, DAN and EPNP.

## Related Peptidases

Acrocylindropepsin is thought to be homologous with other pepsin-type aspartic proteinases.

**References**

Ichihara, S. & Uchino, F. (1975) The specificity of acid proteinase from *Acrocylindrium. Agric. Biol. Chem.* **39**, 423–428.

Takahashi, K. (1987) The amino acid sequence of rhizopuspepsin, an aspartic proteinase from *Rhizopus chinensis. J. Biol. Chem.* **262**, 1468–1478.

Takahashi, K. & Chang, W.-J. (1976) The structure and function of acid proteases. V. Comparative studies on the specific inhibition of acid proteases by diazoacetyl-DL-norleucine methyl ester, 1,2-epoxy-3-(*p*-nitrophenoxy)propane and pepstatin. *J. Biochem (Tokyo)* **80**, 497–506.

Uchino, F., Kurono, Y. & Doi, S. (1967) Purification and some properties of crystalline acid protease from *Acrocylindrium* sp. *Agric. Biol. Chem.* **31**, 428–434.

*Kenji Takahashi*
*School of Life Science,*
*Tokyo University of Pharmacy and Life Science,*
*1432-1 Horinouchi, Hachioji, Tokyo 192-03, Japan*
*Email: kenjitak@ls.toyaku.ac.jp*

# 292. Pycnoporopepsin

## Databanks

*Peptidase classification: clan AX, family A99, MEROPS ID: A9G.010*
*NC-IUBMB enzyme classification: EC 3.4.23.30*
*Chemical Abstracts Service registry number: 77967-78-3*
*Databank codes: no sequence data available*

## Name and History

A study of protein turnover in wood metabolism led to the discovery of an acid proteinase, **pycnoporopepsin**, from a wood-deteriorating basidiomycete fungus *Pycnoporus sanguineus* (formerly designated *Trametes sanguinea* and *Pycnoporus coccineus*; Tomoda & Shimazono, 1964). Forms Ia and Ib (formerly designated carboxyl proteinases Ia and Ib) are recognized.

## Activity and Specificity

The specific activities of highly purified pycnoporopepsin Ia with pI 4.72 and pycnoporopepsin Ib with pI 4.58 were $0.215\,\text{kat}\,\text{kg}^{-1}$ of enzyme and $0.182\,\text{kat}\,\text{kg}^{-1}$ of enzyme, respectively (Ichishima *et al*., 1980). Pycnoporopepsin Ia migrates as a single band on disk gel electrophoresis at pH 9.4 and 2.3. Pycnoporopepsin Ia has an absorbance $A = 15$ at 280 nm, 1% solution, 1 cm path length.

The pH optima are 2.5 against casein (Hammarsten), and 2.3 against hemoglobin (Tomoda & Shimazono, 1964). Pycnoporopepsin Ia at pH 2.7 cleaves at three major points in the oxidized B chain of insulin: Phe24+Phe25, Tyr16+Leu17 and Ala14+Leu15 (Ichishima *et al*., 1980); the relative rates of cleavage are 1.00, 0.92 and 0.81, respectively:

FVNQHLCGSHLVEA+LY+LVCGERGF+FYTPKA

In the experiment, a characteristic small peptide, Leu-Tyr16 is recovered. The peptide map obtained from pycnoporopepsin Ib is identical to that of pycnoporopepsin Ia, and their specificities seem similar. The action on the three bonds of the B chain is similar to the cleavages reported for pig gastricsin (Chapter 276) (Ryle *et al*., 1968). However, pycnoporopepsin Ia could not split the His10+Leu11 or Glu13+Ala14 bonds, which are cleaved by gastricsin.

The specificity of pycnoporopepsin Ia was investigated with oligopeptides at pH 2.7 (Kumagai *et al*., 1981). Pycnoporopepsin Ia hydrolyzes the His10+Leu11 and Ala14+Leu15 bonds in oxidized insulin peptide B1–B16 (Kumagai *et al*., 1981), but does not hydrolyze oxidized insulin peptide B15–B24. The characteristic small dipeptide, Leu15-Tyr16 is released from the B1–B16 peptide, but not from the B15–B24. The results suggest that the susceptibility of peptides to pycnoporopepsin Ia requires the presence of an amino acid residue at the P3 position (Kumagai *et al*., 1981). Pycnoporopepsin Ia exhibits a preference for a hydrophobic amino acid such as leucine or isoleucine in the P1′ site. Thus, the S1′–P′ interactions might be important in increasing enzyme–substrate affinity and turnover rate. The study also showed that the pycnoporopepsin Ia exhibits a preference for a hydrophobic amino acid such as tyrosine or valine in the P2′ position. Furthermore, the present results show that pycnoporopepsin Ia exhibits a lower preference for a hydrophobic amino acid in the P1 position, but that pepsin (Chapter 272) exhibits a strong preference for a hydrophobic amino acid in the P1 position (Ichishima *et al.*, 1980).

Pycnoporopepsin Ia selectively hydrolyzes Ala14┼Leu15, Tyr16┼Leu17 and Phe24┼Phe25 in the B chain of the native insulin molecule at pH 2.7 (Ichishima *et al.*, 1982). No hydrolysis was observed in the A chain of native insulin. Hydrolysis of angiotensin (formerly designated angiotensin II) at pH 2.7 was observed at the Tyr4┼Ile5 bond (Kumagai *et al.*, 1981). Hydrolysis of proangiotensin (formerly designated angiotensin I) at pH 2.7 was also at the Tyr4┼Ile5 and His8┼His9 bonds, but not at the His6-Pro7 bond (Ichishima *et al.*, 1980; Kumagai *et al.*, 1981).

In view of its specificity, pycnoporopepsin is clearly distinguishable from aspergillopepsin I from the fungus *Aspergillus saitoi* (Chapter 294) and mammalian pepsin.

## Structural Chemistry

The sedimentation coefficient ($s_{17,w}$) from Svedberg's equation gave values of 2.95 S for pycnoporopepsin (Tomoda, 1964). Using Svedberg's formula with this $s$ value and a diffusion coefficient of $D_{25}$ of $9 \times 10^{-7}$, an $M_r$ of 34 000 was calculated for the enzyme (Tomoda, 1964). This value agrees quite well with that of 33 000 or 34 000 obtained by Archibald's method (Tomoda & Shimazono, 1964). The pI values of pycnoporopepsin Ia and Ib are 4.72 and 4.58, respectively (Ichishima *et al.*, 1980).

Nothing is known of the gene for pycnoporopepsin. However, the three-dimensional structure and active-site structure of pycnoporopepsin are considered to be similar to those of mammalian pig pepsin. The molecular structure of pig pepsinogen has been determined at 1.8 Å resolution by a combination of molecular replacement and multiple isomorphous phasing techniques (James & Sielecki, 1986; Sielecki *et al.*, 1991). The resulting structure was refined by restrained-parameter least-squares methods. For the enzyme portion of the zymogen, the root-mean-square difference in Ca atom coordinates with the refined pig pepsin structure is 0.90 Å (284 common atoms) and with the Ca atoms of penicillopepsin (Chapter 295) it is 1.63 Å (275 common atoms). The additional 44 N-terminal amino acids of the pro segment (Leu1p–Leu44p, using the letter p after the residue number to distinguish the residues of the pro segment) adopt a relatively compact structure consisting of a long $\beta$ strand followed by two approximately orthogonal $\alpha$ helices and a short 3(10) helix. Intimate contacts, both electrostatic and hydrophobic, are made with residues in the pepsin active site. Electrostatic interactions of Lys36pN and hydrogen-bonding interactions of Tyr37pOH and Tyr9OH with the two catalytic aspartate groups, Asp32 and Asp215, prevent substrate access to the active site of the proenzyme.

## Preparation

Maximum proteolytic activity was attained in a submerged culture of *Pycnoporus sanguineus* after 140 h cultivation. The enzyme was purified approximately 30-fold with about 8% recovery of the original activity (Tomoda & Shimazono, 1964). Pycnoporopepsin was obtained in platelet crystalline form from mycelium-free culture filtrate by

the following successive treatments: acetone precipitation, ion-exchange column chromatography, ammonium sulfate salting-out, dialysis, and crystallization by acetone (Tomoda & Shimazono, 1964). The partially purified enzyme preparation obtained by the method of Tomoda & Shimazono seemed to be homogeneous by free boundary electrophoresis at pH 6.08 and the sedimentation pattern showed it to be monodisperse (Tomoda, 1964). However, the presence of Z-Glu-Tyr hydrolase activity (acid carboxypeptidase activity) in the pycnoporopepsin preparation still remained a possibility since Z-Glu-Tyr is a sensitive substrate for serine carboxypeptidases (Chapter 132) of *Aspergillus* (Ichishima & Arai, 1972, 1973; Arai & Ichishima, 1974; Ichishima, 1991) and *Penicillium* (Yokoyama & Ichishima, 1972; Yokoyama *et al.*, 1975a,b). Further purification of the crystalline enzyme was performed in a chromatographic procedure with DEAE-Sephadex and sulfopropyl (SP)-Sephadex C-50 and isoelectric focusing (Ichishima *et al.*, 1980). The highly purified preparations of pycnoporopepsins Ia and Ib (then termed *P. coccineus* carboxyl proteinases Ia and Ib) migrate as a single band on disc gel electrophoresis at pH 9.4 and 2.3, respectively (Ichishima *et al.*, 1980).

Crude natural enzyme is commercially available from Takeda (see Appendix 2 for full names and addresses of suppliers). No recombinant enzyme of pycnoporopepsin is available.

## Biological Aspects

The enzyme is present in the basidiomycete, *Pycnoporus sanguineus* (Tomoda & Shimazono, 1964); it is an extracellular proteinase and is assumed to be an important digestive enzyme for fungal nutrition in wood metabolism. In its physical and enzymatic properties pycnoporopepsin is analogous to aspergillopepsin I (Ichishima, 1970; Tanaka *et al.*, 1977; Ichishima *et al.*, 1981; Majima *et al.*, 1988; Shintani & Ichishima, 1994; Shintani *et al.*, 1996) of the fungus *Aspergillus saitoi* (now designated *A. phoenicis*) in the imperfect fungi.

## Related Peptidases

An acid serine carboxypeptidase (Chapter 132) is found together with pycnoporopepsin in the culture filtrate of *P. sanguineus* (Ichishima *et al.*, 1983). This carboxypeptidase has a pH optimum at pH 3.4 for Z-Glu-Tyr, a $K_m$ of 0.74 mM and a $k_{cat}$ of 16 s$^{-1}$ with Z-Glu-Tyr. The $K_m$ and $k_{cat}$ values for bradykinin at pH 3.4 and 30°C are 20 mM and 25 s$^{-1}$, respectively. Values for angiotensin at pH 3.4 and 30°C were 0.76 mM and 2.4 s$^{-1}$, respectively. The values of $M_r$ and pI were 54 000 and 4.78, respectively.

## References

Arai, T. & Ichishima, E. (1974) Mode of action on proteins of acid carboxypeptidase from *Aspergillus saitoi*. *J. Biochem.* **76**, 765–769.

Ichishima, E. (1970) Purification and mode of assay for acid proteinase of *Aspergillus saitoi*. *Methods Enzymol.* **19**, 397–406.

Ichishima, E. (1991) Mode of action and application of *Aspergillus* carboxypeptidase. *Comments Agric. Food Chem.* **2**, 279–298.

Ichishima, E. & Arai, T. (1972) Purification and characterization of a new type of acid carboxypeptidase from *Aspergillus*. *Biochim. Biophys. Acta* **258**, 274–288.

Ichishima, E. & Arai, T. (1973) Specificity and mode of action of acid carboxypeptidase from *Aspergillus saitoi. Biochim. Biophys. Acta* **293**, 444–450.

Ichishima, E., Kumagai, H. & Tomoda, K. (1980) Substrate specificity of carboxyl proteinase from *Pycnoporus coccineus*, a wood-deteriorating fungus. *Curr. Microbiol.* **3**, 333–337.

Ichishima, E., Majima, E., Emi, M., Hayashi, K. & Murao, S. (1981) Enzymatic cleavage of the histidyl-prolyl bond of proangiotensin by carboxyl proteinases from *Aspergillus sojae* and *Scytalidium lignicolum. Agric. Biol. Chem.* **45**, 2391–2393.

Ichishima, E., Emi, M., Majima, E., Mayumi, Y., Kumagai, H., Hayashi, K. & Tomoda, K. (1982) Initial sites of insulin cleavage and stereospecificity of carboxyl proteinases from *Aspergillus sojae* and *Pycnoporus coccineus. Biochim. Biophys. Acta* **700**, 247–253.

Ichishima, E., Yoshimura, K. & Tomoda, K. (1983) Acid carboxypeptidase from a wood-deteriorating basidiomycete, *Pycnoporus sanguineus. Phytochemistry* **22**, 825–829.

Ichishima, E., Ito, Y. & Takeuchi, M. (1985) 1,2-α-D-mannosidase from a wood-rotting basidiomycete, *Pycnoporus sanguineus. Phytochemistry* **24**, 2835–2837.

James, M.M.G. & Sielecki, A.R. (1986) Molecular structure of an aspartic proteinase zymogen, porcine pepsinogen, at 1.8 Å resolution. *Nature* **319**, 33–38.

Kumagai, H., Matsue, M., Majima, E., Tomoda, K. & Ichishima, E. (1981) Carboxyl proteinase from the wood-deteriorating basidiomycete *Pycnoporus coccineus*: substrate specificity with oxidized insulin peptide B1–B16 and B15–B24, angiotensin and proangiotensin. *Agric. Biol. Chem.* **45**, 981–985.

Majima, E., Oda, K., Murao, S. & Ichishima, E. (1988) Comparative study on the specificities of several fungal aspartic and acidic proteinases towards the tetradecapeptide of a renin substrate. *Agric. Biol. Chem.* **52**, 787–793.

Ryle, A.P., Leclerc, J. & Falla, F. (1968) The substrate specificity of pepsin C. *Biochem. J.* **110**, 4p.

Shintani, T. & Ichishima, E. (1994) Primary structure of aspergillopepsin I deduced from nucleotide sequence of the gene and aspartic acid-76 is an essential active site of the enzyme for trypsinogen activation. *Biochim. Biophys. Acta* **1204**, 257–264.

Shintani, T., Kobayashi, M. & Ichishima, E. (1996) Characterization of the S1 subsite specificity of aspergillopepsin I by site-directed mutagenesis. *J. Biochem.* **120**, 974–981.

Sielecki, A.R., Fujinaga, M., Read, R.J. & James, M.N.G. (1991) Refined structure of porcine pepsinogen at 1.8 Å resolution. *J. Mol. Biol.* **219**, 671–692.

Tanaka, N., Takeuchi, M. & Ichishima, E. (1977) Purification of an acid proteinase from *Aspergillus saitoi* and determination of peptide bond specificity. *Biochim. Biophys. Acta* **485**, 406–416.

Tomoda, K. (1964) Acid protease produced by *Trametes sanguinea*, a wood-destroying fungus. Part II. Physical and enzymological properties of the enzyme. *Agric. Biol. Chem.* **28**, 774–778.

Tomoda, K. & Shimazono, H. (1964) Acid protease produced by *Trametes sanguinea*, a wood-destroying fungus. Part I. Purification and crystallization of the enzyme. *Agric. Biol Chem.* **28**, 770–773.

Yokoyama, S. & Ichishima, E. (1972) A new type of acid carboxypeptidase of mold of the genus *Penicillium. Agric. Biol. Chem.* **36**, 1259–1261.

Yokoyama, S., Oobayashi, A., Tanabe, O. & Ichishima, E. (1975a) Action of crystalline acid carboxypeptidase from *Penicillium janthinellum. Biochim. Biophys. Acta* **397**, 443–448.

Yokoyama, S., Oobayashi, A., Tanabe, O., Ohata, K., Shibata, Y. & Ichishima, E. (1975b) Kininase and anti-inflammatory activity of acid carboxypeptidase from *Penicillium janthinellum. Experientia* **31**, 1122–1124.

*Eiji Ichishima*
*Laboratory of Molecular Enzymology,*
*Department of Applied Biological Chemistry,*
*Faculty of Agriculture, Tohoku University,*
*1-1, Tsutsumidori-Amamiyamachi,*
*Aoba-ku, Sendai 981, Japan*
*Email: ichisima@t.soka.ac.jp*

# 293. Physaropepsin

## Databanks

*Peptidase classification: clan AX, family A99, MEROPS ID: A9G.012*
*NC-IUBMB enzyme classification: EC 3.4.23.27*
*Chemical Abstracts Service registry number: 94949-28-7*
*Databank codes: no sequence data available*

## Name and History

In the plasmodia of a true slime mold, *Physarum polycephalum*, there are intracellular acid proteinases that have a highly acidic pH optimum (Polanshek *et al.*, 1978; Murakami-Murofushi *et al.*, 1984). The enzyme activity is detected in the particulate fraction and is effectively extracted with a nonionic detergent, indicating the possibility of its localization in a lysosome-like organelle (Murakami-Murofushi *et al.*, 1984). In 1990, this enzyme was purified from the plasmodia to apparent homogeneity and characterized (Murakami-Murofushi *et al.*, 1990). The *Physarum* enzyme was demonstrated to be a novel type of intracellular acid proteinase distinct in several properties from ordinary aspartic enzymes. As this enzyme can cleave a certain synthetic pepsin substrate and is optimally active at pH 1.7 toward hemoglobin as a substrate, it appears to be more similar to pepsin (Chapter 272) than to other aspartic proteinases. The name *physaropepsin* was recommended by the IUBMB in 1992.

## Activity and Specificity

A synthetic peptide, Lys-Pro-Ile-Glu-Phe+Nph-Arg-Leu (Dunn *et al.*, 1986; Matsuzaki & Takahashi, 1988) was cleaved by physaropepsin, but other synthetic pepsin substrates (Tang, 1970) were not hydrolyzed. The protease showed a unique substrate specificity toward oxidized insulin B chain, and the major cleavage sites were the bonds Gly8+Ser9, Leu11+Val12, Cya19+Gly20, Gly20-Glu21 and Phe24-Phe25 (Cya, cysteic acid); of these, the Gly8-Ser9 bond was most susceptible:

FVNQHLCG+SHL+VEALYLVC+G+ERGF+FYTPKA

To our knowledge (Barrett *et al.*, 1977; Matsuzaki & Takahashi, 1988), cleavage of Gly-containing peptide bonds by acid proteinases has never been reported.

The activity of physaropepsin can be assayed with hemoglobin (Murakami-Murofushi *et al.*, 1984, 1990). The pH optimum of the enzyme is about 1.7 and the assay system contains 100 mM HCl–KCl. Physaropepsin is strongly inhibited by DAN (Rajagopalan *et al.*, 1966) in the presence of cupric ions, but other inhibitors tested, including EPNP (Tang, 1971) and pepstatin A (Aoyagi *et al.*, 1971), typical aspartic proteinase inhibitors, showed no significant effects.

## Structural Chemistry

Analysis of the purified enzyme by SDS-PAGE and Sephadex G-100 chromatography showed that it consists of two polypeptide chains, a 31 kDa heavy chain and a 23 kDa light chain, cross-linked by disulfide bond(s), the heavy chain of the enzyme with seven residues of mannose, five residues of glucosamine, and one residue each of fucose and glucose (Murakami-Murofushi *et al.*, 1990). The N-terminal amino acid sequences of the heavy chain and light chains show no significant homology with those of other proteinases found in the protein sequence database.

## Preparation

Physaropepsin is not widely distributed and has been found in the diploid plasmodia and haploid myxoamoebae of a true slime mold, *P. polycephalum* (Murakami-Murofushi *et al.*, 1984). The enzyme was purified by a combination of detergent extraction, acid precipitation, and column chromatographies on DEAE-Sephadex, hydroxyapatite, CM-Sephadex, and Sephadex G-100, and an approximately 142-fold increase in specific activity was attained with an overall yield of 2.4% (Murakami-Murofushi *et al.*, 1990).

## Biological Aspects

The enzyme has been found only in a particulate fraction of a true slime mold, *P. polycephalum*, and its activity changes dramatically during differentiation from haploid myxoamoebae to diploid plasmodia (Murakami-Murofushi *et al.*, 1984). This indicates an important role of this enzyme in the process of cell differentiation.

## Distinguishing Features

The $M_r$ of physaropepsin (54 000–68 000) is larger than that (35 000–45 000) of cathepsin D (Chapter 277) and related aspartic proteinases except for that of cathepsin E (85 000) (Chapter 275), which is composed of two identical subunits (Kageyama & Takahashi, 1980). The *Physarum* enzyme has a two-chain structure (23 and 31 kDa chains), like pig spleen cathepsin D (11 and 26 kDa chains; Huang *et al.*, 1979).

Physaropepsin is thought to be initially synthesized as a single-chain precursor, but the mechanism of its processing to the mature two-chain form remains to be elucidated. The enzyme shows a unique cleavage specificity as mentioned above, particularly the cleavage of Gly-containing bonds. The B chain bonds Leu11+Val12, Leu17+Tyr18 and Phe24+Phe25 are the major sites of cleavage by cathepsin D, cathepsin E and pepsin (Matsuzaki & Takahashi, 1988).

## Further Reading

For further reading, see Murakami-Murofushi *et al.* (1990).

## References

Aoyagi, T., Kunimoto, S., Morishima, H., Takeuchi, T. & Umezawa, H. (1971) Effect of pepstatin on acid proteases. *J. Antibiot. (Tokyo)* **24**, 687–694.

Barrett, A.J. (1977) Cathepsin D and other carboxyl proteinases. In: *Proteinases in Mammalian Cells and Tissues* (Barrett, A.J., ed.). Amsterdam: North-Holland., pp. 209–248.

Dunn, B.M., Jimenez, M., Parten, B.F., Valler, M.J., Rolph, C.E. & Kay, J. (1986) A systematic series of synthetic chromophoric substrates for aspartic proteinases. *Biochem. J.* **237**, 899–906.

Huang, J.S., Huang, S.S. & Tang, J. (1979) Cathepsin D isozymes from porcine spleens. Large scale purification and polypeptide chain arrangements. *J. Biol. Chem.* **254**, 11405–11417.

Kageyama, T. & Takahashi, K. (1980) A cathepsin D-like acid proteinase from human gastric mucosa. Purification and characterization. *J. Biochem. (Tokyo)* **87**, 725–735.

Matsuzaki, O. & Takahashi, K. (1988) Improved purification of slow moving protease from gastric mucosa and its action on the B chain of oxidized bovine insulin. *Biomed. Res.* **9**, 515–523.

Murakami-Murofushi, K., Hiratsuka, A. & Ohta, J. (1984) A novel acid protease from haploid amoebae of *Physarum polycephalum*, and its changes during mating and subsequent differentiation into diploid plasmodia. *Cell Struct. Funct.* **9**, 311–315.

Murakami-Murofushi, K., Takahashi, T., Minowa, Y., Iino, S., Takeuchi, T., Kitagaki-Ogawa, H., Murofushi, H. & Takahashi, K. (1990) Purification and characterization of a novel intracellular acid proteinase from the plasmodia of a true slime mold,

*Physarum polycephalum. J. Biol. Chem.* **265**, 19898–19903.

Polanshek, M.M., Blomquist, J.C., Evans, T.E. & Rusch, H.P. (1978) Aminopeptidase of *Physarum polycephalum* during growth and differentiation. *Arch. Biochem. Biophys.* **190**, 261–269.

Rajagopalan, T.G., Stein, W.H. & Moore, S. (1966) The inactivation of pepsin by diazoacetyl-norleucine methyl ester. *J. Biol. Chem.* **241**, 4295–4297.

Tang, J. (1970) Gastricsin and pepsin. *Methods Enzymol.* **19**, 406–421.

Tang, J. (1971) Specific and irreversible inactivation of pepsin by substrate-like epoxides. *J. Biol. Chem.* **246**, 4510–4517.

***Kimiko Murakami-Murofushi***
*Biochemistry Laboratory, Department of Biology,*
*Faculty of Science,*
*Ochanomizu University,*
*Ohtsuka, Bunkyo-ku, Tokyo 112, Japan*
*Email: murofush@cc.ocha.ac.jp*

# 294. *Aspergillopepsin I*

## Databanks

*Peptidase classification: clan AA, family A1, MEROPS ID: A01.016*
*NC-IUBMB enzyme classification: EC 3.4.23.18*
*Chemical Abstracts Service registry number: 9025-49-4*
*Databank codes:*

| Species | SwissProt | PIR | EMBL (cDNA) | EMBL (genomic) |
|---|---|---|---|---|
| *Aspergillus awamori* | – | PN0090 | – | – |
| *Aspergillus awamori* | P17946 | JU0340 PS0140 | M34454 | – |
| *Aspergillus fumigatus* | – | – | – | X85092: complete gene |
| *Aspergillus niger* | – | JC4052 PC4018 | D45177 | U03507: 5′ region |
| *Aspergillus oryzae* | Q06902 | JN0630 JT0614 PN0534 | D13894 | M92927: complete gene |
| *Aspergillus phoenicis* | P55325 | S42751 | D25318 D49838 | D49839: complete gene |

## Name and History

The crystallization of the serine alkaline proteinase oryzin (formerly designated aspergillopeptidase B) (Chapter 105) from *Aspergillus oryzae* (Crewther & Lennox, 1950) led to the discovery of an acid proteinase, *aspergillopepsin I*, and a correlative study of its enzymatic properties (Yoshida, 1956). This protease, formerly known as *aspergillopeptidase A*, was isolated from *Aspergillus saitoi*, a microorganism used in

fermentation of the traditional Japanese liquors awamori and shochu. The extracellular acid proteinase from *A. saitoi* R-3813 (now designated *A. phoenicis* ATCC 14332) was purified to a high degree by successive chromatographies (Ichishima & Yoshida, 1965a,b). The optimal pH of the enzyme for milk casein digestion is in the pH range of 2.5–3.0 and the proteinase is fairly stable over the range of 2.5–6.0 (Ichishima, 1970).

Aspergillopepsin I has been assigned to the aspartic proteinase family (I) in the pepsin superfamily by Shintani & Ichishima (1994). In this handbook it is placed within clan AA, family A1 (Chapter 271). Aspergillopepsins are found in a variety of species within the genus of imperfect fungi *Aspergillus*, and have been known by a variety of names:

1. *A. awamori* – *awamorin* or *aspergillopepsin A* (Ostoslavskaya *et al.*, 1986; Berka *et al.*, 1990; Thompson, 1990).
2. *A. foetidus* – *aspergillopepsin* (Ostoslavskaya *et al.*, 1976).
3. *A. fumigatus* – *extracellular aspartic protease* (Reichard *et al.*, 1995; Lee & Kolattukudy, 1995).
4. *A. kawachi* – *carboxyl protease* (Yagi *et al.*, 1986).
5. *A. niger* – *extracellular acid protease* (Jarai & Buxton, 1994; Jarai *et al.*, 1994), also known as *proteinase B* or *proctase B* (Morihara & Oka, 1973; Chang *et al.*, 1976; Lu *et al.*, 1995).
6. *A. oryzae* – *trypsinogen kinase* (Davidson *et al.*, 1975; Majima *et al.*, 1988), also known as *acid protease* (Gomi *et al.*, 1993), *aspergillopepsin O* (Berka *et al.*, 1993) and *aspartic proteinase* (Takeuchi *et al.*, 1995).
7. *A. saitoi* – *aspergillopeptidase A* (Ichishima & Yoshida, 1966, 1967a,b), also known as *aspartic proteinase* (Majima *et al.*, 1988) and *aspergillopepsin I* (Shintani & Ichishima, 1994).
8. *A. sojae* – *carboxyl proteinase I* (Ichishima *et al.*, 1981, 1982) or *aspartic proteinase* (Majima *et al.*, 1988).

All these proteases were formerly included in EC 3.4.23.6.

## Activity and Specificity

Aspergillopepsin I from *A. saitoi* primarily hydrolyzes two bonds in the oxidized B chain of insulin: Leu15↓Tyr16 and Phe24↓Phe25 (Tanaka *et al.*, 1977). Minor cleavages are noted at His10↓Leu11, Ala14↓Leu15 and Tyr16↓Leu17:

FVNQHLCGSH↓LVEA↓L↓Y↓LVCGERGF↓FYTPKA

The two primary specificity sites are identical to those of cathepsin D (Chapter 277) from human erythrocytes (Reichelt *et al.*, 1974). Hydrolysis of angiotensin II by aspergillopepsins I from *A. saitoi* (Tanaka *et al.*, 1977) and *A. oryzae* (Majima *et al.*, 1988) is at the Tyr4↓Ile5 bond. Aspergillopepsins I from *A. oryzae*, *A. sojae* and *A. saitoi* cleave Tyr4↓Ile5, His6↓Pro7 and Leu10↓Val11 bonds in the tetradecapeptide renin substrate (DRVY↓IH↓PFHLL↓VYS) at pH 2.7 (Majima *et al.*, 1988). The Xaa-Pro (where Xaa is any amino acid) bond is generally resistant to hydrolysis by proteinases; however, aspergillopepsin I from *A. sojae* (Ichishima *et al.*, 1982) and scytalidopepsin B from *Scytalidium lignicolum* (Chapter 326) (Murao *et al.*, 1972) cleave the His6↓Pro7 bond of proangiotensin (angiotensin I, DRVYIH↓PFHL) at pH 3.0 and 2.2, respectively (Ichishima *et al.*, 1981). This is the first association of the hydrolysis of the Xaa↓Pro bond with the degradation of a hormone.

Generally, aspergillopepsin I from *A. saitoi* favors hydrophobic amino acid residues in P1 and P1′ (Tanaka *et al.*, 1977), but it also accepts Lys in P1, which leads to activation

of trypsinogen at acidic pH (Gabeloteau & Desnuelle, 1960; Abita *et al.*, 1969). The fungal aspartic or acidic proteinases from *Aspergillus*, *Penicillium* (Chapter 295), *Mucor* (Chapter 298) and *Rhizopus* (Chapter 297) have similar specificities towards the tetradecapeptide renin substrate (Majima *et al.*, 1988). Pepsin (Chapter 272), on the other hand, has a different specificity on this substrate (Majima *et al.*, 1988).

Aspergillopepsin I from *A. sojae*, a soy sauce-producing fungus, hydrolyzes primarily at two bonds in the insulin B chain: Leu↓Tyr16 and Phe↓Phe25 (Ichishima *et al.*, 1982):

FVNQHLCGSHLVEAL↓YLVCGERGF↓FYTPKA

Minor cleavage of Tyr16↓Leu17 was also noted. This enzyme favors hydrophobic amino acid residues in P1, P1′ and P2′. Hydrolysis of native undenatured insulin by aspergillopepsin I from *A. sojae* is again at the same two bonds Tyr-Leu and Phe-Phe of the B chain, which are located on the surface of native insulin monomer at pH 3.2, and, secondary cleavage is found in three buried bonds: Leu15↓Tyr16, Ala14↓Leu15 (B chain) and Gln15↓Leu16 (A chain; Ichishima *et al.*, 1982). The pH optimum of aspergillopepsin I from *A. saitoi* is about 2.7 (Yoshida, 1956; Ichishima, 1970; Tanaka *et al.*, 1977).

Aspergillopepsin F from *A. fumigatus*, a pathogenic fungus for human aspergillosis, has an elastolytic activity with elastin Congo red at pH 5 (Lee & Kolattukudy, 1995).

Aspergillopepsin I and other fungal aspartic proteinases are distinct from the mammalian aspartic proteases in their ability to cleave substrates with lysine in the P1 position. Aspergillopepsin I from *A. saitoi* is capable of activating trypsinogen and chymotrypsinogen A at pH 4–4.5 (Gabeloteau & Desnuelle, 1960; Abita *et al.*, 1969). Aspergillopepsin I does not hydrolyze Z-Glu-Tyr or Z-Tyr-Leu (Ichishima, 1972), and does not clot milk (Ichishima, 1970).

## Structural Chemistry

Aspergillopepsin I from *A. saitoi* consists of 325 amino acid residues with an $M_r$ of 34 302 (Shintani & Ichishima, 1994). Although a proenzyme for aspergillopepsin I has not been isolated, the cDNA structure of aspergillopepsin I suggests the existence of a proenzyme (Shintani & Ichishima, 1994). A recombinant experiment shows that activation of the proenzyme of aspergillopepsin I takes place in the medium at a pH less than 5 after expression of the proenzyme in *Saccharomyces cerevisiae* cells (Shintani & Ichishima, 1994). All mammalian aspartic proteinases – pepsin, renin (Chapter 284) and chymosin (Chapter 274) – are synthesized as proenzymes and are subsequently activated.

Aspergillopepsin I from *A. saitoi* is devoid of a complete α helical strand, as judged from the ultraviolet optical rotatory dispersion curves in the 198 and 233 nm spectral zones (Ichishima & Yoshida, 1966, 1967a,b). The infrared result indicates that the deuterium-exchanged aspergillopepsin I exists in the antiparallel β structure (Ichishima & Yoshida, 1966). It is assumed that the characteristic property of acid stability of aspergillopepsin I partially depends upon the backbone conformation of its polypeptide chain (Ichishima & Yoshida, 1967a,b).

The two catalytically essential carboxyl groups, Asp32 and Asp214, of aspergillopepsin I from *A. saitoi* seem to connect to one another by a complex network of hydrogen bonds as in pepsin. The third active site residue, Asp76, is essential for trypsinogen activation at pH 3–4.5 (Shintani & Ichishima, 1994; Shintani *et al.*, 1996). Substitution of Asp76 by Ser or Thr and deletion of Ser78, as in the mammalian aspartic proteinases, pepsin and cathepsin D, causes a drastic decrease in the activities towards substrates containing a basic amino acid residue at P1. In contrast, substrates with hydrophobic residues at P1 are effectively hydrolyzed by each mutant enzyme (Shintani *et al.*, 1996). These results demonstrate that Asp76 and Ser78 residues on the active-site flap play important roles in the recognition of basic amino acid residue at the P1 position (Shintani *et al.*, 1996).

Using the *S. cerevisiae* expression system, Tyr75 on the flap of mucorpepsin (Chapter 298) has been shown to enhance catalytic efficiency of this fungal aspartic proteinase (Beppu *et al.*, 1995).

The *pepO* gene encoding the aspartic proteinase, aspergillopepsin O (PEPO) from *A. oryzae* (Berka *et al.*, 1993), is strikingly similar to the aspergillopepsin A gene *pepA* from *A. niger* var. *awamori* (previously called *A. awamori*) in that both are composed of four exons and three introns with virtually identical lengths, and the positions of the introns are exactly conserved (Berka *et al.*, 1990). From the deduced amino acid sequence it appears that PEPO, like other fungal aspartic proteinases, is synthesized as a zymogen containing a putative N-terminal prepro region of 77 amino acids followed by a mature protein of 327 amino acids.

The aspergillopepsin I gene *apnS* of 1340 bp from *A. saitoi* consists of four exons and three introns (Shintani & Ichishima, 1994). Exon sizes are 320, 278, 249 and 338 bp, and introns are 51, 52 and 52 bp. All intron–exon junction sequences are compatible with the GT...AG rule (Sharp, 1981). Exon 1 encodes the signal sequence and part of the mature enzyme including the active-site Asp32 residue. The predicted amino acid sequence of aspergillopepsin I from *A. saitoi* includes two homologous peptide sequences corresponding to those reported for human pepsin (Sogawa *et al.*, 1983) and containing the two active-site Asp residues. The predicted amino acid sequence of aspergillopepsin I is 32%, 69% and 34%, identical with human pepsin (Sogawa *et al.*, 1983), penicillopepsin (Hsu *et al.*, 1977; James & Sielecki, 1983) and rhizopuspepsin (Suguna *et al.*, 1987a,b), respectively.

Aspergillopepsin A from *A. awamori* is 52% identical to endothiapepsin (Chapter 296) (Choi *et al.*, 1993). The nucleotide sequence of the coding region of the proctase B cDNA from *A. niger* var. *macrosporus* also shares high identity with that of the genes for aspergillopepsins I from *A. awamori* (98%) and *A. saitoi* (95%; Lu *et al.*, 1995). Moreover, the upstream and the downstream untranslated sequences of the cDNA for proctase B are 100% and 93%, respectively, identical with the corresponding sequences in the genomic DNA of *A. awamori* pepsin (Berka *et al.*, 1990). The upstream noncoding sequence in the cDNA for proctase B is also identical with that in the genomic DNA coding for *A. saitoi* pepsin (Shintani & Ichishima, 1994), although their downstream noncoding sequences are not so similar.

The prepro form of *A. oryzae* pepsin, however, is rather different in amino acid sequence from those of the above three aspergillopepsins; there are 9, 28 and 91 residue differences in the signal sequences, pro segments and enzyme parts, respectively, compared with the sequence of preproproctase B (Lu *et al.*, 1995). The nucleotide sequence of the coding region of the *A. oryzae* pepsin gene is 69% identical with that of the proctase B cDNA. Moreover, neither the upstream nor downstream noncoding sequence in the cDNA for proctase B is homologous with those in the genomic DNA for *A. oryzae* pepsin (Berka *et al.*, 1993). The results are consistent with the fact that *A. awamori*, *A. niger* var. *macrosporus* and *A. saitoi* are genetically very close to one another, while *A. oryzae* is more distant. The codon usage, however, is similar among all these aspergillopepsins I.

The nucleotide and amino acid sequences of the prepro-enzyme of the extracellular aspartic proteinase (PEP) from *A. fumigatus* are 70% and 71% identical to the corresponding sequences of the aspergillopepsin from *A. niger* var. *awamori* (Reichard *et al.*, 1995). The amino acid sequence of aspergillopepsin F from *A. fumigatus* is 70, 60 and 67% identical to the sequences of those from *A. oryzae* (Berka *et al.*, 1993), *A. awamori* (Berka *et al.*, 1990), and *A. saitoi* (Shintani & Ichishima, 1994), respectively. The active-site motif (DTG) and the catalytic aspartic residue characteristic of aspartic proteinases are found in these enzymes.

Since the three-dimensional structure of aspergillopepsin I is believed to be similar to that of penicillopepsin (Chapter 295), five of the six and six of the seven differences in amino acids between proctase B and *A. awamori* pepsin and between proctase B and *A. saitoi* pepsin, respectively, are presumed to be in the β sheet V or in the loops between β sheets III and V of the C-terminal domain (James & Seilecki, 1983). Further, there are ten residue differences between *A. awamori* pepsin and *A. saitoi* pepsin. Eight of these ten differences are also presumed to be in the β sheet V or in a loop between β sheets III and V. Therefore some differences in their properties seem partly attributable to the structural differences in or around the β sheet V (Lu *et al.*, 1995).

The polypeptide chain of penicillopepsin folds via an 18-stranded mixed β sheet into two distinct lobes separated by a 30 Å long groove, which forms the extended substrate-binding site. The catalytic residues Asp32 and Asp215 are located in this groove and their carboxyl groups are in intimate contact (Hsu *et al.*, 1977). The structure of rhizopuspepsin has been refined to a crystallographic R-factor of 0.143 at 1.8 Å resolution (Suguna *et al.*, 1987a). The active aspartate residues, Asp35 and Asp218, are involved in similar hydrogen-bonding interactions with neighboring residues and with several water residues in a tightly hydrogen-bonded position. The refinement shows an unambiguous interaction of the highly mobile 'flap', a β hairpin loop region that projects over the binding pocket and makes it easily accessible. The three-dimensional structure of rhizopuspepsin closely resembles that of other aspartic proteinase structures. A detailed comparison with the structure of penicillopepsin showed striking similarities as well as subtle differences in the active site geometry and molecular packing.

The molecular structure of pig pepsinogen at 1.8 Å resolution has been determined by a combination of molecular replacement and multiple isomorphous phasing techniques

(James & Sielecki, 1986; Sielecki *et al.*, 1991). The resulting structure was refined by restrained-parameter least-squares methods. For the enzyme portion of the zymogen, the root-mean-square difference in C$\alpha$ atom coordinates with the refined pig pepsin structure is 0.90 Å (284 common atoms) and with the C$\alpha$ atoms of penicillopepsin it is 1.63 Å (275 common atoms). The additional 44 N-terminal amino acids of the pro segment (Leu1p-Leu44p, using the letter p after the residue number to distinguish the residues of the pro segment) adopt a relatively compact structure consisting of a long $\beta$ strand followed by two approximately orthogonal $\alpha$ helices and a short 3(10) helix. Intimate contacts, both electrostatic and hydrophobic interactions, are made with residues in the pepsin-active site. Electrostatic interactions of Lys36pN and hydrogen-bonding interactions of Tyr37pOH and Tyr9OH with the two catalytic aspartate groups, Asp32 and Asp215, prevent substrate access to the active site of the zymogen.

## Preparation

Crude enzyme preparations from *A. saitoi* are commercially available from Kikkoman, Seisin Pharmaceutical, Sigma (see Appendix 2 for full names and addresses of suppliers). (The Sigma product is listed as Protease XIII: fungal.)

The recombinant proenzyme of aspergillopepsin I was expressed in *S. cerevisiae* cells (Shintani & Ichishima, 1994). After 12 h, the activation of the proenzyme of aspergillopepsin I took place in the medium at a pH of approximately 5. The recombinant enzyme was also expressed in *Escherichia coli* cells (Shintani *et al.*, 1996). The inclusion bodies were solubilized with 8 M urea, and then dialyzed at alkaline pH. The active proenzyme was purified to homogeneity by two chromatography steps on a TOYO-PEARL HW-55F column from Tosoh and a RESOURCE Q column from Pharmacia. The N-terminal sequence of the recombinant aspergillopepsin I, which was expressed in *E. coli* cells, prepared from crude proaspergillopepsin I and acidified at pH 2.7 for 12 h at 25°C, was found to be Glu-Ala-Ala-Ser-Lys-Gly-Ser-Ala- (Shintani *et al.*, 1996). This sequence is three residues longer than that of native aspergillopepsin I from *A. saitoi* (Shintani & Ichishima, 1994).

Expression of calf chymosin (Chapter 274), a milk-clotting aspartic proteinase, has been reported from the filamentous fungus *A. oryzae* (Tsuchiya *et al.*, 1993).

## Biological Aspects

Growth of imperfect fungus *Aspergillus* depends upon having both cell wall-bound and extracellular proteolytic enzymes to satisfy requirements for nutrition, because the fungi feed entirely by absorption, not by photosynthesis or ingestion. Aspergillopepsins are extracellular proteinases secreted by fungal mycelia. The enzymes are of practical importance for fungal nutritional in an acidic environment. Aspergillopepsin I from *A. saitoi* shows two forms of activity at acidic pH, pepsin-like catalytic function and trypsinogen-activating activity like enteropeptidase (Chapter 14) (Shintani & Ichishima, 1994; Shintani *et al.*, 1996).

The major extracellular protease from *A. niger*, which is responsible for 80–85% of the total activity, is aspergillopepsin A, a protein of approximately 43 kDa, the activity of which is inhibited by pepstatin (Mattern *et al.*, 1992). The gene for aspergillopepsin A (*pepA*) is located on chromosome 1 of *A. niger*.

Hydrolysis of structural proteins in the lung by extracellular proteinases secreted by *A. fumigatus* is thought to play a significant role in invasive aspergillosis. The fungus secretes not only elastolytic aspartic proteinase (aspergillopepsin F) but also an elastinolytic serine proteinase, oryzin (Chapter 105) and a metalloproteinase (Chapter 514) (Lee & Kolattukudy, 1995). Aspergillopepsin F can catalyze hydrolysis of the major structural proteins of basement membrane, elastin, collagen and laminin. Immunogold electron microscopy showed that the aspartic proteinase was secreted by *A. fumigatus* invading neutrogenic mouse lung and its secretion was directed towards the germ tubes of penetrating hyphae (Lee & Kolattukudy, 1995).

## Economic Features

*Aspergillus* and *Penicillium* are two of the most economically important genera of fungi (Pitt & Samson, 1990). Much of their economic impact is deleterious, with food spoilage, mycotoxin production and biodeterioration, but their potential for economic use is equally great. Over the past 1200 years the use of hydrolytic enzymes from microorganisms has become more prevalent in traditional Japanese fermentation industries. The mold *Aspergillus* has great practical importance in these fermentation industries and in Japan's culture as a whole (Ichishima, 1986). The extent of the proteolytic breakdown in *Aspergillus* foods such as soy sauce (shoyu) (Fukushima, 1985), Japanese seasoning miso paste (Shurtleff & Aoyagi, 1976, 1977) and Japanese sake or rice wine (Murakami, 1972; Hayashida *et al.*, 1972) can be judged from the production of free amino acids that contribute the 'Umani taste' (O'Mahony & Ishii, 1987) and the flavors typical of Japanese traditional fermented foods (Ichishima, 1989). For the production of soy sauce, miso, sake, shochu and awamori, *A. sojae* and *A. oryzae*, *A. oryzae*, *A. oryzae*, *A. saitoi* and *A. awamori* are used in Japanese industries, respectively. Three species, *A. awamori*, *A. niger* and *A. saitoi*, have black conidia and also have powerful activity for citric acid fermentation.

## Distinguishing Features

Secretory aspartic proteinase from *Candida albicans* (candidapepsin, EC 3.4.23.24) (Chapter 300) (Hube *et al.*, 1991), three distinct secreted aspartic proteinases in *C. albicans* (White *et al.*, 1993), and a fourth secreted aspartic proteinase gene (SAP4) from *C. albicans* (Miyasaki *et al.*, 1994) were isolated and characterized. Aspartic proteinases of the opportunistic pathogen *Candida* species (candidapepsins) have been found to be the major virulence factors in candidiasis (Ross *et al.*, 1990). The proteinase-deficient mutants of *C. albicans* generated by chemical mutagenesis have been shown to be much less virulent than the parent strain, indicating an association between extracellular proteinase and virulence (Kwon-Chung *et al.*, 1985).

*Scytalidium lignicolum* produces three types of acid (carboxyl) proteases: A-1, A-2 (Chapter 328) and B (Chapter 326) (Murao *et al*., 1972). The former two enzymes are insensitive to pepstatin, DAN and 1,2-epoxy-3-(4-azido-2-nitrophenoxy)propane (EPNP), formerly 1,2-epoxy-3-(*p*-nitrophenoxy)propane, all of which are known to be inhibitors specific for aspartic proteinases. The acid protease B is also insensitive to pepstatin and DAN, but is inhibited by EPNP. The complete amino acid sequence of protease B with 204 amino acid residues and an $M_r$ of 21 969 has been established and compared with those of other aspartic proteinases (Maita *et al*., 1984). Unlike the other carboxyl proteinases, one of the catalytic residues of this enzyme is a glutamic acid. The active amino acid residue modified with EPNP was found to be Glu53 (Tsuru *et al*., 1989a). The other catalytic residue was found to be Asp98 by using a new inhibitor, 1-diazo-3-phenyl-2-propanone (Tsuru *et al*., 1989b). The amino acid sequence around Glu53 shows a high similarity with those around the active site Asp215 residue of pig pepsin (Chapter 272) and calf chymosin (Chapter 274). Although the amino acid sequence around Asp98 also shows a high identity with those around the active-site Asp32, there is an insertion of a serine residue between Thr and Gly. This was the first demonstration of a glutamic proteinase.

Acid carboxypeptidases (serine-type carboxypeptidases; EC 3.4.16.1; Chapter 132) from *A. saitoi* and *A. oryzae* were purified from the culture filtrates (Ichishima, 1972). The mode of action and application of *Aspergillus* carboxypeptidase were summarized in a review (Ichishima, 1991). With the aid of the Ouchterlony double immunodiffusion technique good cross-reactions were determined between antiserum to *A. saitoi* acid carboxypeptidase and acid carboxypeptidases from *A. aureus* IAM 2337, *A. awamori* var. *acidus* IAM 2279, *A. inuii* IAM 2258, *A. nakazawai* IAM 2293, *A. niger* IFO 6661 and *A. usamii* IAM 2185, which are strains belonging to black aspergilli (Ushijima *et al*., 1979). The data indicate that acid carboxypeptidases from *A. saitoi, A. aureus, A. awamori* var. *acidus, A. inuii, A. nakazawai, A. niger* and *A. usamii* are immunologically similar, whereas those from *A. oryzae* IAM 2640, *A. oryzae* var. *magnasporus* IAM 2620 and *A. sojae*, which are strains belonging to yellow and green aspergilli, have different immunogenicity from those of black aspergilli.

## Further Reading

For further reading, see Ichishima (1970) and Tang & Wong (1987). For an extensive review of pepstatin-insensitive proteinase, see Takahashi (1995).

## References

Abita, J.P., Delaage, M., Lazdunski, M. & Savrda, J. (1969) The mechanism of activation of trypsinogen. The role of the four N-terminal aspartyl residues. *Eur. J. Biochem.* 8, 314–324.

Beppu, T., Young-Nam, P., Aikawa, J., Nishiyama, M. & Horinouchi, S. (1995) Tyrosine 75 on the flap contributes to enhanced catalytic efficiency of a fungal aspartic proteinase, *Mucor pusillus* pepsin. *Adv. Exp. Med. Biol.* 362, 501–509.

Berka, R.M., Ward, M., Wilson, L.J., Hayenga, K.J., Kodama, K.H., Carlomagno, L.P. & Thompson, S.A. (1990) Molecular cloning and deletion of the gene encoding aspergillopepsin A from *Aspergillus awamori* (published erratum appears in *Gene* (1990) 96, 313). *Gene* 86, 153–162.

Berka, R.M., Carmona, C.L., Hayenga, K.J., Thompson, S.A. & Ward, M. (1993) Isolation and characterization of the *Aspergillus oryzae* gene encoding aspergillopepsin O. *Gene* 125, 195–198.

Chang, W.-J., Horiuchi, S., Takahashi, K., Yamasaki, M. & Yamada, Y. (1976) The structure and function of acid proteases. VI. Effects of acid protease-specific inhibitors on the acid proteases from *Aspergillus niger* var. *macrosporus. J. Biochem.* 80, 975–981.

Choi, G.H., Pawlyk, D.M., Rae, B., Shapira, R. & Nuss, D.L. (1993) Molecular analysis and overexpression of the gene encoding endothiapepsin, an aspartic protease from *Cryphonectria parasitica. Gene* 125, 135–141.

Crewther, W.C. & Lennox, F.G. (1950) Preparation of crystals containing protease from *Aspergillus oryzae. Nature* 165, 680.

Davidson, R., Gertler, A. & Hofmann, T. (1975) *Aspergillus oryzae* acid proteinase. Purification and properties, and formation of $\pi$-chymotrypsin. *Biochem. J.* 147, 45–53.

Fukushima, D. (1985) Fermented vegetable protein and related foods of Japan and China. *Food Rev. Int.* 1, 149–209.

Gabeloteau, C. & Desnuelle, P. (1960) On the activation of beef trypsinogen by a crystallized proteinase of *Aspergillus saitoi. Biochim. Biophys. Acta* 42, 230–237.

Gomi, K., Arikawa, K., Kamiya, N., Kitamoto, K. & Kumagai, C. (1993) Cloning and nucleotide sequence of the acid protease-encoding gene (*pepA*) from *Aspergillus oryzae. Biosci. Biotechnol. Biochem.* 57, 1095–1100.

Hayashida, S., Kamachi, T. & Hongo, M. (1972) Semi-continuous sake fermentation. In: *Fermentation Technology Today* (Terui, G., ed.). Osaka: Soc. Ferm. Technol., pp. 645–649.

Hsu, I.-N., Delbaere, L.T., James, M.N. & Hofmann, T. (1977) Penicillopepsin from *Penicillium janthinellum*: crystal structure at 2.8 Å and sequence homology with porcine pepsin. *Nature* 266, 140–145.

Hube, B., Turver, C.J., Odds, F.C., Eiffert, H., Boulnis, G.J., Kochel, H. & Ruchel, R. (1991) Sequence of the *Candida albicans* gene encoding the secretory aspartate proteinase. *J. Med. Vet. Mycol.* 29, 129–132.

Ichishima, E. (1970) Purification and mode of assay for acid proteinase of *Aspergillus saitoi. Methods Enzymol.* 19, 397–406.

Ichishima, E. (1972) Purification and characterization of a new type of acid carboxypeptidase from *Aspergillus. Biochim. Biophys. Acta* 258, 274–288.

Ichishima, E. (1986) *Aspergillus* enzymes for developing Japanese original biosciences. *Nippon Jozo Gakkaishi* (In Japanese) 81, 756–763, 844–853.

Ichishima, E. (1989) Japanese traditional fermented foods. *Invitation to Fermented Foods: From Food Civilization to New Biotechnology* (In Japanese). Tokyo: Shyokabo, pp. 39–47, 127–143.

Ichishima, E. (1991) Mode of action and application of *Aspergillus* carboxypeptidase. *Comments Agric. Food Chem.* 2, 279–298.

Ichishima, E. & Yoshida, F. (1965a) Molecular weight of acid proteinase of *Aspergillus saitoi. Nature* 207, 525–526.

Ichishima, E. & Yoshida, F. (1965b) Chromatographic purification and physical homogeneity of acid proteinase of *Aspergillus saitoi. Biochim. Biophys. Acta* 99, 360–366.

Ichishima, E. & Yoshida, F. (1966) Conformation of aspergillopeptidase A in aqueous solution. *Biochim. Biophys. Acta* 128, 130–135.

Ichishima, E. & Yoshida, F. (1967a) Ultraviolet optical rotatory dispersion of aspergillopeptidase A. *Agric. Biol. Chem.* 31, 507–510.

Ichishima, E. & Yoshida, F. (1967b) Conformation of aspergillopeptidase A in aqueous solution. II. Ultraviolet optical rotatory dispersion of aspergillopeptidase A. *Biochim. Biophys. Acta* **147**, 341–346.

Ichishima, E., Majima, E., Emi, M., Hayashi, K. & Murao, S. (1981) Enzymatic cleavage of the histydyl-prolyl bond of proangiotensin by carboxyl proteinases from *Aspergillus sojae* and *Scytalidium lignicolumn. Agric. Biol. Chem.* **45**, 2391–2393.

Ichishima, E., Emi, M., Majima, E., Mayumi, Y., Kumagai, H., Hayashi, K. & Tomoda, K. (1982) Initial sites of insulin cleavage and stereospecificity of carboxyl proteinase from *Aspergillus sojae* and *Pycnoporus coccineus. Biochim. Biophys. Acta* **700**, 247–253.

James, M.N.G. & Sielecki, A.R. (1983) Structure refinement of penicillopepsin at 1.8 Å resolution. *J. Mol. Biol.* **163**, 299–361.

James, M.N.G. & Sielecki, A.R. (1986) Molecular structure of an aspartic proteinase zymogen, porcine pepsinogen, at 1.8 Å resolution. *Nature* **319**, 33–38.

Jarai, G. & Buxton, F. (1994) Nitrogen, carbon, and pH regulation of extracellular acidic protease of *Aspergillus niger. Curr. Genet.* **26**, 238–244.

Jarai, G., Van den Homberg, H. & Buxton, F.P. (1994) Cloning and characterization of the gene of *Aspergillus niger* encoding a new aspartic protease and regulation of *pepE* and *pepC. Gene* **145**, 171–178.

Kwon-Chung, K.J., Lehman, D., Good, C. & Marge, P.T. (1985) Genetic evidence for role of extracellular proteinase in virulence of *Candida albicans. Infect. Immun.* **49**, 571–575.

Lee, J.D. & Kolattukudy, P.E. (1995) Molecular cloning of the cDNA and gene for an elastolytic aspartic proteinase from *Aspergillus fumigatus* and evidence of its secretion by the fungus during invasion of the host lung. *Infect. Immun.* **63**, 3796–3803.

Lu, J.-F., Inoue, H., Kimura, T., Makabe, O. & Takahashi, K. (1995) Molecular cloning of a cDNA for proctase from *Aspergillus niger* var. *macrosporus* and sequence comparison with other aspergillopepsins I. *Biosci. Biotechnol. Biochem.* **59**, 954–955.

Maita, T., Nagata, S., Matsuda, G., Murata, S., Oda, K., Murao, S. & Tsuru, D. (1984) Complete amino acid sequence of *Scytalidium lignicolum* acid protease B. *J. Biochem.* **95**, 465–475.

Majima, E., Oda, K., Murao, S. & Ichishima, E. (1988) Comparative study on the specificities of several fungal aspartic and acidic proteinases towards the tetradecapeptide of a renin substrate. *Agric. Biol. Chem.* **52**, 787–793.

Mattern, I.E., van Noort, J.M., van den Berg, P., Archer, D.B., Roberts, I.N. & van den Hondel, C.A. (1992) Isolation and characterization of mutants of *Aspergillus niger* deficient in extracellular proteases. *Mol. Genet.* **234**, 332–336.

Miyasaki, S.H., White, T.C. & Agabian, N. (1994) A fourth secreted proteinase gene (*SAP4*) and a *CARE2* repetitive element are located upstream of the *SAP1* gene in *Candida albicans. J. Bacteriol.* **176**, 1702–1710.

Morihara, K. & Oka, T. (1973) Comparative specificity of microbial acid proteinases for synthetic peptides. III. Relationship with their trypsinogen activating ability. *Arch. Biochem. Biophys.* **157**, 561–572.

Murakami, H. (1972) Some problems in sake brewing. In: *Fermentation Technology Today* (Terui, G., ed.). Osaka: Soc. Ferm. Technol., pp. 639–643.

Murao, S., Oda, K. & Matsushita, Y. (1972) New acid proteases from *Scytalidium lignicolum* M-133. *Agric. Biol. Chem.* **36**, 1647–1650.

O'Mahony, M. & Ishii, R. (1987) The umami concept: implications for the dogma of four basic tastes. In: *Umami: A Basic Taste* (Kawamura, Y. & Kare, R., eds). New York: Marcel Dekker, pp. 75–93.

Ostoslavskaya, V.I., Kotlova, E.K., Stepanov, V.M., Rudenskaya, G.H., Baratova, L.A. & Belyanova, L.P. (1976) Aspergillopepsin F – a carboxylic proteinase from *Aspergillus foetidus. Bioorg. Khim.* **5**, 595–603.

Ostoslavskaya, V.I., Revina, L.P., Kotlova, E.K., Surova, L.A., Levin, E.D., Timokhina, E.A. & Stepanov, V.M. (1986) The primary structure of aspergillopepsin A, aspartic proteinase from *Aspergillus awamori*. IV. Amino acid sequence of the enzyme. *Bioorg. Khim.* **12**, 1030–1047.

Pitt, J.I. & Samson, R.A. (1990) Systematics of *Penicillium* and *Aspergillus* – past, present and future. In: *Modern Concepts in Penicillium and Aspergillus Classification* (Samson, R.A. & Pitt, J.I., eds). New York: Plenum Press, pp. 3–13.

Reichard, U., Monod, M. & Ruchel, R. (1995) Molecular cloning and sequencing of the gene encoding an extracellular aspartic proteinase from *Aspergillus fumigatus. FEMS Microbiol. Lett.* **130**, 69–74.

Reichelt, D., Jacobsohn, E. & Haschen, R.J. (1974) Purification and properties of cathepsin D from human erythrocytes. *Biochim. Biophys. Acta* **341**, 15–26.

Ross, I.K., De Bernardis, F., Emerson, G.W., Cassone, A. & Sullivan, P.A. (1990) The secreted aspartate proteinase of *Candida albicans*: physiology of secretion and virulence of a proteinase-deficient mutant. *J. Gen. Microbiol.* **136**, 687–694.

Sharp, P.A. (1981) Speculations on RNA splicing. *Cell* **23**, 643–646.

Shintani, T. & Ichishima, E. (1994) Primary structure of aspergillopepsin I deduced from nucleotide sequence of the gene and aspartic acid-76 in an essential active site of the enzyme for trypsinogen activation. *Biochim. Biophys. Acta* **1204**, 257–264.

Shintani, T., Kobayashi, M. & Ichishima, E. (1996) Characterization of the S1 subsite specificity of aspergillopepsin I by site-directed mutagenesis. *J. Biochem.* **120**, 974–991.

Shurtleff, W. & Aoyagi, A. (1976) Miso. In: *The Book of Miso*. Hayama, Japan: Autumn Press Inc., pp. 15–44.

Shurtleff, W. & Aoyagi, A. (1977) Miso. In: *The Book of Miso: Miso Production*. Hayama, Japan: Autumn Press Inc., pp. 5–55.

Sielecki, A.R., Fujinaga, M., Read, R.J. & James, M.N.G. (1991) Refined structure of porcine pepsinogen at 1.8 Å resolution. *J. Mol. Biol.* **219**, 671–692.

Sogawa, K., Fujii-Kuriyama, Y., Mizukami, Y., Ichihara, Y. & Takahashi, K. (1983) Primary structure of human pepsinogen gene. *J. Biol. Chem.* **258**, 5306–5311.

Suguna, K., Bott, R.R., Padlan, E.A., Subramanian, E., Sheriff, S., Cohen, G.H. & Davies, D.R. (1987a) Structure and refinement at 1.8 Å resolution of the aspartic proteinase from *Rhizopus chinensis. J. Mol. Biol.* **196**, 877–900.

Suguna, K., Padlan, E.A., Smith, C.W., Carlson, W.D. & Davies, D.R. (1987b) Binding of a reduced peptide inhibitor to the aspartic proteinase from *Rhizopus chinensis*: implication for a mechanism of action. *Proc. Natl Acad. Sci. USA* **84**, 7009–7013.

Takahashi, K. (1995) Proteinase A (aspergillopepsin II, EC 3.4.23.19) from *Aspergillus niger. Methods Enzymol.* **248**, 146–155.

Takeuchi, M., Ogura, K., Hamamoto, K. & Kobayashi, Y. (1995) Molecular cloning and sequence analysis of a gene encoding an aspartic proteinase from *Aspergillus oryzae. Adv. Exp. Med. Biol.* **362**, 577–580.

Tang, J. & Wong, R.N. (1987) Evolution in the structure and function of aspartic protease. *J. Cell Biochem.* **33**, 53–63.

Tanaka, N., Takeuchi, M. & Ichishima, E. (1977) Purification of an acid proteinase from *Aspergillus saitoi* and determination of peptide bond specificity. *Biochim. Biophys. Acta* **485**, 406–416.

Thompson, S.A. (1990) Molecular cloning and deletion of the gene encoding aspergillopepsin A from *Aspergillus awamori. Gene* **96**, 313.

Tsuchiya, K., Gomi, K., Kitamoto, K., Kumagai, C. & Tamura, G. (1993) Secretion of calf chymosin from the filamentous fungus *Aspergillus oryzae. Appl. Microbiol. Biotechnol.* **40**, 327–332.

Tsuru, D., Naotsuka, A., Kobayashi, R., Yoshimoto, T., Oda, K. & Murao, S. (1989a) Inactivation of *Scytalidium lignicolum* acid protease B with 1,2-epoxy-3-(4-azido-2-nitrophenoxy)propane. *Agric. Biol. Chem.* **53**, 2751–2756.

Tsuru, D., Kobayashi, R., Nakagawa, N. & Yoshimoto, T. (1989b) Inhibition of *Scytalidium lignicolumn* acid protease B by 1-diazo-3-phenyl-2-propanone. *Agric. Biol. Chem.* **53**, 1305–1312.

Ushijima, T., Takeuchi, M. & Ichishima, E. (1979) Comparative immunological study of acid carboxypeptidases in genus *Aspergillus. Agric. Biol. Chem.* **43**, 859–860.

White, T.C., Miyasaki, S.H. & Agabian, N. (1993) Three distinct secreted aspartyl proteinases in *Candida albicans. J. Bacteriol.* **175**, 6126–6133.

Yagi, F., Fan, J., Terada, K. & Kobayashi, A. (1986) Purification and characterization of carboxyl proteinase from *Aspergillus kawachii. Agric. Biol. Chem.* **50**, 1029–1033.

Yoshida, F. (1956) Studies on the proteolytic enzymes of black aspergilli. Part I. Investigation of strains producing proteinase yields a black *Aspergillus* and the crystallization of proteolytic enzymes from *Aspergillus saitoi. Bull. Chem. Soc. Jpn* **20**, 252–256.

***Eiji Ichishima***
*Laboratory of Molecular Enzymology,*
*Department of Applied Biological Chemistry,*
*Faculty of Agriculture, Tohoku University,*
*1-1, Tsutsumidori-Amamiyamachi,*
*Aoba-ku, Sendai 981 Japan*
*Email: ichisima@t.soka.ac.jp*

# 295. *Penicillopepsin*

## Databanks

*Peptidase classification: clan AA, family A1, MEROPS ID: A01.011*
*NC-IUBMB enzyme classification: EC 3.4.23.20*
*ATCC entries: CMI 75589 culture*
*Chemical Abstracts Service registry number: 9074-08-2*
*Databank codes:*

| Species | SwissProt | PIR | EMBL (cDNA) | EMBL (genomic) |
|---|---|---|---|---|
| Penicillopepsin | | | | |
| *Penicillium janthinellum* | P00798 | A00991 A38008 | – | – |
| Penicillopepsin-JT2 | | | | |
| *Penicillium janthinellum* | – | – | – | U81483: complete gene |

Brookhaven Protein Data Bank three-dimensional structures:

| Species | ID | Resolution | Notes |
|---|---|---|---|
| *Penicillium janthinellum* | 1APT | 1.8 | complex with isoval-Val-Val-Lysta-OEt |
| | 1APU | 1.8 | complex with isoval-Val-Val-Sta-OEt |
| | 1APV | 1.8 | complex with isoval-Val-Val-hydrated difluorostatone-*N*-methylamide |
| | 1APW | 1.8 | complex with isoval-Val-Val-difluorostatone-*N*-methylamide |
| | 1PPK | 1.8 | complex with isoval-Val-Val-StaP-OEt |
| | 1PPL | 1.7 | complex with isoval-Val-Val-LeuP-O(Phe)-OMe |
| | 1PPM | 1.7 | complex with CBZ-Ala-Ala-LeuP-(O)Phe-OMe |
| | 3APP | 1.8 | |

## Name and History

A proteolytic enzyme with the ability to activate trypsinogen at low pH (pH optimum 3.4) obtained from an unidentified species of a filamentous imperfect fungus of the genus *Penicillium* was first described by Kunitz (1938) who named it **mold kinase**. Twenty years later as part of a study of the mechanism of trypsinogen activation we used an enzyme from *Penicillium janthinellum* (Hofmann, 1960) whose properties were similar to those of Kunitz's enzyme. We called the enzyme **penicillium kinase**. A similar enzyme from *Aspergillus oryzae* was named **trypsinogen kinase** by Nakanishi (1959). The purification and characterization of the enzyme from *Penicillium janthinellum* was later described by Hofmann & Shaw (1964) who renamed it **peptidase A**. This name was changed to **penicillopepsin** after it was found that a heptapeptide isolated after reaction with DAN, an active site-directed inhibitor, had a sequence that was nearly identical with a peptide similarly isolated from pig pepsin (Chapter 272). This suggested that the fungal enzyme was homologous to the mammalian enzyme (Sodek & Hofmann, 1970). The subsequent determination of the complete amino acid sequence (Hsu *et al.*, 1977) showed extensive similarity with that of pig pepsin (Sepulveda *et al.*, 1975). A comparison of the three-dimensional structure (Hsu *et al.*, 1977) with that of pig pepsin, which was determined later (Sielecki *et al.*, 1990; Cooper *et al.*, 1990; Abad-Zapatero *et al.*, 1990), provided clear evidence that the two enzymes were homologous.

During attempts to isolate the gene for penicillopepsin from a genomic library of *P. janthinellum*, we isolated genes for two other aspartic proteinases. The deduced amino acid sequence of one of them differed from the expected sequence by 29% (see below). We suggest therefore that a suffix be added to the name of the original penicillopepsin and that it be called **penicillopepsin-JT1** to distinguish it from these additional two proteases for which we have at present one complete and one partial DNA sequences. We propose the designations penicillopepsin-JT2 and penicillopepsin-JT3 for the proteins for which these genes code. Penicillopepsin-JT2 is described briefly below.

## Activity and Specificity

Penicillopepsin-JT1, a single-chain enzyme of 323 amino acyl residues ($M_r$ 33 422), rapidly activates cattle trypsinogen with a $k_{cat}$ of $420 s^{-1}$ by releasing the same activation peptide, Val-(Asp)$_4$-Lys-OH, as that produced by autocatalytic activation (Hofmann, 1960; Shaw & Hofmann, 1964). Penicillopepsin-JT1 hydrolyzes proteins with broad specificity, but shows preference for hydrophic residues at P1 and P1′, although it also cleaves the Gly20+Glu21 bond in the B chain of insulin (Mains *et al.*, 1971). It also has specificity for lysine in P1 because the ε-ammonium group of the lysyl side chain forms an ionic bond with the side-chain carboxylate of Asp77 (James *et al.*, 1985). This dual specificity is explained by the fact that the methylene groups of the lysyl side chain interact with the hydrophobic residues of the S1 binding site; the ε-ammonium group extends beyond it towards the carboxylate of Asp77 (James *et al.*, 1985). Replacement of Asp77 by threonine in rhizopuspepsin

(EC 3.4.23.21) (Chapter 297), which also has lysine specificity, showed that Asp77 was essential for the interaction of lysine with the enzyme (Lowther *et al.*, 1995). The lysine specificity has been used to study the large rate-enhancing effects of long substrates previously observed with pig pepsin (EC 3.4.23.1) by Fruton (1970) because substrates with a lysyl residue in P1 are very water-soluble unlike substrates for pepsin with two hydrophobic residues in P1 and P1′.

The analysis of the kinetic parameters of the hydrolysis of the series Ac-(Ala)$_m$-Lys+Nph-(Ala)$_n$-amide, where *m* and *n* range from 0 to 3, showed that the addition of alanyl residues in positions P3 and P2′ additively increased $k_{cat}$ about 40-fold each, whereas alanyl residues in the other positions caused only small changes in $k_{cat}$. In contrast, the $K_m$ values remained unchanged for the whole series (Hofmann *et al.*, 1988). Changes in the side chain of P3 affected both $k_{cat}$ and $K_m$ (Hofmann *et al.*, 1988). Similar results have more recently been obtained with endothiapepsin (EC 3.4.23.22) (Chapter 296), rhizopuspepsin (EC 3.4.23.21) and pig pepsin (Balbaa *et al.*, 1993). Occupation of subsite S3 also plays an important role in inhibitor binding to penicillopepsin-JT1. $K_i$ for the pepstatin analog Iva-Val-Sta-OEt, where Iva is isovaleryl and Sta is the γ-amino acid staline, is at least four orders of magnitude larger than that for Iva-Val-Val-Sta-OEt (Blum *et al.*, 1985). These findings suggest strongly that it is not the side chain of P3 that is responsible for the rate-enhancing effect, but the hydrogen bond that exists between the -NH- group of P3 and the OH group of the side chain of Thr217, which is the fourth residue on the C-terminal side of the active-site Asp213. There is a strong possibility that this is a common feature of all aspartic proteinases other than the retroviral ones. In all the known sequences of over 48 aspartic proteinases (listed by Fusek & Vetvicka, 1995) that position is occupied by either threonine or serine, except in those of *Mucor pusillus* and *Mucor miehei* mucorpepsins (Chapter 298) where it is an asparagine whose sidechain carbonyl oxygen can act as a hydrogen acceptor. Furthermore, $k_{cat}$ values for penicillopepsin-JT1 acting on substrates of the type X-Ala-Lys+Ala-Ala-amide, where X is an alanine analog in which the NH group has been replaced by a nonhydrogen bond donating group, such as $CH_3$-CO-$CH_2$-($CH_3$)CH-CO- or $CH_3$-CO-O-($CH_3$)CH-CO-, are significantly lower than that for the regular substrate, where X is Ac-Ala-. The rate enhancements of alanyl residues in P2′ and P3 are probably associated with conformational changes (Allen *et al.*, 1990).

The temperature dependence of peptides without alanine in both P2′ and P3 show linear Arrhenius plots with free energies of activation of $55 kJ mol^{-1}$. Peptides with alanine in P2′ but no substituent in P3, show a sharp break in the Arrhenius plot at 10.5°C and free energies of activation of $90 kJ mol^{-1}$ below the break and $54 kJ mol^{-1}$ above the break. For substrates in which P3 is occupied, the transition is at 14.2°C, with free energies of activation of $66 kJ mol^{-1}$ below and $26–39 kJ mol^{-1}$ above the transition. The breaks in the Arrhenius plots are not due to temperature-dependent changes in the rate constant, but are most probably associated with conformational changes. Evidence for this was obtained from the nonlinear temperature dependence of the 240 nm band in the circular dichroic spectrum of a penicillopepsin-JT1-pepstatin complex; the temperature dependence of the 240 nm band of the free enzyme was linear (Allen *et al.*, 1990).

Further evidence for the influence of binding of P3 in subsite S3 comes from a comparison of the solvent isotope effect on the hydrolysis of Ac-Lys┼Nph-amide (I) and Ac-(Ala)$_2$-Lys┼Nph-(Ala)$_2$-amide (II). The hydrolysis of I shows no solvent isotope effect. In contrast, the isotope effect of the hydrolysis of II has a value of 2.11. Its dependence upon the concentration of D$_2$O in H$_2$O is not linear. This indicates that two or more protons are involved in the rate-limiting step (Cunningham *et al.*, 1990). The absence of a solvent isotope effect with substrate I shows that the rate-limiting step does not involve a bond-breaking step. Detailed discussion of these experiments leads to the tentative conclusion that for substrate I the rate-limiting step occurs before the bond-breaking step(s) and is probably due to distortion of the scissile bond towards a tetrahedral configuration, whereas that for substrate II the rate-limiting step is likely to be a conformational change induced by the hydrogen-bond formation resulting from the occupation of subsites S2′ and S3, as well as other possible protonic reorganizations resulting from the associated conformational change (Cunningham *et al.*, 1990).

A comparison of the three-dimensional structures of endothiapepsin, with and without an inhibitor, shows that a 'rigid-body movement' of the domain comprising residues 190–302 is associated with binding of the inhibitor, which results in changes in the shape of the active-site cleft that are largest around the S3 pocket (Sali *et al.*, 1992). A superposition of the Cα backbones of six structures of other aspartic proteinases, including penicillopepsin-JT1, on those parts of their structures that do not include residues 190–302 (or their corresponding residues) shows that by far the largest differences across the aspartic proteinase family involve the 190–302 domain (Sali *et al.*, 1992).

It has been known for some time that penicillopepsin-JT1 and pig pepsin can catalyze transpeptidation reactions during which an N-terminal amino acid with a large hydrophobic side chain is transferred from a short peptide to form oligomers of itself (Wang & Hofmann, 1976; Antonov *et al.*, 1981). We subsequently studied the action of penicillopepsin-JT1 and of pig pepsin on Nph-Ala-Ala-amide (Blum *et al.*, 1991; Balbaa *et al.*, 1994). The major products in the early stages of the reaction are Ala-Ala-amide and the trimer of Nph (Nph$_3$). One of the major conclusions of these experiments is that the enzymes can noncovalently trap *p*-nitrophenylalanine and its dimer in such a way that they cannot exchange with the free components in the solvent. We propose that Nph and Nph$_2$ are tightly held in the active site by hydrogen bonds and by two strong electrostatic interactions (Blum *et al.*, 1991). The optimum rate of hydrolysis of Ac-(Ala)$_2$-Lys┼Nph-(Ala)$_2$-amide, the standard substrate used for routine assays, is at pH 4.5 (Hofmann *et al.*, 1984). Trypsinogen (EC 3.4.21.4) (Chapter 3) activation, as described by Hofmann (1976), provides a very sensitive assay for the penicillopepsins and also for endothiapepsin and rhizopuspepsin; as little as 5 ng ml$^{-1}$ of the enzymes can be measured accurately (Hofmann, 1976).

## Structural Chemistry

The three-dimensional structure of penicillopepsin-JT1 has been determined, initially at 2.8 Å resolution (Hsu *et al.*, 1977), and subsequently at 1.8 Å resolution (James & Sielecki, 1983). A mode of substrate binding was deduced from the X-ray analysis of complexes with the pepstatin analogs Iva-Val-Val-Sta-OEt and Iva-Val-Val-Lysta-OEt, where Iva is isovaleryl, and Lysta is the lysine analog of statine (James *et al.*, 1982; James & Sielecki, 1985). Deductions about the catalytic pathway for aspartic proteinases are based on structures of penicillopepsin with difluorostatine- and difluorostatone-containing peptides (James *et al.*, 1992) and with peptides, which contain phosphinyl groups as transition state analogs in place of the carbonyl groups of statine in the peptide Iva-Val-Val-StaP-OEt and of leucine in the peptide Iva-Val-Val-LeuP-OPhe-OMe (Fraser *et al.*, 1992). The pI of penicillopepsin-JT1 is less than 3.0.

## Preparation

The large-scale production of penicillopepsin-JT1 in a 250 liter fermenter and the subsequent purification have been described in detail by Hofmann (1976). Yields of up to 1 g of homogeneous protein have been routinely obtained. The method of preparation can easily be scaled down for use in smaller fermenters.

## Biological Aspects

Penicillopepsin-JT1 is produced by the fungus as an extracellular enzyme only after the logarithmic phase of mycelial growth ends and sporulation begins. Its increase approximately parallels the formation of spores (Thangamani & Hofmann, 1966). However, there is no direct correlation between enzyme production and spore formation. When the culture is grown in the presence of acridine orange, enzyme formation is suppressed, but spore formation is only slightly affected. Conversely, 6-ethyl-thiopurine suppresses spore formation, but stimulates enzyme production (Thangamani & Hofmann, 1966). Strains of aspergilli from which the gene for the aspartic proteinase has been deleted, e.g. *Aspergillus awamori* UVK143f, sporulate normally (M. Ward, personal communication). Since *Penicillium* and *Aspergillus* species are closely related it is reasonable to assume that deletion of the aspartic proteinase from a *Penicillium* would not inhibit sporulation. The role of the aspartic proteinase in *Penicillium janthinellum* and related filamentous fungi remains to be elucidated.

In analogy to aspergillopepsin A (Chapter 271) (Berka *et al.*, 1989), mucorpepsin (Tonouchi *et al.*, 1986) and penicillopepsin-JT2, the gene for penicillopepsin-JT1 probably also codes for a leader sequence and an activation peptide. Several attempts at isolating an inactive precursor have, however, been unsuccessful. The conversion of the precursors to the active enzymes most likely occurs during their secretion into the medium. No active penicillopepsin-JT1 has ever been detected in the mycelia, but evidence has been obtained for an intracellular precursor of penicillopepsin-JT1 by the use of an antibody to penicillopepsin-JT1 (Q.-N. Cao & T. Hofmann, unpublished results).

## Distinguishing Features

The two penicillopepsins of *Penicillium janthinellum* strain NRRL 905 share many properties. They are easily distinguished from members of proteinase families other than family A1 because they are inhibited by pepstatin with similar inhibition constants. They are readily separable from each other on a DEAE-Sephadex column (A. Cunningham & T. Hofmann, unpublished results). Thus far, we have been unable to determine whether the gene for penicillopepsin-JT2 is actually expressed by the organism under the growth condition which we have tested (A. Cunningham & T. Hofmann, unpublished results).

## Related Enzymes

### Penicillopepsin-JT2

During a search for the gene of penicillopepsin-JT1, nucleotide sequences for two other aspartic proteinases were identified in a genomic library of *Penicillium janthinellum*, strain NRRL 905. The complete gene for penicillopepsin-JT2 has been sequenced and aligned with that of the homologous preproaspergillopepsin A (Berka *et al.*, 1990). The sequence consists of three exons interrupted by two introns, which are 64 and 67 bp long. They are in the same locations as introns 1 and 3 (length 51 and 59 bp, respectively) of preproaspergillopepsin A (Chapter 294) (Berka *et al.*, 1990). The start codon precedes a sequence of 69 bp which encodes a putative signal peptide of about 20 amino acid and a putative activation peptide of about 49 amino acids. This is analogous to other aspartic proteinases such as rhizopuspepsin (Horiuchi *et al.*, 1988) and endothiapepsin (Razanamparany *et al.*, 1992). This sequence is 66% identical with the nucleotide sequence of the leader sequence and the activation peptide of preproaspergillopepsin A (Chapter 294). The remaining nucleotide sequence, which codes for the active enzyme, is about 71% identical with that of aspergillopepsin A (Q.-N. Cao *et al.*, unpublished results).

The gene was cloned into plasmid pGPT-pyrG1 and transformed into a strain of *Aspergillus niger* var. *awamori*, GAP3-4, from which the glucoamylase and aspartic proteinase genes had been deleted (Berka *et al.*, 1990). Only the active enzyme was found in the growth medium. It was present in high yield and in a highly glycosylated form. It was purified and deglycosylated with endoglycosidase H. Its N-terminal sequence and its amino acid composition correspond to the amino acid sequence deduced from the nucleotide sequence. The amino acid sequence of the active penicillopepsin-JT2 is 71% identical with those of both penicillopepsin-JT1 and aspergillopepsin A; the sequence of penicillopepsin-JT1 is about 69% identical with that of aspergillopepsin A. Penicillopepsin-JT2 ($M_r$ 33 800) is one amino acid longer than penicillopepsin-JT1. The enzymatic properties of the two penicillopepsins are very similar. The action of penicillopepsin-JT2 on the series Ac-(Ala)$_m$-Lys$+$Nph-(Ala)$_n$-amide shows the same rate-enhancing

effects as penicillopepsin-JT1 when alanyl residues are added in positions P3 and P2′, but not in the other positions of the substrate. The $K_m$ values remain the same within experimental error for the whole series. Penicillopepsin-JT2 also activates trypsinogen at low pH (A. Cunningham & T. Hofmann, unpublished results). The genomic DNA sequence and the deduced amino acid sequence have been submitted to Genbank; the gene name is *pepA* and the accession number U81483.

The partial nucleotide sequence of the gene for penicillopepsin-JT3 shows that it has three introns, which are in identical positions to those of preproaspergillopepsin A. It also has a sequence that corresponds to the activation peptide of propenicillopepsin-JT2 and proaspergillopepsin A. The sequence of the 5′ terminal end has not yet been completed.

### Penicillopepsin from Penicillium roqueforti

The isolation and purification of this enzyme ($M_r$ 33 400) were described by Zevaco *et al.* (1973). Its molecular and enzymatic properties are very similar to penicillopepsin-JT1 and an aspartic proteinase from *A. oryzae* (Zevaco *et al.*, 1973).

### Penicillopepsin from Penicillium duponti

The thermophilic fungus *P. duponti* K1014 produces a penicillopepsin with properties that are similar to the enzymes from other penicilliums, except that it is markedly glycosylated (Emi *et al.*, 1976). Its $M_r$ is 41 590; it has a pI of 3.81. It is inactivated by DAN and related diazo-compounds.

### Penicillopepsin from Penicillium camemberti

This enzyme ($M_r$ 33 500) was purified by Chrzanowska *et al.* (1995). The enzyme is comparable to penicillopepsin-JT1 in molecular properties, sensitivity to inhibitors, ability to activate trypsinogen and low specificity (Chrzanowska *et al.*, 1995). At pH 3.3 it specifically cleaves the Leu7$+$Met8 peptide bond of the trypsin inhibitor from *Cucurbita maxima* described by Otlewski *et al.* (1984); at neutral pH this bond can be resynthesized (Chrzanowska *et al.*, 1995).

## References

Abad-Zapatero, C., Rydel, T.J. & Erickson, J. (1990) Revised 2.3 Å structure of porcine pepsin: evidence for a flexible subdomain. *Proteins: Struct. Funct. Genet.* **8**, 62–81.

Allen, B., Blum, M., Cunningham, A., Tu, G.-C. & Hofmann, T. (1990) A ligand-induced, temperature-dependent conformational change in penicillopepsin. *J. Biol. Chem.* **265**, 5060–5065.

Antonov, V.K., Ginodman, L.M., Rumsh, L.D., Kapitannikov, Y.K., Barshevskaya, T.N., Yavashev, L.P., Gurova, A.G. & Volkova, L.I. (1981) Studies on the mechanism of action of proteolytic enzymes using heavy oxygen exchange. *Eur. J. Biochem.* **117**, 195–200.

Balbaa, M., Cunningham, A. & Hofmann, T. (1993) Secondary substrate binding in aspartic proteinases: contributions of subsites S$_2'$ and S$_3$ to $k_{cat}$. *Arch. Biochem. Biophys.* **306**, 297–303.

Balbaa, M., Blum, M. & Hofmann, T. (1994) Mechanism of pepsin-catalyzed aminotranspeptidation reactions. *Int. J. Biochem.* **26**, 35–42.

Berka, R.M., Ward, M., Wilson, L.J., Hayenga, K.J., Kodama, K.H., Carlomagno, L.P. & Thompson, S.A. (1990) Molecular cloning and deletion of the gene encoding aspergillopepsin A from *Aspergillus awamori. Gene* **86**, 153–162.

Blum, M., Cunningham, A., Bendiner, M. & Hofmann, T. (1985). Penicillopepsin, the aspartic proteinase from *Penicillium janthinellum*: substrate-binding effects and intermediates in transpeptidation reactions. *Biochem. Soc. Trans.* **13**, 1044–1047.

Blum, M., Cunningham, A., Pang, H. & Hofmann, T. (1991) Mechanism and pathway of penicillo-pepsin-catalyzed transpeptidation and evidence for noncovalent trapping of amino acid and peptide intermediates. *J. Biol. Chem.* **266**, 9501–9507.

Chrzanowska, J., Kolaczkowska, M., Dryjansky, M., Stachowiak, D. & Polanowski, A. (1995) Aspartic proteinase from *Penicillium camemberti*: purification, properties, and substrate specificity. *Enzyme Microb. Technol.* **17**, 719–724.

Cooper, J.B., Khan, G., Taylor, G., Tickle, I.J. & Blundell, T.L. (1990) X-ray analysis of aspartic proteinases. II. Three-dimensional structure of the hexagonal form of porcine pepsin at 2.3 Å resolution. *J. Mol. Biol.* **214**, 199–220.

Cunningham, A., Hofmann, M.I. & Hofmann, T. (1990) Rate-determining steps in penicillopepsin catalyzed reactions. *FEBS Lett.* **276**, 119–122.

Emi, S., Myers, D.V. & Iacobucci, G.A. (1976) Purification and properties of the thermostable acid protease of *Penicillium duponti. Biochemistry* **15**, 842–848.

Fraser, M.E., Strynadka, N.C., Bartlett, P.A., Hanson, J.E. & James, M.N.G. (1992) Crystallographic analysis of transition-state mimics bound to penicillopepsin: phosphorus-containing peptide analogues. *Biochemistry* **31**, 5201–5214.

Fruton, J.S. (1970) The specificity and mechanism of pepsin action. *Adv. Enzymol. Relat. Areas Mol. Biol.* **33**, 401–443.

Fusek, M. & Vetvicka, V. (eds) (1995) *Aspartic proteinases. Physiology and Pathology.* Boca Raton, FL: CRC Press, pp. 289–302.

Hofmann, T. (1960) On the mechanism of activation of trypsinogen by penicillium kinase. *Bull. Soc. Chim. Biol.* **42**, 1279–1284.

Hofmann, T. (1976). Penicillopepsin. *Methods Enzymol.* **45**, 434–452.

Hofmann, T. & Shaw, R. (1964). Proteolytic enzymes of *Penicillium janthinellum*. I. Purification and properties of a trypsinogen-activating enzyme (peptidase A). *Biochim. Biophys. Acta* **92**, 543–557.

Hofmann, T., Hodges, R.S. & James, M.N.G. (1984). Effect of pH on the activities of penicillopepsin and rhizopuspepsin, and a proposal for the productive substrate binding mode in penicillopepsin. *Biochemistry* **23**, 635–643.

Hofmann, T., Allen, B., Bendiner, M., Blum, M. & Cunningham, A. (1988) The effect of secondary substrate binding in penicillopepsin: the contributions of subsites $S'_2$ and $S_3$ to $k_{cat}$. *Biochemistry* **27**, 1140–1147.

Horiuchi, H., Yanai, K., Okazaki, T., Takagi, M. & Yano, K. (1988) Isolation and sequencing of a genomic clone encoding aspartic proteinase of *Rhizopus niveus. J. Bacteriol.* **170**, 272–278.

Hsu, I.N., Delbaere, L.T.J., James, M.N.G. & Hofmann, T. (1977). Penicillopepsin from *Penicillium janthinellum*: crystal structure at 2.8 Å and sequence homology with porcine pepsin. *Nature (London)* **266**, 140–145.

James, M.N.G. & Sielecki, A.R. (1983) Structure and refinement of penicillopepsin at 1.8 Å resolution. *J. Mol. Biol.* **163**, 299–361.

James, M.N.G. & Sielecki, A.R. (1985) Stereochemical analysis of peptide bond hydrolysis catalyzed by the aspartic proteinase penicillopepsin. *Biochemistry* **24**, 3701–3713.

James, M.N.G., Sielecki, A.R., Salituro, F., Rich, D.H. & Hofmann, T. (1982) Conformational flexibility in the active sites of aspartyl proteinases revealed by a pepstatin fragment binding to penicillopepsin. *Proc. Natl Acad. Sci. USA* **79**, 6137–6141.

James, M.N.G., Sielecki, A.R. & Hofmann, T. (1985). X-ray diffraction studies on penicillopepsin and its complexes: the hydrolytic mechanism. In: *Aspartic Proteinases and Their Inhibitors* (Kostka, V., ed.). Berlin: Walter de Gruyter & Co., pp. 163–177.

James, M.N.G., Sielecki, A.R., Hayakawa, K. & Gelb, M.H. (1992) Crystallographic analysis of transition-state mimics bound to penicillopepsin: difluorostatine- and difluorostatone-containing peptides. *Biochemistry* **31**, 3872–3886.

Kunitz, M. (1938) Formation of trypsin from trypsinogen by an enzyme produced by a mold of the genus *Penicillium. J. Gen. Physiol.* **21**, 601–620.

Lowther, W.T., Majer, P. & Dunn, B.M. (1995) Engineering the substrate specificity of rhizopuspepsin: the role of Asp 77 of fungal aspartic proteinases in facilitating the cleavage of oligopeptide substrates with lysine in P1. *Protein Sci.* **4**, 689–702.

Mains, G., Takahashi, M., Sodek, J. & Hofmann, T. (1971). The specificity of penicillopepsin. *Can. J. Biochem.* **49**, 1134–1149.

Nakanishi, K. (1959) Trypsinogen-kinase in *Aspergillus oryzae*. III. Purification of trypsinogen kinase and its relation to acid-protease. *J. Biochem. (Tokyo)* **46**, 1263–1269.

Otlewski, J., Polanowski, A., Leluk, J. & Wilusz, T. (1984) Trypsin inhibitors in summer squash (*Cucurbita pepo*) seeds. Isolation, purification and partial characterization of three inhibitors. *Acta Biochim. Polon.* **31**, 267–278.

Razanamparany, V., Jara, P., Legoux, R., Delmas, P., Msayeh, F., Kaghad, M. & Loison, G. (1992) Cloning and mutation of the gene encoding endothiapepsin from *Cryphonectria parasitica. Curr. Genet.* **21**, 455–461.

Sali, A., Veerapandian, B., Cooper, J.B., Moss, D.S., Hofmann, T. & Blundell, T.L. (1992) Domain flexibility in aspartic proteinases. *Proteins Struct. Funct. Genet.* **12**, 158–170.

Sepulveda, P., Marciniszyn, J., Liu, D. & Tang, J. (1975) Primary structure of porcine pepsin. III. Amino acid sequence of a cyanogen bromide fragment, CB2A, and the complete structure of porcine pepsin. *J. Biol. Chem.* **250**, 5082–5088.

Shaw, R. & Hofmann, T. (1964) Proteolytic enzymes of *Penicillium janthinellum*. I. Purification and properties of a trypsinogen-activating enzyme (peptidase A). *Biochim. Biophys. Acta* **92**, 543–557.

Sielecki, A.R., Fedorov, A.A., Boodhoo, A., Andreeva, N.S. & James, M.N.G. (1990) Molecular and crystal structure of monoclinic porcine pepsin refined at 1.8 Å resolution. *J. Mol. Biol.* **214**, 143–170.

Sodek, J. & Hofmann, T. (1970). Amino acid sequence around the active site aspartic acid in penicillopepsin. *Can. J. Biochem.* **48**, 1014–1016.

Thangamani, A. & Hofmann, T. (1966) The role of a protease in the sporulation of *Penicillium janthinellum. Can. J. Biochem.* **44**, 579–584.

Tonouchi, N., Shoun, H., Uozumi, T. & Beppu, T. (1986) Cloning and sequencing of a gene for mucor rennin, an aspartic proteinase from *Mucor pusillus. Nucleic Acids Res.* **14**, 7551–7568.

Wang, T.T. & Hofmann, T. (1976) Acyl and amino intermediates in reactions catalysed by pig pepsin. *Biochem. J.* **153**, 691–699.

Zevaco, C., Hermier, J. & Gripon, J.-C. (1973) Le système protéolytique de *Penicillium roqueforti*. II. Purification et propriétés de la protéase acide [The proteolytic system of *Penicillium roqueforti*. II. Purification and properties of the acid protease]. *Biochimie* **55**, 1353–1360.

*Theo Hofmann*
*Department of Biochemistry,*
*University of Toronto,*
*King's College Circle,*
*Toronto, Ontario, Canada M5S 1A8*
*Email: theo@hera.med.utoronto.ca*

# 296. *Endothiapepsin*

## Databanks

*Peptidase classification: clan AA, family A1, MEROPS ID: A01.017*
*NC-IUBMB enzyme classification: EC 3.4.23.22*
*Chemical Abstracts Service registry number: 37205-60-0*
*Databank codes:*

| Species | SwissProt | PIR | EMBL (cDNA) | EMBL (genomic) |
|---|---|---|---|---|
| *Cryphonectria parasitica* | P11838 | JU0143 | X53997 | A17637: unannotated patent |
| | | S00088 | X63351 | A17638: unannotated patent |
| | | S21358 | | A17665: complete gene |
| | | S22136 | | |
| | | S26871 | | |

Brookhaven Protein Data Bank three-dimensional structures:

| Species | ID | Resolution | Notes |
|---|---|---|---|
| *Cryphonectria parasitica* | 1EED | 2 | complex with cyclohexyl renin inhibitor PD125754 |
| | 1ENT | 1.9 | complex with inhibitor PD130328 |
| | 1EPL | 2 | complex with PS1 (Pro-Leu-Glu-PSA-Arg-Leu) |
| | 1EPM | 1.6 | complex with PS2 (Thr-Phe-Gln-Ala-PSA-Leu-Arg-Glu) |
| | 1EPN | 1.6 | complex with CP-80 794 (MOR-Phe-Cys-$CH_3$-NOR) |
| | 1EPO | 2 | complex with CP-81 282 (MOR-Phe-Nle-CHF-NME) |
| | 1EPP | 1.9 | complex with PD-130 693 (MAS-Phe-Lys-MTF-Sta-MBA) |
| | 1EPQ | 1.9 | complex with PD-133 450 (SOT-Phe-Gly-SCC-GCL) |
| | 1EPR | 2.3 | complex with PD-135 040 (TSM-DPH-His-CHF-EMR) |
| | 1ER8 | 2 | complex with H-77 (D-His-Pro-Phe-His-Leu-Leu-Val-Tyr) |
| | 2ER0 | 3 | complex with L364 099 (Iva-His-Pro-Phe-His-CHS-Leu-Phe-$NH_2$) |
| | 2ER6 | 2 | complex with H-256 (Pro-Thr-Glu-Phe-Phe-Arg-Glu) |

*continued overleaf*

| Species | ID | Resolution | Notes |
|---|---|---|---|
| | 2ER7 | 1.6 | complex with H-261 |
| | 2ER9 | 2.2 | complex with L636 564 |
| | 3ER3 | 2 | complex with CP-71 362 |
| | 3ER5 | 1.8 | complex with H-189 |
| | 4APE | 2.1 | |
| | 4ER1 | 2 | complex with PD125967 (BNA-His-CAL-DCI) |
| | 4ER2 | 2 | complex with pepstatin |
| | 4ER4 | 2.1 | complex with Pro-His-Pro-Phe-His-Leu-Val-Ile-His-Lys |
| | 5ER1 | 2 | complex with BW624 (LOL-CH$_2$-Val-Ile-Phe-OMe) |
| | 5ER2 | 1.8 | complex with CP-69 799 (BOC-Phe-His-AHS-Lys-Phe) |

## Name and History

The aspartic proteinase from the chestnut blight fungus (*Cryphonectria* or *Endothia parasitica*) is referred to as **endothiapepsin**. This enzyme can be used as a fungal rennet in cheese production (Sardinas, 1968).

## Activity and Specificity

Like most aspartic proteinases endothiapepsin has an acid pH optimum and a low pI of around 5.5 (Sardinas, 1972). It cleaves protein substrates with a specificity similar to that of pig pepsin A (Chapter 272), preferring hydrophobic residues at P1 and P1'. Williams *et al.* (1972) analyzed the cleavage sites in the oxidized B chain of insulin and found the rates of hydrolysis to be as follows: Phe24⥮Phe25 > Tyr16⥮Leu17 > Gln4⥮His5 >>> Leu17⥮Val18 > Asn3⥮Gln4. The enzyme's milk-clotting activity was initially attributed to cleavage of the same Phe105-Met106 bond of κ-casein, which is cleaved by calf chymosin (Chapter 274) and several other aspartic proteinases. However, Drohse & Foltmann (1989) showed that endothiapepsin cleaves the preceding peptide bond of κ-casein, namely, Ser104⥮Phe105. The proteinase's catalytic activity may be assayed by the use of chromogenic substrates containing nitrophenylalanine, usually at the P1' position (e.g. Hofmann & Hodges, 1982; Dunn & Kay, 1985), which exhibit a change in absorption spectrum on cleavage.

Like many aspartic proteinases, endothiapepsin is potently inhibited by the microbial peptide pepstatin A, which contains the unusual amino acid statine (Cooper *et al.*, 1989; Bailey *et al.*, 1993). The latter contains a main chain -CH(OH)-CH$_2$- group, which is thought to mimic the putative tetrahedral intermediate (-C(OH)$_2$-NH-) of catalysis. Many synthetic analogs such as the reduced bond (-CH$_2$-NH-), hydroxyethylene (-CH(OH)-CH$_2$-) and fluoroketone (-C(OH)$_2$-CF$_2$-) analogs act as potent inhibitors when incorporated into appropriate peptide ligands (Foundling *et al.*, 1987).

## Structural Chemistry

Endothiapepsin is a single-chain proteinase of 330 amino acids and has a molecular mass of 33.8 kDa. The full amino acid sequence was determined chemically by Barkholt (1987) and confirmed by DNA sequencing (Razanamparany *et al.*, 1992; Choi *et al.*, 1993), which established the presence of an 89 residue propeptide. The enzyme possesses a single disulfide bridge between residues 250 and 283 (pig pepsin numbering). The amino acid sequence has signs of an internal repeat relating the two halves of the molecule, their identity being greatest in the vicinity of the catalytic residues, which occur in the two conserved Asp-Thr-Gly sequences. The enzyme was crystallized by Moews & Bunn (1970) using ammonium sulfate as precipitant at pH 4.6. X-ray analysis by multiple isomorphous replacement (Jenkins *et al.*, 1975; Subramanian *et al.*, 1977) culminated in refinement of the crystal structure of the enzyme at 2.1 Å resolution (Blundell *et al.*, 1990) to an R factor of 0.178. The structure possesses the same fold as other aspartic proteinases being largely β sheet and consisting of two topologically related lobes of approximately 170 amino acids each (Figure 296.1). The active site resides in a pronounced cleft between the lobes. The base of this cleft is made of β strands forming two abutting ψ structures that contain the catalytic aspartate residues (32 and 215 in pig pepsin numbering).

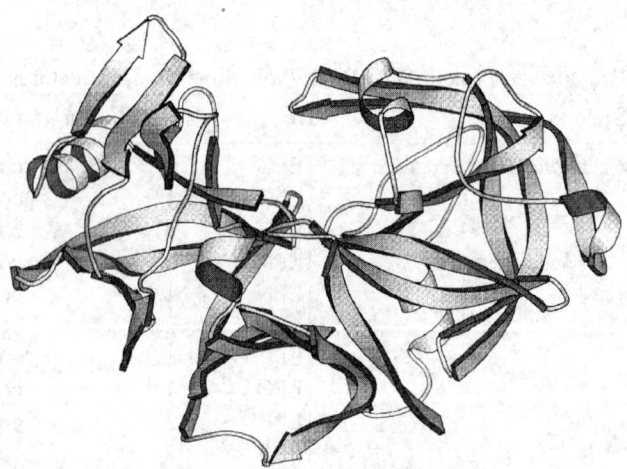

*Figure 296.1* The structure of endothiapepsin at 2.1 Å resolution showing the pronounced active-site cleft and the flap covering the catalytic center.

The side chains of the catalytic aspartates are held coplanar and within hydrogen bonding distance by an intricate arrangement of hydrogen bonds involving main-chain and conserved side-chain groups (Pearl & Blundell, 1984). A solvent molecule bound tightly to both carboxyls by hydrogen bonds is found in all aspartic proteinase crystal structures and in the more distantly related retroviral proteinases (Chapter 310) (Lapatto *et al.*, 1989; Wlodawer *et al.*, 1989). The latter have a broadly similar active-site structure, but consist of two identical subunits with topology similar to each of the two domains of a eukaryotic aspartic proteinase.

Renin (Chapter 284) inhibitors based on synthetic and naturally occurring transition state analogs have been shown by X-ray crystallography to bind to endothiapepsin in extended conformations with up to ten residues of the ligand interacting with the cleft (Blundell *et al.*, 1987; Cooper *et al.*, 1987, 1989, 1992; Foundling *et al.*, 1987; Sali *et al.*, 1989; Veerapandian *et al.*, 1990, 1992; Bailey *et al.*, 1993; Lunney *et al.*, 1993). The hydrogen bonds, which position an inhibitor's main chain in the active-site cleft are largely conserved from one complex to another and indeed, from one aspartic proteinase to another, implying that the main determinants of specificity are the van der Waals contacts between the enzyme and the ligand's side chains. The central regions of the inhibitors except for P2 are almost completely shielded from solvent by the hydrophobic binding pockets and a $\beta$ hairpin covering the active-site cleft (Figures 296.1 and 296.2).

The hydroxyl groups of transition state analogs bind by hydrogen bonds between the catalytic aspartates in the same position as the solvent molecule in the uncomplexed enzyme (Figure 296.2). This water molecule has been implicated in catalysis and Suguna *et al.* (1987) first suggested that it may become partly displaced upon substrate binding and polarized by one of the aspartate carboxyls. The water oxygen could then nucleophilically attack the scissile bond carbonyl carbon thereby forming the noncovalently bound intermediate (-C(OH)$_2$-NH-) of peptide bond hydrolysis. This mechanism has been refined based on the high-resolution structures of *gem*-diol transition state analogs (difluoroketone: -C(OH)$_2$-CF$_2$-) in which both hydroxyls of the putative transition state are mimicked (James *et al.*, 1992; Veerapandian *et al.*, 1992). A detailed comparison of X-ray structures of 21 endothiapepsin inhibitor complexes is given by Bailey & Cooper (1994). The strongly conserved interactions at P4, P3, P1, P1' and P2' compared to the weaker binding and unfavorable geometry of the P2 residue indicate that the enzyme may possibly strain bound substrates at P1 and P2 towards the geometry found in complexes with transition state isosteres. Intriguing effects are also seen in which the side chains of the inhibitors can adopt different conformations to compensate for greater or lesser occupation of the neighboring subsites in different complexes (Figure 296.3). These studies provide much information relevant to design of specific human renin inhibitors with therapeutic applications for treatment of hypertension and congestive heart failure (Blundell *et al.*, 1987).

## Preparation

*Endothia parasitica* culture filtrate is concentrated and ammonium sulfate added to 40% w/v to precipitate the enzyme.

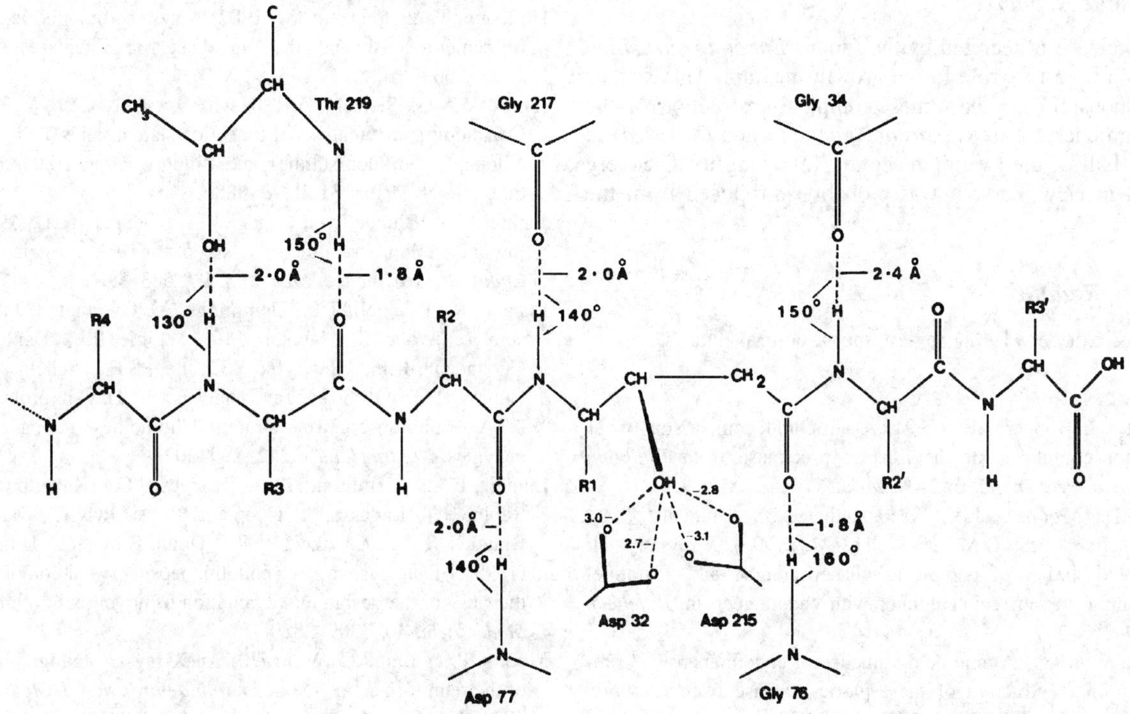

*Figure 296.2* The possible hydrogen-bonding interactions of a statine-containing renin inhibitor (L-363,564) bound to endothiapepsin as defined by X-ray analysis at 2.2 Å. The statine residue spans the central P1–P1' regions of the inhibitor with its hydroxyl located between the catalytic aspartate carboxyl groups (32 and 215)

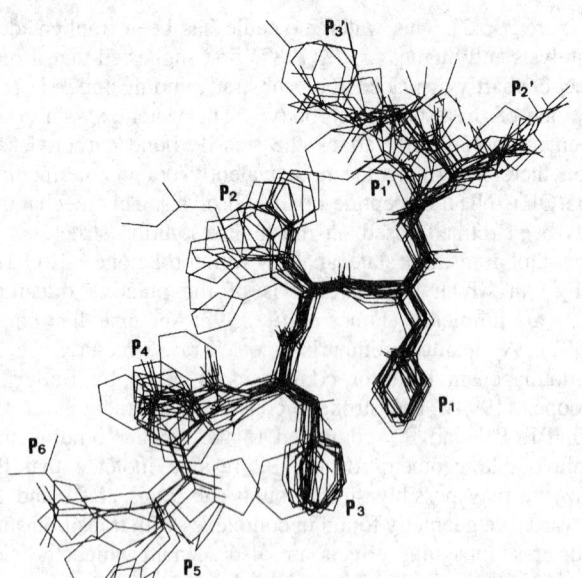

*Figure 296.3* A superposition of 21 inhibitor complexes showing the structural conservation in the central region of the inhibitors except at the P2 position where the side chain appears to adopt a range of conformations depending on the nature of neighboring residues, principally P1'.

Subsequent purification involves conventional gel filtration and ion-exchange chromatography with the enzyme buffered close to pH 4.6.

## Biological Aspects

The proteinase is secreted by the fungus *Endothia parasitica* and has a digestive role in the growth medium. This fungus was responsible for the almost complete destruction of the American chestnut trees (*Castanea dentata* and *C. sativa*) in the first half of the twentieth century following its discovery in 1904 in New York; it was probably introduced from the Far East.

## Further Reading

Bailey & Cooper (1994) review the structural data.

## References

Bailey, D. & Cooper, J.B. (1994) A structural comparison of 21 inhibitor complexes of the aspartic proteinase from *Endothia parasitica*. *Protein Sci.* **3**, 2129–2143.

Bailey, D., Cooper, J.B., Veerapandian, B., Blundell, T.L., Atrash, B., Jones, D.M. & Szelke, M. (1993) X-ray crystallographic studies of complexes of pepstatin A and a statine-containing human renin inhibitor with endothiapepsin. *Biochem. J.* **289**, 363–371.

Barkholt, V. (1987) Amino acid sequence of endothiapepsin. Complete primary structure of the aspartic protease from *Endothia parasitica*. *Eur. J. Biochem.* **167**, 327–338.

Blundell, T.L., Cooper, J.B., Foundling, S.I., Jones, D.M., Atrash, B. & Szelke, M. (1987) On the rational design of renin inhibitors: X-

ray studies of aspartic proteinases complexed with transition state analogues. *Biochemistry* **26**, 5585–5590.

Blundell, T.L., Jenkins, J.A., Sewell, B.T., Pearl, L.H., Cooper, J.B., Tickle, I.J., Veerapandian, B. & Wood, S.P. (1990) X-ray analyses of aspartic proteinases. The three-dimensional structure at 2.1 Å resolution of endothiapepsin. *J. Mol. Biol.* **211**, 919–941.

Choi, G.H., Pawlyk, D.M., Rae, B., Shapira, R. & Nuss, D.L. (1993) Molecular analysis and overexpression of the gene encoding endothiapepsin, an aspartic protease from *Cryphonectria parasitica*. *Gene* **125**, 135–141.

Cooper, J.B., Foundling, S.I., Blundell, T.L., Boger, J., Jupp, R.A. & Kay, J. (1989) X-ray studies of aspartic proteinase–statine inhibitor complexes. *Biochemistry* **28**, 8596–8603.

Cooper, J.B., Foundling, S., Hemmings, A., Blundell, T., Jones, D.M., Hallett, A. & Szelke, M. (1987) The structure of a synthetic pepsin inhibitor complexed with endothiapepsin. *Eur. J. Biochem.* **169**, 215–221.

Cooper, J.B., Quail, W., Frazao, C., Foundling, S.I., Blundell, T.L., Humblet, C., Lunney, E.A., Lowther, W.T. & Dunn, B.M. (1992) X-ray crystallographic analysis of inhibition of endothiapepsin by cyclohexyl renin inhibitors. *Biochemistry* **31**, 8142–8150.·

Drohse, H.B. & Foltmann, B. (1989) Specificity of milk-clotting enzymes towards bovine κ-casein. *Biochim. Biophys. Acta* **995**, 221–224.

Dunn, B.M. & Kay, J. (1985) Design, synthesis and analysis of synthetic substrates for aspartic proteinases. *Biochem. Soc. Trans.* **13**, 1041–1043.

Foundling, S.I., Cooper, J.B., Watson, F.E., Pearl, L.H., Sibanda, B.L., Wood, S.P., Blundell, T.L., Valler, M.J., Norey, C.G., Kay, J., Boger, J., Dunn, B.M., Leckie, B.J., Jones, D.M., Atrash, B., Hallett, A. & Szelke, M. (1987) High resolution X-ray analysis of renin inhibitor aspartic proteinase complexes. *Nature (Lond.)* **327**, 349–352.

Hofmann, T. & Hodges, R.S. (1982) A new chromophoric substrate for penicillopepsin and other fungal aspartic proteinases. *Biochem. J.* **203**, 603–610.

James, M.N.G., Sielecki, A.R., Hayakawa, K. & Gelb, M.H. (1992) Crystallographic analysis of transition state mimics bound to penicillopepsin: difluorostatine- and difluorostatone-containing peptides. *Biochemistry* **31**, 3872–3886.

Jenkins, J.A., Blundell, T.L., Tickle, I.J. & Ungaretti, L. (1975) The low resolution structure analysis of an acid proteinase from *Endothia parasitica*. *J. Mol. Biol.* **99**, 583–590.

Lapatto, R., Blundell, T.L., Hemmings, A., Overington, J., Wilderspin, J., Wood, S., Merson, J.R., Whittle, P.J., Danley, D.E., Geoghegan, K.F., Hawrylik, S.J., Lee, S.E., Scheld, K.G. & Hobart, P.M. (1989) X-ray analysis of HIV-1 proteinase at 2.7 Å resolution confirms structural homology among retroviral enzymes. *Nature (Lond.)* **342**, 299–302.

Lunney, E.A., Hamilton, H.W., Hodges, J.C., Kaltenbrohn, J.S., Repine, J.T., Badasso, M., Cooper, J.B., DeAlwis, C., Wallace, B., Blundell, T.L., Lowther, W.T., Dunn, B.M. & Humblet, C. (1993) The analysis of five endothiapepsin crystal complexes and their use in the design and evaluation of novel renin inhibitors. *J. Med. Chem.* **36**, 3809–3820.

Moews, P. & Bunn, C.W. (1970) An X-ray crystallographic study of the rennin-like enzyme of *Endothia parasitica*. *J. Mol. Biol.* **54**, 395–397.

Pearl, L. & Blundell, T. (1984) The active site structure of aspartic proteinases. *FEBS Lett.* **174**, 96–101.

Razanamparany, V., Jara, P., Legoux, R., Delmas, P., Msayeh, F., Kaghad, M. & Loison, G. (1992) Cloning and mutation of the gene encoding endothiapepsin from *Cryphonectria parasitica*. *Curr. Genet.* **21**, 455–461.

Sali, A., Veerapandian, B., Cooper, J.B., Foundling, S.I., Hoover, D.J. & Blundell, T.L. (1989) High resolution X-ray study of the complex between endothiapepsin and an oligopeptide inhibitor: the analysis of inhibitor binding and description of the rigid body shifts in the enzyme. *EMBO J.* **8**, 2179–2188.

Sardinas, J.L. (1968) Rennin enzyme from *Endothia parasitica*. *Appl. Microbiol.* **16**, 248–255.

Sardinas, J.L. (1972) Microbial rennets. *Adv. Appl. Microbiol.* **15**, 39–73.

Subramanian, E., Swan, I.D.A., Liu, M., Davies, D.R., Jenkins, J.A., Tickle, I.J. & Blundell, T.L. (1977) Homology among acid proteases: comparison of crystal structures at 3.0 Å resolution of acid proteases from *Rhizopus chinensis* and *Endothia parasitica*. *Proc. Natl Acad. Sci. USA* **74**, 556–559.

Suguna, K., Padlan, E.A., Smith, C.W., Carlson, W.D. & Davies, D.

(1987) Binding of a reduced peptide inhibitor to the aspartic proteinase from *Rhizopus chinensis*: implications for a mechanism of action. *Proc. Natl Acad. Sci. USA* **84**, 7009–7013.

Veerapandian, B., Cooper, J., Sali, A. & Blundell, T.L. (1990) Three dimensional structure of endothiapepsin complexed with a transition-state isostere inhibitor of renin at 1.6 Å resolution. *J. Mol. Biol.* **216**, 1017–1029.

Veerapandian, B., Cooper, J.B., Sali, A., Blundell, T.L., Rosatti, R.L., Dominy, B.W., Damon, D.B. & Hoover, D.J. (1992) Direct observation by X-ray analysis of the tetrahedral 'intermediate' of aspartic proteinases. *Protein Sci.* **1**, 322–328.

Williams, D.C., Whitaker J.R. & Caldwell, P.V. (1972) Hydrolysis of peptide bonds of the oxidized B-chain of insulin by *Endothia parasitica* protease. *Arch. Biochem. Biophys.* **149**, 52–61.

Wlodawer, A., Miller, M., Jaskolski, M., Sathyanarayana, B.K., Baldwin, E., Weber, I.T., Selk, L.M., Clawson, L., Schneider, J. & Kent, S. (1989) Conserved folding in retroviral proteinases: crystal structure of synthetic HIV-1 proteinase. *Science* **245**, 616–621.

*Jonathan B. Cooper*
*Department of Biochemistry,*
*School of Biological Sciences,*
*University of Southampton,*
*Bassett Crescent East,*
*Southampton SO16 7PX, UK*
*Email: jbc2@soton.ac.uk*

# 297. *Rhizopuspepsin*

## Databanks

*Peptidase classification: clan AA, family A1, MEROPS ID: A01.012*
*NC-IUBMB enzyme classification: EC 3.4.23.21*
*Chemical Abstracts Service registry number: 9074-09-3*
*Databank codes:*

| Species | Type | SwissProt | PIR | EMBL (cDNA) | EMBL (genomic) |
|---|---|---|---|---|---|
| *Rhizopus chinensis* | I | – | – | L33856 | – |
| *Rhizopus chinensis* | II | P06026 | A26681 | J02651 | L33857: 5′ promoter |
| | | | A26682 | M63451 | L33858: complete gene |
| | | | A40425 | | L33859: complete gene |
| | | | A41415 | | |
| | | | A61330 | | |
| *Rhizopus niveus* | 1 | P10602 | A28672 | M19100 | – |
| *Rhizopus niveus* | 2 | P43231 | – | X56964 | – |

*continued overleaf*

| Species | Type | SwissProt | PIR | EMBL (cDNA) | EMBL (genomic) |
|---------|------|-----------|-----|-------------|----------------|
| *Rhizopus niveus* | 3 | Q03699 | A33223 | – | D00908: unannotated |
| | | | JU0343 | | D13940: unannotated |
| | | | | | X56965: complete gene |
| *Rhizopus niveus* | 4 | Q03700 | – | X56992 | – |
| *Rhizopus niveus* | 5 | P43232 | – | X56993 | – |
| *Rhizopus niveus* | II | – | – | D13939 | – |

Brookhaven Protein Data Bank three-dimensional structures:

| Species | Type | ID | Resolution | Notes |
|---------|------|-----|-----------|-------|
| *Rhizopus chinensis* | II | 2APR | 1.8 | |
| | | 3APR | 1.8 | complex with psi-(CH$_2$-NH)-D-His-Pro-Phe-His-Phe-Phe-Val-Tyr |
| | | 4APR | 2.5 | complex with a pepstatin-like renin inhibitor |
| | | 5APR | 2.1 | complex with a pepstatin-like renin inhibitor |
| | | 6APR | 2.5 | complex with pepstatin |

## Name and History

Although not the first fungal peptidase isolated, the **Rhizopus chinensis acid protease** has provided a very useful model system for the application of a wide variety of experimental and theoretical studies. Fukumoto *et al.* (1967) described the first homogeneous preparation of this enzyme, following earlier reports on isolation of the related enzymes from *Endothia parasitica* (Chapter 296) and *Penicillium janthinellum* (Chapter 295). Following the suggestion by Sodek & Hofmann (1971), Ohtsuru *et al.* (1982) coined the name *rhizopuspepsin*, and this has been used in the ensuing publications.

## Activity and Specificity

The activity maximum of rhizopuspepsin varies from pH 2.9 to 4.5, depending on the substrate under investigation. Classically, this enzyme was initially analyzed for its cleavage properties on protein substrates, where it was seen that cleavages occurred most readily between aromatic or bulky hydrophobic residues (Fukumoto *et al.*, 1967). The oxidized B chain of insulin was used in two studies (Tsuru *et al.*, 1969; Kurono *et al.*, 1971) with the following cleavages noted:

F↓VNQH↓LCGSH↓L↓V↓E↓A↓L↓Y↓LVCG↓ERG↓F↓F↓YTPKA

Rhizopuspepsin exhibits a very broad specificity, a result confirmed in later studies. The data provided in this initial study, however, were not quantitative as the digestions were carried out under forcing conditions of high enzyme and long incubation times.

In a departure from the classic studies using a single oligopeptide such as the oxidized B chain of insulin, Morihara and Oka (Morihara & Oka, 1973; Oka & Morihara, 1974) explored the specificity of enzymes in this class by preparing synthetic oligopeptides of defined and varying sequence. First, following up on the observations by Kunitz (1938) and Sodek & Hofmann (1971), Morihara & Oka (1973) prepared oligopeptides based on the activation point in the conversion of trypsinogen to trypsin. It had been seen that

some acid proteases would cleave at Lys6 of trypsinogen to generate mature trypsin. The peptide, Z-Ala-Ala-Lys↓Ala-Ala-Ala was cleaved at the carbonyl group of Lys with the highest efficiency ($k_{cat}/K_m$ approximately $1000\,M^{-1}\,s^{-1}$) at the pH optimum of 4. Notably, mammalian enzymes such as pig pepsin (Chapter 272) would not cleave such peptides following the Lys residue. These observations were confirmed by Hofmann & Hodges (1982) using a chromogenic substrate. In a follow-up study, Oka & Morihara (1974) used peptides of the type Z-Xaa↓Leu-Ala-Ala and Z-Phe↓Xaa-Ala-Ala, to explore the P1 and P1' specificity of rhizopuspepsin. It was found that aromatic or hydrophobic residues on both sides of the scissile bond provided the best substrates, with the catalytic efficiency, $k_{cat}/K_m$, approaching $2 \times 10^4\,M^{-1}\,s^{-1}$. At the pH optimum of 4, Hofmann *et al.* (1984) determined a $k_{cat}/K_m$ of approximately $2.5 \times 10^7\,M^{-1}\,s^{-1}$ with the substrate Ac-Ala-Ala-Lys-Nph↓Ala-Ala-NH$_2$.

Lowther *et al.* (1991), working with recombinant enzyme produced using the vector of Chen *et al.* (1991), introduced the use of a panel of 28 unique oligopeptides based on the sequence, Lys-Pro-Ala-Lys-Phe↓Nph-Arg-Leu. The other six positions of this sequence, first identified as an excellent substrate for pig pepsin (Pohl & Dunn, 1988), P5, P4, P3, P2, P2' and P3', were substituted with the series, Ala, Ser, Asp, Leu, and Arg. These five residues provide a spectrum of properties intended to probe the interactions within the secondary subsites of the active site cleft (Dunn *et al.*, 1995). The resulting 28 peptides thus provide an internally-consistent set of peptides to analyze the specificity of enzymes in this class, and has been applied to several members of the family. The results with rhizopuspepsin clearly support the conclusion that this enzyme possesses very broad specificity towards oligopeptides (W. T. Lowther & B. M. Dunn, unpublished results). The parent peptide exhibited a value of $k_{cat}/K_m$ of $1.13 \times 10^6\,M^{-1}\,s^{-1}$, at pH 3.5. The specificity constants for the substituted peptides range from 0.38 to $2.39 \times 10^6\,M^{-1}\,s^{-1}$, with two exceptions; with Arg in P4, the value falls to $0.06 \times 10^6\,M^{-1}\,s^{-1}$, and with Asp in P2' the value falls to $0.09 \times 10^6\,M^{-1}\,s^{-1}$. The best substitutions in the six positions were: P5 – Ser, P4 – Leu, P3 – Arg/Leu, P2 – Ala/Leu/Arg, P2' – Ala/Arg, and P3' – Leu. In addition,

Lowther and Dunn have determined the effect of changing the Phe in P1 to Leu, $k_{cat}/K_m$ of $0.86 \times 10^6\,M^{-1}\,s^{-1}$, to Ala, $k_{cat}/K_m$ of $0.17 \times 10^6\,M^{-1}\,s^{-1}$, or to Val, where no cleavage was observed (W. T. Lowther & B. M. Dunn, unpublished results).

In experiments that followed, Lowther evaluated the role of several residues in determining the specificity of rhizopuspepsin, especially the ability of this enzyme to cleave at the peptide bond following Lys (Lowther & Dunn, 1995). The residue at position 77 (pepsin numbering), an Asp in rhizopuspepsin and several other fungal enzymes that can activate trypsinogen, was shown to be critical to the interaction with a Lys residue in P1 of a substrate through mutagenesis and evaluation of the resulting properties using oligopeptide substrates with substitutions in P1.

In a study of processing of a synthetic neurotensin-like peptide of sequence Lys-Pro-Arg-Arg-Pro-Tyr-Ile-Leu+Lys-Arg-Gly-Ser-Tyr-Tyr-Tyr by various aspartic proteinases, rhizopuspepsin was shown (Carraway *et al.*, 1992) to cleave the Leu+Lys bond, as did pepsin, cathepsin D (Chapter 277), and renin (Chapter 284). It was suggested by the authors that it was surprising that other bonds, such as the Tyr-Ile or Ile-Leu, were not cleaved, given the general preference for bulky amino acids in P1 and P1′; however, cleavage at Tyr-Ile would place Pro in the P2 position and cleavage at Ile-Leu would place the $\beta$-branched amino acid in P1, and both are not acceptable.

### Structural Chemistry

The amino acid sequence of rhizopuspepsin isozyme pI 5.1 was determined by Delaney *et al.* (1987). In the same issue of the *J. Biol. Chem.*, Takahashi (1987) reported the sequence analysis of a crystalline preparation containing both the pI 5.1 and 5.8 isozymes, and noted the presence of two divergent amino acid residues at eight positions. The following year, Takahashi (1988) presented the analysis of both isozymes, following separation by isoelectric focusing. The dichotomy at the eight positions was thus resolved as follows: position 15: pI 5.1 = Ile, pI 5.8 = Val; position 61: pI 5.1 = Asn, pI 5.8 = Lys; position 116: pI 5.1 = Ser, pI 5.8 = Asn; position 162: pI 5.1 = Lys, pI 5.8 = Ser; position 230: pI 5.1 = Ile, pI 5.8 = Val; position 241: pI 5.1 = Tyr, pI 5.8 = Ser; position 293: pI 5.1 = Asp, pI 5.8 = Asn; and position 325: pI 5.1 = Glu, pI 5.8 = Gln.

Rhizopuspepsin has served as an excellent model system for many studies of the biochemistry of the acid proteinases, largely due to the work of the Davies laboratory in the determination of the structure of this protein to very high resolution (Suguna *et al.*, 1987a). Of note, the position of the mobile 'flap' segment, a $\beta$ hairpin that partly covers the active-site cleft, was well defined in this structure. The similarities with other members of the aspartic proteinase family were noted. Subsequently, the binding of inhibitors to the enzyme was studied (Suguna *et al.*, 1987b, 1992), providing significant detail on active-site interactions.

Also of interest at the structural level is the report by Fukuda *et al.* (1994) demonstrating that the pro part segment is essential for correct folding and secretion of the mature form of the related enzyme from *Rhizopus niveus*. This effect of a prosegment is believed to be general for enzymes of this class.

Because of the high-resolution structure, a number of authors have used the rhizopuspepsin system as a model to study the pH dependence of activity (Goldblum, 1988), the use of modeling to predict the structures of new proteins (Summers & Karplus, 1989) or aspects of the motion of protein chains (Mao, 1992), as a few examples.

### Preparation

The preferred method of preparation of the native enzyme was described by Ohtsuru *et al.* (1982) following earlier reports by Fukumoto *et al.* (1967) and Kurono *et al.* (1971). The method uses liquid cultures of the mold *R. chinensis*, where the enzyme is secreted into the culture medium. Ammonium sulfate precipitation is employed to concentrate the protein and, following re-solubilization, affinity chromatography on pepstatin-Sepharose yields a preparation consisting of the mixture of isoenzymes. Isoelectric focusing may be used to separate the five species present, with the two major isoforms being of pI 5.1 and 5.8 (Ohtsuru *et al.*, 1982). A more recent report on the preparation of the related enzyme from *R. niveus* has appeared (Horiuchi *et al.*, 1990).

The preparation of the recombinant form of rhizopuspepsinogen, the precursor, has been described by Chen *et al.* (1991). The method requires the expression of the protein in *E. coli*, isolation of inclusion bodies, solubilization, and refolding by dialysis to lower pH. Final purification is provided by ion-exchange and gel filtration chromatography. These methods were used by Lowther *et al.* (1995) in their studies of the properties and specificity of mutant forms of the enzyme.

### Biological Aspects

The mold peptidases are active in the extracellular milieu surrounding the growing fungi. They are believed to be secreted to assist in the degradation of macromolecules in the environment to provide nutrients for the fungal cells.

### Related Peptidases

A report has appeared describing the properties of a related enzyme from *Rhizopus hangchow* (Ichishima *et al.*, 1995). Preparation of the enzyme from *R. niveus* is mentioned above (Horiuchi *et al.*, 1990).

### References

Carraway, R.E., Mitra, S.P. & Salmonsen, R. (1992) Pepsin-mediated processing of synthetic precursor-like sequence yields neurotensin-like peptide. *Peptides* **13**, 319–322.

Chen, Z., Koelsch, G., Han, H.P., Wang, X.J., Lin, X.L., Hartsuck, J.A. & Tang, J. (1991) Recombinant rhizopuspepsinogen. Expression, purification, and activation properties of recombinant rhizopuspepsinogens. *J. Biol. Chem.* **266**, 11718–11725.

Delaney, R., Wong, R.N., Meng, G.Z., Wu, N.H. & Tang, J. (1987) Amino acid sequence of rhizopuspepsin isozyme pI 5. *J. Biol. Chem.* **262**, 1461–1467.

Dunn, B.M., Scarborough, P.E., Lowther, W.T. & Rao-Naik, C. (1995) Comparison of the active site specificity of the aspartic proteinases based on a systematic series of peptide substrates. *Adv. Exp. Med. Biol.* **362**, 1–9.

Fukuda, R., Horiuchi, H., Ohta, A. & Takagi, M. (1994) The pro-sequence of *Rhizopus niveus* aspartic proteinase-I supports correct folding and secretion of its mature part in *Saccharomyces cerevisiae. J. Biol. Chem.* **269**, 9556–9561.

Fukumoto, J., Tsuru, D. & Yamamoto, T. (1967) Studies on mold proteases. Part I. Purification, crystallization and some enzymatic properties of acid protease of *Rhizopus chinensis. Agric. Biol. Chem. (Tokyo)* **31**, 710–717.

Goldblum, A. (1988) Theoretical calculations on the acidity of the active site in aspartic proteinases. *Biochemistry* **27**, 1653–1658.

Hofmann, T. & Hodges, R.S. (1982) A new chromophoric substrate for penicillopepsin and other fungal aspartic proteinases. *Biochem. J.* **203**, 603–610.

Hofmann, T., Hodges, R.S. & James, M.N.G. (1984) Effect of pH on the activities of penicillopepsin and *Rhizopus* pepsin and a proposal for the productive substrate binding mode in penicillopepsin. *Biochemistry* **23**, 635–643.

Horiuchi, H., Ashikari, T., Amachi, T., Yoshizumi, H., Takagi, M. & Yano, K. (1990) High-level secretion of a *Rhizopus niveus* aspartic proteinase in *Saccharomyces cerevisiae. Agric. Biol. Chem.* **54**, 1771–1779.

Ichishima, E., Ojima, M., Yamagata, Y., Hanzawa, S. & Nakamura, T. (1995) Molecular and enzymatic properties of an aspartic proteinase from *Rhizopus hangchow. Phytochemistry* **38**, 27–30.

Kunitz, M. (1938) Formation of trypsin from trypsinogen by an enzyme produced by a mold of the genus *Penicillium. J. Gen. Physiol.* **21**, 601–620.

Kurono, Y., Chidimatsu, M., Horikoshi, K. & Ikeda, Y. (1971) Isolation of a protease from a *Rhizopus* product. *Agric. Biol. Chem.* **35**, 1668–1675.

Lowther, W.T. & Dunn, B.M. (1995) Site-directed mutagenesis of rhizopuspepsin: an analysis of unique specificity. *Adv. Exp. Med. Biol.* **362**, 555–558.

Lowther, W.T., Chen, Z., Lin, X.L., Tang, J. & Dunn, B.M. (1991) Substrate specificity study of recombinant *Rhizopus chinensis* aspartic proteinase. *Adv. Exp. Med. Biol.* **306**, 275–279.

Lowther, W.T., Majer, P. & Dunn, B.M. (1995) Engineering the substrate specificity of rhizopuspepsin: the role of Asp 77 of fungal aspartic proteinases in facilitating the cleavage of oligopeptide substrates with lysine in P1. *Protein Sci.* **4**, 689–702.

Mao, B. (1992) Molecular-dynamics investigation of molecular flexibility in ligand binding. *Biochem. J.* **288**, 109–116.

Morihara, K. & Oka, T. (1973) Comparative specificity of microbial acid proteinases for synthetic peptides. III. Relationship with their trypsinogen activating ability. *Arch. Biochem. Biophys.* **157**, 561–572.

Ohtsuru, M., Tang, J. & Delaney, R. (1982) Purification and characterization of rhisopuspepsin isozymes from a liquid culture of *Rhizopus chinensis. Int. J. Biochem.* **14**, 925–932.

Oka, T. & Morihara, K. (1974) Comparative specificity of microbial acid proteinases for synthetic peptides. Primary specificity with Z-tetrapeptides. *Arch. Biochem. Biophys.* **165**, 65–71.

Pohl, J. & Dunn, B.M. (1988) Secondary enzyme-substrate interactions: kinetic evidence for ionic interactions between substrate side chains and the pepsin active site. *Biochemistry* **27**, 4827–4834.

Sodek, J. & Hofmann, T. (1971) Microbial acid proteinases. *Methods Enzymol.* **19**, 371–397.

Suguna, K., Bott, R.R., Padlan, E.A., Subramanian, E., Sheriff, S., Cohen, G.H. & Davies, D.R. (1987a) Structure and refinement at 1.8 Å resolution of the aspartic proteinase from *Rhizopus chinensis. J. Mol. Biol.* **196**, 877–900.

Suguna, K., Padlan, E.A., Smith, C.W., Carlson, W.D. & Davies, D.R. (1987b) Binding of a reduced peptide inhibitor to the aspartic proteinase from *Rhizopus chinensis*: implications for a mechanism of action. *Proc. Natl Acad. Sci. USA* **84**, 7009–7013.

Suguna, K., Padlan, E.A., Bott, R., Boger, J., Parris, K.D. & Davies, D.R. (1992) Structures of complexes of rhizopuspepsin with pepstatin and other statine-containing inhibitors. *Proteins* **13**, 195–205.

Summers, N.L. & Karplus, M. (1989) Construction of side-chains in homology modelling. Application to the C-terminal lobe of rhizopuspepsin. *J. Mol. Biol.* **210**, 785–811.

Takahashi, K. (1987) The amino acid sequence of rhizopuspepsin, an aspartic proteinase from *Rhizopus chinensis. J. Biol. Chem.* **262**, 1468–1478.

Takahashi, K. (1988) Determination of the amino acid sequences of the two major isozymes of rhizopuspepsin. *J. Biochem. (Tokyo)* **103**, 162–167.

Tsuru, D., Hattori, A., Tsuji, J., Yamamoto, T. & Fukumoto, J. (1969) Studies on mold proteases. Part II. Substrate specificity of acid protease of *Rhizopus chinensis. Agric. Biol. Chem.* **33**, 1419–1426.

*Ben M. Dunn*
*Department of Biochemistry & Molecular Biology,*
*University of Florida College of Medicine,*
*Gainesville, FL 32610-0245, USA*
*Email: bdunn@biochem.med.ufl.edu*

# 298. Mucorpepsin

## Databanks

*Peptidase classification: clan AA, family A1, MEROPS ID: A01.013*
*NC-IUBMB enzyme classification: EC 3.4.23.23*

*Chemical Abstracts Service registry number: 148465-73-0*
*Databank codes:*

| Species | SwissProt | PIR | EMBL (cDNA) | EMBL (genomic) |
|---|---|---|---|---|
| *Rhizomucor miehei* | P00799 | A00992 A26537 A29039 | A02534 M18411 | M15267: complete gene (no introns) |
| *Rhizomucor pusillus* | P09177 | A25767 A34411 S01118 | – | X06219: complete gene (no introns) |

Brookhaven Protein Data Bank three-dimensional structures:

| Species | ID | Resolution | Notes |
|---|---|---|---|
| *Rhizomucor pusillus* | 1MPP | 2 | |

## Name and History

*Mucorpepsin*, one of the microbial aspartic proteinases, was discovered as a milk-clotting enzyme for cheese manufacturing when a shortage of calf chymosin (Chapter 274) arose in the 1960s (Arima *et al*., 1967; Ottesen & Rickert, 1970). Mucorpepsin is now the major microbial milk-clotting enzyme used in industry, although endothiapepsin (see Chapter 271) from *Endothia parasitica*, is also used to a small extent. The enzyme is produced by microorganisms that are zygomycete fungi, originally classified as *Mucor pusillus* and *Mucor miehei*, but now reclassified in the genus *Rhizomucor* (Schipper, 1978). Accordingly, the enzyme has been named with a variety of combinations of these keywords such as **Mucor pusillus pepsin (rennin)**, **Mucor miehei pepsin (rennin)**, and **aspartic proteinases** of *M. pusillus* and *M. miehei*. But these two fungal species are taxonomically very close to each other and the structures of their enzymes are also very similar. The name mucorpepsin has been recommended by IUBMB, and the enzymes from the two species are dealt with as variants.

## Activity and Specificity

The pH optimum of mucorpepsin is around 4, although the exact pH varies with the substrate. Proteolytic activity can be measured with casein, acid-denatured hemoglobin or Leu-Ser-Phe(NO$_2$)+Nle-Ala-Leu-OMe as the substrates (Anson, 1938; Schlamowitz & Peterson, 1959; Somkuti & Babel, 1968; Martin *et al*., 1980). This synthetic substrate is available from Bachem (see Appendix 2 for names and addresses of suppliers).

Rhizomucorpepsin has high milk-clotting activity along with relatively low proteolytic activity, which enables it to be used as a milk coagulant. The enzyme preferentially cleaves the Phe105+Met106 bond of $\kappa$-casein; this destabilizes casein micelles to cause coagulation of milk. Milk-clotting activity can be determined based on the time required for clotting a 10% solution of skim-milk powder (Arima *et al*., 1970). The amount of the enzyme that can clot the milk solution per minute is defined as 400 units. Specific activity is expressed as the milk-clotting activity per absorbance unit of protein at 280 nm.

The enzymatic properties of mucorpepsin are similar to those of chymosin (Richardson *et al*., 1967). Tabulated studies of cleavage of several peptide substrates and oxidized insulin B chain reveals that the enzyme favors hydrophobic residues at P1 and P1′ (Somkuti & Babel, 1968; Rickert, 1970; Sternberg, 1972; Oka *et al*., 1973). The enzyme does not accept Lys at P1 and hence does not activate trypsinogen.

Mucorpepsin is inhibited by general inhibitors for aspartic proteinases such as DAN, 1,2-epoxy-3-(*p*-nitrophenoxy)propane (EPNP) and pepstatin (Takahashi *et al*., 1976).

## Structural Chemistry

Mucorpepsin variants produced by *R. pusillus* and *R. miehei* show 83% sequence identity and are immunologically cross-reactive (Etoh *et al*., 1979; Boel *et al*., 1986; Gray *et al*., 1986; Tonouchi *et al*., 1986). Like the other members of family A1, the enzyme is produced as a zymogen with a propeptide of 44 amino acids in the *R. pusillus* enzyme (Tonouchi *et al*., 1986) and of 47 in the *R. miehei* enzyme (Boel *et al*., 1986; Gray *et al*., 1986). Three and two possible *N*-linked glycosylation sites were found in the amino acid sequences of the proteases from *R. pusillus* (Asn79, Asn113, Asn188) and *R. miehei*, (Asn79 and Asn188), respectively. In the *R. pusillus* enzyme, Asn79 and Asn188 were shown to be glycosylated when expressed in *Saccharomyces cerevisiae* and possibly in the original fungus (Aikawa *et al*., 1990; Murakami *et al*., 1993), although commercial preparations of the *R. pusillus* enzyme contain only traces of carbohydrates. The three-dimensional structure of the *R. pusillus* enzyme has been determined by X-ray crystallographic analyses (Newman *et al*., 1993). Site-directed mutagenesis reveals a contribution of Tyr75 (pepsin numbering) on the flap for the catalytic activity (Park *et al*., 1996).

## Preparation

The enzyme is produced on an industrial scale from filtrates of liquid fermentation broth of *R. miehei* (Sternberg, 1971) or the extracts of solid cultures of *R. pusillus* grown on wheat bran (Iwasaki *et al.*, 1967).

## Biological Aspects

The gene structure of mucorpepsin has been determined (Boel *et al.*, 1986; Gray *et al.*, 1986; Tonouchi *et al.*, 1986). The genes of both *R. pusillus* and *R. miehei* contain no introns. The *R. pusillus* gene encodes a precursor composed of a signal sequence of 22 amino acids for secretion, a propeptide of 44 amino acids, and a mature protease part of 361 amino acids (Tonouchi *et al.*, 1986), whereas the *R. miehei* precursor is composed of a signal sequence of 22 amino acids, a propeptide of 47 amino acids, and a mature protease part of 361 amino acids (Boel *et al.*, 1986; Gray *et al.*, 1986). In both cases, no data are available on the region required for promoter function or on the transcriptional initiation site.

The genes are efficiently expressed in heterologous expression systems: *S. cerevisiae* and *Aspergillus oryzae* (Yamashita *et al.*, 1987; Christensen *et al.*, 1988). In the yeast expression system, the enzyme is excreted into the culture medium in a form of zymogen, which is converted to the mature form mainly by autocatalytic proteolysis (Hiramatsu *et al.*, 1989).

## Further Reading

Beppu *et al.* (1995) report a study using the *R. pusillus* rennin to confirm that the role of Tyr75 is conserved here as in all the aspartic proteinases.

## References

Aikawa, J., Yamashita, T., Nishiyama, M., Horinouchi, S. & Beppu, T. (1990) Effects of glycosylation on the secretion and enzyme activity of *Mucor* rennin, an aspartic proteinase of *Mucor pusillus*, produced by recombinant yeast. *J. Biol. Chem.* **265**, 13955–13959.

Anson, M.I. (1938) Estimation of pepsin, papain and cathepsin with haemoglobin, *J. Gen. Physiol.* **22**, 79–89.

Arima, K., Iwasaki, S. & Tamura, G. (1967) Milk clotting enzyme from microorganisms. Part I. Screening test and the identification of the potent fungus. *Agric. Biol. Chem.* **31**, 540–545.

Arima, K., Yu, J. & Iwasaki, S. (1970) Milk-clotting enzyme from *Mucor pusillus* var. Lindt. *Methods Enzymol.* **19**, 446–459.

Beppu, T., Park, Y.-N., Aikawa, J., Nishiyama, M. & Horinouchi, S. (1995) Tyrosine 75 on the flap contributes to enhance catalytic efficiency of a fungal aspartic proteinase, *Mucor pusillus* pepsin. In: *Aspartic Proteinase: Structure, Function, Biology, and Biomedical Implications* (Takahashi, T., ed.). New York: Plenum Press, pp. 559–563.

Boel, E., Bech, A.M., Randrup, K., Draeger, B., Fiil, N.P. & Foltmann, B. (1986) Primary structure of a precursor to the aspartic proteinase from *Rhizomucor miehei* shows that the enzyme is synthesized as a zymogen. *Proteins* **1**, 363–369.

Christensen, T., Woeldike, H., Boel, E., Mortensen, S.B., Hjortshoej, K., Thim, L. & Hansen, M.T. (1988) High-level expression of recombinant genes in *Aspergillus oryzae*. *Bio/Technology* **6**, 1419–1422.

Etoh, Y., Shoun, H., Beppu, T. & Arima, K. (1979) Physicochemical and immunochemical studies on similarities of acid proteases *Mucor pusillus* rennin and *Mucor miehei* rennin. *Agric. Biol. Chem.* **43**, 209–215.

Gray, G.L., Hayenga, K., Cullen, D., Wilson, L.J. & Norton, S. (1986) Primary structure of *Mucor miehei* aspartyl protease: evidence for a zymogen intermediate. *Gene* **48**, 41–53.

Hiramatsu, R., Aikawa, J., Horinouchi, S. & Beppu, T. (1989) Secretion by yeast of the zymogen form of *Mucor rennin*, an aspartic proteinase of *Mucor pusillus*, and its conversion to the mature form. *J. Biol. Chem.* **264**, 16862–16866.

Iwasaki, S., Tamura, G. & Arima, K. (1967) Milk clotting enzyme from microorganisms. Part II. The enzyme production and the properties of crude enzyme. *Agric. Biol. Chem.* **31**, 546–551.

Martin, P., Raymond, M.-N., Bricas, E. & Ribadeau Ducas, B. (1980) Kinetic studies on the action of *Mucor pusillus*, *Mucor miehei* acid proteases and chymosins A and B on a synthetic chromophoric hexapeptide. *Biochim. Biophys. Acta* **612**, 410–420.

Murakami, K., Aikawa, J., Horinouchi, S. & Beppu, T. (1993) Characterization of an aspartic proteinase of *Mucor pusillus* expressed in *Aspergillus oryzae*. *Mol. Gen Genet.* **241**, 312–318.

Newman, M., Watson, F., Roychowdhury, P., Jones, H., Badasso, M., Cleasby, A., Wood, S.P., Tickle, I.J. & Blundell, T.L. (1993) X-ray analyses of aspartic proteinases. V. Structure and refinement at 2.0 Å resolution of the aspartic proteinase from *Mucor pusillus*. *J. Mol. Biol.* **230**, 260–283.

Oka, T., Ishino, K., Tsuzuki, H., Morihara, K. & Arima, K. (1973) On the specificity of a rennin-like enzyme from *Mucor pusillus*. *Agric. Biol. Chem.* **37**, 1177–1184.

Ottesen, M. & Rickert, W. (1970) The isolation and partial characterization of an acid protease produced by *Mucor miehei*. *C. R. Trav. Lab. Carlsberg* **37**, 301–325.

Park, Y.-N., Aikawa, J., Nishiyama, M., Horinouchi, S. & Beppu, T. (1996) Involvement of a residue at position 75 in the catalytic mechanism of a fungal aspartic proteinase, *Rhizomucor pusillus* pepsin. Replacement of tyrosine 75 on the flap by asparagine enhances catalytic efficiency. *Protein Eng.* **9**, 869–875.

Richardson, G.H., Nelson, J.H., Lubnow, R.E. & Schwarberg, R.L. (1967) Rennin-like enzyme from *Mucor pusillus* for cheese manufacture. *J. Dairy Sci.* **50**, 1066–1072.

Rickert, W. (1970) The degradation of the B-chain of oxidized insulin by *Mucor miehei* protease. *C. R. Trav. Lab. Carlsberg* **38**, 1–17.

Schipper, M.A.A. (1978) On the genera *Rhizomucor* and *Parasitella*. In: *Studies in Mycology*. No. 17. *Inst. Roy. Netherlands Acad. Sci. Lett.* 53–71.

Schlamowitz, M. & Peterson, L.U. (1959) Studies on the optimum pH for the action of pepsin on native and denatured bovine serum albumin and bovine haemoglobin. *J. Biol. Chem.* **234**, 3137–3145.

Somkuti, G.A. & Babel, F.J. (1968) Purification and properties of *Mucor pusillus* acid protease. *J. Bacteriol.* **95**, 1407–1414.

Sternberg, M.Z. (1971) Crystalline milk-clotting protease from *Mucor miehei* and some of its properties. *J. Dairy Sci.* **54**, 150–167.

Sternberg, M. (1972) Bond specificity, active site and milk clotting mechanism of the *Mucor miehei* protease. *Biochim. Biophys. Acta* **285**, 383–392.

Takahashi, K., Chang, W.J. & Arima, K. (1976) The structure and function of acid proteases. IV. Inactivation of the acid protease from *Mucor pusillus* by acid protease-specific inhibitors. *J. Biochem. (Tokyo)* **80**, 61–67.

Tonouchi, N., Shoun, H., Uozumi, T. & Beppu, T. (1986) Cloning and sequencing of a gene for *Mucor* rennin, an aspartate protease from *Mucor pusillus*. *Nucleic Acids Res.* **14**, 7557–7568.

Yamashita, T., Tonouchi, N., Uozumi, T. & Beppu, T. (1987) Secretion of *Mucor* rennin, a fungal aspartic protease of *Mucor pusillus*, by recombinant yeast cells. *Mol. Gen. Genet.* **210**, 462–467.

***Teruhiko Beppu***
*Department of Applied Biological Sciences,*
*College of Bioresource Sciences,*
*Nihon University,*
*Kameino 1866, Fujisawa-shi, Kanagawa 252, Japan*
*Email: beppu@brs.nihon-u.ac.jp*

***Makoto Nishiyama***
*Biotechnology Research Center,*
*The University of Tokyo,*
*Yayoi 1-1-1, Bunkyo-ku, Tokyo 113, Japan*
*Email: umanis@hongo.ecc.u-tokyo.ac.jp*

# 299. *Polyporopepsin*

## Databanks

*Peptidase classification: clan AA, family A1, MEROPS ID: A01.019*
*NC-IUBMB enzyme classification: EC 3.4.23.29*
*Chemical Abstracts Service registry number: 61573-73-7*
*Databank codes:*

| Species | SwissProt | PIR | EMBL (cDNA) | EMBL (genomic) |
|---|---|---|---|---|
| *Polyporus tulipiferae* | P17576 | JU0057 | D00589 | – |

## Name and History

The name **polyporopepsin** denotes a pepsin-like enzyme from *Polyporus tulipiferae* (formerly *Irpex lacteus*). Polyporopepsin was found as a calf rennet substitute among various strains of basidiomycetes and some properties were investigated (Kawai, 1970; Kawai & Mukai, 1970; Kawai, 1971). Polyporopepsin was purified by methods including affinity chromatography with dehydroacetylpepstatin as a ligand (Kobayashi *et al.*, 1983a). It was designated as a milk-clotting enzyme because it showed high milk-coagulating activity compared to other aspartic proteinases.

## Activity and Specificity

Polyporopepsin is most active at pH 2.8–2.9 against hemoglobin and casein. The ratio of milk-clotting activity to caseinolytic activity of the enzyme at pH 6.0 is almost equal to that of mucorpepsin (Chapter 298) and higher than that of endothiapepsin (Chapter 296) (Kobayashi *et al.*, 1983a).

The major cleavage sites of polyporopepsin in the B chain of oxidized insulin are Leu11+Val12, Ala14+Leu15, Phe24+Phe25 and Thr27+Pro28 at pH 3.0; the Ala14+Leu15 bond is preferred.

FVNQHLCGSHL+VEA+LYLVCGERGF+FYT+PKA

The specificity of polyporopepsin is distinct from that of other microbial milk-clotting enzymes and it has a more restricted specificity than chymosin (Chapter 274) and pig pepsin (Chapter 272) (Kobayashi *et al.*, 1983b) on the insulin B chain.

The enzyme cleaves the Phe23+Phe24 bond of $\alpha_{s1}$-casein at pH 6.0 as well as calf chymosin does (Mulvihill *et al.*, 1979a). Microbial aspartic proteinases that activate trypsinogen usually hydrolyze the Lys-X bond (Morihara & Oka, 1973). While the Lys103+Tyr104 bond of $\alpha_{s1}$-casein is cleaved by polyporopepsin, this enzyme does not activate trypsinogen as well as pig pepsin and mucorpepsins (Kobayashi *et al.*, 1985a).

Polyporopepsin hydrolyzes the Phe105+Met106 bond of $\kappa$-casein and subsequent milk coagulation occurs in the same manner as with other milk-clotting enzymes (Lawrence *et al.*, 1969; Kobayashi *et al.*, 1985b). Polyporopepsin and chymosin cleave Leu165+Ser166, Ala189+Phe190 and Tyr192+Glu193 bonds of $\beta$-casein. The differences in specificity between polyporopepsin and chymosin are exhibited by the cleavage at Leu139+Leu140 bond of $\beta$-casein by chymosin and the Ser142+Trp143 bond by polyporopepsin (Mulvihill & Fox, 1979b; Kobayashi *et al.*, 1985b). From these results polyporopepsin is specific for the peptide bonds formed by one or two amino acid(s) with large hydrophobic side chain on $\alpha_{s1}$- and $\beta$-casein. Furthermore, polyporopepsin always requires hydrophobic amino acids in the P4 and/or P3 sites (Kobayashi *et al.*, 1985b).

## Structural Chemistry

The $M_r$ of polyporopepsin calculated from the nucleotide sequence of the cDNA is 35 000; this value is smaller than that found by SDS-PAGE (36 000). Two putative *N*-linked glycosylation sites are found at Asn192 and Asn228 of polyporopepsin. The difference seems to be attributable to glycosylation (Kawai, 1971) because of the affinity for various lectins including concanavalin A, wheatgerm agglutinin and *Ricinus communis* agglutinin (Kobayashi *et al.*, 1989).

The inactivation of polyporopepsin by active-site directed inhibitors of pepsin such as DAN and 1,2-epoxy-(*p*-nitrophenoxy)propane (EPNP) implies that the catalytic mechanism of the enzyme is similar to that of the other aspartic proteinases (Kobayashi *et al.*, 1983a). Comparison of the amino acid sequence of polyporopepsin with those of other aspartic proteinases suggests that Asp32 and Asp212 in polyporopepsin are the active-site residues because the regions Asp32–Ser35 and Asp212–Thr215 are highly conserved among those enzymes.

One striking feature of the amino acid composition of polyporopepsin is the high content of Thr and Ser, as in endothiapepsin. The content of Thr and Ser in polyporopepsin is 30%; for both residues this is double the average values for other proteins (Barkholt, 1987). The other striking feature of polyporopepsin is the absence of cysteine and methionine in its primary structure. This suggests that polyporopepsin is the first aspartic proteinase that contains no disulfide bridge in the molecule (Kobayashi *et al.*, 1989). The number of disulfide bonds is three in vertebrate aspartic proteinases, two in rhizopuspepsin (Chapter 297), mucorpepsin (Chapter 298) and saccharopepsin (Chapter 283), one in endothiapepsin and penicillopepsin (Chapter 295), and none in polyporopepsin. The lack of disulfide bridges might be one of the reasons why polyporopepsin is unstable above neutral pH (Kikuchi *et al.*, 1988).

The pI of the enzyme is 5.3. Single crystals ($1 \times 0.5 \times 0.2$ mm) have been obtained using the hanging drop method with ammonium sulfate as a precipitant. The crystals are monoclinic, space group $P2_1$, with cell dimensions $a = 54.5$ Å, $b = 79.6$ Å, $c = 37.5$ Å and $\beta = 96.8$. The $V$ value is 2.31 Å Da$^{-1}$, assuming one molecule asymmetric unit. The solvent content is 47% by volume. Crystals diffract to beyond 1.9 Å resolution and are generally extremely stable to X-rays (Kobayashi *et al.*, 1992).

## Preparation

A medium containing distillers' solubles was used to produce polyporopepsin (Kobayashi *et al.*, 1989). The enzyme was purified from the culture filtrate by ammonium sulfate precipitation (crude enzyme preparation), chromatographies on dehydroacetylpepstatin-aminohexylagarose and DEAE-cellulose columns, and isoelectric focusing. Eight mg of the purified polyporopepsin was obtained from 1 g of crude enzyme and the enzyme was purified 20-fold from the crude preparation (Kobayashi *et al.*, 1983a).

## Biological Aspects

Nothing is known about the gene for polyporopepsin. The enzyme is secreted into the culture medium (Kawai & Mukai, 1970).

## Distinguishing Features

Polyporopepsin is the least heat stable among milk-clotting enzymes and this is probably due to the lack of disulfide bridges in the molecule. Rabbit polyclonal antibodies have been produced against polyporopepsin (H. Kobayashi, unpublished results).

## Further Reading

Kobayashi (1987) provides an extensive review of properties and substrate specificity of the enzyme.

## References

Barkholt, V. (1987) Amino acid sequence of endothiapepsin. Complete primary structure of the aspartic proteinase from *Endothia parasitica*. *Eur. J. Biochem.* **167**, 327–338.

Kawai, M. (1970) Studies on milk clotting enzymes produced by basidiomycetes. Part II. Some properties of basidiomycete milk clotting enzymes. *Agric. Biol. Chem.* **34**, 164–169.

Kawai, M. (1971) Studies on milk clotting enzymes produced by basidiomycetes. Part III. Partial purification and some properties of the enzyme produced by *Irpex lacteus*. *Agric. Biol. Chem.* **35**, 1517–1525.

Kawai, M. & Mukai, N. (1970) Studies on milk clotting enzyme produced by basidiomycetes. Part. I. Screening tests of basidiomycetes for the production of milk clotting enzyme. *Agric. Biol. Chem.* **34**, 159–163.

Kikuchi, E., Kobayashi, H., Kusakabe, I. & Murakami, K. (1988) Fiber-structured cheese making with *Irpex lacteus* milk-clotting enzyme. *Agric. Biol. Chem.* **52**, 1277–1278.

Kobayashi, H. (1987) Milk-clotting enzyme from *Irpex lacteus*. *Farm Animals* **2**, 1–13.

Kobayashi, H., Kusakabe, I. & Murakami, K. (1983a) Purification and characterization of two milk-clotting enzymes from *Irpex lacteus*. *Agric. Biol. Chem.* **47**, 551–558.

Kobayashi, H., Kusakabe, I. & Murakami, K. (1983b) Substrate specificity of a carboxyl proteinase from *Irpex lacteus*. *Agric. Biol. Chem.* **47**, 1921–1923.

Kobayashi, H., Kusakabe, I. & Murakami, K. (1985a) Substrate specificity of the milk-clotting enzyme from *Irpex lacteus* on αs1-casein. *Agric. Biol. Chem.* **49**, 1611–1619.

Kobayashi, H., Kusakabe, I., Yokoyama, S. & Murakami, K. (1985b) Substrate specificity of the milk-clotting enzyme from *Irpex lacteus* on κ- and β-casein. *Agric. Biol. Chem.* **49**, 1621–1631.

Kobayashi, H., Sekibata, S., Shibuya, H., Yoshida, S., Kusakabe, I. & Murkakami, K. (1989) Cloning and sequence analysis of cDNA for *Irpex lacteus* aspartic proteinase. *Agric. Biol. Chem.* **53**, 1927–1933.

Kobayashi, H., Kasamo, K., Mizuno, H., Kim, H., Kusakabe, I. & Murakami, K. (1992) Crystallization and preliminary X-ray diffraction studies of aspartic proteinase from *Irpex lacteus*. *J. Mol. Biol.* **226**, 1291–1293.

Lawrence, R.C. & Creamer, L.K. (1969) Action of calf rennet and other proteolytic enzymes on κ-casein. *J. Dairy Res.* **36**, 11–20.

Morihara, K. & Oka, T. (1973) Comparative specificity of microbial acid proteinases for synthetic peptides. III. Relationship with their trypsinogen activating ability. *Arch. Biochem. Biophys.* **157**, 561–572.

Mulvihill, O.M. & Fox, P.F. (1979a) Proteolytic specificity of chymosin on bovine αs1-casein. *J. Dairy Res.* **46**, 641–651.

Mulvihill, O.M. & Fox, P.F. (1979b) Proteolytic specificity of chymosins and pepsins on β caseins. *Milchwissenschaft* **34**, 680–683.

*Hideyuki Kobayashi*
*Molecular Function Laboratory,*
*National Food Research Institute,*
*Ministry of Agriculture, Forestry and Fisheries,*
*Tsukuba, Ibaraki 305, Japan*
*Email: hkobayas@nfri.affrc.go.jp*

# 300. *Candidapepsin*

## Databanks

*Peptidase classification: clan AA, family A1, MEROPS ID: A01.014*
*NC-IUBMB enzyme classification: EC 3.4.23.24*
*Chemical Abstracts Service registry number: 69458-91-9*
*Databank codes:*

| Species | Gene | SwissProt | PIR | EMBL (cDNA) | EMBL (genomic) |
|---|---|---|---|---|---|
| *Candida albicans* | *apr1* | P10977 | S03433 | U36754 X13669 | – |
| *Candida albicans* | *sap1* | P28872 | – | L12449 L12450 L12451 L12452 X56867 | – |
| *Candida albicans* | *sap2* | P28871 | A45280 | M83663 X62289 | – |
| *Candida albicans* | *sap3* | P43092 | – | – | L22358: complete gene |
| *Candida albicans* | *sap4* | P43093 | – | L25388 | – |
| *Candida albicans* | *sap5* | P43094 | – | Z30191 | – |
| *Candida albicans* | *sap6* | P43095 | – | Z30192 | – |
| *Candida albicans* | *sap7* | P43096 | – | Z30193 | – |
| *Candida parapsilosis*[a] | *sapp1* | P32951 | S20705 | Z11919 | – |
| *Candida parapsilosis*[a] | *sapp2* | P32950 | S20704 S34581 | Z11918 | – |
| *Candida tropicalis*[a] | *sapt1* | Q00663 | S16971 | X61438 | – |

[a]The last three entries are also included in Chapters 301 and 302.

Brookhaven Protein Data Bank three-dimensional structures:

| Species | Gene | ID | Resolution | Notes |
|---|---|---|---|---|
| *Candida albicans* | *sap2* | 1EAG | 2.1 | complex with A70450 |
| | | 1ZAP | 2.5 | complex with synthetic inhibitor |

## Name and History

Over 30 years ago, the extracellular proteolytic activity of *Candida albicans* was discovered by Staib (1963). The proteolytic activity was later identified as a pepsin-like aspartic proteinase (MacDonald & Odds, 1980a). Subsequently, production of *Candida* proteinase during candidiasis was demonstrated by the appearance of specific antibodies in patients (MacDonald & Odds, 1980b), the detection of proteinase antigen in the blood of patients (Rüchel & Böning, 1983), and the demonstration of proteinase antigen in infected human tissue (Rüchel *et al.*, 1991). Therefore, it was suggested that *Candida* proteinase might play a role in the pathogenesis of candidiasis and in the virulence of these species. Circumstantial evidence supported this view in that highly proteolytic strains were reported to be more pathogenic than less proteolytic strains, chemically induced nonproteolytic mutants were found to be less pathogenic than the wild-type strains and treatment of infected mice with pepstatin, an inhibitor of aspartic proteinases, caused a discrete protective effect (Rüchel *et al.*, 1992). However, in an attempt to establish the role played by *Candida* proteinase, a molecular genetic approach was chosen.

Surprisingly, a whole family of genes *sap1–7* was detected, coding for related *secretory aspartic proteinases* or *SAPs* (Monod *et al.*, 1994). These genes are presumed to encode a multigene family and the regulation and differential expression of these genes is likely to be complex during the life cycle of the yeast. The amino acid sequences of the *C. albicans* SAPs cluster into three separate groups when aligned together. The percentage of identical residues is 68% between SAPs 1–3, 82% between SAPs 4–6, and 25% between SAP7 and any member of either group (Monod *et al.*, 1994).

As far as is known, only four of the seven *sap* genes of *C. albicans* are expressed. The expression of individual *sap* genes is specific during the *C. albicans* life cycle with respect to cell morphology or phenotype (Hube *et al.*, 1994). The *sap6* gene, for example, is expressed only during the virulent hyphal morphology (Hube *et al.*, 1994; White & Agabian, 1995). This phenomenon suggests that *C. albicans* may require stage-specific proteases (Odds, 1994) and that each enzyme may possess specific characteristics, including specificity, to meet the stage-specific functions. Knowledge of the distinct substrate specificity of SAPs from *C. albicans* is of both scientific and clinical interest (Goldman *et al.*, 1995). The name *candidapepsin* has been recommended by IUBMB, and another name that has been used is *candialbicin*, but for convenience, the seven forms of candidapepsin will here be refered to as SAP1–SAP7.

## Activity and Specificity

The SAP enzymes have an extremely broad specificity range. They are implicated in the degradation of host proteins involved in defense and barrier function, and recently *C. albicans* SAP2 has been shown to degrade the Fc portion of IgG, IgA, complement factor C3 and endogenous protease inhibitors in human serum (Kaminishi *et al.*, 1995). These attributes would be ideally suited to fungal virulence factors, which assist in an invasive stage of a disease.

Of the four SAPs secreted, only SAP2 is abundantly expressed in cultures of *C. albicans*. Its specificity has been studied in a number of laboratories and has been partially characterized enzymologically by Fusek *et al.* (1994). Cleavage of oxidized insulin B chain–

$$\text{FV} \!\mid\! \text{NQHLCGSHL} \!\mid\! \text{V} \!\mid\! \text{EA} \!\mid\! \text{LYLVCGERGFF} \!\mid\! \text{YTPKA}$$

and albumin by the SAP2 enzyme shows a low side-chain specificity (Remold *et al.*, 1968), although preferential cleavage between hydrophobic amino acids was observed during short incubation times. This characteristic, together with its acidic pH optimum, indicates its activity is similar to that of other fungal pepsins. A sensitive fluorogenic substrate with the chemical structure, 4-(4-dimethylaminophenylazo)benzoyl-$\gamma$-aminobutyryl-Ile-His-Pro-Phe-His-Leu-Val-Ile-His $\!\mid\!$ Thr-[5-(2-aminoethyl)-amino]naphthalene-1-sulfonic acid was used to monitor the enzyme production, purification and inhibition of SAP2. A $K_m$ of 4.3 mM was measured for this substrate at an optimum pH of 4.5. This substrate was cleaved at the terminal His-Thr bond, which is an unusual specificity for a pepsin-like enzyme (Capobianco *et al.*, 1992). The classic aspartic proteinase inhibitor pepstatin A was found to have an $IC_{50}$ of 27 nM when tested with the purified enzyme.

The SAP2 enzyme was further characterized for specificity by screening a library of synthetic spectrophotometric substrates of the type Lys-Pro-Ala-Leu-Phe $\!\mid\!$ Phe($p$-NO$_2$)-Arg-Leu (Fusek *et al.*, 1994). Highly efficient enzymatic reactions were observed for a variety of modified substrate sequences, which essentially differed only in the chemical identity of a few of the amino acid side chains at secondary specificity subsites adjacent to the scissile Phe $\!\mid\!$ Phe bond. An optimized peptide sequence Lys-Pro-Ala-Arg-Phe $\!\mid\!$ Nph-Arg-Leu was shown to have a $K_m$ of 8 μM at pH 3.25 with the SAP2 enzyme, and with a related substrate, a $K_i$ value of 6 nM for pepstatin A was determined. The results from the specificity screening were used to delineate subtle differences at the level of the primary amino acid sequence between the SAP2 enzyme and other related secreted aspartic proteinases purified from the yeasts *Candida tropicalis* (Chapter 302) and *Candida parapsilosis* (Chapter 301). The overall picture, which is derived from the specificity studies, is that these pepsin-related enzymes show a particularly broad and promiscuous substrate specificity, which may be related to their biological function in fungal virulence.

## Structural Chemistry

Seven gene sequences have been identified for the SAP enzymes. All *C. albicans* SAPs are apparently translated as prepro enzymes with an approximately 60 amino acid precursor part: a 14–21 amino acid secretion signal sequence, followed by a pro sequence (Monod *et al.*, 1994). Dibasic kexin-like cleavage sites are located within the prepro sequence and it is assumed that a kexin-like endopeptidase (Chapter 115) is present in the *C. albicans* cells. The mature enzyme has a mass of approximately 42 kDa and, based on comparisons of sequence identity, is distantly related to other fungal aspartic proteinases in the large aspartic proteinase

*Table 300.1*  General steps used in the purification of *Candida* yeast aspartic proteinases

| Candidapepsin (SAP2) | Canditropsin (SAPT1) | Candiparapsins (SAPP1/SAPP2) |
|---|---|---|
| Sterile supernatant | Sterile supernatant | Sterile supernatant |
| ↓ | ↓ | ↓ |
| Batch adsorption chromatography DEAE Trisacryl | Salt precipitation centrifugation | Batch adsorption chromatography SP-Trisacryl |
| ↓ | ↓ | ↓ |
| Dialysis | Dialysis | Dialysis |
| ↓ | ↓ | ↓ |
| FPLC Mono-Q | Low-pressure column chromatography, DEAE Trisacryl | FPLC Mono-S |
| ↓ | | ↓ |
| FPLC Mono-S | ↓ | Dialysis |
| ↓ | Dialysis | ↓ |
| Dialysis | ↓ | Yield SAPP1: 4.4 mg liter$^{-1}$ |
| ↓ | FPLC Mono-Q | |
| Yield: 4.2 mg liter$^{-1}$ | ↓ | Yield SAPP2: 1.0 mg liter$^{-1}$ |
| | Dialysis | |
| | ↓ | |
| | Yield: 5.1 mg liter$^{-1}$ | |

family (Abad-Zapatero *et al.*, 1996). Within the multigene family *saps* 1–3 are approximately 71–75% identical and 4–6 are more divergent, forming two subgroups. *sap7* is more distantly related with just 25% homology to the *sap1–6* genes. It is not known whether SAP7 is a functional proteinase.

The X-ray crystallographic structure of a SAP2–inhibitor complex has been solved to 2.1 Å by Cutfield *et al.* (1995). A similar structure of a SAP2X was solved as an enzyme–inhibitor complex by Abad-Zapatero *et al.* (1996). The *sap2x* enzyme differed in amino acid sequence from the published SAP2 gene sequence, and likely represents a clinical strain-specific isoform of the *sap2* enzyme. The structures are generally similar and share the typical aspartic proteinase fold, but differ quite dramatically at the active-site binding cleft. The binding cleft of the SAP2/SAP2X proteinase is considerably enlarged compared to pig pepsin (Chapter 272) and related aspartic proteinases in the family.

## *Preparation*

*Candida* yeasts are routinely maintained on Sabouraud agar plates from Difco (see Appendix 2 for full names and addresses of suppliers). *Candida* yeasts are grown typically in liquid culture containing 0.2% serum albumin and 1.2% YCB Difco medium, adjusted to pH 4.0. Fermentation is continued for up to 3–5 days and cells are removed by centrifugation followed by filtration to yield sterile supernatant. Proteinase is secreted into the supernatant to levels of up to 5 mg liter$^{-1}$. Sterile supernatant is used for all subsequent steps in the enzyme isolation. The sterile filtered supernatant can be stored at −70°C for several weeks before use without significant degradation of proteinase.

The isolation of proteolytic enzymes employs standard methods of salting-out or batch adsorption ion-exchange chromatography, low-pressure column chromatography, and FPLC as the final purification step, yielding homogeneous

samples suitable for protein crystallization (Fusek *et al.*, 1995). The schematic chart (Table 300.1) describes the general steps used for the purification of SAP2, SAPT1 and SAPP1 and SAPP2, and is referenced in later chapters dealing with related secreted *Candida* yeast proteinases (Chapters 301 and 302).

## *Biological Aspects*

Yeast species of the genus *Candida* form a large family of about 200 members (Barnett *et al.*, 1991). *Candida* yeasts are asexual yeasts with a variety of shapes and biochemical capabilities, both assimilative and fermentative; however, they lack sexual propagules and carotenoid pigments. The yeast species in the *Candida* genus are dimorphic fungi. The morphology of growth is dependent upon the growth conditions, they are observed to propagate as true budding yeast cells do, grow as buds that elongate and do not separate (pseudohyphae), form true hyphal elements and exist in any combination of these morphologies (Merz, 1990; Scherer & Magee, 1990; Jakab *et al.*, 1991; Rinaldi, 1993).

The predominant human pathogen classified in the genus *Candida* is *Candida albicans*. However, other species classified in this genus are also known to cause human infections. These include the species *C. tropicalis*, *C. parapsilosis* (older name *C. parakrusei*), *C. krusei*, *C. guilliermondii*, *C. lustiniae*, *C. glabrata* and others (Rinaldi, 1993). The yeast *C. albicans* is well adapted to the mucosal milieu of warm-blooded animals and is the major cause of superficial and invasive infections in immunocompromised human hosts, including transplant recipients, patients with HIV, as well as patients undergoing chemotherapy and other immunosuppressive regimens.

The spectrum of candidiasis caused by *C. albicans* as well as by other species includes thrush, vaginitis, skin and nail infections, pulmonary disease, enteritis, esophagitis,

*Table 300.2* Potential virulence factors of *Candida* infections

| Mechanisms | Molecular factors |
|---|---|
| Adherence/cell surface | Extracellular hydrolases |
| Dimorphism (hyphae production) | (proteinases and lipases) |
| Switching (phenotypic variability) | Toxins |
| Persorption | Killer toxins |
| Germ tube formation | Nitrosamines |
| Interference with phagocytosis, immune defenses, complement | Acidic metabolites |
| Synergism with bacteria | |

endocarditis, meningitis, brain abscesses, arthritis, keratomycosis, pyelonephritis, cystitis, septicemia, chronic mucocutaneous disease and other manifestations. Pathologic studies demonstrate three basic varieties of human candidiasis: the first, superficial lesions (e.g., infections of skin or respiratory tracts), the second, locally invasive (often in immunosuppressed patients in the form of pneumonia and other complications), and the third, deep or systemic (the most serious form recognizable by intraparenchymal lesions that usually involves the heart, kidneys, liver, spleen, lung and brain) (Cho & Choi, 1979; Luna & Tortoledo, 1993).

In general, the fungal infections carry a poor prognosis with a mortality rate of approximately 65% in transplant recipients and leukemia patients, and among fungal infections, the *Candida* infections are the most prevalent (Meunier-Carpentier *et al.*, 1981; Verfaillie *et al.*, 1991). The same alarming trend is observed for patients with HIV, where *Candida* infections represent the major fungal infections with extremely high mortality rates (Holmberg & Meyer, 1986; Mandal, 1989; Coleman *et al.*, 1993).

*Candida* species are clearly notorious opportunistic pathogens; that is, they incite disease in the host whose immune system has been impaired, damaged, or is inherently dysfunctional. Perhaps the major factor contributing to *Candida* virulence is the ability to persist on mucosal surfaces (Rüchel, 1990).

A short summary of suggested virulence factors is given in Table 300.2 (Rinaldi, 1993).

## Further Reading

A review of biological aspects is presented by Rüchel *et al.* (1992) and medical aspects are reviewed by Odds (1995).

## References

Abad-Zapatero, C., Goldman, R., Muchmore, S.W., Hutchins, C., Stewart, K., Navaza, J., Payne, C.D. & Ray, T.L. (1996) Structure of a secreted aspartic protease from *C. albicans* complexed with a potent inhibitor: implications for the design of antifungal agents. *Protein Sci.* **5**, 640–652.

Barnett, J.A., Payne, R.W. and Yarrow, D. (1991) *Yeasts – Characteristics and Identification.* New York: Cambridge University Press.

Capobianco, J.O., Lerner, C.G. & Goldman, R.C. (1992) Application of a fluorogenic substrate in the assay of proteolytic activity and in the discovery of a potent inhibitor of *Candida albicans* aspartic proteinase. *Anal. Biochem.* **204**, 96–102.

Cho, S.Y. & Choi, H.Y. (1979) Opportunistic infection among cancer patients. A ten year autopsy study. *Am. J. Clin. Pathol.* **72**, 617–621.

Coleman, D.C., Bennett, D.E., Sullivan, D.J., Gallagher, P.J., Henman, M.C., Shanley, D.B. & Russell, R.J. (1993) Oral *Candida* in HIV infection and AIDS: new perspectives/new approaches. *Crit. Rev. Microbiol.* **19**, 61–82.

Cutfield, S.M., Dodson, E.J., Anderson, B.F., Moody, P.C.E., Marshall, C.J., Sullivan, P.A. & Cutfield, J.F. (1995) The crystal structure of a major secreted aspartic proteinase from *Candida albicans* in complexes with two inhibitors. *Structure* **3**, 1261–1271.

Fusek, M., Smith, E.A., Monod, M., Dunn, B.M. & Foundling, S.I. (1994) Extracellular aspartic proteinases from *Candida albicans, Candida tropicalis,* and *Candida parapsilosis* yeast differ substantially in their specificities. *Biochemistry* **33**, 9791–9799.

Fusek, M., Smith, E. & Foundling, S.I. (1995) Extracellular aspartic proteinases from *Candida* yeasts. In: *Aspartic Proteinases: Structure, Function, Biology, and Biomedical Implications* (Takahashi, K. ed.). New York: Plenum Press, pp. 489–500.

Goldman, R.C., Frost, D.J., Capobianco, J.O., Kadam, S., Rasmussen, R.R. & Abad-Zapatero, C. (1995) Antifungal drug targets: *Candida* secreted aspartyl protease and fungal wall β-glucan synthesis. *Infect. Agents Dis.* **4**, 228–247.

Holmberg, K. & Meyer, R. D. (1986) Fungal infections in patients with AIDS and AIDS-related complex. *Scand. J. Infect. Dis.* **18**, 179–182.

Hube, B., Monod, M., Schofield, D.A., Brown, A.J.P. & Gow, N.A.R. (1994) Expression of seven members of the gene family encoding secretory aspartyl proteinases in *Candida albicans. Mol. Microbiol.* **14**, 87–99.

Jakab, E., Schmidt, T. & Hernadi, F. (1991) Purification and some properties of *Candida albicans* DNA polymerase. *Prep. Biochem.* **21**, 105–123.

Kaminishi, H., Miyaguchi, H., Tamaki, T., Suenaga, N., Hisamatsu, M., Matsumoto, H., Maeda, H. & Hagihara, Y. (1995) Degradation of humoral host defence by *Candida albicans* proteinase. *Infect. Immun.* **63**, 984–988.

Luna, M.A. & Tortoledo, M.E. (1993) Histologic identification and pathologic patterns of disease caused by *Candida*. In: *Candidiasis. Pathogenesis, Diagnosis, and Treatment* (Bodey, O.P., ed.). New York: Raven Press, pp. 21–42.

MacDonald, F. & Odds, F.C. (1980a) Inducible proteinase of *Candida albicans* in diagnostic serology and in pathogenesis of systemic candidiasis. *J. Med. Microbiol.* **13**, 423–435.

MacDonald, F. & Odds, F.C. (1980b) Purified *Candida albicans* proteinase in the serological diagnosis of systemic candidiasis. *J. Am. Med. Assoc.* **243**, 2409–2411.

Mandal, B. (1989) AIDS and fungal infections. *J. Infect.* **19**, 199–205.

Merz, W.G. (1990) *Candida albicans* strain delineation. *Clin. Microbiol. Rev.* **3**, 321–324.

Meunier-Carpentier, F., Kiehn, T.E. & Armstrong, D. (1981) Funginemia in the immunocompromised host. Changing patterns, antigenemia, high mortality. *Am. J. Med.* **17**, 363–370.

Monod, M., Togni, G., Hube, B. & Sanglard, D. (1994) Multiplicity of genes encoding secreted aspartic proteinases in *Candida* species. *Mol. Microbiol.* **13**, 357–368.

Odds, F. (1994) Pathogenesis of *Candida* infections. *J. Am. Acad. Dermatol.* **31**, S2–S5.

Remold, H., Fasold, H. & Staib, F. (1968) Purification and characterization of a proteolytic enzyme from *Candida albicans.*

*Biochim. Biophys. Acta* **167**, 399–406.

Rinaldi, M.G. (1993) Biology and pathogenicity of *Candida* species. In: *Candidiasis: Pathogenesis, Diagnosis, and Treatment* (Bodey, G.P., ed.). New York: Raven Press, pp. 1–20.

Rüchel, R. (1990) Virulence factors of Candida species, in oral candidosis. In: *Oral Candidosis* (Samaranayake, L.P. & MacFarlane, T.W., eds). London: Wright, p. 47.

Rüchel, R. & Böning, B. (1983) Detection of *Candida* proteinase by enzyme immunoassay and interaction of the enzyme with alpha-2-macroglobulin. *J. Immunol. Methods* **61**, 107–114.

Rüchel, R., Zimmerman, F., Böning-Stutzer, B. & Helmchen, U. (1991) Candidiasis visualized by proteinase-directed immunofluorescence. *Virchow's Arch. A: Pathol. Anat. Histopathol.* **419**, 199–202.

Rüchel, R., De Bernardis, F., Ray, T.L, Sullivan, P.A. & Cole, G.T. (1992) *Candida* acid proteinases. *J. Med. Vet. Mycol.* **30** (suppl.1), 123–132.

Scherer, S. & Magee, P.T. (1990) Genetics of *Candida albicans*. *Microbiol. Rev.* **54**, 226–241.

Staib, F. (1963) Serum-proteins as nitrogen source for yeast-like fungi. *Sabouraudia* **4**, 187–193.

Verfaillie, C., Weisdorf, D., Haake, R., Hostetter, M., Ramsey, N.K. & McGlave, P. (1991) *Candida* infections in bone marrow transplant recipients. *Bone Marrow Transplant* **8**, 177–184.

White, T.C. & Agabian, N. (1995) *Candida albicans* secreted aspartyl proteinase: isoenzyme pattern is determined by cell type, and levels are determined by environmental factors. *J. Bacteriol.* **177**, 5215–5221.

***Stephen I. Foundling***
*Siemens Analytical X-ray Systems, Inc.,*
*6300 Enterprise Lane,*
*Madison, WI 53719, USA*
*Email: sfoundling@bruker-axs.com*

# 301. Candiparapsin

## Databanks

*Peptidase classification: clan AA, family A1, MEROPS ID: A01.038*
*NC-IUBMB enzyme classification: 3.4.23.24*
*Databank codes:*

| Species | Gene | SwissProt | PIR | EMBL (cDNA) | EMBL (genomic) |
|---|---|---|---|---|---|
| *Candida parapsilosis* | *sapp1* | P32951 | S20705 | Z11919 | – |
| *Candida parapsilosis* | *sapp2* | P32950 | S20704 S34581 | Z11918 | – |

## Name and History

For almost a decade, only one principal enzyme was believed to be secreted by *Candida parapsilosis* yeast (Rüchel *et al.*, 1986). Some physical chemical properties were known for this enzyme. It had an apparent lower molecular mass, and a higher pI than proteinases from *Candida albicans* (Chapter 300) and *C. tropicalis* (Chapter 302). In 1994, it was demonstrated that two secreted proteinases were made when the yeast was grown in culture media with protein as the sole source of nitrogen (Fusek *et al.*, 1994a). These two enzymes were isolated and purified to homogeneity. Limited chemical sequencing indicated that the enzymes were the products of two independent gene sequences. These two genes had been sequenced earlier by de Viragh *et al.* (1993) and called *ACPR* and *ACPL*. They have now been renamed **SAPP1** and **SAPP2** respectively (for ***secreted aspartic proteinase parapsilosis 1*** and ***secreted aspartic proteinase parapsilosis 2***). Here the further name **candiparapsin** is suggested,

but for convenience the terms SAPP1 and SAPP2 will be used for the two forms.

Evidence for multiple genes coding for multiple SAPs was also obtained by another group. Monod *et al.* (1994) demonstrated that four potential genes existed in a clinical isolate of *C. parapsilosis* E18. This further demonstrates that pathogenic strains of yeast have a high degree of gene complexity and a multigene family similar to *C. albicans*. No further sequencing studies have been carried out to identify these individual gene sequences from the *C. parapsilosis* yeast.

## Activity and Specificity

The SAPP enzymes have a broad substrate specificity, for example, they can degrade keratin, denatured collagen, hemoglobin and serum albumin. When the relative rates of activity were investigated for the two enzymes, acting on macromolecular substrates, both enzymes were found to have

*Table 301.1* Comparison of kinetic properties of proteases from three species of *Candida* with Lys-Pro-Ala-Leu-Phe-Nph-Arg-Leu as substrate

| Protease | $k_{cat}$ (s$^{-1}$) | $K_m$ (μM) | $k_{cat}/K_m$ (s$^{-1}$ M$^{-1}$ × 10$^6$) |
|---|---|---|---|
| SAP2 | 72.5 ± 2.9 | 24.5 ± 9.2 | 2.95 ± 0.25 |
| SAPT | 28.0 ± 1.7 | 17.3 ± 3.7 | 1.60 ± 0.19 |
| SAPP1 | 14.5 ± 3.0 | 24.4 ± 3.9 | 0.59 ± 0.16 |

very similar rates of cleavage and produced similar cleavage patterns, which were visualized by SDS-PAGE analysis. Serum albumin was observed to be completely cleaved to small fragments within several hours of incubation with either enzyme at pH 3.5 and a temperature of 37°C. The apparent pH optimum for both enzymes with serum albumin as substrate was pH 3.5.

A large difference in specific activity was measured for the cleavage of the synthetic substrate Lys-Pro-Ala-Glu-Phe+Phe(NO₂)-Ala-Leu with SAPP1 and SAPP2 at pH 3.5. Initial rates of hydrolysis differed by more than two orders of magnitude with the enzymes. SAPP1 had a measured $K_m$ of 101 mM at pH 3.5, a $k_{cat}$ value of 36.0 s$^{-1}$ and a specificity constant $k_{cat}/K_m$ of 0.350. In comparison, SAPP2 had a measured $K_m$ of 180 mM at pH 3.5 and a $k_{cat}$ value of 0.26 s$^{-1}$ and a specificity constant of 0.0014 (Fusek *et al.*, 1993).

Specificity screening of SAPP1 against a library of synthetic colorimetric substrates showed that similar to the SAP2 (candidapepsin, Chapter 300) and SAPT (canditropsin, Chapter 302) proteinases, the SAPP1 enzyme preferred substrate sequences that had bulky amino acids around the primary cleavage site P1–P1′. Two optimized peptide sequences with the chemical structure Lys-Pro-Ile-Glu-Phe+Nph-Arg-Leu and Lys-Pro-Ala-Glu-Phe+Nph-Ala-Leu were preferred by SAPP1. The SAPP1 enzyme also showed preference for an acidic side chain at the substrate specificity position P2. The substrate Lys-Pro-Ala-Leu-Phe-Nph-Arg-Leu was determined to be the best substrate for all three extracellular *Candida* aspartic proteinases included in the specificity screening of Fusek *et al.* (1994b) (Table 301.1).

The inhibition constant for pepstatin acting on SAPP1 with an optimum substrate is $K_i = 0.6 ± 0.3$ nM at pH 3.25. In a cellular growth inhibition assay, pepstatin A, which was included as a general fungistatic compound, was found to be only moderately effective at retarding the growth of *C. parapsilosis* yeast at pH 4.0 and 7.0, even at 100 nM concentration.

The SAPP enzymes generally show a broad substrate specificity as would be expected for general proteolytic enzymes and virulence factors. SAPP1, in particular, demonstrated a wide tolerance for synthetic substrates, which may be related to its higher pI and its smaller size (relative to SAP2 and SAPT), which could result in a less restricted substrate-binding cleft.

## Structural Chemistry

Two gene sequences have been identified and sequenced for the SAPP enzymes (de Viragh *et al.*, 1993). These are genes *sapp1*ACPR and *sapp2*ACPL: *sapp2* was originally assumed to

```
SAP2   QAIPVTLNNELVSYAADITIGSNKQKFNVIVDTGSSDLW
SAPP1  DSISLSLINEGPSYASKVSVGSNKQQQTVIIDTGSSDFW
SAPP2  SSPSSPLYFEGPSYGIRVSVGSNKQEQQVVLDTGSSDFW
```

*Figure 301.1* Comparison of the sequences for *C. albicans* SAP2 with those for SAPP1 and SAPP2.

be a pseudogene, as no evidence existed that it encoded a secreted product. The studies of Fusek *et al.* (1994a) changed this view and it was shown that the yeast *C. parapsilosis* secretes two aspartic proteinases. Both enzymes are coded by long open reading frames, SAPP1 is coded in 1206 bp, and SAPP2 in 1185 bp, with opposite directions of transcription for the two genes. Both enzymes have 'prepro' regions of approximately 60 amino acids in length. Within the 'pre' region is a classic secretion signal peptide sequence, followed by a putative signal peptidase cleavage site. Upstream of the 'pre' region is the propeptide. This is homologous to propeptides from *C. albicans* and *C. tropicalis*. Pairs of dibasic sites are conserved within the 'pro' region and these conform to the known specificity of a kexin (Chapter 115) processing site. The mature sequence of SAPP1 starts with the sequence DSISLSLINEGP, and the sequence SSPSSPL defines the start of the mature SAPP2 enzyme. Both sequences were obtained by N-terminal chemical sequencing of the enzymes purified to homogeneity from the same culture filtrate and are compared to *C. albicans* in Figure 301.1.

The mature SAPP1 and SAPP2 enzymes run with apparent molecular masses of 38 000, but have quite different pIs as determined by isoelectric focusing.

Sufficient quantities of SAPP1 were available to facilitate both enzymatic studies and X-ray structural studies. The apoenzyme was crystallized by Fusek *et al.* (1995); crystals were of high quality and diffracted to high resolution. The second enzyme, SAPP2, is available in smaller quantities and studies on this protein lag behind those for SAPP1.

## Preparation

*C. parapsilosis* yeast was maintained on Sabouraud agar plates or modified Sabouraud agar plates and replated every 4–6 weeks (Difco; see Appendix 2 for full names and addresses of suppliers). Yeast was grown in liquid culture containing 0.2% serum albumin and 1.5% YCB Difco medium adjusted to pH 4.0 with HCl. Fermentation was carried out for 4–5 days, by which time all serum albumin in the media was completely degraded and yeast cells were removed by centrifugation and sterile filtration. SAPP1 was produced at levels up to 5 mg liter$^{-1}$ of supernatant and SAPP2 was produced at levels up to 1 mg liter$^{-1}$. The isolation of the SAPP enzymes has been described in the previous chapter on candidapepsin (Chapter 300). The final step of purification, on a Mono-S cation-exchange column yields homogeneous SAPP1 and SAPP2 proteinases. These were determined to be pure by SDS-PAGE and amino acid sequencing.

## Biological Aspects

*C. parapsilosis* is an anamorphic yeast (asexual/imperfect) showing typical cell budding or blastoconidia and it can also form pseudohyphae and true filamentous thalli (true hyphae).

The pseudohyphae are curved and relatively short, presenting a 'sagebrush' appearance. The presence of pseudohyphae in a patient's tissues is a reflection of invasion and deep-seated candidiasis. The ability of *C. parapsilosis* to produce large quantities of proteinase in culture may be directly related to its virulence in humans (Rüchel, 1991). Like *C. albicans* and *C. tropicalis*, this organism is a true opportunistic pathogen and ranks third as a clinically significant pathogen (Bodey, 1993). In some hospital situations it has become the most dominant yeast pathogen; however, no epidemiologic data have been provided to explain why it surpassed other species of *Candida* in these instances. This species of yeast is associated with infections consequent to the use of intravenous drugs, indwelling catheters and parenteral hyperalimentation. It is prevalent in patients with solid tumors (Horn *et al.*, 1985). There is a marked association of infections in patients with prosthetic devices (Rinaldi, 1993). Other infections have included fungemia, endocarditis, endophthalmitis, septic arthritis and peritonitis. *C. parapsilosis* is numerically a less significant pathogen and is responsible for less than 2% of all mycoses.

## Related Peptidases

With the increase in numbers of immunocompromised patients, other less common yeasts are also emerging as pathogens. These include the species *Candida krusei*, *Candida lusitaniae*, *Candida guillermondii* and *Candida glabrata*. *Candida krusei* is an opportunistic agent, inciting serious infections in neutropenic patients. It is documented as the agent in diarrhea in infants, it has also been isolated from patients with septicemia and endophthalmitis and more ominous, shows resistance to the antifungal drug fluconazole in AIDS clinics. *C. lusitaniae* is an emerging opportunistic pathogen in the immunocompromised host and is found in blood, lung, kidney and the gastrointestinal tract. It is also noted for developing resistance to amphotericin B, the mainstay of antifungal therapy. *C. guillermondii* is documented as an agent of endocarditis, particularly in intravenous drug users. It causes infections in patients undergoing surgical procedures and in immunocompromised patients. The final strain, *C. glabrata*, is a significant agent of human mycoses. It is emerging as a serious pathogen of the immunocompromised patients with AIDS, and is associated with urinary tract disease.

The literature reports that no proteinase activity is detectable from *in vitro* cultures for many of these yeasts. However, in the author's laboratory, the strains *C. glabrata*, *C. guillermondii* and *C. krusei* have been found to be mildly proteolytic and proteinase has been purified from *C. krusei*. It is interesting to reflect that in the studies of Monod *et al.* (1994), multiple genes encoding aspartic proteinases appear

to be present in *C. guillermondii*. The biology of these less common yeast species is generally poorly understood and virulence factors are inadequately characterized.

## Distinguishing Features

The SAPP1 enzyme has a specificity similar to that of the other *Candida* proteinases, however, there are differences among the three *Candida* species considered in this *Handbook* (Chapters 300 and 301). The most detailed study of these differences is found in Fusek *et al.* (1994b). Perhaps the most noticeable difference is in the greatly reduced ability of SAPP1 to accommodate Arg in position P4 in the substrate Lys-Pro-Ala-Lys-Phe+Nph-Arg-Leu; activity with this substitution is less than 8% of that found for the other two proteases. On the other hand, SAPP1 binds pepstatin ten times tighter than SAP2 or SAPT.

## References

Bodey, G.P. (1993) Hematogenous and major organ candidiasis. In: *Candidiasis: Pathogenesis, Diagnosis and Treatment* (Bodey, G.P., ed.). New York: Raven Press, pp. 279–329.

de Viragh, P.A., Sanglard, D., Togni, G., Falchetto, R. & Monod, M. (1993) Cloning and sequencing of two *Candida parapsilosis* genes encoding acid proteases. *J. Gen. Microbiol.* **139**, 335–342.

Fusek, M., Smith, E.A., Monod, M. & Foundling, S.I. (1994a) *Candida parapsilosis* expresses and secretes two aspartic proteinases. *FEBS Lett.* **327**, 108–112.

Fusek, M., Smith, E.A., Monod, M., Dunn, B.M. & Foundling, S.I. (1994b) Extracellular aspartic proteinases from *Candida albicans*, *Candida tropicalis*, and *Candida parapsilosis* yeast differ substantially in their specificities. *Biochemistry* **33**, 9791–9799.

Fusek, M., Smith, E.A. & Foundling, S.I. (1995) Extracellular aspartic proteinases from *Candida* yeasts. In: *Aspartic Proteinases: Structure, Function, Biology and Biomedical Implications* (Takahashi, K., ed.). New York: Plenum Press, pp. 489–500.

Horn, R., Wong, B. & Kiehn, T.E. (1985) Fungemia in a cancer hospital: changing frequency, earlier onset, and results of therapy. *Rev. Infect. Dis.* **7**, 646–654.

Monod, M., Togni, G., Hube, B. & Sanglard, D. (1994) Multiplicity of genes encoding secreted aspartic proteinases in *Candida* species. *Mol. Microbiol.* **13**, 357–368.

Rinaldi, M.G. (1993) Biology and pathogenicity of *Candida* species. In: *Candidiasis: Pathogenesis, Diagnosis and Treatment* (Bodey, G.P., ed.). New York: Raven Press, pp. 1–20.

Rüchel, R. (1991) Proteinases secreted by *Candida* species. In: *Candida and Candidamycosis* (Tümbay, E., Seeliger, H.P.R. & Ang, O., eds). New York: Plenum Press, pp. 39–42.

Rüchel, R., Boning, B. & Borg, M. (1986) Characterization of a secretory proteinase of *Candida parapsilosis* and evidence for the absence of the enzyme during infection *in vitro*. *Infect. Immun.* **53**, 411–419.

*Stephen I. Foundling*
*Siemens Analytical X-ray Systems, Inc.,*
*6300 Enterprise Lane,*
*Madison, WI 53719, USA*
*Email: sfoundling@bruker-axs.com*

# 302. *Canditropsin*

## Databanks

*Peptidase classification: clan AA, family A1, MEROPS ID: A01.037*
*NC-IUBMB enzyme classification: EC 3.4.23.24*
*Databank codes:*

| Species | Gene | SwissProt | PIR | EMBL (cDNA) | EMBL (genomic) |
|---|---|---|---|---|---|
| *Candida tropicalis* | *sapt1* | Q00663 | S16971 | X61438 | – |

Brookhaven Protein Data Bank three-dimensional structures:

| Species | Type | ID | Resolution | Notes |
|---|---|---|---|---|
| *Candida tropicalis* | | 1PCT | | |

## Name and History

Ample evidence exists that proteolytic strains of *Candida tropicalis* secrete protease(s) in humans during mycosis (Douglas, 1988). High titers of specific antibodies can be monitored in sera of patients with candidiasis, and proteinase-related antigens can be demonstrated in mycotic tissue. Proteinase antigen can be detected by enzyme immunoassay in the sera of patients who are suspected of deep-seated candidiasis (Rinaldi, 1993). In deep-seated mycoses of submucosal tissues, the fungus encounters blood vessels, which are readily invaded. *C. tropicalis* is suspected of causing blood coagulation through activation of coagulation factor X in an as yet unexplained fashion. The protease may also exert a renin-like effect via the bradykinin pathway (Rüchel, 1991).

The principal enzyme secreted by clinical strains of *C. tropicalis* grown in liquid culture was partially characterized by Rüchel *et al*. (1991). The protein chemistry on this enzyme did not progress significantly until the later work of Fusek *et al*. (1994, 1995). Partial N-terminal sequencing of the purified *secreted aspartic proteinase tropicalis (SAPT)* revealed that the predicted gene sequence derived from a cDNA clone coincided with that of the predicted mature terminus of a 334 amino acid enzyme (Togni *et al*., 1991; Fusek *et al*., 1995). The name *canditropsin* was introduced in 1993 (Lin *et al*., 1993). The mature enzyme has a calculated mass of 35.8 kDa and some variation in amino acid sequence has been observed among the different clinical isolates. This indicates that the enzyme may exist as several isoforms. The apparent molecular mass of the purified enzyme determined by SDS-PAGE analysis is closer to 40 kDa and differential staining techniques have indicated that the enzyme is a glycoprotein and this is consistent with its decreased electrophoretic mobility (Togni *et al*., 1994).

The question has been raised of multiple genes coding for multiple secreted aspartic proteinase enzymes for the *C. tropicalis* yeast (Monod *et al*., 1994). *Eco*RI-digested genomic DNAs of *C. tropicalis* were hybridized under low stringency conditions using a *C. albicans* PR1 probe, and probes from *C. tropicalis* and *C. parapsilosis*, which were similar to the PR1 probe. Four *C. tropicalis* bands were detected, indicating that this species may have a level of complexity similar to that of *C. albicans*. No further sequencing studies were done to identify the individual gene sequences from *C. tropicalis*. Like *C. albicans*, *C. tropicalis* has a complicated life cycle that involves morphological switching. It is probable that the gene multiplicity observed for *C. tropicalis* reflects the different roles that proteinase activity plays in the life cycle of these asexual diploid yeasts.

## Activity and Specificity

The SAPT enzyme, like the forms of candidapepsin (Chapter 300), has an extremely broad specificity range. The main studies on enzyme activity and specificity have been done by the groups of Lin *et al*. (1993) (recombinant SAPT or rSAPT) and Fusek *et al*. (1994) (wild-type secreted SAPT). rSAPT produced in *Escherichia coli* hydrolyzes hemoglobin in the pH range 2–6 with an optimum of pH 4.5. It readily degrades the connective tissue proteins keratin and collagen. Using oxidized cattle insulin B chain as substrate one major cleavage site at Ala14$\downarrow$Leu15 was identified and three additional sites were cleaved at greatly reduced rates. The secondary cleavage sites were Val2$\downarrow$Asn3, Tyr16$\downarrow$Leu17 and Phe25$\downarrow$Tyr26:

FV$\downarrow$NQHLCGSHLVEA$\downarrow$LY$\downarrow$LVCGERGFF$\downarrow$YTPKA

One unusual feature was observed during these studies: the autoactivation of the zymogen occurred via cleavage of a Lys$\downarrow$Arg bond to yield the correct N-terminus. A secondary cleavage at Gln$\downarrow$Lys was also observed, which is upstream to the dibasic pair. This demonstrates that the substrate specificity is unusual for this enzyme and is different from that of pepsin. rSAPT was found to be inhibited strongly by pepstatin A with a $K_i$ of $1.75 \times 10^{-8}$ M. The enzyme was also inactivated by DAN and 1,2-epoxy-3-(*p*-nitrophenoxy)propane (EPNP), which chemically modify the catalytic aspartic acids. It is interesting that the serine protease inhibitors leupeptin and antipain competitively inhibited rSAPT with $K_i$ values of $1.7 \times 10^{-4}$ M and $1.5 \times 10^{-5}$ M, respectively.

Specificity screening with wild-type secreted SAPT against a library of synthetic chromogenic substrates showed that the enzyme preferred substrate sequences that had bulky amino acids around the primary cleavage bond. An optimized peptide sequence Lys-Pro-Ala-Leu-Phe$+$Nph-Arg-Leu was shown to have a $K_m$ of 17 µM at pH 3.25. The picture that emerged from this work indicated that the enzyme had a particularly broad and promiscuous substrate specificity, which is likely related to the biological role it plays during fungal virulence and pathogenesis.

## Structural Chemistry

One gene sequence has been identified and sequenced for the enzyme SAPT (Togni *et al.*, 1991). A single long open reading frame of 1182 bp has been identified, which codes for a preproenzyme of 394 amino acids. The Met1 residue suggests the existence of a signal sequence in the protease precursor. The signal sequence of 60 amino acids begins with a hydrophobic core of seven amino acids and 6–10 amino acids downstream is a putative signal peptidase cleavage site. Just upstream of the mature N-terminus is a dibasic **KR** pair (residues 60–61), located within the sequence LIQ**KR**SDVPTT. The mature enzyme starts at position 61, and the dibasic pair is assumed to be a kexin (Chapter 115) cleavage site, although it is worth noting that the proenzyme can autoactivate *in vitro* (see above). A second endopeptidase cleavage site is found further upstream at position 32–33 in the propart. The function of a cleavage here, if any, is unknown. It is likely that other genes exist in the *C. tropicalis* genome that code for related SAPs (Monod *et al.*, 1994). Evidence for this added complexity comes from the studies of Sanglard *et al.* (1992) and Togni *et al.* (1994). These investigators compared the virulence of a pathogenic wild-type strain of *C. tropicalis* with that of a proteinase-deficient strain constructed by gene knockout site-directed mutagenesis. The SAPT gene was knocked out and both *in vitro* assays and a systemic infection murine model indicated that SAPT did not contribute to virulence in systemic infections. In the murine model it was observed that mice infected with the SAPT knockout strain of yeast still produced antibodies that reacted with SAPT. These data, together with those of Monod *et al.* (1994), provide strong evidence that additional aspartic proteinases are secreted *in vivo* in a *C. tropicalis* yeast infection.

The mature SAPT enzyme has been purified to homogeneity and in sufficient quantity to allow both enzymatic studies and X-ray structural studies. The apoenzyme was crystallized by Fusek *et al.*, 1995, and crystals were of suitable quality to resolve the structure to high resolution. Co-ordinates for this structure will be deposited with the Brookhaven Protein Data Bank with the entry code 1PCT. The SAPT three-dimensional structure is homologous to the SAP2 and SAP2X structures, which were discussed in Chapter 300 on *C. albicans* aspartic proteinases.

## Preparation

*C. tropicalis* is maintained on Sabouraud agar plates (Difco; see Appendix 2 for full names and addresses of suppliers),

enriched with 0.02% biotin. Yeasts are grown in liquid culture containing 0.2% serum albumin and 1.2% YCB Difco medium, 0.02% biotin and adjusted to pH 4.0 with HCl. Fermentation is carried out for three days and cells are removed by centrifugation and sterile filtration. SAPT is produced at levels up to 4 mg liter$^{-1}$ of supernatant. The isolation of the SAPT enzyme has been described in Chapter 300. The two-step purification procedure, which uses low-pressure column chromatography, can be reduced to a one-step affinity chromatography procedure on pepstatin A–agarose as was described by Wagner *et al.* (1995). Either procedure results in homogenous enzyme as determined by N-terminal sequencing and samples are suitable for protein crystallization (Fusek *et al.*, 1995).

## Biological Aspects

*C. tropicalis* is an asexual diploid yeast and is a normal commensal of the human gastrointestinal and genito-urinary tract. It is an opportunistic pathogen like *C. albicans* and ranks second only to *C. albicans* as a clinically significant pathogen (Bodey, 1993). In some clinical settings *C. tropicalis* is more prevalent than *C. albicans*. *C. tropicalis* is the major cause of candidiasis in patients with leukemia at some clinics (Winegard *et al.*, 1979). Some strains may produce sucrose assimilation-negative variants, which have been called *C. paratropicalis*. The predisposing factors allowing this *Candida* species to incite disease are the same as those for disease due to *C. albicans*. The onset of pathogenesis in a patient is adherence to, and multiplication on, mucosal surfaces. *C. tropicalis* is known to have a doubling time of slightly less than 1 h, showing a distinct growth dependence on biotin and other vitamins. Pathogenesis is also associated with hyphal production and germ tube production. Hyphal growth is characterized by production of blastoconidia along the entire length of pseudohyphae. Enzyme production and tissue damage and penetration are also associated with pathogenesis, and *C. tropicalis* is noted for a high provocation of inflammation at the site of injury. *C. tropicalis* may be more virulent than *C. albicans* because it is documented to be more likely to cause systemic infection when recovered from fungal surveillance cultures from patients with hematologic malignancies or patients undergoing bone marrow transplantation (Sandford *et al.*, 1980).

## Distinguishing Features

Detailed studies of substrate specificity (Fusek *et al.*, 1994) indicate that the aspartic proteinases from the three species of *Candida* are substantially different. Canditropsin is very sensitive to charged residues in the P2 position; substitution of Asp for Leu at P2 in the substrate Lys-Pro-Ala-Leu-Phe$+$Nph-Arg-Leu causes a 30-fold reduction in the specificity constant. Candidapepsin (Chapter 300) and candiparapsin (Chapter 301) are insensitive to this substitution. Canditropsin is the least sensitive of the three to inhibition by pepstatin.

## References

Bodey, G.P. (1993) Hematogenous and major organ candidiasis. In: *Candidiasis: Pathogenesis, Diagnosis and Treatment* (Bodey, G.P., ed.). New York: Raven Press, pp. 279–329.

Douglas, J. (1988) *Candida* proteinases and candidiasis. *CRC Crit. Rev. Biotechnol.* **8**, 121–129.

Fusek, M., Smith, E.A., Monod, M., Dunn, B.M. & Foundling S.I. (1994) Extracellular aspartic proteinases from *Candida albicans, Candida tropicalis,* and *Candida parapsilosis* yeast differ substantially in their specificities. *Biochemistry* **33**, 9791–9799.

Fusek, M., Smith, E.A. & Foundling, S.I. (1995) Extracellular aspartic proteinases from *Candida* yeasts. In: *Aspartic Proteinases: Structure, Function, Biology and Biomedical implications* (Takahashi, K., ed.). New York: Plenum Press, pp. 489–500.

Lin, X.-L., Tang, J., Koelsch, G., Monod, M. & Foundling, S.I. (1993) Recombinant canditropsin, an extracellular aspartic protease from yeast *Candida tropicalis: Escherichia coli* expression, purification, zymogen activation and enzymic properties. *J. Biol. Chem.* **268**, 20143–20147.

Monod, M., Togni, G., Hube, B. & Sanglard, D. (1994) Multiplicity of genes encoding secreted aspartic proteinases in *Candida* species. *Mol. Microbiol.* **13**, 357–368.

Rinaldi, M.G. (1993) Biology and pathogenicity of *Candida* species. In: *Candidiasis: Pathogenesis, Diagnosis and Treatment* (Bodey, G.P., ed.). New York: Raven Press, pp. 1–20.

Rüchel, R. (1991) Proteinase secreted by *Candida* species. In: *Candida and Candidamycosis* (Tümbay, E., Seeliger, H.P.R. & Ang O., eds). New York: Plenum Press, pp. 39–42.

Sandford, G.R., Merz, W.G., Winegard, J.R., Charache, P. & Saral, R. (1980) The value of fungal surveillance cultures as predictors of systemic fungal infections. *J. Infect. Dis.* **142**, 503–509.

Sanglard, D., Togni, G., de Viragh, P.A. & Monod, M. (1992) Disruption of the gene encoding the secreted acid protease (ACP) in the yeast *C. tropicalis. FEMS Microbiol. Lett.* **95**, 149–156.

Togni, G., Sanglard, D., Falchetto, R. & Monod, M. (1991) Isolation and nucleotide sequence of the extracellular acid protease gene (*ACP*) from the yeast *Candida tropicalis. FEBS Lett.* **286**, 181–185.

Togni, G., Sanglard, D. & Monod, M. (1994) Acid protease secreted by *Candida tropicalis*: virulence in mice of a proteinase negative mutant. *J. Med. Vet. Mycol.* **32**, 257–265.

Wagner, T., Borg von Zepelin, M. & Rüchel, R. (1995) pH-dependent denaturation of extracellular aspartic proteinases from *Candida* species. *J. Med. Vet. Mycol.* **33**, 275–278.

Winegard, J.R., Merz, W.G. & Saral, R. (1979) *Candida tropicalis*: a major pathogen in immunocompromised patients. *Ann. Intern. Med.* **91**, 539–543.

*Stephen I. Foundling*
*Siemens Analytical X-ray Systems, Inc.,*
*6300 Enterprise Lane,*
*Madison, WI 53719, USA*
*Email: sfoundling@bruker-axs.com*

# 303. Barrierpepsin

## Databanks

*Peptidase classification: clan AA, family A1, MEROPS ID: A01.015*
*NC-IUBMB enzyme classification: EC 3.4.23.35*
*Chemical Abstracts Service registry number: 152060-38-3*
*Databank codes:*

| Species | SwissProt | PIR | EMBL (cDNA) | EMBL (genomic) |
|---|---|---|---|---|
| *Saccharomyces cerevisiae* | P12630 | A34084 | J03573 | Z47047: chromosome IX complete sequence |

## Name and History

In the budding yeast *Saccharomyces cerevisiae*, haploid mating-type $\alpha$ (Mat$\alpha$) cells constitutively secrete a diffusible peptide hormone (designated $\alpha$ factor; Duntze *et al.*, 1970) that binds to a G protein-coupled receptor on mating-type **a** (Mat**a**) cells. In the presence of $\alpha$ factor, Mat**a** cells arrest in the $G_1$ stage of the cell cycle, increase production of their own peptide mating hormone (**a**-factor), and undergo a variety of other physiologic events that are preparatory to conjugation between Mat**a** and Mat$\alpha$ cells (see Cross *et al.*, 1988; Herskowitz, 1989; MacKay, 1993 for reviews). Hicks & Herskowitz (1976) reported that Mat**a** cells constitutively secrete another activity that acts as a 'barrier' to the diffusion of $\alpha$-factor through agar medium. Mutants that specifically lack this activity were identified and shown to be supersensitive to $\alpha$ factor (Sprague & Herskowitz, 1981; Chan & Otte, 1982); their defects mapped at a single genetic locus, *bar1*. The *bar1* gene was cloned, determined to be the structural gene for the enzyme, and predicted by homology to be an aspartic protease that contains the two active-site domains characteristic of this family, as well as a third C-terminal domain (MacKay *et al*., 1988). This protease has had various names: ***barrier proteinase, bar proteinase***, but the name recommended by the IUBMB is ***barrierpepsin***.

## Activity and Specificity

There is only one known natural substrate for barrierpepsin – the peptide hormone α factor: Trp-His-Trp-Leu-Gln-Leu┼Lys-Pro-Gly-Gln-Pro-Met-Tyr available from Sigma (see Appendix 2 for full names and addresses of suppliers). The proteinase cleaves between Leu6 and Lys7; Arg can substitute for Lys7. Cleavage of the substrate is readily detected by HPLC analysis on a C18 column (MacKay *et al*., 1991). Like other aspartic proteinases, this one requires four amino acids (P4–P1) on the N-terminal side of scissile bond, but is unusual in requiring at least six (P1′–P6′) on the C-terminal side (MacKay *et al*., 1991; V.L. MacKay & K. Walker, unpublished results). Barrierpepsin is not active on other peptides that are generally cleaved by aspartic proteinases.

Barrierpepsin has a pH optimum of approximately 5.0–5.3, but retains over 50% activity from pH 2.6–pH 6.8. It does not require divalent cations or any cofactors and is resistant to all reversible inhibitors tested, including pepstatin A, EDTA, ε-aminocaproic acid, and a number of serine proteinase inhibitors (MacKay *et al*., 1991). Its resistance to tosyl arginyl methyl ester (TAME) is interesting, since the $G_1$ arrest of Mata cells induced by α factor can be prolonged by the addition of TAME to the culture. This effect was originally thought to be due to inhibition of barrierpepsin by TAME (Ciejek & Thorner, 1979).

## Structural Chemistry

Barrierpepsin is a single-chain protein with a primary translation product of 587 amino acids (MacKay *et al*., 1988), including a 24 residue signal peptide that is cleaved during secretion (Jars *et al*., 1995). Excluding the signal peptide, the primary translation product contains eight cysteine residues and activity is destroyed by boiling with reducing agents (although not by boiling without reducing agents). As secreted by *S. cerevisiae*, it is heavily glycosylated, migrating at over 200 kDa in SDS reducing gels, with both *N*-linked and *O*-linked glycosylation. The first two domains of the protein (205 and 159 amino acids) have substantial homology to the domains of other aspartic proteinases (Chapter 271), particularly around the active-site residues (Asp-Thr-Gly and Asp-Ser-Gly). A third domain of 191 amino acids has no significant homology to members of this family of proteinases or to other proteins in databases. This domain is composed of 33% Ser and Thr, is extensively *O*-glycosylated, and is required for export of the proteinase to the culture medium (MacKay *et al*., 1988, 1991). A thorough analysis of the sites and extent of *O*-linked glycosylation of this domain has been reported (Jars *et al*., 1995).

## Preparation

Barrierpepsin is isolated from the culture supernatants of Mata cells of *S. cerevisiae* (often strain X2180-1A, Yeast Genetics Stock Center), or to obtain more protein, from supernatants of cells that have been transformed with a multicopy plasmid with the *bar1* gene under the control of a strong transcriptional promoter. Active proteinase can be precipitated by mixing the cell-free supernatant with an equal volume of cold 95% ethanol. The concentrated crude fraction can then be purified more than 2000-fold to approximately 90% purity by chromatography on DEAE-Sepharose, Mono-Q, and S-200 resins (MacKay *et al*., 1991).

## Biological Aspects

The function of barrierpepsin may be limited to the cleavage of α factor and the resulting recovery of Mata cells from α factor-induced $G_1$ arrest, since *bar1* deletion mutants, which lack the protein have only the one phenotype of supersensitivity to α factor. Moreover, the *bar1* gene is not transcribed in Matα cells or in the Mata/Matα diploids that result from conjugation. Promoter elements, which restrict *BAR1* transcription to Mata cells and increase its transcription with α factor induction have been described (Kronstad *et al*., 1987).

From pulse-chase labeling studies, it appears that the protein rapidly transits through the secretory pathway and is efficiently exported through the cell wall to the medium (S.K. Welch, M. Zavortink & V.L. MacKay, unpublished results). The enzyme is secreted in active form, unlike many aspartic proteases, which require activation of a zymogen.

## Distinguishing Features

Whereas the primary sequence and pH optimum of barrierpepsin show that it is a member of the aspartic proteinase family A1, its unique substrate specificity, resistance to inhibitors, and third domain indicate its novelty. Other aspartic proteases that degrade α factor cleave at the Leu6-Lys7 bond but also at other sites (V.L. MacKay, unpublished results). A rabbit polyclonal antiserum has been raised against the barrierpepsin expressed as a fusion protein in *Escherichia coli* (MacKay *et al*., 1991), but this antiserum is not commercially available.

## Further Reading

Most aspects of barrierpepsin are discussed in MacKay *et al*. (1991).

## References

Chan, R.K. & Otte, C.A. (1982) Isolation and genetic analysis of *Saccharomyces cerevisiae* mutants super-sensitive to G1 arrest by **a**-factor and α-factor pheromones. *Mol. Cell. Biol.* **2**, 11–20.

Ciejek, E. & Thorner, J. (1979) Recovery of *S. cerevisiae* **a** cells from G1 arrest by α-factor pheromone requires endopeptidase action. *Cell* **18**, 623–635.

Cross, F., Hartwell, L.H., Jackson, C. & Konopka, J.B. (1988) Conjugation in *Saccharomyces cerevisiae*. *Annu. Rev. Cell Biol.* **4**, 429–457.

Duntze, W., MacKay, V. & Manney, T.R. (1970) *Saccharomyces cerevisiae*: a diffusible sex factor. *Science* **168**, 1472–1473.

Herskowitz, I. (1989) A regulatory hierarchy for cell specialization in yeast. *Nature (Lond.)* **342**, 749–757.

Hicks, J.B. & Herskowitz, I. (1976) Evidence for a new diffusible element of mating pheromones in yeast. *Nature (Lond.)* **260**, 246–248.

Jars, M.U., Osborn, S., Forstrom, J. & MacKay, V.L. (1995) *N*- and *O*-glycosylation and phosphorylation of the Bar secretion leader derived from the Barrier protease of *Saccharomyces cerevisiae. J. Biol. Chem.* **270**, 24810–24817.

Kronstad, J.W., Holly, J.A. & MacKay, V.L. (1987) A yeast operator overlaps an upstream activation site. *Cell* **50**, 369–377.

MacKay, V.L. (1993) **a**'s, α's, and shmoos: mating pheromones and genetics. In: *The Early Days of Yeast Genetics* (Hall, M.N. & Linder, P., eds). Cold Spring Harbor, NY: Cold Spring Harbor Press, pp. 273–290.

MacKay, V.L., Welch, S.K., Insley, M.Y., Manney, T.R., Jolly, J., Saari, G.C. & Parker, M.L. (1988) The *Saccharomyces cerevisiae BAR1* gene encodes an exported protein with homology to pepsin. *Proc. Natl Acad. Sci. USA* **85**, 55–59.

MacKay, V.L., Armstrong, J., Yip, C., Welch, S., Walker, K., Osborn, S., Sheppard, P. & Forstrom, J. (1991) Characterization of the Bar proteinase, an extracellular enzyme from the yeast *Saccharomyces cerevisiae.* In: *Proceedings of the Aspartic Proteinase Conference on Structure and Function of the Aspartic Proteinases* (Dunn, B.M., ed.). New York: Plenum Publishing, pp. 161–172.

Sprague, G.F., Jr. & Herskowitz, I. (1981) Control of yeast cell type by the mating type locus. I. Identification and control of expression of the **a**-specific gene, *BAR1. J. Mol. Biol.* **153**, 305–321.

*Vivian L. MacKay*
*ZymoGenetics, Inc.,*
*1201 Eastlake Ave. E,*
*Seattle, WA 98102, USA*
*Email: mackayv@zgi.com*

# 304. Yapsin 1

## Databanks

*Peptidase classification: clan AA, family A1, MEROPS ID: A01.030*
*NC-IUBMB enzyme classification: none*
*Databank codes:*

| Species | SwissProt | PIR | EMBL (cDNA) | EMBL (genomic) |
|---|---|---|---|---|
| *Saccharomyces cerevisiae* | P32329 | S20150 | L31651 | – |

## Name and History

In 1990, an enzyme was cloned from yeast (Egel-Mitani *et al.*, 1990) on the basis of its ability to partially suppress the pro-α-mating factor processing defect in yeast mutants that lacked the natural pro-α-mating factor processing enzyme, the subtilisin-like serine protease, kexin (Chapter 115). Nucleotide sequence analysis identified the new enzyme as an aspartic protease and it was named **yeast aspartic protease 3 (Yap3)** because of its identification as the third aspartic protease from yeast after saccharopepsin (Chapter 283) and barrierpepsin (Chapter 303). The enzyme was independently cloned 3 years later (Bourbonnais *et al.*, 1993) based on its ability to process the heterologously expressed prohormone anglerfish prosomatostatin II to somatostatin-28 by a cleavage after a single arginine residue. Hence, its ability to correctly cleave the basic residue cleavage sites of pro-α-mating factor and prosomatostatin II identified this enzyme as a member of a subgroup of the aspartic proteases with unique specificity.

Since its initial characterization, a C-terminally-truncated soluble form of Yap3 has been overexpressed, purified and further characterized with respect to its physical and chemical properties (see below). Recently, Yap3 has been renamed **yapsin 1**, as the first member of this novel subclass of aspartic proteases that was cloned. Other members in this family include yapsin 2, a Yap3 homolog from *S. cerevisiae* characterized initially as Mkc7 (Chapter 305) (Komano & Fuller, 1995), and yapsin 3, a putative mammalian homolog of Yap3p characterized from cattle pituitary secretory vesicles as pro-opiomelanocortin-converting enzyme (Chapter 306) (PCE; EC 3.4.23.17) (Loh *et al.*, 1985).

## Activity and Specificity

In addition to cleaving pro-α-mating factor and prosomatostatin II *in vivo*, yapsin 1 has been shown to cleave a wide variety of mammalian prohormone substrates at their basic residue cleavage sites *in vitro*, including pro-opiomelanocortin (POMC) (Azaryan *et al.*, 1993), anglerfish prosomatostatin I and II (Cawley *et al.*, 1993) and proinsulin (Cawley *et al.*, 1996a). It is also capable of cleaving shorter peptide substrates such as adrenocorticotropin$_{1-39}$

(ACTH) (Azaryan *et al.*, 1993; Cawley *et al.*, 1996a) and cholecystokinin 33 (Cawley *et al.*, 1996a) and the synthetic peptide Boc-Arg-Val-Arg-Arg+Mca (Azaryan *et al.*, 1993). The catalytic efficiencies, $k_{cat}/K_m$, of yapsin 1 have demonstrated an enhancement of enzymatic activity upon addition of basic residues in the upstream and downstream positions of the cleavage site (Cawley *et al.*, 1996a; Ledgerwood *et al.*, 1996). Hence, the tetra-basic cleavage site of $ACTH_{1-39}$, Lys9+Lys10-Arg-Arg, which contains additional Lys residues at P5 and P6′, has one of the highest $k_{cat}/K_m$ values of $1.3 \times 10^6 \, M^{-1} \, s^{-1}$ (Cawley *et al.*, 1996a).

The optimum pH for enzymatic activity is generally 4.0–4.5 while the pI is approximately 4.5 (Cawley *et al.*, 1995). Pepstatin A is a potent competitive inhibitor of yapsin 1 ($K_i$ approximately 0.4 µM; unpublished results of the author) while all other class-specific protease inhibitors have little or no effect on its activity. Neither calcium nor reducing agents appear to have any effect on the activity. Due to its efficient cleavage of $ACTH_{1-39}$, a standard unit of yapsin 1 activity has recently been defined as the amount of enzyme that generates 0.18 µg of $ACTH_{1-15}$ from 10 µg $ACTH_{1-39}$ in 100 µl of 0.1 M sodium citrate, pH 4.0, at 37°C for 30 min.

## Structural Chemistry

The gene for yapsin 1, which is found on yeast chromosome XII, encodes a protein of 569 amino acids with ten potential *N*-linked glycosylation sites. The two active-site aspartic acid residues characteristic of aspartic proteases are Asp101 and Asp371. The calculated molecular mass is almost 60 kDa; however, due to extensive glycosylation, 70 kDa, 90 kDa and 120–150 kDa forms of the enzyme are found. The enzyme is synthesized as preproyapsin 1 that becomes sequentially processed to remove the signal peptide at Gly21 in the endoplasmic reticulum, followed by autocatalytic processing of its propeptide in an acidic compartment at its cleavage site, Lys66+Arg67. Wild-type yapsin 1 is a membrane-anchored enzyme localized to the extracellular side of the yeast cell membrane (Ash *et al.*, 1995; Cawley *et al.*, 1995). Attachment is through a glycophosphatidylinositol (GPI) membrane anchor situated in the C-terminus. Further processing of the protein into two subunits occurs at Asn144. This cleavage site falls within a loop structure that so far appears to be unique to yapsin 1 and 2 only. The two subunits, designated $\alpha$ and $\beta$, are joined by a disulfide bond. Deglycosylation of the approximately 90 kDa and 120–150 kDa forms by endoglycosidase H results in a decrease of both forms to approximately 65 kDa; however, this band on SDS-PAGE in reducing conditions is still diffuse, probably due to heterogeneous Ser- or Thr-linked glycosylation.

## Preparation

Expression levels of wild-type yapsin 1 from *S. cerevisiae* are too low for efficient purification. A recombinant form of yapsin 1 lacking its GPI anchor has been successfully overexpressed in yeast and purified from both the cell extract (Azaryan *et al.*, 1993) and media (Cawley *et al.*, 1998) yielding approximately 0.1 µg and 1 µg yapsin 1 per g wet yeast, respectively. Proyapsin 1 has also been expressed in insect cells using the baculovirus expression system. However, it has only been partially purified and the yield appears to be lower than that of the yeast expression system. One benefit of the use of the baculovirus expression system is the ability to obtain the zymogen form of the enzyme, since no activation occurs in this system. The zymogen can be activated *in vitro* upon acidification.

## Biological Aspects

The physiologic function of yapsin 1 in yeast is unknown. Although its gene has been determined to be non-essential (Egel-Mitani *et al.*, 1990), its ability to act as an alternative processing enzyme for kexin in the secretory pathway warrants its description as a processing enzyme. Based on the presence of some putative heat-shock elements in the promoter region of the yapsin 1 gene (Olsen, 1994), it has been proposed that the enzyme may be induced in times of cellular stress. In light of the stability of yapsin 1 to temperatures above normal optimal growth temperatures for yeast and a $Q_{10}$ of 1.95, it is possible that its function becomes more evident at elevated temperatures. The fact that yapsin 2 (Chapter 305) was cloned as a suppressor of a cold-sensitive phenotype of a kexin-deficient mutant (Komano & Fuller, 1995), suggests that the yeast yapsins may play an important role as backup enzymes in abnormal growth conditions.

The propeptide of yapsin 1 has been shown to be required for the correct folding of the nascent enzyme and its removal is required for activity. It is unknown where yapsin 1 becomes activated in the cell. The pH optimum for autocatalytic activation is pH approximately 4 (Cawley *et al.*, 1998) and the mechanism appears to be intramolecular (unpublished results of the author). Yapsin 1 has been cotransfected into the pheochromocytoma cell line PC12 cells, with the cDNA of POMC and found to be able to process POMC in the regulated secretory pathway (Cool *et al.*, 1996), indicating that yapsin 1 was targeted to the correct compartment in a mammalian cell, where it was activated and subsequently able to process POMC.

Similar to the homology that exists between the yeast kexin (Chapter 115) and the mammalian prohormone convertases (PCs), it appears likely that mammalian homologs of the yeast yapsins also exist in the capacity of alternative or backup enzymes for the PCs similar to that seen between kexin and yapsins 1 and 2. The existence of mammalian homologs of the yeast yapsins has become more evident from the immunologic characterization of yapsin 1-like proteins in neuropeptide-rich regions of mammalian brain and pituitary (Cawley *et al.*, 1996b).

## Distinguishing Features

One of the dramatic features of this enzyme is the ability of the glycosylated forms to withstand lyophilization without measurable loss of activity. Also, there is no loss of specific activity when purified enzyme is stored for one month at 4°C or for longer periods (more than one year) at −20°C. Lyophilized enzyme appears to be stable indefinitely.

## Further Reading

For further reading, see Loh & Cawley (1995).

## References

Ash, J., Dominguez, M., Bergerson, J.J., Thomas, D.Y. & Bourbonnais, Y. (1995) The yeast proprotein convertase encoded by *YAP3* is a glycophosphatidylinositol-anchored protein that localizes to the plasma membrane. *J. Biol. Chem.* **270**, 20847–20854.

Azaryan, A.V., Wong, M., Friedman, T.C., Cawley, N.X., Estivariz, F.E., Chen, H.C. & Loh, Y.P. (1993) Purification and characterization of a paired basic residue-specific yeast aspartic protease encoded by the YAP3 gene. Similarity to the mammalian pro-opiomelanocortin-converting enzyme. *J. Biol. Chem.* **268**, 11968–11975.

Bourbonnais, Y., Ash, J., Daigle, M. & Thomas, D.Y. (1993) Isolation and characterization of *S. cerevisiae* mutants defective in somatostatin expression: cloning and functional role of a yeast gene encoding an aspartyl protease in precursor processing at monobasic cleavage sites. *EMBO J.* **12**, 285–294.

Cawley, N.X., Noe, B.D. & Loh, Y.P. (1993) Purified yeast aspartic protease 3 cleaves anglerfish pro-somatostatin I and II at di- and monobasic sites to generate somatostatin-14 and -28. *FEBS Lett.* **332**, 273–276.

Cawley, N.X., Wong, M., Pu, L.-P., Tam, W. & Loh, Y.P. (1995) Secretion of yeast aspartic protease 3 is regulated by its carboxy-terminal tail: characterization of secreted YAP3p. *Biochemistry* **34**, 7430–7437.

Cawley, N.X., Chen, H.-C., Beinfeld, M.C. & Loh, Y.P. (1996a) Specificity and kinetic studies on the cleavage of various prohormone mono- and paired-basic residue sites by yeast aspartic protease 3. *J. Biol. Chem.* **271**, 4168–4176.

Cawley, N.X., Pu, L.-P. & Loh, Y.P. (1996b) Immunological identification and localization of yeast aspartic protease 3-like prohormone processing enzymes in mammalian brain and pituitary. *Endocrinology* **137**, 5135–5143.

Cawley, N.X., Olsen V., Zhang, C-F., Chen, H-C., Tan, M. & Loh, Y.P. (1998) Activation and processing of non-anchored yapsin 1 (Yap3p). *J. Biol. Chem.* **273**, 584–591.

Cool, D.R., Louie, D.Y. & Loh, Y.P. (1996) yeast aspartic protease 3 (YAP3p) is sorted to secretory granules and activated to process pro-opiomelanocortin in PC12 cells. *Endocrinology* **137**, 5441–5446.

Egel-Mitani, M., Flygenring, H.P. & Hansen, M.T. (1990) A novel aspartyl protease allowing *KEX2*-independent *MFα* propheromone processing in yeast. *Yeast* **6**, 127–137.

Komano, H. & Fuller, R.S. (1995) Shared functions *in vivo* of a glycosyl-phosphatidylinositol-linked aspartyl protease, Mkc7, and the proprotein processing protease Kex2 in yeast. *Proc. Natl Acad. Sci. USA* **92**, 10752–10756.

Ledgerwood, E.C., Brennan, S.O., Cawley, N.X., Loh, Y.P. & George, P.M. (1996) Yeast aspartic protease 3 (Yap3) prefers substrates with basic residues in the P2, P1 and P2′ positions. *FEBS Lett.* **383**, 67–71.

Loh, Y.P. & Cawley, N.X. (1995) Processing enzymes of pepsin family: yeast aspartic protease 3 and pro-opiomelanocortin converting enzyme. *Methods Enzymol.* **248**, 136–146.

Loh, Y.P., Parish, D.C. & Tuteja, R. (1985) Purification and characterization of a paired basic residue-specific pro-opiomelanocortin converting enzyme from bovine pituitary intermediate lobe secretory vesicles. *J. Biol. Chem.* **260**, 7194–7205.

Olsen, V. (1994) A putative prohormone convertase from *Saccharomyces cerevisiae*. Characterization of biosynthesis and cellular localization of the yap3 gene product. Masters thesis, Laboratory of Cellular and Molecular Physiology, University of Copenhagen.

*Niamh X. Cawley*
*Section on Cellular Neurobiology,*
*Laboratory of Developmental Neurobiology,*
*National Institute of Child Health*
*and Human Development,*
*National Institutes of Health,*
*Bethesda, MD 20892, USA*
*Email: cawley@codon.nih.gov*

*Y. Peng Loh*
*Section on Cellular Neurobiology,*
*Laboratory of Developmental Neurobiology,*
*National Institute of Child Health*
*and Human Development,*
*National Institutes of Health,*
*Bethesda, MD 20892, USA*

# 305. Yapsin 2

## Databanks

*Peptidase classification: clan AA, family A1, MEROPS ID: A01.031*
*NC-IUBMB enzyme classification: none*

*Databank codes:*

| Species | SwissProt | PIR | EMBL (cDNA) | EMBL (genomic) |
|---------|-----------|-----|-------------|----------------|
| *Saccharomyces cerevisiae* | P53379 | – | U14733 | Z50046: chromosome IV cosmid 8358<br>Z54139: chromosome IV cosmid 2943 |

## Name and History

The *MKC7* gene was identified as a multicopy suppressor of the cold-sensitive growth phenotype of a yeast (*Saccharomyces cerevisiae*) kexin mutant (Komano & Fuller, 1995). The closest homolog of *MKC7* is *YAP3*, which was isolated as a multicopy suppressor of the pro-$\alpha$-factor processing defect of a kexin mutant (Egel-Mitani *et al.*, 1990). Yap3p (*YAP3* gene product) and **Mkc7p** (*MKC7* gene product) have recently been termed yapsin 1 (Chapter 304) and **yapsin 2**, respectively.

## Activity and Specificity

Yapsin 2 is a basic residue-specific aspartic protease (Komano & Fuller, 1995), requiring Lys or Arg at P1, but having relaxed specificity at P2 (although Pro is excluded at P2; H. Komano & R.S. Fuller, unpublished results). Yapsin 2 appears to cleave exclusively at the carboxyl side of pairs of basic residues, suggesting exclusion of basic residues from P1'. Although yapsin 2 exhibits a pH optimum ($k_{cat}/K_m$) of 4.5 with a substrate based on a pro-$\alpha$-factor cleavage site, a single-Lys containing substrate with His at P3 and P4 is cleaved with a pH optimum of approximately 6.0. Yapsin 2 exhibits a $k_{cat}$ of approximately $20\,s^{-1}$ with the best internally quenched fluorogenic peptide substrates. The enzyme is sensitive to pepstatin, but only at high concentrations (H. Komano & R.S. Fuller, unpublished results).

## Structural Chemistry

The *MKC7* sequence predicts a preproprotein of 596 residues (a potential propeptide cleavage site exists after Arg65), with an $M_r$ of 64 000 and nine potential Asn-glycosylation sites (Komano & Fuller, 1995). Purified yapsin 2 migrates heterogeneously in SDS-PAGE in the range of 100–200 kDa due to extensive *N*-linked glycosylation (H. Komano & R.S. Fuller, unpublished results). Like yapsin 1 (Cawley *et al.*, 1995; Ash *et al.*, 1995), yapsin 2 is anchored to the yeast plasma membrane via a C-terminal glycosylphosphatidylinositol (GPI) linkage (Komano & Fuller, 1995).

## Preparation

GPI-linked yapsin 2 can be purified by a series of chromatographic steps from a Triton X-100 extract of a yeast strain overproducing the full-length protein (H. Komano & R.S. Fuller, unpublished results).

## Biological Aspects

Knocking out either *MKC7* or *YAP3* gene alone results in no phenotype, but *yap3 mkc7* double mutants exhibit poor growth at 37°C, suggesting functional redundancy between yapsin 1 and yapsin 2 (Komano & Fuller, 1995). Triple mutants (*mkc7 yap3 kex2*) exhibit increased sensitivity of growth to high and low temperatures, suggesting that the physiologic functions of yapsins 1 and 2 significantly overlap that of kexin, despite localization of the yapsins to the external surface of the plasma membrane and kexin to a late Golgi compartment. The phenotypic synergism of the mutations argues further that yapsins 1 and 2 are processing, rather than degradative enzymes. The physiologic substrates of yapsins 1 and 2 are unknown, but localization of these apparent processing enzymes to the plasma membrane suggests secretase-like function.

## Distinguishing Features

Yapsin 2, along with yapsin 1 (Chapter 304), is distinguished from other aspartic proteases by specificity for cleavage at basic residues and by GPI-anchor dependent cell surface localization.

## Acknowledgement

Supported by NIH grant GM39697.

## Further Reading

For further reading, see Komano & Fuller (1995).

## References

Ash, J., Dominguez, M., Bergeron, J.J., Thomas, D.Y. & Bourbonnais, Y. (1995) The yeast proprotein convertase encoded by *YAP3* is a glycophosphatidylinositol-anchored protein that localizes to the plasma membrane. *J. Biol. Chem.* **270**, 20847–20854.

Cawley, N.X., Wong, M., Pu, L.P., Tam, W. & Loh, Y.P. (1995) Secretion of yeast aspartic protease 3 is regulated by its carboxy-terminal tail: characterization of secreted YAP3p. *Biochemistry* **34**, 7430–7437.

Egel-Mitani, M., Flygenring, H.P. & Hansen, M.T. (1990) A novel aspartyl protease allowing *KEX2*-independent *MFα* propheromone processing in yeast. *Yeast* **6**, 127–137.

Komano, H. & Fuller, R.S. (1995) Shared functions *in vivo* of a glycosyl-phosphatidylinositol-linked aspartyl protease, Mkc7, and the proprotein-processing protease Kex2 in yeast. *Proc. Natl Acad. Sci. USA* **92**, 10752–10756.

*Robert S. Fuller*
*Department of Biological Chemistry,*
*University of Michigan Medical School,*
*Ann Arbor, MI 48109-0606, USA*
*Email: bfuller@umich.edu*

# 306. Yapsin 3

## Databanks

*Peptidase classification: clan AA, family A1, MEROPS ID: A01.032*
*NC-IUBMB enzyme classification: EC 3.4.23.17*
*Databank codes: no sequence data available*

## Name and History

The search for prohormone-processing enzymes specific for paired basic residues led to the discovery of **pro-opiomelanocortin converting enzyme** (**PCE**) in cattle pituitary intermediate lobe secretory granules (Loh et al., 1985). This enzyme was given this name since it was first shown to cleave the prohormone pro-opiomelanocortin (POMC). Subsequently, it was purified to apparent homogeneity from cattle intermediate and neural lobe secretory granules and shown to cleave other prohormones as well, such as provasopressin and proinsulin (Loh et al., 1985; Parish et al., 1986). PCE was identified as an aspartic protease, the first enzyme in this class demonstrated to cleave specifically at mono- and paired basic residues. Recently, (Cawley et al., 1996) it was shown to share immunologic identity with another basic residue specific aspartic protease, yapsin 1 (yeast aspartic protease 3) (Chapter 304), based on western blots using an antibody against yapsin 1. PCE has recently been classified as belonging to the yapsin group of aspartic proteinases related to yapsin 1 that have specificity for basic residues. PCE has now been given the name **yapsin 3**.

## Activity and Specificity

Yapsin 3 is highly specific for mono- and di-basic residues of prohormones. However, it cleaves synthetic peptide substrates with paired basic residues very poorly. Furthermore, it only cleaves at selected mono- and di-basic residues and on certain prohormones. The conformation of the substrate appears to be important in determining the selectivity of the substrates and the basic residue sites to be cleaved. Most commonly, the monobasic cleavage site has an additional basic residue upstream from it.

Yapsin 3 has been shown to cleave four prohormones so far. It cleaves POMC at several paired basic residues, both within and on the carboxyl side of the pair, to yield 21–23 kDa ACTH, $ACTH_{1-39}$, $\beta$-lipotropin, $\beta$-endorphin$_{1-31}$, N-POMC$_{1-49}$ and $\gamma^3$-MSH (Birch et al., 1991a; Estivariz et al., 1989; Loh et al., 1985). In addition, yapsin 3 cleaves provasopressin at a Lys-Arg pair to yield vasopressin Lys-Arg (Parish et al., 1986), and at a monobasic residue site to yield neurophysin and a glycopeptide (Parish, 1986). It also cleaves the Lys-Arg and Arg-Arg sites of proinsulin to yield the A and B chains of insulin, respectively (Loh et al., 1985). Recently, it was shown to cleave proenkephalin (Azaryan et al., 1995).

Yapsin 3 has a pH optimum of 4.0–5.5, assayed using mouse POMC as substrate (Loh et al., 1985). The activity is stimulated by calcium, although the extent of stimulation varies with substrate (Estivariz et al., 1989; Birch et al., 1991b). Assays are most conveniently made with $\beta$ lipotropin (Loh, 1986), which is obtainable from the US National Hormone and Pituitary Program, and with other isolated or recombinant prohormones described above. However, there is no short synthetic substrate found that is efficiently cleaved by yapsin 3. Yapsin 3 is inhibited by pepstatin A and DAN (Loh et al., 1985; Loh, 1986).

## Structural Chemistry

Yapsin 3 is a glycoprotein of about 68 kDa in size, with a pI between 3.5 and 4.0 (Loh et al., 1985). Yapsin 3 has been shown to cross-react with antibodies against yapsin 1, indicating structural homology between these two enzymes.

## Preparation

Yapsin 3 has been purified to homogeneity from cattle intermediate lobe and neural lobe secretory granules (Loh et al., 1985; Parish et al., 1986; Birch et al., 1991b) as well as from cattle adrenal medulla chromaffin granules (Azaryan et al., 1995). Conventional procedures such as that used for yapsin 1 (Chapter 295) are recommended when attempting to purify yapsin 3 from tissue or cells. Purified yapsin 3 can be stored at −20°C for 2–3 days.

## Biological Aspects

The enzyme has been found in the pituitary intermediate lobe, anterior lobe and neural lobe of several species of mammals including rats, mice and cows (Loh & Gainer, 1982; Chang & Loh, 1983; Loh et al., 1985; Parish et al., 1986) and in cattle adrenal medulla chromaffin granules (Azaryan et al., 1995). It is present in the secretory granules of these tissues (Loh et al., 1995; Cawley et al., 1996). Yapsin 3 activity is found both in the soluble fraction and associated with the membrane in the secretory granules. The membrane activity is extractable with 1 M NaCl (Chang & Loh, 1984). The activity purified from adrenal chromaffin granules has characteristics identical to yapsin 3 and cleaves recombinant proenkephalin at paired basic residues (Azaryan et al., 1995). Yapsin 3 is secreted from intermediate lobe cells in a coordinate and regulated manner with the processed hormones (Castro et al., 1989). The specificity and subcellular localization of yapsin 3 indicate that it is a prohormone-processing enzyme.

## Further Reading

For further reading, see Loh & Cawley (1995).

## References

Azaryan, A.V., Schiller, M., Mende-Mueller, L. & Hook, V.Y.H. (1995) Characteristics of the chromaffin granule aspartic proteinase involved in proenkephalin processing. *J. Neurochem.* **65**, 1771–1779.

Birch, N.P., Estivariz, F.E., Bennett, H.P.J. & Loh, Y.P. (1991a) Differential glycosylation of N-POMC[1–77] regulates the production of γ3-MSH by purified pro-opiomelanocortin converting enzyme: a possible mechanism for tissue specific processing. *FEBS Lett.* **290**, 191–194.

Birch, N.P., Bennett, H.P.J., Estivariz, F.E. & Loh, Y. (1991b) Effect of calcium ions on the processing of pro-opiomelanocortin by bovine intermediate lobe pro-opiomelanocortin converting enzyme. *Eur. J. Biochem.* **201**, 85–89.

Castro, M.G., Birch, N.P. & Loh, Y.P. (1989) Regulated secretion of pro-opiomelanocortin converting enzyme and an aminopeptidase B-like enzyme from dispersed bovine intermediate lobe pituitary cells. *J. Neurochem.* **52**, 1619–1628.

Cawley, N.X., Pu, L.-P. & Loh, Y.P. (1996) Immunological identification and localization of yeast aspartic protease 3-like prohormone processing enzymes in mammalian brain and pituitary. *Endocrinology* **137**, 5135–5143.

Chang, T.-L. & Loh, Y.P. (1983) Characterization of pro-opiocortin converting activity in rat anterior pituitary granules. *Endocrinology* **112**, 1832–1838.

Chang, T.-L. & Loh, Y.P. (1984) *In vitro* processing of pro-opiocortin by membrane-associated and soluble converting enzyme activities from rat intermediate lobe secretory granules. *Endocrinology* **114**, 2092–2099.

Estivariz, F.E., Birch, N.P. & Loh, Y.P. (1989) Generation of Lys-γ3-melanotropin from pro-opiomelanocortin[1–77] by a bovine intermediate lobe secretory vesicle membrane-associated aspartic protease and purified pro-opiomelanocortin converting enzyme. *J. Biol. Chem.* **264**, 17796–17801.

Loh, Y.P. (1986) Kinetic studies on the processing of β-lipotropin by bovine pituitary intermediate lobe pro-opiomelanocortin converting enzyme. *J. Biol. Chem.* **261**, 11949–11955.

Loh, Y.P. & Cawley, N.X. (1995) Processing enzymes of pepsin family: yeast aspartic protease 3 and pro-opiomelanocortin converting enzyme. *Methods Enzymol.* **248**, 136–146.

Loh, Y.P. & Gainer, H. (1982) Characterization of pro-opiocortin converting activity in purified secretory granules from rat pituitary neurointermediate lobe. *Proc. Natl Acad. Sci. USA* **79**, 108–112.

Loh, Y.P., Parish, D.C. & Tuteja, R. (1985) Purification and characterization of a paired basic residue-specific pro-opiomelanocortin converting enzyme from bovine pituitary intermediate lobe secretory vesicles. *J. Biol. Chem.* **260**, 7194–7205.

Loh, Y.P., Cawley, N.X., Friedman, T.C. & Pu, L.-P. (1995) Yeast and mammalian, basic residue-specific aspartic proteases in prohormone conversion. In: *Aspartic Proteinases* (Takahashi, K., ed.). New York: Plenum Press, pp. 519–527.

Parish, D.C. (1986) Prohormone-converting enzymes. In: *Neural and Endocrine Peptides and Receptors* (Moody, T.W., ed.). New York: Plenum Press, pp. 35–43.

Parish, D.C., Tuteja, R., Altstein, M., Gainer, H. & Loh, Y.P. (1986) Purification and characterization of a paired basic residue specific prohormone converting enzyme from bovine pituitary neural lobe secretory vesicles. *J. Biol. Chem.* **261**, 14392–14397.

*Y. Peng Loh*
*Section on Cellular Neurobiology,*
*Laboratory of Developmental Neurobiology,*
*National Institute of Child Health and Human Development,*
*National Institutes of Health,*
*Bethesda, MD 20892, USA*
*Email: ypl@codon.nih.gov*

# 307. Pregnancy-associated glycoproteins

## Databanks

*Peptidase classification: clan AA, family A1, MEROPS ID: A01.090*
*NC-IUBMB enzyme classification: none*

*Databank codes:*

| Species | Type | SwissProt | PIR | EMBL (cDNA) | EMBL (genomic) |
|---------|------|-----------|-----|-------------|----------------|
| *Bos taurus* | boPAG-1 | – | – | M73962 | L27832: exon 9 and complete CDS |
| | | | | | L27833: complete gene |
| | | | | | L27834: exons 7 and 8 |
| *Bos taurus* | boPAG-2 | – | – | L06151 | – |
| *Equus caballus* | eqPAG-1 | – | – | L38511 | – |
| *Ovis aries* | ovPAG-1 | – | – | M73961 | – |
| *Ovis aries* | ovPAG-2 | – | – | U30251 | – |
| *Sus scrofa* | poPAG-1 | – | – | L34360 | – |
| *Sus scrofa* | poPAG-2 | – | – | L34361 | U39198: promoter |
| | | | | | U39199: exons 3 and 4 |
| | | | | | U39762: exon 7 |
| | | | | | U39763: exons 8 and 9 and complete CDS |
| | | | | | U41421: exon 1 |
| | | | | | U41422: exon 2 |
| | | | | | U41423: exon 5 |
| | | | | | U41424: exon 6 |

## Name and History

Butler *et al*. (1982) described the partial purification of an antigen they named ***pregnancy-specific protein-B (PSP-B)*** from late-term placenta of cows. They first raised an antiserum against total placental homogenates in rabbits and then used tissue and blood from nonpregnant cattle to adsorb out antibodies in the antiserum not specific to pregnancy. The resulting antiserum was employed to monitor the partial purification of a glycoprotein that was reported to have an $M_r$ of approximately 50 000 and a pI of 4.0. Over the next decade this antigen was never thoroughly characterized, and the antiserum was not made generally available.

Zoli *et al*. (1991) reported the purification of a glycoprotein, called ***pregnancy-associated glycoprotein*** (or ***PAG-1***) from cattle placenta using a similar approach to that of Butler *et al*. (1982). Their antiserum was used to track PAG-1 purification through a series of chromatographic steps and led to the isolation of several isoforms (pI range 4.4–5.4) of $M_r$ 67 000. PSP-B and PAG-1 have identical N-terminal sequences and probably represent the same placental product (Roberts *et al*., 1995).

The antiserum generated by Zoli *et al*. (1991) was used to screen cDNA libraries prepared from cattle and sheep placentas (Xie *et al*., 1991). The resulting cDNAs were of length 1.7 kbp and coded for polypeptides of 380 and 382 amino acids, respectively. Curiously, the inferred primary sequences had approximately 50% amino acid identity to mammalian pepsinogens. Each polypeptide had a predicted signal sequence of 15 residues and a propeptide of 38 amino acids, as suggested from the N-terminal sequencing of the purified PAG (Zoli *et al*., 1991; S. Xie, J. Green & R.M. Roberts, unpublished results). Together, these data indicated that the PAGs are related to peptidase family A1.

## Activity and Specificity

Examination of the sequences around the two normally conserved active-site aspartic acid residues within the N-terminal and C-terminal lobes strongly suggested that neither cattle PAG-1 (boPAG-1) nor sheep PAG-1 (ovPAG-1) could be catalytically active as aspartic proteinases, despite their sequence similarity to pepsin (Xie *et al*., 1991). OvPAG-1, for example, lacked one of the two aspartic acid residues and could not therefore be active (Table 307.1). BoPAG-1, in contrast, had an alanine substituted for a normally invariant glycine residue in the N-terminus catalytic center. Molecular modeling has indicated that the presence of the relatively bulky methyl side chain of alanine would cause a displacement of the water molecule that resides symmetrically between the two catalytic aspartic acid residues. More detailed modeling of the entire cleft region of boPAG-1 and ovPAG-1 have reinforced the impression that ovPAG-1 and boPAG-1 cannot be active as proteinases (Guruprasad *et al*., 1996), although they can bind the peptide inhibitor of aspartic proteinases, pepstatin A (J. Green, S. Xie & R.M. Roberts, unpublished results).

The lack of activity of boPAG-1 has been confirmed by the inability of the native protein purified from the placenta to hydrolyze [$^{14}$C]hemoglobin (J. Green, S. Xie & R.M. Roberts, unpublished results). The propeptide is removed before secretion, but it seems likely that this cleavage is achieved through the action of other proteinases rather than by action of boPAG itself.

Recent experiments have indicated that PAG-like molecules are produced by placental tissues of a wide variety of species within the ungulates, including pigs (Szafranska *et al*., 1995) and horses (Green *et al*., 1994) as well as ruminants (Xie *et al*., 1994). Moreover, cattle, sheep and pigs, at least, each produce several different PAGs, which have as little as 50–60% amino acid sequence identity with each other, even within a single species (see Table 307.1). Examination of the sequences comprising the catalytic centers of these different PAGs (see Table 307.1) again suggests that most of them must be inactive, while others retain the consensus sequence required for catalytic activity. At least one PAG, eqPAG-1, is active as a proteinase (Chapter 308).

*Table 307.1*   Comparison of amino acid sequences of pepsin A and selected PAGs

| | Direct comparison[a] | | Overall amino acid sequence identity with[b] | |
|---|---|---|---|---|
| | N-terminal | C-terminal | Human pepsin A | Cattle PAG-1 |
| Human pepsin A | V F D T G S S | I V D T G T S | 100 | 49.5 |
| Cattle PAG-1 | · · · A · · | L · · · · · · | 49.5 | 100 |
| Cattle PAG-2 | · · · · · A | L L · · · · | 50.8 | 57.8 |
| Sheep PAG-1 | · · · · · · | L · G · · · | 49.4 | 70.6 |
| Sheep PAG-2 | · · · · · · | L · · · · · | 50.5 | 60.4 |
| Pig PAG-1 | I · · A · · | · L · S G S A | 48.6 | 48.8 |
| Pig PAG-2 | · · · · · · | · · · · · · | 52.9 | 56.2 |
| Horse PAG-1 | I · · · · A | · · · · · · | 58.6 | 54.9 |

[a]Amino acids that deviate from the sequence of human pepsin A are shown in single letter code.
[b]These comparisons do not include the signal and pro peptide regions.

## Structural Chemistry

Three-dimensional homology models of boPAG-1 and ovPAG-1 have been constructed on the basis of the crystal structures of pig pepsin and cattle chymosin (Guruprasad *et al.*, 1996). The peptide-binding subsites were defined by modeling interactions with pepstatin and the renin inhibitor CH-66. This approach allowed residues likely to affect peptide binding to be identified. Only two (eqPAG-1 and poPAG-2) of the eight PAGs examined retained features of active aspartic proteinases. Results also indicate that the peptide-binding specificities of the PAGs so far characterized differ significantly from each other and from pepsin. In the case of boPAG-1 and ovPAG-1, the data suggest a likely preference for lysine- or arginine-rich peptides.

All of the known PAGs are or have been inferred to be glycoproteins. The high $M_r$ values of some members (>65 000) suggest extensive glycosylation and possibly other forms of post-translational processing, including phosphorylation (Xie *et al.*, 1996).

## Preparation

BoPAG-1 and boPAG-2 have been purified from cattle placenta (Zoli *et al.*, 1991; Xie *et al.*, 1994). Recent experiments indicate that several PAGs can be collectively isolated by binding to pepstatin affinity columns followed by elution at increasing pH values (J. Green, S. Xie & R.M. Roberts, unpublished results). Final purification is achieved by anion-exchange chromatography.

## Biological Aspects

One defining feature of all PAGs so far described is that their expression appears to be restricted to the outer cell layer (trophectoderm or chorion) of the placenta. Certain PAGs (e.g. boPAG-1 and ovPAG-1) are localized to specialized binucleate cells (Xie *et al.*, 1991), while others are more generally expressed throughout the epithelium (Xie *et al.*, 1994; Szafranska *et al.*, 1995).

The gene coding for boPAG-1 consists of nine exons (range 99–281 bp) and eight introns (87–180 bp) organized in a manner very similar to that of proteolytically active mammalian proteinases (Xie *et al.*, 1995). Despite similarities in the transcribed portion of the genes encoding PAG-1, pepsinogen and other mammalian aspartic proteinases, the 1.2 kbp sequence upstream of the transcription start position contains a standard TATA element, but otherwise shows no strong resemblance to the promoter regions of other known genes including pepsinogens.

The function of PAG remains unknown although it is clear that they evolved at least 65 million years ago, most likely from a pepsin-like progenitor (Guruprasad *et al.*, 1996). The fact that many, possibly the majority, are catalytically inactive suggests that they do not function as proteinases. One hypothesis is that the intact substrate-binding cleft allows a PAG to associate with particular peptide sequences on other proteins or peptides. The fact that there appear to be numerous PAGs and that each may have structurally distinct clefts suggests that their binding specificities may differ.

## Distinguishing Features

The PAGs are members of the aspartic proteinase gene family produced as secretory products in the outer cell layer of the placentas of ungulate species. Their presence in other mammalian orders has not been ruled out. In general, they appear to be inactive as proteinases but retain peptide-binding capabilities.

## References

Butler, J.E., Hamilton, W.C., Sasser, R.G., Ruder, C.A., Hass, G.M. & Williams, R.J. (1982) Detection and partial characterization of two bovine pregnancy-specific proteins. *Biol. Reprod.* **26**, 925–933.

Green, J., Xie, S., Newman, A., Szafranska, B., Roberts, R.M., Baker, C.B. & McDowell, K. (1994) Pregnancy-associated glycoproteins of the horse (Abstract). *Biol. Reprod.* **50** (suppl. 1), 152.

Guruprasad, K., Blundell, T.L., Xie, S., Green, J., Szafranska, B., Nagel, R.J., McDowell, K., Baker, C.B. & Roberts, R.M. (1996) Comparative modeling and analysis of amino acid substitutions suggest that the family of pregnancy-associated glycoproteins includes both active and inactive aspartic proteinases. *Protein Eng.* **9**, 849–856.

Roberts, R.M., Xie, X., Nagel, R.J., Low, B., Green, J. & Beckers, J.-F. (1995) Glycoproteins of the aspartyl proteinase gene family secreted by the developing placenta. In: *Aspartic Proteinases: Structure, Function, Biology and Biomedical Implications* (Takahashi, K., ed.). *Adv. Exp. Med. Biol.* **362**. New York: Plenum Press, pp. 231–240.

Szafranska, B., Xie, S., Green, J. & Roberts, R.M. (1995) Porcine pregnancy-associated glycoproteins: new members of the aspartic proteinase gene family expressed in trophectoderm. *Biol. Reprod.* **53**, 21–28.

Xie, S., Low, B.G., Nagel, R.J., Kramer, K.K., Anthony, R.V., Zoli, A.P., Beckers, J.-F. & Roberts, R.M. (1991) Identification of the major pregnancy-specific antigens of cattle and sheep as inactive members of the aspartic proteinase family. *Proc. Natl Acad. Sci. USA* **88**, 10247–10251.

Xie, S., Low, B.G., Nagel, R.J., Beckers, J.-F. & Roberts, R.M. (1994) A novel glycoprotein of the aspartic proteinase gene family expressed in bovine placental trophectoderm. *Biol. Reprod.* **51**, 1145–1153.

Xie, S., Green, J., Beckers, J.-F. & Roberts, R.M. (1995) The gene encoding bovine pregnancy-associated glycoprotein-1, an inactive member of the aspartic proteinase family. *Gene* **159**, 193–197.

Xie, S., Nagel, R.J., Green, J., Beckers, J.-F. & Roberts, R.M. (1996) Trophoblast-specific processing and phosphorylation of pregnancy-associated glycoprotein-1 in day 15 to 25 sheep placenta. *Biol. Reprod.* **54**, 122–129.

Zoli, A.P., Beckers, J.F., Woutters-Ballman, P., Closset, J., Falmagne, P. & Ectors, F. (1991) Purification and characterization of a bovine pregnancy-associated glycoprotein. *Biol. Reprod.* **45**, 1–10.

**R. Michael Roberts**
Departments of Veterinary
Pathobiology, Animal Sciences
and Biochemistry
158 Animal Science Research Center,
University of Missouri, Columbia,
MO 65211-0001, USA
Email: gail_foristal@muccmail.missouri.edu

**Sancai Xie**
Departments of Veterinary
Pathobiology and Biochemistry,
158 Animal Science Research Center,
University of Missouri, Columbia,
MO 65211-0001, USA

**Jonathan Green**
Department of Biochemistry,
158 Animal Science Research Center,
University of Missouri, Columbia,
MO 65211-0001, USA

# 308. Horse placental aspartic proteinase

## Databanks

*Peptidase classification: clan AA, family A1, MEROPS ID: A01.091*
*NC-IUBMB enzyme classification: none*
*Databank codes:*

| Species | Type | SwissProt | PIR | EMBL (cDNA) | EMBL (genomic) |
|---------|------|-----------|-----|-------------|----------------|
| *Equus caballus* | PAG-1 | – | – | L38511 | – |

## Name and History

Pregnancy-associated glycoproteins (PAGs) (Chapter 307) are a large group of proteins secreted by the placenta of cattle, sheep (Xie *et al.*, 1991), pig (Szafranska *et al.*, 1995), and other members of the Artiodactyla order (Wood *et al.*, 1986). PAGs are part of the aspartic proteinase gene family having about 50% amino acid sequence identity with pepsinogens. Analysis of the putative amino acid sequence of several PAG cDNAs revealed that these molecules are not likely to be catalytically active due to changes in critical residues in and around the two catalytic domains (Xie *et al.*, 1991). Both native and recombinant PAG have been tested in proteinase assays with denatured [$^{14}$C]hemoglobin (Hb) as substrate; none demonstrated proteolytic activity. However, these modifications to the active site do not appear to interfere with their ability to bind peptide since both native and recombinant PAG can be isolated by affinity chromatography with pepstatin agarose (J. Green & R.M. Roberts, unpublished results).

To identify additional PAG-related transcripts, an equine d25 placental library was screened for the presence of PAG (see below). A PAG-like transcript (ePAG) was identified that had 50–60% amino acid sequence identity with the PAG family. Unlike other members of this family, **horse pregnancy-associated glycoprotein** is capable of cleaving peptide substrates. In this chapter, the enzyme will be referred to as **equine pregnancy-associated glycoprotein (ePAG)**.

## Activity and Specificity

Proteolytic activity of recombinant ePAG has been demonstrated by using a hemoglobin assay (Lin *et al.*, 1989). The pH optimum is around 4, although some activity remains detectable at pH 7.5 provided the propeptide is removed before the assay. As with most aspartic proteinases, ePAG can catalyze the removal of its own propeptide at low pH (below pH 5.5). The ability of ePAG to cleave its own propeptide as well as its activity toward [$^{14}$C]Hb can be inhibited by pepstatin A.

## Structural Chemistry

Equine PAG is a single-chain secretory protein of about 40 kDa with a theoretical pI of 6.4. The primary sequence has a single putative *N*-linked glycosylation site. However, the native protein does not appear to be glycosylated since the mobility of the native protein on SDS-PAGE is very similar to that of recombinant ePAG expressed in bacteria. Consequently, ePAG is probably a misnomer, indicative of this protein's relationship to the PAG family and not its glycosylation state. The molecule will likely be renamed once its function has been elucidated.

## Preparation

The ePAG transcript was identified by hybridization screening of a horse d25 placental library (kindly provided by K. McDowell, University of Kentucky) with a probe mixture consisting of $^{32}$P-labeled PAG cDNA from cattle, sheep, and pig. Approximately 0.5% of the $10^4$ plaques that were screened were positive. Sequencing of 24 of the positive plaques revealed that they all represented a single transcript (ePAG). Recombinant ePAG can be expressed in *Escherichia coli* as insoluble inclusion bodies, which can be renatured by solubilization with guanidine hydrochloride, followed by stepwise removal of the guanidine by dialysis against Tris buffer. The renatured protein can be purified by successive anion-exchange and gel-filtration chromatography.

## Biological Aspects

The ePAG transcript is detectable by northern analysis in d25–d32 extra-embryonic membranes as well as in the term placenta (J. Green & R.M. Roberts, unpublished results). *In situ* hybridization revealed that the ePAG transcript is restricted to specific cell populations within extraembryonic membranes. No signal could be detected in the embryo or in uterine tissue (B. Szafranska, K. McDowell & J. Green, unpublished results).

## Distinguishing Features

Horse PAG is an aspartic proteinase that seems to be restricted to the placenta. It differs from typical PAG family members in that it is proteolytically active and does not seem to be glycosylated.

## References

Lin, X.-L., Wong, R.N.S. & Tang, J. (1989) Synthesis, purification, and active site mutagenesis of recombinant porcine pepsinogen. *J. Biol. Chem.* **264**, 4482–4489.

Szafranska, B., Xie, S., Green, J. & Roberts, R.M. (1995) Porcine pregnancy-associated glycoproteins: new members of the aspartic proteinase gene family expressed in trophectoderm. *Biol. Reprod.* **53**, 21–28.

Wood, A.K., Short, R.E., Darling, A.E., Dusek G.L., Sasser, R.G. & Ruder, C.A. (1986) Serum assays for detecting pregnancy in mule and white-tailed deer. *J. Wildlife Manag.* **50**, 684–687.

Xie, S., Low, B.G., Nagel, R.J., Kramer, K.K., Anthony, R.V., Zoli, A.P., Beckers, J.-F. & Roberts, R.M. (1991) Identification of the major pregnancy-specific antigens of cattle and sheep as inactive members of the aspartic proteinase family. *Proc. Natl Acad. Sci. USA* **88**, 10247–10251.

*Jonathan Green*
*Department of Biochemistry,*
*158 Animal Sciences Research Center,*
*University of Missouri-Columbia,*
*Columbia, MO 65211, USA*
*Email: gail.foristal@muccmail.missouri.edu*

*Sancai Xie*
*Departments of Veterinary*
*Pathobiology and Animal Sciences,*
*158 Animal Sciences Research Center,*
*University of Missouri-Columbia,*
*Columbia, MO 65211, USA*

*R. Michael Roberts*
*Departments of Veterinary*
*Pathobiology, Animal Sciences*
*and Biochemistry*
*158 Animal Sciences Research Center,*
*University of Missouri-Columbia,*
*Columbia, MO 65211, USA*

# 309. Introduction: other families in clan AA

## Databanks

*MEROPS ID: A2, A3, A9, A10, A11, A12, A13, A14, A15, A16, A17, A18*

**Family A2**
Retropepsin of bovine leukemia virus (Chapter 315)
Retropepsin of equine infectious anemia virus (Chapter 312)
Retropepsin of feline immunodeficiency virus (Chapter 319)
Retropepsin of human immunodeficiency virus 1 (HIV1) (Chapter 310)
Retropepsin of human immunodeficiency virus 2 (HIV2) (Chapter 311)
Retropepsin of human T-cell leukemia virus (HTLV) (Chapter 314)
Retropepsin of Mason–Pfizer leukemia virus (Chapter 316)
Retropepsin of Moloney murine leukemia virus (Chapter 318)
Retropepsin of mouse mammary tumour virus (Chapter 317)
Retropepsin of Rous sarcoma virus (RSV) (Chapter 313)
Others (Chapter 320)

**Family A3**
Cassava vein mosaic virus endopeptidase (Chapter 322)
Cauliflower mosaic virus (caulimovirus) endopeptidase (Chapter 322)
*Commelina* yellow mottle virus putative protease (Chapter 322)

**Family A9**
Spumaretrovirus endopeptidase (see Chapter 320)

**Family A10**
*Schizosaccharomyces* retropepsin-like transposon (Chapter 321)

**Family A11**
*Drosophila* transposon *copia* peptidase (Chapter 321)

**Family A12**
Maize transposon Bs1 peptidase (Chapter 321)

**Family A13**
*Drosophila buzzatii* transposon Osvaldo peptidase (Chapter 321)

**Family A14**
Soybean transposon SIRE-1 peptidase (Chapter 321)

**Family A15**
Rice tungro bacilliform virus endopeptidase (Chapter 322)

**Family A16**
*Ascaris lumbricoides* transposon Tas peptidase (Chapter 321)

**Family A17**
*Bombyx mori* transposon Pao peptidase (Chapter 321)

**Family A18**
*Fusarium oxysporum* transposon skippy peptidase (Chapter 321)

Besides family A1 (Chapter 271), a number of other families have been included in clan AA. Of these, the tertiary structures have been solved only for endopeptidases from family A2. The other families are included because there are similarities in function of the endopeptidases to those in family A2, and there is conservation of the Xaa-Xaa-**Asp**-Xbb-Gly-Xbb motif, in which Xaa is a hydrophobic amino acid and Xbb is either Ser or Thr (see Alignment 309.1 on CD-ROM).

*Family A2* contains polyprotein-processing enzymes from positive-strand RNA viruses. Viral RNA encodes a number of polyproteins, and in the human immunodeficiency virus the genes are known as *gag, pol* and *env*. The *gag* and *env* genes encode structural proteins, whereas the *pol* gene encodes the three enzymes necessary for transcribing the viral RNA to DNA (reverse transcriptase), incorporating it into the host genome (ribonuclease H) and for processing the polyprotein (retropepsin). In some of the viruses, such as Rous sarcoma virus, the retropepsin is part of the Gag polyprotein, whereas in others, for example Mason–Pfizer monkey virus, the retropepsin is encoded by a separate RNA gene known as *vprt*. In many viruses, the *gag* and *pol* genes are transcribed contiguously because of a ribosomal frameshift, and a Gag-Pol polyprotein is synthesized by the infected cells. The retropepsin is required for processing of all three polyproteins, although initial stages of Env polyprotein processing are performed by cellular enzymes.

Tertiary structures of several endopeptidases from family A2 have now been determined, including the retropepsins from human immunodeficiency viruses type 1 and type 2, macaque immunodefiency virus, equine infectious anemia virus, feline immunodeficiency virus, Rous avian sarcoma virus and avian myeloblastosis-associated virus. All the structures are similar, and show a single domain bearing a single catalytic Asp. The endopeptidase is only functional when it dimerizes. The monomer has a similar fold to that of the N-terminal domain of pepsin A, in other words, the retropepsin monomer also carries the specificity site on the 'flap'. A retropepsin dimer therefore has two 'flaps'. The structure of the human immunodeficiency virus type 1 retropepsin is shown in Figure 309.1.

The sequence alignment (Alignment 309.1) shows the sequence around the catalytic Asp. Ala35 replaces the Ser or Thr that occurs in most members of family A1, but is Ala in renin. Both renin and retropepsins act at neutral pH, and Ala35 may be important for a neutral pH optimum (Ido *et al.*, 1991). Besides the motif around the catalytic Asp, a

second motif is also conserved between families A1 and A2. This is known as the 'hydrophobic-hydrophobic-Gly' motif, and is found not only in families A1 and A2, but in other members of clan AA (see Alignment 309.1). In family A1, this motif is found only in the N-terminal domain. Although the function of the hydrophobic-hydrophobic-Gly motif is not understood, in family A2, Arg123, which follows the motif and is invariant, is believed to be important for dimerization (Lapatto *et al.*, 1989).

A subset of retropepsins from oncoviruses, such as Mason–Pfizer virus retropepsin, and avian retroviruses, such as Rous sarcoma virus retropepsin, are larger proteins with N-terminal extensions, which in some representatives are homologous to *Escherichia coli* dUTPase.

An evolutionary tree for family A2 is shown in Figure 309.2. There is considerable divergence in sequence, probably because of the high rate of mutation in RNA viruses (which lack an efficient error-correcting mechanism) (Doolittle *et al.*, 1989).

*Family A3* includes a polyprotein-processing endopeptidase from pararetroviruses; these are double-stranded DNA viruses that infect plants. The genome contains an open reading frame (ORF V) that is analogous to the *pol* gene of retroviruses. ORF V encodes a polyprotein containing a reverse transcriptase homologous to that of retroviruses, and an endopeptidase at the N-terminus. Site-directed mutagenesis confirmed that Asp32 (numbering as in Alignment 309.1) is catalytic in cauliflower mosaic virus endopeptidase, and weak inhibition with pepstatin prevented polyprotein processing (Torruella *et al.*, 1989). A sequence that had been thought to be a transposon known as gypsy from *Drosophila* is now believed to be a retrovirus (Kim *et al.*, 1994); it has a domain homologous to the pararetrovirus endopeptidase, but the Asp has been replaced.

*Family A9* includes the polyprotein-processing enzyme from spumaretroviruses. The endopeptidase is part of the Gag polyprotein in the human foamy virus, but part of the Pol polyprotein in simian foamy virus (Renne *et al.*, 1992) (Table 310.1). The catalytic Asp has been identified by site-specific mutagenesis (Konvalinka *et al.*, 1995). As can be seen from Alignment 309.1, the catalytic Asp is contained in a motif that differs from that of pepsin A, in that residue 30 is not hydrophobic. There is also no hydrophobic-hydrophobic-Gly motif.

*Family A15* includes the polyprotein-processing endopeptidases from rice tungro bacilliform virus and sugarcane bacilliform virus. The bacilliform viruses are plant pararetroviruses from the badnavirus group. The endopeptidase has been shown to be essential for processing, and the catalytic Asp has been identified by site-specific mutagenesis (Laco *et al.*, 1995).

*Families A10, A11, A12, A13, A14, A16, A17* and *A18* include mainly putative endopeptidases from retrotransposons of fungi, plants and animals. Retrotransposons are genetically mobile elements that can copy themselves from one place to another in the genome. A retrotransposon encodes a polyprotein that contains all of the enzymes needed to accomplish the transpositions, including an integrase, an RNAase H, a reverse transcriptase and an endopeptidase that processes the polyprotein. The components of the polyprotein are thus similar to those of a retrovirus Pol polyprotein,

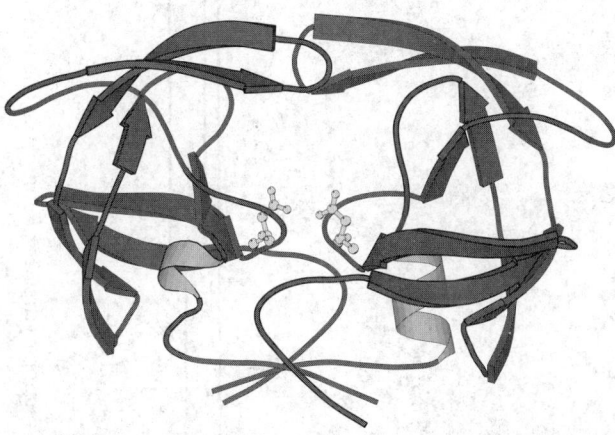

*Figure 309.1*  Richardson diagram of the human immunodefiency type 1 retropepsin dimer. The image was prepared from the Brookhaven Protein Data Bank entry (5HVP) as described in the Introduction (p. xxv). The catalytic residue Asp32 is shown in ball-and-stick representation (one from each monomer; numbering as in Alignment 309.1).

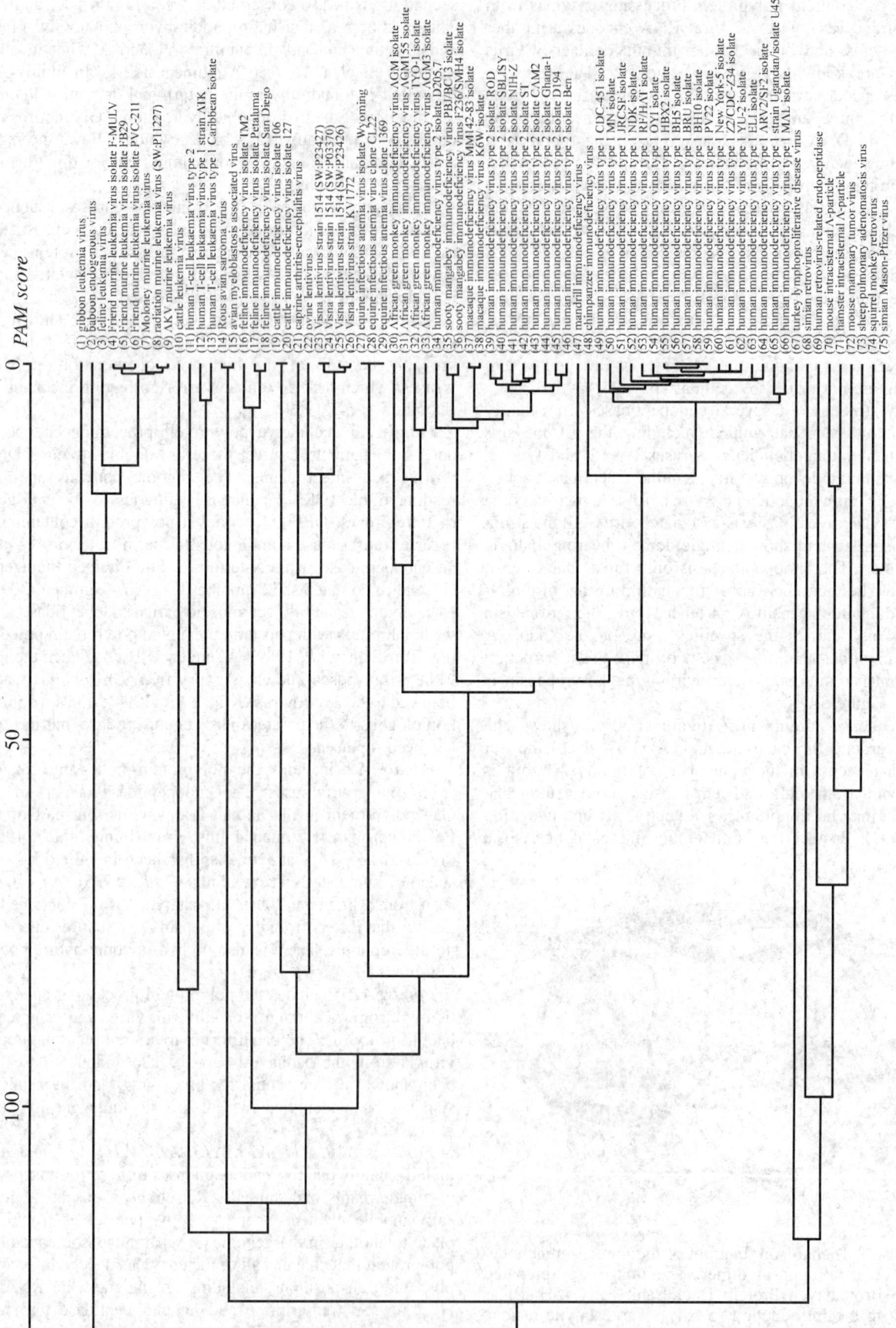

*Figure 309.2* Evolutionary tree for family A2. The tree was constructed as described in the Introduction (p. xxv).

and it has been proposed that one is the ancestor to the other.

Often an endopeptidase has been identified only by the presence of an Asp-Thr-Gly motif, or something similar. In some cases, the hydrophobic-hydrophobic-Gly motif is also present (see Alignment 309.1). In a few cases an endopeptidase has been biochemically characterized, usually by site-directed mutagenesis of the proposed catalytic Asp resulting in abolition of polyprotein processing. Among those endopeptidases that have been so characterized are those from *Drosophila* transposon copia (family A11) (Yoshioka *et al.*, 1990) and yeast Ty3 transposon (family A10) (Kirchner & Sandmeyer, 1993).

### References

Doolittle, R.F., Feng, D.-F., Johnson, M.S. & McClure, M.A. (1989) Origins and evolutionary relationships of retroviruses. *Q. Rev. Biol.* **64**, 1–30.

Ido, E., Han, H., Kezdy, F.J. & Tang, J. (1991) Kinetic studies of human immunodificiency virus type 1 protease and its active-site hydrogen bond mutant A28S. *J. Biol. Chem.* **266**, 24359–24366.

Kim, A., Terzian, C., Santamaria, P., Pelisson, A., Prudhomme, N. & Bucheton, A. (1994) Retroviruses in invertebrates: the gypsy retrotransposon is apparently an infectious retrovirus of *Drosophila melanogaster*. *Proc. Natl Acad. Sci. USA* **91**, 1285–1289.

Kirchner, J. & Sandmeyer, S. (1993) Proteolytic processing of Ty3 proteins is required for transposition. *J. Virol.* **67**, 19–28.

Konvalinka, J., Löchelt, M., Zentgraf, H., Flügel, R.M. & Kräusslich, H.G. (1995) Active foamy virus proteinase is essential for virus infectivity but not for formation of a Pol polyprotein. *J. Virol.* **69**, 7264–7268.

Laco, G.S., Kent, S.B.H. & Beachy, R.N. (1995) Analysis of the proteolytic processing and activation of the rice tungro bacilliform virus reverse transcriptase. *Virology* **208**, 207–214.

Lapatto, R., Blundell, T., Hemmings, A., Overington, J., Wilderspin, A., Wood, S., Merson, J.R., Whittle, P.J., Danley, D.E., Geoghegan, K.F., Hawrylik, S.J., Lee, S.E., Scheld, K.G. & Hobart, P.M. (1989) X-ray analysis of HIV-1 proteinase at 2.7 Å resolution confirms structural homology among retroviral enzymes. *Nature* **342**, 299–302.

Renne, R., Friedl, E., Schweizer, M., Fleps, U., Turek, R. & Neumann-Haefelin, D. (1992) Genomic organization and expression of simian foamy virus type 3 (SFV-3). *Virology* **186**, 597–608.

Torruella, M., Gordon, K. & Hohn, T. (1989) Cauliflower mosaic virus produces an aspartic proteinase to cleave its polyproteins. *EMBO J.* **8**, 2819–2825.

Yoshioka, K., Honma, H., Zushi, M., Kondo, S., Togashi, S., Miyake, T. & Shiba, T. (1990) Virus-like particle formation of *Drosophila* copia through autocatalytic processing. *EMBO J.* **9**, 535–541.

# *310. Human immunodeficiency virus 1 retropepsin*

## Databanks

*Peptidase classification: clan AA, family A2, MEROPS ID: A02.001*
*NC-IUBMB enzyme classification: EC 3.4.23.16*
*Databank codes:*

| Species | Isolate | SwissProt | PIR | EMBL (genomic) |
| --- | --- | --- | --- | --- |
| Human immunodeficiency virus type 1 | ARV2/SF2 | P03369 | A03968 | K02007: – |
| | BH5 | P04587 | – | K02012: – |
| | BH10 | P03366 | A03965 | K02010: – |
| | BRU | P03367 | A03966 | K02013: – |
| | CDC-451 | P05960 | – | M13136: 5′ end of genome |
| | ELI | P04589 | – | A07108: complete genome |
| | | | | K03454: complete genome |
| | HXB2 | P04585 | – | K03455: complete genome |
| | JH3 | P12498 | – | M21137: *gag* gene |
| | JRCSF | P20875 | – | M38429: complete genome |
| | MAL | P04588 | – | A07116: proviral DNA |
| | | | | X04415: complete genome |
| | MN | P05961 | – | M17449: complete genome |
| | NDK | P18802 | JQ0067 | M27323: complete genome |

*continued overleaf*

| Species | Isolate | SwissProt | PIR | EMBL (genomic) |
|---|---|---|---|---|
| | New York-5 | P12497 | – | M19921: complete genome |
| | OYI | P20892 | – | M26727: complete genome |
| | PV22 | P03368 | A03967 | K02083: complete genome |
| | | | | X01762: proviral genome |
| | RF/HAT3 | P05959 | – | M17451: complete genome |
| | U455 | P24740 | – | M62320: complete genome |
| | YU-2 | P35963 | B44001 | M93258: complete genome |
| | Zaire 6 | P04586 | A26192 | K03458: 3′ end of genome |
| | Z2 | P12499 | – | M22639: complete genome |
| Chimpanzee immunodeficiency virus | P17283 | S09984 | X52154 | – |

Brookhaven Protein Data Bank three-dimensional structures:

| Species | Strain | ID | Resolution | Notes |
|---|---|---|---|---|
| Human immunodeficiency virus type 1 | ARV2/SF2 | 1CPI | 2.05 | complex with a cyclic peptide inhibitor |
| | | 3HVP | 2.8 | Cys replaced with $\alpha$-amino-$N$-butyric acid |
| | BH5 | 1HVR | 1.8 | complex with XK263 |
| | | 1HVS | 2.25 | Val82Ala mutant; complex with A77003 |
| | | 1TCX | 2.3 | triple mutant; complex with SB203386 |
| | BH10 | 1AAQ | 2.5 | complex with hydroxyethylene isostere |
| | | 1HBV | 2.3 | complex with SB203238 |
| | | 1HEF | 2.2 | recombinant |
| | | 1HOS | 2.3 | complex with SB204144 |
| | | 1HPS | 2.3 | |
| | | 1HTE | 2.8 | complex with GR123976 |
| | | 1HTF | 2.2 | complex with GR126045 |
| | | 1HTG | 2 | complex with GR137615 |
| | | 1HVI | 1.8 | complex with A77003 |
| | | 1HVJ | 2 | complex with A78791 |
| | | 1HVK | 1.8 | complex with A76928 |
| | | 1HVL | 1.8 | complex with A76889 |
| | | 1HVP | | theoretical model |
| | | 1SBG | 2.3 | complex with SB203386 |
| | | 9HVP | 2.8 | complex with A-747041 |
| | BRU | 1HHP | 2.7 | |
| | | 1HPV | 1.9 | complex with VX-478 |
| | | 1HPX | 2 | complex with inhibitor KNI-272 |
| | HXB2 | 3PHV | 2.7 | recombinant |
| | LAI | 1HVC | 1.8 | complex with A-76928 |
| | New York-5 | 2HVP | 3 | recombinant |
| | | 4PHV | 2 | complex |
| | | 5HVP | 2 | complex with acetyl-pepstatin |
| | PV22 | 1GNM | 2.3 | Val82Asp mutant; complex with U89360E inhibitor |
| | | 1GNN | 2.3 | Val82Asn mutant; complex with U89360E inhibitor |
| | | 1GNO | 2.3 | complex with U89360E inhibitor |
| | synthetic | 4HVP | 2.3 | complex with Ac-Thr-Ile-Nle-PSI(CH$_2$-NH)-Nle-Gln-Arg amide |
| | | 7HVP | 2.4 | complex with ACE-Ser-Leu-Asn-Phe-PSI(CH(OH)-CH$_2$)-Pro-Ile-Val-OMe |
| | | 8HVP | 2.5 | complex with Val-Ser-Gln-Asn-Leu-PSI(CH(OH)-CH$_2$)-Val-Ile-Val |
| | unknown | 1HIV | 2 | |

## Name and History

***Human immunodeficiency virus 1 retropepsin (HIV-1 retropepsin)***, more commonly known as ***HIV-1 protease (HIV-1 PR)***, has become the most thoroughly studied system in the history of proteolytic enzymes. This is remarkable, as the initial discovery of the enzyme dates only from the report of the nucleotide sequence of the viral genome in 1985 (Ratner *et al.*, 1985). The ensuing 12 years have witnessed both a gathering of scientific talent and a steady stream of impressive observations. This has culminated in approval from the Food and Drug Administration (FDA) in the USA for therapeutic antiviral use in humans of four compounds (as of April 1997), based on significant clinical results in reducing viral load in patients with HIV.

Viral proteases have the function of cleaving the viral polyprotein initially formed by translation of the viral mRNA into the constituent functional units such as the matrix, capsid, and nucleocapsid structural proteins of HIV. These are initially connected in a head-to-tail fashion, forming a 55 kDa Gag polyprotein in the case of HIV-1. In addition, translational frameshifting occurs approximately 5% of the time to read through a stop codon producing a Gag-Pol fusion polyprotein of 160 kDa containing, in addition to the Gag structural proteins listed above, the protease, reverse transcriptase and integrase enzymatic proteins. This is further complicated by the necessity in the HIV system of dimerization of the protease sequence in order to form a functional enzyme, structurally similar to the pepsin family (A1) of aspartic proteinases (Chapter 309). In the latter case, one aspartic acid is contributed by the N-terminal (residue 32 in pepsin) and the C-terminal (residue 215 in pepsin) domains to form the catalytic machinery. In the retropepsins, the viral protease protein contributes one of the aspartic acids in the signature-Xaa-Xaa-Asp-Thr-Gly- (where Xaa is a hydrophobic residue) sequence. It requires two of these monomeric units to combine to form the functional enzyme. The necessity for formation of this higher order structure is likely to provide a mechanism for the timing of processing. This follows assembly of the new proteins into a daughter viral particle and budding from the surface of the infected cell.

The retroviruses constitute a unique type of viral system wherein the viral RNA contained within the nucleocapsid is copied into a DNA strand by the action of the viral reverse transcriptase. This activity is not found in mammalian systems, thus providing the first opportunity to create antiviral agents 3'-azido-3'-deoxythymidine (AZT), 2',3'-dideoxyinosine (ddI) and other nucleoside analogs. The DNA copy is eventually incorporated into the genome of the host cell. The name retropepsin was coined to denote the origin in a retroviral system as well as the structural and functional similarity to the pepsin family.

## Activity and Specificity

As discussed above, the activity of the viral enzyme is focused on cleavage of the junctions in the polyprotein to free the individual units from one another. Sequencing efforts from a number of laboratories have provided the details of the sites of cleavage (Sanchez-Pescador *et al.*, 1985; Wain-Hobson *et al.*, 1985; di Marzo Veronese *et al.*, 1986; Lightfoote

*et al.*, 1986; Henderson *et al.*, 1988; Graves *et al.*, 1989; Kräusslich *et al.*, 1989; Graves *et al.*, 1990). The information can be summarized as a list of the sequences surrounding the cleavage sites, presented here in the order in which they appear in the Gag-Pol polyprotein (Table 310.1).

Inspection of these sequences reveals a general division into two types of cleavage sites. One group involves cleavage between an aromatic residue (Tyr or Phe) and a Pro residue (junctions A, E and F). The second type includes two hydrophobic residues, but the amino acid on either side of the cleavage point is different in every case. In considering the residues in the P2 or P2' positions, it can be seen that the aromatic-Pro junctions all have an Asn residue in P2 and an Ile in P2'. In the case of junctions B, C and D, the P2' residue is either a Glu or a Gln residue (Margolin *et al.*, 1990). Examination of amino acid residues in other flanking positions presents a bewildering array of amino acids of varying properties. Thus, it has been difficult to draw conclusions about the precise preferences of the viral enzyme (Darke *et al.*, 1988; Konvalinka *et al.*, 1990; Phylip *et al.*, 1990; Billich & Winkler, 1991; Tözsér *et al.*, 1991a,b, 1992; Griffiths *et al.*, 1992; Dunn *et al.*, 1994; Cameron *et al.*, 1994; Kassel *et al.*, 1995) and it has been termed both a specific as well as a nonspecific enzyme.

The mixture of sequences that are seen at the cleavage points in the HIV system most likely reflects a significant property of the aspartic class of proteolytic enzymes: that they bind their substrates through interactions along an extended active-site cleft, always forcing the cleavage sequence to adopt a β strand conformation. This spreads the interactions along the cleft, typically resulting in the necessity for at least 6–8 amino acids to constitute an efficient oligopeptide substrate. This property also accounts to some degree for the serious problem of resistance development in drug therapy. The viral enzyme is able to mutate one or two residues to create a new enzyme form that still retains adequate binding to substrate polyprotein, but loses significant binding interactions with synthetic inhibitors, which are typically small in size and only interact with two or three subsites of the active-site cleft.

In addition to studies on peptide substrates, several groups have explored the cleavage of general protein substrates in an effort to further define the specificity of the retroviral enzyme (Tomasselli *et al.*, 1990a,b,c, 1991a,b, 1993; Tomasselli & Heinrikson, 1994). This has led to several schemes for prediction of cleavage sites in protein substrates (Poorman *et al.*, 1991; Chou *et al.*, 1996). The potential role of cleavage of

*Table 310.1*   Specificity of polyprotein cleavage

| Site | Sequence | Junction |
|------|----------|----------|
| A | Ser-Gly-Asn-Tyr┼Pro-Ile-Val-Gln | MA/CA |
| B | Ala-Arg-Val-Leu┼Ala-Glu-Ala-Met | CA/p2 |
| C | Ala-Thr-Ile-Met┼Met-Gln-Arg-Gly | p2/NC |
| D' | Arg-Gln-Ala-Asn┼Phe-Leu-Gly-Lys | NC/p6 |
| D | Pro-Gly-Asn-Phe┼Leu-Gln-Ser-Arg | NC/p6X |
| E | Ser-Phe-Asn-Phe┼Pro-Ile-Ser-Pro | p6/PR |
| F | Thr-Leu-Asn-Phe┼Pro-Ile-Ser-Pro | PR/RT |
| G | Ala-Glu-Thr-Phe┼Tyr-Val-Asp-Gly | RT p51/p66 |
| H | Arg-Lys-Ile-Leu┼Phe-Leu-Asp-Gly | RT/IN |

cellular proteins in the cytotoxic effects of HIV has been described by a number of authors who observed degradation by the peptidase of microtubule-associated proteins (Wallin *et al.*, 1990; Ainsztein & Purich, 1992), cytoskeletal proteins (Höner *et al.*, 1991, 1992; Oswald & von der Helm, 1991; Shoeman *et al.*, 1990, 1993; Luftig & Lupo, 1994) and nuclear factor-κB (Rivière *et al.*, 1991; Zhang *et al.*, 1995). Finally, the possible regulation of viral expression through the proteolytic processing of the viral NEF protein has been studied (Freund *et al.*, 1994; Gaedigk-Nitschko *et al.*, 1995; Welker *et al.*, 1996).

## Structural Chemistry

It is in the structural work that the significance of retropepsin becomes obvious. The discovery of the viral enzyme coincided with the efforts of a number of pharmaceutical companies to establish structure-based drug design groups. This approach had been pioneered during the largely unsuccessful attempts to develop renin inhibitors. While pharmacological problems limited the significance of compounds targeted to renin, the process of using crystallographically derived or modeled structures to design new inhibitors became accepted. A number of efforts to produce recombinant protein for crystallography were made, but it was chemical synthesis that yielded sufficient protein for solution of the structure (Wlodawer *et al.*, 1989). This was soon followed by the report of the structure of the enzyme derived from expression (Lapatto *et al.*, 1989). In the eight years following the first structure determinations, several hundred independent crystal structures, many with bound inhibitor molecules, have been established in as many as 50 different laboratories. The Brookhaven Protein Data Bank file designations are provided above for the coordinate sets that have been deposited in an accessible manner, and these may be accessed through a database established by the Wlodawer laboratory (Vondrasek & Wlodawer, 1996; Vondrasek *et al.*, 1997). Unfortunately, not all datasets have been made available to the scientific community, largely for proprietary reasons. Fortunately, the available crystal structures include many with bound inhibitors (Miller *et al.*, 1989; Fitzgerald *et al.*, 1990; Swain *et al.*, 1990; Graves *et al.*, 1991; Jaskolski *et al.*, 1991; Murthy *et al.*, 1992; Thanki *et al.*, 1992; Rutenber *et al.*, 1993; Hoog *et al.*, 1995; Chen *et al.*, 1994; Jhoti *et al.*, 1994; Baldwin *et al.*, 1995a; Priestle *et al.*, 1995; Hong *et al.*, 1996; Rose *et al.*, 1996a; Wang *et al.*, 1996).

Due to the intense interest in the development of antiviral agents, retropepsin has also been used extensively in computational studies to discover methods for structure prediction (Pearl & Taylor, 1987; Weber *et al.*, 1989; Pechik *et al.*, 1989; Weber, 1991; Taylor, 1994), studies of energetics and molecular dynamics (Harte *et al.*, 1990; Gustchina *et al.*, 1994; Collins *et al.*, 1995; Nicholson *et al.*, 1995; Silva *et al.*, 1995; Liu *et al.*, 1996), and drug design (Blundell *et al.*, 1990; Greer *et al.*, 1994). Of particular interest are studies where the role of the 'flap', a β hairpin that hangs over the active-site cleft and can control binding of substrates and inhibitors, is discussed (Gustchina & Weber, 1990).

Variation in the sequence of the enzyme is a major concern due to the development of resistant variants, discussed further below. A number of studies have explored the importance of specific amino acids of the enzyme, beginning with the extensive work of the Swanstrom group (Loeb *et al.*, 1989; Louis *et al.*, 1989; Borders *et al.*, 1994; Moody *et al.*, 1995). Structural work on forms of the viral peptidase resistant to inhibitors has begun recently and will continue as new information from clinical trials becomes available (Chen *et al.*, 1995; Baldwin *et al.*, 1995b; Hong *et al.*, 1996).

As described above, the formation of a dimer is an essential step in creating an active enzyme. Accordingly, a number of studies have explored this through mutagenesis (Guenet *et al.*, 1989; Wondrak & Louis, 1996) and through the creation of a tethered dimer by molecular biology (Cheng *et al.*, 1990b; Bhat *et al.*, 1994, 1995).

## Preparation and Assay

Following the initial descriptions of the activity of the protease discussed above, several groups have reported on methodology for the successful high-yield expression and purification of retropepsin (Danley *et al.*, 1989; Heimbach *et al.*, 1989; Hostomsky *et al.*, 1989; Cheng *et al.*, 1990a; Rittenhouse *et al.*, 1990; Goobar *et al.*, 1991; Margolin *et al.*, 1991; Hui *et al.*, 1993; Stebbins & Debouck, 1994; Gustafson *et al.*, 1995). Improvements in the stability of the enzyme have been achieved by selective mutagenesis to remove internal cleavage sites (Mildner *et al.*, 1994). Finally, the enzyme has been synthesized by totally chemical means, both with the normal L amino acids (Schneider & Kent, 1988; Hoeprich, 1994) and with all D amino acids (Milton *et al.*, 1992).

Assays for the enzyme have been described in several categories, usually using peptides derived from the junction sequences given above. Reliable assays can be conducted by HPLC methods to follow cleavage of defined substrates (Betageri *et al.*, 1993; Pazhanisamy *et al.*, 1995). Accurate assays may be performed by spectrophotometric methods (Nashed *et al.*, 1989; Richards *et al.*, 1990; Tomaszek *et al.*, 1990) or fluorescence methods (Matayoshi *et al.*, 1990; Toth & Marshall, 1990; Krafft & Wang, 1994; Peranteau *et al.*, 1995). Highly sensitive analyses are provided by radiometric procedures (Hyland *et al.*, 1990; Cook *et al.*, 1991; Evans *et al.*, 1992).

## Inhibition

The intense interest in retropepsin in the past decade has centered on the race to design, construct, test, and bring to market compounds that selectively inhibit the viral enzyme and not the related human enzymes. This effort has generated hundreds of literature reports and has served to drive the process of drug discovery forward. The first clue to a way to make a specific inhibitor was derived from the observation that retropepsin was able to cleave peptide bonds involving proline. This is an unusual specificity, especially among aspartic peptidases. An early report that a substrate may be converted into an inhibitor by replacing the Pro with pipecolic acid (Copeland *et al.*, 1990) was followed by the construction of several effective inhibitors. Many are derived from the known binding of the hydroxyl group of statine, an unusual amino acid, to the catalytic aspartic acids of this class of enzyme (Seelmeier *et al.*, 1988; Richards *et al.*, 1989).

These two principles were combined by several companies to produce antiviral compounds that have achieved FDA approval in the USA (Roberts *et al.*, 1990; Dorsey *et al.*, 1994; Varney *et al.*, 1994; Kempf *et al.*, 1995). The guiding principles of the use of structure in the design of inhibitors for retropepsin were clearly detailed in a review by Wlodawer and Erickson (Wlodawer & Erickson, 1993). Furthermore, it should be noted that many thousands of man-years of effort went into the process of searching for effective inhibitors. Space prevents a complete listing of all the citations, however.

Although extremely strong and selective inhibitors have been discovered and demonstrated to be effective in human trials, at least one major consideration remains: the rapid development of forms of the virus that are resistant to the drug. This arises due to the heterogeneity of the virus in the first instance (Fontenot *et al.*, 1992; Barrie *et al.*, 1996; Kozal *et al.*, 1996; Lech *et al.*, 1996; Rose *et al.*, 1996b), as well as to the selective pressure exerted by the drug and the ability of the virus to incorporate mutations in its sequence. The latter phenomena arise due to the error-prone nature of the viral reverse transcriptase. Also of importance is the rapid turnover of the virus (Ho *et al.*, 1995). Studies of the kinetic properties of the variant forms of the peptidase have revealed the quantitative reduction in binding for a number of inhibitors (Otto *et al.*, 1993; Kaplan *et al.*, 1994a; Sardana *et al.*, 1994; Condra *et al.*, 1995; Gulnik *et al.*, 1995; Jacobsen *et al.*, 1995; Lin *et al.*, 1995; Markowitz *et al.*, 1995; Ridky & Leis, 1995; Tisdale *et al.*, 1995; Pazhanisamy *et al.*, 1996; Molla *et al.*, 1996).

## Biological Aspects

As discussed in the first section of this chapter, the biological activity of the retroviral peptidase is to cleave the polyprotein precursor into its constituent units to permit assembly. One remaining question is the timing of this proteolytic processing. Does the viral enzyme function during the budding of new progeny virus? Is the activity a consequence of some property of the new viral particle, such as achieving the correct pH or ionic strength for efficient operation of the proteolytic enzyme? Some answers to these questions are beginning to emerge from studies of the structure of the virus (Ross *et al.*, 1991; Kaplan *et al.*, 1993, 1994b), the processing of the polyprotein (Erickson-Viitanen *et al.*, 1989; Louis *et al.*, 1994; Zybarth *et al.*, 1994; Zybarth & Carter, 1995; Wondrak *et al.*, 1996) and processing of various constructs containing retropepsin (Partin *et al.*, 1991; Kotler *et al.*, 1992; Phylip *et al.*, 1992; Co *et al.*, 1994). A possible role of the peptidase in events taking place early in infection has been disputed by a report using conditional mutants (Kaplan *et al.*, 1996). Finally, the actual form of the enzyme that functions inside the virion is still unclear. A report from Almog *et al.* (1996) reveals that the enzyme is largely in the form of the p6* N-terminal extended form. This question is of importance, as all drug development is typically done with the smallest possible functional unit, the 99 residue enzyme.

## References

Ainsztein, A.M. & Purich, D.L. (1992) Cleavage of bovine brain microtubule associated protein 2 by human immunodeficiency virus proteinase. *J. Neurochem.* **59**, 874–880.

Almog, N., Roller, R., Arad, G., Passi-Even, L., Wainberg, M.A. & Kotler, M. (1996) A p6Pol protease fusion protein is present in mature particles of human immunodeficiency virus type 1. *J. Virol.* **70**, 7228–7232.

Baldwin, E.T., Bhat, T.N., Gulnik, S., Liu, B., Topol, I.A., Kiso, Y., Mimoto, T., Mitsuya, H. & Erickson, J.W. (1995a) Structure of HIV-1 protease with KNI 272, a tight binding transition state analog containing allophenylnorstatine. *Structure* **3**, 581–590.

Baldwin, E.T., Bhat, T.N., Liu, B., Pattabiraman, N. & Erickson, J.W. (1995b) Structural basis of drug resistance for the V82A mutant of HIV-1 proteinase. *Nature Struct. Biol.* **2**, 244–249.

Barrie, K.A., Perez, E.E., Lamers, S.L., Farmerie, W.G., Dunn, B.M., Sleasman, J.W. & Goodenow, M.M. (1996) Natural variation in HIV-1 protease, Gag p7 and p6, and protease cleavage sites within Gag/Pol polyprotein: amino acid substitutions in the absence of protease inhibitors in mothers and children infected by human immunodeficiency virus type 1. *Virology* **219**, 407–416.

Betageri, R., Hopkins, J.L., Thibeault, D., Emmanuel, M.J., Chow, G.C., Skoog, M.T., de Dreu, P. & Cohen, K.A. (1993) Rapid, sensitive and efficient HPLC assays for HIV-1 proteinase. *J. Biochem. Biophys. Methods* **27**, 191–197.

Bhat, T.N., Baldwin, E.T., Liu, B., Cheng, Y.S. & Erickson, J.W. (1994) Crystal structure of a tethered dimer of HIV-1 proteinase complexed with an inhibitor. *Nature Struct. Biol.* **1**, 552–556.

Bhat, T.N., Baldwin, E.T., Liu, B., Cheng, Y.S. & Erickson, J.W. (1995) X ray structure of a tethered dimer for HIV-1 protease. *Adv. Exp. Med. Biol.* **362**, 439–444.

Billich, A. & Winkler, G. (1991) Analysis of subsite preferences of HIV-1 proteinase using MA/CA junction peptides substituted at the P3–P1' positions. *Arch. Biochem. Biophys.* **290**, 186–190.

Blundell, T.L., Lapatto, R., Wilderspin, A.F., Hemmings, A.M., Hobart, P.M., Danley, D.E. & Whittle, P.J. (1990) The 3 D structure of HIV-1 proteinase and the design of antiviral agents for the treatment of AIDS. *Trends Biochem. Sci.* **15**, 425–430.

Borders, C.L., Jr., Broadwater, J.A., Bekeny, P.A., Salmon, J.E., Lee, A.S., Eldridge, A.M. & Pett, V.B. (1994) A structural role for arginine in protein: multiple hydrogen bonds to backbone carbonyl oxygens. *Protein Sci.* **3**, 541–548.

Cameron, C.E., Ridky, T.W., Shulenin, S., Leis, J., Weber, I.T., Copeland, T., Wlodawer, A., Burstein, H., Bizub-Bender, D. & Skalka, A.M. (1994) Mutational analysis of the substrate binding pockets of the Rous sarcoma virus and human immunodeficiency virus 1 proteases. *J. Biol. Chem.* **269**, 11170–11177.

Cheng, Y.S.E., McGowan, M.H., Kettner, C.A., Schloss, J.V., Erickson-Viitanen, S. & Yin, F.H. (1990a) High level synthesis of recombinant HIV-1 protease and the recovery of active enzyme from inclusion bodies. *Gene* **87**, 243–248.

Cheng, Y.S.E., Yin, F.H., Foundling, S., Blomstrom, D. & Kettner, C.A. (1990b) Stability and activity of human immunodeficiency virus protease comparison of the natural dimer with a homologous, single chain tethered dimer. *Proc. Natl Acad. Sci. USA* **87**, 9660–9664.

Chen, Z., Li, Y., Chen, E., Hall, D., Darke, P., Culberson, C., Shafer, J.A. & Kuo, L.A. (1994) Crystal structure at 1.9 Å resolution of human immunodeficiency virus (HIV) II protease complexed with L-735 524, an orally bioavailable inhibitor of the HIV proteases. *J. Biol. Chem.* **269**, 26344–26348.

Chen, Z., Li, Y., Schock, H., Hall, D., Chen, E. & Kuo, L.A. (1995) Three dimensional structure of a mutant HIV-1 protease displaying cross resistance to all protease inhibitors in clinical trials. *J. Biol. Chem.* **270**, 21433–21436.

Chou, K.C., Tomasselli, A.G., Reardon, I.M. & Heinrikson, R.L. (1996) Predicting human immunodeficiency virus protease cleavage sites in proteins by a discriminant function method. *Proteins* **24**, 51–72.

Co, E., Koelsch, G., Lin, Y.Z., Ido, E., Hartsuck, J.A. & Tang, J. (1994) Proteolytic processing mechanisms of a miniprecursor of the aspartic protease of human immunodeficiency virus type 1. *Biochemistry* **33**, 1248–1254.

Collins, J.R., Burt, S.K. & Erickson, J.W. (1995) Flap opening in HIV-1 protease simulated by 'activated' molecular dynamics. *Nature Struct. Biol.* **2**, 334–338.

Condra, J.H., Schleif, W.A., Blahy, O.M., Gabryelski, L.J., Graham, D.J., Quintero, J.C., Rhodes, A., Robins, H.L., Roth, E., Shivaprakash, M., Yang, T., Chodakewitz, J.A., Deutsch, P.J., Leavitt, R.Y., Massari, F.E., Mellors, J.W., Squires, K.E., Steigbigel, R.T., Teppler, H. & Emini, E.A. (1995) *In vivo* emergence of HIV-1 variants resistant to multiple protease inhibitors. *Nature* **374**, 569–571.

Cook, N.D., Jessop, R.A., Robinson, P.S., Richards, A.D. & Kay, J. (1991) Scintillation proximity enzyme assay. A rapid and novel assay technique applied to HIV proteinase. *Adv. Exp. Med. Biol.* **306**, 525–528.

Copeland, T.D., Wondrak, E.M., Tözsér, J., Roberts, M.M. & Oroszlan, S. (1990) Substitution of proline with pipecolic acid at the scissile bond converts a peptide substrate of HIV proteinase into a selective inhibitor. *Biochem. Biophys. Res. Commun.* **169**, 310–314.

Danley, D.E., Geoghegan, K.F., Scheld, K.G., Lee, S.E., Merson, J.R., Hawrylik, S.J., Rickett, G.A., Ammirati, M.J. & Hobart, P.M. (1989) Crystallizable HIV-1 protease derived from expression of the viral pol gene in *Escherichia coli*. *Biochem. Biophys. Res. Commun.* **165**, 1043–1050.

Darke, P.L., Nutt, R.F., Brady, S.F., Garsky, V.M., Ciccarone, T.M., Leu, C.T., Lumma, P.K., Freidinger, R.M., Veber, D.F. & Sigal, I.S. (1988) HIV-1 protease specificity of peptide cleavage is sufficient for processing of gag and pol polyproteins. *Biochem. Biophys. Res. Commun.* **156**, 297–303.

di Marzo Veronese, F., Copeland, T.D., DeVico, A.L., Rahman, R., Oroszlan, S., Gallo, R.C. & Sarngadharan, M.G. (1986) Characterization of highly immunogenic p66/p51 as the reverse transcriptase of hTLV-III/LAV. *Science* **231**, 1289–1291.

Dorsey, B.D., Levin, R.B., McDaniel, S.L., Vacca, J.P., Guare, J.P., Darke, P.L., Zugay, J.A., Emini, E.A., Schleif, W.A., Quintero, J.C., Lin, J.H., Chen, I-W., Holloway, M.K. Fitzgerald, P.M.D., Axel, M.G., Ostovic, D., Anderson, P.S. & Huff, J.R. (1994) L 735 524 the design of a potent and orally bioavailable HIV protease inhibitor. *J. Med. Chem.* **37**, 3443–3451.

Dunn, B.M., Gustchina, A., Wlodawer, A. & Kay, J. (1994) Subsite preferences of retroviral proteinases. *Methods Enzymol.* **241**, 254–278.

Erickson-Viitanen, S., Manfredi, J., Viitanen, P., Tribe, D.E., Tritch, R., Hutchison, C.A. III., Loeb, D.D. & Swanstrom, R. (1989) Cleavage of HIV-1 *gag* polyprotein synthesized *in vitro* sequential cleavage by the viral protease. *AIDS Res. Hum. Retroviruses* **5**, 577–591.

Evans, D.B., Vosters, A.F., McQuade, T.J. & Sharma, S.K. (1992) An ultrasensitive human immunodeficiency virus type 1 protease radioimmuno rate assay with a potential for monitoring blood levels of protease inhibitors in acquired immunodeficiency disease syndrome patients. *Anal. Biochem.* **206**, 288–192.

Fitzgerald, P.M.D., McKeever, B.M., VanMiddlesworth, J.F., Springer, J.P., Heimbach, J.C., Leu, C.T., Herber, W.K., Dixon,

R.A.F. & Darke, P.L. (1990) Crystallographic analysis of a complex between human immunodeficiency virus type 1 protease and acetyl pepstatin at 2.0 Å resolution. *J. Biol. Chem.* **265**, 14209–14219.

Fontenot, G., Johnston, K., Cohen, J.C., Gallaher, W.R., Robinson, J. & Luftig, R.B. (1992) PCR amplification of HIV-1 proteinase sequences directly from lab isolates allows determination of five conserved domains. *Virology* **190**, 1–10.

Freund, J., Kellner, R., Konvalinka, J., Wolber, V., Kräusslich, H.G. & Kalbitzer, H.R. (1994) A possible regulation of negative factor (Nef) activity of human immunodeficiency virus type 1 by the viral protease. *Eur. J. Biochem.* **223**, 589–593.

Gaedigk-Nitschko, K., Schon, A., Wachinger, G., Erfle, V. & Kohleisen, B. (1995) Cleavage of recombinant and cell derived human immunodeficiency virus 1 ( HIV-1) Nef protein by HIV-1 protease. *FEBS Lett.* **357**, 275–278.

Goobar, L., Danielson, U.H., Brodin, P., Grundstrom, T., Oberg, B. & Norrby, E. (1991) High yield purification of HIV-1 proteinase expressed by a synthetic gene in *Escherichia coli*. *Protein Exp. Purif.* **2**, 15–23.

Graves, B.J., Hatada, M.H., Miller, J.K., Graves, M.C., Roy, S., Cook, C.M., Krohn, A., Martin, J.A. & Roberts, N.A. (1991) The three dimensional x ray crystal structure of HIV-1 protease complexed with a hydroxyethylene inhibitor. *Adv. Exp. Med. Biol.* **306**, 455–460.

Graves, M.C., Lim, J.J., Zicopoulos, M.A., Stoller, T.J., Miedel, M.C., Pan, Y.-C.E., Danho, W. & Nalin, C.M. (1989) Expression and characterization of human immunodeficiency virus-1 protease, In: *Proteases of Retroviruses* (Kostka, V., ed.). Berlin: Walter de Gruyter & Co., pp. 83–92.

Graves, M.C., Meidel, M.C., Pan, Y.C., Manneberg, M., Lahm, H.W. & Gruninger-Leitch, F. (1990) Identification of a human immunodeficiency virus-1 protease cleavage site within the 66 000 Dalton subunit of reverse transcriptase. *Biochem. Biophys. Res. Commun.* **168**, 30–36.

Greer, J., Erickson, J.W., Baldwin, J.J. & Varney, M.D. (1994) Application of the three dimensional structures of protein target molecules in structure based drug design. *J. Med. Chem.* **37**, 1035–1054.

Griffiths, J.H., Phylip, L.H., Konvalinka, J., Strop, P., Gustchina, A., Wlodawer, A., Davenport, R.J., Briggs, R., Dunn, B.M. & Kay, J. (1992) Different requirements for productive interaction between the active site of HIV-1 proteinase and substrates containing hydrophobic*hydrophobic or aromatic*pro cleavage sites. *Biochemistry* **31**, 5193–5200.

Guenet, C., Leppik, R.A., Pelton, J.T., Moelling, K., Lovenberg, W. & Harris, B.A. (1989) HIV-1 protease mutagenesis of asparagine 88 indicates a domain required for dimer formation. *Eur. J. Pharmacol.* **172**, 443–451.

Gulnik, S.V., Suvorov, L.I., Liu, B.S., Yu, B., Anderson, B., Mitsuya, H. & Erickson, J.W. (1995) Kinetic characterization and cross resistance patterns of HIV-1 protease mutants selected under drug pressure. *Biochemistry* **34**, 9282–9287.

Gustafson, M.E., Junger, K.D., Foy, B.A., Baez, J.A., Bishop, B.F., Rangwala, S.H., Michener, M.L., Leimgruber, R.M., Houseman, K.A., Mueller, R.A., Matthews, B.K., Olins, P.O., Grabner, R.W. & Hershman, A. (1995) Large scale production of HIV-1 protease from *Escherichia coli* using selective extraction and membrane fractionation. *Protein Exp. Purif.* **6**, 512–518.

Gustchina, A., Sansom, C., Prevost, M., Richelle, J., Wodak, S.Y., Wlodawer, A. & Weber, I.T. (1994) Energy calculations and analysis of HIV-1 protease inhibitor crystal structures. *Protein Eng.* **7**, 309–317.

Gustchina, A. & Weber, I.T. (1990) Comparison of inhibitor binding in HIV-1 protease and in nonviral aspartic protease: the role of the flap. *FEBS Lett.* **269**, 269–272.

Harte, W.J., Swaminathan, S., Mansuri, M.M., Martin, J.C., Rosenberg, I.E. & Beveridge, D.L. (1990) Domain communication in the dynamical structure of human immunodeficiency virus 1 protease. *Proc. Natl Acad. Sci. USA* **87**, 8864–8868.

Heimbach, J.C., Garsky, V.M., Michelson, S.R., Dixon, R.A.F., Sigal, I.S. & Darke, P.L. (1989) Affinity purification of the HIV-1 protease. *Biochem. Biophys. Res. Commun.* **164**, 955–960.

Henderson, L.E., Copeland, T.D., Sowder, R.C., Schultz, A.M. & Oroszlan, S. (1988) Analysis of proteins and peptides purified from sucrose gradient banded HTLV-III. In: *Human Retroviruses, Cancer, and AIDS: Approaches to Prevention and Therapy* (Bolognesi, D., ed.). New York: Alan R. Liss, Inc., pp. 135–147.

Ho, D.D., Neumann, A.U., Perelson, A.S., Chen, L.J. & Markowitz, M. (1995) Rapid turnover of plasma virions and CD3 lymphocytes in HIV-1 infection. *Nature* **373**, 123–126.

Hoeprich, P.J. (1994) Chemical synthesis of the aspartic proteinase from human immunodeficiency virus (HIV). *Methods Mol. Biol.* **36**, 287–304.

Höner, B., Shoeman, R.L. & Traub, P. (1991) Human immunodeficiency virus type 1 protease microinjected into cultured human skin fibroblasts cleaves vimentin and affects cytoskeletal and nuclear architecture. *J. Cell Sci.* **100**, 799–807.

Höner, B., Shoeman, R.L. & Traub, P. (1992) Degradation of cytoskeletal proteins by the human immunodeficiency virus type 1 protease. *Cell Biol. Int. Rep.* **16**, 603–612.

Hong, L., Treharne, A., Hartsuck, J.A., Foundling, S. & Tang, J. (1996) Crystal structures of complexes of a peptidic inhibitor with wild type and two mutant HIV-1 proteases. *Biochemistry* **35**, 10627–10633.

Hoog, S.S., Zhao, B., Winborne, E., Fisher, S., Green, D.W., DesJarlais, R.L., Newlander, K.A., Callahan, J.F., Moore, M.L., Huffman, W.F. & Abdel-Meguid, S.S. (1995) A check on rational drug design crystal structure of a complex of human immunodeficiency virus type 1 protease with a novel gamma turn mimetic inhibitor. *J. Med. Chem.* **38**, 3246–3252.

Hostomsky, Z., Appelt, K. & Ogden, R.C. (1989) High level expression of self processed HIV-1 protease in *Escherichia coli* using a synthetic gene. *Biochem. Biophys. Res. Commun.* **161**, 1056–1063.

Hui, J.O., Tomasselli, A.G., Reardon, I.M., Lull, J.M., Brunner, D.P., Tomich, C.S. & Heinrikson, R.L. (1993) Large scale purification and refolding of HIV-1 protease from *Escherichia coli* inclusion bodies. *J. Protein Chem.* **12**, 323–327.

Hyland, L.J., Dayton, B.D., Moore, M.L., Shu, A.Y.L., Heys, J.R. & Meek, T.D. (1990) A radiometric assay for HIV-1 protease. *Anal. Biochem.* **188**, 408–415.

Jacobsen, H., Yasargil, K., Winslow, D.L., Craig, J.C., Krohn, A., Duncan, I.B. & Mous, J. (1995) Characterization of human immunodeficiency virus type 1 mutants with decreased sensitivity to proteinase inhibitor Ro 31 8959. *Virology* **206**, 527–534.

Jaskolski, M., Tomasselli, A.G., Sawyer, T.K., Staples, D.G., Heinrikson, R.L., Schneider, J., Kent, S.B.H. & Wlodawer, A. (1991) Structure at 2.5 Å resolution of chemically synthesized human immunodeficiency virus type 1 protease complexed with a hydroxyethylene based inhibitor. *Biochemistry* **30**, 1600–1609.

Jhoti, H., Singh, O.M., Weir, M.P., Cooke, R., Murray-Rust, P. & Wonacott, A. (1994) X ray crystallographic studies of a series of penicillin derived asymmetric inhibitors of HIV-1 protease. *Biochemistry* **33**, 8417–8427.

Kaplan, A.H., Zack, J.A., Knigge, M., Paul, D.A., Kempf, D.J., Norbeck, D.W. & Swanstrom, R. (1993) Partial inhibition of the human immunodeficiency virus type 1 protease results in aberrant virus assembly and the formation of noninfectious particles. *J. Virol.* **67**, 4050–4055.

Kaplan, A.H., Michael, S.S., Wehbie, R.F., Knigge, M.S., Paul, D.A., Everitt, L., Kempf, D.J., Norbeck, D.W., Erickson, J.W. & Swanstrom, R. (1994a) Selection of multiple human immunodeficiency virus type 1 variants that encode viral proteases with decreased sensitivity to an inhibitor of the viral protease. *Proc. Natl Acad. Sci. USA* **91**, 5597–5601.

Kaplan, A.H., Manchester, M. & Swanstrom, R. (1994b) The activity of the protease of human immunodeficiency virus type 1 is initiated at the membrane of infected cells before the release of viral proteins and is required for release to occur with maximum efficiency. *J. Virol.* **68**, 6782–6786.

Kaplan, A.H., Manchester, M., Smith, T., Yang, Y.L. & Swanstrom, R. (1996) Conditional human immunodeficiency virus type 1 protease mutants show no role for the viral protease early in virus replication. *J. Virol.* **70**, 5840–5844.

Kassel, D.B., Green, M.D., Wehbie, R.S., Swanstrom, R. & Berman, J. (1995) HIV-1 protease specificity derived from a complex mixture of synthetic substrates. *Anal. Biochem.* **228**, 259–266.

Kempf, D.J., Marsh, K.C., Denissen, J.F., McDonald, E., Vasavanonda, S., Flentge, C.A., Green, B.E., Fino, L., Park, C.H., Kong, X.P., Wideburg, N.E., Saldivar, A., Ruiz, L., Kati, W.M., Shaw, H.L., Robins, T., Stewart, K.D., Hsu, A., Plattner, J.J., Leonard, J.M. & Norbeck, D.W. (1995) ABT 538 is a potent inhibitor of human immunodeficiency virus protease and has high oral bioavailability in humans. *Proc. Natl Acad. Sci. USA* **92**, 2484–2488.

Konvalinka, J., Strop, P., Velek, J., Cerna, V., Kostka, V., Phylip, L.H., Richards, A.D., Dunn, B.M. & Kay, J (1990) Sub site preferences of the aspartic proteinase from the human immunodeficiency virus, HIV-1. *FEBS Lett.* **268**, 35–38.

Kotler, M., Arad, G. & Hughes, S.H. (1992) Human immunodeficiency virus type 1 *gag*-protease fusion proteins are enzymatically active. *J. Virol.* **66**, 6781–6783.

Kozal, M.J., Shah, N., Shen, N.P., Yang, R., Fucini, R., Merigan, T.C., Richman, D.D., Morris, D., Hubbell, E.R., Chee, M. & Gingeras, T.R. (1996) Extensive polymorphisms observed in HIV-1 clade B protease gene using high density oligonucleotide arrays. *Nature Med.* **2**, 753–759.

Krafft, G.A. & Wang, G.T. (1994) Synthetic approaches to continuous assays of retroviral proteases. *Methods Enzymol.* **241**, 70–86.

Kräusslich, H.G., Ingraham, R.H., Skoog, M.T., Wimmer, E., Pallai, P.V. & Carter, C.A. (1989) Activity of purified biosynthetic proteinase of human immunodeficiency virus on natural substrates and synthetic peptides. *Proc. Natl Acad. Sci. USA* **86**, 807–811.

Lapatto, R., Blundell, T., Hemmings, A., Overington, J., Wilderspin, A., Wood, S., Merson, J., Whittle, P.J., Danley, D.E., Geoghegan, K.F., Hawrylik, S.J., Lee, S.E., Scheld, KG. & Hobart, P.M. (1989) X ray analysis of HIV-1 proteinase at 2.7 Å resolution confirms structural homology among retroviral enzymes. *Nature* **342**, 299–302.

Lech, W.J., Wang, G., Yang, Y.L., Chee, Y., Dorman, K., McCrae, D., Lazzeroni, L.C., Erickson, J.W., Sinsheimer, J.S. & Kaplan, A.H. (1996) *In vivo* sequence diversity of the protease of human immunodeficiency virus type 1: presence of protease inhibitor resistant variants in untreated subjects. *J. Virol.* **70**, 2038–2043.

Lightfoote, M.M., Coligan, J.E., Folks, T.M., Fauci, A.S., Martin, M.A. & Venkatesan, S. (1986) Structural characterization of reverse transcriptase and endonuclease polypeptides of the acquired immunodeficiency syndrome retrovirus. *J. Virol.* **60**, 771–775.

Lin, Y., Lin, X., Hong, L., Foundling, S., Heinrikson, R.L., Thaisrivongs, S., Leelamanit, W., Raterman, D., Shah, M., Dunn, B.M. & Tang, J. (1995) Effect of point mutations on the kinetics and the inhibition of human immunodeficiency virus type 1 protease relationship to drug resistance. *Biochemistry* **34**, 1143–1152.

Liu, H., Muller Plathe, F. & van Gunsteren, W.F. (1996) A combined quantum/classical molecular dynamics study of the catalytic mechanism of HIV protease. *J. Mol. Biol.* **261**, 454–469.

Loeb, D.D., Swanstrom, R., Everitt, L., Manchester, M., Stamper, S.E. & Hutchison, C.A. III. (1989) Complete mutagenesis of the HIV-1 protease. *Nature* **340**, 397–400.

Louis, J.M., Smith, C.A.D., Wondrak, E., Mora, P.T. & Oroszlan, S. (1989) Substitution mutations of the highly conserved arginine 87 of HIV-1 protease result in loss of proteolytic activity. *Biochem. Biophys. Res. Commun.* **164**, 30–38.

Louis, J.M., Nashed, N.T., Parris, K.D., Kimmel, A.R. & Jerina, D.M. (1994) Kinetics and mechanism of autoprocessing of human immunodeficiency virus type 1 protease from an analog of the Gag Pol polyprotein. *Proc. Natl Acad. Sci. USA* **91**, 7970–7974.

Luftig, R.B. & Lupo, L.D. (1994) Viral interactions with the host cell cytoskeleton: the role of retroviral proteases. *Trends Microbiol.* **2**, 178–182.

Margolin, N., Heath, W., Osborne, E., Lai, M. & Vlahos, C. (1990) Substitutions at the P2' site of gag p17 p24 affect cleavage efficiency by HIV-1 protease. *Biochem. Biophys. Res. Commun.* **167**, 554–560.

Margolin, N., Dee, A., Lai, M. & Vlahos, C. (1991) Purification of recombinant HIV-1 protease. *Prep. Biochem.* **21**, 163–173.

Markowitz, M., Mo, H., Kempf, D.J., Norbeck, D.W., Bhat, T.N., Erickson, J.W. & Ho, D.D. (1995) Selection and analysis of human immunodeficiency virus type 1 variants with increased resistance to ABT 538, a novel protease inhibitor. *J. Virol.* **69**, 701–706.

Matayoshi, E., Wang, G.T., Krafft, G.A. & Erickson, J.W. (1990) Novel fluorogenic substrates for assaying retroviral proteases by resonance energy transfer. *Science* **247**, 954–958.

Mildner, A.M., Rothrock, D.J., Leone, J.W., Bannow, C.A., Lull, J.M., Reardon, I.M., Sarcich, J.L., Howe, W.J., Tomich, C.S.C., Smith, C.W., Heinrikson, R.L. & Tomasselli, A.G. (1994) The HIV-1 protease as enzyme and substrate: mutagenesis of autolysis sites and generation of a stable mutant with retained kinetic properties. *Biochemistry* **33**, 9405–9413.

Miller, M., Schneider, J., Sathyanarayana, B.K., Toth, M.V., Marshall, G.R., Clawson, L., Selk, L., Kent, S.B.H. & Wlodawer, A. (1989) Structure of complex of synthetic HIV-1 protease with a substrate based inhibitor at 2.3 Å resolution. *Science* **246**, 1149–1152.

Milton, R.C., Milton, S.C. & Kent, S.B.H. (1992) Total chemical synthesis of a D enzyme the enantiomers of HIV-1 protease show reciprocal chiral substrate specificity. *Science* **256**, 1445–1448.

Molla, A., Korneyeva, M., Gao, Q., Vasavanonda, S., Schipper, P.J., Mo, H.M., Markowitz, M., Chernyavskiy, T., Niu, P., Lyons, N., Hsu, A., Granneman, G.R., Ho, D.D., Boucher, C.A.B., Leonard, J.M., Norbeck, D.W. & Kempf, D.J. (1996) Ordered accumulation

of mutations in HIV protease confers resistance to ritonavir. *Nature Med.* **2**, 760–766.

Moody, M.D., Pettit, S.C., Shao, W., Everitt, L., Loeb, D.D., Hutchison, C.A. III. & Swanstrom, R. (1995) A side chain at position 48 of the human immunodeficiency virus type 1 protease flap provides an additional specificity determinant. *Virology* **207**, 475–485.

Murthy, K.H., Winborne, E.L., Minnich, M.D., Culp, J.S. & Debouck, C. (1992) The crystal structures at 2.2 Å resolution of hydroxyethylene based inhibitors bound to human immunodeficiency virus type 1 protease show that the inhibitors are present in two distinct orientations. *J. Biol. Chem.* **267**, 22770–22778.

Nashed, N.T., Louis, J.M., Sayer, J.M., Wondrak, E.M., Mora, P.T., Oroszlan, S. & Jerina, D.M. (1989) Continuous spectrophotometric assay for retroviral proteases of HIV-1 and AMV. *Biochem. Biophys. Res. Commun.* **163**, 1079–1085.

Nicholson, L.K., Yamazaki, T., Torchia, D.A., Grzesiek, S., Bax, A., Stahl, S.J., Kaufman, J.D., Wingfield, P.T., Lam, P.Y.S., Jadhav, P.K., Hodge, C.N., Domaille, P.J. & Chang, C-H. (1995) Flexibility and function in HIV-1 protease (see comments). *Nature Struct. Biol.* **2**, 274–280.

Oswald, M. & von der Helm, K. (1991) Fibronectin is a non-viral substrate for the HIV proteinase. *FEBS Lett.* **292**, 298–300.

Otto, M.J., Garber, S., Winslow, D.L., Reid, C.D., Aldrich, P., Jadhav, P.K., Patterson, C.E., Hodge, C.E. & Cheng, Y.N. (1993) *In vitro* isolation and identification of human immunodeficiency virus (HIV) variants with reduced sensitivity to C 2 symmetrical inhibitors of HIV type 1 protease. *Proc. Natl Acad. Sci. USA* **90**, 7543–7547.

Partin, K., Zybarth, G., Ehrlich, L., DeCrombrugghe, M., Wimmer, E. & Carter, C. (1991) Deletion of sequences upstream of the proteinase improves the proteolytic processing of human immunodeficiency virus type 1. *Proc. Natl Acad. Sci. USA* **88**, 4776–4780.

Pazhanisamy, S., Stuver, C.M. & Livingston, D.J. (1995) Automation of a high performance liquid chromatography based enzyme assay evaluation of inhibition constants for human immunodeficiency virus 1 protease inhibitors. *Anal. Biochem.* **229**, 48–53.

Pazhanisamy, S., Stuver, C.M., Cullinan, A.B., Margolin, N., Rao, B.G. & Livingston, D.J. (1996) Kinetic characterization of human immunodeficiency virus type 1 protease resistant variants. *J. Biol. Chem.* **271**, 17979–17985.

Pearl, L.H. & Taylor, W.R. (1987) A structural model for the retroviral proteases. *Nature* **329**, 351–354.

Pechik, I.V., Gustchina, A.E., Andreeva, N.S. & Fedorov, A.A. (1989) Possible role of sme groups in the structure and function of HIV-1 protease as revealed by molecular modeling studies. *FEBS Lett.* **247**, 118–122.

Peranteau, A.G., Kuzmic, P., Angell, Y., Garcia-Echeverria, C. & Rich, D.H. (1995) Increase in fluorescence upon the hydrolysis of tyrosine peptide: application to proteinase assays. *Anal. Biochem.* **227**, 242–245.

Phylip, L.H., Richards, A.D., Kay, J., Kovalinka, J., Strop, P., Bláha, I., Velek, J., Kostka, V., Ritchie, A.J., Broadhurst, A.V., Farmerie, W.G., Scarborough, P.E. & Dunn, B.M. (1990) Hydrolysis of synthetic chromogenic substrates by HIV-1 and HIV-2 proteinases. *Biochem. Biophys. Res. Commun.* **171**, 439–444.

Phylip, L.H., Mills, J.S., Parten, B.F., Dunn, B.M. & Kay, J. (1992) Intrinsic activity of precursor forms of HIV-1 proteinase. *FEBS Lett.* **314**, 449–454.

Poorman, R.A., Tomasselli, A.G., Heinrikson, R.L. & Kezdy, F.J.

(1991) A cumulative specificity model for proteases from human immunodeficiency virus types 1 and 2, inferred from statistical analysis of an extended substrate data base. *J. Biol. Chem.* **266**, 14554–14561.

Priestle, J.P., Fassler, A., Rosel, J., Blomley, M., Strop, P. & Grutter, M.G. (1995) Comparative analysis of the X ray structures of HIV-1 and HIV-2 proteases in complex with CGP 53820, a novel pseudosymmetric inhibitor. *Structure* **3**, 381–389.

Ratner, L., Haseltine, W., Patarca, R., Livak, K.J., Starcich, B., Josephs, S.F., Dorna, E.R., Rafalski, J.A., Whitehorn, E.A., Baumeister, K., Ivanoff, L., Petteway, S.R., Pearson, M.L., Lauttenberger, J.A., Papas, T.S., Ghrayeb, J., Chang, N.T., Gallo, R.C. & Wong-Staal, F. (1985) Complete nucleotide sequence of the AIDS virus, HTLV-III. *Nature* **313**, 277–285.

Richards, A.D., Roberts, R., Dunn, B.M., Graves, M.C. & Kay, J. (1989) Effective blocking of HIV-1 proteinase activity by characteristic inhibitors of aspartic proteinases. *FEBS Lett.* **247**, 113–117.

Richards, A.D., Phylip, L.H., Farmerie, W.G., Scarborough, P.E., Alvarez, A., Dunn, B.M., Hirel, P.H., Konvalinka, J., Strop, P., Pavlickova, L., Kostka, V. & Kay, J. (1990) Sensitive, soluble chromogenic substrates for HIV-1 proteinase. *J. Biol. Chem.* **265**, 7733–7736.

Ridky, T.W. & Leis, J. (1995) Development of drug resistance to HIV-1 protease inhibitors. *J. Biol. Chem.* **270**, 29621–29623.

Rittenhouse, J., Turon, M.C., Helfrich, R.J., Albrecht, K.S., Weigl, D., Simmer, R.L., Mordini, F., Erickson, J. & Kohlbrenner, W.E. (1990) Affinity purification of HIV-1 and HIV-2 proteases from recombinant *E. coli* strains using pepstatin agarose. *Biochem. Biophys. Res. Commun.* **171**, 60–66.

Rivière, Y., Blank, V., Kourilsky, P. & Israel, A. (1991) Processing of the precursor of NF κ B by the HIV-1 protease during acute infection. *Nature* **350**, 625–626.

Roberts, N.A., Martin, J.A., Kinchington, D., Broadhurst, A.V., Craig, J.C., Duncan, I.B., Galpin, S.A., Handa, B.K., Kay, J., Kröhn, A., Lambert, R.W., Merrett, J.H., Mills, J.S., Parkes, K.E.B., Redshaw, S., Ritchie, A.J., Taylor, D.L., Thomas, G.J. & Machin, P.J. (1990) Rational design of peptide based HIV proteinase inhibitors. *Science* **248**, 358–361.

Rose, R.B., Craik, C.S., Douglas, N.L. & Stroud, R.M. (1996a) Three dimensional structures of HIV-1 and SIV protease product complexes. *Biochemistry* **35**, 12933–12944.

Rose, R.E., Gong, Y.F., Greytok, J.A., Bechtold, C.M., Terry, B.J., Robinson, B.S., Alam, M., Colonno, R.J. & Lin, P.F. (1996b) Human immunodeficiency virus type 1 viral background plays a major role in development of resistance to protease inhibitors. *Proc. Natl Acad. Sci. USA* **93**, 1648–1653.

Ross, E.K., Fuerst, T.R., Orenstein, J.M., O'Neill, T., Martin, M.A. & Venkatesan, S. (1991) Maturation of human immunodeficiency virus particles assembled from the *gag* precursor protein requires *in situ* processing by *gag-pol* protease. *AIDS Res. Hum. Retroviruses* **7**, 475–483.

Rutenber, E., Fauman, E.B., Keenan, R.J., Fong, S., Furth, P.S., Ortiz de Montellano, P.R., Meng, E., Kuntz, I.D., DeCamp, D.L., Salto, R., Rosé, J.R., Craik, C.C. & Stroud, R.M. (1993) Structure of a non peptide inhibitor complexed with HIV-1 protease. Developing a cycle of structure based drug design. *J. Biol. Chem.* **268**, 15343–15346.

Sanchez-Pescador, R., Power, M.D., Barr, P.J., Steimer, K.S., Stempien, M.M., Brown-Shimer, S.L., Gee, W.W., Renard, A., Randolph, A., Levy, J.A., Dino, D. & Luciw, P.A. (1985) Nucleotide sequence and expression of an AIDS associated retrovirus (ARV-2). *Science* **227**, 4686–4849.

Sardana, V.V., Schlabach, A.J., Graham, P., Bush, B.L., Condra, J.H., Culberson, J.C., Gotlib, L., Graham, D.J., Kohl, N.E., LaFemina, R.L., Schneider, C.L., Wolanski, B.S., Wolfgang, J.A. & Emini, E.A. (1994) Human immunodeficiency virus type 1 protease inhibitor: evaluation of resistance engendered by amino acid substitutions in the enzyme's substrate binding site. *Biochemistry* **33**, 2004–2010.

Schneider, J. & Kent, S.B.H. (1988) Enzymatic activity of a synthetic 99 residue protein corresponding to the putative HIV-1 protease. *Cell* **54**, 363–368.

Seelmeier, S., Schmidt, H., Turk, V. & von der Helm, K. (1988) Human immunodeficiency virus has an aspartic type protease that can be inhibited by pepstatin A. *Proc. Natl Acad. Sci. USA* **85**, 6612–6616.

Shoeman, R.L., Höner, B., Stoller, T.J., Kesselmeier, C., Miedel, M.C., Traub, P. & Graves, M.C. (1990) Human immunodeficiency virus type 1 protease cleaves the intermediate filament proteins vimentin, desmin, and glial fibrillary acidic protein. *Proc. Natl Acad. Sci. USA* **87**, 6336–6340.

Shoeman, R.L., Sachse, C., Höner, B., Mothes, E., Kaufmann, M. & Traub, P. (1993) Cleavage of human and mouse cytoskeletal and sarcomeric proteins by human immunodeficiency virus type 1 protease. Actin, desmin, myosin, and tropomyosin. *Am. J. Pathol.* **142**, 221–230.

Silva, A.M., Cachau, R.E., Baldwin, E.T., Gulnik, S., Sham, H.L. & Erickson, J.W. (1995) Molecular dynamics of HIV-1 protease in complex with a difluoroketone containing inhibitor implications for the catalytic mechanism. *Adv. Exp. Med. Biol.* **362**, 451–454.

Stebbins, J. & Debouck, C. (1994) Expression systems for retroviral proteases. *Methods Enzymol.* **241**, 13–16.

Swain, A.L., Miller, M.M., Green, J., Rich, D.H., Schneider, J., Kent, S.B.H. & Wlodawer, A. (1990) X-ray crystallographic structure of a complex between a synthetic protease of human immunodeficiency virus 1 and a substrate based hydroxyethylamine inhibitor. *Proc. Natl Acad. Sci. USA* **87**, 8805–8809.

Taylor, W.R. (1994) Protein structure modelling from remote sequence similarity. *J. Biotechnol.* **35**, 281–291.

Thanki, N., Rao, J.K., Foundling, S.I., Howe, W.J., Moon, J.B., Hui, J.O., Tomasselli, A.G., Heinrikson, R.L., Thaisrivongs, S. & Wlodawer, A. (1992) Crystal structure of a complex of HIV-1 protease with a dihydroxyethylene containing inhibitor: comparisons with molecular modeling. *Protein Sci.* **1**, 1061–1072.

Tisdale, M., Myers, R.E., Maschera, B., Parry, N.R., Oliver, N.M. & Blair, E.D. (1995) Cross resistance analysis of human immunodeficiency virus type 1 variants individually selected for resistance to five different protease inhibitors. *Antimicrob. Agents Chemother.* **39**, 1704–1710.

Tomasselli, A.G. & Heinrikson, R.L. (1994) Specificity of retroviral protease: an analysis of viral and nonviral protein substrates. *Methods Enzymol.* **241**, 279–301.

Tomasselli, A.G., Hui, J.O., Sawyer, T.K., Staples, D.J., Bannow, C.A., Reardon, I.M., Chaudhary, V.K., Fryling, C.M., Pastan, I., Fitzgerald, D.J. & Heinrikson, R.L. (1990a) Proteases from human immunodeficiency virus and avian myeloblastosis virus show distinct specificities in hydrolysis of multidomain protein substrates. *J. Virol.* **64**, 3157–3161.

Tomasselli, A.G., Hui, J.O., Sawyer, T.K., Staples, D.J., FitzGerald, D.J., Chaudhary, V.K., Pastan, I. & Heinrikson, R.L. (1990b)

Interdomain hydrolysis of a truncated *Pseudomonas* exotoxin by the human immunodeficiency virus 1 protease. *J. Biol. Chem.* **265**, 408–413.

Tomasselli, A.G., Hui, J.O., Sawyer, T.K., Staples, D.J., Bannow, C.A., Reardon, I.M., Howe, W.J., DeCamp, D.L., Craik, C.S. & Heinrikson, R.L. (1990c) Specificity and inhibition of proteases from human immunodeficiency viruses 1 and 2. *J. Biol. Chem.* **265**, 14675–14683.

Tomasselli, A.G., Howe, W.J., Hui, J.O., Sawyer, T.K., Reardon, I.M., DeCamp, D.L., Craik, C.S. & Heinrikson, R.L. (1991a) Calcium free calmodulin is a substrate of proteases from human immunodeficiency viruses 1 and 2. *Proteins* **10**, 1–9.

Tomasselli, A.G., Hui, J.O., Adams, L., Chosay, J., Lowery, D.D., Greenberg, B., Yem, A., Deibel, M.R. Jr., Zurcher-Neely, H.A. & Heinrikson, R.L. (1991b) Actin, troponin C, Alzheimer amyloid precursor protein and pro-interleukin 1 beta as substrates of the protease from human immunodeficiency virus. *J. Biol. Chem.* **266**, 14548–14553.

Tomasselli, A.G., Sarcich, J.L., Barrett, L.J., Reardon, I.M., Howe, W.J., Evans, D.B., Sharma, S.K. & Heinrikson, R.L. (1993) Human immunodeficiency virus type 1 reverse transcriptase and ribonuclease H as substrates of the viral protease. *Protein Sci.* **2**, 2167–2176.

Tomaszek, T.A., Magaard, V.W., Bryan, H.G., Moore, M.L. & Meek, T.D. (1990) Chromophoric peptide substrates for the spectrophotometric assay of HIV-1 protease. *Biochem. Biophys. Res. Commun.* **168**, 274–280.

Toth, M.V. & Marshall, G.R. (1990) A simple, continuous fluorometric assay for HIV protease. *Int. J. Pept. Protein Res.* **36**, 544–550.

Tözsér, J., Bláha, I., Copeland, T.D., Wondrak, E.M. & Oroszlan, S. (1991a) Comparison of the HIV-1 and HIV 2 proteinases using oligopeptide substrates representing cleavage sites in Gag and Gag-Pol polyproteins. *FEBS Lett.* **281**, 77–80.

Tözsér, J., Gustchina, A., Weber, I.T., Bláha, I., Wondrak, E.M. & Oroszlan, S. (1991b) Studies on the role of the S4 substrate binding site of HIV proteinases. *FEBS Lett.* **279**, 356–360.

Tözsér, J., Weber, I.T., Gustchina, A., Bláha, I., Copeland, T.D., Louis, J.M. & Oroszlan, S. (1992) Kinetic and modeling studies of S3 S3′ subsites of HIV proteinases. *Biochemistry* **31**, 4793–4800.

Varney, M.D., Appelt, K., Kalish, V., Reddy, M.R., Tatlock, J., Palmer, C.L., Romines, W.H., Wu, B.W. & Musick, L. (1994) Crystal structure based design and synthesis of novel C terminal inhibitors of HIV protease. *J. Med. Chem.* **37**, 2274–2284.

Vondrasek, J. & Wlodawer, A. (1996) New database (letter). *Science* **272**, 337–338.

Vondrasek, J., van Buskirk, C.P. & Wlodawer, A. (1997) Database of three-dimensional structures of HIV proteinases. *Nature Struct. Biol.* **4**, 8.

Wain-Hobson, S., Sonigo, P., Danos, O., Cole, S. & Alizon, M. (1985) Nucleotide sequence of the AIDS virus, LAV. *Cell* **40**, 9–17.

Wallin, M., Deinum, J., Goobar, L. & Danielson, U.H. (1990) Proteolytic cleavage of microtubule associated proteins by retroviral proteinases. *J. Gen. Virol.* **71**, 1985–1991.

Wang, Y.X., Freedberg, D.I., Grzesiek, S., Torchia, D.A., Wingfield, P.T., Kaufman, J.D., Stahl, S.J., Chang, C.H. & Hodge, C.N. (1996) Mapping hydration water molecules in the HIV-1 protease/DMP323 complex in solution by NMR spectroscopy. *Biochemistry* **35**, 12694–12704.

Weber, I.T. (1991) Modeling of structure of human immunodeficiency virus 1 protease with substrate based on crystal structure of Rous sarcoma virus protease. *Methods Enzymol.* **202**, 727–741.

Weber, I.T., Miller, M., Jaskolski, M., Leis, J., Skalka, A.M. & Wlodawer, A. (1989) Molecular modeling of the HIV-1 protease and its substrate binding site. *Science* **243**, 928–931.

Welker, R., Kottler, H., Kalbitzer, H.R. & Kräusslich, H.G. (1996) Human immunodeficiency virus type 1 Nef protein is incorporated into virus particles and specifically cleaved by the viral proteinase. *Virology* **219**, 228–236.

Wlodawer, A. & Erickson, J.W. (1993) Structure based inhibitors of HIV-1 protease. *Annu. Rev. Biochem.* **62**, 543–585.

Wlodawer, A., Miller, M., Jaskolski, M., Sathyanarayana, B.K., Baldwin, E., Weber, I.T., Selk, L.M., Clawson, L., Schneider, J. & Kent, S.B.H. (1989) Conserved folding in retroviral protease: crystal structure of a synthetic HIV-1 protease. *Science* **245**, 616–621.

Wondrak, E.M. & Louis, J.M. (1996) Influence of flanking sequences on the dimer stability of human immunodeficiency virus type 1 protease. *Biochemistry* **35**, 12957–12962.

Wondrak, E.M., Nashed, N.T., Haber, M.T., Jerina, D.M. & Louis, J.M. (1996) A transient precursor of the HIV-1 protease. Isolation, characterization, and kinetics of maturation. *J. Biol. Chem.* **271**, 4477–4481.

Zhang, D., Zhang, N., Wick, M.M. & Byrn, R.A. (1995) HIV type 1 protease activation of NF $\kappa$ B within T lymphoid cells. *AIDS Res. Hum. Retroviruses* **11**, 223–230.

Zybarth, G. & Carter, C. (1995) Domains upstream of the protease (PR) in human immunodeficiency virus type 1 Gag Pol influence PR autoprocessing. *J. Virol.* **69**, 3878–3884.

Zybarth, G., Kräusslich, H.G., Partin, K. & Carter, C. (1994) Proteolytic activity of novel human immunodeficiency virus type 1 proteinase proteins from a precursor with a blocking mutation at the N terminus of the PR domain. *J. Virol.* **68**, 240–250.

*Ben M. Dunn*
*Department of Biochemistry and Molecular Biology,*
*University of Florida College of Medicine,*
*Gainesville, FL 32610-0245, USA*
*Email: bdunn@college.med.ufl.edu*

# 311. Human immunodeficiency virus 2 retropepsin

## Databanks

*Peptidase classification: clan AA, family A2, MEROPS ID: A02.002*
*NC-IUBMB enzyme classification: none*
*Databank codes:*

| Species | Isolate | SwissProt | PIR | EMBL (genomic) |
|---|---|---|---|---|
| Human immunodeficiency virus type 2 | BEN | P18096 | – | M30502: complete genome |
| | CAM2 | P24107 | B38475 | – |
| | D194 | P17757 | S12153 | J04542: complete genome |
| | | | | X52223: proviral genome |
| | D205,7 | P15833 | S08436 | X61240: complete genome |
| | Ghana-1 | P18042 | JS0328 | M30895: complete genome |
| | NIH-Z | P05962 | – | J03654: complete genome |
| | ROD | P04584 | B26262 | M15390: proviral genome |
| | | | | X05291: complete genome |
| | SBLISY | P12451 | – | J04498: complete genone |
| | ST | P20876 | B33943 | M31113: complete genome |

Brookhaven Protein Data Bank three-dimensional structures:

| Species | Strain | ID | Resolution | Notes |
|---|---|---|---|---|
| Human immunodeficiency virus type 2 | ROD | 1HII | 2.3 | complex with inhibitor CGP 53820 |
| | | 1IDA | 1.7 | complex with BILA 1906 |
| | | 1IDB | 2.2 | complex with BILA 2450 |
| | | 1PHV | | theoretical model; complex with Ac-Val-Val-Sta-Ala-Sta |
| | | 2HPE | 2 | Lys57Leu mutant |
| | | 2HPF | 3 | Lys57Leu mutant |
| | | 2MIP | 2.2 | complex with Phe-Val-Phe-PSI(CH$_2$NH)-Leu-Glu-Ile amide inhibitor |
| | | 2PHV | | theoretical model; complex with renin inhibitor H261 |
| | unknown | 1IVP | 2.5 | Lys57Leu mutant; complex with NOA-His-CHA-PSI[CH(OH)CH(OH)]Val-Ile-APY |
| | | 1IVQ | 2.6 | Lys57Leu mutant; complex with QNC-Asn-CHA-PSI[CH(OH)CH(OH)]Val-NPT |

**A**

## Name and History

Shortly after reports of the isolation of the HIV-1 virus, two groups reported the isolation of a new but related virus from patients in West Africa (Clavel *et al.*, 1986a; Kanki *et al.*, 1986; Albert *et al.*, 1987). Kanki *et al.* (1986) reported serologic evidence that demonstrated the new virus to be more closely related to the simian T-lymphotropic virus, STLV-III$_{AGM}$, than to the virus known to be the infectious agent in Europe and the USA, then called HTLV-III/LAV. These authors also isolated the virus and found its growth rate and major proteins similar to those of both simian immunodeficiency virus (SIV) and HIV-1. Clavel *et al.* (1986a) reported on a similar virus isolated from two different patients. Using immunologic criteria, they also demonstrated that the new virus, termed LAV-II by this group, was more closely related to the SIV than to HIV-1. In a follow-up paper, Clavel *et al.* (1986b) described cloning of the 9.5 kilobase genome of the new virus, which had been renamed as HIV-2, and a preliminary analysis of the relationship to HIV-1 and SIV retroviruses. The complete nucleotide sequence was reported by Guyader *et al.* (1987). Based on the similar, but not identical, genetic organization and the limited sequence identity (56% in the *gag* gene and 60% in the *pol* gene, 42% overall), these authors concluded that HIV-1 and HIV-2 are distinct retroviruses and not strains of the same virus. This also led to the conclusion that the retrovirus family diverged well before the current epidemic began, and that the rapid spread of the disease is more likely due to 'modifications of epidemiologic parameters', such as rapid urbanization leading to infection of larger populations. By comparing two separate strains of HIV-2 (HIV-2$_{\text{NIH-Z}}$ and HIV-2$_{\text{ROD}}$), Zagury *et al.* (1988) strengthened the conclusion that HIV-2 is a distinct virus from SIV. Furthermore, they concluded that the variability seen in the HIV-2 nucleotide sequence is similar to that observed in different isolates of HIV-1. This chapter describes the **human immunodeficiency virus 2 retropepsin** (**HIV-2 retropepsin**), more commonly known as **HIV-2 proteinase**.

## Activity and Specificity

Le Grice *et al.* (1989) were the first to report activity of the HIV-2 proteinase, by constructing a hybrid gene in which the processing/release of the reverse transcriptase was catalyzed by HIV-2 rather than HIV-1 retropepsin (Chapter 310). In addition, a second construct was made wherein the processing/release of the Gag p24 protein was again controlled by HIV-2 proteinase. In both cases, when expressed in *Escherichia coli*, the correct processing occurred; however, the rates were slower in the cases of the cleavage by HIV-2 proteinase. This reflects the differences in sequence within the cleavage junctions of the respective polyproteins as well as the differences between the two viral enzymes. Le Grice *et al.* (1989) also demonstrated that pepstatin A, a well-known inhibitor of aspartic proteinases, inhibited HIV-2 proteinase-catalyzed cleavage of a Gag precursor polypeptide more effectively than HIV-1 proteinase cleavage at comparable concentrations.

Richards *et al.* (1989) also studied the activity of HIV-2 proteinase and the inhibition of the viral enzyme by small molecules. However, these authors employed a synthetic peptide substrate, Tyr-Val-Ser-Gln-Asn-Phe+Pro-Ile-Val-Gln-Asn-Arg, based on the MA/CA cleavage site of HIV-1 Gag. The $K_{\text{m}}$ value reported (160 μM at pH 4.7) was about 2.5-fold higher than observed for the same peptide with HIV-1 proteinase. Using this substrate, it was determined that the $K_{\text{i}}$ value for pepstatin A (the isovaleryl form) was 150 nM at pH 4.7, again slightly higher than that observed with HIV-1 proteinase. However, the value of $K_{\text{i}}$ obtained with acetyl-pepstatin was 5 nM, approximately four times lower than that seen for HIV-1 proteinase.

Pichuantes *et al.* (1990) also prepared HIV-2 proteinase by recombinant expression in both yeast and bacteria. They used a myristylated p53-Gag precursor protein to demonstrate catalytic activity of their purified enzyme. In addition, they used two synthetic peptide substrates representing junction sequences in HIV-1 Gag-Pol precursor as substrates. With a peptide based on the MA/CA junction (Ser-Gln-Asn-Tyr+Pro-Ile-Val-Gln), a $k_{\text{cat}}/K_{\text{m}}$ value of 23.2 min$^{-1}$ mM$^{-1}$ was found; with a peptide based on the PR/RT junction (Ala-Thr-Leu-Asn-Phe+Pro-Ile-Ser-Pro-Trp) a $k_{\text{cat}}/K_{\text{m}}$ value of 71.4 min$^{-1}$ mM$^{-1}$ was determined.

A thorough analysis of the specificity of the HIV-1 and -2 retropepsins was provided by Tözsér *et al.* (1991). These authors synthesized 21 oligopeptides spanning all known cleavage sites in both the HIV-1 and HIV-2 Gag-Pol polyproteins. They determined $k_{\text{cat}}$ and $K_{\text{m}}$ values for the 16 of these that were cleaved at measurable rates. As expected, there was a substantial variation in the cleavage rates of the different junctions; however, similar results were obtained for both retroviral enzymes, confirming the earlier studies that reported that HIV-2 proteinase could efficiently process HIV-1 polyprotein junctions. A slightly different conclusion was reached by Tomasselli *et al.* (1990) by exploring comparisons of some analogs of the junction peptides.

Phylip *et al.* (1990), using chromogenic oligopeptides prepared for the analysis of HIV-1 proteinase, measured kinetic parameters for a series of peptides based on junction B of the HIV-1 Gag precursor. Variation in the sequence, Lys-Ala-Arg-Val-Nle+Nph-Glu-Ala-Nle-NH$_2$, with changes in the P1 and P2 residues, resulted in several excellent substrates with values of $k_{\text{cat}}/K_{\text{m}}$ as high as 0.5 s$^{-1}$ μM$^{-1}$ (with either Nle or Tyr in P1).

Tomasselli and colleagues (1990) have also reported on comparative inhibition studies using inhibitors with transition-state analogs replacing the scissile peptide bond. In some cases, HIV-1 proteinase bound compounds 10–100 times more tightly than HIV-2 proteinase.

## Structural Chemistry

With the discovery of the HIV-2 genome sequence coming just before solution of the structure of the HIV-1 retropepsin (for references, see Chapter 310), it was a logical step to use the structure of the type 1 enzyme to model that of the type 2 enzyme. Accordingly, Gustchina and colleagues (Gustchina *et al.*, 1991; Gustchina & Weber, 1991) reported on their modeling work and on the resulting analysis of the active-site cleft. Conservative substitutions were observed within the binding cleft. The validity of this approach was confirmed by subsequent reports (Patterson *et al.*, 1992) that chimeric enzymes could be constructed having one monomer contributed by HIV-1 and one monomer by HIV-2, with little effect on the catalytic activity. A different conclusion was reached by Griffiths *et al.* (1994), using a more detailed analysis of chimeric constructs involving both substrates and inhibitors. A reduction in activity or binding was observed for the mixture of HIV-1 and HIV-2 subunits in a tethered dimer construction. This was attributed to the small but significant differences in the residues making up the active site.

Two papers appeared in 1993 describing direct crystallographic analysis of HIV-2 proteinase with bound inhibitors. Mulichak *et al.* (1993) described the structures of HIV-2 proteinase complexed with both dihydroxyethylene-containing and hydroxyethylene-containing inhibitors. While the overall features of the structure were very similar to those of HIV-1 retropepsin, the presence at position 82 of an Ile residue was noted as important in determination of the precise active-site specificity of the HIV-2 retropepsin. In the second study (Tong *et al.*, 1993), a reduced peptide bond inhibitor was cocrystallized. In this case, the differences between HIV-1 and HIV-2 enzymes was much less apparent, correlating with the similar potency of inhibition observed for the two retropepsins.

Priestle *et al.* (1995) have reported on a comparison of the binding of a pseudosymmetric inhibitor to both HIV-1 and

HIV-2 proteinases. By the analysis of the binding to the two enzymes, they have been able to rationalize differences in the affinity of inhibitor binding reported in other comparisons of the two enzymes. These differences can be accounted for by the small, but important differences in amino acid sequence between the two proteins. Tong *et al.* (1995) have presented the high resolution structures of seven inhibitors in complex with HIV-2 proteinase. In this study, they have found evidence for conformational rearrangement of some side-chains of the enzyme to accommodate substitutions in the inhibitor structure. Both of these studies illustrate the need for high-resolution analysis of carefully matched samples in order to extract the subtle yet vital differences that can explain differences in affinity.

## Preparation and Assay

HIV-2 retropepsin has been prepared by expression in bacterial and yeast recombinant systems (Richards *et al.*, 1989; Pichuantes *et al.*, 1990; Tomasselli *et al.*, 1990; Mulichak *et al.*, 1993) and by chemical synthesis (Wu *et al.*, 1990). The protein from either source can be purified by gel filtration and refolded by dilution from GuHCl solutions into a buffer of pH 5.5–7.5. Final purification can be achieved by an additional gel filtration step.

A number of chromogenic or fluorescent substrates are available for routine assay of HIV-2 retropepsin. As described above, the type 2 enzyme cleaves the same substrates as HIV-1, although frequently at slightly reduced rates. The optimal pH range for activity assays is pH 4.5–5.5.

## Biological Aspects

The role of the retroviral enzymes is to cleave the initially formed polyprotein into smaller functional units. As the genetic organization of HIV-2 is approximately the same as HIV-1, we can conclude that the processing steps are most likely similar. Thus, the polyprotein first assembles at the inner surface of the cell membrane, resulting in budding of the new viral particle from the infected cell. Processing of the cleavage junctions can proceed in the immature virus particle, resulting in structural reorganization to yield the final mature and infectious progeny virus. As in the case of HIV-1 retropepsin, the type 2 enzyme must form a homodimer in order to create a functional active-site. This implies that the polyprotein must dimerize before the first cleavages can happen. Whether there are further structural constraints that lead to preferred steps in the self processing remains to be determined.

## Distinguishing Features

Despite the low identity between HIV-1 and -2 retropepsins (50% sequence identity), the two enzymes are able to cleave most substrates at comparable rates. However, Tomasselli *et al* (1990) have reported two compounds that provide differentiation between the enzymes: the substrate, Val-Ser-Gln-Asn-Tyr┼Val-Ile-Val, is cleaved approximately 35 times faster by HIV-1 retropepsin than by the HIV-2 enzyme; the inhibitor U-71038 (Boc-Pro-Phe-NMeHis-Leu [CH(OH)CH$_2$]Val-Ile-Amp) binds to HIV-1 proteinase with

a $K_i$ of 10 nM, but exhibits a $K_i$ of greater than 1 μM with HIV-2 proteinase.

## References

Albert, J., Bredberg, U., Chiodi, F., Bottiger, B., Fenyo, E.M., Norrby, E. & Biberfeld, G. (1987) A new human retrovirus isolate of West African origin (SBL-6669) and its relationship to HTLV-IV, LAV-II, and HTLV-IIB. *AIDS Res. Hum. Retroviruses* **3**, 3–10.

Clavel, F., Guetard, D., Brun-Vezinet, F., Chamaret, S., Rey, M.A., Santos-Ferreira, M.O., Laurent, A.G., Dauguet, C., Katlama, C., Rouzioux, C., Klatzmann, D., Champalimaud, J.L. & Montagnier, L. (1986a) Isolation of a new human retrovirus from West African patients with AIDS. *Science* **233**, 343–346.

Clavel, F., Guyader, M., Guetard, D., Salle, M., Montagnier, L. & Alizon, M. (1986b) Molecular cloning and polymorphism of the human immune deficiency virus type 2. *Nature* **324**, 691–695.

Griffiths, J.T., Tomchak, L.A., Mills, J.S., Graves, M.C., Cook, N.D., Dunn, B.M. & Kay, J. (1994) Interactions of substrates and inhibitors with a family of tethered HIV-1 and HIV-2 homo- and heterodimeric proteinases. *J. Biol. Chem.* **269**, 4787–4793.

Gustchina, A. & Weber, I.T. (1991) Comparative analysis of the sequences and structures of HIV-1 and HIV-2 proteases. *Proteins: Struct. Funct. Genet.* **10**, 325–339.

Gustchina, A., Weber, I.T. & Wlodawer, A. (1991) Molecular modeling of the HIV-2 protease. *Adv. Exp. Med. Biol.* **306**, 549–553.

Guyader, M., Emerman, M., Sonigo, P., Clavel, F., Montagnier, L. & Alizon, M. (1987) Genome organization and transactivation of the human immunodeficiency virus type 2. *Nature* **326**, 662–669.

Kanki, P.J., Barin, F., M'Boup, S., Allan, J.S., Romet-Lemonne, J.L., Marlink, R., McLane, M.F., Lee, T.-H., Arbeille, B., Denis, F. & Essex, M. (1986) New human T-lymphotropic retrovirus related to simian T-lymphotropic virus type III (STLV-III$_{AGM}$). *Science* **232**, 238–243.

Le Grice, S.F., Ette, R., Mills, J. & Mous, J. (1989) Comparison of the human immunodeficiency virus type 1 and 2 proteases by hybrid gene construction and trans-complementation. *J. Biol. Chem.* **264**, 14902–14908

Mulichak, A.M., Hui, J.O., Tomasselli, A.G., Heinrikson, R.L., Curry, K.A., Tomich, C.S., Thaisrivongs, S., Sawyer, T.K. & Watenpaugh, K.D. (1993) The crystallographic structure of the protease from human immunodeficiency virus type 2 with two synthetic peptidic transition state analog inhibitors. *J. Biol. Chem.* **268**, 13103–13109.

Patterson, C.E., Seetharam, R., Kettner, C.A. & Cheng, Y.S. (1992) Human immunodeficiency virus type 1 and type 2 protease monomers are functionally interchangeable in the dimeric enzymes. *J. Virol.* **66**, 1228–1231.

Phylip, L.H., Richards, A.D., Kay, J., Konvalinka, J., Strop, P., Bláha, I., Velek, J., Kostka, V., Ritchie, A.J., Broadhurst, A.V., Farmerie, W.G., Scarborough, P.E. & Dunn, B.M. (1990) Hydrolysis of synthetic chromogenic substrates by HIV-1 and HIV-2 proteinases. *Biochem. Biophys. Res. Commun.* **171**, 439–444.

Pichuantes, S., Babé, L.M., Barr, P.J., DeCamp, D.L. & Craik, C.S. (1990) Recombinant HIV2 protease processes HIV1 Pr53$^{gag}$ and analogous junction peptides *in vitro*. *J. Biol. Chem.* **265**, 13890–13898.

Priestle, J.P., Fassler, A., Rosel, J., Tintelnot-Blomley, M., Strop, P. & Grutter, M.G. (1995) Comparative analysis of the X-ray structures of HIV-1 and HIV-2 proteases in complex with CGP 53820, a novel pseudosymmetric inhibitor. *Structure* **3**, 381–389.

Richards, A.D., Broadhurst, A.V., Ritchie, A.J., Dunn, B.M. & Kay, J. (1989) Inhibition of the aspartic proteinase from HIV-2. *FEBS Lett.* **253**, 214–216.

Tomasselli, A.G., Hui, J.O., Sawyer, T.K., Staples, D.J., Bannow, C., Reardon, I.M., Howe, W.J., DeCamp, D.L., Craik, C.S. & Heinrikson, R.L. (1990) Specificity and inhibition of proteases from human immunodeficiency viruses 1 and 2. *J. Biol. Chem.* **265**, 14675–14683.

Tong, L., Pav, S., Pargellis, C., Do, F., Lamarre, D. & Anderson, P.C. (1993) Crystal structure of human immunodeficiency virus (HIV) type 2 protease in complex with a reduced amide inhibitor and comparison with HIV-1 protease structures. *Proc. Natl Acad. Sci. USA* **90**, 8387–8391.

Tong, L., Pav, S., Mui, S., Lamarre, D., Yoakim, C., Beaulieu, P. & Anderson, P.C. (1995) Crystal structures of HIV-2 protease in complex with inhibitors containing the hydroxyethylamine dipeptide isostere. *Structure* **3**, 33–40.

Tözsér, J., Bláha, I., Copeland, T.D., Wondrak, E.M. & Oroszlan, S. (1991) Comparison of the HIV-1 and HIV-2 proteinases using oligopeptide substrates representing cleavage sites in Gag and Gag-Pol polyproteins. *FEBS Lett.* **281**, 77–80.

Wu, J.C., Carr, S.F., Jarnagin, K., Kirsher, S., Barnett, J., Chow, J., Chan, H.W., Chen, M.S., Medzihradszky, D.; Yamashiro, D. & Santi, D.V. (1990) Synthetic HIV-2 protease cleaves the GAG precursor of HIV-1 with the same specificity as HIV-1 protease. *Arch. Biochem. Biophys.* **277**, 306–311.

Zagury, J.F., Franchini, G., Reitz, M., Collalti, E., Starcich, B., Hall, L., Fargnoli, K., Jagodzinski, L., Guo, H.-G., Laure, F., Arya, S.K., Josephs, S., Zagury, D., Wong-Staal, F. & Gallo, R.C. (1988) Genetic variability between isolates of human immunodeficiency virus (HIV) type 2 is comparable to the variability among HIV type 1. *Proc. Natl Acad. Sci. USA* **85**, 5941–5945.

*Ben M. Dunn*
*Department of Biochemistry & Molecular Biology,*
*University of Florida College of Medicine,*
*Gainesville, FL 32610-0245, USA*
*Email: bdunn@biochem.med.ufl.edu*

# 312. *Equine infectious anemia virus retropepsin*

## Databanks

*Peptidase classification: clan AA, family A2, MEROPS ID: A02.004*
*NC-IUBMB enzyme classification: none*
*Databank codes:*

| Species | Clone/isolate | SwissProt | PIR | EMBL (genomic) |
| --- | --- | --- | --- | --- |
| Equine infectious anemia virus | 1369 | P11204 | B27842 | M16575: complete genome |
| | CL22 | P32542 | B41991 | M87581: complete genome |
| | pSPEIAV19 | – | – | U01866: complete genome |
| | WSU5 | – | – | L06609: 5′ end of *pol* gene |
| | Wyoming | P03371 | A03970 | M87578: part of *pol* gene |

Brookhaven Protein Data Bank three-dimensional structures:

| Species | ID | Resolution | Notes |
| --- | --- | --- | --- |
| Equine infectious anemia virus | 1EQI | | theoretical model |
| | 1FMB | 1.8 | complex with inhibitor HBY-793 |

EMBL accessions M87572 and M87573 are fragments of the genome lacking the retropepsin; M11337 has been amalgamated into M16575. A note in SwissProt states that the amalgamated EMBL entry M16575 contains sequences from different isolates, including the Wyoming isolate. The retropepsin is coded by residues 81–185 in the Pol polyprotein.

## Name and History

Equine infectious anemia virus (EIAV), causing an episodic wasting disease in horses colloquially known as swamp fever, is classified as a member of the lentivirus group of retroviruses and is closely related to the primate immunodeficiency viruses including human immunodeficiency virus type 1 (HIV-1) (Montelaro, 1994). The location of a retropepsin gene in EIAV genome was first suggested based on sequence homology (Stephens *et al.*, 1986). Protein sequencing of its naturally occurring Gag (Henderson *et al.*, 1987) and Pol (Roberts & Oroszlan, 1990) substrates identified several cleavage sites, and their comparison with those of HIV-1 showed limited conservation and substantial variation in amino acid residues at the scissile bond and the neighboring sequence (Roberts & Oroszlan, 1990). Active retropepsin was found in isolated EIAV capsids capable of *in situ* processing of the nucleocapsid protein (Roberts & Oroszlan, 1989). An oligopeptide substrate representing the cleavage site between the matrix and capsid proteins of HIV-1 was found to be a good substrate of **EIAV retropepsin**, and its P1′ pipecolic acid-substituted derivative specifically inhibited the enzyme (Copeland *et al.*, 1990). Expression of a plasmid containing the EIAV *gag* gene followed by in-frame insertion of the putative retropepsin coding sequence resulted in processing of the Gag polyprotein while out of frame insertion yielded intact Gag proteins (Ruslow *et al.*, 1992). The EIAV retropepsin is commonly referred to as **EIAV protease** (or proteinase).

## Activity and Specificity

The activity of EIAV retropepsin can be conveniently measured using oligopeptide substrates, in the pH range of 4–6, in the presence of 0.3–2.0 M NaCl. Generally, two types of retropepsin cleavage sites exist: type 1 contains aromatic amino acid residue and proline, and type 2 contains hydrophobic residues (excluding Pro) at the site of cleavage (Pettit *et al.*, 1991; Griffiths *et al.*, 1992) (Table 310.1). Assay of type 1 substrates is usually carried out by isolating the cleavage products by reversed-phase HPLC (Tözsér *et al.*, 1993; Weber *et al.*, 1993; Powell *et al.*, 1996). An HIV-1 retropepsin (Chapter 310) substrate suitable to measure the activity of EIAV retropepsin is commercially available (Val-Ser-Gln-Asn-Tyr↓Pro-Ile-Val) from Bachem (see Appendix 2 for full names and addresses of suppliers). Some type 2 substrates were designed to contain the *p*-nitrophenylalanine group at P1′ so that their cleavage can be followed spectrophotometrically (Powell *et al.*, 1996).

The specificity of EIAV retropepsin derived from purified virus or expressed as a recombinant protein was extensively characterized using oligopeptide substrates (Tözsér *et al.*, 1993; Weber *et al.*, 1993; Powell *et al.*, 1996). While HIV-1 and HIV-2 (Chapter 311) retropepsins generally recognize seven residues of the substrates, the substrate binding site of EIAV retropepsin appears to be more extended (Tözsér *et al.*, 1993; Powell *et al.*, 1996). Several specific inhibitors of HIV-1 retropepsin were ineffective on the EIAV retropepsin (Powell *et al.*, 1996). However, Compound 3 (Grobelny *et al.*, 1990) and HBY-793 could be used for active-site titration (Tözsér *et al.*, 1993; Powell *et al.*, 1996) and the latter

was also cocrystallized with the enzyme (Gustchina *et al.*, 1996).

The substrate-binding sites of EIAV retropepsin were characterized by using a series of type 1 (Weber *et al.*, 1993) and type 2 (Powell *et al.*, 1996) substrates. The S1 subsite of EIAV retropepsin seems to be similar to that of HIV. With both series the best P1 substituents were large hydrophobic residues, while peptides containing Val, Ser, Asp or Lys were not hydrolyzed significantly. In contrast, preferences for the S2 and S2′ binding sites were not only different from those of HIV retropepsins, but also different in the two series. In type 1 peptides small hydrophobic or partly hydrophobic residues like Cys, Ala, Thr were preferred while in the type 2 substrates larger residues like Leu or Ile gave the highest specificity constants. Different preferences for these two types of cleavage sites were previously found for HIV (Griffiths *et al.*, 1992) and avian myeloblastosis virus proteinase (Chapter 313) (Tözsér *et al.*, 1996). In the type 1 series, besides the P2 differences, substitutions at P4 also gave results different from those with HIV-1: hydrophobic residues at P4 form better substrates for EIAV than for HIV-1 retropepsin (Weber *et al.*, 1993). In the type 2 series, P2′ substitutions also gave substantially different preferences than those obtained with HIV-1 retropepsin (Powell *et al.*, 1996). Two mutants containing single mutations toward the HIV-1 retropepsin sequence (Thr30Asp and Ile54Gly) were also generated and the specificity of these retropepsins was compared to that of the wild type (Powell *et al.*, 1996).

Activity of the EIAV retropepsin can also be tested by using protein substrates. It can cleave recombinant EIAV Gag polyprotein precursor into proteins with sizes similar to those found in mature viruses (Ruslow *et al.*, 1992; Luukkonen *et al.*, 1995; Powell *et al.*, 1996). It can also cleave recombinant HIV-1 Gag polyprotein, albeit slower than HIV-1 retropepsin (Luukkonen *et al.*, 1995; Powell *et al.*, 1996).

## Structural Chemistry

EIAV retropepsin is a homodimeric enzyme of two 104 amino acid polypeptide chains, each with a calculated $M_r$ of 11 431. The $M_r$ of each subunit was estimated by SDS-PAGE as 10 000 (Ruslow *et al.*, 1992) or 12 000 (Tözsér *et al.*, 1993). The theoretically expected value was also found by mass spectrometry (Powell *et al.*, 1996). EIAV retropepsin shares about 42% sequence similarity with HIV-1 and Rous sarcoma virus (Chapter 313) enzymes and, based on this homology, a molecular model was built (Protein Data Bank code 1EQI, Weber *et al.*, 1993). The crystal structure of EIAV retropepsin containing the Ile54Gly mutation has been solved (Gustchina *et al.*, 1996). As expected, the overall fold is very similar to that of other retroviral proteinases. However, unlike any other retropepsins with solved crystal structure, the EIAV enzyme contains a second helix in the monomer (Gustchina *et al.*, 1996).

## Preparation

EIAV retropepsin has been purified from virus (Wyoming strain) grown in chronically infected dog thymus cells (Cf2Th,

ATCC CRL 1430) by reversed-phase HPLC (Tözsér *et al.*, 1993). The enzyme has been successfully synthesized chemically (Copeland *et al.*, 1991). It was also expressed in *Escherichia coli* and purified by DEAE- and CM-cellulose chromatography, followed by Mono-S separation (Powell *et al.*, 1996).

## Biological Aspects

As in all lentiviruses, the EIAV retropepsin is coded in the 5′ part of the *pol* gene and synthesized as part of the Gag-Pol polyprotein precursor by a ribosomal. frameshift mechanism (Oroszlan & Luftig, 1990). EIAV retropepsin is responsible for cleaving the Gag and Gag-Pol polyprotein precursors into functional proteins during or after assembly and budding of the virions. Protein sequence analysis of the mature viral proteins identified the naturally occurring cleavage sites (Henderson *et al.*, 1987; Roberts & Oroszlan, 1990). The purified EIAV capsids contain retropepsin, and incubation of these capsids results in cleavages in the nucleocapsid protein (Roberts & Oroszlan, 1989). On the basis of these findings, a novel function of the retroviral proteinase in the early phase of viral replication has been suggested (Roberts *et al.*, 1991). Similar to murine leukemia virus retropepsin, the EIAV retropepsin also cleaves at the C-terminal of the transmembrane protein (Rice *et al.*, 1990).

## Distinguishing Features

EAIV is similar to HIV in showing rapid antigenic variation in the infected host (Montelaro, 1994). Furthermore, EIAV is the only lentivirus for which capsid preparation has been developed (Roberts & Oroszlan, 1989), therefore, it is an important model of HIV. Generally, the specificity of EIAV retropepsin is similar to that of HIV-1; however, detailed analysis of subsite interactions also suggests marked differences (Weber *et al.*, 1993; Powell *et al.*, 1996). Most of the tested HIV-1 retropepsin inhibitors are ineffective against the EIAV enzyme (Powell *et al.*, 1996). Polyclonal antiserum towards a peptide having the C-terminal segment of the retropepsin (DIPVTILGRDILQDLGAKLV) was developed by Ruslow *et al.* (1992).

## References

Copeland, T.D., Wondrak, E.M., Tözsér, J., Roberts, M.M. & Oroszlan, S. (1990) Substitution of proline with pipecolic acid at the scissile bond converts a peptide substrate of HIV proteinase into selective inhibitor. *Biochem. Biophys. Res. Commun.* **169**, 310–314.

Copeland, T.D., Bláha, I., Tözsér, J. & Oroszlan, S. (1991) Totally synthetic retroviral proteins: proteases and nucleic acid binding proteins. In: *Peptides: Chemistry and Biology* (Smith, J.A. & Rivier, J.E., eds). Leiden, The Netherlands: ESCOM, pp. 717–718.

Griffiths, J.T., Phylip, L.H., Konvalinka, J., Strop, P., Gustchina, A., Wlodawer, A., Davenport, R.J., Briggs, R., Dunn, B.M. & Kay, J. (1992) Different requirements for productive interaction between the active site of HIV-1 proteinase and substrates

containing-hydrophobic*hydrophobic- or -aromatic*Pro- cleavage sites. *Biochemistry* **31**, 5193–5200.

Grobelny, D., Wondrak, E.M., Galardy, R.E. & Oroszlan, S. (1990) Selective phosphinate transition-state inhibitors of the protease of human immunodeficiency virus. *Biochem. Biophys. Res. Commun.* **169**, 1111–1116.

Gustchina, A., Kervinen, J., Powell, D.J., Zdanov, A., Kay, J. & Wlodawer, A. (1996) Structure of equine infectious anemia virus proteinase complexed with an inhibitor. *Protein Sci.* **5**, 1453–1465.

Henderson, L.E., Sowder, R.C., Smythers, G.W. & Oroszlan, S. (1987) Chemical and immunological characterizations of equine infectious anemia virus *gag*-encoded proteins. *J. Virol.* **61**, 1116–1124.

Luukkonen, B.G.M., Tan, W., Fenyö, E.M. & Schwartz, S. (1995) Analysis of cross reactivity of retrovirus proteases using a vaccinia virus-T7 RNA polymerase-based expression system. *J. Gen. Virol.* **76**, 2169–2180.

Montelaro, R.C. (1994) Equine infectious anemia virus. In: *Encyclopedia of Virology* (Webster, R.G. & Grandoff, A., eds). New York: Academic Press, pp. 430–436.

Oroszlan, S. & Luftig, R.B. (1990) Retroviral proteinases. *Curr. Top. Microbiol. Immunol.* **157**, 153–185.

Pettit, S.C., Simsic, J., Loeb, D.D., Everitt, L., Hutchinson, C.A., III. & Swanstrom, R. (1991) Analysis of retroviral protease cleavage sites reveals two type of cleavage sites and the structural requirements of the P1 amino acid. *J. Biol. Chem.* **266**, 14539–14547.

Powell, D.L., Bur, D., Wlodawer, A., Gustchina, A., Payne, S.L., Dunn, B.M. & Kay, J. (1996) Expression, characterisation and mutagenesis of the aspartic proteinase from equine infectious anemia virus. *Eur. J. Biochem.* **241**, 664–674.

Rice, N.R., Henderson, L.E., Sowder, R.C., Copeland, T.D., Oroszlan, S. & Edwards, J.F. (1990) Synthesis and processing of the transmembrane envelope protein of equine infectious anemia virus. *J. Virol.* **64**, 3770–3778.

Roberts, M.M. & Oroszlan, S. (1989) The preparation and biochemical characterization of intact capsids of equine infectious anemia virus. *Biochem. Biophys. Res. Commun.* **160**, 486–494.

Roberts, M.M. & Oroszlan, S. (1990) The action of retroviral protease in various phases of virus replication. In: *Retroviral Proteases: Control of Maturation and Morphogenesis* (Pearl, L., ed.). London: Macmillan Press, pp. 131–139.

Roberts, M.M., Copeland, T.D. & Oroszlan, S. (1991) *In situ* processing of a retroviral nucleocapsid protein by the viral proteinase. *Protein Eng.* **4**, 695–700.

Ruslow, K., Peng, X.X., Montelaro, R.C. & Shih, D.S. (1992) Expression of the protease gene of equine infectious anemia virus in *Escherichia coli*: formation of the mature processed enzyme and specific cleavage of the Gag precursor. *Virology* **188**, 396–401.

Stephens, R.M., Casey, J.W. & Rice, N.R. (1986) Equine infectious anemia virus *gag* and *pol* genes: relatedness to Visna and AIDS virus. *Science* **231**, 589–594.

Tözsér, J., Friedman, D., Weber, I.T., Bláha, I. & Oroszlan, S. (1993) Studies on the substrate specificity of the proteinase of equine infectious anemia virus using oligopeptide substrates. *Biochemistry* **32**, 3347–3353.

Tözsér, J., Bagossi, P., Weber, I.T., Copeland, T.D. & Oroszlan, S. (1996) Comparative studies on the substrate specificity of avian myeloblastosis virus proteinase and lentiviral proteinases. *J. Biol.*

Chem. **271**, 6781–6788.

Weber, I.T., Tözsér, J., Wu, J., Friedman, D. & Oroszlan, S. (1993) Molecular model of the equine infectious anemia virus proteinase and kinetic measurements for peptide substrates with single amino acid substitutions. *Biochemistry* **32**, 3354–3362.

*József Tözsér*
*Department of Biochemistry,*
*University Medical School of Debrecen,*
*H-4012 Debrecen, Hungary*
*Email: tozser@indi.biochem.dote.hu*

*Luis Menéndez-Arias*
*Centro de Biología Molecular*
*'Severo Ochoa',*
*CSIC-Universidad Autónoma de Madrid,*
*28049 Cantoblanco (Madrid), Spain*
*Email: lmenendez@mvax.cbm.uam.es*

*Stephen Oroszlan*
*Molecular Virology and*
*Carcinogenesis Laboratory,*
*ABL-Basic Research Program,*
*NCI-Frederick Cancer Research and*
*Development Center,*
*Frederick, MD 21702-1201, USA*
*Email: oroszlans@mail.fcrdc.gov*

# 313. Rous sarcoma virus retropepsin and avian myeloblastosis virus retropepsin

## Databanks

*Peptidase classification: clan AA, family A2, MEROPS ID: A02.015*
*NC-IUBMB enzyme classification: none*
*Databank codes:*

| Species | SwissProt | PIR | EMBL (genomic) |
|---|---|---|---|
| Rous avian sarcoma virus | P03322 | A03922 | J02342: complete genome<br>V01197: complete genome |
| Avian myeloblastosis virus | P26315 | – | S78824: retropepsin gene |

Brookhaven Protein Data Bank three-dimensional structures:

| Species | ID | Resolution | Notes |
|---|---|---|---|
| Rous avian sarcoma virus | 2RSP | 2 | |

## Name and History

The Rous sarcoma virus (RSV), identified by Peyton Rous as a tumor-causing agent in 1911 (Rous, 1911) contains an RNA genome which encodes three genes required for replication. These genes, *gag, pol* and *env*, are translated to yield three large precursor polyproteins: Gag, Gag-Pol and Env. During virus maturation, Gag and Gag-Pol are cleaved into eight different mature proteins by a viral specific protease (Vogt *et al.*, 1975; Von der Helm, 1977; Dittmar & Moelling, 1978; Vogt *et al.*, 1979). It does not appear that a host cell protease is required for this maturation process. The viral protease, known originally as the *p15 protein*, is located at the C-terminal region of the 76 kDa Gag polyprotein. The protease is active only as a homodimer and is referred to as **PR** (Leis *et al.*, 1988). RSV belongs to a group of avian retroviruses that includes avian myeloblastosis virus (AMV). The *avian myeloblastosis virus retropepsin* and *Rous sarcoma virus retropepsin* differ by only two amino acid residues and are biochemically indistinguishable.

## Activity and Specificity

Viral replication requires that the protease cleaves nine sites of eight amino acids in the Gag and Gag-Pol polyproteins.

Alignment of the cleavage sites reveals no obvious consensus sequence. Substrate selection is based primarily on optimization of van der Waals interactions between amino acid side chains from the substrate, and enzyme amino acids forming eight individual enzyme subsites. Peptide specificity at each of the subsite positions P4–P4' has been examined using a series of modified peptide substrates based on the naturally occurring cleavage sequence between the viral nucleocapsid and protease subunit (Grinde *et al.*, 1992a; Cameron *et al.*, 1993). This analysis has demonstrated that RSV retropepsin has a preference for large hydrophobic side chains in P1, P1' and P3. Although enzyme subsites act somewhat independently in substrate selection, there exists significant steric interaction between subsites that limits the number of cleavable substrate sequences (Ridky *et al.*, 1996a). For example, large side chains can be accommodated individually into the P1 and P3 positions, but substrates with large side chains in both P1 and P3 are cleaved poorly.

The RSV/AMV retropepsin homodimer is structurally very similar to the HIV-1 retropepsin (Chapter 310), even though it shares only a 30% sequence identity. Despite this structural similarity, the two enzymes have different substrate specificities, which can be altered by mutating key amino acids in their respective substrate binding pockets. Mutant RSV retropepsins with HIV-1 retropepsin-like specificity have been constructed by substituting key specificity-determining residues with amino acids from structurally equivalent positions in HIV-1 PR (Grinde *et al.*, 1992b; Sedlacek *et al.*, 1993; Cameron *et al.*, 1994; Ridky *et al.*, 1996b). While wild-type RSV retropepsin does not cleave the natural HIV-1 retropepsin substrates, an RSV retropepsin mutant containing nine amino acid substitutions (S38T, I42D, I44V, M73V, A100L, V104T, R105P, G106V, S107N) cleaves these substrates efficiently (Ridky *et al.*, 1996b). Additionally, this combinatorial mutant becomes sensitive to HIV-1 specific inhibitors that are ineffective against the wild-type RSV enzyme. Many enzyme positions important for determining specificity in both RSV and HIV-1 protease are those that change in developing resistance to retropepsin inhibitors used with HIV-1 infected patients (Ridky & Leis, 1995).

Recombinant enzymes expressed with two subunits covalently linked have been purified and are at least 30% as active as wild type (Bizub *et al.*, 1991; Ridky *et al.*, 1996b; Xiang *et al.*, 1997). The linked enzymes have allowed for an examination of the importance of enzyme symmetry in substrate selection. Introduction of specificity altering mutations selectively into one subunit has shown that the dimer is functionally symmetric with respect to substrate recognition. That is, each subunit is capable of interacting equally well with both the N- and C-terminals of a peptide substrate (Ridky *et al.*, 1996b).

Protease activity is studied most easily using an *in vitro* peptide cleavage assay (Bizub *et al.*, 1991). Typically, synthetic peptides of 12–14 amino acids are synthesized with the cleavage site after residue 5 or 6, for example PPAVS↓LAMTMRR and PATVL↓TVALRR. The fluorescence assay does not require any unnatural amino acids, but does require that the substrate peptide is synthesized with an N-terminal proline. C-terminal Arg residues are included to improve solubility. After cleavage, a product peptide is detected by the presence of its primary amine, which reacts specifically with fluorescamine. Specificity of the cleavage reaction is confirmed by N-terminal amino acid analysis of the products. One limitation of the assay is that cleavage reactions occurring with a P1' proline are not detected. In a typical assay, nanomolar concentrations of purified protease are incubated with micromolar concentrations of synthetic peptide substrate at a pH optimum of 5.9 and a salt optimum of 2.4 M NaCl. The high salt is required for efficient binding of substrate; $K_m$ values increase dramatically with lower salt concentrations. Unfortunately, the high salt concentrations frequently cause peptide substrates to be insoluble. Solubility can be increased upon addition of no more than 10% DMSO. This does increase $K_m$ values, but usually allows saturating substrate concentrations to be obtained. Other more traditional assay systems based on HPLC separation of products can also be used, but are significantly more time-consuming and not as sensitive.

Pepstatin is inhibitory. Most inhibitors designed for HIV-1 PR are relatively ineffective against RSV PR, in spite of the homology of the two enzymes. The modified peptide Pro-Pro-Cys-Val-Phe-Sta-Ala-Met-Thr-Met is a nanomolar competitive inhibitor (Cameron *et al.*, 1994).

## Structural Chemistry

Functionally active RSV/AMV protease is a symmetric homodimer with two identical subunits of 124 amino acids each. There is one high resolution crystal structure of the native enzyme without substrate or inhibitor (Miller *et al.*, 1989; Jaskolski *et al.*, 1990). The protein has an $M_r$ of 27 000 and contains no disulfide linkages or metal atoms. The active site and substrate binding site are formed by an obvious cleft at the dimer interface. Enzyme subunits are held together through formation of a $\beta$ sheet structure with the N- and C-termini from both subunits. Two highly flexible surface loops known as 'flaps' appear to fold over a peptide substrate to secure it in the binding site. Peptide substrates bind in an extended $\beta$ conformation where adjacent substrate side chains (i.e. P1 and P2) extend in opposite (*trans*) directions into individual subsites within the enzyme. Side chains for alternate substrate positions (i.e. P1 and P3) extend in the same (*cis*) direction.

## Preparation

AMV protease has been purified from whole virus (Alexander *et al.*, 1987). Recombinant RSV wild-type and mutant enzymes were initially expressed in *Escherichia coli* and refolded from inclusion bodies (Bizub *et al.*, 1991). While these preparations are relatively pure (>95% by SDS-PAGE), they refold poorly with maximal folding efficiency of around 20%. These enzyme preparations do not crystallize. Recently RSV PR has been expressed with an N-terminal polyhistidine tag and purified by $Ni^{2+}$ affinity chromatography from the soluble fraction of *E. coli* lysates (Ridky *et al.*, 1996a). This enzyme is over 99% pure by SDS-PAGE, but has only minimal activity, presumably because the histidine extension interferes with dimerization. However, the histidine tag can be removed by cleavage with cattle factor Xa. This leaves the native RSV PR sequence at the N-terminus. This enzyme is as active as PR purified directly from virions and crystallizes with a structure identical to that seen with the viral enzyme. AMV retropepsin is available commercially from Molecular Genetic Resources (see Appendix 2 for full names and addresses of suppliers).

## Biological Aspects

The RSV retropepsin is known to be translated as part of the Gag polyprotein on cytosolic polyribosomes and transported with the polyprotein to sites of virus assembly on the plasma membrane. During the budding process, Gag molecules associate to allow protease dimerization. The active enzyme first cleaves itself free of the polyprotein before proceeding with cleavage at other protein junctions. Studies with covalently linked protease dimers have shown that protease dimerization and initial cleavage at the nucleocapsid–protease junction in Gag is a key regulatory step in the budding process (Xiang *et al.*, 1997). Premature protease activation divorces many of the Gag proteins from membrane targeting signals present at the N-terminus of Gag preventing the viral proteins from reaching the site of assembly. Delayed or inactivated protease does not prevent the actual budding process, but does prevent viral maturation; resulting particles are morphologically immature and non-infectious.

## Distinguishing Features

The 124 amino acid RSV retropepsin monomer is slightly larger than the 99 amino acid HIV-1 retropepsin primarily due to two larger surface loops. One of these loops is known as the highly flexible 'flap' region and helps dictate specificity at the P4 and P4′ position. While HIV-1 retropepsin has a very poorly defined S4 and S4′ subsite on the enzyme surface, the RSV enzyme has a relatively enclosed pocket that allows the protease to accommodate more hydrophobic side chains at these positions. The RSV enzyme is generally several-fold less active than the HIV-1 enzyme; however, this is highly variable and depends upon the substrate sequence used. This lower activity may be compensated by the 20-fold excess of RSV retropepsin relative to HIV-1 retropepsin within the virus. While RSV is expressed as part of Gag, HIV-1 retropepsin is expressed only as part of the larger precursor protein Gag-Pol. This polyprotein results from an inefficient frameshift event that occurs about 5% of the time during translation.

## Further Reading

For further reading see Skalka (1989).

## References

Alexander, F., Leis, J., Soltis, D., Crowl, R., Danho, W., Poonian, M., Pan, Y. & Skalka, A. (1987) Proteolytic processing of avian sarcoma leukosis viruses *pol-endo* recombinant proteins reveals another *pol* gene domain. *J. Virol.* **61**, 534–542.

Bizub, D., Weber, I., Cameron, C., Leis, J. & Skalka, A. (1991) A range of catalytic efficiencies with avian retroviral protease subunits genetically linked to form single polypeptide chains. *J. Biol. Chem.* **266**, 4951–4958.

Cameron, C., Grinde, B., Jacques, P., Jentoft, J. & Leis, J. (1993) Comparison of the substrate binding pockets of the Rous sarcoma virus and human immunodeficiency virus type 1 proteases. *J. Biol. Chem.* **268**, 11711–11720.

Cameron, C.E., Ridky, T.W., Shulenin, S., Leis, J.P., Weber, I.T., Copeland, T.C., Wlodawer, A., Burstein, H., Bizub-Bender, D. & Skalka, A. (1994) Mutational analysis of the substrate binding pockets of the Rous sarcoma virus and human immunodeficiency virus-1 proteases. *J. Biol. Chem.* **269**, 11170–11177.

Dittmar, K.J. & Moelling, K. (1978) Biochemical properties of p15-associated protease in an avian RNA tumor virus. *J. Virol.* **28**, 106–118.

Grinde, B., Cameron, C., Leis, J., Weber, I., Wlodawer, A., Burstein, H. & Skalka, A. (1992a) Analysis of substrate interactions of the Rous sarcoma virus wild type and mutant proteases and human immunodeficiency virus-1 protease using a set of systematically altered peptide substrates. *J. Biol. Chem.* **267**, 9491–9498.

Grinde, B., Cameron, C., Leis, J., Weber, I., Wlodawer, A., Burstein, H., Bizub, D. & Skalka, A. (1992b) Mutations that alter the activity of the Rous sarcoma virus protease. *J. Biol. Chem.* **267**, 9481–9490.

Jaskolski, M., Miller, M., Rao, J., Leis, J. & Wlodawer, A. (1990) Structure of the aspartic protease from Rous sarcoma retrovirus refined at 2-Å resolution. *Biochemistry* **29**, 5889–5898.

Leis, J., Baltimore, D., Bishop, J.M., Coffin, J., Fleissner, E., Goff, S.P., Oroszlan, S., Robinson, H., Skalka, A.M., Temin, H.M. & Vogt, V. (1988) Standardized and simplified nomenclature for proteins common to all retroviruses. *J. Virol.* **62**, 1808–1809.

Miller, M., Jaskolski, M., Rao, J., Leis, J. & Wlodawer, A. (1989) Crystal structure of a retroviral protease proves relationship to aspartic acid protease family. *Nature* **337**, 576–579.

Ridky, T. & Leis, J. (1995) Development of resistance to HIV-1 protease inhibitors. *J. Biol. Chem.* **270**, 29621–29623.

Ridky, T.W., Cameron, C.E., Cameron, J., Leis, J., Weber, I., Wlodawer, A. & Copeland, T. (1996a) HIV-1 protease specificity is limited by interactions between substrate amino acid side chains in adjacent enzyme subsites. *J. Biol. Chem.* **271**, 4709–4717.

Ridky, T.W., Bizub-Bender, D., Cameron, C.E., Weber, I., Wlodawer, A., Copeland, T., Skalka, A. & Leis, J. (1996b) Programming the Rous sarcoma virus protease to cleave new substrate sequences. *J. Biol. Chem.* **271**, 10538–10544.

Rous, P. (1911) A sarcoma of the fowl transmissible by an agent separable from the tumor cells. *J. Exp. Med.* **13**, 397–411.

Sedlacek, J., Fabry, M., Coward, J., Horejsi, M., Strop, P. & Luftig, R. (1993) Myeloblastosis associated virus (MAV) protease site-mutated to be HIV-like has a higher activity and allows production of infectious but morphologically altered virus. *Virology* **192**, 667–672.

Skalka, A.M. (1989) Retrovirus proteases: first glimpses at the anatomy of a processing machine. *Cell* **56**, 911–913.

Vogt, V.M., Eisenman, R. & Diggelmann, H. (1975) Generation of avian myeloblastosis structural proteins by proteolytic cleavage of a precursor polypeptide. *J. Mol. Biol.* **96**, 471–493.

Vogt, V.M., Wight, W. & Eisenman, R. (1979) *In vitro* cleavage of avian retrovirus *gag* proteins by viral protease p15. *Virology* **98**, 154–167.

Von der Helm, K. (1977) Cleavage of Rous sarcoma viral polyprotein precursor into internal structural proteins *in vitro* involves viral p15. *Proc. Natl Acad. Sci. USA* **74**, 911–915.

Xiang, Y., Ridky, T.W., Krishna, N.K. & Leis, J. (1997) Altered Rous sarcoma virus Gag polyprotein processing and its effects on particle formation. *J. Virol.* **71**, 2083–2091.

**Todd W. Ridky**
Case Western Reserve University School of Medicine,
Cleveland, OH 44106, USA
Email: leis@biocserver.bioc.cwru.edu

**Jonathan Leis**
Case Western Reserve University School of Medicine,
Cleveland, OH 44106, USA

# 314. Human T-cell leukemia virus type I retropepsin

## Databanks

*Peptidase classification: clan AA, family A2, MEROPS ID: A02.012*
*NC-IUBMB enzyme classification: none*
*Databank codes:*

| Species | Strain/isolate | SwissProt | PIR | EMBL (genomic) |
| --- | --- | --- | --- | --- |
| Human T-cell leukemia virus I | ATK | P10274 | A24817 | D10033: mid-section of genome |
| | Caribbean | P14074 | B28136 | D13784: complete genome |
| | Om1 | – | – | D11119: retropepsin |
| Human T-cell leukemia virus II | – | P03353 | A03954 | M10060: complete genome |

## Name and History

The identification and sequencing of the entire proviral genome of human T-cell leukemia virus type I (HTLV-I) in 1983 (Seiki *et al.*, 1983) led to the recognition, by similarity to other retroviral sequences, of a putative HTLV-I protease-coding region. This was first recognized in an isolate from an HTLV-I transformed T-cell line MT2 (Nam & Hatanaka, 1986) and verified in an isolate from extra-chromosomal viral DNA in HL60 cells (Hiramatsu *et al.*, 1987). As little protease activity could be detected associated with HTLV-I virions, an analysis of protease activity was done using protease expressed first by vaccinia virus recombinants (Nam *et al.*, 1988) and later in *Escherichia coli* (Kobayashi *et al.*, 1991; Daenke *et al.*, 1994). **HTLV-I protease**, or **HTLV-I retropepsin**, is an aspartic protease and is autoprocessed from a precursor protein expressed in a separate open reading frame within the *Gag-Pol* region of the HTLV-I genome (Nam *et al.*, 1988). Active protease is required for the essential processing of viral structural proteins p15, p19 and p24 and the *pol*-encoded proteins reverse transcriptase, RNAaseH and integrase.

## Activity and Specificity

Cleavage of synthetic peptide substrates by HTLV-I retropepsin shows preference for the 'aromatic+Pro' type substrate sequence described by Griffiths *et al.* (1992), for example p19/p24, Pro-Gln-Val-Leu+Pro-Val-Met-His; Gag/retropepsin, Pro-Ala-Ser-Ile-Leu+Pro-Val-Ile-Leu-Pro; protease/reverse transcriptase, Pro-Pro-Val-Ile-Leu+Pro-Ile-Gln-Ala-Pro. However, some of the native HTLV-I substrates conform to the 'hydrophobic+hydrophobic' substrate pattern, for example p24/p15, Lys-Thr-Lys-Val-Leu+Val-Val-Lys-Pro-Lys. A large diversity of residues is tolerated at the P2–P3 and P2'–P3' positions, with a weak preference for Val at P2'. Deletional mutagenesis of the protease-coding region has shown that the N-terminal residues Pro33–Leu37 are not essential for enzymatic activity. Likewise, the C-terminal residues Pro153–Leu157 are not necessary for proteolytic activity, but may contribute to the stability of the protein (Hayakawa *et al.*, 1992).

Cleavage may be assayed using synthetic peptides incorporating a chromogenic reporter group Nph at P1' (Daenke *et al.*, 1994) or full-length Gag polyprotein precursor molecules (Nam *et al.*, 1988; Hayakawa *et al.*, 1992). The pH optimum is 5–5.5 (Kobayashi *et al.*, 1991) with 2–3 M salt (Ménard *et al.*, 1993). Cleavage is weakly inhibited by pepstatin A ($K_i = 100\,\mu$M) and the HIV-1 Pol transition state mimetic compound Ro31-8959 (Daenke *et al.*, 1994) ($K_i = 3\,\mu$M). Weak inhibition can also be demonstrated in the presence of peptides corresponding to the N- and C-terminal residues of the protease (Daenke *et al.*, 1994) ($K_i = 87\,\mu$M for N-terminal peptide; $K_i = 5$–$70\,\mu$M for C-terminal peptide), presumably by destabilizing the molecule.

## Structural Chemistry

HTLV-I protease resembles the cellular aspartic proteases pepsin (Chapter 272), renin (Chapter 284) and cathepsin D (Chapter 277). In contrast to the single-chain cellular enzymes, HTLV-I protease is a small dimer composed of two identical subunits of 125 amino acids and 14 kDa mass, which are processed from a 234 amino acid precursor protein. The subunits associate noncovalently into the active molecule. Structural similarities shared with other aspartic proteases include the Asp-Thr-Gly active site motif and the sequences flanking the active site. Molecular modeling of the tertiary structure based on the structures of other retroviral proteases suggests an unusually complex association of the C-terminals of each subunit (Hayakawa *et al.*, 1992). The crystal structure of the HTLV-I retropepsin dimer has not been solved.

## Preparation

Active protease can be expressed either from a precursor protein construct or as the mature protein coding sequence in eukaryotic cells from a vaccinia virus recombinant. Purified recombinant HTLV-I retropepsin can be expressed and purified from *E. coli* on a scale of several milligrams.

## Biological Aspects

HTLV-I retropepsin has been shown to be inactive on other retroviral substrates except bovine leukemia virus (BLV) Gag substrate (Luukkonen *et al.*, 1995). However, the activity on cellular proteins is not known. HIV-1 retropepsin is known to cleave cellular vimentin, desmin and glial fibrillary proteins, and this may contribute to the cytotoxic effect observed in HIV-1-infected cells. A similar study with HTLV-I retropepsin has not been undertaken, although its effect on the cell may be limited by sequestration inside the capsid during assembly and maturation of the virion. Probably for the same reason, the protease is not known to be a target for immune attack in immunocompetent infected individuals. As a result of this, the protease sequence is highly conserved between isolates, due to lack of immune selection and strong functional constraints.

## Distinguishing Features

HTLV-I-specific rabbit polyclonal antisera have been raised to the C-terminal peptide 123–140 LVDTKNNWAIIGRDALQQ (Hayakawa *et al.*, 1992) and a N-terminal peptide 38–54 DPARRPVIKAQVDTQTS (Hayakawa *et al.*, 1991). A rabbit antiserum to the protease precursor peptide 221–234 (EPGD-SSTTCGPLTL) has been described (Hayakawa *et al.*, 1992). None of these antisera are available commercially.

## Related Peptidases

The most closely related of the retroviral proteases to HTLV-I protease is that of BLV (32% amino acid identity) (Chapter 315). These proteases are able to cleave their heterologous viral polyprotein substrates (Luukkonen *et al.*, 1995). Proteases from the lentivirus group (HIV, SIV, etc.) are less similar in sequence and size, but share similarity in structure (see above). Cellular aspartic proteases are also similar at the structural level, although these are translated as single-chain molecules with intrinsic activity.

## Further Reading

The cleavage kinetics are given in detail in Daenke *et al.* (1994) and a comparison of the substrate specificities of a number of retroviral proteases is provided by Luukkonen *et al.* (1995).

## References

Daenke, S., Schramm, H.J. & Bangham, C.R.M. (1994) Analysis of substrate cleavage by recombinant protease of human T cell leukaemia virus type 1 reveals preferences and specificity of cleavage. *J. Gen. Virol.* 75, 2233–2239.

Griffiths, J.T., Phylip, L.H., Konvalinka, J., Strop, P., Gustchina, A., Wlodawer, A., Davenport, R.J., Briggs, R., Dunn, B.M. & Kay, J. (1992) Different requirements for productive interaction between the active site of HIV-1 proteinase and substrates containing-hydrophobic*hydrophobic- or -aromatic*Pro- cleavage sites. *Biochemistry* 31, 5193–5200.

Hayakawa, T., Misumi, Y., Kobayashi, M., Ohi, Y., Fujisawa, Y., Kakinuma, A. & Hatanaka, M. (1991) Expression of human T-cell leukemia virus type 1 protease in *Escherichia coli*. *Biochem. Biophys. Res. Commun.* 181, 1281–1287.

Hayakawa, T., Misumi, Y., Kobayashi, M., Yamamoto, Y. & Fujisawa, Y. (1992) Requirement of N- and C-terminal regions for enzymatic activity of human T-cell leukemia virus type 1 protease. *Eur. J. Biochem.* 206, 919–925.

Hiramatsu, K., Nishida, J., Naito, A. & Yoshikura, H. (1987) Molecular cloning of the closed circular provirus of human T cell leukaemia virus type 1: a new open reading frame in the *gag-pol* region. *J. Gen. Virol.* 68, 213–218.

Kobayashi, M., Ohi, Y., Asano, T., Hayakawa, T., Kato, K., Kakinuma, A. & Hatanaka, M. (1991) Purification and characterization of human T-cell leukemia virus type 1 protease produced in *Escherichia coli*. *FEBS Lett.* 293, 106–110.

Luukkonen, B.G.M., Tan, W., Fenyo, E.M. & Schwartz, S. (1995) Analysis of cross reactivity of retrovirus proteases using a vaccinia virus T7 RNA polymerase-based expression system. *J. Gen. Virol.* 76, 2169–2180.

Ménard, A., Mamoun, R.Z., Geoffre, S., Castroviejo, M., Raymond, S., Précigoux, G., Hospital, M. & Guilleman, B. (1993) Bovine leukemia virus: purification and characterization of the aspartic protease. *Virology* 193, 680–689.

Nam, S.H. & Hatanaka, M. (1986) Idenfication of a protease gene of human T-cell leukemia virus type 1 (HTLV-1) and its structural comparison. *Biochem. Biophys. Res. Commun.* 139, 129–135.

Nam, S.H., Kidokoro, M., Shida, H. & Hatanaka, M. (1988) Processing of *gag* precursor polyprotein of human T-cell leukemia virus type 1 by virus-encoded protease. *J. Virol.* **62**, 3718–3728.

Seiki, M., Hattori, S., Hirayama, Y. & Yoshida, M. (1983) Human adult T-cell leukemia virus: complete nucleotide sequence of the provirus genome integrated in leukemia cell DNA. *Proc. Natl Acad. Sci. USA* **80**, 3618–3622.

*S. Daenke*
*Molecular Sciences Division,*
*Nuffield Department of Medicine,*
*University of Oxford,*
*John Radcliffe Hospital,*
*Oxford OX3 9DU, UK*
*Email: sdaenke@molbiol.ox.ac.uk*

# 315. *Bovine leukemia virus retropepsin*

## Databanks

*Peptidase classification: clan AA, family A2, MEROPS ID: A02.013*
*NC-IUBMB enzyme classification: none*
*Databank codes:*

| Species | SwissProt | PIR | EMBL (genomic) |
|---|---|---|---|
| Bovine leukemia virus | P10270 | S29357 | M10987: *gag* and *pol* genes |

*Notes*: EMBL accessions D00647 and K02120 do not include sequence for the *vprt* gene.

## Name and History

Bovine leukemia virus (BLV) is the etiologic agent of enzootic cattle lymphosarcoma and of its preclinical form, persistent lymphocytosis (for review see Burny *et al.*, 1987 and references cited therein). In 1983, a protein of 14 kDa, probably the BLV protease, was first visualized by immuno-precipitation of proteins from purified BLV (Mamoun *et al.*, 1983). The BLV protease gene was identified by Sagata *et al.* (1984a) and by Rice *et al.* (1985). The enzyme was then purified from viral particles by Yoshinaka *et al.* (1986) who carried out the N-terminal amino acid analysis and determined the total amino acid composition, confirming the previously proposed sequence. *Bovine leukemia virus retropepsin* is also named *BLV endopeptidase*, *BLV protease* or *BLV proteinase*. It contains the Asp-Thr-Gly sequence typical of the aspartic proteases (Sagata *et al.*, 1984a, 1985) and is inhibited by pepstatin A, an inhibitor of this class of enzyme (Katoh *et al.*, 1987). BLV retropepsin processes the BLV precursors Gag and Gag-Pro into mature structural proteins and functional protease (Ghysdael *et al.*, 1979; Mamoun *et al.*, 1983; Katoh *et al.*, 1989) enabling formation of new infectious virions. BLV retropepsin is of particular interest because it can be used as a model for the study of the human T-cell leukemia virus type I (HTLV-I) retropepsin (Chapter 314) (Sagata *et al.*, 1984a,b).

## Activity and Specificity

BLV retropepsin is able to hydrolyze its own N- and C-terminals (Andreánsky *et al.*, 1991; Bláha *et al.*, 1992) and to process its precursor Gag into structural proteins of the matrix (MA), capsid (CA) and nucleocapsid (NC) (Yoshinaka *et al.*, 1986; Katoh *et al.*, 1991; Ménard *et al.*, 1993). However, its role in the maturation of reverse transcriptase has yet to be elucidated. The enzyme is also able to cleave murine Moloney leukemia virus (Yoshinaka *et al.*, 1986); Gazdar murine sarcoma virus (Katoh *et al.*, 1987) and HTLV-I (Ménard *et al.*, 1993) Gag precursors into structural proteins. It also recognizes the N-terminal cleavage site of HTLV-I retropepsin and substrates of murine leukemia virus retropepsin, but only a few peptide substrates of the HIV-1 retropepsin (Chapter 310) (Bláha *et al.*, 1992).

The maturation of Gag precursors by the protease is currently studied by western blotting, while the correct hydrolysis of synthetic peptides is determined by HPLC (see references cited above). Assays are most conveniently made with radiolabeled peptide substrates using thin-layer electrophoresis on cellulose plates as described by Ménard *et al.* (1993). The biochemical properties of BLV retropepsin have commonly been investigated using a synthetic decapeptide as substrate (Tyr-Asp-Pro-Pro-Ala-Ile-Leu↓Pro-Ile-Ile), which has a sequence corresponding to the hypothetical BLV

(CA/NC) cleavage site at Leu109+Pro110 in the Gag precursor (Rice *et al.*, 1985), and which is cleaved at the expected site. The protease has been shown to be active from 0 to 80°C with a temperature optimum of 40°C, and from pH 3.5 to pH 7.5 with a maximum activity between pH 5 and 6. Activity is stimulated at high salt concentration (1–2 M NaCl or ammonium sulfate). In contrast, using a natural substrate such as an HTLV-I recombinant Gag precursor, the activity was found to be higher at low salt concentration (0.5 M NaCl) (Ménard *et al.*, 1993).

Hydrolysis of the peptide substrate described above is inhibited by pepstatin A with an $IC_{50}$ in the range of $3–5 \times 10^{-4}$ M and by cerulenin ($IC_{50}$ approximately $5 \times 10^{-3}$ M). The statin residue (Sta), the reduced bond between the two amino acids at the cleavage site, and pipecolic acid and azetidine residues also inhibit BLV retropepsin activity. Potent inhibition is observed with the BLV retropepsin inhibitor Tyr-Asp-Pro-Pro-Ala-Ile-Sta-Ile-Ile ($IC_{50}$ approximately $5 \times 10^{-7}$ M). The HTLV-I retropepsin inhibitors Ac-Pro-Glu-Val-Leu r Pro-Val-Met (r: reduced bond), Ac-Pro-Glu-Val-Leu-pipecolic acid-Val-Met, Ac-Pro-Glu-Val-Leu-azetidine-Val-Met, Pro-Glu-Val-Sta-Ala-Leu and Gly-Val-Leu-Tyr-Sta-Glu-Ala inhibit BLV proteolytic activity with $IC_{50}$ values of 10, 15, 250, 100 and 100 μM, respectively (Ménard *et al.*, 1994). DAN and 1,2-epoxy-3-(*p*-nitrophenoxy)propane (EPNP) inactivate cellular aspartic protease activities. Incubation of BLV retropepsin with DAN leads to around 90% inactivation of enzyme activity in the presence of cupric ions, whereas EPNP does not affect activity (Katoh *et al.*, 1989).

## Structural Chemistry

BLV retropepsin is a dimeric enzyme of about 28 kDa (Katoh *et al.*, 1989) composed of two identical subunits of 126 residues (14 kDa) devoid of cysteine and histidine residues (Sagata *et al.*, 1984a, 1985; Rice *et al.*, 1985) and having a pI of 10.2. The exact three-dimensional structure of this enzyme has yet to be determined. A model of the protease has been proposed in which the structure is similar to that established for other retroviral aspartic proteases. The two identical subunits of the enzyme mirror each other. BLV retropepsin contains essentially β sheets and one α helix. According to this model, the ten C-terminal residues do not appear to be essential for the enzyme activity; they form tails outside the catalytic site. Similar observations have been made on the HTLV-I protease (Précigoux *et al.*, 1993, 1994). As these two retroviruses are closely related (Sagata *et al.*, 1984b, 1985) the role of these ten extra amino acids residues is puzzling. They may perhaps confer a functional advantage on the rate of enzymatic activity or play some part in recognition of a privileged sequence of a precursor polyprotein.

## Preparation

BLV retropepsin was first purified from viral particles under denaturing conditions (60 mg of purified BLV yielded 35 μg of homogeneous protein) (Yoshinaka *et al.*, 1986), while under nondenaturing conditions, 20 mg of virus yielded 40 μg of purified protein (Ménard *et al.*, 1993). The enzyme has also been expressed in *Escherichia coli* as insoluble cytoplasmic inclusion bodies and purified in milligram quantities (Andreánsky *et al.*, 1991). Active enzymes with (Bláha *et al.*, 1992) or without (Précigoux *et al.*, 1993) the ten C-terminal residues have also been prepared by solid-phase peptide synthesis.

## Biological Aspects

The viral protease is encoded in an independent open reading frame located between the *gag* and *pol* genes (Sagata *et al.*, 1984a). A ribosomal frameshift is required for the synthesis of the Gag-Pol precursor polyprotein (Hatfield *et al.*, 1989), which is further cleaved into functional protease. Maturation of BLV reverse transcriptase by the enzyme has yet to be demonstrated. BLV protease is also responsible for the correct processing of its Gag precursor polyprotein into structural proteins of MA (p15), CA (p24) and NC (p12) (Ghysdael *et al.*, 1979; Mamoun *et al.*, 1983). It further processes p15 to generate MA (p10), a short peptide of seven amino acid residues, and p4 (Katoh *et al.*, 1991). To date, no cellular proteins have been identified as substrates of the enzyme.

BLV retropepsin is active as a dimeric complex. Monomers of the enzyme readily reassemble into dimer (Katoh *et al.*, 1989), although the mechanism of dimerization remains to be elucidated. The enzyme also remains active without the ten C-terminal residues indicating that they are not essential for activity.

## Distinguishing Features

BLV retropepsin, like other retroviral proteases, belongs to the group of the, aspartic proteases (clan AA), but being dimeric (Katoh *et al.*, 1989) contrasts with the cellular aspartic proteases, which are monomeric.

Unlike its close relative HIV-1 retropepsin, the BLV enzyme does not contain Cys or His residues (Hatanaka & Nam, 1989; Sagata *et al.*, 1984a,b). To date, no antisera or monoclonal antibodies to the enzyme have been described.

## Further Reading

For a comparative review see Précigoux *et al.* (1994).

## References

Andreánsky, M., Hrusková-Heidingsfeldová, O., Sedlácek, J., Konvalinka, J., Bláha, I., Jecmen, P., Horejsi, M., Strop, P. & Fábry, M. (1991) High-level expression of enzymatically active bovine leukemia virus proteinase in *E. coli. FEBS Lett.* **287**, 129–132.

Bláha, I., Tözér, J., Kim, Y., Copeland, T.D. & Oroszlan, S. (1992) Solid phase synthesis of the proteinase of bovine leukemia virus. *FEBS Lett.* **309**, 389–393.

Burny, A., Cleuter, Y., Kettman, R., Mammerickx, G., Marbaix, G., Portetelle, D., Van der Broeke, A., Willems, L. & Thomas, R. (1987) Bovine leukemia: facts and hypotheses derived from the study of an infectious cancer. *Cancer Surveys* **6**, 139–159.

Ghysdael, J., Kettman, R. & Burny, A. (1979) Translation of bovine leukemia virus virion RNAs in heterologous protein-synthesizing systems. *J. Virol.* **29**, 1087–1098.

Hatanaka, M. & Nam, S.H. (1989) Identification of HTLV-I gag protease and its sequential processing of the gag gene product. *J. Cell. Biochem.* **40**, 15–30.

Hatfield, D., Feng, Y.X., Lee, B.J., Rein, A., Levin, J.G. & Oroszlan, S. (1989) Chromatographic analysis of the aminoacyl-tRNAs which are required for translation of codons at and around the ribosomal frameshift sites of HIV, HTLV-I and BLV. *Virology* **173**, 736–742.

Katoh, I., Yasunaga, T., Ikawa, Y. & Yoshinaka, Y. (1987) Inhibition of retroviral protease activity by an aspartyl proteinase inhibitor. *Nature* **329**, 654–656.

Katoh, I., Ikawa, Y. & Yoshinaka, Y. (1989) Retrovirus protease characterized as a dimeric aspartic proteinase. *J. Virol.* **63**, 2226–2232.

Katoh, I., Kyushiki, H., Sakamoto, Y., Ikawa, Y. & Yoshinaka, Y. (1991) Bovine leukemia virus matrix-associated protein MA(p15): further processing and formation of a specific complex with the dimer of the 5′-terminal genomic RNA fragment. *J. Virol.* **65**, 6845–6855.

Mamoun, R.Z., Astier, T., Guillemain, B. & Duplan, J.F. (1983) Bovine lymphosarcoma: expression of BLV related proteins in cultured cells. *J. Gen. Virol.* **64**, 1895–1905.

Ménard, A., Mamoun, R.Z., Geoffre, S., Castroviejo, M., Raymond, S., Précigoux, G., Hospital, M. & Guillemain, B. (1993) Bovine leukemia virus: purification and characterization of the aspartic protease. *Virology* **193**, 680–689.

Ménard, A., Lénard, R., Llido, S., Geoffre, S., Picard, P., Berteau, F., Précigoux, G., Hospital, M. & Guillemain, B. (1994) Inhibition of activity of the protease from bovine leukemia virus. *FEBS Lett.*

**346**, 268–272.

Précigoux, G., Geoffre, S., Léonard, R., Llido, S., Dautant, A., Langloid, D'Estaintot, B., Picard, P., Ménard, A., Guillemain, B. & Hospital, M. (1993) Modelling, synthesis and biological activity of a BLV proteinase, made of (only) 116 amino acids. *FEBS Lett.* **326**, 237–240.

Précigoux, G, Llido, S., Léonard, R., Guillemain, R., Ménard, A., Candresse, T. & Le-Gall, O. (1994) Biological activity of native BLV proteinase and C-terminal truncated BLV and HTLV-I proteinase. *Curr. Top. Pept. Protein Res.* **1**, 441–452.

Rice, R., Stephens, M., Burny, A. & Gilden, R.V. (1985) The *gag* and *pol* genes of bovine leukemia virus: nucleotide sequence and analysis. *Virology* **142**, 357–377.

Sagata, N., Yasunaga, T. & Ikawa, Y. (1984a) Identification of a potential protease-coding gene in the genomes of bovine leukemia virus and human T-cell leukemia viruses. *FEBS Lett.* **178**, 79–82.

Sagata, N., Yasunaga, T., Oshishi, K., Tsuzuku-Kawamura, J., Onuma, M. & Ikawa, Y. (1984b) Comparison of the entire genomes of bovine leukemia virus and human T-cell leukemia virus and characterization of their unidentified open reading frame. *EMBO J.* **3**, 3231–3237.

Sagata, N., Yasunaga, T., Tsuzuku-Kawamura, J., Ohishi, K., Ogawa, Y. & Ikawa, Y. (1985) Complete nucleotide sequence of the genome of bovine leukemia: its evolutionary relationship to other retroviruses. *Proc. Natl Acad. Sci. USA* **82**, 677–681.

Yoshinaka, Y., Katoh, I., Copeland, T.D., Smythers, G.W. & Oroszlan, S. (1986) Bovine leukemia virus protease: purification, chemical analysis, and *in vitro* processing of *gag* precursor polyproteins. *J. Virol.* **57**, 826–832.

*Armelle Ménard*
*I.N.S.E.R.M. Unité 328, Structures et Fonctions des*
*Rétrovirus Humains,*
*Fondation Bergonié, Université de Bordeaux II,*
*229 cours de l'Argonne, F-33076 Bordeaux Cedex, France*
*Email: u328@bordeaux.inserm.fr*

*Bernard Guillemain*
*I.N.S.E.R.M. Unité 328, Structures et Fonctions des*
*Rétrovirus Humains,*
*Fondation Bergonié, Université de Bordeaux II,*
*229 cours de l'Argonne, F-33076 Bordeaux Cedex, France*

# 316. *Mason–Pfizer monkey virus retropepsin*

## Databanks

*Peptidase classification: clan AA, family A2, MEROPS ID: A02.009*
*NC-IUBMB enzyme classification: none*
*Databank codes:*

| Species | SwissProt | PIR | EMBL (genomic) |
| --- | --- | --- | --- |
| Simian Mason–Pfizer virus | P07570 | B25839 | M12349: complete genome |
| Simian retrovirus SRV-1 | P04024 | A03953 | M11841: complete genome |
| Simian retrovirus SRV-2 | P51518 | – | M16605: complete genome |

## Name and History

Mason–Pfizer monkey virus (M-PMV) encodes a proteinase (PR) that is responsible for the processing of virus-encoded polyprotein precursors into the structural proteins and viral enzymes (Sonigo *et al.*, 1986; Rhee & Hunter, 1987). M-PMV was the first primate retrovirus to be isolated from a spontaneous breast carcinoma of female rhesus monkey (*Macaca mulatta*) by Chopra & Mason (1970) at Pfizer lab. The morphogenetic properties of the virus have led to the classification of M-PMV as a D-type retrovirus (Kramarsky *et al.*, 1971; Fine & Schochetman, 1978). In 1992, M-PMV was classified on the basis of viral serotype as simian retrovirus type 3 (SRV-3) (Sommerfelt *et al.*, 1992).

Mutagenesis of the aspartic acid residue 26 of M-PMV PR demonstrated that the enzyme is an aspartic proteinase (Sommerfelt *et al.*, 1992). The cloning and purification of *Mason–Pfizer monkey virus (M-PMV) retropepsin* (Hrusková-Heidingsfeldová *et al.*, 1995) showed that it exists in two forms, differing in molecular mass. The abbreviations p17 and p12 were used for the 17 kDa and 12 kDa forms of M-PMV proteinase, respectively.

## Activity and Specificity

The analysis of natural cleavage sites in the M-PMV polyprotein precursors has revealed a preference of the proteinase for Met+Ala, Tyr+Trp (Gly, Ala) and Phe+Pro sequences around the scissile bond. In the P2 and P2′ positions, the proteinase prefers small hydrophobic residues. In specificity and activity, M-PMV PR is closer to avian myeloblastosis virus retropepsin (Chapter 313) than to HIV-1 retropepsin (Chapter 310). Statine-type inhibitors (e.g. Phe-Tyr-Val-Phe-Sta-Ala-Met-Thr-NH$_2$) based on the cleavage sites in the MAV Gag polyprotein inhibit the enzyme in the nanomolar range (Hrusková-Heidingsfeldová *et al.*, 1995).

Assay of activity can be performed with the peptide substrate ATPQVY+F(NO$_2$)VRKA as described by Hrusková-Heidingsfeldová *et al.* (1995). This substrate is not commercially available. The pH optimum of the proteinase is about 4.5. The assay system typically includes 2 M sodium chloride and 0.1% ME.

## Structural Chemistry

The proteinase is synthesized as part of the Gag-Pro and Gag-Pro-Pol polyproteins and requires autocatalytic cleavage and dimerization to achieve the proteolytic activity. Each monomer contributes one of the two catalytic Asp (in the motif Asp-Thr-Gly) to the active site of the enzyme. When expressed in bacteria as a 26 kDa precursor, the proteinase is first processed at the N-terminus, yielding a proteolytically active 17 kDa protein. This initial cleavage is followed by a much slower self-processing that leads to emergence of the active 12 kDa form of the retropepsin (Hrusková-Heidingsfeldová *et al.*, 1995). N-terminal processing sites of both forms of M-PMV PR are identical. The X-ray crystallographic structure has not been obtained yet. The pIs of p17 and p12 are 9.15 and 8.05, respectively.

## Preparation

The recombinant M-PMV enzyme has been expressed in *Escherichia coli* as a 26 kDa proteinase precursor accumulated in the insoluble cytoplasmic inclusion bodies. Enzyme was purified on a scale of tens of milligrams (Hrusková-Heidingsfeldová *et al.*, 1995).

## Biological Aspects

M-PMV proteinase is encoded by the second open reading frame (ORF), denoted *pro*, which overlaps gag at its 5′ end and *pol* at its 3′ end. The proteinase is expressed by a frameshifting mechanism. Two frameshifts occur within the *gag/pro* and *pro/pol* overlaps to generate a fused polyprotein Gag-Pro (Pr95) and Gag-Pro-Pol (Pr180) (Sonigo *et al.*, 1986).

M-PMV is characterized by the self-assembly of the Gag-containing precursors into the intracytoplasmic particles (ICAPs) within the infected cell cytoplasm. During or shortly after particle release, the viral proteinase is activated and cleaves the viral polyproteins. This process in M-PMV is tightly controlled, since proteinase precursors preassembled into ICAPs are not activated while the capsids remain in the cytoplasm (Rhee & Hunter, 1987). Heterologously expressed Gag precursors spontaneously assemble *in vitro* within bacteria (Kliková *et al.*, 1995) and in reticulocyte lysates (Sakalian *et al.*, 1996). The ability to produce immature capsid *in vitro* could facilitate the analysis of M-PMV inhibitors of retrovirus replication, including those of proteinase.

M-PMV PR also converts the cell-associated transmembrane protein (TM, gp 22) to a virus-associated gp 20 (Sommerfelt *et al.*, 1992; Strambio-de-Castillia *et al.*, 1992; Brody *et al.*, 1994). While the inactivation of M-PMV PR does not affect the assembly of noninfectious, immature virus particles, the maturation process is abolished and noninfectious immature particles are released (Sommerfelt *et al.*, 1992). In addition to tumor, M-PMV particles were observed in the thymus and lymph nodes of nonhuman primates (Chopra & Mason, 1970; Jensen *et al.*, 1970). M-PMV is not oncogenic. Infection of macaque species with M-PMV causes an acquired immune deficiency syndrome (SAIDS) (Fine *et al.*, 1975; Daniel *et al.*, 1984, 1985; Bryant *et al.*, 1986).

## Distinguishing Features

M-PMV retropepsin specifically cleaves sequences spanning the cleavage sites in viral polyprotein precursors. Rabbit polyclonal antibodies against M-PMV recognize the proteinase. Antibodies have been produced in E. Hunter's laboratory at the University of Alabama Birmingham, USA, but are not commercially available.

## Further Reading

For further reading see Hrusková-Heidingsfeldová *et al.* (1995).

## References

Brody, B.A., Rhee, S.S. & Hunter, E. (1994) Postassembly cleavage of a retroviral glycoprotein cytoplasmic domain removes a necessary incorporation signal and activates fusion activity. *J. Virol.* **7**, 4620–4627.

Bryant, M.L., Gardner, M.B., Marx, P.A., Maul, D.H., Lerche, N.W., Osborn, K.G., Lowenstein, L.J., Bodgen, A., Arthur, L.O. & Hunter, E. (1986) Immunodeficiency in rhesus monkeys associated with the original Mason-Pfizer monkey virus. *J. Natl Cancer Inst.* **77**, 957–965.

Chopra, H.C. & Mason, M.M. (1970) A new virus in a spontaneous mammary tumor of a rhesus monkey. *Cancer Res.* **30**, 2081–2086.

Daniel, M.D., King, N.W., Letwin, N.L., Hunt, R.D., Sehgal, P.K. & Desrosiers, R.C. (1984) A new type D retrovirus isolated from macaques with an immunodeficiency syndrome. *Science* **223**, 602–605.

Daniel, M.D., Letvin, N.L., King, N.W., Kannagi, M., Sehgal, P.K., Hunt, R.D., Kanki, P.J., Essex, M. & Desrosiers, R.C. (1985) Isolation of T-cell tropic HTLV-III like retrovirus from macaques. *Science* **228**, 1201–1204.

Fine, D. & Schochetman, G. (1978) Type D primate retroviruses: a review. *Cancer Res.* **38**, 3123–3139.

Fine, D.L., Landon, J.C., Pienta, R.J., Kubicek, M.T., Valerio, M.J., Loeb, W.F. & Chopra, H.C. (1975) Responses of infant rhesus monkeys to inoculation with Mason–Pfizer monkey virus materials. *J. Natl Cancer Inst.* **54**, 651–658.

Hruskova-Heidingsfeldová, O., Andreansky, M., Fábry, M., Bláha, I., Strop, P. & Hunter, E. (1995) Cloning, bacterial expression, and characterization of the Mason-Pfizer monkey virus proteinase. *J. Biol. Chem.* **25**, 15053–15058.

Jensen, E.M., Zelljadt, I., Chopra, H.C. & Mason, M.M. (1970) Isolation and propagation of a new virus from a spontaneous mammary carcinoma of a rhesus monkey. *Cancer Res.* **30**, 2388–2393.

Kliková, M., Rhee, S.S., Hunter, E. & Ruml, T. (1995) Efficient *in vivo* and *in vitro* assembly of retroviral capsids from Gag precursor proteins expressed in bacteria. *J. Virol.* **69**, 1093–1098.

Kramarsky, B., Sarkar, N.A. & Moore, D.H. (1971) Ultrastructural comparison of a virus from a rhesus-monkey mammary carcinoma with four oncogenic RNA viruses. *Proc. Natl Acad. Sci. USA* **68**, 1603–1697.

Rhee, S.S. & Hunter, E. (1987) Myristylation is required for intracellular transport but not for assembly of D-type retrovirus capsids. *J. Virol.* **61**, 1045–1053.

Sakalian, M., Parker, S.D., Weldon, R.A., Jr. & Hunter, E. (1996) Synthesis and assembly of retrovirus Gag precursors into immature capsids *in vitro*. *J. Virol.* **70**, 3706–3715.

Sommerfelt, M.A., Petteway, S.R., Jr., Dreyer, G.B. & Hunter, E. (1992) Effect of retroviral proteinase inhibitors on Mason-Pfizer monkey virus maturation and transmembrane glycoprotein cleavage. *J. Virol.* **7**, 4220–4227.

Sonigo, P., Barker, C., Hunter, E. & Wain-Hobson, S. (1986) Nucleotide sequence of Mason-Pfizer monkey virus: an immunosuppressive D-type retrovirus. *Cell* **45**, 375–385.

Strambio-de-Castillia, C. & Hunter, E. (1992) Mutational analysis of the major homology region of Mason-Pfizer monkey virus by use of saturation mutagenesis. *J. Virol.* **66**, 7021–7032.

*Iva Pichová*
*Group of the Retroviral Proteinases, Department of Biochemistry,*
*Institute of Organic Chemistry and Biochemistry, Flemingovo n. 2,*
*166 10 Prague 6, Czech Republic*
*Email: iva.pichova@uochb.cas.cz*

# 317. Mouse mammary tumor virus retropepsin

## Databanks

*Peptidase classification: clan AA, family A2, MEROPS ID: A02.010*
*NC-IUBMB enzyme classification: none*
*ATCC entries: 45005 (plasmid p202, including retropepsin coding region)*
*Databank codes:*

| Species | Strain | SwissProt | PIR | EMBL (genomic) |
|---|---|---|---|---|
| Mouse mammary tumor virus | BR6 | P10271 | A45125 B26795 | M15122: complete genome |
| | C3H | – | – | L01464: 3′ end |
| | | – | – | M16766: mid-section of genome |
| | Jiayuguang | – | – | D16249: complete genome |

EMBL accession M30519 is part of the genome only and does not include the retropepsin.

## Name and History

Mouse mammary tumor viruses (MMTV) are endogenous and exogenous retroviruses, mostly milkborne, which often cause mammary carcinomas and sometimes T-cell lymphomas. The presence of a retropepsin in the MMTV genome was suggested by Jacks *et al.* (1987) and Moore *et al.* (1987). They found an open reading frame with the sequence Leu-Asp-Thr-Gly-Ala-Asp, which was homologous to the consensus active site of retroviral and some cellular aspartic proteases (Toh *et al.*, 1985). A viral protein (p13) showing sequence homology to other retropepsins was then purified from virions by HPLC, although the proteolytic activity of this protein was not shown (Hizi *et al.*, 1987). Evidence of MMTV retropepsin activity in lysed virus and partially purified protease-containing viral extracts was later reported (Menéndez-Arias *et al.*, 1992a). Active *mouse mammary tumor virus retropepsin* (p13) was then expressed in *Escherichia coli* and fully characterized (Menéndez-Arias *et al.*, 1992b). The *MMTV retropepsin* is commonly referred to as the *MMTV protease*.

## Activity and Specificity

Maturation sites in Gag and Gag-Pro polyproteins have been inferred by comparison of their primary structure derived from the nucleotide sequence of the MMTV genome (Jacks *et al.*, 1987; Moore *et al.*, 1987) and the N- and C-terminal sequences of proteins purified from the mature virus (Hizi *et al.*, 1987, 1989). Synthetic oligopeptides, mimicking the maturation sites found in the *gag*-related polyproteins are useful substrates to detect MMTV retropepsin activity (Menéndez-Arias *et al.*, 1992a,b). Assays are usually carried out in 0.25 M potassium phosphate buffer, pH 5.0, containing 1–2 M NaCl, 0.2 mM EDTA, 1 mM DTT, 1.25% glycerol and 50 µM substrate, and after incubation at 37°C, products are analyzed by HPLC. Under these conditions, the synthetic dodecapeptide LTFTF↓PVVFMRR, which mimics the cleavage site found at the N-terminal end of the capsid protein in MMTV is apparently the best substrate. Its $k_{cat}$ and $K_m$ values were $0.165 \text{ s}^{-1}$ and 17.0 µM, respectively (Menéndez-Arias *et al.*, 1992b).

The pH optimum of the enzyme is 4.5, although it retains more than 80% activity between pH 4.0 and 6.0. Its activity drops below 15% at pH lower than 3.0 and higher than 7.5 (Menéndez-Arias *et al.*, 1992b). The MMTV retropepsin is more sensitive to salt concentration than other retroviral proteases. Maximum activity is observed at 1.5 M NaCl concentration and drops below 25% when the NaCl concentration is higher than 2.5 M (Menéndez-Arias *et al.*, 1992b).

A specific phosphinate-type inhibitor of HIV-1 retropepsin (Chapter 310) known as Compound 3 (Grobelny *et al.*, 1990) is the most effective inhibitor of MMTV retropepsin, among those tested so far (Menéndez-Arias *et al.*, 1992b, 1993). Its $IC_{50}$ is 1.1 µM in the standard assay conditions. Pepstatin A is a poor inhibitor of MMTV retropepsin ($IC_{50} = 90$ µM).

## Structural Chemistry

MMTV retropepsin is a homodimeric enzyme, formed by two 115 amino acid polypeptide chains (Hizi *et al.*, 1987; Menéndez-Arias *et al.*, 1992a,b). The $M_r$ of each subunit is 13 500, as estimated by SDS-PAGE (Menéndez-Arias *et al.*, 1992b).

## Preparation

The MMTV retropepsin has been partially purified from virus (strain C3H) obtained from culture supernatants of infected cell lines such as MMT 060562 (ATCC CCL51) or Mm5mt/c1, but yields are very low (Menéndez-Arias *et al.*, 1992a). The protease has been expressed in *Escherichia coli* using the vector pGEX-2T (Menéndez-Arias *et al.*, 1992b). A chimeric protein formed by the glutathione *S*-transferase of *Schistosoma japonicum*, a hexapeptide that contains a thrombin cleavage site and the MMTV retropepsin was obtained. Affinity chromatography on a glutathione-Sepharose 4B column was used to isolate the chimeric protein. After thrombin cleavage, the glutathione *S*-transferase and the retropepsin were separated by gel-filtration chromatography on a Sephadex G-75 column. The overall yield of the protease purification procedure was about 1 mg of protease per liter of culture (Menéndez-Arias *et al.*, 1992b).

## Biological Aspects

MMTV maturation involves the accumulation of immature cores (A particles) in the cytoplasm of infected cells. The immature cores are formed by the unprocessed polypeptides Gag (77 kDa), Gag-Pro (110 kDa) and Gag-Pro-Pol (160 kDa), which at a later stage are cleaved by the viral retropepsin to produce an infective mature virion (B particle). Seven retropepsin cleavage sites have been identified in the polyprotein precursors (Hizi *et al.*, 1987, 1989; Menéndez-Arias *et al.*, 1992b). Gag, Gag-Pro and Gag-Pro-Pol derive from translation of three out-of-phase overlapping reading frames: *gag, pro and pol*. The MMTV retropepsin coding region is located in the *pro* open reading frame of the viral genome (Jacks *et al.*, 1987; Moore *et al.*, 1987). Ribosomal frameshifting in the −1 direction renders Gag-Pro and Gag-Pro-Pol after one or two frameshift events, respectively. The efficiency of frameshifting is 23% in the *gag-pro* overlap and 8% in the *pro-pol* overlap (Jacks *et al.*, 1987). Gag-Pro and Gag-Pro-Pol contain the protease sequence. A mature form of the MMTV retropepsin arises from autocatalytic processing of the Gag-Pro precursor, and comprises the 115 amino acids forming the C-terminal portion of Gag-Pro (Hizi *et al.*, 1987). MMTV protease forms derived from Gag-Pro-Pol have not been characterized so far. The mechanism of activation of the MMTV retropepsin is not known.

## Distinguishing Features

The substrate specificity and enzymatic properties of MMTV retropepsin revealed important differences from other retroviral proteases. Thus, synthetic oligopeptides such as HIV-1 matrix protein (MA)/capsid protein (CA), murine leukemia virus p12/CA or human T-cell leukemia virus-1 MA/CA are cleaved more slowly by the MMTV retropepsin than by the retropepsins of HIV-1 (Chapter 310) or Moloney murine leukemia virus (Chapter 318) (Menéndez-Arias *et al.*, 1992b, 1993). In addition, several specific inhibitors of HIV-1 retropepsin were found to be ineffective with the MMTV retropepsin (Menéndez-Arias *et al.*, 1993).

Polyclonal antiserum towards the C-terminal region of the retropepsin was obtained after immunization of rabbits with the HPLC-purified peptide CKDIKVRLMTDSPDDSQDL, coupled to keyhole limpet hemocyanin (Menéndez-Arias et al., 1992c).

### References

Grobelny, D., Wondrak, E.M., Galardy, R.E. & Oroszlan, S. (1990) Selective phosphinate transition-state analogue inhibitors of the protease of human immunodeficiency virus. *Biochem. Biophys. Res. Commun.* **169**, 1111–1116.

Hizi, A., Henderson, L.E., Copeland, T.D., Sowder, R.C., Hixson, C.V. & Oroszlan, S. (1987) Characterization of mouse mammary tumor virus *gag-pro* gene products and the ribosomal frameshift site by protein sequencing. *Proc. Natl Acad. Sci. USA* **84**, 7041–7045.

Hizi, A., Henderson, L.E., Copeland, T.D., Sowder, R.C., Krutzsch, H.C. & Oroszlan, S. (1989) Analysis of *gag* proteins from mouse mammary tumor virus. *J. Virol.* **63**, 2543–2549.

Jacks, T., Townsley, K., Varmus, H.E. & Majors, J. (1987) Two efficient ribosomal frameshifting events are required for synthesis of mouse mammary tumor virus *gag*-related polyproteins. *Proc. Natl Acad. Sci. USA* **84**, 4298–4302.

Majors, J.E. & Varmus, H.E. (1981) Nucleotide sequences at host-proviral junctions for mouse mammary tumour virus. *Nature (Lond.)* **289**, 253–258.

Menéndez-Arias, L., Risco, C. & Oroszlan, S. (1992a) Isolation and characterization of $\alpha_2$-macroglobulin-protease complexes from purified mouse mammary tumor virus and culture supernatants from virus-infected cell lines. *J. Biol. Chem.* **267**, 11392–11398.

Menéndez-Arias, L., Young, M. & Oroszlan, S. (1992b) Purification and characterization of the mouse mammary tumor virus protease expressed in *Escherichia coli*. *J. Biol. Chem.* **267**, 24134–24139.

Menéndez-Arias, L., Risco, C., Pinto da Silva, P. & Oroszlan, S. (1992c) Purification of immature cores of mouse mammary tumor virus and immunolocalization of protein domains. *J. Virol.* **66**, 5615–5620.

Menéndez-Arias, L., Gotte, D. & Oroszlan, S. (1993) Moloney murine leukemia virus protease: bacterial expression and characterization of the purified enzyme. *Virology* **196**, 557–563.

Moore, R., Dixon, M., Smith, R., Peters, G. & Dickson, C. (1987) Complete nucleotide sequence of a milk-transmitted mouse mammary tumor virus: two frameshift suppression events are required for translation of *gag* and *pol*. *J. Virol.* **61**, 480–490.

Toh, H., Ono, M., Saigo, K. & Miyata, T. (1985) Retroviral protease-like sequence in the yeast transposon Ty1. *Nature (Lond.)* **315**, 691.

**Luis Menéndez-Arias**
*Centro de Biología Molecular*
*'Severo Ochoa',*
*CSIC-Universidad Autónoma*
*de Madrid,*
*28049 Cantoblanco (Madrid), Spain*
*Email: lmenendez@mvax.cbm.uam.es*

**József Tözsér**
*Department of Biochemistry,*
*University Medical School*
*of Debrecen,*
*H-4012 Debrecen, Hungary*
*Email: tozser@indi.biochem.dote.hu*

**Stephen Oroszlan**
*Laboratory of Molecular*
*Virology and Carcinogenesis,*
*ABL-Basic Research Program,*
*NCI-Frederick Cancer Research and*
*Development Center,*
*Frederick, MD 21702–1201, USA*
*Email: oroszlans@mail.fcrdc.gov*

# 318. Moloney murine leukemia virus retropepsin

## Databanks

*Peptidase classification: clan AA, family A2, MEROPS ID: A02.008*
*NC-IUBMB enzyme classification: none*
*Databank codes:*

| Species | Strain | SwissProt | PIR | EMBL (genomic) |
| --- | --- | --- | --- | --- |
| AKV murine leukemia virus | – | P03356 | A03957 | J01998: complete genome |
| Avian reticuloendotheliosis virus | – | P03360 | A03959 | K02537: 3' end of genome |
| | | | | X01455: 3' end of genome |
| Baboon endogenous virus | – | P10272 | JT0261 | D00088: complete genome |
| | | | | D10032: complete genome |
| | | | | J02034: *gag* and *pol* genes |

| Species | Strain | SwissProt | PIR | EMBL (genomic) |
|---|---|---|---|---|
| Feline leukemia virus | – | P10273 | – | M16550: complete genome<br>X05470: complete genome<br>K01803: 5′ end of genome<br>L06140: 3′ end of genome<br>M18247: complete genome |
| Gibbon leukemia virus | | P21414 | B32595 | M26927: complete genome |
| Friend murine leukemia virus | 57 | P26810 | – | X02794: complete genome |
| | FB29 | P26809 | – | Z11128: complete genome |
| | pvc-211 | P26808 | S35475 | M93134: complete genome |
| *Mesocricetus auratus* SIGN G-element | – | – | – | U09104: C-type troviral sequence |
| Moloney murine leukemia virus | – | P03355 | A03956 | J02255: complete genome<br>M10413: mid-section of pol gene<br>M87550: complete genome |
| | 3-1R | – | A46311 | M32803: mid-section of pol gene |
| | CAS-BR-E | – | – | X57540: complete genome |
| | type C | – | – | X94150: complete genome |
| | WN1802N | – | – | K01203: gag-pol region |
| | WN1802B | – | – | K01204: gag-pol region |
| | pseudogene | – | – | L08395 |
| Radiation murine leukemia virus | Kaplan | P31795 | A42743 | M93052: 3′ end of genome |
| | VL3 | P11227 | B26183 | K03363: complete genome |
| Rat leukemia virus | – | – | – | M77193: V-ras oncogene |
| Rat sarcoma virus | – | – | – | M76757: transduction protein<br>M77194: complete genome |

The following EMBL accession numbers are fragments of the genomes that do not include the retropepsins: K00021, X59305, J02263, J02266 and V01185.

## Name and History

Early evidence of the proteolytic processing of murine leukemia virus (MLV) Gag polyproteins came from immunoprecipitation studies with Rauscher and Moloney MLV (Arcement *et al*., 1976; Barbacid *et al*., 1976; Jamjoon *et al*., 1977). High molecular mass precursor polyproteins were immunoprecipitated with antisera made against proteins of the mature virus. The presence of a proteolytic factor in detergent-treated extracts of MLV was then reported by Yoshinaka & Luftig (1977a,b), who described the morphological conversion of 'immature' viral cores to a 'mature' form after addition of a partially purified P65-70 proteolytic factor (Yoshinaka & Luftig, 1977b, 1978). This protease was later isolated from Moloney MLV in large quantities and its proteolytic activity was shown using a 65 kDa Gag precursor polyprotein as substrate (Yoshinaka *et al*., 1985a). The N- and C-terminal amino acid sequences of the retropepsin were then determined and aligned with the primary structure deduced from the DNA sequence of Moloney MLV. It was shown that the **Moloney murine leukemia virus retropepsin** was encoded by the *gag-pol* gene and synthesized through suppression of the UAG termination codon found at the end of *gag* (Yoshinaka *et al*., 1985a). This enzyme is also known as **MLV retropepsin** or **MLV protease** (or proteinase).

## Activity and Specificity

Viral precursor polyproteins (e.g. the uncleaved 65 kDa precursor from Gazdar murine sarcoma virus) were used as substrates in assays performed with retropepsin isolated from virions (Yoshinaka *et al*., 1985a,b). Cleavage sites in Gag and Gag-Pol have been identified (Oroszlan *et al*., 1978; Henderson *et al*., 1984). Synthetic peptides mimicking the maturation sites found in those polyproteins were shown to be useful substrates for Moloney MLV retropepsin (Menéndez-Arias *et al*., 1993). Their $K_m$ values ranged from 16.6–98.0 μM, while the corresponding $k_{cat}$ values ranged from 0.07–0.71 s$^{-1}$ (Menéndez-Arias *et al*., 1993). Proteolytic assays were done in 0.25 M potassium phosphate buffer, pH 6.0, containing 3.0 M NaCl and 0.15 mM substrate, and after incubation at 37°C products were analyzed by HPLC. Its optimum pH is 6.0 and its maximum activity is observed at 3.0 M NaCl concentration (Menéndez-Arias *et al*., 1993).

Moloney MLV retropepsin is able to cleave efficiently synthetic peptides representing maturation sites found in polyproteins of other retroviruses (Menéndez-Arias *et al*., 1993). For example, HIV-1 retropepsin (Chapter 310) substrate Val-Ser-Gln-Asn-Tyr+Pro-Ile-Val-Gln is cleaved by the Moloney MLV retropepsin, with $k_{cat}$ and $K_m$ values of 0.55 s$^{-1}$ and 261 μM, respectively. Moloney MLV and HIV retropepsins showed similar substrate specificity as revealed by kinetic analysis using Val-Ser-Gln-Asn-Tyr+Pro-Ile-Val-Gln analogs with single amino acid substitutions in the P4-P3′ positions (Menéndez-Arias *et al*., 1994). Differences between the enzymes were observed with peptides having large hydrophobic residues at P4 or P2, which were much better substrates of the MLV retropepsin. Site-directed mutagenesis studies suggest that His37, Val39 and Ala57 at the substrate-binding pocket of Moloney MLV retropepsin are major determinants of the differences in substrate specificity between MLV and HIV retropepsins (Menéndez-Arias *et al.*, 1995). Shorter analogs of Val-Ser-Gln-Asn-Tyr-Pro-Ile-Val-Gln that are hydrolyzed by Moloney MLV retropepsin (Menendez-Arias *et al.*, 1994) are available from Bachem (see Appendix 2 for full names and addresses of suppliers).

Several HIV-1 retropepsin inhibitors have been reported to be effective on MLV retropepsins, with 50% inhibitory concentrations of less than 1 μM, as demonstrated using purified protease (Menéndez-Arias *et al.*, 1993) as well as in cell culture assays (Black *et al.*, 1993; Lai *et al.*, 1993).

## Structural Chemistry

Moloney MLV retropepsin is a homodimeric enzyme (Yoshinaka & Luftig, 1980) formed by two polypeptides of 125 amino acids with calculated $M_r = 13\,315$ (Yoshinaka *et al.*, 1985a). A molecular model of Moloney MLV retropepsin based on the crystal structure of HIV-1 retropepsin has been described (Menéndez-Arias *et al.*, 1994).

## Preparation

The Moloney MLV retropepsin has been purified from viral particles (Yoshinaka *et al.*, 1985a), although yields were very low – less than 0.3 μg of protease per mg of virus. Moloney MLV retropepsin has been expressed in *Escherichia coli* fused to TrpE (Calkins & Luftig, 1989) or glutathione *S*-transferase of *Schistosoma japonicum* (Menéndez-Arias *et al.*, 1993). The retropepsin was purified to homogeneity from bacterial extracts expressing the chimeric enzyme formed by glutathione *S*-transferase and MLV retropepsin. The purification yield was around 0.2 mg of protease per liter of culture (Menéndez-Arias *et al.*, 1993).

## Biological Aspects

The Moloney MLV retropepsin coding region is located at the 5′ end of the *pol* gene, and its first four amino acids overlap with the 3′ end of the *gag* gene. The fifth amino acid residue is glutamine, which is inserted by suppression of the amber termination codon at the *gag-pol* junction (Yoshinaka *et al.*, 1985a). The efficiency of the translational readthrough is about 4–10% (Jamjoon *et al.*, 1977).

The precursor polyproteins Gag-Pol (180 kDa) and Gag (65 kDa) are cleaved by the Moloney MLV retropepsin to render the mature viral proteins of the virus. Six cleavage sites have been identified in these two polyproteins (see review by Oroszlan & Luftig, 1990). In addition, during the final stage of Env protein maturation, the Moloney MLV retropepsin cleaves the transmembrane intermediate protein Pr15(E) (or TM) to p15(E) and p2(E) (Katoh *et al.*, 1985; Crawford & Goff, 1985; Schultz & Rein, 1985; Menéndez-Arias *et al.*, 1993). This cleavage activates the membrane fusion capability of the MLV Env protein and appears to be essential for infectivity (Rein *et al.*, 1994). The mechanism leading to the activation of Moloney MLV retropepsin is not known, although dimerization of Gag-Pol precursors is likely to play an important role in the process.

## Distinguishing Features

Moloney MLV and HIV retropepsins have similar enzymological properties. However, several specific inhibitors of HIV-1 retropepsin are less effective with the MLV proteases (Black *et al.*, 1993; Menéndez-Arias *et al.*, 1993), and differences in the cleavage rates of various synthetic peptides were reported (Menéndez-Arias *et al.*, 1994).

A polyclonal antiserum against the C-terminal sequence of the Moloney MLV retropepsin has been described (Menéndez-Arias *et al.*, 1993).

## Related Peptidases

MLV retropepsins are highly conserved and most of them share more than 95% identical amino acids with the Moloney MLV protease. Active feline LV retropepsin has been purified from virus (Yoshinaka *et al.*, 1985b). It shares 80% identical amino acids with Moloney MLV retropepsin. All of these enzymes appear to be very similar in terms of substrate specificity as suggested by comparison of the corresponding Gag cleavage site sequences in the different viruses.

## Further Reading

For further reading see Oroszlan & Luftig (1990). This review provides a detailed account of the historical events leading to the discovery of retropepsins, with one section dedicated to MLV retropepsin.

## References

Arcement, L.J., Karshin, W.L., Naso, R.B., Jamjoom, G. & Arlinghaus, R.B. (1976) Biosynthesis of Rauscher leukemia viral proteins: presence of p30 and envelope p15 sequences in precursor polyproteins. *Virology* **69**, 763–774.

Barbacid, M., Stephenson, J.R. & Aaronson, S.A. (1976) *gag* Gene of mammalian type-C RNA tumour viruses. *Nature* **262**, 554–559.

Black, P.L., Downs, M.B., Lewis, M.G., Ussery, M.A., Dreyer, G.B., Petteway, S.R., Jr. & Lambert, D.M. (1993) Antiretroviral activities of protease inhibitors against murine leukemia virus and simian immunodeficiency virus in tissue culture. *Antimicrob. Agents Chemother.* **37**, 71–77.

Calkins, P. & Luftig, R.B. (1989) Bacterial expression and activity of the Moloney murine leukemia virus proteinase. In: *Viral Proteinases as Targets for Chemotherapy* (Kräusslich, H.-G., Oroszlan, S. & Wimmer, E., eds). Cold Spring Harbor: Cold Spring Harbor Laboratory Press, pp. 113–116.

Crawford, S. & Goff, S.P. (1985) A deletion mutation in the 5′ part of the *pol* gene of Moloney murine leukemia virus blocks proteolytic processing of the *gag* and *pol* polyproteins. *J. Virol.* **53**, 899–907.

Henderson, L.E., Sowder, R., Copeland, T.D., Smythers, G. & Oroszlan, S. (1984) Quantitative separation of murine leukemia virus proteins by reversed-phase high-pressure liquid chromatography reveals newly described *gag* and *env* cleavage products. *J. Virol.* **52**, 492–500.

Jamjoom, G.A., Naso, R.B. & Arlinghaus, R.B. (1977) Further characterization of intracellular precursor polyproteins of Rauscher leukemia virus. *Virology* **78**, 11–34.

Katoh, I., Yoshinaka, Y., Rein, A., Shibuya, M., Odaka, T. & Oroszlan, S. (1985) Murine leukemia virus maturation: protease region required for conversion from 'immature' to 'mature' core form and for virus infectivity. *Virology* **145**, 280–292.

Lai, M.-H.T., Tang, J., Wroblewski, V., Dee, A.G., Margolin, N., Vlahos, C., Bowdon, B., Buckheit, R., Colacino, J. & Hui, K.Y. (1993) Impeded progression of Friend disease in mice by an inhibitor of retroviral proteases. *J. Acquir. Immune Defic. Syndr.* **6**, 24–31.

Menéndez-Arias, L., Gotte, D. & Oroszlan, S. (1993) Moloney murine leukemia virus protease: bacterial expression and characterization of the purified enzyme. *Virology* **196**, 557–563.

Menéndez-Arias, L., Weber, I.T., Soss, J., Harrison, R.W., Gotte, D. & Oroszlan, S. (1994) Kinetic and modeling studies of subsites S4–S3′ of Moloney murine leukemia virus protease. *J. Biol. Chem.* **269**, 16795–16801.

Menéndez-Arias, L., Weber, I.T. & Oroszlan, S. (1995) Mutational analysis of the substrate binding pocket of murine leukemia virus protease and comparison with human immunodeficiency virus proteases. *J. Biol. Chem.* **270**, 29162–29168.

Oroszlan, S. & Luftig, R.B. (1990) Retroviral proteinases. *Curr. Top. Microbiol. Immunol.* **157**, 153–185.

Oroszlan, S., Henderson, L.E., Stephenson, J.R., Copeland, T.D., Long, C.W., Ihle, J.N. & Gilden, R.V. (1978) Amino- and carboxyl-terminal amino acid sequences of proteins coded by *gag* gene of murine leukemia virus. *Proc. Natl Acad. Sci. USA* **75**, 1404–1408.

Rein, A., Mirro, J., Haynes, J.G., Ernst, S.M. & Nagashima, K. (1994) Function of the cytoplasmic domain of a retroviral transmembrane protein: p15E-p2E cleavage activates the membrane fusion capability of the murine leukemia virus Env protein. *J. Virol.* **68**, 1773–1781.

Schultz, A. & Rein, A. (1985) Maturation of murine leukemia virus *env* proteins in the absence of other viral proteins. *Virology* **145**, 335–339.

Yoshinaka, Y. & Luftig, R.B. (1977a) Murine leukemia virus morphogenesis: cleavage of P70 *in vitro* can be accompanied by a shift from a concentrically coiled internal strand ('immature') to a collapsed ('mature') form of the virus core. *Proc. Natl Acad. Sci. USA* **74**, 3446–3450.

Yoshinaka, Y. & Luftig, R.B. (1977b) Properties of a P70 proteolytic factor of murine leukemia viruses. *Cell* **12**, 709–720.

Yoshinaka, Y. & Luftig, R.B. (1978) Morphological conversion of 'immature' Rauscher leukaemia virus cores to a 'mature' form after addition of the P65–70 (*gag* gene product) proteolytic factor. *J. Gen. Virol.* **40**, 151–160.

Yoshinaka, Y. & Luftig, R.B. (1980) Physicochemical characterization and specificity of the murine leukaemia virus Pr65$^{gag}$ proteolytic factor. *J. Gen. Virol.* **48**, 329–340.

Yoshinaka, Y., Katoh, I., Copeland, T.D. & Oroszlan, S. (1985a) Murine leukemia virus protease is encoded by the *gag-pol* gene and is synthesized through suppression of an amber termination codon. *Proc. Natl Acad. Sci. USA* **82**, 1618–1622.

Yoshinaka, Y., Katoh, I., Copeland, T.D. & Oroszlan, S. (1985b) Translational readthrough of an amber termination codon during synthesis of feline leukemia virus protease. *J. Virol.* **55**, 870–873.

*Luis Menéndez-Arias*
*Centro de Biología Molecular 'Severo Ochoa',*
*CSIC-Universidad Autónoma de Madrid,*
*28049 Cantoblanco (Madrid), Spain*
*Email: lmenendez@mvax.cbm.uam.es*

*József Tözsér*
*Department of Biochemistry,*
*University Medical School of Debrecen,*
*H-4012 Debrecen, Hungary*
*Email: tozser@indi.biochem.dote.hu*

*Stephen Oroszlan*
*Laboratory of Molecular Virology and Carcinogenesis,*
*ABL-Basic Research Program,*
*NCI-Frederick Cancer Research and Development Center,*
*Frederick, MD 21702–1201, USA*
*Email: oroszlans@mail.fcrdc.gov*

# *319. Feline immunodeficiency virus retropepsin*

## Databanks

*Peptidase classification: clan AA, family A2, MEROPS ID: A02.007*
*NC-IUBMB enzyme classification: none*
*Databank codes:*

| Species | Isolate | SwissProt | PIR | EMBL (genomic) |
|---|---|---|---|---|
| Feline immunodeficiency virus | Petaluma | P16088 | B33543 | M25729: complete genome |
| | San Diego | P19028 | – | M36968: complete genome |
| | TM2 | P31822 | B45557 | M59418: 3′ end of genome |

Brookhaven Protein Data Bank three-dimensional structures:

| Species | Isolate | ID | Resolution | Notes |
|---|---|---|---|---|
| Feline immunodeficiency virus | Petaluma | 1FIV | 2 | complex with acAla-Val-Sta-Glu-Ala-NH$_2$ |

Retropepsin is residues 34–156 of Pol polyprotein.

## Name and History

The feline immunodeficiency virus (FIV), isolated in 1987 by Pedersen *et al.*, is a member of the lentivirus subfamily. The complete nucleotide sequence of the virus was reported by Talbott *et al.* (1989), and these authors identified a sequence at the 5′ end of the *pol* gene that was predicted to code for a protease of 13 493 Da based on homology with HIV. Elder *et al.* (1993) reported the analysis of the sequence and mass spectra of isolated viral proteins, confirming the earlier assignment of the protease based on nucleotide sequence analysis. The first report of the cloning, expression and kinetic activity of *feline immunodeficiency virus retropepsin* (**FIV retropepsin**) appeared in 1991 (Farmerie *et al.*, 1991). This was followed by high-level expression and crystallization in 1994, with the structure determination published the following year (Wlodawer *et al.*, 1995). The enzyme has also been prepared by expression (Slee *et al.*, 1995) and chemical synthesis (Schnölzer *et al.*, 1996).

## Activity and Specificity

The study of Elder *et al.* (1993) on the analysis of the products of polyprotein processing provided the initial indication of the specificity of the enzyme. In exact analogy to the HIV case, cleavage between Tyr┼Pro, linking the MA and CA proteins of the Gag region, is a characteristic feature of the specificity. Thus, the determinants that allow the HIV retropepsin (Chapters 310 and 311) to catalyze the cleavage of the unusual aromatic-Pro junction also must exist in the FIV retropepsin. Table 319.1 lists the cleavage sites derived from the 1993 report (Elder *et al.*, 1993).

Schnölzer *et al.* (1996) utilized a peptide of MA/CA sequence (R)KEEGPPQAY┼PIQTVNGPQY(R), where the Arg residues at the ends are added for increased solubility, to analyze their synthetic enzyme. Complete cleavage at the indicated point is observed, with no other cleavage products apparent. Increasing salt concentration up to 1.5 M NaCl increases enzymatic activity. Maximum activity is obtained in the pH range of 5–6. At pH 5.25, 1 M NaCl, and 37°C, the observed $K_m$ value is 10 mM; however, uncertainty in quantitation of the enzyme concentration prevented determination of $k_{cat}$.

Schnölzer *et al.* (1996) also observed that an eight residue peptide is the minimum size of substrate that can be efficiently cleaved by FIV retropepsin. The cleavage of protected decapeptide MA/CA junction peptides, Ac-Arg-Pro-Gln-Ala-Tyr┼Pro-Ile-Gln-Thr-Arg-NH$_2$ and Ac-Arg-Ser-Gln-Asn-Tyr┼Pro-Ile-Val-Gln-Arg-NH$_2$ to represent the FIV and HIV junctions, respectively, were analyzed with both FIV and HIV retropepsin. The data clearly illustrate a preference of each enzyme for its cognate cleavage junction, with the FIV peptide being cleaved completely by the FIV enzyme and only partially by HIV retropepsin and the HIV peptide being cleaved completely by the HIV enzyme and not at all by FIV retropepsin. Furthermore, in the peptide, Ac-Tyr-Lys-Met-Gln-Leu┼Leu-Ala-Glu-Leu-Thr-Lys-NH$_2$, FIV retropepsin cuts only between the Leu┼Leu residues, while the HIV enzyme cuts at that site as well as at the Leu┼Ala bond. Curiously, FIV retropepsin will cleave modified peptides based on the HIV MA/CA cleavage site, albeit at slower rates. In addition, FIV retropepsin requires a larger residue at the P1 position; in the sequence Lys-Ala-Arg-Val-P1┼Nph-Glu-Ala-Nle-Gly-NH$_2$ the peptides are cleaved in the order P1 = $\beta$-naphthylalanine > Tyr > Phe > Nle > Leu (Farmerie *et al.*, 1991; Zhang & Dunn, 1997). $K_m$ values are in the range of 20–60 mM and $k_{cat}$ values are in the range of 0.3–0.8 s$^{-1}$ for the best three substrates, all at pH 4.7 and 37°C, with ionic strength of 0.3 M. A useful peptide, Lys-Ala-Arg-Val-Nle-Nph-Glu-Ala-Nle-NH$_2$ (H-1048) is available from Bachem (see Appendix 2 for full names and addresses of suppliers).

The development of inhibitors for FIV retropepsin has presented a challenge, despite the similarity of the feline virus enzyme to the human virus enzyme (see below). The explosion of effort in the late 1980s and the early 1990s to create inhibitors targeted against the HIV-1 retropepsin as an additional strategy for antiviral drug therapy has resulted in several compounds achieving Food & Drug Administration (USA) approval for use in humans. However, with rare exception, these compounds do not effectively inhibit FIV retropepsin. In the crystallographic study by Wlodawer *et al.* (1995), a test of over 30 highly potent ($K_i$ < 10 nM) inhibitors of HIV retropepsin revealed none that were inhibitory towards FIV retropepsin at concentrations of 1 mM. A series of general aspartic proteinase inhibitors, loosely based on the sequence of pepstatin, was inhibitory in the range of 200–600 nM, and one of these, Ac-naphthylalanine-Val-Sta-Glu-naphthylalanine-NH$_2$ (Sta (statine): 3-hydroxy,

*Table 319.1* Cleavage of synthetic peptides corresponding to the natural cleavage sites of the polyprotein

| Cleavage site | Sequence |
|---|---|
| MA/CA | Pro-Gln-Ala-Tyr┼Pro-Ile-Gln-Thr |
| CA/NC1 | Lys-Met-Gln-Leu┼Lys-Ala-Glu-Ala |
| CA/NC2 | Thr-Lys-Val-Gln┼Val-Val-Gln-Ser |
| NC/PR | Gly-Phe-Val-Asn┼Tyr-Asn-Lys-Val |
| PR/RT | Arg-Leu-Val-Met┼Ala-Gln-Ile-Ser |
| RT/RNase H | Ala-Glu-Thr-Trp┼Tyr-Ile-Asp-Gly |
| RNase H/DU | Cys-Gln-Thr-Met┼Met-Ile-Ile-Glu |
| DU/IN | Thr-Gly-Val-Phe┼Ser-Ser-Trp-Val |

CA, capsid; DU, dUTPase; IN, integrase; MA, matrix; NC, nucleocapsid; PR, protease.

4-amino,7-methylheptanoic acid), was utilized to prepare a noncovalent complex that was crystallized. The $K_i$ for this compound was determined to be 260 nM for FIV retropepsin, whereas the same compound gave a $K_i$ value of 1.7 nM with HIV retropepsin. This difference of at least 100-fold in potency in favor of HIV retropepsin was also observed in a detailed study of a series of $\alpha$-ketoamides and hydroxyethylamines (Slee *et al*., 1995). The best FIV retropepsin inhibitor reported in this study was Z-PheY[CO-CO]Pro-Ile-Glu-OBu$^t$, with $K_i$ of 29 μM. The corresponding $K_i$ for HIV retropepsin was 154 nM. It was concluded that FIV retropepsin requires additional specificity determining residues, which agrees with the conclusion derived from the lack of cleavage of a six residue peptide noted earlier. It has also been suggested (Wlodawer *et al*., 1995; Slee *et al*., 1995) that FIV retropepsin might be a good model for the resistant forms of HIV retropepsin due to its more stringent specificity and to the substitutions at critical positions noted below.

## Structural Chemistry

The dimeric FIV retropepsin has been crystallized and its structure determined (Wlodawer *et al*., 1995). The enzyme resembles HIV retropepsin in most details; however, the longer length of the FIV enzyme (116 versus 99 amino acid residues) results in three surface loops that are larger in the FIV enzyme. These can be thought of as insertions in the HIV sequence/structure and include residues 48–50, 77–81 and 94–96 of the FIV sequence. This allows the central 'core' components of the FIV enzyme to superimpose exactly on the HIV structure. The now familiar 'flaps' that cover the active site cleft are nearly identical between FIV and HIV retropepsins. Specific details of the active site cleft that lead to the differences in specificity detailed above are most likely related to the following differences in the sequences of FIV and HIV retropepsins. Residues 32 and 50 in the HIV enzyme are Val and Ile; these are replaced by Ile and Val in FIV retropepsin. Although these two changes are reciprocal, the effects upon the structure of the binding site are not neutral. In addition, the replacement of Gly48 in the HIV enzyme by an Ile57 at the structurally homologous location partly compensates for the more open structure around the loop in HIV retropepsin that includes residues 79–82; this is where three residues are inserted in the FIV sequence. The resulting larger loop opens up the binding site to some degree. This will then account for the requirement for larger residues in P1–P3 as well as P1′–P3′ in both substrates and inhibitors.

An additional feature of FIV retropepsin deserves note: the FIV sequence naturally contains substitutions that are identical to the amino acids that arise in HIV retropepsin in response to drug challenge. Specifically, residues Val32, Gly48, Ile50, Asn88 and Leu90 in the HIV sequence are changed to Ile, Val, Val, Asp and Met, respectively, in different drug-resistant forms. Residues in FIV retropepsin that correspond structurally are Ile37, Ile57, Val59, Asp105 and Met107. Thus, the feline virus enzyme has determinants different from the naturally most dominant HIV enzyme sequence, but closer to the drug resistant forms. Future work will have to establish the exact relationship.

## Preparation

The enzyme has been chemically synthesized (Schnölzer *et al*., 1996) and expressed in *Escherichia coli* (Farmerie *et al*., 1991; Wlodawer *et al*., 1995; Slee *et al*., 1995). The recombinant material has been purified by ammonium sulfate precipitation and column chromatography.

## Biological Aspects

The necessity for an enzyme to process the viral polyprotein has become a familiar refrain, and this provides convenient targets for drug development as antiviral agents. Translation of the viral message, either directly in the case of RNA viruses that are not copied into DNA, or as the mRNA produced from transcription of the DNA integrated into the host cell genome in the case of the retroviruses, yields a 'polyprotein' that can be either 50–60 kDa or, with a ribosomal frameshift, 150–200 kDa in size. This larger protein contains a number of individual components connected head-to-tail with the Gag structural proteins typically occurring on the N-terminal end, and the enzymatic functions of the virus occurring on the C-terminal end, after the frameshift position. The retroviral endopeptidase is a component of the larger polyprotein and has the biological function of cutting itself out of this position, as well as separating all the other proteins from one another by selective cleavage. The viral endopeptidase also cleaves the smaller Gag polyprotein of 50–60 kDa, which is usually present in 15–20 times larger amounts. This provides the structural proteins necesary for construction of the new viral particle.

## References

Elder, J.H., Schnölzer, M., Hasselkus, L.C.S., Henson, M., Lerner, D.A., Phillips, T.R., Wagaman, P.C. & Kent, S.B. (1993) Identification of proteolytic processing sites within the Gag and Pol polyproteins of feline immunodeficiency virus. *J. Virol.* **67**, 1869–1876.

Farmerie, W.G., Goodenow, M.M. & Dunn, B.M. (1991) Cloning, expression and kinetic characterization of the feline immunodeficiency virus proteinase. *Adv. Exp. Med. Biol.* **306**, 511–513.

Gustchina, A. (1995) Molecular modeling of the structure of FIV protease. *Adv. Exp. Med. Biol.* **362**, 479–484.

Kakinuma, S., Motokawa, K., Hohdatsu, T., Yamamoto, J.K. & Hashimoto, H. (1995) Nucleotide sequence of feline immunodeficiency virus: classification of Japanese isolates into two subtypes which are distinct from non-Japanese subtypes. *J. Virol.* **69**, 3639–3646.

Pedersen, N.C., Ho, E., Brown, M.L. & Yamamoto, J.K. (1987) Isolation of a T-lymphotropic virus from domestic cats with an immunodeficiency-like syndrome. *Science* **235**, 790–793.

Schnölzer, M., Rackwitz, H.R., Gustchina, A., Laco, G.S., Wlodawer, A., Elder, J.H. & Kent, S.B.H. (1996) Comparative properties of feline immunodeficiency virus (FIV) and human immunodeficiency virus type 1 (HIV-1) proteinases prepared by total chemical synthesis. *Virology* **224**, 268–275.

Slee, D.H., Laslo, K.L., Elder, J.H., Ollmann, I.R., Gustchin, A., Kervinen, J., Zdandov, A., Wlodawer, A. & Wong, C.H. (1995) Selectivity in the inhibition of HIV and FIV protease: inhibitory and mechanistic studies of pyrrolidine-containing $\alpha$-keto amide and core structures. *J. Am. Chem. Soc.* **117**, 11867–11878.

Talbott, R.L., Sparger, E.E., Fitch, M.W., Lovelace, K.M., Pedersen, N.C., Luciw, P.A. & Elder, J.H. (1989) Nucleotide sequence and genomic organization of feline immunodeficiency virus. *Proc. Natl Acad. Sci. USA* **86**, 5743–5747.

Wlodawer, A., Gustchina A., Reshetnikova L., Lubkowski, J., Zdanov, A., Hui, K.Y., Angleton, E.L., Farmerie, W.G., Goodenow, M.M., Bhatt, D., Zhang, L. & Dunn, B.M. (1995) Structure of an inhibitor complex of the proteinase from feline immunodeficiency virus. *Nature Struct. Biol.* **2**, 480–488.

Zhang, L., Pennington, M.W., Baur, P., Byrnes, M.E., de Chastonay, J., Wilson, S.I., Kay, J. & Dunn, B.M. (1997) Binding of mutant HIV-1 proteases with junction B peptides containing methyleneamino isostere replacements. *Protein Pept. Lett.* **4**, 225–235.

***Ben M. Dunn***
*Department of Biochemistry & Molecular Biology,*
*University of Florida College of Medicine,*
*Gainesville, FL 32610-0245, USA*
*Email: bdunn@biochem.med.ufl.edu*

# 320. *Miscellaneous viral retropepsins*

## Databanks

*Peptidase classification: clan AA, families A2 and A9, MEROPS ID: A2, A9*
*NC-IUBMB enzyme classification: none*
*Databank codes:*

| Species | SwissProt | PIR | EMBL (genomic) |
| --- | --- | --- | --- |
| **Family A2** | | | |
| African green monkey immunodeficiency virus | P05895 | B30045 | X07805: complete genome |
| | P12500 | – | M22974: *pol* gene |
| | P12501 | – | M21311: *pol* gene |
| | P27973 | – | M29975: complete genome |
| | P27980 | – | M30931: complete genome |
| | Q02836 | – | M66437: complete genome |
| Bovine immunodeficiency virus | P19560 | – | – |
| | P19561 | B34742 | M32690: complete genome |
| Caprine arthritis-encephalitis virus | P33459 | B45345 | M33677: complete genome |
| *Dendrobates vertrimaculatus* retrovirus | – | – | X95795: *pol* gene |
| | – | – | X95796: *pol* gene |
| Hamster intracisternal A-particle | P04023 | A03952 | – |
| Human retrovirus-related endopeptidase | P10265 | C24483 | – |
| | – | – | U12970 |
| Macaque immunodeficiency virus | P05896 | B28887 | Y00277: complete genome |
| | P05897 | B28873 | Y00295: complete genome |
| Mandrill immunodeficiency virus | P22382 | S28081 | M27470: complete genome |
| Mouse intracisternal A-particle | P11365 | A26787 | – |
| Ovine lentivirus | P16901 | B46335 | M31646: complete genome |
| Sheep pulmonary adenomatosis virus | P31625 | B42740 | M80216 – |
| Sooty mangabey immunodeficiency virus | P12502 | – | X14307: complete genome |
| | P19505 | – | M31325: complete genome |
| Squirrel monkey retrovirus | P21407 | B31827 | M23385 – |
| Turkey lymphoproliferative disease virus | – | – | U09568: complete genome |
| Visna lentivirus | P03370 | A03969 | M10608: complete genome |
| Visna lentivirus | P23426 | – | M60609: complete genome |
| Visna lentivirus | P23427 | – | M60610: complete genome |
| Visna lentivirus | P35956 | B45390 | L06906: complete genome |
| **Family A9** | | | |
| Human spumaretrovirus | P14349 | A28880 | M19427: *gag* and *pol* genes |
| Simian foamy virus (type 1) | P23074 | S15566 | X54482: complete genome |
| | | S18738 | |

Brookhaven Protein Data Bank three-dimensional structures:

| Species | ID | Resolution | Notes |
|---|---|---|---|
| Simian foamy virus (type 3) | P27401 | B40820 | M74895: 3′ end of genome |
| Macaque immunodeficiency virus | 1SIP | 2.3 | |
| | 1SIV | 2.5 | complex with SKF107457 |
| | 1TCW | 2.3 | complex with SB203386 |
| | 2SAM | 2.4 | mutant Ser4His |

## *Name and History*

All retroviruses encode a proteinase (PR) that specifically processes the polyprotein precursors giving rise to the viral structural proteins and replication enzymes. The presence of a conserved Asp-Ser/Thr-Gly sequence in the primary structure of the putative retroviral PRs led to the suggestion that they are related to the aspartic peptidases of the pepsin family (A1) (Toh *et al*., 1985a,b; Pearl & Taylor, 1987a,b) (Chapter 270). It is interesting to note that the first retroviral PRs recognized as of aspartic type were enzymes from less well-known species like yeast retrotransposon or *Drosophila* transposable genetic element. Their classification was later confirmed by mutation of the presumed active-site aspartates and inhibition with pepstatin A leading to inactivation of the enzyme (Kohl *et al*., 1988; Seelmeier *et al*., 1988) (for detailed review, see Chapter 309).

As opposed to the retropepsins from human pathogens like HIV-1 (Chapter 310) and HIV-2 (Chapter 311), the miscellaneous viral retropepsins discussed in this chapter have in most cases never been isolated and/or characterized as chemical species and there is very limited information available regarding their structure, activity and biological relevance. Their existence in the viral genomes has mostly been inferred by analogy with other, more thoroughly studied retropepsins, based on sequence homology or general similarities in genome organization of the particular viruses. Such suggestions are sometimes misleading, however. One example is the gene for a putative aspartic proteinase from vaccinia virus named vaccinia virus protease-like gene (Slabaugh & Roseman, 1989) that seems to encode a dUTPase rather than an aspartic proteinase (Massung *et al*., 1996). Given the lack of experimental data for most of these putative enzymes, we will slightly modify the structure of this chapter giving a brief overview of the limited amount of information available regarding the proteolytic machinery of individual viruses and deal only with the structure and activity of those enzymes for which experimental data are available.

## *Activity, Specificity and Biological Aspects*

### *Bovine Immunodeficiency Virus*

Bovine immunodeficiency virus (BIV) is a recently characterized lentivirus infecting cattle, which is closely related to human and simian immunodeficiency viruses (Chapter 311) (Garvey *et al*., 1990; Baron *et al*., 1995). The structural organization of the BIV genome is similar to that of other members of the lentivirus genus. The putative PR sequence is located in the *pol* open reading frame and the predicted 11 kDa PR protein is synthesized as part of the Gag-Pol precursor, presumably by ribosomal frameshifting (Garvey

*et al*., 1990). The PR has not been immunologically identified and no data are available on its activity and structure.

### *Visna Lentivirus, Ovine Lentivirus, Caprine Arthritis Encephalitis Virus*

Visna ('wasting', in Icelandic) is a slow progressive encephalomyelitis of sheep. The diseased animals suffer from weakness that slowly progresses to paralysis (Braun *et al*., 1987; Andreson *et al*., 1993). Visna virus is the prototypic member of the lentivirus genus within the family Retroviridae. Its genomic organization is similar to that of the primate immunodeficiency viruses in general, but is less complex (Sonigo *et al*., 1985). Two closely related lentiviruses are South African ovine Maedi Visna virus (SA-OMVV) (Querat *et al*., 1990) with homologies at the protein level ranging from 65–91%) and caprine arthritis-encephalitis virus (CAEV) causing leukoencephalitis in young goats and chronic arthritis in adult animals (Saltarelli *et al*., 1990). The *pol* gene products of these viruses are translated as Gag-Pol precursors of 150 kDa via ribosomal frameshifting. The putative PR regions are located at the N-terminals of the Pol polyproteins. At present there are no experimental data regarding polyprotein processing for these viruses and their respective PRs have not been characterized.

### *Squirrel Monkey Retrovirus, Sheep Pulmonary Adenomatosis Virus (Jaagsiekte Retrovirus)*

The jaagsiekte sheep retrovirus (JSRV) (York *et al*., 1992) and the squirrel monkey retrovirus (SMRV-H) (Oda *et al*., 1988; Sun *et al*., 1995) share certain characteristics with D-type and B-type retroviruses. Jaagsiekte is an Afrikaans name for contagious adenocarcinoma of sheep and goats, SMRV-H has been characterized as a type D retrovirus produced in a human lymphoblastoid B cell line. The putative PRs of both viruses are encoded in separate open reading frames, partially overlapping with the *gag* and *pol* reading frames. High-sequence homology to the prototype D-type retrovirus, Mason–Pfizer monkey virus (M-PMV) (Chapter 316) in the case of SMRV and to the mouse mammary tumor virus (MMTV) (Chapter 317), a prototype B retrovirus in the case of the JSRV, suggest that their substrate specificities and activities might be similar to their respective prototypes.

### *Human Endogenous Retroviruses*

The human endogenous retroviruses type K (HERV-K) constitute a group of endogenous retroviruses, which contain *gag*, *prt*, *pol* and *env* open reading frames, but in most cases have multiple stop codons or frameshift mutations within these reading frames. Only the HERV-K10 element appears to contain uninterrupted open reading frames (Ono *et al*., 1986) and can give rise to the release of retrovirus particles, at least in some human teratocarcinoma cell lines (Boller *et al.*, 1993). These particles appear immature by electron microscopy, suggesting that there might be a defect in PR activity or PR activation (for review see Löwer *et al.*, 1996). The biological function of HERV-K10-derived particles remains uncertain.

HERV-K10 PR is encoded in its own reading frame located between the *gag* and *pol* genes. Schommer *et al.* (1996) expressed various constructs containing either the full *prt* domain with flanking regions or truncated versions thereof in *Escherichia coli* under the control of the bacterial *trpE* or T7 promoter. The proteolytic activity of the putative PR was demonstrated by autocatalytic cleavage of the expression product to be an 18 kDa protein, which was recognized by a polyclonal antiserum raised against the PR domain. Bacterially expressed PR also cleaved the *in vitro* translated HERV-K10 Gag polyprotein into major fragments of 39 and 30 kDa. Proteolysis was blocked by the HIV-1 PR-specific inhibitor Ro31-8959 (Roberts *et al.*, 1990). However, much higher concentrations of inhibitor were required in this case compared to inhibition of HIV PR. Expression of an N-terminally deleted version of the putative HERV-K10 PR yielded a 16 kDa protein, which retained proteolytic activity (Schommer *et al.*, 1996).

## Mouse Intracisternal A-type Particle, Hamster Intracisternal A-type Particle

Intracisternal A-particles (IAPs) are defective retroviruses that are present in a variety of tumor cells and in early embryonic cells derived from normal rodents such as mice (Mietz *et al.*, 1987), rats and Syrian hamsters (Ono *et al.*, 1985). These viruses undergo an entirely intracellular life cycle: they assemble and bud into the cisternae of the endoplasmatic reticulum and remain there as immature A-type particles. From a phylogenetic point of view, these viruses are related to MMTV (type B retrovirus) (Chapter 317) and squirrel monkey retrovirus (type D retrovirus, see above). They produce a Gag polyprotein of approximately 70 kDa and exhibit weak, but detectable reverse transcriptase activity (for review see Kuff & Lueders, 1988). However, there appears to be no proteolysis of either the Gag or the putative Gag-Pol polyprotein. The IAP PR coding region has been reported to constitute the 3′ terminal fragment of the *gag* open reading frame, similar to avian C-type retroviruses (Mietz *et al.*, 1987). Recent evidence showed, however, that the IAP PR is encoded in its own reading frame, separate from both *gag* and *pol* (Fehrmann *et al.*, 1997). Thus, IAPs synthesize Gag, Gag-PR, and Gag-PR-Pol polyproteins by two consecutive translational frameshifting events, similar to D-type viruses.

Welker *et al.* (1997) recently analyzed the requirements for intracellular polyprotein transport and PR activation in the case of mouse IAPs. It was shown that deletion or substitution mutants of the presumed targeting signal on the IAP polyprotein led to transport of the respective polyproteins to various intracellular sites. Interestingly, assembly and budding of IAP-like particles occurred at the membrane of the endoplasmic reticulum and at the nuclear membrane, inside the nucleus and at the plasma membrane. All particles consisted of uncleaved polyproteins except for those that had

been assembled at the plasma membrane and were released into the medium. In this case, specific cleavage of the Gag polyprotein by the viral PR occurred, which was abolished by a PR active site mutation. Thus, intracellular localization and transport to a specific site can influence IAP PR activation or PR activity and may trigger virion maturation. The mechanism of PR regulation is currently not known.

In the same study, Welker *et al.* (1997) constructed bacterial expression vectors for the putative IAP PR region and flanking sequences and analyzed the expression products. Upon induction, a major product of 18 kDa was observed in the case of an unmutated PR while a similar construct containing a mutation in the presumed active site of PR yielded a protein of 26 kDa, corresponding to the expected mass of the full-length expression product. Thus, IAP PR can autocatalytically cleave itself from a larger precursor yielding a mature PR molecule of 18 kDa. Incubation of active PR with the IAP Gag polyprotein obtained by *in vitro* translation yielded cleavage products of 35, 24 and 20 kDa, which were not observed in the case of the active-site mutant. The migration of *in vitro* processed IAP Gag proteins was identical with that of Gag proteins in extracellular particles released from the plasma membrane of the cell.

## Simian Foamy Virus Type 1, Simian Foamy Virus Type 3, Human Foamy Virus

Foamy viruses (spumaviruses) are a group of complex retroviruses that have been isolated from various species and at present are not connected to any pathological condition. The amino acid homology between simian foamy virus 1 and 3 and human foamy virus is approximately 80% in the Pol polyprotein (Kupiec *et al.*, 1991; Renne *et al.*, 1992; for references see Löchelt & Flügel, 1995; Saib & de The, 1996). A particular feature of foamy viruses is the fact that their structural proteins remain mostly uncleaved and accumulate inside infected cells with few extracellular virions observed (Kupiec *et al.*, 1991; Renne *et al.*, 1992; for references see Löchelt & Flügel, 1995). Most particles stay immature in morphology and a 1:1 mixture of two *gag*-related proteins of 74 kDa and 78 kDa with little if any further cleavage is observed in cell extracts and particle fractions. In contrast, *pol*-derived proteins of foamy viruses do not remain in the form of polyproteins, but are completely processed to the mature products. The viral PR has been assigned to the 5′ region of the *pol* gene, containing an Asp-Ser-Gly amino acid triplet, and has been reported to correspond to a 10 kDa protein (Kögel *et al.*, 1995).

The putative foamy virus PR (family A9, p. 916) shows very little similarity to other retroviral and cellular aspartic PRs outside of the conserved active site region and Gly-Arg-Lys sequence motif. In particular, no sequence corresponding to the flap region observed in all aspartic proteinases has been detected in this region. Konvalinka *et al.* (1995) showed that PR activity is essential for viral infectivity. Mutation of the putative active-site aspartate residue yielded noninfectious virus containing only the 78 kDa Gag polyprotein. PR appears to cleave a 4 kDa fragment from the C-terminus of Gag with no apparent further processing steps. The PR active-site mutation also abolished processing of the *pol*-derived products to reverse transcriptase and integrase. Surprisingly, however, no Gag-Pol precursor was found in this case, indicating that expression of the *pol* gene may occur by a different mechanism in these viruses. This hypothesis was subsequently confirmed and it was shown that *pol*-derived products are synthesized from a spliced mRNA rather than by translational frameshifting or readthrough of a stop codon as in the other retroviruses (Jordan *et al*., 1996; Löchelt & Flügel, 1996; Yu *et al*., 1996; for reference see Rethwilm, 1996). Luukkonen *et al*. (1995) analyzed cross-reactivity of several retroviral PRs using vaccinia virus-T7 RNA polymerase-mediated expression. Human foamy virus PR only cleaved its cognate Gag polyprotein in *trans*, but did not process the Gag polyproteins of HIV-1 and -2, equine infectious anemia virus or human T cell leukemia virus.

## Structural Chemistry

All viral retropepsins analyzed to date have been symmetrical homodimers with the monomeric subunit encoded as a segment of the viral polyprotein. Although no specific information is available regarding the enzymes discussed in this chapter, they also contain only a single Asp-Thr/Ser-Gly triplet and therefore also need to dimerize to form the active enzyme. Monomeric subunits are generally in the range of 18 kDa, at least in those cases where an active PR species has been identified.

## Preparation

In those cases where active PR species have been analyzed, they had been obtained by recombinant expression of the presumed PR region in bacterial cells. Proteolytic activity contained in virus particles or virus-infected cells has not been analyzed for any of the miscellaneous viral retropepsins nor have the enzymes been purified from natural sources. Purification of the active enzyme in larger quantities following bacterial expression has also not been reported for any of the viral retropepsins discussed in this chapter.

## References

Andreson, O.S., Elser, J.E., Tobin, G.J., Greenwood, J.D., Gonda, M.A., Georgsson, G., Andresdottir, V., Benediktsdottir, E., Carlsdottir, H.M., Mantyla, E.O., Rafnar, B., Palsson, P.A., Casey, J.W. & Petursson, G. (1993) Nucleotide sequence and biological properties of a pathogenic proviral molecular clone of neurovirulent Visna virus. *Virology* **193**, 89–105.

Baron, T., Mallet, F., Polack, B., Betemps, D. & Belli, P. (1995) The bovine immunodeficiency-like virus (BIV) is transcriptionally active in experimentally infected calves. *Arch. Virol.* **140**, 1461–1467.

Boller, K., König, H., Sauter, M., Müller-Lantzsch, N., Löwer, R., Löwer, J. & Kurth, R. (1993) Evidence that HERV-K is the endogenous retrovirus that codes for the human teratocarcinoma-derived retrovirus HDTV. *Virology* **196**, 349–353.

Braun, M.J., Clements, J.E. & Gonda, M.A. (1987) The visna virus genome: evidence for hypervariable site in the *env* gene and sequence homology among lentivirus envelope proteins. *J. Virol.* **61**, 4046–4054.

Fehrmann, F., Welker, R. & Kräusslich H.-G. (1997) Intracisternal A-type particles express their proteinase in a separate reading frame by translational frameshifting similar to D-type retroviruses. *Virology* **235**, 352–359.

Garvey, K.J., Oberste, M.S., Elser, J.E., Braun, M.J. & Gonda, M.A. (1990) Nucleotide sequence and genome organization of biologically active proviruses of the bovine immunodeficiency-like virus. *Virology* **175**, 391–409.

Jordan, I., Enssle, J., Güttler, E., Mauer, B. & Rethwilm, A. (1996) Expression of human foamy virus reverse transcriptase involves a spliced pol mRNA. *Virology* **224**, 314–319.

Kögel, D., Aboud, M. & Flügel, R. (1995) Molecular biological characterization of the human foamy virus reverse transcriptase and ribonuclease H domains. *Virology* **213**, 97–108.

Kohl, N.E., Emini, E.A., Schleif, W.A., Davis, L.J., Heimbach, J.C., Dixon, R.A.F., Scolnick, E.M. & Sigal, I.S. (1988) Active human immunodeficiency virus protease is required for viral infectivity. *Proc. Natl Acad. Sci. USA* **85**, 4686–4690.

Konvalinka, J., Löchelt, M., Zentgraf, H., Flügel, R.M. & Kräusslich, H.-G. (1995) Active foamy virus proteinase is essential for virus infectivity but not for formation of a *pol* polyprotein. *J. Virol.* **69**, 7264–7268.

Kuff, E.L. & Lueders, K.K. (1988) The intracisternal A particle gene family: structure and functional aspects. *Adv. Cancer Res.* **51**, 183–276.

Kupiec, J.-J., Kay, A., Hayat, M., Ravier, R., Peries, J. & Galibert, F. (1991) Sequence analysis of the simian foamy virus type 1 gene. *Gene* **101**, 185–194.

Löchelt, M. & Flügel, R. (1995) The molecular biology of human and primate spuma retroviruses. In: *The Retroviridae*, vol. 4 (Levy, J.A., ed.). New York: Plenum Press, pp. 239–408.

Löchelt, M. & Flügel, R. (1996) The human foamy virus is expressed as a Pro-Pol polyprotein and not as a Gag-Pol fusion protein. *J. Virol.* **70**, 1033–1040.

Löwer, R., Löwer, J. & Kurth, R. (1996) The viruses in all of us: characteristics and biological significance of human endogenous retrovirus sequences. *Proc. Natl Acad. Sci. USA* **93**, 5177–5184.

Luukkonen, B.G., Tan, W., Fenyö, E.M. & Schwartz, S. (1995) Analysis of cross reactivity of retrovirus proteases using a vaccinia virus-T7 RNA polymerase-based expression system. *J. Gen. Virol.* **76**, 2169–2180.

Massung, R.F., Loparev, V., Knight, J.C., Totmenin, A.V., Chizhikov, V.E., Parsons, J.M., Safronov, P.F., Gutorov, V.V., Shchelkunov, S.N. & Esposito, J.J. (1996) Terminal region sequence variations in variola virus DNA. *Virology* **221**, 291–300.

Mietz, J., Grossman, Z., Lueders, K.K. & Kuff, E.L. (1987) Nucleotide sequence of a complete mouse intracisternal A-particle genome: relationship to known aspects of particle assembly and function. *J. Virol.* **61**, 3020–3029.

Oda, T., Ikeda, S., Watanabe, S., Hatsushika, M., Akiyama, K. & Mitsunobu, F. (1988) Molecular cloning, complete nucleotide sequence, and gene structure of the provirus genome of a retrovirus produced in a human lymphoblastoid cell line. *Virology* **167**, 468–476.

Ono, M., Toh, H., Miyata, T. & Awaya, T. (1985) Nucleotide sequence of the Syrian hamster intracisternal A-particle gene: close evolutionary relationship of type A particle gene to types B and D oncovirus genes. *J. Virol.* **55**, 387–394.

Ono, M., Yasunaga, T., Miyata, T. & Ushikubo, H. (1986) Nucleotide sequence of human endogenous retrovirus genome related to the mouse mammary tumor virus genome. *J. Virol.* **60**, 589–598.

Pearl, L.H. & Taylor, W.R. (1987a) Sequence specificity of retroviral proteases. *Nature* **328**, 482.

Pearl, L.H. & Taylor, W.R. (1987b) A structural model for the retroviral proteases. *Nature* **329**, 351–354.

Querat, G., Audoly, G., Sonigo, P. & Vigne, R. (1990) Nucleotide sequence analysis of SA-OMVV, a visna related ovine lentivirus: phylogenetic history of lentiviruses. *Virology* **175**, 434–447.

Renne, R., Friedl, E., Schweizer, M., Fleps, U., Turek, R. & Neuman-Haefelin, D. (1992) Genomic organisation and expression of simian foamy virus type 3 (SFV-3). *Virology* **186**, 597–608.

Rethwilm, A. (1996) Unexpected replication pathways of foamy viruses. *J. Acquir. Immune Defic. Syndr. Hum. Retrovirol.* **13**, S248–S253.

Roberts, N.A., Martin, J.A., Kinchington, D., Broadhurst, A.V., Craig, J.C., Duncan, I.B., Galpin, S.A., Handa, B.K., Kay, J., Kroehn, A., Lambert, R.W., Merret, J.H., Mills, J.S., Parkes, K.E.B., Redshaw, S., Ritchie, A.J., Taylor, D.L., Thomas, G.J. & Machin, P.J. (1990) Rational design of peptide-based HIV proteinase inhibitors. *Science* **248**, 358–361.

Saib, A. & de The, H. (1996) Molecular biology of the human foamy virus. *J. Acquir. Immune Defic. Syndr. Hum. Retrovirol.* **13**, S254–S260

Saltarelli, M., Querat, G., Konings, D.A.M., Vigne, R. & Clements, J.E. (1990) Nucleotide sequence and transcriptional analysis of moelecular clones of CAEV which generate infectious virus. *Virology* **179**, 347–364.

Schommer, S., Sauter, M., Kräusslich, H.-G., Best, B. & Mueller-Lantzsch, N. (1996) Characterization of the human endogenous retrovirus K proteinase. *J. Gen. Virol.* **77**, 375–379.

Seelmeier, S., Schmidt, H., Turk, V. & von der Helm, K. (1988) Human immunodeficiency virus has an aspartic type protease that can be inhibited by pepstatin A. *Proc. Natl Acad. Sci. USA* **85**, 6612–6616.

Slabaugh, M.B. & Roseman, N.A. (1989) Retroviral protease-like gene in the vaccinia virus genome. *Proc. Natl Acad. Sci. USA* **86**, 4152–4155.

Sonigo, P., Aliyon, M., Staskus, K., Klatzman, D., Cole, S., Danos, O., Retzel, E., Tiollais, P., Haase, A. & Wain-Hobson, S. (1985) Nucleotide sequence of the visna lentivirus: relationship to the AIDS virus. *Cell* **42**, 369–382.

Sun, R., Grogan, E., Shedd, D., Bykovsky, A., Kushnaryov, V.M., Grossberg, S.E. & Miller, G. (1995) Transmissible retrovirus in Epstein-Barr virus-producer B95–8 cells. *Virology* **209**, 374–383.

Toh, H., Ono, M., Saigo, K. & Miyata, T. (1985a) Retroviral protease-like sequence in the yeast retrotransposon Ty 1. *Nature* **315**, 691.

Toh, H., Kikuno, R., Hayashida, H., Miyata, T., Kugimiya, W., Inouye, S., Yuki, S. & Saigo, K. (1985b) Close structural resemblance between putative polymerase of a *Drosophila* transposable genetic element 17.6 and pol gene product of Moloney murine leukemia virus. *EMBO J.* **4**, 1267–1272.

Welker, R., Janetzko, A. & Kräusslich, H.-G. (1997) Plasma membrane targeting of chimeric intracisternal A-type particle polyproteins leads to particle release and specific activation of the viral proteinase. *J. Virol.* **71**, 5209–5217.

York, D.F., Vigne, R., Verwoerd, D.W. & Querat, G. (1992) Nucleotide sequence of the Jaagsiekte retrovirus, an exogenous and endogenous type D and B retrovirus of sheep and goats. *J. Virol.* **66**, 4930–4939.

Yu, S.F., Baldwin, D.N., Gwynn, S.R., Yendapalli, S. & Linial, M.L. (1996) Human foamy virus replication: a pathway distinct from that of retroviruses and hepadnaviruses. *Science* **271**, 1579–1582.

*Jan Konvalinka*
*Department of Biochemistry,*
*Institute of Organic Chemistry and Biochemistry,*
*Czech Academy of Science,*
*Flemingovo n.2, 166 10 Praha 6, Czech Republic*
*Email: jan.konvalinka@uochb.cas.cz*

*Hans-Georg Kräusslich*
*Heinrich-Pette Institut für Experimentelle Virologie und Immunologie,*
*Martinistr. 52, D-20251 Hamburg, Germany*
*Email: hgk@hpi.uni-hamburg.de*

# 321. Fungal, plant and animal transposon elements

## Databanks

*Peptidase classification: clan AA, families A10–14, A16–A18, MEROPS ID: A10, A11, A12, A13, A14, A16, A17, A18*
*NC-IUBMB enzyme classification: none*

*Databank codes:*

| Species | Transposon name | SwissProt | PIR | EMBL (genomic) |
|---|---|---|---|---|
| **Family A10** | | | | |
| *Drosophila melanogaster* | 17.6 | P04323 | A03971 | X01472: complete retrotransposon |
| *Drosophila melanogaster* | 297 | P20825 | B24872 | X03431: complete retrotransposon |
| *Drosophila melanogaster* | 412 | P10394 | D29349 | X04132: complete retrotransposon |
| *Drosophila melanogaster* | micropia | – | – | X14037: complete retrotransposon |
| *Saccharomyces cerevisiae* | Ty3-1 | – | – | M34549: *gag3* and *pol3* genes |
| | Ty3-2 | | | M23367: complete retrotransposon |
| *Schizosaccharomyces pombe* | Tf1-107 | – | A36373 | M38526: *pol* gene |
| *Schizosaccharomyces pombe* | Tf2 | Q05654 | – | L10324: *pol* gene |
| *Trichoplusia ni* | TED | – | – | M32662: complete retrotransposon |
| **Family A11** | | | | |
| *Drosophila melanogaster* | copia | P04146 | A03324 | X02599: complete retrotransposon |
| | | | | X04456: complete retrotransposon |
| *Nicotiana otophora* | Tnt 1-94 | P10978 | S04273 | X13777: complete retrotransposon |
| *Saccharomyces cerevisiae* | Ty1-17 | P25384 | B23496 | X03840: complete transposon |
| | | | S53592 | X59720: complete chromosome III |
| *Saccharomyces cerevisiae* | Ty1B | P47098 | – | Z49526: chromosome X ORF |
| *Saccharomyces cerevisiae* | Ty1B | P47100 | – | Z49528: chromosome X ORF |
| *Saccharomyces cerevisiae* | Ty4 | – | – | M94164: 3′ end of retrotransposon |
| | | | | X67284: complete retrotransposon |
| **Family A12** | | | | |
| *Zea mays* | Bs1 | P15718 | S10112 | X16080: complete retrotransposon |
| **Family A13** | | | | |
| *Drosophila buzzatii* | Osvaldo | – | – | S75260: 3′ end of retrotransposon |
| **Family A14** | | | | |
| *Glycine max* | SIRE-1 | – | – | U22103: mid-section of retrotransposon |
| **Family A16** | | | | |
| *Ascaris lumbricoides* | Tas | – | – | Z29712: complete retrovirus-like element |
| *Caenorhabditis elegans* | – | – | – | U41509: chromosome III cosmid |
| **Family A17** | | | | |
| *Bombyx mori* | Pao | – | – | L09635: complete retrotransposon |
| *Drosophila simulans* | ninja | – | – | D83207: complete retrotransposon |
| **Family A18** | | | | |
| *Fusarium oxysporum* | skippy | – | – | L34658: complete retrotransposon |

## *Name and History*

Transposable elements are present in every type of organism from prokaryotes to humans, from archaea to the plants. Aspartic proteinases are encoded by long terminal repeat (LTR) retrotransposons, but not by non-LTR retrotransposons (Eickbush, 1994). Since the LTR retrotransposons are functionally and structurally related to retroviruses, the nomenclature used for retrotransposons and their proteases (PRs) is borrowed from that used in retroviral systems. As shown in Figure 321.1, LTR retrotransposons contain an internal coding region flanked by LTRs. An RNA transcript that initiates in the 5′ LTR and terminates in the 3′ LTR is used as the template for both translation and reverse transcription. The internal coding region encodes two proteins, Gag and a Gag-Pol polyprotein, that are translated from the element mRNA.

The *gag* gene encodes the structural proteins of the virus-like particle (VLP); *pol* encodes the enzymatic proteins: protease (PR), integrase (IN), and reverse transcriptase/ribonuclease H (RT/RH).

Retrotransposon PRs have striking similarities to retroviral PRs (Doolittle *et al.*, 1989). In retrotransposons, PR is found as either the N-terminal protein of the Pol domain, as in Ty1 (Youngren *et al.*, 1988) and Ty3 (Hansen & Sandmeyer, 1990) from *Saccharomyces cerevisiae* and TED from *Trichoplusia ni* (Lerch & Friesen, 1992), or within Gag in copia from *Drosophila melanogaster* (Yoshioka *et al.*, 1990). PR activity is essential for retrotransposition, since it releases Gag from Pol and cleaves the enzymatic proteins within Pol. PR is required for all cleavage events of the precursor proteins. No cellular activity can substitute for its function (Youngren

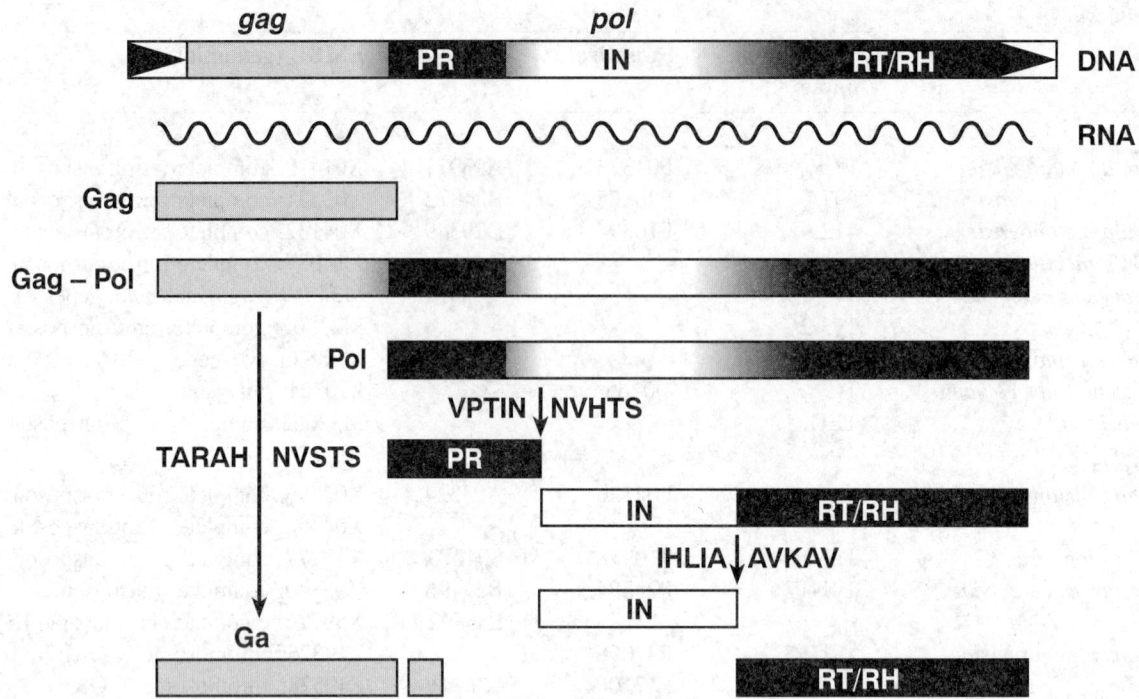

*Figure 321.1* Ty1 protein-processing pathway. The processing of Ty1 proteins is initiated when the Gag-Pol polyprotein is cleaved by Ty1 protease (PR) at the N-terminal of PR to form the Gag and Pol proteins. The Pol polyprotein is subsequently cleaved to form mature PR and an integrase (IN)-reverse transcriptase/ribonuclease H (RT/RH) polyprotein that is further processed into mature IN and RT/RH (Garfinkel *et al.*, 1991). The Gag protein is also processed by PR into the form of Gag present in the mature Ty1 virus-like particle (Merkulov *et al.*, 1996). Boxed arrowheads indicate long terminal repeats. Letters flanking each arrow depict the amino acid residues surrounding each PR cleavage site, (Moore & Garfinkel, 1994; Merkulov *et al.*, 1996).

*et al.*, 1988). Processing of these proteins occurs within VLPs and is necessary for VLP maturation (i.e. the morphological changes in VLP structure that reflect the initial stages of retroviral core assembly) (Garfinkel, 1992).

## Activity and Specificity

Even though numerous LTR retrotransposons have been identified by sequence analysis, the precise mechanism of retrotransposition has only been studied in Ty1, Ty3, *copia*, TED, and Tf1 from *Schizosaccharomyces pombe*. Active PRs have been identified and analyzed from these elements (Youngren *et al.*, 1988; Yoshioka *et al.*, 1990; Kirchner *et al.*, 1992; Lerch & Friesen, 1992; Levin *et al.*, 1993). The location of PR domains within most retrotransposons has usually been determined only by sequence similarity to retroviruses (Eickbush, 1994).

Internal bonds of the Gag-Pol polyprotein are cleaved at specific sites by PR. Cleavages are highly specific (only a few sites are cleaved in the Gag and Gag-Pol precursor proteins), and conservative (only the scissile bond is hydrolyzed at each cleavage site). Despite this specificity, the recognition sites for PRs are diverse in their primary amino acid sequence. The mechanisms by which PR recognizes processing sites in the precursor proteins are poorly understood. Only six amino acid residues, P3 through P3' in the Gag cleavage

site may be sufficient for cleavage of Ty1 Gag by PR, since block substitutions of the flanking sequences do not prevent cleavage at this site (Merkulov *et al.*, 1996). Hydrophobic residues are preferred at the P2 and P2' cleavage sites for Ty1 (Merkulov *et al.*, 1996) and at the P3, P2, and P2' cleavage sites for Ty3 (Kirchner & Sandmeyer, 1993). Analysis of the sites of PR cleavage shows similarities to retroviral PR cleavage sites in that there is no strict amino acid consensus sequence required for cleavage; rather, hydrophobic residues are found flanking the scissile bond (Pettit *et al.*, 1991). Ty1 PR cleaves at the sites shown in Figure 321.1. Ty3 PR cleaves at the following sites (Kirchner & Sandmeyer, 1993):

Gag: Asn-Gly-Tyr-Val-His┼Thr-Val-Arg-Thr-Arg

Gag/PR: Leu-Pro-Ile-Val-His┼Tyr-Ile-Ala-Ile-Pro

PR/RT: Ser-Asn-Val-Val-Ser┼Thr-Ile-Glu-Ser-Val

RT/IN: Ser-Arg-Ala-Val-Tyr┼Thr-Ile-Thr-Pro-Glu

IN: Asp-Pro-Leu-Cys-Ser┼Ala-Val-Leu-Ile-His

Encoded C-terminal to Ty3 PR is the 90 amino acid 'protein X', which has weak amino acid similarity to the PR hydrophobic motif (Kirchner & Sandmeyer, 1993). This protein has been hypothesized to be a vestige of the cellular aspartic proteinase that has acquired mutations and a cleavage site that now separates protein X from the rest of the Ty3 PR coding sequence.

## Structural Chemistry

Little is known about the structure of retrotransposon PRs. Their structures are assumed to be similar to those of retroviral PRs, based on similarities in activities and sequences of amino acids located in the active site. In contrast to the amino acid residues Asp-Thr-Gly present in the PR active site of most retroviruses, the active site motif of Ty1, Ty3, and *copia* PRs contains Asp-Ser-Gly. In Ty1, a nine amino acid domain containing the PR active site is highly similar to that in retroviral PRs (Youngren *et al.*, 1988). In Ty3, this active site is flanked by residues with more similarities to cellular aspartic proteinases (Kirchner & Sandmeyer, 1993). When the Ser in the Asp-Ser-Gly motif of Ty3 PR was changed to Thr, only a modest effect on PR activity occurred *in vivo* with no measurable effect on Ty3 transposition, indicating that having different amino acid residues in the PR active site motif does not distinguish different functional groups of PRs (Kirchner & Sandmeyer, 1993). An Asp to Ile mutation in this site, however, abolished processing in Ty3 PR (Kirchner & Sandmeyer, 1993). Tf1 PR, however, contains Asp-Thr-Gly, and when this site is deleted, proteolysis is blocked (Levin *et al.*, 1993).

## Preparation

Studies of Ty1 and Ty3 retrotransposition have been facilitated by increasing the expression and retrotransposition of these elements by fusing them to the inducible galactose promoter on a high copy plasmid in *S. cerevisiae* (Curcio & Garfinkel, 1992). Galactose induction results in the appearance in the cytoplasm of VLPs that are purified and analyzed for PR and processing activity. Proteolytic activity of PR is detected within the VLP by western analysis with antibodies specific to PR and other Ty proteins (Garfinkel *et al.*, 1991; Hansen *et al.*, 1992). Although the rate of Ty1 Gag processing is increased by galactose induction (Curcio & Garfinkel, 1992), Ty1 PR and Ty3 PR appear to be in low abundance in VLPs. Mature PR is difficult to detect reproducibly in galactose-induced strains by using a variety of antibodies (Garfinkel *et al.*, 1991; Kirchner & Sandmeyer, 1993). TED PR and its processing intermediates have been detected by pulse-chase analysis (Lerch & Friesen, 1992). No retrotransposon PR has been purified to homogeneity.

## Biological Aspects

All retrotransposon PRs appear to undergo autocatalysis. Like retroviral PRs, cleavage of the peptide scissile bond at the N-terminus of the Ty1 PR domain activates PR from its relatively inactive precursor to cleave the other sites in the Gag-Pol polyprotein (Merkulov *et al.*, 1996). The processing cascade of the remaining sites of cleavage occurs subsequently in an ordered pattern (Garfinkel *et al.*, 1991). Whether the initial cleavage of the N-terminus of PR is an intra- or intermolecular reaction is unknown.

The level of protein processing and therefore the level of Ty1 PR appears to limit the rate of Ty1 transposition (Curcio & Garfinkel, 1992). Since PRs from retrotransposons are probably homodimeric, dimerization of PR may be a means of regulating activation (Kirchner & Sandmeyer, 1993). The concentration of mature PR in the cytoplasm is low, even in cells overexpressing the retrotransposon. Concentrating PR within the VLP may facilitate dimerization and allow activity in a compartment that is not harmful to the cell. The inability to produce fully mature Ty1 proteins is a common defect associated with mutant genomic elements, indicating that inactive Ty elements have defects in either PR or cleavage sites (Garfinkel *et al.*, 1991). Plasmid-encoded Ty1 mutants with impaired PR activity are not complemented in *trans* by genomic Ty1 elements (Curcio & Garfinkel, 1992). Genomic Ty1 mutants with impaired PR activity are complemented in *trans* (Curcio & Garfinkel, 1994). Such conflicting results probably reflect the tremendously high levels of PR expressed from galactose-induced Ty1 elements relative to the expression levels from a single genomic Ty1 element (Curcio & Garfinkel, 1994). These data give further evidence that the levels of Ty1 PR are limiting for transposition.

## Acknowledgement

Research sponsored by the National Cancer Institute, Department of Health and Human Services, under contract with ABL. The contents of this publication do not necessarily reflect the views or policies of the Department of Health and Human Services, nor does mention of trade names, commercial products, or organizations imply endorsement by the US Government.

## References

Curcio, M.J. & Garfinkel, D.J. (1992) Posttranslational control of Ty1 retrotransposition occurs at the level of protein processing. *Mol. Cell. Biol.* **12**, 2813–2825.

Curcio, M.J. & Garfinkel, D.J. (1994) Heterogeneous functional Ty1 elements are abundant in the *Saccharomyces cerevisiae* genome. Genetics **136**, 1245–1259.

Doolittle, R.F., Feng, D.F., Johnson, M.S. & McClure, M.A. (1989) Origins and evolutionary relationships of retroviruses. *Q. Rev. Biol.* **64**, 1–29.

Eickbush, T.H. (1994) Origin and evolutionary relationships of retroelements. In: *The Evolutionary Biology of Viruses* (Morse, S.S., ed.). New York: Raven Press, pp. 121–157.

Garfinkel, D.J. (1992) Retroelements in microorganisms. In: *The Retroviridae*, vol. 1 (Levy, J.A., ed.). New York: Plenum Press, pp. 107–158.

Garfinkel, D.J., Hedge, A.M., Youngren, S.D. & Copeland, T.D. (1991) Proteolytic processing of *pol-TYB* proteins from yeast retrotransposon *Ty1*. *J. Virol.* **65**, 4573–4581.

Hansen, L.J. & Sandmeyer, S.B. (1990) Characterization of a transpositionally active Ty3 element and identification of the Ty3 integrase protein. *J. Virol.* **64**, 2599–2607.

Hansen, L.J., Chalker, D.L., Orlinsky, K.J. & Sandmeyer, S.B. (1992) Ty3 *GAG3* and *POL3* genes encode the components of intracellular particles. *J. Virol.* **66**, 1414–1424.

Kirchner, J. & Sandmeyer, S.B. (1993) Proteolytic processing of Ty3 proteins is required for transposition. *J. Virol.* **67**, 19–28.

Kirchner, J., Sandmeyer, S.B. & Forrest, D.B. (1992) Transposition of a Ty3 *GAG3–POL3* fusion mutant is limited by availability of capsid protein. *J. Virol.* **66**, 6081–6092.

Lerch, R.A. & Friesen, P.D. (1992) The baculovirus-integrated retro-transposon TED encodes *gag* and *pol* proteins that assemble into virus-like particles with reverse transcriptase. *J. Virol.* **66**, 1590–1601.

Levin, H.L., Weaver, D.C. & Boeke, J.D. (1993) Novel gene expression mechanism in a fission yeast retroelement: Tf1 proteins are derived from a single primary translation product. *EMBO J.* **12**, 4885–4895.

Merkulov, G.V., Swiderek, K.M., Brachmann, C.B. & Boeke, J.D. (1996) A critical proteolytic cleavage site near the C terminus of the yeast retrotransposon Ty1 Gag protein. *J. Virol.* **70**, 5548–5556.

Moore, S.P. & Garfinkel, D.J. (1994) Expression and partial purification of enzymatically active recombinant Ty1 integrase in *Saccharomyces cerevisiae*. *Proc. Natl Acad. Sci. USA* **91**, 1843–1847.

Pettit, S.C., Simsic, J., Loeb, D.D., Everitt, L., Hutchison, C.A., III. & Swanstrom, R. (1991) Analysis of retroviral protease cleavage sites reveals two types of cleavage sites and the structural requirements of the P1 amino acid. *J. Biol. Chem.* **266**, 14539–14547.

Yoshioka, K., Honma, H., Zushi, M., Kondo, S., Togashi, S., Miyake, T. & Shiba, T. (1990) Virus-like particle formation of *Drosophila copia* through autocatalytic processing. *EMBO J.* **9**, 535–541.

Youngren, S.D., Boeke, J.D., Sanders, N.J. & Garfinkel, D.J. (1988) Functional organization of the retrotransposon Ty from *Saccharomyces cerevisiae*: Ty protease is required for transposition. *Mol. Cell. Biol.* **8**, 1421–1431.

*Lori A. Rinckel*
*Genome Regulation and Chromosome Biology Laboratory,*
*National Cancer Institute, ABL-Basic Research Program,*
*PO Box B, Building 539, Frederick, MD 21702-1201, USA*

*David J. Garfinkel*
*Genome Regulation and Chromosome Biology Laboratory,*
*National Cancer Institute, ABL-Basic Research Program,*
*PO Box B, Building 539, Frederick, MD 21702-1201, USA*
*Email: garfinke@ncifcrf.gov*

# 322. *Cauliflower mosaic virus proteinase*

## Databanks

*Peptidase classification: clan AA, families A3 and A15, MEROPS ID: A03.001*
*NC-IUBMB enzyme classification: none*
*Databank codes:*

| Species | Strain/isolate | SwissProt | PIR | EMBL (genomic) |
| --- | --- | --- | --- | --- |
| **Family A3** | | | | |
| Carnation etched ring virus | | P05400 | S00854 | X04658: complete genome |
| Cassava vein mosaic virus | – | – | – | U20341: complete genome |
| | | | | U59751: complete genome |
| Cauliflower mosaic virus | B29 | – | – | X79465: complete genome |
| | BBC | Q02964 | – | M90542: complete genome |
| | CM-1841 | P03555 | D90799 | V00140: complete genome |
| | CMV-1 | – | – | M90543: complete genome |
| | D/H | P03556 | D90799 | M10376: complete genome |
| | NY8153 | Q00962 | – | M90541: complete genome |
| | Strasbourg | P03554 | D90799 | V00141: complete genome |
| *Commelina* yellow mottle virus | – | P19199 | – | X52938: complete genome |
| Figwort mosaic virus | – | P09523 | S01283 | X06166: complete genome |
| Peanut chlorotic streak virus | – | – | – | U13988: complete genome |
| Soyabean chlorotic mottle virus | – | P15629 | JS0375 | X15828: complete genome |
| Strawberry vein banding virus | – | | | X97304: complete genome |

| Species | Strain/isolate | SwissProt | PIR | EMBL (genomic) |
|---|---|---|---|---|
| **Family A15** | | | | |
| Rice tungro bacilliform virus | – | P27502 | C40785 | M65026: ORF2, ORF3 and ORF4 |
| | | | | X57924: ORFs P24, P12, P194 and P46 |
| Sugar cane bacilliform virus | – | – | – | M89923: complete genome |

Cacao bacilliform virus has no domain homologous to peptidases in A15 (or A3).

## Name and History

Cauliflower mosaic virus (CaMV) contains seven open reading frames (ORFs) located on the 8000-nucleotide-long pregenomic RNA. Sequence comparisons have revealed the relationship of the CaMV ORF V to the retrovirus *pol* genes (Toh *et al.*, 1985). Expression of ORF V in *Escherichia coli* (Gordon *et al.*, 1988; Torruella *et al.*, 1989) revealed that the ORF V polypeptide is cleaved to yield an N-terminal doublet of 20 and 22 kDa and a C-terminal fragment of 56 kDa apparent molecular masses. The doublet includes aspartate protease consensus motifs (i.e. 'Phe-Val-Asp-Thr-Gly-Ala') and 'Ile-Ile-Gly-Asn', which corresponds to the aspartic proteinase flap motif (Wlodawer *et al.*, 1989). The 20 and 22 kDa proteins are active in cleaving the CaMV structural polyprotein and the activity was termed ***cauliflower mosaic virus proteinase***. Mutation of Asp in the aspartic protease consensus sequence to either Glu or Ala destroys the activity and the corresponding mutation in the virus genome is lethal (Torruella *et al.*, 1989).

## Activity and Specificity

So far, the only known substrates are the CaMV polyproteins derived from ORF IV (Gag) and V (Pol). The proteinase activates itself by cleaving the ORF V polypeptide into a protease and a reverse transcriptase. It is not known whether this cleavage is performed in *cis* or in *trans*. Several cleavage sites within ORF IV are probably used by the CaMV proteinase, explaining the variety of ORF IV products detected in virus particles (Al Ani *et al.*, 1979); however, only one of these sites (i.e. the one creating the N-terminus of ORF IV derivative p44) has been identified (Martinez-Izquierdo & Hohn, 1987; Torruella *et al.*, 1989). This cleavage site (Leu+Ala) is within a highly hydrophobic region. Preproteinase produced from the reticulocyte in an *in vitro* translation system yields the active proteinase by self-cleavage upon further incubation.

A convenient assay uses a CaMV ORF IV fragment containing the cleavage site fused to the chloramphenicol acetyl transferase (CAT) coding region. Proteinase activity releases and thereby activates CAT. An uncharacterized inhibitor exists in the wheatgerm extract used for *in vitro* translation (Torruella *et al.*, 1989)

## Structural Chemistry

The CaMV proteinase is assumed to be a dimer in analogy to the related retrovirus proteinases and to contain a similar structure with an active site consisting of the loop with the active aspartate and the flaps (Le Grice *et al.*, 1988; Wlodaver *et al.*, 1989).

## Preparation/Biological Aspects

CaMV proteinase can be prepared by expression of the appropriate clones in *E. coli* or in reticulocyte lysates (Torruella *et al.*, 1989). The enzyme is found in virus particles. It is assumed that its activity contributes to the maturation of the virus after assembly.

## Related Proteinases

The closest-related proteinases (putative, based on sequence homologies) are found in other members of the caulimoviruses (ORF V): figwort virus (Richins *et al.*, 1987), carnation etch ring virus (Hull *et al.*, 1986) and peanut chlorotic streak virus (Richins, 1993); cassava vein mosaic virus (ORF III) (Calvert *et al.*, 1995); and badnaviruses (ORF III): *Commelina* yellow mottle virus (Medberry *et al.*, 1990), rice tungro bacilliform virus (Hay *et al.*, 1991; Qu *et al.*, 1991; Laco *et al.*, 1995), cacao swollen shoot virus (Hagen *et al.*, 1993) and sugar cane bacilliform virus (Bouhida *et al.*, 1993). Retrovirus retropepsins (Chapter 309) are more distantly related. These are usually shorter (i.e. around 100 residues compared to approximately 190 for CaMV proteinase). Non-viral aspartic proteinases (Chapter 271) are much larger (300–350 residues), contain the two Asp residues needed to form the active site and are active as monomers, in contrast to the retrovirus (and probably cauliflower mosaic virus) proteinase, which act as dimers.

## Further Reading

For further reading see Torruella *et al.* (1989).

## References

Al Ani, R., Pfeiffer, P. & Lebeurier, G. (1979) The structure of cauliflower mosaic virus. II: Identity and location of the viral polypeptides. *Virology* **93**, 188–197.

Bouhida, M., Lockhart, B.E.L. & Olszewski, N.E. (1993) An analysis of the complete sequence of sugarcane bacilliform virus genome infectious to banana and rice. *J. Gen. Virol.* **74**, 15–22.

Calvert, L.A., Ospina, M.D. & Shepherd, R.J. (1995) Characterization of cassava vein mosaic virus: a distinct plant pararetrovirus. *J. Gen. Virol.* **76**, 1271–1276.

Gordon, K., Pfeiffer, P., Fütterer, J. & Hohn, T. (1988) *In vitro* expression of cauliflower mosaic virus genes. *EMBO J.* **7**, 309–317.

Hagen, L.S., Jaquemont, M., Lepingle, A., Lot, H. & Tepfer, M. (1993) Nucleotide sequence and genomic organization of cacao swollen shoot virus. *Virology* **196**, 619–628.

Hay, J.M., Jones, M.C., Blackebrough, M.L., Dasgupta, I., Davies, J.W. & Hull, R. (1991) An analysis of the sequence of an infectious clone of rice tungro bacilliform virus, a plant pararetrovirus. *Nucleic Acids Res.* **19**, 2615–2621.

Hull, R., Sadler, J. & Longstaff, M. (1986) The sequence of carnation etch ring virus DNA: comparison with CaMV and retroviruses. *EMBO J.* **12**, 3083–3090.

Laco, G.S., Kent, S.B.H. & Beachy, R.N. (1995) Analysis of the proteolytic processing and activation of the rice tungro bacilliform virus reverse transcription. *Virology* **208**, 207–214.

LeGrice, S.F.J., Mills, J. & Mous, J. (1988) Active site mutagenesis of the AIDS virus protease and its alleviation by trans complementation. *EMBO J.* **7**, 2547–2553.

Martinez-Izquierdo, J. & Hohn, T. (1987) Cauliflower mosaic virus coat protein is phosphorylated *in vitro* by a virion associated protein kinase. *Proc. Natl Acad. Sci. USA* **84**, 1824–1828.

Medberry, S.L., Lockhart, B.E.L. & Olszewski, N.E. (1990) Properties of *Commelina* yellow mottle virus's complete DNA sequence, genomic discontinuities and transcript suggest that it is a pararetrovirus. *Nucleic Acids Res.* **18**, 5505–5512.

Qu, R., Bhattacharyya, M., Laco, G.S., de Kochko, A., Subba Rao, B.L., Kaniewska, M.B., Elmer, J.S., Rochester, D.E.,

Smith, C.E. & Beachy, R.N. (1991) Characterization of the genome of rice tungro bacilliform virus: comparison with commelina yellow mottle virus and caulimoviruses. *Virology* **185**, 354–364.

Richins, R.D. (1993) Organization and expression of the peanut chlorotic streak virus genome. Thesis, University of Kentucky.

Richins, R.D., Scholthof, H.B. & Shepherd, R.J. (1987) Sequence of figwort mosaic virus DNA (caulimovirus group). *Nucleic Acids Res.* **15**, 8451–8466.

Toh, H., Ono, M., Saigo, K. & Miyata, T. (1985) Retroviral protease-like sequence in the yeast transposon Ty1. *Nature* **315**, 691.

Torruella, M., Gordon, K. & Hohn, T. (1989) Cauliflower mosaic virus produces an aspartic proteinase to cleave its polyproteins. *EMBO J.* **8**, 2819–2825.

Wlodawer, A., Miller, M., Jaskolski, M., Sathayanarayana, B.K., Baldwin, E., Weber, I.T., Selk, L.M., Clawson, L., Schneider, J. & Kent, S.B.H. (1989) Conserved folding in retroviral proteases: crystal structure of a synthetic HIV-1 protease. *Science* **249**, 616–621.

***Thomas Hohn***
*Friedrich Miescher Institute,*
*PO Box 3543, CH-4002 Basel, Switzerland*
*Email:hohn@fmi.ch*

# 323. Introduction: clan AB containing the nodavirus coat protein

## Databanks

*MEROPS ID: AB*

**Family A6**
   Nodavirus endopeptidase (Chapter 324)

**Family A21**
   Tetravirus endopeptidase (Chapter 324)

Clan AB includes two families of coat proteins from viruses. *Family A6* contains the coat proteins from nodaviruses, which are single-stranded RNA viruses that infect insects, fish and mammals. The virus produces two RNA molecules, coding for nonstructural proteins (RNA1) and structural proteins (RNA2), respectively. RNA2 encodes the capsid protein, which in the flock house virus is synthesized as a 407 residue precursor. Once the virion has been assembled, and the RNA molecules packaged inside, the capsid protein precursor undergoes an autolytic cleavage that liberates a 44 residue C-terminal fragment. The cleavage is slow ($t_{0.5}$ is 4 h), and not all the coat proteins in the virion shell are cleaved. The bond cleaved is Asn363∔Ala364.

The tertiary structure of the black beetle nodavirus coat protein has been solved (Wery *et al*., 1994). The virion is an icosahedron containing 60 coat protein units, and each unit is triangular in form and consists of three coat protein molecules. The coat protein fold is of two domains, an eight-strand $\beta$ barrel on the surface, and a three-helix bundle on the inner face. The cleavage site precedes helix III, and the presumed catalytic residue, Asp75, is at the end of helix I. Asn363 is also thought to be a component of the catalytic site. Figure 323.1 shows a representation of the coat protein structure.

*Family A21* includes the coat protein from tetraviruses. Like nodaviruses, tetraviruses are single-stranded RNA viruses that infect insects such as the cotton bollworm (*Helicoverpa armigera*). The tertiary structure of the coat protein from the tetravirus that infects the insect *Nudaurelia capensis* has been solved, and, despite being a larger protein (644 residues) with little sequence relationship (see Alignment 323.1 on the CD-ROM), shows a remarkable structural similarity to the nodavirus coat protein (Munshi *et al*, 1996). Like the nodavirus coat protein, there is an autolytic cleavage near the C-terminus at an asparginyl bond (Asn570∔Phe571), and the asparagine is part of a catalytic dyad. The second element of the dyad is a Glu residue, not an Asp as in family A6. Details of the catalytic mechanism have yet to be determined, but it is possible that the tetravirus coat protein is an example of a glutamyl endopeptidase.

The following differences distinguish endopeptidases of clan AB from those of clan AA. The folds show no similarity, and there is no evidence of gene duplication. The coat protein has a potential Asp/Glu, Asn catalytic dyad, whereas in clan

*Figure 323.1*   Richardson diagram of one cleaved molecule of the black beetle virus coat protein. The image was prepared from the Brookhaven Protein Data Bank entry (2BBV) as described in the Introduction (p. xxv). Catalytic residues are shown in ball-and-stick representation: Asp75 and Asn363.

AA there is an Asp, Asp dyad. Conservation of residues around the putative catalytic residues of the coat protein is shown in Alignment 323.1 (on CD-ROM).

## References

Munshi, S., Liljas, L., Cavarelli, J., Bomu, W., McKinney, B., Reddy, V. & Johnson, J.E. (1996) The 2.8 Å structure of a *T*=4 animal virus and its implications for membrane translocation of RNA. *J. Mol. Biol.* **261**, 1–10.

Wery, J.P., Reddy, V.S., Hosur, M.V. & Johnson, J.E. (1994) The refined 3-dimensional structure of an insect virus at 2.8 Å resolution. *J. Mol. Biol.* **235**, 565–586.

# 324. *Nodavirus endopeptidase*

## Databanks

*Peptidase classification: clan AB, family A6, MEROPS ID: A06.001, A21.001*
*NC-IUBMB enzyme classification: none*
*Databank codes:*

| Species | SwissProt | PIR | EMBL (cDNA) | EMBL (genomic) |
|---|---|---|---|---|
| **Family A6** | | | | |
| Black beetle virus | P04329 | A04151 S11036 | X00956 | – |
| Boolarra virus | P12869 | A34011 S11038 | X15960 | – |
| Flock House virus | P12870 | B34011 S11037 | X15959 | – |
| Nodamura virus | P12871 | C34011 S11039 | X15961 | – |
| **Family A21** | | | | |
| *Helicoverpa armigera* stunt virus | – | – | L37299 | – |
| *Nudaurelia capensis* omega virus | – | A43370 | S43937 | |

Brookhaven Protein Data Bank three-dimensional structures:

| Species | ID | Resolution | Notes |
|---|---|---|---|
| Black beetle virus | 2BBV | 2.8 | complex with duplex RNA |

## Name and History

Many animal viruses initially assemble into a noninfectious precursor particle, the provirion, that subsequently matures to the infectious virion. This maturation involves either cellular proteases such as the trypsin-like endoprotease in the case of influenza virus or virally encoded proteases (e.g. the retropepsin of retroviruses) (Chapter 309). In general, the proteases cleave a polypeptide chain associated with the precursor particle and this cleavage results in acquisition of virion infectivity. In some cases it is clear why such a cleavage must occur for infectivity. For example, in influenza virus, cleavage of the hemagglutinin precursor HA0 into HA1 and HA2 makes a fusion peptide available that is required for membrane translocation of the particle. In several small (approximately 30–40 nm diameter), nonenveloped, RNA animal viruses such as picornaviruses, nodaviruses and tetraviruses, the tertiary and quaternary structures of the particle are critical to the mechanism of maturation. For example, autocatalytic cleavage following assembly (Gallagher & Rueckert, 1988) is the final event that produces the infectious virion (Schneemann *et al.*, 1992) of nodaviruses.

There is still debate on the detailed mechanism for cleavage in picornaviruses (Basavappa *et al.*, 1994), but a mechanism for the proteolytic activity of nodaviruses has been proposed on the basis of the black beetle virus (BBV) structure (Hosur *et al.*, 1987; Wery *et al.*, 1994 ) and site-directed mutagenesis has confirmed a critical role of a single aspartic acid residue in the mechanism (Zlotnick *et al.*, 1994). This ***nodavirus endopeptidase*** activity is not a typical enzymatic process in that each active site functions only once to release a 44 amino acid polypeptide from the C-terminal portion of the subunit (Hosur *et al.*, 1987). Thus there is no turnover. Recently the structure of *Nudaurelia capensis* ω virus (N ω V) was reported and the cleavage site in this $T=4$ virus ($T=4$ indicates that the icosahedral particle is composed of 240 copies of one subunit type) is remarkably similar to that of the $T=3$ ($T=3$ indicates that the particle is formed of 180 copies of one subunit type) nodaviruses, with the exception that a glutamic acid residue plays the role attributed to the aspartic acid residue in nodaviruses (Munshi *et al.*, 1996). There is no evidence that there is any similarity between the proposed mechanism of cleavage of particles of the insect viruses and the mammalian picornaviruses.

## Activity and Specificity

The reaction resulting from the endopeptidase activity of nodaviruses is:

$$\alpha(407aa) \rightarrow \beta(363aa) + \gamma(44aa)$$

The maturation proteolysis occurs only after the subunits of nodaviruses or tetraviruses have assembled and it is a slow reaction with a half-life of nearly 4 h at 26°C, the growth temperature of BBV, the only virus for which kinetic data were recorded (Gallagher & Rueckert, 1988). In each of the viruses studied the cleavage occurs following an asparagine residue – Asn363 in nodaviruses (Hosur *et al.*, 1987) and Asn571 in tetraviruses (Agrawal & Johnson, 1992). In nodaviruses the site of cleavage is a conserved Asn┼Ala bond (Kaesberg *et al.*, 1990), while in the N ω V tetravirus it is Asn┼Phe (Agrawal & Johnson, 1992). The reaction is dependent upon divalent metal ions since the presence of EDTA during release and purification of uncleaved precursor particles from infected cells dramatically reduces the rate of maturation (Schneemann *et al*., 1994; Agrawal & Johnson, 1995). It is thought that maturation of provirions *in vivo* is, in fact, dependent upon release of particles from the cell cytoplasm into the growth medium, which contains high concentrations of divalent cations.

### Structural Chemistry

Figure 324.1 illustrates the environment of the cleavage site observed in BBV (Wery *et al*., 1994). The site in N $\omega$ V is very similar. In each case an acidic residue is hydrogen-bonded to the new C-terminal Asn carboxyl group. In the nodaviruses the acidic residue is an aspartic acid, while in

N $\omega$ V it is a glutamic acid. The reaction is intramolecular with the critical acidic residue associated with the same subunit in which the cleavage occurs. The assembly dependence results from (1) the stress placed on the scissile bond by the particle quaternary structure and (2) the stabilization of the hydrophobic pocket in which the critical acid residue is located. The hydrophobic pocket leads to an environmentally elevated p$K$ of the acidic group which becomes protonated and thereby allows formation of the critical hydrogen bond that is still observed in the mature particle. Mutation of the critical Asp75 to Asn in the Flock House nodavirus prevents the reaction (Zlotnick *et al*., 1994).

Figure 324.2 illustrates the proposed mechanism and Figure 324.3 the role of the virus quaternary structure in driving the reaction (Zlotnick *et al*., 1994). The role of divalent metal ions appears to be indirect in that there is no electron density attributable to metal ions in the vicinity

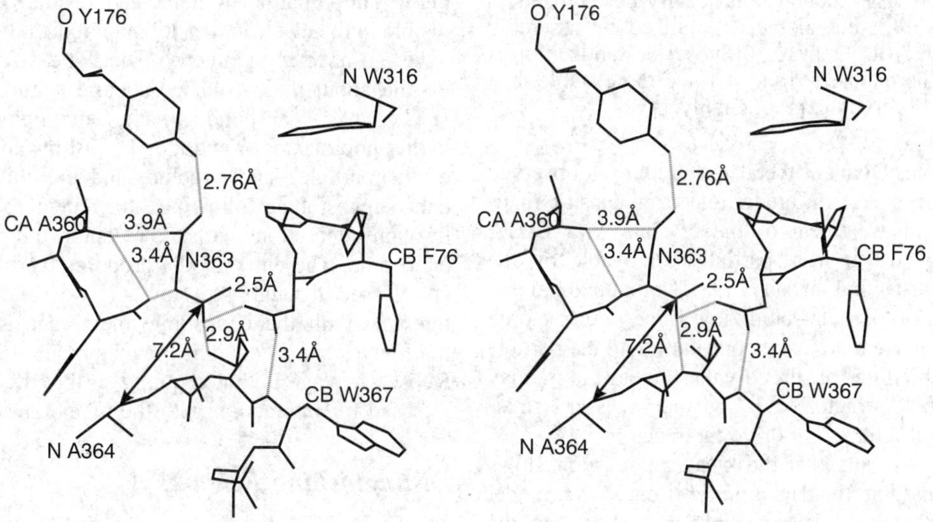

*Figure 324.1*    Stereoview of the active site in BBV showing the configuration of residues in the postcleavage state. Asn363 is the new C-terminus and probably has a position nearly identical to the precleaved state. Hydrogen bonds made by the side chain of this residue direct its carboxyl group (originally at the scissile bond) at Asp75, allowing the hydrogen bond to form. Ala364 is the new N-terminus and moves 7.2 Å following the cleavage.

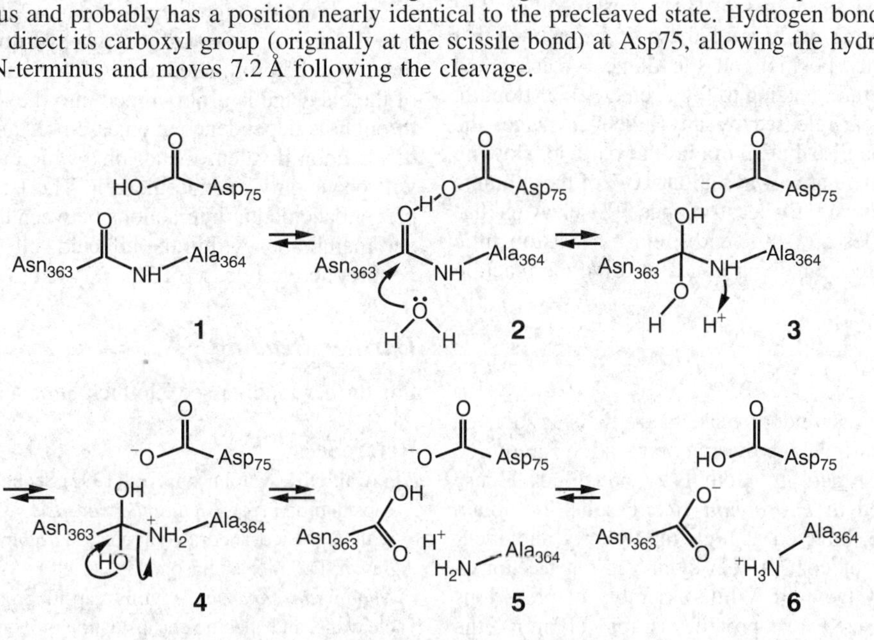

*Figure 324.2*    The proposed chemical mechanism for the acid-catalyzed cleavage of the scissile bond.

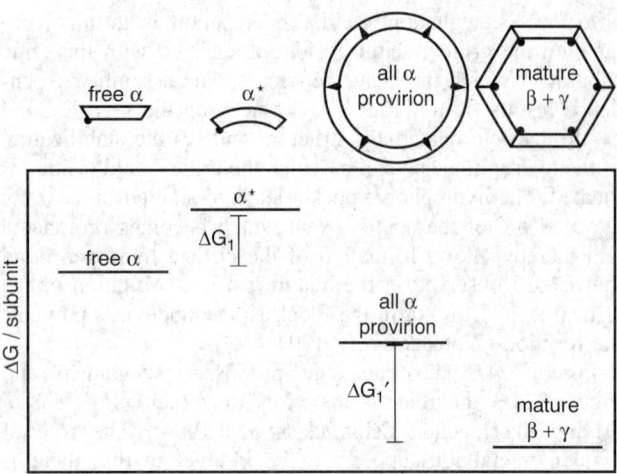

*Figure 324.3* Cartoon showing how the quaternary structure drives the reaction. The subunit α is locally destabilized as the scissile bond (α*), but globally stabilized by assembly into the provirion. After cleavage the local destabilization is relaxed, making the mature (β + γ) particle more stable than the provirions as observed experimentally.

of the cleavage site. Divalent metal ion-binding is observed at subunit interfaces and difference electron density maps computed following treatment of the crystals with EDTA indicate that they are at least partially removable by the chelating agent (Fisher & Johnson, 1993). Computational and chemical studies of particle-stabilizing forces demonstrate that these cations make a dramatic contribution to the particle stability. Thus, binding of the divalent metal ions may 'tighten' the particle structure and add the necessary stress to the scissile bond required to drive the reaction.

It is interesting to note that the cleavage reaction slows down with time and that the data are consistent with a model in which each cleavage in the particle is slower than the one before. This is consistent with the concept of tensegrity observed in the geodesic domes of Buckminster Fuller. The mechanical engineering analysis of these domes shows that stress applied to any position on the dome is distributed throughout the structure, leading to their remarkable strength. Likewise, as stress is released by any given cleavage, the reduction in stress is distributed around the particle, slowing the next reaction. There are always about 10% of the subunits that do not cleave, apparently because once 90% cleavage has occurred, the particles are sufficiently relaxed that too little stress is applied to the remaining bonds to drive the reaction to completion.

## Preparation

Because the nodavirus endoprotease activity is associated with the precursor particle, a provirion preparation is required to study the protease activity. Both BBV and Flock House virus are propagated in *Drosophila* line 1 cells. To obtain provirions, cells are infected at high multiplicity and incubated until the rate of coat protein synthesis reaches maximum levels, usually for 16 h. While assembly of provirions from newly synthesized coat protein is rapid (10 min), the subsequent cleavage reaction is slow ($t_{1/2}$ 4 h) leading to a temporary accumulation of provirions in the infected cells. At 16 h post-infection, cells are lysed in ice-cold buffer containing NP-40 and the lysate is clarified. Provirions in the extracts are pelleted by ultracentrifugation through a 10–30% (w/w) sucrose gradient and resuspended in buffer with or without divalent cations depending on the type of analysis to be performed. The entire purification procedure is performed in the cold and in the presence of EDTA to prevent cleavage of the precursor particles.

## Biological Aspects

The autocatalytic cleavage of nodaviruses releases the C-terminal 44 residues from the subunit. As shown in Figure 324.1, the distance between the new C and N terminals is about 7 Å. The cleaved-off polypeptide is internal and remains associated with the particle following proteolysis. Residues 364–382 of BBV form a canonical, highly amphipathic, α helix and residues 383–407 are not visible in the electron density map, indicating that they do not have icosahedral symmetry. The helices form a pentameric bundle about the 5-fold icosahedral symmetry axes (Cheng *et al.*, 1994). The bundles have a hydrophilic interior and a hydrophobic exterior and they lie just inside the outer surface of the particle. The structure and preliminary biochemical data suggest that following attachment to the cell surface through interactions with the receptor, the particles enter the endosome. The entry model proposes that the helical bundle is released through the (low pH-weakened) particle shell, interacts with the endosomic membrane and forms a pore that may serve as conduit for membrane translocation of the RNA. Cleavage is thus required so that the γ peptide can be released for interaction with the membrane.

## Distinguishing Features

The mechanism for the protease activity described depends upon a single aspartic acid residue and destabilization of the scissile bond through local quaternary structure contacts. The burial of the aspartic acid within a hydrophobic pocket assures that it will be protonated at physiological pH. Thus the timing of the cleavage is programmed into the life cycle of the virus through its dependence upon assembly and through its dependence upon divalent metal ions, which assures that cleavage will occur only outside the cell. The latter requirement may prevent accidental interactions between the infectious particle and membranes within the infected cell, thereby preventing a nonproductive release of RNA.

## Further Reading

For further reading see Zlotnick *et al.* (1994).

## References

Agrawal, D.K. & Johnson, J.E. (1992) Sequence and analysis of the capsid protein of *Nudaurelia capensis* ω virus, an insect virus with *T* = 4 icosahedral symmetry. *Virology* **190**, 806–814.

Agrawal, D.K. & Johnson, J.E. (1995) Assembly of the *T* = 4 *Nudaurelia capensis* ω virus capsid protein, post translational cleavage, and specific encapsidation of its mRNA in a baculovirus expression system. *Virology* **207**, 89–97.

Basavappa, R.,  Syed, R.,  Flore, O.,  Icenogle, J.,  Filman, D.  & Hogle, J. (1994) Role and mechanism of the maturation cleavage of VP0 in poliovirus assembly. *Protein Sci.* **3**, 1651–1669.

Cheng, R.H., Reddy, V.S., Olson, N.H., Fisher, A.J., Baker, T.S. & Johnson, J.E. (1994) Functional implications of quasi-equivalence in a $T=3$ icosahedral animal virus established by cryo-electron microscopy and X-ray crystallography. *Structure* **2**, 271–282.

Fisher, A.J. & Johnson, J.E. (1993) Ordered duplex RNA controls the capsid architecture of an icosahedral animal virus. *Nature* **361**, 176–179.

Gallagher, T.M. & Rueckert, R.R. (1988) Assembly-dependent maturation cleavage in provirions of a small icosahedral insect ribovirus. *J. Virol.* **62**, 3399–3406.

Hosur, M.V., Schmidt, T, Tucker, R.C., Johnson, J.E., Gallagher, T.M., Selling, B.H. & Rueckert, R.R. (1987) Structure of an insect virus at 3.0 Å resolution. *Proteins: Struct. Funct. Genet.* **2**, 167–176.

Kaesberg, P., Dasgupta, R., Sgro, J.-Y., Wery, J.-P., Selling, B.H., Hosur, M.V. & Johnson, J.E. (1990) Structural homology among four nodaviruses as deduced by sequencing and X-ray crystallography. *J. Mol. Biol.* **214**, 423–435.

Munshi, S., Liljas, L., Cavarelli, J., Bomu, W., McKinney, B., Reddy, V. & Johnson, J.E. (1996) The 2.8 Å structure of a $T=4$ animal virus and its implications for membrane translocation of RNA. *J. Mol. Biol.* **261**, 1–10.

Schneemann, A., Zhong, W., Gallagher, T. & Rueckert, R. (1992) Maturation cleavage required for infectivity of a nodavirus. *J. Virol.* **66**, 6728–6734.

Schneemann, A., Gallagher, T. & Rueckert, R. (1994) Reconstitution of Flock house virus provirions: a model system for studying structure and assembly. *J. Virol.* **68**, 4547–4556.

Wery, J.-P., Reddy, V., Hosur, M.V. & Johnson, J.E. (1994) The refined structure of an insect virus at 2.8 Å resolution. *J. Mol. Biol.* **235**, 565–586.

Zlotnick, A., Reddy, V.S., Dasgupta, R., Schneemann, A., Ray, W.J., Rueckert, R.R. & Johnson, J.E. (1994) Capsid assembly in a family of animal viruses primes an autoproteolytic maturation that depends on a single aspartic acid residue. *J. Biol. Chem.* **269**, 13680–13684.

***John E. Johnson***
*Department of Molecular Biology,*
*The Scripps Research Institute,*
*10550 N. Torrey Pines Road,*
*La Jolla, CA 92037, USA*
*Email: jackj@scripps.edu*

***Anette Schneemann***
*Department of Molecular Biology,*
*The Scripps Research Institute,*
*10550 N. Torrey Pines Road,*
*La Jolla, CA 92037, USA*

# 325. Introduction: other families of aspartic endopeptidases

## Databanks

MEROPS ID: A4, A5, A7, A8

| Species | SwissProt | PIR | EMBL (cDNA) | EMBL (genomic) |
|---|---|---|---|---|
| **Family A4** | | | | |
| Aspergillopepsin II (Chapter 327) | | | | |
| Scytalidopepsin B (Chapter 326) | | | | |
| **Family A5** | | | | |
| Thermopsin (Chapter 332) | | | | |
| **Family A7** | | | | |
| Pseudomonapepsin (Chapter 329) | | | | |
| Xanthomonapepsin (Chapter 330) | | | | |
| Others | | | | |
| *Dictyostelium discoideum* | – | – | U27540 | – |
| *Physarum polycephalum* | – | A44869 | – | – |
| **Family A8** | | | | |
| Signal peptidase II (Chapter 333) | | | | |
| Other | | | | |
| *Helicobacter pylori* | P25178 | D38537 | M60398 | – |

There are four families of aspartic-type endopeptidases that cannot be assigned to clans because the catalytic residues have not been identified, and no three-dimensional structures are available. Two of these families contain endopeptidases that are not sensitive to inhibition by pepstatin, a rather general inhibitor of the aspartic peptidases in clan AA. In the absence of identification of the catalytic residues, it may be premature to term these enzymes 'aspartic peptidases' but this is done here for simplicity.

*Family A5* contains thermopsin, an aspartic endopeptidase from the archaean *Sulfolobus acidocaldarius*. Consistent with the fact that the organism is an acidic thermophile, the enzyme is most active at pH 2 and at 70°C. Thermopsin is secreted, remaining covalently attached to the cell wall, and is synthesized with a signal peptide and a propeptide of about 12 residues. Thermopsin is competitively inhibited by pepstatin with a $K_i$ of $2 \times 10^{-7}$ M at 76°C, and is also inhibited by DAN and 1,2-epoxy-3-($p$-nitrophenoxy)propane (Lin & Tang, 1995). Unlike pepsin, thermopsin does not contain an Asp-Thr-Gly (or 'DTG') motif, and there is no evidence from the sequence of internal homology. The mature protein has six aspartates, any one of which could be catalytic.

The other family of pepstatin-sensitive aspartic endopeptidases is *family A8*, containing signal peptidase II from bacteria. Signal peptidase II is responsible for the removal of the signal peptide from the N-terminus of the precursor of murein lipoprotein, one of the most abundant bacterial cell wall proteins. The cysteine residue that will become the new N-terminus of the lipoprotein must be modified by substituting the sulfur with a $CH_2(OOCR_1)CH(OOCR_2)CH_2$-group

before signal peptidase II will cleave it. Signal peptidase II is inhibited by pepstatin (Dev & Ray, 1984) and globomycin (Dev *et al.*, 1985). The enzyme is membrane bound, and has four membrane-spanning domains and two larger, periplasmic regions. *Helicobacter pylori* UreD protein is a distant homolog, but is a protein of unknown function controlled by the urease operon (Labigne *et al.*, 1991).

There are two conserved Asp residues in signal peptidase II. Asp23 is within the first membrane-spanning domain, and Asp123 is periplasmic. Although the catalytic residues are believed to be within periplasmic domains, it may be significant that a natural mutation of Asp23 to Gly produces an enzyme that is globomycin resistant and temperature sensitive. Asn53 is also conserved and is within a periplasmic region. An Asp, Asn catalytic dyad would be similar to that of peptidases from clan AB (Chapter 323). A catalytic dyad of Asn53 and Asp123 would suggest a typical two-domain peptidase, with one catalytic residue per domain and the active site between the domains. An alignment around the potential catalytic residues is shown in Alignment 325.1 (on CD-ROM).

*Family A4* contains the pepstatin-insensitive endopeptidases aspergillopepsin II and scatylidopepsin B. Another homolog is the EapB protein from *Cryphonectria parasitica*. All known members of the family are from fungi. These fungal endopeptidases are active at very low pH, and inhibition by 1,2-epoxy-3-($p$-nitrophenoxy)propane (EPNP), which has been shown to modify a catalytic Asp in pepsin (Tang, 1971), is evidence that they are aspartic-, or at least carboxyl-type peptidases. EPNP has been shown to react with Glu53 in scylidopepsin B (Tsuru *et al.*, 1986). However, aspergillopepsin II and the EapB protein have Gln in this position (Inoue *et al.*, 1991; Jara *et al.*, 1996), as had the originally published amino acid sequence of scytalidopepsin B (Maita *et al.*, 1984). A second Asp (Asp98) predicted to be catalytic in scytalidopepsin B is replaced by Lys in aspergillopepsin II and Leu in the EapB protein.

Aspergillopepsin II is synthesized with a 59 residue pre-propeptide, and activation also involves removal of an internal 11 residue peptide to generate a two-chain form.

The second family of pepstatin-insensitive aspartic-type endopeptidases is *family A7*, which contains the bacterial enzymes pseudomonapepsin and xanthomonapepsin. There are also two members of this family from slime molds: a cell type-specific protein from *Physarum polycephalum* and the vegetative stage-specific V4-7 protein from *Dictyostelium discoideum*. The identified catalytic aspartates are not conserved in the *Dictyostelium* protein, which is probably not a peptidase. Pseudomonapepsin is synthesized with a 200 residue propeptide, and the xanthomonapepsin precursor possesses a C-terminal domain similar to those of other endopeptidases from *Vibrio* and *Achromobacter* that is required for secretion, and is proteolytically removed to activate the enzyme.

### References

Dev, I.K. & Ray, P.H. (1984) Rapid assay and purification of a unique signal peptidase that processes the prolipoprotein from *Escherichia coli B. J. Biol. Chem.* **259**, 11114–11120.

Dev, I.K., Harvey, R.J. & Ray, P.H. (1985) Inhibition of prolipoprotein signal peptidase by globomycin. *J. Biol. Chem.* **260**, 5891–5894.

Inoue, H., Kimura, T., Makabe, O. & Takahashi, K. (1991) The gene and deduced protein sequences of the zymogen of *Aspergillus niger* acid proteinase A. *J. Biol. Chem.* **266**, 19484–19489.

Jara, P., Gilbert, S., Delmas, P., Guillemot, J.C., Kaghad, M., Ferrara, P. & Loison, G. (1996) Cloning and characterization of the *eapB* and *eapC* genes of *Cryphonectria parasitica* encoding two new acid proteinases, and disruption of *eapC. Mol. Gen. Genet.* **250**, 97–105.

Labigne, A., Cussac, V. & Courcoux, P. (1991) Shuttle cloning and nucleotide sequences of *Helicobacter pylori* genes responsible for urease activity. *J. Bacteriol.* **173**, 1920–1931.

Lin, X. & Tang, J. (1995) Thermopsin. *Methods Enzymol.* **248**, 156–168.

Maita, T., Nagata, S., Matsuda, G., Maruta, S., Oda, K., Murao, S. & Tsuru, D. (1984) Complete amino acid sequence of *Scytalidium lignicolum* acid protease B. *J. Biochem. (Tokyo)* **95**, 465–475.

Tang, J. (1971) Specific and irreversible inactivation of pepsin by substrate-like epoxides. *J. Biol. Chem.* **246**, 4510–4517.

Tsuru, D., Shimada, S., Maruta, S., Yoshimoto, T., Oda, K., Murao, S., Miyata, T. & Iwanaga, S. (1986) Isolation and amino acid sequence of a peptide containing an epoxide-reactive residue from the thermolysin-digest of *Scytalidium lignicolum* acid protease B. *J. Biochem. (Tokyo)* **99**, 1537–1539.

**A**

# 326. *Scytalidopepsin B*

### Databanks

*Peptidase classification: clan AX, family A4, MEROPS ID: A04.001*
*NC-IUBMB enzyme classification: EC 3.4.23.32*
*Chemical Abstracts Service registry number: 104781-89-7*
*Databank codes:*

| Species | SwissProt | PIR | EMBL (cDNA) | EMBL (genomic) |
|---|---|---|---|---|
| *Scytalidium lignicolum* | P15369 | A28864 | D83963 | – |

### Name and History

Pepstatin-insensitive carboxyl proteinases were found by testing with a carboxyl proteinase inhibitor, S-PI (Ac-pepstatin) isolated by Murao in 1970 (Murao & Oda, 1985). The structure of S-PI is Ac-Val-Val-AHMHA-Ala-AHMHA, where AHMHA is 4-amino-3-hydroxy-6-methylheptanoic acid (statine). In the same year, Umezawa *et al.* (1970) isolated pepstatin containing isovaleryl as the acyl residue. Since 1970, we focused on the inhibitory spectrum of S-PI, which inhibited all the carboxyl proteinases known at that time. We considered that an S-PI-insensitive carboxyl proteinase, if there was one, should be distinct in the structure of the active site from any other known carboxyl proteinases. We therefore screened for a microorganism that might produce an S-PI-insensitive carboxyl proteinase and succeeded in isolating *Scytalidium lignicolum* in 1972. This strain produces four distinct carboxyl proteinases: A-1, A-2, B and C. None of them is inactivated by S-PI, DAN or 1,2-epoxy-3-(*p*-nitrophenoxy)propane (EPNP), with the exception of carboxyl proteinase B, which is inactivated by EPNP. This enzyme was referred to as **Scytalidium pepstatin-insensitive carboxyl** *proteinase B*. In 1992 the IUBMB recommended the name *scytalidopepsin B*.

## Activity and Specificity

Carboxyl proteinase B is most active at pH 2.0 when casein is used as substrate. The activity is not inhibited by pepstatin, S-PI or DAN, but is inhibited by EPNP. The specific activity is 60% of that of scytalidopepsin A (Chapter 328). This enzyme cleaves the B chain of oxidized insulin at Tyr26↓Thr27 as well as other positions such as Phe24↓Phe25 (Oda & Murao, 1976):

FVNQHLCGSHLVE↓A↓LYLVCGERGF↓FY↓TPKA

No other known carboxyl proteinase cleaves the Tyr26↓Thr27 bond. This enzyme also cleaves the His6↓Pro7 bond of angiotensin I (Majima *et al.*, 1988), another unique characteristic of this enzyme. When Z-tetrapeptides such as Z-Xaa-Leu-Ala-Ala and Z-Phe-Xbb-Ala-Ala (where Xaa and Xbb are various amino acid residues) are used as substrates, scytalidopepsin A (Chapter 328) cleaves these peptides at either the Leu-Ala or Xbb-Ala bond (Oda & Murao, 1991). The $k_{cat}/K_m$ values are between 8 000 and 11 000 M$^{-1}$ s$^{-1}$. Scytalidopepsin B, by comparison, cleaves the C-terminal bonds and its activity is very low ($k_{cat}/K_m = 30$–$40$ M$^{-1}$ s$^{-1}$). This enzyme prefers Lys, Leu or Phe as the Xaa and Tyr or Lys as the Xbb residue. Such S-PI-sensitive carboxyl proteinases as pepsin (Chapter 272) cleave the Xaa-Leu and Phe-Xbb bonds. The $k_{cat}/K_m$ values are between 200 and 1000 M$^{-1}$ s$^{-1}$. Thus the substrate specificity of scytalidopepsin B is distinguishable from that of other microbial carboxyl proteinases sensitive to pepstatin, DAN and EPNP.

## Structural Chemistry

This enzyme is one of the smallest carboxyl proteinases active as a monomer. The pI is pH 3.2. The enzyme is a single polypeptide composed of 204 amino acid residues with a molecular weight of 21 969 Da (Maita *et al.*, 1984). The amino acid sequence is quite different from those of pepstatin-sensitive carboxyl proteinases such as pepsin and penicillopepsin. Furthermore, unlike the other carboxyl proteinases, one of the catalytic residues is a glutamic acid. The active amino acid residue modified with EPNP was found to be Glu53 (Tsuru *et al.*, 1986). The other catalytic residue was found to be Asp98 by using a new inhibitor, 1-diazo-3-phenyl-2-propanone (Tsuru *et al.*, 1989). The amino acid sequence around Glu53 shows similarity to that around the active-site Asp215 residue of pig pepsin (Chapter 272) and other aspartic proteinases. While the amino acid sequence around Asp98 also shows similarity to that around the active Asp32, there is an insertion of a serine residue between Thr

and Gly. This was the first demonstration of a glutamic proteinase. Cloning of the enzyme gene is in progress.

## Preparation

From 1 liter of culture filtrate of *S. lignicolum* ATCC 24568, 60 mg of purified enzyme was obtained.

## Biological Aspects

*S. lignicolum* is a wood-destroying fungus, and the biological function of these enzymes is assumed to be the degradation of protein in the wood of dead trees.

## Distinguishing Features

It is easy to distinguish this enzyme from the other pepstatin-insensitive ones because this enzyme is inhibited by EPNP.

## Further Reading

For further reading see Oda *et al.* (1995).

## References

Maita, T., Nagata, S., Matsuda, G., Maruta, S., Oda, K., Murao, S. & Tsuru, D. (1984) Complete amino acid sequence of *Scytalidium lignicolum* acid protease B. *J. Biochem. (Tokyo)* **95**, 465–475.

Majima, E., Oda, K., Murao, S. & Ichishima, E. (1988) Comparative study on the specificities of several fungal aspartic and acid proteinases toward the tetradecapeptide of a renin substrate. *Agric. Biol. Chem.* **52**, 787–793.

Murao, S. & Oda, K. (1985) Pepstatin-insensitive acid proteinases. In: *Aspartic Proteinases and Their Inhibitors* (Kostka, V., ed.). Berlin: Walter de Gruyter, pp. 379–399.

Oda, K. & Murao, S. (1976) Action of *Scytalidium lignicolum* acid proteases on insulin B-chain. *Agric. Biol. Chem.* **40**, 1221–1225.

Oda, K. & Murao, S. (1991) Pepstatin-insensitive carboxyl proteinases. In: *Structure and Function of The Aspartic Proteinases* (Dunn, B.M., ed.). New York: Plenum Press, pp. 185–201.

Oda, K., Takahashi, S., Shin, T. & Murao, S. (1995) Pepstatin-insensitive carboxyl proteinases. In: *Aspartic Proteinases: Structure, Function, Biology, and Biomedical Implications* (Takahashi, K., ed.). New York: Plenum Press, pp. 529–542.

Tsuru, D., Shimada, S., Maruta, S., Yoshimoto, T., Oda, K., Murao, S., Miyata, T. & Iwanaga, S. (1986) Isolation and amino acid sequence of a peptide containing an epoxide-reactive residue from the thermolysin-digest of *Scytalidium lignicolum* acid protease B. *J. Biochem (Tokyo)* **99**, 1537–1539.

Tsuru, D., Kobayashi, R., Nakagawa, N. & Yoshimoto, T. (1989) Inhibition of *Scytalidium lignicolum* acid protease B by 1-diazo-3-phenyl-2-propane. *Agric. Biol. Chem.* **53**, 1305–1312.

Umezawa, H., Aoyagi, T., Morishima, H., Matsuzaki, M., Hamada, M. & Takeuchi, T. (1970) Pepstatin, a new pepsin inhibitor produced by Actinomycetes. *J. Antibiot.* **23**, 259–265.

*Kohei Oda*
*Department of Applied Biology,*
*Faculty of Textile Science,*
*Kyoto Institute of Technology,*
*Sakyoku, Kyoto 606, Japan*
*Email: bika@ipc.kit.ac.jp*

# 327. *Aspergillopepsin II*

## Databanks

*Peptidase classification: clan AX, family A4, MEROPS ID: A04.002*
*NC-IUBMB enzyme classification: EC 3.4.23.19*
*Databank codes:*

| Species | SwissProt | PIR | EMBL (cDNA) | EMBL (genomic) |
|---|---|---|---|---|
| *Aspergillus niger* | P24665 | A41025 | M68871 | – |
| *Cryphonectria parasitica* (EapB) | – | – | X83998 | – |
| *Cryphonectria parasitica* (EapC) | – | – | X83997 | – |

## Name and History

*Aspergillopepsin II* was discovered in 1964 in the culture filtrate of *Aspergillus niger* var. *macrosporus* together with aspergillopepsin I (Chapter 294) and isolated from the crude powder of the filtrate (commercially named Proctase from Meiji; see Appendix 2 for full names and addresses of suppliers) (Koaze *et al.*, 1964). It was thus named *proctase A* and has also been called *Aspergillus proteinase A*.

## Activity and Specificity

The substrate specificity was examined using performic acid-oxidized cattle insulin B chain (Iio & Yamasaki, 1976):

FVN┼QHLCGSHLVE┼ALYLVCGERGFFY┼TPKA

Some bioactive peptides (Ido *et al.*, 1976; Takahashi, 1995) and performic acid-oxidized cattle pancreatic ribonuclease A (Takahashi, 1995) were also used. These results show that the enzyme has primary specificity at the P1 position for Tyr, Phe, His, Asn, Asp, Gln and Glu and that it has certain additional subsite specificity on both sides of the scissile bonds that has not yet been clearly defined. It does not hydrolyze Mca bonds in synthetic peptide-Mca substrates. The enzyme is thought to have an extended substrate-binding site like ordinary aspartic proteinases. It degrades various proteins and peptides, but less extensively than aspergillopepsin I; the specific activity toward casein is approximately one-tenth of that of aspergillopepsin I. However, it acts on collagen and gelatin more strongly than aspergillopepsin I.

Assay is routinely performed with the protein substrates cattle hemoglobin at pH 3.5 and 37°C (Chang *et al.*, 1976) or casein at pH 2.6 or 1.5 and 30°C (Koaze *et al.*, 1964) by measuring spectrophotometrically the trichloroacetic acid-soluble hydrolysis products. The optimum pH is 1.8 toward cattle hemoglobin (Takahashi, 1995) and 2.0 toward casein (Koaze *et al.*, 1964). The optimum temperature for casein hydrolysis is about 70°C at pH 2.6 or 60°C at pH 1.5 (Koaze *et al.*, 1964).

The enzyme is resistant to the inhibitors of ordinary pepsin-type aspartic proteinases such as pepstatin, DAN and 1,2-epoxy-3-(*p*-nitrophenoxy)propane (EPNP) (Chang *et al.*, 1976), and no naturally occurring inhibitor is known. Therefore, the enzyme is classified as a nonpepsin-type acid proteinase. It is inactivated irreversibly above pH 6.0 (Takahashi *et al.*, 1991a; Fukada *et al.*, 1995).

## Structural Chemistry

The complete amino acid sequence was determined by conventional methods of protein chemistry. It is a two-chain protein of 212 residues, composed of a 39 residue light (L) chain and a 173 residue heavy (H) chain bound noncovalently with each other (Takahashi *et al.*, 1991b). The H chain is blocked at the N-terminus with a pyroglutamic acid and has two intrachain disulfide bonds. The enzyme is a fairly acidic protein and contains only four basic amino acid residues (three Lys, one His and no Arg) while it has 37 free acidic amino acid residues (18 Asp and 19 Glu) per molecule of protein.

The amino acid sequence of the enzyme has no homology with those of typical aspartic or acid proteinases so far sequenced except for scytalidopepsin B (Chapter 326). The sequence identity between aspergillopepsin II and scytalidopepsin B is about 50%. Thus, these two proteinases belong to the same acid proteinase family (A4). However, aspergillopepsin II has a two-chain structure, whereas scytalidopepsin B has a single-chain structure. Further, the former has two disulfide bonds and the latter has three, among which only one disulfide bond is common between them. Aspergillopepsin II is assumed to have catalytic carboxyl group(s) at the active site as judged from its pH activity profile and the results of chemical modification studies (K. Takahashi, Y. Onmura, Y. Sakurai & H. Inoue, unpublished results). Recent site-directed mutagenesis studies indicate that both Asp and Glu may be involved in the catalytic activity (N. Kagami, X.-P. Huang, H. Inoue & K. Takahashi, unpublished results).

The amino acid sequence of the precursor form of the enzyme (i.e. preaspergillopepsinogen II or preproproteinase A), has been deduced from isolation and analysis of the genomic and cDNA clones (Inoue *et al.*, 1991). The encoded protein is a single-chain precursor of 282 residues, and composed of the N-terminal prepropeptide (59 residues), the L chain (39 residues), an intervening sequence (11 residues) and the H chain (173 residues) linked in this order. Therefore, the enzyme is synthesized initially as a single-chain preproprotein and then processed to the mature two-chain form.

The contents of secondary structures were estimated from measurements of the circular dichroism spectra. The estimated contents of $\alpha$ helix, $\beta$ sheet and $\beta$ turn are 4%, 42% and 25%, respectively, at pH 3.0, and 2%, 64% and 16%, respectively, at pH 5.5 (Takahashi *et al*., 1991a; M. Kojima, M. Tanokura & K. Takahashi, unpublished results). The high content of $\beta$-sheet structure is also indicated from the results of two-dimensional NMR spectra (Takahashi *et al*., 1991a; Kojima *et al*., 1995). The NMR data also indicate that the enzyme has a tightly packed core structure (Takahashi *et al*., 1991a; Kojima *et al*., 1995). Three types of the enzyme crystals have been obtained under different conditions in aqueous ammonium sulfate solution containing 3–9% dimethylsulfoxide (Matsuzaki *et al*., 1991; Tanokura *et al*., 1992, 1993) and their diffraction data have been collected to 1.3–1.6 Å resolution (Sasaki *et al*., 1995). They all belong to $P2_12_12_1$ space group. It is notable that the solvent content of type I crystal is 26%, among the smallest values for protein crystals. A $K_2PtCl_4$ derivative has been obtained for type I crystal. The enzyme has an $M_r$ of 22 265 as based on the primary structure (Takahashi *et al*., 1991b). The pI is estimated to be at pH 1.5–2.0.

## Preparation

*A. niger* var. *macrosporus* secretes the enzyme into the culture medium. The enzyme is purified from the commercial enzyme powder Proctase, prepared from the culture filtrate. It has been expressed as the precursor protein in *Escherichia coli* (Kagami *et al*., 1995). The enzyme can be isolated and purified from the Proctase powder by ammonium sulfate fractionation followed by chromatography on sulfoethylcellulose (Koaze *et al*., 1964) or by repeated chromatography on DEAE-Toyopearl (Takahashi *et al*., 1991a).

## Biological Aspects

The enzyme is biosynthesized as a preproprotein, which is converted to the active enzyme through proteolytic processing at four sites. The first processing is the signal peptide cleavage to produce the proenzyme aspergillopepsinogen II, which is followed by cleavages at three sites, i.e. the cleavage of the propeptide to produce the putative one-chain enzyme and the additional cleavages at two sites within it to produce the mature two-chain enzyme with release of an 11 residue internal peptide. The cleavage of the propeptide may occur autocatalytically (Kagami *et al*., 1995), but it is not certain whether any other enzyme(s) is involved in the additional cleavages. The signal peptide cleavage should occur in the endoplasmic reticulum within the cell, but it is unclear where the other cleavages occur. No proform can be isolated from the cells or from the culture medium. The enzyme is an extracellular proteinase and therefore may be implicated in digestion of nutritional and/or harmful foreign proteins and peptides.

The enzyme is fairly stable under mild acidic conditions, whereas it becomes very unstable at around pH 6.0 and above and is inactivated very rapidly with a large conformational change and concomitant dissociation of the two chains (Takahashi *et al*., 1991a; Fukada *et al*., 1995).

## Distinguishing Features

The insensitivity toward the typical inhibitors of ordinary pepsin-type aspartic proteinases such as pepstatin, DAN and EPNP is a distinctive feature of the enzyme.

## Related Peptidases

Two proteinases secreted by *Cryphonectria parasitica*, namely EapB and EapC, have been purified (Jara *et al*., 1996). The products of the EapB and EapC genes as deduced from the nucleotide sequences, are 268 and 269 residues long, respectively, corresponding to the prepro enzymes. After cleavage of the prepro sequences, they yield mature enzymes of 206 and 203 residues, respectively, which share extensive similarities. These enzymes resemble both aspergillopepsin II and scytalidopepsin B (Chapter 326) in sequence, but they are one-chain enzymes like the latter. EapC is unique in that it has no cysteine residue.

## Further Reading

For further reading see Takahashi (1995) and Takahashi *et al*. (1991a).

## References

Chang, W.-J., Horiuchi, S., Takahashi, K., Yamasaki, M. & Yamada, Y. (1976) The structure and function of acid proteases. VI. Effects of acid protease-specific inhibitors on the acid proteases from *Aspergillus niger* var. *macrosporus*. *J. Biochem. (Tokyo)* **80**, 975–981.

Chang, W.-J. & Takahashi, K. (1973) The structure and function of acid proteases. II. Inactivation of bovine rennin by acid protease-specific inhibitors. *J. Biochem. (Tokyo)* **74**, 231–237.

Fukada, H., Takahashi, K., Sorai, M., Kojima, M., Tanokura, M. & Takahashi, K. (1995) Differential scanning colorimetric studies of the thermal unfolding of acid proteinase A from *Aspergillus niger* at various pHs. *Thermochim. Acta* **267**, 373–378.

Ido, E., Saito, T. & Yamasaki, M. (1987) Substrate specificity of acid proteinase A from *Aspergillus niger* var. *macrosporus*. *Agric. Biol. Chem.* **51**, 2855–2856.

Iio, K. & Yamasaki, M. (1976) Specificity of acid proteinase A from *Aspergillus niger* var. *macrosporus* towards B-chain of performic acid oxidized bovine insulin. *Biochim. Biophys. Acta* **429**, 912–924.

Inoue, H., Kimura, T., Makabe, O. & Takahashi, K. (1991) The gene and deduced protein sequences of the zymogen of *Aspergillus niger* proteinase A. *J. Biol. Chem.* **266**, 19484–19489.

Jara, P., Gilbert, S., Delmas, P., Guillemot, J.C., Kaghad, M., Ferrara, P. & Loison, G. (1996) Cloning and characterization of the eapB and eapC genes of *Cryphonectria parasitica* encoding two new acid proteinases, and disruption of eapC. *Mol. Gen. Genet.* **250**, 97–105.

Kagami, N., Inoue, H., Kimura, T., Makabe, O. & Takahashi, K. (1995) Expression in *E. coli* of *Aspergillus niger* var. *macrosporus* proteinase A, a non-pepsin type acid proteinase. In: *Aspartic Proteinases. Structure, Function, Biology, and Biomedical Implications* (Takahashi, K., ed.). New York: Plenum Press, pp. 597–603.

Koaze, Y., Goi, H., Ezawa, K., Yamada, Y. & Hara, T. (1964) Fungal proteolytic enzymes. Part I. Isolation of two kinds of acid-proteases excreted by *Aspergillus niger* var. *macrosporus*. *Agric. Biol. Chem.* **28**, 216–223.

Kojima, M., Tanokura, M., Muto, Y., Miyano, H., Suzuki, E., Hamaya, T., Takizawa, T., Kono, T. & Takahashi, K. (1995) Conformation analysis of non-pepsin-type acid proteinase A from the fungus *Aspergillus niger* by NMR. In: *Aspartic Proteinases: Structure, Function, Biology, and Biomedical Implications* (Takahashi, K., ed.). New York: Plenum Press, pp. 611–615.

Matsuzaki, H., Iwata, S., Hamaya, T., Takizawa, T., Tanokura, M. & Takahashi, K. (1991) Effect of dimethylsulfoxide on the crystallization of *Aspergillus niger* proteinase A. *Proc. Japan Acad.* **67**, 209–212.

Sasaki, H., Tanokura, N., Muramatsu, M., Nakagawa, A., Iwata, S., Hamaya, T., Takizawa, T., Kono, T. & Takahashi, K. (1995) X-ray crystallographic study of a non-pepsin-type acid proteinase, *Aspergillus niger* proteinase A. In: *Aspartic Proteinases: Structure, Function, Biology, and Biomedical Implications* (Takahashi, K., ed.). New York: Plenum Press, pp. 605–609.

Takahashi, K. (1995) Proteinase A from *Aspergillus niger*. *Methods Enzymol.* **248**, 146–155.

Takahashi, K., Tanokura, M., Inoue, H., Kojima, M., Muto, Y.,

Yamasaki, M., Makabe, O., Kimura, T., Takizawa, T., Hamaya, T., Suzuki, E. & Miyano, H. (1991a) Structure and function of a pepstatin-insensitive acid proteinase from *Aspergillus niger* var. *macrosporus*. In: *Structure and Function of the Aspartic Proteinases* (Dunn, B.M., ed.). New York: Plenum Press, pp. 203–211.

Takahashi, K., Inoue, H., Sakai, K., Kohama, T., Kitahara, S., Takishima, K., Tanji, M., Athauda, S.B.P., Takahashi, T., Akanuma, H., Mamiya, G. & Yamasaki, M. (1991b) The primary structure of *Aspergillus niger* acid proteinase A. *J. Biol. Chem.* **266**, 19480–19483.

Tanokura, M., Matsuzaki, H., Iwata, S., Nakagawa, A., Hamaya, T., Takizawa, T. & Takahashi, K. (1992) Crystallization and preliminary X-ray investigation of proteinase A, a non-pepsin-type acid proteinase from *Aspergillus niger* var. *macrosporus*. *J. Mol. Biol.* **223**, 373–375.

Tanokura, M., Sasaki, H., Muramatsu, T., Iwata, S., Hamaya, T., Takizawa, T. & Takahashi, K. (1993) A new crystal form of proteinase A, a non-pepsin-type acid proteinase from *Aspergillus niger* var. *macrosporus*. *J. Biochem. (Tokyo)* **114**, 457–458.

*Kenji Takahashi*
*School of Life Science,*
*Tokyo University of Pharmacy and Life Science,*
*1432-1 Horinouchi, Hachioji,*
*Tokyo 192-03, Japan*
*Email: kenjitak@ls.toyaku.ac.jp*

# 328. Scytalidopepsin A

## Databanks

*Peptidase classification: clan AX, family A99, MEROPS ID: A9G.011*
*NC-IUBMB enzyme classification: EC 3.4.23.31*
*Databank codes: no sequence data available*

## Name and History

The name **scytalidopepsin A** was recommended by the IUBMB in 1992. An earlier name was **Scytalidium sp. pepstatin-insensitive carboxyl proteinase A**. Pepstatin-insensitive carboxyl proteinases were found by testing with a carboxyl proteinase inhibitor, S-PI, isolated by Murao (Murao & Satoi, 1970). The structure of S-PI is Ac-Val-Val-AHMHA-Ala-AHMHA, where AHMHA is 4-amino-3-hydroxy-6-methylheptanoic acid (statine). In the same year, Umezawa *et al*. (1970) isolated pepstatin, which contains isovaleryl as the acyl residue. Since 1970, we have paid attention to the inhibitory spectrum of S-PI, which inhibits all the carboxyl proteinases available. We considered that an S-PI-insensitive carboxyl proteinase, if there was one,

should be distinct in the structure of its active site from any other known carboxyl proteinases. We screened for microorganisms that could produce an S-PI-insensitive carboxyl proteinase and succeeded in isolating one from *Scytalidium lignicolum* in 1972 (Murao *et al*., 1972, 1973). This strain produces four distinct carboxyl proteinases: A-1, A-2, B and C (Oda & Murao, 1974; Oda *et al*., 1986). None of them are inactivated by S-PI, DAN or 1,2-epoxy-3-(*p*-nitrophenoxy)propane (EPNP), with the exception of carboxyl proteinase B, which is inactivated by EPNP. Among these enzymes, carboxyl proteinases A-1 and A-2 are very similar to each other in their enzymatic properties except for a small difference of pI: the pI value for A-1 is 3.6 and for A-2, 3.8.

## Activity and Specificity

Scytalidopepsin A is most active at pH 3.0–3.5 when casein is used as substrate. The activity is not inhibited by pepstatin, S-PI, DAN or EPNP. This enzyme cleaves the B chain of oxidized insulin at the Cya7$\downarrow$Gly8 bond as well as other positions such as Phe24$\downarrow$Phe25 (Oda & Murao, 1976):

FVNQHLC$\downarrow$GSHLVEALYLVCGERGF$\downarrow$FYTPKA

No other carboxyl proteinase reported so far cleaves the Cya$\downarrow$Gly bond. This enzyme cannot cleave the His6$\downarrow$Pro7 bond of angiotensin I, which is cleaved by scytalidopepsin B (Chapter 326). When Z-tetrapeptides such as Z-Xaa-Leu$\downarrow$Ala-Ala and Z-Phe-Xbb$\downarrow$Ala-Ala (Xaa and Xbb: various amino acid residues) are used as substrates, scytalidopepsin A cleaves these peptides at either the Leu-Ala or Xbb-Ala bond. The $k_{cat}/K_m$ values are between 8000 and 11 000 $M^{-1} s^{-1}$. This enzyme prefers Phe, Ala or Tyr as the Xaa residue, and Leu as the Xbb residue. Scytalidopepsin B cleaves the C-terminal bonds, and its activity is very low ($k_{cat}/K_m = 30$–$40 M^{-1} s^{-1}$). Such S-PI-sensitive carboxyl proteinases as pepsin (Chapter 272) cleave the Xaa-Leu and Phe-Xbb bonds. The $k_{cat}/K_m$ values are between 200 and 1 000 $M^{-1} s^{-1}$. Thus the substrate specificity of scytalidopepsin A is distinguishable from that of other microbial carboxyl proteinases, which are sensitive to pepstatin, DAN and EPNP (Murao & Oda, 1985).

## Structural Chemistry

The molecular mass of scytalidopepsin A is 40 000 Da by the sedimentation equilibrium method (Oda *et al.*, 1976). This enzyme is a glycoprotein, containing 3 moles of glucosamine and 10 moles of mannose. Cloning of the enzyme gene is presently in progress.

## Preparation

About 3 mg of A-1 and 6 mg of A-2 were obtained from 1 liter of culture filtrate of *S. lignicolum* ATCC 24568.

## Biological Aspects

*S. lignicolum* belongs to the wood-destroying fungi, and the biological function of these enzymes is assumed to be the degradation of protein in the wood of dead trees.

## Distinguishing Features

There is no specific inhibitor for this enzyme, hence it is difficult to distinguish this enzyme from other pepstatin-insensitive carboxyl proteinases.

## Further Reading

For further reading see Oda & Murao (1991) and Oda *et al.* (1995).

## References

Murao, S. & Satoi, S. (1970) New pepsin inhibitors (S-PI) from *Streptomyces* EF-44-201. *Agric. Biol. Chem.* **34**, 1265–1267.

Murao, S. & Oda, K. (1985) Pepstatin-insensitive acid proteinases. In: *Aspartic Proteinases and Their Inhibitors* (Kostka, V., ed.). Berlin: Walter de Gruyter, pp. 379–399.

Murao, S., Oda, K. & Matsushita, Y. (1972) New acid proteases from *Scytalidium lignicolum* M-133. *Agric. Biol. Chem.* **36**, 1647–1650.

Murao, S., Oda, K. & Matsushita, Y. (1973) Isolation and identification of a microorganism which produces non *Streptomyces* pepsin inhibitor and *N*-diazoacetyl-DL-norleucine methylester sensitive acid proteases. *Agric. Biol. Chem.* **37**, 1417–1421.

Oda, K. & Murao, S. (1974) Purification and some properties of acid protease A and B of *Scytalidium lignicolum* ATCC 24568. *Agric. Biol. Chem.* **38**, 2435–2444.

Oda, K. & Murao, S. (1976) Action of *Scytalidium lignicolum* acid proteases on insulin B-chain. *Agric. Biol. Chem.* **40**, 1221–1225.

Oda, K. & Murao, S. (1991) Pepstatin-insensitive carboxyl proteinases. In: *Structure and Function of the Aspartic Proteinases* (Dunn, B.M., ed.). New York: Plenum Press, pp. 185–201.

Oda, K., Murao, S., Oka, T. & Morihara, K. (1976) Some physico-chemical properties and substrate specificities of acid proteases A-1 and A-2 of *Scytalidium lignicolum* ATCC 24568. *Agric. Biol. Chem.* **40**, 859–866.

Oda, K., Torishima, H. & Murao, S. (1986) Purification and characterization of acid proteinase C of *Scytalidium lignicolum* ATCC 24568. *Agric. Biol. Chem.* **50**, 651–658.

Oda, K., Takahashi, S., Shin, T. & Murao, S. (1995) Pepstatin-insensitive carboxyl proteinases. In: *Aspartic Proteinases: Structure, Function, Biology, and Biomedical Implications* (Takahashi, K., ed.). New York: Plenum Press, pp. 529–542.

Umezawa, H., Aoyagi, T., Morishima, H., Matsuzaki, M., Hamada, M. & Takeuchi, T. (1970) Pepstatin, a new pepsin inhibitor produced by Actinomycetes. *J. Antibiot.* **23**, 259–265.

*Sawao Murao*
*Department of Applied Microbial Technology,*
*Kumamoto Institute of Technology,*
*Ikeda 4-22-1, Kumamoto 860, Japan*
*Email: murao@bio.kumamoto-it.ac.jp*

# 329. *Pseudomonapepsin*

## Databanks

*Peptidase classification: clan AX, family A7, MEROPS ID: A07.001*
*NC-IUBMB enzyme classification: EC 3.4.23.37*
*Databank codes:*

| Species | SwissProt | PIR | EMBL (cDNA) | EMBL (genomic) |
|---|---|---|---|---|
| *Pseudomonas* sp. 101 | P42790 | – | D37970 | – |

## Name and History

The carboxyl proteinases, formerly called acid proteinases, are classified into two groups on the basis of sensitivity to inhibitors: pepstatin-sensitive and -insensitive carboxyl proteinases (Murao & Oda, 1985). Pepstatin-sensitive carboxyl proteinases represented by pig pepsin (Chapter 272) are blocked by such inhibitors as pepstatin, S-PI (acetyl pepstatin), DAN and 1,2-epoxy-3-(*p*-nitrophenoxy)propane (EPNP). Pepstatin-sensitive carboxyl proteinases are called aspartic proteinases.

Murao and Oda found that several pepstatin-insensitive carboxyl proteinases are produced by *Scytalidium lignicolum* ATCC 24568 (Chapters 326 and 328) (Oda & Murao, 1991). None of these are inhibited by pepstatin, S-PI, DAN or EPNP with the exception of carboxyl proteinase B, which is inhibited by EPNP. Furthermore, these enzymes have unique substrate specificities. The primary structure of carboxyl proteinase B shows no similarity to other carboxyl proteinases reported so far. The existence of enzymes having characteristics similar to those of *S. lignicolum* has been reported in fungi, bacteria, and thermophilic bacteria. The first carboxyl proteinase to be isolated from bacteria also belongs to this group: ***Pseudomonas pepstatin-insensitive carboxyl proteinase*** (Oda *et al*., 1987). In 1992 the IUBMB recommended the name ***pseudomonapepsin*** for this protease.

## Activity and Specificity

The substrate specificity of pseudomonapepsin was studied on a series of 28 different oligopeptide substrates with a general structure of P5-P4-P3-P2-Phe┼Nph-P2′-P3′. Pseudomonapepsin hydrolyzes a synthetic substrate, Lys-Pro-Ala-Leu-Phe┼Nph-Arg-Leu, most effectively (Oda *et al*., 1992; Ito *et al*., 1996). The kinetic parameters for this peptide are $K_m = 6.3\,\mu\text{M}$, $k_{cat} = 51.4\,\text{s}^{-1}$, and $k_{cat}/K_m = 8.16\,\text{M}^{-1}\,\text{s}^{-1}$ (pH 3.5, 37°C). Pseudomonapepsin has a more restricted substrate specificity than xanthomonapepsin (Chapter 330), in spite of the high similarity in their primary structure (52% identity).

The pH optimum for the action toward casein is around 3.0. The activity is not inhibited by pepstatin, DAN, and EPNP, but inhibited by tyrostatin (a peptide aldehyde), which was isolated from *Kitasatosporia* ($K_i = 2.6\,\text{nM}$).

## Structural Chemistry

Pseudomonapepsin is synthesized as a proenzyme of 587 amino acid residues, and is processed to a mature enzyme (Oda *et al*., 1994). The prepro part consists of 215 amino acid residues and the mature enzyme consists of 372 amino acid residues and one disulfide bridge (38 kDa). This enzyme has no apparent sequence similarity to other pepstatin-sensitive carboxyl proteinases reported so far, nor to the pepstatin-insensitive carboxyl proteinases except xanthomonapepsin (Chapter 330). Also the characteristic active-site aspartic acid sequence, Asp-Thr-Gly, of aspartic proteinases is absent.

To identify the catalytic residues, the following experiments have been carried out.

1. Kinetic studies indicate that two residues with apparent dissociation constants at pH 2.97 or 4.92 are involved in the catalytic action.
2. By use of a zinc(II)-PAD (pyridine-2-azo-*p*-dimethyl aniline) complex as a probe, evidence was found that this enzyme has two active carboxyl residues participating in the action.
3. By site-directed mutagenesis, Asp170 and Asp328 (mature enzyme numbering) were recently identified as the catalytic residues of this enzyme.

To confirm these results, affinity labeling using a tyrostatin derivative is currently in progress. X-ray crystallographic analysis of the tyrostatin complex is under investigation by a research group at the Rational Drug Design Laboratory, Fukushima, Japan.

## Preparation

*Pseudomonas* sp. 101 was aerobically cultured at 30°C for 25 h. About 105 mg of purified enzyme with a 53% recovery was obtained from a 21 liter culture filtrate. The recombinant enzyme has been expressed in *Escherichia coli* as a soluble protein.

## Biological Aspects

The biological function is not known. It should be noted that this enzyme is secreted in the culture fluid from a gram-negative bacterium that has a double-layered membrane around the cells.

## Distinguishing Features

To distinguish pseudomonapepsin and xanthomonapepsin from other pepstatin-insensitive carboxyl proteinases, inhibition by tyrostatin should be assayed. Tyrostatin (Oda *et al*., 1989) is a specific inhibitor for both enzymes described above.

## Further Reading

For further reading see Oda *et al*. (1995).

## References

Ito, M., Dunn, B.M. & Oda, K. (1996) Substrate specificities of pepstatin-insensitive carboxyl proteinases from Gram-negative bacteria. *J. Biochem. (Tokyo)* **120**, 845–850.

Murao, S. & Oda, K. (1985) Pepstatin-insensitive acid proteinases. In: *Aspartic Proteinases and Their Inhibitors* (Kostka, V., ed.). Berlin: Walter de Gruyter, pp. 379–399.

Oda, K. & Murao, S. (1991) Pepstatin-insensitive carboxyl proteinases. In: *Structure and Function of the Aspartic Proteinases* (Dunn, B.M., ed.). New York: Plenum Press, pp. 185–201.

Oda, K., Sugitani, M., Fukuhara, K. & Murao, S. (1987) Purification and properties of a pepstatin-insensitive carboxyl proteinase from Gram-negative bacterium. *Biochim. Biophys. Acta* **923**, 463–499.

Oda, K., Fukuda, Y., Murao, S., Uchida, K. & Kainosho, M. (1989) A novel proteinase inhibitor, tyrostatin, inhibiting some pepstatin-insensitive carboxyl proteinases. *Agric. Biol. Chem.* **53**, 405–415.

Oda, K., Nakatani, H. & Dunn, B.M. (1992) Substrate specificity and kinetic properties of pepstatin-insensitive carboxyl proteinase from *Pseudomonas* sp. No. 101. *Biochim. Biophys. Acta* **1120**, 208–214.

Oda, K., Takahashi, T., Tokuda, Y., Shibano, Y. & Takahashi, S. (1994) Cloning, nucleotide sequence, and expression of an isovaleryl pepstatin-insensitive carboxyl proteinase gene from *Pseudomonas* sp. 101. *J. Biol. Chem.* **269**, 26518–26524.

Oda, K., Takahashi, S., Shin, T. & Murao, S. (1995) Pepstatin-insensitive carboxyl proteinases. In: *Aspartic Proteinases: Structure, Function, Biology, and Biomedical Implications* (Takahashi, K., ed.). New York: Plenum Press, pp. 529–542.

*Sawao Murao*
*Department of Applied Microbial Technology,*
*Kumamoto Institute of Technology,*
*Ikeda 4-22, Kumamoto 860, Japan*
*Email: murao@bio.kumamoto-it.ac.jp*

# 330. *Xanthomonapepsin*

## Databanks

*Peptidase classification: clan AX, family A7, MEROPS ID: A07.002*
*NC-IUBMB enzyme classification: EC 3.4.23.33*
*Chemical Abstracts Service registry number: 113356-29-9*
*Databank codes:*

| Species | SwissProt | PIR | EMBL (cDNA) | EMBL (genomic) |
|---|---|---|---|---|
| *Xanthomonas* sp. T-22 | – | – | D83740 | – |

## Name and History

A result of the study of pepstatin-insensitive carboxyl proteinases described in Chapter 329 was the discovery of a proteinase homologous with pseudomonapepsin (Chapter 329) in a *Xanthomonas* species. The *Xanthomonas* **pepstatin-insensitive carboxyl proteinase** was the second carboxyl proteinase to be isolated from bacteria (Oda *et al*., 1987). The name **xanthomonapepsin** was recommended by IUBMB in 1992.

## Activity and Specificity

The substrate specificity of xanthomonapepsin was studied by using a series of 28 different oligopeptide substrates with a general structure of P5-P4-P3-P2-Phe+Nph-P2'-P3'. Xanthomonapepsin hydrolyzes a synthetic substrate, Lys-Pro-Ala-Leu-Phe+Nph-Arg-Leu, most effectively (Ito *et al*., 1996). The kinetic parameters for this peptide are $K_m = 3.6\,\mu M$, $k_{cat} = 52.2\,s^{-1}$, and $k_{cat}/K_m = 14.5\,\mu M^{-1}\,s^{-1}$ (pH 3.5, 37°C). Xanthomonapepsin showed a less stringent substrate specificity than that of pseudomonapepsin (Chapter 329), in spite of the similarity in their primary structure (52% identity).

The pH optimum for casein digestion is around 2.7. The activity is not inhibited by pepstatin, DAN or EPNP, but is inhibited by tyrostatin (a peptide aldehyde), which was isolated from *Kitasatosporia* ($K_i = 2.1\,nM$). Assays are most conveniently made by using casein or acid-denatured hemoglobin as substrates. Tyrostatin is not available commercially.

## Structural Chemistry

Xanthomonapepsin is synthesized as a large precursor consisting of three regions: an N-terminal prepro peptide (237 amino acid residues); the mature enzyme (398 residues, 40 kDa); and a C-terminal propeptide (192 residues; Oda *et al*., 1996). This enzyme has no apparent sequence similarity to either pepstatin-sensitive carboxyl proteinases reported to date, nor to the pepstatin-insensitive carboxyl proteinases except pseudomonapepsin. Also the characteristic active site aspartic acid sequence Asp-Thr-Gly of aspartic proteinases is absent.

To identify the catalytic residues, the following experiments have been carried out.

1. By means of a zinc(II)-PAD (pyridine-2-azo-*p*-dimethyl-aniline) complex as a probe, it was deduced that this enzyme has two active carboxyl residues participating in its action.
2. By site-directed mutagenesis, Asp169 and Asp348 were recently identified as the catalytic residues of this enzyme.

To confirm these results, affinity labeling by using a tyrostatin derivatives is currently in progress.

## Preparation

*Xanthomonas* sp. T-22 was aerobically cultured at 30°C for 24 h. About 100 mg of purified enzyme with a 25% recovery was obtained from 60 liters of culture filtrate. The recombinant enzyme has been expressed in *Escherichia coli* as a soluble protein.

## Biological Aspects

We do not know the biological function yet. It should be noted that this enzyme is secreted into the culture fluid from gram-negative bacteria that have a double-layered membrane around the cells.

## Distinguishing Features

To distinguish xanthomonapepsin and pseudomonapepsin from other pepstatin-insensitive carboxyl proteinases, an inhibitory activity against tyrostatin should be assayed. Tyrostatin (Oda *et al*., 1989) is a specific inhibitor for both enzymes described above. Rabbit polyclonal antibodies have been produced against both enzymes.

## Further Reading

For further reading see Oda *et al*. (1995).

## References

Ito, M., Dunn, B.M. & Oda, K. (1996) Substrate specificities of pepstatin-insensitive carboxyl proteinases from Gram-negative bacteria. *J. Biochem. (Tokyo)* **120**, 845–850.

Oda, K., Nakazima, T., Terashita, T., Suzuki, K. & Murao, S. (1987) Purification and properties of an S-PI (Pepstatin Ac)-insensitive carboxyl proteinase from *Xanthomonas* sp. bacterium. *Agric. Biol. Chem.* **51**, 3073–3080.

Oda, K., Fukuda, Y., Murao, S., Uchida, K. & Kainosho, M. (1989) A novel proteinase inhibitor, tyrostatin, inhibiting some pepstatin-insensitive carboxyl proteinases. *Agric. Biol. Chem.* **53**, 405–415.

Oda, K., Takahashi, S., Shin, T. & Murao, S. (1995) Pepstatin-insensitive carboxyl proteinases. In: *Aspartic Proteinases: Structure, Function, Biology, and Biomedical Implications* (Takahashi, K., ed.). New York: Plenum Press, pp. 529–542.

Oda, K., Ito, M., Uchida, K., Shibano, Y., Fukuhara, K. & Takahashi, S. (1996) Cloning and expression of an isovaleryl pepstatin-insensitive carboxyl proteinase gene from *Xanthomonas* sp. T-22. *J. Biochem. (Tokyo)* **120**, 564–572.

*Kohei Oda*
*Department of Applied Biology,*
*Faculty of Textile Science, Kyoto Institute of Technology,*
*Sakyouku, Kyoto 606, Japan*
*Email: bika@ipc.kit.ac.jp*

# 331. Bacillus pepstatin-insensitive acid endopeptidases

## Databanks

*Peptidase classification: clan AX, family A99, MEROPS ID: A9G.015*
*NC-IUBMB enzyme classification: none*
*Databank codes: no sequence data available*

## Name and History

Pepstatin-insensitive acid proteases were originally discovered in fungi such as *Scytalidium lignicolum* (Chapter 326) (Murao & Oda, 1985). Since then, similar proteases have been discovered from other fungi and bacteria. All these proteases have been suggested to belong to the proposed **acetyl-pepstatin-insensitive carboxyl protease** or **S-PI-insensitive carboxyl protease** group (Murao & Oda, 1985). The first was isolated from *Bacillus* st. MN-32 by Murao *et al.* (1988, 1993) from acidic hot springs in Japan. This protease is also known as **kumamolysin** (Murao *et al.*, 1993). Two isolates from New Zealand hot springs, *Bacillus* strains Wai21.A1 (Prescott *et al.*, 1992, 1995) and Wp22.A1 (Toogood *et al.*, 1995) also produce acid proteases. The **MN-32 protease**, **Wai21.A1 protease** and **Wp22.A1 protease** are the subjects of the present article.

## Activity and Specificity

Activity is determined by the detection of cleavage products of casein for MN-32 protease (Murao *et al.*, 1993), hemoglobin for Wai21.A1 (Prescott *et al.*, 1995) and cytochrome *c* for Wp22.A1 proteases (Toogood *et al.*, 1995).

The optimum pH of MN-32 protease is 3.0 at 70°C (Murao *et al.*, 1993). The protease is completely stable in the pH range of 2.0–4.0 at 50°C for 24 h, and in the range of 1–10 at 5°C for 3 days (Murao *et al.*, 1993). Wai21.A1 protease has a pH optimum of 3.0 at 60°C (Prescott *et al.*, 1992). Activity of the enzyme falls to 50% at pH 1.8 and pH 4.1. The protease has a higher stability between pH 4–5 where less than maximal activity occurs (Prescott *et al.*, 1992). The pH optimum for Wp22.A1 protease is 3.5 at 60°C and 3.0 at 45°C (Toogood *et al.*, 1995). Maximal stability is also observed in the pH range of 4–5, where less than maximal activity is observed.

The three proteases are not inhibited by serine or cysteine protease inhibitors. Although all three proteases are insensitive to pepstatin (Table 331.1), only MN-32 protease is insensitive to the active-site-modifying reagents DAN and 1,2-epoxy-3-(*p*-nitrophenoxy)propane (EPNP) (Murao *et al.*, 1988). Inhibition of Wai21.A1 and Wp22.A1 proteases by EPNP is pepsin-like, but inhibition by DAN is only partial (Prescott *et al.*, 1995; Toogood *et al.*, 1995). Further studies showed that this partial inhibition could be due to DAN binding to a carboxyl group at a site other than at the active site, such as in the binding cleft. Alternatively, it may be the result of slow binding at the active-site carboxyl group (Prescott *et al.*, 1995). Both Wp22.A1 and Wai21.A1 proteases are destabilized by the metal chelators EDTA and EGTA, which remove $Ca^{2+}$ ions.

The active-site residues of MN-32 protease were investigated by observing the binding of zinc(II)-pyridine-2-azo-*p*-dimethylaniline (PAD) to the protease. A spectral change was observed at 530 nm (Murao *et al.*, 1993), as seen with the binding of this ligand to the active site aspartate residues of pepsin and other acid proteases. Thus while the MN-32 protease is insensitive to the class-specific inhibitors, the binding of the zinc(II)-PAD complex suggests that it can still be classified as an acid protease.

All three *Bacillus* acid proteases are thermostable. Table 331.1 shows the half-lives of the proteases at 80°C. Wai21.A1 and Wp22.A1 proteases have both been shown to be stabilized by $Ca^{2+}$ ions. In the presence of 5 mM $Ca^{2+}$ ions at 60°C, both the first- and second-order activity loss plots for Wp22.A1 protease approached linearity, suggesting that both thermal denaturation and autolysis contributed to the loss of stability. At lower concentrations of $Ca^{2+}$ ions, only the first order plot was linear, indicating that thermal denaturation was predominant. This suggests that high $Ca^{2+}$ levels stabilize Wp22.A1 protease by preventing thermal

*Table 331.1*   Some properties of *Bacillus* pepstatin-insensitive acid proteases

| Source | $M_r$ (kDa) | Optimum pH | Stability (half-life) | Inhibitors | | |
| --- | --- | --- | --- | --- | --- | --- |
| | | | | Pep[a] | DAN | EPNP |
| *Bacillus* st. MN-32 | 40–41 | 3.0 | 10 min 80°C pH 4.0 | – | – | – |
| *Bacillus* st. Wai21.A1 | 45 | 3.0 | 2 min 80°C pH 3.0 | – | +[b] | + |
| *Bacillus* st. Wp22.A1 | 45 | 3.5 | 21 min 80°C pH 3.5 | – | +[b] | + |

[a]Pepstatin; [b]partial inhibition.
Data from Toogood *et al.* (1995).

denaturation, making autolysis more significant (Toogood *et al.*, 1995). In a similar manner, Wai21.A1 protease is stabilized by 5 mM $Ca^{2+}$ ions against thermal denaturation (Prescott *et al.*, 1995). This stabilizing effect is more marked at higher temperatures (e.g. 80°C), while for Wp22.A1 this effect was seen only at lower temperatures (60°C). The effect of $Ca^{2+}$ ions on the stability of MN-32 protease has not been investigated.

All three proteases showed a narrow substrate specificity with oxidized insulin B chain. MN-32 protease cleaved the bond Leu15┼Tyr16 with high specificity, with a $k_{cat}$ of $71\,s^{-1}$ at 30°C, pH 3.0 (Murao *et al.*, 1993). A second minor cleavage site Phe25┼Tyr26 was also detected. No other cleavage sites were detected.

FVNQHLCGSHLVEAL┼YLVCGERGFF┼YTPKA

Wai21.A1 protease has the same cleavage pattern, though the rates are much slower for the initial cleavage site (Prescott *et al.*, 1995). A few extremely minor sites, such as Gln4┼His5, were also detected. These results suggest that these two proteases have a preference for bulky hydrophobic residues at the P1 and P1' sites of the substrate. Although this is a typical feature of acid proteases, this narrow substrate specificity distinguishes it from many of these proteases, even those that are pepstatin-insensitive (Morihara, 1974).

The cleavage profile is rather different for Wp22.A1 protease:

FV┼NQH┼LCGSHLVEAL┼YLVCGERGFF┼YTPKA

The primary cleavage site is Val2┼Asn3, with 50% of the bonds cleaved in 20 min. After 1 h, the sites His5┼Leu6, Leu15┼Tyr16 and Phe25┼Tyr26 were cleaved (Toogood *et al.*, 1995). In contrast, the other bacterial acid proteases have a high preference for Leu15┼Tyr16, and do not cleave Val2-Asn3 at all. These results suggest that Wp22.A1 protease has a preference for sites where the P1 amino acid is bulky and nonpolar, while the P1' amino acid is preferably bulky and polar (Toogood *et al.*, 1995).

## Structural Chemistry

The $M_r$ of the proteases is 40–45 kDa (Table 331.1). The pIs are 3.8 for Wp22.A1 and Wai21.A1 proteases, and 3.5 for MN-32 protease (Toogood *et al.*, 1995).

## Preparation

*Bacillus* st. MN-32 is cultivated aerobically in a 200 liter fermenter at 65°C (Murao *et al.*, 1993). The protease is purified from the culture supernatant by using DEAE-Sepharose CL-6B, Sephadex G-100 and TSK gel DEAE-5PW. *Bacillus* st. Wai21.A1 is cultivated aerobically at 60°C in a 600 liter fermenter (Prescott *et al.*, 1995). The protease is extracted from the cell exterior by soaking the cells in fresh medium. The protease is purified using S-Sepharose, phenyl-Sepharose and Mono-Q columns (Prescott *et al.*, 1995). *Bacillus* st. Wp22.A1 is cultivated aerobically at 60°C in small batch cultures (Toogood *et al.*, 1995). The protease is extracted from the cell exterior by soaking the cells in HEPES buffer

containing 1 M NaCl. The protease is purified by using phenyl Sepharose (ethanediol gradient), Mono-Q, and phenyl-Sepharose (amino acid gradient).

## Biological Aspects

These proteases are extracellular, and at least Wai21.A1 and Wp22.A1 proteases are cell associated. This suggests that their role is to cleave proteins in the surrounding medium for nutrition.

The compositions of MN-32 and Wai21.A1 proteases have been determined. The compositions and N-terminal sequences of the two proteases are very similar.

## Distinguishing Features

Table 331.1 is a comparison of some of the properties of the three proteases. It shows that differences in features such as molecular weight, pI, pH optimum and thermostability are small. The main differences among the group are inhibitor sensitivities and substrate specificity.

## Related Peptidases

The three proteases are quite different from other acid proteases. Only three other bacterial and archaeal acid proteases have been described. Oda *et al.* (1987a,b) isolated mesophilic acid proteases from *Xanthomonas* sp. strain T-22 (Chapter 330) and *Pseudomonas* sp. strain no. 101 (Chapter 329). The thermophilic protease thermopsin (Chapter 332), from *Solfolobus acidocaldarius*, is significantly different from all other proteases described (Lin & Tang, 1990).

## Further Reading

See Daniel *et al.* (1995) for a review of thermostable proteases, including a section on *Bacillus* acid proteases.

## References

Daniel, R.M., Toogood, H.S. & Bergquist, P.L. (1995) Thermostable proteases. *Biotechnol. Genet. Eng. Rev.* **13**, 51–100.

Lin, X. & Tang, J. (1990) Purification, characterization and gene cloning of thermopsin, a thermostable acid protease from *Sulfolobus acidocaldarius*. *J. Biol. Chem.* **265**, 1490–1495.

Morihara, K. (1974) Comparative specificity of microbial proteases. *Adv. Enzymol.* **41**, 179–243.

Murao, S. & Oda, K. (1985) Pepstatin-insensitive acid proteases. In: *Aspartic Proteinases and Their Inhibitors* (Kostka, V., ed.). Berlin: Walter de Gruyter, pp. 379–399.

Murao, S., Ohkuni, K., Nagao, M., Oda, K. & Shin, T. (1988) A novel thermostable, S-PI (pepstatin Ac)-insensitive acid proteinase from thermophilic *Bacillus* novosp. Strain MN-32. *Agric. Biol. Chem.* **52**, 1629–1631.

Murao, S., Ohkuni, K., Nagao, M., Hirayama, K., Murao, S., Ohkuni, K., Nagao, M., Hirayama, K., Fukuhara, K., Oda, K., Oyama, H. & Shin, T. (1993) Purification and characterization of kumamolysin, a novel thermostable pepstatin-insensitive carboxyl proteinase from *Bacillus* novosp. MN-32. *J. Biol. Chem.* **268**, 349–355.

Oda, K., Sugitani, M., Fukuhara, K. & Murao, S. (1987a) Purification and properties of a pepstatin-insensitive carboxyl proteinase from a Gram-negative bacterium. *Biochim. Biophys. Acta* **923**, 463–469.

Oda, K., Nakazima, T., Terashita, T., Suzuki, K.-I. & Murao, S. (1987b) Purification and properties of an S-PI (pepstatin)-insensitive carboxyl proteinase from a *Xanthomonas* sp. bacterium. *Agric. Biol. Chem.* **51**, 3073–3080.

Prescott, M., Peek, K., Prendergast, E. & Daniel, R.M. (1992) Purification and characterisation of a pepstatin-insensitive acid protease from a bacterium. *Enzyme Eng. XI* **672**, 167–170.

Prescott, M., Peek, K. & Daniel, R.M. (1995) Characterisation of a thermostable pepstatin-insensitive acid proteinase from a *Bacillus* sp. *Int. J. Biochem. Cell Biol.* **27**, 729–739.

Toogood, H.S., Prescott, M. & Daniel, R.M. (1995) A pepstatin-insensitive aspartic proteinase from a thermophilic *Bacillus* sp. *Biochem. J.* **307**, 783–789.

*H.S. Toogood*
*Thermophile Research Unit,*
*University of Waikato,*
*Private Bag 3105, Hamilton, New Zealand*
*Email: Biolsec3@Waikato.ac.nz*

*R.M. Daniel*
*Thermophile Research Unit,*
*University of Waikato,*
*Private Bag 3105, Hamilton, New Zealand*

# 332. *Thermopsin*

## Databanks

*Peptidase classification: clan AX, family A5, MEROPS ID: A05.001*
*NC-IUBMB enzyme classification: EC 3.4.99.43*
*Databank codes:*

| Species | SwissProt | PIR | EMBL (cDNA) | EMBL (genomic) |
|---|---|---|---|---|
| *Sulfolobus acidocaldarius* | P17118 | A35009 | J05184 | – |

## Name and History

**Thermopsin** is an acid protease produced by a thermophilic archaeon, *Sulfolobus acidocaldarius*. This organism was originally isolated from the hot springs of Yellowstone Park and is best cultured in pH 2 at 70°C. Therefore, it is not surprising that thermopsin can withstand extreme acidity and high temperature. The protease was first described with its isolation and cloning (Lin & Tang, 1990; Fusek *et al.*, 1990). Heterologous expression of recombinant thermopsin has also been described (Lin *et al.*, 1992).

## Assay and Specificity

Thermopsin hydrolyzes many proteins, which are all potential substrates for assay. It can be assayed using cattle hemoglobin as substrate (Anson, 1938) only with at least partially purified enzyme, but not with crude cell extract. The assay is greatly improved in sensitivity and accuracy when the substrate is radiolabeled with $^{14}$C-methylation (Lin *et al.*, 1989). In this assay (Lin & Tang, 1995), 0.51% (w/v) substrate and thermopsin are incubated in 0.1 ml of 0.1 M Na formate, pH 3.2, at 80°C for a period of typically 5–30 min. After the addition of 0.1 ml of 10% trichloroacetic acid, the radioactivity in an aliquot of the centrifuged supernate is determined. Thermopsin hydrolyzes a synthetic chromogenic substrate Lys-Pro-Ala-Glu-Phe↓Nph-Ala-Leu, and this substrate can be used for kinetic studies (Fusek *et al.*, 1990).

The specificity of thermopsin has been studied only using the oxidized cattle insulin B chain (Fusek *et al.*, 1990). Four observed bond cleavages are between amino acids Leu11↓Val12, Leu15↓Tyr16, Phe24↓Phe25 and Phe26↓Tyr27:

FVNQHLCGSHL↓VEAL↓YLVCGERGF↓F↓YTPKA

These hydrolytic sites, which are very similar to those produced by other broad specificity aspartic proteases such as pepsin (Chapter 272) and cathepsin D (Chapter 277), suggest thermopsin prefers large hydrophobic residues at both P1 and P1'. No clear preferences of their specificity sites are apparent.

The proteolytic activity of thermopsin on hemoglobin is observed from pH 0–12, although the main activity is at pH 1–5 and the optimum is near pH 2. The enzyme activity increases almost linearly from 40–80°C, the optimum. The activity at 100°C is about 70% of the optimum. Low but real activity can be measured even at 0°C. Using the chromogenic substrate described above, the optimal activity is found between pH 1.9–2.1. The $k_{cat}$ and $K_m$ for this substrate are $14.3\,\text{s}^{-1}$ and $5.3 \times 10^{-5}$ M, respectively. The temperature dependence of $k_{cat}$ and $K_m$ between 26°C and 78°C reveals that $K_m$ increases only slowly from 26°C to about 65°C, then increases rapidly with higher temperature. The $k_{cat}$ increases steadily with the temperature. The highest catalytic efficiency judged by $k_{cat}/K_m$ values, in fact, occurs at 50–65°C. Above that, the decline of catalytic efficiency is mainly due to the $K_m$ increase. Therefore, at the temperature range of optimal bacterial growth, 80–90°C, the enzyme is operating at a reduced efficiency. Its biological function is dependent upon its ability to resist heat denaturation (Fusek *et al.*, 1990).

Thermopsin is inactivated by DAN and 1,2-epoxy-3-(*p*-nitrophenoxy)propane (EPNP), both specific active-site esterification reagents for pepsin and aspartic proteases. The inactivation of thermopsin by these reagents is somewhat slower than that for pepsin and the completely reacted enzyme appears to be partially active. Therefore, it is not clear whether DAN and EPNP esterify the active site of thermopsin as in pepsin. The enzyme is effectively inhibited by mercuric acetate at 50 mM. Iodoacetate and *p*-chloromercuric benzoate, however, have no effect, indicating that the inhibition by mercuric acetate is not mediated by a sulfhydryl group of the enzyme. Pepstatin, a transition-state inhibitor of aspartic proteases, is a competitive inhibitor of thermopsin with the $K_i$ of $4.9 \times 10^{-7}$ M at 37°C, $1.2 \times 10^{-7}$ M at 56°C and $2.0 \times 10^{-7}$ M at 76°C. Thermopsin is also competitively inhibited by urea ($K_i = 0.5$ M), acetamide ($K_i = 0.4$ M) and phenylalaninamide ($K_i = 10$ mM). Inhibition may involve the competitive binding of inhibitors to the amide-recognizing components in the active site (Fusek *et al.*, 1990).

Purified thermopsin cannot be stained by Coomassie Blue or commonly used protein dyes after SDS-PAGE. However, soaking the electrophoresed SDS-gel in acidic cattle hemoglobin solution and incubating at 40°C resulted in a negatively stained band by Coomassie Blue due to the digestion of hemoglobin at the enzyme band. This also indicates that thermopsin resists SDS denaturation. Thermopsin contains an unusually high number (35) of tyrosine residues, which may be related to its stability in SDS or at high temperature.

## Structural Chemistry

The gene that encodes *S. acidocaldarius* thermopsin has been cloned and sequenced. The deduced amino acid sequence shows that thermopsin is synthesized as a 340 residue protein. The position of the N-terminus of the mature enzyme is clearly defined by the chemical sequencing data, which indicate that thermopsin contains 299 residues with a calculated molecular mass of 32 651. There are 41 upstream residues, which contain a signal peptide, probably around 30 residues, and a propeptide of about 11 residues. The peptide bond that is hydrolyzed to yield mature thermopsin is a Leu+Phe bond, which fits the specificity of thermopsin quite well and suggests the autoactivation of thermopsin precursor.

Like the enzymes of pepsin family, thermopsin is apparently an acidic protein. Based on amino acid composition, there are ten net negative charges per molecule. This is advantageous for the approach of positively charged substrates at pH 2. A single cysteine is found at position 237. Judging from the lack of inhibition by two sulfhydryl-modifying reagents, this group is not essential for activity. Many of the 11 potential *N*-glycosylation sites in thermopsin are likely glycosylated since the determined molecular mass (in the range of 46 000–51 000) is considerably larger than the calculated one. In addition, the Edman degradation of thermopsin yielded no positive identification for the phenylthiohydantoin (PTH) amino acid at two potential Asn glycosylation positions 24 and 28 (Lin & Tang, 1990), which frequently indicates the presence of carbohydrate on Asn. Thermopsin appears to be covalently attached to the cell surface. The nature of this linkage is not known.

Thermopsin does not contain a 'signature' active-site sequence, Asp-Thr-Gly, present in all aspartic proteases of the pepsin family. Nor is there any evidence from the amino acid sequence that thermopsin contains internal two-fold structural symmetry as in other aspartic proteases (Tang *et al.*, 1978). Seven Asp residues present in thermopsin are potential active site residues. However, no homology is observed among the sequences around these Asp residues (Lin & Tang, 1990).

## Preparation

The culture medium of *S. acidocaldarius* contains some free thermopsin, but most (about 90%) of the enzymes is associated with the bacterial cells. The cell-attached enzyme is associated with the centrifuged pellet after homogenization. The assay of intact cells produced nearly all the activity, suggesting that the enzyme is located on the outside of the cell surface and only a minor fraction is shed into the media. The attempt to release the enzyme from the cell using detergents and other chemicals had little success, indicating that thermopsin is covalently bound to cells (Lin & Tang, 1990).

Thermopsin is more conveniently purified from the culture medium. The archaeon (from ATCC) needs to be kept in a continuous small-scale culture, from which a large-scale culture is made to produce thermopsin (Lin & Tang, 1995). A typical large-scale culture is 40 liters of ATCC medium 1256, pH 2, for 2 days at 70°C. From 400 liters of culture medium, which contains about 40 mg of thermopsin, the purification is carried out through ultrafiltration and five chromatographic steps (Lin & Tang, 1995) to yield 0.35 mg of pure enzyme. The same procedure can be applied to purify the enzyme released from the cells by formic acid (Lin & Tang, 1990).

Several attempts have been made to produce recombinant thermopsin from *Escherichia coli* and insect cells (Lin *et al.*, 1992; Lin & Tang, 1995). In the former, the thermopsin gene is fused downstream from the cDNA of pig pepsinogen. The resulting inclusion bodies (about 4 mg of fusion protein per liter of culture) are washed and refolded from 8 M urea to generate thermopsin activity. After incubation in acid solution, the pepsinogen moiety is digested and recombinant thermopsin is purified simply from gel filtration. Although the yield is poor, the purified material has a temperature–activity optimum between 70–80°C, in spite of the absence of glycosylation. The production of thermopsin fused to other proteins did not improve the yield (Lin *et al.*, 1992). Insect cells produced active thermopsin. However, the yield was also poor.

### Biological Aspects

The role of thermopsin in *S. acidocaldarius* appears to be digestion of protein substrates in the medium to supply nutrients. The covalent attachment of the enzyme to the cell surface ensures its long-term usage and is physiologically advantageous. The general and somewhat broad proteolytic specificity is consistent with this function. From the $k_{cat}$ and $K_m$ values, it is interesting that thermopsin catalyzes most efficiently between 50–65°C, in which temperature range the bacterium grows only marginally. It can be reasoned that under these adverse conditions, the enzyme needs to respond to low substrate concentration in order to survive. At temperatures higher than 65°C, the bacterium thrives and the enzyme needs to respond only to higher substrate concentration.

In spite of many similarities of enzymic properties between thermopsin and pepsin (Chapter 272), which include pH optima, specificity, inhibitors, kinetic parameters, and activation energies, there is no sequence similarity. This raises the question of whether these enzymes have similar active sites but have evolved through a convergent evolutionary process. Aspartic proteases of the pepsin family exist only in eukaryotes and retroviruses, but not in prokaryotes. As the archaea evolved as a separate branch from the eukaryotes and prokaryotes, the convergent evolutionary route is not unreasonable.

### References

Anson, M.L. (1938) The estimation of pepsin, trypsin, papain, and cathepsin with hemoglobin. *J. Gen. Physiol.* **22**, 79–89.

Fusek, M., Lin, X. & Tang, J. (1990) Enzymic properties of thermopsin. *J. Biol. Chem.* **265**, 1496–1501.

Lin, X. & Tang, J. (1990) Purification, characterization, and gene cloning of thermopsin, a thermostable acid protease from *Sulfolobus acidocaldarius*. *J. Biol. Chem.* **265**, 1490–1495.

Lin, X. & Tang, J. (1995) Thermopsin. *Methods Enzymol.* **248**, 156–168.

Lin, X., Wong, R.N.S. & Tang, J. (1989) Synthesis, purification, and active site mutagenesis of recombinant porcine pepsinogen. *J. Biol. Chem.* **264**, 4482–4489.

Lin, X., Liu, M. & Tang, J. (1992) Heterologous expression of thermopsin, a heat-stable acid proteinase. *Enzyme Microb. Technol.* **14**, 696–701.

Tang, J., James, M.N.G., Hsu, I.N., Jenkins, J.A. & Blundell, T.L. (1978) Structural evidence for gene duplication in the evolution of the acid proteases. *Nature* **271**, 618–621.

*Jordan Tang*
Protein Studies Program,
Oklahoma Medical Research Foundation,
825 NE 13 Street, Oklahoma City, OK 73104, USA
Email: jordan-tang@omrf.uokhsc.edu

*Xinli Lin*
Protein Studies Program,
Oklahoma Medical Research Foundation,
825 NE 13 Street, Oklahoma City, OK 73104, USA

# 333. *Signal peptidase II*

### Databanks

Peptidase classification: clan AX, family A8, MEROPS ID: A08.001
NC-IUBMB enzyme classification: EC 3.4.23.36
Databank codes:

| Species | SwissProt | PIR | EMBL (cDNA) | EMBL (genomic) |
|---|---|---|---|---|
| *Bacillus subtilis* | – | – | U48870 | – |
| *Enterobacter aerogenes* | P13514 | – | M26713 | – |
| *Escherichia coli* | P00804 | A00999 B91325 B93991 S40550 | X00776 | D10483: genome 0–2.4' |
| *Haemophilus influenzae* | P44975 | H64107 | – | U32781 genomic |
| *Lactococcus lactis* | – | – | U63724 | – |
| *Mycobacterium tuberculosis* | Q10764 | – | Z74020 | – |

| Species | SwissProt | PIR | EMBL (cDNA) | EMBL (genomic) |
|---|---|---|---|---|
| *Pseudomonas fluorescens* | P17942 | B37152 | M35366 | – |
| *Staphylococcus aureus* | P31024 | S20433 | M83994 | – |
| *Staphylococcus carnosus* | – | – | X78084 | – |

## Name and History

For the first time in the history of lipoproteins, Hantke & Braun (1973) discovered covalent lipid modification of a major outer membrane protein of *Escherichia coli* (also called Braun's lipoprotein) by demonstrating the presence of Ac-S-*sn*1,2-diacylglyceryl Cys at its N-terminus. The mRNA sequence of this lipoprotein revealed a 20 amino acid extension with the characteristics of a signal peptide at the N-terminal of prolipoprotein (Inouye *et al*., 1977). Soon the study of its biosynthesis, especially the identification of the lipid-modifying enzymes and the putative signal peptidase began. The first clue came in 1978 when Arai and his coworkers found the accumulation of a precursor-form of Braun's lipoprotein when *E. coli* was grown in the presence of globomycin, a cyclic pentapeptide antibiotic (Inukai *et al*., 1978). This precursor was found to contain the intact signal peptide and the diacylglyceryl-modified Cys (Hussain *et al*., 1980). Subsequently, the enzyme involved in the cleavage of the signal peptide in Braun's lipoprotein was identified in crude membrane preparations of *E. coli* (Hussain *et al*., 1982). It was shown by Wu and his coworkers that this **lipoprotein-specific signal peptidase** was distinct from the nonlipoprotein signal peptidase I (Chapter 153) (Tokunaga *et al*., 1984a) and hence it was named **signal peptidase II**.

The gene for this enzyme (*lsp*) was identified and sequenced independently by Innis *et al*. (1984) and Yu *et al*. (1984). From the deduced primary structure it was proposed to be a transmembrane protein and supportive evidence was provided later (Muñoa *et al*., 1991). The enzyme has been purified to homogeneity and shown unambiguously to require the initial diacylglyceryl modification of the prospective N-terminal Cys of the mature protein before it could cleave the signal peptide (Dev & Ray, 1984). The signal peptidase II gene was sequenced from *Enterobacter aerogenes* (Isaki *et al*., 1987), *Pseudomonas fluorescens* (Isaki *et al*., 1990) and *Staphylococcus aureus* (Zhao & Wu, 1992). Comparison of these sequences reveals several conserved regions and suggests that this could be a novel aspartic proteinase (Sankaran & Wu, 1994).

## Activity and Specificity

This enzyme cleaves the signal peptides only of bacterial lipoproteins. The $K_m$ for Braun's diacylglyceryl-modified prolipoprotein is 6 mM (Dev *et al*., 1985). The signal peptides of lipoproteins very much resemble those of nonlipoproteins except that the former end in a lipobox sequence (Inouye *et al*., 1986), which, in a majority of about 132 known lipoproteins reported from a variety of bacteria is -Leu-Ala/Ser-Gly/Ala┼Cys- (amino acid positions −3 to +1; Braun & Wu, 1993). The Cys in this lipobox, has to be modified to diacylglyceryl Cys by phosphatidylglycerol:prolipoprotein diacylglyceryl transferase (Sankaran & Wu, 1994) before the action of signal peptidase II (Dev & Ray, 1984). The −1 position should be Gly, Ala or Ser (Pollitt *et al*., 1986; Gennity *et al*., 1990).

The enzyme is assayed using [$^{35}$S]Met-labeled-diacylglyceryl-modified-Braun's prolipoprotein as the substrate. This substrate is not commercially available, but its preparation has been described in detail by Sankaran & Wu (1995). Upon cleavage of the signal peptide, the substrate is converted to apolipoprotein, which moves faster than the substrate in SDS-PAGE. In a simpler, quicker and quantitative assay developed later, the radioactive signal peptide is recovered in the 80% acetone phase after it is cleaved off from the [$^{35}$S]Met-labeled Braun's prolipoprotein and counted in a liquid scintillation counter (Dev & Ray, 1984).

The enzyme does not require metal ions, thiol compounds, cofactors or phospholipids and is optimally active around neutral pH values (Dev & Ray, 1984). Notable among several protease inhibitors tried, pepstatin, tosyl arginyl methyl ester and mercuric chloride inhibit its activity significantly at concentrations below 1 mM (Dev & Ray, 1984). Globomycin, a cyclic pentapeptide antibiotic, is a potent uncompetitive inhibitor with a $K_i$ of 36 nM (Dev *et al*., 1985). The enzyme requires detergent for its activity; it is active in the presence of Triton X-100 or Nikkol, but loses activity in octylglucoside and Sarkosyl solutions (Tokunaga *et al*., 1984b). It has a good thermal stability, being stable for short exposures at elevated temperatures up to 80°C in the absence of detergent (Hussain *et al*., 1982). It also withstands acidic pH conditions (Dev & Ray, 1984). The optimal pH for its activity in the purified state is 6.0 (Dev & Ray, 1984) and in the crude state it has been reported to be 7.9 (Tokunaga *et al*., 1984b).

## Structural Chemistry

The primary structures of signal peptidase II from *E. coli*, *E. aerogenes, P. fluorescens* and *S. aureus* have been deduced from the corresponding gene sequences. They are all about the same size as the *E. coli* enzyme (164 residues) whose deduced $M_r$ is 18 144; the experimentally-determined $M_r$ is 18 000 (Dev & Ray, 1984). There is close homology among the pro lipoprotein signal peptidases with several well-conserved regions (Sankaran & Wu, 1995). In *E. coli* signal peptidase II is an integral protein of the inner membrane; it contains four transmembrane segments (Muñoa *et al*., 1991). A large and a small loop at the periplasmic side seem to contain the active site. It lacks a signal peptide.

## Preparation

This enzyme has been prepared to apparent homogeneity from *E. coli* inner membrane (Dev & Ray, 1984). Solubilization of the total membrane prepared by the lysozyme-EDTA method with Triton X-100, brief heat treatment at 65°C at pH 4.0, followed by DEAE-cellulose chromatography and chromatofocusing yielded preparations that showed 35 000-fold increase in specific activity and 40% recovery. Yu *et al*. (1984) overexpressed signal peptidase II using the *trp* promoter. Following induction, ammonium sulfate fractionation, DEAE-cellulose chromatography in the presence of sodium EDTA, and gel filtration, yielded apparently homogeneous preparations. Purifications from other bacterial sources have not been reported.

## Biological Aspects

This is an essential enzyme in the biosynthesis of bacterial lipoproteins. When it is defective or inhibited by globomycin, cell growth is stopped and the bacteria die. Since lipoproteins from several different bacteria show the same modifications and processing, it appears that the enzyme is ubiquitous among bacteria. Its gene is present as part of an operon, *x-ileS-lsp-orf149-lytB*, at 0.5 min on the *E. coli* genome (Miller *et al*., 1987). The same organization is also seen in other bacteria but the significance of this organization is not yet clear. Processing of the diacylglyceryl modified Braun's prolipoprotein by signal peptidase II *in vivo* requires functional Sec proteins as well as an intact proton-motive-force, but not the ATPase activity associated with SecA (Kosic *et al*., 1993).

## Distinguishing Features

This is a small (18 kDa) integral protein of the inner membrane of bacteria that specifically cleaves the signal peptides of lipoproteins in bacteria. The signal sequences of unmodified prolipoproteins and other secretory proproteins are not recognized by the enzyme. Requirement of diacylglyceryl-modified Cys and other structural features in the lipobox at the C-terminal end of the signal sequence of prolipoproteins is characteristic of signal peptidase II. Globomycin, a fungal pentapeptide, is a specific and potent uncompetitive inhibitor, and inhibition leads to accumulation of diacylglyceryl prolipoproteins and death of the bacteria. The sequence of signal peptidase II does not show appreciable similarity with other known types of signal peptidases including signal peptidase I (Chapter 153). It appears to be a novel aspartic protease.

## Further Reading

For further reading see Sankaran & Wu (1994, 1995).

## References

Braun, V. & Wu, H.C. (1993) Lipoproteins, structure, function, biosynthesis and model for protein export. In: *New Comprehensive Biochemistry*, vol. 27 (Ghuysen, J.-M. & Hackenbeck, R., eds). Amsterdam: Elsevier Science Publishers, pp. 319–342.

Dev, I.K. & Ray, P.H. (1984) Rapid assay and purification of a unique signal peptidase that processes the prolipoprotein from *Escherichia coli* B. *J. Biol. Chem.* **259**, 11114–11120.

Dev, I.K., Harvey, R.J. & Ray, P.H. (1985) Inhibition of prolipoprotein signal peptidase by globomycin. *J. Biol.Chem.* **260**, 5891–5894.

Gennity, J., Goldstein, J. & Inouye, M. (1990) Signal peptide mutants of *Escherichia coli*. *J. Bioenerg. Biomemb.* **22**, 233–269.

Hantke, K. & Braun, V. (1973) Covalent binding of lipid to protein. *Eur. J. Biochem.* **34**, 284–296.

Hussain, M., Ichihara, S. & Mizushima, S. (1980) Accumulation of glyceride-containing precursor of the outer membrane lipoprotein in the cytoplasmic membrane of *Escherichia coli* treated with globomycin. *J. Biol. Chem.* **255**, 3707–3712.

Hussain, M., Ichihara, S. & Mizushima, S. (1982) Mechanism of signal peptide cleavage in the biosynthesis of the major lipoprotein of the *Escherichia coli* outer membrane. *J. Biol. Chem.* **257**, 5177–5182.

Innis, M.A., Tokunaga, M., Williams, M.E., Loranger, J.M., Chang, S.Y., Chang, S. & Wu, H.C. (1984) Nucleotide sequence of the *Escherichia coli* prolipoprotein signal peptidase (*lsp*) gene. *Proc. Natl Acad. Sci. USA* **81**, 3708–3712.

Inouye, S., Wang, S., Sekizawa, J., Halegoua, S. & Inouye, M. (1977) Amino acid sequence for the peptide extension on the prolipoprotein of the *Escherichia coli* outer membrane. *Proc. Natl Acad. Sci. USA* **74**, 1004–1008.

Inouye, S., Duffaud, G. & Inouye, M. (1986) Structural requirement at the cleavage site for effective processing of the lipoprotein secretory precursor of *Escherichia coli*. *J. Biol. Chem.* **261**, 10970–10975.

Inukai, M., Takeuchi, M., Shimizu, K. & Arai, M. (1978) Mechanism of action of globomycin. *J. Antibiot.* **31**, 1203–1205.

Isaki, L., Kawakami, M., Beers, R., Hom, R. & Wu, H.C. (1987) Cloning and nucleotide sequence of the *Enterobacter aerogenes* signal peptidase II (*lsp*) gene. *J. Bacteriol.* **172**, 469–472.

Isaki, L., Beers, R. & Wu, H.C. (1990) Nucleotide sequence of the *Pseudomonas fluorescens* signal peptidase II gene (*lsp*) and flanking genes. *J. Bacteriol.* **172**, 6512–6517.

Kosic, N., Sugai, M., Fan, C. & Wu, H.C. (1993) Processing of lipid-modified prolipoprotein requires energy and *sec* gene products *in vivo*. *J. Bacteriol.* **19**, 6113–6117.

Miller, K.W., Bouvier, J., Stragier, P. & Wu, H.C. (1987) Identification of the genes in the *Escherichia coli ileS-lsp* operon. *J. Biol. Chem.* **262**, 7391–7397.

Muñoa, F.J., Miller, K.W., Beers, R., Grahman, M. & Wu, H.C. (1991) Membrane topology of *Escherichia coli* prolipoprotein signal peptidase (signal peptidase II). *J. Biol. Chem.* **266**, 17667–17672.

Pollitt, S., Inouye, S. & Inouye, M. (1986) Effect of amino acid substitutions at the signal peptide cleavage site of the *Escherichia coli* major outer membrane lipoprotein. *J. Biol. Chem.* **261**, 1835–1837.

Sankaran, K. & Wu, H.C. (1994) Signal peptidase II – specific signal peptidase for bacterial lipoproteins. In: *Signal Peptidases* (von Heijine, G., ed.). Austin: R.G. Landes Co., pp. 17–29.

Sankaran, K. & Wu, H.C. (1995) Bacterial prolipoprotein signal peptidase. *Methods Enzymol.* **248**, 169–179.

Tokunaga, M., Loranger, J.M. & Wu, H.C. (1984a) A distinct signal peptidase for prolipoprotein in *Escherichia coli*. *J. Cell. Biochem.* **24**, 113–120.

Tokunaga, M., Loranger, J.M. & Wu, H.C. (1984b) Prolipoprotein modification and processing enzymes in *Escherichia coli*. *J. Biol. Chem.* **259**, 3825–3830.

Yu, F., Yamada, H., Daishima, K. & Mizushima, S. (1984)

Nucleotide sequence of the *lspA* gene, the structural gene for lipoprotein signal peptidase of *Escherichia coli. FEBS Lett.* **173**, 264–268.

Zhao, X. & Wu, H.C. (1992) Nucleotide sequence of the *Staphylococcus aureus* signal peptidase II (*lsp*) gene. *FEBS Lett.* **299**, 80–84.

***Krishnan Sankaran***
*Centre for Biotechnology,*
*Anna University,*
*Madras-600 025, India*
*Email: cbiotech@giasmd01.vsnl.net.in*

# METALLO-PEPTIDASES

# 334. Introduction: metallopeptidases and their clans

*MEROPS ID: MA, MB, MC, MD, ME, MF, MG, MH*

Metallopeptidases are among the hydrolases in which the nucleophilic attack on a peptide bond is mediated by a water molecule. This is a characteristic shared with the aspartic peptidases (see Chapter 270), but in the metallopeptidases a divalent metal cation, usually zinc but sometimes cobalt or manganese, activates the water molecule. The metal ion is held in place by amino acid ligands, usually three in number.

Metallopeptidases can be divided into two broad groups depending on the number of metal ions required for catalysis. In many metallopeptidases, only one zinc ion is required, but in some families there are two metal ions that act cocatalytically. All the metallopeptidases in which cobalt or manganese is essential require two metal ions, but there are also families of zinc-dependent metallopeptidases in which two zinc ions are cocatalytic. In the peptidases with cocatalytic metal ions, only five amino acids residues act as ligands, since one of these ligates both metal ions. All metallopeptidases with cocatalytic metal ions so far described are exopeptidases, whereas metallopeptidases with one catalytic metal ion may be exopeptidases or endopeptidases.

The known metal ligands in metallopeptidases are His, Glu, Asp or Lys residues. Besides the metal ligands, at least one other residue is required for catalysis. This is Glu in many metallopeptidases, but Lys and Arg have been implicated in the activity of leucyl aminopeptidase (Chapter 473). The function of the Glu has not been established beyond doubt: initially, it was thought to act as a general base (James, 1993), but recent evidence has suggested an electrophilic role (Mock & Stanford, 1996). In clan MB, a tyrosine acts as general base in astacin (Chapter 405) and serralysin (Chapter 386). This residue may also act as a fourth zinc ligand in astacin, but this is not general in the family.

We have allocated metallopeptidases to eight clans as follows:

*Clan MA* (Chapter 335) includes zinc-dependent metallopeptidases in which the zinc ligands are the two histidines in the **His**-Glu-Xaa-Xaa-**His** ('HEXXH') motif and a Glu 18–72 residues C-terminal to the HEXXH motif. The tertiary structure has been determined only for members of family M4, and shows a two-domain structure with the active site between the domains. The N-terminal domain includes both $\alpha$ helices and $\beta$ sheets, and carries the HEXXH motif. It is this domain that shows most similarity to a domain from peptidases of clan MB. The C-terminal domain is all helical, containing five helices in a closed bundle, and has a fold so far seen only in thermolysin-like peptidases. Both exopeptidases and endopeptidases are found in the clan.

*Clan MB* (Chapter 380) contains zinc-dependent metalloendopeptidases in which the zinc ligands are in the motif **His**-Glu-Xaa-Xaa-**His**-Xaa-Xaa-Gly-Xaa-Xaa-**His/Asp**. The

ligands are the two His residues of the HEXXH motif, together with the final His, or in two families, a corresponding Asp. The tertiary structures have been determined for peptidases from families M7 (Chapter 382), M10 (Chapter 385) and M12 (Chapter 404), and show an N-terminal peptidase domain that includes both $\alpha$ helices and $\beta$ sheets.

*Clan MC* (Chapter 450) includes metallocarboxypeptidases in which the zinc ligands are two His residues and a Glu. One His and the Glu occur in the motif **His**-Xaa-Xaa-**Glu**, and the third ligand is between 103 and 143 residues C-terminal to this motif. The structure consists of an $\alpha + \beta$ sandwich with an antiparallel $\beta$ sheet.

*Clan MD* (Chapter 462) contains the zinc-dependent D-Ala-D-Ala carboxypeptidase from *Streptomyces*. The zinc-ligands are two His residues, and an Asp. The His and Asp are closely spaced, within the sequence **His**-Met-Tyr-Gly-His-Ala-Ala-**Asp**, while the third ligand is 36 residues C-terminal. Again, the tertiary structure shows two domains, with two of the zinc ligands in the C-terminal domain (unlike clans MA and MB, in which two of the zinc ligands are in the N-terminal domain). The secondary structure is that of an $\alpha/\beta$ protein with mainly antiparallel $\beta$ sheets, and the fold of the hedgehog protein is similar (Hall *et al.*, 1995).

*Clan ME* (Chapter 465) includes two families in which two of the zinc ligands are thought to occur in the motif **His**-Xaa-Xaa-Glu-**His** ('HXXEH'), almost the reverse of the motif found in clans MA and MB. The His residues and a Glu C-terminal to this motif have been shown to be zinc ligands by site-directed mutagenesis in pitrilysin (Chapter 467), while the Glu in the HXXEH motif is also important for catalysis (Becker & Roth, 1992). No tertiary structure has been solved for any peptidase in this clan.

*Clan MF* (Chapter 472) contains aminopeptidases that require cocatalytic zinc ions for activity. The tertiary structure determined for leucyl aminopeptidase (Chapter 473) shows a two-domain structure with the active site in the C-terminal domain. Both domains contain $\alpha$ and $\beta$ structure, with a $\beta$ sheet in an $\alpha/\beta/\alpha$ layering. In the N-terminal domain the $\beta$ sheet has five strands, and in the C-terminal domain eight. The fold of the N-terminal domain is unique to leucyl aminopeptidase, but the C-terminal domain shows similarities to the metallocarboxypeptidases of clan MC and *Vibrio* aminopeptidase (Chapter 491) from clan MH (Artymiuk *et al.*, 1992; Chevrier *et al.*, 1994). As discussed above, the positions of the zinc ligands are not conserved between the three clans. In leucyl aminopeptidase, the zinc ligands are three Asp residues, a Lys and a Glu. A Lys and an Arg are believed to be important for catalysis. The motif Asn-Thr-**Asp**-Ala-**Glu**-Gly-**Arg**-Leu, which includes two zinc ligands and the catalytic Arg, is completely conserved in the family.

M

***Clan MG*** (Chapter 475) contains exopeptidases that require cocatalytic ions of cobalt (seen in methionyl aminopeptidase 1; Chapter 476) or manganese (as in X-Pro dipeptidase; Chapter 478). The structure determined for *Escherichia coli* methionyl aminopeptidase 1 shows a two-domain structure with an active site between the domains. The two domains are structurally similar, and it is thought that the structure has arisen from the duplication of an ancestral gene followed by a gene fusion event. The ancestral enzyme was presumably active as a dimer. Such an origin has also been postulated for peptidases related to chymotrypsin (Chapter 76) and pepsin (Chapter 271). The fold includes both $\alpha$ and $\beta$ structure, with an antiparallel $\beta$ sheet, and is similar to the fold of the C-terminal domain of *Pseudomonas putida* creatinase. The ligands in methionyl aminopeptidase 1 are two Asp residues, two Glu residues and a His. These are widely spaced, and are not contained in a well-conserved motif.

***Clan MH*** (Chapter 341) contains the third group of metallopeptidases that require cocatalytic metal ions, which in this clan are all zinc ions. The tertiary structure of *Vibrio* aminopeptidase (Chapter 491) has been determined, and shows an $\alpha/\beta$ structure with a mixed $\beta$ sheet of eight strands. Similar structures occur in carboxypeptidases from clan MC and the C-terminal domain of leucyl aminopeptidase (Chapter 473), but without conservation of the position of zinc ligands (Artymiuk *et al.*, 1992; Chevrier *et al.*, 1994). In the aminopeptidase from *Vibrio*, the zinc ligands are three Asp residues, a Glu and a His. As in clan MG, the ligands are widely separated, and peptidases that belong to the clan cannot be identified by any particular motif.

Tertiary structures are known from at least one member of each of the above clans, apart from clan ME. There is some degree of similarity between the tertiary structures for members of clans MC, MF and MH, but because the relative positions of the metal ligands are not conserved, we do not consider that a common evolutionary origin is possible for the three clans. This is because intermediates would be required that would be unable to bind metal ions and would thus be inactive. Also, clan MC contains peptidases that bind only one zinc ion, whereas peptidases from the other two clans possess cocatalytic zinc ions.

Peptidases from clans MA, MB, MC, MD and ME have only one catalytic zinc ion. In contrast, peptidases from clans MF, MG and MH have cocatalytic metal ions, which in clan MF are zinc or manganese and in clan MG are cobalt or manganese. It was pointed out some time ago that the spacing of metal ligands in metallopeptidases follows a 'short–long' pattern, with the first two ligands close together and the third, towards the C-terminus, more distant (Vallee & Auld, 1990). Since there was no obvious reason why this should be, it is perhaps surprising that despite the subsequent identification of metal ligands in several more metallopeptidase families, the generalization still holds. The short–long pattern is also found in peptidases with cocatalytic metal ions (Vallee & Auld, 1993), as well as in proteins in which the metal atom has a structural rather than a catalytic role.

All metallopeptidases with cocatalytic metal ions are exopeptidases, but of these, clan MH is the only clan to include exopeptidases that attack the C-terminus of the substrate. Clan MF contains only aminopeptidases, and clan MG contains aminopeptidases and dipeptidases.

In clans MA and MB, two of the zinc ligands occur in the sequence: Xaa-Xbb-Xcc-**His**-**Glu**-Xbb-Xbb-**His**-Xbb-Xdd, in which Xaa is hydrophobic or Thr, Xbb is an uncharged residue, Xcc is any amino acid except Pro, and Xdd is hydrophobic. Within this sequence, the two His residues are zinc ligands, and the Glu is a catalytic residue. In clan MA, the third zinc ligand is a Glu 18–72 residues C-terminal to the HEXXH motif. In clan MB, the third zinc ligand is the final His (or Asp) in the motif His-Glu-Xaa-Xaa-His-Xaa-Xaa-Gly-Xaa-Xaa-His/Asp. In both clans, the HEXXH motif occurs on an $\alpha$ helix, as does the third ligand, and a turn is required to bring the ligands together. The presence of the conserved Gly in the consensus sequence for clan MB is essential for this $\beta$ turn to form (Bode *et al.*, 1992). The zinc ligands occur within similar segments of secondary structure in clans MA and MB. Thus the r.m.s. deviation is only 0.76 Å between the helix carrying the HEXXH motif and the third ligand for thermolysin (Chapter 351) in clan MA and the corresponding part of interstitial collagenase (Chapter 389) in clan MB (Lovejoy *et al.*, 1994). However, this could well have resulted from convergence in evolution, and other differences in the structures make it difficult for us to imagine that there could have been a single ancestor for peptidases from both clans. The peptidases from clan MB are distinguished by the presence of a conserved $\beta$ turn that underlies the active site and in which a Met is conserved, leading to the term 'Met-zincins' for these enzymes (Stöcker *et al.*, 1995). Peptidases from clan MA are sometimes known as 'Glu-zincins' because the third zinc ligand is Glu rather than His (Hooper, 1994).

Besides the eight clans outlined above, there are a number of families of metallopeptidases that cannot yet be assigned to clans (see Chapter 496). Some of these contain the HEXXH motif in the sequence, but the metal ligands have not been biochemically identified.

## References

Artymiuk, P.J., Grindley, H.M., Park, J.E., Rice, D.W. & Willett, P. (1992) Three-dimensional structural resemblance between leucine aminopeptidase and carboxypeptidase A revealed by graph-theoretical techniques. *FEBS Lett.* **303**, 48–52.

Becker, A.B. & Roth, R.A. (1992) An unusual active site identified in a family of zinc metalloendopeptidases. *Proc. Natl Acad. Sci. USA* **89**, 3835–3839.

Bode, W., Gomis-Rüth, F.-X., Huber, R., Zwilling, R. & Stöcker, W. (1992) Structure of astacin and implications for activation of astacins and zinc-ligation of collagenases. *Nature* **358**, 164–167.

Chevrier, B., Schalk, C., D'Orchymont, H., Rondeau, J.M., Moras, D. & Tarnus, C. (1994) Crystal structure of *Aeromonas proteolytica* aminopeptidase: a prototypical member of the co-catalytic zinc enzyme family. *Structure* **2**, 283–291.

Hall, T.M.T., Porter, J.A., Beachy, P.A. & Leahy, D.J. (1995) A potential catalytic site revealed by the 1.7-Å crystal structure of the amino-terminal signalling domain of Sonic hedgehog. *Nature* **378**, 212–216.

Hooper, N.M. (1994) Families of zinc metalloproteases. *FEBS Lett.* **354**, 1–6.

James, M.N.G. (1993) Convergence of active-center geometries among the proteolytic enzymes. In: *Proteolysis and Protein Turnover* (Bond, J.S. & Barrett, A.J., eds). London: Portland Press, pp. 1–8.

Lovejoy, B., Cleasby, A., Hassell, A.M., Longley, K., Luther, M.A., Weigl, D., McGeehan, G., McElroy, A.B., Drewry, D., Lambert, M.H. & Jordan, S.R. (1994) Structure of the catalytic domain of fibroblast collagenase complexed with an inhibitor. *Science* **263**, 375–377.

Mock, W.L. & Stanford, D.J. (1996) Arazoformyl dipeptide substrates for thermolysin. Confirmation of a reverse protonation catalytic mechanism. *Biochemistry* **35**, 7369–7377.

Stöcker, W., Grams, F., Baumann, U., Reinemer, P., Gomis-

Rüth, F.-X., McKay, D.B. & Bode, W. (1995) The metzincins – topological and sequential relations between the astacins, adamalysins, serralysins, and matrixins (collagenases) define a superfamily of zinc-peptidases. *Protein Sci.* **4**, 823–840.

Vallee, B.L. & Auld, D.S. (1990) Zinc coordination, function, and structure of zinc enzymes and other proteins. *Biochemistry* **29**, 5647–5659.

Vallee, B.L. & Auld, D.S. (1993) Cocatalytic zinc motifs in enzyme catalysis. *Proc. Natl Acad. Sci. USA* **90**, 2715–2718.

# 335. Introduction: clan MA containing thermolysin and its relatives

## Databanks

*MEROPS ID: MA*

| Species | SwissProt | PIR | EMBL (cDNA) | EMBL (genomic) |
|---|---|---|---|---|
| **Family M1** (Chapter 336) | | | | |
| **Family M2** | | | | |
| Peptidyl-dipeptidase A (Chapter 359) | | | | |
| Peptidyl-dipeptidase A, insect type (Chapter 360) | | | | |
| **Family M4** | | | | |
| Bacillolysin (typically *B. subtilis*) (Chapter 351) | | | | |
| Coccolysin (Chapter 354) | | | | |
| Elastase (*Staphylococcus*) | | | | |
|    *Staphylococcus epidermidis* | P43148 | – | X69957 | – |
| Hemagglutinin proteinase (e.g. *Vibrio, Helicobacter*) (Chapter 356) | | | | |
|    λ Toxin (*Clostridium perfringens*)– | | – | D45904 | – |
| Pseudolysin (Chapter 357) | | | | |
| Thermolysin (typically *Bacillus thermoproteolyticus*) (Chapter 351) | | | | |
| Thermolysin homolog, *Legionella* (Chapter 358) | | | | |
| Thermolysin homolog, *Listeria* (Chapter 353) | | | | |
| Vibriolysin (Chapter 355) | | | | |
| Others | | | | |
|    *Erwinia carotovora* | Q99132 | – | M36651 | – |
|    *Lactobacillus* sp. | – | – | D29673 | – |
|    *Renibacterium salmoninarum* | P55111 | – | X76499 | – |
|    *Serratia marcescens* | Q06517 | – | M59854 | – |
| **Family M5** | | | | |
| Mycolysin (Chapter 361) | | | | |
| **Family M9** | | | | |
| *Clostridium* collagenase (Chapter 368) | | | | |
| Microbial collagenase (*Vibrio*) (Chapter 367) | | | | |
| Others | | | | |
|    *Bacillus cereus* | – | – | M30809 | – |
|    *Bacillus thuringiensis* | – | – | X12952 | – |
| **Family M13** | | | | |
| Endothelin-converting enzyme 1 (Chapter 363) | | | | |
| Endothelin-converting enzyme 2 (Chapter 364) | | | | |
| Kell blood-group protein | | | | |
|    *Homo sapiens* | P23276 | – | S76819 | – |
| Neprilysin (Chapter 362) | | | | |
| Neprilysin homolog | | | | |
|    *Streptococcus gordonii* | P42359 | – | L11577 | – |
| Oligopeptidase O (Chapter 365) | | | | |
| PEX protein | | | | |
|    *Homo sapiens* | – | – | U60475 | – |
|    *Mus musculus* | – | – | U73910 | U73911: exon 3 |
| | | | | U73912: exon 6 |
| | | | | U73913: exon 9 |
| | | | | U73914: exon 19 |
| | | | | U73915: exon 21 |
| Others | | | | |
|    *Caenorhabditis elegans* | – | – | D36217 | – |
|    *Caenorhabditis elegans* | – | – | U41546 | – |
|    *Haemonchus contortus* | – | – | Z75054 | – |

| Species | SwissProt | PIR | EMBL (cDNA) | EMBL (genomic) |
|---|---|---|---|---|
| **Family M30** | | | | |
| Hyicolysin (Chapter 369) | | | | |
| **Family M36** | | | | |
| Fungalysin (*Aspergillus*) (Chapter 514) | | | | |
| **Family M48** | | | | |
| Ste24p protease (yeast) (Chapter 366) | | | | |
| Others | | | | |
| *Arabidopsis thaliana* | – | – | Z34189 | – |
| *Caenorhabditis elegans* | – | – | T00972 | – |
| *Caenorhabditis elegans* | – | – | T02282 | – |
| *Escherichia coli* | – | – | U31523 | – |
| *Escherichia coli* | P23894 | A43659 | M58470 | – |
| *Escherichia coli* | P25894 | A42604 | M32363 | U28377: chromosome 65–68′ |
| *Escherichia coli* | P43674 | – | X82933 | – |
| *Haemophilus influenzae* | P44840 | – | – | U32755: genomic |
| *Homo sapiens* | – | – | Z43273 | – |
| *Methanococcus jannaschii* | – | – | – | U67608: complete genome section 150 of 150 |
| *Schistosoma mansoni* | – | – | W06703 | – |
| *Schistosoma mansoni* | – | – | W06760 | – |
| *Synechocystis* sp. | – | – | D90915 | – |

Clan MA includes all the families of metallopeptidases in which the zinc ligands are the His residues in an His-Glu-Xaa-Xaa-His ('HEXXH') motif, and a Glu at least 14 residues C-terminal to this motif. The clan includes the archetypal metalloendopeptidase thermolysin. The only family in the clan from which tertiary structures of peptidases are known is M4.

Although the catalytic mechanism in thermolysin is probably the best understood of all the metallopeptidases, there have been different views as to which residues perform which roles. It had originally been thought that Glu143, which is part of the HEXXH motif, acts as a proton acceptor during catalysis, but the studies with arazoformyl dipeptide substrates (Mock & Stanford, 1996) has suggested the following interpretation. Besides an activated water molecule, the zinc ion and the zinc ligands, three other residues are important for catalysis. These are Glu143, which acts as an electrophile, Asp226 and His231. Asp226 orientates the imidazolium ring of His231, in much the same way as Asp102 is thought to act in chymotrypsin and its homologs in clan SA (Chapter 76), and His231 acts as proton donor and general base. Another residue that is conserved in most members of the clan is Asp170. In thermolysin, His142 forms a hydrogen bond with Asp170 (Argos *et al.*, 1978), and the equivalent Asp in neprilysin has been shown to be essential for activity, and mutations to Glu, Asn or Ala in this position cause drastic reductions in specific activity (Le Moual *et al.*, 1994). A similar Asp:His interaction has also been postulated for carboxypeptidase A (Argos *et al.*, 1978; Christianson & Alexander, 1990). Tyr157 may aid positioning of the nucleophilic water molecule (Mock & Stanford, 1996).

The zinc ligands have been identified in only five of the families included in clan MA: M1, M2, M4, M5 and M13. In families M2, M4, M5 and M13, the spacing between the third zinc ligand and Asp170 is identical (four amino acids; Alignment 350.1), and the Glu-(Xaa)$_3$-Asp motif provides a means in addition to the HEXXH motif of finding other families of metallopeptidases that may share the thermolysin fold. Three other families can be included in the clan because of the presence of both motifs, families M30, M36 and M48. In family M9 there are HEXXH and a Glu-(Xaa)$_3$-Glu motifs, and we also include this family in clan MA.

Other characteristics common to peptidases in this clan are that (a) the main determinant of substrate specificity is a hydrophobic residue in P1′, and (b) there is strong inhibition by phosphoramidon.

Family M1 is described in more detail in Chapter 336, and the other families of clan MA are described in Chapter 350.

### References

Argos, P., Garavito, R.M., Eventoff, W., Rossman, M.G. & Bränden, C.I. (1978) Similarities in active center geometrics of zinc-containing enzymes, proteases and dehydrogenases. *J. Mol. Biol.* **126**, 141–158.

Christianson, D.W. & Alexander, R.S. (1990) Another catalytic triad? *Nature* **346**, 225.

Le Moual, H., Dion, N., Roques, B.P., Crine, P. & Boileau, G. (1994) Asp650 is crucial for catalytic activity of neutral endopeptidase 24–11. *Eur. J. Biochem.* **221**, 475–480.

Mock, W.L. & Stanford, D.J. (1996) Arazoformyl dipeptide substrates for thermolysin. Confirmation of a reverse protonation catalytic mechanism. *Biochemistry* **35**, 7369–7377.

**M**

# 336. Introduction: family M1 of membrane alanyl aminopeptidase

## Databanks

MEROPS ID: M1

| Species | SwissProt | PIR | EMBL (cDNA) | EMBL (genomic) |
|---|---|---|---|---|
| **Family M1** | | | | |
| Alanyl/arginyl aminopeptidase (*Saccharomyces*) (Chapter 344) | | | | |
| Aminopeptidase II (*Saccharomyces*) (Chapter 344) | | | | |
| Aminopeptidase B (Chapter 348) | | | | |
| Aminopeptidase Ey (Chapter 349) | | | | |
| Aminopeptidase N (*Streptomyces lividans*) (Chapter 346) | | | | |
| Aminopeptidase PS (Chapter 343) | | | | |
| CrylAc toxin-binding protein | | | | |
|   *Heliothis virescens* | – | – | U35096 | – |
|   *Manduca sexta* | – | – | X89081 | – |
| Cystinyl aminopeptidase (Chapter 341) | | | | |
| Glutamyl aminopeptidase (Chapter 339) | | | | |
| Insulin-regulated membrane aminopeptidase (Chapter 342) | | | | |
| Leukotriene $A_4$ hydrolase (Chapter 347) | | | | |
| Lysyl aminopeptidase (*Lactobacillus*) (Chapter 345) | | | | |
| Membrane alanyl aminopeptidase (eukaryote) (Chapters 337, 338) | | | | |
| Pyroglutamyl-peptidase II (Chapter 340) | | | | |
| Others | | | | |
|   *Caenorhabditis elegans* | – | – | – | Z30317: cosmid T16G12 |

Family M1 contains exopeptidases that act at the N-terminus of polypeptides. Most members of the family are aminopeptidases, but the family does include an omega peptidase, pyroglutamyl-peptidase II, which releases N-terminal pyroglutamate from thyrotropin-releasing hormone, Glp┼His-Xaa tripeptides and Glp┼His-Xaa-Gly tetrapeptides. Leukotriene $A_4$ hydrolase possesses two catalytic activities, acting as an arginyl aminopeptidase and as an epoxide hydrolase, converting leukotriene $A_4$ to the inflammatory mediator leukotriene $B_4$. The presence of homologs of this enzyme in yeast and *Dictyostelium* suggests that it has a more fundamental role, however.

The family is widely distributed and examples are known from gram-negative and gram-positive bacteria, cyanobacteria, protozoa, fungi and animals. No complete sequence is known from plants, but there are expressed sequence tag fragments of mRNA from *Arabidopsis thaliana*. The family includes membrane-bound enzymes, such as membrane alanyl aminopeptidase and pyroglutamyl-peptidase II, and cytoplasmic aminopeptidases such as leukotriene $A_4$ hydrolase and lysyl aminopeptidase. In the membrane-bound peptidases, there is an N-terminal cytoplasmic anchor followed by a transmembrane domain, with the remainder of the molecule extracellular and heavily glycosylated. It is no surprise that some of the membrane-bound proteins are also recognized as surface antigens. Membrane alanyl aminopeptidase is also known as the myeloid leukemia marker CD13 (Olsen *et al*., 1989) and glutamyl aminopeptidase is known as a differentiation-related kidney antigen gp160 and an immature B cell marker BP-1/6C3 (Wu *et al*., 1990; Nanus *et al*., 1993). Both alanyl and glutamyl aminopeptidases are homodimers and quantitatively important components of the intestinal brush border membranes, together comprising about 5% of total protein. None of the peptidases in family M1 is synthesized as a proenzyme.

A homolog from *Caenorhabditis elegans*, known only from sequence obtained from the genome-sequencing project, contains two homologous peptidase units. A C-terminal fragment of a homolog from *Dictyostelium* (SW: P52922) is reported to have an HEXH motif rather than HEXXH.

The zinc ligands have been identified in leukotriene $A_4$ hydrolase (Medina *et al*., 1991) by site-directed mutagenesis as His295, His299 and Glu318 (numbered according to Alignment 336.1). The catalytic Glu has been shown to be Glu388 in lysyl aminopeptidase (Vazeux *et al*., 1996). This residue is important for peptidase activity but not for epoxide hydrolysis in leukotriene $A_4$ hydrolase (Minami *et al*., 1992). These residues are the same and in the same order as in thermolysin, and because of this, family M1 is included in clan MA. However, there is no conserved His that could act as the general base in family M1. Tyr476 has been shown to be essential for peptidase activity in leukotriene $A_4$ hydrolase

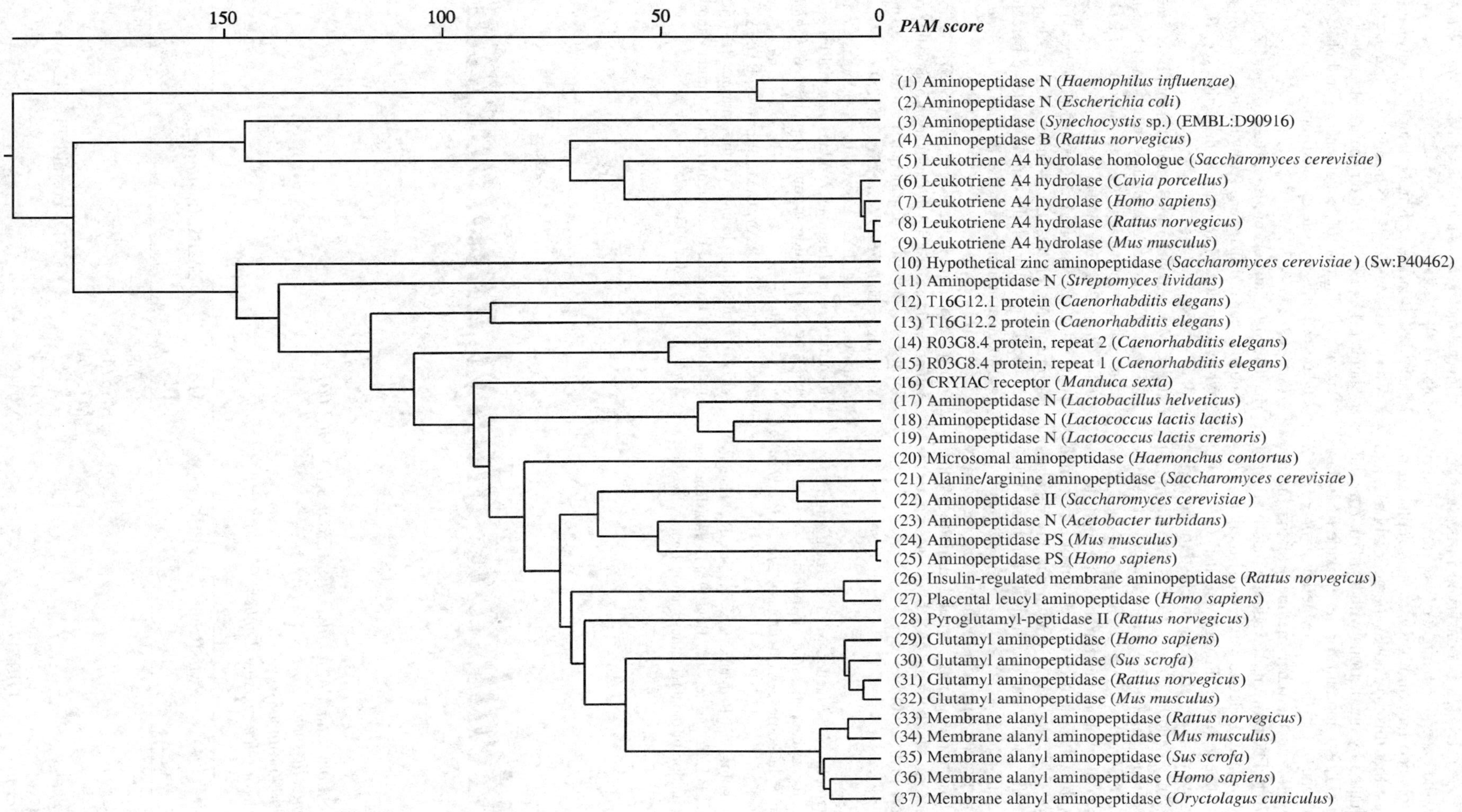

PAM score

(1) Aminopeptidase N (*Haemophilus influenzae*)
(2) Aminopeptidase N (*Escherichia coli*)
(3) Aminopeptidase (*Synechocystis* sp.) (EMBL:D90916)
(4) Aminopeptidase B (*Rattus norvegicus*)
(5) Leukotriene A4 hydrolase homologue (*Saccharomyces cerevisiae*)
(6) Leukotriene A4 hydrolase (*Cavia porcellus*)
(7) Leukotriene A4 hydrolase (*Homo sapiens*)
(8) Leukotriene A4 hydrolase (*Rattus norvegicus*)
(9) Leukotriene A4 hydrolase (*Mus musculus*)
(10) Hypothetical zinc aminopeptidase (*Saccharomyces cerevisiae*) (Sw:P40462)
(11) Aminopeptidase N (*Streptomyces lividans*)
(12) T16G12.1 protein (*Caenorhabditis elegans*)
(13) T16G12.2 protein (*Caenorhabditis elegans*)
(14) R03G8.4 protein, repeat 2 (*Caenorhabditis elegans*)
(15) R03G8.4 protein, repeat 1 (*Caenorhabditis elegans*)
(16) CRY1AC receptor (*Manduca sexta*)
(17) Aminopeptidase N (*Lactobacillus helveticus*)
(18) Aminopeptidase N (*Lactococcus lactis lactis*)
(19) Aminopeptidase N (*Lactococcus lactis cremoris*)
(20) Microsomal aminopeptidase (*Haemonchus contortus*)
(21) Alanine/arginine aminopeptidase (*Saccharomyces cerevisiae*)
(22) Aminopeptidase II (*Saccharomyces cerevisiae*)
(23) Aminopeptidase N (*Acetobacter turbidans*)
(24) Aminopeptidase PS (*Mus musculus*)
(25) Aminopeptidase PS (*Homo sapiens*)
(26) Insulin-regulated membrane aminopeptidase (*Rattus norvegicus*)
(27) Placental leucyl aminopeptidase (*Homo sapiens*)
(28) Pyroglutamyl-peptidase II (*Rattus norvegicus*)
(29) Glutamyl aminopeptidase (*Homo sapiens*)
(30) Glutamyl aminopeptidase (*Sus scrofa*)
(31) Glutamyl aminopeptidase (*Rattus norvegicus*)
(32) Glutamyl aminopeptidase (*Mus musculus*)
(33) Membrane alanyl aminopeptidase (*Rattus norvegicus*)
(34) Membrane alanyl aminopeptidase (*Mus musculus*)
(35) Membrane alanyl aminopeptidase (*Sus scrofa*)
(36) Membrane alanyl aminopeptidase (*Homo sapiens*)
(37) Membrane alanyl aminopeptidase (*Oryctolagus cuniculus*)

*Figure 336.1*  Evolutionary tree for family M1. The tree was prepared as described in the Introduction (p. xxv).

(Blomster *et al.*, 1995). Lys478 is conserved in most members of the family, the exception being Cry1Ac toxin-binding protein from the tobacco hawkmoth where it is replaced by Arg, and we suggest that this may act as the general base, and perhaps be coordinated by interactions with Tyr476. An additional residue, Tyr471, is important for the epoxide hydrolase activity of leukotriene $A_4$ hydrolase (Mueller *et al.*, 1996), and is replaced by Phe in the other members of the family. Mutation of Tyr476 to Phe prevents the suicide inactivation of leukotriene $A_4$ hydrolase by leukotriene $A_4$ (Mueller *et al.*, 1996). It has also been shown through the use of specific inhibitors that arginyl residues are present close to the active center of the enzyme (Mueller *et al.*, 1995). An alignment of the catalytic residues of family M1 is shown in Alignment 336.1 (see CD-ROM).

A motif that is conserved in the family N-terminal to the HEXXH motif, is the Gly-Xaa-Met-Glu-Asn motif, in which Xaa can be Ala or Gly. No role is known for any residues in this motif, but the motif may be helpful for identifying members of the family.

The evolutionary tree (Figure 336.1) shows that the family can be divided into three groups around aminopeptidase N from enterobacteria, leukotriene $A_4$ hydrolase, and the remaining members of the family.

### References

Blomster, M., Wetterholm, A., Mueller, M.J. & Haeggström, J.Z. (1995) Evidence for a catalytic role of tyrosine 383 in the peptidase reaction of leukotriene $A_4$ hydrolase. *Eur. J. Biochem.* **231**, 528–534.

Medina, J.F., Wetterholm, A., Rådmark, O., Shapiro, R., Haeggström, J.Z., Vallee, B.L. & Samuelsson, B. (1991) Leukotriene $A_4$ hydrolase: determination of the three zinc-binding ligands by site-directed mutagenesis and zinc analysis. *Proc. Natl Acad. Sci. USA* **88**, 7620–7624.

Minami, M., Bito, H., Ohishi, N., Tsuge, H., Miyano, M., Mori, M., Wada, H., Mutoh, H., Shimada, S., Izumi, T., Abe, K. & Shimizu, T. (1992) Leukotriene $A_4$ hydrolase, a bifunctional enzyme: distinction of leukotriene $A_4$ hydrolase and aminopeptidase activities by site-directed mutagenesis at Glu-297. *FEBS Lett.* **309**, 353–357.

Mueller, M.J., Samuelsson, B. & Haeggström, J.Z. (1995) Chemical modification of leukotriene $A_4$ hydrolase: indications for essential tyrosyl and arginyl residues at the active site. *Biochemistry* **34**, 3536–3543.

Mueller, M.J., Andberg, M.B., Samuelsson, B. & Haeggström, J.Z. (1996) Leukotriene $A_4$ hydrolase, mutation of tyrosine 378 allows conversion of leukotriene $A_4$ into an isomer of leukotriene $B_4$. *J. Biol. Chem.* **271**, 24345–24348.

Nanus, D.M., Engelstein, D., Gastl, G.A., Gluck, L., Vidal, M.J., Morrison, M., Finstad, C.L., Bander, N.H. & Albino, A.P. (1993) Molecular cloning of the human kidney differentiation antigen gp160: human aminopeptidase A. *Proc. Natl Acad. Sci. USA* **90**, 7069–7073.

Olsen, J., Sjöström, H. & Norén, O. (1989) Cloning of the pig aminopeptidase N gene. Identification of possible regulatory elements and the exon distribution in relation to the membrane-spanning region. *FEBS Lett.* **251**, 275–281.

Vazeux, G., Wang, J.Y., Corvol, P. & Llorens-Cortès, C. (1996) Identification of glutamate residues essential for catalytic activity and zinc coordination in aminopeptidase A. *J. Biol. Chem.* **271**, 9069–9074.

Wu, Q., Lahti, J.M., Air, G.M., Burrows, P.D. & Cooper, M.D. (1990) Molecular cloning of the murine BP-1/6C3 antigen: a member of the zinc-dependent metallopeptidase family. *Proc. Natl Acad. Sci. USA* **87**, 993–997.

# *337. Membrane alanyl aminopeptidase*

### Databanks

*Peptidase classification: clan MA, family M1, MEROPS ID: M01.001*
*NC-IUBMB enzyme classification: EC 3.4.11.2*
*ATCC entries: 106096 (human)*
*Chemical Abstracts Service registry number: 9054-63-1*
*Databank codes:*

| Species | SwissProt | PIR | EMBL (cDNA) | EMBL (genomic) |
|---|---|---|---|---|
| *Arabidopsis thaliana* | – | – | Z48608 | – |
| *Caenorhabditis elegans* | – | – | M75749 M75750 | – |
| *Caenorhabditis elegans* | – | – | – | Z30317: cosmid T16G12 |
| *Caenorhabditis elegans* | – | – | Z69794 | – |
| *Haemonchus contortus* | Q10737 | – | X94187 | – |

| Species | SwissProt | PIR | EMBL (cDNA) | EMBL (genomic) |
|---|---|---|---|---|
| *Homo sapiens* | P15144 | A30325 S01658 | M22324 M55522 X13276 | M55523: exon 1 |
| *Mus musculus* | – | – | U77083 | – |
| *Oryctolagus cuniculus* | P15541 | A25985 S07099 | S68687 X51508 | X89921: promoter |
| *Rattus norvegicus* | P15684 | A32852 | M25073 M26710 | – |
| *Sus scrofa* | P15145 | A53984 | Z29522 | X16088: 5′ end |

## Name and History

The early history of **membrane alanyl aminopeptidase (mAAP)** has been thoroughly summarized in Chapter 338 under its guise as **Cys-Gly dipeptidase**. The enzyme has also been referred to in its earlier days as **aminopeptidase M** (for microsomal or membrane aminopeptidase), reflecting its tight association with a microsomal membrane fraction in pig kidney from which it was purified. The use of the name aminopeptidase M is still occasionally seen today in the literature and the enzyme has also been confused with the cytosolic 'leucine aminopeptidase' because of their overlapping substrate specificities and similar tissue distributions. In 1980 it was suggested that the enzyme should be renamed **aminopeptidase N**, reflecting its preference for action on neutral amino acids (Ferracci & Maroux, 1980) and that terminology is still in common use today. The name membrane alanyl aminopeptidase was introduced to clarify the nature and localization of the enzyme and to distinguish it from its cytosolic counterpart. Much of the original characterization of mAAP was performed on the renal or intestinal enzymes. However, the presence of the enzyme in brain has attracted substantial interest since the discovery that it can participate in the hydrolysis and inactivation of the enkephalins by hydrolysis of the Tyr1+Gly2 bond (Gros *et al.*, 1985; Matsas *et al.*, 1985). mAAP also turns out to be identical with the human **cluster differentiation antigen CD13** expressed on the surface of myeloid progenitors, monocytes, granulocytes and myeloid leukemia cells (Look *et al.*, 1989).

## Activity and Specificity

mAAP has a broad substrate specificity removing N-terminal amino acids (Xaa+Xbb-) from almost all unsubstituted oligopeptides and from an amide or arylamide. It has usually been assayed with derivatives of alanine, e.g. Ala-NHMec or the NHPhNO$_2$ or NNap derivatives, because Ala is the most favored residue. Leu-NHMec and other bulky hydrophobic amino acid derivatives are also good substrates but leucinamide is poorly hydrolyzed. For aminoacyl derivatives, the favored order is reported to be Ala, Phe, Tyr, Leu, Arg, Thr, Trp, Lys, Ser, Asp, His and Val. Pro- and $\alpha$- or $\gamma$-Glu-derivatives are very slowly attacked. When a prolyl residue is preceded by a bulky hydrophobic residue, e.g. Leu, Tyr or Trp, unusual secondary reactions can occasionally arise such that the X-Pro combination is released as an intact dipeptide (see, for example, McDonald & Barrett, 1986). Dipeptides are readily hydrolyzed, e.g. Cys-Gly, as in the original studies on

this activity (Semenza, 1957; see also Chapter 338). Subsite interactions are important and hence chain length greatly affects the rates, although precise rules governing specificity have not been defined.

The pH optimum is around 7.0 although the optimum can rise to 9.0 as the substrate concentration is increased. However, the $K_m$ is lowest in the pH range 7.0–7.5. Metal chelating agents are effective inhibitors, consistent with the metallopeptidase nature of the enzyme, and sulfhydryl reagents are without effect. A comparison of the effects of a range of metallopeptidase inhibitors on membrane aminopeptidases has been carried out by Tieku & Hooper (1992). Amastatin (originally described as an inhibitor of glutamyl aminopeptidase (aminopeptidase A) (Chapter 339) is also a very effective inhibitor of mAAP, with an increase in potency when preincubated with the enzyme, the $K_i$ value decreasing from 20 $\mu$M to 20 nM, i.e. it is a slow, tight-binding inhibitor which involves a conformational change in the enzyme–inhibitor complex (Rich *et al.*, 1984). The kinetics of this reaction have been examined in detail by Rich *et al.* (1984). Probestin is also a potent inhibitor with a reported I$_{50}$ of 50 nM (Tieku & Hooper, 1992). Bestatin is also a well-recognized inhibitor of mAAP although considerably less potent than amastatin or probestin (Stephenson & Kenny, 1987; Tieku & Hooper, 1992). Actinonin (I$_{50}$ = 2 $\mu$M) can be considered a relatively specific inhibitor of mAAP compared with other membrane aminopeptidases (Tieku & Hooper, 1992). The enzyme is only very weakly inhibited by puromycin (see Distinguishing Features, below). Based on such inhibitory data, a selective enzyme assay for mAAP has been devised (Gillespie *et al.*, 1992). More recently, a new range of potent and selective inhibitors of mAAP have been described based on derivatives of 3-amino-2-tetralone (Schalk *et al.*, 1994), some of which exhibit $K_i$ values in the nanomolar range. The proposed mode of binding of these compounds is as bidentate ligands with the amino and carbonyl functions coordinating to the active-site zinc.

## Structural Chemistry

mAAP is a type II integral membrane protein located on the plasma membrane as an ectoenzyme. The pI is approximately 5. The native enzyme exists as a homodimer of subunit $M_r$ 140 000–150 000 in most species, although it is reported to be monomeric in the rabbit (Feracci & Maroux, 1980). It is heavily glycosylated with carbohydrate accounting for at least 20% of the mass of the protein. The polypeptide chain is susceptible to proteolysis, generating two fragments of $M_r$ approximately 90 000 and 45 000 that have been referred to in the earlier literature as the $\beta$ and $\gamma$ subunits respectively (the intact chain being the $\alpha$ subunit). This artifact of preparation led to the suggestion that the native enzyme may be a trimer (Maroux *et al.*, 1973).

The enzyme was originally cloned from a human intestinal cDNA library (Olsen *et al.*, 1988) and subsequently from rat (Watt & Yip, 1989; Malfroy *et al.*, 1989) and rabbit kidney (Yang *et al.*, 1993). The rat enzyme comprises a 966 amino acid polypeptide with a small cytoplasmic domain, a 24 amino acid hydrophobic segment close to the N-terminus which serves as the membrane anchor region and the bulk of the polypeptide chain including the active site present as an ectodomain. The sequence includes nine potential N-linked glycosylation sites and a typical zinc-binding motif (Val-Xaa-Xaa-His-Glu-Xaa-Xaa-His) (Hooper, 1994) in which the two closely spaced histidines represent two of the zinc ligands. The third zinc ligand is a glutamate. The protein contains one $Zn^{2+}$ per subunit. Chemical modification experiments have been used to identify arginyl, histidyl, tyrosyl and aspartyl/glutamyl residues at the active site (Helene *et al.*, 1991). The *Lactococcus lactis pepN* gene encodes an aminopeptidase (Chapter 345) homologous to mAAP with almost 30% identity between the bacterial and mammalian proteins and with particularly high conservation around the active-site region (Tan *et al.*, 1992).

## Preparation

In the kidney, mAAP represents as much as 8% of the brush border membrane protein, thereby providing a convenient and abundant source to initiate purification. It was first isolated from pig kidney as 'cysteinyl-glycinase' (Semenza, 1957) and subsequently as an aminopeptidase (Wachsmuth *et al.*, 1966). The protein can be purified in either hydrophilic or amphipathic form by proteinase (trypsin, papain) treatment or detergent solubilization respectively. Conventional chromatographic procedures can then be used to isolate the enzyme (e.g. Feracci & Maroux, 1980). The pig small intestinal mAAP has also been purified by immunoadsorbent chromatography (Sjöström *et al.*, 1978). A 130 kDa glycoprotein purified from pig kidney brush border membranes by affinity chromatography on immobilized 4-acetamido-4'-isothiocyanostilbene-2,2'-disulfonate (SITS) followed by concanavalin A-Sepharose, turned out serendipitously to be mAAP (See & Reithmeier, 1990), suggesting that the protein possesses an anion-binding site. This procedure provides a convenient purification method for the enzyme which represents the major concanavalin A-binding protein in brush border membranes. mAAP in the larval midgut cell membranes of the silkworm, *Bombyx mori*, is partially sensitive to release by phosphatidylinositol-specific phospholipase C, suggesting that in this species the enzyme may be anchored through a glycolipid anchor rather than a transmembrane domain (Takesue *et al.*, 1992).

## Biological Aspects

mAAP is widely distributed among species and tissues although it is of greatest abundance in brush border membranes of the kidney, mucosal cells of the small intestine and in the liver. It is also present in the lung where it is identical to the p146 type II alveolar epithelial cell antigen (Funkhouser *et al.*, 1991) and is located on endothelial cells in blood vessels. On polarized epithelial cells, mAAP is localized to the apical domain and is targeted there through an apical sorting signal thought to be located in the catalytic head group region of the protein (Vogel *et al.*, 1992). In the kidney, mAAP contributes to the extracellular catabolism of glutathione (Curthoys, 1987). The cysteinyl-glycine generated during the catabolism of glutathione by $\gamma$-glutamyltranspeptidase is hydrolyzed by the two ectoenzymes mAAP and membrane dipeptidase (Chapter 500) contributing approximately equally (McIntyre & Curthoys, 1982). In the intestine, the enzyme functions in the final stages of protein and peptide digestion.

A detailed localization of the enzyme has been carried out in the brain because of its potential involvement in terminating the actions of certain neuropeptides, especially the enkephalins (Solhonne *et al.*, 1987; Barnes *et al.*, 1988, 1994). In addition to being present on endothelial cells and synaptic membranes, mAAP is found on astrocytes and pericytes (Barnes *et al.*, 1994; Kunz *et al.*, 1994). It is abundant in the choroid plexus and can therefore also serve to prevent access to the brain of potentially damaging circulating peptides. On vascular cells, mAAP may serve to metabolize certain vasoactive peptides (Ward *et al.*, 1990). An important location of mAAP is in hematopoietic cells, where it is referred to as CD13 (Look *et al.*, 1989). Here, its expression is restricted primarily to myeloid cells, but it is also found on antigen-presenting cells, melanoma cells and lymphocytes. On granulocytes it may cooperate with neprilysin (Chapter 362) to downregulate responses to chemotactic factors such as formyl-Met-Leu-Phe (Shipp & Look, 1993). More generally in the immune system it may serve to inactivate certain cytokines (Hoffmann *et al.*, 1993; Kanayama *et al.*, 1995). The immunopotentiating and reported antitumor activities of bestatin may relate to inhibition of mAAP (Leyhausen *et al.*, 1983).

The gene locus of human mAAP is 15q13–qter (Kruse *et al.*, 1988). The pig mAAP gene has been cloned (Olsen *et al.*, 1989). Separate promoters control transcription of the human gene in myeloid and intestinal epithelial cells (Shapiro *et al.*, 1991).

A novel feature of mAAP is its ability to serve as a receptor for certain viruses, especially coronavirus 229E, an RNA virus that causes upper respiratory tract infections in humans (Yeager *et al.*, 1992). Mutagenesis studies suggest that the virus-binding site lies close to the active-site region, although enzyme activity is not essential for virus binding. Human mAAP also appears to mediate human cytomegalovirus infection although, again, enzyme activity is not essential for infection (Soderberg *et al.*, 1993). Another coronavirus, transmissible gastroenteritis virus, which causes a fatal diarrhea in newborn pigs, uses intestinal mAAP as its receptor (Delmas *et al.*, 1992). mAAP appears to be the major receptor for the CryIAc toxin of *Bacillus thuringiensis* in *Lymantria dispar* (gypsy moth) (Lee *et al.*, 1996).

mAAP is synthesized in a fully active form. Substance P and bradykinin, which are not substrates for mAAP, have been reported as natural inhibitors of the enzyme with $K_i$

values in the low micromolar range (Xu *et al.*, 1995). However, it is unlikely that they play any physiological role in regulating enzyme activity and the enzyme is therefore probably essentially unregulated at the surface of cells.

## Distinguishing Features

mAAP can be distinguished from the cytosolic leucine aminopeptidase by its membrane association and its poor hydrolysis of leucinamide (see above). It can be distinguished from another membrane aminopeptidase in brain (Chapter 343) capable of hydrolyzing the enkephalins by its relative insensitivity to puromycin ($K_i = 78$ mM compared with 1 mM for the puromycin-sensitive activity). Actinonin is a relatively selective inhibitor. The dipeptidase activity of mAAP can be distinguished from that of the mammalian membrane dipeptidase (Chapter 500) by the sensitivity of the latter to cilastatin (Littlewood *et al.*, 1989).

## Related Peptidases

Several mammalian aminopeptidases with homology to mAAP have recently been cloned, including the puromycin-sensitive aminopeptidase (Chapter 343), which has been implicated in cell growth and viability (Constam *et al.*, 1995), and human placental leucine aminopeptidase/oxytocinase (Chapter 341), which is also a type II integral membrane protein and may play a role in the degradation of oxytocin and vasopressin (Rogi *et al.*, 1996). The major protein present in GLUT4 vesicles in fat and muscle tissues is a glycoprotein of $M_r$ 160 000 that has structural homology to mAAP and exhibits aminopeptidase activity *in vitro* (Kandror *et al.*, 1994) (Chapter 342). The cytosolic leukotriene A$_4$ hydrolase (Chapter 347) also has aminopeptidase activity and belongs to the mAAP family (Toh *et al.*, 1990).

## Further Reading

For a review, see Wang & Cooper (1996).

## References

Barnes, K., Matsas, R., Hooper, N.M., Turner, A.J. & Kenny, A.J. (1988) Endopeptidase-24.11 is striosomally ordered in pig brain and, in contrast to aminopeptidase N and peptidyl dipeptidase A ('angiotensin converting enzyme'), is a marker for a set of striatal efferent fibres. *Neuroscience* **27**, 799–817.

Barnes, K., Kenny, A.J. & Turner, A.J. (1994) Localization of aminopeptidase N and dipeptidyl peptidase IV in pig striatum and in neuronal and glial cell cultures. *Eur. J. Neurosci.* **6**, 531–537.

Constam, D.B., Tobler, A.R., Rensing-Ehl, A., Kemler, I., Hersh, L.B. & Fontana, A. (1995) Puromycin-sensitive aminopeptidase. Sequence analysis, expression, and functional characterization. *J. Biol. Chem.* **270**, 26931–26939.

Curthoys, N.P. (1987) Extracellular catabolism of glutathione. In: *Mammalian Ectoenzymes* (Kenny, A.J. & Turner, A.J., eds). Amsterdam: Elsevier, pp. 249–264.

Delmas, B., Gelfi, J., L'Haridon, R., Vogel, L.K., Sjöström, H., Norén, O. & Laude, H. (1992) Aminopeptidase N is a major receptor for the entero-pathogenic coronavirus TGEV. *Nature* **357**, 417–420.

Feracci, H. & Maroux, S. (1980) Rabbit intestinal aminopeptidase N. Purification and molecular properties. *Biochim. Biophys. Acta* **599**, 448–463.

Funkhouser, J.D., Tangada, S.D., Jones, M., O, S.J. & Peterson, R.D. (1991) p146 type II alveolar epithelial cell antigen is identical to aminopeptidase N. *Am. J. Physiol.* **260**, L274–L279.

Gillespie, T.J., Konings, P.N., Merrill, B.J. & Davis, T.P. (1992) A specific enzyme assay for aminopeptidase M in rat brain. *Life Sci.* **51**, 2097–2106.

Gros, C., Giros, B. & Schwartz, J.-C. (1985) Identification of aminopeptidase M as an enkephalin-inactivating enzyme in rat cerebral membranes. *Biochemistry* **24**, 2179–2185.

Helene, A., Beaumont, A. & Roques, B.P. (1991) Functional residues at the active site of aminopeptidase N. *Eur. J. Biochem.* **196**, 385–393.

Hoffmann, T., Faust, J., Neubert, K. & Ansorge, S. (1993) Dipeptidyl peptidase IV (CD26) and aminopeptidase N (CD13) catalyzed hydrolysis of cytokines and peptides with N-terminal cytokine sequences. *FEBS Lett.* **336**, 61–64.

Hooper, N.M. (1994) Families of zinc metalloproteases. *FEBS Lett.* **354**, 1–6.

Kanayama, N., Kajiwara, Y., Goto, J., el Maradny, E., Maehara, K., Andou, K. & Terao, T. (1995) Inactivation of interleukin-8 by aminopeptidase N (CD13). *J. Leukoc. Biol.* **57**, 129–134.

Kandror, K.V., Yu, L. & Pilch, P.F. (1994) The major protein of GLUT4-containing vesicles, gp160, has aminopeptidase activity. *J. Biol. Chem.* **269**, 30777–30780.

Kruse, T.A., Bolund, L., Grzeschik, K.H., Ropers, H.H., Olsen, J., Sjöström, H. & Norén, O. (1988) Assignment of the human aminopeptidase N (peptidase E) gene to chromosome 15q13-qter. *FEBS Lett.* **239**, 305–308.

Kunz, J., Krause, D., Kremer, M. & Dermietzel, R. (1994) The 140-kDa protein of blood-brain barrier-associated pericytes is identical to aminopeptidase N. *J. Neurochem.* **62**, 2375–2386.

Lee, M.K., You, T.H., Young, B.A., Cotrill, J.A., Valaitis, A.P. & Dean, D.H. (1996) Aminopeptidase N purified from gypsy moth brush border membrane vesicles is a specific receptor for *Bacillus thuringiensis* CryIAc toxin. *Appl. Environ. Microbiol.* **62**, 2845–2849.

Leyhausen, G., Schuster, D.K., Vaith, P., Zahn, R.K., Umezawa, H., Falke, D. & Müller, W.E.G. (1983) Identification and properties of the cell membrane bound leucine aminopeptidase interacting with the potential immunostimulant and chemotherapeutic agent bestatin. *Biochem. Pharmacol.* **32**, 1051–1057.

Littlewood, G.M., Hooper, N.M. & Turner, A.J. (1989) Ectoenzymes of the kidney microvillar membrane. Affinity purification, characterization and localization of the phospholipase C-solubilized form of renal dipeptidase. *Biochem. J.* **257**, 361–367.

Look, A.T., Ashmun, R.A., Shapiro, L.H. & Peiper, S.C. (1989) Human myeloid plasma membrane glycoprotein CD13 (gp150) is identical to aminopeptidase N. *J. Clin. Invest.* **83**, 1299–1307.

McDonald, J.K. & Barrett, A.J. (1986) *Mammalian Proteases: A Glossary and Bibliography*, vol. 2: *Exopeptidases*. London: Academic Press, pp. 59–71.

McIntyre, T.M. & Curthoys, N.P. (1982) Renal catabolism of glutathione: characterization of a particulate rat renal dipeptidase that catalyzes the hydrolysis of cysteinylglycine. *J. Biol. Chem.* **257**, 11915–11921.

Malfroy, B., Kado-Fong, H., Gros, C., Giros, B., Schwartz, J.-C. & Hellmiss, R. (1989) Molecular cloning and amino acid sequence of rat kidney aminopeptidase M: a member of a super family of zinc metallohydrolases. *Biochem. Biophys. Res. Commun.* **161**, 236–241.

Maroux, S., Louvard, D. & Baratti, J. (1973) The aminopeptidase from hog intestinal brush border. *Biochim. Biophys. Acta* **321**, 282–295.

Matsas, R., Stephenson, S.L., Hryszko, J., Kenny, A.J. & Turner, A.J. (1985) The metabolism of neuropeptides: phase separation of synaptic membrane preparations with Triton X-114 reveals the presence of aminopeptidase N. *Biochem. J.* **231**, 445–449.

Olsen, J., Cowell, G.M., Kønigshøfer, E., Danielsen, E.M., Møller, J., Laustsen, L., Hansen, O.C., Welinder, K.G., Engberg, J., Hunziker, W., Spiess, M., Sjöström, H. & Norén, O. (1988) Complete amino acid sequence of human intestinal aminopeptidase N as deduced from cloned cDNA. *FEBS Lett.* **238**, 307–314.

Olsen, J., Sjöström, H. & Norén, O. (1989) Cloning of the pig aminopeptidase N gene. Identification of possible regulatory elements and the exon distribution in relation to the membrane-spanning region. *FEBS Lett.* **251**, 275–281.

Rich, D.H., Moon, B.J. & Harbeson, S. (1984) Inhibition of aminopeptidases by amastatin and bestatin derivatives. Effect of inhibitor structure on slow-binding processes. *J. Med. Chem.* **27**, 417–422.

Rogi, T., Tsujimoto, M., Nakazato, H., Mizutani, S. & Tomoda, Y. (1996) Human placental leucine aminopeptidase/oxytocinase. A new member of type II membrane-spanning zinc metallopeptidase family. *J. Biol. Chem.* **271**, 56–61.

Schalk, C., d'Orchymont, H., Jauch, M.-F. & Tarnus, C. (1994) 3-Amino-2-tetralone derivatives: novel and potent inhibitors of aminopeptidase-M. *Arch. Biochem. Biophys.* **311**, 42–46.

See, H. & Reithmeier, R.A. (1990) Identification and characterization of the major stilbene-disulphonate- and concanavalin A-binding protein of the porcine renal brush-border membrane as aminopeptidase N. *Biochem. J.* **271**, 147–155.

Semenza, G. (1957) Chromatographic purification of cysteinyl-glycinase. *Biochim. Biophys. Acta* **24**, 401–413.

Shapiro, L.H., Ashmun, R.A., Roberts, W.M. & Look, A.T. (1991) Separate promoters control transcription of the human aminopeptidase N gene in myeloid and intestinal epithelial cells. *J. Biol. Chem.* **266**, 11999–12007.

Shipp, M.A. & Look, A.T. (1993) Hematopoietic differentiation antigens that are membrane-associated enzymes: cutting is the key. *Blood* **82**, 1052–1070.

Sjöström, H., Norén, O., Jeppesen, L., Staun, M., Svensson, B. & Christiansen, L. (1978) Purification of different amphiphilic forms of a microvillus aminopeptidase from pig small intestine using immunoadsorbent chromatography. *Eur. J. Biochem.* **88**, 503–511.

Soderberg, C., Giugni, T.D., Zaia, J.A., Larsson, S., Wahlberg, J.M. & Moller, E. (1993) CD13 (human aminopeptidase N) mediates human cytomegalovirus infection. *J. Virol.* **67**, 6576–6585.

Solhonne, B., Gros, C., Pollard, H. & Schwartz, J.-C. (1987) Major localization of aminopeptidase M in rat brain. *Neuroscience* **22**, 225–232.

Stephenson, S.L. & Kenny, A.J. (1987) Metabolism of neuropeptides. Hydrolysis of the angiotensins, bradykinin, substance P and oxytocin by pig kidney microvillar membranes. *Biochem. J.* **241**, 237–247.

Takesue, S., Yokota, K., Miyajima, S., Taguchi, R., Ikezawa, H. & Takesue, Y. (1992) Partial release of aminopeptidase N from larval midgut cell membranes of the silkworm, *Bombyx mori*, by phosphatidylinositol-specific phospholipase C. *Comp. Biochem. Physiol. B* **102**, 7–11.

Tan, P.S., van Alen-Boerrigter, I.J., Poolman, B., Siezen, R.J., de Vos, W.M. & Konings, W.N. (1992) Characterization of the *Lactococcus lactis pepN* gene encoding an aminopeptidase homologous to mammalian aminopeptidase N. *FEBS Lett.* **306**, 9–16.

Tieku, S. & Hooper, N.M. (1992) Inhibition of aminopeptidases N, A and W. A re-evaluation of the actions of bestatin and inhibitors of angiotensin converting enzyme. *Biochem. Pharmacol.* **44**, 1725–1730.

Toh, H., Minami, M. & Shimizu, T. (1990) Molecular evolution and zinc ion binding motif of leukotriene A4 hydrolase. *Biochem. Biophys. Res. Commun.* **171**, 216–221.

Vogel, L.K., Norén, O. & Sjöström, H. (1992) The apical sorting signal on human aminopeptidase N is not located in the stalk but in the catalytic head group. *FEBS Lett.* **308**, 14–17.

Wachsmuth, E.D., Fritze, I. & Pfleiderer, G. (1966) An aminopeptidase occurring in pig kidney. I. An improved method of preparation. Physical and enzymic properties. *Biochemistry* **5**, 169–174.

Wang, J. & Cooper, M.D. (1996) Aminopeptidases: structure and biological function. In: *Zinc Metalloproteases in Health and Disease* (Hooper, N.M., ed.). London: Taylor & Francis, pp. 131–151.

Ward, P.E., Benter, I.F., Dick, L. & Wilk, S. (1990) Metabolism of vasoactive peptides by plasma and purified renal aminopeptidase M. *Biochem. Pharmacol.* **40**, 1725–1732.

Watt, V.M. & Yip, C.C. (1989) Amino acid sequence deduced from a rat kidney cDNA suggests it encodes the Zn-peptidase aminopeptidase N. *J. Biol. Chem.* **264**, 5480–5487.

Xu, Y., Wellner, D. & Scheinberg, D.A. (1995) Substance P and bradykinin are natural inhibitors of CD13/aminopeptidase N. *Biochem. Biophys. Res. Commun.* **208**, 664–674.

Yang, X.F., Milhiet, P.E., Gaudoux, F., Crine, P. & Boileau, G. (1993) Complete sequence of rabbit kidney aminopeptidase N and mRNA localization in rabbit kidney by *in situ* hybridization. *Biochem. Cell Biol.* **71**, 278–287.

Yeager, C.L., Ashmun, R.A., Williams, R.K., Cardellichio, C.B., Shapiro, L.H., Look, A.T. & Holmes, K.V. (1992) Human aminopeptidase N is a receptor for human coronavirus 229E. *Nature* **357**, 420–422.

*Anthony J. Turner*
*Department of Biochemistry and Molecular Biology,*
*University of Leeds,*
*Leeds LS2 9JT, UK*
*Email: a.j.turner@leeds.ac.uk*

# 338. *Cys-Gly dipeptidase*

## Databanks

*Peptidase classification: clan MA, family M1, MEROPS ID: M01.001*
*NC-IUBMB enzyme classification: EC 3.4.13.6 (now replaced by EC 3.4.11.2)*
*Chemical Abstracts Service registry number: 9025-30-3 and 9054-63-1*
*Databank codes: see Chapter 337.*

## Name and History

In the early 1950s, Binkley's group (Olson & Binkley, 1950; Binkley, 1952; Binkley *et al.*, 1957) reported the extensive purification of Cys-Gly dipeptidase (and also of 'leucinamidase' and alkaline phosphatase; Binkley *et al.*, 1957) activities from pig kidney by a procedure which involved two unusually drastic steps – even by the standards of half a century ago: 2 weeks' digestion at room temperature with crude pancreatic proteases and then vigorous repeated shaking with chloroform–octanol (9:1), also at room temperature. The purified preparation(s) were suggested to be protein-free RNAs. An alternative hypothesis for the role of RNAs in protein biosynthesis was put forward (Binkley, 1952), in which the alleged peptidase activities of RNAs would lead to the biosynthesis of polypeptide chains by acting in reverse, step by step.

In 1956, the (then new) chromatographic procedures led to complete separation of **Cys-Gly dipeptidase** activity from the RNA in Binkley's preparations (Semenza, 1957a); the enzyme thereby purified was, by all criteria available at the time, a protein and most likely homogeneous. Research on this enzyme then came essentially to a rest until the late 1970s, when Cys-Gly dipeptidase was localized in the renal brush border membranes (Hughey *et al.*, 1978; Grau *et al.*, 1979); in 1980 it was finally shown to be identical with **aminopeptidase N** (Chapter 337) (Rankin *et al.*, 1980).

## Activity and Specificity

As a substrate, either synthetic Cys⊦Gly or, more conveniently, a partial acid hydrolyzate of glutathione can be used (Olson & Binkley, 1950); still more convenient substrates are *S*-substituted derivatives of Cys-Gly (Hughey *et al.*, 1978) and now, of course, the standard substrates of aminopeptidase N. For the substrate specificity of the enzyme, see Chapter 337.

When Cys-Gly is used as the substrate, advantage can be taken of the fact that Sullivan's reagent ($\beta$-naphthoquinone-4-sulfonate; Sullivan & Hess, 1936) yields different colors with Cys-Gly than with equimolar mixtures of cysteine and glycine (Binkley & Nakamura, 1948; Olson & Binkley, 1950). Alternatively, with Cys-[$^3$H]Gly, the liberation of [$^3$H]Gly can be monitored using organomercurial agarose columns which retain both Cys-Gly and cysteine (Rankin *et al.*, 1980). Cys-Gly dipeptidase activity requires $Mn^{2+}$ (or $Co^{2+}$ or $Fe^{2+}$) (Olson & Binkley, 1950; Binkley, 1952) and is thus inhibited by metal-complexing agents, including 1,10-phenanthroline

(quoted in Rankin *et al.*, 1980) and phosphate. It has a pH optimum of approximately 8. The different colors produced by $\beta$-naphthoquinone-4-sulfonate with Cys-Gly or free cysteine plus glycine also led to a convenient test to detect Cys-Gly dipeptidase activity on paper, e.g., after electrophoresis (Semenza, 1957b).

## Structural Chemistry and Preparation

Cys-Gly dipeptidase, as obtained after 2 weeks' digestion with proteases, and then treatment with chloroform–octanol, both at room temperature (Olson & Binkley, 1950; Binkley, 1952; Binkley *et al.*, 1957) had unusual properties: it was stable towards proteases, to moderately high temperatures, and could be stored at room temperature for at least 5 years without loss of activity (Binkley *et al.*, 1957). No protein could be precipitated with trichloroacetic acid or detected by the biuret test.

The chloroform–octanol treatment deserves a comment. It was first introduced by Sevag *et al.* (1938) and then used extensively by Chargaff's group (e.g. Chargaff *et al.*, 1951) for the preparation of essentially protein-free nucleic acids. Vigorous shaking of the tissue homogenate with a chloroform–octanol (or amylol) mixture, followed by centrifugation, yields three phases: the heaviest consists mainly of chloroform and the alcohol; the middle, gel-like phase is made of denatured, precipitated proteins; the lightest, water-rich phase contains mainly DNA and RNA. Repeated treatment of this phase with chloroform–octanol can reduce the amounts of proteins to nearly undetectable levels. Cys-Gly dipeptidase, 'leucinamidase' and alkaline phosphatase activities of the homogenates (digested for 2 weeks) survive the chloroform–octanol treatment and are recovered with a high specific activity (referred to protein) in the water-rich phase (Binkley *et al.*, 1957). This phase has a UV absorption maximum at approximately 260 nm, and this fact, plus the remarkable stability of Cys-Gly dipeptidase activity towards treatments generally inactivating proteins, led to the suggestion that Cys-Gly dipeptidase could be an RNA.

Yet, even the reported properties of this 'RNA-peptidase' were somewhat unusual for an RNA. Cys-Gly dipeptidase activity was resistant towards RNAase treatment or moderately alkaline pH values. The 260/280 nm ratio in the UV spectrum reported by Binkley (1952) did indicate the presence of residual proteins. Further purification (Semenza, 1957a) by hydroxyapatite and ion-exchange chromatographies of the Cys-Gly dipeptidase prepared according to Binkley's procedures led to complete separation of the enzymatic activity from the RNA or organic phosphorus; the UV absorption spectrum of the purified Cys-Gly dipeptidase now had a maximum at 280 nm, was inactivated by trypsin or chymotrypsin in the presence of urea, or by chloroform–octanol.

It seems clear that during Binkley's 2 week proteolytic digestion Cys-Gly dipeptidase was partially degraded, and the hydrophobic membrane-spanning domain (by which as we now know aminopeptidase N is anchored to the brush border membrane) was also split off from the rest of the protein. The solubilized, partially degraded enzyme still retained its activity. It was protected by nucleic acids (and/or perhaps by other components) against denaturation by chloroform–octanol, and against further proteolytic degradation.

Unfortunately I had too little purified material to further investigate its chemical properties with the methods available in the mid-1950s. Quite likely the purified Cys-Gly peptidase was reasonably homogeneous. It had an unusually fast electrophoretic mobility (on paper).

The identity of Cys-Gly dipeptidase with aminopeptidase M (now called N) (Rankin *et al.*, 1980) now allows us to transfer to the former the properties studied in the latter. However, Binkley's unorthodox procedures may be worth rescuing from total oblivion in some cases: they allowed handling of several pounds of tissue (Binkley *et al.*, 1957) and led to preparations with very high specific activities (referred to protein) and, prior to removal of nucleic acids, endowed with remarkable stability.

## Biological Aspects

Cys-Gly peptidase is an integral protein of the brush border membrane of the kidney tubule (Hughey *et al.*, 1978; Grau *et al.*, 1979) and is identical with aminopeptidase N. We thus know that it is anchored to the membrane via an N-terminal hydrophobic sequence, with 'the body' of the protein protruding in the tubular lumen. In the human it probably contributes to the degradation of peptides in the urine, including Cys-Gly arising from the action of $\gamma$-glutamyltransferase on glutathione.

## References

Binkley, F. (1952) Evidence for the polynucleotide nature of cysteinylglycinase. *Exp. Cell Res.* **2**(suppl), 145–157.

Binkley, F. & Nakamura, K. (1948) Metabolism of glutathione. I. Hydrolysis by tissues of the rat. *J. Biol. Chem.* **173**, 411–421.

Binkley, F., Alexander, V., Bell., R.E. & Lea, C. (1957) Peptidases and alkaline phosphatases of swine kidney. *J. Biol. Chem.* **228**, 559–567.

Chargaff, E., Lipshitz, R., Green, C. & Hodges, M.E. (1951) The composition of the deoxyribonucleic acid of salmon sperm. *J. Biol. Chem.* **192**, 223–230.

Grau, E.M., Marathe, G.V. & Tate, S.S. (1979) Rapid purification of rat kidney brush borders enriched in $\gamma$-glutamyl transpeptidase. *FEBS Lett.* **98**, 91–95.

Hughey, R.P., Rankin, B.B., Elce, J.S. & Curthoys, N.P. (1978) Specificity of a particulate rat renal peptidase and its localization along with other enzymes of mercapturic acid synthesis. *Arch. Biochem. Biophys.* **186**, 211–217.

Olson, C.K. & Binkley, F. (1950) Metabolism of glutathione. III. Enzymatic hydrolysis of cysteinylglycine. *J. Biol. Chem.* **186**, 731–735.

Rankin, B.B., McIntyre, T.M. & Curthoys, N.P. (1980) Brush border membrane hydrolysis of *S*-benzyl-cysteine-nitroanilide, an activity of aminopeptidase M. *Biochem. Biophys. Res. Commun.* **96**, 991–996.

Semenza, G. (1957a) Chromatographic purification of cysteinylglycinase. *Biochim. Biophys. Acta* **24**, 401–413.

Semenza, G. (1957b) Detection of dipeptides and dipeptidase activity on paper. *Experientia* **13**, 166.

Sevag, M.G., Lackman, D.B. & Smolens, J. (1938) The isolation of the components of streptococcal nucleoproteins in serologically active form. *J. Biol. Chem.* **124**, 425–436.

Sullivan, M.X. & Hess, W.C. (1936) The determination of cystine in urine. *J. Biol. Chem.* **116**, 221–232.

*Giorgio Semenza*
*Department of Biochemistry,*
*Swiss Federal Institute of Technology,*
*ETH-Zentrum,*
*CH-8092 Zurich, Switzerland*
*Email: semenza@bc.biol.ethz.ch*

# *339. Glutamyl aminopeptidase*

## Databanks

*Peptidase classification: clan MA, family M1, MEROPS ID: M01.003*
*NC-IUBMB enzyme classification: EC 3.4.11.7*
*ATCC entries: 100767 (human)*

*Chemical Abstracts Service registry number: 9074-83-3*
*Databank codes:*

| Species | SwissProt | PIR | EMBL (cDNA) | EMBL (genomic) |
|---|---|---|---|---|
| *Homo sapiens* | Q07075 | – | L12468 L14721 | – |
| *Mus musculus* | P16406 | – | M29961 | U50280: 5′ flanking region |
| *Rattus norvegicus* | – | – | S81912 | – |
| *Rattus norvegicus* | P50123 | – | S73583 | – |
| *Sus scrofa* | – | – | U66371 | – |

## Name and History

This enzyme was first identified in rat and guinea pig kidney sections as an α-glutamyl peptidase that catalyzed the hydrolysis of N-(α-L-glutamyl)-β-naphthylamide (Glenner & Folk, 1961). Subsequently, the enzyme was found to hydrolyze N-terminal aspartyl residues and to exhibit primarily exopeptidase activity. It was therefore named **aminopeptidase A** (**A**cidic α-amino peptidase) (Glenner *et al.*, 1962). Since α-L-glutamyl derivatives are more efficiently hydrolyzed than are α-L-aspartyl derivatives, the enzyme is now referred to as **glutamyl aminopeptidase (EAP)**.

A broad tissue distribution of the enzyme is indicated by immunohistochemical staining and histoenzymatic assay of EAP activity (Li *et al.*, 1993). It is particularly abundant on the brush borders of intestinal enterocytes and kidney. Preparation of EAP from different sources and the use of different substrates to determine the enzyme activity has led to the assignment of different names to EAP. These include **aspartate aminopeptidase**, **angiotensinase A**, **Ca²⁺-activated glutamate aminopeptidase**, **membrane aminopeptidase II** and the **BP-1/6C3 antigen**.

## Activity and Specificity

The enzyme is activated by $Ca^{2+}$ and inhibited by EDTA. Enzyme activity can be assayed with N-(α-L-glutamyl)-β-naphthylamide or N-(α-L-glutamyl)-p-nitroanilides as described by Glenner *et al.* (1962). These synthetic substrates are available from Bachem (see Appendix 2 for full names and addresses of suppliers). The pH optimum of EAP is approximately 7.2 and the assay system typically includes calcium (1–10 mM) in 50 mM Tris buffer. The reaction is carried out at 37°C and the activity is determined by measuring the optical density at 405 nm. EAP is activated by $Ca^{2+}$ and inhibited by chelating agents such as EDTA, EGTA and 1,10-phenanthroline at 1 mM (Danielsen *et al.*, 1980; Tobe *et al.*, 1980). The enzyme activity is completely inhibited by transitional metal ions such as $Zn^{2+}$, $Ni^{2+}$, $Cu^{2+}$, $Hg^{2+}$ and $Cd^{2+}$ at 1 mM. Amastatin is a potent and relatively specific competitive inhibitor of EAP (Aoyagi *et al.*, 1978).

## Structural Chemistry

EAP has been reported to have molecular weights ranging from 45 000 to 400 000 depending on the tissue source and the method of enzyme purification. In mouse, EAP was found to be identical to the murine BP-1 antigen originally identified as a pre-B/immature B cell-specific marker (Cooper *et al.*,

1986). Immunoprecipitation with the BP-1 antibody reveals the mouse EAP to be a homodimeric, phosphorylated glycoprotein composed of two disulfide-linked 140 kDa subunits. The deduced amino acid sequence of BP-1/EAP cDNA indicates that it is a type II integral membrane protein composed of 945 amino acids (Wu *et al.*, 1990).

EAP contains a highly conserved zinc coordination motif, i.e. HEXXHX₁₈E (Vallee & Auld, 1989; Wu *et al.*, 1990). This zinc-binding motif is shared by many metallopeptidases and has been shown to be critical for enzyme activity. A mutant EAP molecule in which the second His residue was converted to Phe completely lacks enzyme activity, thus demonstrating the importance of the zinc-binding motif (Wang & Cooper, 1993). Although previously atomic absorption spectrophotometry failed to detect $Zn^{2+}$ in the purified enzyme, a recent study indicates that EAP does contain Zn in its zinc-binding motif, thus identifying EAP as a zinc metallopeptidase (Vazeux *et al.*, 1996).

## Preparation

Purification of EAP from various sources has been well described (McDonald & Barrett, 1986). Expression vectors for mouse EAP, both wild type and mutants, are available from either of the undersigned authors. Also available are expression vectors for soluble forms of both wild-type and mutant EAP, generated by replacing the cytoplasmic and the transmembrane domain of EAP cDNA with the human interleukin 4 (IL-4) leader sequence. Recombinant EAP can be expressed by transfecting these vectors into either COS-7 cells transiently or L cells stably, and can be further purified to near homogeneity using a BP-1 antibody column (antibody available from Pharmingen).

## Biological Aspects

Although EAP is expressed in many tissues, its expression is lineage-specific and differentiation stage-specific among hematopoietic cells (Cooper *et al.*, 1986). EAP expression is confined to the pro-B, pre-B and immature B cell stages in the B cell differentiation pathway. EAP is also expressed by bone marrow and thymic stromal cells (Whitlock *et al.*, 1987; Li *et al.*, 1993). Interleukin 7 treatment induces both proliferation of early B-lineage cells and increased expression of EAP, suggesting that this ectoenzyme might be involved in regulating the proliferation of pre-B cells (Welch *et al.*, 1990; Sherwood & Weissman, 1990).

Analysis of genomic clones encoding the gene for EAP, *ENPEP* (Wang *et al.*, 1996), indicates that the gene spans more than 100 kb and contains 20 exons. Except for the first and last exons, the gene is composed of small exons, ranging from 56 to 171 bp in size, separated by introns ranging from less than 100 bp to approximately 10 kb. The zinc-binding motif HEXXH and the glutamic acid residue 19 amino acids downstream that also binds zinc are encoded separately in exons 5 and 6, respectively. Whether this is a common feature for other aminopeptidases is presently unknown. *ENPEP* genomic structure resembles two other peptidase genes, *CD10* (neprilysin; D'Adamio *et al.*, 1989) (Chapter 362) and *CD26* (dipeptidyl-peptidase IV; Bernard *et al.*, 1994) (Chapter 128), in that each contains many small exons separated by introns ranging from 100 bp to tens of kilobases. The promoter region contains a TATA-like element and potential DNA-binding motifs for lymphocyte-specific transcription factors including Ikaros, BSAP, PU.1 and octamer-binding proteins, as well as DNA-binding motifs for several ubiquitous transcription factors. The *ENPEP* gene is localized in the distal region of mouse chromosome 3 in a region that lacks mouse mutations with a phenotype indicating defective B lymphocyte development or aberrant blood pressure regulation (Wang *et al.*, 1996). The distal region of mouse chromosome 3 shares a region of homology with human chromosome 4q. Recent studies (L. Li *et al.*, unpublished results) have localized the *ENPEP* gene to human chromosome 4q25 near the complement factor 1 gene.

A well-known physiological substrate of EAP is angiotensin II, an important regulator of blood pressure. Hence, EAP has also been called angiotensinase (Nagatsu *et al.*, 1965; Sakura *et al.*, 1983). EAP converts angiotensin II to angiotensin III, one of the steps leading to the inactivation of angiotensin II. Angiotensin III may have a function different to that of angiotensin II in certain tissues. For example, angiotensin III appears to be an active neuropeptide in the brain (Song *et al.*, 1993; Healy & Wilk, 1993).

EAP-deficient mice have recently been generated through homologous recombination in embryonic stem cells (Q. Lin *et al.*, unpublished results), and their T and B cell development appears to be normal. While further studies are required to clarify the roles of EAP in the early B cell differentiation pathway, blood pressure homeostasis and other physiological processes, it is possible that EAP function may be compensated by other peptidases in EAP-deficient mice. It will be interesting to cross the BP-1-deficient mice with mice deficient in other peptidases to examine the combined effects.

## Acknowledgments

J.W. was a research associate and M.D.C. is an investigator of the Howard Hughes Medical Institute.

## Further Reading

For reviews, see McDonald & Barrett (1986), Taylor (1996) and Wang & Cooper (1996).

## References

Aoyagi, T., Tobe, H., Kojima, F., Hamada, M., Takeuchi, T. & Umezawa, H. (1978) Amastatin, an inhibitor of aminopeptidase A, produced by actinomycetes. *J. Antibiot.* **31**, 636–638.

Bernard, A.M., Mattei, M.G., Pierres, M. & Marguet, D. (1994) Structure of the mouse dipeptidyl peptidase IV (CD26) gene. *Biochemistry* **33**, 15204–15214.

Cooper, M.D., Mulvaney, D., Coutinho, A. & Cazenave, P.A. (1986) A novel cell surface molecule on early B-lineage cells. *Nature* **321**, 616–618.

D'Adamio, L., Shipp, M.A., Masteller, E.L. & Reinherz, E.L. (1989) Organization of the gene encoding common acute lymphoblastic leukemia antigen (neutral endopeptidase 24.11): multiple miniexons and separate 5' untranslated regions. *Proc. Natl Acad. Sci. USA* **86**, 7103–7107.

Danielsen, E.M., Norén, O., Sjöström, H., Ingram, J. & Kenny, J. (1980) Proteins of the kidney microvillar membrane. Aspartate aminopeptidase: purification by immunoadsorbent chromatography and properties of the detergent- and proteinase-solubilized forms. *Biochem. J.* **189**, 591–603.

Glenner, G.G. & Folk, J.E. (1961) Glutamyl peptidases in rat and guinea pig kidney slices. *Nature* **192**, 338–340.

Glenner, G.G., McMillan, P.J. & Folk, J.E. (1962) A mammalian peptidase specific for the hydrolysis of N-terminal α-L-glutamyl and aspartyl residues. *Nature* **194**, 867.

Healy, D.P. & Wilk, S. (1993) Localization of immunoreactive glutamyl aminopeptidase in rat brain. II. Distribution and correlation with angiotensin II. *Brain Res.* **606**, 295–303.

Li, L., Wu, Q., Wang, J., Bucy, R.P. & Cooper, M.D. (1993) Widespread tissue distribution of aminopeptidase A, an evolutionary conserved ectoenzyme recognized by the BP-1 antibody. *Tissue Antigens* **42**, 488–496.

McDonald, J.K. & Barrett, A.J. (1986) Section 11. Aminopeptidases. In: *Mammalian Proteases*, vol. 2 (McDonald, J.K. & Barrett, A.J., eds). London: Academic Press, pp. 23–100.

Nagatsu, I., Gillespie, L., Folk, J.E. & Glenner G.G. (1965) Serum aminopeptidases, 'angiotensinase', and hypertension. I. Degradation of antiotensin II by human serum. *Biochem. Pharmacol.* **14**, 721–728.

Sakura, H., Kobayashi, H., Mizutani, S., Sakura, N., Hashimoto, T. & Kawashima, Y. (1983) Kinetic properties of placental aminopeptidase A: N-terminal degradation of angiotensin II. *Biochem. Int.* **6**, 609–615.

Sherwood, P.J. & Weissman, I.L. (1990) The growth factor IL-7 induces expression of a transformation-associated antigen in normal pre-B cells. *Int. Immunol.* **2**, 399–406.

Song, L., Wilk, E., Wilk, S. & Healy, D.P. (1993) Localization of immunoreactive glutamyl aminopeptidase in rat brain. I. Association with cerebral microvessels. *Brain Res.* **606**, 286–294.

Taylor, A. (1996) Aminopeptidase: structure and function. *FASEB J.* **7**, 290–298.

Tobe, H., Kojima, F., Aoyagi, T. & Umezawa, H. (1980) Purification by affinity chromatography using amastatin and properties of aminopeptidase A from pig kidney. *Biochim. Biophys. Acta* **613**, 459–468.

Vallee, B.L. & Auld, D.S. (1989) Short and long spacer sequences and other structural features of zinc binding sites in zinc enzymes. *FEBS Lett.* **257**, 138–140.

Vazeux, G., Wang, J., Corvol, P. & Llorens-Cortes, C. (1996) Identification of glutamate residues essential for catalytic activity and zinc coordination in aminopeptidase A. *J. Biol. Chem.* **271**, 9069–9074.

Wang, J. & Cooper, M.D. (1993) Histidine residue in the zinc-binding motif of aminopeptidase A is critical for enzymatic activity. *Proc. Natl Acad. Sci. USA* **90**, 1222–1226.

Wang, J. & Cooper, M.D. (1996) Aminopeptidases: structure and biological function. In: *Zinc Metalloproteases in Health and Disease* (Hooper, N.M., ed.). London: Taylor & Francis, pp. 131–151.

Wang, J., Walker, H., Lin, Q., Jenkins, N., Copeland, N.G., Watanabe, T., Burrows, P.D. & Cooper, M.D. (1996) The mouse BP-1 gene: structure, chromosomal localization, and regulation of expression by type I interferons and interleukin-7. *Genomics* **33**, 167–176.

Welch, P.A., Burrows, P.D., Namen, A.E., Gillis, S. & Cooper, M.D. (1990) Bone marrow stromal cells and interleukin-7 induce coordinate expression of the BP-1/6C3 antigen and pre-B cell growth. *Int. Immunol.* **2**, 697–705.

Whitlock, C.A., Tidmarsh, G.F., Muller-Sieburg, C. & Weissman, I.L. (1987) Bone marrow stromal cell lines with lymphopoietic activity express high levels of pre-B neoplasia-associated molecule. *Cell* **48**, 1009–1021.

Wu, Q., Lahti, J.M., Air, G.M., Burrows, P.D. & Cooper, M.D. (1990) Molecular cloning of the murine BP-1/6C3 antigen: A member of the zinc-dependent metallopeptidase family. *Proc. Natl Acad. Sci. USA* **87**, 993–997.

*Jiyang Wang*
*Medical Institute of Bioregulation,*
*Kyushu University,*
*Fukuoka 812-82, Japan*

*Max D. Cooper*
*Division of Developmental and Clinical Immunology,*
*Departments of Medicine, Pediatrics and Microbiology,*
*University of Alabama at Birmingham and the Howard*
*Hughes Medical Institute,*
*Birmingham, AL 35294, USA*
*Email: max.cooper@ccc.UAB.edu*

# 340. *Pyroglutamyl-peptidase II*

## Databanks

*Peptidase classification: clan MA, family M1, MEROPS ID: M01.008*
*NC-IUBMB enzyme classification: EC 3.4.19.6*
*Databank codes:*

| Species | SwissProt | PIR | EMBL (cDNA) | EMBL (genomic) |
|---|---|---|---|---|
| *Rattus norvegicus* | – | – | X80535 | – |

## Name and History

Thyrotropin-releasing hormone (TRH, thyroliberin; Glp-His-Pro-NH$_2$), the first hypothalamic hypophysiotrophic neuropeptide hormone to be elucidated structurally, stimulates the secretion of the pituitary hormones thyroid-stimulating hormone (TSH), prolactin and, under certain mainly pathological conditions, growth hormone. In addition to these neuroendocrine functions, TRH elicits a variety of biological effects in the central nervous system, suggesting that it may also act as a neuromodulator and/or neurotransmitter. This interpretation is supported by the observation that TRH and TRH receptors are widely distributed throughout the central nervous system.

Studies on the catabolism of TRH (for review see O'Cuinn *et al.*, 1990) revealed that the hydrolysis of the tripeptide amide at the Glp⊣His bond is catalyzed by two enzymes, one cytosolic and one membrane bound. The soluble enzyme exhibits the characteristics of the well-known pyroglutamyl-peptidase, designated pyroglutamyl-peptidase I (Chapter 269) to differentiate it from the membrane-bound TRH-degrading enzyme which has been designated ***pyroglutamyl-peptidase II*** (McDonald & Barrett, 1986) or ***pyroglutamyl aminopeptidase II***. In contrast to the nonspecific soluble activity, the particulate enzyme is not a general pyroglutamyl-peptidase but an enzyme which exhibits an extraordinarily high degree of substrate specificity and thus, this enzyme has also been described as ***membrane-bound TRH-degrading enzyme*** or ***TRH-degrading ectoenzyme***. Because of the strict substrate specificity of the serum form, this enzyme has also been called ***thyroliberinase*** (Bauer *et al.*, 1981).

## Activity and Specificity

Compared to TRH, all TRH analogs and tripeptide derivatives tested (e.g. 3-methyl-TRH, TRH acid, TRH methylamide, Glp-His-pyrrolidine, etc.) are hydrolyzed considerably more slowly or not at all, indicating that the entire entity of the TRH molecule is recognized by the enzyme. With the exception of Glp⌐Phe-Pro-NH$_2$ (Kelly *et al.*, 1997), only tripeptides commencing with the sequence Glp-His are accepted as poor substrates. Neither the TRH-like peptide Glp-Glu-Pro-NH$_2$, recently isolated from various tissues and human semen, nor any of the biologically active Glp-containing peptides such as neurotensin, gastrin, bombesin, etc., are hydrolyzed by this peptidase. Interestingly, the hypothalamic decapeptide amide luteinizing hormone-releasing hormone (LH-RH) containing the sequence Glp-His effectively inhibits the degradation of TRH but is not hydrolyzed (Bauer *et al.*, 1981; O'Connor & O'Cuinn, 1985; Friedman & Wilk, 1986; Elmore *et al.*, 1990).

Enzyme activity can be determined by a radiochemical test using TRH radiolabeled either in the proline residue (available from DuPont NEN; see Appendix 2 for full names and addresses of suppliers) or in the pyroglutamic moiety (Bauer *et al.*, 1981, 1990; Vargas *et al.*, 1987). Friedman & Wilk (1986) developed a coupled enzyme fluorometric assay using Glp⌐His-Pro-NHNap (available from Bachem) and an excess of dipeptidyl-peptidase IV to liberate $\beta$-naphthylamine from the primary cleavage product His-Pro-NHNap. Specificity of the assay is achieved by discriminating against other enzymatic activities. The soluble pyroglutamyl-peptidase I, a cysteine peptidase, is inhibited by alkylating agents such as 2-iodoacetamide or by the enzyme-specific inhibitor pyroglutamyl chloromethane, and the TRH-deamidating proline endopeptidase is inhibited by serine protease inhibitors such as DFP and specifically by Z-Pro-prolinal and Z-Gly-Pro-CH$_2$Cl. Pyroglutamyl-peptidase II is not inhibited by these compounds. As a metallopeptidase, pyroglutamyl-peptidase II is effectively inhibited by 1,10-phenanthroline and slowly inactivated by chelating agents such as EDTA in a time-dependent manner. Specific and highly effective inhibitors are not yet available.

The pH optimum of the enzyme is about pH 7.3. In phosphate or diethylmalonate buffer at pH 7.3 the enzyme is very stable and retains full activity for more than 2 months when stored at 4°C. In Tris and imidazole-containing buffers the activity continuously decreases during storage (O'Connor & O'Cuinn, 1985).

## Structural Chemistry

Subcellular fractionation studies and experiments with various cell culture systems clearly demonstrate that pyroglutamyl-peptidase II is a true ectoenzyme which is located as an integral membrane protein on the surface of neuronal and pituitary cells but not on glial cells (O'Connor & O'Cuinn, 1984; Horsthemke *et al.*, 1984; Garat *et al.*, 1985; Charli *et al.*, 1988; Elmore *et al.*, 1990; Bauer *et al.*, 1990; Cruz *et al.*, 1991).

The cDNA-deduced amino acid sequence correspondingly predicts a type II integral membrane protein consisting of 1025 amino acids. The short, presumably cytoplasmic region at the N-terminus with a potential phosphorylation site is followed by a stretch of hydrophobic amino acids with the characteristics of a membrane-spanning sequence. The large, presumably extracellular domain contains 12 putative glycosylation sites and the HEXXH consensus sequence motif with a second glutamic acid separated by 18 amino acids (Schauder *et al.*, 1994). The sequence data are in full agreement with the biochemical identification of the enzyme as a glycosylated (Bauer, 1994), Zn-dependent metallopeptidase. Furthermore, active-site studies with protein-modifying reagents (O'Conner & O'Cuinn, 1987; Czekay & Bauer, 1993) also indicated that Tyr, His, Arg and an acidic amino acid are constituents of the active site.

Computer searches reveal that pyroglutamyl-peptidase II is homologous with membrane alanyl aminopeptidase (Chapter 337) (34%) and glutamyl aminopeptidase (Chapter 339) (32%). The cDNA-deduced amino acid sequence of the human pyroglutamyl-peptidase II shows a 96% identity when compared to the rat enzyme. While 15 out of 40 amino acid substitutions are found in the presumed stalk region, a domain of 225 amino acids including the HEXXH consensus sequence is completely conserved (L. Schomburg & K. Bauer, unpublished results).

## Preparation

Highest activities of pyroglutamyl-peptidase II are found in brain and throughout the central nervous system. The specific activities vary considerably between different brain regions. While highest activities are found in the olfactory bulb, cerebral cortex and hippocampus, considerably lower activities are detected in cerebellum, spinal cord and posterior pituitary. Significant activities are also found in retina, lung and liver while all other tissues are almost devoid of this peptidase activity (Friedman & Wilk, 1986; Vargas *et al.*, 1987; Wilk *et al.*, 1988; Scharfman & Aratan-Spire, 1991). The activity of the serum enzyme varies considerably among species. High activities are found in some species (e.g. pig, rat) but not in others such as cattle.

After solubilization by trypsin under very mild conditions, the truncated enzyme has been purified 200 000-fold from rat and pig brain (Bauer, 1994). With proteolytically solubilized enzyme preparations a molecular mass of 230 000 Da has been estimated by gel filtration, whereas a single band with an apparent $M_r$ of 116 000 was detected by SDS-PAGE, indicating that the enzyme consists of two identical subunits. Pyroglutamyl-peptidase II has also been purified from pig serum and identified as a glycoprotein. The serum enzyme is recognized by monoclonal and polyclonal antibodies raised against the particulate enzyme. Furthermore, protein sequence data fully support the concept that both enzymes are derived from the same gene (K. Bauer, unpublished results).

## Biological Aspects

The extraordinarily high degree of substrate specificity strongly indicates that pyroglutamyl-peptidase II may serve very specialized functions. In the brain, the high enzymatic activity and the specific localization on neuronal cells suggests that the enzyme could be important for the termination and thus the transmission of TRH signals. In the anterior pituitary the enzyme activity is stringently regulated by peripheral hormones. In rats, the activities and the mRNA levels of the enzyme increase rapidly after administration of triiodothyronine and, conversely, decrease when the animals are rendered hypothyroid (Bauer, 1987; Ponce *et al.*, 1988; Suen & Wilk, 1989; Schomburg & Bauer, 1995). Opposite effects were observed with estradiol (Bauer *et al.*, 1990). The tissue-specific regulation of the adenohypophyseal enzyme supports the concept that the enzyme itself might serve regulatory functions by controlling the responsiveness of adenohypophyseal target cells and thus pituitary hormone secretion.

## Distinguishing Features

Under appropriate assay conditions pyroglutamyl-peptidase II activity can readily be distinguished from the cysteine pyroglutamyl-peptidase I (Chapter 269).

## Further Reading

Detailed reviews of TRH and its catabolism may be found in O'Cuinn *et al.* (1990) and O'Leary & O'Connor (1995). A short review focusing on the hormonal regulation of the adenohypophyseal enzyme is presented in Bauer (1995).

## References

Bauer, K. (1987) Adenohypophyseal degradation of thyrotropin-releasing hormone regulated by thyroid hormones. *Nature* **330**, 375–377.

Bauer, K. (1994) Purification and characterization of the thyrotropin-releasing hormone-degrading ectoenzyme. *Eur. J. Biochem.* **224**, 387–396.

Bauer, K. (1995) Inactivation of thyrotropin-releasing hormone (TRH) by the hormonally regulated TRH-degrading ectoenzyme. A potential regulator of TRH signals? *Trends Endocrinol. Metabol.* **6**, 101–105.

Bauer, K., Nowak, P. & Kleinkauf, H. (1981) Specificity of a serum peptidase-hydrolyzing thyroliberin at the pyroglutamyl-histidine bond. *Eur. J. Biochem.* **118**, 173–176.

Bauer, K., Carmeliet, P., Schulz, M., Baes, M. & Denef, C. (1990) Regulation and cellular localization of the membrane-bound thyrotropin releasing hormone-degrading enzyme in primary cultures of neuronal, glial and adenohypophyseal cells. *Endocrinology* **127**, 1224–1233.

Charli, J.L., Cruz, C., Vargas, M.A. & Joseph-Bravo, P. (1988) The narrow specificity pyroglutamate aminopeptidase degrading TRH in rat brain is an ectoenzyme. *Neurochem. Int.* **13**, 237–242.

Cruz, C., Charli, J.L., Vargas, M.A. & Joseph-Bravo, P. (1991) Neuronal localization of pyroglutamate aminopeptidase II in primary cultures of fetal mouse brain. *J. Neurochem.* **56**, 1594–1601.

Czekay, G. & Bauer, K. (1993) Identification of the thyrotropin-releasing hormone-degrading ectoenzyme as a metallopeptidase. *Biochem. J.* **290**, 921–926.

Elmore, M.A., Griffiths, E.C., O'Connor, B. & O'Cuinn, G. (1990) Further characterization of the substrate specificity of a TRH-hydrolyzing pyroglutamate aminopeptidase from guinea-pig brain. *Neuropeptides* **15**, 31–36.

Friedman, T.C. & Wilk, S. (1986) Delineation of a particulate thyrotropin-releasing hormone-degrading enzyme in rat brain by the use of specific inhibitors. *J. Neurochem.* **46**, 1231–1239.

Garat, B., Miranda, J., Charli, J.L. & Joseph-Bravo, P. (1985) Presence of a membrane-bound pyroglutamyl aminopeptidase degrading thyroliberin-releasing hormone in rat brain. *Neuropeptides* **6**, 27–40.

Horsthemke, B., Leblanc, P., Kordon, C., Wattiaux-De Coninck, S., Wattiaux, R. & Bauer, K. (1984) Subcellular distribution of particle-bound neutral peptidases capable of hydrolyzing gonadotropin, thyroliberin, enkephalin and substance P. *Eur. J. Biochem.* **139**, 315–320.

Kelly, J.A., Loscher, C.E., Gallagher, S. & O'Connor, B. (1997) Degradation of pyroglutamyl-phenylalanine-proline amide by a pyroglutamyl aminopeptidase purified from membrane fractions of bovine brain. *Biochem. Soc. Trans.* **25**, 114S.

McDonald, J.K. & Barrett, A.J. (1986) *Mammalian Proteases: A Glossary and Bibliography*, vol. 2: *Exopeptidases*. New York: Academic Press.

O'Connor, B. & O'Cuinn, G. (1984) Localization of a narrow specificity thyroliberin-hydrolyzing pyroglutamate aminopeptidase in synaptosomal membranes of guinea-pig brain. *Eur. J. Biochem.* **144**, 271–278.

O'Connor, B. & O'Cuinn, G. (1985) Purification of and kinetic studies on a narrow specificity synaptosomal membrane pyroglutamate aminopeptidase from guinea-pig brain. *Eur. J. Biochem.* **150**, 47–52.

O'Connor, B. & O'Cuinn, G. (1987) Active site studies on a narrow specificity thyroliberin-hydrolyzing pyroglutamate aminopeptidase purified from synaptosomal membrane of guinea-pig brain. *J. Neurochem.* **48**, 676–680.

O'Cuinn, G., O'Connor, B. & Elmore, M. (1990) Degradation of thyrotropin-releasing hormone and luteinising hormone-releasing hormone by enzymes of brain tissue. *J. Neurochem.* **54**(1), 1–13.

O'Leary, R. & O'Connor, B. (1995) Thyrotropin-releasing hormone. *J. Neurochem.* **65**, 953–963.

Ponce, G., Charli, J.L., Pasten, J.A., Aceves, C. & Joseph-Bravo, P. (1988) Tissue-specific regulation of pyroglutamate aminopeptidase II activity by thyroid hormones. *Neuroendocrinology* **48**, 211–213.

Scharfman, R. & Aratan-Spire, S. (1991) Ontogeny of two topologically distinct TRH-degrading pyroglutamate aminopeptidase activities in the rat liver. *Regul. Pept.* **32**, 75–83.

Schauder, B., Schomburg, L., Köhrle, J. & Bauer, K. (1994) Cloning of a cDNA encoding an ectoenzyme that degrades thyrotropin-releasing hormone. *Proc. Natl Acad. Sci. USA* **91**, 9534–9538.

Schomburg, L. & Bauer, K. (1995) Thyroid hormones rapidly and stringently regulate the messenger RNA levels of the thyrotropin-releasing hormone (TRH) receptor and the TRH-degrading ectoenzyme. *Endocrinology* **136**, 3480–3485.

Suen, C.S. & Wilk, S. (1989) Regulation of thyrotropin-releasing hormone-degrading enzymes in rat brain and pituitary by L-3,5,3'-triiodothyronine. *J. Neurochem.* **52**, 884–888.

M

Vargas, M., Mendiz, M., Cisneros, M., Joseph-Bravo, P. & Charli, J.L. (1987) Regional distribution of the membrane-bound pyroglutamate aminopeptidase-degrading thyrotropin-releasing hormone in rat brain. *Neurosci. Lett.* **79**, 311–314.

Wilk, S., Suen, C.S. & Wilk, E.K. (1988) Occurrence of pyroglutamyl peptidase II, a specific TRH degrading enzyme in rabbit retinal membranes and in human retinoblastoma cells. *Neuropeptides* **12**, 43–47.

*Karl Bauer*
*Max-Planck-Institut für experimentelle Endokrinologie,*
*Feodor-Lynen-Strasse 7,*
*D-30625 Hannover, Germany*
*Email: 106001,2503@compuserve.com*

# 341. Cystinyl aminopeptidase/oxytocinase

## Databanks

*Peptidase classification: clan MA, family M1, MEROPS ID: M01.011*
*NC-IUBMB enzyme classification: EC 3.4.11.3*
*Chemical Abstracts Service registry number: 9031-41-8*
*Databank codes:*

| Species | SwissProt | PIR | EMBL (cDNA) | EMBL (genomic) |
|---------|-----------|-----|-------------|----------------|
| *Homo sapiens* | – | – | D50810 | – |

## Name and History

Since Fekete first discovered the ability of human pregnancy sera to degrade oxytocin (Fekete, 1930), many investigators have studied this enzyme utilizing bioassays. The enzyme is an aminopeptidase with specificity towards the peptide bonds between the N-terminal cysteine and the adjacent tyrosine residues of oxytocin or vasopressin, hence it has been called *oxytocinase* or *vasopressinase* (Sjöholm & Yman, 1967). Tuppy & Nesbadva (1957) introduced a chemical method for the measurement of this enzyme using a synthetic substrate, L-Cys-di-NHNap. Thus the enzyme is also called *cystine aminopeptidase, cystinyl aminopeptidase* and *CAP*. The *placental leucine aminopeptidase (P-LAP)*, which increases in serum during pregnancy, was purified from retroplacental sera using a synthetic substrate L-Leu⊣NHPhNO$_2$ and shown to be identical with CAP (Tsujimoto *et al.*, 1992). The human cDNA for the enzyme was first cloned after determining the partial amino acid sequences of the purified enzyme (Rogi *et al.*, 1996). A novel insulin-regulated aminopeptidase from GLUT4 vesicles was cloned independently from rat adipocytes (Keller *et al.*, 1995) (Chapter 342) and shown to be the rat homolog of P-LAP/oxytocinase.

## Activity and Specificity

Cystinyl aminopeptidase releases an N-terminal amino acid, Cys⊣Xaa, in which the half-cystine residue is involved in a disulfide loop, notably oxytocin and vasopressin (Sjöholm & Yman, 1967). However, hydrolysis rates on a range of aminoacyl arylamides such as Leu⊣, Arg⊣ and Ala⊣NHNap exceed those of cystinyl derivatives. Almost all arylamides except for Asp-NHNap and Glu-NHNap are hydrolyzed, indicating the broad substrate specificity of the enzyme (Sakura *et al.*, 1981).

While leucyl arylamidase activity normally present in sera and believed to be derived from liver and intestine is inhibited in the presence of 20 mM L-methionine, CAP activity is not. Therefore CAP activity can be determined using L-Leu⊣NHNap or L-Leu⊣NHPhNO$_2$ as a substrate in the presence of 20 mM L-methionine (Mizutani *et al.*, 1976a). This assay is more sensitive than that with Cys-di-NHNap (Mizutani *et al.*, 1976b). The pH optimum of the enzyme is around 7.4 and it is labile after incubation at 60°C for 30 min.

Chelating agents such as 8-hydroxyquinoline and 1,10-phenanthroline (1 mM) inhibit the enzyme, whereas EDTA is less effective (Sakura *et al.*, 1981). Divalent cations (1 mM) such as $Zn^{2+}$, $Cu^{2+}$ and $Cd^{2+}$ are strong inhibitors and $Zn^{2+}$ behaves as a competitive inhibitor ($K_i = 83\,\mu M$; Sakura *et al.*, 1981). Amastatin ($K_i = 1.9\,\mu M$) and bestatin ($K_i = 87\,\mu M$) are also competitive inhibitors (Mizutani *et al.*, 1976a). Moreover, prostaglandins and/or cGMP also inhibit the enzyme (Roy & Karim, 1983).

## Structural Chemistry

Human serum oxytocinase is a glycoprotein and exists in a dimeric form, each subunit containing two tightly bound zinc ions. Depending on the purification scheme, the molecular mass of the enzyme (monomer) is reported to be 14, 17 and 21 kDa (Sakura *et al.*, 1981; Watanabe *et al.*, 1989; Tsujimoto *et al.*, 1992). The precursor form of the enzyme deduced from the cloned cDNA contains three domains, a short (putatively 28 residue) N-terminal cytoplasmic domain, a 23 residue transmembrane domain and an 893 residue C-terminal extracellular domain, indicating that the enzyme is a type II transmembrane protein (Rogi *et al.*, 1996). The N-terminal sequence of the purified serum enzyme was determined to be Ala-Thr-Asn-Gly-Lys-Leu-Phe-Pro-Trp-Ala-Gln-Ile-Arg-Leu-Pro, indicating the presence of a post-translational processing and regulatory mechanism for the appearance of the enzyme in the maternal bloodstream (Rogi *et al.*, 1996; Watanabe *et al.*, 1996). The rat enzyme isolated from low-density microsomal fractions of adipocytes and skeletal muscle is a 165 kDa protein and the precursor form deduced from the cDNA clone shows 87% homology to the human enzyme. Human and rat enzymes carry, respectively, 18 and 17 potential *N*-glycosylation sites, one located in the N-terminal domain in both cases. In the C-terminal domain, both enzymes contain the HEXXH consensus sequence of some zinc metallopeptidases with an additional glutamic acid residue 19 amino acids downstream, which constitute the active site of clan MA metallopeptidases (Keller *et al.*, 1995; Rogi *et al.*, 1996).

## Preparation

The enzyme has been purified from retroplacental serum after normal human delivery (Sakura *et al.*, 1981; Tsujimoto *et al.*, 1992) and also from human (Lampello & Venha-Perttula, 1979) and monkey placenta (Hayashi & Oshima, 1976).

Rabbit polyclonal antibodies to the purified human enzyme (Mizutani *et al.*, 1996) and a synthetic N-terminal amino acid sequence of the enzyme (Watanabe *et al.*, 1996) are available. Antibodies to the N-terminal 108 and C-terminal 16 amino acids have also been produced (Keller *et al.*, 1995).

## Biological Aspects

Although historically the enzyme is known to be found in pregnancy plasma and placenta of human and primates, the presence of the enzyme in the animal hypothalamus has also been reported (Frith & Hooper, 1974). Only trace amounts of the enzyme activity can be detected in the blood of nonpregnant women and none in the blood of men (Mizutani *et al*, 1976a; Watanabe *et al.*, 1996). Northern blot analysis shows that human oxytocinase is expressed not only in placenta but also in several tissues including heart, skeletal muscle and brain. Two forms of the mRNA (3.6 and 10.5 kb transcripts) are expressed (Rogi *et al.*, 1996).

During pregnancy, the placenta releases oxytocinase into the maternal circulation. However, the mechanism of this release is still unknown. While the enzyme activity in serum remains low in the first trimester, it rises progressively during the second and third trimesters to a maximum at about term, then declines following parturition (Mizutani *et al.*, 1976a,b). Oxytocinase hydrolyzes oxytocin and vasopressin *in vitro*. The fetuses produce these factors actively as gestation advances or during fetal acidemia and they act as uterotonic and vasoactive agents, respectively. Therefore it is generally believed that oxytocinase in pregnancy blood serves to prevent a premature onset of uterine contractions by degrading oxytocin. The enzyme may also play an important role in controlling the fetal and maternal blood pressure though the regulation of the concentration of vasoactive peptides in the interface between fetus and mother. In fact, monitoring the changes in maternal oxytocinase activities is useful for predicting the onset of labor (Mizutani *et al.*, 1982), monitoring the preterm labor (Naruki *et al.*, 1995) and predicting pre-eclampsia (a hypertensive disorder peculiar to pregnancy) (Mizutani *et al.*, 1985).

Recently we have shown that oxytocinase degrades somatostatin *in vitro* (Mizutani *et al.*, 1996). Since somatostatin affects the growth of the fetus by regulating insulin and growth hormone secretion, our data suggest that the enzyme may be involved in fetal growth via degradation of somatostatin. Monitoring the feto-placental function and fetal development by medical electronics has largely taken over the role of biochemical methods such as measurement of this enzyme. However, considering all the data, we believe that the clinical use of the enzyme still holds promise.

## Distinguishing Features

Although the mammalian membrane aminopeptidases homologous to oxytocinase are predominantly localized at the cell surface, cystinyl aminopeptidase is recovered mainly from the microsomal fraction of placenta (Mizutani *et al.*, 1995). The rat homolog (vp165) is shown to be concentrated in GLUT4-containing vesicles in basal adipocytes and redistributed to the plasma membrane in response to insulin (Keller *et al.*, 1995; Ross *et al.*, 1996). The N-terminal cytoplasmic domain, which is longer than those of other aminopeptidases, is suggested to play a role in the trafficking of the enzyme. The appearance of the enzyme only in the serum of pregnant women in spite of the mRNA expression in a wide variety tissues warrants future studies to elucidate its biological functions.

## Further Reading

For reviews, see Mizutani & Tomoda (1992, 1996).

## References

Fekete, K. (1930) Beiträge zur Physiologie der Gravidität [Studies on the physiology of pregnancy]. *Endokrinologie* **7**, 364–369.

Frith, D.A. & Hooper, K.C. (1974) The development of an *in vitro* system for estimating oxytocinase in the rabbit hypothalamus and the effects of some protein synthesis inhibitors and estradiol-17β on enzyme activity. *Acta. Endocrinol.* **75**, 443–451.

Hayashi, M. & Oshima, K. (1976) Purification and properties of oxytocinase (cystine aminopeptidase) from monkey placenta. *J. Biochem.* **80**, 389–396.

Keller, S.R., Scott, H.M., Mastick, C.C., Aebersold, R. & Lienhard, G.E. (1995) Cloning and characterization of a novel insulin-regulated membrane aminopeptidase from Glut4 vesicles. *J. Biol. Chem.* **270**, 23612–23618.

Lampello, S. & Vanha-Perttula, T. (1979) Fractionation and characterization of cystine aminopeptidase (oxytocinase) and arylamidase of the human placenta. *J. Reprod. Fert.* **56**, 285–296.

Mizutani, S. & Tomoda, Y. (1992) Oxytocinase: placental cystine aminopeptidase or placental leucine aminopeptidase (P-LAP). *Semin. Reprod. Endocrinol.* **10**, 146–153.

Mizutani, S. & Tomoda, Y. (1996) Effects of placental proteases on maternal and fetal blood pressure in normal pregnancy and preeclampsia. *Am. J. Hypertens.* **9**, 591–597.

Mizutani, S., Yoshino, M. & Oya, M. (1976a) Placental and non-placental leucine aminopeptidases during normal pregnancy. *Clin. Biochem.* **9**, 16–18.

Mizutani, S., Yoshino, M. & Oya, M. (1976b) A comparison of oxytocinase and L-methionine-insensitive leucine aminopeptidase during normal pregnancy. *Clin. Biochem.* **9**, 228.

Mizutani, S., Hayakawa, H., Akiyama, H., Sakura, H., Yoshino, M., Oya, M. & Kawashima, Y. (1982) Simultaneous determinations of plasma oxytocin and placental leucine aminopeptidase (P-LAP) during late pregnancy. *Clin. Biochem.* **15**, 141–145.

Mizutani, S., Akiyama, H., Kurauchi, O., Taira, H., Narita, O. & Tomoda, Y. (1985) Plasma angiotensin I and serum placental leucine aminopeptidase (P-LAP) in pre-eclampsia. *Arch. Gynecol.* **236**, 165–172.

Mizutani, S., Kurauchi, O., Yamada, R., Narita, O. & Tomoda, Y. (1986) Effects of bestatin, its related compounds, puromycin and bacitracin on human oxytocinase and human microsomal placental leucine aminopeptidase. *IRCS Med. Sci.* **14**, 270–271.

Mizutani, S., Safwat, M.A., Goto, K., Tsujimoto, M., Nakazato, H., Itakura, A., Mizuno, M., Kurauchi, O., Kikkawa, F. & Tomoda, Y. (1995) Initiating and responsible enzyme of arginine vasopressin degradation in human placenta and pregnancy serum. *Regul. Pept.* **59**, 371–378.

Mizutani, S., Goto, K., Tsujimoto, M., Nakazato, H., Matsuzawa, K.,

Furuhashi, Y., Arii, K. & Tomoda, Y. (1996) Possible effects of placental aminopeptidase on the regulation of brain-gut hormones in the fetoplacental unit. *Biol. Neonate* **69**, 307–317.

Naruki, M., Mizutani, S., Yamada, R., Itakura, A., Kurauchi, O., Kikkawa, F. & Tomoda, Y. (1995) Changes in maternal oxytocinase activities in preterm labour. *Med. Sci. Res.* **23**, 797–802.

Rogi, T., Tsujimoto, M., Nakazato, H., Mizutani, S. & Tomoda, Y. (1996) Human placental leucine amonopeptidase/oxytocinase: a new member of type II membrane-spanning zinc metallopeptidase family. *J. Biol. Chem.* **271**, 56–61.

Ross, S.A., Scott, H.M., Morris, N.J., Leung, W.-Y., Mao, F., Lienhard, G.E. & Keller, S.R. (1996) Characterization of the insulin-regulated membrane aminopeptidase in 3T3-L1 adipocytes. *J. Biol. Chem.* **271**, 3328–3332.

Roy, A.C. & Karim, M.M. (1983) Significance of the inhibition by prostaglandins and cyclic GMP of oxytocinase activity in human pregnancy and labour. *Prostaglandins* **25**, 55–70.

Sakura, H., Lin, T.Y., Doi, M., Mizutani, S. & Kawashima, Y. (1981) Purification and properties of oxytocinase, a metalloenzyme. *Biochem. Int.* **2**, 173–179.

Sjöholm, I. & Yman, L. (1967) Degradation of oxytocin, lysine vasopressin, angiotensin II and angiotensin II amide by oxytocinase (cystine aminopeptidase). *Acta Pharm. Suecica* **4**, 65–76.

Tsujimoto, M., Mizutani, S., Adachi, H., Kimura, H., Nakazato, H. & Tomoda, Y. (1992) Identification of human placental leucine aminopeptidase as oxytocinase. *Arch. Biochem. Biophys.* **292**, 388–392.

Tuppy, H. & Nesbadva, H. (1957) Über die Aminopeptidaseaktivität des Schwangerenserums und ihre Beziehung zu dessen Vermögen, Oxytocin zu inactiveren [The aminopeptidase acitvity of serum in pregnancy and its relationship to the potential for inactivating oxytocin]. *Monatsh. Chem.* **88**, 977–988.

Watanabe, Y., Kumagai, Y., Kubo, Y., Shimamori, Y. & Fujimoto, Y. (1989) Aminopeptidases in human retroplacental sera: purification and characterization of two enzymes. *Biochem. Med. Metab. Biol.* **41**, 139–148.

Watanabe, Y., Iwaki-Egawa, S., Yokosawa, J., Mizukoshi, H. & Fujimoto, Y. (1996) An antibody specific to N-terminal amino acid sequence of human maternal serum cystine aminopeptidase and its applications. *Biochem. Mol. Biol. Int.* **38**, 653–658.

**Shigehiko Mizutani**
*Department of Obstetrics and Gynecology, Nagoya University School of Medicine, Tsurumai-cho, Showa-ku, Nagoya, 466, Japan*

**Masafumi Tsujimoto**
*Laboratory of Bioorganic Chemistry, The Institute of Physical and Chemical Research (RIKEN), Wako-shi, Saitama, 351-01, Japan*

**Hiroshi Nakazato**
*Suntory Institute for Biomedical Research, Mishima-gun, Osaka, 618, Japan*

# 342. *Insulin-regulated membrane aminopeptidase*

## Databanks

Peptidase classification: clan MA, family M1, MEROPS ID: M01.011
NC-IUBMB enzyme classification: none
Databank codes:

| Species | SwissProt | PIR | EMBL (cDNA) | EMBL (genomic) |
|---|---|---|---|---|
| *Rattus norvegicus* | – | – | U32990 U76997 | – |

## Name and History

This protein was first identified as a major component of specific intracellular vesicles isolated from the low-density microsomes of rat adipocytes that harbor the insulin-responsive glucose transporter isotype GLUT4 (GLUT4 vesicles) and was subsequently purified in denatured form from this source (Kandror & Pilch, 1994a; Mastick *et al.*, 1994). It was named **vp165** for **v**esicle **p**rotein of $M_r$ 165 000 (Mastick *et al.*, 1994), as well as **gp160** for **g**lyco**p**rotein of $M_r$ 160 000 (Kandror & Pilch, 1994b). The sequences of short tryptic peptides derived from the purified vp165 and gp160 showed some homology to sequences found in membrane alanyl aminopeptidase (Chapter 337) and indirect evidence was obtained that vp165/gp160 had aminopeptidase activity (Kandror *et al.*, 1994). With the cloning of its cDNA, as well as the further characterization of the purified protein, it became evident that vp165/gp160 is a novel member of a family containing other zinc-dependent membrane aminopeptidases (Keller *et al.*, 1995). Since, like GLUT4, it redistributed from its intracellular location to the plasma membrane in response to insulin (Kandror & Pilch, 1994a,b; Mastick *et al.*, 1994), we have named it the **insulin-regulated membrane aminopeptidase (IRAP)**.

Recently, when the cloning of the human placental cystinyl aminopeptidase/oxytocinase (Chapter 341) was reported (Rogi *et al.*, 1996), it was discovered that the insulin-regulated membrane aminopeptidase is the rat homolog of this enzyme, which had been previously characterized as a cystinyl aminopeptidase. The amino acid sequence of the human enzyme is 87% identical to that of the insulin-regulated membrane aminopeptidase (S.R. Keller, unpublished results). However, it should be noted that the published amino acid sequence of the human enzyme is incomplete at its N-terminus.

## Activity and Specificity

IRAP cleaves the N-terminal peptide bond of short peptides (Herbst *et al.*, 1997). Four peptide hormones have so far been tested for direct cleavage by IRAP: vasopressin (Cys+Tyr-Phe-Gln-Asn-Cys-Pro-Arg-Gly), lysyl bradykinin (Lys+Arg-Pro-Pro-Gly-Phe-Ser-Pro-Phe-Arg), angiotensin III (Arg+Val-Tyr-Ile-His-Pro-Phe) and angiotensin IV (Val+Tyr-Ile-His-Pro-Phe). The $K_m$ values of these substrates, estimated from their inhibition of the hydrolysis of Leu+NHNap, are 4, 0.6, 0.08 and 0.02 mM, respectively, and the turnover rates of their cleavage at 21–22°C are approximately 80, 12, 2.2 and 1.3 min$^{-1}$, respectively (table 1 in Herbst *et al.*, 1997).

The specificity of IRAP for the N-terminal amino acid has been examined by assay with the aminoacyl $\beta$-naphthylamides as substrates at a single concentration. Leu+NHNap is hydrolyzed most efficiently, followed by lysine at 58%, methionine at 44%, alanine at 28% and arginine at 23% of the rate for the leucine substrate (Keller *et al.*, 1995). The $K_m$ values were determined to be 140 µM for Leu-NHNap and 43 µM for Lys+NHNap (Herbst *et al.*, 1997). Based on these results, IRAP has a broad specificity with regard to the N-terminal amino acid.

Enzyme assays may be performed either at pH 7.0 in 0.1 M Tris–HCl, or at pH 7.4 in 30 mM HEPES, 120 mM NaCl, 4 mM $KH_2PO_4$, 1 mM $MgSO_4$, 1 mM $CaCl_2$, with the detergent octaethyleneglycol dodecylether (0.1%) present in both cases. There is no difference in activity between the two buffers with Leu-NHNap as the substrate (S.R. Keller, unpublished results).

A number of other peptide hormones have also been examined for binding to the enzyme by assaying their inhibitory effect on the hydrolysis of Leu-NHNap, but not for direct cleavage. Their $K_i$ values range from 2 µM for adrenocorticotropic hormone to greater than 200 µM for insulin (Herbst *et al.*, 1997).

Among the well-known aminopeptidase inhibitors (Chan, 1983, Tieku & Hooper, 1992) L-leucinethiol was found to inhibit potently ($K_i = 50$ nM), whereas amastatin inhibits weakly ($K_i = 140$ µM; Herbst *et al.*, 1997). Although we have not examined inhibition by metal-chelating agents carefully, we have observed that 1 mM EDTA inactivates the enzyme. Also, $Zn^{2+}$ at concentrations in the $10^{-5}$ M range has an inhibitory effect on the enzyme (S.R. Keller, unpublished results).

## Structural Chemistry

IRAP is a type II integral membrane protein, with a large extracellular/intraluminal domain, a single transmembrane segment and a cytoplasmic N-terminal tail. The central portion of the extracellular/intraluminal domain contains regions which are highly conserved among the family M1 aminopeptidases and which are essential for catalytic function. This includes the zinc-binding site Ile461-Ile-Ala-His-Glu-Leu-Ala-His-Gln-Trp-Phe-Gly...Leu483-Trp-Leu-Asn-Glu-Gly-Phe characteristic of this group of aminopeptidases (Keller *et al.*, 1995). Although IRAP shares its overall structure with the homologous mammalian membrane aminopeptidases, its cytoplasmic tail is clearly distinct from these. It is considerably longer and contains potential sorting motifs which are also found in GLUT4 and may therefore be responsible for its distinct subcellular localization and regulation by insulin (Keller *et al.*, 1995).

The theoretical $M_r$ for IRAP is 117 317 (value for the larger form, see below). Since IRAP is extensively glycosylated, the molecular mass of the processed protein is larger, 165 kDa in most tissues and 140 kDa in the brain (Keller *et al.*, 1995). Its predicted pI is 5.2.

The cDNA for IRAP was first cloned from rat adipocytes (Keller *et al.*, 1995). The cloning of IRAP from rat skeletal muscle has now also been completed (K.B. Clairmont & J.T. Hart, personal communication). Upon comparison of the two cDNA sequences it was found that they diverged at their 3' ends, and thus encoded different C-termini. The eight amino acids closest to the C-terminus in the adipocyte sequence were absent from the muscle sequence and were replaced by a segment of 117 amino acids (Keller *et al.*, 1995). The mRNA encoding the larger form of the protein has now also been found to be present in rat adipocytes (S.R. Keller, unpublished results). Based on the sequence of the originally described 3.2 kb cDNA and the northern blot analysis (Keller *et al.*, 1995), together with results recently obtained from mapping the mouse IRAP gene (S.R. Keller & T.K. Mahondas, unpublished results), it seems most likely that the smaller size 3.2 kb cDNA is derived from an mRNA of low abundance which is the rare product of differential splicing and processing of the IRAP transcript. If the product is translated, it could potentially yield a truncated protein. The major part of the expressed IRAP protein is most likely the form with the longer C-terminus, since this larger C-terminus contains short segments which are homologous to sequences in the C-termini of other membrane aminopeptidases. However, this conclusion remains to be confirmed experimentally.

## Preparation

A good source of the enzyme, and the only one used for the characterization of its enzymatic activity, is the low-density microsomal fraction isolated from the adipocytes from rat epididymal fat pads (Keller *et al.*, 1995; Herbst *et al.*, 1997). IRAP constitutes about 0.2% of the protein in this fraction and 0.01% of the total protein in adipocytes (Mastick *et al.*, 1994). Approximately 80% of IRAP fractionates in the low-density microsomes and there it is found almost exclusively in GLUT4 vesicles (Kandror & Pilch, 1994b; Mastick *et al.*, 1994). At least 90% of the aminopeptidase activity toward Leu-NHNap in the low-density microsomal fraction and all of this activity in GLUT4 vesicles is due to IRAP (Keller *et al.*, 1995; Herbst *et al.*, 1997). IRAP can be further purified from the low-density microsomes or GLUT4 vesicles by immunoprecipitation (Keller *et al.*, 1995; Herbst *et al.*, 1997).

## Biological Aspects

IRAP has been well characterized in rat and mouse adipocytes and rat muscle cells (Kandror & Pilch, 1994a,b; Kandror *et al.*, 1994, 1995; Mastick *et al.*, 1994; Keller *et al.*, 1995; Ross *et al.*, 1996; Houseknecht *et al.*, 1996; Herbst *et al.*, 1997; Sumitani *et al.*, 1997). In these cells IRAP is sequestered in an intracellular compartment under basal conditions and redistributes from this intracellular location to the plasma membrane in response to insulin. IRAP is also present in human adipocytes (Maianu *et al.*, 1996) and its subcellular distribution under basal and insulin-stimulated conditions is the same as in rat and mouse adipocytes.

In addition, IRAP is also found in other cell types. It has been detected in all the major tissues (adipose tissues, brain, heart, liver, lung, kidney, skeletal muscle, spleen and testis) by immunoblotting (Keller *et al.*, 1995; S.R. Keller, unpublished results). Its abundance (amount per milligram of protein) in each of these tissues is similar to that in adipocytes, with the exception of liver, where it is much less abundant. Preliminary results obtained by immunofluorescence on tissue sections indicate that IRAP is restricted to specific cell types in each tissue and that it is predominantly found in an intracellular location (S.R. Keller, unpublished results). Whether insulin and/or other stimuli can cause translocation of IRAP to the cell surface in cell types other than fat and muscle has not yet been determined.

The intron/exon structure of the mouse IRAP gene has been determined (S.R. Keller, unpublished results) and its human chromosomal localization mapped (S.R. Keller & T.K. Mohandas, unpublished results).

The physiological function for IRAP is unknown. It could participate in peptide hormone processing, either in the intracellular vesicles, where it is found under basal conditions, or at the cell surface, where it moves in response to insulin. In a model for the latter it has been found that insulin treatment of adipocytes enhances the degradation of vasopressin in the medium by 3-fold (Herbst *et al.*, 1997). Which peptide hormones are *in vitro* substrates for IRAP and what the effects of IRAP action are remain to be established.

Insulin-elicited redistribution of IRAP to the cell surface is impaired in adipocytes isolated from insulin-resistant individuals with noninsulin-dependent diabetes mellitus (NIDDM) (Maianu *et al.*, 1996). The consequences of this defect for the manifestations of diseases associated with insulin resistance (NIDDM, obesity and hypertension) (Reaven, 1994) will only become clear once the role of IRAP in insulin action has been elucidated.

## Distinguishing Features

Polyclonal antibodies have been produced against the unique cytoplasmic domain from rat IRAP (Keller *et al.*, 1995), as well as against peptides derived from the extracellular domain of the rat protein (Kandror & Pilch, 1994b; Keller *et al.*, 1995). At present these are not commercially available.

## Further Reading

For a review, see Keller *et al.* (1995).

## References

Chan, W.W.-C. (1983) L-leucinethiol – a potent inhibitor of leucine aminopeptidase. *Biochem. Biophys. Res. Commun.* **116**, 297–302.

Herbst, J.J., Ross, S.R., Scott, H.M., Bobin, S.A., Morris, N.J., Lienhard, G.E. & Keller, S.R. (1997) Insulin stimulates cell surface aminopeptidase acitvity toward vasopressin in adipocytes. *Am. J. Physiol.* **272**, E600–E606.

Houseknecht, K.L., Martin, S., Keller, S.R., James, D.E. & Kahn, B.B. (1996) Overexpression of Glut4 in adipocytes of transgenic mice saturates sorting. *Diabetes* (suppl. 2) **45**, 156A.

Kandror, K. & Pilch, P.F. (1994a) Identification and isolation of glycoproteins that translocate to the cell surface from GLUT4-enriched vesicles in an insulin-dependent fashion. *J. Biol. Chem.* **269**, 138–142.

Kandror, K.V. & Pilch, P.F. (1994b) gp160, a tissue-specific marker for insulin-activated glucose transport. *Proc. Natl Acad. Sci. USA* **91**, 8017–8021.

Kandror, K.V., Yu, L. & Pilch, P.F. (1994) The major protein of GLUT4 containing vesicles, gp 160 has aminopeptidase activity. *J. Biol. Chem.* **269**, 30777–30780.

Kandror, K.V., Coderre, L., Pushkin, A.V. & Pilch, P.F. (1995) Comparison of glucose-transporter-containing vesicles from rat fat and muscle tissues: evidence for a unique endosomal compartment. *Biochem. J.* **307**, 383–390.

Keller, S.R., Scott, H.M., Mastick, C.C., Aebersold, R. & Lienhard, G.E. (1995) Cloning and characterization of a novel insulin-regulated membrane aminopeptidase from GluT4 vesicles. *J. Biol. Chem.* **270**, 23612–23618 and 30236.

Maianu, L., Keller, S.R. & Garvey, W.T. (1996) Fat and muscle tissue express a common abnormality in translocation/trafficking of GLUT4/vp165 containing vesicles in NIDDM. *Diabetes* (suppl. 2) **45**, 157A.

Mastick, C.C., Aebersold, R. & Lienhard, G.E. (1994) Characterization of a major protein in GLUT4 vesicles. Concentration in the vesicles and insulin-stimulated translocation to the plasma membrane. *J. Biol. Chem.* **269**, 6089–6092.

Reaven, G.M. (1994) Syndrome X: 6 years later. *J. Intern. Med.* **736** (suppl.), 13–22.

Rogi, T., Tsujimoto, M., Nakazato, H., Mizutani, S. & Tomoda, Y. (1996) Human placental leucine aminopeptidase/oxytocinase. *J. Biol. Chem.* **271**, 56–61.

Ross, S.A., Scott, H.M., Morris, N.J., Leung, W.-Y., Mao, F., Lienhard, G.E. & Keller, S.R. (1996) Characterization of the insulin-regulated membrane aminopeptidase in 3T3-L1 adipocytes. *J. Biol. Chem.* **271**, 3328–3332.

Sumitani, S., Ramlal, T., Somwar, R., Keller, S.R. & Klip, A. (1997) Insulin regulation and selective segregation with glucose transporter-4 of the membrane aminopeptidase vp165 in rat skeletal muscle cells. *Endocrinology* **138**, 1029–1034.

Tieku, S. & Hooper, N.M. (1992) Inhibition of aminopeptidases N, A and W. *Biochem. Pharmacol.* **44**, 1725–1730.

*Susanna R. Keller*
*Department of Biochemistry,*
*Dartmouth Medical School,*
*7200 Vail Building,*
*Hanover, NH 03755-3844, USA*
*Email: susanna.keller@dartmouth.edu*

# 343. Aminopeptidase PS

## Databanks

*Peptidase classification: clan MA, family M1, MEROPS ID: M01.010*
*NC-IUBMB enzyme classification: EC 3.4.11.14*
*Databank codes:*

| Species | SwissProt | PIR | EMBL (cDNA) | EMBL (genomic) |
|---|---|---|---|---|
| *Homo sapiens* | – | – | Y07701 | – |
| *Mus musculus* | Q11011 | – | U35646 | – |

## Name and History

There have been a number of reports of cytosolic aminopeptidase activities in animal tissues that are inhibited by puromycin. A good many tissues have been reported to contain such activities, but the greater part of the work has been done with brain enzymes, which have been implicated in the metabolism of enkephalins. The biological activity of the enkephalins is eliminated by cleavage of the Tyr1-Gly2 bond (De Souza *et al*., 1991; Hersh & McKelvy, 1981; McDermott *et al*., 1985; Hui *et al*., 1983, Smyth & O'Cuinn, 1994). Dando *et al*. (1997) presented evidence that a single enzyme can account for all of the reports of puromycin-sensitive, cytosolic aminopeptidase activities. Names that have been used in the literature for this enzyme include *alanine aminopeptidase* (Kao *et al*., 1978), *alanyl aminopeptidase* (Flores *et al*., 1996), *enkephalin-degrading aminopeptidase* (Schnebli *et al*., 1979), *neuropeptide-degrading aminopeptidase* (McDermott *et al*., 1985), *aminopeptidase III* (Sharma & Ortwerth, 1986), *thiol aminopeptidase* (Watanabe *et al*., 1987) and *puromycin-sensitive aminopeptidase* (Hui *et al*., 1990; Johnson & Hersh, 1990). Dando *et al*. (1997) proposed the name *aminopeptidase PS* (for 'puromycin-sensitive') for the enzyme, and that is the name that will be used here. However, the official NC-IUBMB name is *cytosol alanyl aminopeptidase*.

## Activity and Specificity

Aminopeptidase PS is maximally active at pH 7.5, and it is activated by thiol compounds and calcium ions (Mantle *et al*., 1983; Dando *et al*., 1997). An optimized enzyme assay procedure was described by Dando *et al*. (1997). Aminopeptidase PS was assayed at 37°C with $10^{-5}$ M (final concentration) Ala-NHMec in 50 mM MOPS/NaOH, pH 7.5, 10% ethanediol, 0.5 mM $CaCl_2$ and 2 mM 2-ME, in a total volume of 2.5 mL. The increase in fluorescence (excitation 360 nm, emission 460 nm) caused by hydrolysis of the substrate was monitored. The characteristic of aminopeptidase PS of being totally inhibited by $25 \times 10^{-5}$ M puromycin was used to distinguish this enzyme from other aminopeptidases in complex mixtures.

Aminopeptidase PS has a broad specificity. Unblocked N-terminal Ala, Leu, Arg, Phe, Tyr and Met residues are hydrolyzed at approximately similar rates from peptides, as well as from aminoacyl-NHNap and -NHMec substrates. Compounds containing Pro in the N-terminal or the subterminal position (P1 or P1') seem to be totally resistant to hydrolysis by aminopeptidase PS, however. Consistent with this, bradykinin (Arg-Pro-Pro-Gly-Phe-Ser-Pro-Phe-Arg) is unaffected by aminopeptidase PS (Hiroi *et al*., 1992; Smyth & O'Cuinn, 1994; Dando *et al*., 1997). Gly is also very unfavorable in P1, as illustrated by the lack of hydrolysis of Gly-NHMec (Mantle *et al*., 1983; Dando *et al*., 1997), and as expected from this, the hydrolysis of [Leu]enkephalin (Tyr-Gly-Gly-Phe-Leu) ceases after removal of the N-terminal Tyr (Hersh, 1981; Hui *et al*., 1988; Mantle *et al*., 1983; Watanabe *et al*., 1987; Dando *et al*., 1997).

Aminopeptidase PS is completely inhibited by EDTA (5 mM) and 1,10-phenanthroline (1 mM). $K_i$ values determined for reversible inhibitors were amastatin 4 nM, bestatin 14 nM and puromycin 186 nM. Less effective were arphamanine B (4.1 µM) and puromycin aminonucleoside (856 µM) (Dando *et al*., 1997). Phosphoramidon (50 µM) was not inhibitory (Constam *et al*., 1995).

The inhibition of aminopeptidase PS by puromycin is strongly reminiscent of the inhibition of another N-terminally acting peptidase, dipeptidyl-peptidase II (Chapter 138), by puromycin and other cationic molecules (McDonald *et al*., 1968). The inhibition of dipeptidyl-peptidase II was shown to be correlated directly with the molecular size of the cation, giving an order of potency of inhibition: puromycin > puromycin aminonucleoside > Tris. We found Tris buffer (pH 7.5) to be totally inhibitory at 150 mM, and Safavi & Hersh (1995) have reported a $K_i$ value of 6.1 mM for Tris.

## Structural Chemistry

cDNA encoding mouse aminopeptidase PS has been cloned and sequenced by Constam *et al*. (1995), and that of the human enzyme by Tobler *et al*. (1997). Sequence identity between the species at the amino acid level is 98%. The N-terminal sequence of the rat liver enzyme is apparently identical to that from brain (Dyer *et al*., 1990; Dando *et al*., 1997). The sequence of aminopeptidase PS shows the enzyme to be a member of peptidase family M1, which contains many other aminopeptidases, all of which depend for activity on a single zinc ion bound at an HEXXH consensus site (see Chapter 336). Like other members of the family, aminopeptidase PS is a single-chain protein of about 100 kDa. The pI is 4.9 (Dando *et al*., 1997).

## Preparation

Aminopeptidase PS has been purified essentially to homogeneity by conventional procedures that typically involve anion-exchange chromatography, gel-permeation chromatography, and chromatography on hydroxyapatite (McDermott *et al*., 1985; Dyer *et al*., 1990; Hiroi *et al*., 1992; Smyth & O'Cuinn, 1994; Dando *et al*., 1997). Dando *et al*. (1997) found that the rat liver enzyme was stabilized and prevented from adsorbing to surfaces by the presence of 10% ethanediol, whereas several detergents were inhibitory. The activity of the purified enzyme is completely destroyed by freezing and thawing (Smyth & O'Cuinn, 1994) or freeze-drying (McDermott *et al*., 1985).

Activity of the isolated enzyme can be detected after gel electrophoresis without denaturation by reaction with Ala-4-methoxy-NHNap, followed by detection of the free naphthylamine by coupling with the diazonium salt Fast Garnet (Dando *et al*., 1997).

## Biological Aspects

Aminopeptidase PS is most abundant in rodent brain, with slightly lower levels in small intestine, skeletal muscle, testis and other tissues (Hiroi *et al*., 1992; Constam *et al*., 1995; Dando *et al*., 1997). Tobler *et al*. (1997) described a similar distribution for the human enzyme. Tissues other than brain that have been found to contain aminopeptidase PS include human skeletal muscle and kidney (Mantle, 1992), pig skeletal muscle (Nishimura *et al*., 1992), human and cattle lenses

(Sharma & Ortwerth, 1986) and rat liver (Hiroi *et al.*, 1992). The enzyme has a very wide distribution throughout animal species (de Souza *et al.*, 1991).

The physiological function of aminopeptidase PS remains unknown. The predominantly cytosolic localization of the enzyme would not be consistent with a primary role in the degradation of extracellular, bioactive peptides. Its broad distribution in rat tissues also argues against a purely neuropeptide-regulative effect. Constam *et al.* (1995) recently suggested that aminopeptidase PS participates in proteolytic events essential for cell growth and viability. Hiroi *et al.* (1992) reported that degradation of acetylated hemoglobin by a purified rat liver neutral endopeptidase was significantly enhanced by the addition of aminopeptidase PS, which by itself did not possess protein-degrading activity. They speculated that the rate-limiting step in the *in vivo* degradation of stable proteins may be a slow aminopeptidase cleavage that exposes a destabilizing residue which then allows further degradation according to the 'N-end rule' (Bachmair *et al.*, 1986).

The turnover of intracellular proteins, believed to be mediated primarily by the proteasome (Coux *et al.*, 1996), leads to the formation of large quantities of oligopeptides, including antigenic peptides to be presented by the MHC class I system. The majority of these peptides are presumably degraded to amino acids by the action of cytosolic oligopeptidases and exopeptidases, including aminopeptidase PS. P.M. Dando & L. Young (unpublished results) have seen synergistic action of aminopeptidase PS with thimet oligopeptidase (Chapter 371) and with neurolysin (Chapter 372) in the degradation of synthetic oligopeptides of 11–17 residues. Aminopeptidase PS may well be the major cytosolic aminopeptidase in mammalian cells (Mantle, 1992).

## Distinguishing Features

Aminopeptidase PS is a metallopeptidase of 100 kDa. The broad-specificity aminopeptidase activity is essentially totally inhibited by $25 \times 10^{-5}$ M puromycin, which has little or no effect on other known aminopeptidases. The thiol activation of aminopeptidase PS is also unusual for a metallopeptidase. The enzyme is relatively insensitive to arphamenine B, a rather specific inhibitor of soluble arginyl aminopeptidase (Belhacene *et al.*, 1993) (Chapter 348).

Polyclonal antisera to aminopeptidase PS have been described by McLellan *et al.* (1988), and monoclonal antibodies have been described by Hui *et al.* (1990). Antibodies are not commercially available at the time of writing.

## References

Bachmair, A., Finley, D. & Varshavsky, A. (1986) *In vivo* half-life of a protein is a function of its amino-terminal residue. *Science* **234**, 179.

Belhacene, N., Mari, B., Rossi, B. & Auberger, P. (1993) Characterization and purification of T lymphocyte aminopeptidase B: a putative marker of T cell activation. *Eur. J. Immunol.* **23**, 1948–1955.

Constam, D.B., Tobler, A.R., Rensing-Ehl, A., Kemler, I., Hersh, L.B. & Fontana, A. (1995) Puromycin-sensitive aminopeptidase – sequence analysis, expression, and functional characterization. *J.*

*Biol. Chem.* **270**, 26931–26939.

Coux, O., Tanaka, K. & Goldberg, A.L. (1996) Structure and functions of the 20S and 26S proteasomes. *Annu. Rev. Biochem.* **65**, 801–847.

Dando, P.M., Young, N.E. & Barrett, A.J. (1997) Aminopeptidase PS: a widely distributed cytosolic peptidase. In: *Proteolysis in Cell Functions. Proceedings of the 11th International Conference on Proteolysis and Protein Turnover* (Hopsu-Havu, V.K., ed.). Amsterdam: IOS Press, pp. 88–95.

de Souza, A.N.C., Bruno, J.A. & Carvalho, K.M. (1991) An enkephalin-degrading aminopeptidase of human brain preserved during the vertebrate phylogeny. *Comp. Biochem. Physiol. [C]* **99C**, 363–367.

Dyer, S.H., Slaughter, C.A., Orth, K., Moomaw, C.R. & Hersh, L.B. (1990) Comparison of the soluble and membrane-bound forms of the puromycin-sensitive enkephalin-degrading aminopeptidases from rat. *J. Neurochem.* **54**, 547–554.

Flores, M., Aristoy, M.C. & Toldra, F. (1996) HPLC purification and characterization of soluble alanyl aminopeptidase from porcine skeletal muscle. *J. Agric. Food Chem.* **44**, 2578–2583.

Hersh, L.B. (1981) Solubilization and characterization of two rat brain membrane-bound aminopeptidases active on Met-enkephalin. *Biochemistry* **20**, 2345–2350.

Hersh, L.B. & McKelvy, J.F. (1981) An aminopeptidase from bovine brain which catalyzes the hydrolysis of enkephalin. *J. Neurochem.* **36**, 171–178.

Hiroi, Y., Endo, Y. & Natori, Y. (1992) Purification and properties of an aminopeptidase from rat-liver cytosol. *Arch. Biochem. Biophys.* **294**, 440–445.

Hui, K.-S., Wang, Y.-J. & Lajtha, A. (1983) Purification and characterization of an enkephalin aminopeptidase from rat brain membranes. *Biochemistry* **22**, 1062–1067.

Hui, K.-S., Hui, M. & Lajtha, A. (1988) Major rat brain membrane-associated and cytosolic enkephalin-degrading aminopeptidases: comparison studies. *J. Neurosci. Res.* **20**, 231–240.

Hui, K.S., Hui, M., Lajtha, A., Saito, M. & Saito, M. (1990) Cellular localization of puromycin-sensitive aminopeptidase isozymes. *Neurochem. Res.* **15**, 1147–1151.

Johnson, G.D. & Hersh, L.B. (1990) Studies on the subsite specificity of the rat brain puromycin-sensitive aminopeptidase. *Arch. Biochem. Biophys.* **276**, 305–309.

Kao, Y.J., Starnes, W.L. & Behal, F.J. (1978) Human kidney alanine aminopeptidase: physical and kinetic properties of a sialic acid containing glycoprotein. *Biochemistry* **17**, 2990–2994.

Mantle, D. (1992) Comparison of soluble aminopeptidases in human cerebral cortex, skeletal muscle and kidney tissues. *Clin. Chim. Acta* **207**, 107–118.

Mantle, D., Hardy, M.F., Lauffart, B., McDermott, J.R., Smith, A.I. & Pennington, R.J.T. (1983) Purification and characterization of the major aminopeptidase from human skeletal muscle. *Biochem. J.* **211**, 567–573.

McDermott, J.R., Mantle, D., Lauffart, B. & Kidd, A.M. (1985) Purification and characterization of a neuropeptide-degrading aminopeptidase from human brain. *J. Neurochem.* **45**, 752–759.

McDonald, J.K., Reilly, T.J., Zeitman, B.B. & Ellis, S. (1968) Dipeptidyl arylamidase II of the pituitary. Properties of lysylalanyl-β-naphthylamide hydrolysis: inhibition by cations, distribution in tissues, and subcellular localization. *J. Biol. Chem.* **243**, 4143–4150.

McLellan, S., Dyer, S.H., Rodriguez, G. & Hersh, L.B. (1988) Studies on the tissue distribution of the puromycin sensitive enkephalin-degrading aminopeptidases. *J. Neurochem.* **51**, 1552–1559.

Nishimura, T., Kato, Y., Rhyu, M.R., Okitani, A. & Kato, H. (1992) Purification and properties of aminopeptidase C from porcine skeletal muscle. *Comp. Biochem. Physiol. [B]* **102**, 129–135.

Safavi, A. & Hersh, L.B. (1995) Degradation of dynorphin-related peptides by the puromycin-sensitive aminopeptidase and aminopeptidase M. *J. Neurochem.* **65**, 389–395.

Schnebli, H.P., Phillips, M.A. & Barclay, R.K. (1979) Isolation and characterization of an enkephalin-degrading aminopeptidase from human brain. *Biochim. Biophys. Acta* **569**, 89–98.

Sharma, K.K. & Ortwerth, B.J. (1986) Isolation and characterization of a new aminopeptidase from bovine lens. *J. Biol. Chem.* **261**, 4295–4301.

Smyth, M. & O'Cuinn, G. (1994) Alanine aminopeptidase of guinea-pig brain: a broad specificity cytoplasmic enzyme capable of hydrolysing short and intermediate length peptides. *Int. J. Biochem.* **26**, 1287–1297.

Tobler, A.R., Constam, D.B., Schmitt-Graff, A., Malipiero, U., Schlapbach, R. & Fontana, A. (1997) Cloning of the human puromycin-sensitive aminopeptidase and evidence for expression in neurons. *J. Neurochem.* **68**, 889–897.

Watanabe, Y., Kumagai, Y., Shimamori, Y. & Fujimoto, Y. (1987) Purification and characterization of thiol aminopeptidase from the cytosolic fraction of human placenta. *Biochem. Med. Metab. Biol.* **37**, 235–245.

*Pam M. Dando*
*Peptidase Laboratory, Department of Immunology,*
*The Babraham Institute, Babraham,*
*Cambridgeshire CB2 4AT, UK*
*Email: pam.dando@bbsrc.ac.uk*

*Alan J. Barrett*
*MRC Peptidase Laboratory, Department of Immunology,*
*The Babraham Institute, Babraham,*
*Cambridgeshire CB2 4AT, UK*
*Email: alan.barrett@bbsrc.ac.uk*

# 344. Yeast aminopeptidases Ape2, Aap1' and Yin7

## Databanks

Peptidase classification: clan MA, family M1, MEROPS ID: M01.006, M01.007, M01.017
NC-IUBMB enzyme classification: none
ATCC entries: 40927 (AAP1)
Databank codes:

| Species | Type | SwissProt | PIR | EMBL (cDNA) | EMBL (genomic) |
|---|---|---|---|---|---|
| *Saccharomyces cerevisiae* | Ape2 | P32454 | S37794 | X63998 | Z28157: chromosome XI ORF |
| *Saccharomyces cerevisiae* | Aap1' | P37898 | – | L12542 | U00062: chromosome VIII cosmid |
| *Saccharomyces cerevisiae* | Yin7 | P40462 | – | – | Z47047: complete chromosome IX sequence |

## Name and History

Three yeast metallo-aminopeptidases, *Ape2 aminopeptidase*, *Aap1' aminopeptidase* and *Yin7 aminopeptidase* are grouped on the basis of their sequence relationships. Although all three show high degrees of similarity and identity (especially in putative catalytic regions), the closest are Ape2 and Aap1', with Yin7 being more divergent.

The first report of enzymatic activity for Ape2 was by Frey & Röhm (1978). Ape2 enzymatic activity was later described by two groups (Achstetter *et al.*, 1983, Trumbly & Bradley, 1983). Achstetter *et al.* (1983) designated the aminopeptidase activity as **yscII** (for **y**east *Saccharomyces* **c**erevisiae aminopeptidase **II**). Due to the specificity of the enzyme for leucine chromogenic substrates, Trumbly & Bradley (1983) designated the enzyme as *Lap1*. Garcia-Alvarez *et al.* (1991) isolated the gene encoding Ape2.

The gene encoding Aap1' was isolated by Caprioglio *et al.* (1993). When we used nondenaturing gel electrophoresis and gel assays with chromogenic substrates, activity of strains with gene disruptions showed loss of an alanine and arginine-specific activity. The gene was named *AAP1* for **a**lanine/**a**rginine **a**mino**p**eptidase. Due to a prior mitochondrial gene designation of *AAP*, the database has designated the alanine/arginine aminopeptidase as *AAP1'*.

*YIN7* encodes a putative metallo-aminopeptidase based on sequence homology of an open reading frame from the

*Saccharomyces* Genome Database. No enzymatic activity has been identified for the gene product.

## Activity and Specificity

Ape2 cleaves internal peptide bonds in di- and tripeptides in *in vitro* assays, with preference for peptides with hydrophobic N-terminal amino acids. Ape2 has less activity with Lys+NHPhNO₂ and Lys+NHNap substrates as compared to similar leucine substrates (Achstetter *et al.*, 1983; Trumbly & Bradley, 1983). Similar results were published by Hirsch *et al.* (1988) and Achstetter *et al.* (1983).

Activity for Ape2 is assayed in most instances with Lys+NHNap, Leu+NHNap, Lys+NHPhNO₂ or Leu+NHPhNO₂ substrates. Assays of activity on di- and tripeptides are also used based on reduced $A_{235}$ after cleavage of the peptide bond. The optimal pH for Ape2 enzymatic activity is 7.5 (Trumbly & Bradley, 1983; Hirsch *et al.*, 1988). Activity is decreased by divalent metal cations $Fe^{2+}$, $Mg^{2+}$, $Zn^{2+}$ and is stimulated by $Co^{2+}$ at concentrations of 1 mM (Trumbly & Bradley, 1983). Hirsch *et al.* (1988) report inhibition with 0.1 mM EDTA or 0.1 mM $HgCl_2$ but 0.1 mM nitrilotriacetic acid has no effect. Bestatin is also reported to be a strong inhibitor (Achstetter *et al.*, 1983).

Aap1 aminopeptidase activity is assayed using β-naphthylamide substrates in nondenaturing PAGE assays at pH 7.5 (Caprioglio *et al.*, 1993). Aap1 activity is detected with alanine and arginine substrates. No activity is seen with Leu-NHNap substrates. Zinc cation at a concentration of 50 μM is required for activity. As of this date, no aminopeptidase activity has been described for Yin7.

## Structural Chemistry

The Ape2 protein is found in two forms, unglycosylated and glycosylated. The size of the unglycosylated form was found by gel filtration to be between 85 000 and 90 000 Da (Trumbly & Bradley, 1983; Hirsch *et al.*, 1988). A glycosylated form of approximately 140 000 Da was identified by Achstetter *et al.* (1983); it is catalytically indistinguishable from the 90 000 Da form. The *APE2* gene sequence predicts a protein of molecular mass of 97 368 Da with a pI of 5.73; it contains two putative N-glycosylation sites (Garcia-Alvarez *et al.*, 1991). The sequence also shows a region of strong sequence similarity to zinc-binding domains.

The *AAP1* gene sequence analysis determines a protein with a molecular mass of 97 634 Da and a pI of 5.11. The sequence also shows four potential N-glycosylation sites and a region with strong homology to zinc-binding regions of other peptidases (Caprioglio *et al.*, 1993). The *YIN7* gene product is predicted to be slightly larger (107 706 Da) with a pI of 6.31 (based on sequence). There are six potential N-glycosylation sites and a region of strong homology to zinc-binding region.

## Preparation

The sources for all of these aminopeptidases (Ape2, Aap1 and Yin7) is the yeast *Saccharomyces cerevisiae*. Ape2 has been purified 331-fold (Trumbly & Bradley, 1983). The

preparation is based on precipitation between 50 and 70% saturation with ammonium sulfate. The resuspended protein is heat-precipitated and then chromatographed on a Sephadex G-150 column. Pooled fractions of activity are separated on a DEAE-cellulose column and eluted at 80 mM NaCl. Final separation is by a diaminobutane-Sepharose column eluted at 40% saturation with ammonium sulfate. Hirsch *et al.* (1988) separated Ape2 activity (Leu+NHNap and Lys+NHNap specific) with a single DEAE-Sepharose column. The Ape2 elutes at 53 mM NaCl. No purification level is given but the enzyme is free of most other aminopeptidase activities.

Aap1 analyses to date have been done on total cell lysate preparations in nondenaturing PAGE.

## Biological Aspects

The gene structure of the *APE2* gene shows several features. There is a strong TATA region 324 bases up from a strong consensus start site AAAATGACC. A putative polyadenylation signal sequence of AAATAAA is present 184 bp downstream from the termination codon. The *APE2* mRNA transcript of 2.7 kb correlates well with the 97 kDa size of the unglycosylated form of the protein (Hirsch *et al.*, 1988). The normal function of the Ape2 aminopeptidase is thought to be in the uptake of hydrophobic peptides, especially leucine N-terminal peptides; this is based on analysis of mutants in Ape2 (Hirsch *et al.*, 1988). This activity may be due to the glycosylated form that is found in the membrane fraction of cell lysates and also in the so-called 'periplasmic' space (Frey & Röhm, 1979). Ape2 enzymatic activity is found in all phases of growth (Achstetter *et al.*, 1983).

The *AAP1* gene structure shows similar start and stop sites to *APE2* (including the polyadenylation site). Although the *AAP1* transcript is also 2.7 kb in length (in line with the 97 kDa predicted molecular weight of its protein), the transcription regulation during growth is different. The mRNA for *AAP1* is detected only during logarithmic growth while the aminopeptidase activity is found throughout growth and into the stationary phase. Disruption of the *AAP1* gene lowers the ability of the strain to accumulate glycogen. A strain with the *AAP1* gene carried on a multicopy plasmid shows increases in Aap1 aminopeptidase activity, glycogen accumulation, and upregulation of the HSP70 gene *SSA3* (Caprioglio *et al.*, 1993).

The *YIN7* gene sequence contains the TATA and polyadenylation sequences. The transcript is slightly larger (3.2 kb), as was the predicted molecular mass of the protein from sequence analysis. The transcript, unlike *APE2* or *AAP1*, is found only during the postdiauxic and stationary phases of growth. Strains with disruptions in the *YIN7* gene have lower glycogen levels, slow growth and are unable to grow on nonfermentable carbon sources. The cells also show a greatly enlarged vacuole.

## Distinguishing Features

The Ape2 aminopeptidase is active on leucine substrates (Leu+NHNap and Leu+NHPhNO₂), whereas Aap1' is not. On nondenaturing PAGE (7.5%) stained for Leu+NHNap activity, Ape2 is usually found as the third-fastest migrating band of activity. The Aap1 activity is the fastest-migrating activity under the same conditions when stained with the Arg+NHNap substrate.

### References

Achstetter, T., Ehmann, C. & Wolf, D.H. (1983) Proteolysis in eucaryotic cells: aminopeptidase and dipeptidyl aminopeptidases of yeast revisited. *Arch. Biochem. Biophys.* **266**, 292–305.

Caprioglio, D.R., Padilla, C. & Werner-Washburne, M. (1993) Isolation and characterization of *AAP1*: a gene encoding an alanine/arginine aminopeptidase in yeast. *J. Biol. Chem.* **268**, 14310–14315.

Frey, J. & Röhm, K.H. (1978) Subcellular localization and levels of aminopeptidases and dipeptidase in *Saccharomyces cerevisiae*. *Biochim. Biophys. Acta* **527**, 31–41.

Frey, J. & Röhm, K.H. (1979) External and internal forms of yeast aminopeptidase II. *Eur. J. Biochem.* **97**, 169–173.

García-Alvarez, N., Cueva, R. & Suárez Rendueles, P. (1991) Molecular cloning of soluble aminopeptidases from *Saccharomyces cerevisiae*: Sequence analysis of aminopeptidase yscII, a putative zinc-metallopeptidase. *Eur. J. Biochem.* **202**, 993–1102.

Hirsch, H.H., Suárez Rendueles, P., Achstetter, T. & Wolf, D.H. (1988) Aminopeptidase yscII of yeast: isolation of mutants and their biochemical and genetic analysis. *Eur. J. Biochem.* **173**, 589–598.

Trumbly, R.J. & Bradley, G. (1983) Isolation and characterization of aminopeptidase mutants of *Saccharomyces cerevisiae*. *J. Bacteriol.* **156**, 36–48.

*Daniel R. Caprioglio*
*Department of Biology,*
*University of Southern Colorado,*
*2200 Bonforte Blvd.,*
*Pueblo, CO 81001, USA*
*Email: dcaprio@uscolo.edu*

# 345. Lysyl aminopeptidase (bacteria)

## Databanks

Peptidase classification: clan MA, family M1, MEROPS ID: M01.002
NC-IUBMB enzyme classification: none
Databank codes:

| Species | SwissProt | PIR | EMBL (cDNA) | EMBL (genomic) |
| --- | --- | --- | --- | --- |
| *Acetobacter turbidans* | Q10736 | – | X94692 | – |
| *Caulobacter crescentus* | P37893 | – | M91449 | – |
| *Escherichia coli* | P04825 | A25058 | M15273 | – |
| | | A27164 | M15676 | |
| | | A29045 | X03709 | |
| | | A91163 | X04020 | |
| | | A91561 | | |
| | | B91163 | | |
| *Haemophilus influenzae* | P45274 | F64132 | – | U32781: genomic section 96 of 163 |
| | | | | U32835: genome section 150 of 163 |
| | | | | Z33502: gene cluster |
| *Lactobacillus delbrueckii* | P37896 | S38364 | Z21701 | – |
| *Lactobacillus helveticus* | Q10730 | JC4054 | U08224 | – |
| | | S47274 | Z30323 | |
| *Lactococcus lactis* | – | JU0191 | D38040 | – |
| *Lactococcus lactis cremoris* | P37897 | JN0324 | M65867 | – |
| | | S23157 | M87840 | |
| | | | S39955 | |
| | | | X61230 | |

## Name and History

The bacterial *lysyl aminopeptidase* is a broad-specificity metallo-aminopeptidase. *Peptidase N (PepN)* is the most commonly used name for the enzyme belonging to family M1, which includes bacterial, fungal and mammalian enzymes (Chapter 336). The name was originally used to describe aminopeptidases hydrolysing aminoacyl β-naphthylamides (the N is derived from naphthylamide). The first *pepN* gene cloned and sequenced was from *Escherichia coli* (Foglino *et al.*, 1986; McCaman & Gabe, 1986).

The databanks currently assign the sequences of bacterial lysyl aminopeptidase (see the Databanks table) to EC 3.4.11.2, for which the recommended name is membrane alanyl aminopeptidase (Chapter 337). This is inappropriate, however, since the bacterial sequences contain no transmembrane or membrane-associated helices or any hydrophobic segments likely to be part of a signal peptide. In contrast, many of the mammalian enzymes, which are thought to play important roles in the hydrolysis of peptides, are membrane-associated glycoproteins. For this reason, the bacterial members are treated separately in this *Handbook*.

A number of bacterial lysyl aminopeptidases have been purified and characterized (Khalid & Marth, 1990; Bockelmann *et al.*, 1992; Exterkate *et al.*, 1992; Tan *et al.*, 1992; Gobbetti *et al.*, 1996). The N-terminal amino acid sequences of the two enzymes purified from *Streptococcus salivarius* subsp. *thermophilus* showed high similarity with that of the lactococcal PepN (Midwinter & Pritchard, 1994; Rul *et al.*, 1994), but only a few bacterial genes have been cloned and sequenced.

## Activity and Specificity

Generally, PepN is capable of hydrolyzing a broad range of peptides, removing the N-terminal amino acid. The lactococcal gene was designated *pepN* since it complemented an *E. coli pepN* mutation. PepN of *Lactococcus lactis* is active on di- and tripeptides and as well on oligopeptides (Baankreis & Exterkate, 1991; Miyakawa *et al.*, 1992; Tan & Konings, 1990). Activity is optimal at 40°C. The enzyme shows a marked preference for substrates containing Arg as the N-terminal residue but, to a lesser extent, is also capable of cleaving other residues such as Lys and Leu. There is a tendency for the activity to increase with the hydrophobicity index of the C-terminal residue of dipeptide substrates. The values detected for $K_m$ and $V$ increase with chain length for oligopeptides of the general formula Lys+Phe-(Gly)$_n$, and the optimal substrate length is a tetramer (Niven *et al.*, 1995). The lactococcal enzyme is completely inactivated by *p*-chloromercuribenzoate, mersalyl, chelating agents and the divalent cations $Cu^{2+}$ and $Cd^{2+}$ (Tan & Konings, 1990).

PepN specificity determined with purified enzyme from *Lactobacillus delbrueckii* subsp. *lactis* was monitored only by chromogenic -NHPhNO$_2$ dipeptide analogs, indicating a preference for bulk or aliphatic N-terminal residues (Lys- > Leu- > Ala- > Phe- > Pro- > Tyr- > Gly-) (Klein *et al.*, 1993). This enzyme has optimum activity at pH 6.5–7 and, in comparison to the lactococcal PepN, is more heat stable (optimum

50°C) and has an 8-fold higher affinity for Lys+NHPhNO$_2$. The enzyme is completely inhibited by 1,10-phenanthroline or EDTA. The activity of the EDTA-treated enzyme can be restored by addition of $Mn^{2+}$ and $Co^{2+}$, while $Zn^{2+}$ or $Mg^{2+}$ have no effect. After 1,10-phenanthroline inhibition, activity can be restored to 50% with $Zn^{2+}$, but not with $Mg^{2+}$ (Klein *et al.*, 1993).

The activity of the *Lactobacillus helveticus* 53/7 enzyme (Varmanen *et al.*, 1994) was tested with chromogenic -NHPhNO$_2$ substrates (Lys+, Leu+, Met+NHPhNO$_2$); the enzyme activity profiles were essentially similar. An aminopeptidase, partially purified from *Lactobacillus helveticus* CNRZ32 (Khalid & Marth, 1990), is likely to be encoded by the sequenced *pepN* gene (Christensen *et al.*, 1995). This enzyme cleaves the following N-terminal amino acids from dipeptides in decreasing order Lys > Arg > Leu > Met > Val > Glu.

Aminopeptidase N purified from *E. coli* HB101 was assayed for arylamidase and peptidase activities with Leu+NHNap and Met+Ala-Ser, as substrates. The two enzyme activities were inseparable throughout the purification. Both the arylamidase and the peptidase were inhibited by 1,10-phenanthroline and *p*-chloromercuribenzoate (Yoshimoto *et al.*, 1988).

All bacterial lysyl aminopeptidases so far characterized have a pH optimum between 6.5 and 7.5.

During whole-genome sequencing of *Haemophilus influenzae* the *pepN* gene was detected by sequence similarity, but the corresponding enzyme has not been identified (Fleischmann *et al.*, 1995).

## Structural Chemistry

Peptidase N is a monomeric protein of about 95–97 kDa. The sequence is well conserved among the lactic acid bacteria. Aminopeptidases N from *Lactococcus lactis* subsp. *cremoris* strains WG2 (Strøman, 1992) and MG1363 (Van Alen-Boerrigter *et al.*, 1991) differ in five amino acids, and the enzymes of *Lactobacillus helveticus* CNRZ32 (Christensen *et al.*, 1995) and 57/3 (Varmanen *et al.*, 1994) in six amino acids. The primary sequence of PepN from *Lactococcus lactis* subsp. *cremoris* is 56% identical to the lysyl aminopeptidase of *Lactococcus lactis* (Pir: JUO191) and 49% identical to the different *Lactobacillus* enzymes (Klein *et al.*, 1993; Varmanen *et al.*, 1994; Christensen *et al.*, 1995). Alignment with other bacterial PepNs displays 22% and 20% identity with PepN of *Haemophilus influenzae* (Fleischmann *et al.*, 1995) and *E. coli* (Foglino *et al.*, 1986; McCaman & Gabe 1986), with conserved regions in and upstream of the presumed active site. Lysyl aminopeptidase contains two consecutive, highly conserved motifs that can be described as FXXGAMENGX and [GSTALIVN]-x(2)-H-E-[LIVMFYW]-{DEHRKP}-H-x-[LIVMFYWGSPQ] by use of the PROSITE conventions, in which the square brackets indicate residues that may occur in this position and the curly brackets indicate residues that cannot occur. The sequence FXX...NGX is confined to family M1, suggesting that it may have functional importance. The two residues of His in the second, 10 residue sequence are thought to be zinc ligands.

**M**

## Preparation

PepN from *Lactococcus* was purified from both wild-type and recombinant strains (Tan & Konings, 1990; Van Alen-Boerrigter *et al.*, 1991; Strøman, 1992). The enzyme was purified from extracts of *Lactobacillus delbrueckii* subsp. *lactis*, but purification from an *E. coli* clone, in which PepN constituted about 50% of total soluble protein, should be easier (Klein *et al.*, 1993). From *Lactobacillus helveticus* CNRZ32, PepN was partially purified (Khalid & Marth, 1990).

## Biological Aspects

The majority of the enzymes involved in the nutritional proteolytic pathway of *Lactococcus lactis* have been purified and biochemically characterized, and the corresponding genes have been analyzed. Breakdown of casein by an extracellular proteinase lactocepin (Chapter 99) is calculated to be sufficient to supply the cell with oligopeptides which are further degraded by intracellular peptidases (Poolman *et al.*, 1995). Consistent with this, bacterial lysyl aminopeptidases do not contain hydrophobic segments that might serve as a signal peptide or membrane-spanning domains, and there are no indications of post-translational modification.

PepN plays an essential role in metabolism since it is involved in nitrogen supply and protein turnover. In studies with peptidase mutants of *Lactococcus lactis* (Mierau *et al.*, 1996) inactivation of PepN alone led to a significant decrease of the growth rate. The biological activity of lysyl aminopeptidase of *E. coli* is regulated at the transcriptional level by anaerobiosis and phosphate starvation (Foglino & Lazdunski, 1987). Little is known about regulation of PepN activity in *Lactococcus*, which appears to be medium-dependent and variable in different strains (Meijer *et al.*, 1996). In *Lactobacillus helveticus* 57/7 a steady level of *pepN* transcription was measured in the exponential growth phase and *pepN* mRNA remained high throughout the entire growth phase (Varmanen *et al.*, 1994).

## Distinguishing Features

No specific substrate is known that permits distinction from other general aminopeptidases. Cleavage studies with purified enzymes as well as transport studies indicate that the general aminopeptidases of *Lactococcus* – PepN, aminopeptidase C (Chapter 216) and peptidase T (Chapter 485) – have overlapping but not identical specificities (Kunji *et al.*, 1996).

Four different monoclonal antibodies against lysyl aminopeptidase from *Lactococcus lactis* were isolated and characterized (Laan *et al.*, 1996). They reacted specifically with their respective antigens in crude cell extracts of *Lactococcus lactis* subsp. *cremoris* and *lactis*. No cross-reaction with proteins from other lactic acid bacteria was observed. Of the four isolated monoclonal antibodies against PepN, only one cross-reacted weakly with a 90 kDa protein of *E. coli*, whereas the others cross-reacted with 80 kDa proteins of *Lactobacillus casei*, *Lactobacillus delbrueckii*, and *Streptococcus bovis*. Immunogold labeling of *Lactococcus lactis* WG2 with the antibodies revealed that PepN is located intracellularly.

## Further Reading

Poolman *et al.* (1995) present an overall perspective of protein degradation in *Lactococcus*. Further details of substrate specificity may be found in Niven *et al.* (1995).

## References

Baankreis, R. & Exterkate, F.A. (1991) Characterization of a peptidase from *Lactococcus lactis* ssp. *cremoris* HP that hydrolyses di- and tripeptides containing proline or hydrophobic residues as the aminoterminal amino acid. *Syst. Appl. Microbiol.* **14**, 317–323.

Bockelmann, W., Schulz, Y. & Teuber, M. (1992) Purification and characterization of an aminopeptidase from *Lactobacillus delbrueckii* subsp. *bulgaricus*. *Int. Dairy J.* **2**, 95–107.

Christensen, J.E., Lin, D.L., Palva, A. & Steele, J.L. (1995) Sequence analysis, distribution and expression of an aminopeptidase N-encoding gene from *Lactobacillus helveticus* CNRZ32. *Gene* **155**, 89–93.

Exterkate, F.A., Jong, M.D., De Veer, G.J.C.M. & Baankreis, R. (1992) Location and characterization of aminopeptidase N in *Lactococcus lactis* subsp. *cremoris* HP. *Appl. Microbiol. Biotechnol.* **37**, 46–54.

Fleischmann, R.D., Adams, M.D., White, O. *et al.* (1995) Whole-genome random sequencing and assembly of *Haemophilus influenzae* Rd. *Science* **269**, 496–512.

Foglino, M. & Lazdunski, A. (1987) Deletion analysis of the promoter region of the *Escherichia coli pepN* gene, a gene subject in vivo to multiple global controls. *Mol. Gen. Genet.* **210**, 523–527.

Foglino, M., Gharbi, S. & Lazdunski, A. (1986) Nucleotide sequence of the *pepN* gene encoding aminopeptidase N of *Escherichia coli*. *Gene* **49**, 303–309.

Gobbetti, M., Smacchi, E. & Corsetti, A. (1996) The proteolytic system of *Lactobacillus sanfrancisco* CB1: purification and characterization of a proteinase, a dipeptidase, and an aminopeptidase. *Appl. Environ. Microbiol.* **62**, 3220–3226.

Khalid, N.M. & Marth, E.H. (1990) Partial purification and characterization of an aminopeptidase from *Lactobacillus helveticus* CNRZ32. *Syst. Appl. Microbiol.* **13**, 3111–3119.

Klein, J.R., Klein, U., Schad, M. & Plapp, R. (1993) Cloning, DNA sequence analysis and partial characterization of *pepN*, a lysyl aminopeptidase from *Lactobacillus delbrueckii* ssp. *lactis* DSM7290. *Eur. J. Biochem.* **217**, 105–114.

Kunji, E.R.S., Mierau, I., Poolman, B., Konings, W., Venema, G. & Kok, J. (1996) Fate of peptides in peptidase mutants of *Lactococcus lactis*. *Mol. Microbiol.* **21**, 123–131.

Laan, H., Haverkort, R.E., Leij, L.D. & Konings, W.N. (1996) Detection and localization of peptidases in *Lactococcus lactis* with monoclonal antibodies. *J. Dairy Res.* **63**, 245–256.

McCaman, M.T. & Gabe, J.D. (1986) The nucleotide sequence of the *pepN* gene and its over-expression in *Escherichia coli*. *Gene* **48**, 145–153.

Meijer, W., Marugg, J.D. & Hugenholtz, J. (1996) Regulation of proteolytic enzyme activity in *Lactococcus lactis*. *Appl. Environ. Microbiol.* **62**, 156–161.

Midwinter, R.G. & Pritchard, G.G. (1994) Aminopeptidase N from *Streptococcus salivarius* subsp. *thermophilus* NCDO 573: purification and properties. *J. Appl. Bacteriol.* **77**, 288–295.

Mierau, I., Kunji, E.R.S., Leenhouts, K.J., Hellendoorn, M.A., Haandrikman, A.J., Poolman, B., Konings, W.N., Venema, G. & Kok, J. (1996) Multiple-peptidase mutants of *Lactococcus lactis* are severely impaired in their ability to grow in milk. *J. Bacteriol.* **178**, 2794–2803.

Miyakawa, H., Kobayashi, S., Shimamura, S. & Tomita, M. (1992) Purification and characterization of an aminopeptidase from *Lactobacillus helveticus* LHE-511. *J. Dairy Sci.* **75**, 27–35.

Niven, G.W., Holder, S.A. & Strøman, P. (1995) A study of the substrate specificity of aminopeptidase N from *Lactococcus lactis* subsp. *cremoris* Wg2. *Appl. Microbiol. Biotechnol.* **44**, 100–105.

Poolman, B., Kunji, E.R.S., Hagting, A., Juillard, V. & Konings, W.N. (1995) The proteolytic pathway of *Lactococcus lactis*. *J. Appl. Bacteriol. Symp. Suppl.* **24**, 65S–75S.

Rul, F., Monnet, V. & Gripon, J. (1994) Purification and characterization of a general aminopeptidase (St-PepN) from *Streptococcus salivarius* ssp. *thermophilus* CNRZ 302. *J. Dairy Sci.* **77**, 2880–2889.

Strøman, P. (1992) Sequence of a gene (*lap*) encoding a 95.3-kDa aminopeptidase from *Lactococcus lactis* ssp. *cremori* Wg2. *Gene* **113**, 107–112.

Tan, P.S.T. & Konings, W.N. (1990) Purification and characterization of an aminopeptidase from *Lactococcus lactis* subsp. *cremoris* Wg2. *Appl. Environ. Microbiol.* **56**, 526–532.

Tan, P.S.T., Van Alen-Boerrigter, I.J., Poolman, B., Siezen, R.J., De Vos, W.M. & Konings, W.N. (1992) Characterization of the *Lactococcus lactis pepN* gene encoding an aminopeptidase homologous to mammalian aminopeptidase N. *FEBS Lett.* **306**, 9–16.

Van Alen-Boerrigter, I.J., Baankreis, R. & De Vos, W.M. (1991) Characterization and overexpression of the *Lactococcus lactis pepN* gene and localization of its product, aminopeptidase N. *Appl. Environ. Microbiol.* **57**, 2555–2561.

Varmanen, P., Vesanto, E., Steele, J. & Palva, A. (1994) Characterization and expression of the *pepN* gene encoding a general aminopeptidase from *Lactobacillus helveticus*. *FEMS Microbiol. Lett.* **124**, 315–320.

Yoshimoto, T., Tamesa, Y., Gushi, K., Murayama, N. & Tsuru, D. (1988) An aminopeptidase N from *Escherichia coli* HB101: purification and demonstration that the enzyme possesses arylamidase and peptidase activities. *Agric. Biol. Chem.* **52**, 217–225.

*Jürgen R. Klein*
*Department of Microbiology, University of Kaiserslautern,*
*PO Box 3049,*
*D-67653 Kaiserslautern, Germany*
*Email: jklein@rhrk.uni-kl.de*

*Bernhard Henrich*
*Department of Microbiology, University of Kaiserslautern,*
*PO Box 3049,*
*D-67653 Kaiserslautern, Germany*

# 346. Aminopeptidase N (Streptomyces lividans)

## Databanks

*Peptidase classification: clan MA, family M1, MEROPS ID: M01.009*
*NC-IUBMB enzyme classification: none*
*Databank codes:*

| Species | SwissProt | PIR | EMBL (cDNA) | EMBL (genomic) |
|---|---|---|---|---|
| *Streptomyces lividans* | Q11010 | – | L23172 | – |

## Name and History

The gene encoding the **Streptomyces lividans aminopeptidase N (PepN)** protein was isolated from a screening experiment (Butler *et al*., 1994b) to characterize the major aminopeptidase activities in the strain. Clones from plasmid genomic libraries were identified by virtue of their ability to hydrolyze either Leu-|-NHNap or Arg-|-NHNap to produce bright red colonies against a background of orange colonies. Nucleotide sequence determination of the cloned genomic DNA fragments revealed a predicted protein of approximately 95 kDa.

Comparison of the predicted protein sequence with databases displayed a strong similarity to aminopeptidase N proteins from several species, with that of the gram-positive bacterium *Lactococcus lactis* (Paris *et al*., 1992) being the closest match.

## Activity and Specificity

The purified PepN hydrolyzes a wide range of different peptide and *p*-nitroanilide substrates with greatest specific activity against Leu+NHPhNO$_2$ followed closely by Lys+NHPhNO$_2$ and Arg+NHPhNO$_2$. Ala-NHPhNO$_2$ and Met-NHPhNO$_2$ are hydrolyzed at about 60% of the rate of Leu-NHPhNO$_2$ while Ser-NHNap and Phe-NHNap are hydrolyzed more slowly (approximately 25% of the rate of Leu-NHPhNO$_2$). Pro-NHPhNO$_2$ and Gly-NHPhNO$_2$ are hydrolyzed at less than 10% of the rate of Leu-NHPhNO$_2$. No activity is detected against Val-NHPhNO$_2$, Asp-NHPhNO$_2$ or Glu-NHPhNO$_2$. The substrate N$\alpha$-benzoyl-L-Arg-NHPhNO$_2$ is not hydrolyzed, indicating that the enzyme is a true aminopeptidase in requiring a free N-terminal amino group for catalysis to occur. The enzyme is a metalloprotease; its activity is inhibited by EDTA and 1,10-phenanthroline but not by PMSF. A zinc-binding motif (HELAH) is found in the predicted protein sequence and this is the region of closest similarity to the *L. lactis* PepN sequence. Furthermore, an adjacent glutamic acid residue is also closely conserved between the two proteins, suggesting that this might provide another ligand to contribute to the binding of zinc at the active site of the enzyme.

### Structural Chemistry and Preparation

The PepN was strongly overexpressed by strains carrying the gene on a multicopy plasmid. SDS-PAGE analysis was consistent with the relatively large molecular mass expected for the predicted protein (95 kDa). No secretion signal sequence was visible at the N-terminus of the predicted protein, suggesting that the PepN is primarily located intracellularly. However, after 40 h of fermentation significant amounts of activity could be detected in the culture medium. SDS-PAGE analysis showed that the PepN protein appeared to be relatively enriched compared to the other intracellular proteins and, therefore, PepN was purified from the medium. The behavior of the protein in gel-filtration chromatography was consistent with a monomeric protein of size approximately 95 kDa. The enzyme was substantially purified by adding ammonium sulfate to 85% saturation. The precipitated material was passed over a Q-Sepharose column and fractions containing enzymatic activity were concentrated for a final gel-filtration step to produce purified PepN protein for characterization as described above.

### Biological Aspects

PepN represents the most active intracellular aminopeptidase removing single amino acid residues from the N-termini of peptides in *S. lividans* strains and, as in other species, is probably important in the turnover of intracellular peptides (Yen *et al.*, 1980). Other peptidases are probably also involved in these processes. X-Pro aminopeptidase (PepP) (Chapter 479) activity has been observed; two genes have been cloned and shown to encode major (Butler *et al.*, 1993) and minor (Butler *et al.*, 1994b) PepP activities, which can be removed by recombinational chromosomal deletion techniques. Similarly, another general aminopeptidase-encoding gene (designated *PepG*) was isolated by screening with the substrate Ser-NHNap (Butler *et al.*, 1994b). Crude cell extracts of strains overproducing this PepG showed fastest hydrolysis of Gly-NHPhNO$_2$, Ser-NHNap, Ala-NHPhNO$_2$ and Leu-NHPhNO$_2$. Thus, the specificities of the PepG and PepN together with the PepPs appear to be complementary and should be adequate to account for the major intracellular peptide turnover capability.

### References

Butler, M.J., Bergeron, A., Soostmeyer, G., Zimny, T. & Malek, L.T. (1993) Cloning and characterisation of an aminopeptidase P-encoding gene from *Streptomyces lividans*. *Gene* **123**, 115–119.

Butler, M.J., Aphale, J.S., DiZonno, M.A., Krygsman, P., Walczyk, E. & Malek, L.T. (1994a) Intracellular aminopeptidases in *Streptomyces lividans* 66. *J. Ind. Microbiol.* **13**, 24–29.

Butler, M.J., Aphale, J.S., Binnie, C., DiZonno, M.A., Krygsman, P., Soltes, G.A., Walczyk, E. & Malek, L.T. (1994b) The aminopeptidase N-encoding *pepN* gene of *Streptomyces lividans* 66. *Gene* **141**, 115–119.

Paris, S.T., Van Alen-Boerrigter, I.J., Poopan, B., Siezen, R.J., de Vos, W.M. & Konings, W.N. (1992) Characterisation of the *Lactococcus lactis pepN* 306: 9–16 gene encoding an animopeptidase homologous to mammalian aminopeptidase N. *FEBS Lett.* **306**, 9–16.

Yen, C., Green, L. & Miller, C.G. (1980) Degradation of intracellular protein in *Salmonella typhimurium* peptidase mutants. *J. Mol. Biol.* **143**, 21–33.

*Michael J. Butler*
*Strangeways Research Laboratory,*
*Cambridge CB1 4RN, UK*
*Email: m.butler@uea.ac.uk*

# 347. *Leukotriene* A4 *hydrolase*

### Databanks

*Peptidase classification: clan MA, family M1, MEROPS ID: M01.004*
*NC-IUBMB enzyme classification: EC 3.3.2.6*

*Databank codes:*

| Species | SwissProt | PIR | EMBL (cDNA) | EMBL (genomic) |
|---|---|---|---|---|
| *Caenorhabditis elegans* | – | – | M88793 | – |
| *Cavia porcellus* | P19602 | S01018 | D16669 | – |
| *Dictyostelium discoideum* | P52922 | – | U27538 | – |
| *Homo sapiens* | P09960 | – | J02959 J03459 | U27275–U27293: complete gene |
| *Mus musculus* | P24527 | – | M63848 | – |
| *Rattus norvegicus* | P30349 | – | – | – |
| *Saccharomyces cerevisiae* | Q10740 | – | Z71321 | – |

## Name and History

The name **leukotriene A$_4$ hydrolase** denotes hydrolysis of the unstable epoxide leukotriene A$_4$ (LTA$_4$, 5S-*trans*-5,6-*oxido*-7,9-*trans*-11,14-*cis*-eicosatetraenoic acid) into the dihydroxy acid leukotriene B$_4$ (LTB$_4$, 5S,12R-dihydroxy-6,14-*cis*-8,10-*trans*-eicosatetraenoic acid). The name leukotriene, in turn, refers to a class of lipid mediators derived from the oxidative metabolism of arachidonic acid via the 5-lipoxygenase pathway (Samuelsson, 1983). These compounds are formed in **leuko**cytes and their structures all contain a conjugated **triene** moiety giving rise to a characteristic UV chromophore. LTA$_4$ hydrolase is distinct from microsomal or soluble xenobiotic epoxide hydrolase (Haeggström *et al.*, 1986). The enzyme is ubiquitous in mammalian tissues and is even found in cells unable to produce LTA$_4$. From sequence comparisons with certain zinc-containing aminopeptidases and proteases, a zinc-binding motif was identified in LTA$_4$ hydrolase (Malfroy *et al.*, 1989; Vallee & Auld, 1990). Further studies showed that the enzyme contains one catalytic zinc and also exhibits a peptide-cleaving activity (Haeggström *et al.*, 1990a,b; Minami *et al.*, 1990). Hence, LTA$_4$ hydrolase is a bifunctional zinc metalloenzyme.

## Activity and Specificity

Both catalytic activities of LTA$_4$ hydrolase are dependent on the catalytic zinc atom and are inhibited by divalent cations with different specificity and potency for each of the two activities (Haeggström *et al.*, 1990a,b; Wetterholm *et al.*, 1994). In addition, the peptidase activity is greatly stimulated by monovalent anions, including chloride (Wetterholm & Haeggström, 1992). Typically, LTA$_4$ hydrolase undergoes suicide inactivation when exposed to its lipid substrate LTA$_4$. During this process, both activities are affected to the same degree (Örning *et al.*, 1992).

The epoxide hydrolase activity of LTA$_4$ hydrolase is very substrate specific and, besides LTA$_4$, the only other known substrates are the double-bond isomers LTA$_5$ and to a lesser extent LTA$_3$ (Ohishi *et al.*, 1987). This activity is measured by incubation with the free acid of LTA$_4$, under conditions compatible with its chemical lability, followed by extraction and HPLC analysis of the product LTB$_4$ (Haeggström, 1990). The peptide-cleaving activity is much less specific and can be assayed with a number of synthetic chromogenic amides, including various nitroanilides and β-napthylamides, in particular alanine, arginine, leucine and proline derivatives (Wetterholm & Haeggström, 1992). Certain opioid peptides can be hydrolyzed by LTA$_4$ hydrolase but with very low efficiency (Griffin *et al.*, 1992). The best substrates described thus far are various arginyl di- and tripeptides (Örning *et al.*, 1994).

LTA$_4$ hydrolase is inhibited by the chelating agent 1,10-phenanthroline (but not EDTA), the general aminopeptidase inhibitor bestatin, and captopril, a classical inhibitor of angiotensin-converting enzyme (Haeggström *et al.*, 1990a; Örning *et al.*, 1991). Several other more potent inhibitors of the enzyme have recently been synthesized (Yuan *et al.*, 1992; Wetterholm *et al.*, 1995).

## Structural Chemistry

LTA$_4$ hydrolase from humans and rodents is a soluble monomeric enzyme composed of 610 amino acids (initial Met excluded) with a calculated $M_r$ of 69 153. The cDNAs encoding human, mouse, rat and guinea pig LTA$_4$ hydrolase have been cloned and sequenced (Funk *et al.*, 1987; Medina *et al.*, 1991a; Makita *et al.*, 1992; Minami *et al.*, 1987, 1995). A zinc-binding motif (HEXXHX$_{18}$E) is present in the primary structure and it contains 1 mol of zinc. Analysis of structure–function relationships by site-directed mutagenesis has identified a number of putative active-site residues (Haeggström *et al.*, 1993). Thus, His295, His299 and Glu318 have been shown to be the zinc-binding ligands (Medina *et al.*, 1991b). Glu296, conserved within the zinc site, is required for the peptidase activity, presumably as a general base (Wetterholm *et al.*, 1992). A 21 residue peptide segment (K21) has been identified to which LTA$_4$ binds during suicide inactivation (Mueller *et al.*, 1995). More specifically, Tyr378 appears to be the primary site for binding of LTA$_4$ and this residue also seems to play a role in the formation of the correct double bond geometry in the enzymatic product LTB$_4$ (Mueller *et al.*, 1996a,b). Tyr383, also located within peptide K21, is required for the peptidase activity, perhaps as a proton donor (Blomster *et al.*, 1995). Taken together, the structural and functional properties of LTA$_4$ hydrolase indicate that the active sites corresponding to the two activities are not identical but overlapping (Figure 347.1). LTA$_4$ hydrolase has been crystallized but no structural information is yet available (Tsuge *et al.*, 1994).

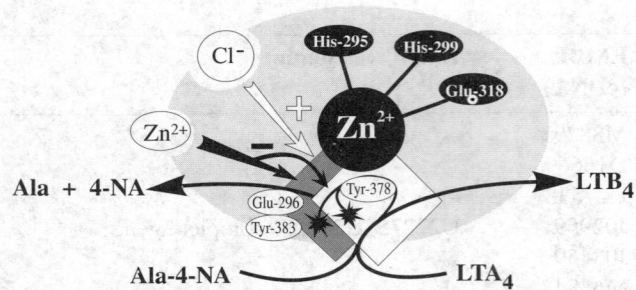

*Figure 347.1* Model of structural and functional elements of LTA$_4$ hydrolase (cf. Haeggström, 1997).

## Preparation

LTA$_4$ hydrolase is a widely distributed enzyme and is particularly abundant in cells derived from the bone marrow as well as epithelial cells of the gut and respiratory tract. It has been purified from a number of human and animal sources, e.g. leukocytes, erythrocytes, lung and liver (Rådmark & Haeggström, 1990). Several methods for purification of LTA$_4$ hydrolase have been described (Haeggström, 1990). Recombinant enzyme has been expressed in *E. coli*, COS cells, and an insect cell/baculovirus system (Minami *et al.*, 1988; Makita *et al.*, 1992; Gierse *et al.*, 1993).

## Biological Aspects

Leukotrienes possess potent biological activities and are implicated in a number of inflammatory and allergic diseases (Lewis *et al.*, 1990; Samuelsson, 1983). The lipid product of LTA$_4$ hydrolase, i.e. LTB$_4$, is one of the most powerful chemotactic agents known to date. Since leukotrienes are believed to be formed almost exclusively in cells derived from the bone marrow, e.g. granulocytes, monocytes, macrophages and mast cells, the widespread occurrence of LTA$_4$ hydrolase has been difficult to rationalize. One explanation has been 'transcellular metabolism', a concept in which LTA$_4$ would be produced by a donor cell, e.g. an activated leukocyte, and further metabolized by a recipient cell containing LTA$_4$ hydrolase. Alternatively, the distribution may be explained by its second catalytic activity such that LTA$_4$ hydrolase exhibits an epoxide hydrolase activity in leukotriene-producing cells and a peptide-cleaving activity in other cell types. The human gene has been cloned and characterized (Mancini & Evans, 1995). It comprises >35 kb of DNA, is divided into 19 exons, and exists as a single copy localized to chromosome 12q22. The regulation of cellular LTA$_4$ hydrolase expression is presently unknown.

## Distinguishing Features

The most exclusive property of LTA$_4$ hydrolase is its ability to convert LTA$_4$ into LTB$_4$. Soluble epoxide hydrolase (EC 3.3.2.3) can also utilize LTA$_4$ as substrate but will convert the epoxide into 5$S$,6$R$-dihydroxy-7,9-*trans*-11,14-*cis*-eicosatetraenoic acid (Haeggström *et al.*, 1986). The substrate LTA$_4$, as well as antibodies against LTA$_4$ hydrolase, are commercially available (Cayman Chemical; see Appendix 2 for full names and addresses of suppliers).

## Related Peptidases

LTA$_4$ hydrolase has a zinc-binding site similar to those of several other zinc aminopeptidases and proteases, e.g. membrane alanyl aminopeptidase M (Chapter 337) and thermolysin (Chapter 351) (Vallee & Auld, 1990). However, the similarity is usually restricted to this short segment of the proteins. In spite of its potent peptidase activity against arginyl di- and tripeptides, LTA$_4$ hydrolase is distinct from arginyl aminopeptidase (aminopeptidase B; Chapter 348). Thus, rat aminopeptidase B has been cloned and the primary structure contained the canonical zinc signature HEXXHX$_{18}$E typical of the M1 family of metallopeptidases and was found to be about 35% identical to LTA$_4$ hydrolase (Fukasawa *et al.*, 1996). Interestingly, *Saccharomyces cerevisiae* contains a homolog of LTA$_4$ hydrolase that is about 40% identical to the mammalian enzymes at the amino acid level (Nasr *et al.*, 1996). The properties of this yeast protein have not yet been reported. A partial sequence with significant homology to LTA$_4$ hydrolase has also been obtained from *Dictyostelium discoideum* (E. Jho & W. Kopachik, unpublished results).

## Further Reading

Haeggström *et al.* (1993) present a review focused on the identification of LTA$_4$ hydrolase as a zinc enzyme and the peptidase activity. Haeggström (1997) provides a broad and updated review covering all aspects of the enzyme.

## References

Blomster, M., Wetterholm, A., Mueller, M.J. & Haeggström, J.Z. (1995) Evidence for a catalytic role of tyrosine 383 in the peptidase reaction of leukotriene A$_4$ hydrolase. *Eur. J. Biochem.* **231**, 528–534.

Fukasawa, K.M., Fukasawa, K., Kanai, M., Fujii, S. & Harada, M. (1996) Molecular cloning and expression of rat liver aminopeptidase B. *J. Biol. Chem.* **271**, 30731–30735.

Funk, C.D., Rådmark, O., Fu, J.Y., Matsumoto, T., Jörnvall, H., Shimizu, T. & Samuelsson, B. (1987) Molecular cloning and amino acid sequence of leukotriene A$_4$ hydrolase. *Proc. Natl Acad. Sci. USA* **84**, 6677–6681.

Gierse, J.K., Luckow, V.A., Askonas, L.J., Duffin, K.L., Aykent, S., Bild, G.S., Rodi, C.P., Sullivan, P.M., Bourner, M.J., Kimack, N.M. & Krivi, G.G. (1993) High-level expression and purification of human leukotriene A$_4$ hydrolase from insect cells infected with a baculovirus vector. *Protein Exp. Purif.* **4**, 358–366.

Griffin, K.J., Gierse, J., Krivi, G. & Fitzpatrick, F.A. (1992) Opioid peptides are substrates for the bifunctional enzyme LTA$_4$ hydrolase/aminopeptidase. *Prostaglandins* **44**, 251–257.

Haeggström, J.Z. (1990) Cytosolic liver enzymes catalyzing hydrolysis of leukotriene A$_4$ into leukotriene B$_4$ and 5,6-dihydroxy-eicosatetraenoic acid. *Methods Enzymol.* **187**, 324–334.

Haeggström, J.Z. (1997) The molecular biology of the leukotriene A$_4$ hydrolase. In: *SRS-A to Leukotrienes* (Holgate, S. & Dahlén, S.-E., eds). Oxford: Blackwell Science, pp. 85–100.

Haeggström, J., Meijer, J. & Rådmark, O. (1986) Leukotriene A$_4$: enzymatic conversion into 5,6-dihydroxy-7,9,11,14-eicosatetraenoic acid by mouse liver cytosolic epoxide hydrolase. *J. Biol. Chem.* **261**, 6332–6337.

Haeggström, J.Z., Wetterholm, A., Shapiro, R., Vallee, B.L. & Samuelsson, B. (1990a) Leukotriene A$_4$ hydrolase: a zinc metalloenzyme. *Biochem. Biophys. Res. Commun.* **172**, 965–970.

Haeggström, J.Z., Wetterholm, A., Vallee, B.L. & Samuelsson, B. (1990b) Leukotriene A₄ hydrolase: an epoxide hydrolase with peptidase activity. *Biochem. Biophys. Res. Commun.* **173**, 431–437.

Haeggström, J.Z., Wetterholm, A., Medina, J.F. & Samuelsson, B. (1993) Leukotriene A₄ hydrolase: structural and functional properties of the active center. *J. Lipid Mediat.* **6**, 1–13.

Lewis, R.A., Austen, K.F. & Soberman, R.J. (1990) Leukotrienes and other products of the 5-lipoxygenase pathway. *N. Engl. J. Med.* **323**, 645–655.

Makita, N., Funk, C.D., Imai, E., Hoover, R.L. & Badr, K.F. (1992) Molecular cloning and functional expression of rat leukotriene A₄ hydrolase using the polymerase chain reaction. *FEBS Lett.* **299**, 273–277.

Malfroy, B., Kado-Fong, H., Gros, C., Giros, B., Schwartz, J.-C. & Hellmiss, R. (1989) Molecular cloning and amino acid sequence of rat kidney aminopeptidase M: a member of a super family of zinc-metallohydrolases. *Biochem. Biophys. Res. Commun.* **161**, 236–241.

Mancini, J.A. & Evans, J.F. (1995) Cloning and characterization of the human leukotriene A₄ hydrolase gene. *Eur. J. Biochem.* **231**, 65–71.

Medina, J.F., Rådmark, O., Funk, C.D. & Haeggström, J.Z. (1991a) Molecular cloning and expression of mouse leukotriene A₄ hydrolase cDNA. *Biochem. Biophys. Res. Commun.* **176**, 1516–1524.

Medina, J.F., Wetterholm, A., Rådmark, O., Shapiro, R., Haeggström, J.Z., Vallee, B.L. & Samuelsson, B. (1991b) Leukotriene A₄ hydrolase: determination of the three zinc-binding ligands by site-directed mutagenesis and zinc analysis. *Proc. Natl Acad. Sci. USA* **88**, 7620–7624.

Minami, M., Ohno, S., Kawasaki, H., Rådmark, O., Samuelsson, B., Jörnvall, H., Shimizu, T., Seyama, Y. & Suzuki, K. (1987) Molecular cloning of a cDNA coding for human leukotriene A₄ hydrolase. *J. Biol. Chem.* **262**, 13873–13876.

Minami, M., Minami, Y., Emori, Y., Kawasaki, H., Ohno, S., Suzuki, K., Ohishi, N., Shimizu, T. & Seyama, Y. (1988) Expression of human leukotriene A₄ hydrolase cDNA in *Escherichia coli*. *FEBS Lett.* **229**, 279–282.

Minami, M., Ohishi, N., Mutoh, H., Izumi, T., Bito, H., Wada, H., Seyama, Y., Toh, H. & Shimizu, T. (1990) Leukotriene A₄ hydrolase is a zinc-containing aminopeptidase. *Biochem. Biophys. Res. Commun.* **173**, 620–626.

Minami, M., Mutoh, H., Ohishi, N., Honda, Z.I., Bito, H. & Shimizu, T. (1995) Amino-acid sequence and tissue distribution of guinea-pig leukotriene A₄ hydrolase. *Gene* **161**, 249–251.

Mueller, M.J., Wetterholm, A., Blomster, M., Jörnvall, H., Samuelsson, B. & Haeggström, J.Z. (1995) Leukotriene A₄ hydrolase: mapping of a heneicosapeptide involved in mechanism-based inactivation. *Proc. Natl Acad. Sci. USA* **92**, 8383–8387.

Mueller, M.J., Blomster, M., Opperman, U.C.T., Jörnvall, H., Samuelsson, B. & Haeggström, J.Z. (1996a) Leukotriene A₄ hydrolase: protection from mechanism-based inactivation by mutation of tyrosine-378. *Proc. Natl Acad. Sci. USA* **93**, 5931–5935.

Mueller, M.J., Blomster, M., Samuelsson, B. & Haeggström, J.Z. (1996b) Leukotriene A₄ hydrolase: mutation of tyrosine-383 allows conversion of leukotriene A₄ into an isomer of leukotriene B₄. *J. Biol. Chem.* **271**, 24345–24348.

Nasr, F., Bécam, A.M. & Herbert, C.J. (1996) The sequence of 12.8 kb from the left arm of chromosome XIV reveals a sigma element, a pro-tRNA and six complete open reading frames, one of which encodes a protein similar to the human leukotriene A₄ hydrolase. *Yeast* **12**, 493–499.

Ohishi, N., Izumi, T., Minami, M., Kitamura, S., Seyama, Y., Ohkawa, S., Terao, S., Yotsumoto, H., Takaku, F. & Shimizu, T. (1987) Leukotriene A₄ hydrolase in the human lung: inactivation of the enzyme with leukotriene A₄ isomers. *J. Biol. Chem.* **262**, 10200–10205.

Örning, L., Krivi, G. & Fitzpatrick, F.A. (1991) Leukotriene A₄ hydrolase: inhibition by bestatin and intrinsic aminopeptidase activity establish its functional resemblance to metallohydrolase enzymes. *J. Biol. Chem.* **266**, 1375–1378.

Örning, L., Gierse, J., Duffin, K., Bild, G., Krivi, G. & Fitzpatrick, F.A. (1992) Mechanism-based inactivation of leukotriene A₄ hydrolase/aminopeptidase by leukotriene A₄. Mass spectrometric and kinetic characterization. *J. Biol. Chem.* **267**, 22733–22739.

Örning, L., Gierse, J.K. & Fitzpatrick, F.A. (1994) The bifunctional enzyme leukotriene A₄ hydrolase is an arginine aminopeptidase of high efficiency and specificity. *J. Biol. Chem.* **269**, 11269–11273.

Rådmark, O. & Haeggström, J. (1990) Properties of leukotriene A₄ hydrolase. *Adv. Prostaglandin Thromboxane Leukot. Res.* **20**, 35–45.

Samuelsson, B. (1983) Leukotrienes: mediators of immediate hypersensitivity reactions and inflammation. *Science* **220**, 568–575.

Tsuge, H., Ago, H., Aoki, M., Furuno, M., Noma, M., Miyano, M., Minami, M., Izumi, T. & Shimizu, T. (1994) Crystallization and preliminary X-ray crystallographic studies of recombinant human leukotriene A₄ hydrolase complexed with bestatin. *J. Mol. Biol.* **238**, 854–856.

Vallee, B.L. & Auld, D.S. (1990) Zinc coordination, function, and structure of zinc enzymes and other proteins. *Biochemistry* **29**, 5647–5659.

Wetterholm, A. & Haeggström, J.Z. (1992) Leukotriene A₄ hydrolase: an anion activated peptidase. *Biochim. Biophys. Acta* **1123**, 275–281.

Wetterholm, A., Medina, J.F., Rådmark, O., Shapiro, R., Haeggström, J.Z., Vallee, B.L. & Samuelsson, B. (1992) Leukotriene A₄ hydrolase: abrogation of the peptidase activity by mutation of glutamic acid-296. *Proc. Natl Acad. Sci. USA* **89**, 9141–9145.

Wetterholm, A., Macchia, L. & Haeggström, J.Z. (1994) Zinc and other divalent cations inhibit purified leukotriene A₄ hydrolase and leukotriene B₄ biosynthesis in human polymorphonuclear leukocytes. *Arch. Biochem. Biophys.* **318**, 263–271.

Wetterholm, A., Haeggström, J.Z., Samuelsson, B., Yuan, W., Munoz, B. & Wong, C.-H. (1995) Potent and selective inhibitors of leukotriene A₄ hydrolase: effects on purified enzyme and human polymorphonuclear leukocytes. *J. Pharmacol. Exp. Ther.* **275**, 31–37.

Yuan, W., Wong, C.-H., Haeggström, J.Z., Wetterholm, A. & Samuelsson, B. (1992) Novel tight-binding inhibitors of leukotriene A₄ hydrolase. *J. Am. Chem. Soc.* **114**, 6552–6553.

**M**

***Jesper Z. Haeggström***
*Department of Medical Biochemistry and Biophysics,*
*Karolinska Institute, S-171 77 Stockholm, Sweden*
*Email: jesper.haeggstrom@mbb.ki.se*

# 348. *Aminopeptidase B*

## Databanks

*Peptidase classification: clan MA, family M1, MEROPS ID: M01.014*
*NC-IUBMB enzyme classification: EC 3.4.11.6*
*Chemical Abstracts Service registry number: 9073-92-1*
*Databank codes:*

| Species | SwissProt | PIR | EMBL (cDNA) | EMBL (genomic) |
|---|---|---|---|---|
| *Rattus norvegicus* | – | – | D87515 | – |

## Name and History

An exopeptidase activity was originally identified in several rat tissues (Hopsu *et al*., 1964), using L-aminoacyl $\beta$-naphthylamides and L-amino acid-7-amido-4-methylcoumarins as substrates. This enzyme was able to remove only basic residues (Arg and Lys) from L-aminoacyl $\beta$-naphthylamides and was consequently called **aminopeptidase B**. The activity was found to be enhanced in the presence of physiological concentrations of chloride (Hopsu *et al*., 1966a). Moreover, aminopeptidase B had the capacity to convert kallidin-10 (Lys-bradykinin) to bradykinin (Hopsu & Makinen, 1966), although this result was disputed (Freitas *et al*., 1979; Kawata *et al*., 1980; Söderling, 1983). This activity was unable to hydrolyze the Arg-Pro bond of bradykinin (Hopsu *et al*., 1966a). A preliminary purification of this enzyme from rat liver gave rise to a 95.5 kDa protein (Hopsu *et al*., 1966b). In the 1967–1993 period, a number of conflicting reports appeared on aminopeptidase B. Indeed, it was reported that aminopeptidase B also exhibited an endopeptidase activity (Söderling & Mäkinen, 1983; Mantle *et al*., 1985; McDermott *et al*., 1988). The enzyme was considered as a $Zn^{2+}$-metallopeptidase (Suda *et al*., 1976; Ocain & Rich, 1987) or as a thiol-protease (Freitas *et al*., 1979; Söderling & Mäkinen, 1983). Although several reports showed an aminopeptidase B with a molecular mass around 70 000 (Mantle *et al*., 1985; Ishiura *et al*., 1987; Flores *et al*., 1993; Belhacène *et al*., 1993; Yamada *et al*., 1994), others described proteins with masses ranging from 43 000 (Freitas *et al*., 1979) to 105 000 (Söderling & Mäkinen, 1983). Finally, aminopeptidase B, also successively called **arylamidase II**, **arginine aminopeptidase**, **Cl⁻-activated arginine aminopeptidase**, **cytosol aminopeptidase IV** and then classified as EC. 3.4.11.6, was at one time suspected to be possibly identical to leukotriene $A_4$ (LTA$_4$) hydrolase (Chapter 347) and/or bleomycin hydrolase (Chapter 215).

However, as discussed below, aminopeptidase B is clearly distinct from LTA$_4$ hydrolase, bleomycin hydrolase and puromycin-sensitive aminopeptidase (Chapter 343).

## Activity and Specificity

Aminopeptidase B removes exclusively basic amino acid residues from L-aminoacyl-NHNap substrates (Arg: $K_m = 20$ mM, $V = 1$ nmol min$^{-1}$; Lys: $K_m = 36$ mM, $V = 0.7$ nmol min$^{-1}$) and from the N-terminus of various peptides like kallidin 10 (Hopsu *et al*., 1966a), [Arg$^0$]-Met$^5$-enkephalin (Gainer *et al*., 1984; Cadel *et al*., 1995: $K_m = 125$ mM, $V = 110$ pmol min$^{-1}$), [Arg$^0$]-Leu$^5$-enkephalin (Cadel *et al*., 1995: $K_m = 111$ mM, $V = 48$ pmol min$^{-1}$), [Arg$^{-1}$, Lys$^0$]-somatostatin-14 (Gluschankof *et al*., 1987; Gomez *et al*., 1988; Cadel *et al*., 1995: $K_m = 5$ mM, $V = 4.16$ pmol min$^{-1}$), [Arg$^0$]-neurokinin A (Cadel *et al*., 1995: $K_m = 58$ mM, $V = 25$ pmol min$^{-1}$), [Arg$^0$]-$\alpha$-atrial natriuretic factor(1–20) (Cadel *et al*., 1995: $K_m = 14$ mM, $V = 3.6$ pmol min$^{-1}$) and thymopentin (Belhacène *et al*., 1993). It was unable to cleave when a proline is adjacent (P1' position) to this basic residue (Hopsu *et al*., 1966a; Gainer *et al*., 1984; Cadel *et al*., 1995). Moreover, aminopeptidase B was unable to cleave non-basic amino acids from the N-terminus of various peptides or from L-aminoacyl $\beta$-naphthylamides.

Assays of the enzyme are most conveniently performed using the chromogenic L-Arg⊣NHNap substrate (Cadel *et al*., 1995). The activity towards L-Arg-NHNap is enhanced by chloride anions in the 150 mM range (Hopsu *et al*., 1966b; Cadel *et al*., 1995) and this effect is variable depending on the peptide (Cadel *et al*., 1995). The activity is functional over a broad range of pH and shows an optimum around pH 7 (Yamada *et al*., 1994; Cadel *et al*., 1995). The apparent pI of the protein is around 4.9 (Cadel *et al*., 1995).

The metallopeptidase character of the enzyme was suggested by significant inhibition in the presence of millimolar concentrations of EDTA, EGTA and 1,10-phenanthroline (Hopsu *et al*., 1966b; Cadel *et al*., 1995). This was confirmed by the ability of low $Zn^{2+}$ concentrations to reactivate aminopeptidase B after inhibition by 1,10-phenanthroline (Cadel *et al*., 1995). Moreover, aminopeptidase B was sensitive to cysteinyl proteinase inhibitors such as *p*-(chloromercuri)-benzenesulfonic acid and DTT, suggesting either a direct, or an indirect, involvement of thiol group(s) in the enzymatic reaction (Hopsu *et al.*, 1966b; Cadel *et al.*, 1995). Neither serine protease inhibitors (PMSF, aprotinin) nor an aspartic protease inhibitor (pepstatin) affected the enzyme activity (Cadel *et al.*, 1995). As expected, aminopeptidase B was sensitive to classical inhibitors of aminopeptidases such as bestatin and arphamenines A and B (Umezawa *et al.*, 1983; Ishiura *et al.*, 1987; Belhacène *et al.*, 1993; Cadel *et al.*, 1995). When tested with [Arg$^0$]-Met$^5$-enkephalin, the activity was also inhibited by amastatin (Cadel *et al.*, 1995).

## Structural Chemistry

Aminopeptidase B is a single-chain protein comprising 648 amino acid residues with a calculated molecular mass of 72 300 Da and a theoretical pI of 5.83. Aminopeptidase B primary structure exhibits an N-terminal putative signal peptide and the consensus $Zn^{2+}$-binding site (HEXXHX$_{18}$E). The protein contains at least one disulfide bond and several potential phosphorylation sites. Neither hydrophobic transmembrane domains nor *N*-glycosylation sites could be predicted from the amino acid sequence. Moreover, aminopeptidase B is not *O*-glycosylated as demonstrated using endo-$\alpha$-*N*-acetylgalactosaminidase. Secondary structure predictions using the Garnier and Robson algorithm indicated that aminopeptidase B might be a $\beta/\alpha$ protein. In the M1 family, the most closely related protein to aminopeptidase B is LTA$_4$ hydrolase, which exhibits 33% identity and 48% similarity. In LTA$_4$ hydrolase (Chapter 347), two tyrosine residues (Tyr379 and Tyr384) were implicated in the suicide inactivation of the enzyme and in catalysis, respectively. Indeed, the first Tyr covalently binds the LTA$_4$ substrate to inactivate the enzyme (Mueller *et al.*, 1996) and the latter is necessary to the peptidase activity as a proton donor in a general base mechanism (Blomster *et al.*, 1995). The conservation of these two residues in aminopeptidase B further supports the evolutionary affinities between these two enzymes. The gene for aminopeptidase B, *Ap-B*, has been localized to the long arm of human chromosome 1 band q32 (Aurich-Costa *et al.*, 1998).

## Preparation

Since quantitative tissue distribution of the activity indicated that the rat testis contains significant amounts of aminopeptidase B activity, it was used as the source of enzyme. Aminopeptidase B was purified from this organ in four main steps as described by Cadel *et al.* (1995). A 1000-fold purification was obtained after the last chromatofocusing step. This factor is probably significantly underestimated since contaminating proteases interfered with the aminopeptidase B assay in the early stages of purification and because a partial loss of enzyme activity was observed after the final purification step.

## Biological Aspects

Aminopeptidase B is present in several rat tissues (cortex, epididymis, heart, kidney, large intestine, liver, lung, muscle, pancreas, small intestine, spleen) (Foulon *et al.*, 1996), in cattle pituitary secretory vesicles (Gainer *et al.*, 1984) and in various cell lines including BE(2)M17 (human neuroblastoma cell line) (Draoui *et al.*, 1997), COS7 (fibroblast-like cell line from kidney of African green monkey) (Foulon *et al.*, 1997), PC12 (rat adrenal pheochromocytoma cell line) (Foulon *et al.*, 1997), Mrc5 (human lung derived cell line) (Foulon *et al.*, 1997), GH3 (rat pituitary tumor-derived cell line) (Foulon *et al.*, 1997), W93 (primary culture of rat pituitary) (Foulon *et al.*, 1997), Jurkat (human lymphoma cell line) (Belhacène *et al.*, 1993) and K562 cells (human chronic myeloid leukemia cells) (Yamada *et al.*, 1994), indicating that the enzyme is ubiquitous, although aminopeptidase B was differently expressed in all these samples.

In testis, aminopeptidase B is mainly expressed in late spermatids during the maturation phase. Observations of seminiferous tubule sections showed that the enzyme is concentrated in the cytoplasm of late spermatids which gives rise to the residual bodies. The absence of diffuse aminopeptidase B specific labeling in the cytoplasm of late spermatids and in residual bodies led to the hypothesis of a link between the exopeptidase and a membrane-derived structure (Cadel *et al.*, 1995). Interestingly, confocal microscopy gives evidence for the presence of aminopeptidase B in interstitial areas of seminiferous tubules, suggesting that it might also be expressed in Leydig cells (Foulon *et al.*, 1997).

Electron microscopic examination of the subcellular localization of aminopeptidase B in germinal cells shows that it is present in the *trans*-Golgi network around the apical pole of the nucleus of round spermatids and that it is concentrated in the proacrosomic granule of these cells. Since the acrosome is generated from saccules of the Golgi apparatus, this confirmed the presence of aminopeptidase B in the secretory apparatus of spermatids during their transformation into spermatozoa.

Considered at first as cytosoluble, aminopeptidase B is also a secreted enzyme. Indeed the protein was found in the culture or incubation medium of dispersed cattle intermediate lobe pituitary cells (Castro *et al.*, 1989), Jurkat T lymphocytes (Belhacène *et al.*, 1993), germinal cells and PC12 cells. Moreover, the enzyme was found to be associated with the membrane of secretory vesicles from cattle pituitary (Gainer *et al.*, 1984), and partially on the membrane surface of both leukemic and normal T cells (Belhacène *et al.*, 1993) as well as on the membrane surface of PC12 cells. Interestingly, aminopeptidase B activity is upregulated during long-term activation of both normal and leukemic T lymphocytes by various stimuli (Belhacène *et al.*, 1993). In the case of PC12 cells, the labeling indicates a membrane association of aminopeptidase B which is observed for only one-fifth of the cells.

The close structural relationships between aminopeptidase B and LTA$_4$ hydrolase is illustrated by the ability of aminopeptidase B to hydrolyze, *in vitro*, LTA$_4$ into LTB$_4$ which is a lipid mediator of inflammation. At the present time, the physiological relevance of the *in vitro* bifunctionality of aminopeptidase B remains an open question. The broad pH dependence of the enzyme and its ubiquitous presence both argue in favor of its adaptability to various cellular subcompartments and its involvement in a broad spectrum of physiological phenomena possibly including inflammatory processes in which some potential peptide substrates of aminopeptidase B such as kallidin, enkephalins and somatostatin might play a central role. Moreover, aminopeptidase B might be associated with processing and regulatory processes at various levels of the membrane network of the producing cells. This might include either post-translational maturation in the *trans*-Golgi network or regulatory processes at the plasma membrane, or both.

## Distinguishing Features and Related Peptidases

Since LTA$_4$ hydrolase (Chapter 347) exhibits an aminopeptidase activity, it was also considered as an arginine aminopeptidase (Örning *et al.*, 1994). These data led to the hypothesis that LTA$_4$ hydrolase and aminopeptidase B might be identical. However, LTA$_4$ hydrolase aminopeptidase activity is not specific for basic residues and exhibits a wide cleavage specificity (Örning *et al.*, 1994) unlike aminopeptidase B, which is strictly selective for Arg and Lys. Moreover, it has been reported that bleomycin hydrolase (Chapter 215), which is involved in bleomycin inactivation, is also able to hydrolyze L-Lys-NHNap and L-Arg-NHNap and is consequently related to aminopeptidase B (Umezawa *et al.*, 1974). In fact, these two activities are due to distinct enzymes (Sebti & Lazo, 1987). Furthermore, bleomycin hydrolase is a cysteine aminopeptidase with an active site related to papain (Sebti *et al.*, 1989; Nishimura *et al.*, 1989; Enenkel & Wolf, 1993; Joshua-Tor *et al.*, 1995).

Finally, aminopeptidase B is also distinct from puromycin-sensitive aminopeptidase (Chapter 343) which cleaves a broad range of L-aminoacyl β-naphthylamides (Constam *et al.*, 1995). Nevertheless, it should be noted that during the purification of aminopeptidase B a second aminopeptidase, able to cleave L-Arg-NHNap and, notably, L-Arg-Arg-NHNap, was observed (Cadel *et al.*, 1995). The latter activity was very sensitive to puromycin (S. Cadel, unpublished results) and might be the puromycin-sensitive aminopeptidase.

## Further Reading

Particular attention should be given to the papers by Hopsu *et al.* (1966a,b), Belhacène *et al.* (1993) and Cadel *et al.* (1995).

## References

Aurich-Costa, J., Cadel, S., Gouzy, C., Foulon, T., Cherif, D. & Cohen, P. (1997) Assignment of the aminopeptidase B gene to human chromosome 1 band q32 by *in situ* hybridization. *Cytogenet. Cell Genet.* **79**, 143–144.

Belhacène, N., Mari, B., Rossi, B. & Auberger, P. (1993) Characterization and purification of T lymphocyte aminopeptidase B: a putative marker of T cell activation. *Eur. J. Immunol.* **23**, 1948–1955.

Blomster, M., Wetterholm, A., Mueller, M.J. & Haeggström, J.Z. (1995) Evidence for a catalytic role of tyrosine 383 in the peptidase reaction of leukotriene A$_4$ hydrolase. *Eur. J. Biochem.* **231**, 528–534.

Cadel, S., Pierotti, A.R., Foulon, T., Créminon, C., Barré, N., Segrétain, D. & Cohen, P. (1995) Aminopeptidase B in the rat testes: isolation, functional properties and cellular localization in the seminiferous tubules. *Mol. Cell. Endocrinol.* **110**, 149–160.

Castro, M.G., Birch, N.P. & Loh, Y.P. (1989) Regulated secretion of pro-opiomelanocortin converting enzyme and an aminopeptidase B-like enzyme from dispersed bovine intermediate lobe pituitary cells. *J. Neurochem.* **52**, 1620–1628.

Constam, D.B., Tobler, A.R., Rensing-Ehl, A., Kemler, I., Hersh, L.B. & Fontana, A. (1995) Puromycin-sensitive aminopeptidase. Sequence analysis, expression, and functional characterization. *J. Biol. Chem.* **270**, 26931–26939.

Draoui, M., Bellincampi, L., Hospital, V., Cadel, S., Foulon, T., Prat, A., Barré, N., Reichert, U., Melino, G. & Cohen, P. (1997) Expression and retinoic modulation of N-arginine dibasic convertase and an aminopeptidase-B in human neuroblastoma cell lines. *J. Neuro-oncol.* **31**, 99–106.

Enenkel, C. & Wolf, D.H. (1993) BHL1 codes for a yeast thiol aminopeptidase, the equivalent of mammalian bleomycin hydrolase. *J. Biol. Chem.* **268**, 7036–7043.

Flores, M., Aristoy, M.C. & Toldra, F. (1993) HPLC purification and characterization of porcine muscle aminopeptidase B. *Biochimie* **75**, 861–867.

Foulon, T., Cadel, S., Chesneau, V., Draoui, M., Prat, A. & Cohen, P. (1996) Two novel metallopeptidases with a specificity for basic residues. Functional properties, structure and cellular distribution. *Ann. N.Y. Acad. Sci.* **780**, 106–120.

Foulon, T., Cadel, S., Prat, A., Chesneau, V., Hospital, V., Segrétain, D. & Cohen, P. (1997) NRD convertase and aminopeptidase B: two putative processing metallopeptidases with a selectivity for basic residues. *Ann. Endocrinol.* **58**, 357–364.

Freitas, J.R., Guimaràes, J.A., Borges, D.R. & Prado, J.L. (1979) Two arylamidases from human liver and their kinin-converting activity. *Int. J. Biochem.* **10**, 81–89.

Gainer, H., Russell, J.T. & Loh, Y.P. (1984) An aminopeptidase activity in bovine pituitary secretory vesicles that cleaves the N-terminal arginine from β-lipotropin. *FEBS Lett.* **175**, 135–139.

Gluschankof, P., Gomez, S., Morel, A. & Cohen, P. (1987) Enzymes that process somatostatin precursors. A novel endoprotease that cleaves before the arginine-lysine doublet is involved in somatostatin-28 convertase activity of the rat brain cortex. *J. Biol. Chem.* **262**, 9515–9520.

Gomez, S., Gluschankof, P., Lepage, A. & Cohen, P. (1988) Relationship between endo- and exopeptidase in a processing enzyme system: activation of an endoprotease by the aminopeptidase B-like activity in somatostatin-28 convertase. *Proc. Natl Acad. Sci. USA* **85**, 5468–5472.

Hopsu, V.K. & Mäkinen, K.K. (1966) Formation of bradykinin from kallidin-10 by aminopeptidase B. *Nature* **212**, 1271–1272.

Hopsu, V.K., Kantonen, U.M. & Glenner, G.G. (1964) A peptidase from rat tissues selectively hydrolysing N-terminal arginine and lysine residues. *Life Sci.* **3**, 1449–1453.

Hopsu, V.K., Mäkinen, K.K. & Glenner, G.G. (1966a) Purification of a mammalian peptidase selective for N-terminal arginine and lysine residues: aminopeptidase B. *Arch. Biochem. Biophys.* **114**, 557–566.

Hopsu, V.K., Mäkinen, K.K. & Glenner, G.G. (1966b) Characterization of aminopeptidase B: substrate specificity and affector studies. *Arch. Biochem. Biophys.* **114**, 567–566.

Ishiura, S., Yamamoto, T., Yamamoto, M., Nojima, M., Aoyagi, T. & Sugita, H. (1987) Human skeletal muscle contains two major aminopeptidases: an anion-activated aminopeptidase B and an aminopeptidase M-like enzyme. *J. Biochem.* **102**, 1023–1031.

Joshua-Tor, L., Xu, H.E., Johnston, S.A. & Rees, D. (1995) Crystal structure of a conserved protease that binds DNA: the bleomycin hydrolase, Gal6. *Science* **269**, 945–950.

Kawata, S., Takayama, S., Ninomiya, K. & Makisumi, S. (1980) Porcine liver aminopeptidase B. *J. Biochem.* **88**, 1601–1605.

Mantle, D., Lauffart, B., McDermott, J.R., Kidd, A.M. & Pennington, R.J.T. (1985) Purification and characterization of two Cl⁻-activated aminopeptidases hydrolysing basic termini from human skelettal muscle. *Eur. J. Biochem.* **147**, 307–312.

McDermott, J.R., Mantle, D., Lauffart, B., Gibson, A.M. & Biggins, J.A. (1988) Purification and characterization of two soluble Cl⁻ activated arginyl aminopeptidases from human brain and their endopeptidase action on neuropeptides. *J. Neurochem.* **50**, 177–182.

Mueller, M.J., Blomster, M., Oppermann, U.C.T., Jörnvall, H., Samuelsson, B. & Haeggström, J.Z. (1996) Leukotriene A4 hydrolase: protection from mechanism-based inactivation by mutation of tyrosine-378. *Proc. Natl Acad. Sci. USA* **93**, 5931–5935.

Nishimura, C., Suzuki, H., Tanaka, N. & Yamaguchi, H. (1989) Bleomycin hydrolase is a unique thiol aminopeptidase. *Biochem. Biophys. Res Commun.* **163**, 788–796.

Ocain, T.D. & Rich, D.H. (1987) L-lysine thiol: a subnanomolar inhibitor of aminopeptidase B. *Biochem. Biophys. Res. Commun.* **145**, 1038–1042.

Örning, L., Gierse, J.K. & Fitzpatrick, F.A. (1994) The bifunctional enzyme leukotriene-A4 hydrolase is an arginine aminopeptidase of high efficiency and specificity. *J. Biol. Chem.* **269**, 11269–11273.

Sebti, S.M. & Lazo, J.S. (1987) Separation of the protective enzyme bleomycin hydrolase from rabbit pulmonary aminopeptidases. *Biochemistry* **26**, 432–437.

Sebti, S.M., Mignami, J.E., Jani, J.P., Srimatkandada, S. & Lazo, J.S. (1989) Bleomycin hydrolase: molecular cloning, sequencing and biochemical studies reveal membership in the cysteine protease family. *Biochemistry* **28**, 6544–6548.

Söderling, E. (1983) Substrates specificities of Cl⁻-activated arginine aminopeptidase from human and rat origin. *Arch. Biochem. Biophys.* **220**, 1–10.

Söderling, E. & Mäkinen, K.K. (1983) Modification of the Cl⁻-activated arginine aminopeptidase from rat liver and human erythrocytes: a comparative study. *Arch. Biochem. Biophys.* **220**, 11–21.

Suda, H., Aoyagi, T., Takeuchi, T. & Umezawa, H. (1976) Inhibition of aminopeptidase B and leucine aminopeptidase by bestatin and its stereoisomer. *Arch. Biochem. Biophys.* **177**, 196–200.

Umezawa, H., Hori, S., Tsutomu, S., Toshioka, T. & Takeuchi, T. (1974) A bleomycin-inactivating enzyme in mouse liver. *J. Antibiot.* **27**, 419–424.

Umezawa, H., Aoyagi, T., Ohuchi, S., Okuyama, A., Suda, H., Takita, T., Hamada, M. & Takeuchi, T. (1983) Arphamenine A and B, new inhibitors of aminopeptidase B, produced by bacteria. *J. Antibiot.* **36**, 1572–1575.

Yamada, M., Sukenaga, Y., Fujii, H., Abe, F. & Takeuchi, T. (1994) Purification and characterization of a ubenimex (Bestatin)-sensitive aminopeptidase B-like enzyme from K562 human chronic myeloid leukemia cells. *FEBS Lett.* **342**, 53–56.

*Thierry Foulon*
*Laboratoire de Biochimie des Signaux Régulateurs Cellulaires et Moléculaires,*
*Unité de Recherche Associée au Centre National de la Recherche Scientifique 1682,*
*Université Pierre et Marie Curie,*
*96 Boulevard Raspail,*
*F-75006 Paris, France*
*Email: foulon@infobiogen.fr*

*Sandrine Cadel*
*Laboratoire de Biochimie des Signaux Régulateurs Cellulaires et Moléculaires,*
*Unité de Recherche Associée au Centre National de la Recherche Scientifique 1682,*
*Université Pierre et Marie Curie,*
*96 Boulevard Raspail,*
*F-75006 Paris, France*
*Email: cadel@infobiogen.fr*

*Paul Cohen*
*Laboratoire de Biochimie des Signaux Régulateurs Cellulaires et Moléculaires,*
*Unité de Recherche Associée au Centre National de la Recherche Scientifique 1682,*
*Université Pierre et Marie Curie,*
*96 Boulevard Raspail,*
*F-75006 Paris, France*
*Email: pcohen@infobiogen.fr*

# 349. *Aminopeptidase Ey*

## Databanks

*Peptidase classification: clan MA, family M1, MEROPS ID: M01.016*
*NC-IUBMB enzyme classification: EC 3.4.11.20*
*Databank codes:*

| Species | SwissProt | PIR | EMBL (cDNA) | EMBL (genomic) |
|---|---|---|---|---|
| *Gallus gallus* | – | – | D87992 | – |

## Name and History

The introduction of a sensitive fluorogenic substrate, Leu-NHMec, and a chromogenic substrate, Leu-NHPhNO$_2$, at pH 7.5 led to the discovery of **aminopeptidase Ey** in hen's egg yolk (Ichishima *et al*., 1989). The enzyme has a broad specificity for N-terminal amino acid residues at the P1 position (Tanaka & Ichishima, 1993b).

## Activity and Specificity

Aminopeptidase Ey has a broad specificity for amino acid residues at the P1 position of substrates (Tanaka & Ichishima, 1993b). The enzyme degrades a variety of peptides having various N-terminal amino acids: hydrophobic, basic and acidic amino acids including proline (Tanaka & Ichishima, 1993b). It rapidly degrades Leu-enkephalin [Tyr+Gly+Gly+Phe+Leu] to the C-terminus (Tanaka & Ichishima, 1993b). Tyrosine sulfate, a post-translationally modified amino acid, is easily liberated from cholecystokinin octapeptide: Asp+Tyr(SO$_3$H)+Met+Gly+Trp+Met-Asp-Phe-NH$_2$ (Tanaka & Ichishima, 1993b).

Aminopeptidase Ey hydrolyzes N-terminal Xaa+Pro bonds in a chicken brain pentapeptide (Leu+Pro-Leu-Arg-Phe-NH$_2$), substance P fragment 1–4 (Arg+Pro-Lys-Pro) and bradykinin fragment 1–5 (Arg+Pro-Pro-Gly-Phe), but does not hydrolyze substance P (Arg-Pro-Lys-Pro-Gln-Gln-Phe-NH$_2$) (Tanaka & Ichishima, 1993b).

The enzyme releases proline from Pro+Phe+Gly-Lys (Tanaka & Ichishima, 1993b), while it is unable to release proline from melanocyte-stimulating hormone release-inhibiting factor (Pro-Leu-Gly-NH$_2$) (MSH-release inhibiting factor), or from schisto FMRE-amide (Pro-Asp-Val-Asp-His-Val-Phe-Leu-Arg-NH$_2$) (Tanaka & Ichishima, 1993b).

The pH optimum of the enzyme for Leu-Leu-Tyr, Leu-NHMec and Leu-NHPhNO$_2$ hydrolysis is about pH 7.5 (Ichishima *et al*., 1989; Tanaka & Ichishima, 1993b). The addition of 1,10-phenanthroline or bestatin (Umezawa *et al*., 1976) at 2 mM or OF 4949-II, an inhibitor of aminopeptidase B from Ehrlich ascites carcinoma cells (Sano *et al*., 1987) at 0.1 mM, results in complete inactivation of the enzyme. It also loses activity upon dialysis against 1 mM EDTA at 4°C for 24 h. The inactive apoenzyme is instantaneously converted to active aminopeptidase Ey not only by the addition of Zn$^{2+}$ but also by the addition of Co$^{2+}$, Cu$^{2+}$, Cd$^{2+}$, Mn$^{2+}$, Ni$^{2+}$ and Ca$^{2+}$ using Leu-NHPhNO$_2$ or Leu-Leu-Tyr as substrates (Tanaka & Ichishima, 1993a). Each metal-reconstituted enzyme contained 1 mol of metal per mol of 150 kDa subunit.

Two values for the activation energy can be calculated from the Arrhenius plot of temperature dependence of Leu-NHPhNO$_2$ hydrolysis by aminopeptidase Ey (Tanaka & Ichishima, 1993a). The higher one of 63 kJ mol$^{-1}$ is observed at temperatures below 45°C, whereas at higher temperatures it diminishes to 37 kJ mol$^{-1}$. A similar result was seen in the Co$^{2+}$-reconstituted enzyme: two values were obtained, 97 kJ mol$^{-1}$ below 32°C and 9.3 kJ mol$^{-1}$ above 32°C.

## Structural Chemistry

Aminopeptidase Ey is a dimeric enzyme with homologous subunits, each having a relative molecular mass of 150 000 Da (Tanaka *et al*., 1993). It contains 1.0 mol of zinc per mol of each subunit (Tanaka *et al*., 1993). The enzyme molecule is seen as a dimer composed of two globular subunits by electron micrography at a magnification of 100 000× (Tanaka *et al*., 1993). The pI of the enzyme is about 2.8 as determined by isoelectric focusing. An asialo form of the enzyme, obtained by treatment with *Arthrobacter* sialidase, has a pI of 4.4 (Tanaka & Ichishima, 1993a). Our CD spectra studies indicate that aminopeptidase Ey requires $Zn^{2+}$ not only for catalytic activity but also for stabilization of molecular conformation (Tanaka & Ichishima, 1993a). The $\alpha$ helix content of native aminopeptidase Ey was calculated as 14%, but the $\alpha$ helix content of apoenzyme drastically diminished to 6% (Tanaka & Ichishima, 1993a); the two forms had the same $\beta$ sheet contents of 68% and 70%, respectively (Tanaka & Ichishima, 1993a). The tertiary structure of aminopeptidase Ey can be assumed to be similar to those of other members of family M1, but none of these has been solved to date.

Analysis of the 3196 bp nucleotide sequence of the cDNA of the aminopeptidase Ey gene, *apdE*, reveals a single open reading frame coding for 967 amino acid residues (T. Midorikawa *et al*., 1998). The coding region of *apdE* occupies 2901 bp of the cDNA. The predicted amino acid sequence of the enzyme is 63.8%, identical to that of aminopeptidase N (Chapter 337) from rabbit (Yang *et al*., 1993). Two histidine residues, His386 and His390, and the Glu409 residue are assumed to correspond to zinc ligands, and Glu387 is also assumed to be an essential catalytic site in the homologous aminopeptidase N.

## Preparation

The highly purified aminopeptidase Ey, having a specific activity of 51 400 μkat kg$^{-1}$ of protein, gave a single band on PAGE (Ichishima *et al*., 1989) and SDS-PAGE (Tanaka *et al*., 1993). The specific activity increased about 13 000-fold higher than that of the initial plasma of hen's egg yolk (Ichishima *et al*., 1989). Recombinant aminopeptidase Ey has not been reported.

## Biological Aspects

Among the many suggested functions of aminopeptidases are protein maturation, terminal degradation of proteins, hormone level regulation, and cell-cycle control (Taylor, 1993). Altered aminopeptidase activity has been associated with a variety of conditions and pathologies including aging, cancers, cataracts, cystic fibrosis and leukemias. It is assumed that aminopeptidase Ey is important for digesting oligopeptides produced by catheptic proteinases in the egg yolk

because of its specificities: broad specificity for amino acid residues at the P1 position of substrate, similar specificity at the P1' position to that of aminopeptidase P (Chapter 479), and preference for smaller peptides of four or five amino acid residues. These specificity characteristics distinguish aminopeptidase Ey from other aminopeptidases (Tanaka & Ichishima, 1993b).

Aminopeptidase Ey was studied for its specificity against N-blocked peptides and hydrolyzed only *N*-formylmethionyl peptides (Tanaka & Ichishima, 1994). *N*-Formyl-Met┼Leu-Phe(fMLP) lost its chemotactic activity towards human neutrophils after incubation with aminopeptidase Ey (Tanaka & Ichishima, 1994). This suggests that the enzyme may offer a biodefense against the infectious microbial product *N*-formyl-peptide.

## Distinguishing Features

The glutamyl aminopeptidase (Chapter 339) from chicken egg white preferentially hydrolyzes bonds of $\alpha$-glutamyl residue at the N-terminal of synthetic substrates and peptides (Petrovic & Vitale, 1990). The glutamyl aminopeptidase is a dimer with an $M_r$ of 320 000 and pI of 4.2.

## References

Ichishima, E., Yamagata, Y., Chiba, H., Sawaguchi, K. & Tanaka, T. (1989) Soluble and bound forms of aminopeptidase in hen's egg yolk. *Agric. Biol. Chem.* **53**, 1867–1872.

Midorikawa, T., Abe, R., Yamagata, Y., Nakajima, T. & Ichisihima, E. (1998) Isolation and characterization of DNA encoding chicken egg yolk aminopeptidase Ey. *Comp. Biochem. Physiol. B. Biochem. Mol. Biol.* (in press).

Petrovič, S. & Vitale, L. (1990) Purification and properties of glutamyl aminopeptidase from chicken egg-white. *Comp. Biochem. Physiol.* **95B**, 589–595.

Sano, S., Ikai, K., Katayama, K., Takesako, K., Nakamura, T., Obayashi, A., Ezure, Y. & Enomoto, H. (1986) OF4949, new inhibitor of aminopeptidase B. II. Elucidation of structure. *J. Antibiot.* **39**, 1685–1696.

Tanaka, T. & Ichishima, E. (1993a) Molecular properties of aminopeptidase Ey as a zinc-metalloenzyme. *Int. J. Biochem.* **25**, 1681–1688.

Tanaka, T. & Ichishima, E. (1993b) Substrate specificity of aminopeptidase Ey from hen's (*Gallus domesticus*) egg yolk. *Comp. Biochem. Physiol.* **105B**, 105–110.

Tanaka, T. & Ichishima, E. (1994) Inactivation of chemotactic peptides by aminopeptidase Ey from hen's (*Gallus gallus domesticus*) egg yolk. *Comp. Biochem. Physiol.* **107B**, 533–538.

Tanaka, T., Oshida, K. & Ichishima, E. (1993) Electron microscopic analysis of dimeric form of aminopeptidase Ey from hen's egg yolk. *Agric. Biol. Chem.* **55**, 2179–2181.

Taylor, A. (1993) Aminopeptidases: towards a mechanism of action. *Trends Biochem. Sci.* **18**, 167–173.

Umezawa, H., Aoyagi, T., Suda, H., Hamada, M. & Takeuchi, T. (1976) Bestatin, an inhibitor of aminopeptidase B, produced by actinomycetes. *J. Antibiot.* **29**, 97–99.

Yang, X.F., Milheit, P.E., Gaudoux, F., Crine, P. & Boileau, G. (1993) Complete sequence of rabbit kidney aminopeptidase N and mRNA localization in rabbit kidney by *in situ* hybridization. *Biochem. Cell Biol.* **71**, 278–287.

***Eiji Ichishima***
*Laboratory of Molecular Enzymology,*
*Department of Applied Biological Chemistry,*
*Faculty of Agriculture, Tohoku University,*
*1–1, Tsutsumidori-Amamiyamachi,*
*Aoba-ku, Sendai 981, Japan*
*Email: ichisima@t.soka.ac.jp*

# 350. Introduction: other families in clan MA

*MEROPS ID: M2, M4, M5, M9, M13, M30, M36, M48*

Clan MA includes a number of families other than family M1 (Chapter 336). The databanks table for these is to be found in Chapter 335. Peptidases from families M2, M5, M9, M13, M30, M36 and M48 are all predicted to share the same tertiary fold as thermolysin from family M4. An alignment around the known catalytic residues and zinc ligands is shown in Alignment 350.1 (on CD-ROM).

**Family M4** is known only from bacteria. Thermolysin (Chapter 351) is active at temperatures up to 60°C, and thermostability has been attributed to the binding of four calcium ions (Matthews *et al.*, 1974). Thermolysin and its precursor lack cysteine residues, but this is not a characteristic of the family, because pseudolysin (Chapter 357) possesses two disulfide bridges (Thayer *et al.*, 1991). The structure of thermolysin is shown in Figure 350.1.

Alignment 350.1 shows that the essential residues in thermolysin are not completely conserved throughout the family. Asp226 may be replaced by Asn in vibriolysin (Chapter 355), which produces a situation analogous to the His159 and Asn175 interaction of papain and other members of clan CA. The spacing between Asp226 and His231 is shorter in *Vibrio cholerae* hemagglutinin protease (Chapter 356), pseudolysin and *Legionella* metalloendopeptidase (Chapter 358) (within an Asp-Xaa-His motif). These three enzymes are contained in the most divergent branch of the evolutionary tree (see Figure 350.2).

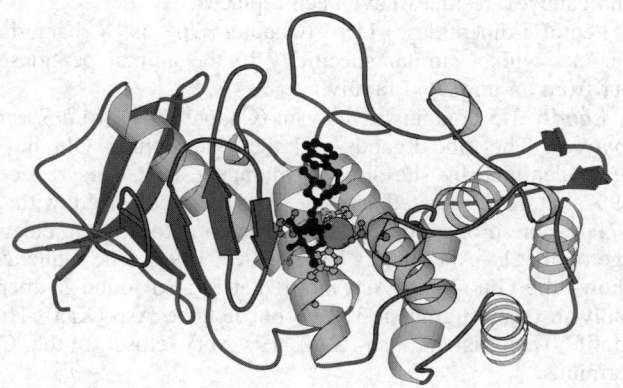

*Figure 350.1* Richardson diagram of the *Bacillus stearothermophilus* thermolysin/phosphoramidon complex. The image was prepared from the Brookhaven Protein Data Bank entry (1TLP) as described in the Introduction (p. xxv). The catalytic zinc ion is shown in CPK representation as a dark gray sphere. The zinc ligands are shown in ball-and-stick representation: His142, His146 and Glu166 (numbering as in Alignment 350.1). The catalytic Glu143 is also shown in ball-and-stick representation. Phosphoramidon is shown in black in ball-and-stick representation.

Members of family M4 are secreted enzymes, and are synthesized as proenzymes. In pseudolysin, which is activated in the periplasm, the propeptide has been shown to be an effective inhibitor (Kessler & Safrin, 1994). As yet, there are no crystal structures of the proenzymes. Most members of the family prefer a substrate with an aromatic residue in P1'. Pseudolysin, which causes tissue damage by degrading collagens, elastin and fibronectin, has a wide active-site cleft (see Figure 350.3), and analysis of the tertiary structures has led to the conclusion that the enzymes undergo hinge-bending motion during catalysis (Holland *et al.*, 1992). Figure 350.3 shows all residues known to be involved in catalysis in ball-and-stick representation, including our prediction that Asp226 orientates the imidazolium ring of His231.

**Family M2** contains the animal peptidyl-dipeptidase A (Chapter 359), which removes C-terminal dipeptides. Its biological function is to process angiotensin I and the peptidase is also known as angiotensin I-converting enzyme. The mammalian enzyme exists in two forms, both derived from the same gene by alternative modes of transcription. The form from lung epithelium contains two homologous peptidase units that have arisen by gene duplication. The structure of the gene reflects this duplication, and the number and size of introns, as well as the phase of the intron–exon junctions, are conserved. As can be seen from the evolutionary tree (Figure 350.4), this represents a recent duplication event, about 415 million years ago, and contrasts with the ancient duplication events seen in the chymotrypsin, pepsin and methionyl aminopeptidase families in which no similarities in intron number, size or position remain between the two halves of the gene. The testis-specific form of the enzyme contains only the C-terminal peptidase unit. There is evidence that the gene promoter was also copied during the duplication event, so that a second promoter exists within intron 12. It is this promoter that is responsible for the testicular transcripts. Because the initiating ATG sequence is within intron 12, the first 72 N-terminal residues of the testicular form are derived from the intron, and show no sequence relationship to the lung enzyme. Intron 12 also includes elements responsive to cAMP and steroid hormones that account for the selective expression in the testis (Hubert *et al.*, 1991).

Because the duplication event occurred so recently in vertebrate evolution, it is no surprise that invertebrate peptidyl-dipeptidase A (Chapter 360) contains only one peptidase unit.

Both peptidase units of mammalian peptidyl-dipeptidase A are catalytically active, but with different kinetic constants. Site-directed mutagenesis has been used to identify residues important for catalysis. The third zinc ligand has been identified as Glu166, and Asp170 is also essential (Williams *et al.*, 1994). There is also an Asp-$(Xaa)_5$-His motif

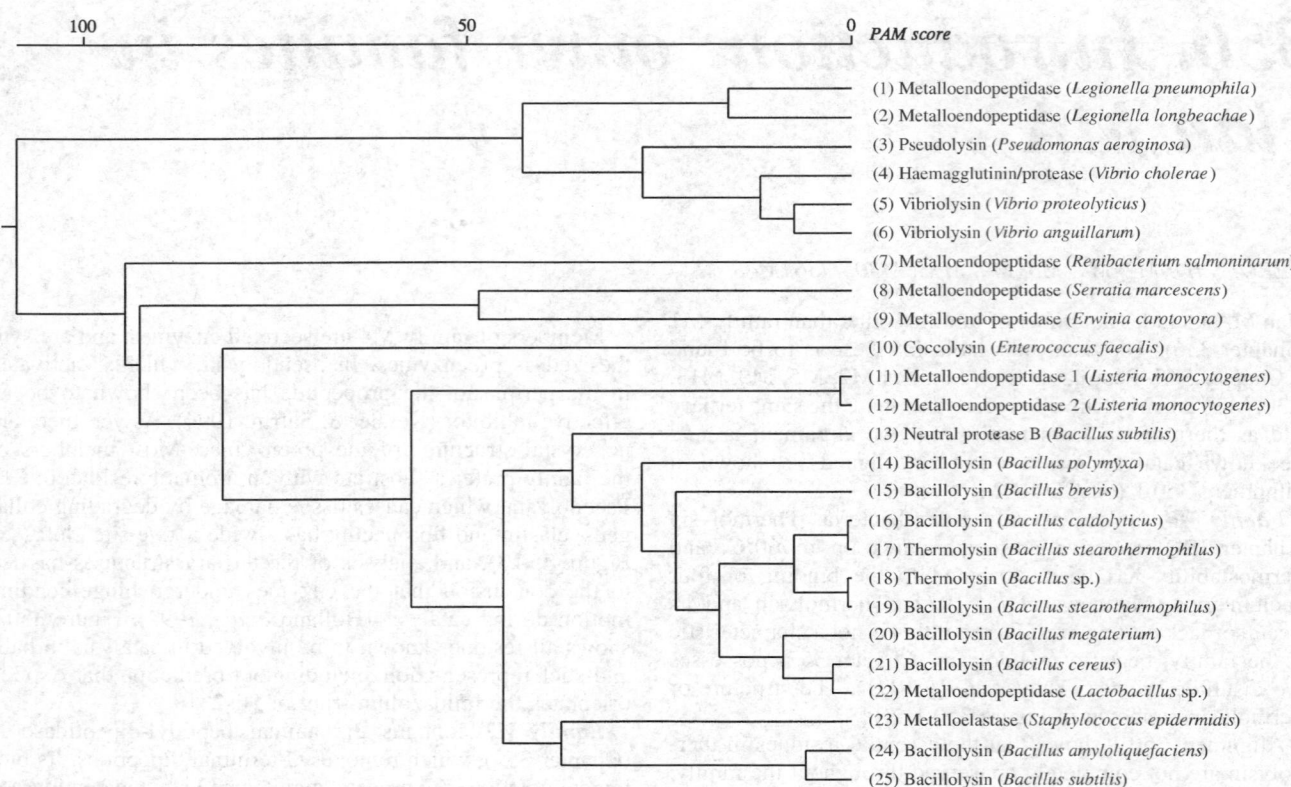

*Figure 350.2* Evolutionary tree for family M4. The tree was prepared as described in the Introduction (p. xxv).

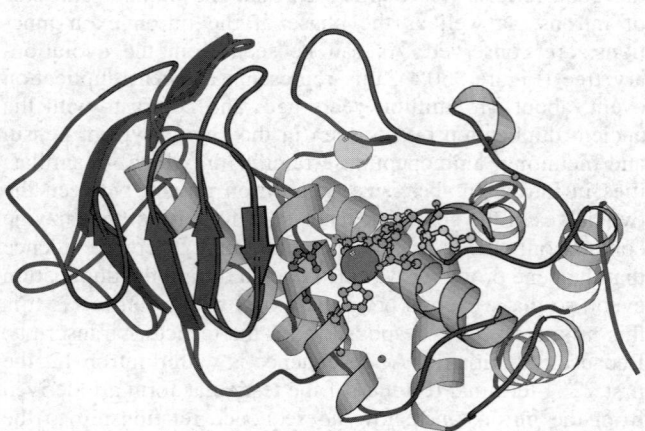

*Figure 350.3* Richardson diagram of pseudolysin. The image was prepared from the Brookhaven Protein Data Bank entry (1EZM) as described in the Introduction (p. xxv). The catalytic zinc ion and the structural calcium ion are shown as dark gray spheres in CPK representation. The zinc ligands are shown in ball-and-stick representation: His142, His146 and Glu166 (numbering as in Alignment 350.1). Other catalytic residues are shown in ball-and-stick representation: Glu143, Asp170, Asp226 and His231.

which may be analogous to the Asp-(Xaa)$_4$-His motif in thermolysin.

The C42D8.5 protein from *Caenorhabditis elegans* (included in EMBL: U56966) is a homolog of peptidyl-dipeptidase A, but is probably not a peptidase because all the zinc ligands and catalytic residues have been replaced.

Peptidyl-dipeptidase Dcp (Chapter 376) is a bacterial enzyme with a similar specificity to the animal peptidase, but from an unrelated family.

**Family M5** contains mycolysin (Chapter 361) from *Streptomyces*. The zinc ligands and the electrophilic Glu have been identified by site-directed mutagenesis (Chang & Lee, 1992). It is a secreted enzyme, and is synthesized with a 171 residue propeptide. Activation is an autocatalytic cleavage of an Ala⊦Ala bond (Chang & Lee, 1992). The sequence shows the Glu-(Xaa)$_3$-Asp motif that is also found in thermolysin (see Alignment 350.1), but not the Asp-(Xaa)$_4$-His motif. There is an Asn-(Xaa)$_4$-His very close to the C-terminus.

**Family M9** contains bacterial collagenases from *Vibrio* and *Clostridium*. There are a number of collagenolytic bacterial endopeptidases, including the metalloendopeptidase cytophagalysin (Chapter 518) from *Cytophaga* in family M43 and a collagenase of unknown catalytic type from *Porphyromonas* in family U32. The *Clostridium* collagenase (Chapter 368) is only distantly related to the *Vibrio* colla-

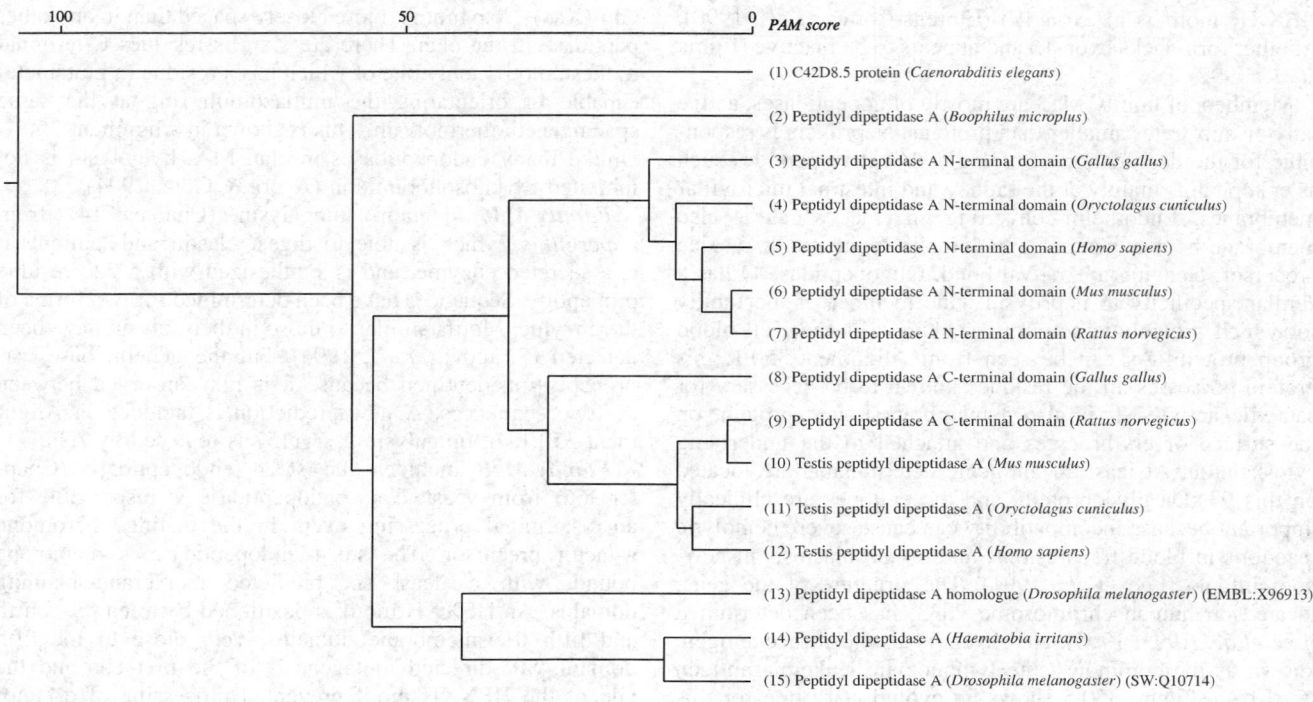

*Figure 350.4*   Evolutionary tree for family M2. The tree was prepared as described in the Introduction (p. xxv).

genase (Chapter 367), and had previously been considered to be the sole member of the now defunct family M31. Both the *Vibrio* and *Clostridium* collagenases are secreted enzymes, and are synthesized as precursors. Both possess the C-terminal secretion domain found in several peptidases from *Vibrio* and its relatives, including *Achromobacter* lysyl endopeptidase (Chapter 85), a *Xanthomonas* extracellular endopeptidase from family S8, vibriolysin (Chapter 355) and *Vibrio* aminopeptidase (Chapter 491). The C-terminus of *Clostridium* collagenase has two 90 residue repeats that are homologous to fragments of two hypothetical open reading frames sequenced in conjunction with the genes for phospholipase C from *Bacillus thuringiensis* (EMBL: X12952) and *B. cereus* (EMBL: M30809).

The zinc ligands have not been biochemically determined for any member of family M9, and Alignment 350.1 shows our prediction for the zinc ligands and catalytic residues. Two of the zinc ligands occur in the HEXXH motif, while the third we predict to be in a **Glu**-(Xaa)$_3$-Glu motif. Because the *Clostridium* and *Vibrio* collagenases are distantly related, sequence alignment outside these two motifs is difficult. There is apparently no conserved His C-terminal to the conserved motifs, and we are unable to predict a residue that could act as the general base.

**Family M13** contains the animal peptidases neprilysin (Chapter 362), endothelin-converting enzymes I (Chapter 363) and II (Chapter 364), and the bacterial oligopeptidase O (Chapter 365). The zinc ligands have been experimentally determined for neprilysin (Le Moual *et al.*, 1993), as have

two other residues essential for catalysis: the His that acts as the general base (Bateman & Hersh, 1987), and an Asp that probably orientates the imidazolium ring of the first His zinc ligand (Le Moual *et al.*, 1994). These residues are shown in Alignment 350.1. In addition, we suggest that the Asp that coordinates the catalytic His corresponds to the more closely spaced Asp226 of pseudolysin. In the homolog from *Haemonchus contortus*, this residue is replaced by Asn. One further important residue, not shown on Alignment 350.1, is an Asn placed 41 residues N-terminally to the HEXXH motif (Asn542, numbered according to human neprilysin) that is important for substrate binding at the P2′ subsite (Dion *et al.*, 1995). Further evidence that neprilysin and thermolysin share a common fold comes from hydrophobic cluster analysis (Dion *et al.*, 1995). Like thermolysin, neprilysin and other members of family M13 are potently inhibited by phosphoramidon.

Neprilysin has a much larger molecule than thermolysin (90 kDa compared to 35 kDa), and is a membrane-bound protein. There is a short cytoplasmic tail at the N-terminus (27 residues), followed by a transmembrane domain (23 residues), and the remainder of the molecule, including the metallopeptidase unit, is extracellular. The cytoplasmic domain includes a conformationally restrained octapeptide that has been proposed to be a 'stop transfer sequence' preventing proteolysis and secretion (Malfroy *et al.*, 1987). There are a number of alternative spliced forms of neprilysin mRNA, including one lacking exons 5–18, thus retaining only the membrane-spanning domain (exon 3) and the peptidase domain (the

HEXXH motif is in exon 19) (Llorens-Cortes *et al*., 1990). Another form lacks exon 16 and appears to be inactive (Iijima *et al*., 1992).

Members of family M13 are mostly oligopeptidases, active only on substrates smaller than proteins. Neprilysin is responsible for the degradation of biologically active peptides such as enkephalins mainly at the kidney and intestinal microvillar membranes. Endothelin-converting enzyme, which is also membrane-bound, generates endothelin from its 38 residue precursor, cleaving a Trp+Val bond. Oligopeptidase O has a similar specificity to neprilysin. One member of the family for which proteolytic activity is unknown is the Kell blood group protein. As can be seen from Alignment 350.1, this protein possesses all the residues known to be necessary for catalytic activity. It is also a membrane-bound protein, on the surface of erythrocytes and attached to the underlying cytoskeleton. At least 24 antigenic determinants are located on this 93 kDa glycoprotein, and the epitopes are clinically important because incompatibility can cause severe hemolytic reactions in blood transfusions, and erythroblastosis in new-born infants (Lee *et al*., 1991). The structure of the gene, located on human chromosome 7q33, has been determined (Lee *et al*., 1993, 1995). Neprilysin is also a surface antigen, known as the common acute lymphocytic leukemia antigen (CALLA). Figure 350.5 shows an evolutionary tree for this family.

*Family M30* contains only hyicolysin (Chapter 369), an endopeptidase from *Staphylococcus hyicus* with thermolysin-like specificity. Like thermolysin, it is active at 55°C (Ayora & Götz, 1994). The sequence includes HEXXH and Glu-(Xaa)$_3$-Asp motifs, more closely spaced than in any other peptidase in the clan. There are six His residues C-terminal to these motifs, only one of which has a residue (a glutamate) capable of orientating the imidazolium ring at the same spacing as in thermolysin. This is shown in Alignment 350.1. Unlike many endopeptidases in clan MA, hyicolysin is not inhibited by phosphoramidon (Ayora & Götz, 1994).

*Family M36* contains fungalysin (Chapter 514) from *Aspergillus*, which is able to digest elastin and laminin. It is a secreted enzyme, and is synthesized with a 227 residue propeptide. Sequences have been determined for two forms of the enzyme. Motifs similar to those in thermolysin have been detected (Sirakova *et al*., 1994), but the general base was probably misidentified because it is not conserved between the two sequences. A new prediction is included in Alignment 350.1. In fungalysin 2, Tyr157 is replaced by Asn.

*Family M48* includes the ste24 endopeptidase (Chapter 366) from yeast. This endopeptidase is responsible for an N-terminal processing event in the mating pheromone **a**-factor precursor. The ste24 endopeptidase is membrane bound, with at least six predicted membrane-spanning domains. An HEXXH motif is positioned between the fourth and fifth transmembrane domains, very close to the fifth domain. Site-directed mutagenesis of the first His and the Glu of the HEXXH motif prevented processing of **a**-factor precursor (Fujimura-Kamada *et al*., 1997). There is a Glu-(Xaa)$_3$-Asp motif closely following the sixth transmembrane domain (see Alignment 350.1).

A potential endoplasmic reticulum retrieval signal is present at the C-terminus of ste24 endopeptidase, implying

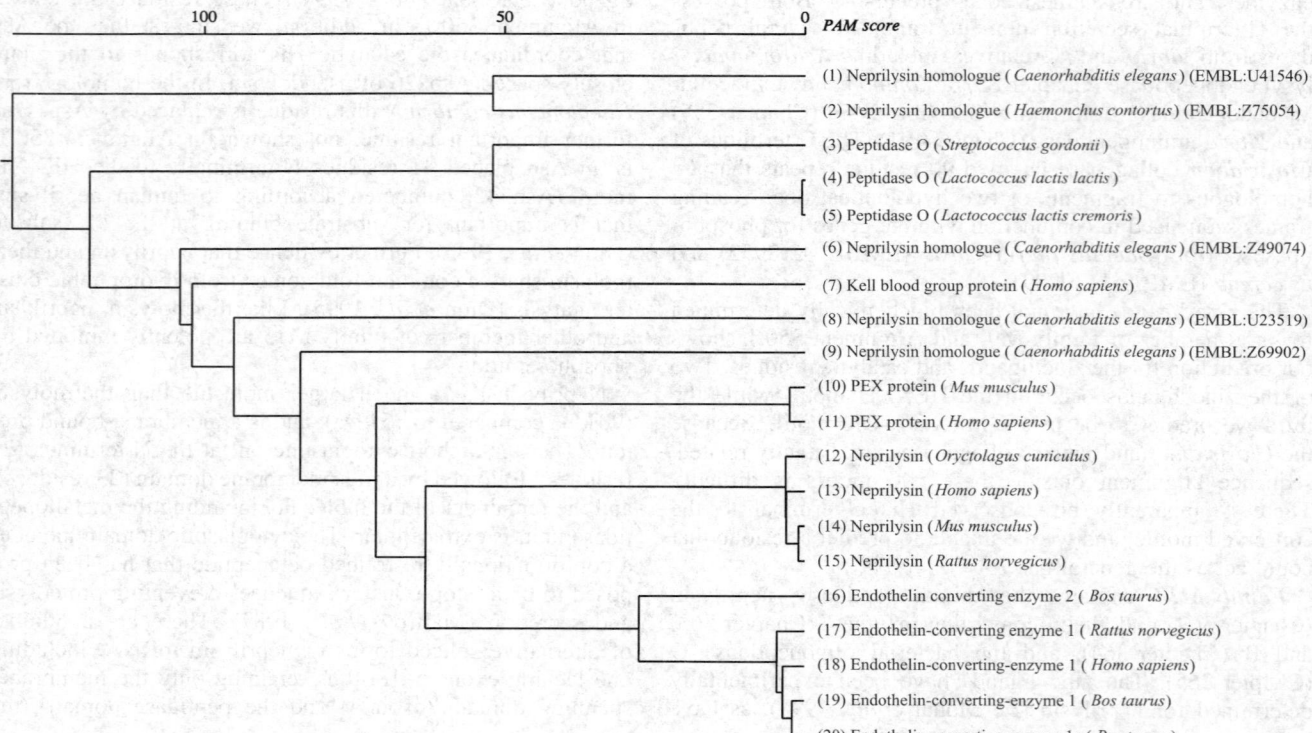

*Figure 350.5*   Evolutionary tree for family M13. The tree was prepared as described in the Introduction (p. xxv).

that the endopeptidase remains in the endoplasmic reticulum membrane, with the active site directed towards the lumen, where processing of the a-factor precursor occurs. A number of homologs of the ste24 endopeptidase are known, including heat-shock protein HtpX from *Escherichia coli* and *Salmonella typhimurium*, and several expressed sequence tag cDNA fragments from *Caenorhabditis elegans, Schistosoma mansoni, Arabidopsis thaliana* and human. The family is clearly widely distributed among organisms, and presumably there is a more general function for the enzymes. No catalytic activity has been demonstrated for any of the homologs. Both the HEXXH and Glu-(Xaa)$_3$-Asp motifs are conserved, as is a His near the C-terminus which may be the general base. However, there is no conserved residue near this that could correspond to Asp226 of thermolysin.

## References

Ayora, S. & Götz, F. (1994) Genetic and biochemical properties of an extracellular neutral metalloprotease from *Staphylococcus hyicus* subsp. *hyicus. Mol. Gen. Genet.* **242**, 421–430.

Bateman, R.C., Jr. & Hersh, L.B. (1987) Evidence for an essential histidine in neutral endopeptidase 24.11. *Biochemistry* **26**, 4237–4242.

Chang, P.-C. & Lee, Y.-H. (1992) Extracellular autoprocessing of a metalloprotease from *Streptomyces cacaoi. J. Biol. Chem.* **267**, 3952–3958.

Dion, N., Le Moual, H., Fournié-Zaluski, M.C., Roques, B.P., Crine, P. & Boileau, G. (1995) Evidence that Asn$^{542}$ of neprilysin (EC 3.4.24.11) is involved in binding of the P$_2'$ residue of substrates and inhibitors. *Biochem. J.* **311**, 623–627.

Fujimura-Kamada, K., Nouvet, F.J. & Michaelis, S. (1997) A novel membrane-associated metalloprotease, Ste24p, is required for the first step of NH$_2$-terminal processing of the yeast a-factor precursor. *J. Cell Biol.* **136**, 271–285.

Holland, D.R., Tronrud, D.E., Pley, H.W., Flaherty, K.M., Stark, W., Jansonius, J.N., McKay, D.B. & Matthews, B.W. (1992) Structural comparison suggests that thermolysin and related neutral proteases undergo hinge-bending motion during catalysis. *Biochemistry* **31**, 11310–11316.

Hubert, C., Houot, A.M., Corvol, P. & Soubrier, F. (1991) Structure of the angiotensin I-converting enzyme gene: Two alternate promoters correspond to evolutionary steps of a duplicated gene. *J. Biol. Chem.* **266**, 15377–15383.

Iijima, H., Gerard, N.P., Squassoni, C., Ewig, J., Face, D., Drazen, J.M., Kim, Y.A., Shriver, B., Hersh, L.B. & Gerard, C. (1992) Exon 16 del: a novel form of human neutral endopeptidase (CALLA). *Am. J. Physiol.* **262**, L725–729.

Kessler, E. & Safrin, M. (1994) The propeptide of *Pseudomonas aeruginosa* elastase acts as an elastase inhibitor. *J. Biol. Chem.* **269**, 22726–22731.

Le Moual, H., Roques, B.P., Crine, P. & Boileau, G. (1993) Substitution of potential metal-coordinating amino acid residues in the zinc-binding site of endopeptidase-24.11. *FEBS Lett.* **324**, 196–200.

Le Moual, H., Dion, N., Roques, B.P., Crine, P. & Boileau, G. (1994) Asp650 is crucial for catalytic activity of neutral endopeptidase 24–11. *Eur. J. Biochem.* **221**, 475–480.

Lee, S., Zambas, E.D., Marsh, W.L. & Redman, C.M. (1991) Molecular cloning and primary structure of Kell blood group protein. *Proc. Natl Acad. Sci. USA* **88**, 6353–6357.

Lee, S., Zambas, E.D., Marsh, W.L. & Redman, C.M. (1993) The human Kell blood group gene maps to chromosome 7q33 and its expression is restricted to erythroid cells. *Blood* **81**, 2804–2809.

Lee, S., Zambas, E., Green, E.D. & Redman, C. (1995) Organization of the gene encoding the human Kell blood group protein. *Blood* **85**, 1364–1370.

Llorens-Cortes, C., Giros, B. & Schwartz, J.-C. (1990) A novel potential metallopeptidase derived from the enkephalinase gene by alternative splicing. *J. Neurochem.* **55**, 2146–2148.

Malfroy, B., Schofield, P.R., Kuang, W.-J., Seeburg, P.H., Mason, A.J. & Henzel, W.J. (1987) Molecular cloning and amino acid sequence of rat enkephalinase. *Biochem. Biophys. Res. Commun.* **144**, 59–66.

Matthews, B.W., Weaver, L.H. & Kester, W.R. (1974) The conformation of thermolysin. *J. Biol. Chem.* **249**, 8030–8044.

Sirakova, T.D., Markaryan, A. & Kolattukudy, P.E. (1994) Molecular cloning and sequencing of the cDNA and gene for a novel elastinolytic metalloproteinase from *Aspergillus fumigatus* and its expression in *Escherichia coli. Infect. Immun.* **62**, 4208–4218.

Thayer, M.M., Flaherty, K.M. & McKay, D.B. (1991) Three-dimensional structure of the elastase of *Pseudomonas aeruginosa* at 1.5-Å resolution. *J. Biol. Chem.* **266**, 2864–2871.

Williams, T.A., Corvol, P. & Soubrier, F. (1994) Identification of two active site residues in human angiotensin I-converting enzyme. *J. Biol. Chem.* **269**, 29430–29434.

# 351. Thermolysin

## Databanks

*Peptidase classification: clan MA, family M4, MEROPS ID: M04.001*
*NC-IUBMB enzyme classification: EC 3.4.24.27*
*ATCC entries: 67500 (B. stearothermophilus npr)*

*Databank codes:*

| Species | Gene | SwissProt | PIR | EMBL (cDNA) | EMBL (genomic) |
|---------|------|-----------|-----|-------------|----------------|
| *Alicyclobacillus acidocaldarius* | – | – | JC4113 | U07824 | – |
| *Bacillus caldolyticus* | *npr* | P23384 | A42464 | M63575 | – |
| *Bacillus caldolyticus* | *npr* | – | – | U25629 | – |
| *Bacillus* sp. | | – | – | U25630 | – |
| *Bacillus stearothermophilus* | *nprS* | P43133 | B36706 | M21663 | – |
| | | | | M34237 | – |
| *Bacillus stearothermophilus* | *nprT* | P06874 | A24924 | M11446 | – |
| | | | A46564 | | |
| *Bacillus thermoproteolyticus* | *npr* | P00800 | A00993 | A20191 | – |
| | | | S41312 | X76986 | – |

Brookhaven Protein Data Bank three-dimensional structures

| Species | ID | Resolution | Notes |
|---------|-----|-----------|-------|
| *Bacillus thermoproteolyticus* | 1HYT | 1.7 | complex with benzylsuccinic acid |
| | 1LNA | 1.9 | complex with cobalt |
| | 1LNB | 1.8 | complex with iron |
| | 1LNC | 1.8 | complex with manganese |
| | 1LND | 1.7 | complex with zinc |
| | 1LNE | 1.7 | complex with cadmium |
| | 1LNF | 1.7 | |
| | 1THL | 1.7 | complex |
| | 1TLP | 2.3 | complex with phosphoramidon |
| | 1TMN | 1.9 | complex with $N$-(1-carboxy-3-phenylpropyl)-L-Leu-L-Trp |
| | 1TRL | | fragment 255–316 |
| | 2TMN | 1.6 | complex with $N$-phosphoryl-L-leucinamide |
| | 3TMN | 1.7 | complex with Val-Trp |
| | 4TLN | 2.3 | complex with L-Leu-hydroxylamine |
| | 4TMN | 1.7 | complex with Z-Phe-P-Leu-Ala |
| | 5TLN | 2.3 | complex with HONH-benzylmalonyl-L-Ala-Gly-NHPhNO$_2$ |
| | 5TMN | 1.6 | complex with Z-Gly-P-Leu-Leu |
| | 6TMN | 1.6 | complex with Z-Gly-P-(O)-Leu-Leu |
| | 7TLN | 2.3 | complex with $CH_2CO(N-OH)Leu-OCH_3$ |
| | 7TMN | | theoretical model; complex with Gly-TPH-Leu-Leu |
| | 8TLN | 1.6 | complex with Val-Lys |

## Name and History

**Thermolysin** is the name first given to an extracellular, 34.6 kDa metalloendopeptidase secreted by the gram-positive thermophilic bacterium *Bacillus thermoproteolyticus* (Endo, 1962). It was the first metalloendopeptidase to be crystallized and to have its structure solved (reviewed in Matthews, 1988).

There is some confusion over the naming of, and the relationships among, the enzymes in metalloproteinase clan MA, family M4. Many of the enzymes have trivial names and are treated as separate entries in this publication (Chapters 351–358). Particularly unclear is the relationship between thermolysin(s) and the enzyme(s) referred to as bacillolysin. Bacillolysin (Chapter 352) was first described as a neutral metalloendopeptidase produced by *Bacillus subtilis*.

Other extracellular metalloendopeptidases of *Bacillus* are variously referred to as thermolysins or bacillolysins, but the justification for assigning one name or the other is unclear. CLUSTALW-based alignment of the mature enzyme sequences yields a family that does not divide strongly into a bacillolysin group and a thermolysin group. Rather, the thermolysins are identifiable as a tightly clustered group of enzymes within the whole set. For the purpose of this chapter and the next, this group of enzymes will be referred to as 'thermolysins', and the other members as 'bacillolysins' (Figure 351.1). However, we suggest that this classification may need to be revisited, although the unevenness of kinetic or physicochemical data suggest that sequence data will be the only useable criterion. One solution may be to combine all of these enzymes under a single EC entry.

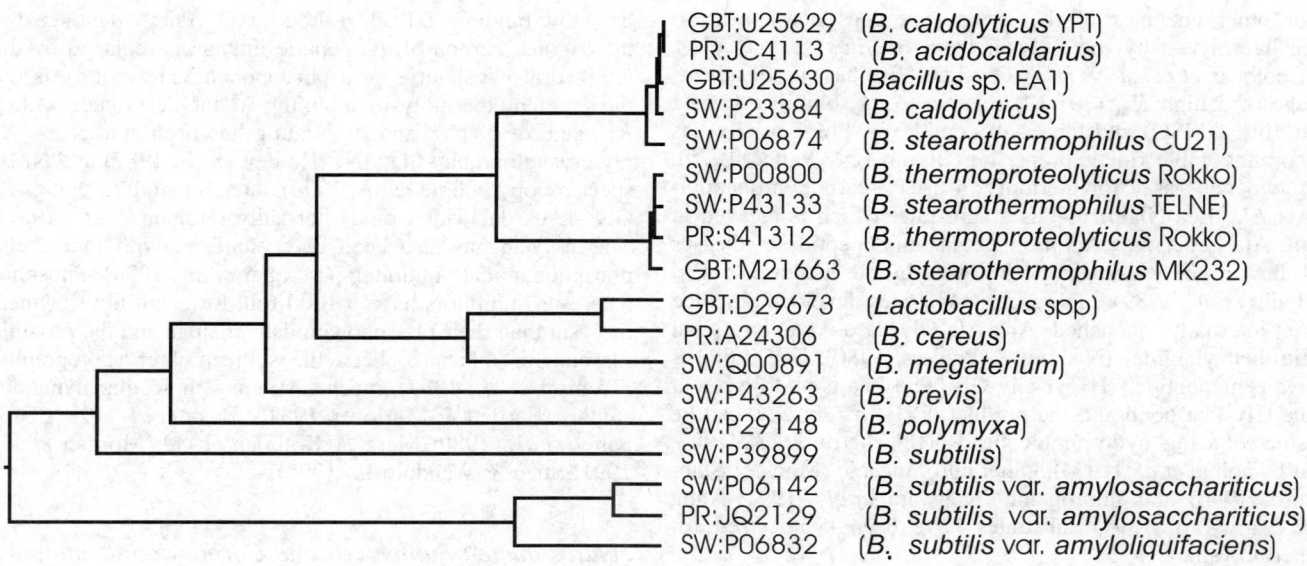

GBT:U25629   (*B. caldolyticus* YPT)
PR:JC4113    (*B. acidocaldarius*)
GBT:U25630   (*Bacillus* sp. EA1)
SW:P23384    (*B. caldolyticus*)
SW:P06874    (*B. stearothermophilus* CU21)
SW:P00800    (*B. thermoproteolyticus* Rokko)
SW:P43133    (*B. stearothermophilus* TELNE)
PR:S41312    (*B. thermoproteolyticus* Rokko)
GBT:M21663   (*B. stearothermophilus* MK232)
GBT:D29673   (*Lactobacillus* spp)
PR:A24306    (*B. cereus*)
SW:Q00891    (*B. megaterium*)
SW:P43263    (*B. brevis*)
SW:P29148    (*B. polymyxa*)
SW:P39899    (*B. subtilis*)
SW:P06142    (*B. subtilis* var. *amylosacchariticus*)
PR:JQ2129    (*B. subtilis* var. *amylosacchariticus*)
SW:P06832    (*B. subtilis* var. *amyloliquifaciens*)

*Figure 351.1*   Evolutionary tree for the thermolysin/bacillolysin group. The tree was produced from a CLUSTAL alignment (DNASTAR). There is no clear segregation of a bacillolysin and thermolysin family, although a group of enzymes known most commonly as 'thermolysins' are most tightly clustered. *Key to enzymes*: U25629, *Bacillus caldolyticus* neutral proteinase (npr) gene; JC4113, neutral protease from *Bacillus* sp.; U25630, *Bacillus* sp. neutral proteinase (*npr*) gene, complete CDS; P23384, bacillolysin precursor (gene *npr*) from *Bacillus caldolyticus*; P06874, thermolysin precursor, *Bacillus stearothermophilus*; S41312, thermolysin from *Bacillus thermoproteolyticus*; X76986, Thermolysin (gene *npr*) from *Bacillus thermoproteolyticus*; P43133, bacillolysin precursor from *Bacillus stearothermophilus* strain TELNE; D29673, hydrolase from *Lactobacillus* sp.; A24306, bacillolysin precursor from *Bacillus cereus*; Q00891, bacillolysin precursor from *Bacillus megaterium*; P43263, bacillolysin precursor from *Bacillus brevis*; P29148, bacillolysin precursor from *Bacillus polymyxa*; P39899, neutral protease precursor from *Bacillus subtilis*; P06142, bacillolysin precursor from *Bacillus subtilis* var. *amylosacchariticus* and *Bacillus mesentericus*; JQ2129, bacillolysin (EC 3.4.24.28) precursor – *Bacillus subtilis* var. *amylosacchariticus*; P06832, bacillolysin precursor from *Bacillus amyloliquefaciens*. PR, PIR; SW, SwissProt; GB, Genbank (translated).

Thermolysin is widely used in protein chemistry as a nonspecific protease to obtain sequence or conformational data (Kanoh *et al*., 1993; De Laureto *et al*., 1995). The enzyme can also act as a peptide synthetase (Wayne & Fruton, 1983; Jakubke & Konnecke, 1987; Clapes *et al*., 1995) and has been used for the production of a precursor of the artificial sweetener, aspartame (Ooshima *et al*., 1985; Murakami *et al*., 1996). Thermolysin has also been used to generate an active-site model for the design of inhibitors for mammalian zinc endopeptidases such as neprilysin (Roques *et al*., 1993) (Chapter 362).

## Activity and Specificity

The pH–activity profile of thermolysin follows a bell-shaped curve with maximal activity at or near pH 7.0 for both proteolysis (Feder & Schuck, 1970; Pangburn & Walsh, 1975; Holmquist & Vallee, 1976; Kunugi *et al*., 1982) and esterolysis (Holmquist & Vallee, 1976). The enzyme is inhibited by phosphate buffers (Feder, 1968) and $CaCl_2$ in the range 1–10 mM is usually included in the buffers to minimize autolysis. High concentrations of neutral salts activate the hydrolysis of certain thermolysin substrates (Holmquist & Vallee, 1976; Inouye, 1992; Yang *et al*., 1994; Inouye *et al*., 1996).

Thermolysin specificity has been investigated using a variety of different peptide and model peptide substrates (Morihara *et al*., 1968; Morihara & Oka, 1968; Morihara & Tsuzuki, 1970; Feder & Schuck, 1970; Morihara & Tsuzuki, 1971; Blumberg & Vallee, 1975; Pozsgay *et al*., 1986; Hersh & Morihara, 1986). The major specificity site of the enzyme is at S1', which accepts large hydrophobic residues. Thermolysin therefore preferentially cleaves peptides and proteins at the N-terminal side of Leu, Phe, Ile and Val, although hydrolysis of bonds with Met, His, Tyr, Ala, Asn, Ser, Thr, Gly, Lys, Glu or Asp at P1' has been observed (Heinrikson, 1977). Substrates can also interact with residues in the S1, S2 and S2' subsites, although these interactions are less important for specificity. A hydrophobic residue is preferred in the P1 position, Ala or Phe is preferred to Gly in P2 and in P2', the order of preference is Leu > Ala > Phe > Gly. Amino acids in the P3 and P3' positions can also affect activity, with Ala or Phe promoting catalysis as compared to Gly in both cases.

Thermolysin activity can be measured using nonspecific protease substrates such as casein (Fujii *et al*., 1983) but more specific substrates are commercially available, of which the furylacryloyl dipeptide FA-Gly⊦Leu-$NH_2$ or FAGLA, is the most commonly used. Hydrolysis of the Gly-Leu bond can be continuously monitored by measuring the decrease in $A_{345}$ (Feder, 1968). FA-Gly-Leu-$NH_2$ has also been used for other enzymes of the group, but does not appear to be hydrolyzed by thermolysin from *Bacillus* sp. var. EA1 (Coolbear *et al.*, 1991). FA-Gly-Leu-NH$_2$ has the disadvantage of a high $K_m$ (30 mM) relative to its solubility (2 mM in 10% DMSO) and the FA tripeptide FA-Phe-Leu-Gly has more favorable kinetic properties (Blumberg & Vallee, 1975). A two-step assay for thermolysin uses 3-carboxypropanoyl-Ala-Ala↓Leu-NHPhNO$_2$ as a substrate, which is cleaved at the Ala-Leu bond. Addition of an aminopeptidase releases 4-nitroaniline which can be followed by the increase in $A_{405}$ (Indig *et al.*, 1989). Fluorescent substrates developed include the internally quenched Abz-Ala-Gly↓Leu-Ala-Nba (Nba: nitrobenzylamide) (Nishino & Powers, 1980). The radioactive pentapeptide [$^3$H]Tyr-Gly-Gly↓Phe-Leu is hydrolyzed at the Gly-Phe bond and the product, [$^3$H]Tyr-Gly-Gly, can be isolated using hydrophobic beads (Benchetrit *et al.*, 1987; O'Donohue *et al.*, 1994). Other chromogenic (Mock & Stanford, 1996) and fluorogenic (Kajiwara *et al.*, 1991; Yang & Van Wart, 1994) substrates have been synthesized for thermolysin.

Thermolysin is reversibly inhibited by millimolar concentrations of zinc-chelating agents such as 1,10-phenanthroline (Holmquist & Vallee, 1974). Inhibition by EDTA is irreversible, as EDTA preferentially chelates the calcium ions, leading to autolysis (Fontana, 1988). The enzyme is also inhibited by high concentrations of zinc >10 μM (Holmquist & Vallee, 1974) which binds to an active-site histidine residue (Holland *et al.*, 1995). More specific thermolysin inhibitors are generally modified di- or tripeptide substrate analogs, with a hydrophobic residue to fit the S1' subsite and containing a strong zinc-binding agent such as a phosphoramidate, phosphonamidate, sulfhydryl, hydroxamate or carboxylate group. The development and synthesis of some of these molecules is discussed in the following references: Nishino & Powers (1978), Kam *et al.* (1979), Nishino & Powers (1979), Holmquist & Vallee (1979), Maycock *et al.* (1981), Bartlett & Marlowe (1983), Blumberg & Tauber (1983), Bartlett & Marlowe (1987a), Grobelny *et al.* (1992), Bohacek & McMartin (1994) and Morgan *et al.* (1994).

## Phosphorus-containing Inhibitors

The commercially available inhibitor phosphoramidon is a naturally occurring phosphoramidate from *Streptomyces tanashiensis* which inhibits thermolysin with a $K_i$ of 30 nM at neutral pH (Suda *et al.*, 1973). Phosphoramidon is a slow binding inhibitor (Kam *et al.*, 1979; Kitagishi & Hiromi, 1984) and its $K_i$ value is highly pH dependent, varying from 1.4 nM at pH 5.0 to 8.5 μM at pH 8.5 (Kitagishi & Hiromi, 1984). The structure of the thermolysin–phosphoramidon complex has been studied by both X-ray crystallographic (Tronrud *et al.*, 1986) and NMR spectroscopy (Gettins, 1988) and the unsubstituted phosphoramidate, *N*-phosphoryl Leu-amide ($K_i = 1.9$ μM) has also been cocrystallized with the enzyme (Tronrud *et al.*, 1986). Both molecules have been proposed to be transition state analogs. A series of phosphonamidate inhibitors have been designated transition state inhibitors due to the strong correlation between their $K_i$ values and the $k_{cat}/K_m$ values for the corresponding substrate analogs (Bartlett & Marlowe, 1983, 1987a). One of these,

the slow-binding Z-Phe$^{P*}$-L-Leu-L-Ala (Phe$^{P*}$ denotes that the trigonal carbon of the peptide linkage is replaced by the tetrahedral phosphorus of a phosphonamidate group), is the most potent thermolysin inhibitor so far described, with a $K_i$ value of 68 pM, and its binding has been studied by X-ray crystallography (4TMN) (Holden *et al.*, 1987) and NMR spectroscopy (Copié *et al.*, 1990) and compared to that of Z-Gly$^P$-Leu-Ala, which binds normally (Holden *et al.*, 1987). Thermolysin has also been co-crystallized with two cyclic phosphonamidate inhibitors (Morgan *et al.*, 1994). Phosphonate ester inhibitors have a 1000-fold lower affinity for thermolysin than their phosphonamidate analogs and the possible reasons for this have been the subject of crystallographic (Tronrud *et al.*, 1987), mechanistic and molecular dynamics studies (Bartlett & Marlowe, 1987b; Bash *et al.*, 1987; Grobelny *et al.*, 1989; Merz & Kollman, 1989; Morgan *et al.*, 1991; Shen & Wendoloski, 1995).

## Hydroxamate-, Sulfhydryl- and Carboxylate-containing Inhibitors

Several hydroxamate-containing thermolysin inhibitors have been synthesized and the enzyme has been co-crystallized with L-Leu-HNOH ($K_i = 190$ μM; Brookhaven PDB: 4TLN) and HONH-benzylmalonyl-L-Ala-Gly-NHPhNO$_2$ ($K_i = 0.43$ μM; Brookhaven PDB: 5TLN) (Holmes & Matthews, 1981). The binding of a hydroxamate inhibitor to thermolysin has also been the subject of a mechanistic study (Izquierdo-Martin & Stein, 1992). Thermolysin has been cocrystallized with three β-thiol inhibitors: (2-benzyl-3-mercaptopropanoyl)-L-alanylglycinamide ($K_i = 0.75$ μM) (Monzingo & Matthews, 1982) *N*-(*S*)-2-(mercaptomethyl)-1-oxo-3-phenylpropylglycine (thiorphan) ($K_i = 1.8$ μM) and ([(*R*)-1-(mercaptomethyl)-2-phenylethyl]amino)-3-oxopropanoic acid (retrothiorphan) ($K_i = 2.3$ μM) (Roderick *et al.*, 1989) and with the carboxylate-containing inhibitors *N*-(*S*)-(1-carboxy-3-phenylpropyl)-(*S*)-leucyl-(*S*)-tryptophan ($K_i = 50$ nM; Brookhaven PDB: 1TMN) (Monzingo & Matthews, 1984), *N*-[1-(2(*R*,*S*)-carboxy-4-phenylbutyl)-cyclopentylcarbonyl]-(*S*)-tryptophan (Holland *et al.*, 1994) and benzylsuccinic acid (Brookhaven PDB: 1HYT) (Hausrath & Matthews, 1994).

## Irreversible Inhibition

Thermolysin is irreversibly inhibited by the D-enantiomer of the active site-directed inhibitor *N*-chloroacetyl-*N*-hydroxyleucine methyl ester (Rasnick & Powers, 1978; Kim & Jin, 1996), which alkylates the active-site glutamate residue (Holmes *et al.*, 1983).

Molecular dynamics studies of the binding of inhibitors to thermolysin can be found in the following references: Ghosh & Edholm (1994), Giessner & Jacob (1989), van Aalten *et al.* (1995), Bohacek & McMartin (1994), Merz & Kollman (1989) and Wasserman & Hodge (1996).

## Structural Chemistry

The first crystallographic structure of thermolysin was reported in 1972 (Colman *et al.*, 1972) and the structure has been refined (Holmes & Matthews, 1982) and analyzed with bound inhibitors since then (Matthews, 1988). However, although the initial structures reported were thought to be those of the free enzyme, it is now believed that a dipeptide was present in the active site (Holland *et al*., 1992). The structure of the holoenzyme is thus not available. This has led to a revision of the previously held idea that thermolysin does not undergo a conformational change on substrate or inhibitor binding (Holland *et al*., 1992, 1995).

The mature enzyme form of thermolysin comprises 316 residues, forming a bilobal structure in which the N-terminal region is predominantly $\beta$-pleated sheet and the C-terminal domain is predominantly $\alpha$ helical (Figure 351.2). The catalytically essential zinc atom is located in a deep active-site cleft that lies between the two lobes. A central helical segment contains the His-Glu-Leu-Thr-His (HELTH) zinc-binding motif (residues 142–146) that is common to this group of enzymes and in which the two histidines are zinc ligands. A Glu166 residue C-terminal to the motif and a water molecule provide the third and fourth ligands for the zinc atom, which is thus bound in an approximate tetrahedral geometry (Figure 351.3). Activity can be restored to the zinc-free enzyme by stoichiometric addition of $Zn^{2+}$ (100%), $Co^{2+}$ (200%) and $Mn^{2+}$ (10%), or a high molar excess of $Fe^{2+}$ (60%) (Holmquist & Vallee, 1974) and the structural changes occurring on zinc replacement have been examined by X-ray crystallography (Holland *et al*., 1995). Thermolysin also binds four calcium atoms, Ca(1) and Ca(2) at a double binding site near the active-site cleft and Ca(3) and Ca(4) at exposed loops in the N- and C-terminal lobes respectively. The calcium ions have no catalytic role but protect the enzyme from autolysis (Roche & Voordouw, 1978; Fontana, 1988); they can also be replaced by a variety of other ions (Matthews & Weaver, 1974; Holland *et al*., 1995). Molecular modeling studies of the enzyme from *Bacillus stearothermophilus* CU21, which has 85% sequence identity with thermolysin, could be reconciled with a very similar structure (Vriend & Eijsink, 1993). Indeed, the alignment of sequences in the context of secondary structure reveals that all enzymes of this group are likely to have similar profiles (Figure 351.4).

## Mechanism of Action

The crystallographic studies on thermolysin, coupled with parallel studies on the zinc endopeptidase carboxypeptidase A (Chapter 451), have led to a general base mechanism of action being proposed for the zinc peptidase family, with the glutamate of the HELTH sequence polarizing the zinc-bound water molecule which subsequently attacks the scissile bond. This is more fully discussed by Hangauer *et al*. (1984), Matthews (1988) and Christianson & Lipscomb (1989). An alternative mechanism has recently been proposed in which an active-site histidine acts as the general base instead of the glutamate residue (Mock & Aksamawati, 1994; Mock & Stanford, 1996).

## Preparation

Commercial preparations of thermolysin can be repurified by crystallization (Matsubara, 1970) or by affinity

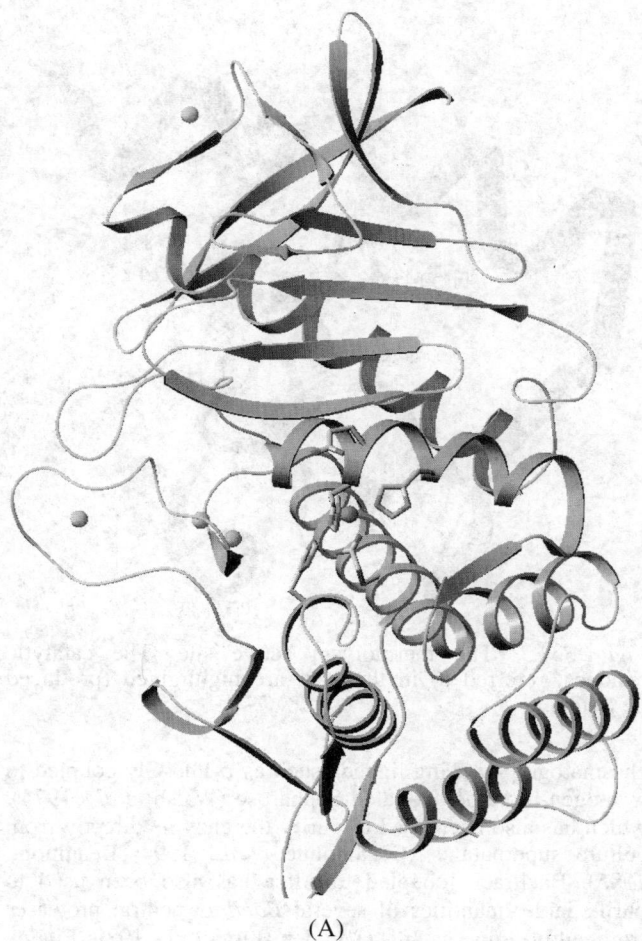

(A)

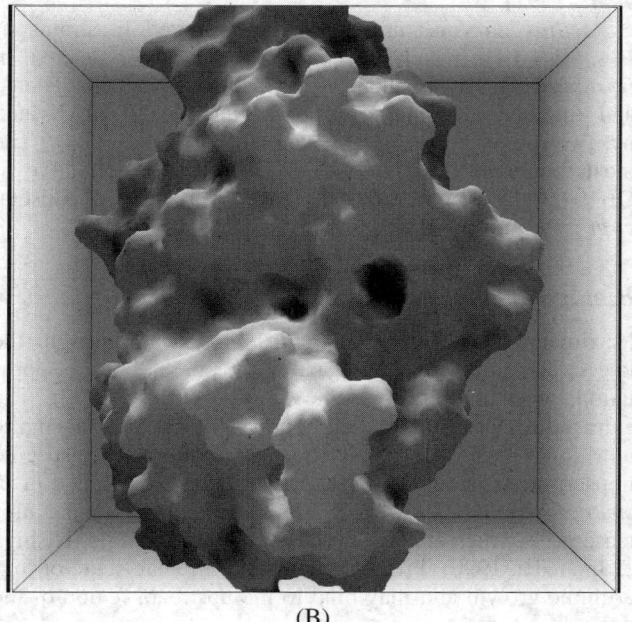

(B)

*Figure 351.2* The structure of thermolysin. (A) The active-site residues, catalytic zinc atom, and the four calcium atoms are highlighted (image produced by Setor; Evans, 1993). (B) A GRASP (Nichols, 1992) image of thermolysin, in approximately the same orientation, highlights the active-site cleft.

M

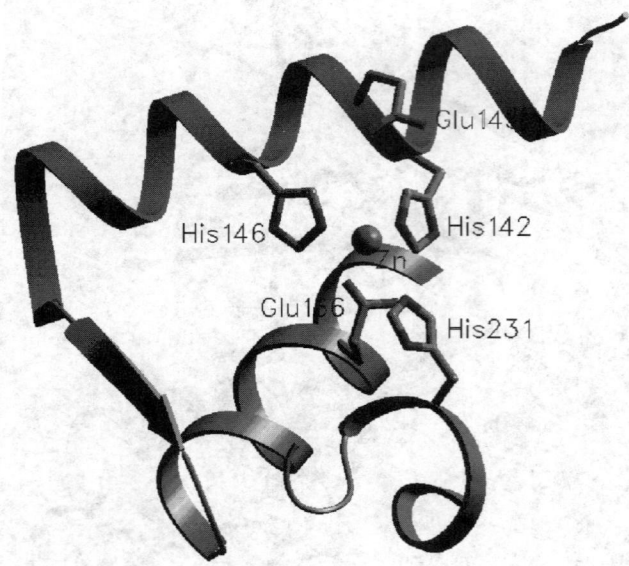

*Figure 351.3* The thermolysin active site. The catalytic residues, referred to in the text, are highlighted (produced by Setor).

chromatography using ligands such as D-Phe-Gly coupled to cyanogen-bromide-activated Sepharose (Walsh *et al.*, 1974), which has also been used to purify the enzyme directly from culture supernatants (O'Donohue *et al.*, 1994; Beaumont, 1995). Bacitracin coupled to silica has also been used to purify large quantities of several *Bacillus* neutral proteases from culture supernatants (Van den Burg *et al.*, 1989; Eijsink *et al.*, 1991).

*Bacillus* enzymes are generally expressed in strains of *B. subtilis* from which other proteases have been deleted. In *B. subtilis* DB104 an alkaline protease gene has been deleted and the neutral protease gene is functionally inactivated by two point mutations (Kawamura & Doi, 1984). The most frequently used, *B. subtilis* DB117, is a derivative of DB104, where both proteases have been deleted (Eijsenk *et al.*, 1990).

### Biological Aspects

Maximal synthesis of the *Bacillus* enzymes occurs in the late exponential and early stationary phase, before sporulation, when nutrients become limiting (Priest, 1977). The extracellular proteases therefore appear to have a scavenging nutritional role for their bacteria. A neutral protease-specific transcriptional activator gene, *prtA*, has been identified in *B. stearothermophilus* strain TELNE, upstream from the protease gene, which enhances protease synthesis 5-fold (Nishiya & Imanaka, 1990). PrtA is produced in the late stage of logarithmic growth and may bind to the upstream region of the neutral protease promoter.

The thermolysins are synthesized as preproteins (Takagi *et al.*, 1985; Kubo & Imanaka, 1988; van den Burg *et al.*, 1991; O'Donohue & Beaumont, 1996; Saul *et al.*, 1996) and the pro sequences, of around 200 residues, are thought to be removed autocatalytically (Kubo *et al.*, 1992; Beaumont

*et al.*, 1995). Although no studies have been reported on the role of these pro sequences *in vivo*, the pro sequence of thermolysin, in *trans*, facilitates refolding of the guanidinium–HCl and acid-denatured mature enzyme *in vitro* and acts as a noncompetitive inhibitor of activity (O'Donohue & Beaumont, 1996)

### Distinguishing Features

Thermolysin was originally identified as a thermostable enzyme (Endo, 1962). In large part, this thermostability is attributable to the bound calcium ions. Enzymes in the thermolysin group are more stable than those in the bacillolysin group that, with the exception of the enzyme from *Bacillus cereus*, lack the third and fourth calcium-binding sites. Thermal inactivation is caused by local unfolding followed by rapid autolysis, and calcium binding protects against this irreversible process. One or two calcium atoms are released from thermolysin during thermal denaturation, depending on the conditions used, and these are thought to be Ca(3) and/or Ca(4) (Dahlquist *et al.*, 1976; Roche & Voorduow, 1978; Fontana, 1988).

Even among the M4 enzymes differences in thermostability exist, the temperature at which they loss 50% of their activity in a 30 min incubation ($T_{50}$) being 82°C, 76.7°C and 68.5°C for the enzymes from *B. thermoproteolyticus*, *B. caldolyticus* and *B. stearothermophilus* CU21 respectively (van den Burg *et al.*, 1991), while the enzyme from *Bacillus* sp. EA1 has been reported to be more stable than that from *B. thermoproteolyticus* (Coolbear *et al.*, 1991). Site-directed mutagenesis studies, mainly concentrating on the enzyme from *B. stearothermophilus* CU21 (reviewed in Vriend & Eijsink, 1993; Eijsink *et al.*, 1995), have identified a cluster of residues in the N-terminal lobe important for thermostability. Multiple mutation of six of these residues, which are situated near the binding site of Ca(3), has led to the production of an enzyme with a $T_{50}$ of 96.9°C (Eijsink *et al.*, 1995).

### Further Reading

Beaumont & Beynon (1998) give a thorough overview of this family of proteases. The broader topic of bacterial metalloprotease is covered by Häse & Finkelstein (1993). The X-ray structure is detailed in Holmes & Matthews (1982). Fontana (1988) deals with structure and stability. Finally, Eijsink *et al.* (1995) provide insight into the heat stability of this group of enzymes.

### References

Bartlett, P.A. & Marlowe, C.K. (1983) Phosphonamidates as transition-state analogue inhibitors of thermolysin. *Biochemistry* **22**, 4618–4623.

Bartlett, P.A. & Marlowe, C.K. (1987a) Possible role for water dissociation in the slow binding of phosphorus-containing transition-state-analogue inhibitors of thermolysin. *Biochemistry* **26**, 8553–8561.

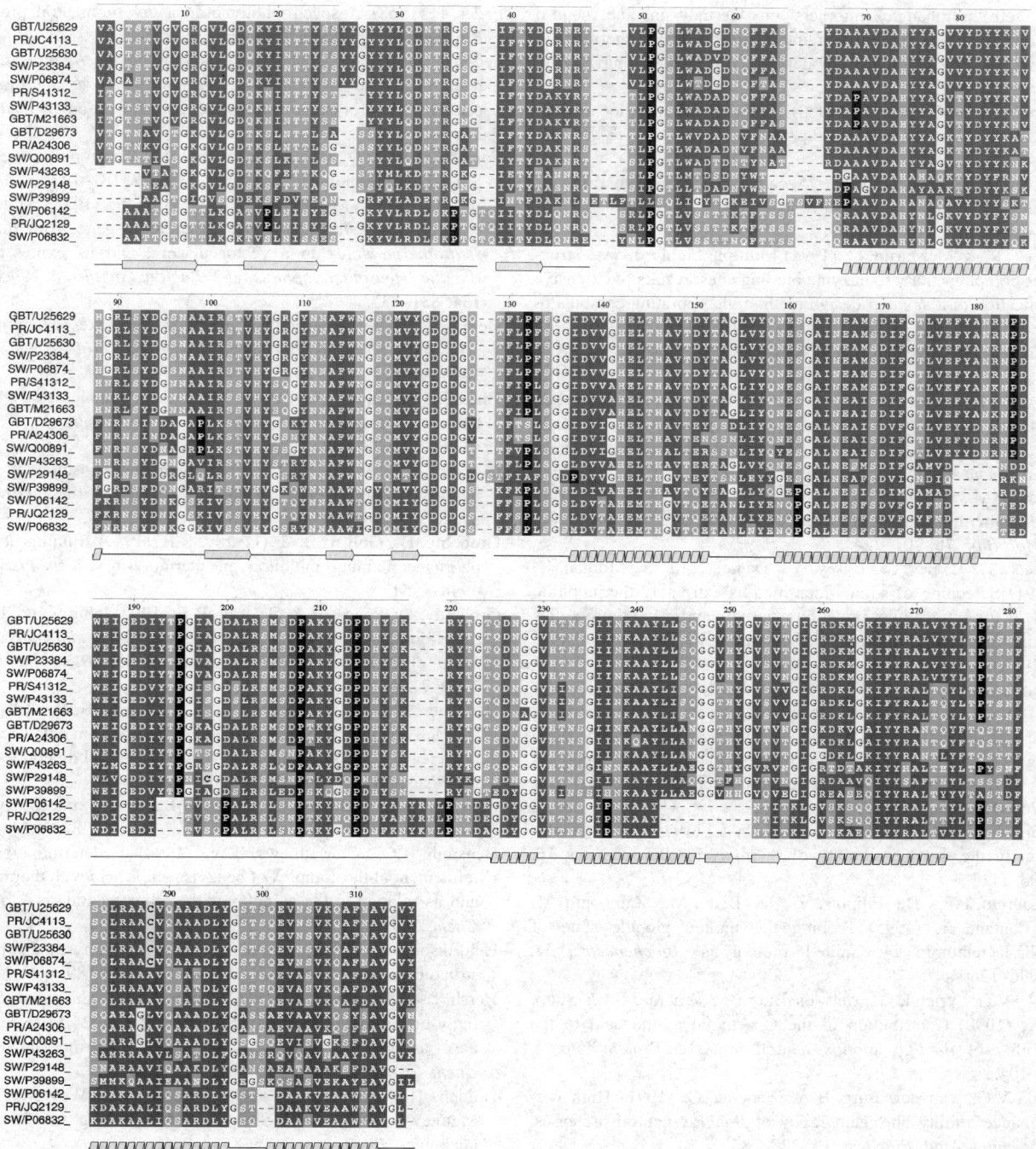

*Figure 351.4* Alignment of the thermolysin/bacillolysin family. The alignment was generated using CLUSTALW and the output was displayed using the program MALPLOT (S.J. Hubbard, personal communication, http://sjh.bi.umist.ac.uk) to include the secondary structure cartoon of thermolysin PDB: 8TMN. *Key to enzymes*: See Figure 351.1.

Bartlett, P. & Marlowe, C. (1987b) Evaluation of intrinsic binding energy from a hydrogen bonding group in an enzyme inhibitor. *Science* **235**, 569–571.

Bash, P., Singh, U., Brown, F., Langridge, R. & Kollman, P. (1987) Calculation of the relative change in binding free energy of a protein-inhibitor complex. *Science* **235**, 574–576.

Beaumont, A. & Beynon, R.J. (1998) The bacterial M4 metallo-endopeptidases. In: *Protein Profiles*. Oxford: IRL Press (in press).

Beaumont, A., O'Donohue, M.J., Paredes, N., Rousselet, N., Assicot, M., Bohuon, C., Fournié-Zaluski, M.-C. & Roques, B.P. (1995) The role of histidine 231 in thermolysin-like enzymes. A site-directed mutagenesis study. *J. Biol. Chem.* **270**, 16803–16808.

Benchetrit, T., Fournié-Zaluski, M.-C. & Roques, B.P. (1987) Relationship between the inhibitory potencies of thiorphan and retrothiorphan enantiomers on thermolysin and neutral endopeptidase 24.11 and their interactions with the thermolysin active site by computer modelling. *Biochem. Biophys. Res. Commun.* **147**, 1034–1040.

Blumberg, S. & Tauber, Z. (1983) Inhibition of metalloendopeptidases by 2-mercaptoacetyl-dipeptides. *Eur. J. Biochem.* **136**, 151–154.

Blumberg, S. & Vallee, B.L. (1975) Superactivation of thermolysin by acylation with amino acid *N*-hydroxysuccinimide esters. *Biochemistry* **14**, 2410–2419.

Bohacek, R. & McMartin, C. (1994) Multiple highly diverse structures complementary to enzyme binding sites: results of extensive application of a *de novo* design method incorporating combinatorial growth. *J. Am. Chem. Soc.* **116**, 5560–5571.

Christianson, D.W. & Lipscomb, W.N. (1989) Carboxypeptidase A. *Acc. Chem. Res.* 22: 62–69.

Clapes, P., Torres, J.L. & Adlercreutz, P. (1995) Enzymatic peptide synthesis in low water content systems: preparative enzymatic synthesis of [Leu]- and [Met]-enkephalin derivatives. *Bioorg. Med. Chem.* **3**, 245–255.

Colman, P.M., Jansonius, J.N. & Matthews, B.W. (1972) The structure of thermolysin: an electron density map at 2.3 Å resolution. *J. Mol. Biol.* **70**, 701–724.

Coolbear, T., Eames, C., Casey, Y., Daniel, R.M. & Morgan, H. (1991) Screening of strains identified as extremely thermophilic *Bacilli* for extracellular proteolytic activity and general properties of the proteinases from two of the strains. *J. Appl. Bacteriol.* **71**, 252–264.

Copié, V., Kolbert, A. C, Drewry, D.H, Bartlett, P.A, Oas, T.G. & Griffin, R.G. (1990) Inhibition of thermolysin by phosphonamidate transition-state analogues: measurement of $^{31}$P-$^{15}$N bond lengths and chemical shifts in two enzyme-inhibitor complexes by solid-state nuclear magnetic resonance. *Biochemistry* **29**, 9176–9184.

Dalquist, F.W., Long, J.W. & Bigbee, W.L. (1976) Role of calcium in the thermal stability of thermolysin. *Biochemistry* **15**, 1103–1111.

De Laureto, P.P., De Filippis, V., Di Bello, M., Zambonin, M. & Fontana, A. (1995) Probing the molten globule state of alpha-lactalbumin by limited proteolysis. *Biochemistry* **34**, 12596–12604.

Eijsink, V.G., Vriend, G., van den Burg, B., Venema, G. & Stulp, B.K. (1990) Contribution of the C-terminal amino acid to the stability of *Bacillus subtilis* neutral protease. *Protein Eng.* **4**, 99–104.

Eijsink, V.G., van den Burg, B. & Venema, G. (1991) High performance affinity chromatography of *Bacillus* neutral proteases. *Biotechnol. Appl. Biochem.* **14**, 275–283.

Eijsink, V.G., Veltman O.R., Aukema W., Vriend, G. & Venema, G. (1995) Structural determinants of the stability of thermolysin-like proteinases. *Nature Struct. Biol.* **2**, 374–379.

Endo, S. (1962) The protease produced by thermophilic bacteria. *Hakko Kogaku Zashi* **40**, 346–347.

Evans, S.V. (1993) SETOR: hardware-lighted three-dimensional solid model representations of macromolecules. *J. Mol. Graphics* **11**, 134–138.

Feder, J. (1968) A spectrophotometric assay for neutral protease. *Biochem. Biophys. Res. Commun.* **32**, 326–332.

Feder, J. & Schuck, J. M. (1970) Studies on the *Bacillus subtilis* neutral-protease and *Bacillus thermoproteolyticus* thermolysin-catalyzed hydrolysis of dipeptide substrates. *Biochemistry* **9**, 2784–2791.

Fontana, A. (1988) Structure and stability of thermophilic enzymes. Studies on thermolysin. *Biophys. Chem.* **29**, 181–193.

Fujii, M., Takagi, M., Imanaka, T. & Aiba, S. (1983) Molecular cloning of a thermostable neutral protease gene from *Bacillus stearothermophilus* in a vector plasmid and its expression in *Bacillus stearothermophilus* and *Bacillus subtilis*. *J. Bacteriol.* **154**, 831–837.

Gettins, P. (1988) Thermolysin-inhibitor complexes examined by $^{31}$P and $^{113}$Cd NMR spectroscopy. *J. Biol. Chem.* **263**, 10208–10211.

Ghosh, I. & Edholm, O. (1994) Molecular dynamics study of the binding of phenylalanine stereoisomers to thermolysin. *Biophys. Chem.* **50**, 237–248.

Giessner, P.C. & Jacob, O. (1989) A theoretical study of Zn$^{++}$ interacting with models of ligands present at the thermolysin active site. *J. Comput. Aided Mol. Des.* **3**, 23–37.

Grobelny, D., Goli, U.B. & Galardy, R.E. (1989) Binding of phosphorus containing inhibitors of thermolysin. *Biochemistry* **28**, 4948–4951.

Grobelny, D., Poncz, L. & Galardy, R.E. (1992) Inhibition of human skin fibroblast collagenase, thermolysin, and *Pseudomonas aeruginosa* elastase by peptide hydroxamic acids. *Biochemistry* **31**, 7152–7154.

Hangauer, D.G., Monzingo, A F. & Matthews, B.W. (1984) An interactive computer graphics study of thermolysin-catalyzed peptide cleavage and inhibition by N-carboxymethyl dipeptides. *Biochemistry* **23**, 5730–5741.

Häse, C.C. & Finkelstein, R.A. (1993) Bacterial extracellular zinc-containing metalloproteases. *Microbiol. Rev.* **57**, 823–827.

Hausrath, A.C. & Matthews, B.W. (1994) Redetermination and refinement of the complex of benzylsuccinic acid with thermolysin and its relation to the complex with carboxypeptidase A. *J. Biol. Chem.* **269**, 18839–18842.

Heinrikson, R. (1977) Applications of thermolysin in protein structural analysis. *Methods Enzymol.* **47**, 175–189.

Hersh, L. & Morihara, K. (1986) Comparison of the subsite specificity of the mammalian neutral endopeptidase 24.11 (enkephalinase) to the bacterial neutral endopeptidase thermolysin. *J. Biol. Chem.* **261**, 6433–6437.

Holden, H.M., Tronrud, D.E., Monzingo, A.F., Weaver, L.H. & Matthews, B.W. (1987) Slow- and fast-binding transition-state analogues. *Biochemistry* **26**, 8542–8553.

Holland, D.R., Tronrud, D.E., Pley, H.W., Flaherty, K.M., Stark, W. Jansonius, J.N., McKay, D.B. & Matthews, B.W. (1992) Structural comparison suggests that thermolysin and related neutral proteases undergo hinge-bending motion during catalysis. *Biochemistry* **31**, 11310–11316.

Holland, D.R., Barclay, P., Danilewicz, J., Matthews, B.W. & James, K. (1994) Inhibition of thermolysin and neutral endopeptidase-24.11 by a novel glutaramide derivative: X-ray structure determination of the thermolysin-inhibitor complex. *Biochemistry* **33**, 51–56.

Holland, D.R., Hausrath, A.C., Juers, D. & Matthews, B.W. (1995) Structural analysis of zinc substitutions in the active site of thermolysin. *Protein Sci.* **4**, 1955–1965.

Holmes, M.A. & Matthews, B.W. (1981) Binding of hydroxamic acid inhibitors to crystalline thermolysin suggests a pentacoordinate zinc intermediate in catalysis. *Biochemistry* **20**, 6912–6920.

Holmes, M.A. & Matthews, B.W. (1982) The structure of thermolysin refined at 1.6 Å resolution. *J. Mol. Biol.* **160**, 623–639.

Holmes, M.A., Tronrud, D.E. & Matthews, B.W. (1983) Structural analysis of the inhibition of thermolysin by an active-site-directed irreversible inhibitor. *Biochemistry* **22**, 236–240.

Holmquist, B. & Vallee, B.L. (1974) Metal substitutions and inhibition of thermolysin: spectra of the cobalt enzyme. *J. Biol. Chem.* **249**, 4601–4607.

Holmquist, B. & Vallee, B.L. (1976) Esterase activity of zinc neutral proteases. *Biochemistry* **15**, 101–107.

Holmquist, B. & Vallee, B. (1979) Metal-coordinating substrate analogs as inhibitors of metalloenzymes. *Proc. Natl Acad. Sci. USA* **76**, 6216–6220.

Indig, F.E., Ben Meir, D., Spungin, A. & Blumberg, S. (1989) Investigation of neutral endopeptidases (EC 3.4.24.11) and of neutral proteinases (EC 3.4.24.4) using a new sensitive two-stage enzymatic reaction. *FEBS Lett.* **255**, 237–240.

Inouye, K. (1992) Effects of salts on thermolysin: activation of hydrolysis and synthesis of *N*-carbobenzoxy-L-aspartyl-L-phenylalanine methyl ester, and a unique change in the absorption spectrum of thermolysin. *J. Biochem.* **112**, 335–340.

Inouye, K., Lee, S.-B. & Tonomura, B. (1996) Effect of amino acid residues at the cleavable site of substrates on the remarkable activation of thermolysin by salts. *Biochem. J.* **315**, 133–138.

Izquierdo-Martin, M. & Stein, R.L. (1992) Mechanistic studies on the inhibition of thermolysin by a peptide hydroxamic acid. *J. Am. Chem. Soc.* **114**, 325–331.

Jakubke, H.D. & Konnecke, A. (1987) Peptide synthesis using immobilized proteases. *Methods Enzymol.* **136**, 178–188.

Kajiwara, K., Kumazaki, T., Sato, E., Kanaoka, Y. & Ishii, S. (1991) Application of bimane-peptide substrates to spectrofluorometric assays of metalloendopeptidases. *J. Biochem.* **110**, 345–349.

Kam, C., Nishino, N. & Powers, J.C. (1979) Inhibition of thermolysin and carboxypeptidase A by phosphoramidates. *Biochemistry* **18**, 3032–3038.

Kanoh, S., Ito, M., Niwa, E., Kawano, Y. & Hartshorne, D.J. (1993) Actin-binding peptide from smooth muscle myosin light chain kinase. *Biochemistry* **32**, 8902–8907.

Kawamura, F. & Doi, R. (1984) Construction of a *Bacillus subtilis* double mutant deficient in extracellular alkaline and neutral proteases. *J. Bacteriol.* **160**, 442–444.

Kim, D.H. & Jin, Y. (1996) Inhibitory stereochemistry of *N*-chloroacetyl-*N*-hydroxyleucine methyl ester for thermolysin. *Bioorg. Med. Chem. Lett.* **6**, 153–156.

Kitagishi, K. & Hiromi, K. (1984) Binding between thermolysin and its specific inhibitor, phosphoramidon. *J. Biochem.* **95**, 529–534.

Kubo, M. & Imanaka, T. (1988) Cloning and nucleotide sequence of the highly thermostable neutral protease gene from *Bacillus*

*stearothermophilus. J. Gen. Microbiol.* **134**, 1883–1892.

Kubo, M., Mitsuda, Y., Takagi, M. & Imanaka, T. (1992) Alteration of specific activity and stability of thermostable neutral protease by site-directed mutagenesis. *Appl. Environ. Microbiol.* **58**, 3779–3783.

Kunugi, S., Hirohara, H. & Ise, N. (1982) pH and temperature dependencies of thermolysin catalysis. Catalytic role of zinc-coordinated water. *Eur. J. Biochem.* **124**, 157–163.

Matsubara, H. (1970) Purification and assay of thermolysin. *Methods Enzymol.* **19**, 642–651.

Matthews, B.W. (1988) Structural basis of the action of thermolysin and related zinc peptidases. *Acc. Chem. Res.* **21**, 333–340.

Matthews, B.W. & Weaver, L.H. (1974) The binding of lanthanides to thermolysin. *Biochemistry* **13**, 1719–1725.

Maycock, A., DeSousa, D., Payne, L., ten Broeke, J., Wu, M. & Patchett, A. (1981) Inhibition of thermolysin by *N*-carboxymethyl dipeptides. *Biochem. Biophys. Res. Commun.* **102**, 963–969.

Merz, K. & Kollman, P. (1989) Free energy perturbation simulations of the inhibition of thermolysin: prediction of the free energy of binding of a new inhibitor. *J. Am. Chem. Soc.* **111**, 5649–5658.

Mock, W.L. & Aksamawti, M. (1994) Binding to thermolysin of phenolate-containing inhibitors necessitates a revised mechanism of catalysis. *Biochem. J.* **302**, 57–68.

Mock, W.L. & Stanford, D.J. (1996) Arazoformyl dipeptide substrates for thermolysin. Confirmation of a reverse protonation catalytic mechanism. *Biochemistry* **35**, 7369–7377.

Monzingo, A.F. & Matthews, B.W. (1982) Structure of a mercaptan-thermolysin complex illustrates mode of inhibition of zinc-proteases by substrate analog mercaptans. *Biochemistry* **21**, 3390–3394.

Monzingo, A.F. & Matthews, B.W. (1984) Binding of *N*-carboxymethyl dipeptide inhibitors to thermolysin determined by X-ray crystallography: a novel class of transition-state analogues for zinc peptidases. *Biochemistry* **23**, 5724–5729.

Morgan, B.P., Scholtz, J.M., Ballinger, M.D., Zipkin, I.D. & Bartlett, P.A. (1991) Differential binding energy: a detailed evaluation of the influence of hydrogen-bonding and hydrophobic groups on the inhibition of thermolysin by phosphorus-containing inhibitors. *J. Am. Chem. Soc.* **113**, 297–307.

Morgan, B.P., Holland, D.R., Matthews, B.W. & Bartlett, P.A. (1994) Structure-based design of an inhibitor of the zinc peptidase thermolysin. *J. Am. Chem. Soc.* **116**, 3251–3260.

Morihara, K. & Oka, T. (1968) The complex active sites of bacterial neutral proteases in relation to their specificities. *Biochem. Biophys. Res. Commun.* **30**, 625–630.

Morihara, K. & Tsuzuki, H. (1970) Thermolysin: kinetic study with oligopeptides. *Eur. J. Biochem.* **15**, 374–380.

Morihara, K. & Tsuzuki, H. (1971) Comparative study of various neutral proteinases from microorganisms: specificity with oligopeptides. *Arch. Biochem. Biophys.* **146**, 291–296.

Morihara, K., Tsuzuki, H. & Oka, T. (1968) Comparison of the specificities of various neutral proteinases from microorganisms. *Arch. Biochem. Biophys.* **123**, 572–588.

Murakami, Y., Hirata, M. & Hirata, A. (1996) Mathematical approach to thermolysin catalyzed synthesis of aspartame precursor. *J Ferment. Bioeng.* **82**, 246–252.

Nichols, A. (1990) *GRASP: Graphical Representation and Analysis of Surface Properties.* New York: Columbia University.

Nishino, N. & Powers, J. (1978) Peptide hydroxamic acids as inhibitors of thermolysin. *Biochemistry* **17**, 2846–2850.

Nishino, N. & Powers, J. (1979) Design of potent reversible inhibitors for thermolysin. Peptides containing zinc coordinating ligands and their use in affinity chromatography. *Biochemistry* **18**, 4340–4347.

Nishino, N. & Powers, J.C. (1980) *Pseudomonas aeruginosa* elastase. Development of a new substrate, inhibitors, and an affinity ligand. *J. Biol. Chem.* **255**, 3482–3486.

Nishiya, Y. & Imanaka, T. (1990) Cloning and nucleotide sequence of the *Bacillus stearothermophilus* neutral protease gene and its transcriptional activator gene. *J. Bacteriol.* **172**, 4861–4869.

O'Donohue, M.J. & Beaumont, A. (1996) The roles of the pro-sequence of thermolysin in enzyme inhibition and folding *in vitro*. *J. Biol. Chem.* **271**, 26477–26481.

O'Donohue, M.J., Roques, B.P. & Beaumont, A. (1994) Cloning and expression in *Bacillus subtilis* of the *npr* gene from *Bacillus thermoproteolyticus* Rokko coding for the thermostable metalloprotease thermolysin. *Biochem. J.* **300**, 599–603.

Ooshima, H., Mori, H. & Harano, Y. (1985) Synthesis of aspartame precursor by thermolysin solid in organic solvent. *Biotechnol. Lett.* **7**, 789–792.

Pangburn, M.K. & Walsh, K.A. (1975) Thermolysin and neutral protease: mechanistic considerations. *Biochemistry* **14**, 4050–4054.

Pozsgay, M., Michaud, C., Liebman, M. & Orlowski, M. (1986) Substrate and inhibitor studies of thermolysin-like neutral metalloendopeptidase from kidney membrane fractions. Comparison with bacterial thermolysin. *Biochemistry* **25**, 1292–1299.

Priest, F. (1977) Extracellular enzyme synthesis in the genus *Bacillus*. *Bacteriol. Rev.* **41**, 711–753.

Rasnick, D. & Powers, J.C. (1978) Active site directed irreversible inhibition of thermolysin. *Biochemistry* **21**, 4363–4369.

Roche, R. & Voorduow, G. (1978) The structural and functional roles of metal ions in thermolysin. *CRC Crit. Rev. Biochem.* **5**, 1–23.

Roderick, S.L., Fournié-Zaluski, M.-C., Roques, B.P. & Matthews, B.W. (1989) Thiorphan and retro-thiorphan display equivalent interactions when bound to crystalline thermolysin. *Biochemistry* **28**, 1493–1497.

Roques, B.P., Noble, F., Daugé, V., Fournié-Zaluski, M.-C. & Beaumont, A. (1993) Neutral endopeptidase 24.11: structure, inhibition, and experimental and clinical pharmacology. *Pharmacol. Rev.* **45**, 87–146.

Saul, D.J., Williams, L.C., Toogood, H.S., Daniel, R.M. & Berquist, P.L. (1996) Sequence of the gene encoding a highly thermostable neutral proteinase from *Bacillus* sp. strain EA1: expression in

*Escherichia coli* and characterisation. *Biochim. Biophys. Acta* **1308**, 74–80.

Shen, J. & Wendoloski, J. (1995) Binding of phosphorus-containing inhibitors to thermolysin studied by the Poisson-Boltzmann method. *Protein Sci.* **4**, 373–381.

Suda, H., Aoyagi, T., Takeuchi, T. & Umezawa, H. (1973) A thermolysin inhibitor produced by Actinomycetes: phosphoramidon. *J. Antibiot.* **26**, 621–623.

Takagi, M., Imanake, T. & Aiba, S (1985) nucleotide sequence and promoter region for the neutral protease gene from *Bacillus stearothermophilus*. *J. Bacteriol.* **163**, 824–831.

Tronrud, D.E., Monzingo, A.F. & Matthews, B.W. (1986) Crystallographic structural analysis of phosphoramidates as inhibitors and transition-state analogs of thermolysin. *Eur. J. Biochem.* **157**, 261–268.

Tronrud, D.E., Holden, H.M. & Matthews, B.W. (1987) Structures of two thermolysin-inhibitor complexes that differ by a single hydrogen bond. *Science* **235**, 571–574.

Van Aalten, D.M.F., Amadei, A., Linssen, A.B.M., Eijsink, V.G.H., Vriend, G. & Berendsen, H.F.C. (1995) The essential dynamics of thermolysin: confirmation of the hinge-bending motion and comparison of simulations in vacuum and water 2. *Proteins Struct. Funct. Genet.* **22**, 43–54.

van den Burg, B., Eijsink, V.G.H., Stulp, B. & Venema, G. (1989) One-step affinity purification of *Bacillus* neutral proteases using Bacitracin-silica. *J. Biochem. Biophys. Methods* **18**, 209–220.

van den Burg, B., Enequist, H., van der Haar, M., Eijsink, V., Stulp, B. & Venema, G. (1991) A highly thermostable neutral protease from *Bacillus caldolyticus*: cloning and expression of the gene in *Bacillus subtilis* and characterization of the gene product. *J. Bacteriol.* **173**, 4107–4115.

Vriend, G. & Eijsink, V (1993) Prediction and analysis of structure, stability and unfolding of thermolysin-like proteases. *J. Comput. Aided Mol. Des.* **7**, 367–396.

Walsh, K.A., Burnstein, Y. & Pangburn, M.K. (1974) Thermolysin and other neutral proteases. *Methods Enzymol.* **34**, 435–440.

Wasserman, Z. & Hodge, C. (1996) Fitting an inhibitor into the active site of thermolysin – a molecular dynamics case study. *Proteins* **24**, 227–237.

Wayne, S.I. & Fruton, J.S. (1983) Thermolysin-catalyzed peptide bond synthesis. *Proc. Natl Acad. Sci. USA* **80**, 3241–3244.

Yang, J.J. & Van Wart, H. (1994) Kinetics of hydrolysis of dansyl peptide substrates by thermolysin: analysis of fluorescence changes and determination of steady-state kinetic parameters. *Biochemistry* **33**, 6508–6515.

Yang, J.J., Artis, D.R. & Van Wart, H.E. (1994) Differential effects of halide ions on the hydrolysis of different dansyl substrates by thermolysin. *Biochemistry* **33**, 6516–6523.

***Robert J Beynon***
*Department of Biochemistry & Applied Molecular Biology,*
*UMIST, PO Box 88,*
*Manchester M60 1QD, UK*
*Email: r.beynon@umist.ac.uk*

***Ann Beaumont***
*Inserm U266 CNRS URA D1500,*
*UFR des Sciences Pharmaceutiques et Biologiques,*
*4, Avenue de l'Observatoire,*
*75270 Paris Cedex 06, France*

# 352. *Bacillolysin*

## Databanks

*Peptidase classification: clan MA, family M4, MEROPS ID: M04.002*
*NC-IUBMB enzyme classification: EC 3.4.24.28*
*ATCC entries: 77217 (B. subtilis nprE)*
*Chemical Abstracts Service registry number: 76774-43-1*
*Databank codes:*

| Species | Gene | SwissProt | PIR | EMBL (cDNA) | EMBL (genomic) |
|---|---|---|---|---|---|
| *Bacillus amyloliquefaciens* | – | – | – | M36723 | – |
| *Bacillus amyloliquefaciens* | – | – | – | M64815 | – |
| *Bacillus amyloliquefaciens* | *npr* | P06832 | – | K02497 | – |
| *Bacillus brevis* | *npr* | P43263 | – | X61286 | – |
| *Bacillus cereus* | *npr* | P05806 | A24306 | M83910 | – |
| *Bacillus megaterium* | *nprM* | Q00891 | A47710 S21934 | X61380 | – |
| *Bacillus polymyxa* | *npr* | P29148 | – | D00861 | – |
| *Bacillus subtilis* | *nprB* | P39899 | | M62845 | – |
| *Bacillus subtilis* | *nprE* | P06142 J02129 | JQ2129 | U30932 | – |
| *Lactobacillus* sp. | | – | – | D29673 | – |

Brookhaven Protein Data Bank three-dimensional structures:

| Species | ID | Resolution | Notes |
|---|---|---|---|
| *Bacillus cereus* | 1ESP | 2.8 | Glu144Ser mutant |
| | 1NPC | 2.0 | |

## Name and History

**Bacillolysin** is one of the extracellular Zn-metalloendo-peptidases produced by bacteria. It was first reported as an extracellular product from cultures of *Bacillus subtilis*, but related enzymes are produced by *B. amyloliquefaciens*, *B. megaterium*, *B. mesentericus*, *B. cereus* and *B. stearother-mophilus*. There is some confusion about the correct classification of this group, particularly in relation to those enzymes entered under the name 'thermolysin' (Chapter 351). The classification is based on enzyme specificity, on thermal stability and on aligned protein sequences. There is, however, a need for the systematic classification of this whole group of enzymes to be revisited – this is discussed in greater detail in Chapter 351. Given the unevenness of catalytic and stability data, it is likely that aligned protein sequences interpreted in terms of known structural features such as calcium-binding sites will provide a more enduring classification in the future. For the present, the entries under bacillolysin will include all the *Bacillus* neutral proteases that do not cluster in the tight, highly thermostable group that includes thermolysin. It is recognized that this is unsatisfactory, and will need to be reconsidered.

## Activity and Specificity

There are insufficient studies comparing hydrolysis of peptides or proteins to define differences in the catalytic activity of this group in detail. Those tested all have the thermolysin-like preference for a hydrophobic residue in P1'. Indeed, it is unlikely that catalytic properties will provide an effective classification of this group of enzymes that will allow them to be discriminated from thermolysins. Pseudolysin (Chapter 357) and the neutral proteases from *Bacillus cereus* (Feder *et al.*, 1971; Holmquist, 1977) and *Bacillus subtilis* have specificities similar to that of thermolysin, although some differences were observed (Beaumont & Beynon, 1998). The three enzymes degrade peptides with Leu or Ile at P1' at comparable rates; however, Phe in this position decreases bacillolysin activity, while promoting pseudolysin-catalyzed hydrolysis. In addition,

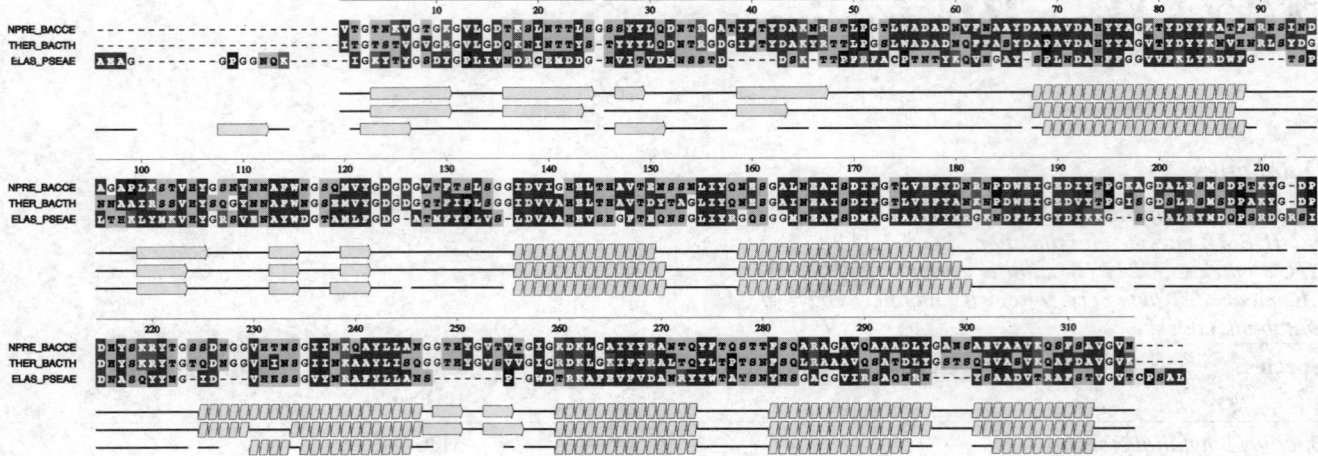

*Figure 352.1* Alignment of three related metalloendopeptidases. The alignment compares three proteinases of known three-dimensional structure. These are thermolysin from *Bacillus thermoproteolyticus* (SW: THER_BACTH, PDB: 8TMN), bacillolysin from *Bacillus cereus* (SW: NPRE_BACCE, PDB: 1NPC) and pseudolysin from *Pseudomonas aeruginosa* (SW: ELAS_PSEAE, PDB: 1E2M). For each protein the aligned sequence is supported by a cartoon representing the secondary structure, arrayed, top to bottom, in the same order as the sequences. This figure was generated using the program MALPLOT (S. Hubbard, personal communication).

pseudolysin is more effective than either thermolysin or bacillolysin at hydrolyzing peptides with Tyr at P1′ (Morihara *et al*., 1968; Feder *et al*., 1971; Morihara & Tsuzuki, 1971). Both thermolysin and subtilisin (Chapter 94) have a similar preference for a hydrophobic residue in P1, whilst pseudolysin prefers Ala to either Gly or Phe (Morihara & Tsuzuki, 1971). Information is also available on the neutral thermolysin/bacillolysin-like neutral proteases from *B. mesentericus* although it should be noted that the mature enzyme has the same sequence as that derived from *B. subtilis* (Stoeva *et al*., 1990). The optimal pH values for the neutral proteinase from *B. subtilis* and thermolysin are both near neutral, but there is a reported difference between the two enzymes with protein substrates: thermolysin has a broader and slightly higher pH optimum with casein (Feder & Schuck, 1970).

## Structural Chemistry

The structure of the enzyme known as bacillolysin, from *Bacillus cereus* (Sidler *et al*., 1986a,b), was first modeled (Pauptit *et al*., 1988) and then solved (Stark *et al*., 1992). The alignment of thermolysin and related enzymes known as bacillolysins and the metalloendopeptidase pseudolysin from *Pseudomonas aeruginosa* (Figure 352.1), particularly in the context of the secondary structure assignments, indicates the extent of the similarity between these three enzymes.

A superposition of the structures of thermolysin (Brookhaven PDB: 8TLN) and bacillolysin (Brookhaven PDB: 1NPC) provides further evidence of the similarity of the fold and active-site geometry of the two enzymes (Figure 352.2). Indeed, the related metalloendopeptidase from *Pseudomonas aeruginosa*, pseudolysin (Thayer *et al*., 1991) has a very similar topology, and is included in Figure 352.2 for comparative purposes. The calcium-binding sites for Ca(3)

and Ca(4) are absent in the enzymes from *B. subtilis* and *B. amyloliquefaciens* (Stoeva *et al*., 1990). Introduction of the Ca(3)-binding loop into the enzyme from *B. subtilis* enhances thermal stability 2-fold over the native enzyme (Toma *et al*., 1991). However, thermal stability also resides in structural features other than the calcium-binding sites (Sidler *et al*., 1986b; Pauptit *et al*., 1988; Stark *et al*., 1992).

## Preparation

A number of one-step affinity purification methods have been described for the different enzymes. One of the most widely used is D-Phe-Gly coupled to cyanogen-bromide-activated Sepharose. This has been used to repurify commercially available thermolysin (Walsh *et al*., 1974) or to purify thermolysins directly from culture supernatants (O'Donohue *et al*., 1994; Beaumont *et al*., 1995). Bacitracin coupled to silica has been used to purify bacillolysin from *B. subtilis* and thermolysin (Van den Burg *et al*., 1989). When the bacitracin-silica method was adapted to an HPLC system, 4.9 mg of bacillolysin were purified from 165 ml of culture supernatant in 1.5 h (Eijsenk *et al*., 1991). Other affinity ligands are *N*-phenylphosphoryl-L-phenylalanine-L-phenylalanine used for the enzyme from *B. cereus* (Holmqist & Vallee, 1976), and D-Phe for thermolysin and bacillolysin (Pangburn *et al*., 1973).

## Biological Aspects

Maximal synthesis of the *Bacillus* enzymes occurs in the late exponential growth and in the early stationary phase before sporulation, when nutrients become limiting (Sidler *et al*., 1986a). The extracellular proteases therefore appear to have a scavenging nutritional role for the bacteria. They may also

*Figure 352.2* Comparison of thermolysin and bacillolysin structures. The structures of thermolysin (PDB: 8TMN, see Chapter 351) and bacillolysin (PDB: 1NPC) were superimposed using the program STALIN (S. Hubbard, personal communication). The aligned structures were then displayed using the program SETOR.

play a role in processing of other secreted enzymes – the neutral proteinase from *B. polymyxa* is able to process the precursor form of α-amylase into multiple active variants (Takekawa *et al.*, 1991). For pathogenic bacteria this may also contribute to their virulence. Deletion of the neutral proteinase from *B. subtilis* is without effect on growth, morphology and sporulation (Yang *et al.*, 1984). The pro sequence of the enzyme from *B. cereus* is essential for correct export and autoactivation (Wetmore *et al.*, 1992).

## Distinguishing Features

Bacillolysin from *B. cereus* is less stable than thermolysin by about 20°C (Sidler *et al.*, 1986b). There are fewer hydrogen bonds than in thermolysin (Chapter 351) (Pauptit *et al.*, 1988; Stark *et al.*, 1992) but the four calcium-binding sites are present. The additional thermal stability of thermolysin might be due to a combination of rigidification by proline residues, hydrogen bonding or salt-bridges (Stark *et al.*, 1992). The thermal stability of this group of enzymes varies considerably (Vriend & Eijsenk, 1993) and multiple factors are likely to contribute. The lack of a consensus method for determination of stability makes comparisons across different studies difficult, but the enzyme from *B. subtilis*, which is one of the least thermally stable, is likely to have lost two of the four calcium-binding sites (calcium ions 3 and 4). Although the calcium ions undoubtedly contribute to prevention of local loop unfolding and autolysis (Vriend & Eijsenk, 1993), they are not the sole factor in thermal stability (Toma *et al.*, 1991).

## Further Reading

For reviews, see Beaumont & Beynon (1998) and Häse & Finkelstein (1993).

## References

Beaumont, A. & Beynon, R.J. (1998) The bacterial M4 metalloendopeptidases. *Protein Profiles*. Oxford: IRL Press (in press).

Beaumont, A., O'Donohue, M.J., Paredes, N., Rousselet, N., Assicot, M., Bohuon, C., Fournié-Zaluski, M.-C. & Roques, B.P. (1995) The role of histidine 231 in thermolysin-like enzymes. A site-directed mutagenesis study. *J. Biol. Chem.* **270**, 16803–16808.

Eijsink, V.G., van den Burg, B. & Venema, G. (1991) High performance affinity chromatography of Bacillus neutral proteases. *Biotechnol. Appl. Biochem.* **14**, 275–283.

Feder, J. & Schuck, J.M. (1970) Studies on the *Bacillus subtilis* neutral-protease and *Bacillus thermoproteolyticus* thermolysin-catalyzed hydrolysis of dipeptide substrates. *Biochemistry* **9**, 2784–2791.

Feder, J., Keay, L., Garrett, L., Cirulis, N., Moseley, M. & Wildi, B. (1971) *Bacillus cereus* neutral protease. *Biochim. Biophys. Acta* **251**, 74–78.

Häse, C. & Finkelstein, R.A. (1993) Bacterial extracellular zinc-containing metalloproteases. *Microbiol. Rev.* **57**, 823–837.

Holmquist, B. (1977) Characterization of the 'microprotease' from *Bacillus cereus*. A zinc neutral endoprotease. *Biochemistry* **16**, 4591–4594.

Holmquist, B. & Vallee, B.L. (1976) Esterase activity of zinc neutral proteases. *Biochemistry* **15**, 101–107.

Morihara, K. & Tsuzuki, H. (1971) Comparative study of various neutral proteinases from microorganisms: specificity with oligopeptides. *Arch. Biochem. Biophys.* **146**, 291–296.

Morihara, K., Tsuzuki, H. & Oka, T. (1968) Comparison of the specificities of various neutral proteinases from microorganisms. *Arch. Biochem. Biophys.* **123**, 572–588.

O'Donohue, M.J., Roques, B.P. & Beaumont, A. (1994) Cloning and expression in *Bacillus subtilis* of the npr gene from *Bacillus thermoproteolyticus* Rokko coding for the thermostable metalloprotease thermolysin. *Biochem. J.* **300**, 599–603.

Pangburn, M., Burstein, P., Morgan, P., Walsh, K. & Neurath, H. (1973) Affinity chromatography of thermolysin and of neutral proteases from *B. subtilis*. *Biochem. Biophys. Res. Commun.* **54**, 371–379.

Pauptit, R.A., Karlsson, R., Picot, D., Jenkins, J.A., Niklaus, R.A. & Jansonius, J.N. (1988) Crystal structure of neutral protease from *Bacillus cereus* refined at 3.0 Å resolution and comparison with the homologous but more thermostable enzyme thermolysin. *J. Mol. Biol.* **199**, 525–537.

Sidler, W., Niederer, E., Suter, F. & Zuber, H. (1986a) The primary structure of *Bacillus cereus* neutral proteinase and comparison with thermolysin and *Bacillus subtilis* neutral proteinase. *Biol. Chem. Hoppe Seyler* **367**, 643–657.

Sidler, W., Kumpf, B., Peterhans, B. & Zuber, H. (1986b) A neutral proteinase produced by *Bacillus cereus* with high sequence homology to thermolysin: production, isolation and characterization. *Appl. Microbiol. Biotechnol.* **25**, 18–24.

Stark, W., Pauptit, R.A., Wilson, K.S. & Jansonius, J.N. (1992) The structure of neutral protease from *Bacillus cereus* at 0.2-nm resolution. *Eur. J. Biochem.* **207**, 781–791.

Stoeva, S., Kleinschmidt, T., Mesrob, B. & Braunitzer, G. (1990) Primary structure of a zinc protease from *Bacillus mesentericus* strain 76. *Biochemistry* **29**, 527–534.

Takekawa, S., Uozumi, N., Tsukagoshi, N & Udaka, S. (1991) Proteases involved in generation of beta and alpha-amylases from a large amylase precursor in *B. polymyxa*. *J. Bacteriol.* **173**, 6820–6825.

Thayer, M.M., Flaherty, K.M. & McKay, D.B. (1991) Three-dimensional structure of the elastase of *Pseudomonas aeruginosa* at 1.5-Å resolution. *J. Biol. Chem.* **266**, 2864–2871.

Toma, S., Campagnoli, S., Margarit, I., Gianna, R., Grandi, G., Bolognesi, M., De Filippis, V. & Fontana, A. (1991) Grafting of a calcium-binding loop of thermolysin to *Bacillus subtilis* neutral protease. *Biochemistry* **30**, 97–106.

Van den Burg, B., Eijsink, V., Stulp, B. & Venema, G. (1989) One-step affinity purification of *Bacillus* neutral proteases using bacitracin-silica. *J. Biochem. Biophys. Methods* **18**, 209–220.

Vriend, G. & Eijsink, V. (1993) Prediction and analysis of structure, stability and unfolding of thermolysin-like proteases. *J. Comput. Aided Mol. Des.* **7**, 367–396.

Walsh, K.A., Burnstein, Y. & Pangburn, M.K. (1974) Thermolysin and other neutral proteases. *Methods Enzymol.* **34**, 435–440.

Wetmore, D.R., Wong, S.L. & Roche, R.S. (1992) The role of the pro-sequence in the processing and secretion of the thermolysin-like neutral protease from *Bacillus cereus. Mol. Microbiol.* **6**, 1593–1604.

Yang, M.Y., Ferrari, E. & Henner, D.J. (1984) Cloning of the neutral protease gene of *B. subtilis* and the use of the cloned gene to create an *in vitro*-derived deletion mutant. *J. Bacteriol.* **160**, 15–21.

**Robert J Beynon**
*Department of Biochemistry and Applied Molecular Biology,
UMIST, PO Box 88,
Manchester, M60 1QD, UK
Email: r.beynon@umist.ac.uk*

**Ann Beaumont**
*Inserm U266 CNRS URA D1500,
UFR des Sciences Pharmaceutiques et Biologiques,
4, Avenue de l'Observatoire,
75270 Paris Cedex 06, France*

# 353. *Listeria metalloprotease Mpl*

## Databanks

*Peptidase classification: clan MA, family M4, MEROPS ID: M04:008*
*NC-IUBMB enzyme classification: none*
*Databank codes:*

| Species | SwissProt | PIR | EMBL (cDNA) | EMBL (genomic) |
|---|---|---|---|---|
| *Listeria monocytogenes* | P23224 | – | X54619 | – |
| *Listeria monocytogenes* | P34025 | – | X60035 | – |

## Name and History

*Listeria monocytogenes* is a gram-positive nonspore-forming, facultative intracellular rod-shaped bacterium which is capable of causing serious infections in human and animals (Farber & Peterkin, 1991). Up to 1991 when the gene encoding the enzyme was cloned and shown by its deduced amino acid sequence to be homologous to proteases of the thermolysin family (M4) (Chapter 350), no proteolytic activity had ever been reported for this species (Domann *et al.*, 1991; Mengaud *et al.*, 1991). The protease was initially designated either as *mpl* or *prtA* following its identification. Nevertheless, a common designation metalloprotease of *Listeria monocytogenes* (*Mpl*) was subsequently agreed upon (Portnoy *et al.*, 1992).

## Activity and Specificity

Protease activity can be detected when using the EnzCheck Assay from Molecular Probes (see Appendix 2 for full names and addresses of suppliers) with fluorescein-labeled casein as substrate. Optimal activity is detected at pH 7.0 but activity can still be observed at pH 5.0 and pH 8.0. The enzyme is inhibited by the metal chelators EDTA, the calcium chelator EGTA, 1,10-phenanthroline and the thermolysin-type protease inhibitor phosphoramidon. Its activity is stimulated by low concentrations (0.1 mM) of $ZnCl_2$, but inhibited by high concentrations (0.5 mM). The specificity of cleavage by the Mpl protease has not been defined.

## Structural Chemistry

The *L. monocytogenes* Mpl protease contains 510 amino acids with a predicted molecular mass of 57.4 kDa. Following cleavage of the N-terminal 24 amino acid signal sequence, the 55 kDa inactive zymogen is secreted to the external medium. Mature active protease has a molecular mass of 36 kDa as a result of further processing of the N-terminal 180 amino acids (Domann *et al.*, 1991; Mengaud *et al.*, 1991). The calculated pI of Mpl is 6.64.

## Preparation

A satisfactory purification scheme for the Mpl protease remains to be established; no such schemes have been published to date. In our hands relatively pure preparations of Mpl have been obtained from supernatant cultures of recombinant *L. monocytogenes* strains that overproduce the protease. One hundred-fold concentrated supernatants are precipitated with 50% ammonium sulfate. Following adsorption to hydroxyapatite, the protease can be eluted as a single peak with 300 mM potassium phosphate buffer. The protease is inactive when produced in *Escherichia coli* (J. Broer, unpublished results).

## Biological Aspects

The gene encoding the Mpl protease is only present in the pathogenic *Listeria* species, *L. monocytogenes* and *L. ivanovii* (Domann *et al*., 1991; Gouin *et al*., 1994). Isogenic *L. monocytogenes* mutants that do not produce active Mpl are attenuated for virulence in a mouse infection model, suggesting a role for the protein in virulence (Raveneau *et al*., 1992). Mpl is responsible for the extracellular degradation and processing of two other listerial factors with which it is located in a transcriptional unit. The listerial ActA protein, an actin-nucleating protein that is extensively degraded in the wild-type strain when grown in complex media, is not degraded in strains where the *mpl* gene has been genetically inactivated (Kuhn & Goebel, 1995). However, this degradation is not seen when bacteria grow within the host cell cytoplasm (Brundage *et al*., 1993). A second listerial protein, the phosphatidylcholine phospholipase C (lecithinase) is synthesized as a 33 kDa precursor and maturation to the 29 kDa active form requires the presence of active Mpl protease (Poyart *et al*., 1993). Hence, Mpl activity in *L. monocytogenes* is best assessed by the level of maturation of lecithinase.

Expression and production of Mpl protease is dependent on growth temperatures and is transcriptionally regulated by the virulence gene regulator PrfA of *L. monocytogenes* (Leimeister-Wächter *et al*., 1992; Chakraborty *et al*., 1992). There are reports to suggest that addition of high salt concentrations to growing cultures increase protease activity (Coffey *et al*., 1996).

## Distinguishing Features

The genetic and biochemical evidence available suggests that Mpl is a neutral protease of the thermolysin family (M4). An unusual feature of Mpl is the presence of several cysteine residues that could create disulfide bridges to stabilize the enzyme. None of the known natural thermolysin-like proteases is known to contain disulfide bridges. Polyclonal antisera and several monoclonal antibodies have been produced against Mpl (J. Broer, unpublished results); these recognize both the mature and processed form of the protease (60 and 35 kDa, respectively).

## References

Brundage, R.A., Smith., G.A., Camilli, A., Theriot, J.A. & Portnoy, D.A. (1993) Expression and phosphorylation of the *Listeria monocytogenes* ActA protein in mammalian cells. *Proc. Natl Acad. Sci. USA* **90**, 11890–11894.

Chakraborty, T., Leimeister-Wächter, M., Domann, E., Hartl, M., Nichterlein, T. & Goebel, W. (1992) Coordinate regulation of virulence genes in *Listeria monocytogenes* requires the product of the *prfA* gene. *J. Bacteriol.* **174**, 568–574.

Coffey, A., Rombouts, F.M. & Abee, T. (1996) Influence of environmental parameters on phosphatidylcholine phospholipase C production in *Listeria monocytogenes*: a convenient method to differentiate *L. monocytogenes* from other *Listeria* species. *Appl. Environ. Microbiol.* **62**, 1252–1256.

Domann, E., Leimeister-Wächter, M., Goebel, W. & Chakraborty, T. (1991) Molecular cloning, sequencing, and identification of a metalloprotease gene from *Listeria monocytogenes* that is species specific and physically linked to the listeriolysin gene. *Infect. Immun.* **59**, 65–72.

Farber, J.M. & Peterkin, P.I. (1991) *Listeria monocytogenes*, a food borne pathogen. *Microbiol. Rev.* **55**, 476–511.

Gouin, E., Mengaud, J. & Cossart, P. (1994) The virulence gene cluster of *Listeria monocytogenes* is also present in *Listeria ivanovii*, an animal pathogen, and *Listeria seeligeri*, a non-pathogenic species. *Infect. Immun.* **63**, 3350–3353.

Kuhn, M. & Goebel, W. (1995) Molecular studies on the virulence of *Listeria monocytogenes*. *Genet. Eng.* **17**, 31–51.

Leimeister-Wächter, M., Domann, E. & Chakraborty, T. (1992) The expression of virulence genes in *Listeria monocytogenes* is thermoregulated. *J. Bacteriol.* **174**, 947–952.

Mengaud, J., Geoffroy, C. & Cossart, P. (1991) Identification of a new operon involved in *Listeria monocytogenes* virulence: its first gene encodes a protein homologous to bacterial metalloproteases. *Infect. Immun.* **59**, 1043–1049.

Portnoy, D.A., Chakraborty, T., Goebel, W. & Cossart, P. (1992) Molecular determinants of *Listeria monocytogenes* pathogenesis. *Infect. Immun.* **60**, 1263–1267.

Poyart, C., Abachin, E., Razafimanantsoa, I. & Berche, P. (1993) The zinc metalloprotease of *Listeria monocytogenes* is required for maturation of phosphatidylcholine phospholipase C: direct evidence obtained by gene complementation. *Infect. Immun.* **61**, 1576–1580.

Raveneau, J., Geoffroy, C., Beretti, J.L., Gaillard, J.L., Alouf, J. & Berche, P. (1992) Reduced virulence of a *Listeria monocytogenes* phospholipase C-deficient mutant obtained by transposon insertion into the zinc metalloprotease gene. *Infect. Immun.* **60**, 916–921.

*Johanna Broer*
*National Center for Biotechnology,*
*Mascheroder Weg 1,*
*D-38124 Braunschweig, Germany*

*Jürgen Wehland*
*National Center for Biotechnology,*
*Mascheroder Weg 1,*
*D-38124 Braunschweig, Germany*

*Trinad Chakraborty*
*Institute for Medical Microbiology,*
*Frankfurterstrasse 107,*
*D-35392 Giessen, Germany*
*Email: trinad.Chakraborty@mikrobio.*
*med.uni-giessen.de*

# 354. Coccolysin

## Databanks

*Peptidase classification: clan MA, family M4, MEROPS ID: M04.007*
*NC-IUBMB enzyme classification: EC 3.4.24.30*
*Databank codes:*

| Species | SwissProt | PIR | EMBL (cDNA) | EMBL (genomic) |
|---|---|---|---|---|
| *Enterococcus faecalis* | – | – | D85393 | – |
| *Enterococcus faecalis* | – | A43580 | M37185 | – |

## Name and History

The first published reports on this enzyme appeared three or four decades ago (Grutter & Zimmerman, 1955; Bleiweis & Zimmerman, 1964; Casas & Zimmerman, 1969; Cano *et al.*, 1971). The cells of *Streptococcus* (currently *Enterococcus*) *faecalis* var. *liquefaciens* were shown to produce an extracellular metalloprotease that liquefied gelatin, hence the enzyme was called **streptococcal gelatinase**. Other previously used names are **metalloendopeptidase II** and **microbial proteinase**. Early studies showed that the enzyme hydrolyzes casein, gelatin and hemoglobin, and clots milk (Grutter & Zimmerman, 1955; Shockman & Cheney, 1969). The enzyme was later reinvestigated, isolated from the human oral strain OG1-10 of *E. faecalis* var. *liquefaciens*, characterized as a zinc-dependent endopeptidase (Mäkinen *et al.*, 1989), and renamed **coccolysin** (EC 3.4.24.30). The nucleotide sequence of the gene was established in 1991 (Su *et al.*, 1991).

## Activity and Specificity

Internal bonds are cleaved in peptides normally containing at least six amino acid residues (the pentapeptide Leu-enkephalin is hydrolyzed at a very low rate). The enzyme is especially active on Azocoll and gelatin; soluble and insoluble collagens are hydrolyzed at a lower rate. Insulin B chain is rapidly hydrolyzed at Phe24↓Phe25, followed by the cleavage of the His5↓Leu6 and other bonds:

FVNQH↓LCGSH↓LVEA↓LYLVCGERGF↓F↓YTPKA

In the best substrates, the scissile bonds involve residues with pronounced hydrophobicity. The residues which tend to be favored in position P1′ are Leu, Phe, Ile and Ala. Examples of peptides that are hydrolyzed are *E. faecalis* pheromone-related peptides (Mäkinen *et al.*, 1989), bradykinin (Arg-Pro-Pro-Gly↓Phe-Ser-Pro↓Phe-Arg), substance P (Arg-Pro-Lys-Pro-Gln-Gln↓Phe↓Phe-Gly↓Leu-Met), human endothelin 1 (primarily at Ser↓Leu and His↓Leu), and angiotensin I (Asp-Arg-Val-Tyr↓Ile-His-Pro↓Phe-His-Leu). Angiotensin II is cleaved only at Tyr↓Ile (Mäkinen *et al.*, 1989; Mäkinen & Mäkinen, 1994). The C-terminal moiety of the scissile bond must be at least two residues long. The enzyme does not hydrolyze small collagenase substrates, such as Pz-Pro-Leu-Gly-Pro-D-Arg and 2-FA-Leu-Gly-Pro-Ala. In general, the specificity profile resembles that of the *Bacillus stearothermophilus* thermolysin (EC 3.4.24.4) (Chapter 351) and the mammalian neprilysin (Chapter 362).

With Azocoll as substrate, the pH optimum is broad (6–8). Shorter peptides hydrolyze readily at pH 7.0–7.2. Chemical modification suggests the importance of histidyl, carboxyl and tyrosyl groups in enzyme activity. *o*-Chloranil, which inactivates bacterial collagenases in an almost equimolar manner (Mäkinen & Mäkinen, 1988), is a near-stoichiometric inhibitor of this enzyme (Mäkinen *et al.*, 1989). Assays have been conveniently made using a homogeneous suspension of Azocoll (the stock mixture must be under continuous stirring) (Mäkinen *et al.*, 1989). Metal chelators inactivate the enzyme. A high molecular mass inflammatory factor is inhibitory (Mäkinen *et al.*, 1989).

## Structural Chemistry

Coccolysin is a single-chain protein of about 31.5 kDa, pI 4.6. The enzyme molecule is strongly hydrophobic; the portion of hydrophobic amino acids is considered to be at least 43% (Mäkinen *et al.*, 1989). The *Staphylococcus aureus* metalloendopeptidase aureolysin may be a closely related enzyme (Chapter 538).

## Biological Aspects

*E. faecalis* is frequently identified as the etiologic agent of several opportunistic infections (such as soft tissue and urinary tract infections, intra-abdominal abscesses and root canal infections), and as the causative agent in several cases of endocarditis, secondary bacteremia, food poisoning, etc. It has been proposed, owing to the specificity profile of the enzyme, that the extracellular production of the enzyme is associated with these clinical conditions (Mäkinen *et al.*, 1989).

## References

Bleiweis, A.S. & Zimmerman, L.N. (1964) Properties of proteinase from *Streptococcus faecalis* var. *liquefaciens*. *J. Bacteriol.* **88**, 653–659.

Cano, F.R., Casas, I.A. & Zimmerman, L.N. (1971) Preparation and characterization of a protease produced by *Streptococcus faecalis* var. *liquefaciens*. *Prep. Biochem.* **1**, 269–282.

Casas, I.A. & Zimmerman, L.N. (1969) Dependence of protease secretion by *Streptococcus faecalis* var. *liquefaciens* on arginine and its possible relation to site of synthesis. *J. Bacteriol.* **97**, 307–312.

Grutter, F.H. & Zimmerman, L.N. (1955) A proteolytic enzyme of *Streptococcus zymogenes*. *J. Bacteriol.* **69**, 728–732.

Mäkinen, P.-L. & Mäkinen, K.K. (1988) Near stoichiometric, irreversible inactivation of bacterial collagenases by *o*-chloranil (3,4,5,6-tetrachloro-1,2-benzoquinone). *Biochem. Biophys. Res. Commun.* **153**, 74–80.

Mäkinen, P.-L. & Mäkinen, K.K. (1994) The *Enterococcus faecalis* extracellular metalloendopeptidase (EC 3.4.24.30; coccolysin)

inactivates human endothelin at bonds involving hydrophobic amino acid residues. *Biochem. Biophys. Res. Commun.* **200**, 981–985.

Mäkinen, P.-L., Clewell, D.B., An, F. & Mäkinen, K.K. (1989) Purification and specificity of a strongly hydrophobic extracellular metalloendopeptidase ('gelatinase') from *Streptococcus faecalis* (strain OG1-10). *J. Biol. Chem.* **264**, 3325–3334.

Shockman, G.D. & Cheney, M.C. (1969) Autolytic enzyme system of *Streptococcus faecalis*. V. Nature of the autolysin-cell wall complex and its relationship to properties of the autolytic enzyme of *Streptococcus faecalis*. *J. Bacteriol.* **98**, 1199–1207.

Su, Y.A., Sulavik, M.C., He, P., Mäkinen, K.K., Mäkinen, P.-L., Fiedler, S., Wirth, R. & Clewell, D.B. (1991) Nucleotide sequence of the gelatinase gene (*gelE*) from *Enterococcus faecalis* subsp. *liquefaciens*. *Infect. Immun.* **59**, 415–420.

*Kauko K. Mäkinen*
*Pirkko-Liisa Mäkinen,*
*Institute of Dentistry,*
*University of Turku,*
*20520 Turku, Finland*
*Email: kmakinen@umich.edu*

# 355. *Vibriolysin*

## Databanks

*Peptidase classification: clan MA, family M4, MEROPS ID: M04.003*
*NC-IUBMB enzyme classification: EC 3.4.24.25*
*ATCC entries: 158338, 53559; bacterial culture: 53559*
*Databank codes:*

| Species | SwissProt | PIR | EMBL (cDNA) | EMBL (genomic) |
|---|---|---|---|---|
| Vibrio proteolyticus | Q00971 | – | M64809 | – |
| Vibrio anguillarum | P43147 | – | L02528 | – |

## Name and History

*Vibrio proteolyticus* (formerly *Aeromonas proteolytica*) is a marine microorganism that was first isolated from the intestine of a small, wood-boring isopod crustacean, *Limnoria tripunctata* (Merkel *et al.*, 1964). During growth of this halotolerant, gram-negative bacterium, an aminopeptidase (Chapter 491) and a neutral protease are secreted into the culture medium (Wilkes & Prescott, 1976). The highly active metalloprotease was first purified and characterized by Griffin & Prescott (1970). Further biochemical characterization of the protease indicated that it was structurally and catalytically similar to thermolysin (Chapter 351) from *Bacillus thermoproteolyticus* (Holmquist & Vallee, 1976; Wilkes & Prescott, 1976; Bayliss *et al.*, 1980). The gene encoding the enzyme

has been cloned, sequenced and expressed in several systems (David *et al.*, 1992).

Industrial as well as biomedical applications for the enzyme have been discovered (Durham, 1990; Bull *et al.*, 1992; Durham *et al.*, 1993). The first reference to the designation of this protease as **vibriolysin** was disclosed by Durham (1989) in the patent literature. Subsequently, this terminology appeared in the scientific literature (David *et al.*, 1992). Earlier literature indicates that the name vibriolysin was applied to a toxin produced by *Vibrio parahaemolyticus* (Goshima *et al.*, 1978), but this terminology is no longer in use. Currently, vibriolysin is formally listed as **aeromonolysin** by the IUBMB, even though the protease is derived from a *Vibrio* rather than an *Aeromonas* species.

## Activity and Specificity

Vibriolysin is an endopeptidase that prefers substrates in which bulky, hydrophobic amino acid residues constitute both the P2 and P1′ sites. Residues at P1′ are significantly more important for catalytic activity than the P1 residue donating the carboxyl portion of the peptide bond (Bayliss *et al.*, 1980). The presence of Leu at the P1′ site results in the highest $k_{cat}/K_m$ ratios for the aliphatic amino acid series. However, Phe at this position was shown to be the most favored residue. In this respect, the specificity of the protease (see below) differs from those of thermolysin (Chapter 351) (Bayliss *et al.*, 1980) and bacillolysin (Chapter 352) (Holmquist & Vallee, 1976). The catalytic efficiency of vibriolysin is governed by three residues on the C-terminal side and two residues on the N-terminal side of the scissile bond (Bayliss *et al.*, 1980). Vibriolysin also exhibits esterase activity with a preference for hydrophobic amino acids adjacent to the cleavage site. The general trend indicates that the protease is more active on peptide bonds compared to ester-peptide pairs (Holmquist & Vallee, 1976).

Optimal proteolytic activity for vibriolysin is at neutral pH and at temperatures between 50 and 60°C (Durham, 1989). Assays for activity with azocasein (Fluka) (see Appendix 2 for full names and addresses of suppliers) as substrate were described by Durham (1990). The enzyme can also be assayed with synthetic peptides, FA-Gly+Ala-NH$_2$ (Wilkes *et al.*, 1988) or FA-Ala+Phe-NH$_2$ (Wilkes & Prescott, 1976; David *et al.*, 1992). Vibriolysin exhibits significant proteolysis of casein, hemoglobin, collagen, elastin and fibrin; the enzyme does not hydrolyze native proteins (Durham *et al.*, 1993).

Vibriolysin has been shown to be superactivated following acylation with *N*-hydroxysuccinimide esters (Holmquist *et al.*, 1976), presumably due to modification of Tyr117 (V.A. David, unpublished results). The enzyme is inactivated at low pH in the presence of chelators by dissociation of the metal ion cofactor (Durham, 1990), by EDTA, DTT, ME, cysteine, reduced glutathione, 1,10-phenanthroline (Wilkes & Prescott, 1976) and phosphoramidon (D.R. Durham, unpublished results). In addition, the protease is reversibly inactivated by substrate analogs and compounds containing aminoacyl hydroxamido or chloroacetyl groups (Wilkes *et al.*, 1988).

## Structural Chemistry

Vibriolysin is secreted into the culture medium as a single polypeptide with an $M_r$ of 44 800. The enzyme undergoes autocatalytic processing during purification, which can be enhanced by heating, to yield a protein with an $M_r$ of 34 800, as determined from sedimentation equilibrium analysis and from the amino acid composition of the purified protease (Griffin & Prescott, 1970). Evaluation of the protein by direct injection electrospray-mass spectrometry (ES-MS) and matrix-assisted laser desorption ionization-time of flight MS reveals molecular masses of 34 131 and 34 127, respectively. The protein contains 1 mol of zinc per mol of enzyme (Griffin & Prescott, 1970), and two disulfide bridges (Cys35-Cys61 and Cys277-Cys306) (David *et al.*, 1992). The pI of the protease is approximately 3.5 (Griffin & Prescott, 1970).

Inhibitor studies by Wilkes *et al.* (1988) demonstrated a protonated histidine group and an unprotonated aspartic/glutamic acid residue in close proximity to the active-site zinc. Based on the tertiary structure of thermolysin, the positions of the residues acting as the zinc ligands and those involved in the catalytic activity are known and are conserved for most prokaryotic metalloproteases (Häse & Finklestein, 1993). Proposed active-site residues for vibriolysin include His230, Glu148, Tyr162 and Asp222, and proposed zinc-binding ligands include His147, His151 and Glu171 (David *et al.*, 1992; Häse & Finklestein, 1993). In addition, vibriolysin has calcium-binding regions within the molecule analogous to those of thermolysin (V.A. David, unpublished results), even though calcium is not required for stability (Wilkes & Prescott, 1976).

## Preparation

Vibriolysin is produced during fermentation of *V. proteolyticus* in a relatively simple culture medium (Wilkes & Prescott, 1976; Durham, 1990). After a 24–28 h fermentation period, gram quantities are secreted into the medium. Expression of the vibriolysin gene (nprV) in *Bacillus subtilis* has been demonstrated by inserting the nucleotide sequence encoding the pro-*nprV* gene downstream from the subtilisin gene (*apr*) promoter and *apr* signal sequence. Transformation of protease-deficient mutants of *B. subtilis* with this DNA construction resulted in transformants that expressed vibriolysin, and further genetic modification of the recombinant strain by insertion of the *SacQ* gene increased expression (David *et al.*, 1992). Recombinant *V. proteolyticus* strains have also been constructed that improve protease titers by 3-fold (D.Z. Fortney & W.J. Jackson, unpublished results).

Purification of vibriolysin is complicated by the secretion of a thermostable aminopeptidase during growth that is physically similar to the metalloprotease with respect to molecular mass and isoelectric point (Griffin & Prescott, 1970). However, the aminopeptidase can be inactivated by incubation under alkaline conditions (pH 11.5), whereas vibriolysin is stable (Durham, 1990). Thus, concentration of the culture medium by ultrafiltration, followed by alkali treatment and anion-exchange chromatography results in a purified protease preparation (Durham, 1990). A recent report indicates that vibriolysin can be purified by hydrophobic interaction chromatography, preceded by inactivation of the aminopeptidase with amastatin (Dickson *et al.*, 1996).

## Biological Aspects

An open reading frame (ORF) specifying 609 amino acids that contains the vibriolysin structural gene has been cloned and sequenced (David *et al.*, 1992). A putative signal sequence has been identified: the first 24 amino acids contain charged residues (Lys3, Arg6 and His7) followed by a series of hydrophobic residues (Ile8 to Leu20), a helix-disrupting residue (Pro21) with an uncharged Ala as the proposed C-terminal amino acid. A prosequence of 172 amino acids was noted between the signal sequence and the beginning of the mature gene. The N-terminus of the mature protein sequence was identified by comparison with the N-terminal sequence of purified vibriolysin. The ORF of the sequence from the mature N-terminus to the stop codon encodes a polypeptide (413 amino acids) with an $M_r$ of 44 800. As mentioned earlier, the protease undergoes additional processing to produce a less hydrophobic protease with an $M_r$ of 34 800. This is presumably due to the removal of a C-terminal portion of the 'mature' protease. Similar C-terminal processing has been reported for bacterial proteases and hemolysins (Häse & Finklestein, 1993).

The protease is not associated with any disease processes. In contrast, vibriolysin has been shown to have potential medical benefits as an effective enzymatic debridement agent for the removal of necrotic tissue from wounds such as burns or cutaneous ulcers (Durham *et al.*, 1993) by virtue of its ability to distinguish between viable and nonviable tissue, to hydrolyze denatured protein components (such as collagen, fibrin and elastin) found in devitalized tissue, to function at physiological pH and temperature and to exhibit compatibility with adjunct therapies. The stability properties of the protease are also attractive for this use, since vibriolysin can be formulated into hydrophilic creams and maintain proteolytic activity for years during storage at ambient temperature (Durham *et al.*, 1993; D.Z. Fortney, unpublished results). In addition, a recent nonclinical study by Nanney *et al.* (1995) demonstrated that the protease stimulated the healing of partial-thickness burn wounds.

Vibriolysin has industrial application for enzyme-mediated synthesis of dipeptides. *V. proteolyticus* neutral protease has been shown, in a water-miscible solvent, to mediate the coupling of *N*-protected aspartic acid and phenylalanine methyl ester to yield *N*-protected aspartylphenylalanine methyl ester, a precursor to the sweetener aspartame (Bull *et al.*, 1992).

## Distinguishing Features

Vibriolysin, unlike most microbial metalloproteases, contains disulfide bonds and is not dependent on calcium ions for stability (V.A. David, unpublished results; Wilkes & Prescott, 1976). In addition, the enzyme exhibits unusual alkaline stability properties for this class of proteases, which closely resemble the alkaline stability properties observed for serine proteases used in laundry detergents (Durham, 1989). The protease is highly active on a wide variety of proteins, and the specificity and catalytic activity differs from the structurally similar proteases, thermolysin (Chapter 351) and pseudolysin (Chapter 357) (Thayer *et al.*, 1991). For example, comparisons of the literature values (Bayliss *et al.*, 1980; Thayer *et al.*, 1991) for the $k_{cat}/K_m$ ratios (mM$^{-1}$ s$^{-1}$) for a series of Z-Gly-S1'-NH$_2$ substrates, where S1' is the variable amino acid, revealed the following specificities and activities, in parenthesis: Phe (1720) > Tyr (32) > Leu (16) for vibriolysin; Leu (2.7) = Phe (2.6) > Tyr (0.1) for thermolysin; and Phe (22.7) > Tyr (8.7) > Leu (3.6) for pseudolysin.

## Related Peptidases

Unlike thermolysin, the *V. proteolyticus* protease is similar to pseudolysin (Thayer *et al.*, 1991) and *V. vulnificus* elastolytic protease (Kothary & Kreger, 1987) in that it contains disulfide bridges. Homologous metalloproteases from *Vibrio* species, as well as the aminopeptidase from *V. proteolyticus* (Chapter 491) undergo autocatalytic processing of the C-terminus to remove small polypeptides (Häse & Finklestein, 1991, 1993; David *et al.*, 1992; Milton *et al.*, 1992; Van Heeke *et al.*, 1992). These C-terminal regions demonstrate significant homology among the metalloproteases and display similarity to the C-terminal region of the serine protease from *V. alginolyticus* that purportedly undergoes similar C-terminal processing (Deane *et al.*, 1989). The role of the C-terminal region in the secretion of vibriolysin has not been established.

## Further Reading

A detailed review of metalloproteases from both gram-negative and gram-positive bacteria can be found in Häse & Finklestein (1993).

## References

Bayliss, M.E., Wilkes, S.H. & Prescott, J.W. (1980) *Aeromonas* neutral protease: specificity toward extended substrates. *Arch. Biochem. Biophys.* **204**, 214–219.

Bull, C., Durham, D.R., Gross, A., Kupper, R.J. & Walter, J.F. (1992) Aspartame synthesis in a miscible organic solvent. In: *Biocatalytic Production of Amino Acids and Derivatives* (Rozzell, J.D. & Wagner, F., eds). Munich: Hanser Publishers, pp. 241–259.

David, V.A., Deutch, A.H., Sloma, A., Pawlyk, D., Ally, A. & Durham, D.R. (1992) Cloning, sequencing and expression of the gene encoding the extracellular neutral protease, vibriolysin, of *Vibrio proteolyticus. Gene* **112**, 107–112.

Deane, S.M., Robb, F.T., Robb, S.M. & Woods, D.R. (1989) Nucleotide sequence of the *Vibrio alginolyticus* calcium-dependent, detergent-resistant alkaline serine exoprotease. *Gene* **76**, 281–288.

Dickson, K.M., Sadler, A.M. & Warneck, J.B.H. (1996) Purified vibriolysin. PCT Patent Application GB96/00258.

Durham, D.R. (1989) Cleaning compositions containing protease produced by *Vibrio* and method of use. US Patent 4,865,983.

Durham, D.R. (1990) The unique stability of *Vibrio proteolyticus* neutral protease under alkaline conditions affords a selective step for purification and use in amino acid coupling reactions. *Appl. Environ. Microbiol.* **56**, 2277–2281.

Durham, D.R., Fortney, D.Z. & Nanney, L.B. (1993) Preliminary evaluation of vibriolysin, a novel proteolytic enzyme composition suitable for the debridement of burn wound eschar. *J. Burn Care Rehabil.* **14**, 544–551.

Goshima, K., Owaribe, K., Yamanaka, H. & Yoshino, S. (1978) Requirement of calcium ions for cell degeneration with a toxin (vibriolysin) from *Vibrio parahaemolyticus. Infect. Immun.* **22**, 821–832.

Griffin, T.B. & Prescott, J.M. (1970) Some physical characteristics of a proteinase from *Aeromonas proteolytica. J. Biol. Chem.* **245**, 1348–1356.

Häse, C.C. & Finklestein, R.A. (1991) Cloning and nucleotide sequence of the *Vibrio cholerae* hemagglutinin/protease (HA/protease) gene and construction an HA/protease-negative strain. *J Bacteriol.* **173**, 3311–3317.

Häse, C.C. & Finklestein, R.A. (1993) Bacterial extracellular zinc-containing metalloproteases. *Microbiol. Rev.* **57**, 823–837.

Holmquist, B. & Vallee, B.L. (1976) Esterase activity of zinc neutral proteases. *Biochemistry* **15**, 101–107.

**M**

Holmquist, B., Blumberg, S. & Vallee, B.L. (1976) Superactivation of neutral proteases: acylation with *N*-hydroxysuccinimide esters. *Biochemistry* **15**, 4675–4680.

Kothary, M.H. & Kreger, A.S. (1987) Purification and characterization of an elastolytic protease of *Vibrio vulnificus. J. Gen. Microbiol.* **133**, 1783–1791.

Merkel, J.R., Traganza, E.D., Mukherjee, B.B., Griffin, T.B. & Prescott, J.M. (1964) Proteolytic activity and general characteristics of a marine bacterium, *Aeromonas proteolytica* sp. N. *J. Bacteriol.* **87**, 1227–1233.

Milton, D.L., Norqvist, A. & Wolf-Watz, H. (1992) Cloning of the metalloprotease gene involved in the virulence mechanism of *Vibrio anguillarum. J. Bacteriol.* **174**, 7235–7244.

Nanney, L.B., Fortney, D.Z. & Durham, D.R. (1995) Effect of vibriolysin, an enzymatic debridement agent, on the healing of partial-thickness burn wounds. *Wound Repair Regen.* **3**, 442–448.

Thayer, M.M., Flaherty, K.M. & McKay, D.B. (1991) Three-dimensional structure of the elastase of *Pseudomonas aeruginosa* at 1.5-Å resolution. *J Biol. Chem.* **266**, 2864–2871.

Van Heeke, G., Denslow, S., Watkins, J.R., Wilson, K.J. & Wagner, F.W. (1992) Cloning and nucleotide sequence of the *Vibrio proteolyticus* aminopeptidase gene. *Biochim. Biophys. Acta* **1131**, 337–340.

Wilkes, S.H. & Prescott, J.M. (1976) *Aeromonas* neutral protease. *Methods Enzymol.* **45**, 404–415.

Wilkes, S.H., Bayliss, M.E. & Prescott, J.M. (1988) Critical ionizing groups in *Aeromonas* neutral protease. *J. Biol. Chem.* **263**, 1821–1825.

*Don R. Durham*
*Bio Science Contract Production Corp.,*
*5901 E. Lombard Street,*
*Baltimore, MD 21224, USA*
*Email: BSCP@erolf.com*

# 356. Hemagglutinin/protease

## Databanks

*Peptidase classification: clan MA, family M4, MEROPS ID: M04.004*
*NC-IUBMB enzyme classification: none*
*Databank codes:*

| Species | SwissProt | PIR | EMBL (cDNA) | EMBL (genomic) |
|---|---|---|---|---|
| *Vibrio cholerae* | P24153 | A40159 A42358 | M59466 | – |
| *Helicobacter pylori* | | S54406 S38665 | Z27239 | |

## Name and History

Cholera is an infectious disease caused by the bacterium *Vibrio cholerae* and characterized by severe vomiting and watery diarrhea. In the late 1940s, Burnet discovered mucinase (EC 3.3.1.35) activity in the culture filtrate of *V. cholerae*, which participated in the desquamation of epithelium from pieces of guinea pig intestine (Burnet & Stone, 1947; Burnet, 1949). In 1982, Finkelstein and his colleagues, who were characterizing agglutination of erythrocytes by this organism (Hanne & Finkelstein, 1982), found and purified a soluble hemagglutinin (Finkelstein & Hanne, 1982). They discovered the soluble hemagglutinin had protease activity against fibronectin, ovomucin and lactoferrin and they proposed the bifunctional protein as a possible pathogenic factor (Finkelstein *et al.*, 1983).

The enzyme had been called **mucinase** of *V. cholerae* at first (Burnet & Stone, 1947; Burnet, 1949). As the protein was studied independently as a hemagglutinin without recognition of its proteolytic activities, it was called **cholera lectin** (Finkelstein & Hanne, 1982) or **soluble hemagglutinin** (Hanne & Finkelstein, 1982) to distinguish it from the many other cell-associated hemagglutinins (Hanne & Finkelstein, 1982). Today the bifunctional protein is more often called **hemagglutinin/protease** or **HA/protease**.

Metalloproteases clearly homologous to the hemagglutinin/protease of *Vibrio cholerae* are found in the fish pathogen *Vibrio anguillarum* (Norqvist *et al.*, 1990; Milton *et al.*, 1992) (see also Chapter 355) and *Helicobacter pylori* (Smith *et al.*, 1994), which is considered to be a causative agent of gastritis, peptic ulcer and gastric carcinoma.

## Activity and Specificity

The pH optimum is between 7.5 and 8.0 (Booth *et al.*, 1983). For full protease activity $Ca^{2+}$ is required and 8-hydoxyquinoline inhibits the activity (Booth *et al.*, 1983). The purified hemagglutinin/protease has several pIs, namely 6.3 and 4.7–5.3, does not stain as a glycoprotein and is relatively hydrophobic (Finkelstein & Hanne, 1982). Hemagglutinating activity is still effective at 4°C, whereas protease activity is not (Finkelstein & Hanne, 1982).

Examples of hemagglutinin/protease cleavage specificity are for El Tor cytolysin, Ala-Ala-Ser-Gly+Phe-Ala-Ser-Pro (Nagamune *et al.*, 1996) and for subunit A of heat-labile enterotoxin of enterotoxigenic *Escherichia coli*, Gly-Asn-Ala-Pro-Arg-Ser+Ser+Met-Ser-Asn-Thr. This cleavage specificity resembles that of *Bacillus thermoproteolyticus* thermolysin (Chapter 351) which is an endopeptidase with rather broad specificity.

Hemagglutinating activity can be measured using a chicken erythrocyte assay (Hanne & Finkelstein, 1982). Protease activity is detected conventionally by using a single-diffusion technique in agar gel containing skim milk as a substrate (Finkelstein *et al.*, 1983). Hemagglutinin/protease is also active on FA-Gly+Leu-NH$_2$, a synthetic substrate for thermolysin and other similar proteases (Booth *et al.*, 1983).

Both hemagglutinating and protease functions are inhibited by chelating agents, including 2-(*N*-hydroxycarboxamido)-4-methyl pentanoyl-L-Ala-Gly-NH$_2$ (Zincov), a hydroxamic acid derivative specifically designed to inhibit zinc metalloproteases (Booth *et al.*, 1983).

## Structural Chemistry

The structural gene (*hap*) for hemagglutinin/protease of *V. cholerae* consists of a 1827 bp encoding a 609 amino acid polypeptide (Häse & Finkelstein, 1991). The deduced protein contains a putative signal sequence followed by a large propeptide. The extracellular hemagglutinin/protease consists of 414 amino acids with a deduced molecular mass of 46 700, which is processed to the 32 kDa form (Häse & Finkelstein, 1991). The deduced amino acid sequence of the mature hemagglutinin/protease shows 61.5% identity with the pseudolysin (Chapter 357) (Häse & Finkelstein, 1991). In a *V. cholerae* non-O1 strain, hemagglutinin/protease was processed from a 34 kDa intermediate to a 32 kDa form by autodigestion at the C-terminus with a resultant increase in proteolytic activity and decrease in hemagglutinating activity (Naka *et al.*, 1992). Electron microscopy shows that purified preparations form long filamentous polymers (Finkelstein & Hanne, 1982; Finkelstein *et al.*, 1983).

Certain residues of hemagglutinin/protease correspond to those in *P. aeruginosa* elastase (Bever & Iglewski, 1988), which is also zinc/calcium dependent. These are His343, His347, Glu367 for zinc liganding and Glu344, Tyr358 and His426 for the active site (Chapter 357) (Häse & Finkelstein, 1990).

Monoclonal antibodies to an extracellular *Pseudomonas cepacia* protease neutralized pseudolysin, *P. pseudomallei* protease and *V. cholerae* hemagglutinin/protease (Kooi *et al.*, 1994).

## Preparation

Natural sources that have yielded essentially homogeneous preparations of hemagglutinin/protease include *Vibrio cholerae* O1 (Finkelstein & Hanne, 1982), O139 (Bengal; Naka *et al.*, 1995) and non-O1 (Honda *et al.*, 1989).

## Biological Aspects

The enzyme is known to be produced by O1 (Finkelstein *et al.*, 1983), O139 (Naka *et al.*, 1995; Jonson *et al.*, 1996) and non-O1 (Honda *et al.*, 1987, 1989) serogroups of *Vibrio cholerae*, *Helicobacter pylori* (Smith *et al.*, 1994) and *Vibrio anguillarum* (Milton *et al.*, 1992). It has been proposed that hemagglutinin/protease may be a virulence factor and protective antigen of *V. cholerae* (Hanne & Finkelstein, 1982; Finkelstein *et al.*, 1983). Specific antibody against hemagglutinin/protease inhibits the attachment of vibrios to intestinal epithelium (Finkelstein & Hanne, 1982; Finkelstein *et al.*, 1983). The enterotoxicity of live *V. cholerae* to rabbits was enhanced by pretreatment of the small intestine with hemagglutinin/protease prior to inoculation (Ichinose *et al.*, 1994). Proteolytic activity of hemagglutinin/protease is likely to play a role in activation of toxins of *V. cholerae* such as cholera toxin by nicking of subunit A (Booth *et al.*, 1984), or El Tor hemolysin by processing to active cytolysin (Nagamune *et al.*, 1996). However, clinical cholera infection gives rise to weak serum and gut mucosal antibody responses to hemagglutinin/protease whereas the responses to cholera toxin or lipopolysaccharide are much stronger (Svennerholm *et al.*, 1984). On the other hand, protease activity may be responsible for detachment of the vibrios from the cells by digestion of several putative receptors for *V. cholerae* adhesins (Finkelstein *et al.*, 1992).

## Distinguishing Features

Polyclonal (Finkelstein & Hanne, 1982) and monoclonal antisera (Honda *et al.*, 1991) against the hemagglutinin/protease from *V. cholerae* have been prepared. The monoclonal antibody neutralizes protease activity but not hemagglutination. These antibodies are not commercially available at the time of writing.

## References

Bever, R.A. & Iglewski, B.H. (1988) Molecular characterization and nucleotide sequence of the *Pseudomonas aeruginosa* elastase structural gene. *J. Bacteriol.* **170**, 4309–4314.

Booth, B.A., Boesman-Finkelstein, M. & Finkelstein, R.A. (1983) *Vibrio cholerae* soluble hemagglutinin/protease is a metalloenzyme. *Infect. Immun.* **42**, 639–644.

Booth, B.A., Boesman-Finkelstein, M. & Finkelstein, R.A. (1984) *Vibrio cholerae* hemagglutinin/protease nicks *cholerae* enterotoxin. *Infect. Immun.* **45**, 558–560.

Burnet, F.M. (1949) Ovomucin as a substrate for the mucinolytic enzymes of *V. cholerae* filtrates. *Aust. J. Exp. Biol. Med. Sci.* **27**, 245–252.

Burnet, F.M. & Stone, J.D. (1947) Desquamation of intestinal epithelium *in vitro* by *V. cholerae* filtrates: characterization of mucinase and tissue disintegrating enzymes. *Aust. J. Exp. Biol. Med. Sci.* **25**, 219–226.

Finkelstein, R.A. & Hanne, L.F. (1982) Purification and characterization of the soluble hemagglutinin (cholera lectin) produced by *Vibrio cholerae*. *Infect. Immun.* **36**, 1199–1208.

Finkelstein, R.A., Boesman-Finkelstein, M. & Holt, P. (1983) *Vibrio cholerae* hemagglutinin/lectin/protease hydrolyzes fibronectin and ovomucin: F.M. Burnet revisited. *Proc. Natl Acad. Sci. USA* **80**, 1092–1095.

Finkelstein, R.A., Boesman-Finkelstein, M., Chang, Y. & Häse, C.C. (1992) *Vibrio cholerae* hemagglutinin/protease, colonial variation, virulence, and detachment. *Infect. Immun.* **60**, 472–478.

Hanne, L.F. & Finkelstein, R.A. (1982) Characterization and distribution of the hemagglutinins produced by *Vibrio cholerae*. *Infect. Immun.* **36**, 209–214.

Häse, C.C. & Finkelstein, R.A. (1990) Comparison of the *Vibrio cholerae* hemagglutinin/protease and the *Pseudomonas aeruginosa* elastase. *Infect. Immun.* **58**, 4011–4005.

Häse, C.C. & Finkelstein, R.A. (1991) Cloning and nucleotide sequence of the *Vibrio cholerae* hemagglutinin/protease (HA/protease) gene and construction of an HA/protease-negative strain. *J. Bacteriol.* **173**, 3311–3317.

Honda, T., Booth, B.A., Boesman-Finkelstein, M. & Finkelstein, R.A. (1987) Comparative study of *Vibrio cholerae* non-O1 protease and soluble hemagglutinin with those of *Vibrio cholerae* O1. *Infect. Immun.* **55**, 451–454.

Honda, T., Lertpocasombat, K., Hata, A., Miwatani, T. & Finkelstein, R.A. (1989) Purification and characterization of a protease produced by *Vibrio cholerae* non-O1 and comparison with a protease of *V. cholerae* O1. *Infect. Immun.* **57**, 2799–2803.

Honda, T., Hata-Naka, A., Lertpocasombat, K. & Miwatani, T. (1991) Production of monoclonal antibodies against a hemagglutinin/protease of *Vibrio cholerae* non-O1. *FEMS Microbiol. Lett.* **62**, 227–230.

Ichinose, Y., Ehara, M., Honda, T. & Miwatani, T. (1994) The effect on enterotoxicity of protease purified from *Vibrio cholerae* O1. *FEMS Microbiol. Lett.* **115**, 265–271.

Jonson, G., Osek, J., Svennerholm, A.-M. & Holmgren, J. (1996) Immune mechanisms and protective antigens of *Vibrio cholerae* serogroup O139 as a basis for vaccine development. *Infect. Immun.* **64**, 3778–3785.

Kooi, C., Cox, A., Darling, P. & Sokol, P.A. (1994) Neutralizing monoclonal antibodies to an extracellular *Pseudomonas cepacia* protease. *Infect. Immun.* **62**, 2811–2817.

Milton, D.L., Norqvist, A. & Wolf-Watz, H. (1992) Cloning of a metalloprotease gene involved in the virulence mechanism of *Vibrio anguillarum*. *J. Bacteriol.* **174**, 7235–7244.

Nagamune, K., Yamamoto, K., Naka, A., Matsuyama, J., Miwatani, T. & Honda, T. (1996) *In vitro* proteolytic processing and activation of the recombinant precursor of El Tor cytolysin/hemolysin (Pro-HlyA) of *Vibrio cholerae* by soluble hemagglutinin/protease of *V. cholerae*, trypsin, and other proteases. *Infect. Immun.* **64**, 4655–4658.

Naka, A., Yamamoto, K., Miwatani, T. & Honda, T. (1992) Characterization of two forms of hemagglutinin/protease produced by *Vibrio cholerae* non-O1. *FEMS Microbiol. Lett.* **77**, 197–200.

Naka, A., Yamamoto, K., Albert, M.J. & Honda, T. (1995) *Vibrio cholerae* O139 produces a protease which is indistinguishable from the haemagglutinin/protease of *Vibrio cholerae* O1 and non-O1. *FEMS Immunol. Med. Microbiol.* **11**, 87–90.

Norqvist, A., Norrman, B. & Wolf-Watz, H. (1990) Identification and characterization of a zinc metalloprotease associated with invasion by the fish pathogen *Vibrio anguillarum*. *Infect. Immun.* **58**, 3731–3736.

Smith, A.W., Chahal, B. & French, G.L. (1994) The human gastric pathogen *Helicobacter pylori* has a gene encoding an enzyme first classified as a mucinase in *Vibrio cholerae*. *Mol. Microbiol.* **13**, 153–160.

Svennerholm, A.M., Levine, M.M. & Holmgren, J. (1984) Weak serum and intestinal antibody responses to *Vibrio cholerae* soluble hemagglutinin in *cholerae* patients. *Infect. Immun.* **45**, 792–794.

*Koichiro Yamamoto*
*Department of Bacterial Infections,*
*Research Institute for Microbial Diseases,*
*Osaka University, Suita, Osaka 565, Japan*
*Email: yamak@biken.osaka-u.ac.jp*

# *357. Pseudolysin*

## *Databanks*

*Peptidase classification: clan MA, family M4, MEROPS ID: M04.005*
*NC-IUBMB enzyme classification: EC 3.4.24.26*
*Databank codes:*

| Species | SwissProt | PIR | EMBL (cDNA) | EMBL (genomic) |
|---|---|---|---|---|
| *Pseudomonas aeruginosa* | P14756 | – | M24531 | – |

Brookhaven Protein Data Bank three-dimensional structures:

| Species | ID | Resolution | Notes |
| --- | --- | --- | --- |
| *Pseudomonas aeruginosa* | 1AKL | 2 | |
| | 1EZM | 1.5 | |

## Name and History

**Pseudolysin** is the current name of the most abundant extracellular endopeptidase of *Pseudomonas aeruginosa* that is commonly called **Pseudomonas aeruginosa elastase**. Pseudo alludes to the source (**Pseudo**monas) and lysin reflects the metalloendopeptidase nature of the enzyme as well as its ability to cause lysis of insoluble elastin (Morihara *et al.*, 1965) and of some connective tissues (Kessler *et al.*, 1977, 1982). Destruction of the arterial elastic laminae in human systemic *P. aeruginosa* infections (Margaretten *et al.*, 1961) was the first evidence that *P. aeruginosa* may secrete an elastinolytic protease. Numerous studies have subsequently established pseudolysin as a major extracellular virulence factor of *P. aeruginosa* (Morihara & Homma, 1985). The name elastase relates to the ability of pseudolysin to degrade elastin and was given by Morihara *et al.* (1965), the first investigators to isolate and characterize this enzyme. In light of its remarkable similarity to microbial neutral metalloendopeptidases of the thermolysin family (M4) (Chapter 350), pseudolysin has also been termed **P. aeruginosa neutral proteinase** (Morihara, 1974), but this name has not been widely used.

## Activity and Specificity

Pseudolysin extensively digests denatured protein substrates such as casein, hemoglobin, ovalbumin, fibrin and oxidized insulin B chain (Morihara *et al.*, 1965; Morihara, 1995). Specific activity against casein is about five times higher than that of trypsin, $\alpha$-chymotrypsin, or subtilisin BPN′, and 10-fold higher than that of aeroginolysin (Chapter 387). Pseudolysin can also degrade proteins of potential biological significance to the host upon infection with *P. aeruginosa*. In addition to elastin, these include collagens III and IV (Heck *et al.*, 1986a), laminin (Heck *et al.*, 1986b), proteoglycans (Kessler *et al.*, 1977, 1982), immunoglobulins G and A (Döring *et al.*, 1981), $\alpha_1$-proteinase inhibitor (Morihara *et al.*, 1979), complement components (Schultz & Miller, 1974), cytokines (Parmely *et al.*, 1990) and coagulation factor XII (Chapter 49) (Holder & Neely, 1989).

Pseudolysin favors hydrophobic or aromatic amino acid residues at the P1′ position (Morihara & Homma, 1985; Morihara, 1974, 1995). In a series of synthetic substrates with the general structure Z-Phe+Xaa-Ala, the order of preference for Xaa is Phe > Leu > Tyr > Val > Ile, i.e. aromatic residues at P1′ seem to be preferred to aliphatic residues. Ala is favored at the P1 and P2′ positions, and elongation of the substrate to the P2 and P2′ positions results in a marked increase in the rate of hydrolysis. In the presence of methanol ($\sim$34%), pseudolysin catalyzes the formation of peptide bonds between Z-Ala and various hydrophobic aminoacyl amides (Pauchon *et al.*, 1993).

Proteolytic activity can be assayed with casein (Morihara, 1995) or azocasein (Kessler *et al.*, 1982) whereas elastinolytic activity is determined with elastin–Congo Red (Rust *et al.*, 1994) or orcein–elastin (Morihara, 1995). Elastin nutrient agar plate assays are used to detect elastase production by *P. aeruginosa* (Rust *et al.*, 1994). Peptidase activity can be determined spectrophotometrically with FA-Gly+Leu-NH$_2$ or FA-Gly+Leu-Ala-NH$_2$ (Morihara, 1995), or with the fluorogenic substrate Abz-Ala-Gly+Leu-Ala-Nba (Abz: *o*-aminobenzoyl; Nba: 4-nitrobenzylamide) (Nishino & Powers, 1980). The latter is available from Enzyme Systems Products (see Appendix 2 for full names and addresses of suppliers).

The pH optimum is 7–8 and Ca$^{2+}$ is required for stability. Activity is inhibited by metal chelators, including EDTA, EGTA, 1,10-phenanthroline and tetraethylene pentamine. Zn$^{2+}$ ($>$0.01 mM) is also inhibitory (Morihara, 1995; E. Kessler, unpublished results). Phosphoramidon, phosphoryl-dipeptides such as phosphoryl-Leu-Phe and phosphoryl-Leu-Trp, and peptides containing thiol or hydroxamate groups such as HSAc-Phe-Leu, HSCH$_2$CH-(CH$_2$C$_6$H$_5$)CO-Ala-Gly-NH$_2$, or HONHCOCH(CH$_2$C$_6$H$_5$)CO-Ala-Gly-NH$_2$), are potent reversible inhibitors ($K_i$ values 0.04 nM to $\sim$1 µM) (Morihara & Tsuzuki, 1978; Nishino & Powers, 1980; Kessler *et al.*, 1982; Poncz *et al.*, 1984). Inhibition by these compounds is not specific to pseudolysin as thermolysin (Chapter 351) and other M4 family members are also inhibited. ClCH$_2$CO-HOLeu-Ala-Gly-NH$_2$ (HOLeu: *N*-hydroxyleucine) is an irreversible inhibitor of pseudolysin (Nishino & Powers, 1980).

## Structural Chemistry

The molecular mass of pseudolysin is 33 kDa and its pI is 5.9 (Bever & Iglewski, 1988; Morihara *et al.*, 1965). It is synthesized as a preproenzyme with 498 amino acid residues and a molecular mass of 53.6 kDa. This larger precursor (see Figure 357.1) has a 2.6 kDa (23 residues) signal peptide that is followed by an 18 kDa (174 residues) propeptide and the 33 kDa mature domain (301 residues) (Bever & Iglewski, 1988; Fukushima *et al.*, 1989; Kessler *et al.*, 1992). Pseudolysin shares considerable sequence homology with thermolysin. This homology is especially pronounced (48% identity) in the region between residues 136 and 180 including part of the active site (Bever & Iglewski, 1988; Fukushima *et al.*, 1989). It contains one zinc atom that is essential for activity and one atom of calcium that is required for stability. A conserved HEXXH sequence at positions 140–144 and a conserved Glu164 are involved in zinc binding, the zinc ligands being His140, His144 and Glu164 (Figure 357.1). Glu141 is essential for catalysis whereas Tyr155, His223 and, apparently Asp221, are involved with substrate binding (Thayer *et al.*, 1991). The calcium atom is ligated by the carboxyl groups of Glu172, Glu175, Asp136 and Asp183, the carbonyl of Leu185, and one molecule of H$_2$O (Thayer *et al.*, 1991). Two disulfide bonds, Cys30-Cys58

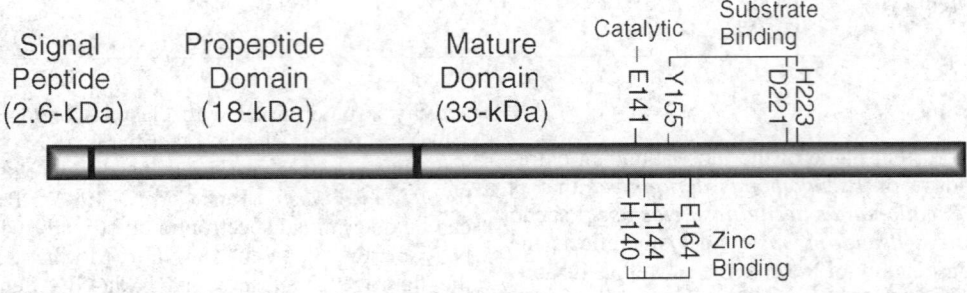

Figure 357.1  Schematic representation of the structure of prepropseudolysin.

and Cys270-Cys297, are also present. The three-dimensional structure of pseudolysin is almost superimposable with that of thermolysin (Chapter 351) (Thayer *et al*., 1991). Both have distinct N-terminal and C-terminal domains that are separated by a cleft, and in both, the zinc-binding residues and the substrate-binding residues are located at the inner surface of the cleft. However, the active-site cleft is significantly more 'open' in pseudolysin than in thermolysin. This may account for the difference in specificity against amino acid residues at the P1′ position (aromatic residues are preferred by pseudolysin as opposed to aliphatic ones favored by thermolysin) and ability to degrade elastin. In both enzymes, the N-terminal domain is constructed primarily of antiparallel $\beta$ strands while the C-terminal domain is $\alpha$ helical, and the two helices which span the active site and the loop connecting them, are identical in length.

## Preparation

*P. aeruginosa* strains IFO 3455 (Morihara *et al*., 1965), FRD2 (McIver *et al*., 1991), PAKS 1 (Wretlind & Wadström, 1977), and Habs serotype 1 (Kessler *et al*., 1977) produce high levels of pseudolysin ($>50\,\mu g\,ml^{-1}$ culture filtrate) and are among the best sources for the enzyme. Pseudolysin has been purified from the culture filtrates of these strains primarily by procedures involving ammonium sulfate precipitation, acetone precipitation, and DEAE-cellulose chromatography (Morihara *et al*., 1965; Kessler *et al*., 1982; Rust *et al*., 1994; Morihara, 1995), but it has also been purified by isoelectric focusing and gel chromatography (Wretlind & Wadström, 1977; Kreger & Gray, 1978). Affinity chromatography has also been used for the isolation of pseudolysin (Morihara & Tsuzuki, 1975; Nishino & Powers, 1980). The enzyme is readily crystallized from concentrated solutions (Morihara *et al*., 1965; Thayer *et al*., 1991).

The *lasB* gene encoding pseudolysin has been expressed in *Escherichia coli* (Fukushima *et al*., 1989; Tanaka *et al*., 1991) or *in vitro* by using a cell-free transcription–translation system (Bever & Iglewski, 1988). An overexpression system in *E. coli* has been described by McIver *et al*. (1991). Pseudolysin produced in *E. coli* is not secreted. To overcome this limitation, an overexpression system in a *lasB*-deleted *P.*

*aeruginosa* strain (FRD740) has also been developed (McIver *et al*., 1995). Since many *P. aeruginosa* strains secrete large amounts of the enzyme, artificial expression systems have no apparent advantage over wild-type *P. aeruginosa* strains as potential sources for pseudolysin. Also, purified pseudolysin is commercially available from Nagase Biochemicals (see Appendix 2 for full names and addresses of suppliers).

## Biological Aspects

The processing of prepropseudolysin and the secretion of active pseudolysin involve two proteolytic steps (Figure 357.2). The signal sequence is apparently removed by signal peptidase during translocation through the inner membrane. The propeptide domain is then rapidly cleaved in the periplasm, but it remains noncovalently associated with the processed 33 kDa periplasmic pseudolysin. The propeptide–pseudolysin complex is inactive and it accumulates temporarily in the periplasm (Kessler & Safrin, 1988a,b). Translocation of pseudolysin through the outer membrane requires a complex extracellular protein export apparatus (*Xcp*) of at least 11 genes which bears homology to the pullulanase (type II) export apparatus of *Klebsiella pneumoniae* (Tommassen *et al*., 1992).

Cleavage of the propeptide is autocatalytic (McIver *et al*., 1991). It may be blocked by exposing *P. aeruginosa* cells to EDTA (Kessler & Safrin, 1988b) or by growing the cells in a medium deficient in zinc and calcium ions (Olson & Ohman, 1992). Also, when *lasB* is expressed in *E. coli*, processing to the active 33 kDa pseudolysin occurs independently of any other *P. aeruginosa* gene product(s). Furthermore, when the codons for active-site residues His223 or Glu141 are changed to encode other amino acids, processing is blocked and the *E. coli* cells accumulate the 51 kDa proenzyme (McIver *et al*., 1991; Kawamoto *et al*., 1993). Intracellular accumulation of such mutant proelastase species is also observed when expressed in *P. aeruginosa* (McIver *et al*., 1993). Thus, propeptide removal is necessary for extracellular localization of pseudolysin in *P. aeruginosa*.

The propeptide plays important roles in the secretion of pseudolysin. Before processing, it limits the activity of

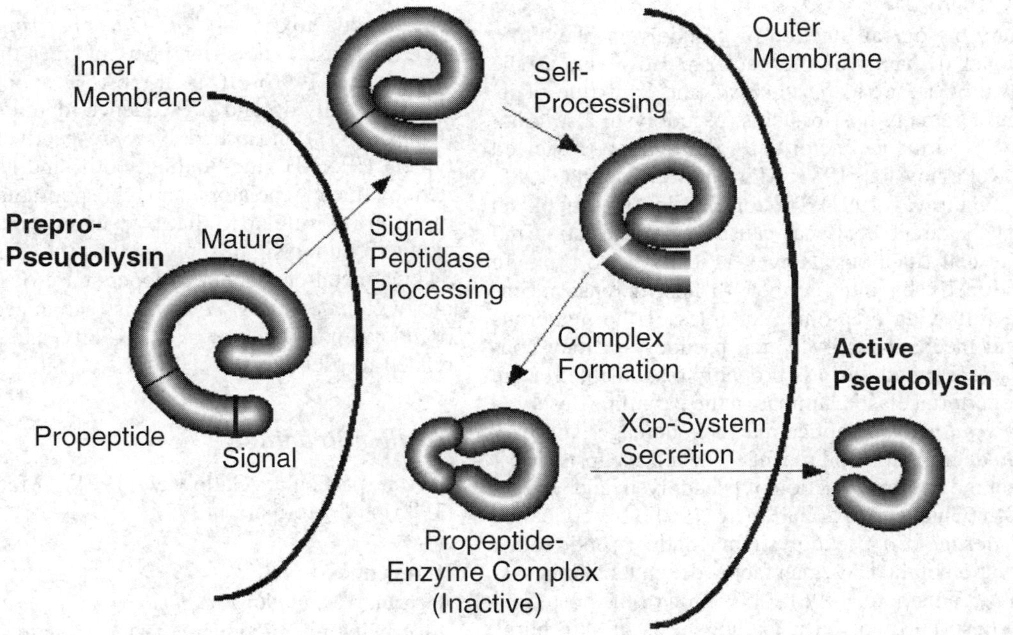

*Figure 357.2*   A scheme illustrating the sequence of events involved in pseudolysin secretion. Expression of *lasB* within the cytoplasm of *P. aeruginosa* yields a 53.6 kDa preproenzyme containing a signal peptide, a propeptide, and the C-terminal mature domain. During translocation across the inner membrane, the signal peptide (2.6 kDa) is removed to generate the 51 kDa proenzyme. In the periplasm, the propeptide (18 kDa) guides proper folding of pseudolysin and is then cleaved rapidly by autoproteolysis. The cleaved propeptide remains noncovalently associated with the processed enzyme (33 kDa) to inhibit its activity while cell bound and maintain the enzyme in a conformation that will ultimately permit activity. The propeptide–enzyme interaction also appears to be required for recognition of pseudolysin by the Xcp export machinery. Mature active pseudolysin is finally secreted through the outer membrane into the extracellular environment.

(pro)pseudolysin in cleaving the His┼Ala peptide bond at the propeptide–mature peptide junction. By binding to the processed periplasmic form of pseudolysin, the propeptide inhibits the activity of the cell-associated enzyme and this may serve to protect the cell from self-destruction by this potent protease (Kessler & Safrin, 1994). The covalently linked propeptide also functions as an intramolecular chaperone (Shinde & Inouye, 1993) that is necessary for correct folding and enzymatic activity of pseudolysin. Expression in *E. coli* of mutant *lasB* alleles lacking the propeptide-coding region results in the formation and accumulation in the cell of an inactive 33 kDa enzyme, while coexpression of such alleles with an independent *lasB* construct that encodes the prepro region can rescue much of the enzyme's activity (McIver *et al.*, 1995; Braun *et al.*, 1996). In *P. aeruginosa* coexpression of pseudolysin and the prepro region from separate plasmids restores activity as well as secretion of the active enzyme (McIver *et al.*, 1995; Braun *et al.*, 1996); this involves direct propeptide–pseudolysin interaction (McIver *et al.*, 1995). Thus, the propeptide functions as an intermolecular as well as intramolecular chaperone and it also plays a role in targeting of pseudolysin into the extracellular environment. The fate of the propeptide once it has fulfilled all of its functions is not known.

The *lasB* gene is located at ~28 min on the 75 min chromosome map of strain PAO1 (Shortridge *et al.*, 1991). A transcriptional start for *lasB* was identified 141 nucleotides upstream of the translational initiation codon, but the promoter DNA upstream from this showed no obvious consensus for a $\sigma$ factor (Rust *et al.*, 1996). Pseudolysin is produced late in the logarithmic phase of growth and the yields of the enzyme are decreased by increasing the concentration of iron, salts such as ammonium sulfate, or glucose. Subinhibitory concentrations of certain antibiotics may also decrease pseudolysin production (reviewed in Galloway, 1991). A complex hierarchy of global regulators controls the expression of *lasB* (as well as genes for other exoenzymes of *P. aeruginosa*) through a cascade of quorum-sensing mechanisms that promote gene expression when cell density is high. There are two such systems in *P. aeruginosa* and both show homology to the bioluminescence regulators LuxR–LuxI in *Vibrio fischeri*. LasR–LasI form one set of autoinducer-responsive transcriptional regulators, where LasR is a positive transcriptional activator of *lasB* (Gambello & Iglewski, 1991) and LasI synthesizes a specific *N*-acylhomoserinelactone which, in combination with LasR, induces transcription of *lasB* (Passador *et al.*, 1993; Pearson *et al.*, 1994). A second autoinducer-responsive system, called RhlR-RhlI, controls expression of *lasB* and other *P. aeruginosa* genes by a similar mechanism (Brint & Ohman, 1995; Ochsner & Reiser, 1995; Pearson *et al.*, 1995). Evidence for a hierarchical linkage between

**M**

the LasR–LasI and RhlR–RhlI systems was recently reported (Latifi *et al*., 1996).

The primary biological function of pseudolysin may be to meet nutritional demands of the bacteria. However, as the predominant protease of *P. aeruginosa*, and by virtue of its ability to degrade many host proteins, pseudolysin also functions as a major virulence factor of *P. aeruginosa* (reviewed in Wretlind & Pavlovskis, 1983; Morihara & Homma, 1985; Holder, 1985; Galloway, 1991). The contribution of pseudolysin to disease may be direct, i.e. it may cause tissue destruction and damage some cell functions. However, it may also promote virulence indirectly by interfering with host defense mechanisms. The following are some examples of the numerous situations and mechanisms by which pseudolysin may contribute to disease. Pseudolysin is probably responsible for the destruction of arterial elastic laminae in the vasculitis observed in cases of *Pseudomonas* septicemia. This damage is largely due to elastin degradation and recent evidence suggests that it may represent the combined action of pseudolysin and staphylolysin (LasA) (Chapter 507) (Galloway, 1991). In septicemia caused by *P. aeruginosa*, pseudolysin may induce septic shock through activation of the Hageman factor-dependent kinin system (Maeda & Yamamoto, 1996). Activation of the host kinin cascade may also be involved in pathogenesis of skin burns (Holder & Neely, 1989). In burns, pseudolysin appears to also support growth and invasiveness of the organisms (Holder, 1985). By rapidly degrading corneal proteoglycans, pseudolysin causes severe corneal destruction during *Pseudomonas* keratitis (Kessler *et al*., 1977; Kreger & Gray, 1978); it may also affect corneal damage indirectly, by activating endogenous corneal proteinases (Twinning *et al*., 1993). In *Pseudomonas* pneumonia pseudolysin may cause lung damage with hemorrhages and necrosis of alveolar septal cells (Gray & Kreger, 1979) and it may destroy alveolar epithelial cell junctions, which increases epithelial permeability to macromolecules (Azghani, 1996). Degradation of immunoglobulins, complement components and $\alpha_1$-proteinase inhibitor by pseudolysin may compromise the host defense mechanisms and impair normal control of the physiological functions of plasma proteases. Pseudolysin seems also to be involved in the chronic disease caused by *P. aeruginosa* in cystic fibrosis patients (Fick, 1989).

## Distinguishing Features

*P. aeruginosa* secretes several endopeptidases, including alkaline proteinase, staphylolysin (Chapter 507), and a lysine-specific serine endopeptidase (c.f. lysyl endopeptidase (Chapter 85)) which may interfere with pseudolysin assays of proteolytic as well as elastolytic activity (Kessler *et al*., 1997). Pseudolysin can be distinguished from these proteases by size, lack of immunological cross-reactivity, inhibition by phosphoramidon, and cleavage of FA-Gly-Leu-Ala-NH$_2$.

Polyclonal antibodies to pseudolysin have been described in many of the studies cited below and are readily obtained by standard procedures. Denatured pseudolysin (e.g. after exposure to SDS) is poorly recognized by antibodies to the native enzyme (Kessler & Safrin, 1988b). Monoclonal antibodies to pseudolysin have been reported (Lagacé & Fréchette, 1991; Yokota *et al*., 1992) but are not available commercially.

## Related Peptidases

The lambda toxin of *Clostridium perfringens* was recently identified as a new member of the thermolysin family (Jin *et al*., 1996). It is a 36 kDa endopeptidase sharing approximately 46 and 27 sequence identity with thermolysin (Chapter 351) and pseudolysin, respectively. It contains the typical HEXXH zinc-binding motif, and its activity is inhibited by metal chelators and phosphoramidon. Its substrate repertoire is similar to that of pseudolysin; however, in contrast to pseudolysin, it shows little or no elastolytic activity. A 34 kDa metal-dependent endopeptidase of *Burkholderia* (formerly *Pseudomonas*) *cepacia* is recognized by antibodies to pseudolysin (McKevitt *et al*., 1989), and thus may also be related to pseudolysin.

## Further Reading

See in particular Galloway (1991), Morihara & Homma (1985) and Morihara (1995).

## References

Azghani, A.O. (1996) *Pseudomonas aeruginosa* and epithelial permeability: role of virulence factors elastase and exotoxin A. *Am. J. Cell Mol. Biol.* **15**, 132–140.

Bever, R.A. & Iglewski, B.H. (1988) Molecular characterization and nucleotide sequence of the *Pseudomonas aeruginosa* elastase structural gene. *J. Bacteriol.* **170**, 4309–4314.

Braun, P., Tommassen, J. & Filloux, A. (1996) Role of the propeptide in folding and secretion of elastase of *Pseudomonas aeruginosa*. *Mol. Microbiol.* **19**, 297–306.

Brint, J. & Ohman, D. (1995) Synthesis of multiple exoproducts in *Pseudomonas aeruginosa* is under the control of RhlR-RhlI, another set of regulators in strain PAO1 with homology to the autoinducer-responsive LuxR-LuxI family. *J. Bacteriol.* **177**, 7155–7163.

Döring, G., Obernesser, J. & Botzenhart, K. (1981) Extracellular toxins of *P. aeruginosa*. II. Effect of two proteases on human immunoglobulins IgG, IgA and secretory IgA. *Zentralbl. Bakteriol. Parasitenkd. Infectionskr. Hyg. Abt. 1 Orig. Reihe A* **249**, 89–98.

Fick, R.B., Jr. (1989) Pathogenesis of the *Pseudomonas* lung lesion in cystic fibrosis. *Chest* **96**, 158–164.

Fukushima, J., Yamamoto, S., Morihara, K., Atsumi, Y., Takeuchi, H., Kawamoto, S. & Okuda, K. (1989) Structural gene and complete amino acid sequence of *Pseudomonas aeruginosa* IFO 3455 elastase. *J. Bacteriol.* **171**, 1698–1704.

Galloway, D.R. (1991) *Pseudomonas aeruginosa* elastase and elastolysis revisited: recent developments. *Mol. Microbiol.* **5**, 2315–2321.

Gambello, M.J. & Iglewski, B.H. (1991) Cloning and characterization of the *Pseudomonas aeruginosa lasR* gene, a transcriptional activator of elastase expression. *J. Bacteriol.* **173**, 3000–3009.

Gray, L. & Kreger, A. (1979) Microscopic characterization of rabbit lung damage production by *Pseudomonas aeruginosa* proteases. *Infect. Immun.* **23**, 150–159.

Heck, L.W., Morihara, K., McRae, W.B. & Miller, E.J. (1986a) Specific cleavage of human type III and IV collagens by *Pseudomonas aeruginosa* elastase. *Infect. Immun.* **51**, 115–118.

Heck, L.W., Morihara, K. & Abrahamson, D.R. (1986b) Degradation of soluble laminin and depletion of tissue-associated basement membrane laminin by *Pseudomonas aeruginosa* elastase and alkaline protease. *Infect. Immun.* **54**, 149–153.

Holder, I.A. (1985) The pathogenesis of infections owing to *Pseudomonas aeruginosa* using the burned mouse model: experimental studies from the Shriners Burn Institute, Cincinnati. *Can. J. Microbiol.* **4**, 393–402.

Holder, I.A. & Neely, A.N. (1989) *Pseudomonas* elastase acts as a virulence factor in burned hosts by Hageman factor-dependent activation of the host kinin cascade. *Infect. Immun.* **57**, 3345–3348.

Jin, F., Matsushita, O., Katayama, S-I., Jin, S., Matsushita, C., Minami, J. & Okabe, A. (1996) Purification, characterization, and primary structure of *Clostridium perfringens* λ-toxin, a thermolysin-like metalloprotease. *Infect. Immun.* **64**, 230–237.

Kawamoto, S., Shibano, Y., Fukushima, J., Ishii, N., Morihara, K. & Okuda, K. (1993) Site-directed mutagenesis of Glu-141 and His-223 in *Pseudomonas aeruginosa* elastase: catalytic activity, processing, and protective activity of the elastase against *Pseudomonas* infection. *Infect. Immun.* **61**, 1400–1405.

Kessler, E. & Safrin, M. (1988a) Partial purification and characterization of an inactive precursor of *Pseudomonas aeruginosa* elastase. *J. Bacteriol.* **170**, 1215–1219.

Kessler, E. & Safrin, M. (1988b) Synthesis, processing, and transport of *Pseudomonas aeruginosa* elastase. *J. Bacteriol.* **170**, 5241–5247.

Kessler, E. & Safrin, M. (1994) The propeptide of *Pseudomonas aeruginosa* elastase acts as an elastase inhibitor. *J. Biol. Chem.* **269**, 22726–22731.

Kessler, E., Kennah, H.E. & Brown, S.I. (1977) *Pseudomonas* protease. Purification, partial characterization, and its effect on collagen, proteoglycan, and rabbit corneas. *Invest. Ophthalmol. Vis. Sci.* **16**, 488–497.

Kessler, E., Israel, M., Landshman, N., Chechick, A. & Blumberg, S. (1982) *In vitro* inhibition of *Pseudomonas aeruginosa* elastase by metal-chelating peptide derivatives. *Infect. Immun.* **38**, 716–723.

Kessler, E., Safrin, M., Peretz, M. & Burstein, Y. (1992) Identification of cleavage sites involved in proteolytic processing of *Pseudomonas aeruginosa* preproelastase. *FEBS Lett.* **299**, 291–293.

Kessler, E., Safrin, M., Abrams, W.R., Rosenbloom, J. & Ohman, D.E. (1997) Inhibitors and specificity of *Pseudomonas aeruginosa* LasA. *J. Biol. Chem.* **272**, 9984–9889.

Kreger, A.S. & Gray, L.D. (1978) Purification of *Pseudomonas aeruginosa* proteases and microscopic characterization of pseudomonal protease-induced rabbit corneal damage. *Infect. Immun.* **19**, 630–648.

Latifi, A., Folingo, M., Tanaka, K., Williams, P. & Lazdunski, A. (1996) A hierarchical quorum-sensing cascade in *Pseudomonas aeruginosa* links the transcriptional activators LasR and RhlR(VsmR) to expression of the stationary-phase sigma factor RpoS. *Mol. Microbiol.* **21**, 1137–1146.

Lagacé, J. & Fréchette, M. (1991) Four epitopes of *Pseudomonas aeruginosa* elastase defined by monoclonal antibodies. *Infect. Immun.* **59**, 712–715.

Maeda, H. & Yamamoto (1996) Pathogenic mechanisms induced by microbial proteases in microbial infections. *Biol. Chem. Hoppe-Seyler* **377**, 217–226.

Margaretten, W., Nakai, H. & Landing, B.H. (1961) Significance of selective vasculitis and the 'bone marrow' syndrome in *Pseudomonas* septicemia. *N. Engl. J. Med.* **265**, 773–776.

McIver, K., Kessler, E. & Ohman, D.E. (1991) Substitution of active-site His-223 in *Pseudomonas aeruginosa* elastase and expression of the mutated *lasB* alleles in *Escherichia coli* show evidence for autoproteolytic processing of proelastase. *J. Bacteriol.* **173**, 7781–7789.

McIver, K.S., Olson, J.C. & Ohman, D.E. (1993) *Pseudomonas aeruginosa lasB1* mutants produce an elastase, substituted at active-site His-223, that is defective in activity, processing, and secretion. *J. Bacteriol.* **175**, 4008–4015.

McIver, K.S., Kessler, E., Olson, J.C. & Ohman, D.E. (1995) The elastase propeptide functions as an intramolecular chaperone required for elastase activity and secretion in *Pseudomonas aeruginosa*. *Mol. Microbiol.* **18**, 877–889.

McKevitt, A.I., Bajaksouzian, S., Klinger, J.D. & Woods, D.E. (1989) Purification and characterization of an extracellular protease from *Pseudomonas cepacia*. *Infect. Immun.* **57**, 771–778.

Morihara, K. (1974) Comparative specificity of microbial proteinases. *Adv. Enzymol.* **41**, 179–242.

Morihara, K. (1995) Pseudolysin and other pathogen endopeptidases of thermolysin family. *Methods Enzymol.* **248**, 242–253.

Morihara, K. & Homma, J.Y. (1985) *Pseudomonas* proteases. In: *Bacterial Enzymes and Virulence* (Holder, I.A., ed.). Boca Raton, FL: CRC Press, pp. 41–79.

Morihara, K. & Tsuzuki, H. (1975) *Pseudomonas aeruginosa* elastase: affinity chromatography and some properties as a metalloneutral proteinase. *Agric. Biol. Chem.* **39**, 1123–1128.

Morihara, K. & Tsuzuki, H. (1978) Phosphoramidon as an inhibitor of elastase from *Pseudomonas aeruginosa*. *Jpn J. Exp. Med.* **48**, 81–84.

Morihara, K., Tsuzuki, H., Oka, T., Inoue, H. & Ebata, M. (1965) *Pseudomonas aeruginosa* elastase. Isolation, crystallization, and preliminary characterization. *J. Biol. Chem.* **240**, 3295–3304.

Morihara, K., Tsuzuki, H. & Oda, K. (1979) Protease and elastase of *Pseudomonas aeruginosa*: inactivation of human plasma $\alpha_1$-proteinase inhibitor. *Infect. Immun.* **24**, 188–193.

Nishino, N. & Powers, J.C. (1980) *Pseudomonas aeruginosa* elastase. Development of a new substrate, inhibitors, and an affinity ligand. *J. Biol. Chem.* **255**, 3482–3486.

Ochsner, U.A. & Reiser, J. (1995) Autoinducer-mediated regulation of rhamnolipid biosurfactant synthesis in *Pseudomonas aeruginosa*. *Proc. Natl Acad. Sci. USA* **92**, 6424–6428.

Olson, J.C. & Ohman, D.E. (1992) Efficient production and processing of elastase and LasA by *Pseudomonas aeruginosa* require zinc and calcium ions. *J. Bacteriol.* **174**, 4140–4147.

Parmely, M., Gale, A., Clabaugh, M., Horvat, R. & Zhou, W.-W. (1990) Proteolytic inactivation of cytokines by *Pseudomonas aeruginosa*. *Infect. Immun.* **58**, 3009–3014.

Passador, L., Cook, J.M., Gambello, M.J., Rust, L. & Iglewski, B.H. (1993) Expression of *Pseudomonas aeruginosa* virulence genes requires cell-to-cell communication. *Science* **260**, 1127–1130.

Pauchon, V., Besson, C., Saulnier, J. & Wallach, J. (1993) Peptide synthesis catalysed by *Pseudomonas aeruginosa* elastase. *Biotechnol. Appl. Biochem.* **17**, 217–221.

Pearson, J.P., Gray, K.M., Passador, L., Tucker, K.D., Eberhard, A., Iglewski, B.H. & Greenberg, E.P. (1994) Structure of the autoinducer required for the expression of *Pseudomonas aeruginosa* virulence genes. *Proc. Natl Acad. Sci. USA* **91**, 197–201.

Pearson, J.P., Passador, L., Iglewski, B.H. & Greenberg, E.P. (1995) A second N-acylhomoserine lactone signal produced by *Pseudomonas aeruginosa*. *Proc. Natl Acad. Sci. USA* **92**, 1490–1494.

Poncz, L., Gerken, T.A., Dearborn, D.G., Grobelny, D. & Galardy, R.E. (1984) Inhibition of the elastase of *Pseudomonas aeruginosa*

**M**

by $N^\alpha$-phosphoryl dipeptides and kinetics of spontaneous hydrolysis of the inhibitors. *Biochemistry* **23**, 2766–2772.

Rust, L., Messing, C.R. & Iglewski, B.H. (1994) Elastase assays. *Methods Enzymol.* **235**, 554–562.

Rust, L., Pesci, E.C. & Iglewski, B.H. (1996) Analysis of the *Pseudomonas aeruginosa* elastase (*lasB*) regulatory region. *J. Bacteriol.* **178**, 1134–1140.

Schultz, D.R. & Miller, K.D. (1974) Elastase of *Pseudomonas aeruginosa*: inactivation of complement components and complement-derived chemotactic and phagocytic factors. *Infect. Immun.* **10**, 128–135.

Shinde, U. & Inouye, M. (1993) Intramolecular chaperones and protein folding. *Trends Biochem. Sci.* **18**, 442–446.

Shortridge, V.D., Pato, M.L., Vasil, A.I. & Vasil, M.L. (1991) Physical mapping of virulence-associated genes in *Pseudomonas aeruginosa* by transverse alternating-field electrophoresis. *Infect. Immun.* **59**, 3596–3603.

Tanaka, E., Kawamoto, S., Fukushima, J., Hamajima, K., Onishi, H., Miyagi, Y., Inami, S., Morihara, K. & Okuda, K. (1991) Detection of elastase production in *Escherichia coli* with the elastase structural gene from several non-elastase producing strains of *Pseudomonas aeruginosa*. *J. Bacteriol.* **173**, 6153–6158.

Thayer, M.M., Flaherty, K.M. & McKay, D.B. (1991) Three-dimensional structure of the elastase of *Pseudomonas aeruginosa* at 1.5-Å resolution. *J. Biol. Chem.* **266**, 2864–2871.

Tommassen, J., Filloux, A., Bally, M., Murgier, M. & Lazdunski, A. (1992) Protein secretion in *Pseudomonas aeruginosa*. *FEMS Microbiol. Rev.* **103**, 73–90.

Twinning, S.S., Kirschner, S.E., Mahnke, L.A. & Frank, D.W. (1993) Effect of *Pseudomonas aeruginosa* elastase, alkaline protease, and exotoxin A on corneal proteinases and proteins. *Invest. Ophthalmol. Vis. Sci.* **34**, 2699–2712.

Wretlind, B. & Pavlovskis, O.R. (1983) *Pseudomonas aeruginosa* elastase and its role in *Pseudomonas* infections. *Rev. Infect. Dis.* **5**(suppl. 5), S998–S1004.

Wretlind, B. & Wadström, T. (1977) Purification of a protease with elastase activity from *Pseudomonas aeruginosa*. *J. Gen. Microbiol.* **103**, 319–327.

Yokota, S., Ohtsuka, H. & Noguchi, H. (1992) Monoclonal antibodies against *Pseudomonas aeruginosa* elastase: a neutralizing antibody which recognizes a conformational epitope related to an active site of elastase. *Eur. J. Biochem.* **206**, 587–593.

*Efrat Kessler*
*Maurice & Gabriela Goldschleger Eye Research Institute,*
*Tel-Aviv University Sackler Faculty of Medicine,*
*Sheba Medical Center,*
*Tel-Hashomer 52621, Israel*
*Email: ekessler@ccsg.tau.ac.il*

*Dennis E. Ohman*
*Department of Microbiology & Immunology,*
*University of Tennessee & Veteran Affairs Medical Center,*
*Memphis, TN 38163, USA*
*Email: dohman@utmem1.utmem.edu*

# 358. *Legionella metalloendopeptidase*

## Databanks

*Peptidase classification: clan MA, family M4, MEROPS ID: M04.006*
*NC-IUBMB enzyme classification: none*
*Databank codes:*

| Species | SwissProt | PIR | EMBL (cDNA) | EMBL (genomic) |
|---|---|---|---|---|
| *Legionella longbeachae* | P55110 | – | X83035 | – |
| *Legionella pneumophila* | P21347 | – | M31884 | – |

## Name and History

An epidemic of pneumonia that broke out among attendees at a convention of the American Legion in Philadelphia in 1976 was ultimately traced to a novel organism, *Legionella pneumophila* (McDade *et al.*, 1977), which grew in the water-cooling units of the air-conditioning system of the convention hotel. The organism is a facultative intracellular parasite in alveolar macrophages. The mechanism of disease production is not fully understood, but early studies showed that an extracellular protease was produced that was cytotoxic and hemolytic (Thompson *et al.*, 1981). The enzyme degraded casein, hide powder and gelatin; activity was inhibited by EDTA and restored by zinc. The enzyme is sometimes referred to as **major secretory protein**, but this does not suggest its proteolytic action. A suitable name for a zinc protease might be legiolysin, but that name is already in use for a hemolytic factor, lacking proteolytic activity, from the same organism (Wintermeyer *et al.*, 1991). Therefore, the recommended name for this enzyme is ***Legionella metalloendopeptidase***.

## Activity and Specificity

Very little is known of the specificity of this enzyme. In addition to the digestion of common protease substrates such as casein, gelatin and hide powder mentioned above, the enzyme cleaves $\alpha_1$-antitrypsin (Conlan *et al.*, 1988a), tumor necrosis factor $\alpha$ (Hell *et al.*, 1993), interleukin 2 and CD4 on human T cell surfaces (Mintz *et al.*, 1993). Berdal *et al.* (1982) noted that crude enzyme preparations could cleave the chromogenic peptide MeOSuc-Arg-Pro-Tyr↓NHPhNO$_2$ (Kabi S2586) (see Appendix 2 for full names and addresses of suppliers). This has been confirmed by McIntyre *et al.* (1991) who also showed that O-Bz-Thr-Pro-Pro-NHPhNO$_2$ (Kabi S2492) and Glp-Gly-Arg-NHPhNO$_2$ (Kabi S2444) are not cleaved. However, the peptide has not been tested with highly purified protease. No information is available on specific peptide bonds that are cleaved.

The enzyme has a pH optimum around neutrality. Activity is inhibited by various chelators such as EDTA, EGTA and 1,10-phenanthroline, but not by DTT (10 mM) (Dreyfus & Iglewski, 1986). Activity can be restored, following EDTA inhibition, by $Zn^{2+}$, $Fe^{2+}$, $Mn^{2+}$ and $Cu^{2+}$, in decreasing order. Potent inhibition is obtained with phosphoramidon and the phosphoramidate analog Z-Gly$^P$(O)Leu-Ala (Black *et al.*, 1990). The zinc-binding motif HEVSH is at the catalytic center; mutation of the Glu378 residue renders the enzyme inactive and abolishes cytotoxicity (Moffat *et al.*, 1994a).

## Structural Chemistry

The cDNA sequence predicts a protein product of 543 amino acid residues and a mass of 60 775 Da (Black *et al.*, 1990). However, the enzyme isolated from cultures typically has masses of 40 000 and 38 000 Da; processing to the mature 40 kDa form takes place in the periplasmic space (Moffat *et al.*, 1994a). The enzyme is most closely related to the elastase of *Pseudomonas aeruginosa*, now known as pseudolysin (Chapter 357). Atomic absorption spectrometry indicates 1 mol of zinc per mol of enzyme (Dreyfus & Iglewski, 1986). The pI is 4.2 and 4.4, for the 40 and 38 kDa forms, respectively.

## Purification

The enzyme was purified 6-fold from culture broth by DEAE-cellulose ion-exchange chromatography (Conlan *et al.*, 1986). A homogeneous preparation was obtained by a 10-fold purification in three steps: DEAE-cellulose, octyl-Sepharose and DEAE-BioGel A (Dreyfus & Iglewski, 1986). Rechnitzer *et al.* (1989a) used FPLC followed by gel filtration to prepare homogeneous protease; an apparent 1500-fold purification appears to be due to a very high level of protein in the starting medium relative to the other two methods cited.

## Biological Aspects

Intranasal administration of the protease in guinea pigs produces hemorrhagic pneumonia; there is evidence for collagenolytic but not elastolytic activity (Baskerville *et al.*, 1986). The amount of protease produced by the organism growing in cells within the guinea pig lung, as assayed by an ELISA method, is similar to the amounts that are lethal when administered intranasally (Conlan *et al.*, 1988b). The protease is cytotoxic to both neutrophils and monocyte/macrophages (Rechnitzer & Kharazmi, 1992) and interferes with the binding of natural killer cells to their target cells (Rechnitzer *et al.*, 1989b). However, it is not completely clear how the protease contributes to the disease process. The enzyme is not required for growth of the bacterium within cultured human macrophages nor for killing the cells, as shown by mutations that block protease production (Szeto & Shuman, 1990). However, when a similar mutant was used in the guinea pig pneumonia model, virulence was greatly attenuated but not completely abolished (Moffat *et al.*, 1994b).

## Further Reading

An extensive review of the virulence factors of members of the Legionellaceae superfamily, including the zinc metalloproteases, is presented by Dowling *et al.* (1992).

## References

Baskerville, A., Conlan, J.W., Ashworth, L.A.E. & Dowsett, A.B. (1986) Pulmonary damage caused by a protease from *Legionella pneumophila*. *Br. J. Exp. Pathol.* **67**, 527–536.

Berdal, B.P., Hushovd, O., Olsvik, O., Odegard, O.R. & Bergan, T. (1982) Demonstration of extracellular proteolytic enzymes from *Legionella* species strains by using synthetic chromogenic peptidase substrates. *Acta Pathol. Microbiol. Immunol. Scand. B* **90**, 119–123.

Black, W.J., Quinn, F.D. & Tompkins, L.S. (1990) *Legionella pneumophila* zinc metalloprotease is structurally and functionally homologous to *Pseudomonas aeruginosa* elastase. *J. Bacteriol.* **172**, 2608–2613.

Conlan, J.W., Baskerville, A. & Ashworth, L.A.E. (1986) Separation of *Legionella pneumophila* proteases and purification of a protease which produces lesions like those of Legionnaires' disease in guinea pig lung. *J. Gen. Microbiol.* **132**, 1565–1574.

Conlan, J.W., Williams, A. & Ashworth, L.A. (1988a) Inactivation of human $\alpha$-1-antitrypsin by a tissue-destructive protease of *Legionella pneumophila*. *J. Gen. Microbiol.* **134**, 481–487.

Conlan, J.W., Williams, A. & Ashworth, L.A. (1988b) *In vivo* production of a tissue-destructive protease by *Legionella pneumophila* in the lungs of experimentally infected guinea-pigs. *J. Gen. Microbiol.* **143**, 143–149.

Dowling, J.N., Saha, A.K. & Glew, R.H. (1992) Virulence factors of the family *Legionellaceae*. *Microbiol. Rev.* **56**, 32–60.

Dreyfus, L.A. & Iglewski, B.H. (1986) Purification and characterization of an extracellular protease of *Legionella pneumophila*. *Infect. Immun.* **51**, 736–743.

Hell, W., Essig, A., Bohnet, S., Gaterman, S. & Marre, R. (1993) Cleavage of tumor necrosis factor-$\alpha$ by *Legionella* exoprotease. *APMIS* **101**, 120–126.

McDade, J.E., Shepard, C.C., Fraser, D.W., Tsai, F.T., Redus, M.A., Dowdle, W.R. & Laboratory Investigation Team (1977) Legionnaires' disease associated with a bacterium and demonstration of its role in other respiratory disease. *N. Engl. J. Med.* **297**, 1197–1203.

M

McIntyre, M., Quinn, F.D., Fields, P.I. & Berdal, B.P. (1991) Rapid identification of *Legionella pneumophila* zinc metalloprotease using chromogenic detection. *APMIS* **99**, 316–320.

Mintz, C.S., Miller, R.D., Gutgsell, N.S. & Malek, T. (1993) *Legionella pneumophila* protease inactivates interleukin-2 and cleaves CD4 on human T cells. *Infect. Immun.* **61**, 3416–3421.

Moffat, J.F., Black, W.J. & Tompkins, L.S. (1994a) Further molecular characterization of the cloned *Legionella pneumophila* zinc metalloprotease. *Infect. Immun.* **62**, 751–753.

Moffat, J.F., Edelstein, P.H., Regula, D.P., Jr., Cirilllo, J.D. & Tompkins, L.S. (1994b) Effects of an isogenic Zn-metalloprotease-deficient mutant of *Legionella pneumophila* in a guinea-pig pneumonia model. *Mol. Microbiol.* **12**, 693–705.

Rechnitzer, C. & Kharazmi, A. (1992) Effect of *Legionella pneumophila* cytotoxic protease on human neutrophil and monocyte function. *Microb. Pathog.* **12**, 115–125.

Rechnitzer, C., Tvede, M. & Döring, G. (1989a) A rapid method for purification of homogeneous *Legionella pneumophila* cytotoxic protease using fast protein liquid chromatography. *FEMS Microbiol. Lett.* **59**, 39–44.

Rechnitzer, C., Diamant, M. & Pedersen, B.K. (1989b) Inhibition of human natural killer cell activity by *Legionella pneumophila* protease. *Eur. J. Clin. Microbiol. Infect. Dis.* **8**, 989–992.

Szeto, L. & Shuman, H.A. (1990) The *Legionella pneumophila* major secretory protein, a protease, is not required for intracellular growth or cell killing. *Infect. Immun.* **58**, 2585–2592.

Thompson, M.R., Miller, R.D. & Iglewski, B.H. (1981) *In vitro* production of an extracellular protease by *Legionella pneumophila*. *Infect. Immun.* **34**, 299–302.

Wintermeyer, E., Rdest, U., Ludwig, B., Debes, A. & Hacker, J. (1991) Characterization of *legiolysin (lly)*, responsible for haemolytic activity, colour production and fluorescence of *Legionella pneumophila*. *Mol. Microbiol.* **5**, 1135–1143.

*J. Fred Woessner*
*Department of Biochemistry & Molecular Biology,*
*University of Miami School of Medicine,*
*PO Box 016960, Miami, FL 33101, USA*
*Email: fwoessne@mednet.med.miami.edu*

# 359. *Peptidyl-dipeptidase A/angiotensin I-converting enzyme*

## Databanks

*Peptidase classification: clan MA, family M2, MEROPS ID: M02.001*
*NC-IUBMB enzyme classification: EC 3.4.15.1*
*Chemical Abstracts Service registry number: 9015-82-1*
*Databank codes:*

| Species | SwissProt | PIR | EMBL (cDNA) | EMBL (genomic) |
|---|---|---|---|---|
| Somatic form | | | | |
| *Bos taurus* | P12820 | A26376 | – | – |
| *Canis familiaris* | – | – | U67199 | – |
| *Gallus gallus* | Q10751 | JC2489 | L40175 | – |
| *Homo sapiens* | P22966 | A33979 | M26657 | – |
| | | S05238 | M26658 | |
| | | | X16295 | |
| *Mus musculus* | P09470 | A34171 | J03940 | – |
| | | | J04946 | |
| | | | J04947 | |
| | | | M34433 | |
| *Oryctolagus cuniculus* | P12822 | A49726 | – | – |
| | | S35484 | | |

| Species | SwissProt | PIR | EMBL (cDNA) | EMBL (genomic) |
|---|---|---|---|---|
| *Rattus norvegicus* | P47820 | JC2038 | L36664 U03708 U03734 | – |
| Testicular form | | | | |
| *Homo sapiens* | P12821 | A31759 | A00914 J04144 | – |
| *Mus musculus* | P22967 | A35655 | M55333 | M61904: exons 11, 12, 12A and 13 |
| *Oryctolagus cuniculus* | P22968 | A34402 A36232 A60724 C18700 | J05041 M58580 | – |

## Name and History

*Angiotensin I-converting enzyme* (*ACE*; *peptidyl-dipeptidase A*) is a zinc metallopeptidase which cleaves the C-terminal dipeptide from angiotensin I to produce the potent vasopressor octapeptide angiotensin II (Skeggs *et al.*, 1956) and inactivates bradykinin by the sequential removal of two C-terminal dipeptides (Yang *et al.*, 1970). In addition to these two main physiological substrates, which are involved in blood pressure regulation and water and salt metabolism, ACE cleaves C-terminal dipeptides from various oligopeptides with a free C-terminus. ACE is also able to cleave a C-terminal dipeptide amide. Clearly, ACE displays a wide substrate specificity and none of the designations, such as *kininase II*, *dipeptidyl carboxypeptidase I*, *carboxycathepsin*, *dipeptide hydrolase* and *peptidase P*, assigned to this enzyme adequately describe its varied actions. ACE has also been implicated in a range of physiological processes unrelated to blood pressure regulation, such as immunity, reproduction and neuropeptide regulation, due to its localization and/or the *in vitro* cleavage of a range of biologically active peptides.

The role of ACE in blood pressure control and water and salt metabolism has been defined mainly by the use of highly specific ACE inhibitors developed during the last decade (Waeber *et al.*, 1990). These drugs clearly show their efficacy in the treatment of hypertension, congestive heart failure and diabetic nephropathy. Several reviews have been published on the biochemistry of ACE (Erdös & Skidgel, 1987; Ehlers & Riordan, 1989; Hooper, 1991), on its main enzymatic characteristics and on ACE inhibitors (Lawton *et al.*, 1992). The present chapter will emphasize new aspects of the structure and function of ACE as revealed by molecular biology.

## Tissue Distribution

### Somatic ACE

ACE is an ectoenzyme anchored to the plasma membrane with the bulk of its mass exposed at the extracellular surface of the cell. There are two ACE isoforms: a somatic form of around 150–180 kDa found in endothelial, epithelial and neuronal cells and a smaller isoform (90–110 kDa) present in germinal cells. Somatic ACE is found in the plasma membrane of vascular endothelial cells, particularly in the

lung, and oriented to metabolize circulating substrates (Oparil & Haber, 1974a,b). Early work indicated that the predominant conversion of angiotensin I to II takes place in the lung (Ng & Vane, 1967, 1968). However, more recently this concept has been challenged by the discovery of tissue renin–angiotensin systems (see Chapter 284) which may be the major sites of angiotensin I conversion.

ACE is also expressed in the brush borders of absorptive epithelia of the small intestine and the kidney proximal convoluted tubule (Bruneval *et al.*, 1986), in mononuclear cells (such as monocytes after macrophage differentiation), T lymphocytes and fibroblasts (Friedland *et al.*, 1978; Costerousse *et al.*, 1993). *In vitro* autoradiography, employing radiolabeled specific ACE inhibitors, and immunohistochemical studies have mapped the principal locations of ACE in brain (Defendini *et al.*, 1983; Strittmatter *et al.*, 1984; Barnes *et al.*, 1988). ACE is found primarily in the choroid plexus, which may be the source of ACE in cerebrospinal fluid, ependyma, subfornical organ, basal ganglia (caudate putamen and globus pallidus), substantia nigra and pituitary. The colocalization of ACE and the mammalian tachykinin peptide substance P to the striatonigral neuronal pathway led to suggestions of a physiological role for ACE in substance P hydrolysis in brain (Strittmatter *et al.*, 1985; Strittmatter & Snyder, 1987). However, another mammalian ectoenzyme neprilysin (Chapter 362) also present in the same neuronal pathway is more likely to be involved in the *in vivo* degradation of this neuropeptide (Matsas *et al.*, 1984; Oblin *et al.*, 1984; Barnes *et al.*, 1988). Neuronal ACE is smaller in size than the more widely distributed form of the enzyme, probably due to differences in glycosylation (Hooper & Turner, 1987). Neuronal ACE and the endothelial form of the enzyme display the same neuropeptide specificities and do not represent distinct isoenzymes (Williams *et al.*, 1991).

### Soluble ACE

A soluble form of ACE is present in many biological fluids, such as serum (Das *et al.*, 1977), seminal fluid (El-Dorry *et al.*, 1983), amniotic fluid (Yasui *et al.*, 1984) and cerebrospinal fluid (Schweisfurth & Schioberg-Schiegnitz, 1984). It appears to be derived from the membrane-bound form of the enzyme in endothelial cells and will be discussed later.

M

## Germinal ACE

The smaller isoenzyme of ACE (germinal ACE) is found exclusively in the testes (El-Dorry *et al.*, 1982; Velletri, 1985); it is found in spermatids but not in immature forms of germinal cells (Berg *et al.*, 1986; Brentjens *et al.*, 1986). The maximum expression of ACE occurs during the acrosome phase in murine species. ACE is exclusively produced in haploid germ cells and belongs to the group of proteins whose expression during definite maturation steps of spermiogenesis appears to be correlated with the unique process of germ cell differentiation (Sibony *et al.*, 1994). The inactivation of the ACE gene by homologous recombination leads to homozygous male mice with markedly reduced blood pressure, severe renal abnormalities and severely reduced fertility (Krege *et al.*, 1995; Esther *et al.*, 1996). The effect on male fertility is attributed to the inactivation of the germinal form of ACE, although the physiological substrates of this ACE isoform have not been identified.

## Activity and Specificity

ACE was shown to be a metalloprotease by Skeggs *et al.* (1956); subsequent studies using atomic absorption spectroscopy showed that the associated metal was zinc (Bünning & Riordan, 1985; Ehlers & Riordan, 1991; Williams *et al.*, 1992). Zinc is essential for the catalytic activity of ACE (Bünning & Riordan, 1985). The zinc ion functions directly in the catalytic step by ionizing or polarizing the zinc-bound water molecule, which thus becomes activated to initiate the nucleophilic attack on the substrate carbonyl scissile bond (Vallee & Auld, 1990).

### Substrate Specificity

The primary action of ACE is to cleave C-terminal dipeptides from oligopeptide substrates with a free C-terminus and the absence of a penultimate Pro residue (Hooper, 1991). Bradykinin (RPPGFSP+FR) is the most favorable substrate for ACE with a $k_{cat}/K_m$ of 3900–5000 mM$^{-1}$ s$^{-1}$, which is considerably higher than that of angiotensin I (DDVYIHPF+HL) hydrolysis (147–189 mM$^{-1}$ s$^{-1}$) (Ehlers & Riordan, 1990). *In vitro* ACE hydrolyzes a wide range of substrates including neurotensin (Skidgel *et al.*, 1984), [Met$^5$]-enkephalin, [Met$^5$]-enkephalin-Arg$^6$-Gly$^7$-Leu$^8$, $\beta$-neoendorphin, dynorphins (Skidgel & Erdös, 1987) and the insulin B chain (Igic *et al.*, 1972). The specificity of ACE is not restricted to its C-terminal peptidyl dipeptidase action; it acts as an endopeptidase on certain substrates blocked at the C-terminus. ACE cleaves the C-terminal tripeptide amides from substance P (RPKPQQFF+GLM-NH$_2$) (Yokosawa *et al.*, 1983; Skidgel *et al.*, 1984) and luteinizing hormone-releasing hormone (LH-RH) (pEHWSYGL+RPG-NH$_2$) (Skidgel & Erdös, 1985) and it cleaves des-Arg$^9$-bradykinin, which contains a penultimate Pro residue (Inokuchi & Nagamatsu, 1981). It cleaves the C-terminal dipeptide amides from cholecystokinin-8 (DYMGWM+DF-NH$_2$) and various gastrin analogs (Dubreuil *et al.*, 1989). In addition, and perhaps more surprisingly, ACE is also able to cleave the N-terminal tripeptide from LH-RH as well as the blocked C-terminal tripeptide (Skidgel & Erdös, 1985). However, it should be noted that none of these hydrolyses have been observed *in vivo* except for those of angiotensin I and bradykinin.

ACE has recently been implicated in the *in vivo* hydrolysis of the tetrapeptide Ac-Ser-Asp-Lys-Pro, which is involved in the control of hematopoietic stem cell proliferation by preventing their recruitment into S phase (Rieger *et al.*, 1993). The *acute* administration of captopril, an ACE inhibitor, produces a 7-fold increase in the plasma concentration of Ac-Ser-Asp-Lys-Pro in normal volunteers, which demonstrates the *in vivo* role of the enzyme in the metabolism of this substrate (Azizi *et al.*, 1996). The increase in Ac-Ser-Asp-Lys-Pro that results from *chronic* ACE inhibitor treatment is particularly marked during renal failure (M. Azizi *et al.*, unpublished results) and may account for the rare cases of anemia and/or leukopenia that have been observed in these circumstances.

## Chloride Activation

The requirement of chloride for the hydrolysis of angiotensin I by ACE has long been recognized (Skeggs *et al.*, 1956); in fact, this substrate cannot be cleaved in the absence of chloride (Bünning & Riordan, 1983). However, chloride is not essential for the hydrolysis of all substrates: bradykinin is hydrolyzed in the absence of chloride, but the maximal rate occurs only at 20 mM chloride (Dorer *et al.*, 1974; Bünning & Riordan, 1983). The kinetics of chloride activation are rather complex, being both substrate and pH dependent (Shapiro *et al.*, 1983; Ehlers & Kirsch, 1988). It is proposed that chloride binding induces a subtle change in the conformation of the active site which facilitates substrate binding (Ehlers & Riordan, 1990). Recent studies, discussed later, demonstrate the different chloride dependence of the two active sites of ACE, which may display different substrate specificities. Thus, different conditions of chloride and pH could alter the relative activities of the two active sites of somatic ACE and, as such, determine the substrate specificity of the enzyme. Reductive methylation of rabbit pulmonary ACE by formaldehyde and sodium cyanoborohydride has implicated a single Lys residue located at or near the active site of ACE as the chloride-binding site (Shapiro & Riordan, 1983). Confirmation of this and identification of the specific Lys residue awaits further study.

## Structural Chemistry

### Somatic ACE

Complete cDNAs for the somatic form of ACE were first isolated from human endothelial and from mouse kidney cDNA libraries (Soubrier *et al.*, 1988; Bernstein *et al.*, 1988). Human endothelial ACE mRNA is 4.3 kb and in rabbit the pulmonary ACE mRNA is 5.0 kb (Soubrier *et al.*, 1988; Thekkumkara *et al.*, 1992). In mice two ACE mRNAs have been detected in kidney and in lung of 4.9 and 4.15 kb, respectively (Bernstein *et al.*, 1989). A high level of sequence similarity, 80–90% in both the nucleotide and amino acid sequences, has been found between the different somatic mammalian ACE cDNA sequences cloned to date from human, mouse, rabbit, cattle and rat ACE (Soubrier *et al.*, 1988; Bernstein *et al.*, 1989; Thekkumkara *et al.*, 1992; Shai et al., 1992; Koike et al., 1994). Human endothelial ACE comprises 1306 amino acid residues, of which 14 are cysteine residues, and contains 17 potential N-linked glycosylation sites but no Ser/Thr-rich region indicative of O-glycosylation (Soubrier et al., 1988). The primary sequence of ACE reveals two hydrophobic regions: a sequence of 29 amino acids at the N-terminus is absent in the mature enzyme and, therefore, constitutes the cleaved signal peptide, and a second hydrophobic sequence of 17 amino acids is located near the C-terminus of the enzyme which comprises the membrane anchor (Figure 359.1).

Molecular cloning of the somatic form of ACE demonstrated that the enzyme is composed of two homologous domains, called hereafter the N- and C-domains, indicative of the duplication of an ancestral gene (Soubrier et al., 1988; Bernstein et al., 1989). The germinal form contains only one of these domains, presumably the ancestral form of the gene. The overall sequence similarity in human ACE is 60% in the two sequences, but increases to 89% in a sequence of 40 amino acids in each domain containing essential residues of the active site. Both domains contain a putative catalytic site based on the zinc-binding motif HEXXH widely found in metalloproteases (Vallee & Auld, 1990). The two His residues provide two zinc-coordinating ligands and the third has been identified in both thermolysin (Chapter 351) and neprilysin (Chapter 362) as a Glu residue which is separated, on the C-terminal side, from the second His zinc-binding ligand by 19 and 58 amino acid residues, respectively (Kester & Matthews, 1977; LeMoual et al., 1991). The putative Glu third zinc-coordinating residues in ACE have been proposed as Glu389 and Glu987 in the N- and C-domains, respectively (Soubrier et al., 1988) and this has been confirmed by site-directed mutagenesis (see below).

The ACE gene appears to be duplicated in all mammalian species from which ACE has been cloned to date. Furthermore, a peptidyl-dipeptidase was purified by affinity chromatography from the electric organ of *Torpedo marmorata* (Turner et al., 1987). The molecular mass of the *Torpedo* ACE was 190 kDa compared to 170 kDa of the human somatic form of ACE and, therefore, it would appear that *Torpedo* ACE is also transcribed from a duplicated

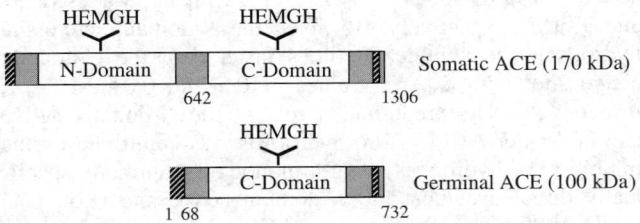

*Figure 359.1* Schematic representation of somatic and germinal ACE. Somatic ACE comprises two highly similar domains called N- and C-domains, each of which contain the HEXXH consensus sequence for zinc binding. A hydrophobic sequence in the C-terminal region (shaded box on the right) comprises the membrane-spanning domain. Germinal ACE comprises a single catalytically active domain identical to the C-domain of somatic ACE except for an N-terminal 67 residue germinal ACE-specific sequence.

gene. ACE-like enzymes have been identified in the housefly, *Musca domestica* (Lamango & Isaac, 1993), and in *Drosophila melanogaster* (Cornell et al., 1995). These invertebrate ACEs are discussed in Chapter 360.

### Germinal ACE

The primary structure of the germinal form of ACE found in testes has been deduced from the cDNA sequence from human (Lattion et al., 1989; Ehlers et al., 1989), rabbit (Kumar et al., 1989) and mouse (Howard et al., 1990). Germinal ACE is composed of 732 amino acids, compared to 1306 of somatic ACE, and corresponds to the C-domain of somatic ACE and therefore contains a single active site (Figure 359.1). In fact, amino acid residues 68–732 of human germinal ACE are identical to residues 642–1306 of human somatic ACE, which means that germinal ACE contains the same hydrophobic membrane-anchoring domain located near the C-terminus of somatic ACE. The N-terminal 67 amino acid sequence of germinal ACE is specific to this form of the enzyme and contains the signal peptide and a Ser/Thr-rich region for O-glycosylation. Germinal ACE is indeed heavily O-glycosylated, with 90% of the O-linked carbohydrates added to a region covering 36 amino acids in the N-terminal germinal ACE-specific sequence (Ehlers et al., 1992). The function of the O-linked carbohydrates has not been determined; however, it has been demonstrated that there is no relation between the O-linked glycosylation and enzyme activity or stability (Ehlers et al., 1992).

### Active-site Residues

In order to establish that both putative active sites in somatic ACE are functional, a series of ACE mutants containing only a single intact domain was constructed (Wei et al., 1991a). These mutants contained point mutations of putative critical active-site residues in either the N- or C-domain, or contained a deletion of either the N- or C-domain to form a truncated mutant. Both the N- and C-domains displayed catalytic activity hydrolyzing the synthetic ACE substrate Bz-Gly+His-Leu and the physiological substrate angiotensin I. Each domain displayed a similar $K_m$ for substrate binding, but the N-domain exhibited a 9- and 3-fold lower $k_{cat}$ compared to the C-domain for the hydrolysis of Bz-Gly-His-Leu and angiotensin I, respectively. In addition, both the N- and C-domain active sites possessed an absolute zinc requirement for activity and appeared to function independently. Therefore, this study demonstrated conclusively that both putative active sites in ACE were functional. In agreement with these findings, the zinc content of somatic ACE has been determined as two atoms of zinc bound per molecule of enzyme and each domain binds a specific ACE inhibitor (Ehlers & Riordan, 1991; Williams et al., 1992; Perich et al., 1992).

Sequence alignment of both ACE domains with other zinc metalloproteases indicated a Glu residue that could represent the third zinc ligand, and an Asp residue that might be indirectly involved in zinc interaction. This hypothesis was confirmed by site-directed mutagenesis of the ACE C-domain which showed that substitution of Glu987 (mature human ACE numbering) by a Val residue completely abolished the activity of this ACE domain, and its replacement by an Asp residue resulted in a 300-fold decrease in $k_{cat}$ (Williams et al.,

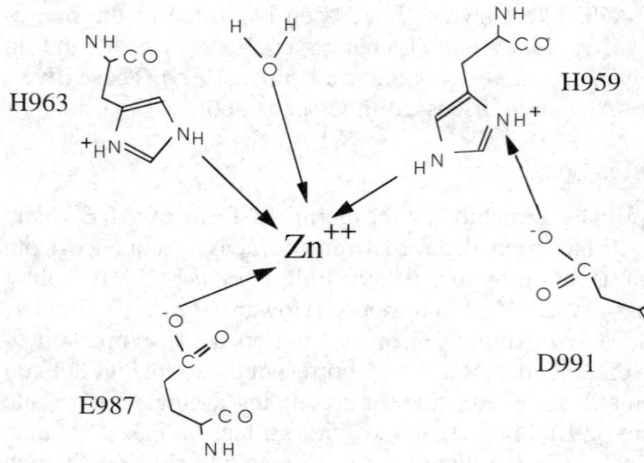

*Figure 359.2* Zinc ligands in the C-domain active site of ACE. The amino acid residues which coordinate the active-site zinc ion are shown with their corresponding amino acid positions. The water molecule provides the fourth zinc ligand. The Asp residue (D991), which may function in the precise positioning of the first His zinc ligand, is also shown. The other arrows represent zinc-coordinating bands. This is a hypothetical model based on the known active site of thermolysin (Colman *et al.*, 1972). (Modified from Williams *et al.*, 1994.)

1994). In the same report, Asp991 was found to be important for catalytic activity, probably playing a structural role in positioning the first His zinc ligand (Figure 359.2).

A Tyr and a Lys residue have been identified as essential active-site residues in the C-domain of rabbit pulmonary (the somatic form) and also in rabbit testicular (germinal form) ACE (Bünning *et al.*, 1990; Chen & Riordan, 1990). These residues were identified by a chemical modification using 1-fluoro-2,4-dinitrobenzene, which inactivated both forms of the enzyme. Amino acid analysis and sequencing of tryptic digests of the modified enzymes defined the amino acid positions of the Lys and Tyr residues as Lys118 and Tyr200 (mature rabbit testicular ACE numbering), respectively (Chen & Riordan, 1990). The functional role of the Tyr residue in human testicular ACE was investigated by mutating this residue to Phe (Chen *et al.*, 1992). The mutant enzyme displayed a decreased specific activity for the hydrolysis of the synthetic substrate FA-Phe⊣Gly-Gly and angiotensin I. The change in the $K_m$ for the binding of these substrates was quite minor and the $K_a$ for chloride activation was similar to that of the wild-type enzyme. In contrast, the mutant displayed a 100-fold decrease in affinity for the specific ACE inhibitor lisinopril, which is believed to act partly as a transition state analog (Patchett & Cordes, 1985). Thus, these results indicated that the Tyr residue was not involved in substrate or chloride binding and was not crucial for catalysis but could influence the rate of catalysis by stabilizing the transition state complex, in a manner analogous to Tyr198 in carboxypeptidase A (Chapter 451) (Gardell *et al.*, 1987). However, a subsequent study on rabbit testicular ACE produced quite different results. In this study the Lys and Tyr residues previously identified by chemical labeling with 1-fluoro-2,4-dinitrobenzene were mutated to produce two single mutants (Sen *et al.*, 1993). These mutated enzymes exhibited

the same enzymatic properties as the wild-type enzyme, indicating that the single replacement of either residue can be tolerated without disturbing the usual functional properties of the enzyme.

## Comparison of the Two Catalytic Sites in Somatic ACE

### Chloride Dependency

The N-domain of somatic ACE appears to be less chloride dependent than the C-domain; activity is observed in the absence of chloride and optimal activity on angiotensin I is achieved at 10 mM chloride. In contrast, the C-domain is essentially inactive in the absence of chloride and higher chloride concentrations are required for maximal activity (30 mM for angiotensin I hydrolysis). The chloride dependence of the C-domain is particularly marked for synthetic substrates such as Bz-Gly⊣His-Leu, which requires 800 mM Cl⁻ for maximal enzymatic activity.

### Hydrolytic Activity

Somatic ACE has been reported to perform the endopeptidase cleavage from the N-terminus of LH-RH (Trp3⊣Ser4) at a faster rate than testicular ACE, thus implying that it is the N-domain active site that is primarily responsible for this cleavage *in vitro* (Ehlers & Riordan, 1991). ACE mutants containing only a single intact N-domain cleave the Trp3⊣Ser4 bond of LH-RH 30 times faster than the C-domain active site (Jaspard *et al.*, 1993). This same study also reported that both active sites in ACE exhibit the same kinetic parameters for the hydrolysis of bradykinin although only the C-domain activity is stimulated by chloride. In addition, both domains of ACE hydrolyze substance P to produce the C-terminal dipeptide and tripeptide amides. For the hydrolysis of this substrate the C-domain active site is more readily activated by chloride and hydrolyzes substance P at a faster rate than the N-domain.

As mentioned before, the peptide Ac-Ser-Asp-Lys-Pro is the first highly specific substrate for the N-active site of ACE, with kinetic constants in the range of physiological substrates. Ac-Ser-Asp-Lys-Pro hydrolysis by ACE was studied using wild-type recombinant ACE and two full-length mutants containing a single functional site. Both the N- and C-domain active sites of ACE exhibit peptidyl dipeptidase activity towards Ac-Ser-Asp-Lys-Pro, with $K_m$ values of 31 and 39 μM, respectively. However, the N-domain active site hydrolyzes the peptide 50 times faster than the C-domain active site, with $k_{cat}/K_m$ values of 0.5 and 0.01 μM$^{-1}$s$^{-1}$, respectively. The predominant role of the N-domain active site in Ac-Ser-Asp-Lys-Pro hydrolysis was confirmed by the inhibition of hydrolysis using a monoclonal antibody specifically directed against the N-domain active site (Rousseau *et al.*, 1995). This suggests that ACE might be involved via its N-terminal active site in the *in vivo* regulation of the local concentration of this hemoregulatory peptide.

### ACE Inhibitors

The binding of various ACE inhibitors has been used to probe possible structural differences between the N- and C-domain active sites (Perich *et al.*, 1992). The binding of ACE inhibitors to purified rat lung and testis ACE was studied and the binding parameters for the two inhibitor-binding sites in lung ACE were determined from the displacement of the ACE inhibitor [$^{125}$I]Ro31-8472, with either Ro31-8472 or 351A, an analog of lisinopril. Similar $K_d$ values for Ro31-8472 were determined for the two binding sites in lung ACE; however, with 351A, the $K_d$ for the N-domain was more than 100-fold higher than that for the C-domain. Since the main structural difference between these two inhibitors is the length of the inhibitor side chain, which interacts with the active-site zinc atom in each domain of ACE, it was proposed that the N-domain of ACE contains a deeply recessed active site and, therefore, the N-domain active site zinc atom is less accessible to the inhibitor 351A than the C-domain zinc atom.

Wei *et al.* (1992) reported that both ACE domains contain a single high-affinity binding site for the highly potent inhibitor trandolaprilat, and that chloride enhances inhibitor potency by stabilizing the enzyme–inhibitor complex and slowing the dissociation rate. Interestingly, this effect of chloride was more marked for the C-domain than for the N-domain. Furthermore, the order of potency of the specific ACE inhibitors captopril, enalaprilat and lisinopril was lisinopril > enalaprilat > captopril for the C-domain, whereas this order was reversed for the N-domain. A. Michaud *et al.* (unpublished results) evaluated the substrate dependence of ACE inhibition. They showed that the $K_i$ of captopril was 16-fold lower with Ac-Ser-Asp-Lys-Pro than with angiotensin I as substrate, when using wild-type ACE. This raises the interesting possibility of using a specific N-domain ACE inhibitor for enhancing plasma Ac-Ser-Asp-Lys-Pro levels during chemotherapy to specifically protect hematopoietic stem cells.

## Biological Aspects

### Gene Structure

The complete intron–exon structure of the human ACE gene was determined from the restriction mapping of genomic clones and by sequencing the intron–exon boundaries (Hubert *et al.*, 1991). The human ACE gene contains 26 exons: the somatic ACE mRNA is transcribed from exon 1 to 26, but exon 13 is removed by splicing from the primary RNA transcript; the germinal ACE mRNA is transcribed from exon 13 to 26 (Figure 359.3). The structure of the human ACE gene provides further support for the hypothesis of the duplication of an ancestral gene. Exons 4–11 and 17–24, which encode the two homologous domains of the enzyme, are highly similar in size and in sequence. In contrast, the intron sizes are not conserved.

The human somatic and germinal forms of ACE mRNA are transcribed from a single gene (Hubert *et al.*, 1991) and the two species of ACE mRNA are generated either by differential splicing of a common primary RNA transcript or by the initiation of transcription from alternative start sites under the control of separate promoters. There are two functional promoters in the ACE gene. The somatic ACE promoter is located on the 5′ side of the first exon of the gene. Positive regulatory elements are found inside the 132 bp region upstream from the transcription start site and there are negative regulatory elements between positions −132 and −343 and between −472 and −754 (Testut *et al.*, 1993).

The germinal ACE promoter is located on the 5′ side of the 5′ end of the germinal ACE mRNA. Primer extension and RNAase protection assays on mouse, rabbit and human germinal RNA demonstrated that the transcription initiation site was located inside the ACE gene (Howard *et al.*, 1990; Hubert *et al.*, 1991; Kumar *et al.*, 1991). Thus intron 12, which corresponds to the genomic sequence 5′ to the germinal specific exon 13, as deduced from the complete analysis of the ACE gene in humans, was proposed as the putative germinal ACE promoter. Unequivocal evidence for the promoter function of this sequence was obtained by using intron 12 to drive the transcription of a reporter gene in a germinal-specific fashion, in elongating spermatozoa (Langford *et al.*, 1991). In another series of transgenic mice, a 91 bp fragment of intron 12 of the ACE gene was used as the promoter and was sufficient to confer a germinal cell-restricted pattern of transcription to the transgene (Howard *et al.*, 1993). Further mapping of the elements controlling transcription was achieved by DNAase footprinting and gel mobility shift assays which demonstrated that a sequence between positions −42 and −62 specifically binds nuclear factors from testes extracts and contains a consensus cAMP responsive element (Howard *et al.*, 1993). The two alternative promoters of the ACE gene exhibit highly contrasted cell specificities since the somatic ACE promoter is active in several cell types in contrast to the germinal ACE promoter, which is only active in a stage-specific manner in male germinal cells (Nadaud *et al.*, 1992).

### Mechanism of ACE Anchorage and Solubilization

Hooper *et al.* (1987) showed that ACE was inserted into the membrane via a sequence located at the C-terminus of the enzyme since the N-terminal ends of the enzyme released by trypsin digestion or by solubilization with detergent were identical. The primary structure of ACE contains a highly

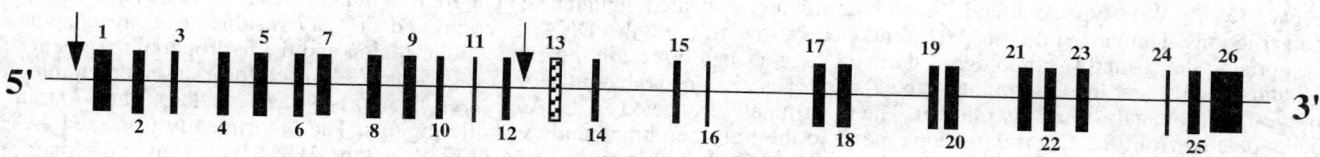

*Figure 359.3*  Organization of the human ACE gene. The relative locations of the 26 exons (vertical bars) are shown. Exon 13 (hatched bar) is specific to the testicular ACE mRNA. The two promoters are indicated by vertical arrows. (Modified from Hubert *et al.*, 1991.)

hydrophobic sequence located near the C-terminal end of the enzyme (Soubrier *et al.*, 1988). This sequence must constitute the membrane anchor since it is the only region of high hydrophobicity capable of spanning the lipid bilayer.

Further evidence for a C-terminal anchor was obtained by site-directed mutagenesis studies on human endothelial ACE and human testicular ACE (Wei *et al.*, 1991a). Transfection of Chinese hamster ovary (CHO) cells with the wild-type recombinant ACE cDNA resulted in 95% of the newly synthesized enzyme bound to the cell surface and the remaining 5% secreted into the culture medium. However, when a stop codon was introduced immediately before the putative membrane anchor, all newly synthesized ACE was liberated into the culture medium. Similarly, the expression of mutants of testicular ACE with a deletion of the putative membrane anchor also results in the production of a secreted form of the enzyme (Ehlers *et al.*, 1991). In addition, the soluble form of the enzyme purified from human plasma is not recognized by

an antibody raised against the C-terminal cytoplasmic domain of the enzyme (Wei *et al.*, 1991b).

Amino acid C-terminal microsequencing of the spontaneously secreted recombinant human somatic ACE from CHO cells established that the post-translational cleavage occurred at the Arg1137-Leu1138 bond (Beldent *et al.*, 1993) (Figure 359.4). In addition, the C-terminus of the circulating form of ACE in human plasma indicates the same cleavage; thus it seems that a similar mechanism for ACE secretion exists in human vascular cells and in CHO cells. Transfection of mouse epithelial cells with rabbit testicular ACE cDNA demonstrated that the cleavage process is enhanced by phorbol esters (Ramchandran *et al.*, 1994) and occurs exclusively at the cell surface; the cleavage site was identified as Arg663-Ser664 (rabbit testicular ACE numbering). The enzyme involved in the secretion of ACE from pig kidney microvillar membranes has been partially characterized (Oppong & Hooper, 1993) and bears some resemblance to

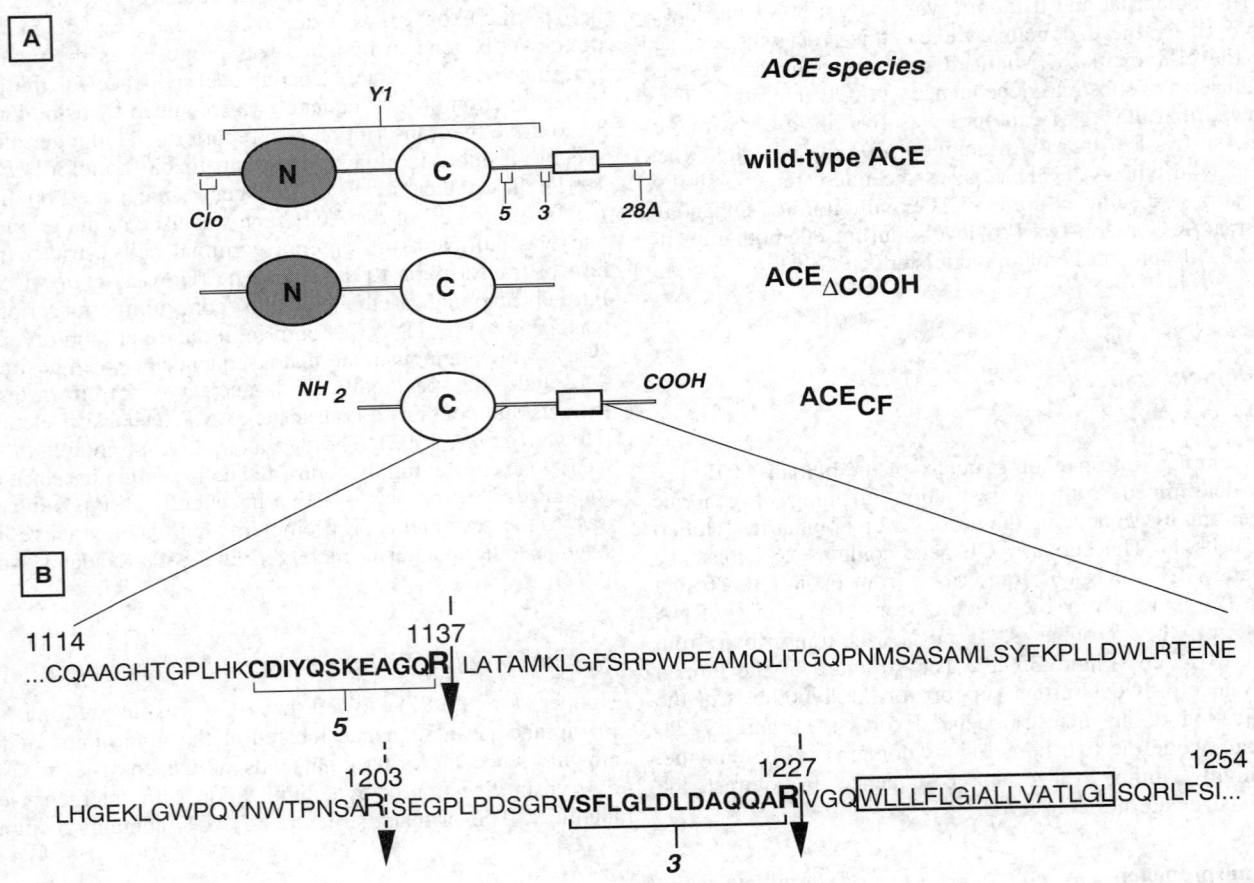

*Figure 359.4* Wild-type ACE and C- and N-terminally truncated mutants. (A) Diagram of human ACEs encoded by the cDNA constructions. Human wild-type ACE, encoded by the full-length cDNA, is composed of 1277 residues (mature enzyme), comprising the N- and C-homologous domains (egg-shaped boxes) in the extracellular region and the hydrophobic segment of 17 amino acids near the C-terminus (boxed). Epitopes recognized by antisera (Y1, Clo and 28A) used for immunoprecipitation or western blot analysis are also shown. The C-terminally truncated ACE (ACE$_{\Delta COOH}$) comprises 1230 residues. The 47 amino acids deleted from the C-terminus correspond to the transmembrane and cytosolic regions. The N-terminally truncated ACE (ACE$_{CF}$) comprises residues 1–4 of the N-terminus followed by residues 572–1277 of mature ACE. (B) C-terminal sequence of ACE from Cys1114 to Ile1254. Cleavage sites of human endothelial ACE (R1137), rabbit testicular ACE (R1203) and ACE$_{CF}$ (R1227) are shown by an arrow. Corresponding regions of the two antibody epitopes 5 and 3 are indicated. The hydrophobic domain is boxed. (Reproduced with permission from Beldent *et al.*, 1995.)

tumor necrosis factor $\alpha$ (TNF$\alpha$) convertase (Chapter 449). The ACE-secreting enzyme is tightly associated with the membrane and is EDTA-sensitive, with the activity being substantially restored by the addition of $Mg^{2+}$ and to a lesser extent by $Zn^{2+}$ and $Mn^{2+}$. Ramchandran & Sen (1995) recently found that both somatic and testicular rabbit ACE were solubilized by an integral membrane metalloprotease which was specifically inhibited by a synthetic hydroxamic acid derivative called compound 3.

The subcellular localization of ACE proteolysis was recently established by pulse-chase experiments, cell surface immunolabeling and biotinylation of radiolabeled mature proteins (Beldent *et al.*, 1995). The proteolysis of ACE takes place primarily at the plasma membrane. The solubilization of ACE is less than 2% within 1 h, is increased 2.4-fold by phorbol esters, but is not influenced by ionophores. An ACE mutant lacking the transmembrane domain and the cytosolic part is secreted at a faster rate without a C-terminal cleavage, and phorbol esters or ionophores have no effect on its rate of production in the medium. Therefore, the proteolysis of ACE is dependent on the presence of the membrane anchor and the present data suggest that the secretase(s) involved is also membrane-associated. An ACE mutant lacking the N-domain (ACE$_{CF}$) is secreted 10-fold faster than wild-type ACE. The solubilization of ACE$_{CF}$ also occurs at the plasma membrane and is stimulated 2.7-fold by phorbol esters, and the cleavage site is at Arg1227$\downarrow$Val1228. The N-domain of ACE slows down the proteolysis and seems to act as a 'conformational inhibitor' of the proteolytic cleavage, possibly via interactions with the 'stalk' of ACE and the secretase(s) itself. It will be of considerable interest to elucidate the nature of the secretase involved in the process of ACE solubilization because it may involve a category of enzyme(s) implicated in the solubilization of other ectoproteins such as transforming growth factor $\alpha$ receptor, $\beta$-amyloid precursor protein and colony-stimulating factor.

## Further Reading

Reviews of the biochemistry of ACE may be found in Ehlers & Riordan (1989), Erdös & Skidgel (1987) and Hooper (1991). A review of ACE inhibitors is presented by Lawton *et al.* (1992).

## References

Azizi, M., Rousseau, A., Ezan, E., Guyene, T.-T., Michelet, S., Grognet, J.-M., Lenfant, M., Corvol, P. & Ménard, J. (1996) Acute angiotensin-converting enzyme inhibition increases the plasma level of the natural stem cell regulator *N*-acetyl-seryl-aspartyl-lysyl-proline. *J. Clin. Invest.* **97**, 839–844.

Barnes, K., Matsas, R., Hooper, N.M., Turner, A.J. & Kenny, A.J. (1988) Endopeptidase 24.11 is striosomally ordered in pig brain and, in contrast to aminopeptidase N and dipeptidyl dipeptidase, an angiotensin converting enzyme, is a marker for a set of striatal efferent fibers. *Neuroscience* **27**, 799–817.

Beldent, V., Michaud, A., Wei, L., Chauvet, M.-T. & Corvol, P. (1993) Proteolytic release of human angiotensin-converting enzyme. Localization of the cleavage site. *J. Biol. Chem.* **268**, 26428–26433.

Beldent, V., Michaud, A., Bonnefoy, C., Chauvet, M.-T. & Corvol, P. (1995) Cell surface localization of proteolysis of human endothelial angiotensin I-converting enzyme. *J. Biol. Chem.* **270**, 28962–28969.

Berg, T., Sulner, J., Lai, C.Y. & Soffer, R.L. (1986) Immunohistochemical localization of two angiotensin I-converting isoenzymes in the reproductive tract of the male rabbit. *J. Histochem. Cytochem.* **34**, 753–760.

Bernstein, K.E., Martin, B.M., Bernstein, E.A., Linton, J., Striker, L. & Striker, G. (1988) The isolation of angiotensin converting enzyme cDNA. *J. Biol. Chem.* **263**, 11021–11024.

Bernstein, K.E., Martin, B.M., Edwards, A.S. & Bernstein, E.A. (1989) Mouse angiotensin I-converting enzyme is a protein composed of two homologous domains. *J. Biol. Chem.* **264**, 11945–11951.

Brentjens, J.R., Matsuo, S., Andres, G.A., Caldwell, P.R.B. & Zamboni, L. (1986) Gametes contain angiotensin converting enzyme (kininase II). *Experientia* **42**, 399–402.

Bruneval, P., Hinglais, N., Alhenc-Gelas, F., Tricottet, V., Corvol, P., Ménard, J., Camilleri, J.-P. & Bariety, J. (1986) Angiotensin I converting enzyme in human intestine and kidney. Ultrastructural immunohistochemical localization. *Histochemistry* **85**, 73–80.

Bünning, P. & Riordan, J.F. (1983) Activation of angiotensin converting enzyme by monovalent anions. *Biochemistry* **22**, 110–116.

Bünning, P. & Riordan, J.F. (1985) The functional role of zinc in angiotensin converting enzyme. Implication for the enzyme mechanism. *J. Inorg. Biochem.* **24**, 183–198.

Bünning, P., Kleemann, S.G. & Riordan, J.F. (1990) Essential residues in angiotensin converting enzyme: modification with 1-fluoro-2,4-dinitrobenzene. *Biochemistry* **29**, 10488–10492.

Chen, Y.N.P. & Riordan, J.F. (1990) Identification of essential tyrosine and lysine residues in angiotensin converting enzyme: evidence for a single active site. *Biochemistry* **29**, 10493–10498.

Chen, Y.N.P., Ehlers, M.R.W. & Riordan, J.F. (1992) The functional role of tyrosine-200 in human testis angiotensin-converting enzyme. *Biochem. Biophys. Res. Commun.* **184**, 306–309.

Colman, P.M., Jansonius, J.N. & Matthews, B.W. (1972) The structure of thermolysin: an electron density map at 2.3 Å resolution. *J. Mol. Biol.* **70**, 701–724.

Cornell, M.J., Williams, T.A., Lamango, N.S., Coates, D., Corvol, P., Soubrier, F., Hoheisel, J., Lehrach, H. & Isaac, R.E. (1995) Cloning and expression of an evolutionary conserved single-domain angiotensin converting enzyme from *Drosophila melanogaster*. *J. Biol. Chem.* **270**, 13613–13619.

Costerousse, O., Allegrini, J., Lopez, M. & Alhenc-Gelas, F. (1993) Angiotensin I-converting enzyme in human circulatory mononuclear cells. Genetic polymorphism in T-lymphocytes. *Biochem. J.* **290**, 33–40.

Das, M., Hartley, J.L. & Soffer, R.L (1977) Serum angiotensin-converting enzyme. Isolation and relationship to the pulmonary enzyme. *J. Biol. Chem.* **252**, 1316–1319.

Defendini, R., Zimmerman, E.A., Weare, J.A., Alhenc-Gelas, F. & Erdös, E.G. (1983) Angiotensin-converting enzyme in epithelial and neuroepithelial cells. *Neuroendocrinology* **37**, 32–40.

Dorer, F.E., Kahn, J.R., Lentz, K.E., Levine, M. & Skeggs, L.T. (1974) Hydrolysis of bradykinin by angiotensin-converting enzyme. *Circ. Res.* **34**, 824–827.

Dubreuil, P., Fulcrand, P., Rodriguez, M., Fulcrand, H., Laur, J. & Martinez, J. (1989) Novel activity of angiotensin-converting enzyme hydrolysis of cholecystokinin and gastrin analogues with release of the amidated carboxy-terminal dipeptide. *Biochem. J.* **262**, 125–130.

M

Ehlers, M.R.W. & Kirsch, R.E. (1988) Catalysis of angiotensin I hydrolysis by human angiotensin-converting enzyme: effect of chloride and pH. *Biochemistry* **27**, 5538–5544.

Ehlers, M.R.W. & Riordan, J.F. (1989) Angiotensin-converting enzyme: new concepts concerning its biological role. *Biochemistry* **28**, 5311–5318.

Ehlers, M.R.W. & Riordan, J.F. (1990) Angiotensin-converting enzyme. Biochemistry and molecular biology. In: *Hypertension Pathophysiology, Diagnosis and Management* (Laragh, J.H. & Brenner, B.M., eds). New York: Raven Press, pp. 1217–1231.

Ehlers, M.R.W., Fox, E.A., Strydom, D.J. & Riordan, J.F. (1989) Molecular cloning of human testicular angiotensin-converting enzyme: the testis isozyme is identical to the C-terminal half of endothelial angiotensin-converting enzyme. *Proc. Natl Acad. Sci. USA* **86**, 7741–7745.

Ehlers, M.R.W. & Riordan, J.F. (1991) Angiotensin-converting enzyme: zinc- and inhibitor-binding stoichiometries of the somatic and testis isozymes. *Biochemistry* **30**, 7118–7126.

Ehlers, M.R.W., Chen, Y.N.P. & Riordan, J.F. (1991) Spontaneous solubilization of membrane-bound human testis angiotensin-converting enzyme expressed in Chinese hamster ovary cells. *Proc. Natl Acad. Sci. USA* **88**, 1009–1013.

Ehlers, M.R.W., Chen, Y.N.P. & Riordan, J.F. (1992) The unique N-terminal sequence of testis angiotensin-converting enzyme is heavily *O*-glycosylated and unessential for activity or stability. *Biochem. Biophys. Res. Commun.* **183**, 199–205.

El-Dorry, H., Iwata, K., Thornberg, N.A., Cordes, E.H. & Soffer, R.L. (1982) Molecular and catalytic properties of rabbit testicular dipeptidyl carboxypeptidase. *J. Biol. Chem.* **257**, 14128–14133.

El-Dorry, H.A., MacGregor, J.S. & Soffer, R.L. (1983) Dipeptidyl carboxypeptidase from seminal fluid resembles the pulmonary rather than the testicular isoenzyme. *Biochem. Biophys. Res. Commun.* **115**, 1096–1100.

Erdös, E.G. & Skidgel, R.A. (1987) The angiotensin I-converting enzyme. *Lab. Invest.* **56**, 345–348.

Esther, C.R., Jr, Howard, T.E., Marino, E.M., Goddard, J.M., Capecchi, M.R. & Bernstein, K.E. (1996) Mice lacking angiotensin-converting enzyme have low blood pressure, renal pathology and reduced male fertility. *Lab. Invest.* **74**, 953–965.

Friedland, J., Setton, C. & Silverstein, E. (1978) Induction of angiotensin-converting enzyme in human monocytes in culture. *Biochem. Biophys. Res. Commun.* **83**, 843–849.

Gardell, S.J., Hilvert, D., Barnett, J., Kaiser, E.T. & Rutter, W.J. (1987) Use of directed mutagenesis to probe the role of tyrosine 198 in the catalytic mechanism of carboxypeptidase A. *J. Biol. Chem.* **262**, 576–582.

Hooper, N.M. (1991) Angiotensin-converting enzyme. Implications from molecular biology for its physiological functions. *Int. J. Biochem.* **23**, 641–647.

Hooper, N.M. & Turner, A.J. (1987) Isolation of two differentially glycosylated forms of peptidyl-dipeptidase A (angiotensin-converting enzyme) from pig brain: a re-evaluation of their role in neuropeptide metabolism. *Biochem. J.* **241**, 625–633.

Hooper, N.M., Keen, J., Pappin, D.J.C. & Turner, A.J. (1987) Pig kidney angiotensin-converting enzyme. Purification and characterization of amphiphatic and hydrophilic forms of the enzyme establishes C-terminal anchorage to the plasma membrane. *Biochem. J.* **247**, 85–93.

Howard, T.E., Shai, S.Y., Langford, K.G., Martin, B.M. & Bernstein, K.E. (1990) Transcription of testicular angiotensin-converting enzyme (ACE) is initiated within the 12th intron of the somatic ACE gene. *Mol. Cell. Biol.* **10**, 4294–4302.

Howard, T., Balogh, R., Overbeek, P. & Bernstein, K.E. (1993) Sperm-specific expression of angiotensin-converting enzyme (ACE) is mediated by a 91-base-pair promoter containing a CRE-like element. *Mol. Cell. Biol.* **13**, 1–27.

Hubert, C., Houot, A.-M., Corvol, P. & Soubrier, F. (1991) Structure of the angiotensin I-converting enzyme gene. Two alternate promoters correspond to evolutionary steps of a duplicated gene. *J. Biol. Chem.* **266**, 15377–15383.

Igic, R., Erdös, E.G., Yen, H.S.G., Sorrells, K. & Nakajima, T. (1972) Angiotensin I converting enzyme of the lung. *Circ. Res.* **31**(suppl 2), 51–61.

Inokuchi, J.J. & Nagamatsu, A. (1981) Tripeptidyl carboxypeptidase activity of kininase II (angiotensin converting enzyme). *Biochim. Biophys. Acta* **662**, 300–307.

Jaspard, E., Wei, L. & Alhenc-Gelas, F. (1993) Differences in properties and enzymatic specificities between the two active sites of angiotensin I-converting enzyme. Studies with bradykinin and other natural peptides. *J. Biol. Chem.* **268**, 9496–9503.

Kester, W.R. & Matthews, B.W. (1977) Crystallographic study of the binding of dipeptide inhibitors to thermolysin: implications for the mechanism of catalysis. *Biochemistry* **16**, 2506–2516.

Koike, G., Krieger, J.E., Jacob, H.J., Mukoyama, M., Pratt, R.E. & Dzau, V.J. (1994) Angiotensin-converting enzyme and genetic hypertension: cloning of rat cDNAs and characterization of the enzyme. *Biochem. Biophys. Res. Commun.* **198**, 380–386.

Krege, J.H., John, S.W.M., Laugenbach, L.L., Hodgin, J.B., Hagaman, J.R., Bachman, E.S., Jennette, J.C., O'Brien, D.A. & Smithies, O. (1995) Male-female differences in fertility and blood pressure in ACE-deficient mice. *Nature* **375**, 146–148.

Kumar, R.S., Kusari, J., Roy, S.N., Soffer, R.L. & Sen, G.C. (1989) Structure of testicular angiotensin-converting enzyme. A segmental mosaic isozyme. *J. Biol. Chem.* **264**, 16754–16758.

Kumar, R.S., Thekumkara, T.J. & Sen, G. (1991) The mRNAs encoding the two angiotensin-converting isozymes are transcribed from the same gene by a tissue-specific choice of alternative transcription initiation sites. *J. Biol. Chem.* **266**, 3854–3862.

Lamango, N. & Isaac, R.E. (1993) Identification of an ACE-like peptidyl dipeptidase activity in the housefly, *Musca domestica*. *Biochem. Soc. Trans.* **21**, 245S.

Langford, K.G., Shai, S.Y., Howard, T.E., Kovac, M.J., Overbeek, P.A. & Bernstein, K.E. (1991) Transgenic mice demonstrate a testis-specific promoter for angiotensin-converting enzyme. *J. Biol. Chem.* **266**, 15559–15562.

Lattion, A.L., Soubrier, F., Allegrini, J., Hubert, C., Corvol, P. & Alhenc-Gelas, F. (1989) The testicular transcript of the angiotensin I-converting enzyme encodes for the ancestral, non-duplicated form of the enzyme. *FEBS Lett.* **252**, 99–104.

Lawton, G., Paciorek, P.H. & Waterfall, J.F. (1992) The design and biological profile of ACE inhibitors. In: *Advances in Drug Research* (Testa, B., ed.), vol. 23. New York: Academic Press, pp. 161–220.

Le Moual, H., Devault, A., Roques, B.P., Crine, P. & Boileau, G. (1991) Identification of glutamic acid 616 as a zinc-coordinating residue in endopeptidase-24.11. *J. Biol. Chem.* **266**, 15670–15674.

Matsas, R., Kenny, A.J. & Turner, A.J. (1984) The metabolism of neuropeptides. The hydrolysis of peptides including enkephalins, tachykinins and their analogues by endopeptidase 24.11 (E.C. 3.4.24.11). *Biochem. J.* **223**, 433–440.

Nadaud, S., Houot, A.., Hubert, C., Corvol, P. & Soubrier, F. (1992) Functional study of the germinal angiotensin I-converting enzyme promoter. *Biochem. Biophys. Res. Commun.* **189**, 134–140.

Ng, K.K.F. & Vane, J.R. (1967) Conversion of angiotensin I to angiotensin II. *Nature* **216**, 762–766.

Ng, K.K.F. & Vane, J.R. (1968) Fate of angiotensin I in the circulation. *Nature* **218**, 144–150.

Oblin, A., Danse, M.J. & Zivkovic, B. (1988) Degradation of substance P by membrane peptidases in the rat substantia nigra: effect of selective inhibitors. *Neurosci. Lett.* **84**, 91–96.

Oparil, S. & Haber, E. (1974a) The renin angiotensin system. *N. Engl. J. Med.* **291**, 389–401.

Oparil, S. & Haber, E. (1974b) The renin angiotensin system. *N. Engl. J. Med.* **291**, 446–457.

Oppong, S.Y. & Hooper, N.M. (1993) Characterization of a secretase acitivity which releases angiotensin-converting enzyme from the membrane. *Biochem. J.* **292**, 597–603.

Patchett, A.A. & Cordes, E.H. (1985) The design and properties of of *N*-carboxyalkyldipeptide inhibitors of angiotensin converting enzyme. *Adv. Enzymol.* **57**, 1–84.

Perich, R.B., Jackson, B., Rogerson, F., Mendelsohn, F.A.O., Paxton, D. & Johnston, C.I. (1992) Two binding sites on angiotensin-converting enzyme: evidence from radioligand binding studies. *Mol. Pharmacol.* **42**, 286–293.

Ramchandran, R. & Sen, I. (1995) Cleavage processing of angiotensin-converting enzyme by a membrane-associated metalloprotease. *Biochemistry* **34**, 12645–12652.

Ramchandran, R., Sen, G.C., Misono, K. & Sen, I. (1994) Regulated cleavage-secretion of the membrane-bound angiotensin-converting enzyme. *J. Biol. Chem.* **269**, 2125–2130.

Rieger, K.J., Saez-Servent, N., Papet, M.P., Wdzieczak-Bakala, J., Morgat, J.L., Thierry, J., Woelter, W. & Lenfant, M. (1993) Involvements of human plasma angiotensin converting enzyme in the degradation of the hemoregulatory peptide *N*-acetyl-seryl-aspartyl-lysyl-proline. *Biochem. J.* **296**, 373–378.

Rousseau, A., Michaud, A., Chauvet, M.-T., Lenfant, M. & Corvol, P. (1995) The hemoregulatory peptide *N*-acetyl-Ser-Asp-Lys-Pro is a natural and specific substrate of the N-terminal active site of human angiotensin-converting enzyme. *J. Biol. Chem.* **270**, 3656–3661.

Schweisfurth, H. & Schioberg-Schiegnitz, S. (1984) Assay and biochemical characterization of angiotensin I-converting enzyme in cerebrospinal fluid. *Enzyme* **32**, 12–19.

Sen, I., Kasturi, S., Jabbar, M.A. & Sen, G.C. (1993) Mutations in two specific residues of testicular angiotensin-converting enzyme change its catalytic properties. *J. Biol. Chem.* **268**, 25748–25754.

Shai, S.Y., Fishel, R.S., Martin, B.M., Berk, B.C. & Bernstein, K.B. (1992) Bovine angiotensin converting enzyme cDNA cloning and regulation. Increased expression during endothelial cell growth arrest. *Circ. Res.* **70**, 1274–1281.

Shapiro, R. & Riordan, J.F. (1983) Critical lysine residue at the chloride binding site of angiotensin converting enzyme. *Biochemistry* **22**, 5315–5321.

Shapiro, R., Holmquist, B. & Riordan, J.F. (1983) Anion activation of angiotensin-converting enzyme: dependence on nature of substrate. *Biochemistry* **22**, 3850–3857.

Sibony, M., Segretain, D. & Gasc, J.-M. (1994) Angiotensin-converting enzyme in murine testis: step-specific expression of the germinal isoform during spermiogenesis. *Biol. Reprod.* **50**, 1015–1026.

Skeggs, L.T., Kahn, J.R. & Shumway, N.P. (1956) The preparation and function of the hypertensin-converting enzyme. *J. Exp. Med.* **103**, 295–299.

Skidgel, R.A. & Erdös, E.G. (1985) Novel activity of human angiotensin I converting enzyme: release of the $NH_2$- and COOH-terminal tripeptides from the luteinizing hormone-releasing hormone. *Proc. Natl Acad. Sci. USA* **82**, 1025–1029.

Skidgel, R.A. & Erdös, E.G. (1987) The broad substrate specificity of human angiotensin I converting enzyme. *Clin. Exp. Hypertens.* **A9**, 243–259.

Skidgel, R.A., Engelbrecht, S., Johnson, A.R. & Erdös, E.G. (1984) Hydrolysis of substance P and neurotensin by converting enzyme and neutral endopeptidase. *Peptides* **5**, 769–776.

Soubrier, F., Alhenc-Gelas, F., Hubert, C., Allegrini, J., John, M., Tregear, G. & Corvol, P. (1988) Two putative active centers in human angiotensin I-converting enzyme revealed by molecular cloning. *Proc. Natl Acad. Sci. USA* **85**, 9386–9390.

Strittmatter, S.M. & Snyder, S.H. (1987) Angiotensin-converting enzyme immuno-histochemistry in rat brain and pituitary gland: correlation of isozyme type with cellular localization. *Neuroscience* **21**, 407–420.

Strittmatter, S.M., Lo, M.M.S., Javitch, J.A. & Snyder, S.H. (1984) Autoradiographic visualization of angiotensin-converting enzyme in rat brain with [$^3$H]captopril: localization to a striatonigral pathway. *Proc. Natl Acad. Sci. USA* **81**, 1599–1603.

Strittmatter, S.M., Thiele, E.A., Kapiloff, M.S. & Snyder, S.H. (1985) A rat brain isozyme of angiotensin-converting enzyme. *J. Biol. Chem.* **260**, 9825–9832.

Testut, P., Corvol, P. & Hubert, C. (1993) Functional analysis of the somatic angiotensin I converting enzyme promoter. *Biochem. J.* **293**, 843–848.

Thekkumkara, T.J., Livingston, I.W., Kumar, R.S. & Sen, G.C. (1992) Use of alternative polyadenylation sites for tissue-specific transcription of two angiotensin-converting enzyme mRNAs. *Nucleic Acids Res.* **20**, 683–687.

Turner, A.J., Hryszko, J., Hooper, N.M. & Dowdall, M.J. (1987) Purification and characterization of a peptidyl dipeptidase resembling angiotensin converting enzyme from the electric organ of *Torpedo marmorata*. *J. Neurochem.* **48**, 910–916.

Vallee, B.L. & Auld, D.S. (1990) Active-site zinc-ligands and activated $H_2O$ of zinc enzymes. *Proc. Natl Acad. Sci. USA* **87**, 220–224.

Velletri, P.A. (1985) Testicular angiotensin I-converting enzyme (E.C. 3.4.15.1). *Life Sci.* **36**, 1597–1608.

Waeber, B., Nussberger, J. & Brunner, HS. (1990) Angiotensin-converting enzyme inhibitors in hypertension. In: *Hypertension: Pathophysiology, Diagnosis and Management* (Laragh, J.H. & Brenner, B.M., eds). New York: Raven Press, pp. 2209–2232.

Wei, L., Alhenc-Gelas, F., Corvol, P. & Clauser, E. (1991a) The two homologous domains of the human angiotensin I-converting enzyme are both catalytically active. *J. Biol. Chem.* **266**, 9002–9008.

Wei, L., Alhenc-Gelas, F., Soubrier, F., Michaud, A., Corvol, P. & Clauser, E. (1991b) Expression and characterization of recombinant human angiotensin I-converting enzyme. Evidence for a C-terminal transmembrane anchor and for a proteolytic processing of the secreted recombinant and plasma enzymes. *J. Biol. Chem.* **266**, 5540–5546.

Wei, L., Clauser, E., Alhenc-Gelas, F. & Corvol, P. (1992) The two homologous domains of human angiotensin I-converting enzyme interact differently with competitive inhibitors. *J. Biol. Chem.* **267**, 13398–13405.

Williams, T.A., Hooper, N.M. & Turner, A.J. (1991) Characterization of neuronal and endothelial forms of angiotensin converting enzyme in pig brain. *J. Neurochem.* **57**, 193–199.

Williams, T.A., Barnes, K., Kenny, A.J., Turner, A.J. & Hooper, N.M. (1992) A comparison of the zinc content and substrate specificity of the endothelial and testicular forms of porcine angiotensin converting enzyme and the isolation of isoenzyme specific antisera. *Biochem. J.* **288**, 878–881.

Williams, T.A., Corvol, P. & Soubrier, F. (1994) Identification of two active site residues in human angiotensin I-converting enzyme. *J. Biol. Chem.* **269**, 29430–29434.

Yang, H.Y.T., Erdös, E.G. & Levin, Y. (1970) A dipeptidyl carboxypeptidase that converts angiotensin I and inactivates bradykinin. *Biochim. Biophys. Acta* **214**, 374–376.

Yasui, T., Alhenc-Gelas, F., Corvol, P. & Menard, J. (1984) Angiotensin I-converting enzyme in amniotic fluid. *J. Lab. Clin. Med.* **104**, 741–751.

Yokosawa, H., Endo, S., Ogura, Y. & Ishii, S. (1983) A new feature of angiotensin-converting enzyme in the brain: hydrolysis of substance P. *Biochem. Biophys. Res. Commun.* **116**, 735–742.

*Pierre Corvol*
*INSERM U36,*
*Collège de France,*
*3 rue d'Ulm,*
*75005 Paris, France*

*Tracy A. Williams*
*INSERM U36,*
*Collège de France,*
*3 rue d'Ulm,*
*75005 Paris, France*

# 360. Peptidyl-dipeptidase A (invertebrate)

## Databanks

*Peptidase classification: clan MA, family M2, MEROPS ID: M02.002*
*NC-IUBMB enzyme classification: EC 3.4.15.1*
*Chemical Abstracts Service registry number: 9015-82-1*
*Databank codes:*

| Species | SwissProt | PIR | EMBL (cDNA) | EMBL (genomic) |
|---|---|---|---|---|
| *Boophilus microplus* | – | – | U62809 | – |
| *Caenorhabditis elegans* | – | – | – | U56966: cosmid |
| *Drosophila melanogaster* | Q10714 | – | U25344 U34599 | – |
| *Drosophila melanogaster* | – | – | X96913 | – |
| *Haematobia irritans* | Q10715 | – | L43965 | – |

## Name and History

The first reports of an ***invertebrate peptidyl-dipeptidase A*** came from a study of the metabolism of Tyr-D-Ala-Gly+Leu-Phe by a membrane preparation from the heads of the housefly, *Musca domestica* (Lamango & Isaac, 1993, 1994). This hydrolysis was partly inhibited by 10 μM captopril, a potent inhibitor of mammalian peptidyl-dipeptidase A. Enzymatic studies and structural information from cDNA sequences shows that the invertebrate enzyme is closely related to mammalian peptidyl-dipeptidase A, which is often called angiotensin I-converting enzyme (ACE) (Chapter 359), after its important role in the renin–angiotensin system. ACE is a frequently used abbreviation for the invertebrate enzyme also, but because the acetylcholinesterase gene of *Drosophila melanogaster* is already known as *Ace*, the acronym *Ance* is used for the *D. melanogaster* peptidyl-dipeptidase A gene (Cornell *et al.*, 1995).

Peptidyl-dipeptidase A has been isolated from the insects *D. melanogaster* (Cornell *et al.*, 1995), *M. domestica* (Lamango *et al.*, 1996) and *Haematobia irritans* (Wijffels *et al.*, 1996), from an acarine, *Boophilus microplus* (Jarmey *et al.*, 1995) and from an annelid, *Theromyzon tessulatum* (Laurent & Salzet, 1996).

## Activity and Specificity

The insect enzyme, like mammalian peptidyl-dipeptidase A, has a broad substrate specificity and can function as an endopeptidase by cleaving dipeptide amides and tripeptide amides from oligopeptides with an amidated C-terminus. For example, Phe-Leu-NH$_2$ is cleaved from [Leu$^5$]-enkephalinamide (Tyr-Gly-Gly+Phe-Leu-NH$_2$) and Phe-Met-NH$_2$ from [Met$^5$]-enkephalinamide (Tyr-Gly-Gly+Phe-Met-NH$_2$) and both Met-Leu-NH$_2$ and Gly-Met-Leu-NH$_2$ are primary products from the hydrolysis of substance P (Lamango *et al*., 1996; Wijffels *et al*., 1996). Bradykinin, Arg-Pro-Pro-Gly-Phe-Ser-Pro+Phe-Arg, is a favored *in vitro* substrate (Lamango *et al*., 1996). *In vivo* substrates have not yet been identified, although a number of insect peptide hormones are hydrolyzed by insect peptidyl-dipeptidase A (Lamango *et al*., 1997). A convenient assay employs Bz-Gly+His-Leu as substrate and reverse-phase HPLC with detection at 214 nm to measure the production of Bz-Gly. The pH optimum is 7.5–8.2. Chloride or sulfate ions (0.6 M) are required for optimum activity and zinc ions (10 μM) are often included in the reaction buffer. Other substrates sometimes require different assay conditions for optimal activity. Mammalian peptidyl-dipeptidase A inhibitors (for example, captopril, lisinopril, enalopril) are also potent inhibitors of the insect enzyme with captopril displaying greatest potency (IC$_{50}$ 50 nM). Chelators of bivalent metal ions (EDTA and 1,10-phenanthroline) are inhibitory.

## Structural Chemistry

Peptidyl-dipeptidase A from *D. melanogaster*, *M. domestica* and *H. irritans* has an $M_r$ of 70 000, whereas the *B. microplus* peptidyl-dipeptidase A is larger ($M_r$ 86 500), due to a greater degree of glycosylation. The existence of a membrane-bound form of the enzyme has not been confirmed. The *T. tessulatum* enzyme has an $M_r$ of around 120 000. No two-domain form of peptidyl-dipeptidase A, equivalent to mammalian somatic peptidyl-dipeptidase A, has been found in invertebrates. *N*-Glycosylation is not required for expression of enzymically active *D. melanogaster* peptidyl-dipeptidase A in yeast (*Pichia pastoris*) but does contribute to the thermal stability of the recombinant protein (Williams *et al*., 1996).

## Biological Aspects

The enzyme is particularly concentrated in the midgut, testes, ovaries and blood of adult insects and significant activity is found in the central nervous system (Wijffels *et al*., 1996; Isaac *et al*., 1998). Expression is also seen in early embryonic development and increases during larval stages, reaching a peak during pupation (Tatei *et al*., 1995; Isaac *et al*., 1998). In *H. irritans* expression is restricted to the midgut, testes and brain of adult males, but is absent from ovaries of the adult female fly (Wijffels *et al*., 1996). Peptidyl-dipeptidase A is found in neurosecretory cells of the brain and subesophageal ganglion of insects and is likely to be involved in the intracellular processing of insect peptide hormones, such as members of the *Locusta* myotropin/pheromone biosynthesis-activating peptide family (Isaac *et al*., 1998). Some peptidergic endocrine cells of the locust midgut also contain peptidyl-dipeptidase A. *D. melanogaster* with mutations in the *Ance* gene die during larval/pupal development and male

transheterozygotes of two mutant *Ance* alleles are infertile (Tatei *et al*., 1995).

## Distinguishing Features

The *D. melanogaster* and *H. irritans* enzymes differ from mammalian forms of the enzyme in that they have a much lower carbohydrate content and that they are not bound to the plasma membrane (Cornell *et al*., 1995; Wijffels *et al*., 1996). Captopril is equipotent at inhibiting the insect and mammalian enzymes but lisinopril, enalaprilat, trandolaprilat and fosinoprilat have higher $K_i$ values for inhibition of *D. melanogaster* peptidyl-dipeptidase A than for the inhibition of mammalian peptidyl-dipeptidase A (Chapter 359) (Williams *et al*., 1996). Antibodies to the pig kidney peptidyl-dipeptidase A do not cross-react with the insect enzyme.

## Further Reading

See Isaac *et al*. (1998) for a recent review.

## References

Cornell, M.J., Williams, T.A., Lamango, N.S., Coates, D., Corvol, P., Soubrier, F., Hoheisel, J., Lehrach, H. & Isaac, R.E. (1995) Cloning and expression of an evolutionary conserved single-domain angiotensin-converting enzyme from *Drosophila melanogaster*. *J. Biol. Chem.* **270**, 13613–13619.

Isaac, R.E., Coates, D., Williams, T.A. & Schoofs, L. (1998) Insect angiotensin-converting enzyme: comparative biochemistry and evolution. In: *Arthropod Endocrinology: Perspectives and Recent Advances* (Coast, G.M. & Webster, S.G., eds). Cambridge: Cambridge University Press, pp. 357–377.

Jarmey, J.M., Riding, G.A., Pearson, R.D., McKenna, R.V. & Willadsen, P. (1995) Carboxypeptidase from *Boophilus microplus*: a concealed antigen with similarity to angiotensin-converting enzyme. *Insect Biochem. Mol. Biol.* **25**, 969–974.

Lamango, N.S. & Isaac, R.E. (1993) Identification of an ACE-like peptidyl dipeptidase activity in the housefly, *Musca domestica*. *Biochem. Soc. Trans.* **21**, 245S.

Lamango, N.S. & Isaac, R.E. (1994) Identification and properties of a peptidyl dipeptidase in the housefly, *Musca domestica*, that resembles mammalian angiotensin-converting enzyme. *Biochem. J.* **299**, 651–657.

Lamango, N.S., Sajid, M. & Isaac, R.E. (1996) The endopeptidase activity and the activation by Cl$^-$ of angiotensin-converting enzyme is evolutionarily conserved: purification and properties of an angiotensin-converting enzyme from the housefly, *Musca domestica*. *Biochem. J.* **314**, 639–646.

Lamango, N.S., Nachman, R.J., Hayes, T.K., Strey, A. & Isaac, R.E. (1997) Hydrolysis of insect neuropeptides by an angiotensin-converting enzyme from the housefly, *Musca domestica*. *Peptides* **18**, 47–52.

Laurent, V. & Salzet, M. (1996) Biochemical properties of the angiotensin-converting enzyme-like enzyme from the leech *Theromyzon tessulatum*. *Peptides* **17**, 737–745.

Tatei, K., Cai, H., Ip, T. & Levine, M. (1995) Race – a *Drosophila* homolog of the angiotensin-converting enzyme. *Mech. Dev.* **51**, 157–168.

Wijffels, G., Fitzgerald, C., Gough, J., Riding, G., Elvin, C., Kemp, D. & Willadsen, P. (1996) Cloning and characterisation of angiotensin-converting enzyme from the dipteran species,

**M**

*Haematobia irritans exigua*, and its expression in the maturing male reproductive system. *Eur. J. Biochem.* **237**, 414–423.

Williams, T.A., Michaud, A., Houard, X., Chauvet, M.T., Soubrier, F. & Corvol, P. (1996) *Drosophila melanogaster* angiotensin

I-converting enzyme expressed in *Pichia pastoris* resembles the C domain of the mammalian homologue and does not require glycosylation for secretion and enzymic activity. *Biochem. J.* **318**, 125–131.

***R.E. Isaac***
*Department of Biology,*
*University of Leeds,*
*Leeds LS2 9JT, UK*
*Email: r.e.isaac@leeds.ac.uk*

***D. Coates***
*Department of Biology,*
*University of Leeds,*
*Leeds LS2 9JT, UK*
*Email: d.coates@leeds.ac.uk*

# 361. Mycolysin

## Databanks

*Peptidase classification: clan MA, family M5, MEROPS ID: M05.001*
*NC-IUBMB enzyme classification: EC 3.4.24.31*
*Chemical Abstracts Service registry number: 153190-34-2*
*Databank codes:*

| Species | SwissProt | PIR | EMBL (cDNA) | EMBL (genomic) |
|---|---|---|---|---|
| *Streptomyces cacaoi* | P20910 | JQ0530 | M37055 | – |

## Name and History

Various microbes belonging to Streptomycetes have been reported to produce metallo-endopeptidases. Among them, the enzymes with a molecular mass of 35–37 kDa (or more) are named **mycolysin**. Those with a mass of about 16 kDa are classified as the *Streptomyces* small neutral endopeptidase (Chapter 382). Mycolysin was first purified from a culture filtrate of *Streptomyces naraensis* (Hiramatsu, 1967a). *S. griseus* mycolysin was then isolated from a commercially available protease preparation 'pronase' (Narahashi *et al.*, 1968). At least two molecular species seem to be present in this mycolysin (Narahashi *et al.*, 1968; Löfqvist, 1974; Tsuyuki *et al.*, 1991). Purification and characterization have also been described for mycolysins from *S. violaceorectus* (Nakamura *et al.*, 1972; Inoue *et al.*, 1984a), *S. griseoruber* (Inoue *et al.*, 1984b) and *S. mauvecolor* (Inoue *et al.*, 1985). Chang *et al.* (1990) have reported on the cloning, sequencing and expression of a gene encoding prepromycolysin of *S. cacaoi*.

## Activity and Specificity

Cleavage of internal bonds was observed in an *N*-ethylsuccinimide derivative of cytochrome *c* (Hiramatsu, 1967b), oxidized insulin B chain (Morihara *et al.*, 1968)–

F┼VNQH┼LCGSH┼LVEA┼LY┼L┼VCGERG┼F┼F┼YTPKA

and various synthetic peptides (Morihara *et al.*, 1968; Inoue *et al.*, 1984a,b, 1985; Kajiwara *et al.*, 1991a). The analytical results indicate that all the mycolysins examined favor hydrophobic residues in the P1' position, as in the case of thermolysin (Chapter 351). Mycolysin shows higher preference for aromatic residues than for other hydrophobic residues at P1'. Hydrophobic residues are also favored in P1 and P2'.

Activity has been routinely determined with casein (Hiramatsu, 1967a; Narahashi *et al.*, 1968), Z-Gly┼Phe-NH$_2$ (Inoue *et al.*, 1984a,b, 1985; Tsuyuki *et al.*, 1991) or FA-Gly┼Leu-NH$_2$ (Blumberg & Tauber, 1983) as substrates. The assay is more conveniently made with the quenched fluorescence substrate, Bim-SCH$_2$-Phe┼Trp-Leu-OH (Bim: 1,7-dioxo-2,5,6-trimethyl-1*H*,7*H*-pyrazolo[1,2-α]pyrazol-3-yl-methyl) (Kajiwara *et al.*, 1991a). The pH optimum for these substrates is generally about 7. The only exception is *S. violaceorectus* mycolysin which shows a pH optimum for casein at 9.0–9.5. Even with this enzyme, however, the value of $k_{cat}/K_m$ for Z-Gly-Phe-NH$_2$ at pH 7.0 is 9 times higher than that at pH 9.0, due to the much higher $k_{cat}$ at pH 7.0 (Inoue *et al.*, 1984a).

EDTA is inhibitory, and the suppressed activity is almost completely restored by Zn$^{2+}$ (Hiramatsu, 1967a; Inoue *et al.*, 1984b, 1985). The effects of various active site-directed inhibitors were compared between *S. griseus* mycolysin and thermolysin (Chapter 351). Thus, phosphoramidon was found to be almost equally potent for the two ($K_i$ about 0.01 M at pH 7.0) (Tsuyuki *et al*., 1991). The $K_i$ value at pH 7.5 of 2-mercaptoacetyl-Phe-Leu is 0.07 μM for *S. griseus* mycolysin and 4.2 μM for thermolysin (Blumberg & Tauber, 1983). The second-order rate constant for inactivation by ClCH$_2$CO-D,L-(N-OH)Leu-Ala-Gly-NH$_2$ is 8.9 M$^{-1}$ s$^{-1}$ for the former enzyme and about one-fourth of that for the latter (Kumazaki *et al*., 1994). Mycolysins from *S. griseus* and from *S. violaceorectus* are inhibited by *Streptomyces* protein protease inhibitors (formerly believed to be specific to serine proteases) such as plasminostreptin, alkaline protease inhibitor-2c′, and *Streptomyces* subtilisin inhibitor (Kajiwara *et al*., 1991b). Activities of thermolysin (Chapter 351) and pseudolysin (Chapter 357) are not susceptible to these inhibitors.

## Structural Chemistry

Mycolysins are single-chain proteins of about 37 kDa, with the exception of those from *S. griseoruber* (52 kDa: Inoue *et al*., 1984b) and from *S. mauvecolor* (59 kDa; Inoue *et al*., 1985). The amino acid sequence deduced from the gene encoding a precursor of *S. cacaoi* mycolysin consists of a signal peptide with 34 residues, a propeptide with 171 residues, and the protease portion with 345 residues (Chang *et al*., 1990). Site-specific mutations in a series of gene expression studies have demonstrated the importance of His202, Glu203, His206 and Glu240 (residue numbers for the mature enzyme) as putative zinc ligands and a catalytic residue (Chang & Lee, 1992). The propeptide, which is autocatalytically processed by cleavage between Ala residues, plays an essential role in maturation and secretion of the protease (Chang *et al*., 1994). The mature protease seems to be a basic protein, because it has 21 residues of Asp, 7 Glu, 11 Lys, 21 Arg and 8 His. *S. naraensis* mycolysin has been reported to have a pI of 4.2, however (Hiramatsu & Ouchi, 1972). Mycolysins from *S. naraensis, S. griseus* (Tsuyuki *et al*., 1991) and *S. cacaoi* are known to contain Cys (2, 2 and 7 residues, respectively). No information is available on disulfide bridges. Atomic absorption spectrophotometry showed the presence of 1 mol of zinc per mol of enzyme in these three mycolysins.

## Preparation

A commercial heterogeneous protease preparation pronase is the most convenient source of *S. griseus* mycolysin. For mycolysins of the other *Streptomyces* spp., ammonium sulfate or acetone precipitates obtained from supernatants of the microbe culture broth are generally used as the enzyme sources. Purification has been achieved by a series of chromatographic steps with gel filtration and ion-exchange columns in early studies (Narahashi *et al*., 1968; Hiramatsu & Ouchi, 1972; Nakamura *et al*., 1972). The introduction of affinity chromatography offered more efficient ways for the purification (Inoue *et al*., 1984a,b, 1985; Tsuyuki *et al*., 1991). The gene encoding a precursor of *S. cacaoi* mycolysin was expressed in *S. lividans*, and the

mature enzyme secreted in the medium was purified (Chang *et al*., 1990).

## Biological Aspects

Mycolysins are secretory proteins and are considered to act in the digestion of extracellular food proteins. Some of them have been reported to show anti-inflammatory activity against carrageenan-induced edema in rats (Nakamura *et al*., 1972; Inoue *et al*., 1985).

## References

Blumberg, S. & Tauber, Z. (1983) Inhibition of metalloendopeptidases by 2-mercaptoacetyl-dipeptides. *Eur. J. Biochem.* **136**, 151–154.

Chang, P.C. & Lee, Y.-H.W. (1992) Extracellular autoprocessing of a metalloprotease from *Streptomyces cacaoi. J. Biol. Chem.* **267**, 3952–3958.

Chang, P.C., Kuo, T.-C., Tsugita, A. & Lee, Y.-H.W. (1990) Extracellular metalloprotease gene of *Streptomyces cacaoi*: structure, nucleotide sequence and characterization of the cloned gene product. *Gene* **88**, 87–95.

Chang, S.-C., Chang, P.-C. & Lee, Y.-H.W. (1994) The roles of propeptide in maturation and secretion of Npr protease from *Streptomyces. J. Biol. Chem.* **269**, 3548–3554.

Hiramatsu, A. (1967a) A neutral proteinase from *Streptomyces naraensis*. I. Purification and some properties. *J. Biochem. (Tokyo)* **62**, 353–363.

Hiramatsu, A. (1967b) A neutral proteinase from *Streptomyces naraensis*. II. Its mode of action on baker's yeast cytochrome c. *J. Biochem. (Tokyo)* **62**, 364–372.

Hiramatsu, A. & Ouchi, T. (1972) A neutral proteinase from *Streptomyces naraensis*. Improved purification and some physicochemical properties. *J. Biochem. (Tokyo)* **71**, 767–781.

Inoue, Y., Kawaguchi, Y. & Nakamura, S. (1984a) Affinity chromatography of alkinonase A on *N*-carbobenzoxy-glycyl-leucyl-aminohexyl-Sepharose. *Chem. Pharm. Bull.* **32**, 2333–2339.

Inoue, Y., Kawaguchi, Y. & Nakamura, S. (1984b) Affinity chromatography of neutral metalloendopeptidase produced by *Streptomyces griseoruber* on *N*-benzyloxycarbonyl-glycyl-leucyl-aminohexyl-Sepharose. *Chem. Pharm. Bull.* **32**, 4532–4538.

Inoue, Y., Kawaguchi, Y. & Nakamura, S. (1985) Affinity chromatography of neutral metalloendopeptidase produced by *Streptomyces mauvecolor* on *N*-benzyloxycarbonyl-glycyl-D-leucyl-aminohexyl-Sepharose. *Chem. Pharm. Bull.* **33**, 1544–1551.

Kajiwara, K., Kumazaki, T., Sato E., Kanaoka, Y. & Ishii, S. (1991a) Application of bimane-peptide substrates to spectrofluorometric assays of metalloendopeptidases. *J. Biochem. (Tokyo)* **110**, 345–349.

Kajiwara, K., Fujita, A., Tsuyuki, H., Kumazaki, T. & Ishii, S. (1991b) Interaction of *Streptomyces* serine-protease inhibitors with *Streptomyces griseus* metalloendopeptidase II. *J. Biochem. (Tokyo)* **110**, 350–354.

Kumazaki, T., Ishii, S. & Yokosawa, H. (1994) Inhibition of *Streptomyces griseus* metallo-endopeptidase II by active-site-directed inhibitors. *J. Biochem. (Tokyo)* **115**, 532–535.

Löfqvist, B. (1974) Zinc enzymes in commercial pronase-P. Further characterization of the heterogeneity of the proteolytic enzymes of the K-1 strain of *Streptomyces griseus. Acta Chem. Scand.* **B28**, 1013–1023.

Morihara, K., Tsuzuki, H. & Oka, T. (1968) Comparison of the specificities of various neutral proteinases from microorganisms. *Arch. Biochem. Biophys.* **123**, 572–588.

Nakamura, S., Fukuda, H., Yamamoto, T., Ogura, M., Hamada, M., Matsuzaki, M. & Umezawa, H. (1972) Alkinonase A and AF, new alkaline proteinases produced by *Streptomyces violaceorectus.* *Chem. Pharm. Bull.* **20**, 385–390.

Narahashi, Y., Shibuya, K. & Yanagita, M. (1968) Studies on proteolytic enzymes (pronase) of *Streptomyces griseus* K-1. *J. Biochem. (Tokyo)* **64**, 427–437.

Tsuyuki, H., Kajiwara, K., Fujita, A., Kumazaki, T. & Ishii, S. (1991) Purification and characterization of *Streptomyces griseus* metalloendopeptidase I and II. *J. Biochem. (Tokyo)* **110**, 339–344.

*Shin-ichi Ishii*
*Department of Biochemistry,*
*Faculty of Pharmaceutical Sciences,*
*Hokkaido University,*
*Kita-ku, Sapporo, 060, Japan*
*Email: ishiisn@mxm.meshnet.or.jp*

*Takashi Kumazaki*
*Department of Biochemistry,*
*Faculty of Pharmaceutical Sciences,*
*Hokkaido University,*
*Kita-ku, Sapporo, 060, Japan*

# 362. *Neprilysin*

## Databanks

*Peptidase classification: clan MA, family M13, MEROPS ID: M13.001*
*NC-IUBMB enzyme classification: EC 3.4.24.11*
*ATCC entries: 251182, 252559, 252809, 254846 (human); 760846, 903486 (mouse)*
*Chemical Abstracts Service registry number: 82707-54-8*
*Databank codes:*

| Species | SwissProt | PIR | EMBL (cDNA) | EMBL (genomic) |
|---|---|---|---|---|
| Caenorhabditis elegans | – | – | U23519 | – |
| Caenorhabditis elegans | – | – | Z49074 | – |
| Caenorhabditis elegans | – | – | Z69902 | – |
| Homo sapiens | – | – | A30431 | – |
| Homo sapiens | P08473 | A41387 | J03779 | L08103: exons 1, 2a and 2b |
| | | | X07166 | M26605: exon 1 |
| | | | Y00811 | M26606: exon 2 |
| | | | | M26607: exon 3 |
| | | | | M26608: exon 4 |
| | | | | M26609: exon 5 |
| | | | | M26610: exon 6 |
| | | | | M26611: exon 7 |
| | | | | M26612: exon 8 |
| | | | | M26613: exon 9 |
| | | | | M26614: exon 10 |
| | | | | M26615: exon 11 |
| | | | | M26616: exon 12 |
| | | | | M26617: exon 13 |
| | | | | M26618: exon 14 |
| | | | | M26619: exon 15 |
| | | | | M26620: exon 16 |
| | | | | M26621: exon 17 |
| | | | | M26622: exon 18 |
| | | | | M26623: exon 19 |

| Species | SwissProt | PIR | EMBL (cDNA) | EMBL (genomic) |
|---|---|---|---|---|
| | | | | M26624: exon 20 |
| | | | | M26625: exon 21 |
| | | | | M26626: exon 22 |
| | | | | M26627: exon 23 |
| | | | | M26628: exon 24 |
| *Mus musculus* | – | – | M81591 | – |
| *Oryctolagus cuniculus* | P08049 | – | M16593 X05338 | – |
| *Rattus norvegicus* | P07861 | – | M15944 | – |
| *Sus scrofa* | P19621 | – | – | – |

## Name and History

The first report of neprilysin was as an activity in brush border membranes of rat kidney capable of hydrolyzing [$^{125}$I]iodoinsulin B chain at neutral pH (Wong-Leung & Kenny, 1968) and the enzyme was hence referred to as **kidney brush border neutral proteinase**. The enzyme was shown to be a metallopeptidase and was subsequently purified to homogeneity from rabbit kidney (Kerr & Kenny, 1974a). Neprilysin constitutes approximately 4% of the protein of the renal brush border membrane facilitating enzyme purification.

Preliminary characterization of the purified enzyme suggested that it was a unique mammalian membrane endopeptidase which resembled in its properties the microbial metalloproteinase thermolysin (Chapter 351). This similarity was reinforced when it was shown that the thermolysin inhibitor phosphoramidon also inhibited neprilysin with a similar potency ($K_i$ 2 nM; Kenny, 1977). Neprilysin was rediscovered independently as a brain membrane enzyme responsible for the inactivation of the enkephalins (Malfroy *et al*., 1978), which led to the use of the trivial name **enkephalinase**. However, the subsequent discovery of a range of other biologically active peptide substrates for the enzyme meant that this terminology was inappropriate and the more general name **endopeptidase-24.11** was suggested (Matsas *et al*., 1983). Neprilysin is also commonly referred to as **neutral endopeptidase** or **NEP** in the literature.

Another twist in the history of neprilysin came with the cloning and sequencing of the enzyme (Devault *et al*., 1987) and the recognition that it was identical with a leukocyte cell surface antigen, the **common acute lymphoblastic leukemia antigen** (**CALLA** or **CD10**) (LeTarte *et al*., 1988). To this day the substrate(s) for neprilysin and its functions in the immune system are unknown. However, a number of other cell surface peptidases have been identified as cluster differentiation (CD) antigens, e.g. membrane alanyl aminopeptidase (CD13; Chapter 337), and a fuller discussion of these activities is to be found in Turner (1993). The conflicting nomenclature for the phosphoramidon-sensitive neutral endopeptidase-24.11 led the IUBMB to recommend the name **neprilysin** in 1992.

## Activity and Specificity

Neprilysin is essentially an oligopeptidase hydrolyzing peptides up to about 40 amino acids in length, although generally the efficiency of hydrolysis declines with increasing length of peptide. One of the most efficiently hydrolyzed substrates is the undecapeptide substance P (Matsas *et al*., 1984). The substrate specificity of neprilysin has been explored with a wide range of synthetic and natural peptide substrates, although those for which neprilysin has been established to have a physiological role in metabolism are limited. The principal substrates *in vivo* appear to be the enkephalins, tachykinins such as substance P, and the atrial natriuretic peptide family. Thus inhibitors of neprilysin may have potential in the treatment of pain, inflammation and hypertension (Roques & Beaumont, 1990).

The primary specificity requirement is a bulky hydrophobic residue in the P1' position, the enzyme normally functioning as an endopeptidase. Leu in the P1' position gives the highest rates of hydrolysis whereas Phe exhibits the lowest $K_m$ values. The smallest identified substrate is the chemotactic peptide formyl-Met-Leu-Phe, cleavage occurring at the Met┼Leu bond (Connelly *et al*., 1985). The subsite specificity of neprilysin has been explored in some detail (Pozsgay *et al*., 1986; Hersh & Morihara, 1986), revealing that the enzyme has an extended active site accommodating at least four amino acid side chains. A consensus sequence for efficient hydrolysis by neprilysin was suggested as: -Phe-Phe-Gly┼Phe-Leu-(Ala)-. With some substrates, neprilysin displays peptidyl-dipeptidase character, for example in the hydrolysis of [Leu]-enkephalin: Tyr-Gly-Gly┼Phe-Leu ($k_{cat}/K_m$ 43.9 min$^{-1}$ mM$^{-1}$). The C-terminally amidated derivative of [Leu]-enkephalin is hydrolyzed much less efficiently ($k_{cat}/K_m$ 1.7 min$^{-1}$ mM$^{-1}$), reflecting the interaction of the free C-terminal carboxylate group in [Leu]-enkephalin with an arginyl residue (Arg102) in the active site of the enzyme (Bateman *et al*., 1989). Examples of cleavage positions in other biologically active substrates are for substance P, Arg-Pro-Lys-Pro-Gln-Gln┼Phe┼Phe-Gly┼Leu-Met-NH$_2$, the most rapid hydrolysis occurring at the Gly┼Leu bond, consistent with the consensus sequence for hydrolysis. In neurotensin the following hydrolysis sites are seen: Glp-Leu-Tyr-Glu-Asn-Lys-Pro-Arg-Arg-Pro┼Tyr┼Ile-Leu, and in cholecystokinin-8, Asp-Tyr-Met-Gly┼Trp-Met-Asp┼Phe-NH$_2$. $K_m$ values for synthetic substrates and natural peptides are usually in the range 10 μM–1 mM.

There are many fluorometric and colorimetric assays available for neprilysin utilizing synthetic peptides and [D-Ala$^2$,

Leu⁵]-enkephalin provides a convenient substrate for HPLC analysis (see, for example, Turner *et al*., 1989; Li & Hersh, 1995). A particularly convenient and sensitive two-stage assay for neprilysin, which can be adapted for microtiter plate assay, involves the substrate Suc-Ala-Ala⌐Leu-NHPhNO₂ in the presence of *S. griseus* aminopeptidase (Chapter 490) as coupling enzyme (Indig *et al*., 1989). The release of Leu-NHPhNO₂ by neprilysin is followed by aminopeptidase release of the 4-nitroaniline leaving group which can be monitored at 405 nm. In tissue preparations it is necessary to check the specificity of the reaction with phosphoramidon.

Neprilysin exhibits a pH optimum of 6.0 (Kerr & Kenny, 1974b). It is inhibited by zinc-chelating reagents and a variety of synthetic inhibitors of the enzyme have been designed, of which the first was thiorphan ([DL-3-mercapto-2-benzylpropanoyl]-glycine) ($K_i = 2.3$ nM; Roques *et al*., 1980). Potent inhibitors of neprilysin are based on the general design of zinc-peptidase inhibitors and, in particular, thiorphan was modeled on the design of inhibitors of carboxypeptidase A (Chapter 451). Mercapto-, phosphonamidate-, carboxyalkyl- and hydroxamate-based inhibitors have all been reported (see, for example, Roques *et al*., 1993 for discussion).

## Structural Chemistry

Neprilysin is a type II integral membrane protein of the plasma membrane; it exists as an ectoenzyme with the bulk of the protein, including the active site, facing the extracellular space. It contains 1 mol of zinc per subunit and no other metal ions have been found associated with the protein. In most species it appears to exist as a noncovalently associated homodimer (Fulcher & Kenny, 1983) although in the rabbit it is reported to exist as a monomer (Kerr & Kenny, 1974b). Intrachain disulfide bonds are important for the maintenance of structure and activity (Tam *et al*., 1985). There are 12 cysteine residues, four of which are clustered into the first 32 residues immediately following the transmembrane anchor. This is a feature also seen in the closely related proteins endothelin-converting enzyme (Chapter 363) and the Kell blood group protein (see below), as well as in γ-glutamyltransferase and sucrase-isomaltase. The $M_r$ of

neprilysin ranges from about 85 000 to 100 000 depending on tissue source, the variation being attributable to differences in glycosylation (Relton *et al*., 1983). Subsequent cDNA cloning revealed rabbit (Devault *et al*., 1987), rat (Malfroy *et al*., 1987) and human (Malfroy *et al*., 1988) enzymes to consist of 750, 742 and 742 amino acid polypeptides, respectively. A schematic representation of the rabbit neprilysin molecule, highlighting the various domains, *N*-linked glycosylation sites and cysteine residues is shown in Figure 362.1.

The C-terminal domain of neprilysin contains the HEXXH zinc-binding motif typical of many zinc peptidases. The histidines in this motif (His583 and 587 in the rabbit sequence) constitute two of the zinc ligands with a glutamate at position 646 acting as the third zinc ligand. Glu584 is essential for catalysis. A number of other residues essential for substrate or inhibitor binding, or catalysis, have been identified by comparisons with thermolysin coupled with site-directed mutagenesis studies of neprilysin (see Roques *et al*., 1993 for review). A closer bacterial homolog to neprilysin than thermolysin is the recently cloned lactococcal oligopeptidase O (Chapter 365) (Mierau *et al*., 1993) which is the product of the *pepO* gene of *Lactococcus lactis* and which has an overall identity of 27.1% with human neprilysin. To date there is no crystal structure available for neprilysin.

## Preparation

The purification of neprilysin has been much facilitated by the application of immunoaffinity chromatography employing a monoclonal antibody after solubilization of the enzyme from the membrane by detergent treatment. Such a procedure permitted the first isolation of the enzyme from brain, an overall purification factor of approximately 23 000 being achieved (Relton *et al*., 1983). More conventional chromatographic procedures have also been applied in the purification of neprilysin from a wide range of mammalian species and tissues (Almenoff & Orlowski, 1983; Li & Hersh, 1995). More recently, large-scale expression and purification of recombinant neprilysin has been achieved from a baculovirus-infected insect cell line (Fossiez *et al*., 1992).

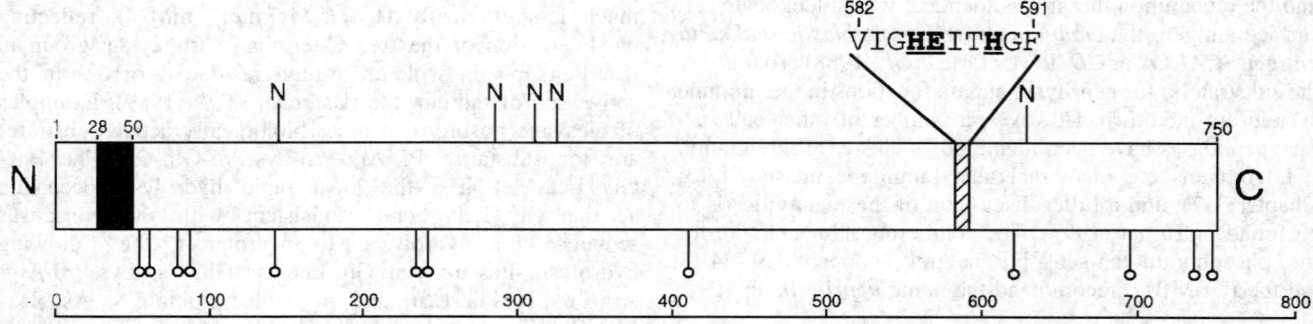

*Figure 362.1* Schematic model of the primary structure of rabbit neprilysin (Devault *et al*., 1987) which consists of an intracellular domain (residues 1–27), amphipathic membrane-spanning domain (residues 28–50: black box) and an extracellular domain (residues 51–750). The figure also indicates the zinc-binding motif (striped box), putative *N*-linked glycosylation sites (indicated by N), and Cys residues (indicated by 'lollipops'). The scale shows the number of amino acid residues.

## Biological Aspects

Neprilysin is a widely distributed enzyme and phosphoramidon-sensitive activities with the appropriate specificity have been identified in diverse species including mammals, insects (Isaac, 1988), *Aplysia* (Bawab *et al.*, 1993) and *Mytilus edulis* (Shipp *et al.*, 1990). In mammals the enzyme is abundant in the brush border membranes of kidney, intestine, placenta and choroid plexus, and it is also found on reticular cells in the immune system (Howell *et al.*, 1991). It is present in much lower abundance in many other tissues and cell types, including the brain where it is located on neuronal cells, especially in the striatonigral pathway (see, for example, Barnes *et al.*, 1988). A close correspondence has been observed between the distribution in brain of neprilysin and of opioid receptors (Waksman *et al.*, 1986) and colocalization of neprilysin and the neuropeptide substrate substance P has been observed (Barnes *et al.*, 1993). The enzyme has also been found on Schwann cells in the peripheral nervous system and it is present in the genitourinary system (e.g. testis and ovary).

The discovery that neprilysin is identical with the common acute lymphoblastic leukemia antigen (LeTarte *et al.*, 1988) has revealed its transient expression on the surface of certain hematopoietic cells and increased levels are found on mature lymphocytes in certain disease states (see Le Bien & McCormack, 1989, for review). The enzyme is an important diagnostic marker in certain forms of childhood leukemia. Much attention has focused on the role of neprilysin in inactivating the natriuretic peptides *in vivo* (see, for example, Kenny & Stephenson, 1988) and it has been shown that a neprilysin inhibitor can markedly increase the amount of immunoreactive atrial natriuretic peptide in the circulation (Wilkins, 1993). Neprilysin inhibitors may be clinically useful in certain forms of cardiovascular and renal disease and are complementary to the use of angiotensin-converting enzyme inhibitors.

Neprilysin may participate in bone metabolism both by its presence on the surface of human osteoblasts and by its ability to metabolize calcitonin (Howell *et al.*, 1993). Both calcitonin and 1,25-dihydroxyvitamin $D_3$ were able to upregulate neprilysin expression at physiologically relevant doses, suggesting that it might play a role in $Ca^{2+}$ homeostasis in this tissue (Howell *et al.*, 1993). Neprilysin in the lung may play a role in the airways by modulating responses to inflammatory neuropeptides, for example substance P, which constricts the airway smooth muscle (Borson, 1991).

Compatible with its exclusive occurrence as a plasma membrane ectoenzyme, neprilysin is heavily glycosylated with five or six *N*-linked glycosylation sites depending on the species. There appear to be no other post-translational modifications of the enzyme nor have any natural, circulating inhibitors been identified. It is not inhibited by matrix metalloproteinase inhibitors such as TIMP. Thus, the biological activity of neprilysin appears to be essentially unregulated. The neprilysin gene, located on chromosomal region 3q21–q27 (Barker *et al.*, 1989) exists in a single copy, spans more than 80 kb, is composed of 24 exons and is highly conserved among mammalian species (D'Adamio *et al.*, 1989). Three distinct neprilysin mRNAs have been identified in human and rat which differ only in their 5′ noncoding regions

(D'Adamio *et al.*, 1989; Li *et al.*, 1995). A gene knockout of neprilysin in mice has been reported in which the animals appeared developmentally normal but the neprilysin null mice were, surprisingly, highly sensitive to endotoxic shock (Lu *et al.*, 1995). This observation may reflect a general role for neprilysin in the metabolism of proinflammatory peptides.

## Distinguishing Features

Neprilysin is characterized by its membrane association, its substrate specificity and particularly by its sensitivity to inhibition by phosphoramidon and thiorphan at nanomolar levels. Although the closely related endothelin-converting enzyme is inhibited by phosphoramidon, it is only sensitive at micromolar concentrations of the inhibitor and is not affected by thiorphan. Phosphoramidon-sensitive hydrolysis of the Gly-Phe bond in [D-Ala$^2$,Leu$^5$]-enkephalin therefore typifies neprilysin action. Polyclonal and monoclonal antisera to human neprilysin (anti-CALLA or anti-CD10 antisera) are available from a number of commercial sources.

## Related Peptidases

In addition to endothelin-converting enzyme which is described elsewhere (Chapters 363 and 364), two other related mammalian gene products have been reported. The Kell blood group protein is located on the surface of erythrocytes and incompatibility involving Kell antigens can cause severe hemolytic reactions to blood transfusions. Molecular cloning of Kell cDNA revealed it to be a type II integral membrane glycoprotein related in sequence to neprilysin (Lee *et al.*, 1991). Although it retains the zinc-binding domain and certain other residues presumed to be important for substrate binding or catalysis, no peptidase activity has yet been detected for Kell protein.

The *PEX* gene is associated with X-linked hypophosphatemic rickets. A positional cloning approach was used to isolate a candidate gene for the disease from human patients. A composite partial sequence of the PEX protein showed considerable homology with the neprilysin sequence and the overall similarity between the PEX, neprilysin and Kell sequences was between 50 and 60% (HYP Consortium, 1995). The putative protein has a potential membrane-spanning segment near the N-terminus followed by a cluster of cysteine residues, as well as a zinc-binding motif towards the C-terminus. Although the complete coding sequence of the PEX protein was not obtained, it has the hallmarks of being a membrane-bound zinc metallopeptidase.

## Further Reading

For a review, see Erdös & Skidgel (1989). An extensive review of all aspects of the enzyme is presented by Roques *et al.* (1993).

## References

Almenoff, J. & Orlowski, M. (1983) Membrane-bound kidney neutral metalloendopeptidase: interaction with synthetic substrates, natural peptides, and inhibitors. *Biochemistry* **22**, 590–599.

Barker, P.E., Shipp, M.A., D'Adamio, L., Masteller, E.L. & Reinherz, E.L. (1989) The common acute lymphoblastic leukemia antigen maps to chromosomal region 3(q21-q27). *J. Immunol.* **142**, 283–287.

Barnes, K., Matsas, R., Hooper, N.M., Turner, A.J. & Kenny, A.J. (1988) Endopeptidase-24.11 is striosomally ordered in pig brain and, in contrast to aminopeptidase N and peptidyl dipeptidase A ('angiotensin converting enzyme') is a marker for a set of striatal efferent fibres. *Neuroscience* **27**, 799–817.

Barnes, K., Turner, A.J. & Kenny, A.J. (1993) An immunoelectron microscopic study of pig substantia nigra shows co-localisation of endopeptidase-24.11 with substance P. *Neuroscience* **53**, 1073–1082.

Bateman, R.C., Jackson, D., Slaughter, C.A., Unnithan, S., Chai, Y.G., Moomaw, C.A. & Hersh, L.B. (1989) Identification of the active site arginine in rat neutral endopeptidase 24.11 (enkephalinase) as Arg 102 and analysis of a glutamine 102 mutant. *J. Biol. Chem.* **264**, 6151–6157.

Bawab, W., Aloyz, R.S., Crine, P., Roques, B.P. & DesGroseillers, L. (1993) Identification and characterization of a neutral endopeptidase activity in *Aplysia californica*. *Biochem. J.* **296**, 459–465.

Borson, D.B. (1991) Roles of neutral endopeptidase in airways. *Am. J. Physiol.* **260**, L212–L225.

Connelly, J.C., Skidgel, R.A., Schulz, W.W., Johnson, A.R. & Erdös, E.G. (1985) Neutral endopeptidase 24.11 in human neutrophils: cleavage of chemotactic peptide. *Proc. Natl Acad. Sci. USA* **82**, 8737–8741.

D'Adamio, L., Shipp, M.A., Masteller, E.L. & Reinherz, E.L. (1989) Organization of the gene encoding common acute lymphoblastic leukemia antigen (neutral endopeptidase 24.11): multiple miniexons and separate 5-prime untranslated regions. *Proc. Natl Acad. Sci. USA* **86**, 7103–7107.

Devault, A., Lazure, C., Nault, C., Le Moual, H., Seidah, N.G., Chrétien, M., Kahn, P., Powell, J., Mallet, J., Beaumont, A., Roques, B.P., Crine, P. & Boileau, G. (1987) Amino acid sequence of rabbit kidney neutral endopeptidase-24.11 (enkephalinase) deduced from a complementary DNA. *EMBO J.* **6**, 1317–1322.

Erdös, E.G. & Skidgel, R.A. (1989) Neutral endopeptidase 24.11 (enkephalinase) and related regulators of peptide hormones. *FASEB J.* **3**, 145–151.

Fossiez, F., Lemay, G., Labonte, N., Parmentier-Lesage, F., Boileau, G. & Crine, P. (1992) Secretion of a functional soluble form of neutral endopeptidase-24.11 from a baculovirus-infected insect cell line. *Biochem. J.* **284**, 53–59.

Fulcher, I.S. & Kenny, A.J. (1983) Proteins of the kidney microvillar membrane. The amphipathic forms of endopeptidase purified from pig kidneys. *Biochem. J.* **211**, 743–753.

Hersh, L.B. & Morihara, K. (1986) Comparison of subsite specificity of the mammalian neutral endopeptidase-24.11 (enkephalinase) to the bacterial neutral endopeptidase, thermolysin. *J. Biol. Chem.* **261**, 6433–6437.

Howell, S., Murray, H., Turner, A.J. & Kenny, A.J. (1991) A highly sensitive elisa for endopeptidase-24.11, the common acute lymphoblastic leukemia antigen (CALLA, CD-10), applicable to material of porcine and human origin. *Biochem. J.* **278**, 417–421.

Howell, S., Caswell, A.M., Kenny, A.J. & Turner, A.J. (1993) Membrane peptidases on human osteoblast-like cells in culture: hydrolysis of calcitonin and hormonal regulation of endopeptidase-24.11. *Biochem. J.* **290**, 159–164.

HYP Consortium (1995) A gene (*PEX*) with homologies to endopeptidases is mutated in patients with X-linked hypophosphatemic rickets. *Nature Genet.* **11**, 130–136.

Indig, F.E., Ben-Meir, D., Spungin, A. & Blumberg, S. (1989) Investigation of neutral endopeptidase (EC 3.4.24.11) and of neutral proteinases (EC 3.4.24.4) using a new two-stage enzymatic reaction. *FEBS Lett.* **255**, 237–240.

Isaac, R.E. (1988) Neuropeptide-degrading endopeptidase activity of locust (*Schistocerca gregaria*) synaptic membranes. *Biochem. J.* **255**, 843–847.

Kenny, A.J. (1977) Proteinases associated with cell membranes. In: *Proteinases in Mammalian Cells and Tissues* (Barrett, A.J., ed.). Amsterdam: Elsevier, pp. 393–444.

Kenny, A.J. & Stephenson, S.L. (1988) Role of endopeptidase-24.11 in the inactivation of atrial natriuretic peptide. *FEBS Lett.* **232**, 1–8.

Kerr, M.A. & Kenny, A.J. (1974a) The purification and specificity of a neutral endopeptidase from rabbit kidney brush border. *Biochem. J.* **137**, 477–488.

Kerr, M.A. & Kenny, A.J. (1974b) The molecular weight and properties of a neutral metalloendopeptidase from rabbit kidney brush border. *Biochem. J.* **137**, 489–495.

Le Bien, T.W. & McCormack, R.T. (1989) The common acute lymphoblastic leukaemia antigen (CD10) – emancipation from a functional enigma. *Blood* **73**, 625–634.

Lee, S., Zambas, E.D., Marsh, W.L. & Redman, C.M. (1991) Molecular cloning and primary structure of Kell blood group protein. *Proc. Natl Acad. Sci. USA* **88**, 6353–6357.

LeTarte, M., Vera, S., Tran, R., Addis, J.B.L., Onizuka, R.J., Quackenbush, E.J., Jongeneel, C.V. & McInnes, R.R. (1988) Common acute lymphocytic leukemia antigen is identical to neutral endopeptidase. *J. Exp. Med.* **168**, 1247–1253.

Li, C. & Hersh, L.B. (1995) Neprilysin: assay methods, purification and characterization. *Methods Enzymol.* **248**, 253–263.

Li, C., Booze, R.M. & Hersh, L.B. (1995) Tissue-specific expression of rat neutral endopeptidase (neprilysin) mRNAs. *J. Biol. Chem.* **270**, 5723–5728.

Lu, B., Gerard, N.P., Kolakowski, L.F., Jr., Bozza, M., Zurakowski, D., Finco, O., Carroll, M.C. & Gerard, C. (1995) Neutral endopeptidase modulation of septic shock. *J. Exp. Med.* **181**, 2271–2275.

Malfroy, B., Swerts, J.P., Guyon, A., Roques, B.P. & Schwartz, J.-C. (1978) High-affinity enkephalin-degrading peptidase is increased after morphine. *Nature* **276**, 523–526.

Malfroy, B., Schofield, P.R., Kuang, W.J., Seeburg, P.H., Mason, A.J. & Henzel, W.J. (1987) Molecular cloning and amino acid sequence of rat enkephalinase. *Biochem. Biophys. Res. Commun.* **144**, 59–66.

Malfroy, B., Kuang, W.J., Seeburg, P.H., Mason, A.J. & Schofield, P.R. (1988) Molecular cloning and amino acid sequence of human enkephalinase (neutral endopeptidase). *FEBS Lett.* **229**, 206–210.

Matsas, R., Fulcher, I.S., Kenny, A.J. & Turner, A.J. (1983) Substance P and [Leu]enkephalin are hydrolyzed by an enzyme in pig caudate synaptic membranes that is identical with the endopeptidase of kidney microvilli. *Proc. Natl Acad. Sci. USA* **80**, 3111–3115.

Matsas, R., Turner, A.J. & Kenny, A.J. (1984) The metabolism of neuropeptides: the hydrolysis of peptides, including enkephalins, tachykinins and their analogues by endopeptidase-24.11. *Biochem. J.* **223**, 433–440.

Mierau, I., Tan, P.S.T., Haandrikman, A.J., Kok, J., Leenhouts, K.J., Konings, W.N. & Venema, G. (1993) Cloning and sequencing of the gene for a lactococcal endopeptidase, an enzyme with sequence similarity to mammalian enkephalinase. *J. Bacteriol.* **175**, 2087–2096.

Pozsgay, M., Michaud, C., Liebman, M. & Orlowski, M. (1986) Substrate and inhibitor studies of thermolysin-like neutral metalloendopeptidase from kidney membrane fractions. Comparison with bacterial thermolysin. *Biochemistry* **25**, 1292–1299.

Relton, J.M., Gee, N.S., Matsas, R., Turner, A.J. & Kenny, A.J. (1983) Purification of endopeptidase-24.11 ('enkephalinase') from pig brain by immunoadsorbent chromatography. *Biochem. J.* **215**, 519–523.

Roques, B.P. & Beaumont, A. (1990) Neutral endopeptidase-24.11 inhibitors: from analgesics to antihypertensives. *Trends Pharmacol. Sci.* **11**, 245–249.

Roques, B.P., Fournié-Zaluski, M.-C., Soroca, E., Lecomte, J.M., Malfroy, B., Llorens, C. & Schwartz, J.-C. (1980) The enkephalinase inhibitor thiorphan shows antinociceptive activity in mice. *Nature* **288**, 286–288.

Roques, B.P., Noble, F., Daugé, V., Fournié-Zaluski, M.-C. & Beaumont, A. (1993) Neutral endopeptidase 24.11. Structure, inhibition, and experimental and clinical pharmacology. *Pharmacol. Rev.* **45**, 87–146.

Shipp, M.A., Stefano, G.B., D'Adamio, L., Switzer, S.N., Howard, F.D., Sinisterra, J., Scharrer, B. & Reinherz, E.L. (1990) Downregulation of enkephalin-mediated inflammatory responses by CD10/neutral endopeptidase 24.11. *Nature* **347**, 394–396.

Tam, L.T., Engelbrecht, S., Talent, J.M., Gracy, R.W. & Erdös, E.G. (1985) The importance of disulfide bridges in human endopeptidase (enkephalinase) after proteolytic cleavage. *Biochem. Biophys. Res. Commun.* **133**, 1187–1192.

Turner, A.J. (1993) Membrane peptidases of the nervous and immune systems. *Adv. Neuroimmunol.* **3**, 163–170.

Turner, A.J., Hooper, N.M. & Kenny, A.J. (1989) Neuropeptide-degrading enzymes. In: *Neuropeptides: A Methodology* (Fink, G. & Harmar, A.J., eds). Chichester: John Wiley, pp. 189–223.

Waksman, G., Hamel, E., Fournié-Zaluski, M.-C. & Roques, B.P. (1986) Autoradiographic comparison of the distribution of the neutral endopeptidase 'enkephalinase' and of $\mu$ and $\delta$ opioid receptors in rat brain. *Proc. Natl Acad. Sci. USA* **83**, 1523–1527.

Wilkins, M.R. (1993) Clinical potential of endopeptidase-24.11 inhibitors in cardiovascular disease. *Biochem. Soc. Trans.* **21**, 673–678.

Wong-Leung, Y.L. & Kenny, A.J. (1968) Some properties of a microsomal peptidase in rat kidney. *Biochem. J.* **110**, 5P.

*Anthony J. Turner*
*Department of Biochemistry & Molecular Biology,*
*University of Leeds,*
*Leeds LS2 9JT, UK*
*Email: a.j.turner@leeds.ac.uk*

# 363. *Endothelin-converting enzyme 1*

## Databanks

*Peptidase classification: clan MA, family M13, MEROPS ID: M13.002*
*NC-IUBMB enzyme classification: EC 3.4.24.71*
*Databank codes:*

| Species | SwissProt | PIR | EMBL (cDNA) | EMBL (genomic) |
|---|---|---|---|---|
| *Bos taurus* | P42891 | S47268 | S73774 Z35306 | – |
| *Bos taurus* | – | A54667 | – | – |
| *Homo sapiens* | P42892 | JC2521 JC4136 | D49471 S74573 Z35307 | X91922: exons 1 and 2 |
| *Rattus norvegicus* | P42893 | A53679 | D29683 | – |

## Name and History

Endothelin 1 (ET-1) was first discovered in 1988 by Yanagisawa and coworkers in the culture media of pig aortic endothelial cells and shown to be a 21 amino acid peptide with the most potent vasoconstricting activity known (Yanagisawa *et al.*, 1988). Subsequently, three distinct genes encoding three closely related peptides, ET-1, ET-2 and ET-3, were identified (Inoue *et al.*, 1989). Endothelins are produced from peptide precursors of approximately 200 amino acid residues. They are first processed by prohormone-processing enzyme(s) (Seidah *et al.*, 1993) into biologically inactive, 38–41 amino acid intermediates called big ET-1, -2 and -3. Big endothelins are then cleaved at the Trp21+Val22/Ile22 to yield 21 amino acid endothelins by **endothelin-converting enzyme (ECE)** (Yanagisawa *et al.*, 1988).

Evidence indicates that the physiologically relevant ECE is inhibited by phosphoramidon, a nonselective peptidic metalloprotease inhibitor. Phosphoramidon has been shown to inhibit the pressor and airway contractile effects of big ET-1 *in vivo* (Fukuroda *et al.*, 1990; Matsumura *et al.*, 1990) and to suppress the secretion of ET-1 from cultured endothelial cells (Ikegawa *et al.*, 1990; Sawamura *et al.*, 1991). ECE from endothelial cells as well as other tissue sources has been found to be a membrane-bound metalloprotease. It has also been shown that ECE is insensitive to inhibitors of other metalloproteases such as thiorphan (neprilysin: Chapter 362) and captopril (angiotensin-converting enzyme: Chapter 359) (Ahn *et al.*, 1992, 1995, 1996; Ohnaka *et al.*, 1993; Takahashi *et al.*, 1993; Shimada *et al.*, 1994; Xu *et al.*, 1994).

One enzyme which possesses these qualifying characteristics, ECE-1, has been purified to homogeneity from rat lung (Takahashi *et al.*, 1993) and pig aortic endothelium (Ohnaka *et al.*, 1993). Partial amino acid sequence data obtained from the purified enzyme rapidly led to the cloning of the gene for the rat and cattle enzymes (Shimada *et al.*, 1994; Xu *et al.*, 1994). Subsequently, cDNA for the human enzyme was also obtained (Schmidt *et al.*, 1994; Shimada *et al.*, 1995a). Since then, another ECE family member was cloned and termed ECE-2. ECE-2 was shown to have an overall 59% amino acid identity to ECE-1 (Emoto & Yanagisawa, 1995) and will be discussed further in a separate chapter (Chapter 364). Additionally, two isoforms of ECE-1 with distinct N-terminal tails have been identified and termed ECE-1a and ECE-1b. ECE-1a and ECE-1b have been shown to be encoded by the same gene by way of two distinct promoters (Shimada *et al.*, 1995b; Valdenaire *et al.*, 1995). The amino acid sequences of the ECE-1 enzymes were shown to be homologous among mammalian species. Human ECE-1a has 93 and 95% amino acid identity to rat and cattle ECE-1a, respectively. Similarly, human ECE-1b has 96 and 95% amino acid identity to rat and cattle ECE-1b, respectively (Schmidt *et al.*, 1994; Shimada *et al.*, 1994, 1995a,b; Valdenaire *et al.*, 1995; Xu *et al.*, 1994).

Because the vasoconstrictor activity of big ET-1 is over 100-fold less than that of ET-1 (Kimura *et al.*, 1989), the conversion from big ET-1 to ET-1 appears to be essential for biological activity. Therefore, specific and potent ECE inhibitors will be an invaluable tool in understanding the pathophysiological role of endothelin and in evaluating the therapeutic potential of ECE inhibition, and many pharmaceutical companies are pursuing development of such inhibitors. The biochemistry and molecular pharmacology of ECE-1 have been reviewed (Opgenorth *et al.*, 1992; Turner & Murphy, 1996).

## Activity and Specificity

Assay for ECE-1 activity can be performed with big ET-1 (1–38) as substrate and the ET-1 generated can be quantitated by radioimmunoassay (RIA) (Ahn *et al.*, 1992; Ohnaka *et al.*, 1993), enzyme immunoassay (Xu *et al.*, 1994), enzyme-linked immunosorbant assay (ELISA) (Ahn *et al.*, 1998), or ET-1 scintillation proximity assay (SPA) (Takahashi *et al.*, 1993). Kits for ET-1 ELISA and SPA suitable for the measurement of ET-1 after the ECE-1 reaction are available from Amersham (see Appendix 2 for full names and addresses of suppliers). Polyclonal antibody against ET-1, with low cross-reactivity for big ET-1, is available from Biodesign and can be used for RIA. Synthetic ETs and big ETs are available from Peptides International and American Peptide.

ECE-1 has a very sharp pH optimum at 7.0 (Ahn *et al.*, 1992; Ohnaka *et al.*, 1990; Takahashi *et al.*, 1993; Xu *et al.*, 1994). Only substrate specificity and kinetic data obtained from ECE-1 which was either purified to homogeneity or cloned and expressed will be discussed here, thus most of the earlier and most inconsistent data are omitted. ECE-1 converts big ET-1 to ET-1 without further degradation of the products (Takahashi *et al.*, 1993; Xu *et al.*, 1994). No other biological peptides have been identified as substrates for ECE-1. There is a large variation in the reported ability of ECE-1 to process big ET isopeptides. With purified rat lung ECE-1, the relative conversion rates for big ET-1, ET-2 and ET-3 were reported to be 4:2:1 (Takahashi *et al.*, 1993). For ECE-1 purified from pig aortic endothelium, big ET-1 was a much preferred substrate over big ET-2 (12.5:1). However, it did not convert big ET-3 (Ohnaka *et al.*, 1993). Cattle ECE-1 expressed in CHO cells processed all three big ET isopeptides, with the conversion rates, big ET-1 > big ET-3 > big ET-2 (Xu *et al.*, 1994). These discrepancies in the reported relative rates of big ET conversion might reflect variations in the specificity of several isotypes of ECE-1 for substrate. However, it has been shown that there is no substrate specificity difference at least between ECE-1a and ECE-1b of either human or rat (Shimada *et al.*, 1995b). Greater attention needs to be given to careful determination of exact peptide concentration in these experiments. Most researchers rely on commercial sources of these peptides, supplied as preweighed lyophilized powders. The stated weight may not always be accurate or the peptide may not be fully soluble.

The reported $K_m$ values of ECE-1 for big ET-1 range widely from 0.2 µM (purified from rat lung; Takahashi *et al.*, 1993) to 23 µM (recombinant human ECE-1 expressed in CHO cells; Schmidt *et al.*, 1994). $V$ values of 3.1 and 410 nmol min$^{-1}$ mg$^{-1}$ protein were reported for the purified ECE-1 from rat lung and pig aortic endothelium, respectively (Ohnaka *et al.*, 1993; Takahashi *et al.*, 1993). These large variations in kinetic data might likewise be due partly to errors in measurement of big ET concentration.

More detailed study of substrate specificity was carried out with truncated forms of big ET-1 using the purified enzyme (Ohnaka *et al.*, 1993). The removal of the N-terminal disulfide loop increases the hydrolysis rates by 2.7- or 2.5-fold compared to native big ET-1 (1–38) as shown with big ET-1 (16–37) or (18–34) as substrates, respectively. When Val22 in the P1' position is replaced by Ala22 in big ET-1 (18–34), the hydrolysis rate decreases by 11.2-fold compared to big ET-1 (18–34), whereas substitution by Phe22 increases the rate by 4.8-fold. The C-terminal extension of the substrate at residues 32–37 seems to be essential since big ET-1 (1–31) and big ET-1 (17–26) are not cleaved.

The reported IC$_{50}$ values for phosphoramidon with ECE-1 range from 0.35 µM (rat lung ECE-1; Takahashi *et al.*, 1993) to 8 µM (recombinant human ECE-1; Schmidt *et al.*, 1994), approximately three orders of magnitude less potent than with neprilysin. Other peptidic and nonpeptidic ECE-1 inhibitors have been reported. The peptidic inhibitors include phosphinic acid-based inhibitors (Chackalamannil *et al.*, 1996; McKittrick *et al.*, 1996) and thiol-based inhibitors (Deprez *et al.*, 1996; Kukkola *et al.*, 1996). However, selectivity of these inhibitors against other metalloproteases needs further study. A nonpeptidic inhibitor, CGS 26303, was shown to be 1000-fold more potent against neprilysin (IC$_{50}$ of 0.1 nM) than ECE-1 (IC$_{50}$ of 1.1 µM) (De Lombaert *et al.*, 1994). Nonpeptidic inhibitors from fungal broth and marine sponge have been reported (Tsurumi *et al.*, 1995a,b; Patil *et al.*, 1996). PD 069185, a trisubstituted quinazoline, was recently shown to be a highly selective ECE-1 inhibitor with an IC$_{50}$ value of 0.9 µM. The closely related enzyme, ECE-2 (Chapter 364), is not inhibited at up to 100 µM PD 069185 (Ahn *et al.*, 1998).

## Structural Chemistry

ECE-1 is a type II integral membrane protein with a short N-terminal cytoplasmic tail, a transmembrane hydrophobic domain and a large extracellular domain containing the catalytic site and a zinc-binding motif (HEXXH). ECE-1 shows significant sequence similarities, particularly in the C-terminal region, to neprilysin (Chapter 362) and the human Kell group protein (Chapter 362). ECE-1 has 10 sites predicted to be *N*-glycosylated and a high glycosylation level is evident from the apparent molecular mass of 120–130 kDa estimated from SDS-PAGE (Schmidt *et al.*, 1994; Shimada *et al.*, 1994; Xu *et al.*, 1994). The predicted molecular mass based on amino acid content alone is 86 kDa. ECE-1 was reported to form a disulfide-linked dimer because ECE-1 runs as a 120–130 or 250–300 kDa protein on SDS-PAGE under reducing or nonreducing conditions, respectively (Schmidt *et al.*, 1994; Xu *et al.*, 1994; Takahashi *et al.*, 1995). Studies relying on site-directed mutagenesis have indicated the involvement of Cys412 (numbered according to mature human ECE-1) in dimer formation. The third ligand for zinc was shown to be Glu651. Two mutants, Glu592Gln and His716Gln, show a complete loss of ECE-1 activity. However, mutants containing substitutions at Arg129 and Gly752 in ECE-1, which correspond to Arg109 and Arg747 which have been shown to be responsible for substrate and inhibitor binding in neprilysin, have activity comparable to the wild type (Shimada *et al.*, 1996). A pI of 4.1 was reported for the pig enzyme (Ohnaka *et al.*, 1993).

## Preparation

ECE-1 has been purified 6300- or 12 000-fold to homogeneity from a microsomal fraction of rat lung or from the crude membrane fraction of pig aortic endothelium, respectively (Takahashi *et al.*, 1993; Ohnaka *et al.*, 1993). The recombinant enzyme has been expressed in COS and CHO cells (Schmidt *et al.*, 1994; Shimada *et al.*, 1994, 1995a; Xu *et al.*,

1994). Phosphoramidon at 100 µM has been shown to increase the level of ECE-1 expression in endothelial cells by up to 10-fold, but no corresponding change in mRNA levels for ECE-1 was observed (Schmidt *et al.*, 1994; Barnes *et al.*, 1996). The mechanism of this phenomenon still needs to be elucidated.

ECE-1 has been shown to be localized either on the cell surface or in the Golgi depending on the tissue source of the enzyme (Gui *et al.*, 1993; Corder *et al.*, 1995; Takahashi *et al.*, 1995). It has been suggested that different isoforms of ECE-1 are responsible for determining the localization of the enzyme.

## Biological Aspects

Recent studies using endothelin receptor antagonists suggest that endothelins may play important roles in a number of pathological conditions including stroke (Patel *et al.*, 1995), chronic heart failure (Kiowski *et al.*, 1995) and hypertension (Clozel *et al.*, 1993). In addition, disruptions of ET-1 (Kurihara *et al.*, 1994), ET-3 (Baynash *et al.*, 1994), and ET$_B$ receptor genes (Puffenberger *et al.*, 1994) have demonstrated important roles for endothelins in the development of neural crest-derived tissues. The increase in ECE-1 expression after balloon angioplasty suggests that ECE-1 plays an active role in the pathogenesis of neointima formation (Wang *et al.*, 1996).

Northern analysis shows that ECE-1 mRNA is expressed in most of the cattle tissues tested, with the highest expression in the ovary and testis, followed by the adrenal gland (Xu *et al.*, 1994). Widespread expression of ECE-1 in human tissues has also been observed using RT-PCR (Rossi *et al.*, 1995). The two isoforms of ECE-1 display dissimilar tissue distribution by northern analysis. ECE-1b is expressed more abundantly than ECE-1a in pancreas, peripheral blood leukocytes, prostate, testis and colon, whereas lung, spleen, placenta and small intestine have higher expression of ECE-1a (Valdenaire *et al.*, 1995). In part contrary to these findings, ECE-1b mRNA was shown to be more abundant than ECE-1a mRNA in several rat and human tissues (Shimada *et al.*, 1995b). The gene of human ECE-1 is composed of 19 exons that span more than 68 kb and has been mapped to the 1p36 band of the human genome (Valdenaire *et al.*, 1995).

## Distinguishing Features

A very narrow neutral pH optimum for ECE-1 is in contrast to the acidic pH optimum of ECE-2 (Chapter 364). The potency of phosphoramidon for these two enzymes differs by 250-fold (IC$_{50}$ 1 µM and 4 nM for ECE-1 and ECE-2, respectively) (Emoto & Yanagisawa, 1995). ECE-1 appears to have much higher substrate selectivity than neprilysin, which is known to hydrolyze many biologically active peptides including atrial natriuretic peptide, enkephalins, substance P, big ET-1 and ET-1 (Vijayaraghavan *et al.*, 1990; Roques *et al.*, 1993; Murphy *et al.*, 1994). ECE-1 purified from pig aortic endothelium does not cleave human adrenocorticotropic hormone (1–24), rat atrial natriuretic peptide (1–28), human angiotensin II (1–8), ET-1, ET-2 or ET-3 (Ohnaka *et al.*, 1993). So far as presently known, ECE-1 distinguishes itself from all other proteases in its rigid specificity for processing big ET-1. Other proteases do not restrict themselves to the Trp21┼Val22/Ile22 bond nor refrain from further degradation of the generated products with the exception of cathepsin E (Chapter 275) (Lees *et al*., 1990), which in any case is an aspartic- not a metalloprotease. ECE-1 has been shown to be localized both in Golgi and on the cell surface, this is in marked contrast to the absence of neprilysin from the Golgi (Roques *et al*., 1993). Both polyclonal and monoclonal antibodies against ECE-1 have been reported (Shimada *et al*., 1994; Emoto & Yanagisawa, 1995), but are not commercially available at the time of writing.

## Further Reading

A good overview is provided by Turner & Murphy (1996).

## References

Ahn, K., Beningo, K., Olds, G. & Hupe, D. (1992) The endothelin-converting enzyme from human umbilical vein is a membrane-bound metalloprotease similar to that from bovine aortic endothelial cells. *Proc. Natl Acad. Sci. USA* **89**, 8606–8610.

Ahn, K., Pan, S., Beningo, K. & Hupe, D. (1995) A permanent human cell line (EA.hy926) preserves the characteristics of endothelin converting enzyme from primary human umbilical vein endothelial cells. *Life Sci.* **56**, 2331–2341.

Ahn, K., Pan, S.M., Zientek, M.A., Guy, P.M. & Sisneros, A.M. (1996) Characterization of endothelin converting enzyme from intact cells of a permanent human endothelial cell line, EA.hy926. *Biochem. Mol. Biol. Int.* **39**, 573–580.

Ahn, K., Sisneros, A.M., Herman, S.B., Pan, S.M., Hupe, D., Lee, D., Nikam, S., Cheng, X.-M., Doherty, A.M., Schroeder, R.L., Haleen, S.J., Kaw, S., Emoto, N. & Yanigisawa, M. (1998) Novel selective quinazoline inhibitors of endothelin converting enzyme-1. *Biochem. Biophys. Res. Commun.* **243**, 184–190.

Barnes, K., Shimada, K., Takahashi, M., Tanzawa, K. & Turner, A.J. (1996) Metallopeptidase inhibitors induce an up-regulation of endothelin-converting enzyme levels and its redistribution from the plasma membrane to an intracellular compartment. *J. Cell Sci.* **109**, 919–928.

Baynash, A.G., Hosoda, K., Giaid, A., Richardson, J.A., Emoto, N., Hammer, R.E. & Yanagisawa, M. (1994) Interaction of endothelin-3 with endothelin-B receptor is essential for development of epidermal melanocytes and enteric neurons. *Cell* **79**, 1277–1285.

Chackalamannil, S., Chung, S., Stamford, A.W., McKittrick, B.A., Wang, Y., Tsai, H., Cleven, R., Fawzi, A. & Czarniecki, M. (1996) Highly potent and selective inhibitors of endothelin converting enzyme. *Bioorg. Med. Chem. Lett.* **6**, 1257–1260.

Clozel, M., Breu, V., Burri, K., Cassal, J.-M., Fischli, W., Gray, G.A., Hirth, G., Löffler, B.-M., Müller, M., Neidhart, W. & Ramuz, H. (1993) Pathophysiological role of endothelin revealed by the first orally active endothelin receptor antagonist. *Nature* **365**, 759–761.

Corder, R., Khan, N. & Harrison, V.J. (1995) A simple method for isolating human endothelin converting enzyme free from contamination by neutral endopeptidase 24.11. *Biochem. Biophys. Res. Commun.* **207**, 355–362.

De Lombaert, S., Ghai, R.D., Jeng, A.Y., Trapani, A.J. & Webb, R.L. (1994) Pharmacological profile of a non-peptidic dual inhibitor of neutral endopeptidase 24.11 and endothelin-converting enzyme.

*Biochem. Biophys. Res. Commun.* **204**, 407–412.

Deprez, P., Guillaume, J., Dumas, J. & Vevert, J.-P. (1996) Thiol inhibitors of endothelin-converting enzyme. *Bioorg. Med. Chem. Lett.* **6**, 2317–2322.

Emoto, N. & Yanagisawa, M. (1995) Endothelin-converting enzyme-2 is a membrane-bound, phosphoramidon-sensitive metalloprotease with acidic pH optimum. *J. Biol. Chem.* **270**, 15262–15268.

Fukuroda, T., Noguchi, K., Tsuchida, S., Nishikibe, M., Ikemoto, F., Okada, K. & Yano, M. (1990) Inhibition of biological actions of big endothelin-1 by phosphoramidon. *Biochem. Biophys. Res. Commun.* **172**, 390–395.

Gui, G., Xu, D., Emoto, N. & Yanagisawa, M. (1993) Intracellular localization of membrane-bound endothelin-converting enzyme from rat lung. *J. Cardiovasc. Pharmacol.* **22**(suppl 8), S53–S56.

Ikegawa, R., Matsumura, Y., Tsukahara, Y., Takaoka, M. & Morimoto, S. (1990) Phosphoramidon, a metalloproteinase inhibitor, suppresses the secretion of endothelin-1 from cultured endothelial cells by inhibiting a big endothelin-1 converting enzyme. *Biochem. Biophys. Res. Commun.* **171**, 669–675.

Inoue, A., Yanagisawa, M., Kimura, S., Kasuya, Y., Miyauchi, T., Goto, K. & Masaki, T. (1989) The human endothelin family: three structurally and pharmacologically distinct isopeptides predicted by three separate genes. *Proc. Natl Acad. Sci. USA* **86**, 2863–2867.

Kimura, S., Kasuya, Y., Sawamura, T., Shinmi, O., Sugita, Y., Yanagisawa, M., Goto, K. & Masaki, T. (1989) Conversion of big endothelin-1 to 21-residue endothelin-1 is essential for expression of full vasoconstrictor activity: structure-activity relationships of big endothelin-1. *J. Cardiovasc. Pharmacol.* **13**(suppl 5), S5–S7.

Kiowski, W., Sütsch, G., Hunziker, P., Müller, P., Kim, J., Oechslin, E., Schmitt, R., Jones, R. & Bertel, O. (1995) Evidence for endothelin-1-mediated vasoconstriction in severe chronic heart failure. *Lancet* **346**, 732–736.

Kukkola, P.J., Bilci, N.A., Kozak, W.X., Savage, P. & Jeng, A.J. (1996) Optimization of retro-thiorphan for inhibition of endothelin converting enzyme. *Bioorg. Med. Chem. Lett.* **6**, 619–624.

Kurihara, Y., Kurihara, H., Suzuki, H., Kodama, T., Maemura, K., Nagai, R., Oda, H., Kuwaki, T., Cao, W-H., Kamada, N., Jishage, K., Ouchi, Y., Azuma, S., Toyoda, Y., Ishikawa, T., Kumada, M. & Yazaki, Y. (1994) Elevated blood pressure and craniofacial abnormalities in mice deficient in endothelin-1. *Nature* **368**, 703–710.

Lees, W.E., Kalinka, S., Meech, J., Capper, S.J., Cook, N.D. & Kay, J. (1990) Generation of human endothelin by cathepsin E. *FEBS Lett.* **273**, 99–102.

Matsumura, Y., Hisaki, K., Takaoka, M. & Morimoto, S. (1990) Phosphoramidon, a metalloproteinase inhibitor, suppresses the hypertensive effect of big endothelin-1. *Eur. J. Pharmacol.* **185**, 103–106.

McKittrick, B.A., Stamford, A.W., Weng, X., Ma, K., Chackalamannil, S., Czarniecki, M., Cleven, R.M. & Fawzi, A.B. (1996) Design and synthesis of phosphinic acids that triply inhibit endothelin converting enzyme, angiotensin converting enzyme and neutral endopeptidase 24.11. *Bioorg. Med. Chem. Lett.* **6**, 1629–1634.

Murphy, L.J., Corder, R., Mallet, A.I. & Turner, A.J. (1994) Generation by the phosphoramidon-sensitive peptidases, endopeptidase-24.11 and thermolysin, of endothelin-1 and C-terminal fragment from big endothelin-1. *Br. J. Pharmacol.* **113**, 137–142.

Ohnaka, K., Takayanagi, R., Yamauchi, T., Okazaki, H., Ohashi, M., Umeda, F. & Nawata, H. (1990) Identification and characterization of endothelin converting activity in cultured bovine endothelial cells. *Biochem. Biophys. Res. Commun.* **168**, 1128–1136.

Ohnaka, K., Takayanagi, R., Nishikawa, M., Haji, M. & Nawata, H. (1993) Purification and characterization of a phosphoramidon-sensitive endothelin-converting enzyme in porcine aortic endothelium. *J. Biol. Chem.* **268**, 26759–26766.

Opgenorth, T.J., Wu-Wong, J.R. & Shiosaki, K. (1992) Endothelin-converting enzymes. *FASEB J.* **6**, 2653–2659.

Patel, T.R., Galbraith, S.L., McAuley, M.A., Doherty, A.M., Graham, D.I. & McCulloch, J. (1995) Therapeutic potential of endothelin receptor antagonists in experimental stroke. *J. Cardiovasc. Pharmacol.* **26**(suppl 3), S412–S415.

Patil, A.D., Freyer, A.J., Breen, A., Carte, B. & Johnson, R.K. (1996) Halistanol disulfate B, a novel sulfated sterol from the sponge *Pachastrella* sp.: inhibitor of endothelin converting enzyme. *J. Nat. Prod.* **59**, 606–608.

Puffenberger, E.G., Hosoda, K., Washington, S.S., Nakao, K., deWit, D., Yanagisawa, M. & Chakravarti, A. (1994) A missense mutation of the endothelin-B receptor gene in multigenic Hirschsprung's disease. *Cell* **79**, 1257–1266.

Roques, B.P., Noble, F., Daugé, V., Fournié-Zaluski, M.-C. & Beaumont, A. (1993) Neutral endopeptidase 24.11: structure, inhibition, and experimental and clinical pharmacology. *Pharmacol. Rev.* **45**, 87–146.

Rossi, G.P., Albertin, G., Franchin, E., Sacchetto, A., Cesari, M., Palú, G. & Pessina, A.C. (1995) Expression of the endothelin-converting enzyme gene in human tissues. *Biochem. Biophys. Res. Commun.* **211**, 249–253.

Sawamura, T., Kasuya, Y., Matsushita, Y., Suzuki, N., Shinmi, O., Kishi, N., Sugita, Y., Yanagisawa, M., Goto, K., Masaki, T. & Kimura, S. (1991) Phosphoramidon inhibits the intracellular conversion of big endothelin-1 to endothelin-1 in cultured endothelial cells. *Biochem. Biophys. Res. Commun.* **174**, 779–784.

Schmidt, M., Kröger, B., Jacob, E., Seulberger, H., Subkowski, T., Otter, R., Meyer, T., Schmalzing, G. & Hillen, H. (1994) Molecular characterization of human and bovine endothelin converting enzyme (ECE-1). *FEBS Lett.* **356**, 238–243.

Seidah, N.G., Day, R., Marcinkiewicz, M. & Chrétien, M. (1993) Mammalian paired basic amino acid convertases of prohormones and proproteins. *Ann. N.Y. Acad. Sci.* **680**, 135–146.

Shimada, K., Takahashi, M. & Tanzawa, K. (1994) Cloning and functional expression of endothelin-converting enzyme from rat endothelial cells. *J. Biol. Chem.* **269**, 18275–18278.

Shimada, K., Matsushita, Y., Wakabayashi, K., Takahashi, M., Matsubara, A., Iijma, Y. & Tanzawa, K. (1995a) Cloning and functional expression of human endothelin-converting enzyme cDNA. *Biochem. Biophys. Res. Commun.* **207**, 807–812.

Shimada, K., Takahashi, M., Ikeda, M. & Tanzawa, K. (1995b) Identification and characterization of two isoforms of an endothelin-converting enzyme-1. *FEBS Lett.* **371**, 140–144.

Shimada, K., Takahashi, M., Turner, A.J. & Tanzawa, K. (1996) Rat endothelin-converting enzyme-1 forms a dimer through Cys(412) with a similar catalytic mechanism and a distinct substrate binding mechanism compared with neutral endopeptidase-24.11. *Biochem. J.* **315**, 863–867.

Takahashi, M., Matsushita, Y., Iijima, Y. & Tanzawa, K. (1993) Purification and characterization of endothelin-converting enzyme from rat lung. *J. Biol. Chem.* **268**, 21394–21398.

Takahashi, M., Fukuda, K., Shimada, K., Barnes, K., Turner, A.J., Ikeda, M., Koike, H., Yamamoto, Y. & Tanzawa, K. (1995) Localization of rat endothelin-converting enzyme to vascular endothelial cells and some secretory cells. *Biochem. J.* **311**, 657–665.

Tsurumi, Y., Fujie, K., Nishikawa, M., Kiyoto, S. & Okuhara, M. (1995a) Biological and pharmacological properties of highly selective new endothelin converting enzyme inhibitor WS79089B isolated from *Streptosporangium roseum* No. 79089. *J. Antibiot.* **48**, 169–174.

Tsurumi, Y., Ueda, H., Hayashi, K., Takase, S., Nishikawa, M., Kiyoto, S. & Okuhara, M. (1995b) WS75624 A and B, new endothelin converting enzyme inhibitors isolated from *Saccharothrix* sp. No. 75624. *J. Antibiot.* **48**, 1066–1072.

Turner, A.J. & Murphy, L.J. (1996) Molecular pharmacology of endothelin converting enzymes. *Biochem. Pharmacol.* **51**, 91–102.

Valdenaire, O., Rohrbacher, E. & Mattei, M.-G. (1995) Organization of the gene encoding the human endothelin-converting enzyme (ECE-1). *J. Biol. Chem.* **270**, 29794–29798.

Vijayaraghavan, J., Scicli, A.G., Carretero, O.A., Slaughter, C., Moomaw, C. & Hersh, L.B. (1990) The hydrolysis of endothelins by neutral endopeptidase 24.11 (enkephalinase). *J. Biol. Chem.* **265**, 14150–14155.

Wang, X., Douglas, S.A., Louden, C., Vickery-Clark, L.M., Feuerstein, G.Z. & Ohlstein, E.H. (1996) Expression of endothelin-1, endothelin-3, endothelin-converting enzyme-1, and endothelin-A and endothelin-B receptor mRNA after angioplasty-induced neointimal formation in the rat. *Circ. Res.* **78**, 322–328.

Xu, D., Emoto, N., Giaid, A., Slaughter, C., Kaw, S., deWit, D. & Yanagisawa, M. (1994) ECE-1: a membrane-bound metalloprotease that catalyzes the proteolytic activation of big endothelin-1. *Cell* **78**, 473–485.

Yanagisawa, M., Kurihara, H., Kimura, S., Tomobe, Y., Kobayashi, M., Mitsui, Y., Yazaki, Y., Goto, K. & Masaki, T. (1988) A novel potent vasoconstrictor peptide produced by vascular endothelial cells. *Nature* **332**, 411–415.

***Kyunghye Ahn***
*Department of Biochemistry,*
*Parke-Davis Pharmaceutical Research,*
*Division of Warner-Lambert Company,*
*Ann Arbor, MI 48105, USA*
*Email: ahnk@aa.wl.com*

M

# 364. Endothelin-converting enzyme 2

## Databanks

*Peptidase classification: clan MA, family M13, MEROPS ID: M13.003*
*NC-IUBMB enzyme classification: none*
*Databank codes:*

| Species | SwissProt | PIR | EMBL (cDNA) | EMBL (genomic) |
|---------|-----------|-----|-------------|----------------|
| *Bos taurus* | Q10711 | – | U27341 | – |

## Name and History

Since the discovery of three distinct genes encoding three closely related endothelin (ET) peptides, ET-1, ET-2 and ET-3 (Inoue *et al.*, 1989), it has been anticipated that there are several endothelin-converting enzymes (ECEs) which generate these isopeptides. ECE-1 was shown to cleave big ET-1 more efficiently than big ET-2 or -3 (Ohnaka *et al.*, 1993; Takahashi *et al.*, 1993; Xu *et al.*, 1994), suggesting that there are additional ECE(s) that would preferentially convert big ET-2 and/or big ET-3. Further, although ECE-1 was shown to be widely expressed in diverse tissues and cell types, some, such as neurons that produce ETs, lacked ECE-1 expression (Xu *et al.*, 1994). This suggested the existence of a different ECE in these cells. The gene for another isoenzyme, ***endothelin-converting enzyme 2 (ECE-2)***, was obtained by RT-PCR against cattle adrenal cortex mRNA using highly degenerate primers based on a peptide microsequence from purified cattle ECE-1 followed by screening of a cDNA library with the cloned ECE-2 RT-PCR product (Emoto & Yanagisawa, 1995). More general history of ECE is provided in a separate entry for ECE-1 (Chapter 363).

## Activity and Specificity

Assay for ECE-2 can be performed with big ET-1 (1–38) as substrate by the same method as that described for ECE-1. ECE-2 has a narrow acidic pH optimum at pH 5.5 and is virtually inactive at neutral pH (Emoto & Yanagisawa, 1995), suggesting that ECE-2 acts as an intracellular enzyme responsible for the conversion of endogenously synthesized big ET-1 at the *trans*-Golgi network, acidic compartment of the secretory pathway. Rather surprisingly, recombinant cattle ECE-2 expressed in CHO cells showed a similar substrate specificity for big ET isopeptides, a strong preference toward big ET-1 compared to big ET-2 and big ET-3. That these substrates are indeed cleaved exclusively at the Trp21╀Val/Ile22 bond still needs to be verified. A $K_m$ value of approximately 1–2 μM for big ET-1 has been reported with cattle ECE-2 expressed in CHO cells. ECE-2 is inhibited by phosphoramidon with nanomolar potency, similar to neprilysin (Chapter 362), but is not inhibited by thiorphan or captopril as is ECE-1. FR 901533 has been shown to inhibit both ECE-1 and ECE-2 with similar potency (Emoto & Yanagisawa, 1995).

## Structural Chemistry

Cattle ECE-2 cDNA encodes a 787 amino acid peptide. It is a type II integral membrane protein with a short N-terminal cytoplasmic tail (82 residues), a transmembrane domain (23 residues), and a large extracellular C-terminal domain (682 residues). The extracellular region includes the putative catalytic domain which contains the HEXXH zinc-binding motif. ECE-2 has 10 sites in the extracellular domain predicted to be *N*-glycosylated. Its highly glycosylated nature, like that of ECE-1, is evident from immunoblot analysis, which shows that ECE-2 is expressed as an approximately 130 kDa protein. The predicted molecular mass based on amino acid content alone is 89 kDa. ECE-1, neprilysin (Chapter 362), and human Kell group protein (Chapter 362) show significant sequence similarity to ECE-2. The amino acid sequence of ECE-2 is most similar to that of ECE-1, with an overall identity of 59%. The sequence similarity is especially high within the C-terminal one-third of the putative extracellular domain. Within this region (amino acids 582–787 of ECE-2), the identities of ECE-2 with ECE-1, neprilysin, and Kell are 71, 44 and 40%, respectively (Emoto & Yanagisawa, 1995).

## Preparation

Purification of ECE-2 has not been reported to date. Recombinant cattle ECE-2 has been expressed in CHO cells. ECE-2 is mainly localized intracellularly, as shown by conversion of big ET-1 generated endogenously in CHO cells into which genes for both ECE-2 and prepro-ET-1 have been transfected (Emoto & Yanagisawa, 1995). The acidic pH optimum of ECE-2 precisely matches the luminal pH of the *trans*-Golgi network, which has been directly measured at 5.5–5.7. These data suggest that the C-terminal catalytic domain of ECE-2 is probably facing the luminal side of secretory vesicles, where it encounters the substrate big ET-1.

## Biological Aspects

Northern analysis of cattle tissues showed the highest expression of ECE-2 mRNA in neural tissues, i.e. cerebral cortex, cerebellum and adrenal medulla. However, in all the cattle tissues tested, ECE-2 mRNA was shown to be expressed at much lower levels than ECE-1 mRNA. Even in the brain, where ECE-1 mRNA expression is comparatively low, the ECE-1 mRNA signals were several-fold higher than that for ECE-2 mRNA. In cultured endothelial cells, ECE-2 mRNA is expressed at levels only 1–2% of that of ECE-1 mRNA (Emoto & Yanagisawa, 1995). The predominance of ECE-2 mRNA in neural tissues suggests that ECE-2 may be the major ECE in neurons, glia, and certain neuroendocrine cells. Little is known about the ability of ECE-2 to process other biologically active peptide hormones.

## Distinguishing Features

ECE-2 resembles ECE-1 (Chapter 363) in that it is inhibited by FR 901533 with similar potency or by phosphoramidon albeit with a 250-fold difference in potency ($IC_{50}$ 1 μM and 4 nM for ECE-1 and ECE-2, respectively). But other metalloprotease inhibitors such as thiorphan (neprilysin; Chapter 362) or captopril (peptidyl-dipeptidase A, Chapter 359) do not inhibit either enzyme. ECE-2, like ECE-1, converts big ET-1 much more efficiently than big ET-2 or big ET-3. However, ECE-2 is very different from ECE-1 in that ECE-2 has an acidic pH optimum in contrast to the neutral pH optimum for ECE-1 (Emoto & Yanagisawa, 1995). PD 069185 selectively inhibits ECE-1, but not ECE-2 (K. Ahn *et al.*, 1998). The very similar substrate specificity for big ET isopeptides shown by both ECE-1 and ECE-2 suggests that there may be still other isoenzymes of ECE which would cleave big ET-2 and/or big ET-3 more efficiently.

## Further Reading

See Emoto & Yanagisawa (1995) for an overview of this enzyme.

## References

Ahn, K., Sisneros, A.M., Herman, S.B., Pan, S.M., Hupe, D., Lee, D., Nikam, S., Cheng, X.-M., Doherty, A.M., Schroeder, R.L., Haleen, S.J., Kaw, S., Emoto, N. & Yanigisawa, M. (1998) Novel selective quinazoline inhibitors of endothelin converting enzyme-1. *Biochem. Biophys. Res. Commun.* **243**, 184–190.

Emoto, N. & Yanagisawa, M. (1995) Endothelin-converting enzyme-2 is a membrane-bound, phosphoramidon-sensitive metalloprotease with acidic pH optimum. *J. Biol. Chem.* **270**, 15262–15268.

Inoue, A., Yanagisawa, M., Kimura, S., Kasuya, Y., Miyauchi, T., Goto, K. & Masaki, T. (1989) The human endothelin family: three structurally and pharmacologically distinct isopeptides predicted by three separate genes. *Proc. Natl Acad. Sci. USA* **86**, 2863–2867.

Ohnaka, K., Takayanagi, R., Nishikawa, M., Haji, M. & Nawata, H. (1993) Purification and characterization of a phosphoramidon-sensitive endothelin-converting enzyme in porcine aortic endothelium. *J. Biol. Chem.* **268**, 26759–26766.

Takahashi, M., Matsushita, Y., Iijima, Y. & Tanzawa, K. (1993) Purification and characterization of endothelin-converting enzyme from rat lung. *J. Biol. Chem.* **268**, 21394–21398.

Xu, D., Emoto, N., Giaid, A., Slaughter, C., Kaw, S., deWit, D. & Yanagisawa, M. (1994) ECE-1: a membrane-bound metalloprotease that catalyzes the proteolytic activation of big endothelin-1. *Cell* **78**, 473–485.

*Kyunghye Ahn*
*Department of Biochemistry,*
*Parke-Davis Pharmaceutical Research,*
*Division of Warner-Lambert Company,*
*Ann Arbor, MI 48105, USA*
*Email: ahnk@aa.wl.com*

# 365. *Oligopeptidase O*

## Databanks

*Peptidase classification: clan MA, family M13, MEROPS ID: M13.004*
*NC-IUBMB enzyme classification: none*
*Databank codes:*

| Species | SwissProt | PIR | EMBL (cDNA) | EMBL (genomic) |
|---|---|---|---|---|
| *Lactococcus lactis lactis* | Q07744 | – | L18760 | – |
| *Lactococcus lactis cremoris* | Q09145 | – | L04938 | – |
| | | | U09553 | |

## Name and History

In the search for peptidases in the dairy lactic acid bacterium *Lactococcus lactis* which could participate in the degradation of casein-derived peptides, a necessary process for growth of the organism in milk and for cheese ripening, several authors isolated an endopeptidase. This enzyme was either given no specific name (Tan *et al.*, 1991) or was named **NOP** (*neutral thermolysin-like oligoendopeptidase*; Baankreis *et al.*, 1995). The present, generally used name is **PepO** (*oligoendopeptidase*; Mierau *et al.*, 1993) following the recommendation of Tan *et al.* (1993). [The present editors suggest that *oligopeptidase O* is a suitable name for general use.]

## Activity and Specificity

PepO cleaves various neuropeptides, which have been used as model substrates, and several peptides derived from β- and κ-casein. The substrate size ranges between five amino acid residues ([Met]enkephalin) to at least 30 residues (oxidized insulin B chain). PepO does not hydrolyze di-, tri- and tetrapeptides or intact α-, β- or κ-casein (Tan *et al.*, 1991; Pritchard *et al.*, 1994; Baankreis *et al.*, 1995; Stepaniak & Fox, 1995). Analysis of peptide cleavage patterns shows that PepO has a preference for peptide bonds with a hydrophobic amino acid in the P1' position. Frequently either Phe or Leu is found in this position and, less frequently, Ile, Val or Tyr (Pritchard *et al.*, 1994; Baankreis *et al.*, 1995; Stepaniak & Fox, 1995). Kinetic measurements with peptides of increasing size showed that PepO has higher affinity for larger peptides. (Lian *et al.*, 1996).

The temperature optimum of the enzyme was reported to be 50°C (at pH 5.5) according to one study (Baankreis *et al.*, 1995), which used $\alpha_{S1}$-CN(f1–23) as a substrate, and between 30 and 38°C according to others (Tan *et al.*, 1991; Stepaniak & Fox, 1995 using substrates [Met]-enkephalin and $\alpha_{S1}$-CN(f1–23), respectively). The pH optimum is 6.0 and 6.5 for [Met]-enkephalin and $\alpha_{S1}$-CN(f1–23), respectively (Tan *et al.*, 1991; Stepaniak & Fox, 1995). The enzyme is inhibited by phosphoramidon ($K_i \sim 4$ nM), 1,10-phenanthroline (1 mM) and thiorphan ($K_i \sim 6000$ nM) (Tan *et al.*, 1991; Pritchard *et al.*, 1994; Baankreis *et al.*, 1995; Stepaniak & Fox, 1995; Lian *et al.*, 1996). EDTA was also found to inhibit PepO but had to be added at high concentrations (10 mM and 50 mM, in Baankreis *et al.*, 1995, Lian *et al.*, 1996, respectively) otherwise the inhibition was only partial (Pritchard *et al.*, 1994). After EDTA inactivation, activity could be restored by 50–300 μM $Co^{2+}$ (Tan *et al.*, 1991) or by 50 μM $Zn^{2+}$ or 50 μM $Co^{2+}$ (Baankreis *et al.*, 1995). $Mn^{2+}$ was found to restore 50% of the enzyme activity after inhibition with 10 mM EDTA (Baankreis *et al.*, 1995). PepO activity was completely inhibited by 1 mM $Cu^{2+}$ and 1 mM $Zn^{2+}$. No effect on enzyme activity was found after addition of 1 mM $Ca^{2+}$, $Mg^{2+}$, $Fe^{2+}$, $Co^{2+}$ or $Mn^{2+}$ (Tan *et al.*, 1991). Furthermore, inhibitory effects of PMSF (1 mM), 2-ME (1%) and diethylpyrocarbonate have also been observed (Tan *et al.*, 1991; Stepaniak & Fox, 1995; Lian *et al.*, 1996). The substrate specificity and the effect of the various inhibitors suggest a functional similarity of PepO with thermolysin (Chapter 351) and with neprilysin (Chapter 362) (Pritchard *et al.*, 1994; Stepaniak & Fox, 1995; Baankreis *et al.*, 1995; Lian *et al.*, 1996).

PepO activity can be determined with various peptide substrates and by thin-layer chromatography (Tan *et al.*, 1991), HPLC (Pritchard *et al.*, 1994; Baankreis *et al.*, 1995) or capillary electrophoresis (Bockelmann *et al.*, 1995). Also, fluorescent substrates (Glt(or Suc)-Ala-Ala↓Phe-4-methoxy-NHNap (or -NHMec) and Suc-Arg-Pro-Phe-His-Leu↓Leu-Val-Tyr-NHMec) (Lian *et al.*, 1996) (peptides available from Sigma) (see Appendix 2 for full names and addresses of suppliers) and various chromogenic substrates (Stepaniak & Fox, 1995) have been used to determine PepO activity.

## Structural Chemistry

PepO is a monomeric enzyme of 71.5 kDa (Mierau *et al.*, 1993) with a pI of 4.3 (Tan *et al.*, 1991; Stepaniak & Fox, 1995). A pI of 4.62 was calculated from the derived amino acid sequence of PepO (Mierau *et al.*, 1993). PepO has a zinc-binding motif HEISH (Mierau *et al.*, 1993). Furthermore, homology comparison with the well-characterized neprilysin (Chapter 362) revealed other residues which are also involved in zinc binding (Glu535) or the formation of the active site: Arg28, Asn434, Ala435, Val472, His 587 and Lys625. The functional relevance in the lactococcal enzyme of Arg28 was shown by studies with modified substrates and that of Glu535 and His587, by site-directed mutagenesis (Lian *et al.*, 1996).

## Preparation

PepO has been purified from different *Lactococcus* strains with purification factors ranging from 147-fold (Pritchard *et al.*, 1994) to 678-fold (Tan *et al.*, 1991). In *Lactobacillus delbrueckii* subsp. *bulgaricus* and in *Streptococcus salivarius* subsp. *thermophilus* 70 kDa proteins have been found which reacted with PepO-specific antibodies, indicating that PepO is also present in species other than *Lactococcus* (Tan *et al.*, 1991). The *pepO* gene has been expressed in *Escherichia coli* with and without an N-terminal hexahistidine tag. Expression of His-tagged PepO reached 10 mg liter$^{-1}$ culture and the active PepO fusion-protein could be isolated and purified in a two-step process with an Ni-NTA-agarose column followed by a Mono-Q column. With this method a purification factor of 127-fold and a yield of 47% was reached (~4 mg enzyme liter$^{-1}$ culture) (Lian *et al.*, 1996).

## Biological Aspects

The gene of PepO, *pepO*, has been cloned and its nucleotide sequence has been determined (Mierau *et al.*, 1993). *pepO* is the last gene of an operon that encodes an oligopeptide transport system (*opp*). Expression of *pepO* is either directed by a promoter at the beginning of the operon and/or by a putative promoter upstream of the preceding *oppA* gene. Insertional inactivation of *oppA* leads to inactivation of *pepO* expression (Mierau *et al.*, 1993; Tynkkynen *et al.*, 1993). The physiological function of PepO is only partially understood. Both genetic and biochemical studies indicate that PepO is located in the cytoplasm (Tan *et al.*, 1992; Mierau *et al.*,

1993; Pritchard *et al*., 1994). Inactivation of PepO did not influence growth rate or acid production in milk or in complex medium (Mierau *et al*., 1993). Inactivation of PepO in a strain in which the aminopeptidase N (Chapter 345) had already been deleted caused an increase in the doubling time in milk from 60 min to 120 min, indicating that PepO, at least under these circumstances, has a role in the degradation of casein-derived peptides for the delivery of essential amino acids (Mierau *et al*., 1996). The physiological meaning of the coupling of *pepO* to the *opp* operon is not yet understood.

Apart from growth of lactococci in milk, peptidases also play a role in cheese ripening. Milk protein degradation is one major aspect in this process. At the end of growth, lactococcal cells lyse easily in the cheese matrix and release their enzyme contents. In young Gouda cheese, casein-derived peptides have been found that are indicative of PepO activity in cheese. Thus, a function of this endopeptidase in cheese fermentation has been proposed (Baankreis *et al*., 1995).

PepO is structurally and functionally highly similar to mammalian neprilysin (NEP) (Chapter 362), a glycoprotein which plays a pivotal role in regulating the physiological action of a large number of neuro- and hormone peptides (Mierau *et al*., 1993). Currently, the possibility of using PepO as a model for NEP is being assessed. PepO can easily be isolated and manipulated and provide valuable information complementing that obtained from thermolysin (Chapter 351), the current NEP model (Lian *et al*., 1996).

## Distinguishing Features

Thus far, PepO has been the only representative of its group in the bacterial kingdom (Mierau *et al*., 1993). Recently however, a gene was identified in one *Lactococcus lactis* subsp. *cremoris* strain which encodes a protein (PepO2) with 88% amino acid sequence identity to PepO. The gene of PepO2 has an entirely different genetic context and did not evolve by a simple duplication of the *pepO* region (I. Mierau & M.A. Hellendoorn, unpublished results). Furthermore, from *Streptococcus gordonii* PK488 a gene has been partially characterized that could encode a truncated protein with 58% amino acid sequence identity over a stretch of 219 amino acids surrounding the $Zn^{2+}$-binding site of PepO (Kolenbrander *et al*., 1994).

## Further Reading

For a review, see Monnet (1995).

## References

Baankreis, R., van Schalkwijk, S., Alting, A.C. & Exterkate, F.A. (1995) The occurrence of two intracellular oligoendopeptidases in *Lactococcus lactis* and their significance for peptide conversion in cheese. *Appl. Microbiol. Biotechnol.* **44**, 386–392.

Bockelmann, W., Hoppe-Seyler, T. & Heller, K.J. (1995) Quantitative determination of endopeptidase activity from lactic acid bacteria with capillary electrophoresis. *Milchwissenschaft* **50**, 13–17.

Kolenbrander, P.E., Andersen, R.N. & Ganeshkumar, N. (1994) Nucleotide sequence of the *Streptococcus gordonii* PK488 coaggregation adhesin gene, *scaA*, and ATP-binding cassette. *Infect. Immun.* **62**, 4469–4480.

Lian, W., Wu, D., Konings, W.N., Mierau, I. & Hersh, L.B. (1996) Heterologous expression and characterization of recombinant *Lactococcus lactis* neutral endopeptidase (neprilysin). *Arch. Biochem. Biophys.* **333**, 121–126.

Mierau, I., Tan, P.S.T., Haandrikmaan, A.J., Kok, J., Leenhouts, K.J., Konings, W.N. & Venema, G. (1993) Cloning and sequencing of the gene for a lactococcal endopeptidase, an enzyme with sequence similarity to mammalian enkephalinase. *J. Bacteriol.* **175**, 2087–2096.

Mierau, I., Kunji, E.R.S., Leenhouts, K.J., Hellendoorn, M.A., Haandrikman, A.J., Poolman, B., Konings, W.N., Venema, G. & Kok, J. (1996). Multiple peptidase mutants of *Lactococcus lactis* are severely impaired in their ability to grow in milk. *J. Bacteriol.* **10**, 2794–2803.

Monnet, V. (1995). Oligopeptidases from *Lactococcus lactis*. *Methods Enzymol.* **248**, 579–592.

Pritchard, G.G., Freebairn, A.D. & Coolbear, T. (1994) Purification and characterization of an endopeptidase from *Lactococcus lactis* subsp. *lactis* SK11. *Microbiology* **140**, 923–930.

Stepaniak, L. & Fox, P.F. (1995) Characterization of the principal intracellular endopeptidase from *Lactococcus lactis* subsp. *lactis* MG1363. *Int. Dairy J.* **5**, 699–713.

Tan, P.S.T., Pos, K.M., & Konings, W.N. (1991) Purification and characterization of an endopeptidase from *Lactococcus lactis* subsp. *cremoris* Wg2. *Appl. Environ. Microbiol.* **57**, 3593–3599.

Tan, P.S.T., Chapot-Chartier, M.-P., Pos, K.M., Rousseaud, M., Boquien, C.-Y., Gripon, J.-C. & Konings, W.N. (1992) Localization of peptidases in *Lactococci*. *Appl. Environ. Microbiol.* **58**, 285–290.

Tan, P.S.T., Poolman, B. & Konings, W.N. (1993) Proteolytic enzymes of *Lactococcus lactis*. *J. Dairy Res.* **60**, 269–286.

Tynkkynen, S., Buist, G., Kunji, E., Kok, J., Poolman, B., Venema, G. & Haandrikman, A. (1993) Genetic and biochemical characterization of the oligopeptide transport system of *Lactococcus lactis*. *J. Bacteriol.* **175**, 7523–7532.

*Igor Mierau*
*Department of Genetics,*
*Groningen Biomolecular Sciences and*
*Biotechnology Institute,*
*University of Groningen,*
*Kerklaan 30,*
*9751 NN Haren, The Netherlands*

*Jan Kok*
*Department of Genetics,*
*Groningen Biomolecular Sciences and*
*Biotechnology Institute,*
*University of Groningen,*
*Kerklaan 30,*
*9751 NN Haren, The Netherlands*
*Email: kokj@biol.rug.nl*

# 366. *Ste24 protease*

## *Databanks*

*Peptidase classification: clan MA, family M48, MEROPS ID: M48.001*
*NC-IUBMB enzyme classification: none*
*Databank codes:*

| Species | | SwissProt | PIR | EMBL (cDNA) | EMBL (genomic) |
|---|---|---|---|---|---|
| *Arabidopsis thaliana* | | EST | – | – | Z34189 | – |
| *Caenorhabditis elegans* | | ESTs | – | – | C07677 | – |
| | | | | | T00972 | |
| | | | | | T00973 | |
| | | | | | T02282 | |
| *Drosophila melanogaster* | | ESTs | – | – | AA439020 | – |
| | | | | | AA201578 | |
| *Escherichia coli* | *htpX* | P23894 | A43659 | M32363 | U28377: chromosome section 65–68' |
| | | | A42604 | M58470 | |
| | *ycaL* | P43674 | – | X82933 | – |
| | *aroA* | – | – | U31523 | – |
| *Haemophilus influenzae* | *htpX* | P44840 | – | – | U32755 |
| *Homo sapiens* | | ESTs | – | – | AA018886 | – |
| | | | | | AA210930 | |
| | | | | | AA295448 | |
| | | | | | AA296746 | |
| | | | | | AA359058 | |
| | | | | | F11310 | |
| | | | | | N76181 | |
| | | | | | R54272 | |
| | | | | | T12172 | |
| | | | | | T35312 | |
| | | | | | Z43273 | – |
| *Methanococcus jannaschii* | *MJ1682* | Q59076 | – | – | U67608: genome section 150 of 150 |
| *Saccharomyces cerevisiae* | *ste24* | P47154 | – | U77137 | Z49617: chromosome X ORF |
| *Schistosoma mansoni* | | ESTs | – | – | T14623 | |
| | | | | | W06703 | |
| | | | | | W06760 | |
| *Schizosaccharomyces pombe* | *yan5* | Q10071 | – | Z68144 | – |

## *Name and History*

In the yeast *Saccharomyces cerevisiae*, mutants defective in mating are designated sterile (*ste*). Mating in yeast occurs between two haploid cell types (*MAT*a and *MAT*α). The *STE24* gene was identified in a genetic screen for yeast sterile mutants that exhibited a decrease in *MAT*a-specific mating (Fujimura-Kamada *et al.*, 1997). The *STE24* gene encodes a protease, Ste24p. This **Ste24 protease** is necessary for the production of the **a**-factor mating pheromone secreted by *MAT*a haploid cells. Mature, bioactive **a**-factor is a prenylated, carboxylmethylated peptide (12-mer) derived from a precursor (36-mer or 38-mer) containing a C-terminal CAAX motif (C is Cys; A is aliphatic, **not** necessarily alanine, and X is any amino acid) and an N-terminal extension. Maturation of the **a**-factor precursor involves a series of C-terminal CAAX modifications, followed by two sequential

N-terminal endoproteolytic cleavages (Chen *et al.*, 1997). *STE24* mutants are defective at the first N-terminal processing step of **a**-factor biogenesis (Fujimura-Kamada *et al.*, 1997) (see Figure 366.1). The gene *AFC1*, for **a**-factor-converting enzyme, was subsequently identified in an independent genetic screen for mutants defective in **a**-factor production and was found to be identical to *STE24* (Boyartchuk *et al.*, 1997). Here the *STE24* designation will be used to be consistent with the standard nomenclature for genes involved in mating. In the *Saccharomyces* Genome Database, the *STE24* open reading frame is designated YJR117w.

## *Activity and Specificity*

Ste24p is required for efficient **a**-factor production. Fujimura-Kamada *et al.* (1997) demonstrated that Ste24p is necessary

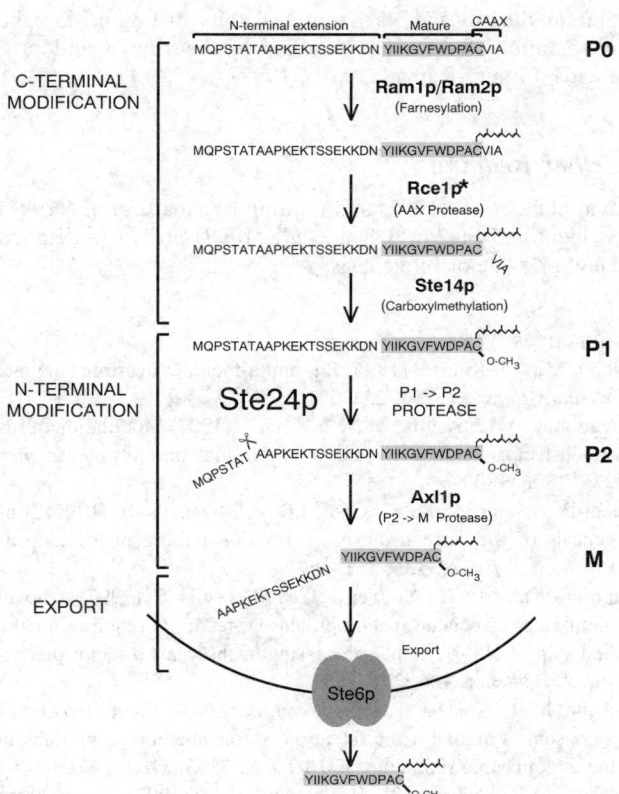

*Figure 366.1* The role of Ste24p in the biogenesis of the mating pheromone **a**-factor. The mature portion of the **a**-factor precursor (**P0**) is shaded in light gray, and is flanked by a C-terminal CAAX motif and an N-terminal extension. The **a**-factor precursor undergoes a series of C-terminal modifications (farnesylation, cleavage of AAX and carboxylmethylation) followed by two N-terminal cleavages; subsequently, mature **a**-factor is exported out of the cell by the Ste6p transporter. The enzymes involved at each step of **a**-factor processing are listed and their respective activities are indicated. The **a**-factor precursor and its biosynthesis intermediates (**P0, P1, P2, M**) can be separated with SDS-PAGE (Chen *et al.*, 1997). Ste24p is necessary for the N-terminal P1→P2 cleavage step of **a**-factor processing (Fujimura-Kamada *et al.*, 1997). (*) Ste24p also appears to perform a redundant function with Rce1p as the AAX protease (Boyartchuk *et al.*, 1997). [It is notable that the action of the AAX protease is similar to that of the mammalian isoprenylated protein peptidase (Chapter 260)–eds.]

for **a**-factor processing (see Figure 366.1) at the first N-terminal cleavage (Thr7┼Ala8; also designated the P1 → P2 cleavage site), since *STE24* mutants accumulate the P1 precursor of **a**-factor. (Note: P1 is here defined as the **a**-factor precursor in which the C-terminal modifications are complete and the N-terminus is intact; P2 is the precursor which has an intermediate cleavage of the N-terminus.) Furthermore, point mutations in the **a**-factor precursor at or near the P1 → P2 cleavage site lead to accumulation of the P1 form, just as observed in a *STE24* mutant (Fujimura-Kamada *et al.*, 1997; F.J. Nouvet, A. Kistler & S. Michaelis, unpublished results). Taken together, these findings are consistent with the role

of Ste24p in the first N-terminal processing step of the **a**-factor precursor; when P1 → P2 processing is blocked either by a *STE24* mutation or the appropriate **a**-factor mutation, neither subsequent maturation nor secretion of **a**-factor can proceed. Boyartchuk *et al.* (1997) reported that Ste24p also functions at an earlier step of **a**-factor processing. In this study, Ste24p apparently performs a redundant function with another enzyme, Rce1p, as a CAAX prenyl protease to cleave the last three residues, AAX, of the **a**-factor precursor and other prenylated proteins. Although our data indicate the primary role of Ste24p is in the P1 → P2 cleavage of the **a**-factor precursor, the two proteolytic activities of Ste24p may not be mutually exclusive and future experiments will determine the specificity of Ste24p function.

An *in vitro* assay is not yet available for the P1 → P2 cleavage activity of Ste24p. Ashby & Rine (1995) have described an *in vitro* assay for AAX cleavage using yeast membrane extracts and a synthetic substrate, the farnesylated and tritiated peptide KWWDPA(*S-trans-trans*-farnesyl)CV[4,5-$^3$H]IA. In this assay, the AAX activity of Ste24p is inhibited by the zinc chelator 1,10-phenanthroline (Boyartchuk *et al.*, 1997).

## Structural Chemistry

Ste24p is a 52 kDa membrane protein characterized by several distinctive features. The first and most notable feature of Ste24p is a putative zinc metalloprotease motif, HEXXH. Mutations in the conserved His and Glu residues within the HEXXH motif disrupt Ste24p function (Boyartchuk *et al.*, 1997; Fujimura-Kamada *et al.*, 1997). In *Escherichia coli*, overexpression of the Ste24p bacterial homolog, HtpX, results in increased degradation of abnormal proteins, lending further confirmation to the suggestion that Ste24p and its homologs function as proteases (Kornitzer *et al.*, 1991).

A second notable feature of Ste24p is that it is predicted to be an integral membrane protein with multiple transmembrane spans (Fujimura-Kamada *et al.*, 1997). Additional characteristics of Ste24p include the endoplasmic reticulum (ER) retrieval motif KKXX (Jackson *et al.*, 1990, 1993) at the C-terminus and several potential *N*-linked glycosylation sites. Ste24p resides in the ER membrane, as shown by subcellular fractionation and indirect immunofluorescence (K. Fujimura-Kamada, W.K. Schmidt & S. Michaelis, unpublished results). Since the ER retrieval motif KKXX is expected to face the cytoplasm, the HEXXH protease motif, which is separated from KKXX by a pair of potential membrane spans, is also predicted to lie in a cytosolic loop. Thus, Ste24p substrates are all likely to reside on the cytosolic side of the ER membrane.

## Biological Aspects

In yeast, Ste24p is not essential for viability but is necessary for efficient mating. To initiate the mating process, two haploid cells (*MAT***a** and *MAT*α) secrete their respective pheromones, **a**-factor and α-factor. The sterile defect of an *STE24* mutant is specific for the *MAT***a** cell type, since **a**-factor (but not α-factor) requires Ste24p activity for maturation (Fujimura-Kamada *et al.*, 1997). Residual mating occurs in a *MAT***a** *STE24* mutant and a small amount of mature

**a**-factor can be detected; presumably, another cellular protease can also carry out the function of Ste24p to process the **a**-factor precursor within the N-terminal extension, albeit at a low level.

*STE24* is expressed in both haploid mating types, as well as in the diploid (Fujimura-Kamada *et al.*, 1997). This ubiquitous expression strongly suggests that Ste24p has other substrates in addition to the **a**-factor precursor, whose expression is restricted to the *MAT***a** haploid mating type.

As listed in the Databanks table, Ste24p has homologs in bacteria and *Schizosaccharomyces pombe*. All of these putative peptidases have multiple transmembrane spans; importantly, the HEXXH motif is in an analogous position relative to the C-terminal membrane span for each homolog. Furthermore, sequences that are strikingly similar to *STE24* have also been identified in the Database of Expressed Sequence Tags (dbEST), which contains ESTs from cDNA libraries derived from diverse organisms including trypanosome, nematode, mouse, human. In humans, the *STE24* ESTs originated from cDNA libraries of the brain, heart, retina and prostate. This finding predicts that expression of the human Ste24p homolog is distributed in a wide variety of tissues.

## Distinguishing Features

Ste24p has several distinctive features, as described above: (a) a zinc metalloprotease motif, HEXXH, (b) six predicted transmembrane spans, (c) an ER localization motif, KKXX, and (d) three domains (I, II and III) that are highly conserved in a wide range of organisms. The three conserved domains are located in the C-terminal half of Ste24p and are predicted to be cytosolically disposed. Domain I includes the HEXXH motif. The conserved domains I, II and III are only found in Ste24p and its homologs, thus distinguishing the Ste24p family from other families of the zinc metalloproteases (Fujimura-Kamada *et al.*, 1997).

## Further Reading

Boyartchuk *et al.* (1997) and Fujimura-Kamada *et al.* (1997) first identified Ste24p. Chen *et al.* (1997) present a detailed review of **a**-factor biogenesis.

## References

Ashby, M.N. & Rine, J. (1995) Ras and **a**-factor converting enzyme. *Methods Enzymol.* **250**, 235–251.

Boyartchuk, V.L., Ashby, M.N. & Rine, J. (1997) Modulation of ras and **a**-factor function by carboxyl-terminal proteolysis. *Science* **275**, 1796–1800.

Chen, P., Sapperstein, S.K., Choi, J.D. & Michaelis, S. (1997) Biogenesis of the *Saccharomyces cerevisiae* mating pheromone **a**-factor. *J. Cell Biol.* **136**, 251–269.

Fujimura-Kamada, K., Nouvet, F.J. & Michaelis, S. (1997) A novel membrane-associated metalloprotease, Ste24p, is required for the first step of $NH_2$-terminal processing of the yeast **a**-factor precursor. *J. Cell Biol.* **136**, 271–285.

Jackson, M.R., Nilsson, T. & Peterson, P.A. (1990) Identification of a consensus motif for the retention of transmembrane proteins in the endoplasmic reticulum. *EMBO J.* **9**, 3153–3162.

Jackson, M.R., Nilsson, T. & Peterson, P.A. (1993) Retrieval of transmembrane proteins to the endoplasmic reticulum. *J. Cell Biol.* **121**, 317–333.

Kornitzer, D., Teff, D., Altuvia, S. & Oppenheim, A.B. (1991) Isolation, characterization, and sequence of an *Escherichia coli* heat shock gene, *htpx*. *J. Bacteriol.* **173**, 2944–2953.

*Amy Tam*
*Department of Cell Biology and Anatomy,*
*The Johns Hopkins University School of Medicine,*
*725 North Wolfe Street,*
*Baltimore, MD 21205, USA*

*Konomi Fujimura-Kamada*
*Department of Cell Biology and Anatomy,*
*The Johns Hopkins University School of Medicine,*
*725 North Wolfe Street,*
*Baltimore, MD 21205, USA*

*Susan Michaelis*
*Department of Cell Biology and Anatomy,*
*The Johns Hopkins University School of Medicine,*
*725 North Wolfe Street,*
*Baltimore, MD 21205, USA*
*Email: susan_michaelis@qmail.bs.jhu.edu*

# 367. *Vibrio collagenase*

## Databanks

*Peptidase classification: clan MA, family M9, MEROPS ID: M09.001*
*NC-IUBMB enzyme classification: EC 3.4.24.3*
*Chemical Abstracts Service registry number: 9001-12-1*

*Databank codes:*

| Species | SwissProt | PIR | EMBL (cDNA) | EMBL (genomic) |
|---|---|---|---|---|
| *Vibrio alginolyticus* | P43154 | – | X62635 | – |
| *Vibrio parahaemolyticus* | – | – | Z46782 | – |

## Name and History

Collagenolytic activity in hide bacteria was first reported by Woods *et al*. (1972) and one of these bacteria was first identified as *Achromobacter iophagus* (Welton & Woods, 1973). Microbiological study later showed that this bacterium belongs to the largely marine genus *Vibrio* and accordingly it was renamed as *Vibrio alginolyticus* chemovar. *iophagus* (Emod *et al*., 1983). This collagenolytic enzyme, quoted frequently under its historical name as **Achromobacter collagenase**, and more recently as **Vibrio collagenase**, is one of the most thoroughly studied bacterial collagenases (Fukushima *et al*., 1990).

## Activity and Specificity

*Vibrio* collagenase cleaves native collagen much more rapidly than does vertebrate interstitial collagenase (Chapter 389). In the fist step of degradation, this enzyme acts on collagen in a manner similar to the vertebrate collagenase, i.e. attacking at a point three-quarters of the way from the N-terminus, although the bond preferentially cleaved is different: Xaa┼Gly instead of Gly┼Leu or Gly┼Ile (Lecroisey & Keil, 1979). Collagenase activity can be measured colorimetrically using a synthetic substrate (Pz-Pro-Leu┼Gly-Pro-D-Arg from Fluka; see Appendix 2 for full names and addresses of suppliers), and the assay system typically includes Tris–HCl buffer (pH 7.2; 50 mM), $CaCl_2$ (1 mM) and NaCl (0.4 M). *Vibrio* and *Clostridium* collagenase (Chapter 368) cleave the same bond in the synthetic peptide Pz-Pro-Leu┼Gly-Pro-D-Arg; however their specific activities are different: the purest *Clostridium* collagenase has a specific activity about 300 nkat mg$^{-1}$ and *Vibrio* collagenase 1800 nkat mg$^{-1}$ (Lecroisey *et al*., 1975; Keil-Dlouha, 1976).

It is believed that a bacterial collagenase degrades exclusively collagen-like structures. Nevertheless, numerous proteins have short sequence stretches analogous to those in collagen. In fact, this collagenase has been successfully applied to produce highly specific cleavages in β-casein (Gilles & Keil, 1976), myosin A, prolactin and adenylate kinase (Keil, 1992). The corresponding substrate subsites display a consistent characteristic pattern: Pro is either in position P2 or P2′ and a small neutral residue (Gly, Ala) in P1′ (Keil, 1992). Inhibitor analysis using synthetic peptides further reveals that the S3′ subsite is important for substrate binding and the charged groups in the P3′ position play a key role in the interaction of the inhibitors with enzyme, e.g. HS-CH$_2$-CH$_2$-CO-Pro-Arg was a much more potent inhibitor ($K_i$ 0.5 µM) than HS-CH$_2$-CH$_2$-CO-Pro-Asp ($K_i$ 70 µM) (Yiotakis & Dive, 1986).

The enzyme activity is inhibited by EDTA, cysteine and histidine.

## Structural Chemistry

*Vibrio* collagenase is a protein of 81 875 Da and 739 amino acid residues as deduced from the cDNA sequence analysis (Takeuchi *et al*., 1992). Collagenase protein is synthesized in precursor form (preprocollagenase), then the signal peptide and pro region are removed and the resulting mature form is secreted. The active enzyme contains 1 mol of zinc per mol enzyme (Keil-Dlouha, 1976) and there are two residues of His in the sequence -His-Glu-Tyr-Val-His- that may well be zinc ligands. The calculated pI from the amino acid composition is 4.4. Secondary structure comparison of this enzyme with thermolysin (Chapter 351) and clostridial collagenase (Chapter 368) reveals that *Vibrio* collagenase contains a higher fraction of α helix than the *Clostridium* enzyme (Heindl *et al*., 1980).

## Preparation

The enzyme protein is prepared from the culture medium and purified by DEAE-cellulose chromatography (Lecroisey *et al*., 1975) or followed by preparative gel electrophoresis (Tong *et al*., 1986). These reports indicate that there are several molecular forms of enzyme in the preparation; these are considered to be autodigestion products from a single polypeptide chain (Keil-Dlouha, 1976).

## Biological Aspects

This nonpathogenic bacterium was isolated from cured hides and shown to lyse collagen rapidly under aerobic conditions. The primary role of bacterial collagenase is the attack upon the host, destruction of the collagen barrier and provision of nutritional resources for bacteria. Collagenase is only induced by the addition of collagen (Keil-Dlouha *et al*., 1976) or peptone (Reid *et al*., 1980) in the culture medium. There is a bacterial receptor for detecting the collagen or peptides derived from collagen. This receptor is identified in membranes by a radioiodination reaction (Keil-Dlouha *et al*., 1983). Production of the enzyme is also regulated by temperature and oxygen (Hare *et al*., 1981).

The use of *Vibrio* collagenase in tissue cell dispersions, elastin purification and selective cleavages of gene products in biotechnology is attractive. However, more important applications in human therapy are in the removal of necrotic tissues from burns, ulcers and decubitus ulcers because of its strong and specific activity against native collagen.

## Distinguishing Features

A quantitative difference exists between *Vibrio* and *Clostridium* (Chapter 368) collagenase in the action on a synthetic peptide, Pz-Pro-Leu+Gly-Ala-D-Arg, which is cleaved only by the *Vibrio* enzyme (Keil, 1992). Polyclonal antisera against the enzyme have been described (Tong *et al.*, 1988; Fukushima *et al.*, 1990), but are not commercially available at this time.

### Further Reading

An extensive review of many aspects of this enzyme is presented by Keil (1992).

### References

Emod, I., Soubigou, P., Tong, N.T., Keil, B. & Richard, G. (1983) Assignment of *Achromobacter iophagus* strain I.029 to *Vibrio alginolyticus* chemovar *iophagus*. *Int. J. Syst. Bacteriol.* **33**, 451–459.

Fukushima, J., Takeuchi, H., Tanaka, E., Hamajima, K., Sato, Y., Kawamoto, S., Morihara, K., Keil, B. & Okuda, K. (1990) Molecular cloning and partial DNA sequencing of the collagenase gene of *Vibrio alginolyticus*. *Microbiol. Immunol.* **34**, 977–984.

Gilles, A.M. & Keil, B. (1976) Cleavage of β-casein by collagenase from *Achromobacter iophagus* and *Clostridium histolyticum*. *FEBS Lett.* **65**, 369–372.

Hare, P., Long, S., Robb, F.T. & Woods, D.R. (1981) Regulation of exoprotease production by temperature and oxygen in *Vibrio alginolyticus*. *Arch. Microbiol.* **130**, 276–280.

Heindl, M.C., Fermandjian, S. & Keil, B. (1980) Circular dichroism comparative studies of two bacterial collagenases and thermolysin. *Biochim. Biophys. Acta* **624**, 51–59.

Keil, B. (1992) *Vibrio alginolyticus* ('*Achromobacter*') collagenase: biosynthesis, function and application. *Matrix Suppl.* **1**, 127–133.

Keil-Dlouha, V. (1976) Chemical characterization and study of the autodigestion of pure collagenase from *Achromobacter iophagus*. *Biochim. Biophys. Acta* **429**, 239–251.

Keil-Dlouha, V., Misrahi, R. & Keil, B. (1976) The induction of collagenase and neutral proteinase by their high molecular weight substrates in *Achromobacter iophagus*. *J. Mol. Biol.* **107**, 293–305.

Keil-Dlouha, V., Emod, I., Soubigou, P., Bagilet, L.K. & Keil, B. (1983) Cell-surface collagen-binding protein in the procaryote *Achromobacter iophagus*. *Biochim. Biophys. Acta* **727**, 115–121.

Lecroisey, A. & Keil, B. (1979) Differences in the degradation of native collagen by two microbial collagenases. *Biochem. J.* **179**, 53–58.

Lecroisey, A., Keil-Dlouha, V., Woods, D.R., Perrin, D. & Keil, B. (1975) Purification, stability and inhibition of the collagenase from *Achromobacter iophagus*. *FEBS Lett.* **59**, 167–172.

Reid, G.C., Woods, D.R. & Robb, F.T. (1980) Peptone induction and rifampin-insensitive collagenase production by *Vibrio alginolyticus*. *J. Bacteriol.* **142**, 447–454.

Takeuchi, H., Shibano, Y., Morihara, K., Fukushima, J., Inami, S., Keil, B., Gilles, A.M., Kawamoto, S. & Okuda, K. (1992). Structural gene and complete amino acid sequence of *Vibrio alginolyticus* collagenase. *Biochem. J.* **281**, 703–708.

Tong, N.T., Tsugita, A. & Keil-Dlouha, V. (1986) Purification and characterization of two high molecular forms of *Achromobacter iophagus*. *Biochim. Biophys. Acta* **874**, 296–303.

Tong, N.T., Dumas, J. & Keil-Dlouha, V. (1988) New *Achromobacter* collagenase and its immunological relationship with a vertebrate collagenase. *Biochim. Biophys. Acta* **955**, 43–49.

Welton, R.L. & Woods, D.R. (1973) Halotolerant collagenolytic activity of *Achromobacter iophagus*. *J. Gen. Microbiol.* **75**, 191–196.

Woods, D.R., Welton, R.L., Thomson, J.A. & Cooper, D.R. (1972) Collagenolytic activity of cured hide bacteria. *J. Appl. Bacteriol.* **35**, 123–128.

Yiotakis, A. & Dive, V. (1986) New thiol inhibitor of *Achromobacter iophagus* collagenase: specificity of the enzyme's S3′ subsite. *Eur. J. Biochem.* **160**, 413–418.

*J. Fukushima*
*Department of Bacteriology,*
*Yokohama City University School of Medicine,*
*3–9 Fukuura, Kanazawa-ku, Yokohama 236, Japan*
*Email: jfukusim@med.yokohama-cu.ac.jp*

*K. Okuda*
*Department of Bacteriology,*
*Yokohama City University School of Medicine,*
*3–9 Fukuura, Kanazawa-ku, Yokohama 236, Japan*

# 368. Clostridium collagenases

### Databanks

*Peptidase classification: clan MA, family M9, MEROPS ID: M09.002*
*NC-IUBMB enzyme classification: EC 3.4.24.3*

*ATCC entries: 6282, 8034, 17859, 17860, 19401, 21000, 25770 (Clostridium histolyticum cell line); 69334 (clone encoding a 68 kDa type I collagenase)*
*Chemical Abstracts Service registry number: 9001-12-1*
*Databank codes:*

| Species | SwissProt | PIR | EMBL (cDNA) | EMBL (genomic) |
|---|---|---|---|---|
| *Clostridium histolyticum* | – | – | D29981 | – |
| *Clostridium perfringens* | P43153 | – | D13791 | – |

## Name and History

The collagenases produced by *Clostridium histolyticum* were the first collagenases to be discovered and studied in any detail. Although other microorganisms are now known to produce collagenases, the terms **bacterial collagenase** and **microbial collagenase** have become synonymous with **clostridium collagenase**. The culture filtrate of *Clostridium histolyticum* contains a mixture of collagenases and other proteinases that exhibits potent hydrolytic activity toward connective tissue. Accordingly, crude preparations derived from this mixture have become the reagents of choice for tissue dissociation experiments (Seglen, 1976) and are among the largest-selling enzymatic products.

Mandl and coworkers first attempted to isolate 'the collagenase' from *Clostridium histolyticum* (Mandl *et al.*, 1953, 1958). This preparation has subsequently been referred to as **clostridiopeptidase A**. Subsequent studies established that the culture filtrate contained multiple collagenases (Grant & Alburn, 1959; Mandl *et al.*, 1964; Yoshida & Noda, 1965; Kono, 1968; Harper & Kang, 1970; Lwebuga-Mukasa *et al.*, 1976; Bond & Van Wart, 1984a,b). Seven collagenases with molecular masses ranging from 68 to 130 kDa have been purified to homogeneity (Bond & Van Wart, 1984a) and characterized (Bond & Van Wart, 1984b,c). These have been designated as class I ($\alpha$, $\beta$, $\gamma$ and $\eta$) and class II ($\delta$, $\varepsilon$ and $\zeta$) collagenases based on a variety of criteria including their relative activities toward collagen versus synthetic peptide substrates. The two classes of collagenases are probably derived from two distinct gene products with the lower molecular mass members of each class derived from a larger precursor by proteolysis. The collagenase I (Yoshida & Noda, 1965) or B (Seifter & Harper, 1970) described earlier is probably predominantly $\beta$-collagenase, while collagenases II (Yoshida & Noda, 1965) and A (Seifter & Harper, 1970) are probably mixtures of individual class I and II collagenases.

## Activity and Specificity

The clostridial collagenases are distinguished by their ability to digest native, triple-helical types I, II and III collagens into a mixture of small peptides under physiological conditions. A variety of collagen-based assays is available to quantify these activities (see Mallya *et al.*, 1986, and references cited therein). Clostridial collagenases also digest other types of collagen, but these reactions have not been characterized in detail. The initial proteolytic events in the hydrolysis of type I, II and III collagens by the class I and II collagenases have been delineated (French *et al.*, 1987, 1992) and the kinetic parameters for these reactions measured (Mallya *et al.*, 1992). The class I and II enzymes initially attack all three collagens at distinct hyper-reactive sites whose sequences have been identified (French *et al.*, 1992). All of the cleavages are at Yaa+Gly bonds in the repeating Gly-X-Y collagen sequence. The hyper-reactivity of these cleavage sites appears to be more related to some local conformational feature of the collagen fold than to the surrounding sequence. The two classes of collagenases are complementary in their patterns of attack on these collagens.

Spectrophotometric assays based on substrates containing either the 2-furanacryloyl (FA) or cinnamoyl groups in subsite P2 (Van Wart & Steinbrink, 1981) or the 4-nitrophenylalanine group in subsite P1 (Van Wart & Steinbrink, 1985) have been developed. The peptide FA-Leu+Gly-Pro-Ala has become a standard substrate for these collagenases and is available commercially. This assay has largely replaced an earlier one based on hydrolysis of the chromogenic peptide P3-Pro-Leu+Gly-Pro-D-Arg (Wünsch & Heidrich, 1963). The specificity of both the class I and II collagenases toward peptides with collagen-like sequences has been extensively investigated (Mookhtiar *et al.*, 1985; Steinbrink *et al.*, 1985; Van Wart & Steinbrink, 1985). The influence of both the size of the peptide and the identity of the residues in the P3–P3' subsites has been determined. Both classes of collagenases have a strong preference for Gly in subsites P1' and P3, and for aromatic residues in subsite P1. The class II collagenases have a preference for Leu in P1'. Both classes prefer Pro or Ala in subsites P2 and P2', and Hyp, Ala or Arg in subsite P3' (Hyp: hydroxyproline). The class II enzymes have a broader specificity, but are less active toward substrates with Hyp in subsites P1 and P3'. Both classes of collagenases are true endopeptidases whose best peptide substrates have the P3–P3' subsites occupied. Interestingly, however, these collagenases strongly prefer substrates with an unblocked carboxyl group in P3', thus acting as peptidyl-tripeptidase (Mookhtiar *et al.*, 1985).

Based on the specificity of the class I and II collagenases, a number of substrate analog inhibitors have been synthesized (Schwartz & Van Wart, 1992). These have centered on the use of mercaptan, ketone and phosphonamidate moieties

M

as scissile bond surrogates. Selective inhibition of class II enzymes has been achieved by optimizing the residues placed in subsites P1 and P3′.

## Structural Chemistry

All of the clostridial collagenases are single polypeptide chains with molecular masses that vary from 68 kDa for α-collagenase to 130 kDa for η-collagenase (Bond & Van Wart, 1984b). The individual collagenases contain between 2 and 5 Cys residues per protein chain, but it is not known whether there are any disulfide bonds. The pI values of all of the collagenases fall into the 5.35–6.20 range. Two genes that encode clostridial collagenases have been cloned. Yoshihara *et al*. (1994) have cloned the *col*H gene from strain JCM 1403 (ATCC 19401). It encodes a 116 kDa protein consisting of 1021 amino acids of which the first 40 are believed to be a signal sequence. There is no propeptide and the mature enzyme is produced directly on secretion as the signal peptide is removed. The same gene has been cloned by Ambrosius *et al*. (1995) who have characterized the expressed enzyme and concluded that it corresponds to a class II collagenase. A second distinct gene has been cloned from *Clostridium histolyticum* strain ATCC 21000 (Lin & Lei, 1993). The full-length gene encodes a 936 residue protein with a molecular mass of 110 kDa while a partial clone (ATCC 69334) encodes a 68 kDa truncated form. Expression of these genes in *E. coli* produces enzymes with the characteristics of the class I collagenases (Hun-Chi Lin, personal communication). Other than the assignment of these two genes as coding for collagenases from the two known classes, the relationship between them and the individual species purified by Bond & Van Wart (1984a) remains uncertain.

All of the clostridial collagenases isolated by Bond & Van Wart (1984a,b) contain approximately 1 atom of zinc per protein chain. The single zinc atom is essential for activity (Angleton & Van Wart, 1988a) and presumably resides at the active site. Both of the clones described above contain an HEYTH zinc-binding motif (Yoshihara *et al*., 1994; Hun-Chi Lin, personal communication). The zinc atom of γ-collagenase (class I) can be exchanged with cobalt, copper and nickel with retention of activity, but only the cobalt- and nickel-substituted forms of ζ-collagenase (class II) are active (Angleton & Van Wart, 1988a,b). Cadmium and mercury do not restore activity to the apo forms of either enzyme. The clostridial collagenases also contain between 1.9 (α-collagenase) and 6.8 (β-collagenase) atoms of calcium per molecule that probably function to stabilize the structure of the protein, as observed for thermolysin (Chapter 351).

No reports of crystallization trials of any clostridial collagenase have appeared and very little is known about the tertiary structure of these enzymes. Circular dichroism studies reveal that the secondary structures of the two classes are somewhat different. The class I enzymes contain 45% α helix and 25% β structure while the class II enzymes contain 28% α helix and 42% β structure (Bond & Van Wart, 1984c).

## Preparation

The clostridial collagenases are isolated from the culture filtrate of the bacterium. A number of commercial suppliers (Sigma, Worthington, Advanced Biofactures, Boehringer Mannheim, etc.; see Appendix 2 for full names and addresses of suppliers) sell both crude and partially purified preparations that vary widely in their composition (Bond & Van Wart, 1984a). The crude preparations contain the collagenases as well as a brown pigment, clostripain (Chapter 257), an aminopeptidase and several neutral proteinases. These crude preparations can serve as a source for the purification of the individual class I and II collagenases, but the purification scheme is quite long (Bond & Van Wart, 1984a). As noted above, other purification schemes (Yoshida & Noda, 1965; Seifter & Harper, 1970) most likely produce mixtures of the individual collagenases. A recombinant collagenase produced in *E. coli* using the class I collagenase gene isolated by Lin & Lei (1993) is sold by Sigma.

## Biological Aspects

*Clostridium histolyticum* is a pathogenic anaerobe that causes gas gangrene. All strains of the bacterium elaborate a collagenase, but the amount depends on the strain and the culture medium (Mandl *et al*., 1953). The bacterium presumably uses the collagenases as a means to invade the host and possibly to degrade host protein for nutritional purposes. It is not clear whether all strains produce the same collagenases. Yoshihari *et al*. (1994) have concluded from southern hybridization experiments that strain JCM 1403 from which the *col*H gene was isolated contains no similar genes. However, the gene encoding a collagenase from another class could escape detection if it were distantly related.

Clostridial collagenase preparations continue to be very useful for tissue dissociation experiments. This includes the isolation of hepatocytes, adipocytes, and other cells for research purposes (Seglen, 1976), as well as medical procedures involving the preparation of vascular endothelial cells for seeding vascular prostheses (Sharefkin *et al*., 1987) and isolating pancreatic islets for transplantation (Wolters *et al*., 1995). The use of *Clostridium* collagenase preparations for these purposes has been plagued by reproducibility problems that are undoubtedly related to their variable proteolytic compositions. Although attempts have been made to develop defined preparations that are optimized for different purposes, the use of clostridial preparations for cell isolation is still largely empirical. A clostridial collagenase preparation called Santyl is produced by Biospecifics Technologies and approved for use in the treatment of bed sores. The direct use of clostridial collagenase in the treatment of herniated discs is also under investigation (Hedtmann *et al*., 1992).

## Distinguishing Features

Although the clostridial collagenases share with vertebrate collagenases (Chapters 389 and 390) the ability to hydrolyze interstitial (types I, II and III) collagens, the mode of attack is distinctly different. The vertebrate collagenases initiate collagenolysis by making a single scission across all three $\alpha$ chains at a distinct locus three-quarters from the N-terminus, after which subsequent attack on the $\alpha$ chains is very limited. Subsequent collagenolysis is carried out by gelatinases and other synergistic proteinases only after denaturation of the collagen triple helix. In contrast, the clostridial collagenases not only initiate hydrolysis of interstitial collagens by making multiple scissions within the collagen triple helix, but also complete the digestion by hydrolyzing these initial fragments into a mixture of small peptides. The clostridial collagenases also appear able to digest most if not all of the other collagen types, while the vertebrate collagenases cannot.

## Related Proteinases

The clostridial collagenase that corresponds to the *col*H gene isolated by Yoshihara *et al.* (1994) is homologous to the *Clostridium perfringens* (Matsushita *et al.*, 1994) and *Vibrio alginolyticus* collagenases (Takeuchi *et al.*, 1992) (Chapter 367). All three contain the HEXXH zinc-binding motif.

## Further Reading

For a review of the earlier literature on clostridial collagenase, see Seifter & Harper (1971). For a more recent review of the properties of these collagenases, see Mookhtiar & Van Wart (1992).

## References

Ambrosius, D., Hesse, F. & Burtscher, H. (1995) Rekombinante Kollagenase Typ II aus *Clostridium histolyticum* und ihre Verwendung zur Isolierung von Zellen und Zellverbanden. European Patent 0 677 586 A1. [Recombinant type II collagenase from *C. histolyticum* and its application for the isolation of cells and cell assemblies.]

Angleton, E.L. & Van Wart, H.E. (1988a) Preparation and reconstitution with divalent metal ions of class I and class II *Clostridium histolyticum* apocollagenases. *Biochemistry* **27**, 7406–7412.

Angleton, E.L. & Van Wart, H.E. (1988b) Preparation by direct metal exchange and kinetic study of active site metal substituted class I and class II *Clostridium histolyticum* collagenases. *Biochemistry* **27**, 7413–7418.

Bond, M.D. & Van Wart, H.E. (1984a) Purification and separation of the individual collagenases of *Clostridium histolyticum* using red dye ligand chromatography. *Biochemistry* **23**, 3077–3085.

Bond, M.D. & Van Wart, H.E. (1984b) Characterization of individual collagenases from *Clostridium histolyticum*. *Biochemistry* **23**, 3085–3091.

Bond, M.D. & Van Wart, H.E. (1984c) Relationship between the individual collagenases of *Clostridium histolyticum*: evidence for evolution by gene duplication. *Biochemistry* **23**, 3092–3099.

French, M.F., Mookhtiar, K.A. & Van Wart, H.E. (1987) Limited proteolysis of type I collagen at hyperreactive sites by class

I and II *Clostridium histolyticum* collagenases: complementary digestion patterns. *Biochemistry* **26**, 681–687.

French, M.F., Bhown, A. & Van Wart, H.E. (1992) Identification of *Clostridium histolyticum* collagenase hyperreactive sites in type I, II and III collagens: lack of correlation with local triple helical stability. *J. Protein Chem.* **11**, 83–97.

Grant, N.H. & Alburn, H.E. (1959) Studies on the collagenase of *Clostridium histolyticum. Arch. Biochem. Biophys.* **82**, 245–255.

Harper, E. & Kang, A.H. (1970) Studies on the specificity of bacterial collagenase. *Biochem. Biophys. Res. Commun.* **41**, 482–487.

Hedtmann, A., Fett, H., Steffen, R. & Kramer, J. (1992) Chemonukleolyse mit Chymopapain und Kollagenase. 3-Jahres-Ergebnisse einer prospektiv-randomisierten Studie [Chemonucleosis with chymopapain and collagenase, results of a three-year prospective randomized study]. *Z. Orthop. Grenz.* **130**, 36–44.

Kono, T. (1968) Purification and partial characterization of collagenolytic enzymes from *Clostridium histolyticum. Biochemistry* **7**, 1106–1114.

Lin, H.-C. & Lei, S.-P. (1993) Molecular cloning of the genes responsible for collagenase production from *Clostridium histolyticum*. US Patent 5,177,017.

Lwebuga-Mukasa, J.S., Harper, E. & Taylor, P. (1976) Collagenase enzymes from *Clostridium*: characterization of individual enzymes. *Biochemistry* **15**, 4736–4741.

Mallya, S.K., Mookhtiar, K.A. & Van Wart, H.E. (1986) Accurate, quantitative assays for the hydrolysis of soluble type I, II and III $^3$H-acetylated collagens by bacterial and tissue collagenases. *Anal. Biochem.* **158**, 334–345.

Mallya, S.K., Mookhtiar, K.A. & Van Wart, H.E. (1992) Kinetics of hydrolysis of type I, II and III collagens by the class I and class II *Clostridium histolyticum* collagenases. *J. Protein Chem.* **11**, 99–107.

Mandl, I., MacLennon, J.D. & Howes, E.L. (1953) Isolation and characterization of proteinase and collagenase from *Cl. histolyticum. J. Clin. Invest.* **32**, 1323–1329.

Mandl, I., Zipper, H. & Ferguson, L.T. (1958) *Clostridium histolyticum* collagenase: its purification and properties. *Arch. Biochem. Biophys.* **74**, 465–475.

Mandl, I., Keller, S. & Manahan, J. (1964) Multiplicity of *Clostridium histolyticum* collagenases. *Biochemistry* **3**, 1737–1741.

Matsushita, O., Yoshihara, K., Katayama, S.-I., Minami, J. & Okabe, A. (1994) Purification and characterization of a *Clostridium perfringens* 120-kilodalton collagenase and nucleotide sequence of the corresponding gene. *J. Bacteriol.* **176**, 149–156.

Mookhtiar, K.A. & Van Wart, H.E. (1992) *Clostridium histolyticum* collagenases: a new look at some old enzymes. *Matrix Suppl.* **1**, 116–126.

Mookhtiar, K.A., Steinbrink, D.R. & Van Wart, H.E. (1985) Mode of hydrolysis of collagen-like peptides by class I and class II clostridium collagenases: evidence for both endopeptidase and tripeptidylcarboxypeptidase activities. *Biochemistry* **24**, 6527–6533.

Schwartz, M.A. & Van Wart, H.E. (1992) Synthetic inhibitors of bacterial and mammalian interstitial collagenases. *Prog. Med. Chem.* **29**, 271–334.

Seifter, S. & Harper, E. (1970) The collagenases. *Methods Enzymol.* **19**, 613–635.

Seifter, S. & Harper, E. (1971) The collagenases. In: *The Enzymes*, 3rd edn (Boyer, P.D., ed.). New York: Academic Press, pp. 649–697.

**M**

Seglen, P. (1976) Preparation of isolated rat liver cells. *Methods Cell Biol.* **13**, 29–83.

Sharefkin, J.B., Van Wart, H.E. & Williams, S.K. (1987) Enzymatic harvesting of adult human endothelial cells for use in autogenous endothelial vascular prosthetic seeding. In: *Endothelial Seeding in Vascular Surgery* (Herring, M.D. & Glover, J.L., eds). Philadelphia: Grune & Stratton, pp. 79–101.

Steinbrink, D.R., Bond, M.D. & Van Wart, H.E. (1985) Substrate specificity of β-collagenase from *Clostridium histolyticum*. *J. Biol. Chem.* **260**, 2771–2776.

Takeuchi, H., Shibano, Y., Morihara, K., Fukushima, J., Inami, S., Keil, B., Gilles, A.-M., Kawamoto, S. & Okuda, K. (1992) Structural gene and complete amino acid sequence of *Vibrio alginolyticus* collagenase. *Biochem. J.* **281**, 703–708.

Van Wart, H.E. & Steinbrink, D.R. (1981) A continuous spectrophotometric assay for *Clostridium histolyticum* collagenase. *Anal. Biochem.* **113**, 356–365.

Van Wart, H.E. & Steinbrink, D.R. (1985) Complementary substrate specificities of class I and class II collagenases from *Clostridium histolyticum*. *Biochemistry* **24**, 6520–6526.

Wolters, G.H., Vos-Scheperkeuter, G.H., Lin, H.-C. & van Schilfgaarde, R. (1995) Different roles of class I and II *Clostridium histolyticum* collagenase in rat pancreatic islet isolation. *Diabetes* **44**, 227–233.

Wünsch, E. & Heidrich, H.G. (1963) Zur quantitativen Bestimmung der Kollagenase [On the quantitative determination of collagenase]. *Z. Physiol. Chem.* **333**, 149–151.

Yoshida, E. & Noda, H. (1965) Isolation and characterization of collagenase I and II from *Clostridium histolyticum*. *Biochim. Biophys. Acta* **105**, 562–574.

Yoshihara, K., Matsushita, O., Minami, J. & Okabe, A. (1994) Cloning and nucleotide sequence analysis of the *col*H gene from *Clostridium histolyticum* encoding a collagenase and a gelatinase. *J. Bacteriol.* **176**, 6489–6496.

*Harold E. Van Wart*
*Inflammatory Diseases Unit,*
*Roche Bioscience, Mailstop S3-1,*
*3401 Hillview Avenue,*
*Palo Alto, CA 94304, USA*
*Email: harold.vanwart@roche.com*

# 369. *Hyicolysin*

## Databanks

*Peptidase classification: clan MA, family M30, MEROPS ID: M30.001*
*NC-IUBMB enzyme classification: none*
*Databank codes:*

| Species | SwissProt | PIR | EMBL (cDNA) | EMBL (genomic) |
|---|---|---|---|---|
| *Staphylococcus hyicus* | Q08002 | – | X73315 | – |

## Name and History

Most *Staphylococcus aureus* strains secrete proteolytic enzymes; three different types of proteases have been studied in detail. A serine protease from *S. aureus* V8, commonly referred to as V8 protease or protease I (Chapter 79), is an endoprotease which cleaves peptide bonds on the C-terminal side of glutamic acid and to a lesser extent of aspartic acid (Drapeau *et al.*, 1972). Similar serine proteases with the same cleavage preference are also found in other *S. aureus* strains; however, they differ from the above described enzyme in their size and sensitivity to DFP (Arvidson *et al.*, 1973; Ryden *et al.*, 1974). The amino acid sequence of a 27 kDa serine protease of *S. aureus* has been determined (Drapeau, 1978).

Another protease of *S. aureus* V8, a metalloprotease called aureolysin (Chapter 538) (Arvidson *et al.*, 1973), is completely inactivated by EDTA and is insensitive to sulfhydryl reagents and DFP. This metalloprotease has a rather broad proteolytic specificity, cleaving preferentially at the N-terminal side of hydrophobic residues. A metalloprotease with a molecular weight of 38 000 and a pH optimum of 7.0 has been isolated from a mutant of *S. aureus* V8 (Björklind & Jörnvall, 1974).

*Staphylococcus hyicus* subsp. *hyicus* also shows strong proteolytic and cytotoxic activity (Devriese *et al.*, 1978; Allaker *et al.*, 1991), but little is known about the proteases from this strain. A protease was partially characterized by Takeuchi *et al*. (1985). We have some evidence that in *S. hyicus* subsp. *hyicus* an exoprotease is involved in the propeptide processing of the extracellular prolipase (Wenzig *et al.*, 1990; Demleitner & Götz, 1994). The preferential cleavage site of this protease should be at Thr┼Val; this substrate specificity is unique among the proteases described to date (Ayora *et al.*, 1994). Here we focus on the cloning and sequencing of the gene (*shpI*) encoding a metalloprotease from *S. hyicus* subsp. *hyicus*, and the biochemical characterization of the purified hyicolysin (ShpI). The name **hyicolysin** is suggested to conform to the concept of having metalloproteinase names end in lysin. Earlier names include **Staphylococcus neutral protease** and **ShpI**.

### Activity and Specificity

On the basis of the presence of the zinc-binding motif of neutral proteases in the sequence, hyicolysin is predicted to be a zinc-dependent neutral protease (Ayora & Götz, 1994). To confirm this, hyicolysin was assayed with several inhibitors. The enzyme was resistant to all inhibitors for serine, cysteine and aspartic proteases and could only be inhibited by the metal chelator EDTA, the calcium chelator EGTA and the zinc-specific chelator 1,10-phenanthroline. Phenanthroline could inhibit even in the presence of 10 mM $CaCl_2$, indicating that zinc is the catalytic metal of hyicolysin, and calcium is required for enzyme stability. Both enzyme preparations from *S. hyicus* subsp. *hyicus* and *S. carnosus* (pCAshp1) behaved similarly. Confirmation of hyicolysin as a zinc metalloenzyme was accomplished by removal of $Zn^{2+}$, followed by reactivation with several metals. Monovalent cations were ineffective; chloride salts of zinc and cobalt could reactivate the enzyme completely, whereas $Mg^{2+}$, $Ca^{2+}$ and $Cu^{2+}$ were only partially effective.

The optimal pH for hyicolysin is between 7.4 and 8.5. The effect of pH values below 6.5 could not be determined because of the precipitation of the azocasein substrate. The temperature maximum is 55°C with a rapid decrease of activity at temperatures above 55°C; however, hyicolysin is unstable at temperatures higher than 45°C.

The purified protease does not cleave elastin nor the synthetic substrates FA-Gly-Leu-amide (FAGLA) (a substrate for neutral proteinases) and FA-Leu-Gly-Pro-Ala (FALGPA) (a substrate for collagenase). The preferred cleavage sites of hyicolysin were determined by using the oxidized insulin B chain, glucagon and melittin as substrates. Some peptide bonds such as His10┼Leu11 and Tyr16┼Leu17 from the insulin B chain, Asp15┼Ser16, Arg18┼Ala19, Gln28┼Trp29 from glucagon and Lys7┼Val8, Ala15┼Leu16, Arg22┼Lys23 from melittin are readily cleaved, whereas some bonds like Asn3┼Gln4 from insulin, Ser8┼Asp9, Lys12┼Tyr13 from glucagon, Ala4┼Val5, Leu16┼Ile17, Ile17┼Ser18 and Ser19┼Trp20 from melittin are slowly cleaved (Ayora & Götz, 1994).

### Structural Chemistry

The gene encoding the extracellular neutral metalloprotease hyicolysin from *S. hyicus* subsp. *hyicus* has been cloned (Ayora & Götz, 1994). DNA sequencing revealed an ORF of 1317 nucleotides encoding a 438 amino acid protein with an $M_r$ of 49 698. When the cloned gene was expressed in *S. carnosus*, a 42 kDa protease was found in the culture medium. The protease was purified from both *S. carnosus* (pCAshp1) and *S. hyicus* subsp. *hyicus*. The N-terminal amino acid sequences of the two proteases revealed that hyicolysin is organized as a preproenzyme with a proposed 26 amino acid signal peptide, a 75 amino acid hydrophilic pro region, and a 337 amino acid extracellular mature form with a calculated $M_r$ of 38 394 (Ayora & Götz, 1994). The N-termini showed microheterogeneity in both host strains.

The deduced amino acid sequence of this hyicolysin was compared with other known protease sequences in the Microgenie (Beckman), PC/GENE (Intelli Genetics, Inc.) and the EMBL databanks. The protein showed no significant homology to other sequences. Only a short pentapeptide in the sequence HEYQH corresponds to a zinc-binding motif found in a number of zinc-dependent metalloproteases of both gram-positive and gram-negative bacteria (Jongeneel *et al.*, 1989).

### Preparation

Hyicolysin was purified from the culture supernatant of *S. hyicus* subsp. *hyicus* (Ayora & Götz, 1994). Ammonium sulfate precipitation of crude culture supernatants gave poor recoveries of activity, probably due to denaturation and digestion of the protein. DEAE-chromatography was effective in separating hyicolysin from other proteins; however, only after AcA-44 gel filtration was it possible to separate hyicolysin from a dominant protease, ShpII (Ayora & Götz, 1994), present in the *S. hyicus* subsp. *hyicus* culture supernatants. Purity of the final products was assessed by SDS-PAGE; the molecular mass of hyicolysin was calculated from this gel to be 42 kDa. Hyicolysin was also purified from the culture supernatant of *S. carnosus* (pCAshp1).

### Biological Aspects

A putative zinc-binding motif for metalloproteases was found in the deduced amino acid sequence of hyicolysin. In the thermolysin-like neutral proteinases (and using thermolysin numbering), His142 and His146 are the zinc ligands and Glu143 is thought to be involved in catalysis (Matthews *et al.*, 1972). Moreover, the metalloproteinases of the thermolysin family also have secondary ligand sites, including a conserved amino acid sequence located in an α-helix structure, Ile-Asn-Glu-Ala-Ile-Ser-Asp, with the glutamic acid located 19 amino acids downstream of the last histidine of the pentapeptide, and a region, Asp-Asn-Gly-Gly-Val-His-Ile-Asn-Ser-Gly, of which the His231 is an active residue.

Hyicolysin does not contain amino acid sequences resembling these additional active-site residues and there is only limited similarity in the α-helix region surrounding a Glu at position 269 which could be the third ligand for zinc. Whether (in hyicolysin numbering) His242, His246 and Glu269 are zinc ligands and Glu243 is an active-site residue must still be determined. It is interesting to note that some bacterial metalloproteases such as the neutral protease from *Streptomyces* sp. strain C5 have the zinc-binding motif but the third ligand is different from that in thermolysin, and has not yet been identified. The deduced sequence of hyicolysin shows little similarity with that of a thermolysin homologue from *S. epidermidis* (Teufel & Götz, 1993).

We found that hyicolysin production is correlated with growth both in *S. hyicus* subsp. *hyicus* and in *S. carnosus* (pCAshp1). The only marked difference is that in the *S. carnosus* clone, hyicolysin activity remained constant in the stationary phase, while in *S. hyicus* subsp. *hyicus* protease activity declined rapidly at the end of the exponential growth phase (Ayora & Götz, 1994). The reason for this is very likely the production or activation of other proteases in this host in the stationary growth phase; one further protease, ShpII, was already identified in the course of this study (Ayora *et al.*, 1994).

A remarkable finding was that the specific activity of the purified hyicolysin from *S. hyicus* subsp. *hyicus* was three times higher than that of the *Staphylococcus carnosus* clone. One reason for this difference could be that hyicolysin is processed differently in the two species. The variations in the N-terminus of the mature protease isolated from *S. hyicus* subsp. *hyicus* and from *S. carnosus* (pCAshp1) suggest that the processing of hyicolysin is autocatalytic and that other proteases may participate by facilitating the propeptide processing. However, such a protease was not detected in protease activity gels.

The apparent lack of homology between hyicolysin and the active site of thermolysin, as well as the lack of inhibition of hyicolysin by phosphoramidon, suggests that hyicolysin is distinct from the thermolysin class of metalloproteases (see Chapters 350 and 351). The fact that FAGLA was not hydrolyzed by hyicolysin also indicates that this protease differs from the thermolysin family of proteases. Since elastin and FALGPA were not hydrolyzed by this enzyme, hyicolysin also differs from pseudolysin (Chapter 357) (Bever & Iglewski, 1988) and the collagenase family (Chapter 389) (Vallee & Auld, 1990) which also contain the zinc-binding motif in the sequence.

## Distinguishing Features

Two different extracellular proteases from *Staphylococcus hyicus* subsp. *hyicus*, hyicolysin and ShpII, have been characterized. Hyicolysin is a neutral metalloprotease with a broad substrate specificity; the gene has been cloned and sequenced. While ShpII was mainly produced in the late logarithmic growth phase, hyicolysin is mainly produced in the late stationary growth phase. The molecular mass of ShpII, estimated from SDS-PAGE, was 34 kDa; the temperature optimum was 55°C and the pH optimum 7.4. ShpII,

a neutral metalloprotease, was strongly inhibited by zinc and calcium chelators (Ayora *et al.*, 1994). The substrate specificity of ShpII was similar to that of thermolysin-like proteases, with the exception that ShpII also recognized aromatic amino acids. We demonstrated *in vitro* that ShpII, but not hyicolysin, cleaves the 86 kDa *S. hyicus* subsp. *hyicus* prolipase at Thr245+Val246 to generate the mature 46 kDa lipase (Ayora *et al.*, 1994). Thus, the two proteases differ not only in their molecular mass but also in substrate specificity. While ShpII reacts with both azocasein and FAGLA as a substrate, hyicolysin shows activity only on azocasein.

## References

Allaker, R.P., Whitlock, M. & Lloyd, D.H. (1991) Cytotoxic activity of *Staphylococcus hyicus*. *Vet. Microb.* **26**, 161–166.

Arvidson, S., Holme, T. & Lindholm, B. (1973) Studies on extracellular proteolytic enzymes from *Staphylococcus aureus*. I. Purification and characterization of one neutral and one alkaline protease. *Biochim. Biophys. Acta* **302**, 135–148.

Ayora, S. & Götz, F. (1994) Genetic and biochemical properties of an extracellular neutral metalloprotease from *Staphylococcus hyicus* subsp. *hyicus*. *Mol. Gen. Genet.* **242**, 421–430.

Ayora, S., Lindgren, P.-E. & Götz, F. (1994) Biochemical properties of a novel metalloprotease from *Staphylococcus hyicus* subsp. *hyicus* involved in extracellular lipase processing. *J. Bacteriol.* **176**, 3218–3223.

Bever, R.A. & Iglewski, B.H. (1988) Molecular characterization and nucleotide sequence of the *Pseudomonas aeruginosa* elastase structural gene. *J. Bacteriol.* **170**, 4309–4314.

Björklind, A. & Jörnvall, A. (1974) Substrate specificity of three different extracellular proteolytic enzymes from *Staphylococcus aureus*. *Biochim. Biophys. Acta* **370**, 524–529.

Demleitner, G. & Götz, F. (1994) Evidence for importance of the *Staphylococcus hyicus* lipase pro-peptide in lipase secretion, stability, and activity. *FEMS Microbiol. Lett.* **121**, 189–198.

Devriese, L.A., Hajek, V., Oeding, P., Meyer, S.A. & Schleifer, K.H. (1978) *Staphylococcus hyicus* (Sompolinsky, 1953) comb. nov. and *Staphylococcus hyicus* subsp. *chromogenes* subsp. nov. *Int. J. Syst. Bacteriol.* **28**, 482–490.

Drapeau, G.R. (1978) The primary structure of staphylococcal protease. *Can. J. Biochem.* **56**, 534–544.

Drapeau, G.R., Boily, Y. & Houmard, J. (1972) Purification and properties of an extracellular protease of *Staphylococcus aureus*. *J. Biol. Chem.* **247**, 6720–6726.

Jongeneel, C.V., Bouvier, J. & Bairoch, A. (1989) A unique signature identifies a family of zinc-dependent metallopeptidases. *FEBS Lett.* **242**, 7859–7864.

Matthews, B.W., Jansonius, J.N., Colman, P.M., Schoenborn, B.P. & Duporque, D. (1972) Three dimensional structure of thermolysin. *Nature (Lond.) New Biol.* **238**, 37–41.

Ryden, A., Ryden, L. & Philipson, L. (1974) Isolation and properties of a staphylococcal protease preferentially cleaving glutamoyl-peptide bonds. *Eur. J. Biochem.* **44**, 105–114.

Takeuchi, S., Kobayashi, Y. & Morozumi, T. (1985) Purification and some properties of protease produced by *Staphylococcus hyicus* subsp. *hyicus* strain III. *Jpn J. Vet. Sci.* **47**, 769–775.

Teufel, P. & Götz, F. (1993) Characterization of an extracellular metallo protease with elastase activity of *Staphylococcus epidermidis*. *J. Bacteriol.* **175**, 4218–4224.

Vallee, B.L. & Auld, D.S. (1990) Zinc coordination, function and structure of zinc-enzymes and other proteins. *Biochemistry* **29**, 5647–5657.

Wenzig, E., Lottspeich, F., Verheij, B., de Haas, G.H. & Götz, F. (1990) Extracellular processing of the *Staphylococcus hyicus* lipase. *Biochem. (Life Sci. Adv.)* **9**, 47–56.

*Friedrich Götz*
*Microbial Genetics,*
*University of Tübingen,*
*D-72076 Tübingen, Germany*
*Email: friedrich.goetz@uni-tuebingen.de*

**M**

# 370. Introduction: family M3 of thimet oligopeptidase

## Databanks

*MEROPS ID: M3*

| Species | SwissProt | PIR | EMBL (cDNA) | EMBL (genomic) |
|---|---|---|---|---|
| **Family M3** | | | | |
| Mitochondrial intermediate peptidase (Chapter 377) | | | | |
| Neurolysin (Chapter 372) | | | | |
| Oligopeptidase A (Chapter 375) | | | | |
| Oligopeptidase F (Chapter 378) | | | | |
| Oligopeptidase MepB (Chapter 374) | | | | |
| Oligopeptidase PepB (Chapter 379) | | | | |
| Peptidyl-dipeptidase Dcp (Chapter 376) | | | | |
| Saccharolysin (Chapter 373) | | | | |
| Thimet oligopeptidase (Chapter 371) | | | | |
| Others | | | | |
| *Bacillus licheniformis* | – | – | D88209 | – |
| *Saccharomyces cerevisiae* | P35999 | – | – | – |

*Family M3* is the largest and most complex family of metallopeptidases in which zinc ligands have not definitely been identified. The family includes intracellular oligopeptidases from mammals (thimet oligopeptidase and neurolysin), fungi (saccharolysin, oligopeptidase MepB) and bacteria (oligopeptidase A, oligopeptidase F and oligopeptidase PepB), that only degrade peptides of certain lengths (between 4 and 17 residues, in the case of thimet oligopeptidase). Another endopeptidase in the family is mitochondrial intermediate peptidase, which removes a second targeting signal (an N-terminal octapeptide) from cytoplasmically synthesized proteins that are imported into the mitochondrion. The family also includes the bacterial exopeptidase peptidyl-dipeptidase Dcp, which releases C-terminal dipeptides, and has a specificity similar to that of the animal peptidyl-dipeptidase A (Chapters 359 and 360) from family M2.

The peptidases of family M3 contain the 'HEXXH' motif, suggesting possible relationship to clan MA or to clan MB. As can be seen from the alignment (Alignment 370.1 on CD-ROM), there are several regions that resemble motifs in thermolysin, including a Glu501-$(Xaa)_6$-Glu that could correspond to the Glu-$(Xaa)_3$-Asp motif that includes the third zinc ligand in clan MA, and an Asp560-$(Xaa)_{3-4}$-His that could correspond to the Asp-$(Xaa)_{1-5}$-His motif in clan MA that includes the general base. Mutation of His472, Glu473, His476 and Glu501 eliminated catalytic activity of mitochondrial intermediate peptidase, implying thermolysin-like zinc binding, but no zinc-binding studies were made (Chew *et al.*, 1996). If a thermolysin-like fold is assumed for peptidases in family M3, then His564 would be the general base, but this residue is replaced by Tyr and Lys in oligopeptidase A from two species of *Mycoplasma*.

The catalytic mechanism in clan MB has not been completely determined, and so comparing regions of conservation with family M3 is of limited use. However, one feature conserved in clan MB is the presence of a 'Met-turn'. The absence of a conserved Met C-terminal to the HEXXH motif in family M3 makes it unlikely that thimet oligopeptidase and its homologs will possess a fold similar to that of endopeptidases in clan MB. There are also no exopeptidases in clan MB, in contrast to family M3.

The possibility thus remains that thimet oligopeptidase and its homologs have a fold that is quite unlike that of other metallopeptidases. With this in mind, and because Tyr has been implicated in the catalytic mechanisms of leukotriene $A_4$ hydrolase, astacin and serralysin, we draw attention to a region of conserved tyrosine residues in Alignment 370.1 (see CD-ROM).

An evolutionary tree showing relationships between the sequences is shown in Figure 370.1. The family can be seen as having three branches, the most divergent containing oligopeptidase F, the second containing mitochondrial intermediate peptidase, and the remainder of the family forming the third branch.

## Reference

Chew, A., Rollins, R.A., Sakati, W.R. & Isaya, G. (1996) Mutations in a putative zinc-binding domain inactivate the mitochondrial intermediate peptidase. *Biochem. Biophys. Res. Commun.* **226**, 822–829.

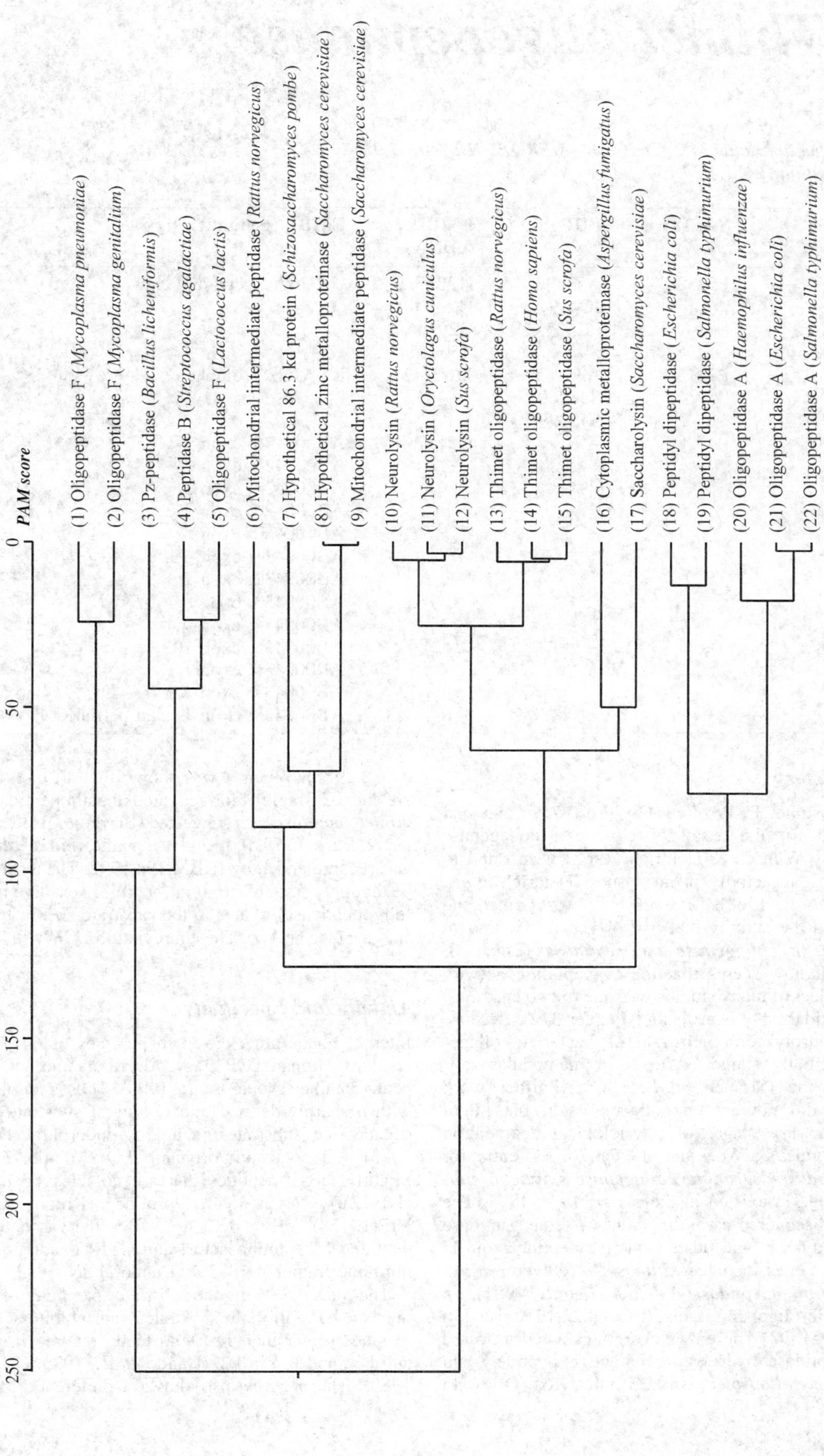

**PAM score**

(1) Oligopeptidase F (*Mycoplasma pneumoniae*)
(2) Oligopeptidase F (*Mycoplasma genitalium*)
(3) Pz-peptidase (*Bacillus licheniformis*)
(4) Peptidase B (*Streptococcus agalactiae*)
(5) Oligopeptidase F (*Lactococcus lactis*)
(6) Mitochondrial intermediate peptidase (*Rattus norvegicus*)
(7) Hypothetical 86.3 kd protein (*Schizosaccharomyces pombe*)
(8) Hypothetical zinc metalloproteinase (*Saccharomyces cerevisiae*)
(9) Mitochondrial intermediate peptidase (*Saccharomyces cerevisiae*)
(10) Neurolysin (*Rattus norvegicus*)
(11) Neurolysin (*Oryctolagus cuniculus*)
(12) Neurolysin (*Sus scrofa*)
(13) Thimet oligopeptidase (*Rattus norvegicus*)
(14) Thimet oligopeptidase (*Homo sapiens*)
(15) Thimet oligopeptidase (*Sus scrofa*)
(16) Cytoplasmic metalloproteinase (*Aspergillus fumigatus*)
(17) Saccharolysin (*Saccharomyces cerevisiae*)
(18) Peptidyl dipeptidase (*Escherichia coli*)
(19) Peptidyl dipeptidase (*Salmonella typhimurium*)
(20) Oligopeptidase A (*Haemophilus influenzae*)
(21) Oligopeptidase A (*Escherichia coli*)
(22) Oligopeptidase A (*Salmonella typhimurium*)

*Figure 370.1*   Evolutionary tree for family M3. The tree was prepared as described in the Introduction (p. xxv).

# 371. Thimet oligopeptidase

*Peptidase classification: clan MX, family M3, MEROPS ID: M03:001*
*Enzyme classification: EC 3.4.24.15*
*Databank codes:*

| Species | SwissProt | PIR | EMBL (cDNA) | EMBL (genomic) |
|---|---|---|---|---|
| *Homo sapiens* | P52888 | A53633 JC4197 PC4053 | I12383 I12393 U29366 Z50115 | U29367: 5′ end and exon 1 |
| *Rattus norvegicus* | P24155 | A36165 S38760 S55999 | M61142 | RNAJ66: promoter |
| *Sus scrofa* | P47788 | – | D21871 | AB000426: exon 1 AB000427: exon 2 AB000428: exon 3 AB000429: exon 4 AB000430: exon 5 AB000431: exon 6 AB000432: exon 7 AB000433: exon 8 AB000434: exon 9 AB000435: exon 10 AB000436: exon 11 AB000437: exon 12 AB000438: exon 13 and partial CDS |

## Name and History

A synthetic substrate, $P_3$-Pro-Leu$+$Gly-Pro-D-Arg, that had been introduced for the assay of clostridial collagenase (Chapter 368) by Wünsch & Heidrich (1963) was found to be hydrolyzed by an activity in rat tissues (Heidrich *et al.*, 1973; Strauch *et al.*, 1968). The substrate was termed the 'Pz-peptide', and the activity that hydrolyzed it was known as *Pz-peptidase* or *collagenase-like peptidase* (Lukac & Koren, 1977). It now seems that the Pz-peptidase activity that was the subject of many studies was due to two enzymes, thimet oligopeptidase and neurolysin (Chapter 372). Because Pz-peptidase hydrolyzed a substrate of bacterial collagenase, it was initially suggested to be a mammalian collagenolytic enzyme (Strauch *et al.*, 1968). Further work showed that it did not act on collagen itself, but might contribute to the late stages of physiological degradation of collagen (Morales & Woessner, 1977). An EC entry for *tissue endopeptidase-degrading collagenase synthetic substrate* (EC 3.4.99.31) existed in the period 1978–1992. The enzyme was rediscovered as an apparent cysteine peptidase hydrolyzing bradykinin and other bioactive peptides, and in this connection names included *kininase A* (Oliveira *et al.*, 1976) and *endo-oligopeptidase A* (Coelho *et al.*, 1981); an EC entry for endo-oligopeptidase A (EC 3.4.22.19) existed in the period 1989–1992. The enzyme was again rediscovered as a metallopeptidase hydrolyzing bioactive peptides, and termed *soluble metalloproteinase* (EC 3.4.24.15) (Orlowski

*et al.*, 1983). When it was shown that all of these activities are due to a single, thiol-activated, metallopeptidase, the name *thimet peptidase* was suggested (Barrett & Brown, 1990; Tisljar & Barrett, 1990a), and the name *thimet oligopeptidase* was recommended by IUBMB in 1992. The term 'thimet' can be thought of as an acronym for '**thi**ol-sensitive **met**allo', and 'oligopeptidase' relates to the substrate-size restriction of the enzyme (see below). The abbreviation *TOP* will be used here.

## Activity and Specificity

Internal bonds are cleaved in peptides of 6–17 amino acid residues. Human TOP cleaved hexa-alanine, but not tetra- or penta-alanine (Dando *et al.*, 1993). TOP is an oligopeptidase with a sharply-defined upper limit of substrate size. Heptadecapeptide substrates include $\gamma$-endorphin (Dando *et al.*, 1993) and nociceptin/orphanin FQ (Montiel *et al.*, 1997). Slightly larger peptides that are not cleaved include (Gly-Hyp-Leu)$_6$ (Hyp: hydroxyproline) (Knight *et al.*, 1995), VIP(10–28), VIP and CLIP (VIP is the vasoactive intestinal peptide of 28 amino acids, and CLIP is adrenocorticotropic hormone fragment 18–39) (Dando *et al.*, 1993).

The rules governing the substrate specificity of TOP remain unclear, but with both substrates and inhibitors, hydrophobic residues (other than Ile) tend to be favored in positions P1 and P3′, and Pro in P2′ (Dando *et al.*, 1993; Orlowski *et al.*, 1988). The enzyme also shows a preference for cleaving a bond three to six residues from the C-terminus (Knight & Barrett, 1991; Knight *et al.*, 1995). Examples of cleavage positions are, in bradykinin, Arg-Pro-Pro-Gly-Phe+Ser-Pro-Phe-Arg, and in neurotensin, Glp-Leu-Tyr-Glu-Asn-Lys-Pro-Arg+Arg-Pro-Tyr-Ile-Leu.

Maximal activity of TOP is seen at about pH 7.8 (Barrett *et al.*, 1995). Activity is stimulated by low concentrations of thiol compounds (e.g. 0.1 mM DTT, 2 mM 2-ME), but inhibited at higher concentrations (Barrett & Brown, 1990). Evidence has been put forward that the low activity of TOP under oxidizing conditions is due to disulfide-mediated formation of inactive oligomers of the enzyme, and that the reducing effect of a low concentration of thiol causes dissociation of these oligomers to the active monomer (Shrimpton *et al.*, 1997). The inhibition of TOP at high concentrations of thiol compounds is attributable to binding of the thiol to the catalytic zinc ion, and is not seen with the manganese form of the enzyme (Barrett & Brown, 1990).

A spectrophotometric assay for TOP activity has been described (Orlowski *et al.*, 1988), but assays are more conveniently made with the quenched fluorescence substrate, QF02, Mcc-Pro-Leu+Gly-Pro-D-Lys(Dnp). This substrate, discovered by Tisljar *et al.* (1990), permits continuous assays, and is available from Calbiochem/Novabiochem (see Appendix 2 for full names and addresses of suppliers). The assay methods are fully described by Barrett *et al.* (1995). The substrates available for TOP are also hydrolyzed by neurolysin, so specific assays depend upon the inclusion of selective inhibitors such as antibodies and Pro-Ile (Serizawa *et al.*, 1995).

TOP is inhibited by the chelating agent 1,10-phenanthroline, but (especially from some species) is not readily inhibited by EDTA, and for a time this obscured the metallopeptidase character of the enzyme, leading to controversy. Potent inhibition (low nanomolar $K_i$ values) is shown by Cpp-Ala-Ala-Phe-pAb and Cpp-Ala-Pro-Tyr-pAb (Knight & Barrett, 1991; Orlowski *et al.*, 1988), as well as by *N*-[(2*R*,4*R*)-2-(2-hydroxyphenyl)-3-(3-mercaptopropionyl)-4-thiazolidinecarbonyl]-L-phenylalanine (SA898) (Ukai *et al.*, 1996). Combinatorial chemistry has been used to develop exceptionally powerful and selective inhibitors, of general structure Z-(L,D)Phe-$\psi$(PO$_2$CH$_2$)(L,D)Ala-Arg-Xaa. The most effective of these ($K_i$ 70 pM) has Xaa = Met, whereas the most selective compound for TOP over neurolysin contains Xaa = Phe, and exhibits a $K_i$ value of 0.160 nM, three orders of magnitude less than for neurolysin (Jirácek *et al.*, 1995).

Amongst compounds that might occur physiologically, TOP is strongly inhibited by dynorphin A(1–13) (Dando *et al.*, 1993), but is not affected by $\alpha_2$-macroglobulin. The failure to cleave the bait region of the macroglobulin, which is sensitive to the action of the great majority of endopeptidases of all catalytic types (Barrett & Starkey, 1973), is a clear demonstration of the restricted, oligopeptidase specificity of TOP.

## Structural Chemistry

TOP is a single-chain protein of about 78.5 kDa, pI 5.1, which does not contain disulfide bonds (Barrett *et al.*, 1995). The recombinant enzyme contained about two atoms of zinc per molecule (McKie *et al.*, 1995), and there are two residues

of histidine in the sequence -**His**-Glu-Phe-Gly-**His**- that are almost certainly zinc ligands (see also Chapter 370).

There are no indications of post-translational modifications in the biosynthesis of TOP, or of a proteolytically activatable proenzyme. Nor is there any obvious hydrophobic segment that might serve as a signal peptide or a membrane-spanning domain.

Directed mutagenesis has been used to identify amino acid residues important for the catalytic activity of TOP (J.-M. Chen & A.J. Barrett, unpublished results). PCR-based site-directed mutagenesis was used, and the recombinant proteins were expressed in *E. coli*. First, conservative replacements were made in the HEXGH (472–476) sequence predicted by analogy with thermolysin to form part of the catalytic site. The mutants His472Arg, Glu473Gln and His476Arg were catalytically inactive, consistent with the prediction that His472 and His476 are zinc ligands, and Glu473 is catalytic. Candidates for the expected third zinc ligand were Asp, Glu and His residues conserved in an alignment of the sequences: Asp413, His479, Glu486, Asp498, Glu501, Glu508, His523, Asp560 and His564. The recombinant TOP from mutants Glu486Asp, Glu508Gln, His523Gln and His564Ala remained active, but mutants His479Asp, Glu501Ala, and His564Arg were inactive, suggesting that these residues are crucial either for the catalytic activity of the enzyme or for maintaining the conformation of the molecule (the latter being more likely for His523, since His523Ala had been found to be active). Binding of $^{65}$Zn was used in an attempt to identify the third zinc ligand of TOP. As predicted, mutants His472Arg and His476Arg did not bind zinc, whereas Glu473Gln, although inactive, did bind the metal. Several of the other inactive mutants of TOP including His479Asp, Glu501Ala and His564Arg were found to bind zinc, eliminating these residues as candidates for the third ligand. We hypothesize that one of the two conserved Asp residues, Asp498 and Asp560, may represent the third ligand. Cys482 was shown not to be responsible for the thiol dependence of TOP, contrary to a previous suggestion, since Cys482Ala and Cys482Ser mutants retained thiol dependence.

## Preparation

Natural sources that have yielded essentially homogeneous preparations of TOP (with the corresponding purification factors) include chicken embryo (400-fold) (Morales & Woessner, 1977), chicken liver (1600-fold) (Barrett & Brown, 1990), rat testis (270-fold) (Orlowski *et al.*, 1989; Dando *et al.*, 1993), and human erythrocytes (40 000-fold) (Dando *et al.*, 1993). The recombinant rat testis enzyme has also been expressed in *Escherichia coli* as a soluble protein (McKie *et al.*, 1995; Pierotti *et al.*, 1990). TOP is separated from neurolysin in column chromatography on hydroxyapatite (Checler *et al.*, 1995; Serizawa *et al.*, 1995) and also on DEAE-cellulose (Serizawa *et al.*, 1997).

## Biological Aspects

TOP is present in birds and mammals, and almost certainly in lower animals (Tisljar, 1993). It is present in cells of most animal tissues, but is particularly abundant in mature

**M**

testis (Orlowski *et al.*, 1989). In tissue homogenates, the activity is predominantly not sedimentable, and the enzyme is generally thought to be a component of the soluble phase of the cytoplasm. A minor fraction (5–20%) is reported to be membrane associated, however (Acker *et al.*, 1987; Molineaux & Ayala, 1990), and an immunofluorescence study has shown TOP in endosome-like organelles (Chen *et al.*, 1995). In biochemical work on the intracellular localization of TOP it is important to be aware of possible interference by neurolysin (Chapter 372) (Serizawa *et al.*, 1995).

The gene locus of human TOP is 19q13.3 (Meckelein *et al.*, 1996). Sequence data are available for the pig and human genes for TOP (see the Databanks table).

As Pz-peptidase or collagenase-like peptidase, TOP was thought to be involved in the physiological catabolism of collagen, but any such role now seems likely to be minor (Barrett *et al.*, 1995; Knight *et al.*, 1995). In the test-tube, TOP efficiently hydrolyzes many peptides that have important biological activities, and experiments with the inhibitor Cpp-Ala-Ala-Phe-pAb initially suggested that TOP acts in this way *in vivo* also. Peptides studied in these experiments included enkephalin-containing peptides, luliberin and bradykinin (Lasdun *et al.*, 1989; Molineaux & Ayala, 1990; Schriefer & Molineaux, 1993). However, it has been found that neprilysin (Chapter 362) cleaves the -Ala┼Phe- bond in the inhibitor, to generate Cpp-Ala-Ala, which is a potent inhibitor of peptidyl-dipeptidase A (Chapter 359), and it may well be that some of the effects obtained were due to such indirect inhibition of peptidyl-dipeptidase A (Cardozo & Orlowski, 1993; Lew *et al.*, 1996; Yang *et al.*, 1994). Experiments with another inhibitor have given results consistent with a role for TOP in the inactivation of nociceptin/orphanin FQ (Noble & Roques, 1997). It has also been suggested that the abundant TOP in testis may function to destroy luliberin and maintain an appropriate hormonal balance (Pierotti *et al.*, 1991).

It has been proposed that TOP may contribute to the $\beta$-secretase activity thought to give rise to the amyloidogenic $\beta$-amyloid peptide in Alzheimer's disease, although such activity would require that the enzyme make an endopeptidase cleavage in a large protein, i.e. would not be an oligopeptidase. The proposals have been based on the ability of TOP to cleave oligopeptide substrates mimicking the site hydrolyzed by the $\beta$-secretase, such as Ac-Glu-Val-Lys-Met┼Asp-Ala-Glu-Phe-NH$_2$, and on the fact that the locus of the human TOP gene is within the linkage region for late-onset Alzheimer's disease (McDermott *et al.*, 1992; Papastoitsis *et al.*, 1994; Thompson *et al.*, 1995). However, the results of more recent investigations indicate that TOP is distinct from $\beta$-secretase (Marks *et al.*, 1995; Brown *et al.*, 1996; Chevallier *et al.*, 1997). Low activity in the Alzheimer's brain has also been suggested to lead to a decreased half-life of substance P in the affected brain (Waters & Davis, 1995).

TOP is mostly located inside cells, not being secreted or presented on the plasma membrane to any great extent. If the indications of endosomal localization mentioned above are confirmed, the intracellular enzyme may still act on exogenous substrates, but there are also many potential substrates formed within cells, and the thiol dependence of TOP would suit the enzyme to acting in the cytoplasmic environment. Intracellular substrates would include the oligopeptide products of action of the 20S and 26S proteasomes (Chapters 172 and 173) amongst which are the antigenic peptides of the MHC class I system.

The biological activity of TOP seems to be essentially unregulated. The protein is synthesized in fully active form, and there is little evidence of natural inhibitors, apart from possibly dynorphin A(1–13).

## Distinguishing Features

In birds and mammals, any enzyme hydrolyzing QF02, and inhibited by Cpp-Ala-Ala-Phe-pAb, is likely to be either TOP or neurolysin (Chapter 372). Neurolysin differs in its point of hydrolysis of neurotensin (Glp-Leu-Tyr-Glu-Asn-Lys-Pro-Arg-Arg-Pro┼Tyr-Ile-Leu), in lack of thiol activation, and in its predominantly mitochondrial localization in most tissues (Serizawa *et al.*, 1995). Inhibitors can be used to distinguish the two activities, as described above.

Polyclonal antisera against TOP from various species have been described (Toffoletto *et al.*, 1988; Orlowski *et al.*, 1989; Dando *et al.*, 1993), but are not commercially available at the time of writing.

## Further Reading

For further reading see Barrett *et al.* (1995).

## References

Acker, G.R., Molineaux, C. & Orlowski, M. (1987) Synaptosomal membrane-bound form of endopeptidase-24.15 generates Leu-enkephalin from dynorphin$_{1-8}$, $\alpha$- and $\beta$-neoendorphin, and Met-enkephalin from Met-enkephalin-Arg$^6$-Gly$^7$-Leu$^8$. *J. Neurochem.* **48**, 284–292.

Barrett, A.J. & Brown, M.A. (1990) Chicken liver Pz-peptidase, a thiol-dependent metallo-endopeptidase. *Biochem. J.* **271**, 701–706.

Barrett, A.J. & Starkey, P.M. (1973) The interaction of $\alpha_2$-macroglobulin with proteinases: characteristics and specificity of the reaction and a hypothesis concerning its molecular mechanism. *Biochem. J.* **133**, 709–724.

Barrett, A.J., Brown, M.A., Dando, P.M., Knight, C.G., McKie, N., Rawlings, N.D. & Serizawa, A. (1995) Thimet oligopeptidase and oligopeptidase M. *Methods Enzymol.* **248**, 529–556.

Brown, A.M., Tummolo, D.M., Spruyt, M.A., Jacobsen, J.S. & Sonnenberg-Reines, J. (1996) Evaluation of cathepsins D and G and EC 3.4.24.15 as candidate $\beta$-secretase proteases using peptide and amyloid precursor protein substrates. *J. Neurochem.* **66**, 2436–2445.

Cardozo, C. & Orlowski, M. (1993) Evidence that enzymatic conversion of N-[1(R,S)-carboxy-3-phenylpropyl]-Ala-Ala-Phe-p-aminobenzoate, a specific inhibitor of endopeptidase 24.15, to N-[1(R,S)-carboxy-3-phenylpropyl]-Ala-Ala is necessary for inhibition of angiotensin converting enzyme. *Peptides* **14**, 1259–1262.

Checler, F., Barelli, H., Dauch, P., Dive, V., Vincent, B. & Vincent, J.P. (1995) Neurolysin: purification and assays. *Methods Enzymol.* **248**, 593–614.

Chen, J.-M., Changco, A., Brown, M.A. & Barrett, A.J. (1995) Immunolocalization of thimet oligopeptidase in chicken embryonic fibroblasts. *Exp. Cell Res.* **216**, 80–85.

Chevallier, N., Jirácek, J., Vincent, B., Baur, C.P., Spillantini, M.G., Goedert, M., Dive, V. & Checler, F. (1997) Examination of the role of endopeptidase 3.4.24.15 in Ab secretion by human transfected cells. *Br. J. Pharmacol.* **121**, 556–562.

Coelho, H.L.L., Cicilini, M.A., Carvalho, K.M., Carvalho, I.F. & Camargo, A.C.M. (1981) Inhibition of rabbit tissue kininase by anti-(endo-oligopeptidase A) antibodies. *Biochem. J.* **197**, 85–93.

Dando, P.M., Brown, M.A. & Barrett, A.J. (1993) Human thimet oligopeptidase. *Biochem. J.* **294**, 451–457.

Heidrich, H.-G., Kronschnabl, O. & Hannig, K. (1973) Eine Endopeptidase aus der Matrix von Rattenlebermitochondrien. Isolierung, Reinigung und Charakterisierung des Enzymes. [An endopeptidase from the matrix of rat liver mitochondria. Isolation, purification and characterization of the enzyme.] *Z. Physiol. Chem.* **354**, 1399–1404.

Jirácek, J., Yiotakis, A., Vincent, B., Lecoq, A., Nicolaou, A., Checler, F. & Dive, V. (1995) Development of highly potent and selective phosphinic peptide inhibitors of zinc endopeptidase 24–15 using combinatorial chemistry. *J. Biol. Chem.* **270**, 21701–21706.

Knight, C.G. & Barrett, A.J. (1991) Structure/function relationships in the inhibition of thimet oligopeptidase by carboxyphenylpropylpeptides. *FEBS Lett.* **294**, 183–186.

Knight, C.G., Dando, P.M. & Barrett, A.J. (1995) Thimet oligopeptidase specificity: evidence of preferential cleavage near the C-terminus and product inhibition from kinetic analysis of peptide hydrolysis. *Biochem. J.* **308**, 145–150.

Lasdun, A., Reznik, S., Molineaux, C.J. & Orlowski, M. (1989) Inhibition of endopeptidase 24.15 slows the *in vivo* degradation of luteinizing hormone-releasing hormone. *J. Pharmacol. Exp. Ther.* **251**, 439–447.

Lew, R.A., Tomoda, F., Evans, R.G., Lakat, L., Boublik, J.H., Pipolo, L.A. & Smith, A.I. (1996) Synthetic inhibitors of endopeptidase EC 3.4.24.15: potency and stability *in vitro* and *in vivo*. *Br. J. Pharmacol.* **118**, 1269–1277.

Lukač, J. & Koren, E. (1977) The metalloenzymic nature of collagenase-like peptidase of the rat testis. *J. Reprod. Fertil.* **49**, 95–99.

Marks, N., Berg, M.J., Sapirstein, V.S., Durrie, R., Swistok, J., Makofske, R.C. & Danho, W. (1995) Brain cathepsin B but not metalloendopeptidases degrade rAPP[751] with production of amyloidogenic fragments – comparison with synthetic peptides emulating β- and γ-secretase sites. *Int. J. Pept. Protein Res.* **46**, 306–313.

McDermott, J.R., Biggins, J.A. & Gibson, A.M. (1992) Human brain peptidase activity with the specificity to generate the N-terminus of the Alzheimer β-amyloid protein from its precursor. *Biochem. Biophys. Res. Commun.* **185**, 746–752.

McKie, N., Dando, P.M., Brown, M.A. & Barrett, A.J. (1995) Rat thimet oligopeptidase: large-scale expression in *Escherichia coli* and characterization of the recombinant enzyme. *Biochem. J.* **309**, 203–207.

Meckelein, B., de Silva, H.A.R., Roses, A.D., Rao, P.N., Pettenati, M.J., Xu, P.T., Hodge, R., Glucksman, M.J. & Abraham, C.R. (1996) Human endopeptidase (THOP1) is localized on chromosome 19 within the linkage region for the late-onset Alzheimer disease AD2 locus. *Genomics* **31**, 246–249.

Molineaux, C.J. & Ayala, J.M. (1990) An inhibitor of endopeptidase-24.15 blocks the degradation of intraventricularly administered dynorphins. *J. Neurochem.* **55**, 611–618.

Montiel, J.L., Cornille, F., Roques, B.P. & Noble, F. (1997) Nociceptin/orphanin FQ metabolism: role of aminopeptidase and endopeptidase 24.15. *J. Neurochemistry* **68**, 354–361.

Morales, T.I. & Woessner, J.F., Jr. (1977) PZ-peptidase from chick embryos. Purification, properties, and action on collagen peptides. *J. Biol. Chem.* **252**, 4855–4860.

Noble, F. & Roques, B.P. (1997) Association of aminopeptidase N and endopeptidase 24.15 inhibitors potentiate behavioral effects mediated by nociceptin/orphanin FQ in mice. *FEBS Lett.* **401**, 227–229.

Oliveira, E.B., Martins, A.R. & Camargo, A.C.M. (1976) Isolation of brain endopeptidases: influence of size and sequence of substrates structurally related to bradykinin. *Biochemistry* **15**, 1967–1974.

Orlowski, M., Michaud, C. & Chu, T.G. (1983) A soluble metalloendopeptidase from rat brain. Purification of the enzyme and determination of specificity with synthetic and natural peptides. *Eur. J. Biochem.* **135**, 81–88.

Orlowski, M., Michaud, C. & Molineaux, C.J. (1988) Substrate-related potent inhibitors of brain metalloendopeptidase. *Biochemistry* **27**, 597–602.

Orlowski, M., Reznik, S., Ayala, J. & Pierotti, A.R. (1989) Endopeptidase 24.15 from rat testes. Isolation of the enzyme and its specificity toward synthetic and natural peptides, including enkephalin-containing peptides. *Biochem. J.* **261**, 951–958.

Papastoitsis, G., Siman, R., Scott, R. & Abraham, C.R. (1994) Identification of a metalloprotease from Alzheimer's disease brain able to degrade the β-amyloid precursor protein and generate amyloidogenic fragments. *Biochemistry* **33**, 192–199.

Pierotti, A., Dong, K.-W., Glucksman, M.J., Orlowski, M. & Roberts, J.L. (1990) Molecular cloning and primary structure of rat testes metalloendopeptidase EC 3.4.24.15. *Biochemistry* **29**, 10323–10329.

Pierotti, A.R., Lasdun, A., Ayala, J.M., Roberts, J.L. & Molineaux, C.J. (1991) Endopeptidase-24.15 in rat hypothalamic/pituitary/gonadal axis. *Mol. Cell. Endocrinol.* **76**, 95–103.

Schriefer, J.A. & Molineaux, C.J. (1993) Modulatory effect of endopeptidase inhibitors on bradykinin-induced contraction of rat uterus. *J. Pharmacol. Exp. Ther.* **266**, 700–706.

Serizawa, A., Dando, P.M. & Barrett, A.J. (1995) Characterization of a mitochondrial metallopeptidase reveals neurolysin as a homologue of thimet oligopeptidase. *J. Biol. Chem.* **270**, 2092–2098.

Serizawa, A., Dando, P.M. & Barrett, A.J. (1997) Oligopeptidase M (neurolysin). Targeting to mitochondria and cytosol in rat tissues. In: *Proteolysis in Cell Functions* (Hopsu-Havu, V.K., Järvinen, M. & Kirschke, H., eds). Amsterdam: IOS Press, pp. 248–255.

Shrimpton, C.N., Glucksman, M.J., Lew, R.A., Tullai, J.W., Margulies, E.H., Roberts, J.L. & Smith, A.I. (1997) Thiol activation of endopeptidase EC 3.4.24.15 – a novel mechanism for the regulation of catalytic activity. *J. Biol. Chem.* **272**, 17395–17399.

Strauch, L., Vencelj, H. & Hannig, K. (1968) Kollagenase in Zellen höher entwickelter Tiere. [Collagenase in cells of higher animals.] *Z. Physiol. Chem.* **349**, 171–178.

Thompson, A., Huber, G. & Malherbe, P. (1995) Cloning and functional expression of a metalloendopeptidase from human brain with the ability to cleave a β-APP substrate peptide. *Biochem. Biophys. Res. Commun.* **213**, 66–73.

Tisljar, U. (1993) Thimet oligopeptidase – a review of a thiol dependent metallo-endopeptidase also known as Pz-peptidase endopeptidase 24.15 and endo-oligopeptidase. *Biol. Chem. Hoppe-Seyler* **374**, 91–100.

Tisljar, U. & Barrett, A.J. (1990) Thiol-dependent metallo-endopeptidase characteristics of Pz-peptidase in rat and rabbit. *Biochem. J.* **267**, 531–533.

Tisljar, U., Knight, C.G. & Barrett, A.J. (1990) An alternative quenched fluorescence substrate for Pz-peptidase. *Anal. Biochem.* **186**, 112–115.

**M**

Toffoletto, O., Metters, K.M., Oliveira, E.B., Camargo, A.C.M. & Rossier, J. (1988) Enkephalin is liberated from metorphamide and dynorphin A₁₋₈ by endo-oligopeptidase A, but not by metallo-endopeptidase EC 3.4.24.15. *Biochem. J.* **252**, 35–38.

Ukai, Y., Li, Q.L., Ito, S. & Mita, S. (1996) A novel synthetic inhibitor of endopeptidase-24.15. *J. Enzym. Inhib.* **11**, 39–49.

Waters, S.M. & Davis, T.P. (1995) Alterations of substance P metabolism and neuropeptidases in Alzheimer's disease. *J.*

*Gerontol. [A]* **50A**, B315–B319.

Wünsch, E. & Heidrich, H.G. (1963) Zur quantitativen Bestimmung der Kollagenase. [On the quantitative determination of collagenase.] *Z. Physiol. Chem.* **333**, 149–151.

Yang, X.-P., Saitoh, S., Scicli, A.G., Mascha, E., Orlowski, M. & Carretero, O.A. (1994) Effects of a metalloendopeptidase-24.15 inhibitor on renal hemodynamics and function in rats. *Hypertension* **23**(suppl.), I235–I239.

**Alan J. Barrett**
*MRC Peptidase Laboratory,*
*The Babraham Institute,*
*Cambridge CB2 4AT, UK*
*Email: alan.barrett@bbsrc.ac.uk*

**Jinq-May Chen**
*MRC Peptidase Laboratory,*
*The Babraham Institute,*
*Cambridge CB2 4AT, UK*

# 372. Neurolysin

*Peptidase classification: clan MX, family M3, MEROPS ID: M03.002*
*NC-IUBMB enzyme classification: EC 3.4.24.16*
*Chemical Abstracts Service registry number: 149371-24-4*
*Databank codes:*

| Species | SwissProt | PIR | EMBL (cDNA) | EMBL (genomic) |
|---|---|---|---|---|
| *Oryctolagus cuniculus* | P42675 | A45985 | D13310 | – |
| *Rattus norvegicus* | P42676 | – | X87157 | – |
| *Sus scrofa* | Q02038 | A43411 | D11336 | – |

## Name and History

The name **neurolysin** recommended in the IUBMB enzyme nomenclature is reminiscent of the fact that much of the research on this enzyme has been motivated by interest in its capacity to hydrolyze neurotensin. Some scientists have chosen to use a term based on the EC number, **endopeptidase 24.16** (Kato *et al.*, 1997), and several other names have been used during the long history of the enzyme.

In the 1960s, a synthetic substrate, P₃-Pro-Leu┤Gly-Pro-D-Arg, that had been introduced for the assay of clostridial collagenase (Chapter 368) by Wünsch & Heidrich (1963), was found to be hydrolyzed by an activity in rat tissues (Heidrich *et al.*, 1973; Strauch *et al.*, 1968). The substrate was termed the 'Pz-peptide', and the activity that hydrolyzed it was known as **Pz-peptidase** or **collagenase-like peptidase** (Lukac & Koren, 1977). It now seems that the Pz-peptidase activity that was the subject of many studies was due to two enzymes, thimet oligopeptidase (Chapter 371) and neurolysin. Because Pz-peptidase hydrolyzed a substrate of bacte-

rial collagenase, it was initially suggested to be a mammalian collagenolytic enzyme (Strauch *et al.*, 1968). Further work showed that it did not act on collagen itself, but might contribute to the late stages of physiological degradation of collagen (Morales & Woessner, 1977). An EC entry for **tissue endopeptidase-degrading collagenase synthetic substrate** (EC 3.4.99.31) existed in the period 1978–1992. In many of the studies on Pz-peptidase, it is difficult to say how much of the activity was due to thimet oligopeptidase and how much to neurolysin, but a mitochondrial form of Pz-peptidase was detected that was clearly neurolysin. The study in which Heidrich *et al.* (1969) detected Pz-peptidase activity in highly purified rat liver mitochondria was followed up by the study of Tisljar & Barrett (1990) that showed the enzyme to be similar to thimet oligopeptidase, but not identical. This in turn led to the definitive demonstration that neurolysin is predominantly mitochondrial in liver of pig and rat (Serizawa *et al.*, 1995), and also in other tissues (P.M. Dando, unpublished results). The enzyme obtained from mitochondria was first termed **thimet oligopeptidase II**

(Tisljar & Barrett, 1990), and then *oligopeptidase M* (Serizawa *et al*., 1995), prior to its conclusive identification as neurolysin.

The N-terminal amino acid sequence and recognition by a monoclonal antibody also showed oligopeptidase M to be identical to a *soluble angiotensin II-binding protein* that had been described by Soffer and coworkers (Soffer *et al*., 1991). The complete amino acid sequence of the pig soluble angiotensin II-binding protein was already known, and this was so similar to that of a rat enzyme that had been termed *microsomal endopeptidase* (Kawabata *et al*., 1993) as to leave no doubt that again these seemingly distinct proteins are species variants of neurolysin. The cloning and sequencing of the authentic rat neurolysin (Dauch *et al*., 1995) finally confirmed this chain of identities.

## Activity and Specificity

All known substrates of neurolysin contain 17 or fewer amino acids (as is also the case with thimet oligopeptidase). The hydrolysis of dynorphin A(1–17) is of particular interest, however, because this peptide is an inhibitor and not a substrate of thimet oligopeptidase. The activity of neurolysin is also readily distinguished from that of thimet oligopeptidase with neurotensin as substrate, in that neurolysin cleaves the peptide solely at the Pro-Tyr bond (Vincent *et al*., 1996). There is no evidence that neurolysin has activity on protein substrates, as has occasionally been suggested (Kawabata & Davie, 1992).

Neurolysin is not appreciably activated by the low concentrations of thiol compounds that enhance the activity of thimet oligopeptidase, but is inhibited by EDTA and 1,10-phenanthroline, like other metallopeptidases. The dipeptide Pro-Ile is a readily available inhibitor that serves to distinguish neurolysin from thimet oligopeptidase and immunoinhibition also has been used to render assays selective for neurolysin (Serizawa *et al*., 1995).

Several potent and selective inhibitors of neurolysin have been described. These (with the respective $K_i$ values and ratios of $K_i$ for thimet oligopeptidase/neurolysin) include Pro-Phe-$\psi$(PO$_2$CH$_2$)Gly-Pro (4 nM, 2000) (Jirácek *et al*., 1996), Pro-Phe-$\psi$(PO$_2$CH$_2$)Leu-Pro-NH$_2$ (12 nM, 5540) (Vincent *et al*., 1997) and HONHCO-CH$_2$-CH(CH$_2$-C$_6$H$_5$)CO-Ile-Ala (12 nM, 800) (Bourdel *et al*., 1996).

## Structural Chemistry

The complete amino acid sequence of the pig soluble angiotensin II-binding protein was already known when Serizawa *et al*. (1995) showed that the angiotensin-binding protein is identical with neurolysin, and this amounted to the discovery of the primary structure of pig neurolysin. The sequence is closely related to that of thimet oligopeptidase, so that the enzymes belong together in family M3. At the same time, it became clear that the sequence of rabbit neurolysin had already been determined in work on an enzyme thought to be a microsomal endopeptidase (Kawabata *et al*., 1993). The molecular cloning and sequencing of rat brain neurolysin were described by Dauch *et al*. (Dauch *et al*., 1995).

## Preparation

Neurolysin has been purified from a number of sources, including rat kidney (Barelli *et al*., 1993), human brain (Vincent *et al*., 1996) and rat liver mitochondria (Barrett *et al*., 1995). Care is needed in separating the enzyme from thimet oligopeptidase, but the activities are usually well resolved by chromatography on hydroxyapatite and DEAE-cellulose (Checler *et al*., 1995; Serizawa *et al*., 1997).

## Biological Aspects

More contrasting subcellular localizations have been described for neurolysin than for most cellular peptidases. Results obtained in the authors' laboratory show that in tissues of rat and pig, at least, the enzyme is predominantly mitochondrial and cytosolic, with the ratio between these two compartments varying with tissue and species (Table 372.1). A report that neurolysin (as 'microsomal metalloendopeptidase' ) might be located in the endoplasmic reticulum, and contribute to the post-translational processing of secreted proteins (Kawabata & Davie, 1992) has been revised (Nakagawa *et al*., 1997). The belief that neurolysin contributes significantly to the physiological destruction of neurotensin and perhaps other bioactive peptides has led to emphasis

*Table 372.1* Subcellular distribution of neurolysin in tissues of rat and pig

| Tissue | Rat | | | Pig | | |
|--------|--------|--------------|-------|---------|--------------|-------|
|        | Cytosol | Mitochondria | Ratio | Cytosol | Mitochondria | Ratio |
| Brain  | 0.30 | 0.51 | 0.59 | 0.17 | 0.34 | 0.50 |
| Heart  | 0.12 | 0.14 | 0.86 | n/d | n/d | n/d |
| Kidney | 0.28 | 0.48 | 0.59 | n/d | n/d | n/d |
| Liver  | 0.48 | 1.13 | 0.42 | n/d | n/d | n/d |
| Lung   | 0.16 | 0.48 | 0.34 | n/d | n/d | n/d |
| Spleen | 0.38 | 1.72 | 0.22 | 0.14 | 0.15 | 0.93 |
| Testis | 0.00 | 0.90 | <0.01 | 0.23 | 0.25 | 0.89 |

The cytosol and mitochondria fractions were prepared as described by Serizawa *et al*. (1995). Values are in milliunits of activity per mg protein, and the ratio shown is the value for cytosol divided by that for mitochondria.

on the evidence that a fraction of neurolysin may have an extracellular location. The data in support of this have been presented by Vincent *et al.* (1997) who found that an inhibitor of neurolysin protected exogenous neurotensin from degradation in rat brain, and also potentiated the neurotensin-induced antinociception of mice.

The targeting of neurolysin to cytosol and to mitochondria is apparently due to the existence of two forms of the protein that differ only in the length of the N-terminus. The longer form contains a cleavable mitochondrial targeting sequence (first recognized by Serizawa *et al.*, 1995) at the N-terminus, and is directed to mitochondria, whereas the shorter form, lacking the targeting sequence, remains in the cytosol. Two explanations have been put forward for the existence of the two forms of the protein. Kato *et al.* (1997) consider that multiple mRNAs are produced by alternative promoter usage, and mRNA splicing. In contrast, Serizawa *et al.* (1997) were unable to detect multiple mRNAs, and concluded that the two forms of the protein probably arise from the use of alternative initiation codons.

Despite the evidence for a role of neurolysin in the physiological degradation of neurotensin, the bulk of the neurolysin in most tissues, located in mitochondria and cytosol, is not well placed to mediate the degradation of extracellular peptides, and most likely contributes to the hydrolysis of oligopeptides that are intermediates in the turnover of cellular proteins.

## Distinguishing Features

As is described above, neurolysin is most likely to be confused with thimet oligopeptidase, but it is distinguished by its lack of activation by thiol compounds, inhibition by millimolar concentrations of Pro-Ile, cleavage of neurotensin at the Pro-Tyr bond, and different mobility in column chromatography on hydroxyapatite and DEAE-cellulose. A monoclonal antibody to neurolysin has been raised in the authors' laboratory, and is available for academic use.

## Further Reading

Full reviews of neurolysin have been provided by Barrett *et al.* (1995) and Checler *et al.* (1995).

## References

Barelli, H., Vincent, J.-P. & Checler, F. (1993) Rat kidney endopeptidase 24.16. Purification, physico-chemical characteristics and differential specificity towards opiates, tachykinins and neurotensin-related peptides. *Eur. J. Biochem.* **211**, 79–90.

Barrett, A.J., Brown, M.A., Dando, P.M., Knight, C.G., McKie, N., Rawlings, N.D. & Serizawa, A. (1995) Thimet oligopeptidase and oligopeptidase M. *Methods Enzymol.* **248**, 529–556.

Bourdel, E., Doulut, S., Jarretou, G., Labbe-Jullie, C., Fehrentz, J.A., Doumbia, O., Kitabgi, P. & Martinez, J. (1996) New hydroxamate inhibitors of neurotensin-degrading enzymes – synthesis and enzyme active-site recognition. *Int. J. Pept. Protein Res.* **48**, 148–155.

Checler, F., Barelli, H., Dauch, P., Dive, V., Vincent, B. & Vincent, J.P. (1995) Neurolysin: purification and assays. *Methods Enzymol.* **248**, 593–614.

Dauch, P., Vincent, J.P. & Checler, F. (1995) Molecular cloning and expression of rat brain endopeptidase 3.4.24.16. *J. Biol. Chem.* **270**, 27266–27271.

Heidrich, H.G., Prokopova, D. & Hannig, K. (1969) The use of synthetic substrates for the determination of mammalian collagenases. Is collagenolytic activity present in mitochondria? *Z. Physiol. Chem.* **350**, 1430–1436.

Heidrich, H.-G., Kronschnabl, O. & Hannig, K. (1973) Eine Endopeptidase aus der Matrix von Rattenlebermitochondrien. Isolierung, Reinigung und Charakterisierung des Enzymes [An endopeptidase from the matrix of rat liver mitochondria. Isolation, purification and characterization of the enzyme.]. *Z. Physiol. Chem.* **354**, 1399–1404.

Jirácek, J., Yiotakis, A., Vincent, B., Checler, F. & Dive, V. (1996) Development of the first potent and selective inhibitor of the zinc endopeptidase neurolysin using a systematic approach based on combinatorial chemistry of phosphinic peptides. *J. Biol. Chem.* **271**, 19606–19611.

Kato, A., Sugiura, N., Saruta, Y., Hosoiri, T., Yasue, H. & Hirose, S. (1997) Targeting of endopeptidase 24.16 to different subcellular compartments by alternative promoter usage. *J. Biol. Chem.* **272**, 15313–15322.

Kawabata, S. & Davie, E.W. (1992) A microsomal endopeptidase from liver with substrate specificity for processing proproteins such as the vitamin K-dependent proteins of plasma. *J. Biol. Chem.* **267**, 10331–10336.

Kawabata, S., Nakagawa, K., Muta, T., Iwanaga, S. & Davie, E.W. (1993) Rabbit liver microsomal endopeptidase with substrate specificity for processing proproteins is structurally related to rat testes metalloendopeptidase 24.15. *J. Biol. Chem.* **268**, 12498–12503.

Lukač, J. & Koren, E. (1977) The metalloenzymic nature of collagenase-like peptidase of the rat testis. *J. Reprod. Fertil.* **49**, 95–99.

Morales, T.I. & Woessner, J.F., Jr. (1977) PZ-peptidase from chick embryos. Purification, properties, and action on collagen peptides. *J. Biol. Chem.* **252**, 4855–4860.

Nakagawa, K., Kawabata, S., Nakashima, Y., Iwanaga, S. & Sueishi, K. (1997) Tissue distribution and subcellular localization of rabbit liver metalloendopeptidase. *J. Histochem. Cytochem.* **45**, 41–47.

Serizawa, A., Dando, P.M. & Barrett, A.J. (1995) Characterization of a mitochondrial metallopeptidase reveals neurolysin as a homologue of thimet oligopeptidase. *J. Biol. Chem.* **270**, 2092–2098.

Serizawa, A., Dando, P.M. & Barrett, A.J. (1997) Oligopeptidase M (neurolysin). Targeting to mitochondria and cytosol in rat tissues. In: *Proteolysis in Cell Functions* (Hopsu-Havu, V.K., Järvinen, M. & Kirschke, H., eds). Amsterdam: IOS Press, pp. 248–255.

Soffer, R.L., Kiron, M.A.R., Mitra, A. & Fluharty, S.J. (1991) Soluble angiotensin II-binding protein. *Methods Neurosci.* **5**, 192–203.

Strauch, L., Vencelj, H. & Hannig, K. (1968) Kollagenase in Zellen höher entwickelter Tiere. [Collagenase in cells of higher animals.] *Z. Physiol. Chem.* **349**, 171–178.

Tisljar, U. & Barrett, A.J. (1990) A distinct thimet peptidase from rat liver mitochondria. *FEBS Lett.* **264**, 84–86.

Vincent, B., Vincent, J.P. & Checler, F. (1996) Purification and characterization of human endopeptidase 3.4.24.16. Comparison with the porcine counterpart indicates a unique cleavage site on neurotensin. *Brain Res.* **709**, 51–58.

Vincent, B., Jiracek, J., Noble, F., Loog, M., Roques, B., Dive, V.,

Vincent, J.P. & Checler, F. (1997) Effect of a novel selective and potent phosphinic peptide inhibitor of endopeptidase 3.4.24.16 on neurotensin-induced analgesia and neuronal inactivation. *Br. J.*

*Pharmacol.* **121**, 705–710.

Wünsch, E. & Heidrich, H.G. (1963) Zur quantitativen Bestimmung der Kollagenase. *Z. Physiol. Chem.* **333**, 149–151.

***Alan J. Barrett***
*MRC Peptidase Laboratory,*
*The Babraham Institute,*
*Cambridge CB2 4AT, UK*
*Email: alan.barrett@bbsrc.ac.uk*

***Pam M. Dando***
*MRC Peptidase Laboratory,*
*The Babraham Institute,*
*Cambridge CB2 4AT, UK*

# 373. *Saccharolysin*

## Databanks

*Peptidase classification: clan MX, family M3, MEROPS ID: M03.003*
*NC-IUBMB enzyme classification: EC 3.4.24.37*
*Databank codes:*

| Species | SwissProt | PIR | EMBL (cDNA) | EMBL (genomic) |
|---|---|---|---|---|
| *Saccharomyces cerevisiae* | P25375 | S19387 | X59720 X76504 | – |

## Name and History

**Saccharolysin** was discovered in a systematic search for proteolytic activities from the yeast *Saccharomyces cerevisiae* using chromogenic peptide substrates (Achstetter *et al.*, 1984). It was identified by its ability to cleave the standard substrate Bz-Pro+Phe-Arg-NHPhNO$_2$ although the release of the nitroanilide chromogen was dependent on the addition of membrane alanyl aminopeptidase (Chapter 337) (Achstetter *et al.*, 1985). Furthermore, it was shown to be a metalloproteinase. The proteinase was named **yscD** according to the nomenclature introduced by Achstetter *et al.* (1984) for *S. cerevisiae* proteinases and an EC entry as a cysteine peptidase (EC 3.4.22.22) existed in the period 1989–1992. The name *saccharolysin* (EC 3.4.24.37) was recommended by IUBMB in 1992. The cloning of the *S. cerevisiae PRD1* gene encoding saccharolysin showed that it is highly similar to mammalian thimet oligopeptidase (Chapter 371) and its ability to degrade typical peptide substrates of thimet oligopeptidase was demonstrated (Büchler *et al.*, 1994). Saccharolysin was therefore classified in the metalloproteinase M3 family of Rawlings & Barrett (1995).

## Activity and Specificity

The activity of saccharolysin has been tested on various chromogenic peptide substrates. The following peptides are cleaved: Bz-Pro+Phe-Arg-NHPhNO$_2$ ($K_m$ 60 μM, $k_{cat}$ 1.5 × 10$^4$ s$^{-1}$), Ac-Ala+Ala-Pro-Met-NHPhNO$_2$ ($K_m$ 190 μM, $k_{cat}$ 6 × 10$^4$ s$^{-1}$), MeOSuc-Ala+Ala-Pro-Met-NHPhNO$_2$ ($K_m$ 35 μM, $k_{cat}$ 3.8 × 10$^4$ s$^{-1}$), Ac-Ala+Ala-Pro-Phe-NHPhNO$_2$ ($K_m$ 85 μM, $k_{cat}$ 3.8 × 10$^4$ s$^{-1}$), Pz-Pro-Leu+Gly-Pro-D-Arg, Mcc-Pro-Leu+Gly-Pro-D-Lys(Dnp) ($K_m$ 6 μM) and Dnp-Pro-Leu+Gly-Pro-Trp-D-Lys ($K_m$ 29 μM) (Achstetter *et al.*, 1985; Büchler *et al.*, 1994). Bradykinin is cleaved at the Phe+Ser bond (Büchler *et al.*, 1994). Peptides with a free N-terminus are not cleaved (Achstetter *et al.*, 1985). The primary specificity of the enzyme is likely to depend on the presence of hydrophobic amino acids in position P1 but the nature of the N-terminal blocking group has a strong influence on the activity as do hydrophobic residues at position P3′ and Pro in P2′. The enzyme cannot hydrolyze azocasein or [$^3$H]methylcasein.

The pH optimum is about 7.9 with Bz-Pro-Phe-Arg-NHPhNO$_2$, about 5.9 with Ac-Ala-Ala-Pro-Met-NHPhNO$_2$ and about 5.5 with MeOSuc-Ala-Ala-Pro-Met-NHPhNO$_2$ (Achstetter *et al.*, 1985). The activity is inhibited by EDTA (50% at 1 mM, 90% at 10 mM) and can be efficiently reactivated by the addition of Zn$^{2+}$, Co$^{2+}$ and Mn$^{2+}$ (Achstetter *et al.*, 1985; Büchler *et al.*, 1994). The enzyme is strongly inhibited by Cpp-Ala-Ala-Phe-*p*-aminobenzoic acid, an inhibitor specially developed for thimet oligopeptidase (Chapter 371) (Orlowski *et al.*, 1988; Büchler *et al.*, 1994). Inhibition is obtained with mercurials (100% inhibition with 0.01 mM HgCl$_2$ and 0.1 mM 4-hydroxymercuribenzoate) suggesting that a sulfhydryl residue is necessary for its activity (Achstetter *et al.*, 1985). However, in contrast to thimet oligopeptidase, saccharolysin is not affected by DTT, *N*-ethylmaleimide, iodoacetic acid or iodoacetamide, suggesting that it is not thiol dependent (Büchler *et al.*, 1994).

Saccharolysin is best assayed with Bz-Pro-Phe-Arg-NHPhNO$_2$ or Ac-Ala-Ala-Pro-Met-NHPhNO$_2$ in the presence of membrane alanyl aminopeptidase (Chapter 337) as described in Achstetter *et al.* (1985). Other assays are described in Büchler *et al.* (1994).

## Structural Chemistry

Saccharolysin is a single-chain protein of 712 amino acids with a molecular mass of about 81.8 kDa and a pI of 4.75 as determined by chromatofocusing (Achstetter *et al*, 1985; Büchler *et al.*, 1994). The deduced amino acid sequence of the *PRD1* gene shows significant similarity to *Aspergillus fumigatus* MepB oligopeptidase (Chapter 374) and to mammalian thimet oligopeptidase (Chapter 371) and neurolysin (Chapter 372) (Büchler *et al.*, 1994; Ibrahim-Granet & d'Enfert, 1997). The main region of homology is in the vicinity of two amino acids motifs (**His**-Glu-Leu-Gly-**His** (HELGH) at position 501 and Asp-Phe-Val-**Glu**-Ala-Pro at position 528) that correspond to two motifs found in thermolysin (Chapter 351) and shown to contribute to the catalytic site through two zinc-binding His residues and a Glu residue (Matthews *et al.*, 1972).

## Preparation

*Saccharolysin* has been purified from stationary phase cells of the yeast *S. cerevisiae* using standard chromatographic procedures. A purification factor of more than 250-fold was obtained (Achstetter *et al.*, 1985). It can also be purified with the method of Tisljar & Barrett (1990) that was developed for the purification of thimet oligopeptidase (Chapter 371) from rat testis cytosol.

## Biological Aspects

Saccharolysin has been classified as a nonvacuolar proteinase (Emter & Wolf, 1984) and appears to be localized mostly in the cytosol. However, some activity (3–5%) purifies with the intermembrane space of mitochondria. This localization is probably due to the occurrence of a stretch of 26 predominantly hydrophobic and hydroxylated amino acid residues at the N-terminal end of the protein that may form a mitochondrial targeting signal (Büchler *et al.*, 1994).

*S. cerevisiae* mutants devoid of saccharolysin activity have been obtained following chemical mutagenesis and screening of the mutant cell for the inability to degrade the chromogenic substrate Ac-Ala-Ala-Pro-Met-NHPhNO$_2$ *in situ* (*prd1–6*; Garcia-Alvarez *et al.*, 1987) or following gene replacement at the *PRD1* locus with an inactive allele (*prd1::URA3*; Büchler *et al.*, 1994). *prd1–6* mutant cells have thermolabile saccharolysin activity while *prd1::URA3* mutant cells are devoid of the cytoplasmic and mitochondrial activities. Both mutant

strains do not show any obvious phenotype, suggesting that the enzyme is not involved in a proteolytic process that is essential in yeast (Garcia-Alvarez *et al.*, 1987; Büchler *et al.*, 1994). However, although protein degradation is not impaired in the mutant cells, the accumulation of peptides that may result from protein degradation has been observed in the *prd1::URA3* strain (Büchler *et al.*, 1994). It is therefore likely that saccharolysin is the major intracellular oligopeptidase responsible for the degradation of peptides resulting from nonvacuolar proteolysis. A similar function has been proposed for thimet oligopeptidase (Chapter 371) and related proteinases (Orlowski, 1990).

## Distinguishing Features

Saccharolysin can be distinguished from the other members of the thimet oligopeptidase family (Chapter 370) by its ability to cleave the chromogenic peptide substrate Bz-Pro-Phe-Arg-NHPhNO$_2$. In contrast to thimet oligopeptidases, it is not thiol dependent, a feature that is shared with neurolysin (Chapter 372), a close homolog of thimet oligopeptidase that has been localized in the mitochondria (Serizawa *et al*, 1995). This might be related to the low cysteine content of saccharolysin (4 residues), which contrasts with that of thimet oligopeptidase (14 residues).

## Further Reading

See the overview by Büchler *et al.* (1994).

## References

Achstetter, T., Ehmann, C., Osaki, A. & Wolf, D.H. (1984) Proteolysis in eukaryotic cells. Proteinase yscE, a new yeast peptidase. *J. Biol. Chem.* **259**, 13344–13348.

Achstetter, T., Ehmann, C. & Wolf, D.H. (1985) Proteinase yscD. Purification and characterization of a new yeast peptidase. *J. Biol. Chem.* **260**, 4585–4590.

Büchler, M., Tisljar, U. & Wolf, D.H. (1994) Proteinase yscD (oligopeptidase yscD) structure, function and relationship of the yeast enzyme with mammalian thimet oligopeptidase (metalloendopeptidase, EP 24.15). *Eur. J. Biochem.* **219**, 627–639.

Emter, O. & Wolf, D.H. (1984) Vacuoles are not the sole compartment of proteolytic enzymes in yeast. *FEBS Lett.* **166**, 321–325.

Garcia-Alvarez, N., Teichert, U. & Wolf, D.H. (1987) Proteinase yscD mutants of yeasts. Isolation and characterization. *Eur. J. Biochem.* **163**, 339–346.

Ibrahim-Granet, O. & d'Enfert, C. (1997) The *Aspergillus fumigatus mepB* gene encodes an 82 kDa intracellular metalloproteinase structurally related to mammalian thimet oligopeptidases. *Microbiology* **143**, 2247–2253.

Matthews, B.W., Jansonius, J.N., Colman, P.M., Schoenborn, B.P. & Duporque, D. (1972) Three-dimensional structure of thermolysin. *Nature New Biol.* **238**, 37–41.

Orlowski, M. (1990) The multicatalytic proteinase complex, a major extralysosomal proteolytic system. *Biochemistry* **29**, 10289–10297.

Orlowski, M., Michaud, C. & Wolf, D.H. (1988) Substrate related potent inhibitors of brain metalloendopeptidase. *Biochemistry* **27**, 597–602.

Rawlings, N.D. & Barrett, A.J. (1995) Evolutionary families of metallopeptidases. *Methods Enzymol.* **248**, 183–228.

Serizawa, A., Dando, P.M. & Barrett, A.J. (1995) Characterization of a mitochondrial metallopeptidase reveals neurolysin as a homologue of thimet oligopeptidase. *J. Biol. Chem.* **270**, 2092–2098.

Tisljar, U. & Barrett, A.J. (1990) Thiol-dependent metallo-endopeptidase characteristics of Pz-peptidase in rat and rabbit. *Biochem. J.* **267**, 531–533.

***Christophe d'Enfert***
*Laboratoire des Aspergillus, Institut Pasteur,*
*25 rue du Docteur Roux,*
*75724 Paris Cedex 15, France*

***Oumaïma Ibrahim-Granet,***
*Laboratoire des Aspergillus, Institut Pasteur,*
*25 rue du Docteur Roux,*
*75724 Paris Cedex 15, France*
*Email: ogranet@pasteur.fr*

# 374. Oligopeptidase MepB

## Databanks

*Peptidase classification: clan MX, family M3, MEROPS ID: M03.009*
*NC-IUBMB enzyme classification: none*
*Databank codes:*

| Species | SwissProt | PIR | EMBL (cDNA) | EMBL (genomic) |
|---|---|---|---|---|
| *Aspergillus fumigatus* | – | – | – | U85769: complete gene |

## Name and History

*Aspergillus fumigatus*, an airborne pathogenic fungus, is the main causative agent of pulmonary invasive aspergillosis. Diagnosis of aspergillosis relies on immunological assays for antibodies against *Aspergillus* antigens (Fisher *et al.*, 1981) or the presence of antigens in the patient's body fluid (Stynen *et al.*, 1995). *A. fumigatus* antigens may be proteinases. It is known that *A. fumigatus* produces several extracellular proteolytic enzymes of the serine, aspartic and metalloprotease classes. Kolattukudy *et al.* (1993) have suggested that an elastinolytic serine proteinase is involved in aspergillosis pathogenesis. However, genetically engineered mutants of *A. fumigatus* that are deficient for oryzin (Chapter 105) (the ALP gene product) and/or for fungalysin (Chapter 514) (the MEP gene product) behave like wild type in a mouse model of invasive aspergillosis (Jaton-Ogay *et al.*, 1994). One of the residual proteolytic activities identified in the double mutant ALP⁻MEP⁻ was attributed to an 82 kDa intracellular metalloproteinase. MepB was the name given to this first cytosolic enzyme of *A. fumigatus* (Ibrahim-Granet *et al.*, 1994). The cloning of the *mepB* gene showed that it is highly similar to mammalian thimet oligopeptidase (Chapter 371) and it was therefore classified in the M3 family of Rawlings & Barrett (1995). The suggested name is ***oligopeptidase MepB***.

## Activity and Specificity

MepB, like all the members of the thimet oligopeptidase family M3, degrades the Pz-peptide substrate Pz-Pro-Leu+Gly-Pro-D-Arg (Pz-PLGPR) (Wünsch & Heidrich, 1963). The enzyme was also reported to cleave native type I collagen (Ibrahim-Granet *et al.*, 1994). On the other hand, azocasein and azocoll were not hydrolyzed by MepB nor was FA-Leu+Gly-Pro-Ala (FALGPA) a specific substrate for bacterial collagenolytic enzymes (Van Wart & Steinbrink, 1981).

The optimum pH for degrading the Pz-PLGPR substrate is 7.1. The kinetic constants are $K_m$ $7 \times 10^{-5}$ M; $k_{cat}$ 232 min⁻¹; $k_{cat}/K_m$ $3.32 \times 10^6$ M⁻¹ min⁻¹. The most efficient inhibitor is 1,10-phenanthroline while EDTA had no effect at 1 mM and produced 74% inhibition at 100 mM. MepB is also inhibited by 1 mM DFP, β-ME and DTT (Ibrahim-Granet *et al.*, 1994).

## Structural Chemistry

MepB is a single-chain protein of 82 kDa, pI 5.6. The amino acid sequence is closely similar to that of saccharolysin (Chapter 373) (Büchler *et al.*, 1994) and to mammalian thimet oligopeptidase (Chapter 371) (Pierotti *et al.*, 1990), neurolysin (Chapter 372) (Serizawa *et al.*, 1995), oligopeptidase A (Chapter 375) and peptidyl-dipeptidase (Chapter 376) (Conlin & Miller, 1992; Conlin *et al.*, 1992). The most conserved region is found in the putative catalytic domain which contains two amino acid motifs, **His-Glu**-Leu-Gly-**His** (HELGH), where the two His residues are assumed to be zinc ligands and the Glu residue may polarize the water molecule involved in nucleophilic attack on the scissile peptide bond,

**M**

and Asp-Phe-Val-**Glu**-Ala-Pro, which contains a second Glu that may participate in zinc binding (Matthews *et al.*, 1972).

## Preparation

MepB was purified 30-fold from the aqueous extract of *A. fumigatus* by dye-binding chromatography (Ibrahim-Granet *et al.*, 1994).

## Biological Aspects

The gene structure of MepB has been determined. An open reading frame (ORF) of 2061 bp interrupted by two introns was identified. As shown by the amino acid sequence, MepB is not synthesized as a precursor. The absence of a signal peptide confirms the cytoplasmic localization. Despite the occurrence of two potential *N*-glycosylation sites in MepB, the purified protein appears not to be *N*-glycosylated since it does not bind to concanavalin A.

An *A. fumigatus* mutant lacking the MepB protein has been constructed by gene disruption. This mutant is devoid of Pz-peptidase activity, confirming that MepB is the only proteinase responsible for the Pz-peptidase activity present in the cytoplasm of *A. fumigatus*. The mutant does not show an altered phenotype. Moreover, the same mortality rate as produced by the wild type was obtained in an animal model. Since MepB is very similar to the other members of the thimet oligopeptidase family, a possible role of MepB is the intracellular degradation of small peptides resulting from protein degradation or possessing bioactivity (Büchler *et al.*, 1994).

## Distinguishing Features

MepB can be distinguished from the other members of M3 family of Rawlings & Barrett (1995) by its ability to hydrolyze *in vitro* native, as well as denatured, collagen. One clue to this property is the presence of a Ser residue very close to the catalytic site. However, the evidence that MepB is a true collagenase is not strong. Given its intracellular location and the fact that other members of the (M3) oligopeptidase family are not truly collagenolytic (Rawlings & Barrett, 1995), additional evidence will be needed to confirm that MepB can hydrolyze collagen *in vivo* and how it may act on other exogenous substrates.

Another distinctive feature of MepB is the occurrence of nonconserved amino acid stretches on both side of the putative catalytic domain. These amino acids may act as hinges between the catalytic domain and the N- and C-terminal domains. MepB, like saccharolysin (Chapter 373) (Büchler *et al.*, 1994) and in contrast to thimet oligopeptidases (Chapter 371) is not thiol dependent. This lack of thiol activation was probably related to the absence of Cys residues, located at position 246 and 253 in thimet oligopeptidases, that are strong candidates for explaining the thiol dependence. Comparison of the amino acid sequence of MepB with those of two *A. fumigatus* metalloendopeptidases, deuterolysin (Chapter 514) (Jaton-Ogay *et al.*, 1994) and fungalysin (Chapter 514) (Ramesh *et al.*, 1995) did not reveal any significant similarity.

## Related Peptidases

Several metalloproteinases with chromatographic properties similar to those of MepB have been identified in the aqueous cellular extract of different filamentous fungi. One of these proteins, purified from *Trichophyton schoenleinii*, was shown by peptide sequencing to have homology with *Aspergillus* oligopeptidase MepB and with thimet oligopeptidase (Chapter 371) as well with saccharolysin (Chapter 373) (Ibrahim-Granet *et al.*, 1996).

## Further Reading

For a review, see Ibrahim-Granet *et al.* (1994).

## References

Büchler, M., Tisljar, U. & Wolf, D.H. (1994) Proteinase yscD (oligopeptidase yscD) structure, function and relationship of the yeast enzyme with mammalian thimet oligopeptidase (metalloendopeptidase, EP 24.15). *Eur. J. Biochem.* **219**, 627–639.

Conlin, C.A. & Miller, C.G. (1992) Cloning and nucleotide sequence of *opdA*, the gene encoding oligopeptidase A in *Salmonella typhimurium. J. Bacteriol.* **174**, 1631–1640.

Conlin, C.A., Trun, N.J., Silhavy, T.J. & Miller, C.G. (1992) *Escherichia coli prlC* encodes an endopeptidase and is homologous to the *Salmonella typhimurium opdA* gene. *J. Bacteriol.* **174**, 5881–5887.

Fisher, B.D., Armstrong, D., Yu, B. & Gold, J.W.M. (1981) Invasive aspergillosis. Progress in early diagnosis and treatment. *Am. J. Med.* **71**, 571–577.

Ibrahim-Granet, O., Bertrand, O., Debeaupuis, J.-P., Planchenault, T. & Dupont, B. (1994) *Aspergillus fumigatus* metalloproteinase that hydrolyses native collagen: purification by dye-binding chromatography. *Protein Exp. Purif.* **5**, 84–88.

Ibrahim-Granet, O., Hernandez, F.H., Chevrier, G. & Dupont, B. (1996) Expression of PZ-peptidases by cultures of several pathogenic fungi. Purification and characterization of a collagenase from *Trychophyton schoenleini. J. Med. Vet. Mycol.* **34**, 83–90.

Jaton-Ogay, K., Paris, S., Huerre, M., Quadroni, M., Falchetto, R., Togni, G., Latgé, J.-P. & Monod, M. (1994) Cloning and disruption of the gene encoding an extracellular metalloprotease of *Aspergillus fumigatus. Mol. Microbiol.* **14**, 917–928.

Kolattukudy, P.E., Lee, J.D., Rogers, L.M., Zimmerman, P., Ceselki, S., Fox, B., Stein, B. & Copelan, E. (1993) Evidence for possible involvement of an elastolytic serine protease in aspergillosis. *Infect. Immun.* **61**, 2357–2368.

Matthews, B.W., Jansonius, J.N., Colman, P.M., Schoenborn, B.P. & Duporque, D. (1972) Three-dimensional structure of thermolysin. *Nature New Biol.* **238**, 37–41.

Pierotti, A., Dong, K.W., Glucksman, M.J., Orlowski, M. & Roberts, J.L. (1990) Molecular cloning and primary structure of rat testes metalloendopeptidase. *Biochemistry* **29**, 10323–10329.

Ramesh, M.V., Sirakova, T.D. & Kolattukudy, P.E. (1995) Cloning and characterization of the cDNAs and genes (*mep20*) encoding homologous metalloproteinases from *Aspergillus flavus* and *Aspergillus fumigatus. Gene* **165**, 121–125.

Rawlings, N.D., & Barrett, A.J. (1995) Evolutionary families of metallopeptidases. *Methods Enzymol.* **248**, 183–228.

Serizawa, A., Dando, P.M. & Barrett, A.J. (1995) Characterization of a mitochondrial metallopeptidase reveals neurolysin as a homologue of thimet oligopeptidase. *J. Biol. Chem.* **270**, 2092–2098.

Stynen, D., Goris, A., Sarfati, J. & Latgé, J.-P. (1995) A new sensitive sandwich enzyme-linked immunosorbent assay to detect galactofuran in patients with invasive aspergillosis. *J. Clin. Microbiol.* **33**, 497–500.

Van Wart, H.E. & Steinbrink, D.R. (1981) A continuous spectrophotometric assay for *Clostridium histolyticum* collagenase. *Anal.*

*Biochem.* **113**, 356–365.

Wünsch, E. & Heidrich, H.G. (1963) Darstellung von Prolinpeptiden. III. Ein neues Substrat zur Bestimmung der Kollagenase [Preparation of proline peptides. III. A new substrate for the determination of collagenase]. *Z. Physiol. Chem.* **332**, 300–304.

*Oumaïma Ibrahim-Granet*
*Laboratoire des Aspergillus, Institut Pasteur,*
*25, rue du Dr Roux,*
*75724 Paris Cedex 15, France*
*Email: ogranet@pasteur.fr*

*Christophe d'Enfert*
*Laboratoire des Aspergillus, Institut Pasteur,*
*25, rue du Dr Roux,*
*75724 Paris Cedex 15, France*

# 375. *Oligopeptidase A*

## Databanks

*Peptidase classification: clan MX, family M3, MEROPS ID: M03.004*
*NC-IUBMB enzyme classification: EC 3.4.24.70*
*Databank codes:*

| Species | SwissProt | PIR | EMBL (cDNA) | EMBL (genomic) |
| --- | --- | --- | --- | --- |
| *Escherichia coli* | P27298 | S47718 | M93984 | U00039: chromosome region from 76.0 to 81.5′ |
| *Haemophilus influenzae* | P44573 | C64055 | – | U32706: complete genome section 21 of 163 |
| *Salmonella typhimurium* | P27237 | A42298 | M84574 | – |

## Name and History

*Oligopeptidase A* (*OpdA*) was first recognized as an Ac-(Ala)$_4$-hydrolyzing activity present in extracts of a *Salmonella typhimurium dcp* (peptidyl-dipeptidase Dep, Chapter 376) mutant (Vimr & Miller, 1983; Vimr *et al.*, 1983). The gene locus was originally called *optA* but the name was changed to *opdA* to avoid a nomenclatural conflict. Mutations affecting the *Escherichia coli* homolog (*prlC*) of *opdA* were isolated earlier as suppressors of the protein-localization defect conferred by certain signal sequence mutations (Emr *et al.*, 1981) but the identification of *prlC* with *opdA* was not made until much later (Conlin *et al.*, 1992).

## Activity and Specificity

Although no metal analysis has been reported, the presence of a conserved HEXXH motif makes it likely that OpdA contains tightly bound $Zn^{2+}$. The enzyme can be purified in active form using buffers without added metal ions but its activity can be stimulated by $Co^{2+}$ (0.1–0.5 mM) and to a lesser extent by $Mn^{2+}$. $Zn^{2+}$, $Ni^{2+}$ and $Cu^{2+}$ (all 1 mM) and EDTA all inhibit activity. OpdA is active at pH 8.

OpdA hydrolyzes *N*-blocked peptides containing at least four amino acids but is unable to hydrolyze unblocked peptides with fewer than five amino acids (Vimr *et al.*, 1983): Ac(Ala)$_4$ and (Ala)$_5$ are substrates but (Ala)$_4$ is not. A study of a rather small group of peptides led to the conclusion that Gly or Ala on either side of the scissile bond is permissive for hydrolysis. More distant residues clearly affect hydrolysis however, since Z-Gly-Pro-Gly+Gly-Pro-Ala is cleaved between the Gly residues, but Z-(Gly)$_5$ is not a substrate. OpdA is the major soluble activity able to hydrolyze the lipoprotein signal peptide (Novak *et al.*, 1986; Novak & Dev, 1988). The 20 amino acid signal peptide is attacked only after it is cleaved from the precursor protein, so OpdA is a signal peptide peptidase, not a signal peptidase. The signal peptide is cleaved at six sites each involving a Gly or Ala on one or the other side of the cleaved bond. OpdA also catalyzes the processing of the bacteriophage P22 gp7 protein. The 229 amino acid gp7 precursor is cleaved by OpdA at a Glu+Lys bond to remove 20 N-terminal amino acids.

## Structural Chemistry

The nucleotide sequence of *S. typhimurium* OpdA predicts a 680 amino acid, 77 kDa, pI 5.0 protein (Conlin & Miller,

1992). OpdA is a member of the thimet oligopeptidase family of metallopeptidases (M3) (Chapter 370) (Barrett *et al.*, 1995; Rawlings & Barrett, 1995). The putative HEXXH zinc-binding motif is in a region that is strongly conserved among members of the family. The active enzyme is monomeric.

## Preparation

OpdA has been purified from extracts of *S. typhimurium* strains carrying the *opdA* gene on a pBR328 plasmid (Conlin & Miller, 1992). The purification factor is approximately 3000-fold relative to the level of enzyme in extracts of strains carrying the gene in single copy and the enzyme was estimated to be >95% pure.

## Biological Aspects

The *opdA* gene is regulated as part of the heat shock regulon and its transcription is dependent on $\sigma^{32}$ (Conlin & Miller, 1995). Null mutations in the gene are not lethal under normal growth conditions nor do such mutations confer temperature sensitivity. OpdA can clearly function as a specific processing enzyme in the maturation of bacteriophage P22 gp7 precursor (Conlin & Miller, 1992). No cellular proteins are known to be processed by OpdA. *In vitro* OpdA can degrade the lipoprotein signal peptide (Novak *et al.*, 1986; Novak & Dev, 1988) but there is little evidence to indicate that it participates in this process in the cell (Miller & Conlin, 1994). The fact that certain *opdA* alleles suppress the secretion defect of some signal sequence mutations (Emr *et al.*, 1981; Trun & Silhavy, 1987, 1989) suggests, however, that OpdA may play some role in the protein localization process. The mechanism of this suppression is not understood.

The recognition by sequence of an *opdA* gene in *Haemophilus influenzae* suggests that the enzyme will be present in most gram-negative bacteria. A similar sequence does not seem to be present in *Methanococcus janaschii*, the single archaean species for which a genome sequence is currently available.

## Distinguishing Features

In crude extracts of *E. coli* and *S. typhimurium*, dipeptidyl-peptidase Dep (Chapter 376) and OpdA are the only enzymes that hydrolyze Ac-(Ala)$_4$ (Conlin & Miller, 1995). Both of these enzymes are members of family M3, both are inhibited by metal chelators and $Zn^{2+}$ and both are stimulated by $Co^{2+}$. The two enzymes are almost exactly the same size.

Dcp requires substrates with a free C-terminus from which it releases dipeptides sequentially and only Dcp will attack Ac-(Ala)$_3$. Unlike OpdA, Dcp hydrolysis is inhibited by captopril.

## Further Reading

Conlin & Miller (1995) have provided a summary of the current status of our understanding of OpdA.

## References

Barrett, A.J., Brown, M.A., Dando, P.M., Knight, C.G., McKie, N., Rawlings, N.D. & Serizawa, A. (1995) Thimet oligopeptidase and oligopeptidase M or neurolysin. *Methods Enzymol.* **248**, 529–556.

Conlin, C.A. & Miller, C.G. (1992) Cloning and nucleotide sequence of opdA, the gene encoding oligopeptidase A in *Salmonella typhimurium*. *J. Bacteriol.* **174**, 1631–1640.

Conlin, C.A. & Miller, C.G. (1995) Dipeptidyl carboxypeptidase and oligopeptidase A from *Escherichia coli* and *Salmonella typhimurium*. *Methods Enzymol.* **248**, 567–579.

Conlin, C.A., Trun, N.J., Silhavy, T.J. & Miller, C.G. (1992) *Escherichia coli prlC* encodes an endopeptidase and is homologous to the *Salmonella typhimurium opdA* gene. *J. Bacteriol.* **174**, 5881–5887.

Emr, S.D., Hanley-Way, S. & Silhavy, T.J. (1981) Suppressor mutations that restore export of a protein with a defective signal sequence. *Cell* **23**, 79–88.

Miller, C.G. & Conlin, C.A. (1994) Signal peptide hydrolases. In: *Signal Peptidases* (von Heijne, G., ed.). Austin, TX: R.G. Landes, pp. 49–57.

Novak, P. & Dev, I.K. (1988) Degradation of a signal peptide by protease IV and oligopeptidase A. *J. Bacteriol.* **170**, 5067–5075.

Novak, P., Ray, P.H. & Dev, I.K. (1986) Localization and purification of two enzymes from *Escherichia coli* capable of hydrolyzing a signal peptide. *J. Biol. Chem.* **261**, 420–427.

Rawlings, N. & Barrett, A.J. (1995) Evolutionary families of metallopeptidases. *Methods Enzymol.* **248**, 183–228.

Trun, N.J. & Silhavy, T.J. (1987) Characterization and *in vivo* cloning of prlC, a suppressor of signal sequence mutations in *Escherichia coli* K-12. *Genetics* **116**, 513–521.

Trun, N.J. & Silhavy, T.J. (1989) PrlC, a suppressor of signal sequence mutations in *Escherichia coli*, can direct the insertion of the signal sequence into the membrane. *J. Mol. Biol.* **205**, 665–676.

Vimr, E.R. & Miller, C.G. (1983) Dipeptidyl carboxypeptidase-deficient mutants of *Salmonella typhimurium*. *J. Bacteriol.* **153**, 1252–1258.

Vimr, E.R., Green, L. & Miller, C.G. (1983) Oligopeptidase-deficient mutants of *Salmonella typhimurium*. *J. Bacteriol.* **153**, 1259–1265.

*Charles G. Miller*
*Department of Microbiology,*
*B103 Chemical and Life Sciences Laboratory,*
*University of Illinois at Champaign-Urbana,*
*Urbana, IL 61801, USA*
*Email: charlesm@uiuc.ed*

# 376. *Peptidyl-dipeptidase Dcp*

## Databanks

*Peptidase classification: clan MX, family M3, MEROPS ID: M03.005*
*NC-IUBMB enzyme classification: EC 3.4.15.5*
*Databank codes:*

| Species | SwissProt | PIR | EMBL (cDNA) | EMBL (genomic) |
|---|---|---|---|---|
| *Escherichia coli* | P24171 | A49931 | X57947 | – |
| *Salmonella typhimurium* | P27236 | A42297 | M84575 | – |
| Other related sequences | | | | |
| *Arabidopsis thaliana* | – | – | Z35217 | – |
| *Caenorhabditis elegans* | – | – | D33493 | – |
| *Trypanosoma brucei* | – | – | T26735 | – |
| *Zea mays* | – | – | T18736 | – |

## Name and History

The name **peptidyl-dipeptide hydrolase** was initially used to denote enzymes which cleave C-terminal dipeptides from longer peptide substrates. In 1972, only one activity of this type was formally recognized by the IUB (EC 3.4.15.1) (Chapter 359). This entry referred to the mammalian angiotensin I-converting enzyme (ACE) known to hydrolyze the peptides angiotensin I and bradykinin involved in the regulation of blood pressure (Corvol *et al.*, 1995). In 1978, an enzyme from *Escherichia coli* with catalytic properties very similar to those of ACE (Yaron *et al.*, 1972) was included into the same entry (EC 3.4.15.1) which then was renamed **dipeptidyl carboxypeptidase**. The *E. coli* enzyme was separated out in 1981 and moved to a new entry (EC 3.4.15.3) termed peptidyl-dipeptidase B to distinguish it from EC 3.4.15.1 which accordingly was renamed peptidyl-dipeptidase A. In the period 1984–1988 the names **dipeptidyl carboxypeptidase II** and dipeptidyl-carboxypeptidase I, and in the period 1989–1992 the terms **peptidyl dipeptidase II** and peptidyl dipeptidase I were recommended for the *E. coli* enzyme and ACE, respectively. In 1992, entry EC 3.4.15.3 was closed and reincluded into EC 3.4.15.1 which since then was referred to as **peptidyl-dipeptidase A** and covered enzymes of both prokaryotic and eukaryotic origin until 1996. On the basis of recent sequencing studies (Hamilton & Miller, 1992; Henrich *et al.*, 1993), the enzymes from *E. coli* and *Salmonella typhimurium* were separated out again and transferred to the new entry 3.4.15.5. They are now referred to as **peptidyl-dipeptidase Dcp**.

By use of the synthetic substrate Bz-Gly┼His-Leu, metalloenzymes with peptidyl-dipeptidase activity have also been detected in other enteric bacterial species and in the opportunistic human pathogen *Pseudomonas maltophilia* (Stevens *et al.*, 1990) as well as in the gram-positive bacteria *Corynebacterium equi* (Lee *et al.*, 1971), *Bacillus pumilus* (Nagamori *et al.*, 1991) and *Streptomyces* (Miyoshi *et al.*, 1992). Each of them however is distinguished from Dcp of *E. coli* and *S. typhimurium* by some features such as molecular mass, substrate spectrum, susceptibility to EDTA and ACE inhibitors, or salt requirement. Well-founded EC assignments of these enzymes may require more information on their molecular relationships.

## Activity and Specificity

The penultimate peptide bond is hydrolyzed in $\alpha$-*N*-blocked tripeptides, free tetrapeptides and higher peptides; free tripeptides are not cleaved. Peptide bonds in which the nitrogen is donated by a Pro residue or which link two consecutive Gly residues are refractory to Dcp. Peptides with a blocked C-terminus or with a D-amino acid in the C-terminal position are not accepted as substrates. As an example, the nonapeptide bradykinin is cleaved at two positions: Arg-Pro-Pro-Gly-Phe┼Ser-Pro┼Phe-Arg. $K_m$ values towards several selected tetrapeptides range between 0.6 and 6.1 mM (Yaron, 1976).

Activity is conventionally determined with the specific substrate Ac-Ala┼Ala-Ala, and the reaction products are detected by the ninhydrin method (Conlin & Miller, 1995) or by HPLC (Hamilton & Miller, 1992). A more convenient assay relies on the photometric detection of Bz-Gly released from the synthetic peptidyl-dipeptidase A (ACE) substrate Bz-Gly┼His-Leu (Cushman & Cheung, 1971). Since this substrate is not cleaved by any other peptidases of *E. coli* (Henrich *et al.*, 1993), it is also suitable for the assay of Dcp in crude cell extracts. Both substrates may be purchased from Bachem (see Appendix 2 for full names and addresses of suppliers). pH optima of 8.2 and 7.5 have been determined for Dcp purified from *E. coli* B (Yaron *et al.*, 1972) and *E. coli* K-12 (Henrich *et al.*, 1993), respectively. The temperature of highest activity is 42°C. The enzyme is slightly activated by 1 mM $Mn^{2+}$, $Mg^{2+}$, $Ca^{2+}$ or $Co^{2+}$. A 5- to 8-fold stimulation by 0.05 mM $Co^{2+}$ observed for Dcp from *E. coli* B (Yaron *et al.*, 1972) is not observed for the *E. coli* K-12 enzyme (Henrich *et al.*, 1993). $Cu^{2+}$, $Ni^{2+}$ and $Zn^{2+}$ are potent inhibitors. NaCl is not required for optimal activity and

**M**

is inhibitory at a concentration of 100 mM. EDTA, even at high concentrations (10 mM), has only a minor effect whereas virtually no activity is detectable in the presence of 1,10-phenanthroline. Dcp of both *E. coli* (Deutch & Soffer, 1978) and *S. typhimurium* (Vimr & Miller, 1983) is highly sensitive to captopril, a chemical inhibitor of ACE (Cushman *et al.*, 1977). Of several other protease inhibitors, only chymostatin (0.2 mM) clearly inactivates the *E. coli* enzyme (Henrich *et al.*, 1983).

### Structural Chemistry

The proteins predicted from the genes for Dcp from *E. coli* and *S. typhimurium* are both 680 amino acids in length and share 79% sequence identity. They are quite similar (32% identity) to the oligopeptidase A enzymes of both bacterial species, which also fall into the thimet oligopeptidase family (Chapter 375) (Hamilton & Miller, 1992; Henrich *et al.*, 1993).

Dcp contains the potential zinc-binding signature HEXXH (Jongeneel *et al.*, 1989); however, catalytic residues or zinc ligands have not been experimentally determined. This motif is embedded into a larger region, suspected to cover the active center, which is highly conserved among a number of proteases from eubacteria, fungi and animals. Some of these enzymes in addition share blocks of clear sequence similarity near their N- and C-termini (Conlin & Miller, 1995). Dcp is most probably active as a monomer of 77.5 kDa. For the *E. coli* enzyme a pI of 5.2 has been determined.

### Preparation

Due to the lack of sequencing information for functionally similar enzymes known from various organisms, the natural distribution of peptidyl-dipeptidase Dcp is largely unknown. The enzyme has been purified to homogeneity from *E. coli* B carrying only the chromosomal copy of the *dcp* gene (1200-fold; Yaron, 1976) and from transformants of *S. typhimurium*, in which *dcp* was 50-fold overexpressed from a plasmid (Conlin & Miller, 1995). Since heavy overproduction of Dcp, in contrast, is not tolerated by *E. coli* K-12, in this case the enzyme was enriched 80-fold from transformants carrying a recombinant low copy-number plasmid variant (Henrich *et al.*, 1993).

### Biological Aspects

Dcp is the only C-terminal exopeptidase known from *E. coli* or *S. typhimurium* apart from the D-Ala-D-Ala carboxypeptidases, Chapters 143 and 144). Starting at well-identified promoters, the corresponding gene is transcribed as a single species of monocistronic mRNA in both bacterial species, and there is no evidence for regulation at the transcriptional level (Hamilton & Miller, 1992; Henrich *et al.*, 1993). Codon preferences suggest that Dcp is present in less than 100 copies per cell. Apart from the removal of N-terminal *N*-formyl-Met (which may be incomplete for the *S. typhimurium* enzyme), no post-translational modifications are known.

The hydropathic properties of the enzyme are indicative of a soluble protein with cytoplasmic location. Earlier observations, however, indicated that about 10% of the Dcp activity is located in the periplasmic space (Deutch & Soffer, 1978). Although the enzyme can function in the utilization of peptides supplied in the medium (Deutch & Soffer, 1978; Vimr & Miller, 1983), its major role appears to be in the breakdown of intracellular proteins (Vimr *et al.*, 1983), since most Dcp substrates are expected to be poorly transported into the cell (Payne, 1980). Dcp closely resembles the mammalian glycoprotein ACE (Chapter 359) in its substrate specificity and in its susceptibility to the antihypertensive drug captopril, but, except for a thermolysin-type zinc-binding motif, there are no significant homologies between the amino acid sequences of these proteins.

Enzymatic activities, cleaving C-terminal dipeptides from oligopeptides, have been detected in a variety of bacterial species. The significance of their potential to act on vasoactive peptides in mammalian hosts however remains unknown. In the case of *P. maltophilia*, it has been speculated that the bacteria may promote their own dissemination in infected endothelial tissue by modulating the vascular smooth muscle contraction via the hydrolysis of angiotensin I and bradykinin (Stevens *et al.*, 1990).

### Distinguishing Features

The structural dissimilarity of Dcp and ACE (now obvious from the lack of sequence homology) was previously recognized by Das & Soffer (1976). They found that antibodies raised against rabbit ACE did not affect the activity of the *E. coli* enzyme. Unlike ACE, Dcp is not activated by chloride.

Although oligopeptidase A is able to hydrolyze some of the same small peptide substrates as Dcp (e.g. Ac-(Ala)₄), the specificity and inhibitor characteristics of both enzymes are clearly distinguishable. Dcp is a strict exopeptidase, and oligopeptidase A is an oligopeptidase. Oligopeptidase A, in contrast to Dcp, is not inhibited by captopril (Conlin & Miller, 1992) and is required to support the propagation of phage P22 (Conlin *et al.*, 1992). In view of the discrepancy between structural similarity and functional distinctness of Dcp and oligopeptidase A, it seems inappropriate to assess the enzymatic properties of similar proteins purely on the basis of amino acid homologies. Therefore, the assignment of some predicted proteins, such as a putative oligopeptidase of *Haemophilus influenzae* (Fleischmann *et al.*, 1995) and a potential metalloendopeptidase (MEP) of *Schizophyllum commune* (Stankis *et al.*, 1992), which share significant homology with Dcp (see also Chapter 377), remains unclear as long as no biochemical data are available.

### Further Reading

See Conlin & Miller (1995) for further information on this enzyme and oligopeptidase A.

### References

Conlin, C.A. & Miller, C.G. (1992) Cloning and nucleotide sequence of *opdA*, the gene encoding oligopeptidase A in *Salmonella typhimurium*. *J. Bacteriol.* **174**, 1631–1640.

Conlin, C.A. & Miller, C.G. (1995) Oligopeptidase A and peptidyl-dipeptidase of *Escherichia* and *Salmonella*. *Methods Enzymol.* **248**, 567–579.

Conlin, C.A., Vimr, E.R. & Miller, C.G. (1992) Oligopeptidase A is required for normal phage P22 development. *J. Bacteriol.* **174**, 5869–5880.

Corvol, P., Williams, T.A. & Soubrier, F. (1995) Peptidyl dipeptidase A: angiotensin I-converting enzyme. *Methods Enzymol.* **248**, 283–305.

Cushman, D.W. & Cheung, H.S. (1971) Spectrophotometric assay and properties of the angiotensin-converting enzyme of rabbit lung. *Biochem. Pharmacol.* **20**, 1637–1648.

Cushman, D.W., Cheung, H.S., Sabo, E.F. & Ondetti, M.A. (1977) Design of potent competitive inhibitors of angiotensin-converting enzyme. Carboxyalkanoyl and mercaptoalkanoyl amino acids. *Biochemistry* **16**, 5484–5491.

Das, M. & Soffer, R.L. (1976) Pulmonary angiotensin-converting enzyme antienzyme antibody. *Biochemistry* **15**, 5088–5094.

Deutch, C.E. & Soffer, R.L. (1978) *Escherichia coli* mutants defective in dipeptidyl carboxypeptidase. *Proc. Natl Acad. Sci. USA* **75**, 5998–6001.

Fleischmann, R.D., Adams, M.D., White, O. *et al.* (1995) Whole-genome random sequencing and assembly of *Haemophilus influenzae* Rd. *Science* **269**, 496–512.

Hamilton, S. & Miller, C.G. (1992) Cloning and nucleotide sequence of *Salmonella typhimurium dcp* gene encoding dipeptidyl carboxypeptidase. *J. Bacteriol.* **174**, 1626–1630.

Henrich, B., Becker, S., Schroeder, U. & Plapp, R. (1993) *dcp* Gene of *Escherichia coli*: cloning, sequencing, transcript mapping, and characterization of the gene product. *J. Bacteriol.* **175**, 7290–7300.

Jongeneel, C.V., Bouvier, J. & Bairoch, A. (1989) A unique signature identifies a family of zinc-dependent metallopeptidases.

*FEBS Lett.* **242**, 211–214.

Lee, H.-J., Larue, J.N. & Wilson, I.B. (1971) Dipeptidyl carboxypeptidase from *Corynebacterium equi*. *Biochim. Biophys. Acta* **250**, 608–613.

Miyoshi, S., Nomura, G., Suzuki, M., Fukui, F., Tanaka, H. & Maruyama, S. (1992) Purification and characterization of a novel dipeptidyl carboxypeptidase from a *Streptomyces* species. *J. Biochem.* **112**, 253–257.

Nagamori, Y., Kusaka, K., Fujishima, N. & Okada, S. (1991) Enzymatic properties of dipeptidyl carboxypeptidase from *Bacillus pumilus*. *Agric. Biol. Chem.* **55**, 1695–1699.

Payne, J.W. (1980) Transport and utilization of peptides by bacteria. In: *Microorganisms and Nitrogen Sources* (Payne, J.W., ed.). New York: John Wiley, pp. 211–256.

Stankis, M.M., Specht, C.A., Yang, H., Giasson, L., Ullrich, R.C. & Novotny, C.P. (1992) The *Aα* mating locus of *Schizophyllum commune* encodes two dissimilar multiallelic homeodomain proteins. *Proc. Natl Acad. Sci. USA* **89**, 7169–7173.

Stevens, J., Fanburg, B.L. & Lanzillo, J.J. (1990) Determination of peptidyl dipeptidase activity in 24 bacterial species. *Can. J. Microbiol.* **36**, 56–59.

Vimr, E.R. & Miller, C.G. (1983) Dipeptidyl carboxypeptidase-deficient mutants of *Salmonella typhimurium*. *J. Bacteriol.* **153**, 1252–1258.

Vimr, E.R., Green, L. & Miller, C.G. (1983) Oligopeptidase-deficient mutants of *Salmonella typhimurium*. *J. Bacteriol.* **153**, 1259–1265.

Yaron, A. (1976) Dipeptidyl carboxypeptidase from *Escherichia coli*. *Methods Enzymol.* **50**, 599–610.

Yaron, A., Mlynar, D. & Berger, A. (1972) A dipeptidyl carboxypeptidase from *E. coli*. *Biochem. Biophys. Res. Commun.* **47**, 897–902.

*Bernhard Henrich*
*Department of Microbiology,*
*University of Kaiserslautern,*
*PO Box 3049,*
*D-67653 Kaiserslautern, Germany*
*Email: henrich@rhrk.uni-kl.de*

*Jürgen R. Klein*
*Department of Microbiology,*
*University of Kaiserslautern,*
*PO Box 3049,*
*D-67653 Kaiserslautern, Germany*

**M**

# *377. Mitochondrial intermediate peptidase*

## Databanks

*Peptidase classification: clan MX, family M3, MEROPS ID: M03.006*
*NC-IUBMB enzyme classification: EC 3.4.24.59*
*ATCC entries: 973145 (mouse)*

*Databank codes:*

| Species | SwissProt | PIR | EMBL (cDNA) | EMBL (genomic) |
|---|---|---|---|---|
| *Rattus norvegicus* | Q01992 | A46273 S23380 | M96633 | – |
| *Saccharomyces cerevisiae* | P51980 | – | U10243 | – |
| *Schizophyllum commune* | P37932 | – | – | L43072: complete gene M97179: 3′ end M97180: 3′ end M97181: 3′ end |
| *Schizosaccharomyces pombe* | Q10415 | – | Z70690 | – |

## Name and History

The name *mitochondrial intermediate peptidase* (*MIP*; formerly *P2* or *MPP-II* ) denotes cleavage of intermediate-sized mitochondrial proteins to the mature form. The acronyms *HMIP*, *RMIP*, *SMIP* and *YMIP* are used to designate the MIP proteins of human, rat, the basidiomycete fungus *Schizophyllum commune*, and the budding yeast *Saccharomyces cerevisiae*, respectively. The acronym *MIP1* denotes the *S. cerevisiae* gene for the YMIP polypeptide, and the acronym *MIPEP* the human gene for the HMIP polypeptide.

Early studies on mitochondrial protein trafficking showed that a group of nuclear-encoded precursor polypeptides are processed to the mature form in two sequential steps via formation of an intermediate with an octapeptide targetting signal (Hurt *et al.*, 1985; Hartl *et al.*, 1986; Sztul *et al.*, 1987). Contrary to suggestions that the mitochondrial processing peptidase (MPP; EC 3.4.99.41) (Chapter 469) was probably responsible for both cleavages, studies of rat liver mitochondria showed that the intermediate-processing activity had requirements different from those exhibited by MPP (Kalousek *et al.*, 1988). Subsequently, MIP was purified to homogeneity from rat liver mitochondrial matrix (Kalousek *et al.*, 1992), and a full-length RMIP cDNA was isolated (Isaya *et al.*, 1992b).

## Activity and Specificity

MIP is involved in two-step processing of mitochondrial protein precursors which contain the motif Arg-Xaa-(Phe/Leu/Ile)-Xaa-Xaa-(Ser/Thr/Gly)-(Xaa)$_4$+ at the C-terminus of their leader peptide. During or upon import to the mitochondrial matrix, MPP cleaves the precursor after Arg-Xaa. This initial cleavage yields a processing intermediate with a typical octapeptide at the N-terminus; the octapeptide is subsequently cleaved by MIP to form mature protein. Positioning of the octapeptide at the protein N-terminus is necessary but not sufficient for cleavage, as the mature portion of the intermediate is also required for substrate recognition (Isaya *et al.*, 1991, 1992a).

The general characteristics of the MIP cleavage sites have been identified by sequence analysis of mitochondrial protein precursors (Hendrick *et al.*, 1989; Gavel & von Heijne, 1990; Branda & Isaya, 1995). A Phe residue is found at P8 more frequently than Leu or Ile. Gly is common at P5 in mammalian octapeptides, while Ser or Thr are preferred at this position in *S. cerevisiae* octapeptides. While Ser is common at P7, no particular amino acids are preferred at

any of the remaining five positions within the octapeptide. Known octapeptide sequences have similar overall amino acid compositions, with a predominance of the six smallest amino acids – Ser, Thr, Ala, Val, Pro and Gly. In particular, Ser or Thr are normally found at two or more positions and are often clustered together. Positively but not negatively charged residues may be present. No particular amino acids are preferred at P1′ or P2′ nor any other positions within the first ten amino acids of the mature protein.

MIP activity can be measured by incubation of *in vitro* translated octapeptide-containing precursors with isolated mitochondria (Isaya & Kalousek, 1995). Under such conditions, the cleavage catalyzed by MIP is at the end of a mitochondrial protein import reaction which also involves outer membrane receptors, outer and inner membrane translocation complexes, molecular chaperones and MPP. If the precursor is incubated directly with mitochondrial matrix or purified enzyme, initial cleavage by MPP is still required to observe processing of the intermediate by MIP. This requirement is circumvented when MIP activity is determined using an intermediate protein as the substrate. Intermediates that are translated *in vitro* from a methionine artificially placed at the octapeptide N-terminus can be processed to the mature form by MIP independent of the presence of MPP (Isaya *et al.*, 1992a). Proteolytic reactions are carried out at 27°C in the presence of 10 mM HEPES-KOH, pH 7.4, containing 1 mM MnCl$_2$ and 1 mM DTT, and analyzed by SDS-PAGE and fluorography. MIP activity is determined by the relative amounts of intermediate processed to the mature form, which is measured by densitometric analysis of fluorograms (Isaya & Kalousek, 1995).

## Structural Chemistry

Active MIP is a monomer of approximately 75 kDa in mammals (Kalousek *et al.*, 1992; Chew *et al.*, 1997) and 80 kDa in fungi (Isaya *et al.*, 1994, 1995). HMIP and RMIP are homologous (87% identity) and present over 50% identity to YMIP and SMIP. HMIP, RMIP and SMIP have been expressed in *S. cerevisiae* and shown to be able to functionally replace YMIP. Over 40% identity is found between the MIP sequences and thimet oligopeptidase (Chapter 371).

MIP is activated by divalent cations and inactivated by thiol-blocking agents (Kalousek *et al.*, 1992). Amino acids potentially involved in the catalytic mechanism have been identified on the basis of sequence analysis and site-directed mutagenesis (Chew *et al*., 1996). All known MIP sequences contain a highly conserved domain: Phe-His-Glu-Xaa-Gly-His-(Xaa)$_2$-His-(Xaa)$_{12}$-Gly-(Xaa)$_5$-Asp-(Xaa)$_2$-Glu-Xaa-Pro-Ser-(Xaa)$_3$-Glu, which is also shared by thimet oligopeptidase. In addition to a zinc-binding motif, this domain includes a His and two Glu residues which could participate in catalysis. In RMIP, HMIP, YMIP and SMIP there are 18, 16, 17 and 8 Cys residues, respectively, but only two of them, Cys131 and Cys581, are conserved in all four proteins. The His and Glu residues in the HEXGH motif, and a Glu 25 residues from the second His in the motif are essential for YMIP function *in vivo*. In contrast, Cys131 and Cys581 are not required for activity *in vivo* or *in vitro*. Ser or Val substitutions for Cys131 and/or Cys581 reduce enzyme stability *in vitro* and at the same time increase its sensitivity to thiol-blocking agents. While these observations are consistent with MIP being a metallopeptidase, enzyme inactivation by thiol-blocking agents seems to result from derivatization of multiple thiol groups and loss of protein stability without reflecting a role for cysteine residues in catalysis.

## Preparation

Native RMIP has been purified 2250-fold from rat liver mitochondrial matrix with a final yield of about 2% (Kalousek *et al*., 1992). Expression of recombinant enzyme has been achieved in *S. cerevisiae* (Chew *et al*., 1996). In this system, the YMIP precursor is synthesized in the cytoplasm, imported into the mitochondrial matrix, and finally processed to the active form. To achieve high levels of expression, the *MIP1* gene is cloned into a yeast expression vector, pG-3, downstream of a strong constitutive promoter, the yeast glyceraldehyde-3-phosphate dehydrogenase (GPD) promoter. This plasmid, pG3-*MIP1*, also carries a 2-μm origin of replication, which results in an average plasmid copy number of 10–20 per cell, and the yeast *TRP1* gene as a selectable marker. A c-*myc* tag fused to the C-terminus of the recombinant YMIP polypeptide enables protein detection by immunoblotting as well as rapid protein purification by affinity chromatography. Mitochondria are isolated from yeast cells, and matrix fractions are prepared by sonication of mitochondria and ultracentrifugation. Recombinant YMIP-myc is affinity-purified from total matrix using an agarose-bead conjugate of a mouse monoclonal anti-c-*myc* (9E10) antibody, which is commercially available. This system yields approximately 2 μg purified active YMIP per g of yeast cells.

## Biological Properties

MIP is encoded in the nucleus and initially synthesized in the cytoplasm as a large precursor molecule carrying a mitochondrial leader peptide; upon import into the mitochondrion, the precursor is processed to the mature form by MPP (Isaya *et al*., 1992b). In both rat and yeast mitochondria, the active enzyme is a soluble monomer localized to the matrix compartment. The HMIP gene is differentially expressed in human tissues and very high mRNA levels are detected in the heart and skeletal muscle. The HMIP locus has been mapped to 13q12 (Chew *et al*., 1997).

A number of natural substrates have been identified in *S. cerevisiae* (Branda & Isaya, 1995) and man (Chew *et al*., 1997). YMIP cleaves proteins required for maintenance of the oxidative phosphorylation system, including respiratory enzyme subunits and factors mediating replication and expression of mitochondrial DNA. When YMIP is genetically inactivated, these protein are no longer processed to the mature form, which results in a complex respiratory-deficient phenotype with loss of mtDNA. Proteins cleaved by HMIP include respiratory enzyme subunits but no components of the human mitochondrial genetic system. Given that only a handful of proteins involved in replication and expression of mtDNA are known in humans, the identification of a larger sample of human mitochondrial precursor polypeptides will be required to precisely define the nature of the HMIP-cleaved proteins.

## Distinguishing Features

The *S. cerevisiae* gene *PRD1* (formerly ORF YCL57w) encodes the oligopeptidase yscD (Chapter 373) (Büchler *et al*., 1994), an enzyme located to both the cytoplasm and the mitochondria with 24% identity and 47% similarity to YMIP. Although the *PRD1* gene product had initially been proposed as the yeast ortholog of mammalian MIP (Isaya *et al*., 1992b), a *prd1* knockout mutant showed normal mitochondrial function (Büchler *et al*., 1994) and normal processing of octapeptide-containing precursors to the mature form (Isaya *et al*., 1994). As a double *prd1*–*mip1* mutant did not present a more severe phenotype than a single *mip1* mutant, it was concluded that YMIP and yscD do not have overlapping biologic roles.

## Further Reading

See the review by Isaya & Kalousek (1995).

## References

Branda, S.S. & Isaya, G. (1995) Prediction and identification of new natural substrates of the yeast mitochondrial intermediate peptidase. *J. Biol. Chem.* **270**, 27366–27373.

Büchler, M., Tisljar, U. & Wolf, D.H. (1994) Proteinase yscD (oligopeptidase yscD). Structure, function, and relationship of the yeast enzyme with mammalian thimet oligopeptidase (metallopeptidase, EP 24.15). *Eur. J. Biochem.* **219**, 627–639.

Chew, A., Rollins, R.A., Sakati, W.R. & Isaya, G. (1996) Mutations in a putative zinc-binding domain inactivate the mitochondrial intermediate peptidase. *Biochem. Biophys. Res. Commun.* **226**, 822–829.

Chew, A., Buck, E.A., Peretz, S., Sirugo, G., Rinaldo, P. & Isaya, G. (1997) Cloning, expression, and chromosomal assignment of the human mitochondrial intermediate peptidase (MIPEP)s. *Genomics* **40**, 493–496.

Gavel, Y. & von Heijne, G. (1990) Cleavage-site motifs in mitochondrial targeting peptides. *Protein Eng.* **4**, 33–37.

Hartl, F.U., Schmidt, B., Wachter, E., Weiss, H. & Neupert, W. (1986) Transport into mitochondria and intramitochondrial sorting of the Fe/S protein of ubiquinol-cytochrome *c* reductase. *Cell* **47**, 939–951.

Hendrick, J.P., Hodges, P.E. & Rosenberg, L.E. (1989) Survey of amino-terminal proteolytic cleavage sites in mitochondrial

M

precursor proteins: leader peptides cleaved by two matrix protease share a three-amino acid motif. *Proc. Natl Acad. Sci. USA* **86**, 4056–4060.

Hurt, E.C., Muller, U. & Schatz, G. (1985) The first twelve amino acids of a yeast mitochondrial outer membrane protein can direct a nuclear-encoded cytochrome *c* oxidase subunit to the mitochondrial inner membrane. *EMBO J.* **4**, 2061–2068.

Isaya, G. & Kalousek, F. (1995) Mitochondrial intermediate peptidase. *Methods Enzymol.* **248**, 556–567.

Isaya, G., Kalousek, F., Fenton, W.A. & Rosenberg, L.E. (1991) Cleavage of precursors by the mitochondrial processing peptidase requires a compatible mature protein or an intermediate octapeptide. *J. Cell Biol.* **113**, 65–76.

Isaya, G., Kalousek, F. & Rosenberg, L.E. (1992a) Amino-terminal octapeptides function as recognition signals for the mitochondrial intermediate peptidase. *J. Biol. Chem.* **267**, 7904–7910.

Isaya, G., Kalousek, F. & Rosenberg, L.E. (1992b) Sequence analysis of rat mitochondrial intermediate peptidase: similarity to zinc metallopeptidases and to a putative yeast homologue. *Proc. Natl Acad. Sci. USA* **89**, 8317–8321.

Isaya, G., Miklos, D. & Rollins, R.A. (1994) *MIP1*, a new yeast gene homologous to rat mitochondrial intermediate peptidase, is required for oxidative metabolism in *Saccharomyces cerevisiae*. *Mol. Cell. Biol.* **14**, 5603–5616.

Isaya, G., Sakati, W.R., Rollins, R.A., Shen, G.P., Hanson, L.E., Ullrich, R.C. & Novotny, C.P. (1995) Mammalian mitochondrial intermediate peptidase: structure/function analysis of a new homologue from *Schizophyllum commune* and relationship to thimet oligopeptidases. *Genomics* **28**, 450–461.

Kalousek, F., Hendrick, J.P. & Rosenberg, L.E. (1988) Two mitochondrial matrix proteases act sequentially in the processing of mammalian matrix enzymes. *Proc. Natl Acad. Sci. USA* **85**, 7536–7540.

Kalousek, F., Isaya, G. & Rosenberg, L.E. (1992) Rat liver mitochondrial intermediate peptidase (MIP): purification and initial characterization. *EMBO J.* **11**, 2803–2809.

Sztul, E.S., Hendrick, J.P., Kraus, J.P., Wall, D., Kalousek, F. & Rosenberg, L.E. (1987) Import of rat ornithine transcarbamylase precursor into mitochondria: two-step processing of the leader peptide. *J. Cell Biol.* **105**, 2631–2639.

*Grazia Isaya*
*Department of Genetics,*
*Yale University School of Medicine,*
*New Haven, CN 06510, USA*
*Email: Grazia.Isaya@Yale.edu*

# *378. Oligopeptidase F*

## Databanks

*Peptidase classification: clan MX, family M3, MEROPS ID: M03.007*
*NC-IUBMB enzyme classification: none*
*Databank codes:*

| Species | Type | SwissProt | PIR | EMBL (cDNA) | EMBL (genomic) |
|---|---|---|---|---|---|
| *Lactococcus lactis* | pepF1 | P54124 | A55485 S49150 | X99798 Z32522 | – |
| *Lactococcus lactis* | pepF2 | – | – | X99710 | – |
| *Mycoplasma genitalium* | – | P47429 | – | U02198 | U39695: complete genome section 17 of 56 |
| *Mycoplasma pneumoniae* | – | P54125 | – | – | U34795: cosmid pcosMPGT9 fragment |

## Name and History

The presence of **oligopeptidase F** (**PepF**) was probably first reported in *Lactococcus lactis* under the name **LEPI** (Yan *et al.*, 1987) and then as **alkaline endopeptidase** (Baankreis *et al.*, 1995). More recently, two homologous proteins with the same oligopeptidase activity were purified and genetic studies revealed that the gene coding for the oligopeptidase F is duplicated (Monnet *et al.*, 1994). The two copies of the gene were named *pepF1* (located on a large plasmid with the genes necessary to the utilization of lactose and casein) and *pepF2* (located on the chromosome) (M. Nardi *et al.*, unpublished results). The corresponding oligopeptidases are called **PepF1** and **PepF2** according to the nomenclature established for the peptidases from lactic acid bacteria (Tan *et al.*, 1993).

## Activity and Specificity

Oligopeptidase F is able to cleave peptides ranging in length from 7 to 23 amino acids with a broad specificity (Yan *et al.*, 1987; Monnet *et al.*, 1994; Baankreis *et al.*, 1995), different from that observed with the other lactococcal oligopeptidase PepO (Chapter 365) (Monnet, 1995). Although a low breakdown of larger peptides after a long incubation period with oligopeptidase F was reported (Baankreis *et al.*, 1995), it seems that the oligopeptidase is a substrate-size-recognizing endopeptidase which is not able to hydrolyze either very short peptides or proteins (Yan *et al.*, 1987; Monnet *et al.*, 1994; Baankreis *et al.*, 1995).

The best substrate among the peptides tested is bradykinin Arg-Pro-Pro-Gly-Phe+Ser-Pro-Phe-Arg. The fluorogenic substrate Mcc-Pro-Leu-Gly-Pro-D-Lys (Dnp) introduced for the assay of thimet oligopeptidase (Chapter 371) (Tisljar *et al.*, 1990; Vincent *et al.*, 1995) is also a substrate for oligopeptidase F (V. Monnet, unpublished results).

The optimum pH for oligopeptidase F activity is rather alkaline (7–9) and the optimal temperature is around 40°C. Oligopeptidase F is clearly a metalloenzyme, inhibited by 1 mM EDTA or 1,10-phenanthroline. After inactivation by EDTA, oligopeptidase F activity is restored with $Mn^{2+}$ and $Co^{2+}$ (Yan *et al.*, 1987; Monnet *et al.*, 1994). The presence of 1 M NaCl decreases oligopeptidase F activity by 25% (Baankreis *et al.*, 1995).

## Structural Chemistry

Oligopeptidase F is described as a monomeric enzyme with a molecular mass of 90 000 Da (Monnet *et al.*, 1994) or 98 000 (Yan *et al.*, 1987) estimated by gel filtration and 70 000 Da from the size of the corresponding gene (Monnet *et al.*, 1994). The characteristic HGXXH zinc-binding motif is found in the oligopeptidase F sequence (Monnet *et al.*, 1994).

## Preparation

Oligopeptidase F can be purified by chromatography from intracellular extracts of *Lactococcus lactis*. It is also present in intracellular extracts of another lactic acid bacterium, *Streptococcus thermophilus* (Rul & Monnet, 1997).

## Biological Aspects

Oligopeptidase F is found in the cytoplasm of *Lactococcus lactis*, which is widely used in cheese-making technology. However, it does not seem to play a significant role in the hydrolysis of peptides during cheese ripening (Baankreis *et al.*, 1995). It is most probably involved in a function essential for the bacteria such as protein turnover (M. Nardi *et al.*, unpublished results). The selective advantage *Lactococcus lactis* get from oligopeptidase F amplification by gene duplication remains to be evaluated.

## Distinguishing Features

A sequence of 33 amino acids in the oligopeptidase F sequence, not present in the other peptidases from the M3 family, shows similarity (up to 48% identity) with creatine and arginine kinases (Monnet *et al.*, 1994).

## Related Enzymes

PepB (Chapter 379), an oligopeptidase 66% identical to oligopeptidase F and exhibiting a similar specificity, was recently isolated from *Streptococcus agalactiae* (Lin *et al.*, 1996). In addition, the sequencing of the complete genome of *Mycoplasma genitalium* (Fraser *et al.*, 1995) revealed the existence in this organism of a gene coding for a protein which is 30% identical to oligopeptidase F.

## References

Baankreis, R., van Schalkwijk, S., Alting, A.C. & Exterkate, F.A. (1995) The occurrence of two intracellular oligoendopeptidases in *Lactococcus lactis* and their significance for peptide conversion in cheese. *Appl. Microbiol. Biotechnol.* **44**, 386–392.

Fraser, C.M., Gocayne, J.D., White, O. *et al.* (1995) The minimal gene complement of *Mycoplasma genitalium*. *Science* **270**, 397–403.

Lin, B., Averett, W.F., Novak, J., Chatam, W.W., Hollingshead, S.K., Coligan, J.E., Egan, M.L. & Pritchard, D.G. (1996) Characterization of PepB, a group B streptococcal oligopeptidase. *Infect. Immun.* **64**, 3401–3406.

Monnet, V. (1995) Oligopeptidases from *Lactococcus lactis*. *Methods Enzymol.* **248**, 579–592.

Monnet, V., Nardi, M., Chopin, A., Chopin, M.C. & Gripon, J.C. (1994) Biochemical and genetic characterization of PepF, an oligopeptidase from *Lactococcus lactis*. *J. Biol. Chem.* **269**, 32070–32076.

Rul, F. & Monnet, V. (1997) Presence of additional peptidases in *Streptococcus thermophilus* CNRZ302 compared to *Lactococcus lactis*. *J. Appl. Bacteriol.* **82**, 695–704.

Tan, P.S.T., Poolman, B. & Konings, W.N. (1993) Proteolytic enzymes of *Lactococcus lactis*. *J. Dairy Res.* **60**, 269–286.

Tisljar, U., Knight, C.G. & Barrett, A.J. (1990) An alternative quenched fluorescence substrate for Pz-peptidase. *Anal. Biochem.* **186**, 112–115.

Vincent, B., Dive, V., Yiotakis, A., Smadja, C., Maldonado, R., Vincent, J.P. & Checler, F. (1995) Phosphorus-containing peptides as mixed inhibitors of endopeptidase 3.4.24.15 and 3.4.24.16: effect on neurotensin degradation *in vitro* and *in vivo*. *Br. J. Pharmacol.* **115**, 1053–1063.

Yan, T.R., Azuma, N., Kaminogawa, S. & Yamauchi, K. (1987) Purification and characterization of substrate-size-recognizing metalloendopeptidase from *Streptococcus cremoris* H61. *Appl. Environ. Microbiol.* **53**, 2296–2302.

*Véronique Monnet*
*INRA, Laboratoire de Biochimie et Structure des Protéines,*
*Domaine de Vilvert,*
*78352 Jouy en Josas Cedex, France*
*Email: veronique.monnet@jouy.inra.fr*

# 379. Oligopeptidase PepB

## Databanks

*Peptidase classification: clan MX, family M3, MEROPS ID: M03.008*
*NC-IUBMB enzyme classification: none*
*Databank codes*

| Species | SwissProt | PIR | EMBL (cDNA) | EMBL (genomic) |
|---------|-----------|-----|-------------|----------------|
| *Streptococcus agalactiae* | – | – | U49821 | – |

## Name and History

Oligopeptidase PepB from group B *Streptococcus* was originally thought to be a collagenase (Jackson *et al.*, 1994) since it hydrolyzed the synthetic collagen-like substrate FA-Leu+Gly-Pro-Ala (FALGPA). Subsequently, Lin and coworkers (1996) cloned the gene for the enzyme and expressed it in *Escherichia coli*. Neither the highly purified native nor the recombinant enzyme, however, were capable of solubilizing a film of reconstituted rat tail collagen, an activity regarded as an obligatory criterion of a true collagenase. The enzyme also did not degrade gelatin or any other protein tested. Its high amino acid sequence similarity to oligopeptidase F from *Lactococcus lactis* (Chapter 378) suggested that it might really be an oligopeptidase. This was subsequently confirmed when the enzyme was shown to degrade a variety of small bioactive peptides, including bradykinin, neurotensin and peptide fragments of substance P and adrenocorticotropic hormone (Lin *et al.*, 1996). The enzyme is best designated *oligopeptidase PepB* in order to distinguish it from oligopeptidase B (Chapter 126), an unrelated serine protease.

## Activity and Specificity

Oligopeptidase PepB cleaves small basic peptides at positions that do not appear to be determined by the amino acids immediately flanking the sensitive bonds (Lin *et al.*, 1996). The exact rules governing which bonds are cleaved have not yet been elucidated but, as was suggested in the case of thimet oligopeptidase (Chapter 371) (Camargo *et al.*, 1994); the length, flexibility and tertiary structure of the peptide are likely to be important. Oligopeptidase PepB cleaves bradykinin, Arg-Pro-Pro-Gly-Phe+Ser-Pro-Phe-Arg, at the same position as several other related peptidases, including thimet oligopeptidase and neurolysin (Chapter 372) (Serizawa *et al.*, 1995). Oligopeptidase PepB, also cleaves neurotensin, Glp-Leu-Tyr-Glu-Asn-Lys-Pro-Arg+Arg-Pro+Tyr-Ile-Leu. In this peptide, PepB cleaves not only the Arg+Arg bond that is cleaved by thimet oligopeptidase, but also the Pro-Tyr bond that is the specific cleavage site of neurolysin.

Oligopeptidase PepB is conveniently assayed using FA-Leu+Gly-Pro-Ala (obtained from Sigma; see Appendix 2 for full names and addresses of suppliers) as a substrate (Van Wart & Steinbrink, 1981). A 10 ml sample is added to 300 ml of an enzyme assay mixture composed of 0.45 mg ml$^{-1}$ FAL-GPA, 0.24 M NaCl, 50 mM imidazole, 12 mM CaCl$_2$ and 2.4 mM ZnCl$_2$, pH 7.0. The rate of decrease in absorbance at 345 nm is followed in a recording spectrophotometer at room temperature.

The pH optimum for FALGPA hydrolysis is 7.4. No increase in enzyme activity is observed in the presence of DTT (0.2–5 mM). Enzyme activity is completely inhibited in the presence of EDTA or 1,10-phenanthroline (10 mM). A low concentration (2 μM) of Zn$^{2+}$ ions activates purified enzyme preparations. Excess Zn$^{2+}$ ions (10 μM), however, strongly inhibit, whereas the same concentration of Co$^{2+}$ ions activates the enzyme.

## Structural Chemistry

Oligopeptidase PepB consists of a single 69.6 kDa polypeptide chain of 601 amino acids. It has a pI of 4.90. A typical zinc-binding motif, HETGH, is present at positions 385–389.

## Preparation

Oligopeptidase PepB has been purified to homogeneity from cell lysates of serotype III strain 3502 group B *Streptococcus* by a combination of ion-exchange and hydrophobic interaction chromatography (Lin *et al.*, 1996). Much higher yields of the enzyme were subsequently obtained for the recombinant enzyme purified from cell lysates of *Escherichia coli* using essentially the same procedure.

## Biological Aspects

The gene for oligopeptidase PepB has been cloned and expressed in *E. coli* (Lin *et al.*, 1996). The start codon is preceded by a putative ribosomal binding sequence GGAGG 7 bp before the ATG. A 10 bp inverted repeat, probably a transcription termination sequence, is present 23 bp after the stop codon TAA. The deduced amino acid sequence shows 66.4% identity to oligopeptidase F from *Lactococcus lactis* (Chapter 378).

The biological function of oligopeptidase PepB is not known. The intracellular location of the enzyme, however, suggests the possibility that it functions in concert with a peptide transport system that brings in peptides from outside the bacterial cell. Many of the known peptide substrates of oligopeptidase PepB, including bradykinin and substance P, are important mediators of normal inflammatory responses to microorganisms. It has recently been found that group B streptococci rapidly deplete a culture medium of added bradykinin, even in the presence of high concentrations of all 20 amino acids (D.G. Pritchard, unpublished results). Uptake and degradation of these bioactive peptides, therefore, may have the effect of attenuating an inflammatory response to the bacteria and thereby facilitating long-term colonization of mucosal surfaces.

## References

Camargo, A.C.M., Gomes, M.D., Toffoletto, O., Ribeiro, M.J.F., Ferro, E.S., Fernandes, B.L., Suzuki, K., Sasaki, Y. & Juliano, L. (1994) Structural requirements of bioactive peptides for interaction with endopeptidase 22.19. *Neuropeptides* **26**, 281–287.

Jackson, R.J., Dao, M.L. & Lim, D.V. (1994) Cell-associated collagenolytic activity by group B streptococci. *Infect. Immun.* **62**, 5647–5651.

Lin, B., Averett, W.F., Novak, J., Chatham, W.W., Hollingshead, S.K., Coligan, J.E., Egan, M.L. & Pritchard, D.G. (1996) Characterization of PepB, a group B streptococcal oligopeptidase. *Infect. Immun.* **64**, 3401–3406.

Serizawa, A., Dando, P.M. & Barrett, A.J. (1995) Characterization of a mitochondrial metallopeptidase reveals neurolysin as a homologue of thimet oligopeptidase. *J. Biol. Chem.* **270**, 2092–2098.

Van Wart, H.E. & Steinbrink, D.R. (1981) A continuous spectrophotometric assay for *Clostridium histolyticum* collagenase. *Anal. Biochem.* **113**, 356–365.

***David G. Pritchard***
*Department of Biochemistry and Molecular Genetics,*
*University of Alabama at Birmingham,*
*Birmingham, AL 35294, USA*
*Email: dpritchard@biochem_cellbio.bhs.uab.edu*

M

# 380. Introduction: clan MB containing 'metzincins'

## Databanks

MEROPS ID: M6, M7, M8, M10, M11, M12

| Species | SwissProt | PIR | EMBL (cDNA) | EMBL (genomic) |
|---|---|---|---|---|
| **Family M6** | | | | |
| Immune inhibitor A (*Bacillus thuringiensis*) (Chapter 381) | | | | |
| prtV protease | | | | |
|   *Vibrio cholerae* | – | – | – | Y00557 |
| **Family M7** | | | | |
| Snapalysin (Chapter 382) | | | | |
| **Family M8** | | | | |
| Leishmanolysin (Chapter 383) | | | | |
| Others | | | | |
|   *Homo sapiens* | – | – | T08083 | – |
|   *Trypanosoma brucei* | – | – | W04127 | – |
| **Family M10** (Chapter 385) | | | | |
| **Family M11** | | | | |
| Gametolysin (Chapter 384) | | | | |
| **Family M12** (Chapter 404) | | | | |
| **Others** | | | | |
| Pregnancy-associated plasma protein A | | | | |
|   *Homo sapiens* | – | – | U28727 | – |
| Putative zinc metallopeptidase | | | | |
|   *Saccharomyces cerevisiae* | P40483 | S48464 | – | Z47047: chromosome IX complete sequence |

Clan MB is the second group of metallopeptidases in which two of the three zinc ligands are His residues in an 'HEXXH' motif. Unlike clan MA, all the members of the clan are endopeptidases, and the third zinc ligand is a third His or an Asp. In all families in the clan except family M8, the zinc ligands occur in the extended motif **HEXXHXXGXX(H/D)** (Alignment 380.1 on CD-ROM). The Glu is predicted to have the same role in catalysis as Glu142 of thermolysin (Chapter 335), and the conserved Gly allows the formation of the $\beta$ turn that brings the zinc ligands together (Gomis-Rüth *et al.*, 1993). The endopeptidases from clan MB are also known as met-zincins, because there is a conserved Met (Met145 in astacin, Chapter 405) in a turn that underlies the active site. Tertiary structures are known for endopeptidases from three families in the clan.

Clan MB contains the families M6, M7, M10, M11 and M12, and provisionally, M8. In families M10, M11 and M12 the third zinc ligand is His, whereas in families M6 and M7 it is Asp. Families M10 and M12 are each divided into two well-defined subfamilies. One of the subfamilies of family M10 contains the mammalian endopeptidases that degrade extracellular matrix proteins, known as 'matrix metalloproteinases' (MMPs) or 'matrixins', and the other contains the bacterial serralysins. Family M12 contains astacins and closely related enzymes in one subfamily, and the reprolysins

of snake venoms and mammals in the other. Family M8 is provisionally included in clan MB primarily because of the mechanism of inhibition in the enzyme precursor (see below). The HEXXHXXGXXH motif is also found in the hypothetical YIL108W protein from *Saccharomyces cerevisiae* and human pregnancy-associated plasma protein A, which are unrelated to each other or other proteins in the clan, and if shown to be peptidases would be considered the sole representatives of two new families.

The catalytic mechanism of peptidases from clan MB is not as well understood as it is for thermolysin. In thermolysin, two residues are essential for catalysis, the Glu of the HEXXH motif, and His231. The Glu acts as an electrophile and the His as a proton donor/general base. In astacin (family M12), a tyrosine residue (Tyr149) occupies the same position as His231 of thermolysin, and this was originally thought to be a fourth zinc ligand. However, it has now been shown that this residue moves away from the zinc ion when a transition-state analog inhibitor is bound, forming a hydrogen bond with the inhibitor (Gomis-Rüth *et al.*, 1993). Although astacin in its apoenzyme state binds zinc in a novel trigonal-bipyramidal geometry, the zinc ion is tetrahedrally coordinated in the complex with the inhibitor. The phenolate anion of Tyr149 is insecurely bound to the pentacoordinate zinc ion in the apoenzyme as is demonstrated by the small $pK_a$ perturbation of five units. In contrast, the water molecule acidified by the tetracoordinated zinc ion in thermolysin can experience a 10 unit shift in $pK_a$. Therefore, Tyr149 could function as a general base at pH values >5 (Mock & Yao, 1997). Tyr149 is on the Met-turn, but although it is conserved in the astacin and serralysin subfamilies, it is replaced by Pro in the matrix in and reprolysin subfamilies and plays no part in the catalytic mechanism. It is clear from the structure of adamalysin II (Chapter 421) that Glu93 cannot act as a general base, because there is no residue close enough to stabilize the negative charge of a deprotonated carboxylate group (Gomis-Rüth *et al*., 1994). Alignment 380.1 shows an alignment around the zinc ligands and also the Met-turn for families other than M8. Leishmanolysin (family M8) does not contain the complete **HEXXHXXGXX(H/D)** active-site motif, because residue 102 cannot be a zinc ligand and there is presumably an insert between the conserved Gly99 and the third ligand. There are one conserved His and three conserved Glu residues C-terminal to the HEXXH motif, and it is not possible to predict which is the third ligand. There are three conserved methionines C-terminal to the HEXXH, and again it is not possible to predict which forms the Met-turn.

The endopeptidases of families M7, M8 and M11 are secreted enzymes, but there are also integral membrane proteins in families M10 and M12. These families also contain mosaic proteins, and gene structures from M10 suggest that nonpeptidase domains have been acquired by phase 1 exon shuffling. A similar genetic mechanism is thought to apply to family S1, the trypsin family (Chapter 2), and domains found in family M12 are also found in S1 (Rawlings & Barrett, 1990). One difference is that the extra domains are appended to the C-terminus of the peptidase unit in family M12, but to the N-terminus in family S1.

Several peptidases in clan MB, including astacin, meprin (Chapter 406 and 407) and serralysin (Chapter 386), are active on arylamide substrates such as peptide nitroanilides and peptide aminomethylcoumarins. These compounds are hydrolyzed by many serine- and cysteine-type endopeptidases, but not by many metalloendopeptidases. The enzymes of clan MB are scarcely inhibited by phosphoramidon, unlike many in clan MA.

Endopeptidases in clan MB are synthesized as inactive precursors. In some members of the clan, the propeptide inhibits the enzyme because a conserved cysteine interacts with the catalytic zinc ion to preventing binding to a water molecule. This mechanism is found in members of families M8, M10, M11 and M12 (see Alignment 380.2 on CD-ROM).

The endopeptidases in clan MB almost invariably contain either the cysteine-switch activation mechanism or a catalytic Tyr149, but not both; the one known exception is fragilysin (Chapter 403) from family M10, which has neither.

**Family M6** contains a metalloendopeptidase from *Bacillus thuringiensis* known as immune inhibitor A. The bacterium infects lepidoptera and is used in their control. The endopeptidase degrades host antibacterial proteins attacin and cecropin, and is inhibited by 1,10-phenanthroline, but not by phosphoramidon (Lövgren *et al*., 1990). It is synthesized as a preproenzyme. The mature enzyme lacks cysteine residues. A homolog exists in *Vibrio cholerae*.

**Family M7** contains snapalysin from *Streptomyces*. The tertiary structure has been solved for the enzyme from

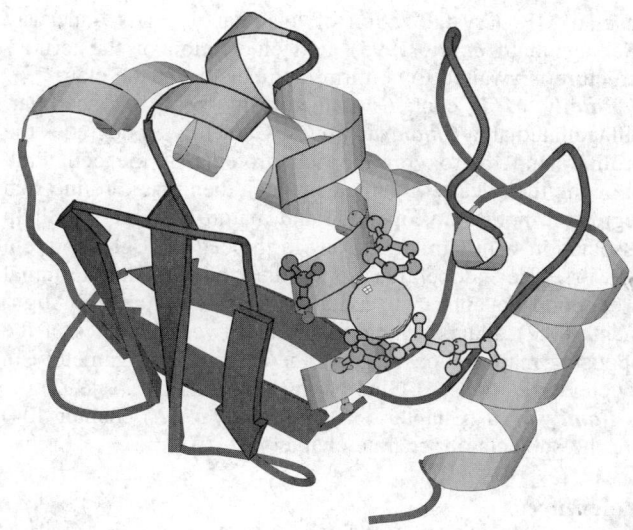

*Figure 380.1*   Richardson diagram of snapalysin. The image was prepared from the Brookhaven Protein Data Bank entry (1KUH) as described in the Introduction (p. xxv). The catalytic zinc ion is shown as a dark gray sphere in CPK representation. The zinc ligands are shown in ball-and-stick representation: His92, His96 and Asp102 (numbering as in Alignment 380.1). The catalytic Glu93 is also shown in ball-and-stick representation.

*Streptomyces caespitosus* and shows a similar tertiary fold to astacin (Kurisu *et al*., 1997). The structure of snapalysin is shown in Figure 380.1. Snapalysin is one of the smallest monomeric peptidases known, the mature enzyme consisting of only 132 residues, 16 kDa. This is reflected in the structure, with a much smaller C-terminal domain than other members of the clan. It is not inhibited by phosphoramidon.

**Family M8** contains leishmanolysin, the major cell surface protein in the promastigote stage of *Leishmania* and other flagellated, parasitic protozoa. As a cell surface protein, the endopeptidase is also known as gp63, and is attached to the cell membrane by a glycosylphosphatidylinositol anchor and heavily glycosylated. Site-directed mutagenesis has shown that all three potential *N*-linked glycosylation sites are utilized, and the last of these (Asn577) is required for attachment of the glycosylphosphatidylinositol moiety (McGwire & Chang, 1996). The enzyme must be present on the parasite cell surface for the organism to invade host macrophages (Chakrabarty *et al*., 1996) and is essential for the survival of the amastigotes in the host phagolysosomes (Seay *et al*., 1996). Leishmanolysin has an acidic pH optimum, which is very unusual for a metallopeptidase, but consistent with the environment of the parasite in host lysosomes. The His and Glu residues in the HEXXH motif have been shown to be essential for catalytic activity, but not for processing of the precursor, implying that another peptidase is responsible for the activation of leishmanolysin (McGwire & Chang, 1996). Activation of the endopeptidase is by the cysteine-switch mechanism, and Cys92 (numbering as in Alignment 380.2) is conserved throughout the family (Macdonald *et al*., 1995). As explained above, the extended motif typical of the clan MB active site is not present, and it is because a cysteine switch is present that family M8 is provisionally included

**M**

in clan MB. Crystallization of leishmanolysin is underway (Schlagenhauf *et al.*, 1995), and elucidation of the tertiary structure is awaited to confirm membership of the clan.

**Family M11** contains gametolysin from the acellular, biflagellated alga *Chlamydomonas*. Gametolysin degrades the proline- and hydroxyproline-rich proteins of the cell wall, allowing the release of gametes, which then fuse. Proline-rich regions in both the propeptide and mature enzyme may help association with similar regions in the cell wall glycoprotein network. The endopeptidase is synthesized with an N-terminal propeptide that appears to contain a cysteine switch (see Alignment 380.2). Flagellar agglutination between gametes of the opposing mating types triggers the activation of gametolysin by cleavage at a Lys3-Glu4 bond (Kinoshita *et al.*, 1992).

**Family M10** (Chapter 385) and *family M12* (Chapter 404) are the subjects of separate chapters.

## References

Chakrabarty, R., Mukherjee, S., Lu, H.G., McGwire, B.S., Chang, K.P. & Basu, M.K. (1996) Kinetics of entry of virulent and avirulent strains of *Leishmania donovani* into macrophages: a possible role of virulence molecules (gp63 and LPG). *J. Parasitol.* **82**, 632–635.

Gomis-Rüth, F.-X., Stöcker, W., Huber, R., Zwilling, R. & Bode, W. (1993) Refined 1.8 Å X-ray crystal structure of astacin, a zinc-endopeptidase from the crayfish *Astacus astacus* L. Structure determination, refinement, molecular structure and comparison with thermolysin. *J. Mol. Biol.* **229**, 945–968.

Gomis-Rüth, F.-X., Kress, L.F., Kellermann, J., Mayr, I., Lee, X., Huber, R. & Bode, W. (1994) Refined 2.0 Å X-ray crystal structure of the snake venom zinc-endopeptidase adamalysin II. Primary and tertiary structure determination, refinement, molecular structure and comparison with astacin, collagenase and thermolysin. *J. Mol. Biol.* **239**, 513–544.

Kinoshita, T., Fukuzawa, H., Shimada, T., Saito, T. & Matsuda, Y. (1992) Primary structure and expression of a gamete lytic enzyme in *Chlamydomonas reinhardtii*: similarity of functional domains to matrix metalloproteases. *Proc. Natl Acad. Sci. USA* **89**, 4693–4697.

Kurisu, G., Kinoshita, T., Sugimoto, A., Nagara, A., Kai, Y., Kasai, N. & Harada, S. (1997) Structure of the zinc endoprotease from *Streptomyces caespitosus*. *J. Biochem. (Tokyo)* **121**, 304–308.

Lövgren, A., Zhang, M., Engström, A., Dalhammar, G. & Landén, R. (1990) Molecular characterization of immune inhibitor A, a secreted virulence protease form *Bacillus thuringiensis*. *Mol. Microbiol.* **4**, 2137–2146.

Macdonald, M.H., Morrison, C.J. & McMaster, W.R. (1995) Analysis of the active site and activation mechanism of the *Leishmania* surface metalloproteinase GP63. *Biochim. Biophys. Acta* **1253**, 199–207.

McGwire, B.S. & Chang, K.P. (1996) Posttranslational regulation of a *Leishmania* HEXXH metalloprotease (gp63) – the effects of site-specific mutagenesis of catalytic, zinc binding, *N*-glycosylation, and glycosyl phosphatidylinositol addition sites on N-terminal end cleavage, intracellular stability, and extracellular exit. *J. Biol. Chem.* **271**, 7903–7909.

Mock, W.L. & Yao, J. (1997) Kinetic characterization of the seralysins: a divergent catalytic mechanism pertaining to astacin-type metalloproteases. *Biochemistry* **36**, 4949–4958.

Rawlings, N.D. & Barrett, A.J. (1990) Bone morphogenetic protein 1 is homologous in part with calcium-dependent serine proteinase. *Biochem. J.* **266**, 622–624.

Schlagenhauf, E., Etges, R. & Metcalf, P. (1995) Crystallization and preliminary X-ray diffraction studies of leishmanolysin, the major surface metalloproteinase from *Leishmania major*. *Proteins: Struct. Funct. Genet.* **22**, 58–66.

Seay, M.B., Heard, P.L. & Chaudhuri, G. (1996) Surface Zn-proteinase as a molecule for defense of *Leishmania mexicana amazonensis* promastigotes against cytolysis inside macrophage phagolysosomes. *Infect. Immun.* **64**, 5129–5137.

# 381. Immune inhibitor A

## Databanks

*Peptidase classification: clan MB, family M6, MEROPS ID: M06.001*
*NC-IUBMB enzyme classification: none*
*Databank codes:*

| Species | SwissProt | PIR | EMBL (cDNA) | EMBL (genomic) |
|---|---|---|---|---|
| *Bacillus thuringiensis* | P23382 | S12399 | X55436 | – |

## Name and History

*Bacillus thuringiensis* is widely used in agriculture for the control of insect pests. The bacterium produces an endotoxin which is found in a crystalline inclusion. In addition to the toxic factor, other proteins are involved in invasion of the host and blocking the host's immune defenses. Edlund *et al.* (1976) isolated two protein materials that could block the immune factors of the hemolymph of saturniid pupae and named them ***immune inhibitor*** A and B. Inhibitor A is heat sensitive and is able to block the lysis of *Escherichia coli* by hemolymph taken from pupae of *Hyalophora cecropia*. The immune inhibitor A was purified by Sidén *et al.* (1979); the homogeneous preparation still retained some proteolytic activity. The proteolytic nature of this factor was demonstrated clearly in a study by Dalhammar & Steiner (1984) and more recently the gene (*ina*) has been cloned and sequenced (Lövgren *et al.*, 1990).

## Activity and Specificity

The purified factor was first tested on casein as a substrate; activity was weak and was blocked by EDTA (Sidén *et al.*, 1979). Subsequently, it was shown that immune inhibitor A was quite active in digesting two classes of antibacterial humoral factors from insects: cecropins and attacins. The specificity of the enzyme against synthetic cecropin A (1–33) was determined by HPLC separation of the peptides and amino acid composition studies (Dalhammar & Steiner, 1984). A number of bonds were cleaved:

KWKL┼FKK┼IEK┼CG┼Q┼N┼IRDG┼IIKA┼GPAVA┼V┼VGQAT

There is no distinctive pattern and the authors note that cecropins are largely in a random coil conformation so the enzyme may make a broad general attack. On the other hand, the protease action on folded proteins such as myoglobin or ribonuclease is weak unless they are first reduced and alkylated to open their structure.

The enzyme is stabilized by divalent cations such as $Ca^{2+}$ or $Mg^{2+}$, but high (10 mM) concentrations of $Ca^{2+}$ are inhibitory. The enzyme is inhibited by EDTA but not by inhibitors of serine or aspartic proteases. It is also inhibited by 1,10-phenanthroline at 0.5 mM, but is insensitive to phosphoramidon at concentrations of 250 µg ml$^{-1}$ (Lövgren *et al.*, 1990).

## Structural Chemistry

The *ina* gene was cloned from *B. thuringiensis* serotype 3a,b and found to code for a protein of 647 residues and a mass of 72 kDa (Lövgren *et al.*, 1990). The sequence HEYGH is the presumptive zinc-binding sequence, the third zinc ligand may be Glu287 or Asp276 (Chapter 380). The protease appears to be distinct from the metalloprotease reported earlier for *B. thuringiensis* (Li & Yousten, 1975) because it is twice as large.

## Purification

Sidén *et al.* (1979) purified the enzyme to homogeneity. Culture filtrate (16 liters) was concentrated on an ultrafiltration membrane to 15 ml, fractionated with 35–60% saturated ammonium sulfate and eluted from a hydroxyapatite column. The protein was homogeneous upon electrophoresis. However, subsequent studies of isoelectric focusing revealed three bands at pI 4.9, 5.1 and 5.3; all had identical substrate specificity on cecropin A (Dalhammar & Steiner, 1984).

## Biological Aspects

The immune inhibitor A appears to be an important participant in the insecticidal properties of *B. thuringiensis*. It rapidly degrades cecropins (37 residue peptides) and attacins (23 kDa proteins) from several species of insect, inhibiting the ability of the insect immune system to kill invading bacteria. It is suggested, on the basis of its ability to digest hide powder, that this enzyme may have a hemorrhagic activity as well (Dalhammar & Steiner, 1984).

It is not clear why digesting antibacterial proteins could be lethal to the insect. Lövgren *et al.* (1990) postulated that immune inhibitor A may digest small peptides within the larvae, e.g. a neuropeptide. Larvae injected with the protease turn black, possibly due to destruction of juvenile hormones.

## References

Dalhammar, G. & Steiner, H. (1984) Characterization of inhibitor A, a protease from *Bacillus thuringiensis* which degrades attacins and cecropins, two classes of antibacterial proteins in insects. *Eur. J. Biochem.* **139**, 247–252.

Edlund, T., Sidén, I. & Boman, H.G. (1976) Evidence for two immune inhibitors from *Bacillus thuringiensis* interfering with the humoral defense system of Saturniid pupae. *Infect. Immun.* **14**, 934–941.

Li, E. & Yousten, A.A. (1975) Metalloprotease from *Bacillus thuringiensis*. *Appl. Microbiol.* **30**, 354–361.

Lövgren, A., Zhang, M., Engstrom, A., Dalhammar, G. & Landén, R. (1990) Molecular characterization of immune inhibitor A, a secreted virulence protease from *Bacillus thuringiensis*. *Mol. Microbiol.* **4**, 2137–2146.

Sidén, I., Dalhammar, G., Telander, B., Boman, H.G. & Somerville, H. (1979) Virulence factors in *Bacillus thuringiensis*: purification and properties of a protein inhibitor of immunity in insects. *J. Gen. Microbiol.* **114**, 45–52.

**M**

*J. Fred Woessner*
*Department of Biochemistry and Molecular Biology,*
*University of Miami School of Medicine,*
*PO Box 016960,*
*Miami, FL 33101, USA*
*Email: fwoessne@mednet.med.miami.edu*

# 382. *Snapalysin*

## Databanks

*Peptidase classification: clan MB, family M7, MEROPS ID: M07.001*
*NC-IUBMB enzyme classification: none*
*Databank codes:*

| Species | SwissProt | PIR | EMBL (cDNA) | EMBL (genomic) |
|---|---|---|---|---|
| *Streptomyces coelicolor* | P43164 | – | Z11929 | – |
| *Streptomyces lividans* | P43162 | – | D00670 | – |
| | | | M81703 | |
| | | | M89476 | |
| *Streptomyces* sp. | P43163 | – | M86606 | – |

Brookhaven Protein Data Bank three-dimensional structures

| Species | ID | Resolution | Notes |
|---|---|---|---|
| *Streptomyces caespitosus* | 1 KUH | 1.6 | |

## Name and History

The name *snapalysin*, derived from *Streptomyces* **n**eutral **p**roteinase **A**, is proposed here for a small neutral proteinase species, the gene for which has been cloned from three independent *Streptomyces* species. The *S. lividans* gene was cloned independently by three groups and the protein named **SnpA** (Butler *et al.*, 1992) or **Prt** (Lichenstein *et al.*, 1992). A gene encoding a very similar protein (also designated SnpA) was isolated from *Streptomyces* sp. C5 (Lampel *et al.*, 1992). The corresponding gene cloned from *Streptomyces coelicolor* 'Muller' was named *MprA* (Dammann & Wohlleben, 1992).

These genes were cloned by screening genomic libraries from each *Streptomyces* strain that had been used to transform protoplasts of *Streptomyces lividans*, which were regenerated on agar plates containing skimmed milk. Positive colonies were replated on LB agar containing skimmed milk to confirm their phenotype (since the agar medium used for regeneration was slightly turbid, which made zone visualization difficult). Zones of clearing around colonies carrying the cloned DNA indicated the presence of the secreted SnpA protease.

## Activity and Specificity

The milk hydrolysis activity of the C5 enzyme was assayed using dry milk (Carnation™) in HEPES buffer at 37°C. The decrease in $A_{580}$ was monitored and the activity defined as that required to decrease the $A_{580}$ of the milk solutions by 0.001 absorbance units $min^{-1}$. The purified enzyme was tested for its ability to cleave synthetic *p*-nitroanilides (Suc-Ala-Ala-Pro-Phe+NHPhNO₂, Bz-Arg+NHPhNO₂, Z-Gly-Gly-Leu+NHPhNO₂ and Leu+NHPhNO₂ representing substrates for chymotrypsin, trypsin, subtilisin and leucine aminopeptidase, respectively.

The milk-hydrolyzing activity is not inhibited by DFP but is completely inhibited by 10 mM 1,10-phenanthroline. No activity is observed with the synthetic substrates. The enzyme appears during exponential growth and shows a pH optimum of 7.0, temperature optimum of 55°C and optimal zinc concentration of 20 mM. It is also inhibited by DTT, iodoacetamide and $HgCl_2$. The *S. lividans* enzyme has a significant hydrolytic activity against azocasein whereas the C5 enzyme does not. This may be due to the presence of additional N-terminal amino acid residues in the *S. lividans* enzyme.

## Structural Chemistry

The *S. lividans* protein was observed as a 24 kDa species on SDS-PAGE analysis after 19 h growth in tryptone soya broth at 30°C. This protein appeared to have a blocked N-terminus. On further growth, however, lower bands were seen at approximately 23 and 20 kDa. Both these smaller species yielded N-terminal protein sequences consistent with the predicted protein sequence derived from the cloned DNA. The purified C5 SnpA behaved as a monomeric species when analyzed by gel filtration.

## Preparation

The C5 SnpA was purified from a 10 liter fermentation culture of an *S. lividans* 66 strain carrying the cloned C5 DNA on the multicopy plasmid vector used in the cloning experiment. Dried milk was included in the fermentation medium to induce maximal protease activity. The conditioned medium was collected using a membrane filtration unit and the protein precipitated by acetone. The protein was collected by centrifugation and resuspended in 50 mM HEPES buffer (pH 6.5). The active enzyme was purified by carboxymethyl-Sepharose chromatography followed by Sepharose G-75 gel-filtration chromatography.

## Biological Aspects

The SnpA is secreted early during liquid fermentation of *S. lividans* cultures overexpressing the *snpA* gene. The putative signal peptide does not appear to be removed, possibly because the single positively charged residue at the N-terminus is not sufficient to retain the protein at the cell membrane for signal peptidase processing. Since proteolytic activity is observed even when the protein is essentially unprocessed (i.e. blocked N-terminus) it seems that it is not produced as a latent proenzyme, although N-terminal processing does occur.

The C5 enzyme is directly processed to the 15–16 kDa species whereas the *S. lividans* enzyme is seen as 24, 23 and finally 20 kDa species. These differences presumably reflect the differences in the amino acid sequences of the two proteins in their N-terminal regions. A divergent gene was identified and designated *snpR*. The predicted protein sequence of this gene showed homology to members of the LysR family of transcriptional regulators (Viale *et al.*, 1991). Both direct and inverted nucleotide repeat sequences were noticed in the intergenic DNA sequence consistent with the LysR family control proteins, which act as transcriptional activators for a relevant structural gene but repressors for their own transcription. Consistent with this hypothesis was the observation that the purified MprR was shown to bind directly to the intergenic DNA in *S. coelicolor* 'Muller' DNA.

## References

Butler, M.J., Davey, C.C., Krygsman, P., Walczyk, E. & Malek, L.T. (1992) Cloning of genetic loci involved in endoprotease activity in *Streptomyces lividans 66*: a novel neutral protease with an adjacent putative regulatory gene. *Can. J. Microbiol.* **38**, 912–920.

Dammann, T. & Wohlleben, W. (1992) A metalloprotease gene from *Streptomyces coelicolor* 'Muller' and its transcriptional activator, a member of the LysR family. *Mol. Microbiol.* **6**, 2267–2278.

Lampel, J.S., Aphale, J.S., Lampel, K.A. & Strohl, W.R. (1992) Cloning and sequencing of a gene encoding a novel extracellular neutral proteinase from *Streptomyces* sp. strain C5 and expression of the gene in *Streptomyces lividans* 1326. *J. Bacteriol.* **174**, 2797–2808.

Lichenstein, H.S., Busse, L.A., Smith, G.A., Nahri, L.O., McGinley, M.O., Rhode, M.F., Katzowitz, J.L. & Sukowski, M.M. (1992) Cloning and characterization of a gene encoding extracellular metalloprotease from *Streptomyces lividans. Gene* **111**, 125–130.

Viale, A.M., Kobayashi, H., Akazama, T. & Henikoff, S. (1991) *rcbR*, a gene coding for a member of the LysR family of transcriptional regulators, is located upstream of the expressed set of ribulose 1,5-biphosphate carboxylase/oxygenase genes in the photosynthetic bacterium *Chromatium vinosum. J. Bacteriol.* **173**, 5224–5229.

*Michael J. Butler*
*Strangeways Research Laboratory,*
*Cambridge CB1 4RN, UK*
*Email: m.butler@uea.ac.uk*

# 383. *Leishmanolysin*

## Databanks

*Peptidase classification: clan MB, family M8, MEROPS ID: M08.001*
*NC-IUBMB enzyme classification: EC 3.4.24.36*
*Databank codes:*

| Species | Gene/clone | SwissProt | PIR | EMBL (cDNA) | EMBL (genomic) |
|---|---|---|---|---|---|
| *Crithidia fasciculata* | *gp63* | Q06031 | A60961 | M94364 | M94365: 5′ end |
| *Leishmania amazonensis* | *gp63* | – | – | L46798 | – |
| *Leishmania chagasi* | *gp63* | P15706 | A44951 | M28527 M80669 M80672 | – |
| *Leishmania donovani* | *gp63* | P23223 | A45621 | M60048 M80671 | – |
| *Leishmania donovani* | *mspS4* | – | – | L19562 | – |
| *Leishmania donovani* | *mspS2* | – | – | L19563 | – |
| *Leishmania guyanensis* | *gp63* | Q00689 | – | M85203 | – |
| *Leishmania guyanensis* | Lg63c7 | – | – | L16776 | – |
| *Leishmania guyanensis* | Lg63c8 | – | – | L16777 | – |
| *Leishmania guyanensis* | Lg63c9 | – | – | L16778 | – |
| *Leishmania guyanensis* | Lg63c10 | – | – | L16779 | – |
| *Leishmania major* | *gp63* | P08148 | PL0221 | Y00647 | – |
| *Leishmania mexicana* | *gp63-c1* | P43150 | A48564 | X64394 | – |

## Name and History

**Leishmanolysin** is a membrane-bound metalloproteinase expressed at the surface of the promastigote of the protozoan parasite *Leishmania*. This enzyme was first identified by surface radioiodination of promastigotes in several species of *Leishmania*. These studies revealed an abundant membrane protein of about 63 kDa (Lepay *et al.*, 1983; Bouvier *et al.*, 1985), which was named p63, or **gp63**. Peptide-mapping analyses showed that these proteins were structurally related and conserved among the recognized species of *Leishmania* (Etges *et al.*, 1985; Colomer-Gould *et al.*, 1985; Bouvier *et al.*, 1987; Etges, 1992). The major surface glycoprotein, or 'surface antigen', of the promastigote was then shown to be a proteolytic enzyme (Etges *et al.*, 1986b; Chaudhuri & Chang, 1988) designated the **promastigote surface protease** (**PSP**), and later characterized as a zinc endopeptidase (Bouvier *et al.*, 1989; Chaudhuri *et al.*, 1989). Although leishmanolysin appeared to be expressed at low levels by the amastigote from an abundant, constitutively transcribed mRNA (Button *et al.*, 1989; Medina-Acosta *et al.*, 1993a,b), the results of efforts to demonstrate the presence of the active protease at the surface of amastigotes have been negative or inconclusive. It was, however, shown that the amastigotes of *L. mexicana* produce significant amounts (0.3–0.4% of the total cellular protein) of a soluble, intracellular metalloproteinase, localized within the parasite's lysosomal compartment, the megasome (Ilg *et al.*, 1993). Since the enzyme is neither surface oriented, nor membrane bound in the amastigote, nor promastigote specific, the designation promastigote surface proteinase (PSP) was clearly no longer tenable, it was therefore proposed to call the metalloproteinases of *Leishmania*, the leishmanolysins (Rawlings & Barrett, 1993; Bouvier *et al.*, 1995). This term also applies to the homologous metalloproteinases produced by other kinetoplastid protozoa. Indeed, surface metalloprotease activity was shown to be not only a highly conserved feature of the genus *Leishmania*, but also to occur at the surface of the monogenetic trypanosomatids *Crithidia fasciculata* and *Herpetomonas samuelpessoai* (Etges, 1992; Schneider & Glaser, 1993; Inverso *et al.*, 1993).

## Activity and Specificity

The peptide bond specificity of leishmanolysin from *L. major* was investigated using synthetic peptide substrates (Figure 383.1); it appears to be defined essentially by the P' subsites of the substrate. Leishmanolysin shows a clear preference for hydrophobic amino acids at the P1' site, although this is not as strict as that of other metalloproteinases since it can also accommodate hydrophilic residues like Ser or Thr (Bouvier *et al.*, 1990; Ip *et al.*, 1990). In one instance, when the DO$\beta$-HLA peptide is hydrolyzed, leishmanolysin exhibits a dipeptidyl-peptidase activity. Further, the fact that the peptide bond Tyr+Leu is often cleaved but not always at the same rate, and in glucagon not at all, clearly indicates the importance of the amino acids flanking the scissile bond. The frequent presence of tyrosine at the P1 site and of basic amino acids at the P2' and P3' sites of susceptible peptides suggests that some specificity exists at these subsites (Bouvier *et al.*, 1990).

| Substrates | Cleavage sites and initial velocity ($v_0$) of hydrolysis |
|---|---|

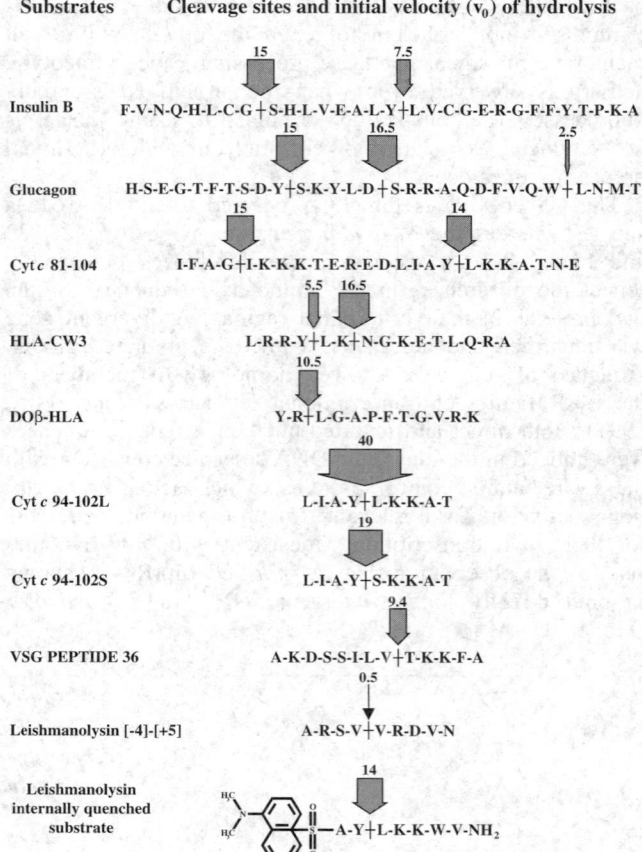

*Figure 383.1* Proteolytic activity of purified leishmanolysin from *Leishmania major* and *L. mexicana* on peptide substrates. Time course experiments were performed using 250 µM of each peptide and 8 nM of leishmanolysin from *L. major* at 37°C in Tris–buffer saline at pH 7.5. Sites of hydrolysis are indicated by arrows whose width is proportional to the initial rate of hydrolysis. The numbers indicate the specific activity of leishmanolysin on each peptide in moles of peptide bond cleaved per second per mole of leishmanolysin (Bouvier *et al.*, 1990). The internally quenched fluorogenic dansyl heptapeptide was treated as described (Bouvier *et al.*, 1993). Hydrolysis of VSG peptide 36 with leishmanolysin from *L. mexicana* was conducted as described by Ip *et al.* (1990).

In an independent study, Ip *et al.* (1990) described the specific cleavage by leishmanolysin from *L. mexicana* of a synthetic peptide with Val in the P1 site, Thr in the P1' site, and lysines in both the P2' and P3' sites (Figure 383.1). These findings strongly suggest that the latter two sites may contribute more to the actual peptide bond specificity of leishmanolysin than those forming the scissile peptide bond in the P1 and P1' sites. The synthetic peptide cytochrome *c* 94–102, which has a Tyr at the P1 site, a Leu at the P1' site, and the basic amino acid Lys in both the P2' and P3' sites, was used as a model substrate to determine a $k_{cat}/K_m$ ratio of $1.8 \times 10^6 \text{ s}^{-1}\text{M}^{-1}$. The fluorogenic internally quenched dansyl heptapeptide (Bouvier *et al.*, 1993) derived from these studies presented a smaller $k_{cat}$ of $14.1 \text{ s}^{-1}$ compared to that for the cytochrome *c* peptide 94–102 ($k_{cat} = 40 \text{ s}^{-1}$), but its substantially greater $K_m$ ($1.6 \times 10^{-6}$ M) resulted in a $k_{cat}/K_m$ ratio of $8.8 \times 10^6$ s$^{-1}$ M$^{-1}$.

Promastigotes from several species of *Leishmania* were shown to cleave azocasein optimally at neutral to alkaline pH (Bouvier *et al.*, 1987), while oxidized insulin B chain was more susceptible at pH 5–6 depending upon the species (Tzinia & Soteriadou, 1991). In assays with defined synthetic peptide substrates, where conformational constraints of the substrates are unlikely to interfere with access to susceptible peptide bonds, both enzymes from *L. mexicana* and *L. major* show a neutral optimum pH (Ip *et al.*, 1990; Bouvier *et al.*, 1990). A more acidic optimum pH was reported for leishmanolysin from *L. amazonensis* (Chaudhuri & Chang, 1988; Chaudhuri *et al.*, 1989). Little is known concerning the activity of the metalloproteinase in the amastigotes.

Leishmanolysin is inhibited by metal-chelating agents like 1,10-phenanthroline, but EDTA and phosphoramidon are totally inactive. The chelating peptide, Z-Tyr-Leu-hydroxamate inhibits leishmanolysin in the high micromolar range (17 µM), which is much lower than the millimolar concentrations needed for inhibition with 1,10-phenanthroline (Bouvier *et al.*, 1990). Purified leishmanolysin is inhibited in solution by human $\alpha_2$-macroglobulin. This inhibition does not occur at the surface of promastigotes, however, indicating that the active site of the membrane-bound leishmanolysin is inaccessible to the relatively large $\alpha_2$-macroglobulin *in vivo* (Heumann *et al.*, 1989), due either to its orientation relative to the membrane or steric interference caused by the abundant surface glycoconjugate of the promastigote, lipophosphoglycan (Pimenta *et al.*, 1994).

## Structural Chemistry

Leishmanolysin is a glycosylated, membrane-bound, zinc-containing metalloproteinase. The enzyme occurs as a dimer of 63 kDa monomers at the surface of the promastigote and in detergent solution. Each monomer of leishmanolysin contains one atom of zinc, as shown by atomic emission and atomic absorption spectroscopy, as well as by biosynthetic labeling with [$^{65}$Zn]Cl$_2$ (Bouvier *et al.*, 1989). An extended substrate-binding site in leishmanolysin was suggested on the basis of sequence similarity between the putative active site of leishmanolysin and the active sites of several metalloendopeptidases (Bouvier *et al.*, 1989; Chaudhuri *et al.*, 1989). Site-directed mutagenesis of the gene coding for leishmanolysin from *L. major* showed that its active site is similar to those of matrix metalloproteinases (Chapter 380) (MacDonald *et al.*, 1995; McGwire & Chang, 1996).

Deglycosylation of the purified enzyme with trifluoromethanesulfonic acid, by treatment with endoglycosidases, or by synthesis by promastigotes treated with the glycosylation inhibitor tunicamycin, showed a decreased molecular mass of approximately 54 kDa (Bouvier *et al.*, 1985; Chang *et al.*, 1986; Funk *et al.*, 1994). These findings were in agreement with the molecular mass of 53 kDa predicted from the genes encoding the mature polypeptides of *L. major* and *L. chagasi* (Button & McMaster, 1988; Miller *et al.*, 1990). The predicted amino acid sequence contains two or three potential *N*-glycosylation sites. Structural analysis

of the *N*-linked oligosaccharides of leishmanolysin from *L. mexicana* revealed glycans consisting of four related biantennary oligomannoses with a unique terminal glucopyranosyl residue on the $\alpha 1$–3 arm in one case (Olafson *et al.*, 1990), representing a rare occurrence of glucose in the *N*-linked glycans of a mature surface glycoprotein.

Leishmanolysin is bound to the membrane of the parasite by a glycosylphosphatidylinositol (GPI) anchor (Etges *et al.*, 1986a; Schneider *et al.*, 1990) similar to that which attaches a wide variety of proteins to membranes of higher eukaryotes (Cross, 1990; McConville & Ferguson, 1993). The enzyme can be solubilized in detergent solution by the action of phosphatidylinositol-specific phospholipases C (PI-PLC) from bacteria or trypanosomes (Bordier *et al.*, 1986; Schneider *et al.*, 1990), revealing the cross-reacting determinant (CRD), which is common to many PI-PLC-solubilized GPI-anchored proteins. All glycoprotein GPI anchors so far characterized have an identical carbohydrate core structure that is completely conserved in leishmanolysin. In *Leishmania*, the anchoring lipid consists of a 1-*O*-alkyl-2-*O*-acyl glycerol with a marked preference for a fully saturated 24 carbon alkyl chain, in contrast to the diacylglycerol of the variable surface glycoprotein (VSG) from *T. brucei*, which contains uniquely the 14 carbon myristic acid. The mature C-terminal asparagine (Asn577) of leishmanolysin to which the GPI anchor is attached is located 25 residues before the predicted C-terminus of the protein (Schneider *et al.*, 1990). Substitution of this Asn residue causes release of all mutant products, indicative of its specificity as a GPI addition site for membrane anchoring of leishmanolysin (McGwire & Chang, 1996).

The secondary structure of leishmanolysin, determined by Raman spectroscopy (Jähnig & Etges, 1988) and circular dichroism (Bouvier *et al.*, 1989) is approximately 50% antiparallel $\beta$ sheet and less than 20% $\alpha$-helical structure. Purified leishmanolysin from *L. major* has recently been crystallized in its mature form (Schlagenhauf *et al.*, 1995). Two crystal forms have been obtained by the vapor diffusion method. One tetragonal crystal form belongs to the space group $P4_12_12$ or the enantiomorph $P4_32_12$, with unit cell parameters of $a = b = 63.6$ Å, $c = 251.4$ Å and containing one molecule per asymmetric unit. The second crystal form is monoclinic, belonging to the space group $C2$, with unit cell dimensions $a = 107.2$ Å, $b = 90.6$ Å, $c = 70.6$ Å, $\beta = 110.6°$, and also containing one molecule per asymmetric unit. Both crystal forms diffract X-rays beyond 2.6 Å resolution. Native diffraction data sets have been collected and the structure determination of leishmanolysin, using a combination of the isomorphous replacement and the molecular replacement methods, is in progress.

## Preparation

Promastigotes of *Leishmania* can be grown axenically to high density. The abundance of leishmanolysin (up to 1% of the total cellular protein in *L. major*) led to its purification to homogeneity (Bouvier *et al.*, 1985, 1995) and crystallization (Schlagenhauf *et al.*, 1995). In a typical purification, starting with $2.3 \times 10^{12}$ promastigotes containing 184 mg of leishmanolysin, the overall yield of amphiphilic leishmanolysin was about 33% (61 mg), and that of the hydrophilic form after treatment with the phospholipase C was about 23% (43 mg). Recombinant protein has been expressed in the baculovirus insect cell expression system, leading to the secretion of a latent enzyme with a yield of 1 mg liter$^{-1}$. Treatment with HgCl$_2$ results in appearance of full proteinase activity and a concomitant loss in $M_r$ (Button *et al.*, 1993).

## Biological Aspects

Leishmanolysin from *L. major*, like human fibroblast collagenase, is synthesized as an inactive precursor protein of 602 amino acid residues with a conventional signal sequence followed by a pro sequence of 100 amino acids (Button & McMaster, 1988). The capacity of leishmanolysin from *L. major* to self-activate by cleavage of its propeptide was investigated using as a substrate a synthetic peptide spanning the Val+Val cleavage site (Ala97-Arg-Ser-**Val**+**Val**-Arg-Asp-Val-Asn105; residues numbered according to Button & McMaster, 1988). The enzyme hydrolyzes this synthetic peptide uniquely at the Val-Val peptide bond, however, with a rate of only 0.46 mol s$^{-1}$, or about 90 times less than that found with cytochrome *c* 94–102L (40 mol s$^{-1}$; Figure 383.1). Recent studies, however, show that recombinant leishmanolysin is not able to self-activate (McGwire & Chang, 1996); activity only appears after treatment with organomercurial compounds that result in removal of the propeptide, ultimately generating the mature N-terminus. This processing, included the removal of a highly conserved cysteine residue, occurs by a *cis* mechanism, since the addition of active enzyme did not enhance latent leishmanolysin activity (MacDonald *et al.*, 1995). Interestingly, the positions of all 18 cysteine residues in the mature protein sequence and the single cysteine residue in the propeptide sequence, which may act as the 'cysteine switch' mechanism proposed for matrix collagenases (discussed in Chapter 380), are absolutely conserved in all *Leishmania* spp.

Genes encoding leishmanolysin are linked at a single chromosome locus in *L. major* and occur in five repeats of 3.1 kb (with 1.8 kb of open reading frame) plus an additional gene 8 kb away, all of which display conserved restriction maps (Button *et al.*, 1989). Similar tandemly linked genes occur in *L. chagasi* (Miller *et al.*, 1990). The genomic organization and expression of the leishmanolysin loci of *L. mexicana* (Medina-Acosta *et al.*, 1993a,b) and *L. chagasi* (Ramamoorthy *et al.*, 1995) have been analyzed in detail.

In spite of the numerous investigations devoted to the characterization of leishmanolysin, the role of the enzyme and the identity of its substrate(s) *in vivo* remain unknown and the subject of intense speculation. Leishmanolysin has been shown to be proteolytically active at the surface of live and fixed promastigotes (Bouvier *et al.*, 1987). The enzyme hydrolyzes a wide range of denatured polypeptides, but fails to cleave synthetic chromogenic or fluorogenic substrates, including those designed for thermolysin. It was recently shown, however, that leishmanolysin from *L. major* and *L. donovani* cleaves CD4 molecules at the surface of human T cells (Hey *et al.*, 1994) as well as protecting the promastigotes from lysis by complement (Brittingham *et al.*, 1995), confirming the possible role of leishmanolysin as a virulence factor (Chang *et al.*, 1990).

Leishmanolysin's involvement in the early phases of infection as a ligand for the mannosyl-fucosyl receptor, as an acceptor for C3b deposition, or as the major surface antigen (Russell, 1987; Puentes *et al.*, 1989) has led to considerable effort to use the surface metalloproteinase as a molecularly defined vaccine. Indeed, intraperitoneal vaccination of inbred mice with purified leishmanolysin from *L. mexicana* with Freund's complete adjuvant conferred significant protection to CBA mice upon challenge infection (Russell & Alexander, 1988); however, attempts to protect BALB/c mice with recombinant leishmanolysin from *L. major* were unsuccessful (Handman *et al.*, 1990). The inability of T cells from human cutaneous leishmaniasis patients to recognize purified leishmanolysin *in vitro* suggests that leishmanolysin or leishmanolysin-derived peptides alone are inadequate to vaccinate human populations (Jaffe *et al.*, 1990). Further, the ability of T lymphocytes from different strains of mice immunized with purified leishmanolysin to recognize the same antigen *in vitro* was shown not to correspond at all to the susceptibility or resistance of the strain of mouse to infection with *Leishmania* promastigotes (Lopez *et al.*, 1991). Nevertheless, a recent report claims significant protection against leishmaniasis in mice injected with naked DNA plasmid encoding leishmanolysin (Xu & Liew, 1995).

## Further Reading

A detailed overview is given by Bouvier *et al.* (1995).

## References

Bordier, C., Etges, R.J., Ward, J., Turner, M.J. & Cardoso de Almeida, M.L. (1986) *Leishmania* and *Trypanosoma* surface glycoproteins have a common glycophospholipid membrane anchor. *Proc. Natl Acad. Sci. USA* **83**, 5988–5991.

Bouvier, J., Etges, R.J. & Bordier, C. (1985) Identification and purification of membrane and soluble forms of the major surface protein of *Leishmania* promastigotes. *J. Biol. Chem.* **260**, 15504–15509.

Bouvier, J., Etges, R. & Bordier, C. (1987) Identification of the promastigote surface protease in seven species of *Leishmania*. *Mol. Biochem. Parasitol.* **24**, 73–79.

Bouvier, J., Bordier, C., Vogel, H., Reichelt, R. & Etges, R. (1989) Characterization of the promastigote surface protease of *Leishmania* as a membrane-bound zinc endopeptidase. *Mol. Biochem. Parasitol.* **37**, 235–246.

Bouvier, J., Schneider, P., Etges, R. & Bordier, C. (1990) Peptide bond specificity of the membrane-bound metalloprotease of *Leishmania*. *Biochemistry* **29**, 10113–10119.

Bouvier, J., Schneider, P. & Malcolm B. (1993) A fluorescent peptide substrate for the surface metalloproteases of *Leishmania*. *Exp. Parasitol.* **76**, 146–155.

Bouvier, J., Schneider, P. & Etges, R. (1995) Leishmanolysin: the surface metalloproteinase of *Leishmania*. *Methods Enzymol.* **248**, 614–628.

Brittingham, A., Morrison, C.J., McMaster, W.R., McGwire, B.S., Chang, K.-P. & Mosser, D.M. (1995) Role of the *Leishmania* surface protease gp63 in complement fixation, cell adhesion, and resistance to complement-mediated lysis. *J. Immunol.* **155**, 3102–3111.

Button, L.L. & McMaster, W.R. (1988) Molecular cloning of the major surface antigen of *Leishmania*. *J. Exp. Med.* **167**, 724–729 (published erratum appeared in *J. Exp. Med.* **171**, 589).

Button, L.L., Russell, D.G., Klein, H.L., Medina-Acosta, E., Karess, R.E. & McMaster, W.R. (1989) Genes encoding the major surface glycoprotein in *Leishmania* are tandemly linked at a single chromosomal locus and are constitutively transcribed. *Mol. Biochem. Parasitol.* **32**, 271–284.

Button, L.L., Wilson, G., Astell, C.R. & McMaster, W.R. (1993) Recombinant *Leishmania* surface glycoprotein GP63 is secreted in the baculovirus expression system as a latent metalloproteinase. *Gene* **134**, 75–81.

Chang, C.S., Inserra, J.T., Kink, J.A., Fong, D. & Chang, K.-P. (1986) Expression and size heterogeneity of a 63 kDa membrane glycoprotein during growth and transformation of *Leishmania mexicana amazonensis*. *Mol. Biochem. Parasitol.* **18**, 197–210.

Chang, K.-P., Chaudhuri, G. & Fong, D. (1990) Molecular determinants of *Leishmania* virulence. *Annu. Rev. Microbiol.* **44**, 499–529.

Chaudhuri, G. & Chang, K.-P. (1988) Acid protease activity of a major surface membrane glycoprotein (gp63) from *Leishmania mexicana* promastigotes. *Mol. Biochem. Parasitol.* **27**, 43–52.

Chaudhuri, G., Chaudhuri, M., Pan, A. & Chang, K.-P. (1989) Surface acid proteinase (gp63) of *Leishmania mexicana*: a metalloenzyme capable of protecting liposome-encapsulated proteins from phagolysosomal degradation by macrophages. *J. Biol. Chem.* **264**, 7483–7489.

Colomer-Gould, V., Quintao, L.G., Keithly, J. & Nogueira, N. (1985) A common major surface antigen on amastigotes and promastigotes of *Leishmania* species. *J. Exp. Med.* **162**, 902–916.

Cross, G.A.M. (1990) Glycolipid anchoring of plasma membrane proteins. *Annu. Rev. Cell Biol.* **6**, 1–10.

Etges, R. (1992) Identification of a surface metalloproteinase on 13 species of *Leishmania* isolated from humans, *Crithidia fasciculata*, and *Herpetomonas samuelpessoai*. *Acta Tropica* **50**, 205–217.

Etges, R.J., Bouvier, J., Hoffman, R. & Bordier, C. (1985) Evidence that the major surface proteins of three *Leishmania* species are structurally related. *Mol. Biochem. Parasitol.* **14**, 141–149.

Etges, R., Bouvier, J. & Bordier, C. (1986a) The major surface protein of *Leishmania* promastigotes is anchored in the membrane by a myristic acid-labeled phospholipid. *EMBO J.* **3**, 597–601.

Etges, R., Bouvier, J. & Bordier, C. (1986b) The major surface protein of *Leishmania* promastigotes is a protease. *J. Biol. Chem.* **261**, 9098–9101.

Funk, V.A., Jardim, A. & Olafson, R.W. (1994) An investigation into the significance of the N-linked oligosaccharides of *Leishmania* gp63. *Mol. Biochem. Parasitol.* **63**, 23–35.

Handman, E., Button, L.L. & McMaster, R.W. (1990) *Leishmania major*: production of recombinant gp63, its antigenicity and immunogenicity in mice. *Exp. Parasitol.* **70**, 427–435.

Heumann, D., Burger, D., Vischer, T., de Colmenares, M., Bouvier, J. & Bordier, C. (1989) Molecular interactions of *Leishmania* promastigote surface protease with human α2-macroglobulin. *Mol. Biochem. Parasitol.* **33**, 67–72.

Hey, A.S., Theander, T.G., Hviid, L., Hazrati, S.M., Kemp, M. & Kharazmi, A. (1994) The major surface glycoprotein (gp63) from *Leishmania major* and *Leishmania donovani* cleaves CD4 molecules on human T cells. *J. Immunol.* **152**, 4542–4548.

Ilg, T., Harbecke, D. & Overath, P. (1993) The lysosomal gp63-related protein in *Leishmania mexicana* amastigotes is a soluble metalloproteinase with an acidic pH optimum. *FEBS Lett.* **327**, 103–107.

Inverso, J.A., Medina-Acosta, E., O'Connor, J., Russell, D.G. & Cross, G.A.M. (1993) *Crithidia fasciculata* contains a transcribed

leishmanial surface proteinase (gp63) gene homologue. *Mol. Biochem. Parasitol.* **57**, 47–54.

Ip, H.S., Orn, A., Russell, D.G. & Cross, G.A.M. (1990) *Leishmania mexicana mexicana* gp63 is a site-specific neutral endopeptidase. *Mol. Biochem. Parasitol.* **40**, 163–172.

Jaffe, C.L., Shor, R., Trau, H. & Passwell, J.H. (1990) Parasite antigens recognized by patients with cutaneous leishmaniasis. *Clin. Exp. Immunol.* **80**, 77–82.

Jähnig, F. & Etges, R. (1988) Secondary structure of the promastigote surface protease of *Leishmania*. *FEBS Lett.* **241**, 79–82.

Lepay, D.A., Nogueira, N. & Cohn, Z. (1983) Surface antigens of *Leishmania donovani* promastigotes. *J. Exp. Med.* **157**, 1562–1572.

Lopez, J.A., Reins, H.-A., Etges, R.J., Button, L.L., McMaster, W.R., Overath, P. & Klein, J. (1991) Genetic control of the immune response in mice to *Leishmania mexicana* surface protease. *J. Immunol.* **146**, 1328–1334.

MacDonald, M.H., Morrison, C.J. & McMaster, W.R. (1995) Analysis of the active site and activation mechanism of the *Leishmania* surface metalloproteinase GP63. *Biochim. Biophys. Acta* **1253**, 199–207.

McConville, M.J. & Ferguson, M.A.J. (1993) The structure, biosynthesis and function of glycosylated phosphatidylinositols in the parasitic protozoa and higher eukaryotes. *Biochem. J.* **294**, 305–324.

McGwire, B.S. & Chang, K.-P. (1996) Posttranslational regulation of a *Leishmania* HEXXH metalloproteinase (gp63). *J. Biol. Chem.* **271**, 7903–7909.

Medina-Acosta, E., Karess, R.E. & Russell, D.G. (1993a) Structurally distinct genes for the surface protease of *Leishmania mexicana* are developmentally regulated. *Mol. Biochem. Parasitol.* **57**, 31–45.

Medina-Acosta, E., Beverley, S.M. & Russell, D.G. (1993b) Evolution and expression of the *Leishmania* surface proteinase (gp63) gene locus. *Infect. Agents Dis.* **2**, 25–34.

Miller, R.A., Reed, S.G. & Parsons, M. (1990) *Leishmania* gp63 molecule implicated in cellular adhesion lacks an Arg-Gly-Asp sequence. *Mol. Biochem. Parasitol.* **39**, 267–274.

Olafson, R.W., Thomas, J.R., Ferguson, M.A.J., Dwek, R.A., Chaudhuri, M., Chang, K.-P. & Rademacher, T.W. (1990) Structures of the N-linked oligosaccharides of gp63, the major surface glycoprotein from *Leishmania mexicana amazonensis*. *J. Biol. Chem.* **265**, 12240–12247.

Pimenta, P.F., Pinto da Silva, P., Rangarajan, D., Smith, D.F. & Sacks, D.L. (1994) *Leishmania major*: association of the differentially expressed gene B protein and the surface lipophosphoglycan as revealed by membrane capping. *Exp. Parasitol.* **79**, 468–479.

Puentes, S.M., Dwyer, D.M., Bates, P.A. & Joiner, K.A. (1989) Binding and release of C3 from *Leishmania donovani* promastigotes during incubation in normal human serum. *J. Immunol.* **143**, 3743–3749.

Ramamoorthy, R., Swihart, K.G., McCoy, J.J., Wilson, M.E. & Donelson, J.E. (1995) Intergenic regions between tandem gp63 genes influence the differential expression og gp63 RNAs in *Leishmania chagasi* promastigotes. *J. Biol. Chem.* **270**, 12133–12139.

Rawlings, N.D. & Barrett, A.J. (1993) Evolutionary families of peptidases. *Biochem. J.* **290**, 205–218.

Russell, D.G. (1987) The macrophage-attachment glycoprotein gp63 is the predominant C3-acceptor site on *Leishmania mexicana* promastigotes. *Eur. J. Biochem.* **164**, 213–221.

Russell, D.G. & Alexander, J. (1988) Effective immunization against cutaneous leishmaniasis with defined membrane antigens reconstituted into liposomes. *J. Immunol.* **140**, 1274–1279.

Schlagenhauf, E., Etges, R. & Metcalf, P. (1995) Crystallization and preliminary X-ray diffraction studies of leishmanolysin, the major surface metalloproteinase from *Leishmania major*. *Proteins* **22**, 58–66.

Schneider, P. & Glaser, T.A. (1993) Characterization of a surface metalloproteinase from *Herpetomonas samuelpessoai* and comparison with *Leishmania major* promastigote surface protease. *Mol. Biochem. Parasitol.* **58**, 277–282.

Schneider, P., Ferguson, M.A.J., McConville, M.J., Mehlert, A., Homans, S.W. & Bordier, C. (1990) Structure of the glycosyl-phosphatidylinositol membrane anchor of the *Leishmania major* promastigote surface protease. *J. Biol. Chem.* **265**, 16955–16964.

Tzinia, A.K. & Soteriadou, K.P. (1991) Substrate-dependent pH optima of gp63 purified from seven strains of *Leishmania*. *Mol. Biochem. Parasitol.* **47**, 83–89.

Xu, D. & Liew, F.Y. (1995) Protection against leishmaniasis by injection of DNA encoding a major surface glycoprotein, gp63, of *L. major*. *Immunology* **84**, 173–176.

*Jacques Bouvier*
*Biochemical Parasitology Unit, Animal Health Sector,*
*Novartis Inc.,*
*CH-1566 St. Aubin, Switzerland*
*Email: jacques.bouvier@chsa.mhs.ciba.com*

# 384. Gametolysin

## Databanks

*Peptidase classification: clan MB, family M11, MEROPS ID: M11.001*
*NC-IUBMB enzyme classification: EC 3.4.24.38*
*Databank codes:*

| Species | SwissProt | PIR | EMBL (cDNA) | EMBL (genomic) |
|---|---|---|---|---|
| *Chlamydomonas reinhardtii* | P31178 | A45287 | D10542 | – |

## Name and History

This enzyme is a matrix metalloproteinase (Kinoshita *et al.*, 1992) produced by a unicellular green alga, *Chlamydomonas*. It was first documented (Claes, 1971) as a gamete enzyme (*autolysin*) responsible for removal of cell walls of gametes of both mating types during mating as a necessary prelude to cell fusion. The enzyme was then purified from the medium of mating gametes (Matsuda *et al.*, 1984) and termed *cell wall lytic enzyme*. The names *gamete autolysin* (Jaenicke *et al.*, 1987), *lysin* (Buchanan & Snell, 1988) and *g-lysin* (Adair & Snell, 1990) were also used in papers. The name *gamete lytic enzyme* (*GLE*) was proposed after determination of its primary structure (Kinoshita *et al.*, 1992). However, this name does not convey the idea of proteolysis or of a metalloprotease. Taking the naming history for this enzyme into consideration, a new name *gametolysin*, which denotes a metalloprotease (*-lysin*) that acts on *gametes*, is proposed here.

## Activity and Specificity

Gametolysin cleaves α-neoendorphin (YGGF↓LRKYPK), dynorphin$_{1-13}$ (YGGF↓LRRIRPKLK), neurotensin (pELYE-NKPRRP↓YIL) and mastoparan (INLKALA↓A↓LAKK), indicating that it preferentially splits the peptide bond between consecutive hydrophobic residues. The presence of an amino acid at the P2′ position is required for cleavage (Matsuda *et al.*, 1990).

Activity can be assayed using walled cells. Bioassays of enzyme activity measure either the formation of protoplasts from walled cells (Snell, 1982) or the liberation of daughter cells from sporangia (Tamaki *et al.*, 1981), the latter assay being much more sensitive than the former (Matsuda, 1988) (Figure 384.1). The pH optimum is about 7.5 and the temperature optimum is 35°C. Activity is slightly stimulated by divalent cations (0.75–1 mM Ca$^{2+}$, Mg$^{2+}$; Matsuda *et al.*, 1984). EDTA, EGTA and 1,10-phenanthroline are inhibitory (Snell, 1982; Matsuda *et al.*, 1984; Jaenicke *et al.*, 1987).

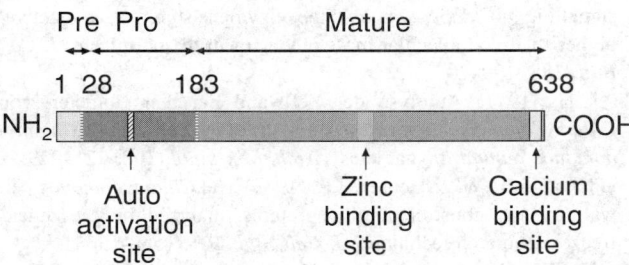

**Figure 384.1**  Bioassay of gametolysin activity by measuring the degradation of the cell walls of mother (sporangial) cells (*left*) and the release of daughter cells (*right*). Scale bar, 10 μm.

Pre Pro          Mature

1  28          183                              638

NH₂ ☐▨░░░░░░░░▨░░░░░░░░░░░░░░░░░░░░☐ COOH

Auto          Zinc          Calcium
activation    binding       binding
site          site          site

**Figure 384.2**  Schematic representation of the prepropeptide structure of gametolysin. Numbers indicate positions of amino acid residues from the N-terminus. (Adapted from Kinoshita *et al.*, 1992).

Inhibition is also shown by $\alpha_2$-macroglobulin, SH-blocking agents such as *p*-chloromercuribenzoic acid and $HgCl_2$ (Matsuda *et al.*, 1984), and by phosphoramidon at a high concentration (Matsuda *et al.*, 1985).

## Structural Chemistry

Gametolysin consists of a 28 amino acid signal peptide, a 155 amino acid propeptide and a 455 amino acid mature peptide ($M_r$ 49 633) (Figure 384.2). A potential site for autocatalytic activation (cysteine switch) is contained in the middle of the propeptide and a zinc-binding site is found within the mature peptide; both sites are similar to those in mammalian collagenases. A putative calcium-binding site is present in the near C-terminal region of the mature peptide (Kinoshita *et al.*, 1992). The active enzyme secreted into the mating medium is a glycoprotein of 60–62 kDa (Matsuda *et al.*, 1984; Buchanan & Snell, 1988), pI 6.5 (Matsuda *et al.*, 1995). A proenzyme with a slightly higher molecular mass than the active enzyme is found in gametes (Buchanan *et al.*, 1989) and also in vegetative cells (Y. Matsuda *et al.*, unpublished results). The proenzyme (65 kDa), which has been purified from both vegetative and gametic cells, contains a 25 amino acid propeptide at the N-terminus of the mature peptide (Y. Matsuda *et al.*, unpublished results). It has therefore lost the putative cysteine switch motif, and can cleave several model peptides (e.g. mastoparan) but is

unable to digest the cell wall (Y. Matsuda *et al.*, unpublished results).

## Preparation

A crude form of active enzyme is prepared by mixing gametes of opposite mating types (Harris, 1989). A relatively simple method to prepare the mating mixture from plate cultures is available from our laboratory (http://www.biol.kobe-u.ac.jp/labs/matsuda/GLE-e.html). The enzyme has been purified about 140-fold (Matsuda *et al.*, 1984), 30-fold (Jaenicke *et al.*, 1987) and 60-fold (Buchanan & Snell, 1988) by different methods. The proenzyme can be released from gametes by freeze-thawing (Buchanan *et al.*, 1989) and purified to homogeneity (Y. Matsuda *et al.*, unpublished results).

## Biological Aspects

The enzyme is present in *C. reinhardtii* and many other species of *Chlamydomonas* whose gametes shed their cell walls during mating as a necessary prelude to cell fusion (Matsuda, 1988; Harris, 1989). Gametolysin from *C. reinhardtii* can digest the cell walls of related species of *Chlamydomonas*, *Gonium* and *Astrephomene* (Matsuda *et al.*, 1987a). It can degrade not only the gametic cell walls but also the cell walls of vegetative cells and those of mother (sporangial) cells (Claes, 1971). Therefore, gametolysin is now becoming very important to prepare protoplasts from the walled *Chlamydomonas* cells for transforming genomes (Kindle, 1990).

In *C. reinhardtii*, the proenzyme is stored in the periplasm of gametes (Matsuda *et al.*, 1987b; Millikin & Weiss, 1984) (Figure 384.3). Activation of proenzyme occurs by the signal of flagellar agglutination between $mt^+$ and $mt^-$ gametes (Snell *et al.*, 1989) or by the exogenous presentation of dibutyryl-cAMP to gametes (Pasquale & Goodenough, 1987). The active enzyme acts specifically on the framework proteins of the cell wall (Goodenough & Heuser, 1985; Matsuda *et al.*, 1985; Imam & Snell, 1988). In normal mating, gametes release all of their enzyme into the medium almost immediately after mixing (Snell, 1982; Buchanan *et al.*, 1989).

Conversion of proenzyme to active enzyme is due to a second enzyme, designated *p*-lysinase (Snell *et al.*, 1989). The *p*-lysinase is a serine protease of about 300 kDa and can activate the proenzyme *in vitro*. De-walled gametes release *p*-lysinase into the medium after dibutyryl-cAMP treatment, suggesting that it is released into the periplasm as a consequence of sexual signaling (Snell *et al.*, 1989).

The proenzyme of gametolysin exists not only in gametes but also in the periplasm of vegetative cells (Matsuda *et al.*, 1987b). Steady-state levels of mRNA (Kinoshita *et al.*, 1992) and proenzyme (Y. Matsuda *et al.*, unpublished results) increase markedly during growth and mitotic cell division in the vegetative cell cycle. The significance of proenzyme in vegetative cells is not clear. One possible explanation for its presence is that it might be required for cell wall expansion during vegetative growth (Adair & Snell, 1990; Kinoshita *et al.*, 1992). Matsuda *et al.* (1987b) reported that when cells are disrupted in a French pressure cell, the proenzyme from vegetative cells is not activated while that from

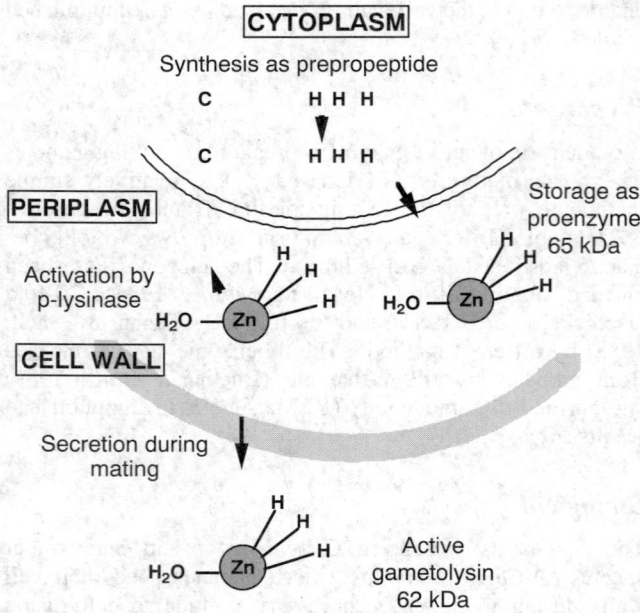

**CYTOPLASM**

Synthesis as prepropeptide

C        H H H

C        H H H

**PERIPLASM**

Storage as proenzyme 65 kDa

Activation by p-lysinase

CELL WALL

Secretion during mating

Active gametolysin 62 kDa

*Figure 384.3* Schematic diagram of the synthesis, storage and secretion of gametolysin in *Chlamydomonas reinhardtii*.

gametes is activated. Proenzyme from vegetative cells can be activated by sonication or by freeze-thawing prior to cell disruption. Although they suggested that these differences are due to the storage form of gametolysin in the two cell types, it is also possible that the mechanism of enzyme activation by *p*-lysinase is different in vegetative and gametic cells (Y. Matsuda *et al.*, unpublished results).

The gene encoding gametolysin maps to linkage group 19 approximately 34 cM from the centromere, between the PF10 and UNI1 loci (C.D. Silflow *et al.*, unpublished results).

### Distinguishing Features

A polyclonal antibody against purified enzyme has been described (Buchanan *et al.*, 1989). However, this antibody reacts with multiple bands in western blots due to shared carbohydrate determinants (Adair, 1985). Therefore, absorption of the antibody with cell walls is necessary to raise its specificity (Buchanan *et al.*, 1989). Recently, antisera specific to the propeptide and mature peptide portions of the enzyme were prepared. Interestingly, the antipropeptide antibody recognizes two forms of proenzyme in vegetative cells: a major 65 kDa form and a minor 72 kDa form (Y. Matsuda *et al.*, unpublished results).

### Related Peptidases

During the mating reaction, two novel endopeptidases, each of which can digest the B chain of cattle insulin, are released into the mating medium, together with gametolysin (Matsuda *et al.*, 1994). The function *in vivo* of these endopeptidases is not known. In the vegetative cell cycle, a serine endopeptidase, named sporangial autolysin (Jaenicke *et al.*, 1987;

Schlösser 1976) or vegetative lytic enzyme (VLE) (Matsuda *et al.*, 1995) is responsible for liberation of daughter cells arising from mitosis. For this enzyme, I propose a new name sporangin in this *Handbook*.

### References

Adair, W.S. (1985) Characterization of *Chlamydomonas* sexual agglutinins. *J. Cell Sci.* **2**(suppl.), 233–260.

Adair, W.S. & Snell, W.J. (1990) The *Chlamydomonas reinhardtii* cell wall: structure, biochemistry and molecular biology. In: *Organization and Assembly of Plant and Animal Extracellular Matrix* (Mecham, R.P. & Adair, W.S., eds). New York: Academic Press, pp. 15–84.

Buchanan, M.J. & Snell, W.J. (1988) Biochemical studies on lysin, a cell wall degrading enzyme released during fertilization in *Chlamydomonas*. *Exp. Cell Res.* **179**, 181–193.

Buchanan, M.J., Imam, S.H., Eskue, W.A. & Snell, W.J. (1989) Activation of the cell wall degrading protease, lysin, during sexual signalling in *Chlamydomonas*: the enzyme is stored as an inactive, higher relative molecular mass precursor in the periplasm. *J. Cell Biol.* **108**, 199–207.

Claes, H. (1971) Autolyse der Zellwand bei den Gameten von *Chlamydomonas reinhardtii* [Autolysis of the cell wall of *Chlamydomonas reinhardtii* gametes]. *Arch. Mikrobiol.* **108**, 221–229.

Goodenough, U.W. & Heuser, J.E. (1985) The *Chlamydomonas* cell wall and its constituent glycoproteins analyzed by the quick-freeze, deep-etch technique. *J. Cell Biol.* **101**, 1550-1568.

Harris, E.H. (1989) The *Chlamydomonas Sourcebook*. San Diego: Academic Press.

Imam, S.H. & Snell, W.J. (1988) The *Chlamydomonas* cell wall degrading enzyme, lysin, acts on two substrates within the framework of the wall. *J. Cell Biol.* **106**, 2211–2221.

Jaenicke, L., Kuhne, W., Spessert, R., Wahle, U. & Waffenschmidt, S. (1987) Cell-wall lytic enzymes (autolysins) of *Chlamydomonas reinhardtii* are (hydroxy)proline-specific proteases. *Eur. J. Biochem.* **170**, 485–491.

Kindle, K.L. (1990) High-frequency nuclear transformation of *Chlamydomonas reinhardtii*. *Proc. Natl Acad. Sci. USA* **87**, 1228–1232.

Kinoshita, T., Fukuzawa, H., Shimada, T., Saito, T. & Matsuda, Y. (1992) Primary structure and expression of a gamete lytic enzyme in *Chlamydomonas reinhardtii*: similarity of functional domains to matrix metalloproteases. *Proc. Natl Acad. Sci. USA* **89**, 4693–4697.

Matsuda, Y. (1988) The *Chlamydomonas* cell walls and their degrading enzymes. *Jpn J. Phycol.* **36**, 246–264.

Matsuda, Y., Yamasaki, A., Saito, T. & Yamaguchi, T. (1984) Purification and characterization of cell wall lytic enzyme released by mating gametes of *Chlamydomonas reinhardtii*. *FEBS Lett.* **166**, 293–297.

Matsuda, Y., Saito, T., Yamaguchi, T. & Kawase, H. (1985) Cell wall lytic enzyme released by mating gametes of *Chlamydomonas reinhardtii* is a metalloprotease and digests the sodium perchlorate-insoluble component of cell wall. *J. Biol. Chem.* **260**, 6373–6377.

Matsuda, Y., Musgrave, A., van den Ende, H. & Roberts, K. (1987a) Cell walls of algae in the Volvocales; their sensitivity to a cell wall lytic enzyme and labeling with an anti-cell wall glycopeptide of *Chlamydomonas reinhardtii*. *Bot. Mag. Tokyo* **100**, 373–384.

Matsuda, Y., Saito, T., Yamaguchi, T., Koseki, M. & Hayashi, K. (1987b) Topography of cell wall lytic enzyme in *Chlamydomonas reinhardtii*: form and location of the stored enzyme in vegetative cell and gamete. *J. Cell Biol.* **104**, 321–329.

Matsuda, Y., Uzaki, T., Iwasawa, N., Tanaka, T. & Saito, T. (1990) Proteolytic activity against model peptides of the cell wall lytic enzyme from *Chlamydomonas*. *Plant Cell Physiol.* **31**, 717–720.

Matsuda, Y., Saito, T. & Taketoshi, T. (1994) Two novel endopeptidases released into the medium during mating of gametes of *Chlamydomonas reinhardtii*. *Plant Cell Physiol.* **35**, 957–961.

Matsuda, Y., Koseki, M., Shimada, T. & Saito, T. (1995) Purification and characterization of a vegetative lytic enzyme responsible for liberation of daughter cells during the proliferation of *Chlamydomonas reinhardtii*. *Plant Cell Physiol.* **36**, 681–689.

Millikin, B.E. & Weiss, R.L. (1984) Distribution of concanavalin A binding carbohydrates during mating in *Chlamydomonas*. *J. Cell Sci.* **66**, 223–239.

Pasquale, S.M. & Goodenough, U.W. (1987) Cyclic AMP functions as a primary sexual signal in gametes of *Chlamydomonas reinhardtii*. *J. Cell Biol.* **105**, 2279–2292.

Schlösser, U.W. (1976) Entwicklungsstadien- und sippenspezifische Zellwand-Autolysine bei der Freisetzung von Fortpflanzungszellen in der Gattung *Chlamydomonas* [Developmental stage- and lineage-specific cell wall autolysins in the release of gametes in the genus *Chlamydomonas*]. *Berl. Deutsch. Bot. Ges.* **89**, 1–56.

Snell, W.J. (1982) Study of the release of cell wall degrading enzymes during adhesion of *Chlamydomonas* gametes. *Exp. Cell Res.* **138**, 109–119.

Snell, W.J., Eskue, W.A. & Buchanan, M.J. (1989) Regulated secretion of a serine protease that activates an extracellular matrix-degrading metalloprotease during fertilization in *Chlamydomonas*. *J. Cell Biol.* **109**, 1689–1694.

Tamaki, S., Matsuda, Y. & Tsubo, Y. (1981) The isolation and properties of the lytic enzyme of the cell wall released by mating gametes of *Chlamydomonas reinhardtii*. *Plant Cell Physiol.* **22**, 127–133.

*Yoshihiro Matsuda*
*Department of Biology,*
*Faculty of Science, Kobe University,*
*Nada, Kobe 657, Japan*
*Email: matsuda@kobe-u.ac.jp*

M

# 385. Introduction: family M10 of interstitial collagenase (clan MB)

## Databanks

*MEROPS ID: M10*

| Species | SwissProt | PIR | EMBL (cDNA) | EMBL (genomic) |
|---|---|---|---|---|
| **Subfamily A: Matrixins** | | | | |
| Collagenase 3 (Chapter 391) | | | | |
| Collagenase 4 (Chapter 392) | | | | |
| Enterotoxin of *Bacteroides fragilis* (Chapter 403) | | | | |
| Envelysin (Chapter 400) | | | | |
| Gelatinase A (Chapter 401) | | | | |
| Gelatinase B (Chapter 402) | | | | |
| Interstitial collagenase (Chapter 389) | | | | |
| Macrophage elastase (Chapter 395) | | | | |
| Matrilysin (Chapter 396) | | | | |
| Membrane-type matrix metalloproteinase (Chapter 399) | | | | |
| Neutrophil collagenase (Chapter 390) | | | | |
| Soybean metalloproteinase (Chapter 398) | | | | |
| Stromelysin 1 (Chapter 393) | | | | |
| Stromelysin 2 (Chapter 394) | | | | |
| Stromelysin 3 (Chapter 397) | | | | |
| Others | | | | |
|    *Caenorhabditis elegans* | – | – | U00038 | – |
|    *Homo sapiens* | – | – | D83647 | – |
|    *Mus musculus* | – | A32963 | – | – |
| **Subfamily B: Serralysins** | | | | |
| Aeruginolysin (Chapter 387) | | | | |
| Mirabilysin (Chapter 388) | | | | |
| Serralysin (Chapter 386) | | | | |

Family M10 is one of the larger families in clan MB. The family contains secreted bacterial and eukaryotic metallo-endopeptidases that are so diverse in sequence that the family is divided into two subfamilies, subfamily A containing the eukaryotic enzymes also known as matrixins, and subfamily B containing the bacterial serralysins. The deep divergence between the subfamilies is very clear from the evolutionary tree (Figure 385.1). As yet, no members of the family are known from fungi, protozoa or archaea.

Both subfamilies contain mosaic proteins. Tertiary structures have been determined for endopeptidases from both subfamilies, at least for the catalytic domains. An alignment around the zinc ligands and known catalytic residues is shown in Alignment 380.1. Although enzymes from both subfamilies bind a single catalytic zinc ion by three His residues, and Glu93 is a known catalytic residue, one fundamental difference is that Tyr149 acts as the general base in the serralysins, but there is no equivalent residue in the matrixins. As we shall see, another significant differences between the subfamilies is the mechanism of activation.

The subfamily B includes serralysin from *Serratia*, aeruginolysin from *Pseudomomas*, and mirabilysin from *Proteus*. The serralysins are synthesized without signal peptides, and instead have a special domain appended to the C-terminus that contains six glycine-rich repeats (Gly-Gly-Xaa-Gly-Asn-Asp) important for binding calcium ions. Similar domains are found in the nonproteolytic hemolysins. These domains are required for secretion by the ATP-binding cassette family of transmembrane transport proteins, and mutational studies have shown that a C-terminal 29 residue segment acts as a translocation signal, and that the C-terminal Asp-Xaa-Xaa-Xaa motif (where Xaa is any hydrophobic amino acid), which is conserved throughout the subfamily, is essential for secretion. This C-terminal domain is not proteolytically removed from serralysins on secretion to the medium. Activation is by removal of an N-terminal propeptide. The structure of aeruginolysin is shown in Figure 385.2. Serralysin is inhibited by a 10 kDa inhibitor produced by *Serratia*, but inhibition by the vertebrate tissue inhibitors of metalloproteinases (TIMPs) has not been shown.

Subfamily A includes a number of animal endopeptidases that are involved in the degradation of extracellular matrix proteins such as collagens and gelatins. Degradation is required for tissue remodeling and repair. Like the serralysins,

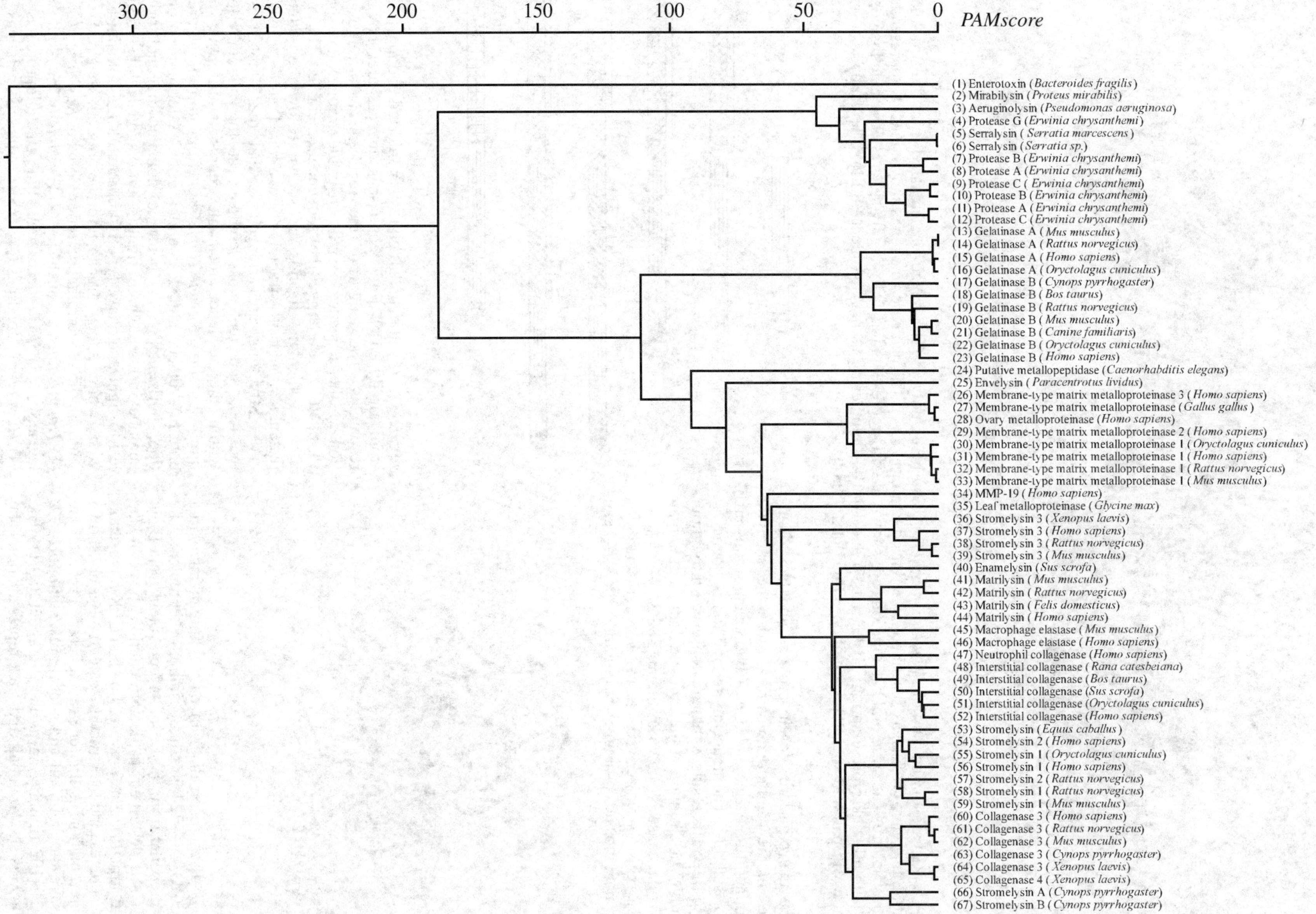

*Figure 385.1*  Evolutionary tree for family M10. The tree was prepared as described in the Introduction (p. xxv). Subfamily A (matrixins) is represented by sequences (13)–(65), and subfamily B (serralysins) by sequences (2)–(12).

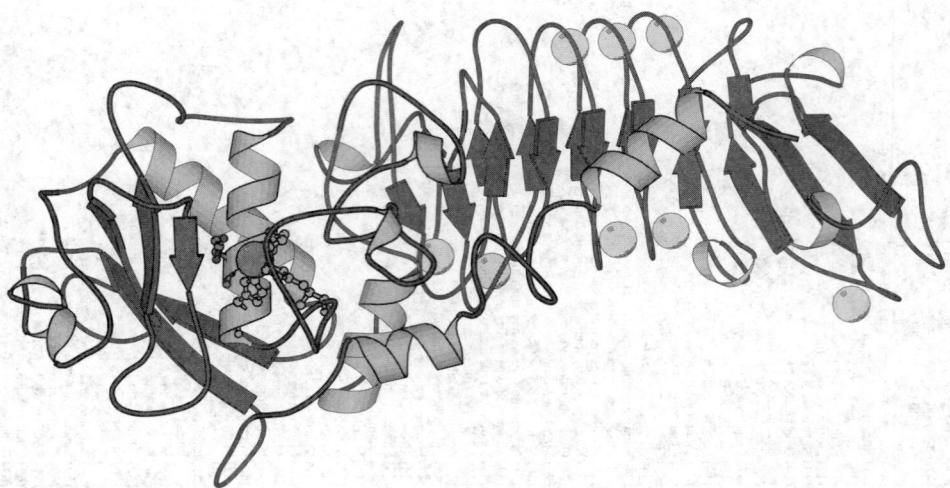

*Figure 385.2*   Richardson diagram of *Pseudomonas aeruginosa* aeruginolysin. The image was prepared from the Brookhaven Protein Data Bank entry (1AKL) as described in the Introduction (p. xxv). The catalytic zinc ion is shown in CPK representation as a dark gray sphere. The structural calcium ions are shown in CPK representation as light gray spheres. The zinc ligands are shown in ball-and-stick representation: His92, His96 and His102 (numbering as in Alignment 380.1). The catalytic residues are also shown in ball-and-stick representation: Glu93 and Tyr149.

most of the matrixins are mosaic proteins and lack disulfide bridges in the catalytic domain. Besides the catalytic zinc ion, an additional zinc and one or two calcium ions are required for stability. The noncatalytic zinc ion is bound by one Asp and three His residues that are conserved throughout the subfamily. The structure of the catalytic domain of human interstitial collagenase is shown in Figure 385.3.

The matrixins are derived from simpler enzymes that were not mosaic proteins, resembling the soybean

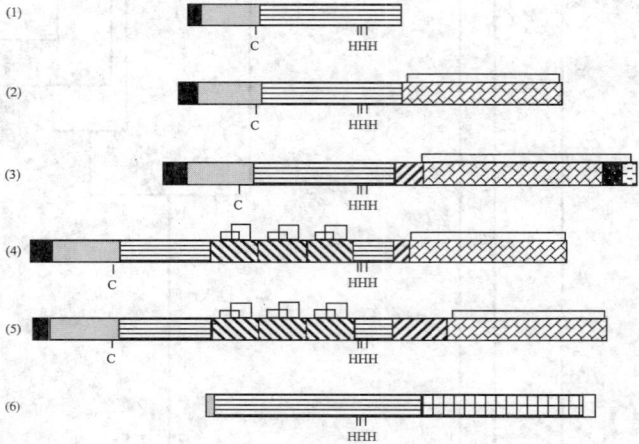

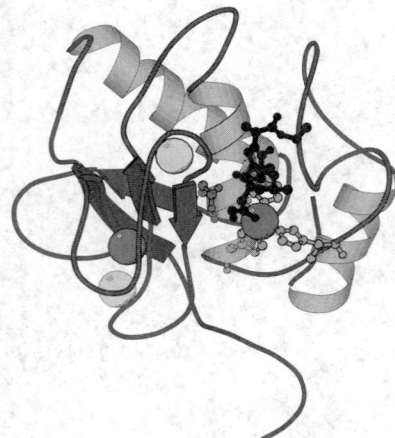

*Figure 385.3*   Richardson diagram of the human interstitial collagenase/RO 31-4724 synthetic inhibitor complex. The image was prepared from the Brookhaven Protein Data Bank entry (2TCL) as described in the Introduction (p. xxv). The catalytic and structural zinc ions are shown in CPK representation as gray spheres, and the structural calcium ions are shown in CPK representation as lighter gray spheres. The catalytic zinc ligands are shown in ball-and-stick representation: His92, His96 and His102 (numbering as in Alignment 380.1). The catalytic Glu93 is shown in ball-and-stick representation. The inhibitor is shown in ball-and-stick representation in black.

*Figure 385.4*   Domain structures for selected endopeptidases from family M10. Endopeptidase sequences are shown as rectangles. The length of the rectangle is a representation of the sequence length. Disulfide bridges are shown as open boxes above the sequence rectangle. The positions of the zinc ligands are shown below the sequence rectangles, and all structures are aligned at the zinc ligands. Structures are arranged so that the simplest is at the top, and the most complex at the bottom. Boxes within the sequence rectangle represent different structural domains. *Key to domains* (from top of figure to bottom): black, signal peptide; gray, N-terminal propeptide; horizontal lines, catalytic domain; diagonal bricks, hemopexin-like domain; right-to-left diagonals, hinge region; left-to-right diagonals, type II fibronectin-like domain; black with white spots, transmembrane region; hatched, calcium-binding C-terminal secretion domain. *Key to sequences:* (1) *Homo sapiens* matrilysin, (2) *H. sapiens* interstitial collagenase, (3) *H. sapiens* membrane metalloproteinase 1, (4) *H. sapiens* gelatinase A, (5) *H. sapiens* gelatinase B, (6) aeruginolysin.

metalloendopeptidase. During the course of evolution, additional domains have been added to the basic peptidase precursor unit, and Figure 385.4 shows a diagrammatic representation of this increasing sophistication. Most matrixins possess at least a C-terminal domain homologous to hemopexin and vitronectin that may help bond the enzyme to the extracellular matrix. A 'hinge' region, rich in Pro, exists between the peptidase unit and the hemopexin-like domain. The hemopexin-like domain was evidently acquired early in animal evolution, because the sea-urchin enzyme envelysin has such a domain that shows most similarity to limunectin, an invertebrate equivalent of vitronectin. Animal enzymes such as interstitial collagenase, neutrophil collagenase, macrophage elastase and stromelysin 1 possess only the hemopexin-like domain in addition to the peptidase precursor. Remarkably, the gelatinases have acquired three additional domains inserted within the peptidase unit; these domains are homologous to type II segments of fibronectin and appear to have been acquired by exon shuffling followed by two tandem duplication events. The size difference between gelatinase A and gelatinase B is attributable to a longer hinge region in gelatinase B. A transmembrane domain has been inserted towards the end of the hemopexin-like domain in membrane-type matrix metalloproteinase 1.

Gene structures are known for several members of the subfamily, and a single exon codes for that region of the protein bearing the zinc ligands. This exon is bounded by phase 1 junctions in all the gene structures so far determined, so it is believed that insertion of extra exons in an ancestral matrixin gene was by phase 1 exon shuffling. Matrilysin is not a mosaic protein, and has a structure similar to that of soybean metalloproteinase 1. As can be seen from the evolutionary tree (Figure 385.1), this simple structure must result from loss of the hemopexin-like domain.

Proteolytic activation of the matrixins involves proteolytic removal of an N-terminal propeptide. The propeptide inhibits by means of a 'cysteine switch' mechanism, with Cys92 of the propeptide acting as the fourth zinc ligand (see Alignment 380.2). The structure of the stromelysin 1 precursor has been determined (Figure 385.5).

The matrixins are inhibited by animal protein inhibitors known as TIMPs (tissue inhibitors of metalloproteinases) which are not known to inhibit any other metalloproteinases.

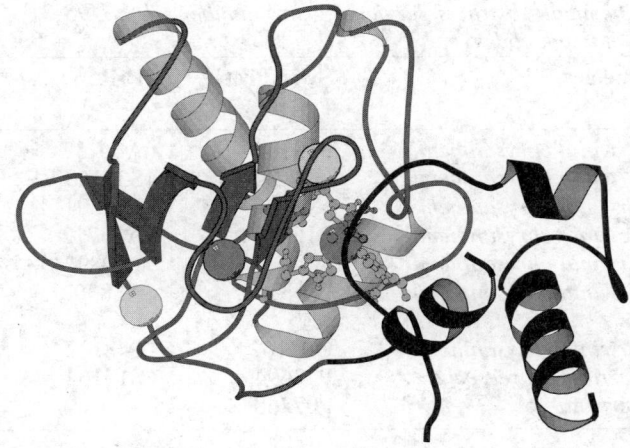

*Figure 385.5* Richardson diagram of a truncated human stromelysin 1 precursor lacking the hemolysin-like domain. The image was prepared from the Brookhaven Protein Data Bank entry (1SLM) as described in the Introduction (p. xxv). The image has been rotated in the Y direction with respect to Figures 385.2 and 385.3. The catalytic and structural zinc ions are shown in CPK representation as gray spheres, and the structural calcium ions are shown in CPK representation as lighter gray spheres. The catalytic zinc ligands are shown in ball-and-stick representation: His92, His96 and His102 (numbering as in Alignment 380.1) and Cys92P (numbering as in Alignment 380.2). The catalytic Glu93 is shown in ball-and-stick representation. The propeptide is shown in black. Because no electron density corresponded to the first 15 residues of the propeptide and 10 residues between the first two helices of the propeptide, these are not shown in the diagram.

An oddity in family M10 is fragilysin. This is an enterotoxin from *Bacteroides fragilis* that causes diarrhea in some strains of cattle, degrading the tight junctions and basement membrane of intestinal epithelial cells. As can be seen from the alignment (Alignment 380.1), fragilysin is more similar to the matrixins than to the serralysins, especially around the Met-turn. Because it lacks Tyr149, the mechanism of action must also be more like that of matrixins. However, it lacks a propeptide, and thus the cysteine-switch activation mechanism, a feature more in keeping with the serralysins.

# 386. Serralysin

## Databanks

*Peptidase classification: clan MB, family M10, MEROPS ID: M10.051*
*NC-IUBMB enzyme classification: EC 3.4.24.40*

*Chemical Abstracts Service registry number: 70851-98-8*
*Databank codes:*

| Species | SwissProt | PIR | EMBL (cDNA) | EMBL (genomic) |
|---|---|---|---|---|
| *Erwinia chrysanthemi* | P19144 | A36137 | M37390 | – |
| *Erwinia chrysanthemi* | Q07295 | S30160 | X70011 | – |
| *Erwinia chrysanthemi* | – | JN0891 | – | – |
| *Erwinia chrysanthemi* | P16316 | A33712 | J04736 | – |
| *Erwinia chrysanthemi* | – | JN0892 | – | – |
| *Erwinia chrysanthemi* | P16317 | A38307 | M59229 B33712 | – |
| *Erwinia chrysanthemi* | Q07162 | S48132 | X71365 | – |
| *Serratia marcescens* | P23694 | S12164 | X55521 | – |
| *Serratia* sp. | P07268 | – | X04127 | – |

Brookhaven Protein Data Bank three-dimensional structures:

| Species | ID | Resolution | Notes |
|---|---|---|---|
| *Serratia marcescens* | 1SAT | 1.75 | |
| *Serratia* sp. | 1SRP | 2 | complex with zinc |

## Name and History

**Serralysin** was first discovered in the culture medium of *Serratia* sp. E-15 (Miyata *et al.*, 1970) and was named from the genus name *Serratia* + *lysin*. Similar proteases have been found to be secreted by other gram-negative bacteria, e.g. *Pseudomonas aeruginosa* (Morihara & Oda, 1992) and *Erwinia chrysanthemi* (Delepelaire & Wandersman, 1989, 1990; Ghigo & Wandersman, 1992a,b). These proteases are quite similar in their physicochemical properties and are grouped together in the serralysin subfamily. Later, it was established on evidence from three-dimensional structures that serralysins, together with matrix metalloproteases, astacins and snake venom proteinases, belong to clan MB (the metzincins) (Bode *et al.*, 1993).

## Activity and Specificity

Serralysin from *Serratia* shows a broad specificity with a preference for small- to medium-sized and hydrophobic residues in P1′ position (notably Gly and Ala). Furthermore, hydrophobic residues in the P2 and P2′ positions are very favorable. The length of the substrate polypeptide chain is important, and peptides of more than four residues are preferred over shorter ones (Morihara *et al.*, 1973). Serralysin also hydrolyzes peptide bonds after Arg in *p*-nitroanilides or methylcoumarylamides (Shibuya *et al.*, 1991; Semba *et al.*, 1992; Tanaka *et al.*, 1992). The alkaline protease from *Pseudomonas aeruginosa* (aeruginolysin, Chapter 387) shows a similar, though slightly different specificity pattern.

The pH optimum is around pH 8, but the range is rather broad, ranging roughly from pH 6 to 10. Activity is inhibited by EDTA, tetramethylenepentamine and 1,10-phenanthroline, but can be regained by addition of $Co^{2+}$, $Zn^{2+}$ and, to a lesser extent, $Cu^{2+}$, $Fe^{2+}$ and $Mn^{2+}$ ions. The serralysins are not inhibited by the classical phosphoramidons which are thermolysin inhibitors. On the other hand, most serralysins form inactive complexes with a 10 kDa protein inhibitor produced by some strains of *Serratia*, *Pseudomonas* or *Erwinia chrysanthemi*. The $K_i$ value of these inhibitors is in the 0.1 µM range and they cross-react with serralysins from different species (Létoffé *et al.*, 1989).

Activity can be assayed by caseinolysis or, more conveniently, with peptidyl-methylcoumarylamide or *p*-nitroanilide substrates. These assays are described by Maeda & Morihara (1995).

## Structural Chemistry

Serralysin is a single-chain protein with a molecular mass of about 55 kDa and a pI between 4.5 and 5.5. All serralysins bind one zinc atom which is required for catalysis and 7–8 calcium ions.

The primary structures of serralysin from *Serratia marcescens* SM6 and *Serratia* sp. E-15 have been determined (Nakahama *et al.*, 1986; Braunagel & Benedik, 1990) as well as the sequences of homologous proteases from *Pseudomonas aeruginosa*, *Erwinia chrysanthemi* and *Proteus mirabilis*. The sequence identity between the various species is about 50%.

The X-ray crystal structures for arginolysin (Baumann *et al.*, 1993; Miyatake *et al.*, 1995), serralysin from *Serratia marcescens* SM6 (Baumann, 1994), *Serratia* sp. E-15 (Hamada *et al.*, 1996) and protease C from *Erwinia chrysanthemi* (U. Baumann *et al.*, unpublished results) have been determined. Also the structure of a complex between *S. marcescens* SM6 protease and the 10 kDa inhibitor Inh from *E. chrysanthemi* has been solved by X-ray crystallography (Baumann *et al.*, 1995). These structures show for the proteolytical domains (approximately residues 20–220) a typical Met-zincin fold, as is found in astacin (Chapter 405), adamalysin (Chapter 421) and various matrix metalloproteases (Stöcker *et al.*, 1995). The catalytically active $Zn^{2+}$ is bound to the three histidines of the HEXXHXXGXXH motif. Tyr216 is sometimes found to be a distant zinc ligand in uncomplexed proteases and is presumably involved in substrate binding and/or stabilizing the tetrahedral transition state (Baumann *et al.*, 1993, 1995; Grams *et al.*, 1996). The C-terminal domain consists of an extended β-sheet sandwich where the GGXGXDX(L/I/F/V)X motifs form a parallel β roll which binds five calcium ions. Two other calcium ions are found in this domain. There is little α-helical structure within this domain, especially in the C-terminal secretion signal, which is entirely folded into a β-sheet structure.

## Preparation

Homogeneous preparations of serralysins can be obtained in the milligram range from the culture supernatant of the relevant bacterial strains (see Maeda & Morihara, 1995, for detailed procedures). The *Pseudomonas aeruginosa* (Duong *et al.*, 1992) and *Erwinia chrysanthemi* (Delepelaire & Wandersman, 1989) proteins have been produced as active proteins in *E. coli* where the parent translocation machinery has to be present in order to obtain soluble protein.

## Biological Aspects

The physiological function of serralysins is not clear. Presumably, they play a role in nutrient digestion/uptake. Most studies of serralysins have focused on their peculiar secretion mechanism (Delepelaire & Wandersman, 1989, 1990; Létoffé *et al.*, 1990; Duong *et al.*, 1992). This pathway, the so-called hemolysin pathway, is shared by a number of proteins, mostly cell toxins, which are secreted into the medium by gram-negative bacteria and are grouped as RTX toxins (repeats in toxins) (Wandersman, 1989; Welch, 1991; Wandersman *et al.*, 1992). Characteristic of the hemolysin pathway is the absence of an N-terminal signal peptide and a one-step instead of the usual two-step secretion mechanism. RTX toxins possess a C-terminal secretion signal and are secreted in one step by the aid of only three membrane proteins through both membranes into the medium. The essential secretion signal is located in the C-terminal 50–100 residues and is preceded by a number of glycine-rich tandem repeats with the consensus sequence GGXGXDX(L/I/F/V)X. The function of these repeats is not well understood, though their number is correlated with the size of the protein and they are thought to enhance secretion efficiency. Serralysins possess 4–6 of these repeats.

All serralysins are synthesized and secreted as inactive zymogens and activated autocatalytically in the medium in the presence of divalent cations, such as $Zn^{2+}$ or $Ca^{2+}$ (Delepelaire & Wanderman, 1989). The propeptide, which is cleaved off during activation, consists of an N-terminal extension of 10–20 residues in length. There must be a major conformational rearrangement occurring during the activation process since the mature N-terminus and the catalytic site are over 45 Å apart in the active enzyme, a distance too large to be bridged by 20 residues.

Serralysin is considered as one of the virulence factors produced during *Serratia* or *Pseudomonas* infection, though its importance seems to be less than that of other toxins. Aeruginolysin from *P. aeruginosa* (Chapter 357) was found to enhance the binding of this organism to mouse corneal epithelium (Gupta *et al.*, 1996). *Serratia* metalloprotease mimics the action of the endogenous shedding protease (Chapter 449) which releases several membrane-bound cytokines and cytokine receptors (Vollmer *et al.*, 1996).

## Distinguishing Features

Serralysins can be distinguished from the thermolysins by inhibition with certain monoclonal antibodies (Kooi & Sokol, 1996).

## Further Reading

For a review, see Maeda & Morihara (1995).

## References

Baumann, U. (1994) Crystal structure of the 50 kDa metalloprotease from *Serratia marcescens*. *J. Mol. Biol.* **242**, 244–251.

Baumann, U., Wu, S., Flaherty, K.M. & McKay, D.B. (1993) Three-dimensional structure of the alkaline protease of *Pseudomonas aeruginosa*: a two-domain protein with a calcium binding parallel beta roll motif. *EMBO J.* **12**, 3357–3364.

Baumann, U., Bauer, M., Létoffé, S., Delepelaire, P. & Wandersman, C. (1995) Crystal structure of a complex between *Serratia marcescens* metalloprotease and an inhibitor from *Erwinia chrysanthemi*. *J. Mol. Biol.* **248**, 653–661.

Bode, W., Gomis-Rüth, F.X. & Stöcker, W. (1993). Astacins, serralysins, snake venom and matrix metalloproteinases exhibit identical zinc-binding environments (HEXXHXXGXXH and Met-turn) and topologies and should be grouped into a common family, the 'metzincins'. *FEBS Lett.* **331**, 134–140.

Braunagel, S.C. & Benedik, M.J. (1990) The metalloprotease gene from *Serratia marcescens* strain SM6. *Mol. Gen. Genet.* **222**, 446–451.

Delepelaire, P. & Wandersman, C. (1989) Protease secretion by *Erwinia chrysanthemi*. *J. Biol. Chem.* **264**, 9083–9089.

Delepelaire, P. & Wandersman, C. (1990) Protein secretion in Gram-negative bacteria. *J. Biol. Chem.* **265**, 17118–17125.

Duong, F., Lazdunski, A., Cami, B. & Murgier, M. (1992) Sequence of a cluster of genes controlling synthesis and secretion of alkaline protease in *Pseudomonas aeruginosa*: relationship to other secretory pathways. *Gene* **121**, 47–54.

Ghigo, J.M. & Wandersman, C. (1992a) A fourth metalloprotease gene in *Erwinia chrysanthemi*. *Res. Microbiol.* **143**, 857–867.

Ghigo, J.M. & Wandersman, C. (1992b) Cloning, nucleotide sequence and characterization of the gene encoding the *Erwinia chrysanthemi* B374 PrtA metalloprotease: a third metalloprotease secreted via a C-terminal secretion signal. *Mol. Gen. Genet.* **236**, 135–144.

Grams, F., Dive, V., Yiotakis, A., Yiallouros, I., Vassiliou, S., Zwilling, R., Bode, W. & Stöcker, W. (1996) Structure of astacin with a transition-state analog inhibitor. *Nature Struct. Biol.* **3**, 671–675.

Gupta, S.K., Masinick, S.A., Hobden, J.A., Berks, R.S. & Hazlett, L.D. (1996) Bacterial proteases and adherence of *Pseudomonas aeruginosa* to mouse cornea. *Exp. Eye Res.* **62**, 641–649.

Hamada, K., Hata, Y., Katsuya, Y., Hiramatsu, H., Fujiwara, T., & Katsube, Y. (1996) Crystal structure of Serratia proteinase from *Serratia* sp. E-15, containing a beta-sheet coil motif at 2.0 Å resolution. *J. Biochem.* **119**, 844–851.

Kooi, C. & Sokol, P.A. (1996) Differentiation between thermolysins and serralysins by monoclonal antibodies. *J. Med. Microbiol.* **45**, 219–225.

M

---

Létoffé, S., Delepelaire, P. & Wandersman, C. (1989) Characterization of a protein inhibitor of extracellular proteases produced by *Erwinia chrysanthemi*. *Mol. Microbiol.* **3**, 79–86.

Létoffé, S., Delepelaire, P. & Wandersman, C. (1990) Protease secretion by *Erwinia chrysanthemi*: the specific secretion functions are analogous to those of *Escherichia coli* A-haemolysin. *EMBO J.* **9**, 1375–1382.

Maeda, H. & Morihara, K. (1995) Serralysin and related bacterial proteases. *Methods Enzymol.* **248**, 395–413.

Miyata, K., Maejima, K., Tomoda, K. & Isono, M. (1970). Serratia protease. I. Purification and general properties of the enzyme. *Agric. Biol. Chem.* **34**, 310–318.

Miyatake, H., Hata, Y., Fujii, T., Hamada, K., Morihara, K. & Katsube, Y. (1995). Crystal structure of the unliganded alkaline protease from *Ps. aeruginosa* IFO3080 and its conformational changes upon ligand binding. *J. Biochem.* **118**, 474–479.

Morihara, K. & Oda, K. (1992) Microbial degradation of proteins. In: *Microbial Degradation of Natural Products* (Winkelmann, G., ed.). Weinheim: Verlag Chemie, pp. 293–364.

Morihara, K., Tsuzuki, H. & Oka, T. (1973) On the specificity of *Pseudomonas aeruginosa* alkaline proteinase with synthetic peptides. *Biochim. Biophys. Acta* **309**, 414–429.

Nakahama, K., Yoshimura, K., Marumoto, R., Kikuchi, M., Lee, I.S., Hase, T. & Matsubara, H. (1986) Cloning and sequencing of *Serratia* protease gene. *Nucleic Acids Res.* **14**, 5843–5855.

Semba, U., Yamamoto, T., Kunisada, T., Shibuya, Y., Tanase, S., Kambara, T. & Okabe, H. (1992) Primary structure of guinea-pig Hageman factor: sequence around the cleavage site differs from the human molecule. *Biochim. Biophys. Acta* **1159**, 113–121.

Shibuya, Y., Yamamoto, T., Morimoto, T., Nishino, N., Kambara, T. & Okabe, H. (1991) *Pseudomonas aeruginosa* alkaline protease might share a biological function with plasmin. *Biochim. Biophys. Acta* **1077**, 316–324.

Stöcker, W., Grams, F., Baumann, U., Reinemer, P., Gomis-Rüth, F.X., McKay, D.B. & Bode, W. (1995) The metzincins – topological and sequential relations between astacins, adamalysins, serralysins and matrixins define a superfamily of zinc-peptidases. *Protein Sci.* **4**, 823–840.

Tanaka, H., Yamamoto, T., Shibuya, Y., Nishino, N., Tanase, S., Miyauchi, Y. & Kambara, T. (1992) Activation of human prekallikrein by *Pseudomonas aeruginosa* elastase. II. Kinetic analysis and identification of scissile bond of prekallikrein in the activation. *Biochim. Biophys. Acta* **1138**, 243–250.

Vollmer, P., Walev, I., Rosejohn, S. & Bhakdi, S. (1996) Novel pathogenic mechanism of microbial metalloproteases – liberation of membrane-anchored molecules in biologically active form. *Infect. Immun.* **64**, 3646–3651.

Wandersman, C. (1989) Secretion, processing and activation of bacterial extracellular proteases. *Mol. Microbiol.* **3**, 1825–1831.

Wandersman, C., Delepelaire, P., Létoffé, S. & Ghigo, J.M. (1992), A signal peptide-independent protein secretion pathway. *Antonie van Leeuwenhoek* **61**, 111–113.

Welch, R.A. (1991) Pore-forming cytolysins of Gram-negative bacteria. *Mol. Microbiol.* **5**, 521–528.

*Ulrich Baumann*
*The Institute of Cancer Research,*
*Cotswold Road,*
*Sutton, Surrey SM2 5NG, UK*
*Email: u.baumann@icr.ac.uk*

# 387. Aeruginolysin

## Databanks

*Peptidase classification: clan MB, family M10, MEROPS ID: M10.056*
*NC-IUBMB enzyme classification: EC 3.4.24.40*
*Databank codes:*

| Species | SwissProt | PIR | EMBL (cDNA) | EMBL (genomic) |
|---|---|---|---|---|
| *Pseudomonas aeruginosa* | Q03023 | A41463 S26699 | D87921 X64558 | – |

Brookhaven Protein Data Bank three-dimensional structures:

| Species | ID | Resolution | Notes |
|---|---|---|---|
| *Pseudomonas aeruginosa* | 1AKL | 2.0 | |
| | 1KAP | 1.64 | complex with Gly-Ser-Asn-Ser |

## Name and History

Morihara (1957) isolated **aeruginolysin** from the cultural supernatant of *Pseudomonas myxogenes* which was cultured by shaking in semisynthetic medium containing high concentration of glucose. Later, the identification of the organism was revised as *Pseudomonas aeruginosa* (Morihara, 1960, 1962). The proteases, whether isolated from various strains of *P. aeruginosa* such as IFO 3080, IFO 3455, T 30 (stock cultures at the Institute for Fermentation of Osaka, Japan) (Morihara, 1963, 1964), or from 18 strains isolated from patients (Morihara & Tsuzuki, 1977), were indistinguishable from one another by immunospecificity, molecular mass and pI. These proteases were at first called **Pseudomonas aeruginosa alkaline proteinases** because of their pH optimum.

## Activity and Specificity

Although the specificity has been studied using the oxidized insulin B chain$N$

FVN┼QHLC┼G┼SHLVE┼ALY┼LVCG┼ER┼G┼FF┼Y┼TPKA

and various synthetic peptides as substrates (Maeda & Morihara, 1995), the rules governing specificity remain unclear. Among various oligopeptides tested Z-Gly-Leu┼Gly-Gly-Ala, Z-Ala-Gly┼Gly-Leu-Ala and Abz-Gly-Phe-Arg┼Leu-Leu-4-nitrobenzylamide are most susceptible to the enzyme. The molecular size of substrates seems more important for their susceptibility to the enzyme than the kinds of amino acid residues on either side of the cleavage site.

Assay of activity can be done with casein or fluorescein isothiocyanate (FITC)-gelatin as described by Maeda & Morihara (1995). Aeruginolysin is inhibited by chelators such as EDTA and 1,10-phenanthroline as well as by peptidyl-mercaptoanilides such as Bz-Phe-Arg-SH ($K_i = 7.5\,\mu M$), but is insensitive to phosphoramidon and Zincov which inhibit many metalloproteinases of clan MA (Maeda & Morihara, 1995).

## Structural Chemistry

The molecular mass is 49 500 Da and the pI is 4.1 (Maeda & Morihara, 1995). The amino acid sequence (470 amino acid residues) is closely similar to that of serralysin (Chapter 386) and related proteinases (Okuda *et al.*, 1990; Duong *et al.*, 1992). The sequence is 55% identical to that of the serralysin from *Serratia* sp. E-15. The consensus sequence that is responsible for zinc binding in astacin, namely HEXXHX-UGUXH (in which X represents any amino acid and U is a bulky hydrophobic residue), can be seen in aeruginolysin.

The tertiary structure has been solved by X-ray analysis to a resolution of 1.6 Å (Baumann *et al.*, 1993) and 2.0 Å (Miyatake *et al.*, 1995). The enzyme has two distinct structural domains, the N- and C-terminal domains. The N-terminal domain is the proteolytic domain; it has an overall tertiary fold and active-site zinc ligation similar to that of astacin (Chapter 405). The C-terminal domain consists of a 21-strand $\beta$ sandwich. Within this domain is a novel so-called $\beta$-roll structure, in which successive $\beta$ strands are wound in a right-handed spiral, and in which $Ca^{2+}$ is bound within the turns between strands by a repeated GGXGXD sequence motif.

## Preparation

The production of aeruginolysin is negligible when *P. aeruginosa* is cultured in a complex medium such as bouillon or nutrient medium (Morihara, 1964). $Ca^{2+}$ is essential for the production in synthetic or semisynthetic medium containing glucose (Morihara, 1964). The highest production was observed in the medium containing 7% glucose, 1% $(NH_4)_2HPO_4$, 1% $Na_2HPO_4.12H_2O$, 0.2% $KH_2PO_4$, 0.05% $MgSO_4.7H_2O$, 0.2% yeast extract, and 2.5% $CaCO_3$, pH 7.0. The organism is cultured with shaking (1.3 Hz, 10 cm amplitude) in a 500 ml flask containing 100 ml of the medium for 3–4 days at 28°C. Purification from the culture filtrate involves ammonium sulfate precipitation (0.3–0.6 saturation), acetone fractionation (30–60%, v/v), and column chromatography on DEAE-cellulose. The enzyme is eluted in 0.02 M phosphate buffer (pH 8) with a 0.2–0.4 M gradient of NaCl. The enzyme is crystallized in the presence of acetone (Maeda & Morihara, 1995).

## Biological Aspects

Aeruginolysin is synthesized as an inactive precursor with short N-terminal extensions (nine amino acid residues), which do not resemble a classical signal peptide (Duong *et al.*, 1992). Secretion across the two membranes probably takes place in one step by the action of a translocation apparatus formed from three accessory proteins, designated aprD, aprE and aprF (Guzzo *et al.*, 1991).

Aeruginolysin can cause a wide range of pathogenic effects in hosts infected with *P. aeruginosa*, including tissue degradation, spreading of infection and septicemia, and inactivation of defense-oriented proteins including immunoglobulins, lysozyme and transferrin (Morihara & Homma, 1985). Thus, a multicomponent *P. aeruginosa* vaccine contains a toxoid form of aeruginolysin as one component (Homma & Tanimoto, 1988).

## Related Peptidases

Serralysin (Chapter 386) and similar enzymes produced by *Pseudomonas fluorescens* (Kumura *et al.*, 1993), *Pseudomonas tolaasii* (Baral *et al.*, 1995) and *Pseudomonas* sp.AFT-36 (Matta *et al.*, 1994) are related to aeruginolysin.

## Further Reading

For a review, see Maeda & Morihara (1995).

## References

Baral, A., Fox, P.F. & O'Connor, T.P. (1995) Isolation and characterization of an extracellular proteinase from *Pseudomonas tolaasii*. *Phytochemistry* **39**, 757–762.

Baumann, U., Wu, S., Flaherty, K.M. & McKay, D.B. (1993) Three-dimensional structure of the alkaline protease of *Pseudomonas aeruginosa*: a two-domain protein with a calcium binding parallel beta roll motif. *EMBO J.* **12**, 3357–3364.

Duong, F., Lazdunski, A., Cami, B. & Murgier, M. (1992) Sequence of a cluster of genes controlling synthesis and secretion of alkaline protease in *Pseudomonas aeruginosa*: relationship to other secretory pathways. *Gene* **121**, 47–54.

Guzzo, J., Pages, J.-M., Duong, F., Lazdunski, A. & Murgier, M. (1991) *Pseudomonas aeruginosa* alkaline protease: evidence for secretion gene and study of secretion mechanism. *J. Bacteriol.* **173**, 5290–5297.

Homma, J.Y. & Tanimoto, H. (1988) A multicomponent *Pseudomonas aeruginosa* vaccine consisting of toxoids protease, elastase, exotoxin A and a common protective antigen (OEP): basic concept of vaccination and prospect of clinical application. *Kitasato Arch. Exp. Med.* **61**, 81–93.

Kumura, H., Mikawa, K. & Saito, Z. (1993) Purification and some properties of proteinase from *Pseudomonas fluorescens* No. 33. *J. Dairy Res.* **60**, 229–237.

Maeda, H. & Morihara, K. (1995) Serralysin and the related bacterial proteinases. *Methods Enzymol.* **248**, 395–413.

Matta, H., Punj, V. & Kalra, M.S. (1994) Isolation and partial characterization of a heat-stable extracellular protease from *Pseudomonas* sp. AFT-36. *Milchwissenschaft-Milk Sci. Int.* **49**, 186–189.

Miyatake, H., Hata, Y., Fujii, T., Hamada, K., Morihara, K. & Katsube, Y. (1995) Crystal structure of the unliganded alkaline protease from *Pseudomonas aeruginosa* IFO 3080 and its conformational changes on ligand binding. *J. Biochem. (Tokyo)* **118**, 474–479.

Morihara, K. (1957) Studies on the protease of *Pseudomonas*. II. Crystallization of the protease and its physicochemical and general properties. *Bull. Agric. Chem. Soc. Japan* **21**, 11–17.

Morihara, K. (1960) Studies on the protease of *Pseudomonas*. VII. An immunological study of the crystalline protease. *Bull. Agric. Chem. Soc. Japan* **24**, 467–473.

Morihara, K. (1962) Studies on the protease of *Pseudomonas*. VIII. Proteinase production of various *Pseudomonas* species, especially *Ps. aeruginosa*. *Agric. Biol. Chem.* **26**, 842–847.

Morihara, K. (1963) *Pseudomonas aeruginosa* proteinase. I. Purification and general properties. *Biochim. Biophys. Acta* **73**, 113–124.

Morihara, K. (1964) Production of elastase and proteinase by *Pseudomonas aeruginosa*. *J. Bacteriol.* **88**, 745–757.

Morihara, K. & Homma, J.Y. (1985) *Pseudomonas* proteases. In: *Bacterial Enzymes and Virulence* (Holder, I.A., ed.). Boca Raton: CRC Press, pp. 41–79.

Morihara, K. & Tsuzuki, H. (1977) Production of protease and elastase by *Pseudomonas aeruginosa* strains isolated from patients. *Infect. Immun.* **15**, 679–685.

Okuda, K., Morihara, K. Atsumi, Y., Takeuchi, H., Kawamoto, S., Kawasaki, H., Suzuki, K. & Fukushima, J. (1990) Complete nucleotide sequence of the structural gene for alkaline proteinase from *Pseudomonas aeruginosa* IFO 3455. *Infect. Immun.* **58**, 4083–4088.

*Kazuyuki Morihara*
*University of East Asia,*
*Graduate School,*
*Institute for Applied Life Science,*
*2-1 Ichinomiya-gakuencho, Shimonoseki,*
*Yamaguchi 751, Japan*
*Email: morihara@po.pios.cc.toua-u.ac.jp*

# 388. Mirabilysin

## Databanks

*Peptidase classification: clan MB, family M10, MEROPS ID: M10.057*
*NC-IUBMB enzyme classification: none*
*Databank codes:*

| Species | SwissProt | PIR | EMBL (cDNA) | EMBL (genomic) |
|---|---|---|---|---|
| Proteus mirabilis | Q11137 | – | U25950 | – |

## Name and History

Among the virulence components known to be expressed by the urinary tract pathogen *Proteus mirabilis* is a metallo-protease referred to as **mirabilysin**. This is an extracellular protease that can be isolated when *P. mirabilis* is grown on media containing a suitable substrate such as skim milk agar (Wassif *et al.*, 1995). Many strains of *P. mirabilis* produce this protease, also referred to as an IgA protease because of its specificity for both serum and secretory forms of IgA1 (immunoglobulin A1) and IgA2, as well as IgG (Loomes *et al.*, 1990, 1992, 1993; Senior *et al.*, 1991).

## Activity and Specificity

The purified enzyme is stable for long periods at 4°C at pH 8.0 and is unaffected by heating at 60°C for 5 min (Loomes *et al.*, 1992). Enzyme activity is detected from pH 6.0 to 10.0, with an optimum activity at pH 8.0 (Loomes *et al.*, 1992). The addition of either $Mg^{2+}$ or $Ca^{2+}$ increases activity, while EDTA is inhibitory. The finding of activity in the absence of added divalent cations (ca. 65% of maximal activity) suggests that the purified protein has divalent metal ions already bound to it, in agreement with the observations of Senior *et al.* (1987, 1988) and Loomes *et al.* (1992).

In addition to EDTA, 1,10-phenanthroline and $\alpha,\alpha'$-dipyridyl both inhibit the activity of mirabilysin, while incubation in the presence of either DFP or iodoacetamide has no effect. Phosphoramidon, an inhibitor of some metalloproteases, has no effect on mirabilysin (Loomes *et al.*, 1992), a result that is typical of peptidases in family M10.

The *zapA* gene that encodes mirabilysin has been cloned and its gene product overexpressed (Wassif *et al.*, 1995). SDS-PAGE was used to examine the substrate specificity of the recombinant mirabilysin. Both human serum IgA1 and human IgG were digested by the recombinant metalloprotease. Recombinant mirabilysin degrades human IgA1 in a time-dependent manner resulting in complete digestion of the IgA1 substrate into numerous smaller fragments. Mouse IgA is also a substrate for mirabilysin, an unusual finding because the action of IgA proteases is normally limited to IgA from human and related primates (Kornfeld & Plaut, 1981 (cf. Chapter 508). Recombinant mirabilysin also digested human serum IgA2 and secretory IgA, as well as casein and azocasein, in agreement with the observations of others (Senior *et al.*, 1991; Loomes *et al.*, 1993). The enzyme does not have activity against serum albumin, cytochrome *c*, *P. mirabilis* flagellin, ovalbumin, phosphorylase *b* or other proteins, indicating that the specificity of mirabilysin is limited to immunoglobulins (Senior *et al.*, 1991; Loomes *et al.*, 1993; Wassif *et al.*, 1995).

## Structural Chemistry

The amino acid sequence of mirabilysin has been deduced from the nucleotide sequence of the *zapA* gene. A 1473 bp open reading frame provides convincing evidence that this is the structural gene of the secreted metalloprotease. The deduced amino acid sequence of mirabilysin predicts an acidic protein (pI 4.30) composed of 491 residues with a total molecular mass of 54 000 Da. Based on gel electrophoresis analysis, recombinant mirabilysin is composed of a single protein of 55 kDa. The enzyme is rich in glycine (11.2 mol %).

Computerized protein homology searches comparing mirabilysin to other protein sequences from both prokaryotes and eukaryotes revealed that mirabilysin is homologous to members of the serralysin subgroup of family M10, which includes the proteases of *Serratia marcescens* (serralysin: Chapter 386), *Erwinia chrysanthemi*, and *Pseudomonas aeruginosa* (aeruginolysin: Chapter 387). Four distinctive protein signature motifs are found in the mirabilysin sequence that support this idea. The first signature motif, encompassing Asn178 to Gly203, is homologous to the zinc-binding region of this protein family. The location of Glu187 is appropriate for this residue to act as the catalytic base (Baumann, 1994). Additionally, the three histidines in this region, His186, His190 and His196, are correctly located to function as a putative zinc-binding site. The second motif found in the serralysin subgroup of proteases is the Met-turn of metzincins (Baumann, 1994), which in mirabilysin is located at Thr222 to Tyr226, with Met224 being the conserved residue in this motif.

The proteases of the serralysin subgroup are secreted by the ATP-binding cassette (ABC) superfamily of prokaryotic and eukaryotic transporters (Higgins, 1992; Pugsley, 1992; Wandersman, 1992; see also Chapters 386, 387). In these systems, three transport proteins probably combine to form zones of adhesion between the inner and outer membranes, through which the proteins are secreted (Delepelaire & Wandersman, 1991; Holland *et al.*, 1990). The proteins secreted by this system do not possess an N-terminal signal sequence, but they do contain a C-terminal targeting signal that is essential for secretion (Delepelaire & Wandersman, 1990). This sequence ends with the conserved four amino acid motif DXXX (Ghigo & Wandersman, 1994). In the case of mirabilysin and serralysin this sequence is DFIV (Nakahama *et al.*, 1986). A further characteristic of this group of secreted proteins is that the member proteins contain 4–13 repeats of the consensus sequence GGXGXD near the C-terminal secretion signal. This motif produces a $\beta$-roll conformation that serves as a $Ca^{2-}$-binding site (Baumann, 1994). Mirabilysin has three of these sites located at Gly343 to Asp348, Gly361 to Asp366, and Gly379 to Asp384. A fourth $Ca^{2+}$-binding motif may also exist (Gly388 to Asn393), although it lacks consensus at the last residue with asparagine substituted for aspartate.

## Preparation

Mirabilysin is purified by hydrophobic chromatography on phenyl-Sepharose using the procedure described by Wassif *et al.* (1995). This purification scheme typically provides between 200 and 500 μg of protein per liter of culture supernatant. The relative protease activity of the recombinant enzyme is ca. 15–17 units per μg of protein as measured by azocasein digestion.

## Biological Aspects

Mirabilysin is similar to the proteases of *P. aeruginosa* (Chapter 387) and *S. marcescens* (Chapter 386), both of which degrade IgA and IgG (Döring *et al.*, 1981; Molla *et al.*, 1988), and to the proteases of *E. chrysanthemi* (Létoffé *et al.*, 1990). Although the role of mirabilysin in the virulence of *P. mirabilis* has not been directly confirmed, immunoblotting of urine from patients who had *P. mirabilis* urinary tract infections showed that 64% of the specimens with IgA contained IgA heavy-chain fragments identical in size to those formed when purified IgA was degraded by pure protease (Senior *et al.*, 1991). Thus, it is likely that the protease is produced and is active upon urinary tract IgA during *Proteus* infections.

## Further Reading

For reviews, see Hooper (1994) and Maeda & Morihara (1995).

## References

Baumann, U. (1994) Crystal structure of the 50 kDa metallo protease from *Serratia marcescens*. *J. Mol. Biol.* **242**, 244–251.

Delepelaire, P. & Wandersman, C. (1990) Protein secretion in gram-negative bacteria. The extracellular metalloprotease B from *Erwinia chrysanthemi* contains a C-terminal secretion signal analogous to that of *Escherichia coli* α-hemolysin. *J. Biol. Chem.* **265**, 17118–17125.

Delepelaire, P. & Wandersman, C. (1991) Characterization, localization and transmembrane organization of the three proteins PrtD, PrtE and PrtF necessary for protease secretion by the gram-negative bacterium *Erwinia chrysanthemi*. *Mol. Microbiol.* **5**, 2427–2434.

Döring, G., Obernesser, H.-J. & Botzenhart, K. (1981) Extracellular toxins of *Pseudomonas aeruginosa*. II. Effect of two proteases on human immunoglobulins IgG, IgA and secretory IgA. *Zentralbl. Bakteriol. Parasitenkd. Infektionskr. Hyg. Abt. 1 Orig. Reihe A* **249**, 89–98.

Ghigo, J.M. & Wandersman, C. (1994) A carboxyl-terminal four-amino acid motif is required for secretion of the metalloprotease PrtG through the *Erwinia chrysanthemi* protease secretion pathway. *J. Biol. Chem.* **269**, 8979–8985.

Higgins, C. (1992) ABC-transporters: from microorganisms to man. *Annu. Rev. Cell Biol.* **8**, 67–113.

Holland, I., Kenny, B. & Blight, M. (1990) Haemolysin secretion from *E. coli*. *Biochimie* **72**, 131–141.

Hooper, N.M. (1994) Families of zinc metalloproteases. *FEBS Lett.* **354**, 1–6.

Kornfeld, S.J. & Plaut, A.G. (1981) Secretory immunity and the bacterial IgA proteases. *Rev. Infect. Dis.* **3**, 521–534.

Létoffé, S., Delepelaire, P. & Wandersman, C. (1990) Protease secretion by *Erwinia chrysanthemi*: the specific secretion functions are analogous to those of *Escherichia coli* α-haemolysin. *EMBO J.* **9**, 1375–1382.

Loomes, L.M., Senior, B.W. & Kerr, M.A. (1990) A proteolytic enzyme secreted by *Proteus mirabilis* degrades immunoglobulins of the immunoglobulin A1 (IgA1), IgA2, and IgG isotypes. *Infect. Immun.* **58**, 1979–1985.

Loomes, L.M., Senior, B.W. & Kerr, M.A. (1992) Proteinases of *Proteus* spp.: purification, properties, and detection in urine of infected patients. *Infect. Immun.* **60**, 2267–2273.

Loomes, L.M., Kerr, M.A. & Senior, B.W. (1993) The cleavage of immunoglobulin G *in vitro* and *in vivo* by a proteinase secreted by the urinary tract pathogen *Proteus mirabilis*. *J. Med. Microbiol.* **39**, 225–232.

Maeda, H. & Morihara, K. (1995) Serralysin and related bacterial proteinases. *Methods Enzymol.* **248**, 395–413.

Molla, A., Kagimoto, T. & Maeda, H. (1988) Cleavage of immunoglobulin G (IgG) and IgA around the hinge region by proteases from *Serratia marcescens*. *Infect. Immun.* **56**, 916–920.

Nakahama, K., Yoshimura, K., Marumoto, R., Kikuchi, M., Lee, I., Hase, T. & Matsubara, H. (1986) Cloning and sequencing of *Serratia* protease gene. *Nucleic Acids Res.* **14**, 5843–5855.

Pugsley, A. (1992) Superfamilies of bacterial transport systems with nucleotide binding components. *Symp. Soc. Gen. Microbiol.* **47**, 223–248.

Senior, B.W., Albrechtsen, M. & Kerr, M.A. (1987) *Proteus mirabilis* strains of diverse type have IgA protease activity. *J. Med. Microbiol.* **24**, 175–180.

Senior, B.W., Albrechtsen, M. & Kerr, M.A. (1988) A survey of IgA protease production among clinical isolates of Proteeae. *J. Med. Microbiol.* **25**, 27–31.

Senior, B.W, Loomes, L. & Kerr, M. (1991) The production and activity *in vivo* of *Proteus mirabilis* IgA protease in infections of the urinary tract. *J. Med. Microbiol.* **35**, 203–207.

Wandersman, C. (1992) Secretion across the bacterial outer membrane. *Trends Genet.* **8**, 317–321.

Wassif, C., Cheek, D. & Belas, R. (1995) Molecular analysis of a metalloprotease from *Proteus mirabilis*. *J. Bacteriol.* **177**, 5790–5798.

***Robert Belas***
*Center of Marine Biotechnology,*
*The University of Maryland Biotechnology Institute,*
*701 East Pratt Street,*
*Baltimore, MD 21202, USA*
*Email: belas@umbi.umd.edu*

# 389. *Interstitial collagenase*

## Databanks

*Peptidase classification: clan MB, family M10, MEROPS ID: M10.001*
*NC-IUBMB enzyme classification: EC 3.4.24.7*
*ATCC entries: 57684, 57685, 79062, 79063 (human)*
*Chemical Abstracts Service registry number: 9001-12-1*
*Databank codes:*

| Species | SwissProt | PIR | EMBL (cDNA) | EMBL (genomic) |
|---|---|---|---|---|
| *Bos taurus* | P28053 | S14654 S14655 S20336 | X58256 | – |
| *Homo sapiens* | P03956 | A00996 A37308 A44518 B60964 D29157 S06132 S22766 | M13509 M15996 X05231 X54925 | D26110: upstream region M16567: 5′ end U78045: complete gene |
| *Oryctolagus cuniculus* | P13943 | A27500 B27500 | M25663 | M17820: exon 1 M17821: exons 2, 3 and 4 M17822: exons 5 and 6 M17823: exons 9 and 10 and complete CDS M19240: exons 7 and 8 |
| *Rana catesbeiana* | Q11133 | – | S75623 | – |
| *Sus scrofa* | P21692 | S13597 S15986 | X54724 | – |

Brookhaven Protein Data Bank three-dimensional structures:

| Species | ID | Resolution | Notes |
|---|---|---|---|
| *Homo sapiens* | 1CGE | 1.9 | recombinant peptidase domain |
|  | 1CGF | 2.1 | catalytic domain binary complex |
|  | 1CGL | 2.4 | recombinant catalytic domain |
|  | 1HFC | 1.56 | recombinant |
|  | 1TCL | 2.2 | catalytic domain; complex with inhibitor |
| *Sus scrofa* | 1FBL | 2.5 | |

## Name and History

The name collagenase was first used to describe a bacterial enzyme that degraded collagen, but Gross & Lapière (1962) identified an enzyme produced by resorbing tadpole tail in culture that degraded triple-helical collagen at neutral pH. This discovery provoked an important shift in understanding concerning the susceptibility of collagen to attack by endogenous proteinases. The term *interstitial collagenase* was used to describe this enzyme as it specifically cleaved triple-helical collagen across all three α chains at a single point three-quarters of the way from the N-terminal end of the interstitial collagens types I, II and III. The human α1(I) chain is cleaved between Gly-Pro-Gln-Gly775⊢Ile-Ala-Gly-Gln. The enzyme belongs to the matrixin family and has been designated *matrix metalloproteinase 1* (**MMP-1**) (Nagase *et al.*, 1992). It has also been called *vertebrate collagenase, mammalian collagenase* or *fibroblast collagenase*. More recently it has been called *collagenase 1* (*Col1*) to distinguish it from enzymes subsequently described from the matrixin family that cleave interstitial collagens in the same way, namely neutrophil collagenase (MMP-8; collagenase 2) (Chapter 390), collagenase 3 (MMP-13) (Chapter 391) and collagenase 4 (MMP-18) (Chapter 392). However, the name collagenase 1 could still be confused with some of the bacterial collagenases. Thus, in order to avoid any ambiguity, the best name to use is MMP-1. Gelatinase A (MMP-2) (Chapter 401) also cleaves interstitial collagen at the three-quarters/one-quarter site (Aimes &

Quigley, 1995) as does MT1-MMP (MMP-14) (Chapter 399) (Ohuchi *et al.*, 1997).

## Activity and Specificity

MMP-1 cleaves collagen types I, II, III, VII and X (Gadher *et al.*, 1989; Seltzer *et al.*, 1989). Highly cross-linked collagen is slowly attacked by MMP-1 but the rate increases if other proteinases, such as stromelysin (Chapter 393), are present. Gelatin and the core protein of proteoglycan can be cleaved (both to a limited extent). The activity against gelatin is low compared to that of collagenase 3/MMP-13 (Chapter 391), which has a broad proteolytic activity (Knäuper *et al.*, 1996). MMP-1 cleaves itself within the polypeptide linker between the N- and C-terminal domains (Clark & Cawston, 1989) at Pro269┼Ile270-Gly in human MMP-1. The corresponding Pro-Ser-Gly sequence in pig is resistant to cleavage but Ala238┼Ile239 is slowly cleaved at high concentration in pig MMP-1 (Clark *et al.*, 1995). A human enzyme with these residues mutated to the pig linker sequence is stable (O'Hare *et al.*, 1995). When the C-terminal domain is lost then the N-terminal domain containing the catalytic zinc retains the ability to act as a proteinase (and it can also act as an activator of the proenzyme) but it is unable to cleave triple-helical collagen. This implicates the C-terminal domain in collagen binding. This autoprocessing may be physiologically relevant but could represent an artifact of isolation.

$\alpha_2$-Macroglobulin is cleaved in a susceptible sequence Gly-Pro-Glu-Gly679┼Leu680-Arg-Val-Gly adjacent to the bait region (Mortensen *et al.*, 1981) and this cleavage proceeds much more rapidly than for any other protein substrate so far studied for this enzyme. MMP-1 also cleaves $\alpha_1$-proteinase inhibitor at two sites (Met-Phe┼Leu-Glu and Ile-Pro┼Met-Ser), $\alpha_1$-antichymotrypsin at one site (Ser-Ala┼Leu-Val) (Desrochers *et al.*, 1991) and serum amyloid A (SAA) at some point in the sequence Gly-Gly-Val-Trp-Ala-Ala-Glu-Val. The cleavage of $\alpha_2$-macroglobulin allows MMP-1 to be rapidly inhibited, serpin cleavage is likely to have a physiological function and as SAA is known to induce collagenase secretion, cleavage of this protein may act to limit tissue destruction (Mitchell *et al.*, 1993).

Netzel-Arnett *et al.* (1991b) measured the rate of hydrolysis of 60 oligopeptides covering the P4 to the P5′ subsites of the substrate. They modeled these peptides on the known cleavage sites in collagens as well as on other protein substrates such as $\alpha_2$-macroglobulin. The amino acids in subsites P4 to P4′ all influence the rate of hydrolysis. Some substitutions not found in collagens led to a higher rate of hydrolysis, particularly at subsites P1, P1′ and P2′. Ala is preferred in subsite P1 and Trp or Phe in subsite P2′. MMP-1 does not tolerate aromatic residues in subsite P1′. Of peptides tested the best sequence appears to be Ala-Leu-Ala┼Leu-Arg-Val-Thr. Tyr and Met could be easily substituted at P2; His, Phe Pro, Gln, Val and Met in P1; Val, Met and Phe in P1′; Glu and Leu in P1′; Ser, Met and Ala in P3′ and Pro in P4′. Only Val in subsite P1 and Phe in subsite P1′ are predicted to be bad substitutions. It is not clear why only one site in each collagen $\alpha$ chain is cleaved whilst many potentially cleavable sites are unaffected.

Fields (1991) suggests that the cleavage site in the interstitial collagens is characterized by a region of tightly coiled triple helix preceding the cleavage site, with a loosely coiled region immediately after. Arg is the only charged residue, always found at P5′ or P8′, in a 25 amino acid sequence that is found around the collagen cleavage site. There must not be an imino acid adjacent to the Gly┼Ile/Leu bond cleaved. Careful examination of 31 identified sites with the required sequence, but that remain uncleaved, reveal that none fulfilled all of the above properties. As the C-terminal domain is critical to allow collagen binding and is known to bind collagen, then its role may be to correctly position the N-terminal active-site zinc exactly over the cleavage site. Triple-helical peptides can compete with collagen and the C-terminal domain could contain a triple-helix recognition site (Netzel-Arnett *et al.*, 1994).

Collagen cleavage is unusual as three peptide bonds must be cleaved. It is likely that MMP-1 binds to collagen, cleaves one peptide bond, then the second and third prior to disengaging (Welgus *et al.*, 1985). The active-site cleft is too small and rigid to accommodate all three $\alpha$ chains of the collagen molecule at one time. The binding of the C-terminal domain to collagen may perturb the triple helix, perhaps where it is loosely coiled, to extend the first chain into the active-site cleft. Subsequent cleavages could then presumably occur with increasing ease. It has been proposed that the C-terminal domain could interact with collagen on the opposite side of the helix to the active site, sandwiching and trapping the substrate at the active-site cleft (Gomis-Rüth *et al.*, 1996). Alternatively the C-terminal domain may bind to a point along the helix distant from the cleavage point and the length of the linker may then position the active site at the cleavage point.

Assay of activity uses radiolabeled collagen at pH 7.6 (the pH optimum of the enzyme), and calcium (1–10 mM) is also included to confer stability at 37°C. Collagen is allowed to form fibrils and at the end of the assay period undigested collagen is removed by centrifugation (Cawston & Barrett, 1979). Other assay systems have been employed using 96-well plate formats with collagen bound to each well. Care needs to be taken that collagen is not denatured to ensure that these assays just measure collagenase activity. Trypsin resistance is usually taken as a reliable measure of collagen integrity. To definitively show that collagenase cleavage at the three-quarters/one-quarter position occurs involves cleavage followed by separation by SDS-PAGE to identify the three-quarters and one-quarter products. Different assay systems and the labeling methods for collagen are described by Dioszegi *et al.* (1995). Activity assays do not distinguish among various enzymes that cleave collagen and for reliable assays that just measure MMP-1, ELISAs are used (Cooksley *et al.*, 1990; Clark *et al.*, 1992; Zhang *et al.*, 1993). Some immunoassays can recognize neoepitopes revealed after specific cleavage of collagen by MMP-1 and other collagenases and these assays can be used to demonstrate collagenase activity *in vivo* and *in vitro* (Hollander *et al.*, 1994).

Quenched fluorescent substrates can also be used (Stack & Gray, 1989; Knight, 1995; Beekman *et al.*, 1996) and although substrates can be made that are preferentially cleaved by MMP-1, these are never totally specific and other MMPs can usually cleave but at a lower rate. The substrate Dnp-Pro-Leu-Ala┼Leu-Trp-Ala-Arg was developed as a specific substrate for MMP-1 (Netzel-Arnett *et al.*, 1991a) and Knight *et al.* (1992) have reported a more sensitive assay for MMP-1 using Mca-Pro-Leu-Gly┼Leu-Dpa-Ala-Arg-NH₂ (Dpa: *N*-3-(2,4-dinitrophenyl)-L-2,3-diaminopropionyl).

MMP-1 is inhibited stoichiometrically by TIMP-1 (tissue inhibitor of metalloproteinases 1) (Cawston *et al.*, 1983) and TIMP-2 (DeClerck *et al.*, 1991). It is rapidly inhibited by $\alpha_2$-macroglobulin (Cawston & Mercer, 1986) in preference to TIMP-1 and inhibited by chelators such as 1,10-phenanthroline (2 mM) and EDTA (10 mM). Specific inhibitors with hydroxamate or chelating groups coupled to peptides that mimic the cleavage sequence are effective in the nanomolar range (Henderson *et al.*, 1990; Cawston, 1996). These inhibitors can be modeled on cleavage sites in substrates, propeptide cleavage sites, propeptide sequences that interact with zinc, autolytic cleavage sites or sequences cleaved in other inhibitors such as serpins or $\alpha_2$-macroglobulin. Some specificity is seen with these inhibitors for individual enzymes. Tetracyclines are also reported to inhibit MMP-1 both *in vivo* and *in vitro* (Greenwald, 1994; Greenwald *et al.*, 1987).

### Structural Chemistry

MMP-1 is synthesized as a preproenzyme of 469 amino acids with a prepropeptide (19 amino acids), a propeptide domain (81 amino acids), a catalytic domain (162 amino acids), a linking peptide (16 amino acids) and the C-terminal domain (192 amino acids) as shown in Figure 389.1. Its sequence is closely related to those of neutrophil collagenase (Chapter 390) and collagenase 3 (Chapter 391) as well as other matrix metalloproteinases. The proenzyme of $M_r$ 51 929 is activated to an $M_r$ of 42 570. This activation is accomplished by organomercurials, proteolysis by trypsin, plasmin or stromelysin 1 (Chapter 393) and by chaotropic or chemical agents (Springman *et al.*, 1990). Activation depends on disrupting the interaction between the catalytic zinc and the conserved cysteine in the propeptide sequence Pro-Arg-Cys92-(Val/Asn)-Pro-Asp-(Val/Leu)-(Ala/Gly) with subsequent proteolysis to remove the propeptide (Windsor *et al.*, 1991). The final step to give maximal activation is a cleavage at Gln99┼Phe100 (Suzuki *et al.*, 1990).

A number of X-ray structures are published for the N-terminal domain of MMP-1 (Borkakoti *et al.*, 1994; Lowry *et al.*, 1992; Spurlino *et al.*, 1994; Lovejoy *et al.*, 1994). This domain consists of three $\alpha$ helices with five $\beta$ sheets; two zinc atoms are found, one in the catalytic site and one in a structural role. Some biochemical studies have suggested that native MMP-1 contains only one zinc ion (Willenbrock *et al.*, 1995) and it is possible that the inclusion of zinc in refolding mixtures for recombinant proteins may introduce a second ion into the molecule. The structural studies suggest that the second zinc ion is tightly held and performs a structural role although other studies suggest that the second zinc may not be an absolute requirement or could be replaced by other similar ions (Springman *et al.*, 1995). Two to three calcium ions are also present in this domain. The full-length structure of the pig MMP-1 (Fig. 389.2) shows the N-terminal domain linked by an unstructured proline-rich linker peptide which is highly exposed (thus explaining the susceptibility to autocatalytic cleavage). This links to the C-terminal domain which has a unique four-bladed $\beta$-propeller structure stabilized by a calcium ion and a disulfide bond that links the first unit of $\beta$ sheet to the fourth unit at the C-terminus of the protein (Li *et al.*, 1995). In Figure 389.2, a peptide-based inhibitor, cocrystallized with the protein, is shown binding to the active-site zinc atom. The other zinc and calcium ions are also shown. The position of the two domains, relative to each other, will change in solution as there is little interaction between the domains and little structure within the linking peptide. Knowledge of the structure of the catalytic site (Figure 389.3) will allow design of new inhibitors that prevent the action of MMP-1 and act as therapeutic molecules preventing collagen turnover.

The structure of the N-terminus of neutrophil collagenase (MMP-8) (Chapter 390) is well-ordered if Phe is present as the N-terminal amino acid but disordered if other residues are at the N-terminus of the protein (Reinemer *et al.*, 1994). The authors suggest that this is likely to also be the case for MMP-1. An ammonium group from the N-terminal Phe100 forms a salt-link with the side-chain carboxylate group of the strictly conserved Asp252. Thus stabilization of the catalytic site could be conferred by strong hydrogen bonds made by the strictly conserved Asp252 with the characteristic conserved Met236 which lies underneath the active-site residues. These structural observations explain

**M**

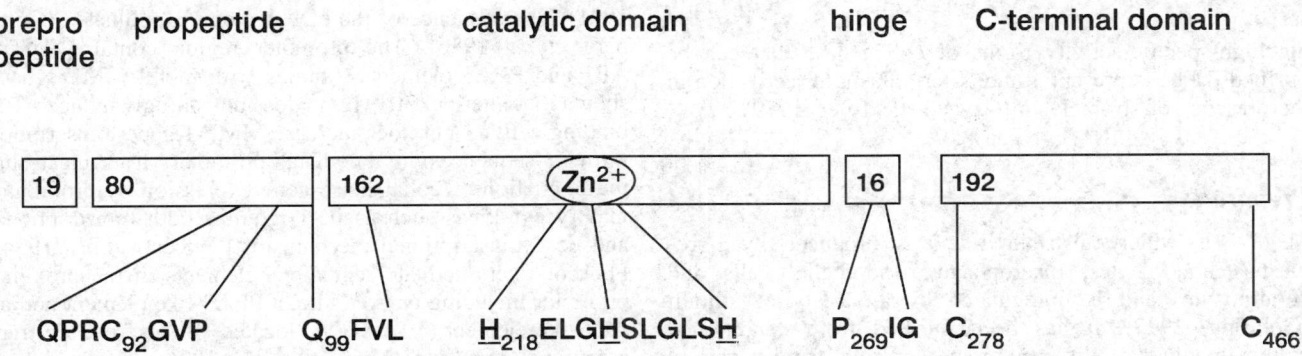

*Figure 389.1*  Domain structure of MMP-1.

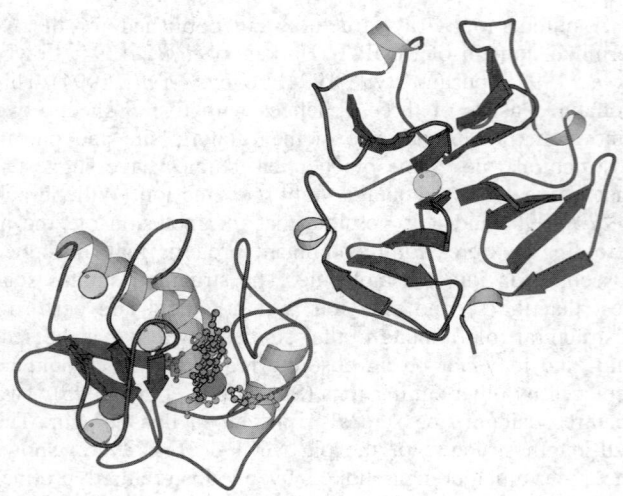

*Figure 389.2* The structure of MMP-1 showing the N-terminal domain which contains two zinc ions and three calcium ions. A highly exposed linker peptide joins to the C-terminal domain which has a four-bladed β-propeller structure. The first blade of the propeller is joined to the fourth by a disulfide bond and a calcium ion is found in the central channel where the four blades meet.

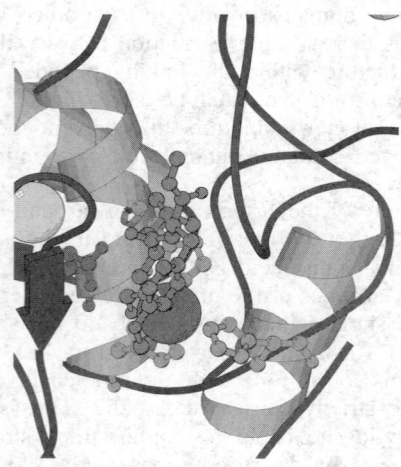

*Figure 389.3* A high-power view of the active-site zinc with a hydroxamate peptide inhibitor bound to the zinc and filling the active site.

the high-specific-activity forms of collagenase that can be purified if Phe is present at the N-terminus after activation in the presence of stromelysin (Cawston & Tyler, 1979).

## Preparation

MMP-1 is widely distributed and is produced by fibroblasts, chondrocytes, macrophages, endothelial cells and keratinocytes and is upregulated by a variety of stimuli (Goldring, 1993). It has been purified from conditioned culture medium of different tissues or cells from a variety of species (early references listed in Barrett & McDonald,

1980; Woessner, 1992). Standard chromatographic procedures can be used to successfully purify this enzyme from these sources (Cawston & Murphy, 1981). Commonly used chromatographic steps include zinc-chelate affinity chromatography, heparin-Sepharose or CM-Sepharose and gel filtration. Zinc chelate affinity chromatography is a useful first step as it binds the enzyme but allows TIMP-1 (often found in cell-conditioned media) to pass through the column without binding. Antibody affinity chromatography has also been used. Other methods involve the use of specific inhibitors coupled to Sepharose (Moore & Spilburg, 1986). Coupling of these inhibitors to activated Sepharose should be allowed to proceed for 24–48 h to ensure good coupling efficiency if proline is the amino acid at the N-terminus of the peptide-hydroxamate. These peptide-hydroxamate matrices only purify active forms of the enzyme and are rarely totally specific, so that MMP-1 preparations may be contaminated with other MMPs. The enzyme is relatively stable to exposure to either high (9.5) or low (4.5) pH providing this is kept to short time periods. The inclusion of calcium ions and a detergent such as Brij 35 is essential for maximum yields.

MMP-1s from rabbit, pig and human have been cloned. MMP-1 has been expressed in COS cells (Murphy *et al.*, 1987), *E. coli* as inclusion bodies (Windsor *et al.*, 1991) and also in *E. coli* as a fusion protein (O'Hare *et al.*, 1992). Recombinant proteins can be purified as described above or fusion proteins can be expressed and specific purification strategies used for the fusion partner followed by subsequent cleavage of MMP-1. Refolding of MMP-1 is best performed by slowly removing denaturant and many protocols include low levels of zinc ions in addition to calcium ions. When selecting a vector for expression of active enzyme, care needs to be taken to ensure that the correct N-terminal Phe is expressed, providing enzyme of the highest specific activity. The enzyme has been purified 135-fold from human skin fibroblasts (Stricklin *et al.*, 1977), 235-fold from pig synovial tissue conditioned media (Cawston & Tyler, 1979) and 8500-fold from cattle nasal cartilage extract (Boulka, 1990). Large yields have been obtained from gingival fibroblasts in culture stimulated with interleukin 1β yielding 5 mg liter$^{-1}$ of conditioned culture medium (Lark *et al.*, 1990).

## Biological Aspects

The structure of the MMP-1 gene contains ten exons and nine introns in 8–12 kbp of DNA and is located on the long arm of chromosome 11 in the human (Huhtala *et al.*, 1991). The sequence of the cDNA clone is published (Goldberg *et al.*, 1986). The promoter region contains TATA, AP1 and PEA3 elements. Studies with rabbit MMP-1 have shown that interleukin-1 (IL-1) does not strongly induce AP1-binding activity but does increase MMP-1 gene transcription through sequences in the distal promoter. It is clear that the regulation of collagenase gene expression by proinflammatory cytokines such as IL-1 requires both transcriptional and post-transcriptional mechanisms (Vincenti *et al.*, 1994). Phorbol esters activate regulatory elements in the proximal promoter including an AP1 site, a PEA3-like element and an AP1-like element (TTAATCA) located 186 bp from the transcriptional start site (Vincenti *et al.*, 1996). These authors have shown that activation of rabbit collagenase transcription by proinflammatory cytokines involves the activation of an src-related tyrosine kinase (Vincenti *et al.*, 1996). Maximal activation of collagenase transcription can occur when the cytokine oncostatin M and IL-1 are added together (Korzus *et al.*, 1997). Both AP1- and STAT-binding sites in the promoter are necessary to confer oncostatin M responsiveness in the oncostatin M response element and synergistic cooperation is required to accomplish maximum rates of transcription. In cartilage cultures oncostatin M and IL-1 synergize to promote cartilage collagen loss through upregulation of MMP-1, demonstrating that this mechanism could be important in disease (Cawston *et al.*, 1995). Thus MMP-1 is induced by the proinflammatory cytokines IL-1 and tumor necrosis factor $\alpha$ (TNF$\alpha$), various growth factors such as EGF, PDGF, basic FGF and oncostatin M.

Other agents can also regulate collagenase and these include serotonin, leukoregulin, calcium ionophore, lipopolysaccharide, substance P and UV irradiation. MMP-1 is downregulated by a variety of agents and these include IL-4, transforming growth factor $\beta$ (TGF$\beta$), interferon $\gamma$ (IFN$\gamma$), retinoic acid and glucocorticoids. Heparin and also calmodulin have been shown to affect the production of collagenase in some systems. The response to individual cytokines can be specific to individual cell types (Goldring, 1993) and can be altered by cell–matrix interactions. It can be modulated by several hormones such as progesterone and 1,25-dihydroxyvitamin D$_3$ (Delaissé *et al.*, 1988). Induction of expression can occur with physical rather than chemical stimuli, such as cell shape change, phagocytosis of particles such as crystals, heat shock, treatment with cytochalasin B, altered cell–matrix interactions or direct contact with inflammatory cells. Cell–cell contact between activated T cell clones (or cell membranes prepared from T cells) and either monocytes or dermal fibroblasts upregulates MMP-1 production (Miltenburg *et al.*, 1995).

MMP-1 is implicated in a wide variety of physiological and pathological processes where collagen degradation occurs. These include rheumatoid arthritis, osteoarthritis, periodontal disease, tumor invasion, angiogenesis, corneal ulceration, tissue remodeling, inflammatory bowel disease and atherosclerosis, aneurysm and restenosis (Birkedal-Hansen *et al.*, 1993 for review). The role of MMP-1 in bone resorption is not completely clear. It is produced by osteoblasts and was thought to be responsible for removing the layer of osteoid to allow osteoclasts to then bind to the exposed bone surface (Birkedal-Hansen *et al.*, 1993). Recent studies also suggest that MMP-1 can be found in osteoclasts and may be left on the bone surface as these cells move over the bone. Once the cell has moved on and as the pH rises then collagen cleavage occurs (Okada *et al.*, 1995).

Activation of MMP-1 is a neglected control point and it is still not clear exactly how this occurs *in vivo*. Plasminogen activators are produced by many cells that also produce MMP-1 and in the presence of plasminogen, plasmin (Chapter 59) can be generated and MMP-1 is activated. Other studies have shown that stromelysin (Chapter 393) is able to activate and also other MMPs such as gelatinase. There is some evidence from tissue-based studies that thiol proteinases may also sometimes be involved (Buttle *et al.*, 1993). In many biological systems MMP-1 can be upregulated and secreted from the cell but no collagen breakdown occurs as the enzyme is not activated. Thus activation may be the limiting control point in these systems (van der Zee *et al.*, 1996). Some studies have suggested that collagenolytic activity can be associated with the plasma membrane of certain cells (O'Grady *et al.*, 1982). It is not yet clear if this activity is a membrane-associated form of one of the recognized collagenases or if this is one of the membrane-associated metalloproteinases of the matrixin family known as MT-MMPs. It is known that MT1-MMP (Chapter 399) can degrade interstitial collagens at the three-quarters/one-quarter cleavage site (Ohuchi *et al.*, 1997).

## Distinguishing Features

As substrate, MMP-1 prefers type III collagen, neutrophil collagenase (MMP-8) (Chapter 390) prefers type I and collagenase 3 (MMP-13) (Chapter 391) prefers type II collagen. Care must be taken when comparing different collagenases to ensure that the high-specific-activity forms of each enzyme are being compared. Collagenase 3 (MMP-13) is able to cleave type I collagen at an additional amino telopeptide locus (Krane *et al.*, 1996). The sequence of amino acids cleaved in peptide substrates is very similar for MMP-1 and neutrophil collagenase (MMP-8). MMP-1 is smaller in size than neutrophil collagenase (MMP-8) and differs only slightly in size (approximately 3 kDa) from collagenase 3 (MMP-13). Although both MMP-1 and collagenase 3 (MMP-13) bind to many of the same chromatographic matrices during purification, it is reported that Sephadex QAE does not bind MMP-1 but binds collagenase 3 (MMP-13) (Van der Stappen *et al.*, 1992). If specific antibodies are available, the sizes of the three enzymes can be shown to be different by western blotting. The collagenases are differentially regulated with neutrophil collagenase (MMP-8) produced constitutively and stored, MMP-1 is downregulated by retinoic acid and collagenase 3 (MMP-13) is upregulated by this agent. Rabbit and sheep polyclonal antibodies have been produced for a wide variety of species and monoclonal antibodies to the human enzyme and are available from a range of suppliers that include Cambio and Oncogene Research Products. A commercial ELISA kit for MMP-1 is available from Amersham (see Appendix 2 for full names and addresses of suppliers). A polyclonal antibody to pig MMP-1 is available from Chemicon International. Collagenase 3 (MMP-13) has a broad proteolytic activity. Gelatinase A (MMP-2) (Chapter 401) which can also cleave collagen, can be separated from MMP-1 by gelatin-Sepharose.

## Further Reading

Extensive reviews may be found in Birkedal-Hansen *et al.* (1993), Woessner (1991) and Dioszegi *et al.* (1995).

## References

Aimes, R.T. & Quigley, J.P. (1995) Matrix metalloproteinase-2 is an interstitial collagenase. *J. Biol. Chem.* **270**, 5872–5876.

Barrett, A.J. & McDonald, J.K. (1980) Vertebrate collagenase. In: *Mammalian Proteases: A Glossary and Bibliography*, vol. 1. London: Academic Press, pp. 359–378.

Beekman, B., Drijfhout, J.W., Bloemhoff, W., Ronday, K.H., Tak, P.P. & te Koppele, J.M. (1996) Convenient fluorometric assay for

M

matrix metalloproteinase activity and its application in biological media. *FEBS Lett.* **390**, 221–225.

Birkedal-Hansen, H., Moore, W.G.I., Bodden, M.K., Windsor, L.J., Birkedal-Hansen, B., DeCarlo, A. & Engler, J.A. (1993) Matrix metalloproteinases: a review. *Crit. Rev. Oral Biol. Med.* **4**, 197–250.

Borkakoti, N., Winkler, F.K., Williams, D.H., D'Arcy, A., Broadhurst, M.J., Brown, P.A., Johnson, W.H. & Murray, E.J. (1994) Structure of the catalytic domain of human fibroblast collagenase complexed with an inhibitor. *Struct. Biol.* **1**, 106–110.

Boukla, A. (1990) Purification and properties of bovine nasal hyaline cartilage collagenase. *Int. J. Biochem.* **11**, 1273–1282

Buttle, D.J., Handley, C.J., Ilic, M.Z., Saklatvala, J., Murata, M. & Barrett, A.J. (1993) Inhibition of cartilage proteoglycan release by a specific inactivator of cathepsin B and an inhibitor of matrix metalloproteinases: evidence for two converging pathways of chondrocyte–mediated proteoglycan degradation. *Arthritis Rheum.* **36**, 1709–1717.

Cawston, T.E. (1996) Metalloproteinase inhibitors and the prevention of connective tissue breakdown. *Pharmacol. Ther.* **70**, 163–182.

Cawston, T.E. & Barrett, A.J. (1979) A rapid and reproducible assay for collagenase using [1-$^{14}$C]acetylated collagen. *Anal. Biochem.* **99**, 340–345.

Cawston, T.E. & Mercer, E. (1986) Preferential binding of collagenase to $\alpha_2$-macroglobulin in the presence of the tissue inhibitor of metalloproteinases. *FEBS Lett.* **209**, 9–12.

Cawston, T.E. & Murphy, G. (1981) Mammalian collagenases. *Methods Enzymol.* **80**, 711–722.

Cawston, T.E. & Tyler, J.A. (1979) Purification of pig synovial collagenase to high specific activity. *Biochem. J.* **183**, 647–656.

Cawston, T.E., Murphy, G., Mercer, E., Galloway, W.A., Hazleman, B.L. & Reynolds, J.J. (1983) The interaction of purified rabbit bone collagenase with purified rabbit bone metalloproteinase inhibitor. *Biochem. J.* **211**, 313–318.

Cawston, T.E., Ellis, A.J., Humm, G., Lean, E., Ward, D. & Curry, V. (1995) Interleukin-1 and oncostatin M in combination promote the release of collagen fragments from bovine nasal cartilage in culture. *Biochem. Biophys. Res. Commun.* **215**, 377–385.

Clark, I.M. & Cawston, T.E. (1989) Fragments of human fibroblast collagenase: purification and characterization. *Biochem. J.* **263**, 201–206.

Clark, I.M., Wright, J.K., Cawston, T.E. & Hazleman, B.L. (1992) Polyclonal antibodies against human fibroblast collagenase and the design of an enzyme-linked immunosorbent assay to measure TIMP-collagenase complex. *Matrix* **12**, 108–115.

Clark, I.M., Mitchell, R.E., Powell, L.K. & Bigg, H.F. (1995) Recombinant porcine collagenase: purification and autolysis. *Arch. Biochem. Biophys.* **316**, 123–127.

Cooksley, S., Hipkiss, J.B., Tickle, S.P., Holmes-Leavers, E., Docherty, A.J.P., Murphy, G. & Lawson, A.D.G. (1990) Immunoassays for the detection of human collagenase, stromelysin, tissue inhibitor of metalloproteinases (TIMP) and enzyme-inhibitor complexes. *Matrix* **10**, 285–291.

DeClerck, Y.V., Yean, T.-D., Lu, H.S., Ting, J. & Langley, K.E. (1991) Inhibition of autoproteolytic activation of interstitial procollagenase by recombinant metalloproteinase inhibitor MI/TIMP-2. *J. Biol. Chem.* **266**, 3893–3899.

Delaissé, J.M., Eeckhout, Y. & Vaes, G. (1988) Bone resorbing agents affect the production and distribution of procollagenase as well as the activity of collagenase in bone tissue. *Endocrinology* **123**, 264–276.

Desrochers, P.E., Jeffrey, J.J. & Weiss, S.J. (1991) MMP-1 (matrix metalloproteinase-1) expresses serpinase activity. *J. Clin. Invest.* **87**, 2258–2265.

Dioszegi, M., Cannon, P. & Van Wart, H.E. (1995) Vertebrate collagenases. *Methods Enzymol.* **248**, 413–431

Fields, G.B. (1991) A model for interstitial collagen catabolism by mammalian collagenases. *J. Theor. Biol.* **153**, 585–602.

Gadher, S.J., Schmid, T.M., Heck, L.W. & Woolley, D.E. (1989) Cleavage of collagen type X by human synovial collagenase and neutrophil elastase. *Matrix* **9**, 109–115.

Goldberg, G.I., Wilhelm, S.M., Kronberger, A., Bauer, E.A., Grant, G.A. & Eisen, A.Z. (1986) Human fibroblast collagenase: complete primary structure and homology to an oncogene transformation-induced rat protein. *J. Biol. Chem.* **261**, 6600–6605.

Goldring, M.B. (1993) Degradation of articular cartilage in culture: regulatory factors. In: *Joint Cartilage Degradation* (Woessner, J.F. & Howell, D.S., eds). New York: Marcel Dekker, pp. 281–345.

Gomis-Rüth, F.X., Gohlke, U., Betz, M., Knäuper, V., Murphy, G., López-Otin, C. & Bode, W. (1996) The helping hand of collagenase-3 (MMP-13): 2.7 Å crystal structure of its C-terminal haemopexin-like domain. *J. Mol. Biol.* **264**, 556–566.

Greenwald, R.A. (1994) Treatment of destructive arthritic disorders with MMP inhibitors. Potential role of tetracyclines. *Ann. N.Y. Acad. Sci.* **732**, 181–198.

Greenwald, R.A., Golub, L.M., Lavietes, B., Ramamurthy N.S., Gruber, B., Laskin, R.S. & NcNamara, T.F. (1987) Tetracyclines inhibit human synovial collagenase *in vivo* and *in vitro*. *J. Rheumatol.* **14**, 28–32.

Gross, J. & Lapière, C.M. (1962) Collagenolytic activity in amphibian tissues: a tissue culture assay. *Proc. Natl Acad. Sci. USA* **48**, 1014–1022.

Henderson, B., Docherty, A.J.P. & Beetey, N.R.A. (1990) Design of inhibitors of articulor cartilage destruction. *Drugs Future* **15**, 490–508.

Hollander, A.P., Heathfield, T.F., Webber, C., Iwata, Y., Bourne, R., Rorabeck, C. & Poole, A.R. (1994) Increased damage to type II collagen in osteoarthritic cartilage detected by a new immunoassay. *J. Clin. Invest.* **93**, 1722–1732.

Huhtala, P., Tuuttila, A., Chow, L.T., Lohi, J., Keski-Oja, J. & Tryggvason, K. (1991) Complete structure of the human gene for 92-kDa type IV collagenase. *J. Biol. Chem.* **266**, 16485–16490.

Knäuper, V., López-Otin, C., Smith, B., Knight, G. & Murphy, G. (1996) Biochemical characterization of human collagenase-3. *J. Biol. Chem.* **271**, 1544–1550.

Knight, C.G. (1995) Fluorimetric assays of proteolytic enzymes. *Methods Enzymol.* **248**, 18–34.

Knight, C.G., Willenbrock, F. & Murphy, G. (1992) A novel coumarin-labelled peptide for sensitive continuous assays of the matrix metalloproteinases. *FEBS Lett.* **296**, 263–266.

Korzus, E., Nagase, H., Rydell, R. & Travis, J. (1997) The mitogen-activated protein kinase and JAK-STAT signaling pathways are required for an oncostatin M-responsive element-mediated activation of matrix metalloproteinase 1 gene expression. *J. Biol. Chem.* **272**, 1188–1196.

Krane, S.M., Byrne, M.H., Lemaitre, V., Henriet, P., Jeffrey, J.J., Witter, J.P., Liu, X., Wu, H., Jaenisch, R. & Eeckhout, Y. (1996) Different collagenase gene products have different roles in degradation of type 1 collagen. *J. Biol. Chem.* **45**, 28509–28515.

Lark, M.W., Walakovits, L.A., Shah, T.K., Vanmiddlesworth, J., Cameron, P.M. & Lin, T.-Y. (1990) Production and purification of prostromelysin and procollagenase from IL-1 beta-stimulated human gingival fibroblasts. *Connect. Tiss. Res.* **25**, 49–65.

Li, J., Brick, P., O'Hare, M.C., Skarzynski, T., Lloyd, L.F., Curry, V.A., Clark, I.M., Bigg, H.F., Hazleman, B.L., Cawston, T.E. & Blow, D.M. (1995) Structure of full-length porcine synovial collagenase reveals a C-terminal domain containing a calcium-linked, four-bladed β-propeller. *Structure* **3**, 541–549.

Lovejoy, B., Cleasby, A., Hassell, A.M., Longley, K., Luther, M.A., Weigle, D., McGeehan, G., McElroy, A.B., Drewry, D., Lambert, M.H. & Jordan, S.R. (1994) Structure of the catalytic domain of fibroblast collagenase complexed with an inhibitor. *Science* **263**, 375–377.

Lowry, C.L., McGeehan, G. & LeVine, H. (1992) Metal ion stabilization of the conformation of a recombinant 19-kDa catalytic fragment of human fibroblast collagenase. *Proteins: Struct. Funct. Genet.* **12**, 42–48.

Miltenburg, A.M.M., Lacraz, S., Welgus, H.G. & Dayer, J.-M. (1995) Immobilized anti-CD3 antibody activates T cell clones to induce the production of MMP-1, but not tissue inhibitor of metalloproteinases, in monocytic THP-1 cells and dermal fibroblasts. *J. Immunol.* **154**, 2655–2667.

Mitchell, T.I., Jeffrey, J.J., Palmiter, R.D. & Brinckerhoff, C.E. (1993) The acute phase reactant serum amyloid A (SAA3) is a novel substrate for degradation by the metalloproteinases collagenase and stromelysin. *Biochim. Biophys. Acta* **1156**, 245–254.

Moore, W.M. & Spilburg, C.A. (1986) Purification of fibroblast collagenase with a peptide-hydroxamic acid affinity column. *Biochemistry* **25**, 5189–5195.

Mortensen, S.B., Sottrup-Jensen, L., Hansen, H.F., Petersen, T.E. & Magnusson, S. (1981) Primary and secondary cleavage sites in the bait region of α2-macroglobulin. *FEBS Lett.* **135**, 295–300.

Murphy, G., Cockett, M.I., Stephens, P.E., Smith, B.J. & Docherty, A.J.P. (1987) Stromelysin is an activator of procollagenase: a study with natural and recombinant enzymes. *Biochem. J.* **248**, 265–268.

Nagase, H., Barrett, A.J. & Woessner, J.F. (1992) Nomenclature and glossary of the matrix metalloproteinases. *Matrix* **suppl. 1**, 1–8.

Netzel-Arnett, S., Mallya, S.K., Nagase, H., Birkedal-Hansen, H. & Van Wart, H.E. (1991a) Continuously recording fluorescent assays optimized for five human matrix metalloproteinases. *Anal. Biochem.* **195**, 86–92.

Netzel-Arnett, S., Fields, G., Birkedal-Hansen, H. & Van Wart, H.E. (1991b) Sequence specificities of human fibroblast and neutrophil collagenases. *J. Biol. Chem.* **266**, 6747–6755.

Netzel-Arnett, S., Salari, A., Goli, U.B. & Van Wart, H.E. (1994) Evidence for a triple helix recognition site in the hemopexin-like domains of human fibroblast and neutrophil interstitial collagenases. *Ann. N.Y. Acad. Sci.* **732**, 22–30.

O'Grady, R.L., Harrop, P.J. & Cameron, D.A. (1982) Collagenolytic activity by malignant tumours. *Pathology* **14**, 135–138.

O'Hare, M.C., Clarke, N.J. & Cawston, T.E. (1992) Production by *Escherichia coli* of porcine type-I collagenase as a fusion protein with β-galactosidase. *Gene* **111**, 245–248.

O'Hare, M.C., Curry, V.A., Mitchell, R.E. & Cawston, T.E. (1995) Stabilization of purified human collagenase by site-directed mutagenesis. *Biochem. Biophys. Res. Commun.* **216**, 329–337.

Ohuchi, E., Imai, K., Fujii, Y. & Sato, H. (1997) Membrane type 1 matrix metalloproteinase digests interstitial collagens and other extracellular matrix macromolecules. *J. Biol. Chem.* **272**, 2446–2451.

Okada, Y., Naka, K., Kawamura, K., Matsumoto, T., Nakanishi, I., Fujimoto, N., Sato, H. & Seiki, M. (1995) Localization of matrix metalloproteinase 9 (92-kilodalton gelatinase/type IV collagenase = gelatinase B) in osteoclasts: implications for bone resorption. *Lab. Invest.* **72**, 311–322.

Reinemer, P., Grams, F., Huber, R., Kleine, T., Schnierer, S., Piper, M., Tschesche, H. & Bode, W. (1994) Structural implications for the role of the N terminus in the 'superactivation' of collagenases. *FEBS Lett.* **338**, 227–233.

Seltzer, J.L., Eisen, A.Z., Bauer, E.A., Morris, N.P., Glanville, R.W. & Burgeson, R.E. (1989) Cleavage of type VII collagen by MMP-1 and type IV collagenase (gelatinase) derived from human skin. *J. Biol. Chem.* **264**, 3822–3826.

Springman, E.B., Angleton, E.L., Birkedal-Hansen, H. & Van Wart, H.E. (1990) Multiple modes of activation of latent human fibroblast collagenase: evidence for the role of $Cys^{73}$ active-site zinc complex in latency and a 'cysteine switch' mechanism for activation. *Proc. Natl Acad. Sci. USA* **87**, 364–368.

Springman, E.B., Nagase, H., Birkedal-Hansen, H. & Van Wart, H.E. (1995) Zinc content and function in human fibroblast collagenase. *Biochemistry* **34**, 15173–15720.

Spurlino, J.C., Smallwood, A.M., Carlton, D.D., Banks, T.M., Vavra, K.J., Johnson, J.S., Cook, E.R., Falvo, J., Wahl, R.C., Pulvino, T.A., Wendoloski, J.J. & Smith, D.L. (1994) 1.56 Å structure of mature truncated human fibroblast collagenase. *Proteins: Struct. Funct. Genet.* **19**, 98–109.

Stack, M.S. & Gray, R.D. (1989) Comparison of vertebrate collagenase and gelatinase using a new fluorogenic substrate peptide. *J. Biol. Chem.* **264**, 4277–4281.

Stricklin, G.P., Bauer, E.A., Jeffrey, J.A. & Eisen, A.Z. (1977) Human skin collagenase: isolation of precursor and active forms from both fibroblast and organ cultures. *Biochemistry* **16**, 1607–1615.

Suzuki, K., Enghild, J.J., Morodomi, T., Salvesen, G. & Nagase, H. (1990) Mechanisms of activatin of tissue procollagenase by matrix metalloproteinase 3 (stromelysin). *Biochemistry* **29**, 10261–10270.

Van der Stappen, J.W.J., Hendriks, T. & de Man, B.M. (1992) Collagenases from human and rat skin fibroblasts purified on a zinc chelating column reveal marked differences in latency as a result of serum culture conditions. *Int. J. Biochem.* **24**, 725–735.

van der Zee, E., Everts, V. & Beertsen, W. (1996) Cytokine-induced endogenous procollagenase stored in the extracullar matrix of soft connective tissue results in a burst of collagen breakdown following its activation. *J. Periodont. Res.* **31**, 483–488.

Vincenti, M.P., Coon, C.I., Lee, O. & Brinckerhoff, C.E. (1994) Regulation of collagenase gene expression by IL-1β requires transcriptional and post-transcriptional mechanisms. *Nucleic Acids Res.* **22**, 4818–4827.

Vincenti, M.P., Coon, C.I., White, L.A., Barchowsky, A. & Brinckerhoff, C.E. (1996) *src*-Related tyrosine kinases regulate transcriptional activation of the interstitial collagenase gene, MMP-1, in interleukin-1-stimulated synovial fibroblasts. *Arthritis Rheum.* **39**, 574–582.

Welgus, H.G., Jeffrey, J.J., Eisen, A.Z., Roswit, W.T. & Stricklin, G.P. (1985) Human skin fibroblast collagenase: interaction with substrate and inhibitor. *Coll. Rel. Res.* **5**, 167–179.

Willenbrock, F., Murphy, G., Phillips, I.R. & Brocklehurst, K. (1995) The second zinc atom in the matrix metalloproteinase

M

catalytic domain is absent in the full-length enzymes: a possible role for the C-terminal domain. *FEBS Lett.* **358**, 189–192.

Windsor, J.L., Birkedal-Hansen, H., Birkedal-Hansen, B. & Engler, J.A. (1991) An internal cysteine plays a role in the maintenance of the latency of human fibroblast collagenase. *Biochemistry* **30**, 641–647.

Woessner, J.F., Jr. (1991) Matrix metalloproteinases and their inhibitors in connective tissue remodeling. *FASEB J.* **5**, 2145–2154.

Woessner, J.F. (1992) Literature on vertebrate matrix metalloproteinases and their tissue inhibitors. *Matrix* **suppl. 1**, 425–501.

Zhang, J., Fujimoto, N., Iwata, K., Sakai, T., Okada, Y. & Hayakawa, T. (1993) A one-step sandwich enzyme immunoassay for human matrix metalloproteinase 1 (interstitial collagenase) using monoclonal antibodies. *Clin. Chim. Acta* **219**, 1–14.

*Tim E. Cawston*
*Department of Rheumatology,*
*University of Newcastle,*
*The Medical School,*
*Newcastle Upon Tyne NE2 4HH, UK*
*Email: T.E.Cawston@newcastle.ac.uk*

# 390. *Neutrophil collagenase*

## Databanks

*Peptidase classification: clan MB, family M10, MEROPS ID: M10.002*
*NC-IUBMB enzyme classification: EC 3.4.24.34*
*Chemical Abstracts Service registry number: 9001-12-1*
*Databank codes:*

| Species | SwissProt | PIR | EMBL (cDNA) | EMBL (genomic) |
|---|---|---|---|---|
| *Homo sapiens* | P22894 | A36230 A37073 A61175 S09680 S11026 S19576 S27225 | J05556 | – |

Brookhaven Protein Data Bank three-dimensional structures:

| Species | ID | Resolution | Notes |
|---|---|---|---|
| *Homo sapiens* | 1JAN | 2.5 | catalytic domain; complex with Pro-Leu-Gly-hydroxylamine |
| | 1JAO | 2.4 | catalytic domain; complex with 3-mercapto-2-benzyl-propanoyl-Ala-Gly-NH$_2$ |
| | 1JAP | 1.82 | catalytic domain; complex with Pro-Leu-Gly-hydroxylamine |
| | 1JAQ | 2.25 | catalytic domain; complex with 1-hydroxylamine-2-isobutylmalonyl-Ala-Gly-NH$_2$ |
| | 1MNC | 2.1 | catalytic domain |

## Name and History

The name collagenase denotes an enzyme capable of cleaving triple-helical collagen. A collagenase isolated from human polymorphonuclear leukocytes (PMNLs) was first described by Lazarus *et al.* (1968). Later it was shown to be localized to the specific granules (Murphy *et al.*, 1977). Because of its origin the enzyme was given the name **PMNL-collagenase** or **neutrophil collagenase**. As a member of the matrixin family the neutrophil collagenase is also referred to as **matrix metalloproteinase 8 (MMP-8)** (Nagase *et al.*, 1992). In 1990 the cDNA encoding human neutrophil collagenase was cloned and sequenced using a λgt11 cDNA library constructed from mRNA extracted from the peripheral leukocytes of a patient with chronic granulocytic leukemia (Hasty *et al.*, 1990). Until recently, human neutrophil collagenase was thought to be expressed exclusively by neutrophil leukocytes, but the enzyme and its mRNA were recently found in normal human articular chondrocytes (Cole *et al.*, 1996). Following the discovery of a third form of collagenase in humans (MMP-13; Chapter 391), the neutrophil collagenase is frequently referred to as *collagenase 2*.

## Activity and Specificity

Neutrophil collagenase is stored intracellularly as a latent proenzyme in the specific granules of polymorphonuclear leukocytes (Murphy *et al.*, 1977). Activation of the procollagenase occurs in the extracellular space after secretion. The active enzyme is capable of cleaving types I, II and III triple-helical collagen into the characteristic one-quarter and three-quarters fragments by hydrolyzing the Gly775$\dagger$Leu776 and Gly775$\dagger$Ile776 peptide bonds of collagen α chains (Miller *et al.*, 1976; Hasty *et al.*, 1987; Mallya *et al.*, 1990; Tschesche *et al.*, 1992; Van Wart, 1992). Besides this triple-helical cleavage activity and self-degradation during autocatalytic activation (Sorsa *et al.*, 1985; Hasty *et al.*, 1986; Knäuper *et al.*, 1990a; Mookhtiar & Van Wart, 1990; Tschesche *et al.*, 1992) and autolysis (Knäuper *et al.*, 1993a), the neutrophil collagenase has more general proteolytic properties. These include the hydrolysis of natural substrates such as gelatin peptides, fibronectin, proteoglycans (Tschesche *et al.*, 1992), cartilage aggrecan (Fosang *et al.*, 1993), serpins such as human C1-inhibitor or α$_1$-proteinase inhibitor (Desrochers & Weiss, 1988; Knäuper *et al.*, 1990b, 1991; Michaelis *et al.*, 1992), β-casein (Fields *et al.*, 1990), and peptides like angiotensin and substance P (Diekmann & Tschesche, 1994). To investigate the sequence specificity of human neutrophil collagenase, the rate of hydrolysis of 52 octapeptides was measured (Netzel-Arnett *et al.*, 1991). The preference for Ala in subsite P1 and Tyr or Phe in subsite P1′ is noteworthy, whereas Gly-Pro-Gln-Gly$\dagger$Ile-Trp-Gly-Gln was the best substrate.

Synthetic substrates such as Dnp-Pro-Gln-Gly$\dagger$Ile-Ala-Gly-Gln-D-Arg-OH, Dnp-Pro-Leu-Gly$\dagger$Leu-Trp-Ala-D-Arg-NH$_2$ or Mca-Pro-Leu-Gly$\dagger$Leu-Dpa-Ala-Arg-NH$_2$ (Dpa: *N*-3-(2,4-dinitrophenyl)-L-2,3-diaminopropionyl) are available from Bachem (see Appendix 2 for full names and addresses of suppliers). They are suitable for easy and rapid activity tests but are also cleaved by many other members of the matrix metalloproteinase family. The assay works best using a buffer system which contains $Zn^{2+}$ ions (0.5 mM), $Ca^{2+}$ ions (5 mM) and sodium chloride (100 mM) at a pH of about 7.0.

Enzymatic activity of neutrophil collagenase is inhibited by the family of TIMPs (tissue inhibitor of metalloproteinases) (Stetler-Stevenson *et al.*, 1989; Docherty *et al.*, 1985; Pavloff *et al.*, 1992; Kleine *et al.*, 1993) and by α$_2$-macroglobulin (Sottrup-Jensen, 1989). In addition to the tetracyclines, especially doxycyclines (Golub *et al.*, 1983, 1990; Lauhio *et al.*, 1994, 1995) and chelating agents like 1,10-phenanthroline and EDTA, a broad spectrum of small synthetic inhibitors have been developed. These are based on short substrate analog peptide

sequences linked to chelating moieties such as hydroxamate, thiol, carboxylate or phosphinic groups (Nishino & Powers, 1979; Moore & Spilburg, 1986; Powers & Harper, 1986; Johnson *et al.*, 1987).

## Structural Chemistry

Human neutrophil preprocollagenase consists of a signal peptide, a propeptide containing a sequence motif responsible for maintaining the enzyme in the latent form, a catalytic domain and a hemopexin-like domain responsible for binding native collagen. Conversion of the latent proenzyme to the active form is accompanied by a molecular weight reduction to an apparent $M_r$ of 64 000. Depending on the type of activator, different intermediate forms are observed (Tschesche *et al.*, 1992).

Unfortunately, no three-dimensional structure of the full-length enzyme of neutrophil collagenase has been obtained so far, because of the rapid autocatalytic cleavage of the hemopexin-like domain. However, the recombinant catalytic domains, residues Met80–Gly242 and residues Phe79–Gly242, the latter also referred to as the 'superactivated' form (Knäuper *et al.*, 1993b), were cloned and expressed in *E. coli* (Schnierer *et al.*, 1993). Both have been cocrystallized with the substrate analog inhibitor Pro-Leu-Gly-NHOH (Bode *et al.*, 1994; Reinemer *et al.*, 1994). The structure displays a spherical molecule divided by a relatively flat active-site cleft into a small 'lower' part and an 'upper' main body. The lower part consists of multiple turns and ends in a long C-terminal α helix. The upper part is formed mainly by a central, highly twisted, five-stranded β-pleated sheet and two long α helices. The catalytic zinc ion is coordinated by the three histidine residues – His197, His201 and His207 – of the conserved HEXXHXXGXXH sequence motif. The tight 1,4-turn Ala213-Leu-Met-Tyr216, also called the Met-turn, is conserved among all metzincins (Bode *et al.*, 1993, 1996; Stöcker *et al.*, 1995) and provides a hydrophobic base for the three zinc-liganding histidine residues. Besides the catalytic zinc ion, the neutrophil collagenase catalytic domain harbors a second structural zinc ion and two calcium ions. In the X-ray structure of the Met80–Gly242 form six N-terminal residues are disordered, with the chain position only being defined from Pro86 onward. In contrast, in the 'superactivated' Phe79–Gly242 form the N-terminus is fixed to the C-terminal helix by numerous contacts and by the free amino group of the Phe79 forming a salt-bridge with the carboxylate moiety of the strictly conserved Asp232. This seems to lead to stabilization of the active site by the neighboring Asp233 that forms a hydrogen bond to the Met-turn and, hence, could lead to more favorable transition states, explaining the higher activity (Reinemer *et al.*, 1994) (Figure 390.1).

In order to understand the protein–inhibitor interactions and to discover new inhibitor lead structures the MMP-8 catalytic domain was also cocrystallized with various other peptide thiols and peptide hydroxamate inhibitors, including batimastat (Grams *et al.*, 1995a,b).

## Preparation

Human neutrophil collagenase was found in PMNLs (Lazarus *et al.*, 1968) and has been purified to homogeneity by

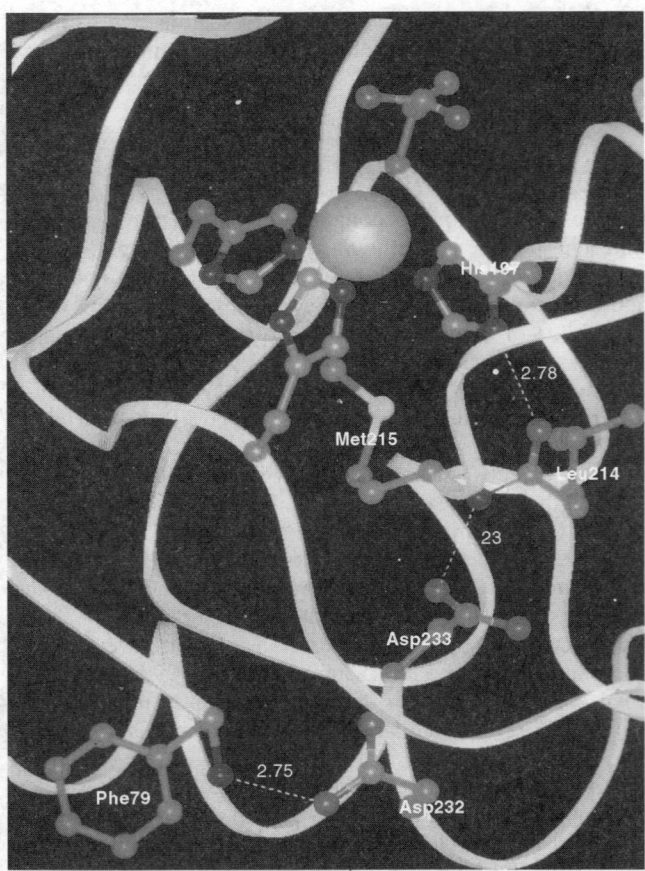

*Figure 390.1* Detailed view of the hydrogen network between the N-terminal Phe79 and the Met-turn. All relevant amino acids are labeled. The catalytic zinc ion is displayed as a large sphere.

several methods (Christner *et al.*, 1982; Sorsa *et al.*, 1985; Callaway *et al.*, 1986; Hasty *et al.*, 1986; Mookhtiar *et al.*, 1990). One rapid and reproducible method includes affinity chromatography on zinc-chelate Sepharose, ion-exchange chromatography on Q-Sepharose fast flow, followed by affinity chromatography on orange Sepharose and finally a gel-permeation step on Sephacryl S-300 (Knäuper *et al.*, 1990a). Neutrophil collagenase has also been detected in human articular chondrocytes, but the enzyme has not yet been purified (Cole *et al.*, 1996).

The coding regions for full-length human neutrophil collagenase and for procollagenase lacking the hemopexin-like domain were cloned and expressed in *E. coli* and purified to homogeneity (Schnierer *et al.*, 1993; Tschesche, 1995). The full-length enzyme was also cloned and expressed in COS-7 cells (Hasty *et al.*, 1990).

## Biological Aspects

The gene for human neutrophil collagenase is located on the long arm of chromosome 11 and mapped to 11q21–q22 by *in situ* hybridization (Yang-Feng *et al.*, 1991).

In addition to their primary function of phagocytosis, human PMNLs have a high capacity for infiltrating connective tissue. This is often associated with a breakdown of the extracellular matrix, especially during pathological processes such as inflammation, rheumatoid arthritis or osteoarthritis, and this is initiated by human neutrophil collagenase. The enzyme stored in the specific granules is released from the neutrophils as latent procollagenase upon various stimuli, e.g. interleukin 1 and 8 (Luger *et al.*, 1983; Yoshimura *et al.*, 1987), tumor necrosis factor α (TNFα) (Klebanoff *et al.*, 1986), chemotactic formylpeptides (Schettler *et al.*, 1991), human C5a anaphylatoxins (Ward & Hill, 1970), fibrinogen and fibrin-derived products (Kay *et al.*, 1973), granulocyte-macrophage colony-stimulating factor (GM-CSF) (Yuo *et al.*, 1989) and calcium ionophore A23187 (Hibbs *et al.*, 1984).

The precise pathway of *in vivo* activation after secretion still remains unclear, although several different agents initiating *in vitro* activation according to the 'cysteine switch' activation mechanism have been reported (Van Wart & Birkedal-Hansen, 1990). These include proteases such as trypsin, chymotrypsin, cathepsin G, tissue kallikrein (Knäuper *et al.*, 1990a; Tschesche *et al.*, 1992), and stromelysin 1 or 2 (Lindy *et al.*, 1986; Van Wart, 1992; Knäuper *et al.*, 1993b). Activation can also be achieved by autocatalysis (Hasty *et al.*, 1986; Tschesche *et al.*, 1992; Van Wart, 1992), mercury compounds (Springman *et al.*, 1990; Bläser *et al.*, 1991; Tschesche *et al.*, 1992), oxygen radicals, hydrogen peroxide or hypochlorite (Weiss *et al.*, 1985; Springman *et al.*, 1990), or sodium gold thiomalate (Lindy *et al.*, 1986; Springman *et al.*, 1990).

## Distinguishing Features

The cDNA sequences show that neutrophil collagenase and interstitial collagenase are coded by two different genes (Hasty *et al.*, 1990). Nevertheless the three highly homologous human collagenases – neutrophil collagenase, fibroblast collagenase and the recently discovered collagenase 3 (Freije *et al.*, 1994) – share more than 50% sequence identity (Knäuper *et al.*, 1996). Neutrophil collagenase and fibroblast collagenase even display 57% identity and 72% chemical similarity (Hasty *et al.*, 1990), but they differ in immunological cross-reactivity (Woolley *et al.*, 1976; Hasty *et al.*, 1987). The former enzyme is highly glycosylated, with about one-third of its molecular mass arising from carbohydrate (Tschesche *et al.*, 1992).

All three enzymes cleave type I, II and III triple-helical collagen into the characteristic one-quarter and three-quarter fragments, but with different substrate specificity. Human neutrophil collagenase preferentially hydrolyzes type I collagen (Horwitz *et al.*, 1977; Hasty *et al.*, 1987; Mallya *et al.*, 1990; Tschesche *et al.*, 1992), whereas fibroblast collagenase prefers type III (Mallya *et al.*, 1990; Tschesche *et al.*, 1992) and collagenase 3, type II collagen (Knäuper *et al.*, 1996). Studies with synthetic oligopeptides also demonstrate similar but distinct substrate specificities (Netzel-Arnett *et al.*, 1991).

In contrast to the fibroblast collagenase which is constitutively transcribed and expressed *de novo* by human skin or synovial cells, the neutrophil collagenase is stored in subcellular specific granules of the neutrophils. It is assumed to be synthesized during maturation in the bone marrow (Bainton *et al.*, 1971). Monoclonal antibodies to this enzyme are commercially available from various sources including Fuji Chemical Industries and Oncogene Research Products.

## Further Reading

A short review is given by Tschesche (1995). The first cloned and sequenced cDNA encoding the human neutrophil collagenase was reported by Hasty *et al.* (1990) and the first X-ray structure was published by Bode *et al.* (1994).

## References

Bainton, D.F., Ullyot, J.L. & Farquhar, M.G. (1971) The development of neutrophilic polymorphonuclear leukocytes in human bone marrow. *J. Exp. Med.* **134**, 907–934.

Bläser, J., Knäuper, V., Osthues, A., Reinke, H. & Tschesche, H. (1991) Mercurial activation of human polymorphonuclear leukocyte procollagenase. *Eur. J. Biochem.* **202**, 1223–1230.

Bode, W., Gomis-Rüth, F.-X. & Stöcker, W. (1993) Astacins, serralysins, snake venom and matrix metalloproteinases exhibit identical zinc-binding environments (HEXXHXXGXXH and Met-turn) and topologies and should be grouped into a common family, the 'metzincins'. *FEBS Lett.* **331**, 134–140.

Bode, W., Reinemer, P., Huber, R., Kleine, T., Schnierer, S. & Tschesche, H. (1994) The X-ray crystal structure of the catalytic domain of human neutrophil collagenase inhibited by a substrate analogue reveals the essentials for catalysis and specificity. *EMBO J.* **13**, 1263–1269.

Bode, W., Grams, F., Reinemer, P., Gomis-Rüth, F.-X., Baumann, U., McKay, D.B. & Stöcker, W. (1996) The metzincin-superfamily of zinc-peptidases. *Adv. Exp. Med. Biol.* **389**, 1–11.

Callaway, J.E., Garcia, J.A., Hersch, C.L., Yeh, R.K. & Gilmore-Mebert, M. (1986) Use of lectin affinity chromatography for the purification of collagenase from human polymorphonuclear leukocytes. *Biochemistry* **25**, 4757–4762.

Christner, P., Damato, D., Reinhart, M. & Abrams, W. (1982) Purification of human neutrophil collagenase and production of a monospecific antiserum. *Biochemistry* **21**, 6005–6011.

Cole, A.A., Chubinskaya, S., Schumacher, B., Huch, K., Cs-Szabo, G., Yao, J., Mikecz, K., Hasty, K.A. & Kuettner, K.E. (1996) Chondrocyte matrix metalloproteinase-8: human articular chondrocytes express neutrophil collagenase *J. Biol. Chem.* **271**, 11023–11026.

Desrochers, D.E. & Weiss, S.J. (1988) Proteolytic inactivation of $\alpha_1$-proteinase inhibitor by a neutrophil metalloproteinase. *J. Clin. Invest.* **81**, 1646–1650.

Diekmann, O. & Tschesche, H. (1994) Degradation of kinins, angiotensins and substance P by polymorphonuclear matrix metalloproteinases MMP-8 and MMP-9. *Braz. J. Med. Biol. Res.* **27**, 1877.

Docherty, A.J.P., Lyons, A., Smith, B.J., Wright, E.M., Stephens, P.E. & Harris, T.J.R. (1985) Sequence of human tissue inhibitor of metalloproteinase and its identity to erythroid-potentiating activity. *Nature* **318**, 66–69

Fields, G.B., Netzel-Arnett, S.J., Windsor, S.J., Engler, L.J., Birkedal-Hansen, H. & Van Wart, H.E. (1990) Proteolytic activity of human fibroblast collagenase: hydrolysis of a broad range of substrates at a single active site. *Biochemistry* **29**, 6670–6677.

Fosang, A.J., Last, K., Knäuper, V., Neame, P.J., Murphy, G., Hardingham, T.E., Tschesche, H. & Hamilton, J.A. (1993) Fibroblast and neutrophil collagenase cleave at two sites in the cartilage aggrecan interglobular domain. *Biochem. J.* **295**, 273–276.

Freije, J.M.P., Diez-Itza, L., Balbin, L., Sanchez, M., Blasco, L.M.,

Tolivia, R.T. & López-Otín, C. (1994) Molecular cloning and expression of collagenase-3, a novel human matrix metallo-proteinase produced by breast carcinomas. *J. Biol. Chem.* **269**, 16766–16773.

Golub, L.M., Lee, H.M., Lehrer, G., Nemiroff, A., McNamara, T.F., Kaplan, R. & Ramamurthy, N.S. (1983) Minocycline reduces gingival collagenolytic activity during diabetes: preliminary observations and a proposed new mechanism of action. *J. Periodontol. Res.* **18**, 515–525.

Golub, L.M., Ciancio, S., Ramamurthy, N.S., Leung, M. & McNamara, T.F. (1990) Low-dose doxycycline therapy: effect on gingival and crevicular fluid collagenase activity in humans. *J. Periodontol.* **25**, 321–330.

Grams, F., Reinemer, P., Powers, J.C., Kleine, T., Pieper, M., Tschesche, H., Huber, R. & Bode, W. (1995a) X-ray structures of human neutrophil collagenase complexed with peptide hydroxamate and peptide thiol inhibitors. *Eur. J. Biochem.* **228**, 830–841.

Grams, F., Crimmin, M., Hinnes, L., Huxley, P., Pieper, M., Tschesche, H. & Bode, W. (1995b) Structure determination and analysis of human neutrophil collagenase complexed with a hydroxamate inhibitor. *Biochemistry* **34**, 14012–14020.

Hasty, K.A., Hibbs, S.M., Kang, A.H. & Mainardi, C.I. (1986) Secreted forms of human neutrophil collagenase. *J. Biol. Chem.* **261**, 5645–5650.

Hasty, K.A., Jeffrey, J.J., Hibbs, M.S. & Welgus, H.G. (1987) The collagen substrate specificity of human neutrophil collagenase. *J. Biol. Chem.* **262**, 10048–10052.

Hasty, K.A., Poumotabbed, T.F., Goldberg, G.I., Thompson, J.P., Spinella, D.G., Stevens, R.M. & Mainardi, C.L. (1990) Human neutrophil collagenase: a distinct gene product with homology to other matrix metalloproteinases. *J. Biol. Chem.* **265**, 11421–11424.

Hibbs, M.S., Hasty, K.A., Kang, A.H. & Mainardi, C.L. (1984) Secretion of collagenolytic enzymes by human polymorphonuclear leukocytes. *Collagen Rel. Res.* **4**, 467–477.

Horwitz, A.J., Hance, A.J. & Crystal, R.G. (1977) Granulocyte collagenase: selective digestion of type I relative to type II collagen. *Proc. Natl Acad. Sci. USA* **74**, 894–901.

Johnson, W.H., Roberts, N.A. & Borkakoti, N. (1987) Collagenase inhibitors: their design and potential therapeutic use. *J. Enzyme Inhib.* **2**, 1–22.

Kay, A.B., Pepper, D.S. & Ewart, M.R. (1973) Generation of chemotactic activity for leukocytes by the action of thrombin on human fibrinogen. *Nature New Biol.* **243**, 56.

Klebanoff, S.J., Vadas, M.A., Harlan, J.M., Sparkes, L.H., Gamble, J.R., Agosti, J.M. & Waltersdorph, A.M. (1986) Stimulation of neutrophils by tumor necrosis factor. *J. Immunol.* **136**, 4220–4225.

Kleine, T., Bartsch, S., Bläser, J., Schnierer, S., Triebel, S., Valentin, M., Gote, T. & Tschesche, H. (1993) Preparation of active recombinant TIMP-1 from *Escherichia coli* inclusion bodies and complex formation with the recombinant catalytic domain of PMNL collagenase. *Biochemistry* **32**, 14125–14131.

Knäuper, V., Krämer, S., Reinke, H. & Tschesche, H. (1990a) Characterization and activation of procollagenase from human polymorphonuclear leukocytes. *Eur. J. Biochem.* **189**, 295–300.

Knäuper, V., Reinke, H. & Tschesche, H. (1990b) Inactivation of human plasma $\alpha_1$-proteinase inhibitor by human PMN-leukocyte collagenase. *FEBS Lett.* **263**, 355–357.

Knäuper, V., Triebel, S., Reinke, H. & Tschesche, H. (1991) Inactivation of human plasma C1-inhibitor by human PMN-leukocyte matrix metalloproteinase. *FEBS Lett.* **290**, 99–102.

Knäuper, V., Osthues, A., DeClerck, Y.A., Langley, K.E., Bläser, J.

& Tschesche, H. (1993a) Fragmentation of human polymorpho-nuclear-leukocyte collagenase. *Biochem. J.* **291**, 847–854.

Knäuper, V., Wilhelm, S.M., Seperack, P.K., DeClerck, Y.A., Langley, K.E., Osthues, A. & Tschesche, H. (1993b) Direct activation of human neutrophil collagenase by recombinant stromelysin. *Biochem. J.* **295**, 581–586.

Knäuper, V., López-Otín, C., Smith, B., Knight, G. & Murphy, G. (1996) Biochemical characterization of human collagenase-3. *J. Biol. Chem.* **271**, 1544–1550.

Lauhio, A., Salo, T., Ding, Y., Konttinen, Y.T., Nordström, D., Tschesche, H., Lähdevirta, J., Golub, L.M. & Sorsa, T. (1994) *In vivo* inhibition of neutrophil collagenase (MMP-8) activity during long-term combination therapy of doxycycline and non-steroidal anti-inflammatory drugs (NSAID) in acute reactive arthritis. *Clin. Exp. Immunol.* **98**, 21–28.

Lauhio, A., Konttinen, Y.T., Salo, T., Tschesche, H., Nordström, D., Lähdevirta, J., Golub, L.M. & Golub, L.M. (1995) The *in vivo* effect of doxycycline treatment on matrix metalloproteinases in reactive arthritis. *Ann. N.Y. Acad. Sci.* **732**, 431–432.

Lazarus, G.S., Brown, R.S., Daniels, J.R. & Fullmer, H.M. (1968) Human granulocyte collagenase. *Science* **159**, 1483–1485.

Lindy, S., Sorsa, T., Suomalainen, K. & Turto, H. (1986) Gold sodium thiomalate activates latent human leukocyte collagenase. *FEBS Lett.* **208**, 23–25.

Luger, T.A., Charon, J.A., Colot, M., Micksche, M. & Oppenheim, J.J. (1983) Chemotactic properties of partially purified human epidermal cell derived thymocyte-activation factor (ETAF) for polymorphonuclear and mononuclear cells. *J. Immunol.* **131**, 816–820.

Mallya, S.K., Mookhtiar, K.A., Gao, Y., Brew, K., Dioszegi, M., Birkedal-Hansen, H. & Van Wart, H.E. (1990) Characterization of 58-kilodalton human neutrophil collagenase: comparison with human fibroblast collagenase. *Biochemistry* **29**, 10628–10634.

Michaelis, J., Vissers, M.C.M. & Winterbourn, C.C. (1992) Cleavage of $\alpha$-1-antitrypsin by human neutrophil collagenase. *Matrix* **suppl. 1**, 80–81.

Miller, E.J., Harris, E.D., Jr., Chung, E., Finch, J.E., Jr., McCroskery, P.A. & Butler, W.T. (1976) Cleavage of type II and type III collagens with mammalian collagenase: site of cleavage and primary structure at the NH$_2$-terminal portion of the smaller fragment released from both collagenases. *Biochemistry* **15**, 787–792.

Mookhtiar, K.A. & Van Wart, H.E. (1990) Purification to homogeneity of latent and active 58 kilodalton forms of human neutrophil collagenase. *Biochemistry* **29**, 10620–10627.

Moore, W.M. & Spilburg, C.A. (1986) Purification of human collagenases with a hydroxamic acid affinity column. *Biochemistry* **25**, 5189–5195.

Murphy, G., Reynolds, J.J., Bretz, U. & Baggiolini, M. (1977) Collagenase is a component of the specific granules of human neutrophil leukocytes. *Biochem. J.* **162**, 195–197.

Nagase, H., Barrett, A.J. & Woessner, J.F., Jr. (1992) Nomenclature and glossary of matrix metalloproteinases. *Matrix* **suppl. 1**, 421–424.

Netzel-Arnett, S., Fields, G., Birkedal-Hansen, H. & Van Wart, H.E. (1991) Sequence specificities of human fibroblast and neutrophil collagenases. *J. Biol. Chem.* **266**, 6747–6755.

Nishino, N. & Powers, J.C. (1979) Design of potent reversible inhibitors for thermolysin. Peptides containing zinc coordinating ligands and their use in affinity chromatography. *Biochemistry* **18**, 4340–4347.

Pavloff, N., Staskus, P.W., Kishnani, N.S. & Hawkes, S.P. (1992) A new inhibitor of metalloproteinases from chicken: ChIMP-3. *J.*

*Biol. Chem.* **267**, 17321–17326.

Powers, J.C. & Harper, J.W. (1986) Inhibitors of metalloproteinases. In: *Proteinase Inhibitors* (Barrett, A.J. & Salvesen, G., eds). Amsterdam: Elsevier, pp. 219–298.

Reinemer, P., Grams, F., Huber, R., Kleine, T., Schnierer, S., Pieper, M., Tschesche, H. & Bode, W. (1994) Structural implications for the role of the N-terminus in the 'superactivation' of collagenases; a crystallographic study. *FEBS Lett.* **338**, 227–233.

Schettler, A., Thorn, H., Jokusch, B.M. & Tschesche, H. (1991) Release of proteinases from stimulated polymorphonuclear leukocytes. *Eur. J. Biochem.* **197**, 197–202.

Schnierer, S., Kleine, T., Gote, T., Hillemann, A., Knäuper, V. & Tschesche, H. (1993) The recombinant catalytic domain of human neutrophil collagenase lacks type I collagen substrate specificity. *Biochem. Biophys. Res. Commun.* **191**, 319–326.

Sorsa, T., Suomalainen, K., Turto, H. & Lindy, S. (1985) Partial purification and characterization of latent human leukocyte collagenase. *Med. Biol.* **63**, 66–77.

Sottrup-Jensen, L. (1989) $\alpha$-Macroglobulins: structure, shape, and mechanism of proteinase complex formation. *J. Biol. Chem.* **264**, 11539–11542.

Springman, E.B., Angleton, E.L., Birkedal-Hansen, H. & Van Wart, H.E. (1990) Multiple models of activation of latent human fibroblast collagenase: Evidence for the role of Cys[73] active site zinc complex in latency and a 'cysteine-switch' mechanism for activation. *Proc. Natl Acad. Sci. USA* **87**, 364–368.

Stetler-Stevenson, W.G., Krutzsch, H.C. & Liotta, L.A. (1989) Tissue inhibitor of metalloproteinase (TIMP-2). A new member of the metalloproteinase inhibitor family. *J. Biol. Chem.* **264**, 17374–17378.

Stöcker, W., Grams, F., Baumann, U., Reinemer, P., Gomis-Rüth, F.-X., McKay, D.B. & Bode, W. (1995) The metzincins: topological and sequential relations between the astacins, adamalysins, serralysins, and matrixins (collagenases) define a superfamily of zinc-peptidases. *Protein Sci.* **4**, 823–840.

Tschesche, H. (1995) Human neutrophil collagenase *Methods Enzymol.* **248**, 431–449.

Tschesche, H., Knäuper, V., Krämer, S., Michaelis, J., Oberhoff, R. & Reinke, H. (1992) Latent collagenase and gelatinase from human neutrophils and their activation. *Matrix* **suppl. 1**, 245–255.

Van Wart, H.E. (1992) Human neutrophil collagenase. *Matrix* **suppl. 1**, 31–36.

Van Wart, H.E. & Birkedal-Hansen, H. (1990) The cysteine-switch: a principle of regulation of metalloproteinase activity with potential applicability to the entire matrix metalloproteinase gene family. *Proc. Natl Acad. Sci. USA* **87**, 5578–5582.

Ward, P.A. & Hill, J.H. (1970) C5-chemotactic fragments produced by an enzyme in lyososomal granules of neutrophils. *J. Immunol.* **104**, 535–543.

Weiss, S.J., Peppin, G., Ortiz, X., Ragsdale, C. & Test, S.T. (1985) Oxidative autoactivation of latent collagenase by human neutrophils. *Science* **227**, 747–749.

Woolley, D.E., Crossley, M.J. & Evanson, J.M. (1976) Antibody to rheumatoid synovial collagenase. *Eur. J. Biochem.* **69**, 421–428.

Yang-Feng, T.L., Berliner, N., Deverajan, P. & Johnston, J. (1991) Assignment of two human neutrophil secondary granule protein genes, transcolbalmin I and neutrophil collagenase to chromosome 11. *Cytogenet. Cell Genet.* **58**, 1974.

Yoshimura, T., Matsushima, K., Tanaka, S., Robinson, E.A., Apella, E., Oppenheim, J.J. & Leonard, E.J. (1987) Purification of a human monocyte-derived neutrophil chemotactic factor that has peptide sequence similarity to other host defence cytokines. *Proc. Natl Acad. Sci. USA* **84**, 9233–9237.

Yuo, A., Kitagawa, S., Ohsaka, A., Ohta, M., Miyazona, K., Okabe, T., Urabe, A., Saito, M & Takabu, F. (1987) Recombinant human granulocyte colony-stimulating factor as an activator of human granulocytes: potentiation of responses triggered by receptor-mediated agonists and stimulation of C3bi receptor expression and adherence. *Blood* **74**, 2144–2149.

***Harald Tschesche***
*Faculty of Chemistry, Department of Biochemistry,*
*University of Bielefeld,*
*Universitätsstrasse 25,*
*D-33615 Bielefeld, Germany*
*Email: tschesche@chema.uni-bielefeld.de*

***Michael Pieper***
*Faculty of Chemistry, Department of Biochemistry,*
*University of Bielefeld,*
*Universitätsstrasse 25,*
*D-33615 Bielefeld, Germany*

# 391. *Collagenase 3*

## Databanks

*Peptidase classification: clan MB, family M10, MEROPS ID: M10.013*
*NC-IUBMB enzyme classification: none*
*ATCC entries: 101581 (human)*
*Databank codes:*

| Species | SwissProt | PIR | EMBL (cDNA) | EMBL (genomic) |
|---|---|---|---|---|
| *Cynops pyrrhogaster* | – | – | D82055 | – |
| *Homo sapiens* | P45452 | – | X75308 | – |
| | | | X81334 | |
| *Mus musculus* | P33435 | S29243 | X66473 | – |
| *Rattus norvegicus* | P23097 | A23685 | M60616 | – |
| *Xenopus laevis* | Q10835 | – | L49412 | – |
| *Xenopus laevis* | – | – | U41824 | – |

## Name and History

Human *collagenase 3* was cloned from a breast carcinoma cDNA library (Freije *et al.*, 1994). The protein was expressed, purified by these workers and shown to be an interstitial collagenase by virtue of its ability to cleave native type I collagen. This finding resulted in the suggested assignment of the names ***collagenase 3*** and ***matrix metalloproteinase 13 (MMP-13)*** to this new matrix metalloproteinase, along with collagenase 1 (human MMP-1) (Chapter 389) (Goldberg *et al.*, 1986) and neutrophil collagenase (collagenase 2, human MMP-8) (Chapter 390) (Hasty *et al.*, 1990). Although collagenase 3 clearly belongs in the matrix metalloproteinase subfamily, its percentage identity to matrix metalloproteinase 1 (MMP-1) is quite low – on the order of 45% (Freije *et al.*, 1994). On the other hand, the identity of human collagenase 3 to the interstitial collagenases purified from both rat (Roswit *et al.*, 1983; Quinn *et al.*, 1990) and mouse (Henriet *et al.*, 1992) is very high – approximately 90%. The enzymes from the rat and the mouse are nearly 100% identical (Quinn *et al.*, 1990; Henriet *et al.*, 1992). For some time the low level of similarity between the known rodent interstitial collagenases and human MMP-1 represented a biological conundrum, and a marked departure from the similarities displayed between the interstitial collagenases of other species (reviewed by Sang & Douglas, 1996). After the discovery of human collagenase 3, it was of considerable interest to determine whether the genome of all mammalian species contains both forms of interstitial collagenase. Probing genomic DNA from a wide variety of species has revealed that, while most species so examined contained homologs of both collagenase 1 and collagenase 3, mouse and rat genomic DNA contained only a collagenase 3 homolog. No evidence for a collagenase 1 enzyme could be found in these species (J.E. Hambor, P.G. Mitchell, S.A. Wilcox & J.J. Jeffrey, unpublished results). Similarly, there appears to be no collagenase 2 (neutrophil collagenase, MMP-8) in the rat (S. Wilhelm, personal communication). These findings, aside from evolutionary considerations, lead to a significant nomenclature problem: while it is proper, indeed essential, to refer to this gene product as collagenase 3 in the human, the same nomenclature is meaningless in rodentia, where only one interstitial collagenase appears to exist in the genome.

Thus, while the term collagenase 3 is appropriate in the human, the term *rodent interstitial collagenase* is the term I prefer for the rat and mouse.

## Activity and Specificity

The detailed study of the activity and specificity of human collagenase 3 has just begun. Based upon the high level of identity that exists between this enzyme and its rodent homologs, it was expected that, at least to a first approximation, the activity of collagenase 3 would be found to be very similar to that of rat interstitial collagenase. The substrate specificity of this latter enzyme has been described in some detail by Welgus *et al*. (1983). These studies showed that the enzyme displays very similar affinities for both monomeric and fibrillar collagen: $K_m$ and $K_d$ values of approximately $1 \times 10^{-6}$ M were observed for both substrates, values very similar to those determined for human collagenase 1 (MMP-1) (Welgus *et al*., 1981) (Chapter 389). Again, in common with MMP-1, the values of $k_{cat}$ on native substrates were quite low: values of approximately $25\,h^{-1}$ were observed. It should also be noted that the same cleavages are catalyzed in native type I collagen molecules by the two enzymes: Gly-Pro-Gln-Gly775+Ile776-Ala-Gly-Gln. Thus, in many ways the two homologs are enzymatically quite similar. Nevertheless, major differences have been found to exist between rat interstitial collagenase and human MMP-1. First, the rate of cleavage of native type II collagen by the rat enzyme was approximately equal to that observed with type I collagen as a substrate (Welgus *et al*., 1983), a rate much higher than that displayed by human MMP-1. So slow was the cleavage of type II collagen by human MMP-1 that Welgus *et al*. (1981) speculated that a separate enzyme might exist whose biological role was the cleavage of this substrate and, potentially, the management of cartilage connective tissue turnover and repair. Recently, two reports have suggested that human collagenase 3 cleaves type II collagen considerably faster than type I (Mitchell *et al*., 1996; Knäuper *et al*., 1996). This finding would be consistent with the speculation of Welgus and coworkers, as noted above.

A second major difference observed between MMP-1 and rodent collagenase was the ability of the rodent enzyme to degrade denatured collagen. On a rate basis, rat collagenase displayed a higher capacity to degrade gelatin than native collagen (Welgus *et al*., 1985); the reverse was true for MMP-1 (Welgus *et al*., 1982). The two enzymes catalyze the cleavage of the same bonds in gelatin (Gly+Leu, Gly+Ile, Gly+Phe, Gly+Ala), but at markedly different rates. Thus, the rodent enzyme could be viewed as playing a dual enzymatic role – as a true collagenase and as a gelatinase – possibly in specific, as yet undefined, biological settings. This aspect of the activity of collagenase 3 has not yet been explored but, interestingly, a recent report (Fosang *et al*., 1996) indicates that collagenase 3 demonstrates considerable activity in digesting the cartilage proteoglycan aggrecan, suggesting the possibility that a wide substrate specificity may exist for this enzyme as well. Further studies are clearly necessary to shed light on this potentially important issue.

Assay of these enzymes, particularly in impure biological samples, is most reliably performed using strictly native collagen, preferably in fibrillar form, since many other matrix metalloproteinases degrade virtually every other substrate introduced to date for the assay of interstitial collagenase. A variety of techniques have been utilized to label collagen *in vitro* (see Cawston & Barrett, 1979 for an example). This methodology has been very useful, but caution must be exercised to insure that the chemically modified collagen is truly native. This is usually done by assessing the amount of degradation of the radiolabeled fibrillar substrate by high concentrations of trypsin. In addition, a number of synthetic peptides have been utilized to assess the activity of pure enzyme (Bickett *et al*., 1993; Weingarten & Feder, 1985).

## Structural Chemistry

Collagenase 3, in common with its rat and mouse homologs, is synthesized as a zymogen with an approximate $M_r$ of 51 500. The three homologous proenzymes are characterized by a distinctly acidic propiece. The rat and mouse enzymes, for example, have a sequence of six contiguous aspartates (at residues 19–25); the human sequence is quite similar as well. All three proenzymes are converted to active enzymes by trypsin or the organomercurial compound aminophenylmercuric acetate (APMA) in a two-step process, resulting in active forms displaying an approximate $M_r$ of 42 300, all of which have tyrosine at the N-terminus (Tyr85 in human procollagenase 3). As an example of the acidic nature of the propiece of these enzymes, conversion from proenzyme to the active species raises the pI of the converted molecule by nearly 0.75 units (Sang & Douglas, 1996). In contrast, the analogous conversion of MMP-1 homologs decreases the pI of the active forms modestly. The reason for the acidic nature of the propeptides of collagenase 3 and its homologs is unknown. In essentially all other respects the three enzymes are extremely similar. All contain identical 'cysteine switch' sequences (Van Wart & Birkedal-Hansen, 1990) and identical sequences in the catalytic zinc-binding domain. The observation that human collagenase 3 displays a markedly increased rate of degradation of type II collagen relative to type I, whereas the rat enzyme does not, suggests that some subtle differences do exist. Knäuper *et al*. (1996) report that trypsin, when used for proenzyme activation, rapidly degrades the C-terminal domain of human collagenase 3, whereas APMA does not. This situation is essentially reversed for rat interstitial collagenase: trypsin appears reliably to result in full activation, whereas substantial enzyme activity is lost when APMA is used for activation (W.T. Roswit & J.J. Jeffrey, unpublished results).

Both human collagenase 3 (Knäuper *et al*., 1996) and rat interstitial collagenase (Roswit *et al*., 1992) are inhibited stoichiometrically and with high affinity by TIMP.

## Preparation

Human collagenase 3 has been prepared by expression of the recombinant form of the enzyme using expression vectors containing full-length cDNA (Freije *et al*., 1994; Mitchell *et al*., 1996). Naturally occurring rodent collagenase has been purified from the medium of uterine smooth muscle cells in culture (Roswit *et al*., 1983). Milligram amounts of enzyme can be obtained from this source. The purification of this enzyme is made quite convenient by the extremely high avidity of both the pro- and active forms for heparin conjugated to a solid matrix (Roswit *et al.*, 1983). Interestingly, this affinity has no obvious basis in the overall chemistry of the protein, nor do there appear to be any of the several previously identified heparin-binding motifs present in the molecule. Indeed, proenzyme and active species as well, after elution from the negatively charged heparin matrix, bind rather tightly to positively charged DEAE in the next step of the purification. Thus, the possibility exists that this affinity of this set of collagenases has biological significance; no information on this possibility is currently available.

## Biological Aspects

Human collagenase 3 appears, as of this writing, to be considerably restricted in its tissue expression. Clearly, its presence in human breast carcinomas suggests significant importance in this pathological setting, but further evidence for the nature of its involvement is currently unavailable. Subsequent reports have established, however, that normal chondrocytes are a potentially major source of collagenase 3. This finding would make a great deal of biological sense, as discussed above, in that the cells of cartilaginous tissues produce an interstitial collagenase whose specificity is directed toward type II collagen. This finding opens the way for a better understanding of the pathological degradation of collagen in both rheumatoid arthritis and osteoarthritis.

In rodents, the homologous enzyme has been identified in numerous organs, including the uterus, the skin, bone and a variety of other cellular sources. Thus, as restricted as the expression of collagenase 3 is in the human, its rodent homolog appears to be ubiquitous in these species, and to be the sole source of interstitial collagenase activity in these animals. In common with numerous mammalian collagenases, rodent interstitial collagenase is induced by interleukin 1 and tumor necrosis factor $\alpha$ (TNF$\alpha$). Similar cytokine-dependent inductions of human collagenase 3 have recently been reported (Reboul *et al.*, 1996). In addition, rat interstitial collagenase has been reported to be induced by serotonin (5-hydroxytryptamine) in myometrial smooth muscle cells (Jeffrey *et al.*, 1991). This induction is at least cell selective, and has provided a paradigm for collagenase gene activation in smooth muscle cells. It will be of interest to determine whether human collagenase 3 is under the same control in uterine smooth muscle.

## Distinguishing Features

The set of collagenase 3 homologs, as a group, displays catalytic selectivity for type II collagen, of potentially major significance in the biology of cartilage collagen turnover as noted above. Chemically, all of the members of the set contain highly acidic proenzyme domains, in contrast to the MMP-1s. In addition, the proenzymes of the collagenase 3 set appear, in one way or another, less stable, as probed by various reagents utilized for their activation. Finally, although they are highly homologous in their overall sequence, it appears that small differences in sequence between the molecules result in rather major differences in substrate specificity.

## Further Reading

In view of the recent discovery of this enzyme in humans, insufficient information has been developed which would warrant extensive review. Thus, the reader is referred to the following articles which should provide an appropriate overview of the enzyme: Mitchell *et al.* (1996), Freije *et al.* (1994) and Sang & Douglas (1996).

## References

Bickett, D.M., Green, M.D., Berman, J., Dezube, M., Howe, A.S., Brown, P.J., Roth, J.T. & McGeehan, G.M. (1993) A high throughput fluorogenic substrate for interstitial collagenase (MMP-1) and gelatinase (MMP-9). *Anal. Biochem.* **212**, 58–64.

Cawston, T.E. & Barrett, A.J. (1979) A rapid and reproducible assay for collagenase using [1-$^{14}$C]-acetylated collagen. *Anal. Biochem.* **99**, 340–345.

Fosang, A.J., Last, K., Knäuper, V., Murphy, G. & Neame, P.T. (1996) Degradation of cartilage aggrecan by collagenase 3 (MMP-13). *FEBS Lett.* **380**, 17–20.

Freije, J.M.P., Díez-Itza, I., Balbín, M., Sánchez, L.M., Blasco, R., Tolivia, J. & López-Otin, C. (1994) Molecular cloning and expression of collagenase 3, a novel human matrix metalloproteinase produced by breast carcinomas. *J. Biol. Chem.* **269**, 16766–16773.

Goldberg, G.I., Wilhelm, S.M., Kronberger, A., Bauer, E.A., Grant, G.A. & Eisen, A.Z. (1986) Human fibroblast collagenase. Complete primary structure and homology to an oncogene transformation-induced rat protein. *J. Biol. Chem.* **261**, 6600–6605.

Hasty, K.A., Pourmotabbed, T.F., Goldberg, G.I., Thompson, S.P., Spinella, D.G., Stevens, R.M. & Mainardi, C.L. (1990) Human neutrophil collagenase. A distinct gene product with homology to other matrix metalloproteinases. *J. Biol. Chem.* **265**, 11421–11424.

Henriet, P., Rousseau, G.G. & Eeckhout, Y. (1992) Cloning and sequencing of mouse collagenase cDNA. Divergence of mouse and rat collagenases from the other mammalian collagenases. *FEBS Lett.* **310**, 175–178.

Jeffrey, J.J., Ehlich, L.S. & Roswit, W.T. (1991) Serotonin: an inducer of collagenase in myometrial smooth muscle cells. *J. Cell. Physiol.* **146**, 399–406.

Knäuper, V., López-Otin, C., Smith, B., Knight, G. & Murphy, G. (1996) Biochemical characterization of human collagenase 3. *J. Biol. Chem.* **271**, 1544–1550.

Mitchell, P.G., Magna, H.A., Reeves, L.M., Lopresti-Morrow, L.L., Yocum, S.A., Rosner, P.J., Geoghegen, K.F. & Hambor, J.E. (1996) Cloning, expression and type II collagenolytic activity of matrix metalloproteinase-13 from human osteoarthritic cartilage. *J. Clin. Invest.* **97**, 761–768.

Quinn, C.O., Scott, D.K., Brinckerhoff, C.E., Matrisian, L.M., Jeffrey, J.J. & Partridge, N.C. (1990) Rat collagenase. Cloning, amino acid sequence comparison and parathyroid hormone regulation in osteoblastic cells. *J. Biol. Chem.* **265**, 22343–22347.

Reboul, P., Pelletier, J.-P., Tardif, G., Cloutier, J.M. & Martel-Pelletier, J. (1996) The new collagenase, collagenase 3, is expressed and synthesized by human chondrocytes, but not by synoviocytes. A role in osteoarthritis. *J. Clin. Invest.* **97**, 2011–2019.

Roswit, W.T., Halme, J. & Jeffrey, J.J. (1983) The purification and properties of rat uterine procollagenase. *Arch. Biochem. Biophys.* **225**, 285–295.

Roswit, W.T., Partridge, N.C., Kahn, A.J. & Jeffrey, J.J. (1992) Purification of two tissue inhibitors of metalloproteinases (TIMP) from the rat. *Arch. Biochem. Biophys.* **292**, 402–410.

Sang, Q.A. & Douglas, D.A. (1996) Computational sequence analysis of matrix metalloproteinases. *J. Protein Chem.* **15**, 137–160.

Van Wart, H.E. & Birkedal-Hansen, H. (1990) The cysteine switch: a principle of regulation of metalloproteinase activity with potential applicability to the entire matrix metalloproteinase gene family. *Proc. Natl Acad. Sci. USA* **87**, 5578–5582.

Welgus, H.G., Jeffrey, J.J. & Eisen, A.Z. (1981) The collagen substrate specificity of human skin fibroblast collagenase. *J. Biol. Chem.* **256**, 9511–9516.

Welgus, H.G., Jeffrey, J.J., Stricklin, G.P. & Eisen, A.Z. (1982) The gelatinolytic activity of human skin fibroblast collagenase. *J. Biol. Chem.* **257**, 11534–11539.

Welgus, H.G., Kobayashi, D.K. & Jeffrey, J.J. (1983) The collagen substrate specificity of rat uterus collagenase. *J. Biol. Chem.* **258**, 14162–14165.

Welgus, H.G., Sacchettini, J., Grant, G.A., Roswit, W.T. & Jeffrey, J.J. (1985) The gelatinolytic activity of rat uterus collagenase. *J. Biol. Chem.* **260**, 13601–13606.

Weingarten, H. & Feder, J. (1985) Spectrophotometric assay for vertebrate collagenase. *Anal. Biochem.* **147**, 437–440.

*John J. Jeffrey*
*Department of Biochemistry,*
*Albany Medical College,*
*Albany, NY 12208-3479, USA*
*Email: JohnJJeff@aol.com*

# 392. Collagenase 4

## Databanks

*Peptidase classification: clan MB, family M10, MEROPS ID: M10.018*
*Enzyme classification: none*
*Databank codes:*

| Species | SwissProt | PIR | EMBL (cDNA) | EMBL (genomic) |
|---|---|---|---|---|
| Xenopus laevis | – | – | L76275 | – |

## Name and History

The name **collagenase 4 (Col4)** refers to the fourth collagenase of the vertebrate type I collagenase group. The other three collagenases are the interstitial collagenase (MMP-1) (Chapter 389), neutrophil collagenase (MMP-8) (Chapter 390) and collagenase 3 (MMP-13) (Chapter 391) (Alexander & Werb, 1991; Woessner, 1991; Matrisian, 1992; Birkedal-Hansen *et al.*, 1993; Sang & Douglas, 1996). Col4 was first discovered in the amphibian *Xenopus laevis* (Stolow *et al.*, 1996). Stolow *et al.* were interested in cloning matrix metalloproteinases (MMPs) involved in amphibian metamorphosis. They used a human stromelysin 1 cDNA probe (Wilhelm *et al.*, 1987) to screen a cDNA library made of tadpole intestinal RNA under reduced stringency conditions. The *Xenopus* Col4 cDNA was thus cloned. Its deduced amino acid sequence shows homology to a number of different MMPs. Its collagenase activity is similar to that of human Col1 but it is not the *Xenopus* homolog of any known MMPs based on sequence analysis (Stolow *et al.*, 1996). Thus, it was given the name Col4. Within the matrixin subfamily it is known as **matrix metalloproteinase 18 (MMP-18)**.

## Activity and Specificity

Native collagenase 4 has not been purified from any species. Stolow *et al.* (1996) have studied the biochemical properties of *Xenopus* Col4 (xCol4). Col4 has been overexpressed in *Escherichia coli* as a fusion protein of the propeptide containing a short histidine tag at the N-terminus. It can be activated by $HgCl_2$ treatment. The activated xCol4 can hydrolyze rat tail tendon type I collagen to produce the specific TC$^A$ and TC$^B$ fragments produced by human Col1.

Sequence analysis of the major TC$^B$ fragment shows that human Col1 and xCol4 cleave the $\alpha1(I)$ chain at the same Gly+Ile site to generate the TC$^B$ fragment with the N-terminal sequence of Ile-Ala-Gly-Gln-Arg-Gly. Furthermore, both Col1 and xCol4 can digest gelatin. However, they produce slightly different profiles of digested gelatin products. In addition, casein zymography shows that xCol4 fails to digest casein while human Col1 can, under the conditions tested. Thus, xCol4 and Col1 appear to have distinct but overlapping substrate specificities. (A caution is that the differences between xCol4 and human Col1 in gelatin digestion may be in part contributed by the presence of small amounts of partially degraded xCol4 generated during purification.) The xCol4 can also digest rat type II and human type III collagens, human $\alpha_1$-antitrypsin, and human $\alpha_2$-macroglobulin (D.D. Bauzon *et al.*, unpublished results).

The xCol4 overexpressed in *E. coli* can undergo auto-degradation during purification (Stolow *et al.*, 1996). This autodegradation can be inhibited by 100 mM of the peptide Pro-Leu-Gly-NHOH.HCl. A closely related peptide Z-Pro-Leu-Gly-NHOH is known to inhibit Col1 with an $IC_{50}$ of about 40 mM (Moore & Spilburg, 1986). Thus, both xCol4 and Col1 (Chapter 389) appear to be similarly inhibited by such peptide inhibitors. Kinetic analysis of the digestion of rat tail tendon type I collagen by xCol4 and human Col1 in the presence of 3 mM of Pro-Leu-Gly-NHOH.HCl at 25°C has shown that xCol4 and human Col1 have similar properties. The $K_m$ values of 1.6 and 2.3 mM for xCol4 and human Col1, respectively, are comparable to those previously reported for human Col1 (Welgus *et al.*, 1981; Aimes & Quigley, 1995) and human neutrophil collegenase (Mallya *et al.*, 1990). The values for $k_{cat}$ (2.9 h$^{-1}$ and 4.1 h$^{-1}$ for xCol4 and human Col1, respectively) and specific activity (85 and 91 mg min$^{-1}$ mg$^{-1}$ for xCol4 and human Col1, respectively) are lower than those reported earlier for collagenases (Welgus *et al.*, 1981; Mallya *et al.*, 1990; Mookhtiar *et al.*, 1990; Aimes & Quigley, 1995). This is probably due to the presence of the peptide inhibitor under the assay conditions (xCol4 was renatured in the presence of the inhibitor to prevent autodegradation).

### Structural Chemistry

Collagenase 4 consists of a single peptide of 467 amino acids (Stolow *et al.*, 1996). Like other collagenases, it has a signal peptide that is predicted to target the protein for secretion and a propeptide, the removal of which is believed to be required for enzymatic activation. The predicted molecular masses are 53 000, 51 000 and 42 000 Da for the full-length peptide, proenzyme and mature enzyme, respectively. The proenzyme, with pI of 6.2, can be activated by HgCl$_2$ to produce the predicted mature enzyme with a pI of 6.1.

No structural studies have been carried out for xCol4. The primary sequence shares about 50% identity with collagenases and stromelysins (Stolow *et al.*, 1996). The amino acid sequence of xCol4 is slightly more similar to human Col1 (Chapter 389) (54% identity) than to neutrophil collagenase (Chapter 390) and Col3 (Chapter 391) but lacks the last three residues present in Col1 and neutrophil collagenase. The xCol4 has all the features commonly present in other metalloproteinases, i.e. the zinc-binding sites, the highly conserved sequence PRCGVPD in the propeptide with the exception that the second Pro is replaced with a Tyr in xCol4. Nevertheless, the propeptide functions like that of other MMPs and can be activated by HgCl$_2$ treatment.

### Preparation

Collagenase 4 has thus far been reported only in *Xenopus laevis*. Its homologs in other animal species, if any, are yet to be cloned. The *xCol4* gene is transiently expressed at high levels toward the end of embryogenesis in *X. laevis*. In metamorphosing tadpoles, its mRNA is present in the intestine, tail and hindlimb, the only three organs analyzed (Stolow *et al.*, 1996). The protein levels or enzymatic activities have not been determined in any tissues at the present time.

The xCol4 protein has not been purified. It has been overexpressed as a fusion protein of the pro-xCol4 with a short N-terminal histidine tag. The tag allows purification with a metal chelation resin (Ni-column) to about 85% purity (Stolow *et al.*, 1996). The purified proenzyme can be activated with HgCl$_2$ and cleaves type I collagen.

### Biological Aspects

The *xCol4* gene is regulated in an organ-dependent manner during *Xenopus* metamorphosis (Stolow *et al.*, 1996). The xCol4 mRNA is present at high levels in the tail toward the end of metamorphosis as the tail resorbs. In the intestine, the mRNA level is also elevated during the stages when intestinal remodeling takes place, i.e. the transformation of the simple tubular tadpole intestine to a complex frog intestine with multiple epithelial folds through larval cell apoptosis, proliferation and differentiation of adult cells. Similarly, the gene is expressed at higher levels, compared to at other stages, during early stages of hindlimb development when morphogenesis (digit formation) takes place. These results point to a role of xCol4 in facilitating larval tissue degeneration and adult organogenesis during the tadpole-to-frog transformation.

Metamorphosis is a highly regulated process. The controlling agent is thyroid hormone. Treatment of premetamorphic tadpoles with thyroid hormone can cause precocious metamorphosis. In agreement with the proposed role of xCol4 in metamorphosis, thyroid hormone treatment of premetamorphic tadpoles leads to the activation of the *xCol4* gene in the tail. The activation appears to be indirect as it requires several days of hormone treatment (Stolow *et al.*, 1996).

### Distinguishing Features

The xCol4 has enzymatic properties very similar to those of Col1, but its sequence lacks the C-terminal three amino acids that are present in Col1 (Chapter 389). The PRCGVYD sequence in the propeptide is unique in that the Tyr residue replaces the Pro found in all other known MMPs.

### Related Peptidases

A collagenase has been purified from the amphibian *Rana catesbeiana* (Oofusa & Yoshizato, 1991). Using an antibody against this enzyme to screen an expression cDNA library, a *Rana* cDNA clone has been isolated which encodes a polypeptide of 384 amino acids (Oofusa *et al.*, 1994). The peptide shares over 80% identity with human Col1 but is much shorter than the human Col1, which is 469 amino acid in length (Goldberg *et al.*, 1986). Although no biochemical studies have been performed with the cloned *Rana* collagenase, the sequence homology suggests that it is a type I collagenase. However, its much shorter length compared to other collagenases suggests that it may represent a novel collagenase, instead of the *Rana* homolog of Col1.

Two closely related *Xenopus* homologs of collagenase 3 (Chapter 391) have been cloned (Brown *et al.*, 1996). The regulation of *Xenopus* collagenase 3 genes during metamorphosis is similar to that of the *xCol4* gene (Brown *et al.*, 1996; Stolow *et al.*, 1996). In particular the collagenase 3 genes are expressed at extremely high levels during tail resorption.

## References

Aimes, R.T. & Quigley, J.P. (1995) Matrix metalloproteinase-2 is an interstitial collagenase: inhibitor-free enzyme catalyzes the cleavage of collagen fibrils and soluble native type I collagen generating the specific 3/4- and 1/4-length fragments. *J. Biol. Chem.* **270**, 5872–5876.

Alexander, C.M. & Werb, Z. (1991) Extracellular matrix degradation. In: *Cell Biology of Extracellular Matrix*, 2nd edn (Hay, E.D., ed.). New York: Plenum Press, pp. 255–302.

Birkedal-Hansen, H., Moore, W.G.I., Bodden, M.K., Windsor, L.J., Birkedal-Hansen, B., DeCarlo, A. & Engler, J.A. (1993) Matrix metalloproteinases: a review. *Crit. Rev. Oral Biol. Med.* **4**(2), 197–250.

Brown, D.D., Wang, Z., Furlow, J.D., Kanamori, A., Schwartzman, R.A., Remo, B.F. & Pinder, A. (1996) The thyroid hormone-induced tail resorption program during *Xenopus laevis* metamorphosis. *Proc. Natl Acad. Sci. USA* **93**, 1924–1929.

Goldberg, G.I., Wilhelm, S.M., Kronberger, A., Bauer, E.A., Grant, G.A., & Eisen, A.Z. (1986) Human fibroblast collagenase: complete primary structure and homology to an oncogene transformation-induced rat protein. *J. Biol. Chem.* **261**, 6600–6605.

Mallya, S.K., Mooktiar, K.A., Gao, Y., Brew, K., Dioszegi, M., Birkedal-Hansen, H., van Wart, H. (1990) Characterization of 58-kilodalton human neutrophil collagenase: comparison with human fibroblast collagenase. *Biochemistry* **29**, 10628–10634.

Matrisian, L.M. (1992) The matrix-degrading metalloproteinases. *Bioessays* **14**, 455–463.

Mookhtiar, K.A. & Van Wart, H.E. (1990) Purification to homogeneity of latent and active 58-kilodalton forms of human neutrophil collagenase. *Biochemistry* **29**, 10620–10627.

Moore, W.M. & Spilburg, C.A. (1986) Purification of human collagenases with a hydroxamic acid affinity column. *Biochemistry* **25**, 5189–5195.

Oofusa, K. & Yoshizato, K. (1991) Biochemical and immunological characterization of collagenase in tissues of metamorphosing bullfrog tadpoles. *Dev. Growth Differ.* **33**, 329–339.

Oofusa, K., Yomori, S. & Yoshizato, K. (1994) Regionally and hormonally regulated expression of genes of collagen and collagenase in the anuran larval skin. *Int. J. Dev. Biol.* **38**, 345–350.

Sang, Q.A. & Douglas, D.A. (1996) Computational sequence analysis of matrix metalloproteinases. *J. Protein Chem.* **15**, 137–160.

Stolow, M.A., Bauzon, D.D., Li, J., Sedgwick, T., Liang, V.C.-T., Sang, Q.A. & Shi, Y.-B. (1996) Identification and characterization of a novel collagenase in *Xenopus laevis*: possible roles during frog development. *Mol. Biol. Cell* **7**, 1471–1483.

Welgus, H.G., Jeffrey, J.J. & Eisen, A.Z. (1981) The collagen substrate specificity of human skin fibroblast collagenase. *J. Biol. Chem.* **256**, 9511–9515.

Wilhelm, S.M., Collier, I.E., Kronberger, A., Eisen, A.Z., Marmer, B.L., Grant, G.A., Bauer, E.A. & Goldberg, G.I. (1987) Human skin fibroblast stromelysin: structure, glycosylation, substrate specificity, and differential expression in normal and tumorigenic cells. *Proc. Natl Acad. Sci. USA* **84**, 6725–6729.

Woessner, J.F., Jr (1991) Matrix metalloproteinases and their inhibitors in connective tissue remodeling. *FASEB J.* **5**, 2145–2154.

*Yun-Bo Shi*
*UMM/LME/NICHD/NIH,*
*Bldg. 18T, Rm. 106,*
*Bethesda, MD 20892-5431, USA*
*Email: Shi@helix.nih.gov*

*QingXiang Amy Sang*
*Department of Chemistry and Institute of*
*Molecular Biophysics,*
*Florida State University,*
*Tallahassee, FL 32306, USA*
*Email: qxs@Kyoko.chem.fsu.edu*

# 393. Stromelysin 1

## Databanks

*Peptidase classification: clan MB, family M10, MEROPS ID: M10.005*
*NC-IUBMB enzyme classification: EC 3.4.24.17*
*ATCC entries: 63083 (rat, 1.70 kb)*
*Chemical Abstracts Service registry number: 79955-99-0*

*Databank codes:*

| Species | SwissProt | PIR | EMBL (cDNA) | EMBL (genomic) |
|---|---|---|---|---|
| *Canis familiaris* | – | S23518 | – | – |
| *Cynops pyrrhogaster* | – | – | D82053 | – |
| *Cynops pyrrhogaster* | – | – | D82054 | – |
| *Equus caballus* | – | – | U62529 | – |
| *Homo sapiens* | P08254 | A28156 A28399 A60964 C29157 S15427 | X05232 | U78045: complete gene |
| *Mus musculus* | P28862 | JC1476 S18867 S33139 | X63162 X66402 | – |
| *Oryctolagus cuniculus* | P28863 | A29157 A37306 | A12061 M25664 | – |
| *Rattus norvegicus* | P03957 | A00997 PS0150 S22767 | M65233 X02601 | – |

Brookhaven Protein Data Bank three-dimensional structures:

| Species | ID | Resolution | Notes |
|---|---|---|---|
| *Homo sapiens* | 1SLM | 1.9 | precursor |
|  | 1SLN | 2.27 | catalytic domain; complex with *N*-carboxyl-alky inhibitor L-702,842 |
|  | 2SRT |  | catalytic domain; complex with *N*-(*R*-carboxyethyl)-α-(*S*)-(2-phenyl-ethyl)-glycyl-L-arginine-*N*-phenylamide |

## Name and History

The name *stromelysin* denotes a stromal cell-derived metallo-proteinase that hydrolyzes extracellular matrix (Chin *et al.*, 1985). The enzymic activity was first reported in 1974 as a proteoglycan-degrading metalloproteinase from cartilage (Sapolsky *et al.*, 1974) and a neutral proteinase from rabbit fibroblasts (Werb & Reynolds, 1974). The enzyme was then purified from the rabbit bone culture medium and called *proteoglycanase* (Galloway *et al.*, 1983). Later, it was renamed stromelysin (Chin *et al.*, 1985). Okada *et al.* (1986) purified two isoforms of the enzyme of 45 kDa and 28 kDa from human rheumatoid synoviocytes in culture and referred to it as *matrix metalloproteinase 3* (*MMP-3*) to distinguish it from interstitial collagenase (MMP-1) (Chapter 389) and gelatinase A (MMP-2) (Chapter 401) (Nagase *et al.*, 1992). The rat stromelysin was first discovered as a viral transformation- or epidermal growth factor-induced mRNA without known function (Matrisian *et al.*, 1985) and the protein was named *transin* (Matrisian *et al.*, 1986). A metalloproteinase that enhances procollagenase activation was purified from rabbit fibroblasts and called *procollagenase activator* (Vater *et al.*, 1983). A similar activator called *collagenase activator protein* was purified from cattle cartilage (Treadwell *et al.*, 1986). These activators have been identified as stromelysin. A second stromelysin called stromelysin 2 (Chapter 394) was discovered by cDNA cloning (Muller *et al.*, 1988). To distinguish the two closely related enzymes, the name *stromelysin 1* was recommended for stromelysin by the IUBMB in 1992. The enzyme is a member of the matrixin subfamily of family M12. An *acid metalloproteinase* purified from human cartilage was first designated as matrix metalloproteinase 6 (Azzo & Woessner, 1986), but later proven to be stromelysin 1 (Wilhelm *et al.*, 1993). '*Type I collagen telopeptidase*' from skin fibroblasts (Goldberg & Scott, 1986) was referred to as matrix metalloproteinase 4 by Overall & Sodek (1987), but this enzyme is also likely to be stromelysin 1.

## Activity and Specificity

Nagase (1995) has tabulated studies of the cleavage of various proteins; the enzyme shows a strong preference for hydrophobic residues at P1′, but the P1 site is not specified. The catalytic efficiency depends on the length of the peptide substrate: a peptide with three residues in the P site and two residues at the P′ site is not hydrolyzed unless the P3 and the P2′ sites are blocked (Niedzwiecki *et al.*, 1992). Preferred residues are large hydrophobic groups at P2 and P2′ and Pro at P3 (Nagase & Fields, 1996).

Assay of activity can be done with reduced *S*-carboxymethylated transferrin (Nagase, 1995), azocoll, casein, Mca-Pro-Leu-Gly↓Leu-[3,(2,4-dinitrophenyl)-L-2,3-diaminopropionyl]-Ala-Arg-NH₂ (available from Bachem) (Knight *et al.*, 1992) or Mca-Arg-Pro-Lys-Pro-Val-Glu↓norv-aline-Trp-Arg-Lys(Dnp)-NH₂ (available from Bachem; Peptide Institute; Peptides International) (Nagase, 1995) (see Appendix 2 for full names and addresses of suppliers). The latter Mca-peptide is the most susceptible synthetic substrate for stromelysin 1 so far described. No significant hydrolysis of this substrate is detected with interstitial collagenase (MMP-1) (Chapter 389) or gelatinase A (MMP-2) (Chapter 401) (Nagase & Fields, 1996). Stromelysin 1 exhibits optimal activity around pH 5.5–6.0 with a shoulder of activity from pH 7.5–8.0 for most substrates (Harrison *et al*., 1992; Wilhelm *et al*., 1993), but the optimal activity against *S*-carboxymethylated transferrin is shifted to pH 7.5 with a shoulder of activity around pH 6.0–6.5 (Okada *et al*., 1986). Calcium is required to maintain the active conformation of the enzyme. The assay system typically includes calcium (1–10 mM), sodium chloride (0.1–0.4 M) and a detergent Brij-35 (Nagase, 1995).

Stromelysin 1 is inhibited by tissue inhibitor of metalloproteinases 1 (TIMP-1), TIMP-2 (Murphy & Willenbrock, 1995), $\alpha_2$-macroglobulin and ovostatin (ovomacroglobulin) (Enghild *et al*., 1989) in a stoichiometric fashion. The enzyme is strongly inhibited by chelators such as 1,10-phenanthroline, EDTA, EGTA, cysteine and DTT. Various hydroxamates and phosphonamidate compounds are effective inhibitors, but are not currently available in forms highly specific to stromelysin 1.

### Structural Chemistry

Stromelysin 1 shares structural homology with other matrixins. The human prepro-enzyme consists of a signal peptide (17 amino acids), a propeptide (82 amino acids), a catalytic domain (165 amino acids), a proline-rich hinge region (25 amino acids) and a C-terminal domain (188 amino acids) that shows a sequence similarity to hemopexin and vitronectin. The C-terminal domain is often called the 'hemopexin-like domain'. The enzyme is synthesized as a preproenzyme and secreted from the cell as a proenzyme of 57 kDa. The pI of rabbit prostromelysin 1 is 5.48 (Vater *et al*., 1983). About 20% of prostromelysin 1 is *N*-linked glycosylated and detected as a 59 kDa form. The catalytic domain of the enzyme contains two atoms of zinc: one at the active site where it is bound to side chains of three His residues (His201, His205 and His211, numbered according to human prostromelysin 1) and one structural zinc that interacts with side chains of Asp153, His151, His166 and His179 (Becker *et al*., 1995). The crystal structure of the C-terminal domain-truncated prostromelysin 1 indicates that the pro domain forms a separate folding unit with three $\alpha$ helices and an extended peptide that includes the Pro73-Arg-Cys-Gly-Val-Pro-Asp79 sequence, a motif common to all matrixins (Becker *et al*., 1995). Residues Lys71 to Val77 occupy the active site of the enzyme in a manner similar to a substrate by forming several $\beta$ structure-like hydrogen bonds, but the direction of the propeptide is opposite to that of the substrate (Figure 393.1B,C). Cys75 interacts with the catalytic zinc atom to maintain the latency of the proenzyme. The catalytic domain consists of three $\alpha$ helices, a five-stranded $\beta$ sheet and bridging loops (Becker *et al*., 1995). The overall topology of the catalytic domain is similar to those of interstitial (Chapter 389) and neutrophil (Chapter 390) collagenases, but stromelysin 1 has a larger and deeper S1′ pocket than those collagenases (Figure 393.1A). The structure of the catalytic domain of stromelysin 1 has also been determined by NMR (Gooley *et al*., 1994; Van Doren *et al*.,

1995). Two cysteines in the C-terminal domain are disulfide bonded.

The activation of prostromelysin 1 can be accomplished by organomercurials such as 4-aminophenylmercuric acetate, by heat, or by proteinases (trypsin: Chapter 3; chymotrypsin:

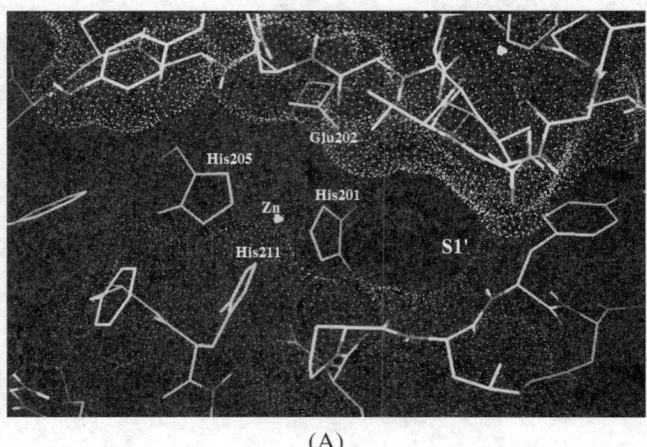

(A)

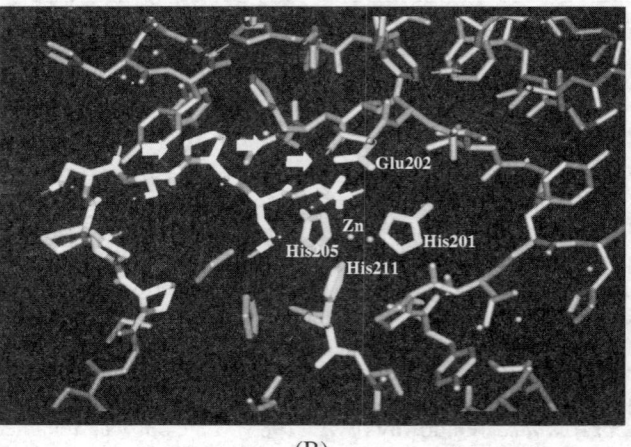

(B)

*Figure 393.1*  Binding of a peptide substrate and the propeptide in the stromelysin 1 active site. (A) Free active site. His201, His205 and His211 coordinate the catalytic zinc. Glu202 is involved in catalysis. The large S1′ pocket is at the bottom next to His203. The substrate binds from left to right. (B) Location of the C-terminal peptide Pro248-Pro-Pro-Asp-Ser-Pro-Glu-Thr of the neighboring stromelysin molecule (catalytic domain) bound to the active-site groove in a substrate-like manner. This interaction occurred during crystallization of the catalytic domain. The C-terminal peptide is shown with white carbon atoms. His201, His205 and His211 and Glu202 are labeled. Arrows indicate the direction of the C-terminal peptide. (C) Interaction of the cysteine switch region with the active-site groove. Lys72-Pro-Arg-Cys-Gly-Val-Pro-Asp-Val-Gly-His (the propeptide segment) is shown with white carbon atoms. The side chains of Arg74 and Asp79 form a salt-bridge. The arrows indicate the direction in which the cysteine switch region of the propeptide binds to the active site groove. (Courtesy of Dr Krzysztof Appelt.)

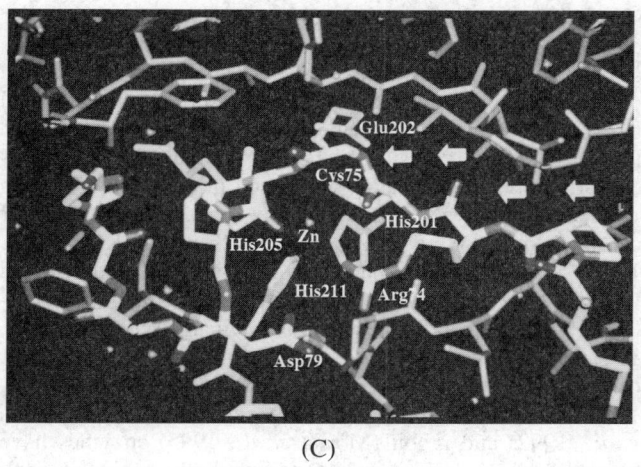

(C)

*Figure 393.1   (continued).*

Chapter 8; tryptase: Chapter 19; chymases: Chapter 18; plasmin: Chapter 59; plasma kallikrein: Chapter 50; leukocyte elastase: Chapter 15; cathepsin G: Chapter 16; pseudolysin: Chapter 357; and thermolysin: Chapter 351) (Nagase, 1995). The common feature of activation is that it proceeds in a stepwise manner (Nagase *et al.*, 1990). The activation by organomercurials is initiated by perturbation of the proenzyme rather than by interaction with Cys75 of the propeptide (Chen

*et al.*, 1993). This treatment generates several intermediates and the first is considered to be formed by an intramolecular reaction because the reaction follows first-order kinetics and it is not inhibited by an active-site inhibitor of MMPs (Okada *et al.*, 1988; Cameron *et al.*, 1995) (Figure 393.2). The activation by proteinases is initiated by proteolytic attacks on the residues near the middle of the propeptide located between the first and second $\alpha$ helices and the generation of intermediates (Figure 393.2). The final process is an intermolecular autolytic cleavage by intermediates or an active stromelysin 1 at the His82$\dashv$Phe83 bond, resulting in an active 45 kDa enzyme (Nagase *et al.*, 1990). The 45 kDa form can be further processed to an active 28 kDa form by removing the C-terminal domain by autolysis (Okada *et al.*, 1988). Both 45 kDa and 28 kDa forms exhibit indistinguishable substrate specificity and specific activity (Okada *et al.*, 1986, 1988). When trypsin is used to activate prostromelysin 1, the N-terminus of the resulting species is Thr85 (Figure 393.2). This form of the enzyme has only about 20% of the activity of the 45 kDa enzyme generated by stromelysin 1 (Benbow *et al.*, 1996).

## Preparation

Stromelysin 1 is not widely distributed nor readily detected in normal cells in culture. The enzyme has been purified from human cartilage (Gunja-Smith *et al.*, 1989). Fibroblasts treated with interleukin 1, tumor necrosis factor $\alpha$, phorbol

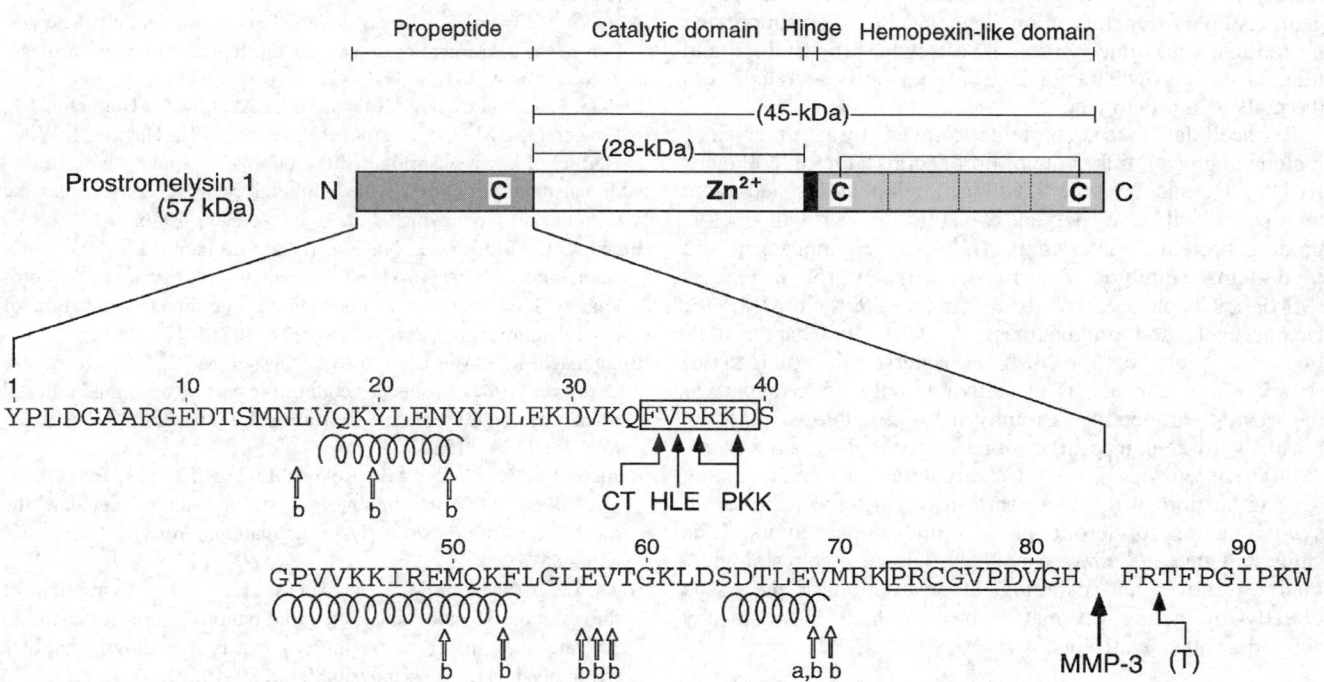

*Figure 393.2*   Cleavage sites in the prodomain of prostromelysin 1 during activation by proteinases and aminophenylmercuric acetate (APMA). Initial cleavage sites of the propeptide by chymotrypsin (CT), human leukocyte elastase (HLE), plasma kallikrein (PKK) are shown by arrows. The sites cleaved during APMA treatment are indicated by open arrows (a, after Nagase *et al.*, 1990; b, after Cameron *et al.*, 1995). The final cleavage of the His82-Phe83 bond is by the action of stromelysin 1 (MMP-3) intermediates, forming a 45 kDa stromelysin 1. The 45 kDa form undergoes autolysis to the 28 kDa MMP-3. Trypsin (T) cleaves the Arg84-Thr85 bond and produces a 45 kDa enzyme, but the activity of this species is about 20% of the fully active stromelysin 1 with Phe83 at the N-terminus (Benbow *et al.*, 1996).

or the activated macrophage-conditioned medium in culture provide a good source for the enzyme purification (Chin *et al.*, 1985; Okada *et al.*, 1986). The recombinant full-length prostromelysin 1 has been expressed in mammalian cells (Murphy *et al.*, 1992; Housley *et al.*, 1993) and in *Escherichia coli* as an insoluble form which has been subsequently refolded (Rosenfeld *et al.*, 1994), and the C-terminal-truncated prostromelysin 1, in *E. coli* as a soluble protein in high yield (Marcy *et al.*, 1991). The catalytic domain has been expressed in *E. coli* in both soluble and insoluble forms; the latter has been refolded to the active form in high yield (Ye *et al.*, 1992).

## Biological Aspects

The human stromelysin 1 gene is located on chromosome 11q22–q23 (Formstone *et al.*, 1993), in which other matrixin genes including stromelysin 2, interstitial collagenase, collagenase 3, macrophage elastase and matrilysin are also located. The genes for human (Sirum & Brinckerhoff, 1989) and rat (Breathnach *et al.*, 1987) have been isolated. The rat gene consists of ten exons (Breathnach *et al.*, 1987) whose arrangement corresponds to that of other matrixin genes. The promoter region contains TATA, AP1 and PEA3 elements. The enzyme may be expressed in fibroblasts, chondrocytes, endothelial cells, macrophages, vascular smooth muscle cells, osteoblasts and keratinocytes in response to appropriate stimuli. The inductive agents include inflammatory cytokines (interleukin 1, tumor necrosis factor $\alpha$), epidermal growth factor, platelet-derived growth factor, phorbol and oncogenic cellular transformation. The synthesis is suppressed by retinoic acid, glucocorticoids, estrogen, progesterone and transforming growth factor $\beta$. The enzyme is secreted from the cells as a proenzyme.

Extracellular matrix proteins cleaved by stromelysin 1 include aggrecan, link protein, fibronectin, laminin, collagens III, IV, IX and X, type I collagen telopeptides and large tenascin C. It also degrades insulin-like growth factor-binding protein 3, interleukin 1$\beta$, $\alpha_1$-antichymotrypsin and $\alpha_1$-proteinase inhibitor. The enzyme activates procollagenases (MMP-1: Chapter 389; MMP-8: Chapter 390; and MMP-13: Chapter 391) and progelatinase B (MMP-9: Chapter 402). For activation of proMMP-1 (interstitial collagenase) stromelysin 1 can act as an activator only when a part of the MMP-1 propeptide is removed by proteinases (plasmin, plasma kallikrein, trypsin) or organomercurials (Suzuki *et al.*, 1990). The enzyme is found in involuting mammary gland, cycling endometrium, osteoarthritic cartilage, rhematoid synovium, atherosclerotic plaques and some tumors. It is suggested that the enzyme may participate in physiological matrix turnover and pathological destruction of the tissue directly by acting on matrix proteins and indirectly by activating other matrixins.

## Distinguishing Features

Stromelysin 1 (MMP-3) and interstitial collagenase (MMP-1) have similar molecular masses and they are often secreted from the same cells. Stromelysin 1 can be distinguished from MMP-1 (Chapter 389) by its failure to cleave types I and II collagen. Stromelysin 1 and stromelysin 2 (Chapter 394) share 78% identity in amino acid sequence and have similar enzymic properties, but the pH optimum of stromelysin 1 is around pH 5.5–6.0, whereas that of stromelysin 2 is around pH 7.5–8.0 (Nagase, 1995). Prostromelysin 2 is not activated by neutrophil elastase (Chapter 15) while prostromelysin 1 is. The cellular expression of the two stromelysins are different: stromelysin 1 is readily expressed in mesenchymal cells whereas stromelysin 2 is expressed in keratinocytes under certain stimuli (Windsor *et al.*, 1993; Saarialho-Kere *et al.*, 1994). Stromelysin 3 (Chapter 397) is more distantly related to these two stromelysins in structure and enzymic properties. Polyclonal antibodies against human (Wilhelm *et al.*, 1987; Okada *et al.*, 1989), rabbit (Murphy *et al.*, 1986), cattle (DenBesten *et al.*, 1989) and rat (Sanchez-Lopez *et al.*, 1988), and monoclonal antibodies against human (Obata *et al.*, 1992) and rabbit (Morita *et al.*, 1995) enzymes have been described. Rabbit polyclonal antibody against human stromelysin 1 (Wilhelm *et al.*, 1987) fails to react with stromelysin 2 (Saarialho-Kere *et al.*, 1994). A commercial ELISA kit for human stromelysin 1 and monoclonal antibody reactive with human, mouse and rat stromelysin 1 are available from Amersham and from Oncogene Research Products, respectively.

## Further Reading

A detailed description of the properties and purification of this enzyme may be found in Nagase (1995).

## References

Azzo, W. & Woessner, J.F. Jr. (1986) Purification and characterization of an acid metalloproteinase from human articular cartilage. *J. Biol. Chem.* **261**, 5434–5441.

Becker, J.W., Marcy, A.I., Rokosz, L.L., Axel, M.G., Burbaum, J.J., Fitzgerald, P.M.D., Cameron, P.M., Esser, C.K., Hagmann, W.K., Hermes, J.D. & Springer, J.P. (1995) Stromelysin-1: three-dimensional structure of the inhibited catalytic domain and the C-truncated proenzyme. *Protein Sci.* **4**, 1966–1976.

Benbow, U., Buttice, G., Nagase, H. & Kurkinen, M. (1996) Characterization of the 46-kDa intermediates of matrix metalloproteinase 3 (stromelysin 1) obtained by site-directed mutation of phenylalanine 83. *J. Biol. Chem.* **271**, 10715–10722.

Breathnach, R., Matrisian, L.M., Gesnel, M.-C., Staub, A. & Leroy, P. (1987) Sequences coding for part of oncogene-induced transin are highly conserved in a related rat gene. *Nucleic Acids Res.* **15**, 1139–1151.

Cameron, P.M., Marcy, A.I., Rokosz, L.L. & Hermes, J.D. (1995) Use of an active-site inhibitor of stromelysin to elucidate the mechanism of prostromelysin activation. *Bioorg. Chem.* **23**, 415–426.

Chen, L.C., Noelken, M.E. & Nagase, H. (1993) Disruption of the cysteine-75 and zinc ion coordination is not sufficient to activate the precursor of human matrix metalloproteinase 3 (stromelysin 1). *Biochemistry* **32**, 10289–10295.

Chin, J.R., Murphy, G. & Werb, Z. (1985) Stromelysin, a connective tissue-degrading metalloendopeptidase secreted by stimulated rabbit synovial fibroblasts in parallel with collagenase. Biosynthesis, isolation, characterization, and substrates. *J. Biol. Chem.* **260**, 12367–12376.

DenBesten, P.K., Heffernan, L.M., Treadwell, B.V. & Awbrey, B.J. (1989) The presence and possible functions of the matrix metalloproteinase collagenase activator protein in developing enamel matrix. *Biochem. J.* **264**, 917–920.

Enghild, J.J., Salvesen, G., Brew, K. & Nagase, H. (1989) Interaction of human rheumatoid synovial collagenase (matrix metalloproteinase 1) and stromelysin (matrix metalloproteinase 3) with human $\alpha_2$-macroglobulin and chicken ovostatin. *J. Biol. Chem.* **264**, 8779–8785.

Formstone, C.J., Byrd, O.J., Ambrose, H.J., Riley, J.H., Hernandez, D., McConville, C.M. & Taylor, A.M. (1993) The order and orientation of a cluster of metalloproteinase genes, stromelysin 2, collagenase, and stromelysin, together with D11S385, on chromosome 11q22-q23. *Genomics* **16**, 289–291.

Galloway, W.A., Murphy, G., Sandy, J.D., Gavrilovic, J., Cawston, T.E. & Reynolds, J.J. (1983) Purification and characterization of a rabbit bone metalloproteinase that degrades proteoglycan and other connective tissue components. *Biochem. J.* **209**, 741–752.

Goldberg, H.A. & Scott, P.G. (1986) Isolation from cultured porcine gingival explants of a neutral proteinase with collagen telopeptidase activity. *Connect. Tissue Res.* **15**, 209–219.

Gooley, P.R., O'Connell, J.F., Marcy, A.I., Cuca, G.C., Salowe, S.P., Springer, J.B. & Johnson, B.A. (1994) The NMR structure of the inhibited catalytic domain of human stromelysin-1. *Nature Struct. Biol.* **1**, 111–118.

Gunja-Smith, Z., Nagase, H. & Woessner, J.F., Jr. (1989) Purification of the neutral proteoglycan-degrading metalloproteinase from human articular cartilage tissue and its identification as stromelysin matrix metalloproteinase-3. *Biochem. J.* **258**, 115–119.

Harrison, R.K., Chang, B., Niedzwiecki, L. & Stein, R.L. (1992) Mechanistic studies on the human matrix metalloproteinase stromelysin. *Biochemistry* **31**, 10757–10762.

Housley, T.J., Baumann, A.P., Braun, I.D., Davis, G., Seperack, P.K. & Wilhelm, S.M. (1993) Recombinant Chinese hamster ovary cell matrix metalloprotease-3 (MMP-3, stromelysin-1). Role of calcium in promatrix metalloprotease-3 (pro-MMP-3, prostromelysin-1) activation and thermostability of the low mass catalytic domain of MMP-3. *J. Biol. Chem.* **268**, 4481–4487.

Knight, C.G., Willenbrock, F. & Murphy, G. (1992) A novel coumarin-labelled peptide for sensitive continuous assays of the matrix metalloproteinases. *FEBS Lett.* **296**, 263–266.

Marcy, A.I., Eiberger, L.L., Harrison, R., Chan, H.K., Hutchinson, N.I., Hagmann, W.K., Cameron, P.M., Boulton, D.A. & Hermes, J.D. (1991) Human fibroblast stromelysin catalytic domain: expression, purification, and characterization of a C-terminally truncated form. *Biochemistry* **30**, 6476–6483.

Matrisian, L.M., Glaichenhaus, N., Gesnel, M.-C. & Breathnach, R. (1985) Epidermal growth factor and oncogenes induce transcription of the same cellular mRNA in rat fibroblasts. *EMBO J.* **4**, 1435–1440.

Matrisian, L.M., Bowden, G.B., Krieg, P., Furstenberger, G., Briand, J.-P., Leroy, P. & Breathnach, R. (1986) The mRNA coding for the secreted protease transin is expressed more abundantly in malignant than benign tumors. *Proc. Natl Acad. Sci. USA* **83**, 9413–9417.

Morita, Y., Igeta, K., Azumano, E., Kunisawa, T., Iwata, K., Okada, Y., Ito, A., Mori, Y. & Nagase, H. (1995) A two-step sandwich enzyme immunoassay for rabbit proMMP-3 using monoclonal antibodies. *Connective Tissue (Japan)* **27**, 183–190.

Muller, D., Quantin, B., Gesnel, M.C., Millon-Collard, R., Abecassis, J. & Breathnach, R. (1988) The collagenase gene family in humans consists of at least four members. *Biochem. J.* **253**, 187–192.

Murphy, G. & Willenbrock, F. (1995) Tissue inhibitors of matrix metalloproteinases. *Methods Enzymol.* **248**, 496–510.

Murphy, G., Hembry, R.M. & Reynolds, J.J. (1986) Characterization of a specific antiserum to rabbit stromelysin and demonstration of the synthesis of collagenase and stromelysin by stimulated rabbit articular chondrocytes. *Collagen Relat. Res.* **6**, 351–364.

Murphy, G., Allan, J.A., Willenbrock, F., Cockett, M.I., O'Connell, J.P. & Docherty, A.J.P. (1992) The role of the C-terminal domain in collagenase and stromelysin specificity. *J. Biol. Chem.* **267**, 9612–9618.

Nagase, H. (1995) Stromelysins 1 and 2. *Methods Enzymol.* **248**, 449–470.

Nagase, H. & Fields, G.B. (1996) Human matrix metalloproteinase specificity studies using collagen sequence-based synthetic peptides. *Biopolymers* **40**, 399–416.

Nagase, H., Enghild, J.J., Suzuki, K. & Salvesen, G. (1990) Stepwise activation mechanisms of the precursor of matrix metalloproteinase 3 (stromelysin) by proteinases and (4-aminophenyl)mercuric acetate. *Biochemistry* **29**, 5783–5789.

Nagase, H., Barrett, A.J. & Woessner, J.F. Jr. (1992) Nomenclature and glossary of the matrix metalloproteinases. *Matrix* **suppl. 1**, 421–424.

Niedzwiecki, L., Teahan, J., Harrison, R.K. & Stein, R.L. (1992) Substrate specificity of the human matrix metalloproteinase stromelysin and the development of continuous fluorometric assays. *Biochemistry* **31**, 12618–12623.

Obata, K., Iwata, K., Okada, Y., Kohrin, Y., Ohuchi, E., Yoshida, S., Shinmei, M. & Hayakawa, T. (1992) A one-step sandwich enzyme immunoassay for human matrix metalloproteinase 3 (stromelysin-1) using monoclonal antibodies. *Clin. Chim. Acta* **211**, 59–72.

Okada, Y., Nagase, H. & Harris, E.D., Jr. (1986) A metalloproteinase from human rheumatoid synovial fibroblasts that digests connective tissue matrix components. Purification and characterization. *J. Biol. Chem.* **261**, 14245–14255.

Okada, Y., Harris, E.D., Jr. & Nagase, H. (1988) The precursor of a metalloendopeptidase from human rheumatoid synovial fibroblasts. Purification and mechanisms of activation by endopeptidases and 4-aminophenylmercuric acetate. *Biochem. J.* **254**, 731–741.

Okada, Y., Takeuchi, N., Tomita, K., Nakanishi, I. & Nagase, H. (1989) Immunolocalisation of matrix metalloproteinase 3 (stromelysin) in rheumatoid synovioblasts (B cells): correlation with rheumatoid arthritis. *Ann. Rheum. Dis.* **48**, 645–653.

Overall, C.M. & Sodek, J. (1987) Initial characterization of a neutral metalloproteinase, active on native 3/4-collagen fragments, synthesized by ROS 17/2.8 osteoblastic cells, periodontal fibroblasts, and identified in gingival crevicular fluid. *J. Dent. Res.* **66**, 1271–1282.

Rosenfeld, S.A., Ross, O.H., Corman, J.I., Pratta, M.A., Blessington, D.L., Feeser, W.S. & Freimark, B.D. (1994) Production of human matrix metalloproteinase 3 (stromelysin) in *Escherichia coli*. *Gene* **139**, 281–286.

Saarialho-Kere, U.K., Pentland, A.P., Birkedal-Hansen, H., Parks, W.G. & Welgus, H.G. (1994) Distinct populations of basal keratinocytes express stromelysin-1 and stromelysin-2 in chronic wounds. *J. Clin. Invest.* **94**, 79–88.

Sanchez-Lopez, R., Nicholson, R., Gesnel, M.C., Matrisian, L.M. & Breathnach, R. (1988) Structure–function relationships in the collagenase family member transin. *J. Biol. Chem.* **263**, 11892–11899.

Sapolsky, A.I., Howell, D.S. & Woessner, J.F., Jr. (1974) Neutral proteases and cathepsin D in human articular cartilage. *J. Clin. Invest.* **53**, 1044–1053.

**M**

Sirum, K.L. & Brinckerhoff, C.E. (1989) Cloning of the genes for human stromelysin and stromelysin 2: differential expression in rheumatoid synovial fibroblasts. *Biochemistry* **28**, 8691–8698.

Suzuki, K., Enghild, J.J., Morodomi, T., Salvesen, G. & Nagase, H. (1990) Mechanisms of activation of tissue procollagenase by matrix metalloproteinase 3 (stromelysin). *Biochemistry* **29**, 10261–10270.

Treadwell, B.V., Neidel, J., Pavia, M., Towle, C.A., Trice, M.E. & Mankin, H.J. (1986) Purification and characterization of collagenase activator protein synthesized by articular cartilage. *Arch. Biochem. Biophys.* **251**, 715–723.

Van Doren, S.R., Kurochkin, A.V., Hu, W., Ye, Q., Johnson, L.L., Hupe, D.J. & Zuiderweg, E.R.P. (1995) Solution structure of the catalytic domain of human stromelysin complexed with a hydrophobic inhibitors. *Protein Sci.* **4**, 2487–2498.

Vater, C.A., Nagase, H. & Harris, E.D., Jr. (1983) Purification of an endogenous activator of procollagenase from rabbit synovial fibroblast culture medium. *J. Biol. Chem.* **258**, 9374–9382.

Werb, Z. & Reynolds, J.J. (1974) Stimulation by endocytosis of the secretion of collagenase and neutral proteinase from rabbit synovial fibroblasts. *J. Exp. Med.* **140**, 1482–1497.

Wilhelm, S.M., Collier, I.E., Kronberger, A., Eisen, A.Z., Marmer, B.L., Grant, G.A., Bauer, E.A. & Goldberg, G.I. (1987) Human skin fibroblast stromelysin: structure, glycosylation, substrate specificity and differential expression in normal and tumorgenic cells. *Proc. Natl Acad. Sci. USA* **84**, 6725–6729.

Wilhelm, S.M., Shao, Z.H., Housley, T.J., Seperack, P.K., Baumann, A.P., Gunja-Smith, Z. & Woessner, J.F. Jr. (1993) Matrix metalloproteinase-3 (stromelysin-1). Identification as the cartilage acid metalloprotease and effect of pH on catalytic properties and calcium affinity. *J. Biol. Chem.* **268**, 21906–21913.

Windsor, L.J., Grenett, H., Birkedal-Hansen, B., Bodden, M.K., Engler, J.A. & Birkedal-Hansen, H. (1993) Cell type-specific regulation of SL-1 and SL-2 genes. Induction of the SL-2 gene but not the SL-1 gene by human keratinocytes in response to cytokines and phorbolesters. *J. Biol. Chem.* **268**, 17341–17347.

Ye, Q.-Z., Johnson, L.L., Hupe, D.J. & Baragi, V. (1992) Purification and characterization of the human stromelysin catalytic domain expressed in *Escherichia coli. Biochemistry* **31**, 11231–11235.

*Hideaki Nagase*
*Department of Biochemistry and Molecular Biology,*
*University of Kansas Medical Center,*
*3901 Rainbow Boulevard,*
*Kansas City, KS 66160, USA*
*Email: hnagase@kumc.edu*

# 394. Stromelysin 2

## Databanks

*Peptidase classification: clan MB, family M10, MEROPS ID: M10.006*
*NC-IUBMB enzyme classification: EC 3.4.24.22*
*ATCC entries: 63083 (rat)*
*Chemical Abstracts Service registry number: 140610-48-6*
*Databank codes:*

| Species | SwissProt | PIR | EMBL (cDNA) | EMBL (genomic) |
|---|---|---|---|---|
| *Homo sapiens* | P09238 | A28816 A47496 | X07820 | – |
| *Mus musculus* | – | – | X76537 | – |
| *Rattus norvegicus* | P07152 | A41775 B26403 S26498 | M65253 X05083 | – |

## Name and History

**Stromelysin 2** is so named as a result of its similarity in primary sequence and substrate specificity to stromelysin 1 (Chapter 393). 'Stromelysin' refers to the ability of these enzymes to degrade multiple components of the extracellular matrix, or 'stroma'. The existence of a second stromelysin was originally indicated by the presence of an additional band in Southern blot analysis of rat genomic DNA using rat stromelysin 1/transin as a probe. The complete genomic structure and cDNA sequence for rat stromelysin 2 was determined and referred to as **transin-2** (Breathnach *et al*., 1987). The human stromelysin 2 cDNA was identified by screening a cDNA library generated from a pool of 11 human cancers with rat stromelysin 1 cDNA as a probe (Muller *et al*., 1988). Stromelysin 2 has proteoglycanase and collagenase activator activity that cannot be distinguished from stromelysin 1, suggesting that early enzyme preparations may have contained either or both enzymes. Stromelysin 2 is also referred to as **matrix metalloproteinase 10 (MMP-10)**.

## Activity and Specificity

Stromelysin 2 activity can be measured using synthetic substrates such as Mca-Pro-Leu-Gly+Leu-Dpa-Ala-Arg-NH$_2$ (Dpa: *N*-3-(2,4-dinitrophenyl)-L-2,3-diaminopropionyl) (Knäuper *et al*., 1996) and [$^{14}$C]casein (Nicolson *et al*., 1989), although these substrates are cleaved by other matrix metalloproteinases as well. Stromelysin 2 cleaves collagen types III, IV and V, fibronectin, gelatins, aggrecan, and pig cartilage proteoglycans in a manner similar to stromelysin 1 (Nicolson *et al*., 1989; Fosang *et al*., 1991; Nguyen *et al*., 1993). Some differences are observed in the cleavage of human cartilage link protein, in that stromelysin 2 can cleave between Leu25+Leu26, while this cleavage product is not observed with stromelysin 1 (Chapter 393) (Nguyen *et al*., 1993). Stromelysin 2, like stromelysin 1, can 'superactivate' the zymogen forms of interstitial collagenase (Chapter 389) (Nicolson *et al*., 1989; Windsor *et al*., 1993) and neutrophil collagenase (Chapter 390) (Knäuper *et al*., 1996). In studies in which chimeric proteins containing domains of collagenase and stromelysin 2 were tested for substrate specificity, it was determined that exon 5 of stromelysin 2, the exon encoding the zinc-binding domain, confers the cleavage specificity of this metalloproteinase (Sanchez-Lopez *et al*., 1993).

Stromelysin 2, like other matrix metalloproteinases, is inhibited by chelating agents such as EDTA, EGTA and 1,10-phenanthroline, and in a stoichiometric fashion by the tissue inhibitor of metalloproteinases (TIMP-1) (Nicolson *et al*., 1989; Windsor *et al*., 1993). Stromelysin 2 forms a binary, SDS-resistant complex with TIMP-1 and TIMP-2 similar to that described for other matrix metalloproteinases (Windsor *et al*., 1993).

## Structural Chemistry

The deduced amino acid sequence of stromelysin 2 reveals that this enzyme contains a signal prepropeptide, propeptide, catalytic domain and C-terminal hemopexin/vitronectin-like domain. Prostromelysin 2 purified from the conditioned medium of phorbol ester-induced human keratinocytes is a 54 kDa protein with an N-terminal sequence consistent with the start of the propeptide, and is converted to a 44 kDa form following activation and the removal of the pro domain (Windsor *et al*., 1993). The enzyme is presumed to be glycosylated based on differences between the deduced and observed sizes and the existence of an *N*-glycosylation site.

## Preparation

Human stromelysin 2 has been purified from phorbol ester-induced human keratinocyte cultures by affinity chromatography with antistromelysin 1 antibodies (Windsor *et al*., 1993). Recombinant human stromelysin 2 has been expressed in myeloma cells (Fosang *et al*., 1991) and Chinese hamster ovary cells as a protein A fusion protein (Nicholson *et al*., 1989).

## Biological Aspects

The most significant differences between stromelysin 1 (Chapter 393) and stromelysin 2 are in their differential patterns of expression, suggesting that these two enzymes represent isozymes of each other. Although the promoters for these two enzymes have AP1 and PEA3 response elements in common, they are differentially regulated in rabbit and human rheumatoid synovial cells and human foreskin fibroblasts in that stromelysin 2 is much less responsive to growth factors and cytokines than stromelysin 1 (Brinckerhoff *et al*., 1992). Stromelysin 2 expression in normal human tissues has been reported in T cells (Conca & Willmroth, 1994) and menstrual endometrium (Rodgers *et al*., 1994). The stromelysin 2 transcript has been detected in a limited number of human tumor samples as reviewed in Powell & Matrisian (1996). Both stromelysin 1 and stromelysin 2 have been identified as differentially expressed genes in cultured rat cells transformed with v-*mos* (Chan *et al*., 1992) and v-K-*ras* (DeVouge & Mukherjee, 1991; Sreenath *et al*., 1992). A striking example of differential regulation of these two enzymes is that stromelysin 2, but not stromelysin 1, is expressed in phorbol ester-induced keratinocytes, but the opposite is true for mucosal fibroblasts (Windsor *et al*., 1993). During wound healing, distinct populations of basal keratinocytes express stromelysin 1 and stromelysin 2 (Saarialho-Kere *et al*., 1994). In the mouse, stromelysin 2 transcripts are found in abundance in heart and kidney with lower levels in many other tissues, while stromelysin 1 expression is restricted to heart, lung and muscle (Gack *et al*., 1994). The specific function of these enzymes in normal tissues has not been elucidated but is believed to involve maintenance and turnover of extracellular matrix components.

## Distinguishing Features

Stromelysin 2 mRNA can be distinguished from stromelysin 1 (Chapter 393) transcripts with specific subcloned sequences, and differences in protein size have been noted (Windsor *et al*., 1993; Sreenath *et al*., 1993). The human stromelysin 2 gene is closer to the centromere on chromosome 11q22.3 and is separated from the stromelysin 1 gene by that of interstitial collagenase (Formstone *et al*., 1993).

## Further Reading

Windsor *et al*. (1993) provide the most complete information on the comparison of human stromelysin 2 with stromelysin 1.

## References

Breathnach, R., Matrisian, L.M., Gesnel, M.C., Staub, A. & Leroy, P. (1987) Sequences coding for part of oncogene-induced transin are highly conserved in a related rat gene. *Nucleic Acids Res.* **15**, 1139–1151.

Brinckerhoff, C.E., Sirum-Connolly, L.K., Karmilowicz, M.J. & Auble, D.T. (1992) Expression of stromelysin and stromelysin-2 in rabbit and human fibroblasts. *Matrix* **suppl. 1**, 165–175.

Conca, W. & Willmroth, F. (1994) Human T lymphocytes express a member of the *matrix metalloproteinase* gene family. *Arthritis Rheum.* **37**, 951–956.

Chan, J.C., Scanlon, M., Zhang, H.-Z., Jia, L.-B., Yu, D., Hung, M.-C., French, M. & Eastman, E.R. (1992) Molecular cloning and characterization of v-mos-activated transformation-associated proteins. *J. Biol. Chem.* **267**, 1099–1103.

DeVouge, M.W. & Mukherjee, B.B. (1992) Transformation of normal rat kidney cells by v-K-*ras* enhances expression of transin 2 and an S-100-related calcium-binding protein. *Oncogene* **7**, 109–119.

Formstone, C.J., Byrd, P.J., Ambrose, H.J., Riley, J.H., Hernandez, D., McConville, C.M. & Taylor, A.M.R. (1993) The order and orientation of a cluster of metalloproteinase genes, stromelysin 2, collagenase, and stromelysin, together with D11S385, on chromosome 11q22-q23. *Genomics* **16**, 289–291.

Fosang, A.J., Neame, P.J., Hardingham, T.E., Murphy, G. & Hamilton, J.A. (1991) Cleavage of cartilage proteoglycan between G1 and G2 domains by stromelysins. *J. Biol. Chem.* **266**, 15579–15582.

Gack, S., Vallon, R., Schaper, J., Ruther, U. & Angel, P. (1994) Phenotypic alterations in fos-transgenic mice correlate with changes in Fos/Jun-dependent collagenase type I expression. Regulation of mouse metalloproteinases by carcinogens, tumor promoters, cAMP, and Fos oncoprotein. *J. Biol. Chem.* **269**, 10363–10369.

Knäuper, V., Murphy, G. & Tschesche, H. (1996) Activation of human neutrophil procollagenase by stromelysin 2. *Eur. J. Biochem.* **235**, 187–191.

Muller, D., Quantin, B., Gesnel, M., Millon-Collard, R., Abecassis, J. & Breathnach, R. (1988) The collagenase gene family in humans consists of at least four members. *Biochem. J.* **253**, 187–192.

Nguyen, Q., Murphy, G., Hughes, C.E., Mort, J.S. & Roughley, P.J. (1993) Matrix metalloproteinases cleave at two distinct sites on human cartilage link protein. *Biochem. J.* **295**, 595–598.

Nicholson, R., Murphy, G. & Breathnach, R. (1989) Human and rat malignant-tumor-associated mRNAs encode stromelysin-like metalloproteinases. *Biochemistry* **28**, 5195–5203.

Powell, W.C. & Matrisian, L.M. (1996) Complex roles of matrix metalloproteinases in tumor progression. In: *Attempts to Understand Metastasis Formation. I. Metastasis-related Molecules* (Gunthert, U. & Birchmeier, W., eds). Berlin: Springer-Verlag, pp. 1–21.

Rodgers, W.H., Matrisian, L.M., Giudice, L.C., Dsupin, B., Cannon, P., Svitek, C., Gorstein, F. & Osteen, K.G. (1994) Patterns of matrix metalloproteinase expression in cycling endometrium imply differential functions and regulation by steroid hormones. *J. Clin. Invest.* **94**, 946–953.

Saarialho-Kere, U.K., Pentland, A.P., Birkedal-Hansen, H., Parks, W.C. & Welgus, H.G. (1994) Distinct populations of basal keratinocytes express stromelysin-1 and stromelysin-2 in chronic wounds. *J. Clin. Invest.* **94**, 79–88.

Sanchez-Lopez, R., Alexander, C.M., Behrendtsen, O., Breathnach, R. & Werb, Z. (1993) Role of zinc-binding- and hemopexin domain-encoded sequences in the substrate specificity of collagenase and stromelysin-2 as revealed by chimeric proteins. *J. Biol. Chem.* **268**, 7238–7247.

Sreenath, T., Matrisian, L.M., Stetler-Stevenson, W., Gattoni-Celli, S. & Pozzatti, R.O. (1992) Expression of matrix metalloproteinase genes in transformed rat cell lines of high and low metastatic potential. *Cancer Res.* **52**, 4942–4947.

Windsor, L.J., Grenett, H., Birkedal-Hansen, B., Bodden, M.K., Engler, J.A. & Birkedal-Hansen, H. (1993) Cell type-specific regulation of SL-1 and SL-2 genes. Induction of the SL-2 gene but not the SL-1 gene by human keratinocytes in response to cytokines and phorbolesters. *J. Biol. Chem.* **268**, 17341–17347.

*Lynn M. Matrisian*
*Department of Cell Biology, MCN C-2310,*
*Vanderbilt University School of Medicine,*
*Nashville, TN 37232, USA*
*Email: Lynn.Matrisian@MCMAIL.Vanderbilt.edu*

# 395. Macrophage elastase

## Databanks

*Peptidase classification: clan MB, family M10, MEROPS ID: M10.009*
*NC-IUBMB enzyme classification: EC 3.4.24.65*

*Databank codes:*

| Species | SwissProt | PIR | EMBL (cDNA) | EMBL (genomic) |
|---------|-----------|-----|-------------|----------------|
| *Homo sapiens* | P39900 | – | L23808 | U25346: promoter<br>U78045: 3' end |
| *Mus musculus* | P34960 | – | M82831 | – |

## Name and History

The name ***macrophage elastase*** incorporates both the principal cellular source of this enzyme and the fact that it is able to degrade insoluble elastin. The name ***macrophage metalloelastase*** is also frequently used to emphasize the distinction between this enzyme and the serine elastases. The enzyme was first identified in the conditioned media of mouse macrophages that had been obtained after intraperitoneal stimulation with thioglycollate (Werb & Gordon, 1975). Further characterization and purification from mouse macrophage-conditioned media established that the enzyme is a metalloproteinase rather than a serine proteinase (White *et al.*, 1980; Banda & Werb, 1981). However, the status of the enzyme as a distinct member of the matrix metalloproteinase subfamily remained unsettled until 1992 when the cDNA was cloned and found to have characteristic matrix metalloproteinase features, but only 33–48% amino acid homology with other matrix metalloproteinases (Shapiro *et al.*, 1992). Subsequently, the human ortholog of mouse macrophage elastase was found in human alveolar macrophages (Shapiro *et al.*, 1993). The cDNAs for human and mouse macrophage metalloelastase have 74% identity; there is 64% identity between the enzymes at the amino acid level.

Macrophage elastase is a member of the matrix metalloproteinase subfamily; hence its alternate name, ***matrix metalloproteinase 12 (MMP-12)***. Because macrophages can express several elastolytic proteinases other than macrophage elastase, including gelatinase B (Chapter 402) and cathepsins S (Chapter 211) and L (Chapter 210), macrophage elastase activity and macrophage elastase should not be considered interchangeable. In this context, rabbit and guinea pig alveolar macrophages in culture release elastolytic activity (Werb, 1978; Banda *et al.*, 1985), presumably due to elastase, but the enzymatic basis of the elastase activity from these sources has not, in fact, been established.

## Activity and Specificity

As with the enzymatic activities of other matrix metalloproteinases, macrophage elastase has a pH optimum at 8.0 with considerable activity at pH 7.4. Calcium is required for activity; addition of $Ca^{2+}$ to a final concentration of 5 mM in calcium-free Tris buffer restores full elastase activity.

Cleavage fragments of elastin (Banda & Werb, 1981) and the insulin B chain (Kettner *et al.*, 1981):

FVNQHLCGSHLVEA↓LY↓LVCGERGFFYTPKA

produced by mouse macrophage elastase show leucine in the P1' position. A single cleavage in $\alpha_1$-proteinase inhibitor ($\alpha_1$PI) occurs between Pro357↓Met358 after it is exposed to macrophage elastase. However, if Met358 is oxidized, the cleavage occurs at Phe352↓Leu353 (Banda *et al.*, 1987a). Interestingly, human macrophage elastase cleaves $\alpha_1$-PI at Phe352↓Leu353 as well as Glu199↓Val200, and cleaves $\alpha_1$-PI with at least one order of magnitude more efficiency than any other matrix metalloproteinase (Gronski *et al.*, 1997).

Macrophage elastase activity is typically quantified against radiolabeled insoluble elastin at pH 7.4 in 0.05 M Tris–HCl, 5 mM $CaCl_2$ (Banda *et al.*, 1987b). For determination of the presence of elastase activity at the appropriate molecular mass, $\kappa$-elastin zymography can be used (Senior *et al.*, 1991). Casein zymography is also a convenient means of documenting the presence of the enzyme in samples known to contain it. Using synthetic octapeptides it was shown that human metalloelastase has a preference for Leu in the P1' position but can tolerate a variety of large and small residues in that position (Gronski *et al.*, 1997). This is consistent with the prediction that this enzyme should have a deep S1' pocket based on its predicted amino acid sequence.

TIMP-1 and $\alpha_2$-macroglobulin inhibit macrophage elastase (Shapiro *et al.*, 1992). Because enzymatic activity requires zinc in the catalytic domain, the enzyme can be reversibly inactivated by zinc chelators such as 1,10-phenanthroline. Various hydroxamates are also effective inhibitors, including the peptide hydroxamate BB94 and other nonselective matrix metalloproteinase inhibitors (S.D. Shapiro *et al.*, unpublished results).

## Structural Chemistry

Mouse macrophage elastase cDNA codes for a predicted proenzyme of 53 kDa (Shapiro *et al.*, 1992). The proenzyme has the typical signal peptide, propeptide segment and catalytic domain, as well as the C-terminal hemopexin-like domain found in all matrix metalloproteinases except matrilysin (Chapter 396). During purification, both the propeptide and the C-terminal hemopexin-like domain are removed, leading to an active enzyme of 22 kDa. Human macrophage elastase codes for a protein of 54 kDa and, like the mouse enzyme, it undergoes processing to an active 22 kDa form during purification (Shapiro *et al.*, 1993). The activation occurring upon isolation explains the marked increase in total elastase activity recovered after purification in early studies (Banda & Werb, 1981), and eliminates the likelihood that the enzyme is dissociated from an endogenous inhibitor in the course of purification.

The human macrophage elastase gene is 13 kb and consists of 10 exons and 9 introns. It is located on chromosome 11q22.2–22.3, physically linked to the interstitial collagenase (Chapter 389) and stromelysin 1 (Chapter 393) genes (Belaaouaj *et al.*, 1995). Mouse macrophage elastase is on chromosome 9 (Shapiro *et al.*, 1992).

## Preparation

Mouse macrophage elastase is present in conditioned medium of cultures of normal peritoneal macrophages and mouse macrophage cell lines (Werb *et al.*, 1978), for example P388D1 cells (available from the ATCC; see Appendix 2 for full names and addresses of suppliers). Gel filtration and ion-exchange chromatography (Banda & Werb, 1981), affinity chromatography over α-elastin linked to agarose (White *et al.*, 1980), and heparin-agarose chromatography followed by gel filtration over ACA-54 (Shapiro *et al.*, 1992) have been used as purification procedures. Recombinant mouse and human macrophage elastases can be generated in *Escherichia coli* after transformation with full-length macrophage elastase cDNAs that have been ligated into pET vectors (Shapiro *et al.*, 1992, 1993). Purified mouse macrophage elastase is available commercially as the proenzyme (Elastin Products Co.).

## Biological Aspects

Macrophage elastase cleaves a number of proteins besides elastin, including basement membrane components, entactin, laminin, fibronectin, various proteoglycans, pepsinogen, type IV collagen, $\alpha_1$-PI (Banda *et al.*, 1980; R.M. Senior, unpublished results), fibrinogen, myelin basic protein, casein (Banda & Werb, 1981), the insulin B chain (Kettner *et al.*, 1981), mouse immunoglobulins (Banda *et al.*, 1983), latent tumor necrosis factor α and plasminogen yielding angiostatin (L.A. Cornelius *et al.*, unpublished results). Considering that it can cleave both extracellular matrix components and other proteins, one might question naming the enzyme for its elastolytic activity. Mice completely lacking macrophage elastase as a result of a targeted disruption of the gene demonstrate marked impairment in capacity to recruit macrophages into inflammatory lesions and, unlike macrophages from wild-type animals, macrophages from these animals do not penetrate reconstituted basement membrane or degrade insoluble elastin (Shipley *et al.*, 1996). These findings point to roles for macrophage elastase in macrophage responses that do not necessarily involve elastin degradation.

By *in situ* hybridization, human placenta (Belaaouaj *et al.*, 1995), carotid atherosclerotic plaques (Halpert *et al.*, 1996) and breast carcinoma (Heppner *et al.*, 1996) demonstrate macrophage elastase mRNA. In the placenta, macrophages and stromal cells are positive, whereas in carcinomatous breast tissue the elastase mRNA is confined to macrophages.

One may postulate that macrophage elastase is appropriately expressed during the host inflammatory response to remodel extracellular matrix, enhance macrophage recruitment and perhaps limit tumor angiogenesis. If elastase is excessively or inappropriately expressed it may cause tissue destruction. Mice that lack elastase as a result of targeted mutagenesis fail to develop cigarette smoke-induced emphysema (Hautemaki *et al.*, 1997).

## Distinguishing Features

Antibody raised to a peptide that represents the first 12 amino acids of the N-terminus of active human macrophage elastase recognizes both the proenzyme and active forms of the enzyme, but does not cross-react with other matrix metalloproteinases (Shapiro *et al.*, 1993).

## References

Banda, M.J. & Werb, Z. (1981) Mouse macrophage elastase. Purification and characterization as a metalloproteinase. *Biochem. J.* **193**, 589–605.

Banda, M.J., Clark, E.J. & Werb, Z. (1980) Limited proteolysis by macrophage elastase inactivates human alpha-1-proteinase inhibitor. *J. Exp. Med.* **152**, 1563–1570.

Banda, M.J., Clark, E.J. & Werb, Z. (1983) Selective proteolysis of immunoglobulins by mouse macrophage elastase. *J. Exp. Med.* **157**, 1184–1196.

Banda, M.J., Clark, E.J. & Werb, Z. (1985) Regulation of alpha-1-proteinase inhibitor function by rabbit alveolar macrophages. Evidence for proteolytic rather than oxidative inactivation. *J. Clin. Invest.* **75**, 1758–1762.

Banda, M.J., Clark, E.J., Sinha, S. & Travis, J. (1987a) Interaction of mouse macrophage elastase with native and oxidized human alpha-1-proteinase inhibitor. *J. Clin. Invest.* **79**, 1314–1317.

Banda, M.J., Werb, Z. & McKerrow, J.H. (1987b) Elastin degradation. *Methods Enzymol.* **144**, 288–305.

Belaaouaj, A., Shipley, J.M., Kobayashi, D.K., Zimonjic, D.B., Popescu, N., Silverman, G. & Shapiro, S.D. (1995) Human macrophage metalloelastase: genomic organization, chromosomal location & tissue specific expression. *J. Biol. Chem.* **270**, 14568–14575.

Gronski, T.J. Jr, Martin, R.L., Kobayashi, D.K., Walsh, B.C., Holman, M.C., Huber, M., Van Wart, H.E. & Shapiro, S.D. (1997) Hydrolysis of a broad spectrum of extracellular matrix proteins by human macrophage elastase. *J. Biol. Chem.* **272**, 12189–12194.

Halpert, I., Sires, U.I., Roby, J.D., Potter-Perigo, S., Wight, T.N., Shapiro, S.D., Welgus, H.G., Wickline, S.A. & Parks, W.C. (1996) Matrilysin is expressed by lipid-laden macrophages at sites of potential rupture in atherosclerotic lesions and localizes to areas of versican deposition, a proteoglycan substrate for the enzyme. *Proc. Natl Acad. Sci. USA* **92**, 9748–9753.

Hautemaki, R.D., Kobayashi, D.K., Senior, R.M. & Shapiro, S.D. (1997) Requirement for macrophage elastase for cigarette smoke-induced emphysema in mice. *Science* **277**, 2002–2004.

Heppner, K.J., Matrisian, L.M., Jensen, R.A. & Rodgers, W.H. (1996) Expression of most matrix metalloproteinase family members in breast cancer represents a tumor-induced host response. *Am. J. Pathol.* **149**, 273–282.

Kettner, C., Shaw, E., White, R. & Janoff, A. (1981) The specificity of macrophage elastase on the insulin B-chain. *Biochem. J.* **195**, 369–372.

Senior, R.M., Griffin, G.L., Fliszar, C.J., Shapiro, S.D., Goldberg, G.I. & Welgus, H.G. (1991) Human 92- and 72-kilodalton type IV collagenases are elastases. *J. Biol. Chem.* **266**, 7870–7875.

Shapiro, S.D., Griffin, G.L., Gilbert, D.J., Jenkins, N.A., Copeland, N.G., Welgus, H.G., Senior, R.M. & Ley, T.J. (1992) Molecular cloning, chromosomal localization, and bacterial expression of a murine macrophage metalloelastase. *J. Biol. Chem.* **267**, 4664–4671.

Shapiro, S.D., Kobayashi, D.K. & Ley, T.J. (1993) Cloning and characterization of a unique elastolytic metalloproteinase produced by human alveolar macrophages. *J. Biol. Chem.* **268**, 23824–23829.

Shipley, J.M., Wesselschmidt, R.L., Kobayashi, D.K., Ley, T.J. & Shapiro, S.D. (1996) Metalloelastase is required for macrophage-mediated proteolysis and matrix invasion in mice. *Proc. Natl Acad. Sci. USA* **93**, 3942–3946.

Werb, Z. (1978) Biochemical actions of glucocorticoids on macrophages in culture. Specific inhibition of elastase, collagenase, and plasminogen activator secretion and effects on other metabolic functions. *J. Exp. Med.* **147**, 1695–1712.

Werb, Z. & Gordon, S. (1975) Elastase secretion by stimulated

macrophages. *J. Exp. Med.* **142**, 361–377.

Werb, Z., Foley, R. & Munck, A. (1978) Glucocorticoid receptors and glucocorticoid-sensitive secretion of neutral proteinases in a macrophage line. *J. Immunol.* **121**, 115–121.

White, R.R., Norby, D., Janoff, A. & Dearing, R. (1980) Partial purification and characterization of mouse peritoneal exudative macrophage elastase. *Biochim. Biophys. Acta* **612**, 233–244.

*Robert M. Senior*
*Department of Medicine,*
*Barnes-Jewish Hospital (North Campus),*
*216 South Kingshighway,*
*St. Louis, MO 63110, USA*
*Email: rsenior@imgate.wustl.edu*

*Stephen D. Shapiro*
*Department of Medicine,*
*Barnes-Jewish Hospital (North Campus),*
*216 South Kingshighway,*
*St. Louis, MO 63110, USA*

# 396. Matrilysin

## Databanks

*Peptidase classification: clan MB, family M10, MEROPS ID: M10.008*
*NC-IUBMB enzyme classification: EC 3.4.24.23*
*ATCC entries: 107942 (173 bp of human matrilysin)*
*Chemical Abstracts Service registry number: 141256-52-2*
*Databank codes:*

| Species | SwissProt | PIR | EMBL (cDNA) | EMBL (genomic) |
|---|---|---|---|---|
| *Felis catus* | P55032 | – | U04444 | – |
| *Homo sapiens* | P09237 | – | X07819 Z11887 | L22524: exon 6 and complete CDS |
| *Mus musculus* | Q10738 | – | L36244 | L36238: exon 6 and complete CDS L36239: exon 5 L36240: exon 4 L36241: exon 3 L36242: exon 2 L36243: exon 1 |
| *Rattus norvegicus* | P50280 | – | L24374 | – |

Brookhaven Protein Data Bank three-dimensional structures:

| Species | ID | Resolution | Notes |
|---|---|---|---|
| *Homo sapiens* | 1MMP | 2.3 | complex with carboxylate inhibitor |
| | 1MMQ | 1.9 | complex with hydroxamate inhibitor |
| | 1MMR | 2.3 | complex with sulfodiimine inhibitor |

## Name and History

The name **matrilysin** denotes lysis of (extracellular) matrix. The Latin root 'matrix' has the further meaning of uterus; this provides an allusion to the role of this proteinase in the postpartum involution of the rat uterus (Sellers & Woessner, 1980). This rat enzyme was subsequently purified (Woessner & Taplin, 1988) and referred to as **uterine metalloproteinase**

**(UMP)**. The human enzyme was first discovered by cDNA cloning (Muller *et al.*, 1988) and was given the name **putative metalloproteinase 1 (pump-1)**. Later the abbreviation was taken to stand for **punctuated metalloproteinase**, alluding to the lack of a C-terminal hemopexin domain. The name **matrin** was used in one paper (Miyazaki *et al.*, 1990) and the name matrilysin was earlier applied to gelatinase A in a review (Alexander & Werb, 1991). The enzyme is a member of the matrixin family and has been designated *matrix metalloproteinase 7 (MMP-7)* (Nagase *et al*., 1992). This designation should not be confused with MMP-7ase (muscle metalloproteinase 7), a cytosolic proteinase not in the matrixin family and not covered by a separate entry in this *Handbook*.

## Activity and Specificity

Matrilysin is capable of digesting a large series of proteins of the extracellular matrix including gelatins I, III, IV, V, fibronectin, collagen IV, laminin, entactin/nidogen, aggrecan, cartilage link protein, elastin and tenascin-C (tabulated in Wilson & Matrisian, 1996) as well as fibulin 1 and 2 (Sasaki *et al*., 1996), vitronectin (Imai *et al*., 1995) and versican (Halpert *et al*., 1996). Matrilysin also cleaves a variety of other proteins including tumor necrosis factor $\alpha$ precursor, u-plasminogen activator precursor (pro-uPA) (Chapter 57), $\alpha_1$-proteinase inhibitor ($\alpha_1$PI), transferrin, casein and insulin B chain (Wilson & Matrisian, 1996) as well as myelin basic protein (Chandler *et al*., 1995) and alpha 2HS glycoprotein (Ochieng & Green, 1996). It has only limited action on type I gelatin (Abramson *et al*., 1995).

Woessner (1995) has tabulated studies of cleavage of various matrix proteins; the major feature is the presence of a large hydrophobic group at P1'. The denatured $\alpha2$ chain of type I collagen is split at several points including the Gly-Ile bond cleaved in intact collagen by interstitial collagenase (Chapter 389); the collagenase substrate Dnp-Pro-Leu-Gly+Ile-Ala-Gly-D-Arg is also cleaved. In a series of 58 octapeptides, Gly-Pro-Gln-Ala+Ile-Ala-Gly-Gln was best (Netzel-Arnett *et al*., 1993). More recently, a bacteriophage display method was employed to generate 200 million hexapeptide recombinants which were then screened to find sequences that were cleaved at high rates (Smith *et al*., 1995). Synthesis of blocked hexapeptides based on these sequences showed that the optimum substrate was Ac-Pro-Leu-Glu+Leu-Arg-Ala-NH$_2$ with a $k_{cat}/K_m$ of 177 000 M$^{-1}$ s$^{-1}$. Pro was clearly preferred at the P3 position, Phe or Leu at P2 and Leu or Ile at P1'. A review of specificity distinctions among MMP-1, -2, -3, -7, -8, -9 and -10 is given by Nagase & Fields (1996).

Assay of activity can be done with azocoll, transferrin or Mca-Pro-Leu-Gly+Leu-Dpa-Ala-Arg-NH$_2$ (Dpa: *N*-3-(2,4-dinitrophenyl)-L-2,3-diaminopropionyl) as described by Woessner (1995). This synthetic substrate is available from Bachem (see Appendix 2 for full names and addresses of suppliers). It is cleaved by many other matrix metalloproteases and therefore is suitable only for assay of the purified enzyme. The pH optimum of the protease is about 7.0 and the assay system typically includes calcium (1–10 mM), sodium chloride (0.1–0.2 M) and a detergent such as Brij-35 (Woessner, 1995). Commercial ELISA kits for measuring human MMP-7 are available from Amersham and from Fuji Chemical Industries; the latter is based on the sandwich method of Ohuchi *et al*. (1996).

Matrilysin is inhibited in stoichiometric fashion by the tissue inhibitor of metalloproteinases (TIMP-1) and also by TIMP-2 (Ward *et al*., 1991). The weakened binding of these inhibitors to MMP-7 relative to their binding to other matrixins may be attributed to the lack of the C-terminal hemopexin domain on MMP-7 (Baragi *et al*., 1994). Other natural inhibitors include $\alpha_2$-macroglobulin, $\alpha$1I3 (Zhu & Woessner, 1991) and ovostatin (Woessner & Taplin, 1988). The enzyme is strongly inhibited by chelators such as 1,10-phenanthroline and DTT, but only poorly by phosphoramidon. Various hydroxamates and phosphinate compounds are effective inhibitors, but are not yet available in forms highly specific for matrilysin (Woessner, 1995). An X-ray structure comparison of MMP-7 complexed to hydroxamate, carboxylate and sulfodiimine inhibitors of closely similar geometry emphasizes the role of binding to the catalytic zinc in determining the efficacy of the inhibitor; the hydroxamate had a $K_i$ about 100 times smaller than the sulfodiimine compound (Browner *et al*., 1995). A combinatorial library method is currently being developed for the identification of matrilysin inhibitors (Schullek *et al*., 1997).

## Structural Chemistry

Matrilysin is the smallest member of the matrixin subfamily; the precursor consists of a signal peptide (cleaved prior to secretion), a propeptide and a catalytic domain. Its sequence is closely similar to that of other matrix metalloproteinases. It is synthesized as a proenzyme of $M_r$ 28 000 and is activated to an $M_r$ of 19 000. The activation can be accomplished by organomercurials such as aminophenylmercuric acetate, by heat or by proteolysis by trypsin, plasmin and other proteases. Several intermediates are observed, but the final step is autolytic cleavage at Glu77+Tyr78 (Crabbe *et al*., 1992). The active enzyme contains two atoms of zinc (Soler *et al*., 1994): one atom at the active center where it is bound to three His residues and one structural zinc. X-ray crystallographic structures have been obtained for the active enzyme complexed to various inhibitors (Browner *et al*., 1995); the best resolution is 1.9 Å. The structure closely resembles that of the other matrixins and fits into the metzincin group of structures (Chapter 380). In addition to the zinc ions there are two calcium ions that help to stabilize the structure. The pI for the rat enzyme is 5.9.

## Preparation

As pointed out in the next section, matrilysin is not widely distributed and is found only in glandular epithelial cells. This means there are few tissues that are suitable for the usual methods of enzyme purification. The enzyme is greatly elevated in the postpartum involuting uterus of the rat (Sellers & Woessner, 1980) and this source has been employed for the extraction and purification (7000-fold) of matrilysin by classical protein purification methods (Woessner & Taplin, 1988; Woessner, 1995). However, yields are measured in micrograms.

Therefore, it is desirable to prepare the enzyme from cultured cells. It has been expressed in COS cells (Quantin *et al*., 1989) and Chinese hamster ovary cells (Barnett *et al*., 1994) and purified from the medium. The proenzyme has been expressed in *E. coli* as an N-terminal fusion protein with ubiquitin (Welch *et al*., 1995) or as a His-tagged protein (Itoh *et al*., 1996). These methods offer the facile preparation of large amounts of correctly folded protein. A baculovirus method has also been developed (López de Turiso *et al.*, 1996).

## Biological Aspects

The human matrilysin gene is localized to chromosome 11q21–q22; the gene structure has been elucidated by Gaire *et al.* (1994). The first five exons match those highly conserved in the various matrixins; the sixth exon is atypical. The promoter region contains TATA, AP1, PEA3 and TIE elements. The last appears to be particularly important in the regulation of this enzyme by progesterone in the menstrual cycle; progesterone acts on stromal cells to induce the synthesis of transforming growth factor $\beta$ (TGF$\beta$), which then diffuses to the epithelium and downregulates matrilysin through interaction with TIE (TGF$\beta$-inhibitory element) (Bruner *et al.*, 1995). The enzyme is induced by epidermal growth factor and phorbol.

Matrilysin has a very restricted tissue distribution: it is produced only by glandular epithelial cells in uterine endometrium, pancreas, skin and kidney mesangium as well as in promonocyte-like cells and blood monocytes (Busiek *et al.*, 1995). It has also been found in epithelium of mammary gland, liver bile duct, prostate and parotid glands (Wilson & Matrisian, 1996). Activity is generally not high unless there is an unusual stimulus as in the postpartum involution of the uterine endometrium, the involuting rat ventral prostate following castration (Powell *et al.*, 1996) and the antral gland epithelium in the vicinity of gastric carcinomas (Honda *et al.*, 1996). Considerable attention has been given to the role of this enzyme in tumors because of the epithelial origin of many tumor cells (Wilson & Matrisian, 1996). It appears to provide a mechanism for the cells to degrade basement membranes (Witty *et al.*, 1994); increasing cell enzyme levels by transfection or reducing them by the use of antisense clones supports such a role. A knockout mouse lacking matrilysin had suppressed tumorigenesis (Wilson *et al.*, 1997).

A considerable number of proteins of the extracellular matrix are cleaved by matrilysin, including collagen type IV, gelatin types I, III, IV, V, fibronectin, vitronectin, elastin, laminin, aggrecan, versican, link protein, fibulin and entactin/nidogen. However, it is not certain which of these, if any, are the natural substrates of the enzyme (Woessner, 1995). There is no direct evidence for the existence in tissue of degraded matrix molecules displaying termini that would be expected based on the cleavage specificity of matrilysin. Certain molecules such as vitronectin and $\alpha_1$-PI are cleaved much more efficiently by matrilysin than by any of the other matrixins. In atherosclerotic lesions, versican was rapidly degraded in the basement membrane at sites where MMP-7 was being expressed by lipid-laden macrophages (Halpert *et al.*, 1996).

The enzyme can convert pro-uPA to uPA (Marcotte *et al.*, 1992) and can activate latent procollagenase 1 (MMP-1) (Chapter 389), progelatinase A (MMP-2) (Chapter 401) (Crabbe *et al.*, 1994), and progelatinase B (MMP-9) (Chapter 402) (Sang *et al.*, 1995). It can also degrade $\alpha_1$-PI, cleaving at Pro357$+$Met358 (Zhang *et al.*, 1994); again, it is much more efficient at this than other matrixins (Sires *et al.*, 1994). These effects point to a direct role in matrix degradation as well as an indirect role through activation of other proteases. The

finding of matrilysin in situations of extensive matrix breakdown in uterine involution, endometrial cycling and malignancies of various sorts is congruent with such functions.

## Distinguishing Features

Although matrilysin is the smallest of the matrix metalloproteinases, the active form of 19 000 Da could be confused with active macrophage elastase (Chapter 395). However, zymography on transferrin gels (Woessner, 1995) would reveal that matrilysin originates from a 28 kDa zymogen whereas elastase starts as a 55 kDa form. Rabbit polyclonal antibodies have been produced against both rat (J.F. Woessner, unpublished results) and human (Rodgers *et al.*, 1993) matrilysin. Monoclonal antibodies against the human enzyme are commercially available from several suppliers including Fuji Chemical Industries and Oncogene Research Products.

## Further Reading

An extensive review of all aspects of this enzyme is presented by Wilson & Matrisian (1996).

## References

Abramson, S.R., Conner, G.A., Nagase, H., Neuhaus, I. & Woessner, J.F., Jr. (1995) Characterization of rat uterine matrilysin and its cDNA: relationship to human pump-1 and activation of procollagenase. *J. Biol. Chem.* **270**, 16016–16022.

Alexander, C.M. & Werb, Z. (1991) Extracellular matrix degradation. In: *Cell Biology of the Extracellular Matrix*, 2nd edn (Hay, E.D., ed.). New York: Plenum Press, pp. 255–302.

Baragi, V.M., Fliszar, C.J., Conroy, M.C., Ye, Q.Z., Shipley, J.M. & Welgus, H.G. (1994) Contribution of the C-terminal domain of metalloproteinases to binding by tissue inhibitor of metalloproteinases. C-terminal truncated stromelysin and matrilysin exhibit equally compromised binding affinities as compared to full-length stromelysin. *J. Biol. Chem.* **269**, 12692–12697.

Barnett, J., Straub, K., Nguyen, B., Chow, J., Suttman, R., Thompson, K., Tsing, S., Benton, P., Schatzman, R., Chen, M. & Chan, H. (1994) Production, purification and characterization of human matrilysin (pump) from recombinant Chinese hamster ovary cells. *Protein Exp. Purif.* **5**, 27–36.

Browner, M.F., Smith, W.W. & Castelhano, A.L. (1995) Matrilysin-inhibitor complexes: common themes among metalloproteases. *Biochemistry* **34**, 6602–6610.

Bruner, K.L., Rodgers, W.H., Gold, L.I., Korc, M., Hargrove, J.T., Matrisian, L.M. & Osteen, K.G. (1995) Transforming growth factor beta mediates the progesterone suppression of an epithelial metalloproteinase by adjacent stroma in the human endometrium. *Proc. Natl Acad. Sci. USA* **92**, 7362–7366.

Busiek, D.F., Baragi, V., Nehring, L.C., Parks, W.C. & Welgus, H.G. (1995) Matrilysin expression by human mononuclear phagocytes and its regulation by cytokines and hormones. *J. Immunol.* **154**, 6484–6491.

Chandler, S., Coates, R., Gearing, A., Lury, J., Wells, G. & Bone, E. (1995) Matrix metalloproteinases degrade myelin basic protein. *Neurosci. Lett.* **201**, 223–226.

Crabbe, T., Willenbrock, F., Eaton, D., Hynds, P., Carne, A.F., Murphy, G. & Docherty, A.J.P. (1992) Biochemical characterization of matrilysin. Activation conforms to the stepwise mechanisms proposed for other matrix metalloproteinases. *Biochemistry* **31**, 8500–8507.

Crabbe, T., Smith, B., O'Connell, J. & Docherty, A. (1994) Human progelatinase A can be activated by matrilysin. *FEBS Lett.* **345**, 14–16.

Gaire, M., Magbanua, Z., McDonnell, S., McNeil, L., Lovett, D.H. & Matrisian, L.M. (1994) Structure and expression of the human gene for the matrix metalloproteinase matrilysin. *J. Biol. Chem.* **269**, 2032–2040.

Halpert, I., Sires, U.I., Roby, J.D., Potter-Perigo, S., Wight, T.N., Shapiro, S.D., Welgus, H.G., Wickline, S.A. & Parks, W.C. (1996) Matrilysin is expressed by lipid-laden macrophages at sites of potential rupture in atherosclerotic lesions and localizes to areas of versican deposition, a proteoglycan substrate for the enzyme. *Proc. Natl Acad. Sci. USA* **93**, 9748–9753.

Honda, M., Mori, M., Ueo, H., Sugimachi, K. & Akiyoshi, T. (1996) Matrix metalloproteinase-7 expression in gastric carcinoma. *Gut* **39**, 444–448.

Imai, K., Shikata, H. & Okada, Y. (1995) Degradation of vitronectin by matrix metalloproteinases-1, -2, -3, -7 and -9. *FEBS Lett.* **369**, 249–251.

Itoh, M., Masuda, K., Ito, Y., Akizawa, T., Yoshioka, M., Imai, K., Okada, Y., Sato, H. & Seiki, M. (1996) Purification and refolding of recombinant human proMMP-7 (pro-matrilysin) expressed in *Escherichia coli* and its characterization. *J. Biochem.* **119**, 667–673.

López de Turiso, J., Fernández, P., Barbacid, M.M., Mira, E., Quesada, A.R., Márquex, G. & Aracil, M. (1996) Expression and purification of human matrilysin produced in baculovirus-infected insect cells. *J. Biotechnol.* **46**, 235–241.

Marcotte, P.A., Kozan, I.M., Dorwin, S.A. & Ryan, J.M. (1992) The matrix metalloproteinase Pump-1 catalyzes formation of low molecular weight (pro)urokinase in cultures of normal human kidney cells. *J. Biol. Chem.* **267**, 13803–13806.

Miyazaki, K., Hattori, Y., Umenishi, F., Yasumitsu, F., Yasumitsu, H. & Umeda, M. (1990) Purification and characterization of extracellular matrix-degrading metalloproteinase, matrin (pump-1), secreted from human rectal carcinoma cell line. *Cancer Res.* **50**, 7758–7764.

Muller, D., Quantin, B., Gesnel, M.C., Millon-Collard, R., Abecassis, J. & Breathnach, R. (1988) The collagenase gene family in humans consists of at least four members. *Biochem. J.* **253**, 187–192.

Nagase, H. & Fields, G.B. (1996) Human matrix metalloproteinase specificity studies using collagen sequence-based synthetic peptides. *Biopolymers* **40**, 399–416.

Nagase, H., Barrett, A.J. & Woessner, J.F., Jr. (1992) Nomenclature and glossary of the matrix metalloproteinases. In: *Matrix Metalloproteinases and Inhibitors* (Birkedal-Hansen, H. *et al.*, eds). Stuttgart: Gustav Fischer, pp. 421–424.

Netzel-Arnett, S., Sang, Q.X., Moore, W.G.I., Navre, M., Birkedal-Hansen, H. & Van Wart, H.E. (1993) Comparative sequence specificities of human 72-kDa and 92-kDa gelatinases (type IV collagenases) and PUMP (matrilysin). *Biochemistry* **32**, 6427–6432.

Ochieng, J. & Green, B. (1996) The interactions of alpha 2HS glycoprotein with metalloproteinases. *Biochem. Mol. Biol. Int.* **40**, 13–20.

Ohuchi, E., Azumano, I., Yoshida, S., Iwata, K & Okada, Y. (1996) A one-step sandwich enzyme immunoassay for human matrix metalloproteinase 7 (matrilysin) using monoclonal antibodies.

*Clin. Chem. Acta* **244**, 181–198.

Powell, W.C., Domann, F.E. Jr, Mitchen, J.M., Matrisian, L.M., Nagle, R.B. & Bowden, G.T. (1996) Matrilysin expression in the involuting rat ventral prostate. *Prostate* **29**, 159–168.

Quantin, B., Murphy, G. & Breathnach, R. (1989) Pump-1 cDNA codes for a protein with characteristics similar to those of classical collagenase family members. *Biochemistry* **28**, 5327–5334.

Rodgers, W.H., Osteen, K.G., Matrisian, L.M., Navre, M., Giudice, L.C. & Gorstein, F. (1993) Expression and localization of matrilysin, a matrix metalloproteinase, in human endometrium during the reproductive cycle. *Am. J. Obstet. Gynecol.* **168**, 253–260.

Sang, Q.-X., Birkedal-Hansen, H. & Van Wart, H.E. (1995) Proteolytic and non-proteolytic activation of human neutrophil progelatinase B. *Biochim. Biophys. Acta* **1251**, 99–108.

Sasaki, T., Mann, K., Murphy, G., Chu, M.-L. & Timpl, R. (1996) Different susceptibilities of fibulin-1 and fibulin-2 to cleavage by matrix metalloproteinases and other tissue proteases. *Eur. J. Biochem.* **240**, 427–434.

Schullek, J.R., Butler, J.H., Ni, Z.J., Chen, D. & Yuan, Z.Y. (1997) A high-density screening format for encoded combinatorial libraries: assay miniaturization and its application to enzymatic reactions. *Anal. Biochem.* **246**, 20–29.

Sellers, A. & Woessner, J.F., Jr. (1980) The extraction of a neutral metalloproteinase from the involuting rat uterus, and its action on cartilage proteoglycan. *Biochem. J.* **189**, 521–531.

Sires, U.I., Murphy, G., Baragi, V.M., Fliszar, C.J., Welgus, H.G. & Senoir, R.M. (1994) Matrilysin is much more efficient than other matrix metalloproteinases in the proteolytic inactivation of $\alpha_1$-antitrypsin. *Biochem. Biophys. Res. Commun.* **204**, 613–620.

Smith, M.M., Shi, L. & Navre, M. (1995) Rapid identification of highly active and selective substrates for stromelysin and matrilysin using bacteriophage peptide display libraries. *J. Biol. Chem.* **270**, 6440–6449.

Soler, D., Nomizu, T., Brown, W.E., Chen, M., Ye, Q.Z., Van Wart, H.E. & Auld, D.S. (1994) Zinc content of promatrilysin, matrilysin and the stromelysin catalytic domain. *Biochem. Biophys. Res. Commun.* **201**, 917–913.

Ward, R.V., Hembry, R.M., Reynolds, J.J. & Murphy, G. (1991) The purification of tissue inhibitor of metalloproteinases-2 from its 72 kDa progelatinase complex. *Biochem. J.* **278**, 179–187.

Welch, A.R., Holman, C.M., Browner, M.F., Gehring, M.R., Kan, C.-C. & Van Wart, H.E. (1995) Purification of human matrilysin produced in *Escherichia coli* and characterization using a new optimized fluorogenic peptide substrate. *Arch. Biochem. Biophys.* **324**, 59–64.

Wilson, C.L. & Matrisian, L.M. (1996) Matrilysin: an epithelial matrix metalloproteinase with potentially novel functions. *Int. J. Biochem. Cell Biol.* **28**, 123–136.

Wilson, C.L., Heppner, K.J., Labosky, P.A., Hogan, B.L.M. & Matrisian, L.M. (1997) Intestinal tumorigenesis is suppressed in mice lacking the metalloproteinase matrilysin. *Proc. Natl Acad. Sci. USA* **94**, 1402–1407.

Witty, J.P., McDonnell, S., Newell, K.J., Cannon, P., Navre, M., Tressler, R.J. & Matrisian, L.M. (1994) Modulation of matrilysin levels in colon carcinoma cell lines affects tumorigenicity in vivo. *Cancer Res.* **54**, 4805–4812.

Woessner, J.F., Jr. (1995) Matrilysin. *Methods Enzymol.* **248**, 485–495.

Woessner, J.F., Jr. & Taplin, C.J. (1988) Purification and properties of a small latent matrix metalloproteinase of the rat uterus. *J. Biol. Chem.* **263**, 16918–16925.

Zhang, Z., Winyard, P.G., Chidwick, K., Murphy, G., Wardell, M., Carrell, R.W. & Blake, D.R. (1994) Proteolysis of human native and oxidised alpha-1-proteinase inhibitor by matrilysin and stromelysin. *Biochim. Biophys. Acta* **1199**, 224–228.

Zhu, C. & Woessner, J.F., Jr. (1991) A tissue inhibitor of metalloproteinases (TIMP) and α-macroglobulins in the ovulating rat ovary: possible regulators of collagen matrix breakdown. *Biol. Reprod.* **45**, 334–342.

*J. Fred Woessner*
*Department of Biochemistry and Molecular Biology,*
*University of Miami School of Medicine,*
*PO Box 016960,*
*Miami, FL 33101, USA*
*Email: fwoessne@mednet.med.miami.edu*

# 397. Stromelysin 3

## Databanks

*Peptidase classification: clan MB, family M10, MEROPS ID: M10.007*
*NC-IUBMB enzyme classification: none*
*ATCC entries: 124711 (human)*
*Databank codes:*

| Species | SwissProt | PIR | EMBL (cDNA) | EMBL (genomic) |
|---|---|---|---|---|
| *Homo sapiens* | P24347 | I38250 S13423 S58912 | X57766 | X84664: 5′ end |
| *Mus musculus* | Q02853 | A44399 | Z12604 | – |
| *Rattus norvegicus* | – | – | U46034 | – |
| *Xenopus laevis* | Q11005 | S38623 | Z27093 | – |

## Name and History

**Stromelysin 3** was named originally on the basis of the similarity of its domain structure to the stromelysin subgroup (i.e. stromelysin 1: Chapter 393 and stromelysin 2: Chapter 394) as well as its stromal pattern of tissue distribution. The stromelysin 3 cDNA was cloned by differential screening between malignant and benign breast tumor samples (Basset *et al.*, 1990). It was the first MMP to be specifically associated with the invasive phase of breast cancer where its distribution was primarily confined to the surrounding stromal cells by *in situ* hybridization (Basset *et al.*, 1990). The mouse homolog of stromelysin 3 was cloned by screening mouse cDNA libraries with human cDNA as a probe (Lefebvre *et al.*, 1992). In *Xenopus*, it was cloned as a thyroid hormone-response gene from metamorphosing intestines and subsequently shown to have the highest sequence homology to human stromelysin 3 (Patterton *et al.*, 1995). The enzyme belongs to the matrixin subfamily and has been designated **matrix metalloproteinase 11 (MMP-11)**.

## Activity and Specificity

Although originally proposed to be involved in extracellular matrix degradation during breast cancer cell invasion, stromelysin 3 has the most restricted substrate specificity of all known MMP family members. While human stromelysin 3 has not been found to degrade extracellular matrix components, the ~45–47 kDa mature form of the enzyme specifically cleaved a single protein in the conditioned media of the MCF7 breast cancer cell line. The stromelysin 3 substrate was subsequently identified as the serine proteinase inhibitor (serpin), α1-proteinase inhibitor. Proteolysis occurred at a single site (Ala350⫫Met351) within the reactive-site loop of the inhibitor, which resulted in a complete loss of antiproteolytic activity (Pei *et al.*, 1994). Human stromelysin 3 can also cleave additional members of the serpin family including α2-antiplasmin and plasminogen activator inhibitor 2 (Pei *et al.*, 1994).

A 28 kDa C-terminally truncated form of stromelysin 3 was isolated as the major active species when the full-length mouse cDNA was expressed in mammalian cells (Murphy *et al.*, 1993). The truncated enzyme (but not the mature full-length proteinase) expressed weak matrix-degrading activity against laminin as well as type IV collagen (Murphy *et al.*, 1993). A similarly truncated human enzyme, however, did not express matrix-degrading activity (Noël *et al.*, 1995). Differences between the mouse and human enzymes have been attributed to an unusual Pro235 → Ala 'substitution' that is distinct from human stromelysin 3 (Noël *et al.*, 1995). Further experiments are needed to define the substrate specificity of the mature and C-terminally truncated forms of stromelysin 3. Currently, stromelysin 3 enzymic activity can be assayed by using either $\alpha_1$-proteinase inhibitor, $\alpha_2$-macroglobulin or $\alpha$-casein as substrates (Murphy *et al.*, 1993; Pei *et al.*, 1994; Noël *et al.*, 1995). No synthetic substrate has been employed or designed for the specific characterization of stromelysin 3 activity. While the optimal reaction conditions for stromelysin 3 have not been fully explored, assays are typically performed at pH 7.6, with 5 mM $CaCl_2$ and 150 mM NaCl (Pei *et al.*, 1994). Like other MMPs, mouse or human stromelysin 3 activity can be blocked by either natural inhibitors (tissue inhibitors of metalloproteases: TIMP-1, -2 or -3, and $\alpha_2$-macroglobulin), synthetic inhibitors (e.g. the hydroxamate compound BB94) or metal chelators (e.g. EDTA or 1,10-phenanthroline) (Murphy *et al.*, 1993; Pei *et al.*, 1994; Pei & Weiss, 1995; Noël *et al.*, 1995).

## Structural Chemistry

The overall domain structure for preprostromelysin 3 is arranged from its N-terminal end to its C-terminal end as follows: signal peptide – pro domain – catalytic domain – hemopexin-like domain (Basset *et al.*, 1990). Unlike other MMPs, prostromelysin 3 cannot be activated by organomercurials or detected by standard zymographic techniques. This unusual behavior has been attributed to structural characteristics peculiar to both the pro and catalytic domains of the proteinase (Pei *et al.*, 1994; Santavicca *et al.*, 1996). Furthermore, while MMPs are normally synthesized and secreted as inactive zymogens, stromelysin 3 can be recovered in culture media from normal as well as transiently or stably transfected cells as an active enzyme (Pei *et al.*, 1994; Santavicca *et al.*, 1995; Noël *et al.*, 1995). Initially synthesized as a ~54–56 kDa precursor, the proteinase is then modified in the *trans*-Golgi network to a ~62–65 kDa form, presumably following the glycosylation of serine/threonine residues in the pro domain (Pei & Weiss, 1995; Santavicca *et al.*, 1996). Subsequently, the ~45–47 kDa mature species is generated intracellularly following the removal of the stromelysin 3 pro domain. This activation process is mediated by furin (Chapter 117), a *trans*-Golgi network-associated proprotein convertase, which cleaves prostromelysin 3 at the C-terminal end of a unique RXRXKR motif sandwiched between the pro and catalytic domains (Pei & Weiss, 1995; Santavicca *et al.*, 1996). In the absence of metalloproteinase inhibitors, the 45 kDa species is unstable and undergoes autocatalytic processing into smaller, C-terminal truncation products (Pei *et al.*, 1994; Pei & Weiss, 1995).

## Preparation

Natural sources for stromelysin 3 are generated during a diverse array of extracellular matrix remodeling events such as uterine and mammary gland involution, tumor invasion and metastasis (e.g. Rodgers *et al.*, 1994; Lefebvre *et al.*, 1995; Anderson *et al.*, 1995). However, these *in vivo* sources are not practical for large-scale purification of the enzyme. Instead, stromelysin 3 has been expressed in mammalian cell lines such as COS, mouse myeloma, MCF7, HT1080 and 293 cells (Murphy *et al.*, 1993; Pei *et al.*, 1994) or in prokaryotes (Noël *et al.*, 1995). In the presence of the reversible MMP inhibitor BB94, a ~45–47 kDa active species can be purified in large scale by a combination of dye-affinity, heparin-affinity and gel-filtration chromatographies (Pei *et al.*, 1994). In the absence of BB94, a ~28 kDa C-terminally truncated species is obtained (Murphy *et al.*, 1993).

## Biological Aspects

Stromelysin 3 has been localized to a novel locus on chromosome 22q11.2 (Levy *et al.*, 1992) which differs from that reported for stromelysin 1 and 3 on chromosome 11. The genomic structure for stromelysin 3 has also been determined in humans (Anglard *et al.*, 1995) and mice (D. Pei *et al.*, unpublished results). The first four exons match those of other classic MMPs while the remaining four exons are organized differently: (a) the hinge region in stromelysin 3 is an extension of exon 5 which encodes a portion of the catalytic domain; (b) the first Cys residue for the conserved disulfide bond in the hemopexin domain is not part of the exon encoding the hinge region as in other MMPs, but part of the first exon of the hemopexin-like domain; (c) a conserved intron located between the second and third exons of the hemopexin domains present in all other MMPs is absent in stromelysin 3. These differences in exon/intron organization support phylogenetic analyses which place stromelysin 3 outside of the main MMP cluster between the mammalian MMPs and the bacterial metalloproteinases (Murphy *et al.*, 1991, see also Fig. 385.1).

Stromelysin 3 expression can be stimulated by phorbol esters and growth factors including epidermal growth factor and platelet-derived growth factor (Basset *et al.*, 1990; Anderson *et al.*, 1995). Unlike many MMP promoters, no consensus sequences for AP1-binding sites were found, but a specific *cis*-acting retinoic acid responsive element of the DR1-type was characterized (Anglard *et al.*, 1995). However, while transient transfection experiments demonstrated that the human stromelysin 3 promoter can be transactivated by retinoic acid receptors in the presence of retinoic acid (Anglard *et al.*, 1995), stromelysin 3 expression appears to be downregulated by retinoic acid in normal fibroblasts (Anderson *et al.*, 1995). The promoter elements responsible for growth factor-dependent stimulation as well as tissue or tumor stroma-specific expression have not been defined.

Despite interesting correlations between stromelysin 3 expression, active matrix remodeling events and rates of breast cancer recurrence (Wolf *et al.*, 1993; Engel *et al.*, 1994), the precise role of stromelysin 3 in tumor invasion and metastasis remains unclear. Stromelysin 3 expression can promote tumor take in nude mice, but increases in local invasiveness or metastasis were not observed in either nude mice studies or *in vitro* (Noël *et al*., 1996). The mechanism by which stromelysin 3 enhances cancer cell survival requires further study.

Studies of stromelysin 3 processing led to the discovery that furin (Chapter 117), a member of the proprotein convertase family that can specifically cleave after Lys/Arg-Arg or Arg-Xaa-Arg/Lys-Arg motifs (Steiner *et al*., 1992), mediates the intracellular processing of the zymogen to its active form (Pei & Weiss, 1995; Santavicca *et al*., 1996). Similar motifs have recently been identified between the pro and catalytic domains of all four members of the membrane-type matrix metalloproteinases (MT-MMPs) (Chapter 399) (Sato *et al*., 1994; Takino *et al*., 1995; Will & Hinzmann, 1995; Puente *et al*., 1996). Indeed, furin can process the activation of a soluble, transmembrane-deleted form of MT1-MMP (Pei & Weiss, 1996), but the role of this or related proprotein convertases in the processing of the wild-type enzyme remains controversial (Cao *et al*., 1996). Since members of the MT-MMP subgroup have been implicated in the activation of MMPs lacking RXKR motifs, such as gelatinase A (Chapter 401) (Sato *et al*., 1994) and collagenase 3 (Chapter 391) (Knäuper *et al*., 1996; D. Pei *et al*., unpublished results), furin or related proprotein convertases (Seidah *et al*., 1994) may play a broader role in extracellular matrix remodeling events.

## Further Reading

For a review, see Rouyer *et al*. (1995).

## References

Anderson, I.C., Sugarbaker, D.J., Ganju, R.K., Tsarwhas, D.G., Richards, W.G., Sunday, M., Kobzik, L. & Shipp, M.A. (1995) Stromelysin 3 is overexpressed by stromal elements in primary non-small cell lung cancers and regulated by retinoic acid in pulmonary fibroblasts. *Cancer Res.* **55**, 4120–4126.

Anglard, P., Melot, T., Guerin, E., Thomas, G. & Basset, P. (1995) Structure and promoter characterization of the human stromelysin 3 gene. *J. Biol. Chem.* **270**, 20337–20344.

Basset, P., Bellocq, J.P., Wolf, C., Stoll, I., Hutin, P., Limacher, J.M., Podhajcer, O.L., Chenard, M.P., Rio, M.C. & Chambon, P. (1990) A novel metalloproteinase gene specifically expressed in stromal cells of breast carcinomas. *Nature* **348**, 699–704.

Cao, J., Rehemtulla, A., Bahou, W. & Zucker, S. (1996) Membrane type matrix metalloproteinase 1 activates pro-gelatinase A without furin cleavage of the N-terminal domain. *J. Biol. Chem.* **271**, 30174–30180.

Engel, G., Heselmeyer, K., Auer, G., Bäckdahl, M., Eriksson, E. & Linder, S. (1994) Correlation between stromelysin 3 mRNA level and outcome of human breast cancer. *Int. J. Cancer* **58**, 830–835.

Knäuper, V., Will, H., López-Otín, C., Smith, B., Atkinson, S.J., Stanton, H., Hembry, R.M. & Murphy, G. (1996) Cellular mechanisms for human procollagenase-3 (MMP-13) activation. *J. Biol. Chem.* **271**, 17124–17131.

Lefebvre, O., Wolf, C., Limacher, J.M., Hutin, P., Wendling, C., LeMeur, M., Basset, P. & Rio, M. (1992) The breast cancer-associated stromelysin 3 gene is expressed during mouse mammary gland apoptosis. *J. Cell Biol.* **119**, 997–1002.

Lefebvre, O., Régnier, C., Chenard, M.P., Wendling, C., Chambon, P., Basset, P. & Rio, M.C. (1995) Developmental expression of mouse stromelysin 3 mRNA. *Development* **121**, 947–955.

Levy, A., Zucman, J., Delattre, O., Mattei, M.G., Rio, M.C. & Basset, P. (1992) Assignment of the human stromelysin 3 (STMY3) gene to the q11.2 region of chromosome 22. *Genomics* **13**, 881–883.

Murphy, G.J.P., Murphy, G. & Reynold, J.J. (1991) The origin of matrix metalloproteinases and their relationships. *FEBS Lett.* **289**, 4–7.

Murphy, G., Segain, J.P., O'Shea, M., Cockett, M., Ioannou, C., Lefebvre, O., Chambon, P. & Basset, P. (1993) The 29-kDa N-terminal domain of mouse stromelysin 3 has the general properties of a weak metalloproteinase. *J. Biol. Chem.* **268**, 15435–15441.

Noël, A., Santavicca, M., Stoll, I., L'Hoir, C., Staub, A., Murphy, G., Rio, M.-C. & Basset, P. (1995) Identification of structural determinants controlling human and mouse stromelysin 3 proteolytic activities. *J. Biol. Chem.* **270**, 22866–22872.

Noël, A.C., Lefebvre, O., Maquoi, E., VanHoorde, L., Chenard, M.P., Mareel, M., Foidart, J.M., Basset, P. & Rio, M.-C. (1996) Stromelysin expression promotes tumor take in nude mice. *J. Clin. Invest.* **97**, 1924–1930.

Patterton, D., Hayes, W.P. & Shi, Y.B. (1995) Transcriptional activation of the matrix metalloproteinase gene stromelysin 3 coincides with thyroid hormone-induced cell death during frog metamorphosis. *Dev. Biol.* **167**, 252–262.

Pei, D. & Weiss, S.J. (1995) Furin-dependent intracellular activation of the human stromelysin 3 zymogen. *Nature* **375**, 244–247.

Pei, D. & Weiss, S.J. (1996) Transmembrane-deletion mutants of the membrane-type matrix metalloproteinase-1 process progelatinase A and express intrinsic matrix-degrading activity. *J. Biol. Chem.* **271**, 9135–9140.

Pei, D., Majmudar, G. & Weiss, S.J. (1994) Hydrolytic inactivation of a breast carcinoma cell-derived serpin by human stromelysin 3. *J. Biol. Chem.* **269**, 25849–25855.

Puente, X.S., Pendás, A.M., Llano, E., Velasco, G. & López-Otín, C. (1996) Molecular cloning of a novel membrane-type matrix metalloproteinase from a human breast carcinoma. *Cancer Res.* **56**, 944–949.

Rodgers, W.H., Matrisian, L.M., Giudice, L.C., Dsupin, B., Cannon, P., Sviteck, C., Gorstein, F. & Osteen, K.G. (1994) Patterns of matrix metalloproteinase expression in cycling endometrium imply differential functions and regulation by steroid hormones. *J. Clin. Invest.* **94**, 946–953.

Rouyer, N., Wolf, C., Chenard, M.P., Rio, M.C., Chambon, P., Bellocq, J.P. & Basset, P. (1995) Stromelysin 3 gene expression in human cancer: an overview. *Invasion Metastasis* **14**, 269–275.

Santavicca, M., Noël, A., Chenard, M.P., Lutz, Y., Stoll, I., Segain, J.P., Rouyer, N., Rio, M.C., Wolf, C., Bellocq, J.P. & Basset, P. (1995) Characterization of monoclonal antibodies against stromelysin 3 and their use to evaluate stromelysin 3 levels in breast carcinoma by semi-quantitative immunohistochemistry. *Int. J. Cancer* **64**, 336–341.

Santavicca, M., Noël, A.C., Angliker, H., Stoll, I., Segain, J.P., Anglard, P., Chretien, M., Seidah, N. & Basset, P. (1996) Characterization of structural determinants and molecular mechanisms involved in pro-stromelysin 3 activation by 4-aminophenylmercuric acetate and furin-type convertase. *Biochem. J.* **315**, 953–958.

Sato, H., Takino, T., Okada, Y., Cao, J., Shinagawa, A., Yamamoto, E. & Seiki, M. (1994) A matrix metalloproteinase expressed on the surface of invasive tumour cells. *Nature* **370**, 61–65.

M

Seidah, H.G., Chrétien, M. & Day, R. (1994) The family of subtilisin/kexin-like proprotein and pro-hormone convertases: divergent or shared functions. *Biochimie (Paris)* **76**, 197–209.

Steiner, D.F., Smeekens, S.P., Ohagi, S. & Chan, S.J. (1992) The new enzymology of precursor processing endoproteases. *J. Biol. Chem.* **267**, 23435–23438.

Takino, T., Sato, H., Shinagawa, A. & Seiki, M. (1995) Identification of the second membrane-type matrix metalloproteinase (MT-MMP-2) gene from a human placenta cDNA library. *J. Biol.*

*Chem.* **270**, 23013–23020.

Will, H. & Hinzmann, B. (1995) cDNA sequence and mRNA tissue distribution of a novel human matrix metalloproteinase with a potential transmembrane segment. *Eur. J. Biochem.* **231**, 602–608.

Wolf, C., Rouyer, N., Lutz, Y., Adida, C., Loriot, M., Bellocq, J.P., Chambon, P. & Basset, P. (1993) Stromelysin 3 belongs to a subgroup of proteinases expressed in breast carcinoma fibroblastic cells and possibly implicated in tumor progression. *Proc. Natl Acad. Sci. USA* **90**, 1843–1847.

***Duanqing Pei***
*Division of Hematology/Oncology,*
*University of Michigan,*
*5220 MSRB III,*
*1150 W. Medical Center Drive,*
*Ann Arbor, MI 48109-0640, USA*
*Email: jrohr@vs2.im.med.umich.edu*

***Stephen J. Weiss***
*Division of Hematology/Oncology,*
*University of Michigan,*
*5220 MSRB III,*
*1150 W. Medical Center Drive,*
*Ann Arbor, MI 48109-0640, USA*
*Email: jrohr@vs2.im.med.umich.edu*

# 398. *Soybean metalloproteinase 1*

## Databanks

*Peptidase classification: clan MB, family M10, MEROPS ID: M10.012*
*NC-IUBMB enzyme classification: none*
*Databank codes:*

| Species | SwissProt | PIR | EMBL (cDNA) | EMBL (genomic) |
|---|---|---|---|---|
| *Arabidopsis thaliana* | – | – | – | AC002062: chromosome 1 BAC F20P5 |
| *Glycine max* | P29136 | – | – | – |

## Name and History

**Soybean metalloproteinase 1 (SMEP1)** was initially described as one of two Azocoll-digesting activities found in crude soluble extracts of soybean leaves (Ragster & Chrispeels, 1979). These two activities are referred to as Azocollase A (EDTA sensitive) and Azocollase B (PCMB sensitive) and it was suggested that these activities fell into the metallo and cysteine classes of proteinases, respectively. These findings aided in the subsequent characterization of Azocollase A. The protein was purified to homogeneity and shown to be a 19 kDa metalloproteinase (Graham *et al*., 1991). The complete chemical sequencing of the polypeptide revealed that the plant enzyme shared many biochemical features with the vertebrate matrix metalloproteinases (MMPs) (McGeehan *et al*., 1992). These included the conserved zinc-binding motif, collagenase-like cleavage specificity for peptide substrates and stoichiometric inhibition by the tissue inhibitor of metalloproteinase 1 (TIMP-1). These findings suggest that the MMP family of enzymes is not restricted to vertebrates and that they are more widely present than originally thought. In addition, there is the possibility that the TIMP-like inhibitors may exist in plants and regulate the activity of these enzymes in a system analogous to that in vertebrates.

## Activity and Specificity

SMEP1 was originally characterized as an Azocoll-cleaving activity. The activity can be monitored by performing a limited digest (25 mM Tris, 1 mg ml$^{-1}$ Azocoll, pH 9.0) at 37°C. Reaction mixtures can be centrifuged and product formation monitored at 520 nm. The enzyme has pH optimum of 8.0–9.0 using Azocoll as a substrate.

The high homology between SMEP1 and the MMP sequences suggested that the enzymes are functionally related. The enzyme has no activity against fibrillar collagens. However, the fluorogenic substrate Dnp-Pro-Leu-Gly+Leu-Trp-Ala-D-Arg-NH$_2$ (Stack & Gray, 1989) and the chromogenic substrate Ac-Pro-Leu-Gly-thioester+Leu-Leu-Gly-OEt (Weingarten *et al.*, 1985) are specifically cleaved by SMEP1 at the Gly-Leu scissile bond. The pH optimum is about 8.0 and the enzyme can be assayed in the standard assay buffers used for MMP assays (150 mM NaCl, 50 mM Tris, 5 mM Ca$^{2+}$, 20 μM Zn$^{2+}$, 0.05% Brij-35, pH 7.6).

SMEP1 is inhibited by several known MMP inhibitors (McGeehan *et al.*, 1992). The tissue inhibitor of metalloproteinases (TIMP-1) binds to the enzyme with an affinity similar to that of TIMP-1 for the catalytic 19 kDa fragment of collagenase (24 nM versus 14 nM). In addition, the synthetic hydroxamate inhibitor SC43937 (Dickens *et al.*, 1986) is also a potent inhibitor of SMEP1 activity (IC$_{50}$ = 41 nM) with binding affinity similar to that reported for collagenase. The enzyme is inhibited by zinc chelators such as 1,10-phenanthroline, but it is only poorly inhibited by phosphoramidon, very similar to what is observed for the MMPs.

## Structural Chemistry

SMEP1 is one of the smallest members of the matrixin family. It consists of a signal peptide, a propeptide and a catalytic domain. The catalytic domain primary sequence is ~40% identical to that of the other vertebrate MMPs (McGeehan *et al.*, 1992), which is similar to the identity among the vertebrate MMPs, as well. The enzyme is expected to have two zinc atoms, one at the catalytic site and one at the structural site. The amino acids making Zn$^{2+}$ contacts at the catalytic site and at the structural site in interstitial collagenase (Chapter 389) (Lovejoy *et al.*, 1994) are completely conserved in the SMEP1 sequence as well.

## Preparation

SMEP1 was originally characterized from mature leaves of soybean (Graham *et al.*, 1991). The enzyme was purified 1160-fold in a three-step sequence: (1) DEAE-cellulose chromatography, (2) ammonium sulfate precipitation and (3) size-exclusion chromatography, to give the 19 kDa catalytically active fragment. The enzyme could be stored at 4°C in assay buffer (150 mM NaCl, 50 mM Tris, 5 mM Ca$^{2+}$, 20 mM Zn$^{2+}$, 0.05% Brij-35, pH 7.6) for an extended period of time with only marginal loss in activity.

## Biological Aspects

A cDNA clone coding for SMEP1 has been obtained by PCR and rapid amplification of cDNA ends (RACE) (Pak *et al.*, 1997). The open reading frame encodes a 305 amino acid polypeptide of calculated $M_r$ 34 042. SMEP1 has a signal sequence which is probably cleaved between Ala28 and His29

to afford a 31 kDa proenzyme. It contains the conserved 'cysteine switch' sequence, Pro-Arg-Cys-Gly-Val-Pro-Asp, which has been shown to play a role in maintaining zymogen latency (Van Wart & Birkedal-Hansen, 1990).

Digestion of genomic DNA shows only a single band, indicating that the *smep1* gene is present as a single copy (Pak *et al.*, 1997). In addition, initial studies by PCR indicate that the gene lacks introns.

Data from northern and western blot analyses indicate that SMEP1 expression is regulated in a developmental- and tissue-specific manner. The gene appears to be under tight transcriptional control. Message and protein begin to accumulate 10 days after leaf emergence and remain at steady-state levels until leaf senescence. The physiological role of the enzyme is still not clarified.

## Distinguishing Features

SMEP1, like matrilysin (Chapter 396) (Quantin *et al.*, 1989) represents a distinct subset of the MMP family that is characterized by the lack of the C-terminal hemopexin domain. A SMEP1 homolog has recently been characterized in *Arabidopsis*, suggesting that this enzyme may be widespread in the plant kingdom (J.S. Graham, unpublished results). The interesting finding that TIMP-1 binds to this enzyme suggests that regulation of SMEP1 activity in plants may be analogous to vertebrate systems and efforts to elucidate such activities are underway.

## References

Dickens, J.P., Donald, D.K., Keen, G. & McKay, W.R. (1986) US Patent 4599361, 8 July 1986.

Graham, J.S., Xiong, J. & Gillikin, J.W. (1991) Purification and developmental analysis of a metalloendoproteinase from the leaves of *Glycine max*. *Plant Physiol.* **97**, 786–792.

Lovejoy, B., Cleasby, A., Hassell, A., Longley, K., Luther, M.A., Weigl, D., McGeehan, G.M., McElroy, A.B., Drewry, D., Lambert, M. & Jordan, S.R. (1994) Structure of the catalytic domain of fibroblast collagenase complexed with an inhibitor. *Science* **263**, 375–377.

McGeehan, G., Burkhart, W., Anderegg, R., Becherer, J.D., Gillikin, J.W. & Graham, J.S. (1992) Sequencing and characterization of a soybean leaf metalloproteinase. *Plant Physiol.* **99**, 1179–1183.

Pak, J.H., Liu, C.Y. Huangpu, J. & Graham, J.S. (1997) Construction and characterization of the soybean leaf metalloproteinase cDNA. *FEBS Lett.* **404**, 283–288.

Quantin, B., Murphy, G. & Breathnach, R. (1989) Pump-1 encodes for a protein with characteristics similar to those of classical collagenase family members. *Biochemistry* **28**, 5327–5334.

Ragster, L. & Chrispeels, M.J. (1979) Azocoll digesting proteinases in soybean leaves. Characteristics and changes during leaf maturation and senescence. *Plant Physiol.* **64**, 857–862.

Stack, M.S. & Gray, R.D. (1989) Comparison of vertebrate collagenase and gelatinase using a new fluorogenic peptide substrate. *J. Biol. Chem.* **264**, 4277–4281.

Van Wart, H.E. & Birkedal-Hansen, H. (1990) The cysteine switch: a principle of regulation of metalloproteinase activity with potential applicabililty to the entire matrix metalloproteinase family. *Proc.* *Natl Acad. Sci. USA* **87**, 5578–5582.

Weingarten, H., Martin, R.J. & Feder, J. (1985) Synthetic substrates of vertebrate collagenase. *Biochemistry* **24**, 6730–6734.

***Gerard McGeehan***
*New Leads Generation,*
*Rhone-Poulenc Rorer,*
*500 Arcola Road,*
*Collegeville, PA 19426, USA*
*Email: mcgeegm@rpr.rpna.com*

***John S. Graham***
*Department of Biological Sciences,*
*Bowling Green State University,*
*Bowling Green, OH 43403, USA*
*Email: jgraham@opie.bgsu.edu*

# 399. Membrane-type matrix metalloproteinase 1

## Databanks

Peptidase classification: clan MB, family M10, MEROPS ID: M10.014
NC-IUBMB enzyme classification: none
Databank codes:

| Species | SwissProt | PIR | EMBL (cDNA) | EMBL (genomic) |
|---|---|---|---|---|
| Membrane-type matrix metalloproteinase 1 | | | | |
| *Homo sapiens* | P50281 | – | D26512 | – |
| | | | U41078 | |
| | | | X83535 | |
| | | | Z48481 | |
| *Mus musculus* | P53690 | – | X83536 | – |
| *Oryctolagus cuniculus* | – | – | U73940 | – |
| *Rattus norvegicus* | Q10739 | – | X83537 | – |
| | | | X91785 | |
| Membrane-type matrix metalloproteinase 2 | | | | |
| *Homo sapiens* | P51511 | – | Z48482 | |
| Membrane-type matrix metalloproteinase 3 | | | | |
| *Gallus gallus* | – | – | U66463 | – |
| *Homo sapiens* | P51512 | – | D50477 | – |
| | | | D83646 | – |
| Membrane-type matrix metalloproteinase 4 | | | | |
| *Homo sapiens* | | – | X89576 | – |

## Name and History

***Membrane-type matrix metalloproteinase 1 (MT1-MMP)*** was discovered by cDNA cloning using RT-PCR and degenerate oligonucleotide primers derived from the conserved sequences of enzymes of the matrix metalloproteinase (MMP) family (Sato *et al*., 1994; Takino *et al*., 1995b; Okada *et al*., 1995). MT1-MMP contains a C-terminal proximal transmembrane domain and a short cytoplasmic tail in addition to the domains shared with other MMPs. The enzyme was named membrane-type matrix metalloproteinase (MT-MMP) since it was the first integral membrane protein found among MMPs (Sato *et al.*, 1994). Three other MT-MMPs encoded by different genes were reported subsequently: MT2-MMP (Will & Hinzmann, 1995), MT3-MMP (Takino *et al.*, 1995a) and MT4-MMP (Puente *et al.*, 1996). Two MT-MMP-2s were reported simultaneously by different groups and one of these was renamed to MT-MMP-3 (Takino *et al.*, 1995a). Recently, the names MT1-, MT2-, MT3- and MT4-MMP have become more common and are more appropriate to avoid a possible confusion with other MMP family members such as collagenase 1 (MMP-1), which is not bound to the plasma membrane. According to the MMP nomenclature (Okada *et al.*, 1986), the four MT-MMPs correspond to *MMP-14* through MMP-17, respectively.

## Activity and Specificity

MT1-MMP was first reported as an activator of progelatinase A (proMMP-2) (Chapter 401) (Sato *et al.*, 1994). Upon expression of MT1-MMP, progelatinase A binds to the cell surface and is activated via MT-MMP-dependent cleavage of its propeptide (Atkinson *et al.*, 1995; Strongin *et al.*, 1995; Pei & Weiss, 1996; Sato *et al.*, 1996a). MT1-MMP cleaves the Asn66+Leu67 peptide bond in gelatinase A, generating an intermediate form which is converted into a fully activated enzyme by an autoproteolytic mechanism (Kinoshita *et al.*, 1996; Will *et al.*, 1996). Activation of progelatinase A on the cell surface has been reported to depend on its hemopexin-like domain (Murphy *et al.*, 1992; Strongin *et al.*, 1993) and a small amount of the tissue inhibitor of metalloproteinases 2 (TIMP-2) (Strongin *et al.*, 1995).

TIMP-2 is thought to mediate the binding of progelatinase A to the MT1-MMP on the cell surface. TIMP-2 binds to the activated MT1-MMP and at the same time to the hemopexin-like domain of progelatinase A through its N-terminal and C-terminal domains respectively. Strongin *et al.* (1995) indeed demonstrated the formation of such a trimolecular complex *in vitro*. Binding of progelatinase A to the cells is thought to facilitate its activation by increasing local concentration of the substrate and presenting it to the neighboring free MT1-MMP. An excessive amount of TIMP-2, however, inhibits the binding and the activation of progelatinase A. Thus, local concentration of TIMP-2 on the cell surface is a critical factor controlling the progelatinase A activation. Integrins such as $\alpha_v\beta_3$ may also participate in this process as a part of a, or an additional, gelatinase A receptor (Brooks *et al.*, 1996). It is also possible that progelatinase A binds to TIMP-2 which is forming a complex with a molecule other than MT1-MMP (Emmert-Buck *et al.*, 1995). Although MT1-MMP does not activate progelatinase B, it can activate procollagenase 3 (Chapter 391) (Knäuper *et al.*, 1996). A similar cell surface mechanism may be responsible for this activation.

MT1-MMP degrades extracellular matrix macromolecules such as fibronectin, vitronectin, B chain of laminin and dermatan sulfate proteoglycan (Ohuchi *et al.*, 1996; Pei & Weiss, 1996). The enzyme also degrades gelatin, casein and elastin; a synthetic peptide can be used to monitor the catalytic activity (Imai *et al.*, 1996; Pei & Weiss, 1996; Will *et al.*, 1996). The enzyme also shows activity against interstitial collagens (Ohuchi *et al.*, 1996). MT1-MMP is specifically inhibited by TIMP-2 and TIMP-3, but not by TIMP-1 (Kinoshita *et al.*, 1996; Will *et al.*, 1996).

## Structural Chemistry

ProMT1-MMP has three characteristic insertions when it is aligned with the sequence of other MMPs (Sato *et al.*, 1994; Takino *et al.*, 1995b). These are: (a) 11 amino acid residues between propeptide and catalytic domains (INS-1: Ala101–Arg111), (b) 8 amino acid residues in the catalytic domain (INS-2: Pro163–Gly170) and (c) 66 amino acid residues at the C-terminus of the hemopexin-like domain (INS-3: Pro509–Val582). INS-1 contains a consensus sequence (Arg-Arg-Lys-Arg) known to be cleaved by furin (Chapter 117), a processing enzyme in the *trans*-Golgi apparatus (Hosaka *et al.*, 1991). A similar furin-sensitive site is present in other proMT-MMPs and in prostromelysin 3 (proMMP-11) (Chapter 397). INS-3 is composed of a 24 amino acid residues long hydrophobic stretch (Ala539-Phe562) that functions as a transmembrane domain and a short cytoplasmic tail. The function of the INS-2 is not known. As expected, furin activates proMT-MMP by cleaving the propeptide domain downstream from the INS-1 recognition sequence (at Arg111-Tyr112) (Pei & Weiss, 1996; Sato *et al.*, 1996b). Recombinant proMT1-MMP can also be activated *in vitro* by trypsin that cleaves at the same site (Will *et al.*, 1996). The molecular mass of the translated protein is 66 kDa and 53.7 kDa for the activated enzyme, and the pI is 5.67.

## Preparation

MT1-MMP is expressed by various tumor cell lines (HT1080, MDA-MB-231, etc.) and fibroblasts. Treatment of these cells with concanavalin A or phorbol (TPA) enhances activation of progelatinase A. MT1-MMP was purified from a crude plasma membrane fraction of the human fibrosarcoma cell line HT1080 treated with TPA using affinity chromatography column with immobilized C-terminal domain of gelatinase A (Strongin *et al.*, 1995). TIMP-2 was copurified in a trimolecular complex with the enzyme. A soluble form of MT1-MMP was also purified as a complex with TIMP-2 from the culture supernatant of a breast carcinoma cell line MDA-MB-231 (Imai *et al.*, 1996). Recombinant enzyme was expressed in cultured mammalian cells (Imai *et al.*, 1996; Pei & Weiss, 1996) and in *Escherichia coli* (Kinoshita *et al.*, 1996; Will *et al.*, 1996), retaining its proteolytic activity against the propeptide sequence of gelatinase A (Chapter 401).

## Biological Aspects

The appearance of the activated form of gelatinase A (Chapter 401) in tumor tissue has been reported to associate with tumor spread of breast (Brown *et al.*, 1993b; Davies *et al.*, 1993) and lung carcinomas (Brown *et al.*, 1993a). MT1-MMP is expressed in various types of human tumors. MT1-MMP was immunolocalized in carcinoma cells of lung (Sato *et al.*, 1994; Tokuraku *et al.*, 1995; Polette *et al.*, 1996), cervical (Gilles *et al.*, 1996), gastric (Nomura *et al.*, 1995), colon (Ohtani *et al.*, 1996) and brain tumors (Yamamoto *et al.*, 1996). Stromal cells adjacent to tumors were also MT1-MMP positive. However, the results of *in situ* hybridization using MT1-MMP-specific probes are not in agreement with the observations made by the immunohistochemistry, especially in the case of breast carcinomas. The *in situ* hybridization signal is weak in tumor cells compared to the surrounding stroma (Okada *et al.*, 1995; Heppner *et al.*, 1996; Polette *et al.*, 1996) while the reverse is observed by immunostaining (Polette *et al.*, 1996). The major producer cells in the tissue observed by *in situ* hybridization vary depending on the tumor types (Okada *et al.*, 1995; Ohtani *et al.*, 1996; Polette *et al.*, 1996). The level of MT1-MMP mRNA correlated well with the activation of progelatinase A in lung (Tokuraku *et al.*, 1995) and gastric (Nomura *et al.*, 1995) carcinomas, supporting the idea that MT1-MMP is an activator of gelatinase A.

MT1-MMP is also expressed during mouse embryogenesis (Kinoh *et al.*, 1996). Expression of MT1-MMP mRNA increases during the gestation period and decreases with maturation after birth. *In situ* hybridization and immunohistochemistry localize MT1-MMP mRNA and protein to the cells of ossifying tissues where gelatinase A was also expressed. The amount of activated gelatinase A appears to be proportional to the expression of MT1-MMP. Thus activation of gelatinase A by MT1-MMP is thought to be a physiological system utilized in tissue remodeling.

## Related Peptidases

Three additional related MT-MMPs have been reported (above). Amino acid sequences of MT1-, MT2- and MT3-MMPs are closely related to each other, showing 50% identity, while MT4-MMP shows only 30% identity (Puente *et al.*, 1996). MT3-MMP also activates progelatinase A (Takino *et al.*, 1995a). The molecular masses expected for the activated enzymes are 61.2, 55.4 and 57.8 kDa for MT2-, MT3- and MT4-MMP, respectively. Monoclonal antibodies that recognize MT1- and MT3-MMP have been described (Sato *et al.*, 1994; Takino *et al.*, 1995a). Monoclonal antibodies to human MT1-MMP are commercially available from Fuji Chemical Industries and Oncogene Research Products (see Appendix 2 for full names and addresses of suppliers).

## References

Atkinson, S.J., Crabbe, T., Cowell, S., Ward, R.V., Butler, M.J., Sato, H., Seiki, M., Reynolds, J.J. & Murphy, G. (1995) Intermolecular autolytic cleavage can contribute to the activation of progelatinase A by cell membranes. *J. Biol. Chem.* **270**, 30479–30485.

Brooks, P.C., Strömblad, S., Sanders, L.C., von Schalscha, T.L., Aimes, R.T., Stetler-Stevenson, W.G., Quigley, J.P. & Cheresh, D.A. (1996) Localization of matrix metalloproteinase MMP-2 to the surface of invasive cells by interaction with integrin $\alpha v \beta 3$. *Cell* **85**, 683–693.

Brown, P.D., Bloxidge, R.E., Stuart, N.S.A., Gatter, K.C. & Carmichael, J. (1993a) Association between expression of activated 72-kilodalton gelatinase and tumor spread in non-small-cell lung carcinoma. *J. Natl Cancer Inst.* **85**, 574–578.

Brown, P.D., Bloxidge, R.E., Anderson, E. & Howell, A. (1993b) Expression of activated gelatinase in human invasive breast carcinoma. *Clin. Exp. Metastasis.* **11**, 183–189.

Davies, B., Miles, D.W., Happerfield, L.C., Naylor, M.S., Bobrow, L.G., Rubens, R.D. & Balkwill, F.R. (1993) Activity of type IV collagenases in benign and malignant breast disease. *Br. J. Cancer* **67**, 1126–1131.

Emmert-Buck, M.R., Emonard, H.P., Corcoran, M.L., Krutzsch, H.C., Foidart, J.M. & Stetler-Stevenson, W.G. (1995) Cell surface binding of TIMP-2 and pro-MMP-2/TIMP-2 complex. *FEBS Lett.* **364**, 28–32.

Gilles, C., Polette, M., Piette, J., Munaut, C., Thompson, E.W., Birembaut, P. & Foidart, J.M. (1996) High level of MT-MMP expression is associated with invasiveness of cervical cancer cells. *Int. J. Cancer* **65**, 209–213.

Heppner, K.J., Matrisian, L.M., Jensen, R.A. & Rodgers, W.H. (1996) Expression of most metalloproteinase family members in breast cancer represents a tumor-induced host response. *Am. J. Pathol.* **149**, 273–282.

Hosaka, M., Nagahama, M., Kim, W.S., Watanabe, T., Hatsuzawa, K., Ikemizu, J., Murakami, K. & Nakayama, K. (1991) Arg-X-Lys/Arg-Arg motif as a signal for precursor cleavage catalyzed by furin within the constitutive secretory pathway. *J. Biol. Chem.* **266**, 12127–12130.

Imai, K., Ohuchi, E., Aoki, T., Nomura, H., Fujii, Y., Sato, H., Seiki, M. & Okada, Y. (1996) Membrane-type matrix metalloproteinase 1 is a gelatinolytic enzyme and is secreted in a complex with tissue inhibitor of metalloproteinases 2. *Cancer Res.* **56**, 2707–2710.

Kinoh, H., Sato, H., Tsunezuka, Y., Takino, T., Kawashima, A., Okada, Y. & Seiki, M. (1996) MT-MMP, the cell surface activator of proMMP-2 (pro-gelatinase A), is expressed with its substrate in mouse tissue during embryogenesis. *J. Cell Sci.* **109**, 953–959.

Kinoshita, T., Sato, H., Takino, T., Itoh, M., Akizawa, T. & Seiki, M. (1996) Processing of a precursor of 72-kilodalton type IV collagenase/gelatinase A by a recombinant membrane-type 1 matrix metalloproteinase. *Cancer Res.* **56**, 2535–2538.

Knäuper, V., Will, H., López-Otin, C., Smith, B., Atkinson, S.J., Stanton, H., Hembry, R.M. & Murphy, G. (1996) Cellular mechanisms for human procollagenase-3 (MMP-3) activation. *J. Biol. Chem.* **271**, 17124–17131.

Murphy, G., Willenbrock, F., Ward, R.V., Cockett, M.I., Eaton, D. & Docherty, A.J. (1992) The C-terminal domain of 72 kDa gelatinase A is not required for catalysis, but is essential for membrane activation and modulates interactions with tissue inhibitors of metalloproteinases. *Biochem. J.* **283**, 637–641.

Nomura, H., Sato, H., Seiki, M., Mai, M. & Okada, Y. (1995) Expression of membrane-type matrix metalloproteinase in human gastric carcinomas. *Cancer Res.* **55**, 3263–3266.

Ohtani, H., Motohashi, H., Sato, H., Seiki, M. & Nagura, M. (1996) Dual overexpression pattern of membrane-type metalloproteinase-1 in cancer and stromal cells in human gastrointestinal carcinoma revealed by in situ hybridization and immunoelectron microscopy. *Int. J. Cancer* **68**, 565–570.

Ohuchi, E., Imai, K., Fujii, Y., Sato, H., Seiki, M. & Okada, Y. (1996) Membrane-type 1 matrix metalloproteinase digests interstitial collagens and other extracellular matrix macromolecules. *J. Biol. Chem.* **272**, 2446–2451.

Okada, A., Bellocq, J.P., Rouyer, N., Chenard, M.P., Rio, M.C., Chambon, P. & Basset, P. (1995) Membrane-type matrix metalloproteinase (MT-MMP) gene is expressed in stromal cells of human colon, breast, and head and neck carcinomas. *Proc. Natl Acad. Sci. USA* **92**, 2730–2734.

Okada, Y., Nagase, H. & Harris, E.J. (1986) A metalloproteinase from human rheumatoid synovial fibroblasts that digests connective tissue matrix components. Purification and characterization. *J. Biol. Chem.* **261**, 14245–14255.

Pei, D. & Weiss, S. (1996) Transmenbrane-deletion mutants of the membrane-type matrix metalloproteinase-1 process progelatinase A and express intrinsic matrix-degrading activity. *J. Biol. Chem.* **271**, 9135–9140.

Polette, M., Nawrocki, B., Gilles, C., Sato, H., Seiki, M., Tournier, J.M. & Birembaut, P. (1996) MT-MMP expression and localization in human lung and breast cancers. *Virchows Arch.* **428**, 29–35.

Puente, X.S., Pendas, A.M., Llano, E., Velasco, G. & López-Otin, C. (1996) Molecular cloning of a novel membrane-type matrix metalloproteinase from a human breast carcinoma. *Cancer Res.* **56**, 944–949.

Sato, H., Takino, T., Okada, Y., Cao, J., Shinagawa, A., Yamamoto, E. & Seiki, M. (1994) A matrix metalloproteinase expressed on the surface of invasive tumor cells. *Nature* **370**, 61–65.

Sato, H., Takino, T., Kinoshita, T., Imai, K., Okada, Y., Stetler-Stevenson, W.G. & Seiki, M. (1996a) Cell surface binding and activation of gelatinase A induced by expression of membrane-type-1-matrix metalloproteinase (MT1-MMP). *FEBS Lett.* **385**, 238–240.

Sato, H., Kinoshita, T., Takino, T., Nakayama, K. & Seiki, M. (1996b) Activation of a recombinant membrane type 1-matrix metalloproteinase (MT1-MMP) by furin and its interaction with tissue inhibitor of metalloproteinases (TIMP)-2. *FEBS Lett.* **393**, 101–104.

Strongin, A.Y., Marmer, B.L., Grant, G.A. & Goldberg, G.I. (1993) Plasma membrane-dependent activation of the 72-kDa type IV collagenase is prevented by complex formation with TIMP-2. *J. Biol. Chem.* **268**, 14033–14039.

Strongin, A.Y., Collier, I., Bannikov, G., Marmer, B.L., Grant, G.A. & Goldberg, G.I. (1995) Mechanism of cell surface activation of 72 kDa type IV collagenase: isolation of the activated form of the membrane metalloproteinase. *J. Biol. Chem.* **270**, 5331–5338.

Takino, T., Sato, H., Shinagawa, A. & Seiki, M. (1995a) Identification of the second membrane-type matrix metalloproteinase (MT-MMP-2) gene from a human placenta cDNA library. *J. Biol. Chem.* **270**, 23013–23021.

Takino, T., Sato, H., Yamamoto, E. & Seiki, M. (1995b) Cloning of a human gene potentially encoding a novel matrix metalloproteinase having a C-terminal transmembrane domain. *Gene* **155**, 293–298.

Tokuraku, M., Sato, H., Murakami, S., Okada, Y., Watanabe, Y. & Seiki, M. (1995) Activation of the precursor of gelatinase A/72-kDa type IV collagenase/MMP-2 in lung carcinomas correlates with the expression of membrane-type matrix metalloproteinase (MT-MMP) and with lymph node metastasis. *Int. J. Cancer* **64**, 355–359.

Will, H. & Hinzmann, B. (1995) cDNA sequence and mRNA tissue distribution of a novel human matrix metalloproteinase with a potential transmembrane segment. *Eur. J. Biochem.* **231**, 602–608.

Will, H., Atkinson, S.J., Butler, G.S., Smith, B. & Murphy, G. (1996) The soluble catalytic domain of membrane type I matrix metalloproteinase cleaves the propeptide of progelatinase A and initiates autocatalytic activation. *J. Biol. Chem.* **271**, 17119–17123.

Yamamoto, M., Mohanam, S., Sawaya, R., Fuller, G.N., Seiki, M., Sato, H., Gokaslan, Z.L., Liotta, L.A., Nicolson, G.L. & Rao, J.S. (1996) Differential expression of membrane-type matrix metalloproteinase and its correlation with gelatinase A activation in human malignant brain tumors *in vivo* and *in vitro*. *Cancer Res.* **56**, 384–392.

*Motoharu Seiki*
*University of Tokyo, Institute of Medical Science,*
*Department of Cancer Cell Research,*
*4-6-1 Shirokane-dai, Minato-ku,*
*Tokyo 108, Japan*
*Email: mseiki@ims.u-tokyo.ac.jp*

# *400. Envelysin*

## *Databanks*

*Peptidase classification: clan MB, family M10, MEROPS ID: M10.010*
*NC-IUBMB enzyme classification: EC 3.4.24.12*
*Chemical Abstracts Service registry number: 50812-13-0*
*Databank codes:*

| Species | SwissProt | PIR | EMBL (cDNA) | EMBL (genomic) |
|---|---|---|---|---|
| *Paracentrotus lividus* | P22757 | S12805 | X53598 S41409 | X65722: complete gene |

## Name and History

Eggs and embryos from most animal species are enclosed in protective coats from which the embryo must escape, at a specific stage of development, through a process referred to as hatching. In order to hatch, many vertebrate and invertebrate species rely on enzymatic degradation of their protective envelopes. The term 'hatching enzyme' was apparently first used by Needham (1931) to refer to this type of activity. The *sea urchin hatching enzyme* was discovered and named by Ishida (1936), who showed that a proteolytic enzyme present in the culture supernatant of hatching embryos was able to digest the fertilization envelope of early embryos. The name recommended by the NC-IUBMB is *envelysin*.

The first attempts to purify envelysin were only partially successful and thus its nature and origin were long debated. Eventually, the enzyme was purified (Lepage & Gache, 1989) and its cDNA and gene were cloned (Lepage & Gache, 1990; Ghiglione *et al.*, 1994), allowing studies at the molecular level.

## Activity and Specificity

There is no convenient and specific assay for envelysin. The natural substrate is a huge and complex extracellular matrix which can hardly be used in a quantitative assay, and no specific synthetic substrate is available. Two methods should be used in parallel. Proteolytic activity can be measured by the method of Barrett & Edwards (1976) in which *N,N*-dimethylcasein is the substrate and the liberated amino groups are detected by reaction with fluorescamine. Activity on the fertilization envelope can be checked by the method of Barrett & Angelo (1969). Two-cell stage embryos surrounded by their fertilization envelope and lightly fixed in ethanol served as substrate, the reaction being monitored under the microscope. The time taken to digest the fertilization envelope is a semiquantitative, comparative, but specific estimation of the hatching activity (Lepage & Gache, 1989). Both assays are carried out in a medium which resembles seawater (500 mM NaCl, 10 mM CaCl$_2$, 10 mM EPPS (*N*-[2-hydroxyethyl]piperazine-*N'*-[3-propanesulfonic acid]), pH 8.3). It is important to use high ionic strength as degraded forms of envelysin and proteases unrelated to the hatching process are unable to work under this condition.

Envelysin specificity has been reported only for the Japanese species *Hemicentrotus pulcherrimus* (Nomura *et al.*, 1991) using a short proteolyzed form lacking the C-terminal part. This form of envelysin preferentially cleaves peptide bonds on the amino side of hydrophobic bulky residues (Leu, Ile, Phe, Tyr). The oxidized B chain of insulin is cleaved at:

FVNQ↓HLCGSHLVEA↓LY↓LVCGERGFF↓YTPKA

The hatching enzyme requires Ca$^{2+}$ for full activity with a $K_{0.5}$ of about 60 µM (Lepage & Gache, 1989). It is completely inhibited by 20 mM EDTA. This inhibition is not reversible by addition of Ca$^{2+}$, probably because Zn$^{2+}$, which is expected to be present at the active center, is also chelated.

Jaspisin ((*E*)-5,6-dihydroxystyryl sulfate) inhibits envelysin activity in the assay of Barrett & Angelo (1969) with an IC$_{50}$ of about 9 µg ml$^{-1}$ (Ikegami *et al.*, 1994). A synthetic peptide whose sequence (Ac-Pro-Arg-Cys-Gly-Val-Pro-Asp-NH$_2$) is derived from the highly conserved activation sequence involved in the cysteine switch (Van Wart & Birdekal-Hansen, 1990) inhibits hatching of live embryos with an IC$_{50}$ of 0.1–0.5 mM (Nomura & Suzuki, 1993).

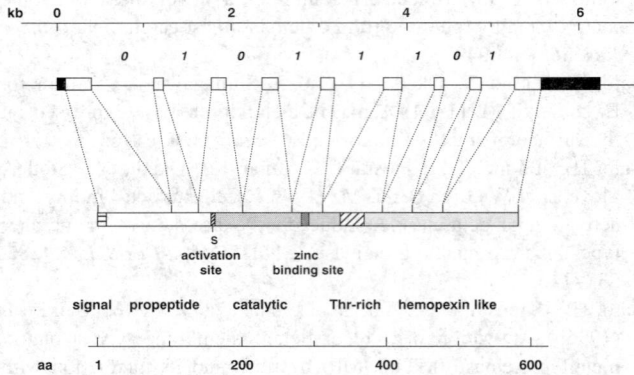

*Figure 400.1* Organization of envelysin gene locus and envelysin protein. *Top*: Hatching enzyme locus. The transcription-start site is taken as the origin. Introns are represented by lines, exons by boxes. Open boxes correspond to coding sequences, black boxes to 5′ and 3′ UTR. Italic numbers indicate phase of introns. *Bottom*: Envelysin protein. Characteristic regions are shown as marked areas. The activation sequence is just upstream of the N-terminal of the mature enzyme. Lower scale refers to amino acid residue numbers.

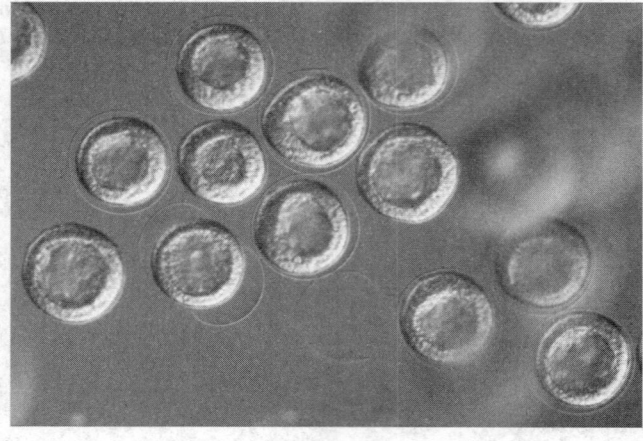

*Figure 400.2* Embryos at the blastula stage during hatching. Successive stages of the process can be seen. Some blastulae are still surrounded by their fertilization envelopes, which appear spherical and continuous but thinner than the intact fertilization envelopes. Other fertilization envelopes are locally very thin, torn and fragmented. Motile blastulae escape broken envelopes before complete digestion. Free blastulae swimming rapidly and out of focus appear as blurred spheres. Empty, partially digested fertilization envelopes are left behind and will completely disappear later.

## Structural Chemistry

Envelysin is synthesized from a 3 kb mRNA as a preproenzyme (587 amino acids, mass 65.2 kDa; Lepage & Gache, 1990). The proenzyme (569 amino acids, 63.2 kDa, pI 5.6) is activated by removal of an N-terminal propeptide (148 amino acids) to give the mature form which consists of 421 amino acids and has a molecular mass of 47.7 kDa and a calculated pI of 9.3. The zymogen activation thus implies a large decrease in mass (15.5 kDa) and a large increase in net charge from about $-14$ to about $+6$ at the pH of seawater. The active form is probably a monomer of a single polypeptide chain. Envelysin is a glycoprotein with little if any $N$-linked carbohydrate, but there are possibly carbohydrates $O$-linked on a cluster of Thr residues. The enzyme is predicted to have five cysteines and thus at least one free SH group.

Envelysin has all the hallmarks of a matrix metalloproteinase (Figure 400.1) (cf. Chapter 380). The secreted preproenzyme comprises five domains: signal peptide, propeptide, catalytic, Thr-repeat and hemopexin-like. The hatching enzyme catalytic domain sequence resembles those of the MMPs (e.g. 47% identity and 76% similarity with human stromelysin 1: Chapter 393). In addition, two key sequences are almost identical to their counterparts in vertebrate enzymes: the activation sequence, located in the propeptide just upstream of the N-terminal of the mature enzyme, and the active-site sequence with the catalytic and zinc-binding residues.

The transcription unit of the envelysin gene spans 6.3 kb and comprises nine exons (Figure 400.1) (Ghiglione *et al.*, 1994). A symmetrical 1–1 exon whose sequence, length and limits are very conserved among MMPs codes for a short domain comprising the active center. The Thr-repeat domain which is not found in other MMPs, is coded for by a 1–1 exon unique to the hatching enzyme gene. The first pair of hemopexin repeats is also flanked by phase 1 introns. These

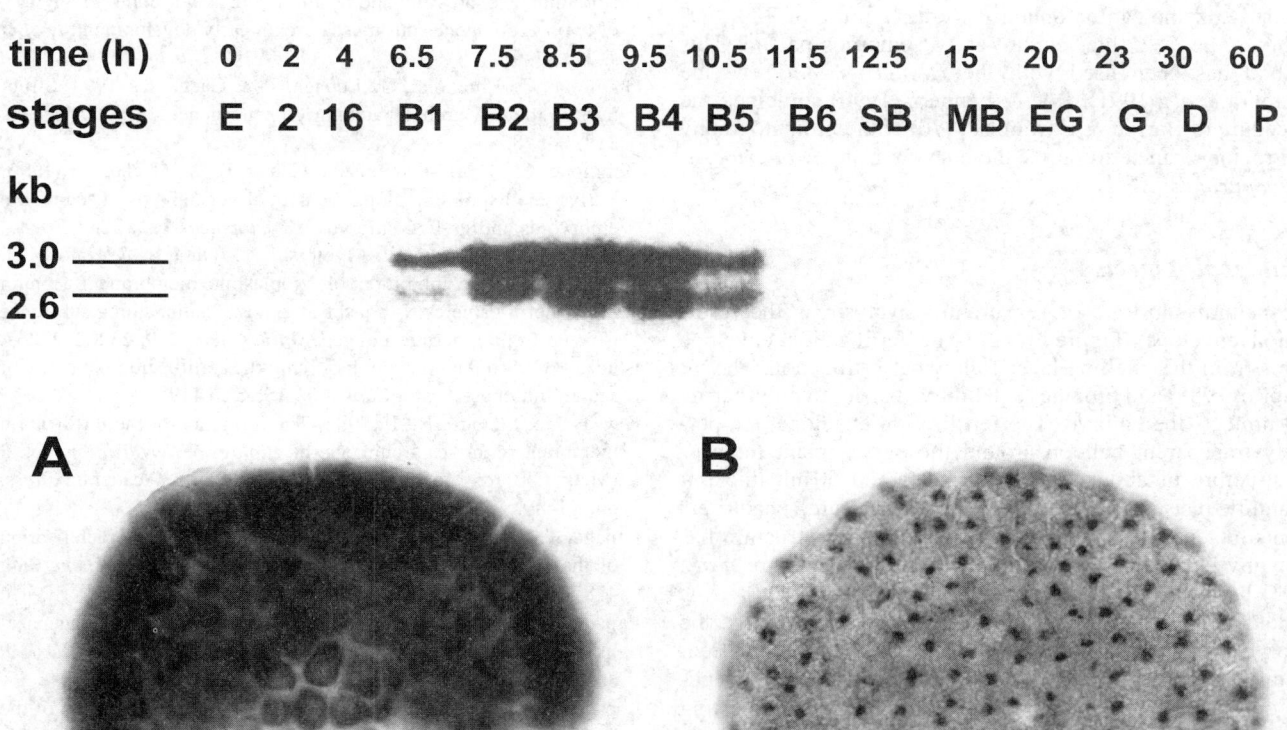

*Figure 400.3* Temporal and spatial pattern of envelysin gene expression during development. The envelysin gene is expressed transiently during early stages, in a restricted territory. *Top*: Envelysin transcript accumulation as seen by northern blot. Time and stages are indicated. E, unfertilized egg; 2, two-cell stage; 16, 16-cell stage; B1–B6, prehatching blastula stages; SB, swimming blastula; MB, mesenchyme blastula; EG, early gastrula; G, gastrula; D, prism; P, pluteus. *Bottom*: Localization of envelysin transcripts and envelysin protein. (A) Unlabeled blastula. (B) Whole mount *in situ* hybridization. Side views with the animal pole near the top.

structures might be considered as exchangeable modules and their occurrence probably indicates that exon shuffling played a role in the evolution of the MMPs.

## Preparation

Embryo culture supernatants are the only source for sea urchin envelysin, but are difficult to fractionate as the envelysin seems to remain tightly bound to digested fertilization envelope fragments (Barrett & Edwards, 1976; Takeuchi *et al.*, 1979). An efficient purification method of the hatching enzyme from *Paracentrotus lividus* was based on three important features (Lepage & Gache, 1989): (a) fertilization envelopes are removed before hatching, using envelysin itself to ensure a specific and complete removal; (b) the enzyme is handled at high salt concentration (1 M NaCl, i.e. about twice that of seawater), in the presence of detergent (0.5% CHAPS); and (c) concentrated supernatant is fractionated by hydrophobic chromatography on Procion-agarose. About 100 µg of purified enzyme can be obtained from a culture of $2.5 \times 10^8$ embryos, i.e. a 25 liter culture at $10^4$ embryos per ml. This method has been used with the *H. pulcherrimus* enzyme (Nomura *et al.*, 1991). Roe & Lennarz (1990) combined the procedure of Barrett & Edwards (1976) with chromatography on Procion-agarose to purify the enzyme from *Strongylocentrotus purpuratus*.

## Biological Aspects

The natural substrate of sea urchin envelysin is the fertilization envelope (Figure 400.2). The fertilization envelope forms from the vitelline layer following fertilization (Kay & Shapiro, 1985) and provides a definitive barrier to polyspermy and protects the embryo. The fertilization envelope is a perfectly transparent balloon around the egg, a giant molecular structure of about 100 µm diameter and 50 nm thick. It is highly resistant to extreme pH, temperature, chaotropic agents, detergents and proteolysis. Its complex structure has been investigated by electron microscopy (Mozingo *et al.*, 1993) but is poorly known at the molecular level.

Hatching is a landmark in early development as the embryo switches from a nonmotile encased form to a free-swimming form. About 12–20 h after fertilization (depending on species), the hatching enzyme is secreted, digesting the fertilization envelope and freeing the swimming blastula (Figure 400.2).

The expression of the envelysin gene (Figure 400.3) is strictly zygotic and controlled at the transcriptional level. Envelysin transcripts transiently accumulate during the cleavage stages (Lepage & Gache, 1990; Reynolds *et al.* 1992). The gene is activated around the eight-cell stage, peaks at the 128-cell stage and is completely turned off well before hatching (Lepage & Gache, 1990). This activation/inactivation pattern is unaltered in dissociated blastomeres, suggesting an autonomous expression (Ghiglione *et al.*, 1993). Envelysin gene expression is spatially restricted to the animal-most two-thirds of the blastula, a large area which corresponds approximately to presumptive ectoderm and whose limit is perpendicular to the animal–vegetal axis (Lepage *et al.*, 1992). This spatial pattern is apparently unaltered by strongly inductive cells implanted at ectopic positions and is probably independent of interaction and contact between cells. It is affected by Li$^+$, a well-known teratogenic reagent, which shifts the border of the envelysin expression domain towards the animal pole (Ghiglione *et al.*, 1993, 1996). The envelysin gene is thus the earliest zygotic gene activated in the sea urchin embryo and it is spatially restricted. Its temporal and spatial expression pattern shows that its regulation is closely linked to the maternally defined animal–vegetal axis.

## References

Barrett, D. & Angelo, G.M. (1969) Maternal characteristics of hatching enzymes in hybrid sea urchin embryos. *Exp. Cell Res.* **57**, 159–166.

Barrett, D. & Edwards, B. F. (1976) Hatching enzyme of the sea urchin *Strongylocentrotus purpuratus*. *Methods Enzymol.* **45**, 354–373.

Ghiglione, C., Lhomond, G., Lepage, T. & Gache, C. (1993) Cell-autonomous expression and position-dependent repression by Li$^+$ of two zygotic genes during sea urchin early development. *EMBO J.* **12**, 87–96.

Ghiglione, C., Lhomond, G., Lepage, T. & Gache, C. (1994) Structure of the sea urchin hatching enzyme gene. *Eur. J. Biochem.* **219**, 845–854.

Ghiglione, C., Emily-Fenouil, F., Chang, P. & Gache, C. (1996) Early gene expression along the animal-vegetal axis in sea urchin embryoids and grafted embryos. *Development* **122**, 3067–3074.

Ikegami, S., Kobayashi, H., Myotoishi, Y., Ohta, S. & Kato, K.H. (1994) Selective inhibition of exoplasmic membrane fusion in echinoderm gametes with jaspisin, a novel antihatching substance isolated from a marine sponge. *J. Biol. Chem.* **269**, 23262–23267.

Ishida, J. (1936) An enzyme dissolving the fertilization envelope of sea urchin eggs. *Annot. Zool. Jap.* **15**, 453–459.

Kay, E.S. & Shapiro, B.M. (1985) The formation of the fertilization membrane of the sea urchin egg. In: *Biology of Fertilization*, vol. 3 (Metz, C.B. & Monroy, A., eds). New York: Academic Press, pp. 45–80.

Lepage, T. & Gache, C. (1989) Purification and characterization of the sea urchin embryo hatching enzyme. *J. Biol. Chem.* **264**, 4787–4793.

Lepage, T. & Gache, C. (1990) Early expression of a collagenase-like hatching enzyme gene in the sea urchin embryo. *EMBO J.* **9**, 3003–3012.

Lepage, T., Sardet, C. & Gache, C. (1992) Spatial expression of the hatching enzyme gene in the sea urchin embryo. *Dev. Biol.* **150**, 23–32.

Mozingo, N.M., Hollar, L.R. & Chandler, D.E. (1993) Degradation of an extracellular matrix: sea urchin hatching enzyme removes cortical granule-derived proteins from the fertilization envelope. *J. Cell Sci.* **104**, 929–938.

Needham, J. (1931) *Chemical Embryology*. London: Cambridge University Press.

Nomura, K. & Suzuki, N. (1993) Stereo-specific inhibition of sea urchin envelysin (hatching enzyme) by a synthetic autoinhibitor peptide with a cysteine-switch consensus sequence. *FEBS Lett.* **321**, 84–88.

Nomura, K., Tanaka, H., Kikkawa, Y., Yamaguchi, M. & Suzuki, N. (1991) The specificity of sea urchin hatching enzyme (envelysin) places it in the mammalian matrix metalloproteinase family. *Biochemistry* **30**, 6115–6123.

Reynolds, S.D., Angerer, L.M., Palis, J., Nasir, A. & Angerer, R.C. (1992) Early mRNAs, spatially restricted along the animal-vegetal axis of sea urchin embryos, include one encoding a protein related to tolloid and BMP-1. *Development* **114**, 769–786.

Roe, J.L. & Lennarz, W.J. (1990). Biosynthesis and secretion of the hatching enzyme during sea urchin embryogenesis. *J. Biol. Chem.* **265**, 8704–8711.

Takeuchi, K., Yokosawa, H. & Hoshi, M. (1979) Purification and characterization of hatching enzyme of *Strongylocentrotus inter-medius. Eur. J. Biochem.* **100**, 257–265.

Van Wart, H.E. & Birdekal-Hansen, H. (1990) The cysteine switch: a principle of regulation of metalloproteinase activity with potential applicability to the entire matrix metalloproteinase gene family. *Proc. Natl Acad. Sci. USA* **87**, 5578–5582.

***Christian Gache***
*URA 671 CNRS/Université Paris VI,*
*Station Marine,*
*06230 Villefranche-sur-Mer, France*
*Email: gache@ccrv.obs-vlfr.fr*

***Thierry Lepage***
*URA 671 CNRS/Université Paris VI,*
*Station Marine,*
*06230 Villefranche-sur-Mer, France*

***Christian Ghiglione***
*URA 671 CNRS/Université Paris VI,*
*Station Marine,*
*06230 Villefranche-sur-Mer, France*

***Françoise Emily-Fenouil***
*URA 671 CNRS/Université Paris VI,*
*Station Marine,*
*06230 Villefranche-sur-Mer, France*

***Guy Lhomond***
*URA 671 CNRS/Université Paris VI,*
*Station Marine,*
*06230 Villefranche-sur-Mer, France*

# *401.* *Gelatinase A*

## Databanks

*Peptidase classification: clan MB, family M10, MEROPS ID: M10.003*
*NC-IUBMB enzyme classification: EC 3.4.24.24*
*ATCC entries: 65016, 65017 (human); 79064, 79065, 79066, 79067 (human)*
*Databank codes:*

| Species | SwissProt | PIR | EMBL (cDNA) | EMBL (genomic) |
|---|---|---|---|---|
| *Gallus gallus* | – | S46492 | U07775 | – |
| *Homo sapiens* | P08253 | A28153 | J03210 | M33789: exon 1 |
| | | A31480 | | M55582: exon 2 |
| | | A34202 | | M55583: exon 3 |
| | | A42225 | | M55584: exon 4 |
| | | A60187 | | M55585: exon 5 |
| | | S13858 | | M55586: exon 6 |
| | | S39436 | | M55587: exon 7 |
| | | | | M55588: exon 8 |
| | | | | M55589: exon 9 |
| | | | | M55590: exon 10 |
| | | | | M55591: exon 11 |
| | | | | M55592: exon 12 |
| | | | | M55593: exon 13 |
| | | | | M58552: exon 1 |
| *Mus musculus* | P33434 | A42496 | M84324 | – |
| *Oryctolagus cuniculus* | P50757 | – | D63579 | – |

M

Brookhaven Protein Data Bank three-dimensional structures:

| Species | ID | Resolution | Notes |
|---|---|---|---|
| *Homo sapiens* | 1GEN | 2.15 | C-terminal domain |
| | 1RTG | 2.6 | hemopexin domain |

## Name and History

The human enzyme was first described as a gelatin-degrading activity from the culture media of rheumatoid synovial tissue (Harris & Krane, 1972) and was purified from cultures of human skin (Seltzer *et al*., 1981), rabbit bone (Murphy *et al*., 1985) and human gingiva (Nakano & Scott, 1986). *Gelatinase A (GelA)* was cloned from a human fibroblast cDNA library (Collier *et al*., 1988) and named *72 kDa gelatinase* or *type IV collagenase*. In human the gene is located on chromosome 16 at q21 (Collier *et al*., 1991). The cDNA/protein sequences are extremely highly conserved across species (mouse protein, 96% similar; chick protein, 84% similar). The enzyme is a member of the matrixin subfamily and has been designated as *matrix metalloproteinase 2 (MMP-2)* (Nagase *et al*., 1992). It is a distinct gene product from the 92 kDa gelatinase, now known as gelatinase B (Chapter 402).

## Activity and Specificity

Seltzer *et al*. (1990) have studied the cleavage specificity of gelatinase A using denatured type I collagen. Cleavages between glycine and hydrophobic residues as well as Gly+Glu, Gly+Asn and Gly+Ser cleavages were identified. It was noted that hydroxyproline always occurred at P5' and often at P5. A systematic study of the subsite requirements of gelatinase A was completed by Netzel-Arnett *et al*. (1993)

using synthetic peptide substrates. Putting these two studies together gives the following optimal residues:

Hyp-Xaa-Pro-Leu-Ala+Met-Phe-Gly-Xaa-Hyp

where P1 can be Ala or Gly and P1' can be Met, Ile or Tyr. The enzyme readily degrades the denatured form of all collagens that have been studied. In some instances the cleavage of solubilized native collagens has been described, including type I collagen (Aimes & Quigley, 1995), type V collagen (Okada *et al*., 1990) and type VII collagen (Seltzer *et al*., 1989). The ability to degrade type IV collagen varies with the extent of exposure of the latter to reducing agents and the temperature, i.e. with the extent of loss of native state by the preparation under study (Eble *et al*., 1996). It is not at all clear that this enzyme can effectively cleave native type IV collagen *in vivo*. Numerous other extracellular matrix components are degraded, including elastin, laminin, fibronectin and aggrecan. In some cases the cleavage site(s) have been defined; whereas the degradation of gelatins conforms to the peptide studies, activity against other substrates yields many variations (Table 401.1).

Gelatinase A may be assayed using radiolabeled gelatin ($k_{cat}/K_m$ $4.47 \times 10^5$ $M^{-1}$ $s^{-1}$; Murphy & Crabbe, 1995) or fluorimetrically using Mca-Pro-Leu-Gly+Leu-Dpa-Ala-Arg-NH$_2$ (Dpa: *N*-3-(2,4-dinitrophenyl)-L-2,3-diaminopropionyl) ($k_{cat}/K_m$ $6.29 \times 10^5$ $M^{-1}$ $s^{-1}$) (Knight *et al*., 1992). This

*Table 401.1*   Protein sequences cleaved by gelatinase A

| | P4 | P3 | P2 | P1 | + | P1' | P2' | P3' | P4' |
|---|---|---|---|---|---|---|---|---|---|
| Progelatinase A[a] | Asp | Val | Ala | Asn | | Tyr | Ser | Leu | Phe |
| Procollagenase 3[b] | Asp | Val | Gly | Glu | | Tyr | Asn | Val | Phe |
| Aggrecan[c] | Ile | Pro | Glu | Asn | | Phe | Phe | Gly | Val |
| Insulin-like growth factor[d] | Leu | Arg | Ala | Tyr | | Leu | Leu | Pro | Ala |
| α1(I)collagen (denatured)[e] | Gly | Ala | Hyp | Gly | | Leu | Glx | Gly | His |
| | Gly | Pro | Gln | Gly | | Val | Arg | Gly | Glu |
| | Gly | Pro | Ser | Gly | | Leu | Hyp | Gly | Pro |
| | Gly | Pro | Ala | Gly | | Phe | Ala | Gly | Pro |
| | Gly | Pro | Ile | Gly | | Asn | Val | Gly | Ala |
| | Gly | Pro | Hyl | Gly | | Ser | Arg | Gly | Ala |
| | Gly | Pro | Ala | Gly | | Glu | Arg | Gly | Ser |
| | Gly | Pro | Ala | Gly | | Glx | Asp | Gly | Pro |
| | Gly | Pro | Ala | Gly | | Val | Gln | Gly | Pro |
| | Gly | Ala | Lys | Gly | | Leu | Thr | Gly | Ser |
| CB3 α1(IV)collagen (denatured)[f] | Gly | Pro | Pro | Gly | | Ile | Val | Ile | Gly |
| | Gly | Pro | Leu | Gly | | Glu | Lys | Gly | Asp |
| | Gly | Pro | Ser | Gly | | Arg | Asp | Gly | Leu |
| | Gly | Pro | Pro | Gly | | Asp | Gln | Gly | Pro |
| | Asp | Ile | Asp | Gly | | Tyr | Arg | Gly | Pro |
| | Gly | Thr | Ala | Gly | | Leu | Ile | Gly | Gln |

Data taken from: [a]Stetler-Stevenson *et al*. (1989); [b]Knäuper *et al*. (1996); [c]Fosang *et al*. (1992); [d]Fowlkes *et al*. (1994); [e]Seltzer *et al*. (1981); [f]Eble *et al*. (1996). Hyp, hydroproline; Hyl, hydroxylysine.

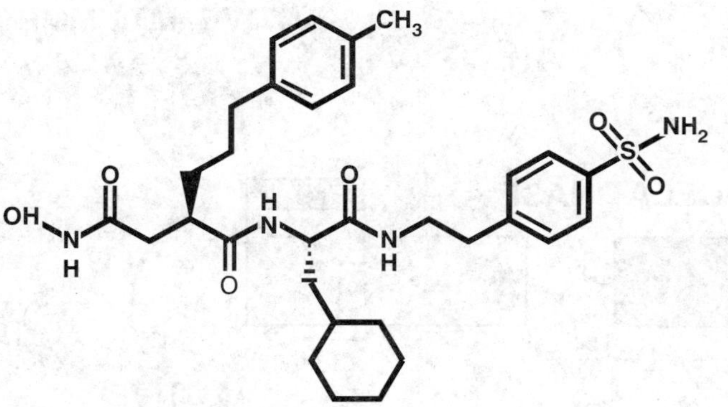

*Figure 401.1* Structure of a typical hydroxamic acid inhibitor of gelatinase A (C0T989-00) $N4$-hydroxy-$N1$-[1-(s)-(4-aminosulfonyl)phenylethylaminocarboxyl-2-cyclohexylethyl)-2$R$-[4-methylphenylpropyl] succinamide. Gelatinase A: $K_i < 0.01$ nM (2.98 nM for stromelysin 1 and 329 nM for collagenase 1), $k_{on}$ $7.13 \times 10^6$ M$^{-1}$ s$^{-1}$, $k_{off}$ $3.67 \times 10^{-5}$ s$^{-1}$ (J. O'Connell, unpublished results; Porter *et al.*, 1994).

synthetic substrate is available from Calbiochem-Novabiochem or Bachem (see Appendix 2 for full names and addresses of suppliers). It is not specific for gelatinase A but has a markedly greater $k_{cat}/K_m$ value for the gelatinases than the other matrix metalloproteinases, allowing the detection of as little as 10 pM enzyme. The pH optimum is about 8.5 and the assay system should contain 2–10 mM Ca$^{2+}$ ions, 0.1 M NaCl and a detergent such as Brij-35. The active enzyme is exceedingly unstable below pH 6.5 but can be partially stabilized by the presence of $>10\,\mu$M Zn$^{2+}$ ions. As little as 1 pg of enzyme is detectable by gelatin zymography, which allows the identification of latent and active forms. With care the technique can be quantitative (Kleiner & Stetler-Stevenson, 1994).

Gelatinase A is inhibited in a 1:1 stoichiometric fashion by the tissue inhibitors of metalloproteinases, TIMP-1, TIMP-2 and TIMP-3, which are all tight binding ($K_i \sim 2$ pM). Binding of the enzyme is extremely rapid, being facilitated by interactions between the C-terminal domains of both gelatinase and TIMP (Willenbrock *et al.*, 1993). This is most marked for TIMP-2 and to a lesser extent for TIMP-3. The C-domain interactions between progelatinase A and TIMP-2 and TIMP-3 are sufficiently strong to form stable complexes in the absence of the N-domain binding observed for active enzyme (Howard & Banda, 1991; Kleiner *et al.*, 1993). Other natural inhibitors include $\alpha_2$-macroglobulin.

Gelatinase A is inhibited by 1,10-phenanthroline and DTT as well as by peptide inhibitors incorporating a chelating moiety (hydroxamates, phosphinates and carboxylates). Some of the peptide inhibitors are relatively specific for gelatinase A (Porter *et al.*, 1994) (Figure 401.1), but are not available commercially.

## Structural Chemistry and Biochemistry

Gelatinase A precursor is made up of the typical matrixin arrangement of a signal peptide (cleaved prior to secretion), a propeptide, a catalytic unit and a C-terminal domain. The proenzyme has an $M_r$ of 72 000 and an active form $M_r$ of about 62 000. It should be noted that these $M_r$ values will be considerably reduced (by about 6000) when the enzyme is analyzed by gelatin zymography under nonreducing conditions. Loss of the C-terminal domain can occur by autolysis to give a species of $M_r$ 42 000. The free C-terminal domain has an $M_r$ of 31 000 and accumulates in autolysing preparations. Both gelatinase A and B (Chapter 402) contain a further domain inserted into the catalytic unit, which takes the form of three repeats of the type II module found in fibronectin and cattle seminal fluid protein PDC-109 b domain. Expression of these modules in isolation has been used to demonstrate their ability to interact with denatured type I collagen (Banyai & Patthy, 1991; Collier *et al.*, 1992). They also bind to denatured type IV and V collagens, elastin and native type I collagen (Steffensen *et al.*, 1995). Deletion of this domain from gelatinase A leaves a proteinase that is catalytically competent for the cleavage of peptide substrates but has a much reduced ability to cleave macromolecules such as denatured collagens (Murphy *et al.*, 1994).

The C-terminal domain structure of gelatinase A has been determined using X-ray crystallography (Libson *et al.*, 1995; Gohlke *et al.*, 1996). Like other members of the hemopexin family, it has a disc-like shape with the chain folded into a $\beta$ propeller structure that has pseudo 4-fold symmetry. The four blades of the propeller are made up of antiparallel four-stranded $\beta$ sheets arranged around a tunnel harboring a number of cations (calcium) and anions (chloride). The first and fourth blades are linked by a disulfide bond which is conserved throughout the matrixins and must be intact to retain many of the properties of the domain.

The C-terminal domain of gelatinase A is important for docking interactions with the C-terminal three loops of the TIMPs. In the case of TIMP-2 these C-domain interactions with gelatinase A are particularly strong ($K_d$ about 50 pM; Willenbrock *et al.*, 1993), such that progelatinase A can be complexed to TIMP-2. The biological significance of this property is not clear. Progelatinase A–TIMP-2 complexes retain all the inhibitory capacity of free TIMP-2 (Ward *et al.*, 1991; Kolkenbrock *et al.*, 1991). If purified complexes are treated with the organomercurial 4-aminophenylmercuric acetate (APMA), the propeptide–catalytic domain interactions are perturbed, such that the bound TIMP-2 can interact with the active-site cleft without propeptide processing. This new complex has no gelatinolytic capacity (Itoh *et al.*, 1995).

The progelatinase A–TIMP-2 interactions are thought to account for the binding of progelatinase to the surface of the cells via the membrane-type MMP (MT1-MMP) (Chapter 399) (Sato *et al.*, 1994). Cell-based studies have indicated that such an association of progelatinase A with the cell membrane is a prerequisite for propeptide processing to generate the active form of the enzyme (Murphy *et al.*, 1992; Strongin *et al.*, 1993, 1995). Although it is clear that both MT1-MMP and gelatinase cleavages lead to gelatinase A activation, the precise mechanism and particularly the role of TIMP-2 is not known (Atkinson *et al.*, 1995; Kinoshita *et al.*, 1996; Imai *et al.*, 1996; Will *et al.*, 1996). Other studies have shown that progelatinase A can efficiently self-activate at high concentrations or by concentration on heparin molecules (Crabbe *et al.*, 1993). It may therefore be postulated that active MT1-MMP at the cell

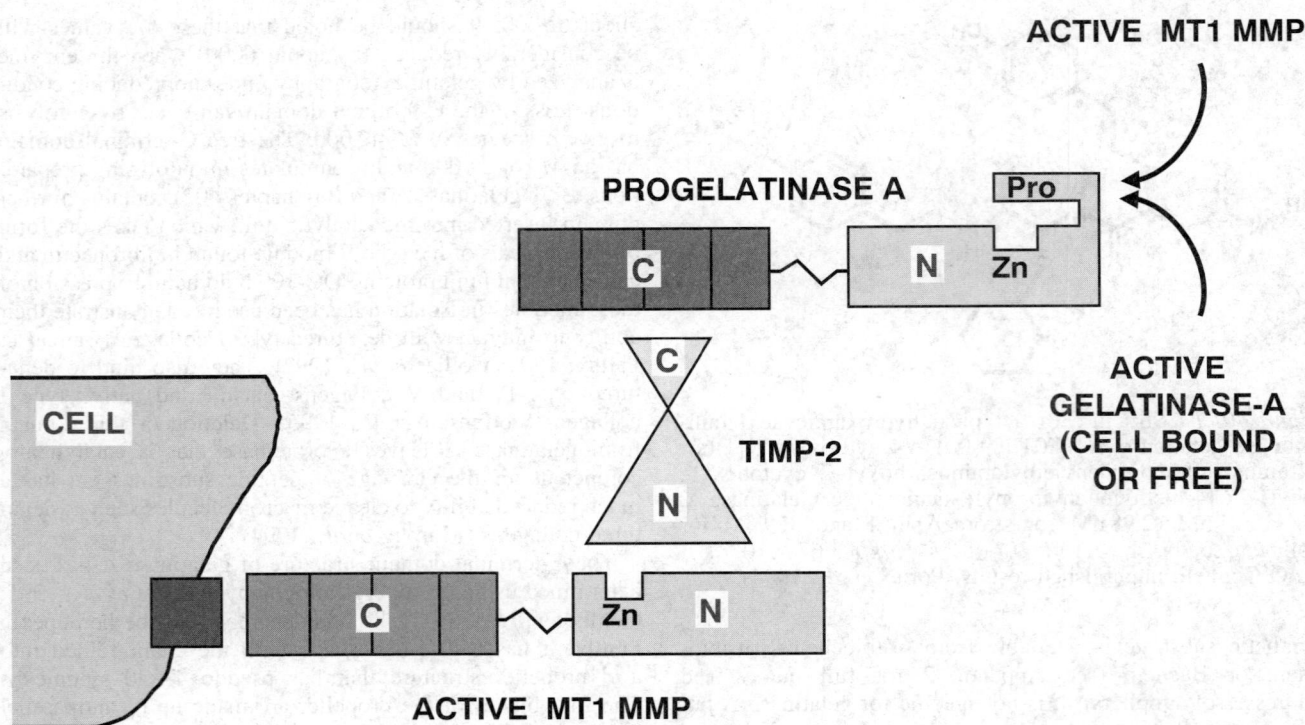

*Figure 401.2*   Hypothetical mechanism for the binding of progelatinase A to cell surfaces and its subsequent activation. Expression of active MT1-MMP at the cell surface induces TIMP-2 binding by N-domain interactions to give a functionally inactive complex. A ternary complex with progelatinase A is formed by C-domain interactions with TIMP-2. This is an effective mechanism for the concentration of progelatinase A on the plasma membrane which facilitates the initiation of propeptide processing by active MT1-MMP (TIMP-2 free). Alternatively, cell-bound or free active progelatinase A may initiate and complete propeptide cleavage to yield active progelatinase A. The level of TIMP-2 in the system is critical in the determination of the extent of progelatinase A activation and final activity.

surface binds TIMP-2 (by N-domain interactions), such that progelatinase A can also be associated in a ternary complex (Figure 401.2). If MT1-MMP molecules are clustered at the cell membrane, progelatinase A will be effectively concentrated such that cleavage could occur either by the action of active MT1-MMP activity and/or gelatinase A activity. This mechanism is supported by the studies of Itoh *et al.* (1995) using stromelysin 1 (Chapter 393) in progelatinase A–TIMP-2 activation studies. The level of TIMP-2 in the system would be critical – sufficient to permit gelatinase binding but insufficient to inhibit all the active MMPs in the system.

## Preparation

Gelatinase A is very widely expressed, notably by cells in culture, although levels may be very low. It has been found to be associated with the stroma of invasive tumors with a high content of stroma (Pyke *et al.*, 1993; Polette *et al.*, 1994; MacDougall & Matrisian, 1995) and is present (mRNA) in mesenchymal tissues of the developing embryo (Reponen *et al.*, 1992) and in plasma (Zucker *et al.*, 1992). It can be purified from the conditioned media of fibroblasts from many tissue sources. The recombinant form of the enzyme has also been expressed from eukaryotic cells (Murphy *et al.*, 1992; Strongin *et al.*, 1993) and the fibronectin-like domain has been expressed in *Escherichia coli* (Banyai & Patthy, 1991; Collier *et al.*, 1992; Steffensen *et al.*, 1995). Human

progelatinase A is now available commercially (TCS Biologicals or Novus Molecular).

The purification of progelatinase A is extremely simple, using gelatin-Sepharose as an affinity matrix. A proportion of the enzyme is often associated with TIMP-2, which binds to both pro and active forms (Murphy & Crabbe, 1995). The only other significant contaminants are likely to be fibronectin and gelatinase B (Chapter 402). Progelatinase may be activated by treatment with 4-aminophenylmercuric acetate, interstitial collagenase (Chapter 389), matrilysin (Chapter 396) or the membrane-type (MT) MMPs (Chapter 399). As discussed above, all these routes involve gelatinase A self-cleavage at some stage. At very high concentrations the enzyme will efficiently self-activate, particularly in the presence of heparin (Crabbe *et al.*, 1993). The N-terminal active sequence of the active enzyme is Tyr81-Asn-Phe-Phe-Pro-Arg-Lys. Although trypsin cleaves progelatinase A it does not result in efficient activation of the enzyme.

## Biological Aspects

The gene structure of human gelatinase A has been published (Huhtala *et al.*, 1990; Collier *et al.*, 1991). It has 13 exons, of which 1–4 and 8–12 correspond to those of collagenase 1 and stromelysin 1, indicating the close structural relationship of these genes. Exons 5–7 are small, each encoding one of the type II modules of the fibronectin-like domain. The 5′ flanking region has no TATA box or AP1 site but has two SPl-binding sites. There is a putative AP2 site in the first exon. In some cell types gelatinase A can be induced by transforming growth factor $\beta$ (Overall *et al*., 1991), but in general, expression levels are little affected by the action of growth factors and cytokines. Many cells in culture make low levels of gelatinase A, but it is not clear whether it is so widely expressed by cells *in vivo*. The precise role of gelatinase A is not clear. It has been proposed that the membrane-bound enzyme may be involved in normal collagen turnover, fragmenting fibrils prior to phagocytosis (Everts *et al*., 1996). The localization of gelatinase A on the membrane at the invading front (invadopodia) of some tumor cell lines had also been described and linked to cellular invasiveness (Chen, 1996).

## Distinguishing Features

Gelatinase A may be distinguished by its extreme efficiency in the degradation of gelatin, e.g. in zymography. Gelatinase B has a similar activity but should be identifiable by its higher $M_r$ value. Care is needed if the C-terminal domain of the latter is lost as it will then migrate at a similar $M_r$ to gelatinase A. Gelatinase A can also easily lose its C-terminal domain and will retain its ability to degrade gelatin, making it distinguishable as a band at $M_r$ 45 000 on nonreducing PAGE zymography. Gelatinase A is unique in its ability to form complexes with TIMP-2 when it is in the proenzyme form; the complex may be the predominant form in many cell culture media. Monoclonal antibodies to human gelatinase A are commercially available from Oncogene Research Products and Fuji Chemical Industries, as well as a rabbit polyclonal antibody from Chemicon International.

## Further Reading

For a review, see Murphy & Crabbe (1995).

## References

Aimes, R.T. & Quigley, J.P. (1995) Matrix metalloproteinase-2 is an interstitial collagenase. Inhibitor-free enzyme catalyzes the cleavage of collagen fibrils and soluble native type I collagen generating the specific 3/4- and 1/4-length fragments. *J. Biol. Chem.* **270**, 5872–5876.

Atkinson, S.J., Crabbe, T., Cowell, S., Ward, R.V., Butler, M.J., Sato, H., Seiki, M., Reynolds, J.J. & Murphy, G. (1995) Intermolecular autolytic cleavage can contribute to the activation of progelatinase A by cell membranes. *J. Biol. Chem.* **270**, 30479–30485.

Banyai, L. & Patthy, L. (1991) Evidence for the involvement of type II domains in collagen binding by 72 kDa type IV procollagenase. *FEBS Lett.* **282**, 23–25.

Chen, W.T. (1996) Proteases associated with invadopodia, and their role in degradation of extracellular matrix. *Enzyme Protein* **49**, 59–71.

Collier, I.E., Wilhelm, S.M., Eisen, A.Z., Marmer, B.L., Grant, G.A., Seltzer, J.L., Kronberger, A., He, C., Bauer, E.A. & Goldberg, G.I. (1988) H-ras oncogene-transformed human bronchial epithelial cells (TBE-1) secrete a single metalloprotease capable of degrading basement membrane collagen. *J. Biol. Chem.* **263**, 6579–6587.

Collier, I.E., Bruns, G.A.P., Goldberg, G.I. & Gerhard, G.S. (1991) On the structure and chromosome location of the 72- and 92-kDa human type IV collagenase genes. *Genomics* **9**, 429–434.

Collier, I.E., Krasnov, P.A., Strongin, A.Y., Birkedal-Hansen, H. & Goldberg, G.I. (1992) Alanine scanning mutagenesis and functional analysis of the fibronectin-like collagen-binding domain from human 92-kDa type IV collagenase. *J. Biol. Chem.* **267**, 6776–6781.

Crabbe, T., Ioannou, C. & Docherty, A.J.P. (1993) Human progelatinase A can be activated by autolysis at a rate that is concentration-dependent and enhanced by heparin bound to the C-terminal domain. *Eur. J. Biochem.* **218**, 431–438.

Eble, J.A., Ries, A., Lichy, A., Mann, K., Stanton, H., Gavrilovic, J., Murphy, G. & Kühn, K. (1996) The recognition sites of the integrins $\alpha_1\beta_1$ and $\alpha_2\beta_1$ within collagen IV are protected against gelatinase A attack in the native protein. *J. Biol. Chem.* **271**, 30964–30970.

Everts, V., Van der Zee, E., Creemers, L. & Beertsen, W. (1996) Phagocytosis and intracellular digestion of collagen, its role in turnover and remodelling. *Histochem. J.* **28**, 229–245.

Fosang, A.J., Neame, P.J., Last, K., Hardingham, T.E., Murphy, G. & Hamilton, J.A. (1992) The interglobular domain of cartilage aggrecan is cleaved by PUMP, gelatinases, and cathepsin B. *J. Biol. Chem.* **267**, 19470–19474.

Fowlkes, J.L., Enghild, J.J., Suzuki, K. & Nagase, H. (1994) Matrix metalloproteinases degrade insulin-like growth factor-binding protein-3 in dermal fibroblast cultures. *J. Biol. Chem.* **269**, 25742–25746.

Gohlke, U., Gomis-Rüth, F.X., Crabbe, T., Murphy, G., Docherty, A.J.P. & Bode, W. (1996) The C-terminal (haemopexin-like) domain structure of human gelatinase A (MMP-2): structural implications for its function. *FEBS Lett.* **378**, 126–130.

Harris, E.D. & Krane, S.M. (1972) An endopeptidase from rheumatoid synovial tissue culture. *Biochim. Biophys. Acta* **258**, 566–576.

Howard, E.W. & Banda, M.J. (1991) Binding of tissue inhibitor of metalloproteinases 2 to two distinct sites on human 72-kDa gelatinase. Identification of a stabilization site. *J. Biol. Chem.* **266**, 17972–17977.

Huhtala, P., Chow, L.T. & Tryggvason, K. (1990) Structure of the human type IV collagenase gene. *J. Biol. Chem.* **265**, 11077–11082.

Imai, K., Ohuchi, E., Aoki, T., Nomura, H., Fujii, Y., Sato, H., Seiki, M. & Okada, Y. (1996) Membrane-type matrix metalloproteinase 1 is a gelatinolytic enzyme and is secreted in a complex with tissue inhibitor of metalloproteinases 2. *Cancer Res.* **56**, 2707–2710.

Itoh, Y., Binner, S. & Nagase, H. (1995) Steps involved in activation of the complex of pro-matrix metalloproteinase 2 (progelatinase A) and tissue inhibitor of metalloproteinases (TIMP)-2 by 4-aminophenylmercuric acetate. *Biochem. J.* **308**, 645–651.

Kinoshita, T., Sato, H., Takino, T., Itoh, M., Akizawa, T. & Seiki, M. (1996) Processing of a precursor of 72-kilodalton type IV collagenase/gelatinase A by a recombinant membrane-type 1 matrix metalloproteinase. *Cancer Res.* **56**, 2535–2538.

Kleiner, D.E., Tuuttila, A., Tryggvason, K. & Stetler-Stevenson, W.G. (1993) Stability analysis of latent and active 72-kDa type IV collagenase: the role of tissue inhibitor of metalloproteinases-2 (TIMP-2). *Biochemistry* **32**, 1583–1592.

**M**

Kleiner, D.E. & Stetler-Stevenson, W.G. (1994) Quantitative zymography: detection of picogram quantities of gelatinases. *Anal. Biochem.* **218**, 325–329.

Knäuper, V., Will, H., López-Otin, C., Smith, B., Atkinson, S.J., Stanton, H., Hembry, R.M. & Murphy, G. (1996) Cellular mechanisms for human procollagenase-3 (MMP-13) activation – evidence that MT1-MMP (MMP-14) and gelatinase A (MMP-2) are able to generate active enzyme. *J. Biol. Chem.* **271**, 17124–17131.

Knight, C.G., Willenbrock, F. & Murphy, G. (1992) A novel coumarin-labelled peptide for sensitive continuous assays of the matrix metalloproteinases. *FEBS Lett.* **296**, 263–266.

Kolkenbrock, H., Orgel, D., Hecker-Kia, A., Noack, W. & Ulbrich, N. (1991) The complex between a tissue inhibitor of metalloproteinases (TIMP-2) and 72 kDa progelatinase is a metalloproteinase inhibitor. *Eur. J. Biochem.* **198**, 775–781.

Libson, A.M., Gittis, A.G., Collier, I.E., Marmer, B.L., Goldberg, G.I. & Lattman, E.E. (1995) Crystal structure of the haemopexin-like C-terminal domain of gelatinase A. *Nature Struct. Biol.* **2**, 938–942.

MacDougall, J.R. & Matrisian, L.M. (1995) Contributions of tumor and stromal matrix metalloproteinases to tumor progression, invasion and metastasis. *Cancer Metastasis Rev.* **14**, 351–362.

Murphy, G. & Crabbe, T. (1995) Gelatinase A and B. *Methods Enzymol.* **248**, 470–495.

Murphy, G., McAlpine, C.G., Poll, C.T. & Reynolds, J.J. (1985) Purification and characterization of a bone metalloproteinase that degrades gelatin and types IV and V collagen. *Biochim. Biophys. Acta* **831**, 49–58.

Murphy, G., Willenbrock, F., Ward, R.V., Cockett, M.I., Eaton, D. & Docherty, A.J.P. (1992) The C-terminal domain of 72 kDa gelatinase A is not required for catalysis, but is essential for membrane activation and modulates interactions with tissue inhibitors of metalloproteinases. *Biochem. J.* **283**, 637–641.

Murphy, G., Nguyen, Q., Cockett, M.I., Atkinson, S.J., Allan, J.A., Knight, C.G., Willenbrock, F. & Docherty, A.J.P. (1994) Assessment of the role of the fibronectin-like domain of gelatinase A by analysis of a deletion mutant. *J. Biol. Chem.* **269**, 6632–6636.

Nagase, H., Barrett, A.J. & Woessner, J.F., Jr. (1992) Nomenclature and glossary of the matrix metalloproteinases. *Matrix* **suppl. 1**, 421–424.

Nakano, T. & Scott, P.G. (1986) Purification and characterization of a gelatinase produced by fibroblasts from human gingiva. *Biochem. Cell Biol.* **64**, 387–393.

Netzel-Arnett, S., Sang, Q.-X., Moore, W.G.I., Navre, M., Birkedal-Hansen, H. & Van Wart, H.E. (1993) Comparative sequence specificities of human 72- and 92-kDa gelatinases (type IV collagenases) and PUMP (matrilysin). *Biochemistry* **32**, 6427–6432.

Okada, Y., Morodomi, T., Enghild, J.J., Suzuki, K., Yasui, A., Nakanishi, I., Salvesen, G. & Nagase, H. (1990) Matrix metalloproteinase 2 from human rheumatoid synovial fibroblasts. Purification and activation of the precursor and enzymic properties. *Eur. J. Biochem.* **194**, 721–730.

Overall, C.M., Wrana, J.L. & Sodek, J. (1991) Transcriptional and post-transcriptional regulation of 72-kDa gelatinase/type IV collagenase by transforming growth factor $\beta$1 in human fibroblasts. Comparisons with collagenase and tissue inhibitor of matrix metalloproteinase gene expression. *J. Biol. Chem.* **266**, 14064–14071.

Polette, M., Gilbert, N., Stas, I., Nawrocki, B., Noel, A., Remacle, A., Stetler-Stevenson, W.G., Birembaut, P. & Foidart, M. (1994) Gelatinase A expression and localization in human breast cancers. An *in situ* hybridization study and immunohistochemical detection using confocal microscopy. *Virchows Arch.* **424**, 641–645.

Porter, J.R., Beeley, R.A., Boyce, B.A., Mason, B., Millican, A., Millar, K., Leonard, J., Morphy, R. & O'Connell, J.P. (1994) Potent and selective inhibitors of gelatinase A. 1. Hydroxamic acid derivatives. *Biomed. Chem. Lett.* **4**, 2741–2746.

Pyke, C., Ralfkiaer, E., Huhtala, P., Hurskainen, T., Danø, K. & Tryggvason, K. (1992) Localization of messenger RNA for Mr 72,000 and 92,000 type IV collagenases in human skin cancers by *in situ* hybridization. *Cancer Res.* **52**, 1336–1341.

Pyke, C., Ralfkiaer, E., Tryggvason, K. & Danø, K. (1993) Messenger RNA for two type IV collagenases is located in stromal cells in human colon cancer. *Am. J. Pathol.* **142**, 359–365.

Reponen, P., Sahlberg, C., Huhtala, P., Hurskainen, T., Thesleff, I. & Tryggvason, K. (1992) Molecular cloning of murine 72-kDa type IV collagenase and its expression during mouse development. *J. Biol. Chem.* **267**, 7856–7862.

Sato, H., Takino, T., Okada, Y., Cao, J., Shinagawa, A., Yamamoto, E. & Seiki, M. (1994) A matrix metalloproteinase expressed on the surface of invasive tumour cells. *Nature* **370**, 61–65.

Seltzer, J.L., Adams, S.A., Grant, G.A. & Eisen, A.Z. (1981) Purification and properties of a gelatin-specific neutral protease from human skin. *J. Biol. Chem.* **256**, 4662–4668.

Seltzer, J.L., Eisen, A.Z., Bauer, E.A., Morris, N.P., Glanville, R.W. & Burgeson, R.E. (1989) Cleavage of type VII collagen by interstitial collagenase and type IV collagenase (gelatinase) derived from human skin. *J. Biol. Chem.* **264**, 3822–3826.

Seltzer, J.L., Akers, K.T., Weingarten, H., Grant, G.A., McCourt, D.W. & Eisen, A.Z. (1990) Cleavage specificity of human skin type IV collagenase (gelatinase). Identification of cleavage sites in type I gelatin, with confirmation using synthetic peptides. *J. Biol. Chem.* **265**, 20409–20413.

Steffensen, B., Wallon, M. & Overall, C.M. (1995) Extracellular matrix binding properties of recombinant fibronectin type II-like modules of human 72 kDa gelatinase/type IV collagenase. *J. Biol. Chem.* **270**, 11555–11566.

Stetler-Stevenson, W.G., Krutzsch, H.C., Wacher, M.P., Margulies, I.M.K. & Liotta, L.A. (1989) The activation of human type IV collagenase proenzyme. Sequence identification of the major conversion product following organomercurial activation. *J. Biol. Chem.* **264**, 1353–1356.

Strongin, A.Y., Marmer, B.L., Grant, G.A. & Goldberg, G.I. (1993) Plasma membrane-dependent activation of the 72-kDa type IV collagenase is prevented by complex formation with TIMP-2. *J. Biol. Chem.* **268**, 14033–14039.

Strongin, A.Y., Collier, I., Bannikov, G., Marmer, B.L., Grant, G.A. & Goldberg, G.I. (1995) Mechanism of cell surface activation of 72-kDa type IV collagenase. Isolation of the activated form of the membrane metalloprotease. *J. Biol. Chem.* **270**, 5331–5338.

Ward, R.V., Hembry, R.M., Reynolds, J.J. & Murphy, G. (1991) The purification of tissue inhibitor of metalloproteinases-2 from its 72 kDa progelatinase complex. Demonstration of the biochemical similarities of tissue inhibitor of metalloproteinases-2 and tissue inhibitor of metalloproteinases-1. *Biochem. J.* **278**, 179–187.

Will, H., Atkinson, S.J., Butler, G.S., Smith, B. & Murphy, G. (1996) The soluble catalytic domain of membrane type 1 matrix metalloproteinase cleaves the propeptide of progelatinase A and initiates autoproteolytic activation – regulation by TIMP-2 and TIMP-3. *J. Biol. Chem.* **271**, 17119–17123.

Willenbrock, F., Crabbe, T., Slocombe, P.M., Sutton, C.W., Docherty, A.J.P., Cockett, M.I., O'Shea, M., Brocklehurst, K., Phillips, I.R. & Murphy, G. (1993) The activity of the tissue inhibitors of metalloproteinases is regulated by C-terminal domain interactions: a kinetic analysis of the inhibition of gelatinase A.

*Biochemistry* **32**, 4330–4337.

Zucker, S., Lysik, R.M., Gurfinkel, M., Zarrabi, M.H., Stetler-Stevenson, W., Liotta, L.A., Birkedal-Hansen, H. & Mann, W. (1992) Immunoassay of type-IV collagenase/gelatinase (MMP-2) in human plasma. *J. Immunol. Methods* **148**, 189–198.

*Gillian Murphy*
The School of Biological Sciences,
University of East Anglia,
Norwich NR4 7TL, UK
Email: g.murphy@uea.ac.uk

# 402. Gelatinase B

## Databanks

*Peptidase classification: clan MB, family M10, MEROPS ID: M10.004*
*NC-IUBMB enzyme classification: EC 3.4.24.35*
*ATCC entries: 215917 (human)*
*Databank codes:*

| Species | SwissProt | PIR | EMBL (cDNA) | EMBL (genomic) |
|---|---|---|---|---|
| *Bos taurus* | P52176 | S43112 | X78324 | – |
| *Canis familiaris* | – | – | U68533 | – |
| *Cynops pyrrhogaster* | – | – | D82052 | – |
| *Homo sapiens* | P14780 | A34458 | D10051 | M68350: exon 8 |
| | | A41166 | J05070 | M68351: exon 9 |
| | | A42253 | X58968 | M68352: exon 10 |
| | | A45114 | | M68353: exon 11 |
| | | A61385 | | M68354: exon 12 |
| | | B48417 | | M68355: exon 13 |
| | | S16097 | | |
| *Mus musculus* | P41245 | JC1456 | D12712 | X72794: complete gene |
| | | S38654 | X72795 | |
| | | | Z27231 | |
| *Oryctolagus cuniculus* | P41246 | A53796 | D26514 | – |
| | | A55398 | | |
| *Rattus norvegicus* | P50282 | JC4364 | U24441 | – |
| | | | U36476 | |

## Name and History

**Gelatinase B (GelB)** was initially identified as a secretion product of alveolar macrophages (Mainardi *et al.*, 1984; Hibbs *et al.*, 1987), phorbol ester-treated monocytic leukemia cells U937 and granulocytes (Hibbs *et al.*, 1995). The fact that it has been given a multitude of names, including *92 kDa type IV collagenase, matrix metalloprotease 9 (MMP-9), 92 kDa gelatinase, type V collagenase*, reflects the controversy regarding the nature of its physiological substrate.

## Activity and Specificity

### Activation

GelB is secreted as a proenzyme and has to be activated for enzymologycal studies. All known modes of MMP activation result in a proteolytic removal of the N-terminal propeptide (Grant *et al.*, 1987) involving a cyteine switch mechanism (Van Wart & Birkedal-Hansen, 1990); GelB is no exception (Wilhelm *et al.*, 1989). Activation of GelB *in vitro* can be initiated by treatment with organomercurials to produce an active enzyme with Met75 at the N-terminus (Wilhelm *et al.*, 1989; Sang *et al.*, 1995), leaving the conserved propeptide sequence (PRC) uncleaved. Trypsin is a poor activator of GelB, perhaps due to the absence of three basic amino acid residues in the GelB propeptide compared to interstitial collagenase (Chapter 389) (Wilhelm *et al.*, 1989). Among activators that may be relevant physiologically, stromelysin (Chapter 393) is perhaps the most efficient (Ogata *et al.*, 1992; Goldberg *et al.*, 1992). Human matrilysin (Chapter 396) and interstitial collagenase can activate GelB by a mechanism that is very similar to that of stromelysin (Sang *et al.*, 1995). Ultimately, treatment with stromelysin, matrilysin, collagenase, plasmin or trypsin results in active GelB with Phe88 at the N-terminus. Tissue kallikrein (Desrivières *et al.*, 1993) (Chapter 29) is a more effective activator of GelB than plasmin (Chapter 59) or elastase (Chapter 15). Activation by gelatinase A (GelA) (Chapter 401) (Fridman *et al.*, 1995) has been also reported.

*Peptide Substrates*

Although GelA and GelB are similar gelatinolytic enzymes and can degrade a nearly identical battery of macromolecular matrix components, some enzymatic distinctions have been established. The peptide substrate sequence specificities of human GelA (Chapter 401) and GelB are similar but distinct (Netzel-Arnett *et al.*, 1993). They tolerate only small amino acids such as Gly and Ala in subsite P1 and prefer hydrophobic, aliphatic residues in subsite P1′. The P3′ subsite specificities of these two gelatinases are also very similar. The best substrate for GelB is Gly-Pro-Gln-Gly+Ile-Phe-Gly-Gln. In a study by Xia *et al.* (1996a) cleavage of the series of dodecylpeptides by GelB showed a different kinetic profile from that of GelA. Also, GelA digested type I gelatin about 2.5 times faster than GelB and SDS-PAGE analysis of gelatin cleavage products showed different patterns of peptides for the two enzymes. The substrate specificity of both gelatinases is closely related to those of other MMPs (Nagase *et al.*, 1994).

*Biological Substrates*

Gelatin is by far the most favored substrate of GelB *in vitro*. Since abundance of gelatin as well as the physiological role of its proteolysis in tissues are not well documented, the substrate specificity of GelB in tissues remains a subject of controversy. The latter has been a major contributor to the fact that multiple names have been given to the enzyme. *In vitro*, activated GelB can cleave collagens types III, IV and V as well as gelatins, as has been documented by many laboratories (for review see Birkedal-Hansen, 1995). Elastin (Senior *et al.*, 1991; Katsuda *et al.*, 1994) and the cross-link-containing N-terminal telopeptides of the α2 chain of type I collagen (Y. Okada *et al.*, 1995) have been reported as substrates for GelB. The interglobular domain of cartilage aggrecan was shown to be cleaved at the Asn+Phe site by GelB (Fosang *et al.*, 1993). Both GelA and GelB were shown to cleave a Pro-Gly-rich N-terminal domain of the galactoside-binding protein galectin 3 (Ochieng *et al.*, 1994). Limited cleavage by GelB of neonatal human proteoglycan link protein between His16+Ile17 was observed (Nguyen

*et al.*, 1993). GelB was found to cleave human myelin basic protein (MBP) (Proost *et al.*, 1993; Gijbels *et al.*, 1993). The positions of the cleavage sites in human MBP were such that at least one peptide coincided with a documented major MBP autoantigen, suggesting a possible role for the enzyme in the pathogenesis of demyelinating diseases such as multiple sclerosis.

## Structural Chemistry

We have reported the primary structure of human GelB isolated from SV40-transformed human lung fibroblasts and determined it to be identical to that of the 92 kDa metalloprotease secreted by normal human alveolar macrophages, phorbol ester-differentiated monocytic leukemia U937 cells, fibrosarcoma HT1080 cells, cultured human keratinocytes (Wilhelm *et al.*, 1989) and the enzyme released by polymorphonuclear cells (Wilhelm *et al.*, 1989; Devarajan *et al.*, 1992). Since then, GelB from several different species has been cloned (Masure *et al.*, 1993; A. Okada *et al.*, 1995; Xia *et al.*, 1996b; Graubert *et al.*, 1993) and its genomic structure has been determined (Huhtala *et al.*, 1991; Collier *et al.*, 1991; Munaut *et al.*, 1994; Fini *et al.*, 1994).

The GelB preproenzyme is synthesized as a polypeptide with a predicted $M_r$ of 78 426 containing a 19 amino acid signal peptide and secreted as a single 92 kDa glycosylated proenzyme. The GelB protein contains 17 cysteine residues (Wilhelm *et al.*, 1989; Goldberg *et al.*, 1992). The propeptide contains one conserved Cys99 that plays an essential role in the maintenance of the proenzyme state by interacting with the zinc ion of the active center, and is cleaved off upon enzyme activation (see above).

The gelatin-binding domain of GelB consists of three contiguous fibronectin-derived type II homology units (T2HU), T2HU-1, T2HU-2 and T2HU-3 and is responsible for the gelatin binding of the proenzyme (Collier *et al.*, 1992).

Four Cys residues are found in each of the three T2HU repeats and are conserved in GelA: Cys230, Cys244, Cys256, Cys271 in T2HU-1; Cys288, Cys302, Cys314, Cys329 in T2HU-2; and Cys347, Cys361, Cys373, Cys388 in T2HU-3. Gelatin-binding assays demonstrate that the three repeats differ significantly in their capacity to bind gelatin, with T2HU-1 and T2HU-3 having less binding activity than T2HU-2 (Collier *et al.*, 1992). There is preliminary evidence that similar differences obtain for the T2HU repeats in GelA (Steffensen *et al.*, 1995). Amino acid residues Arg307, Asp309, Asn319, Tyr320, Asp323 as well as intact disulfide bridges in the T2HU-2 are critical for gelatin binding (Collier *et al.*, 1992). The role of fibronectin domain-mediated gelatin binding in the hydrolysis of gelatin by GelB is still a matter of controversy (Collier *et al.*, 1992; Murphy *et al.*, 1994).

The gelatin-binding domain is followed by the zinc-binding part of the catalytic domain which is in turn connected to the C-terminal hemopexin-like domain by an extended proline-rich hinge region. Within the C-terminal domain are Cys516 and Cys704, conserved in all extracellular matrix metalloproteases that contain a C-terminal hemopexin-like domain, and Cys468 and Cys647, two additional residues unique to GelB. The residue Cys468 is situated in the hinge region.

The tertiary structure of GelB has not been determined, but the structures of its gelatin-binding, catalytic and C-terminal domains can be modeled with high confidence since structures of these domains from other members of the subfamily have been solved (Constantine *et al.*, 1992; Stöcker *et al.*, 1995; Libson *et al.*, 1995). The relative orientation of these domains in solution, however, is not known.

In contrast to collagenase 1 (MMP-1) (Chapter 389), which requires activation to form a complex with inhibitor, GelB forms a specific complex with tissue inhibitor of metalloproteinases 1 (TIMP-1) in a proenzyme form (Wilhelm *et al.*, 1989; Goldberg *et al.*, 1989, 1992). An intact C-terminal domain of GelB is required for the formation of this complex (Goldberg *et al.*, 1992). Additionally, GelB can form a covalent reduction-sensitive homodimer or a complex with MMP-1 in the absence of TIMP-1. Formation of the GelB/TIMP-1 complex prevents dimerization, formation of the MMP-1 complex and activation of the GelB proenzyme by stromelysin 1 (Goldberg *et al.*, 1992). The GelB proenzyme homodimer cannot form a complex with TIMP-1. Activation of the GelB/MMP1 complex yields a complex active against both gelatin and fibrillar type I collagen, suggesting a possible mechanism for cooperative action of two enzymes in reducing collagen fibrils to small peptides under physiological conditions.

## Preparation

GelB is secreted into the media by a great variety of normal and tumorigenic cells in culture; these can serve as an adequate source of the enzyme. Only a few such sources permit purification of proenzyme TIMP-1 free. Recombinant enzyme has been expressed in mammalian cells from which the enzyme monomer, dimer and complex with TIMP-1 were purified (Goldberg *et al.*, 1992). Purification of the recombinant enzyme from *Escherichia coli* must be accompanied by multiple tests to assess the correct folding, which cannot be guaranteed *a priori*. A simple purification procedure that is based on the ability of the proenzyme to bind to red dye and gelatin affinity columns has been described in detail (Wilhelm *et al.*, 1989; Goldberg *et al.*, 1992) and currently is accepted as the method of choice. Separation of different forms of enzyme is achieved by developing the gelatin affinity column with a gradient of 0–10% DMSO (Wilhelm *et al.*, 1989; Goldberg *et al.*, 1992). For most of the enzyme sources step elution of gelatin-affinity column with DMSO should be avoided.

## Biological Aspects

GelB belongs to the subfamily of matrix metalloproteases (MMPs) that play an important role in tissue remodeling in normal and pathological processes. Various *in vitro* and animal models of tumorigenesis, arthritis and several other diseases are being actively investigated to determine the role of these proteases in pathogenesis and to develop new therapies based on protease inhibitor drug development (Davies *et al.*, 1993). Regulation of gene transcription and tissue-specific expression of these enzymes in normal and diseased states are also a major focus of attention and these measures of enzyme metabolism clearly distinguish GelB

from GelA. GelB is produced by osteoclasts and plays an important role in bone resorption (Reponen *et al.*, 1994; Wucherpfennig *et al.*, 1994; Blavier & Delaissé, 1995; Witty *et al.*, 1996). GelB is also a major secretion product of activated monocytes (Welgus *et al.*, 1991 ) and a component of the specific granules of human neutrophils (Kjeldsen *et al.*, 1992). The enzyme is also produced by T cells and is induced by T cell activation (Weeks *et al.*, 1993; Leppert *et al.*, 1995). As a consequence, the enzyme is present in most cases of inflammatory response. Human skin tumors immunostained with anti-GelB-specific antibody showed no detectable enzyme in the epithelial component of tumor. Rather, GelB-positive cells were found within the connective tissue stroma of the tumors in the inflammatory infiltrates (Pyke *et al.*, 1992; Karelina *et al.*, 1993). GelB was confined to polymorphonuclear leukocytes in colorectal carcinomas (Gallegos *et al.*, 1995). *In situ* hybridization showed that GelB expression was restricted to macrophages infiltrating the human primary prostate tumors (Knox *et al.*, 1996). Evidence linking GelB to cancer progression has been reviewed in detail (Himelstein *et al.*, 1994). The role of MMPs in pathogenesis of arthritis has been under intense investigation and remains controversial. GelB has been detected in synovial fluid and cartilage of arthritis patients (Mohtai *et al.*, 1993; Koolwijk *et al.*, 1995; Okada *et al.*, 1995). The elaboration of metalloproteinases, including GelB, subsequent to the recruitment of inflammatory cells into the adventitia may contribute to the rapid growth and rupture of larger aortic aneurysms (Newman *et al.*, 1994; Freestone *et al.*, 1995; McMillan *et al.*, 1995). GelB appearance in the cerebrospinal fluid is associated with inflammatory diseases of the central nervous system (Norga *et al.*, 1995). GelB was found to be present only in samples from mice developing experimental autoimmune encephalomyelitis and correlated with the cytosis (Gijbels *et al.*, 1993). Gelatinases were reported to contribute to proteolytic disruption of basement membranes that opens the blood–brain barrier after ischemic and hemorrhagic brain injury, leading to secondary vasogenic brain edema (Rosenberg, 1995; Rosenberg *et al.*, 1995; Gottschall *et al.*, 1995). Finally, several studies indicate a possible role of the enzyme in wound healing of human respiratory epithelium (Buisson *et al.*, 1996) and human skin (Oikarinen *et al.*, 1993).

Explants of human endometrium release large amounts of MMPs including GelB (Marbaix *et al.*, 1992). The release was totally abolished under physiological concentrations of progesterone. GelB is expressed in specific organs in a precise temporal sequence during embryo implantation (Shimonovitz *et al.*, 1994; Behrendtsen *et al.*, 1992; Reponen *et al.*, 1995) and development (Canete-Soler *et al.*, 1995). At the time of implantation, MMP-9 mRNA was localized to the invading trophoblast cells and yolk sac. On day 11 it was seen in the central nervous system (Soler *et al.*, 1995). By day 15 strong signals were seen in the liver, in the developing bronchial epithelium and primordial alveoli of lungs, in the epithelium of the thyroid gland, in the thymus, in the endochondral plates of the bone, and in neural cells. Expression of GelB is controlled by a host of factors including cytokines, signal transduction through integrin receptors and cell shape (Huhtala *et al.*, 1990; Sato *et al.*, 1993; Larjava *et al.*, 1993; Fini *et al.*, 1994; Sato & Seiki, 1993; Gum *et al.*, 1996; Huhtala *et al.*, 1995). In a human melanoma cell line alteration in cell shape induced by plating on poly(HEMA) or by an inhibitor of actin polymerization, cytochalasin D, resulted in a specific loss of the constitutive production of GelB (MacDougall & Kerbel, 1995). RGD peptides induced synthesis of GelB in chondrocytes (Arner & Tortorella 1995).

## Distinguishing Features

One of the most distinctive features of GelB is its tissue-specific pattern of expression. On the protein level the formation of proenzyme complex with TIMP-1 and dimerization of the enzyme are unique for this protease among MMPs. Rabbit polyclonal antibodies to human GelB are available from Chemicon International and monoclonal antibodies are supplied by Fuji Chemical Industries and Oncogene Research Products (see Appendix 2 for full names and addresses of suppliers).

## Further Reading

GelB is among the many matrix metalloproteinases reviewed by Birkedal-Hansen (1995).

## References

Arner, E.C. & Tortorella, M.D. (1995) Signal transduction through chondrocyte integrin receptors induces matrix metalloproteinase synthesis and synergizes with interleukin-1. *Arthritis Rheum.* **38**, 1304–1314.

Behrendtsen, O., Alexander, C.M. & Werb, Z. (1992) Metalloproteinases mediate extracellular matrix degradation by cells from mouse blastocyst outgrowths. *Development* **114**, 447–456.

Birkedal-Hansen, H. (1995) Proteolytic remodeling of extracellular matrix. *Curr. Opin. Cell Biol.* **7**, 728–735.

Blavier, L. & Delaissé, J.M. (1995) Matrix metalloproteinases are obligatory for the migration of preosteoclasts to the developing marrow cavity of primitive long bones. *J. Cell Sci.* **108**, 3649–3659.

Buisson, A.C., Zahm, J.M., Polette, M., Pierrot, D., Bellon, G., Puchelle, E., Birembaut, P. & Tournier, J.M. (1996) Gelatinase B is involved in the in vitro wound repair of human respiratory epithelium *J. Cell. Physiol.* **166**, 413–426.

Canete-Soler, R., Gui, Y.H., Linask, K.K. & Muschel, R.J. (1995) Developmental expression of MMP-9 (gelatinase B) mRNA in mouse embryos. *Dev. Dyn.* **204**, 30–40.

Collier, I.E., Bruns, G.A., Goldberg, G.I. & Gerhard, D.S. (1991) On the structure and chromosome location of the 72- and 92-kDa human type IV collagenase genes. *Genomics* **9**, 429–434.

Collier, I.E., Krasnov, P.A., Strongin, A.Y., Birkedal-Hansen, H. & Goldberg, G.I. (1992) Alanine scanning mutagenesis and functional analysis of the fibronectin-like collagen-binding domain from human 92-kDa type IV collagenase. *J. Biol. Chem.* **267**, 6776–6781.

Constantine, KL., Madrid, M., Banyai, L., Trexler, M., Patthy, L., Llinas, M. (1992) Refined solution structure and ligand-binding properties of PDC-109 domain b. A collagen-binding type II domain. *J. Mol. Biol.* **223**, 281–298

Davies, B., Brown, P.D., East, N., Crimmin, M.J. & Balkwill, F.R. (1993) A synthetic matrix metalloproteinase inhibitor decreases tumor burden and prolongs survival of mice bearing human ovarian carcinoma xenografts. *Cancer Res.* **53**, 2087–2091.

Desrivières, S., Lu, H., Peyri, N., Soria, C., Legrand, Y. & Ménashi, S. (1993) Activation of the 92 kDa type IV collagenase by tissue kallikrein. *J. Cell Physiol.* **157**, 587–593.

Devarajan, P., Johnston, J.J., Ginsberg, S.S., Van Wart, H.E. & Berliner, N. (1992) Structure and expression of neutrophil gelatinase cDNA. Identity with type IV collagenase from HT1080 cells. *J. Biol. Chem.* **267**, 25228–25232.

Fini, M.E., Bartlett, J.D., Matsubara, M., Rinehart, W.B., Mody, M.K., Girard, M.T. & Rainville, M. (1994) The rabbit gene for 92-kDa matrix metalloproteinase. Role of AP1 and AP2 in cell type-specific transcription. *J. Biol. Chem.* **269**, 28620–28628.

Fosang, A.J., Last, K., Knäuper, V., Neame, P.J., Murphy, G., Hardingham, T.E., Tschesche, H. & Hamilton, J.A. (1993) Fibroblast and neutrophil collagenases cleave at two sites in the cartilage aggrecan interglobular domain. *Biochem. J.* **295**, 273–276.

Freestone, T., Turner, R.J., Coady, A., Higman, D.J., Greenhalgh, R.M. & Powell, J.T. (1995) Inflammation and matrix metalloproteinases in the enlarging abdominal aortic aneurysm. *Arterioscler. Thromb. Vasc. Biol.* **15**, 1145–1151.

Fridman, R., Toth, M., Pena, D. & Mobashery, S. (1995) Activation of progelatinase B (MMP-9) by gelatinase A (MMP-2). *Cancer Res.* **55**, 2548–2555.

Gallegos, N.C., Smales, C., Savage, F.J., Hembry, R.M. & Boulos, P.B. (1995) The distribution of matrix metalloproteinases and tissue inhibitor of metalloproteinases in colorectal cancer. *Surg. Oncol.* **4**, 111–119.

Gijbels, K., Proost, P., Masure, S,. Carton, H., Billiau, A. & Opdenakker, G. (1993) Gelatinase B is present in the cerebrospinal fluid during experimental autoimmune encephalomyelitis and cleaves myelin basic protein. *J. Neurosci. Res.* **36**, 432–440.

Goldberg, G.I., Marmer, B.L., Grant, G.A., Eisen, A.Z., Wilhelm, S.M. & He, C. (1989) Human 72-kilodalton type IV collagenase forms a complex with a tissue inhibitor of metalloproteases designated TIMP-2. *Proc. Natl Acad. Sci. USA* **86**, 8207–8211.

Goldberg, G.I., Strongin, A., Collier, I.E., Genrich, L.T. & Marmer, B.L. (1992) Interaction of 92-kDa type IV collagenase with the tissue inhibitor of metalloproteinases prevents dimerization, complex formation with interstitial collagenase, and activation of the proenzyme with stromelysin. *J. Biol. Chem.* **267**, 4583–4591.

Gottschall, P.E., Yu, X. & Bing, B. (1995) Increased production of gelatinase B (matrix metalloproteinase-9) and interleukin-6 by activated rat microglia in culture. *J. Neurosci. Res.* **42**, 335–342.

Grant, G.A., Eisen, A.Z., Marmer, B.L., Roswit, W.T. & Goldberg, G.I. (1987) The activation of human skin fibroblast procollagenase: Sequence identification of the major conversion products. *J. Biol. Chem.* **262**, 5886–5889.

Graubert, T., Johnston, J. & Berliner, N. (1993) Cloning and expression of the cDNA encoding mouse neutrophil gelatinase: demonstration of coordinate secondary granule protein gene expression during terminal neutrophil maturation. *Blood* **82**, 3192–3197.

Gum, R., Lengyel, E., Juarez, J., Chen, J.H., Sato, H., Seiki, M. & Boyd, D. (1996) Stimulation of 92-kDa gelatinase B promoter activity by ras is mitogen-activated protein kinase kinase 1-independent and requires multiple transcription factor binding sites including closely spaced PEA3/ets and AP-1 sequences. *J. Biol. Chem.* **271**, 10672–10680.

Hibbs, M.S., Hasty, K.A., Seyer, J.M., Kang, A.H. & Mainardi, C.L. (1985) Biochemical and immunological characterization of the secreted forms of human neutrophil gelatinase. *J. Biol. Chem.* **260**, 2493–2500.

Hibbs, M.S., Hoidal, J.R. & Kang, A.H. (1987) Expression of a met-alloproteinase that degrades native type V collagen and denatured collagens by cultured human alveolar macrophages. *J. Clin. Invest.* **80**, 1644–1650

Himelstein, B.P., Canete-Soler, R., Bernhard, E.J., Dilks, D.W. & Muschel, R.J. (1994) Metalloproteinases in tumor progression – the contribution of MMP-9. *Invasion Metastasis* **14**, 246–258.

Huhtala, P., Chow, L.T. & Tryggvason, K. (1990) Structure of the human type IV collagenase gene. *J. Biol. Chem.* **265**, 11077–11082.

Huhtala, P., Tuuttila, A., Chow, L.T., Lohi, J., Keski-Oja, J. & Tryggvason, K. (1991) Complete structure of the human gene for 92-kDa type IV collagenase. Divergent regulation of expression for the 92- and 72-kilodalton enzyme genes in HT-1080 cells. *J. Biol. Chem.* **266**, 16485–16490.

Huhtala, P., Humphries, M.J., McCarthy, J.B., Tremble, P.M., Werb, Z. & Damsky, C.H. (1995) Cooperative signaling by $\alpha$ 5 $\beta$ 1 and $\alpha$ 4 $\beta$ 1 integrins regulates metalloproteinase gene expression in fibroblasts adhering to fibronectin. *J. Cell Biol.* **129**, 867–879.

Karelina, T.V., Hruza, G.J., Goldberg, G.I. & Eisen, A.Z. (1993) Localization of 92 kDa type IV collagenase in human skin tumors. Comparison with normal human fetal and adult skin. *J. Invest. Dermatol.* **100**, 159–165.

Katsuda, S., Okada, Y., Okada, Y., Imai, K. & Nakanishi, I. (1994) Matrix metalloproteinase-9 (92-kd gelatinase/type IV collagenase equals gelatinase B) can degrade arterial elastin. *Am. J. Pathol.* **145**, 1208–1218.

Kjeldsen, L., Bjerrum, O.W., Askaa, J. & Borregaard, N. (1992) Subcellular localization and release of human neutrophil gelatinase, confirming the existence of separate gelatinase-containing granules. *Biochem. J.* **287**, 603–610

Knox, J.D., Wolf, C., McDaniel, K., Clark, V., Loriot, M., Bowden, G.T. & Nagle, R.B. (1996) Matrilysin expression in human prostate carcinoma *Mol. Carcinogen.* **15**, 57–63.

Koolwijk, P., Miltenburg, A.M., van Erck, M.G., Oudshoorn, M., Niedbala, M.J., Breedveld, F.C. & van Hinsbergh, V.W. (1995) Activated gelatinase-B (MMP-9) and urokinase-type plasminogen activator in synovial fluids of patients with arthritis. Correlation with clinical and experimental variables of inflammation. *J. Rheumatol.* **22**, 385–393.

Larjava, H., Lyons, J.G., Salo, T., Makela, M., Koivisto, L., Birkedal-Hansen, H., Akiyama, S.K., Yamada, K.M. & Heino, J. (1993) Anti-integrin antibodies induce type IV collagenase expression in keratinocytes. *J. Cell Physiol.* **157**, 190–200.

Leppert, D., Waubant, E., Galardy, R., Bunnett, N.W. & Hauser, S.L. (1995) T cell gelatinases mediate basement membrane transmigration in vitro. *J. Immunol.* **154**, 4379–4389.

Libson, A.M., Gittis, A.G., Collier, I.E., Marmer, B.L., Goldberg, G.I. & Lattman, E.E. (1995) Crystal structure of the haemopexin-like C-terminal domain of gelatinase A. *Nature Struct. Biol.* **2**, 938–942.

MacDougall, J.R. & Kerbel, R.S. (1995) Constitutive production of 92-kDa gelatinase B can be suppressed by alterations in cell shape. *Exp. Cell Res.* **218**, 508–515.

Mainardi, C.L., Hibbs, M.S., Hasty, K.A. & Seyer, J.M. (1984) Purification of a type V collagen degrading metalloproteinase for rabbit alveolar macrophages. *Collagen Rel. Res.* **4**, 479–492.

Marbaix, E., Donnez, J., Courtoy, P.J. & Eeckhout, Y. (1992) Progesterone regulates the activity of collagenase and related gelatinases A and B in human endometrial explants. *Proc. Natl Acad. Sci. USA* **89**, 11789–11793.

Masure, S., Nys, G., Fiten, P., Van Damme, J. & Opdenakker, G. (1993) Mouse gelatinase B. cDNA cloning, regulation of expression and glycosylation in WEHI-3 macrophages and gene organisation. *Eur. J. Biochem.* **218**, 129–141.

McMillan, W.D., Patterson, B.K., Keen, R.R., Shively, V.P., Cipollone, M. & Pearce, W.H. (1995) In situ localization and quantification of mRNA for 92-kD type IV collagenase and its inhibitor in aneurysmal, occlusive, and normal aorta. *Arterioscler. Thromb. Vasc. Biol.* **15**, 1139–1144.

Mohtai, M., Smith, R.L., Schurman, D.J., Tsuji, Y., Torti, F.M., Hutchinson, N.I., Stetler-Stevenson, W.G. & Goldberg, G.I. (1993) Expression of 92-kD type IV collagenase/gelatinase (gelatinase B) in osteoarthritic cartilage and its induction in normal human articular cartilage by interleukin 1. *J. Clin. Invest.* **92**, 179–185.

Munaut, C., Reponen, P., Huhtala, P., Kontusaari, S., Foidart, J.M. & Tryggvason, K. (1994) Structure of the mouse 92-kDa type IV collagenase gene. In vitro and in vivo expression in transient transfection studies and transgenic mice. *Ann. N.Y. Acad. Sci.* **732**, 369–371.

Murphy, G., Nguyen, Q., Cockett, M.I., Atkinson, S.J., Allan, J.A., Knight, G.C., Willenbrock, F. & Docherty, A.J.P. (1994) Assessment of the role of the fibronectin-like domain of gelatinase A by analysis of a deletion mutant. *J. Biol. Chem.* **269**, 6632–6636.

Nagase, H., Fields, C.G. & Fields, G.B. (1994) Design and characterization of a fluorogenic substrate selectively hydrolyzed by stromelysin 1 (matrix metalloproteinase-3). *J. Biol. Chem.* **269**, 20952–20957.

Netzel-Arnett, S., Sang, Q.X., Moore, W.G., Navre, M., Birkedal-Hansen, H. & Van Wart, H.E. (1993) Comparative sequence specificities of human 72- and 92-kDa gelatinases (type IV collagenases) and PUMP (matrilysin). *Biochemistry* **32**, 6427–6432.

Newman, K.M., Ogata, Y., Malon, A.M., Irizarry, E., Gandhi, R.H., Nagase, H. & Tilson, M.D. (1994) Identification of matrix metalloproteinases 3 (stromelysin-1) and 9 (gelatinase B) in abdominal aortic aneurysm. *Arterioscler. Thromb.* **14**, 1315–1320.

Nguyen, Q., Murphy, G., Hughes, C.E., Mort, J.S. & Roughley, P.J. (1993) Matrix metalloproteinases cleave at two distinct sites on human cartilage link protein. *Biochem. J.* **295**, 595–598.

Norga, K., Paemen, L., Masure, S., Dillen, C., Heremans, H., Billiau, A., Carton, H., Cuzner, L., Olsson, T., Van Damme, J. & Opdenakker, G. (1995) Prevention of acute autoimmune encephalomyelitis and abrogation of relapses in murine models of multiple sclerosis by the protease inhibitor D-penicillamine *Inflammation Res.* **44**, 529–534.

Ochieng, J., Fridman, R., Nangia-Makker, P., Kleiner, D.E., Liotta, L.A., Stetler-Stevenson, W.G. & Raz, A. (1994) Galectin-3 is a novel substrate for human matrix metalloproteinases-2 and -9. *Biochemistry* **33**, 14109–14114.

Ogata, Y., Pratta, M.A., Nagase, H. & Arner, E.C. (1992) Matrix metalloproteinase 9 (92-kDa gelatinase/type IV collagenase) is induced in rabbit articular chondrocytes by cotreatment with interleukin 1 beta and a protein kinase C activator. *Exp. Cell. Res.* **201**, 245–249.

Oikarinen, A., Kylmaniemi, M., Autio-Harmainen, H., Autio, P. & Salo, T. (1993) Demonstration of 72-kDa and 92-kDa forms of type IV collagenase in human skin: variable expression in

M

various blistering diseases, induction during re-epithelialization, and decrease by topical glucocorticoids. *J. Invest. Dermatol.* **101**, 205–210.

Okada, A., Santavicca, M. & Basset, P. (1995) The cDNA cloning and expression of the gene encoding rat gelatinase B. *Gene* **164**, 317–321.

Okada, Y., Naka, K., Kawamura, K., Matsumoto, T., Nakanishi, I., Fujimoto, N., Sato, H. & Seiki, M. (1995) Localization of matrix metalloproteinase 9 (92-kilodalton gelatinase/type IV collagenase = gelatinase B) in osteoclasts: implications for bone resorption. *Lab. Invest.* **72**, 311–322.

Proost, P., Van Damme, J. & Opdenakker, G. (1993) Leukocyte gelatinase B cleavage releases encephalitogens from human myelin basic protein. *Biochem. Biophys. Res. Commun.* **192**, 1175–1181.

Pyke, C., Ralfkiaer, E., Huhtala, P., Hurskainen, T., Danø, K. & Tryggvason, K. (1992) Localization of messenger RNA for Mr 72,000 and 92,000 type IV collagenases in human skin cancers by in situ hybridization. *Cancer Res.* **52**, 1336–1341.

Reponen, P., Sahlberg, C., Munaut, C., Thesleff, I. & Tryggvason, K. (1994) High expression of 92-kDa type IV collagenase (gelatinase) in the osteoclast lineage during mouse development. *Ann. N.Y. Acad. Sci.* **732**, 472–475.

Reponen, P., Leivo, I., Sahlberg, C., Apte, S.S., Olsen, B.R., Thesleff, I. & Tryggvason, K. (1995) 92-kDa type IV collagenase and TIMP-3, but not 72-kDa type IV collagenase or TIMP-1 or TIMP-2, are highly expressed during mouse embryo implantation. *Dev. Dyn.* **202**, 388–396.

Rosenberg, G.A. (1995) Matrix metalloproteinases in brain injury. *J. Neurotrauma* **12**, 833–842.

Rosenberg, G.A., Estrada, E.Y., Dencoff, J.E. & Stetler-Stevenson, W.G. (1995) Tumor necrosis factor-alpha-induced gelatinase B causes delayed opening of the blood-brain barrier – an expanded therapeutic window. *Brain Res.* **703**, 151–155.

Sang, Q.X., Birkedal-Hansen, H. & Van Wart, H.E. (1995) Proteolytic and non-proteolytic activation of human neutrophil progelatinase B. *Biochim. Biophys. Acta* **1251**, 99–108.

Sato, H. & Seiki, M. (1993) Regulatory mechanism of 92 kDa type IV collagenase gene expression which is associated with invasiveness of tumor cells. *Oncogene* **8**, 395–405.

Sato, H., Kita, M. & Seiki, M. (1993) v-Src activates the expression of 92-kDa type IV collagenase gene through the AP-1 site and the GT box homologous to retinoblastoma control elements. A mechanism regulating gene expression independent of that by inflammatory cytokines. *J. Biol. Chem.* **268**, 23460–23468.

Senior, R.M., Griffin, G.L., Fliszar, C.J., Shapiro, S.D., Goldberg, G.I. & Welgus, H.G. (1991) Human 92- and 72-kilodalton type IV collagenases are elastases. *J. Biol. Chem.* **266**, 7870–7875.

Shimonovitz, S., Hurwitz, A., Dushnik, M., Anteby, E., Geva-Eldar, T. & Yagel, S. (1994) Developmental regulation of the expression of 72 and 92 kD type IV collagenases in human trophoblasts: a possible mechanism for control of trophoblast invasion. *Am. J. Obstet. Gynecol.* **171**, 832–838.

Soler, R.C., Gui, Y.H., Linask, K.K. & Muschel, R.J. (1995) MMP-9 (Gelatinase B) mRNA is expressed during mouse neurogenesis and may be associated with vascularization. *Brain Res.* **88**, 37–52.

Steffensen B., Wallon U.M. & Overall, C.M. (1995) Extracellular matrix binding properties of recombinant fibronectin type II-like modules of human 72-kDa gelatinase/type IV collagenase. *J. Biol. Chem.* **270**, 11555–11566.

Stöcker, W., Grams, F., Baumann, U., Reinemer, P., Gomis-Rüth, F.X., Mckay, D.B. & Bode, W. (1995) The metzincins – topological and sequential relations between the astacins, adamalysins, serralysins, and matrixins (collagenases) define a superfamily of zinc-peptidases *Protein Sci.* **4**, 823–840.

Van Wart, H.E. & Birkedal-Hansen, H. (1990) The cysteine switch: a principle of regulation of metalloproteinase activity with potential applicability to the entire matrix metalloproteinase gene family. *Proc. Natl Acad. Sci. USA* **87**, 5578–5582.

Weeks, B.S., Schnaper, H.W., Handy, M., Holloway, E. & Kleinman, H.K. (1993) Human T lymphocytes synthesize the 92 kDa type IV collagenase (gelatinase B). *J. Cell. Physiol.* **157**, 644–649.

Welgus, H.G., Campbell, E.J., Cury, J.D., Eisen, A.Z., Senior, R.M., Wilhelm, S.M. & Goldberg, G.I. (1991) Neutral metalloproteinases produced by human mononuclear phagocytes. Enzyme profile, regulation, and expression during cellular development. *J. Clin. Invest.* **86**, 1496–1502.

Wilhelm, S.M., Collier, I.E., Marmer, B.L., Eisen, A.Z., Grant, G.A. & Goldberg, G.I. (1989) SV-40-transformed human lung fibroblasts secrete a 92 kDa type IV collagenase which is identical to that secreted by normal human macrophages. *J. Biol. Chem.* **264**, 17213–17221.

Witty, J.P., Foster, S.A., Stricklin, G.P., Matrisian, L.M. & Stern, P.H. (1996) Parathyroid hormone-induced resorption in fetal rat limb bones is associated with production of the metalloproteinases collagenase and gelatinase B. *J. Bone Miner. Res.* **11**, 72–78.

Wucherpfennig, A.L., Li, Y.P., Stetler-Stevenson, W.G., Rosenberg, A.E. & Stashenko, P. (1994) Expression of 92 kD type IV collagenase/gelatinase B in human osteoclasts. *J. Bone Miner. Res.* **9**, 549–556.

Xia, T., Akers, K.T., Eisen, A.Z., Seltzer, J.L. (1996a) Comparison of cleavage site specificity of gelatinases A and B using collagenous peptides. *Biochim. Biophys. Acta* **1293**, 259–266.

Xia, Y., Garcia, G., Chen, S., Wilson, C.B. & Feng, L. (1996b) Cloning of rat 92-kDa type IV collagenase and expression of an active recombinant catalytic domain. *FEBS Lett.* **382**, 285–288.

*Ivan E. Collier*
*Department of Medicine and Molecular*
*Biochemistry and Biophysics,*
*Washington University School of Medicine,*
*St. Louis, MO 63130, USA*

*Greg I. Goldberg*
*Department of Medicine and Molecular*
*Biochemistry and Biophysics,*
*Washington University School of Medicine,*
*St. Louis, MO 63130, USA*
*Email: goldberg@medicine.wustl.edu*

# 403. *Fragilysin*

## Databanks

*Peptidase classification: clan MB, family M10, MEROPS ID: M10.020*
*NC-IUBMB enzyme classification: EC 3.4.24.74*
*Databank codes:*

| Species | SwissProt | PIR | EMBL (cDNA) | EMBL (genomic) |
|---|---|---|---|---|
| *Bacteroides fragilis* | P54355 | – | S75941 U67735 | – |

## Name and History

**Fragilysin**, the *Bacteroides fragilis* enterotoxin, is a metallo-proteinase belonging to the metzincin clan. It was first discovered in the mid-1980s by Myers *et al*. (1984), who reported that some strains of *B. fragilis* caused a diarrheal disease in livestock and that these strains appeared to produce an enterotoxin. We now know that these pathogenic strains produce the fragilysin protease. *Bacteroides fragilis* is an anaerobic bacterium that is a part of the normal flora found in the large intestines of humans and most other mammals at about the same concentration as *Escherichia coli* (Moore *et al*., 1978). Like *E. coli*, some strains of *B. fragilis* can produce toxin; approximately 10% of *B. fragilis* strains produce this enterotoxin (Shoop *et al*., 1990). Since Myers' findings, enterotoxic *B. fragilis* have been reported as the cause of diarrhea in several animal species (Border *et al*., 1985; Collins *et al*., 1989; Myers *et al*., 1985, 1987a, 1989, 1991; Myers & Shoop, 1987). Fragilysin has also been implicated as a cause of diarrhea in humans (Myers *et al*., 1987b; Sack *et al*., 1992, 1994; Pantosti *et al*., 1994a,b; San Joaquin *et al*., 1995; Meisel-Mikulajczyk *et al*., 1994).

Despite more than a decade of research by several laboratories, the mechanism of action of the enterotoxin remained unknown until Moncrief *et al*. (1995) showed that the enterotoxin gene codes for a signature zinc-binding motif belonging to the metzincin clan. Obiso *et al*. (1995) went on to show that the purified fragilysin causes damage to the intestinal epithelium resulting in the secretion of fluid into the lumen of the intestine. This is the first reported example of a protease causing diarrhea.

Fragilysin causes proteolytic degradation of the tight junctions and basement membranes (the extracellular matrix) of tissue-cultured epithelial cells, and this is probably the major mechanism by which the intestinal epithelium is altered *in vivo* (Obiso *et al*., 1997a). Fragilysin also may activate host proteases and inactivate protease inhibitors, leading to exacerbation of the destruction (Obiso *et al*., 1997a).

## Activity and Specificity

Fragilysin cleaves collagen type IV, gelatin and fibrinogen as well as actin, tropomyosin, myosin, human complement C3 and $\alpha_1$-proteinase inhibitor (Moncrief *et al*., 1995).

Several researchers have detected fragilysin in clinical samples by using the tissue culture assay developed by Weikel *et al*. (1992). Fragilysin causes the cells of the human colon cell line HT-29 (ATCC HTB-38) to become rounded and disaggregated. Fragilysin degrades the tight junction proteins that hold these cells together which results in the rounding of the individual cells and increases permeability of the epithelial layer. This increased permeability also occurs when tissue cultured cells from nonintestinal epithelia are exposed to the protease, but the cells do not become completely detached and rounded (unpublished results of the author). Fragilysin is active on the synthetic substrate Pz-Pro-Leu+Gly-Pro-D-Arg. It has also been shown to cleave at Cys+Leu, Ser+Leu, Thr+Leu, Gly+Leu, or Leu+Gly peptide bonds in a number of proteins (Moncrief *et al*. 1995).

The pH optimum is about 6.5. Activity is stimulated by low concentrations of zinc (e.g., 1 µM), but inhibited by higher concentrations. 1,10-Phenanthroline and EDTA are inhibitory, as are Zincov, diethyl pyrocarbonate and diethylenetriaminepentaacetic acid (Moncrief *et al*., 1995).

## Structural Chemistry

Fragilysin is a single-chain protein of about 20 600 Da, with a pI of 4.5, and does not contain disulfide bonds. The activity is stable from pH 4 to 10, at −20°C and 4°C, and upon freeze-drying, but is unstable and rapidly autodigests above 37°C (van Tassell *et al*., 1992). The enzyme contains one atom of zinc per molecule, which is required for its activity (Moncrief *et al*., 1995). There are three histidine residues in the active-site sequence – Met-Ala-**His**-Glu-Leu-Gly-**His**-Ile-Leu-Gly-Ala-Glu-**His** – that may serve as zinc ligands (Moncrief *et al*., 1995). Fragilysin has between 12 and 20% sequence identity with other metzincins (Obiso *et al*., 1997b). This may seem low, but metzincins do not show a high sequence similarity, but rather, have conserved secondary structures and topologies (Bode *et al*., 1993; Stöcker *et al*., 1995). There is obvious sequence identity at the active site, as well as at a conserved methionine turn region. Fragilysin is predicted to contain three $\alpha$ helices and five $\beta$ sheets that show similar size and spacing among all the metzincins, but especially so with adamalysin (Chapter 421) (Obiso *et al*., 1997b).

**M**

## Preparation

The only natural source of fragilysin is from enterotoxigenic strains of *B. fragilis*, which contain the toxin gene. The enterotoxin is produced when these strains are grown in anaerobic brain–heart infusion supplemented with hemin, vitamin K and sodium bicarbonate (Obiso *et al*., 1995). The enzyme can be purified (2000–3000-fold) from the culture filtrates by sequential ammonium sulfate precipitation, ion-exchange chromatography on Q-Sepharose, hydrophobic chromatography on phenyl-agarose, and high-resolution ion-exchange chromatography using a Mono-Q column (van Tassell *et al*., 1992). The recombinant fragilysin gene also has been cloned into *Escherichia coli*, but the protein was expressed as inclusion bodies and was not soluble or active (Obiso *et al*., 1997a).

## Biological Aspects

The enzyme is only produced by certain strains of *Bacteroides fragilis*. Other proteases are produced by these and other strains but only this protease has been shown to be involved in pathogenicity (Gibson & MacFarlane, 1988). In culture, the enterotoxic activity is secreted. The enzyme is also attached to the outer membrane in a proenzyme form that must be cleaved by proteolysis to form the mature form (van Tassell *et al*., 1994).

Data from the amino acid sequence for fragilysin reveal similarity to other metzincins, especially adamalysin (Chapter 421) from *Crotalus adamanteus*. Secondary structural predictions and molecular modeling of fragilysin show that these two enzymes are very similar, not only at the active site, but with overall similarity (Obiso *et al*., 1997).

Fragilysin hydrolyzes a number of proteins in the extracellular matrix of epithelial cells, which is probably how this enzyme causes its pathological effects in the intestine (Moncrief *et al*., 1995). It has been shown in tissue culture experiments that the enzyme contributes to the invasion of intracellular bacterial species into epithelial cells by the proteolytic degradation of the extracellular matrix (Wells *et al*., 1996). The intestinal tract has to be resistant to the action of proteases in general because it is constantly bathed in proteases. How this protease can have such an unusual effect is not at all clear.

The regulation of the enzyme is unknown. The protein is produced as a preproenzyme, and may be attached to the outer membrane of the microbial cell through a lipoprotein attachment (Obiso *et al*., 1997a).

## Further Reading

For reviews, see Stöcker *et al*. (1995) and Häse & Finkelstein (1993).

## References

Bode, W., Gomis-Rüth, F.X., & Stöcker, W. (1993) Astacins, serralysins, snake venom and matrix metalloproteases exhibit identical zinc-binding environments (HEXXHXXGXXH and Met-turn) and topologies and should be grouped into a common family, 'the metzincins'. *FEBS Lett.* **331**, 134–140.

Border, M.M., Firehammer, B.D., Shoop, D.S. & Myers, L.L. (1985) Isolation *of Bacteroides fragilis* from the feces of diarrheic calves and lambs. *J. Clin. Microbiol.* **21**, 472–473.

Collins, J.E., Bergeland, M.E., Myers, L.L. & Shoop, D.S. (1989) Exfoliating colitis associated with enterotoxigenic *Bacteroides fragilis* in a piglet. *J. Vet. Diagn. Invest.* **1**, 349–351.

Gibson, S.A. & MacFarlane, G.T. (1988) Studies on the proteolytic activity of *Bacteroides fragilis*. *J. Gen. Microbiol.* **134**, 19–27 and 2231–2240.

Häse, C.C. & Finkelstein, R.A. (1993) Bacterial extracellular zinc-containing proteases. *Microbiol. Rev.* **57**, 823–837.

Meisel-Mikulajczyk, F., Sebald, M. Torbicka, E, Rafalowska, K. & Zielinska, U. (1994) Isolation of enterotoxigenic *B. fragilis* strains in Poland. *Acta Microbiol. Pol.* **43**, 389–392.

Moncrief, J.S., Obiso, R., Jr., Barroso, L.A., Kling, J.J. Wright, R.L., van Tassell, D.M., Lyerly, D.M. & Wilkins, T.D. (1995) The enterotoxin of *Bacteroides fragilis* is a matrix metalloprotease. *Infect. Immun.* **63**, 175–181.

Moore, W.E.C., Cato, E.P. & Holdeman, L.V. (1978) Some current concepts in intestinal bacteriology. *Am. J. Clin. Nutr.* **31**, S33–S42.

Myers, L.L. & Shoop, D.S. (1987) Association of enterotoxigenic *Bacteroides fragilis* with diarrheal disease in young pigs. *Am. J. Vet. Res.* **48**, 774–775.

Myers, L.L., Firehammer, B.D., Shoop, D.S. & Border, M.M. (1984) *B. fragilis*: a possible cause of acute diarrheal disease in newborn lambs. *Infect. Immun.* **44**, 241–244.

Myers, L.L., Shoop, D.S., Firehammer, B.D. & Border, M.M. (1985) Association of enterotoxigenic *Bacteroides fragilis* with diarrheal disease. *J. Infect. Dis.* **152**, 1344–1347.

Myers, L.L., Shoop, D.S. & Byars, T.B. (1987a) Diarrhea associated with enterotoxigenic *Bacteroides fragilis* in foals. *Am. J. Vet. Res.* **48**, 1565–1567.

Myers, L.L., Shoop, D.W., Stackhouse, L.L., Newman, F.S., Flaherty, R.J., Letson, G.W. & Sack, R.B. (1987b) Isolation of enterotoxigenic *Bacteroides fragilis* from humans with diarrhea. *J. Clin. Microbiol.* **25**, 2330–2333.

Myers, L.L., Shoop, D.S., Collins, J.E. & Bradbury, W.C. (1989) Diarrheal disease caused by enterotoxigenic *Bacteroides fragilis* in infant rabbits. *J. Clin. Microbiol.* **27**, 2025–2030.

Myers, L.L. Collins, J.E. & Shoop, D.S. (1991) Ultrastructural lesions of enterotoxigenic *Bacteroides fragilis*. *Vet. Pathol.* **28**, 336–338.

Obiso, R.J., Jr., Lyerly, D.M., van Tassell, R.L. & Wilkins, T.D. (1995) Proteolytic activity of the *Bacteroides fragilis* enterotoxin causes fluid accumulation and intestinal damage *in vivo*. *Infect. Immun.* **63**, 3820–3826.

Obiso, R.J., Jr, Azghani, A.O. & Wilkins, T.D. (1997a) The *Bacteroides fragilis* toxin fragilysin disrupts the paracellular barrier function of epithelial cells. *Infect. Immun.* **65**, 1431–1439.

Obiso, R.J., Jr., Bevan, D.R. & Wilkins, T.D. (1997b) Molecular modeling and analysis of the *Bacteroides fragilis* toxin. *Clin. Infect. Dis.* **25**(suppl. 2), S153–S155.

Pantosti, A., Piersimoni, C. & Perissi, G. (1994a) Detection of *Bacteroides fragilis* enterotoxin in the feces of a child with diarrhea. *Clin. Infect. Dis.* **19**, 809–810.

Pantosti, A., Cerquetti, M. Colongel, R. & D'Ambrosio, F. (1994b) Detection of intestinal and extraintestinal strains of enterotoxigenic *B. fragilis* by the HT-29 cytotoxicity assay. *J. Med. Microbiol.* **41**, 191–196.

Sack, R.B., Myers, L.L., Aleido-Hill, J., Shoop, D.S., Bradbury, W.C., Reidand, R. & Santosham, M. (1992) Enterotoxigenic

*Bacteroides fragilis* epidemiological studies of its role as a human pathogen. *J. Diarr. Dis. Res.* **10**, 4–9.

Sack, B.S., Albert, M.J. Alan, K., Neogi, P.K.B. & Akbar, M.S. (1994) Isolation of enterotoxigenic *Bacteroides fragilis* from Bangladeshi children with diarrhea: a controlled. study. *J. Clin. Microbiol.* **32**, 960–963.

San Joaquin, V.H., Griffis, J.C., Lee, C. & Sears, C.L. (1995) Association of *Bacteroides fragilis* with childhood diarrhea. *Scand. J. Infect. Dis.* **27**, 211–215.

Shoop, G.L., Myers, L.L. & LeFever, J.B. (1990) Enumeration of enterotoxigenic *Bacteroides fragilis* in municipal sewage. *Appl. Environ. Microbiol.* **56**, 2243–2244.

Stöcker, W.L., Grams, F., Bauman, U., Reinemer, P., Gomis-Rüth, F.X., McKay, D.B. & Bode, W. (1995) The metzincins – topological and sequential relations between the astacins, adamalysins, serralysins, and matrixins (collagenases) define a superfamily of zinc peptidases. *Protein Sci.* **4**, 823–840.

van Tassell, R.L., Lyerly, D.M. & Wilkins, T.D. (1992) Purification and characterization of an enterotoxin from *Bacteroides fragilis*. *Infect. Immun.* **60**, 1343–1350.

van Tassell, R.L., Lyerly, D.M. & Wilkins, T.D. (1994) Characterization of the enterotoxigenic *Bacteroides fragilis* by a toxin specific enzyme linked immunosorbant assay. *Clin. Diag. Lab. Immun.* **1**, 578–587.

Weikel, C., Grieco, F., Reuben, J., Myers, L.L. & Sack, R.B. (1992) Human colonic epithelial cells HT29/C1, treated with crude *Bacteroides fragilis* enterotoxin dramatically alter their morphology. *Infect. Immun.* **60**, 321–327.

Wells, C.L., Van de Westerlo, E.M.A., Jechorek, R.P., Feltis, B.A., Wilkins, T.D. & Erlandsen, S.L. (1996) *Bacteroides fragilis* enterotoxin modulates epithelial permeability and bacterial internalization by HT-29 enterocytes. *Gastroenterology* **110**, 1429–1437.

*Richard J. Obiso, Jr*
*Fralin Biotechnology Center,*
*Virginia Polytechnic Institute and State University,*
*Department of Biochemistry*
*and Anaerobic Microbiology,*
*Blacksburg, Virginia 24061-0346, USA*

*Tracy D. Wilkins*
*Fralin Biotechnology Center,*
*Virginia Polytechnic Institute and State University,*
*Department of Biochemistry*
*and Anaerobic Microbiology,*
*Blacksburg, Virginia 24061-0346, USA*
*Email: tracyw@.vt.edu*

**M**

# 404. Introduction: family M12 of astacin (clan MB)

## Databanks

*MEROPS ID: M12*

| Species | SwissProt | PIR | EMBL (cDNA) | EMBL (genomic) |
|---|---|---|---|---|
| **Subfamily A: Astacins** | | | | |
| Astacin (Chapter 405) | | | | |
| Choriolysin H (Chapter 410) | | | | |
| Choriolysin L (Chapter 409) | | | | |
| Flavastacin (Chapter 412) | | | | |
| Meprin (human) (Chapter 408) | | | | |
| Meprin B (Chapter 407) | | | | |
| Meprin A (Chapter 406) | | | | |
| Other astacin homologs (Chapter 413) | | | | |
| Procollagen N-endopeptidase (Chapter 414) | | | | |
| Procollagen C-endopeptidase (Chapter 411) | | | | |
| Protease 10, sea urchin blastula (*Paracentrotus*) (Chapter 413) | | | | |
| SpAN gene product (sea urchin) (Chapter 413) | | | | |
| Tolkin (Chapter 413) | | | | |
| Tolloid (Chapter 413) | | | | |
| **Subfamily B: Reprolysins** (Chapter 416) | | | | |
| Acutolysin (Chapter 431) | | | | |
| ADAM metalloproteinases (Chapter 447) | | | | |
| Adamalysin (Chapter 421) | | | | |
| Atrolysin A (Chapter 418) | | | | |
| Atrolysin B (Chapter 420) | | | | |
| Atrolysin C (Chapter 423) | | | | |
| Atrolysin E (Chapter 430) | | | | |
| Atroxase (Chapter 419) | | | | |
| Basilysin (Chapter 442) | | | | |
| Bothrolysin (Chapter 439) | | | | |
| Ecarin (Chapter 432) | | | | |
| Fibrolase (*Agkistrodon contortrix*) (Chapter 417) | | | | |
| *Agkistrodon halys* venom metalloproteinases (Chapter 428) | | | | |
| Horrilysin (Chapter 443) | | | | |
| Jararafibrase II | | | | |
|    *Bothrops jararaca* | – | A42766 | – | – |
| Jararhagin (Chapter 429) | | | | |
| Kistomin (Chapter 434) | | | | |
| Lebetase | | | | |
|    *Vipera lebetina* | – | – | X97894 | – |
| Mutalysins (Chapter 424) | | | | |
| Mucrolysin (Chapter 426) | | | | |
| Myelin-associated metalloproteinase/ADAM 10 (Chapter 448) | | | | |
| Ruberlysin (Chapter 422) | | | | |
| Russellysin (Chapter 433) | | | | |
| Tumor necrosis factor $\alpha$-converting enzyme (Chapter 449) | | | | |
| Trimerelysin I (Chapter 425) | | | | |
| Trimerelysin II (Chapter 427) | | | | |
| Other venom reprolysins | | | | |
|    *Agkistrodon contortrix* | – | – | U18233 | – |
|    *Agkistrodon contortrix* | – | – | U18234 | – |
|    *Crotalus atrox* | – | – | U21003 | – |
|    *Echis pyramidum* | – | – | X78970 | – |
|    *Echis pyramidum* | – | – | X78971 | – |
|    *Trimeresurus flavoviridis* | P14530 | JX0074 | – | – |
|    *Trimeresurus gramineus* | P15503 | – | X51530 | – |
|    *Trimeresurus mucrosquamatus* | – | – | X91190 | – |

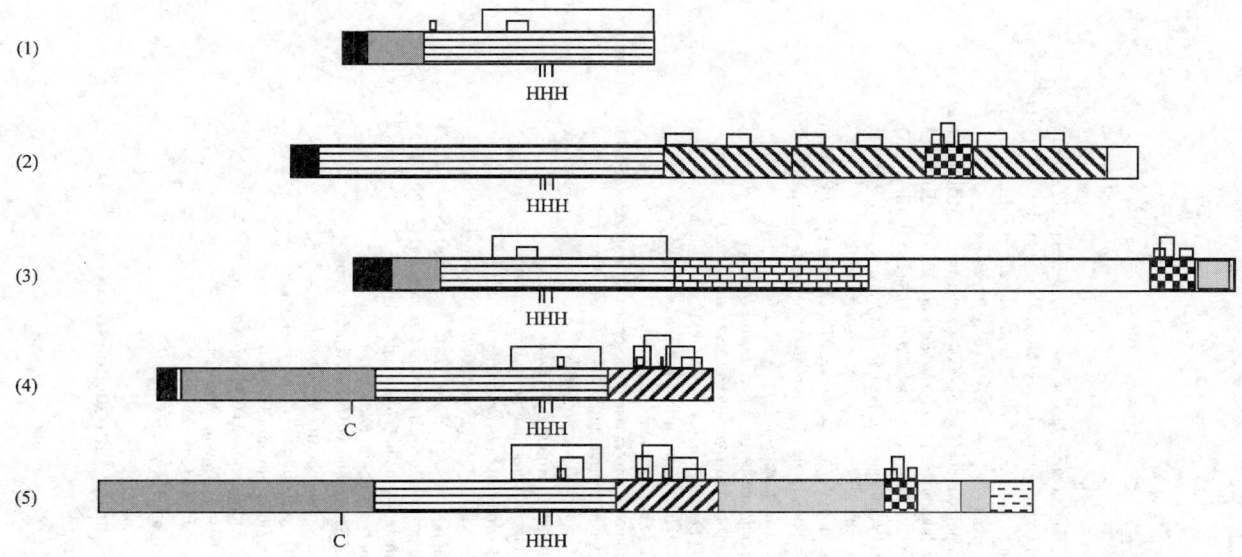

*Figure 404.1*  Domain structures for selected endopeptidases from family M12. Endopeptidase sequences are shown as rectangles. The length of the rectangle is a representation of the sequence length. Disulfide bridges are shown as open boxes above the sequence rectangle. The position of the zinc ligands is shown below the sequence rectangles, and all structures are aligned at the zinc ligands. Structures are arranged so that the simplest is at the top, and the most complex at the bottom. Shaded boxes within the sequence rectangle represent different structural domains. *Key to domains* (from top of figure to bottom): black, signal peptide; grey, N-terminal propeptide; horizontal stripes, catalytic domain; left-to-right diagonals, CUB domain; checker, epidermal growth factor-like domain; bricks, MAM domain; black with spots, transmembrane region; right-to-left diagonals, disintegrin domain; light grey, cysteine-rich region; short horizontal lines, cytoplasmic anchor. *Key to sequences:* (1) *Oryzias latipes* choriolysin H1, (2) *Homo sapiens* procollagen C-endopeptidase, (3) *Mus musculus* meprin α subunit, (4) *Crotalus atrox* atrolysin E, (5) *Cavia porcellus* ADAM 1 protein.

Family M12 is included in clan MB. The enzymes in the family are all endopeptidases dependent on one catalytic zinc ion that is bound by three His residues. The tertiary structure of the crayfish digestive enzyme astacin was the first in the clan to be solved.

Family M12 shares many similarities with family M10 (Chapter 385), in addition to the obvious structural resemblances. Like family M10, M12 is also divided into two subfamilies because of the divergence in sequence, and as also seen in family M10, only one subfamily has a catalytic Tyr on the Met-turn (subfamily A, the astacin subfamily), and only one subfamily possesses the cysteine-switch activation mechanism (subfamily B, the reprolysin subfamily). The pattern of disulfide bonds is different between the subfamilies, and both subfamilies include mosaic proteins, as is shown in Figure 404.1.

The structure of astacin is shown in Figure 404.2. The catalytic mechanism as presently understood involves three His zinc ligands (His93, His96 and His102), a Tyr that is a fourth zinc ligand in the apoenzyme state and acts as a general base during catalysis (Tyr149), and a Glu that acts as an electrophile to stabilize the transition state complex (Glu93).

Astacin represents the simplest enzyme in the subfamily. It is a small endopeptidase (only 200 residues) and is not a mosaic protein. There are analogous enzymes in the matrixin subfamily, for example matrilysin and soybean metalloproteinase 1 (Chapters 396, 398). Besides astacin, other simple members of the astacin subfamily include choriolysin L and choriolysin H. Other members of the

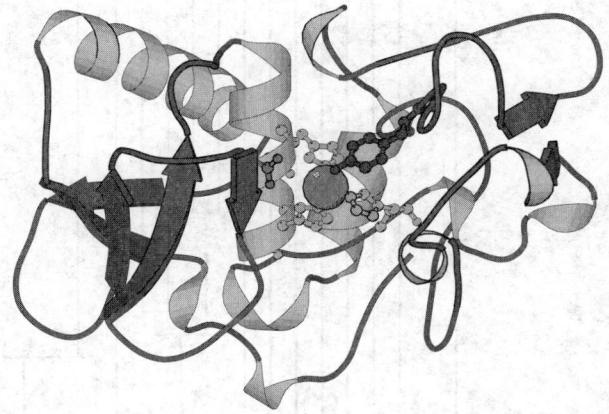

*Figure 404.2*  Richardson diagram of *Astacus astacus* astacin. The image was prepared from the Brookhaven Protein Data Bank entry (1AST) as described in the Introduction (p. xxv). The catalytic zinc ion is shown in CPK representation as a dark gray sphere. The zinc ligands are shown in ball-and-stick representation: His92, His96 and His102 (numbering as in Alignment 380.1). The catalytic residues are also shown in ball-and-stick representation: Glu93 and Tyr149.

subfamily are more complex, with nonpeptidase domains appended to the C-terminus of the peptidase unit. Human procollagen C-endopeptidase has three CUB domains that are similar to regions of the complement subcomponent C1r from family S1 (Chapter 43), as well as an epidermal

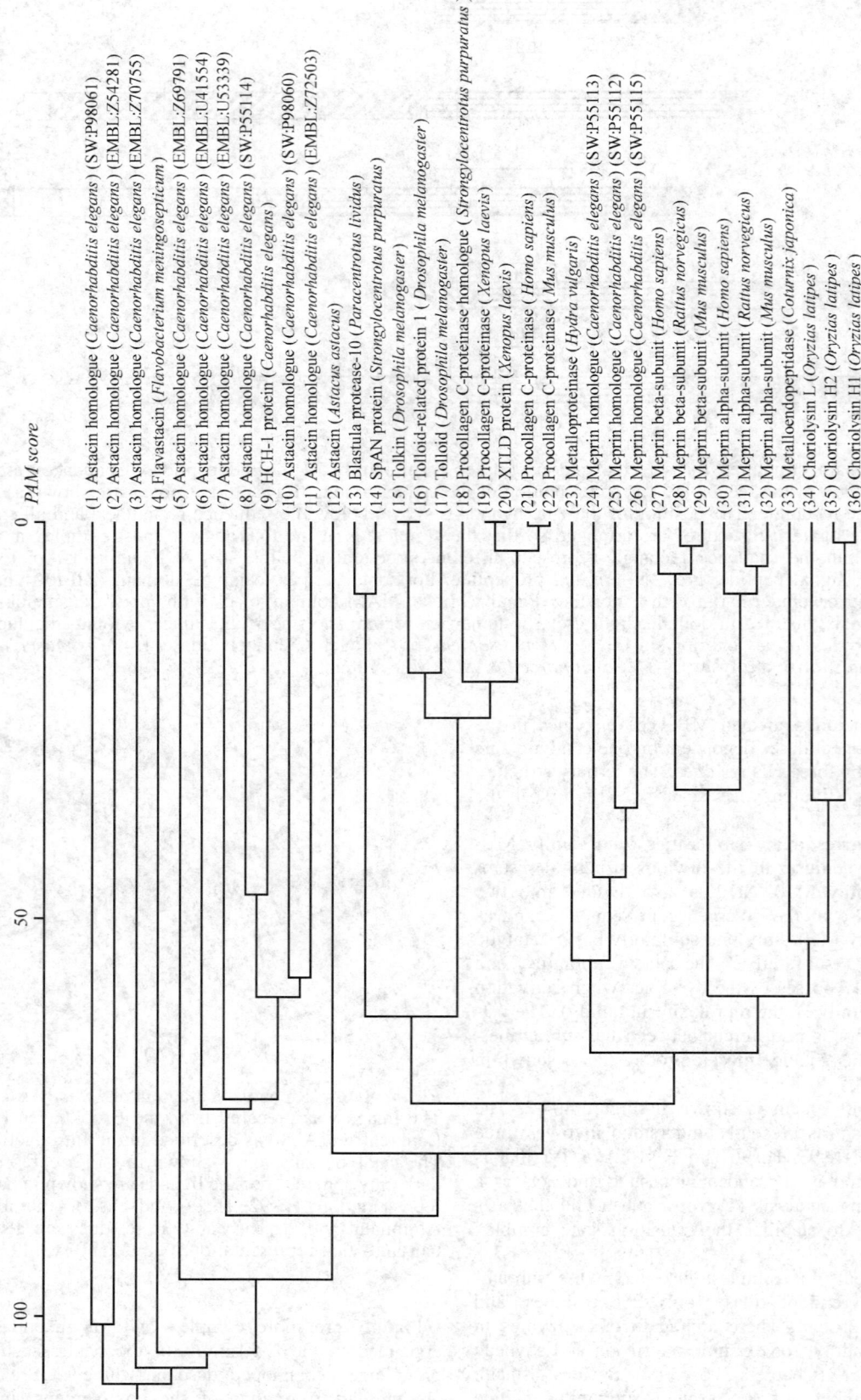

*PAM score*

0

50

100

(1) Astacin homologue (*Caenorhabditis elegans*) (SW:P98061)
(2) Astacin homologue (*Caenorhabditis elegans*) (EMBL:Z54281)
(3) Astacin homologue (*Caenorhabditis elegans*) (EMBL:Z70755)
(4) Flavastacin (*Flavobacterium meningosepticum*)
(5) Astacin homologue (*Caenorhabditis elegans*) (EMBL:Z69791)
(6) Astacin homologue (*Caenorhabditis elegans*) (EMBL:U41554)
(7) Astacin homologue (*Caenorhabditis elegans*) (EMBL:U53339)
(8) Astacin homologue (*Caenorhabditis elegans*) (SW:P55114)
(9) HCH-1 protein (*Caenorhabditis elegans*)
(10) Astacin homologue (*Caenorhabditis elegans*) (SW:P98060)
(11) Astacin homologue (*Caenorhabditis elegans*) (EMBL:Z72503)
(12) Astacin (*Astacus astacus*)
(13) Blastula protease-10 (*Paracentrotus lividus*)
(14) SpAN protein (*Strongylocentrotus purpuratus*)
(15) Tolkin (*Drosophila melanogaster*)
(16) Tolloid-related protein 1 (*Drosophila melanogaster*)
(17) Tolloid (*Drosophila melanogaster*)
(18) Procollagen C-proteinase homologue (*Strongylocentrotus purpuratus*)
(19) Procollagen C-proteinase (*Xenopus laevis*)
(20) XTLD protein (*Xenopus laevis*)
(21) Procollagen C-proteinase (*Homo sapiens*)
(22) Procollagen C-proteinase (*Mus musculus*)
(23) Metalloproteinase (*Hydra vulgaris*)
(24) Meprin homologue (*Caenorhabditis elegans*) (SW:P55113)
(25) Meprin homologue (*Caenorhabditis elegans*) (SW:P55112)
(26) Meprin homologue (*Caenorhabditis elegans*) (SW:P55115)
(27) Meprin beta-subunit (*Homo sapiens*)
(28) Meprin beta-subunit (*Rattus norvegicus*)
(29) Meprin beta-subunit (*Mus musculus*)
(30) Meprin alpha-subunit (*Homo sapiens*)
(31) Meprin alpha-subunit (*Rattus norvegicus*)
(32) Meprin alpha-subunit (*Mus musculus*)
(33) Metalloendopeptidase (*Coturnix japonica*)
(34) Choriolysin L (*Oryzias latipes*)
(35) Choriolysin H2 (*Oryzias latipes*)
(36) Choriolysin H1 (*Oryzias latipes*)

*Figure 404.3* Evolutionary tree for subfamily A (the astacins) of family M12. The tree was prepared as described in the Introduction (p. xxv)

growth factor (EGF)-like domain which is also present in the complement subcomponent. In invertebrate forms of procollagen C-endopeptidase, such as the tolloid protein from *Drosophila melanogaster*, the number of CUB domains increases to five, with two EGF-like domains interspersed between them. The tolloid protein is important for dorsal–ventral patterning in the developing fly.

The membrane-bound kidney endopeptidases, the meprins, are dimers of two astacin homologs. Meprin B is a homodimer of meprin $\beta$ subunits, whereas meprin A is a heterodimer of meprin $\alpha$ and $\beta$ subunits. Both the $\alpha$ and $\beta$ subunits are mosaic proteins, each containing a MAM domain, an EGF-like domain, a transmembrane domain and a cytoplasmic anchor. MAM domains are believed to aid cell surface adhesion, and similar domains are found in other membrane-bound proteins, such as the receptor protein tyrosine phosphatase $\mu$, which is involved in cytosolic dephosphorylation, and the A5 protein from the visual center of the *Xenopus* brain, which also contains C1r-like repeats.

The gene structure of mouse meprin $\alpha$ has been determined, and the region encoding the peptidase domain is followed by a series of phase 1 intron–exon junctions (Jiang & Flannery, 1997). Thus, phase 1 exon shuffling in an ancestral astacin would explain the acquisition of C-terminal domains.

The mechanism of activation in astacin is analogous to that of family S1, the trypsin family (see Chapter 2). After proteolytic removal of the N-terminal propeptide, the new N-terminal Ala is buried in a water-filled cavity with the -$NH_3^+$ group forming a salt bridge to the carboxylate of Glu99, which is the residue next to the third zinc ligand, His98. Presumably, there is a large conformational change when the proenzyme is processed.

The astacin subfamily includes a bacterial endopeptidase, flavastacin. This enzyme is secreted by *Flavobacterium* and cleaves Asp+ bonds in peptides. It is synthesized with an N-terminal propeptide, but does not contain the disulfide bonds thought to be structurally important in astacin. Figure 404.3 shows an evolutionary tree for the astacin subfamily.

Subfamily B represents the reprolysins, which were discovered in snake venoms. Homologs are now known from a variety of animals, although not all are peptidases. The human enzyme fertilin is a heterodimer of two reprolysins, one of which is inactive because the zinc ligands and catalytic residues have been replaced. The non-snake venom reprolysins have recently been renamed the ADAM metalloproteinases. Physiological functions are mostly unknown for the ADAM proteins, the exception being tumor necrosis factor $\alpha$-converting enzyme, which releases tumor necrosis factor $\alpha$ from its precursor.

Reprolysins are derived from complex precursors that are mosaic proteins. As can be seen from Figure 404.1, the precursors contain a signal peptide, an N-terminal propeptide and a number of domains appended to the C-terminus, which may include a disintegrin domain, an EGF-like domain, a cysteine-rich region and a transmembrane domain. The mechanism of inhibition by the propeptide is thought to be a cysteine-switch mechanism (see Alignment 380.2).

Some C-terminal domains are cleaved off, the extent of which determines the size of the mature enzyme. Ruberlysin

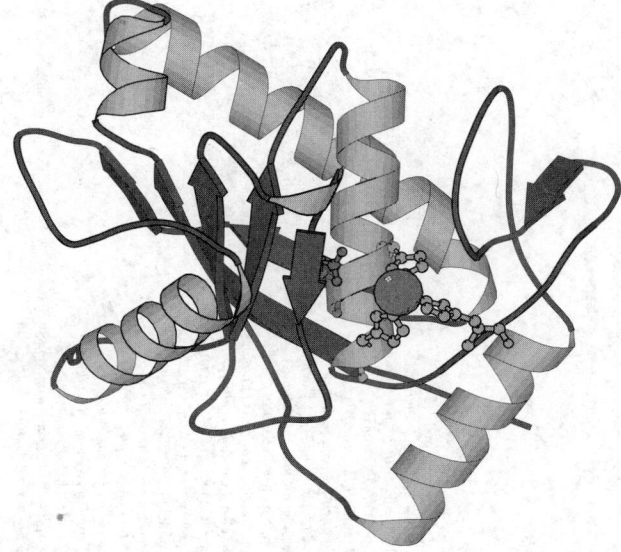

*Figure 404.4*    Richardson diagram of adamalysin. The image was prepared from the Brookhaven Protein Data Bank entry (1IAG) as described in the Introduction (p. xxv). The catalytic zinc ion is shown in CPK representation as a light gray sphere. The zinc ligands are shown in ball-and-stick representation: His92, His96 and His102 (numbering as in Alignment 380.1). The catalytic Glu93 is also shown in ball-and-stick representation.

is 25 kDa, whereas trimerelysin I, which retains the disintegrin domain, is 60 kDa. The disintegrin domain from the green habu snake reprolysin is known as trigramin, and binds to platelet fibrinogen receptors preventing platelet aggregation. Association with the receptor does not appear to be dependent on the presence of an Arg-Asp-Gly (RGD) motif.

There are a considerable number of snake venom reprolysins (see overview in Chapter 416) showing different specificities for insulin B chain. The tertiary structure of adamalysin has been determined and is shown diagrammatically in Figure 404.4. Unlike astacin, but like interstitial collagenase, adamalysin lacks Tyr149, which is believed to act as the general base in catalysis (Mock & Yao, 1997). There is no obvious corresponding residue, and the general base has not been identified.

The evolutionary tree (Figure 404.5) shows that the ADAM proteins are the more ancient and that the snake venom enzymes are a specialized subset of them. The most divergent ADAM proteins are tumor necrosis factor $\alpha$ (ADAM 17) and the myelin-associated metalloendopeptidase, ADAM 10.

## References

Jiang, W.P. & Flannery, A.V. (1997) Correlation of the exon/intron organization to the secondary structures of the protease domain of mouse meprin $\alpha$ subunit. *Gene* **189**, 65–71.

Mock, W.L. & Yao, J. (1997) Kinetic characterization of the serralysins: a divergent catalytic mechanism pertaining to astacin-type metalloproteases. *Biochemistry* **36**, 4949–4958.

M

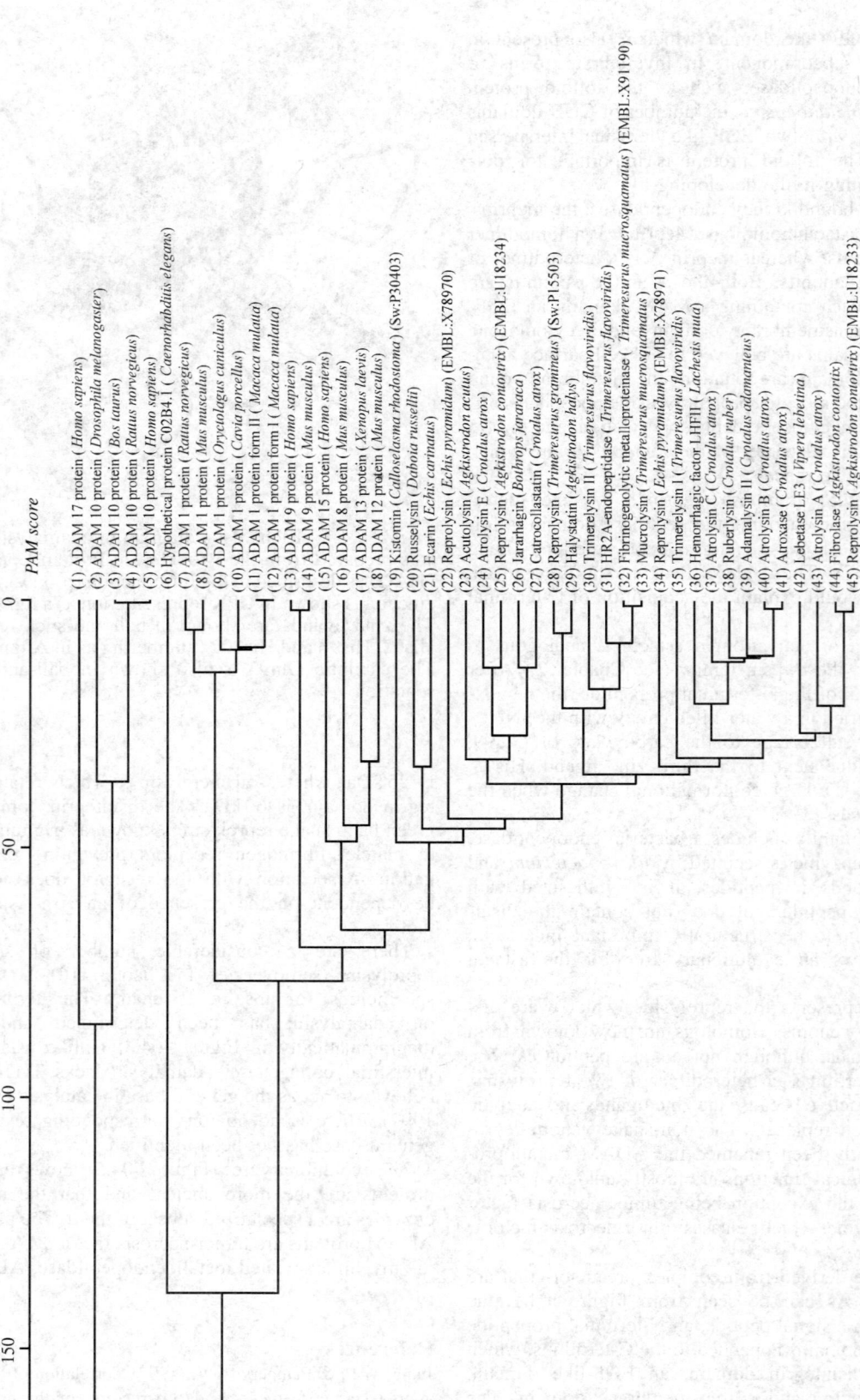

*PAM score*

(1) ADAM 17 protein ( *Homo sapiens* )
(2) ADAM 10 protein ( *Drosophila melanogaster* )
(3) ADAM 10 protein ( *Bos taurus* )
(4) ADAM 10 protein ( *Rattus norvegicus* )
(5) ADAM 10 protein ( *Homo sapiens* )
(6) Hypothetical protein C02B4.1 ( *Caenorhabditis elegans* )
(7) ADAM 1 protein ( *Rattus norvegicus* )
(8) ADAM 1 protein ( *Mus musculus* )
(9) ADAM 1 protein ( *Oryctolagus cuniculus* )
(10) ADAM 1 protein ( *Cavia porcellus* )
(11) ADAM 1 protein form II ( *Macaca mulatta* )
(12) ADAM 1 protein form I ( *Macaca mulatta* )
(13) ADAM 9 protein ( *Homo sapiens* )
(14) ADAM 9 protein ( *Mus musculus* )
(15) ADAM 15 protein ( *Homo sapiens* )
(16) ADAM 8 protein ( *Mus musculus* )
(17) ADAM 13 protein ( *Xenopus laevis* )
(18) ADAM 12 protein ( *Mus musculus* )
(19) Kistomin ( *Calloselasma rhodostoma* ) (Sw:P30403)
(20) Russelysin ( *Daboia russellii* )
(21) Ecarin ( *Echis carinatus* )
(22) Reprolysin ( *Echis pyramidum* ) (EMBL:X78970)
(23) Acutolysin ( *Agkistrodon acutus* )
(24) Atrolysin E ( *Crotalus atrox* )
(25) Reprolysin ( *Agkistrodon contortrix* ) (EMBL:U18234)
(26) Jararhagin ( *Bothrops jararaca* )
(27) Catrocollastatin ( *Crotalus atrox* )
(28) Reprolysin ( *Trimeresurus gramineus* ) (Sw:P15503)
(29) Halystatin ( *Agkistrodon halys* )
(30) Trimerelysin II ( *Trimeresurus flavoviridis* )
(31) HR2A-endopeptidase ( *Trimeresurus flavoviridis* )
(32) Fibrinogenolytic metalloproteinase ( *Trimeresurus mucrosquamatus* ) (EMBL:X91190)
(33) Mucrolysin ( *Trimeresurus mucrosquamatus* )
(34) Reprolysin ( *Echis pyramidum* ) (EMBL:X78971)
(35) Trimerelysin 1 ( *Trimeresurus flavoviridis* )
(36) Hemorrhagic factor LHFII ( *Lachesis muta* )
(37) Atrolysin C ( *Crotalus atrox* )
(38) Ruberlysin ( *Crotalus ruber* )
(39) Adamalysin II ( *Crotalus adamanteus* )
(40) Atrolysin B ( *Crotalus atrox* )
(41) Atroxase ( *Crotalus atrox* )
(42) Lebetase LE3 ( *Vipera lebetina* )
(43) Atrolysin A ( *Crotalus atrox* )
(44) Fibrolase ( *Agkistrodon contortrix* )
(45) Reprolysin ( *Agkistrodon contortrix* ) (EMBL:U18233)

*Figure 404.5*   Evolutionary tree for subfamily B (the reprolysins) of family M12. The tree was prepared as described in the Introduction (p. xxv).

# 405. *Astacin*

## Databanks

*Peptidase classification: clan MB, family M12, MEROPS ID: M12.001*
*NC-IUBMB enzyme classification: EC 3.4.24.21*
*Databank codes:*

| Species | SwissProt | PIR | EMBL (cDNA) | EMBL (genomic) |
|---|---|---|---|---|
| *Astacus astacus* | P07584 | A25829 | – | X95684: complete gene |
| *Chionoecetes opilio* | P34156 | – | – | – |

Brookhaven Protein Data Bank three-dimensional structures:

| Species | ID | Resolution | Notes |
|---|---|---|---|
| *Astacus astacus* | 1AST | 1.8 | |
| | 1IAA | 1.9 | zinc replaced with copper |
| | 1IAB | 1.79 | zinc replaced with cobalt |
| | 1IAC | 2.1 | zinc replaced with mercury |
| | 1IAD | 2.3 | zinc removed |
| | 1IAE | 1.83 | zinc replaced with nickel |

## Name and History

In 1967, Zwilling and colleagues reported the purification of an endopeptidase of $M_r$ 11 000 from the digestive tract of the freshwater crayfish, *Astacus astacus* L. (syn. *Astacus fluviatilis* Fabr.) (Pfleiderer *et al.*, 1967). The apparent small size of the enzyme led to the designation 'low molecular weight protease' or **crayfish small molecule proteinase** (Zwilling *et al.*, 1981; Zwilling & Neurath, 1981). When the protein sequence had been established, the name was altered to **Astacus protease** (Titani *et al.*, 1987); the originally determined low molecular mass could be attributed to an anomalous retardation in gel filtration. Subsequently, the enzyme was shown to be zinc dependent (Stöcker *et al.*, 1988), and the designation **astacin** was introduced following the recommendation of the NC-IUBMB (Stöcker *et al.*, 1991a; Dumermuth *et al.*, 1991).

Astacin is the prototype for a family of extracellular zinc endopeptidases termed the astacin family (M12) (Stöcker *et al.*, 1991a, 1993, 1995; Dumermuth *et al.*, 1991; Bond & Beynon, 1995). Many members of this family are involved in developmental processes. The astacins are found throughout the animal kingdom and also in bacteria. Examples are the bone morphogenetic protein (BMP-1), which is identical with procollagen C-proteinase (Chapter 411), an enzyme critical for the assembly of collagen fibers (Kessler *et al.*, 1996; Li *et al.*, 1996; Suzuki *et al.*, 1996), and the meprins (Chapters 406 and 407), which are anchored to the surface of epithelial cells where they are thought to process biologically active peptides (Bond & Beynon, 1995). Most of these proteins have a multidomain structure including one astacin-like catalytic domain which exhibits at least about 30% sequence identity with the crayfish proteinase (Stöcker *et al.*, 1993;

Bond & Beynon, 1995). The designation 'astacin' had also been used earlier for $\beta$-carotene derivatives of crustacean origin (Kuhn & Lederer, 1933).

## Activity and Specificity

The 71 cleavage sites of astacin in the denatured chains of $\alpha$- and $\beta$-tubulin revealed an extended substrate-binding region with preference for Ala in P1′, Pro in P2′ and P3, hydrophobic residues in P3′ and P4′, and Lys, Arg, Asn or Tyr in P1 and P2 (Krauhs *et al.*, 1982; Stöcker *et al.*, 1990; Stöcker & Zwilling, 1995). Optimal astacin substrates contain five or more amino acid residues.

The pH optimum is between pH 6 and 8 depending on the substrate (Stöcker *et al.*, 1991a). Astacin activity is assayed with denatured casein, with synthetic nitroanilides like Suc-Ala-Ala-Ala-NHPhNO$_2$ or with more sensitive fluorescent substrates (Stöcker *et al.*, 1990; Wolz, 1994; Stöcker & Zwilling, 1995; Wolz & Bond, 1995). Most convenient are quenched fluorescent substrates like Dns-Pro-Lys-Arg+Ala-Pro-Trp-Val (Stöcker *et al.*, 1990) and Abz-Arg-Pro-Ile-Phe+Ser-Pro-Nph-Arg (Wolz, 1994), which were the best substrates out of a series of tubulin-derived heptapeptides and bradykinin-derived octapeptides, respectively.

Astacin and other members of the astacin family are not inhibited by TIMP-1 (tissue inhibitor of metalloproteinases) or by phosphoramidon. The only potent natural inhibitor is $\alpha_2$-macroglobulin (Stöcker *et al.*, 1991b). Other protein inhibitors have been identified in seeds of the Mediterranean plant *Ecballium elaterium* and in potato tubers (Herkert *et al.*, 1990; Stöcker & Zwilling, 1995). Powerful astacin inhibitors are phosphinic transition state analog pseudopeptides like FMOC-Pro-Lys-Phe-$\psi$(PO$_2$-CH$_2$)-Ala-Pro-Leu-Val-OH ($K_i = 42$ nM) (Grams *et al.*, 1996; I. Yiallouros, S. Vassiliou, A. Yiotakis, R. Zwilling, W. Stöcker & V. Dive, unpublished results).

Metal chelating inhibitors include 1,10-phenanthroline, EDTA (only slowly inhibitory), dipicolinic acid, 8-hydroxyquinoline-5-sulfonic acid, 2,2'-bipyridyl, amino acid hydroxamates and thiol compounds (Stöcker *et al.*, 1988; Wolz & Zwilling, 1989; Wolz *et al.*, 1990; Wolz, 1994). The slow inhibition by EDTA was the reason for the late identification of astacin as a zinc enzyme. The commercially available hydroxamate Pro-Leu-Gly-NHOH (Sigma) (see Appendix 2 for full names and addresses of suppliers) ($K_i = 14$ μM; Grams *et al.*, 1996) is used for affinity purification of recombinant astacin (S. Reyda, E. Jacob, R. Zwilling & W. Stöcker, unpublished results).

## Structural Chemistry

The covalent structure of the mature form of astacin comprises a single chain of 200 residues, corresponding to an $M_r$ of 22 614; two disulfide bridges cross-link Cys42-Cys198 and Cys64-Cys84 (Titani *et al.*, 1987). From the nucleotide sequences of both the cDNA and the entire astacin gene, an N-terminal prepro sequence of 49 amino acid residues and two additional residues at the C-terminus have been deduced (Geier *et al.*, 1997; S. Reyda, E. Jacob, R. Zwilling & W. Stöcker, unpublished results). The pI is 3.5 for the native (undenatured) protein and 4.8 for the urea-denatured astacin (Stöcker & Zwilling, 1995).

The X-ray crystal structure has been solved to 1.8 Å resolution (Figure 405.1) (Bode *et al.*, 1992; Gomis-Rüth *et al.*, 1993; Stöcker *et al.*, 1993; Grams *et al.*, 1996). The catalytic zinc is located at the bottom of the long and deep active-site cleft, ligated by three His residues, a Tyr residue and a water molecule in a trigonal bipyramidal coordination sphere. The histidine zinc ligands are part of the consensus motif **HEXXHXXGXXH**, whose glutamic acid residue, in analogy to that of thermolysin (Matthews, 1988), is thought to polarize a water molecule for nucleophilic attack on the scissile peptide bond. The tyrosine zinc ligand is embedded in another conserved motif (**SXMHY**) beneath the active-site zinc. In the complex of astacin with the transition state analog inhibitor Cbz-Pro-Lys-Phe-$\psi$(PO$_2$-CH$_2$)-Ala-Pro-OMe (Figure 405.1),

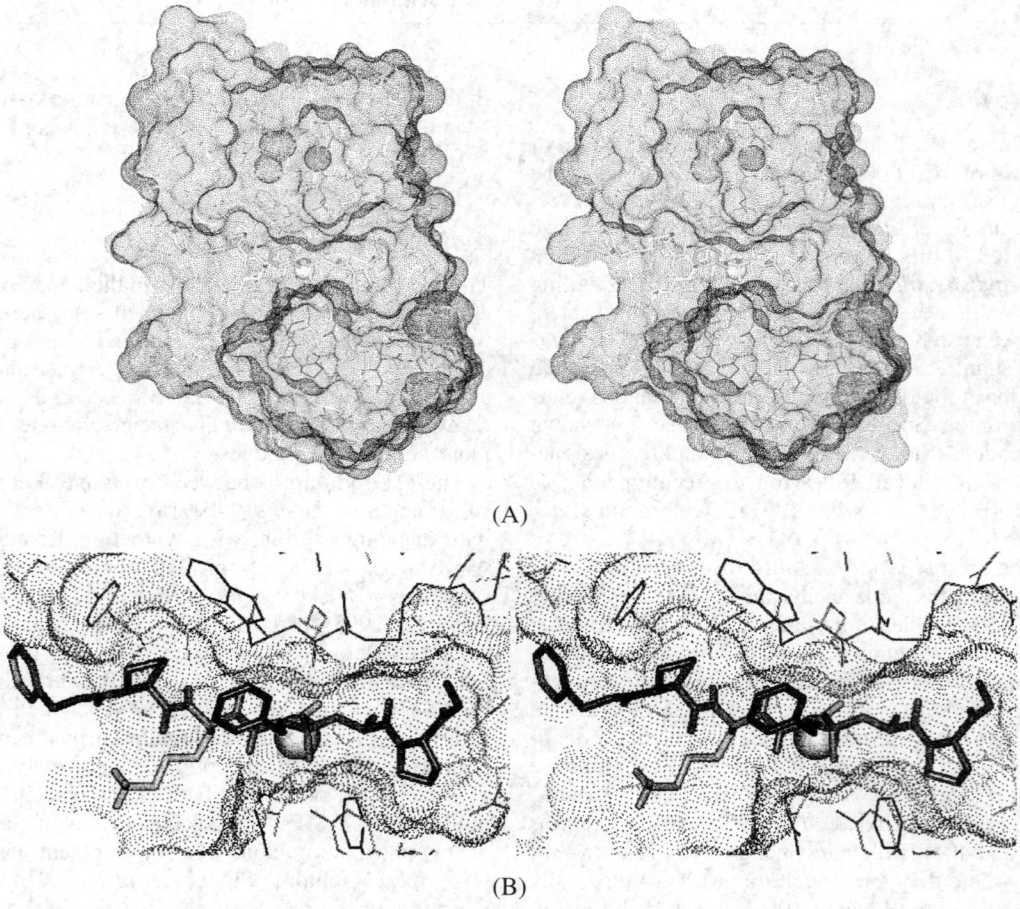

(A)

(B)

*Figure 405.1* (A) Model of astacin overlaid with the corresponding Connolly dot surface (rolling sphere of 1.4 Å) complexed with the inhibitor Z-Pro-Lys-Phe-$\psi$(PO$_2$-CH$_2$)-Ala-Pro-OMe (Grams *et al.*, 1996). (B) Close-up of the active site with the bound inhibitor.

this tyrosine side chain is shifted and hydrogen-bonded with the $PO_2$ group that mimics a water-attacked peptide bond (Grams *et al.*, 1996). This unique 'tyrosine switch' is a special feature of the astacin-like proteinases and presumably also of the serralysins (Stöcker *et al.*, 1995).

The zinc ion of astacin has been replaced by $Cu^{2+}$, $Co^{2+}$, $Hg^{2+}$ and $Ni^{2+}$ (Stöcker *et al.*, 1988). The copper- and cobalt-enzymes are enzymatically active; they contain trigonal bipyramidally coordinated metal just like the zinc enzyme. By contrast, mercury- and nickel-astacin are inactive, which is probably due to their distorted tetrahedral and octahedral metal coordination, respectively (Gomis-Rüth *et al.*, 1994). In the mature enzyme, the N-terminus is buried and salt-bridged with Glu103, the direct neighbor of the zinc ligand His102. This indicates an activation mechanism, since N-terminally extended proforms could not exhibit such a conformation (Bode *et al.*, 1992).

## Preparation

Astacin can be isolated from the digestive juice of the crayfish *Astacus astacus* (Zwilling & Neurath, 1981; Stöcker *et al.*, 1991a; Stöcker & Zwilling, 1995). The recombinant enzyme has been expressed in bacteria and purified from urea-dissolved inclusion bodies (S. Reyda, E. Jacob, R. Zwilling & W. Stöcker, unpublished results).

## Biological Aspects

The astacin gene, obtained from a crayfish genomic library, spans 2616 bp. These are distributed over a pattern of five exons and four introns that correlates with the secondary structure on the protein level (Geier *et al.*, 1997). The first intron disrupts the sequence encoding the signal peptide. Introns 2 and 3 are in positions equivalent to those in other astacins, whereas intron 4 is unique. In the promoter region a TATA box has been identified (Geier *et al.*, 1997).

The enzyme is synthesized in the midgut gland as a pre-proenzyme. The mode of proenzyme activation is unclear. However, it has been shown for the closely related meprins that activation involves proteolytic (tryptic) removal of the propeptide (Bond & Beynon, 1995). The proform of astacin does not accumulate and the mature proteinase is stored extra-cellularly in the stomach (Vogt *et al.*, 1989). As a prerequisite, astacin is remarkably stable against self-digestion.

Astacin degrades the triple helix of type I collagen, gelatin and other proteins at neutral pH (Stöcker & Zwilling, 1995). This most likely reflects the physiological function in the crayfish stomach, which does not contain an acidic barrier.

## Distinguishing Features

The astacins share the consensus zinc-binding motif **HEXXHXXGXXH** with the matrixins, the adamalysins/re-prolysins and the serralysins (Stöcker *et al.*, 1995). These four subfamilies of zinc peptidases form a clan which has been called the 'metzincins' clan MB (Chapter 380) (Bode *et al.*, 1993; Stöcker & Bode, 1995; Stöcker *et al.*, 1995). The metzincins are characterized by this zinc-binding motif,

by a conserved methionine participating in a 1,4-$\beta$-turn (Met-turn) containing the tyrosine zinc ligand in case of the astacins and serralysins, and by equivalent topologies of their three-dimensional structures. Most interestingly, on the genomic level, there is a conserved intron position shared by the astacins and the matrixins (intron 2 of astacin) that is inserted between the sequences encoding the N-terminal framework and the catalytic apparatus (Geier *et al.*, 1997). There is also partial topological similarity between astacin and the archetypical zinc proteinase thermolysin (Chapter 351) (Gomis-Rüth *et al.*, 1993) including, for example, the general consensus motif **HEXXH** present in many metalloproteinases.

## Further Reading

A comprehensive account of the function and structure of the zinc endopeptidase astacin is presented by Stöcker & Zwilling (1995). A review of the various members of the astacin family is found in Bond & Beynon (1995). Stöcker *et al.* (1995) present a detailed report on the metzincin clan.

## References

Bode, W., Gomis-Rüth, F.-X., Huber, R., Zwilling, R. & Stöcker, W. (1992) Structure of astacin and implications for activation of astacins and zinc ligation of collagenases. *Nature* **358**, 164–167.

Bode, W., Gomis-Rüth, F.-X. & Stöcker, W. (1993) Astacins, serralysins, snake venom and matrix metalloproteinases exhibit identical zinc-binding environments (HEXXHXXGXXH and Met-turn) and topologies and should be grouped into a common family, the 'metzincins'. *FEBS Lett.* **331**, 134–140.

Bond, J.S. & Beynon, R.B. (1995) The astacin family of metallo-endopeptidases. *Protein Sci.* **4**, 1247–1261.

Dumermuth, E., Sterchi, E.E., Jiang, W., Wolz, R.L., Bond, J.S., Flannery, A.V. & Beynon, R.J. (1991) The astacin family of metalloendopeptidases. *J. Biol. Chem.* **266**, 21381–21385.

Geier, G., Jacob, E., Stöcker, W. & Zwilling, R. (1997) Genomic organization of the zinc-endopeptidase astacin. *Arch. Biochem. Biophys.* **337**, 300–307.

Gomis-Rüth, F.-X., Stöcker, W., Huber, R., Zwilling, R., & Bode, W. (1993) The refined 1.8 Å X-ray crystal structure of astacin, a zinc-endopeptidase from the crayfish *Astacus astacus* L. Structure determination, refinement, molecular structure, and comparison to thermolysin. *J. Mol. Biol.* **229**, 945–968.

Gomis-Rüth, F.-X., Grams, F., Yiallouros, I., Nar, H., Küsthardt, U., Zwilling, R., Bode, W. & Stöcker, W. (1994) Crystal structures, spectroscopic features and catalytic properties of cobalt (II)-, copper (II)-, nickel (II)- and mercury (II)-derivatives of the zinc-endopeptidase astacin. A correlation of structure and proteolytic activity. *J. Biol. Chem.* **269**, 17111–17117.

Grams, F., Dive, V., Yiotakis, A., Yiallouros, I., Vassilou, S., Zwilling, R., Bode, W. & Stöcker, W. (1996) Structure of astacin with a transition state analogue inhibitor. *Nature Struct. Biol.* **3**, 671–675.

Herkert, M., Stöcker, W. & Zwilling, R. (1990) Structure, function and localization of a proteinase inhibitor from potato tubers (*Solanum tuberosum*). *Biol. Chem. Hoppe-Seyler* **371**, 760–761.

Kessler, E., Takahara, K., Biniaminov, L., Brusel, M. & Greenspan, D.S. (1996) Bone morphogenetic protein-1: the type I procollagen C-proteinase. *Science* **271**, 360–362.

**M**

Krauhs, E., Dörsam, H., Little, M., Zwilling, R. & Ponstingl, H. (1982) A protease from *Astacus fluviatilis* as an aid in protein sequencing. *Anal. Biochem.* **119**, 153–157.

Kuhn, R. & Lederer, E. (1933) Über die Farbstoffe des Hummers (*Astacus gammarus* L.) und ihre Stammsubstanz, das Astacin [On the pigments of the lobster (*Astacus gammarus* L.) and their progenitor, astacin]. *Berichte der Chemischen Gesellschaft* **66**, 488–495.

Li, S.-W., Sieron, A.L., Fertala, A., Hojima, Y., Arnold, W.V. & Prockop, D.J. (1996) The C-proteinase that processes procollagens to fibrillar collagen is identical to the protein previously identified as bone morphogenetic protein-1. *Proc. Natl Acad. Sci. USA* **93**, 5127–5130.

Matthews, B.W. (1988) Structural basis of the action of thermolysin and related zinc peptidases. *Accts Chem. Res.* **21**, 333–340.

Pfleiderer, G., Zwilling, R. & Sonneborn, H.H. (1967) Zur Evolution der Endopeptidasen III. Eine Protease vom Molekulargewicht 11000 und eine trypsinähnliche Fraktion aus *Astacus fluviatilis* [On the evolution of endopeptidases. III. A protease of molecular weight 11,000 and a trypsin-like fraction from *Astacus fluviatilis*]. *Z. Physiol. Chem.* **348**, 1319.

Stöcker, W. & Bode, W. (1995) Structural features of a superfamily of zinc-endopeptidases: the metzincins. *Curr. Opin. Struct. Biol.* **5**, 383–390.

Stöcker, W. & Zwilling, R. (1995) Astacin. *Methods Enzymol.* **248**, 305–325.

Stöcker, W., Wolz, R.L., Zwilling, R., Strydom, D.J. & Auld, D.S. (1988) *Astacus* protease – a zinc-metalloenzyme. *Biochemistry* **27**, 5026–5032.

Stöcker, W., Ng, M. & Auld, D.S. (1990) Fluorescent oligopeptide substrates for kinetic characterization of the specificity of *Astacus* protease. *Biochemistry* **29**, 10418–10425.

Stöcker, W., Sauer, B. & Zwilling, R. (1991a) Kinetics of nitroanilide cleavage by astacin. *Biol. Chem. Hoppe-Seyler* **372**, 385–392.

Stöcker, W., Breit, S., Sottrup-Jensen, L. & Zwilling, R. (1991b) $\alpha_2$-Macroglobulin from the hemolymph of the freshwater crayfish *Astacus astacus*. *Comp. Biochem. Physiol.* **98B**, 501–509.

Stöcker, W., Gomis-Rüth, F.-X., Bode, W. & Zwilling, R. (1993)

Implications of the three-dimensional structure of astacin for the structure and function of the astacin family of zinc-endopeptidases. *Eur. J. Biochem.* **214**, 215–231.

Stöcker, W., Grams, F., Baumann, U., Reinemer, P., Gomis-Rüth, F.-X., McKay, D.B. & Bode, W. (1995) The metzincins – topological and sequential relations between the astacins, adamalysins, serralysins, and matrixins (collagenases) define a superfamily of zinc-peptidases. *Protein Sci.* **4**, 823–840.

Suzuki, N., Labosky, P.A., Furuta, Y., Hargett, L., Dunn, R., Fogo, A.B., Takahara, K., Peters, D.M.P., Greenspan, D.S. & Hogan, B.L.M. (1996) Failure of ventral body wall closure in mouse embryos lacking a procollagen C-proteinase encoded by BMP1, a mammalian gene related to *Drosophila* tolloid. *Development* **122**, 3587–3595.

Titani, K., Torff, H.-J., Hormel, S., Kumar, S., Walsh, K.A., Rödl, J., Neurath, H. & Zwilling, R. (1987) Amino acid sequence of a unique protease from the crayfish *Astacus fluviatilis*. *Biochemistry* **26**, 222–226.

Vogt, G., Stöcker, W., Storch, V. & Zwilling, R. (1989) Biosynthesis of *Astacus* protease a digestive enzyme from crayfish. *Histochemistry* **91**, 373–381.

Wolz, R.L. (1994) A kinetic comparison of the homologous proteases astacin and meprin A. *Arch. Biochem. Biophys.* **310**, 144–151.

Wolz, R.L. & Bond, J.S. (1995) Meprins A and B. *Methods Enzymol.* **248**, 325–345.

Wolz, R.L. & Zwilling, R. (1989) Kinetic evidence for cooperative binding of two ortho-phenanthroline molecules to *Astacus* protease during metal removal. *J. Inorg. Biochem.* **35**, 157–167.

Wolz, R.L., Zeggaf, C., Stöcker, W. & Zwilling, R. (1990) Thiol containing compounds and amino acid hydroxamates as reversible inhibitors of *Astacus* protease. *Arch. Biochem. Biophys.* **281**, 275–281.

Zwilling, R. & Neurath, H. (1981) Invertebrate proteases. *Methods Enzymol.* **80**, 633–664.

Zwilling, R., Dörsam, H., Torff, H.-J. & Rödl, J. (1981) Low molecular mass protease: evidence for a new family of proteolytic enzymes. *FEBS Lett.* **127**, 75–78.

*Walter Stöcker*
*Westfälische Wilhelms-Universität Münster,*
*Institut für Zoophysiologie,*
*Molekulare Physiologie, Hindenburgplatz 55,*
*D-48143 Münster, Germany*
*Email: wst@uni-muenster.de*

# 406. Meprin A

## Databanks

*Peptidase classification: clan MB, family M12, MEROPS ID: M12.002*
*NC-IUBMB enzyme classification: EC 3.4.24.18*

*ATCC entries: 222190, 229926, 265565, 729126, 1058809 (mouse, α subunit); 977145 (mouse, β subunit)*
*Chemical Abstracts Service registry number: 148938-24-3*
*Databank codes:*

| Species | SwissProt | PIR | EMBL (cDNA) | EMBL (genomic) |
|---------|-----------|-----|-------------|----------------|
| α subunit | | | | |
| *Mus musculus* | P28825 | – | M74897 | – |
| *Rattus norvegicus* | – | S24134 | S43408 | – |
| β subunit | | | | |
| *Mus musculus* | – | – | L15193 | |
| *Rattus norvegicus* | P28826 | – | M88601 | – |

Brookhaven Protein Data Bank three-dimensional structures:

| Species | ID | Resolution | Notes |
|---------|-----|-----------|-------|
| α subunit | | | |
| *Mus musculus* | 1IAF | | theoretical model |

## Name and History

Meprin was first purified by Beynon *et al.* (1981) from mouse (BALB/c) kidneys as a membrane-bound metalloproteinase with azocasein-degrading activity. Purified preparations were found to have disulfide-bridged subunits of approximately 85 kDa constituting oligomers of 320 kDa. Further investigations revealed an inherited 'deficiency' of meprin in kidney of certain inbred mouse strains (C stock mice, such as C3H/He, and CBA; Beynon & Bond, 1983). The observed polymorphism, mice with 'high' or 'low' kidney meprin activity, led to the localization of the gene responsible for the deficiency and to the discovery of a latent form of meprin in the kidney of mice with 'low' activity (Bond *et al.* 1984; Butler & Bond, 1988). It is now clear that meprins are oligomers that consist of one or two evolutionarily related subunits (the α and β subunits). Mice with 'high' kidney meprin express both subunits, while the 'low' or 'deficient' meprin strains express only the latent β subunit. Meprin A (EC 3.4.24.18) is defined as those isoforms that contain the α subunit; this includes membrane-bound heterotetrameric forms, and a secreted meprin of homo-oligomeric form. Meprin B (Chapter 407) is the homo-oligomeric form of β subunits. **Meprin A** from rat kidney (formerly called **endopeptidase-2**, or **endopeptidase 24.18**), and from human intestine (also called **PABA-peptide hydrolase**) (Chapter 408) are enzymatically similar to the mouse enzyme; for example, they all hydrolyze bradykinin, azocasein and gelatin (Wolz & Bond, 1995).

Cloning and sequencing of the protease domain of mouse and human meprin subunits led to the identification of the astacin family of metalloendopeptidases (M12) (Dumermuth *et al.*, 1991) (Chapter 405). The structural studies have clarified the relation of astacin family enzymes to other metalloendopeptidases (Jiang & Bond, 1992; Hooper, 1994; Rawlings & Barrett, 1995).

## Activity and Specificity

Meprin A is a zinc metalloendopeptidase, displaying activity against a variety of biologically active peptides (Wolz & Bond, 1995) such as parathyroid hormone (Yamaguchi *et al.*, 1991), angiotensins and bradykinin (Butler *et al.*, 1987), α-melanocyte stimulating hormone (Wolz *et al.*, 1991), luliberin, substance P and neurotensin (Kenny & Ingram, 1987). The protein substrates azocasein (Beynon *et al.*, 1981) and gelatin (Beynon *et al.*, 1996) are commonly used to assay meprin activity. Degradation of type IV collagen by meprin A has also been observed (Kaushal *et al.*, 1994). Studies using synthetic peptide substrates revealed that meprin A also possesses arylamidase activity (Sterchi *et al.*, 1982; Wolz, 1994), a property rare for metalloendopeptidases. The peptide bond specificity of the enzyme has been examined using the oxidized insulin B chain (Butler *et al.*, 1987)–

$$\text{FVNQH}+\text{L}+\text{CG}+\text{SH}+\text{LVEA}+\text{L}+\text{YL}+\text{VCG}+\text{ERGF}+\text{F}+\text{YTPKA}$$

and analogs of bradykinin (Wolz *et al.*, 1991) as substrates. These studies indicate that optimal meprin A substrates have aromatic P1 side chains and small, uncharged P1′ side chains. However, peptides with virtually any side chain in the P1 and P1′ positions can serve as substrates for meprin A.

Meprin A activity is optimal between the pH values of 7.5 and 9.5, depending on the substrate, with an optimal ionic strength of 0.2 (Beynon *et al.*, 1981; Kenny & Ingram, 1987). Therefore, meprin activity is generally assayed at neutral or alkaline pH values. The enzyme is routinely assayed at pH 9.5 using azocasein (casein that has been derivatized with diazotized sulfanilamide). Activity is quantitated by the absorbance at 340 nm of released azopeptides soluble in 4% trichloroacetic acid (Beynon & Kay, 1978; Reckelhoff *et al.*, 1985). The quenched fluorescent peptide Abz-Arg-Pro-Pro-Gly-Phe+Ser-Pro-Phe-Arg-Lys(Dnp)-Gly-OH is also used to assay meprin A. Hydrolysis of the Phe+Ser bond is monitored spectrofluorimetrically using excitation and emission wavelengths of 320 and 417 nm, respectively (Marchand *et al.*, 1996). Commercially available, inexpensive peptides have also been used to assay the arylamidase activity of meprin A. Substrates include Abz-Ala-Ala-Phe+NHPhNO₂, Glt-Ala-Ala-Phe+NHNapOMe, Suc-Ala-Ala-Ala+NHPhNO₂ and Z-Phe-Arg-NHMec. These substrates are not specific for meprin

A, but Z-Phe-Arg-NHMec has been used successfully to monitor the purification of the enzyme from rat kidneys (Kenny & Ingram, 1987; Johnson & Hersh, 1992).

Meprin A is completely inhibited by metal chelators such as EDTA and 1,10-phenanthroline (Beynon *et al.*, 1981). The thiol-containing compounds 2-ME and cysteine are also inhibitory. Aminoacyl and peptidyl hydroxamates also inhibit (Wolz *et al.*, 1991). The peptidyl hydroxamate actinonin, with a $K_i$ of 0.15 mM (Wolz, 1994), is selective for the inhibition of meprin A, but is also known to inhibit other metallopeptidases, particularly aminopeptidases. There are no other known naturally occurring inhibitors of meprin A, and no specific synthetic inhibitors have yet been described. Meprin is not inhibited by thiorphan or phosphoramidon (neprilysin inhibitors, Chapter 362), captopril (an effective inhibitor of angiotensin I-converting enzyme, Chapter 359), or TIMPs (tissue inhibitors of metalloproteinases, inhibitors of matrixins, Chapter 385).

## Structural Chemistry

Meprin A is a glycosylated, oligomeric metalloendopeptidase that can be composed solely of $\alpha$, or both $\alpha$ and $\beta$ subunits (Gorbea *et al.*, 1991; Bond & Beynon, 1995). Secreted and plasma membrane-associated forms are known to exist (Flannery *et al.*, 1990; Marchand *et al.*, 1994). The secreted form of meprin A contains only $\alpha$ subunits (Bond & Beynon, 1995; Beynon *et al.*, 1996). The bulk of the plasma membrane form of meprin A is extracellular, making it an ectoenzyme. The meprin A ectoenzyme is particularly abundant on the brush border membranes of the kidney and intestine of mice, rats and humans. The individual subunits range in molecular mass from 82 kDa to 110 kDa (Wolz & Bond, 1995), depending on the species and degree of post-translational modification. The basic structural unit of meprin A is a dimer of disulfide-linked subunits, either $\alpha/\alpha$ homodimers or $\alpha/\beta$ heterodimers, with higher order oligomers forming through noncovalent interactions between subunits (Gorbea *et al.*, 1991; Johnson & Hersh, 1992). Because the $\beta$ subunit remains an integral membrane protein (see below), meprin oligomers containing $\beta$ subunits are found exclusively on the cell surface; meprin $\alpha$ is found at the cell surface when associated with $\beta$, but secreted meprin $\alpha$ homo-oligomers are found in mouse urine (Beynon *et al.* 1996).

The primary structure of meprin A has been deduced through the cloning and sequencing of cDNAs encoding both the $\alpha$ and $\beta$ subunits. Meprin $\alpha$ subunit cDNAs have been sequenced from mouse (Jiang *et al.*, 1992), rat (Corbeil *et al.*, 1992) and human (Dumermuth *et al.*, 1993). Meprin $\beta$ subunit cDNAs have been cloned for mouse (Gorbea *et al.*, 1993) and rat (Johnson & Hersh, 1992). The mouse $\alpha$ and $\beta$ subunits are 42% identical in amino acid sequence and share a similar arrangement of structural domains (Figure 406.1).

Meprin cDNAs encode an N-terminal signal sequence (S), followed by a pro sequence, a zinc metalloprotease domain (protease domain) homologous to that of astacin, a MAM domain (**m**eprin, **A**-5 protein, receptor protein tyrosine phosphatase $\mu$; Beckmann & Bork, 1993), a MATH domain (**m**eprin **a**nd **TRAF h**omology; Uren & Vaux, 1996), a domain of unknown function called the AM domain (**a**fter**MATH**), an epidermal growth factor (EGF)-like domain,

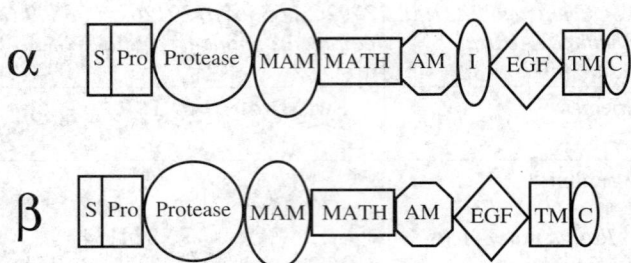

*Figure 406.1*   Deduced domain structure of meprin subunits. See text for description of the domains.

a putative C-terminal transmembrane (TM) domain, and a cytoplasmic (C) domain. The $\alpha$ subunit differs from the $\beta$ subunit in that it has a 56 amino acid domain ('I') inserted between the AM and EGF-like domains; this domain has been shown to be sufficient and essential for the C-terminal processing of the $\alpha$ subunit during biosynthesis (Marchand *et al.*, 1995). Of these domains, the structure and function of the protease domain is the best understood.

The X-ray crystal structure of astacin has been solved (Bode *et al.*, 1992), and molecular modeling studies of the mouse meprin $\alpha$ subunit protease domain indicate that all the residues essential for zinc coordination and peptide bond hydrolysis in astacin are conserved in the meprin protease domain (Stöcker *et al.*, 1993). Cysteine residues of meprin MAM domains are essential for the formation of intersubunit disulfide linkages (Chevallier *et al.*, 1996; Marchand *et al.*, 1996). There is some evidence that the oligomeric structure of meprins is essential for stability against heat and proteases, and for activity against large proteins (Marchand *et al.*, 1996). The MAM domain is proposed to be an adhesion domain, and this domain on receptor protein tyrosine phosphatase $\mu$ and $\kappa$ has been shown to mediate specific homophilic cell–cell interactions in transfected insect cells (Zondag *et al.*, 1995). MATH domains may contribute to noncovalent interactions between meprin subunits, and possibly to interactions with other proteins; MATH domains in cytosolic proteins are essential for homodimerization and heterologous interactions with receptor proteins in processes such as acute phase responses, lymphocyte activation and nerve cell growth (e.g. Cheng *et al.*, 1995).

## Preparation

Meprin A is abundant in the kidneys of mouse strains such as ICR, C57BL/6 and in Sprague–Dawley rats, so these sources are generally employed for the purification of the enzyme (Kenny & Ingram, 1987; Kounnas *et al.*, 1991; Johnson & Hersh, 1992). Meprin A is less abundant in human kidneys, but the enzyme has been purified from this source (Yamaguchi *et al.*, 1994), and from human intestine (Sterchi *et al.*, 1988) (see Chapter 408). Using rodent kidneys as the source, meprin A is released from the brush border membranes by treatment with papain, and then purified as a soluble protein using ammonium sulfate precipitation, gel filtration and anion-exchange chromatography. Purification of the enzyme is readily monitored by assay with azocasein or Z-Phe-Arg-NHMec.

## Biological Aspects

The $\alpha$ and $\beta$ meprin subunits are initially biosynthesized as type I integral membrane proteins with single membrane-spanning regions near their C-termini. However, heterologous expression of meprin subunits in mammalian systems (COS-1, human embryonic kidney 293, and Madin Darby canine kidney cells) has revealed differences in the post-translational modification of the meprin $\alpha$ and $\beta$ subunits. While the mature $\beta$ subunit remains a type I integral membrane protein, the $\alpha$ subunit was found to be secreted into the culture medium. Coexpression of both subunits resulted in the localization of both subunits to the cell surface, as is seen in the rodent kidney (Johnson & Hersh, 1994; Marchand *et al.*, 1994; Milhiet *et al.*, 1994). Secretion of the $\alpha$ subunit is now known to be a result of proteolytic processing in the I-domain that cleaves the transmembrane domain from the rest of the $\alpha$ subunit (Marchand *et al.*, 1995). Both subunits were found to be synthesized as zymogens in cultured cells, retaining their N-terminal pro sequences. Removal of the pro sequences by treatment with trypsin resulted in the appearance of meprin activity, indicating that pro sequence removal is essential for activation of the astacin-like protease domain (Johnson & Hersh, 1994; Milhiet *et al.*, 1994). It was noted that both $\alpha$ and $\beta$ activated subunits are able to degrade azocasein, but that only the $\alpha$ subunit possesses significant arylamidase activity (Johnson & Hersh, 1994).

The $\alpha$ subunit is very abundant in the mouse kidney brush border membranes when it is expressed; it has been estimated to represent 5% of the brush border membrane protein in ICR mice (Bond & Beynon, 1986). Studies of the rate of incorporation of radiolabeled amino acids into immunoprecipitable forms of the $\alpha$ and $\beta$ subunits indicated that the rate of synthesis of the $\alpha$ subunits is about twice that of the $\beta$ subunit (Hall *et al.*, 1993). The half-lives of the subunits *in vivo* (approximately 10 h) are similar. The higher rate of synthesis of the $\alpha$ subunit is, therefore, responsible for the greater abundance of this protein compared to $\beta$ in microvillus membranes, and for secreted forms of meprin $\alpha$ which exceed the quantity of the $\beta$ anchor.

Expression of meprin subunits appears to be highly regulated and tissue specific. Kidney and intestinal expression has been observed in rats, mice and humans (Craig *et al.*, 1987; Sterchi *et al.*, 1988; Barnes *et al.*, 1989; Yamaguchi *et al.*, 1994). In adult mouse intestine, meprin B (the $\beta$ homo-oligomeric form) (Chapter 407) is the predominant form of the enzyme, although meprin A is found in the ileum (Bankus & Bond, 1996). Meprins have been detected in salivary glands, thyroid tissue and neuroepithelial cells in the inner ear, nasal conchae, and choroid plexus in rat embryos (Craig *et al.*, 1991; Spencer-Dene *et al.*, 1994). Glycosylation of the subunits is affected by estrogens (Stroupe *et al.*, 1991).

The structural genes for the $\alpha$ and $\beta$ meprin subunits have been localized in the mouse and human genomes by interspecific backcrossing methods and by the use of radiation and somatic cell hybrids (Jiang *et al.*, 1993; Gorbea *et al.*, 1993; Bond *et al.*, 1984, 1995). The $\alpha$ subunit is located on mouse chromosome 17 and human chromosome 6p, both linked to the histocompatibility complex; the $\beta$ structural gene is on mouse and human chromosomes 18. The structures of the genes are currently under investigation.

The role of meprin A in renal disease has been studied in rodent models. Meprin activities were found to be elevated in streptozocin-induced diabetes, and decreased in reaction to the toxin puromycin aminonucleoside (Trachtman *et al.*, 1993). Proteinurea was elevated in the mice administered the toxin, and it was suggested that the decreased meprin activity could contribute to the progressive renal injury. When different strains of mice were administered glycerol, or subjected to bilateral renal artery occlusion, the severity of the acute renal failure was much higher in 'high' compared to 'low' meprin mice (Trachtman *et al.*, 1995). Thus, meprin A is a potential mediator of cellular damage in response to toxic injury. Additionally, a recent study has demonstrated an abrupt downregulation of meprin expression during experimentally induced hydronephrosis (Ricardo *et al.*, 1996). It was proposed that the downregulation of meprins, which could protect cells from damage, could also contribute to the onset of fibrosis in the diseased kidney.

## Distinguishing Features

The complex oligomeric structure of meprins in which homo- or hetero- disulfide-linked dimers associate noncovalently with other dimers to form tetramers distinguishes meprins from other proteases. While many cell surface proteins exist in oligomeric forms, meprin's structure is thus far unique. Other membrane-bound metalloproteases can be distinguished from meprin in their responses to inhibitors; for example, neprilysin (Chapter 362) and endothelin-converting enzyme (Chapter 363) are inhibited by phosphoramidon (unlike meprin), angiotensin I-converting enzyme (Chapter 359) by captopril, aminopeptidases by bestatin, and membrane-type matrix metalloproteinases (Chapter 399) are inhibited by TIMPs. Less is known about the enzymatic properties of the disintegrins (e.g. fertilin, metargidin), membrane-bound metalloproteinases of the reprolysin subfamily (Chapter 447), however, these proteases do not have disulfide-linked subunits, and can be distinguished from meprins in this way.

Antibodies have been raised against meprin by several laboratories (Barnes *et al.*, 1989; Craig *et al.*, 1991; Johnson & Hersh, 1992; Marchand *et al.*, 1995). These antibodies have been used for immunoblotting, immunoprecipitation, purification and immunohistochemical localization.

## Further Reading

The structure and function of the astacin family enzymes, with emphasis on meprins, are reviewed in Bond & Beynon (1995). The purification and assay of meprins is detailed in Wolz & Bond (1995). The structure and biosynthesis of meprins are reviewed in Marchand & Bond (1996).

## References

Bankus, J. & Bond, J.S. (1996) Expression and distribution of meprin protease subunits in mouse intestine. *Arch. Biochem. Biophys.* **331**, 87–94.

Barnes, K., Ingram, J. & Kenny, A.J. (1989) Structural and immuno-chemical properties of rat endopeptidase-2 and its immunohisto-chemical localization in tissues of rat and mouse. *Biochem. J.* **264**, 335–346.

Beckmann, G. & Bork, P. (1993) An adhesive domain detected in functionally diverse receptors. *Trends Biochem. Sci.* **18**, 40–41.

Beynon, R.J. & Bond, J.S. (1983) Deficiency of a kidney metallo-proteinase activity in inbred mouse strains. *Science* **219**, 1351–1353.

Beynon, R.J. & Kay, J. (1978) The inactivation of native enzymes by a neutral proteinase from rat intestinal muscle. *Biochem. J.* **173**, 291–298.

Beynon, R.J., Shannon, J.D. & Bond, J.S. (1981) Purification and characterization of a metallo-endopeptidase from mouse kidney. *Biochem. J.* **199**, 591–598.

Beynon, R.J., Oliver, S. & Robertson, D.H.L. (1996) Characteriza-tion of the soluble, secreted form of urinary meprin. *Biochem. J.* **315**, 461–466

Bode, W., Gomis-Rüth, F.X., Huber, R., Zwilling, R. & Stöcker, W. (1992) Structure of astacin and implications for activation and zinc-ligation of collagenases. *Nature* **358**, 164–167.

Bond, J.S. & Beynon, R.J. (1986) Meprin: a membrane-bound metalloendo-peptidase. *Curr. Top. Cell. Regul.* **28**, 263–290.

Bond, J.S. & Beynon, R.J. (1995) The astacin family of metalloen-dopeptidases. *Protein Sci.* **4**, 1247–1261.

Bond, J.S., Beynon, R.J., Reckelhoff, J.F. & David, C.S. (1984) *Mep-1* gene controlling a kidney metalloendopeptidase is linked to the major histocompatibility complex in mice. *Proc. Natl Acad. Sci. USA* **81**, 5542–5545.

Bond, J.S., Rojas, K., Overhauser, J, Zoghbi, H.Y. & Jiang, W. (1995) The structural genes *MEP1A* and *MEP1B*, for the α and β subunits of the metalloproteinase meprin map to human chromo-some 6p and 18q, respectively. *Genomics* **25**, 300–303.

Butler, P.E. & Bond, J.S. (1988) A latent proteinase in mouse kidney membranes. *J. Biol. Chem.* **263**, 13419–13426.

Butler, P.E., McKay, M.J. & Bond, J.S. (1987) Characterization of meprin, a membrane-bound metalloendopeptidase from mouse kidney. *Biochem. J.* **241**, 229–235.

Cheng, G., Cleary, A.M., Ye, Z.S., Hong, D.I., Lederman, S. & Bal-timore, D. (1995) Involvement of CRAF1, a relative of TRAF, in CD40 signaling. *Science* **267**, 1494–1498.

Chevallier, S., Ahn, J., Boileau, G. & Crine, P. (1996) Identification of the cysteine residues implicated in the formation of α₂ and α/β dimers of rat meprin. *Biochem. J.* **317**, 731–738.

Corbeil, D., Gaudoux, F., Wainwright, S., Ingram, J., Kenny, A.J., Boileau, G. & Crine, P. (1992) Molecular cloning of the α-subunit of rat endopeptidase-24.18 (endopeptidase-2) and colocalization with endopeptidase-24.11 in rat kidney by in situ hybridization. *FEBS Lett.* **309**, 203–208.

Craig, S.S., Reckelhoff, J.F. & Bond, J.S. (1987) Distribution of meprin in kidneys from mice with high- and low-meprin activity. *Am. J. Physiol. Cell Physiol.* **253**, C535–540.

Craig, S.S., Mader, C. & Bond, J.S. (1991) Immunohistochemical localization of the metalloproteinase meprin in salivary glands of male and female mice. *J. Histochem. Cytochem.* **39**, 123–129.

Dumermuth, E., Sterchi, E.E., Jiang, W., Wolz, R.L., Bond, J.S., Flannery, A.V. & Beynon, R. J. (1991) The astacin family of metalloendopeptidases. *J. Biol. Chem.* **266**, 21381–21385.

Dumermuth, E., Eldering, J.A., Grünberg, J., Jiang, W. & Ster-chi, E.E. (1993) Cloning of the PABA peptide hydrolase α subunit (PPH-α) from human small intestine and its expression in COS-1

cells. *FEBS Lett.* **335**, 367–375.

Flannery, A.V., Dalzell, G.N. & Beynon, R.J. (1990) Proteolytic activity in mouse urine: relationship to the kidney metallo-endopeptidase, meprin. *Biochim. Biophys. Acta* **1041**, 64–70.

Gorbea, C.M., Flannery, A.V. & Bond, J.S. (1991) Homo- and het-erotetrameric forms of the membrane-bound metalloendopepti-dases meprin A and B. *Arch. Biochem. Biophys.* **290**, 549–553.

Gorbea, C.M., Marchand, P., Jiang, W., Copeland N.G., Gilbert, D.J., Jenkins, N.A. & Bond, J.S. (1993) Cloning, expression, and chro-mosomal localization of the mouse meprin β subunit. *J. Biol. Chem.* **268**, 21035–21043.

Hall, J.L., Sterchi, E.E. & Bond, J.S. (1993) Biosynthesis and degradation of meprins, kidney brush border proteinases. *Arch. Biochem. Biophys.* **307**, 73–77.

Hooper, N.M. (1994) Families of zinc metalloproteases. *FEBS Lett.* **354**, 1–6.

Jiang, W. & Bond, J.S. (1992) Families of metalloendopeptidases and their relationships. *FEBS Lett.* **312**, 110–114.

Jiang, W., Gorbea, C.M., Flannery, A.V., Beynon, R.J., Grant, G.A. & Bond, J.S. (1992) The a subunit of meprin A. Molecular cloning and sequencing, differential expression in inbred mouse strains, and evidence for divergent evolution of the α and β subunits. *J. Biol. Chem.* **267**, 9185–9193.

Jiang, W., Sadler, P.M., Jenkins, N.A., Gilbert, D.J., Copeland, N.G. & Bond, J.S. (1993) Tissue-specific expression and chromosomal localization of the α subunit of mouse meprin A. *J. Biol. Chem.* **268**, 10380–10385.

Johnson, G.D. & Hersh, L.B. (1992) Cloning a rat meprin cDNA reveals the enzyme is a heterodimer. *J. Biol. Chem.* **267**, 13505–13512. (Erratum *J. Biol. Chem.* 1993: **268**, 17647.)

Johnson, G.D. & Hersh, L.B. (1994) Expression of meprin subunit precursors. Membrane anchoring through the β subunit and mech-anism of zymogen activation. *J. Biol. Chem.* **269**, 7682–7688.

Kaushal, G.P., Walker, P.D. & Shah, S.V. (1994) An old enzyme with a new function: purification and characterization of a distinct matrix-degrading metalloproteinase in rat kidney cortex and its identification as meprin. *J. Cell Biol.* **126**, 1319–1327.

Kenny, A.J. & Ingram, J. (1987) Proteins of the kidney microvillar membrane. Purification and properties of the phosphoramidon-insensitive endopeptidase (endopeptidase-2) from rat kidney. *Biochem. J.* **245**, 515–524.

Kounnas, M.Z., Wolz, R.L., Gorbea, C.M. & Bond, J.S. (1991) Meprin-A and -B. Cell surface endopeptidases of the mouse kid-ney. *J. Biol. Chem.* **266**, 17350–17357.

Marchand, P. & Bond, J.S. (1996) Structure and biosynthesis of meprins. In: *Intracellular Protein Catabolism* (Suzuki, K. & Bond, J.S., eds). New York: Plenum Press, pp. 13–22.

Marchand, P., Tang, J. & Bond, J.S. (1994) Membrane association and oligomeric organization of the α and β subunits of mouse meprin A. *J. Biol. Chem.* **269**, 15388–15393.

Marchand, P., Tang, J., Johnson, G.D. & Bond, J.S. (1995) COOH-terminal proteolytic processing of secreted and membrane forms of the a subunit of the metalloprotease meprin A. Requirement of the I domain for processing in the endoplasmic reticulum. *J. Biol. Chem.* **270**, 5449–5456.

Marchand, P., Volkmann, M. & Bond, J.S. (1996) Cysteine muta-tions in the MAM domain result in monomeric meprin and alter stability and activity of the proteinase. *J. Biol. Chem.* **271**, 24236–24241.

Milhiet, P.E., Corbeil, D., Simon, V., Kenny, A.J., Crine, P. & Boileau, G. (1994) Expression of rat endopeptidase-24.18 in

COS-1 cells: membrane topology and activity. *Biochem. J.* **300**, 37–43.

Rawlings, N. & Barrett, A.J. (1995) Evolutionary families of metallopeptidases. *Methods Enzymol.* **248**, 183–228.

Reckelhoff, J.F., Bond, J.S., Beynon, R.J., Savarirayan, S. & David, C.S. (1985) Proximity of the *Mep-1* gene to H-2D on chromosome 17 in mice. *Immunogenetics* **22**, 617–623

Ricardo, S.D., Bond, J.S., Johnson, G.D., Kaspar, J. & Diamond, J.R. (1996) Expression of subunits of the metalloendopeptidase meprin in renal cortex in experimental hydronephrosis. *Am. J. Physiol.* **270**, F669–F676.

Spencer-Dene, B., Torogood, P., Nair, S., Kenny, A.J., Harris, M. & Henderson, B. (1994) Distribution of, and a putative role for, the cell-surface neutral metalloendopeptidases during mammalian craniofacial development. *Development* **120**, 3213–3226.

Sterchi, E.E., Green, J.R. & Lentze, M.J. (1982) Non-pancreatic hydrolysis of *N*-benzoyl-L-tyrosyl-*p*-aminobenzoic acid (PABA-peptide) in the human small intestine. *Clin. Sci.* **62**, 557–560.

Sterchi, E.E., Naim, H.Y., Lentze, M.J., Hauri, H.-P. & Fransen, J.A.M. (1988) *N*-benzoyl-L-tyrosyl-*p*-aminobenzoic acid hydrolase: a metalloendopeptidase of the human intestinal microvillus membrane which degrades biologically active peptides. *Arch. Biochem. Biophys.* **265**, 105–118.

Stöcker, W., Gomis-Rüth, F.X., Bode, W. & Zwilling, R. (1993) Implications of the three-dimensional structure of astacin for the structure and function of the astacin family of zinc-endopeptidases. *Eur. J. Biochem.* **214**, 215–231.

Stroupe, S.T., Craig, S.S., Gorbea, C.M. & Bond, J.S. (1991) Sex-related differences in meprin-A, a membrane-bound mouse kidney proteinase. *Am. J. Physiol.* **261**, E354–E361.

Trachtman, H., Greenwald, R., Moak, S., Tang, J. & Bond, J.S. (1993) Meprin activity in rats with experimental renal disease. *Life Sci.* **53**, 1339–1344.

Trachtman, H., Valderrama, E., Dietrich, J.M. & Bond, J.S. (1995) The role of meprin A in the pathogenesis of acute renal failure. *Biochem. Biophys. Res. Commun.* **208**, 498–505.

Uren, A.G. & Vaux, D.L. (1996) TRAF proteins and meprins share a conserved domain. *Trends Biochem. Sci.* **21**, 244–245.

Wolz, R.L. (1994) A kinetic comparison of the homologous proteases astacin and meprin A. *Arch. Biochem. Biophys.* **310**, 144–151.

Wolz, R.L. & Bond, J.S. (1995) Meprins A and B. *Methods Enzymol.* **248**, 325–345.

Wolz, R.L., Harris, R.B. & Bond, J.S. (1991) Mapping the active site of meprin-A with peptide substrates and inhibitors. *Biochemistry* **30**, 8488–8493.

Yamaguchi, T., Kido, H., Fukase, T., Fujita, T. & Katunuma, N. (1991) A membrane-bound metallo-endopeptidase from rat kidney hydrolyzing parathyroid hormone. Purification and characterization. *Eur. J. Biochem.* **200**, 563–571.

Yamaguchi, T., Fukase, M., Sugimoto, T., Kido, H. & Chihara, K. (1994) Purification of meprin from human kidney and its role in parathyroid hormone degradation. *Biol. Chem. Hoppe-Seyler* **375**, 821–824

Zondag, G.C.M., Koningstein, G.M., Jiang, Y.-P., Sap, J., Moolenaar, W.H. & Gebbink, M.F.B.G. (1995) Homophilic interactions mediated by receptor tyrosine phosphatases μ and κ. *J. Biol. Chem.* **270**, 14247–14250.

***Gary D. Johnson***
*Department of Biochemistry and Molecular Biology,*
*Penn State University College of Medicine,*
*Hershey, PA 17033-0850, USA*
*Email: jbond@bcmic.hmc.psu.edu*

***Judith S. Bond***
*Department of Biochemistry and Molecular Biology,*
*Penn State University College of Medicine,*
*Hershey, PA 17033-0850, USA*
*Email: jbond@bcmic.hmc.psu.edu*

# *407. Meprin B*

## Databanks

*Peptidase classification: clan MB, family M12, MEROPS ID: M12.004*
*NC-IUBMB enzyme classification: EC 3.4.24.63*
*ATCC entries: 977145 (mouse, β subunit)*
*Databank codes:*

| Species | SwissProt | PIR | EMBL (cDNA) | EMBL (genomic) |
|---|---|---|---|---|
| *Mus musculus* | – | – | L15193 | – |
| *Rattus norvegicus* | P28826 | – | M88601 | – |

## Name and History

**Meprin B** (EC 3.4.24.63) is found in the kidney in a number of mouse strains (e.g. C3H/He) in which there is little or no expression of meprin α subunits (Reckelhoff *et al.*, 1985). Meprin B, which contains only β subunits, has also been identified in the duodenum of all mouse strains tested (Flannery *et al.*, 1991; Gorbea *et al.*, 1993; Bankus & Bond, 1996). It was initially identified as an isoform of meprin with low azocaseinase activity and described as a latent proteinase with activity that could be dramatically increased by activation with trypsin (Butler & Bond, 1988). The basis of this latency became evident with the cloning and expression of meprin β subunits, as it was realized that mouse meprin B is composed of β subunits that retain their N-terminal pro sequences in the kidney, rendering the protease domain inactive (Gorbea *et al.*, 1993; Johnson & Hersh, 1994; Milhiet *et al.*, 1994).

## Activity and Specificity

Meprin B is assayed with azocasein under the same conditions as those used for meprin A (Chapter 406). Trypsin-activated meprin B has a specific activity for azocasein comparable to that of meprin A (Wolz & Bond, 1995). When the oxidized insulin B chain was used as a substrate, it was found that meprin A and B have some cleavage sites in common, however, meprin B hydrolyzes only a subset of the peptide bonds hydrolyzed by meprin A and cleaves one distinct bond, Cya19↓Gly20 (Kounnas *et al.*, 1991):

FVNQH↓L↓CGSHLVEA↓LYLVC↓GERGFFYTPKA

Meprin B also has little activity against shorter peptides that are commonly used to assay meprin A, including bradykinin analogs and arylamides (Kounnas *et al.*, 1991; Wolz & Bond, 1995).

## Structural Chemistry

Meprin B is similar to meprin A in that it is a zinc metalloendopeptidase of the astacin family (M12), is highly glycosylated, and is composed of disulfide-linked dimers which form higher order oligomers (Gorbea *et al.*, 1991). By definition, meprin B contains only β subunits. Unlike α subunits, β subunits lack the I-domain, and are therefore not proteolytically processed to remove the C-terminal transmembrane domain during biosynthesis (Marchand *et al.*, 1994, 1995). Therefore, meprin B is always found as an integral membrane protein.

## Preparation

Meprin B is isolated from the kidneys of C3H/He mice, which do not express meprin α subunits, by the method of Kounnas *et al.* (1991). The initial stages of the purification are identical to those employed for meprin A; two additional steps are required to remove the leucyl aminopeptidase and angiotensin I-converting enzyme (ACE) that tend to copurify with meprin B. The first is an additional chromatography step using Mono-Q anion-exchange resin, the second step is affinity chromatography on lisinopril-Sepharose-6B (Ehlers *et al.*, 1989). Lisinopril (Nα-[(S)-1-carboxy-3-phenylpropyl]-L-lysyl-L-proline) is an active site-directed inhibitor of ACE, so it is effective in removing the ACE that copurifies with meprin B in the other purification steps.

## Biological Aspects

Some mouse strains are deficient in the expression of meprin α, but there are no known strains lacking meprin β (Flannery *et al.*, 1991). The significance of this is not clear, but it may indicate that the meprin β subunit has an indispensable function in these animals. Because meprin B is found in mice as a latent proteinase, it is possible that its activity is highly regulated in response to specific events. However, meprin B could also perform important functions independent of its protease activity. The mature β subunit possesses several extracellular domains that are either known to be, or postulated to be, involved in protein–protein interactions. These are the MAM, MATH and EGF-like domains (as defined in Chapter 406), and their presence in meprin B may allow this protein to function as a cell surface receptor for one or more as yet unidentified proteins or peptides or to organize other cell surface proteins into functional units via protein–protein interactions.

A novel meprin β subunit mRNA, called β′, has been isolated from embryonal carcinoma and human colon adenocarcinoma cell lines (Dietrich *et al.*, 1996). This mRNA encodes a meprin β subunit that differs from the β subunit only in the region of the signal sequence and the propeptide. The functional consequences of these changes have yet to be determined, but it is possible that meprin β may be involved in the development or maintenance of cancer.

## Distinguishing Features

Meprin B is distinguished from other integral membrane proteins by its complex oligomeric structure (disulfide-linked homodimers associating noncovalently with similar dimers to form tetramers), and by its inhibitor profile which is similar to that of meprin A (Chapter 406). Meprin B differs from meprin A in that it does not degrade small peptides (less than 10 amino acids, including bradykinin and acylamides), and displays some differences in peptide bond specificity using the oxidized insulin B chain as substrate (Kounnas *et al.*, 1991).

Specific polyclonal antibodies are available to the meprin β protein isolated from C3H/He mouse and rat kidney, to recombinant meprin mouse β subunit protease domain, and to the cytoplasmic domain of the mouse β subunit (Johnson & Hersh, 1992; Marchand *et al.*, 1994, 1995).

## Further Reading

The structure and function of the astacin family, with emphasis on meprins, is reviewed in Bond & Beynon (1995). The structure and biosynthesis of meprins is reviewed in Marchand & Bond (1996). Details of purification and assay may be found in Wolz & Bond (1995).

## References

Bankus, J. & Bond, J.S. (1996) Expression and distribution of meprin protease subunits in mouse intestine. *Arch. Biochem. Biophys.* **331**, 87–94.

Bond, J.S. & Beynon, R.J. (1995) The astacin family of metallo-endopeptidases. *Protein Sci.* **4**, 1247–1261.

Butler, P.E. & Bond, J.S. (1988) A latent proteinase in mouse kidney membranes. *J. Biol. Chem.* **263**, 13419–13426.

Dietrich, J.M., Jiang, W. & Bond, J.S. (1996) A novel meprin β' mRNA in mouse embryonal carcinoma and human colon carcinoma cells. *J. Biol. Chem.* **271**, 2271–2278.

Ehlers, M.R., Fox, E.A., Strydom, D.J. & Riordan, J.F. (1989) Molecular cloning of human testicular angiotensin-converting enzyme: the testis isozyme is identical to the C-terminal half of endothelial angiotensin-converting enzyme. *Proc. Natl Acad. Sci. USA* **86**, 7741–7745.

Flannery, A.V., Macadam, G.C. & Beynon, R.J. (1991) Immunological characterization of different meprin species in mice. *Biochim. Biophys. Acta* **1079**, 119–122.

Gorbea, C.M., Flannery, A.V. & Bond, J.S. (1991) Homo- and heterotetrameric forms of the membrane-bound metalloendopeptidases meprin A and B. *Arch. Biochem. Biophys.* **290**, 549–553.

Gorbea, C.M., Marchand, P., Jiang, W., Copeland, N.G., Gilbert, D.J., Jenkins, N.A. & Bond, J.S. (1993) Cloning, expression, and chromosomal localization of the mouse meprin β subunit. *J. Biol. Chem.* **268**, 21035–21043.

Johnson, G.D. & Hersh, L.B. (1992) Cloning a rat meprin cDNA reveals the enzyme is a heterodimer. *J. Biol. Chem.* **267**, 13505–13512. (Erratum *J. Biol. Chem.* 1993: **268**, 17647.)

Johnson, G.D. & Hersh, L.B. (1994) Expression of meprin subunit precursors. Membrane anchoring through the β subunit and mechanism of zymogen activation. *J. Biol. Chem.* **269**, 7682–7688.

Kounnas, M.Z., Wolz, R.L., Gorbea, C.M. & Bond, J.S. (1991) Meprin-A and -B. Cell surface endopeptidases of the mouse kidney. *J. Biol. Chem.* **266**, 17350–17357.

Marchand, P. & Bond, J.S. (1996) Structure and biosynthesis of meprins. In: *Zinc Metalloproteases in Health and Disease* (Hooper, N.M, ed.). London: Taylor & Frances, pp. 23–45.

Marchand, P., Tang, J. & Bond, J.S. (1994) Membrane association and oligomeric organization of the α and β subunits of mouse meprin A. *J. Biol. Chem.* **269**, 15388–15393.

Marchand, P., Tang, J., Johnson, G.D. & Bond, J.S. (1995) COOH-terminal proteolytic processing of secreted and membrane forms of the α subunit of the metalloprotease meprin A. Requirement of the I domain for processing in the endoplasmic reticulum. *J. Biol. Chem.* **270**, 5449–5456.

Milhiet, P.E., Corbeil, D., Simon, V., Kenny, A.J., Crine, P. & Boileau, G. (1994) Expression of rat endopeptidase-24.18 in COS-1 cells: membrane topology and activity. *Biochem. J.* **300**, 37–43.

Reckelhoff, J.F, Bond, J.S., Beynon, R.J., Savarirayan, S. & David, C.S. (1985) Proximity of the *Mep-1* gene to H-2D on chromosome 17 in mice. *Immunogenetics* **22**, 617–623.

Wolz, R.L. & Bond, J.S. (1995) Meprins A and B. *Methods Enzymol.* **248**, 325–345.

***Gary D. Johnson***
*Department of Biochemistry and Molecular Biology,*
*Penn State University College of Medicine,*
*Hershey, PA 17033-0850, USA*

***Judith S. Bond***
*Department of Biochemistry and Molecular Biology,*
*Penn State University College of Medicine,*
*Hershey, PA 17033-0850, USA*
*Email: jbond@bcmic.hmc.psu.edu*

# 408. *Human meprin*

## Databanks

*Peptidase classification: clan MB, family M12, MEROPS ID: M12.004*
*NC-IUBMB enzyme classification: none*
*ATCC entries: 982943 (human, β subunit)*
*Databank codes:*

| Species | SwissProt | PIR | EMBL (cDNA) | EMBL (genomic) |
|---|---|---|---|---|
| α subunit | | | | |
| *Homo sapiens* | Q16819 | – | M82962 | – |
| β subunit | | | | |
| *Homo sapiens* | Q16820 | – | X81333 | – |

## Name and History

Bz-Tyr-*p*-aminobenzoic acid (Paba-peptide) is a substrate used to assess exocrine pancreas function (Imondi *et al*., 1972). After oral administration of the substrate, *p*-aminobenzoate is released in the small intestine by pancreatic chymotrypsin (Chapter 8) and excreted in the urine. Residual recovery of *p*-aminobenzoate observed in urine of patients with severe or total pancreatic insufficiency and in urine of bile duct-ligated rats led to the discovery of a **Paba-peptide hydrolase (PPH)** activity in the mucosa of small intestine (Sterchi *et al*., 1982, 1983). PPH was subsequently isolated from microvillus membranes of intestinal epithelial cells and partially characterized (Sterchi *et al*., 1988a,b). Cloning of PPH revealed a polypeptide very similar to mouse meprin (Chapter 406), human procollagen C-endopeptidase (Chapter 411), and UVS2 (Chapter 413) from *Xenopus laevis* and led to the definition of the astacin family (M12) of zinc metalloendopeptidases (Dumermuth *et al*., 1991).

The primary structure of rodent meprins (Corbeil *et al*., 1992; Jiang *et al*., 1992; Johnson & Hersh, 1992; Gorbea *et al*., 1993) and PPH (Dumermuth *et al*., 1993) were subsequently elucidated. Rodent meprin and human PPH consist of two subunits, α and β, with extensive sequence similarities. Thus, PPH is presently best referred to as **human meprin** (Bond & Beynon, 1995), but PPH will be used here for convenience.

## Activity and Specificity

PPH is synthesized as an inactive zymogen and requires proteolytic processing of the N-terminal propeptide to gain enzymatic activity. In the human gut propeptide removal is most likely catalyzed by luminal trypsin. PPH isolated from gut mucosa cleaves a number of biologically active peptides and proteins including angiotensin I and II, bradykinin, substance P, insulin B chain and azocasein. Hydrolysis of these substrates occurs preferentially after an aromatic amino acid. PPH is assayed using 20 mM *N*-Bz-L-Tyr-*p*-aminobenzoate in 50 mM Tris–HCl/1 mM MgCl$_2$, pH 7.5. Studies with transfected mammalian cells indicate that PPHα is primarily responsible for hydrolysis of small peptide substrates while PPHβ prefers larger polypeptides or proteins such as azocasein. Rodent meprin has been studied extensively using different substrates (Butler *et al*., 1987; Kenny & Ingram, 1987; Wolz & Bond, 1990; Wolz *et al*., 1991), and degrades extracellular matrix proteins (Kaushal *et al*., 1994) and protein kinase A (Chestukhin *et al*., 1996). An enzyme from human kidney with almost identical properties to rat meprin degrades parathyroid hormone (Yamaguchi *et al*., 1994). Extensive comparative studies of substrate specificities of the rodent and human enzymes are needed to shed more light on the function of these enzymes.

## Structural Chemistry

PPH consists of two distinct but related subunits, α and β, with an overall sequence identity of over 40% and a similarity score of about 60%. Both subunits are composed of several distinct domains starting with an N-terminal leader sequence and a propeptide sequence preceding the protease domain. The catalytic astacin domain is followed by a MAM or adhesive domain (Beckman & Bork, 1993), a MATH (**m**eprin **a**nd **T**RAF **h**omology) domain (Uren & Vaux, 1996), an intervening sequence, an epidermal growth factor (EGF)-like domain, a transmembrane sequence and a C-terminal cytosolic tail. Overall similarity is considerably less in the C-terminal third of the molecule. The α subunit has an additional inserted sequence (I-domain) of 58 amino acids preceding the EGF-like domain, which appears essential for proteolytic processing (see section on Distinguishing Features). The two subunits also differ in their transmembrane sequence and the cytosolic domain. The cytosolic domain of PPHβ is 24 amino acids long compared to the six amino acids of PPHα and contains a consensus sequence for phosphorylation by protein kinase C.

The protease domain of PPH, in common with all astacin family members, is approximately 200 residues long and contains a zinc-binding site motif (HEXXH) within an extended highly conserved sequence of HEXXHXXGFXHE, which clearly distinguishes the astacins from metalloendopeptidases outside clan MB. A comparison of the mouse meprin protease domain with the X-ray structure of astacin provides some structural insights (Stöcker *et al*., 1993). Astacin has a kidney-shaped structure consisting of an N-terminal and a C-terminal domain separated by a deep active-site cleft. The zinc ion is located at the bottom of this cleft and is bound to three histidines, a tyrosine and a glutamic acid residue via a water molecule bound to its carboxyl side group in a trigonal bipyramidal arrangement.

The gene (*MEP1A*) for the α subunit is on human chromosome 6p21.2–p21.1 telomeric to *GSTA2* and closely linked to the HLA loci encoding proteins of the major histocompatibility complex. The structural gene (*MEP1B*) for the β subunit was mapped to region 18q12.2–q12.3 proximal to the *TTR/PALB* gene on human chromosome 18 (Jiang *et al*., 1995).

## Preparation

PPH has been isolated from human small intestinal mucosa using subcellular fractionation and immunoaffinity batch chromatography. The average yield was 6.8%, the enrichment factor 765-fold (Sterchi *et al*., 1988b). Inactive zymogen forms of the two subunits may be immunoisolated from transfected MDCK or COS-1 cells.

## Biological Aspects

PPH is present throughout the human small intestine with an enzyme gradient increasing from proximal duodenum to distal ileum (Sterchi *et al*., 1988b). PPH mRNA has also been detected in human colon and colon tumor tissue (Sterchi *et al*., 1994) and recently a different mRNA has been detected in a subclone of a human intestinal adenocarcinoma cell line (Dietrich *et al*., 1996). A metalloendopeptidase cross-reacting with antibodies raised against rat meprin has been purified from human kidney (Yamaguchi *et al*., 1994).

Biosynthesis of PPH has been investigated in cultured small intestinal explants (Sterchi *et al*., 1988a) and in transfected mammalian cell lines (Grünberg *et al*., 1993). The two subunits are processed differently in that the α subunit requires C-terminal proteolytic processing as an early post-translational event leading to a truncated form lacking the transmembrane and cytosolic domains. The truncated α subunit is completely secreted from cells transfected with PPHα cDNA. The I-domain, which is only present in the α subunit, appears essential for this processing. The β subunit predominantly retains the transmembrane domain and is inserted into the apical microvillus membrane of MDCK cells. When both subunits are expressed together the α subunit is bound to the β subunit via disulfide bridges and also retained at the apical membrane of epithelial cells. In contrast to rodent meprin, a proportion of human PPHβ is also proteolytically processed. This processing is different to PPHα (Eldering *et al*., 1997). The significance of this difference in β subunit processing is interesting and suggests that the human and the rodent enzymes may be regulated differently. The physiological function of human PPH is presently not known. It may serve in digestion processes or it may be involved in processing of biologically active peptides such as growth factors and thus influence proliferation and differentiation of small intestinal epithelial cells.

## Further Reading

For a review, see Bond & Beynon (1995).

## References

Beckman, G. & Bork, P. (1993) An adhesive domain detected in functionally diverse receptors. *Trends Biochem. Sci.* **18**, 40–41.

Bond, J.S. & Beynon, R.J. (1995) The astacin family of metalloendopeptidases. *Protein Sci.* **4**, 1247–1261.

Butler, P.E., McKay, M.J. & Bond, J.S. (1987) Characterization of meprin, a membrane-bound metalloendopeptidase from mouse kidney. *Biochem. J.* **241**, 229–235.

Chestukhin, A., Muradov, K., Litovchick, L. & Shaltiel, S. (1996) The cleavage of protein kinase A by the kinase-splitting membranal protease is reproduced by meprin B. *J. Biol. Chem.* **271**, 30272–30280.

Corbeil, D., Gaudoux, F., Wainwright, S., Ingram, J., Kenny, A.J., Boileau, G. & Crine, P. (1992) Molecular cloning of the alpha-subunit of rat endopeptidase-24.18 (endopeptidase-2) and co-localization with endopeptidase-24.11 in rat kidney by *in situ* hybridization. *FEBS Lett.* **309**, 203–208.

Dietrich, J., Jiang, W. & Bond, J. (1996) A novel meprin beta′ mRNA in mouse embryonal and human colon carcinoma cells. *J. Biol. Chem.* **271**, 2271–2278.

Dumermuth, E., Sterchi, E.E., Jiang, W.P., Wolz, R.L., Bond, J.S., Flannery, A.V. & Beynon, R.J. (1991) The astacin family of metalloendopeptidases. *J. Biol. Chem.* **266**, 21381–21385.

Dumermuth, E., Eldering, J.A., Grünberg, J., Jiang, W. & Sterchi, E.E. (1993) Cloning of the PABA peptide hydrolase alpha subunit (PPH alpha) from human small intestine and its expression in COS-1 cells. *FEBS Lett.* **335**, 367–375.

Eldering, J.A., Grunberg, J., Hahn, D., Croes, H.J., Fransen, J.A. & Sterchi, E.E. (1997) Polarised expression of human intestinal N-benzoyl-L-tyrosyl-p-aminobenzoic acid hydrolase (human meprin) alpha and beta subunits in Madin–Darby canine kidney cells. *Eur. J. Biochem.* **247**, 920–932.

Gorbea, C.M., Marchand, P., Jiang, W., Copeland, N.G., Gilbert, D.J., Jenkins, N.A. & Bond, J.S. (1993) Cloning, expression, and chromosomal localization of the mouse meprin beta subunit. *J. Biol. Chem.* **268**, 21035–21043.

Grünberg, J., Dumermuth, E., Eldering, J.A. & Sterchi, E.E. (1993) Expression of the alpha subunit of PABA peptide hydrolase (EC 3.4.24.18) in MDCK cells. Synthesis and secretion of an enzymatically inactive homodimer. *FEBS Lett.* **335**, 376–379.

Imondi, A., Stradley, R. & Wolgemuth, R. (1972) Synthetic peptides in the diagnosis of exocrine pancreatic insufficiency in animals. *Gut* **13**, 726–731.

Jiang, W., Gorbea, C.M., Flannery, A.V., Beynon, R.J., Grant, G.A. & Bond, J.S. (1992) The alpha subunit of meprin A. Molecular cloning and sequencing, differential expression in inbred mouse strains, and evidence for divergent evolution of the alpha and beta subunits. *J. Biol. Chem.* **267**, 9185–9193.

Jiang, W., Dewald, G., Brundage, E., Mucher, G., Schildhaus, H.U., Zerres, K. & Bond, J.S. (1995) Fine mapping of *MEP1A*, the gene encoding the alpha subunit of the metalloendopeptidase meprin, to human chromosome 6P21. *Biochem. Biophys. Res. Commun.* **216**, 630–635.

Johnson, G.D. & Hersh, L.B. (1992) Cloning a rat meprin cDNA reveals the enzyme is a heterodimer. *J. Biol. Chem.* **267**, 13505–13512.

Kaushal, G.P., Walker, P.D. & Shah, S.V. (1994) An old enzyme with a new function: purification and characterization of a distinct matrix-degrading metalloproteinase in rat kidney cortex and its identification as meprin. *J. Cell Biol.* **126**, 1319–1327.

Kenny, A.J. & Ingram, J. (1987) Proteins of the kidney microvillar membrane. Purification and properties of the phosphoramidon-insensitive endopeptidase ('endopeptidase-2') from rat kidney. *Biochem. J.* **245**, 515–524.

Sterchi, E.E., Green, J.R. & Lentze, M.J. (1982) Non-pancreatic hydrolysis of N-benzoyl-L-tyrosyl-p-aminobenzoic acid (PABA-peptide) in the human small intestine. *Clin. Sci.* **62**, 557–60.

Sterchi, E.E., Green, J.R. & Lentze, M.J. (1983) Nonpancreatic hydrolysis of N-benzoyl-L-tyrosyl-p-aminobenzoic acid (PABA peptide) in the rat small intestine. *J. Pediatr. Gastroenterol. Nutr.* **2**, 539–547.

Sterchi, E.E., Naim, H.Y. & Lentze, M.J. (1988a) Biosynthesis of N-benzoyl-L-tyrosyl-p-aminobenzoic acid hydrolase: disulfide-linked dimers are formed at the site of synthesis in the rough endoplasmic reticulum. *Arch. Biochem. Biophys.* **265**, 119–127.

Sterchi, E.E., Naim, H.Y., Lentze, M.J., Hauri, H.P. & Fransen, J.A. (1988b) N-benzoyl-L-tyrosyl-p-aminobenzoic acid hydrolase: a metalloendopeptidase of the human intestinal microvillus membrane which degrades biologically active peptides. *Arch. Biochem. Biophys.* **265**, 105–118.

Sterchi, E., Dumermuth, E., Eldering, J. & Grünberg, J. (1994) Paba-peptide hydrolase (PPH): structure and expression of a metalloendopeptidase from human small intestinal epithelial cells. In: *Mammalian Brush Border Membrane Proteins II* (Lentze, M.J., Naim, H.Y. & Grand, R., eds). Stuttgart: Georg Thième, pp. 141–151.

Stöcker, W., Gomis-Rüth, F.X., Bode, W. & Zwilling, R. (1993) Implications of the three-dimensional structure of astacin for the structure and function of the astacin family of zinc-endopeptidases. *Eur. J. Biochem.* **214**, 215–231.

Uren, A. & Vaux, D. (1996) TRAF proteins and meprins share a conserved domain. *Trends Biochem. Sci.* **21**, 244–245.

Wolz, R.L. & Bond, J.S. (1990) Phe5(4-nitro)-bradykinin: a chromogenic substrate for assay and kinetics of the metalloendopeptidase meprin. *Anal. Biochem.* **191**, 314–320.

Wolz, R.L., Harris, R.B. & Bond, J.S. (1991) Mapping the active site of meprin-A with peptide substrates and inhibitors. *Biochemistry* **30**, 8488–8493.

Yamaguchi, T., Fukase, M., Sugimoto, T., Kido, H. & Chihara, K. (1994) Purification of meprin from human kidney and its role in parathyroid hormone degradation. *Biol. Chem. Hoppe-Seyler* **375**, 821–824.

*Erwin E. Sterchi*
*University of Berne,*
*Institute of Biochemistry and Molecular Biology,*
*Buehlstrasse 28,*
*CH-3012 Berne, Switzerland*
*Email: sterchi@mci.unibe.ch*

# *409. Choriolysin L*

## Databanks

*Peptidase classification: clan MB, family M12, MEROPS ID: M12.006*
*NC-IUBMB enzyme classification: EC 3.4.24.66*
*Databank codes:*

| Species | SwissProt | PIR | EMBL (cDNA) | EMBL (genomic) |
|---------|-----------|-----|-------------|----------------|
| *Oryzias latipes* | P31579 | – | M96169 | D83949: complete gene |

## Name and History

This enzyme was discovered and partly characterized as one of two constituent proteases of the hatching enzyme of a teleost fish, the medaka (*Oryzias latipes*), and has been called **low choriolytic enzyme** (**LCE**) (Yasumasu *et al.*, 1988, 1989a,b). The fish hatching enzyme, initially discovered by Moriwaki (1910) and Wintrebert (1912) in salmoid fishes, is synthesized in and secreted from the hatching gland cell of embryos during development and dissolves the inner layer of the egg envelope (chorion) to facilitate hatching of the embryos (Yamagami, 1988). The medaka hatching enzyme was first found by Ishida (1944) and partly purified by Yamagami (1972), but has recently been shown to consist of two constituent proteases as mentioned above. The fish hatching enzyme was tentatively called **chorionase** (Kaighn, 1964; Yamagami, 1973), since it digests 'chorion'.

Among the hatching enzymes of various animal species, the medaka enzyme is the only one that is confirmed to be an enzyme system consisting of two component proteases: **choriolysin L** (low choriolytic enzyme, LCE) and choriolysin H (high choriolytic enzyme, HCE) (Chapter 410). Although the name choriolysin L has been adopted following the recommendation of the NC-IUBMB for a component of the medaka hatching enzyme with low choriolytic activity, the more general term choriolysin might be used for other fish hatching proteases such as the zebrafish choriolysin, as all of these digest chorion.

## Activity and Specificity

Choriolysin L displays very low choriolytic activity when examined by the turbidimetric method (Yamagami, 1970; Yasumasu *et al.*, 1988). It hardly digests the intact inner layer of egg envelope, but efficiently digests the swollen inner layer of egg envelope prepared by a limited digestion by choriolysin H. The specificity (swollen chorion-digesting activity/caseinolytic activity) of choriolysin L is 8000 and 10 000 times greater than those of trypsin and thermolysin, respectively. The hatching of medaka embryos results from a two-step digestion of the two enzymes; choriolysin H swells the inner layer of egg envelope by limited digestion, and choriolysin L solubilizes the swollen part of it completely. Choriolysin L also hydrolyzes casein and some synthetic substrates such as Suc-Leu-Leu-Val-Tyr+NHMec (Yasumasu *et al.*, 1989b).

The pH optimum is about 8.6 for caseinolytic activity. The activity is inhibited by EDTA at a concentration of 1 mM, but not inhibited by DFP, iodoacetamide or iodoacetic acid. EDTA-inactivated apoenzyme is reactivated by a low concentration of zinc ion and partly reactivated by a high concentration of copper ion (Yasumasu *et al*., 1989b).

## Structural Chemistry

Choriolysin L is a single-chain protein of about 25.5 kDa and pI 9.8. The enzyme is synthesized as a preproenzyme and the mature enzyme consists of 200 amino acids (Yasumasu *et al*., 1992b). The activation is inhibited by EDTA. According to the similarity of the amino acid sequence to that of astacin (Chapter 405), whose three-dimensional structure is well elucidated (Bode *et al*., 1992), three residues of His in the sequence **HELLHALGFYHE** and a Tyr in the sequence **IMHYG** may be the zinc ligands.

## Preparation

Choriolysin L is purified from the hatching liquid, a filtrate of culture medium after larvae have hatched out, through three repeats of a gel-filtration column (Toyopearl HW50SF) chromatography and HPLC with a CM-300 cation-exchange column (Yasumasu *et al*., 1989b).

## Biological Aspects

Choriolysin L is colocalized with choriolysin H in the secretory granules of the hatching gland cells which are located in the inner wall of the pharyngeal cavity of prehatching medaka embryos. Immunocytochemical study reveals that the intragranular arrangements of choriolysin L and choriolysin H are different: choriolysin L is localized at the periphery of the secretory granules, while choriolysin H is evenly distributed in the same granules (Yasumasu *et al*., 1992a).

The gene for choriolysin L comprises eight exons and seven introns. The genes for choriolysin L and choriolysin H are expressed synchronously. A putative promoter region of the gene contains the TATA box sequence, but not the CAT box nor GC box sequences (Yasumasu *et al*., 1996).

## Distinguishing Features

Though choriolysin L and choriolysin H play distinct roles in hydrolysis of the egg envelope and are different in gene organization, their amino acid sequences are similar to each other, with a sequence identity of 55% (Yasumasu *et al*., 1992b, 1994).

## Further Reading

For a review, see Yamagami *et al*. (1992).

## References

Bode, W., Gomis-Rüth, F.X., Huber, R., Zwilling, R. & Stöcker, W. (1992) Structure of astacin and implications for activation of astacins and zinc-ligation of collagenases. *Nature* **358**, 164–167.

Ishida, J. (1944) Hatching enzyme in the fresh-water fish, *Oryzias latipes. Annot. Zool. Jpn* **22**, 137–154.

Kaighn, M.E. (1964) A biochemical study of the hatching process in *Fundulus heteroclitus. Dev. Biol.* **9**, 56–80.

Moriwaki, I. (1910) The mechanism of escape of the fry out of the egg chorion in dog salmon. *3rd Report Hokkaido Fish. Res. Stn* 524–531.

Wintrebert, P. (1912) Le mécanisme de l'éclosion chez la truite arc-en-ciel [The mechanism of hatching of the rainbow trout]. *C.R. Seances Soc. Biol. Fil.* **72**, 724–727.

Yamagami, K. (1970) A method for rapid and quantitative determination of the hatching enzyme (chorionase) activity of the medaka, *Oryzias latipes. Annot. Zool. Jpn* **43**, 1–9.

Yamagami, K. (1972) Isolation of choriolytic enzyme (hatching enzyme) of the teleost, *Oryzias latipes. Dev. Biol.* **29**, 343–348.

Yamagami, K. (1973) Some enzymological properties of a hatching enzyme (chorionase) isolated from the fresh water teleost, *Oryzias latipes. Comp. Biochem. Physiol.* **46B**, 603–616.

Yamagami. K. (1988) Mechanism of hatching in fish. In: *Fish Physiology*, vol. XIA (Hoar, W.S. & Randall, D.J., eds). San Diego: Academic Press, pp. 447–499.

Yamagami, K., Hamazaki, T.S., Yasumasu, S., Masuda K. & Iuchi, I. (1992) Molecular and cellular basis of formation, hardening, and breakdown of the egg envelope in fish, *Int. Rev. Cytol.* **136**, 51–92

Yasumasu, S., Iuchi, I. & Yamagami, K. (1988) Medaka hatching enzyme consists of two kinds of proteases which act cooperatively. *Zool. Sci.* **5**, 191–195.

Yasumasu, S., Iuchi, I. & Yamagami, K. (1989a) Purification and partial characterization of high choriolytic enzyme, a component of the hatching enzyme of the teleost, *Oryzias latipes. J. Biochem.* **105**, 204–211.

Yasumasu, S., Iuchi, I. & Yamagami, K. (1989b) Isolation and some properties of low choriolytic enzyme (LCE), a component of the hatching enzyme of the teleost, *Oryzias latipes. J. Biochem.* **105**, 212–218.

Yasumasu, S., Katow, S., Hamazaki, T.S., Iuchi, I. & Yamagami, K. (1992a) Two constituent proteases of a teleostean hatching enzyme: concurrent syntheses and packaging in the same secretory granules in discrete arrangement. *Dev. Biol.* **149**, 349–356.

Yasumasu, S., Yamada, K., Akasaka, K., Mitsunaga, K., Iuchi, I., Shimada, H. & Yamagami, K. (1992b) Isolation of cDNAs for LCE and HCE, two constituent proteases of the hatching enzyme of *Oryzias latipes*, and concurrent expression of their mRNAs during development. *Dev. Biol.* **153**, 250–258.

Yasumasu, S., Iuchi, I. & Yamagami, K. (1994) cDNAs and the genes of HCE and LCE, two constituents of the medaka hatching enzyme. *Dev. Growth Differ.* **36**, 241–250.

Yasumasu, S., Shimada, H., Inohaya, K., Yamazaki, K., Iuchi, I., Yasumasu, I. & Yamagami, K. (1996) Different exon–intron organizations of the genes for two astacin-like proteases, high choriolytic enzyme (choriolysin H) and low choriolytic enzyme (choriolysin L), the constituents of the fish hatching enzyme. *Eur. J. Biochem.* **237**, 752–758.

*Shigeki Yasumasu*
*Life Science Institute,*
*Sophia University,*
*7-1 Kioi-cho, Chiyoda-ku, Tokyo 102, Japan*
*Email: y-sigeki@hoffman.cc.sophia.ac.jp*

*Kenjiro Yamagami*
*Life Science Institute,*
*Sophia University,*
*7-1 Kioi-cho, Chiyoda-ku, Tokyo 102, Japan*

# 410. *Choriolysin H*

## Databanks

Peptidase classification: clan MB, family M12, MEROPS ID: M12.007
NC-IUBMB enzyme classification: EC 3.4.24.67
Databank codes:

| Species | Type | SwissProt | PIR | EMBL (cDNA) | EMBL (genomic) |
|---|---|---|---|---|---|
| *Oryzias latipes* | H1 | P31580 | – | M96170 | D83950 complete gene |
| *Oryzias latipes* | H2 | P31581 | – | M96171 | – |

## Name and History

This enzyme was first discovered and partly characterized as one of two constituent proteases of the hatching enzyme of a teleost fish, the medaka (*Oryzias latipes*), and has been called **high choriolytic enzyme (HCE)** (Yasumasu *et al.*, 1988, 1989a,b). The fish hatching enzyme, initially discovered by Moriwaki (1910) and Wintrebert (1912) in salmoid fishes, is synthesized in, and secreted from, the hatching gland cell of embryos during development, and dissolves the inner layer of the egg envelope (chorion) to facilitate hatching of the embryos (Yamagami, 1988). The medaka hatching enzyme was found by Ishida (1944) and partly purified by Yamagami (1972). However, it has recently been shown to consist of two constituent proteases, as mentioned above. The fish hatching enzyme was tentatively called **chorionase** (Kaighn, 1964; Yamagami, 1973), since it digests 'chorion'. Among the hatching enzymes of various animal species, the medaka enzyme is the only one that is confirmed to be an enzyme system consisting of two component proteases, **choriolysin H** (high choriolytic enzyme, HCE) and choriolysin L (low choriolytic enzyme, LCE) (Chapter 409). Although the name choriolysin H has been adopted according to the recommendation of the NC-IUBMB for the high choriolytic hatching activity of the medaka, the more general name choriolysin might be used for other fish hatching proteases such as zebrafish choriolysin, as they also digest chorion.

## Activity and Specificity

Choriolysin H hydrolyzes the inner layer of egg envelope and releases unique proline-rich polypeptides from it. As a result of hydrolysis by choriolysin H, the inner layer of egg envelope is swollen. The released polypeptides consist of repeats of Pro-Xaa-Yaa, mainly Pro-Glu-Yaa, which are present in ZI-1,2, one of the major subunit proteins of the egg envelope (Lee *et al.*, 1994). Choriolysin H also hydrolyzes casein and some synthetic substrates such as Suc-Leu-Leu-Val-Tyr+NHMec (Yasumasu *et al.*, 1989a).

The pH optimum is about 8.0 and 8.7 for caseinolytic activity and choriolytic activity, respectively. Both activities are inhibited by EDTA at a concentration of 1 mM, but are not inhibited by DFP, iodoacetamide and iodoacetic acid (Yamagami, 1973; Yasumasu *et al.*, 1989a). EDTA-inactivated apoenzyme is reactivated by low concentration of zinc ion and high concentration of copper ion. The enzyme contains zinc, calcium and magnesium as revealed by metal analysis.

Choriolysin H tends to bind tightly to the egg envelope (Yasumasu *et al.*, 1989c). One of the antichoriolysin H monoclonal antibodies (mAbs) does not affect caseinolytic activity, but inhibits both choriolytic activity and binding to the egg envelope. The analysis using mAbs suggests that a binding site and a catalytic site are present separately on the molecule of choriolysin H (Yasumasu, 1993).

## Structural Chemistry

Choriolysin H is a single-chain protein of about 24 kDa and pI 10.2 (Yasumasu *et al.*, 1989a). The enzyme is synthesized as a preproenzyme and the mature enzyme consists of 200 amino acids (Yasumasu *et al.*, 1992b). Activation of the proenzyme is inhibited by EDTA. The active enzyme contains about one atom of zinc per molecule. According to the similarity with the amino acid sequence of astacin (Chapter 405), which has a known three-dimensional structure (Bode *et al.*, 1992), three residues of His in the sequence HELNHAL-GFQHE and a Tyr in the sequence IMHYG may be the zinc ligands.

## Preparation

Choriolysin H was purified from the hatching liquid, a filtrate of culture medium after the larvae hatch out, through cation-exchange (CM-cellulose) column chromatography, gel-filtration column (Toyopearl HW50SF) chromatography and HPLC with a CM-300 cation-exchange column (Yasumasu *et al.*, 1989a). Choriolysin H consists of two isoforms (HCE-1 and HCE-2), which are fractionated by HPLC using a cation-exchange (CM-300) column but not distinguished by other fractionation procedures such as native PAGE, SDS-PAGE or gel-filtration chromatography. The cDNA clones for choriolysin H (HCE23 and HCE21), which are 95.5% identical, seem to correspond to the two proteins (Yasumasu *et al.*, 1994).

## Biological Aspects

Choriolysin H is colocalized with choriolysin L in the secretory granules of the hatching gland cells which are located in the inner wall of the pharyngeal cavity of pre-hatching medaka embryos (Yasumasu *et al*., 1992a).

The choriolysin H genes are multicopy genes and lack introns. Six of seven HCE genes cloned are situated in a cluster and arranged tandemly. A putative promoter region of each gene contains a TATA box sequence, but not CAT box nor GC box sequences (Yasumasu *et al*., 1996). Gene expression of the enzyme starts early in the embryonic development, i.e. at the late gastrula stage, as compared with the time of actual hatching (Inohaya *et al*., 1995).

## Related Peptidases

The hatching enzymes of other animals such as mammals, amphibians and sea urchins have been studied, and the sequence relationships of the enzymes are especially interesting. Amphibian, *Xenopus laevis*, hatching enzyme also belongs to the astacin family, but possesses an additional two repeats of the CUB domain following the astacus protease domain (Sato & Sargent, 1990; C. Katagiri, unpublished results). QuCAM-1, which is considered to be a hatching-related protease from quail, also belongs to the astacin family (Elaroussi & DeLuca, 1994). However, the sea urchin hatching enzyme envelysin (Chapter 400) belongs to the matrix metalloprotease family (M10) (LePage & Gache, 1990). The proteolytic properties of mammalian hatching enzymes are not yet clear, but they are suggested to be trypsin-like enzymes (Perona & Wassarman, 1986; Sawada *et al*., 1990). The hatching enzyme of another teleost, the pike (*Esox licius*), is also a zinc protease and its molecular mass is 24 kDa (Shoots & Denuce, 1981).

## Further Reading

For reviews, see Yamagami *et al*. (1992, 1993).

## References

Bode, W., Gomis-Rüth, F.X., Huber, R., Zwilling, R. & Stöcker, W. (1992) Structure of astacin and implications for activation of astacins and zinc-ligation of collagenases. *Nature* **358**, 164–167.

Elaroussi, M.A. & DeLuca, H.F. (1994) A new member to the astacin family of metallopeptidases: A novel 1,25-dihydroxy-vitamin D-3-stimulated mRNA from chorioallantoic membrane of quail. *Biochim. Biophys. Acta* **1217**, 1–8.

Inohaya, K., Yasumasu, S., Ishimaru, S., Ohyama, A., Iuchi, I. & Yamagami, K. (1995) Temporal and spatial patterns of gene expression for the hatching enzyme in the teleost embryo, *Oryzias latipes. Dev. Biol.* **171**, 374–385.

Ishida, J. (1944) Hatching enzyme in the fresh-water fish, *Oryzias latipes. Annot. Zool. Jpn* **22**, 137–154.

Kaighn, M.E. (1964) A biochemical study of the hatching process in *Fundulus heteroclitus. Dev. Biol.* **9**, 56–80.

Lee, K.S., Yasumasu, S., Nomura, K. & Iuchi, I. (1994) HCE, a constituent of the hatching enzyme of *Oryzias latipes* embryos, releases unique proline-rich polypeptides from its natural substrate, the hardened chorion. *FEBS Lett.* **339**, 281–284.

LePage, T. & Gache, C. (1990) Early expression of a collagenase-like hatching enzyme gene in the sea urchin embryo. *EMBO J.* **9**, 3003–3012.

Moriwaki, I. (1910) The mechanism of escape of the fry out of the egg chorion in dog salmon. *3rd Report Hokkaido Fish. Res. Stn* 524–531.

Perona, R.M. & Wassarman, P.M. (1986) Mouse blastocysts hatch *in vitro* by using a trypsinlike proteinase associated with cells of mural trophectoderm. *Dev. Biol.* **114**, 42–52.

Sato, S.M. & Sargent, T.D. (1990) Molecular approach to dorso-anterior development in *Xenopus laevis. Dev. Biol.* **137**, 135–141.

Sawada, H., Yamazaki, K. & Hoshi, M. (1990) Trypsin-like hatching protease from mouse embryos: evidence for the presence in culture medium and its enzymatic properties. *J. Exp. Zool.* **254**, 83–87.

Shoots, A.F.M. & Denuce, J.M. (1981) Purification and characterization of hatching enzyme of pike, *Esox lucius. Int. J. Biochem.* **13**, 591–602.

Wintrebert, P. (1912) Le mécanisme de l'éclosion chez la truite arc-en-ciel [The hatching mechanism of the rainbow trout]. *C.R. Seances Soc. Biol. Fil.* **72**, 724–727.

Yamagami, K. (1972) Isolation of choriolytic enzyme (hatching enzyme) of the teleost, *Oryzias latipes. Dev. Biol.* **29**, 343–348.

Yamagami, K. (1973) Some enzymological properties of a hatching enzyme (chorionase) isolated from the fresh water teleost, *Oryzias latipes. Comp. Biochem. Physiol.* **46B**, 603–616.

Yamagami, K. (1988) Mechanism of hatching in fish. In: *Fish Physiology*, vol. XIA (Hoar, W.S. & Randall, D.J., eds). San Diego: Academic Press, pp. 447–499.

Yamagami, K., Hamazaki, T.S., Yasumasu, S., Masuda K. & Iuchi, I. (1992) Molecular and cellular basis of formation, hardening, and breakdown of the egg envelope in fish. *Int. Rev. Cytol.* **136**, 51–92.

Yamagami, K., Yasumasu, S. & Iuchi, I. (1993) Choriolysis by the medaka hatching enzyme, an enzyme system. In: *Physiological and Biochemical Aspects of Fish Development* (Walther, B.T. & Fyhn, H.J., eds). Bergen: University of Bergen, pp. 104–111.

Yasumasu, S. (1993) Gene structure of the hatching enzymes of the medaka and molecular basis for their action. *Fish Biol. J. Medaka* **5**, 23–28.

Yasumasu, S., Iuchi, I. & Yamagami, K. (1988) Medaka hatching enzyme consists of two kinds of proteases which act cooperatively. *Zool. Sci.* **5**, 191–195.

Yasumasu, S., Iuchi, I. & Yamagami, K. (1989a) Purification and partial characterization of high choriolytic enzyme, a component of the hatching enzyme of the teleost, *Oryzias latipes. J. Biochem.* **105**, 204–211.

Yasumasu, S., Iuchi, I. & Yamagami, K. (1989b) Isolation and some properties of low choriolytic enzyme (LCE), a component of the hatching enzyme of the teleost, *Oryzias latipes. J. Biochem.* **105**, 212–218.

Yasumasu, S., Katow, S., Umino, Y., Iuchi, I. & Yamagami, K. (1989c) A unique proteolytic action of HCE, a constituent protease of a fish hatching enzyme: tight binding to its natural substrate, egg envelope. *Biochem. Biophys. Res. Commun.* **162**, 54–63.

Yasumasu, S., Katow, S., Hamazaki, T.S., Iuchi, I. & Yamagami, K. (1992a) Two constituent proteases of a teleostean hatching enzyme: concurrent syntheses and packaging in the same secretory granules in discrete arrangement. *Dev. Biol.* **149**, 349–356.

Yasumasu, S., Yamada, K., Akasaka, K., Mitsunaga, K., Iuchi, I., Shimada, H. & Yamagami, K. (1992b) Isolation of cDNAs for LCE and HCE, two constituent proteases of the hatching enzyme

**M**

of *Oryzias latipes*, and concurrent expression of their mRNAs during development. *Dev. Biol.* **153**, 250–258.

Yasumasu, S., Iuchi, I. & Yamagami, K. (1994) cDNAs and the genes of HCE and LCE, two constituents of the medaka hatching enzyme. *Dev. Growth Differ.* **36**, 241–250.

Yasumasu, S., Shimada, H., Inohaya, K., Yamazaki, K., Iuchi, I.,

Yasumasu, I. & Yamagami, K. (1996) Different exon-intron organizations of the genes for two astacin-like proteases, high choriolytic enzyme (choriolysin H) and low choriolytic enzyme (choriolysin L), the constituents of the fish hatching enzyme. *Eur. J. Biochem.* **237**, 752–758.

*Shigeki Yasumasu*
*Life Science Institute,*
*Sophia University,*
*7-1 Kioi-cho, Chiyoda-ku, Tokyo 102, Japan*
*Email: y-sigeki@hoffman.cc.sophia.ac.jp*

*Kenjiro Yamagami*
*Life Science Institute,*
*Sophia University,*
*7-1 Kioi-cho, Chiyoda-ku, Tokyo 102, Japan*

# 411. Procollagen C-endopeptidase

## Databanks

*Peptidase classification: clan MB, family M12, MEROPS ID: M12.005*
*NC-IUBMB enzyme classification: EC 3.4.24.19*
*ATCC entries: 40295, 40311 (human)*
*Databank codes:*

| Species | SwissProt | PIR | EMBL (cDNA) | EMBL (genomic) |
|---|---|---|---|---|
| *Homo sapiens* | P13497 | A37278 | L35278 L35279 M22488 U50330 | – |
| *Mus musculus* | P98063 | I49540 | L24755 L35280 L35281 | – |
| *Strongylocentrotus purpuratus* | P98069 | – | L23838 | – |
| *Xenopus laevis* | P98070 | – | L12249 | – |
| *Xenopus laevis* | – | – | D83476 | – |

## Name and History

The name **procollagen C-endopeptidase**, recommended by the NC-IUBMB, reflects the role this enzyme plays in procollagen processing by removal of the carboxyl propeptides from procollagens. This activity was first noted in cultures of human and mouse fibroblasts (Goldberg *et al*., 1975) and in chick embryo calvaria and tendon (Davidson *et al*., 1975; Fessler *et al*., 1975; Leung *et al*., 1979). Early names for the enzyme included **procollagen C-peptidase, carboxyl-procollagen peptidase** and **carboxy-terminal procollagen peptidase**; the name **procollagen C-proteinase (PCP)** is used in this chapter. Though cathepsin-like acidic proteinases have been implicated in C-terminal processing of procollagen (Davidson *et al*., 1979; Helseth & Veis, 1984), specific PCPs which act at neutral pH were ultimately identified (Hojima *et al*., 1985; Kessler *et al*., 1986). Recently, PCP was found

to be the same as **bone morphogenetic protein-1 (BMP-1)** (Kessler *et al*., 1996; Li *et al*., 1996), a bone protein discovered by its bone-inducing activity (Wozney *et al*., 1988). An alternatively spliced form of BMP-1 was named **mammalian tolloid** (mTld) (Takahara *et al*., 1994b) because it has the same structure as *Drosophila* tolloid (Chapter 413) (Shimell *et al*., 1991). (At present, the names BMP-1 and PCP are used interchangeably.)

## Activity and Specificity

PCP cleaves the carboxyl propeptides of procollagens I, II and III as well as those of the respective pC-collagens (procollagen processing intermediates containing the C- but not the N-propeptides). The cleavage site in the pro- (or pC) α1(I), α2(I) and α1(II) chains is an Ala↓Asp bond at the

C-telopeptide/C-propeptide junctions in these chains (Hojima *et al.*, 1985; Kessler *et al.*, 1986). The human and chick pro α1(III) chains are cleaved at the respective Gly┼Asp and Arg┼Asp bonds. Substitution of the Asp at position P1′ in pro α2(I) blocks cleavage (Lee *et al.*, 1990) but heat-denatured procollagen is cleaved. Pro-lysyl oxidase is cleaved at the predicted processing site Met-Val-Gly┼Asp-Asp-Pro-Tyr-Asn-Arg (Panchenko *et al.*, 1996) and the prolaminin γ2 chain is cleaved at Tyr-Ser-Gly┼Asp-Glu-Asn-Pro-Asp-Ile-Thr-Gly (R. Burgeson, personal communication). Thus, PCP activity is not restricted to folded procollagen trimers and seems to depend on Asp at the P1′ position (as does that of the bacterial flavastacin, Chapter 412).

Activity can be assayed with [$^3$H]tryptophan-labeled type I procollagen, in which almost all of the radioactivity is in the C-propeptide (Kessler & Goldberg, 1978; Kessler & Adar, 1989), or procollagen labeled throughout the pro α chains with a mixture of [$^{14}$C]amino acids (Hojima *et al.*, 1985; Kadler & Watson, 1995). In a rapid assay, radioactivity in the free C-propeptide is measured after ethanol precipitation of undigested and partially processed procollagen. In an electrophoretic assay, reaction products are separated by SDS-PAGE and the relative amount of [$^{14}$C]-labeled pNα1(I) chain (processing intermediate containing the N- but not the C-propeptide) is determined by fluorography of the dried gel.

The pH optimum of PCP is 8–8.5 and the enzyme requires calcium for activity (Hojima *et al.*, 1985; Kessler *et al.*, 1986). The assay system includes 0.15 M NaCl and 5 mM CaCl$_2$. Serum albumin (1 mg ml$^{-1}$) and Brij-35 (0.005–0.05%) may be added. Activity is stimulated in the presence of the procollagen C-proteinase enhancer glycoprotein (PCPE) (Adar *et al.*, 1986; Kessler & Adar, 1989) or when procollagen is aggregated with dextran sulfate or polyethylene glycol (Hojima *et al.*, 1994a). Inhibitors of the enzyme include chelators such as EDTA, EGTA, and 1,10-phenanthroline, basic amino acids (e.g. Arg, Lys; 5–10 mM), DTT, *N*-ethylmaleimide (>20 mM), concanavalin A and

α$_2$-macroglobulin (Hojima *et al.*, 1985; Kessler *et al.*, 1986). CdCl$_2$, ZnCl$_2$ and CuCl$_2$ (10–100 μM) are also inhibitory (Hojima *et al.*, 1985, 1994b). The peptides Tyr-Tyr-Arg-Ala-Asp-Asp-Ala and Ac-Met-Val-Gly-Asp-Asp-Pro-Tyr-Asn-NH$_2$, which mimic the sequences around the cleavage site in pro α1(I) and prolysyl oxidase, respectively, are weak inhibitors (Njieha *et al.*, 1982; Panchenko *et al.*, 1996).

## Structural Chemistry

PCP belongs to the astacin family (M12) of metalloendopep-tidases (Chapter 404) (Bond & Beynon, 1995). The enzyme consists of a single multidomain polypeptide chain and at least two alternatively spliced forms exist (Takahara *et al.*, 1994b; Li *et al.*, 1996). The short variant is identical to BMP-1 (Wozney *et al.*, 1988; Kessler *et al.*, 1996) and the longer one, mTld (Takahara *et al.*, 1994b), has the same domain structure as that of *Drosophila* tolloid (Chapter 413) (Shimell *et al.*, 1991). The PCPs are synthesized as preproen-zymes that in human (hu) BMP-1 and mTld, are 730 and 987 residues long, respectively. In both, the prepro sequence (22 and 98 residues, respectively) is followed by a 201 residue astacin-like zinc metalloprotease domain, two CUB pro-tein–protein interactions domains (Bork & Beckmann, 1993), a Ca-binding epidermal growth factor (EGF)-like domain, and a third CUB domain (Figure 411.1). In mTld, CUB3 is fol-lowed by another EGF-like domain and two additional CUB domains. Both, BMP-1 and mTld also have short (23 and 11 residues, respectively) nonhomologous sequences at their C-termini. Each CUB domain comprises ~110 residues and four conserved cysteine residues; the EGF-like domains are ~40 residues long and each contains six conserved cysteine residues. The cysteine residues presumably form intradomain disulfide bonds.

The catalytic domain contains an 18 residue signature zinc-binding sequence, **HEXXHXXGFXHEXXRXDR**, and another conserved sequence, SXMHY, that forms a critical

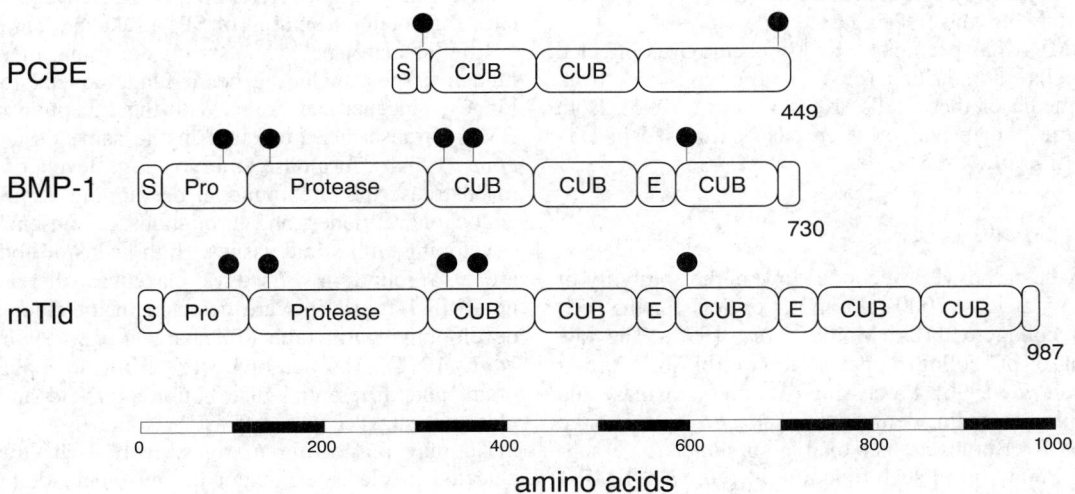

*Figure 411.1*  Domain structure of the two procollagen C-proteinase variants, BMP-1 and mTld, and of the procollagen C-proteinase enhancer protein, PCPE. S, signal peptide; Pro, propeptide domain; CUB, the CUB (**C**1r/s complement, **U**egf, **B**MP-1) module (Bork & Beckmann, 1993); E, EGF-like domain; blank box, nonhomologous region; ●, potential *N*-glycosylation sites; number at right, residue number in the human protein.

'methionine turn' beneath the active-site metal. This places PCP in the metzincin clan of zinc metalloendopeptidases (Stöcker *et al*., 1995). By similarity to other metzincins, the three His residues, the intermediate Glu and the Tyr residue in the Met-turn, are involved in zinc binding. BMP-1 and mTld contain five conserved $N$-glycosylation sites at positions (human/mouse) 91/96 142/147, 332/337, 363/368 and 599/604 of the respective preproenzyme sequences, and the native mouse BMP-1 is highly glycosylated (Takahara *et al*., 1994b; Kessler *et al*., 1996). The sequences of mature human and mouse BMP-1 and mTld are 98% identical but those of the respective prepro regions show only 50% identity. A conserved Arg-Arg-Ser-Arg sequence at the junction between the pro region and the protease domain suggests intracellular processing by dibasic endopeptidases (Barr, 1991). The predicted molecular masses of prepro-BMP-1 (human) and mTld (mouse) are 82 899 and 111 607 Da, respectively (Wozney *et al*., 1988; Fukagawa *et al*., 1994) and the values derived by SDS-PAGE for the mature, $N$-glycosylated, mouse and chick enzymes are ~80 and 100 kDa, respectively (Hojima *et al*., 1985; Kessler & Adar, 1989). The sea urchin (*Strongylocentrotus purpuratus*) BMP-1 gene (*suBMP-1*) encodes a 639 residue, 72 000 Da protein, which shows ~55% sequence identity with huBMP-1 (Hwang *et al*., 1994), whereas that of *Xenopus laevis* encodes a 707 residue, 80 673 Da protein that in an overlapping region of 672 residues, shares 82% identity with huBMP-1 (Maéno *et al*., 1993).

## Preparation

The level of PCP in natural sources is very low (Hojima *et al*., 1985; Kessler & Adar, 1989). Microgram amounts of the enzyme have been purified from media of organ cultures of chick embryo tendons (Hojima *et al*., 1985) and media of cultured mouse 3T6 fibroblasts (Kessler & Adar, 1989) with corresponding purification factors of 21 000 and 18 000. The enzyme isolated from chick apparently represents the mTld variant, whereas the mouse fibroblast enzyme corresponds to BMP-1. The purified enzyme is stored at −70°C and is stable to repeated thawing and freezing.

Human BMP-1 has been expressed in baculovirus-infected Sf21insect cells and purified (on a microgram scale) from the culture media of these cells (Kessler *et al*., 1996). Both variants of the human enzyme were also expressed in HT-1080 cells (Li *et al*., 1996).

## Biological Aspects

Cleavage of the carboxyl propeptides lowers the solubility of procollagen by at least 1000-fold and is critical for the self-assembly of collagen fibrils (Mould *et al*., 1990). The rate of processing of pC-collagen appears to control fibril formation (Kadler *et al*., 1996). Processing of prolysyl oxidase and the prolaminin γ2 chain, demonstrated *in vitro* (Panchenko *et al*., 1996; R. Burgeson, personal communication), suggests that PCP may fulfill such functions also *in vivo*. BMP-1 can induce ectopic bone formation when implanted in rats (Wozney *et al*., 1988). In bone extracts, BMP-1 is associated with transforming growth factor β (TGFβ)-like molecules and may serve as a processing enzyme for growth factors belonging to the TGFβ superfamily (Wozney *et al*., 1988). The

alternatively spliced products of the BMP-1 gene (Takahara *et al*., 1994b) may have different activities and specificities for various TGFβ-like molecules, different types of procollagens, and other noncollagenous matrix molecules. Thus, PCPs may govern matrix deposition at several levels: through activation of TGFβ-like molecules they may stimulate matrix production and then facilitate matrix deposition by processing precursors of various matrix macromolecules.

*In vitro*, procollagen processing by PCP is stimulated by PCPE, a glycoprotein that contains two CUB domains (Takahara *et al*., 1994a) (Figure 411.1), binds to the C-propeptide of procollagen type I (Kessler & Adar, 1989), and is abundant in connective tissues rich in fibrous collagens (Kessler *et al*., 1990; Takahara *et al*., 1994a). PCPE binding of the C-propeptide is mediated by the CUB domains and appears to affect both the $K_m$ and the $V$ of the PCP reaction on procollagen I (Adar *et al*., 1986). PCPE may be an important regulator of PCP activity.

PCP belongs to an evolutionarily conserved family of proteases implicated in pattern formation during development (Bond & Beynon, 1995). The mouse *BMP1* gene is required for development. Although homozygous mutant mouse embryos lacking the PCPs encoded by *BMP1* appear to have normal skeletons, they show reduced ossification of skull bones, have herniation of the gut, and do not survive beyond birth (Suzuki *et al*., 1996). The morphology of the collagen fibrils in the amnion is abnormal and consistent with the incorporation of partially processed procollagen molecules.

The locus of the human PCP/BMP-1 gene is 8p21, 706 bp downstream of the polyadenylation site of the SFTP2 surfactant gene (Yoshiura *et al*., 1993; Takahara *et al*., 1995). A single gene encodes alternatively spliced transcripts for BMP-1, mTld, and a third form in which a histidine-rich domain is inserted near the BMP-1 terminus (Takahara *et al*., 1994b). The protein product of the latter transcript is unknown. The 46 kb gene contains 22 exons. The promoter region, like those of 'housekeeping genes' and a number of growth control-related genes, lacks a TATA box, is GC rich, and has abundant sites for potential binding of SP1 (Takahara *et al*., 1995).

Low levels of mTld transcripts are found in various adult human tissues, including heart, lung, skeletal muscle, liver, kidney, placenta and brain. With the exception of the brain, BMP-1 transcripts are found in the same tissues (Takahara *et al*., 1994b). In mouse embryos, high levels of BMP-1 and mTld transcripts are found in developing membranous and endochondral bones, and throughout the mesenchyme of the developing embryonic tissues. High levels of both transcripts are also found in 17.5 days placenta, whereas mTld but no BMP-1 transcripts are detected in the floor plate of the developing neural tube (Takahara *et al*., 1994b; Fukagawa *et al*., 1994). Absence of BMP-1 from floor plate and adult brain, and differential distribution of mTld and BMP-1 in placenta, suggest that at least in some tissues, BMP-1 and mTld may play different roles. BMP-1/His transcripts are detected in placenta but not in any other adult tissues, and low-level expression is also evident in embryonic placenta and developing bone (Takahara *et al*., 1994b).

*Xenopus laevis* BMP-1 RNA is detected in blastula and early gastrula embryos and in hatched tadpole, but little or no expression is observed in morula and late gastrula embryos (Maéno *et al*., 1993). Sea urchin BMP-1 mRNA is found in both ectodermal and primary mesenchyme cells in hatched blastula stage embryos. Maximal expression is observed at mesenchyme blastula, just before the onset of primitive skeleton (spicule) formation. Sea urchin BMP-1 antigen is detected in all cell types, in the cytoplasm, on the surface of the cells, and within the blastocoel in late gastrula stage embryos (Hwang *et al*., 1994).

## Distinguishing Features

The region surrounding the PCP cleavage site in procollagen is sensitive to other endopeptidases, including cathepsin D (Chapter 277) (Helseth & Veis, 1984), a cell-associated enzyme called 'telopeptidase' (Bateman *et al*., 1987; Lee *et al*., 1990), and an unidentified secreted endopeptidase(s) (Kessler & Adar, 1989). These can be differentiated from PCP by inhibition properties, in particular, lack of inhibition by arginine, pH optima (~8.5 for PCP; neutral or acidic for the other endopeptidases), and SDS-PAGE analysis of reaction products. Nonspecific cleavages often produce α (or pNα) chains and C-propeptides with abnormal electrophoretic mobilities.

## Further Reading

The astacin family is reviewed by Bond & Beynon (1995). The CUB domain is discussed by Bork & Beckmann (1993). Further details of PCP purification and assay are given by Kadler & Watson (1995).

## References

Adar, R., Kessler, E. & Goldberg, B. (1986) Evidence for a protein that enhances the activity of type I procollagen C-proteinase. *Collagen Rel. Res.* **6**, 267–277.

Bateman, J.F., Pillow, J.J., Mascara, T., Medvedec, S., Ramshaw, J.A.M. & Cole, W.G. (1987) Cell-layer-associated proteolytic cleavage of the telopeptides of type I procollagen in fibroblast culture. *Biochem. J.* **245**, 677–682.

Barr, P.J. (1991) Mammalian subtilisins: the long-sought dibasic processing endoproteases. *Cell* **66**, 1–3.

Bond, J.S. & Beynon, R.J. (1995) The astacin family of metallo-endopeptidases. *Protein Sci.* **4**, 1247–1261.

Bork, P. & Beckmann, G. (1993) The CUB domain. A widespread module in developmentally regulated proteins. *J. Mol. Biol.* **231**, 539–545.

Davidson, J.M., McEneany, L.S.G. & Bornstein, P. (1975) Intermediates in the limited proteolytic conversion of procollagen to collagen. *Biochemistry* **14**, 5188–5194.

Davidson, J.M., McEneany, L.S.G. & Bornstein, P. (1979) Procollagen processing. Limited proteolysis of COOH-terminal extension peptides by a cathepsin-like protease secreted by tendon fibroblasts. *Eur. J. Biochem.* **100**, 551–558.

Fessler, L.I., Morris, N.P. & Fessler, J.H. (1975) Procollagen: biological scission of amino and carboxyl extension peptides. *Proc. Natl Acad. Sci. USA* **72**, 4905–4909.

Fukagawa, M., Suzuki, N., Hogan, B.L.M. & Jones, C.M. (1994) Embryonic expression of mouse bone morphogenetic protein-1 (BMP-1), which is related to the *Drosophila* dorsoventral gene *tolloid* and encodes a putative astacin metalloendopeptidase. *Dev. Biol.* **163**, 175–183.

Goldberg, B., Taubman, M.B. & Radin, A. (1975) Procollagen peptidase: its mode of action on the native substrate. *Cell* **4**, 45–50.

Helseth, D.L., Jr. & Veis, A. (1984) Cathepsin D-mediated processing of procollagen: lysosomal enzyme involvement in secretory processing of procollagen. *Proc. Natl Acad. Sci. USA* **81**, 3302–3306.

Hojima, Y., van der Rest, M. & Prockop, D.J. (1985) Type I procollagen carboxyl-terminal proteinase from chick embryo tendons. Purification and characterization. *J. Biol. Chem.* **260**, 15996–16003.

Hojima, Y., Behta, B., Romanic, A.M. & Prockop D.J. (1994a) Cleavage of type I procollagen by C- and N-proteinases is more rapid if the substrate is aggregated with dextran sulfate or polyethylene glycol. *Anal. Biochem.* **223**, 173–180.

Hojima, Y., Behta, B., Romanic, A.M. & Prockop, D.J. (1994b) Cadmium ions inhibit procollagen C-proteinase and cupric ions inhibit procollagen N-proteinase. *Matrix Biol.* **14**, 113–120.

Hwang, S.-P.L., Partin, J.S. & Lennarz, W.J. (1994) Characterization of a homolog of human bone morphogenetic protein 1 in the embryo of the sea urchin, *Strongylocentrotus purpuratus*. *Development* **120**, 559–568.

Kadler, K.E. & Watson, R.B. (1995) Procollagen C-peptidase: procollagen C-proteinase. *Methods Enzymol.* **248**, 771–781.

Kadler, K.E., Holmes, D.F., Trotter, J.A. & Chapman, J.A. (1996) Collagen fibril formation. *Biochem. J.* **316**, 1–11.

Kessler, E. & Adar, R. (1989) Type I procollagen C-proteinase from mouse fibroblasts. Purification and demonstration of a 55 kDa enhancer glycoprotein. *Eur. J. Biochem.* **186**, 115–121.

Kessler, E. & Goldberg, B. (1978) A method for assaying the activity of the endopeptidase which excises the nonhelical carboxy-terminal extensions from type I procollagen. *Anal. Biochem.* **86**, 463–469.

Kessler, E., Adar, R., Goldberg, B. & Niece, R. (1986) Partial purification of a procollagen C-proteinase from the culture medium of mouse fibroblasts. *Collagen Rel. Res.* **6**, 249–266.

Kessler, E., Mould, P.A. & Hulmes, D.J.S. (1990) Procollagen type I C-proteinase enhancer is a naturally occurring connective tissue glycoprotein. *Biochem. Biophys. Res. Commun.* **173**, 81–86.

Kessler, E., Takahara, K., Biniaminov, L., Brusel, Marina & Greenspan, D.S. (1996) Bone morphogenetic protein-1: the type I procollagen C-proteinase. *Science* **271**, 360–362.

Lee, S.-T., Kessler, E. & Greenspan, D.S. (1990) Analysis of site-directed mutations in human pro-α2(I) collagen which block cleavage by the C-proteinase. *J. Biol. Chem.* **265**, 21992–21996.

Leung, M.K.K., Fessler, L.I., Greenberg, D.B. & Fessler, J.H. (1979) Separate amino and carboxyl procollagen peptidases in chick embryo tendon. *J. Biol. Chem.* **254**, 224–232.

Li, S.-W., Sieron, A.L., Fertala, A., Hojima, Y., Arnold, W.V. & Prockop, D.J. (1996) The C-proteinase that processes procollagens to fibrillar collagens is identical to the protein previously identified as bone morphogenetic protein-1. *Proc. Natl Acad. Sci. USA* **93**, 5127–5130.

Maéno, M., Xue, Y., Wood, T.I., Ong, R.C. & Kung, H.F. (1993) Cloning and expression of cDNA encoding *Xenopus laevis* bone morphogenetic protein-1 during early embryonic development. *Gene* **134**, 257–261.

Mould, A.P., Hulmes, D.J.S., Holmes, D.F., Cummings, C., Sear, C.H.J. & Chapman, J.A. (1990) D-periodic assemblies of type I procollagen. *J. Mol. Biol.* **211**, 581–594.

Njieha, F.K., Morikawa, T., Tuderman, L. & Prockop, D.J. (1982) Partial purification of a procollagen C-proteinase. Inhibition by synthetic peptides and sequential cleavage of type I procollagen. *Biochemistry* **21**, 757–764.

Panchenko, M.V., Stetler-Stevenson, W.G., Trubetskoy, O.V., Gacheru, S.N. & Kagan, H.M. (1996) Metalloproteinase activity secreted by fibrogenic cells in the processing of prolysyl oxidase. Potential role of procollagen C-proteinase. *J. Biol. Chem.* **271**, 7113–7119.

Shimell, M.J., Ferguson, E.L., Childs, S.R. & O'Connor, M.B. (1991) The *Drosophila* dorsal-ventral patterning gene *tolloid* is related to human bone morphogenetic protein-1. *Cell* **67**, 469–481.

Stöcker, W., Grams, F., Baumann, U., Reinemer, P., Gomis-Rüth, F.-X., McKay, D.B. & Bode, W. (1995) The metzincins – topological and sequential relations between the astacins, adamalysins, serralysins, and matrixins (collagenases) define a superfamily of zinc-peptidases. *Protein Sci.* **4**, 823–840.

Suzuki, N., Labosky, P.A., Furuta, Y., Hargett, L., Dunn, R., Fogo, A.B., Takahara, K., Peters, D.M.P., Greenspan, D.S. & Hogan, B.L.M. (1996) Failure of ventral body wall closure in mouse embryos lacking a procollagen C-proteinase encoded by *BMP-1*, a mammalian gene related to *Drosophila tolloid*. *Development* **122**, 3587–3595.

Takahara, K., Kessler, E., Biniaminov, L., Brusel, M., Eddy, R.L., Jani-Sait, S., Shows, T.B. & Greenspan, D.S. (1994a) Type I procollagen COOH-terminal proteinase enhancer protein: identification, primary structure, and localization of the cognate human gene (*PCOLCE*). *J. Biol. Chem.* **269**, 26280–26285.

Takahara, K., Lyons, G.E. & Greenspan, D.S. (1994b) Bone morphogenetic protein-1 and a mammalian tolloid homologue (mTld) are encoded by alternatively spliced transcripts which are differentially expressed in some tissues. *J. Biol. Chem.* **269**, 32572–32578.

Takahara, K., Lee, S., Wood, S. & Greenspan, D.S. (1995) Structural organization and genetic localization of the human bone morphogenetic protein 1/mammalian tolloid gene. *Genomics* **29**, 9–15.

Wozney, J.M., Rosen, V., Celeste, A.J., Mitsock, L.M., Whitters, M.J., Kriz, R.W., Hewick, R.M. & Wang, E.A. (1988) Novel regulators of bone formation: molecular clones and activities. *Science* **242**, 1528–1534.

Yoshiura K., Tamura, T., Hong, H.-S., Ohta, T., Soejima, H., Kishino, T., Jinno, Y. & Niikawa, N. (1993) Mapping of the bone morphogenetic protein 1 gene (BMP1) to 8p21: removal of BMP1 from candidacy for the bone disorder in Langer-Giedion syndrome. *Cytogenet. Cell Genet.* **64**, 208–209.

***Efrat Kessler***
*Maurice & Gabriela Goldschleger Eye Research Institute,*
*Tel-Aviv University Sackler Faculty of Medicine,*
*Sheba Medical Center,*
*Tel-Hashomer 52621, Israel*
*Email: ekessler@ccsg.tau.ac.il*

# 412. Flavastacin

## Databanks

*Peptidase classification: clan MB, family M12*
*NC-IUBMB enzyme classification: none*
*Databank codes*

| Species | SwissProt | PIR | EMBL (cDNA) | EMBL (genomic) |
|---|---|---|---|---|
| *Flavobacterium meningosepticum* | – | – | L37784 | – |

## Name and History

***Flavastacin*** is a zinc metalloendopeptidase secreted by the bacterium *Flavobacterium meningosepticum* and is structurally related to astacin (Chapter 405). Flavastacin was previously referred to as ***P40*** (Plummer *et al.*, 1995; Reinhold *et al.*, 1995) before its function was determined by molecular cloning (Tarentino *et al.*, 1995) and its relationship to the eukaryotic metalloendopeptidases of family M12 was known. It is debated whether flavastacin demonstrates that the zinc-binding motif of the astacin subfamily appeared long ago in evolution before the divergence of prokaryotes and eukaryotes, or whether it is the result of a horizontal transfer.

## Activity and Specificity

Flavastacin is a neutral endopeptidase of limited specificity that hydrolyzes peptides on the N-terminal side of aspartic acid. A series of pentapeptides of the type FA-Leu-Ala↓Asp-Ala-Ser-NH$_2$ was synthesized and used to determine the relative specificity of the enzyme. The Asp-containing pentapeptide was cleaved completely to FA-Leu-Ala and Asp-Ala-Ser-NH$_2$. The asparagine and glutamine analogs were not hydrolyzed, but at high enzyme concentrations the corresponding glutamic acid analog was cleaved slowly (Tarentino *et al.*, 1995). A suitable substrate for this enzyme is not commercially available.

## Structural Chemistry

Flavastacin is synthesized as a single polypeptide with a 91 amino acid prepro region that is removed during secretion. The mature protein contains 352 amino acids with a theoretical mass of 38 843 Da. The mass determined experimentally is actually 40 087 Da, because of the presence of an unusual *O*-linked heptasaccharide (1244 Da) attached at a consensus site (Plummer *et al.*, 1995; Reinhold *et al.*, 1995). Flavastacin is about 20% identical to astacin. Its protease domain contains the highly conserved extended zinc-binding signature, HEXXHXXGXXHEXXRXDRD, and the characteristic downstream 'Met-turn' sequence, SVMMY (Jiang & Bond, 1992) that places this bacterial enzyme in the astacin subfamily (Bond & Beynon, 1995). Flavastacin, however, does not contain the four cysteine residues thought to be important for maintaining the three-dimensional structure of this subfamily of enzymes (Stöcker *et al.*, 1993).

## Preparation

Flavastacin is purified to homogeneity from the culture filtrate of *F. meningosepticum* in essentially one step using hydrophobic interaction chromatography (Tarentino *et al.*, 1995). The enzyme corresponds to peak C as reported earlier by Grimwood *et al.* (1994). About 4 mg of pure flavastacin is obtained from 1 liter of growth medium. To avoid slow autoproteolysis the enzyme is best purified in the presence of 10 mM EDTA. Activation is then accomplished by dialysis against 1 mM zinc acetate. An overexpression system fusing the synthesis of flavastacin to the *Escherichia coli* maltose-binding protein is in development.

## Related Peptidases

Flavastacin is similar in specificity to a *Pseudomonas fragi* peptidyl-Asp metalloendopeptidase (Chapter 539) sold by Boehringer Mannheim (see Appendix 2 for full names and addresses of suppliers) for protein sequencing under the trade name endoproteinase Asp:N (Drapeau, 1980). No other bacterial sources for this type of enzyme are known.

## References

Bond, J.S. & Beynon, R.J. (1995) The astacin family of metalloendopeptidases. *Protein Sci.* **4**, 1247–1261.

Drapeau, G.R. (1980) Substrate specificity of a proteolytic enzyme isolated from a mutant of *Pseudomonas fragi. J. Biol. Chem.* **255**, 839–840.

Grimwood, B.G., Plummer, T.H. Jr. & Tarentino, A.L. (1994) Purification and characterization of a neutral zinc endopeptidase secreted by *Flavobacterium meningosepticum. Arch. Biochem. Biophys.* **311**, 127–132.

Jiang, W. & Bond, J.S. (1992) Families of metalloendoproteases and their relationships. *FEBS Lett.* **312**, 110–114.

Plummer, T.H. Jr., Tarentino, A.L. & Hauer, C.R. (1995) Novel specific *O*-glycosylation of secreted *Flavobacterium meningosepticum* proteins. *J. Biol. Chem.* **270**, 13192–13196.

Reinhold, B.B., Hauer, C.R., Plummer, T.H. & Reinhold, V. (1995) Detailed structural analysis of a novel specific *O*-linked glycan from *Flavobacterium meningosepticum. J. Biol. Chem.* **270**, 13197–13203.

Stöcker, W., Gomis-Rüth, F.X., Bode, W. & Zwilling, R. (1993) Implications of the three dimensional structure of astacin for the structure and function of the astacin family of zinc-endopeptidases. *Eur. J. Biochem.* **214**, 215–231.

Tarentino, A.L., Quinones, G., Grimwood, B.G., Hauer, C.R. & Plummer, T.H. Jr. (1995) Molecular cloning and sequence analysis of flavastacin: an *O*-glycosylated prokaryotic zinc metallo-endopeptidase. *Arch. Biochem. Biophys.* **319**, 281–285.

*Anthony L. Tarentino*
*New York State Department of Health,*
*Wadsworth Center, PO Box 509,*
*Albany, NY 12201-0509, USA*
*Email: tarentin@wadsworth.org*

# 413. Miscellaneous astacin homologs

## Databanks

*Peptidase classification: clan MB, family M12, MEROPS ID: M12.009, M12.010, M12.011*
*NC-IUBMB enzyme classification: none*

*Databank codes:*

| Species | SwissProt | PIR | EMBL (cDNA) | EMBL (genomic) |
|---|---|---|---|---|
| Tolkin protein | | | | |
| *Drosophila melanogaster* | – | S58984 | U34777 | – |
| Tolloid protein[a] | | | | |
| *Caenorhabditis elegans* | – | – | – | U46668: complete gene |
| *Drosophila melanogaster* | P25723 | A39288 | M76976 | U04239: complete gene |
| *Drosophila melanogaster* | – | – | U12634 | – |
| *Homo sapiens* | – | – | L35279 | – |
| *Mus musculus* | – | – | L35281 U34042 | – |
| Miscellaneous homologs | | | | |
| *Caenorhabditis elegans* | P55112 | – | U00048 | – |
| *Caenorhabditis elegans* | P55113 | – | U13072 | – |
| *Caenorhabditis elegans* | P55114 | – | U39666 | – |
| *Caenorhabditis elegans* | P55115 | – | U41274 | – |
| *Caenorhabditis elegans* | P98060 | – | U00036 | – |
| *Caenorhabditis elegans* | P98061 | – | U10414 | – |
| *Caenorhabditis elegans* | – | – | D85744 | – |
| *Caenorhabditis elegans* | – | – | M75746 | – |
| *Caenorhabditis elegans* | – | – | Z72503 | – |
| *Caenorhabditis elegans* | – | – | – | U41554: complete gene |
| *Caenorhabditis elegans* | – | – | – | U53339: complete gene |
| *Caenorhabditis elegans* | – | – | – | Z54281: complete gene |
| *Caenorhabditis elegans* | – | – | – | Z69791: complete gene |
| *Caenorhabditis elegans* | – | – | – | Z70755: complete gene |
| *Chionoecetes opilio* | P34156 | – | – | – |
| *Coturnix coturnix* | P42662 | – | U12642 | – |
| *Hydra vulgaris* | – | – | U22380 | – |
| *Paracentrotus lividus* | P42674 | A44880 | S99978 X56224 | X65721: complete gene |
| *Strongylocentrotus purpuratus* | P98068 | – | M84144 | – |

[a] 'Tolloid' is an alternatively spliced variant of procollagen C-proteinase and the database accession numbers are repeated from the Table in Chapter 411.

## Name and History

The 'astacin family of metalloendopeptidases' was recognized in 1991 when the protease domains of mouse and human meprins were cloned and sequenced (Dumermuth *et al.*, 1991). At that time three other proteins were identified from the databanks to have sequence similarity with the meprins; they are the crayfish astacin, human bone morphogenetic protein 1 (BMP-1), and UVS.2, a developmental protein from *Xenopus laevis* embryos (Titani *et al.*, 1987; Wozney *et al.*, 1988; Sato & Sargent, 1990). Now there are approximately 50 sequences in the databanks that contain domains that are homologs of astacin. Several chapters in this book are devoted to members of this family that have been enzymatically characterized, i.e. astacin (Chapter 405), meprins A and B (Chapters 406 and 407), human meprin A (Chapter 408), fish enzymes chorolysin L and H (Chapters 409 and 410), procollagen C-proteinase (which appears to be identical to BMP-1) (Chapter 411) and the bacterial flavastacin (Chapter 412). Most of the members of the family have been identified from mRNA or DNA sequences but have not been enzymatically characterized. These include nine nematode sequences, a crab-beetle hepatopancreas protein, sea urchin embryonic proteins (BP10, Span), *tolloids* (from *Drosophila*, mouse and human), *tolloid-like* proteins (*tolloid-related-1*, *Tlr-1* or *tolkin* from *Drosophila* and mammalian *tolloid* (*mtld*)-like or *mTll* from mouse), a quail protein CAM-1, and a hydra protein HMP-1 (Jiang & Bond, 1996; Takahara *et al.*, 1996).

## Activity and Specificity

Most of the 'miscellaneous astacin homologs' have not been purified and studied for enzymatic activity against defined substrates. Some functional studies that have been conducted with the hydra metalloproteinase 1 (HMP-1) indicate that the enzyme can activate transforming growth factor $\beta_1$ (Yan *et al.*, 1995). *Tolloid* and *tolloid-related-1* in *Drosophila*, and SpAN and BP10 in sea urchins, and UVS.2 in *Xenopus laevis* have been implicated in dorsal ventral patterning and early embryo development, but their specific substrates are unknown (Sato & Sargent, 1990; Shimell *et al.*, 1991; Lepage *et al.*, 1992; Reynolds *et al.*, 1992; Nguyen *et al.*, 1994). CAM-1 of quail embryos is implicated in degradation of eggshell matrix proteins in preparation for hatching (Elaroussi & DeLuca, 1994).

## Structural Chemistry

Little is known about the structure of these enzymes. Amino acid sequences deduced from the cDNA predict that they are secreted proteins and contain a pro sequence upstream of the protease domain. The pro sequence may be important for latency/activation of these proteases. CAM-1 is an exception; no signal peptide or pro sequence was reported for this protein. A majority of these proteins contain domains C-terminal to the protease domain such as epidermal growth factor (EGF)-like and/or CUB (complement components C1r/C1s, embryonic sea urchin protein Uegf, BMP-1) domains in one or more copies (Bond & Beynon, 1995). These domains may be involved in calcium-binding and protein-protein or enzyme-substrate interactions. SpAN and BP10 contain a Ser/Thr-rich region, a unique feature for these two members of the astacin family. This region could be the site of *O*-glycosylation.

## Preparation

The hydra HMP-1 metalloproteinase has been purified using standard methods such as homogenization, ammonium sulfate precipitation, and Sephacryl-300 gel filtration (Yan *et al.*, 1995). Purification of the enzyme is monitored by gelatin zymography.

## Biological Aspects

The proposed developmental functions of these proteins correlate well with their expression patterns in embryos. Expression of their mRNA is temporally and spatially controlled, consistent with specific roles in embryonic development. For example, the *tolloid* RNA is expressed dorsally at the blastoderm stage, which agrees with its function as the dorsal–ventral patterning gene in *Drosophila* embryos (Shimell *et al.*, 1991). Expression of homologous proteins (showing overall sequence similarities in addition to the protease domain) in the same species is coordinately and differentially controlled, indicating complementary and independent functions of these counterparts. These pairs are *tolloid* and *Tlr-1* or *tolkin* (*Drosophila*), SpAN and BP10 (sea urchin), and *mTld* and *mTll* (mouse) (Nguyen *et al.*, 1994; Finelli *et al.*, 1995; Reynolds *et al.*, 1992; Lepage *et al.*, 1992; Takahara *et al.*, 1995, 1996). The genes for *tolloid* and *Tlr-1* are closely linked in the *Drosophila* genome but their sizes differ greatly (3 versus 30 kb). The human *mTld* gene (46 kb) and mouse *mTll* gene are located on human and mouse chromosome 8, respectively.

## Distinguishing Features

All members of the astacin subfamily contain a signature sequence, HEXXHXXGFXHEXXRXDR, which can be used to distinguish them from other metalloendopeptidases of the metzincin type such as serralysins, reprolysins and matrixins (Jiang & Bond, 1992; Bode *et al.*, 1993). Multidomain structures of these proteins indicate complex protein–protein and enzyme–substrate interactions. Temporal, spatial, coordinate and differential expression of these proteases in embryos indicates they play specific roles during embryonic development.

Antibodies have been generated against the purified HMP-1 (Yan *et al.*, 1995)

## Further Reading

The astacin family of enzymes is reviewed in Bond & Beynon (1995) and Jiang & Bond (1996).

## References

Bode, W., Gomis-Rüth, F.X. & Stöcker, W. (1993) Astacins, serralysins, snake venom and matrix metalloproteinases exhibit identical zinc-binding environments (HEXXHXXGXXH and Met-turn) and topologies and should be grouped into a common family, the 'metzincins'. *FEBS Lett.* **331**, 134–140.

Bond, J.S. & Beynon, R.J. (1995) The astacin family of metallo-endopeptidases. *Protein Sci.* **4**, 1247–1261.

Dumermuth, E., Sterchi, E.E., Jiang, W., Wolz, R.L., Bond, J.S., Flannery, A.V. & Beynon, R.J. (1991) The astacin family of metalloendopeptidases. *J. Biol. Chem.* **266**, 21381–21385.

Elaroussi, M.A. & DeLuca, H.F. (1994) A new member of the astacin family of metalloendopeptidases: a novel 1,25-dihydroxy-vitamin D-3-stimulated mRNA from chorioallantoic membrane of quail. *Biochim. Biophys. Acta* **1217**, 1–8.

Finelli, A.L., Xie, T., Bossie, C.A., Blackman, R.K. & Padgett, R.W. (1995) The *tolkin* gene is a *tolloid*/BMP-1 homologue that is essential for *Drosophila* development. *Genetics* **141**, 271–281.

Jiang, W. & Bond, J.S. (1992) Families of metalloendopeptidases and their relationships. *FEBS Lett.* **312**, 110–114.

Jiang, W. & Bond, J.S. (1996) The astacin family of metallo-endoproteases. In: *Zinc Metalloproteases in Health and Disease* (Hooper, N.M., ed.). London: Taylor & Frances, pp. 23–45.

Lepage, T., Sardet, C. & Gache, C. (1992) Spatial and temporal expression pattern during sea urchin embryogenesis of a gene coding for a protease homologous to the human protein BMP-1 and to the product of the *Drosophila* dorsal-ventral patterning gene *tolloid*. *Development* **114**, 147–164.

Nguyen, T., Jamal, J., Shimell, M.J., Arora, K. & O'Connor, M.B. (1994) Characterization of *tolloid-related-1*: a BMP-1-like product that is required during larval and pupal stages of *Drosophila* development. *Dev. Biol.* **166**, 569–586.

Reynolds, S.D., Angerer, L.M., Palis, J., Nasir, A. & Angerer, R.C. (1992) Early mRNAs, spatially restricted along the animal-vegetal axis of sea urchin embryos, include one encoding a protein related to *tolloid* and BMP-1. *Development* **114**, 769–785.

Sato, S.M. & Sargent, T.D. (1990) Molecular approach to dorso-anterior development in *Xenopus laevis*. *Dev. Biol.* **137**, 135–141.

Shimell, M.J., Ferguson, E.L., Childs, S.T. & O'Connor, M.B. (1991) The *Drosophila* dorsal-ventral patterning gene *tolloid* is related to human bone morphogenetic protein-1. *Cell* **67**, 469–481.

Takahara, K., Lee, S., Wood, S. & Greenspan, D.S. (1995) Structural organization and genetic localization of the human bone morphogenetic protein 1/mammalian tolloid gene. *Genomics* **29**, 9–15.

Takahara, K., Brevard, R., Hoffman, G.G., Suzuki, N. & Greenspan, D.S. (1996) Characterization of a novel gene product (mammalian tolloid-like) with high sequence similarity to mammalian tolloid/bone morphogenetic protein-1. *Genomics* **34**, 157–165.

Titani, K., Torff, H.-J., Hormel, S., Kumar, S., Walsh, K.A., Rödl, J., Neurath, H. & Zwilling, R. (1987) Amino acid sequence of a unique protease from the crayfish *Astacus fluviatilis*. *Biochemistry* **26**, 222–226.

M

Yan, L., Pollock, G.H., Nagase, H. & Sarras, M.P., Jr. (1995) A 25.7 × 10³$M_r$ hydra metalloproteinase (HMP1), a member of the astacin family, localizes to the extracellular matrix of *Hydra vulgaris* in a head-specific manner and has a developmental function. *Development* **121**, 1591–1602.

Wozney, J.M., Rosen, V., Celeste, A.J., Mitsock, L.M., Whitters, M.J., Kriz, R.W., Hewick, R.M. & Wang, E.A. (1988) Novel regulators of bone formation: molecular clones and activities. *Science* **242**, 1528–1534.

*Weiping Jiang*
Department of Biochemistry and Molecular Biology,
Penn State University College of Medicine,
Hershey, PA 17033-0850, USA

*Judith S. Bond*
Department of Biochemistry and Molecular Biology,
Penn State University College of Medicine,
Hershey, PA 17033-0850, USA
Email: jbond@bcmic.hmc.psu.edu

# 414. Procollagen N-endopeptidase

## Databanks

Peptidase classification: clan MB, family M12, MEROPS ID: M12.301
NC-IUBMB enzyme classification: EC 3.4.24.14
Databank codes:

| Species | SwissProt | PIR | EMBL (cDNA) | EMBL (genomic) |
|---|---|---|---|---|
| *Bos taurus* | – | – | X96389 | – |
| *Caenorhabditis elegans* | – | – | Z69361 | – |

## Name and History

*Procollagen N-endopeptidase* was initially identified with the help of a genetic disease in cattle and sheep, dermatosparaxis (Lenaers *et al*., 1971). This genetic defect is characterized by the presence in the skin of the animals of a longer collagen α chain, pN-collagen, which contains the uncleaved N-propeptide. Tissue extracts of dermatosparactic calves did not cleave pN-collagen, whereas extracts of normal tissues cleaved the protein (Lapière *et al*., 1971). The defect appeared therefore to be a deficiency of procollagen N-protease, and subsequently the enzyme activity was partially purified from calf tissues (Lapière *et al*., 1971) and chick embryos (Fessler *et al*., 1975, Tuderman *et al*., 1978), and later to near homogeneity from cattle skin (Colige *et al*., 1995) and tendons (Hojima *et al*., 1994a). The enzyme has also been known as *procollagen N-protease*, *procollagen N-proteinase* and *procollagen I N-proteinase*.

## Activity and Specificity

Procollagen N-endopeptidase removes the N-terminal propeptide from procollagen I and procollagen II prior to deposition of the collagen molecule in fibrils in the extracellular space. The peptide bonds cleaved are Asn-Phe-Ala-Pro+Gln-Leu-Ser-Tyr in the α1(I) chain, Pro+Gln in the α1(II) chain and Ala+Gln in the α2(I) chain. The enzyme only cleaves the procollagens in native conformation, no cleavage is observed after heat denaturation of the substrate (Tuderman *et al*., 1978). Procollagen N-endopeptidase does not act on procollagen III or collagen XIV (Tuderman *et al*., 1978; Colige *et al*., 1995). Other members of the collagen protein family have not been tested as substrates. Aggregation of procollagen I with dextran sulfate increases the rate of cleavage about 4-fold (Hojima *et al*., 1994b). The enzyme activity requires calcium. Under standard assay conditions (pH 7.5, 2 mM CaCl₂, 0.2 M NaCl, 35°C) the $K_m$ is 435 nM and the $V$ is 39 nM h⁻¹, but both vary significantly with pH and ionic strength (Colige *et al*., 1995). Metal chelators, serum, acidic polymers (heparin, heparan sulfate), concanavalin A, SDS, copper, cadmium and zinc ions (Hojima *et al*., 1994a,c) are potent inhibitors of the enzymic activity, whereas serine proteinase inhibitors are not (Hojima *et al*., 1994a; Colige *et al*., 1995).

## Structural Chemistry

The enzyme is presumably a single polypeptide of about 107 kDa that contains intrachain disulfide bridges. Its primary structure is not known, except for a few short peptide sequences (Colige *et al*., 1995). The pI of a partially purified enzyme preparation was determined to be 3.8 (Hojima *et al*., 1994a).

## Preparation

The enzyme has been extensively purified from cattle skin (Colige *et al*., 1995) and tendons (Hojima *et al*., 1994a). The first procedure included several steps, con A- and heparin-Sepharose chromatography (Tuderman & Prockop, 1982) followed by type XIV collagen affinity chromatography and a second heparin-Sepharose step. It yielded over 1000-fold purification of an enzyme with a mass of about 107 kDa. The second procedure utilized chromatography on concanavalin A- and heparin-Sepharose, Sephacryl S-300 and CM-Sephadex C-50. With this methodology, a 500 kDa form of the enzyme was purified over 16 000-fold. This preparation contained unreduced polypeptide chains of 58, 125, 170 and 190 kDa on SDS-PAGE. The relation of the two enzyme preparations is not yet clear, however, the 170 and 190 kDa peptides in the cattle tendon enzyme preparation might represent collagen XIV which is known to bind to the enzyme with high affinity.

## Biological Aspects

Systematic studies of the tissue distribution of procollagen N-endopeptidase have not been carried out; however, the enzyme activity is present in tissues rich in procollagen I, such as skin, tendons and bone. There is no evidence for a proenzyme, and the only known function is the processing of procollagen.

The enzyme is implicated in the pathogenesis of Ehlers–Danlos syndrome type VII, a heritable human connective tissue disorder comparable to dermatosparaxis in cattle and sheep and characterized by the accumulation of collagen precursors in connective tissues. Ehlers–Danlos VIIA and B are caused by mutations in the genes encoding procollagen $\alpha$ chains that result in the disruption of the cleavage site of procollagen N-endopeptidase. Recently, reduced activity of procollagen N-endopeptidase was observed in patients with Ehlers–Danlos syndrome VIIC (Smith *et al*., 1992; Nusgens *et al*., 1992). The disease is characterized by laxity and fragility of the skin, altered collagen fibrils seen as hieroglyphic pictures with electron microscopy, accumulation of pN-$\alpha$1 and pN-$\alpha$2 collagen polypeptides in the dermis and absence of processing of the procollagen in fibroblast cultures.

## Distinguishing Features

Procollagen N-endopeptidase cleaves only native procollagens I and II, but not III.

## Note Added in Proof

After the manuscript for this chapter was submitted, the complete cDNA sequence of procollagen N-endopeptidase was published (Colige *et al*., 1997). The cDNA codes for a polypeptide chain of 1205 amino acid residues. The enzyme contained zinc-binding sequences and four repeats that are homologous to domains found in thrombospondins and in properdin. When expressed as a recombinant protein, the enzyme actively cleaved type I procollagen. The same investigators have also recently disclosed mutations in the gene for procollagen N-endopeptidase in families suffering from Ehlers–Danlos syndrome type VIIC (B. Nusgens, personal communication).

## References

Colige, A., Beschin, A., Samyn, B., Goebels, Y., Van Beeumen, J., Nusgens, B.V. & Lapière, C.M. (1995) Characterization and partial amino acid sequencing of a 107-kDa procollagen I N-proteinase purified by affinity chromatography on immobilized type XIV collagen. *J. Biol. Chem.* **270**, 16724–16730.

Colige, A., Li, S.-W., Sieron, A.L., Nusgens, B.V., Prockop, D.J. & Lapière, C.M. (1997) cDNA cloning and expression of bovine procollagen I N-proteinase: a new member of the superfamily of zinc-metalloproteinases with binding sites for cells and other matrix components. *Proc. Natl Acad. Sci. USA* **94**, 2374–2379.

Fessler, L.I., Morris, N.P. & Fessler, J.H. (1975) Procollagen: biological scission of amino and carboxyl extension peptides. *Proc. Natl Acad. Sci. USA* **72**, 4905–4909.

Hojima, Y., Mörgelin, M.M., Engel, J., Boutillon, M.M., van der Rest, M., McKenzie, J., Chen, G.C., Rafi, N., Romanic, A.M. & Prockop, D.J. (1994a) Characterization of a type I procollagen N-proteinases from fetal bovine tendon and skin. Purification of the 500 kDa form of the enzyme from bovine tendon. *J. Biol. Chem.* **269**, 11381–11390.

Hojima, Y., Behta, B., Romanic, A.M. & Prockop, D.J. (1994b) Cleavage of type I procollagen by C- and N-proteinases is more rapid if the substrate is aggregated with dextran sulfate or polyethylene glycol. *Anal. Biochem.* **223**, 173–180.

Hojima, Y., Behta, B., Romanic, A.M. & Prockop, D.J. (1994c) Cadmium ions inhibit procollagen by C-proteinase and cupric ions inhibit procollagen by N-proteinase. *Matrix Biol.* **14**, 113–120.

Lapière, C.M., Lenaers, A. & Kohn, L.D. (1971) Procollagen peptidase: an enzyme excising the coordination peptides of procollagen. *Proc. Natl Acad. Sci. USA* **68**, 3054–3058.

Lenaers, A., Ansay, M., Nusgens, B. & Lapière, C.M. (1971) Collagen made of extended alpha-chains, procollagen, in genetically defective dermatosparactic calves. *Eur. J. Biochem.* **23**, 533–543.

Nusgens, B.V., Verellen-Dumoulin, C., Hermanns-Le, T., De Paepe, A., Nuytinck, L., Pierard, G.E. & Lapière, C. (1992) Evidence for a relationship between Ehlers-Danlos type VII C in humans and bovine dermatosparaxis. *Nature Genet.* **1**, 214–217.

Smith, L.T., Wertelecki, W., Milstone, L.M., Petty, E.M., Seashore, M.R., Braverman, I.M., Jenkins, T.G. & Byers, P.H. (1992) Human dermatosparaxis: a form of Ehlers–Danlos syndrome that results from failure to remove the amino-terminal propeptide of type I procollagen. *Am. J. Hum. Genet.* **51**, 235–244.

Tuderman, L. & Prockop, D.J. (1982) Procollagen N-proteinase. Properties of the enzyme purified from chick embryo tendons. *Eur. J. Biochem.* **125**, 545–549.

Tuderman, L., Kivirikko, K.I. & Prockop, D.J. (1978) Partial purification and characterization of a neutral protease which cleaves the N-terminal propeptides from procollagen. *Biochemistry* **17**, 2948–2954.

M

*Leena Bruckner-Tuderman*
*Department of Dermatology,*
*University of Münster, 48149 Münster, Germany*
*Email: tuderma@uni-muenster.de*

# 415. Procollagen III N-endopeptidase

## Databanks

*Peptidase classification: clan MX, family M99, MEROPS ID: M9G.004*
*NC-IUBMB enzyme classification: none*
*Databank codes: no sequence data available*

## Name and History

After the enzyme that processes procollagens I and II (Chapter 414) had been identified in tendon and skin extracts and in fibroblast cultures, a similar enzyme activity was predicted to cleave the N-propeptide from procollagen III. Nusgens *et al*. (1980) identified ***procollagen III N-endopeptidase*** activity in calf tendon fibroblasts, and Halila & Peltonen (1984) characterized it in cattle aortic smooth muscle cells *in vitro*.

## Activity and Specificity

Procollagen III N-endopeptidase removes the N-terminal propeptide from native procollagen III, but not from procollagens I or IV, nor from denatured procollagen III. Cleavage is at the Tyr-Ser-Pro+Gln-Tyr-Glu-Ala bond. The activity requires calcium, but the specificity with which the enzyme cleaves procollagen III has not been established definitively. A pH optimum of approximately 7.5 and an apparent $K_m$ of 300 nM are observed (Nusgens *et al*., 1980). Metal chelators and reducing agents are potent inhibitors of the enzymic activity, whereas serine proteinase inhibitors, lysine, or serum are not (Nusgens *et al*., 1980; Halila & Peltonen, 1984).

## Structural Chemistry

No structural details are known.

## Preparation

The enzyme has been partially purified from cultured calf tendon fibroblasts (Nusgens *et al*., 1980) and from cattle aortic smooth muscle cells (Halila & Peltonen, 1984) using a two-step purification procedure with concanavalin A chromatography and affinity chromatography with pN-collagen III-Sepharose. With this system, an approximately 400-fold purification of the enzyme activity was achieved. On gel-filtration columns, the activity eluted with an apparent molecular mass of 72 kDa.

## Distinguishing Features

Procollagen III N-endopeptidase cleaves only procollagen III, but not procollagen I or IV. In contrast, procollagen N-endopeptidase (Chapter 414) removes the propeptide from procollagens I and II, but not III.

## References

Halila, R. & Peltonen, L. (1984) Neutral protease cleaving the N-terminal propeptide of type III procollagen: partial purification and characterization of the enzyme from smooth muscle cells of bovine aorta. *Biochemistry* **23**, 1251–1256.

Nusgens, B.V., Goebels, Y., Shinkai, H. & Lapière, C. (1980) Procollagen type III N-terminal endopeptidase in fibroblast culture. *Biochem. J.* **191**, 699–706.

*Leena Bruckner-Tuderman*
*Department of Dermatology,*
*University of Münster,*
*48149 Münster, Germany*
*Email: tuderma@uni-muenster.ge*

# 416. Introduction to the reprolysins

*MEROPS ID: M12*

## Classification and Diversity

The term **reprolysin** has been introduced (Bjarnason & Fox, 1995) to refer to a subfamily of the family M12 metalloproteinases in which the peptidase units are more closely related to each other than they are to astacin, and the proteins contain a characteristic C-terminal extension composed of a series of non-peptidase domains, commonly including a disintegrin domain. The word reprolysin reflects the fact that much of the work on these enzymes has been done with examples from snake (**rept**ile) venoms (which we shall here term **snake venon reprolysins, SVR**s), but also that members of the subfamily are known from mammalian tissues, especially **repr**oductive tissues. The mammalian reprolysins are often called *ADAM* metalloproteinases (see also Chapter 447).

To date there are reports of at least 102 snake venom reprolysins (SVRs) from 36 species of snakes (Bjarnason & Fox, 1995). However, this is not to say that there are over 100 distinct and different SVRs in the sense that trypsin and chymotrypsin are different enzymes whereas trypsins from different species are forms of the same enzyme, or orthologues. On the contrary, they are probably much fewer in number, probably fewer than ten.

Some reprolysins are now known to be isoenzymes from the same species of snake. For example, astrolysin C (EC 3.4.24.42) (Chapter 423) is a trivial name for two hemorrhagic toxin isoenzymes, HT-c and HT-d, from *Crotalus atrox* venom. These enzymes have a difference of only one amino acid (one charge) out of a total of 203 (Bjarnason & Fox, 1987; Hite *et al.*, 1993). The same seems to apply to the isoenzymes proteinase I and proteinase II, both termed adamalysin (Chapter 421) from the venom of *Crotalus adamanteus* (Kurecki *et al.*, 1978; Gomis-Rüth *et al.*, 1993). There also appear to exist more distantly related isoenzymes from the same species of snake. Thus, atrolysin B (Chapter 420) should probably be regarded as a cationic isoenzyme of the anionic atrolysin C (HT-c/d) isoenzymes. They share identical cleavage sites on the insulin B chain, as well as a 78% sequence identity (Fox *et al.*, 1986; Bjarnason *et al.*, 1988; Hite *et al.*, 1993), while, for comparison, the anionic and cationic forms of cattle trypsinogen share 65% sequence identity (LeHueron *et al.*, 1990). Atroxase (Chapter 419) from *Crotalus atrox* venom should probably be classified as an isoenzyme of atrolysin B since the two enzymes are almost identical (98% sequence identity). These two isoenzymes were probably the major constituents of the preparation from *Crotalus atrox* venom previously termed α-protease (Zwilling & Pfleiderer, 1967; Bjarnason & Fox, 1995). Thus, if *Crotalus atrox* venom is used as an example of the diversity of the snake venom reprolysins, one can make the following

observations. There are, to date, reports on 13 reprolysins from *Crotalus atrox* venom. These are the six atrolysins (A, B, Cc, Cd, E and F), HT-g, atroxase, protease I, protease IV, collagenase and new α-protease. These 13 metalloproteinases can apparently be reduced to six or fewer distinct and disparate enzymes, i.e. the atrolysin A group containing atrolysin F (Chapter 441) and HT-g, in addition to atrolysin A (Chapter 418); the atrolysin C group with HT-c and -d (both termed atrolysin C), atrolysin B, protease I, atroxase and new α-protease; atrolysin E (Chapter 430); protease IV; collagenase; and fibrinogenase (Fox & Bjarnason, 1995). Furthermore, atrolysin B and protease I appear to be the same enzyme, as do atroxase and new α-protease (Fox & Bjarnason, 1995).

As might be expected, orthologous enzymes appear to exist in the venoms of most of the different species of snakes, similar to the general distribution of trypsins in the digestive tract of most organisms. Thus, atrolysin C and adamalysin appear to be orthologous enzymes from the venoms of *Crotalus atrox* and *Crotalus adamanteus* respectively, as previously suggested (Bjarnason & Fox, 1988–89). They have similar amino acid compositions, similar insulin B chain specificities and their X-ray crystal structures show identical topologies thought to be characteristic of the two disulfide-bond reprolysins (Gomis-Rüth *et al.*, 1994; Zhang *et al.*, 1994). Adamalysin has not been sequenced as such, but the amino acid sequence of atrolysin C, its homolog from *Crotalus atrox* venom, was used to assist in making structural assignments in the determination of the X-ray crystal structure of adamalysin. It therefore seems plausible that all of the enzymes found in the *Crotalus atrox* venom may have orthologous enzymes in other venoms. Furthermore, there may be some additional enzyme groups (such as russellysin, Chapter 433) in other venoms not yet found in *Crotalus atrox* venom. Thus, as previously suggested, there may be fewer than ten distinct reprolysins in snake venom. Clearly, more data are needed on most of the reprolysins, in particular sequence and specificity data, and preferably structural information, before reliable determinations of relationships and group assignments can be made.

The snake venom metalloendopeptidases have been purified and investigated from standpoints of a diversity of activities and biological manifestations. Thus they have been studied simply as proteolytic enzymes, or as hemorrhagic toxins, fibrinogenases, fibrinolytic enzymes, collagenases, inactivators of serpins, clotting factor X activators, etc. This has caused some confusion in the field, and at times emphasized diversity over similarity amongst these enzymes. The issue of hemorrhagic activity versus nonhemorrhagic activity has also been a source of some controversy. Most proteinases will cause some sort of hemorrhagic effect if injected in high amounts.

This means that homologous enzymes from different venoms may be classified as hemorrhagic or nonhemorrhagic, depending on the discretion of the investigators. A case in point would be the orthologous enzymes atrolysin C (HT-c and d) from *Crotalus atrox* venom and adamalysin (proteinases I and II) from *Crotalus adamanteus*. HT-c and HT-d were found to be very weakly hemorrhagic, about three orders of magnitude less than atrolysin A from the same venom, and could probably have been classified as nonhemorrhagic proteinases. They were, however, presented in the literature as hemorrhagic toxins, and probably make a relevant synergistic contribution to the hemorrhagic effect with the other toxins, rather than as single purified proteins. In contrast, their orthologs from *Crotalus adamanteus* venom, proteinases I and II (adamalysin), were initially presented as collagenases I and II with hemorrhagic activity (Kurecki & Laskowski, 1978) in a proceedings abstract, but were then described in the literature as proteinase I and II with serpin inactivation activity (Kurecki *et al.*, 1978). The consequence of this is that many investigators in the field of reprolysins have considered these enzymes, atrolysin C and adamalysin, to be unrelated enzymes with dissimilar activities, when in fact they are homologous enzymes, with weak hemorrhagic activities, and they probably both interact with serpins. Another subject of some confusion is the issue of fibrinogenolytic activities of the SVRs. Apparently all the SVRs, hemorrhagic or not, that have been assayed for fibrinogenase activity have been found to be active, sometimes forming a fibrin precipitate, but a precipitate that is different from the thrombin-generated one (Bjarnason & Fox, 1988–89). Many SVRs have also been found to be fibrinolytic.

It has been suggested that the reprolysins should be classified into four main categories based on their sizes (molecular masses) and more recently on sequences published to date (Bjarnason & Fox, 1988–89, 1995) (c.f. Fig. 416.1). These categories are:

- class I, the small reprolysins, having molecular masses of 20–30 kDa and containing only the protease domain;
- class II, the medium-size enzymes with molecular masses of 30–50 kDa, having also a disintegrin-like domain;
- class III, the reprolysins and most potent hemorrhagic toxins, having molecular masses of 50–80 kDa and a third 'high-cysteine' domain; and
- class IV of molecular mass 80–100 kDa, containing of a fourth lectin-like domain in addition to the aforementioned three.

Of the 102 reprolysins known to date about half, or 54 enzymes, belong to class I – the small reprolysins, containing only a protease domain; nine appear to fall into class II – the medium size proteins; 30 are placed in class III – the three-domain enzymes; seven are assigned to class IV; and two cannot be assigned to any class due to lack of molecular mass data (Bjarnason & Fox, 1995). It must be clearly stated that these assignments are based solely on the molecular masses of the reprolysins except for those with known sequences (14 cases). Furthermore, of the 102 reprolysins counted so far, most of them, or 69, have been found to cause hemorrhage, another 20 have been determined to be nonhemorrhagic and 13 have not been assayed for hemorrhagic activity.

The first reprolysins to be clearly purified to homogeneity and shown to be zinc-dependent metalloproteinases were leucolysin (Chapter 436) from *Agkistrodon piscivorus leucostoma* (Spiekerman *et al.*, 1973) and five hemorrhagic metalloproteinases (atrolysins A–E) from *C. atrox* venom (Bjarnason & Tu, 1978). The atrolysins have been quite extensively studied since this first report appeared. The zinc ion of atrolysin E has been exchanged for cobaltous ion, with no observed structural perturbations and only a minor decrease in proteolytic activity (Bjarnason & Fox, 1983). Their specificities on the B chain of insulin have been determined (Bjarnason & Fox, 1983, 1988–89; Fox *et al.*, 1986) and their substrate-binding sites have been probed using a multitude of synthetic substrates and inhibitors (Fox *et al.*, 1986). Their cleavage of basement membrane matrix components has been described (Baramova *et al.*, 1989, 1990a, 1991; Bjarnason *et al.*, 1988, 1993). Their interaction with serum inhibitors has been studied (Baramova *et al.*, 1990b), and their cDNA and amino acid sequences have been determined (Shannon *et al.*, 1989; Hite *et al.*, 1992, 1993). Most recently the X-ray crystal structure of atrolysin C has been analyzed (Zhang *et al.*, 1994), as well as that of adamalysin from *C. adamanteus* venom (Chapter 421) (Gomis-Rüth *et al.*, 1994).

## Classification Based on Domain Composition

Considerable data are now available on cDNA sequences of some venom reprolysins (Hite *et al.*, 1992, 1993; Paine *et al.*, 1992) as well as protein sequences, and these also can be used to assign SURs to four classes (Bjarnason & Fox, 1995). The first part of the scheme (N-I to N-IV) is based on the nucleotide cDNA sequences of the venom metalloproteinases (Figure 416.1). In the first class, termed N-I (nucleotide class I), the cDNA (mRNA) codes for a signal sequence, pro sequence and metalloproteinase sequence. The second nucleotide class, N-II, has signal, pro, proteinase and disintegrin-like cDNA sequences. N-III is the third nucleotide class and these sequences represent information for a signal, pro, proteinase, disintegrin-like and cysteine-rich domains. The cDNA sequence of atrolysin A (HT-a) is a member of this class (Hite *et al.*, 1993). We also propose a putative N-IV class which has sequence information for a signal, pro, proteinase, disintegrin-like, cysteine-rich and lectin-like domains, based on the protein sequence of RVV-X, a high molecular mass metalloproteinase from *Vipera russelli* venom (Takeya *et al.*, 1992).

By use of protein sequence data for the venom metalloproteinases one can propose a parallel and interrelated classification based on the structure of the mature protein as it occurs after protein purification. The P-I class (protein class I as above) has only a proteinase domain in its mature form, i.e. following proteolytic processing of the signal and zymogen structures (Figure 416.1). No venom metalloproteinase has yet been sequenced that can be classified as a P-II proteinase, but there are venom proteinases of which there are data to suggest that they belong to the P-II class (table 2 of Bjarnason & Fox, 1995).

There are at least three members of the P-III class with known protein or cDNA sequences: atrolysin A from *C. atrox*

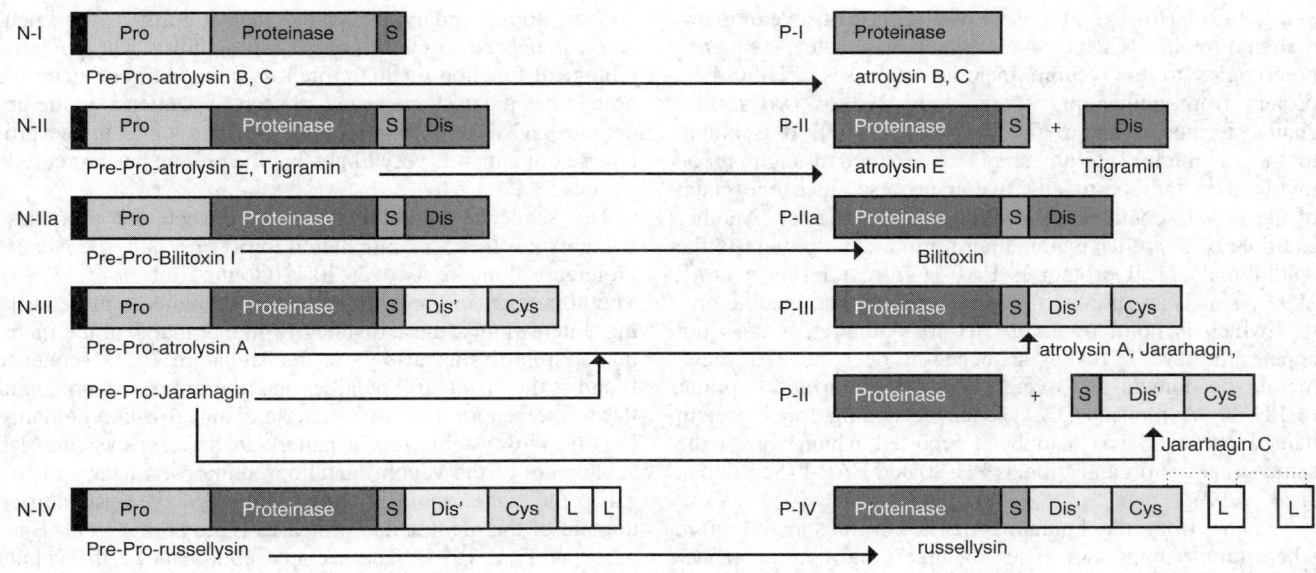

*Figure 416.1* mRNA and processed protein classes of snake venom reprolysins. S, spacer region; Dis, disintegrin domain; Dis′, disintegrin-like domain; Cys, cysteine-rich domain; L, lectin domain. The light-colored proteinase domains have two disulfides and the darker proteinase domains have three.

(Figure 416.1); trimerelysin I or HR-1B from *Trimeresurus flavoviridis* (Chapter 425); and jararhagin (Chapter 429) from *Bothrops jararaca* (Takeya *et al*., 1990; Paine *et al*., 1992; Hite *et al*., 1993). In their mature form they all comprise a proteinase domain, a disintegrin-like domain and a cysteine-rich domain. All of these proteinases are potent hemorrhagic toxins, more so than the lower molecular mass toxins, strongly suggesting the importance of the disintegrin-like and the high-cysteine domains to the hemorrhagic activity of these toxins.

The final protein class, P-IV, has one member definably identified to date by amino acid sequence data, russellysin (Chapter 433) from *Vipera russelli* (Takeya *et al*., 1992), a 79 kDa protein comprised of two disulfide-bonded protein chains (59 kDa and 21 kDa) which functions as a coagulation factor X-activating enzyme (Figure 416.1). The structure of the heavy chain is similar to that of the high molecular mass hemorrhagic toxins such as HT-a and HR-1B although russellysin is not hemorrhagic. The light chain is disulfide bonded to the heavy chain through an extra cysteine residue present in the second subdomain of the high cysteine domain. Mucrotoxin A (Chapter 426) is a hemorrhagic metalloproteinase from *T. mucrosquamatus* (Sugihara *et al*., 1983) with a molecular mass of approximately 94 kDa. This toxin has 39 cysteinyl residues, which is similar to the number found in russellysin (44 cysteines). There will probably be more examples of the P-IV class reprolysins appearing with time in the literature.

## The Disintegrin-like Domain

A short discussion on disintegrin peptides and disintegrin-like domains of reprolysins is appropriate here. Disintegrins are peptides of 49–83 amino acids in length which are isolated from crotalid and viperid venoms (Gould *et al*., 1990;

Scarborough *et al*., 1993). The disintegrins have been demonstrated to bind the platelet integrin GPIIb/IIIa (aIIb/b3) and inhibit platelet aggregation (Dennis *et al*., 1989). Essentially all of the disintegrins have an Arg-Gly-Asp integrin-binding consensus sequence. Recent cDNA sequence studies suggest that the disintegrins are proteolytically processed from venom metalloproteinase precursors, rhodostomin being an example of such a phenomenon (Neeper & Jacobson, 1990; Au *et al*., 1991). The majority of the large SVRs (P-II, III and IV) and all the ADAM metalloproteinases (Chapter 447) have a disintegrin-like domain. These are distinguished by the term disintegrin-like domain, since they do not have the Arg-Gly-Asp (RGD) consensus sequence of the disintegrins proper, but instead a wide variety of other nonconserved residues in this position. However, the SVRs and ADAMs have a highly conserved Arg residue in the first position of the RGD loop, next to the 12th Cys residue of the domain, and an Asp residue in the 8th position of the RGD loop, on the C-terminal side of extra Cys residue of the disintegrin-like domain. Furthermore, they have a conserved Glu in the 11th position of the loop. Two of these three residues of the disintegrin-like domains, i.e. R and D or R and E, must be considered strong candidates for the role of ligands for integrin binding in place of R and D of the RGD triad in the disintegrins. The disintegrin-like domains also have a different disulfide bond arrangement from the disintegrins proper with at least two additional disulfide bonds present; one probably connecting to the proteinase domain and the other (at the RGD loop region) leading to the cysteine-rich domain.

## Non-venom Reprolysins

Until a few years ago the snake venom metalloproteinases appeared to be representatives of a new family of metalloproteinases with no homologs of nonvenomous origin (Hite *et al.*, 1992). However, there are now reports of several mammalian proteins (Chapter 447) which have clear sequence homologies to the venom metalloproteinases. PH-30 is a protein from guinea pig sperm composed of two similar chains, termed $\alpha$ and $\beta$ (Blobel *et al.*, 1992). It is bound to the membrane of guinea pig sperm and thought to be involved in the sperm–egg fusion process since one chain of the protein contains a viral-like fusion sequence. Another example of a similar mammalian reproductive protein is the epididymal apical protein I (EAP-I) from rat (Perry *et al.* 1992). This is an androgen-induced, epididymus-specific protein which is homologous to PH-30. Although it does not appear to have a fusion sequence it does have a transmembrane domain and overall domain organization similar to PH-30. A mouse cDNA sequence coding for a protein named cyritestin has also been reported which shows the same structural organization as PH-30 and EAP-I (Senftleben *et al.* 1992).

A gene from the human 17q21.3 chromosomal region, where tumor suppressor genes for breast and ovarian cancer are thought to reside, was isolated and sequenced (Emi *et al.*, 1993). From the DNA sequence, Emi *et al.* predicted a protein of 524 amino acids with domain homology to the SVR class P-III proteins and PH-30. The DNA sequence coded for **m**etalloproteinase, **d**isintegrin-like and **c**ysteine-rich domains, thus termed MDC protein. As in the case of PH-30b, cyritestin and EAP I the zinc-binding consensus sequence of MDC protein has been lost. From careful consideration of this DNA sequence, it appears that additional open reading frames may be present in the clone. If a potential sequencing error is taken into account, sequence coding for two additional domains is observed: a transmembrane domain and a cytoplasmic domain. Confirmation of this possibility will require further studies. Although the function of this protein is unknown, the authors demonstrated that rearrangements in both tumor types involve multiple exons and disrupt the coding region of the MDC gene.

A sequence from mouse macrophages was identified as a protein containing a pre sequence, extracellular domain, transmembrane domain and a cytoplasmic domain by use of subtractive cDNA cloning techniques (Yoshida *et al.*, 1990) Although it was not immediately recognized, the extracellular region is homologous to the SVRs and the ADAM metalloproteinases. It shares the pro, proteinase, disintegrin-like and cysteine-rich domains with the other reprolysin groups. The proteinase domain has a complete zinc-binding consensus sequence and thus may have proteolytic activity. The authors did demonstrate that the expression of this protein could be upregulated by macrophage stimulators. The function of the protein in monocytic cell lines is unclear, but one may speculate that proteolytic activity may be important in the migration of these cells through matrix or perhaps in the processing of bioactive proteins such as cytokines.

A sequence has been identified from cDNA transcribed from human myeloblast mRNA (cell line KG-1) which is homologous to ADAM metalloproteinases (Nomura *et al.*, 1994). The cDNA sequence codes for pre, pro, proteinase, disintegrin-like and cysteine-rich domains. The cDNA clone has 3541 nucleotides which suggests that it may, in fact, code for a longer protein sequence which would encompass

transmembrane and cytoplasmic domains. Further experimentation is needed to confirm such a possibility. The potential biological function of this protein is unknown, but since the protein has the cysteine switch consensus sequence in the pro domain and a valid zinc-binding consensus site in the proteinase domain, it is very likely that the protein has proteolytic activity.

The sequence homology shared between the proteinase domains of these mammalian reprolysins and the venom proteinase domains is over 30% (Gomis-Rüth *et al.*, 1993). The homology in the disintegrin-like domains is more striking. Interestingly, these disintegrin-like domains in the mammalian reprolysins also lack the signature RGD sequence found in the disintegrin peptides and, therefore, are more similar to the venom metalloproteinase disintegrin-like domains. Finally, the cysteine repeat pattern in the first cysteine-rich subdomain of the venom metalloproteinases is nearly identical to the same region at the beginning of the cysteine-rich domain of the mammalian proteases (Hite *et al.*, 1993; Bjarnason & Fox, 1994). The second subdomain of the venom cysteine-rich domain is different from those of the mammalian proteases, suggesting a different function.

Based on the structural similarities of the venom metalloproteinases and the mammalian reproductive proteins, it is tempting to speculate on the evolutionary relationship of their functions. Proteolytic activity for most of the mammalian proteins has yet to be demonstrated, but the myelin-associated metalloproteinase (Chapter 448) and tumor necrosis factor $\alpha$ converting enzyme (Chapter 449) illustrate that at least some are active. It is likely that the proteinase function and thus the biological activity of these related venom and mammalian enzymes is modulated by the domains following the proteinase domain. For example, in the case of the hemorrhagic toxins the disintegrin-like domain may function to alter platelet aggregation via interaction with the GPIIb IIIa receptor and also target the enzyme to regions where platelets may adhere (i.e. disruptions in the capillary basement membrane). In the PH-30 chains the disintegrin-like domains may also serve to target the protein and therefore the sperm to receptors present on the egg. In the case of PH-30$\alpha$ chain, the second high-cysteine subdomain contains within it the viral-like fusion sequence and one can speculate that this subdomain is important in presenting the fusion sequence in a recognizable form to a cell membrane. With the venom protein russellysin the second subdomain serves as the connecting site for the light chain (C lectin-like) (Chapter 433).

It is thus clear that the venom metalloproteinases, regardless of size and function, are structurally related via the classification scheme based on domain structure described above, as well as their nucleotide and amino acid sequences. Each new sequence of a snake venom metalloproteinase appearing in the scientific literature has fallen into the class of reprolysins and we expect that to continue. Furthermore, these venom proteinases, particularly of the P-III and P-IV class, are structurally related to the mammalian reproductive proteins EAP-I, PH-30 and cyritestin and others. It is expected that additional proteins will be found from snake, mammals, as well as other organisms, which will strengthen the foundation for the proposal of an encompassing class of unique proteins forming a subfamily of metalloendopeptidases related by the structural homologies outlined above, the reprolysins.

called the reprolysins, reflecting the reptilian and reproductive tract origin of these proteins.

## X-Ray Crystal Structure of Reprolysins

The most recent and informative developments in the field of reprolysin research has been the analysis of the X-ray crystal structures of adamalysin (proteinase II) and atrolysin C (HT-d), homologous enzymes from the venoms of *C. adamanteus* and *C. atrox* respectively (Gomis-Rüth *et al.*, 1994; Zhang *et al.*, 1994) (see Fig. 404.4). Their structures are identical, revealing an oblate ellipsoidal molecule with an active-site cleft separating a fairly irregularly folded small 'lower' subdomain from the 'upper' main domain, which is composed of an open twisted $\alpha$-$\beta$ structure having a five-stranded $\beta$ sheet and four $\alpha$ helices. The peptide chain of each of these enzymes starts at the lower subdomain surface and forms the regularly folded main domain with residues 6–150 (Figure 416.2). The topology of this upper domain begins with a $\beta$-$\alpha$-$\beta$ motif followed by an $\alpha$-helix link of the C helix to a $\beta$-$\beta$-$\beta$ motif and ending in the D helix, which is terminated by Gly149. The lower subdomain, consisting of the last 50 residues, is organized in multiple turns cross-connected by a disulfide bridge linking Cys157 and Cys164. The chain ends in a long C-terminal $\alpha$ helix and

an extended segment clamped to the upper domain with a disulfide bridge connecting Cys117 to Cys197. These disulfide bridges are typical of the two-disulfide reprolysins. The majority of the reprolysins, however, appear to be three-disulfide proteinases, having disulfide connections between Cys157-Cys181 and Cys159-Cys164 in the lower subdomain, as well as the Cys117-Cys197 connection.

The catalytic zinc ion in the active-site cleft is coordinated by His142, His146, His152 and a water molecule anchored to Glu143 in a nearly tetrahedral manner as predicted for atrolysin E by spectroscopic studies of cobaltous HT-e (Bjarnason & Fox, 1983). Residues His142, Glu143 and His146 are part of the long active-site D helix which extends to Gly149, where it turns sharply toward His152. The substrate-binding site is bordered by the Met-turn (Met166) which forms a 'basement' of the Zn locus and thus helps to anchor the essential 168–172 strand. This strand forms one 'wall' of the extended binding site. The segment between residues 109 and 114 of the antiparallel $\beta$ strand of the upper main domain defines the upper wall of the extended binding site. A calcium ion is observed liganded by the side chains of Glu9 and Asp93 of the main upper domain and Asn200 and the carbonyl group of Cys197 of C-terminal 'tail' as well as by internal and external water molecules.

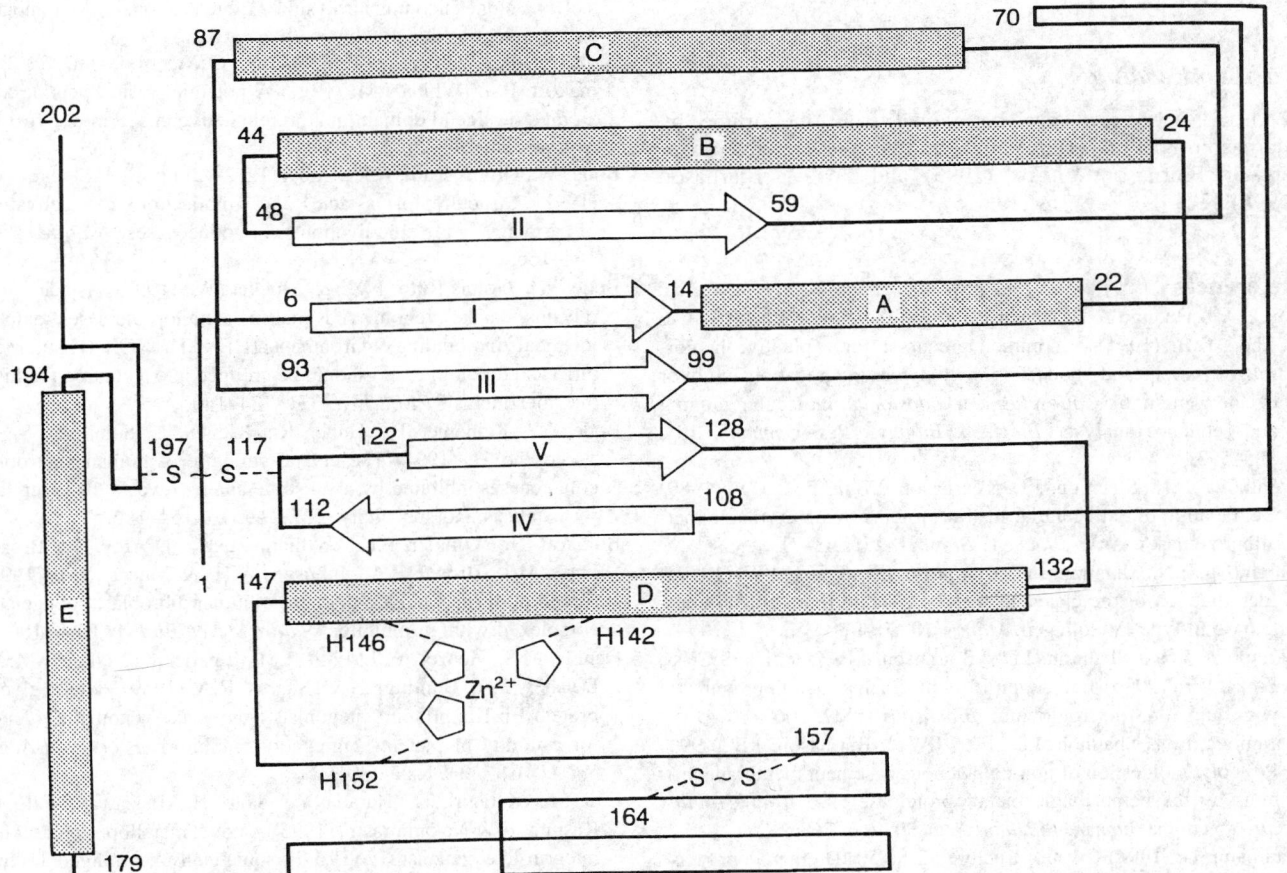

*Figure 416.2* Diagram of the structure of adamalysin/atrolysin C. Arrows represent $\beta$ sheet structure and shaded rectangles are $\alpha$ helices. The main domain extends from residue 6 to 150 and the minor domain from 153 to 202.

In addition to those of the reprolysins, several crystal structures of metalloendopeptidases from clan MB have been reported in recent years, in particular those of astacin (EC 3.4.24.27) (Chapter 405), serralysin (Chapter 386), aeruginolysin (Chapter 387) (Bode *et al*., 1992; Baumann *et al*., 1993; Baumann, 1994) and several of the matrixins, namely interstitial collagenases (Chapter 389) (Borakoti *et al*., 1994; Lovejoy *et al*., 1994a,b; Spurlino *et al*., 1994), human neutrophil collagenase (Chapter 390) (Bode *et al*., 1994; Stams *et al*., 1994), and human stromelysin 1 (Chapter 393) (Gooley *et al*., 1994). These metalloendopeptidases have a similar overall tertiary fold and active-site zinc ligation to the reprolysins. The three histidines in the HEXXHXXGXXH motif are zinc ligands and the fourth ligand is a water or a carboxylate group, conferring a tetrahedral structure on the zinc ion of the reprolysins and the matrixins. However, members of the astacin and serralysin subfamilies of metalloendopeptidases have a tyrosine as a fifth ligand to the zinc ion, conferring a penta coordinated trigonal bipyramidal geometry on the metal ion. On the other hand, all the above mentioned metalloendopeptidases share an identical active-site basement formed by a common Met-turn (Bode *et al*., 1993). Based on these common structural motifs of these four families or subfamilies, a clan or superfamily designated as clan MB or metzincins has been proposed (Rawlings & Barrett, 1995; Stöcker *et al*., 1995; Hooper, 1995) (see also Chapters 380, 385, 404).

## Further Reading

A further overview may be gained from the article by Bjarnason & Fox (1994) and by the two review chapters by Bjarnason & Fox (1995) and Fox & Bjarnason (1996).

## References

Au, L.-C., Huang, Y.-B., Huang, T.-F., Teh, G.-W., Lin, H.-H. & Choe, K.B. (1991) A common precursor for a putative hemorrhagic protein and rhodostomin, a platelet aggregation inhibitor of the venom of *Calloselasma rhodostoma*: molecular cloning and sequence analysis. *Biochem. Biophys. Res. Commun.* **181**, 585–593.

Baramova, E.N., Shannon, J.D., Bjarnason, J.B. & Fox, J.W. (1989) Degradation of extracellular matrix proteins by hemorrhagic metalloproteinases. *Arch. Biochem. Biophys.* **275**, 63–71.

Baramova, E.N., Shannon, J.D., Bjarnason, J.B. & Fox, J.W. (1990a) Identification of the cleavage sites by a hemorrhagic metalloproteinase in type IV collagen. *Matrix* **10**, 91–97.

Baramova, E.N., Shannon, J.D., Bjarnason, J.B., Gonias, S.C. & Fox, J.W. (1990b) Interaction of hemorrhagic metalloproteinases with human $\alpha_2$-macroglobulin. *Biochemistry* **29**, 1069–1074.

Baramova, E.N., Shannon, J.D., Fox, J.W. & Bjarnason, J.B. (1991) Proteolytic digestion of non-collagenous basement membrane proteins by the hemorrhagic metalloproteinase Ht-e from *Crotalus atrox* venom. *Biomed. Biochim. Acta* **50**, 763–768.

Baumann, U. (1994) Crystal structure of the 50 kDa metalloprotease from *Serratia marcescens*. *J. Mol. Biol.* **242**, 244–251.

Baumann, U., Wu, S., Flaherty, K.M. & McKay, D.B. (1993) Three-dimensional structure of the alkaline protease of *Pseudomonas aeruginosa*: a two-domain protein with a calcium binding parallel-beta roll motif. *EMBO J.* **12**, 3357–3364.

Bjarnason, J.B. & Fox, J.W. (1983) Proteolytic specificity and cobalt exchange of hemorrhagic toxin e, a zinc protease isolated from the venom of the Western diamondback rattlesnake *(Crotalus atrox)*. *Biochemistry* **22**, 3770–3778.

Bjarnason, J.B. & Fox, J.W. (1987) Characterization of two hemorrhagic zinc proteinases, toxin c and toxin d, from Eastern diamondback rattlesnake *(Crotalus atrox)* venom. *Biochim. Biophys. Acta* **911**, 356–363.

Bjarnason, J.B. & Fox, J.W. (1988–89) Hemorrhagic toxins from snake venoms. *J. Toxic. Toxin Rev.* **7**, 121–209.

Bjarnason, J.B & Fox, J.W. (1994) Hemorrhagic metalloproteinases from snake venoms. *Pharmacol. Ther.* **62**, 325–372.

Bjarnason, J.B. & Fox, J.W. (1995) Snake venom metalloproteinases: reprolysins. *Methods Enzymol.* **248**, 345–368.

Bjarnason, J.B. & Tu, A.T. (1978) Hemorrhagic toxins from Western diamondback rattlesnake *(Crotalus atrox)* venom: isolation and characterization of five toxins and the role of zinc in hemorrhagic toxin e. *Biochemistry* **17**, 3395–3404.

Bjarnason, J.B., Hamilton, D. & Fox, J.W. (1988) Studies on the mechanism of hemorrhage production by free proteolytic hemorrhagic toxin from *Crotalus atrox* venom. *Biol. Chem. Hoppe-Seyler* **369**, 121–129.

Bjarnason, J.B., Hite, L.A., Shannon, J.D., Tapaninaho, S.J. & Fox, J.W. (1993) Atrolysin E (Hte, E.C. 3.4.24.44) and the non-hemorrhagic $\alpha$-protease – comparative properties and sequence of a cDNA clone encoding atrolysin E. Evidence for signal, zymogen and disintegrin-like structures. *Toxicon* **31**, 513.

Blobel, C.P., Wolfsberg, T.G., Turck, C.W., Myles, D.G., Primakoff, P. & White, J.M. (1992) A potential fusion peptide and an integrin ligand domain in a protein active in sperm-egg fusion. *Nature* **356**, 248–252.

Bode, W., Gomis-Rüth, F.X., Huber, R., Zwilling, R. & Stöcker, W. (1992) Structure of astacin and implication for activation of astacins and zinc-ligation of collagenases. *Nature* **358**, 164–166.

Bode, W., Gomis-Rüth, F.X. & Stöcker, W. (1993) Astacin, serralysins, snake venom and matrix metalloproteinases exhibit identical zinc-binding environments (HEXXHXXGXXH and Met-turn) and topologies and should be grouped into a common family, the 'metzincins'. *FEBS Lett.* **331**, 134–140.

Bode, W., Reinemer, P., Huber, R., Kleine, T., Schnierer, S. & Tschesche, H. (1994) The crystal structure of human neutrophil collagenases inhibited by a substrate analog reveals the essentials for catalysis and specificity. *EMBO J.* **13**, 164–166.

Borakoti, N., Winkler, F.K., Williams, D.H., D'Arcy, A., Broadhurst, M.J., Brown, P.A., Johnson, W.H. & Murray, E.J. (1994) Structure of the catalytic domain of human fibroblast collagenase complexed with an inhibitor. *Nature Struct. Biol.* **1**, 106–110.

Dennis, M.S., Henzel, W.J., Pitti, R.M., Lipari, M.T., Napier, M.A., Deisher, T.A., Bunting, S. & Lazarus, R.A. (1989) Platelet glycoprotein IIb-IIIa protein antagonists from snake venoms: Evidence for a family of platelet aggregation inhibitors. *Proc. Natl Acad. Sci. USA* **87**, 2471–2475.

Emi, M., Katagiri, T., Harada, Y., Saito, H., Inazawa, J., Ito, I., Kasumi, F. & Nakamura, Y. (1993) A novel metalloproteinase/disintegrin-like gene at 17q21.3 is somatically rearranged in two primary breast cancers. *Nature Genet.* **5**, 151–157.

Fox, J.W. & Bjarnason, J.B. (1995) Atrolysins: metalloproteinases from *Crotalus atrox* venom. *Methods Enzymol.* **248**, 368–387.

Fox, J.W. & Bjarnason, J.B. (1996) The reprolysins: a family of metalloproteinases defined by snake venom and mammalian metalloproteinases. In: *Zinc Metalloproteinases in Health and Disease* (Hooper, N.M., ed.). London: Taylor & Francis, pp. 47–83.

Fox, J.W., Campbell, R., Beggerly, L. & Bjarnason, J.B. (1986) Substrate specificities and inhibition of two hemorrhagic zinc proteases Ht-c and Ht-d from *C. atrox* venom. *Eur. J. Biochem.* **156**, 65–72.

Gomis-Rüth, F.-X., Kress, L.F. & Bode, W. (1993) First structure of a snake venom metalloproteinase: a prototype for matrix metalloproteinases/collagenases. *EMBO J.* **12**, 4151–4157.

Gomis-Rüth, F.X., Kress, L.F. & Bode, W. (1994) Refined 2.0 Å X-ray crystal structure of the snake venom zinc endopeptidase adamalysin II. Primary and tertiary structure determination, refinement, molecular structure and comparison with astacin, collagenase and thermolysin. *J. Mol. Biol.* **239**, 513–544.

Gooley, P.R., O'Connell, J.F., Marcy, A.I., Cuca, G.C., Salowe, S.P., Bush, B.L., Hermes, J.D., Esser, C.K., Hagmann, W.K., Springer, J.P. & Johnson, B.A. (1994) The NMR structure of the inhibited catalytic domain of human stromelysin-1. *Nature Struct. Biol.* **1**, 111–118.

Gould, R.J., Polokoff, M.A., Friedman, P.A., Huang, T.-F., Holt, J.C., Cook, J.J. & Niewiarowski, S. (1990) Disintegrins: a family of integrin inhibitory proteins from viper venoms. *Proc. Soc. Exp. Biol. Med.* **195**, 168–171.

Hite, L.A., Shannon, J.D., Bjarnason, J.B. & Fox, J.W. (1992) Sequence of a cDNA clone encoding zinc metalloproteinase hemorrhagic toxin e from *Crotalus atrox*: evidence for signal, zymogen and disintegrin-like structures. *Biochemistry* **31**, 6203–6211.

Hite, L.A., Jia, L.-G., Bjarnason, J.B. & Fox, J.W. (1993) cDNA sequence for four snake venom metalloproteinases: structure, classification and their relationship to mammalian reproductive proteins. *Arch. Biochem. Biophys.* **308**, 182–191.

Hooper, N.M. (1995) The biological roles of zinc and zinc metalloproteinases In: *Zinc Metalloproteinases in Health and Disease* (Hooper, N.M., ed.). London: Taylor & Francis, pp. 1–21.

Kurecki, T. & Laskowski, M., Sr. (1978) Purification of collagenases from *Crotalus adamanteus* venom. In: *Toxicon Supplement No. 1, Toxins, Animals, Plant and Microbial.* Elmsford, New York: Pergamon Press, p. 311.

Kurecki, T., Laskowski, M., Sr. & Kress, L.F. (1978) Purification and some properties of two proteinases from *Crotalus adamanteus* venom that inactive human α1-proteinase inhibitor. *J. Biol. Chem.* **253**, 8340–8345.

LeHueron, I., Wicker, C., Guilloteau, P., Toullec, R. & Puigserver, A. (1990) Isolation and nucleotide sequence of cDNA clone for bovine pancreatic anionic trypsinogen. Structural identity within the trypsin family. *Eur. J. Biochem.* **193**, 767–773.

Lovejoy, B., Cleasby, A., Hassell, A.M., Longley, K., Luther, M.A., Weigl, D., McGeehan, G., McElroy, A.B., Drewry, D., Lambert, M.H. & Jordan, S.R. (1994a) Structure of the catalytic domain of fibroblast collagenase complexed with an inhibitor. *Science* **263**, 375–377.

Lovejoy, B., Hassell, A.M., Luther, M.A., Weigl, D. & Jordan, S.R. (1994b) Crystal structure of recombinant 19-kDa human fibroblast collagenase complexed to itself. *Biochemistry* **33**, 8207–8217.

Neeper, M.P. & Jacobson, M.A. (1990) Sequence of a cDNA encoding the platelet aggregation inhibitor trigramin. *Nucleic Acids Res.* **18**, 4255–4261.

Nomura, N., Miyajima, N., Sazuka, T., Tanaka, S., Kawarabayasi, Y., Sato, S., Nagase, T., Seki, N., Ishikawa, K. & Tabata, S. (1994) Prediction of the coding sequence of unidentified human genes. I. The coding sequences of 40 new genes (KIAA0001–KIAA0040) deduced by analysis of randomly sampled cDNA clones from human immature myeloid cell line KG-1. *DNA Res.* **1**, 27–35.

Paine, M.J.I., Desond, H.P., Theakston, R.D.G. & Crampton, J.M. (1992) Purification, molecular characterization of a high molecular weight hemorrhagic metalloproteinase, jararhagin, from *Bothrops jararaca* venom. *J. Biol. Chem.* **247**, 22869–22876.

Perry, A.C.F., Jones, R., Barker, P.J. & Hall, L. (1992) A mammalian epididymal protein with remarkable sequence similarity to snake venom hemorrhagic peptides. *Biochem. J.* **286**, 671–675.

Rawlings, N.D. & Barrett, A.J. (1995) Evolutionary families of metallopeptidases. *Methods Enzymol.* **248**, 183–228.

Scarborough, R.M., Rose, J.W., Naughton, M.A., Philips, D.R., Nannizzi, C., Arfsten, A., Campbell, A.M. & Charo, I.F. (1993) Characterization of the integrin specificities of disintegrins isolated from American pit viper venoms. *J. Biol. Chem.* **268**, 1058–1065.

Senftleben, A., Wallat, S., Lemaire, L. & Heinlein, U.A. (1992) Pre and postmeiotic germ cell specific expression of TAZ83, a gene encoding a putative, cysteine-rich transmembrane protein. GenBank Accession Number X64227 (no other publication).

Shannon, J.D., Baramova, E.N., Bjarnason, J.B. & Fox, J.W. (1989) Amino acid sequence of a *Crotalus atrox* venom metalloproteinase which cleaves type IV collagen and gelatin. *J. Biol. Chem.* **264**, 11575–11583.

Spiekerman, A.M., Fredericks, K.K., Wagner F.W. & Prescott, J.M. (1973) Leucostoma peptidase A: a metalloprotease from snake venom. *Biochim. Biophys. Acta* **293**, 464–472.

Spurlino, J.C., Smallwood, A.M., Carlton, D.D., Banks, T.M., Vavra, K.J., Johnson, J.J., Cook, E.R., Falvo, J., Wahl, R.C., Pulvino, T.A., Wendolski, J.J. & Smith, D.L. (1994) 1.56A structure of mature truncated human fibroblast collagenase. *Proteins Struct. Funct. Genet.* **19**, 98–109.

Stams, T., Spurlino, J.C. Smith, D.L., Wahl, R.C., Ho, T.F., Qorronfleh, M.W., Banks, T.M. & Rubin, B. (1994) Structure of human neutrophil collagenase reveals large S1′ specificity pocket. *Nature Struct. Biol.* **1**, 119–123.

Stöcker, W., Grams, F., Baumann, U., Reinemer, P., Gomis-Rüth, F.-X., McKay, D.B. & Bode, W. (1995) The metzincins – topological and sequential relations between the astacins, adamalysins, serralysins, and matrixins (collagenases) define a superfamily of zinc-peptidases. *Protein Sci.* **4**, 823–840.

Sugihara, H., Moriura, M. & Nikai, T. (1983) Purification and properties of a lethal, hemorrhagic protein, 'mucrotoxin A', from the venom of the Chinese habu snake (*Trimeresurus mucrosquamatus*). *Toxicon* **21**, 247–255.

Takeya, H., Oda, K., Miyata, T., Omori-Satoh, T. & Iwanaga, S. (1990) The complete amino acid sequence of the high molecular mass hemorrhagic protein HR1B isolated from the venom of *Trimeresurus flavoviridis. J. Biol. Chem.* **265**, 16068–16073.

Takeya, H., Nishida, S., Miyata, T., Kawada, S.-I., Saisaka, Y., Morita, T. & Iwanaga, S. (1992) Coagulation factor X activating enzyme from Russell's viper venom (RVV-X). *J. Biol. Chem.* **267**, 14109–14117.

Yoshida, S., Setoguchi, M., Higuchi, Y., Akizuki, S. & Yamamoto, S. (1990) Molecular cloning of cDNA encoding MS2 antigen, a novel cell surface antigen strongly expressed in murine monocytic lineage. *Int. Immunol.* **2**, 585–591.

Zhang, D., Botos, I., Gomis-Rüth, F.X., Doll, R., Blood, C., Njoroge, F.G., Fox, J.W., Bode, W. & Meyer, E. (1994) Structural interaction of natural and synthetic inhibitors with the venom metalloproteinase, atrolysin C (Ht-d). *Proc. Natl Acad. Sci. USA* **91**, 8447–8451.

Zwilling, R. & Pfleiderer, G. (1967) Eigenschaften der $\alpha$-protease aus dem Gift von *Crotalus atrox* [Properties of the $\alpha$-protease from the venom of *Crotalus atrox*]. *Z. Physiol. Chem.* **38**, 519–528.

*Jón B. Bjarnason*
Science Institute,
University of Iceland,
Reykjavík, Iceland
Email: jonbragi@raunvis.hi.is

*Jay W. Fox*
Department of Microbiology,
University of Virginia Health Sciences Center,
Charlottesville, VA 22908, USA
Email: jwf8x@virginia.edu

# 417. Fibrolase

## Databanks

*Peptidase classification: clan MB, family M12, MEROPS ID: M12.133*
*NC-IUBMB enzyme classification: EC 3.4.24.72*
*Databank codes:*

| Species | SwissProt | PIR | EMBL (cDNA) | EMBL (genomic) |
|---|---|---|---|---|
| *Agkistrodon contortrix* | P28891 | A41827 | – | – |

## Name and History

Fibrolase is a metalloproteinase with fibrinolytic activity that was isolated from the venom of *Agkistrodon contortrix contortrix* (southern copperhead snake). Didisheim & Lewis (1956) identified a number of venoms that possessed fibrinolytic protease activity and were among the first to suggest a clinical use for this activity as an agent for the dissolution of blood clots. Kornalik (1966) reported fibrinolytic activity in the venom of *A. c. contortrix*. In this work he showed that proteinases from *A. c. contortrix* venom acted directly on fibrin and did not function as plasminogen activators. Bajwa *et al.* (1982) described a ***fibrinolytic protease*** from *A. c. contortrix* venom with a molecular mass of approximately 25 000; the enzyme is now known as ***fibrolase***. Amino acid sequence analysis resulted in the determination of the molecular mass as 23 000 (Randolph *et al.*, 1992). Markland *et al.* (1993) reported that two different isoforms of fibrolase exist and can be separated by weak cation-exchange HPLC.

## Activity and Specificity

Fibrolase is a direct-acting fibrinolytic enzyme cleaving the A$\alpha$ chain of fibrinogen initially at position Lys413$\dagger$Leu414; it also cleaves the B$\beta$ chain, but at a slower rate. Fibrolase has no effect on the $\gamma$ chain. Retzios & Markland (1988) described the lack of activity of fibrolase on a number of artificial chromogenic substrates commonly cleaved by serine proteinases. Their findings indicate that fibrolase is not a serine protease. Fibrolase differs from the clinically used thrombolytic agents tissue plasminogen activator (t-PA, Chapter 58) and streptokinase in that it does not activate plasminogen (Retzios & Markland, 1988). After incubation of fibrolase with plasminogen there is no evidence of plasmin formation. Subsequent incubation of the fibrolase–plasminogen mixture with streptokinase yields plasmin, indicating that fibrolase neither activates, degrades nor inactivates plasminogen.

While fibrolase degrades fibrin and fibrinogen *in vitro*, pretreatment of fibrolase with plasma inhibits fibrinolytic activity (Markland *et al.*, 1988). Further studies indicated that incubation of fibrolase with $\alpha_2$-macroglobulin results in loss of fibrolase activity. SDS-PAGE showed that the 180 kDa $\alpha_2$-macroglobulin is cleaved into two 90 kDa fragments indicating binding of fibrolase to the proteinase inhibitor. Recent studies have shown that fibrolase interacts with $\alpha_2$-macroglobulin rapidly and with high affinity. *In vivo* experiments have shown that fibrolase lyses experimental thrombi in both arteries and veins (Markland, 1996).

Fibrolase activity is completely and rapidly inhibited by the addition of EDTA, tetraethylenepentamine and 1,10-phenanthroline (Markland *et al.*, 1988). Inhibition by agents that chelate zinc supports the conclusion from atomic absorption spectroscopy (Guan *et al.*, 1991) that fibrolase is a zinc metalloproteinase and not a serine or cysteine proteinase.

Fibrolase activity *in vitro* can be measured by several different methods. Two of the most commonly used procedures are the degradation of azocasein and dissolution of fibrin clots in a multiwell plate (Retzios & Markland, 1992). The azocasein assay is not specific, as is the fibrin plate assay, but the fibrin plate assay requires an incubation of from several hours to overnight and is not very sensitive. A more sensitive and rapid method of determination of fibrinolytic activity has recently been utilized, the cleavage of a synthetic octapeptide (Abz-His-Thr-Glu-Lys+Leu-Val-Thr-Ser-Dnp) containing the fibrolase scissile Lys413+Leu414 bond found in the $\alpha$ chain sequence of fibrin, an N-terminal fluorophore and a C-terminal quenching group (F.S. Markland, unpublished results). In the uncleaved peptide, the quenching moiety blocks the emission of a fluorescent signal, but when the peptide is cleaved, the two groups become separated, allowing the emission of a fluorescent signal. Cleavage of the peptide can be used both to confirm proteolytic activity and to monitor the kinetics of fibrolase activity. This method is rapid and highly sensitive.

The pH optimum for fibrolase activity in both the azocasein and fibrin plate assays is 7.4 and the pI of the enzyme, as determined by isoelectric focusing in an immobilized pH gradient, is 6.8 (Guan & Markland, 1988).

## Structural Chemistry

Fibrolase is a 23 kDa single-chain protein containing 1 mol of zinc per mol of protein (Guan et al., 1991). Structurally, the protein comprises five $\alpha$-helical regions and four parallel and one antiparallel $\beta$ sheets (Pretzer et al., 1992). The structure is stabilized by three disulfide bonds (Randolph et al., 1992; Manning, 1995). As evidenced by separation of fibrolase by weak cation-exchange HPLC (CM300) there are at least two distinct isoforms of the enzyme. The difference between the two isolated isoforms lies in their N-termini, one of the isoforms possesses two Gln residues while the other has only one (Loayza et al., 1994). This alteration in the primary sequence of the protein has no apparent effect on enzymatic activity, as both isoforms possess identical activity against different substrates and both isoforms are equally inhibited when incubated with agents known to inhibit fibrolase activity.

Fibrolase shares a high degree ($\sim$60%) of amino acid sequence similarity with other enzymes in the adamalysin/reprolysin subfamily (Bode et al., 1993). Crystals of fibrolase have been grown from polyethylene glycol (Markland et al., 1993) but the structure of fibrolase has not yet been solved. A three-dimensional model of the fibrolase structure has been created based on the structures of other members of the adamalysin subfamily of proteinases (Bolger et al., 1996). The energy-minimized model shows an identical structure in the active-site groove and allows for the three disulfide bonds in fibrolase compared to two disulfide bonds in adamalysin (Chapter 421; Gomis-Rüth et al., 1994) and atrolysin C (Chapter 423; Zhang et al. 1994), both members of the adamalysin subfamily with known three-dimensional structures.

Sequence similarity in the active-site region of the enzymes, as shown in Figure 417.1, is close to 100%. The conserved amino acids between residues 140 and 166 include the three histidines involved in zinc ligation and the Met-turn motif, the hallmarks of this group of enzymes.

```
Fibrolase:  141 AHELGHNLGM NHDGNQCHCG ANSCVMAAML 170
Atrolysin:  141 AHELGHNLGM EHDGKDCLRG ASLCIMRPGL 170
Adamalysin: 141 AHELGHNLGM EHDGKDCLRG ASLCIMRPGL 170
Consensus:  141 AHELGHNLGM HDG  C G A C M  L 170
```

Figure 417.1 Comparison of amino acid sequences in the active site region of three different venom metalloproteinases: fibrolase (Randolph et al., 1992), atrolysin C (Zhang et al., 1994) and adamalysin (Gomis-Rüth et al., 1994). Both fibrolase and adamalysin are nonhemorrhagic while atrolysin C is hemorrhagic. Adamalysin degrades plasma serpins such as antithrombin III (Kress, 1988) and $\alpha_1$-proteinase inhibitor. The active-site histidine residues (142, 146, 152) and Met166 of the Met-turn are invariant; differences lie in the residues surrounding the active site.

## Preparation

Fibrolase has been purified by a number of different procedures, but the method which was developed most recently provides a high yield and is fairly simple. This method involves a three-step HPLC purification (Loayza et al., 1994). In this method hydrophobic interaction HPLC is followed by hydroxyapatite chromatography. The fractions which possess fibrinolytic activity from the hydroxyapatite HPLC column contain one major band when analyzed by SDS-PAGE. The final purification step by cation-exchange HPLC (CM300) eliminates the minor contamination and allows for separation of two isoforms of fibrolase. When starting from crude Agkistrodon contortrix contortrix venom the expected yield of fibrolase is approximately 15–30 mg g$^{-1}$ of crude venom. Other methods of purification employ open column chromatography and result in an enzyme identical to that purified by the HPLC procedure (Markland et al., 1988).

## Biological Aspects

Fibrolase is present in the venom of the southern copperhead snake. It is assumed that the role of this enzyme upon envenomation is to remove fibrin and fibrinogen from the blood to maintain an anticoagulant state that will enable the rapid spread of the other venom components throughout the circulatory system. Fibrolase shares a high degree of sequence similarity with a number of other venom metalloproteinases. However, fibrolase has a unique fibrin(ogen)olytic activity. A number of the enzymes sharing sequence identity with fibrolase are hemorrhagic. Fibrolase does not exhibit hemorrhagic activity and as stated previously cleaves the $\alpha$ and $\beta$ chains of fibrin and A$\alpha$ and B$\beta$ chains of fibrinogen.

Fibrolase is inhibited in vivo by $\alpha_2$-macroglobulin, but not by any of the serine proteinase inhibitors. Fibrolase does not activate or degrade plasminogen nor does it possess thrombin-like activity (Retzios & Markland, 1988).

A cDNA copy of fibrolase has been cloned into a yeast expression vector and the protein produced has been observed to possess activity identical to that of the native enzyme (Loayza et al., 1994). Other than the expression of fibrolase as a preproenzyme in the snake venom gland, nothing else is known about the gene coding for or the expression of fibrolase in A. c. contortrix.

## Distinguishing Features

Fibrolase has been used in experimental systems as a thrombolytic agent against artificially created thrombi in both rabbit and canine models. Upon gross or histological examination of the treated animals following lysis of thrombi, no side-effects could be detected. Fibrolase shows promise as a therapeutic agent as it differs from thrombolytic agents available today. Fibrolase acts directly on the thrombus, while agents such as t-PA and urokinase act indirectly by converting plasminogen to plasmin.

Fibrolase, while high in sequence similarity, differs in activity from other protease members of the adamalysin subfamily (Chapter 416). The primary activity of fibrolase is the degradation of fibrin(ogen). Other members of the subfamily possess different activities including hemorrhagic activity. While past experimentation has proven fibrolase to be an effective and unique thrombolytic agent, decreasing its inhibition by $\alpha_2$-macroglobulin would make the enzyme even more effective.

## Related Peptidases

Enzymes with similar physicochemical properties and biological activities have been isolated from venom of a number of different types of snakes. Nonhemorrhagic enzymes with fibrinolytic activity that have been purified and characterized and shown to have sequence homology with fibrolase and that are not described in this *Handbook* include the proteinase from *Vipera lebetina* venom, lebetase (Siigur & Siigur, 1991).

*A. c. contortrix* venom possesses four additional proteinases with activity toward fibrinogen as described by Dyr *et al.* (1983). Two of these enzymes are procoagulant and two are fibrinogenolytic. These four proteinases range in size from 25 000 to 68 000. Unlike fibrolase, all of the enzymes, including the one with a molecular mass of 25 000, possess activity toward chromogenic substrates, suggesting they are serine proteinases.

## Further Reading

An overview of fibrolase is found in Markland (1996).

## References

Bajwa, S.S., Russell, F.E. & Markland, F.S. (1982) Thrombolytic agents in snake venoms. *Proc. West. Pharmacol. Soc.* **25**, 353–356.

Bode, W., Gomis-Rüth, F.-X. & Stöcker, W. (1993) Astacins, serralysins snake venom and matrix metalloproteinases exhibit identical zinc-binding environments (HEXXHXXGXXH and Met turn) and topologies and should be grouped into a common family, the 'metzincins'. *FEBS Lett.* **331**, 134–140.

Bolger, M., Swenson, S. & Markland, F.S. (1996) Three-dimensional structure of fibrolase, the fibrinolytic enzyme from Southern copperhead venom, modeled from the x-ray structure of adamalysin II. *Pharmacol. Res.* **13**, S144 (abstr.).

Didisheim, P. & Lewis, J.H. (1956) Fibrinolytic and coagulant activities of certain snake venoms and proteases. *Proc. Soc. Exptl. Biol. Med.* **93**, 10–13.

Dyr, J.E., Blomback, B. & Kornalik, F. (1983) The fibrinogenolytic and procoagulant activity of Southern copperhead venom enzymes. *Thromb. Res.* **30**, 185.

Gomis-Rüth, F.X., Kress, L.F., Kellermann, J., Mayr, I., Lee, X., Huber, R. & Bode, W. (1994) Refined 2.0 Å X-ray crystal structure of the snake venom zinc-endopeptidase adamalysin II. Primary and tertiary structure determination, refinement, molecular structure and comparison with astacin, collagenase and thermolysin. *J. Mol. Biol.* **239**, 513–544.

Guan, A.L. & Markland, F.S. (1988) Isoelectric focusing in immobilized pH gradients of a snake venom fibrinolytic enzyme. *J. Biochem. Biophys. Methods* **16**, 215–226.

Guan, A.L., Retzios, A.D., Henderson, G.N. & Markland, F.S. (1991) Purification and characterization of a fibrinolytic enzyme from the venom of the Southern copperhead snake (*Agkistrodon contortrix contortrix*). *Arch. Biochem. Biophys.* **289**, 197–207.

Kornalik, F. (1966) The influence of snake venoms of fibrinogen conversion and fibrinolysis. *Mem. Inst. Butantan* **33**, 179.

Kress, L.F. (1988) The action of snake venom metalloproteinase on plasma proteinase inhibitors. In: *Hemostasis and Animal Venoms* (Pirkle, H. & Markland, F.S., eds). New York: Marcel Dekker, pp. 335–348.

Loayza, S.L., Trikha, M., Markland, F.S., Riquelme, P. & Kuo, J. (1994) Resolution of isoforms of natural and recombinant fibrolase, the fibrinolytic enzyme from *Agkistrodon contortrix contortrix* snake venom, and comparison of their EDTA sensitivities. *J. Chromatogr. B.* **662**, 227–243.

Manning, M.C. (1995). Sequence analysis of fibrolase, a fibrinolytic metalloproteinase from *Agkistrodon contortrix contortrix*. *Toxicon* **33**, 1189–1200.

Markland, F.S. (1996). Fibrolase, an active thrombolytic enzyme in arterial and venous thrombosis model systems. In: *Natural Toxins II. Adv. Exp. Med. Biol.*, vol. 391 (Singh, B.R. & Tu, A.T., eds). New York: Plenum Press, pp. 427–438.

Markland, F.S., Reddy, K.N.N. & Guan, L. (1988) Purification and characterization of a direct-acting-fibrinolytic enzyme from southern copperhead venom. In: *Hemostasis and Animal Venoms* (Pirkle, H. & Markland, F.S., eds). New York: Marcel Dekker, pp. 173–189.

Markland, F.S., Morris, S., Deschamps, J.R. & Ward, K.B. (1993) Resolution of isoforms of natural and recombinant fibrinolytic snake venom enzyme using high performance capillary electrophoresis. *J. Liquid Chromatogr.* **16**, 2189–2201.

Pretzer, D.B.S., Vander Velde, D.G., Smith, C.D., Mitchell, J.W. & Manning, M.C. (1992) Effect of zinc binding on the structure and stability of fibrolase, a fibrinolytic protein from snake venom. *Pharm. Res.* **9**, 870–877.

Randolph, A., Chamberlain, S.H., Chu, C., Retzios, A.D., Markland, F.S. & Masiarz, F.R. (1992) Amino acid sequence of fibrolase, a direct-acting fibrinolytic enzyme from *Agkistrodon contortrix contortrix* venom. *Protein Sci.* **1**, 590–600.

Retzios, A.D. & Markland, F.S. (1988) A direct-acting fibrinolytic enzyme from the venom of *Agkistrodon contortrix contortrix*: effects on various components of the human blood coagulation and fibrinolytic systems. *Thromb. Res.* **52**, 541–552.

Retzios, A.D. & Markland, F.S. (1992) Purification, characterization, and fibrinogen cleavage sites of three fibrinolytic enzymes from the venom of *Crotalus basiliscus basiliscus*. *Biochemistry* **31**, 4547–4557.

Siigur, E. & Siigur, J. (1991) Purification and characterization of lebetase, a fibrinolytic enzyme from *Vipera lebetina* (snake)

venom. *Biochim. Biophys. Acta* **1074**, 223–229.
Zhang, D., Botos, I., Gomis-Rüth, F.X., Doll, R., Blood, C., Njoroge, F.G., Fox, J.W., Bode, W. & Meyer, E.F. (1994) Structural inter-

action of natural and synthetic inhibitors with the venom metalloproteinase, atrolysin C (form d). *Proc. Natl Acad. Sci. USA* **91**, 8447–8451.

**Stephen Swenson**
*Department of Biochemistry and Molecular Biology,*
*University of Southern California, School of Medicine,*
*Cancer Research Laboratory no. 104,*
*1303 N. Mission Road,*
*Los Angeles, CA 90033, USA*
*Email: sswenson@hsc.usc.edu*

**Francis S. Markland, Jr**
*Department of Biochemistry and Molecular Biology,*
*University of Southern California, School of Medicine,*
*Cancer Research Laboratory no. 106,*
*1303 N. Mission Road,*
*Los Angeles, CA 90033, USA*
*Email: markland@hsc.usc.edu*

# *418. Atrolysin A*

## Databanks

*Peptidase classification: clan MB, family M12, MEROPS ID: M12.142*
*NC-IUBMB enzyme classification: EC 3.4.24.1*
*Chemical Abstracts Service registry number: 37288-82-7*
*Databank codes:*

| Species | SwissProt | PIR | EMBL (cDNA) | EMBL (genomic) |
|---|---|---|---|---|
| *Crotalus atrox* | – | S41607 | U01234 | – |

## Name and History

**Atrolysin A**, formerly known as **hemorrhagic toxin a (HT-a)**, is isolated from the venom of the western diamondback rattlesnake, *Crotalus atrox* (Bjarnason & Tu, 1978). This zinc metalloproteinase is the most potent hemorrhagic toxin isolated to date from *C. atrox* venom. It is a member of the P-III class of snake venom metalloproteinases (Chapter 416). The classification of atrolysin A as EC 3.4.24.1 stems from the original classification of a DEAE-chromatography fraction of *C. atrox* venom as α-proteinase (Pfleiderer & Krauss, 1965). This α-proteinase fraction is heterogeneous and comprises the metalloproteinases atroxase (Chapter 419) (Willis & Tu, 1988) and atrolysin B (Chapter 420) (Kruzel & Kress, 1985; Bjarnason & Fox, 1995). α-Proteinase should no longer be used as a name for a *C. atrox* metalloproteinase.

## Activity and Specificity

The peptidase activity of atrolysin A can be assayed with the quenched fluorescence substrate Abz-Ala-Gly+Leu-Ala-nitrobenzylamide (Shimokawa *et al.*, 1996) or by using protein substrates such as dimethylcasein (Bjarnason & Tu, 1978). Proteolytic activity is stabilized by low concentrations of calcium (e.g. 2 mM; Bjarnason & Tu, 1978).

The proteolytic and hemorrhagic activity of atrolysin E is inhibited by EDTA and 1,10-phenanthroline (Bjarnason & Tu, 1978). Unlike the lower molecular weight atrolysins, atrolysin A is not significantly inhibited by $\alpha_2$-macroglobulin (Baramova *et al.*, 1990). Atrolysin A does, nevertheless, cleave the bait region of $\alpha_2$-macroglobulin at positions His-Ala-Arg696+Leu697-Val-His and Arg-Gly-His694+Ala695-Arg-Leu. The poor inhibition of atrolysin A by $\alpha_2$-macroglobulin may indirectly contribute to the observed high hemorrhagic activity for this snake venom metalloproteinase. The serum protein oprin from the opossum can only partially inhibit the proteolytic and hemorrhagic activity of atrolysin A (Catanese & Kress, 1992). However, an additional, uncharacterized fraction from opossum serum can completely inhibit these activities of atrolysin A (Catanese & Kress, 1992). Although the mechanism of inhibition is uncertain, a noncovalent, inhibitor/inactive enzyme complex is formed.

The sites of cleavage of the oxidized B chain of insulin by atrolysin A according to Bjarnason *et al.* (1988) are:

FVN+QH+LCGSH+LVEA+LY+LVCGERGFFYTPKA

This corrects the earlier report of Tu *et al.* (1981) which did not note the cleavages at Asn3+Gln4 and His5+Leu6.

Of the atrolysins A, E, B and D, atrolysin A is most active in the production of soluble peptides from a basement membrane preparation from Englebreth–Holm–Swarm tumor (Bjarnason *et al*., 1988). This proteolytic activity on the basement membrane reflects the fact that atrolysin A is the most potent hemorrhagic atrolysin. Atrolysin A can degrade fibrinogen to unclottable products with the A$\alpha$ chain being rapidly degraded followed by the degradation of the B$\beta$ and $\gamma$ chains (Bjarnason *et al*., 1988). Fibrin is not significantly digested by atrolysin A. The basement membrane proteins – collagen type IV, laminin, fibronectin and nidogen – are readily degraded by atrolysin A (Bjarnason *et al*., 1988; Baramova *et al*., 1989). The interstitial collagen types I and III and type V collagen are not digested by atrolysin A. Gelatin is rapidly digested by atrolysin A (Baramova *et al*., 1989).

## Structural Chemistry

Atrolysin A as isolated from the venom of *Crotalus atrox* has an $M_r$ of 68 000 by SDS-PAGE with a pI value in the acidic pH range (Bjarnason & Tu, 1978). There is one zinc atom per molecule along with a variable number of calcium ions. The cDNA sequence analysis of atrolysin A establishes that it is a member of the P-III class of snake venom reprolysins (Hite *et al*., 1994). The proteinase domain has three disulfide bonds compared to two observed for the P-I class of snake venom metalloproteinases. The protein has potential glycosylation sites at Asn263, Asn517, Asn530 and Asn533. From Edman sequence analysis of atrolysin A, Asn517 and Asn533 appear to be glycosylated, but Asn533 is not (Hite *et al*., 1994).

## Preparation

Isolation of atrolysin A begins with the dialysis of lyophilized crude *Crotalus atrox* venom followed by ion-exchange chromatography on DEAE-cellulose (Bjarnason & Tu, 1978). Four additional chromatography steps (gel filtration on Sephadex G-75; ion exchange on DEAE-cellulose and two gel-filtration steps on AcA44) of fraction A-1 from the initial DEAE chromatography step yield homogeneous atrolysin A. A typical isolation yields approximately 32 mg of pure protein from 10 g of crude venom.

## Biological Aspects

Atrolysin A is the most potent hemorrhagic toxin isolated to date from *Crotalus atrox* venom; it has a minimum hemorrhagic dose (MHD) of 0.04 $\mu$g (Fox & Bjarnason, 1995). Based on the amount and activity of atrolysin A, it is estimated that it contributes approximately 16% of the total hemorrhagic activity of *Crotalus atrox* venom. When purified atrolysin A is injected into the muscle tissue of mice, a rapid appearance of ecchymotic hemorrhage occurs along with degradation of vessels and erythrocytes present in the surrounding stroma (Ownby *et al*., 1978). Under electron microscopic observation, atrolysin A appears to produce lysis of endothelial plasma membrane and loss of adjacent basement membrane (Ownby *et al*., 1978).

Atrolysin A has three domains in the mature protein; the proteinase, disintegrin-like and cysteine-rich domains. The protein has platelet-aggregation inhibitory activity which can be attributed to the disintegrin-like domain (Jia *et al*., 1997). The cysteine-rich domain appears to function by binding to collagen type I. There is a possibility for the proteolytic release of the spacer/disintegrin-like/cysteine-rich portion of the molecule. A protein of similar composition, jararhagin-C, from the P-III snake venom metalloproteinase jararhagin (Chapter 429) has been isolated from the venom of *Bothrops jararaca jararaca* (Usami *et al*., 1994). The free spacer/disintegrin-like/cysteine-rich portion of catrocollastatin, termed catrocollastatin-C, has also been isolated from *Crotalus atrox* venom (Shimokawa *et al*., 1997). It appears that the disintegrin-like and cysteine-rich domains of the atrolysin A may function to modulate the overall pathological effect of the toxin by inhibiting platelet aggregation and localizing the toxin to the subendothelium by attachment to collagen (Jia *et al*., 1997). The functional properties of these nonproteinase domains in atrolysin A, as well as in other P-III hemorrhagic toxins, may be one of the most important features contributing to the potent hemorrhagic activity observed for this class of toxins as compared to the P-I toxins (Bjarnason & Fox, 1995).

## Related Peptidases

Similar P-III class snake venom metalloproteinases have been isolated from the venom of the prairie rattlesnake, *Crotalus viridis viridis* (54 kDa; Komori *et al*., 1994), the red rattlesnake, *Crotalus ruber ruber* (hemorrhagic toxin, HT-1, 60 kDa; Mori *et al*., 1987) (Chapter 422) and *Crotalus atrox* (catrocollastatin, 50 kDa; Zhou *et al*., 1995). The lower molecular masses of these three proteinases compared to atrolysin A probably reflect differences in glycosylation. The toxins from *C. viridis* and *C. ruber*, like atrolysin A, can degrade fibrinogen without causing clot formation. *Crotalus ruber* HT-1 cleaves the oxidized B chain of insulin with a similar pattern to that of atrolysin A except that it does not cleave the Asn3+Gln4 bond and does cleave the Phe24+Phe25 bond (Komori *et al*., 1994).

## Distinguishing Features

The distinguishing feature of atrolysin A is its very potent hemorrhagic activity as compared to the other atrolysins.

## Further Reading

An overall review of venoms and related mammalian proteases, including atrolysin A, may be found in Fox & Bjarnason (1996).

## References

Baramova, E.N., Shannon, J.D., Bjarnason, J.B., Gonias, S.L. & Fox, J.W. (1989) Degradation of extracellular matrix proteins by hemorrhagic metalloproteinases. *Arch. Biochem. Biophys.* **275**, 63–71.

Baramova, E.N., Shannon, J.D., Bjarnason, J.B., Gonias, S.L. & Fox, J.W. (1990) Interaction of hemorrhagic metalloproteinases with human $\alpha_2$-macroglobulin. *Biochemistry* **29**, 1158–1164.

Bjarnason, J.B. & Fox, J.W. (1995) Snake venom metallo-endopepti-dases. *Methods Enzymol.* **248**, 345–368.

Bjarnason, J.B. & Tu, A.T. (1978) Hemorrhagic toxins from Western diamondback rattlesnake (*Crotalus atrox*) venom: isolation and characterization of five toxins and the role of zinc in hemorrhagic toxin e. *Biochemistry* **17**, 3395–3404.

Bjarnason, J.B., Hamilton, D.& Fox, J.W. (1988) Studies on the mechanism of hemorrhagic production by five proteolytic hem-orrhagic toxins from *Crotalus atrox* venom. *Biol. Chem. Hoppe-Seyler* **369**, 121–129.

Catanese, J.J. & Kress, L.F. (1992) Isolation from opossum serum of a metalloproteinase inhibitor homologous to human $\alpha$1B-glycoprotein. *Biochemistry* **31**, 410–418.

Fox, J.W. & Bjarnason, J.R. (1995) Atrolysins – metalloproteinases from *Crotalus atrox* venom. *Methods Enzymol.* **248**, 369–387.

Fox, J.W. & Bjarnason, J.B. (1996) Reprolysins: snake venom and mammalian reproductive metalloproteinases. In: *Zinc Metallopro-teinases in Health and Disease* (Hooper, N.M., ed.). London: Tay-lor & Francis, pp. 47–81.

Hite, L.A., Jia, L.-G., Bjarnason, J.B. & Fox, J.W. (1994) cDNA sequences for four snake venom metalloproteinases: structure, classification and their relationship to mammalian reproductive proteins. *Arch. Biochem. Biophys.* **308**, 182–191.

Jia, L.G., Wang, X.-M., Shannon, J.D., Bjarnason, J.B. & Fox, J.W. (1997) Function of disintegrin-like/cysteine-rich domains of atrolysin A – inhibition of platelet aggregation by recombinant protein and peptide antagonists. *J. Biol. Chem.* **272**, 13094–13102.

Komori, Y., Nikai, T., Sekido, C., Fuwa, M.& Sugihara, H. (1994) Biochemical characterization of hemorrhagic toxin from *Crotalus viridis viridis* (prairie rattlesnake) venom. *Int. J. Biochem.* **26**, 1411–1418.

Kruzel, M. & Kress, L.F. (1985) Separation of *Crotalus atrox* (West-ern diamondback rattlesnake) $\alpha$-proteinase from serine proteinase and hemorrhagic factor activities. *Anal. Biochem.* **151**, 471–478.

Mori, N., Nikai, T., Sugihara, H. & Tu, A.T. (1987) Biochemical characterization of hemorrhagic toxins with fibrinogenase activ-ity isolated from *Crotalus ruber ruber* venom. *Arch. Biochem. Biophys.* **253**, 108–121.

Ownby, C.L., Bjarnason, J.B. & Tu, A.T. (1978) Hemorrhagic toxins from rattlesnake (*Crotalus atrox*) venom: pathogenesis of hemorr-hage induced by three purified toxins. *Am. J. Pathol.* **93**, 201–218.

Pfleiderer, G. & Krauss, A. (1965) Specificity of snake venom proteases (*Crotalus atrox*). *Biochem. Z.* **342**, 85–94.

Shimokawa, K., Jia, L.-G., Wang, X.-M. & Fox, J.W. (1996) Expres-sion, activation and processing of the recombinant snake venom metalloproteinase, pro-atrolysin E. *Arch. Biochem. Biophys.* **335**, 283–294.

Shimokawa, K., Jia, L.-G., Shannon, J.D. & Fox, J.W. (1997) Sequence and biological activity of catrocollastatin C – a disintegrin-like/cysteine-rich two-domain protein from *Crotalus atrox* venom. *Arch. Biochem. Biophys.* **343**, 35–43.

Tu, A.T., Nikai, T. & Baker, J.O. (1981) Proteolytic specificity of hemorrhagic toxin a isolated from Western diamondback rat-tlesnake (*Crotalus atrox*) venom. *Biochemistry* **20**, 7004–7009.

Usami, Y., Fujimura, Y., Miura, S., Shima, H., Yoshida, E., Yosh-ioka, A, Hirano, K., Suzuki, M. & Titani, K. (1994) A 28 kDa-protein with disintegrin-like structure (jarahagin-C) purified from *Bothrops jararaca* venom inhibits collagen and ADP-induced platelet aggregation. *Biochem. Biophys. Res. Commun.* **201**, 331–339.

Willis, T.W. & Tu, A.T. (1988) Purification and biochemical char-acterization of atroxase, a nonhemorrhagic fibrinolytic protease from Western diamondback rattlesnake venom. *Biochemistry* **27**, 4769–4777.

Zhou, Q., Smith, J.B. & Grossman, M.H. (1995) Molecular cloning and expression of catrocollastin, a snake-venom protein from *Crotalus atrox* (Western diamondback rattlesnake) which inhibits platelet adhesion to collagen. *Biochem. J.* **307**, 411–417.

*Jay W. Fox*
*Department of Microbiology,*
*University of Virginia Health Sciences Center,*
*Charlottesville, VA 22908, USA*
*Email: jwf8x@virginia.edu*

*Jón B. Bjarnason*
*Science Institute,*
*University of Iceland,*
*Reykjavik, Iceland*
*Email: jonbragi@raunvis.hi.is*

# 419. Atroxase

## Databanks

*Peptidase classification: clan MB, family M12, MEROPS ID: M12.147*
*NC-IUBMB enzyme classification: EC 3.4.24.43*
*Databank codes:*

| Species | SwissProt | PIR | EMBL (cDNA) | EMBL (genomic) |
| --- | --- | --- | --- | --- |
| *Crotalus atrox* | – | – | S77086 | – |

## Name and History

This enzyme was named *atroxase* because it was isolated from the venom of western diamondback rattlesnake, *Crotalus atrox* (Willis & Tu, 1988). As it was known that rattlesnake bite significantly affects coagulation of the victim's blood, it was proposed that if an anticoagulant principle could be isolated from the venom, it might have potential medical use. The enzyme isolated was found to have fibrinolytic activity as well as fibrinogenolytic activity. Other names for this enzyme are *Crotalus atrox fibrinogenase* and *Crotalus atrox nonhemorrhagic metalloendopeptidase* (Willis *et al*., 1989).

## Activity and Specificity

The enzyme hydrolyzes the oxidized insulin B chain:

FVNQH┼LCGS┼H┼LVEA┼LY┼LVCGERGFFYTPKA

The A$\alpha$ and B$\beta$ chains of fibrinogen are readily hydrolyzed, but the $\gamma$ chain is resistant to hydrolysis. With fibrin, both $\alpha$ and $\beta$ chains are hydrolyzed, but the $\gamma$-$\gamma$ chain is not affected. The synthetic substrate S-2238 for thrombin is not hydrolyzed.

## Structural Chemistry

The cDNA sequence encoding atroxase has been determined (Baker *et al*., 1995; Baker & Tu, 1996). Nucleotides 1–440 are the 5′-untranslated region, 441–1053 are the translated region, and 1054–1586 constitute the 3′-untranslated region. The cDNA encodes a 23 500 Da protein. The sequence of atroxase is similar to that of lebetase, another fibrinolytic enzyme isolated from a snake venom (Siigur *et al*., 1996). The enzyme contains 1 mol of zinc per mol of protein and 0.3 mol of calcium per mol of protein.

## Preparation

Atroxase was prepared from *C. atrox* venom by DEAE-cellulose, G-75 superfine resin, and CM-cellulose column chromatography (Willis & Tu, 1988).

## Biological Aspects

*In vitro*, fibrinolytic activity was tested on artificial thrombi induced in the posterior vena cava of Sprague–Dawley rats. Thrombolysis was then characterized by angiographic techniques over a period of 3 h. Intravenous administration of the protease, at a dosage of 6.0 mg kg$^{-1}$, resulted in thrombolysis within 1 h followed by recanalization of the originally occluded vein within 2 h. Fibrinogenolytic activity resulted in a 60% decrease in the plasma fibrinogen level of the rat. Histological examination of kidney, liver, heart and lung tissue showed no necrosis or hemorrhage. These results are the first step in evaluating the thrombolytic potential of anticoagulant proteases within *C. atrox* venom using laboratory animals.

## Distinguishing Features

Atroxase is generally similar to other venom proteases, but the sequence Asp64-Gln-Asp-Phe-Ile-Thr-Val70 is not found in other proteases. Like atroxase, M4 fibrinolytic protease from *C. molossus molossus* venom hydrolyzed $\alpha$ and $\beta$ chains of fibrin (Rael *et al*., 1992).

## Further Reading

For reviews, see Baker & Tu (1996) and Marsh (1994).

## References

Baker, B.J. & Tu, A.T. (1996) Atroxase – a fibrinolytic enzyme isolated from the venom of western diamondback rattlesnake. In: *Natural Toxins II* (Singh, B.R. & Tu, A.T., eds). New York: Plenum Press, pp. 203–211.

Baker, B.J., Wongvibulsin, S., Nyborg, J. & Tu, A.T. (1995) Nucleotide sequence encoding the snake venom fibrinolytic enzyme atroxase obtained from a *Crotalus atrox* venom gland cDNA library. *Arch. Biochem. Biophys.* **317**, 357–364.

Marsh, N. (1994) Inventory of haemorrhagic factors from snake venoms: on behalf of the registry of exogenous hemostatic factors (K. Stocker, Chairman) of the Scientific and Standardisation Committee of the International Society on Thrombosis and Haemostasis. *Thromb. Haemost.* **71**, 793–797.

Rael, E.D., Martinez, M. & Molina, O. (1992) Isolation of a fibrinolytic protease, M4, from venom of *Crotalus molossus molossus* (northern blacktail rattlesnake). *Haemostasis* **22**, 41–49.

Siigur, E., Aaspõllu, A., Tu, A.T. & Siigur, J. (1996) cDNA cloning and deduced amino acid sequence of fibrinolytic enzyme (Lebetase) from *Vipera lebetina* snake venom. *Biochem. Biophys. Res. Commun.* **224**, 229–236.

Willis, T.W. & Tu, A.T. (1988) Purification and biochemical characterization of atroxase, a nonhemorrhagic fibrinolytic protease from western diamondback rattlesnake venom. *Biochemistry* **27**, 4769–4777.

Willis, T.W., Tu, A.T. & Miller, C.W. (1989) Thrombolysis with a snake venom protease in a rat model of venous thrombosis. *Thromb. Res.* **53**, 19–29.

*Anthony T. Tu*
*Department of Biochemistry and Molecular Biology,*
*Colorado State University,*
*Ft. Collins, CO 80523-1870, USA*
*Email: atu@vines.colostate.edu*

# 420. *Atrolysin B*

## Databanks

*Peptidase classification: clan MB, family M12, MEROPS ID: M12.143*
*NC-IUBMB enzyme classification: EC 3.4.24.41*
*Databank codes:*

| Species | SwissProt | PIR | EMBL (cDNA) | EMBL (genomic) |
|---|---|---|---|---|
| *Crotalus atrox* | – | S41608 | U01235 | – |

## Name and History

**Atrolysin B**, formerly known as **hemorrhagic toxin b (HT-b)**, is isolated from the venom of the western diamondback rattlesnake, *Crotalus atrox* (Bjarnason & Tu, 1978). The proteinase was one of two snake venom metalloproteinases (the other being atroxase, Chapter 419) found in a DEAE-chromatographic fraction of the venom termed α-proteinase (Pfleiderer & Krauss, 1965).

## Activity and Specificity

The peptidase activity of atrolysin B can be assayed with the quenched fluorescence substrate Abz-Ala-Gly-Leu+Ala-nitrobenzylamide (Shimokawa *et al.*, 1996) or using protein substrates such as dimethylcasein (Bjarnason & Tu, 1978). The proteolytic and hemorrhagic activities of atrolysin B are inhibited by EDTA and 1,10-phenanthroline (Bjarnason & Tu, 1978). Both the hemorrhagic and proteolytic activities of atrolysin B are inhibited by the opossum serum protein oprin (Catanese & Kress, 1992).

The sites of cleavage of the oxidized B chain of insulin by atrolysin B are given by Bjarnason *et al.* (1988):

FVNQH+LCGSH+LVEA+LY+LVCGERG+FFYTPKA

This specificity is identical to that of atrolysin C, forms c and d (Chapter 423). Interestingly, atrolysin B was demonstrated to have a low activity, similar to that observed for atrolysin C, for soluble peptide production from basement membrane preparations (Bjarnason *et al.*, 1988). Atrolysin B can degrade fibrinogen to produce soluble, nonclotting products. No significant degradation of fibrin by atrolysin B occurs. Gelatin is readily digested by atrolysin B (Bjarnason *et al.*, 1988).

## Structural Chemistry

Atrolysin B as isolated from the venom of *Crotalus atrox* has a mass of 24 kDa by SDS-PAGE with a pI value in the basic pH range (Bjarnason & Tu, 1978). There is one zinc atom per molecule along with a variable number calcium ions. Based on the cDNA sequence analysis of atrolysin B, it is classified as a member of the P-I class of snake venom metalloproteinases (Hite *et al.*, 1994). The proteinase domain has two disulfide bonds. No glycosylation consensus sequence sites are present.

## Preparation

Atrolysin B is isolated from the lyophilized crude venom of *Crotalus atrox* by an initial ion-exchange chromatography step on DEAE-cellulose followed by three additional chromatographies: gel filtration on Sephadex G-75, ion-exchange chromatography on CM-cellulose and an additional ion-exchange step on CM-cellulose (Bjarnason & Tu, 1978). A typical isolation yields approximately 370 mg of proteinase from 10 g of crude venom.

## Biological Aspects

Atrolysin B is one of the less potent hemorrhagic toxins in *Crotalus atrox* venom; it has a minimum hemorrhagic dose of 3 μg (Bjarnason & Tu, 1978). However, due to the relatively larger amount of this proteinase present in the venom, it is calculated to contribute approximately 16% of the hemorrhagic activity of the venom. When atrolysin B is injected into muscle tissue, the typical degradation of the basement membrane structure surrounding capillaries is observed (Ownby *et al.*, 1978). Hemorrhage into the stroma is *per rhexis*, with the intracellular junctions of the endothelial cells remaining intact. As expected, these pathologies appear more slowly than with the more potent hemorrhagic toxins atrolysins A and E.

## Distinguishing Features

A distinguishing feature of atrolysin B is that, in addition to the usual consequences of injection of hemorrhagic toxins into muscle tissue, localized necrosis of the muscle (i.e. myonecrosis) is observed (Ownby *et al.*, 1978). This is not a typical pathology associated with other atrolysins and the basis for this difference is unknown.

## Further Reading

A broad review of the venom metalloproteases may be found in Fox & Bjarnason (1996).

## References

Bjarnason, J.B. & Tu, A.T. (1978) Hemorrhagic toxins from Western diamondback rattlesnake (*Crotalus atrox*) venom: isolation and characterization of five toxins and the role of zinc in hemorrhagic toxin e. *Biochemistry* **17**, 3395–3404.

Bjarnason, J.B., Hamilton, D. & Fox, J.W. (1988) Studies on the mechanism of hemorrhagic production by five proteolytic hemorrhagic toxins from *Crotalus atrox* venom. *Biol. Chem. Hoppe-Seyler* **369**, 121–129.

Catanese, J.J. & Kress, L.F. (1992) Isolation from opossum serum of a metalloproteinase inhibitor homologous to human α1B-glycoprotein. *Biochemistry* **31**, 410–418.

Fox, J.W. & Bjarnason, J.B. (1996) Reprolysins: snake venom and mammalian reproductive metalloproteinases. In: *Zinc Metalloproteinases in Health and Disease* (Hooper, N.M., ed.). London: Taylor & Francis, pp. 47–81.

Hite, L.A., Jia, L.-G., Bjarnason, J.B. & Fox, J.W. (1994) cDNA sequences for four snake venom metalloproteinases: structure, classification and their relationship to mammalian reproductive proteins. *Arch. Biochem. Biophys.* **308**, 182–191.

Ownby, C.L., Bjarnason, J.B. & Tu, A.T (1978) Hemorrhagic toxins from rattlesnake (*Crotalus atrox*) venom: pathogenesis of hemorrhage induced by three purified toxins. *Am. J. Pathol.* **93**, 201–218.

Pfleiderer, G. & Krauss, A. (1965) Die Wirkungsspezificität von Schlagengift-Proteasen (*Crotalus atrox*) [Specificity of snake venom proteases (*Crotalus atrox*.)] *Biochem. Z.* **342**, 85–94.

Shimokawa, K., Jia, L.-G., Wang, X.-M. & Fox, J.W. (1996) Expression, activation and processing of the recombinant snake venom metalloproteinase, pro-atrolysin E. *Arch. Biochem. Biophys.* **335**, 283–294.

*Jay W. Fox*
*Department of Microbiology,*
*University of Virginia Health Sciences Center,*
*Charlottesville, VA 22908, USA*
*Email: jwf8x@virginia.edu*

*Jón B. Bjarnason*
*Science Institute,*
*University of Iceland,*
*Reykjavik, Iceland*
*Email: jonbragi@raunvis.hi.is*

# *421. Adamalysin*

## Databanks

*Peptidase classification: clan MB, family M12, MEROPS ID: M12.141*
*NC-IUBMB enzyme classification: EC 3.4.24.46*
*Chemical Abstracts Service registry number: 74812-51-4*
*Databank codes:*

| Species | SwissProt | PIR | EMBL (cDNA) | EMBL (genomic) |
|---------|-----------|-----|-------------|----------------|
| *Crotalus adamanteus* | P34179 | – | – | – |

Brookhaven Protein Data Bank three-dimensional structures:

| Species | ID | Resolution | Notes |
|---------|-----|-----------|-------|
| *Crotalus adamanteus* | 1IAG | 2 | |

## Name and History

The recommended name **adamalysin** is based upon the enzyme's source, the venom of the eastern diamondback rattlesnake, *Crotalus adamanteus*. Adamalysin (or **adamalysin II** in some of the references cited herein) was previously named **proteinase II**, with a related form designated proteinase I, in the initial report describing their purification and characterization (Kurecki *et al.*, 1978). An alternative name is **Crotalus adamanteus metalloendopeptidase**. Proteinase II is the more abundant of the two and has been the most thoroughly investigated (Gomis-Rüth *et al.*, 1994; Kress, 1986). Unless otherwise noted, the name adamalysin in this chapter refers to proteinase II.

Adamalysin has been grouped according to active-site environment and folding topology in the metzincin clan (MB) of zinc peptidases (Chapter 380) (Stöcker *et al.*, 1995), and according to sequence homology in the reprolysin subfamily (Chapter 416) of the M12 family of metalloproteinases (Bjarnason & Fox, 1995). The crystal structure of adamalysin (Gomis-Rüth *et al.*, 1994) is a prototype for the metalloproteinase-like domains of proteins expressed by the ADAM gene family (Wolfsberg *et al.* 1995). Adamalysin and proteinase I have unfortunately been recently included with the hemorrhagic metalloproteinases from snake venoms (Bjarnason & Fox, 1994, 1995). However, previous reports had indicated that neither adamalysin nor proteinase I elicited a hemorrhagic response in mice or rabbits, and that 100% of the hemorrhagic activity in *C. adamanteus* venom was due to another metalloproteinase, proteinase H, from that venom (Kurecki & Kress, 1985; Gomis-Rüth *et al.*, 1993; Marsh, 1994).

## Activity and Specificity

Adamalysin enzymatically inactivates human plasma serine proteinase inhibitors (serpins) by cleavage of a single bond in the inhibitor reactive site loop without formation of a stoichiometric enzyme–inhibitor complex. The cleavages are as follows: $\alpha_1$-proteinase inhibitor, Ala350$\downarrow$Met351, and antithrombin III, Ala375$\downarrow$Ser376 (Kress, 1986). Adamalysin cleaves the oxidized insulin B chain (Kurecki *et al.*, 1978) as follows:

$$F\downarrow VNQH\downarrow LCGSH\downarrow LVEA\downarrow L\downarrow Y\downarrow LVCGERGFFYTPKA$$

No activity is found on native type I collagen or elastin. Denatured casein and Congo red elastin are poor substrates. Adamalysin is assayed using hide powder azure or casein at pH 10.0 as described by Kurecki *et al.* (1978). Proteolytic activity is lost in the presence of EDTA or 1,10-phenanthroline. Adamalysin activity is inhibited ($K_i = 3.4\,\mu M$) by the hexapeptide Pro-Lys-Met-Cys-Gly-Val-NH$_2$, which corresponds to the conserved sequence portion in the putative propeptides of several snake venom zinc endopeptidases (Grams *et al.*, 1993). Adamalysin is also partially inhibited by a metalloproteinase inhibitor isolated from opossum serum (Catanese & Kress, 1992).

## Structural Chemistry

Adamalysin is a single-chain 24 kDa zinc endopeptidase containing two disulfide bonds and no free sulfhydryl groups. The adamalysin sequence exhibits 47–83% identity with other venom metalloproteinases, and 27–42% identity with the metalloproteinase-like domains of the mammalian multidomain proteins (ADAMs) (Chapter 447) reported by Wolfsberg *et al.* (1995).

X-Ray crystallographic analysis (Gomis-Rüth *et al.*, 1994) has revealed that adamalysin is an ellipsoidal molecule with a relatively flat active-site cleft separating a 'lower' subdomain from the 'upper' main body (see Fig. 404.4). The regularly folded N-terminal upper domain consists essentially of a central, highly twisted five-stranded $\beta$-pleated sheet flanked by long and short surface-located helices on its convex side, and by two long helices, one of which represents the central 'active-site helix' on its concave side. The lower subdomain, comprising the last 50 residues, is organized in multiple turns, with the chain ending in a long C-terminal helix and an extended segment clamped to the upper domain via a disulfide bridge.

The catalytic zinc ion, located at the bottom of the active-site cleft, is tetrahedrally coordinated by His142, His146, His152, and a water molecule anchored to Glu143. His142, Glu143 and His146 are part of the long active-site helix, which extends up to Gly149 where it turns sharply away towards His152. Asp153, which is strictly conserved in snake venom and mammalian reproductive tract metalloproteinases, is buried in the subdomain and seems to stabilize the hydrophobic active-site basement. A calcium ion localized close to the surface of adamalysin opposite its active site is liganded by seven oxygen atoms donated by the side chains of the strongly conserved residues Asp93, Asn200 and Glu9, by the carbonyl group of Cys197, and by two water molecules.

The catalytic domain of adamalysin and other snake venom metalloproteinases and the corresponding domains of the ADAMs can be aligned with only a few single-residue insertions and deletions. All substitutions, including some cysteine residues engaged in additional disulfide bridges, are in agreement with the adamalysin structure, which can, therefore, be considered as a prototype.

Adamalysin exhibits a strong overall topological equivalence with the crayfish digestive metalloproteinase astacin (Chapter 405). In particular, both enzymes possess a virtually identical zinc-binding consensus segment and a methionine-containing turn (the 'Met-turn'), with the methionine side chain providing a hydrophobic basement for the three zinc-liganding histidine side chains. Serralysins (Chapter 386) and vertebrate matrixins share these characteristic structural features. The adamalysins, the astacins, the serralysins and the matrixins are therefore suggested to be grouped into a clan called the 'metzincins' (Gomis-Rüth *et al.*, 1993), clan MB in the present Handbook.

## Preparation

Adamalysin is isolated from lyophilized venom of *Crotalus adamanteus* (Miami Serpentarium) (see Appendix 2 for full names and addresses of suppliers) by gel filtration of the venom on BioGel P-150 followed by chromatography on SP-Sephadex to separate the two forms, and rechromatography on SP-Sephadex to achieve electrophoretic homogeneity of the two forms (Kurecki *et al.*, 1978).

## Biological Aspects

The biological significance of the enzymatic inactivation of mammalian serpins by adamalysin has not been fully established. Metalloproteinases exhibiting this activity are present in representative venoms of the Crotalid, Viperid and Colubrid families of poisonous snakes (Kress, 1988). Adamalysin itself does not elicit hemorrhage. However, a role in the sequence of proteolytic events which culminate in localized hemorrhage, necrosis and edema following rattlesnake (Crotalid) envenomation is suggested by the fact that the opossum (*Didelphis virginiana*) is resistant to these effects, and its serum contains metalloproteinase inhibitors which block the activity of adamalysin and related venom metalloproteinases (Catanese & Kress, 1992).

M

## References

Bjarnason, J.B. & Fox, J.W. (1994) Hemorrhagic metalloproteinases from snake venoms. *Pharmacol. Ther.* **62**, 325–372.

Bjarnason, J.B. & Fox, J.W. (1995) Snake venom metalloendopeptidases: reprolysins. *Methods Enzymol.* **248**, 345–368.

Catanese, J.J. & Kress, L.F. (1992) Isolation from opossum serum of a metalloproteinase inhibitor homologous to human α1B-glycoprotein. *Biochemistry* **31**, 410–418.

Gomis-Rüth, F.-X., Kress, L.F. & Bode, W. (1993) First structure of a snake venom metalloproteinase: a prototype for matrix metalloproteinases/collagenases. *EMBO J.* **12**, 4151–4157.

Gomis-Rüth, F.X., Kress, L.F., Kellermann, J., Mayr, I., Lee, X., Huber, R. & Bode, W. (1994) Refined 2.0 Å X-ray crystal structure of the snake venom zinc-endopeptidase adamalysin II. Primary and tertiary structure determination, refinement, molecular structure and comparison with astacin, collagenase and thermolysin. *J. Mol. Biol.* **239**, 513–544.

Grams, F., Huber, R., Kress, L.F., Moroder, L. & Bode, W. (1993) Activation of snake venom metalloproteinases by a cysteine-like switch mechanism. *FEBS Lett.* **335**, 76–80.

Kress, L.F. (1986) Inactivation of human plasma serine proteinase inhibitors (serpins) by limited proteolysis of the reactive site loop with snake venom and bacterial metalloproteinases. *J. Cell. Biochem.* **32**, 51–58.

Kress, L.F. (1988) The action of snake venom metalloproteinases on plasma proteinase inhibitors. In: *Hemostasis and Animal Venoms* (Pirkle, H. & Markland, F.S., Jr., eds). New York: Marcel Dekker, pp. 335–348.

Kurecki, T. & Kress, L.F. (1985) Purification and partial characterization of the hemorrhagic factor from the venom of *Crotalus adamanteus* (eastern diamondback rattlesnake). *Toxicon* **23**, 657–668.

Kurecki, T., Laskowski, M., Sr. & Kress, L.F. (1978) Purification and some properties of two proteinases from *Crotalus adamanteus* venom that inactivate human α1-proteinase inhibitor. *J. Biol. Chem.* **253**, 8340–8345.

Marsh, N. (1994) Inventory of haemorrhagic factors from snake venoms. *Thromb. Haemost.* **71**, 793–797.

Stöcker, W., Grams, F., Baumann, U., Reinemer, P., Gomis-Rüth, F.-X., McKay, D.B. & Bode, W. (1995) The metzincins – topological and sequential relations between the astacins, adamalysins, serralysins, and matrixins (collagenases) define a superfamily of zinc-peptidases. *Protein Sci.* **4**, 823–840.

Wolfsberg, T.G., Straight, P.D., Gerena, R.L., Huovila, A.-P.J., Primakoff, P., Myles, D.G. & White, J.M. (1995) ADAM, a widely distributed and developmentally regulated gene family encoding membrane proteins with A Disintegrin And Metalloprotease domain. *Dev. Biol.* **169**, 378–383.

***Lawrence F. Kress***
*Department of Molecular and Cellular Biology,*
*Roswell Park Cancer Institute,*
*Buffalo, NY 14263, USA*
*Email: lkress@mcbio.med.buffalo.edu*

***Joseph J. Catanese***
*Department of Molecular and Cellular Biology,*
*Roswell Park Cancer Institute,*
*Buffalo, NY 14263, USA*

# *422. Ruberlysin*

## *Databanks*

*Peptidase classification: clan MB, family M12, MEROPS ID: M12.150*
*NC-IUBMB enzyme classification: EC 3.4.24.48*
*Databank codes:*

| Species | SwissProt | PIR | EMBL (cDNA) | EMBL (genomic) |
|---|---|---|---|---|
| *Crotalus ruber* | P20897 | – | – | – |

## *Name and History*

The venom of *Crotalus ruber ruber* (red rattlesnake) contains three hemorrhagic toxins with fibrinogenolytic activity (Mori *et al.*, 1987). The snake is found in southern California and has a very limited range of distribution. Among three hemorrhagic proteases, **HT-1**, **HT-2** and **HT-3**, the amino acid sequence of HT-2 has been determined (Takeya *et al.*, 1990). HT-2 is also called **ruberlysin**, **red rattlesnake metalloendopeptidase**, and **hemorrhagic toxin II**. All three enzymes will be described under this entry.

## Activity and Specificity

The proteolytic specificity of HT-1, HT-2 and HT-3 was analyzed using angiotensin I, luteinizing hormone releasing hormone (LH-RH) and [Met]enkephalin. HT-1, HT-2 and HT-3 all cleaved angiotensin I at Pro7↓Phe8, LH-RH at Gly6↓Leu7 and [Met]enkephalin at Gly3↓Phe4. HT-2 and HT-3 cut additionally on angiotensin I at His6↓Pro7, and on LH-RH at Trp3↓Ser4:

Angiotensin I:   Asp-Arg-Val-Tyr-Ile-His↓Pro↓Phe-His-Leu
LH-RH:           Pyr-His-Trp↓Ser-Tyr-Gly↓Leu-Arg-Pro-
                 Gly-NH$_2$
[Met]enkephalin: Tyr-Gly-Gly↓Phe-Met

With the B chain of insulin, the cleavage points of HT-2 are as follows:

FVNQHLCGSH↓LVEA↓LY↓LVCGERG↓FFYTPKA

All three hemorrhagic proteases hydrolyzed A$\alpha$ and B$\beta$ chains of fibrinogen, but the $\gamma$ chain was resistant to hydrolysis by all three hemorrhagic proteases. These proteases are zinc-containing enzymes and the protease activity is abolished by EDTA (Mori *et al.*, 1987). A trypsin inhibitor, benzamidine, has no effect.

## Structural Chemistry

HT-2 has a molecular mass of 25 000, a pI of 5.2, and binds 1.1 mol of zinc and 1.1 mol of calcium per mol of protein. The secondary structure of HT-2 as studied by circular dichroism was 27.5% $\alpha$ helix, 30.8% antiparallel $\beta$ sheet, 10.2% parallel $\beta$ sheet, 18.7% $\beta$ turn and 13.1% random coil. By Raman spectroscopy, the $\alpha$ helix content was 30.7%, 26.9% $\beta$ sheet, and 42.4% random coil (Mori *et al.*, 1987). The primary structure of HT-2 has been elucidated; it contains 202 amino acid residues.

## Preparation

Isolation of HT-2 was achieved by a combination of gel filtration (G-75) and ion-exchange chromatography (CM-cellulose) from the venom of *Crotalus ruber ruber* (Mori *et al.*, 1987). The venom contains 12% of HT-2. The enzyme hydrolyzed casein, dimethylcasein and hide powder azure.

## Biological Aspects

The HT-2 protease has hemorrhagic activity with an MHD (minimum hemorrhagic dose) of 0.27 µg per mouse. Comparable MHD values for HT-1 and HT-3 are 0.17 and 1.43 µg per mouse, respectively. Unfractionated crude venom has an MHD of 0.5 µg per mouse, indicating that purification of the HT-2 increased the potency of hemorrhagic activity 2-fold. The protease HT-2 also possesses lethal activity with an LD$_{50}$ of 7.4 µg g$^{-1}$ in mice. The LD$_{50}$ values for HT-1 and HT-3 are 4.0 and 13.9 µg g$^{-1}$, respectively.

## Distinguishing Features

The molecular masses of hemorrhagic proteases lie between 25 000 and 100 000 Da. HT-2 has a molecular mass of 25 000 and is a zinc metalloenzyme. HT-1 is a high molecular mass protein of 60 000, and HT-3 has a much smaller molecular mass of 25 500.

## Further Reading

Marsh (1994) summarized all hemorrhagic factors from snake venoms. Siigur *et al.* (1996) cloned nonhemorrhagic protease from *Vipera lebetina* venom.

## References

Marsh, N. (1994) Inventory of haemorrhagic factors from snake venoms: on behalf of the registry of oxogenous hemostatic factors (K. Stöcker, Chairman) of the Scientific and Standardisation Committee of the International Society on Thrombosis and Haemostasis. *Thromb. Haemost.* **71**, 793–797.

Mori, N., Nikai, T., Sugihara, H. & Tu, A.T. (1987) Biochemical characterization of hemorrhagic toxins with fibrinogenase activity isolated from *Crotalus ruber ruber* venom. *Arch. Biochem. Biophys.* **253**, 108–121.

Siigur, E., Aaspõllu, A., Tu, A.T. & Siigur, J. (1996) cDNA cloning and deduced amino acid sequence of fibrinolytic enzyme (Lebetase) from *Vipera lebetina* snake venom. *Biochem. Biophys. Res. Commun.* **224**, 229–236.

Takeya, H., Onikura, A., Nikai, T., Sugihara, H. & Iwanaga, S. (1990) Primary structure of a hemorrhagic metalloproteinase, HT-2, isolated from the venom of *Crotalus ruber ruber*. *J. Biochem.* **108**, 711–719.

*Anthony T. Tu*
*Department of Biochemistry and Molecular Biology,*
*Colorado State University,*
*Ft. Collins, CO 80523-1870, USA*
*Email: atu@vines.colostate.edu*

# 423. Atrolysin C

## Databanks

*Peptidase classification: clan MB., family M12, MEROPS ID: M12.144*
*NC-IUBMB enzyme classification: EC 3.4.24.42*
*Databank codes:*

| Species | SwissProt | PIR | EMBL (cDNA) | EMBL (genomic) |
|---|---|---|---|---|
| Form | | | | |
| *Crotalus atrox* | P15167 | S41609 | U01236 | – |
| Form | | | | |
| *Crotalus atrox* | – | S41610 | U01237 | – |

Brookhaven Protein Data Bank three-dimensional structures:

| Species | ID | Resolution | Notes |
|---|---|---|---|
| Form | | | |
| *Crotalus atrox* | 1ATL | 1.8 | |

## Name and History

**Atrolysin C** is the general term now used to describe the two weakly hemorrhagic zinc metalloproteinase isoenzymes atrolysin C, form C (formerly **hemorrhagic toxin C, HT-c**) and atrolysin C, form D (formerly **hemorrhagic toxin D, HT-d**) isolated from *Crotalus atrox* venom (Bjarnason & Tu, 1978). From both protein and cDNA sequence analyses it was discovered that the C and D forms differ from one another by only one amino acid substitution: the Asp181 residue in form D is an Ala residue in form C (Bjarnason & Fox, 1987; Shannon *et al.*, 1989; Hite *et al.*, 1994). For the remainder of this chapter, unless specifically noted, the use of the term atrolysin C denotes both forms C and D. Fortunately, the chemical and biological activities of the two forms are essentially identical and this allows for the general designation of atrolysin C for the two forms.

## Activity and Specificity

*In vitro*, the peptidase activity of atrolysin C can be assayed with the quenched fluorescence substrate Abz-Ala-Gly-Leu+-Ala-nitrobenzylamide (Fox *et al.*, 1986) or using protein substrates such as dimethylcasein (Bjarnason & Tu, 1978). The proteolytic and hemorrhagic activity of atrolysin C is inhibited by EDTA, 1,10-phenanthroline and $\alpha_2$-macroglobulin (Bjarnason & Tu, 1978; Fox *et al.*, 1986; Bjarnason & Fox, 1987; Baramova *et al.*, 1990). Atrolysin C cleaves the bait region of $\alpha_2$-macroglobulin at His-Ala-Arg696+Leu697-Val-His (Baramova *et al.*, 1990). Various amino acid hydroxamates, chloromethyl esters and the thermolysin inhibitor phosphoramidon can inhibit atrolysin C with $K_i$ values in the millimolar to micromolar range (Fox *et al.*, 1986). A naturally occurring peptide in the venom of *Crotalus atrox* has been observed in the active site of atrolysin C, form D, and acts as an inhibitor (Zhang *et al.*, 1994).

The peptide bond specificity of atrolysin C has been probed using the oxidized B chain of insulin and short synthetic peptides. The insulin B chain peptide bonds cleaved are (Fox *et al.*, 1986):

FVNQH+LCGSH+LVEA+LY+LVCGERG+FFYTPKA

These sites are identical to those observed for the other weakly hemorrhagic toxin, atrolysin B (Chapter 420; Bjarnason *et al.*, 1988). Using synthetic peptides, it was determined that there is an extended substrate-binding site in atrolysin C (Fox *et al.*, 1986). Furthermore, based on the kinetics of peptide bond hydrolysis using a panel of synthetic peptides, there appears to be a general specificity preference for an Xaa+Leu or Xaa+aliphatic at the S1–S1′ substrate positions. Peptides of four residues or less are not hydrolyzed (Fox *et al.*, 1986).

Atrolysin C is capable of degrading basement membrane preparations, although less rapidly than the high-potency hemorrhagic toxins, atrolysins A (Chapter 418) and E (Chapter 430). Atrolysin C cleaves fibrinogen to nonclottable peptides. The pattern of degradation of fibrinogen by atrolysin C is similar to that of the other weak hemorrhagic toxin, atrolysin B (Bjarnason *et al.*, 1988). There is no significant degradation of fibrin by atrolysin C.

## Structural Chemistry

Atrolysin C, forms C and D, are class P-I snake venom metalloproteinases (Chapter 416) comprising 202 amino acid residues (Bjarnason & Tu, 1978; Hite *et al.*, 1994). From tryptic peptide mapping and cDNA sequence analysis, the two forms differ by only one amino acid: Asp181 in form D is Ala181 in form C (Bjarnason & Fox, 1987; Hite *et al.*, 1994). Based on SDS-PAGE, atrolysin C has a molecular mass of 24 kDa (Bjarnason & Tu, 1978). There is one zinc ion per molecule and a variable number of calcium ions (Bjarnason & Tu, 1978). The pI of the protein is in the acidic pH range.

The X-ray crystal structure of atrolysin C, form D, has been solved (Zhang *et al.*, 1994). Analysis of the metalloproteinase domain structures of atrolysin C indicates an oblate ellipsoidal molecule with an active-site cleft dividing an 'upper' domain from a smaller, 'lower' domain. The larger domain is composed of a twisted $\alpha/\beta$ structure having a five-stranded $\beta$ sheet and four $\alpha$ helices. The smaller subdomain is somewhat irregularly folded. The active-site region of atrolysin C has the catalytic zinc ion which is coordinated by His146, His142, His152 and a water molecule bound to Glu143, thus forming a tetrahedral coordination sphere. These residues lie on the long active-site helix extending to Gly149. At Gly149, the peptide backbone turns towards His152 with the conserved Met-turn structure (Met166) forming the 'basement' of the zinc-binding site. The S1' primary specificity binding pocket of atrolysin is particularly deep (10 Å) and can accommodate fairly large hydrophobic moieties as has been observed with various synthetic peptides (Botos *et al.*, 1995). Furthermore, the S1' pocket can readily hold the pyroglutamate structure of the peptide Glp-Asn-Trp which thereby blocks the S1' pocket from interaction with other substrates.

## Preparation

Atrolysin C, forms C and D, are isolated from *Crotalus atrox* venom by an initial DEAE ion-exchange chromatography step followed by gel filtration on Sephadex G-75 and two additional DEAE ion-exchange steps. Ten grams of crude venom yields 17 mg of atrolysin C, form C and 31 mg of form D (Bjarnason & Tu, 1978).

## Biological Aspects

Atrolysin C, forms C and D are only very weakly hemorrhagic with minimum hemorrhagic doses (MHD) of 8 and 11 µg, respectively (Bjarnason & Tu, 1978). Based on these MHD values atrolysin C, form C and form D contribute 0.04 and 0.06%, respectively, of the hemorrhagic activity of the venom. Given the very low hemorrhagic potency of the atrolysin C forms, it is questionable whether these toxins are significant producers of hemorrhage in envenomation by *Crotalus atrox*.

## Distinguishing Features

The rather unusual point about these two isoenzymes is the fact that they differ by only one amino acid in their protein sequences and, based on cDNA sequence data, they are different gene products (Hite *et al.*, 1994).

## Related Peptidases

We postulate, based on the sequence similarity of ADAM 1 (Chapter 447) to the snake venom metalloproteinases, that the active-site core region of the metalloproteinase domain of fertilin $\alpha$ is likely to be folded in a manner nearly identical to that of the snake venom metalloproteinases. Furthermore, modeling studies in our laboratory indicate that the peptide backbone of ADAM 1 can readily be superimposed on the tertiary structure of atrolysin C with little distortion (unpublished results of the authors).

## Further Reading

General overviews of the venom metalloproteases are given in Fox & Bjarnason (1996) and Bjarnason & Fox (1994).

## References

Baramova, E.N., Shannon, J.D., Bjarnason, J.B., Gonias, S.L. & Fox, J.W. (1990) Interaction of hemorrhagic metalloproteinases with human $\alpha_2$-macroglobulin. *Biochemistry* **29**, 1158–1164.

Bjarnason, J.B. & Fox, J.W. (1987) Characterization of two hemorrhagic zinc proteinases, toxin c and toxin d, from western diamondback rattlesnake (*Crotalus atrox*) venom. *Biochim. Biophys. Acta* **911**, 356–363.

Bjarnason, J.B. & Fox, J.W. (1994) Hemorrhagic metalloproteinases from snake venoms. *J. Pharmacol. Therapeut.* **62**, 325–372.

Bjarnason, J.B. & Tu, A.T. (1978) Hemorrhagic toxins from western diamondback rattlesnake (*Crotalus atrox*) venom: isolation and characterization of five toxins and the role of zinc in hemorrhagic toxin e. *Biochemistry* **17**, 3395–3404.

Bjarnason, J.B., Hamilton, D. & Fox, J.W. (1988) Studies on the mechanism of hemorrhagic production by five proteolytic hemorrhagic toxins from *Crotalus atrox* venom. *Biol. Chem. Hoppe-Seyler* **369**, 121–129.

Botos, I., Scapozza, L., Shannon, J.D., Fox, J.W. & Meyer, E.F. (1995) Structure-based analysis of inhibitor binding to Ht-d. *Acta Crystallogr.* **D51**, 597–604.

Fox, J.W. & Bjarnason, J.B. (1996) Reprolysins: snake venom and mammalian reproductive metalloproteinases. In: *Zinc Metalloproteinases in Health and Disease* (Hooper, N.M., ed.). London: Taylor & Francis, pp. 47–81.

Fox, J.W., Campbell, R., Beggerly, L. & Bjarnason, J.B. (1986) Substrate specificity and inhibition of two hemorrhagic zinc proteases Ht-c and Ht-d from *Crotalus atrox* venom. *Eur. J. Biochem.* **156**, 65–72.

Hite, L.A., Jia, L-G., Bjarnason, J.B. & Fox, J.W. (1994) cDNA sequences for four snake venom metalloproteinases: structure, classification and their relationship to mammalian reproductive proteins. *Arch. Biochem. Biophys.* **308**, 182–191.

Shannon, J.D., Baramov, E.N., Bjarnason, J.B. & Fox, J.W. (1989) Amino acid sequence of a *Crotalus atrox* venom metalloproteinase which cleaves type IV collagen and gelatin. *J. Biol. Chem.* **264**, 11575–11583.

Zhang, D., Botos, I., Gomis-Rüth, F.-X., Doll, R., Blood, C., Njoroge, G., Fox, J.W., Bode, W. & Meyer, E. (1994) Structural interaction of natural and synthetic inhibitors with the matrix metalloproteinase, atrolysin C (Ht-d), *Proc. Natl Acad. Sci. USA* **91**, 8447–8451.

**M**

**Jay W. Fox**
*Department of Microbiology,*
*University of Virginia Health Sciences Center,*
*Charlottesville, VA 22908, USA*
*Email: jwf8x@virginia.edu*

**Jón B. Bjarnason**
*Science Institute,*
*University of Iceland,*
*Reykjavik, Iceland*
*Email: jonbragi@raunvis.hi.is*

# 424. *Mutalysins*

## Databanks

Peptidase classification: clan MB, family M12, MEROPS ID: M12.162
NC-IUBMB enzyme classification: none
Databank codes:

| Species | SwissProt | PIR | EMBL (cDNA) | EMBL (genomic) |
|---|---|---|---|---|
| *Lachesis muta* LHF-II | P22796 | – | – | – |

## Name and History

The *mutalysins* are zinc metalloendopeptidases originally isolated from the venom of the bushmaster snake, *Lachesis muta muta*. The pioneering research of Brazil & Vellard (1928), Houssay (1930) and others has demonstrated the existence of several enzymes in bushmaster venom which cause severe bleeding in pit viper-envenomed human beings and other mammals. Based on that hemorrhagic activity, two hemorrhagic factors were purified to a homogeneous state. These enzymes have been designated *Lachesis hemorrhagic factor I* and *II* (*LHF-I* and *LHF-II*) (Sanchez *et al.*, 1987, 1991a).

## Activity and Specificity

The small $M_r$ mutalysin II has a low hemorrhagic effect and a high proteolytic activity on commonly used protein substrates: casein, dimethylcasein and hide powder azure. In contrast, the large $M_r$ mutalysin I has a restricted specificity and potent hemorrhagic effect and shows low proteolytic activity on these substrates (Sanchez *et al.*, 1987, 1991a, 1995a,b). Similar patterns occurred when peptides and small chromogenic substrates were used. Hydrolysis of the oxidized insulin B chain has been used by other investigators for analysis of specificity and also to obtain kinetic data (Civello *et al.*, 1983; Fox *et al.*, 1986; Sanchez *et al.*, 1995a). Like other reprolysins, mutalysin II cleaves the Ala14+Leu15 bond very rapidly and the Phe24+Phe25, His10+Leu11 and His5+Leu6 bonds more slowly.

FVNQH+LCGSH+LVEA+LYLVCGERGF+FYTPKA

In contrast, mutalysin I cleaves only the Ala14+Leu15 bond at a slower rate (specific activity 1/1000 that of mutalysin II). In addition, hydrolysis of the collagenase substrate Abz-Pro-Leu-Gly-Leu+Leu-Gly-Arg-EDDnp by mutalysin II was 7060-fold higher than by mutalysin I (Sanchez *et al.*, 1995a).

The pH optimum for both reprolysins is about 8. Activity is stimulated by low concentrations of $Ca^{2+}$, $Mg^{2+}$ and $Ba^{2+}$ (1–10 mM). Furthermore, calcium ions appear to stabilize the proteinases in aqueous solution. The enzymes are strongly inhibited by chelators (e.g. EDTA above 1 mM). Other inhibitors include $\beta$-mercaptoethanol, cysteine and polyvalent anti-snake serum (Sanchez *et al.*, 1991a, 1995b).

## Structural Chemistry

Mutalysin II is a member of the reprolysin class P-1 (Hite *et al.*, 1994; Bjarnason & Fox, 1995) (Chapter 416). It is a single polypeptide chain of 200 amino acid residues (22.5 kDa, pI 6.6) containing no carbohydrate and two disulfide bonds (Sanchez *et al.*, 1991b). The enzyme contains 1 mol of zinc and 2 mol of calcium per mol of protein (Sanchez *et al.*, 1991a). The proteinase domain contains a zinc-binding HEXXHXXGXXH-consensus motif and a strictly-conserved Met-turn (Cys163-Ile-Met-Pro) bordering the substrate-binding site, which are characteristic features of the metzincin clan of metalloproteinases (Stöcker *et al.*, 1995). Mutalysin I is a 100 kDa glycoprotein, pI 4.7, consisting of two subunits and containing 1 mol of zinc and 1.2 mol of calcium per mol of protein (Sanchez *et al.*, 1987, 1995b). At this time the sequence data for mutalysin I are not available. However, based on the most recent structural data of the large reprolysins, russellysin from *Vipera russelli* (Chapter 433) (Takeya *et al.*, 1992) and mucrolysin (Chapter 426) from *Trimeresurus mucrosquamatus* (Kishida *et al.*, 1985; Bjarnason & Fox, 1994, 1995), the large hemorrhagic metalloproteinase from the bushmaster falls into class P-IV (Chapter 416). According to this model, mutalysin I should consist of metalloproteinase, disintegrin-like, cysteine-rich and lectin-domains.

## Preparation

The mutalysins were purified to homogeneity from two fractions, P1 and P5, obtained by gel filtration of the crude venom on Sephadex G-100. Mutalysin I was isolated from P1 by ion-exchange chromatography on DEAE-cellulose and two cycles of gel filtration on Sephacryl S-300. Mutalysin II was purified from P5 by ion exchange on CM-Sepharose CL-6B and two cycles of gel filtration on Sephadex G-50. The content of each protein in the crude venom was approximately 0.3 and 1.6% for mutalysin I and II, respectively. In accordance with the literature, mutalysin I expresses ~32-fold higher hemorrhagic activity than the low-mass mutalysin II, while the latter shows a 27-fold higher caseinolytic activity than mutalysin I (Sanchez *et al.*, 1987, 1991a).

## Biological Aspects

Hemorrhage in victims bitten by the bushmaster snake is related to the metalloproteinase content of the venom (Sanchez *et al*., 1995a). Mutalysin II disrupts the pericellular basement membrane of capillary walls through proteolytic activity. As observed microscopically, these effects are similar to those of other reprolysins from Viperidae venoms (Ohsaka *et al*., 1973; Ownby *et al*., 1978; Baramova *et al*., 1989). Mutalysin I and II showed fibrin(ogeno)lytic activity. The enzymes degrade the A$\alpha$ chain of fibrinogen much faster than the B$\beta$ chain. With fibrin as substrate, mutalysin II showed greater $\alpha$- than $\beta$-fibrinolytic activity. Interestingly, mutalysin I selectively degraded the $\alpha$ chain of fibrin as revealed by SDS-PAGE analysis (Sanchez *et al*., 1991a, 1995b). However, indirect effects may ensue such as enhanced fibrinolysis following release of tissue plasminogen activator from endothelial cells damaged by the hemorrhagic toxins (Hutton & Warrel, 1993) and activation of endogenous proteinases. In addition, the reprolysins do not possess procoagulant activity.

## Distinguishing Features

Snake venom metalloproteinases have structural similarities and also striking differences. In this regard, substrate specificity on insulin B chain by several hemorrhagic and nonhemorrhagic proteinases are quite similar. Furthermore, cross-reactions between mutalysins I and II using monoclonal antibodies against mutalysin II and rabbit polyclonal antisera against mutalysin I clearly indicated that both reprolysins are related but not identical proteins. Evidence such as proteolytic specificity and immunological reactivity suggest that each enzyme is unique among snake venom metalloendopeptidases (Sanchez *et al*., 1995a,b).

## Further Reading

For a review, see Bjarnason & Fox (1995).

## References

Baramova, E.N., Shannon, J.D., Bjarnason, J.B. & Fox, J.W. (1989) Degradation of extracellular matrix proteins by hemorrhagic metalloproteinases. *Arch. Biochem. Biophys.* **275**, 63–71.

Bjarnason, J.B. & Fox, J.W. (1994) Hemorrhagic metalloproteases from snake venoms. *Pharmac. Theor.* **62**, 325–372.

Bjarnason, J.B. & Fox, J.W. (1995) Snake venom metalloendopeptidases: reprolysins. *Methods Enzymol.* **248**, 345–387.

Brazil, V. & Vellard, J. (1928) Action coagulante et anti-coagulante des venins [Coagulant and anti-coagulant activity of venoms]. *Ann. Inst. Pasteur, Paris* **42**, 403–451.

Civello, D.J., Moran, J.B. & Geren, C.R. (1983) Substrate specificity of a hemorrhagic proteinase from timber rattlesnake venom.

*Biochemistry* **22**, 755–762.

Fox, J.W., Campbell, R., Beggerly, L. & Bjarnason, J.B. (1986) Substrate specificities and inhibition of two hemorrhagic zinc proteases Ht-c and Ht-d from *C. atrox* venom. *Eur. J. Biochem.* **156**, 65–72.

Hite, L.A., Jia, L.G., Bjarnason, J.B. & Fox, J.W. (1994) cDNA sequence for a four snake venom metalloendoproteinases: structure, classification and their relationship to mammalian reproductive proteins. *Arch. Biochem. Biophys.* **38**, 182–191.

Houssay, B.A. (1930) Classification des actions de venins de serpentes sur l'organisme animal [Classification of the activities of snake venoms on the animal organism]. *C.R. Soc. Biol. Paris* **105**, 308–310.

Hutton, R.A. & Warrell, D.A. (1993) Action of snake venom components on the haemostatic system. *Blood Rev.* **7**, 176–189.

Kishida, M., Nikai, T., Mori, N., Kohmura, K. & Sugihara, H. (1985) Characterization of mucrotoxin A from venom of *Trimeresurus mucrosquamatus* (the Chinese habu snake). *Toxicon* **23**, 637–645.

Ohsaka, A., Just, M. & Haberman, E. (1973) Action of snake venom hemorrhagic principles on isolated glomular basement membrane. *Biochim. Biophys. Acta* **323**, 415–428

Ownby, C.L., Bjarnason, J.B. & Tu, A.T. (1978) Hemorrhagic toxin from rattlesnake (*Crotalus atrox*) venom, pathogenesis of hemorrhage induced by three purified toxins. *Am. J. Pathol.* **3**, 201–218

Sanchez, E.F., Magalhaes, A. & Diniz, C.R. (1987) Purification of a hemorrhagic factor (LHF-1) from the venom of the bushmaster snake, *Lachesis muta muta. Toxicon* **25**, 611–619.

Sanchez, E.F., Magalhaes, A; Mandelbaum, F.R. & Diniz, C.R. (1991a) Purification and characterization of the hemorrhagic factor II from the venom of the bushmaster snake (*Lachesis muta muta*). *Biochim. Biophys. Acta* **1074**, 347–356.

Sanchez, E.F., Diniz, C.R. & Richardson, M. (1991b) The complete amino acid sequence of the hemorrhagic factor LHF-II, a metalloproteinase isolated from the venom of the bushmaster shake (*Lachesis muta muta*). *FEBS Lett.* **282**, 178–182.

Sanchez, E.F., Cordeiro, M.N., Oliveria, E.B., Juliano, L., Prado, E.S. & Diniz, C.R. (1995a) Proteolytic specificity of two hemorrhagic factors LHF-I and LHF-II, isolated from the venom of the bushmaster snake (*Lachesis muta muta*). *Toxicon* **33**, 1061–1069.

Sanchez, E.F., Costa, M.I.E., Chavez-Olortegui, C., Assakura, M.T., Mandelbaum, F.R. & Diniz, C.R. (1995b) Characterization of a hemorrhagic factor LHF-I, isolated from the bushmaster snake (*Lachesis muta muta*) venom. *Toxicon* **33**, 1653–1667.

Stöcker, W., Grams, F., Baumann, V., Reinemer, P., Gomis-Rüth, F.-X., McKay, D.B. & Bode, W. (1995) The metzincins – topological and sequential relations between the astacins, adamalysins, serralysins, and matrixins (collagenases) define a superfamily of zinc-peptidases. *Protein Sci.* **4**, 823–840.

Takeya, H., Nishida, S., Miyata, T., Kawada, S.I. Saisaka, Y., Morita, T. & Iwanaga, S. (1992) Coagulation factor X activating enzyme from Russell's viper venom (RVV-X). *J. Biol. Chem.* **267**, 14109–14117.

*Eladio F. Sanchez*
*Centro de Pesquisa e Desenvolvimento,*
*Fundação Ezequil Diez,*
*30510-010 Belo Horizonte, MG, Brazil*

**M**

# 425. *Trimerelysin I*

## Databanks

*Peptidase classification: clan MB, family M12, MEROPS ID: M12.154*
*NC-IUBMB enzyme classification: EC 3.4.24.52*
*Chemical Abstracts Service registry number: 151125-16-5*
*Databank codes:*

| Species | Form | SwissProt | PIR | EMBL (cDNA) | EMBL (genomic) |
|---------|------|-----------|-----|-------------|----------------|
| *Trimeresurus flavoviridis* | HR1B | P20164 | A37877 | – | – |

## Name and History

Venoms of snakes belonging to the families Viperidae and Crotalidae contain large amounts of hemorrhagic and non-hemorrhagic metalloproteinases (Iwanaga & Suzuki, 1979). In the 1970s, four hemorrhagic metalloproteinases, HR1A, HR1B, HR2a and HR2b, and one nonhemorrhagic proteinase, $H_2$-proteinase, were purified from the venom of the habu snake, *Trimeresurus flavoviridis* (Ohsaka, 1979). HR1A and HR1B have high $M_r$ values 60 000 and the other three have low $M_r$ values 23 000 (see Chapter 416 for further details of size classification). The two large proteinases express 10–25 times higher hemorrhagic activities than do the two small enzymes HR2a and HR2b (Chapter 427). The $H_2$-protease, trimerelysin II (Chapter 427) has no hemorrhagic action.

**HR1A** has been designated ***trimerelysin I*** by IUBMB; the related HR1B has not yet been classified. The amino acid sequences of HR1B (Takeya *et al.*, 1990a) and HR2a (Miyata *et al.*, 1989) have been determined and because of similarities between HR1A and HR1B, we anticipate that HR1A will also prove to belong to family M12. In this chapter, the two enzymes will be treated together, but the name trimerelysin I should be used only for HR1A.

## Activity and Specificity

Studies on the substrate specificity of these venom metalloproteinases using the oxidized insulin B chain show that they both have an identical primary specificity, favoring bulky hydrophobic residues such as Leu and Met at P1′. Recently, intramolecularly quenched fluorogenic peptide substrates have been developed, based on sequences around autoproteolyic cleavage sites in high molecular mass metalloproteinases and the sites of fibrinogen cleaved by HR2a and trimerelysin II (Takeya *et al.*, 1993). A number of these are cleaved by HR1A (HR1B has not been tested) as shown in Table 425.1.

The hemorrhagic and proteolytic activities of HR1A and B are both inhibited by EDTA or 1,10-phenanthroline.

## Structural Chemistry

No structural studies have been performed on HR1A (trimerelysin I). HR1B is a mosaic protein composed of 416 amino acid residues containing four Asn-linked oligosaccharide chains and the N-terminal half (residues 1–203) contains a metalloproteinase domain with sequence similarity

*Table 425.1*   Cleavage of synthetic substrates by trimerelysin I

| Substrates | | | | | | | | | $k_{cat}/K_m$ (M$^{-1}$ s$^{-1}$) |
|---|---|---|---|---|---|---|---|---|---|
| P4 | P3 | P2 | P1 | | P1′ | P2′ | P3′ | P4′ | |
| | Abz | Ser | Pro | + | Met | Leu | Dna | | 180 000 |
| Abz | Thr | Glu | Lys | + | Leu | Val | Dna | | 130 000 |
| Abz | Phe | Ser | Pro | + | Met | Leu | Dna | | 92 000 |
| Abz | Gly | Phe | Arg | + | Leu | Leu | Dna | | 18 000 |
| | Abz | Ala | Gly | + | Leu | Ala | Dna | | 2 800 |
| Abz | Asn | Ala | Pro | + | Leu | Ala | Dna | | 1 500 |
| Abz | Asn | Ala | Pro | | Ser | Ala | Dna | | not hydrolyzed |
| Abz | Ala | Thr | Asp | | Ile | Val | Dna | | not hydrolyzed |
| | Abz | Gly | Phe | | Arg | Leu | Dna | | not hydrolyzed |
| Abz | Gly | Pro | Leu | | Gly | Pro | Dna | | not hydrolyzed |

Dna: 2,4-dinitroanilinoethylamide.

to HR2a (62% identity) and atrolysin C (52% identity), (Chapter 423) (Takeya *et al.*, 1990a). The middle region of HR1B (residues 204–300) shows sequence similarity to disintegrins, platelet aggregation inhibitors with a potential integrin-binding sequence of Arg-Gly-Asp. Furthermore, the middle region (residues 213–336) shows 30% sequence identity to residues 1543–1656 of human von Willebrand factor.

## Preparation

HR1A and HR1B are purified from the venom of *Trimeresurus flavoviridis* (Ohsaka, 1979; Omori-Satoh & Sadahiro, 1979).

## Biological Aspects

The whole venom causes severe hemorrhage, necrosis and edema, and often also induces marked alterations or changes in blood coagulation. The hemorrhagic metalloproteinases cause localized hemorrhage through the degradation of basement membrane containing type IV collagen, laminin, nidogen and fibronectin (Baramova *et al.*, 1989). Hemorrhage is the most common occurrence in a victim bitten by crotalid or viperid snakes.

## References

Baramova, E.N., Shannon, J.D., Bjarnason, J.B. & Fox, J.W. (1989) Degradation of extracellular matrix proteins by hemorrhagic metalloproteinases. *Arch. Biochem. Biophys.* **275**, 63–71.

Iwanaga, S. & Suzuki, T. (1979) Enzymes in snake venom. In: *Handbook of Experimental Pharmacology*, vol. 52 (Lee, G.-Y., ed.). New York: Springer-Verlag, pp. 61–158.

Miyata, T., Takeya, H., Ozeki, Y., Arakawa, M., Tokunaga, F., Iwanaga, S. & Omori-Satoh, T. (1989) Primary structure of the hemorrhagic protein, HR2a, isolated from the venom of *Trimeresurus flavoviridis*. *J. Biochem. (Tokyo)* **105**, 847–853.

Ohsaka, A. (1979) Hemorrhagic, necrotizing and edema forming effects of snake venoms. In: *Handbook of Experimental Pharmacology*, vol. 52 (Lee, G.-Y., ed.). New York: Springer-Verlag, pp. 480–546.

Omori-Satoh, T. & Sadahiro, S. (1979) Resolution of the major hemorrhagic component of *Trimeresurus flavoviridis* venom into two parts. *Biochim. Biophys. Acta* **580**, 392–404.

Takeya, H., Oda, K., Miyata, T., Omori-Satoh, T. & Iwanaga, S. (1990) The complete amino acid sequence of the high molecular mass hemorrhagic protein HR1B isolated from the venom of *Trimeresurus flavoviridis*. *J. Biol. Chem.* **265**, 16068–16073.

Takeya, H., Nishida, S., Nishino, N., Makinose, Y., Omori-Satoh, T., Nikai, T., Sugihara, H. & Iwanaga, S. (1993) Primary structure of platelet aggregation inhibitor (disintegrins) autoproteolytically released from snake venom hemorrhagic metalloproteinases and new fluorogenic peptide substrates for these enzymes. *J. Biochem. (Tokyo)* **113**, 473–483.

*Shun-ichiro Kawabata*
*Department of Biology,*
*Faculty of Science,*
*Kyushu University,*
*Fukuoka, 812-81 Japan*
*Email: skawascb@mbox.nc.kyushu-u.ac.jp*

*Sadaaki Iwanaga*
*Department of Biology,*
*Faculty of Science,*
*Kyushu University,*
*Fukuoka, 812-81 Japan*

# 426. Mucrolysin

## Databanks

*Peptidase classification: clan MB, family M12, MEROPS ID: M12.157*
*NC-IUBMB enzyme classification: EC 3.4.24.54*
*Databank codes:*

| Species | SwissProt | PIR | EMBL (cDNA) | EMBL (genomic) |
|---|---|---|---|---|
| *Trimeresurus mucrosquamatus* | – | – | X77089 | – |

## Name and History

**Mucrolysin** is the trivial name given to the hemorrhagic toxin **mucrotoxin A** classified as a zinc metalloproteinase isolated from the venom of *Trimeresurus mucrosquamatus* (Chinese habu snake) found in Taiwan (Sugihara *et al.*, 1983).

## Activity and Specificity

Proteinase activity is assayed by the method of Murata *et al.* (1963) using casein as a substrate. The pH optimum is about 8.5. Fibrinogenase activity is measured according to Ouyang & Teng (1976). Dimethylcasein hydrolytic activity is assayed by the method of Lin *et al.* (1969).

Mucrolysin possesses lethal and hemorrhagic activities. However, no caseinolytic activity is observed. Mucrolysin possesses proteolytic activity hydrolyzing the following bonds of the insulin B chain:

FVNQHLCGS┼H┼LVEA┼L┼Y┼LVCGERGFFYTPKA

It also hydrolyses cattle fibrinogen, the A$\alpha$ chain first and then the B$\beta$ chain. Mucrolysin hydrolyzes dipeptides of Gly┼Pro, Leu┼Gly, Leu┼Tyr, Tyr┼Leu, Ala┼Leu, Gly┼Met, His┼Leu, Gly┼Tyr and Gly┼Gly (Kishida *et al.*, 1985). The lethal and hemorrhagic activities are inhibited by EDTA, 1,10-phenanthroline or antivenin, but not by DFP, *p*-chloromercuribenzoate or soybean trypsin inhibitor. The activities are lost upon heating the toxin at 55°C for 10 min at pH 7.0 (Sugihara *et al.*, 1983).

## Structural Chemistry

Mucrotoxin is composed of 862 amino acid residues and of these, 39 are cysteinyl residues. Its molecular mass is 94 kDa as determined by SDS-PAGE and its pI is 4.3. The enzyme contains 2 mol of zinc and 3 mol of calcium per mol of protein.

Metalloproteinases in snake venoms are classified into four groups (Hite *et al.*, 1994) (see Chapter 416); mucrolysin belongs to class IV.

## Preparation

Mucrolysin is isolated from *Trimeresurus mucrosquamatus* venom from Taiwan by a combination of gel-filtration and ion-exchange column chromatographies. The final preparation is homogeneous as judged by native-PAGE, SDS-PAGE, isoelectric focusing and immunodiffusion. By these procedures, 14 mg of mucrolysin is obtained from 1 g of crude venom (Sugihara *et al.*, 1983).

## Biological Aspects

Mucrolysin has dimethylcasein hydrolytic activity (0.04 units mg$^{-1}$), lethal activity (LD$_{50}$ = 12 µg per mouse) and hemorrhagic activity (minimum hemorrhagic dose = 2.31 µg). The pathologic changes in the mice injected with mucrolysin are characterized by bleeding into various organs, especially the heart and stomach. Mucrolysin induces coagulation necrosis, and creatine phosphokinase activity is 72.5 units ml$^{-1}$, compared to 15.1 units ml$^{-1}$ for the control (Sugihara *et al.*, 1983; Kishida *et al.*, 1985).

## Related Peptidases

Hemorrhagic metalloproteinases a and b are purified from *Trimeresurus mucrosquamatus* venom. The molecular masses for these toxins are 15 kDa and 27 kDa, and their pI values are 4.72, and 8.90, respectively. Both hemorrhagic toxins have no caseinolytic activity. Hemorrhagic factor a hydrolyzes the following bonds of the oxidized insulin B chain:

FVNQHLCGSH┼LVEALY┼LVCGER┼GFFYTPKA

It also hydrolyzes the His9┼Leu10 bond of angiotensin I, and the B$\beta$ chain of cattle fibrinogen. Hemorrhagic factor b hydrolyzes only the Ala14┼Leu15 bond of the oxidized insulin B chain, Pro7┼Phe8 bond of angiotensin I, and A$\alpha$ chain of cattle fibrinogen. Hemorrhagic factors a and b produce severe hemorrhage in the heart and stomach. Hemorrhagic factor b also produces severe hemorrhage in the liver. Hemorrhagic activity is inhibited by EDTA, 1,10-phenanthroline or *p*-chloromercuribenzoate, but not by soybean trypsin inhibitor or DFP (Nikai *et al.*, 1985). Hemorrhagic factors a and b belong to the class I venom metalloproteases (see Chapter 416).

Three metalloproteinases with A$\alpha$ fibrinogenase activity have been purified from *Trimeresurus mucrosquamatus* venom, and a complete cDNA sequence for the 24 kDa protease was determined by Huang *et al.* (1995). These proteinases are inhibited by EDTA or 1,10-phenanthroline, but not by PMSF.

## Further Reading

For a review, see Bjarnason & Fox (1995).

## References

Bjarnason, J.B. & Fox, J.W. (1995) Snake venom metalloendopeptidases: reprolysins. *Methods Enzymol.* **248**, 345–387.

Huang, K.-F., Huang, C.-C., Pan, F.-M., Chow, L.-P., Tsugita, A. & Chiou, S.-H. (1995) Characterization of multiple metalloproteinases with fibrinogenolytic activity from the venom of Taiwan habu (*Trimeresurus mucrosquamatus*): protein microsequencing coupled with cDNA sequence analysis. *Biochem. Biophys. Res. Commun.* **216**, 223–233.

Kishida, M., Nikai, T., Mori, N., Kohmura, S. & Sugihara, H. (1985) Characterization of mucrotoxin A from the venom of *Trimeresurus mucrosquamatus* (the Chinese habu snake). *Toxicon* **23**, 637–645.

Lin, Y., Means, G.E. & Feeney, R.E. (1969) The action of proteolytic enzymes on *N,N*-dimethyl proteins. *J. Biol. Chem.* **244**, 789–793.

Murata, Y., Satake, M. & Suzuki, T. (1963) Studies on snake venom. XII. Distribution of proteinase activities among Japanese and Formosan snake venoms. *J. Biochem. (Tokyo)* **53**, 431–437.

Nikai, T., Mori, N., Kishida, M., Kato, Y., Takenaka, C., Murakami, T., Shigezane, S. & Sugihara, H. (1985) Isolation and characterization of hemorrhagic factors a and b from the venom of the Chinese habu snake (*Trimeresurus mucrosquamatus*). *Biochim. Biophys. Acta* **838**, 122–131.

Ouyang, C. & Teng, C.M. (1976) Fibrinogenolytic enzymes of *Trimeresurus mucrosquamatus* venom. *Biochim. Biophys. Acta* **420**, 298–308.

Sugihara, H., Moriura, M. & Nikai, T. (1983) Purification and properties of a lethal protein, 'Mucrotoxin A', from the venom of the Chinese habu snake (*Trimeresurus mucrosquamatus*). *Toxicon* **21**, 247–255.

*Toshiaki Nikai*
*Department of Microbiology, Faculty of Pharmacy, Meijo University,*
*Yagotoyama 150, Tenpaku-Ku,*
*Nagoya, 468, Japan*
*Email: nikai@meijo-u.ac.jp*

# 427. *Trimerelysin II*

## Databanks

*Peptidase classification: clan MB, family M12, MEROPS ID: M12.155*
*NC-IUBMB enzyme classification: EC 3.4.24.53*
*Chemical Abstracts Service registry number: 151125-15-4*
*Databank codes:*

| Species | Type | SwissProt | PIR | EMBL (cDNA) | EMBL (genomic) |
|---|---|---|---|---|---|
| *Trimeresurus flavoviridis* | H2 | P20165 | JU0037 | – | – |
| *Trimeresurus flavoviridis* | HR2a | P14530 | JX0074 | – | – |

## Name and History

As was described in Chapter 425, four hemorrhagic metalloproteases named HR1A, HR1B, HR2a and HR2b, and one **nonhemorrhagic metalloprotease**, named **H2-proteinase**, have been purified from the venom of *Trimeresurus flavoviridis*. The nonhemorrhagic H2-proteinase protease has been assigned the name **trimerelysin II**. Trimerelysin II is very similar to the hemorrhagic protein HR2a in properties such as molecular mass, pH and heat stability, and susceptibility to EDTA, but it is completely free from hemorrhagic activity. HR2a, which is not in the EC listing at this time, will also be dealt with briefly in this chapter.

## Activity and Specificity

Intramolecularly quenched fluorogenic peptide substrates (Abz-peptide-Dna) (Dna: 2,4-dinitroanilinoethylamide) have been developed, based on sequences in close proximity to the autoproteolytic cleavage sites in high-molecular mass metalloproteinases and the sites of fibrinogen cleaved by HR2a and H2-proteases (Takeya *et al.*, 1993). Results with these substrates are shown in Table 427.1.

## Structural Chemistry

H2-Proteinase is a nonglycosylated metalloproteinase consisting of 201 amino acid residues with an N-terminal pyroglutamic acid. It contains the typical zinc-binding sequence HEXXH and shows sequence similarity to HR2a (74% identity; Takeya *et al.*, 1989). HR2a is also blocked by N-terminal pyroglutamate and consists of 202 residues. It also has 62% identity to the N-terminal protease domain (residues 1–203) of trimerelysin I (Chapter 425) (Miyata *et al.*, 1989).

## Preparation

H2-Proteinase is purified from the venom of *Trimeresurus flavoviridis* (Takahashi & Ohsaka, 1970). HR2a is purified at the same time.

## Biological Aspects

HR2a causes hemorrhaging but trimerelysin II does not. The difference may be due to a middle portion of the sequences (residues 51–130 in trimerelysin II) that is only 54% identical between the two proteins, and contains a free Cys in trimerelysin II. This part of the HR2a proteinase may

*Table 427.1*  Hydrolysis of synthetic substrates by trimerelysin II

| Substrates | | | | | | | | | $k_{cat}/K_m$ (M$^{-1}$ s$^{-1}$) | |
| --- | --- | --- | --- | --- | --- | --- | --- | --- | --- | --- |
| P4 | P3 | P2 | P1 | | P1′ | P2′ | P3′ | P4′ | H$_2$-protease | HR2a |
| | Abz | Ser | Pro | + | Met | Leu | Dna | | 20 000 | 340 000 |
| Abz | Thr | Glu | Lys | + | Leu | Val | Dna | | 32 000 | 120 000 |
| Abz | Gly | Phe | Arg | + | Leu | Leu | Dna | | 7 800 | 12 000 |
| Abz | Asn | Ala | Pro | | Ser | Ala | Dna | | not hydrolyzed | |
| Abz | Ala | Thr | Asp | | Ile | Val | Dna | | not hydrolyzed | |
| Abz | Gly | Phe | Arg | | Leu | Dna | | Dna | not hydrolyzed | |
| Abz | Gly | Pro | Leu | | Gly | Pro | Dna | | not hydrolyzed | |

Dna: 2,4-dinitroanilinoethylamide.

bind to basement membrane, and so cause the hemorrhagic activity.

## Related Peptidases

Takahashi & Ohsaka (1970) isolated a second hemorrhagic protease (HR2b) from the venom of *Trimeresurus flavoviridis*. This enzyme has close similarity to HR2a in molecular mass and susceptibility to heat, acidic pH and EDTA. Both proteins lack caseinolytic and collagenolytic activities, but digest various proteins including azocasein, azoalbumin, dimethylcasein and hide powder azure (Nikai *et al*., 1987).

## References

Miyata, T., Takeya, H., Ozeki, Y., Arakawa, M., Tokunaga, F., Iwanaga, S. & Omori-Satoh, T. (1989) Primary structure of hemorrhagic protein, HR2a, isolated from the venom of *Trimeresurus flavoviridis*. *J. Biochem. (Tokyo)* **105**, 847–853.

Nikai, T., Niikawa, M., Komori, Y., Sekoguchi, S. & Sugihara, H. (1987) Proof of proteolytic activity of hemorrhagic toxins, HR-2a and HR-2b, from *Trimeresurus flavoviridis* venom. *Int. J. Biochem.* **19**, 221–226.

Takahashi, T. & Ohsaka, A. (1970) Purification and some properties of two hemorrhagic principles (HR2a and HR2b) in the venom of *Trimeresurus flavoviridis*: complete separation of the principles from proteolytic activity. *Biochim. Biophys. Acta* **207**, 65–75.

Takeya, H., Arakawa, M., Miyata, T., Iwanaga, S. & Omori-Satoh, T. (1989) Primary structure of H$_2$-proteinase, a non-hemorrhagic metalloproteinase, isolated from venom of the habu snake, *Trimeresurus flavoviridis*. *J. Biochem. (Tokyo)* **106**, 151–157.

Takeya, H., Nishida, S., Nishino, N., Makinose, Y., Omori-Satoh, T., Nikai, T., Sugihara, H. & Iwanaga, S. (1993) Primary structure of platelet aggregation inhibitor (disintegrins) autoproteolytically released from snake venom hemorrhagic metalloproteinases and new fluorogenic peptide substrates for these enzymes. *J. Biochem. (Tokyo)* **113**, 473–483.

*Shun-ichiro Kawabata*
*Department of Biology,*
*Faculty of Science,*
*Kyushu University,*
*Fukuoka, 812-81 Japan*
*Email: skawascb@mbox.nc.kyushu-u.ac.jp*

*Sadaaki Iwanaga*
*Department of Biology,*
*Faculty of Science,*
*Kyushu University,*
*Fukuoka, 812-81 Japan*

# 428. *Agkistrodon halys venom metalloproteinases*

## Databanks

*Peptidase classification: clan MB, family M12, MEROPS ID: M12.134*
*NC-IUBMB enzyme classification: none*

*Databank codes:*

| Species | SwissProt | PIR | EMBL (cDNA) | EMBL (genomic) |
|---|---|---|---|---|
| *Agkistrodon halys blomhoffii* | – | – | D28870 | – |

## Name and History

Four proteinases are isolated from the venom of *Agkistrodon halys blomhoffii* (mamushi), found in Japan. Hemorrhagic toxin is the name given to proteinase b (HR-II); proteinase a and c are also present in this venom but are nonhemorrhagic proteinases. A hemorrhagic toxin I (HR-I) that does not hydrolyze casein is a second hemorrhagic toxin isolated from this venom. The Databanks table includes the sequence of a potential metalloendopeptidase from *Agkistrodon halys* that is the precursor of the disintegrin halystatin. It is not known if this putative metalloendopeptidase corresponds to proteinases a, b or c, or HR-I.

## Activity and Specificity

Proteinase activity is assayed by the method of Murata *et al.* (1963) using casein as a substrate. Azocasein and azoalbumin hydrolytic activities are measured by the method of Charney & Tomarelli (1947); hide powder azure hydrolytic activity according to Civello *et al.* (1983); fibrinogenase activity according to Ouyang & Teng (1976); and dimethylcasein hydrolytic activity by the method of Lin *et al.* (1969).

All four of the *A. halys* venom metalloproteinases hydrolyse the B chain of oxidized insulin, with different specificities:

Proteinase a (Iwanaga & Suzuki, 1979)

FVNQHLCG↓SHLVEA↓LYLVCGERG↓FFYTPKA

Proteinase b (Iwanaga & Suzuki, 1979)

FVNQH↓LCGS↓H↓LVEA↓LYLVCGERG↓FFYTPKA

Proteinase c (Satake *et al.*, 1963a)

FVNQHLCGSH↓LVEA↓LYLVCGERG↓FFYTPKA

HR-I

FVNQHLCGSH↓LVEA↓LY↓LVCGERGFFYTPKA

All three of the proteinases also hydrolyze casein, with similar specific activities of 68.1 units/mg (proteinase a), 70.5 units/mg (proteinase b) and 81.3 units/mg (proteinase c). Proteinase b has kininase activity, cleaving the Gly4↓Phe5 bond of bradykinin (Suzuki & Iwanaga, 1970), but it does not hydrolyze Tos-Arg-OMe or clot fibrinogen (Iwanaga & Suzuki, 1979). Proteinase c cleaves five bonds in glucagon: Thr5↓Phe6, Thr7↓Ser8, Asp15↓Ser16, Asp21↓Phe22 and Tyr25↓Leu26 (Satake *et al.*, 1963), but does not cleave the Gly-Phe bond of bradykinin (Suzuki, 1970), although it cleaves the Gly23↓Phe24 bond in the insulin B chain.

The pH optima are 10.5, 9.8 and 8.9 for proteinase a, b and c, respectively (Satake *et al.*, 1963b). All these proteinases are inhibited by EDTA, cysteine, $Mn^{2+}$, $Cu^{2+}$, $Zn^{2+}$, $Co^{2+}$, $Hg^{2+}$, $Cd^{2+}$ or antivenin. The inhibition by EDTA is found to be irreversible.

HR-I has azocasein hydrolytic (11.57 units $mg^{-1}$), azoalbumin hydrolytic (3.81 units $mg^{-1}$), dimethylcasein hydrolytic (5.80 units $mg^{-1}$), hide powder azure hydrolytic (4.28 units $mg^{-1}$).

It also hydrolyses the Ser4↓Tyr5, Tyr5↓Gly6, Gly6↓Leu7 and Pro9↓Gly10 bonds of luteinizing hormone-releasing hormone, and the A$\alpha$ and B$\beta$ chains of cattle fibrinogen. HR-I is inhibited by EDTA, EGTA, tetraethylenepentamine, 1,10-phenanthroline, cysteine, and antivenin, and inactivated by $\alpha$-chymotrypsin or elastase but not by trypsin (Oshima *et al.*, 1972; Nikai *et al.*, 1986).

## Structural Chemistry

Proteinase a has a molecular mass of 50 kDa comprising 82.54 g residues per 100 g protein and the pI is 6.0 (Oshima *et al.*, 1968a). Proteinase b is composed of 78.84 g residues per 100 g protein of 95 kDa with a pI of 4.18 (Oshima *et al.*, 1968b). The molecular mass of proteinase c is 70 kDa and there are 77.8 g residues per 100 g protein and the pI is 3.85 (Oshima *et al.*, 1968a). These three proteinases are glycoproteins. Proteinase b contains 2 mol of calcium and 1 mol of -SH per mol of protein. The ultraviolet difference spectrum of proteinase b treated with EDTA has been described (Oshima *et al.*, 1971). HR-I is composed of 84.56 g residues per 100 g protein and is a glycoprotein of 85 kDa with a pI of 4.7 (Oshima *et al.*, 1972).

Metalloproteinases in snake venoms are classified into four groups (Hite *et al*, 1994): proteinase a, b, c and HR-I belong to classes II, IV, III and IV, respectively (Chapter 416).

## Preparation

The two hemorrhagic proteinases, HR-I and proteinase b, and two nonhemorrhagic proteinases, proteinase a and c, are isolated from the venom of *Agkistrodon halys blomhoffii* by using zone-electrophoresis, gel-filtration and ion-exchange column chromatographies. By these procedures, 0.05% of HR-I, 0.42% of proteinase a, 10.46% of proteinase b, and 8.2% of proteinase c are obtained from the total protein in the venom (Oshima *et al.*, 1968a,b, 1972).

## Biological Aspects

Proteinase b has lethal ($LD_{50}$: 4.96 µg/kg in mouse), and hemorrhagic (MHD 0.19 µg) activities (Iwanaga *et al.*, 1965). It causes severe hemorrhage under the skin of mice and in muscle when injected intravenously (Suzuki, 1970).

Proteinase c has edema-forming activities (Suzuki 1970).

HR-I has lethal ($LD_{50}$: 0.36 µg $kg^{-1}$ mouse), and hemorrhagic (MHD = 0.0012 µg) activities. HR-I produces systemic hemorrhage in internal organs such as the stomach, intestine, kidney and heart of mice by intravenous injection (Suzuki, 1970; Oshima *et al*., 1972; Nikai *et al*., 1986).

## Related Peptidases

A nonhemorrhagic metalloproteinase named protease L4 is isolated from *Agkistrodon halys brevicaudus* (Chinese mamushi). Protease L4 is composed of 173 amino acid residues and is a protein of 22 kDa. It is stable at pH 5.0–9.0 having an optimum pH 8.5. Protease L4 possesses casein hydrolytic activity; it hydrolyzes the A$\alpha$ chain of cattle fibrinogen, the Tyr5↓Gly6 bond of luteinizing hormone-releasing hormone, the Ser12↓Leu13 and Tyr14↓Gln15 bonds of oxidized insulin A chain, and the following bonds of oxidized insulin B chain:

FVNQ↓HLCGS↓H↓LVEA↓LY↓LVCGERGFFYTPKA

The proteolytic activity of protease L4 is inhibited by $Zn^{2+}$, $Co^{2+}$, $Cu^{2+}$, $Hg^{2+}$, EDTA, EGTA, 1,10-phenanthroline or cysteine, but not by phosphoramidon. The proteolytic activity is activated by $Ca^{2+}$, $Ba^{2+}$ and $Sr^{2+}$ (Fujimura *et al*., 1995). Protease L4 belongs to the class I venom metalloproteinases (Chapter 416).

A snake in the Viperidae, *Vipera berus berus*, has a hemorrhagic metalloprotease that has specificity on the oxidized B chain of insulin identical to that of HR-1 and that also cleaves the A$\alpha$ chain of fibrinogen (Samel & Siigur, 1990). Its $M_r$ of 56 000 places it in class III, whereas HR-1 is in class IV.

## Further Reading

For a review, see Tu (1977).

## References

Charney, J. & Tomarelli, R.M. (1947) A colorimetric method for the determination of the proteolytic activity of duodenal juice. *J. Biol. Chem.* **131**, 501–503.

Civello, D.J., Duong, H.L. & Geren, C.R. (1983) Isolation and characterization of a hemorrhagic proteinase from timber rattlesnake venom. *Biochemistry* **22**, 749–755.

Fujimura, S., Rikimaru, T., Baba, S., Hori, J., Hao, X.-O., Terada, S. & Kimoto, E. (1995) Purification and characterization of a non-hemorrhagic metalloproteinase from *Agkistrodon halys brevicaudus* venom. *Biochim. Biophys. Acta* **1243**, 94–100.

Hite, L.A., Jia, L.-G., Bjarnason, J.B. & Fox, J.W. (1994) cDNA sequences for snake venom metalloproteinases: structure, classification, and their relationship to mammalian reproductive proteins. *Arch. Biochem. Biophys.* **308**, 182–191.

Iwanaga, S., Omori, T., Oshima, G. & Suzuki, T. (1965) Studies on snake venoms. XVI. Demonstration a proteinase with hemorrhagic activity in the venom of *Agkistrodon halys blomhoffii*. *J. Biochem. (Tokyo)* **57**, 392–401.

Iwanaga, S. & Suzuki, T. (1979) Enzymes in snake venom. In: *Handbook of Experimental Pharmacology*, vol. 52 (Lee, C.Y., ed.). Berlin: Springer-Verlag, pp. 61–144.

Lin, Y., Means, G.E. & Feeney, R.E. (1969) The action of proteolytic enzymes on N,N-dimethyl proteins. *J. Biol. Chem.* **244**, 789–793.

Murata, Y., Satake, M. & Suzuki, T. (1963) Studies on snake venom. XII. Distribution of proteinase activities among Japanese and Formosan snake venoms. *J. Biochem. (Tokyo)* **53**, 431–437.

Nikai, T., Oguri, E., Kishida, M. Sugihara, H., Mori, N. & Tu, A.T. (1986) Reevaluation of hemorrhagic toxin, HR-I, from *Agkistrodon halys blomhoffii venom*: proof of proteolytic enzyme. *Int. J. Biochem.* **18**, 103–108.

Oshima, G., Iwanaga, S. & Suzuki, T. (1971) Some properties of proteinase b in the venom of *Agkistrodon halys blomhoffii*. *Biochim. Biophys. Acta* **250**, 416–427.

Oshima, G., Matsuo, Y., Iwanaga, S. & Suzuki, T. (1968a) Studies on snake venoms. XIX. Purification and some properties of proteinase a and c from the venom of *Agkistrodon halys blomhoffii*. *J. Biochem. (Tokyo)* **64**, 227–238.

Oshima, G., Iwanaga, S. & Suzuki, T. (1968b) Studies on snake venoms. XVIII. An improved method for purification of the proteinase b from the venom of *Agkistrodon halys blomhoffii* and its physicochemical properties. *J. Biochem. (Tokyo)* **64**, 215–225.

Oshima, G., Omori-Sato, T., Iwanaga, S. & Suzuki, T. (1972) Studies on snake venom hemorrhagic factor I (HR-I) in the venom of *Agkistrodon halys blomhoffii*. Its purification and biological properties. *J. Biochem. (Tokyo)* **72**, 1483–1494.

Ouyang, C. & Teng, C.M. (1976) Fibrinogenolytic enzymes of *Trimeresurus mucrosquamatus* venom. *Biochim. Biophys. Acta* **420**, 298–308.

Samel, M. & Siigur, J. (1990) Isolation and characterization of hemorrhagic metalloproteinase from *Vipera berus berus* (common viper) venom. *Comp. Biochem. Physiol.* **97C**, 109–214.

Satake, M., Omori, T., Iwanaga, S. and Suzuki, T. (1963a) Studies on snake venoms. XIV. Hydrolysis of insulin B chain and glucagon by proteinase c from *Agkistrodon halys blomhoffii* venom. *J. Biochem. (Tokyo)* **54**, 8–16.

Satake, M., Murata, Y. & Suzuki, T. (1963b) Studies on snake venoms. XIII. Chromatographic separation and properties of three proteinases from *Agkistrodon halys blomhoffii* venom. *J. Biochem. (Tokyo)* **53**, 438–447.

Suzuki, T. (1970) Studies on snake venom enzymes, centering around *Agkistrodon halys blomhoffii* venom. *The Snake* **2**, 75–94.

Suzuki, T. & Iwanaga, S. (1970) Bradykinin, kallidin and kallikrein, snake venoms. In: *Handbook der experimentelle Pharmakologie*, vol. XXV (Erdos, E.G. & Wilde, A.F., eds). Berlin: Springer-Verlag, pp. 193–212.

Tu, A.T. (1977) Proteolytic enzymes In: *Venoms: Chemistry and Molecular Biology*. New York: John Wiley & Sons, pp. 104–126.

*Toshiaki Nikai*
*Department of Microbiology, Faculty of Pharmacy, Meijo University,*
*Yagotoyama 150, Tenpaku-Ku,*
*Nagoya,468, Japan*
*Email: nikai@meijo-u.ac.jp*

# 429. *Jararhagin*

## Databanks

*Peptidase classification: clan MB, family M12, MEROPS ID: M12.138*
*NC-IUBMB enzyme classification: EC 3.4.24.73*
*Databank codes:*

| Species | SwissPort | PIR | EMBL (cDNA) | EMBL (genomic) |
|---|---|---|---|---|
| *Bothrops jararaca* | P30431 | – | X68251 | – |

## Name and History

The Brazilian pit viper or jararaca, *Bothrops jararaca*, produces a highly toxic venom which effects hemostasis, causing consumptive coagulopathy and local and systemic hemorrhage (Kamiguti *et al.*, 1996). Hemorrhagic activity is associated with venom proteases and a high molecular mass (52 000) endopeptidase with hemorrhagic activity has been purified from *B. jararaca* venom by Paine *et al.* (1992). The amino acid sequence determined by cDNA cloning identified the enzyme as a member of the snake venom metalloprotease/disintegrin family (M12) (Paine *et al.*, 1992). The name **jararhagin** was derived from the snake species (**jarar-**) and the hemorrhagic activity (**-hagin**) of the enzyme. Several hemorrhagic proteases have been isolated from *B. jararaca* venom (Mandelbaum *et al.*, 1982; Assakura *et al.*, 1986; Maruyama *et al.*, 1992a,b, 1993). Of these, **HF₂-proteinase** (also known as **HF₂** or **hemolytic factor-2**; Mandelbaum *et al.*, 1976) and **jararafibrase 1** (also known as JF 1; Maruyama *et al.*, 1992b) share similar biophysical and biochemical characteristics with jararhagin and are considered synonymous.

## Activity and Specificity

Jararhagin has a minimum hemorrhagic dose of approximately 20 µg (Paine *et al.*, 1992). Assays are carried out by measuring the area of hemorrhage following intradermal injection into the dorsal skin of rats or mice (Theakston & Reid, 1983).

The enzyme is fibrinolytic, cleaving the C-terminal portion of the $\alpha$ chain of fibrinogen (Maruyama *et al.*, 1992a,b; Kamiguti *et al.*, 1996). Other coagulation factors cleaved by jararhagin include the platelet collagen receptor integrin $\alpha_2\beta_1$ and von Willebrand factor (Kamiguti *et al.*, 1996). Jararhagin cleaves type IV collagen, gelatin and casein (Assakura *et al.*, 1986; Maruyama *et al.*, 1992a). It hydrolyzes the following bonds in insulin B chain (Mandelbaum *et al.*, 1976):

FVNQHLCGSH┼LVEA┼L┼YLVCGERGF┼FYTPKA

Enzyme activity is stable within the pH range 6–9 (Assakura *et al.*, 1986). It is inhibited by EDTA, 1,10-phenanthroline and DTT (Maruyama *et al.*, 1992b) but not effectively by $\alpha_2$-macroglobulin (Kamiguti *et al.*, 1994a).

## Structural Chemistry

Jararhagin is a single-chain hemorrhagic endopeptidase with a molecular mass of approximately 52 kDa and a pI of 4.5 (Paine *et al.*, 1992). The enzyme is a class III reprolysin (Bjarnason & Fox, 1995) (see Chapter 416) comprising a metalloprotease domain and a C-terminal disintegrin domain. A propeptide sequence is predicted in the cDNA, indicating that the enzyme is synthesized as an inactive zymogen. In the disintegrin domain the Arg-Gly-Asp sequence is replaced by Glu-Cys-Asp. A high degree of overall sequence similarity exists with other members of the reprolysin family (Chapter 416).

## Preparation

Jararhagin is synthesized and stored in the venom glands of *Bothrops jararaca*. It is a major component of the venom and has been purified to homogeneity (Mandelbaum *et al.*, 1976; Maruyama *et al*, 1992a; Paine *et al.*, 1992; De Luca *et al.*, 1995). Fresh venom may be obtained by milking live snakes, or may be purchased commercially from Sigma (see Appendix 2 for full names and addresses of suppliers).

## Biological Aspects

The natural function of venom is to immobilize and kill prey. The broad substrate specificity and hemorrhagic nature of jararhagin indicates that it probably contributes substantially to the lethal activity of *B. jararaca* venom and may play a role in prey digestion. With respect to humans, jararhagin is considered to be one of the major components responsible for local and systemic hemorrhage following envenomation by the Brazilian pit viper, *B. jararaca* (Rosenfeld, 1971; Maruyama *et al.*, 1990; Kamiguti *et al.*, 1991). The role of jararhagin in causing hemorrhage has been recently reviewed by Kamiguti *et al.* (1996). Jararhagin inhibits collagen- and ristocetin-induced platelet aggregation (Kamiguti *et al.*, 1995), thus interfering with platelet function. Flow cytometry experiments have shown the loss of the platelet collagen receptor $\alpha_2\beta_1$, and the enzyme cleaves von Willebrand factor (Kamiguti *et al.*, 1996). Since jararhagin is not effectively inhibited by plasma protease inhibitors (Kamiguti *et al.*, 1994a), prolonged access to coagulation factors in the bloodstream may be an important factor in the bleeding manifestations of envenomation. Jararhagin may also contribute to local necrosis and hemorrhage through degradation of extracellular matrix proteins as do other hemorrhagic proteases (Bjarnason & Fox, 1994).

It is well known that the hemorrhagic activity of large venom metalloprotease tends to be greater than that of low molecular weight forms (Bjarnasson & Fox, 1994). Jararhagin contains a 'non-RGD' disintegrin similar to mammalian disintegrins involved in cell–cell and cell–matrix interactions (Chapter 416), and it has been suggested that it might contribute to hemorrhagic toxicity by acting as a ligand attaching the molecule to cellular integrin receptors (Takeya *et al.*, 1990; Paine *et al.*, 1992).

Jararhagin-C, a 28 kDa peptide with structural identity to the disintegrin domain of jararhagin (residues Ile240–Tyr421), has recently been purified from *B. jararaca* venom (Usami *et al.*, 1994). Thus, following activation, jararhagin undergoes further processing through proteolysis or autoproteolysis. Jararhagin-C inhibits both collagen and ADP-induced platelet aggregation. De Luca *et al.* (1995) have reported the isolation of a processed disintegrin domain fragment with N-terminal sequence identity to jararhagin-C, which they have termed jaracetin. In contrast to jararhagin-C, the peptide forms a dimer and blocks $\alpha_2\beta_1$-dependent platelet adhesion to collagen. The reasons for these functional differences are unclear but may be due to differences in their internal sequence. To date, the cleaved metalloprotease domain fragment has not been identified in venom.

## Distinguishing Features

*Bothrops jararaca* venom contains several proteases with similar biochemical characteristics (Assakura *et al.*, 1986; Mandelbaum *et al.*, 1982; Maruyama *et al.*, 1992a,b) (Chapter 440). Jararhagin may be distinguished by its size (52 kDa), strong hemorrhagic activity, and its effective inhibition of collagen-induced platelet aggregation (Kamiguti *et al.*, 1994b, 1996). Polyclonal antibodies have been raised against jararhagin (Paine *et al.*, 1992), and cross-reaction is to be expected against immunologically related proteases in the *Bothrops* genus.

## Related Peptidases

Jararafibrases II–IV, low molecular mass (21.4 kDa, 20.4 kDa and 21.2 kDa respectively) fibrinolytic/hemorrhagic peptidases isolated from *B. jararaca* venom (Maruyama *et al.*, 1992b, 1993), share similar substrate specificities with jararhagin. The N-terminal sequence of jararafibrase II is distinct to that of jararhagin. However, there are no sequence data for jararafibrase III and I and there is a possibility they may be isoenzymes corresponding to the metalloprotease region of jararhagin. The non-hemorrhagic bothrolysin from *B. jararaca* venom is considered in Chapter 439.

## Further Reading

Paine *et al.* (1992) review the cloning and characterization of jararhagin; its hemorrhagic effects are considered by Kamiguti *et al.* (1996).

## References

Assakura, M.T., Reichl, A.P. & Mandelbaum, F.R. (1986) Comparison of immunological, biochemical and biophysical properties of three hemorrhagic factors isolated from the venom of *Bothrops jararaca* (jararaca). *Toxicon* **24**, 943–946.

Bjarnason, J.B. & Fox, J.W. (1994) Hemorrhagic metalloproteinases from snake venoms. *Pharmacol. Ther.* **62**, 325–372.

Bjarnason, J.B. & Fox, J.W. (1995) Snake venom metalloendopeptidases: reprolysins. *Methods Enzymol.* **248**, 345–368.

De Luca, L.M., Ward, C.M., Ohmori, K., Andrews, R.K. & Berndt, M.C. (1995) Jararhagin and jaracetin: novel snake venom inhibitors of the integrin collagen receptor, $\alpha_2\beta_1$. *Biochem. Biophys. Res. Commun.* **206**, 570–576.

Kamiguti, A.S., Cardoso, J.L., Theakston, R.D., Sano, M.I., Hutton, R.A., Rugman, F.P., Warrell, D.A. & Hay, C.R. (1991) Coagulopathy and haemorrhage in human victims of *Bothrops jararaca* envenoming in Brazil. *Toxicon* **29**, 961–972.

Kamiguti, A.S., Desmond, H.P., Theakston, R.D., Hay, C.R. & Zuzel, M. (1994a) Ineffectiveness of the inhibition of the main haemorrhagic metalloproteinase from *Bothrops jararaca* venom by its only plasma inhibitor, $\alpha_2$-macroglobulin. *Biochim. Biophys. Acta* **1200**, 307–314.

Kamiguti, A.S., Slupsky, J.R., Zuzel, M. & Hay, C.R. (1994b) Properties of fibrinogen cleaved by Jararhagin, a metalloproteinase from the venom of *Bothrops jararaca*. *Thromb. Haemost.* **72**, 244–249.

Kamiguti, A.S., Hay, C.R.M. & Zuzel, M. (1995) Selective proteolysis of platelet $\alpha_2\beta_1$ integrin (gpIa/IIa) by jararhagin, a hemorrhagic metalloprotease from *Bothrops jararaca* venom. *Br. J. Haematol.* **89**, 8.

Kamiguti, A.S., Hay, R.M., Theakston, R.D.G. & Zuzel, M. (1996) Insights into the mechanism of haemorrhage by snake venom metalloproteases. *Toxicon* **34**, 627–642.

Mandelbaum, F.R., Reichl, A.P. & Assakura, M.T. (1976) Some physical and biochemical characteristics of $HF_2$, one of the hemorrhagic factors in the venom of *Bothrops jararaca*. In: *Animal, Plant and Microbial Toxins* (Ohsaka, A., Hayashi, K. & Sawai, Y., eds). New York: Plenum Press, pp. 111–121.

Mandelbaum, F.R., Reichl, A.P. & Assakura, M.T. (1982) Isolation and characterization of a proteolytic enzyme from the venom of the snake *Bothrops jararaca* (jararaca). *Toxicon* **20**, 955–960.

Maruyama, M., Kamiguti, A.S., Cardoso, J.L., Sano-Martins, I.S., Chudzinski, A.M., Santoro, M.L., Morena, P., Tomy, S.C., Antonio, L.C., Mihara, H., & Kelen, E.M.A. (1990) Studies on blood coagulation and fibrinolysis in patients bitten by *Bothrops jararaca* (jararaca). *Thromb. Haemost.* **63**, 449–453.

Maruyama, M., Sugiki, M., Yoshida, E., Shimaya, K. & Mihara, H. (1992a) Broad substrate specificity of snake venom fibrinolytic enzymes: possible role in haemorrhage. *Toxicon* **30**, 1387–1397.

Maruyama, M., Sugiki, M., Yoshida, E., Mihara, H. & Nakajima, N. (1992b) Purification and characterization of two fibrinolytic enzymes from *Bothrops jararaca* (jararaca) venom. *Toxicon* **30**, 853–864.

Maruyama, M., Tanigawa, M., Sugiki, M., Yoshida, E. & Mihara, H. (1993) Purification and characterization of low molecular weight fibrinolytic/hemorrhagic enzymes from snake (*Bothrops jararaca*) venom. *Enzyme Protein* **47**, 124–135.

Paine, M.J., Desmond, H.P., Theakston, R.D. & Crampton, J.M. (1992) Purification, cloning, and molecular characterization of a high molecular weight hemorrhagic metalloprotease, jararhagin, from *Bothrops jararaca* venom. Insights into the disintegrin gene family. *J. Biol. Chem.* **267**, 22869–22876.

Rosenfeld, G. (1971) Symptomology, pathology and treatment of snake bites in South America. In: *Venomous Animals and their Venoms* (Bucherl, W., Buckley, E. & Deulofeu, V., eds). New York: Academic Press, pp. 345–384.

Takeya, H., Oda, K., Miyata, T., Omori, S.T. & Iwanaga, S. (1990)

The complete amino acid sequence of the high molecular mass hemorrhagic protein HR1B isolated from the venom of *Trimeresurus flavoviridis*. *J. Biol. Chem.* **265**, 16068–16073.

Theakston, R.D. & Reid, H.A. (1983) Development of simple standard assay procedures for the characterization of snake venom. *Bull. World Health Organ.* **61**, 949–956.

Usami, Y., Fujimura, Y., Miura, S., Shima, H., Yoshida, E., Yoshioka, A., Hirano, K., Suzuki, M. & Titani, K. (1994) A 28 kDa-protein with disintegrin-like structure (jararhagin-C) purified from *Bothrops jararaca* venom inhibits collagen- and ADP-induced platelet aggregation. *Biochem. Biophys. Res. Commun.* **201**, 331–339.

*Mark J.I. Paine*
*University of Dundee,*
*Biomedical Research Centre,*
*Ninewells Hospital and Medical School,*
*Dundee DD1 9SY, Scotland*
*Email: mjpaine@dux.dundee.ac.uk*

# 430. Atrolysin E

## Databanks

*Peptidase classification: clan MB, family M12, MEROPS ID: M12.145*
*NC-IUBMB enzyme classification: EC 3.4.24.44*
*Databank codes:*

| Species | SwissProt | PIR | EMBL (cDNA) | EMBL (genomic) |
|---|---|---|---|---|
| *Crotalus atrox* | P34182 | A43296 | M89784 | – |

## Name and History

**Atrolysin E** was originally termed **hemorrhagic toxin e (HT-e)** (Bjarnason & Tu, 1978). This zinc metalloproteinase was isolated from the venom of the western diamondback rattlesnake, *Crotalus atrox*, and was determined to be one of the most hemorrhagic metalloproteinases of the P-II class (Chapter 416) in the venom (Bjarnason & Tu, 1978).

## Activity and Specificity

*In vitro*, the peptidase activity of atrolysin E can be assayed with the quenched fluorescence substrate Abz-Ala-Gly-Leu+Ala-nitrobenzylamide (Shimokawa *et al.*, 1996) or using protein substrates such as dimethylcasein (Bjarnason & Fox, 1983). Proteolytic activity is stabilized by low concentrations of calcium (e.g. 2 mM; Bjarnason & Tu, 1978). The proteolytic and hemorrhagic activities of atrolysin E are inhibited by EDTA, 1,10-phenanthroline and $\alpha_2$-macroglobulin (Bjarnason & Tu, 1978; Bjarnason & Fox, 1985; Baramova

*et al.*, 1990; Fox & Bjarnason, 1995). Atrolysin E cleaves the bait region of $\alpha_2$-macroglobulin at Ser-Asp-Val689+Met690-Gly-Arg and Gly-Arg-Gly693+His694-Ala-Arg (Baramova *et al.*, 1990). Synthetic carboxyalkyl peptides and venom-derived pyroglutamyl-containing peptides are also effective inhibitors of atrolysin E (Robeva *et al.*, 1991). Antihemorrhagins from the serum of *Crotalus atrox* can inhibit the proteolytic and hemorrhagic activities of atrolysin E (Weissenberg *et al.*, 1992).

The sites of peptide bond hydrolysis by atrolysin E of the oxidized B chain of insulin are typical for venom metalloproteinases. The first site cleaved of the insulin B chain is the Ala14+Leu15 bond followed by the Ser9+His10 and Asn3+Gln4 bonds (Bjarnason & Fox, 1983, 1995):

FVN+QHLCGS+HLVEA+LYLVCGERGFFYTPKA

Using a basement membrane preparation from Engelbreth–Holm–Swarm tumor, atrolysin E was demonstrated to have the second greatest proteolytic activity (following atrolysin A) on this substrate compared to other *Crotalus atrox* atrolysins (Bjarnason *et al*., 1988). This is in concordance with the hemorrhagic activities of the atrolysins.

Atrolysin E can digest fibrinogen by rapidly cleaving the A$\alpha$ chain followed by the B$\beta$ and $\gamma$ chains, but does not produce a fibrin clot (Bjarnason *et al*., 1988). Fibrin is not cleaved by atrolysin E to any significant extent. Atrolysin E digests collagen type IV, laminin and nidogen at specific sites. In the case of collagen type IV, atrolysin cleaves the $\alpha$1(IV) chains at Ala258$+$Gln259 and the $\alpha$2(IV) chain at Gly191$+$Leu192 (Baramova *et al*., 1990). Nidogen is also cleaved by atrolysin E (Baramova *et al*., 1991). However, when the nidogen/laminin complex is digested by atrolysin E, the digestion pattern is different from that of nidogen alone. In both situations laminin was cleaved at position 2666 in the A chain and 1238 in the B2 chain (Baramova *et al*., 1989, 1991).

## Structural Chemistry

Atrolysin E is a single-chain polypeptide with a molecular mass of 25.7 kDa based on SDS-PAGE. There is one zinc atom per molecule along with a variable number of calcium ions (Bjarnason *et al*., 1978). The pI of the mature protein is 5.6. From Edman sequence studies on atrolysin E, the mature protein has 202 residues (Hite *et al*., 1992b). Based on cDNA cloning and sequence analysis, the protein is produced as a zymogen with a signal sequence (18 residues), latency domain (168 residues), metalloproteinase domain (202 residues), spacer (21 residues) and disintegrin domain (66 residues), thereby making it a member of the P-II class (Chapter 416) of snake venom metalloproteinases (Hite *et al*., 1992b, 1994). The nascent protein is proteolytically processed by venom metalloproteinases (which distinguishes it from P-IIa class members) to give rise to the mature protein comprising only the metalloproteinase domain (Shimokawa *et al*., 1996). The disintegrin domain, which can be isolated intact from the venom, has platelet-aggregation inhibitory activity (Shimokawa *et al*., 1997).

The zinc ion in the active site of atrolysin E can be replaced by a cobaltous ion (Bjarnason & Fox, 1983). The presence of the cobaltous ion in the active site does not significantly alter the secondary structure of the protein as evidenced by the lack of change in the circular dichroic spectra of the proteins in the aromatic or peptide region. Based on the absorption spectrum of cobaltous atrolysin E, the metal is in a distorted tetrahedral coordination (Bjarnason & Fox, 1983). Replacement of the zinc ion with cobalt ion does not significantly alter the proteolytic activity or hemorrhagic activity of atrolysin E.

## Preparation

Isolation of atrolysin E begins with dialysis of lyophilized crude *Crotalus atrox* venom, followed by ion-exchange chromatography on DEAE-cellulose (Bjarnason & Tu, 1978). Fraction A-4 from the DEAE chromatography step is further chromatographed by gel filtration on Sephadex G-75, followed by ion-exchange chromatography on DE-32. A typical isolation yields approximately 16 mg of purified atrolysin E from 10 g of crude venom. Recombinant proatrolysin E

with an intact disintegrin domain has been produced with a CMV expression vector in human embryonal kidney 293 cells (Shimokawa *et al*., 1996).

## Biological Aspects

Atrolysin E is the most hemorrhagic P-II hemorrhagic toxin isolated from *Crotalus atrox*, having a minimum hemorrhagic dose (MHD) of approximately 1 µg (Bjarnason & Tu, 1978). Based on the amount and activity of atrolysin E, it is estimated that it contributes nearly 3.2% of the hemorrhagic activity of *Crotalus atrox* venom. The purified toxin, when injected into muscle tissue, causes extensive hemorrhage and, as viewed by light microscopy, extensive degeneration of capillaries (Ownby *et al*., 1978). At the level of electron microscopic observation, capillaries appear in various stages of degeneration with gaps formed in the capillary wall. The basement membrane surrounding the endothelial cells is degraded or lost as is, to a certain extent, the surrounding stroma (Ownby *et al*., 1978). As with most snake venom hemorrhagic metalloproteinases, atrolysin E is thought to manifest its hemorrhagic activity by degradation of basement membrane proteins and surrounding stroma, allowing the escape of capillary contents into the stroma (Fox & Bjarnason, 1996). The structural basis for its relatively potent hemorrhagic activity, as compared to the class P-I toxins is not clear, but may reside in a greater turnover number for peptide bonds in the substrate proteins in the basement membranes.

Atrolysin E is synthesized with a disintegrin domain following the proteinase domain (Hite *et al*., 1992a). Recently, it has been demonstrated, using a recombinant form of proatrolysin E in which both the pro domain and the disintegrin domain are present, that the activation of the latent zymogen occurs by proteolytic processing of the pro domain. This is followed by removal of the disintegrin domain by metalloproteinases present in the crude venom (Shimokawa *et al*., 1996).

## Distinguishing Features

Atrolysin E is distinguished by the presence of a disintegrin domain following the proteinase domain (Hite *et al*., 1992a). The disintegrin domain is proteolytically processed from the protein, resulting in the proteinase toxin and the disintegrin. Atrolysin E is one of the most potent class P-II toxins. Its free disintegrin domain, although lacking the classic RGD sequence, also possesses platelet-aggregation inhibitory activity and therefore probably impacts on the overall pathology associated with the presence of atrolysin E in the venom of *C. atrox* (Shimokawa *et al*., 1997).

## Further Reading

For reviews, see Fox & Bjarnason (1996) and Bjarnason & Fox (1994).

## References

Baramova, E.N., Shannon, J.D., Bjarnason, J.B. & Fox, J.W. (1989) Degradation of extracellular matrix proteins by hemorrhagic metalloproteinases. *Arch. Biochem. Biophys.* **275**, 63–71.

Baramova, E.N., Shannon, J.D., Bjarnason, J.B., Gonias, S.L. & Fox, J.W. (1990) Interaction of hemorrhagic metalloproteinases with human $\alpha_2$-macroglobulin. *Biochemistry* **29**, 1158–1164.

Baramova, E.N., Shannon, J.D., Fox, J.W. & Bjarnason, J.B. (1991) Proteolytic digestion of non-collagenous basement membrane proteins by the hemorrhagic metalloproteinase Ht-e from *Crotalus atrox* venom. *Biomed. Biochim. Acta* **50**, 763–768.

Bjarnason, J.B. & Fox, J.W. (1983) Proteolytic specificity and cobalt exchange of hemorrhagic toxin e, a zinc protease isolated from the venom of the western diamondback rattlesnake (*Crotalus atrox*). *Biochemistry* **22**, 3770–3778.

Bjarnason, J.B. & Fox, J.W. (1985) Role of zinc in the structure and function of five hemorrhagic zinc proteases from *C. atrox* venom. In: *Zinc Proteases* (Bertini, I., ed.). Boston: Birkhauser, pp. 249–264.

Bjarnason, J.B. & Fox, J.W. (1994) Hemorrhagic metalloproteinases from snake venoms. *J. Pharmacol. Ther.* **62**, 325–372.

Bjarnason, J.B. & Tu. A.T. (1978) Hemorrhagic toxins from western diamondback rattlesnake (*Crotalus atrox*) venom: isolation and characterization of five toxins and the role of zinc in hemorrhagic toxin e. *Biochemistry* **17**, 3395–3404.

Bjarnason, J.B., Hamilton, D. & Fox, J.W. (1988) Studies on the mechanism of hemorrhagic production by five proteolytic hemorrhagic toxins from *Crotalus atrox* venom. *Biol. Chem. Hoppe-Seyler* **369**, 121–129.

Fox, J.W. & Bjarnason, J.B. (1995) Atrolysins: metalloproteinases from *Crotalus atrox* venom. *Methods Enzymol.* **248**, 369–387.

Fox, J.W. & Bjarnason, J.B. (1996) Reprolysins: snake venom and mammalian reproductive metalloproteinases. In: *Zinc Metalloproteinases in Health and Disease* (Hooper, N.M., ed.). London: Taylor & Francis, pp. 47–81.

Hite, L.A., Fox, J.W. & Bjarnason, J.B. (1992a) A new family of proteinases is defined by several snake venom metalloproteinases. *Biol. Chem. Hoppe-Seyler* **373**, 381–385.

Hite, L.A., Shannon, J.D., Bjarnason, J.B. & Fox, J.W. (1992b) Sequence of a cDNA clone encoding the zinc metalloproteinase hemorrhagic toxin e from *Crotalus atrox*: Evidence for signal, zymogen, and disintegrin-like structures. *Biochemistry* **31**, 6203–6211.

Hite, L.A., Jia, L.-G., Bjarnason, J.B. & Fox, J.W. (1994) cDNA sequences for four snake venom metalloproteinases: structure, classification and their relationship to mammalian reproductive proteins. *Arch. Biochem. Biophys.* **308**, 182–191.

Ownby, C.L., Bjarnason, J.B. & Tu, A.T. (1978) Hemorrhagic toxins from rattlesnake (*Crotalus atrox*) venom: pathogenesis of hemorrhage induced by three purified toxins. *Am. J. Pathol.* **93**, 201–218.

Robeva, A., Politi, V., Shannon, J.D., Bjarnason, J.B. & Fox, J.W. (1991) Synthetic and endogenous inhibitors of snake venom metalloproteinases. *Biomed. Biochim. Acta* **50**, 769–773.

Shimokawa, K., Jia, L.-G., Wang, X.-M. and Fox, J.W. (1996) Expression, activation and processing of the recombinant snake venom metalloproteinase, pro-atrolysin E. *Arch. Biochem. Biophys.* **335**, 283–294.

Shimokawa, K., Shannon, J.D., Jia, L.-G. & Fox, J.W. (1997) Sequence and biological activity of catrocollastatin-C – a disintegrin-like/cysteine-rich two-domain protein from *Crotalus atrox* venom. *Arch. Biochem. Biophys.* **343**, 35–43.

Weissenberg, S., Ovadia, M. & Kochva, K. (1992) Inhibition of the proteolytic activity of hemorrhagin-e from *Crotalus atrox* venom by antihemorrhagins from homologous serum. *Toxicon* **30**, 591–597.

**Jay W. Fox**
*Department of Microbiology,*
*University of Virginia Health Sciences Center,*
*Charlottesville, VA 22908, USA*
*Email: jwf8x@virginia.edu*

**Jón B. Bjarnason**
*Science Institute,*
*University of Iceland,*
*Reykjavik, Iceland*
*Email: jonbragi@raunvis.hi.is*

**M**

# 431. Acutolysin

## Databanks

*Peptidase classification: clan MB, family M12, MEROPS ID: M12.131*
*NC-IUBMB enzyme classification: none*
Databank codes:

| Species | SwissProt | PIR | EMBL (cDNA) | EMBL (genomic) |
|---|---|---|---|---|
| *Agkistrodon acutus* | – | JC2550 | – | – |

## Name and History

*Acutolysin* is the trivial name given to the hemorrhagic toxin *Ac1-proteinase* classified as a zinc metalloproteinase isolated from the venom of *Agkistrodon acutus* (hundred-pace snake) found in Taiwan and China (Nikai *et al*., 1977).

## Activity and Specificity

Proteinase activity is assayed by the method of Murata *et al*. (1963) using casein as a substrate. The pH optimum is about 8.5. Azocasein and azoalbumin hydrolytic activities are measured by the method of Charney & Tomarelli (1947). Hide powder azure hydrolytic activity is determined according to Rinderknecht *et al*. (1968). Fibrinogenase activity is measured according to Ouyang & Teng (1976).

Acutolysin possesses proteolytic activity hydrolyzing the following bonds of oxidized insulin B chain:

FVNQHLCGSHLVEA↓LY↓LVCGERGFFYTPKA

It also cleaves the A$\alpha$ chain of fibrinogen, casein (0.157 units mg$^{-1}$), azoalbumin (0.060 units mg$^{-1}$), and hide powder azure (0.205 units mg$^{-1}$). However, this toxin does not have Tos-Arg-OMe hydrolytic, clotting, 5′-nucleotidase, ATPase, phosphodiesterase or phospholipase A$_2$ activities. The proteolytic activity is inhibited by EDTA, 1,10-phenanthroline or cysteine but not by soybean trypsin inhibitor or DFP (Nikai *et al*., 1977, 1991).

## Structural Chemistry

Acutolysin is composed of 202 amino acid residues and is a single-chain protein of 22944 Da (from sequence data) with a pI of 4.7 (Nikai *et al*., 1977, 1995). The HEXXHXXGXXGH sequence of acutolysin contains the putative zinc-chelating His142, His146, His152 residues and the catalytic Glu143 residue (Nikai *et al*., 1995). This enzyme contains 1 mol of zinc per mol of toxin, and a drastic change is observed in the low-frequency region of the Raman spectrum if zinc is removed from acutolysin (Nikai *et al*., 1982).

Metalloproteinases in snake venoms are classified into four groups (Hite *et al*., 1994); acutolysin belongs to class I (see Chapter 416).

## Preparation

Acutolysin is isolated from *Agkistrodon acutus* venom by a combination of gel-filtration and ion-exchange column chromatographies. By these procedures, 38 mg of acutolysin is obtained from 1 g of crude venom (Nikai *et al*., 1977; Komori *et al*., 1984).

## Biological Aspects

Acutolysin has lethal (LD$_{50}$ = 55–106 µg per mouse) and hemorrhagic (minimum hemorrhagic dose: 0.223 µg) activities. Acutolysin induces hemorrhage in the kidney and stomach, as well as necrosis. Creatine phosphokinase activity (209 U/ml) appears in mouse blood (Nikai *et al*., 1977, 1991; Homma *et al*., 1980; Sakurai *et al*., 1986).

## Related Peptidases

Three main hemorrhagic proteinases (Ac1-, Ac2- and Ac3-proteinases) are isolated from *Agkistrodon acutus* venom with molecular masses of 24.5 kDa, 25 kDa and 57 kDa as determined by SDS-PAGE, respectively. The pI values of these proteinases are determined to be 4.7, 4.9 and 4.7, respectively. Both Ac2- and Ac3-proteinases have casein hydrolytic activity (Nikai *et al*., 1977; Sugihara *et al*., 1978, 1979). Ac3-proteinase (Chapter 435) possesses proteolytic activity hydrolyzing the His10↓Leu11, Ala14↓Leu15, Tyr16↓Leu17 and Phe24↓Phe25 bonds of oxidized insulin B chain:

FVNQHLCGSH↓LVEA↓LY↓LVCGERGF↓FYTPKA

and the A$\alpha$ and B$\beta$ chains of fibrinogen (Yagihashi *et al*., 1986). Acutolysin, Ac2-proteinase and Ac3-proteinase produce hemorrhage in kidney, intestine and lung, respectively (Homma *et al*., 1980). Acutolysin in the venom from Taiwan is identical to that from China as judged by native-PAGE, isoelectric focusing and immunodiffusion (Komori *et al*., 1984).

## Further Reading

For a review, see Tu (1983).

## References

Charney, J. & Tomarelli, R.M. (1947) A colorimetric method for the determination of the proteolytic activity of duodenal juice. *J. Biol. Chem.* **131**, 501–503.

Hite, L.A., Jia, L.-G., Bjarnason, J.B. & Fox, J.W. (1994) cDNA sequences for snake venom metalloproteinases: structure, classification, and their relationship to mammalian reproductive proteins. *Arch. Biochem. Biophys.* **308**, 182–191.

Homma, M., Kubota, F., Nikai, T. & Sugihara, H. (1980) Pathologic changes produced by 100-pace snake venom and its purified proteinases; with special reference to hemorrhagic lesion. *The Kitakanto Med. J.* **30**, 485–494.

Komori, Y., Nikai, T. & Sugihara, H. (1984) Comparative study of *Agkistrodon acutus* venoms from China and Taiwan. *Comp. Biochem. Physiol.* **79**, 51–57.

Murata, Y., Satake, M. & Suzuki, T. (1963) Studies on snake venom. XII. Distribution of proteinase activities among Japanese and Formosan snake venoms. *J. Biochem. (Tokyo)* **53**, 431–437.

Nikai, T., Sugihara, H. & Tanaka, T. (1977) Enzymochemical studies on snake venoms. II. Purification of lethal protein Ac1-proteinase in the venom of *Agkistrodon acutus*. *J. Pharm. Soc. Japan* **97**, 507–514.

Nikai, T., Ishizaki, H., Tu, A.T. & Sugihara, H. (1982) Presence of zinc in proteolytic hemorrhagic toxin isolated from *Agkistrodon acutus* venom. *Comp. Biochem. Physiol.* **72**, 103–106.

Nikai, T., Kato, C., Komori, Y., Sugihara, H. & Homma, M. (1991) Characterization of Ac1 proteinase from the venom of *Agkistrodon acutus* (hundred-pace snake) from Taiwan. *Int. J. Biochem.* **23**, 311–315.

Nikai, T., Kato, C., Komori, Y., Nodani, H. & Sugihara, H. (1995) Primary structure of Ac1 proteinase of *Deinagkistrodon acutus* (hundred-pace snake) from Taiwan. *Biol. Pharm. Bull.* **18**, 631–633.

Ouyang, C. & Teng, C.M. (1976) Fibrinogenolytic enzymes of *Trimeresurus mucrosquamatus* venom. *Biochim. Biophys. Acta* **420**, 298–308.

Rinderknecht, H., Gokas, M.C., Silverman, P. & Haverback, B.J. (1968) A new ultrasensitive method for the determination of proteolytic activity. *Clin. Chim. Acta* **21**, 197–203.

Sakurai, N., Sugimoto, K., Sugihara, H., Muro, H., Kaneko, M., Nikai, T. & Shibata, K. (1986) Glomerular injury in mice induced by *Agkistrodon* venom. *Am. J. Pathol.* **122**, 240–251.

Sugihara, H., Nikai, T., Kawaguchi, T. & Tanaka, T. (1978) Enzymochemical studies on snake venoms. IV. Purification of lethal protein Ac2-Proteinase in the venom of *Agkistrodon acutus*. *J. Pharm. Soc. Japan* **98**, 1523–1529.

Sugihara, H., Nikai, T., Umeda, H. & Tanaka, T. (1979) Enzymochemical studies on snake venoms V. Purification of lethal protein Ac3-Proteinase in the venom of *Agkistrodon acutus*. *J. Pharm. Soc. Japan* **99**, 1161–1167.

Tu., A.T. (1983) The role of zinc in snake toxins. In: *Metal Ions in Biological Systems*, vol. 15 (Sigel, H., ed.). New York & Basel: Marcel Dekker, pp. 193–211.

Yagihashi, S., Nikai, T., Mori, N., Kishida, M. & Sugihara, H. (1986) Characterization of Ac3-proteinase from the venom of *Agkistrodon acutus* (hundred-pace snake). *Int. J. Biochem.* **18**, 885–892.

***Toshiaki Nikai***
*Department of Microbiology, Faculty of Pharmacy, Meijo University,*
*Yagotoyama 150, Tenpaku-Ku,*
*Nagoya,468, Japan*
*Email: nikai@meijo-u.ac.jp*

# 432. *Ecarin*

## Databanks

Peptidase classification: clan MB, family M12, MEROPS ID: M12.151
NC-IUBMB enzyme classification: none
Databank codes:

| Species | SwissProt | PIR | EMBL (cDNA) | EMBL (genomic) |
|---|---|---|---|---|
| *Echis pyramidum* | – | – | D32212 | – |

## Name and History

The saw scaled or carpet viper, *Echis carinatus*, produces a highly toxic venom responsible for a large number of fatalities each year (Warrell & Arnett, 1976). Incoagulable blood is a distinguishing feature in bite victims (Warrel *et al.*, 1977) and a protein showing calcium-independent coagulation of citrated plasma was first isolated from *E. carinatus* venom by Kornalik *et al.* (1969). The procoagulant factor was initially referred to as the ***prothrombin-activating principle*** (Schieck *et al.*, 1972a), but later given the name *ecarin* (Kornalik & Blomback, 1975). Early work describes ecarin as being fibrinogenolytic; however, when purified to homogeneity by Morita & Iwanaga (1978), the enzyme showed strict substrate specificity for prothrombin (Chapter 55) and no fibrinogenolytic activity. Ecarin was thus found to differ substantially from other venom metalloproteases, nearly all of which are fibrino(geno)lytic. Ecarin was cloned from a cDNA library derived from the venom glands of the Kenyan snake *Echis pyramidum leakyi* (previously called *Echis carinatus leakyi*) (Nishida *et al.*, 1995), which provided the first structural information. Close sequence similarities were found with the metalloprotease/disintegrin subfamily of viperine proteases, or reprolysins (Chapter 416).

## Activity and Specificity

Ecarin catalyzes the conversion of prothrombin to $\alpha$-thrombin via the formation of a meizothrombin intermediate (Morita *et al.*, 1976a; Rhee *et al.*, 1982; Briet *et al.*, 1982). In cattle prothombin, ecarin cleaves a single Arg+Ile site linking thrombin A and B chains within the zymogen molecule (Morita *et al.*, 1976a). $K_m$ values towards prothrombin and prethrombin 1 are of the order 1 mM (Morita & Iwanaga, 1978). In human prothrombin there are two cleavage sites, Gly158+Ser159 and Arg232+Ile323 (Briet *et al.*, 1982). Caseinolytic, fibrinogenolytic or hemorrhagic activities are not usually detected when using highly pure enzyme preparations (Morita & Iwanaga, 1978), however, low-level fibrinogenolytic activity may be manifested following prolonged incubation (Fortova *et al*, 1983).

The pH optimum is between 8.0 and 8.5 (Morita & Iwanaga, 1978). Activity is inhibited by chelating agents such as EDTA and 1,10-phenanthroline; by the sulfhydryl blocking agents 2-ME, DTT, glutathione, ascorbate and L-cysteine (Morita & Iwanaga, 1978; Rhee *et al*., 1982); and by the alkylating agent iodoacetamide (Rhee *et al*., 1982). Ecarin is not affected by serine protease inhibitors.

Biological activity may be assessed by measuring clotting times of citrated human plasma in the presence of enzyme (Schieck *et al*., 1972a) or by direct determination of meizothrombin activation with appropriate chromogenic substrates. A chromogenic assay using Tos-Gly-L-Pro-L-Arg-NHPhNO$_2$ (Chromozyme TH) is described by Stöcker *et al*. (1986). Ecarin in partially pure form is commercially available from Pentapharm (see Appendix 2 for full names and addresses of suppliers).

## Structural Chemistry

Ecarin is a single-chain glycoprotein of around 55 kDa, pI 4.5 (Morita & Iwanaga, 1978; Rhee *et al*., 1982). Higher molecular masses of 72 kDa (Nishida *et al*., 1995), and 84–86 kDa (Schieck *et al*., 1972b; Fortova *et al*., 1983) have been reported, possibly reflecting differences in glycosylation, post-translational processing or subspecies differences. A signal sequence and propeptide sequence are found in the cDNA sequence, indicating that ecarin is synthesized as a zymogen molecule. The enzyme is a class III reprolysin (Chapter 416) comprising a metalloprotease and a disintegrin domain. In the disintegrin domain, the Arg-Gly-Asp (RGD) sequence is replaced by Arg-Asp-Asp, and there is strong sequence homology with other members of the reprolysin subfamily.

The total carbohydrate composition of ecarin has been calculated as 16.6%, with equal distribution of hexose, hexosamine and sialic acid (Rhee *et al*., 1982). Five potential *N*-glycosylation sites (Asn-Xaa-Ser/Thr) are located in the enzyme (Nishida *et al*., 1995). When analyzed by atomic absorption spectroscopy, zinc ion was found to be present at very low levels (less than 0.01 mol mol$^{-1}$; Rhee *et al*., 1982).

## Preparation

Ecarin is produced and stored in the venom glands of *E. carinatus*. Homogeneous preparations (approximately 50-fold purification) have been described by several workers (Morita *et al*., 1978; Rhee *et al*., 1982; Fortova *et al*., 1983; Nishida *et al*., 1995). The simplest purification, taking advantage of the glycoprotein nature of ecarin, is a two-step process involving affinity chromatography with wheatgerm agglutinin-Sepharose 4B and ion-exchange chromatography on DEAE-Sephacel (Rhee *et al*., 1982). Purified preparations can contain two protein bands of similar size which may be isoforms containing different carbohydrate composition (Nishida *et al*., 1995). Geographical and taxonomic variation in *E. carinatus* venom occurs (Stöcker *et al*., 1986; Schaeffer, 1987), thus different isoforms of ecarin probably exist among snakes from different regions. Venom can be purchased commercially from Latoxan and Sigma.

## Biological Aspects

Ecarin plays an important role with respect to coagulation defects associated with envenomation by *E. carinatus*. In humans, incoagulable blood following bites by *Echis* is associated with defibrination and a decrease in clotting factors as a result of prothrombin activation (Kornalik & Blomback, 1975; Warrell *et al*., 1977; Edgar *et al*., 1980).

The mechanism of prothrombin activation has been well characterized (Franza *et al*., 1975; Kornalik & Blomback, 1975; Morita *et al*., 1976b; Rhee *et al*., 1982; Briet *et al*., 1982). Cleavage of prothrombin yields an active prothrombin *Echis carinatus* venom (ECV) derivative (Morita *et al*., 1976a), and α-thrombin is formed following autocatalytic conversion. The process of prothrombin activation differs significantly from activation by factor Xa. Unlike normal physiological activators, ecarin does not require calcium ions or cofactors for activation and can generate thrombin from abnormal decarboxythrombin (Schieck *et al*., 1972a; Franza *et al*., 1975). It is therefore commonly used in the detection of low serum prothrombin levels (Franza & Aronson, 1976) and the study of clotting disorders including the selective quantification of acarboxyprothrombin in patients treated with vitamin K antagonists, or those with hepatic diseases (Kornalik & Vorlova, 1988; Rosing & Tans, 1992). Ecarin may be used for confirmation of lupus anticoagulants in patient sera (Triplett *et al*., 1993; Bokorewa *et al*., 1995). More recently it has been shown that the stable intermediate product of ecarin-induced prothrombin conversion, meizothrombin, can effectively inhibit hirudin in blood (Nowak & Bucha, 1995), thus providing an effective neutralizing agent against toxic hirudin levels.

## Distinguishing Features

Ecarin is distinguishable from other venom metalloproteases by its strict substrate specificity towards prothrombin. Prothrombin is converted into three products with amidolytic activity, meizothrombin, meizothrombin-1 and α-thrombin. Thus, meizothrombin activity can be measured using human plasma in the presence of inhibitors of α-thombin (e.g. heparin) with a suitable chromogenic substrate such as Chromozyme TH (Stöcker *et al*., 1986). Polyclonal antibodies against partially purified ecarin have been described (Taylor *et al*., 1986), but are not commercially available.

## Related Peptidases

Two metalloprotease genes, EcH-I and EcH-II, have been described from *E. pyramidum leakyi* (Paine *et al*., 1994). The cDNA sequences (Chapter 404) predict two enzymes containing a signal sequence, propeptide domain, metalloprotease domain and disintegrin domain, thus sharing close similarity with class III reprolysins (Chapter 416). EcH-I and -II and ecarin were all derived from the same cDNA library constructed by Paine *et al*. (1994), although the functional relationship between them is as yet unknown. In keeping with the current nomenclature for proteolytic enzymes it is proposed to refer to EcH-I and EcH-II as echilysins-1 and -2 respectively, a reference to the species and proteolytic nature of the molecules.

*Echis carinatus* venom contains echistatin, a 5.4 kDa RGD-containing disintegrin peptide (Gan *et al*., 1988), which is presumably the product of proteolytic cleavage from a larger metalloprotease precursor. To date the protease precursor has not been characterized.

## Further Reading

Molecular characterization of ecarin is described by Nishida *et al*. (1995); its biochemical properties are detailed in Morita & Iwanaga (1978).

## References

Bokarewa, M.I., Bremme, K., Falk, G., Sten, L.M., Egberg, N. & Blomback, M. (1995) Studies on phospholipid antibodies, APC-resistance and associated mutation in the coagulation factor V gene. *Thromb. Res.* **78**, 193–200.

Briet, E., Noyes, C.M., Roberts, H.R. & Griffith, M.J. (1982) Cleavage and activation of human prothrombin by *Echis carinatus* venom. *Thromb. Res.* **27**, 591–600.

Edgar, W., Warrell, M.J., Warrell, D.A. & Prentice, C.R. (1980) The structure of soluble fibrin complexes and fibrin degradation products after *Echis carinatus* bite. *Br. J. Haematol.* **44**, 471–481.

Fortova, H., Dyr, J.E., Vodrazka, Z. & Kornalik, F. (1983) Isolation of the prothrombin-converting enzyme from fibrinogenolytic enzymes of *Echis carinatus* venom by chromatographic and electrophoretic methods. *J. Chromatogr.* **259**, 473–479.

Franza, B.J. & Aronson, D.L. (1976) Detection and measurement of low levels of prothrombin. Use of a procoagulant from *Echis carinatus* venom. *Thromb. Res.* **8**, 329–336.

Franza, B.J., Aronson, D.L. & Finlayson, J.S. (1975) Activation of human prothrombin by a procoagulant fraction from the venom of *Echis carinatus*. Identification of a high molecular weight intermediate with thrombin activity. *J. Biol. Chem.* **250**, 7057–7068.

Gan, Z.R., Gould, R.J., Jacobs, J.W., Friedman, P.A. & Polokoff, M.A. (1988) Echistatin. A potent platelet aggregation inhibitor from the venom of the viper, *Echis carinatus*. *J. Biol. Chem.* **263** 19827–19832.

Kornalik, F. & Blomback, B. (1975) Prothrombin activation induced by ecarin – a prothrombin converting enzyme from *Echis carinatus* venom. *Thromb. Res.* **6**, 57–63.

Kornalik, F. & Vorlova, Z. (1988) Ecarin test in diagnosis of dicoumarol therapy, liver diseases and DIC. *Folia Haematol. Int. Mag. Klin. Morphol. Blutforsch.* **115**, 483–487.

Kornalik, F., Schieck, A. & Habermann, E. (1969) Isolation, biochemical and pharmacologic characterization of a prothrombin-activating principle from *Echis carinatus* venom. *Naunyn Schmiedebergs Arch. Pharmacol.* **264**, 259–260.

Morita, T. & Iwanaga, S. (1978) Purification and properties of prothrombin activator from the venom of *Echis carinatus*. *J.*

Biochem. (Tokyo) **83**, 559–570.

Morita, T., Iwanaga, S. & Suzuki, T. (1976a) The mechanism of activation of bovine prothrombin by an activator isolated from *Echis carinatus* venon and characterization of the new active intermediates. *J. Biochem. (Tokyo)* **79**, 1089–1108.

Morita, T., Iwanaga, S. & Suzuki, T. (1976b) Activation of bovine prothrombin by an activator isolated from *Echis carinatus* venom. *Thromb. Res.* **8**(suppl.), 59–65.

Nishida, S., Fujita, T., Kohno, N., Atoda, H., Morita, T., Takeya, H., Kido, I., Paine, M.J., Kawabata, S. & Iwanaga, S. (1995) cDNA cloning and deduced amino acid sequence of prothrombin activator (ecarin) from Kenyan *Echis carinatus* venom. *Biochemistry* **34**, 1771–1778.

Nowak, G. & Bucha, E. (1995) Prothrombin conversion intermediate effectively neutralizes toxic levels of hirudin. *Thromb. Res.* **80**, 317–325.

Paine, M.J., Moura, D.S.A., Theakston, R.D. & Crampton, J.M. (1994) Cloning of metalloprotease genes in the carpet viper (*Echis pyramidum leakeyi*). Further members of the metalloprotease/disintegrin gene family. *Eur. J. Biochem.* **224**, 483–488.

Rhee, M.J., Morris, S. & Kosow, D.P. (1982) Role of meizothrombin and meizothrombin-(des F1) in the conversion of prothrombin to thrombin by the *Echis carinatus* venom coagulant. *Biochemistry* **21**, 3437–4343.

Rosing, J. & Tans, G. (1992) Structural and functional properties of snake venom prothrombin activators. *Toxicon* **30**, 1515–1527.

Schaeffer, R.J. (1987) Heterogeneity of *Echis* venoms from different sources. *Toxicon* **25**, 1343–1346.

Schieck, A., Habermann, E. & Kornalik, F. (1972a) The prothrombin-activating principle from *Echis carinatus* venom. II. Coagulation studies *in vitro* and *in vivo*. *Naunyn Schmiedebergs Arch. Pharmacol.* **274**, 7–17.

Schieck, A., Kobnalik, F. & Habermann, E. (1972b) The prothrombin-activating principle from *Echis carinatus* venom. I. Preparation and biochemical properties. *Naunyn Schmiedebergs Arch. Pharmacol.* **272**, 402–416.

Stöcker, K., Fischer, H. & Brogli, M. (1986) Chromogenic assay for the prothrombin activator ecarin from the venom of the saw-scaled viper (*Echis carinatus*). *Toxicon* **24**, 81–89.

Taylor, D., Iddon, D., Sells, P., Semoff, S. & Theakston, R.D. (1986) An investigation of venom secretion by the venom gland cells of the carpet viper (*Echis carinatus*). *Toxicon* **24**, 651–659.

Triplett, D.A., Stocker, K.F., Unger, G.A. & Barna, L.K. (1993) The textarin/ecarin ratio: a confirmatory test for lupus anticoagulants. *Thromb. Haemost.* **70**, 925–931.

Warrell, D.A. & Arnett, C. (1976) The importance of bites by the saw-scaled or carpet viper (*Echis carinatus*): epidemiological studies in Nigeria and a review of the world literature. *Acta Trop.* **33**, 307–341.

Warrell, D.A., Davidson, N., Greenwood, B.M., Ormerod, L.D., Pope, H.M., Watkins, B.J. & Prentice, C.R. (1977) Poisoning by bites of the saw-scaled or carpet viper (*Echis carinatus*) in Nigeria. *Q. J. Med.* **46**, 33–62.

**M**

*Mark J.I. Paine*
*University of Dundee,*
*Biomedical Research Centre,*
*Ninewells Hospital and Medical School,*
*Dundee DD1 9SY, Scotland*
*Email: mjpaine@dux.dundee.ac.uk*

# 433. Russellysin

## Databanks

*Peptidase classification: clan MB, family M12, MEROPS ID: M12.158*
*NC-IUBMB enzyme classification: EC 3.4.24.58*
*Chemical Abstracts Service registry number: 79393-92-3*
*Databank codes:*

| Species | SwissProt | PIR | EMBL (cDNA) | EMBL (genomic) |
|---|---|---|---|---|
| *Daboia russellii* | – | A42972 | – | – |

## Name and History

Russell's viper venom contains two well-known proteases: a serine protease designated Russell's viper venom factor V activator (Chapter 72) and a metalloprotease named **Russell's viper venom coagulation factor X-activating enzyme (RVV-X)**. This latter enzyme, known more simply as *russellysin*, specifically activates coagulation factor X (Chapter 54) as a result of a single cleavage at the Arg┼Ile bond, the same point cleaved by coagulation factors IXa and VIIa (Chapters 52, 53) (Furie & Furie, 1976; Kisiel *et al*., 1976; Lindquist *et al*., 1978).

## Activity and Specificity

In addition to the cleavage of factor X at Val-Arg51┼Ile52-Val-Gly-Gly, the enzyme also cleaves coagulation factor IX and protein C (Chapter 56) at Arg┼Xaa bonds. It has no action on the B chain of insulin. RVV-X contains $Ca^{2+}$ and $Zn^{2+}$ essential for proteolytic activity and is inhibited by EDTA (Amphlett *et al*., 1982).

Intramolecularly quenched fluorogenic peptide substrates (Abz-peptide-2,4-dinitroanilinoethylamide) have been developed, based on sequences around autoproteolytic cleavage sites in high $M_r$ metalloproteinases and the sites in fibrinogen cleaved by HR2a and trimerelysin II (Chapter 427) (Takeya *et al*., 1993). Russellysin hydrolyzes one of these substrates, Abz-Gly-Phe-Arg┼Leu-Leu-2,4-dinitroanilinoethylamide ($k_{cat}/K_m = 76\,000\,M^{-1}\,s^{-1}$).

## Structural Chemistry

Russellysin is composed of three disulfide-linked chains, a heavy chain of 59 kDa and two light chains of 18 kDa and 21 kDa (Gowda *et al*., 1994). The heavy chain (427 amino acid residues) consists of metalloproteinase, disintegrin and cysteine-rich domains, and its overall domain organization is the same as the high molecular mass hemorrhagic metalloproteinase HR1B (Chapter 425) with 53% sequence identity (Takeya *et al*., 1992). One of two light chains consists of 123 amino acid residues and shows sequence similarity to $Ca^{2+}$-dependent lectin (C-type lectin).

## Preparation

Russellysin is purified from the venom of *Daboia* (formerly *Vipera*) *russelli* by the method of Kisiel *et al*. (1976).

## Biological Aspects

The middle region of the russellysin heavy chain (residues 212–301) shows high sequence similarity to the RGD sequence-containing disintegrins, such as bitistatin (60% identity), trigramin (59% identity), barbourin (55% identity) and echistatin (38% identity). Russellysin is able to inhibit platelet aggregation but the RGD sequence is replaced by Arg-Asp-Glu in the corresponding region (Takeya *et al*., 1992). This is similar to the case of *Trimeresurus* protease HR1B, which has a Glu-Ser-Glu replacement (Takeya *et al*., 1990).

## Related Peptidases

There is a serine protease(s) in Russell's viper venom that activates factor V (Chapter 72). Another serine protease, designated VRH-1, acts as a hemorrhagic toxin and cleaves dimethylcasein (Chakrabarty *et al*., 1993). This has an $M_r$ of 22 000. These are both readily distinguished from russellysin, which is metal dependent.

## References

Amphlett, G.W., Byrne, R. & Castellino, F.J. (1982) Cation binding properties of the multiple subforms of RVV-X, the coagulant protein from *Vipera russelli*. *Biochemistry* **21**, 125–132.

Chakrabarty, D., Bhattacharyya, D., Sarkar, H.S. & Lahiri, S.C. (1993) Purification and partial characterization of a haemorrhagin (VRH-1) from *Vipera russelli russelli* venom. *Toxicon* **31**, 1601–1614.

Furie, B.C. & Furie, B. (1976) Coagulation protein of Russell's viper venom. *Methods Enzymol*. **45**, 191–205.

Gowda, D.C., Jackson, C.M., Hensley, P. & Davidson, E.A. (1994) Factor X-activating glycoprotein of Russell's viper venom: polypeptide composition and characterization of the carbohydrate moieties. *J. Biol. Chem.* **269**, 10644–10650.

Kisiel, W., Hermodson, M.A. & Davie E.W. (1976) Factor X activating enzyme from Russell's viper venom: isolation and characterization. *Biochemistry* **15**, 4901–4906.

Lindquist, P.A., Fujikawa, K. & Davie, E.W. (1978) Activation of bovine factor IX (Christmas factor) by factor XIa (activated plasma thromboplastin antecedent) and a protease from Russell's viper venom. *J. Biol. Chem.* **253**, 1902–1909.

Takeya, H., Oda, K., Miyata, T., Omori-Satoh, T. & Iwanaga, S. (1990) The complete amino acid sequence of the high molecular mass hemorrhagic protein HR1B isolated from the venom of *Trimeresurus flavoviridis*. *J. Biol. Chem.* **265**, 16068–16073.

Takeya, H., Nishida, S., Miyata, T., Kawada, S., Saisaka, Y., Morita, T. & Iwanaga, S. (1992) Coagulation factor X activating enzyme from Russell's viper venom (RVV-X): a novel metalloproteinase with disintegrin (platelet aggregation inhibitor)-like and C-type lectin-like domains. *J. Biol. Chem.* **267**, 14109–14117.

Takeya, H., Nishida, S., Nishino, N., Makinose, Y., Omori-Satoh, T., Nikai, T., Sugihara, H. & Iwanaga, S. (1993) Primary structure of platelet aggregation inhibitor (disintegrins) autoproteolytically released from snake venom hemorrhagic metalloproteinases and new fluorogenic peptide substrates for these enzymes. *J. Biochem. (Tokyo)* **113**, 473–483.

*Shun-ichiro Kawabata*
*Department of Biology,*
*Faculty of Science,*
*Kyushu University,*
*Fukuoka, 812-81 Japan*
*Email: skawascb@mbox.nc.kyushu-u.ac.jp*

*Sadaaki Iwanaga*
*Department of Biology,*
*Faculty of Science,*
*Kyushu University,*
*Fukuoka, 812-81 Japan*

# 434. Kistomin

## Databanks

*Peptidase classification: clan MB, family M12, MEROPS ID: M12.161*
*NC-IUBMB enzyme classification: none*
*Databank codes:*

| Species | SwissProt | PIR | EMBL (cDNA) | EMBL (genomic) |
|---|---|---|---|---|
| *Calloselasma rhodostoma* | P30403 | – | L08780 | – |

## Name and History

General hemorrhagic syndrome is one of the characteristic symptoms associated with envenomation by snakes from the family Viperidae and the hemorrhagic activity of the venom has been ascribed to the metalloproteinases known as hemorrhagins (Bjarnason & Fox, 1989). In very severe envenomation, systemic hemorrhage which eventually causes death due to internal bleeding has been reported (Bjarnason & Fox, 1983). Esnouf & Tunnah (1967) reported the presence of a hemorrhagic fraction during their isolation of the thrombin-like enzyme from *Calloselasma rhodostoma* (Malayan pit viper, earlier known as *Ancistrodon rhodostoma* or *Agkistrodon rhodostoma*) venom. Subsequent investigations indicated the presence of several hemorrhagins in the venom. **Kistomin**, the major hemorrhagin of the venom, has been purified to electrophoretic homogeneity (Ponnudurai *et al.*, 1993) and is also termed ***rhodostoxin***. Kistomin is a 203 amino acid residue protein belonging to class P-I of snake venom hemorrhagins (Chapter 416).

Cloning of the gene for rhodostomin, the disintegrin from *C. rhodostoma* venom, has shown the presence of a metalloproteinase precursor sequence in the disintegrin transcript (Au *et al.*, 1991). Au *et al.* therefore predicted the existence of a hemorrhagin in the venom and that the putative hemorrhagin shares a common precursor with rhodostomin. A comparison of the amino acid sequence of rhodostoxin with the translated sequence of the putative hemorrhagin indicates that kistomin is the putative hemorrhagin (Chung *et al.*, 1996).

## Activity and Specificity

Kistomin exhibits strong proteolytic activity of 61 units mg$^{-1}$ when casein is used as substrate. The minimum hemorrhage dose (MHD) of the purified toxin in mice was determined to be 0.13 µg using the method described by Kondo *et al.* (1960). Treatment of kistomin with EDTA (1.25 mM, 4 h, 26°C, pH 7.4) eliminated both the proteolytic and hemorrhagic activities completely. When the oxidized B chain of cattle insulin was used as substrate, kistomin cleaved a total of six peptide bonds: the most rapid cleavage occurring at the Ala14╎Leu15 bond (Tan *et al.*, 1997):

FVNQHLCGS╎H╎LVEA╎LY╎LVCG╎ERGF╎FYTPKA

The Ala-Leu cleavage site is common to many proteases. Kistomin retained full activity after storage in frozen form in 20 mM Tris–HCl, pH 7.4, buffer for 4 weeks, but lost 30% of the activity when stored at 4°C for the same period.

## Structural Chemistry

Amino acid sequencing shows that kistomin is a single polypeptide with 203 amino acid residues and four disulfide bonds. The polypeptide has a calculated $M_r$ of 23 438. Kistomin contains 15% carbohydrate and the molecular mass of the two oligosaccharides (*vide infra*) was calculated to be 4704, thus yielding a calculated molecular mass of 28 142 for kistomin. Gel-filtration chromatography of kistomin yielded a molecular mass of 30 000. Kistomin contains 1 mol of zinc per mol of protein and the pI is 5.3.

The sequence determined by Edman microsequencing confirmed the deduced amino acid sequence of the putative hemorrhagic protein encoded by the prorhodostomin cDNA. However, the N-terminal residue should be asparagine instead of valine, and residue 99 should be threonine instead of methionine. These differences could be due to a minor error in cDNA sequence determination or possibly a point mutation. The amino acid sequence of kistomin is homologous to that of the low molecular mass hemorrhagins as well as the metalloproteinase domains of high molecular mass hemorrhagins isolated from crotalid venoms (Chung *et al.*, 1996): kistomin has the highest identity (54%) with the metalloproteinase domain of atrolysin E from *Crotalus atrox* venom (Hite *et al.*, 1992a) (Chapter 430), followed by a 50% homology with the metalloproteinase domains of the high molecular mass HR1B from *Trimeresurus flavoviridis* venom (Chapter 425) (Takeya *et al.*, 1989) and jararhagin from *Bothrops jararaca* venom (Paine *et al.*, 1992) (Chapter 429). Based on sequence homology data (Jongeneel *et al.*, 1989; Murphy *et al.*, 1991), it appears that His142 and His146 chelate zinc ions and Glu143 is one of the catalytic residues. The sequence includes the extended consensus active-site sequence of Met-Xaa-His-Glu-Xaa-Gly-His-Asn-Leu-Gly-Xaa-Xaa-His-Asp (residues 140–153) (Hite

*et al.*, 1992b) and the predicted secondary structure in the zinc-binding region of the hemorrhagin is similar to that of other snake venom metalloproteinases and thermolysin (Chapter 351). However, it is interesting to note that the phylogenetic tree of snake venom metalloproteinases (constructed using the Higgins–Sharp multiple alignment program) indicates that kistomin is least related to the other metalloproteinases, exhibiting only 33.7% identity (Ponnudurai, 1995) (see also Fig. 404.5).

Kistomin is glycosylated with two *N*-linked glycopeptides attached to residues 91 and 181 (Chung *et al.*, 1996). The glycan chain is a complex type of carbohydrate structure with novel 2,3-linked sialic acids and Gal(1,3)GlcNAc linkages as shown in Figure 434.1. It is interesting to note that the oligosaccharide structure of kistomin is very similar to the carbohydrate chain of the thrombin-like serine protease ancrod (Chapter 70) isolated from the same venom (Pfeiffer *et al.*, 1992) but differs from that of the carbohydrate chain of batroxobin (Chapter 70) isolated from *Bothrops atrox moojeni* venom (Tanaka *et al.*, 1992).

Kini & Evans (1992) suggested that the carbohydrate moiety in hemorrhagin from *Trimeresurus flavoviridis* reduces the susceptibility of the protein to proteolysis. Li *et al.* (1993) reported that in their studies on the oxidation of the sugar moiety and *N*-glycosidase treatment of hemorrhagins from *Crotalus viridis viridis* venom that the carbohydrate moieties of hemorrhagins may help to maintain the stability of the hemorrhagins. Tan *et al.* (1997), however, reported that deglycosylation of kistomin results in an *apparent increase* in the stability of the hemorrhagin as well as an increased specificity: the deglycosylated rhodostoxin cleaved the oxidized cattle insulin B chain primarily at Arg22╎Gly23. It appears therefore that the glyco moiety of hemorrhagins play vastly different roles, ranging from nonspecific roles such as maintaining stability, to specific involvement including altering the substrate specificity of the protease.

## Preparation

Kistomin is the major hemorrhagin of the *C. rhodostoma* venom. The *C. rhodostoma* snake belongs to the family Viperidae and is found in Thailand, Cambodia, Laos, Vietnam, Burma, Java, southern Sumatra and northern Peninsular Malaysia. Venoms of *C. rhodostoma* collected from all these different geographical locations exhibited hemorrhagic activity in mice with minimum hemorrhagic doses of between 9 and 24 µg (Daltry *et al.*, 1996). *C. rhodostoma* venoms from Vietnam exhibited the highest hemorrhagic activity.

Kistomin can be purified to electrophoretic homogeneity from the venom using high-performance MA7Q strong

NeuAcα2–3Galβ1–3GlcNAcβ1–2Manα1

NeuAcα2–3Galβ1–3GlcNAcβ1–2Manα1

Fucα1

Manβ1–4GlcNAcβ1–4GlcNAc–

*Figure 434.1*  Structure of the glycan chain of kistomin.

anion-exchange chromatography following Sephadex G-200 gel-filtration chromatography of the crude venom (Ponnudurai *et al.*, 1993). Kistomin constitutes less than 4% of the venom protein.

## Biological Aspects

The minimum hemorrhage doses (MHD) of *C. rhodostoma* venom and kistomin in mice were determined to be 6.0 μg and 0.13 μg, respectively. Kistomin is a potent hemorrhagin with MHD comparable to the venom hemorrhagins acutolysin (from *Agkistrodon acutus*) (Chapter 431), HR-II (from *Agkistrodon halys blomhoffii*) (Chapter 428) and HR-2 (from *Vipera palaestinae*). The hemorrhagin was not lethal to mice at a dosage of $6 \mu g\, g^{-1}$ (i.v.), indicating that it is not a major lethal factor of *C. rhodostoma* venom, which has an intravenous $LD_{50}$ of approximately 6 μg. This, however, does not imply that kistomin is not important in the toxic action of the venom as hemorrhagin may act synergistically with other venom components that affect the blood-clotting mechanism. Indeed, the fact that kistomin shares a common precursor with rhodostomin, the platelet aggregation inhibitor, strongly suggests that the two proteins are synergistic in action.

Kistomin also exhibited potent edema-inducing activity with a minimum edema dose of 0.02 μg. Time course study indicates that rhodostoxin induces the 'delayed' type edema where the edema reaches a peak 4–6 h after injection (Bonta, 1969). It has been suggested that the release of prostaglandin and bradykinin are involved in this type of edema.

It has been observed that hemorrhagins from venoms of different snake genera share common epitopes (Mandelbaum *et al.*, 1989). Indeed, polyclonal antibodies to kistomin yielded strong indirect ELISA cross-reactions with venoms of *Agkistrodon contortrix contortrix*, *Crotalus adamanteus* and *Crotalus atrox*, and moderately to low against the *Trimeresurus* snake venoms. The antibodies could neutralize effectively the hemorrhagic activities of *Calloselasma rhodostoma*, *Trimeresurus flavoviridis* and *Trimeresurus sumatranus* venoms, but not those of *Agkistrodon*, *Bothrops*, *Crotalus* and *Vipera* venoms. Epitope mapping using polyclonal antibodies located four antigenic sites, corresponding to residues 46–76, 81–88, 111–120 and 165–177, and showed that the oligosaccharide moieties of kistomin do not function as antigenic determinants (Ponnudurai, 1995).

## Distinguishing Features

Rhodostoxin is the first four-disulfide proteinase reported among all known venom metalloproteinases, which are either of the two-disulfide or three-disulfide type.

## Further Reading

For a review, see Chung *et al.* (1996).

## References

Au, L.C., Huang, Y.B., Huang, T.F., Teh, G.W., Lin, H.H. & Choo, K.B. (1991) A common precursor for a putative hemorrhagic protein and rhodostomin, a platelet aggregation inhibitor of the venom of *Calloselasma rhodostoma*: molecular cloning and sequence analysis. *Biochem. Biophys. Res. Commun.* **181**, 585–593.

Bjarnason, J.B. & Fox, J.W. (1983) Proteolytic specificity and cobalt exchange of hemorrhagic toxin-e, a zinc protease isolated from the Western diamondback rattlesnake (*Crotalus atrox*). *Biochemistry* **22**, 3770–3778.

Bjarnason, J.B. & Fox, J.W. (1989) Hemorrhagic toxins from snake venoms. *J. Toxicol. Toxin Rev.* **7**, 121–209.

Bonta, I.L. (1969) Microvascular lesions as a target of anti-inflammatory and certain other drugs. *Acta Physiol. Pharmacol. Neerl.* **15**, 188–222.

Chung, M.C.M., Ponnudurai, G., Kataoka, M., Shimizu, S. & Tan, N.H. (1996) Structural studies of a major hemorrhagin (rhodostoxin) from the venom of *Calloselasma rhodostoma* (Malayan pit viper). *Arch. Biochem. Biophys.* **325**, 199–208.

Daltry, J.C., Ponnudurai, G., Chai, K.S., Tan, N.H., Thorpe, R.S. & Wuster, W. (1996) Electrophoretic profiles and biological activities: intraspecific variation in the venom of the Malayan pit viper (*Calloselasma rhodostoma*). *Toxicon* **34**, 67–79.

Esnouf, M.P. & Tunnah, G.W. (1967) The isolation and properties of the thrombin-like activity from *Ancistrodon rhodostoma* venom. *Br. J. Haematol.* **13**, 581–590.

Hite, L.A., Shannon, J.D., Bjarnason, J.B. & Fox, J.W. (1992a) Sequence of a cDNA clone encoding the zinc metalloproteinase hemorrhagic toxin e from *Crotalus atrox*: evidence for signal, zymogen and disintegrin-like structures. *Biochemistry* **31**, 6203–6211.

Hite, L.A., Fox, J.W. & Bjarnason, J.B. (1992b) A new family of proteinases is defined by several snake venom metalloproteinases. *Biol. Chem. Hoppe-Seyler* **373**, 381–385.

Jongeneel, C.V., Bouvier, J. & Bairoch, A. (1989) A unique signature identifies a family of zinc-dependent metalloproteinases. *FEBS Lett.* **242**, 211–214.

Kini, R.M. & Evans, H.J. (1992) Structural domains in venom proteins: evidences that metalloproteinases and nonenzymatic platelet aggregation inhibitors (disintegrins) from snake venoms are derived by proteolysis from a common precursor. *Toxicon* **30**, 265–293.

Kondo, H., Kondo, S., Ikezawa, H., Murata, R. & Ohsaka, A. (1960) Studies on the quantitative method for determination of hemorrhagic activity of habu snake venom. *Jpn J. Med. Sci. Biol.* **13**, 43–51.

Li, Q., Colberg, T.R. & Ownby, C.L. (1993) Purification and characterization of two high molecular weight hemorrhagic toxins from *Crotalus viridis viridis* venom using monoclonal antibodies. *Toxicon* **31**, 711–722.

Mandelbaum, F.R., Serrano, S.M.T., Sakurada, J.K., Rangel, A. & Assakura, M.T. (1989) Immunological comparison of hemorrhagic principles present in venoms of the Crotalinae and Viperinae subfamilies. *Toxicon* **27**, 169–177.

Murphy, G.J.P., Murphy, G. & Reynolds, J.J. (1991) The origin of matrix metalloproteinases and their familial relationships. *FEBS Lett.* **289**, 4–7.

Paine, M.J.I., Desmond, H.P., Theakston, R.D.G. & Crampton, J.M. (1992) Purification, cloning and molecular characterization of a high molecular weight hemorrhagic metalloproteinase, jararhagin, from *Bothrops jararaca* venom. *J. Biol. Chem.* **267**, 22869–22876.

**M**

Pfeiffer, G., Dabrowski, U., Dabrowski, J., Stirm, S., Strube, K.H. & Geyer, R. (1992) Carbohydrate structure of a thrombin-like serine protease from *Agkistrodon rhodostoma. Eur. J. Biochem.* **205**, 961–978.

Ponnudurai, G. (1995) Biochemical and immunological studies on Malayan pit viper (*Calloselasma rhodostoma*) venom hemorrhagin. PhD thesis, University of Malaya, Kuala Lumpur, Malaysia.

Ponnudurai, G., Chung, M.C.M. & Tan, N.H. (1993) Isolation and characterization of a hemorrhagin from the venom of *Calloselasma rhodostoma* (Malayan pit viper). *Toxicon* **31**, 997–1005.

Takeya, H., Arakawa, M., Miyata, T., Iwanaga, S. & Omori-Satoh, T. (1989) Primary structure of $H_2$-proteinase, a non-hemorrhagic metalloproteinase, isolated from the venom of the Habu snake, *Trimeresurus flavoviridis. J. Biochem.* **106**, 151–157.

Tan, N.H., Ponnudurai, G. & Chung, M.C.M. (1997) Proteolytic specificity of rhodostoxin, the major hemorrhagin of *Calloselasma rhodostoma* (Malayan pit viper) venom. *Toxicon* **35**, 979–984.

Tanaka, N., Nakada, H., Itoh, N., Mizuno, Y., Takanishi, M., Kawasaki, T., Tate, S.I., Inagaki, F. & Yamashina, I. (1992) Novel structure of the *N*-acetylgalactosamine-containing *N*-glycosidic carbohydrate chain of batroxobin a thrombin-like snake venom enzyme. *J Biochem.* **112**, 68–74.

*Nget Hong Tan*
*Department of Biochemistry,*
*Faculty of Medicine, University of Malaya,*
*Kuala Lumpur, Malaysia*
*Email: tannh@medicine.med.um.edu.my*

# *435. Bilitoxin*

## Databanks

*Peptidase classification: clan MB, family M12, MEROPS ID: M12.132*
*NC-IUBMB enzyme classification: none*
*Databank codes: no sequence data available*

## Name and History

**Bilitoxin** is the name given to the hemorrhagic toxin classified as a zinc metalloproteinase isolated from the venom of *Agkistrodon bilineatus* (common cantil) found in southern Mexico and Nicaragua.

## Activity and Specificity

Proteinase activity is assayed by the method of Murata *et al*. (1963) using casein as a substrate. The pH optimum is about 8.5. Fibrinogenase activity is measured according to Ouyang & Teng (1976). Bilitoxin possesses proteolytic activity hydrolyzing the casein (18.9 U mg$^{-1}$), the Aα and Bβ chains of cattle fibrinogen and the following bonds of the insulin B chain:

FVN┼QHLCGSH┼LVEA┼LY┼LVCGER┼GF┼FYTPKA

This toxin does not have Tos-Arg-OMe hydrolytic, clotting, 5'-nucleotidase, ATPase or phosphomonoesterase activities. The casein hydrolytic, lethal and hemorrhagic activities are inhibited by EDTA and EGTA, but not by cysteine or soybean trypsin inhibitor. Proteolytic and hemorrhagic activities are lost upon heating the toxin at 70°C for 10 min at pH 7.2 (Imai *et al*., 1989).

## Structural Chemistry

Bilitoxin is composed of 432 amino acid residues, and is a glycoprotein of 48 kDa. The pI for this enzyme is 4.2. Bilitoxin contains 1 mol of zinc and 2 mol of calcium per mol of toxin. Metalloproteinases in snake venoms are classified into four groups (Hite *et al*., 1994): bilitoxin belongs to class II (Chapter 416).

## Preparation

Bilitoxin is isolated from *Agkistrodon bilineatus* venom from southern Mexico by a combination of gel-filtration and ion-exchange column chromatographies. By these procedures, 32.8 mg of bilitoxin is obtained from 700 mg of crude venom (Imai *et al*., 1989).

## Biological Aspects

Bilitoxin has lethal (LD$_{50}$ 9.9 µg per mouse) and hemorrhagic (MHD: 0.008 µg) activities (Imai *et al*., 1989). Light microscopically, hemorrhage is recognized in the connective tissue surrounding muscle cells and fibrin is present both intravascularly and extravascularly. Electron microscopically, hemorrhage is observed without damage to the intercellular junctions. Bilitoxin induces myonecrosis upon intramuscular injection (Ownby *et al.*, 1990).

## Related Peptidases

Bilitoxin is enzymatically most similar to the Ac3-proteinase in *Agkistrodon acutus* venom from Taiwan (Chapter 431). Ac3-Proteinase hydrolyzes casein, azocasein, dimethylcasein and hide powder azure and possesses lethal and hemorrhagic activities. These activities are inhibited by EDTA, EGTA, 1,10-phenanthroline, tetraethylenpentamine, phosphoramidon or 2-ME. Its molecular mass is 57 kDa and the pI is 4.7. Ac3-Proteinase contains 1 mol of zinc and 5 mol of calcium. It possesses proteolytic activity hydrolyzing the Ser12+Leu13 and Gln15+Leu16 bonds of oxidized insulin A chain; the Pro7+Phe8 bonds of angiotensin I, and the A$\alpha$ and B$\beta$ chains of fibrinogen. It cleaves the oxidized insulin B chain at the following bonds:

FVNQHLCGSH+LVEA+LY+LVCGERGF+FYTPKA

Ac3-Proteinase induces a coagulate type of necrosis and creatine phosphokinase activity is 405 units ml$^{-1}$ (Sugihara *et al.*, 1979; Komori *et al.*, 1984; Yagihashi *et al.*, 1986).

Bilitoxin-2 in *Agkistrodon bilineatus* venom has a molecular mass of 50 kDa and possesses proteolytic and hemorrhagic activities. Both activities are inhibited by EDTA, EGTA or *N*-bromosuccinimide. This enzyme possesses hydrolytic activity towards casein, azocasein, azoalbumin, azocoll, dimethylcasein, hide powder azure and the A$\alpha$, B$\beta$ and $\gamma$ chains of cattle fibrinogen. It cleaves the oxidized B chain of insulin at the following bonds:

FVN+QHLCGSH+LVEA+LY+LVCGERGFFYTPKA

This toxin contains 1 mol of zinc per mol of protein (Nikai *et al.*, 1996). Ac3-Proteinase and bilitoxin-2 belong to the class III and II groups of Hite *et al.* (1994), respectively.

## Further Reading

For a review, see Bjarnason & Fox (1995).

## References

Bjarnason, J.B. & Fox, J.W. (1995) Snake venom metalloendopeptidases: Reprolysins. *Methods Enzymol.* **248**, 345–387.

Hite, L.A., Jia, L.-G., Bjarnason, J.B. & Fox, J.W. (1994) cDNA sequences for snake venom metalloproteinases: structure, classification, and their relationship to mammalian reproductive proteins. *Arch. Biochem. Biophys.* **308**, 182–191.

Imai, K., Nikai, T., Sugihara, H. & Ownby, C.L. (1989) Hemorrhagic toxin from the venom of *Agkistrodon bilineatus* (common cantil). *Int. J. Biochem.* **21**, 667–673.

Komori, Y., Nikai, T. & Sugihara, H. (1984) Comparative study of *Agkistrodon acutus* venoms from China and Taiwan. *Comp. Biochem. Physiol.* **79**, 51–57.

Murata, Y., Satake, M. & Suzuki, T. (1963) Studies on snake venom. XII. Distribution of proteinase activities among Japanese and Formosan snake venoms. *J. Biochem. (Tokyo)* **53**, 431–437.

Nikai, T., Taniguchi, K., Komori, Y., Sugihara, H. & Homma, M. (1996) Hemorrhagic toxin, bilitoxin-2, from *Agkistrodon bilineatus* venom. *J. Nat. Tox.* **5**, 95–106.

Ouyang, C. & Teng, C. M. (1976) Fibrinogenolytic enzymes of *Trimeresurus mucrosquamatus* venom. *Biochim. Biophys. Acta* **420**, 298–308.

Ownby, C.L., Nikai, T., Imai, K. & Sugihara, H. (1990) Pathogenesis of hemorrhage induced by bilitoxin, a hemorrhagic toxin isolated from the venom of the common cantil (*Agkistrodon bilineatus*). *Toxicon* **28**, 837–846.

Sugihara, H., Nikai, T., Umeda, H. & Tanaka, T. (1979) Enzymochemical studies on snake venoms. V. Purification of lethal protein Ac3-proteinase in the venom of *Agkistrodon acutus. J. Pharm. Soc. Jpn* **99**, 1161–1167.

Yagihashi, S., Nikai, T., Mori, N., Kishida, M. & Sugihara, H. (1986) Characterization of Ac3-proteinase from the venom of *Agkistrodon acutus* (hundred-pace snake). *Int. J. Biochem.* **18**, 885–892.

*Toshiaki Nikai*
*Department of Microbiology, Faculty of Pharmacy, Meijo University,*
*Yagotoyama 150, Tenpaku-Ku, Nagoya, 468, Japan*
*Email: nikai@meijo-u.ac.jp*

# *436. Leucolysin*

## Databanks

*Peptidase classification: clan MB, family M12, MEROPS ID: M12.136*
*NC-IUBMB enzyme classification: EC 3.4.24.6*
*Chemical Abstracts Service registry number: 72561-59-1*
*Databank codes: no sequence data available*

## Name and History

The comparatively high level of proteolytic activity in the venom of *Agkistrodon piscivorus leucostoma* (cottonmouth water moccasin) was the criterion for selecting it from a group of venoms from North American snakes for vigorous biochemical characterization (Wagner & Prescott, 1966). All of the information on the major protease (leucolysin) in this venom originated from John Prescott's laboratory at Texas A & M University and was performed prior to the development of recombinant molecular biology techniques or microprotein sequencing techniques. Thus, considerable information is known about its enzymatic specificity and certain molecular properties, but its sequence and ultrastructural properties have not been determined.

Leucolysin was the first venom protease purified to a high degree of homogeneity and characterized as a metalloendopeptidase of defined substrate specificity. The trivial name **leucostoma peptidase A** was chosen to distinguish it from other proteases present in this venom. The name **leucolysin** was assigned by the IUBMB, and it was one of the first two venom proteases catalogued by that agency.

## Activity and Specificity

Solutions of crude *Agkistrodon piscivorus leucostoma* venom hydrolyze a multiplicity of ester and peptide substrates (Wagner & Prescott, 1966). The homogeneous leucolysin, however, hydrolyzes only peptides, separating it from the venom esterases. The enzyme contains 1 mol zinc and 2 mol calcium per mol of protein and its activity is completely inhibited by EDTA. It is also inhibited with 5 mM 1,10-phenanthroline (a poor chelating agent for $Ca^{2+}$) in the presence of 10 mM $Ca^{2+}$ (Spiekerman *et al*., 1973). The metal chelator-inhibited enzyme can be restored to full activity by the addition of $Zn^{2+}$. Presumably, the two calcium ions have structural requirements (Spiekerman *et al*., 1973). Thus, leucolysin was the first well-characterized snake venom metallopeptidase and is still among the few venom proteases properly characterized as a zinc metalloendopeptidase.

Leucolysin was isolated as the major protease active toward hemoglobin in *Agkistrodon piscivorus leucostoma* venom (Wagner *et al*., 1968). The purified enzyme possesses none of the other peptidase or esterase activities detectable in crude venom, e.g. activity toward Bz-L-Arg-OEt or L-Leu-NHPhNO$_2$, substrates commonly used to assay trypsin-like enzymes and aminopeptidases, respectively (Wagner *et al*., 1968). Its hemorrhagic and fibrogenolytic activities have not been characterized. Leucolysin will hydrolyze N-blocked dipeptides such as Z-Gly┼Arg-OH, Z-Gly┼PheNH$_2$ but not Z-Gly-Leu. Using the B chain of oxidized insulin as a substrate, cleavage specificity is towards Leu and Phe residues in the P1′ position (residues 6, 11, 15, 24 and 25):

FVNQHL┼CGSHL┼VEAL┼YLVCGERGF┼F┼YTPKA

It does not cleave bonds containing basic residues in the P1′ site as is the case with certain N-blocked dipeptides (Spiekerman *et al*., 1973).

## Structural Chemistry

The purification and characterization of leucolysin was performed prior to 1974. It has not subsequently been studied using the methods of molecular biology, X-ray crystallography, NMR or MALDI mass spectrometry, nor has it been sequenced.

Leucolysin is a monomeric metallopeptidase with an $M_r$ of 22 500, as determined by sedimentation equilibrium and separate measurements of $s_{0,20,w}$ and $D_{0,20,w}$ (Wagner *et al*., 1968). It has a pI of pH 6.5 and a well-defined amino acid composition. It possesses six Cys residues (Wagner *et al*., 1968), but it is not known if these are all in disulfide linkages, a criterion that differentiates group I-A from I-B venom proteases.

## Preparation

Leucolysin is not commercially available and must be prepared from crude venom, utilizing a five-step procedure described by Wagner *et al*. (1968) and modified by Spiekerman *et al*. (1973). The final preparation lacks detectable protein contaminants when subjected to native PAGE analysis and is devoid of activity toward esters, dipeptides and Leu-NHPhNO$_2$.

## Biological Aspects

Leucolysin is clearly a class I venom metalloendopeptidase based on its molecular mass and metal ion content (Bjarnason & Fox, 1995) (Chapter 416). The natural substrate(s) of leucolysin have not been identified; thus, its contribution to the toxicity of venom is not known.

## Further Reading

A review of the classification and structure and function of this and other venom proteases is given by Bjarnason & Fox (1995).

## References

Bjarnason, J.B. & Fox, J.W. (1995) Snake venom metalloendopeptidases: reprolysins. *Methods Enzymol.* **248**, 345–368.

Spiekerman, A.M., Fredericks, K.K., Wagner, F.W. & Prescott, J.M. (1973) *Leucostoma* peptidase A: a metalloprotease from snake venom. *Biochim. Biophys. Acta* **293**, 464–475.

Wagner, F.W. & Prescott, J.M. (1966) A comparative study of proteolytic activities in the venoms of some North American snakes. *Comp. Biochem. Physiol.* **17**, 191–201.

Wagner, F.W., Spiekerman, A.M. & Prescott, J.M. (1968) *Leucostoma* peptidase A: isolation and physical properties. *J. Biol. Chem.* **243**, 4486–4493.

***Fred W. Wagner***
*BioNebraska, Inc.,*
*3820 NW 46th Street,*
*Lincoln, NE 68524, USA*
*Email: fwagner@unlinfo.unl.edu*

# 437. *Bitis arietans* hemorrhagic proteinases

## Databanks

*Peptidase classification: clan MB, family M12, MEROPS ID: M12.137*
*NC-IUBMB enzyme classification: none*
*Databank codes: no sequence data available*

## Name and History

Hemorrhagic, proteolytic and aminoacyl esterase activities in the venom of the viperous snake *Bitis arietans* (puff adder) have been reported by several workers (Ohsaka *et al.*, 1966; Delpierre, 1968; Soto *et al.*, 1988). Omori-Satoh *et al.* (1995) and Yamakawa *et al.* (1995) have purified two hemorrhagins (**BHRa** and **BHRb**) from the venom and characterized them as metalloproteinases. The name BHR is derived from *Bitis arietans* **h**emo**r**rhagic principles (hemorrhagins).

## Activity and Specificity

The minimum hemorrhagic dose (MHD) producing a hemorrhagic spot of 10 mm in diameter in rabbits is as low as 10 ng for BHRa and 15 ng for BHRb. They are stable at higher pH (8–9) and lose their hemorrhagic activity at pH 6. The optimum pH for hide powder azure hydrolyzing activity of hemorrhagins is around 9. Specific activities of hemorrhagins toward casein and hide powder azure are 10–20 times lower than those of trypsin. At concentrations as high as 25 μg ml$^{-1}$, the hemorrhagins show no arginine esterase activity on Bz-Arg-OEt. Both BHRa and BHRb hydrolyze angiotensin I (Asp-Arg-Val-Tyr-Ile-His-Pro+Phe-His-Leu) but they do not hydrolyze angiotensin II (Asp-Arg-Val-Tyr-Ile-His-Pro-Phe). BHRa hydrolyzes one peptide bond in luteinizing hormone-releasing hormone (Glp-His-Trp-Ser-Tyr+Gly-Leu-Arg-Pro-Gly-NH$_2$), however BHRb hydrolyzes four peptide bonds in the peptide (Glp-His+Trp-Ser+Tyr+Gly+Leu-Arg-Pro-Gly-NH$_2$). BHRa and BHRb cleave the oxidized insulin B chain at 11 and 10 positions, respectively. Five of these positions are common to both hemorrhagins:

FVNQHLCGSHLVEA+LY+LVCG+ERG+F+FYTPKA

Cleavage of Gly20+Glu21 is unusual among reprolysins; the only other reprolysin reported to cleave this bond being leucolysin from *Agkistrodon piscivorus leucostoma* (Chapter 436) (Spiekerman *et al.*, 1973). BHRb hydrolyses the Phe25+Tyr26 bond, and proteinase A from *Bothrops jararaca* venom is the only other reprolysin known to cleave this bond (Mandelbaum *et al.*, 1967). Neither Z-Gly-Pro-Leu-Gly-Pro, a substrate for bacterial collagenase (Nagai *et al.*, 1960), nor Dnp-Pro-Leu-Gly-Ile-Ala-Gly-Arg-NH$_2$, a substrate for animal collagenase (Matsui *et al.*, 1977), was hydrolyzed by either hemorrhagin. None of the di-, tri-, or tetrapeptides such as Gly-Gly, Gly-Ala, Gly-Leu, Z-Gly-Leu, Z-Gly-Pro, Z-Gly-Leu-NH$_2$, Z-Gly-Phe-NH$_2$, Gly-Gly-Gly, Z-Gly-Pro-Leu and Z-Gly-Pro-Leu-Gly are hydrolyzed by either hemorrhagin. Both the hemorrhagic and proteolytic activities of hemorrhagins are strongly inhibited by EDTA and cysteine, but not by PMSF.

## Structural Chemistry

The apparent molecular mass values estimated by SDS-PAGE are 68 000 Da and 75 000 Da, for BHRa and BHRb, respectively. Considering the molecular masses and high cysteine contents of hemorrhagins, they are class III reprolysins (Bjarnason & Fox, 1995) (Chapter 416). BHRb adsorbed on cation-exchange resin (Bio-Rex 70) at pH 7.0 but BHRa did not, indicating that BHRa is an acidic protein whereas BHRb is considered to be a basic protein.

## Preparation

BHRa and BHRb were coeluted near the void volume of the Sephadex G-200 column. Complete separation of BHRa and BHRb was achieved by Bio-Rex 70 cation-exchange chromatography; BHRa was not retained on the column at 50 mM phosphate buffer pH 7.0 and eluted in a pass-through fraction. BHRa was further purified by anion-exchange (DEAE-) chromatography on cellulofine A-500 and gel filtration on cellulofine GCL-1000. BHRb was eluted from the Bio-Rex 70 column with around 0.2 M NaCl concentration and further purified by adsorption chromatography on hydroxyapatite (Omori-Satoh et al., 1995).

## Biological Aspects

The hemorrhagins hydrolyze all gelatins prepared from type I, II, III and IV collagens, although only native type IV collagen was hydrolyzed by either hemorrhagin. BHRa preferentially hydrolyzed the $\alpha 2$ chain of collagen, whereas BHRb hydrolyzed both $\alpha 1$ and $\alpha 2$ chains simultaneously. Hydrolysis of native type IV collagen is consistent with reports (Baramova et al., 1989) on four hemorrhagic metalloproteinases, atrolysin A (Chapter 418), -C, -D (both in Chapter 423), and -E (Chapter 430), from the venom of western diamondback rattlesnake (*Crotalus atrox*). Since type IV collagen is the major component of the basement membranes (Martin & Timpl, 1987), a plausible explanation has been proposed for the mechanism of hemorrhagic reaction: disruption of basement membrane of capillary vessels is the first event of hemorrhage (Ohsaka, 1979).

## Distinguishing Features

BHRa and BHRb formed independent precipitin lines in an immunodiffusion test using the horse polyvalent antisnake bite serum against *Bitis*, *Dendroaspis*, *Hemachates* and *Naja* venoms. The serum is available from the South Africa Institute for Medical Research (see Appendix 2 for full names and addresses of suppliers).

## Related Peptidases

A class I reprolysin named protease A, whose molecular mass is 24 300 (by ultracentrifugation) has been purified and characterized from the venom of *B. arietans* by Van der Walt & Joubert (1971, 1972a,b) and Van der Walt (1972). The N-terminal sequence of protease A (Strydom et al., 1986) shows significant homology to those of reprolysins (Takeya et al., 1993). Protease A does not hydrolyze Tos-Arg-OMe (TAME) and Ac-Tyr-OEt (ATEE). Protease A may not have significant hemorrhagic activity, since the activity in the venom of *B. arietans* was eluted only in the high molecular weight fraction by gel filtration (Omori-Satoh et al., 1995).

Beside reprolysins, two kallikrein-like proteinases (venombins) have been purified from the venom of *B. arietans*. Both proteinases are DFP sensitive and increase capillary permeability in rabbits by intradermal injection. Kinin-releasing enzyme (Sekoguchi et al., 1986) is a single-chain protein of about 45 kDa (SDS-PAGE), pI 5.1, which contains 3% of carbohydrate. The enzyme is specific to arginine esters and hydrolyzes Bz-Arg-OEt (BAEE) ($K_m$ $5 \times 10^{-2}$ M, pH 8.5, 44.8 U mg$^{-1}$), TAME (16.8 U mg$^{-1}$), Pro-Phe-Arg-NHMec, Z-Phe-Arg-NHMec. The enzyme does not hydrolyze casein but hydrolyzes insulin B chain at the positions of Leu6↓Cya7, His10↓Leu11 and Ala14↓Leu15:

FVNQHL↓CGSH↓LVEA↓LYLVCGERGFFYTPKA

The enzyme also hydrolyzes A$\alpha$ and B$\beta$ chain of fibrinogen but the $\gamma$ chain is unaffected.

Kallidin-releasing enzyme (Nikai et al., 1993) is also a single-polypeptide protein of about 58 kDa (SDS-PAGE), pI 4.4 and hydrolyzes TAME ($K_m$ = $6.3 \times 10^{-3}$ M, 32.8 U mg$^{-1}$) much faster than BAEE (2.9 U mg$^{-1}$). The N-terminal sequence of the enzyme resembles those of crotalase (Chapter 70) from *C. adamanteus*, E1 and E2 from *C. atrox*, kininogenin from *Vipera ammodytes* and kallikrein-like enzyme from *C. viridis viridis* and the $\alpha$ chain of pig pancreatic kallikrein (Chapter 29). The enzyme hydrolyzes the oxidized insulin B chain at the positions of Gly8↓Ser9, Ala14↓Leu15, Tyr16↓Leu17 and Phe25↓Tyr26. The enzyme preferentially hydrolyzes A$\alpha$ chain of fibrinogen; however, prolonged incubation resulted in the hydrolysis of both B$\beta$ and $\gamma$ chains.

## References

Baramova, E.N., Shannon, J.D., Bjarnason, J.B. & Fox, J.W. (1989) Degradation of extracellular matrix proteins by hemorrhagic metalloproteinases. *Arch. Biochem. Biophys.* **275**, 63–71.

Bjarnason, L.B. & Fox, J.W. (1995) Snake venom metalloendopeptidases: reprolysins. *Methods Enzymol.* **248**, 345–368.

Delpierre, G.R. (1968) Studies on African snake venoms. II. Differentiation between proteinase and amino-acid esterase activities of some African Viperidae venoms. *Toxicon* **6**, 103–108.

Mandelbaum, F.R., Carrillo, M. & Henriques, S.B. (1967) Proteolytic activity of *Bothrops* protease A on the B chain of oxidized insulin. *Biochim. Biophys. Acta* **132**, 508–510.

Martin, G.R. & Timpl, R. (1987) Laminin and other basement membrane components. *Annu. Rev. Cell Biol.* **3**, 57–85.

Masui, Y., Takemoto, T., Sakakibara, S., Hori, H. & Nagai, Y. (1977) Synthetic substrates for vertebrate collagenase. *Biochem. Med.* **17**, 215–221.

Nagai, Y. Sakakibara, S., Noda, H. & Akabori, S. (1960) Hydrolysis of synthetic peptides by collagenase. *Biochim. Biophys. Acta* **37**, 567–569.

Nikai, T., Momose, M., Okumura, Y., Ohara, A., Komori, Y. & Sugihara, H. (1993) Kallidin-releasing enzyme from *Bitis arietans* (puff adder) venom. *Arch. Biochem. Biophys.* **307**, 304–310

Ohsaka, A. (1979) Hemorrhagic, necrotic and edema-forming effects of snake venoms. In: *Handbook of Experimental Pharmacology*, vol. 52 (Lee, C.Y., ed.). Berlin: Springer, pp. 480–546.

Ohsaka, A., Omori-Satoh, T., Kondo, H., Kondo, S., & Murata, R. (1966) Biochemical and pathological aspects of hemorrhagic principles in snake venoms with special reference to habu (*Trimeresurus flavoviridis*) venom. *Mem. Inst. Butantan* **33**, 193–205.

Omori-Satoh, T., Yamakawa, Y. & Mebs, D. (1995) Hemorrhagic principles in the venom of *Bitis arietans*, a viperous snake. I. Purification and characterization. *Biochim. Biophys. Acta* **1246**, 61–66.

Sekoguchi, S., Nikai, T., Suzuki, Y. & Sugihara, H. (1986) Kinin-releasing enzyme from the venom of *Bitis arietans* (puff adder). *Biochim. Biophys. Acta* **884**, 502–509.

Soto, J.G., Perez, J.C. & Minton, S.A. (1988) Proteolytic and hemorrhagic and hemolytic activities of snake venoms. *Toxicon* **26**, 875–882

Spiekerman, A.M., Fredericks, K.K., Wagner, F.W. & Prescott, J.M. (1973) *Leucostoma* peptidase A: a metalloproteinase from snake venom. *Biochim. Biophys. Acta* **293**, 464–475.

Strydom, D.J., Joubert, F.J. & Howard, N.L. (1986) Chemical studies on protease A of *Bitis arietans* (puff adder) venom. *Toxicon* **24**, 247–257.

Takeya, H., Nishida, S., Nishino, N., Makinose, Y., Omori-Satoh, T., Nikai, T., Sugihara, H. & Iwanaga, S. (1993) Primary structures of platelet aggregation inhibitors (disintegrins) autoproteolytically released from snake venom hemorrhagic metalloproteinases and new fluorogenic peptide substrates for these enzymes. *J. Biochem. (Tokyo)* **113**, 473–483.

Van der Walt, S.J. (1972) Studies on puff adder (*Bitis arietans*) venom. IV. Association of protease A. *Z. Physiol. Chem.* **353**, 1217–1227.

Van der Walt, S.J. & Joubert, F.J. (1971) Studies on puff adder (*Bitis arietans*) venom. I. Purification and properties of protease A. *Toxicon* **9**, 153–161.

Van der Walt, S.J. & Joubert, F.J. (1972a) Studies on puff adder (*Bitis arietans*) venom. II. Specificity of protease A. *Toxicon* **10**, 341–349.

Van der Walt, S.J. & Joubert, F.J. (1972b) Studies on puff adder (*Bitis arietans*) venom. III. Ultracentrifuge and ORD studies on protease A. *Toxicon* **10**, 351–356.

Yamakawa, Y., Omori-Satoh, T. & Mebs, D. (1995) Hemorrhagic principles in the venom of *Bitis arietans*, a viperous snake. II. Enzymatic properties with special reference to substrate specificity. *Biochim. Biophys. Acta* **1247**, 17–23.

*Yoshio Yamakawa*
*Department of Biochemistry and Cell Biology,*
*National Institute of Health Japan,*
*Toyama 1-23-1, Shinjuku-ku, Tokyo, 162, Japan*
*Email: yamakawa@nih.go.jp*

# 438. Bothrops asper hemorrhagic proteinases BaH1 and BaP1

## Databanks

*Peptidase classification: clan MB, family M12, MEROPS ID: M12.166*
*NC-IUBMB enzyme classification: none*
*Databank codes: no sequence data available*

## Name and History

**BaH1** and **BaP1** are hemorrhagic metalloproteinases from the venom of the snake *Bothrops asper* (family Viperidae, subfamily Crotalinae; common name terciopelo). Their isolation and partial characterization were described by Borkow *et al*. (1993) and Gutiérrez *et al*. (1995a), respectively. BaH1 is the most important hemorrhagic component in this venom, whereas BaP1 is a metalloproteinase with low hemorrhagic activity. In addition to these two proteinases, Borkow *et al*. (1993) described the isolation of hemorrhagic metalloproteinases BH2 and BH3, and Aragón-Ortiz & Gubensek (1987) characterized a metalloproteinase devoid of hemorrhagic activity from the same venom.

## Activity and Specificity

BaH1 hydrolyzes Azocoll (Borkow *et al*., 1993) and BaP1 has proteolytic activity on hide powder azure, casein, fibrinogen, fibronectin, collagen type I, collagen type IV and laminin (Gutiérrez *et al.*, 1995a; Rucavado, 1996). Upon incubation with fibrinogen, BaP1 causes rapid degradation of the A$\alpha$ chain and, more slowly, of the B$\beta$ chain, whereas no degradation of the $\gamma$ chain is observed (Gutiérrez *et al.*, 1995a). Hydrolysis of hide powder azure by BaP1 is optimal at pH 8.0 (Gutiérrez *et al.*, 1995a). Both BaH1 and BaP1 are thermolabile, losing their activity after incubation at 60°C (Borkow *et al.*, 1993; Gutiérrez *et al.*, 1995a).

BaH1 and BaP1 are inhibited by chelating agents (EDTA salts and 1,10-phenanthroline), but not by soybean trypsin inhibitor, PMSF, pepstatin A or phosphoramidon (Borkow *et al.*, 1993; Gutiérrez *et al.*, 1995a). In addition, BaH1 is inhibited by NtAH and BaSAH, antihemorrhagic factors isolated from the serum of the snakes *Natrix tessellata* and *Bothrops asper*, respectively (Borkow *et al.*, 1994, 1995b).

## Structural Chemistry

According to the proposed classification of snake venom hemorrhagic metalloproteinases (Bjarnason & Fox, 1995) (Chapter 416), BaH1 belongs to class III since it is a 64 kDa protein, probably having protease, disintegrin-like and high-cysteine domains. BaP1 is a 24 kDa protein belonging to class I hemorrhagic toxins, probably having only the protease domain. BaH1 is an acidic glycoprotein (pI 4.5; Borkow *et al.*, 1993) and BaP1 is a basic protein (pI $\approx$8.5; A. Rucavado & J.M. Gutiérrez, unpublished results). The amino acid composition of BaP1 reveals six Cys and large numbers of Asp, Leu, Ser and Glu residues (Gutiérrez *et al.*, 1995a).

## Preparation

Both BaH1 and BaP1 have been purified from venom obtained from adult specimens of *Bothrops asper* collected in the Pacific region of Costa Rica. BaH1 is purified by gel filtration on Sephacryl S-200, ion-exchange chromatography on DEAE-Sepharose, metal chelate affinity chromatography and hydrophobic interaction chromatography (Borkow *et al.*, 1993). BaP1 is purified by ion-exchange chromatography on CM-Sephadex and gel filtration on Sephacryl S-200 (Gutiérrez *et al.*, 1995a). An additional step (Affi-Blue gel affinity chromatography) has been recently introduced in the purification of BaP1 (A. Rucavado & J.M. Gutiérrez, unpublished results). The method described for the isolation of BaP1 was used with venom samples of snakes collected in the Pacific region of Costa Rica. When venom from specimens collected in the Atlantic region of this country are fractionated by CM-Sephadex, there are variations in the chromatographic pattern, especially in the position where BaP1 elutes (Gutiérrez *et al.*, 1995a).

## Biological aspects

BaH1 and BaP1 induce hemorrhage in mice. BaH1 has higher hemorrhagic activity than BaP1 (Borkow *et al.*, 1993; Gutiérrez *et al.*, 1995a). Both proteinases induce a qualitatively similar pattern of tissue damage upon intramuscular injection in mice, with hemorrhage developing within the first 5 min. In the case of BaH1, capillary vessels are drastically

affected, with degradation of basal lamina, development of blebs, thinning of endothelial cells and detachment of these cells from the basal lamina (Moreira *et al.*, 1994). Eventually, gaps are formed within endothelial cells and extravasation occurs *per rhexis* (Moreira *et al.*, 1994). In addition, some endothelial cells were swollen, with dilated endoplasmic reticulum. Intravital microscopy in mouse cremaster muscle reveals the onset of microbleedings of an explosive character in capillaries several minutes after application of BaP1 (Rucavado, 1996).

In cell culture, BaH1 and BaP1 are not cytotoxic to endothelial cells of various origins, but induce detachment of these cells from the substratum (Lomonte *et al.*, 1994; Borkow *et al.*, 1995a; Rucavado, 1996). It has been proposed that proteolytic degradation of basal lamina components in capillary vessels is a key element in the pathogenesis of hemorrhage induced by BaH1 and BaP1, and that endothelial cell damage observed *in vivo* is a consequence of the disruption in the interaction between basal lamina and endothelial cells (Moreira *et al.*, 1994; Rucavado, 1996). In addition to hemorrhage, BaH1 and BaP1 induce muscle necrosis in mice, probably as a consequence of ischemia (Gutiérrez *et al.*, 1995b; Rucavado, 1996). A poor skeletal muscle regeneration is observed after myonecrosis induced by these hemorrhagic metalloproteinases (Gutiérrez *et al.*, 1995b; Rucavado, 1996), owing to the alterations induced in the microvasculature, since an adequate blood supply is a key requisite for skeletal muscle regeneration. In addition to hemorrhage, BaP1 also induces edema in the mouse foot pad assay (Gutiérrez *et al.*, 1995a).

## Distinguishing Features

Rabbit polyclonal antibodies have been raised against BaH1 and BaP1, but are not available commercially (Borkow *et al.*, 1993; Rucavado *et al.*, 1995). These proteinases do not cross-react immunologically by gel immunodiffusion, western blotting and neutralization (Rucavado *et al.*, 1995). Cross-reacting components have been found in a variety of viperine and crotaline snake venoms (Rucavado *et al.*, 1995). A rabbit antiserum against BaH1 neutralizes the hemorrhagic activity of the venoms of *Bothrops asper*, *B. atrox*, *B. jararaca*, *Crotalus atrox*, *C. durissus durissus*, *Echis carinatus* and *Trimeresurus flavoviridis* (Rucavado *et al.*, 1995).

## Related Peptidases

Many viperine and crotaline snake venoms contain components that cross-react with anti-BaH1 and anti-BaP1 polyclonal antibodies (Rucavado *et al.*, 1995). Several hemorrhagic metalloproteinases have been isolated from *Bothrops* snake venoms (see Bjarnason & Fox, 1995, for a review). BaP1 has strong similarities with bothropasin (Chapter 440), purified from the venom of *Bothrops moojeni* (Assakura *et al.*, 1985; Reichl & Mandelbaum, 1993).

## References

Aragón-Ortiz, F. & Gubensek, F. (1987) Characterization of a metallo-proteinase from *Bothrops asper* (terciopelo) snake venom. *Toxicon* **25**, 759–766.

Assakura, M.T., Reichl, A.P., Asperti, M.C.A. & Mandelbaum, F.R. (1985) Isolation of the major proteolytic enzyme from the venom of the snake *Bothrops moojeni* (caissaca). *Toxicon* **23**, 691–706.

Bjarnason, J.B. & Fox, J.W. (1995) Snake venom metalloendopeptidases: reprolysins. *Methods Enzymol.* **248**, 345–368.

Borkow, G., Gutiérrez, J.M. & Ovadia, M. (1993) Isolation and characterization of synergistic hemorrhagins from the venom of the snake *Bothrops asper*. *Toxicon* **31**, 1137–1150.

Borkow, G., Gutiérrez, J.M. & Ovadia, M. (1994) A potent antihemorrhagin in the serum of the non-poisonous water snake *Natrix tessellata*: isolation, characterization and mechanism of neutralization. *Biochim. Biophys. Acta* **120**, 482–490.

Borkow, G., Gutiérrez, J.M. & Ovadia, M. (1995a) *In vitro* activity of BaH1, the main hemorrhagic toxin of *Bothrops asper* snake venom on bovine endothelial cells. *Toxicon* **33**, 1387–1391.

Borkow, G., Gutiérrez, J.M. & Ovadia, M. (1995b) Isolation, characterization and mode of neutralization of a potent antihemorrhagic factor from the serum of the snake *Bothrops asper*. *Biochim. Biophys. Acta* **1245**, 232–238.

Gutiérrez, J.M., Romero, M., Díaz, C., Borkow, G. & Ovadia, M. (1995a) Isolation and characterization of a metalloproteinase with weak hemorrhagic activity from the venom of the snake *Bothrops asper* (terciopelo). *Toxicon* **33**, 19–29.

Gutiérrez, J.M., Romero, M., Núñez, J., Chaves, F., Borkow, G. & Ovadia, M. (1995b) Skeletal muscle necrosis and regeneration after injection of BaH1, a hemorrhagic metalloproteinase isolated from the venom of the snake *Bothrops asper* (terciopelo). *Exp. Mol. Pathol.* **62**, 28–41.

Lomonte, B., Gutiérrez, J.M., Borkow, G., Ovadia, M., Tarkowski, A. & Hanson, L. (1994) Activity of hemorrhagic metalloproteinase BaH1 and myotoxin II from *Bothrops asper* snake venom on capillary endothelial cells *in vitro*. *Toxicon* **32**, 505–510.

Moreira, L., Borkow, G., Ovadia, M. & Gutiérrez, J.M. (1994) Pathological changes induced by BaH1, a hemorrhagic metalloproteinase isolated from *Bothrops asper* (terciopelo) snake venom, on mouse capillary blood vessels. *Toxicon* **32**, 977–987.

Reichl, A. & Mandelbaum, F.R. (1993) Proteolytic specificity of *moojeni* protease A isolated from the venom of *Bothrops moojeni*. *Toxicon* **31**, 187–194.

Rucavado, A. (1996) Efectos patológicos inducidos por BaP1 y caracterización inmunológica de BaH1 y BaP1, dos hemorraginas del veneno de *Bothrops asper* (terciopelo) [Pathologic effects induced by BaP1 and immunologic characterization of BaH1 and BaP1, two hemorrhagic factors from the venom of *Bothrops asper* (terciopelo)]. MSc thesis, University of Costa Rica.

Rucavado, A., Borkow, G., Ovadia, M. & Gutiérrez, J.M. (1995) Immunological studies on BaH1 and BaP1, two hemorrhagic metalloproteinases from the venom of the snake *Bothrops asper*. *Toxicon* **33**, 1103–1106.

*José María Gutiérrez*
*Instituto Clodomiro Picado,*
*Facultad de Microbiología,*
*Universidad de Costa Rica,*
*San José, Costa Rica*
*Email: jgutierr@cariari.ucr.ac.cr*

*Michael Ovadia*
*George S. Wise Faculty of Life Sciences,*
*Tel Aviv University,*
*Tel Aviv 69978, Israel*
*Email: movadia@ccsg.tau.ac.il*

# *439. Bothrolysin*

## *Databanks*

*Peptidase classification: clan MB, family M12, MEROPS ID: M12.139*
*NC-IUBMB enzyme classification: EC 3.4.24.50*
*Databank codes:*

| Species | SwissProt | PIR | EMBL (cDNA) | EMBL (genomic) |
|---------|-----------|-----|-------------|----------------|
| *Bothrops jararaca* | P20416 | – | – | – |

Note: This is an N-terminal fragment only.

## *Name and History*

Venom of snakes from the Crotalinae and Viperinae subfamilies contain a number of proteolytic enzymes. These enzymes have been characterized and identified to be either serine or metalloproteases (Iwanaga & Suzuki, 1979). Metalloproteases from these venoms are of two types according to their substrate specificity: enzymes with high substrate specificity that are hemorrhagic factors and enzymes with broad substrate specificity that do not induce hemorrhage. The enzyme from *Bothrops jararaca* known as ***J protease*** is a low molecular weight metalloprotease that does not have hemorrhagic action (Tanizaki *et al.*, 1989). The name recommended by the IUBMB for this enzyme is **bothrolysin**.

## Activity and Specificity

Enzyme activity is usually measured on casein as substrate at pH 8.8. Calcium increases the reaction velocity and the concentration for half-maximal activation is 0.4 mM. The enzyme is totally inhibited by metal chelators such as EDTA (0.1 mM) and 1,10-phenanthroline (0.1 mM), but not by phosphoramidon (0.5 mM). DTNB has an inhibitory effect which is partially prevented by addition of calcium. However, J protease is not inhibited by E-64 and its derivatives.

Using the oxidized insulin B chain as substrate, the following bonds are cleaved:

$$\text{FVNQ}\dagger\text{HLCGS}\dagger\text{HLVEA}\dagger\text{LYLVCGERGFFYTPKA}$$

and with angiotensin I as substrate, cleavage is at the Pro7$\dagger$Phe8 bond: Asp-Arg-Val-Tyr-Ile-His-Pro$\dagger$Phe-His-Leu.

## Structural Chemistry

Bothrolysin is a single-chain protein with a mass of about 23 kDa. The protein structure contains 1 mol each of zinc and calcium per mol of enzyme. Since calcium is an activator of bothrolysin, and a value of $4 \times 10^{-4}$ M can be taken as an apparent dissociation constant for calcium binding, the enzyme might have two calcium-binding sites: one with low affinity, related to activation and other one, a tight-binding site related to enzyme structure and stability.

Amino acid sequencing was performed by gas phase sequenator after protein cleavage by lysyl peptidase, cyanogen bromide and V8 protease. The C-terminus was determined with carboxypeptidase. About 80% of the total sequence has been determined. By homology with other venom metalloproteinases, HR2a from *Trimeresurus flavoviridis* (Miyata *et al.*, 1989) (Chapter 427) and Ht-d from *Crotalus atrox* (Shannon *et al.*, 1989) (Chapter 423), the sequence shown in Figure 439.1 was suggested (Tanizaki *et al.*, 1991a).

## Preparation

This enzyme was purified from the venom of *Bothrops jararaca* on four chromatographic columns: DEAE-Sephacel, Sephacryl S-200, CM-cellulose and Sephadex G-75.

## Biological Aspects

A protein inhibitor which inhibits venom proteinases was isolated from the blood plasma of *Bothrops jararaca* (Tanizaki *et al.*, 1991b). This natural inhibitor shows a broad specificity on venom proteinases inhibiting metalloproteinases. Thus, hemorrhagic factors, bothropasin (Tanizaki *et al.*, 1991b) (Chapter 440) and prothrombin-activating enzyme (De Olivera & Tanizaki, 1992) as well as bothrolysin (Tanizaki *et al.*, 1991b), are completely inhibited.

## Distinguishing Features

In venom metalloproteinases, blocked N-terminal amino acids are common, as a result of cyclization of glutamine residues to form the pyroglutamyl amino acid (Zhou *et al.*, 1995). Bothrolysin and atrolysin E (Bjarnason & Fox, 1995) (Chapter 420) are exceptions, since their N-terminal residues are Thr and Gln, respectively.

## Further Reading

For a review, see Tanizaki *et al.* (1989).

## References

Bjarnason, J.B. & Fox, J.W. (1995) Snake venom metalloendopeptidases: reprolysin. *Methods Enzymol.* **248**, 345–387.

De Oliveira, E. & Tanizaki, M.M. (1992) Effect of a proteinase inhibitor from the plasma of *Bothrops jararaca* on coagulant and myotoxic activities of *Bothrops* venoms. *Toxicon* **30**, 123–128.

Iwanaga, S. & Suzuki, T. (1979) Enzymes in snake venom. In: *Handbook of Experimental Pharmacology*, vol. 52 (Lee, C.-Y., ed.). Berlin: Springer, p. 61.

Miyata, T., Takeya, H., Ozeki, Y., Arakawa, M., Tokunaga, F., Iwanaga, S. & Omori-Satoh, T. (1989) Primary structure of hemorrhagic protein, HR2a, isolated from the venom of *Trimeresurus flavoviridis*. *J. Biochem.* **105**, 847–853.

Shannon, J.D., Baramova, E.N., Bjarnason, J.B. & Fox, J.W. (1989) Amino acids sequence of a *Crotalus atrox* venom metalloproteinases which cleaves type IV collagen and gelatin. *J. Biol. Chem.* **264**, 11575–11583.

Tanizaki, M.M., Zingali, R.B., Kawazaki, H., Imajoh, S., Yamazaki, S. & Suzuki, K. (1989) Purification and some characterisitics of a zinc metalloprotease from the venom of *Bothrops jararaca* (jararaca). *Toxicon* **27**, 747–755.

Tanizaki, M.M., Kawazaki, H. & Suzuki, K. (1991a) Primary structure of J protease, a metalloprotease from the venom of *Bothrops jararaca*. *Annual Meeting of the Brazilian Society of Biochemistry and Molecular Biology*, Caxambu, Brazil.

```
TPEHQYIELF  LVVDSGMFMK  YNGNSDKIRR  RIHQMVNIMG  SEYRP-----  50
------IWSN   KDMINVQPAA  PQTLDSFGEW  RKTDLLNRKS  HDNAQLLTST  100
DFHR------  -VGSMCDPKR  STAVI----E  TDLLVAVTMA  HELGHNLGIR  150
HDTGSHS-GG  YSIMP-----  ----YFSDCS  YIQCSDFIIM  KDNPQCILNK  200
Q(E)
```

*Figure 439.1*   Suggested partial sequence for bothrolysin. The putative zinc-binding site is shown in bold.

Tanizaki, M.M., Kawazaki, H., Suzuki, K. & Mandelbaum, F.R. (1991b) Purification of a proteinase inhibitor from the plasma of *Bothrops jararaca (jararaca)*. *Toxicon* **29**, 673–681.

Zhou, Q., Smith, J.B. & Grossman, M.H. (1995). Molecular cloning and expression of catrocollastin, a snake venom protein from *Crotalus atrox* (Western diamondback rattlesnake) which inhibits platelet adhesion to collagen. *Biochem. J.* **307**, 411–417.

***Martha M. Tanizaki***
*Centro de Biotecnologia,*
*Instituto Butantan,*
*Av. Vital Brasil 1500,*
*CEP 05503-900, São Paulo, Brazil*

# 440. Bothropasin

## Databanks

*Peptidase classification: clan MB, family M12, MEROPS ID: M12.140*
*NC-IUBMB enzyme classification: EC 3.4.24.49*
*Databank codes: no sequence data are available*

## Name and History

In the isolation of the hemorrhagic factors from the venom of *Bothrops jararaca* (Mandelbaum *et al*., 1976) an endopeptidase with caseinolytic activity was found, coeluting with jararhagin (Chapter 429). Further purification led to the complete isolation of **bothropasin**, a metalloendopeptidase responsible for 15% of the caseinolytic activity of the crude venom (Mandelbaum *et al*., 1982). Another related metalloenzyme, named MPB, has been isolated from the venom of *Bothrops moojeni* (Serrano *et al*., 1993).

## Activity and Specificity

Bothropasin is active on casein. The pH optimum of the endopeptidase is 8.8 and the assay system includes 4 mM $CaCl_2$. It causes hemorrhage on rabbit skin with a minimum hemorrhagic dose (MHD) of 1 μg (Mandelbaum & Assakura, 1988). Sites of cleavage on B chain of oxidized insulin are:

FVNQH+LCGSH+LVEA+LY+LVCGERGF+FYTPKA

Bothropasin is inhibited by 2 mM EDTA, EGTA, 1,10-phenanthroline and DTT. Similarly to bothropasin, MPB is active on casein but it is devoid of hemorrhagic activity. Sites of its cleavage of the B chain of oxidized insulin are (Reichl *et al*., 1993):

FVNQHLCGS+HLVEA+LY+LVCGERGF+FYTPKA

The enzyme degrades fibronectin, fibrinogen, fibrin, type I collagen and gelatin.

## Structural Chemistry

Bothropasin is an acidic (pI 4.8) glycoprotein that belongs to class II of reprolysins (Bjarnason & Fox, 1995) (Chapter 416). This metalloendopeptidase is a medium-sized enzyme of 48 kDa estimated by SDS-PAGE. By sedimentation equilibrium a relative molecular mass of 49 870 was calculated. The native tertiary structure of the protein is dependent on disulfide and metal bonds. In the presence of EDTA, the denatured and reduced protein shows a relative molecular mass of 37 300 by sedimentation equilibrium, confirmed by SDS-PAGE. MPB presents two protein bands corresponding to $M_r$ values of 65 000 and 55 000 which stain with Schiff's reagent. On PAGE at pH 4.3, MPB appears as a single diffuse protein band.

## Preparation

Bothropasin has been purified 7-fold from *Bothrops jararaca* venom fraction by precipitation between 40 and 50% with ammonium sulfate followed by chromatographies on DEAE-cellulose, DEAE-Sephadex A-50 and Sephadex G-100. $CaCl_2$ (1 mM) is usually added to all preparation buffers to stabilize the enzyme. MPB is isolated from the venom of *B. moojeni* by chromatographies on Sephadex G-100, DEAE-Sephacel, SP-Sephadex C-50 and Sepharose 12.

M

## Biological Aspects

Bothropasin is one of the proteolytic enzymes present in *B. jararaca* venom that plays a role in the local tissue damage after snakebite. It causes hemorrhage followed by myonecrosis and arterial necrosis after intramuscular injection of doses of 20 µg in mice (Queiroz *et al.*, 1985). Myonecrosis is still observed with lower doses down to 1 µg. However, doses of 50 ng of jararhagin produce similar lesions (Chapter 429).

## Distinguishing Features

Bothropasin differs immunologically from hemorrhagic factors isolated from the same venom (Mandelbaum & Assakura, 1988). Polyclonal antibodies raised in rabbits against the enzyme neither give a precipitin line with hemorrhagic factors (HF$_1$ and HF$_2$) nor neutralize their hemorrhagic activity.

## Further Reading

For a review, see Bjarnason & Fox (1995).

## References

Bjarnason, J.B. & Fox, J.W. (1995). Snake venom metalloendopeptidases: reprolysins. *Methods Enzymol.* **248**, 345–368.

Mandelbaum, F.R. & Assakura, M.T. (1988) Antigenic relationship of hemorrhagic factors and proteases isolated from the venoms of three species of *Bothrops* snakes. *Toxicon* **26**, 379–385.

Mandelbaum, F.R., Reichl, A.P. & Assakura, M.T. (1976) Some physical and biochemical characteristics of HF$_2$, one of the hemorrhagic factors in the venom of *Bothrops jararaca*. In: *Animal, Plant and Microbial Toxins* (Ohsaka, A., Hayashi, K. & Sawai, Y, eds). New York: Plenum Publishing, pp. 111–121.

Mandelbaum, F.R., Reichl, A.P. & Assakura, M.T. (1982) Isolation and characterization of a proteolytic enzyme from the venom of the snake *Bothrops jararaca* (jararaca). *Toxicon* **20**, 955–972.

Queiroz, L.S., Santo Neto, H., Assakura, M.T., Reichl, A.P. & Mandelbaum, F.R. (1985) Pathological changes in muscle caused by haemorrhagic and proteolytic factors from *Bothrops jararaca* snake venom. *Toxicon* **23**, 341–345.

Reichl, A.P., Serrano, S.M.T., Sampaio, C.A.M. & Mandelbaum, F.R. (1993) Hydrolytic specificity of three basic proteinases isolated from the venom of *Bothrops moojeni* for the B-chain of oxidized insulin. *Toxicon* **31**, 1479–1482.

Serrano, S.M.T., Sampaio, C.A.M. & Mandelbaum, F.R. (1993) Basic proteinases from *Bothrops moojeni* (caissaca) venom. II. Isolation of the metalloproteinase MPB. Comparison of the proteolytic activity on natural substrates by MPB, MSP 1 and MSP 2. *Toxicon* **31**, 483–492.

*Fajga R. Mandelbaum*
*Laboratório de Bioquímica e Biofísica,*
*Instituto Butantan,*
*Av. Vital Brasil 1500,*
*CEP 05503-900, São Paulo-SP, Brazil*
*Email: butlbioq@eu.ansp.br*

*Marina T. Assakura*
*Laboratório de Bioquímica e Biofísica,*
*Instituto Butantan,*
*Av. Vital Brasil 1500,*
*CEP 05503-900, São Paulo-SP, Brazil*

*Antonia P. Reichl*
*Laboratório de Bioquímica e Biofísica,*
*Instituto Butantan,*
*Av. Vital Brasil 1500,*
*CEP 05503-900, São Paulo-SP, Brazil*

*Solange M.T. Serrano*
*Laboratório de Bioquímica e Biofísica,*
*Instituto Butantan, Av. Vital Brasil 1500,*
*CEP 05503-900, São Paulo-SP, Brazil*

# 441. Atrolysin F

## Databanks

*Peptidase classification: clan MB, family M12, MEROPS ID: M12.146*
*NC-IUBMB enzyme classification: EC 3.4.24.43*
*Databank codes: no sequence data available*

## Name and History

**Atrolysin F** was originally termed *hemorrhagic toxin f* (*HT-f*; Nikai *et al.*, 1984). It is a zinc metalloproteinase that was isolated from the venom of the western diamondback rattlesnake, *Crotalus atrox*. This is the second P-III class snake venom metalloproteinase (Chapter 416) isolated to date from *C. atrox* venom, the other being atrolysin A (Chapter 418).

## Activity and Specificity

*In vitro*, the proteolytic activity of atrolysin F can be assayed by either a modification of the Kunitz method or using dimethylcasein (Nikai *et al.*, 1984). The proteolytic activity of atrolysin F is inhibited by EDTA and 1,10-phenanthroline. The sites of peptide bond hydrolysis by atrolysin F in the oxidized B chain of insulin are:

FV┼NQ┼HL┼CGSH┼LVEA┼LY┼LVCGERGFFYTPKA

This cleavage pattern is very similar to that observed for the other P-III atrolysin, atrolysin A (Chapter 418) (Bjarnason *et al.*, 1988). Fibrinogen is cleaved by atrolysin F, but does not produce clottable peptides (Nikai *et al.*, 1984).

## Structural Chemistry

Atrolysin F is a single polypeptide chain with a molecular weight of 64 000 based on SDS-PAGE. There are 1.15 mol of zinc per mol of enzyme. From amino acid analysis, the protein comprises 572 amino acids with 45 half-cystines (Nikai *et al.*, 1984). The amino acid composition data tentatively place atrolysin F in the P-III category. Antibodies generated against atrolysin F do not cross-react with other atrolysins.

## Preparation

Isolation of atrolysin F from the venom of *Crotalus atrox* employs a five-step chromatographic procedure. The first step is ion-exchange chromatography on DEAE followed by gel filtration on Sephadex G-75, ion exchange on Whatman DE 32 and two additional CM-cellulose ion-exchange steps

(Nikai *et al.*, 1984). A typical isolation yields approximately 1.8 mg of purified atrolysin F from 1 g of crude venom.

## Biological Aspects

The minimum hemorrhagic dose of atrolysin F is 0.53 μg (Nikai *et al.*, 1984). This places atrolysin F as the second-most potent hemorrhagic toxin of all the atrolysins. The lethal dose 50% (LD$_{50}$) for atrolysin F is 9.8 μg g$^{-1}$. When atrolysin F is injected i.v. into mice, systemic hemorrhage is observed in the lung, kidney, liver and heart, all tissues rich in basement membranes. Intravenous injection of atrolysin F also increases the serum levels of creatinine phosphokinase, suggesting muscle tissue destruction by the toxin (Nikai *et al.*, 1984).

## Further Reading

For reviews, see Fox & Bjarnason (1996) and Bjarnason & Fox (1994).

## References

Bjarnason, J.B. & Fox, J.W. (1994) Hemorrhagic metalloproteinases from snake venoms. *J. Pharmacol. Ther.* **62**, 325–372.

Bjarnason, J.B., Hamilton, D. & Fox, J.W. (1988) Studies on the mechanism of hemorrhagic production by five proteolytic hemorrhagic toxins from *Crotalus atrox* venom. *Biol. Chem. Hoppe-Seyler* **369**, 121–129.

Fox, J.W. & Bjarnason, J.B. (1996) Reprolysins: snake venom and mammalian reproductive metalloproteinases. In: *Zinc Metalloproteinases in Health and Disease* (Hooper, N.M., ed.). London: Taylor & Francis, pp. 47–81.

Nikai, T., Mori, N., Kishida, M., Sugihara, H. & Tu, A.T. (1984) Isolation and biochemical characterization of hemorrhagic toxin f from the venom of *Crotalus atrox* (western diamondback rattlesnake). *Arch. Biochem. Biophys.* **231**, 309–319.

*Jay W. Fox*
*Department of Microbiology,*
*University of Virginia Health Sciences Center,*
*Charlottesville, VA 22908, USA*
*Email: jwf8x@virginia.edu*

*Jón B. Bjarnason*
*Science Institute,*
*University of Iceland,*
*Reykjavik, Iceland*
*Email: jonbragi@raunvis.hi.is*

# 442. Basilysin

## Databanks

*Peptidase classification: clan MB, family M12, MEROPS ID: M12.148*
*NC-IUBMB enzyme classification: none*
*Databank codes: An N-terminal fragment has been deposited in GenPept (accession number 1000402).*

## Name and History

Basilysin is the fibrinolytic enzyme from *Crotalus basiliscus basiliscus* (Mexican west coast rattlesnake) venom. One of the earliest references to fibrinolytic activity in *C. b. basiliscus* venom was by Didisheim & Lewis (1956) who reported that of the 11 venoms tested with fibrinolytic activity, *C. b. basiliscus* was the only one that was devoid of thrombic, hemolytic and hemagglutinating activities. They suggested that the isolation of a fibrinolytic enzyme with clinical potential might be possible from this, or other, venoms. Some years later three fibrinolytic, nonhemorrhagic enzymes were purified from this venom and designated **Cbfib1.1**, **Cbfib1.2** and **Cbfib2** (Retzios & Markland, 1992). These enzymes were later referred to as **basiliscusfibrases 1**, **2** and **3**, respectively (Retzios & Markland, 1994). On the basis of structural and functional activities, it appears that basiliscusfibrases 1 and 2 are isoforms with molecular masses of approximately 23 500. Basiliscusfibrase 3, with a molecular mass of approximately 22 500, appears to be a different enzyme.

Datta *et al*. (1995) screened a number of snake venoms for fibrinolytic activity using a fibrin plate assay (Brakman & Astrup, 1971). They found that *C. b. basiliscus* venom contained the highest level of activity and they isolated a fibrin(ogen)olytic enzyme from this venom. These authors reported that the enzyme had a molecular mass by mass spectrometry of 23 171. The enzyme was fibrinolytic and nonhemorrhagic and was named **basilase** (Datta *et al*., 1995). On the basis of almost identical physicochemical properties and similarity of fibrinogen digestion profiles, basilase is probably identical to one of the basiliscusfibrase isozymes. For future reference the enzyme will be called **basilysin**.

## Activity and Specificity

Internal bonds are cleaved in the $\alpha$ and $\beta$ chains of fibrin and fibrinogen; there is little or no digestion of the $\gamma$ chains or the $\gamma$-$\gamma$ dimer (Retzios & Markland, 1992; Datta *et al*., 1995). The enzyme acts directly on fibrin(ogen) and does not rely on the activation of plasminogen. The enzyme degrades the A$\beta$ and B$\beta$ chains of fibrinogen more rapidly than the $\alpha$ and $\beta$ chains of fibrin. With fibrinogen the rate of digestion of the A$\alpha$ and B$\beta$ chains is similar. This feature distinguishes these enzymes from atroxase (Chapter 419) (Willis & Tu, 1988) and fibrolase (Chapter 417) (Retzios & Markland, 1988), both of which degrade the A$\alpha$ chain more rapidly than the B$\beta$ chain. With fibrin, basilysin digests the $\alpha$ chain more rapidly than the $\beta$ chain. During the initial stages of fibrinogen digestion, basiliscusfibrases 1 and 2 cleave the $\alpha$ chain at Lys413$\downarrow$Leu414, the primary cleavage site for fibrolase (Retzios & Markland, 1988), Ser505$\downarrow$Thr506 and Tyr560$\downarrow$Ser561. Another cleavage site was found in the $\beta$ chain approximately 80–90 residues from the N-terminus, but the bond cleaved was not identified (Retzios & Markland, 1992).

Peptide bond cleavage specificity was also examined for basilysin using oxidized insulin B chain and several synthetic octapeptides (Retzios & Markland, 1992). These studies revealed that the enzyme cleaves internal Ala$\downarrow$Leu, Tyr$\downarrow$Leu and Lys$\downarrow$Leu bonds in these peptides and that cleavage specificity appears to be dictated by the amino acid sequence rather than a specific peptide bond *per se*.

Although the pH optimum has not been determined, the enzyme is active in the pH range from 7.0 to 8.4 (Retzios & Markland, 1992; Datta *et al*., 1995). Treatment with EDTA causes rapid loss of activity. Assays for enzyme activity are performed using the fibrin plate method (Retzios & Markland, 1992; Datta *et al*., 1995) or azocasein hydrolysis (Retzios & Markland, 1992). However, both of these assays are rather insensitive and not very accurate. An alternate more sensitive and more specific procedure is to use HPLC to analyze the hydrolysis of one of the peptides described above (Retzios & Markland, 1994). However, this method assumes that the investigator has access both to HPLC and to a peptide synthesizer to prepare the octapeptide His-Thr-Glu-Lys$\downarrow$Leu-Val-Thr-Ser, which reproduces the sequence around the Lys413$\downarrow$Leu414 cleavage site in the $\alpha$ chain of fibrin.

## Structural Chemistry

Basilysin is a metalloproteinase and a member of the adamalysin/reprolysin subfamily of the metzincins (clan MB) (Stöcker *et al*., 1995). Metal analysis by atomic emission spectroscopy reveals that basilysin contains 1 mol of zinc and 2 mol of calcium per mol of enzyme (Datta *et al*., 1995). The enzyme is a single-chain protein of approximately 23 kDa with a pI of approximately 4.5. It contains at least one disulfide bond (Retzios & Markland, 1992; Datta *et al*., 1995). Based on infrared spectroscopy, the enzyme contains approximately 24% $\alpha$ helix and 31% $\beta$ sheet (Datta *et al*., 1995). These findings are in fairly good agreement with the 40% $\alpha$ helix and 20% $\beta$ sheet found in the related venom metalloproteinase, adamalysin (Chapter 421), following X-ray crystal structure determination (Gomis-Rüth *et al*., 1994). The N-terminus is blocked by a pyroglutamic acid (Datta *et al*., 1995), a common finding in this class of enzymes. After enzymatic removal of the N-terminal residue, the sequence of the first 25 amino acids was determined. When these results were combined with the sequence of a cyanogen bromide peptide, the sequence of the first 58 residues of basilysin was established (Datta *et al*., 1995). This sequence shows extensive similarity to two other snake venom fibrinolytic proteinases, atroxase (Chapter 419) (Baker *et al*., 1995) and fibrolase (Chapter 417) (Randolph *et al*., 1992).

## Preparation

The enzyme has been purified by two different methods from crude *C. b. basiliscus* venom. In one procedure a three-step method was employed that included hydrophobic interaction HPLC, hydroxyapatite HPLC and weak anion-exchange HPLC. This procedure resulted in a combined recovery for the isoforms of >50% with an increase in specific activity of slightly more than 3-fold (Retzios & Markland, 1992). In the second procedure the enzyme was purified by a two-step method employing molecular sieve and anion-exchange chromatography. The enzyme was purified 3-fold with high recovery (Datta *et al*., 1995).

## Biological Aspects

The enzyme is present in the venom of the Mexican west coast rattlesnake and is a metalloproteinase that is inhibited by EDTA (Retzios & Markland, 1992; Datta *et al.*, 1995). The enzyme is very similar structurally to fibrinolytic enzymes of approximately 23 kDa (class I venom metalloproteinases; see Chapter 416) from the venom of other snake species. The enzyme does not possess hemorrhagic activity when injected subcutaneously in mice.

The enzyme is inhibited partially by $\alpha_2$-macroglobulin but does not inactivate $\alpha_2$-macroglobulin (Datta *et al.*, 1995). The enzyme does not induce platelet aggregation and it does not possess thrombin-like or factor X-like activities, nor protein C or plasminogen-activating activities (Datta *et al.*, 1995).

Nothing is known at the present time about the gene for basilysin. There are no indications of post-translational modifications (Retzios & Markland, 1992).

## Distinguishing Features

Retzios & Markland (1992, 1994) have purified and characterized two isoforms of basilysin. The enzymes, named basiliscusfibrases 1 and 2, have molecular masses of 23 500. Basiliscusfibrases 1 and 2 appear to be isoforms on the basis of amino acid analyses, comparative HPLC analyses of the tryptic digestion profiles of the two isoforms, and peptide bond cleavage specificity of the enzymes using fibrinogen as substrate. The isoforms have pI values of 4.1 and 4.7, respectively. The enzymes do not appear to be glycoproteins. Datta *et al.* (1995) separately reported the isolation of basilase which appears to be identical to one of the basiliscusfibrase isoforms. These authors reported the presence of 1 mol of zinc and 2 mol of calcium per mol of protein.

Despite being structurally very similar to fibrolase (Chapter 417) (Randolph *et al.*, 1992) and atroxase (Chapter 419) (Willis & Tu, 1988), the *C. b. basiliscus* enzyme has different bond cleavage specificity in the (presumably) natural substrate fibrin(ogen) (Retzios & Markland, 1992, 1994). It is interesting that some of the basilysin-susceptible bonds in fibrinogen, Ser+Thr and Tyr+Ser, are not those expected for metalloproteinases, where cleavage preference for the N-terminal side of a hydrophobic amino acid has been demonstrated (Bjarnason & Fox, 1989, 1994; Takeya & Iwanaga, 1994). Basilysin is distinguished by its low general proteolytic activity and relatively high fibrinolytic activity (Retzios & Markland, 1994). Thus, it may prove interesting as a fibrinolytic enzyme for *in vivo* use.

## Related Peptidases

Fibrinolytic enzymes with similar physicochemical properties and biological activities have been isolated from viperid and crotalid snake venoms. Nonhemorrhagic enzymes with fibrinolytic activity that have been purified and characterized and shown to have sequence homology with basilysin; those that are not described in this *Handbook* include lebetase from *Vipera lebetina* venom (Siigur & Siigur, 1991).

Other metalloproteinases have also been isolated from *Crotalus basiliscus* venom, but they are clearly different based on their activity profiles. Two hemorrhagic metalloproteinases

that also possess fibrinolytic activity have molecular masses of 27 000–27 500 (Molina *et al.*, 1990). Three metalloproteinases with molecular masses in the 23 500–24 200 range were also isolated from the venom (Svoboda *et al.*, 1995). These metalloenzymes inactivate and degrade $\alpha_2$-antiplasmin and inactivate $\alpha_2$-macroglobulin. They also degrade the $\alpha$ and $\beta$ chains of fibrinogen and induce plasma extravasation following intradermal injection into the abdominal skin of test animals.

## Further Reading

A good review of this enzyme may be found in Datta *et al.* (1995).

## References

Baker, B.J., Wongvibgulsin, S., Nyborg, J. & Tu, A.T. (1995) Nucleotide sequence encoding the snake venom fibrinolytic enzyme atroxase obtained from a *Crotalus atrox* venom gland cDNA library. *Arch. Biochem. Biophys.* **317**, 357–364.

Bjarnason, J.B. & Fox, J.W. (1989) Hemorrhagic toxins from snake venoms. *Toxin Rev.* **7**, 121–209.

Bjarnason, J.B. & Fox, J.W. (1994) Hemorrhagic metalloproteinases from snake venoms. *Pharmacol. Ther.* **62**, 325–372.

Brakman, P. & Astrup, T. (1971) The fibrin plate method for assay of fibrinolytic agents. In: *Thrombosis and Bleeding Disorders Theory and Methods* (Bang, N.U., Beller, F.K., Deutsch, E. & Mammen E.F., eds). New York: Academic Press, pp. 332–336.

Datta, G., Dong, A., Witt, J. & Tu, A.T. (1995) Biochemical characterization of basilase, a fibrinolytic enzyme from *Crotalus basiliscus basiliscus*. *Arch. Biochem. Biophys.* **317**, 365–373.

Didisheim, P. & Lewis, J.H. (1956) Fibrinolytic and coagulant activities of certain snake venoms and proteases. *Proc. Soc. Exp. Biol. Med.* **93**, 10–13.

Gomis-Rüth, F.X., Kress, L.F., Kellermann, J., Mayr, I., Lee, X., Huber, R. & Bode, W. (1994) Refined 2.0 Å X-ray crystal structure of the snake venom zinc-endopeptidase adamalysin II. Primary and tertiary structure determination, refinement, molecular structure and comparison with astacin, collagenase and thermolysin. *J. Mol. Biol.* **239**, 513–544.

Molina, O., Seriel, R.K., Martinez, M., Sierra, M.L., Varela-Ramirez, A. & Rael, E.D. (1990) Isolation of two hemorrhagic toxins from *Crotalus basiliscus basiliscus* (Mexican west coast rattlesnake) venom and their effect on blood clotting and complement. *Int. J. Biochem.* **22**, 253–261.

Randolph, A., Chamberlain, S.H., Chu, C., Retzios, A.D., Markland, F.S. & Masiarz, F.R. (1992) Amino acid sequence of fibrolase, a direct-acting fibrinolytic enzyme from *Agkistrodon contortrix contortrix* venom. *Protein Sci.* **1**, 590–600.

Retzios, A.D. & Markland, F.S. (1988) A direct-acting fibrinolytic enzyme from the venom of *Agkistrodon contortrix contortrix*: effects on various components of the human blood coagulation and fibrinolytic systems. *Thromb. Res.* **52**, 541–552.

Retzios, A.D. & Markland, F.S. (1992) Purification, characterization, and fibrinogen cleavage sites of three fibrinolytic enzymes from the venom of *Crotalus basiliscus basiliscus*. *Biochemistry* **31**, 4547–4557.

Retzios, A.D. & Markland, F.S. (1994) Fibrinolytic enzymes from the venoms of *Agkistrodon contortrix contortrix* and *Crotalus basiliscus basiliscus*: cleavage site specificity towards the $\alpha$-chain of fibrin. *Thromb. Res.* **74**, 355–367.

M

Siigur, E. & Siigur, J. (1991) Purification and characterization of lebetase, a fibrinolytic enzyme from *Vipera lebetina* (snake) venom. *Biochim. Biophys. Acta* **1074**, 223–229.

Stöcker, W., Grams, F., Baumann, U., Reinemer, P., Gomis-Rüth, F.X., McKay, D.B. & Bode, W. (1995) The metzincins: topological and sequential relations between the astacins, adamalysins, serralysins, and matrixins (collagenases) define a superfamily of zinc-peptidases. *Protein Sci.* **4**, 823–840.

Svoboda, P., Meier, J. & Freyvogel, T.A. (1995) Purification and characterization of three alpha 2-antiplasmin and alpha 2-macroglobulin inactivating enzymes from the venom of the Mexican west coast rattlesnake (*Crotalus basiliscus*). *Toxicon* **33**, 1331–1346.

Takeya, H. & Iwanaga, S. (1994) Snake venom hemorrhagic and non-hemorrhagic metalloproteinases: their structure and function relationships. In: *Biological Functions of Proteinases and Inhibitors* (Katunuma, M., ed.). Basel: Karger, pp. 231–252.

Willis, T.W. & Tu, A.T. (1988) Purification and biochemical characterization of atroxase, a nonhemorrhagic fibrinolytic protease from western diamondback rattlesnake venom. *Biochemistry* **27**, 4769–4777.

**Francis S. Markland, Jr**
*Department of Biochemistry and Molecular Biology,*
*University of Southern California, School of Medicine,*
*Cancer Research Laboratory no. 106,*
*1303 N. Mission Road,*
*Los Angeles, CA 90033, USA*
*Email: markland@hsc.usc.edu*

**Stephen Swenson**
*Department of Biochemistry and Molecular Biology,*
*University of Southern California, School of Medicine,*
*Cancer Research Laboratory no. 104,*
*1303 N. Mission Road,*
*Los Angeles, CA 90033, USA*
*Email: sswenson@hsc.usc.edu*

# *443. Horrilysin*

## Databanks

*Peptidase classification: clan MB, family M12, MEROPS ID: M12.149*
*NC-IUBMB enzyme classification: EC 3.4.24.47*
Databank codes:

| Species | SwissProt | PIR | EMBL (cDNA) | EMBL (genomic) |
|---|---|---|---|---|
| *Crotalus ruber* | P20897 | – | – | – |

## Name and History

*Horrilysin* was originally termed **hemorrhagic proteinase IV** (**HP-IV**; Civello *et al*., 1983a,b) and is a zinc metalloendopeptidase purified from the venom of the timber rattlesnake, *Crotalus horridus horridus*. This is the only hemorrhagic toxin purified from this venom to date and appears to belong to the P-III class of snake venom metalloendopeptidases (Bjarnason & Fox, 1994, 1995). It was found in all commercial preparations of timber rattlesnake venom examined and constitutes approximately 25% of the total venom protein.

## Activity and Specificity

*In vitro*, the proteolytic activity of horrilysin can be assayed by a modification of the method of Rinderknecht *et al*. (1968) using hide powder azure as a substrate (Civello *et al*., 1983a). Horrilysin also catalyzes the hydrolysis of cow hide powder that does not contain covalently bound dye, although only at 20% of the rate at which hide powder azure is hydrolyzed. Dansylation of the hydrolysis fragments of cow hide produces six new N-terminal residues (Civello *et al*., 1983b). Horrilysin exhibits activity on type I collagen from cattle achilles tendon at 37°C, pH 7.4, albeit much slower and more limited in quantity than *Clostridium* collagenase (Chapter 368) (Civello *et al*., 1983b). Since the experiment was performed at 37°C, one can conclude that horrilysin has gelatinase activity but probably not true collagenase type I activity, similar to the atrolysins (Baramova *et al*., 1989; Fox & Bjarnason, 1995).

Incubation of horrilysin with fibrinogen solutions causes rapid digestion of the A$\alpha$ chain followed by a slower degradation of the B$\beta$ chain. This is accompanied by formation of a precipitate which can easily be removed from solution by centrifugation. No firm clot forms even with incubation up to 20 h, nor does thrombin induce a clot with horrilysin-treated fibrinogen (Civello *et al*., 1983b). Furthermore, fibrin clots formed by thrombin in the presence of horrilysin have a supernatant absorbance at 280 nm of 1.17 as compared to 1.10 for clots formed in the absence of horrilysin. This corresponds to a 55% increase in soluble protein. Electrophoresis shows that horrilysin degrades the α chain of fibrin. When monitored by the fibrin plate and plasma clot assay, horrilysin causes significant lysis within 20 h and completely dissolves each clot within 72 h. Horrilysin also hydrolyzes hemoglobin and denatured casein but at a considerably slower rate than Pronase. However, horrilysin catalyzes the complete solubilization of glomerular basement membrane in the presence of 10 mM calcium chloride at a rate 605 times as fast as an equal weight of *Clostridium* collagenase. The authors concluded that horrilysin exhibits little activity toward most protein substrates (Civello *et al*., 1983b).

Only one peptide is cleaved by horrilysin in each of the oxidized A and B chains of insulin: the Val-Glu-Ala15┼Leu16-Tyr-Leu bond of the B chain and Val-Cys-Ser12┼Leu13-Tyr-Gly bond of the A chain. Bee venom melittin is cleaved at the Ile2┼Gly3, Pro14┼Ala15 and Ser18┼Trp19 bonds (Civello *et al*., 1983b):

GI┼GAVLKVLTTGLP┼ALIS┼WIKRKRQQ

Dansylation of the hydrolysis fragments of cowhide shows the appearance of six new N-terminal residues, namely Tyr, Leu, Met, Trp, Pro and hydroxylysine. Various unblocked dipeptides and the doubly blocked dipeptides Z-Ser-Leu-NH$_2$, Z-Ala-Leu-NH$_2$, Z-Ile-Gly-NH$_2$ were not cleaved.

## Structural Chemistry

Horrilysin has a single polypeptide chain devoid of carbohydrates with a molecular mass of 57 kDa based on SDS-PAGE. The enzyme contains 1 mol of zinc ion per mol of protein and 2–3 mol of calcium. Amino acid analysis indicates that the protein comprises 507 amino acids, of which 41 are half-cystines (Civello *et al*., 1983a). Thus, horrilysin is most likely a three-domain snake venom metalloendopeptidase and a member of the P-III group of reprolysins (Chapter 416) (Bjarnason & Fox, 1994, 1995).

## Preparation

Horrilysin is purified by a two-step column chromatographic procedure. The first column step is DEAE-cellulose anion-exchange separation, yielding fraction IV upon elution by a continuous gradient of sodium acetate. Fraction IV is purified further on a series of two Waters Associates I-125 protein analysis HPLC columns (Civello *et al*., 1983a).

## Biological Aspects

Horrilysin possesses hemorrhagic and lethal activities. An LD$_{50}$ value of 10.0 μg g$^{-1}$ of body weight in mice is consistently demonstrated in all preparations of horrilysin as compared to 2.84 μg g$^{-1}$ for crude timber rattlesnake venom. A minimum hemorrhagic dose is 4.0 μg.

## References

Baramova, E.N., Shannon, J.D., Bjarnason, J.B. & Fox, J.W. (1989) Degradation of extracellular matrix proteins by hemorrhagic metalloproteinases. *Arch. Biochem. Biophys.* **275**, 63–71.

Bjarnason, J.B & Fox, J.W. (1994) Hemorrhagic metalloproteinases from snake venoms. *Pharmacol. Ther.* **62**, 325–372.

Bjarnason, J.B & Fox, J.W. (1995) Snake venom metalloproteinases: reprolysins. *Methods Enzymol.* **248**, 345–368.

Civello, D.J., Duong, H.L. & Geren, C.R. (1983a) Isolation and characterization of a hemorrhagic proteinase from timber rattlesnake venom. *Biochemistry* **22**, 749–755.

Civello, D.J., Moran, J.B. & Geren, C.R. (1983b) Substrate specificity of a hemorrhagic proteinase from timber rattlesnake venom. *Biochemistry* **22**, 755–762.

Fox, J.W. & Bjarnason, J.B. (1995) Atrolysins: metalloproteinases from *Crotalus atrox* venom. *Methods Enzymol.* **248**, 368–387.

Rinderknecht, H., Geokas, M.C., Silverman, P. & Haverback, B.J. (1968) A new ultrasensitive method for the determination of proteolytic activity. *Clin. Chim. Acta* **21**, 197–203.

*Jón B. Bjarnason*
*Science Institute,*
*University of Iceland,*
*Reykjavík, Iceland*
*Email: jonbragi@raunvis.hi.is*

*Jay W. Fox*
*Department of Microbiology,*
*University of Virginia Health Sciences Center,*
*Charlottesville, VA 22908, USA*
*Email: jwf8x@virginia.edu*

# 444. Najalysin

## Databanks

*Peptidase classification: clan MB, family M12, MEROPS ID: M12.167*
*NC-IUBMB enzyme classification: none*
*Databank codes: no sequence data available*

## Name and History

The name *najalysin* refers to peptidases isolated from venoms of the *Naja* genus of snakes, commonly called cobras. Various reports have confirmed the presence of peptidases in *Naja* venoms, despite early indications that these venoms had little or no peptidase activity, based on their inability to cleave Ac-Tyr-OEt (ATEE), Tos-Arg-OMe (TAME) and Bz-Arg-OMe (BAEE) (Murata *et al.*, 1963; Tu *et al.*, 1965), and their limited ability to cleave casein and kininogen (Murata *et al.*, 1963; Mebs, 1968; Oshima *et al.*, 1969; Robertson *et al.*, 1969; Kocholaty *et al.*, 1971).

Screening of the venom of *Naja nigricollis* (spitting cobra) for its effects on various components of the coagulation cascade led to the discovery of an enzyme able to cause limited cleavage of the Aα chain, but with no apparent effect on the Bβ and γ chains of fibrinogen, as monitored by SDS-PAGE (Evans, 1981). The enzyme was subsequently purified from the venom of *N. nigricollis*, shown to require $Zn^{2+}$ for activity, and named *proteinase F1* (Evans, 1984). The name was derived from the fact that this was the first proteinase isolated from a cobra venom, and it acted on fibrinogen. Another cobra venom metallopeptidase was recently isolated from the venom of the Mozambiquan cobra, *N. mocambique mocambique*, and named **mocarhagin**. The name in this case was derived from the source venom. This enzyme was shown to cleave a specific neutrophil receptor, P-selectin glycoprotein ligand 1 (PSGL-1), which is involved with the rolling of neutrophils along the endothelial cells of blood vessel walls (De Luca *et al.*, 1995).

## Activity and Specificity

Treatment of PSGL-1 with *N. mocambique mocambique* najalysin gives a specific cleavage at Tyr10↓Asp11, near the N-terminal end. The enzyme also cleaves the α chain of the platelet glycoprotein GPIb–V–IX complex at Glu282↓Asp283. The common features of these two specificities are the Asp residue at P1′ and the presence of three potentially sulfated Tyr residues in a negative cluster on the N-terminal side of the scissile bond (De Luca *et al.*, 1995).

The timed digestion of fibrinogen with *N. nigricollis* najalysin indicates that this enzyme also cleaves relatively few peptide bonds. The first cleavage occurs at about residue 520 of the 610 in the Aα chain, or about 90 residues from the C-terminus of the chain. The C-terminal 10 kDa peptide is released, whereas the remainder of the chain is held by disulfides to the rest of the molecule. Prolonged treatment of fibrinogen results in cleavage at other sites farther from the C-terminus of the Aα chain, releasing one or more additional peptides. Fibrinogen treated with najalysin remains clottable by thrombin (Chapter 55), indicating that the N-terminal region of the molecule is unaffected (Evans & Barrett, 1988). The fibrinogen Aα chain contains very few Tyr residues in the region of cleavage, suggesting the specificity may differ from that of *N. mocambique mocambique* najalysin. The *N. nigricollis* enzyme, however, inhibits platelet aggregation by a process not involving fibrinogen degradation (Kini & Evans, 1991), consistent with the action of the *N. mocambique mocambique* enzyme on the platelet glycoprotein GPIb–V–IX complex (De Luca *et al.*, 1995).

Both najalysins can be assayed based on their effects on their specific substrates, either by SDS-PAGE analysis of cleavage, or loss of their biological effect. An assay based on the inactivation of $\alpha_2$-macroglobulin can also be used (Evans & Guthrie, 1984). The pH optimum for the *N. nigricollis* najalysin is about 9.0 (Evans, 1984). The activity is inhibited by EDTA and $\alpha_2$-macroglobulin.

## Structural Chemistry

The two examples of najalysins that have been purified to homogeneity include proteinase F1, from *N. nigricollis* venom, with a single 58 kDa polypeptide chain (Evans, 1984) and mocarhagin from the venom of *N. m. mocambique*. The structural details and characterization of the latter are not yet published, but it is indicated to consist of a single 55 kDa polypeptide chain (De Luca *et al.*, 1995). Because of their similarity in mass, their apparently narrow specificity, and the relative paucity of peptidases in cobra venoms, these two enzymes are probably closely related metallopeptidases. These two African cobra venoms have shown similar protein compositions in the past, e.g. with anticoagulant phospholipases that differ by one out of 118 amino acid residues (Kini & Evans, 1987). However, the substrate preferences of the two najalysins have not yet been compared.

## Preparation

The enzyme responsible for the activity of *N. nigricollis* venom on the Aα chain of fibrinogen was purified by a combination of Bio-Rex 70 chromatography and phenyl-Sepharose chromatography (Evans, 1984). The *N. mocambique mocambique* enzyme was isolated by chromatography on a heparin-Sepharose CL-6B column, concentration by ultrafiltration and rechromatography on a CL-6B column (De Luca *et al.*, 1995).

## Biological Aspects

Najalysin is apparently widely distributed in cobras of the genus *Naja*. Several crude cobra venoms show activity on hide powder azure and the ability to inactivate $\alpha_2$-macroglobulin, including those of *N. nivea*, *N. nigricollis* (both East and West African), *N. nigricollis crawshawii*, *N. naja atra* and *N. naja kaouthia* (Evans & Guthrie, 1984). All of the these venoms, as well as those from *N. nigricollis nigricollis* and *N. nigricollis pallida*, cleave the Aα chain, while showing no effect on the Bβ and γ chains of fibrinogen (Evans & Barrett, 1988). The crude venoms of *N. nigricollis*, *N. haje*, *N. melanoleuca* and *N. nigricollis pallida* cleave and inactivate $\alpha_1$-antichymotrypsin (Kress & Hufnagel, 1984). The purified *N. nigricollis* enzyme cleaves fibronectin, in addition to activity on the Aα chain of fibrinogen and $\alpha_2$-macroglobulin (Evans, 1984), but has no visible effect on denatured hemoglobin, casein, serum albumin or type I collagen (Evans, 1984; Evans & Barrett, 1988).

The normal function(s) of najalysin are unknown. The enzyme from *N. nigricollis* venom has very narrow specificity as described above, acting on such protein substrates as fibrinogen, fibronectin, hide powder azure (but not type I collagen), and $\alpha_2$-macroglobulin. It also inhibits platelet aggregation in a process not involving fibrinogen degradation (Kini

& Evans, 1991). It is inactive on the denatured protein substrates casein, hemoglobin and serum albumin. The enzyme from *N. m. mocambique* venom cleaves single peptide bonds in two receptor molecules, the glycoprotein GPIb–V–IX complex of platelets and PSGL-1 on neutrophils. Trials of other substrates for this peptidase have not yet been published.

## Distinguishing Features

Two other peptidases, a metallopeptidase hemorrhagin (Chapter 445) (Tan & Saifuddin, 1990) and a factor X-activating serine peptidase (Lee *et al.*, 1995) have been isolated and characterized from *Ophiophagus hannah* (king cobra) venom. The king cobra and *Naja* cobras are all in the Elapidae family of snakes, and all members of the family have relatively few peptidases. The hemorrhagin from king cobra can lyse casein, whereas the najalysins cannot.

## References

De Luca, M., Dunlop, L.C., Andrews, R.K., Flannery, J.V., Jr., Ettling, R., Cumming, D.A., Veldman, G.M. & Berndt, M.C. (1995) A novel cobra venom metalloproteinase, mocarhagin, cleaves a 10-amino acid peptide from the mature N terminus of P-selectin glycoprotein ligand receptor, PSGL-1, and abolishes P-selectin binding. *J. Biol. Chem.* **270**, 26734–26737.

Evans, H.J. (1981) Cleavage of the Aα-chain of fibrinogen and the α-polymer of fibrin by the venom of spitting cobra (*Naja nigricollis*). *Biochim. Biophys. Acta* **660**, 219–226.

Evans, H.J. (1984) Purification and properties of a fibrinogenase from the venom of *Naja nigricollis*. *Biochim. Biophys. Acta* **802**, 49–54.

Evans, H.J. & Barrett, A.J. (1988) The action of proteinase F1 from *Naja nigricollis* venom on the Aα-chain of human fibrinogen. In: *Hemostasis and Animal Venoms* (Pirkle, H. & Markland, F.S., eds). New York: Marcel Dekker, pp. 213–222.

Evans, H.J. & Guthrie, V.H. (1984) Proteolytic activities of cobra venoms based on inactivation of α2-macroglobulin. *Biochim. Biophys. Acta* **784**, 97–101.

Kini, R.M. & Evans, H.J. (1987) Structure-function relationships in phospholipases: the anticoagulant region of phospholipases A2. *J. Biol. Chem.* **262**, 14402–14407.

Kini, R.M. & Evans, H.J. (1991) Inhibition of platelet aggregation by a fibrinogenase from *Naja nigricollis* venom is independent of fibrinogen degradation. *Biochim. Biophys. Acta* **1095**, 117–121.

Kocholaty, W.F., Ledford, E.B., Daly, J.G. & Billings, T.A. (1971) Toxicity and some enzymatic properties and activities in the venoms of Crotalidae, Elapidae and Viperidae. *Toxicon* **9**, 131–138.

Kress, L.F. & Hufnagel, M.E. (1984) Enzymatic inactivation of human α1-antichymotrypsin by metalloproteinases in snake venoms of the family Elapidae. *Comp. Biochem. Physiol.* **77B**, 431–436.

Lee, W.H., Zhang, Y., Wang, W.Y., Xiong, Y.L. & Gao, R. (1995) Isolation and properties of a blood coagulation factor X activator from the venom of king cobra (*Ophiophagus hannah*). *Toxicon* **33**, 1263–1276.

Mebs, D. (1968) Vergleichende Enzymuntersuchungen an Schlangengiften unter besonderer Berücksichtigung ihrer casein-spaltenden Proteasen. [Comparative enzyme studies on snake venoms with special consideration of their casein-splitting proteases]. *Z. Physiol. Chem.* **349**, 1115–1125.

Murata, Y., Satake, M. & Suzuki, T. (1963) Studies on snake venom. XII. Distribution of proteinase activities among Japanese and Formosan snake venoms. *J. Biochem. (Tokyo)* **53**, 431–437.

Oshima, G., Sato-Ohmori, T. & Suzuki, T. (1969) Proteinase, arginine ester hydrolase and a kinin releasing enzyme in snake venoms. *Toxicon* **7**, 229–233.

Robertson, S.S.D., Steyn, K. & Delpierre, G.R. (1969) Studies on African snake venoms. III. The caseinase activity of some African Elapidae venoms. *Toxicon* **6**, 243–245.

Tan, N.H. & Saifuddin, M.N. (1990) Isolation and characterization of a hemorrhagin from the venom of *Ophiophagus hannah* (king cobra). *Toxicon* **28**, 385–392.

Tu, A.T., James, G.P. & Chua, A. (1965) Some biochemical evidence in support of the classification of venomous snakes. *Toxicon* **3**, 5–8.

*Herbert J. Evans*
*Department of Biochemistry and Molecular Biophysics,*
*Virginia Commonwealth University,*
*School of Medicine, PO Box 980614,*
*Richmond, VA 23298-0614, USA*
*Email: hjevans@gems.vcu.edu*

**M**

# 445. *Ophiolysin*

## Databanks

*Peptidase classification: clan MB, family M12, MEROPS ID: M12.152*
*NC-IUBMB enzyme classification: none*
*Databank codes: no sequence data available*

## Name and History

Although there are many descriptions of snake venom metalloproteinases (reprolysins), most investigations are concerned with those from the venoms of Crotalidae and Viperidae (Bjarnason & Fox, 1995) (Chapter 416). King cobra (*Ophiophagus hannah*) is the only known elapid whose venom contains hemorrhagic as well as protease activity (Ohsaka *et al.*, 1966; Soto *et al.*, 1988). Yamakawa & Omori-Satoh (1988) have purified a nonhemorrhagic proteinase from the venom of king cobra. The enzyme is now known as **ophiolysin** (Bjarnason & Fox, 1995). A hemorrhagin (hemorrhagic proteinase) termed hannahtoxin was also purified from the same venom and characterized by Tan & Safuddin (1990).

## Activity and Specificity

Ophiolysin is a nonhemorrhagic reprolysin and shows a specific activity approximately 1/25 of that of crystalline trypsin when casein is used as a substrate. Ophiolysin hydrolyzes the insulin B chain at the following bonds:

FVN↓Q↓HLCGSH↓LVEA↓LY↓LVCGERGFFYTPKA

The cleavage of the Asn3↓Gln4 bond is unusual among reprolysins; few reprolysins, such as a basic proteinase (Yamakawa *et al.*, 1987), trimerelysin II (Chapter 427) (Takahashi *et al.*, 1970) from *Trimeresurus flavoviridis*, atrolysins A and E (Chapter 418 and 430) from *Crotalus atrox* and AaH-III from *C. acutus* are known to cleave the peptide bond (Bjarnason & Fox, 1995). The proteolytic activity of ophiolysin is strongly inhibited by EDTA and cysteine.

## Structural Chemistry

The apparent molecular mass of ophiolysin estimated by SDS-PAGE is 70 000 Da. A similar value of 65 000 Da has been obtained for the native enzyme by gel filtration, indicating that ophiolysin belongs to type III reprolysins (Bjarnason & Fox 1995) (Chapter 416). Ophiolysin seems to be a basic protein since the enzyme is adsorbed on a cation-exchange resin (CM-) at neutral pH.

## Preparation

Ophiolysin has been purified to an electrophoretically homogeneous state by successive chromatographies on Sephadex G-100 superfine, DEAE-cellulose, hydroxyapatite (HPLC) and CM-Asahipak (HPLC) columns (Yamakawa & Omori-Satoh, 1988).

## Distinguishing Features

Only proteinase c from *Agkistrodon halys blomhoffii* (Chapter 428) (Ohshima *et al.*, 1968), a collagenase from *Crotalus atrox* (Hong, 1982) and ophiolysin from *Ophiophagus hannah* are nonhemorrhagic metalloproteinases among class III reprolysins (Chapter 416) (Bjarnason & Fox, 1995).

## Related Peptidases

A hemorrhagic metalloproteinase termed hannahtoxin has been purified from the venom of the king cobra by Tan & Safuddin (1990). Hannahtoxin is a single-chain protein of about 66 kDa (SDS-PAGE), pI 5.3, which contains 1 mol of zinc per mol of protein and 12% carbohydrate (glucose equivalent). The minimum hemorrhagic doses for hannahtoxin are 0.7 mg and 75 mg, respectively, in rabbit and in mice. Such species-specific sensitivity towards this hemorrhagin of *Ophiophagus hannah* was also reported by Weissenberg *et al.* (1987). Treatment of hannahtoxin with EDTA and 1,10-phenanthroline eliminated both the proteolytic and hemorrhagic activities completely. Rabbit antisera against hannahtoxin neutralize hemorrhage caused by snake venoms of Viperidae and Crotalidae families (Tan *et al.*, 1990).

A serine proteinase that is a blood coagulation factor X activator has been purified from the venom of king cobra by Lee *et al.* (1995). The enzyme is a single-chain protein of 64.5 kDa (SDS-PAGE), pI 5.6. The enzyme activates factor X *in vitro* in the presence of $Ca^{2+}$, although it cannot activate prothombin nor have any effect on fibrinogen. The procoagulant activity is inhibited by serine protease inhibitors such as PMSF, TPCK and soybean trypsin inhibitor but not by EDTA.

## Further Reading

An extensive review of snake venom metalloendopepeptidase or reprolysins is given in Bjarnason & Fox (1995).

## References

Bjarnason, L.B. & Fox, J.W. (1995) Snake venom metalloendopeptidases: reprolysins. *Methods Enzymol.* **248**, 345–368.

Hong, B.-S. (1982) Isolation and identification of a collagenolytic enzyme from the venom of the western diamondback rattlesnake (*Crotalus atrox*). *Toxicon* **20**, 535–545.

Lee, W.-H., Zhang, Y., Wang, W.-Y., Xiong, Y.-L. & Gao, R. (1995) Isolation and properties of a blood coagulation factor X activator from the venom of King cobra (*Ophiophagus hannah*). *Toxicon* **33**, 1263–1276.

Ohsaka, A., Omori-Satoh, T., Kondo, H., Kondo, S. & Murata, R. (1966) Biochemical and pathological aspects of hemorrhagic principles in snake venoms with special reference to habu (*Trimeresurus flavoviridis*) venom. *Mem. Inst. Butantan* **33**, 193–205.

Ohshima, G., Matsuo, Y., Iwanaga, S. & Suzuki, T. (1968) Studies on snake venom. XIX. Purification and some physicochemical properties of proteinases a and c from the venom of *Agkistrodon halys blomhoffii*. *J. Biochem. (Tokyo)* **64**, 227–238.

Soto, J.G., Perez, J.C. & Minton, S.A. (1988) Proteolytic and hemorrhagic and hemolytic activities of snake venoms. *Toxicon* **26**, 875–882.

Takahashi, T & Ohsaka, A. (1970) Purification and characterization of a proteinases in the venom of *Trimeresurus flavoviridis*. Complete separation of the enzyme from hemorrhagic activity. *Biochim. Biophys. Acta* **198**, 293–307.

Tan, N.-H. & Safuddin, N. (1990) Isolation and characterization of hemorrhagin from the venom of *Ophiophagus hannah* (king cobra). *Toxicon* **28**, 385–392.

Tan, N.-H., Safuddin, N. & Nik Jaafar, M.I. (1990) Preparation of antibodies to king cobra (*Ophiophagus hannah*) venom hemorrhagin and investigation of their cross-reactivity. *Toxicon* **28**, 1355–1359.

Weissenberg, S. Ovadia, M. & Kochva, E. (1987) Species specific sensitivity toward the hemorrhagin of *Ophiophagus hannah* (Elapidae). *Toxicon* **25**, 475–481.

Yamakawa, Y. & Omori-Satoh, T. (1988) A protease in the venom of king cobra (*Ophiophagus hannah*): purification, characterization and substrate specificity on oxidized insulin B-chain. *Toxicon* **26**, 1145–1155.

Yamakawa, Y., Omori-Satoh, T. & Sadahiro, H. (1987) Purification and substrate specificity of a basic proteinase in the venom of habu (*Trimeresurus flavoviridis*). *Biochim. Biophys. Acta* **925**, 124–132.

*Yoshio Yamakawa*
*Department of Biochemistry and Cell Biology,*
*National Institute of Health Japan,*
*Toyama 1-23-1, Shinjuku-ku, Tokyo, 162, Japan*
*Email: yamakawa@nih.go.jp*

# 446. *Philodryas venom metalloproteinases*

## Databanks

*Peptidase classification: clan MB, family M12, MEROPS ID: M12.153*
*NC-IUBMB enzyme classification: none*
*Databank codes: no sequence data available*

## Name and History

*Philodryas olfersii*, a Colubridae xenodontinae, is an opisthoglyphous snake that has a well-developed Duvernoy's gland connected with a grooved tooth. From the venom of this snake five fibrin(ogen)olytic enzymes were isolated (Assakura *et al.*, 1992, 1994). The names **PofibC$_1$**, **C$_2$**, **C$_3$**, **H** and S indicate the source and activities of these enzymes. PofibH denotes the **h**emorrhagic activity of this enzyme. PofibS indicates that this enzyme is a **s**erine endopeptidase. The other three are metalloendopeptidases active on **c**asein.

## Activity and Specificity

PofibC$_1$, C$_2$ and C$_3$ show proteolytic activity on casein, PofibC$_2$ being the most active. They are free of hemorrhagic activity as tested with doses up to 6 µg of protein. The hemorrhagic fibrinogenase PofibH has a minimum hemorrhagic dose (MHD) of 0.2 µg and a caseinolytic activity 60% lower than that of the crude venom. The amidolytic fibrinogenase PofibS shows no caseinolytic activity, as tested with doses up to 12 µg. Its activity on Ac-Phe-Arg-NHPhNO$_2$ is 245 µmol min$^{-1}$ mg$^{-1}$ protein. PofibC$_1$, C$_2$ and C$_3$ hydrolyze the insulin B chain at the following bonds:

FVNQHLCGSH+LVEA+LY+LVCGERGFFYTPKA

PofibH, in addition to these bonds, also hydrolyzes the His5+Leu6 and Phe24+Phe25 positions.

FVNQH+LCGSH+LVEA+LY+LVCGERGF+FYTPKA

PofibS hydrolyzes only the Arg22+Gly23 peptide bond in insulin B chain. PofibC$_1$, C$_2$, C$_3$ and H are metalloendopeptidases inhibited by EDTA or 1,10-phenanthroline. PofibS is a serine endopeptidase inhibited by PMSF. All five enzymes are inactivated by DTT or DTE.

## Structural Chemistry

The five enzymes are constituted by single polypeptide chains. PofibC$_1$ and C$_2$ present the same relative molecular mass of 47 000 and pI of 6.2. PofibC$_3$ is a basic endopeptidase of pI 8.5 and $M_r$ 45 000. The hemorrhagic endopeptidase PofibH has an $M_r$ of 58 000 and pI 4.6. PofibS has an $M_r$ of 36 000 and pI 4.5.

## Preparation

*P. olfersii* endopeptidases have been prepared by fractionation of the venom on Sephadex G-100 followed by Mono-Q FPLC. PofibC$_1$, C$_2$ and C$_3$ are isolated by chromatography of the Mono-Q nonadsorbed material on SP-Sephadex C-50. Mono-Q fractions containing hemorrhagic and amidolytic activities are chromatographed separately on DEAE-3SW to yield PofibH and PofibS, respectively.

## Biological Aspects

The five endopeptidases degrade fibrin and fibrinogen. PofibC$_1$, C$_2$ C$_3$ and H cleave preferentially A$\alpha$ chains while PofibS breaks concomitantly A$\alpha$ and B$\beta$ chains of fibrinogen. None of these endopeptidases cleaves the $\gamma$ chain of fibrinogen. All of them digest only $\alpha$ polymer and $\alpha$ chains of fibrin. Although they degrade fibrinogen, each one promotes a different delay of the clotting time of fibrinogen by thrombin. The amount of protein that causes a shift of 20–60 s on the thrombin clotting time is 19.2, 5.6 and 0.9 $\mu$g for PofibC$_2$, C$_3$ and S, respectively. PofibH shows almost no activity and PofibC$_1$ shows no effect on thrombin clotting time. They promote *in vivo* a loss of the circulant plasma fibrinogen (as observed in rats injected with the venom).

## References

Assakura, M.T., Salomão, M.G., Puorto, G. & Mandelbaum, F.R. (1992). Hemorrhagic, fibrinogenolytic and edema-forming activites of the venom of the colubrid snake *Philodryas olfersii* (green snake). *Toxicon* **30**, 427–438.

Assakura, M.T., Reichl, A.P. & Mandelbaum, F.R. (1994). Isolation and characterization of five fibrin(ogen)olytic enzymes from the venom of *Philodryas olfersii* (green snake). *Toxicon* **32**, 819–831.

*Fajga R. Mandelbaum*
*Laboratório de Bioquímica e Biofísica,*
*Instituto Butantan,*
*Av. Vital Brasil 1500 CEP 05503-900,*
*São Paulo-SP, Brazil*
*Email: butlbioq@eu.ansp.br*

*Antonia P. Reichl*
*Laboratório de Bioquímica e Biofísica,*
*Instituto Butantan,*
*Av. Vital Brasil 1500 CEP 05503-900,*
*São Paulo-SP, Brazil*

*Marina T. Assakura*
*Laboratório de Bioquímica e Biofísica,*
*Instituto Butantan,*
*Av. Vital Brasil 1500 CEP 05503-900,*
*São Paulo-SP, Brazil*

*Solange M.T. Serrano*
*Laboratório de Bioquímica e Biofísica,*
*Instituto Butantan,*
*Av. Vital Brasil 1500 CEP 05503-900,*
*São Paulo-SP, Brazil*

# 447. ADAM metalloproteinases

## Databanks

*Peptidase classification: clan MB, family M12, MEROPS ID: M12.200*
*NC-IUBMB enzyme classification: none*
*Databank codes:*

| Species | SwissProt | PIR | EMBL (cDNA) | EMBL (genomic) |
|---|---|---|---|---|
| **Endopeptidases** | | | | |
| ADAM 1 protein (PH-30$\alpha$; fertilin $\alpha$) | | | | |
| *Cavia porcellus* | – | – | Z11719 | – |
| *Macaca mulatta* | – | – | X79808 | – |
| *Macaca mulatta* | – | – | X79809 | – |
| *Mus musculus* | – | – | U22056 | – |
| *Oryctolagus cuniculus* | – | – | U46069 | – |
| *Rattus norvegicus* | – | – | Y08616 | – |
| ADAM 8 protein (MS2) | | | | |
| *Mus musculus* | Q05910 | – | X13335 | – |
| ADAM 9 protein (MDC9) | | | | |
| *Homo sapiens* | – | – | U41766 | – |
| | | | D14665 | |

| Species | SwissProt | PIR | EMBL (cDNA) | EMBL (genomic) |
|---|---|---|---|---|
| **Endopeptidases** (*continued*) | | | | |
| *Mus musculus* | – | – | D50410 | – |
| *Mus musculus* | – | – | D50412 | – |
| | | | U41765 | |
| ADAM 10 protein (MADM; kuzbanian) | | | | |
| *Bos taurus* | – | – | Z21961 | – |
| *Drosophila melanogaster* | – | – | U60591 | – |
| *Homo sapiens* | – | – | Z48579 | – |
| *Rattus norvegicus* | – | – | Z48444 | |
| ADAM 12 protein (meltrin $\alpha$) | | | | |
| *Mus musculus* | – | – | D50411 | – |
| ADAM 13 protein | | | | |
| *Xenopus laevis* | – | – | U66003 | – |
| ADAM 15 protein (metargidin; MDC 15) | | | | |
| *Homo sapiens* | – | – | U41767 | – |
| | | | U46005 | – |
| ADAM 17 protein (TNF-$\alpha$ converting enzyme; TACE) | | | | |
| *Homo sapiens* | – | – | U69611 | – |
| | | | U86755 | |
| Others | | | | |
| *Caenorhabditis elegans* | – | – | Z50006 | – |
| **Non-proteolytic homologs** | | | | |
| ADAM 2 protein (PH-30$\beta$; fertilin $\beta$) | | | | |
| *Cavia porcellus* | – | – | Z11720 | – |
| *Homo sapiens* | – | – | U52370 | |
| | | | X99374 | |
| *Macaca mulatta* | – | – | U33959 | |
| | | | X77653 | |
| *Mus musculus* | – | – | U16242 | |
| | | | U22057 | |
| *Oryctolagus cuniculus* | – | – | U46070 | |
| *Rattus norvegicus* | – | – | X99794 | |
| ADAM 3 protein (cyritestin; tMDC I) | | | | |
| *Macaca fuscata* | – | – | X76637 | |
| *Mus musculus* | – | – | X64227 | – |
| *Rattus norvegicus* | – | – | Y07903 | – |
| ADAM 4 protein | | | | |
| *Mus musculus* | – | – | U22058 | – |
| ADAM 5 protein (tMDC II) | | | | |
| *Cavia porcellus* | – | – | U22060 | – |
| *Macaca fuscata* | – | – | X77619 | – |
| *Mus musculus* | – | – | U22059 | – |
| ADAM 6 protein (tMDC IV) | | | | |
| *Cavia porcellus* | – | – | U22061 | – |
| *Macaca fascicularis* | – | – | X87205 | – |
| *Macaca fascicularis* | – | – | X87206 | – |
| *Macaca fascicularis* | – | – | X87207 | – |
| *Rattus norvegicus* | – | – | Y09111 | – |
| ADAM 7 protein (EAP I) | | | | |
| *Macaca fascicularis* | – | – | X66139 | – |
| *Rattus norvegicus* | – | – | X66140 | – |
| ADAM 11 protein (MDC) | | | | |
| *Homo sapiens* | – | – | D17390 | D31872: 3' end |
| ADAM 14 protein (adm-1) | | | | |
| *Caenorhabditis elegans* | – | – | U68185 | – |
| ADAM 18 protein (tMDC III) | | | | |
| *Macaca fascicularis* | – | – | Y08617 | – |

M

## Name and History

The first two members of the **ADAM** (proteins with **a** **d**isintegrin **a**nd **m**etalloprotease domain) gene family were called **PH-30** $\alpha$ and $\beta$ because of their localization to the posterior head of the guinea pig sperm surface (Primakoff *et al*., 1987). The names were changed to *fertilin* $\alpha$ and $\beta$ as evidence accumulated suggesting that this heterodimeric complex of tightly associated transmembrane proteins is involved in sperm–egg membrane fusion, and thus, in the process of fertilization (Blobel *et al*., 1990, 1992; Myles *et al*., 1994). Fertilin $\alpha$ and $\beta$ contain a pro, metalloprotease-like, disintegrin-like and cysteine-rich domain, followed by an epidermal growth factor (EGF)-like repeat, transmembrane domain and cytoplasmic tail (Wolfsberg *et al*., 1993). They are thus similar in sequence and domain organization to the snake venom reprolysins (Bjarnason & Fox, 1995) (see Chapter 416) or adamalysins (Stöcker *et al*., 1995). These proteases, together with the astacins (Chapter 405), matrix metalloproteases and serralysins (Chapter 386), make up a clan (MB) of zinc peptidases termed the metzincins. All metzincins share a similar topology of a five-stranded $\beta$ sheet and three $\alpha$ helices, a characteristic active-site sequence, and a conserved methionine residue which helps to stabilize the active site.

The metalloprotease domains of ADAMs 1, 8–10, 12, 13, 15 and 17, as well as all the snake venom reprolysins, contain the extended active-site sequence HEXGHXXGXXHD (Wolfsberg *et al*., 1995). The crystal structure of adamalysin (Chapter 421) shows that the three histidine residues (underlined) ligand the zinc, the glutamic acid (bold) is the catalytic base, and the glycine (italicized) allows an important structural turn (Stöcker *et al*., 1995). The snake venom enzymes have been shown to cleave a variety of extracellular matrix molecules (Fox & Bjarnason, 1995) as well as the precursor to TNF$\alpha$ (tumor necrosis factor $\alpha$) (Moura-da Silva *et al*., 1996). Thus, we predict that similar molecules will serve as substrates for the ADAM 1, 8–10, 12, 13, 15 and 17 proteases. ADAMs 2–7, 11, 14 and 18, on the other hand, contain different residues at the position corresponding to the active site and are likely to be catalytically inactive.

Proteolytic activity has been demonstrated for cattle ADAM 10 (MADM) (Chapter 448) (Howard & Glynn, 1995) and human ADAM 17 (tumor necrosis factor $\alpha$ (TNF$\alpha$)-converting enzyme; TACE) (Chapter 449) (Black *et al*., 1997; Moss *et al*., 1997). ADAM 10, a single-chain, *N*-glycosylated 58 kDa metalloprotease, cleaves myelin basic protein at Pro73$\downarrow$Gln74 in an *in vitro* assay. Proteolytic activity is inhibited by 1,10-phenanthroline and by DTT, but not by phosphoramidon. ADAM 10 is active between pH 7 and pH 9, and its activity is not significantly changed by millimolar amounts of $Ca^{2+}$. The identity of the natural substrate of ADAM 10 is not known. ADAM 17, a $\sim$85 kDa metalloprotease, cleaves the TNF$\alpha$ precursor at Ala76$\downarrow$Val77. Proteolytic activity is inhibited by a peptide hydroxamate (Immunex compound 3) (Black *et al*., 1997) as well as by other compounds (Moss *et al*., 1997). The natural substrate of TACE is thought to be TNF$\alpha$.

## Structural Chemistry

All ADAMs are multidomain proteins consisting of a signal sequence, pro-, metalloprotease-like, disintegrin-like and cysteine-rich domains, followed by an EGF-like repeat and transmembrane and cytoplasmic tail domains. Full-length guinea pig fertilin $\alpha$ and $\beta$ (ADAMs 1 and 2) have molecular masses of 105 and 85 kDa, respectively, when analyzed on nonreducing SDS gels (Blobel *et al*., 1990). However, many of the snake venom reprolysins are cleaved at interdomain boundaries, and the activity of certain ADAM domains may be regulated by proteolytic processing (Wolfsberg & White, 1996). For example, proteolytically processed forms of ADAMs 1–3, 10, 12, 13 and 17 have been detected. Furthermore, a number of ADAMs may be cleaved at the di- or tetrabasic residues between their pro- and metalloprotease-like or metalloprotease-like and disintegrin-like domains. All ADAMs have sites for *N*-linked glycosylation in their extracellular domains, and guinea pig fertilin $\alpha$ and $\beta$ are known to be glycosylated (Blobel *et al*., 1990).

## Preparation

The proteolytically active form of ADAM 10 (MADM) has been purified from cattle kidney by DEAE-Sephacel and CM-Sepharose chromatography (Howard & Glynn, 1995). TACE has been affinity purified from pig spleen using a biotinylated inhibitor of TNF$\alpha$ release (Moss *et al*., 1997) and from a human monocytic cell line by several chromotography steps (Black *et al*., 1997).

## Biological Aspects

The sequences of 17 full-length members of the ADAM gene family are currently available. Although most have been cloned from mammals, recent reports describe ADAMs from a frog (Alfandari *et al*., 1997), fruit fly (Fambrough *et al*., 1996; Rooke *et al*., 1996), and nematode (Podbilewicz, 1996). Some ADAMs are expressed in many tissues, while others display a more limited distribution (Wolfsberg & White, 1996).

ADAMs are probably unique among membrane proteins in possessing both a proteolytic and an adhesive (disintegrin) domain. Snake venom disintegrins are known to interact with integrins, and we have predicted that ADAM disintegrin domains will also interact with integrins or other cell surface receptors. Such interactions could serve not only to bring ADAM-expressing cells into close contact with other cells or extracellular matrix components, but also to target ADAM metalloprotease domains to their sites of action. Recent experiments suggest that sperm fertilin $\beta$ (ADAM 2) may mediate sperm–egg binding by interacting with an integrin on the egg surface (Almeida *et al*., 1995). Other ADAMs may also play important roles during development. For example, mouse ADAM 12 (meltrin $\alpha$) has been implicated in myoblast fusion (Yagami-Hiromasa *et al*., 1995), *Drosophila* ADAM 10 (kuzbanian) is involved in neuronal cell fate determination (Rooke *et al*., 1996) and is required for axonal extension (Fambrough *et al*., 1996), *Caenorhabditis elegans* ADAM 14 may play a role in somatic and gamete cell fusions (Podbilewicz, 1996), human ADAM 17 releases soluble TNF$\alpha$ from its membrane-bound precursor (Black *et al.*, 1997; Moss *et al.*, 1997) and *Xenopus* ADAM 13 may be involved in neural crest cell adhesion and migration and in myoblast differentiation (Alfandari *et al.*, 1997).

## Related Peptidases

As described above, the ADAM proteases are most similar in sequence ($\sim$30% identical) to the group of metalloproteases isolated from snake venoms (Chapter 416).

## Further Reading

For a recent, comprehensive review on the ADAM gene family, see Wolfsberg & White (1996). See also Chapters 416, 448 and 449.

## References

Alfandari, D., Wolfsberg, T.G., White, J.M. & DeSimone, D.W. (1997) ADAM 13: a novel ADAM expressed in somatic mesoderm and neural crest cells during *Xenopus laevis* development. *Dev. Biol.* **182**, 314–330.

Almeida, E.A., Huovila, A.P., Sutherland, A.E. *et al.* (1995) Mouse egg integrin alpha 6 beta 1 functions as a sperm receptor. *Cell* **81**, 1095–1104.

Bjarnason, J.B. & Fox, J.W. (1995) Snake venom metalloendopeptidases: reprolysins. *Methods Enzymol.* **248**, 345–368.

Black, R.A., Rauch, C.T., Kozlosky, C.J. *et al.* (1997) A metalloproteinase disintegrin that releases tumour-necrosis factor-alpha from cells. *Nature* **385**, 729–733.

Blobel, C.P., Myles, D.G., Primakoff, P. & White, J.M. (1990) Proteolytic processing of a protein involved in sperm-egg fusion correlates with acquisition of fertilization competence. *J. Cell Biol.* **111**, 69–78.

Blobel, C.P., Wolfsberg, T.G., Turck, C.W., Myles, D.G., Primakoff, P. & White, J.M. (1992) A potential fusion peptide and an integrin ligand domain in a protein active in sperm-egg fusion [see comments]. *Nature* **356**, 248–252.

Fambrough, D., Pan, D., Rubin, G.M. & Goodman, C.S. (1996) The cell surface metalloprotease/disintegrin Kuzbanian is required for axonal extension in *Drosophila*. *Proc. Natl Acad. Sci. USA* **93**, 13233–13238.

Fox, J.W. & Bjarnason, J.B. (1995) Atrolysins: metalloproteinases from *Crotalus atrox* venom. *Methods Enzymol.* **248**, 368–387.

Howard, L. & Glynn, P. (1995) Membrane-associated metalloproteinase recognized by characteristic cleavage of myelin basic protein: assay and isolation. *Methods Enzymol.* **248**, 388–395.

Moss, M.L., Jin, S.L., Milla, M.E. *et al.* (1997) Cloning of a disintegrin metalloproteinase that processes precursor tumour-necrosis factor-alpha. *Nature* **385**, 733–736.

Moura-da Silva, A.M., Laing, G.D., Paine, M.J.I., Dennison, J.M.T.J., Politi, V., Crampton, J.M. & Theakston, R.D.G. (1996) Processing of pro-tumor necrosis factor-$\alpha$ by venom metalloproteinases: a hypothesis explaining local tissue damage following snake bite. *Eur. J. Immunol.* **26**, 2000–2005.

Myles, D.G., Kimmel, L.H., Blobel, C.P., White, J.M. & Primakoff, P. (1994) Identification of a binding site in the disintegrin domain of fertilin required for sperm-egg fusion. *Proc. Natl Acad. Sci. USA* **91**, 4195–4198.

Podbilewicz, B. (1996) ADM-1, a protein with metalloprotease- and disintegrin-like domains, is expressed in syncytial organs, sperm, and sheath cells of sensory organs in *Caenorhabditis elegans*. *Mol. Biol. Cell* **7**, 1877–1893.

Primakoff, P., Hyatt, H. & Tredick-Kline, J. (1987) Identification and purification of a sperm surface protein with a potential role in sperm-egg membrane fusion. *J. Cell Biol.* **104**, 141–149.

Rooke, J., Pan, D., Xu, T. & Rubin, G.M. (1996) KUZ, a conserved metalloprotease-disintegrin protein with two roles in *Drosophila* neurogenesis. *Science* **273**, 1227–1231.

Stöcker, W., Grams, F., Baumann, U., Reinemer, P., Gomis-Rüth, F.X., McKay, D.B. & Bode, W. (1995) The metzincins – topological and sequential relations between the astacins, adamalysins, serralysins, and matrixins (collagenases) define a superfamily of zinc-peptidases. *Protein Sci.* **4**, 823–840.

Weskamp, G. & Blobel, C.P. (1994) A family of cellular proteins related to snake venom disintegrins. *Proc. Natl Acad. Sci. USA* **91**, 2748–2751.

Wolfsberg, T.G. & White, J.M. (1996) ADAMs in fertilization and development. *Dev. Biol.* **180**, 389–401.

Wolfsberg, T.G., Bazan, J.F., Blobel, C.P., Myles, D.G., Primakoff, P. & White, J.M. (1993) The precursor region of a protein active in sperm–egg fusion contains a metalloprotease and a disintegrin domain: structural, functional, and evolutionary implications. *Proc. Natl Acad. Sci. USA* **90**, 10783–10787.

Wolfsberg, T.G., Straight, P.D., Gerena, R.L., Huovila, A.P., Primakoff, P., Myles, D.G. & White, J.M. (1995) ADAM, a widely distributed and developmentally regulated gene family encoding membrane proteins with a disintegrin and metalloprotease domain. *Dev. Biol.* **169**, 378–383.

Yagami-Hiromasa, T., Sato, T., Kurisaki, T., Kamijo, K., Nabeshima, Y. & Fujisawa-Sehara, A. (1995) A metalloprotease-disintegrin participating in myoblast fusion. *Nature* **377**, 652–656.

**M**

***Tyra G. Wolfsberg***
*National Center for Biotechnology Information,*
*National Library of Medicine, National Institutes of Health,*
*Building 38A/Room 8N805,*
*Bethesda, MD 20894, USA*
*Email: wolfsberg@ncbi.nlm.nih.gov*

***Judith M. White***
*Department of Cell Biology,*
*University of Virginia Health Sciences Center,*
*Box 439,*
*Charlottesville, VA 22908, USA*
*Email: jw7g@virginia.edu*

# 448. Myelin-associated metalloproteinase/ADAM 10

## Databanks

*Peptidase classification: clan MB, family M12, MEROPS ID: M12.210*
*NC-IUBMB enzyme classification: none*
*Databank codes:*

| Species | SwissProt | PIR | EMBL (cDNA) | EMBL (genomic) |
|---|---|---|---|---|
| *Bos taurus* | – | – | Z21961 | – |
| *Drosophila melanogaster* | – | – | U60591 | – |
| *Homo sapiens* | – | – | Z48579 | – |
|  |  |  | Z21961 |  |
| *Rattus norvegicus* | – | – | Z48444 | – |

## Name and History

Brain myelin membrane preparations contain a **myelin-associated metalloproteinase** capable of mediating a characteristic limited cleavage of myelin basic protein (MBP) (Chantry *et al.*, 1988). The enzyme responsible was isolated from cattle brain (Chantry *et al.*, 1989) and has recently been cloned from cDNA libraries of cattle, rat and human brain and human U937 histiocytic lymphoma cells (Howard *et al.*, 1996). The deduced protein sequence indicates that this enzyme is a member of the reprolysin subfamily which includes snake venom disintegrin metalloproteases and a number of cell surface mammalian proteins (Bjarnason & Fox, 1995) (Chapter 416). In recognition of this similarity the enzyme was named **MADM** for **ma**mmalian **d**isintegrin **m**etalloprotease (Howard *et al.*, 1996). The mammalian reprolysins have been named ADAM (**a d**isintegrin **a**nd **m**etalloprotease) (Chapter 447) and MADM has been designated **ADAM 10** (Wolfsberg *et al.*, 1995).

## Activity and Specificity

Detection of the characteristic MBP cleavage products is, at present, the only assay for MADM's proteolytic activity (Howard & Glynn, 1995). The enzyme purified from brain cleaves a number of bonds in 18 kDa MBP but two fragments are always prominent (Chantry *et al.*, 1989); these result from a single cleavage at Gly-Ser-Leu-Pro73┼Gln74-Lys-Ser-Gln, generating an 8 kDa N-terminal fragment and a 10 kDa C-terminal fragment which has an N-terminal pyroglutamyl residue resulting from cyclization of the newly exposed glutamine (Groome *et al.*, 1988). This cleavage of MBP occurs between pH 7 and 9, and is inhibited by 1,10-phenanthroline and DTT but not phosphoramidon; degradation of other substrates has not been investigated in any detail (Chantry *et al.*, 1989).

## Structural Chemistry

MADM isolated from cattle brain is a single-chain, N-glycosylated polypeptide of about 60 kDa (Chantry *et al.*, 1989). Cattle MADM cDNA predicts a multidomain type I transmembrane glycoprotein with: (a) a prepro domain with a single cysteine and a potential furin (Chapter 117) cleavage site; (b) a metalloprotease domain with the active-site signature HEXHXXGXXHD including the D which is characteristic of the reprolysin subfamily; (c) disintegrin and cysteine-rich domains with putative integrin-binding capability; (d) a putative transmembrane helix and a C-terminal cytoplasmic tail rich in proline and basic residues (Howard *et al.*, 1996).

## Preparation

Similar low levels (<0.01% of total protein) are found in a wide range of mammalian tissues and cell lines (Chantry & Glynn, 1990; Howard *et al*, 1996). The enzyme may be isolated by standard chromatographic procedures following Triton X-100-solubilization of crude membrane fractions from bulk tissue such as cattle kidney (Howard & Glynn, 1995).

## Biological Aspects

An activity mediating the characteristic cleavage of MBP is present in brain tissue of all vertebrates (Chantry *et al.*, 1992), although recent immunohistochemical studies indicate that MADM is localized in white matter microglia (S. Mitchell & P. Glynn, unpublished results) rather than in oligodendrocytes as suggested in the earlier work. The characteristic 10 kDa C-terminal fragment formed by MADM-mediated cleavage of MBP has been detected immunohistochemically within microglia in human brain white matter (Li *et al.*, 1993). Contrary to the predictions of the primary sequence data, MADM in microglia and a variety of other cells appears to have an intracellular location, although cell surface MADM has been observed on rodent and human myeloma cell lines (S. Mitchell, X. Lu & P. Glynn, unpublished results).

The very striking conservation of sequence in the mature MADM protein between cattle, rat and human (97% identity) suggests preservation for an important biological function. In evolutionary terms, MADM diverged early from other vertebrate members of the reprolysin subfamily (Rawlings & Barrett, 1995; see also Fig. 404.5). While MADM shows only moderate homology (about 25% identity) to other mammalian ADAM proteins, it bears substantially greater similarity (43% identity) to KUZ, a recently described *Drosophila* protein (Chapter 447) involved in relatively early stages of neural development (Rooke *et al.*, 1996). This suggests that the search for the normal substrate of MADM, its putative cognate integrins, and its physiological role might best be directed towards a study of early development in vertebrates.

## Distinguishing Features

Whether any other membrane-associated metalloproteases can mediate the same characteristic cleavage of MBP is unknown. A mouse monoclonal antibody, CG4, specific for cattle MADM (Chantry & Glynn, 1990), and rabbit antisera (against two different synthetic peptides) that appear to cross-react with all mammalian MADMs on western blots, have been produced in this laboratory but are not commercially available.

## References

Bjarnason, J.B. & Fox, J.W. (1995) Snake venom metalloendoproteinases: reprolysins. *Methods Enzymol.* **248**, 345–368.

Chantry, A. & Glynn, P. (1990) A novel metalloproteinase originally isolated from brain myelin membranes is present in many tissues. *Biochem. J.* **268**, 245–248.

Chantry, A., Earl, C., Groome, N. & Glynn, P. (1988) Metallo-endoproteinase cleavage of 18.2 and 14.1-kilodalton basic proteins dissociating from rodent myelin membranes generates 10.0 and 5.1-kilodalton C-terminal fragments. *J. Neurochem.* **50**, 688–694.

Chantry, A., Gregson, N.A. & Glynn, P. (1989) A novel metalloprotease associated with brain myelin membranes: isolation and characterization. *J. Biol. Chem.* **264**, 21603–21607.

Chantry, A., Gregson, N.A. & Glynn, P. (1992) Degradation of myelin basic protein by a membrane-associated metalloprotease: neural distribution of the enzyme. *Neurochem. Res.* **17**, 861–868.

Groome, N., Chantry, A., Earl, C., Newcombe, J., Keen, J., Findlay, J. & Glynn, P. (1988) A new epitope on human myelin basic protein arising from cleavage by a metalloprotease associated with brain myelin membranes *J. Neuroimmunol.* **19**, 77–88.

Howard, L. & Glynn, P. (1995) Membrane-associated metalloproteinase recognized by characteristic cleavage of myelin basic protein: assay and isolation. *Methods Enzymol.* **248**, 388–395.

Howard, L., Lu, X., Mitchell, S., Griffiths, S. & Glynn, P. (1996) Molecular cloning of MADM: a catalytically-active mammalian disintegrin-metalloproteases expressed in various cell types. *Biochem. J.* **317**, 45–50.

Li, H., Newcombe, J., Groome, N.P. & Cuzner, M.L. (1993) Characterization and distribution of phagocytic macrophages in multiple sclerosis plaques. *Neuropathol. Appl. Neurobiol.* **19**, 214–233.

Rawlings, N.D. & Barrett, A.J. (1995) Evolutionary families of metallopeptidases. *Methods Enzymol.* **248**, 183–226.

Rooke, J. Pan, D., Xu, T. & Rubin, G.M. (1996) KUZ, a conserved metalloprotease-disintegrin protein with two roles in *Drosophila* neurogenesis. *Science* **273**, 1227–1231.

Wolfsberg, T.G., Primakoff, P., Myles, D.G. & White, J.M. (1995) ADAM, a novel family of membrane proteins containing A Disintegrin And Metalloprotease domain: multipotential functions in cell–cell and cell–matrix interactions. *J. Cell Biol.* **131**, 275–278.

*Paul Glynn*
*Medical Research Council Toxicology Unit,*
*University of Leicester, Leicester LE1 9HN, UK*
*Email: pg8@leicester.ac.uk*

# 449. Tumor necrosis factor α-converting enzyme

## Databanks

*Peptidase classification: clan MB, family M12, MEROPS ID: M12.217*
*NC-IUBMB enzyme classification: none*

**Databank codes:**

| Species | SwissProt | PIR | EMBL (cDNA) | EMBL (genomic) |
|---------|-----------|-----|-------------|----------------|
| *Homo sapiens* | – | – | U69611 U86755 | – |

## Name and History

This enzyme is called the ***tumor necrosis factor α-converting enzyme (TACE)*** because it was identified by its ability to cleave 26 kDa tumor necrosis factor α (TNFα) at the physiological processing site (Mohler *et al.*, 1994; Black *et al.*, 1997; Moss *et al.*, 1997). In cells, this cleavage releases the soluble, 17 kDa form of TNFα, which induces inflammation. A concerted effort was therefore made to find the responsible enzyme, using recombinant 26 kDa TNFα as substrate (Mohler *et al.*, 1994; Black *et al.*, 1997). In an independent approach, hydroxamate-based inhibitors of the matrix metalloproteinases were found to block the release of TNFα from a variety of cell types (McGeehan *et al.*, 1994), and a purification effort ensued to identify the physiologically relevant enzyme by applying the criterion of hydroxamate-sensitivity to enzymes with the appropriate processing activity (Moss *et al.*, 1997). Cloning of the enzyme revealed that it is a member of the reprolysin/ADAM subfamily (Chapters 416, 447), and it is designated as ***ADAM 17***. The extracellular domains of various proteins other than TNFα are 'shed' from cells, but it is not known whether TACE is responsible for any of these phenomena (see discussion below).

## Activity and Specificity

As indicated above, a defining characteristic of TACE is its specific cleavage of 26 kDa TNFα at the physiological processing site, Pro-Leu-Ala-Gln-Ala+Val-Arg-Ser-Ser-Ser. The same specificity is observed with 12 and 20 residue peptides (acetylated at the N-terminus and amidated at the C-terminus) spanning the scissile bond (Mohler *et al.*, 1994; Black *et al.*, 1996), and with a Dnp-labeled substrate (Dnp-Ser-Pro-Leu-Ala-Gln-Ala+Val-Arg-Ser-Ser-Ser-Arg-amide; Moss *et al.*, 1997). Changing the P1 amino acid in the 12-residue peptide to Ile eliminates cleavage, while changing it to Phe does not (Mullberg *et al.*, 1997), but nothing else has been published regarding substrate preferences. Various mutations around the cleavage site do not prevent the release of TNFα from cells (Perez *et al.*, 1990), but it is not known whether TACE is the enzyme that cleaves the mutant forms of 26 kDa TNFα.

The enzyme has been assayed by following cleavage of the peptide substrates described above by HPLC (Black *et al.*, 1996; Moss *et al.*, 1997). In addition, the peptide substrates can be biotinylated and tritiated to allow TACE activity to be monitored by scintillation proximity assay in a 96-well plate format (Moss *et al.*, 1997).

TACE is inhibited by EDTA and a variety of peptide hydroxamates (Gearing *et al.*, 1994; McGeehan *et al.*, 1994; Mohler *et al.*, 1994; Moss *et al.*, 1997). TIMP-1 and phosphoramidon do not cause significant inhibition (Mohler *et al.*, 1994). A hydroxamate inhibitor (INH-3850-PI) is available from Peptides International (see Appendix 2 for full names and addresses of suppliers).

## Structural Chemistry

The protein encoded by the TACE cDNA is a typical reprolysin/ADAM in that it contains a signal peptide, an apparent pro domain with a putative cysteine switch and furin cleavage sequence, a catalytic domain with the expected zinc-binding site and Met-turn, a region with a disintegrin and EGF-like domain, a transmembrane domain, and a cytoplasmic tail (Black *et al.*, 1997; Moss *et al.*, 1997). However, the amino acid identity with other ADAMs is low: the closest member, ADAM 10 (Chapter 448), is identical at only 29% of the residues (see Fig. 404.5). The catalytic domains of the matrix metalloproteinases, as defined by the crystal structures, show only 10% identity with the corresponding region of TACE. The cytoplasmic domain contains potential phosphorylation sites. The active form of the enzyme runs as an 85 kDa protein in SDS gels under reducing conditions.

## Preparation

TACE has been purified from membrane preparations from human monocytic cells (Black *et al.*, 1997) and pig spleen (Moss *et al.*, 1997). The membranes were prepared following protocols designed to yield plasma membrane in the first case and microsomal membrane in the second case. In both procedures, proteins were then solubilized with detergent, and TACE was purified either by a sequence of standard chromatographic steps (Black *et al.*, 1997) or by a combination of conventional chromatography and hydroxamate-based affinity chromatography (Moss *et al.*, 1997).

The extracellular domain of TACE has been expressed in both 293/EBNA cells (Black *et al.*, 1997) and insect cells (Moss *et al.*, 1997), purified from the medium, and shown to be active. The N-terminus of the active form of TACE purified from the insect cell medium began with Arg215, implicating a furin-like enzyme in the removal of the pro domain.

## Biological Aspects

The role of TACE in releasing TNFα has been confirmed by inactivating the gene in mouse T cells and monocytic cells (Black *et al.*, 1997). Such cells released about 85% less TNFα than wild-type cells, and the residual release was not inhibited by peptide hydroxamates. Surface biotinylation experiments indicate that TACE resides on the cell exterior, although the purification from microsomal preparations (see above) suggests that there may also be an intracellular pool. The biotinylation experiments show the apparently active form of TACE on the surface of several cell types, in the presence or absence of various stimuli, but the precise location of TNFα processing remains to be determined. Northern analyses indicate that TACE mRNA is present in virtually every tissue.

## Distinguishing Features

Only one other ADAM, ADAM 10 (Chapter 448), has actually been shown to have a proteolytic activity (Howard *et al.*, 1996; Lunn *et al.*, 1997), and no comparisons between it and TACE have been published. ADAM 10 does also cleave 26 kDa TNFα (Lunn *et al.*, 1997). TACE and ADAM 10 differ from other ADAMs in having three fewer cysteines between the catalytic domain and the transmembrane domain. TACE is distinguished from the matrix metalloproteinases by its insensitivity to TIMP-1 (Mohler *et al.*, 1994) and its greater specificity in cleaving TNFα-processing site peptides (Black *et al.*, 1996).

## Related Peptidases

The shedding of several proteins besides TNFα is inhibited by peptide hydroxamates, at concentrations similar to that required to inhibit TNFα release (for a thorough review, see Hooper *et al.*, 1997). These proteins include L-selectin, transforming growth factor α, the p75 TNF receptor, the interleukin 6 (IL-6) receptor, angiotensin-converting enzyme (Chapter 359) and the amyloid-β precursor protein. Whether these shedding phenomena involve TACE or one or more related enzymes is unknown. Virtually all the published studies rely on intact cell systems, and none of the responsible enzymes have been purified. However, evidence for a common mechanism, beyond the inhibition by hydroxamates, is that phorbol myristic acid stimulates release in most if not all cases; a mutant CHO cell line has been isolated that is deficient in shedding several of these proteins (Arribas *et al.*, 1996); and alkylating agents stimulate several shedding events. On the other hand, some of the cleavage sites are highly divergent in sequence; the number of residues between the transmembrane domain and the processing site varies from a few to at least 40; the topologies represented include type I single-spanning proteins, type II single-spanning proteins, a seven-spanning protein, and a GPI-linked protein; and TACE appears to be highly specific in its cleavage of peptide substrates. Determining the ability of TACE-deficient cells to shed proteins other than TNFα should help to establish whether multiple 'sheddases' exist.

## References

Arribas, J., Coodly, L., Vollmer, P., Kishimoto, T.K., Rose-John, S. & Massague, J. (1996) Diverse cell surface protein ectodomains are shed by a system sensitive to metalloprotease inhibitors. *J. Biol. Chem.* **271**, 11376–11382.

Black, R.A., Durie, F.H., Otten-Evans, C., Miller, R., Slack, J.L., Lynch, D.H., Castner, B., Mohler, K.M., Gerhart, M., Johnson, R.S., Itoh, Y., Okada, Y. & Nagase, H. (1996) Relaxed specificity of matrix metalloproteinases (MMPs) and TIMP insensitivity of tumor necrosis factor-α (TNF-α) production suggest the major TNF-α converting enzyme is not an MMP. *Biochem. Biophys. Res. Commun.* **225**, 400–405.

Black, R.A., Rauch, C.T., Kozlosky, C.J., Peschon, J.J., Slack, J.L., Wolfson, M.F., Castner, B.J., Stocking, K.L., Reddy, P.L., Srinivasan, S., Nelson, N., Boiani, N., Schooley, K.A., Gerhart, M., Davis, R., Fitzner, J.N., Johnson, R.S., Paxton, R.T., March, C.J. & Cerretti, D.P. (1997) A metalloproteinase disintegrin that releases tumour-necrosis factor-α from cells. *Nature* **385**, 729–733.

Gearing, A.J.H., Beckett, P., Christodoulou, M., Churchill, M., Clements, J.M., Crimmin, M., Davidson, A.H., Drummond, A.H., Galloway, W.A., Gilbert, R., Gordon, J.L., Leber, T.M., Mangan, M., Miller, K., Nayee, P., Owen, K., Patel, S., Thomas, W., Gells, G., Wood, L.M. and Woolley, K. (1994) Processing of tumour necrosis factor-alpha precursor by metalloproteinases. *Nature* **370**, 555–557 (1994).

Hooper, N.M., Karran, E.H. & Turner, A.J. (1997) Membrane protein secretases. *Biochem. J.* **321**, 265–279.

Howard, L., Lu, X., Mitchell, S., Griffiths, S. & Glynn, P. (1996) Molecular cloning of MADM: a catalytically active mammalian disintegrin-metalloprotease expressed in various cell types. *Biochem J* **317**, 45–50.

Lunn, C., Fan, X., Dalie, B., Miller, K., Zavodny, P.J., Narula, S.K. & Lundell, D. (1997) Purification of ADAM 10 from bovine spleen as a TNFα convertase. *FEBS Lett.* **400**, 333–335.

McGeehan, G.M., Becherer, J.D., Bast, R.C., Jr, Boyer, C.M., Champion, B., Connolly, K.M., Conway, J.G., Furdon, P., Karp, S., Kidao, S., McElroy, A.B., Nichols, J., Pryzwansky, K.M., Schoenen, F., Sekut, L., Truesdale, A., Verghese, M., Warner, J., & Ways, J.P. (1994) Regulation of tumour necrosis factor-alpha processing by a metalloproteinase inhibitor. *Nature* **370**, 558–561.

Mohler, K.M., Sleath, P.R., Fitzner, J.N., Cerretti, D.P., Alderson, M., Kerwar, S.S., Torrance, D.S., Otten-Evans, C., Greenstreet, T., Weerawarna, K., Kronheim, S.R., Petersen, M., Gerhart, M., Kozlosky, C.J., March, C.J. & Black, R.A. (1994) Protection against a lethal dose of endotoxin by an inhibitor of tumor necrosis factor processing. *Nature* **370**, 218–220.

Moss, M.L., Jin, S.L.C., Milla, M.E., Burkhart, W., Carter, H.L., Chen, W.J., Clay, W.C., Didsbury, J.R., Hassler, D., Hoffman, C.R., Kost, T.A., Lambert, M.H., Leesnitzer, M.A., McCauley, P., McGeehan, G., Mitchell, J., Moyer, M., Pahel, B., Rocque, W., Overton, L.K., Schoenen, F., Seaton, T., Su, J.L., Warner, J., Willard, D. and Becherer, J.D. (1997) Cloning of a disintegrin metalloproteinase that processes precursor tumour-necrosis factor-α. *Nature* **385**, 733–736.

Mullberg, J., Rauch, C.T., Wolfson, M.F., Castner, B., Fitzner, J.N., Otten-Evans, C., Mohler, K.M., Cosman, D. & Black, R.A. (1997) Further evidence for a common mechanism for shedding of cell surface proteins. *FEBS Lett.* **401**, 235–238.

Perez, C., Albert, I., DeFay, K., Zachariades, N., Gooding, L. & Kriegler, M. (1990) A nonsecretable cell surface mutant of tumor necrosis factor (TNF) kills by cell-to-cell contact. *Cell* **63**, 251–258.

**M**

*Roy A. Black*
*Department of Protein Chemistry,*
*Immunex Corp.,*
*51 University Street,*
*Seattle, WA 98101, USA*
*Email: rblack@immunex.com*

*J. David Becherer*
*Department of Molecular Biochemistry,*
*Glaxo Wellcome Research and Development Inc.,*
*5 Moore Drive,*
*Research Triangle Park, NC 27709, USA*
*Email: becherer~jd@glaxo.com*

# 450. Introduction: clan MC containing metallocarboxypeptidases

## Databanks

MEROPS ID: MC

| Species | SwissProt | PIR | EMBL (cDNA) | EMBL (genomic) |
|---|---|---|---|---|
| **Family M14** | | | | |
| **Subfamily A** | | | | |
| Carboxypeptidase A (Chapter 451) | | | | |
| Carboxypeptidase A2 (Chapter 452) | | | | |
| Carboxypeptidase B (Chapter 455) | | | | |
| Carboxypeptidase T (Chapter 456) | | | | |
| Carboxypeptidase U (Chapter 453) | | | | |
| Mast cell carboxypeptidase A (Chapter 454) | | | | |
| Others | | | | |
| *Saccharomyces cerevisiae* | P38836 | – | U10398 | – |
| *Simulium vittatum* | – | – | U22453 | – |
| **Subfamily B** | | | | |
| Carboxypeptidase H (Chapter 458) | | | | |
| Carboxypeptidase M (Chapter 460) | | | | |
| Lysine carboxypeptidase (Chapter 459) | | | | |
| Metallocarboxypeptidase D (Chapter 461) | | | | |
| Transcription repressor AEBP1 | | | | |
| *Mus musculus* | – | – | X80478 | – |
| **Subfamily C** | | | | |
| γ-D-Glutamyl-meso-(L)-diaminopimelate peptidase I (Chapter 457) | | | | |

Clan MC comprises only family M14, which contains the animal enzymes carboxypeptidase A and carboxypeptidase B, as well as the bacterial carboxypeptidase T and γ-D-glutamyl-(L)-meso-diaminopimelate peptidase I.

In this clan, there is one catalytic zinc ion tetrahedrally coordinated by a water molecule, two histidines and the glutamate. One of the histidines and the Glu occur in the motif **His**-Xaa-Xaa-**Glu**, and the third zinc ligand is 103–143 residues C-terminal to this motif. The tertiary structures have been solved for three members of the family, and the fold is unlike that of thermolysin or astacin, but shows some similarity to those of two peptidases in which two zinc atoms are essential for activity: leucyl aminopeptidase (family M17; Chapter 472) and *Vibrio* aminopeptidase (family M28; Chapter 491). Because the zinc ligands are not conserved between the three families, however, these enzymes are considered representatives of different clans.

The current hypothesis for the catalytic mechanism of carboxypeptidase A involves two residues in addition to the zinc ligands His69, Glu72 and His 196. These are Arg127, which is an electrophile that helps stabilize the 'oxyanion hole', and Glu270, which is thought to be the general base for catalysis (Rees *et al.*, 1983). Other residues are important for substrate binding: Tyr248 hydrogen bonds the P1 amino group, and Arg71 the P2 carbonyl oxygen (Sebastian *et al.*, 1996). Three further residues, Asn144 and Arg145, which bind the C-terminal carboxylate group of the substrate (Rees & Lipscomb, 1983; Osterman *et al.*, 1992), and Tyr198, are also important for substrate binding. An alignment around these residues is shown in Alignment 450.1 (on CD-ROM).

The S1' subsite binds the C-terminus of the substrate and is the primary recognition site for carboxypeptidases. In rat mast cell carboxypeptidase A2, which releases bulky C-terminal residues, Thr268 is replaced by Ala and Leu203 by Met, enlarging the specificity pocket (Faming *et al.*, 1991). Alignment 450.1 also shows a conserved region, the function of which is unknown. This is the Asn112-Xaa-Asp-Gly motif, in which Xaa is predominantly Pro, which is situated some distance from the active site and may take part in shaping the substrate binding cavity (Osterman *et al.*, 1992).

Mallard (*Anas platyrhynchos*) metallocarboxypeptidase D unusually possesses three carboxypeptidase A-like units, but only the first two can be active. As can be seen in Alignment 450.1, the zinc ligand Glu72 is replaced by Ala in the third domain. In the mouse transcriptional repressor AEBP1, the third zinc ligand His196 is replaced by Asn, which is not known to coordinate metal ions in any metallopeptidase.

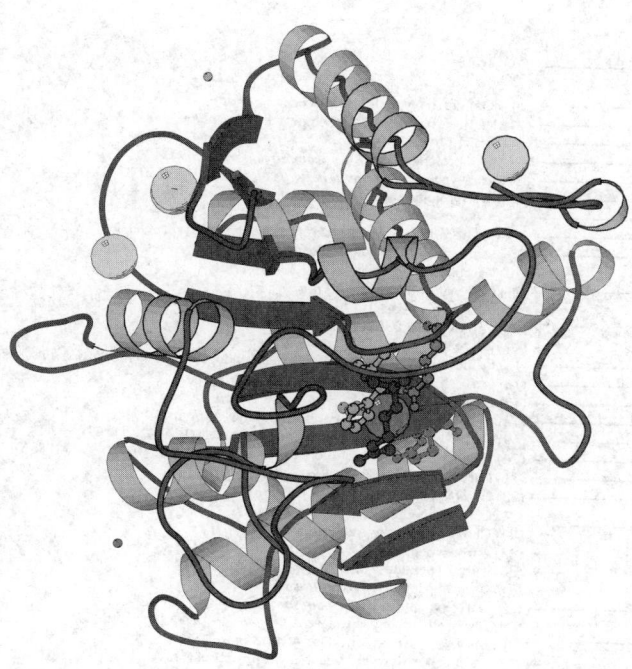

*Figure 450.1* Richardson diagram of *Thermoactinomyces vulgaris* carboxypeptidase T. The image was prepared from the Brookhaven Protein Data Bank entry (1OBR) as described in the Introduction (p. xxv). The catalytic zinc ion and three structural calcium ions are shown in CPK representation as light spheres. The zinc ligands are shown in ball-and-stick representation: His69, Glu72 and His196 (numbering as in Alignment 450.1). The catalytic Arg 127 and Glu270 are also shown in ball-and-stick representation.

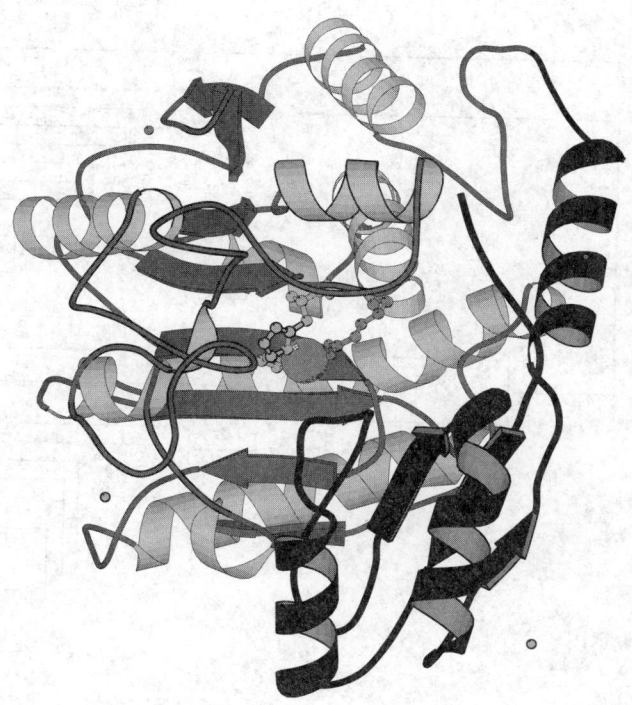

*Figure 450.3* Richardson diagram of cattle procarboxypeptidase A. The image was prepared from the Brookhaven Protein Data Bank entry (1PCA) as described in the Introduction (p. xxv). The catalytic zinc ion is shown in CPK representation as a light gray sphere. The zinc ligands are shown in ball-and-stick representation: His69, Glu72 and His196 (numbering as in Alignment 450.1). The propeptide is shown in black.

Thus, the repressor may not be a peptidase. The hypothetical yeast protein YHR132C may also not be a peptidase, because the catalytic Glu270 is replaced by Lys.

As can be seen from Alignment 450.1 (and Fig. 450.4), family M14 contains subfamilies that can be distinguished by conservation of sequence around the zinc ligands and catalytic residues. In subfamily A, the zinc ligands occur within the motifs **His**-Xaa-Arg-**Glu**-Xbb, in which Xaa is Ser or Ala, and Xbb is Trp or His; and Xaa-**His**-Xbb-Tyr-Ser-Xcc, in which Xaa is a hydrophobic residue, Xbb is Ser or Thr, and Xcc is Gln or Glu. In subfamily B, the motifs are **His**-Gly-Xaa-**Glu**-Xbb, in which Xaa is Asp or Asn, and

**M**

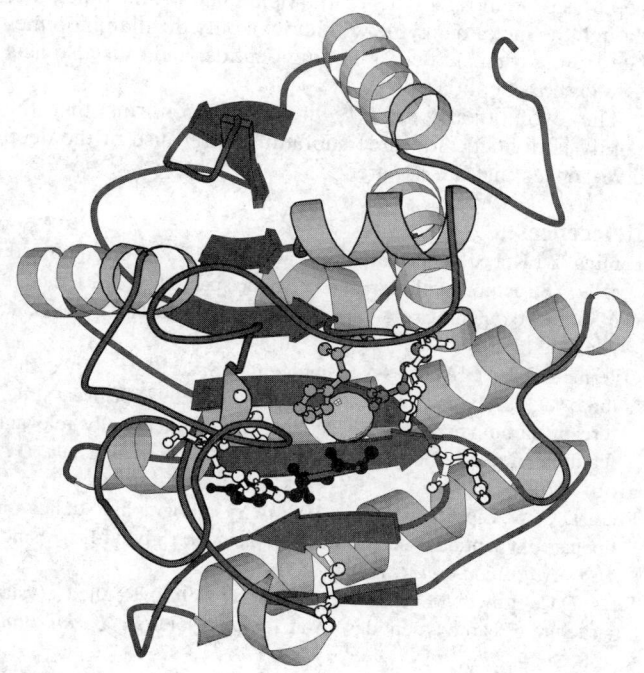

*Figure 450.2* Richardson diagram of the cattle carboxypeptidase A/glycyl-L-tyrosine complex. The image was prepared from the Brookhaven Protein Data Bank entry (3CPA) as described in the Introduction (p. xxv). The catalytic zinc ion is shown in CPK representation as a light gray sphere. The zinc ligands are shown in ball-and-stick representation: His69, Glu72 and His196 (numbering as in Alignment 450.1). The substrate binding residues Arg71, Asn144, Arg145, Tyr198 and Tyr247 are shown in ball-and-stick representation in white. The substrate is shown in black in ball-and-stick representation.

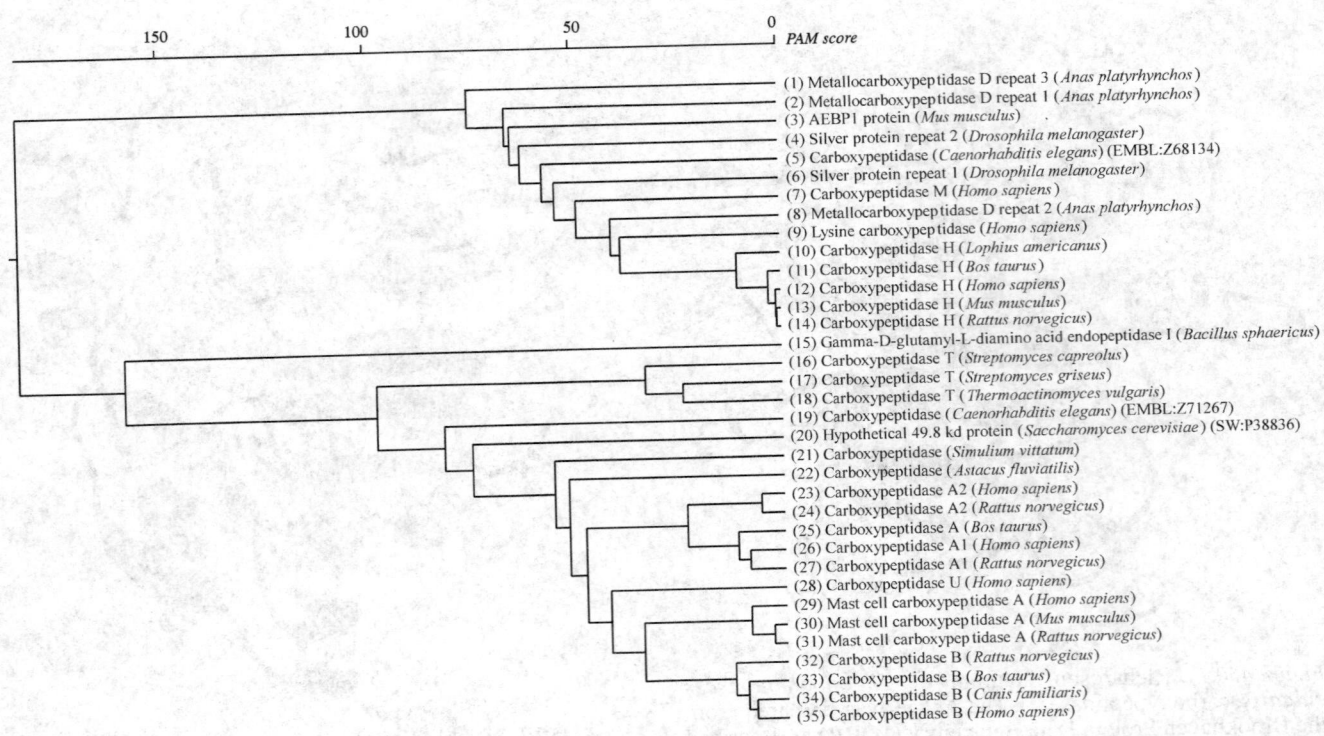

*Figure 450.4* Evolutionary tree for family M14. The tree was prepared as described in the Introduction (p. xxv).

Xbb is uncharged, and Xaa-**His**-Gly-Gly-Xaa-Xbb, in which Xaa is any small amino acid, and Xbb is hydrophobic or Arg. Members of subfamily B are also longer at the C-terminus.

Subfamily A contains not only animal carboxypeptidases, but also carboxypeptidase T from actinomycetes. Carboxypeptidases A and B are found in the pancreas, and are involved in food protein digestion. Hence subfamily A is also known as the 'digestive' subfamily (Osterman *et al.*, 1992).

The structure and catalytic residues of *Thermoactinomyces* carboxypeptidase T are shown in Figure 450.1. Figure 450.2 shows the structure of a cattle carboxypeptidase A/substrate complex, with selected substrate-binding residues shown in ball-and-stick representation.

Most members of family M14 are soluble, but carboxypeptidase M is membrane bound via a glycosylphosphatidylinositol anchor. Members of family M14 are synthesized with an N-terminal propeptide. In the proenzyme, the propeptide inhibits by shielding the catalytic site without making contact with it, and the substrate-binding pocket is blocked by specific contacts. This is shown in Figure 450.3.

Subfamily B includes a number of carboxypeptidases that are important for activation or inactivation of biological mediators and peptide hormones, hence subfamily B is also known as the 'regulatory' subfamily (Osterman *et al.*, 1992). Carboxypeptidase H processes bioactive peptides such as enkephalin, and lysine carboxypeptidase inactivates bradykinin and anaphylotoxins in blood plasma.

Subfamily C includes the γ-glutamyl-(L)-meso-diamino-pimelate peptidase I from *Bacillus sphaericus*. This is one of many enzymes involved in turnover and lysis of bacterial cell

walls, cleaving the cross-linking peptide. An enzyme with a similar specificity, dipeptidyl-peptidase VI (Chapter 265), is known from family C40. Both enzymes act as dipeptidyl-peptidases, removing the diaminopimelate-D-Ala dipeptide, though the metalloenzyme will remove only the diaminopimelate from a truncated cross-linking peptide, and thus also acts as a carboxypeptidase.

The evolutionary tree (Figure 450.4) confirms that the family is divided into three subfamilies, because of the deep divergences that are apparent.

## References

Faming, Z., Kobe, B., Stewart, C.-B., Rutter, W.J. & Goldsmith, E.J. (1991) Structural evolution of an enzyme specificity. The structure of rat carboxypeptidase A2 at 1.9-Å resolution. *J. Biol. Chem.* **266**, 24606–24612.

Osterman, A.L., Grishin, N.V., Smulevitch, S.V., Matz, M.V., Zagnitko, O.P., Revina, L.P. & Stepanov, V.M. (1992) Primary structure of carboxypeptidase T: delineation of functionally relevant features in Zn-carboxypeptidase family. *J. Protein Chem.* **11**, 561–570.

Rees, D.C. & Lipscomb, W.N. (1983) Crystallographic studies on apocarboxypeptidase A and the complex with glycyl-L-tyrosine. *Proc. Natl Acad. Sci. USA* **80**, 7151–7154.

Rees, D.C., Lewis, M. & Lipscomb, W.N. (1983) Refined crystal structure of carboxypeptidase A at 1.54 Å resolution. *J. Mol. Biol.* **168**, 367–387.

Sebastian, J.F., Liang, G.Q., Jabarin, A., Thomas, K. & Wu, H.B. (1996) Effect of enzyme-substrate interactions away from the reaction site on carboxypeptidase A catalysis. *Bioorg. Chem.* **24**, 290–303.

# 451. Carboxypeptidase A

## Databanks

*Peptidase classification: clan MC, family M14, MEROPS ID: M14.001*
*NC-IUBMB enzyme classification: EC 3.4.17.1*
*ATCC entries: 61300, 61301 (human); 105693 (human)*
*Chemical Abstracts Service registry number: 11075-17-5*
*Databank codes:*

| Species | SwissProt | PIR | EMBL (cDNA) | EMBL (genomic) |
|---|---|---|---|---|
| *Bos taurus* | P00730 | A00910 | M61851 | – |
| | | A31406 | Z33906 | |
| | | A38834 | | |
| | | A90355 | | |
| | | JN0126 | | |
| | | JT0440 | | |
| | | S00059 | | |
| *Homo sapiens* | P15085 | A34205 | X67318 | – |
| | | S02810 | | |
| | | S02811 | | |
| | | S08253 | | |
| | | S29127 | | |
| *Rattus norvegicus* | P00731 | A00911 | J00713 | M23960: exon 1 |
| | | | | M23985–M23990: exons 2–10 |
| | | B32129 | V01232 | |
| *Sus scrofa* | P09954 | A25833 | – | – |

Brookhaven Protein Data Bank three-dimensional structures:

| Species | ID | Resolution | Notes |
|---|---|---|---|
| *Bos taurus* | 1BAV | 1.6 | complex with 2-benzyl-3-iodo-propanoic acid |
| | 1CBX | 2 | complex with L-benzylsuccinate inhibitor |
| | 1CPS | 2.25 | complex with [L-2-carboxy-3-phenylpropyl] methyl-sulfodiimine |
| | 2CTB | 1.5 | |
| | 2CTC | 1.4 | complex with L-phenyl lactate |
| | 3CPA | 2 | complex with glycyl-L-tyrosine |
| | 4CPA | 2.5 | complex with potato carboxypeptidase A inhibitor |
| | 5CPA | 1.54 | |
| | 6CPA | 2 | complex with phosphonate |
| | 7CPA | 2 | complex with Bz-Phe-Val-PO-Phe |
| | 8CPA | 2 | complex with Bz-Ala-Gly-PO-Phe |
| | 8CPA | 2 | complex with Bz-Ala-Gly-(P)-(O)-Phe |

## Name and History

**Carboxypeptidase A (CPD A)** gets its name from the fact that it prefers peptide and protein substrates with an **a**romatic or branched-chained C-terminal amino acid. CPD A was isolated in pure form in 1937 (Anson, 1937) and was one of the first proteolytic enzymes for which considerable effort was made to understand the complexity of its kinetics towards acyl amino acids, dipeptides and their ester analogs. It was also the first metalloprotease and second zinc enzyme to be identified (Vallee & Neurath, 1954).

## Activity and Specificity

Carboxypeptidase A cleaves the C-terminal peptide or ester bond of peptides or depsipeptides that have a free C-terminal carboxyl group. Acylated -amino and -hydroxy carboxylic acids are also substrates. The substrates conform to the general structure:

$$R' - \underset{\underset{Y}{\parallel}}{C} - X - \underset{\underset{R}{\mid}}{CH} - COOH$$

where X is O, NH or S and Y is O or S. The rate of peptide substrate hydrolysis is enhanced if the side chain R is aromatic or branched aliphatic, e.g. Phe, Tyr, Trp, Leu or Ile. Early kinetic studies of the hydrolysis of acyl amino acids, dipeptides and their ester analogs by CPD A were characterized by varying degrees of substrate activation and inhibition making the determination of meaningful kinetic constants difficult (Auld & Vallee, 1987, and references therein). The use of tri- and tetrapeptides and their depsipeptide analogs obviated this problem (Auld & Vallee, 1970a).

Analysis of the individual $k_{cat}$ and $K_m$ profiles are supportive of a three protonation state model $EH_2 \rightleftharpoons EH \rightleftharpoons E$ (Auld & Vallee, 1970b). According to this scheme, when the enzyme is in its $EH_2$ form below the acidic $pK_{EH2}$ it can bind substrates but not hydrolyze them. The ionization of the group $EH_2$ with a $pK_a$ of 6.0 transforms the enzyme to its active form $EH$. Further ionization of the enzyme to the E form, occurring with a $pK_{EH}$ of 9.0, markedly reduces substrate binding and therefore catalysis. The enzyme is typically assayed in 1 M NaCl, 20–50 mM HEPES or Tris buffer at pH 7.5. The use of FA-L-Phe-L-Phe (FAPP) for routine assays has been described (Riordan & Holmquist, 1984).

## Structural Chemistry

The amino acid sequences of members of the carboxypeptidase A family (M14) are between about 60 and 80% conserved. The putative His and Glu ligands to the zinc, Cys residues that should form disulfides and other amino acids proposed to be important to specificity and catalysis in the CPD A family, e.g. Glu270, Arg145, Arg127, Tyr248 and Tyr198, are all conserved.

### Tertiary Structure

X-ray crystallographic analysis of cattle carboxypeptidase A and its substrate and transition state analog complexes in conjunction with amino acid sequences of several carboxypeptidases has led to the assignment of a number of active-site residues in catalysis (see Fig. 450.2). Thus Arg145 is postulated to be the site of interaction for the free $\alpha$-carboxyl group of the substrate and Glu270 is the principle catalytic moiety (Quiocho & Lipscomb, 1971; Hartsuck & Lipscomb, 1971; Rees *et al.*, 1983) assisted by Arg127 as an electrophile (Christianson & Lipscomb, 1989). Two conserved Tyr residues, Tyr198 and Tyr248, have been assigned roles in substrate binding. An unusual feature is the presence of a *cis* peptide bond between Ser197 and Tyr198 (Rees *et al.*, 1983). Since Tyr198 has been assigned a role in peptide binding and His196 is a ligand to the zinc, the energetics of the interaction of the substrate with Tyr198 may be directly relayed to the catalytic zinc.

The catalytic zinc site is comprised of His69 ($N^{\delta 1}$), Glu72 ($O^{\epsilon 1}$ and $O^{\epsilon 2}$), His196 ($N^{\delta 1}$) and a water molecule. The first two ligands, separated by a short spacer of two, are at the ends of a reverse turn in an $\alpha$ helix while His196 is the last residue in a $\beta$-pleated sheet extending from amino acids 191 to 196 (Rees *et al.*, 1983). This site is highly conserved throughout the extended carboxypeptidase family (Chapter 451) (Vallee & Auld, 1990).

X-ray diffraction analyses have included high-resolution studies of the interaction of the cross-linked crystals with a number of inhibitors. The earliest studies were performed on the very slowly hydrolyzed substrate Gly-L-Tyr, the potato inhibitor and several aldehydes and ketones at 1.7–2.5 Å (Christianson & Lipscomb, 1989, and references therein). More recently the structural studies have been devoted to a number of inhibitors that could be mimics of the transition state in peptide hydrolysis. These have included the phosphonates (Kaplan & Bartlett, 1991), Z-Phe-Val$^P$(O)Phe, Z-Ala-Gly$^P$(O)Phe, Z-Ala-Ala$^P$(O)Phe at 2.0 Å (where the scissile amide is replaced by PO) (Kim & Lipscomb, 1991) and the sulfodiimine, (L-(−)-2-carboxy-3-phenylpropyl)methylsulfodiimine (Mock & Tsay, 1989) at 2.25 Å (Cappalonga *et al.*, 1992). A common feature in all of these complexes is the interaction of Arg127 and Glu270 with atoms of the inhibitor equivalent to those of the expected tetrahedral intermediate in peptide hydrolysis (phosphonate oxygen or sulfoimine NH). In the case of the phosphonates, Arg127 interacts by a possible hydrogen bond with the oxygen (bond distance of 2.8–3.0 Å) that is coordinated to the zinc (bond distance of 1.9–2.2 Å) while the second oxygen forms a strong interaction with Glu270 (2.2–2.4 Å) and a weaker interaction with the zinc (3.0–3.2 Å) (Kim & Lipscomb, 1991). On the other hand the sulfodiimine forms a unidentate (N1) complex with the zinc (2.2 Å) and a possible hydrogen bond to Glu270 (N1, 3.0) while the N2 imine may be hydrogen bonded to Arg127 (3.2 Å; Cappalonga *et al.*, 1992).

### Mechanistic Studies

Being one of the first isolated zinc enzymes, cattle carboxypeptidase has been investigated by a great number of structure/function approaches. These have included high-resolution X-ray diffraction, NMR, X-ray absorption fine structure, XAFS, chemical modification and cryokinetic and cryospectroscopy techniques.

Cryospectrokinetics of zinc and cobalt cattle carboxypeptidase (CoCPD) at −20°C disclose two intermediates in the hydrolysis of dansylated tripeptides and their depsipeptide analogs and furnish all the rate and equilibrium constants for the reaction scheme $E + S \longleftrightarrow ES_1 \longleftrightarrow ES_2 \rightarrow E + P$ (Galdes *et al.*, 1983). The chemical and kinetic data indicate that neither of the ES intermediates is an acyl intermediate. Rapid-scanning visible spectroscopy at low temperatures indicates the $ES_2$ complexes are different for the tripeptides and their depsipeptide ester analogs (Geoghegan *et al.*, 1983b). Chemical quenching of $ES_2$ intermediates indicates that the scissile bond of peptides is not broken before the rate-determining step whereas the bond is broken in the corresponding ester analogs (Galdes *et al.*, 1986). Among the many differences noted in peptide and ester hydrolysis the different rate-determining steps may explain why the metal affects the $k_{cat}$ values for peptides and $K_m$ values for esters

(Auld, 1987; Auld & Vallee, 1987). The metal may function in the cleavage of the scissile bond of both peptides and esters. However, for esters this has occurred in the $K_m$ 'time domain' while for peptides it is the rate-determining step (Auld, 1987).

Active-site residues that have been proposed to participate in different aspects of catalysis are Glu270, Arg145, Arg127 and Tyr248. Several reasons led to the assignment of the apparent $pK_a$ that controls activity on the acid side of the profile, $pK_{EH2}$, to the ionization of the carboxyl group of Glu270 and its subsequent interaction with the water ligand of the zinc, stabilizing the active-site structure (EH$_2$ $\longleftrightarrow$ EH). It should be noted that ionization of the metal-bound water followed by an interaction with the protonated Glu270 residue would give the equivalent enzyme species. This assignment is consistent with crystallographic studies that show the interatomic distance between the zinc and the carboxylate is 4.5 Å, a distance consistent with a hydrogen bond between the metal-bound water molecule and Glu270 (Lipscomb, 1973). Chemical modification of Glu270 with 1-cyclohexyl-3-(2-morpholinoethyl)-carbodiimide metho-*p*-toluene sulfonate inactivates the enzyme (Nau & Riordan, 1975). Temperature-jump studies provide evidence for a conformational change coupled to the formation of the EH$_2$ form of the enzyme which might reflect movement of Glu270 away from the metal (French *et al.*, 1974). Spectrokinetic and NMR studies on CoCPD (Geogehegan *et al.*, 1983a; Bicknell *et al.*, 1988; Auld *et al.*, 1992) and X-ray absorption fine structure (XAFS) studies of ZnCPD (Zhang & Auld, 1995) demonstrate that the binding of anions to the EH$_2$ form is reasonable since the protonation of Glu270 breaks its interaction with the water allowing its displacement. On the other hand, the enzyme is inhibited by zinc hydroxide binding to the EH form of the enzyme (Larsen & Auld, 1989, 1991). This is also consistent with this model since the positive ZnOH$^+$ can bind to the negative Glu270 carboxylate oxygen and displace the metal-bound water forming a Zn-O-Zn bridge.

Chemical modification studies early on demonstrated that butanedione in borate buffer modifies a single Arg residue with concomitant loss of peptidase activity and increase in esterase activity (Riordan, 1973). The site of this modification has generally been believed to be Arg145. The C-terminal carboxyl group of Gly-L-Tyr and inhibitor analogs of peptides interacts with the guanidinium group of Arg145 (Quiocho & Lipscomb, 1971; Christianson & Lipscomb, 1989), suggesting that this is an anchoring group for substrate binding. More recently both crystallographic (Kim & Lipscomb, 1991; Cappalonga *et al.*, 1992) and mutagenic studies (Phillips *et al.*, 1990, 1992) have suggested a role for Arg127 in stabilizing the transition state in peptide hydrolysis. The crystallographic studies suggest a hydrogen bond is made between Arg127 and an oxygen of a putative tetrahedral intermediate (see section above). The mutagenesis studies have shown conversion of Arg127 to Lys, Met or Ala in the rat A1 enzyme all have profound effects on substrate catalysis and transition-state analog-inhibitor binding. Thus the $k_{cat}$ value for the hydrolysis of Z-Gly-Gly-Phe is reduced 50-, 1100- and 1700-fold for the Lys, Met and Ala replacements, respectively (Phillips *et al.*, 1990). The $K_m$ value is increased about 10-fold for all three mutants.

Moreover, there is a direct correlation in the increase in $K_i$ values for the potent phosphonate analogs of tripeptides (Kaplan & Bartlett, 1991) and the increase in $K_m/k_{cat}$ values for the corresponding peptides (Phillips *et al.*, 1992). Such behavior is suggestive of a role for Arg127 in the rate-determining step in catalysis, possibly by stabilizing the transition state.

Tyr248 was suggested to play a role in peptide catalysis by donating a proton from its phenolic group to the scissile peptide bond (Hartsuck & Lipscomb, 1971; Quiocho & Lipscomb, 1971). However, this role was ruled out by mutagenic studies on the rat A1 enzyme where the conversion of Tyr248Phe led to only a 3-fold decrease in $k_{cat}$ and 5-fold increase in $K_m$ for Z-Gly-Gly-Phe hydrolysis (Hilvert *et al.*, 1986). In addition the alkaline $pK_{EH}$ was not greatly influenced by the modification, suggesting that Tyr248 was not responsible for this $pK_a$. The pH dependence of $K_m$ for peptide hydrolysis does depend on the $pK_a$ of the phenolic group of *o*-nitro Tyr248 (Auld *et al.*, 1986). However, the $pK_a$ for Tyr248 in the native enzyme may be 10 or higher (Mock & Tsay, 1988; Larsen & Auld, 1989). Mutagenic studies on Tyr198Phe also demonstrate that its phenolic group is not obligatory for catalysis (Gardell *et al.*, 1987). A role in stabilizing the transition state was suggested. However, the effect on $k_{cat}$ was much less here than for the mutants of the Arg127.

The examination of the XAFS spectrum of the zinc enzyme over the pH range 7–11 reveals a near-edge feature that titrates with pH to give a $pK_a$ identical to the kinetic $pK_{EH}$ of 9.0 (Zhang & Auld, 1993). XAFS structural analysis indicates that the average first shell ligand distance decreases as the pH increases. Outer-shell analyses further indicate that the conformation of the histidine ligands remains unchanged over this pH region. Water is the one ligand which of course could ionize and whose distance from the zinc might be expected to decrease upon forming the anion hydroxide; thus favoring the assignment of the ionization of the metal-ligated water as the group responsible for $pK_{EH}$ (EH $\longleftrightarrow$ E).

Addition of azide to ZnCPD•L-Phe complex at pH 7 markedly changes the zinc coordination sphere from 4 N/O atoms at $2.021 \pm 0.006$ Å and $1.4 \pm 0.5$ N/O atoms at $2.54 \pm 0.5$ Å to 3.9 N/O atoms at $1.995 \pm 0.006$ Å (Zhang & Auld, 1995). The decrease in the average inner shell distance, $R$, of about 0.03 Å in both the Zn- and CoCPD•L-Phe•N$_3^-$ complexes is probably due to the ligand exchange from a neutral water to an anion. Outer-shell analysis agrees with this assignment. The XAFS spectra of the ternary Zn- and CoCPD•L-Phe•N$_3^-$ complexes are pH independent from 7 to 9 in agreement with the ionization of the water being the source of the spectral changes in the free enzyme and its binary L-Phe complex. The enzyme•azide•L-Phe complex is probably bound in a manner analogous to that expected for a post-transition state in a biproduct complex for peptide hydrolysis, i.e. the carboxylate anion of the peptide bound to the Zn and the protonated form of L-Phe hydrogen bonded to the catalytic Glu270 carboxylate (Auld, 1997).

The zinc-bound water therefore plays a role in both the acid and alkaline $pK_a$ values of carboxypeptidase A. It needs to be in its protonated state ([ZnL$_3$(H$_2$O)]$^+$) to be catalytically active. In this way it can poise the ionized carboxylate Glu270 for catalysis. Ionization of the water decreases the charge on the zinc, making it a poorer Lewis acid for interaction with the substrate. This would be an example of polarization-assisted catalysis (Vallee & Auld, 1990).

## Preparation

Three different forms of cattle carboxypeptidase A have been identified, $A_\alpha$, $A_\beta$ and $A_\gamma$ that consist of 307, 305 and 300 amino acids respectively (Neurath *et al.*, 1970). These enzymes arise from the action of proteases during the activation of the zymogen since the changes are all at the N-terminus of the protein. The form of the enzyme depends on the activation method (Pétra, 1970). Thus the enzyme prepared according to the procedure of Cox *et al.* (1964) is predominately $A_\alpha$, while that prepared according to the methods of Allen and Anson are mainly $A_\beta$ and $A_\gamma$, respectively (Anson, 1937; Allen *et al.*, 1964). In addition, Neurath and collaborators have shown that each of these enzymes exists in two distinct forms that differ in three positions, Val/Leu305, Ile/Val179 and Ala/Glu228 (Pétra *et al.*, 1969). The 'Val' and 'Leu' allotypic forms have remarkably different physical properties but their kinetic characteristics are virtually identical (Pétra, 1970).

Serine protease activity has been detected in the preparations of 3x-crystallized CPD A and in PMSF-treated preparations (Bicknell *et al.*, 1985). The use of such preparations of CPD A can lead to artifacts in mechanistic and sequencing studies. Affinity purification of the enzyme on [*N*-(*ε*-aminocaproyl)-*p*-aminobenzyl]succinyl-Sepharose 4B can remove these enzymes (Bazzone *et al.*, 1979).

Human procarboxypeptidase A1 has been expressed in *Saccharomyces cerevisiae* (Laethem *et al.*, 1996). The zymogen was purified by a combination of Q Sepharose and S-200 Sephadex chromatography. Trypsin activation followed by chromatography on a Sephadex G-50 column yielded pure enzyme.

## Biological Aspects

Zinc-containing exopeptidases such as carboxypeptidase A and B are important in the degradation of food proteins leading to the formation of amino acids. The protein macromolecules must first be broken down to smaller polypeptides by the action of endoproteases before the exopeptidases can act. Pepsin (Chapter 272), trypsin (Chapter 3) and chymotrypsin (Chapter 8) serve this purpose. Carboxypeptidase B complements the specificity of trypsin by liberating Lys and Arg. In the adult human, lysine is an essential amino acid since it can not be synthesized *de novo* from nonprotein sources (Riordan, 1974). Carboxypeptidase A complements the actions of chymotrypsin and pepsin by allowing the production of essential amino acids such as Phe and Trp.

Proteases are generally stored in an inactive precursor form called a zymogen and activated by proteolytic processing of the N-terminal peptide when their activity is required by the organism. Carboxypeptidases are secreted as zymogens from acinar pancreatic cells and activated by trypsin in the duodenum. A 94 amino acid peptide is lost from the N-terminus of proCPD A during the activation process (Vendrell *et al.*, 1992). Both trypsin and the released carboxypeptidase A participate in the degradation of the severed activation regions.

In addition to the pancreatic forms of carboxypeptidase from human, cattle, crayfish, rodent and pig sources (Auld & Vallee, 1987) carboxypeptidase activity has been identified in such diverse sources as orange peel, yeast, fungi, molds, barley, spleen, kidney and connective tissue (Riordan, 1974). Several new carboxypeptidases have been identified by gene sequencing and their activity in various tissues. Thus the human mast cell (Chapter 454), E (Chapter 458), M (Chapter 460), and lysine (Chapter 459) carboxypeptidases are believed to be involved in immune/inflammatory and hormone processing (Vallee & Auld, 1990, and references therein).

## Further Reading

Since carboxypeptidase A is one of the most-studied enzymes by many different types of structure–function analyses, it is not surprising that no single review can do justice to the entire picture. For overviews of the crystallographic studies the reader is directed to the early review by Hartsuck & Lipscomb (1971) and the more recent one by Christianson & Lipscomb (1989). The use of chemical modification techniques, visible spectroscopy and kinetics is covered in the reviews by Vallee *et al.* (1983) and Auld & Vallee (1987). For the use of cryokinetics and XAFS in mechanistic studies of the enzyme the reader is directed to the reviews by Auld (1987 and 1997, respectively).

## References

Allen, B.J., Keller, P.J. & Neurath, H. (1964) Procedures for the isolation of crystalline bovine pancreatic carboxypeptidase A. Isolation from acetone powders of pancreas glands. *Biochemistry* **3**, 40–43.

Anson, M.L. (1937) The preparation of crystalline carboxypeptidase. *J. Gen. Physiol.* **20**, 663–669.

Auld, D.S. (1987) Acyl group transfer-metalloproteinases. In: *Enzyme Mechanisms* (Page, M. & Williams, A., eds). London: Royal Society of Chemistry, pp. 240–258.

Auld, D.S. (1997) Zinc catalysis in metalloproteases. In: *Metal Sites in Proteins and Models: Phosphatases, Lewis Acids and Vanadium* (Hill, H.A.O., Sadler, P.H. & Thomson, A.J., eds). Heidelberg: Springer-Verlag, pp. 29–50.

Auld, D.S. & Vallee, B.L. (1970a) Kinetics of carboxypeptidase A. II. Inhibitors of the hydrolysis of oligopeptides. *Biochemistry* **9**, 602–609.

Auld, D.S. & Vallee, B.L. (1970b) Kinetics of carboxypeptidase A. The pH dependence of tripeptide hydrolysis catalyzed by zinc, cobalt and manganese enzymes. *Biochemistry* **9**, 4352–4359.

Auld, D.S. & Vallee, B.L. (1987) Carboxypeptidase A. In: *Hydrolytic Enzymes* (Neuberger, A. & Brocklehurst, K., eds). Amsterdam: Elsevier, pp. 201–255.

Auld, D.S., Larsen, K.S. & Vallee, B.L. (1986) Active site residues of carboxypeptidase A. In: *Zinc Enzymes* (Bertini, I., Luchinat, C., Maret, W. & Zeppezauer, M., eds). Boston: Birkhauser, pp. 133–154.

Auld, D.S, Bertini, B., Donaire, A., Messori, L. & Moratal, J.M. (1992) pH-dependent properties of cobalt(II) carboxypeptidase A-inhibitor complexes. *Biochemistry* **31**, 3840–3846.

Bazzone, T.J., Sokolovsky, M., Cueni, L.B. & Vallee, B.L. (1979) Single step isolation and resolution of pancreatic carboxypeptidase A and B. *Biochemistry* **18**, 4362–4366.

Bicknell, R., Schäffer, A., Auld, D.S., Riordan, J.F., Monnanni, R. & Bertini, I. (1985) Protease susceptibility of zinc- and apo-carboxypeptidase A. *Biochem. Biophys. Res. Commun.* **133**, 787–793.

Bicknell, R., Schäffer, A., Bertini, I., Luchinat, C., Vallee, B.L. & Auld, D.S. (1988) Interactions of anions with the active site of carboxypeptidase A. *Biochemistry* **27**, 1050–1056.

Cappalonga, A.M., Alexander, R.S. & Christianson, D.W. (1992) Structural comparison of sulfodiimine and sulfonamide inhibitors in their complexes with zinc enzymes. *J. Biol. Chem.* **267**, 19192–19197.

Christianson, D.W. & Lipscomb, W.N. (1989) Carboxypeptidase A. *Acct. Chem. Res.* **22**, 62–69.

Cox, D.J., Bovard, F.C., Bargetzi, J.-P., Walsh, K.A. & Neurath, H. (1964) Procedures for the isolation of crystalline bovine pancreatic carboxypeptidase A. II. Isolation of carboxypeptidase A from procarboxypeptidase A. *Biochemistry* **3**, 44–48.

French, T.C., Yu, N.-T. & Auld, D.S. (1974) Relaxation spectra of proteinases. Isomerizations of carboxypeptidase A (Cox) and (Anson). *Biochemistry* **13**, 2877–2882.

Galdes, A., Auld, D.S. & Vallee, B.L. (1983) Cryokinetic studies of the intermediates in the mechanism of carboxypeptidase A. *Biochemistry* **22**, 1888–1893.

Galdes, A., Auld, D.S. & Vallee, B.L. (1986) Elucidation of the chemical nature of the steady-state intermediates in the mechanism of carboxypeptidase A. *Biochemistry* **25**, 646–651.

Gardell, S.J., Hilvert, D., Barnett, J., Kaiser, E.T. & Rutter, W.J. (1987) Use of directed mutagenesis to probe the role of tyrosine 198 in the catalytic mechanism of carboxypeptidase A. *J. Biol. Chem.* **262**, 576–582.

Geoghegan, K.F., Holmquist, B., Spilburg, C.A. & Vallee, B.L (1983a) Spectral properties of cobalt carboxypeptidase A. Interaction of the metal atom with anions. *Biochemistry* **22**, 1847–1852.

Geoghegan, K.F., Galdes, A., Martinelli, R.A., Holmquist, B., Auld, D.S. & Vallee, B.L. (1983b) Cryospectroscopy of intermediates in the mechanism of carboxypeptidase A. *Biochemistry* **22**, 2255–2262.

Hartsuck, J.A. & Lipscomb, W.N. (1971) Carboxypeptidase A. In: *The Enzymes*, 3rd edn, vol. 3 (Boyer, P., ed.). New York: Academic Press, pp. 1–56.

Hilvert, D., Gardell, S.J., Rutter, W.J. & Kaiser, E.T. (1986) Evidence against a crucial role for the phenolic hydroxyl of Tyr-248 in peptide and ester hydrolysis catalyzed by carboxypeptidase A: comparative studies of the pH dependencies of the native and Phe-248-mutant forms. *J. Am. Chem. Soc.* **108**, 5298–5304.

Kaplan, A.P. & Bartlett, P.A. (1991) Synthesis and evaluation of an inhibitor of carboxypeptidase A with a $K_i$ value in the femtomolar range. *Biochemistry* **30**, 8165–8170.

Kim, H. & Lipscomb, W.N. (1991) Comparison of the structures of three carboxypeptidase A-phosphonate complexes determined by x-ray crystallography. *Biochemistry* **30**, 8171–8180.

Laethem, R.M., Blumenkopf, T.A., Corey, M., Elwell, L., Moxham, C.P., Ray, P.H., Walton, L.M. & Smith, G.K. (1996) Expression and characterization of human pancreatic prepro-carboxypeptidase A1 and preprocarboxypeptidase A2. *Arch. Biochem. Biophys.* **332**, 8–18.

Larsen, K.S. & Auld, D.S. (1989) Carboxypeptidase A: mechanism of zinc inhibition. *Biochemistry* **28**, 9620–9625.

Larsen, K.S. & Auld, D.S. (1991) Characterization of an inhibitory metal binding site in carboxypeptidase A. *Biochemistry* **30**, 2614–2618.

Lipscomb, W.N. (1973) Enzymatic activities of carboxypeptidase As in solution and in crystals. *Proc. Natl Acad. Sci. USA* **70**, 3797–3801.

Mock, W.L. & Tsay, J.-T. (1989) Sulfoximine and sulfodiimine transition-state analogue inhibitors for carboxypeptidase A. *J. Am. Chem. Soc.* **111**, 4467–4472.

Nau, H. & Riordan, J.F. (1975) Gas-chromatography-mass spectrometry for probing the structure and mechanism of action of enzyme active sites. The role of Glu-270 in carboxypeptidase A. *Biochemistry* **14**, 5285–5294.

Neurath, H., Bradshaw, R.A., Petra, P.H. & Walsh, K.A. (1970) Bovine carboxypeptidase A-activation, chemical structure and molecular heterogeneity. *Philos. Trans. R. Soc. Lond. B* **257**, 159–176.

Pétra, P.H. (1970) Bovine procarboxypeptidase and carboxypeptidase A. *Methods Enzymol.* **19**, 460–503.

Pétra, P.H., Bradshaw, R.A., Walsh, K.A. & Neurath, H. (1969) Identification of the amino acid replacements characterizing the allotypic forms of bovine carboxypeptidase A. *Biochemistry* **8**, 2762–2768.

Phillips, M.A., Fletterick, R. & Rutter, W.J. (1990) Arginine 127 stabilizes the transition state in carboxypeptidase. *J. Biol. Chem.* **265**, 20692–20698.

Phillips, M.A., Kaplan, A.P., Rutter, W.J. & Bartlett, P.A. (1992) A new approach combining inhibitor analogues and variation in enzyme structure. *Biochemistry* **31**, 959–963.

Quiocho, F.A. & Lipscomb, W.N. (1971) Carboxypeptidase A: a protein and an enzyme. *Adv. Protein Chem.* **25**, 1–78.

Rees, D.C., Lewis, M. & Lipscomb, W.N. (1983) Refined crystal structure of carboxypeptidase A at 1.54 Å resolution. *J. Mol. Biol.* **168**, 367–387.

Riordan, J.F. (1973) Functional arginyl residues in carboxypeptidase A. Modification with butanedione. *Biochemistry* **12**, 3915–3923.

Riordan, J.F. (1974) Metal-containing exopeptidases. In: *Food Related Enzymes, Adv. Chem. Ser.*, vol. 136 (Whitaker, J.R., ed.). Washington, DC: Amer. Chem. Soc., pp. 220–240.

Riordan, J.F. & Holmquist, B. (1984) Carboxypeptidase A. In: *Methods of Enzymatic Analysis*, 3rd edn, vol. V: *Peptidases, Proteinases and Their Inhibitors* (Bergmeyer, H., ed.). Weinheim: Verlag Chemie, pp. 43–60.

Vallee, B.L. & Auld, D.S. (1990) Zinc coordination, function, and structure of zinc enzymes and other proteins. *Biochemistry* **29**, 5647–5659.

Vallee, B.L. & Neurath, H. (1954) Carboxypeptidase, a zinc metalloprotein. *J. Am. Chem. Soc.* **76**, 5006.

Vallee, B.L., Galdes, A., Auld, D.S. & Riordan, J.F. (1983) Carboxypeptidase A. In: *Zinc Enzymes* (Spiro, T.G., ed.). New York: John Wiley, pp. 26–75.

Vendrell, J., Guasch, A., Coll, M., Villegas, V., Billeter, M., Wider, G., Huber, R., Wüthrich, K. & Avilés, F.X. (1992) Pancreatic procarboxypeptidases: their activation processes related to the structural features of the zymogens and activation segments. *Biol. Chem. Hoppe-Seyler* **373**, 387–392.

**M**

Zhang, K. & Auld, D.S. (1993) XAFS studies of carboxypeptidase A: detection of a structural alteration in the zinc coordination sphere coupled to the catalytically important alkaline pKa. *Biochemistry* **32**, 13844–13851.

Zhang, K. & Auld, D.S. (1995) Structures of binary and ternary inhibitor complexes of native zinc carboxypeptidase A determined by XAFS. *Biochemistry* **34**, 16306–16312.

*David S. Auld*
*Department of Pathology &*
*Center for Biochemical, Biophysical Sciences and Medicine,*
*Harvard Medical School, Boston, MA 02115, USA*
*Email: dauld@warren.med.harvard.edu*

# 452. Carboxypeptidase A2

## Databanks

*Peptidase classification: clan MC, family M14, MEROPS ID: M14.002*
*NC-IUBMB enzyme classification: EC 3.4.17.15*
Databank codes:

| Species | SwissProt | PIR | EMBL (cDNA) | EMBL (genomic) |
|---|---|---|---|---|
| *Caenorhabditis elegans* | – | – | Z71267 | – |
| *Homo sapiens* | P48052 | A56171 S02809 | U19977 | – |
| *Rattus norvegicus* | P19222 | A32128 | – | M23714: exon 1 |
| | | | | M23715: exon 2 |
| | | | | M23716: exons 3 and 4 |
| | | | | M23717: exon 5 |
| | | | | M23718: exon 6 |
| | | | | M23719: exons 7, 8 and 9 |
| | | | | M23720: exon 10 |
| | | | | M23721: exon 11 and complete CDS |

## Name and History

Until recently, just two pancreatic carboxypeptidases were recognized, A and B. Carboxypeptidase A (CPD A) prefers peptide and protein substrates with an aromatic or branched-chain C-terminal while carboxypeptidase B (CPD B) prefers basic side chain amino acids such as Arg and Lys. The discovery of three carboxypeptidase genes in the rat and the characterization of their gene products led to a suggested change in the nomenclature (Gardell *et al*., 1988; Clauser *et al*., 1988). Since two of the proteins possessed activity towards branched-chained amino acids they were termed A1 (see Chapter 451) and A2 while the third displayed the specificity typical of the B enzyme (Chapter 455). There is no known counterpart to the rat and human *carboxypeptidase A2* enzymes found in cattle.

## Activity and Specificity

Only cursory studies of the substrate specificity of the CPD A2 enzymes have been performed. The most extensive studies were performed on the rat A2 enzyme (Gardell *et al*., 1988). The kinetic parameters for the hydrolysis of some carbobenzoxy di- and tripeptides catalyzed by rat and cattle CPD A were compared to those for rat CPD A2. The use of dipeptides for obtaining kinetic parameters for the cattle A enzyme is difficult due to bi- and triphasic kinetic plots while tripeptides generally obey Michaelis–Menten kinetics (Auld & Vallee, 1970). No information was given on the presence or absence of such behavior in the rat A2 enzyme or over what substrate range the kinetic parameters were obtained (Gardell *et al*., 1988). If one compares the results on the two tripeptides, Z-Gly-Gly┼Phe and Z-Gly-Gly┼Leu, the decrease in $k_{cat}/K_m$ values going from Phe to Leu is 14-, 6.0- and 137-fold for the cattle A, rat A1 and rat A2 enzymes (Gardell *et al*., 1988). In addition the rat A1 enzyme displays negligible activity towards Z-Gly┼Trp while the rat A2 enzyme catalyzes its hydrolysis about 10 times better than the cattle enzyme. The combination of these results has led to the suggestion that the A2 enzyme should be more effective than the A1 enzymes towards substrates containing C-terminal amino acids with bulky side chains. A recent study of the human A1- and A2-catalyzed hydrolysis of methotrexate-α-(1-naphthyl)alanine is in agreement with this deduction. The $k_{cat}/K_m$ value for the A2 enzyme is 1000-fold greater than for the A1 enzyme (Laethem *et al.*, 1996). In general, the substrate preference of the rat A1 enzyme is skewed towards smaller amino acids while the A2 enzyme prefers bulkier amino acids when compared to the cattle A enzyme (Gardell *et al.*, 1988). Carboxypeptidases are typically assayed in 1 M NaCl, 20–50 mM HEPES or Tris buffer at pH 7.5. The use of FA-L-Phe-L-Phe for routine assays has been described (Riordan & Holmquist, 1984).

## Structural Chemistry

The amino acid sequences of cattle CPD A (Chapter 451) and rat CPD A2 are 62% identical while rat CPD A1 and A2 enzymes are 63% identical (Gardell *et al.*, 1988). All Cys residues that are involved in disulfide formation in cattle CPD A are conserved in the rat A2 CPD. In addition the ligands to the zinc and other amino acids proposed to be important to specificity and catalysis in the CPD A family (e.g. Glu270, Arg145, Arg127, Tyr248, Tyr198 and Ile255; numbered according to mature cattle carboxypeptidase A) are conserved in CPD A2. The presence of an Ile or Asp residue is believed to reflect the differences in specificity of the A and B groups of carboxypeptidases (Titani *et al.*, 1975).

The structure of rat carboxypeptidase A2 has been reported at a resolution of 1.9 Å (Faming *et al.*, 1991). In addition, the sequence of the human CPD A2 that shows an 89% identity to rat CPD A1 enzyme has been modeled to the three-dimensional structures of the cattle and pig CPD A enzymes (Catasús *et al.*, 1995). The catalytic zinc-binding site characteristic of the carboxypeptidase family (Vallee & Auld, 1990), consisting of a His and Glu residue separated by two amino acids and a third His 123 amino acids C-terminal, is conserved in the CPD A2 enzymes. The conformation of the zinc ligands and the proposed catalytic residue Glu270 are very similar in rat CPD A2 and cattle CPD A (Rees *et al.*, 1983; Faming *et al.*, 1991). Conformational differences in the binding residues Asn144, Arg145, Arg127 occur in the side chains only. One marked difference is the position of Tyr248. The phenolic ring of Tyr248 in the rat CPD A2 is in a conformation similar to that found in the cattle CPD A•glycyl-L-tyrosine complex (Quiocho & Lipscomb, 1971). As such, the phenolic group would need to rotate out of the active site in order for the substrate to bind. This structure is in agreement with stopped-flow and temperature jump studies on cattle arsanilazotyrosine-248 CPD A in solution (Harrison *et al.*, 1975). The stopped-flow studies indicated that Tyr248 was in the active site in the resting state but was rapidly displaced by substrates.

Cattle CPD A and the rat and human CPD A2 differ by only two amino acid replacements in the binding cavity, only one of which should directly affect substrate binding. The amino acid replacements, Thr268Ala and Leu203Met, both belong to the core β sheet (Faming *et al.*, 1991). In addition, a new disulfide between cysteines 210 and 244 is present in the rat enzyme. This disulfide may stabilize the conformation of a surface loop of the binding pocket. The differences in the active-site surface loop together with the replacement of Thr268Ala in the core β sheet appear to enlarge the specificity pocket relative to that of the cattle A enzyme (Faming *et al.*, 1991). The change in amino acid stereochemistry in the binding pocket is consistent with the preference of the CPD A2 enzyme for substrates having C-terminal amino acids with bulky side chains such as Trp.

## Preparation

The rat A2, A1 and B enzymes have been isolated from water-soluble acetone powders of rat pancreas (Gardell *et al.*, 1988). The extract was applied to an affinity column containing immobilized glycyl-L-tyrosyl-azo-benzyl-succinate (Cueni *et al.*, 1980). Stepwise elution with increasing concentrations of NaCl resolved the three different carboxypeptidases. Further chromatography on a Mono-Q anion-exchange column purified the proteins to homogeneity as judged by SDS-PAGE. Carboxypeptidases in general have been affinity purified on [*N*-(ε-aminocaproyl)-*p*-aminobenzyl]succinyl-Sepharose 4B (CABS-Sepharose) (Bazzone *et al.*, 1979; Cueni *et al.*, 1980). Human CPD A2 has been purified from an aqueous extract of human pancreatic powder by HPLC chromatography using a DEAE anion-exchange column (Pascual *et al.*, 1989). The human procarboxypeptidase A2 has been expressed in *Saccharomyces cerevisiae* (Laethem *et al.*, 1996). The zymogen was partially purified by a combination of phenyl-Sepharose and Mono-Q chromatography. Trypsin activation followed by chromatography on Mono-Q and Superose-12 columns yielded pure enzyme.

## Biological Aspects

Carboxypeptidases are secreted from pancreatic acinar cells as zymogens that are activated by trypsin (Chapter 3) in the intestines. Of the 19 proteins identified in human pancreatic juice by two-dimensional isoelectric focusing and SDS-PAGE, two were CPD B and two were CPD A forms (Scheele *et al.*, 1981). Evolutionary tree analysis of the amino acid sequences of the cattle and rat carboxypeptidases has been used to delineate the genealogy of the pancreatic carboxypeptidase family (Gardell *et al.*, 1988). These studies suggest that a gene duplication event gave rise to the A2 enzyme subsequent to the divergence of the CPD B lineage (see also Fig. 450.4).

## Distinguishing Features

The human A1 and A2 enzymes show clear differences in electrophoretic mobility in SDS-PAGE gels, isoelectric points (4.9 and 5.1, respectively) and rate of thermal inactivation (A2 much more stable) and the kinetics of the proteolytic activation process (Pascual *et al.*, 1989).

## References

Auld, D.S. & Vallee, B.L. (1970) Kinetics of carboxypeptidase A. II. Inhibitors of the hydrolysis of oligopeptides. *Biochemistry* **9**, 602–609.

Bazzone, T.J., Sokolovsky, M., Cueni, L.B. & Vallee, B.L. (1979) Single step isolation and resolution of pancreatic carboxypeptidase A and B. *Biochemistry* **18**, 4362–4366.

**M**

Catasús, L., Vendrell, J., Avilés, F.X., Carreira, S., Puigserver, A. & Billeter, M. (1995) The sequence and conformation of human pancreatic procarboxypeptidase A2. *J. Biol. Chem.* **270**, 6651–6657.

Clauser, E., Gardell, S.J., Craik, C.S., MacDonald, R.J. & Rutter, W.J. (1988) Structural characterization of the rat carboxypeptidase A1 and B genes. Comparative analysis of the rat carboxypeptidase gene family. *J. Biol. Chem.* **263**, 17837–17845.

Cueni, L.B., Bazzone, T.J., Riordan, J.F. & Vallee, B.L. (1980) Affinity chromatographic sorting of carboxypeptidase A and its chemically modified derivatives. *Anal. Biochem.* **107**, 341–349.

Faming, Z., Kobe, B., Stewart, C-B., Rutter, W.J. & Goldsmith, E.J. (1991) Structural evolution of an enzyme specificity. The structure of rat carboxypeptidase A2 at 1.9 Å resolution. *J. Biol. Chem.* **266**, 24606–24612.

Gardell, S.J., Craik, C.S., Clauser, E., Goldsmith, E.J., Stewart, C.-B., Graff, M. & Rutter, W.J. (1988) A novel rat carboxypeptidase, CPA2: characterization, molecular cloning, and evolutionary implications on substrate specificity in the carboxypeptidase gene family. *J. Biol. Chem.* **263**, 17828–17836.

Harrison, L.W., Auld, D.S. & Vallee, B.L. (1975) Intramolecular arsanilazotyrosine-248 Zn complex of carboxypeptidase A: a monitor of catalytic events. *Proc. Natl Acad. Sci. USA* **72**, 3930–3933.

Laethem, R.M., Blumenkopf, T.A., Corey, M., Elwell, L., Moxham, C.P., Ray, P.H., Walton, L.M. & Smith, G.K. (1996) Expression and characterization of human pancreatic prepro-carboxypeptidase

A1 and preprocarboxypeptidase A2. *Arch. Biochem. Biophys.* **332**, 8–18.

Pascual, R., Burgos, F.J., Salva, M., Soriano, F., Mendez, E. & Avilés, F.X. (1989) Purification and properties of five different forms of human procarboxypeptidases. *Eur. J. Biochem.* **179**, 609–616.

Quiocho, F.A. & Lipscomb, W.N. (1971) Carboxypeptidase A: a protein and an enzyme. *Adv. Protein Chem.* **25**, 1–78.

Rees, D.C., Lewis, M. & Lipscomb, W.N. (1983) Refined crystal structure of carboxypeptidase A at 1.54 Å resolution. *J. Mol. Biol.* **168**, 367–387.

Riordan, J.F. & Holmquist, B. (1984) Carboxypeptidase A. In: *Methods of Enzymatic Analysis*, 3rd edn, vol. V: *Peptidases, Proteinases and Their Inhibitors* (Bergmeyer, H., ed.). Berlin: Verlag Chemie, pp. 43–60.

Scheele, G., Bartelt, D. & Bieger, W. (1981) Characterization of human exocrine pancreatic proteins by two dimensional isoelectric focusing/sodium dodecyl sulfate gel electrophoresis. *Gastroenterology* **80**, 461–473.

Titani, K., Ericsson, L.H., Walsh, K.A. & Neurath, H. (1975) Amino acid sequence of bovine carboxypeptidase B. *Proc. Natl Acad. Sci. USA* **72**, 1666–1670.

Vallee, B.L. & Auld, D.S. (1990) Zinc coordination, function, and structure of zinc enzymes and other proteins. *Biochemistry* **29**, 5647–5659.

*David S. Auld*
*Department of Pathology & Center for Biochemical,*
*Biophysical Sciences and Medicine,*
*Harvard Medical School,*
*Boston, MA 02115, USA*
*Email: dauld@warren.med.harvard.edu*

# 453. *Carboxypeptidase U*

## Databanks

*Peptidase classification: clan MC, family M14, MEROPS ID: M14.009*
*NC-IUBMB enzyme classification: EC 3.4.17.20*
*Databank codes:*

| Species | SwissProt | PIR | EMBL (cDNA) | EMBL (genomic) |
|---|---|---|---|---|
| *Homo sapiens* | – | – | M75106 | – |

## Name and History

Hendriks & coworkers (1988, 1989a,b) reported on the presence of a labile carboxypeptidase activity in fresh human serum which interfered with the assay of lysine carboxypeptidase (Chapter 459). This novel carboxypeptidase activity was undetectable in human plasma, but its activity appeared upon coagulation of blood. This unstable basic carboxypeptidase activity was separated from plasma lysine carboxypeptidase by means of ion-exchange chromatography. Due to the marked instability of this novel enzyme, they named it **carboxypeptidase U** (CPU, where 'U' indicates **u**nstable) (Hendriks *et al*., 1989a,b, 1990). CPU was characterized as a carboxypeptidase B-like entity able to cleave off basic C-terminal amino acids arginine and lysine. Campbell & coworkers (1989, 1990) confirmed these findings and called the enzyme **carboxypeptidase R** (CPR, where 'R' stands for arginine). Then Eaton *et al*. (1991) discovered it as a contaminant in preparations of $\alpha_2$-antiplasmin. They cloned the CPU cDNA from human liver, deduced the amino acid sequence and describe the activation of the proenzyme to its active form. They designated the proenzyme **pCPB (plasma carboxypeptidase B)**, since the active enzyme behaves very much like pancreatic carboxypeptidase B (Chapter 455) in terms of enzyme activity. Later on, they changed the name to **plasma procarboxypeptidase B (pro-pCPB)** and retained the name pCPB for the active form of the enzyme (Tan & Eaton, 1995). These names, however, are rather confusing, since plasma carboxypeptidase B has been used already for a long time as a synonym for plasma lysine carboxypeptidase (Skidgel, 1988).

Wang *et al*. (1994) described the identity between procarboxypeptidase U (proCPU) and plasma procarboxypeptidase B. In addition, the group of Bajzar *et al*. (1995) independently found this enzyme and showed that it can also be activated by thrombin (Chapter 55), and that upon activation, it can inhibit fibrinolysis. Consequently, they named it **TAFI (thrombin activatable fibrinolysis inhibitor)**. Another synonym recently used for this enzyme is '**inducible carboxypeptidase activity**', since it circulates in the bloodstream as an inactive precursor and its activity can be 'induced' during coagulation (Redlitz *et al*., 1996). Another designation is **arginine carboxypeptidase**.

## Activity and Specificity

CPU specifically removes C-terminal basic amino acids arginine and lysine from a number of peptides. Its activity has been demonstrated on the small synthetic peptides Bz-Gly+Lys, Bz-Gly+Arg, FA-Ala+Lys, FA-Ala+Arg (Campbell & Okada, 1989; Hendriks *et al*., 1989b, 1990; Tan & Eaton, 1995), as well as on a number of hexapeptide enkephalins and on bradykinin (Hendriks *et al*., 1992; Shinohara *et al*., 1994; Tan & Eaton, 1995).

The pH optimum is between pH 7.5 and 7.8 (Hendriks *et al*., 1989b). Activity is stimulated by $Cd^{2+}$ (Campbell & Okada, 1989) but inhibited by $Co^{2+}$ (Hendriks *et al*., 1989b). CPU activity is inhibited by 1,10-phenanthroline, EDTA, 2-ME, the lysine carboxypeptidase inhibitors GEMSA (guanidinoethyl-mercaptosuccinic acid) and MERGETPA (D,L-2-mercaptomethyl-3-guanidino-ethylthiopropanoic acid) and potato carboxypeptidase inhibitor (Eaton *et al*., 1991; Hendriks *et al*., 1989b, 1990; Redlitz *et al*., 1995; Tan & Eaton, 1995; Wang *et al*., 1994). Assays are most conveniently performed with the substrate Bz-Gly+Arg either in a spectrophotometric assay (Bajzar *et al*., 1995) or in an HPLC-assisted assay (Hendriks *et al*., 1985).

## Structural Chemistry

The primary structure of proCPU has been deduced from a cDNA sequence. The primary translation product consists of 423 amino acids containing a 22 amino acid signal peptide, a 92 amino acid activation peptide and a 309 amino acid catalytic domain (Eaton *et al*., 1991). ProCPU has a molecular mass of 60 kDa (SDS-PAGE). Proteolytic cleavage of the 92 amino acid activation peptide, which is heavily glycosylated, results in the formation of the nonglycosylated 36 kDa catalytic unit. Further cleavage at Arg330 inactivates CPU (Tan & Eaton, 1995).

CPU exhibits high sequence similarity (around 50% identity) to both pancreatic carboxypeptidases A and B and thus is presumed to have a fold similar to that of the well-known carboxypeptidase A structure (Fig. 450.2): a zinc-containing metalloenzyme having a central eight-strand $\beta$ sheet over which eight $\alpha$ helices pack on both sides to form a globular molecule. Amino acids important in the catalytic mechanism, substrate binding and coordination of the active-site zinc atom of these carboxypeptidases are conserved in CPU. The pI for the human CPU is 5.0 (Eaton *et al*., 1991).

## Preparation

Procarboxypeptidase U is synthesized in the liver and released as a glycosylated zymogen into the circulation (Tan & Eaton, 1995). ProCPU has been purified to homogeneity from human plasma (14 300-fold) using affinity chromatography on plasminogen-Sepharose as a key step in the purification scheme (Eaton *et al*., 1991; Wang *et al*., 1994; Bajzar *et al*., 1995). It is present in plasma at a concentration of 2–5 mg liter$^{-1}$. CPU activity is undetectable in human blood cells (Hendriks *et al*., 1989b). The proCPU/CPU system also has been described in canine plasma (Redlitz *et al*., 1996).

## Biological Aspects

It was postulated by Hendriks *et al*. (1990) that CPU, generated during the coagulation cascade, could play a role in the fibrinolytic system, where it could act on fibrin partially degraded by plasmin. This hypothesis was substantiated when it was demonstrated that proCPU exhibits a strong affinity for plasminogen and that it can be activated by plasmin (Chapter 59) and thrombin (Chapter 55) (Eaton *et al*., 1991). The thrombin-thrombomodulin complex is the most likely physiological activator, since thrombomodulin increases the catalytic efficiency of CPU activation by thrombin by a factor of 1250 (Bajzar *et al*., 1996). Plasmin degradation of fibrin exposes C-terminal lysine residues which are essential for the high-affinity binding of plasminogen to fibrin. Thus, activated CPU at the site could control the rate of fibrinolysis by cleaving off some of the C-terminal lysine residues. Recent data indicate that CPU could indeed play a role in plasminogen activation since it is able to delay t-plasminogen activator-induced clot lysis *in vitro* (Bajzar *et al*., 1995, 1996; Redlitz *et al*., 1995).

Another possible function of proCPU could be in the regulation of peptide hormone activity, as described for other human basic carboxypeptidases (Skidgel *et al*., 1988) (see Chapters 458 and 459). An intricate interrelationship exists between the processes of coagulation, fibrinolysis, complement activation and the kallikrein/kinin pathway. ProCPU could be activated at the exact site where a lot of its potential substrates are being released during the activation of these different cascades. CPU could act on bradykinin (transforming it to des-Arg$^9$-bradykinin) (Hendriks *et al*., 1992; Shinohara *et al*., 1994), on released anaphylatoxins C3a and C5a, or on $\alpha_2$-antiplasmin.

The gene locus of proCPU is 13q14.11 (Tsai & Drayna, 1992; Vanhoof *et al.*, 1996).

## Further Reading

See Skidgel (1988) for an extensive overview of other basic carboxypeptidases.

## References

Bajzar, L., Manuel, R. & Nesheim, M. (1995) Purification and characterization of TAFI, a thrombin-activatable fibrinolysis inhibitor. *J. Biol. Chem.* **270**, 14477–14484.

Bajzar, L., Morser, J. & Nesheim, M. (1996) TAFI, or plasma procarboxypeptidase B, couples the coagulation and fibrinolytic cascades through the thrombin-thrombomodulin complex. *J. Biol. Chem.* **271**, 16603–16608.

Campbell, W. & Okada, H. (1989) An arginine specific carboxypeptidase generated in blood during coagulation or inflammation which is unrelated to carboxypeptidase N or its subunits. *Biochem. Biophys. Res. Commun.* **162**, 933–939.

Campbell, W., Yonezu, K., Shinohara, T. & Okada, H. (1990) An arginine carboxypeptidase generated during coagulation is diminished or absent in patients with rheumatoid arthritis. *J. Lab. Clin. Med.* **115**, 610–612.

Eaton, D.L., Malloy, B.E., Tsai, S.P., Henzel, W. & Drayna, D. (1991) Isolation, molecular cloning, and partial characterization of a novel carboxypeptidase B from human plasma. *J. Biol. Chem.* **266**, 21833–21838.

Hendriks, D., Scharpé, S. & van Sande, M. (1985) Assay of carboxypeptidase N is serum by liquid chromatographic determination of hippuric acid. *Clin. Chem.* **31**, 1936–1939.

Hendriks, D., Scharpé, S. & van Sande, M. (1988) Partial purification and characterization of a new arginine carboxypeptidase from human serum. *J. Clin. Chem. Clin. Biochem.* **26**, 305.

Hendriks, D., Scharpé, S., van Sande, M. & Lommaert, M.P. (1989a) A labile enzyme in fresh human serum interferes with the assay of carboxypeptidase N. *Clin. Chem.* **35**, 177.

Hendriks, D., Scharpé, S., van Sande, M. & Lommaert, M.P. (1989b) Characterisation of a carboxypeptidase in human serum distinct from carboxypeptidase N. *J. Clin. Chem. Clin. Biochem.* **27**, 277–285.

Hendriks, D., Wang W., Scharpé, S., Lommaert, M.P. & van Sande, M. (1990) Purification and characterization of a new arginine carboxypeptidase in human serum. *Biochim. Biophys. Acta* **1034**, 86–92.

Hendriks, D., Wang, W., van Sande, M. & Scharpé, S (1992) Human serum carboxypeptidase U: a new kininase? *Agents Actions* **38/I**, 407–413.

Redlitz, A., Tan, A.K., Eaton, D.L. & Plow, E.F. (1995) Plasma carboxypeptidases as regulators of the plasminogen system. *J. Clin. Invest.* **96**, 2534–2538.

Redlitz, A., Nicolini, F.A., Malycky, J.L., Topol, E.J. & Plow, E.F. (1996) Inducible carboxypeptidase activity: a role in clot lysis in vivo. *Circulation* **93**, 1328–1330.

Shinohara, T., Sakurada, C., Suzuki, T., Campbell, W., Ikeda, S., Okada, N. & Okada, H. (1994) Pro-carboxypeptidase R cleaves bradykinin following activation. *Int. Arch. Allergy Immunol.* **103**, 400–404.

Skidgel, R.A. (1988) Basic carboxypeptidases: regulators of peptide hormone activity. *Trends Pharmacol. Sci.* **9**, 299–304.

Tan, A.K. & Eaton, D.L. (1995) Activation and characterization of procarboxypeptidase B from human plasma. *Biochemistry* **34**, 5811–5816.

Tsai, S.P. & Drayna, D. (1992) The gene encoding human plasma carboxypeptidase B (CPB2) resides on chromosome 13. *Genomics* **14**, 549–550.

Vanhoof, G., Wauters, J., Schatteman, K., Hendriks, D., Goossens, F., Bossyt, P. & Scharpé, S. (1996) The gene for human carboxypeptidase U (CPU) – a proposed novel regulator of plasminogen activation – maps to 13q14.11. *Genomics* **38**, 454–455.

Wang, W., Hendriks, D. & Scharpé, S. (1994) Carboxypeptidase U: a new plasma carboxypeptidase with high affinity for plasminogen. *J. Biol. Chem.* **269**, 15937–15944.

*Dirk F. Hendriks*
*Department of Medical Biochemistry,*
*University of Antwerp,*
*Universiteitsplein 1,*
*B-2610 Wilrijk, Belgium*
*Email: hendriks@uia.ua.ac.be*

# 454. Mast cell carboxypeptidase

## Databanks

*Peptidase classification: clan MC, family M14, MEROPS ID: M14.010*
*NC-IUBMB enzyme classification: EC 3.4.17.1*
*Chemical Abstracts Service registry number: 11075-17-5*

*Databank codes:*

| Species | SwissProt | PIR | EMBL (cDNA) | EMBL (genomic) |
|---------|-----------|-----|-------------|----------------|
| *Homo sapiens* | P15088 | A43929 | M27717 | M73716: exons 1 and 2 |
| | | S02809 | S40234 | M73717: exon 3 |
| | | S02810 | | M73718: exons 4–10 |
| | | S02811 | | M73719: exon 10 |
| | | | | M73720: exon 11 and complete CDS |
| *Mus musculus* | P15089 | A34487 | J05118 | M76557: promoter and 5′ flank |
| *Rattus norvegicus* | P21961 | A33118 | – | – |
| | | A38395 | | |

## Name and History

**Mast cell carboxypeptidase (MC-CP)**, also known as **mast cell carboxypeptidase A (MC-CPA)**, was originally identified as a carboxypeptidase A-like activity in the granules of rat mast cells (Everitt & Neurath, 1980) and has subsequently been found in the secretory granules of human and mouse mast cells (Goldstein *et al.*, 1987; Serafin *et al.*, 1987). There is some division on the appropriate naming of this enzyme, since it has enzymatic specificity similar to, but distinct from, pancreatic carboxypeptidase A (Chapter 451) and has apparently evolved more directly from an ancestor of pancreatic carboxypeptidase B (Chapter 455) (Goldstein *et al.*, 1989; Reynolds *et al.*, 1989a,b, Fig. 450.4).

## Activity and Specificity

MC-CP is most similar in activity to carboxypeptidase A, cleaving C-terminal residues with hydrophobic side chains (Phe, Tyr, Leu), whereas C-terminal Arg, Lys and Glu are not cleaved (Goldstein *et al.*, 1987). Sensitive assays for MC-CP activity have been developed which use either angiotensin I or Bz-Gly⊣Phe as the substrate and reverse-phase HPLC for identification and quantification of cleavage products (Serafin *et al.*, 1987; Goldstein *et al.*, 1989). MC-CP is most active at neutral to slightly basic pH, 7.0–9.5 (Everitt & Neurath, 1980; Goldstein *et al.*, 1987; Serafin *et al.*, 1987). MC-CP activity is inhibited by the metal-chelating agents 1,10-phenanthroline, 8-hydroxyquinoline and EDTA but not by PMSF or DFP (Everitt & Neurath, 1980; Goldstein *et al.*, 1987; Serafin *et al.*, 1987). Potato tuber carboxypeptidase inhibitor (PCI) is a potent inhibitor of MC-CP, and because of its specific reversible tight-binding and commercial availability, it has been utilized effectively as an affinity purification ligand (described below) (Everitt & Neurath, 1980; Goldstein *et al.*, 1989). No endogenous inhibitors are known at present for MC-CP in rodents or humans.

## Structural Chemistry

MC-CP is believed to be a zinc metallohydrolase of the same structural family (M14) as carboxypeptidase A and B (Reynolds *et al.*, 1989a,b; Cole *et al.*, 1991; Springman *et al.*, 1995). The unique 32 kb human gene has been sequenced; it contains 11 exons and has been localized to chromosome 3 (Reynolds *et al.*, 1989b, 1992). The 1.2 kb coding sequences for mouse and human MC-CP are known and encode a 417 amino acid preproprotein (Reynolds *et al.*, 1989a,b). By identity with its pancreatic homologs, MC-CP is thought to contain a single catalytic zinc ion coordinated by residues His67, Glu70 and His195 (Reynolds *et al.*, 1989a,b; Springman *et al.*, 1995).

The MC-CP zymogen is a single-chain polypeptide of 402 amino acids ($M_r$ 52 000) with two distinct globular domains (Dikov *et al.*, 1994; Springman *et al.*, 1995). Activation of the proenzyme occurs by proteolytic cleavage in the connecting loop between the two domains and dissociation of the N-terminal propeptide, yielding a single-chain mature form of 308 residues ($M_r$ 36 000) (Dikov *et al.*, 1994; Springman *et al.*, 1995).

MC-CP is stored in secretory granules predominantly in its mature form in complexes with proteoglycan (Schwartz *et al.*, 1982; Serafin *et al.*, 1986). Mature MC-CP is highly positively charged (calculated pI 9.5) at pH 7.4 and its complexes with proteoglycan are believed to be stabilized primarily by electrostatic interactions (Reynolds *et al.*, 1989a,b; Springman *et al.*, 1995). Although the nature of the protein–proteoglycan complex differs between human and rodent mast cells, MC-CP appears to reside in the same complexes as chymase (Chapter 18), which are larger in size than the complexes in which tryptase (Chapter 19) resides (Goldstein *et al.*, 1992).

## Preparation

MC-CP can be substantially purified from most sources by a single affinity chromatography step (Everitt & Neurath, 1980; Goldstein *et al.*, 1989). Potato carboxypeptidase inhibitor (PCI) is immobilized at 5–10 mg ml$^{-1}$ activated resin. A crude extract can be applied directly to the PCI resin in the presence of appropriate inhibitors and after washing, MC-CP is eluted at high pH (0.1 M Na$_2$CO$_3$, pH 11.4) and the fractions neutralized immediately (Everitt & Neurath, 1980; Goldstein *et al.*, 1989). Human, mouse and rat skin, cultured mouse bone marrow-derived mast cells, Kirsten sarcoma virus-immortalized mouse mast cell lines, and rat basophil leukemia line are rich sources of MC-CP. Several attempts have been made to develop a heterologous expression system for MC-CP using *Escherichia coli*, mammalian cell lines and yeast (Dikov *et al.*, 1994;

**M**

M.M. Dikov, E.B. Springman & W.E. Serafin, unpublished results). Although these systems yielded MC-CP protein, they failed to yield active MC-CP.

## Biological Aspects

MC-CP expression is regulated transcriptionally by the GATA-1 promoter (Zon *et al.*, 1991). Newly translated preproprotein is routed through the endoplasmic reticulum to the secretory granules. Presumably, the 15 amino acid signal sequence is removed cotranslationally, but the propeptide remains attached until MC-CP has reached the *trans*-Golgi or secretory granule (Dikov *et al.*, 1994; Springman *et al.*, 1995). The enzymes responsible for the proteolytic processing of MC-CP are currently not known but appear to be different from those responsible for activation of tryptase and chymase (Dikov *et al.*, 1994; Springman *et al.*, 1995).

In humans, MC-CP is found in mast cells that contain tryptase and chymase as well as those that contain only chymase, but not in those that contain only tryptase (Irani *et al.*, 1991; Weidner & Austen, 1993). Human skin mast cells contain substantial amounts of MC-CP but those of the intestinal mucosa do not (Goldstein *et al.*, 1989). Similarly in rodents, MC-CP is found in abundance in skin mast cells, designated connective tissue mast cells, but is not present in mucosal or immature bone marrow mast cells (Serafin *et al.*, 1987, 1991; Cole *et al.*, 1991). Ultrastructural examination of human mast cells indicates that the secretory granules containing MC-CP and chymase are qualitatively different from those containing tryptase alone (Craig *et al.*, 1988).

The biological substrates and normal or pathological roles of MC-CP have not been delineated, but it is known that MC-CP can cleave a number of peptide hormones *in vitro*. *In vivo*, it is likely that MC-CP acts in concert with chymase, since the two are colocalized and chymase cleavage results in an exposed hydrophobic C-terminal residue that is a potential substrate for MC-CP.

## Further Reading

For reviews, see Goldstein (1995) and Springman & Serafin (1995).

## References

Cole, K.R., Kumar, S., Le Trong, H., Woodbury, R.G., Walsh, K.A. & Neurath, H. (1991) Rat mast cell carboxypeptidase: amino acid sequence and evidence of enzyme activity within mast cell granules. *Biochemistry* **30**, 648–655.

Craig, S.S., Schechter, N.M. & Schwartz, L.B. (1988) Ultrastructural analysis of human T and TC mast cells identified by by immunoelectron microscopy. *Lab. Invest.* **58**, 682–691.

Dikov, M.M., Springman, E.B., Yeola, S. & Serafin, W.E. (1994) Processing of procarboxypeptidase A and other zymogens in murine mast cells. *J. Biol. Chem.* **269**, 25897–25904.

Everitt, M.T. & Neurath, H. (1980) Rat peritoneal mast cell carboxypeptidase: localization, purification and enzymatic properties. *FEBS Lett.* **110**, 292–296.

Goldstein, S.M. (1995) Mast cell carboxypeptidase: structure and regulation of gene expression. In: *Mast Cell Proteases in Immunology and Biology* (Caughey, G.A., ed.). New York: Marcel Dekker, pp. 109–126.

Goldstein, S.M., Kaempfer, C.E., Proud, D., Schwartz, L.B. & Wintroub, B.U. (1987) Detection and partial characterization of a human mast cell carboxypeptidase. *J. Immunol.* **139**, 2724–2729.

Goldstein, S.M., Kaempfer, C.E., Kealey, J.T. & Wintroub, B.U. (1989) Human mast cell carboxypeptidase: purification and characterization. *J. Clin. Invest.* **83**, 1630–1636.

Goldstein, S.M., Leong, J., Schwartz, L.B., & Cooke, D. (1992) Protease composition of exocytosed human skin skin mast cell protease-proteoglycan complexes: tryptase resides in a complex distinct from chymase and carboxypeptidase. *J. Immunol.* **148**, 2475–2482.

Irani, A.-M.A., Goldstein, S.M., Wintroub, B.U., Bradford, T. & Schwartz, L.B. (1991) Human mast cell carboxypeptidase: selective localization to MCTC cells. *J. Immunol.* **147**, 247–253.

Reynolds, D.S., Stevens, R.L., Gurley, D.S., Lane, W.S., Austen, K.F. & Serafin, W.E. (1989a) Isolation and molecular cloning of mast cell carboxypeptidase A: a novel member of the carboxypeptidase gene family. *J. Biol. Chem.* **264**, 20094–20099.

Reynolds, D.S., Gurley, D.S., Stevens, R.L., Sugarbaker, D.J., Austen, K.F. & Serafin, W.E. (1989b) Cloning of cDNAs that encode human mast cell carboxypeptidase A, and comparison of the protein with mouse mast cell carboxypeptidase A and rat pancreatic carboxypeptidases. *Proc. Natl Acad. Sci. USA* **86**, 9480–9484.

Reynolds, D.S., Gurley, D.S. & Austen, K.F. (1992) Cloning and characterization of the novel gene for mast cell carboxypeptidase A. *J. Clin. Invest.* **89**, 273–282.

Schwartz, L.B., Riedel, B., Schratz, J. & Austen, K.F. (1982) Localization of carboxypeptidase A to the macromolecular heparin protein-proteoglycan complex in secretory granules of rat serosal mast cells. *J. Immunol.* **128**, 1128–1133.

Serafin, W.E., Katz, H.R., Austen, K.F. & Stevens, R.L. (1986) Complexes of heparin proteoglycans, chondroitin sulfate E proteoglycans, and [3]H-diisopropyl-fluorophosphate-binding proteins are exocytosed from activated mouse bone marrow-derived mast cells. *J. Biol. Chem.* **261**, 15017–15021.

Serafin, W.E., Dayton, E.T., Gravallese, P.M., Austen, K.F. & Stevens, R.L. (1987) Carboxypeptidase A in mouse mast cells: identification, characterization and use as a differentiation marker. *J. Immunol.* **139**, 3771–3776.

Serafin, W.E., Sullivan, T.P., Conder, G.A., Ebrahimi, E., Marcham, P.M., Johnson, S.S., Austen, K.F. & Reynolds, D.S. (1991) Cloning of the cDNA and gene for mouse mast cell protease-4: demonstration of its late transcription in mast cell subclasses and analysis of its homology to subclass-specific neutral proteases of the mouse and rat. *J. Biol. Chem.* **266**, 1934–1941.

Springman, E.B. & Serafin, W.E. (1995) Secretory endo- and exopeptidases of mouse mast cells: structure, genetics and regulation of expression. In: *Mast Cell Proteases in Immunology and Biology* (Caughey, G.A., ed.). New York: Marcel Dekker, pp. 169–201.

Springman, E.B., Dikov, M.M. & Serafin, W.E. (1995) Mast cell procarboxypeptidase A: molecular modeling and biochemical characterization of its processing within secretory granules. *J. Biol. Chem.* **270**, 1300–1307.

Weidner, N. & Austen, K.F. (1993) Heterogeneity of mast cells at multiple body sites: fluorescent determination of avidin binding and immunofluorescent determination of chymase, tryptase, and carboxypeptidase content. *Pathol. Res. Pract.* **189**, 156–162.

Zon, L.I., Gurish, M.F., Stevens, R.L. Mather, C., Reynolds, D.S.,

Austen, K.F. & Orkin, S.H. (1991) GATA-binding transcription factors in mast cells regulate the promoter of the mast cell carboxypeptidase A gene. *J. Biol. Chem.* **266**, 22948–22953.

*Eric B. Springman*
*Department of Biochemistry and Enzymology,*
*Arris Pharmaceuticals,*
*180 Kimball Way,*
*South San Francisco, CA 94080, USA*
*Email: eric_springman@arris.com*

# 455. *Carboxypeptidase B*

## Databanks

*Peptidase classification: clan MC, family M14, MEROPS ID: M14.003*
*NC-IUBMB enzyme classification: EC 3.4.17.2*
*ATCC entries: 105720 (human)*
*Chemical Abstracts Service registry number: 9025-24-5*
*Databank codes:*

| Species | SwissProt | PIR | EMBL (cDNA) | EMBL (genomic) |
|---|---|---|---|---|
| Astacus fluviatilis | P04069 | A05141 | – | – |
| Bos taurus | P00732 | A14178 A00912 A92150 A93797 | – | – |
| Canis familiaris | P55261 | – | D78348 | – |
| Homo sapiens | P15086 | A42332 S02812 S02813 S08254 | M81057 | – |
| Protopterus aethiopicus | P19628 | A26212 B26212 | – | – |
| Rattus norvegicus | P19223 | A32129 | – | M23947–M23959: complete gene |
| Sus scrofa | P09955 | A38354 | – | – |

Brookhaven Protein Data Bank three-dimensional structures:

| Species | ID | Resolution | Notes |
|---|---|---|---|
| Bos taurus | 1CPB | 2.8 | fraction II |
|  | 1PBA |  | procarboxypeptidase activation domain (NMR) |

## Name and History

**Carboxypeptidase B** was discovered in autolyzed cattle and pig pancreas as an enzyme which releases C-terminal lysine and arginine from substrates such as Bz-Gly┼Lys or Bz-Gly┼Arg (Folk *et al.*, 1960). It was initially designated **basic carboxypeptidase** because of its specificity and was given the letter 'B' to distinguish it from Anson's carboxypeptidase A (Chapter 451). The name **protaminase** had also been used previously to describe its ability to release free arginine from protamines. It has also been called **tissue carboxypeptidase B** to distinguish it from plasma carboxypeptidase B (carboxypeptidase U: Chapter 453) (Eaton *et al.*, 1991). The use of 'tissue carboxypeptidase B' may lead to confusion since a number of tissue carboxypeptidases other than carboxypeptidase B show a basic carboxypeptidase activity (Skidgel, 1988).

## Activity and Specificity

Carboxypeptidase B is highly specific for excising C-terminal Lys and Arg residues from peptides and proteins with a preference for arginine (Tan & Eaton, 1995). It also acts at a slower rate on C-terminal Val, Leu, Ile, Asn, Gly or Gln (Villegas *et al*., 1995; Nishihira *et al*., 1995). A preference for Ala in position P2 of synthetic substrates has been observed (McKay *et al*., 1979).

Typical assay methods use either Bz-Gly┼Arg (Folk *et al*., 1960) or FA-Ala┼Arg (Tan & Eaton, 1995) as substrates. Assays are carried out at pH 7.5–7.8 and 0.1–0.15 M sodium chloride.

A number of synthetic inhibitors of basic carboxypeptidases are termed 'biproduct analogs'. Guanidinoethyl-mercaptosuccinic acid (McKay *et al*., 1979) and 2-mercaptomethyl-3-guanidinoethyl-propanoic acid (Plummer & Ryan, 1981) are relevant examples, with $K_i$ in the micromolar range. 2-Mercaptomethyl-5-guanidinopentanoic acid (Ondetti *et al*., 1979) shows a $K_i$ in the nanomolar range. Another powerful inhibitor tested is the naturally occurring potato carboxypeptidase inhibitor (Hass & Ryan, 1981), with a $K_i$ in the nanomolar range.

## Structural Chemistry

Depending on the tissue of origin, carboxypeptidase B varies in length between 306 residues (pancreas) and 309 (plasma), with an $M_r$ of about 36 000 (Avilés *et al*., 1993). The propeptides or activation segments vary between 92 and 95 residues. The $M_r$ of the pancreatic zymogens is about 47 000. The enzyme contains 1 mol of zinc per mol of protein.

Titani *et al*. (1975) completed the amino acid sequence determination of carboxypeptidase B. The sequences of the pig proenzyme activation segment (Burgos *et al*., 1991) and the complete rat proenzyme (Clauser *et al*., 1988) are also known. An early unrefined structure at 2.8 Å resolution (Schmid & Herriot, 1976) indicated a strong structural similarity between carboxypeptidases B and A. This was confirmed and detailed in a refined structure of procarboxypeptidase B (Coll *et al*., 1991). The zinc-coordinating residues are conserved: His69, Glu72, His196 and a water molecule (the numbering is that of the reference protein of this family, carboxypeptidase A). This is also true for other important residues for catalysis like Glu270 and Asp127. Differences in specificity between both carboxypeptidases are explained by the presence of Ser207, Glu243 and, particularly, Asp255 at positions occupied by Gly, Ile and Ile in carboxypeptidase A (Coll *et al*., 1991; Avilés *et al*., 1993).

Pig pancreatic procarboxypeptidase B folds in two distinguishable domains: the activation domain (residues 1–80 in the proenzyme) and the enzyme domain (residues 96–401), linked by a connecting region (residues 81–95). There is a close coincidence between the NMR structure of the isolated activation domain in solution and the structure it adopts within the proenzyme in the crystallographic form (Billeter *et al*., 1992).

## Preparation

Carboxypeptidase B has been isolated from the pancreas of several species including human (Pascual *et al*., 1989), cattle (Reeck *et al*., 1971), pig (Burgos *et al*., 1989), ostrich (Bradley *et al*., 1996) and others, as well as human serum (Yamamoto *et al*., 1992) and longitudinal muscle layer of cattle small intestine (Kase *et al*., 1987). A stable enzyme can be purified from pancreatic acetone powder (Folk *et al*., 1960; Burgos *et al*., 1989; Pascual *et al*., 1989). Internal proteolysis and degradation may occur if the selected source is pancreatic secretion (Brodrick *et al*., 1976). The purified enzyme is stable for years in the precipitated or frozen states.

## Biological Aspects

The rat preprocarboxypeptidase B gene spans 11 exons dispersed throughout 34 kb of genomic DNA (Clauser *et al*., 1988). The gene is organized very similar to rat CPA1 and CPA2 genes and conserved sequences in the 5′ flanking regions of the three genes have been identified; evolutionary trees for the pancreatic carboxypeptidases have been constructed (Gardell *et al*., 1988; Avilés *et al*., 1993; see also Fig. 450.4).

Trypsin (Chapter 3) is the only pancreatic proteinase capable of generating carboxypeptidase B from the zymogen *in vitro* (Burgos *et al*., 1991). The generated enzyme participates in the degradation of the released activation peptide (Villegas *et al*., 1995). The zymogen has no intrinsic carboxypeptidase activity, in contrast to the zymogen of carboxypeptidase A, and this is explained by the presence of a salt-bridge between Asp41 in the activation segment and Arg145 in the enzyme moiety, the residue that binds the C-terminal carboxyl group of substrate molecules in the free enzyme form (Coll *et al*., 1991). Tryptic activation of procarboxypeptidase B is much faster than that of the A form, and this can also be explained by comparison of the three-dimensional structures of both zymogens (Vendrell *et al*., 1993). Procarboxypeptidase B was characterized as a serum marker for acute pancreatitis and pancreatic graft rejection (Yamamoto *et al*., 1992).

## Related Peptidases

The term 'plasma carboxypeptidase B' has been used for carboxypeptidase U (Chapter 453), which has similar specificity to carboxypeptidase B (Eaton *et al*., 1991). Glycosylation of the activation peptide of procarboxypeptidase U distinguishes it from other procarboxypeptidases and may stabilize and increase the half-life of the circulating proenzyme (Eaton *et al*., 1991). This activation peptide may mediate binding to plasminogen (Tan & Eaton, 1995) and it has been suggested that carboxypeptidase U can modulate plasminogen binding to cells and control the rate of fibrinolysis (Redlitz *et al*., 1995). The active enzyme has been shown to bind to $\alpha_2$-macroglobulin and pregnancy zone protein in plasma (Valnicokva *et al*., 1996).

## References

Avilés, F.X., Vendrell, J., Guasch, A., Coll, M. & Huber, R. (1993) Advances in metallo-procarboxypeptidases. *Eur. J. Biochem.* **211**, 381–389.

Billeter, M., Vendrell, J., Wider, G., Avilés, F.X., Coll, M., Guasch, A., Huber, R. & Wüthrich, K. (1992) Comparison of the NMR structure with the X-ray structure of the activation domain from procarboxypeptidase B. *J. Biomol. NMR* **2**, 1–10.

Bradley, G., Naude, R.J., Muramoto, K., Yamauchi, F. & Oelofsen, W. (1996) Ostrich (*Struthio camelus*) carboxypeptidase B: purification, kinetic properties and characterization of the pancreatic enzyme. *Int. J. Biochem. Cell Biol.* **28**, 521–529.

Brodrick, J.W., Geokas, M.C. & Largman, C. (1976) Human carboxypeptidase B. II. Purification of the enzyme from pancreatic tissue and comparison with the enzymes present in pancreatic secretion. *Biochim. Biophys. Acta* **97**, 39–43.

Burgos, F.J., Pascual, R., Vendrell, J., Cuchillo, C.M. & Avilés, F.X. (1989) The separation of pancreatic procarboxypeptidases by high-performance liquid chromatography. *J. Chromatogr.* **481**, 233–243.

Burgos, F.J., Salvà, M., Villegas, V., Soriano, F., Méndez, E. & Avilés, F.X. (1991) Analysis of the activation process of porcine procarboxypeptidase B and determination of the sequence of its activation segment. *Biochemistry* **30**, 4082–4089.

Clauser, E., Gardell, S.J., Craik, C.S., MacDonald, R.J. & Rutter, W.J. (1988) Structural characterization of the rat carboxypeptidase A1 and B genes. Comparative analysis of the rat carboxypeptidase gene family. *J. Biol. Chem.* **263**, 17837–17845.

Coll, M., Guasch, A., Avilés, F.X. & Huber, R. (1991) Three-dimensional structure of porcine procarboxypeptidase B: a structural basis for its inactivity. *EMBO J.* **10**, 1–9.

Eaton, D.L., Malloy, B.E., Tsai, S.P., Henzel, W. & Drayna, D. (1991) Isolation, molecular cloning, and partial characterization of a novel carboxypeptidase B from human plasma. *J. Biol. Chem.* **266**, 21833–21838.

Folk, J.E., Piez, K.A., Carroll, W.R. & Gladner, J.A. (1960) Carboxypeptidase B. IV. Purification and characterization of the porcine enzyme. *J. Biol. Chem.* **235**, 2272–2277.

Gardell, S.J., Craick, C.S., Clauser, E., Goldsmith, E.J., Stewart, C.-B., Graf, M. & Rutter, W.J. (1988) A novel rat carboxypeptidase, CPA2: characterization, molecular cloning, and evolutionary implications on substrate specificity in the carboxypeptidase gene family. *J. Biol. Chem.* **263**, 17828–17836.

Hass, G.M. & Ryan, C.A. (1981) Carboxypeptidase inhibitor from potatoes. *Methods Enzymol.* **80**, 778–791.

Kase, R., Hideshima, C. & Hazato, T. (1987) Separation of carboxypeptidase A and B from longitudinal muscle layer of bovine small intestine: their properties regarding hydrolysis of enkephalins and enkephalin-analogs. *Biochem. Int.* **14**, 889–896.

McKay, T.L., Phelan, A.W. & Plummer, T.H., Jr. (1979) Comparative studies on human carboxypeptidases B and N. *Arch. Biochem. Biophys.* **2**, 487–492.

Nishihira, J., Hibiya, Y., Sakai, M., Nishi, S., Kumazaki, T., Ohki, S. & Sakamoto, W. (1995) The C-terminal region, Arg$^{201}$-Gln$^{209}$, of glutathione S-transferase P contributes to stability of the active-site conformation. *Biochim. Biophys. Acta* **1252**, 233–238.

Ondetti, M.A., Condon, M.E., Reid, J., Sabo, E.F., Chemy, H.S. & Cushman, D.W. (1979) Design of specific inhibitors for carboxypeptidase A and B. *Biochemistry* **18**, 1427–1430.

Pascual, R., Burgos, F.J., Salvà, M., Soriano, F., Méndez, E. & Avilés, F.X. (1989) Purification and properties of five different forms of human procarboxypeptidases. *Eur. J. Biochem.* **179**, 609–616.

Plummer, T.H., Jr. & Ryan, T.J. (1981) A potent mercapto bi-product analogue inhibitor for human carboxypeptidase N. *Biochem. Biophys. Res. Commun.* **98**, 448–454.

Redlitz, A., Tan, A.K., Eaton, D.L. & Plow, E.F. (1995) Plasma carboxypeptidases as regulators of the plaminogen system. *J. Clin. Invest.* **96**, 2534–2538.

Reeck, G.R., Walsh, K.A. & Neurath, H. (1971) Isolation and characterization of carboxypeptidases A and B from activated pancreatic juice. *Biochemistry* **10**, 4690–4698.

Schmid, M.F. & Herriot, J.R. (1976) Structure of carboxypeptidase B at 2.8 Å resolution. *J. Mol. Biol.* **103**, 175–190.

Skidgel, R.A. (1988) Basic carboxypeptidases: regulators of peptide hormone activity. *Trends Pharmacol. Sci.* **9**, 299–304.

Tan, A.K. & Eaton, D.L. (1995) Activation and characterization of procarboxypeptidase B from human plasma. *Biochemistry* **34**, 5811–5816.

Titani, K., Ericsson, L.H., Walsh, K.A. & Neurath, H. (1975) Amino-acid sequence of bovine carboxypeptidase B. *Proc. Natl Acad. Sci. USA* **72**, 1666–1670.

Valnickova, Z., Thogersen, I.B., Christensen, S., Chu, C.T., Pizzo, S.V. & Enghild, J.J. (1996) Activated human plasma carboxypeptidase B is retained in the blood by binding to $\alpha_2$-macroglobulin and pregnancy zone protein. *J. Biol. Chem.* **271**, 12937–12943.

Vendrell, J., Catasús, L., Oppezzo, O., Ventura, S., Villegas, V. & Avilés, F.X. (1993) Activation of pancreatic procarboxypeptidases. In: *Innovations in Proteases and their Inhibitors* (Avilés, F.X., ed.). Berlin: Walter de Gruyter, pp. 279–297.

Villegas, V. Vendrell, J. & Avilés, F.X. (1995) The activation pathway of procarboxypeptidase B from porcine pancreas: participation of the active enzyme in the proteolytic processing. *Protein Sci.* **4**, 1792–1800.

Yamamoto, K.K., Pousette, A., Chow, P., Wilson, H. el Shami, S. & French, C.K. (1992) Isolation of a cDNA encoding a human serum marker for acute pancreatitis. Identification of pancreas-specific protein as pancreatic procarboxypeptidase B. *J. Biol. Chem.* **267**, 2575–2581.

***Francesc X. Avilés***
*Departament de Bioquímica i Biologia Molecular,
Unitat de Ciències,
Universitat Autònoma de Barcelona,
08193 Cerdanyola del Vallès. Spain
Email: ibfn0@cc.uab.es*

***Josep Vendrell***
*Departament de Bioquímica i Biologia Molecular,
Unitat de Ciències,
Universitat Autònoma de Barcelona,
08193 Cerdanyola del Vallès. Spain
Email: Josep.Vendrell@blues.uab.ed*

# 456. Carboxypeptidase T

## Databanks

Peptidase classification: clan MC, family M14, MEROPS ID: M14.007
NC-IUBMB enzyme classification: EC 3.4.17.18
Databank codes:

| Species | SwissProt | PIR | EMBL (cDNA) | EMBL (genomic) |
|---|---|---|---|---|
| Streptomyces capreolus | P39041 | – | U00619 | – |
| Streptomyces griseus | P18143 | – | – | – |
| Thermoactinomyces vulgaris | P29068 | S17571 | X56901 | – |

Brookhaven Protein Data Bank three-dimensional structures

| Species | ID | Resolution | Notes |
|---|---|---|---|
| Thermoactinomyces vulgaris | 1OBR | 2.3 | |

## Name and History

The first isolation of a bacterial metallocarboxypeptidase was reported in 1976 by Seber *et al*. (1976), who purified and partially characterized the enzyme from *Streptomyces griseus* (Pronase preparation). This enzyme, named **carboxypeptidase SG**, has a dual specificity pattern that combines the features peculiar to mammalian carboxypeptidases A (Chapter 451) and B (Chapter 455) (Breddam *et al*., 1979). Later, a similar metallocarboxypeptidase was isolated from the culture filtrate of *Thermoactinomyces vulgaris*, a bacterium that shares a number of biochemical traits with bacilli (Osterman *et al*., 1984). The enzyme was thoroughly studied, its structural gene cloned and sequenced (Smulevich *et al*., 1991) and the tertiary structure determined by X-ray crystallography (Teplyakov *et al*., 1992) (see Fig. 450.1). All these data clearly demonstrate the homology of this enzyme with mammalian carboxypeptidases (Osterman *et al*., 1992), in particular with carboxypeptidase A. The letter T was added to its name to indicate its production by *Thermoactinomycetes*, hence, **carboxypeptidase T**.

## Activity and Specificity

The enzyme specificity pattern is typical for metallocarboxypeptidases. It is an exopeptidases and cleaves off the C-terminal L-amino acid residues from peptides. As shown by its tertiary structure, the limitation of the specificity to C-terminal residues with a free carboxyl group is determined by the latter binding to an arginine guanidino group at the active center.

From model peptides carboxypeptidase T cleaved off the C-terminal Arg, Lys, His, Phe, Tyr, Trp, Leu, Met, Ala. A free α-amino group hampers the hydrolysis, hence, only acylated dipeptides are hydrolyzed. The amino acid residue preceding the C-terminal one (P1) influences the rate of hydrolysis.

Thus, glycine and glutamic acid at the penultimate position retard the cleavage, whereas proline blocks it completely. The enzyme hydrolyzes substrates optimally at pH 7–8. It is completely inhibited by the chelators EDTA or 1,10-phenanthroline. The activity of the enzyme increases with temperature up to 60–70°C.

Activity of carboxypeptidase T is measured against chromogenic substrates: Dnp-Ala-Ala+Arg or Dnp-Ala-Ala-Leu+Arg or their analogs. This method is also suitable for assay of carboxypeptidase B or other carboxypeptidases capable of cleaving the C-terminal arginine from the substrates. Splitting of the C-terminal arginine changes the solubility of the Dnp moiety; the chromophore can be extracted into ethyl acetate from acidified aqueous solution, whereas the non-cleaved substrate remains in the aqueous phase. An even simpler protocol, especially suitable for carboxypeptidase T assay in multiple samples, is based on an ion-exchange separation of the colored hydrolysis product from the excess of the substrate. Dnp-Ala-Ala-Arg in weakly acid solution is bound by a cationic-sulfoethyl or sulfopropyl-Sephadex resin due to the positively charged guanidino group of the arginine residue. In contrast, the product Dnp-Ala-Ala falls through and can be measured spectrophotometrically (Osterman *et al*., 1984). The substrate is available from Institute of Genetics and Selection of Industrial Microorganisms (see Appendix 2 for full names and addresses of suppliers).

## Structural Chemistry

The molecular mass of the single-chain enzyme is 36 928 Da based on the amino acid sequence, and its pI is 5.3. The primary structure of carboxypeptidase T determined by DNA sequencing of the cloned structural gene agreed with the N-terminal sequence of 31 amino acid residues established by automated Edman degradation of the mature enzyme. The mature enzyme sequence consists of 326 amino acid residues, preceded by 98 residues that comprise the prepro region. The N-terminal stretch of the latter sequence starts with the motif characteristic for a signal peptide: three positively charged residues followed by a prolonged hydrophobic stretch. The boundary between pre and pro sequences cannot be specified yet. The pro part of the carboxypeptidase T precursor contains about 70 amino acid residues and does not show any homology with the propeptides of mammalian metallocarboxypeptidases.

The crystal structure of carboxypeptidase T was determined by X-ray diffraction at 2.35 Å resolution (Teplyakov *et al.*, 1992) (see Fig. 450.1). The core of the molecule is composed of a twisted eight-stranded $\beta$ sheet and six $\alpha$ helices antiparallel to the $\beta$ strands. The most prominent feature of the enzyme specificity – its capacity to digest peptides with hydrophobic C-terminal residues (like carboxypeptidase A; Chapter 451) as well as to cleave C-terminal arginine or lysine (like carboxypeptidase B; Chapter 455) is explained by the structure of the primary specificity (S1′) pocket of carboxypeptidase T. The presence of Asp255 in carboxypeptidase B and Asp253 in carboxypeptidase T (carboxypeptidase A numbering) explains qualitatively the specificity of these two enzymes toward positively charged arginine or lysine residues at the C-termini of the peptide substrates.

Kinetic parameters for hydrolysis of a series of tripeptide substrates by carboxypeptidases A, B and T indicate that the hydrophobic amino acids phenylalanine and leucine are cleaved by carboxypeptidases A and T with comparable efficiency (Stepanov, 1995). On the other hand, carboxypeptidase B splits off the C-terminal arginine and lysine much better than carboxypeptidase T. $K_m$ values for the latter enzyme indicate a rather loose binding of substrates, only partially compensated by relatively high $k_{cat}$ values. It appears that this loss of efficiency represents a functionally acceptable cost for the broad specificity of carboxypeptidase T.

The enzyme contains 1 mol of zinc per mol of protein, bound in the same manner as in carboxypeptidase A. One important feature that distinguishes the structure of carboxypeptidase T from those of carboxypeptidases A and B is that it contains four binding sites for $Ca^{2+}$ ions. The latter seems to be important for thermostability.

## Preparation

The procedure is based on affinity chromatography on bacitracin-silochrome to isolate the proteolytic enzymes contained in *Thermoactinomyces vulgaris*, strain INMI, culture filtrate. Further chromatography on a bacitracin-Sepharose 4B column permits elution of carboxypeptidase T specifically due to its inhibition with EDTA. The activity is then restored by extensive dialysis of the preparation against calcium chloride solution (Osterman *et al.*, 1984). The adsorption of carboxypeptidase T proceeds apparently due to the interaction of the active-site with the bacitracin peptide chain, which might be supplemented by $Zn^{2+}$ interaction with the cyclopeptide (Stepanov *et al.*, 1981; Stepanov & Rudenskaya, 1983). Good results on the final purification step are also obtained using Sepharose 4B with covalently bound *p*-aminobenzylsuccinate that serves as a substrate analog ('CABS-Sepharose').

## Biological Aspects

Sequence alignment of carboxypeptidase T and 15 other metallocarboxypeptidases has shown that the enzyme belongs to a subfamily of 'short' metallocarboxypeptidases referred to as 'digestive' enzymes due to their apparently simple degradative function (Osterman *et al.*, 1992). Carboxypeptidase SG from *Streptomyces griseus* (the nearest homolog of carboxypeptidase T) has 64% identical residues (Narahachi, 1990), although some differences exist among the residues involved in the substrate binding. Metallocarboxypeptidases homologous to carboxypeptidase T have been found in *Streptomyces capreolus* (A.S. Thiara & E. Cundliffe, SwissProt Accession P39041) and *Streptomyces spheroides* (Rudenskaya *et al.*, 1986). There is 27% identity between carboxypeptidase T and carboxypeptidase A (Chapter 451) and 26% with carboxypeptidase B (Chapter 455), whereas their tertiary folds and the structural elements responsible for the catalytic mechanisms are remarkably similar. Superposition of carboxypeptidases A and T tertiary structures gave an average deviation 0.8 Å for 235 C-$\alpha$ atoms compared. Certain differences were detected in the flexible regions of the molecule distant to the active site.

According to the gene structure, carboxypeptidase T should be secreted as a proenzyme, although the latter was not observed in the culture filtrate – a phenomenon common for many bacterial proteinases. The mature enzyme sequence starts from an aspartic acid residue, and a Gln-Asp bond must be hydrolyzed to convert the precursor into the mature carboxypeptidase T. Hence, a proteinase with rather unusual specificity pattern, capable of hydrolyzing the Gln-Asp peptide bond, seems to be involved in procarboxypeptidase T activation. Homologous carboxypeptidases from *Streptomyces griseus* and *Streptomyces capreolus* also possess Asp as the N-terminal amino acid.

Breddam *et al.* (1979) suggested that the common ancestor of carboxypeptidases A and B might have had a broad specificity similar to that of carboxypeptidese SG. However, because the aspartate crucial to the specificity of carboxypeptidase T is not equivalent to that in carboxypeptidase B, the capacity to hydrolyse basic residues seems to have developed independently in these two enzymes.

Although the physiological role of carboxypeptidase T and related bacterial enzymes remains to be elucidated, it seems plausible that these exopeptidases are involved in degradation of protein substrates. Their unusually broad specificity would be consistent with this.

## Distinguishing Features

As a rule, identification and isolation of carboxypeptidase T is not complicated, although care must be taken to exclude the possibility of metalloendoproteinases that can hydrolyze the Ala-Leu bond in Dnp-Ala-Ala-Leu+Arg, one of the chromogenic substrates used for carboxypeptidase assessment. It is advisable to check the products of the substrate hydrolysis.

## Further Reading

For a review, see Stepanov (1995).

## References

Breddam, K., Bazzone, T.Y., Holmquist, B. & Vallee, B.L. (1979) Carboxypeptidase of *Streptomyces griseus*. Implications of its characteristics. *Biochemistry* **18**, 1563–1570.

Narahachi, Y. (1990) The amino acid sequence of zinc-carboxypeptidase from *Streptomyces griseus*. *J. Biochem.* **107**, 879–886.

Osterman, A.L., Stepanov, V.M., Rudenskaya, G.N., Khodova, O.M., Tsaplina, I.A., Yakovleva, M.B. & Loginova, L.G. (1984) Carboxypeptidase T: extracellular carboxypeptidase of thermoactinomycetes – a distant analog of animal carboxypeptidases. *Biokhimiya* **49**, 292–301.

Osterman, A.L., Grishin, N.V., Smulevitch, S.V., Matz, M.V., Zagnitko, O.P., Revina, L.P. & Stepanov, V.M. (1992) Primary structure of carboxypeptidase T: delineation of functionally relevant features in Zn-carboxypeptidase family. *J. Protein Chem.* **11**, 561–570.

Rudenskaya, G.N., Kreier, V.G., Landau, N.S., Tarasova, N.I., Timokhina, E.A., Egorov, N.S. & Stepanov, V.M. (1986) Isolation and properties of carboxypeptidase from *Streptomyces spheroides*, strain 35. *Biokhimiya* **52**, 2002.

Seber, Y.F., Toomey, T.P., Powell, Y.T., Brew, K. & Awad, W.M. (1976) Proteolytic ezymes of the K-1 strain of *Streptomyces griseus* obtained from a commercial preparation (Pronase).

Purification and characterization of the carboxypeptidase. *J. Biol. Chem.* **251**, 204–208.

Smulevich, S.V., Osterman, A.L., Galperina, O.V., Matz, M.V., Zagnitko, O.P., Kadyrov, R.M., Tsaplina, I.A., Grishin, N.V., Chestukhina, G.G. & Stepanov, V.M. (1991) Molecular cloning and primary structure of *Thermoactinomyces vulgaris* carboxypeptidase T endowed with double substrate specificity. *FEBS Lett.* **291**, 75–78.

Stepanov, V.M. (1995) Carboxypeptidase T. *Methods Enzymol.* **248**, 675–683.

Stepanov, V.M. & Rudenskaya, G.N. (1983) Proteinase affinity chromatography on bacitracin-Sepharose. *J. Appl. Biochem.* **2**, 420–428.

Stepanov, V.M., Rudenskaya, G.N., Gaida, A.V. & Osterman, A.L. (1981) Affinity chromatography of proteolytic enzymes on silica-based biospecific sorbent. *J. Biochem. Biophys. Methods* **5**, 177–186.

Teplyakov, A., Polyakov, K., Obmolova, G., Strokopytov, B., Kuranova, I., Osterman, A., Grishin, N., Smulevich, S., Zagnitko, O., Galperina, O., Matz, M. & Stepanov, V. (1992) Crystal structure of carboxypeptidase T from *Thermoactinomyces vulgaris*. *Eur. J. Biochem.* **208**, 281–288.

*Valentin M. Stepanov*
*Protein Chemistry Laboratory,*
*Institute of Genetics and Selection of Industrial Microorganisms,*
*1 Dorozhny pr., 1, 113545, Moscow, Russia*
*Email: ost@vnigen.msk.su*

# 457. γ-D-Glutamyl-(L)-meso-diaminopimelate peptidase I

## Databanks

*Peptidase classification: clan MC, family M14, MEROPS ID: M14.008*
*NC-IUBMB enzyme classification: EC 3.4.19.11*
Databank codes:

| Species | SwissProt | PIR | EMBL (cDNA) | EMBL (genomic) |
|---|---|---|---|---|
| *Bacillus sphaericus* | Q03415 | – | X69507 | – |
| *Bacillus subtilis* | – | – | – | Z99116: complete genome section 13 of 21 |

## Name and History

*Bacillus sphaericus* NCTC 9602 produces two sporulation-related *γ-D-glutamyl-(L)-diamino acid-hydrolyzing peptidases*, named *I* and II (Guinand *et al*., 1974; Vacheron *et al*., 1979) that cleave selectively the peptide sequence of the peptidoglycans of chemotype A$_{1g}$ (Schleifer & Kandler, 1972). They differ with respect to cellular localization (Guinand *et al*., 1979; Vacheron *et al*., 1979), molecular mass and catalytic mechanism (Garnier *et al*., 1985; Bourgogne *et al*., 1992; Hourdou *et al*., 1992, 1993). From their specificity profiles (Arminjon *et al.*, 1977; Valentin *et al.*, 1983), they are now identified as a carboxypeptidase/peptidyl-dipeptidase (see below) and as a dipeptidyl-peptidase (Chapter 265), respectively.

## Activity and Specificity

Substrates are either: (a) peptides or monomeric units of glycopeptides, derived from the bacterial peptidoglycan precursors (Guinand *et al.*, 1974; Arminjon *et al.*, 1976) or (b) dimeric or polymeric fragments of the *Escherichia coli* peptidoglycan (Valentin *et al.*, 1983). The reaction products can be separated by cellulose thin-layer chromatography and quantified by radioactivity measurements. The enzyme removes the C-terminal *meso*-diaminopimelic acid (*ms*-A$_2$pm) or the C-terminal dipeptide *ms*-A$_2$pm-D-Ala from the peptides L-Ala-$γ$-D-Glu⊣(L)*ms*-A$_2$pm or L-Ala-$γ$-D-Glu⊣(L)*ms*-A$_2$pm(L)-D-Ala, respectively. Activity is reduced on peptides containing D-isoGln versus D-Glu, but is enhanced when the L-Ala amino group is substituted by either an *N*-acetylmuramoyl or an *N*-acetylglucosaminyl-$β$-1,4-*N*-acetylmuramoyl moiety. But there is no activity with peptides containing L-Lys instead of *ms*-A$_2$pm. Unsubstituted amino and carbonyl groups on the D-center of *ms*-A$_2$pm in P1′ and unsubstituted C-terminal groups of D-Ala in P2′ are a strict requirement for activity (Arminjon *et al.*, 1977; Valentin *et al.*, 1983). Thus, activity on peptidoglycans requires a prior hydrolysis of the D-Ala-(D)*ms*-A$_2$pm cross-linkages (Valentin *et al.*, 1983; Morel *et al.*, 1986a). In consequence, the enzyme is a *meso*-diaminopimelic acid carboxypeptidase/peptidyl-dipeptidase.

A $K_m$ value of 0.57 mM and a specific activity of 8.3 μmol min$^{-1}$ mg$^{-1}$ were derived from kinetic studies with Mur-*N*-Ac-L-Ala-$γ$-D-Glu⊣(L)*ms*-A$_2$pm(L)-D-Ala as substrate (Garnier *et al.*, 1985); the pH optimum is about 8. This peptidase is a heat-stable protein with an apparent inactivation temperature of 80°C. Moreover, anionic, cationic or zwitterionic detergents enhance the specific activity up to 20-fold (Garnier *et al.*, 1985; Hourdou *et al.*, 1993). The enzyme is inhibited by 1,10-phenanthroline and EDTA, and is reactivated by zinc, cobalt and manganese ions (Garnier *et al.*, 1985). Activity is quickly lost in *N*-2-hydroxyethyl piperazine-*N*′-2-ethane sulfonic acid buffer (pH 8) (Guinand *et al.*, 1974); *ms*-A$_2$pm is a competitive inhibitor ($K_i = 0.02$ mM) and dicarboxylic D-amino acids also are inhibitors depending on the length of the side chain, the maximum inhibition being observed with D-$α$-aminopimelic acid (Arminjon *et al.*, 1977). The $γ$-L-glutamyl peptidase of *Bacillus cereus* (Cheng *et al.*, 1973) and an enzyme of *Bacillus subtilis* implicated in the biosynthesis of $γ$-glutamyl peptides (Williams *et al.*, 1955), are unrelated enzymes.

## Structural Chemistry

This *B. sphaericus* peptidyl-dipeptidase has been analyzed by proton-induced X-ray emission. It contains 1 mol of zinc per mol of protein. As determined by gene cloning and sequencing, it is a two module protein of $M_r$ 44 724 (pI 5.4) with no cysteine residue (Hourdou *et al.*, 1993). A 100 amino acid residue N-terminal domain is linked to a 296 amino acid residue C-terminal catalytic domain which belongs to the carboxypeptidase A family (Chapter 450). By similarity, the catalytic domain possesses the zinc-binding triad His162, Glu165, His307 and the active site residues Tyr347, Glu366. The N-terminal domain consists of two tandem segments which have 41% identities; it has similarity with the C-terminal repeats of various peptidoglycan hydrolases and is probably involved in the recognition of a particular moiety of the bacterial cell envelope. Polyclonal antisera to the purified enzyme have been prepared.

## Preparation

Two forms of the enzyme have been identified: an extracellular one in the sporulation medium (Garnier *et al.*, 1985) and a membrane-bound one in the integument of the forespores and the spores (Guinand *et al.*, 1979). In this latter case, the enzyme may be bound to its cortex substrate. A bacterial culture was grown in an industrial fermenter of 1600 liters total capacity and the enzyme was purified either in small amount from the sporulation medium (5200-fold) or in larger amount from the integuments of the spores in the presence of Brij-58 (2400-fold; Baji-Kourda *et al.*, 1989). Overexpression of the gene has not been attempted.

## Biological Aspects

A similar activity is found in *B. subtilis* (Guinand *et al.*, 1976, 1978). It also exists in gram-negative bacteria such as *Escherichia coli* W7 (Goodell & Schwarz, 1985) or *Proteus mirabilis* (Gmeiner & Kroll, 1981). In *B. sphaericus*, this enzyme appears at stage IV of sporulation and its period of synthesis closely precedes and parallels spore cortex formation (Guinand *et al.*, 1974). This synthesis is prevented in the presence of netropsin, a drug which selectively inhibits sporulation, and is reduced in a *B. subtilis* asporogenous mutant (Guinand *et al.*, 1978). The cortex peptidoglycan of *B. sphaericus* and other bacilli differs from the vegetative peptidoglycan by the nature of the diamino acid in the peptide units (*ms*-A$_2$pm versus L-Lys) and by the presence, in the glycan chains, of an appreciable proportion of unsubstituted muramyl lactams and *N*-acetyl-muramic acids bearing a single L-alanine residue (Warth & Strominger, 1972). These structural features require the coordinated and sequential action of several peptidoglycan hydrolases, including the present enzyme. The action of $γ$-D-Glu-*ms*-A$_2$pm peptidase I excludes any polymerization of the peptide side chains of new cortex units via transpeptidation, since the $ε$-amino group of *ms*-A$_2$pm is the essential acceptor group for the transpeptidation reaction. Apart from its unique action on $γ$-D-glutamyl linkages, interest in peptidase I arises from its use in the preparation of potential immunostimulants (Morel *et al.*, 1986b).

## References

Arminjon, F., Guinand, M., Michel, G., Coyette, J. & Ghuysen, J.-M. (1976) Préparation enzymatique de peptide du type L-Ala-$γ$-D-Glu-(L)*meso*-diaminopimélyl(L)-D-Ala ($^{14}$C) par réaction d'échange entre les peptides correspondants non radioactifs et la D-alanine ($^{14}$C) [Enzymatic preparation of a peptide of the type L-Ala-γ-D-Glu-(L)*meso*-diaminopimelic(L)-D-Ala($^{14}$C) by an exchange reaction between the corresponding nonradioactive peptides and D-alanine($^{14}$C)]. *Biochimie* **58**, 1167–1172.

Arminjon, F., Guinand, M., Vacheron, M.-J. & Michel. G. (1977) Specificity profiles of the membrane bound γ-D-glutamyl-(L)*meso*-diaminopimelate endopeptidase and LD-carboxypeptidase from *Bacillus sphaericus* 9602. *Eur. J. Biochem.* **73**, 557–565.

Baji-Kourda, F., Guinand, M., Vacheron, M.-J. & Michel, G. (1989) An improved preparation of purified γ-glutamyl-(L)-*meso*-diaminopimelate endopeptidase I from *Bacillus sphaericus* 9602. *Biotechnol. Appl. Biochem.* **11**, 169–175.

Bourgogne, T., Vacheron, M.-J., Guinand, M. & Michel, G. (1992) Purification and partial characterization of the γ-D-glutamyl-L-di-amino acid endopeptidase II from *Bacillus sphaericus*. *Int. J. Biochem.* **24**, 471–476.

Cheng, M.M., Aronson, A.I. & Holt, S.C. (1973) Role of glutathione in the morphogenesis of the bacterial spore coat. *J. Bacteriol.* **113**, 1134–1143.

Garnier, M., Vacheron, M.-J., Guinand, M. & Michel, G. (1985) Purification and partial characterization of the extracellular γ-D-glutamyl-(L)*meso*-diaminopimelate endopeptidase I, from *Bacillus sphaericus* NCTC 9602. *Eur. J. Biochem.* **148**, 539–543.

Gmeiner, J. & Kroll, H.P. (1981) *N*-acetylglucosaminyl-*N*-acetylmuramyl-dipeptide, a novel murein building block formed during the cell division cycle of *Proteus mirabilis. FEBS Lett.* **129**, 142–144.

Goodell, E.W. & Schwarz, U. (1985) Release of cell wall peptides into culture medium by exponentially growing *E. coli. J. Bacteriol.* **162**, 391–397.

Guinand, M., Michel, G. & Tipper, D.J. (1974) Appearance of a γ-D-glutamyl-(L)*meso*-diaminopimelate peptidoglycan hydrolase during sporulation in *Bacillus sphaericus. J. Bacteriol.* **120**, 173–184.

Guinand, M., Michel, G. & Balassa, G. (1976) Lytic enzymes in sporulating *Bacillus subtilis. Biochem. Biophys. Res. Commun.* **68**, 1287–1293.

Guinand, M., Vacheron, M.-J. & Michel, G. (1978) Relation between inhibition of *Bacillus* sporulation and synthesis of lytic enzymes. *Biochem. Biophys. Res. Commun.* **80**, 429–434.

Guinand, M., Vacheron, M.-J., Michel, G. & Tipper, D.J. (1979) Location of peptidoglycan lytic enzymes in *Bacillus sphaericus. J. Bacteriol.* **138**, 126–132.

Hourdou, M.-L., Duez, C., Joris, B., Vacheron, M.-J., Guinand, M., Michel, G. & Ghuysen, J.-M. (1992) Cloning and nucleotide sequence of the gene encoding the γ-D-glutamyl-L-diaminoacid endopeptidase II of *Bacillus sphaericus. FEMS Microbiol. Lett.* **91**, 165–170.

Hourdou, M.-L., Guinand, M., Vacheron, M.-J., Michel, G., Denoroy, L., Duez, C., Englebert, S., Joris, B., Weber, G. & Ghuysen, J.-M. (1993) Characterization of the sporulation-related γ-D-glutamyl-(L)*meso*-diaminopimelic-acid-hydrolysing peptidase I of *Bacillus sphaericus* NCTC 9602 as a member of the metallo (zinc) carboxypeptidase A family. Modular design of the protein. *Biochem. J.* **292**, 563–570.

Morel, P., Guinand, M., Vacheron, M.-J. & Michel, G. (1986a) Biologically active glycopeptides from *Actinomadura* R 39. I. Continuous preparation of glycotri- and glyco-tetrapeptides with immobilized DD-carboxypeptidase from *Streptomyces albus* G. *Biotechnol. Appl. Biochem.* **8**, 404–413.

Morel, P., Guinand, M., Vacheron, M.-J. & Michel, G. (1986b) Biologically active glycopeptides from *Actinomadura* R 39. II. Continuous preparation of glycopeptides with immobilized endopeptidase I from *B. sphaericus* NCTC 9602. *Biotechnol. Appl. Biochem.* **8**, 414–422.

Schleifer, K.H. & Kandler, O. (1972) Peptidoglycan types of bacterial cell walls and their taxonomic implications. *Bacteriol. Rev.* **36**, 407–477.

Vacheron, M.-J., Guinand, M., Françon, A. & Michel, G. (1979) Caractérisation d'une nouvelle endopeptidase spécifique des liaisons γ-D-glutamyl-L-lysine et γ-D-glutamyl-(L)*meso*-diaminopimélate de substrats peptidoglycaniques chez *Bacillus sphaericus* 9602 au cours de la sporulation [Characterization of a novel endopeptidase specific for the bonds γ-D-glutamyl-L-lysine and γ-D-glutamyl-diaminopimelate of peptidoglycan substrates in *Bacillus sphaericus* 9602 during sporulation]. *Eur. J. Biochem.* **100**, 189–196.

Valentin, C., Vacheron, M.-J., Martinez, C., Guinand, M. & Michel, G. (1983) Action d'endopeptidases de *Bacillus sphaericus* sur les peptidoglycanes bactériens et sur des fragments peptidoglycaniques [Action of endopeptidases of *Bacillus sphaericus* on bacterial peptidoglycans and on peptidoglycan fragments]. *Biochimie* **65**, 239–245.

Warth, A.D. & Strominger, J.L. (1972) Structure of the peptidoglycan from spores of *Bacillus subtilis. Biochemistry* **11**, 1389–1395.

Williams, W.J., Litwin, J. & Thorne, C.B. (1955) Further studies on the biosynthesis of γ-glutamyl peptides by transfer reactions. *J. Biol. Chem.* **212**, 427–438.

*Micheline Guinand*
*Laboratoire de Biochimie Analytique,*
*Université Claude Bernard Lyon 1,*
*43 Boulevard du 11 Novembre 1918,*
*F-69622 Villeurbanne Cedex, France*

# 458. Carboxypeptidase E/H

## Databanks

*Peptidase classification: clan MC, family M14, MEROPS ID: M14.005*
*NC-IUBMB enzyme classification: EC 3.4.17.10*
*ATCC entries: 78805, 78805D (human); 103457 (human)*
*Chemical Abstracts Service registry number: 81876-95-1*
*Databank codes:*

| Species | SwissProt | PIR | EMBL (cDNA) | EMBL (genomic) |
|---|---|---|---|---|
| *Bos taurus* | P04836 | – | X04411 | – |
| *Caenorhabditis elegans* | – | – | – | Z68134: cosmid T27A8 |
| *Homo·sapiens* | P16870 | S09489 | X51405 | – |
| *Lophius americanus* | P37892 | – | U01909 | – |
| *Mus musculus* | Q00493 | – | U23184 X61232 | – |
| *Rattus norvegicus* | P15087 | A40469 | J04625 M31602 X51406 | L07273–L07281: complete gene |

## Name and History

*Carboxypeptidase E (CPE)* was originally discovered in the cattle adrenal medulla where it is localized to the enkephalin-containing chromaffin vesicles (Fricker & Snyder, 1982). For publication, the name *enkephalin convertase* was used (Fricker & Snyder, 1982). An earlier report on insulin processing presumably described the same enzyme, but this insulin-processing carboxypeptidase was not well characterized at the time (Zuhlke *et al*., 1977). Subsequently, CPE has been given a variety of names, the most commonly used being *carboxypeptidase H*, and this is the name recommended by the NC-IUBMB. CPE has been purified from several neuroendocrine tissues (Fricker & Snyder, 1983; Davidson & Hutton, 1987).

For many years CPE was thought to be the only carboxypeptidase involved in peptide processing. However, the recent finding that *fat/fat* mice lack CPE activity (due to a point mutation in a coding region) but still retain the ability to partially process peptides has raised the possibility that additional carboxypeptidases participate in peptide processing (Naggert *et al*., 1995). One candidate peptide-processing enzyme is metallocarboxypeptidase D (Chapter 461). However, peptide processing in the *fat/fat* mouse is greatly affected by the lack of CPE activity, which is strong evidence that CPE is the major carboxypeptidase for many peptides and that the other enzymes play a minor role.

## Activity and Specificity

CPE removes C-terminal basic residues (Arg, Lys ≫ His) from a variety of substrates, with no detectable activity towards nonbasic residues (Fricker & Snyder, 1983; Smyth *et al*., 1989). For all the substrates examined to date, Ala is preferred in the penultimate (P1) position (Fricker & Snyder, 1983). All residues are tolerated in the P1 position, although the Pro┼Arg bond is cleaved several orders of magnitude more slowly than substrates with other residues in P1 (Smyth *et al*., 1989). Amongst N-terminally blocked small peptides, CPE shows lower activity with dipeptides than with tri- or tetrapeptides.

The optimal pH for CPE activity is 5.0–5.5, and activity falls off dramatically with increasing pH. Kinetic analysis shows that this pH effect is due to a change in $K_m$, and not a change in $V$ (Greene *et al*., 1992). The sensitivity of CPE to pH over the range 5.5–7 may serve as a mechanism to keep CPE inactive in the Golgi (which has a neutral internal pH) and allow for the activation of CPE in mature secretory vesicles (which have an internal pH of 5–5.5).

As found with other metallocarboxypeptidases, CPE is stimulated by millimolar concentrations of $Co^{2+}$ and inhibited by 1,10-phenanthroline. However, unlike other metallocarboxypeptidases, CPE is potently inhibited by some thiol-directed reagents. The purified enzyme is inhibited >95% by 0.1 mM *p*-chloromercuriphenylsulfonate or by 0.01 mM $HgCl_2$ (Song & Fricker, 1995).

A variety of substrates for CPE has been described (Fricker & Snyder, 1982; Hook & Loh, 1984; Stack *et al*., 1984; Grimwood *et al*., 1988; Fricker & Devi, 1990), although none of them are currently commercially available. Several of the assays make use of the solubility change incurred when the basic residue is removed from the C-terminus of the substrate, and follow the product either by fluorescence (Dns-Phe-Ala┼Arg) or by radioactivity ([$^3$H]Bz-Phe-Ala┼Arg or [$^{125}$I]Ac-Tyr-Ala-Arg). After removal of the basic residue, the products readily partition into organic solvents such as chloroform when the aqueous phase is first acidified. This extraction permits the rapid analysis of a large number of samples. Assays for CPE using HPLC or thin-layer chromatography to separate the product from substrate have also been described (Hook & Loh, 1984; Grimwood *et al.*, 1988).

None of the assays for CPE are specific for this enzyme. The other major enzyme with CPE-like properties is metallocarboxypeptidase D (Chapter 461), not to be confused with the serine carboxypeptidase D (Chapter 134), which can only be distinguished from CPE by partial purification on a substrate affinity column (Song & Fricker, 1995). CPE elutes from a *p*-aminobenzoyl-Arg-Sepharose resin when the pH is raised to 8, whereas carboxypeptidase D remains bound at this pH and does not elute until a competitive ligand such as Arg is added (Song & Fricker, 1995).

## Structural Chemistry

CPE is a single-chain protein of 52–56 kDa, with a pI around 5. It is likely that CPE contains 1 mol of zinc per mol of protein based on the homology between CPE and other metallocarboxypeptidases (Fricker *et al.*, 1986). In addition to a carboxypeptidase domain of about 390 amino acids, CPE contains a C-terminal domain of about 40 amino acids; this domain is not conserved with other members of the metallocarboxypeptidase family (Fricker *et al.*, 1986). The C-terminal domain directs the pH-dependent peripheral association of CPE with membranes and contributes to the sorting of this protein into the secretory pathway (Varlamov & Fricker, 1996). Variation in the C-terminal region has been detected and presumably gives rise to the forms of the protein that are soluble. In addition to the C-terminal variability, a form of CPE with a longer N-terminal region has been isolated (Parkinson, 1990). This form corresponds to proCPE with the 27 residue signal peptide removed, but with a 14 residue pro region remaining on the N-terminus. Interestingly, proCPE is enzymatically active (Parkinson, 1990). CPE is glycosylated on the two Asn-Xaa-Ser/Thr sites found in the protein. Glycosylation does not appear to affect enzyme activity.

## Preparation

CPE has been purified to homogeneity in microgram quantities from cattle pituitary, rat brain, and several other tissues. The soluble form can be purified directly, while the membrane forms require extraction with a combination of 1% Triton X-100 and 1 M NaCl prior to purification (Fricker *et al.*, 1990). From tissues in which CPE is moderately abundant (pituitary, brain) purification to homogeneity is accomplished by a single step on a *p*-aminobenzoyl-Arg-Sepharose affinity resin. The synthesis of this noncommercially available resin has been described (Plummer & Hurwitz, 1978). The commercially available Arg-Sepharose does not bind CPE with high affinity and is not suitable for the purification of this enzyme. When purifying CPE from tissues with lower amounts of this protein, an additional ion-exchange step is usually sufficient to provide homogeneous enzyme (Fricker & Snyder, 1983).

CPE (rat) has been expressed in the baculovirus system in high levels, with approximately 1 mg of enzyme purified from 1 liter of cultured cells (Varlamov *et al.*, 1996). The properties of the baculovirus-produced CPE are indistinguishable from those of CPE purified from brain or pituitary. Bacterial expression of CPE does not produce active enzyme.

## Biological Aspects

CPE functions in the biosynthesis of many peptide hormones and neurotransmitters. Most bioactive peptides are initially produced as precursors that require proteolytic processing at specific sites. In many cases, the processing sites are pairs of basic amino acids (Lys, Arg). The initial cleavage occurs to the C-terminus of the basic amino acids and is catalyzed by one of several enzymes, including prohormone convertases 1 and 2 (Chapters 118 and 119) (Steiner *et al.*, 1992). Removal of the C-terminal basic residues is necessary before many peptides are biologically active and is an absolute requirement for peptides that undergo C-terminal amidation, which require a free C-terminal Gly for the reaction catalyzed by peptidylglycine-$\alpha$-amidating monooxygenase.

The distributions of CPE mRNA and protein are consistent with a broad role for this enzyme in the production of numerous biologically active peptides (Zheng *et al.*, 1994; Schafer *et al.*, 1994). CPE is found in high levels in brain, pituitary and pancreatic islets, and in low to moderate levels in many other neuroendocrine tissues. In several tissues, subcellular fractionation has shown that CPE is enriched in peptide-containing secretory vesicles (Fricker & Snyder, 1982; Docherty & Hutton, 1983; Hook & Loh, 1984). Also, CPE is secreted from neuroendocrine cells via the regulated secretory pathway, consistent with the colocalization of CPE and peptides in the same vesicles (Mains & Eipper, 1984; Vilijn *et al.*, 1989).

The CPE gene consists of nine exons, of which the first eight share homology with other members of the metallocarboxypeptidase gene family (Jung *et al.*, 1991). The CPE gene has been mapped to human chromosome 4 and mouse chromosome 8 (Hall *et al.*, 1993; Naggert *et al.*, 1995). The *fat* mutation, a spontaneously arising mouse mutation discovered in 1973, maps to the CPE locus on chromosome 8 (Naggert *et al.*, 1995). A point mutation is present within the coding region of CPE that changes Ser202 into a Pro. This mutation causes CPE to be completely inactive and to be degraded in the early secretory pathway, prior to packaging into secretory vesicles (Varlamov *et al.*, 1996). Consistent with the lack of CPE activity, mice homozygous for the *fat* mutation have levels of correctly processed peptides that are 10–50% of the levels in control mice (Naggert *et al.*, 1995; Rovere *et al.*, 1996; Fricker *et al.*, 1996). In addition, levels of the C-terminally extended peptides are greatly increased in the *fat/fat* mice. Mutations of CPE in humans have not been reported.

## Distinguishing Features

Of all the known metallocarboxypeptidases, CPE has the lowest pH optimum (5.0–5.5) and is the most sensitive to thiol reagents like $HgCl_2$. However, carboxypeptidase D (CPD) (Chapter 461) has considerable activity at pH 5.0–5.5 and is partially sensitive to thiol reagents (Song & Fricker, 1995). Since CPD is present in all tissues that contain CPE activity, it is not possible to specifically measure CPE in crude homogenates. As described above, the two enzymes can be physically separated on a substrate affinity resin. The physical properties of CPE and D are much different; CPE is 52–56 kDa whereas CPD is primarily 180 kDa in cattle pituitary and 100–180 kDa in various rat tissues (Song & Fricker, 1995, 1996).

The size of CPE is comparable to the size of carboxypeptidase M (Chapter 460) and lysine carboxypeptidase (Chapter 459), which share approximately 50% amino acid identity within the carboxypeptidase domain. Other proteins which contain carboxypeptidase domains with approximately 40–50% amino acid identity to CPE include CPD (Xin *et al.*, 1997), a putative transcription factor designated AEBP1 (He *et al.*, 1995), and a novel protein designated carboxypeptidase Z (L. Song & L.D. Fricker, unpublished results). Carboxypeptidases M, N and Z are maximally active at neutral pH. The properties of AEBP1 have not been reported, but the baculovirus-expressed protein does not appear to have much enzymatic activity (L.D. Fricker, unpublished results). All of these related proteins only share the carboxypeptidase domain, and so antisera directed against the N- or C-termini of CPE are quite specific for this protein (Fricker *et al.*, 1990; Parkinson, 1990).

## Further Reading

The properties and distribution of CPE have been reviewed (Fricker, 1988, 1991). A detailed description of the assay has recently been published (Fricker, 1995), but this assay was written prior to the discovery of CPD so the discussion regarding the specificity of the assay is not current.

## References

Davidson, H.W. & Hutton, J.C. (1987) The insulin secretory granule carboxypeptidase H: purification and demonstration of involvement in proinsulin processing. *Biochem. J.* **245**, 575–582.

Docherty, K. & Hutton, J.C. (1983) Carboxypeptidase activity in the insulin secretory granule. *FEBS Lett.* **162**, 137–141.

Fricker, L.D. (1988) Carboxypeptidase E. *Annu. Rev. Physiol.* **50**, 309–321.

Fricker, L.D. (1991) Peptide processing exopeptidases: amino- and carboxypeptidases involved with peptide biosynthesis. In: *Peptide Biosynthesis and Processing* (Fricker, L.D., ed.). Boca Raton: CRC Press, pp. 199–230.

Fricker, L.D. (1995) Methods for studying carboxypeptidase E. *Methods Neurosci.* **23**, 237–250.

Fricker, L.D. & Devi, L. (1990) Comparison of a spectrophotometric, a fluorometric, and a novel radiometric assay for carboxypeptidase E (EC 3.4.17.10) and other carboxypeptidase B-like enzymes. *Anal. Biochem.* **184**, 21–27.

Fricker, L.D. & Snyder, S.H. (1982) Enkephalin convertase: purification and characterization of a specific enkephalin-synthesizing carboxypeptidase localized to adrenal chromaffin granules. *Proc. Natl Acad. Sci. USA* **79**, 3886–3890.

Fricker, L.D. & Snyder, S.H. (1983) Purification and characterization of enkephalin convertase, an enkephalin-synthesizing carboxypeptidase. *J. Biol. Chem.* **258**, 10950–10955.

Fricker, L.D., Evans, C.J., Esch, F.S. & Herbert, E. (1986) Cloning and sequence analysis of cDNA for bovine carboxypeptidase E. *Nature (Lond.)* **323**, 461–464.

Fricker, L.D., Das, B. & Angeletti, R.H. (1990) Identification of the pH-dependent membrane anchor of carboxypeptidase E (EC 3.4.17.10). *J. Biol. Chem.* **265**, 2476–2482.

Fricker, L.D., Berman, Y.L., Leiter, E.H. & Devi, L.A. (1996) Carboxypeptidase E activity is deficient in mice with the *fat* mutation: effect on peptide processing. *J. Biol. Chem.* **271**, 30619–30624.

Greene, D., Das, B. & Fricker, L.D. (1992) Regulation of carboxypeptidase E: effect of pH, temperature, and Co++ on kinetic parameters of substrate hydrolysis. *Biochem. J.* **285**, 613–618.

Grimwood, B.G., Tarentino, A.L. & Plummer, T.H.J. (1988) High performance liquid chromatographic quantitation of carboxypeptidase activity secreted by human Hep G2 cells. *Anal. Biochem.* **170**, 264–268.

Hall, C., Manser, E., Spurr, N.K. & Lim, L. (1993) Assignment of the human carboxypeptidase E (CPE) gene to chromosome 4. *Genomics* **15**, 461–463.

He, G.P., Muise, A., Li, A.W. & Ro, H.S. (1995) A eukaryotic transcriptional repressor with carboxypeptidase activity. *Nature* **378**, 92–96.

Hook, V.Y.H. & Loh, Y.P. (1984) Carboxypeptidase B-like converting enzyme activity in secretory granules of rat pituitary. *Proc. Natl Acad. Sci. USA* **81**, 2776–2780.

Jung, Y.K., Kunczt, C.J., Pearson, R.K., Dixon, J.E. & Fricker, L.D. (1991) Structural characterization of the rat carboxypeptidase-E gene. *Mol. Endocrinol.* **5**, 1257–1268.

Mains, R.E. & Eipper, B.A. (1984) Secretion and regulation of two biosynthetic enzyme activities, peptidyl-glycine alpha-amidating monooxygenase and a carboxypeptidase, by mouse pituitary corticotropic tumor cells. *Endocrinology* **115**, 1683–1690.

Naggert, J.K., Fricker, L.D., Varlamov, O., Nishina, P.M., Rouille, Y., Steiner, D.F., Carroll, R.J., Paigen, B.J. & Leiter, E.H. (1995) Hyperproinsulinemia in obese *fat/fat* mice associated with a point mutation in the carboxypeptidase E gene and reduced carboxypeptidase E activity in the pancreatic islets. *Nature Genet.* **10**, 135–142.

Parkinson, D. (1990) Two soluble forms of bovine carboxypeptidase H have different NH2-terminal sequences. *J. Biol. Chem.* **265**, 17101–17105.

Plummer, T.H.J. & Hurwitz, M.Y. (1978) Human plasma carboxypeptidase N: isolation and characterization. *J. Biol. Chem.* **253**, 3907–3912.

Rovere, C., Viale, A., Nahon, J. & Kitabgi, P. (1996) Impaired processing of brain proneurotensin and promelanin-concentrating hormone in obese *fat/fat* mice. *Endocrinology* **137**, 2954–2958.

Schafer, M.K.H., Day, R., Cullinan, W.E., Chrétien, M. & Watson, S.J. (1994) Gene expression of prohormone and proprotein convertases in the rat CNS: a comparative *in situ* hybridization analysis. *J. Neurosci.* **13**, 1258–1279.

Smyth, D.G., Maruthainar, K., Darby, N.J. & Fricker, L.D. (1989) C-terminal processing of neuropeptides: involvement of carboxypeptidase H. *J. Neurochem.* **53**, 489–493.

Song, L. & Fricker, L.D. (1995) Purification and characterization of carboxypeptidase D, a novel carboxypeptidase E-like enzyme, from bovine pituitary. *J. Biol. Chem.* **270**, 25007–25013.

Song, L. & Fricker, L.D. (1996) Tissue distribution and characterization of soluble and membrane-bound forms of metallocarboxypeptidase D. *J. Biol. Chem.* **271**, 28884–28889.

Stack, G., Fricker, L.D. & Snyder, S.H. (1984) A sensitive radiometric assay for enkephalin convertase and other carboxypeptidase B-like enzymes. *Life Sci.* **34**, 113–121.

Steiner, D.F., Smeekens, S.P., Ohagi, S. & Chan, S.J. (1992) The new enzymology of precursor processing endoproteases. *J. Biol. Chem.* **267**, 23435–23438.

Varlamov, O. & Fricker, L.D. (1996) The C-terminal region of carboxypeptidase E involved in membrane binding is distinct from

M

the region involved with intracellular routing. *J. Biol. Chem.* **271**, 6077–6083.

Varlamov, O., Leiter, E.H. & Fricker, L.D. (1996) Induced and spontaneous mutations at Ser202 of carboxypeptidase E: effect on enzyme expression, activity, and intracellular routing. *J. Biol. Chem.* **271**, 13981–13986.

Vilijn, M.H., Das, B., Kessler, J.A. & Fricker, L.D. (1989) Cultured astrocytes and neurons synthesize and secrete carboxypeptidase E, a neuropeptide-processing enzyme. *J. Neurochem.* **53**, 1487–1493.

Xin, X., Varlamov, O., Day, R., Dong, W., Bridgett, M.M., Leiter, E.H. & Fricker, L.D. (1996) Cloning and sequence analysis of

cDNA encoding rat carboxypeptidase D. *DNA Cell Biol.* **16**, 897–909.

Zheng, M., Streck, R.D., Scott, R.E.M., Seidah, N.G. & Pintar, J.E. (1994) The developmental expression in rat of proteases furin, PC1, PC2, and carboxypeptidase E: implications for early maturation of proteolytic processing capacity. *J. Neurosci.* **14**, 4656–4673.

Zuhlke, H., Kohnert, K.D., Jahr, H., Schmidt, S., Kirschke, H. & Steiner, D.F. (1977) Proteolytic and transhydrogenolytic activities in isolated pancreatic islets of rats. *Acta Biol. Med. Germ.* **36**, 1695–1700.

*Lloyd D. Fricker*
*Department of Molecular Pharmacology,*
*Albert Einstein College of Medicine,*
*1300 Morris Park Avenue,*
*Bronx, NY 10461, USA*
*Email: fricker@aecom.yu.edu*

# 459. *Lysine carboxypeptidase*

## Databanks

*Peptidase classification: clan MC, family M14, MEROPS ID: M14.004*
*NC-IUBMB enzyme classification: EC 3.4.17.3*
*ATCC entries: 102806 (human); 103173 (human); 40927 (human)*
*Chemical Abstracts Service registry number: 9013-89-2*
*Databank codes:*

| Species | SwissProt | PIR | EMBL (cDNA) | EMBL (genomic) |
|---|---|---|---|---|
| 50 kDa catalytic subunit | | | | |
| *Homo sapiens* | P15169 | – | J05158 X14329 | – |
| 83 kDa noncatalytic subunit | | | | |
| *Homo sapiens* | P22792 | A34901 | J05158 | – |

## Name and History

Studies on bradykinin metabolism by Erdös and Sloane in the early 1960s led to the discovery of a plasma enzyme that inactivated the peptide and it was named *kininase I* (Erdös & Sloane, 1962). Because the enzyme removed the C-terminal Arg of bradykinin, it was called *carboxypeptidase N (CPN)* (Erdös, 1979). Although the Enzyme Commission originally named it *arginine carboxypeptidase*, it preferentially cleaves C-terminal lysine and for this reason the designation was recently changed to *lysine carboxypeptidase*. CPN has been rediscovered and renamed many times, acquiring numerous aliases including *anaphylatoxin inactivator, creatine kinase conversion factor, protaminase, carboxypeptidase K, plasma carboxypeptidase B* and *serum carboxypeptidase B* (Skidgel, 1996).

## Activity and Specificity

CPN cleaves C-terminal Arg or Lys from a variety of synthetic and naturally occurring peptides. In general, the enzyme cleaves Lys faster than Arg (due to a higher $k_{cat}$) and Ala is the preferred penultimate residue in most cases (Oshima *et al.*, 1975; Skidgel, 1995). Enzyme activity can be conveniently measured using FA-Ala↓Lys in a spectrophotometric assay (Plummer & Erdös, 1981) or Dns-Ala↓Arg in a fluorometric assay (Skidgel, 1991). CPN contains zinc as a required cofactor and is therefore inhibited by EDTA and 1,10-phenanthroline (Plummer & Hurwitz, 1978; Erdös, 1979). High-affinity biproduct analog inhibitors include guanidinoethylmercaptosuccinic acid ($K_i$ 1 µM) and DL-2-mercaptomethyl-3-guanidinoethylthiopropanoic acid ($K_i$ 2 nM)

(Plummer & Ryan, 1981; Plummer & Erdös, 1981). $Co^{2+}$ activates and $Cd^{2+}$ inhibits the enzyme (Erdös, 1979). The pH optimum of CPN is in the neutral range but added $Co^{2+}$ at pH 5.5 enhances the activity to 156% of that measured at pH 7.5 without $Co^{2+}$ (Deddish *et al.*, 1989).

## Structural Chemistry

Human CPN is a tetrameric enzyme of 280 kDa consisting of a dimer of heterodimers. Each heterodimer contains one catalytic subunit, with a reported molecular mass ranging from 44 to 55 kDa in SDS-PAGE (referred to here as 50 kDa), and one noncatalytic 83 kDa subunit (Plummer & Hurwitz, 1978; Levin *et al.*, 1982). The 83 kDa subunit is heavily glycosylated (about 28% by weight) whereas the active subunit lacks carbohydrate (Levin *et al.*, 1982; Plummer & Hurwitz, 1978). The function of the 83 kDa subunit is to bind, protect and stabilize the active subunit at 37°C in the circulation (Levin *et al.*, 1982; Skidgel, 1988). The cDNA sequence of the 50 kDa catalytic subunit (Gebhard *et al.*, 1989) encodes a mature protein of 438 amino acids with low sequence identity to human pancreatic carboxypeptidase A (21%; Chapter 451) or B (17%; Chapter 455) but higher identity with human carboxypeptidase M (41%; Chapter 460) and carboxypeptidase E/H (51%; Chapter 458). The cDNA sequence of the 83 kDa subunit encodes a 59 kDa protein with no sequence similarity to any known carboxypeptidases (Tan *et al.*, 1990). The most striking feature is the presence of 12 leucine-rich tandem repeats of 24 amino acids each, a motif found in a variety of other proteins whose only common property is to bind to other macromolecules (Tan *et al.*, 1990; Skidgel & Tan, 1992). The leucine-rich repeat region is probably involved in the interaction of the 83 kDa subunit with the 50 kDa active subunit.

## Preparation

CPN can be readily purified from outdated human plasma by a relatively simple two-step procedure involving DE-52 ion-exchange chromatography and affinity chromatography on arginine-Sepharose (Oshima *et al.*, 1974; Plummer & Hurwitz, 1978; Levin *et al.*, 1982; Skidgel, 1995). By using batchwise adsorption for the chromatographic steps, the purification can be completed in 2–3 days (Skidgel, 1995). The yield from 2 liters of plasma is on the order of 10–15 mg. The individual 83 kDa and 50 kDa subunits of CPN can be isolated by treating the enzyme with 3 M guanidine followed by separation on a gel-filtration column (Levin *et al.*, 1982; Skidgel, 1995). Under these conditions, the 50 kDa subunits remain active and retain most of the enzymatic properties of the intact 280 kDa tetramer (Levin *et al.*, 1982).

## Biological Aspects

The 50 kDa subunit of CPN does not contain an N-terminal propeptide as do the pancreatic carboxypeptidases and is thus active once synthesized. However, treatment of the 50 kDa subunit with trypsin, plasmin or kallikrein results in a 41–84% increase in peptidase activity (Levin *et al.*, 1982), possibly caused by changes in conformation.

CPN is synthesized in the liver and released into the circulation where it is present at a relatively high concentration of approximately $30 \mu g \, ml^{-1}$ ($10^{-7}$ M; Erdös, 1979; Skidgel, 1988). Although liver is the site of synthesis, only low levels of CPN have been found there (Oshima *et al.*, 1975), probably due to the fact it is not stored, but secreted soon after synthesis. Northern blot analysis (F. Tan & R.A. Skidgel, unpublished results) does not indicate synthesis in other cells or organs.

The high concentration of CPN in blood and the fact that no person has been found who lacks the enzyme (Erdös *et al.*, 1965; Mathews *et al.*, 1980) suggest an important protective role in the body. Genetically determined low blood levels (about 20% of normal), caused by decreased hepatic synthesis, were associated with repeated attacks of angioedema in one patient (Mathews *et al.*, 1980, 1986). Conditions that affect hepatic plasma protein synthesis also alter plasma CPN levels: cirrhosis of the liver decreases it whereas pregnancy elevates it (Erdös *et al.*, 1965). Although a wide variety of diseases (e.g. cardiovascular disease, diabetes, allergic conditions) do not cause altered CPN levels (Erdös *et al.*, 1965; Mathews *et al.*, 1980), higher activities have been noted in certain types of cancer and in the blood and synovial fluid of arthritic patients (Erdös *et al.*, 1965; Mathews *et al.*, 1980; Skidgel, 1996).

The most likely function of CPN is to protect the organism from the actions of vasoactive peptides (e.g. kinins, anaphylatoxins C3a, C4a and C5a) that may be released in the circulation and enter the extravascular space (Skidgel, 1988). Evidence for this role was provided in studies on guinea pigs. These animals, when pretreated with a potent carboxypeptidase inhibitor, died within 5 min after complement activation (Huey *et al.*, 1983). In humans, low CPN activity may be a factor in the protamine-reversal syndrome. Protamine, when given to neutralize the effects of heparin after extracorporeal circulation, can trigger a catastrophic reaction in some patients, resulting in pulmonary vasoconstriction, bronchoconstriction and systemic hypotension (Lowenstein *et al.*, 1983). As protamine is a potent inhibitor of CPN (Tan *et al.*, 1989), possibly the decrease in anaphylatoxin and kinin inactivation contributes to the protamine reversal syndrome.

The activity of CPN in plasma can be relevant in the diagnosis of myocardial infarction where the blood level of creatine kinase (CK) released from heart muscle is measured. This is because CPN cleaves the C-terminal lysine residue from either the M or B subunit of CK to yield different MM or MB isoforms that can be separated electrophoretically (Michelutti *et al.*, 1987; Prager *et al.*, 1992). A modified assay was recently developed that relies on measuring the ratio of the CK-MB$_2$ subform (containing C-terminal lysine) to that of CK-MB$_1$ (des-lysine form), which significantly increases the specificity and sensitivity of this diagnostic procedure (Puleo *et al.*, 1994). A variation in blood CPN activity would have obvious relevance to the use of the CK-MB isoform ratio (Erdös & Skidgel, 1995).

Finally, CPN may regulate plasminogen binding to cells. It is known that C-terminal lysine residues on cell surface proteins act as plasminogen 'receptors' and thereby control its activation (Miles *et al.*, 1991). A major cellular plasminogen

M

receptor was identified as $\alpha$-enolase (Miles *et al.*, 1991), a known substrate of CPN (Skidgel, 1996). Indeed, CPN in plasma, as well as purified CPN, effectively reduces plasminogen binding to cells (Redlitz *et al.*, 1995), indicating that this constitutively active enzyme may help to regulate fibrinolysis.

## Distinguishing Features

CPN can be distinguished from other B-type carboxypeptidases by the fact that it is constitutively active in the circulation, has a much higher molecular mass than related enzymes and is the only member of this family to have multiple subunits.

## Related Peptidases

The other known mammalian carboxypeptidases that specifically cleave C-terminal Arg or Lys are pancreatic carboxypeptidase B (Chapter 455), carboxypeptidase U (Chapter 453), carboxypeptidase M (Chapter 460), carboxypeptidase E (or H) (Chapter 458), and the newly described metallocarboxypeptidase D (Chapter 461) (Skidgel, 1996; Song & Fricker, 1995).

## Further Readings

Purification and assay are outlined in Skidgel (1995). The enzyme's relationship to other metallocarboxypeptidases is reviewed in Skidgel (1996).

## References

Deddish, P.A., Skidgel, R.A. & Erdös, E.G. (1989) Enhanced $Co^{2+}$ activation and inhibitor binding of carboxypeptidase M at low pH. Similarity to carboxypeptidase H (enkephalin convertase). *Biochem. J.* **261**, 289–291.

Erdös, E.G. (1979) Kininases. In: *Bradykinin, Kallidin and Kallikrein. Handbook of Experimental Pharmacology*, vol. 25 suppl. (Erdös, E.G., ed.). Heidelberg: Springer-Verlag, pp. 427–448.

Erdös, E.G. & Skidgel, R. (1995) More on the subforms of creatine kinase MB [Letter]. *N. Engl. J. Med.* **333**, 390.

Erdös, E.G. & Sloane, E.M. (1962) An enzyme in human blood plasma that inactivates bradykinin and kallidins. *Biochem. Pharmacol.* **11**, 585–592.

Erdös, E.G., Wohler, I.M., Levine, M.I. & Westerman, P. (1965) Carboxypeptidase in blood and other fluids. Values in human blood in normal and pathological conditions. *Clin. Chim. Acta* **11**, 39–43.

Gebhard, W., Schube, M. & Eulitz, M. (1989) cDNA cloning and complete primary structure of the small, active subunit of human carboxypeptidase N (kininase 1). *Eur. J. Biochem.* **178**, 603–607.

Huey, R., Bloor, C.M., Kawahara, M.S. & Hugli, T.E. (1983) Potentiation of the anaphylatoxins *in vivo* using an inhibitor of serum carboxypeptidase N (SCPN). I. Lethality and pathologic effects on pulmonary tissue. *Am. J. Pathol.* **112**, 48–60.

Levin, Y., Skidgel, R.A. & Erdös, E.G. (1982) Isolation and characterization of the subunits of human plasma carboxypeptidase N (kininase I). *Proc. Natl Acad. Sci. USA* **79**, 4618–4622.

Lowenstein, E., Johnston, E.W., Lappas, D.G., D'Ambra, M.N., Schneider, R.C., Daggett, W.M., Akins, C.W. & Philbin, D.M.

(1983) Catastrophic pulmonary vasoconstriction associated with protamine reversal of heparin. *Anesthesiology* **59**, 470–473.

Mathews, K.P., Pan, P.M., Gardner, N.J. & Hugli, T.E. (1980) Familial carboxypeptidase N deficiency. *Ann. Intern. Med.* **93**, 443–445.

Mathews, K.P., Curd, J.G. & Hugli, T.E. (1986) Decreased synthesis of serum carboxypeptidase N (SCPN) in familial SCPN deficiency. *J. Clin. Immunol.* **6**, 87–91.

Michelutti, L., Falter, H., Certossi, S., Marcotte, B. & Mazzuchin, A. (1987) Isolation and purification of creatine kinase conversion factor from human serum and its identification as carboxypeptidase N. *Clin. Biochem.* **20**, 21–29.

Miles, L.A., Dahlberg, C.M., Plescia, J., Felez, J., Kato, K. & Plow, E.F. (1991) Role of cell-surface lysines in plasminogen binding to cells: Identification of $\alpha$-enolase as a candidate plasminogen receptor. *Biochemistry* **30**, 1682–1691.

Oshima, G., Kato, J. & Erdös, E.G. (1974) Subunits of human plasma carboxypeptidase N (kininase I; anaphylatoxin inactivator). *Biochim. Biophys. Acta* **365**, 344–348.

Oshima, G., Kato, J. & Erdös, E.G. (1975) Plasma carboxypeptidase N, subunits and characteristics. *Arch. Biochem. Biophys.* **170**, 132–138.

Plummer, T.H., Jr. & Erdös, E.G. (1981) Human carboxypeptidase N. *Methods Enzymol.* **80**, 442–449.

Plummer, T.H., Jr. & Hurwitz, M.Y. (1978) Human plasma carboxypeptidase N. Isolation and characterization. *J. Biol. Chem.* **253**, 3907–3912.

Plummer, T.H., Jr. & Ryan, T.J. (1981) A potent mercapto biproduct analogue inhibitor for human carboxypeptidase N. *Biochem. Biophys. Res. Commun.* **98**, 448–454.

Prager, N.A., Suzuki, T., Jaffe, A.S., Sobel, B.E. & Abendschein, D.R. (1992) Nature and time course of generation of isoforms of creatine kinase, MB fraction *in vivo*. *J. Am. Coll. Cardiol.* **20**, 414–419.

Puleo, P.R., Meyer, D., Wathen, C., Tawa, C.B., Wheeler, S., Hamburg, R.J., Ali, N., Obermueller, S.D., Triana, J.F., Zimmerman, J.L., Perryman, M.B. & Roberts, R. (1994) Use of a rapid assay of subforms of creatine kinase MB to diagnose or rule out acute myocardial infarction. *N. Engl. J. Med.* **331**, 561–566.

Redlitz, A., Tan, A.K., Eaton, D.L. & Plow, E.F. (1995) Plasma carboxypeptidases as regulators of the plasminogen system. *J. Clin. Invest.* **96**, 2534–2538.

Skidgel, R.A. (1988) Basic carboxypeptidases: regulators of peptide hormone activity. *Trends Pharmacol. Sci.* **9**, 299–304.

Skidgel, R.A. (1991) Assays for arginine/lysine carboxypeptidases: carboxypeptidase H (E; Enkephalin convertase), M and N. In: *Methods in Neurosciences: Peptide Technology*, vol. 6 (Conn, P.M., ed.). Orlando: Academic Press, pp. 373–385.

Skidgel, R.A. (1995) Human carboxypeptidase N (lysine carboxypeptidase). *Methods Enzymol.* **248**, 653–663.

Skidgel, R.A. (1996) Structure and function of mammalian zinc carboxypeptidases. In: *Zinc Metalloproteases in Health and Disease*. (Hooper, N.M., ed.). London: Taylor & Francis, pp. 241–283.

Skidgel, R.A. & Tan, F. (1992) Structural features of two kininase I-type enzymes revealed by molecular cloning. *Agents Actions* (suppl.) **38/I**, 359–367.

Song, L. & Fricker, L.D. (1995) Purification and characterization of carboxypeptidase D, a novel carboxypeptidase E-like enzyme, from bovine pituitary. *J. Biol. Chem.* **270**, 25007–25013.

Tan, F., Jackman, H., Skidgel, R.A., Zsigmond, E.K. & Erdös, E.G. (1989) Protamine inhibits plasma carboxypeptidase N, the inactivator of anaphylatoxins and kinins. *Anesthesiology* **70**, 267–275.

Tan, F., Weerasinghe, D.K., Skidgel, R.A., Tamei, H., Kaul, R.K., Roninson, I.B., Schilling, J.W. & Erdös, E.G. (1990) The deduced protein sequence of the human carboxypeptidase N high molecular weight subunit reveals the presence of leucine-rich tandem repeats. *J. Biol. Chem.* **265**, 13–19.

***Randal A. Skidgel***
*Laboratory of Peptide Research,*
*Departments of Pharmacology and Anesthesiology,*
*University of Illinois College of Medicine at Chicago,*
*835 S. Wolcott, Chicago, IL 60612, USA*
*Email: rskidgel@uic.edu*

***Ervin G. Erdös***
*Laboratory of Peptide Research,*
*Departments of Pharmacology and Anesthesiology,*
*University of Illinois College of Medicine at Chicago,*
*835 S. Wolcott, Chicago, IL 60612, USA*

# 460. *Carboxypeptidase M*

## Databanks

*Peptidase classification: clan MC, family M14, MEROPS ID: M14.006*
*NC-IUBMB enzyme classification: EC 3.4.17.12*
*Chemical Abstracts Service registry number: 120038-28-0*
*Databank codes:*

| Species | SwissProt | PIR | EMBL (cDNA) | EMBL (genomic) |
|---|---|---|---|---|
| *Homo sapiens* | P14384 | A32619 | J04970 | – |

## Name and History

Until the early 1980s, only two mammalian B-type carboxy-peptidases were known: pancreatic carboxypeptidase B (Chapter 455) and human plasma carboxypeptidase N (lysine carboxypeptidase; Chapter 459) (Erdös, 1979; Skidgel, 1996). Although a few reports hinted that a 'tissue' carboxypeptidase B-type enzyme might exist, the activity was never well characterized (Erdös, 1979). We discovered a membrane-bound carboxypeptidase that was present in a variety of tissues including kidney, lung, placenta and blood vessels, and in cultured endothelial cells and fibroblasts (Skidgel *et al.*, 1984b; Skidgel, 1988, 1996). Purification of the enzyme from human placenta (Skidgel *et al.*, 1989) and cloning and sequencing of its cDNA (Tan *et al.*, 1989) proved it to be unique. We named it ***carboxypeptidase M*** to denote the fact that it is membrane bound. The enzyme was recently found to be identical with the ***MAX.1 monocyte/macrophage differentiation antigen*** (Rehli *et al.*, 1995).

## Activity and Specificity

Carboxypeptidase M is activated by cobalt and inhibited by 1,10-phenanthroline, DL-2-mercaptomethyl-3-guanidinoethyl-thiopropanoic acid (MGTA), guanidinoethylmercaptosuccinic acid (GEMSA) and cadmium acetate, demonstrating that it is a metallopeptidase (Skidgel *et al.*, 1984b, 1989). It has a neutral pH optimum (Skidgel *et al.*, 1984b, 1989), but at lower pH values it can be more efficiently activated by cobalt and inhibited by GEMSA (Deddish *et al.*, 1989). Carboxypeptidase M cleaves only C-terminal Arg or Lys from a variety of synthetic and naturally occurring peptide substrates such as bradykinin, Arg$^6$- and Lys$^6$-enkephalins and dynorphin A$_{1-13}$ (Skidgel *et al.*, 1989). The enzyme prefers C-terminal arginine over lysine, however the penultimate residue can also dramatically affect the rate of hydrolysis (Skidgel *et al.*, 1989). Enzyme activity is routinely measured using Dns-Ala+Arg in a fluorometric assay (Skidgel, 1991; Tan *et al.*, 1995).

## Structural Chemistry

Carboxypeptidase M gives a single band of molecular mass 62 kDa with or without reduction in SDS-PAGE, showing that it is a single-chain protein (Skidgel *et al.*, 1989). In gel filtration in the presence of CHAPS detergent, it has a slightly higher molecular mass of 73 kDa (Skidgel *et al.*, 1989). Carboxypeptidase M is a glycoprotein with 23% carbohydrate content by weight as determined by chemical deglycosylation which reduced the molecular mass to 47.6 kDa (Skidgel *et al.*, 1989). A carboxypeptidase M clone was isolated from a human placental cDNA library coding for a mature protein of 426 residues (after removal of the signal peptide) (Tan *et al.*, 1989). The deduced sequence contains six potential glycosylation sites and a C-terminal hydrophobic sequence that acts as a signal for glycosylphosphatidylinositol (GPI) attachment (Tan *et al.*, 1989). The deduced amino acid sequence has significant identity with carboxypeptidase E/H (43%; Chapter 458) and the lysine carboxypeptidase active subunit (41%; Chapter 459) but much lower (15%) with pancreatic carboxypeptidase A (Chapter 451) or B (Chapter 455) (Tan *et al.*, 1989).

## Preparation

Carboxypeptidase M can be purified from a microvillous membrane fraction of human placenta (Skidgel *et al.*, 1989; Tan *et al.*, 1995). After solubilization by either bacterial phosphatidylinositol-specific phospholipase C (PI-PLC) or CHAPS detergent, the enzyme is purified by ion-exchange chromatography on a Q-Sepharose column followed by affinity chromatography on arginine-Sepharose (Skidgel *et al.*, 1989; Tan *et al.*, 1995). An additional one or two steps of purification can be performed (e.g. gel filtration), depending on the purity required (Tan *et al.*, 1995). Recombinant carboxypeptidase M has also been successfully expressed in baculovirus-infected insect cells (F. Tan & R.A. Skidgel, unpublished results).

## Biological Aspects

Carboxypeptidase M is a widely distributed enzyme, present in a variety of tissues and cells where it is anchored via GPI to the plasma membrane. The highest levels are found in lung and placenta, but significant amounts are also present in kidney, blood vessels, intestine, brain and in peripheral nerves (Nagae *et al.*, 1992, 1993; Skidgel *et al.*, 1984b; Skidgel, 1988, 1996). Carboxypeptidase M is also present in soluble form in various body fluids including amniotic fluid, seminal plasma and urine (McGwire & Skidgel, 1995; Skidgel *et al.*, 1984a,b; Skidgel, 1996). It was recently discovered to be a differentiation-dependent cell surface antigen on white blood cells (Rehli *et al.*, 1995; de Saint-Vis *et al.*, 1995).

Because of its distribution and subcellular localization, carboxypeptidase M may participate in a variety of processes, such as control of peptide hormone activity at the cell surface, in degradation of extracellular proteins and peptides, and in extracellular prohormone processing, if incompletely processed peptides are released from secretory granules (Skidgel, 1988, 1996). For example, carboxypeptidase M might either inactivate or alter the specificity of kinins which interact with two types of receptors: the B2 receptor, which binds the native peptide, and the B1 receptor, which binds to kinins lacking the C-terminal arginine (Bhoola *et al.*, 1992). Thus, carboxypeptidase M can produce an agonist for the B1 receptor or inactivate the agonist for the B2 receptor. This may play an important role in inflammatory or pathological responses. Carboxypeptidase M can alter the activity of other inflammatory mediators as well. For instance, removal of the C-terminal Arg of human anaphylatoxin C5a abolishes its spasmogenic and histamine-releasing activity whereas about 10% of the chemotactic activity is retained (Marceau *et al.*, 1987).

Carboxypeptidase M may also participate in the metabolism of growth factors, several of which contain a C-terminal Arg or Lys. These include epidermal growth factor (EGF; urogastrone) and EGF-like peptides, nerve growth factor, amphiregulin, hepatocyte growth factor, erythropoietin, macrophage-stimulating protein, brain-derived neurotrophic factor, vascular endothelial growth factor, Müllerian-inhibiting substance, hemoregulatory peptide and oncostatin M. Indeed, our recent investigations showed that purified carboxypeptidase M readily converts EGF to des-Arg$^{53}$-EGF and that CPM is the primary extracellular EGF-metabolizing enzyme in Madin–Darby canine kidney cells, urine and amniotic fluid (McGwire & Skidgel, 1995).

In the kidney, carboxypeptidase M may control the activity of kinins which are synthesized and released in the distal tubules. Carboxypeptidase M may also have important functions in the lung as indicated by the high activity present in all species tested (Chodimella *et al.*, 1991; Nagae *et al.*, 1993). The pulmonary type I epithelial cells exhibit strong immunochemical staining for carboxypeptidase M (Nagae *et al.*, 1993) and comprise 93% of the total surface area, indicating that the enzyme might play a protective role here. Carboxypeptidase M may also be readily mobilized from the cell surface. For example, bronchoalveolar lavage fluid contains significant carboxypeptidase M activity which may be elevated almost 5-fold in patients with pneumocystic or bacterial pneumonia or lung cancer (Dragovic *et al.*, 1995). Carboxypeptidase M was also released into the edema fluid of rat lung in an experimental model of lung injury (Dragovic *et al.*, 1993) and the carboxypeptidase M inhibitor MGTA enhanced the noncholinergic bronchoconstrictor response to capsaicin and vagus nerve stimulation in guinea pig lungs (Desmazes *et al.*, 1992).

## Distinguishing Features

Carboxypeptidase M can be distinguished from other B-type carboxypeptidases by the following properties: it is membrane bound, has a neutral pH optimum and can be released from membranes with bacterial PI-PLC.

## Related Peptidases

The other known mammalian carboxypeptidases that specifically cleave C-terminal Arg or Lys are pancreatic carboxypeptidase B (Chapter 455), carboxypeptidase U (or plasma carboxypeptidase B; Chapter 453), carboxypeptidase N (lysine carboxypeptidase; Chapter 459), carboxypeptidase E/H (Chapter 458), and the newly described metallocarboxypeptidase D (Chapter 461) (Song & Fricker, 1995; Skidgel, 1996).

## Further Reading

Skidgel (1996) presents an extensive review of metallocarboxypeptidases. Further details of carboxypeptidase M may be found in Tan *et al.* (1995).

## References

Bhoola, K.D., Figueroa, C.D. & Worthy, K. (1992) Bioregulation of kinins: kallikreins, kininogens, and kininases. *Pharmacol. Rev.* **44**, 1–80

Chodimella, V., Skidgel, R.A., Krowiak, E.J. & Murlas, C.G. (1991) Lung peptidases, including carboxypeptidase, modulate airway reactivity to intravenous bradykinin. *Am. Rev. Respir. Dis.* **144**, 869–874

Deddish, P.A., Skidgel, R.A. & Erdös, E.G. (1989) Enhanced $Co^{2+}$ activation and inhibitor binding of carboxypeptidase M at low pH. Similarity to carboxypeptidase H (enkephalin convertase). *Biochem. J.* **261**, 289–291

de Saint-Vis, B., Cupillard, L., Pandrau-Garcia, D., Ho, S., Renard, N., Grouard, G., Duvert, V., Thomas, X., Galizzi, J.P., Banchereau, J. & Saeland, S. (1995) Distribution of carboxypeptidase-M on lymphoid and myeloid cells parallels the other zinc-dependent proteases CD10 and CD13. *Blood* **86**, 1098–1105

Desmazes, N.A., Lockhart, A., Lacroix, H. & Dusser, D.J. (1992) Carboxypeptidase M-like enzyme modulates the noncholinergic bronchoconstrictor response in guinea pigs. *Am. J. Respir. Cell Mol. Biol.* **7**, 477–484

Dragovic, T., Igic, R., Erdös, E.G. & Rabito, S.F. (1993) Metabolism of bradykinin by peptidases in the lung. *Am. Rev. Respir. Dis.* **147**, 1491–1496

Dragovic, T., Schraufnagel, D.E., Becker, R.P., Sekosan, M., Votta-Velis, E.G. & Erdös, E.G. (1995) Carboxypeptidase M activity is increased in bronchoalveolar lavage in human lung disease. *Am. J. Respir. Crit. Care Med.* **152**, 760–764

Erdös, E.G. (1979) Kininases. In: *Handbook of Experimental Pharmacology*, vol. 25 suppl. (Erdös, E.G., ed.). Heidelberg: Springer-Verlag, pp. 427–448

Marceau, F., Lundberg, C. & Hugli, T.E. (1987) Effects of the anaphylatoxins on circulation. *Immunopharmacology* **14**, 67–84

McGwire, G.B. & Skidgel, R.A. (1995) Extracellular conversion of epidermal growth factor (EGF) to [des-$Arg^{53}$]EGF by carboxypeptidase M. *J. Biol. Chem.* **270**, 17154–17158

Nagae, A., Deddish, P.A., Becker, R.P., Anderson, C.H., Abe, M., Skidgel, R.A. & Erdös, E.G. (1992) Carboxypeptidase M in brain and peripheral nerves. *J. Neurochem.* **59**, 2201–2212

Nagae, A., Abe, M., Becker, R.P., Deddish, P.A., Skidgel, R.A. &

Erdös, E.G. (1993) High concentration of carboxypeptidase M in lungs: presence of the enzyme in alveolar type I cells. *Am. J. Respir. Cell Mol. Biol.* **9**, 221–229

Rehli, M., Krause, S.W., Kreutz, M. & Andreesen, R. (1995) Carboxypeptidase M is identical to the MAX.1 antigen and its expression is associated with monocyte to macrophage differentiation. *J. Biol. Chem.* **270**, 15644–15649

Skidgel, R.A. (1988) Basic carboxypeptidases: regulators of peptide hormone activity. *Trends Pharmacol. Sci.* **9**, 299–304

Skidgel, R.A. (1991) Assays for arginine/lysine carboxypeptidases: carboxypeptidase H (E; enkephalin convertase), M and N. In: *Methods in Neurosciences: Peptide Technology*, vol. 6. (Conn, P.M., ed.). Orlando: Academic Press, pp. 373–385

Skidgel, R.A. (1996) Structure and function of mammalian zinc carboxypeptidases. In: *Zinc Metalloproteases in Health and Disease* (Hooper, N.M., ed.). London: Taylor & Francis, pp. 241–283

Skidgel, R.A., Davis, R.M. & Erdös, E.G. (1984a) Purification of a human urinary carboxypeptidase (kininase) distinct from carboxypeptidase A, B, or N. *Anal. Biochem.* **140**, 520–531

Skidgel, R.A., Johnson, A.R. & Erdös, E.G. (1984b) Hydrolysis of opioid hexapeptides by carboxypeptidase N. Presence of carboxypeptidase in cell membranes. *Biochem. Pharmacol.* **33**, 3471–3478

Skidgel, R.A., Davis, R.M. & Tan, F. (1989) Human carboxypeptidase M: purification and characterization of a membrane-bound carboxypeptidase that cleaves peptide hormones. *J. Biol. Chem.* **264**, 2236–2241

Song, L. & Fricker, L.D. (1995) Purification and characterization of carboxypeptidase D, a novel carboxypeptidase E-like enzyme, from bovine pituitary. *J. Biol. Chem.* **270**, 25007–25013

Tan, F., Chan, S.J., Steiner, D.F., Schilling, J.W. & Skidgel, R.A. (1989) Molecular cloning and sequencing of the cDNA for human membrane-bound carboxypeptidase M: comparison with carboxypeptidases A, B, H and N. *J. Biol. Chem.* **264**, 13165–13170

Tan, F., Deddish, P.A. & Skidgel, R.A. (1995) Human carboxypeptidase M. *Methods Enzymol.* **248**, 663–675

*Randal A. Skidgel*
*Laboratory of Peptide Research,*
*Departments of Pharmacology and Anesthesiology,*
*University of Illinois College of Medicine at Chicago,*
*835 S. Wolcott, Chicago, IL 60612, USA*
*Email: rskidgel@uic.edu*

M

# 461. *Metallocarboxypeptidase D*

## Databanks

*Peptidase classification: clan MC, family M14, MEROPS ID: M14.011*
*NC-IUBMB enzyme classification: EC 3.4.17.22*

*Databank codes:*

| Species | SwissProt | PIR | EMBL (cDNA) | EMBL (genomic) |
|---|---|---|---|---|
| *Anas platyrhynchos* | – | – | U25126 | – |
| *Drosophila melanogaster* | P42787 | – | U03883 | – |
| *Drosophila melanogaster* | – | – | U29591 | – |
| *Rattus norvegicus* | – | – | U62897 | – |

## Name and History

*Metallocarboxypeptidase D* was independently discovered by three separate groups working in distinct areas. Song & Fricker (1995) identified and characterized cattle *carboxypeptidase D (CPD)* based on its enzymatic properties. The duck homolog of CPD was named *gp180* by Kuroki *et al*. (1994) and was identified by its ability to bind avian hepatitis B virus particles. When the cDNA encoding gp180 was cloned and sequenced, it became apparent that this protein represented a 180 kDa member of the metallo-carboxypeptidase family (Kuroki *et al*., 1995). The recent cloning of rat CPD showed a high degree of amino acid sequence similarity between the rat CPD and duck gp180 (Xin *et al*., 1997). A third group identified a *Drosophila* homolog of CPD as the *silver* gene (Settle *et al*., 1995). The *silver* mutation, which was initially discovered in 1918, causes altered pigmentation, wing differentiation, behavioral responses to light, and viability. The *silver* gene was recently isolated and sequenced; this analysis showed the *silver gene product* to be a member of the metallocarboxypeptidase gene family, with highest similarity to CPD (Settle *et al*., 1995). It should be mentioned that CPD, which is a metalloenzyme, is distinct from serine carboxypeptidase D (Chapter 134).

## Activity and Specificity

CPD removes C-terminal Arg residues from a variety of substrates, with no detectable activity towards nonbasic residues (Song & Fricker, 1995). It prefers an Ala in the penultimate position. In comparing N-terminally blocked small peptides, CPD shows lower activity with dipeptides than with tripeptides. It is likely that it also cleaves C-terminal Lys residues, although this has not been directly tested.

The optimal pH for CPD activity is 5.5–6.5. CPD is stimulated by millimolar concentrations of $Co^{2+}$, and inhibited by 1,10-phenanthroline. The enzyme is partially inhibited by some thiol-directed reagents such as *p*-chloromercuriphenyl-sulfonate or $HgCl_2$ (Song & Fricker, 1995).

CPD has been assayed with substrates previously developed for carboxypeptidase E/H (CPE) (Song & Fricker, 1995; see Chapter 458). This assay make use of the solubility change that occurs when the basic residue is removed from the C-terminus of the substrate (Dns-Phe-Ala+Arg). After removal of the basic residue, the product readily partitions into organic solvents such as chloroform from an acidified aqueous phase. The assay for CPD is not specific for this enzyme, and will also detect other metallocarboxypeptidases if they are present. The assay can be performed at pH 5.5 to minimize the contribution from the neutral pH optimal carboxypeptidases, but this will then detect CPE. CPD and CPE can be distinguished by partial purification on a substrate affinity column (Song & Fricker, 1995). CPE elutes from a *p*-aminobenzoyl-Arg-Sepharose resin when the pH is raised to 8, whereas CPD remains bound at this pH and can subsequently be eluted with a competitive ligand such as Arg (Song & Fricker, 1995).

## Structural Chemistry

CPD is a single-chain protein that exists in several forms, with the major form being 180 kDa (Song & Fricker, 1995, 1996). It is likely that CPD contains 2–3 mol of zinc per mol of protein based on the homology between CPD and other metallocarboxypeptidases (Kuroki *et al*., 1995). CPD contains three carboxypeptidase domains of about 390 amino acids, which are followed by a transmembrane domain and a short 60 residue sequence that forms the cytosolic tail (Kuroki *et al*., 1995). The first two metallocarboxypeptidase domains have 40–50% amino acid identity with other members of the CPE (Chapter 458) subfamily. The third domain shows less homology to the other enzymes. Most importantly, residues thought to be critical for catalytic activity are not conserved in the third domain of either rat or duck CPD (Kuroki *et al*., 1995; Xin *et al*., 1997).

The C-terminal cytosolic tail of CPD appears to direct the retention of this protein in the trans-Golgi network (O. Varlamov & L.D. Fricker, unpublished results). Variation in the C-terminal region has been detected, and gives rise to the forms of the protein that are soluble (Song & Fricker, 1996).

## Preparation

CPD has been purified to homogeneity in microgram quantities from bovine pituitary, rat brain and several other tissues. The major 180 kDa membrane form requires extraction with a combination of 1% Triton X-100 and 1 M NaCl prior to purification (Song & Fricker, 1995). Both soluble and membrane forms are purified on a *p*-aminobenzoyl-Arg-Sepharose affinity resin. The synthesis of this noncommercially available resin has been described (Plummer & Hurwitz, 1978). From tissues with abundant CPD (pituitary, brain) purification to homogeneity is accomplished by the single affinity column step (Song & Fricker, 1995). When purifying CPD from tissues with lower amounts of this protein, an additional ion-exchange step is usually sufficient to provide homogeneous enzyme (Song & Fricker, 1996).

CPD (duck) has been expressed in the baculovirus system in high levels (1 mg liter$^{-1}$ of cultured cells) (L.D. Fricker, unpublished results)). The properties of the baculovirus-produced CPD are indistinguishable from those of CPD purified from brain or pituitary.

## Biological Aspects

The distribution of CPD mRNA and protein are consistent with a broad role for this enzyme in the processing of numerous proteins that transit the secretory pathway (Song & Fricker, 1996; Xin *et al.*, 1997). Many proteins that are transported through the secretory pathway require proteolytic processing at specific sites containing multiple basic residues (Lys, Arg). Examples include the receptors for insulin and growth factors, and many of the molecules that serve as extracellular messengers (growth factors, hormones, neurotransmitters). The initial cleavage of the precursors is thought to occur to the C-terminus of the basic amino acids, and a carboxypeptidase would then function to remove these basic residues from the C-terminus of the intermediate. The distribution of CPD is much broader than that of CPE. Whereas CPE (Chapter 458) is primarily found in cell types with a regulated secretory pathway and is associated with neuropeptide biosynthesis, CPD is found in virtually all tissues examined (Song & Fricker, 1996). Also, CPD has recently been found to be enriched in the *trans*-Golgi network, suggesting a role in the processing of many proteins that transit either the regulated or constitutive secretory pathways (O. Varlamov & L.D. Fricker, unpublished results).

The CPD gene has been mapped to mouse chromosome 11 and is not associated with any known mutations (Xin *et al.*, 1997).

## Distinguishing Features

Of all the known metallocarboxypeptidases, CPD is by far the largest in size. CPD is primarily 180 kDa in cattle pituitary and 100–180 kDa in various rat tissues (Song & Fricker, 1995, 1996). In contrast, all other reported metallocarboxypeptidases are smaller than 100 kDa, and are typically 30–60 kDa.

The pH optimum of CPD (5.5–6.5) is also unusual, but is fairly similar to that of CPE (5.0–5.5) so it is difficult to specifically measure CPD activity in crude homogenates. As described above, CPD and CPE can be physically separated on a substrate affinity resin.

The carboxypeptidase domains of CPD have approximately 40–50% amino acid identity with CPE and other related proteins. In contrast, the C-terminal region of CPD is not conserved with other metallocarboxypeptidases. An antiserum directed against the C-terminus of CPD is specific for this protein (Song & Fricker, 1996).

## Further Reading

See Song & Fricker (1995) for a more detailed overview of this enzyme.

## References

Kuroki, K., Cheung, R., Marion, P.L. & Ganem, D. (1994) A cell surface protein that binds avian hepatitis B virus particles. *J. Virol.* **68**, 2091–2096.

Kuroki, K., Eng, F., Ishikawa, T., Turck, C., Harada, F. & Ganem, D. (1995) gp180, a host cell glycoprotein that binds duck hepatitis B virus particles, is encoded by a member of the carboxypeptidase gene family. *J. Biol. Chem.* **270**, 15022–15028.

Plummer, T.H.J. & Hurwitz, M.Y. (1978) Human plasma carboxypeptidase N: isolation and characterization. *J. Biol. Chem.* **253**, 3907–3912.

Settle, S.H., Jr., Green, M.M. & Burtis, K.C. (1995) The silver gene of *Drosophila melanogaster* encodes multiple carboxypeptidases similar to mammalian prohormone-processing enzymes. *Proc. Natl Acad. Sci. USA* **92**, 9470–9474.

Song, L. & Fricker, L.D. (1995) Purification and characterization of carboxypeptidase D, a novel carboxypeptidase E-like enzyme, from bovine pituitary. *J. Biol. Chem.* **270**, 25007–25013.

Song, L. & Fricker, L.D. (1996) Tissue distribution and characterization of soluble and membrane-bound forms of metallocarboxypeptidase D. *J. Biol. Chem.* **271**, 20884–20889.

Xin, X., Varlamov, O., Day, R., Dong, W., Bridgett, M.M., Leiter, E.H. & Fricker, L.D. (1997) Cloning and sequence analysis of cDNA encoding rat carboxypeptidase D. *DNA Cell Biol.* **16**, 897–909.

*Lloyd D. Fricker*
*Professor, Department of Molecular Pharmacology,*
*Albert Einstein College of Medicine,*
*1300 Morris Park Avenue,*
*Bronx, NY 10461, USA*
*Email: fricker@aecom.yu.edu*

# 462. Introduction: clan MD containing zinc D-Ala-D-Ala carboxypeptidase

## Databanks

**MEROPS ID: MD**

| Species | Gene | SwissProt | PIR | EMBL (cDNA) | EMBL (genomic) |
|---|---|---|---|---|---|
| **Family M15** | | | | | |
| Zinc D-Ala-D-Ala carboxypeptidase (Chapter 463) | | | | | |
| **Hedgehog protein family** | | | | | |
| Hedgehog protein | | | | | |
| *Brachydanio rerio* | *twhh* | – | – | L27585 | – |
| | *vhh-1* | – | – | U30710 | – |
| *Cynops pyrrhogaster* | *Shh* | – | – | D63339 | |
| *Drosophila melanogaster* | *HH* | Q02936 | – | L02793 | Z11840: unannotated |
| | | | | L05404 | |
| *Gallus gallus* | *Shh* | – | – | L28099 | – |
| *Homo sapiens* | *lhh* | – | – | L38517 | – |
| | *Shh* | – | – | L38518 | – |
| *Mus musculus* | *lhh* | – | – | X76290 | – |
| | *Shh* | – | – | X76291 | – |
| *Rattus norvegicus* | *vhh-1* | – | – | L27340 | – |
| *Xenopus laevis* | *vhh-1* | – | – | L35248 | – |
| | | | | L39213 | |
| | *hh4* | – | – | U26350 | – |
| | *cephalic* | – | – | U26349 | – |

Brookhaven Protein Data Bank three-dimensional structures:

| Species | ID | Resolution | Notes |
|---|---|---|---|
| Sonic hedgehog protein | | | |
| *Mus musculus* | 1VHH | 1.7 | N-terminal domain complexed with zinc |

This clan consists of only one family of peptidases, namely M15, which contains the zinc D-Ala-D-Ala carboxypeptidase from *Streptomyces* that is involved in the biosynthesis of bacterial cell walls. The complex polymer of glycoproteins that constitutes the cell wall requires a cross-linking peptide that contains D-amino acids. The cross-linking peptide is synthesized as a precursor with an additional D-Ala at the C-terminus. There are a number of different D-Ala-D-Ala carboxypeptidases that process the cross-linking peptide precursor, including two serine-type carboxypeptidases, D-Ala-D-Ala carboxypeptidase PBP5 from family S11 (Chapter 142) and D-Ala-D-Ala carboxypeptidase from family S12 (Chapter 145), plus the VanY D-Ala-D-Ala carboxypeptidase of unknown catalytic type (Chapter 554), as well as this zinc-dependent carboxypeptidase.

Immediately following the release of the cross-linking peptide from its precursor, a transpeptidation reaction occurs to link the new C-terminal D-Ala to the *meso*-diaminopimelic acid. This is usually performed by the D-Ala-D-Ala carboxypeptidase, but in *Streptomyces* the processing and transpeptidation reactions are carried out by different enzymes. The zinc D-Ala-D-Ala carboxypeptidase releases the cross-linking peptide from its precursor, and a serine-type D-Ala-D-Ala transpeptidase performs the transpeptidation.

The tertiary structure of the *Streptomyces albus* zinc D-Ala-D-Ala carboxypeptidase has now been determined at greater resolution (1.8 Å), allowing a more correct determination of the zinc ligands than had previously been possible (see Figure 462.1). The zinc ligands have been identified as His154, Asp161 and His197. This conforms to the short–long spacing identified by Vallee & Auld (1990) for metal ligands generally. A previous prediction of the zinc ligands from a lower resolution crystal structure implied a long–short spacing (Dideberg *et al.*, 1982).

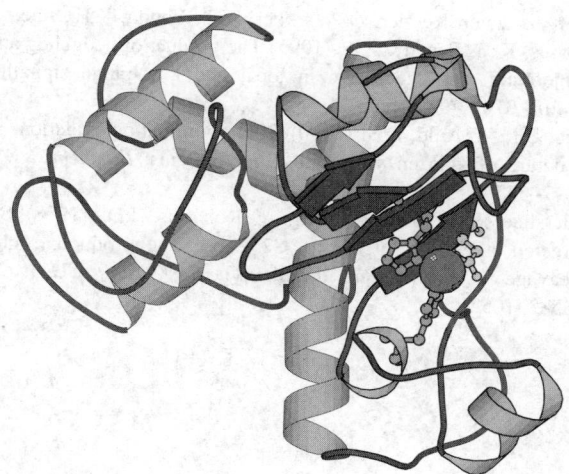

**Figure 462.1** Richardson diagram of *Streptomyces albus* zinc D-Ala-D-Ala carboxypeptidase. The image was prepared from the Brookhaven Protein Data Bank entry (1LBU) as described in the Introduction (p. xxv). The catalytic zinc ion is shown in CPK representation as a light gray sphere. The zinc ligands are shown in ball-and-stick representation: His154, Asp161 and His197.

The zinc D-Ala-D-Ala carboxypeptidase has a two-domain structure, with the zinc ligands and active site in the C-terminal domain. The N-terminal domain (80 residues) is homologous to a C-terminal domain of *N*-acylmuramoyl-Ala amidase (EC 3.5.1.28) from *Bacillus* species. This is an enzyme involved in bacterial cell wall turnover, breaking the bond between the acetylmuramoyl of the glycopeptide and the L-Ala of the cross-linking peptide. This bond is also broken by the predatory bacterium *Lysobacter* by $\beta$-lytic endopeptidase (Chapter 506).

The tertiary structure of zinc D-Ala-D-Ala carboxypeptidase is similar to the fold of the N-terminal domain of the sonic hedgehog protein from mouse (see Figure 462.2). Hedgehog proteins are involved in embryonic patterning; the *sonic hedgehog* gene is expressed in the notochord and is responsible for the local and long-range induction of ventral cell types within the neural tube and somites. The sonic hedgehog protein undergoes an autocatalytic reaction, releasing a N-terminal domain that is responsible for the signaling activity, and a C-terminal domain that carries the autoprocessing activity. The N-terminal domain contains a zinc ion coordinated by His141, Asp148 and His183 and a water molecule (Hall *et al.*, 1995); the nature and spacing of the zinc ligands is almost identical to that of the zinc D-Ala-D-Ala carboxypeptidase. The ligands are conserved in all sonic hedgehog homologs, except for *Drosophila melanogaster* hedgehog protein, in which Asp148 is replaced by Thr and His183 is replaced by Tyr (see Alignment 462.1 on CD-ROM).

Other residues that may be important for catalysis are Glu127, His135 and His181. These are also conserved in all sonic hedgehog homologs, except for rat hedgehog protein in which His181 is replaced by Arg (Hall *et al.*, 1995). As can be seen from Alignment 462.1, only His181 is conserved in the zinc D-Ala-D-Ala carboxypeptidases.

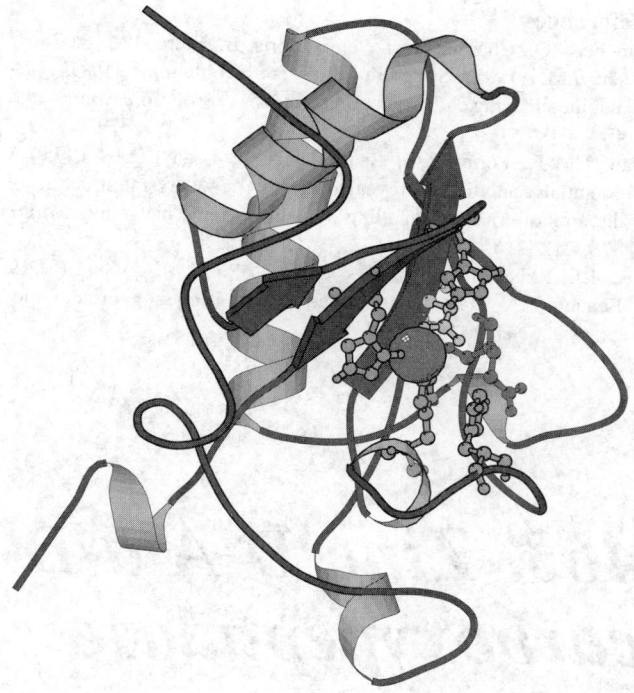

**Figure 462.2** Richardson diagram of mouse sonic hedgehog protein. The image was prepared from the Brookhaven Protein Data Bank entry (1VHH) as described in the Introduction (p. xxv). The catalytic zinc ion is shown in CPK representation as a light gray sphere. The zinc ligands are shown in ball-and-stick representation: His141, Asp148 and His183. Additionally, three residues that may be important for any catalytic activity are also shown in ball-and-stick representation: Glu127, His135 and His181.

Because no proteolytic activity has been shown for the hedgehog proteins, they have not been given a peptidase family designation. However, the possibility that they are peptidases exists.

One catalytic activity that is known for the hedgehog protein is its autoactivation. Autoactivation involves cleavage of the Gly258⊣Cys259 bond in *Drosophila* hedgehog protein, and these residues are conserved throughout the family. Two fragments are generated: an 18 kDa N-terminal fragment that possesses the signalling activity, and a C-terminal 25 kDa C-terminal fragment that is not known to possess any further activity. Mutational studies have shown that Cys258 and His329 are essential for autoactivation in *Drosophila* hedgehog protein (Lee *et al.*, 1994). The autoactivation is not inhibited by any standard peptidase inhibitors, but requires 50 mM DTT *in vitro*. The cleavage mechanism is believed to be similar to that of the activation of prohistidine decarboxylase (Vanderslice *et al.*, 1988), involving activation of Cys258 by His329 for a nucleophilic attack on the carbonyl of Gly257. There is a motif around the cleavage site which is similar to that found at the N-terminal splice junction of self-splicing proteins (Chapter 569), and splicing may involve a thioester intermediate (Porter *et al.*, 1995).

**M**

## References

Dideberg, O., Charlier, P., Dive, G., Joris, B., Frère, J.M. & Ghuysen, J.M. (1982) Structure of a $Zn^{2+}$-containing D-alanyl-D-alanine-cleaving carboxypeptidase at 2.5 Å resolution. *Nature* **299**, 469–470.

Hall, T.M.T., Porter, J.A., Beachy, P.A. & Leahy, D.J. (1995) A potential catalytic site revealed by the 1.7-A crystal structure of the amino-terminal signalling domain of Sonic hedgehog. *Nature* **378**, 212–216.

Lee, J.J., Ekker, S.C., von Kessler, D.P., Porter, J.A., Sun, B.I. & Beachy, P.A. (1994) Autoproteolysis in hedgehog protein biogenesis. *Science* **266**, 1528–1537.

Porter, J.A., von Kessler, D.P., Ekker, S.C., Young, K.E., Lee, J.J., Moses, K. & Beachy, P.A. (1995) The product of hedgehog autoproteolytic cleavage active in local and long-range signalling. *Nature* **374**, 363–366.

Vallee, B.L. & Auld, D.S. (1990) Zinc coordination, function, and structure of zinc enzymes and other proteins. *Biochemistry* **29**, 5647–5659.

Vanderslice, P., Copeland, W.C. & Robertus, J.D. (1988) Site-directed alteration of serine 82 causes nonproductive chain cleavage in prohistidine decarboxylase. *J. Biol. Chem.* **263**, 10583–10586.

# 463. Zinc D-Ala-D-Ala carboxypeptidase

## Databanks

*Peptidase classification: clan MD, family M15, MEROPS ID: M15.001*
*NC-IUBMB enzyme classification: EC 3.4.17.14*
Databank codes:

| Species | SwissProt | PIR | EMBL (cDNA) | EMBL (genomic) |
|---|---|---|---|---|
| *Streptomyces albus* | P00733 | A00913 | X55794 | – |

Brookhaven Protein Data Bank three-dimensional structures:

| Species | ID | Resolution | Notes |
|---|---|---|---|
| *Streptomyces albus* | 1LBU | 1.8 | |

*Note*: There used to be a sequence from *Streptomyces griseus* in SwissProt, which has disappeared.

## Name and History

Bacteriolytic enzymes isolated from culture supernatants of *Streptomyces albus* G and selected for their specificity on critical linkages were useful in establishing the structure of the bacterial cell wall peptidoglycan (Ghuysen, 1968). The **KM endopeptidase** has the specificity of a carboxypeptidase (Ghuysen *et al.*, 1969, 1970). It selectively hydrolyzes bacterial walls in which the peptidoglycan interpeptide bonds extend between a D-alanine residue and another D-amino acid residue in α-position to a free carboxylate; it cleaves the bis-disaccharide peptide dimer (obtained by muramidase treatment of the peptidoglycan of *Escherichia coli*) into monomers by hydrolyzing the C-terminal D-alanyl-D-*meso*-diaminopimelic acid interpeptide bond; and it hydrolyzes the C-terminal D-alanyl-D-alanine linkage of the precursor UDP-*N*-acetylmuramoyl-L-alanyl-γ-D-glutamyl-(L)-*meso*-diaminopimelyl-(L)-D-alanyl-D-alanine. Subsequently, the KM endopeptidase was identified as a zinc enzyme (Dideberg *et al.*, 1980) and has been given the name **zinc D-Ala-D-Ala carboxypeptidase**. It contains 1 mol of zinc per 18 000 Da protein molecule, the apoprotein binds $Zn^{2+}$ with a $K_a$ of $2 \times 10^{14}$ M$^{-1}$ and the $Zn^{2+}$ cofactor is required for activity.

## Specificity and Activity

The specificity of the zinc D-Ala-D-Ala carboxypeptidase was studied on peptides of the form R3-R2-R1-COOH (Leyh-Bouille *et al.*, 1970a). There is a strict requirement that the C-terminal amino acid R1 should have a free carboxylate.

R1 may be glycine, leucine, alanine, ornithine, lysine or diaminopimelic acid, but if the amino acid is not glycine, then the asymmetric carbon must be in the D configuration. The side chain of R1 may be very bulky. High substrate activity is compatible with the presence of polypeptide substituents on the $\varepsilon$-amino group of D-lysine, the $\delta$-amino group of D-ornithine, and the amino group on the L-carbon of *meso*-diaminopimelic acid.

There is a strict requirement that the R2 side chain be either a hydrogen atom or, much better, a methyl group, in which case the asymmetric carbon must be D. Replacement of the penultimate D-Ala in $N^{\alpha}, N^{\varepsilon}$-diacetyl-L-Lys-D-Ala-D-Ala by a Gly does not modify the $V$ value but drastically decreases the $K_m$ value. The length, structure and polarity of the R3 side chain is of great importance. The presence at this position of Gly, L-Ala, L-Tyr, L-homoserine, $N^{\gamma}$-acetyl-L-diaminobutyric acid, $N^{\delta}$-acetyl-L-ornithine and $N^{\varepsilon}$-acetyl-L-Lys is paralleled by a progressive and large increase of the rate of hydrolysis of the terminal D-Ala-D-Ala linkage. The presence of an acyl substituent on the $\omega$-amino group of the R3 side chain is another requirement. The $N^{\alpha}, N^{\varepsilon}$-diacetyl-L-Lys-D-Ala-D-Ala is a much better substrate than the $N^{\alpha}$-monoacetyl derivative. Also, the transformation of the $\varepsilon$-amino group of L-Lys to an $\alpha$-amino group by introduction of a carboxyl group in $\alpha$-position (as occurs when L-Lys is replaced by diaminopimelic acid) and its substitution by a pentaglycine sequence result in a large increased substrate activity of the corresponding peptides.

## Preparation

The wild-type strain produces about 100–200 ng enzyme per liter of culture. The enzyme can be purified using classical chromatographic procedures (Duez *et al.*, 1978). Gene overexpression in *Streptomyces lividans* (Duez *et al.*, 1990) results in the export of the synthesized enzyme (yield: 10 mg liter$^{-1}$).

## Structural Chemistry

The sequence of the mature 213 amino acid residue zinc-containing D-Ala-D-Ala carboxypeptidase was determined by chemical degradation (Joris *et al.*, 1983) and firmly established by gene sequencing (Duez *et al.*, 1990). The enzyme is synthesized as a 255 residue precursor that contains a cleavable signal peptide of 42 amino acids. The three-dimensional structure of the enzyme was established by X-ray diffraction at high resolution (Dideberg *et al.*, 1982; Wéry, 1987) (see Fig. 462.1). The molecular structure in solution was studied by small-angle X-ray scattering (Labischinski *et al.*, 1984). The model compares well with that of the crystal structure. Radius of gyration: $182 \pm 0.05$ nm; largest diameter: $5.9 \pm 0.2$ mm; relative molecular mass: $17\,000 \pm 200$; volume: $35 \pm 2$ mm$^3$; degree of hydration: $0.25 \pm 0.02$ g water g$^{-1}$ protein.

## Bimodular Design

The fold adopted by the zinc D-Ala-D-Ala carboxypeptidase, consisting of two globular modules connected by a single link, is unprecedented. The catalytic 132 amino acid residue C-terminal module possesses three $\alpha$ helices and a five-stranded $\beta$ sheet that defines a cleft in which the Zn$^{2+}$ ion is ligated by one water molecule, two His and one Asp residues (Wéry, 1987). The open cleft can accommodate extended structures, a feature which is related to the specificity profile of the enzyme. Arg138 is probably concerned with the binding, by charge pairing, of the carboxylate substrate, and His and Asp residues probably play the required functions of proton abstraction-donation.

Fused to the catalytic module, the noncatalytic, 81 amino acid residue, N-terminal module possesses three $\alpha$ helices and an elongated crevice defined by a loop-helix-loop-helix motif consisting of two repeats, each 16 amino acid residues long, connected by a heptapeptide (Ghuysen *et al.*, 1994). Similar motifs are borne by the C-terminal regions of the N-acetylmuramoyl-peptide amidases of *Bacillus subtilis* and *Bacillus licheniformis*, and the lysozyme of *Clostridium acetobutylicum*. As a common feature of these exocellular enzymes is their substrate, it is likely that their noncatalytic modules are involved in the binding to the insoluble wall peptidoglycan.

## Active-site-directed Inactivators

The zinc D-Ala-D-Ala carboxypeptidase can be inactivated by acyclic compounds possessing both a C-terminal carboxylate and, at the other end of the molecule, a thiol, hydroxamate or carboxylate (Charlier *et al.*, 1984). 3-Mercaptopropionate (racemic) and 3-mercaptoisobutyrate (L-isomer) are competitive inhibitors ($K_i \approx 5–10 \times 10^{-9}$ M). $6\beta$-Iodopenicillinate binds to the active site in front of the Zn$^{2+}$ cofactor, close to His190 and causes permanent inactivation of the enzyme (Charlier *et al.*, 1984). Classical $\beta$-lactam compounds are very weak inhibitors (Leyh-Bouille *et al.*, 1970b). Enzyme inhibition may be competitive or noncompetitive, in which case (as observed with cephalothin and cephalosporin C) binding causes disruption of the protein crystal lattice (Charlier *et al.*, 1984) and drastically alters the scattering behavior of the protein in solution (Labischinski *et al.*, 1984).

In summary, in spite of differences in stereospecificity, polypeptide folding and function, the zinc D-Ala-D-Ala carboxypeptidase is mechanistically similar to carboxypeptidase A (Chapter 451) and thermolysin (Chapter 351). A similar mechanism, dictated by common catalytic requirements, has evolved from different starting tertiary structures by converging evolution. The zinc D-Ala-D-Ala carboxypeptidase and other peptidoglycan hydrolases have similar binding-site-shaped repeated sequences. The acquisition of a noncatalytic substrate-binding module is an evolutionary advantage for exocellular enzymes that hydrolyze bonds in an insoluble polymer.

## References

Charlier, P., Dideberg, O., Jamoulle, J.C., Frère, J.-M., Ghuysen, J.-M., Dive, G. & Lamotte-Brasseur, J. (1984) Active-site directed inactivators of the Zn$^{++}$ D-alanyl-D-alanine-cleaving carboxypeptidase of *Streptomyces albus* G. *Biochem. J.* **219**, 763–772.

Dideberg, O., Joris, B., Frère, J.-M., Ghuysen, J.-M., Weber, G., Robaye, R., Delbrouck, J.-M. & Roelandts, I. (1980) The

M

exocellular DD-carboxypeptidase of *Streptomyces albus* G: a metallo (Zn$^{++}$) enzyme. *FEBS Lett.* **117**, 215–218.

Dideberg, O., Charlier, P., Dive, G., Joris, B., Frère, J.-M. & Ghuysen, J.-M. (1982) Structure at 2.5 Å resolution of a Zn$^{++}$-containing D-alanyl-D-alanine-cleaving carboxypeptidase. *Nature* **299**, 469–470.

Duez, C., Frère, J.-M., Geurts, F., Ghuysen, J.-M., Dierickx, L. & Delcambe, L. (1978) The exocellular DD-carboxypeptidase-endopeptidase from *Streptomyces albus* strain G. Purification and chemical properties. *Biochem. J.* **175**, 793–800.

Duez, C., Lakaye, B., Houba, S., Dusart, J. & Ghuysen, J.-M. (1990) Cloning, nucleotide sequence and amplified expression of the gene encoding the extracellular metallo (Zn) DD-peptidase of *Streptomyces albus* G. *FEMS Microbiol. Lett.* **71**, 215–220.

Ghuysen, J.-M. (1968) Use of bacteriolytic enzymes in determination of wall structure and their role in cell metabolism. *Bacteriol. Rev.* **32**, 425–464.

Ghuysen, J.-M., Dierickx, L., Coyette, J., Leyh-Bouille, M., Guinand, M. & Campbell, J.N. (1969) An improved technique for the preparation of Streptomyces peptidases and *N*-acetylmuramyl-L-alanine amidase active on bacterial wall peptidoglycans. *Biochemistry* **8**, 213–220.

Ghuysen, J.-M., Leyh-Bouille, M., Bonaly, R., Nieto, M., Perkins, H.R., Schleifer, K.H. & Kandler, O. (1970) Isolation of DD-carboxypeptidase from *Streptomyces albus* G culture filtrates. *Biochemistry* **9**, 2955–2961.

Ghuysen, J.-M., Lamotte-Brasseur, J., Joris, B. & Shockman, G.D. (1994) Binding site-shaped repeated sequences of *Streptomyces albus* G DD-peptidase, *Bacillus subtilis* CwlA amidase, *Corynebacterium acetobutylicum* lysozyme and *Bacillus* sp ORF-L3 gene product. *FEBS Lett.* **342**, 23–28.

Joris, B., Van Beeumen, J., Casagrande, F., Gerday, Ch., Frère, J.-M. & Ghuysen, J.-M. (1983) The complete amino acid sequence of the Zn$^{++}$-containing D-alanyl-D-alanine-cleaving carboxypeptidase of *Streptomyces albus* G. *Eur. J. Biochem.* **130**, 53–69.

Labischinski, H., Giesbrecht, P., Fischer, E., Barnickel, G., Bradaczek, H., Frère, J.-M., Houssier, Cl., Charlier, P., Dideberg, O. & Ghuysen, J.-M. (1984) Study of the Zn-containing DD-carboxypeptidase of *Streptomyces albus* G by small angle X-ray scattering in solution. *Eur. J. Biochem.* **138**, 83–87.

Leyh-Bouille, M., Ghuysen, J.-M., Bonaly, R., Nieto, M., Perkins, H.R., Schleifer, K.H. & Kandler, O. (1970a) Substrate requirements of the *Streptomyces albus* G DD-carboxypeptidase. *Biochemistry* **9**, 2961–2971.

Leyh-Bouille, M., Ghuysen, J.-M., Nieto, M., Perkins, H.R., Schleifer, K.H. & Kandler, O. (1970b) On the *Streptomyces albus* G DD-carboxypeptidase. Mechanism of action of penicillin, vancomycin and ristocetin. *Biochemistry* **9**, 2971–2975.

Wéry, J.P. (1987) Applications de méthodes de détermination de phases à la résolution de deux structures de protéine [Application of the methods of phase determination to the resolution of two protein structures]. PhD thesis, University of Liège.

***Jean-Marie Ghuysen***
*Centre d'Ingénierie des Protéines,*
*Université de Liège,*
*Institut de Chimie, B6,*
*Sart Tilman, B-4000 Liège, Belgium*

# 464. *N-Acetylmuramoyl-L-alanine amidase*

## Databanks

*Peptidase classification: clan MD, not a peptidase, related to family M15, MEROPS ID: M15.960*
*NC-IUBMB enzyme classification: EC 3.5.1.28 & EC 3.4.19.10 (discontinued)*
*Chemical Abstracts Service registry number: 37288-68-9*
*Databank codes:*

| Species | Gene | SwissProt | PIR | EMBL (cDNA) | EMBL (genomic) |
|---|---|---|---|---|---|
| Homologs of peptidases in family M15 | | | | | |
| *Bacillus licheniformis* | *cwlL* | P36550 | S39916 | D13377 | – |
| | *orfL3* | Q99125 | D49754 | M63942 | |

| Species | Gene | SwissProt | PIR | EMBL (cDNA) | EMBL (genomic) |
|---------|------|-----------|-----|-------------|----------------|
| *Bacillus subtilis* | *cwlA* | P24808 | C44816 | X51424 | – |
| | *lytC* | Q02114 | B41322 | D10388 | – |
| | | | | M81324 | |
| | | | | M87645 | |
| Others | | | | | |
| *Staphylococcus aureus* | *atl* | P52081 | – | L41499 | – |
| *Streptococcus pneumoniae* | *lytA* | P06653 | A25634 | M13812 | – |

## Name and History

Peptidoglycan is the major structural polymer in most bacterial cell walls and consists of glycan chains of repeating *N*-acetylglucosamine and *N*-acetylmuramic acid residues cross-linked via peptide side chains. Peptidoglycan hydrolases are produced by many bacteria, bacteriophages and eukaryotes. **N-Acetylmuramoyl-L-alanine amidases** are peptidoglycan hydrolases with the specificity to cleave the amide bond between the lactyl group of the muramic acid and the $\alpha$-amino group of L-alanine and are able to cause dissolution of the peptidoglycan structure and thus do not exhibit peptidase activity (Shockman & Höltje, 1994). Amidases are properly included at EC 3.5.1.28 (Margot *et al.*, 1991) and in the future it is appropriate that this citation be used exclusively as the enzymes are specific for peptidoglycan or derived muropeptides. The EC 3.4.19.10 listing is now discontinued. Bacterial amidases which are able to hydrolyze their own peptidoglycan are called autolysins and are produced by many gram-positive and gram-negative species (Shockman & Höltje, 1994). Lytic enzymes with amidase activity produced by bacteriophages and eukaryotes are numerous (Harz *et al.*, 1990; Foster, 1993). Only examples of amidases illustrating specific points are cited here.

## Activity and Specificity

Most amidases are able to act on intact cell wall peptidoglycan, either associated with living bacteria or as isolated peptidoglycan, resulting in a loss of absorbance by the insoluble substrate. A general assay for peptidoglycan hydrolases by zymogram activity has been developed (Foster, 1992, 1995). The assay for amidase activity is the hydrolysis of peptidoglycan substrate followed by the identification of new L-alanine N-termini (Margot *et al.*, 1991). The amidases show varying substrate specificities and biochemical properties. The LytA autolytic amidase of *Streptococcus pneumoniae* requires cell walls containing choline for activity (Höltje & Tomasz, 1976). Both the autolysin LytC and the cryptic prophage lytic enzyme CwlA of *Bacillus subtilis* are able to hydrolyze purified peptidoglycan although they have a higher specific activity on cell wall peptidoglycan containing teichoic acids (Herbold & Glaser, 1975; Foster, 1992). This may be due to weaker binding to the teichoic acid free substrate. They have a broad pH range for activity with optima at around pH 8.0. The activity of both enzymes is greatly enhanced by the presence of divalent cations, particularly $Mg^{2+}$, $Ca^{2+}$ or $Mn^{2+}$ (20 mM optimum).

## Structural Chemistry

A number of amidases have been purified and their structural genes identified. The major vegetative autolytic amidase of *B. subtilis*, LytC, is a monomer of 49.9 kDa with a pI of 10.6 (Herbold & Glaser, 1975; Lazarevic *et al.*, 1992). The enzyme is translated with a signal sequence which is proteolytically processed on secretion. Like many autolysins, it is a two-domain protein, having an N-terminal domain which is involved in binding to its cell wall substrate. The cell wall-binding domain is characterized by the presence of three imperfect repeats of approximately 59 amino acids separated by 36–39 residues.

CwlA is an amidase of proposed *B. subtilis* cryptic prophage origin which has been expressed in *E. coli* (Foster, 1991, 1993). The *cwlA* gene encodes a protein of 29.9 kDa, which is processed to a 21 kDa form which still exhibits activity.

The Atl autolysin of *Staphylococcus aureus* is unusual in that it is a three-domain protein with a central wall-binding region containing three imperfect direct repeats, each between 140 and 164 residues long, showing 31% identity among the three repeats (Foster, 1995). Atl is a bifunctional protein, the N-terminal domain has amidase activity and C-terminal *N*-acetylglucosaminidase activity (Oshida *et al.*, 1995). Atl is produced as a 137.4 kDa proform which is proteolytically processed into at least six active forms of different molecular mass and activities.

## Preparation

Several amidases from various origins have been purified to apparent homogeneity (Foster, 1993; Herbold & Glaser, 1975; Höltje & Tomasz, 1976). Naturally produced enzymes are found associated with the bacterial cell wall peptidoglycan or in the environment. Cell wall binding is ionic and amidases can be generally extracted from whole gram-positive cells with 4 M LiCl. The sporulation-specific amidase (CwlC) of *B. subtilis* and CwlA were purified by single-step reactive red dye affinity chromatography (Foster, 1993; Smith & Foster, 1995). A number of bacterial enzymes have been purified from culture supernatants and a human amidase from sera (Harz *et al.*, 1990).

## Biological Aspects

The roles of bacterial and phage amidases have begun to be elucidated. The *cwlA* gene of *B. subtilis* is a silent gene, probably of cryptic prophage origin (Foster, 1993). CwlA has 50% amino acid identity with XlyA which is a PBSX prophage-encoded amidase of *B. subtilis*. XlyA is involved in host cell lysis to allow release of mature phage particles from the host cell (Longchamp *et al.*, 1994).

The *lytC* gene of *B. subtilis* is part of a tricistronic operon with two promoters (Lazarevic *et al.*, 1992). Approximately 75 and 25% of transcription of the *lytABC* operon are under the control of the sigma factor, $\sigma^D$ (which controls the expression of the flagellar, chemotaxis and motility regulon) and $\sigma^A$ (the major housekeeping sigma factor), respectively. The two other genes in the operon encode a putative 11.2 kDa lipoprotein (*lytA*) and a 76.7 kDa modifier protein (*lytB*). The modifier protein associates with the LytC amidase enhancing its activity 3-fold (Herbold & Glaser, 1975). LytC and LytB are found predominantly wall-associated in an active form. LytC is the major enzyme involved in release of cell wall material and in generalized post mortem cell lysis in response to metabolic inhibitors and antibiotics (Kuroda & Sekiguchi, 1991; Margot & Karamata, 1992). It accounts for 97% of the total cell-associated amidase activity during vegetative growth.

CwlC is a sporulation-specific amidase of *B. subtilis* which is under the control of the late mother cell sigma factor, $\sigma^K$ (Kuroda *et al.*, 1993). The enzyme is expressed only late during sporulation and is found associated with the mother cell wall in an active form prior to mother cell lysis (Foster, 1992). CwlC has a role in mother cell lysis but its activity can be compensated for by LytC (Smith & Foster, 1995). A *lytC cwlC* mutant has mother cells which are unable to lyse whereas both of the single mutants lyse.

Atl of *S. aureus* is involved in cell division, peptidoglycan release at the cell surface and generalized cell lysis (Foster, 1995). The LytA amidase is the major autolysin of *Streptococcus pneumoniae* (Garcia *et al.*, 1985). A *lytA* mutant shows reduced virulence in a mouse model and a vaccine based on LytA gives good protection against infection (Berry *et al.*, 1989).

## Distinguishing Features

Amidases from various origins can easily be confused with a whole range of other peptidoglycan hydrolases. The zymogram technique and hydrolysis of peptidoglycan substrates are not specific for amidases. Amidases have diverse properties and relative molecular masses and thus can only be distinguished by the specific amidase assay. It is essential to determine that new N-terminal groups are attributable to L- and not D-alanine. New D-alanine N-terminals would be due to the action of an endopeptidase.

## Further Reading

A general review of peptidoglycan hydrolases, including the major amidases is found in Shockman & Höltje (1994).

## References

Berry, A.M., Lock, R.A., Hansman, D. & Paton, J.C. (1989) Contribution of autolysin to virulence of *Streptococcus pneumoniae*. *Infect. Immun.* **57**, 2324–2330.

Foster, S.J. (1991) Cloning, sequence analysis and biochemical characterization of an autolytic amidase of *Bacillus subtilis* 168 *trpC2*. *J. Gen. Microbiol.* **137**, 1987–1998.

Foster, S.J. (1992) Analysis of the autolysins of *Bacillus subtilis* 168 during vegetative growth and differentiation by using renaturing gel electrophoresis. *J. Bacteriol.* **174**, 464–470.

Foster, S.J. (1993) Analysis of *Bacillus subtilis* 168 prophage-associated lytic enzymes; identification and characterization of CWLA-related prophage proteins. *J. Gen. Microbiol.* **139**, 3177–3184.

Foster, S.J. (1995) Molecular characterization and functional analysis of the major autolysin of *Staphylococcus aureus* 8325/4. *J. Bacteriol.* **177**, 5723–5725.

Garcia, E., Garcia, J.L., Ronda, C., Garcia, P. & Lopez, R. (1985) Cloning and expression of the pneumococcal autolysin gene in *Escherichia coli*. *Mol. Gen. Genet.* **201**, 225–230.

Harz, H., Burgdorf, K. & Höltje, J.V. (1990) Isolation and separation of the glycan strands from murein of *Escherichia coli* by reversed-phase high-performance liquid-chromatography. *Anal. Biochem.* **190**, 120–128.

Herbold, D.R. & Glaser, L. (1975) *Bacillus subtilis N*-acetylmuramic acid L-alanine amidase. *J. Biol. Chem.* **250**, 1676–1682.

Höltje, J.V. & Tomasz, A. (1976) Purification of the pneumococcal the *N*-acetylmuramyl-L-alanine amidase to biochemical homogeneity. *J. Biol. Chem.* **251**, 4199–4207.

Jacobs, C., Joris, B., Jamin, M., Klarsov, K., Van Beeumen, J., Mengin-Lecreulx, D., van Heijenoort, J., Park, J.T., Normark, S. & Frere, J.-M. (1995) AmpD, essential for both β-lactamase regulation and cell wall recycling, is a novel cytosolic *N*-acetyl muramyl-L-alanine amidase. *Mol. Microbiol.* **15**, 553–559.

Kuroda, A. & Sekiguchi, J. (1991) Molecular cloning and sequencing of a major *Bacillus subtilis* autolysin gene. *J. Bacteriol.* **173**, 7304–7312.

Kuroda, A., Asami, Y. & Sekiguchi, J. (1993) Molecular cloning of a sporulation-specific cell wall hydrolase gene of *Bacillus subtilis*. *J. Bacteriol.* **175**, 6260–6268.

Lazarevic, V., Margot, P., Soldo, B. & Karamata, D. (1992) Sequencing and analysis of the *Bacillus subtilis lytRABC* divergon: a regulatory unit encompassing the structural genes of the *N*-acetyl muramoyl-L-alanine amidase and its modifier. *J. Gen. Microbiol.* **138**, 1949–1961.

Longchamp, P.F., Mauël, C. & Karamata, D. (1994) Lytic enzymes associated with defective prophages of *Bacillus subtilis*: sequencing and characterisation of the region comprising the *N*-acetyl-muramoyl-L-alanine amidase gene of prophage PBSX. *Microbiology* **140**, 1855–1867.

Margot, P. & Karamata, D. (1992) Identification of the structural genes for *N*-acetyl muramoyl-L-alanine amidase and its modifier in *Bacillus subtilis* 168: inactivation of these genes by insertional mutagenesis has no effect on growth or cell separation. *Mol. Gen. Genet.* **232**, 359–366.

Margot, P., Roten, C.-A.H. & Karamata, D. (1991) *N*-Acetyl muramoyl-L-alanine amidase assay based on specific radioactive labeling of muropeptide L-alanine: quantitation of the enzyme activity in the autolysin deficient *Bacillus subtilis* 168, *flaD* strain. *Anal. Biochem.* **198**, 15–18.

Oshida, T., Sugai, M., Komatsuzawa, H., Hong, Y.M., Suginaka, H. & Tomasz, A. (1995) A *Staphylococcus aureus* autolysin that has an *N*-acetylmuramoyl-L-alanine amidase domain and an endo-β-*N*-acetylglucosaminidase domain: cloning, sequence analysis and characterization. *Proc. Natl Acad. Sci. USA* **92**, 285–289.

Shockman, G.D. & Höltje, J.V. (1994) Microbial peptidoglycan (murein) hydrolases. In: *New Comprehensive Biochemistry: Bacterial Cell Wall* (Ghuysen, J.M. & Hakenbeck, R., eds). Amsterdam: Elsevier, pp. 131–166.

Smith, T.J. & Foster, S.J. (1995) Characterization of the involvement of two compensatory autolysins in mother cell lysis during sporulation of *Bacillus subtilis* 168. *J. Bacteriol.* **177**, 3855–3862.

***Simon J. Foster***
*Department of Molecular Biology and Biotechnology,*
*University of Sheffield,*
*Firth Court, Western Bank,*
*Sheffield S10 2TN, UK*
*Email: S.Foster@sheffield.ac.uk*

**M**

# 465. Introduction: clan ME containing pitrilysin and its relatives

## Databanks

MEROPS ID: ME

| Species | SwissProt | PIR | EMBL (cDNA) | EMBL (genomic) |
|---|---|---|---|---|
| **Family M16** | | | | |
| AXL1 gene product | | | | |
|   *Saccharomyces cerevisiae* | P40851 | – | D17787 | – |
| Chloroplast stromal processing peptidase (Chapter 470) | | | | |
| Insulysin (Chapter 466) | | | | |
| Mitochondrial processing peptidase (Chapter 469) | | | | |
| Nardilysin (Chapter 468) | | | | |
| Nonpeptidase homologs | | | | |
|   *Caenorhabditis elegans* | P98080 | – | | U13644: cosmid F56D2 |
|   *Salmonella typhimurium* | P50335 | – | X91397 | – |
| Pitrilysin (Chapter 467) | | | | |
| PQQ-like protein | | | | |
|   *Klebsiella pneumoniae* | P27508 | – | X58778 | – |
|   *Methylbacterium extorquens* | – | – | L43135 | – |
|   *Pseudomonas fluorescens* | P55174 | – | X87299 | – |
| Others | | | | |
|   *Arabidopsis thaliana* | – | – | Z17458 | – |
|   *Bacillus subtilis* | Q04805 | – | L08471 | – |
| | Q04805 | – | U27560 | – |
|   *Caenorhabditis elegans* | P98080 | – | – | U13644: cosmid F56D2 |
|   *Eimeria bovis* | P42789 | – | M98842 | – |
|   *Escherichia coli* | P31828 | – | – | |
| | P37648 | – | – | U00039: chromosome region 76.0–81.5′ |
|   *Haemophilus influenzae* | P45181 | – | – | U32816: complete genome section 131 of 163 |
|   *Plasmodium falciparum* | – | – | X56851 | – |
|   *Synechocystis* sp. | – | – | D64001 | – |
| **Family M44** | | | | |
| Vaccinia virus metalloendopeptidase (Chapter 471) | | | | |

Clan ME contains peptidases in which the zinc ligands are His, His and Glu, as in clan MA (Chapter 335), but with the His ligands in a **His**-Xaa-Xaa-Glu-**His** (HXXEH) motif, rather than the HEXXH motif of clan MA. Site-directed mutagenesis has shown that in insulysin the Glu is important for activity (Perlman *et al*., 1993). No tertiary structure has been solved for any member of the clan. There are two families in clan ME, both possessing the HXXEH motif. Peptidases from family M16 are found in bacteria and eukaryotes, whereas family M44 contains an endopeptidase from vaccinia virus.

*Family M16* can be divided into two subfamilies. Subfamily A contains oligopeptidases such as insulysin and nardilysin from animals, and pitrilysin from bacteria. The peptidases in this subfamily are large proteins, with the catalytic unit restricted to the N-terminus. The function of the C-terminal domain is unknown.

Pitrilysin with 962 residues is a very large protein for a bacterium. It is located in the periplasmic space and is synthesized with a signal peptide. The zinc ligands are His65, His69 and Glu139 (numbering according to mature pitrilysin). Besides the zinc ligands, site-directed mutagenesis has shown that Glu68 and Glu146 (Becker & Roth, 1993) are also important for activity. Mutants with Glu146 replaced by Gln show a 20% loss of activity, and it is possible that this residue orientates an imidazolium ring of one the histidines, rather than playing a direct role in catalysis. These residues are conserved in all the proteolytically active members of the family. Another residue completely conserved among active members of the family is Asn96 (see Alignment 465.1 on CD-ROM). Tyr203 is conserved among all active members except *Saccharomyces cerevisiae* AXL1 protease.

Insulysin is thiol-activated. In insulysin, the thiol sensitivity was at one time attributed to Cys67, but site-directed

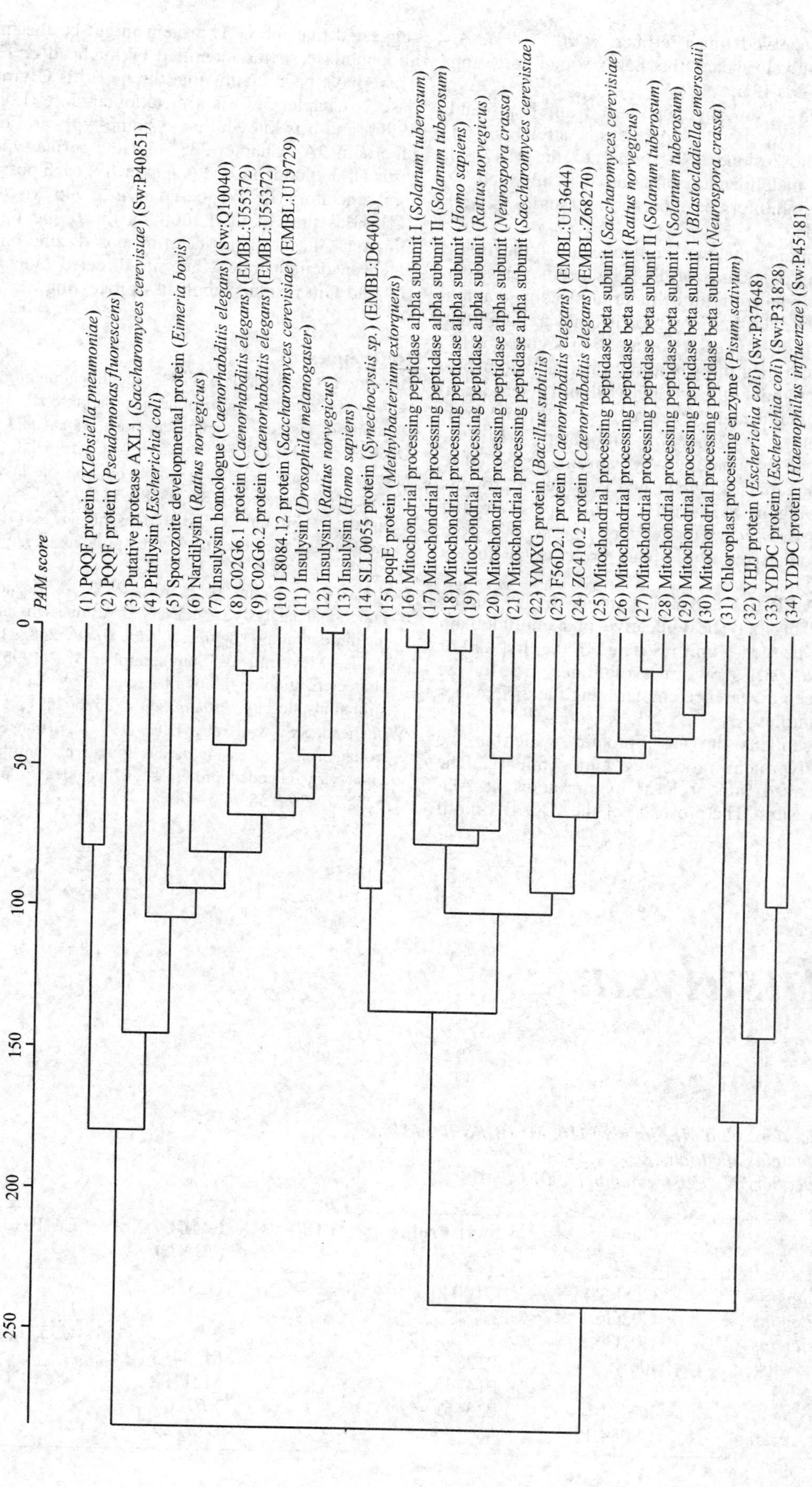

*PAM score*

(1) PQQF protein (*Klebsiella pneumoniae*)
(2) PQQF protein (*Pseudomonas fluorescens*)
(3) Putative protease AXL1 (*Saccharomyces cerevisiae*) (Sw:P40851)
(4) Pitrilysin (*Escherichia coli*)
(5) Sporozoite developmental protein (*Eimeria bovis*)
(6) Nardilysin (*Rattus norvegicus*)
(7) Insulysin homologue (*Caenorhabditis elegans*) (Sw:Q10040)
(8) C02G6.1 protein (*Caenorhabditis elegans*) (EMBL:U55372)
(9) C02G6.2 protein (*Caenorhabditis elegans*) (EMBL:U55372)
(10) L8084.12 protein (*Saccharomyces cerevisiae*) (EMBL:U19729)
(11) Insulysin (*Drosophila melanogaster*)
(12) Insulysin (*Rattus norvegicus*)
(13) Insulysin (*Homo sapiens*)
(14) SLL0055 protein (*Synechocystis sp.*) (EMBL:D64001)
(15) pqqE protein (*Methylbacterium extorquens*)
(16) Mitochondrial processing peptidase alpha subunit I (*Solanum tuberosum*)
(17) Mitochondrial processing peptidase alpha subunit II (*Solanum tuberosum*)
(18) Mitochondrial processing peptidase alpha subunit (*Homo sapiens*)
(19) Mitochondrial processing peptidase alpha subunit (*Rattus norvegicus*)
(20) Mitochondrial processing peptidase alpha subunit (*Neurospora crassa*)
(21) Mitochondrial processing peptidase alpha subunit (*Saccharomyces cerevisiae*)
(22) YMXG protein (*Bacillus subtilis*)
(23) F56D2.1 protein (*Caenorhabditis elegans*) (EMBL:U13644)
(24) ZC410.2 protein (*Caenorhabditis elegans*) (EMBL:Z68270)
(25) Mitochondrial processing peptidase beta subunit (*Saccharomyces cerevisiae*)
(26) Mitochondrial processing peptidase beta subunit (*Rattus norvegicus*)
(27) Mitochondrial processing peptidase beta subunit II (*Solanum tuberosum*)
(28) Mitochondrial processing peptidase beta subunit I (*Solanum tuberosum*)
(29) Mitochondrial processing peptidase beta subunit 1 (*Blastocladiella emersonii*)
(30) Mitochondrial processing peptidase beta subunit (*Neurospora crassa*)
(31) Chloroplast processing enzyme (*Pisum sativum*)
(32) YHJJ protein (*Escherichia coli*) (Sw:P37648)
(33) YDDC protein (*Escherichia coli*) (Sw:P31828)
(34) YDDC protein (*Haemophilus influenzae*) (Sw:P45181)

*Figure 465.1*   Evolutionary tree for family M16. The tree was prepared as described in the Introduction (p. xxv).

mutagenesis has disproved this (Perlman *et al*., 1993). At its C-terminus, insulysin bears the peroxisomal targeting sequence (Ala/Ser)-Lys-Leu.

Nardilysin, which is an oligopeptidase from brain and testis cleaving Xaa┼Arg bonds, possesses a stretch of 71 predominantly acidic residues N-terminal to the HXXEH motif. Unlike most metalloendopeptidases, it is inhibited by the aminopeptidase inhibitors bestatin and amastatin, as well as *N*-ethylmaleimide.

Subfamily B contains the two components of the mitochondrial processing peptidase (MPP), which removes an N-terminal targeting signal from mitochondrial proteins that are synthesized in the cytoplasm. Many proteins that are imported into the mitochondrion possess two targeting signals; MPP removes the first, and then mitochondrial intermediate peptidase (Chapter 377) removes the second. MPP is a heterodimer of an inactive α subunit and an active β subunit, and both are pitrilysin homologs. In *Neurospora* and potato, β-MPP is also a component of the ubiquinol–cytochrome *c* reductase complex, and in mammals, *Euglena* and yeast, core proteins 1 and 2 are homologous to β-MPP. A chloroplast stromal processing peptidase, which removes targeting signals from cytoplasmically synthesized chloroplast proteins, is known from pea.

*Neurospora* α-MPP has a Ser-rich insert in the middle of the sequence. This may be indicative of a multidomain structure, with the peptidase unit restricted to the first 250 or so residues. Figure 465.1 shows an evolutionary tree for the family, with the deep divergence emphasizing the differences between the subfamilies.

*Family M44* contains the G1L protein, which is the vaccinia virus polyprotein processing endopeptidase. The polyprotein is cleaved into at least five proteins at Ala-Gly┼Xaa cleavage sites. The presence of the HXXEH motif suggested that the G1L protein might be the first viral metallopeptidase so far identified (Whitehead & Hruby, 1994). There have been claims that the hepatitis C virus endopeptidase 2 (Chapter 567) is a metallopeptidase (Lohmann *et al*., 1996), and it is known that cysteine-type viral endopeptidase picornain 2A (Chapter 238) and the serine-type hepatitis C virus NS3 endopeptidase (Chapter 89) each possesses a structural zinc atom (De Francesco *et al*., 1996; Voss *et al*., 1995).

Besides the HXXEH motif, a third zinc ligand within a Glu-Asn-Glu motif has been proposed; zinc binding has not yet been demonstrated, but site-directed mutagenesis of the His and Glu residues abolishes processing.

### References

Becker, A.B. & Roth, R.A. (1993) Identification of glutamate-169 as the third zinc-binding residue in proteinase III, a member of the family of insulin-degrading enzymes. *Biochem. J.* **292**, 137–142.

De Francesco, R., Urbani, A., Nardi, M.C., Tomei, L., Steinkühler, C. & Tramontano, A. (1996) A zinc binding site in viral serine proteinases. *Biochemistry* **35**, 13282–13287.

Lohmann, V., Koch, J.O. & Bartenschlager, R. (1996) Processing pathways of the hepatitis C virus proteins. *J. Hepatol.* **24** (suppl. 2), 11–19.

Perlman, R.K., Gehm, B.D., Kuo, W.-L. & Rosner, M.R. (1993) Functional analysis of conserved residues in the active site of insulin-degrading enzyme. *J. Biol. Chem.* **268**, 21538–21544.

Voss, T., Meyer, R. & Sommergruber, W. (1995) Spectroscopic characterization of rhinoviral protease 2A: Zn is essential for the structural integrity. *Protein Sci.* **4**, 2526–2531.

Whitehead, S.S. & Hruby, D.E. (1994) A transcriptionally controlled *trans*-processing assay – putative identification of a vaccinia virus-encoded proteinase which cleaves precursor protein p25k. *J. Virol.* **68**, 7603–7608.

# 466. *Insulysin*

## Databanks

*Peptidase classification: clan ME, family M16, MEROPS ID: M16.002*
*NC-IUBMB enzyme classification: EC 3.4.24.56*
*Chemical Abstracts Service registry number: 9013-83-6*
*Databank codes:*

| Species | Gene | SwissProt | PIR | EMBL (cDNA) | EMBL (genomic) |
|---|---|---|---|---|---|
| *Caenorhabditis elegans* | C28F5.4 | Q10040 | – | U23180 | – |
| *Caenorhabditis elegans* | C02G6.1 | – | – | – | U55372: cosmid C02G6 |
| *Caenorhabditis elegans* | C02G6.2 | – | – | – | U55372: cosmid C02G6 |
| *Drosophila melanogaster* | ide | P22817 | – | M58465 | – |
| *Homo sapiens* | ide | P14735 | – | M21188 | – |
| *Rattus norvegicus* | ide | P35559 | – | X67269 | – |
| *Saccharomyces cerevisiae* | L8084.12 | – | – | U19729 | – |

## Name and History

Soon after the discovery of insulin it was found that it had a very short half-life after injection into animals. Subsequent studies showed that insulin is rapidly proteolyzed both in whole animals and with isolated cells. Indeed, every tissue and cell that responds to insulin can also degrade the insulin molecule. However, it has been very difficult to definitively establish which protease is responsible for mediating this process.

In the whole animal, the liver and kidney are responsible for the major share of insulin degradation. Studies of Mirsky and colleagues in the 1940s established that extracts of liver have an enzymatic activity which can degrade the insulin molecule (Mirsky & Broh-Kahn, 1949). They called this enzyme *insulinase*. One study at this time suggested that this activity was at least partly responsible for the normal turnover of insulin, as mice with elevated levels of insulinase were found to be more resistant to elevated levels of insulin than normal mice (Beyer, 1955). Further attempts to purify this enzyme over the next 30 years by several groups were largely unsuccessful because of the low levels of the enzyme and its poor stability (Duckworth, 1988). These groups called the enzyme *insulin-degrading enzyme* and it was renamed *insulysin* by the IUBMB in 1992.

A major advance in the study of this enzyme was the development of several monoclonal antibodies to the molecule (Shii & Roth, 1986). These antibodies allowed workers to immunocapture the protease to study its enzymatic properties. They were also capable of depleting more than 90% of the insulin-degrading activity from extracts of liver, demonstrating that this enzyme was responsible for most of the insulin proteolytic activity in these cell lysates (Shii & Roth, 1986). The antibodies were also capable of immunoblotting the enzyme and precipitating a labeled band from metabolically labeled cells, allowing the demonstration that the enzyme is composed of only a single polypeptide of 110 000 Da (Shii *et al.*, 1986). Affinity columns composed of these antibodies also allowed a sufficient amount of the protein to be purified so that some of its peptide sequence could be determined, thereby allowing the isolation of a cDNA clone encoding the enzyme (Affholter *et al.*, 1988). Finally, the loading of cells with a pool of these monoclonal antibodies to the enzyme was found to partially inhibit insulin degradation by these cells, suggesting that this enzyme may be partly responsible for the insulin degradation (Shii *et al.*, 1986).

The demonstration that the molecular mass of insulysin was approximately 110 000, as well as some of its other properties, suggested that it might be related to a high molecular mass metalloendoprotease that had been purified from the cytosol of human red blood cells by Kirschner & Goldberg (1983). Both enzymes were inhibited by chelating agents and hence appeared to be metalloproteases. However, a number of differences in their properties were noted. First, the metalloprotease of Kirschner & Goldberg was reported to have a molecular mass of 300 000 on nondenaturing columns whereas the insulin-degrading activity appeared to run closer to 200 000 (Shii *et al.*, 1986). Second, the metalloprotease had a pH optimum of 8.5 whereas that of the insulin-degrading enzyme was around 7.0 (Kirschner

& Goldberg, 1983; Shii *et al.*, 1986). Finally, the cytosolic metalloprotease was reported to only poorly degrade the intact insulin molecule and have a much greater activity on insulin B chain (Kirschner & Goldberg, 1983). This latter difference appears to be due to the assays used in the different studies. The proteolysis of intact insulin by this enzyme results in only a limited number of cleavage sites being generated in the hormone molecule. Thus, assays for this enzyme with the intact insulin molecule by trichloroacetic acid (TCA) precipitation are a very insensitive way to monitor the activity of this protease (Shii *et al.*, 1986). It is now accepted that the two enzymes are, in fact, a single molecule.

It should be noted that a number of prior publications on the insulin-degrading enzyme have reported other discrepancies. For example, a number of studies have suggested that the molecular mass of the protein on SDS gels was about 50 kDa. This could be due to different tissues having different species. However, it appears more likely that these reports were incorrect due to either proteolysis of the molecule or insufficient purification of the enzyme since the 110 000 Da insulin-degrading enzyme has been identified in a variety of tissues (Shii *et al.*, 1985).

The isolation of a cDNA which encodes insulysin established that this enzyme is homologous to a bacterial protease called pitrilysin (Chapter 467) (Affholter *et al.*, 1988). There were three regions of these two molecules which were conserved more than 50%. These two enzymes appear to be representative of a new clan of metalloproteases in that they have an inversion of the traditional active site of metalloproteases (Becker & Roth, 1992). That is, instead of HEXXH, they have HXXEH at the active site. Other members of this family include a yeast enzyme (called STE23) as well as an enzyme in *Caenorhabditis elegans*, although it should be noted that these proteins have not been demonstrated to actually have insulin-degrading activities. More distant members of this family include the mitochondrial-processing endopeptidases (Chapter 469), a sporozoite development protein (*Eimeria*) and nardilysin (Chapter 468) (Rawlings & Barrett, 1991, 1995; Pierotti *et al.*, 1994) (see Fig. 465.1).

## Activity and Specificity

As noted above, it has been very difficult to purify insulysin and hence many of the studies of the enzymatic activity of this molecule may actually represent mixed populations of proteases. Also, it is very labile, so studies of the homogeneous enzyme may not be representative of the fully active molecules. It is clear that the enzyme can cleave intact insulin as well as a number of other molecules. These other molecules include glucagon (Kirschner & Goldberg, 1983), insulin B chain (Kirschner & Goldberg, 1983)–

FVNQHLCGSHLVE↓ALY↓LVCGERGFFYTPKA

oxidatively damaged hemoglobin (Fagan & Waxman, 1992), atrial natriuretic factor (Muller *et al.*, 1991), β-amyloid peptide (Kurochkin & Goto, 1994), transforming growth factor α (Gehm & Rosner, 1991) and, most recently, dynorphin $A_{1-13}$

(Safavi *et al.*, 1996):

YGGFLR+RIRPKLK

It is very difficult to compare the relative activities of insulysin for these different peptides since the activity, as noted above, is highly dependent upon the assay utilized. For example, intact insulin is cleaved at multiple sites by the enzyme (Duckworth, 1988). Assays which monitor a loss in TCA-precipitability clearly underestimate the amount of insulin cleaved since the initial cleavage of insulin results in the formation of a molecule which is still precipitated by TCA. It is even possible that this product may inhibit the enzyme. This would explain why it is more difficult to measure the degradation of high concentrations of insulin than lower concentrations of insulin.

Insulysin is also unusual in its high affinity for its substrates. The enzyme appears to have the highest affinity for intact insulin, a $K_m$ of approximately 100 nM (Duckworth, 1988). This high affinity has allowed the cross-linking of labeled insulin to the protease as well as measurements of the direct binding of insulin to the protease (Shii *et al.*, 1985; Hari *et al.*, 1987; Affholter *et al.*, 1990). The affinity of the enzyme for other substrates is generally less. However, the catalytic rate for several other peptides is higher than that observed with insulin. For example, insulin B chain, glucagon and β-endorphin have been reported to be cleaved more rapidly than native insulin (Kirschner & Goldberg, 1983; Safavi *et al.*, 1996). Again, such results will depend on the method used to assay the cleavage. The specificity constants ($k_{cat}/K_m$) for different substrates are however quite similar (Safavi *et al.*, 1996).

No synthetic substrates have been described for assaying insulysin. However, it is likely that the synthetic substrate QF27 (Mca-Nle-Ala-Val-Lys-Lys-Tyr+Leu-Asn-Ser-Lys(Dnp)-Leu-Asp-D-Lys) which has been described for pitrilysin will serve as a substrate for insulysin (Anastasi *et al.*, 1993). Both enzymes appear to cleave on the amino side of hydrophobic residues. However, the enzyme appears to be influenced by adjoining residues and it is difficult to generalize too much from the known cleavage sites of the different substrates. Indeed, it may be that its specificity is influenced by the overall conformation of the peptide.

The pH optimum of insulysin has been reported to range from 7.0 to 8.5 (Kirschner *et al.*, 1983; Shii *et al.*, 1986; Safavi *et al.*, 1996). The particular substrates used may influence this. Its activity is inhibited by chelating agents such as EDTA and 1,10-phenanthroline, although phenanthroline is the more potent inhibitor (Kirschner & Goldberg, 1983; Shii *et al.*, 1986; Duckworth, 1988; Safavi *et al.*, 1996). The enzyme is also potently inhibited by alkylating agents. The particular cysteine that is required for activity has not yet been reported. Inhibition of the enzyme has also been observed with the antibiotic bacitracin with an $IC_{50}$ of about 1.0 mg ml$^{-1}$ (Shii *et al.*, 1986).

The cleavage of the different substrates by insulysin has been monitored by loss in TCA solubility, HPLC analysis and radioreceptor assays for the substrates (Becker & Roth, 1995). Each assay has its advantages and disadvantages.

## Structural Chemistry

Insulysin is a single-chain metalloprotease of about 110 kDa on SDS gels (Kirschner & Goldberg, 1983; Shii *et al.*, 1986). The calculated molecular mass of the protein (assuming that the second methionine is the preferred start site) is 113 000 and its calculated pI is 5.9 while an experimentally determined pI (by chromatofocusing) was 5.3 (Shii *et al.*, 1986). On nondenaturing columns, the enzyme behaves as a dimer, migrating on most columns with a molecular mass of about 200 000 although some studies have reported larger size complexes (Kirschner & Goldberg, 1983; Shii *et al.*, 1986; Safavi *et al.*, 1996). Attempts to identify insulysin-associated proteins by immunoprecipitation from metabolically labeled cells have been unsuccessful (Ding *et al.*, 1992) although it has been reported that under certain conditions the enzyme is associated with the proteasome (Duckworth *et al.*, 1994). It has also been reported that the enzyme can associate with androgen and glucocorticoid receptors (Kupfer *et al.*, 1994). However, the majority of insulysin in the cell appears to be in the form of a homodimer and this dimer appears to be required for enzymatic activity (Ding *et al.*, 1992). Indeed, the lability of the purified enzyme appears in part due to the breakdown of the enzyme into catalytically inactive monomers.

Insulysin is a metalloprotease in that it is inhibited by chelating agents and its enzymatic activity can be restored by the addition of metals. One study has demonstrated that zinc is present in insulysin preparations although the exact stoichiometry is not clear from these studies (Ebrahim *et al.*, 1991). Manganese was also detected in these preparations although the amount was about 10-times lower than zinc and this metal could have been present as a contaminant. Cells labeled with radioactive zinc also contain radioactivity in the insulysin immunoprecipitates (Perlman & Rosner, 1994). Site-directed mutagenesis studies have clearly established the role of His112 in the binding of zinc to the enzyme since a mutant His112Gln did not contain as much labeled zinc as the native enzyme (Perlman *et al.*, 1993). This mutant enzyme did not have insulin-degrading activity although it still bound to insulin. Other mutant insulysin molecules have also been described. Mutation of His108, Glu111 and Glu189 all resulted in a loss in enzymatic activity with a retention of insulin-binding activity (Perlman *et al.*, 1993). By analogy with pitrilysin, it is therefore likely that His108 and glutamate are also involved in binding zinc although this has not been directly tested.

As mentioned above, insulysin is also inactivated by alkylating agents. A cysteine near the residues involved in zinc binding (Cys110) was mutated and found not to be required for enzymatic activity (Kuo *et al.*, 1994). Thus, the cysteine which is critical for the enzymatic activity of insulysin has not yet been determined.

The sequence of the human cDNA encoding insulysin had two potential AUG start sites at its 5′ end (Affholter *et al.*, 1988). The second start site is in a sequence context that better fits a consensus Kozak sequence. Attempts to produce insulysin from the first start site (by changing the nucleotides surrounding this site to a consensus Kozak sequence) demonstrated that the molecule produced from the first start site is slightly larger than the endogenous molecule. This result would suggest that the majority of the endogenous enzyme is produced with the second start site. Interestingly, analyses of the sequence following the first start site suggest the possibility of a partial signal sequence. It therefore remains a possibility that two forms of the enzyme are produced and directed to different compartments in the cells.

The deduced sequences for the human and rat insulysin contain the peroxisomal targeting sequence Ala/Ser-Lys-Leu at the C-terminus (Baumeister *et al.*, 1993). Immunolocalization studies of the overexpressed enzyme also indicate that a fraction of the enzyme is located in this organelle (Kuo *et al.*, 1994; Authier *et al.*, 1995).

## Preparation

The enzyme has been purified to homogeneity from the cytosol of human erythrocytes, rat brain and, most recently, from the media of a thymoma cell line (Kirschner *et al.*, 1983; Shii *et al.*, 1986; Muller *et al.*, 1991; Safavi *et al.*, 1996). All the purification schemes are quite laborious and require multiple steps. In the case of the erythrocyte enzyme, a purification of 50 000-fold was required. During such a lengthy purification, it is likely that much of the enzymatic activity of the protease is lost, considering that the enzyme is quite labile, especially at low concentrations. Attempts to overproduce the mammalian enzyme in bacteria or yeast have not resulted in an active molecule although the enzyme has been expressed in mammalian cells. However, the amounts produced in mammalian cells are not very high. Thus, at the present time, the easiest way to measure the properties of a purified preparation of the enzyme is to immunocapture the protease from lysates (Becker & Roth, 1995).

## Biological Aspects

The biological role of insulysin has been extremely controversial (Becker *et al.*, 1996). Although it has been established that insulin is readily degraded by many cells and tissues, the enzyme responsible for this degradation is not so clear. Insulysin is the primary protease which degrades insulin in cell lysates (Shii *et al.*, 1986). However, in intact cells and tissues, it is possible that the insulin may not come in contact with insulysin. Cells bind insulin via a specific receptor protein (Becker *et al.*, 1996). The insulin molecule is then internalized in an endosome (Levy & Olefsky, 1990). Acidification of the endosome results in the release of the insulin molecule from its receptor. Exactly where the insulin then goes and when it is degraded is still a matter of dispute. Some studies have suggested that a portion of the insulin molecule may be released from endosomes. Insulin in the cytoplasm of cells clearly could be degraded by insulysin. Insulin in endosomes could also be degraded by insulysin if, as suggested by some studies, insulysin is present in this organelle (Hamel *et al.*, 1991).

Several observations are consistent with a role for insulysin in the degradation of insulin. First, the degradation products of insulin from intact cells are nearly identical with those observed with purified insulysin (Duckworth, 1988). Second,

inhibitors of this enzyme also inhibit insulin degradation by intact cells (Duckworth, 1988). Third, pinocytotic loading of cells with antibodies directed against the enzyme partially inhibit insulin degradation (Shii *et al.*, 1986). Fourth, insulin added to intact cells and internalized by a receptor-mediated process can be cross-linked to the enzyme (Hamel *et al.*, 1991). And finally, overexpression of insulysin has been shown to increase the rate of insulin degradation by intact cells (Kuo *et al.*, 1994).

Evidence against a role for this enzyme in the normal processing of insulin by cells comes from studies suggesting that a distinct enzyme is present in endosomes which can degrade insulin (Authier *et al.*, 1994). Unlike insulysin, this protease was reported to have a pH optimum of 4.5–5.5. Also unlike insulysin, this activity was reported not to be inhibited by 1 mM of the alkylating agent *N*-ethylmaleimide, although it was almost completely inhibited by a different alkylating agent. Similar to insulysin, the activity was only weakly inhibited by the chelating agent EDTA and more potently inhibited by phenanthroline. Although the authors suggested that the molecular mass of this protease on SDS gels may be 80 or 66 kDa, no evidence was presented which establishes these bands as part of the enzyme. It will be very important to purify this enzyme to homogeneity and to determine whether it is related to insulysin or belongs to another protease family.

The sequence of insulysin, as noted above, was found to contain a C-terminal sequence which directs proteins to the peroxisomes. Subsequent immunostaining studies have further shown that a portion of overexpressed enzyme is localized to this organelle. In addition, insulysin was found to be capable of degrading the cleaved leader peptide present in a peroxisomal protein (thiolase) (Authier *et al.*, 1995). Insulysin was not, however, capable of removing the leader peptide from thiolase. It has therefore been suggested that a biological role for insulysin would be to cleave leader peptides that are removed from peroxisomal proteins after import of these proteins into the peroxisomes.

Other roles for insulysin are suggested by its different activities *in vitro*. As noted above, insulysin is capable of degrading a variety of molecules, including oxidatively damaged hemoglobin (Fagan & Waxman, 1992). Thus, it is possible that insulysin may be involved in degrading damaged proteins in cells. Such a role would be consistent with the data suggesting an association of this enzyme with the multicatalytic proteasome (Duckworth *et al.*, 1994). The ability of insulysin to degrade atrial natriuretic factor, β-amyloid peptide, transforming growth factor α- and β-endorphin might indicate a role for this enzyme in the regulation of these peptides. The high level of the enzyme in the testis also suggests that it has some role in this tissue (Baumeister *et al.*, 1993). Thus, the present data do not allow one to definitively state what the biological role of insulysin is in the body, however its widespread distribution and ability to cleave a variety of molecules suggest a number of possibilities.

## Distinguishing Features

Insulysin appears to be responsible for the majority of the insulin-degrading activity in most cell lysates. It is inhibited by both chelating agents and alkylating agents. Because of its high affinity for insulin, it is possible to test whether a particular substrate may be cleaved by this enzyme by determining whether insulin inhibits its degradation. Both polyclonal antisera and a monoclonal antibody against the enzyme have been described (Shii *et al.*, 1986; Kuo *et al.*, 1994), but are not commercially available.

## Further Reading

For a review, see Becker *et al.* (1996).

## References

Affholter, J.A., Fried, V.A. & Roth, R.A. (1988) Human insulin-degrading enzyme shares structural and functional homologies with *E. coli* protease III. *Science* **242**, 1415–1418.

Affholter, J.A., Cascieri, M.A., Bayne, M.L., Brange, J., Casaretto, M. & Roth, R.A. (1990) Identification of residues in the insulin molecule important for binding to insulin-degrading enzyme. *Biochemistry* **29**, 7727–7733.

Anastasi, A., Knight, C.G. & Barrett, A.J. (1993) Characterization of the bacterial metalloendopeptidase pitrilysin by use of a continuous fluorescence assay. *Biochem. J.* **290**, 601–607.

Authier, F., Rachubinski, R.A., Posner, B.I. & Bergeron, J.J.M. (1994) Endosomal proteolysis of insulin by an acidic thiol metalloprotease unrelated to insulin degrading enzyme. *J. Biol. Chem.* **269**, 3010–3016.

Authier, F., Bergeron, J.J., Ou, W.J., Rachubinski, R.A., Posner, B.I. & Walton, P.A. (1995) Degradation of the cleaved leader peptide of thiolase by peroxisomal proteinase. *Proc. Natl Acad. Sci. USA* **92**, 3859–3863.

Baumeister, H., Muller, D., Rehbein, M. & Richter, D. (1993) The rat insulin-degrading enzyme; molecular cloning and characterization of tissue-specific transcripts. *FEBS Lett.* **317**, 250–254.

Becker, A.B. & Roth, R.A. (1992) An unusual active site identified in a family of zinc metalloendopeptidases. *Proc. Natl Acad. Sci. USA* **89**, 3835–3839.

Becker, A.B. & Roth, R.A. (1995) Insulysin and pitrilysin: insulin-degrading enzymes of mammals and bacteria. *Methods Enzymol.* **248**, 693–703.

Becker, A.B., Ding, L. & Roth, R.A. (1996) Insulin degradation. In: *Diabetes Mellitus: A Fundamental and Clinical Text* (LeRoith, D., Olefsky, J. & Taylor, S., eds). Philadelphia/New York: Lippincott-Raven, pp. 242–247.

Beyer, R.E. (1955) A study of insulin metabolism in an insulin-tolerant strain of mouse. *Acta Endocrinol.* **19**, 309–332.

Ding, L., Becker, A.B., Suzuki, A. & Roth, R.A. (1992) Comparison of the enzymatic and biochemical properties of human insulin-degrading enzyme and *Escherichia coli* protease III. *J. Biol. Chem.* **267**, 2414–2420.

Duckworth, W.C. (1988) Insulin degradation: mechanisms, products, and significance. *Endocrine Rev.* **9**, 319–345.

Duckworth, W.C., Bennett, R.G. & Hamel, F.G. (1994) A direct inhibitory effect of insulin on a cytosolic proteolytic complex containing insulin-degrading enzyme and multicatalytic proteinase. *J. Biol. Chem.* **269**, 24575–24580.

Ebrahim, A., Hamel, F.G., Bennett, R.G. & Duckworth, W.C.

(1991) Identification of the metal associated with the insulin-degrading enzyme. *Biochem. Biophys. Res. Commun.* **181**, 1398–1406.

Fagan, J.M. & Waxman, L. (1992) The ATP-independent pathway in red blood cells that degrades oxidant-damaged hemoglobin. *J. Biol. Chem.* **267**, 23015–23022.

Gehm, B.D. & Rosner, M.R. (1991) Regulation of insulin, epidermal growth factor, and transforming growth factor-α levels by growth factor-degrading enzymes. *Endocrinology* **128**, 1603–1610.

Hamel, F.G., Mahone, M.J. & Duckworth, W.C. (1991) Degradation of intraendosomal insulin by insulin-degrading enzyme without acidification. *Diabetes* **40**, 436–443.

Hari, J., Shii, K. & Roth, R.A. (1987) *In vivo* association of $^{125}$I-insulin with a cytosolic insulin-degrading enzyme: detection by covalent cross-linking and immunoprecipitation with a monoclonal antibody. *Endocrinology* **120**, 829–831.

Kirschner, R.J. & Goldberg, A.L. (1983) A high molecular weight metalloendoprotease from the cytosol of mammalian cells. *J. Biol. Chem.* **258**, 967–976.

Kuo, W.-L., Gehm, B.D., Rosner, M.R., Li, W. & Keller, G. (1994) Inducible expression and cellular localization of insulin-degrading enzyme in a stably transfected cell line. *J. Biol. Chem.* **269**, 22599–22606.

Kupfer, S.R., Wilson, E.M. & French, F.S. (1994) Androgen and glucocorticoid receptors interact with insulin-degrading enzyme. *J. Biol. Chem.* **269**, 20622–20628.

Kurochkin, I. & Goto, S. (1994) Alzheimer's β-amyloid peptide specifically interacts with and is degraded by insulin-degrading enzyme. *FEBS Lett.* **345**, 33–37.

Levy, J.R. & Olefsky, J.M. (1990) Receptor-mediated internalization and turnover. In: *Handbook of Experimental Pharmacology: Insulin* (Cuatrecasas, P. & Jacobs, S., eds). Berlin/Heidelberg: Springer-Verlag, pp. 237–266.

Mirsky, I.A. & Broh-Kahn, R.H. (1949) The inactivation of insulin by tissue extracts. I. The distribution and properties of insulin inactivating extracts (insulinase). *Arch. Biochem. Biophys.* **20**, 1–9.

Muller, D., Baumeister, H., Buck, F. & Richter, D. (1991) Atrial natriuretic peptide (ANP) is a high-affinity substrate for rat insulin-degrading enzyme. *Eur. J. Biochem.* **202**, 285–292.

Perlman, R. & Rosner, M. (1994) Identification of zinc ligands of the insulin-degrading enzyme. *J. Biol. Chem.* **269**, 33140–33145.

Perlman, R.K., Gehm, B.D., Kuo, W.-L. & Rosner, M.R. (1993) Functional analysis of conserved residues in the active site of insulin-degrading enzyme. *J. Biol. Chem.* **268**, 21538–21544.

Pierotti, A.R., Prat, A., Chesneau, V., Gaudoux, F., Leseney, A.-M., Foulon, T. & Cohen, P. (1994) N-arginine dibasic convertase, a metalloendopeptidase as a prototype of a class of processing enzymes. *Proc. Natl Acad. Sci. USA* **91**, 6078–6082.

Rawlings, N.D. & Barrett, A.J. (1991) Homologues of insulinase, a new superfamily of metalloendopeptidases. *Biochem. J.* **275**, 389–391.

Rawlings, N.D. & Barrett, A.J. (1995) Evolutionary families of metallopeptidases. *Methods Enzymol.* **248**, 183–228.

Safavi, A., Miller, B.C., Cottam, L. & Hersh, L.B. (1996) Identification of γ-endorphin generating enzyme as insulin-degrading enzyme. *Biochemistry* **35**, 14318–14325.

Shii, K. & Roth, R.A. (1986) Inhibition of insulin degradation by hepatoma cells after microinjection of monoclonal antibodies to a specific cytosolic protease. *Proc. Natl Acad. Sci. USA* **83**, 4147–4151.

Shii, K., Baba, S., Yokono, K. & Roth, R.A. (1985) Covalent linkage of $^{125}$I-insulin to a cytosolic insulin-degrading enzyme. *J. Biol. Chem.* **260**, 6503–6506.

Shii, K., Yokono, K., Baba, S. & Roth, R.A. (1986) Purification and characterization of an insulin-degrading enzyme from human erythrocytes. *Diabetes* **35**, 675–683.

*Richard A. Roth*
*Department of Molecular Pharmacology,*
*Stanford University School of Medicine,*
*Stanford, CA 94305, USA*
*Email: roth@cmgm.stanford.edu*

# 467. *Pitrilysin*

## Databanks

*Peptidase classification: clan ME, family M16, MEROPS ID: M16.001*
*NC-IUBMB enzyme classification: EC 3.4.24.55*
*Databank codes:*

| Species | SwissProt | PIR | EMBL (cDNA) | EMBL (genomic) |
|---|---|---|---|---|
| *Escherichia coli* | P05458 | A29093 | M17095 X04581 X06227 | U29581: chromosome 63–64' |

## Name and History

Cheng & Zipser (1979) identified an endoproteolytic enzyme in *Escherichia coli* based on its ability to cleave fragments of β-galactosidase (Cheng & Zipser, 1979). This enzyme, called **protease III**, was purified and extensively characterized. In addition to degrading fragments of β-galactosidase, they showed that it could degrade the B chain of insulin. Swamy & Goldberg (1981) fractionated *E. coli* extracts and identified a cytosolic (called Ci) and periplasmic (called Pi) activities which could cleave the B chain of insulin. Subsequently, they were able to show that the periplasmic activity (Pi) was due to protease III since their enzymatic activity was missing in bacteria lacking this enzyme (Swamy & Goldberg, 1982). The cytoplasmic enzyme is clearly a distinct molecule but it is not known yet whether the two enzymes are related (Kim *et al.*, 1995). The name **pitrilysin** was introduced for the periplasmic enzyme by the IUBMB in 1992 since it is the product of the *ptr* gene in *E. coli*. The *ptr* gene was identified fortuitously since it was physically located between the *recB* and *recC* genes. When these genes were overexpressed, an increase in pitrilysin activity was observed (Dykstra *et al.*, 1984).

Interest in pitrilysin has been stimulated more recently by the finding that it is homologous with the human enzyme, insulysin (Chapter 466) (Affholter *et al.*, 1988). Moreover, these two enzymes are representatives of a new clan of proteases (MD) with an 'inverted' active site (called an HXXEH; Becker & Roth, 1992). More distant members of this family include the mitochondrial-processing endopeptidases (Chapter 469), a sporozoite development protein and nardilysin (Chapter 468) (Rawlings & Barrett, 1991, 1995; Pierotti *et al.*, 1994) (see Fig. 465.1).

## Activity and Specificity

Pitrilysin does not exhibit aminopeptidase or carboxypeptidase activity (Cheng & Zipser, 1979; Anastasi *et al.*, 1993). The enzyme preferentially degrades fragments of β-galactosidase of less than 7000 Da. It preferentially cleaves insulin B chain between residues 16 and 17 (the Ala-Leu-Tyr↓Leu-Val-Cya peptide bond). A variety of other peptides have been examined by Anastasi *et al.* (1993) for cleavage by pitrilysin. Thus, pig vasoactive intestinal peptide fragment (10–28) is cleaved at multiple points:

YTRL↓R↓KQMAVKKYL↓NS↓ILN

The smallest peptides readily cleaved by the enzyme were substance P (11 residues) and a 13 residue fragment of vasoactive intestinal peptide. Using the sequence of the latter peptide, a synthetic substrate (Mca-Nle-Ala-Val-Lys-Lys-Tyr↓Leu-Asn-Ser-Lys(Dnp)-Leu-Asp-D-Lys called QF27) was synthesized and shown to yield a fluorescent product upon incubation with pitrilysin. The $K_m$ of pitrilysin for QF27 was 7.7 mM and the $k_{cat}$ was 2.9 s$^{-1}$ (Anastasi *et al.*, 1993).

In early studies of the enzyme, insulin degradation was assayed in the presence of reducing agent which could have generated insulin B chain during these assays (Swamy & Goldberg, 1981). Subsequent studies in the absence of reducing agent have shown that the enzyme can degrade intact insulin and insulin-like molecules such as insulin-like growth factor II (Ding *et al.*, 1992). The enzyme has a higher affinity for intact insulin than insulin B chain but has a higher catalysis rate with the insulin B chain.

The pH optimum of the enzyme is about 7.4 (Cheng & Zipser, 1979). Its activity is inhibited by chelating agents such as 1 mM EDTA and 1,10-phenanthroline (Cheng & Zipser, 1979; Ding *et al.*, 1992; Anastasi *et al.*, 1993). High concentrations of DTT also can inhibit the enzyme, presumably by binding to the catalytic zinc (Anastasi *et al.*, 1993). Inhibition of the enzyme has also been observed with the antibiotic bacitracin with an IC$_{50}$ of about 0.1 mg ml$^{-1}$ (Ding *et al.*, 1992; Anastasi *et al.*, 1993).

A variety of different assays for pitrilysin have been reported. As noted above, the first assay for the enzyme made use of the ability of the enzyme to cleave small fragments of the N-terminus of $\beta$-galactosidase called auto $\alpha$. These fragments, when intact, would complement mutant forms of $\beta$-galactosidase (Cheng & Zipser, 1979). Their cleavage would thus decrease the amount of active $\beta$-galactosidase activity generated in this reaction. Degradation of intact insulin and insulin B chain can be assessed by the use of radioactive insulin (readily available because of its use in radioimmunoassay) and trichloroacetic acid precipitation assays (Swamy & Goldberg, 1981; Ding *et al.*, 1992). However, the sensitivity of such assays with intact insulin is quite low because the protease only makes a limited number of cleavages in the insulin molecule. A much more sensitive way to assess degradation of the intact insulin molecule is to use a receptor-binding assay (Ding *et al.*, 1992; Becker & Roth, 1995).

The recent introduction of a fluorometric substrate for pitrilysin (called QF27) allows one to continuously monitor its enzymatic activity as well as to avoid the use of radioactivity (Anastasi *et al.*, 1993; Anastasi & Barrett, 1995). The sensitivity of this assay has been reported to be similar to that with insulin as substrate.

## Structural Chemistry

Pitrilysin is a single-chain metalloprotease of about 110 kDa on SDS gels (Cheng & Zipser, 1979; Ding *et al.*, 1992; Anastasi *et al.*, 1993). The calculated molecular mass of the mature protein is 105 000 and its pI is 5.7. Pitrilysin appears to be a monomer under physiological conditions (Ding *et al.*, 1992). The protein contains only a single cysteine and hence cannot contain any disulfide bonds (Finch *et al.*, 1986). The recombinant enzyme contains almost one atom of zinc/molecule (Ding *et al.*, 1992; Becker & Roth, 1993). Mutagenesis studies have implicated the two His residues in the sequence His-Tyr-Leu-Glu-His and one Glu residue 76 residues from the second His in the binding of the zinc (Becker & Roth, 1992, 1993). These residues are His88, His92 and Glu169 (numbered according to prepropitrilysin).

In addition, the Glu adjacent to the second His, Glu91, is required for enzymatic activity but not zinc binding. Thus the active site of the enzyme is HXXEH(X)$_{76}$E.

## Preparation

The enzyme was originally purified from native *E. coli* in a purification scheme which required almost a 10 000-fold enrichment (Cheng & Zipser, 1979). The finding that the enzyme was a periplasmic protein and the genetic engineering of bacteria to overproduce the enzyme have resulted in more simplified procedures for the purification of the molecule (Ding *et al.*, 1992; Anastasi *et al.*, 1993; Becker & Roth, 1993). The extent of purification of the overexpressed periplasmic form of the enzyme required to attain homogeneity was approximately 100- to 200-fold. About 0.5 mg of the purified enzyme can be obtained from 1 liter of cells.

## Biological Aspects

The enzyme is present in the periplasmic space of *E. coli* (Swamy & Goldberg, 1982); however, its function is not known. Strains deficient in the enzyme appear to be normal (Baneyx & Georgiou, 1991). Such strains have been found to less readily degrade a protein A–$\beta$-lactamase chimeric molecule. This hybrid protein has a much shorter half-life than either of the two parent molecules (Baneyx & Georgiou, 1991). Thus, it may be that pitrilysin is involved in the turnover of abnormally folded proteins.

## Distinguishing Features

The major periplasmic insulin-degrading activity in *E. coli* can be ascribed to pitrilysin (Swamy & Goldberg, 1981). This activity is inhibitable by chelating agents. The active-site sequence of HXXEH in pitrilysin (as well as its surrounding sequence) has been used in part to define related enzymes. However, another region of pitrilysin may be even more suitable to further define related molecules. This sequence (NQLRTEEQLGY) is 81% identical to the comparable region of the human insulin-degrading enzyme, insulysin (Chapter 466) (Affholter *et al.*, 1988). This region is conserved in the known insulin-degrading enzyme from *Drosophila* as well as in the yeast Ste23 gene product. Additional studies are required to define the role of this region in the enzymatic activity of pitrilysin. A test of whether this region can better define closely related members of this family will be to examine the insulin-degrading activities of other members of this family. Pitrilysin also appears to be unique among all the known insulin-degrading enzymes in that it clearly contains a signal peptide to direct it to the periplasmic region of *E. coli* (Swamy & Goldberg, 1982).

Both polyclonal antisera and a monoclonal antibody against the *E. coli* enzyme have been described (Ding *et al.*, 1992; Anastasi *et al.*, 1993), but are not commercially available.

## Related Peptidases

With the increase in sequences available from different organisms, many homologous molecules have been identified (Chapter 465). Such homologs have been identified in other bacteria (*B. subtilis*, SwissProt entry YMXG_BACSU), yeast (Ste23 and *axl1*), *Caenorhabditis elegans* and humans (insulysin; Chapter 466). Only in the yeast have the biological functions for these proteins been conclusively demonstrated. The pitrilysin homolog Axl1p has been shown to function in both the processing of one of the yeast mating factors (α-factor) and in axial bud site selection (Fujita *et al.*, 1994; Adames *et al.*, 1995). The proteolytic activity of this molecule appears, however, to be required only for the processing of the mating factor and not for bud site selection (Adames *et al.*, 1995). The other yeast homolog of pitrilysin (Ste23) also appears to play a role in both of these processes although to a lesser extent then Axl1p (Adames *et al.*, 1995). That is, yeast with a mutation in *STE23* but wild-type Axl1p appear normal in mating factor processing and bud site selection. However, yeast with a defective Axl1p shows a further decrease in propheromone processing but no change in bud site selection, suggesting that Ste23 is involved in the former function but not the latter. It should be noted that neither of these yeast proteins has been expressed and directly assayed for its protease activity.

## Further Reading

For a review, see Anastasi & Barrett (1995).

## References

Adames, N., Blundell, K., Ashby, M.N. & Boone, C. (1995) Role of yeast in insulin-degrading enzyme homologs in propheromone processing and bud site selection. *Science* **270**, 464–467.

Affholter, J.A., Fried, V.A. & Roth, R.A. (1988) Human insulin-degrading enzyme shares structural and functional homologies with *E. coli* protease III. *Science* **242**, 1415–1418.

Anastasi, A. & Barrett, A.J. (1995) Pitrilysin. *Methods Enzymol.* **248**, 684–694.

Anastasi, A., Knight, C.G. & Barrett, A.J. (1993) Characterization of the bacterial metalloendopeptidase pitrilysin by use of a continuous fluorescence assay. *Biochem. J.* **290**, 601–607.

Baneyx, F. & Georgiou, G. (1991) Construction and characterization of *Escherichia coli* strains deficient in multiple secreted proteases: protease III degrades high-molecular-weight substrates *in vivo*. *J. Bacteriol.* **173**, 2696–2703.

Becker, A.B. & Roth, R.A. (1992) An unusual active site identified in a family of zinc metalloendopeptidases. *Proc. Natl Acad. Sci. USA* **89**, 3835–3839.

Becker, A.B. & Roth, R.A. (1993) Identification of glutamate 169 as the third zinc binding residue in protease III, a member of the family of insulin-degrading enzymes. *Biochem. J.* **292**, 137–142.

Becker, A.B. & Roth, R.A. (1995) Insulysin and pitrilysin: insulin-degrading enzymes of mammals and bacteria. *Methods Enzymol.* **248**, 693–703.

Cheng, Y.-S. & Zipser, D. (1979) Purification and characterization of protease III from *Escherichia coli*. *J. Biol. Chem.* **254**, 4698–4706.

Ding, L., Becker, A.B., Suzuki, A. & Roth, R.A. (1992) Comparison of the enzymatic and biochemical properties of human insulin-degrading enzyme and *Escherichia coli* protease III. *J. Biol. Chem.* **267**, 2414–2420.

Dykstra, C., Prasher, D. & Kushner, S. (1984) Physical and biochemical analysis of the cloned recB and recC genes of *Escherichia coli* K-12. *J. Bacteriol.* **157**, 21–27.

Finch, P.W., Wilson, R.E., Brown, K., Hickson, I.D. & Emmerson, P.T. (1986) Complete nucleotide sequence of the *Escherichia coli ptr* gene encoding protease III. *Nucleic Acids Res.* **14**, 7695–7703.

Fujita, A., Oka, C., Arikawa, Y., Katagai, T., Tonouchi, A., Kuhara, S. & Misumi, Y. (1994) A yeast gene necessary for bud-site selection encodes a protein similar to insulin-degrading enzymes. *Nature* **372**, 567–570.

Kim, K.I., Baek, S.H., Hong, Y.-M., Kang, M.-S., Ha, D.B., Goldberg, A.L. & Chung, C.H. (1995) Purification and characterization of protease Ci, a cytoplasmic metalloendoprotease in *Escherichia coli*. *J. Biol. Chem.* **270**, 29799–29805.

Pierotti, A.R., Prat, A., Chesneau, V., Gaudoux, F., Leseney, A.-M., Foulon, T. & Cohen, P. (1994) *N*-arginine dibasic convertase, a metalloendopeptidase as a prototype of a class of processing enzymes. *Proc. Natl Acad. Sci. USA* **91**, 6078–6082.

Rawlings, N.D. & Barrett, A.J. (1991) Homologues of insulinase, a new superfamily of metalloendopeptidases. *Biochem. J.* **275**, 389–391.

Rawlings, N.D. & Barrett, A.J. (1995) Evolutionary families of metallopeptidases. *Methods Enzymol.* **248**, 183–228.

Swamy, K.H.S. & Goldberg, A.L. (1981) *E. coli* contains eight soluble proteolytic activities, one being ATP-dependent. *Nature* **292**, 652–654.

Swamy, K.H.S. & Goldberg, A.L. (1982) Subcellular distribution of various proteases in *Escherichia coli*. *J. Bacteriol.* **149**, 1027–1033.

M

*Richard A. Roth*
*Department of Molecular Pharmacology,*
*Stanford University School of Medicine,*
*Stanford, CA 94305, USA*
*Email: roth@cmgm.stanford.edu*

# 468. *Nardilysin*

## Databanks

*Peptidase classification: clan ME, family M16, MEROPS ID: M16.005*
*NC-IUBMB enzyme classification: EC 3.4.24.61*
*Databank codes:*

| Species | SwissProt | PIR | EMBL (cDNA) | EMBL (genomic) |
|---|---|---|---|---|
| *Homo sapiens* | – | – | Z21675 | – |
| *Rattus norvegicus* | P47245 | – | L27124 | – |

## Name and History

The search for the processing enzyme(s) responsible for the production of somatostatin-14 by cleavage of somatostatin-28 at the Arg-Lys site led to the discovery of **NRD convertase** (Chesneau *et al.*, 1994). However, there is no evidence to date for its implication in the *in vivo* processing of somatostatin-28. By using a synthetic peptide mimicking the maturation site, rat brain cortex extracts were found to contain a 'somatostatin-28 convertase' activity (Gluschankof *et al.*, 1984, 1987). Further purification of this activity revealed that the *in vitro* processing of somatostatin-28 was achieved by the successive intervention of an endopeptidase cleaving at the N-terminus of the Arg residue, and an aminopeptidase of the B-type (Chapter 348) trimming off the two basic residues (Gluschankof *et al.*, 1987). Initially purified from rat brain cortex, the endopeptidase was then purified from the testis where it is abundant (Chesneau *et al.*, 1994). Its cleavage specificity when assayed on various peptides revealed the selectivity of the enzyme for the N-terminal side (**N**) of Arg (**R**) residues in dibasic (**D**) sites; it was thus named NRD convertase. More recently, the IUBMB has adopted **nardilysin** as the recommended name.

## Activity and Specificity

Optimal activity of the enzyme is found at pH 8.85, with a minor peak at pH 6.5. The zinc metallopeptidase character of nardilysin is shown by its high sensitivity to metal chelators, such as EDTA and 1,10-phenanthroline, and to $Zn^{2+}$ ions, which exhibit the strongest inhibitory and reactivating effects. The possible implication of a cysteine in the catalytic process is suggested by the fact that *N*-ethylmaleimide completely blocks activity at low concentration. The best inhibitor of nardilysin is amastatin, a commonly used aminopeptidase inhibitor (Chesneau *et al.*, 1994).

The substrate specificity of the enzyme was studied using synthetic peptides mimicking dibasic maturation sites of prohormonal sequences, and physiologically important peptides (Chesneau *et al.*, 1994). Cleavage occurred strictly at the N-terminus of arginine residues in dibasic sites. This specificity was independent of the nature of the basic doublet: somatostatin-related peptides were cleaved upstream of the Arg-Lys moiety, the major cleavage of opioid substrates occurred within the Arg┼Arg sequence, and preproneurotensin-(154–170) Lys┼Arg doublet and ANF-(1–28) Arg┼Arg doublet were both cleaved within the dibasic pair. For some of the opioid peptides a second, minor, cleavage was seen upstream of the Arg-Arg site. The only exception was α-neoendorphin, where the cleavage occurred within the Arg┼Lys doublet. However, the corresponding $K_m$ value suggests that it is an unlikely specific substrate for the enzyme. The lower $K_m$ values and the higher cleavage efficiencies were obtained with the larger substrates: 43 µM for somatostatin-28 and 6.45 µM for dynorphin-A.

## Structural Chemistry

A full-length cDNA was cloned from a rat testis cDNA library (Pierotti *et al.*, 1994). The open reading frame of 1161 codons corresponds to a protein of 133 kDa which exhibits 35% and 48% similarity with pitrilysin (Chapter 467) (Finch *et al.*, 1986) and rat or human insulysin (Chapter 466) (Affholter *et al.*, 1988; Baumeister *et al.*, 1993), respectively. Moreover, the presence of the HXXEH motif clearly classifies nardilysin as a member of the pitrilysin family (M16) of zinc metalloendopeptidases (Rawlings & Barrett, 1993). Interestingly, several members of this family have been shown to be involved in polypeptide processing. For example, the yeast Ste23 endopeptidase is responsible for the cytoplasmic cleavage of the mating α-factor precursor (Adames *et al.*, 1995).

The nardilysin protein sequence contains a distinctive additional feature consisting of a 71 acidic amino acid stretch composed of 79% Glu and Asp. Located 35 residues upstream of the HXXEH pentapeptide, this stretch interrupts this most conserved region among the M16 family members. Although this domain might represent a binding region for divalent cations like $Ca^{2+}$, its physiological significance remains to be determined.

In addition to their sequence homology (up to 47% identity in the 220 residue region containing the HXXEH motif), insulysin and nardilysin share several biochemical characteristics such as their sensitivity to some sulfhydryl reagents (Müller *et al.*, 1992). The unique conserved cysteine in the alignment of nardilysin (Cys959) and the three (human, rat and *Drosophila*) insulysin sequences (Affholter *et al.*, 1988; Kuo *et al.*, 1990; Baumeister *et al.*, 1993) may thus be responsible for the thiol sensitivity. Moreover, this Cys is absent in pitrilysin which is not sensitive to sulfhydryl reagents (Cheng & Zipser, 1979).

## Preparation

Initially prepared from rat brain cortex, the enzyme was later extracted from rat testis, where it is present in high concentration (Chesneau *et al.*, 1994). Briefly, nardilysin was purified in a five-step protocol. Eight rat testes were homogenized and the supernatant precipitated at pH 4.7 in ammonium acetate buffer. After the pH of the supernatant was readjusted to pH 7.5, it was loaded on a Sephadex G150 superfine column and the relevant collected fractions were then applied to a DEAE-Trisacryl M ion-exchange column. Nardilysin activity was eluted with a linear salt gradient and finally purified on a hydroxyapatite column. About 200 μg of pure enzyme were generally obtained.

## Biological Aspects

Although present in brain and several other tissues, nardilysin is particularly abundant in testis. In seminiferous tubules, its expression is restricted to germ cell and reaches a maximum during the late steps of spermiogenesis. By electron microscopy, nardilysin immunoreactivity was localized in the cytoplasm of elongated spermatids, with a noticeable concentration at the level of two microtubular structures, i.e. the manchette and the axoneme. Nardilysin may thus be involved in cytoplasmic processing events, potentially associated with the morphological transformations occurring during spermiogenesis (Chesneau *et al.*, 1996).

In addition to its cytoplasmic subcellular localization in late spermatids, nardilysin was also shown to be present in the endoplasmic reticulum of neuroendocrine and endocrine cells. This latter localization is in agreement with the presence of a putative signal peptide at the N-terminus of its deduced protein sequence. The molecular basis of the two distinct subcellular localizations observed for the enzyme remains to be elucidated.

## Related Peptidases

Recently, Hersh and collaborators reported the purification of a secreted arginine-specific dibasic cleaving enzyme (Csuhai *et al.*, 1995) which shares its main biochemical properties with nardilysin. This enzyme, purified from the supernatant of cultured mouse thymoma cells (EL-4) as a dynorphin-converting enzyme (Chapter 542), may well be related to nardilysin.

## Further Reading

For reviews, see articles by Chesneau *et al.* (1995, 1996) and Pierotti *et al.* (1994).

## References

Adames, N., Blundell, K., Ashby, M.N. & Boone, C. (1995) Role of yeast insulin-degrading enzyme homologs in prepheromone processing and bud site selection. *Science* **270**, 464–467.

Affholter, J.A., Fried, V.A. & Roth, R.A. (1988) Human insulin-degrading enzyme shares structural and functional homologies with *E. coli* protease III. *Science* **242**, 1415–1418.

Baumeister, H., Müller, D., Rehbein, M. & Richter, D. (1993) The rat insulin-degrading enzyme. Molecular cloning and characterization of tissue-specific transcripts. *FEBS Lett.* **317**, 250–254.

Cheng, Y.-S.E. & Zipser, D. (1979) Purification and characterization of protease III from *Escherichia coli*. *J. Biol. Chem.* **254**, 4698–4706.

Chesneau, V., Pierotti, A.R., Barré, N., Créminon, C., Tougard, C. & Cohen, P. (1994) Isolation and characterization of a dibasic selective metalloendopeptidase from rat testes that cleaves at the amino terminus of arginine residues. *J. Biol. Chem.* **269**, 2056–2061.

Chesneau, V., Prat, A., Segretain, D., Hospital, V., Dupaix, A., Foulon, T., Jégou, B. & Cohen, P. (1996) NRD convertase: a putative processing endoprotease associated with the axoneme and the manchette in late spermatids *J. Cell Sci.* **109**, 2737–2745.

Csuhai, E., Safavi, A. & Hersh, L.B. (1995) Purification and characterization of a secretes arginine-specific dibasic cleaving enzyme from EL-4 cells. *Biochemistry* **34**, 12411–12419.

Finch, P.W., Wilson, R.E., Brown, K., Hickson, I.D. & Emmerson, P.T. (1986) Complete nucleotide sequence of the *Escherichia coli ptr* gene encoding protease III. *Nucleic Acids Res.* **14**, 7695–7703.

Gluschankof, P., Morel, A., Gomez, S., Nicolas, P., Fahy, C. & Cohen, P. (1984) Enzymes processing somatostatin precursors: an Arg-Lys esteropeptidase from the rat brain cortex converting somatostatin-28 into somatostatin-14. *Proc. Natl Acad. Sci. USA* **81**, 6662–6666.

**M**

Gluschankof, P., Gomez, S., Morel, A. & Cohen, P. (1987) Enzymes that process somatostatin precursors. A novel endoprotease that cleaves before the arginine-lysine doublet is involved in somatostatin-28 convertase activity of rat brain cortex. *J. Biol. Chem.* **262**, 9615–9620.

Kuo, W.-L., Gehn, B.D. & Rosner, M.R. (1990) Cloning and expression of the cDNA for *Drosophila* insulin-degrading enzyme. *Mol. Endocrinol.* **4**, 1580–1591.

Müller, D., Schulze, C., Baumeister, H., Buck, F. & Richter, D.

(1992) Rat insulin-degrading enzyme : cleavage pattern of the natriuretic peptide hormone ANP, BNP, and CNP revealed by HPLC and mass spectrometry. *Biochemistry* **31**, 11138–11143.

Pierotti, A.R., Prat, A., Chesneau, V., Gaudoux, F., Leseney, A.-M., Foulon, T. & Cohen, P. (1994) N-Arginine dibasic convertase, a metalloendopeptidase as a prototype of a class of processing enzymes. *Proc. Natl Acad. Sci. USA* **91**, 6078–6082.

Rawlings, N.D. & Barrett, A.J. (1993) Evolutionary families of peptidases. *Biochem. J.* **290**, 205–218.

*Annik Prat*
Laboratoire de Biochimie des Signaux Régulateurs
Cellulaires et Moléculaires,
Université Pierre et Marie Curie,
Unité de Recherches Associée au Centre National de la
Recherche Scientifique 1682, 96,
boulevard Raspail, F-75006 Paris, France
Email: prat@infobiogen.fr

*Thierry Foulon*
Laboratoire de Biochimie des Signaux Régulateurs
Cellulaires et Moléculaires,
Université Pierre et Marie Curie,
Unité de Recherches Associée au Centre National de la
Recherche Scientifique 1682, 96,
boulevard Raspail, F-75006 Paris, France

*Valérie Chesneau*
Laboratoire de Biochimie des Signaux Régulateurs
Cellulaires et Moléculaires,
Université Pierre et Marie Curie,
Unité de Recherches Associée au Centre National de la
Recherche Scientifique 1682, 96,
boulevard Raspail, F-75006 Paris, France

*Paul Cohen*
Laboratoire de Biochimie des Signaux Régulateurs
Cellulaires et Moléculaires,
Université Pierre et Marie Curie,
Unité de Recherches Associée au Centre National de la
Recherche Scientifique 1682, 96,
boulevard Raspail, F-75006 Paris, France
Email: pcohen@infobiogen.fr

# *469. Mitochondrial processing peptidase*

## Databanks

*Peptidase classification: clan ME, family M16, MEROPS ID: M16.003*
*NC-IUBMB enzyme classification: EC 3.4.24.64*
*ATCC entries: 71207 (yeast cosmid); 763893, 982223, 992568 (mouse, α subunit)*
*Databank codes:*

| Species | SwissProt | PIR | EMBL (cDNA) | EMBL (genomic) |
|---|---|---|---|---|
| **α subunit** | | | | |
| *Homo sapiens* | Q10713 | – | D21064 | –<br>D50913 |
| *Neurospora crassa* | P23955 | A36442 | J05484<br>S36362 | – |
| *Oryza sativa* | – | – | D25241 | – |
| *Rattus norvegicus* | P20069 | A36205 | M57728<br>B36205<br>S36361 | – |

| Species | SwissProt | PIR | EMBL (cDNA) | EMBL (genomic) |
|---|---|---|---|---|
| *Saccharomyces cerevisiae* | P11914 | B38734 | M36596 S05738 S36360 S46778 | – X13455 X14105 |
| α1 subunit | | | | |
| *Solanum tuberosum* | P29677 | S23558 | X66284 | – |
| α2 subunit | | | | |
| *Solanum tuberosum* | – | – | X80236 | – |
| β subunit | | | | |
| *Blastocladiella emersonii* | – | – | U41300 | – |
| *Caenorhabditis elegans* | – | – | Z68270 | – |
| *Neurospora crassa* | P11913 | A29881 | M20928 B29881 S03968 | – |
| *Rattus norvegicus* | Q03346 | PC1229 | D13907 S36390 S36391 | – L12965 |
| *Saccharomyces cerevisiae* | P10507 | A38734 | X07649 S00552 S68479 | – |
| β1 subunit | | | | |
| *Solanum tuberosum* | – | A48529 | X80237 | – |
| β2 subunit | | | | |
| *Solanum tuberosum* | – | B48529 | X80235 | – |

## Name and History

The activity of the **mitochondrial processing peptidase (MPP)** was discovered in 1980 when Schatz and coworkers found that extracts of the mitochondrial matrix cleave off the pre sequences of *in vitro* synthesized mitochondrial precursor proteins (Böhni *et al.*, 1980). These presequences are N-terminal signals for targeting nuclear encoded proteins to mitochondria and to the correct intramitochondrial location. Although MPP was the first component of the mitochondrial protein import apparatus to be identified, its isolation from fungi and rat took nearly another decade (Hawlitschek *et al.*, 1988, Yang *et al.*, 1988; Ou *et al.*, 1989). Even later, when MPP was purified from potato, it turned out that in plants the enzyme is not localized in the mitochondrial matrix but in a protein complex of the inner membrane (Braun *et al.*, 1992). This finding effected a new naming of the enzyme as mitochondrial processing peptidase instead of *matrix processing peptidase* (Emmermann *et al.*, 1993).

To discriminate between MPP and another matrix-localized processing peptidase which removes an octapeptide at the C-terminus of some mitochondrial presequences, Kalousek *et al.* (1988) proposed the nomenclature **MPP-1** for the enzyme acting on mitochondrial targeting signals and MPP-II for the octapeptidase. This nomenclature is no longer in use as in 1992 the mitochondrial octapeptidase was purified and termed 'mitochondrial intermediate peptidase' (MIP) (Chapter 377) (Kalousek *et al.*, 1992). MPP is a dimeric protein; the subunits are non-identical but homologous. In different organisms different names were given to the subunits of MPP (the smaller subunit is always given first): **PEP** (*processing enhancing protein*) and **MPP** in *Neurospora*

(Hawlitschek *et al.*, 1988), **Mas1** and **Mas2** in yeast (Yaffe & Schatz, 1984; Witte *et al.*, 1988) and **P-52** and **P-55** in rat (Ou *et al.*, 1989; Kleiber *et al.*, 1990).

According to the new and generally accepted nomenclature the components of the enzyme are designated **α-MPP** (larger subunit) and **β-MPP** (smaller subunit) (Kalousek *et al.*, 1993). From the early 1980s it was assumed that MPP is a metal-dependent endopeptidase (Böhni *et al.*, 1980; McAda & Douglas, 1982) but Ou *et al.* (1989) were the first to show that the pre sequences of mitochondrial precursors are indeed removed in a single cleavage step. In 1991 both MPP subunits were classified as members of a new family of metalloendoproteases (M16) (Rawlings & Barrett, 1991).

## Activity and Specificity

The signals for mitochondrial protein import processed by MPP are at the same time matrix targeting signals and either represent the complete pre sequence or the N-terminal part of bipartite pre sequences. Matrix-targeting signals are characterized by a high content of basic amino acid residues and often form α-helical amphipathic structures (von Heijne, 1986; Roise *et al.*, 1986). Otherwise, pre sequences are rather diverse without significant sequence similarity. A common theme of the cleavage sites is an arginine residue located mostly at the P2 or less often at the P3 position (von Heijne *et al.*, 1989; Hendrick *et al.*, 1989). A proline and/or glycine residue further away from the scissile bond toward the N-terminus and an aromatic (or sometimes hydrophobic) residue at the P1' position have also been suggested as critical determinants for specific cleavage of the substrates (Ogishima *et al.*, 1995). Mutational analysis of matrix targeting signals and MPP cleavage sites suggests that not the amphiphilic $\alpha$ helices but rather other parts of pre sequences are recognized by MPP (Brunner *et al.*, 1994).

Assays are most often made with radiolabeled substrates or more recently with fluorogenic ones (Ogishima *et al.*, 1995). The activity of MPP is strictly dependent on the presence of both subunits, $\alpha$-MPP and $\beta$-MPP (Géli *et al.*, 1990; Saavedra-Alanis *et al.*, 1994). MPP from fungi and mammals is maximally active between pH 7 and pH 8 while the plant enzyme has a broader optimum up to pH 9 (Emmermann & Schmitz, 1993; Brunner *et al.*, 1994). In contrast to MPP from fungi and mammals the complex-integrated processing activity of plant mitochondria is highly salt resistant (Emmermann *et al.*, 1993).

1,10-Phenanthroline and EDTA completely block processing. Pre sequence peptides are competitive inhibitors of MPP but have a much lower affinity than precursors (Brunner *et al.*, 1994). MPP from *Neurospora* is inhibited by *N*-ethylmaleimide (Schneider *et al.*, 1990) while MPP from other organisms are insensitive toward inhibitors of cysteine, serine or aspartic proteases. Although being integrated into the $bc_1$ complex in plants, MPP activity is not affected by inhibitors of electron transfer like antimycin A or myxothiazol (Emmermann & Schmitz, 1993).

## Structural Chemistry

MPP consists of two nonidentical subunits which are structurally related (Pollock *et al.*, 1988; Jensen & Yaffe, 1988). The larger subunit, $\alpha$-MPP, has a molecular mass of 53–57 kDa, the smaller subunit, $\beta$-MPP, of 48–52 kDa. Only in plants is the $\beta$-MPP subunit (53–55 kDa) larger than $\alpha$-MPP (51 kDa). Both MPP subunits are encoded in the nucleus and synthesized with pre sequences which are cleaved off by the holoenzyme. Zinc is supposed to be the favored ligand but $Mn^{2+}$ also enhances activity. Only $\beta$-MPP contains an intact HXXEH zinc binding site motif, and was shown to be the catalytically active subunit (Braun & Schmitz, 1995; Kitada *et al.*, 1995; Striebel *et al.*, 1996). In $\alpha$-MPP, which initially was thought to be the catalytic subunit (Hawlitschek *et al.*, 1988), there is only an incomplete zinc-binding site motif. In potato both subunits of MPP are present in two isoforms, termed $\alpha$1-MPP, $\alpha$2-MPP, $\beta$1-MPP and $\beta$2-MPP (Emmermann *et al.*, 1993, 1994).

## Preparation

The enzyme was purified from *Neurospora* (9000-fold purification from total cell protein; Hawlitschek *et al.*, 1988), yeast (200-fold from mitochondrial matrix; Yang *et al.*, 1988), rat (2000-fold from mitochondrial matrix; Ou *et al.*, 1989; Kleiber *et al.*, 1990) and plants (40-fold from potato, spinach and wheat mitochondria; Braun *et al.*, 1992, 1995; Eriksson *et al.*, 1994). In plants the enzyme is more abundant as it is integrated into the cytochrome $bc_1$ complex. Both MPP subunits from *Neurospora* (Arretz *et al.*, 1994), yeast (Géli, 1993), rat (Saavedra-Alanis *et al.*, 1994) and potato (Emmermann *et al.*, 1993) have been expressed in *E. coli*, but the recombinant potato subunits are not active most probably because of being separated from the $bc_1$ complex.

## Biological Aspects

In autotrophic and heterotrophic organisms MPP is located in different submitochondrial compartments: the enzyme from yeast and mammals forms a heterodimer in the mitochondrial matrix while in plants the two MPP subunits substitute the so-called 'core' proteins of the $bc_1$ complex of the inner mitochondrial membrane. *Neurospora* represents an intermediate situation with the $\beta$-MPP subunit being identical to the 'core' I protein of the $bc_1$ complex and the $\alpha$-MPP subunit being present in the mitochondrial matrix (Schulte *et al.*, 1989). An evolutionary model has been proposed according to which the 'core' proteins are relics of an ancient processing peptidase (Braun & Schmitz, 1995).

Little is known about the genes encoding the MPP subunits. In potato it was shown that the genes encoding two isoforms of $\alpha$-MPP are differentially expressed in all tissues analyzed, but the transcript levels of each isoform do not vary between tissues (Emmermann *et al.*, 1994).

## Distinguishing Features

MPP is not related to the other known mitochondrial processing enzymes: MIP (Chapter 377) (Kalousek *et al.*, 1992); IMP1 (Chapter 155) (Schneider *et al.*, 1991) or IMP2 (Chapter 155) (Nunnari *et al.*, 1993). There is significant sequence identity between MPP and the stromal processing peptidase (Chapter 470) from chloroplasts which also belongs to the pitrilysin family (Chapter 465) (VanderVere *et al.*, 1995). Some chloroplast import signals are processed by MPP but not at the authentic cleavage sites (Brink *et al.*, 1994). Antisera show cross-reactions between MPP and the 'core' proteins of the $bc_1$ complex, which are not unexpected due to the phylogenetic relationship of the two proteins.

Polyclonal antisera against the enzyme from *N. crassa*, yeast *S. cerevisiae*, rat and potato have been described (Hawlitschek *et al.*, 1988; Yang *et al.*, 1988; Kleiber *et al.*, 1990; Braun & Schmitz, 1992; Jänsch *et al.*, 1995), but are not commercially available at the time of writing.

## Further Reading

Reviews may be found in Braun & Schmitz (1995) and in Brunner & Neupert (1995).

## References

Arretz, M., Schneider, H., Guiard, B., Brunner, M. & Neupert, W. (1994) Characterization of the mitochondrial processing peptidase of *Neurospora crassa*. *J. Biol. Chem.* **269**, 4959–4967.

Böhni P., Gasser S., Leaver C. & Schatz, G. (1980) A matrix-localized mitochondrial protease processing cytoplasmically-made precursors to mitochondrial proteins. In: *The Organisation and Expression of the Mitochondrial Genome* (Kroon, A.M. & Saccone, C., eds). Amsterdam: Elsevier Science Publishers, pp. 423–433.

Braun, H.P. & Schmitz, U.K. (1992) Affinity purification of cytochrome *c* reductase from potato mitochondria. *Eur. J. Biochem.* **208**, 761–767.

Braun, H.P. & Schmitz, U.K. (1995) Are the 'core' proteins of the mitochondrial $bc_1$ complex evolutionary relics of a processing protease? *Trends Biochem. Sci.* **20**, 171–175.

Braun, H.P., Emmermann, M., Kruft, V. & Schmitz, U.K. (1992) The general mitochondrial processing peptidase from potato is an integral part of cytochrome *c* reductase of the respiratory chain. *EMBO J.* **11**, 3219–3227.

Braun, H.P., Emmermann, M., Kruft, V., Bödicker, M. & Schmitz, U.K. (1995) The general mitochondrial processing peptidase from wheat is integrated into the cytochrome bc₁ complex of the respiratory chain. *Planta* **195**, 396–402.

Brink, S., Flügge, U.-I., Chaumont, F., Boutry, M., Emmermann, M., Schmitz, U.K., Becker, K. & Pfanner, N. (1994) Preproteins of chloroplast envelope inner membrane contain targeting information for receptor-dependent import into fungal mitochondria. *J. Biol. Chem.* **269**, 16478–16485.

Brunner, M. & Neupert, W. (1995) Purification and characterization of mitochondrial processing peptidase of *Neurospora crassa*. *Methods Enzymol.* **248**, 717–728.

Brunner, M., Klaus, C. & Neupert, W. (1994) The mitochondrial processing peptidase. In: *Signal Peptidases* (von Heijne, G., ed.). Austin: Landes Co., pp. 73–86.

Emmermann, M. & Schmitz, U.K. (1993) The cytochrome *c* reductase integrated processing peptidase from potato mitochondria belongs to a new class of metalloendoproteases. *Plant Physiol.* **103**, 615–620.

Emmermann, M., Braun, H.P., Arretz, M. & Schmitz, U.K. (1993) Characterization of the bifunctional cytochrome *c* reductase/processing peptidase complex from potato mitochondria. *J. Biol. Chem.* **268**, 18936–18942.

Emmermann, M., Braun, H.P. & Schmitz, U.K. (1994) The mitochondrial processing peptidase from potato: a self-processing enzyme encoded by two differentially expressed genes. *Mol. Gen. Genet.* **245**, 237–245.

Eriksson, A.C., Sjöling, S. & Glaser, E. (1994) The ubiquinol cytochrome *c* oxidoreductase complex of spinach leaf mitochondria is involved in both respiration and protein processing. *Biochim. Biophys. Acta* **1186**, 221–231.

Géli, V. (1993) Functional reconstitution in *Escherichia coli* of the yeast mitochondrial matrix peptidase from its two inactive subunits. *Proc. Natl Acad. Sci. USA* **90**, 6247–6251.

Géli, V., Yang, M., Suda, K., Lustig, A. & Schatz, G. (1990) The *Mas*-encoded processing protease of yeast mitochondria. *J. Biol. Chem.* **265**, 19216–19222.

Hawlitschek, G., Schneider, H., Schmidt, B., Tropschug, M., Hartl, F.-U. & Neupert, W. (1988) Mitochondrial protein import: identification of processing peptidase and of PEP, a processing enhancing protein. *Cell* **53**, 795–806.

Hendrick, J.P., Hodges, P.E. & Rosenberg, L.E. (1989) Survey of amino-terminal proteolytic cleavage sites in mitochondrial precursor proteins: leader peptides cleaved by two matrix proteases share a three amino acid motif. *Proc. Natl Acad. Sci. USA* **86**, 4056–4060.

Jänsch, L., Kruft, V., Schmitz, U.K. & Braun, H.P. (1995) Cytochrome *c* reductase from potato does not comprise three core proteins but contains an additional low-molecular mass subunit. *Eur. J. Biochem.* **228**, 878–885.

Jensen, R.E. & Yaffe, M.P. (1988) Import of proteins into yeast mitochondria: the nuclear *Mas2* gene encodes a component of the processing protease that is homologous to the *Mas1*-encoded subunit. *EMBO J.* **7**, 3863–3871.

Kalousek, F., Hendrick, J.P. & Rosenberg, L.E. (1988) Two mitochondrial matrix proteases act sequentially in the processing of mammalian matrix enzymes. *Proc. Natl Acad. Sci. USA*

**85**, 7536–7540.

Kalousek, F., Isaya, G. & Rosenberg, L.E. (1992) Rat liver mitochondrial intermediate peptidase (MIP): purification and initial characterization. *EMBO J.* **11**, 2803–2809.

Kalousek, F., Neupert W., Omura T., Schatz, G. & Schmitz, U.K. (1993) Uniform nomenclature for the mitochondrial proteases cleaving precursors of mitochondrial proteins. *Trends Biochem. Sci.* **18**, 249.

Kitada, S., Shimokata, K., Niidome, T., Ogishima, T. & Ito, A. (1995) A putative metal-binding site in the β-subunit of rat mitochondrial processing peptidase is essential for its catalytic activity. *J. Biochem.* **117**, 1148–1150.

Kleiber, J., Kalousek, F., Swaroop, M. & Rosenberg, L.E. (1990) The general mitochondrial matrix processing protease from rat liver: structural characterization of the catalytic subunit. *Proc. Natl Acad. Sci. USA* **87**, 7978–7982.

McAda, P.C. & Douglas, M.G. (1982) A neutral metallo endoprotease involved in the processing of an F₁-ATPase subunit precursor in mitochondria. *J. Biol. Chem.* **257**, 3177–3182.

Nunnari, J., Fox, T. & Walter, P. (1993) A mitochondrial protease with two catalytic subunits of nonoverlapping specificities. *Science* **262**, 1997–2003.

Ogishima, T., Niidome, T., Shimokata, K., Kitada, S. & Ito, A. (1995) Analysis of elements in the substrate required for processing by mitochondrial processing peptidase. *J. Biol. Chem.* **270**, 30322–30326.

Ou, W.-J., Ito, A., Okazaki, H. & Omura, T. (1989) Purification and characterization of a processing protease from rat liver mitochondria. *EMBO J.* **8**, 2605–2612.

Pollock, R.A., Hartl, F.-U., Cheng, M.Y., Ostermann, J., Horwich, A. & Neupert, W. (1988) The processing peptidase of yeast mitochondria: the two co-operating components MPP and PEP are structurally related. *EMBO J.* **7**, 3493–3500.

Rawlings, N.D. & Barrett, A.J. (1991) Homologues of insulinase, a new superfamily of metalloendopeptidases. *Biochem. J.* **275**, 389–391.

Roise, D., Horvath, S.J., Tomich, J.M., Richards, J.H. & Schatz, G. (1986) A chemically synthesized presequence of an imported mitochondrial protein can form an amphiphilic helix and perturb natural and artificial phospholipid bilayers. *EMBO J.* **5**, 1327–1334.

Saavedra-Alanis, V.M., Rysavy, P., Rosenberg, L.E. & Kalousek, F. (1994) Rat liver mitochondrial processing peptidase. *J. Biol. Chem.* **269**, 9284–9288.

Schneider, H., Arretz, M., Wachter, E. & Neupert, W. (1990) Matrix processing peptidase of mitochondria. *J. Biol. Chem.* **265**, 9881–9887.

Schneider, A., Behrens, M., Scherer, P., Pratje, E., Michaelis, G. & Schatz, G. (1991) Inner membrane protease I, an enzyme mediating intramitochondrial protein sorting in yeast. *EMBO J.* **10**, 247–254.

Schulte, U., Arretz, M., Schneider, H., Tropschug, M., Wachter, E., Neupert, W. & Weiss, H. (1989) A family of mitochondrial proteins involved in bioenergetics and biogenesis. *Nature* **339**, 147–149.

Striebel, H.-M., Rysavy, P., Spizek, J. & Kalousek, F. (1996) Mutational analysis of both subunits from rat mitochondrial processing peptidase. *Arch. Biochem. Biophys.* **335**, 211–218.

VanderVere, P.S., Bennett, T.M., Oblong, J.E. & Lamppa, G.K. (1995) A chloroplast processing enzyme involved in precursor maturation shares a zinc-binding motif with a recently recognized

**M**

family of metalloendopeptidases. *Proc. Natl Acad. Sci. USA* **92**, 7177–7181.

von Heijne, G. (1986) Mitochondrial targeting sequences may form amphiphilic helices. *EMBO J.* **5**, 1335–1342.

von Heijne, G., Stepphuhn, J. & Herrmann, R.G. (1989) Domain structure of mitochondrial and chloroplast targeting peptides. *Eur. J. Biochem.* **180**, 535–545.

Witte, C., Jensen, R.E., Yaffe, M.P. & Schatz, G. (1988) Import of proteins into yeast mitochondria: the nuclear *Mas2* gene encodes a component of the processing protease that is homologous to the

Mas1-encoded subunit. *EMBO J.* **7**, 1439–1447.

Yaffe, M.P. & Schatz, G. (1984) Two nuclear mutations that block mitochondrial protein import in yeast. *Proc. Natl Acad. Sci. USA* **81**, 4819–4823.

Yang, M., Jensen, R.E., Yaffe, M.P., Oppliger, W. & Schatz, G. (1988) Import of proteins into yeast mitochondria: the purified matrix processing protease contains two subunits which are encoded by the nuclear MAS 1 and MAS 2 genes. *EMBO J.* **7**, 3857–3862.

**Hans-Peter Braun**
*Institut für Angewandte Genetik,*
*Universität Hannover,*
*Herrenhäuserstrasse 2, 30419 Hannover, Germany*

**Udo K. Schmitz**
*Institut für Angewandte Genetik,*
*Universität Hannover,*
*Herrenhäuserstrasse 2, 30419 Hannover, Germany*

# 470. *Chloroplast stromal processing peptidase*

## Databanks

*Peptidase classification: clan ME, family M16, MEROPS ID: M16.004*
*NC-IUBMB enzyme classification: none*
*Databank codes:*

| Species | SwissProt | PIR | EMBL (cDNA) | EMBL (genomic) |
| --- | --- | --- | --- | --- |
| *Pisum sativum* | – | – | U25111 | – |

## Name and History

Proteins targeted to the chloroplast are typically synthesized in the cytosol as precursors with N-terminal extensions called transit peptides, which are removed upon import. A proteolytic activity that cleaves the transit peptide releasing the mature protein was initially identified in whole-cell lysates from *Chlamydomonas reinhardtii* (Dobberstein *et al.*, 1977) and subsequently in a soluble extract from isolated pea chloroplasts (Smith & Ellis, 1979; Robinson & Ellis, 1984). Both assays used as a substrate the precursor for the small subunit (preS) of ribulose-1,5-bisphosphate carboxylase/oxygenase (Rubisco), which is located in the stromal compartment. Precursors for the light-harvesting chlorophyll-binding proteins (LHCP), integral membrane proteins of the thylakoids, are also cleaved by a soluble protease that was originally referred to as a *chloroplast processing enzyme* (Lamppa & Abad, 1987; Clark & Lamppa, 1992). Cumulative evidence now indicates that a general *chloroplast stromal processing peptidase (SPP)* with the properties of a metalloendopeptidase carries out these reactions, cleaving the transit peptides of proteins destined for different compartments and biosynthetic pathways of the chloroplast (Robinson & Ellis, 1984; Lamppa & Abad, 1987; Abad *et al.*, 1991; VanderVere *et al.*, 1995). SPP is most likely involved in the maturation of proteins not only required for photosynthesis, but for the synthesis of fatty acids, aromatic and branched amino acids, starch production, terpenoids, phytohormones, tetrapyrroles, as well as for the reduction of nitrogen and sulfur. Hence, SPP is a key component of the import machinery and may be essential for overall chloroplast biogenesis and function. Interestingly, proteins that enter the thylakoid lumen, such as plastocyanin, are synthesized with a bipartite transit peptide which is first processed to an intermediate form by SPP, and then cleaved by a distinct signal peptidase-like endoprotease located in the thylakoid membrane (Hageman *et al.*, 1986).

## Activity and Specificity

The rules for cleavage by SPP are not readily apparent from a comparison of transit peptides or cleavage sites of different precursors. Transit peptides show little sequence similarity other than being slightly basic and rich in the hydroxylated amino acids serine and threonine; furthermore, they vary substantially in length, e.g. from 34 to more than 100 amino acids (for review see Keegstra & Olsen, 1989). However, processing can occur before the mature protein completely translocates across the chloroplast envelope, an indication that the determinants for cleavage reside in the transit peptide and/or at the N-terminus of the mature protein (Schnell & Blobel, 1993). The consensus motif (Val/Ile)-Xaa-(Ala/Cys)+Ala has been suggested as the preferred site for cleavage, but it is found in only one-third of chloroplast precursors (Gavel & von Heijne, 1990). Mutational studies indicate that the C-terminus of the transit peptide is important for processing, which usually contains a basic residue within 10 amino acids of the scissile bond. Substitution of the basic residues with a neutral or acidic residue can lower the efficiency of cleavage although this has only been demonstrated thus far for the precursor of the major LHCP (preLHCP) of photosystem II (Clark & Lamppa, 1991) and preS (Archer & Keegstra, 1993). The preLHCP contains two cleavage sites that are independently recognized during import. However, only one, the secondary site Ala-Lys+Ala-Lys, is processed in an organelle-free reaction (see below) presumably because precursor folding prevents access to the primary site. Site-directed mutagenesis showed that the sequence Ala-Lys-Ala-Lys is essential for processing preLHCP, yet only a four amino acid insertion at the junction of the transit peptide and mature protein abolished secondary site cleavage. If the Ala-Lys+Ala-Lys motif is introduced in the correct structural context into other precursors, such as preS, it is recognized and cleaved by SPP (Clark & Lamppa, 1991).

An *in vitro* organelle-free processing assay using soluble extracts of lysed chloroplasts has been optimized; it requires a pH of 7–9 at 26°C. The metal chelators EDTA and 1,10-phenanthroline inhibit the reaction (Robinson & Ellis, 1984; Abad *et al.*, 1989), which is ATP-independent. SPP processes a large diversity of imported substrates, which can be synthesized by *in vitro* translation in either a wheatgerm or reticulocyte lysate, or recovered from *Escherichia coli*. In the latter case, the recombinant precursor is often sequestered in inclusion bodies and thus must be solubilized, and the denaturant removed, before it is cleavable by SPP (Abad *et al.*, 1991). The source of the precursor may influence the efficiency of processing.

## Structural Chemistry

Gel filtration, sucrose gradient centrifugation and native gel electrophoresis analyses indicate that SPP from pea has a molecular mass of 150–190 kDa (Robinson & Ellis, 1984; Oblong & Lamppa, 1992). Two proteins of 145 and 143 kDa copurify in an apparent 1:1 ratio with the processing activity that cleaves preLHCP in the organelle-free assay. Immunodepletion experiments indicate that they are also required for the processing of the precursors of the small subunit of Rubisco and acyl carrier protein, required for carbon fixation

and fatty acid synthesis, respectively. The 145 and 143 kDa proteins are antigenically related (Oblong & Lamppa, 1992). Given the estimated size of native SPP, it is likely that the 145 and 143 kDa proteins are isoenzymes, function as monomers, and originate either from two related genes or by post-translational modification of a single gene product. Chromatofocusing of the active enzyme indicates a pI of 5.0, and the polypeptide deduced from a cDNA (see below) has a predicted pI of 5.8.

A full-length cDNA coding for a 140 kDa polypeptide was reconstructed from two clones isolated from a pea expression library using antibodies raised against the 145/143 kDa doublet. The 140 kDa polypeptide contains a functional transit peptide of ~7 kDa. An HXXEH zinc-binding motif is present that is conserved in the family (M16) of metalloendopeptidases which includes pitrilysin (Chapter 467), insulysin (Chapter 466), and subunit β of the mitochondrial processing peptidase (Chapter 469) (Becker & Roth, 1992). Following the transit peptide and concentrated near the N-terminus of the mature protein, 25–30% identity is found with these proteases. A short acidic domain (27 residues) immediately precedes the conserved region. In addition, the protein contains a cysteine residue within a potential lipid-modification site. The predicted secondary structure of the peptidase is rich in α helices (VanderVere *et al.*, 1995).

SPP activity from *Chlamydomonas* elutes during gel filtration as a 90 kDa protein. This activity is insensitive to divalent cation chelators (Creighton *et al.*, 1993; Su & Boschetti, 1993).

## Preparation

Using preLHCP as a substrate, SPP was originally purified from chloroplast soluble extracts prepared from leaves of pea (dicotyledonous) (Oblong & Lamppa, 1992). SPP activity is also present in chloroplasts from other higher plants including wheat (monocotyledonous) although its molecular composition remains to be established. Preparation from monocots is more difficult due to the lower recovery of intact chloroplasts. Affinity chromatography, using preLHCP as a ligand, was a key step in purification from pea, allowing rapid enrichment of the labile processing activity and identification of the coeluting 145 and 143 kDa proteins. Preparative gel electrophoresis was used to isolate the 145/143 kDa doublet for the production of antibodies that recognize the native enzyme. These antibodies detect only the 145/143 kDa doublet in pea chloroplast extracts. In wheat extracts, however, two additional prominent species were also identified, equal to ~100 kDa and 110 kDa. The origin of these latter proteins and their relationship to the 145 and 143 proteins remains to be investigated, but the observation raises the possibility that some SPP heterogeneity may exist among higher plants.

## Biological Aspects

SPP activity is found in leaf chloroplasts from higher plants, e.g. pea, spinach, *Arabidopsis* (dicots) and wheat (a monocot), as well as in the alga *Chlamydomonas*. The algal enzyme differs from that of higher plants in its size and insensitivity to divalent-cation chelating agents (Creighton *et al.*,

1993; Su & Boschetti, 1993). In pea, the 145/143 kDa doublet and enzymatic activity are present in both dark- and light-grown plants, allowing one to conclude that gene expression is not light dependent. During greening the level of SPP increases in parallel with the massive synthesis of other stromal proteins, i.e. Rubisco, and membrane proteins such as LHCP, needed for photosynthesis. In contrast to pea, a study using barley (a monocot) showed that precursor cleavage does not occur using organelles from dark-grown plants, but rather activity was only found after plants were exposed to light for 5 h (Chitnis *et al.*, 1988). Hence, there may be plant-specific (dicot versus monocot) developmental regulation of the processing enzyme in leaves. Significantly, SPP is synthesized and active in pea roots (VanderVere *et al.*, 1995), where the nonphotosynthetic plastid – the generic name for the organelle in different plant tissues – performs many other biosynthetic functions (see section on Name and History).

Characterization of an SPP cDNA revealed that it contains a poly(A) tail; thus SPP is encoded by the nuclear genome. In addition, the presence of a functional transit peptide shows that SPP itself undergoes post-translational proteolytic processing upon import. The availability of the SPP cDNA and SPP-specific antibodies now makes it possible to conduct a comprehensive study on the regulation and distribution of SPP in different organs, both during plant development and in response to environmental stimuli that influence the biosynthetic pathways carried out by the plastid.

## Distinguishing Features

The substrates recognized by SPP all contain transit peptides, and hence, the presence of SPP activity can be established by incubation of an extract with a precursor targeted to the chloroplast in an organelle-free assay. Recombinant precursors synthesized in *E. coli* can be used as affinity ligands to enrich for SPP. Polyclonal antibodies raised against the 145/143 kDa doublet will immunodeplete the processing activity, and recognize only these two proteins in plastid extracts from leaves and roots. They are not yet commercially available.

## Related Peptidases

Related peptidases include a yeast protein required for bud site selectin and propheromone processing (Adames *et al.*, 1995), and nardilysin from rat (Chapter 468) (Pierotti *et al.*, 1994).

## Further Reading

See the articles by Oblong & Lamppa (1992) and VanderVere *et al.* (1995).

## References

Abad, M., Clark, S. & Lamppa, G. (1989) Properties of a chloroplast enzyme that cleaves the chlorophyll *a/b* binding protein precursor. *Plant Physiol.* **90**, 117–124.

Abad, M.S., Oblong, J.E., & Lamppa, G.K. (1991) Soluble chloroplast enzyme cleaves preLHCP made in *Escherichia coli* to a mature form lacking a basic N-terminal domain. *Plant Physiol.* **96**, 1220–1227.

Adames, N., Blundell, K., Ashby, M. & Boone, C. (1995) Role of yeast insulin-degrading enzyme homologs in propheromone processing and bud site selection. *Science* **270**, 464–467.

Archer, E. & Keegstra, K. (1993) Analysis of chloroplast transit peptide function using mutations in the carboxyl-terminal region. *Plant Mol. Biol.* **23**, 1105–1115.

Becker, A.B. & Roth, R.A. (1992) An unusual active site identified in a family of zinc metalloendopeptidases. *Proc. Natl Acad. Sci. USA* **89**, 3835–3839.

Chitnis P., Morishige, D., Nechushtai, R. & Thornber, J.P. (1988) Assembly of the barley light-harvesting chlorophyll a/b proteins in barley etiochloroplasts involves processing of the precursor on thylakoids. *Plant Mol. Biol.* **11**, 95–107.

Clark, S.E. & Lamppa, G.K. (1991) Determinants for cleavage of the chlorophyll *a/b*-binding protein precursor: a requirement for a basic residue that is not universal for chloroplast imported proteins. *J. Cell Biol.* **114**, 681–688.

Clark, S.E. & Lamppa, G.K. (1992) Processing of the precursors for the light-harvesting chlorophyll-binding proteins of photosystem II and photosystem I during import and in an organelle-free assay *Plant Physiol.* **98**, 595–601

Creighton, A.M., Bassham, D. & Robinson, C. (1993) The stromal processing peptidase activities from *Chlamydomonas reinhardtii* and *Pisum sativum*: unexpected similarities in reaction specificity. *Plant Mol. Biol.* **23**, 1291–1296.

Dobberstein, B., Blobel, G. & Chua, N.-H. (1977) In vitro synthesis and processing of a putative precursor for the small subunit of ribulose-1,5-bisphosphate carboxylase of *Chlamydomonas reinhardtii*. *Proc. Natl Acad. Sci. USA* **74**, 1082–1085.

Gavel, Y. & von Heijne, G. (1990) A conserved cleavage-site motif in chloroplast transit peptides. *FEBS Lett.* **261**, 455–458.

Hageman, J., Robinson, C. Smeekens, S. & Weisbeek, P. (1986) A thylakoid processing protease is required for complete maturation of the lumen protein plastocyanin. *Nature* **324**, 567–569.

Keegstra, K. & Olsen, L.J. (1989) Chloroplastic precursors and their transport across the envelope membranes. *Annu. Rev. Plant Physiol. Plant Mol. Biol.* **40**, 471–501.

Lamppa, G. & Abad, M. (1987) Processing of a wheat light-harvesting chlorophyll a/b protein precursor by a soluble enzyme from higher plant chloroplasts. *J. Cell Biol.* **105**, 2641–2648.

Oblong, J.E. & Lamppa, G.K. (1992) Identification of two structurally related proteins involved in proteolytic processing of precursors targeted to the chloroplast. *EMBO J.* **11**, 4401–4409.

Pierotti, A., Prat, A., Chesneau, V., Gaudoux, F., Leseny, A.-M., Foulon, T. & Cohen, P. (1994) N-Arginine dibasic convertase, a metalloendopeptidase as a prototype of a class of processing enzymes. *Proc. Natl Acad. Sci. USA* **91**, 6078–6082.

Robinson, C. & Ellis, R.J. (1984) Transport of proteins into chloroplasts. Partial purification of a chloroplast protease involved in the processing of imported precursor polypeptides. *Eur. J. Biochem.* **142**, 337–342.

Schnell, D.J. & Blobel, G. (1993) Identification of intermediates in the pathway of protein import into the chloroplasts and their localization to envelope contact sites. *J. Cell Biol.* **120**, 103–105.

Smith, S.M. & Ellis, R.J. (1979) Processing of small subunit precursor of ribulose bisphosphate carboxylase and its assembly into whole enzyme are stromal events. *Nature* **278**, 662–664.

Su, Q. & Boschetti, A. (1993) Partial purification and properties of enzymes involved in the processing of a chloroplast import protein from *Chlamydomonas reinhardtii. Eur. J. Biochem.* **217**, 1039–1047.

VanderVere, P., Bennett, T., Oblong, J. & Lamppa, G. (1995) A chloroplast processing enzyme involved in precursor maturation

shares a zinc-binding motif with a recently recognized family of metalloendopeptidases. *Proc. Natl Acad. Sci. USA* **92**, 7177–7181.

*Gayle K. Lamppa*
*Department of Molecular Genetics and Cell Biology,*
*University of Chicago,*
*920 E. 58th Street,*
*Chicago, IL 60637-1432, USA*
*Email: gklamppa@midway.uchicage.edu*

# 471. Vaccinia virus proteinase

## Databanks

*Peptidase classification: clan ME, family M44, MEROPS ID: M44.001*
*NC-IUBMB enzyme classification: none*
*Databank codes:*

| Species | Strain | SwissProt | PIR | EMBL (genomic) |
|---|---|---|---|---|
| Fowlpox virus | – | – | H48563 | – |
| Molluscum contagiosum virus | subtype 1 | – | – | U60315: complete genome |
| | | | | U86909: mid-section |
| Vaccinia virus | WR | P16713 | C38497 | J03399: genome fragment |
| Vaccinia virus | Copenhagen | P21022 | F42511 | M35027: complete genome |
| Variola virus | Bangladesh-1975 | – | – | L22579: complete genome |
| | Garcia-1966 | – | – | X76267: complete genome |
| | India-1967 | P32991 | S33078 | X67119: genome fragment |
| | | | | X69198: complete genome |

## Name and History

The involvement of a proteinase activity in the infection cycle of vaccinia virus (VV) was first indicated when Holowczak & Joklik (1967), surveying the structural proteins of VV, noted differences between the apparent molecular masses of radiolabeled proteins present in VV-infected cells and those found in purified virions. It has since become appreciated that proteolytic processing is a crucial event in VV particle maturation. Indeed, the three most abundant proteins in the VV particle, 4a, 4b and 25K, have all been demonstrated to be derived from higher molecular mass precursors (Moss & Rosenblum, 1973; Silver & Dales, 1982; Yang *et al.*, 1988).

To determine the identity of the responsible *vaccinia virus proteinase*, our laboratory went about genetically mapping the proteinase activity *in vivo*, systematically assaying the entire VV genome (Goebel *et al.*, 1990) for proteinase activity. The capability to rescue proteolysis of a test substrate was determined to reside exclusively within the G1L open reading frame (ORF) of the VV genome (Whitehead &

Hruby, 1994a). Computer analysis of this coding region reveals the sequence HLLEH, separated from the tripeptide ENE by a gap of 67 amino acid residues. The HLLEH sequence is a direct inversion of the HEXXH motif present in all members of the thermolysin family (Chapter 350) of metalloendopeptidases (Jiang & Bond, 1992) and is similar to the HXXEH motif found in the zinc-dependent insulysin family (M16) enzymes (Finch *et al.*, 1986; Affholter *et al.*, 1988) (Chapter 465). The downstream ENE triplet in G1L provides both the forward and reverse orientation of the NE doublet that is present in the thermolysin and IDE families. Mutational analysis revealed that both histidine residues and all three glutamate residues were each absolutely necessary to facilitate proteolysis of the test substrate. It was therefore proposed that the proteolytic activity that is observed in VV-infected cells is provided by a proteinase encoded by the G1L ORF, with active-site requirements similar to those of known zinc-dependent metalloendopeptidases. It should be noted that no other VV ORF contains any sequence resembling the active sites of any known proteinase.

## Activity and Specificity

The vaccinia virus proteinase cleaves its substrates at the site Ala-Gly+Xaa (VanSlyke *et al*., 1991a,b; Whitehead & Hruby, 1994b; Bersani *et al*., 1998). Four variants of the P1′ residue (alanine, serine, threonine and lysine) have been observed in natural substrates. However, *in vivo* processing assays in which the cleavage site of a VV-encoded substrate was systematically altered revealed considerable plasticity at the P1′ site (Lee & Hruby, 1994): out of 12 P1′ mutants tested, only proline prevented cleavage of the substrate. The P1 and P2 sites, on the other hand, exhibit more stringent specificity, and few substitutions at these sites can support proteolysis.

The majority of known cleavage events catalyzed by the VV proteinase take place relatively close to the N-terminus of the substrate, releasing small peptides which are predicted to be of markedly higher acidity than the liberated mature product. The observation of stable cleavage intermediates (Whitehead *et al*., 1995) and mature internal fragments of precursors bearing discrete N-termini (Bersani *et al*., 1998) strongly suggest that the VV proteinase has an endoproteolytic activity. The VV proteinase cannot perform cleavage *in vitro*. The enzyme is evidently foiled by a lack of proper substrate context, which will be discussed further in the section on Biological Aspects.

## Structural Chemistry

The gene product of the G1L ORF is predicted to be 591 amino acid residues in length, with a predicted molecular mass of 68 kDa and a theoretical pI of 6.49. Binding of metal ions has not yet been demonstrated, as no pure preparation of the VV proteinase has been obtained at this writing. The results of mutational analysis of the putative active site of G1L (Whitehead & Hruby, 1994a), however, are consistent with the requirements of the catalytic centers of metalloenzymes (Jiang & Bond, 1992).

## Biological Aspects

The nucleotide sequence immediately upstream of the G1L ORF is typical of a VV late gene promoter (Davison & Moss, 1989), and its activity can be observed in a tightly regulated environment which allows only late transcription (Whitehead & Hruby, 1994a).

The VV proteinase has proven to be extremely selective with respect to its conditions for cleavage. The accumulated data suggest that, beyond the presence of the tripeptide cleavage motif, there are contextual requirements which must be satisfied if cleavage of the substrate is to take place. Specifically, substrates of the VV proteinase must be in the proper temporal and conformational configuration in order to undergo cleavage. This context seems to be provided only in the environment provided by the maturing virus particle (Katz & Moss, 1970; Stern *et al*., 1977; Silver & Dales, 1982; VanSlyke *et al*., 1991a). The coordinate linkage of virion maturation and proteolytic processing can

be observed by isolating from infected cells discrete populations of VV particles at intermediate stages of maturity. Each type of particle exhibits a ratio of precursor protein to mature product that decreases with increasing condensation of the virion (VanSlyke *et al*., 1993). These characteristics of proteolysis in the VV replication cycle support the model of morphogenic, rather than formative proteolysis (Hellen & Wimmer, 1992) as the primary function of the VV proteinase.

One aspect of the VV proteinase which must be addressed is that, to date, only one cleavage site in one test substrate (Ala-Gly+Ser in precursor P25K) has been demonstrated *in vivo* to be processed by G1L (Whitehead & Hruby, 1994b). The fact that this event can be observed under the conditions of the *trans*-processing assay, i.e. in the absence of virion assembly, indicates that although this site is a natural variant of the AGX motif, it is not subject to the same contextual requirements as those that have been demonstrated to be coupled with particle maturation. It remains a challenge to determine how this AGS site escapes the contextual requirements of the VV proteinase, and to observe more directly the contextual cleavage of the other precursors.

## Related Peptidases

Homologs of G1L are encoded by poxviruses other than vaccinia virus. The predicted amino acid sequence of the G1L ORF of molluscum contagiosum virus (MCV) (Senkevich *et al*., 1996) reveals 47.5% identity and 72% weighted similarity to VV G1L. The variola major H1L ORF (Massung *et al*., 1993) shares 98% identity and 100% weighted similarity with the VV proteinase. In each case, both the HLLEH and the ENE motifs are absolutely conserved. There have been no attempts to demonstrate proteinase activity in these homologs.

No virus has yet been unequivocally demonstrated to encode a metalloproteinase. The existence of such an enzyme is therefore an exciting prospect. Investigations into the several proteolytic activities of hepatitis C virus (HCV) have also revealed evidence of a virus-encoded metalloendopeptidase (Hijikata *et al*., 1993). The region of the HCV genome which codes for the Cpro-1 proteinase (Chapter 567) activity does not contain any sequence elements resembling active sites of known metalloproteinases. However, of several proteinase inhibitors tested, only EDTA resulted in inhibition of proteolysis. This inhibition could be reversed by the adding of $ZnCl_2$, but not by any other metal ion. Furthermore, mutational analysis revealed a histidine and a cysteine residue that were required for activity. Together, these data strongly implicate a zinc-dependent activity in the HCV replication cycle.

## Further Reading

For a review, see Hellen & Wimmer (1992).

## References

Affholter, J.A., Fried, V.A. & Roth, R.A. (1988) Human insulin-degrading enzyme shares structural and functional homologies with *E. coli* protease III. *Science* **242**, 1415–1418.

Bersani, N.A., Whitehead, S.S. & Hruby, D.E. (1998) Multistep proteolytic maturation of vaccinia virus A12L protein: utilization of a novel variant of a conserved cleavage motif. *Virology* (in press).

Davison, A.J. & Moss, B. (1989) Structure of vaccinia virus late promoters. *J. Mol. Biol.* **210**, 771–784.

Finch, P.W., Wilson, R.E., Brown, K., Hickson, I.D. & Emmerson, P.T. (1986) Complete nucleotide sequence of the *Escherichia coli ptr* gene encoding protease III. *Nucleic Acids Res.* **14**, 7695–7703.

Goebel, S.J., Johnson, G.P., Perkus, M.E., Davis, S.W., Winslow, J.P. & Paoletti, E. (1990) The complete DNA sequence of vaccinia virus. *Virology* **179**, 247–266.

Hellen, C.U.T. & Wimmer, E. (1992) The role of proteolytic processing in the morphogenesis of virus particles. *Experientia* **48**, 210–215.

Hijikata, M., Mizushima, H., Akagi, T., Mori, S., Kakiuchi, N., Kato, N., Tanaka, T., Kimura, K. & Shimotono, K. (1993) Two distinct proteinase activities required for the processing of a putative nonstructural precursor protein of hepatitis C virus. *J. Virol.* **67**, 4665–4675.

Holowczak, J.A. & Joklik, W.K. (1967) Studies on the structural protein of vaccinia virus. Structural protein of virions and cores. *Virology* **33**, 717–725.

Jiang, W. & Bond, J.S. (1992) Families of metalloendopeptidases and their relationships. *FEBS Lett.* **312**, 110–114.

Katz, E. & Moss, B. (1970) Formation of a vaccinia virus polypeptide from a higher molecular weight precursor: inhibition by rifampicin. *Proc. Natl Acad. Sci. USA* **66**, 677–684.

Lee, P. & Hruby, D.E. (1994) Proteolytic cleavage of vaccinia virus virion proteins: mutational analysis of the specificity determinants. *J. Biol. Chem.* **269**, 8616–8622.

Massung, R.F., Esposito, J.J., Liu, L.-I., Qi, J., Utterback, T.R., Knight, J.C., Aubin, L., Yuran, T.E., Parsons, J.M., Loparev, V.N., Selivanov, N.A., Cavallaro, K.F., Kerlavage, A.R., Mahy, B.W.J. & Venter, J.C. (1993) Potential virulence determinants in terminal regions of variola smallpox virus genome. *Nature* **366**, 748–751.

Moss, B. & Rosenblum, E.N. (1973) Protein cleavage and poxvirus morphogenesis: tryptic peptide analysis of core precursors accumulated by blocking assembly with rifampicin. *J. Mol. Biol.* **81**, 267–279.

Senkevich, T.G., Bugert, J.J., Sisler, J.R., Koonin, E.V., Darai, G. & Moss, B. (1996) Genome sequence of a human tumorigenic poxvirus: prediction of specific host response evasion genes. *Science* **273**, 813–816.

Silver, M. & Dales, S. (1982) Biogenesis of vaccinia: interrelationship between posttranslational cleavage, virus assembly and maturation. *Virology* **117**, 341–356.

Stern, W., Pogo, B.G.T. & Dales, S. (1977) Biogenesis of poxviruses: analysis of the morphogenetic sequence using a conditional-lethal mutant defective in envelope self-assembly. *Proc. Natl Acad. Sci. USA* **74**, 2162–2166.

VanSlyke, J.K., Franke, C.A. & Hruby, D.E. (1991a) Proteolytic maturation of vaccinia virus core proteins: identification of a conserved motif at the N-termini of the 4b and 25K virion proteins. *J. Gen. Virol.* **72**, 411–416.

VanSlyke, J.K., Whitehead, S.S., Wilson, E.M. & Hruby, D.E. (1991b) The multistep proteolytic maturation pathway utilized by vaccinia virus P4a protein: a degenerate conserved cleavage motif within core proteins. *Virology* **183**, 467–478.

VanSlyke, J.K., Lee, P., Wilson, E.M. & Hruby, D.E. (1993) Isolation and analysis of vaccinia virus previrions. *Virus Genes* **7**, 311–324.

Whitehead, S.S. & Hruby, D.E. (1994a) A transcriptionally-controlled *trans*-processing assay: identification of a vaccinia virus-encoded proteinase which cleaves precursor protein P25K. *J. Virol.* **68**, 7603–7608.

Whitehead, S.S. & Hruby, D.E. (1994b) Differential utilization of a conserved motif for the proteolytic maturation of vaccinia virus proteins. *Virology* **200**, 154–161.

Whitehead, S.S., Bersani, N.A. & Hruby, D.E. (1995) Physical and molecular genetic analysis of the multistep proteolytic maturation pathway utilized by vaccinia virus P4a protein. *J. Gen. Virol.* **76**, 717–721.

Yang, W., Kao, S. & Bauer, W.R. (1988) Biosynthesis and post-translational cleavage of vaccinia virus structural protein VP8. *Virology* **167**, 585–590.

*Neil A. Bersani*
*Department of Biochemistry and Biophysics,*
*Oregon State University,*
*Corvallis, OR 97331, USA*
*Email: bersani@ava.bcc.orst.edu*

*Dennis E. Hruby*
*Department of Microbiology,*
*Oregon State University,*
*Corvallis, OR 97331, USA*
*Email: hrubyd@bcc.orst.edu*

**M**

# 472. Introduction: clan MF containing co-catalytic leucyl aminopeptidases

## Databanks

*MEROPS ID: MF*

| Species | SwissProt | PIR | EMBL (cDNA) | EMBL (genomic) |
|---|---|---|---|---|
| **Family M17** | | | | |
| Leucyl aminopeptidase (bacteria) (Chapter 474) | | | | |
| Leucyl aminopeptidase (animal and plant) (Chapter 473) | | | | |
| Others | | | | |
| *Caenorhabditis elegans* | P34629 | – | – | – |
| *Caenorhabditis elegans* | – | – | Z68320 | |
| *Escherichia coli* | – | – | D84499 | – |
| *Haemophilus influenzae* | – | – | – | U32715: genomic |
| *Schizosaccharomyces pombe* | Q09735 | – | – | Z54096: chromosome I cosmid |

Clan MF contains only one family, family M17, which is a family of leucyl aminopeptidases from eukaryotes and bacteria. The crystal structure of cattle leucyl aminopeptidase has been solved, revealing a two-domain structure and the presence of two catalytic zinc ions (Fig. 472.1). The N- and C-terminal domains are structurally unrelated, unlike the domains of methionyl aminopeptidase in family M24. The bacterial leucyl aminopeptidase, also known as aminopeptidase A, has two manganese ions in the active site. Cattle leucyl aminopeptidase exists as a homohexamer, with the active sites located in the interior (Kim & Lipscomb, 1993).

The N-terminal domain of cattle leucyl aminopeptidase shows no relationship to any other known protein structure. The zinc ligands and catalytic residues are located in the C-terminal domain, which shows some structural similarity to carboxypeptidase A and carboxypeptidase T from clan MC (Chapter 450), and to *Vibrio* aminopeptidase from clan MH (Chapter 482). The structures are regarded here as representatives of different clans, however, because the zinc ligands are not in similar positions, and carboxypeptidases A and T contain only one zinc ion.

The zinc ions are pentahedrally coordinated in the free leucyl aminopeptidase and when substrate is bound. By contrast, in clans MA, MB and MC the single zinc ion is tetrahedrally coordinated. In leucyl aminopeptidase, one zinc ion (Zn1) is bound by Asp255 $O^{\delta 1}$, Asp332 $O^{\delta 2}$ and Glu334 $O^{\varepsilon 2}$, and the other (Zn2) by Asp255 $O^{\delta 1}$, Lys250 $NH_2$, Asp273 $O^{\delta 1}$ and Glu334 $O^{\varepsilon 1}$. Together, the zinc ions bind a water molecule that is activated to act as the nucleophile. Besides the zinc ligands, Lys262 and Arg336 are also important for catalysis. Lys262 is an electrophile, hydrogen bonding to the polarized carbonyl group of the substrate. Arg336 acts as a proton donor, polarizing a second water molecule that contributes

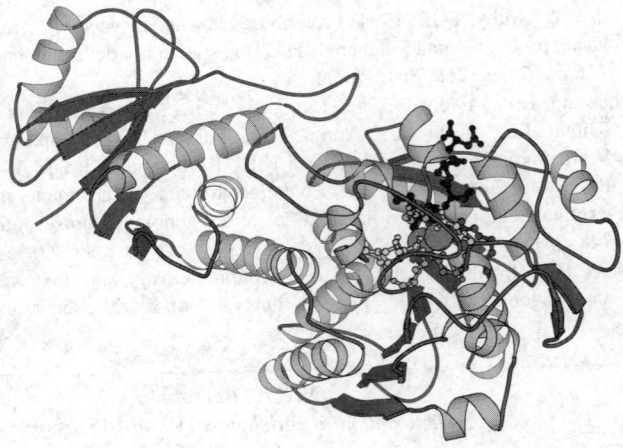

*Figure 472.1* Richardson diagram of the cattle leucyl aminopeptidase/amastatin complex. The image was prepared from the Brookhaven Protein Data Bank entry (1BLL) as described in the Introduction (p. xxv). The catalytic zinc ions are shown in CPK representation as light gray spheres. The zinc ligands are shown in ball-and-stick representation: Lys250, Asp255, Asp273, Asp332 and Glu334 (numbering as in Alignment 472.1). The catalytic residues are also shown in ball-and-stick representation: Lys262 and Arg336. Amastatin is shown in ball-and-stick representation in black.

a proton to the leaving group as the *gem*-diolate intermediate breaks down (Sträter & Lipscomb, 1995). An alignment around these residues is shown in Alignment 472.1 (on CD-ROM). Unusually for a metallopeptidase, no histidines are implicated as zinc ligands or catalytic residues.

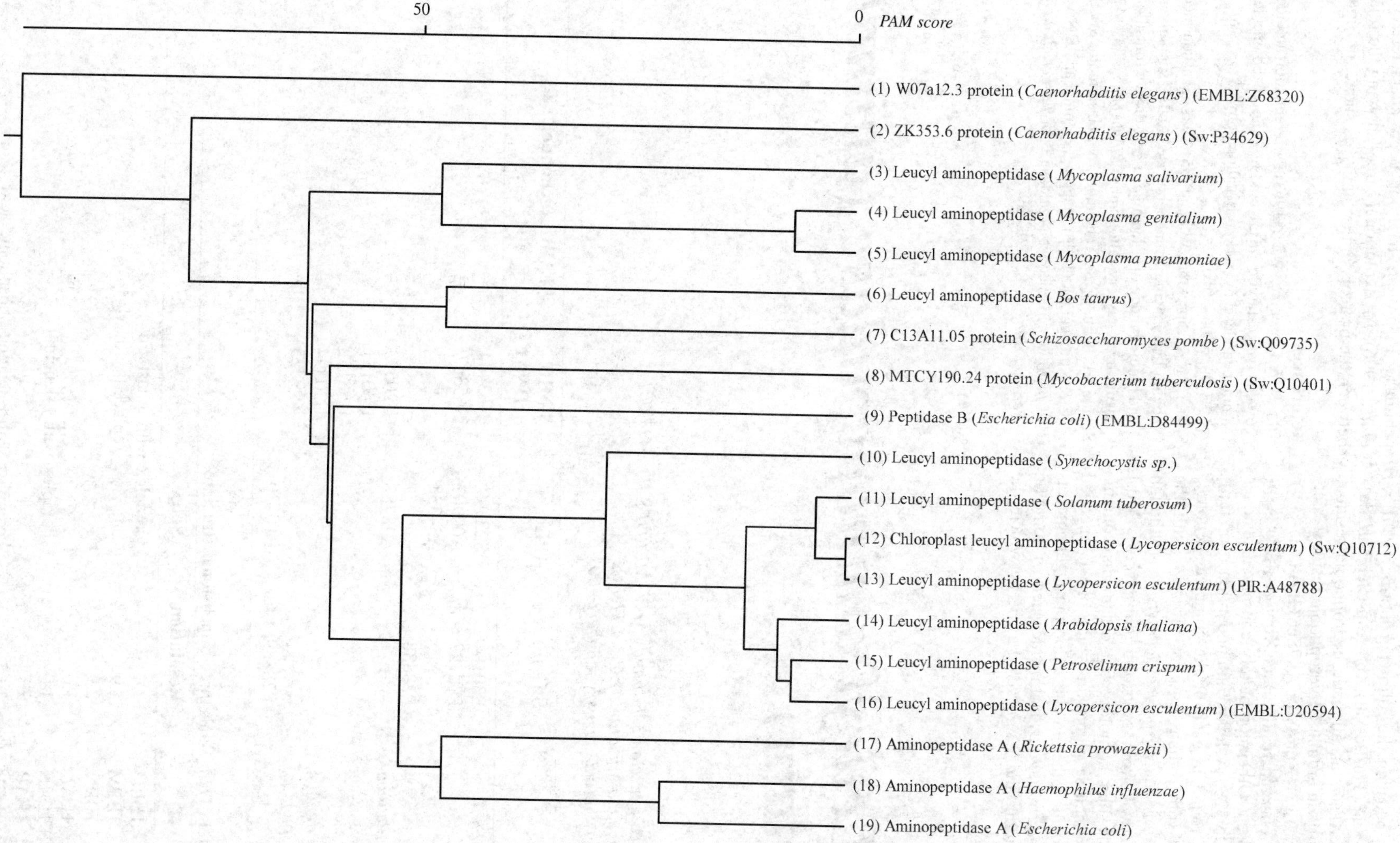

0 PAM score

50

(1) W07a12.3 protein (*Caenorhabditis elegans*) (EMBL:Z68320)

(2) ZK353.6 protein (*Caenorhabditis elegans*) (Sw:P34629)

(3) Leucyl aminopeptidase (*Mycoplasma salivarium*)

(4) Leucyl aminopeptidase (*Mycoplasma genitalium*)

(5) Leucyl aminopeptidase (*Mycoplasma pneumoniae*)

(6) Leucyl aminopeptidase (*Bos taurus*)

(7) C13A11.05 protein (*Schizosaccharomyces pombe*) (Sw:Q09735)

(8) MTCY190.24 protein (*Mycobacterium tuberculosis*) (Sw:Q10401)

(9) Peptidase B (*Escherichia coli*) (EMBL:D84499)

(10) Leucyl aminopeptidase (*Synechocystis sp.*)

(11) Leucyl aminopeptidase (*Solanum tuberosum*)

(12) Chloroplast leucyl aminopeptidase (*Lycopersicon esculentum*) (Sw:Q10712)

(13) Leucyl aminopeptidase (*Lycopersicon esculentum*) (PIR:A48788)

(14) Leucyl aminopeptidase (*Arabidopsis thaliana*)

(15) Leucyl aminopeptidase (*Petroselinum crispum*)

(16) Leucyl aminopeptidase (*Lycopersicon esculentum*) (EMBL:U20594)

(17) Aminopeptidase A (*Rickettsia prowazekii*)

(18) Aminopeptidase A (*Haemophilus influenzae*)

(19) Aminopeptidase A (*Escherichia coli*)

*Figure 472.2*    Evolutionary tree for family M17. The tree was prepared as described in the Introduction (p. xxv).

The zinc ligands occur in the following motifs: (Val/Ile)-Gly-**Lys**-(Gly/Ser)-Xaa-(Thr/Ile/Val)-Xbb-**Asp**-(Ser/Thr/Ala)-Gly-Gly, in which Xaa is an aliphatic hydrophobic residue, Xbb is predominantly an aromatic hydrophobic residue; Met-Xaa- Xbb-**Asp**-(Met/Lys)- Xbb-Gly-(Ala/Ser/Gly)-(Ala-Gly), in which Xaa is a charged residue and Xbb; Asn-Thr-**Asp**-Ala-**Glu**-Gly-Arg-Leu. Other completely conserved residues are Asn305 and Gly323. A leucyl aminopeptidase of family M17 has been found in each of the complete genomes sequenced to date (other than those of viruses), with the notable exception of *Saccharomyces cerevisiae* (although an example is known from *Schizosaccharomyces*). Most genomes have only one homolog, but organisms containing two homologs include plants, in which one is encoded in the nucleus and the other in the chloroplast genome, as well as *Caenorhabditis elegans*, from which neither homolog is known as protein, and *Escherichia coli*, in which the peptidase B protein has not been characterized (Suzuki *et al*., 1996). An evolutionary tree, showing the relationships between the proteins in the family, is shown in Figure 472.2.

### References

Kim, H. & Lipscomb, W.N. (1993) Differentiation and identification of the two catalytic metal binding sites in bovine lens leucine aminopeptidase by x-ray crystallography. *Proc. Natl Acad. Sci. USA* **90**, 5006–5010.

Sträter, N. & Lipscomb, W.N. (1995) Two-metal ion mechanism of bovine lens leucine aminopeptidase: Active site solvent structure and binding mode of L-leucinal, a *gem*-diolate transition state analogue, by X-ray crystallography. *Biochemistry* **34**, 14792–14800.

Suzuki, H., Kim, E.S., Yamamoto, N., Hashimoto, W., Yamamoto, K. & Kumagai, H. (1996) Mapping, cloning, and DNA sequencing of *pepB* which encodes peptidase B of *Escherichia coli* K-12. *J. Ferment. Bioeng.* **82**, 392–397.

# 473. Leucyl aminopeptidase (animal and plant)

## Databanks

*Peptidase classification: clan MF, family M17, MEROPS ID: M17.001*
*NC-IUBMB enzyme classification: EC 3.4.11.1*
*Chemical Abstracts Service registry number: 9001-61-0*
Databank codes:

| Species | SwissProt | PIR | EMBL (cDNA) | EMBL (genomic) |
|---|---|---|---|---|
| *Arabidopsis thaliana* | P30184 | S22399 | X63444 | – |
| *Bos taurus* | P00727 | A00907 A54338 | S65367 | – |
| *Homo sapiens* | P28838 | – | – | – |
| *Lycopersicon esculentum* | – | A48788 | U20593 U50151 U50152 | – |
| *Lycopersicon esculentum* | – | – | U20594 | – |
| *Petroselinum crispum* | – | – | X99825 | – |
| *Sus scrofa* | P28839 | PT0430 | – | – |

Brookhaven Protein Data Bank three-dimensional structures:

| Species | ID | Resolution | Notes |
|---|---|---|---|
| *Bos taurus* | 1BLL | 2.4 | complex with amastatin |
| | 1BPM | 2.9 | one $Zn^{2+}$ replaced with $Mg^{2+}$ |
| | 1BPN | 2.9 | |
| | 1LAM | 1.6 | |
| | 1LAN | 1.9 | complex with L-leucinal |
| | 1LAP | 2.7 | |
| | 1LCP | 1.65 | complex with L-leucine phosphonic acid |

## Name and History

Linderstrøm-Lang (1929) reported the presence of an enzyme in erepsin (extracts of pig intestinal mucosa) that cleaved leucylglycine about 20 times faster than it cleaved glycylglycine. This peptidase was provisionally named *dipeptidase II*, but later referred to as *aminoleucylpeptidase* in a Danish review article (Holter, 1979). Johnson *et al.* (1936) partially purified the enzyme, then called *leucylpeptidase*. Although it is now clear that *leucyl aminopeptidase (LAP)* hydrolyzes a wide variety of peptides and amides, this name has persisted. The activation of LAP by magnesium and manganese was shown by Johnson *et al.* (1936) and Berger & Johnson (1939). Studies by Spackman *et al.* (1955) and Smith & Spackman (1955) on purified LAP from swine kidney revealed many chemical and physical properties of the enzyme. The zinc metalloenzyme nature of LAP was determined by Himmelhoch (1969).

LAP was also among the seven crystalline proteins first obtained in Sumner's laboratory (Dounce & Allen, 1988). However, the identity of these protein crystals was unknown at that time and the protein was named the 'football protein', based on the resemblance of the crystals to an American football. Fifty years later the football protein was finally identified as beef liver LAP (Dounce & Allen, 1987). LAP was the first dizinc enzyme for which a crystal structure at atomic resolution was determined (Burley *et al.*, 1990). Some other names for the enzyme are *cytosol aminopeptidase* and *peptidase S*.

## Activity and Specificity

Pig kidney LAP is maximally active between pH 9 and 9.5 and cleaves a variety of amino acid amides, dipeptides and other compounds (Spackman *et al.*, 1955; Smith & Spackman, 1955; Delange & Smith, 1971). All tested substances with an N-terminal L-amino acid (or glycine) are hydrolyzed; however, compounds that have L-leucyl residues in the N-terminal position are the preferred substrates. Peptides having proline in P1′ (Xaa-Pro-) are not cleaved by LAP. Esters are also substrates of the enzyme, although they are cleaved at about 10% of the rate of the corresponding amides. A comparison of the activities of cattle lens and pig kidney LAP on amino acid amides, aminoacyl-$\beta$-naphthylamides, and aminoacyl-$p$-nitroanilides showed that the general specificities of the two enzymes are very similar (Hanson *et al.*, 1967). For activity assays L-leucyl amide or the chromogenic substrate L-Leu+NHPhNO$_2$ are usually employed.

LAP is inhibited by the naturally occurring inhibitors bestatin (Figure 473.1) and amastatin with $K_i$ values of 1.3 nM (cattle lens enzyme) and 0.2 µM (pig kidney LAP), respectively (Rich *et al.*, 1984; Taylor *et al.*, 1993). Compared to a true peptide substrate these inhibitors have an extra carbon atom that bears a hydroxyl group. Bestatin also inhibits other dizinc aminopeptidases. Several transition state analog inhibitors have been synthesized which resemble the *gem*-diol intermediate after attack of a water nucleophile on the peptide bond. These include aminophosphonates (Lejczak *et al.*, 1989; Giannousis & Bartlett, 1987), aminoaldehydes (Andersson *et al.*, 1985), chloromethanes (Birch *et al.*, 1972),

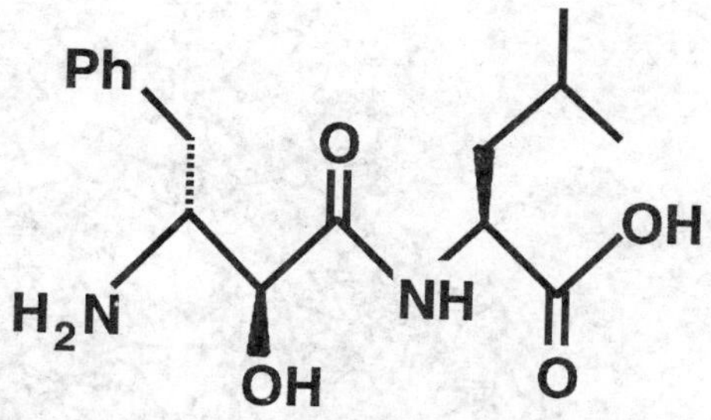

Figure 473.1  Molecular structure of the natural LAP inhibitor bestatin.

and peptides containing a ketomethylene amide bond replacement (Harbeson & Rich, 1989). Also, amino acid hydroxamates (Chan *et al.*, 1982) and L-leucylthiol (Chan, 1983) are strong inhibitors of LAP.

## Structural Chemistry

Cattle lens LAP is a homohexamer of 324 kDa molecular mass (Figure 473.2). Its primary structure was determined by chemical sequencing (Cuypers *et al.*, 1982) and for cattle kidney LAP, from the cDNA sequence (Wallner *et al.*, 1993). The monomer has a mixed $\alpha + \beta$ structure and consists of an N-terminal domain (160 amino acid residues) and a catalytic C-terminal domain (327 amino acid residues; Figure 473.2) (Burley *et al.*, 1991, 1992). The six active sites are located in the interior of the hexamer, where they line a disk-shaped cavity of radius 15 Å and thickness 10 Å. Access to this cavity is provided by solvent channels. This structural feature explains why LAP does not cleave longer peptides or proteins (Kim *et al.*, 1974).

The active site contains two five-coordinated zinc ions which are 3.0 Å apart (Figure 473.3) (Burley *et al.*, 1991, 1992; Sträter & Lipscomb, 1995b). One water ligand bridges the two zinc ions symmetrically. Besides this water molecule the carboxylate side chains of Asp273, Asp255, Asp332 and Glu334, the amino group of Lys250, and the backbone carbonyl group of Asp332 coordinate the dimetal center. The specificity pocket for the N-terminal side chain is formed by Met270, Thr359, Gly362, Ala451 and Met454. Crystal structures of LAP complexed with the inhibitors bestatin (Burley *et al.*, 1991, 1992; Kim *et al.*, 1993), amastatin (Kim & Lipscomb, 1993a), leucylphosphonic acid (Sträter & Lipscomb, 1995a), and leucinal (Sträter & Lipscomb, 1995b) provide models for the binding mode of the *gem*-diolate intermediate of the hydrolysis reaction. Leucinal binds with its amino group to Zn2. One of the *gem*-diol oxygens bridges the two zinc ions and one oxygen is terminally coordinated to Zn1.

The two metal ions in LAP show different exchange behavior: the site 1 metal ion is readily exchanged against other metal ions, including Mn$^{2+}$, Mg$^{2+}$ and Co$^{2+}$ (Carpenter & Vahl, 1973; Thompson & Carpenter, 1976a,b). Exchange of Zn$^{2+}$ against Mn$^{2+}$ or Mg$^{2+}$ in site 1 activates LAP. Zn2 is much more tightly bound and can only be replaced by Co$^{2+}$ (Allen *et al.*, 1983). Substitutions at both metal sites in cattle lens LAP affect both $K_m$ and $k_{cat}$. The unambiguous identification of the two metal-binding sites in LAP was

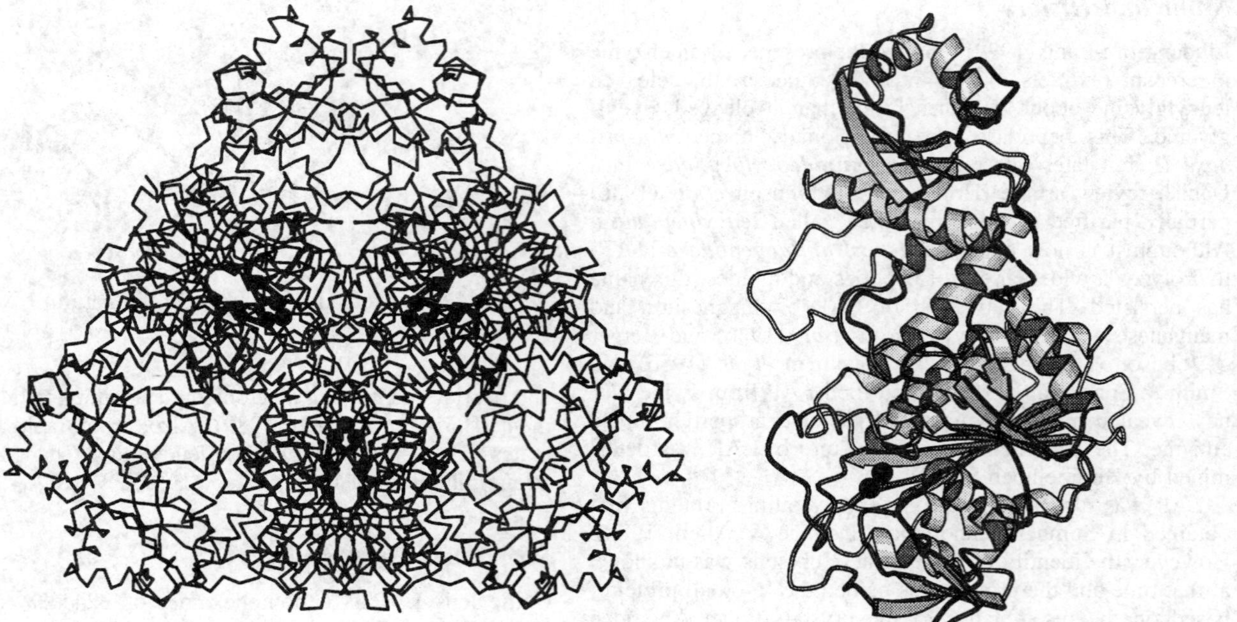

*Figure 473.2   Left*: Hexameric structure of LAP (Cα-trace) viewed along the 3-fold molecular axis. The zinc ions are shown as filled circles. The hexamer has a large solvent cavity in its interior, which contains the six active sites. *Right*: Structure of a cattle lens LAP monomer. The N-terminal domain is in the upper part of the molecule and the larger catalytic domain with the two zinc ions (filled circles) is shown below.

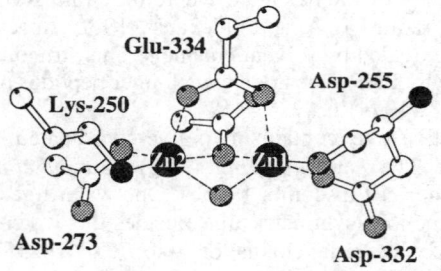

*Figure 473.3*   Structure of the dizinc cluster in the active site of LAP.

achieved by a structure determination of the Mg-Zn enzyme, in which $Mg^{2+}$ occupied the readily exchangeable site 1 (Kim & Lipscomb, 1993b).

## Preparation

Procedures for the purification of LAP from cattle lens (Allen *et al.*, 1983) and pig kidney (for reference to different procedures see Delange & Smith, 1971) to homogeneity have been published. Cytosolic pig kidney LAP is also available from commercial sources (Sigma) (see Appendix 2 for full names and addresses of suppliers). This commercial preparation can be purified as described by Van Wart & Lin (1981). The plant LAPs from *Arabidopsis thaliana* (Bartling & Weiler, 1992), potato (Hildmann *et al.*, 1992) and tomato (Pautot *et al.*, 1993) are less well-studied and have not been purified.

## Biological Aspects

The precise function of LAP in animal organisms is not yet known. Since LAP cannot act on protein substrates, a role in the degradation of peptides, including hormonal peptides, appears likely. Altered LAP activity has been implicated in certain pathological conditions such as human eye cataracts (Taylor *et al.*, 1984). Changes in LAP levels have been used for several clinical assays, including prenatal diagnosis of cystic fibrosis (Buffone *et al.*, 1988). Quantitative immunological techniques indicated that within a species LAPs from different tissues are indistinguishable (Taylor *et al.*, 1984). The plant LAPs from the tomato and the potato have been implicated in wound healing and plant defense response (Hildmann *et al.*, 1992; Pautot *et al.*, 1993). mRNAs encoding tomato LAP were barely detectable in healthy tomato leaves and accumulated to high levels in response to infection by *Pseudomonas syringae*, insect infestation or mechanical wounding.

LAPs are also found in bacteria: *Escherichia coli* aminopeptidase A (PepA) (Chapter 474) shows significant homology to cattle lens LAP (Stirling *et al.*, 1989). PepA has a role in Xer site-specific recombination in *E. coli* and the aminopeptidase activity is not required for this function (McCulloch *et al.*, 1994). The homology between the LAPs from different kingdoms (animals, plants, bacteria) is about 30–40% for the whole protein and higher in the C-terminal domain. The active-site residues involved in metal binding and catalysis are conserved between all LAPs characterized so far.

Based on biochemical and structural studies on the well-studied LAPs from pig and cattle, a model for the catalytic

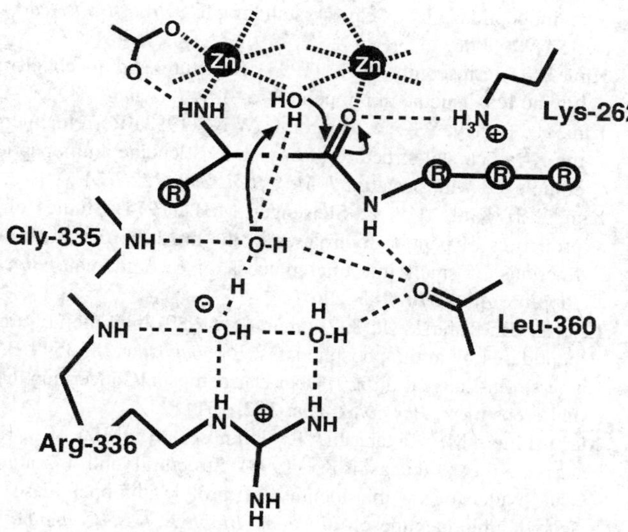

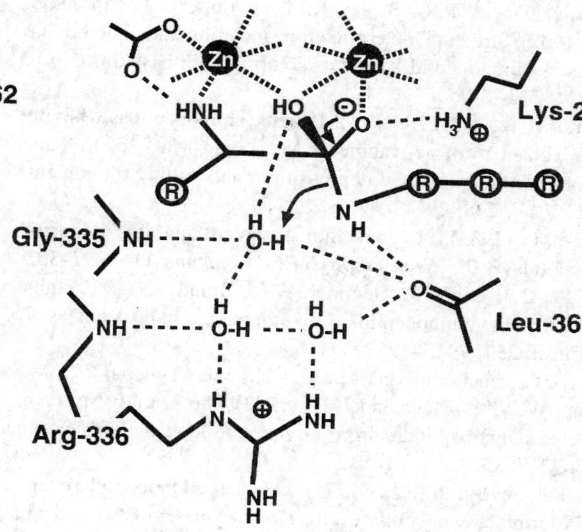

*Figure 473.4*   Proposed reaction mechanism of LAP.

function of this dizinc peptidase has been proposed (Sträter & Lipscomb, 1995b). The substrate probably binds with its terminal amino-group to Zn2 and with the peptide carbonyl group to Zn1 (Figure 473.4). The zinc-bridging water molecule is then in an excellent position for a nucleophilic attack on the peptide bond. The *gem*-diolate intermediate is stabilized by both zinc ions and Lys262. Water molecules between Arg336 and the substrate might also stabilize the intermediate or protonate the leaving group.

## Distinguishing Features

Other aminopeptidases have similar substrate specificities as LAP and have to be distinguished by their molecular properties. The bacterial aminopeptidases from *Vibrio proteolyticus* (Chapter 474) and *Streptomyces griseus* (Chapter 490) also contain a dizinc center and have a preference for peptides with a N-terminal leucyl residue. However, these peptidases are smaller monomeric proteins and the active-site structures are different from LAP. A very distant structural similarity of the protein fold relates the C-terminal domain of LAP, the other bacterial leucyl aminopeptidases, carboxypeptidase A (Chapter 451) and carboxypeptidase $G_2$ (Chapter 483), indicating that they might have arisen from a common ancestor peptidase. However, this similarity does not extend to the location or structure of the active site. Another group of zinc aminopeptidases that should not be confused with LAP are the membrane alanyl aminopeptidases (Chapter 337) which are also sometimes named leucyl aminopeptidase. These have a subunit molecular mass around 110 kDa.

## Related Peptidases

Pig intestinal prolyl aminopeptidase and human liver prolyl aminopeptidase, formerly at EC 3.4.11.5, have been shown to be identical to leucyl aminopeptidase by their kinetic properties and N-terminal sequences (Turzynski & Mentlein,

1990; Matsushima *et al*., 1991). However, other known prolyl aminopeptidases which are distinct from LAP are now assigned to EC 3.4.11.5 (Chapter 139).

## Further Reading

An early survey may be found in Delange & Smith (1971). For other overviews, see Hanson & Frohne (1976), Kim & Lipscomb (1994) and Taylor (1993).

## References

Allen, M.P., Yamada, A.H. & Carpenter, F.H. (1983) Kinetic parameters of metal-substituted leucine aminopeptidase from bovine lens. *Biochemistry* **22**, 3778–3783.

Andersson, L., MacNeela, J. & Wolfenden, R. (1985) Use of secondary isotope effects and varying pH to investigate the mode of binding of inhibitory amino aldehydes by leucine aminopeptidase. *Biochemistry* **24**, 330–333.

Bartling, D. & Weiler, E.W. (1992) Leucine aminopeptidase from *Arabidopsis thaliana*. Molecular evidence for a phylogenetically conserved enzyme of protein turnover in higher plants. *Eur. J. Biochem.* **205**, 425–431.

Berger, J. & Johnson, M.J. (1939) Metal activation of peptidases. *J. Biol. Chem.* **130**, 641–654.

Birch, P.L., El-Obeid, H.A. & Akhtar, M. (1972) The preparation of chloromethylketone analogues of amino acids: inhibition of leucine aminopeptidase. *Arch. Biochem. Biophys.* **148**, 447–451.

Buffone, G.J., Spence, J.E., Fernbach, S.D., Curry, M.R., O'Brien, W.E. & Beaudet, A.L. (1988) Prenatal diagnosis of cystic fibrosis: microvillar enzymes and DNA analysis compared. *Clin. Chem.* **34**, 933–937.

Burley, S.K., David, P.R., Taylor, A. & Lipscomb, W.N. (1990) Molecular structure of leucine aminopeptidase at 2.7-Å resolution. *Proc. Natl Acad. Sci. USA* **87**, 6878–6882.

Burley, S.K., David, P.R. & Lipscomb, W.N. (1991) Leucine aminopeptidase: bestatin inhibition and a model for enzyme-catalyzed peptide hydrolysis. *Proc. Natl Acad. Sci. USA* **88**, 6916–6920.

**M**

Burley, S.K., David, P.R., Sweet, R.M., Taylor, A. & Lipscomb, W.N. (1992) Structure determination and refinement of bovine lens leucine aminopeptidase and its complex with bestatin. *J. Mol. Biol.* **224**, 113–140.

Carpenter, F.H. & Vahl, J.M. (1973) Leucine aminopeptidase (bovine lens). Mechanism of activation by $Mg^{2+}$ and $Mn^{2+}$ of the zinc metalloenzyme, amino acid composition, and sulfhydryl content. *J. Biol. Chem.* **248**, 294–304.

Chan, W.W.-C. (1983) L-Leucinethiol – a potent inhibitor of leucine aminopeptidase. *Biochem. Biophys. Res. Commun.* **116**, 297–302.

Chan, W.W.-C., Dennis, P., Demmer, W. & Brand, K. (1982) Inhibition of leucine aminopeptidase by amino acid hydroxamates. *J. Biol. Chem.* **257**, 7955–7957.

Cuypers, H.T., van Loon-Klaassen, L.A.H., Vree Egberts, W.T.M., de Jong, W.W. & Bloemendal, H. (1982) The primary structure of leucine aminopeptidase from bovine eye lens. *J. Biol. Chem.* **257**, 7077–7085.

Delange, R.J. & Smith, E.L. (1971) Leucine aminopeptidase and other N-terminal exopeptidases. In: *The Enzymes*, vol. III, 3rd edn (Boyer, P.D., ed.). London: Academic Press, pp. 81–118.

Dounce, A.L. & Allen, P.Z. (1987) Identification of the FTBL protein of Sumner and Dounce as a leucine aminopeptidase. *Arch. Biochem. Biophys.* **257**, 13–16.

Dounce, A.L. & Allen, P.Z. (1988) Fifty years later: recollections of the early days of protein crystallization. *Trends Biochem. Sci.* **13**, 317–320.

Giannousis, P.P. & Bartlett, P.A. (1987) Phosphorus amino acid analogues as inhibitors of leucine aminopeptidase. *J. Med. Chem.* **30**, 1603–1609.

Hanson, H. & Frohne, M. (1976) Crystalline leucine aminopeptidase from lens ($\alpha$-aminoacyl-peptide hydrolase; EC 3.4.11.1). *Methods Enzymol.* **45**, 504–521.

Hanson, H., Glässer, D., Ludewig, M., Mannsfeldt, H.-G., John, M. & Nesvadba, H. (1967) Struktur- und Wirkungsidentität der Leucinaminopeptidase aus Schweinenieren und Rinderaugenlinsen und Vergleich mit der Partikelaminopeptidase aus Schweinenieren. [Structure- and activity identity of leucine aminopeptidase from pig kidney and bovine lens; comparison with the particulate aminopeptidase from pig kidney]. *Z. Physiol. Chem.* **348**, 689–704.

Harbeson, S.L. & Rich, D.H. (1989) Inhibition of aminopeptidases by peptides containing ketomethylene and hydroxyethylene amide bond replacements. *J. Med. Chem.* **32**, 1378–1392.

Hildmann, T., Ebneth, M., Pena-Cortes, H., Sanchez-Serrano, J.J., Willmitzer, L. & Prat, S. (1992) General roles of abscisic and jasmonic acids in gene activation as a result of mechanical wounding. *Plant Cell* **4**, 1157–1170.

Himmelhoch, S.R. (1969) Leucine aminopeptidase: a zinc metalloenzyme. *Arch. Biochem. Biophys.* **134**, 597–602.

Holter, H. (1979) 50 years ago. Linderstrøm-Lang and leucylaminopeptidase. *Trends Biochem. Sci.* **4**, 239–240.

Johnson, M.J., Johnson, G.H. & Peterson, W.H. (1936) The magnesium-activated leucyl peptidase of animal erepsin. *J. Biol. Chem.* **116**, 515–526.

Kim, H. & Lipscomb, W.N. (1993a) X-ray crystallographic determination of the structure of bovine lens leucine aminopeptidase complexed with amastatin: formulation of a catalytic mechanism featuring a *gem*-diolate transition state. *Biochemistry* **32**, 8465–8478.

Kim, H. & Lipscomb, W.N. (1993b) Differentiation and identification of the two catalytic metal binding sites in bovine lens leucine aminopeptidase by X-ray crystallography. *Proc. Natl Acad. Sci. USA* **90**, 5006.

Kim, H. & Lipscomb, W.N. (1994) Structure and mechanism of bovine lens leucine aminopeptidase. *Adv. Enzymol.* **68**, 153–213.

Kim, H., Burley, S.K. & Lipscomb, W.N. (1993) Re-refinement of the X-ray crystal structure of bovine lens leucine aminopeptidase complexed with bestatin. *J. Mol. Biol.* **230**, 722–724.

Kim, Y.S., Kim, Y.W. & Sleisenger, M.H. (1974) Studies on the properties of peptide hydrolases in the brush-border and soluble fractions of small intestinal mucosa of rat and man. *Biochim. Biophys. Acta* **370**, 283–296.

Lejczak, B., Kafarski, P. & Zygmunt, J. (1989) Inhibition of aminopeptidases by aminophosphonates. *Biochemistry* **28**, 3549–3555.

Linderstrøm-Lang, K. (1929) Über Darmerepsin [Concerning intestinal erepsin]. *Z. Physiol. Chem.* **182**, 151–174.

Matsushima, M., Takahashi, T., Ichinose, M., Miki, K., Kurokawa, K. & Takahashi, K. (1991) Structural and immunological evidence for the identity of prolyl aminopeptidase with leucyl aminopeptidase. *Biochem. Biophys. Res. Commun.* **178**, 1459–1464.

McCulloch, R., Burke, M.E. & Sherratt, D.J. (1994) Peptidase activity of *Escherichia coli* aminopeptidase A is not required for its role in Xer site-specific recombination. *Mol. Microbiol.* **12**, 241–251.

Pautot, V., Holzer, F.M., Reisch, B. & Walling, L.L. (1993) Leucine aminopeptidase: an inducible component of the defense response in *Lycopersicon esculentum* (tomato). *Proc. Natl Acad. Sci. USA* **90**, 9906–9910.

Rich, D.H., Moon, B.J. & Harbeson, S. (1984) Inhibition of aminopeptidases by amastatin and bestatin derivatives. Effect of inhibitor structure on slow-binding processes. *J. Med. Chem.* **27**, 417–422.

Smith, E.L. & Spackman, D.H. (1955) Leucine aminopeptidase. V. Activation, specificity, and mechanism of action. *J. Biol. Chem.* **212**, 271–299.

Spackman, D.H., Smith, E.L. & Brown, D.M. (1955) Leucine aminopeptidase. IV. Isolation and properties of the enzyme from swine kidney. *J. Biol. Chem.* **212**, 255–269.

Stirling, C.J., Colloms, S.D., Collins, J.F., Szatmari, G. & Sherratt, D.J. (1989) xerB, an *Escherichia coli* gene required for plasmid ColE1 site-specific recombination, is identical to pepA, encoding aminopeptidase A, a protein with substantial similarity to bovine lens leucine aminopeptidase. *EMBO J.* **8**, 1623–1627.

Sträter, N. & Lipscomb, W.N. (1995a) Transition state analogue L-leucinephosphonic acid bound to bovine lens leucine aminopeptidase: X-ray structure at 1.65 Å resolution in a new crystal form. *Biochemistry* **34**, 9200–9210.

Sträter, N. & Lipscomb, W.N. (1995b) Two-metal ion mechanism of bovine lens leucine aminopeptidase: active site solvent structure and binding mode of L-leucinal, a *gem*-diolate transition state analogue, by X-ray crystallography. *Biochemistry* **34**, 14792–14800.

Taylor, A. (1993) Aminopeptidases: structure and function. *FASEB J.* **7**, 290–298.

Taylor, A., Surgenor, T., Thomson, D.K.R., Graham, R.J. & Oettgen, H.C. (1984) Comparison of leucine aminopeptidase from human lens, beef lens and kidney, and hog lens and kidney. *Exp. Eye Res.* **38**, 217–229.

Taylor, A., Peltier, C.Z., Torre, F.J. & Hakamian, N. (1993) Inhibition of bovine lens leucine aminopeptidase by bestatin: number of binding sites and slow binding of this inhibitor. *Biochemistry* **32**, 784–790.

Thompson, G.A. & Carpenter, F.H. (1976a) Leucine aminopeptidase (bovine lens). Effect of pH on the relative binding of $Zn^{2+}$ and $Mg^{2+}$ to and on activation of the enzyme. *J. Biol. Chem.* **251**, 53–60.

Thompson, G.A. & Carpenter, F.H. (1976b) Leucine aminopeptidase (bovine lens). The relative binding of cobalt and zinc to leucine aminopeptidase and the effect of cobalt substitution on specific activity. *J. Biol. Chem.* **251**, 1618–1624.

Turzynski, A. & Mentlein, R. (1990) Prolyl aminopeptidase from rat brain and kidney. Action on peptides and identification as leucyl

aminopeptidase. *Eur. J. Biochem.* **190**, 509–515.

Van Wart, H.E. & Lin, S.H. (1981) Metal binding stoichiometry and mechanism of metal ion modulation of the activity of porcine kidney leucine aminopeptidase. *Biochemistry* **20**, 5682–5689.

Wallner, B.P., Hession, C., Tizard, R., Frey, A.Z., Zuliani, A., Mura, C., Jahngen-Hodge, J. & Taylor, A. (1993) Isolation of bovine kidney leucine aminopeptidase cDNA: comparison with the lens enzyme and tissue-specific expression of two mRNAs. *Biochemistry* **32**, 9296–9301.

***Norbert Sträter***
*Department of Chemistry and Chemical Biology,*
*Harvard University, 12 Oxford Street,*
*Cambridge, MA 02138, USA*

***William N. Lipscomb***
*Department of Chemistry and Chemical Biology,*
*Harvard University, 12 Oxford Street,*
*Cambridge, MA 02138, USA*
*Email: lipscomb@chemistry.harvard.edu*

# *474. Leucyl aminopeptidase (bacteria)*

## Databanks

Peptidase classification: clan MF, family M17, MEROPS ID: M17.003
NC-IUBMB enzyme classification: EC 3.4.11.10
Chemical Abstracts Service registry number: 37288-67-8
Databank codes:

| Species | SwissProt | PIR | EMBL (cDNA) | EMBL (genomic) |
|---|---|---|---|---|
| *Chlamydia trachomatis* | P38019 | – | M86605 | – |
| *Escherichia coli* | P11648 | S04462 | X15130 | U14003: chromosomal region from X86443 92.8 to 00.1′ |
| *Haemophilus influenzae* | P45334 | C64137 | – | U32843: genome section 158 of 163 |
| *Mycobacterium tuberculosis* | Q10401 | – | Z70283 | – |
| *Mycoplasma genitalium* | P47631 | – | U39724 | – |
| *Mycoplasma salivarium* | P47707 | – | D17450 | – |
| *Rickettsia prowazekii* | P27888 | A40631 | M68966 | – |

## Name and History

The purification of a bacterial enzyme with properties similar to those of mammalian leucyl aminopeptidase (Chapter 473) was first reported by Vogt (1970). This *Escherichia coli* aminopeptidase, designated **aminopeptidase I**, had a similar size and exhibited comparable peptide specificity to that of the mammalian enzyme. A notable characteristic of this peptidase that distinguished it from other *E. coli* peptidase activities was its heat stability. Subsequently, peptidase mutants (*pepA*) of *Salmonella typhimurium* (Miller & Mackinnon, 1974) and *E. coli* (Miller & Schwartz, 1978) were isolated that lacked a broad-specificity peptidase activity. This missing enzyme was named **peptidase A**, and recognized to be equivalent, in substrate specificity and heat stability, to the aminopeptidase

I described by Vogt (1970). The enzyme was rediscovered in a different context when the *E. coli xerB* gene, whose product is required for plasmid ColE1 site-specific recombination, was characterized (Stirling *et al.*, 1989). Sequencing of *xerB* revealed that the deduced amino acid sequence shared 31% of its residues with cattle lens leucyl aminopeptidase (Chapter 473). In the C-terminal portion of the proteins, 52% of the residues were identical. This sequence similarity, the fact that the *xerB* product possesses a heat-stable leucyl aminopeptidase activity and the absence of peptidase A (PepA) activity in *E. coli xerB* mutants, led to the conclusion that **xerB** is equivalent to *pepA*. The multifunctional nature of *E. coli* PepA was further expanded when it was discovered that the *E. coli* **carP** gene product, involved in regulation of the carbamoylphosphate synthetase operon, was identical to XerB/PepA (Charlier *et al*., 1995b). Finally, the designation **aminopeptidase My** has been given to an aminopeptidase of *Mycoplasma salivarium* that is composed of two proteins, one of which shares significant amino acid identity with other leucyl aminopeptidases (Shibata *et al*., 1995). This group of enzymes will be treated here as **bacterial leucyl aminopeptidase** or **pepA aminopeptidase**.

## Activity and Specificity

The PepA aminopeptidases of the closely related *E. coli* and *S. typhimurium* are the best characterized of the bacterial enzymes and exhibit peptidase activities similar to their mammalian counterparts. They are enzymes with broad substrate specificity that catalyze the release of N-terminal amino acids from peptides, especially leucyl and methionyl substrates (Vogt, 1970; Miller & Mackinnon, 1974). However, unlike the zinc-dependent mammalian enzymes, the bacterial enzymes are $Mn^{2+}$ dependent and are inhibited by $Zn^{2+}$ (Vogt, 1970; Stirling *et al*., 1989). The *E. coli* enzyme is irreversibly inactivated below pH 6.5 but remains completely active to pH 10.5 (Vogt, 1970). Although the specificity of PepA overlaps with other *S. typhimurium* broad-specificity aminopeptidases (Miller, 1996), there is some unique specificity as evidenced by the fact that PepA appears to be the only *Salmonella* aminopeptidase capable of hydrolyzing the toxic peptide alafosfalin (Gibson *et al*., 1984). Studies with *S. typhimurium* peptidase mutants indicate that PepA is unable to remove N-terminal amino acids if they are adjacent to a proline residue (Miller & Green, 1983). The *M. salivarium* aminopeptidase My has been hypothesized to be an enzyme complex expressing two activities, since it exhibits a high specificity for N-terminal L-arginine as well as L-leucine. This aminopeptidase is most active at pH 8.5, is inhibited by bestatin and EDTA and is stimulated by $Mn^{2+}$ (Shibata & Watanabe, 1987).

A sensitive assay for PepA involves the release of *p*-nitroaniline from *p*-nitroanilide substrates (Tuppy *et al*., 1962). The increase in absorbance at 405 nm as *p*-nitroaniline is generated from L-leucine-*p*-nitroanilide has been used to measure leucyl aminopeptidase activity present in crude bacterial lysates as well as to quantitate the activity of purified enzyme (Stirling *et al*., 1989; Wood *et al*., 1993; Shibata *et al*., 1995).

## Structural Chemistry

*E. coli* PepA consists of 503 amino acids with an $M_r$ of 55 300 (Stirling *et al*., 1989). The active enzyme, like that of the mammalian enzyme, appears to be a hexamer with an $M_r$ of 323 000 (Vogt, 1970). Gene analogs (see the Databanks table), identified through genomic sequencing projects or direct isolation, vary in size from 491 to 520 amino acids. The C-terminal portion of the *Chlamydia trachomatis pepA* gene has been sequenced and most of the *Mycobacterium tuberculosis* coding region can be identified. The full-length analogs and the *M. tuberculosis* partial sequence all contain the leucyl aminopeptidase signature sequence motif (NTDAEGRL) (Bairoch & Bucher, 1994). In addition, the active-site residues identified for cattle lens leucyl

aminopeptidase (Lys250, Asp255, Lys262, Asp 273, Asp332, Glu334, Arg336) (Burley *et al*., 1992) are conserved in all of the bacterial proteins. Deduced secondary structures and loop regions are also conserved between the *E. coli* and cattle lens aminopeptidases (Burley *et al*., 1992; Taylor, 1993). The *M. salivarium* aminopeptidase My differs from the other characterized bacterial enzymes in that it is tightly membrane bound, exhibits a larger molecular mass (390 kDa) and appears to be composed of two heterologous subunits of 46 and 50 kDa (Shibata & Watanabe, 1987). Sequencing of the gene coding for the 46 kDa protein revealed a deduced protein sequence of 520 amino acids that shared significant amino acid identity to leucyl aminopeptidase (Shibata *et al*., 1995).

## Preparation

PepA has been purified from *E. coli* (Stirling *et al*., 1989; Vogt, 1970). Induction of an *E. coli* recombinant plasmid carrying *pepA* resulted in expression at a level of 5–10% of total protein (Stirling *et al*., 1989). The *M. salivarium* aminopeptidase My has also been purified following Triton X-100 solubilization and papain digestion to release the enzyme from the membrane (Shibata & Watanabe, 1987).

## Biological Aspects

In *E. coli* and *S. typhimurium* PepA functions as an aminopeptidase to hydrolyze exogenous peptides as a source of amino acids (Miller & Mackinnon, 1974), to promote protein turnover during starvation (Yen *et al*., 1980) and to degrade abnormal proteins (Miller & Green, 1981). At least for *E. coli* PepA, additional functions, apparently independent of aminopeptidase activity, have also been identified. These include participation in site-specific recombination at *cer* sites and in the regulation of the *carAB* operon. Site-specific recombination at *cer* sites, which are 220 bp regions present in the plasmid ColE1, is necessary in order to maintain stable plasmid monomers rather than plasmid multimers within the bacterial cell. Although PepA provides only an accessory function and its specific role in the recombinational events at *cer* sites is unknown, PepA is necessary for recombination at *cer* to occur. However, PepA aminopeptidase activity is not required for recombination. Using a mutant with an active-site amino acid substitution that eliminated aminopeptidase activity, McCullough *et al*. (1994) demonstrated that PepA-dependent recombination at *cer* was unaffected. However, not all bacterial enzymes are capable of complementing this function. The *R. prowazekii* PepA, while expressing heat-stable leucine aminopeptidase activity, is unable to complement the recombination of *cer* sites in an *E. coli* host (Wood *et al*., 1993).

The multifunctional properties of the *E. coli* and *S. typhimurium* PepA proteins also includes a role as regulator of gene transcription in the expression of the *carAB* operon. Charlier *et al*. (1995a,b) demonstrated that this regulation involves the specific binding of purified PepA/CarP to two sites in the control region of both the *E. coli* and *S. typhimurium carAB* operons. The binding of PepA and integration host factor (IHF) is required for the repression of the *carAB* P1 promoter by pyrimidines, possibly by inducing conformational changes that are required for pyrimidine interaction. Interestingly, these authors also demonstrated that PepA binds to a site within its own control region suggesting an autoregulatory role. Three promoters for *pepA* have been identified with the extreme upstream promoter (P1) the putative target for PepA autoregulation.

## Distinguishing Features

Like several bacterial aminopeptidases, *E. coli* PepA is heat stable (70°C for 5 min). In *E. coli* it appears this is the only peptidase activity stable at high temperature (Vogt, 1970). The enzyme activity of *R. prowazekii* PepA was also demonstrated to be stable following heat treatment (Wood *et al.*, 1993). Inhibition by $Zn^{2+}$ appears to be characteristic of the bacterial enzymes (Stirling *et al.*, 1989; Vogt, 1970).

## Further Reading

Recent articles on metallopeptidases (Rawlings & Barrett, 1995), bacterial aminopeptidases (Gonzales & Robert-Baudouy, 1996) and *E. coli* and *S. typhimurium* protein degradation (Miller, 1996) provide comprehensive reviews of these topics and include information on bacterial PepA and its relationship to other peptidases.

## References

Bairoch, A. & Bucher, P. (1994) PROSITE: recent developments. *Nucleic Acids Res.* **22**, 3583–3589.

Burley, S.K., David, P.R., Sweet, R.M., Taylor, A. & Lipscomb, W.N. (1992) Structure determination and refinement of bovine lens leucine aminopeptidase and its complex with bestatin. *J. Mol. Biol.* **224**, 113–140.

Charlier, D., Gigot, D., Huysveld, N., Roovers, M., Piérard, A. & Glansdorff, N. (1995a) Pyrimidine regulation of the *Escherichia coli* and *Salmonella typhimurium carAB* operons: CarP and integration host factor (IHF) modulate the methylation status of a GATC site present in the control region. *J. Mol. Biol.* **250**, 383–391.

Charlier, D., Hassanzadeh, G., Kholti, A., Gigot, D., Piérard, A. & Glansdorff, N. (1995b) *carP*, involved in pyrimidine regulation of the *Escherichia coli* carbamoylphosphate synthetase operon encodes a sequence-specific DNA-binding protein identical to XerB and PepA, also required for resolution of ColE1 multimers. *J. Mol. Biol.* **250**, 392–406.

Gibson, M.M., Price, M. & Higgins, C.F. (1984) Genetic characterization and molecular cloning of the tripeptide permease *tpp* genes of *Salmonella typhimurium. J. Bacteriol.* **160**, 122–130.

Gonzales, T. & Robert-Baudouy, J. (1996) Bacterial aminopeptidases: properties and functions. *FEMS Microbiol. Rev.* **18**, 319–344.

McCulloch, R., Burke, M.E. & Sherratt, D.J. (1994) Peptidase activity of *Escherichia coli* aminopeptidase A is not required for its role in Xer site-specific recombination. *Mol. Microbiol.* **12**, 241–251.

Miller, C.G. (1996) Protein degradation and proteolytic modification. In: Escherichia coli *and* Salmonella *Cellular and Molecular Biology* (Neidhardt, F.C., Curtiss, R., III, Ingraham, J.L., Lin, E.C.C., Brooks Low, K., Jr., Magasanik, B., Reznikoff, W.S., Riley, M., Schaechter, M. & Umbarger, H.E., eds). Washington, DC: ASM Press, pp. 938–954.

Miller, C.G. & Green, L. (1981) Degradation of abnormal proteins in peptidase-deficient mutants of *Salmonella typhimurium. J. Bacteriol.* **147**, 925–930.

Miller, C.G. & Green, L. (1983) Degradation of proline peptides in peptidase-deficient strains of *Salmonella typhimurium. J. Bacteriol.* **153**, 350–356.

Miller, C.G. & Mackinnon, K. (1974) Peptidase mutants of *Salmonella typhimurium. J. Bacteriol.* **120**, 355–363.

Miller, C.G. & Schwartz, G. (1978) Peptidase-deficient mutants of *Escherichia coli. J. Bacteriol.* **135**, 603–611.

Rawlings, N.D. & Barrett, A.J. (1995) Evolutionary families of metallopeptidases. *Methods Enzymol.* **248**, 183–228.

Shibata, K.-I. & Watanabe, T. (1987) Purification and characterization of an aminopeptidase from *Mycoplasma salivarium. J. Bacteriol.* **169**, 3409–3413.

Shibata, K., Tsuchida, N. & Watanabe, T. (1995) Cloning and sequence analysis of the aminopeptidase My gene from *Mycoplasma salivarium. FEMS Microbiol. Lett.* **130**, 19–24.

Stirling, C.J., Colloms, S.D., Collins, J.F., Szatmari, G. & Sherratt, D.J. (1989) *xerB*, an *Escherichia coli* gene required for plasmid ColE1, site-specific recombination, is identical to *pepA*, encoding aminopeptidase A, a protein with substantial similarity to bovine lens leucine aminopeptidase. *EMBO J.* **8**, 1623–1627.

Taylor, A. (1993) Aminopeptidases: structure and function. *FASEB J.* **7**, 290–298.

Tuppy, H., Wiesbauer, U. & Wintersberger, E. (1962) Aminosäure-*p*-nitroanilide als Substrate für Aminopeptidasen und andere proteolytische Fermente [Amino acid-*p*-nitroanilides as substrates for aminopeptidases and other proteolytic enzymes]. *Z. Physiol. Chem.* **329**, 278–288.

Vogt, V.M. (1970) Purification and properties of an aminopeptidase from *Escherichia coli. J. Biol. Chem.* **245**, 4760–4769.

Wood, D.O., Solomon, M.J., & Speed, R.R. (1993) Characterization of the *Rickettsia prowazekii pepA* gene encoding leucine aminopeptidase. *J. Bacteriol.* **175**, 159–165.

Yen, C., Green, L., & Miller, C.G. (1980) Degradation of intracellular protein in *Salmonella typhimurium* peptidase mutants. *J. Mol. Biol.* **143**, 21–33.

**M**

*David O. Wood*
*Laboratory of Molecular Biology,*
*Department of Microbiology and Immunology,*
*College of Medicine,*
*University of South Alabama,*
*Mobile, AL 36688, USA*
*Email: wood@sungcg.usouthal.edu*

# 475. Introduction: clan MG containing cobalt and zinc aminopeptidases

## Databanks

*MEROPS ID: MG*

| Species | SwissProt | PIR | EMBL (cDNA) | EMBL (genomic) |
|---|---|---|---|---|
| **Family M24** | | | | |
| Methionyl aminopeptidase type I (Chapter 476) | | | | |
| Methionyl aminopeptidase type II (Chapter 477) | | | | |
| X-Pro aminopeptidase (eukaryote) (Chapter 480) | | | | |
| X-Pro aminopeptidase (*Lactococcus*) (Chapter 481) | | | | |
| X-Pro aminopeptidase (prokaryote) (Chapter 479) | | | | |
| X-Pro dipeptidase (Chapter 478) | | | | |
| Others | | | | |
| *Alteromonas* sp. | – | – | U29240 | – |
| *Bacillus subtilis* | P54518 | – | – | D84432: genome section |
| *Homo sapiens* | P53582 | – | D42084 | – |
| *Micrococcus lysodeikticus* | P33111 | – | X17524 | – |
| *Mycobacterium tuberculosis* | – | – | X59509 | – |
| *Mycobacterium tuberculosis* | Q10698 | – | Z73966 | – |
| *Mycoplasma capricolum* | – | – | Z33034 | – |
| *Saccharomyces cerevisiae* | P40051 | – | U18839 | – |
| *Saccharomyces cerevisiae* | P43590 | – | – | D50617: chromosome VI complete |
| *Schizosaccharomyces pombe* | Q10439 | – | Z70721 | – |
| *Synechococcus* sp. | – | – | D26161 | – |

Clan MG contains only family M24. Like clan MF (Chapter 472) and clan MH (Chapter 482), clan MG contains peptidases in which two metal ions are catalytic. The family contains several very different exopeptidases, two methionyl aminopeptidases (the cytosolic methionyl aminopeptidase II and the mitochondrial/bacterial methionyl aminopeptidase I), X-Pro aminopeptidase and X-Pro dipeptidase (Rawlings & Barrett, 1993).

The tertiary structure of *Escherichia coli* methionyl aminopeptidase I has been solved, and shows a two-domain structure, with each domain possessing a similar fold (see Figure 475.1). This situation is similar to that of the two-domain structures of chymotrypsin and pepsin, and is believed to be due to ancient gene duplication and fusion events. The ancestral methionyl aminopeptidase was probably active as a homodimer. The fold of methionyl aminopeptidase is also similar to that of *Pseudomonas* creatinase, even though this enzyme is not metal dependent, and the metal ligands have been replaced.

The two cobalt ions are bound by ligands on five amino acid residues, with two ligands from each domain. One cobalt ion (Co1) is coordinated by Asp108, His171 and Glu204, and the second (Co2) by Asp97, Asp108 and Glu235. Asp97 and Glu204 are in equivalent positions in the two domains, but Asp108 is displaced two residues in the N-terminal domain from Glu235 in the C-terminal domain. His171 is also in the C-terminal domain but is in a part of the domain that cannot

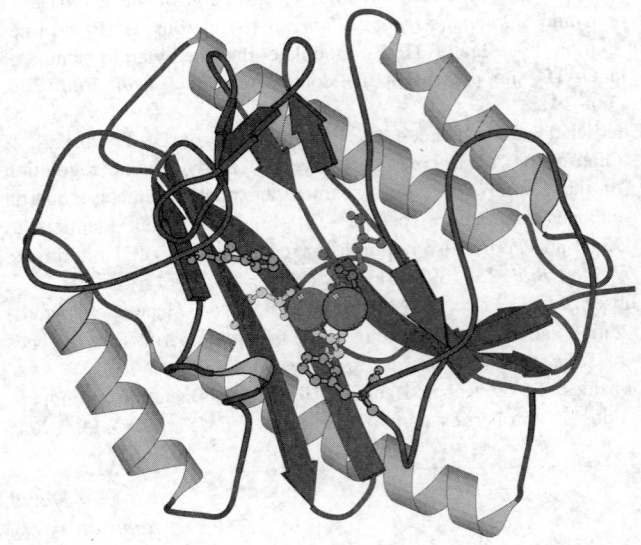

*Figure 475.1* Richardson diagram of *Escherichia coli* methionyl aminopeptidase type I. The image was prepared from the Brookhaven Protein Data Bank entry (1MAT) as described in the Introduction (p. xxv). The catalytic cobalt ions are shown in CPK representation as spheres. The cobalt ligands are shown in ball-and-stick representation: Asp97, Asp108, His171, Glu204 and Glu235 (numbering as in Alignment 475.1). The catalytic His97 is also shown in ball-and-stick representation.

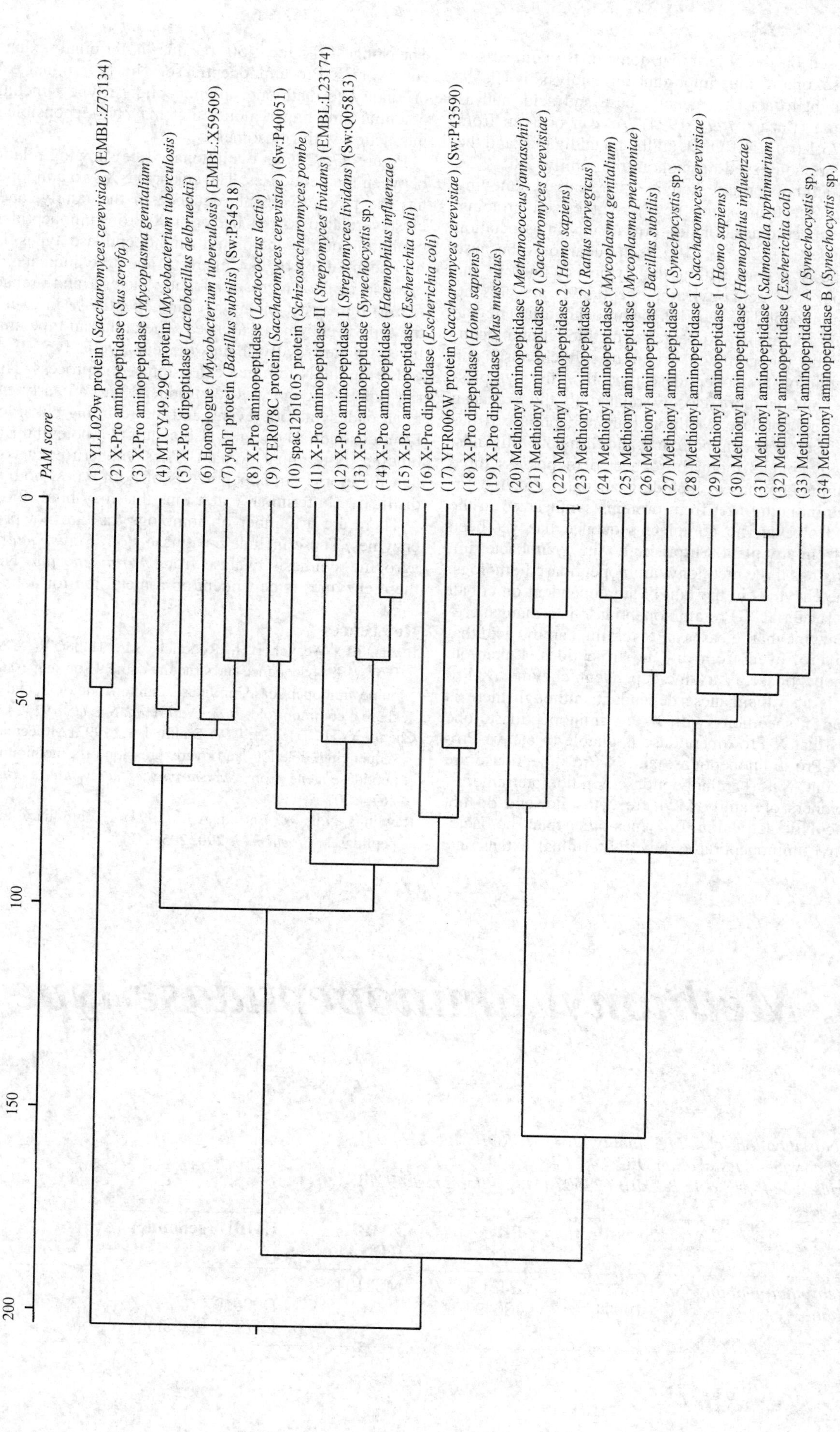

PAM score

(1) YLL029w protein (*Saccharomyces cerevisiae*) (EMBL:Z73134)
(2) X-Pro aminopeptidase (*Sus scrofa*)
(3) X-Pro aminopeptidase (*Mycoplasma genitalium*)
(4) MTCY49.29C protein (*Mycobacterium tuberculosis*)
(5) X-Pro dipeptidase (*Lactobacillus delbrueckii*)
(6) Homologue (*Mycobacterium tuberculosis*) (EMBL:X59509)
(7) yqhT protein (*Bacillus subtilis*) (Sw:P54518)
(8) X-Pro aminopeptidase (*Lactococcus lactis*)
(9) YER078C protein (*Saccharomyces cerevisiae*) (Sw:P40051)
(10) spac12b10.05 protein (*Schizosaccharomyces pombe*)
(11) X-Pro aminopeptidase II (*Streptomyces lividans*) (EMBL:L23174)
(12) X-Pro aminopeptidase I (*Streptomyces lividans*) (Sw:Q05813)
(13) X-Pro aminopeptidase (*Synechocystis sp.*)
(14) X-Pro aminopeptidase (*Haemophilus influenzae*)
(15) X-Pro aminopeptidase (*Escherichia coli*)
(16) X-Pro dipeptidase (*Escherichia coli*)
(17) YFR006W protein (*Saccharomyces cerevisiae*) (Sw:P43590)
(18) X-Pro dipeptidase (*Homo sapiens*)
(19) X-Pro dipeptidase (*Mus musculus*)
(20) Methionyl aminopeptidase (*Methanococcus jannaschii*)
(21) Methionyl aminopeptidase 2 (*Saccharomyces cerevisiae*)
(22) Methionyl aminopeptidase 2 (*Homo sapiens*)
(23) Methionyl aminopeptidase 2 (*Rattus norvegicus*)
(24) Methionyl aminopeptidase (*Mycoplasma genitalium*)
(25) Methionyl aminopeptidase (*Mycoplasma pneumoniae*)
(26) Methionyl aminopeptidase (*Bacillus subtilis*)
(27) Methionyl aminopeptidase C (*Synechocystis sp.*)
(28) Methionyl aminopeptidase 1 (*Saccharomyces cerevisiae*)
(29) Methionyl aminopeptidase 1 (*Homo sapiens*)
(30) Methionyl aminopeptidase (*Haemophilus influenzae*)
(31) Methionyl aminopeptidase (*Salmonella typhimurium*)
(32) Methionyl aminopeptidase (*Escherichia coli*)
(33) Methionyl aminopeptidase A (*Synechocystis sp.*)
(34) Methionyl aminopeptidase B (*Synechocystis sp.*)

*Figure 475.2*  Evolutionary tree for family M24. The tree was prepared as described in the Introduction (p. xxv).

be superimposed on the N-terminal domain. By comparison with creatinase, one residue important for catalysis is His79, which is thought to act as a general base and acid, and as a proton shuttle (Bazan *et al.*, 1994). As can be seen from Alignment 475.1 (on CD-ROM), all the metal ligands and the catalytic His are conserved throughout the family.

Methionyl aminopeptidase activity is essential, removing the initiating Met from some newly synthesized proteins. The enzyme acts cotranslationally, presumably in association with ribosomes (Chang *et al.*, 1992). In eukaryotes, because protein synthesis occurs within the cytoplasm and in the mitochondrion, two methionyl aminopeptidases are present. Methionyl aminopeptidase type I is the mitochondrial enzyme and shows most similarity to bacterial methionyl aminopeptidase, reflecting the common origin of mitochondria and bacteria. Methionyl aminopeptidase type II is the cytoplasmic enzyme, and is more closely related to the archaean methionyl aminopeptidase. Methionyl aminopeptidase II has a 65 residue insert between the fourth and fifth cobalt ligands.

Theoretically, plants might be expected to possess three methionyl aminopeptidases, including an extra enzyme for removing the initiating Met from proteins synthesized in the chloroplast. However, no complete sequence has yet been determined from any plant. Surprisingly, the cyanobacterium *Synechocystis* has three methionyl aminopeptidase homologs.

Not all the enzymes in the family are dependent on cobalt for activity. Bacterial X-Pro aminopeptidase is a manganese-dependent enzyme that releases an N-terminal amino acid that is linked to Pro, even from di- and tripeptides. Eukaryote X-Pro aminopeptidase is a zinc-dependent enzyme. X-Pro dipeptidase is also manganese dependent; although there is some overlap in specificity with X-Pro aminopeptidase, one difference is that X-Pro dipeptidase is unable to cleave Pro-Pro. Both X-Pro aminopeptidase and X-Pro dipeptidase are thiol dependent. X-Pro aminopeptidase is a homotetramer.

Most members of family M24 are cytosolic, and do not require proteolytic activation. Enzymes other than the bacterial methionyl aminopeptidases have N-terminal extensions.

The longest is the 450 residue N-terminal extension in eukaryote X-Pro aminopeptidase. The N-terminal extension of eukaryote methionyl aminopeptidase type I includes two potential zinc-finger motifs that may be responsible for the interaction with ribosomes.

Figure 475.2 is an evolutionary tree showing relationships between the peptidases. The eukaryote X-Pro aminopeptidase has the most divergent sequence in the family, and is not closely related to the bacterial X-Pro aminopeptidase. The evolution of the methionyl aminopeptidase types I and II has been of particular interest. The tree indicates that an enzyme in the precursor of both bacteria and archaea gave rise to the type I methionyl aminopeptidase in bacteria and the type II enzyme of archaea. The archaean-type aminopeptidase in turn evolved into the eukaryotic type II enzyme, acquiring the N-terminal extension in the process. The bacterial type I enzyme was reintroduced into an early eukaryote in an endosymbiotic event, and the resulting eukaryotic type I (mitochondrial) enzyme also acquired a potential ribosome-binding, N-terminal domain. Thus, modern eukaryotes contain methionyl aminopeptidases of both types, both with the distinctive N-terminal extensions that may bind ribosomes.

There are a number of homologs that are not proteolytic enzymes. These include creatinase from *Flavobacterium* and agropine synthesis cyclase from *Agrobacterium*. Neither of these enzymes is dependent on a metal ion for activity.

## References

Bazan, J.F., Weaver, L.H., Roderick, S.L., Huber, R. & Matthews, B.W. (1994) Sequence and structure comparison suggest that methionine aminopeptidase, prolidase, aminopeptidase P, and creatinase share a common fold. *Proc. Natl Acad. Sci. USA* **91**, 2473–2477.

Chang, Y.-H., Teichert, U. & Smith, J.A. (1992) Molecular cloning, sequencing, deletion, and overexpression of a methionine aminopeptidase gene from *Saccharomyces cerevisiae*. *J. Biol. Chem.* **267**, 8007–8011.

Rawlings, N.D. & Barrett, A.J. (1993) Evolutionary families of peptidases. *Biochem. J.* **290**, 205–218.

# *476. Methionyl aminopeptidase type I*

## *Databanks*

*Peptidase classification: clan MG, family M24, MEROPS ID: M24.001*
*NC-IUBMB enzyme classification: EC 3.4.11.18*
*ATCC entries: 53245 (E. coli, 3.2 kb); 623507 (M. genitalium, 1.49 kb)*
*Databank codes:*

| Species | SwissProt | PIR | EMBL (cDNA) | EMBL (genomic) |
|---------|-----------|-----|-------------|----------------|
| *Bacillus stearothermophilus* | P28617 | C42196 | M88104 | – |
| *Bacillus subtilis* | P19994 | JS0493 | – | D00619 |
| | | | | L47971: spc-alpha region |

| Species | SwissProt | PIR | EMBL (cDNA) | EMBL (genomic) |
|---|---|---|---|---|
| *Clostridium perfringens* | P50614 | – | X86486 | – |
| *Escherichia coli* | P07906 | A27761 S45233 | M15106 | D26562: chromosome 2.4–4.1' |
| *Haemophilus influenzae* | P44421 | C64138 | – | U32845: genome section 160 of 163 |
| *Homo sapiens* | P53582 | – | D42084 | – |
| *Klebsiella pneumoniae* | P41392 | – | S43195 | |
| *Mycoplasma genitalium* | P47418 | – | – | U39695: complete genome section 17 of 56 |
| *Mycoplasma pneumoniae* | Q11132 | – | – | U34795: cosmid pcosMPGT9 fragment |
| *Saccharomyces cerevisiae* | Q01662 | S31268 S32590 S59390 | M77092 | U20865: chromosome XII cosmid 9672 |
| *Salmonella typhimurium* | P10882 | S03562 | X55778 | – |
| *Synechocystis* sp. | P53579 | – | – | D64003: complete genome part 22 |
| *Synechocystis* sp. | P53580 | – | D64005 | |
| *Synechocystis* sp. | P53581 | – | – | D64003: complete genome part 22 |

Brookhaven Protein Data Bank three-dimensional structures:

| Species | ID | Resolution | Notes |
|---|---|---|---|
| *Escherichia coli* | 1MAT | 2.4 | |

## Name and History

Two classes of cobalt-containing aminopeptidases have been isolated that are apparently involved in the nonprocessive cleavage of the initiator methionine of protein synthesis. These are generally termed *methionine aminopeptidases* in the current literature, but are recommended by IUBMB to be called *methionyl aminopeptidases* to reflect the fact that a methionyl bond is hydrolysed. The abbreviation *MetAP* is used here. Both classes of MetAPs have the same well-defined specificity based on the penultimate residue (Arfin *et al.*, 1995). Using this profile, MetAPs can be distinguished from a number of other aminopeptidases, commonly containing zinc ion cofactors, that catalyze the release of several hydrophobic amino acids in addition to methionine (Grdisa & Vitale, 1991; Ben-Meir *et al.*, 1993; Chevrier *et al.*, 1994) or are involved in other specific actions, such as the processing of bioactive peptides. Aminopeptidase M (Chapter 337), a membrane-bound extracellular enzyme, also zinc-containing (Olsen *et al.*, 1988; Malfroy *et al.*, 1989), is an example of the latter type. No other MetAPs specific for the nonprocessive hydrolysis of methionine from any substrate are presently known. The *methionyl aminopeptidase type I* (*MetAP I*) enzymes, which are discussed in this chapter, are found in both prokaryotes and eukaryotes, but not archaebacteria. The type II aminopeptidase is described in the next chapter (Chapter 477).

## Activity and Specificity

Although virtually all protein synthesis is initiated with methionine, most proteins do not appear to retain it in their mature forms (Jackson & Hunter, 1970; Driessen *et al.*, 1985). Proteins that are transported across membranes generally have an N-terminal signal peptide that is removed by a specific peptidase (see Chapters 153, 154, 155 and 156), resulting in the loss of the initiator methionine along with several other N-terminal residues. However, cytoplasmic proteins must rely on specific MetAPs to remove this methionine. In eukaryotes, this is a cotranslational process (Arfin & Bradshaw, 1988); in prokaryotes, it may be both co- and post-translational, and deformylation must occur before the methionine is removed (Pine, 1969). Importantly, not all initiator methionine residues are cleaved; the P1' residue determines whether cleavage occurs. Sherman *et al.* (1985), using mutants of iso-cytochrome *c* from *S. cerevisiae*, correctly predicted that the size of the adjacent residue was the principal specifier of susceptibility to cleavage by MetAP. Subsequently, it has been shown *in vivo* using site-directed mutagenesis that if the P1' residue has a side chain with a low radius of gyration (Gly, Ala, Ser, Thr, Pro, Val or Cys), the methionine will be removed, otherwise it is retained (Huang *et al.*, 1987; Boissel *et al.*, 1988; Hirel *et al.*, 1989; Moerschell *et al.*, 1990). This correlates with the N-end rule of protein stability, which predicts that large or charged residues at the N-termini of proteins are destabilizing and lead to ubiquitin-dependent degradation via an E3 (ligase) mechanism (Ciechanover, 1994). It appears that the specificity of MetAP is such that when the penultimate residue of a protein would be destabilizing, the initiator methionine is retained to protect the protein.

Although the N-end rule provides an explanation for the retention of methionine, it does not provide insight into why the seven smallest amino acids act as cleavage signals. There are several possible explanations as to why initiator methionines are cleaved at all. One possibility is that some proteins require a residue other than methionine at their N-terminus to obtain an active structure (Meinnel *et al.*, 1993). For example, an N-terminal cysteine is apparently required for the activity of several glutamine amidotransferases (Weng & Zalkin, 1987). A second reason is that some post-translational modifications such as myristoylation require removal of the methionine, since this alteration has an absolute requirement for N-terminal glycine (plus additional specific residues;

M

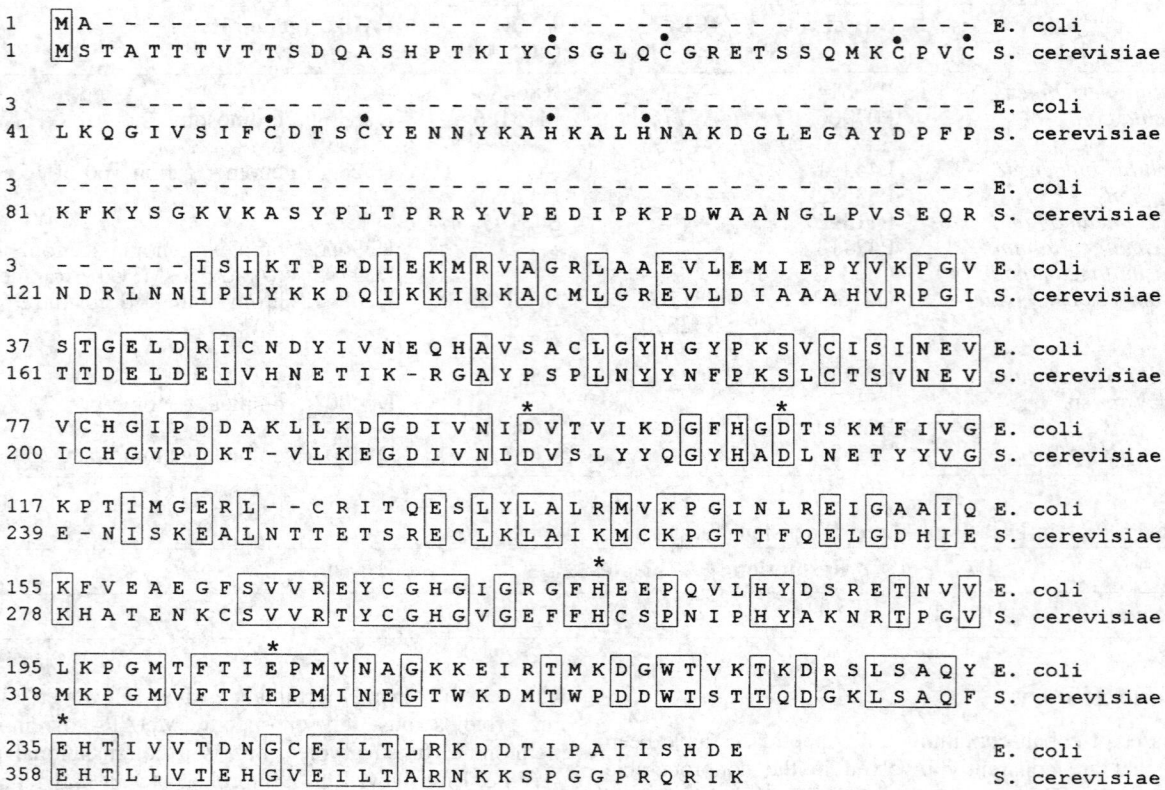

*Figure 476.1* Sequence alignment of *E. coli* and *S. cerevisiae* type I MetAPs. Dots identify putative zinc ion-binding residues in the N-terminal domain, while the asterisks identify the cobalt liganding residues in the catalytic domain.

Duronio *et al*. 1989; Gordon *et al*., 1991). A third, and highly likely, explanation is that methionine cleavage may help maintain the pool of free methionine which is one of the energetically most expensive amino acids to synthesize, requiring sulfur which can be limiting in some environments (Old *et al*., 1991).

All type I MetAPs investigated to date are strongly inhibited by the presence of EDTA, indicating a requirement for divalent metal ions (Miller *et al*., 1987; Chang *et al*., 1990). It has been proposed that cobalt may be the essential metal, but it has not been conclusively demonstrated that it is the ion present *in vivo*. In support of the role of cobalt in MetAP activity, it has been observed that cobalt ions can restore significant activity (37%) of *S. cerevisiae* MetAP I treated with EDTA, while magnesium, manganese, copper, iron and zinc ions restore less than 1% of the activity (Chang *et al*., 1990). Furthermore, *S. typhimurium* MetAP is activated by cobalt 9-fold while manganese, magnesium and zinc have no significant effect (Miller *et al*., 1987).

## Structural Chemistry

The type I MetAPs are found in both prokaryotes and eukaryotes; however, of the nine type I MetAPs putatively identified, MetAP activity has only been verified for the *E. coli*, *S. typhimurium* and *S. cerevisiae* enzymes (Miller *et al*., 1987; Hirel *et al*., 1989; Chang *et al*., 1990). The remaining type I enzymes have been assigned only on the basis of amino acid sequence. As shown in Figure 476.1, which compares the

*E. coli* and yeast MetAP (type I) sequences, the prokaryotic forms have a calculated molecular mass of 27–29 kDa, while the human and yeast enzymes have molecular masses of 44 and 43 kDa, respectively (the human sequence lacks the 5′ initiation codon but by comparison to the yeast sequence, it is likely to be missing only a few amino acids). Gel-filtration studies of the yeast enzyme suggest it is normally found as a monomer (Chang *et al*., 1990).

A high-resolution structure of *E. coli* MetAP has been solved by single crystal X-ray diffraction (Roderick & Matthews, 1993; see Figure 476.2). The structure has been described as a 'pita-bread' fold where the two metal ions are sandwiched between two β sheets, which are surrounded by four α helices yielding a structure with pseudo-2-fold symmetry. Although this is a novel fold for a protease, the structure is similar to that of *Pseudomonas putida* creatinase, which does not possess metal ions in its active site (Bazan *et al*., 1994). The two metal ions of MetAP are only 2.9 Å apart and are liganded by two aspartic acids, two glutamic acids and one histidine. Since a structure of MetAP with bound substrate is not available, it is not known if the metal ions are required for enzyme structure, substrate binding and/or general base catalysis. For the same reason, other details concerning active-site architecture are also not available.

The type I MetAPs show different degrees of similarity typical of species variations. All except *Mycoplasma* MetAP have more than 39% amino acid identity to *E. coli* MetAP (Table 476.1). As expected, the relatedness in the prokaryotic forms is highest, with *E. coli* and *S. typhimurium* showing

*Table 476.1*    Amino acid sequence identity comparisons for the catalytic domain of type I methionine aminopeptidases. The entire *E. coli* enzyme was used to define the catalytic domain of all other proteins. Below and to the left of the identity diagonal is a ratio of the number of identical residues to the total number of residue positions compared. Above and to the right of the identity diagonal is the percent identity of the pairwise comparisons

|  | *E. coli* | *S. typhimurium* | *H. influenzae* | *Synechocystis* | *B. subtilis* | *M. genitalium* | *S. cerevisiae* | *H. sapiens* |
|---|---|---|---|---|---|---|---|---|
| *E. coli* | – | 89.4 | 62.9 | 49.8 | 43.5 | 23.4 | 39.2 | 44.8 |
| *S. typhimurium* | 236/264 | – | 64.0 | 51.4 | 44.0 | 23.8 | 38.8 | 44.4 |
| *H. influenzae* | 170/264 | 173/264 | – | 44.3 | 39.1 | 23.4 | 41.4 | 39.4 |
| *Synechocystis* | 129/250 | 135/250 | 117/252 | – | 47.6 | 25.3 | 46.2 | 52.2 |
| *B. subtilis* | 114/249 | 115/249 | 105/251 | 120/243 | – | 34.3 | 38.7 | 39.1 |
| *M. genitalium* | 60/210 | 61/213 | 63/251 | 69/249 | 89/251 | – | 23.4 | 21.4 |
| *S. cerevisiae* | 111/259 | 110/259 | 115/266 | 123/250 | 97/219 | 57/240 | – | 56.8 |
| *H. sapiens* | 121/262 | 120/262 | 110/253 | 136/247 | 95/242 | 56/247 | 150/249 | – |

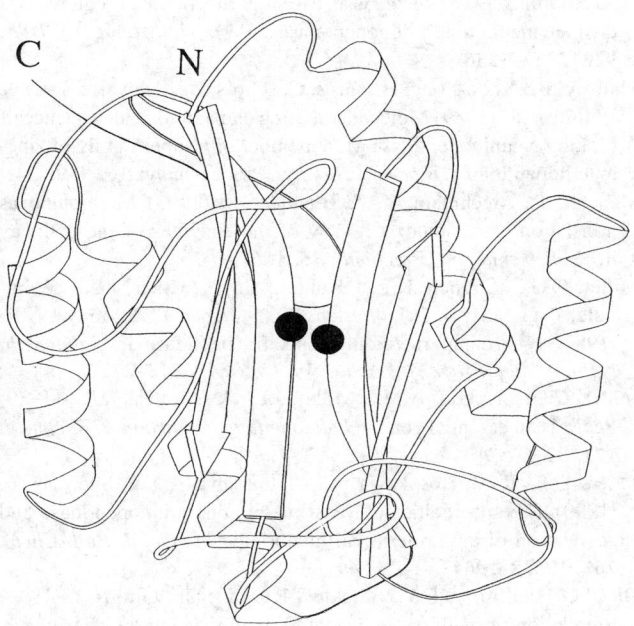

*Figure 476.2*    Ribbon diagram showing the 'pita-bread' fold of *E. coli* MetAP. The view direction is essentially parallel to the local 2-fold axis of symmetry and the active site is marked by the two cobalt ions shown as solid circles. (Reproduced with permission from Bazan *et al*., 1994.)

89% sequence identity. All five metal-binding residues in the entire MetAP I family are completely conserved, suggesting the importance of metal binding for MetAP activity. The yeast and human enzymes have a 125- and 134-amino acid extension, respectively, on their N-termini, indicating that the eukaryotic enzymes have a second domain. Accordingly, the eukaryotic forms are designated type Ib and the prokaryotic forms, type Ia. These N-terminal regions possess putative zinc finger consensus sequences of the type C2C2 and C2H2 (see Figure 476.1). Although the exact role of these zinc finger domains is not known, it has been suggested that they may be involved in ribosome binding given that such motifs are known to bind to nucleic acids (Chang *et al*., 1992). Their absence on the prokaryotic forms suggests that the bacterial enzymes do not associate with ribosomes but probably act as soluble entities.

## Biological Aspects

MetAP activity has been assumed to be associated with ribosomes (in eukaryotes) to ensure that methionine removal occurs efficiently as a cotranslational process. Consistent with this hypothesis, replacement of the wild-type MetAP I in yeast with a MetAP I containing either a deletion or mutations in the N-terminal domain does not complement the slow growth phenotype of null mutants for MetAP I, even though these altered enzymes show full catalytic activity *in vitro* (Zuo *et al*., 1995). However, direct evidence that this alteration in the putative zinc fingers affects ribosome binding has not been obtained. The importance of MetAP to the cell has been demonstrated by deletion of the MetAP genes in *E. coli*, *S. typhimurium* and *S. cerevisiae*, which results in a lethal phenotype (Chang *et al*., 1989; Miller *et al*., 1989; Li & Chang, 1995). In yeast, enzymes of both types must be deleted for lethality.

Although the MetAP enzymes of both prokaryotes and eukaryotes have the same methionine cleavage specificity, production of recombinant proteins can result in methionine cleavage problems in expression systems. For instance, over-production of recombinant proteins in *E. coli* can result in saturation of the methionine processing machinery which can prevent cleavage (Devlin *et al*., 1988; Yasueda *et al*., 1991; Shen *et al*., 1993; Proudfoot *et al*., 1996). This often results in inactive products, or products that are highly immunogenic. Furthermore, production of a normally secreted protein in the cytoplasm of a host cell can result in methionine retention on the N-terminus, due to the presence of a noncleaving penultimate residue, even if the native protein would normally have a nonmethionine N-terminal residue after cleavage with signal peptidase. Thus, expression problems provide additional reasons for better understanding of N-terminal processing systems, particularly with respect to the role of MetAPs.

## Distinguishing Features

At present, no antisera for any MetAP I is available. Identification is based on sequence alignment with the *E. coli* protein and includes conservation of the metal-binding side chains. In eukaryotes, the N-terminal extension with the zinc-finger motifs adds another feature, which may be generally characteristic. The unique specificity of MetAP (type I), which it shares with the type II enzymes, is the most compelling means of identification.

**M**

## Further Reading

For reviews, see Arfin & Bradshaw (1995) and Bradshaw & Arfin (1996).

## References

Arfin, S.M. & Bradshaw, R.A. (1988) Cotranslational processing and protein turnover in eukaryotic cells. *Biochemistry* **27**, 7979–7984.

Arfin, S.M. & Bradshaw, R.A. (1995) Mechanisms of regulated intracellular protein degradation, In: *The Encyclopedia of Molecular Biology: Fundamentals and Applications* (Kendrew, J., ed.). Weinheim: VCH Verlag, pp. 346–354.

Arfin, S.M., Kendall, R.L., Hall, L., Weaver, L.H., Stewart, A.E., Matthews, B.W. & Bradshaw, R.A. (1995) Eukaryotic methionyl aminopeptidases: two classes of cobalt-dependent enzymes. *Proc. Natl Acad. Sci. USA* **92**, 7714–7718.

Bazan, J.F., Weaver, L.H., Roderick, S.L., Huber, R. & Matthews, B.W. (1994) Sequence and structure comparison suggest that methionine aminopeptidase, prolidase, aminopeptidase P, and creatinase share a common fold. *Proc. Natl Acad. Sci. USA* **91**, 2473–2477.

Ben-Meir, D., Spungin, A., Ashkenazi, R. & Blumberg, S. (1993) Specificity of *Streptomyces griseus* aminopeptidase and modulation of activity by divalent metal ion binding and substitution. *Eur. J. Biochem.* **212**, 107–112.

Boissel, J.P., Kasper, T.J. & Bunn, H.F. (1988) Cotranslational amino-terminal processing of cytosolic proteins. *J. Biol. Chem.* **263**, 8443–8449.

Bradshaw R.A. & Arfin, S.M. (1996) Methionine aminopeptidase: structure and function, In: *The Aminopeptidases*, vol. 6 (Taylor, A., ed.). Georgetown: R.G. Landes Co., pp. 91–106.

Chang, S.Y.P., McGary, E.C. & Chang, S. (1989) Methionine aminopeptidase gene of *Escherichia coli* is essential for cell growth. *J. Bacteriol.* **171**, 4071–4072.

Chang, Y.H., Teichert, U. & Smith, J.A. (1990) Purification and characterization of a methionine aminopeptidase from *Saccharomyces cerevisiae. J. Biol. Chem.* **265**, 19892–19897.

Chang, Y.H., Teichert, U. & Smith, J.A. (1992) Molecular cloning, sequencing, deletion, and overexpression of a methionine aminopeptidase gene from *Saccharomyces cerevisiae. J. Biol. Chem.* **267**, 8007–8011.

Chevrier, B., Schalk, C., D'Orchymont, H., Rondeau, J.M., Moras, J.M. & Tarnus, C. (1994) Crystal structure of *Aeromonas* proteolytic aminopeptidase: a prototypical member of the co-catalytic zinc enzyme family. *Structure* **2**, 283–291.

Ciechanover, A. (1994) The ubiquitin-proteasome proteolytic pathway. *Cell* **79**, 13–21.

Devlin, P.E., Drummond, R.J., Toy, P., Mark, D.F., Watt, K.W.K. & Devlin, J.J. (1988) Alteration of amino-terminal codons of human granulocyte-colony-stimulating factor increases expression levels and allows efficient processing by methionine aminopeptidase in *Escherichia coli. Gene* **65**, 13–22.

Driessen, H.P.C., de Jong, W.W., Tesser, G.I. & Bloemendal, H. (1985) The mechanism of N-terminal acetylation of proteins. *CRC Crit. Rev. Biochem.* **18**, 281–325.

Duronio, R.J., Towler, D.A., Heuckeroth, R.O. & Gordon, J.I. (1989) Disruption of the yeast N-myristoyl transferase gene causes recessive lethality. *Science* **243**, 796–800.

Gordon, J.I., Duronio, R.J., Rudnick, D.A., Adams, S.P. & Gokel, G.W. (1991) Protein N-myristoylation. *J. Biol. Chem.* **266**, 8647–8650.

Grdisa, M. & Vitale, L. (1991) Types and localization of aminopeptidases in different human blood cells. *Int. J. Biochem.* **23**, 339–345.

Hirel, P.H., Schmitter, J.M., Dessen, P., Fayat, G. & Blanquet, S. (1989) Extent of N-terminal methionine excision from *Escherichia coli* proteins is governed by the side-chain length of the penultimate amino acid. *Proc. Natl Acad. Sci. USA* **86**, 8247–8251.

Huang, S., Elliott, R.C., Liu, P.S., Koduri, R.K., Weickmann, J.L., Lee, J.H., Blair, L.C., Ghosh-Dastidar, P., Bradshaw, R.A., Bryan, K.M., Einarson, B., Kendall, R.L., Kolacz, K.H. & Saito, K. (1987) Specificity of cotranslational amino-terminal processing of proteins in yeast. *Biochemistry* **26**, 8242–8246.

Jackson, R. & Hunter, T. (1970) Role of methionine in the initiation of haemoglobin synthesis. *Nature* **227**, 672–676.

Li, X. & Chang, Y.H. (1995) Amino-terminal protein processing in *Saccharomyces cerevisiae* is an essential function that requires two distinct methionine aminopeptidases. *Proc. Natl Acad. Sci USA.* **92**, 12357–12361.

Malfroy, B., Kado-Fong, H., Gros, C., Giros, B., Schwartz, J.-C. & Hellmiss, R. (1989) Molecular cloning and amino acid sequence of rat kidney aminopeptidase M: a member of a super family of zinc-metallopeptidases. *Biochem. Biophys. Res. Commun.* **161**, 236–241.

Meinnel, T., Mechulam, Y. & Blanquet, S. (1993) Methionine as translation start signal: a review of the enzymes of the pathway in *Escherichia coli. Biochimie* **75**, 1061–1075.

Miller, C.G., Strauch, K.L., Kukral, A.M., Miller, J.L., Wingfield, P.T., Mazzei, G.J., Werlen, R.C., Graber, P. & Movva, N.R. (1987) N-terminal methionine-specific peptidase in *Salmonella typhimurium. Proc. Natl Acad. Sci. USA* **84**, 2718–2722.

Miller, C.G., Kukral, A.M., Miller, J.L. & Movva, N.R. (1989) PepM is an essential gene in *Salmonella typhimurium. J. Bacteriol.* **171**, 5215–5217.

Moerschell, R.P., Hosokawa, Y., Tsunasawa, S. & Sherman, F. (1990) The specificities of yeast methionine aminopeptidase and acetylation of amino-terminal methionine *in vivo. J. Biol. Chem.* **265**, 19638–19643.

Old, I.G., Phillips, S.E.V., Stockley, P.G. & Saint Girons, I. (1991) Regulation of methionine biosynthesis in the *Enterobacteriaceae. Prog. Biophys. Mol. Biol.* **56**, 145–185.

Olsen, J., Cowell, G.M., Kønigshøfer, E., Danielsen, E.M., Møller, J., Laustsen, L., Hansen, O.C., Welinder, K.G., Engberg, J., Hunziker, W., Spiess, M., Sjöström, H. & Norén, O. (1988) Complete amino acid sequence of human aminopeptidase N as deduced from cloned cDNA. *FEBS Lett.* **238**, 307–314.

Pine, M. (1969) Kinetics of maturation of the amino termini of the cell proteins of *Escherichia coli. Biochem. Biophys. Acta* **174**, 359–372.

Proudfoot, A.E.I., Power, C.A., Hoogewerf, A.J., Montjovent, M.O., Borlat, F., Offord, R.E. & Wells, T.N.C. (1996) Extension of recombinant human RANTES by the retention of the initiating methionine produces a potent antagonist. *J. Biol. Chem.* **271**, 2599–2603.

Roderick, S.L. & Matthews, B.W. (1993) Structure of the cobalt-dependent methionine aminopeptidase from *Escherichia coli*: a new type of proteolytic enzyme. *Biochemistry* **32**, 3907–3912.

Shen, T.J., Ho, N.T., Simplaceanu, V., Zou, M., Green, B.N., Tam, M.F. & Ho, C. (1993) Production of unmodified human adult hemoglobin in *Escherichia coli. Proc. Natl Acad. Sci. USA* **90**, 8108–8112.

Sherman, F., Stewart, J. & Tsunasawa, S. (1985) Methionine or not methionine at the beginning of a protein. *BioEssays* **3**, 27–31.

Weng, M. & Zalkin, H. (1987) Structural role for a conserved region in the CTP synthetase glutamine amide transfer domain. *J. Bacteriol.* **169**, 3023–3028.

Yasueda, H., Kikuchi, Y., Kojima, H. & Nagase, K. (1991) *In vivo* processing of the initiator methionine from recombinant methionyl human interleukin-6 synthesized in *Escherichia coli*

overproducing aminopeptidase-P. *Appl. Microbiol. Biotechnol.* **36**, 211–215.

Zuo, S., Guo, Q., Ling, C. & Chang, Y.H. (1995) Evidence that two zinc fingers in the methionine aminopeptidase from *Saccharomyces cerevisiae* are important for normal growth. *Mol. Gen. Genet.* **246**, 247–253.

**Kenneth W. Walker**
*Department of Biological Chemistry,*
*College of Medicine,*
*University of California, Irvine,*
*CA 92687, USA*
*Email: rablab@uci.edu*

**Stuart M. Arfin**
*Department of Biological Chemistry,*
*College of Medicine,*
*University of California, Irvine,*
*CA 92687, USA*

**Ralph A. Bradshaw**
*Department of Biological Chemistry,*
*College of Medicine,*
*University of California, Irvine,*
*CA 92687, USA*

# 477. *Methionyl aminopeptidase type II*

## Databanks

*Peptidase classification: clan MG, family M24, MEROPS ID: M24.002*
*NC-IUBMB enzyme classification: EC 3.4.11.18*
*Databank codes:*

| Species | SwissProt | PIR | EMBL (cDNA) | EMBL (genomic) |
|---|---|---|---|---|
| *Arabidopsis thaliana* | – | – | Z34850 | – |
| *Caenorhabditis elegans* | P50581 | – | – | Z68134: cosmid T27A8 |
| *Caenorhabditis elegans* | – | – | X54129 | – |
| *Homo sapiens* | P50579 | – | U13261 U29607 | – |
| *Methanococcus jannaschii* | – | – | – | U67573: complete genome section 115 of 150 |
| *Methanothermus fervidus* | P22624 | PS0039 | M26978 | – |
| *Pyrococcus furiosus* | P56218 | – | – | |
| *Rattus norvegicus* | P38062 | – | L10652 | |
| *Saccharomyces cerevisiae* | P38174 | S45411 | U17437 Z35852 | X79489: chromosome II segment |
| *Sulfolobus solfataricus* | P95963 | – | Y08257 | – |

Brookhaven Protein Data Bank three-dimensional structures

| Species | ID | Resolution | Notes |
|---|---|---|---|
| *Pyrococcus furiosus* | 1XGM | 2.8 | dimer |
| | 1XGN | 2.9 | dimer |
| | 1XGO | 3.5 | monomer |
| | 1XGS | 1.75 | dimer |

## Name and History

The enzymes best termed **methionyl aminopeptidases** (**MetAP**), but also commonly called **methionine aminopeptidases** that are active in processing the N-termini of proteins, particularly cotranslationally, are cobalt-containing enzymes. They are divided into two types, designated type I and II. The **methionine aminopeptidase type II** enzymes were first identified in a higher eukaryote (pig; Kendall & Bradshaw, 1992), but are now known to occur in yeast and, in modified form, in archaebacteria (Li & Chang, 1995; Keeling & Doolittle, 1996; Bult *et al.*, 1996). As with type I MetAPs (see Chapter 476), methionine is removed from peptide and protein substrates in which the second residue is one of the seven smallest amino acids and is not cleaved when this position is occupied by the 13 largest amino acids. This spectrum of activity is diagnostic for cotranslational processing, which has been determined by direct measurements (Huang *et al*., 1987; Boissel *et al*., 1988) and by comparison to known sequences in the database (Flinta *et al*., 1986).

### Activity and Specificity

At present, the MetAP classes are not distinguishable in terms of specificity. The pig liver enzyme, the only member of this subfamily to be isolated and characterized (Kendall & Bradshaw, 1992), is active in removing methionine from peptides and proteins where the adjacent (P1′) residue is Gly, Ala, Ser, Thr, Pro, Cys and Val. It shows a very low activity with penultimate Asp substrates but is inactive against substrates with the remaining residues as determined using a peptide series of the structure Met-Xaa-Ser-(Gly)$_5$-(Leu)$_3$. It is also inactive with the substrate Met-Val-Pro-Thr-Leu-Pro-Glu-Glu but is fully active with Met-Val-His-Thr-Leu-Pro-Glu-Glu, indicating that proline is inhibitory in the P2′ position. This is in keeping with the abnormal processing of a human hemoglobin variant with this substitution (Blouquit *et al*., 1984; Barwick *et al*., 1985).

The addition of $Co^{2+}$ (0.5 mM) increases the activity approximately 2-fold, while $Mn^{2+}$, $Zn^{2+}$, 2-ME and EDTA (5.0 mM) are inhibitory. $Mg^{2+}$ (5.0 mM) is without effect. The enzyme shows a broad pH optimum from 6.0 to 8.0 (Kendall & Bradshaw, 1992).

### Structural Chemistry

The pig enzyme is monomeric with an apparent molecular mass of 67 000 (as judged by gel filtration and SDS-PAGE). Partial sequence analysis of peptides derived from Lys-C

```
1   - - A L F Q R A G S M A A V E T R V C E T D - - - G C S S E A - - - - - - - - -   H. sapiens I
1   M A G V E E V A A S G S H L N G D L D P D D R E E G A A S T A E E A A K K K R R   H. sapiens II

27  - K L Q C P T C I K L G I Q G S Y F C S Q E C F K G S W A T H K L L H K K A K D   H. sapiens I
41  K K K K S K G P S A A G E Q E P - - - D K E S G A S V D E V A R Q L E R S A L E   H. sapiens II

66  E K A K R E V S S W T V E G D I N T D P W A G Y R Y T G K L R P H Y P L M P T R   H. sapiens I
78  D K E R D E D D E - - - D G D G D G D G A T G K K K K K K K K K R G P K V Q T D   H. sapiens II

106 P - - V P S Y - - - - - - - I Q R P D Y A D H P L G M S E S E Q A L K G T S Q I   H. sapiens I
115 P P S V P I C D L Y P N G V F P K G Q E C E Y P P T Q D G R T A A W R T T S E E   H. sapiens II

137 K L L - - - S S E D I - E G M R L V C R L A R E V L D V A A G M I K P G V T T E   H. sapiens I
155 K K A L D Q A S E E I W N D F R E A A E A H R Q V R K Y V M S W I K P G M T M I   H. sapiens II

173 E I D H A V H L A C I A R N C Y P S P L N Y - Y N F P K S C C T S V N E V I C H   H. sapiens I
195 E I C E K L E - D C S R K L I K E N G L N A G L A F P T G C - - S L N N C A A H   H. sapiens II

212 G I P - - - D R R P L Q E G D I V N V D I T L Y R N G Y H G D - - L N E T F - -   H. sapiens I
232 Y T P N A G D T T V L Q Y D D I C K I D F G T H I S G R I I D C A F T V T F N P   H. sapiens II

245 - - - - F V G E V D D G A R K L V Q T T Y E C L M Q A I D A V K P G V R Y R E L   H. sapiens I
272 K Y D T L L K A V K D A T - - - - N T G I K C A G I D V R L C D V G E A I Q E V   H. sapiens II

281 G N I I Q K H A Q A N G F S V - - V R S Y C G H G I H K L - F H T A P N V P H Y   H. sapiens I
308 M E S Y E V E I D G K T Y Q V K P I R N L N G H S I G Q Y R I H A G K T V P - I   H. sapiens II

318 A K N K A V G V M K S G H V F T I E P - - - - - - - - - - - - - M I C - - - - - -   H. sapiens I
347 V K G G E A T R M E E G E V Y A I E T F G S T G K G V V H D D M E C S H Y M K N   H. sapiens II

340 - E G G - - - - - - - - - - - - - - - - - - - - - - - - - - - W Q D E - - - -   H. sapiens I
387 F D V G H V P I R L P R T K H L L N V I N E N F G T L A F C R R W L D R L G E S   H. sapiens II

347 - - - - - - - - - - - - - T W P D G W T A V - T R D G K R S A Q F E H T L L V T D   H. sapiens I
427 K Y L M A L K N L C D L G I V D P Y P P L C D I K G S Y T A Q F E H T I L L R P   H. sapiens II

374 T G C E I L T R R L D S A R P H F M S Q F                                         H. sapiens I
467 T - C K E V V S R G D - - - - - - - - - D Y                                       H. sapiens II
```

*Figure 477.1* Sequence alignment of *H. sapiens* type I and *H. sapiens* type II MetAPs. Dots identify putative zinc ion-binding residues in the N-terminal domain, while the asterisks identify the cobalt liganding residues in the catalytic domain.

peptidase cleavage allowed the preparation of suitable oligo-nucleotides and the subsequent isolation of a partial clone from a human brain cDNA library (Arfin *et al.*, 1995). The sequence of this clone, as well as the peptide sequence data, revealed that the type II MetAP is very similar to the sequence of a previously determined rat cDNA for a protein associated with eukaryotic initiation factor 2 (eIF2 AP, also designated p67) (Wu *et al.*, 1993). p67 is reported to block the phosphorylation of eIF2, a key subunit in the translation initiation complex (Chakraborty *et al.*, 1994). Bazan *et al.* (1994) suggested independently that this protein, along with several other enzymes (based on partial or full sequences), formed a related family and predicted that it would have exopeptidase activity (by comparison to the MetAP type I sequences).

The full-length human cDNA of MetAP II (2569 nucleotides) predicts a protein sequence of 478 amino acids with a calculated molecular mass of 52 832 Da (Arfin *et al.*, 1995). This value is substantially less than the observed molecular mass (67 000), which presumably arises either from postribosomal modification or from aberrant behavior of the protein on the electrophoretic analyses. Interestingly, Datta *et al.* (1989) have suggested that the p67 protein contains monomeric glycosylation sites, but a considerable number of such modifications would be required to explain the molecular mass discrepancy. Alternatively, proteins which contain strings of basic and acidic residues, as are found in the

N-terminal extension of human MetAP (type II) (see below) have been previously reported to show migratory deviations on SDS-PAGE (Fuller *et al.*, 1989; Benton *et al.*, 1994).

The comparison of the amino acid sequences of human MetAP I and II is shown in Figure 477.1. While both contain N-terminal extensions (relative to the prokaryotic type I MetAP), there is no significant amino acid sequence identity in this region. However, the putative catalytic domains of both proteins do show sequence relatedness, as indicated by the boxed residues, that establishes the homologous relationship of the two family branches. The most notable difference in this domain is the fairly lengthy insert found in the type II enzymes. However, as indicated by the asterisks in Figure 477.1, the five cobalt liganding residues (as judged by the *E. coli* MetAP I structure; Roderick & Matthews, 1993) are absolutely conserved, which is a characteristic of all of the type I and II MetAPs (Arfin *et al.*, 1995; Keeling & Doolittle, 1996). Furthermore, model-building studies indicate that the MetAP II catalytic domain can be incorporated into the three-dimensional structure of the type I enzyme (from *E. coli*) with the inserted segment probably forming a surface loop structure (Arfin *et al.*, 1995).

Although the type I N-terminal segment contains a putative zinc-finger region, the type II N-terminal extension does not contain any recognizable domain or motif. However, it is characterized by clusters of acidic and basic residues, as noted above, the function of which is presently unknown. It is likely

**Table 477.1** Amino acid sequence identity comparisons for the catalytic domain of type II methionine aminopeptidases. The entire *E. coli* enzyme was used to define the catalytic domain of all other proteins. Below and to the left of the identity diagonal is a ratio of the number of identical residues to the total number of residue positions compared. Above and to the right of the identity diagonal is the percent identity of the pairwise comparisons

|  | *H. sapiens* | *R. norvegicus* | *S. cerevisiae* | *M. jannaschii* | *M. fervidus* |
|---|---|---|---|---|---|
| *H. sapiens* | – | 95.0 | 61.0 | 31.6 | 25.0 |
| *R. norvegicus* | 306/308 | – | 58.2 | 31.3 | 25.0 |
| *S. cerevisiae* | 208/337 | 198/322 | – | 32.0 | 25.0 |
| *M. jannaschii* | 106/331 | 104/323 | 105/345 | – | 43.6 |
| *M. fervidus* | 53/206 | 53/206 | 51/217 | 98/198 | – |

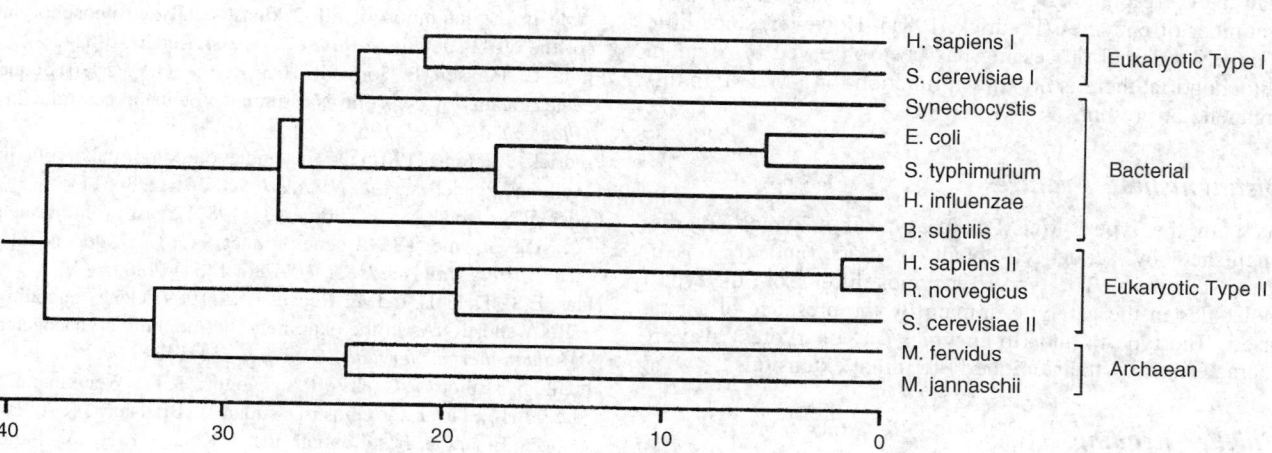

*Figure 477.2*  Phylogenetic tree of likely methionine aminopeptidases. Assignments based on a PAM 250 substitution matrix.

that this portion of MetAP II also functions to form a complex with other entities involved in protein translation, possibly another portion of the ribosome structure, as suggested for eukaryotic MetAP I.

Although the eubacteria apparently contain only type I MetAP (at our present level of knowledge), archaea contain only a type II enzyme. This separation was first observed with a partial sequence fragment from *Methanothermus fervidus* (Haas *et al.*, 1989) and has been confirmed by full-length sequences from *Sulfolobus sulfataricus* (Keeling & Doolittle, 1996) and *Methanococcus jannaschii* (Bult *et al.*, 1996). A cladogram shown in Figure 477.2 depicts the possible evolutionary relationships of the type I and type II MetAPs (as discussed in Chapter 475). It is interesting that the bacterial type I enzymes and the archaean type II enzymes both lack the N-terminal extensions that are present in the respective eukaryotic enzymes. The lack of an N-terminal domain, which could interact with ribosomes, in the archaean enzymes may indicate that these enzymes act in a post-translational manner as has been suggested for the bacterial MetAP.

## Biological Aspects

It is interesting to speculate why eukaryotes have retained both type I and II enzymes while the eubacteria and archaea have not. Several possible explanations present themselves. In the first place, the MetAP system may simply be redundant and both enzymes function in a cotranslational fashion in order to insure the efficiency of the process. Another explanation would suggest that one form of the protein may function in association with the ribosome and the second may be a soluble form (despite the presence of the N-terminal extension) that can carry out these modifications postribosomally as opposed to coribosomally. If this is the case, it is unclear which type functions in which fashion at present. A third possibility, as has been suggested by Keeling & Doolittle (1996), is that the enzymes may exist in different locales in eukaryotic cells. The close similarity of the eukaryotic type I enzyme with the prokaryotic form is consistent with the possibility that this enzyme might be found in the mitochondria. Similar distributions of isozymes, in which the mitochondrial form shows a much closer similarity to the prokaryotic form than the corresponding cytosolic one, have been reported for a number of enzymes (Fredrick, 1981). However, since little if any N-terminal processing has been observed to occur in mitochondria, there is no known function for a MetAP in this organelle at present.

## Distinguishing Features

As with the type I MetAPs, the type II enzymes are distinguished by sequence, cobalt-cofactors, and their substrate specificity. The type II enzymes differ from the type I molecules in the catalytic domain by the presence of a long insert. The type Ib and IIb enzymes (in eukaryotes) also are characterized by their unique N-terminal extensions.

## Further Reading

For reviews, see Arfin & Bradshaw (1995) and Bradshaw & Arfin (1996).

## References

Arfin, S.M. & Bradshaw, R.A. (1995) Mechanisms of regulated intracellular protein degradation. In: *The Encyclopedia of Molecular Biology: Fundamentals and Applications*. Weinheim: VCH Verlag, pp. 346–354.

Arfin, S.M., Kendall, R.L., Hall, L., Weaver, L.H., Steward, A.E., Matthews, B.W. & Bradshaw, R.A. (1995) Eukaryotic methionyl aminopeptidases: two classes of cobalt dependent enzymes. *Proc. Natl Acad. Sci. USA* **92**, 7714–7718.

Barwick, R.C., Jones, R.T., Head, C.G., Shih, M.F.C., Prchal, J.T. & Shih, D.T.B. (1985) Hb long island: a hemoglobin variant with a methionyl extension at the NH2 terminus and a prolyl substitution for the normal histidyl residue 2 of the chain. *Proc. Natl Acad. Sci. USA* **82**, 4602–4605.

Bazan, J.F., Weaver, L.H., Roderick, S.L., Huber, R. & Matthews, B.W. (1994) Sequence and structure comparison suggest that methionine aminopeptidase, prolidase, aminopeptidase P, and creatinase share a common fold. *Proc. Natl Acad. Sci. USA* **91**, 2473–2477.

Benton, B.M., Zang, J.H. & Thorner, J. (1994) A novel FK506- and rapamycin-binding protein (FPR3 gene product) in the yeast *Saccharomyces cerevisiae* is a proline rotamase localized to the nucleolus. *J. Cell Biol.* **127**, 623–639.

Blouquit, Y., Arous, N., Lena, D., Delanoe-Garin, L.J., Lacombe, C., Bardakdjian, J., Vovan, L., Orsini, A. & Galacteros, F. (1984) Hb Marseille [$\alpha 2 \beta 2$ *N* methionyl-2 (NA2)His→Pro]: a new $\beta$ chain variant having an extended N-terminus. *FEBS Lett.* **178**, 315–318.

Boissel, J.P., Kasper, T.J. & Bunn, H.F. (1988) Cotranslational amino-terminal processing of cytosolic proteins. *J. Biol. Chem.* **263**, 8443–8449.

Bradshaw, R.A. & Arfin, S.M. (1996) Methionine aminopeptidase: structure and function. In: *The Aminopeptidases*, vol. 6 (Taylor, A., ed.). Georgetown: R.G. Landes Co., pp. 91–106.

Bult, C.J., White, O., Olsen, G.J. *et al.* (1996) Complete genome sequence of the methanogenic archaeon, *Methanococcus jannaschii. Science* **273**, 1058–1073.

Chakraborty, A., Saha, D., Bose, A., Hileman, R.E., Chatterjee, M. & Gupta, N.K. (1994) Mechanism of action of an eukaryotic initiation factor-2 (eIF-2) associated 67 kDa glycoprotein (p67) and eIF-2 kinase (dsI). *Indian J. Biochem. Biophys.* **31**, 236–242.

Datta, B., Ray, M.K., Chakrabarti, D., Wylie, D.E. & Gupta, N.K. (1989) Glycosylation of eukaryotic peptide chain initiation factor 2 (eIF-2)-associated 67-kDa polypeptide (p67) and its possible role in the inhibition of eIF-2 kinase-catalyzed phosphorylation of the eIF-2-subunit. *J. Biol. Chem.* **264**, 20620–20624.

Flinta, C., Persson, B., Jornvall, H. & Heijne, G.V. (1986) Sequence determinants of cytosolic N-terminal protein processing. *Eur. J. Biochem.* **154**, 193–196.

Fredrick, J.F. (ed.) (1981) Origins and evolution of eukaryotic intracellular organelles. *Ann. NY Acad. Sci.* **361**, 273–511.

Fuller, R.S., Brake, A. & Thorner, J. (1989) Yeast prohormone processing enzyme (KEX2 gene product) is a $Ca^{2+}$-dependent protease. *Proc. Natl Acad. Sci. USA* **86**, 1434–1438.

Haas, E.S., Daniels, C.J. & Reeve, J.N. (1989) Genes encoding 5S rRNA and tRNAs in the extremely thermophilic archaebacterium *Methanothermus fervidus. Gene* **77**, 253–263.

Huang, S., Elliott, R.C., Liu, P.S., Koduri, R.K., Weickmann, J.L., Lee, J.H., Blair, L.C., Ghosh-Dastidar, P., Bradshaw, R.A., Bryan, K.M., Einarson, B., Kendall, R.L., Kolacz, K.H. & Saito, K. (1987) Specificity of cotranslational amino-terminal processing of proteins in yeast. *Biochemistry* **26**, 8242–8246.

Keeling, P.J. & Doolittle, W.F. (1996) Methionine aminopeptidase-1: the MAP of the mitochondrion? *Trends Biochem. Sci.* **21**, 285–286.

Kendall, R.L. & Bradshaw, R.A. (1992) Isolation and characterization of the methionine aminopeptidase from porcine liver responsible for the co-translational processing of proteins. *J. Biol. Chem.* **267**, 20667–20673.

Li, X. & Chang, Y.H. (1995) Amino-terminal protein processing in *Saccharomyces cerevisiae* is an essential function that requires two distinct methionine aminopeptidases. *Proc. Natl Acad. Sci.*

*USA* **92**, 12357–12361.

Roderick, S.L. & Matthews, B.W. (1993) Structure of the cobalt-dependent methionine aminopeptidase from *Escherichia coli*: a new type of proteolytic enzyme. *Biochemistry* **32**, 3907–3912.

Wu, S., Gupta, S., Chatterjee, N., Hileman, R.E., Kinzy, T.G., Denslow, N.D., Merrick, W.C., Chakrabarti, D., Osterman, J.C. & Gupta, N.K. (1993) Cloning and characterization of complementary DNA encoding the eukaryotic initiation factor 2-associated 67-kDa protein (p67). *J. Biol. Chem.* **268**, 10796–10801.

**Ralph A. Bradshaw**
*Department of Biological Chemistry,
College of Medicine,
University of California, Irvine,
CA 92697, USA
Email: rablab@uci.edu*

**Stuart M. Arfin**
*Department of Biological Chemistry,
College of Medicine,
University of California, Irvine,
CA 92697, USA*

**Kenneth W. Walker**
*Department of Biological Chemistry,
College of Medicine,
University of California, Irvine,
CA 92697, USA*

# 478. X-Pro dipeptidase

## Databanks

*Peptidase classification: clan MG, family M24, MEROPS ID: M24.003
NC-IUBMB enzyme classification: EC 3.4.13.9
Chemical Abstracts Service registry number: 9025-32-5
Databank codes:*

| Species | SwissProt | PIR | EMBL (cDNA) | EMBL (genomic) |
|---|---|---|---|---|
| *Caenorhabditis elegans* | – | – | M88837 | – |
| *Escherichia coli* | P21165 | JQ0753 S30738 | X52837 X54687 | M87049: chromosome 84.5–86.5′ |
| *Homo sapiens* | P12955 | – | J04605 | M35496: intron M36549: exon 1 |
| *Lactobacillus delbrueckii*[a] | P46545 | A32454 | Z34896 Z54205 | – |
| *Mus musculus* | Q11136 | – | U51014 | |
| *Rattus norvegicus* | – | – | H32298 | |

[a]The *Lactobacillus* sequence is also included in Chapter 481.

## Name and History

A dipeptidase showing specificity for C-terminal proline was originally obtained from mammalian intestine and designated as **prolidase** (Bergmann & Fruton, 1937). Several years later, it was isolated, shown to be a metalloenzyme, and characterized for substrate preference (Davis & Smith, 1957; Sjöström, 1974). The enzyme has been found in a variety of tissues, as well as in bacteria, and is variously known as **Xaa-Pro dipeptidase, X-Pro dipeptidase (XPD), proline dipeptidase, imidodipeptidase** or **peptidase D**.

## Activity and Specificity

Recently, more extensive specificity studies have been carried out with pig kidney XPD, employing a continuous spectrophotometric assay (Mock *et al.*, 1990; Mock & Green, 1990; Mock & Liu, 1995a). The prototypical substrate is Gly┼Pro ($K_m$ 0.7 mM at pH 7.6). A specificity requirement for L-Pro in the P1′ position prevails: a thiazolidinecarboxylic acid residue is accommodated, but C-terminal hydroxyproline and sarcosine are cleaved only ~0.02 as fast, and primary amino acid residues there yield similarly insusceptible substrates. Most other amino acids of L-configuration may substitute for Gly in the P1 position, a picolinic acid residue is accepted there, and RSCH$_2$CO+Pro or BrCH$_2$CO+Pro are cleaved as well, but Ac-Pro is inactive except as an inhibitor. XPD is also inhibited by 1,2-cyclopentanedicarboxylic acid ($K_i < 0.1\,\mu$M at pH 6). The normal catalytic activity range with dipeptides is pH 6–9, with the upper limit depending on the basicity of the amino group within substrates.

## Structural Chemistry

XPD from a number of sources has been shown to exist in solution as a dimer of 55 kDa subunits, each of which consists of an identical single chain (Manao *et al.*, 1972; Sjöström & Norén, 1974; Endo *et al.*, 1982; Yoshimoto *et al.*, 1983; Browne & O'Cuinn, 1983; Richter *et al.*, 1989). An essential Arg has been identified at the active site by chemical modification (Mock & Zhuang, 1991). The human enzyme has been sequenced (Endo *et al.*, 1989; Tanoue *et al.*, 1990) and primary structures are also known for the mouse and bacterial counterparts. Although a crystallographic structure for prolidase is unavailable, sequence homology has been demonstrated with *Escherichia coli* methionyl aminopeptidase (Chapter 476), a metalloenzyme of established three-dimensional structure, as well as with aminopeptidase P (Chapter 479), which also preferentially hydrolyzes an acyl-proline linkage (Roderick & Matthews, 1993; Bazan *et al.*, 1994; Denslow *et al.*, 1994).

In common with a number of other aminopeptidases, the structurally characterized enzyme from *E. coli* contains a *pair* of closely linked metal ions at the active site. Because the metal ion-binding region of the latter species appears to be conserved in prolidase, and because of similarity in catalytic specificities, the enzyme is likely to be a member of the dual-metal aminopeptidases, of which leucyl aminopeptidase (Chapter 473) is the unrelated prototype. The enzymic active-site metal ligands in prolidase would then be Asp (2), Glu (2) and His, located within a trough region composed of $\beta$ sheet. An assumption found in the literature is that Mn$^{2+}$ is involved in the mechanism, based on a catalytically stimulatory response to incubation of the enzyme with that metal ion at nonphysiological concentrations. However, the latter phenomenon depends upon the mode of protein isolation, and is absent in minimally processed enzyme, so there is no reason to think that Zn$^{2+}$ is not ordinarily the prevalent ion, although it is probably Co$^{2+}$ in the methionyl aminopeptidases.

## Preparation

A convenient source of XPD is pig kidney (Manao *et al.*, 1972). The steps involve (a) formation of an acetone powder from the filtrate of homogenized kidneys, (b) ammonium sulfate and acetone fractionation followed by lyophilization, and (c) final chromatographic purification. In the first step ethanolic PMSF (150 mg liter$^{-1}$) should be incorporated in the homogenization mixture as a serine protease inhibitor (Mock & Liu, 1995a). Without this precaution the final 55 kDa enzyme is found by SDS-PAGE to be contaminated with proteins of $M_r$ of approximately 20 000 and 35 000. A commercially obtained specimen of prolidase was found to consist mostly of the latter mixture. The apparent chain-nicking

has minimal kinetic consequences, except in competitive inhibition studies, as well as at the low pH end of the catalytic efficacy range, where the evidently fragmented enzyme loses activity that the protected protein does not. Also, the intact enzyme was found not to benefit kinetically or in terms of stability from the aqueous Mn$^{2+}$ incubation procedure that has been stipulated for XPD. Step (c) in the preparation, chromatographic purification, may be carried out by HPLC with a polymeric hydroxyethyl methacrylate, macroporous, amine-type ion-exchange column, yielding high-activity XPD as a single band with an $M_r$ of 55 000 according to SDS-PAGE, and with an $M_r$ of approximately 110 000 (dimer) when the electrophoresis is carried out without protein denaturation (Mock & Liu, 1995a).

## Biological Aspects

Apparently the enzyme is required in humans, because many other peptidases are unable to cleave the Xaa+Pro linkage. As a result, these dipeptides (commonly produced in collagen recycling) build to toxic concentrations in individuals with a genetic deficiency of XPD (Myara *et al.*, 1994).

## Distinguishing Features

XPD has been shown to cleave only the *trans* rotamer of the acyl-proline peptide linkage within its substrates (Lin & Brandts, 1979; King *et al.*, 1986). The active site evidently contains catalytic metal ions that are especially Lewis-acidic (sufficient to induce a p$K_a$ of 6.6 for a water molecule that is ligated in the absence of substrate). Activation of the dipeptide scissile linkage by chelative binding of the substrate N-terminus to one of the metal ions is indicated from specificity studies. Depending on substrate structure, either that step or subsequent hydrolytic cleavage of the amide linkage may become rate-limiting (Mock & Liu, 1995a). Cooperativity between protein monomers within the enzyme dimer is manifested kinetically, and appears to be of physiological significance with regard to a biphasic competitive inhibition pattern noted with XPD for the ubiquitous intermediary metabolite phosphoenolpyruvate (Mock & Green, 1990; Mock & Liu, 1995b). A chemical catalytic amide-cleavage mechanism for XPD has been advanced, which may be general for the dual-metal aminopeptidases; it involves bridging of a transient carbonyl-hydration adduct between the two active-site metal ions (Mock & Liu, 1995a). The potency of a ketonic substrate-analog inhibitor is in accord with that mechanism (Kumaravel *et al.*, 1995).

## Further Reading

For a review, see Der Kaloustian *et al.* (1982).

## References

Bazan, J.F., Weaver, L.H., Roderick, S.L., Huber, R. & Matthews, B.W. (1994) Sequence and structure comparison suggest that methionine aminopeptidase, prolidase, aminopeptidase-P, and creatinase share a common fold. *Proc. Natl Acad. Sci. USA* **91**, 2473–2477.

Bergmann, M. & Fruton, J.S. (1937) Proteolytic enzymes. XII. Specificity of aminopeptidases and carboxypeptidase. A new type of enzyme in the intestinal tract. *J. Biol. Chem.* **117**, 189–202.

Browne, P. & O'Cuinn, G. (1983) The purification and characterization of a proline dipeptidase from guinea pig brain. *J. Biol. Chem.* **258**, 6147–6154.

Davis, N.C. & Smith, E.L. (1957) Purification and properties of prolidase of the kidney. *J. Biol. Chem.* **224**, 261–275.

Denslow, N.D., Ryan, J.W. & Nguyen, H.P. (1994) Guinea pig membrane-bound aminopeptidase P is a member of the proline peptidase family. *Biochem. Biophys. Res. Commun.* **205**, 1790–1795.

Der Kaloustian, V.M., Freij, B.J. & Kurban, A.K. (1982) Prolidase deficiency – an inborn error of metabolism with major dermatological manifestations. *Dermatologica* **164**, 293–304.

Endo, F., Matsuda, I., Ogata, A. & Tanaka, S. (1982) Human erythrocyte prolidase (EC 3.4.13.9) and prolidase deficiency. *Pediat. Res.* **16**, 227–231.

Endo, F., Tanoue, A., Nakai, H., Hata, A., Indo, Y., Titani, K. & Matsuda, I. (1989) Primary structure and gene localization of human prolidase. *J. Biol. Chem.* **264**, 4476–4481.

King, G.F., Middlehurst, C.R. & Kuchel, P.W. (1986) Direct NMR evidence that prolidase is specific for the trans isomer of iminopeptide substrates. *Biochemistry* **25**, 1054–1062.

Kumaravel, G., Boettcher, B.R., Shapiro, M.J. & Petter, R.C. (1995) Peptide mimics of glycylproline as inhibitors of prolidase. *Bioorg. Med. Chem. Lett.* **5**, 2825–2828.

Lin, L.-N. & Brandts, J.F. (1979) Evidence suggesting that some proteolytic enzymes may cleave only the trans form of the peptide bond. *Biochemistry* **18**, 43–47.

Manao, G., Nassi, P., Capuggi, G., Camici, G. & Ramponi, G. (1972) Swine kidney prolidase. Assay, isolation procedure, and molecular properties. *Physiol. Chem. Physics* **4**, 75–87.

Mock, W.L. & Green, P.C. (1990) Mechanism and inhibition of prolidase. *J. Biol. Chem.* **265**, 19606–19610.

Mock, W.L. & Liu, Y. (1995a) Hydrolysis of picolinylprolines by prolidase. *J. Biol. Chem.* **270**, 18437–18446.

Mock, W.L. & Liu, Y. (1995b) Inhibition of prolidase is biphasic. Avoidance of endogenous-metabolite inactivation by cooperativity within an enzyme dimer. *Bioorg. Med. Chem. Lett.* **5**, 627–630.

Mock, W.L. & Zhuang, H. (1991) Chemical modification locates guanidinyl and carboxylate groups within the active site of prolidase. *Biochem. Biophys. Res. Commun.* **180**, 401–406.

Mock, W.L., Green, P.C. & Boyer, K.D. (1990) Specificity and pH dependence for acylproline cleavage by prolidase. *J. Biol. Chem.* **265**, 19600–19605.

Myara, I., Cosson, C., Moatti, N. & Lemonnier, A. (1994) Prolidase and prolidase deficiency. *Int. J. Biochem.* **26**, 207–214.

Richter, A.M., Lancaster, G.L., Choy, F.Y.M. & Hechtman, P. (1989) Purification and characterization of activated erythrocyte prolidase. *Biochem. Cell. Biol.* **67**, 34–41.

Roderick, S.L. & Matthews, B.W. (1993) Structure of the cobalt-dependent methionine aminopeptidase from *Escherichia coli*: a new type of proteolytic enzyme. *Biochemistry* **32**, 3907–3912.

Sjöström, H. (1974) Enzymic properties of pig intestinal proline dipeptidase. *Acta Chem. Scand., Ser. B* **28**, 802–808.

Sjöström, H. & Norén, O. (1974) Structural properties of pig intestinal proline dipeptidase. *Biochem. Biophys. Acta* **359**, 177–185.

Tanoue, A., Endo, F. & Matsuda, I. (1990) Structural organization of the gene for human prolidase (peptidase D) and demonstration of a partial gene deletion in a patient with prolidase deficiency. *J. Biol. Chem.* **265**, 11306–11311.

Yoshimoto, T., Matsubara, F., Kawano, E. & Tsuru, D. (1983) Prolidase from bovine intestine: purification and characterization. *J. Biochem.* **94**, 1889–1896.

***William L. Mock***
*Department of Chemistry,*
*University of Illinois at Chicago,*
*Chicago, IL 60607-7061, USA*
*Email: wlmock@uic.edu*

# 479. X-Pro aminopeptidase (prokaryote)

## Databanks

*Peptidase classification: clan MG, family M24, MEROPS ID: M24.004*
*NC-IUBMB enzyme classification: EC 3.4.11.9*
*ATCC entries: 623559 (M. genitalium)*
*Chemical Abstracts Service registry number: 37288-66-7*

*Databank codes:*

| Species | Form | SwissProt | PIR | EMBL (cDNA) | EMBL (genomic) |
|---|---|---|---|---|---|
| *Escherichia coli* | – | P15034 | B47020 JQ0843 JX0067 | D00398 D90281 | U28377: genome 65 to 68′ |
| *Haemophilus influenzae* | – | P44881 | B64096 | – | U32764: genome section 79 of 163 |
| *Lactococcus lactis* | – | – | – | Y08842 | – |
| *Mycoplasma genitalium* | – | P47566 | – | U39715 | – |
| *Streptomyces lividans* | P1 | Q05813 | JN0491 PN0466 | M91546 | – |
| *Streptomyces lividans* | P2 | – | – | L23174 | – |
| *Synechocystis* sp. | – | – | – | – | D90915: genome section 17 of 27 |

## Name and History

Enzyme activity which specifically hydrolyzes N-terminal Xaa+Pro- bonds was first purified from the soluble fraction of *Escherichia coli* B on the basis of its ability to release proline from poly-L-proline (Yaron & Berger, 1970; Yaron & Mlynar, 1968). This activity, originally named *aminopeptidase P*, has been found not only in *E. coli* (two forms; Yoshimoto *et al.*, 1988b), but also in *Salmonella typhimurium* (McHugh & Miller, 1974), *Neisseria gonorrhoeae* (Chen & Buchanan, 1980), and *Streptomyces lividans* (two forms; Butler *et al.*, 1993, 1994). The activity has also been referred to as *amino-acylproline aminopeptidase, peptidase P (PepP)*, and the currently accepted designation, *X-Pro aminopeptidase*.

## Activity and Specificity

Prokaryotic X-Pro aminopeptidase removes the N-terminal residue (P1) from peptides which have a proline residue in the penultimate position (P1′) (Chen & Buchanan, 1980; Yaron & Berger, 1970). The enzyme can accommodate a variety of amino acids including proline in both the P1 and P2′ positions (Yoshimoto *et al.*, 1988b, 1994). It is stereospecific and requires an L-amino acid in P1 and L-proline in P1′. X+Pro dipeptides are cleaved more slowly than larger peptides. Only the *trans* form of the X-Pro bond is hydrolyzed (Lin & Brandts, 1979).

Bacterial X-Pro aminopeptidase has been assayed at pH 7.5–8.2 using poly-L-proline, Phe($p$-NO$_2$)+Pro-Pro-NH-CH$_2$-CH$_2$-NH-Abz, Gly+Pro-NHNap (or -NHPhNO$_2$ or -NHMec), each coupled with prolyl aminopeptidase (Chapter 139), as well as Arg+Pro-Pro or Phe+Pro (reviewed by Yaron & Naider, 1993). Activity is inhibited by EDTA; the addition of Mn$^{2+}$ or Co$^{2+}$ is required for optimal activity. Pro-Phe, Pro-Tyr and apstatin (available from Sigma; see Appendix 2 for full names and addresses of suppliers) inhibit the enzyme in the low micromolar range (Yoshimoto *et al.*, 1994; Prechel *et al.*, 1995).

## Structural Chemistry

The nucleotide and deduced amino acid sequences have been determined for the X-Pro aminopeptidase genes from *Escherichia coli* (type II; Yoshimoto *et al.*, 1989) and *Streptomyces lividans* (two forms; Butler *et al.*, 1993, 1994). The sequences for hypothetical X-Pro aminopeptidase genes have been reported for *Haemophilus influenzae* (Fleischmann *et al.*, 1995), *Mycoplasma genitalium* (Fraser *et al.*, 1995) and *Synechocystis* sp. (Kaneko *et al.*, 1996). The enzyme subunit has an $M_r$ of about 50 000 and aggregates to form a dimer (*Streptomyces lividans* PepP1) or either a dimer or tetramer depending on enzyme concentration (*E. coli* type II). Conserved metal-binding ligands characteristic of the methionyl aminopeptidase family (M24) are present. Purified recombinant *E. coli* enzyme has 0.2 mol zinc per subunit and is activated by Zn$^{2+}$ or Mn$^{2+}$.

## Preparation

*E. coli* X-Pro aminopeptidase (type II) has been purified from strain B (Yaron & Berger, 1970). The gene from *E. coli* HB101 has been cloned and overexpressed in *E. coli* strains DH1 and JM83, followed by purification of the recombinant enzyme (Yoshimoto *et al.*, 1988a, 1989).

## Biological Aspects

X-Pro aminopeptidase is possibly involved in intracellular protein turnover by hydrolyzing X-Pro and particularly X-Pro-Y peptides resistant to most other peptidases. Mutants of *Salmonella typhimurium* lacking this enzyme have a much reduced ability to produce free proline during starvation-induced protein breakdown (McHugh & Miller, 1974; Miller & Green, 1983).

## Distinguishing Features

The structurally related X-Pro dipeptidase (prolidase) (Chapter 478) has also been found in bacteria and is specific for Xaa-Pro dipeptides. In contrast, X-Pro aminopeptidase cleaves Xaa-Pro-Y peptides more readily than Xaa-Pro dipeptides. The structural features giving rise to these different specificities are not yet known. Assignment of nucleotide sequences to one of the two enzyme activities should be considered tentative (Stucky *et al.*, 1995).

Prolyl aminopeptidase (proline iminopeptidase) (Chapter 139), like prokaryote X-Pro aminopeptidase, is capable of cleaving poly-L-proline but is specific for an N-terminal Pro-Xaa bond.

## Further Reading

A review of proline-specific peptidases is found in Yaron & Naider (1993).

## References

Butler, M.J., Bergeron, A., Soostmeyer, G., Zimny, T. & Malek, L.T. (1993) Cloning and characterization of an aminopeptidase P-encoding gene from *Streptomyces lividans. Gene* **123**, 115–119.

Butler, M.J., Aphale, J.S., DiZonno, M.A., Krygsman, P., Walczyk, E. & Malek, L.T. (1994) Intracellular aminopeptidases in *Streptomyces lividans* 66. *J. Indust. Microbiol.* **13**, 24–29.

Chen, K.C.S. & Buchanan, T.M. (1980) Hydrolases from *Neisseria gonorrhoeae*. The study of gonocosin, an aminopeptidase P, a proline iminopeptidase, and an asparaginase. *J. Biol. Chem.* **255**, 1704–1710.

Fleischmann, R.D., Adams, M.D., White, O. *et al.* (1995) Whole-genome random sequencing and assembly of *Haemophilus influenzae* Rd. *Science* **269**, 496–512.

Fraser, C.M., Gocayne, J.D., White, O. *et al.* (1995) The minimal gene complement of *Mycoplasma genitalium. Science* **270**, 397–403.

Kaneko, T., Sato, S., Kotani, H. *et al.* (1996) Sequence analysis of the genome of the unicellular cyanobacterium *Synechocystis sp.* strain PCC803. II. Sequence determination of the entire genome and assignment of potential protein-coding regions. *DNA Res.* **3**, 109–136.

Lin, L.-N. & Brandts, J.F. (1979) Role of *cis-trans* isomerism of the peptide bond in protease specificity. Kinetic studies on small proline-containing peptides and on polyproline. *Biochemistry* **18**, 5037–5042.

McHugh, G.L. & Miller, C.G. (1974) Isolation of proline peptidase mutants of *Salmonella typhimurium. J. Bacteriol.* **120**, 364–374.

Miller, C.G. & Green, L. (1983) Degradation of proline peptides in peptidase-deficient strains of *Salmonella typhimurium. J. Bacteriol.* **153**, 350–356.

Prechel, M.M., Orawski, A.T., Maggiora, L.L. & Simmons, W.H. (1995) Effect of a new aminopeptidase P inhibitor, apstatin, on bradykinin degradation in the rat lung. *J. Pharmacol. Exp. Ther.* **275**, 1136–1142.

Stucky, K., Klein, J.R., Schüller, A., Matern, H., Henrich, B. & Plapp, R. (1995) Cloning and DNA sequence of *pepQ*, a prolidase gene from *Lactobacillus delbrueckii* subsp. *lactis* DSM7290 and partial characterization of its product. *Mol. Gen. Genet.* **247**, 494–500.

Yaron, A. & Berger, A. (1970) Aminopeptidase P. *Methods Enzymol.* **19**, 521–534.

Yaron, A. & Mlynar, D. (1968) Aminopeptidase P. *Biochem. Biophys. Res. Commun.* **32**, 658–663.

Yaron, A. & Naider, F. (1993) Proline-dependent structural and biological properties of peptides and proteins. *Crit. Rev. Biochem. Mol. Biol.* **28**, 31–81.

Yoshimoto, T., Murayama, N., Honda, T., Tone, H. & Tsuru, D. (1988a) Cloning and expression of aminopeptidase P gene from *Escherichia coli* HB101 and characterization of expressed enzyme. *J. Biochem.* **104**, 93–97.

Yoshimoto, T., Murayama, N. & Tsuru, D. (1988b) A novel assay method for aminopeptidase P and partial purification of two types of the enzyme in *Escherichia coli. Agric. Biol. Chem.* **52**, 957–963.

Yoshimoto, T., Tone, H., Honda, T., Osatomi, K., Kobayashi, R. & Tsuru, D. (1989) Sequencing and high expression of aminopeptidase P gene from *Escherichia coli* HB101. *J. Biochem.* **105**, 412–416.

Yoshimoto, T., Orawski, A.T. & Simmons, W.H. (1994) Substrate specificity of aminopeptidase P from *Escherichia coli*: comparison with membrane-bound forms from rat and bovine lung. *Arch. Biochem. Biophys.* **311**, 28–34.

*William H. Simmons*
*Department of Molecular and Cellular Biochemistry,*
*Loyola University Chicago Stritch School of Medicine,*
*2160 S. First Avenue,*
*Maywood, IL 60153, USA*
*Email: WSIMMON@wpo.it.luc.edu*

# 480. X-Pro aminopeptidase (eukaryote)

## Databanks

*Peptidase classification: clan MG, family M24, MEROPS ID: M24.005*
*NC-IUBMB enzyme classification: EC 3.4.11.9*
*Chemical Abstracts Service registry number: 37288-66-7*

*Databank codes:*

| Species | SwissProt | PIR | EMBL (cDNA) | EMBL (genomic) |
|---|---|---|---|---|
| *Cavia porcellus* | – | PC2307 | – | – |
| | – | PC2308 | – | – |
| | – | PC2309 | – | – |
| | – | PC2310 | – | – |
| | – | PC2311 | – | – |
| | – | PC2312 | – | – |
| | – | PC2313 | – | – |
| | – | PC2314 | – | – |
| *Saccharomyces cerevisiae* | – | – | – | Z73134: chromosome XII ORF |
| *Sus scrofa* | – | – | U55039 | – |

## Name and History

Enzyme activity hydrolyzing Gly+Pro-Hyp (in which Hyp is hydroxyproline) was first described in both the membrane and cytosolic fractions of pig kidney homogenates by Dehm & Nordwig (1970). The membrane-bound enzyme was partially purified and called *X-prolyl-aminopeptidase*. The currently accepted name is *X-Pro aminopeptidase*. The activity has also been referred to as *aminopeptidase P, aminoacylproline hydrolase* and *aminoacylprolyl-peptide hydrolase*.

## Activity and Specificity

X-Pro aminopeptidase removes the N-terminal residue (P1) from peptides which have a proline residue in the penultimate position (P1'). The cytosolic and membrane-bound X-Pro aminopeptidases appear to be distinct proteins with different specificities. Both enzymes have a broad specificity for the P1 residue. The cytosolic enzyme also has a broad specificity for the P2' residue and can readily cleave Xaa+Pro dipeptides (Harbeck & Mentlein, 1991; Rusu & Yaron, 1992). In contrast, the membrane-bound enzyme prefers peptides of three amino acids or longer which have a small amino acid side chain in the P2' position (Yoshimoto *et al.*, 1994). The nonapeptide bradykinin (N-terminus: Arg+Pro-Pro-) is a good substrate for both isozymes.

Routine assays for X-Pro aminopeptidase have been carried out at pH 6.8–8.0 utilizing one of the following substrates (reviewed by Yaron & Naider, 1993): Gly+Pro-Hyp, Arg+Pro-Pro, Lys($\varepsilon$-Dnp)+Pro-Pro-NH-CH$_2$-CH$_2$-NH-Abz, Arg+Pro-Pro-[$^3$H]benzylamide, Gly+Pro-Pro-NHNap (or -NHPhNO$_2$; coupled with dipeptidyl-peptidase IV; Chapter 128). Activity of both isozymes is inhibited by 1,10-phenanthroline. Mn$^{2+}$ stimulates activity toward some, but not all, substrates (Orawski & Simmons, 1995; Lloyd *et al.*, 1996). Certain sulfhydryl-type angiotensin-converting enzyme inhibitors can inhibit the membrane-bound enzyme, while carboxyalkyl-types inhibit only in the presence of millimolar concentrations of Mn$^{2+}$. A selective synthetic inhibitor called apstatin (Prechel *et al.*, 1995) is available from Sigma (see Appendix 2 for full names and addresses of suppliers).

## Structural Chemistry

The complete amino acid sequence of membrane-bound X-Pro aminopeptidase from pig kidney has been determined (Vergas Romero *et al.*, 1995), as has the sequence of the full-length cDNA derived from the same source (Hyde *et al.*, 1996). Sequences of fragments of the guinea pig lung and kidney (Denslow *et al.*, 1994) and rat and cattle lung (Orawski & Simmons, 1995) membrane-bound enzymes have been reported. The pig kidney-derived cDNA encodes a protein of 673 amino acids with a calculated $M_r$ of 75 755 which includes an N-terminal signal peptide and a C-terminal recognition sequence for the attachment of a glycosyl-phosphatidylinositol membrane anchor. The mature protein is larger ($M_r$ 90 000) due to glycosylation at six sites in the N-terminal half of the protein. The C-terminal half presumably contains the active site since it has conserved metal ion-binding ligands characteristic of the methionyl aminopeptidase family (M24) (Chapter 475). The enzyme contains 1 mol zinc per mol of protein (Hooper *et al.*, 1992). Oligomers with $M_r$ values of 217 000–360 000 have been reported, depending on the species and experimental conditions. Cytosolic X-Pro aminopeptidase has a $M_r$ of 71 000 and exists as a dimer or trimer.

## Preparation

Membrane-bound X-Pro aminopeptidase has been purified from pig kidney (Hooper *et al.*, 1990; Vergas Romero *et al.*, 1995; Lloyd *et al.*, 1996); cattle lung (Simmons & Orawski, 1992); guinea pig lung, kidney and serum (Ryan *et al.*, 1992, 1994b); and rat lung (Orawski & Simmons, 1995). The pig kidney enzyme has been expressed in COS-1 cells (Hyde *et al.*, 1996). Cytosolic X-Pro aminopeptidase has been purified to homogeneity from rat brain (Harbeck & Mentlein, 1991), human platelets (Vanhoof *et al.*, 1992) and human leukocytes (Rusu & Yaron, 1992).

## Biological Aspects

Membrane-bound X-Pro aminopeptidase is proposed to have a role in the degradation of bradykinin, a potent vasodepressor and cardioprotective peptide. Inhibition of X-Pro aminopeptidase blocks the cleavage of the Arg1+Pro2 bond of bradykinin in the pulmonary and coronary circulations of the rat (Pesquero *et al.*, 1992; Prechel *et al.*, 1995; Ersahin & Simmons, 1996) and potentiates the blood pressure-lowering effects of this peptide (Ryan *et al.*, 1994a; Kitamura *et al.*, 1995). The enzyme is present on the surface of human aortic endothelial cells (Ryan *et al.*, 1996). Membrane-bound X-Pro aminopeptidase may also play a role in neuropeptide Y metabolism in the brain (Medeiros & Turner, 1996).

## Distinguishing Features

Cytosolic X-Pro aminopeptidase can cleave X-Pro dipeptides and therefore can be confused with X-Pro dipeptidase (prolidase) (Chapter 478) which is specific for X-Pro dipeptides. The assay of X-Pro aminopeptidase should therefore utilize substrates such as Arg┼Pro-Pro which are resistant to X-Pro dipeptidase as well as to dipeptidyl-peptidases and carboxypeptidases. In contrast to prokaryote X-Pro aminopeptidase (Chapter 479), the mammalian enzymes cannot cleave poly-L-proline.

## References

Dehm, P. & Nordwig, A. (1970) The cleavage of prolyl peptides by kidney peptidases. Partial purification of a 'X-prolyl-aminopeptidase' from swine kidney microsomes. *Eur. J. Biochem.* **17**, 364–371.

Denslow, N.D., Ryan, J.W. & Nguyen, H.P. (1994) Guinea pig membrane-bound aminopeptidase P is a member of the proline peptidase family. *Biochem. Biophys. Res. Commun.* **205**, 1790–1795.

Ersahin, C. & Simmons, W.H. (1996) Bradykinin metabolism in the rat coronary circulation: role of aminopeptidase P and angiotensin converting enzyme. *FASEB J.* **10**, A1017 (abstr.).

Harbeck, H.-T. & Mentlein, R. (1991) Aminopeptidase P from rat brain: purification and action on bioactive peptides. *Eur. J. Biochem.* **198**, 451–458.

Hooper, N.M., Hryszko, J. & Turner, A.J. (1990) Purification and characterization of pig kidney aminopeptidase P: a glycosyl-phosphatidylinositol-anchored ectoenzyme. *Biochem. J.* **267**, 509–515.

Hooper, N.M., Hryszko, J., Oppong, S.Y. & Turner, A.J. (1992) Inhibition by converting enzyme inhibitors of pig kidney aminopeptidase P. *Hypertension* **19**, 281–285.

Hyde, R.J., Hooper, N.M. & Turner, A.J. (1996) Molecular cloning and expression in COS-1 cells of pig kidney aminopeptidase P. *Biochem. J.* **319**, 197–201.

Kitamura, S., Carbini, L.A., Carretero, O.A., Simmons, W.H. & Scicli, G. (1995) Potentiation by aminopeptidase P [inhibitor] of blood pressure response to bradykinin. *Br. J. Pharmacol.* **114**, 6–7.

Lloyd, G.S., Hryszko, J., Hooper, N.M. & Turner, A.J. (1996) Inhibition and metal ion activation of pig kidney aminopeptidase P: dependence on nature of substrate. *Biochem. Pharmacol.* **52**, 229–236.

Medeiros, M.S. & Turner, A.J. (1996) Metabolism and functions of neuropeptide Y. *Neurochem. Res.* **21**, 1125–1132.

Orawski, A.T. & Simmons, W.H. (1995) Purification and properties of membrane-bound aminopeptidase P from rat lung. *Biochemistry* **34**, 11227–11236.

Pesquero, J.B., Jubilut, G.N., Lindsey, C.J. & Paiva, A.C.M. (1992) Bradykinin metabolism pathway in the rat pulmonary circulation. *J. Hypertens.* **10**, 1471–1478.

Prechel, M.M., Orawski, A.T., Maggiora, L.L. & Simmons, W.H. (1995) Effect of a new aminopeptidase P inhibitor, apstatin, on bradykinin degradation in the rat lung. *J. Pharmacol. Exp. Ther.* **275**, 1136–1142.

Rusu, I. & Yaron, A. (1992) Aminopeptidase P from human leukocytes. *Eur. J. Biochem.* **210**, 93–100.

Ryan, J.W., Valido, F., Berryer, P., Chung, A.Y.K. & Ripka, J.E. (1992) Purification and characterization of guinea pig serum aminoacylproline hydrolase (aminopeptidase P). *Biochim. Biophys. Acta* **1119**, 140–147.

Ryan, J.W., Berryer, P., Chung, A.Y.K. & Sheffy, D.H. (1994a) Characterization of rat pulmonary vascular aminopeptidase P *in vivo*: role in the inactivation of bradykinin. *J. Pharmacol. Exp. Ther.* **269**, 941–947.

Ryan, J.W., Denslow, N.D., Greenwald, J.A. & Rogoff, M.A. (1994b) Immunoaffinity purifications of aminopeptidase P from guinea pig lungs, kidney and serum. *Biochem. Biophys. Res. Commun.* **205**, 1796–1802.

Ryan, J.W., Papapetropoulos, A., Ju, H., Denslow, N.D., Antonov, A., Virmani, R., Kolodgie, F.D., Gerrity, R.G. & Catravas, J.D. (1996) Aminopeptidase P is disposed on human endothelial cells. *Immunopharmacology* **32**, 149–152.

Simmons, W.H. & Orawski, A.T. (1992) Membrane-bound aminopeptidase P from bovine lung: its purification, properties, and degradation of bradykinin. *J. Biol. Chem.* **267**, 4897–4903.

Vanhoof, G., De Meester, I., Goossens, F., Hendriks, D., Scharpé, S. & Yaron, A. (1992) Kininase activity in human platelets: cleavage of the Arg[1]-Pro[2] bond of bradykinin by aminopeptidase P. *Biochem. Pharmacol.* **44**, 479–487.

Vergas Romero, C., Neudorfer, I., Mann, K. & Schäfer, W. (1995) Purification and amino acid sequence of aminopeptidase P from pig kidney. *Eur. J. Biochem.* **229**, 262–269.

Yaron, A. & Naider, F. (1993) Proline-dependent structural and biological properties of peptides and proteins. *Crit. Rev. Biochem. Mol. Biol.* **28**, 31–81.

Yoshimoto, T., Orawski, A.T. & Simmons, W.H. (1994) Substrate specificity of aminopeptidase P from *Escherichia coli*: comparison with membrane-bound forms from rat and bovine lung. *Arch. Biochem. Biophys.* **311**, 28–34.

**M**

*William H. Simmons*
*Department of Molecular and Cellular Biochemistry,*
*Loyola University Chicago Stritch School of Medicine,*
*2160 S. First Avenue,*
*Maywood, IL 60153, USA*
*Email: WSIMMON@wpo.it.luc.edu*

# 481. X-Pro aminopeptidase (Lactococcus)

## Databanks

Peptidase classification: clan MG, family M24, MEROPS ID: M24.006
NC-IUBMB enzyme classification: none
Databank codes:

| Species | Gene | SwissProt | PIR | EMBL (cDNA) | EMBL (genomic) |
|---------|------|-----------|-----|-------------|----------------|
| *Lactobacillus delbrueckii*[a] | *pepQ* | P46545 | A32454 | Z34896 Z54205 | – |
| *Lactococcus lactis cremoris* | *pepP* | – | – | Y08842 | – |

[a]The *Lactobacillus* peptidase has also been included in Chapter 479.

## Name and History

The presence of an X-Pro aminopeptidase was reported both in lactococci (Mou *et al.*, 1975; Booth *et al.*, 1990) and in lactobacilli (Hickey *et al.*, 1983). More recently, the enzyme was purified from *Lactococcus lactis* and characterized (Mars & Monnet, 1995). The corresponding gene, named *pepP*, was cloned and sequenced (J. Matos *et al.*, unpublished results). In the NC-IUBMB nomenclature, the enzyme is recommended to be called X-Pro aminopeptidase, the present form may be referred to as **X-Pro aminopeptidase (Lactococcus)**. The name in common use is **aminopeptidase P (PepP)**.

## Activity and Specificity

PepP is optimally active at pH 8 and 37°C. It is clearly a metalloenzyme inhibited by 1 mM EDTA at pH 7 or by 1,10-phenanthroline. DTT (1 mM) and the fragment 2–7 of bradykinin (1 mM) are also inhibitors.

The lactococcal PepP seems to be highly specific for X+Pro-Pro N-terminal sequences since bradykinin fragments (Arg+Pro-Pro, Arg+Pro-Pro-Gly-Phe) and Leu+Pro-Pro-Ser-Arg are the best substrates among those tested. Some peptides with Xaa-Pro-Yaa- N-termini are hydrolyzed also, but at a slower rate. The optimal size of the substrate is around five residues. No Xaa-Pro hydrolysis was observed with PepP (Mars & Monnet, 1995).

## Structural Chemistry

PepP is described as a monomeric enzyme with a molecular mass of 43 000 Da estimated by gel filtration (Mars & Monnet, 1995), which fits well with the size (1059 bp) of the corresponding gene (J. Matos *et al.*, unpublished results). The sequences surrounding the cobalt ligands and the catalytic residues of the methionyl aminopeptidase family (M24) (Chapter 475) are well conserved also for the lactococcal enzyme.

## Preparation

PepP was purified by chromatography from intracellular extracts of *Lactococcus lactis*. It is also present in intracellular extracts of another lactic acid bacterium, *Streptococcus thermophilus* (Rul & Monnet, 1997).

## Biological Aspects

PepP is found in the cytoplasm of *Lactococcus lactis*, which is widely used in cheese-making technology. Its role in the bacterial nutrition and during cheese ripening remains unknown.

## Related Enzymes

PepP shows similarity to other X-Pro aminopeptidases (Chapter 479) (27–30% identity) and also with X-Pro dipeptidase (Chapter 478) from lactobacilli (31% identity), creatinases (24–26% identity) and methionine aminopeptidases (Chapter 476) (22–24% identity). Interestingly, the sequencing of the complete genome of *Haemophilus influenzae* (Fleischmann *et al.*, 1995) and *Mycoplasma genitalium* (Fraser *et al.*, 1995) revealed the existence in these organisms of genes coding for proteins which are respectively 33 and 31% identical to PepP (see Chapter 479 for databanks codes).

## References

Booth, M., Donnelly, W.J., Fhaolain, I.N., Jennings, P.V. & O'Cuinn, G. (1990) Proline-specific peptidases of *Streptococcus cremoris* AM2. *J. Dairy Res.* **57**, 79–88.

Fleischmann, R.D., Adams, M.D., White, O. *et al.* (1995) Whole-genome random sequencing and assembly of *Haemophilus influenzae* Rd. *Science* **269**, 496–512.

Fraser, C.M., Gocayne, J.D., White, O. *et al.* (1995) The minimal gene complement of *Mycoplasma genitalium*. *Science* **270**, 397–403.

Hickey, M.W., Hillier, A.J. & Jago, G.R. (1983) Peptidase activities in lactobacilli. *Austr. J. Dairy Technol.* **38**, 118–123.

Mars, I. & Monnet, V. (1995) An aminopeptidase P from *Lactococcus lactis* with original specificity. *Biochim. Biophys. Acta* **1243**, 209–215.

Mou, L., Sullivan, J.J. & Jago, G.R. (1975) Peptidase activities in group N streptococci. *J. Dairy Res.* **42**, 147–155.

Rul, F. & Monnet, V. (1997) Presence of additional peptidases in *Streptococcus thermophilus* CNRZ302 compared to *Lactococcus lactis. J. Appl. Microbiol.* **82**, 695–704.

***José Matos***
*INRA, Laboratoire de Biochimie et Structure des Protéines,*
*Domaine de Vilvert,*
*78352 Jouy en Josas Cedex, France*

***Véronique Monnet***
*INRA, Laboratoire de Biochimie et Structure des Protéines,*
*Domaine de Vilvert,*
*78352 Jouy en Josas Cedex, France*
*Email: monnet@jouy.imra.fr*

M

# 482. Introduction: clan MH containing varied co-catalytic metallopeptidases

## Databanks

MEROPS ID: MH

| Species | SwissProt | PIR | EMBL (cDNA) | EMBL (genomic) |
|---|---|---|---|---|
| **Family M18** | | | | |
| Aminopeptidase I (Chapter 496) | | | | |
| Others | | | | |
| *Arabidopsis thaliana* | – | – | H36142 | – |
| *Borrelia burgdorferi* | – | – | X78708 | – |
| *Caenorhabditis briggsae* | – | – | R03805 | – |
| | | | R05220 | |
| *Homo sapiens* | – | – | AA085961 | – |
| | | | AA402037 | |
| | | | AA402156 | |
| *Leishmania major* | – | – | AA060740 | – |
| | | | T93455 | |
| *Mus musculus* | – | – | AA137322 | – |
| | | | AA212374 | |
| | | | AA265191 | |
| | | | AA272238 | |
| | | | AA286608 | |
| | | | W18310 | |
| *Mycobacterium leprae* | – | – | – | U15182: cosmid B2266 |
| *Saccharomyces cerevisiae* | P38821 | – | U00059 | – |
| **Family M20** | | | | |
| Dipeptidase | | | | |
| *Lactococcus lactis* | – | – | U78036 | – |
| Glutamate carboxypeptidase (Chapter 483) | | | | |
| Gly-X carboxypeptidase (Chapter 484) | | | | |
| Peptidase T (Chapter 485) | | | | |
| Peptidase V (Chapter 486) | | | | |
| Others | | | | |
| *Bacillus stearothermophilus* | – | S43914 | L13418 | – |
| *Bacillus subtilis* | P39635 | S39732 | X73124 | – |
| *Bacillus subtilis* | P54542 | – | – | D84432: multiple entry |
| *Saccharomyces cerevisiae* | P38149 | S44543 | – | X76053: chromosome II 32 420 bp |
| | | | | Z36150: chromosome II ORF ybr281c |
| **Family M25** | | | | |
| X-His dipeptidase bacteria (Chapter 487) | | | | |
| Others | | | | |
| *Caenorhabditis elegans* | – | – | – | U00040: gene C18H2.4 |
| **Family M28** | | | | |
| Alkaline phosphatase isozyme conversion protein | | | | |
| *Escherichia coli* | P10423 | A28382 | M18270 | U29579: genome 61–62′ |
| | | | M74586 | |

Aminopeptidase Y (Chapters 488 and 489)
Glutamate carboxypeptidase II (Chapter 492)
Pteroylpoly-$\gamma$-glutamate carboxypeptidase (Chapter 493)
*Streptomyces griseus* aminopeptidase (Chapter 490)
*Vibrio* aminopeptidase (Chapter 491)

| Species | SwissProt | PIR | EMBL (cDNA) | EMBL (genomic) |
|---|---|---|---|---|
| **Others** | | | | |
| *Bacillus subtilis* | P25152 | – | – | X52480: multiple entry |
| | | | | X73124: genome 325–333 |
| *Mycobacterium tuberculosis* | – | – | – | Z84724: cosmid SCY22G10 |
| *Saccharomyces cerevisiae* | P38244 | – | X76294 | Z35943: chromosome II ORF |
| *Saccharomyces cerevisiae* | P47161 | – | Z49626 | – |
| **Family M40** | | | | |
| *Sulfolobus* carboxypeptidase (Chapter 494) | | | | |
| Others | | | | |
| *Agrobacterium tumefaciens* | – | – | L38609 | – |
| *Arabidopsis thaliana* | P54968 | – | U23794 | – |
| *Arabidopsis thaliana* | P54969 | – | U23795 | – |
| *Arabidopsis thaliana* | P54970 | – | U23796 | – |
| *Arabidopsis thaliana* | – | – | Z37250 | – |
| *Bacillus subtilis* | P54955 | – | D45912 | – |
| *Haemophilus influenzae* | P44765 | – | – | U32740: complete genome section 55 of 163 |
| *Homo sapiens* | – | – | Z13316 | – |
| *Oryza sativa* | – | – | D22895 | – |
| *Oryza sativa* | – | – | D23741 | – |
| *Oryza sativa* | – | – | D24956 | – |
| *Synechocystis* sp. | P54984 | – | – | D64004: complete genome part 23 |
| **Family M42** | | | | |
| *Bacillus* aminopeptidase I (Chapter 499) | | | | |
| Glutamyl aminopeptidase (*Lactococcus*) (Chapter 495) | | | | |
| Others | | | | |
| *Clostridium thermocellum* | – | – | L13461 | – |
| *Escherichia coli* | P32153 | – | L19201 | – |
| *Escherichia coli* | P39366 | – | – | – |
| *Homo sapiens* | – | – | R47269 | – |

Clan MH is the third clan of metallopeptidases that require two metal ions for catalytic activity. The other clans are clan MF (Chapter 472) and clan MG (Chapter 475), and each consists of just one family. Clan MH contains six families, and the conservation of metal ligands among the families was discovered by Neuwald *et al*. (1997).

Some of the families in clan MH contain sequences so divergent that they were placed in separate families until linking sequences revealed the true relationships. Gly-X carboxypeptidase was originally placed in a separate family to that of glutamate carboxypeptidase, but the sequence of aminoacylase provided the link (Rawlings & Barrett, 1995), and both carboxypeptidases are now included in family M20. Family M28 was originally considered to represent several different families, but a *Streptomyces* aminopeptidase provided a link, and yeast aminopeptidase Y (formerly family M33) is now included in family M28 (Rawlings & Barrett, 1997). We report here that the alkaline phosphatase isozyme conversion peptidase from *Escherichia coli*, formerly the sole member of family U2, is also a member of M28, having statistically significant sequence similarity with the ipa-8r protein from *Bacillus subtilis* ($z = 17.95$) and yeast aminopeptidase Y. The possibility of such a relationship had been noted previously (Rawlings & Barrett, 1995).

All known metallopeptidases with co-catalytic metal ions are exopeptidases. Clan MF contains only aminopeptidases, and clan MG contains aminopeptidases and dipeptidases. Clan MH is the only clan of the three to contain exopeptidases that attack the C-terminus of the substrate: carboxypeptidases are known from families M20, M28 and M40.

Tertiary structures have been solved for peptidases from families M28 and M20. These include *Vibrio* aminopeptidases and *Streptomyces* aminopeptidase from M28, and *Pseudomonas* glutamate carboxypeptidase from family M20. There is some similarity to the structures of carboxypeptidase A (Chapter 451) and the C-terminal domain of leucyl aminopeptidase (Chapter 473), but this is thought to be the result of convergent rather than divergent evolution, because the positions of the metal ligands and catalytic residues are not conserved. The structure of *Vibrio* aminopeptidase is shown in Figure 482.1.

The zinc ligands in *Vibrio* aminopeptidase are Asp117, Glu152 and His256 for one zinc ion (Zn1) and His97, Asp117 and Asp179 for the other (Zn2). Asp117, Glu152 and His256 are conserved in all the catalytically active members of the clan. His97 can be replaced by Asp, and Asp179 can be replaced by Glu (see Alignment 482.1 on CD-ROM). In families M20 and M40 there are no conserved aspartates between Glu152 and His256. In family M20, this region is acid-rich, and predicting the fourth metal ligand is not possible. In family M40 there are two conserved His residues in this region that could serve as metal ligands. The distance between the fourth and fifth zinc ligands is very variable, being 104 residues in *Vibrio* aminopeptidase, and 343 residues in the rocB protein from *Bacillus subtilis*.

**M**

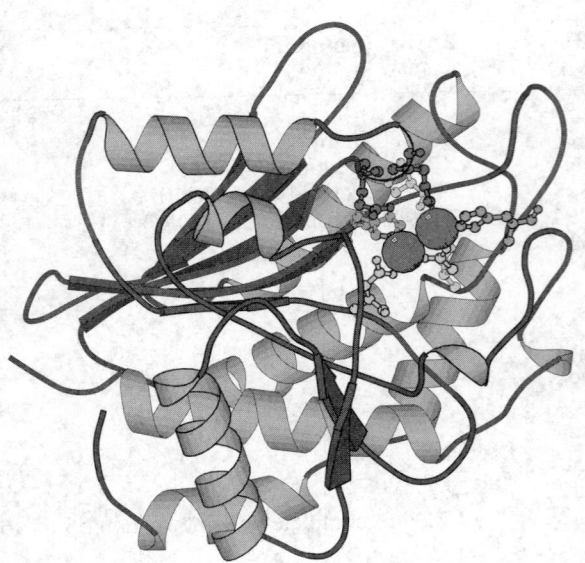

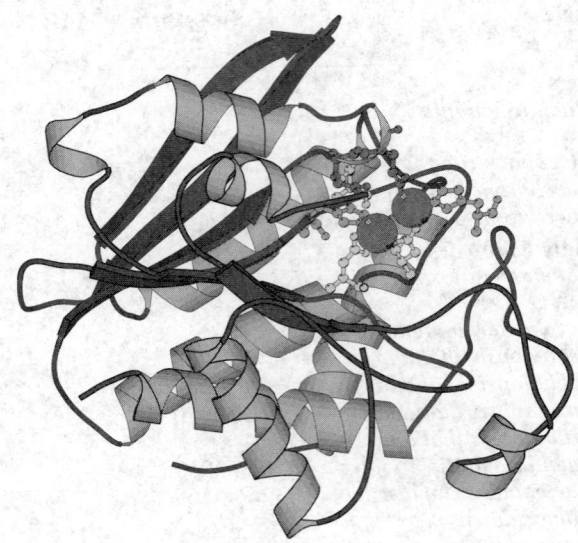

*Figure 482.1*   Richardson diagram of *Vibrio* aminopeptidase. The image was prepared from the Brookhaven Protein Data Bank entry (1AMP) as described in the Introduction (p. xxv). The catalytic zinc ions are shown in CPK representation as dark gray spheres. The zinc ligands are shown in ball-and-stick representation: His97, Asp117, Glu152, Asp179 and His256 (numbering as in Alignment 482.1). The putative catalytic Asp99 and Glu151 are also shown in ball-and-stick representation.

*Figure 482.2*   Richardson diagram of *Streptomyces* aminopeptidase. The image was prepared from the Brookhaven Protein Data Bank entry (1XJO) as described in the Introduction (p. xxv). The catalytic zinc ions are shown in CPK representation as dark gray spheres. The zinc ligands are shown in ball-and-stick representation: His97, Asp117, Glu152, Asp179 and His256 (numbering as in Alignment 482.1). The putative catalytic Asp99 and Glu151 are also shown in ball-and-stick representation.

Glu151 is conserved throughout all the peptidases of the clan and has been shown to interact with the hydroxamate group of a *p*-iodo-D-phenylalanine hydroxamate inhibitor, suggesting the role of general base in catalysis (Chevrier *et al.*, 1996). Asp99 is also conserved in all the active enzymes in the clan and, as can be seen from Figure 482.1, it is ideally placed to be involved in coordination of the imidazolium ring of His97.

Most families in the clan have homologs that are not peptidases. There are also a considerable number of putative proteins from genome sequencing projects for which there is no biochemistry; these may represent peptidases or enzymes with other hydrolytic activities.

**Family M28** includes aminopeptidases from *Vibrio, Streptomyces, Escherichia* and *Saccharomyces*. The *Vibrio* and *Streptomyces* aminopeptidases have been described as 'leucyl aminopeptidases', but actually release a variety of N-terminal amino acids. The *Escherichia* enzyme is an arginyl aminopeptidase, probably membrane bound, that either converts one form of alkaline phosphatase to another by successively releasing N-terminal arginines, or activates the phosphatase (Ishino *et al.*, 1987). The yeast aminopeptidase Y is a vacuolar enzyme that preferentially releases Arg and Lys among other amino acids. The aminopeptidases enter secretory pathways and are synthesized with N-terminal propeptides. Cerevisin (Chapter 104) activates yeast aminopeptidase Y (Nishizawa *et al.*, 1994; Yasuhara *et al.*, 1994). The structure of *Streptomyces* aminopeptidase is shown in Figure 482.2.

Family M28 also includes a carboxypeptidase. Glutamate carboxypeptidase II is a mammalian, membrane-bound enzyme that has activities that were, until very recently, attributed to separate enzymes, and as a result, two chapters are devoted to it here (Chapters 492 and 493). In brain, it degrades the neuropeptide Ac-Asp-Glu, whereas in the intestinal mucosa it converts folylpoly-$\gamma$-glutamate to pteroylglutamate (folate) which is then available for intestinal uptake. Glutamate carboxypeptidase II is also a marker antigen for prostate cancer, as prostate-specific membrane antigen, and it is suggested that an intracellular form of the enzyme may contribute to resistance of tumor cells to methotrexate (Heston, 1997; Rawlings & Barrett, 1997). There is only one other family that includes both aminopeptidases and carboxypeptidases: family S12, which contains D-Ala-D-Ala carboxypeptidase (Chapter 145) and D-stereospecific aminopeptidase (Chapter 146).

Nonpeptidase homologs of family M28 include *N*-acetylpuromycin *N*-acetylhydrolase from *Streptomyces alboniger*, which removes acetate from the antibiotic puromycin, and the mammalian transferrin receptor, which cannot be a peptidase because the zinc ligands and putative catalytic residue have been replaced.

**Family M20** includes carboxypeptidases from *Pseudomonas* and yeast, a bacterial tripeptidase, a bacterial dipeptidase and a number of enzymes that are hydrolases but not peptidases. The *Pseudomonas* glutamate carboxypeptidase is a homodimer, and the crystal structure has been solved (see Figure 482.3). The structure shows that each monomer consists of two domains, a catalytic N-terminal domain and a C-terminal domain that links the monomers. *Pseudomonas* glutamate carboxypeptidase is a periplasmic

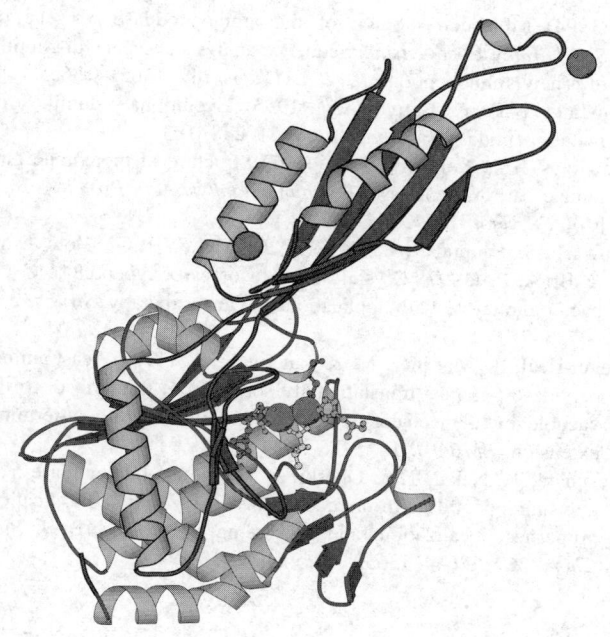

*Figure 482.3* Richardson diagram of one subunit of *Pseudomonas* glutamate carboxypeptidase. Atomic coordinates were kindly provided by Dr P. Brick. This image was prepared as described in the Introduction (p. xxv). The catalytic and structural zinc ions are shown in CPK representation as light gray spheres. The zinc ligands are shown in ball-and-stick representation: His97, Asp117, Glu152, Asp179 and His256 (numbering as in Alignment 482.1).

enzyme and is synthesized with a signal peptide. The carboxypeptidase has an application in cancer therapy through destruction of folate and methotrexate (Rowsell *et al.*, 1997). Yeast Gly-X carboxypeptidase is a vacuolar enzyme that possesses an N-terminal propeptide and is activated by cerevisin.

Peptidase T is a bacterial aminopeptidase that acts only on tripeptide substrates, and for this reason has been called a 'tripeptidase'. Peptidase V is primarily a dipeptidase.

There are a number of other enzymes that are not peptidases but are homologous to members of family M20. These are termed the ArgE. The majority are enzymes that cleave carbon–nitrogen bonds other than peptide bonds, usually to release a moiety substituted at the amino group. These include aminoacylase, which releases a carboxylate group from *N*-acyl-amino acids; acetylornithine deacetylase, which releases acetate from acetylornithine; and succinyl-diaminopimelate desuccinylase, which releases succinate from succinyl-diaminopimelate. The acetylornithine deacetylase from *Dictyostelium* has the zinc ligand Glu152 replaced by Asn.

There are a number of homologs for which no biochemical data are available. Among them is the hypothetical YBR281C protein from yeast which is predicted to have a 450 residue N-terminal extension compared to other members of the family.

**Family M25** contains only X-His dipeptidase. This periplasmic bacterial enzyme breaks down the animal muscle dipeptide carnosine. The dipeptidase is known only from enterobacteria, which may use carnosine as a nutrient source. The predicted fifth zinc ligand (His256 in Alignment 482.1)

is 310 residues from the fourth ligand, and a region between the two ligands presumably corresponds to an insert relative to other members of the clan. This insert is homologous to the C18H2.4 protein from *Caenorhabditis elegans*.

**Family M40** includes a thermostable carboxypeptidase of broad specificity from the archaean *Sulfolobus*. The carboxypeptidase is stable to 80°C, and probably exists as a tetramer. The carboxypeptidase is homologous to a variety of other proteins, none of which is known to be a peptidase. Most of the homologs cleave nonpeptidase carbon–nitrogen bonds, releasing moieties substituted on the amino group of amino acids, including aminoacylases from *Bacillus stearothermophilus*, *Haemophilus influenzae* and *Campylobacter jejuni* (the latter two known as hippuricases); an *N*-carbamoyl-L-amino acid amidohydrolase from *Bacillus stearothermophilus*, which releases $CO_2$ and $NH_3$ from a substituted amino acid; and indole-3-acetic acid-amino acid hydrolases from *Arabidopsis thaliana*, which release the growth regulator indole-3-acetic acid from the conjugate (Bartel & Fink, 1995). These activities are similar to those of the nonpeptidase homologs of family M20.

**Family M42** includes aminopeptidases from gram-positive bacteria. A glutamyl aminopeptidase (Chapter 495) from *Lactococcus lactis* has a superficially similar specificity to a glutamyl aminopeptidase (Chapter 339) from family M1, but is unusual in that it acts also as a carboxyamidase, for example hydrolysing *N*-succinyl *p*-nitroanilide (Chapter 499).

The thermostable aminopeptidase I (Chapter 499) from *Bacillus stearothermophilus* is reported to contain twelve subunits of two non-identical but homologous types designated $\alpha$ and $\beta$. Only short N-terminal fragments are known for the subunits, but both show significant similarity to the ysdC protein from *Bacillus subtilis*. The yscD protein is not known to be a peptidase, but is homologous to *Lactococcus* glutamyl aminopeptidase. Like other families in the clan, family M42 includes enzymes that are not peptidases, such as an endoglucanase from *Clostridium thermocellum*, which degrades carboxymethylcellulose (Kobayashi *et al.*, 1993). An homologous human expressed sequence tag exists.

**Family M18** contains aminopeptidase I from yeast. Aminopeptidase I is a vacuolar aminopeptidase, and is synthesized as a proprotein that is activated at a Leu+Glu bond by cerevisin (Chapter 104). The precursor is synthesized without a signal peptide, and targeting to the lysosome is independent of the secretory pathway (Seguí-Real *et al.*, 1995). The mature enzyme is a nondisulfide-linked dodecamer, and each monomer contains two zinc ions. Other homologs are known from *Mycobacterium leprae* and the Lyme disease spirochete *Borrelia burgdorferi*. Although zinc ligands have not been determined for any member of the family, the only conserved His, Asp and Glu residues follow the pattern of zinc ligands in clan MH (see Alignment 482.1) and we predict that aminopeptidase I and its homologs will have a similar tertiary fold to the *Vibrio* aminopeptidase. A number of expressed sequence tags from human, mouse, *Caenorhabditis*, *Leishmania* and *Arabidopsis* show homology to yeast aminopeptidase I.

## References

Bartel, B. & Fink, G.R. (1995) Ilr1, an amidohydrolase that releases active indole-3-acetic-acid from conjugates. *Science*

**268**, 1745–1748.

Chevrier, B., D'Orchymont, H., Schalk, C., Tarnus, C. & Moras, D. (1996) The structure of the *Aeromonas proteolytica* aminopeptidase complexed with a hydroxamate inhibitor – involvement in catalysis of Glu151 and two zinc ions of the co-catalytic unit. *Eur. J. Biochem.* **237**, 393–398.

Heston, W.D.W. (1997) Characterization and glutamyl preferring carboxypeptidase function of prostate specific membrane antigen: a novel folate hydrolase. *Urology* **49**, 104–112.

Ishino, Y., Shinagawa, H., Makino, K., Amemura, M. & Nakata, A. (1987) Nucleotide sequence of the *iap* gene, responsible for alkaline phosphatase isozyme conversion in *Escherichia coli*, and identification of the gene product. *J. Bacteriol.* **169**, 5429–5433.

Kobayashi, T., Romaniec, M.P.M., Barker, P.J., Gerngross, U.T. & Demain, A.L. (1993) Nucleotide-sequence of gene *celM* encoding a new endoglucanase (celM) of *Clostridium thermocellum* and purification of the enzyme. *J. Ferment. Bioeng.* **76**, 251–256.

Neuwald, A.F., Liu, J.S., Lipman, D.J. & Lawrence, C.E. (1997) Extracting protein alignment models from the sequence database. *Nucleic Acids Res.* **25**, 1665–1677.

Nishizawa, M., Yasuhara, T., Nakai, T., Fujiki, Y. & Ohashi, A. (1994) Molecular cloning of the aminopeptidase Y gene of *Saccharomyces cerevisiae*. Sequence analysis and gene disruption of a new aminopeptidase. *J. Biol. Chem.* **269**, 13651–13655.

Rawlings, N.D. & Barrett, A.J. (1995) Evolutionary families of metallopeptidases. *Methods Enzymol.* **248**, 183–228.

Rawlings, N.D. & Barrett, A.J. (1997) Structure of membrane glutamate carboxypeptidase. *Biochim. Biophys. Acta Protein Struct. Mol. Enzymol.* **1339**, 247–252.

Rowsell, S., Pauptit, R.A., Tucker, A.D., Melton, R.G., Blow, D.M. & Brick, P. (1997) Crystal structure of carboxypeptidase G₂, a bacterial enzyme with applications in cancer therapy. *Structure* **5**, 337–347.

Seguí-Real, B., Martinez, M. & Sandoval, I.V. (1995) Yeast aminopeptidase I is post-translationally sorted from the cytosol to the vacuole by a mechanism mediated by its bipartite N-terminal extension. *EMBO J.* **14**, 5476–5484.

Yasuhara, T., Nakai, T. & Ohashi, A. (1994) Aminopeptidase Y, a new aminopeptidase from *Saccharomyces cerevisiae*. Purification, properties, localization, and processing by protease B. *J. Biol. Chem.* **269**, 13644–13650.

# *483. Glutamate carboxypeptidase*

## *Databanks*

*Peptidase classification: clan MH, family M20, MEROPS ID: M20.001*
*NC-IUBMB enzyme classification: EC 3.4.17.11*
Databank codes:

| Species | SwissPort | PIR | EMBL (cDNA) | EMBL (genomic) |
|---------|-----------|-----|-------------|----------------|
| *Pseudomonas* sp. | P06621 | A24955 | A06774 M12599 | – |

Brookhaven Protein Data Bank three-dimensional structures:

| Species | ID | Resolution | Notes |
|---------|-----|-----------|-------|
| *Pseudomonas* sp. | 1CG2 | 2.5 | homotetramer |

## *Name and History*

Exopeptidases of the **carboxypeptidase G** class are defined by their specificity of release of C-terminal glutamate residues from a wide range of *N*-acylating moieties, including peptidyl, aminoacyl, benzoyl, benzyloxycarbonyl, folyl and pteroyl groups. They were first isolated from pseudomonads 30 years ago (Levy & Goldman, 1967; Goldman & Levy, 1967). Subsequently, enzymes with similar activity were identified in a *Flavobacterium* sp. (Albrecht *et al.*, 1978) and *Acinetobacter* sp. (Albrecht *et al.*, 1976), but the enzyme best characterized at that time was from *Pseudomonas stutzeri*, designated **carboxypeptidase G₁** (McCullough *et al.*, 1971).

The prime interest in the enzyme was due to its ability to cleave the glutamate residue from folic (pteroylglutamic) acid and more significantly folate analogs, such as the chemotherapeutic agent methotrexate. This property provided the opportunity to assess the enzyme as an antitumor agent in its own right through depletion of reduced folates, essential cofactors in DNA synthesis (Chabner *et al.*, 1972a; Kalghatgi & Bertino, 1981) and as a rescue agent against methotrexate (inhibitor of dihydrofolate reductase) toxicity, particularly encountered in patients exhibiting renal failure (Chabner *et al.*, 1972b; Howell *et al.*, 1978). Progress was limited through availability of the enzyme until isolation

*Table 483.1*  Comparison of the physicochemical properties and substrate affinities of carboxypeptidases G$_1$ and G$_2$

| | Carboxypeptidase | |
| --- | --- | --- |
| | G$_1$ | G$_2$ |
| Molecular mass (Da) | $2 \times 46\,000$ | $2 \times 41\,800$ |
| pH optimum | 6.3–7.3 | 7.0–7.5 |
| Isoelectric point | – | 7.8 |
| $K_m$ values (μM) in Tris–HCl pH 7.3 at 37°C: | | |
| Pteroylglutamate (folate) | 1.1 | 4 |
| 5-Formyltetrahydrofolate | 18.1 | 120 |
| 5-Methyltetrahydrofolate | 12.9 | 34 |
| 4-Amino-$N^{10}$-methylpteroyl-glutamate (methotrexate) | 3.9 | 8 |
| 4-Amino-$N^{10}$-methylpteroyl-aspartate | 580 | – |
| 4-Aminopteroylasparate | 104 | – |
| Aminopterin | 8.3 | – |

of **carboxypeptidase G$_2$** from *Pseudomonas* sp. strain RS-16 (Sherwood *et al.*, 1985). The enzyme was subsequently cloned into *Escherichia coli* (Minton *et al.*, 1983) and *Saccharomyces cerevisiae* (Clarke *et al.*, 1985) and fully sequenced (Minton *et al.*, 1984). The name recommended by the NC-IUBMB is **glutamate carboxypeptidase**.

## Activity and Specificity

Glutamate carboxypeptidases are characterized as $Zn^{2+}$-requiring exopeptidases with specificity for glutamate; this specificity is not absolute as carboxypeptidase G$_1$ was reported as exhibiting activity against pteroyl compounds with C-terminal aspartic acid (McCullough *et al.*, 1971). Physicochemical properties and substrate affinities ($K_m$ values) for folic acid and a variety of analogs are compared for the G$_1$ and G$_2$ enzymes in Table 483.1.

There is a strict requirement to have an unsubstituted α-carboxyl group for glutamate release, but substitutions at the γ-carboxyl group (P. Burke, personal communication) and sites on the adjacent benzoyl ring (Springer *et al.*, 1994) are well tolerated by carboxypeptidase G$_2$. In addition, a series of glutamyl-containing enzyme-activated substrate inhibitors based on phenylenediamine and *p*-hydroxyaniline mustards have been shown to be high-affinity substrates (Blakey *et al.*, 1995; Dowell *et al.*, 1996).

Assays are most conveniently performed by a spectrophotometric method based on change in absorbance at 320 nm using methotrexate as substrate (Sherwood *et al.*, 1985). 1,10-Phenanthroline is inhibitory, but only after prolonged exposure. Full activity can be restored by addition of $Zn^{2+}$ ions ($k_{cat}$ for G$_2 = 9 \times 10^{-7}$ M; $k_{cat}$ for G$_1 = 2 \times 10^{-6}$ M). Prior addition of $Co^{2+}$, $Hg^{2+}$, $Cu^{2+}$, $Ni^{2+}$ or $Mn^{2+}$ to 'metal-free' enzyme inhibits reactivation (Chabner & Bertino, 1972).

## Structural Chemistry

Carboxypeptidase G$_2$ is a dimeric protein of 83 600 kDa containing two atoms of zinc per subunit. There are no disulfide bonds and the complete nucleotide and amino acid sequence were resolved following cloning of the gene into *Escherichia*

*coli* (Minton *et al.*, 1984). In both native and recombinant form the enzyme is located in the periplasmic space, targeted by a 22 amino acid signal peptide. Carboxypeptidase G$_2$ has been crystallized (Lloyd *et al.*, 1991; Tucker *et al.*, 1996) and the crystal structure determined at 2.5 Å resolution (Rowsell *et al.*, 1997) (see Fig. 482.3). Each subunit of the molecular dimer consists of a larger catalytic domain containing two zinc ions at the active site, and a separate smaller domain which forms the dimer interface. The two active sites in the dimer are more than 60 Å apart and are presumed to be independent; each contains a symmetric distribution of carboxylate and histidine ligands around two zinc ions separated by 3.2 Å.

## Preparation

Essentially homogeneous preparations of G$_2$ can be obtained from *Pseudomonas* sp. or via recombinant DNA expression in *E. coli* (gram quantities) by a combination of ion exchange, size exclusion and affinity chromatography, based on metal ion-mediated binding to triazine dye (Sherwood *et al.*, 1985). The purification factor is close to 2000-fold from *Pseudomonas* sp. with a 40% yield of enzyme with a specific activity of >500 units mg$^{-1}$. Expression levels in *E. coli* are 10–20%, requiring a 5–10-fold purification with 30–40% yield.

## Biological Aspects

There is no known mammalian equivalent to glutamate carboxypeptidase and it was recognized that the high affinity of the bacterial enzymes for folic acid provided an opportunity to develop anticancer regimes based on depletion of reduced folates, key intermediates in DNA synthesis. Although antitumor activity was demonstrated with both G$_1$ and G$_2$ enzymes (Chabner *et al.*, 1972a; R.F. Sherwood, unpublished results) the use of enzyme alone was considered insufficient to warrant full clinical evaluation. The enzyme has, however, been developed for clinical use in two settings.

In the first, it is used as a rescue agent during high dose (3–30 g m$^{-2}$) methotrexate therapy, against a range of cancers. The principle of 'rescue' is well established using leucovorin (5-formyltetrahydrofolate) to rapidly restore cellular levels of reduced folates depleted by the inhibitory action of methotrexate on dihydrofolate reductase. Removal of methotrexate then relies on patient hydration and renal function. If the latter is impaired then high circulating levels of methotrexate can result in bone marrow and other toxicities. Glutamate carboxypeptidase can be used to rapidly eliminate methotrexate from the circulation. The principle was originally developed using carboxypeptidase G$_1$ (Chabner *et al.*, 1972b; Howell *et al.*, 1978), but all subsequent methodology has been established using carboxypeptidase G$_2$ initially in primate models (Adamson *et al.*, 1992) and subsequently in patients (Widemann *et al.*, 1995; DeAngelis *et al.*, 1996). Rescue protocols have been extended to include intrathecal, as opposed to systemic, methotrexate therapy (Adamson *et al.*, 1991; O'Marcaigh *et al.*, 1996). Combination therapy with folate antagonists, which lack a terminal glutamic acid (e.g. trimetrexate) has also been considered (Romanini *et al.*, 1989; Searle *et al.*, 1990).

The second approach is to use targeted carboxypeptidase G$_2$ to remove the glutamate residue from prodrugs, releasing a highly cytotoxic agent at tumor sites. This method has been given the acronym ADEPT (**a**ntibody **d**irected **e**nzyme **p**rodrug **t**herapy). Targeting is achieved by covalent linkage of carboxypeptidase $G_2$ to tumor-associated antibodies and monoclonal antibody fragments, including antihuman chorionic gonadotrophin (Searle *et al*., 1986; Melton *et al*., 1990), antihuman carcinoembryonic antigen (Melton *et al*., 1993; Michael *et al*., 1996) and antihuman c-*erb*B2 protooncogene product (Eccles *et al*., 1994). Once delivered to the tumor site, the enzyme is used to activate prodrugs, predominantly based on glutamyl derivatives of nitrogen mustard compounds and notably 4-[(2-mesyloxyethyl)(2-chloroethyl)amino] benzoyl glutamic acid (Bagshawe *et al*., 1988; Springer *et al*., 1990, 1991) and 4-[*N,N*,-bis(2-iodoethyl)amino]phenol linked to glutamic acid (Blakey *et al*., 1996). Clinical studies (Bagshawe *et al*., 1991, 1995) are ongoing.

Carboxypeptidase $G_2$ has also been used to facilitate the synthesis of methotrexate analogs (McGuire *et al*., 1991) and thymidylate synthase inhibitors (Bissett *et al*., 1992).

## Distinguishing Features

A monoclonal antibody termed SB43 has been raised and shown to inactivate carboxypeptidase $G_2$ *in vitro* and *in vivo* (Sharma *et al*., 1990). In the context of ADEPT, SB43 is used to accelerate the clearance of enzyme from plasma and is administered in galactosylated form to avoid inactivation of carboxypeptidase $G_2$ at tumor sites and to facilitate the clearance of complexes via carbohydrate receptors in the liver (Sharma *et al*., 1994). Carboxypeptidase $G_2$ is competitively inhibited by D-amino acid analogs of glutamyl prodrugs and sulfur-linked enzyme-activated glutamyl prodrugs (P. Burke & R. Melton, personal communication).

A mutant form of carboxypeptidase $G_2$ with a threonine replacing the alanine at the N-terminal position was constructed and used to produce antibody–enzyme conjugates by a technique based on reverse proteolysis rather than conventional chemical linkage using hetero-bifunctional cross-linking agents (Werlen *et al*., 1994).

## Related Peptidases

An exopeptidase designated as carboxypeptidase $G_3$ has been isolated (Chapter 535), which also cleaves acidic amino acids, including D-forms, but shows affinity for the long-chain fatty acyl group, not the benzoyl group.

Carboxypeptidase $G_2$ does not exhibit any significant sequence homology with other zinc-requiring metallopeptidases, such as carboxypeptidases A and B, but the catalytic domain has structural homology with other zinc-dependent exopeptidases, both those with a single zinc ion or a pair of zinc ions in the active site. The closest structural homology is with the aminopeptidase from *Vibrio proteolyticus* (Chapter 491) where the similarity includes superimposable active site residues despite a different substrate.

A variety of mammalian and bacterial aminoacylases do have related cDNA sequences (human aminoacylase-1, EC 3.5.1.14; *Escherichia coli* and *Bacillus stearothermophilus* aminoacylases, EC 3.5.1.16), plus *E. coli* and *Haemophilus influenzae* succinyl diaminopimelate desuccinylase (EC 3.5.1.18). It is relevant to note that these are also zinc-containing enzymes.

A catalytic antibody (abzyme) with glutamate carboxypeptidase activity has been described (Wentworth *et al*., 1996). This was generated using a phosphonamidate transition state analog of a phenylcarbamoyl mustard prodrug known to be a substrate. The abzyme has a $K_m$ for the mustard prodrug of 201 µM and a $k_{cat}$ of 1.88 min$^{-1}$, compared with figures of about 1 µM and 32 s$^{-1}$ respectively for carboxypeptidase $G_2$, but is nevertheless capable of causing marked reduction in viability of cultured human colon carcinoma cells relative to appropriate controls.

## Further Reading

For review of metal ion-promoted binding of carboxypeptidase $G_2$ and other enzymes to dye ligands, see Hughes & Sherwood (1987). For a review of ADEPT using carboxypeptidase $G_2$, carboxypeptidase A and other enzymes see Melton & Sherwood (1996).

## References

Adamson, P.C., Balis, F.M., McCully, C.L., Godwin, K.S., Bacher, J.D., Walsh, T.J. & Poplack, D.G. (1991) Rescue of experimental intrathecal methotrexate overdose with carboxypeptidase-$G_2$. *J. Clin. Oncol.* **9**, 670–674.

Adamson, P.C., Balis, F.M., McCully, C.L., Godwin, K.S. & Poplack, D.G. (1992) Methotrexate pharmacokinetics following administration of recombinant carboxypeptidase-$G_2$ in rhesus monkeys. *J. Clin. Oncol.* **10**, 1359–1364.

Albrecht, A.M., Boldizsar, E. & Hutchinson, D.J. (1976) Folate- and antifolate-degradation by an *Acinetobacter* enzyme. *Fedn Proc.* **35**, 787.

Albrecht, A.M., Boldizsar, E. & Hutchinson, D.J. (1978) Carboxypeptidase displaying differential velocity in hydrolysis of methotrexate, 5-methyltetrahydrofolic acid, and leucovorin. *J. Bacteriol.* **134**, 506–513.

Bagshawe, K.D., Springer, C.J., Searle, F., Antoniw, P., Sharma, S.K., Melton, R.G. & Sherwood, R.F. (1988) A cytotoxic agent can be generated selectively at cancer sites. *Br. J. Cancer* **58**, 700–703.

Bagshawe, K.D., Sharma, S.K., Springer, C.J., Antoniw, P., Boden, J.A., Rogers, G.T., Burke, P., Melton, R.G. & Sherwood, R.F. (1991) Antibody directed enzyme prodrug therapy (ADEPT): clinical report. *Disease Markers* **9**, 233–238.

Bagshawe, K.D., Sharma, S.K., Springer, C.J. & Antoniw, P. (1995) Antibody directed enzyme prodrug therapy: a pilot-scale clinical trial. *Tumor Targeting* **1**, 17–29.

Bissett, G.M.F., Pawelczak, K., Jackman, A.L., Calvert, A.H. & Hughes, L.R. (1992) Syntheses and thymidylate synthase inhibitory activity of the poly-γ-glutamyl conjugates of *N*-[5-[*N*-(3,4-dihydro-2-methyl-4-oxoquinazolin-6-ylmethyl)-*N*-methylamino]-2-thenoyl]-l-glutamic acid (ICI D1694) and other quinazoline antifolates. *J. Med. Chem.* **35**, 859–866.

Blakey, D.C., Davies, D.H., Dowell, R.I., East, S.J., Burke, P.J., Sharma, S.K., Springer, C.J., Mauger, A.B. & Melton, R.G. (1995) Anti-tumour effects of an antibody-carboxypeptidase $G_2$ conjugate in combination with phenol mustard prodrugs. *Br. J. Cancer* **72**, 1083–1088.

Blakey, D.C., Burke, P.J., Davies, D.H., Dowell, R.I., East, S.J.,

Eckersley, K.P., Fitton, J.E., McDaid, J., Melton, R.G., Niculescu-Duvaz, I.A., Pinder, P.E., Sharma, S.K., Wright, A.F. & Springer, C.J. (1996) ZD2767, an improved system for antibody-directed enzyme prodrug therapy that results in tumour regression in colorectal tumor xenografts. *Cancer Res.* **56**, 3287–3292.

Chabner, B.A. & Bertino, J.R. (1972) Activation and inhibition of carboxypeptidase $G_1$ by divalent cations. *Biochim. Biophys. Acta* **276**, 234–240.

Chabner, B.A., Chello, P.L. & Bertino, J.R. (1972a) Antitumour activity of a folate-cleaving enzyme, carboxypeptidase $G_1$. *Cancer Res.* **32**, 2114–2119.

Chabner, B.A., Johns, D.G. & Bertino, J.R. (1972b) Enzymatic cleavage of methotrexate provides a method for prevention of drug toxicity. *Nature (Lond.)* **239**, 395–397.

Clarke, L.E., Gibson, R.K., Sherwood, R.F. & Minton, N.P. (1985) Expression of the *Pseudomonas* gene coding for carboxypeptidase $G_2$ in *Saccharomyces cerevisiae. J. Gen. Microbiol.* **131**, 897–903.

DeAngelis, L.M., Tang, W.P., Silan, L., Fleisher, M. & Bertino, J.R. (1996) Carboxypeptidase $G_2$ rescue after high-dose methotrexate. *J. Clin. Oncol.* **14**, 2145–2149.

Dowell, R.I., Springer, C.J., Davies, D.H., Hadley, E.M., Burke, P.J., Boyle, F.T., Melton, R.G., Connors, T.A., Blakey, D.C. & Mauger, A.B. (1996) New mustard prodrugs for antibody-directed enzyme prodrug therapy: alternatives to the amide link. *J. Med. Chem.* **39**, 1100–1105.

Eccles, S.A., Court, W.J., Box, G.A., Dean, C.J., Melton, R.G. & Springer, C.J. (1994) Regression of established breast carcinoma xenografts with antibody-directed enzyme prodrug therapy against c-*erb*B2 p185[1]. *Cancer Res.* **54**, 5171–5177.

Goldman, P. & Levy, C.C. (1967) Carboxypeptidase G: purification and properties. *Proc. Natl Acad. Sci. USA* **58**, 1299–1306.

Howell, S.B., Blair, H.E., Uren, J. & Frei, E. (1978) Hemodialysis and enzymatic cleavage of methotrexate in man. *Eur. J. Cancer* **14**, 787–792.

Hughes, P. & Sherwood, R.F. (1987) Metal ion-promoted dye-ligand chromatography. In: *Reactive Dyes in Protein and Enzyme Technology* (Clonis, Y., Atkinson, A., Bruton, C. & Lowe, C., eds). London: Macmillan Press, pp. 125–160.

Kalghatgi, K.K. & Bertino, J.R. (1981) Folate-degrading enzymes: a review with special emphasis on carboxypeptidase G. In: *Enzymes as Drugs* (Holcenberg, J.S. & Roberts, J., eds). New York: John Wiley & Sons, pp. 77–102.

Levy, C.C. & Goldman, P. (1967) The enzymatic hydrolysis of methotrexate and folic acid. *J. Biol. Chem.* **242**, 2933–2938.

Lloyd, L.F., Collyer, C.A. & Sherwood, R.F. (1991) Crystallisation and preliminary crystallographic analysis of carboxypeptidase $G_2$ from *Pseudomonas sp.* strain RS-16. *J. Mol. Biol.* **220**, 17–18.

McCullough, J.L., Chabner, B.A. & Bertino, J.R. (1971) Purification and properties of carboxypeptidase $G_1$. *J. Biol. Chem.* **246**, 7207–7213.

McGuire, J.J., Bolanowska, W.E., Coward, J.K., Sherwood, R.F., Russell, C.A. & Felschow, D.M. (1991) Biochemical and biological properties of methotrexate analogs containing D-glutamic acid or D-erythro, threo-4-fluoroglutamic acid. *Biochem. Pharmacol.* **42**, 2400–2403.

Melton, R.G. & Sherwood, R.F. (1996) Antibody–enzyme conjugates for cancer therapy. *J. Natl Cancer Inst.* **88**, 153–165.

Melton, R.G., Searle, F., Sherwood, R.F., Bagshawe, K.D. & Boden, J.A. (1990) The potential of carboxypeptidase $G_2$: antibody conjugates as antitumour agents. II. *In vivo* localisation and clearance properties in a choriocarcinoma model. *Br. J. Cancer* **61**, 420–424.

Melton, R.G., Boyle, J.M.B., Rogers, G.T., Burke, P.J., Bagshawe, K.D. & Sherwood, R.F. (1993) Optimisation of small-scale coupling of $A_5B_7$ monoclonal antibody to carboxypeptidase $G_2$. *J. Immunol. Methods* **158**, 49–56.

Michael, N.P., Chester, K.A., Melton, R.G., Robson, L., Nicholas, W., Boden, J.A., Pedley, R.B., Begent, R.H.J., Sherwood, R.F. & Minton, N.P. (1996) *In vitro* and *in vivo* characterization of a recombinant carboxypeptidase $G_2$::anti-CEA scFv fusion protein. *Immunotechnology* **2**, 47–57.

Minton, N.P., Atkinson, T. & Sherwood, R.F. (1983) Molecular cloning of the *Pseudomonas* carboxypeptidase $G_2$ gene and its expression in *Escherichia coli* and *Pseudomonas putida. J. Bacteriol.* **156**, 1222–1227.

Minton, N.P., Atkinson, T., Bruton, C.J. & Sherwood, R.F. (1984) The complete nucleotide sequence of the *Pseudomonas* gene coding for carboxypeptidase $G_2$. *Gene* **31**, 31–38.

O'Marcaigh, A.S., Johnson, C.M., Smithson, W.A., Patterson, M.C., Widemann, B.C., Adamson, P.C. & McManus, M.J. (1996) Successful treatment of intrathecal methotrexate overdose by using ventriculolumbar perfusion and intrathecal instillation of carboxypeptidase $G_2$. *Mayo Clin. Proc.* **71**, 161–165.

Romanini, A., Sobrero, A.F., Chou, T.-C., Sherwood, R.F. & Bertino, J.R. (1989) Enhancement of trimetrexate cytotoxicity *in vitro* and *in vivo* by carboxypeptidase $G_2$. *Cancer Res.* **49**, 6019–6023.

Rowsell, S., Pauptit, R.A., Tucker, A.D., Melton, R.G., Blow, D.M. & Brick, P. (1997) Crystal structure of carboxypeptidase G2, a bacterial enzyme with applications in cancer therapy. *Structure* **5**, 337–347.

Searle, F., Bier, C., Buckley, R.G., Newman, S., Pedley, R.B., Bagshawe, K.D., Melton, R.G., Alwan, S.M. & Sherwood, R.F. (1986) The potential of carboxypeptidase $G_2$–antibody conjugates as antitumour agents. I. Preparation of antihuman chorionic gonadotrophin: carboxypeptidase $G_2$ and cytotoxicity against JAR choriocarcinoma cells *in vitro. Br. J. Cancer* **53**, 377–384.

Searle, F., Bagshawe, K.D., Pedley, R.B., Bradshaw, T., Melton, R.G. & Sherwood, R.F. (1990) Carboxypeptidase $G_2$ and trimetrexate cause growth delay of human colonic cancer cells *in vitro. Biochem. Pharmacol.* **39**, 1787–1791.

Sharma, S.K., Bagshawe, K.D., Burke, P.J., Boden, R.W. & Rogers, G.T. (1990) Inactivation and clearance of an anti-CEA carboxypeptidase $G_2$ conjugate in blood after localisation in a xenograft model. *Br. J. Cancer* **61**, 659–662.

Sharma, S.K., Bagshawe, K.D., Burke, P.J., Boden, J.A., Rogers, G.T., Springer, C.J., Melton, R.G. & Sherwood, R.F. (1994) Galactosylated antibodies and antibody–enzyme conjugates in antibody-directed enzyme prodrug therapy. *Cancer* **73**, 1114–1120.

Sherwood, R.F., Melton, R.G., Alwan, S.M. & Hughes, P. (1985) Purification and properties of carboxypeptidase $G_2$ from *Pseudomonas sp.* strain RS-16. *Eur. J. Biochem.* **148**, 447–453.

Springer, C.J., Antoniw, P., Bagshawe, K.D., Searle, F., Bissett, G.M.F. & Jarman, M. (1990) Novel prodrugs which are activated to cytotoxic alkylating agents by carboxypeptidase $G_2$. *J. Med. Chem.* **33**, 677–681.

Springer, C.J., Bagshawe, K.D., Sharma, S.K., Searle, F., Boden, J.A., Antoniw, P., Burke, P.J., Rogers, G.T., Sherwood, R.F. & Melton, R.G. (1991) Ablation of human choriocarcinoma xenografts in nude mice by antibody-directed enzyme prodrug therapy (ADEPT) with three novel compounds. *Eur. J. Cancer* **27**, 1361–1366.

Springer, C.J., Niculescu-Duvaz, I. & Pedley, R.B. (1994) Novel prodrugs of alkylating agents derived from 2-fluoro- and

**M**

3-fluorobenzoic acids for antibody-directed enzyme prodrug therapy. *J. Med. Chem.* **37**, 2361–2370.

Tucker, A.D., Rowsell, S., Melton, R.G. & Pauptit, R.A. (1996) A new crystal form of carboxypeptidase G2 from *Pseudomonas sp.* strain RS-16 which is more amenable to structure determination. *Acta Crystallogr. D* **51**, 890–892.

Wentworth, P., Datta, A., Blakey, D., Boyle, T., Partridge, L.J. & Blackburn, G.M. (1996) Toward antibody-directed 'abzyme prodrug therapy', ADAPT: carbamate prodrug activation by a catalytic antibody and its *in vitro* application to human tumour cell killing. *Proc. Natl Acad. Sci. USA* **93**, 799–803.

Werlen, R.C., Lankinen, M., Rose, K., Blakey, D.C., Shuttleworth, H., Melton, R.G. & Offord, R.E. (1994) Site-specific conjugation of an enzyme and an antibody fragment. *Bioconjug. Chem.* **5**, 411–417.

Widemann, B.C., Hetherington, M.L., Murphy, R.F., Balis, F.M. & Adamson, P.C. (1995) Carboxypeptidase-G2 rescue in a patient with high dose methotrexate-induced nephrotoxicity. *Cancer* **76**, 521–526.

*Roger F. Sherwood*
*Duramed Europe Ltd.,*
*Magdalen Centre,*
*The Oxford Science Park,*
*Oxford OX4 4GA, UK*
*Email: roger@dura-uk.demon.co.uk*

*Roger G. Melton*
*Centre for Applied Microbiology and Research,*
*Porton Down, Salisbury,*
*Wiltshire SP4 0JG, UK*
*Email: 100316,3556@compuserve.com*

# 484. *Gly-X carboxypeptidase*

## Databanks

*Peptidase classification: clan MH, family M20, MEROPS ID: M20.002*
*NC-IUBMB enzyme classification: EC 3.4.17.4*
*Chemical Abstracts Service registry number: 9025-25-6*
*Databank codes:*

| Species | SwissProt | PIR | EMBL (cDNA) | EMBL (genomic) |
|---|---|---|---|---|
| *Saccharomyces cerevisiae* | P27614 | S16693 S16881 | X57316 X63068 | Z49447: chromosome X ORF |

## Name and History

The introduction of the synthetic substrate Z-Gly┼Leu led to the discovery and biochemical characterization of an enzyme that was initially termed *peptidase α* (Felix & Brouillet, 1966). This enzyme was rediscovered later in a *Saccharomyces cerevisiae* strain deficient in carboxypeptidase Y (Chapter 132) activity and was called *carboxypeptidase S* because of its origin in *Saccharomyces* (Wolf & Weiser, 1977). The name *Gly-X carboxypeptidase* recommended by IUBMB relates to the substrate specificity of the peptidase.

## Activity and Specificity

Gly-X carboxypeptidase is a true carboxypeptidase releasing C-terminal aliphatic, aromatic, acid or basic residues from N-blocked peptides as long as glycine is in the penultimate (P1) position (Felix & Brouillet, 1966). Its pH optimum is between 6.0 and 7.0 (Felix & Brouillet, 1966; Wolf & Weiser, 1977) and assays are most conveniently made with Z-Gly┼Leu as substrate as described by Wolf & Weiser (1977). This synthetic substrate is available from Bachem (see Appendix II for full names and addresses of suppliers). Z-Gly┼Leu is also cleaved by *S. cerevisiae* carboxypeptidase Y (Chapter 132) and therefore, when using crude enzyme preparations, the assay has to be performed in the presence of PMSF or DFP, which are strong inhibitors of the enzyme that do not affect Gly-X carboxypeptidase.

Gly-X carboxypeptidase is strongly inhibited by metal chelators such as 1,10-phenanthroline and EDTA and can be reactivated by a number of divalent metal ions. Some thiol reagents are also inhibitory (Felix & Brouillet, 1966; Wolf & Weiser, 1977).

## Biological Aspects

Most of the enzyme activity resides within the vacuole in *S. cerevisiae* (Emter & Wolf, 1984). The gene structure of Gly-X carboxypeptidase S (*CPS1*) has been elucidated

(Spormann *et al*., 1991; Bordallo *et al*., 1991). Five putative *N*-glycosylation sites are present in the protein sequence and specific antibodies recognize two glycoprotein species with three and two glycosyl residues added respectively (Spormann *et al*., 1992). Biosynthesis of this enzyme depends on the secretory pathway. The N-terminal region contains a single hydrophobic stretch of 19 residues which seems to be responsible for maintaining the precursor in a membrane-associated form during transit to the vacuole and that is cleaved by cerevisin (Chapter 104) upon delivery (Spormann *et al*., 1992).

Carboxypeptidase S activity levels are strongly regulated by the availability of nutrients and the regulation is brought about at the transcriptional level (Wolf & Ehmann, 1978; Bordallo & Suárez-Rendueles, 1993). A drastic increase in both *CPS1* mRNA and enzymatic activity is observed when two metabolic conditions are met: nitrogen deprivation and sugar fermentation (Bordallo & Suárez-Rendueles, 1993). Several upstream regulatory regions that bind proteins from a yeast extract have been located within the *CPS1* promoter. The sequence located between positions −644 and −591 was found to be responsible for transcriptional repression of the *CPS1* gene in yeast cells grown on rich nitrogen sources (Bordallo & Suárez-Rendueles, 1995).

### References

Bordallo, J., Bordallo, C., Gascón, S. & Suárez-Rendueles, P. (1991) Molecular cloning and sequencing of genomic DNA encoding yeast vacuolar carboxypeptidase yscS. *FEBS Lett.* **283**, 27–32.

Bordallo, J. & Suárez-Rendueles, P. (1993) Control of *Saccharomyces cerevisiae* carboxypeptidase S (*CPS1*) gene expression under nutrient limitation. *Yeast* **9**, 339–349.

Bordallo, J. & Suárez-Rendueles, P. (1995) *Cis* and *trans*-acting regulatory elements required for regulation of the *CPS1* gene in *Saccharomyces cerevisiae. Mol. Gen. Genet.* **246**, 580–589.

Emter, O. & Wolf, D.H. (1984) Vacuoles are not the sole compartment of proteolytic enzymes in yeast. *FEBS Lett.* **166**, 321–325.

Felix, F. & Brouillet, N. (1966) Purification et propietés de deux peptidases de levure de brasserie [Purification and properties of two peptidases from brewer's yeast]. *Biochim. Biophys. Acta* **122**, 127–144.

Spormann, D.O., Heim, J. & Wolf, D.H. (1991) Carboxypeptidase yscS: gene structure and function of the vacuolar enzyme. *Eur. J. Biochem.* **197**, 399–405.

Spormann, D.O., Heim, J. & Wolf, D.H. (1992) Biogenesis of the yeast vacuole. The precursor forms of the soluble hydrolase carboxypeptidase yscS are associated with the vacuolar membrane. *J. Biol. Chem.* **267**, 8021–8029.

Wolf, D.H. & Ehmann, C. (1978) Carboxypeptidase S from yeast: Regulation of its activity during vegetative growth and differentiation. *FEBS Lett.* **92**, 59–62.

Wolf, D.H. & Weiser, U. (1977) Studies on a carboxypeptidase Y mutant of yeast and evidence for a second carboxypeptidase activity. *Eur. J. Biochem.* **73**, 553–556.

*Paz Suárez-Rendueles*
*Departamento de Bioquímica y Biología Molecular,*
*Facultad de Medicina. Universidad de Oviedo,*
*E-33071 Oviedo, Spain*

*Javier Bordallo*
*Departamento de Bioquímica y Biología Molecular,*
*Facultad de Medicina. Universidad de Oviedo,*
*E-33071 Oviedo, Spain*

# 485. *Peptidase T*

## Databanks

Peptidase classification: clan MH, family M20, MEROPS ID: M20.003
NC-IUBMB enzyme classification: none
Databank codes:

| Species | SwissProt | PIR | EMBL (cDNA) | EMBL (genomic) |
|---|---|---|---|---|
| *Bacillus subtilis* | P55179 | – | X99339 | – |
| *Escherichia coli* | P29745 | – | M64519 | – |
| *Haemophilus influenzae* | P45172 | – | U32814 | – |
| *Lactococcus lactis* | P42020 | – | L27596 | – |
| *Salmonella typhimurium* | P26311 | – | M62725 | – |

## Name and History

*Peptidase T* was discovered as an activity in extracts of *Salmonella typhimurium* that removes the N-terminal amino acid from tripeptides (Strauch & Miller, 1983). *S. typhimurium* mutants overexpressing peptidase T were isolated by selection for peptide utilization in a *pepN pepA pepB*-deficient *S. typhimurium* strain (Strauch *et al.*, 1983). Studies of the regulation of *pepT* showed that the gene is transcriptionally regulated and is induced under anaerobic growth conditions as part of the *fnr* (originally called *oxrA* in *S. typhimurium*) regulon (Strauch *et al.*, 1985). The relationship of this enzyme to the mammalian tripeptide aminopeptidase (Chapter 520) is not yet clear.

## Activity and Specificity

Peptidase T removes the N-terminal amino acid from a variety of tripeptides. Met+Gly-Gly, Met+Gly-Met, Met-Ala-Ser, Phe+Gly-Gly, Ala+Ala-Ala are all substrates and all are hydrolyzed to produce the free N-terminal amino acid and the C-terminal dipeptide (Strauch & Miller, 1983). Phe-Gly-Gly-$NH_2$, Met-Gly-Met-Met and (Ala)$_4$ are not substrates. Based on *in vitro* studies of peptide utilization in strains dependent on peptidase T for tripeptide hydrolysis, N-terminal Met tripeptides with Ala, Thr, Gly, Met, Leu and Phe in the P1′ position are all hydrolyzed. Leu+Gly-Gly and Leu+Leu-Leu are also substrates but Pro-Gly-Gly is not. Based on this rather small set of peptides, peptidase T appears to be relatively specific for tripeptides but relatively nonspecific for the amino acid composition of the peptide.

Peptidase T can be assayed conveniently by HPLC analysis of reaction mixtures derivatized with trinitrobenzenesulfonyl chloride (Strauch & Miller, 1983). Met-Gly-Gly has been used for this purpose but other substrates should work equally well. No specific substrates that are hydrolyzed by PepT but not by other peptidases present in crude extracts of *S. typhimurium* have been identified. The enzyme is active at pH 7.5. The activity can be conveniently detected in crude extracts after electrophoresis in nondenaturing polyacrylamide gels using a coupled peptidase activity stain (Strauch *et al.*, 1983).

Peptidase T is inhibited by EDTA as expected of a metallopeptidase (Strauch & Miller, 1983). In crude extracts activity is stimulated by $Zn^{2+}$ (0.1 mM) and less efficiently by $Mn^{2+}$ (1 mM). $Mg^{2+}$ and $Co^{2+}$ (0.1–1 mM) neither stimulate nor inhibit. Peptidase T presumably contains a tightly bound divalent cation since it can be purified in active form using buffers which do not contain metal ions (Miller *et al.*, 1991). The identity of this ion has not been determined.

## Structural Chemistry

Metal-binding residues have not been identified for this family but the existence of conserved residues that might serve this function has been noted (Miller *et al.*, 1991, Rawlings &

Barrett, 1995, see also Alignment 482.1 on the CD-ROM). The region between amino acids 134 and 179 (*S. typhimurium* sequence) contains several strongly conserved amino acids that might be involved in metal ion binding. Several of these residues are also conserved in other peptidases that are not tripeptide specific (*Escherichia coli* peptidase D (Chapter 478) and aminopeptidase iap and *Pseudomonas* carboxypeptidase G$_2$ (Chapter 483). All of the peptidase T homologs (*S. typhimurium, E. coli, Haemophilus influenzae, Lactococcus lactis, Bacillus subtilis*) have molecular masses of 45 kDa (409–412 amino acids) and all have acidic calculated isoelectric points (4.6–5.6). None of the *pepT* open reading frames show any evidence of a signal peptide and all are presumably cytoplasmic enzymes.

## Preparation

Peptidase T has been purified from an *S. typhimurium* strain carrying the gene on a high copy number plasmid (Miller *et al.*, 1991).

## Biological Aspects

The *S. typhimurium pepT* gene has two promoters: one allows low-level, constitutive expression under aerobic growth conditions and the other allows high-level, Fnr-dependent expression under anaerobic conditions (Strauch *et al.*, 1985; Miller *et al.*, 1991; Lombardo *et al.*, 1997). Fnr is a transcriptional regulator required for high-level transcription of nitrate reductase, fumarate reductase and other genes involved in anaerobic respiratory pathways. The sequence of the *pepT* promoter region of *E. coli* is identical to that of *S. typhimurium*, indicating that *pepT* regulation is the same in the two organisms. The physiological role of peptidase T is not fully understood. The dependence of *pepT* expression on Fnr in *E. coli* and *S. typhimurium* suggests that the enzyme may play a role in anaerobic respiration. It has been suggested that peptidase T might help to liberate amino acids that can either act directly or indirectly as electron donors or acceptors for energy generation. Aspartate or asparagine, for example, can be converted to fumarate which can function as an electron acceptor in anaerobic respiration. The enzyme from *Lactococcus lactis* (Mierau *et al.*, 1994) may be one of a number of peptidases in this organism that allow it to grow on peptides generated from casein in milk (Mierau *et al.*, 1996). The enzymes from *H. influenzae* and *B. subtilis* have been identified by sequence only and nothing is known about their physiological functions.

## Distinguishing Features

The specificity of peptidase T for tripeptides distinguishes it from any other enzyme known to be present in extracts of *E. coli* or *S. typhimurium*. It is also the only peptidase in these organisms that is present at elevated levels in anaerobically grown cells. Unlike most of the other enzymes that hydrolyze small peptides in these organisms, peptidase T is not inhibited by $Zn^{2+}$.

## References

Lombardo, M.J., Lee, A.A., Knox, T.M. & Miller, C.G. (1997) Regulation of the *Salmonella typhimurium pepT* gene by cyclic AMP receptor protein (CRP) and FNR acting at a hybrid CRP-FNR site. *J. Bacteriol.* **179**, 1909–1917.

Mierau, I., Haandrikman, A.J., Velterop, O., Tan, P.S., Leenhouts, K.L., Konigs, W.N., Venema, G. & Kok, J. (1994) Tripeptidase gene (*pepT*) of *Lactococcus lactis*: molecular cloning and nucleotide sequencing of *pepT* and construction of a chromosomal deletion mutant. *J. Bacteriol.* **176**, 2854–2861.

Mierau, I., Kunji, E.R., Leenhouts, K.J., Hellendoorn, M.A., Haan-

drikman, A.J., Poolman, B., Konings, W.N., Venema, G. & Kok, J. (1996) Multiple-peptidase mutants of *Lactococcus lactis* are severely impaired in their ability to grow in milk. *J. Bacteriol.* **178**, 2794–2803.

Miller, C.G., Miller, J.L. & Bagga, D.A. (1991) Cloning and nucleotide sequence of the anaerobically regulated *pepT* gene of *Salmonella typhimurium. J. Bacteriol.* **173**, 3554–3558.

Rawlings, N. & Barrett, A.J. (1995) Evolutionary families of metallopeptidases. *Methods Enzymol.* **248**, 183–228.

Strauch, K.L. & Miller, C.G. (1983) Isolation and characterization of *Salmonella typhimurium* mutants lacking a tripeptidase (peptidase T). *J. Bacteriol.* **154**, 763–771.

Strauch, K.L., Carter, T.H. & Miller, C.G. (1983) Overproduction of *Salmonella typhimurium* peptidase T. *J. Bacteriol.* **156**, 743–751.

Strauch, K.L., Lenk, J.B., Gamble, B.L. & Miller, C.G. (1985) Oxygen regulation in *Salmonella typhimurium. J. Bacteriol.* **161**, 673–680.

*Charles G. Miller*
*Department of Microbiology,*
*B103 Chemical and Life Sciences Laboratory,*
*University of Illinois at Champaign-Urbana,*
*Urbana, IL 61801, USA*
*Email: charlesm@uiuc.ed*

# 486. Peptidase V

## Databanks

*Peptidase classification: clan MH, family M20, MEROPS ID: M20.004*
*NC-IUBMB enzyme classification: none*
*Databank codes:*

| Species | SwissProt | PIR | EMBL (cDNA) | EMBL (genomic) |
|---|---|---|---|---|
| *Lactobacillus delbrueckii* | P45494 | S57902 | Z31377 | – |

## Name and History

Several di- or di/tripeptidases of lactic acid bacteria have been characterized (Wohlrab & Bockelmann, 1992; Montel *et al.*, 1995; Tan *et al.*, 1995; Gobbetti *et al.*, 1996) or cloned and sequenced recently (Dudley *et al.*, 1996; Mierau *et al.*, 1994; Vesanto *et al.*, 1996) but **peptidase V** is so far known only from *Lactobacillus delbrueckii* subsp. *lactis* DSM7290 (Vongerichten *et al.*, 1994). The original EMBL nucleotide

submission of *pepV* describes the corresponding enzyme precisely. It is a carnosinase, cleaving the unusual peptide β-Ala-His (carnosine) and is also active on di- and tripeptides. The eukaryotic carnosinase and the bacterial dipeptidase PepD, both hydrolyzing β-Ala-His, are dealt with in Chapters 528 and 487. The name PepV has also been assigned to a dipeptidase from *Lactococcus lactis* NCDO712 (P. Strøman, unpublished results), which has significant amino acid identity with the *Lb. delbrueckii* subsp. *lactis* enzyme. The *pepV* gene was screened by complementation in *Escherichia coli*. Recombinant clones harboring *pepV* allowed the utilization of carnosine as a source of histidine by an *E. coli* mutant (*pepD⁻*, *hisG⁻*).

## Activity and Specificity

The substrate specificity of peptidase V was determined with crude cell extracts of recombinant *E. coli* by native disk gel electrophoresis (Davis, 1964) with subsequent histochemical staining (method of Sugiura *et al.*, 1977, modified by Vongerichten *et al.*, 1994) with a variety of different substrates. The hydrolyzed amino acids were detected either in a reaction with amino acid oxidase or in an L-alanine dehydrogenase/NAD coupled reaction. This method allows the determination and comparison of substrate specificities without the necessity of protein purification. A band with an $R_F$ value of 0.9 indicates PepV activity. The enzyme cleaves a variety of dipeptides, notably those with an N-terminal β-alanyl residue, and some tripeptides (Met┼Ala-Ser, Met┼Met-Met, Phe┼Gly-Gly, Phe┼Phe-Phe).

The enzyme is a metalloprotease: 1,10-phenanthroline or EDTA give full inhibition. Inhibitors of other mechanistic classes have no significant effect on enzyme activity.

## Structural Chemistry

The *pepV* gene encodes a protein with a calculated molecular mass of 51 998 Da. Comparison with a still unpublished dipeptidase from *Lactococcus lactis* NCDO712 (P. Strøman, personal communication) revealed 47% identity.

## Preparation

PepV has not yet been isolated, but purification should be facilitated by the use of recombinant plasmids allowing heavy overexpression (25% of the soluble protein) in *E. coli* (Vongerichten *et al.*, 1994).

## Biological Aspects

The 5′ end of the *pepV* mRNA was determined by primer extension analysis; the upstream region revealed two conserved sequences TTGcCA and TAgAAT, separated by 17 bp, which closely resemble the −35 and −10 boxes of the *E. coli* consensus promoter. Regulation of *pepV* expression has not yet been investigated. The hydrophilicity plot of PepV does not indicate any obvious membrane-spanning domains, nor does the N-terminus reveal any putative signal peptide sequence. This suggests that PepV is intracellularly located.

The physiological role of PepV, as an enzyme which exclusively degrades peptides, is not certain. The cleavage of unusual β-alanyl peptides and the homology with non-peptidase homologues of family M20 might suggest participation in metabolic pathways having nothing in common with casein degradation.

## Distinguishing Features

The specificity of PepV determined with crude cell extracts of *Lactobacillus delbrueckii* subsp. *lactis* by native disc gel electrophoresis distinguishes the enzyme from other general aminopeptidases and defines β-Ala-dipeptides as specific substrates.

## Related Peptidases

A somewhat similar di/tripeptidase (DTP) from *Lactobacillus helveticus* CNRZ32 was described by Nowakowski *et al.* (1993), but the enzyme was not further characterized. DTP and PepV have some overlapping substrate specificities, but some substrates are hydrolyzed specifically by only one of the two peptidases. PepV cleaves Phe┼Leu, but is not active on Leu-Leu-Leu, which is a substrate of DTP. On the other hand, the PepV substrate Phe-Leu is not hydrolyzed by DTP. However, since both peptidases are isolated from closely related *Lactobacillus* strains, homology might be expected.

## Further Reading

For a review, see Vongerichten *et al.* (1994).

## References

Davis, B.J. (1994) Disk electrophoresis II. *Ann. NY Acad. Sci.* **121**, 404–427.

Dudley, E.G., Husgen, A.C., He, W. & Steele, J.L. (1996) Sequencing, distribution, and inactivation of the dipeptidase A gene (*pepDA*) from *Lactobacillus helveticus* CNRZ32. *J. Bacteriol.* **178**, 701–704.

Gobbetti, M., Smacchi, E. & Corsetti, A. (1996) The proteolytic system of *Lactobacillus sanfrancisco* CB1: purification and characterization of a proteinase, a dipeptidase, and an aminopeptidase. *Appl. Environ. Microbiol.* **62**, 3220–3226.

Mierau, I., Haandrikman, A.J., Velterop, O., Tan, P.S., Leenhouts, K.L., Konings, W.N., Venema, G. & Kok, J. (1994) Tripeptidase gene (pepT) of *Lactococcus lactis*: molecular cloning and nucleotide sequencing of pepT and construction of a chromosomal deletion mutant. *J. Bacteriol.* **176**, 2854–2861.

Montel, M.C., Seronie, M.P., Talon, R. & Hebraud, M. (1995) Purification and characterization of a dipeptidase from *Lactobacillus sake*. *Appl. Environ. Microbiol.* **61**, 837–839.

Nowakowski, C.M., Bhowmik, T.K. & Steele, J.L. (1993) Cloning of peptidase genes from *Lactobacillus helveticus* CNRZ 32. *Appl. Microbiol. Biotechnol.* **39**, 204–210.

Sugiura, M., Ito, Y., Hirano, K. & Sawaki, S. (1977) Detection of dipeptidase and tripeptidase activities on poly acrylamide gel and cellulose acetate gel by reduction of tetrazolium salts. *Anal. Biochem.* **81**, 481–484.

Tan, P.S.T., Sasaki, M., Bosman, B.W. & Iwasaki, T. (1995)

Purification and characterization of a dipeptidase from *Lactobacillus helveticus* SBT 2171. *Appl. Environ. Microbiol.* **61**, 3430–3435.

Vesanto, E., Peltoniemi, K., Purtsi, T., Steele, J.L. & Palva, A. (1996) Molecular characterization, over-expression and purification of a novel dipeptidase from *Lactobacillus helveticus. Appl. Microbiol. Biotechnol.* **45**, 638–645.

Vongerichten, K.F., Klein, J.R., Matern, H. & Plapp, R. (1994) Cloning and DNA sequence analysis of *pepV*, a carnosinase gene from *Lactobacillus delbrueckii* subsp. *lactis* DSM7290 and partial characterization of the enzyme. *Microbiology* **140**, 2591–2600.

Wohlrab, Y. & Bockelmann, W. (1992) Purification and characterization of a dipeptidase from *Lactobacillus delbrueckii* subsp. *bulgaricus. Int. Dairy J.* **2**, 345–361.

**Jürgen R. Klein**
*Department of Microbiology, University of Kaiserslautern,*
*PO Box 3049,*
*D-67653 Kaiserslautern, Germany*
*Email: jklein@rhrk.uni-kl.de*

**Bernhard Henrich**
*Department of Microbiology, University of Kaiserslautern,*
*PO Box 3049,*
*D-67653 Kaiserslautern, Germany*

# 487. X-His dipeptidase (bacteria)

## Databanks

*Peptidase classification: clan MH, family M25, MEROPS ID: M25.001*
*NC-IUBMB enzyme classification: none*
*Chemical Abstracts Service registry number: 9027-21-8*
*Databank codes:*

| Species | SwissProt | PIR | EMBL (cDNA) | EMBL (genomic) |
|---|---|---|---|---|
| *Escherichia coli* | P15288 | JU0300 | M34034 X14790 | – |
| *Haemophilus influenzae* | P44817 | – | U32750 | – |

## Name and History

Bacterial enzymes acting on X-His dipeptides were initially detected as **carnosinases**. This name refers to the cleavage of a basic dipeptide, $\beta$-Ala-L-His, which is also termed carnosine to indicate its natural occurrence in the skeletal muscle of animals (*carne*, meat). Carnosine has been suspected to be involved in physiological processes such as the production of histamine, the complexing of cations or the release of neurotransmitters. Tissues and body fluids of various animals as well as bacteria have been investigated for the presence of peptidases which split carnosine into $\beta$-Ala and L-His. Animal carnosinase was first detected by Hanson & Smith (1949). Early growth experiments with bacteria indicated that carnosine can be utilized as a source for essential amino acids by *Corynebacterium* (Mueller, 1938), *Aerobacter aerogenes* (Magasanik, 1955), *Salmonella typhimurium* (Hartman, 1956) and *Escherichia coli* (Payne, 1973). However, it was not until 1974 that the first direct proof of a carnosine-hydrolyzing activity in bacteria was obtained with *Pseudomonas aeroginosa* (van der Drift & Ketelaars, 1974). This activity was assigned to the EC entry for **aminoacylhistidine dipeptidase** (EC 3.4.13.3) (Chapter 529) which initially applied to carnosinases of both eukaryotic and prokaryotic origin. An earlier entry (EC 3.4.3.3) for the same enzyme was transferred to EC 3.4.13.3 before 1972. In 1989, EC 3.4.13.3 was restricted to aminoacylhistidine dipeptidases which do not act on the meat peptides homocarnosine and homoanserine, in order to distinguish these enzymes from a separate entry (EC 3.4.13.13) which, since 1981, had been allocated to homocarnosine from mammalian kidney and uterus (Lenney *et al.*, 1977). On the basis of biochemical similarities (inhibition by metal chelators, activation by thiols) this distinction was abolished in 1992, and EC 3.4.13.13 was closed and transferred to EC 3.4.13.3, which since then refers to *X-His dipeptidase* from the cytosol of mammalian cells. The mammalian enzyme is covered by a separate entry (Chapter 529) in this *Handbook*.

Carnosine-cleaving enzymes are known from a number of bacterial species (Bersani *et al.*, 1980), but only PepD of *S. typhimurium* (Kirsh *et al.*, 1978) and *E. coli* (Schroeder *et al.*, 1994), and PepV of *Lactobacillus delbrueckii* subsp. *lactis* (Vongerichten *et al.*, 1994) have been biochemically and genetically characterized. Also, the existence of a PepD homolog in *Haemophilus influenzae* has been inferred from nucleotide sequence data (Fleischmann *et al.*, 1995). Since these enzymes exhibit rather broad substrate specificities, the designation of some of them as X-His dipeptidases (EC 3.4.13.3), found in recent database entries, is misleading. We therefore recommend the name *dipeptidase PepD* for bacterial carnosinases of the *E. coli* type. Due to the lack of sequence homology with *E. coli* PepD, the *Lb. delbrueckii* subsp. *lactis* enzyme has been separated out and moved to Chapter 486.

The properties of dipeptidase PepD are similar to those of the so-called cytosol nonspecific dipeptidase (EC 3.4.13.18) (Chapter 527), a broad-specificity metallopeptidase which, in addition to mammals and yeast, has also been detected in *Mycobacterium* (Plancot & Han, 1972), *Streptococcus* (Rabier & Desmazeaud, 1973), and *E. coli* B (Patterson *et al.*, 1973). Whether these bacterial enzymes are identical with PepD of *S. typhimurium* and *E. coli* remains unclear.

## Activity and Specificity

Although PepD is the only peptidase of *S. typhimurium* and *E. coli* that cleaves carnosine, the enzyme is active towards a number of other dipeptides (Kirsh *et al.*, 1978; Klein *et al.*, 1986; Schroeder *et al.*, 1994). Further data on specificity are available only for the purified *E. coli* enzyme (Schroeder *et al.*, 1994): a wide variety of unmodified dipeptides (including L-Pro-Xaa peptides) are hydrolyzed, irrespective of the character of the amino acids constituting them. Tripeptides are not cleaved. Certain nonprotein amino acids (D-Ala, $\beta$-Ala, sarcosine) are accepted in the P1 but not in the P1′ position. Hydrolysis of dipeptides is abolished by blocking the N-termini with acyl or urethane groups, whereas some C-protected dipeptides (methyl esters, dipeptide amines) are cleaved. Deformylase activity towards several formyl amino acids has also been detected. $K_m$ values towards several unique substrates range between 2 and 7 mM.

Activity can be determined by a fluorometric assay with carnosine (Kirsh *et al.*, 1978), or more conveniently by a redox assay with $\beta$-Ala-L-Ala (Henrich *et al.*, 1989). Both substrates are available from Bachem (see Appendix II for full names and addresses of suppliers). The pH optimum of purified PepD is about 9.0, and the temperature of highest activity is 37°C. Enzyme activity is stimulated by $Zn^{2+}$ and $Co^{2+}$ at concentrations of 0.1 mM, whereas $Cu^{2+}$ and $Mn^{2+}$ are potent inhibitors. Exposure to chelators such as EDTA and 1,10-phenanthroline leads to complete loss of activity. Moderate inhibitory effects are also observed in the presence of several other protease inhibitors (Schroeder *et al.*, 1994).

## Structural Chemistry

The gene for the *E. coli* enzyme specifies a protein of 52 kDa (Henrich *et al.*, 1990), but the native molecular masses of 158 kDa for *S. typhimurium* (Kirsh *et al.*, 1978) and 100 kDa for *E. coli* (Klein *et al.*, 1986) indicate that PepD may be active as a di- or trimer. The *E. coli* enzyme has a pI of 4.7. Its deduced amino acid sequence is neither similar to that of any other protein nor does it display any common metal-binding motif.

## Preparation

PepD has been enriched 20-fold from *S. typhimurium* (Kirsh *et al.*, 1978). The *E. coli* enzyme was purified to electrophoretic homogeneity from a clone in which PepD, due to heavy overexpression from a recombinant plasmid, made up more than 50% of total soluble protein (Schroeder *et al.*, 1994).

## Biological Aspects

Bacterial peptidases play essential roles in the utilization of substrates supplied in the medium as well as in the breakdown of intracellular proteins, especially under conditions of nutritional starvation (Miller & Schwartz, 1978; Yen *et al.*, 1980). As carnosinase apparently is not present among intestinal peptidases of animals, carnosine may occur extensively in the usual environment of enteric microorganisms. Enzymes with carnosinase activity may be active in various gram-negative and gram-positive bacteria (Payne, 1973; Bersani *et al.*, 1980), but assigning them to the PepD type seems untimely since appropriate biochemical and genetic data are missing.

The gene encoding the *E. coli* enzyme has been mapped and sequenced (Henrich *et al.*, 1989, 1990; Henrich & Plapp, 1991). Transcription is initiated with low efficiency at two adjacent promoters and increases 5-fold in response to phosphate limitation. This effect is independent of the regulatory genes of the *pho* regulon (Henrich *et al.*, 1992). N-Formyl-Met is removed from the nascent *E. coli* PepD, but further post-translational modifications are not known. Being a typical soluble protein, the enzyme is mainly located in the cytoplasm, but under conditions of heavy overexpression, a significant portion of activity seems to occur in the periplasmic space (Henrich *et al.*, 1990). Inducibility by $\beta$-Ala has been observed with a carnosinase from *P. aeruginosa* (van der Drift & Ketelaars, 1974).

## Distinguishing Features

Considering substrate patterns and other biochemical properties, PepD may easily be confused with other broad-specificity carnosine-cleaving metallopeptidases of bacterial origin. This applies especially to the carnosinase peptidase V (Chapter 486) of *Lb. delbrueckii* subsp. *lactis*, which is in peptidase family M20 (Chapter 482) and is distinguished from PepD mainly on the basis of amino acid sequence comparisons. The assignment of bacterial carnosinases will certainly be facilitated as soon as more information on the cleavage preferences and the molecular relationships of individual enzymes becomes available.

## Further Reading

For a review, see Schroeder (1994).

## References

Bersani, C., Bianchi, M.A., Cantoni, C. & Baretta, G. (1980) Carnosinase: its presence in bacteria. *Arch. Vet. Ital.* **31** (suppl. 1–2), 1–3.

Fleischmann, R.D., Adams, M.D., White, O. *et al.* (1995) Whole-genome random sequencing and assembly of *Haemophilus influenzae* Rd. *Science* **269**, 496–512.

Hanson, H.T. & Smith, E.L. (1949) Carnosinase: an enzyme of swine kidney. *J. Biol. Chem.* **179**, 789–801.

Hartman, P.E. (1956) Linked loci in the control of consecutive steps in the primary pathway of histidine synthesis in *Salmonella typhimurium*. In: *Genetic Studies with Bacteria*. Washington: Carnegie Institute of Washington, pp. 35–62.

Henrich, B. & Plapp, R. (1991) Locations of the genes from *PepD* through *proA* on the physical map of the *Escherichia coli* chromosome. *J. Bacteriol.* **173**, 7407–7408.

Henrich, B., Schroeder, U., Frank, R.W. & Plapp, R. (1989) Accurate mapping of the *Escherichia coli* PepD gene by sequence analysis of its 5′ flanking region. *Mol. Gen. Genet.* **215**, 369–373.

Henrich, B., Monnerjahn, U. & Plapp, R. (1990) PepD gene (*PepD*) of *Escherichia coli* K-12: nucleotide sequence, transcript mapping, and comparison with other peptidase genes. *J. Bacteriol.* **172**, 4641–4651.

Henrich, B., Backes, H., Klein, J.R. & Plapp, R. (1992) The promoter region of the *Escherichia coli* PepD gene: deletion analysis and control by phosphate concentration. *Mol. Gen. Genet.* **232**, 117–125.

Kirsh, M., Dembinski, D.R., Hartman, P.E. & Miller, C. (1978) *Salmonella typhimurium* peptidase active on carnosine. *J. Bacteriol.* **134**, 361–374.

Klein, J., Henrich, B. & Plapp, R. (1986) Cloning and expression of the *PepD* gene of *Escherichia coli*. *J. Gen. Microbiol.* **132**, 2337–2343.

Lenney, J.F., Kan, S.-C., Siu, K. & Sugiyama, G.H. (1977) Homocarnosinase: a hog kidney dipeptidsae with a broader specificity than carnosinase. *Arch. Biochem. Biophys.* **184**, 257–266.

Magasanik, B. (1955). The metabolic control of histidine assimilation and dissimilation in *Aerobacter aerogenes*. *J. Biol. Chem.* **213**, 557–569.

Miller, C.G. & Schwartz, G. (1978) Peptidase-deficient mutants of *Escherichia coli*. *J. Bacteriol.* **135**, 603–611.

Mueller, J.H. (1938) The utilization of carnosine by the diphtheria *Bacillus*. *J. Biol. Chem.* **123**, 421–423.

Patterson, E.K., Gatmaitan, J.S. & Hayman, S. (1973) Substrate specificity and pH dependence of dipeptidases purified from *Escherichia coli* B and from mouse ascites tumor cells. *Biochemistry* **12**, 3701–3709.

Payne, J.W. (1973) Peptide utilization in *Escherichia coli*: studies with peptides containing β-alanyl residues. *Biochim. Biophys. Acta* **298**, 469–478.

Plancot, M.-T. & Han, K.-K. (1972) Purification and characterization of an intracellular dipeptidase from *Mycobacterium phlei*. *Eur. J. Biochem.* **28**, 327–333.

Rabier, D. & Desmazeaud, M.J. (1973) Inventory of different intracellular peptidase activities of *Streptococcus thermophilus*. *Biochimie* **55**, 389–404.

Schroeder, U., Henrich, B., Fink, J. & Plapp, R. (1994) PepD of *Escherichia coli* K-12, a metallopeptidase of low substrate specificity. *FEMS Microbiol. Lett.* **123**, 153–160.

van der Drift, C. & Ketelaars, H.C.J. (1974) Carnosinase: its presence in *Pseudomonas aeruginosa*. *Antonie van Leeuwenhoek* **40**, 377–384.

Vongerichten, K.F., Klein, J.R., Matern, H. & Plapp, R. (1994) Cloning and nucleotide sequence analysis of *pepV*, a carnosinase gene from *Lactobacillus delbrueckii* subsp. *lactis* DSM7290, and partial characterization of the enzyme. *Microbiology* **140**, 2591–2600.

Yen, C., Green, L. & Miller, C.G. (1980) Degradation of intracellular protein in *Salmonella typhimurium* peptidase mutants. *J. Mol. Biol.* **143**, 21–33.

*Bernhard Henrich*
*Department of Microbiology, University of Kaiserslautern,*
*PO Box 3049,*
*D-67653 Kaiserslautern, Germany*
*Email: henrich@rhrk.uni-kl.de*

*Jürgen R. Klein*
*Department of Microbiology, University of Kaiserslautern,*
*PO Box 3049,*
*D-67653 Kaiserslautern, Germany*

# 488. Aminopeptidase Co/Y

## Databanks

*Peptidase classification: clan MH, family M28, MEROPS ID: M28.001*
*NC-IUBMB enzyme classification: 3.4.11.15*

*Databank codes:*

| Species | SwissProt | PIR | EMBL (cDNA) | EMBL (genomic) |
|---|---|---|---|---|
| *Saccharomyces cerevisiae* | P37302 | A54134<br>S39142<br>S44548 | L31635<br>Z36155 | X76053: chromosome II fragment |

## Name and History

This enzyme was discovered in 1982 in the yeast *Saccharomyces cerevisiae* as an activity preferentially cleaving the basic amino acids lysine and arginine from the respective aminoacyl-*p*-nitroanilides (Achstetter *et al.*, 1982). In crude extracts, activity against these substrates was only visible in the presence of $Co^{2+}$ ions and the enzyme was therefore named **aminopeptidase Co** (Achstetter *et al.*, 1982). Aminopeptidase Co was localized to the yeast lysosome, the vacuole (Emter & Wolf, 1984). More than 10 years later a vacuolar aminopeptidase was purified which exhibited the same characteristics, a strong $Co^{2+}$ activation of the cleavage of aminoacyl derivatives of fluorogenic compounds (aminoacyl-NHMec) and a preference for basic amino acids. The authors named this enzyme **aminopeptidase Y** (Yasuhara *et al.*, 1994). The molecular masses of aminopeptidase Co and aminopeptidase Y are 100 and 80 kDa, respectively, as determined by gel filtration. The fact that both enzymes were localized to the vacuole and that only one aminopeptidase, strongly activated by $Co^{2+}$ and splitting basic aminoacyl-*p*-nitroanilides, could be found in yeast cell extracts (Achstetter *et al.*, 1982; Yasuhara *et al.*, 1994) led to the conclusion that aminopeptidase Co and aminopeptidase Y are one and the same enzyme. The name originally recommended by IUBMB for this enzyme was **lysyl aminopeptidase**, but this has now been revised to **aminopeptidase Y**.

## Activity and Specificity

The enzyme protein is synthesized as a 74 kDa precursor which can be cleaved by the vacuolar proteinase cerevisin (Chapter 104) to a 70 kDa mature enzyme. Both enzyme forms are active, the matured 70 kDa form, however, exhibits an about 5-fold enhanced activity over the precursor form (Yasuhara *et al.*, 1994). The enzyme cleaves the N-terminal amino acid from aminoacyl arylamides (Achstetter *et al.*, 1982; Yasuhara *et al.*, 1994) as well as of peptides. Peptide derivatives are far more rapidly hydrolyzed than aminoacyl derivatives. Of aminoacyl-$NHPhNO_2$ or -NHMec derivatives the lysyl and arginyl derivatives are the best substrates. Activity then decreases in the order Leu, Met, Ala, Ser, Phe, Tyr, Pro, Gly, Gln as N-terminal amino acid. The activities against the lysyl derivative and the glutamyl-derivative differ by a factor of more than 100 (Achstetter *et al.*, 1982, Yasuhara *et al.*, 1994). While $Co^{2+}$ ions have a dramatic activating effect of 3- to 26-fold on the cleavage of aminoacyl derivatives of *p*-nitroaniline (Achstetter *et al.*, 1982) and 4-methylcoumaryl-7-amine (Yasuhara *et al.*, 1994), they exhibit an inhibitory effect on the cleavage of most of the peptides and peptidyl derivatives measured (Yasuhara *et al.*, 1994). Met-Gly-Gly, Leu-Leu, Arg-Val, Phe-Ala and Val-Gly are

exceptions. Their cleavage by the enzyme is $Co^{2+}$ activated (Yasuhara *et al.*, 1994). The saturating $Co^{2+}$ concentration for activation is above 0.25 mM (Yasuhara *et al.*, 1994). The pH optimum of the enzyme was determined to be between 7.5 and 8.5 (Achstetter *et al.*, 1982; Yasuhara *et al.*, 1994). The enzyme is inhibited by metal chelating agents (Achstetter *et al.*, 1982; Yasuhara *et al.*, 1994), DTT, amastatin, triphosphate and pyrophosphate (Yasuhara *et al.*, 1994). $Cu^{2+}$ and $Zn^{2+}$ ions are inhibitory (Achstetter *et al.*, 1982; Yasuhara *et al.*, 1994).

## Structural Chemistry

Aminopeptidase Y is synthesized as a preproenzyme containing 537 amino acids. As an enzyme imported into the vacuole of the yeast cell the presumably 21 amino acid residue pre and the 35 amino acid residue pro sequences are cleaved off, resulting in the mature enzyme protein of 481 amino acids (molecular mass 52 900 Da; Nishizawa *et al.*, 1994). The enzyme contains carbohydrate and is synthesized as a 74 kDa precursor, which is converted into the mature form of 70 kDa. Also a mature enzyme of 75 kDa is found which contains a higher degree of glycosylation (Yasuhara *et al.*, 1994; Nishizawa *et al.*, 1994). One zinc atom per molecule of enzyme was found. However, none of the known zinc-binding motifs is present in aminopeptidase Y.

## Preparation

Aminopeptidase Y can be purified to homogeneity from commercial baker's yeast (*Saccharomyces cerevisiae*) (Yasuhara *et al.*, 1994).

## Biological Aspects

Up to now the enzyme has only been found in the yeast *Saccharomyces cerevisiae*. The nucleotide sequence of the enzyme (1611 nucleotides) is localized on chromosome II (Feldmann *et al.*, 1994; Nishizawa *et al.*, 1994). Chromosomal deletion of aminopeptidase Y resulted in viable cells. However, the aminopeptidase activity against substrates containing N-terminal lysine or alanine was dramatically reduced in vacuoles of the deletion mutant (Nishizawa *et al.*, 1994). As the proteolytic capacity of the vacuole of yeast is necessary only under nutritional stress conditions (Teichert *et al.*, 1989; Rupp & Wolf, 1995), one may expect the function of aminopeptidase Y to be important under starvation conditions to provide amino acids for energy production and new protein synthesis (Teichert *et al.*, 1989). As expected, the level of the active enzyme increases in cells grown in minimal medium or deprived of nitrogen (Herrera-Camacho & Suárez-Rendueles, 1996).

## Editors' Note

*The editors agree with the conclusion of Dr Wolf that this enzyme is the same as that described in the following entry (Chapter 489). The name aminopeptidase Y should be applied in both instances.*

## References

Achstetter, T., Ehmann, C. & Wolf, D.H. (1982) Aminopeptidase Co, a new yeast peptidase. *Biochem. Biophys. Res. Commun.* **109**, 341–347.

Emter, O. & Wolf, D.H. (1984) Vacuoles are not the sole compartments of proteolytic enzymes in yeast. *FEBS Lett.* **166**, 321–325.

Feldmann, H. Aigle, M., Aljinovic, G. *et al.* (1994) Complete DNA sequence of yeast chromosome II. *EMBO J.* **13**, 5795–5809.

Herrera-Camacho, I. & Suárez-Rendueles, P. (1996) Regulation of the yeast vacuolar aminopeptidase Y gene (*APY1*) expression. *FEMS Microbiol. Lett.* **139**, 127–132.

Nishizawa, M., Yasuhara, T., Nakai, T., Fujiki, Y. & Ohashi, A. (1994) Molecular cloning of the aminopeptidase Y gene of *Saccharomyces cerevisiae*. Sequence analysis and gene disruption of a new aminopeptidase. *J. Biol. Chem.* **269**, 13651–13655.

Rupp, S. & Wolf, D.H. (1995) Biogenesis of the yeast vacuole (lysosome). The use of active-site mutants of proteinase yscA to determine the necessity of the enzyme for vacuolar proteinase maturation and proteinase yscB stability. *Eur. J. Biochem.* **231**, 115–125.

Teichert, U., Mechler, B., Müller, H. & Wolf, D.H. (1989) Lysosomal (vacuolar) proteinases of yeast are essential catalysts for protein degradation, differentiation and cell survival. *J. Biol. Chem.* **264**, 16037–16045.

Yasuhara, T., Nakai, T. & Ohashi, A. (1994) Aminopeptidase Y, a new aminopeptidase from *Saccharomyces cerevisiae*. Purification, properties, localization and processing by protease B. *J. Biol. Chem.* **269**, 13644–13650.

*Dieter H. Wolf*
*Institut für Biochemie, Universität Stuttgart,*
*Pfaffenwaldring 55,*
*D-70569 Stuttgart, Germany*
*Email: dieter.wolf@po.uni-stuttgart.de*

# 489. Aminopeptidase Y

## Databanks

*Peptidase classification: clan MH, family M28, MEROPS ID: M28.001*
*NC-IUBMB enzyme classification: EC 3.4.11.15*
*Databank codes:*

| Species | SwissProt | PIR | EMBL (cDNA) | EMBL (genomic) |
|---|---|---|---|---|
| *Saccharomyces cerevisiae* | P37302 | A54134 S39142 S44548 | L31635 Z36155 | X76053: chromosome II fragment |

*Note*: This entry is identical to that in Chapter 488.

## Name and History

The name **aminopeptidase Y** denotes aminopeptidase of yeast, a counterpart of carboxypeptidase Y (Chapter 132) which has been known as the major carboxypeptidase localized in the vacuole/lysosome of yeast *Saccharomyces cerevisiae*. In spite of a number of reports on aminopeptidases in yeast (Jones, 1984), aminopeptidase Y was not identified until recently (Yasuhara *et al.*, 1994). The reason for this may due to its relative low activity toward dipepeptide-like substrates such as aminoacyl-NHMec. However, a finding that cobalt enhances the aminoacyl-NHMec-hydrolyzing activity led to the purification of this enzyme. Aminopeptidase Y is a highly potent aminopeptidase in the yeast vacuole.

## Activity and Specificity

Aminopeptidase Y hydrolyzes dipeptidyl-NHMec or oligopeptides far more rapidly than aminoacyl-NHMec:

Lys┼Ala-NHMec/Lys┼NHMec = 350-fold, Arg┼Arg-NH-Mec/Arg┼NHMec = 150-fold. Hydrolysis of the dipeptidyl-NHMec to amino acid and aminoacyl-NHMec can be assayed conveniently by making use of its quenching feature in the hydrolysis (Yasuhara *et al.*, 1994).

Arg-NHMec or Lys-NHMec are the most sensitive amino acyl-NHMec substrates. Leu-, Met- and Ala-NHMec are next to the basic aminoacyl-NHMec in susceptibility to the hydrolysis, followed by Ser-, Phe-, Tyr- and Pro-NHMec. In the assay, cobalt (0.25 mM) is added as an effective activator and 25-fold enhancement is observed in the hydrolysis of Lys- or Arg-NHMec. Interestingly, although cobalt acts as activator on the hydrolysis of aminoacyl-NHMec or dipeptides, hydrolysis of substrates longer than tripeptide or dipeptidyl-NHMec is inhibited by cobalt with an $IC_{50}$ value of about 0.1 mM. NHMec substrates are available from Bachem and Peptide Institute (see Appendix II for full names and addresses of suppliers).

The activity is strongly inhibited by chelators such as 1,10-phenanthroline, EDTA, DTT and the aminopeptidase-specific inhibitors bestatin and amastatin. The pH optimum is about 7.5 in the presence of cobalt.

## Structural Chemistry

Aminopeptidase Y is a 70 kDa single-chain protein. The molecule consists of 53 kDa protein and 17 kDa high mannose type sugar chains. From commercial baker's yeast, a more heavily glycosylated form, 75 kDa, was purified as well. Aminopeptidase Y contains about one atom of zinc per molecule. The calculated pI value is 5.2.

## Preparation

Aminopeptidase Y has been purified from commercial baker's yeast cells. The enzyme may account for about 0.1% of vacuolar soluble protein.

## Biological Aspects

ABYS1 mutant yeast cells from which the four vacuolar proteases have been deleted, contain a 74 kDa precursor of aminopeptidase Y. The proaminopeptidase Y is converted to the 70 kDa mature form by purified yeast cerevisin (Chapter 104). This processing results in activation of the proaminopeptidase Y.

The aminopeptidase Y gene encodes a 56 residue prepro sequence (Nishizawa *et al.*, 1994). The first 21 residues form a hydrophobic stretch that may function as a signal

sequence for the endoplasmic reticulum, and the remaining 35 residue segment accounts for the 4 kDa difference between the proform and mature aminopeptidase Y.

Anti-aminopeptidase Y antibody adsorbs over half the aminopeptidase activity in vacuolar extracts. Disruption of the aminopeptidase Y gene also results in reduction of the major part of aminopeptidase activity in the organelle, but no distinctive phenotype was observed in the cell growth. The gene is localized on chromosome II.

## Distinguishing Features

The effects of cobalt on its activity is a distinctive characteristic of aminopeptidase Y.

## Related Peptidases

Activity of aminopeptidase Co (Chapter 488) reported in the ABYS1 mutant cells and enhanced by cobalt (Achstetter *et al.*, 1982) might be due to the proform of aminopeptidase Y. The yeast chromosome II contains a homolog gene as well (Feldmann *et al.*, 1994). *Bacillus subtilis* also contains a homolog gene, *ipa-8r* (Glaser *et al.*, 1993).

## Editors' Note

*The editors consider the enzyme activity described in this chapter to be identical to that described in the preceding entry (Chapter 488). The IUBMB recommended name 'aminopeptidase Y' should be applied to both.*

## References

Achstetter, T., Ehmann, C. & Wolf, D.H. (1982) Aminopeptidase Co, a new yeast peptidase. *Biochem. Biophys. Res. Commun.* **109**, 341–347.

Feldmann, H., Aigle, M., Aljinovic, G. *et al.* (1994) Complete DNA sequence of yeast chromosome II. *EMBO J.* **13**, 5795–5809.

Glaser, P., Kunst, F., Arnaud, M. *et al.* (1993) *Bacillus subtilis* genome project: cloning and sequencing of the 97 kb region from 325° to 333°. *Mol. Microbiol.* **10**, 371–384.

Jones, E.W. (1984) The synthesis and function of proteases in *Saccharomyces*: genetic approaches. *Annu. Rev. Genet.* **18**, 233–270.

Nishizawa, M., Yasuhara, T., Nakai, T., Fujiki, Y. & Ohashi, A. (1994) Molecular cloning of the aminopeptidase Y gene of *Saccharomyces cerevisiae*. *J. Biol. Chem.* **269**, 13651–13655.

Yasuhara, T., Nakai, T. & Ohashi A. (1994) Aminopeptidase Y, a new aminopeptidase from *Saccharomyces cerevisiae*. *J. Biol. Chem.* **269**, 13644–13650.

*Toshimasa Yasuhara*
*Research Center Kyoto, Bayer Yakuhin, Ltd.,*
*Kizu-cho, Soraku-gun, Kyoto 619-02, Japan*

# 490. *Streptomyces griseus aminopeptidase*

## Databanks

*Peptidase classification: clan MH, family M28, MEROPS ID: M28.003*
*NC-IUBMB enzyme classification: none*
*Databank codes:*

| Species | SwissProt | PIR | EMBL (cDNA) | EMBL (genomic) |
|---|---|---|---|---|
| *Streptomyces griseus* | P80561 | – | – | – |

Brookhaven Protein Data Bank three-dimensional structures:

| Species | ID | Resolution | Notes |
|---|---|---|---|
| *Streptomyces griseus* | 1XJO | 1.75 | |

## Name and History

Pronase, a commercially available mixture of extracellular enzymes derived from the K-1 strain of *Streptomyces griseus*, contains both exo- and endopeptidases as demonstrated by its ability to yield a large amount of free amino acids in the digestion of protein substrates (Nomoto *et al.*, 1960). Preliminary characterization of a partially purified ***Streptomyces griseus aminopeptidase*** preparation indicated that the enzyme is dependent upon metal ions for activity; cobaltous or calcium ions confer the highest activity (Narahashi & Yanagita, 1967; Narahashi *et al.*, 1968). Following this work, two components were purified to homogeneity (Vosbeck *et al.*, 1973).

## Activity and Specificity

The activity of the enzyme can be followed by the hydrolysis of Leu-NHPhNO₂. Substrate specificity was explored on a wide variety of peptide substrates (Vosbeck *et al.*, 1975). A preference is found for hydrophobic residues at the ultimate (P1) and penultimate (P1′) positions. Charged residues are released at a slower rate. D-Amino acid residues at the P1 or P1′ positions reduce the activity substantially; no activity can be demonstrated against D-Leu-D-Leu. No prolidase activity is demonstrable, but N-terminal prolyl residues are easily released. Studies with the $\alpha$ and $\beta$ polypeptide chains of human hemoglobin revealed the enzyme to be functionally pure in that hydrolysis occurs sequentially and stops completely when a proline residue presents at a penultimate position.

The enzyme is completely inhibited by 10 mM EDTA. Maximal activity is gained by the addition of calcium chloride. At a low calcium concentration (50 µM), a broad range of activity is seen between pH values of 6.3 and 11.5 with the highest activity seen at pH 10. At a high concentration of calcium (50 mM) the activity is restricted to pH values between 6.3 and 9.5. There is no significant difference in the activity at pH 8 over a very broad range of calcium concentrations; however, at pH 10 there is virtually no activity at high calcium concentration (Vosbeck *et al.*, 1978). In these studies, free amino acids were found to be inhibitors in a noncompetitive manner against Leu-NHPhNO₂, with histidine showing the strongest inhibition. The enzyme shows relatively high thermal stability (to 70°C) over a broad pH range (6–10.5).

## Structural Chemistry

*S. griseus* aminopeptidase is a zinc enzyme (Vosbeck *et al.*, 1978) with 2 mol of the metal per mol of enzyme (Spungin & Blumberg, 1989). A variety of metal ions can reactivate the protein after treatment with EDTA (Vosbeck *et al.*, 1973; Vosbeck 1974; Ben-Meir *et al.*, 1993). In fact, calcium superactivates the enzyme, but as noted above, the effect of calcium is complex. Some metals bind to the model substrate Leu-NHPhNO₂; however, this apparently is not true for calcium (Wu & Lin, 1995). It has been claimed that the metal-free enzyme shows low activity (Wu & Lin, 1995). However, it is not clear whether the protein in that study was entirely devoid of metal or metal was removed only from one of the two sites. Zinc and calcium stabilized the enzyme as demonstrated by circular dichroism (CD) studies. The protein in the presence of either molar guanidinium chloride or EDTA shows little change in the CD spectrum with a calculated helical content of 32–37% (Vosbeck, 1974). However, in the presence of both EDTA and denaturant the helical ellipticity band is lost.

Complete sequence studies show that the protein is a single polypeptide chain with 284 residues and a calculated molecular mass of 29 723 (Maras *et al.*, 1996). Only two half-cystine residues are present; they are in the C-terminal part of the protein and are separated by four residues. In these studies, a comparison was made with the sequences of a *Bacillus subtilis* protease, the leucine aminopeptidase from *Vibrio proteolyticus* (formerly *Aeromonas proteolytica*) (*Vibrio* aminopeptidase) (Chapter 491) and aminopeptidase Y (Chapter 489) from *Saccharomyces cerevisiae*, where 26.6, 29.6, and 34.4% identities were found. Since the tertiary structure for the *V. proteolyticus* enzyme had been determined with knowledge of those residues involved in metal coordination, the corresponding residues of His85, Asp97, Glu132, Asp160 and His247 were identified in the *S. griseus* protein.

Crystallographic studies (Almog *et al*., 1993; Greenblatt *et al*., 1997) with a synchrotron radiation source have defined the tertiary structure (see Fig. 482.2) and confirmed the roles of the putative zinc-binding residues. The two zinc atoms are separated by 3.6 Å. One zinc site can be isomorphously replaced by mercury. The apoenzyme structure shows little difference from that of the holoenzyme, suggesting that the zincs are required more for activity and less for stability. Though there is only modest (24%) homology with the *Vibrio* enzyme, the overall structures are very similar. A separate calcium site has been identified and is distant (25 Å) from the zinc atoms. The structure of the enzyme is quite different from that of cattle lens leucyl aminopeptidase (Chapter 473).

Chemical modification has been undertaken in two laboratories (Vosbeck *et al*., 1975; Yang *et al*., 1994). Reaction of the protein with acetic anhydride in aqueous buffers generates inactive derivatives; however, in the presence of 50% (v/v) glycerol, acetylation yields a homogeneous product in which the N-terminal Ala is completely acetylated without loss of activity. Complete reaction with Lys residues is not observed under these conditions. Modification of His residues by diethyl pyrocarbonate identified two residues, one of which reduces the activity when reacted, whereas the other is essential for activity. The reaction of these histidines with diethyl pyrocarbonate is inhibited by calcium (Yang *et al*., 1994). Modification of amino groups with maleic anhydride does not affect activity nor does modification of either Tyr residue by tetranitromethane or Arg residues by cyclohexanedione affect catalysis by the enzyme. Only two Tyr residues are accessible to tetranitromethane (Yang *et al*., 1994).

## Comment

All studies have been with a crude commercial preparation, where previous limited proteolysis may have occurred and accounted for the finding of two or more molecular species of the aminopeptidase (Vosbeck *et al*., 1973; Yang *et al*., 1994). A significant impediment in the study of this enzyme is the inaccessibility of the parent microorganism. If it were made available, the recent resolution of the tertiary structure could provide opportunities for structure–function studies, chromosomal analysis and molecular genetic manipulations.

## References

Almog, O., Greenblatt, H.M., Spungin, A., Ben-Meir, D., Blumberg, S. & Shoham, G. (1993) Crystallization and preliminary analysis of *Streptomyces griseus* aminopeptidase. *J. Mol. Biol.* **57**, 342–344.

Ben-Meir, D., Spungin, A., Ashkenazi, R. & Blumberg, S. (1993) Specificity of *Streptomyces griseus* aminopeptidase and modulation of activity by divalent metal ion binding and substitution. *Eur. J. Biochem.* **212**, 107–112.

Greenblatt, H.M., Almog, O., Maras, B., Spungin-Bialik, A., Barra, D., Blumberg, S. & Shoham, G. (1997) *Streptomyces griseus* aminopeptidase: X-ray crystallographic structure at 1.75 Å resolution. *J. Mol. Biol.* **265**, 620–636.

Maras, B., Greenblatt, H.M., Shoham, G., Spungin-Bialik, A., Blumberg, S. & Barra, D. (1996) Aminopeptidase from *Streptomyces griseus*, primary structure and comparison with other zinc-containing aminopeptidases. *Eur. J. Biochem.* **236**, 843–846.

Narahashi, Y. & Yanagita, M. (1967) Studies on proteolytic enzymes (Pronase) of *Streptomyces griseus* K-1. Nature and properties of the proteolytic enzyme system. *J. Biochem. (Tokyo)* **62**, 633–641.

Narahashi, Y., Shibuya, K. & Yanagita, M. (1968) Studies on proteolytic enzymes (Pronase) of *Streptomyces griseus* K-1. I. Separation of exo- and endopeptidases of pronase. *J. Biochem (Tokyo)* **64**, 427–437.

Nomoto, M., Narahashi, Y. & Murakami, M. (1960) A proteolytic enzyme of *Streptomyces griseus*. VI. Hydrolysis of protein by *Streptomyces griseus* protease. *J. Biochem. (Tokyo)* **48**, 593–602.

Spungin, A. & Blumberg, S. (1989) *Streptomyces griseus* aminopeptidase is a calcium-activated zinc metalloprotein. Purification and properties of the enzyme. *Eur. J. Biochem.* **183**, 471–477.

Vosbeck, K.D. (1974) Pronase aminopeptidases. PhD thesis, University of Miami, Florida.

Vosbeck, K.D., Chow, K.F. & Awad, W.M., Jr. (1973) The proteolytic enzymes in the K-1 strain of *Streptomyces griseus* obtained from obtained from a commercial preparation (Pronase). Purification and characterization of the aminopeptidases. *J. Biol. Chem.* **248**, 6029–6034.

Vosbeck, K.D., Greenberg, B.D. & Awad, W.M., Jr. (1975) The proteolytic enzymes of the K-l strain of *Streptomyces griseus* obtained from a commercial preparation (Pronase). Specificity and immobilization of aminopeptidase. *J. Biol. Chem.* **250**, 3981–3987.

Vosbeck, K.D., Greenberg, B.D., Ochoa, M.S., Whitney, P.L. & Awad, W.M., Jr. (1978) Proteolytic enzymes of the K-1 strain of *Streptomyces griseus* obtained from a commercial preparation (Pronase). Effect of pH, metal ions, and amino acids on aminopeptidase activity. *J. Biol. Chem.* **253**, 257–260.

Wu, C.-H. & Lin, W.-Y. (1995) Effects of metal ions on the catalytic and thermodynamic properties of the aminopeptidase isolated from Pronase. *J. Inorg. Biochem.* **57**, 79–89.

Yang, S.-H., Wu, C.-H & Lin, W.-Y. (1994) Chemical modification of aminopeptidase isolated from Pronase. *Biochem. J.* **302**, 595–600.

*William M. Awad, Jr.*
*Departments of Medicine and Biochemistry and Molecular Biology,*
*University of Miami School of Medicine,*
*PO Box 016960,*
*Miami, FL 33101, USA*
*Email: wawad@mednet.med.miami.edu*

# 491. *Vibrio aminopeptidase*

## Databanks

*Peptidase classification: clan MH, family M28, MEROPS ID: M28.002*
*NC-IUBMB enzyme classification: EC 3.4.11.10*
*ATCC entries: 15338*
*Chemical Abstracts Service registry number: 37288-67-8*
*Databank codes:*

| Species | SwissProt | PIR | EMBL (cDNA) | EMBL (genomic) |
|---|---|---|---|---|
| *Vibrio proteolyticus* | Q01693 | S21684 | M85159 | – |
| | | S24314 | Z11993 | |

Brookhaven Protein Data Bank three-dimensional structures:

| Species | ID | Resolution | Notes |
|---|---|---|---|
| *Vibrio proteolyticus* | 1AMP | 1.8 | |
| | 1IGB | 2.3 | complex with *para*-iodo-D-Phe hydroxamate |

## Name and History

A bacterial species with high proteolytic activity was obtained from the alimentary canal of the marine borer *Limnoria*, and was assumed to be a new *Aeromonas* species, to which the name *A. proteolytica* was given (Merkel *et al.*, 1964). The proteolytic enzyme isolated from this organism was identified as an aminopeptidase for which the trivial name **Aeromonas aminopeptidase** was proposed (Prescott & Wilkes, 1966). Furthermore it was also characterized as a two-zinc metalloprotease (Prescott *et al.*, 1971). As *Aeromonas proteolytica* was recently reclassified as *Vibrio proteolyticus*, the name **Vibrio aminopeptidase** is now more appropriate to designate this enzyme. Other synonyms are **Aeromonas proteolytica aminopeptidase** and **bacterial leucyl aminopeptidase** (IUBMB).

## Activity and Specificity

The enzyme exhibits a broad specificity in releasing free N-terminal residues of oligopeptides and polypeptides. Its preference for large hydrophobic residues was assessed on amino acid amide and dipeptide substrates (Wagner *et al.*, 1972). Basic amino acids and proline are also susceptible to hydrolysis; in contrast aspartyl, glutamyl and cysteic acid residues are not cleaved (Wilkes *et al.*, 1973).

Bestatin and amastatin are slow tight-binding inhibitors (Wilkes & Prescott, 1985). Amino acid hydroxamic acids have been extensively studied as inhibitors. The *Vibrio* aminopeptidase is inhibited to a greater extent by D-isomers than by the L-enantiomers; in the case of D-Leu-NHOH and D-Val-NHOH, the L-isomers were bound 150 times less tightly (Baker *et al.*, 1983; Wilkes & Prescott, 1983). The X-ray crystallographic structure of the *Vibrio* aminopeptidase complexed with the synthetic inhibitor *p*-iodo-D-phenylalanine hydroxamate suggests that the two zinc ions and a vicinal glutamic acid could play a key role in the catalytic mechanism of the aminopeptidase (Chevrier *et al.*, 1996).

## Structural Chemistry

In contrast to a number of mammalian aminopeptidases which are oligomeric structures of considerable size (McDonald & Schwabe, 1977), *Vibrio* aminopeptidase is a monomeric metalloprotease of about 30 kDa. The crystal structure is available at 1.8 Å resolution (Chevrier *et al.*, 1994) (see Fig. 482.1). The protein consists of a single peptide comprising 291 amino acids, while the primary sequence determinations predict 297 amino acids (Guenet *et al.*, 1992) or 299 amino acids (Van Heeke *et al.*, 1992). The protein is folded into a single $\alpha/\beta$ globular domain. The active site contains two zinc ions, 3.5 Å apart, with shared ligands and symmetrical coordination spheres. The zincs are bridged by the carboxylate oxygens of Asp117 and by a water molecule. Zinc-1 is also coordinated to a nitrogen of His256 and one carboxylate oxygen of Glu152, whereas zinc-2 is coordinated to a nitrogen of His97 and one carboxylate oxygen of Asp179.

There is a well-defined hydrophobic pocket adjacent to the metal-binding site with approximate dimensions $7 \times 7 \times 13$ Å. It is lined by Met180, Ile193, Cys223, Tyr225, Cys227, Met242, Phe244, Phe248, Tyr251 and Ile255. The hydrophobic character of this subsite reflects the substrate preference of the protein for hydrophobic N-terminal residues (Wilkes *et al.*, 1973). The catalytic pocket is located at the surface of the protein and is accessible in the crystalline state via a solvent channel, consistent with the remaining catalytic activity found in the crystal (Schalk *et al.*, 1992): The only two cysteine residues present in the structure form a disulfide bridge.

## Preparation

The enzyme was isolated from the marine bacterial species *Vibrio proteolyticus* (then *Aeromonas proteolytica*) (Prescott & Wilkes, 1966). It is an unusually stable extracellular two-zinc aminopeptidase that can be obtained in high yield from culture filtrates (Prescott & Wilkes, 1976). In order to eliminate two major contaminants in the culture filtrate, amber pigments and a zinc-dependent neutral proteinase, two chromatographic steps based on hydrophobic interactions and ion-exchange were performed (Schalk *et al.*, 1992). Heat treatment at a temperature of 70°C for several hours provides a specific and rapid purification for this thermostable enzyme (Prescott & Wilkes, 1976; Schalk *et al.*, 1992).

## Biological Aspects

Genes coding for 43 kDa and 54 kDa proteins were cloned and sequenced (Guenet *et al.*, 1992; Van Heeke *et al.*, 1992). The deduced amino acid sequence of the prepro-enzyme encodes a 21-residue predicted signal peptide, 84-residue propeptide, 299-residue mature protein and 100-residue *C*-terminal propeptide. It was suggested that the active and thermosensitive protein precursor is rapidly transformed to thermostable forms of 30 and 32 kDa (Guenet *et al.*, 1992).

## Further Reading

Further details of the crystal structure may be found in Chevrier *et al.* (1994) and the properties and preparation are reviewed by Prescott & Wilkes (1976).

## References

Baker, J.O., Wilkes, S.H., Bayliss, M.E. & Prescott, J.M. (1983) Hydroxamates and aliphatic boronic acids: marker inhibitors for aminopeptidase. *Biochemistry* 22, 2098–2103.

Chevrier, B., Schalk, C., D'Orchymont, H., Rondeau, J.-M., Moras, D. & Tarnus, C. (1994) Crystal structure of *Aeromonas proteolytica* aminopeptidase: a prototypical member of the co-catalytic zinc enzyme family. *Structure* 2, 283–291.

Chevrier, B., D'Orchymont, H., Schalk, C., Tarnus, C. & Moras, D. (1996) The structure of the *Aeromonas proteolytica* aminopeptidase complexed with a hydroxamate inhibitor: involvement in catalysis of Glu151 and two zinc ions of the co-catalytic unit. *Eur. J. Biochem.* 237, 393–398.

Guenet, C., Lepage, P. & Harris, B. (1992) Isolation of the leucine aminopeptidase gene from *Aeromonas proteolytica. J. Biol. Chem.* 267, 8390–8395.

McDonald, J.K. & Schwabe, C. (1977) Intracellular exopeptidases. In: *Proteinases in Mammalian Cells and Tissues* (Barrett, A.J., ed.). Amsterdam: Elsevier, pp. 311–392.

Merkel, J.R., Traganza, E.D., Mukherjee, B.B., Griffin, T.B. & Prescott, J.M. (1964) Proteolytic activity and general characteristics of a marine bacterium, *Aeromonas proteolytica* sp. N. *J. Bacteriol.* 87, 1227–1233.

Prescott, J.M. & Wilkes, S.H. (1966) *Aeromonas* aminopeptidase: purification and some general properties. *Arch. Biochem. Biophys.* 117, 328–336.

Prescott, J.M. & Wilkes, S.H. (1976) *Aeromonas* aminopeptidase. *Methods Enzymol.* 44, 530–543.

Prescott, J.M., Wilkes, S.H., Wagner, F.W. & Wilson, K.J. (1971) *Aeromonas* aminopeptidase. Improved isolation and some physical properties. *J. Biol. Chem.* 246, 1756–1764.

Schalk, C., Remy, J.-M., Chevrier, B., Moras, D. & Tarnus, C. (1992) Rapid purification of the *Aeromonas proteolytica* aminopeptidase: crystallization and preliminary X-ray data. *Arch. Biochem. Biophys.* 294, 91–97.

Van Heeke, G.V., Denslow, S., Watkins, J.R., Wilson, K.J. & Wagner, F.W. (1992) Cloning and nucleotide sequence of the *Vibrio proteolyticus* aminopeptidase gene. *Biochim. Biophys. Acta* 1131, 337–340.

Wagner, F.W., Wilkes, S.H. & Prescott, J.M. (1972) Specificity of *Aeromonas* aminopeptidase toward amino acid amides and dipeptides. *J. Biol. Chem.* 247, 1208–1210.

Wilkes, S.H. & Prescott, J.M. (1983) Stereospecificity of amino acid hydroxamate inhibition of aminopeptidases. *J. Biol. Chem.* 258, 13517–13521.

Wilkes, S.H. & Prescott, J.M. (1985) The slow, tight binding of bestatin and amastatin to aminopeptidases. *J. Biol. Chem.* 260, 13154–13162.

Wilkes, S.H., Bayliss, M.E. & Prescott, J.M. (1973) Specificity of *Aeromonas* aminopeptidase toward oligopeptides and polypeptides. *Eur. J. Biochem.* 34, 459–466.

*Bernard Chevrier*
*UPR 9004 du CNRS,*
*Biologie Structurale, IGBMC,*
*1 rue Laurent Fries – BP 163,*
*67404 Illkirch Cedex, France*
*Email: chevrier@igbmc.u-strasbg.fr*

*Hugues D'Orchymont*
*Marion Merrell Research Institute,*
*16 rue d'Ankara,*
*67080 Strasbourg Cedex, France*
*Email: huguesdorchymont@hmri.com*

# 492. *Glutamate carboxypeptidase II*

## Databanks

*Peptidase classification: clan MH, family M28, MEROPS ID: M28.010*
*NC-IUBMB enzyme classification: EC 3.4.17.21*

*Databank codes:*

| Species | SwissProt | PIR | EMBL (cDNA) | EMBL (genomic) |
| --- | --- | --- | --- | --- |
| *Caenorhabditis elegans* | – | – | M79569 | – |
| *Homo sapiens* | Q04609 | – | M99487 | – |
| *Rattus norvegicus* | – | – | U75973 | – |

## Name and History

This enzyme was first characterized in the rat nervous system by its hydrolysis of the neuropeptide *N*-acetylaspartylglutamate (NAAG; Robinson *et al*. 1987) and was termed *N-acetylated-alpha-linked acidic dipeptidase (NAALADase)*. The prostate cancer marker known as the *prostate-specific membrane antigen (PSM, PSMA)* also possesses NAALADase activity (Carter *et al*., 1996). (Note that PSM is quite different from prostate specific antigin, PSA, which is a term used for semenogelase, Chapter 31). In view of the enzyme's specificity for Glu residues, the name *glutamate carboxypeptidase II* is now recommended.

## Activity and Specificity

Rat brain glutamate carboxypeptidase II demonstrates a high apparent affinity for Ac-Asp⊦Glu ($K_m = 140$ nM; Slusher *et al*., 1990) and has also been shown to hydrolyze other acidic dipeptides, specifically Asp⊦Glu, Glu⊦Glu and $\gamma$-Glu⊦Glu (Serval *et al*., 1990). The PSM form of the enzyme has been shown to catalyze Ac-Asp⊦Glu hydrolysis (Carter *et al*., 1996) and to possess processive omega peptidase activity against the folate pteroyl-$\gamma$-pentaglutamate and the antifolate methotrexate-$\gamma$-triglutamate (Pinto *et al*., 1996). The spectrum of the enzyme's activity and specificity have not been defined entirely, although inferences have been made from competition studies (Robinson *et al*., 1987; Slusher *et al*., 1990; Pinto *et al*., 1996).

A radioenzymatic assay using the substrate Ac-Asp-[$^3$H]Glu, available from Dupont NEN, is the most common method used to analyze activity of glutamate carboxypeptidase II (Robinson *et al*., 1987) (see Appendix II for full names and addresses of suppliers). The rat brain enzyme displays peak Ac-Asp-Glu hydrolyzing activity in the pH range of 6.0–7.4, appears to require monovalent anions such as Cl$^-$ for activity, and is inhibited by phosphate or sulfate anions. Specificity of hydrolysis is determined by inhibition by quisqualic acid (Robinson *et al*., 1987; Serval *et al*., 1990), which is available commercially from multiple sources. Glutamate carboxypeptidase II is also inhibited by Ac-Asp-Glu analogs and *N*-acylated glutamate species (Serval *et al*., 1990, 1992; Subasinghe *et al*., 1990). The most potent known inhibitors of glutamate carboxypeptidase II are substituted phosphonylglutamates (Jackson *et al*., 1996).

## Structural Chemistry

The purified rat brain enzyme has an isoelectric point of 9.0. The rat brain and the PSM forms of glutamate carboxypeptidase II are both glycoproteins and display apparent molecular masses of 94–100 kDa with corresponding deglycosylated polypeptides of approximately 84 kDa. Both of these forms of the enzyme are membrane-bound species and may homodimerize (Slusher *et al*., 1990; Israeli *et al*., 1993; Troyer *et al*., 1995). Analysis of the primary amino acid sequence deduced from the PSM cDNA (Israeli *et al*., 1993) predicts that glutamate carboxypeptidase II is a type II membrane protein, with amino acids 16, 17 and 19 serving as a basic cytosolic anchor, and residues 20–43 serving as a putative transmembrane domain. Asparagine-linked glycosylation consensus sites occur at residues 76, 121, 140, 153, 195, 336, 459, 476 and 638. Homology with the transferrin receptor has also been noted.

Through sequence alignment with other members of the M28 evolutionary family whose structures have been characterized crystallographically, Rawlings & Barrett (1997) have drawn conclusions about the structure and catalytic domain of glutamic carboxypeptidase II. Zinc ligands for human glutamate carboxypeptidase II are predicted to be Asp387, Glu425 and His553 for the first zinc atom and His377, Asp387 and Asp453 for the second. The basic residues Arg463, Lys499, Arg536 and Lys545 may be determinants in the selection of negatively charged substrates by glutamate carboxypeptidase II.

## Preparation

Purified preparations of glutamate carboxypeptidase II are not available commercially. One form of rat brain glutamate carboxypeptidase II has been purified to homogeneity using conventional chromatographic methods (Slusher *et al*., 1990). The PSM form of glutamate carboxypeptidase II has also been purified by immunoaffinity techniques (Israeli *et al*., 1993; Troyer *et al*., 1995).

## Biological Aspects

The expression of glutamate carboxypeptidase II in rat tissues has been examined by radioenzymatic assay and immunodetection, with highest levels observed in the rat nervous system and kidney, and lesser but significant levels in several other tissues (Berger *et al*., 1995; Robinson *et al*., 1987; Slusher *et al*., 1990, 1992). Subcellular fractionation of brain reveals enrichment of the enzyme in neurosynaptosomal complexes (Blakely *et al*., 1988). Electron microscopy localizes glutamate carboxypeptidase II immunoreactivity to the outer surface of the plasma membrane (Berger *et al*., 1995).

PSM was characterized originally as the ligand of the monoclonal antibody 7E11.C5, which exhibited immunohistochemical staining restricted among human tissues to the prostate epithelium (Horoszewicz *et al*., 1987). Subsequent analyses have reported 7E11.C5 immunoreactivity and/or positive RNAase protection assays with PSM-derived probes in human salivary gland, brain, small intestine, seminal plasma and blood serum (Israeli *et al.*, 1994b; Troyer *et al.*, 1995; Murphy *et al.*, 1996). Two gene loci closely related to PSM have been identified on human chromosome 11. These map specifically to 11q14 and 11p11.1–13 (or 11cen–p12; Leek *et al.*, 1995; Rinker-Schaeffer *et al.*, 1995).

The distribution of glutamate carboxypeptidase II in the mammalian nervous system appears to correlate with the distribution of its putative endogenous substrate NAAG in most regions. Intact NAAG and its glutamate carboxypeptidase II-derived metabolite glutamate exhibit differential effects via members of the glutamate receptor family; whereas glutamate conveys excitatory signals, NAAG has negative modulatory effects (Wroblewska *et al.*, 1993; Grunze *et al.*, 1996). Thus, the regulation of glutamate carboxypeptidase II activity may govern whether excitatory or inhibitory effects predominate following NAAG release in neural systems. Further supporting this hypothesis, perturbations in human glutamate carboxypeptidase II activity have been observed in neurologic and psychiatric disorders which involve dysregulation of glutamatergic neurotransmission (Tsai *et al.*, 1995). The role of glutamate carboxypeptidase II as a folate hydrolase in the nervous system has not been explored.

The PSM form of glutamate carboxypeptidase II is expressed in normal and neoplastic prostate epithelial cells, and PSM has been examined for utility in the diagnosis, assessment and treatment of prostatic carcinoma (Israeli *et al.*, 1994a; Murphy *et al.*, 1996). A radioconjugated derivative of the 7E11.C5 anti-PSM antibody, [$^{121}$I]CYT-356, has shown particular promise as an imaging ligand for prostatic tumor metastases (Abdel-Nabi *et al.*, 1992).

### Distinguishing Features

Polyclonal antisera against the rat brain enzyme (Slusher *et al.*, 1990) and monoclonal antibodies against PSM (Horoszewicz *et al.*, 1987; Murphy *et al.*, 1996) have been described, but none are available commercially.

### Further Reading

Determination of PSM/glutamate carboxypeptidase II activity and an overview of the enzyme's relevance to both nervous system function and prostate cancer may be found in Carter *et al.* (1996). The recent article of Heston (1997) is also of interest.

### References

Abdel-Nabi, H., Wright, G.L., Gulfo, J.V., Petrylak, D.P., Neal, C.E., Texter, J.E., Begun, F.P., Tyson, I., Heal, A., Mitchell, E., Purnell, G. & Harwood, S.J. (1992) Monoclonal antibodies and radioimmunoconjugates in the diagnosis and treatment of prostate cancer. *Semin. Urol.* **10**, 45–54.

Berger, U.V., Carter, R.E., McKee, M. & Coyle, J.T. (1995) *N*-Acetylated alpha-linked acidic dipeptidase is expressed by non-myelinating Schwann cells in the peripheral nervous system. *J. Neurocytol.* **24**, 99–109.

Blakely, R.D., Robinson, M.B., Thompson, R.C. & Coyle, J.T. (1988) Hydrolysis of the brain dipeptide *N*-acetyl-L-aspartyl-L-glutamate: subcellular and regional distribution, ontogeny & the effects of lesions on *N*-acetylated-alpha-linked acidic dipeptidase activity. *J. Neurochem.* **50**, 1200–1209.

Carter, R.E., Feldman, A.R. & Coyle, J.T. (1996) Prostate-specific membrane antigen is a hydrolase with substrate and pharmacologic characteristics of a neuropeptidase. *Proc. Natl Acad. Sci. USA* **93**, 749–753.

Grunze, H.C.R., Rainnie, D.G., Hasselmo, M.E., Barkai, E., Hearn, E.F., McCarley, R.W. & Greene, R.W. (1996) NMDA-dependent modulation of CA1 local circuit inhbition. *J. Neurosci.* **16**, 2034–2043.

Heston, W.D.W. (1997) Characterization and glutamyl preferring carboxypeptidase function of prostate specific membrane antigen: a novel folate hydrolase. *Urology* **49** (3A Suppl.): 104–112.

Horoszewicz, J.S., Kawinski, E. & Murphy, G.P. (1987) Monoclonal antibodies to a new antigenic marker in epithelial prostatic cells and serum of prostatic cancer patients. *Anticancer Res.* **7**, 927–936.

Israeli, R.S., Powell, C.T., Fair, W.R. & Heston, W.D.W. (1993) Molecular cloning of a complementary DNA encoding a prostate-specific membrane antigen. *Cancer Res.* **53**, 227–230.

Israeli, R.S., Miller, W.H., Su, S.L., Powell, C.T., Fair, W.R., Samadi, D.S., Huryk, R.F., DeBlasio, A., Edwards, E.T., Wise, G.J. & Heston, W.D.W. (1994a) Sensitive nested reverse transcription polymerase chain reaction detection of circulating prostatic tumor cells: comparison of prostate-specific membrane antigen and prostate-specific antigen-based assays. *Cancer Res.* **54**, 6306–6310.

Israeli, R.S., Powell, C.T., Corr, J.G., Fair, W.R. & Heston, W.D.W. (1994b) Expression of the prostate-specific membrane antigen. *Cancer Res.* **54**, 1807–1811.

Jackson, P.F., Cole, D.C., Slusher, B.S., Stetz, S.L., Ross, L.E., Donzanti, B.A. & Trainor, D.A. (1996) Design, synthesis & biological activity of a potent inhibitor of the neuropeptidase N-acetylated α-linked acidic dipeptidase. *J. Med. Chem.* **39**, 619–622.

Leek, J., Lench, N., Maraj, B., Bailey, A., Carr, I.M., Andersen, S., Cross, J., Whelan, P., MacLennan, K.A., Meredith, D.M. & Markham, A.F. (1995) Prostate-specific membrane antigen: evidence for a second related human gene. *Br. J. Cancer* **72**, 583–588.

Murphy, G.P., Tino, W.T., Holmes, E.H., Boynton, A.L., Erickson, S.J., Bowes, V.A., Barren, R.J., Tjoa, B.A., Misrock, S.L., Ragde, H. & Kenny, G.M. (1996) Measurement of PSMA (prostate specific membrane antigen) in the serum with a new antibody. *Prostate* **28**, 266–271.

Pinto, J.T., Suffoletto, B.P., Berzin, T.M., Qiao, C.H., Lin, S., Tong, W.P., May, F., Mukherjee, B. & Heston, W.D.W. (1996) Prostate specific membrane (PSM) antigen: a novel member of folate hydrolase in human prostatic carcinoma cells. *Clin. Cancer Res.* **2**, 1445–1451.

Rawlings, N.D. & Barrett, A.J. (1997) Structure of membrane glutamate carboxypeptidase. *Biochim. Biophys. Acta* **1339**, 247–252.

Rinker-Schaeffer, C.W., Hawkins, A.L., Su, S.L., Israeli, R.S., Griffin, C.A., Isaacs, J.T. & Heston, W.D.W. (1995) Location and physical mapping of the prostate-specific membrane antigen (PSM) gene to human chromosome 11. *Genomics* **30**, 105–108.

Robinson, M.B., Blakely, R.D., Couto, R. & Coyle, J.T. (1987) Hydrolysis of the brain dipeptide *N*-acetyl-L-aspartyl-L-glutamate. *J. Biol. Chem.* **262**, 14498–14506.

Serval, V., Barbeito, L., Pittaluga, A., Cheramy, A., Lavielle, S. & Glowinski, J. (1990) Competitive inhibition of *N*-acetylated-alpha-linked acidic dipeptidase activity by *N*-acetyl-L-aspartyl-β-linked-L-glutamate. *J. Neurochem.* **55**, 39–46.

Serval, V., Galli, T., Cheramy, A., Glowinski, J. & Lavielle. S. (1992) *In vitro* and *in vivo* inhibition of *N*-acetyl-L-aspartyl-L-glutamate catabolism by *N*-acylated L-glutamate analogs. *J.*

*Pharmacol. Exp. Ther.* **260**, 1093–1100.

Slusher, B.S., Robinson, M.B., Tsai, G., Simmons, M., Richards, S.S. & Coyle, J.T. (1990) Rat brain N-acetylated alpha-linked acidic dipeptidase activity: purification and immunologic characterization. *J. Biol. Chem.* **265**, 21297–21301.

Slusher, B.S., Tsai, G., Yoo, G. & Coyle, J.T. (1992) Immunocytochemical localization of the *N*-acetyl-aspartyl-glutamate (NAAG) hydrolyzing enzyme *N*-acetylated α-linked acidic dipeptidase (NAALADase). *J. Comp. Neurol.* **315**, 217–229.

Subasinghe, N., Schulte, M., Chan, M.Y.-M., Roon, R.J. & Koerner, J.F. (1990) Synthesis of acyclic and dehydroaspartic acid analogues of Ac-Asp-Glu-OH and their inhibition of rat brain N-acetylated α-linked acidic dipeptidase (NAALA dipeptidase). *J.*

*Med. Chem.* **33**, 2734–2744.

Troyer, J.K., Beckett, M.L. & Wright, G.L. (1995) Detection and characterization of the prostate-specific membrane antigen (PSMA) in tissue extracts and body fluids. *Int. J. Cancer* **62**, 552–558.

Tsai, G., Passani, L.A., Slusher, B.S., Carter, R., Kleinman, J.E. & Coyle, J.T. (1995) Abnormal excitatory neurotransmitter metabolism in schizophrenic brains. *Arch. Gen. Psychiatry* **52**, 829–836.

Wroblewska, B., Wroblewski, J.T., Saab, O.H. & Neale, J.H. (1993) *N*-Acetylaspartylglutamate inhibits forskolin-stimulated cyclic AMP levels via a metabotropic glutamate receptor in cultured cerebellar granule cells. *J. Neurochem.* **61**, 946–948.

***Ruth E. Carter***
*Consolidated Department of Psychiatry,*
*Harvard Medical School,*
*Massachusetts General Hospital – East, Rm. 2510,*
*Belmont, MA 02178-1048, USA*
*Email: carter@helix.mgh.harvard.edu*

***Joseph T. Coyle***
*Consolidated Department of Psychiatry,*
*Harvard Medical School,*
*McLean Hospital, 115 Mill Street,*
*Charlestown, MA 02129, USA*
*Email: jcoyle@warren.med.harvard.edu*

# 493. *Pteroylpoly-γ-glutamate carboxypeptidase*

## Databanks

Peptidase classification: clan MH, family M28, MEROPS ID: M28.010
NC-IUBMB enzyme classification: EC 3.4.17.21
Databank codes:

| Species | SwissProt | PIR | EMBL (cDNA) | EMBL (genomic) |
|---|---|---|---|---|
| *Caenorhabditis elegans* | – | – | M79569 | – |
| *Homo sapiens* | Q04609 | – | M99487 | – |
| *Rattus norvegicus* | – | – | U75973 | – |

*Note:* The same table appears in Chapter 492.

## Name and History

The name *pteroylpoly-γ-glutamate carboxypeptidase* denotes the progressive removal of γ-glutamate residues from the C-terminus of pteroylpoly-γ-glutamates (PteGlu$_n$). Other names used for this enzyme have included *intestinal folate conjugase, pteroylpoly-glutamate hydrolase (PPH)* or *brush border PPH*. The observation that dietary folates exist as a mixture of PteGlu$_n$ (Butterworth *et al.*, 1963) led to a search for a specific PPH capable of digestive removal of terminal glutamates prior to intestinal transport of the pteroylglutamate (PteGlu) derivative. In early studies of intestinal PPH,

PteGlu$_n$ was converted to PteGlu during incubation in everted gut sacs of the rat (Rosenberg *et al.*, 1969) and a rat intestinal PPH was purified and identified as an intracellular soluble enzyme (Elsenhans *et al.*, 1984). In subsequent studies, intermediate chain length PteGlu$_n$ and PteGlu end-products were identified in the contents of the intestinal lumen during perfusion of PteGlu$_7$ in the jejunum of human volunteers, a finding that was consistent with a hydrolytic reaction on the brush border surface of the intestine (Halsted *et al.*, 1975). These studies were followed by the discovery of two distinct PPH activities in human jejunal mucosa, one in the soluble intracellular fraction and the other localized to the brush border fraction (Reisenauer *et al.*, 1977). Using an enzyme assay that differentiates the two PPH activities by different pH and inhibitor conditions, both the intracellular and the brush border PPH activities were identified in pig and human jejunal mucosa, whereas only the intracellular activity was found in rat intestine (Wang *et al.*, 1985). Following its purification, human jejunal brush border PPH was found to have properties that were distinct from those of partly pure human jejunal intracellular PPH (Chandler *et al.*, 1986a; Wang *et al.*, 1986), whereas two independent groups found that human and pig jejunal brush border PPH shared similar biochemical properties (Chandler *et al.*, 1986b; Gregory *et al.*, 1987).

## Activity and Specificity

Purified human and pig jejunal brush border PPH is an exopeptidase that progressively releases intermediate chain length $PteGlu_n$ and end-product PteGlu during timed incubation with $PteGlu_7$. The enzyme has a pH optimum of 5.5 but is most stable at pH 6.5. The $K_m$ of 0.6 mM is identical for the alternate substrates $PteGlu_2$, $PteGlu_3$ and $PteGlu_7$, and substrate specificity requires the complete pteroyl moiety and $Glu_n$ chain in $\gamma$-peptide linkage. Enzyme activity is unaffected by *p*-hydroxymercuribenzoate, and postdialysis activity is restored by the addition of $Zn^{2+}$ or $Cu^{2+}$ (Chandler *et al.*, 1986a,b). These observations established the conditions for adapting a PPH radioassay (Krumdieck & Baugh, 1970) to measure the specific activity of brush border PPH per protein concentration in jejunal mucosal biopsies or surgical tissue, using end-labeled $PteGlu_3$ substrate at pH 6.2 in the presence of $Zn^{2+}$ activator and of *p*-hydroxymercuribenzoate, a potent inhibitor of intracellular PPH that has no effect on the brush border enzyme (Halsted *et al.*, 1986).

## Structural Chemistry

The structural chemistry of this enzyme has not been established (but see Related Peptidases).

## Preparation

Human and pig jejunal brush border PPH were purified 5000- and 9700-fold, respectively, by a sequence of brush border isolation from fresh intestinal mucosa, followed by organomercurial affinity chromatography to remove intracellular PPH, DEAE column chromatography and Bio-Gel 1.5m gel filtration (Chandler *et al.*, 1986a,b, 1991). Denaturing gel electrophoresis was followed by immunoblot with specific monoclonal antibody. The purified pig intestinal brush border PPH was identified as a 120 kDa protein (Chandler *et al.*, 1991) that may be a subunit of a larger 237 kDa protein dimer found by others who used nondenaturing gel electrophoresis (Gregory *et al.*, 1987).

## Biological Aspects

### Distribution in Tissues

Pig intestinal brush border PPH activity was found in the duodenal and proximal jejunal mucosa and was absent in the ileum in contrast to the uniform distribution of intracellular PPH activity throughout the pig intestine. Immunohistochemistry with specific monoclonal antibody localized PPH to the pig jejunal brush border surface and confirmed its absence in the ileum. The distribution of activity and immunoreactivity of brush border PPH in the jejunum corresponds with the site of *in vivo* hydrolysis of $PteGlu_3$ in the pig, consistent with the functional specificity of jejunal brush border PPH in folate digestion (Chandler *et al.*, 1991).

### Normal Function

In the human and pig, intestinal brush border PPH accomplishes the progressive hydrolysis of dietary $PteGlu_n$ in the jejunum. The functional role of this enzyme was shown by the similar progressive appearance of hydrolytic intermediate chain length and end-product PteGlu during *in vivo* perfusion of $PteGlu_7$ in the human jejunum (Halsted *et al.*, 1975) or during timed incubation of the purified human jejunal brush border PPH with $PteGlu_7$ (Chandler *et al.*, 1986a). Calculations of the rates of *in vivo* hydrolysis and luminal disappearance of $PteGlu_7$ during its perfusion in the human jejunum indicated that the hydrolytic conversion of $PteGlu_7$ to its end-product PteGlu is rate limiting to the overall intestinal absorption of $PteGlu_7$ (Halsted *et al.*, 1977). Comparisons of data from young adult and very old human volunteers indicate that the aging process affects neither the *in vivo* hydrolysis of perfused $PteGlu_7$ nor the activity of jejunal brush border PPH (Bailey *et al.*, 1984; Halsted *et al.*, 1986).

### Effects of Diseases

Patients with the intestinal mucosal injury of celiac disease demonstrated marked reduction in both the *in vivo* jejunal hydrolysis of perfused $PteGlu_7$ and the specific activity of brush border PPH in jejunal biopsies (Halsted *et al.*, 1977, 1986). Two clinical studies demonstrated the adaptive response of brush border PPH to diet and to excision of the distal small intestine. In one study, brush border PPH was measured in mucosal samples obtained from patients undergoing surgical revision of previous jejunal ileal bypass operation for severe obesity. The specific activity of brush border PPH was reduced in the bypass limb of intestine that was excluded from diet but was increased in the hypertrophied jejunal mucosa that remained in continuity with the duodenum and was thus exposed to a high concentration of dietary nutrients (Reisenauer *et al.*, 1985). In the other study, the functional adaptation of brush border PPH was shown by its increased specific activity in jejunal biopsies from patients who had undergone surgical resection of the distal ileum for inflammatory bowel disease (Halsted *et al.*, 1986).

### Inhibitory Effects of Dietary Constituents and Drugs

Extracts of legumes, tomatoes and orange juice, but not of cereal grains or dairy products, were found to inhibit the activity of both human and pig brush border PPH to a degree sufficient to potentially impair the bioavailability of dietary folates (Bhandari & Gregory, 1990). Drugs known to be associated with clinical folate deficiency include the anticonvulsant phenytoin (Dilantin), the intestinal anti-inflammatory drug salicylazosulfapyridine (Azulfidine or Sulfasalazine), and ethanol when abused chronically (Halsted, 1990). Each drug has been tested as a potential inhibitor of intestinal PPH. Two studies using whole intestinal mucosal homogenates provided conflicting results on the inhibitory effect of phenytoin (Rosenberg *et al.*, 1968; Baugh & Krumdieck, 1969), whereas a pharmacologic concentration of phenytoin had no effect on the activity of human brush border PPH (Reisenauer & Halsted, 1981). By contrast, pharmacologic concentrations of the anti-inflammatory drug sulfasalazine were found to inhibit both the activity of human jejunal brush border PPH (Reisenauer & Halsted, 1981) as well as the *in vivo* hydrolysis of $PteGlu_7$ perfused in the jejunum of patients with ulcerative colitis (Halsted *et al.*, 1981). In a study of control and ethanol-fed pigs, both acute and chronic exposure to ethanol reduced the activity of jejunal brush border PPH (Naughton *et al.*, 1989).

## Distinguishing Features

Intestinal brush border PPH is distinguished by its exopeptidase mechanism of activity, neutral pH optimum, $K_m$, and concentration in the human and pig jejunal mucosa appropriate and adequate for complete hydrolysis of dietary $PteGlu_n$, and substrate requirement for the full pteroyl moiety and $\gamma$-$Glu_n$ linkages of $PteGlu_n$ (Chandler *et al.*, 1986a,b; Reisenauer & Halsted, 1987; Chandler *et al.*, 1991). Both polyclonal and monoclonal antibodies to purified pig jejunal brush border PPH have been produced and were used to localize the enzyme to the jejunal site of hydrolysis of $PteGlu_n$ (Chandler *et al.*, 1991). Both pancreatic PPH and soluble intracellular intestinal PPH have been proposed as alternate folate digestive enzymes. Pig pancreatic PPH is a 29 kDa endopeptidase that liberates Glu, $Glu_n$ and PteGlu in random fashion, has a pH optimum of 4.5, and is stimulated by feeding (Bhandari *et al.*, 1990b). Since the calculated postprandial capacity of pig pancreatic PPH is less than 1% of that of fasting pig jejunal mucosal PPH (Chandler *et al.*, 1991), pancreatic PPH may serve to initiate but not complete the digestion of dietary $PteGlu_n$. Rat intestinal PPH is a soluble 80 kDa enzyme with optimal pH 4.5, and endopeptidase mechanism, liberating PteGlu as the initial cleavage product of $PteGlu_7$ (Elsenhans *et al.*, 1984). Human jejunal intracellular PPH is found in the lysosomal fraction and has a molecular size, pH optimum and endopeptidase mechanism similar to rat intestinal PPH (Wang *et al.*, 1986). Although human jejunal intracellular PPH is nearly twice as abundant as jejunal brush border PPH (Reisenauer *et al.*, 1977), it is unaffected by celiac disease which is known to impair the *in vivo* hydrolysis of $PteGlu_n$ perfused in the jejunum and the activity of brush border PPH in jejunal biopsies (Halsted *et al.*, 1977, 1986). Furthermore, intracellular PPH is equally active throughout the length of the pig small intestine in contrast to the exclusive jejunal site of hydrolysis of $PteGlu_n$ in this species (Chandler *et al.*, 1991). Thus, it is unlikely that intracellular PPH plays a role in the digestion of dietary $PteGlu_n$, although this enzyme may serve other functions in folate metabolism in the absorbing intestinal enterocytes (Halsted, 1991).

## Related Peptidases

Recently, two independent laboratories described a highly conserved mammalian protein that exhibits PPH activity. Summarizing, the human prostate-specific membrane antigen (PSMA) from the human prostate cancer cell line LNCaP (Israeli *et al.*, 1993) is a 94 kDa class II hydrolytic membranous glycoprotein with fully characterized gene that shares 87% nucleotide and amino acid sequence similarity with *N*-acetylated $\alpha$-linked acidic dipeptidase (NAALADase from rat brain) (Chapter 492) (Carter *et al.*, 1996). The potential identity of this gene to human intestinal brush border PPH was suggested by the capacity of noncancer cells transfected with the PSMA gene to hydrolyze $PteGlu_7$ with similar exopeptidase, pH, substrate specificity and inhibitory conditions as previously described for purified human intestinal brush border PPH (Chandler *et al.*, 1986a; Pinto *et al.*, 1996), and by immunohistochemical staining of human duodenal brush border by monoclonal antibody to PSMA (Silver *et al.*, 1997).

In other studies, the gene sequence of a $\gamma$-glutamyl hydrolase (Chapter 264) from rat H35 hepatoma cells was used to establish the cDNA structure of a 67% homologous human enzyme from the expressed tag database (Yao *et al.*, 1996a,b). Studies with methotrexate polyglutamate substrates identified rat $\gamma$-glutamyl hydrolase as an endopeptidase and the human enzyme as an exopeptidase (Yao *et al.*, 1996b). Other comparisons with human jejunal brush border PPH remain to be established.

## Note added in proof

The cDNA of pig jejunal folylpoly-$\gamma$-glutamate carboxypeptidase has been isolated (Genbank AF050502) (Halstead *et al.*, 1998). It encodes a 751 amino acid polypeptide that shares >85% identity with human prostate-specific membrane antigen and rat brain glutamate carboxypeptidase II. It hybridizes to pig and human jejunum, rat and human brain, and human prostate LNCaP cancer cells. The three proteins may represent the expression of a single gene for glutamate carboxypeptidase II (EC 3.4.17.21) in different tissues and species.

## Further Reading

For an extensive review of the discovery, biochemical, and immunological characterization of the enzyme, see Halsted (1991), and for a review that places brush border PPH in the physiological context of folate digestion and absorption, see Halsted (1990).

## References

Bailey, L.B., Cerda, J.J., Bloch B.S., Busby, M.J., Vargas, L., Chandler, C.J. & Halsted, C.H. (1984) Effect of age on poly- and monoglutamyl folacin absorption in human subjects. *J. Nutr.* **114**, 1770–1776.

Baugh, C.M. & Krumdieck, C.L. (1969) Effects of phenytoin on folic acid conjugases in man. *Lancet* **2**, 519–521.

Bhandari, S.D. & Gregory, J.F. (1990) Inhibition by selected food components of human and porcine intestinal pteroylpolyglutamate hydrolase activity. *Am. J. Clin. Nutr.* **51**, 87–94.

**M**

Bhandari, S.D., Gregory, J.F., Renuart, D.R. & Merritt, A.M. (1990b) Properties of pteroylpolyglutamate hydrolase in pancreatic juice of the pig. *J. Nutr.* **120**, 467–475.

Butterworth, C.E., Jr, Santini, R. & Frommeyer, W.B. (1963) The pteroylglutamate components of American diets as determined by chromatographic fractionation. *J. Clin. Invest.* **42**, 1929–1939.

Carter, R.E., Feldman, A.R., & Coyle, J.T. (1996) Prostate-specific membrane antigen is a hydrolase with substrate and pharmacologic characteristics of a neuropeptidase. *Proc. Natl Acad. Sci. USA* **93**, 749–753.

Chandler, C.J., Wang, T.T.Y. & Halsted, C.H. (1986a) Pteroylpolyglutamate hydrolase from human jejunal brush borders. Purification and characterization. *J. Biol. Chem.* **261**, 928–933.

Chandler, C.J., Wang, T.T.Y. & Halsted, C.H. (1986b) Brush border pteroyl-polyglutamate hydrolase in pig jejunum. In: *Chemistry and Biology of Pteridines* (Cooper, B.A. & Whitehead, V.M., eds). New York: Walter de Gruyter, pp. 539–542.

Chandler, C.J., Harrison, D.A., Buffington, C.A., Santiago, N.A. & Halsted, C.H. (1991) Functional specificity of jejunal brush-border pteroylpolyglutamate hydrolase in pig. *Am. J. Physiol.* **260**, G865–G872.

Elsenhans, B., Ahmad, O. & Rosenberg, I.H. (1984) Isolation and characterization of pteroylpolyglutamate hydrolase from rat intestinal mucosa. *J. Biol. Chem.* **259**, 6364–6368.

Gregory, J.F., Ink, S.L. & Cerda, J.J. (1987) Comparison of pteroyl-polyglutamate hydrolase (folate conjugase) from porcine and human intestinal brush border membrane. *Comp. Biochem. Physiol.* **88B**, 1135–1141.

Halsted, C.H. (1990) Intestinal absorption of dietary folates. In: *Folic Acid Metabolism in Health and Disease* (Picciano, M.F., Stokstad, E.L.R. & Gregory, J.F., eds). New York: Wiley-Liss, pp. 23–45.

Halsted, C.H. (1991) Jejunal brush-border folate hydrolase. A novel enzyme. *West. J. Med.* **155**, 605–609.

Halsted, C.H., Baugh, C.M. & Butterworth, C.E., Jr. (1975) Jejunal perfusion of simple and conjugated folates in man. *Gastroenterology* **68**, 261–269.

Halsted, C.H., Reisenauer, A.M., Romero, J.J., Cantor, D.S. & Ruebner B. (1977) Jejunal perfusion of simple and conjugated folates in celiac sprue. *J. Clin. Invest.* **59**, 933–940.

Halsted, C.H., Gandhi, G. & Tamura, T. (1981) Sulfasalazine inhibits the absorption of folates in ulcerative colitis. *N. Engl. J. Med.* **305**, 1513–1517.

Halsted, C.H., Beer, W.H., Chandler, C.J., Ross, K., Wolfe, B.M., Bailey, L. & Cerda, J.J. (1986) Clinical studies of intestinal folate conjugases. *J. Lab. Clin. Med.* **107**, 228–232.

Halsted, C.H., Ling, E.-H., Luthi-Carter, R., Villanueva, J.A., Gardner, J.M. & Coyle, J.T. (1998) Folylpoly-γ-glutamate carboxypeptidase from pig jejunum: molecular characterization and relation to glutamate carboxypeptidase II. *J. Biol. Chem.* In Press.

Israeli, R.S., Powell, C.T., Fair, W.R. & Heston, W.D.W. (1993) Molecular cloning of a complementary DNA encoding a prostate-specific membrane antigen. *Cancer Res.* **53**, 227–230.

Krumdieck, C.L. & Baugh, C.M. (1970) Radioactive assay of folic acid polyglutamate conjugase(s). *Anal. Biochem.* **35**, 123–129.

Naughton, C.A., Chandler, C.J., Duplantier, R.B. & Halsted, C.H. (1989) Folate absorption in alcoholic pigs: in vitro hydrolysis and transport at the brush border membrane. *Am. J. Clin. Nutr.* **50**, 1436–1441.

Pinto, J.T., Suffoletto, B.P., Berzin, T.M., Qiao, C.H., Lin, S. Tong, W.P., May, F., Mukherjee, B. & Heston, W.D.W. (1996) Prostate specific membrane antigen: a novel folate hydrolase in human prostatic carcinoma cells. *Clin. Cancer Res.* **2**, 1445–1451.

Reisenauer, A.M. & Halsted, C.H. (1981) Human brush border folate conjugase: characteristics and inhibition by salicylazosulfapyridine. *Biochim. Biophys. Acta* **659**, 62–69.

Reisenauer, A.M. & Halsted, C.H. (1987) Issues and opinions in nutrition. Human folate requirements. *J. Nutr.* **117**, 600–602.

Reisenauer, A.M., Krumdieck, C.L. & Halsted, C.H. (1977) Folate conjugase: two separate activities in human jejunum. *Science* **198**, 196–197.

Reisenauer, A.M., Halsted, C.H., Jacobs, L.R. & Wolfe, B.M. (1985) Human intestinal folate conjugase: adaptation after jejunoileal bypass. *Am. J. Clin. Nutr.* **42**, 660–665.

Rosenberg, I.H., Godwin, H.A., Striff, R.R. & Castle, W.B. (1968) Impairment of intestinal deconjugation of dietary folate: a possible explanation of megaloblastic anemia associated with phenytoin therapy. *Lancet* **2**, 530–532.

Rosenberg, I.H., Streiff, R.R., Godwin, H.A., & Castle, W.B. (1969) Absorption of polyglutamic folate: participation of deconjugating enzymes of the intestinal mucosa. *N. Engl. J. Med.* **280**, 985–988.

Silver, D.A., Pellicer, I., Fair, W.R., Heston, W.D.W. & Cordon-Cardo, C. (1997) Prostate-specific membrane antigen expression in normal and malignant human tissues. *Clin. Cancer Res.* **3**, 81–85.

Wang, T.T.Y., Reisenauer, A.M. & Halsted C.H. (1985) Comparison of folate conjugase activities in human, pig. rat, and monkey intestine. *J. Nutr.* **115**, 814–819.

Wang, T.T.Y., Chandler, C. J. & Halsted, C.H. (1986) Intracellular pteroylpolyglutamate hydrolase from human jejunal mucosa. Isolation and characterization. *J. Biol. Chem.* **261**, 13551–13555.

Yao, R., Nimec, Z., Ryan, T.J. & Galivan, J. (1996a) Identification, cloning. and sequencing of a cDNA coding for rat γ-glutamyl hydrolase. *J. Biol. Chem.* **271**, 8525–8528.

Yao, R. Schneider, E., Ryan, T.J. & Galivan, J. (1996b) Human γ-glutamyl hydrolase: cloning and characterization of the enzyme expressed *in vitro*. *Proc. Natl Acad. Sci. USA* **93**, 10134–10138.

*Charles H. Halsted*
*Division of Clinical Nutrition and Metabolism,*
*T.B. 156, School of Medicine, University of California Davis,*
*Davis, CA 95616, USA*
*Email: chhalsted@ucdavis.edu*

# 494. *Sulfolobus carboxypeptidase*

## Databanks

*Peptidase classification: clan MH, family M40, MEROPS ID: M40.001*
*NC-IUBMB enzyme classification: none*
*Databank codes:*

| Species | SwissProt | PIR | EMBL (cDNA) | EMBL (genomic) |
|---|---|---|---|---|
| *Sulfolobus solfataricus* | P80092 | S23180 | Z48497 | – |

## Name and History

Thermostable carboxypeptidase activity was first detected in cell extracts of the thermoacidophilic archaean *Sulfolobus solfataricus* using Z+Asp as an artificial substrate (Fusi *et al.*, 1991). The **Sulfolobus carboxypeptidase** was subsequently purified to electrophoretic homogeneity and extensively characterized. In particular, specificity studies confirmed that it was able to cleave an N-blocked tripeptide, and, at least partially, whole proteins (Colombo *et al.*, 1992).

## Activity and Specificity

The enzyme is inhibited by EDTA and 1,10-phenanthroline; furthermore, dialysis against EDTA leads to a complete loss of activity which can be restored by addition of $Zn^{2+}$ ions in the micromolar range. This led to its identification as a zinc metalloprotease (Colombo *et al.*, 1992). Also, it is endowed with an unusually broad substrate specificity as shown by its ability to release basic, acidic, aromatic and aliphatic amino acids from the respective Bz-Gly- and Z-amino acids, in decreasing order of catalytic efficiency. Only proline and tryptophan are not released. The specificity constants ($k_{cat}/K_m$) are comparable to those reported for mammalian carboxypeptidases. pH/activity profiles obtained using Z-Arg and Z-Phe at almost saturating concentrations and 70°C display broad pH optima ranging over 5.5–9.0; when using Z-Asp, the pH optimum ranges over 5.5–7.0. The temperature dependence of enzyme activity at saturating Z-Arg concentration shows a linear increase in an Arrhenius plot with no break-point in the temperature range 40–85°C and an apparent activation energy of 30–35 kJ mol$^{-1}$, irrespective of the substrate used. Because of thermal inactivation, the activity declines above 85°C. As a consequence of the relatively small activation energy, *S. solfataricus* carboxypeptidase retains a significant fraction of its maximal activity at lower temperatures, e.g. 20% at 40°C. Furthermore, it is also endowed with esterase activity, as shown by its ability to hydrolyze ethyl and methyl esters of amino acids and dipeptides blocked at the N-terminus (Colombo *et al.*, 1992).

## Structural Chemistry

Based on the nucleotide sequence of the coding gene, a primary structure of 393 amino acids with a molecular mass of 43 068 Da was deduced (Colombo *et al.*, 1995), in good agreement with the value of 42 kDa determined in SDS-PAGE. However, a molecular mass of 170 kDa was assessed under nondenaturing conditions, which points to a tetrameric structure. An isoelectric point of 5.9 was also determined, in good agreement with a value of 6.0 deduced from the primary structure. No carbohydrate content was found in the enzyme.

It has been suggested (see Chapter 482) that the *Sulfolobus* carboxypeptidase is a member of peptidase clan MH, and that the ligands of catalytic metal ions in the active site are as shown in Alignment 482.1.

The *S. solfataricus* enzyme is completely stable up to 80°C, whereas at higher temperatures it undergoes first-order irreversible thermal inactivation, with half-lives of 131 and 13.5 min at 85°C and 90°C, respectively (Villa *et al.*, 1993). First-order kinetics demonstrates that autolytic cleavage is not responsible for the inactivation, but that it is more a case of thermal unfolding, as also supported by the fact that no proteolyzed form of the enzyme can be detected by native gel or SDS-PAGE. At 70°C *S. solfataricus* carboxypeptidase is completely stable in the pH range 5.8–8.0, whereas it is abruptly destabilized beyond these values. At 40°C the enzyme withstands 50% methanol, 50% acetonitrile and 99% ethanol, whereas at 70°C it undergoes extensive inactivation in the presence of the organic solvents. When subjected to pressure in the range 200–400 MPa, it is further stabilized against thermal inactivation (Bec *et al.*, 1996) unlike most proteins which tend instead to unfold around 400–500 MPa (Mozhaev *et al.*, 1996). This capacity to withstand extreme pressure and solvent conditions might make it suitable as a catalyst for organic syntheses under these unusual physiochemical conditions.

Nevertheless, the structural basis of this resistance to extreme conditions is not completely understood. A role in enzyme stabilization was assigned to the zinc ion by comparing the thermodynamic parameters of thermal inactivation of holo- and metal-depleted enzyme. In particular, metal removal leads to a drop in the activation energy value of the first-order inactivation from 494.4 to 205.6 kJ mol$^{-1}$. As a consequence, thermal inactivation of metal-depleted enzyme is very rapid at 80°C (half-life of 14 min), whereas the holo-enzyme is completely stable at the same temperature (Villa *et al*., 1993). Furthermore, the carboxypeptidase is strongly destabilized by high ionic strength, although this effect is abolished at the extremes of the pH stability curve: this pattern points to a major stabilizing role for intramolecular ionic bonds (Villa *et al*., 1993). The following evidence suggests that the enzyme is also endowed with a high surface hydrophobicity. First, in spite of its thermostability, it loses most of its activity in few days when stored in an aqueous medium at room temperature. Second, the loss of activity is prevented by nonionic detergents. Third, based on its amino acid composition a mean hydrophobicity value of 1.12 kcal mol$^{-1}$ was calculated, which is distinctly above average. Nevertheless, the enzyme does not behave as an integral membrane protein, since little, if any, activity is found in the membrane fraction, both before and after extraction with nonionic detergents (Colombo *et al*., 1992).

## Preparation

*S. solfataricus* carboxypeptidase is obtained from strain MT-4 (ATCC 49155) using a 5-step procedure. A 76-fold enrichment is sufficient to obtain pure enzyme, which indicates that the molecule is abundant in its natural source. Expression of the recombinant molecule was achieved in *E. coli* strain BL21(DE3)[pLysE] using the plasmid pT7-7 as an expression vector (Colombo *et al*., 1995). At the time of writing, optimization of the expression system is underway.

## Biological Aspects

Little is known regarding the physiological role(s) of *S. solfataricus* carboxypeptidase. Its intracellular level is 3- to 4-fold higher in cells grown in a medium containing yeast extract instead of glucose as the sole carbon source, which suggests an involvement in the digestion of extracellularly available peptides after their internalization (Fusi *et al*., 1991). On the other hand, this protease is not likely to be secreted into the medium, since typical pH values of the extracellular environment (2–3) are incompatible with enzyme stability. Furthermore, the N-terminal sequence of the purified carboxypeptidase corresponds to the 5′ end of the coding gene, which allows us to rule out N-terminal processing of a precursor form.

## Distinguishing Features

The unusually broad substrate specificity represents an almost unique feature of *S. solfataricus* carboxypeptidase. Only the 56 kDa metallocarboxypeptidase from the thermophilic eubacterium *Thermus aquaticus*, carboxypeptidase Taq (Chapter 504) (Lee *et al*., 1994) displays a comparably broad substrate specificity. However, there is no significant sequence similarity to *S. solfataricus* carboxypeptidase.

## Related Peptidases

Slight sequence similarities are found between *S. solfataricus* carboxypeptidase and a set of hydrolytic enzymes from eubacterial sources (Chapter 482). The latter mainly catalyze hydrolytic removal of small moieties such as succinate (Bouvier *et al*., 1992; Fleischmann *et al*., 1995), β-alanine (Henrich *et al*., 1990), acyl groups (Sakanyan *et al*., 1993) and carbamate (Watabe *et al*., 1992) from the α-amino group of amino acid residues. Carboxypeptidase G$_2$ (Chapter 483), which specifically removes glutamate residues from folate, is also a member of the same clan (Boyen *et al*., 1992). These metalloenzymes share similar molecular masses, mostly around 42 kDa, and are either dimeric or tetrameric.

## Further Reading

For a review, see Colombo *et al*. (1995).

## References

Bec, N., Villa, A., Tortora, P., Mozhaev, V.V., Balny, C. & Lange, R. (1996) Enhanced stability of carboxypeptidase from *Sulfolobus solfataricus* at high pressure. *Biotechnol. Lett.* **18**, 483–488.

Boyen, A., Charlier, D., Charlier, J., Sakanyan, V., Mett, I. & Glansdorff, N. (1992) Acetylornithine deacetylase, succinyldiaminopimelate desuccinylase and carboxypeptidase G2 are evolutionarily related. *Gene* **116**, 1–6.

Bouvier, J., Richaud, C., Higgins, W., Bögler, O. & Stragier, P. (1992) Cloning, characterization and expression of the *dapE* gene of *Escherichia coli*. *J. Bacteriol.* **174**, 5265–5271.

Colombo, S., D'Auria, S., Fusi, P., Zecca, L., Raia, C.A. & Tortora, P. (1992) Purification and characterization of a thermostable carboxypeptidase from the extreme thermophilic archaebacterium *Sulfolobus solfataricus*. *Eur. J. Biochem.* **206**, 349–357.

Colombo, S., Toietta, G., Zecca, L., Vanoni, M. & Tortora, P. (1995) Molecular cloning, nucleotide sequence and expression of a carboxypeptidase-encoding gene from the archaebacterium *Sulfolobus solfataricus*. *J. Bacteriol.* **177**, 5561–5566.

Fleischmann, R.D., Adams, M.D., White, O. *et al*. (1995) Whole-genome random sequencing and assembly of *Haemophilus influenzae* Rd. *Science* **269**, 496–512.

Fusi, P., Burlini, N., Villa. M., Tortora, P. & Guerritore, A. (1991) Intracellular proteases from the extreme thermophilic archaebacterium *Sulfolobus solfataricus*. *Experientia* **47**, 1057–1060.

Henrich, B., Monnerjahn, U. & Plapp, R. (1990) Peptidase D gene (*pep D*) of *Escherichia coli* K-12: nucleotide sequence, transcript mapping, and comparison with other peptidase genes. *J. Bacteriol.* **172**, 4641–4651.

Lawrence, C.E., Altschul, S.F., Boguski, M.S., Liu, J.S., Neuwald, A.F. & Wotton, J.C. (1993) Detecting subtle sequence signals: a Gibbs sampling strategy for multiple alignments. *Science* **262**, 208–214.

Lee, S.-H., Taguchi, H., Yoshimura, E., Minagawa, E., Kaminogawa, S, Ohta, T. & Matsuzawa, H. (1994) Carboxypeptidase *Taq*, a thermostable zinc enzyme from *Thermus aquaticus* YT-1: molecular cloning, sequencing and expression of the encoding gene in *Escherichia coli*. *Biosci. Biotechnol. Biochem.* **58**, 1490–1495.

Mozhaev, V.V., Heremans, K., Frank, J., Masson, P. & Balny, C. (1996) High pressure effects on protein structure and function.

*Proteins* **24**, 81–91.

Sakanyan, V., Desmarez, L., Legrain, C., Charlier, D., Mett, I., Kochikyan, A., Savchenko, A., Boyen, A., Falmagne, P., Pierard, A. & Glansdorff, N. (1993) Gene cloning, sequence analysis, purification, and characterization of a thermostable aminoacylase from *Bacillus stearothermophilus*. *Appl. Environ. Microbiol.* **59**, 3878–3888.

Schuler, G.D., Altschul, S.F. & Lipman, D.J. (1991) A workbench for multiple alignment and analysis. *Proteins* **9**, 180–190.

Smulevitch, S.V., Osterman, A.L., Galperina, O.V., Matz, M.V., Zagnitko, O.P., Kadyrov, R.M., Tsaplina, I.A. Grishin, N.V., Chestukhina, G.G. & Stepanov, V.M. (1991) Molecular cloning

and primary structure of *Thermoactinomyces vulgaris* carboxypeptidase T. A metalloenzyme endowed with dual substrate specificity. *FEBS Lett.* **291**, 75–78.

Villa, A., Zecca, L., Fusi, P., Colombo, S., Tedeschi, G. & Tortora, P. (1993) Structural features responsible for kinetic thermal stability of a carboxypeptidase from the archaebacterium *Sulfolobus solfataricus*. *Biochem. J.* **295**, 827–831.

Watabe, K., Ishikawa, T., Mukohara, Y. & Nakamura, H. (1992) Cloning and sequencing of the genes involved in the conversion of 5-substituted hydantoins to the corresponding L-amino acids from the native plasmid of *Pseudomonas* sp. strain NS671. *J. Bacteriol.* **174**, 962–969.

***Paolo Tortora***
*Dipartimento di Fisiologia e Biochimica Generali,*
*Via Celoria 26,*
*I-20133 Milano, Italy*
*Email: emmevi@imiucca.csi.unimi.it*

***Marco Vanoni***
*Dipartimento di Fisiologia e Biochimica Generali,*
*Via Celoria 26,*
*I-20133 Milano, Italy*

# 495. *Glutamyl aminopeptidase (Lactococcus)*

## Databanks

*Peptidase classification: clan MH, family M42, MEROPS ID: M42.001*
*NC-IUBMB enzyme classification: none*
*Databank codes:*

| Species | SwissProt | PIR | EMBL (cDNA) | EMBL (genomic) |
|---|---|---|---|---|
| *Lactococcus lactis* | – | – | X81089 | – |

## Name and History

The peptidases of lactic acid bacteria are of particular interest because of their role in the provision of amino acids for growth during milk fermentations and the ripening process of hard cheeses. It is now known that the proteolytic system of lactococci includes approximately 10–12 peptidases, but it has only been since the 1980s that individual enzymes have been purified and characterized. The activity that was later identified as glutamyl aminopeptidase was first observed in lactococci during screening experiments using various substrates to determine the general peptidolytic capacities of cheese starter bacteria. For example, Mou *et al.* (1975) reported the hydrolysis of the synthetic aminopeptidase substrate Glu+NHNap by *Streptococcus cremoris*. This was most likely due to glutamyl aminopeptidase, but at that time the agents responsible for individual activities had not been

identified. Exterkate (1984) was perhaps first to carry out studies that dealt with the glutamyl aminopeptidase as a discrete component of the peptidolytic system.

Glutamyl aminopeptidase was purified from *Lactococcus lactis* subsp. *cremoris* (previously *Streptococcus cremoris*) by Exterkate & de Veer (1987) and Bacon *et al.* (1994), and from *L. lactis* subsp. *lactis* by Niven (1991). In these publications, the enzyme was referred to as ***aminopeptidase A***, as is commonly the case for enzymes with this specificity (c.f. Chapter 339). There are no indications from the gene sequences or physicochemical characteristics that the lactococcal enzyme is related to other aminopeptidases, from eukaryotes or other bacteria, with similar specificity. The grouping together of these enzymes from such diverse sources is a result of their similar substrate specificities, which tend to be restricted to acidic N-terminal amino acid residues.

In a nomenclature of lactococcal aminopeptidases put forward by Tan *et al*. (1993), the name **glutamyl aminopeptidase (GAP)** was initially suggested. This system also recommended that enzymes that have been genetically characterized be given a single letter code following the abbreviation 'Pep'. Since the cloning and sequencing of the gene by l'Anson *et al*. (1995), the names **aminopeptidase A** and **PepA** have become most appropriate, according to this nomenclature. However, in conformity with the more systematic terminology, this chapter is titled **glutamyl aminopeptidase (Lactococcus)**.

## Activity and Specificity

As the name suggests, glutamyl aminopeptidase will remove glutamyl residues from the N-terminus of peptide substrates, but it is also effective against aspartyl and, to a lesser extent, seryl residues. Niven (1991) compared the ability of the enzyme to hydrolyze 18 aminoacyl-Ala dipeptides at 1 mM and found relative activities of: Glu┼Ala, 100%; Asp┼Ala, 74% and Ser┼Ala, 31%. The activity against all other aminoacyl-Ala dipeptides was less than 2% of the rate of Glu-Ala hydrolysis. $K_m$ values of 0.21 mM, 0.28 mM and 27.3 mM were obtained for Asp, Glu and Ser, respectively. No systematic study has been carried out of the effect of the C-terminal residue on activity, although Bacon *et al*. (1994) observed the hydrolysis of glutamyl or aspartyl dipeptides with C-terminal Lys, Gly, Tyr, Leu and Phe. They also reported the cleavage of N-terminal acidic amino acid residues from oligopeptides up to 10 residues in length, the largest substrate tested. The tetrapeptide Glu-Pro-Glu-Trp was not hydrolyzed, which may indicate that the enzyme cannot cleave aminoacyl-proline bonds.

Exterkate & de Veer (1987) carried out a qualitative analysis of the hydrolysis of several substrates in which the N-terminal residues were not $\alpha$-amino acids. Their observation of the cleavage of glutaryl- and succinyl-Phe-*p*-nitroanilides indicated that the presence of an N-terminal amine is not an absolute substrate requirement. As $\gamma$-glutamyl residues were not cleaved, they suggested that a free $\gamma$-carboxy group was essential, rather than a free $\alpha$-amino group. It has since been reported that substrates containing N-terminal seryl residues are hydrolysed (Niven, 1991; Bacon *et al*., 1994) and so a free $\gamma$-carboxy group is not essential. However, acidic groups in the $\gamma$ position may be an important feature for substrate binding in the active site. Bacon *et al*. (1994) reported no cleavage of glutaryl- or succinyl-substrates in their assay. In the absence of a quantitative comparative study this matter remains unresolved.

Glutamyl aminopeptidase is not active against substrates with a blocked N-terminal amine such as Z-Glu-Tyr, and will not cleave pyroglutamyl residues (Exterkate & de Veer, 1987). It has no activity against phosphoserine residues (F. Mulholland, personal communication). The activity of the enzyme against the synthetic chromogenic substrates Glu-NHNap and Glu-NHPhNO$_2$ and their Asp-equivalents is commonly used in enzyme assays.

## Structural Chemistry

All three of the main published studies of this enzyme conclude that glutamyl aminopeptidase is likely to be a metalloenzyme, based on its inhibition by chelating agents (Exterkate & de Veer, 1987; Niven, 1991; Bacon *et al*., 1994). The actual metal has not yet been identified, although $Co^{2+}$ stimulates activity or overcomes EDTA inhibition. $Zn^{2+}$ also had this effect in two of these studies, but Niven (1991) found $Zn^{2+}$ to be inhibitory. The gene sequence reported by l'Anson *et al*. (1995) contained no conserved regions that could be identified as known metal-binding sites (but see Alignment 482.1).

The gene sequence suggests a subunit molecular mass of 38.1 kDa, and values of 40–43 kDa have been obtained by SDS-PAGE of the purified enzyme. Gel filtration chromatography under nondenaturing conditions carried out by Niven (1991) and Bacon *et al*. (1994) suggested that this is a hexamer of approximately 240 kDa. Values of 130 kDa (Exterkate & de Veer, 1987) and 440–520 kDa (Baankreis, 1992) have also been reported for the native enzyme. A pI of 4.35 was determined by Exterkate & de Veer (1987) and activity optima are at approximately pH 8 and 50–65°C.

## Preparation

Glutamyl aminopeptidase has been purified from cell-free extracts of the following strains of *Lactococcus lactis* (the catalogue number of the equivalent strain from the National Collection of Dairy Organisms being given): NCDO 607 (634-fold: Exterkate & de Veer, 1987; 144-fold: Baankreis, 1992); NCDO 712 (727-fold: Niven, 1991), and NCDO 1968 (387-fold: Bacon *et al*., 1994).

## Biological Aspects

Glutamyl aminopeptidase is generally regarded as a component of the multienzyme proteolytic system of lactic acid bacteria, the biological function of which is to obtain amino acids for growth. It is also of technological interest as peptide hydrolysis by enzymes released from lysed starter bacteria are also thought to be involved in the development of flavor during cheese ripening. As similar activities are known from various other organisms that do not have primarily proteolytic nutrition, it is feasible that this enzyme is also involved in other cellular functions.

Growth of *Lactococcus* in milk is not dependent on glutamyl aminopeptidase as the growth rates of deletion mutants are similar to control strains (l'Anson *et al*., 1995). Similar observations have been made for several other peptidases by Mierau *et al*. (1996) who showed that no single peptidase of those tested was essential, but that multiple deletions had an accumulatively detrimental effect on growth. In the case of glutamyl aminopeptidase, this may be because other enzymes, such as aminopeptidase C (Chapter 216) (Neviani *et al*., 1989) and the dipeptidase (Hwang *et al*., 1982) have low levels of activity against acidic amino acid residues which may be sufficient to sustain growth of deletion mutants.

There remain some doubts as to the subcellular localization of this enzyme. It was originally described as 'weakly associated with the membrane' by Exterkate (1984) who detected activity in the membrane fraction of lysed protoplasts, and on the basis of the activities of cells treated with organic solvents. In addition, immunogold labeling of glutamyl aminopeptidase by Baankreis (1992) showed the enzyme to be associated with the outer cell surface. Baankreis (1992) and Bacon *et al*. (1994) detected some activity in the membrane component of fractionated cells, although most activity was detected in the cytoplasm in these studies. A cytoplasmic location is suggested by the gene sequence which contains no identified signal sequences or transmembrane domains (l'Anson *et al*., 1995). In each reported purification, the enzyme has been isolated from the soluble fraction of cell-free extracts. Glutamyl aminopeptidase appears highly hydrophobic during hydrophobic interaction chromatography (Niven, 1991), and also has a strong tendency to bind to ultrafiltration membranes (unpublished results of the author). It is therefore possible that the enzyme associates nonspecifically with the membrane *in vivo* or during purification procedures.

## Distinguishing Features

The lactococcal glutamyl aminopeptidase has a high degree of thermal stability. Niven (1991) reported that 85% of activity remained after incubation of the purified enzyme from subsp. *lactis* for 60 min at 70°C. Bacon *et al*. (1994) found the enzyme from subsp. *cremoris* to be stable for 120 min at 50°C and that 54% of activity remained after 120 min at 70°C.

The possible carboxyamidase activity reported for this enzyme by Exterkate & de Veer (1987) is an unusual feature among aminopeptidases. However, this finding was not replicated by Bacon *et al*. (1994) with similar substrates (e.g. Suc-Phe-NHPhNO$_2$). Rul *et al*. (1995) observed cleavage of substrates with an N-terminal malic acid group by a similar enzyme from *Streptococcus thermophilus*. Systematic and quantitative studies are required to characterize this feature further.

## Related Peptidases

A glutamyl aminopeptidase has been purified from the lactic acid bacterium *Streptococcus thermophilus* (Rul *et al*., 1995). This appears to have very similar activity and physicochemical characteristics to the enzyme from *Lactococcus lactis*. An aminopeptidase A purified from *Staphylococcus chromogenes* has similar subunit molecular mass to the enzyme from *Lactococcus*, has high temperature and alkaline activity optima and is relatively thermostable (Yoshpe-Besançon *et al*., 1993). The ability to cleave N-terminal malic and lactic acid residues was also reported for this enzyme. It is reported in Chapter 482 that the thermostable *Bacillus* aminopeptidase I (Chapter 499) is a homolog of the *Lactococcus* glutamyl aminopeptidase.

## Further Reading

For general reviews of the proteolytic system of lactic acid bacteria, see Tan *et al*. (1993) and Law & Mulholland (1995).

## References

Baankreis, R. (1992) The role of lactococcal peptidases in cheese ripening. PhD thesis, University of Amsterdam.

Bacon, C.L., Jennings, P.V., Ni Fhaolain, I. & O'Cuinn, G. (1994) Purification and characterisation of an aminopeptidase A from cytoplasm of *Lactococcus lactis* subsp. *cremoris* AM2. *Int. Dairy J.* **4**, 503–519.

Exterkate, F.A. (1984) Location of peptidases outside and inside the membrane of *Streptococcus cremoris*. *Appl. Environ. Microbiol.* **47**, 177–183.

Exterkate, F.A. & de Veer, G.J.C.M. (1987) Purification and some properties of a membrane-bound aminopeptidase A from *Streptococcus cremoris*. *Appl. Environ. Microbiol.* **53**, 577–583.

Hwang, I.-K., Kaminogawa, S. & Yamauchi, K. (1982) Kinetic properties of a dipeptidase from *Streptococcus cremoris*. *Agric. Biol. Chem.* **46**, 3049–3053.

l'Anson, K.J.A., Movahedi, S., Griffin, H.G., Gasson, M.J. & Mulholland, F. (1995) A non-essential glutamyl aminopeptidase is required for optimal growth of *Lactococcus lactis* MG 1363 in milk. *Microbiology* **141**, 2873–2881.

Law, B.A. & Mulholland, F. (1995) Enzymology of lactococci in relation to flavour development from milk proteins. *Int. Dairy J.* **5**, 833–854.

Mierau, I., Kunji, E.R.S., Leenhouts, K.J., Hellendoorn, M.A., Haandrikman, A.J., Poolman, B., Konings, W.N., Venema, G. & Kok, J. (1996) Multiple-peptidase mutants of *Lactococcus lactis* are severely impaired in their ability to grow in milk. *J. Bacteriol.* **178**, 2794–2803.

Mou, L., Sullivan, J.J. & Jago, G.R. (1975) Peptidase activities in group N streptococci. *J. Dairy Res.* **42**, 147–155.

Neviani, E., Boquien, C.Y., Monnet, V., Phan Thanh, L. & Gripon, J.-C. (1989) Purification and characterization of an aminopeptidase from *Lacticoccus lactis* subsp. *cremoris* AM2. *Appl. Environ. Microbiol.* **55**, 2308–2314.

Niven, G.W. (1991) Purification and characterization of aminopeptidase A from *Lactococcus lactis* subsp. *lactis* NCDO 712. *J. Gen. Microbiol.* **137**, 1207–1212.

Rul, F., Gripon, J.-C. & Monnet, V. (1995) St-PepA, a *Streptococcus thermophilus* aminopeptidase with high specificity for acidic residues. *Microbiology* **141**, 2281–2287.

Tan, P.S.T., van Alen-Boorrigter, I.J., Poolman, B., Siezen, R.J., de Vos, W.M. & Konings, W.M. (1992) Characterization of the *Lactococcus lactis pepN* gene encoding an aminopeptidase homologous to mammalian aminopeptidase N. *FEBS Lett.* **306**, 9–16.

Tan, P.S.T., Poolman, B. & Konings, W.N. (1993) Proteolytic enzymes of *Lactococcus lactis*. *J. Dairy Res.* **60**, 269–286.

Yoshpe-Besançon, I., Auriol, D., Paul, F., Monsan, P., Gripon, J.-C. & Ribadeau-Dumas, B. (1993) Purification and characterization of an aminopeptidase A from *Staphylococcus chromogenes* and its use for the synthesis of amino-acid derivatives and dipeptides. *Eur. J. Biochem.* **211**, 105–110.

**M**

*Gordon W. Niven*
*Institute of Food Research,*
*Reading Laboratory,*
*Earley Gate, Whiteknights Road,*
*Reading RG6 6BZ, UK*
*Email: gordon.niven@bbsrc.ac.uk*

# 496. Aminopeptidase I

## Databanks

*Peptidase classification: clan MH, family M18, MEROPS ID: M18.001*
*NC-IUBMB enzyme classification: EC 3.4.11.22*
*Chemical Abstracts Service registry number: 9001-61-0*
*Databank codes:*

| Species | SwissProt | PIR | EMBL (cDNA) | EMBL (genomic) |
|---|---|---|---|---|
| *Saccharomyces cerevisiae* | P14904 | A33879 S39101 | – | M25548: complete gene X71133: 17 kb fragment of chromosome XI Y07522: complete gene |
|  |  |  | Z28103 |  |

## Name and History

Aminopeptidases in yeast were first described by Grassmann & Dyckerhoff in 1928. Thereafter Johnson (1941) isolated and characterized a high molecular mass protein called *aminopolypeptidase*. This enzyme was subsequently purified (Metz & Röhm, 1976; Metz *et al*., 1977) and referred to as *aminopeptidase I*, because it was the first peptidase characterized in some detail. It proved to be identical to the *aminopeptidase III* of Matile *et al*. (1971) detected after separating crude extracts from yeast by starch gel electrophoresis. The name *leucine aminopeptidase IV (LAP IV)* was used in a genetic characterization of the enzyme (Trumbly & Bradley, 1983). Finally, Achstetter *et al*. (1984) introduced the name *aminopeptidase yscI*, the suffix 'ysc' indicating its origin in the yeast *Saccharomyces cerevisiae*. IUBMB now recommends the name aminopeptidase I (Roman numeral one).

## Activity and Specificity

Aminopeptidase I belongs to family M18 (Chapter 482). All kinds of aminoacyl and peptidyl derivatives are attacked provided they contain a free $\alpha$-amino group; this includes amino acid amides and esters. As a rule, compounds bearing leucine or another hydrophobic N-terminal residue are among the best substrates whereas glycine or a charged amino acid in the P1 position are split much more slowly. Leu+NHPhNO$_2$, although routinely used to monitor enzyme activity, is a rather poor substrate and it is also cleaved by other yeast aminopeptidases. The enzyme is strongly and specifically activated by $Zn^{2+}$ and $Cl^-$ and inactivated by metal-chelating agents (Metz & Röhm, 1976).

## Structural Chemistry

The smallest active form of aminopeptidase I is a dodecamer with an $M_r$ of 640 000 which does not contain disulfide bonds, assembled from a single type of subunit of $M_r$ 53 000 which is in equilibrium at neutral pH with inactive dimeric and hexameric forms. The active enzyme contains 2 mol of zinc per mol of protein and the pI is 4.7 (Metz & Röhm, 1976;

Metz *et al*., 1977). It is synthesized as an inactive zymogen which is processed in a PEP4-dependent manner by cerevisin (Chapter 104) (Hemmings *et al*., 1981; Trumbly & Bradley, 1983; Oda *et al*., 1996).

## Preparation

Aminopeptidase I is localized inside the yeast vacuole (Matile *et al*., 1971; Frey & Röhm, 1978). It has been purified 900-fold from yeast autolysates (Metz & Röhm, 1976).

## Biological Aspects

The structure of the *APE1* gene encoding aminopeptidase I has been elucidated by Cueva *et al*. (1989) and Chang & Smith (1989). The precursor form contains an N-terminal 45 amino acid propeptide with an unusual helix-turn-helix structure and lacks a consensus signal sequence (Cueva *et al*., 1989; Chang & Smith, 1989; Klionsky *et al*., 1992; Oda *et al*., 1996). The precursor is not glycosylated, remains in the cytoplasm, and does not enter the secretory pathway; instead the sorting signal within the predicted first $\alpha$ helix directs it to the vacuole (Klionsky *et al*., 1992; Oda *et al*., 1996).

Transcription of the *APE1* gene is regulated by the growth phase and by the carbon source used for yeast growth, responding to carbon catabolite repression (Frey & Röhm, 1978; Distel *et al*., 1983; Cueva *et al*., 1989; Bordallo *et al*., 1995). An upstream activating element consisting of at most 18 bp has been shown to be involved in carbon repression and to bind a protein of 78 kDa and a heterodimer of 48 kDa (Bordallo & Suárez-Rendueles, 1995).

## References

Achstetter, T., Ehmann, C., Osaki, A. & Wolf, D.H. (1984) Proteolysis in eukaryotic cells. Proteinase yscE a new yeast peptidase. *J. Biol. Chem.* **259**, 13344–13348.

Bordallo, J. & Suárez-Rendueles, P. (1995) Identification of regulatory proteins that might be involved in carbon catabolite repression of the aminopeptidase I gene of the yeast *Saccharomyces cerevisiae*. *FEBS Lett.* **376**, 120–124.

Bordallo, J., Cueva, R. & Suárez-Rendueles, P. (1995) Transcriptional regulation of the yeast vacuolar aminopeptidase yscI encoding gene (*APE1*) by carbon sources. *FEBS Lett.* **364**, 13–16.

Chang, Y.H. & Smith, J.A. (1989) Molecular cloning and sequencing of genomic DNA encoding aminopeptidase I from *Saccharomyces cerevisiae*. *J. Biol. Chem.* **264**, 6979–6983.

Cueva, R., García-Alvarez, N. & Suárez-Rendueles, P. (1989) Yeast vacuolar aminopeptidase yscI. Isolation and regulation of the APE1 (LAP4) structural gene. *FEBS Lett.* **259**, 125–129.

Distel, B., Al, E.J.M., Tabak, H.F. & Jones, E.W. (1983) Synthesis and maturation of the yeast vacuolar enzymes carboxypeptidase Y and aminopeptidase I. *Biochim. Biophys. Acta* **741**, 128–135.

Frey, J. & Röhm, K.H. (1978) Subcellular localization and levels of aminopeptidases and dipeptidase in *Saccharomyces cerevisiae*. *Biochim. Biophys. Acta* **527**, 31–41.

Grassmann, W. & Dyckerhoff, H. (1928) Über die proteinase und die polypeptidase der Hefe [On the proteinases and polypeptidases of yeast]. *Z. Physiol. Chem.* **179**, 41–78.

Hemmings, B.A., Zubenko, G.S., Hasilik, A. & Jones, E.W. (1981) Mutant defective in processing of an enzyme located in the lysosome-like vacuole of *Saccharomyces cerevisiae*. *Proc. Natl Acad. Sci. USA* **78**, 435–439.

Johnson, M.J. (1941) Isolation and properties of a pure yeast polypeptidase. *J. Biol. Chem.* **137**, 575–586.

Klionsky, D.J., Cueva, R. & Yaver, D.S. (1992) Aminopeptidase I of *Saccharomyces cerevisiae* is localized to the vacuole independent of the secretory pathway. *J. Cell. Biol.* **119**, 287–299.

Matile, P., Wiemken, A. & Guyer, W. (1971) A lysosomal aminopeptidase isozyme in differentiating yeast cells and protoplasts. *Planta* **96**, 43–53.

Metz, G. & Röhm, K.H. (1976) Yeast aminopeptidase I. Chemical composition and catalytic properties. *Biochim. Biophys. Acta* **429**, 933–949.

Metz, G., Marx, R. & Röhm, K.H. (1977) The quaternary structure of yeast aminopeptidase I. *Z. Naturforsch.* **32c**, 929–937.

Oda, M.N., Scott, S.V., Hefner-Gravink, A., Caffarelli, A.D. & Klionsky, D.J. (1996) Identification of a cytoplasm to vacuole targeting determinant in aminopeptidase I. *J. Cell Biol.* **132**, 999–1010.

Trumbly, R.J. & Bradley, G. (1983) Isolation and characterization of aminopeptidase mutants of *Saccharomyces cerevisiae*. *J. Bacteriol.* **156**, 36–48.

*Paz Suárez-Rendueles*
*Departamento de Bioquímica y Biología Molecular,*
*Facultad de Medicina,*
*Universidad de Oviedo,*
*E-33071 Oviedo. Spain*

*Javier Bordallo*
*Departamento de Bioquímica y Biología Molecular,*
*Facultad de Medicina,*
*Universidad de Oviedo,*
*E-33071 Oviedo. Spain*

*Rosario Cueva*
*Departamento de Microbiología,*
*Facultad de Ciencias,*
*Universidad de Extremadura,*
*E-06071 Badajoz, Spain*

**M**

# 497. Introduction: other families of metallopeptidases

## Databanks

MEROPS ID: *M19, M22, M23, M26, M27, M29, M32, M34, M35, M36, M37, M38, M41, M43, M45*

| Species | SwissProt | PIR | EMBL (cDNA) | EMBL (genomic) |
|---|---|---|---|---|
| **METALLOPEPTIDASES WITH HEXXH MOTIF** | | | | |
| **Family M3** (Chapter 370) | | | | |
| **Family M26** | | | | |
| IgA-specific metalloendopeptidase (Chapter 508) | | | | |
| **Family M27** | | | | |
| Bontoxilysin (Chapter 510) | | | | |
| Tentoxilysin (Chapter 509) | | | | |
| **Family M32** | | | | |
| Carboxypeptidase Taq (Chapter 504) | | | | |
| Others | | | | |
|    *Homo sapiens* | – | – | Z15449 | – |
| **Family M34** | | | | |
| Anthrax toxin lethal factor (Chapter 511) | | | | |
| **Family M35** | | | | |
| Deuterolysin (Chapter 513) | | | | |
| Penicillolysin (Chapter 512) | | | | |
| **Family M36** | | | | |
| Fungalysin (Chapter 514) | | | | |
| **Family M41** | | | | |
| FtsH endopeptidase (Chapter 516) | | | | |
| FtsH endopeptidase homolog, chloroplast | | | | |
|    *Arabidopsis thaliana* | – | – | X99808 | – |
|    *Capsicum annuum* | – | – | X90472 | – |
|    *Odontella sinensis* | P49825 | – | – | Z67753: chloroplast complete genome |
|    *Porphyra purpurea* | P51327 | – | U38804 | – |
|    *Synechocystis* sp. | – | – | – | D64000: complete genome part 19 |
| *i*-AAA protease (Chapter 517) | | | | |
| *m*-AAA protease (Chapter 517) | | | | |
| **Family M43** | | | | |
| Cytophagalysin (Chapter 518) | | | | |
| **Family M47** | | | | |
| Metallopeptidase PRSM1 | | | | |
|    *Homo sapiens* | – | – | D38554 U58048 | – |
| **OTHER FAMILIES OF METALLOPEPTIDASES** | | | | |
| **Family M19** | | | | |
| *Acinetobacter* dipeptidase (Chapter 501) | | | | |
| Membrane dipeptidase (Chapter 500) | | | | |
| **Family M22** | | | | |
| *O*-Sialoglycoprotein endopeptidase (Chapter 505) | | | | |

| Species | SwissProt | PIR | EMBL (cDNA) | EMBL (genomic) |
|---|---|---|---|---|
| **Family M23** | | | | |
| β-Lytic metalloendopeptidase (Chapter 506) | | | | |
| Staphylolysin (Chapter 507) | | | | |
| **Family M29** | | | | |
| Aminopeptidase T (Chapter 498) | | | | |
| *Bacillus* aminopeptidase I (Chapter 499) | | | | |
| **Family M37** | | | | |
| Lysostaphin (Chapter 515) | | | | |
| Others | | | | |
| Bacteriophage 80 α | – | – | U72397 | – |
| Bacteriophage φg1e | – | – | – | X98106: complete genome |
| *Escherichia coli* | P24204 | – | M77039 | – |
| | | | U38702 | |
| *Haemophilus influenzae* | P44693 | – | – | U32724: genome section 39 of 163 |
| *Sinorhizobium meliloti* | – | – | U81296 | – |
| *Staphylococcus capitis* | – | – | D86328 | – |
| *Streptococcus zooepidemicus* | – | – | U50357 | – |
| *Synechococcus elongatus* | – | – | – | D13173: phycocyanin operon genes |
| *Synechocystis* sp. | – | – | – | D90915: genome section 17 of 27 |
| | – | – | – | D90907: genome section 9 of 27 |
| *Vibrio cholerae* | – | JC2569 | U07173 | – |
| **Family M38** | | | | |
| β-Aspartyl dipeptidase (Chapter 502) | | | | |
| **Family M45** | | | | |
| VanX D,D-dipeptidase (Chapter 503) | | | | |

There are 17 families of metallopeptidases that cannot yet be assigned to clans. Of these, ten possess the HEXXH motif that includes two zinc ligands and the catalytic Glu that is found in clan MA (Chapter 335) and clan MB (Chapter 380). In these families, the additional zinc ligands have not been biochemically characterized, and the other motifs that are characteristic of clans MA and MB are absent. Seven families do not possess the HEXXH motif, nor any motifs similar to those in any clans of metallopeptidases.

The families with the HEXXH motif will be dealt with first. An alignment comparing sequence around the zinc ligands of thermolysin is shown in Alignment 497.1 (on CD-ROM).

*Family M3* is the subject of a separate group of chapters (Chapters 370, 371, 372, 373, 374, 375, 376, 377, 378 and 379) that immediately follow the section on clan MA because there are some indications of particularly close connections between these groups.

*Family M26* contains IgA-specific metalloendopeptidase from *Streptococcus*. The enzyme cleaves the Pro227↓Thr228 bond in the Pro-rich hinge region of the heavy chain of immunoglobulin A, rendering the IgA molecule ineffective. His142 and Glu143 (numbering as in Alignment 497.1) in the HEXXH motif have been shown to be essential for activity by site-directed mutagenesis. Glu166 is equally spaced from the HEXXH in both thermolysin and the IgA endopeptidase.

The IgA endopeptidase is synthesized as a large precursor of 1669 amino acids. At the C-terminus there are ten tandem repeats of an 18–20 residue sequence, VXPXQVXXXPEYXGXXXGAX, that is possibly important for secretion. It is not known whether this domain is proteolytically removed. Serralysin (Chapter 386) has a C-terminal domain with repeating elements that binds calcium and is important for secretion.

There are a number of IgA endopeptidases from a variety of bacterial pathogens, including those from *Neisseria* and *Haemophilus* which are serine-type IgA endopeptidases (Chapter 87). These are also large proteins with a C-terminal secretion domain, but cleave at a different Pro↓Thr bond in the IgA1 hinge region.

*Family M27* contains the neurotoxins from *Clostridium* species that block acetylcholine release at neuromuscular junctions causing motor paralysis in the diseases tetanus and botulism. Bontoxilysin is the toxin in botulism, and is synthesized as a 150 kDa precursor. The precursor is proteolytically activated into a disulfide-linked heterodimer, with an N-terminal light chain (50 kDa) and a C-terminal heavy chain (100 kDa). The light chain includes the endopeptidase active site, whereas the heavy chain has a high and specific affinity for neurons. The toxin accumulates in the bacterial cell until it lyses, and is then internalized by neuron cells into endosome-like vesicles. The toxin passes to the cytoplasm where the free light chain is able to degrade synaptobrevin, a component of the neuroexocytosis apparatus, at the Gln60╫Lys61 bond. Tentoxilysin cleaves synaptobrevin 2 at the Gln76╫Phe77 bond.

X-ray spectroscopy has suggested that the zinc ion is coordinated by three residues with aromatic rings, and besides the histidines of the HEXXH motif, Tyr152 (numbering as in Alignment 497.1) is suggested to be a zinc ligand (Morante *et al.*, 1996). Tyrosine is known to be a zinc ligand and the general base in some members of clan MB.

**Family M32** contains carboxypeptidase Taq from the thermophilic bacterium *Thermus aquaticus*. The carboxypeptidase contains an HEXXH motif: mutation of the His residues diminished activity and zinc binding, and mutation of the Glu diminished activity but did not affect zinc binding. This implies a catalytic mechanism similar to that of clans MA and MB, but because mutagenesis did not abolish activity, there must still be some doubt that carboxypeptidase Taq is the first carboxypeptidase with an HEXXH zinc-binding motif. A homolog is now known from *Bacillus subtilis*, which is shorter at the C-terminus. A human expressed sequence tag also shows sequence similarity.

**Family M34** contains the anthrax lethal factor from *Bacillus anthracis*. The anthrax toxin is a complex of three proteins, of which lethal factor is one. The mode of action of the anthrax toxin is not known, but like the tetanus and botulism toxins, it enters the endosomes of host cells, which are macrophages for anthrax toxin. Inhibitors of the vacuolar ATPase proton pump prevent macrophage toxicity. Lethal factor has an HEXXH motif, and site-directed mutagenesis of His142 and His146 fully inactivated the protein. Moreover, bestatin and captopril block toxicity. The effects of the metallopeptidase inhibitors strongly suggest that the lethal factor is a peptidase, and similarity to leukotriene $A_4$ hydrolase (Chapter 347) has been suggested (Ménard *et al.*, 1996), but no substrate has yet been found.

**Family M35** includes the fungal metalloendopeptidases deuterolysin from *Aspergillus*, which preferentially degrades nuclear proteins such as histones and protamine, and penicillolysin from *Penicillium*, which is an extracellular enzyme able to cleave a wide range of substrates. Both enzymes are synthesized with N-terminal propeptides.

**Family 36** contains fungalysin (Chapter 514). The family has now been assigned to clan MA, but the change was too recent to be reflected in the present arrangement of chapters.

**Family M41** includes the FtsH endopeptidase, which is known from bacteria, eukaryote mitochondria and chloroplasts. The FtsH protein from bacteria is active during cell division and has been suggested to be an ATP-dependent metallopeptidase involved in the degradation of $\sigma^{32}$ factor. The protein contains a peptidase unit with an HEXXH motif, and an ATPase domain. The ATPase is homologous to many other ATP-hydrolyzing proteins, including regulatory subunits of the eukaryote 26S proteasome (Chapter 173) and the clpX subunit of endopeptidase Clp (Chapter 177).

Under normal conditions of growth in *E. coli*, most promoters respond to $\sigma^{70}$ factor. When the temperature is raised above 40°C, the heat-shock response is initiated, which allows the bacterium to grow as normally as possible under the new temperature conditions. The heat-shock response is inversely related to the cellular levels of $\sigma^{70}$. During heat shock, some 17 proteins are synthesized, and these have promoters that respond to $\sigma^{32}$ rather than $\sigma^{70}$. Among these proteins are the serine-type endopeptidases protease Do (Chapter 83), which is membrane bound, and endopeptidase La (Chapter 178), which is cytoplasmic and ATP-dependent. The gene for one of the subunits of $\sigma^{70}$ also responds to $\sigma^{32}$, acting as a negative feedback to switch off the heat-shock response. The FtsH endopeptidase gene responds to both $\sigma$ factors.

The FtsH endopeptidase is a membrane-bound protein attached to the inner membrane. There is an N-terminal cytoplasmic tetrapeptide, a membrane-spanning region, a periplasmic domain, a second membrane-spanning region, and a large C-terminal cytoplasmic portion containing the ATPase and the peptidase units. Besides degradation of $\sigma^{32}$, the endopeptidase has also been implicated in the lysogenic decision of bacteriophage λ by degradation of the cII protein, and in degradation of uncomplexed SecY, a component of a complex that permits translocation of proteins across the plasma membrane.

The eukaryote homologs of the FtsH endopeptidase include the mitochondrial *m*- and *i*-AAA proteases. In yeast, there are three homologs of the FtsH endopeptidase: the YME1 protein is the *i*-AAA protease, and the AFG3 and RCA1 proteins form a complex known as the *m*-AAA protease. *m*-AAA protease has a chaperone function important for the correct assembly of protein complexes in the mitochondrion, including elements of the respiratory chain and ATP-dependent enzymes, and both endopeptidases are essential for the breakdown of uncomplexed components. Mutations in any of the three genes lead to mitochondrial disfunction, including leakage of mitochondrial DNA into the cytoplasm.

Homologs of the FtsH endopeptidase are also known from the cyanobacterium *Synechocystis* and the thylakoid membranes of algal and plant chloroplasts (Lindahl *et al.*, 1996).

**Family M43** contains cytophagalysin, a secreted microbial collagenase from the soil bacterium *Cytophaga*.

**Family M47** contains the 33 kDa human putative metalloendopeptidase PRSM1, known only as a sequence from a placenta library and located on chromosome 16q24.3. The library was immunoscreened with antibodies to the procollagen III N-endopeptidase (Chapter 415), a 70 kDa endopeptidase also known from placenta, but as yet unsequenced, raising the possibility that the two endopeptidases are either homologous or identical.

The remaining seven families of metallopeptidases do not have the HEXXH motif in the protein sequence.

**Family M19** includes membrane dipeptidase. This is a renal brush border membrane protein from mammals that is anchored to the membrane by a glycophosphatidylinositol anchor. The dipeptidase is synthesized as a precursor with a signal peptide and C-terminal hydrophobic domain, both of which are removed in post-translational processing. The active enzyme is a homodimer with Cys361 providing an intersubunit disulfide bond (Keynan *et al.*, 1996). There is one zinc ion per monomer. Site-directed mutagenesis has been used to show that His20, Glu125, His198 and His219 are essential for catalysis, though it is not known which of these are zinc ligands, and His152 is required for substrate binding (Keynan *et al.*, 1997). Unusually, the dipeptidase also acts as a β-lactamase and is inhibited by cilastatin (Keynan *et al.*, 1995), and the enzyme can also hydrolyze dipeptides in which the C-terminal residue is a D-amino acid.

The family also includes bacterial dipeptidases from *Acinetobacter* and *Klebsiella*. The *Acinetobacter* dipeptidase has a preference for a D-amino acid as the C-terminal residue in the dipeptide substrate (Adachi & Tsujimoto, 1995; Watanabe *et al.*, 1996).

**Family M22** includes *O*-sialoglycoprotein endopeptidase from the causative agent of cattle shipping fever, *Pasteurella haemolytica*. The endopeptidase cleaves only proteins that are heavily *O*-sialoglycosylated, such as glycophorin A from erythrocytes and leukocyte surface antigens CD34, CD43, CD44 and CD45. The enzyme is inhibited by EDTA and prolonged treatment with 1,10-phenanthroline, but not by phosphoramidon. A His111-Met-Glu-Gly-His sequence has been suggested to be a zinc-binding site (Abdullah *et al.*, 1991), but the Glu is not conserved throughout the family. Other potential ligands or catalytic residues are three aspartates: Asp159, Asp167 and Asp305.

There are homologs in bacteria, the archaean *Haloarcula marismortui*, the cyanobacterium *Synechocystis*, *Saccharomyces*, and a human expressed sequence tag.

**Family M23** includes the bacteriolytic enzyme β-lytic endopeptidase from *Lysobacter*. The enzyme cleaves the *N*-acylmuramoyl-L-Ala bond between the cell wall glycoproteins and the cross-linking peptide, a reaction also performed by the *N*-acylmuramoyl-L-Ala amidase. β-Lytic endopeptidase is not just an amidase, because it also cleaves the insulin B chain at Val18+Cya19 and Gly23+Phe24. The α-lytic endopeptidase (Chapter 82) from the same predatory bacterium cleaves within the cross-linking peptide.

A preference for cleaving glycyl bonds is a feature of most members of the family. Staphylolysin is a bacteriolytic endopeptidase from *Pseudomonas* that breaks down the pentaglycine cross-links in *Staphylococcus* and also cleaves pentaglycine and hexaglycine peptides as well as elastin at Gly+ bonds. The fibrinogenolytic endopeptidase from *Aeromonas* cleaves a -Gly-Gly+Ala bond located near the cross-link site in fibrin.

The 171 residue propeptide of β-lytic endopeptidase contains several clusters of Arg residues, as do the propeptides of α-lytic endopeptidase, lysyl endopeptidase (Chapter 85) and subtilisin (Chapter 94). It has been suggested that the basic residues assist in the correct folding of the enzyme by providing hydrophilic and hydrophobic regions, and they may help in transporting the mature enzyme through the outer membrane of the gram-negative bacterium (Li *et al.*, 1990).

It is possible that endopeptidases from family M23 have a similar tertiary fold to the zinc D-Ala-D-Ala carboxypeptidase in clan MD (Chapter 462). Enzymes from both families M23 and M15 hydrolyze bacterial cell walls, and both possess a His-Xaa-His motif. His120 has been shown to be essential for lytic activity in staphylolysin (Gustin *et al.*, 1996). However, it is difficult to align the other zinc ligands between the two families, so family M23 has not been included in clan MD. Other potential catalytic residues and zinc ligands in family M23 are His22, Asp36, His71 and His81 (see Alignment 497.2 on CD-ROM).

**Family M37** includes another bacteriolytic enzyme, lysostaphin from *Staphylococcus*. Similarly to β-lytic endopeptidase, lysostaphin also cleaves the cross-linking peptide at Gly+Gly bonds and contains one ion of zinc per molecule. Lysostaphin is a secreted enzyme and is synthesized as a precursor; proteolytic activation occurs extracellularly. The propeptide contains 13 repeats of the sequence Ala-Glu-Val-Glu-Thr-Ser-Lys-(Ala/Pro)-Pro. At the C-terminus of the mature enzyme there is a cell-attachment domain homologous to one in *N*-acetylmuramoyl-L-Ala amidase from *Staphylococcus aureus*.

There are a number of homologs from other bacteria and bacteriophages. As in families M23 and M15, there is a conserved His-Xaa-His motif, but it is not possible to align other potential catalytic residues and zinc ligands among the families. Alignment 497.3 (on CD-ROM) is an alignment showing predicted catalytic residues or zinc ligands for selected members of family M37.

**Family M38** contains β-Aspartyl dipeptidase from *Escherichia coli*. Asp and Asn residues can undergo a spontaneous post-translational cyclization reaction, which in the case of Asp produces an isoaspartate residue. This residue forms peptide bonds through its β-carboxyl group rather than the α-carboxyl in normal peptide bonds. β-Aspartyl dipeptidase is an omega peptidase that releases the isoaspartyl residue from the N-terminus of a dipeptide. In *E. coli*, the peptidase is a cytosolic enzyme, probably existing as a homo-oligomer. An unsequenced mammalian enzyme with a similar specificity exists, releasing an isoaspartyl residue from carnosine (Chapter 528). *E. coli* β-Asp-peptidase is homologous to bacterial dihydroorotase, which generates *N*-carbamoyl-L-aspartate from (*S*)-dihydroorotate. In the dihydroorotases, there is a conserved His-Xaa-His motif that has been suggested to be important for zinc binding (Souciet *et al.*, 1989). This motif is also conserved in β-Asp-peptidase, suggesting a possible structural relationship with clan MD (Chapter 462), and families M23 and M37 above. All the peptidases in these families cleave unusual peptide bonds, which may be a further indication of relatedness.

**Family M29** contains aminopeptidase T from *Thermus aquaticus* and aminopeptidase II from *Bacillus stearothermophilus*. In family M29 the metal ion is probably cobalt; other cobalt-dependent aminopeptidases are found in clan MG (Chapter 475). Two metal ions are bound per catalytic subunit, suggesting cocatalytic cobalt ions. Both aminopeptidase T and aminopeptidase II are homodimers not stabilized by disulfide bonds.

Although it has been suggested that enzymes from family M29 may have a tertiary fold similar to peptidases of clan MH (Chapter 482), such as *Vibrio* aminopeptidase (Chapter 491), all members of that clan are zinc-dependent peptidases. Because there are only three complete sequences, and these are very similar to one another, it is not possible to identify metal ligands and catalytic residues.

**Family M45** contains *Lactococcus* D-Ala-D-Ala dipeptidase, an enzyme involved in the final stages of breakdown of cell wall cross-link precursors. The dipeptidase is encoded by a plasmid gene (*vanX*). The dipeptidase is one of several genes important for plasmid-borne resistance to the antibiotic vancomycin. Unlike families M23 and M37, there is no His-Xaa-His motif, and metal ligands have not been identified.

Likely candidates occur in the **Glu**181-Trp-Trp-**His** sequence, however.

## References

Abdullah, K.M., Lo, R.Y.C. & Mellors, A. (1991) Cloning, nucleotide sequence, and expression of the *Pasteurella haemolytica* A1 glycoprotease gene. *J. Bacteriol.* **173**, 5597–5603.

Adachi, H. & Tsujimoto, M. (1995) Cloning and expression of dipeptidase from *Acinetobacter calcoaceticus* ATCC 23055. *J. Biochem. (Tokyo)* **118**, 555–561.

Gustin, J.K., Kessler, E. & Ohman, D.E. (1996) A substitution at His-120 in the lasA protease of *Pseudomonas aeruginosa* blocks enzymatic activity without affecting propeptide processing or extracellular secretion. *J. Bacteriol.* **178**, 6608–6617.

Keynan, S., Hooper, N.M., Felici, A., Amicosante, G. & Turner, A.J. (1995) The renal membrane dipeptidase (dehydropeptidase I) inhibitor, cilastatin, inhibits the bacterial metallo-β-lactamase enzyme CphA. *Antimicrob. Agents Chemother.* **39**, 1629–1631.

Keynan, S., Habgood, N.T., Hooper, N.M. & Turner, A.J. (1996) Site-directed mutagenesis of conserved cysteine residues in porcine membrane dipeptidase. Cys 361 alone is involved in disulfide-linked dimerization. *Biochemistry* **35**, 12511–12517.

Li, S.L., Norioka, S. & Sakiyama, F. (1990) Molecular cloning and nucleotide sequence of the β-lytic protease gene from *Achromobacter lyticus. J. Bacteriol.* **172**, 6506–6511.

Lindahl, M., Tabak, S., Cseke, L., Pichersky, E., Andersson, B. & Adam, Z. (1996) Identification, characterization, and molecular cloning of a homologue of the bacterial FtsH protease in chloroplasts of higher plants. *J. Biol. Chem.* **271**, 29329–29334.

Ménard, A., Papini, E., Mock, M. & Montecucco, C. (1996) The cytotoxic activity of *Bacillus anthracis* lethal factor is inhibited by leukotriene A₄ hydrolase and metallopeptidase inhibitors. *Biochem. J.* **320**, 687–691.

Morante, S., Furenlid, L., Schiavo, G., Tonello, F., Zwilling, R. & Montecucco, C. (1996) X-ray absorption spectroscopy study of zinc coordination in tetanus neurotoxin, astacin, alkaline protease and thermolysin. *Eur. J. Biochem.* **235**, 606–612.

Souciet, J.L., Nagy, M., Le, G.M., Lacroute, F. & Potier, S. (1989) Organization of the yeast URA2 gene: identification of a defective dihydroorotase-like domain in the multifunctional carbamoylphosphate synthetase-aspartate transcarbamylase complex. *Gene* **79**, 59–70.

Watanabe, T., Kera, Y., Matsumoto, T. & Yamada, R.H. (1996) Purification and kinetic properties of a D-amino-acid peptide hydrolyzing enzyme from pig kidney cortex and its tentative identification with renal membrane dipeptidase. *Biochim. Biophys. Acta Protein Struct. Mol. Enzymol.* **1298**, 109–118.

# 498. *Aminopeptidase T*

## Databanks

*Peptidase classification: clan MX, family M29, MEROPS ID: M29.001*
*NC-IUBMB enzyme classification: none*
*Databank codes:*

| Species | SwissProt | PIR | EMBL (cDNA) | EMBL (genomic) |
|---|---|---|---|---|
| *Bacillus stearothermophilus* | P24828 | – | D13385 | – |
| *Bacillus subtilis* | P39762 | – | D37799 | U51911: ampS-nprE gene region |
| *Thermus aquaticus* | P23341 | JN0087 | D00814 | – |
| *Thermus thermophilus* | P42778 | – | D13386 | – |

## Name and History

Zuber and his collaborators studied the thermophilic aminopeptidases of *Bacillus stearothermophilus* in detail (Roncari & Zuber, 1976). *B. stearothermophilus* produces at least three aminopeptidases: a highly thermostable aminopeptidase, designated API, and two more thermolabile and mesophilic aminopeptidases, designated APII and APIII. All three of these enzymes are metallopeptidases activated by cobalt ion. API is a membrane-associated enzyme. It is a member of peptidase family M42 in clan MH (see Chapter 482) and is described in Chapter 499 under the name *Bacillus* aminopeptidase I. APII and APIII are low molecular mass enzymes located in the cytoplasm and each is composed of two identical subunits of $M_r$ 46 000 and 47 500, respectively (Balerna & Zuber, 1974; Stoll *et al.*, 1976). APII and APIII are included here with the *Thermus* peptidase.

*Thermus aquaticus* YT-1, an extremely thermophilic bacterium which can grow at temperatures above 70°C also possesses a thermostable aminopeptidase activity. This enzyme was purified and characterized for the first time by Minagawa *et al*. (1988) and named *aminopeptidase T (AP-T)*, the letter 'T' denoting *Thermus* or thermophilic. The nucleotide sequence of the gene encoding AP-T was subsequently determined (Motoshima *et al*., 1990). The N-terminal sequence of AP-T showed similarity to that of APII of *B. stearothermophilus* as determined by Stoll *et al*. (1976). The *aminopeptidase II (APII)* gene has been cloned but its sequence is published only in the databases (see the Databanks table). The sequence of APII is similar throughout to that of AP-T. Therefore, AP-T and APII constitute an aminopeptidase T family (M29) (Rawlings & Barrett, 1995). The *B. subtilis* and *T. thermophilus* sequences are also closely related.

## Activity and Specificity

AP-T releases any amino acids, including proline, from the N-terminus of peptides with broad substrate specificity. Its substrate specificity has been well tabulated (Minagawa *et al*., 1988). The enzyme preferentially hydrolyzes peptides containing hydrophobic residues such as Leu, Val, Phe or Tyr at the N-terminus. However, peptides in which proline occupies a penultimate (P1') position are not hydrolyzed, as in the case of leucyl aminopeptidase (Chapter 473). AP-T hydrolyzes dipeptides, tripeptides and longer oligopeptides. The substrate specificity of APII appears to be similar to that of AP-T, but detailed data are lacking. It has been demonstrated that AP-T efficiently degrades bitter peptides (e.g. the peptide of positions 91–100 in $\alpha_{s1}$-casein) in trypsin hydrolyzates of casein and distinctly decreases their bitterness (Minagawa *et al*., 1989).

The hydrolytic activity of AP-T with various peptides as substrates is measured by the ninhydrin method (Matheson & Tattrie, 1964), this includes peptides having an N-terminal proline (Yaron & Mlyner, 1968). The activity with aminoacyl-$\beta$-naphthylamides as substrates is measured by the colorimetric method of Goldberg & Rutenburg (1958), and in the case of aminoacyl-NHPhNO$_2$ the activity is measured by the spectrophotometric method of Fossy (1978). The optimal temperature for activity of AP-T is 75–80°C. This assay is conveniently done with Leu$+$NHPhNO$_2$ as the substrate at 70°C by monitoring the change in absorbance at 405 or 410 nm. The optimum pH for activity is 8.5–9.0. This enzyme is metal dependent; it is completely inhibited by EDTA (1 mM) or 1,10-phenanthroline, and considerably inhibited by amastatin (1 µM). It is also inhibited by PCMB, 2-ME, 10 mM bestatin, or cysteine. Activity of the native AP-T is not affected by $Co^{2+}$ or $Mg^{2+}$. It is weakly inhibited by $Cu^{2+}$ and $Ca^{2+}$, and strongly inhibited by $Fe^{2+}$, $Sn^{2+}$, $Zn^{2+}$ and $Mn^{2+}$ (1 mM each). The inhibited activity of the apoenzyme is restored by $Co^{2+}$ or $Mg^{2+}$, but not by $Zn^{2+}$ (only tested at 1 mM).

## Structural Chemistry

AP-T is a homodimeric enzyme. The $M_r$ of the enzyme is 108 000 as determined by gel filtration, and 48 000 as indicated by SDS-PAGE. No disulfide bonds are present. Nucleotide sequence analysis reveals that the gene encodes 408 amino acid residues, and the calculated molecular mass is 44 820 Da for each subunit. Comparison of the sequences of AP-T and APII suggested that one of the most conserved sequences (Asp306-Thr-Asp-Glu-Gly-Ala-Arg312) of AP-T is similar to the leucyl aminopeptidase signature (Asn-Thr-Asp-Ala-Glu-Arg-Leu) in the PROSITE database (Bairoch & Bucher, 1994). If this is taken to indicate that family M29 belongs in clan MF (Chapter 472), then we could postulate that the metal ligand residues of AP-T may be Asp308 and Glu249. The positively charged (basic) amino acid Arg312 of AP-T is also conserved, and this may act as a catalytic residue. According to the data concerning the three-dimensional structure of cattle lens leucyl aminopeptidase (Chapter 473) and bestatin complex (Burley *et al*., 1992), Thr336 and Leu337 are also conserved as residues involved in stabilizing the enzyme–bestatin complex. The other potential metal ligand of AP-T may be Glu249, and an additional active-site residue may be Lys237.

It has not been determined yet whether the native AP-T contains $Zn^{2+}$ or $Co^{2+}$. APII binds 2 mol of cobalt per mol of subunit (Stoll *et al*., 1976), and two different types of metal ions are bound by the enzyme (Myrin & Hofsten, 1974). Therefore, AP-T may contain two metal ions per subunit, as in the case of APII, and the same number is present in leucyl aminopeptidase. As mentioned above, the tertiary structure of AP-T may resemble the C-terminal domain of leucyl aminopeptidase, and this enzyme may belong to the group of Zn-dependent exopeptidases (Murzin *et al*., 1995).

## Preparation

AP-T was purified about 640-fold from the cell-free extract of *Thermus aquaticus* (Minagawa *et al*., 1988). The gene has been expressed in *Escherichia coli*. The expressed enzyme had a thermostability equivalent to the authentic enzyme, and was easily purified by heat treatment (80°C for 20 min) and centrifugation (Motoshima *et al*., 1990). The recombinant AP-T is commercially available from Seikagaku Corp. or Wako Pure Chemicals (see Appendix 2 for full names and addresses of suppliers).

## Biological Aspects

The AP-T gene of *Thermus thermophilus* HB8 has been cloned. The location of the AP-T gene in the physical map of the genome of *Thermus thermophilus* HB27 has been determined (Tabata & Hoshino, 1996). It seems that the AP-T gene might be common in *Thermus* spp. and it seems to be a constitutive enzyme in the cell. This enzyme does not have a prepro sequence and is located in the cytoplasm. APII also is produced in various strains of *Bacillus stearothermophilus* (Zuber & Roncari, 1967). Recently, the product of the aminopeptidase S gene (*ampS*) of *Bacillus subtilis* was reported as a putative member of this family of proteins based on the homology of *ampS* to the APII gene. This protein has a high similarity to APII. Thus, the AP-T family is probably ubiquitous in bacteria other than thermophiles.

## Distinguishing Features

It is easy to distinguish AP-T from other exopeptidases in *Thermus* spp. because this enzyme seems to be the only component active against Leu-NHPhNO$_2$ in the cytoplasm of *Thermus* spp. It remains to be examined whether the API type (i.e. high molecular mass, membrane bound) enzyme exists in *Thermus* spp.

## Further Reading

For a review, see Minagawa *et al*. (1988).

## References

Bairoch, A. & Bucher, P. (1994) PROSITE: recent developments. *Nucleic Acids Res.* **22**, 3583–3589.

Balerna, M. & Zuber, H. (1974) Thermophilic aminopeptidase from *Bacillus stearothermophilus*. IV. *Int. J. Protein Res.* **6**, 499–514.

Burley, S.K., David, P.R., Sweet, R.M., Taylor, A. & Lipscomb, W.N. (1992) Structure determination and refinement of bovine lens leucine aminopeptidase and its complex with bestatin. *J. Mol. Biol.* **224**, 113–140.

Fossy, H. (1978) Aminopeptidase from *Brevibacterium linens*: production and purification. *Milchwissenschaft* **33**, 221–223.

Goldberg, J.A. & Rutenberg, A.M. (1958) The colorimetric determination of leucine aminopeptidase in urine and serum of normal subjects and patients with cancer and other diseases. *Cancer* **11**, 283.

Matheson, A.T. & Tattrie, B.L. (1964) A modified Yemm and Cocking ninhydrin reagent for peptidase assay. *Can. J. Biochem.* **42**, 95–103.

Minagawa, E., Kaminogawa, S., Matsuzawa, H., Ohta, T. & Yamauchi, K. (1988) Isolation and characterization of a thermostable aminopeptidase (aminopeptidase T) from *Thermus aquaticus* YT-1, an extremely thermophilic bacterium. *Agric. Biol. Chem.* **52**, 1755–1763.

Minagawa, E., Kaminogawa, S., Tsukasaki, F. & Yamauchi, K. (1989) Debittering mechanism in bitter peptides of enzymatic hydrolysates from milk casein by aminopeptidase T. *J. Food Sci.* **54**, 1225–1229.

Motoshima, H., Azuma, N., Kaminogawa, S., Ono, M., Minagawa, E., Matsuzawa, H., Ohta, T. & Yamauchi, K. (1990) Molecular cloning and nucleotide sequence of the aminopeptidase T gene of *Thermus aquaticus* YT-1 and its high-level expression in *Escherichia coli*. *Agric. Biol. Chem.* **54**, 2385–2392.

Murzin, A., Brenner, S.E., Hubbard, T.J.P. & Chothia, C. (1995) SCOP: a structural classification protein database for the investigation of sequences and structures. *J. Mol. Biol.* **247**, 536–540.

Myrin, P. & Hofsten, B.V. (1974) Purification and metal ion activation of an aminopeptidase (aminopeptidase II) from *Bacillus stearothermophilus*. *Biochim. Biophys. Acta* **350**, 13–25.

Rawlings, D. & Barrett, A.J. (1995) Evolutionary families of metallopeptidases. *Methods Enzymol.* **248**, 183–228.

Roncari, G. & Zuber, H. (1969) Thermophilic aminopeptidase from *Bacillus stearothermophilus*. I. Isolation, specificity, and general properties of the thermostable aminopeptidase I. *Int. J. Protein Res.* **1**, 45–61.

Roncari, G. & Zuber, H. (1976) Thermophilic aminopeptidase I. *Methods Enzymol.* **45**, 522–530.

Stoll, E., Lowell, H., Ericsson, H. & Zuber, H. (1973) The function of the two subunits of thermophilic aminopeptidase I. *Proc. Natl Acad. Sci. USA* **70**, 3781–3784.

Stoll, E., Weder, H.-G. & Zuber, H. (1976) Aminopeptidase II from *Bacillus stearothermophilus*. *Biochim. Biophys. Acta* **438**, 212–220.

Tabata, K. & Hoshino, T. (1996) Mapping of 61 genes on the refined physical map of the chromosome of *Thermus thermophilus* HB27 and comparison of genome organization with that of *T. thermophilus* HB8. *Microbiology* **142**, 401–410.

Yaron, A. & Mlyner, D. (1968) Aminopeptidase-P. *Biochem. Biophys. Res. Commun.* **32**, 658–663.

Zuber, H. & Roncari, G. (1967) Thermophilic and mesophilic aminopeptidases from *Bacillus stearothermophilus*. *Angew. Chem. Int. Edit.* **6**, 880–881.

*Hidemasa Motoshima*
*Research Center,*
*Yotsuba Milk Products Co., Ltd,*
*465-1, Wattsu, Kitahiroshima, Hokkaido, 061-12 Japan*
*Email: motosima@ppp.bekkoame.or.jp*

*Shuichi Kaminogawa*
*Department of Applied Biochemistry,*
*The University of Tokyo,*
*1-1-1 Yayoi, Bunkyouku, Tokyo 113, Japan*
*Email: akamino@hongo.ecc.u-tokyo.ac.jp*

# *499. Bacillus aminopeptidase I*

*Peptidase classification: clan MH, family M42, MEROPS ID: M42.002*
*NC-IUBMB enzyme classification: none*

*Databank codes:*

| Species | SwissProt | PIR | EMBL (cDNA) | EMBL (genomic) |
|---|---|---|---|---|
| *Bacillus stearothermophilus* | P00728 | – | – | – |
| *Bacillus subtilis* | P94521 | – | – | Z75208: 89 kb sequence |

## Name and History

Zuber & Roncari (1967) purified and characterized three different aminopeptidases from *Bacillus stearothermophilus*, a moderately thermophilic bacterium. These enzymes were named aminopeptidase I (API) (Roncari & Zuber, 1969), aminopeptidase II (APII) (Stoll *et al.*, 1976), and aminopeptidase III (APIII) (Balerna & Zuber, 1974). These enzymes differ in $M_r$, heat stability and substrate specificity. Although all three of these aminopeptidases are produced simultaneously in the same strain, only the ***Bacillus aminopeptidase I (API)*** is thermophilic, and this was at one time designated ***thermophilic aminopeptidase*** (EC 3.4.11.12) by IUBMB. *B. stearothermophilus* APII and APII are unrelated, and are included in Chapter 498. Aminopeptidases biochemically similar to API have been reported from fungi, *Talaromyces duponti* and *Mucor* sp. (Chapuis & Zuber, 1970) but there are still no sequence data available for these enzymes.

## Activity and Specificity

The substrate specificity of API in reactions with various synthetic peptides has been tabulated (Zuber & Roncari, 1967; Roncari & Zuber, 1969). It is an aminopeptidase of broad specificity that releases not only neutral (preferentially aliphatic and aromatic), but also acidic and basic amino acids including proline from the N-terminus. It is more active against peptides than against Leu-NHPhNO$_2$. API consists of 12 subunits of two different types ($\alpha$ and $\beta$, with identical molecular mass). The activity of API as mentioned above refers to that of API hybrids (containing $\alpha$ and $\beta$). The substrate specificity of the $\alpha$ subunit differs from that of the $\beta$ subunit. The $\beta$ subunit does not hydrolyze Leu-NHPhNO$_2$, or does so only at a very slow rate compared with the $\alpha$ subunit (Stoll *et al.*, 1972). In experiments concerning the recombination of API subunits, it was revealed that only the $\alpha$ subunit is needed for degradation of neutral peptides, whereas dipeptides having an N-terminal aspartic or glutamic acid residue are substantially hydrolyzed only by forms of the enzyme containing the $\beta$ subunit as well (Stoll *et al.*, 1973).

The methods for determining the activity of this enzyme have been described in detail (Roncari *et al.*, 1976). Conveniently, the assay is performed by monitoring the hydrolysis of Leu-NHPhNO$_2$ at 65°C by spectrophotometric determination of *p*-nitroaniline at 410 nm. The optimal temperature for activity of API is 90°C in assays of 10 min duration. The optimal pH for activity of API using Gly-Leu-Tyr as the substrate at 65°C is pH 9.2–9.5. Using Leu-|-NHPhNO$_2$, optimum activity is observed at pH 7.5–8.0.

API is a metallo-aminopeptidase, and it requires Co$^{2+}$ ions for maximum activity. Since the metal ions are strongly bound in the metal-enzyme complex, this enzyme is stable in the presence of EDTA at concentrations up to 10 mM at pH 8.1. Under pH conditions of less than pH 6.0, the enzyme is converted into the apoenzyme following dialysis against 10 mM EDTA (Moser *et al.*, 1970). The activity of the apoenzyme can be restored by treatment with various bivalent cations. It is most effectively reactivated by Co$^{2+}$ (1 mM). The different metal–enzyme complexes display varying activities against different substrates depending on the type of metal incorporated (Roncari & Zuber, 1969). In cobalt-free buffer the thermostability of the enzyme is markedly reduced, and the apoenzyme is also thermolabile. In 8 M urea, the activity of API in hydrolysis of Gly-Leu-Tyr decreases by only 7%, but its activity against Leu-NHPhNO$_2$ is markedly reduced by 75%. The enzyme remains stable for several hours even at 80°C and its activity increases by about 20% after 30 min at this temperature. API is also activated by the presence of a tertiary butanol (10%), its activity is increased by about 20% within the temperature range of 20–90°C.

## Structural Chemistry

The $M_r$ of API is estimated to be $400\,000 \pm 45\,000$ based on the results of equilibrium centrifugation. In the presence of 8 M urea and 10 mM EDTA at pH 5.6, API dissociates into its subunits. API is composed of two different subunit types, $\alpha$ and $\beta$ with similar $M_r$ of $36\,500 \pm 4000$. The N-terminal sequences of the $\alpha$ and $\beta$ subunits have been determined, demonstrating that they are homologous (67% identity in 30 amino acid residues) (Stoll *et al.*, 1972, 1973). The two subunits can combine in varying ratios, resulting in the hybrids $\alpha_6\beta_6$ (API A), $\alpha_8\beta_4$ (API B) and $\alpha_{10}\beta_2$ (API C). The API C type hybrid is a major component (70% of the total API hybrids) when API is isolated from *B. stearothermophilus* strain NCIB8924 by the described methods. In experiments concerning subunit rearrangement, $\alpha_8\beta_4$ hybrids were found to interconvert to $\alpha_{12}$ enzyme and $\alpha_{10}\beta_2$ hybrids under special conditions. The $\beta_{12}$ enzyme could not be obtained, because the purified $\beta$ subunits precipitated during the reactivation procedure and the activity could not be restored (Stoll *et al.*, 1972, 1973; Stoll & Zuber, 1974). API contains a total of about 23 metal atoms per molecule, corresponding to an average distribution of two metal atoms per subunit (Roncari *et al.*, 1972). It is not known whether native API is present in the cell as a zinc enzyme or as a cobalt–zinc enzyme. The ratio of cobalt and zinc in the purified enzyme may reflect the concentration of Co$^{2+}$ and Zn$^{2+}$ in the cell homogenate. The Zn enzyme has only 5% of the activity displayed by the Co enzyme toward the substrate Gly-Leu-Tyr. The Zn atoms are more strongly bound than the Co atoms.

**M**

## Preparation

A detailed method of preparation of API from *B. stearothermophilus* has been described by Roncari & Zuber (1969, 1970). For the preparation of larger amounts of API, another method has been described by Roncari *et al.* (1976). This enzyme has been purified about 19-fold on the basis of its activity in hydrolysis of Leu-NHPhNO$_2$, and the low yield of activity in purification was due to the separation of APII fraction which was more active against Leu-NHPhNO$_2$ during the purification steps. When the bacterial cells were treated with lysozyme, only APII and APIII were released, whereas API remained associated with the cells and was almost completely membrane bound. Therefore, it is clear that API is a membrane-bound or membrane-associated enzyme. Another API-type enzyme which is even more tightly bound to the membrane than API has been purified from an obligately mesophilic variant of *B. stearothermophilus*. This membrane-bound aminopeptidase (APIm) is very similar to API, but differs slightly in thermostability and substrate specificity (Balerna & Zuber, 1974).

## Biological Aspects

The amounts of the three aminopeptidases in *Bacillus stearothermophilus* vary according to the strain. Obligately thermophilic strains (temperature optimum, 55°C; no growth at 37°C) produce substantial amounts of API and APII but little APIII, whereas obligately mesophilic strains (temperature optimum, 37°C; no growth at 55°C) produce substantial amounts of APIII but very little API or APII. Facultative strains (growth at 37°C and 55°C) produce all three of these aminopeptidases in comparable amounts (Zuber & Roncari, 1967). The gene for API has not been cloned, and the reasons for such temperature-dependent expression of the API gene are still unknown. The enzyme activity may be regulated both by a variable Co/Zn ratio and by certain metabolites or membrane (especially hydrophobic) components (Deranleau & Zuber, 1977).

## Distinguishing Features

Both API and APII (Chapter 498) hydrolyze Leu⊣NHPhNO$_2$, however, in the course of purification, the high molecular mass API is readily separated from APII and from APIII, which are low molecular mass enzymes, by Sephadex G150 gel filtration (Roncari *et al.*, 1976).

## Further Reading

For a review, see Roncari *et al.* (1976).

## References

Balerna, M. & Zuber, H. (1974) Thermophilic aminopeptidase from *Bacillus stearothermophilus*. IV. *Int. J. Protein Res.* **6**, 499–514.

Chapuis, R. & Zuber, H. (1970) Thermophilic aminopeptidase I from *Talaromyces duponti. Methods Enzymol.* **19**, 552–555.

Deranleau, D.A. & Zuber, H. (1977) Thermophilic aminopeptidase IV. Cooperative effects in ANS binding by the thermophilic aminopeptidase I from *B. stearothermophilus. Int. J. Pept. Protein Res.* **9**, 258–268.

Moser, P., Roncari, G. & Zuber, H. (1970) Thermophilic aminopeptidases from *B. stearothermophilus*. II. Aminopeptidase I (AP I): physico-chemical properties; thermostability and activation; formation of the apoenzyme and subunits. *Int. J. Protein Res.* **2**, 191–207.

Rawlings, D. & Barrett, A.J. (1995) Evolutionary families of metallopeptidases. *Methods Enzymol.* **248**, 183–228.

Roncari, G. & Zuber, H. (1969) Thermophilic aminopeptidase from *Bacillus stearothermophilus*. I. Isolation, specificity, and general properties of the thermostable aminopeptidase I. *Int. J. Protein Res.* **1**, 45–61.

Roncari, G. & Zuber, H. (1970) Thermophilic aminopeptidase: AP I from *Bacillus stearothermophilus. Methods Enzymol.* **19**, 544–552.

Roncari, G., Zuber, H. & Wyttenbach, A. (1972) Thermophilic aminopeptidases from *B. stearothermophilus*. III. Determination of the cobalt and zinc content in aminopeptidase I by neutron activation analysis. *Int. J. Pept. Protein Res.* **4**, 267–271.

Roncari, G., Stoll, E. & Zuber, H. (1976) Thermophilic aminopeptidase I. *Methods Enzymol.* **45**, 522–530.

Stoll, E. & Zuber, H. (1974) Interconversion of the different hybrids of aminopeptidase I. *FEBS Lett.* **40**, 210–212.

Stoll, E., Hermodson, M.A., Ericsson, L.H. & Zuber, H. (1972) Subunit structure of the thermophilic aminopeptidase I. *Biochemistry* **11**, 4731–4735.

Stoll, E., Ericsson, L.H. & Zuber, H. (1973) The function of the two subunits of thermophilic aminopeptidase I. *Proc. Natl Acad. Sci. USA* **70**, 3781–3784.

Stoll, E., Weder, H.-G. & Zuber, H. (1976) Aminopeptidase II from *Bacillus stearothermophilus. Biochim. Biophys. Acta* **438**, 212–220.

Zuber, H. & Roncari, G. (1967) Thermophilic and mesophilic aminopeptidases from *Bacillus stearothermophilus. Angew. Chem. Int. Edit.* **6**, 880–881.

*Hidemasa Motoshima*
*Research Center,*
*Yotsuba Milk Products Co., Ltd,*
*465-1, Wattsu, Kitahiroshima, Hokkaido, 061-12 Japan*
*Email: motosima@ppp.bekkoame.or.jp*

*Shuichi Kaminogawa*
*Department of Applied Biochemistry,*
*The University of Tokyo,*
*1-1-1 Yayoi, Bunkyouku, Tokyo 113 Japan*
*Email: akamino@hongo.ecc.u-tokyo.ac.jp*

# 500. *Membrane dipeptidase*

## Databanks

*Peptidase classification: clan MX, family M19, MEROPS ID: M19.001*
*NC-IUBMB enzyme classification: EC 3.4.13.19*
*Chemical Abstracts Service registry number: 9031-99-6*
*Databank codes:*

| Species | SwissProt | PIR | EMBL (cDNA) | EMBL (genomic) |
|---------|-----------|-----|-------------|----------------|
| *Homo sapiens* | P16444 | A35467 | D13138 | D13128: exon 1 |
| | | JS0756 | J05257 | D13129: exon 2 |
| | | JS0757 | | D13130: exon 3 |
| | | PX0021 | | D13131: exon 4 |
| | | S08193 | | D13132: exon 5 |
| | | S29848 | | D13133: exon 6 |
| | | | | D13134: exon 7 |
| | | | | D13135: exon 8 |
| | | | | D13136: exon 9 |
| | | | | D13137: exon 10 and complete CDS |
| | | | | S70329: segment 1 |
| | | | | S70330: segment 2 |
| *Mus musculus* | P31428 | S33757 | D13139 | – |
| | | | | U48387: type A; 3′ untranslated region[a] |
| | | | | U48388: type B; 3′ untranslated region[a] |
| | | | | U48389: type I; 5′ untranslated region[a] |
| | | | | U48390: type II; 5′ untranslated region[a] |
| *Oryctolagus cuniculus* | P31429 | – | X61503 | – |
| *Ovis aries* | P43477 | – | L27113 | – |
| *Rattus norvegicus* | P31430 | – | L07315 | – |
| | | | M94056 | |
| *Sus scrofa* | P22412 | S08194 | D13142 | – |
| | | | D13143 | |
| | | | X53730 | |

[a]The EMBL entries U48387–90 are not genomic; they are from the mRNA, but they are listed here to allow space for the comments.

## Name and History

Membrane dipeptidase was originally described as a dehydropeptidase activity capable of hydrolyzing Gly+dehydroPhe in crude extracts of a number of animal tissues (Bergmann & Schleich, 1932). The enzyme was subsequently purified to homogeneity from pig kidney and shown to hydrolyze a wide range of dipeptides (Robinson *et al.*, 1953). The enzyme was further extensively characterized by Campbell and colleagues in the late 1960s through to the 1980s (Campbell, 1970; Armstrong *et al.*, 1974). At this time it was called *dehydropeptidase I*, or *renal* or *microsomal dipeptidase* and classified under EC 3.4.13.11. The enzyme was then rediscovered by Merck when it was found that the β-lactam antibiotics thienamycin and imipenem were hydrolyzed by the enzyme in the kidney (Kropp *et al.*, 1982), leading to the development of the selective inhibitor cilastatin (Kahan *et al.*, 1983). The enzyme was again rediscovered when it was shown to be a glycosylphosphatidylinositol (GPI)-anchored ectoenzyme of both pig and human kidney (Hooper *et al.*, 1987, 1990), and

subsequently renamed *membrane dipeptidase* (EC 3.4.13.19) to reflect its membrane location.

## Activity and Specificity

The activity of membrane dipeptidase, as its name implies, is restricted to dipeptides with free N- and C-termini (Campbell, 1970). The enzyme will hydrolyze a wide range of such dipeptides, and is unique in that activity is shown against dipeptides in which the C-terminal amino acid is in either the L- or the D-configuration. The pH optimum is around 7.5. Most assays exploit the unique specificity of the enzyme, either utilizing dehydropeptides, such as Gly+dehydroPhe which produce a decrease in absorbance at 275 nm upon hydrolysis (Campbell, 1970), or dipeptides containing a D-amino acid, such as Gly+D-Phe. The released D-Phe is quantified by reverse-phase HPLC (Hooper *et al.*, 1987) or a coupled fluorimetric assay with D-amino acid oxidase (Heywood & Hooper, 1995).

The enzyme is inhibited somewhat poorly by metal chelators (50% inhibition by 1 mM 1,10-phenanthroline or 10 mM EDTA; Littlewood *et al.*, 1989). DTT causes substantial inhibition (95% at 1 mM; Littlewood *et al.*, 1989). The selective inhibitor cilastatin exhibits an $IC_{50}$ of 0.1 mM (Kahan *et al.*, 1983). Negligible inhibition is seen with a range of other zinc metallopeptidase inhibitors (Littlewood *et al.*, 1989).

## Structural Chemistry

Membrane dipeptidase is a disulfide-linked homodimer with a subunit of $M_r$ 45 000 in pig and 59 000 in humans (Littlewood *et al.*, 1989; Hooper *et al.*, 1990). The difference in size between the two species is due entirely to more extensive *N*-linked glycosylation of the human enzyme (Hooper *et al.*, 1990). The subunits are held together by one interchain disulfide bond, and there are two intrachain disulfide bonds (Keynan *et al.*, 1996a). The enzyme contains one atom of zinc per subunit (Armstrong *et al.*, 1974), but does not contain any of the known zinc-binding signatures in its amino acid sequence. Residues potentially involved in catalysis have been identified by site-directed mutagenesis and expression (Adachi *et al.*, 1993; Keynan *et al.*, 1994). The structure of the GPI anchor and its site of attachment have been determined in both the pig and human enzymes (Brewis *et al.*, 1995).

## Preparation

The enzyme has been purified from a number of natural sources. With the observation that it was anchored in the membrane by a GPI moiety, we were able to exploit the ability of bacterial phosphatidylinositol-specific phospholipase C to release the enzyme in a soluble form and we then purified it in essentially a single step by affinity chromatography using the inhibitor cilastatin as ligand (Littlewood *et al.*, 1989; Hooper *et al.*, 1990). Membrane dipeptidase has been expressed in CHO and COS cells (Adachi *et al.*, 1993; Keynan *et al.*, 1994).

## Biological Aspects

Membrane dipeptidase is present as a GPI-anchored ectoenzyme in a number of tissues including kidney, lung, intestine and pancreas. In the kidney the enzyme has been implicated in the extracellular catabolism of glutathione (McIntyre & Curthoys, 1982). After removal of the glutamate residue by $\gamma$-glutamyl transpeptidase, the resulting Cys┼Gly dipeptide is metabolized either by membrane dipeptidase or by membrane alanyl aminopeptidase (EC 3.4.11.2) (Chapter 337), each enzyme contributing equally to this process. However, membrane dipeptidase is the principal activity hydrolyzing the oxidized dipeptide, Cys-*bis*-Gly.

In the lungs the major physiological substrate for the enzyme appears to be the peptidyl leukotriene, leukotriene $D_4$, which is converted into the inactive leukotriene $E_4$ (Campbell *et al.*, 1990). Leukotriene $D_4$ is one of the major components of the slow-reacting substances of anaphylaxis, a potent bronchoconstrictor and vasoconstrictor, and involved as a mediator in bronchial asthma and inflammation. Therefore, the action of membrane dipeptidase is essentially one of inactivation.

Membrane dipeptidase is the only known example of a mammalian $\beta$-lactamase. It readily hydrolyzes the carbapenem class of antibiotics, such as thienamycin and imipenem, but is unable to cleave the *cis*-conformation of the $\beta$-lactam ring seen in the classical penicillins and cephalosporins (Kropp *et al.*, 1982; Kahan *et al.*, 1983).

The gene for membrane dipeptidase is located on human chromosome 16q24.3 (Nakagawa *et al.*, 1992), a region of the genome known to contain a tumor suppressor gene. Loss of alleles in this region of chromosome 16 has been described in the embryonic, renal malignancy, Wilms' tumor, and in a number of carcinomas. In particular, patients with Wilms' tumor have a loss of membrane dipeptidase mRNA (Austry *et al.*, 1993), and the enzyme appears to be a transformation-sensitive protein whose expression is actively repressed by DNA viral oncogenes (Keynan *et al.*, 1997), suggesting that membrane dipeptidase is a tumor-suppressor gene.

## Distinguishing Features

Membrane dipeptidase is the only known mammalian peptidase capable of hydrolyzing substrates containing a D-amino acid, and of metabolizing $\beta$-lactam antibiotics. The enzyme is selectively inhibited by cilastatin and, due to its GPI anchor, is susceptible to release from the membrane by bacterial phosphatidylinositol-specific phospholipase C. Polyclonal antisera against the pig and human enzymes have been described (Littlewood *et al.*, 1989).

## Related Peptidases

A dipeptidase from *Acinetobacter calcoaceticus* has been cloned and sequenced (Chapter 501). It shows 49% sequence similarity (25% sequence identity) to membrane dipeptidase (Adachi & Tsujimoto, 1995). However, the bacterial enzyme is not inhibited by cilastatin and does not hydrolyze dehydropeptides.

## Further Reading

For an extensive review of all aspects of the enzyme see Keynan *et al.* (1996b).

## References

Adachi, H. & Tsujimoto, M. (1995) Cloning and expression of dipeptidase from *Acinetobacter calcoaceticus* ATCC 23055. *J. Biochem.* **118**, 555–561.

Adachi, H., Katayama, T., Nakazato, H. & Tsujimoto, M. (1993) Importance of Glu-125 in the catalytic activity of human renal dipeptidase. *Biochim. Biophys. Acta* **1163**, 42–48.

Armstrong, D.J., Mukhopadhyay, S.K. & Campbell, B.J. (1974) Physicochemical characterization of renal dipeptidase. *Biochemistry* **13**, 1745–1750.

Austry, E., Cohen-Salmon, M., Antignac, C., Béroud, C., Henry, I., Van Cong, N., Brugières, L., Junien, C. & Jeanpierre, C. (1993) Isolation of kidney complementary DNAs down-expressed in Wilms' tumour by a subtractive hybridization approach. *Cancer Res.* **53**, 2888–2894.

Bergmann, M. & Schleich, H. (1932) Uber die enzymatische Spaltung dehydrierter Peptide. Auffindung einer Dehydropeptidase [On the enzymatic cleavage of dehydropeptides. Discovery of a dehydropeptidase]. *Z. Physiol. Chem.* **205**, 65–82.

Brewis, I.A., Ferguson, M.A.J., Mehlert, A., Turner, A.J. & Hooper, N.M. (1995) Structures of the glycosyl-phosphatidylinositol anchors of porcine and human membrane dipeptidase. Interspecies comparison of the glycan core structures and further structural studies on the porcine anchor. *J. Biol. Chem.* **270**, 22946–22956.

Campbell, B.J. (1970) Renal dipeptidase. *Methods Enzymol.* **19**, 722–729.

Campbell, B.J., Baker, S.F., Shukla, S.D., Forrester, L.J. & Zahler, W.L. (1990) Bioconversion of leukotriene D4 by lung dipeptidase. *Biochim. Biophys. Acta* **1042**, 107–112.

Heywood, S.P. & Hooper, N.M. (1995) Development and application of a fluorometric assay for mammalian membrane dipeptidase. *Anal. Biochem.* **226**, 10–14.

Hooper, N.M., Low, M.G. & Turner, A.J. (1987) Renal dipeptidase is one of the membrane proteins released by phosphatidylinositol-specific phospholipase C. *Biochem. J.* **244**, 465–469.

Hooper, N.M., Keen, J.N. & Turner, A.J. (1990) Characterization of the glycosyl-phosphatidylinositol-anchored human renal dipeptidase reveals that it is more extensively glycosylated than the pig enzyme. *Biochem. J.* **265**, 429–433.

Kahan, F.M., Kropp, H., Sundelof, J.G. & Birnbaum, J. (1983) Thienamycin: development of imipenem-cilastatin. *J. Antimicrob. Chemother.* **12**(suppl. D), 1–35.

Keynan, S., Hooper, N.M. & Turner, A.J. (1994) Directed mutagenesis of pig renal membrane dipeptidase. His[219] is critical but the DHXXH motif is not essential for zinc binding or catalytic activity. *FEBS Lett.* **349**, 50–54.

Keynan, S., Habgood, N.T., Hooper, N.M. & Turner, A.J. (1996a) Site-directed mutagenesis of conserved cysteine residues in porcine membrane dipeptidase. Cys 361 alone is involved in disulfide-linked dimerization. *Biochemistry* **35**, 12511–12517.

Keynan, S., Hooper, N.M. & Turner, A.J. (1996b) Molecular and functional aspects of membrane dipeptidase. In: *Zinc Metalloproteases in Health and Disease* (Hooper, N.M., ed.). London: Taylor & Francis, pp. 285–309.

Keynan, S., Asipu, A., Hooper, N.M., Turner, A.J. & Blair, G.E. (1997) Stable and temperature-sensitive transformation of baby rat kidney cells by SV40 suppresses expression of membrane dipeptidase. *Oncogene* **15**, 1241–1245.

Kropp, H., Sundelof, J.G., Hajdu, R. & Kahan, F.M. (1982) Metabolism of thienamycin and related carbapenum antibiotics by the renal dipeptidase, dehydropeptidase-I. *Antimicrob. Agents Chemother.* **22**, 62–70.

Littlewood, G.M., Hooper, N.M. & Turner, A.J. (1989) Ectoenzymes of the kidney microvillar membrane. Affinity purification, characterization and localization of the phospholipase C-solubilized form of renal dipeptidase. *Biochem. J.* **257**, 361–367.

McIntyre, T. & Curthoys, N.P. (1982) Renal catabolism of glutathione. Characterization of a particulate rat renal dipeptidase that catalyzes the hydrolysis of cysteinylglycine. *J. Biol. Chem.* **257**, 11915–11921.

Nakagawa, H., Inazawa, J., Inoue, K., Misawa, S., Kashima, K., Adachi, H., Nakazato, H. & Abe, T. (1992) Assignment of the human renal dipeptidase gene (DPEP1) to band q24 of chromosome 16. *Cytogenet. Cell. Genet.* **59**, 258–260.

Robinson, D.S., Birnbaum, S.M. & Greenstein, J.P. (1953) Properties of an aminopeptidase from kidney cellular particulates. *J. Biol. Chem.* **202**, 1–26.

**Nigel M. Hooper**
*Department of Biochemistry and Molecular Biology,*
*The University of Leeds,*
*Leeds LS2 9JT, UK*
*Email: n.m.hooper@leeds.ac.uk*

**M**

# 501. *Acinetobacter dipeptidase*

## Databanks

*Peptidase classification: clan MX, family M19, MEROPS ID: M19.003*
*NC-IUBMB enzyme classification: none*

*Databank codes:*

| Species | SwissProt | PIR | EMBL (cDNA) | EMBL (genomic) |
| --- | --- | --- | --- | --- |
| *Acinetobacter calcoaceticus* | P07783 | – | X06452 | – |
| | – | – | D50330 | – |
| *Klebsiella pneumoniae* | P27509 | – | X58778 | – |

## Name and History

When the cDNAs for mammalian membrane dipeptidases (Chapter 500) were cloned from several species, it was found that these enzymes showed extensive similarity to each other (Adachi *et al*., 1990, 1992; Rached *et al*., 1990; Igarashi & Karniski, 1991). Availability of cDNA of the enzyme and recombinant protein made it possible to characterize the enzyme, and amino acid residues essential for the catalytic activity were identified (Adachi *et al*., 1993; Keynan *et al*., 1994). In some prokaryotic cells such as *Acinetobacter calcoaceticus* and *Klebsiella pneumoniae*, a gene which shows significant homology to mammalian dipeptidase is coded near the *pqq* genes involved in pyrrolo-quinoline-quinone (PQQ) biosynthesis (Goosen *et al*., 1989; Meulenberg *et al*., 1992). This homologous gene is designated as R or orfX for these two organisms, respectively. The gene *acdp*, encoding dipeptidase from the prokaryote *A. calcoaceticus* ATCC 23055, was cloned and sequenced completely (Adachi & Tsujimoto, 1995). The ***Acinetobacter dipeptidase (ACDP)*** conserved the same critical amino acid residues identified in mammalian dipeptidases (Keynan *et al*., 1997).

## Activity and Specificity

The specific activity for purified recombinant dipeptidase from *E. coli* toward Leu+D-Leu is 103 nmol min$^{-1}$ mg$^{-1}$. The enzyme does not hydrolyze tripeptides. Furthermore, there are no D-aminoacylase nor aminopeptidase activities, indicating that the enzyme is, indeed, a dipeptide-specific enzyme.

When hydrolytic activities against the four possible enantiomers of Leu+Leu are compared, it is evident that Leu-D-Leu is preferentially hydrolyzed. When the N-terminal residue of substrates is an L-amino acid, some activity is detected regardless of the configuration of the C-terminal residue. However, if the N-terminal amino acid is in the D-configuration, little activity is detected. These results suggested that a stereospecific interaction between substrate and enzyme is required for activity. The importance of the D-configuration at the C-terminus is also observed with other sets of substrates. In these cases, the enzyme hydrolyzes substrates having D-amino acids at the C-terminus. On the other hand, human membrane dipeptidase expressed in Chinese hamster ovary cells showed similar activities toward these substrates.

The effect of various divalent metal ions on the hydrolytic activity toward Leu+D-Leu has been determined. In the absence of exogeneously added metal ions, a specific activity of 7.72 nmol min$^{-1}$ mg$^{-1}$ is measured. In the presence of EDTA, a decrease in the enzyme activity is observed, suggesting that metal ions are required for maximum activity. Addition of metal ions such as $Zn^{2+}$, $Co^{2+}$ and $Ni^{2+}$ results in an increase in the activity, while $Ca^{2+}$ or $Mn^{2+}$ show little effect on the activity. $Co^{2+}$ is the most effective ion tested, resulting in a specific activity of 92 nmol min$^{-1}$ mg$^{-1}$.

## Structural Chemistry

The recombinant preparation shows a single band at about 40 kDa on SDS-PAGE analysis. The recombinant enzyme contains about one atom of zinc per molecule (Adachi & Tsujimoto, 1995).

## Preparation

The recombinant enzyme has been expressed in *E. coli* as a soluble protein and purified (Adachi & Tsujimoto, 1995).

## Biological Aspects

It has been reported that in *Acinetobacter calcoaceticus* and *Klebsiella pneumoniae*, the genes are coded near the *pqq* genes. The physiological function of the enzyme awaits further investigation.

## References

Adachi, H. & Tsujimoto, M. (1995) Cloning and expression of dipeptidase from *Acinetobacter calcoaceticus* ATCC 23055. *J. Biochem.* **118**, 555–561.

Adachi, H., Tawaragi, Y., Inuzuka, C., Kubota, I., Tsujimoto, M., Nishihara, T. & Nakazato, H. (1990) Primary structure of human microsomal dipeptidase deduced from molecular cloning. *J. Biol. Chem.* **265**, 3992–3995.

Adachi, H., Ishida, N. & Tsujimoto, M. (1992) Primary structure of rat renal dipeptidase and expression of its mRNA in rat tissues and COS-1 cells. *Biochim. Biophys. Acta* **1132**, 311–314.

Adachi, H., Katayama, T., Nakazato, H. & Tsujimoto, M. (1993) Importance of Glu-125 in the catalytic activity of human renal dipeptidase. *Biochim. Biophys. Acta* **1163**, 42–48.

Goosen, N., Horsman, H.P.A., Huinen, R.G.M. & van de Putte, P. (1989) *Acinetobacter calcoaceticus* genes involved in biosynthesis of the coenzyme pyrrolo-quinoline-quinone: nucleotide sequence and expression in *Escherichia coli* K-12. *J. Bacteriol.* **171**, 447–455.

Igarashi, P. & Karniski, L.P. (1991) Cloning of cDNAs encoding a rabbit renal brush border membrane protein immunologically related to band 3. Sequence similarity with microsomal dipeptidase. *Biochem. J.* **280**, 71–78.

Keynan, S., Hooper, N.M. & Turner, A.J. (1994) Directed mutagenesis of pig renal membrane dipeptidase: His[219] is critical but the DHXXH motif is not essential for zinc binding or catalytic activity. *FEBS Lett.* **349**, 50–54.

Keynan, S., Hooper, N.M. & Turner, A.J. (1997) Identification by site-directed mutagenesis of three essential histidine residues in membrane dipeptidase, a novel mammalian zinc peptidase. *Biochem. J.* **326**, 47–51.

Meulenberg, J.J.M., Sellink, E., Riegman, N.H. & Postma, P.W. (1992) Nucleotide sequence and structure of the *Klebsiella pneumoniae* pqq operon. *Mol. Gen. Genet.* **232**, 284–294.

Rached, E., Hooper, N.M., James, P., Semenza, G., Turner, A.J. & Mantei, N. (1990) cDNA cloning and expression in *Xenopus laevis* oocyte of pig renal dipeptidase, a glycosyl-phosphatidylinositol-anchored ectoenzyme. *Biochem. J.* **271**, 755–760.

*Hideki Adachi*
*Laboratory of Bioorganic Chemistry,*
*The Institute of Physical and Chemical Research (RIKEN),*
*Wako, Saitama 351-01, Japan*
*Email: adachih@postman.riken.go.jp*

# 502. β-Aspartyl dipeptidase

## Databanks

*Peptidase classification: clan MX, family M38, MEROPS ID: M38.001*
*NC-IUBMB enzyme classification: none*
*Databank codes:*

| Species | SwissProt | PIR | EMBL (cDNA) | EMBL (genomic) |
|---|---|---|---|---|
| *Escherichia coli* | P39377 | B55889 | U15029 | U14003: chromosome 92.8 to 00.1′ |

## Name and History

Isoaspartyl residues arise by the spontaneous post-translational alteration of aspartyl and asparaginyl residues in proteins (see Figure 502.1). This abnormal residue forms the peptide bond linkage through its side chain β-carboxyl group rather than its α-carboxyl group. Several early studies using isoaspartyl-containing peptides showed that the isoaspartyl β-carboxyl bond was generally resistant to proteolysis. For example, extracts of yeast and mammalian cells that could hydrolyze the dipeptide aspartyl-glycine did not hydrolyze the corresponding isoaspartyl-glycine dipeptide (Grassmann & Schneider, 1934). Rat liver or kidney extracts demonstrated no hydrolytic activity towards an isoaspartyl-alanine dipeptide (Greenstein & Price, 1949). Finally, leucyl aminopeptidase (Chapter 473) was shown to hydrolyze the aspartyl-containing angiotensin II (Asp-Arg-Val-Tyr-Val-Ala-His-Pro-Phe) but not the corresponding isoaspartyl form (Riniker & Schwyzer, 1964). The first evidence for hydrolysis of a β-carboxyl linkage was the cleavage of an isoaspartyl-histidine dipeptide by a partially purified preparation of pig kidney carnosinase (β-alanyl-histidine hydrolase; Hanson & Smith, 1949).

Consistent with the general resistance of isoaspartyl β-linkages to proteolysis, a number of isoaspartyl-containing peptides were found in normal human urine (Buchanan *et al.*, 1962). These peptides arise from endogenous metabolism as well as from dietary sources (Dorer *et al.*, 1966). The finding that the levels of various isoaspartyl peptides did not vary in the same proportions with food consumption (Dorer *et al.*, 1966) suggested the presence of a proteolytic mechanism that minimized the accumulation of isoaspartyl peptides. Interestingly, when rats were given radiolabeled isoaspartyl di- and tripeptides by stomach tube, only a small portion of the radioactivity was rapidly excreted, suggesting that these peptides could be metabolized by one or more enzymes that cleaved the isoaspartyl linkage (Dorer *et al.*, 1968). This rat enzyme is classified as EC 3.4.19.5 and its mechanistic class has never been determined. It is discussed separately in Chapter 551. An enzyme from *Escherichia coli* was partially purified by Haley (1968) and has been more recently purified to homogeneity (Gary & Clarke, 1995). Unlike the mammalian enzyme, this bacterial enzyme does not hydrolyze tripeptides and is therefore called **β-aspartyl dipeptidase** or **isoaspartyl dipeptidase**.

## Activity and Specificity

Isoaspartyl peptidase activity was first characterized in rat liver extracts (Dorer *et al.*, 1968) (Chapter 551). This activity, assayed by the hydrolysis of an L-isoaspartyl⊣glycine

**L-Asx**      **L-succinimidyl**      **L-isoaspartyl**

*Figure 502.1*   Aspartyl (R = -OH) and asparaginyl (R = -NH$_2$) residues can undergo a spontaneous intramolecular cyclization reaction in which the peptide-backbone nitrogen atom of the following residue attacks the carbonyl of the Asx side chain to form an L-succinimidyl residue (*center*). Upon hydrolysis, this cyclic product opens to give either the L-aspartyl residue or an L-isoaspartyl residue that now has the peptide bond connecting through the side-chain β-carboxyl group, resulting in a kink in the peptide backbone.

*Table 502.1*   Dipeptidase activity

| Substrate | % of maximal activity | | |
|---|---|---|---|
| | Partially purified rat liver preparation[a] | Partially purified E. coli preparation[b] | Homogeneous E. coli enzyme[c] |
| β-Asp-Gly | 100 | 0 | 8 |
| β-Asp-Met | 82 | 68 | |
| β-Asp-Leu | 65 | 100 | 100 |
| β-Asp-Ser | 56 | 82 | |
| β-Asp-Ala | 51 | 33 | |
| β-Asp-Ile | 37 | 19 | |
| β-Asp-Thr | 29 | 18 | |
| β-Asp-Val | 2 | 56 | |
| β-Asp-Gln | | 48 | |
| β-Asp-Phe | | 38 | |
| β-Asp-Asn | | 10 | |
| β-Asp-His | | 0 | |
| β-Asp-Gly-Gly | 95 | 0 | |
| β-Asp-Gly-Ala | 55 | 0 | |
| β-Asp-Gly-Val | 13 | | |
| β-Asp-Leu-Gly | | 0 | |
| Asp-Gly | 3 | | |
| Asp-Leu | 9 | 0 | 13 |
| γ-Glu-Leu | 0 | 0 | 0 |
| γ-Glu-Gly | | | 0 |
| γ-Glu-His | | | 0 |
| γ-Glu-Cys | | | 0 |

[a]Dorer *et al*. (1968); [b]Haley (1968); [c]Gary & Clarke (1995).

dipeptide, was purified approximately 15-fold over crude cytosol and was maximally active in phosphate buffer at pH values between 7.5 and 8.0, falling off sharply below 6 and above 9 (Dorer *et al*., 1968).

An analogous isoaspartyl dipeptidase activity was found shortly thereafter in the bacterium *Escherichia coli* (Haley, 1968). This activity was initially characterized using a preparation that was enriched 110-fold over crude cytosol (Haley, 1968) and has more recently been purified to homogeneity (Gary & Clarke, 1995). The pH activity profile of this bacterial preparation is similar to that of the originally characterized rat enzyme (Haley, 1968). The *E. coli* enzyme, however, has little or no activity towards the isoaspartyl-glycine dipeptide (Haley, 1968; Gary & Clarke, 1995) (see Table 502.1). Instead, maximal activity is observed with an isoaspartyl–leucine dipeptide substrate which has a $K_m$ of 0.81 mM (Haley, 1968). As with the mammalian enzyme, normal aspartyl and γ-glutamyl dipeptides are either not substrates or are much poorer substrates (Table 502.1) (Haley, 1968; Gary & Clarke, 1995). Interestingly, the *E. coli* enzyme is unable to utilize isoaspartyl tripeptides as substrates (Haley, 1968). Another distinguishing feature of the *E. coli* enzyme activity is its sensitivity to $Mn^{2+}$, $Co^{2+}$ and $Zn^{2+}$, which inhibit the enzyme 42, 90 and 100%, respectively (Haley, 1968). On the other hand, the bacterial enzyme activity is unaffected by iodoacetamide (5 mM), *o*-iodosobenzoate (1 mM), and ammonium persulfate (0.4 mM), but is inhibited 17% by 2 mM *p*-hydroxymercuribenzoate (Haley, 1968).

For comparison, the partially purified rat liver enzyme was able to hydrolyze a number of isoaspartyl dipeptides; isoaspartyl⊣glycine was the best substrate tested (Table 502.1). In addition to hydrolyzing these dipeptides, the enzyme preparation also cleaved the N-terminal isoaspartyl peptide bond present in several tripeptides (Dorer *et al.*, 1968). The activity of this peptidase seems to be specific for isoaspartyl residues, because normal aspartyl- and isoglutamyl (γ-glutamyl)-containing dipeptides were not substrates (Dorer *et al.*, 1968). The rat liver isoaspartyl peptidase was neither greatly inhibited nor strongly activated by various divalent metal ions ($Mg^{2+}$, $Mn^{2+}$, $Ca^{2+}$ and $Co^{2+}$) at a concentration of 1 mM, the greatest effect being a 46% inhibition by 1 mM $Zn^{2+}$ (Dorer *et al.*, 1968). Sodium or potassium ions were required for maximal activity (Dorer *et al.*, 1968). Activity was unaffected by sodium EDTA, 2-ME and iodoacetamide at 1 mM concentrations; however, 1 mM *p*-hydroxymercuribenzoate inhibited the activity by 70% (Dorer *et al.*, 1968).

On the basis of these differences, it seems best to retain two separate entries for the mammalian and bacterial peptidases. The mammalian enzyme cannot be definitively assigned to a class and so is treated with the unclassified peptidases in this *Handbook* (Chapter 551). The bacterial enzyme is treated more fully in the following sections.

## Structural Chemistry

The *E. coli* isoaspartyl dipeptidase is encoded by the gene *iadA*, located in the 98 min region of the *E. coli* chromosome (Gary & Clarke, 1995). The IadA protein is cytosolic and contains 390 amino acids with a calculated pI of 5.02. The enzyme has a molecular mass of 41 kDa as predicted from its DNA sequence; this size was confirmed by SDS-PAGE analysis (Gary & Clarke, 1995). The isoaspartyl dipeptidase activity elutes as a species of approximately 120 kDa on a Sephadex G200 gel-filtration column (Haley, 1968; cf. Gary & Clarke, 1995), suggesting that the enzyme may exist in a multimeric complex.

## Preparation

Two bacterial isoaspartyl dipeptidase purifications have been reported. Haley (1968) partially purified (110-fold) the dipeptidase from *E. coli* strain B by size-exclusion and anion-exchange chromotography, achieving a specific activity of 770 nmol min$^{-1}$ mg$^{-1}$ (using L-isoaspartyl-L-leucine as the substrate) with a 34% yield. Gary & Clarke (1995) included an additional phenyl Sepharose chromatography step to achieve a homogeneous preparation (>3000-fold purification) of the enzyme from the wild-type *E. coli* strain MC1000, as well as from strain JDG100, which overexpresses IadA at

least 40-fold over MC1000. Both preparations gave a specific activity of 1900 nmol min$^{-1}$ mg$^{-1}$ (using L-isoaspartyl-L-leucine as the substrate) and a 15% yield.

## Biological Aspects

L-Isoaspartyl residues in proteins arise spontaneously over time from L-aspartyl and L-asparaginyl residues through nonenzymatic, intramolecular isomerization and deamidation reactions (see Figure 502.1; Stephenson & Clarke, 1989; Wright 1991; Clarke *et al.* 1992). Protein function may be impaired due to the resulting kink in the polypeptide backbone. If this damaged protein is not adequately repaired by the L-isoaspartyl/D-aspartyl methyltransferase, it may be degraded (for a review see Visick & Clarke, 1995). Isoaspartyl-containing peptides can arise from the incomplete degradation of these damaged proteins, because most cellular proteases do not recognize the isoaspartyl linkage (Haley *et al.*, 1966; Murray & Clarke, 1984; Johnson & Aswad, 1990).

In *E. coli*, which has a minimal doubling time of 20 min, the accumulation of isoaspartyl residues may be insignificant during exponential growth. However, after a prolonged period in stationary phase the level of these damaged residues may increase along with their isoaspartyl dipeptide degradation products. The isoaspartyl dipeptidase may prevent the accumulation of these dipeptides, which may be toxic to the bacteria or may result in the depletion of the pool of amino acids necessary for survival in stationary phase (Mandelstam, 1958, 1960; Reeve *et al.*, 1984a,b). These hypotheses were tested in a strain of *E. coli* lacking the *iadA* gene, JDG11000 (Gary & Clarke, 1995). The null mutant strain did not exhibit any obvious phenotypes; exponential growth and stationary phase survival were similar to the parent strain (Gary & Clarke, 1995). In addition, radiolabeling studies did not reveal any accumulation of isoaspartyl-leucine dipeptides (Gary & Clarke, 1995). However, an assay of the null mutant cytosol did show the presence of a secondary isoaspartyl dipeptidase activity that accounts for approximately 31% of the total isoaspartyl dipeptidase activity in crude extracts of the parent strain (Gary & Clarke, 1995). It is possible that the remaining activity is sufficient to compensate for the loss of IadA activity; only a strain with both enzymes deleted would then show a phenotype due to isoaspartyl dipeptide accumulation. Preliminary results from a partial purification of this second isoaspartyl dipeptidase activity indicate that its specificity is very similar to that of IadA: it hydrolyzes isoaspartyl-leucine but uses neither aspartyl-leucine nor isoaspartyl-glycine as substrates (J.D. Gary, unpublished results).

## Distinguishing Features

A database search for protein sequences similar to the *E. coli* isoaspartyl dipeptidase amino acid sequence (performed in July 1996) using the NCBI BLAST program (Altschul *et al.*, 1990) did not identify any related proteases or peptidases. However, significant amino acid sequence similarity to the family of bacterial dihydroorotases was found. The region of highest similarity is within the first 100 amino acids (Figure 502.2) and centers around a ten amino acid region of

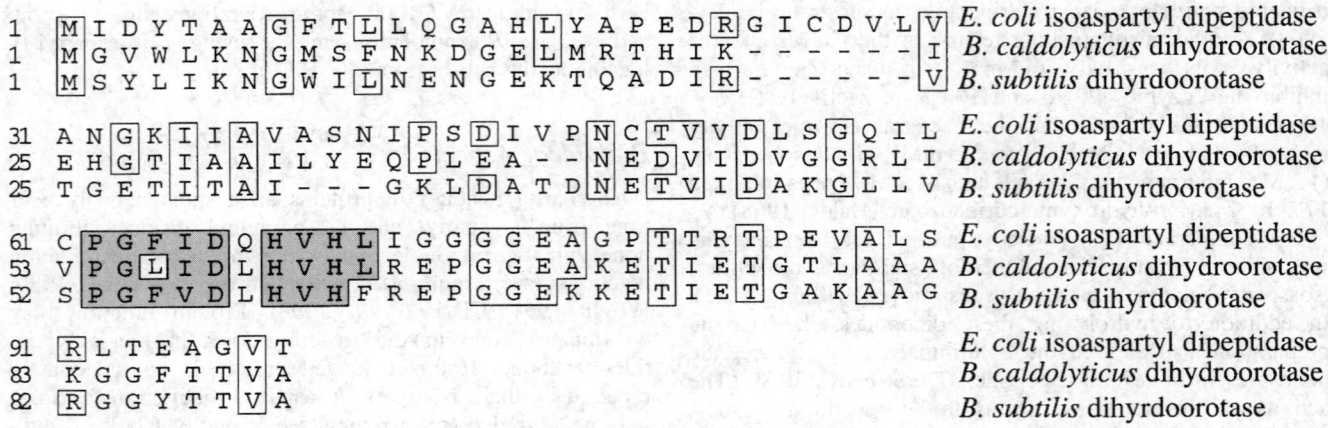

```
 1  M I D Y T A A G F T L L Q G A H L Y A P E D R G I C D V L V    E. coli isoaspartyl dipeptidase
 1  M G V W L K N G M S F N K D G E L M R T H I K - - - - - - I    B. caldolyticus dihydroorotase
 1  M S Y L I K N G W I L N E N G E K T Q A D I R - - - - - - V    B. subtilis dihyrdoorotase

31  A N G K I I A V A S N I P S D I V P N C T V V D L S G Q I L    E. coli isoaspartyl dipeptidase
25  E H G T I A A I L Y E Q P L E A - - N E D V I D V G G R L I    B. caldolyticus dihydroorotase
25  T G E T I T A I - - - G K L D A T D N E T V I D A K G L L V    B. subtilis dihyrdoorotase

61  C P G F I D Q H V H L I G G G G E A G P T R T P E V A L S      E. coli isoaspartyl dipeptidase
53  V P G L I D L H V H L R E P G G E A K E T I E T G T L A A A    B. caldolyticus dihydroorotase
52  S P G F V D L H V H F R E P G G E K K E T I E T G A K A A G    B. subtilis dihyrdoorotase

91  R L T E A G V T                                               E. coli isoaspartyl dipeptidase
83  K G G F T T V A                                               B. caldolyticus dihydroorotase
82  R G G Y T T V A                                               B. subtilis dihyrdoorotase
```

*Figure 502.2*   The N-terminal region of the *E. coli* isoaspartyl dipeptidase sequence aligned with dihydroorotases from *Bacillus caldolyticus* and *Bacillus subtilis*. Residues identical to the dipeptidase are boxed, and the region of highest similarity is shaded. Over the entire protein sequence the dipeptidase is 13% identical in sequence to the dihydroorotases, but this N-terminal region demonstrates 25% identity.

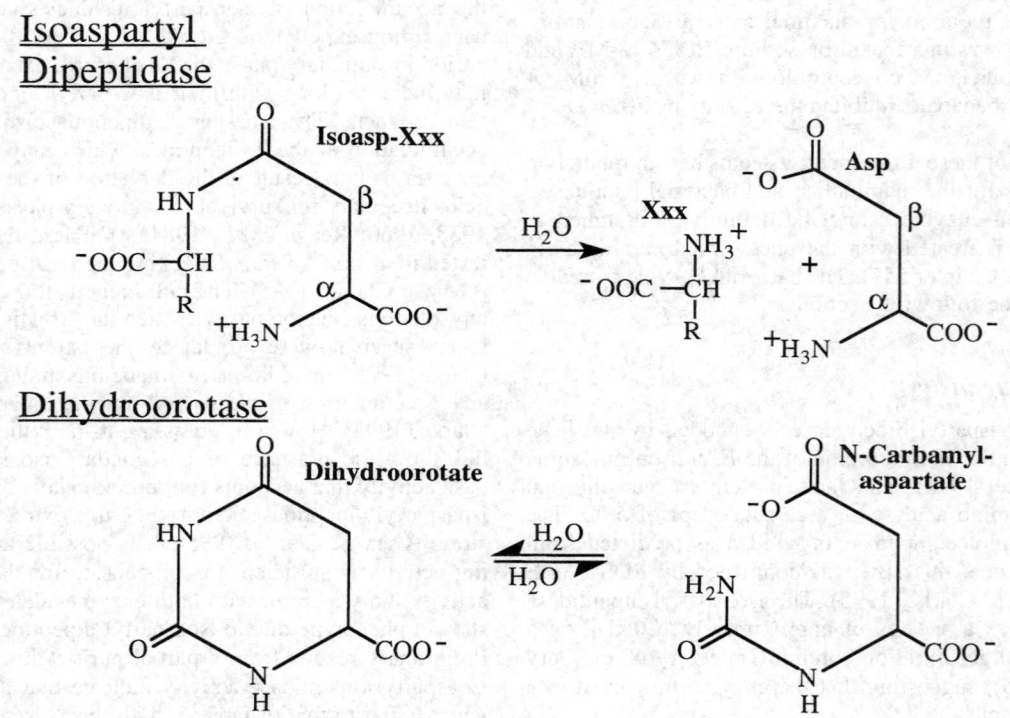

*Figure 502.3*   Comparison of the reactions catalyzed by the isoaspartyl dipeptidase (*top*) and by dihydroorotase (*bottom*), showing the similarity of the reactions and the structural similarity of the substrates.

the dipeptidase with the sequence PGFIDQHVHL (see shaded region in Figure 502.2). Studies on the *E. coli* dihydroorotase suggest that this region is a zinc-binding motif, with the two conserved histidines serving as ligands for a catalytic zinc ion (Washabaugh & Collins, 1984. 1986; Brown & Collins, 1991). The observed sequence similarity between the isoaspartyl dipeptidase and dihydroorotases is consistent with a similarity in their substrates and in the chemistry of the reaction catalyzed (Figure 502.3). Isoaspartyl dipeptides and

dihydroorotate have very similar molecular geometries and functional groups, and hydrolysis occurs at a similar position on each molecule. The *E. coli* isoaspartyl dipeptidase, however, has no dihydroorotase activity (Gary & Clarke, 1995). It appears that in order to cleave an abnormal type of peptide bond, *E. coli* cells have evolutionarily derived a unique dipeptidase from an unrelated metabolic enzyme rather than modifying an existing peptidase from an α- to a β-carboxyl specificity.

# References

Altschul, S.F., Gish, W., Miller, W., Myers, E.W. & Lipman, D.J. (1990) Basic local alignment search tool. *J. Mol. Biol.* **21**, 403–410.

Brown, D.C. & Collins, K.D. (1991) Dihydroorotase from *Escherichia coli*. Substitution of Co(II) for the active site Zn(II). *J. Biol. Chem.* **266**, 1597–1604.

Buchanan, D.L, Haley, E.E., Markiw, R.T. & Peterson, A.A. (1962) Occurrence of β-aspartyl and γ-glutamyl oligopeptides in human urine. *Biochemistry* **1**, 612–620.

Clarke, S., Stephenson, R.C. & Lowenson, J.L. (1992) Lability of asparagine and aspartic-acid residues in proteins and peptides: spontaneous deamidation and isomerization. In: *Stability of Protein Pharmaceuticals, Part A* (Ahern, T.J. & Manning, M.C., eds). New York: Plenum, pp. 1–29.

Dorer, F.E., Haley, E.E. & Buchanan, D.L. (1966) Quantitative studies of urinary beta-aspartyl oligopeptides. *Biochemistry* **5**, 3236–3240.

Dorer, F.E., Haley, E.E. & Buchanan, D.L. (1968) The hydrolysis of β-aspartyl peptides by rat tissue. *Arch. Biochem. Biophys.* **127**, 490–495.

Gary, J.D. & Clarke, S. (1995) Purification and characterization of an isoaspartyl dipeptidase from *Escherichia coli*. *J. Biol. Chem.* **270**, 4076–4087.

Grassmann, W. & Schneider, F. (1934) Zur Spezifität der Dipeptidase. Enzymatisches Verhalten von Asparaginsaure- und Glutaminsaurepeptiden [On the specificity of dipeptidases. Enzymatic action on asparagine and glutamine peptides]. *Biochem. Z.* **273**, 452–465.

Greenstein, J.P. & Price, V.E. (1949) α-Keto acid-activated glutaminase and asparaginase. *J. Biol. Chem.* **178**, 695–705.

Haley, E.E. (1968) Purification and properties of a β-aspartyl peptidase from *Escherichia coli*. *J. Biol. Chem.* **243**, 5748–5752.

Haley, E.E., Corcoran, B.J., Dorer, F.E. & Buchanan, D.L. (1966) β-aspartyl peptides in enzymatic hydrolyzates of protein. *Biochemistry* **5**, 3229–3235.

Hanson, H.T. & Smith, E.L. (1949) Carnosinase: an enzyme of swine kidney. *J. Biol. Chem.* **179**, 789–801.

Johnson, B.A. & Aswad, D.W. (1990) Fragmentation of isoaspartyl peptides and proteins by carboxypeptidase Y: release of isoaspartyl dipeptides as a result of internal and external cleavage. *Biochemistry* **29**, 4373–4380.

Mandelstam, J. (1958) Turnover of protein in growing and non-growing populations of *Escherichia coli*. *Biochem. J.* **69**, 110–119.

Mandelstam, J. (1960) The intracellular turnover of protein and nucleic acids and its role in biochemical differentiation. *Bacteriol. Rev.* **24**, 289–308.

Murray, E.D., Jr. & Clarke, S. (1984) Metabolism of a synthetic L-isoaspartyl-containing hexapeptide in erythrocyte extracts: enzymatic methyl esterification is followed by nonenzymatic succinimide formation. *J. Biol. Chem.* **259**, 10722–10732.

Reeve, C.A., Bockman, A.T. & Matin, A. (1984a) Role of protein degradation in the survival of carbon-starved *Escherichia coli* and *Salmonella typhimurium*. *J. Bacteriol.* **157**, 758–763.

Reeve, C.A., Amy, P.S. & Matin, A. (1984b) Role of protein synthesis in the survival of carbon-starved *Escherichia coli* K-12. *J. Bacteriol.* **160**, 1041–1046.

Riniker, B. & Schwyzer, R. (1964) Synthetische Analoge des Hypertensins. α-L-, β-L-, α-D- und β-D-Asp-Val-Hypertensin II; desamino-Val-Hypertensin II [Synthetic analogs of the hypertensins. α-L-, β-L-, α-D- and β-D-Asp-Val-Hypertensin II; desamino-Val-Hypertensin II]. *Helv. Chim. Acta* **47**, 2357–2374.

Stephenson, R.C. & Clarke, S. (1989) Succinimide formation from aspartyl and asparaginyl peptides as a model for the spontaneous degradation of proteins. *J. Biol. Chem.* **264**, 6164–6170.

Visick, J.E. & Clarke, S. (1995) Repair, refold, recycle: how bacteria can deal with spontaneous and environmental damage to proteins. *Mol. Microbiol.* **16**, 835–845.

Washabaugh, M.W. & Collins, K.D. (1984) Dihydroorotase from *Escherichia coli*: purification and characterization. *J. Biol. Chem.* **259**, 3293–3298.

Washabaugh, M.W. & Collins, K.D. (1986) Dihydroorotase from *Escherichia coli*: sulfhydryl group-metal ion interactions. *J. Biol. Chem.* **261**, 5920–5929.

Wright, H.T. (1991) Nonenzymatic deamidation of asparaginyl and glutaminyl residues in proteins. *Crit. Rev. Biochem. Mol. Biol.* **26**, 1–52.

*Jonathan D. Gary*
*Molecular Biology Institute and*
*Department of Chemistry and Biochemistry,*
*University of California, Los Angeles,*
*Los Angeles, CA 90095-1569, USA*

*Steven Clarke*
*Molecular Biology Institute and*
*Department of Chemistry and Biochemistry,*
*University of California, Los Angeles,*
*Los Angeles, CA 90095-1569, USA*
*Email: clarke@ewald.mbi.ucla.edu*

**M**

# 503. *VanX D-, D-dipeptidase*

## Databanks

*Peptidase classification: clan MX, family M45, MEROPS ID: M45.001*
*NC-IUBMB Enzyme classification: none*

*Databank codes:*

| Species | SwissProt | PIR | EMBL (cDNA) | EMBL (genomic) |
|---|---|---|---|---|
| *Enterococcus faecalis* | – | – | U35369 | – |
| *Enterococcus faecium* | Q06241 | – | M97927 | – |
| *Synechocystis* sp. | – | – | D90913 | – |

## Name and History

The last defense against gram-positive pathogenic infections is currently the glycopeptide antibiotic vancomycin. Clinical cases of resistance to vancomycin have risen dramatically from 0.4% in 1989 to an alarming 13% in 1993 (Swartz, 1994). Resistance to vancomycin is mediated by a plasmid-borne transposon, first isolated by Courvalin and coworkers from the blood of a patient who died of an enterococcal infection which was resistant to both β-lactam and glycopeptide antibiotics (Arthur & Courvalin, 1993). Courvalin's group sequenced the *van* genes (*van* for resistance to **van**comycin) within the transposon and found seven open reading frames (*van*S, R, H, A, X, Y, Z). They proved by insertional mutagenesis that five genes, *van*S, R, H, A, X, were both necessary and sufficient to induce a high level of antibiotic resistance (Figure 503.1). The *vanY* gene product is a D-Ala-D-Ala carboxypeptidase of unknown catalytic type (Chapter 554). From the DNA sequence of the five genes, Courvalin and co-workers inferred putative functions for four of the gene products, VanR, S, H and A (Table 503.1). The deduced amino acid sequence of VanX provided no clue to its probable function, and therefore its role remained obscure until Reynolds *et al.* (1994) found evidence that **VanX** could hydrolyze D-Ala+D-Ala, a crucial intermediate in cell wall biosynthesis, and the target of vancomycin high-affinity binding (Barna & Williams, 1984). In addition, they also observed that VanX failed to hydrolyze D-Ala-D-lactate, the terminal depsipeptide moiety which replaces the D-Ala-D-Ala portion of the pentapeptide intermediate in vancomycin-resistant enterococci and which binds vancomycin with 1000-fold lower affinity (Bugg *et al.*, 1991b). Walsh and colleagues soon after purified **VanX D-, D-dipeptidase** from *Escherichia coli* harboring an overproduction vector, confirmed the dipeptidase activity, showed a zinc dependence and quantitated the selectivity for hydrolysis of D-Ala-D-Ala compared to D-Ala-D-lactate (Wu *et al.*, 1995; Wu & Walsh, 1996).

## Activity and Specificity

VanX is a D-, D-dipeptidase, catalyzing the hydrolysis of the bacterial cell wall biosynthesis intermediate D-Ala+D-Ala into 2 molar equivalents of D-alanine (Reynolds *et al.*, 1994). Substrate specificity appears to be limited to dipeptides of the D-, D-configuration with unmodified N- or C-termini, for example H-D-Ala+D-Ala-OH is hydrolyzed but not Ac-D-Ala-D-Ala-OH or H-D-Ala-D-Ala-NH$_2$. The enzyme neither hydrolyzes tripeptides, nor dipeptides of L-, L- or mixed diastereomeric configuration (L-, D- and D-, L-). However, within these apparently strict requirements for D-, D-dipeptide substrates is some side-chain residue tolerance at sites P1 and P1'. Dipeptides with D-Ala or D-Ser at position P1, and Gly, D-Ser, D-Phe, D-Val, and D-Asn at position P1' are substrates for VanX (Wu *et al.*, 1995).

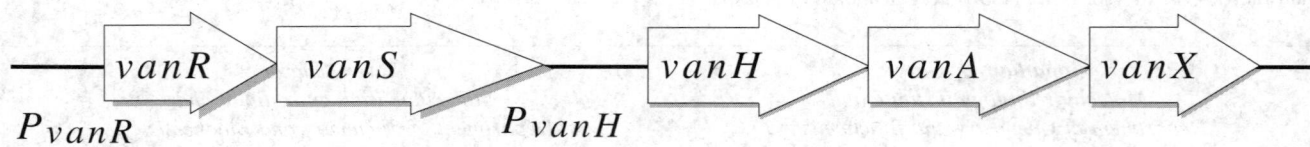

**Figure 503.1** Organization of the *van* gene cluster in *Enterococcus faecium* transposon Tn1546 (Arthur *et al.*, 1993).

**Table 503.1** Activities and functions of the protein products of the five *van* genes which are necessary and sufficient to confer vancomycin resistance

| Protein | Activity | Function |
|---|---|---|
| VanS | Transmembrane histidine kinase | Sensor protein that initiates signal transduction pathway |
| VanR | Two-domain response regulator | Accepts PO$_3$$^{2-}$ group from phospho-VanS, activates *van*H, A, X transcription |
| VanH | D-Specific α-keto acid reductase | Generates D-lactate required for VanA action |
| VanA | Depsipeptidase ligase for D-Ala-D-lactate | Generates an ester D-Ala-D-lactate in competition with normal amide D-Ala-D-Ala |
| VanX | Zn$^{2+}$-dependent D-Ala-D-Ala dipeptidase | Selective removal of D-Ala-D-Ala allows accumulation of D-Ala-D-lactate for addition to growing UDP-muramyl tripeptide |

Perhaps the most intriguing observation is that VanX does not catalyze the hydrolysis of the depsipeptide D-Ala-D-lactate, the ester analog of D-Ala-D-Ala (Reynolds *et al.*, 1994; Wu *et al.*, 1995). The exact mechanism by which VanX preferentially hydrolyzes the amide substrate versus its kinetically and thermodynamically more favorable ester analog has not yet been determined and is the subject of current research efforts. It is also not clear whether this lack of esterase activity is broad based and extends to other esters or thioesters.

VanX has a pH optimum of 8.0 (Wu *et al.*, 1995). Dipeptidase and depsipeptidase activity is determined by measuring the release of free amino acid with the modified cadmium-ninhydrin method (Doi *et al.*, 1981). Hydrolysis of D-Ala-containing substrates may also be detected using D-amino acid oxidase (Messer & Reynolds, 1992).

VanX is resistant to the action of $\beta$-lactam antibiotics (Reynolds *et al.*, 1994) but is inhibited by transition-state substrate analogs or by chelating agents. Phosphinate analogs of D-, D-dipeptides are micromolar affinity inhibitors which exhibit time-dependent slow-binding kinetics (Wu & Walsh, 1995). Dithiols such as DTT and dithioerythritol are also potent time-dependent micromolar inhibitors whose mechanism of inactivation is postulated to occur by the formation of an enzyme–metal–dithiol ternary complex (Wu & Walsh, 1996). Other bidentate chelating agents such as 1,10-phenanthroline inhibit the enzyme by removing the active-site zinc, as evidenced by loss of activity and metal content upon exposure to 10 mM 1,10-phenanthroline and restoration of activity upon stoichiometric addition of zinc (D.G. McCafferty, I.A.D. Lessard & C.T. Walsh, unpublished results). However, higher concentrations (>5 mM) of zinc are inhibitory.

## Structural Chemistry

VanX is a 202 amino acid polypeptide (23.5 kDa) existing in solution as a homodimer. It contains up to 1.0 mol of bound zinc per mol of polypeptide (Wu & Walsh, 1995). There is little sequence similarity between VanX and other known zinc-dependent peptidases and as a result the zinc-coordinating residues are not readily apparent by direct sequence comparison. Mutational analysis of the enzyme is underway to locate the zinc-coordinating ligands (D.G. McCafferty, I.A.D. Lessard & C.T. Walsh, unpublished results). There are two Cys residues in the free thiol form which have been shown by chemical modification not to be required for catalysis (Wu *et al.*, 1995).

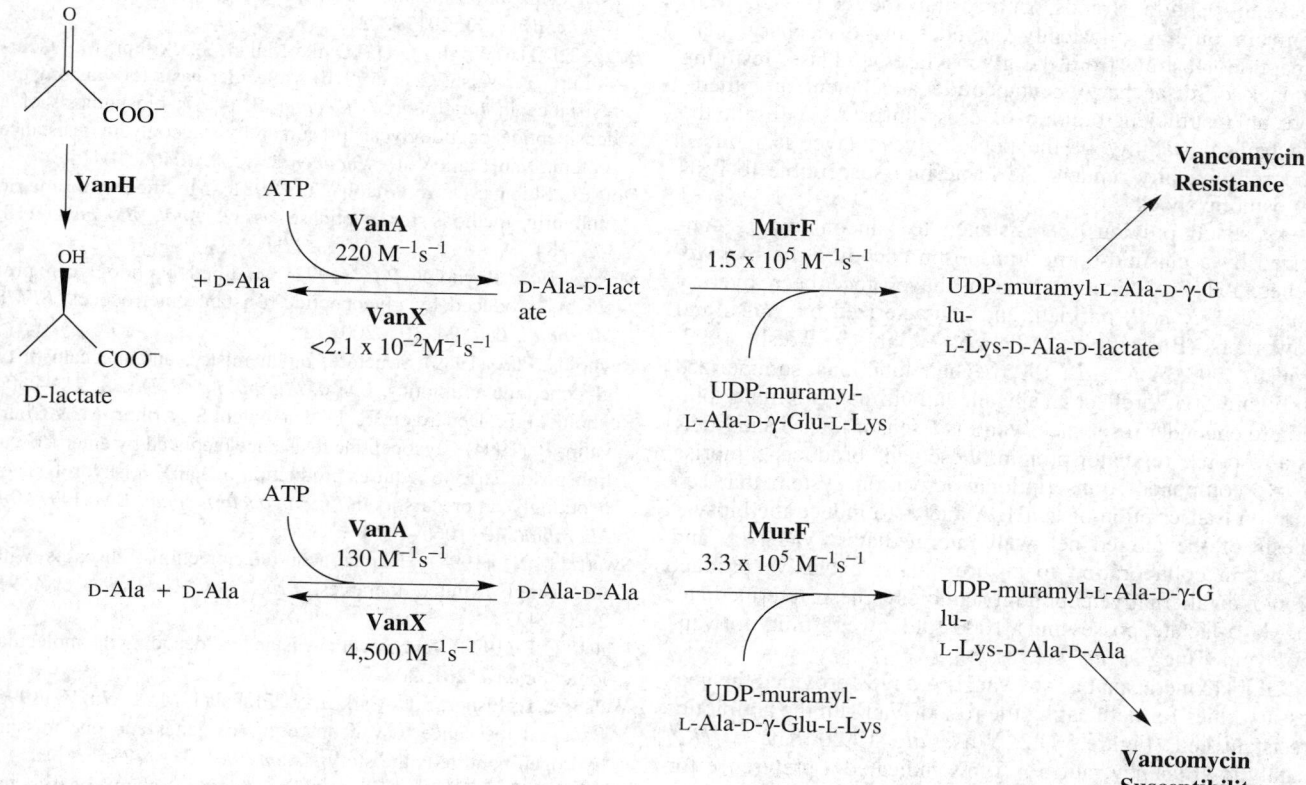

*Figure 503.2* Global kinetic scheme of the VanH, A, X-catalyzed biosynthetic pathway leading to the net production of an alternative peptidoglycan intermediate terminating in D-Ala-D-lactate. Reduction of pyruvate by VanH leads to the production of D-lactate, which is a substrate for the ATP-dependent dipeptide/depsipeptide ligase VanA. VanA produces D-Ala-D-lactate and D-Ala-D-Ala with comparable catalytic efficiency ratios ($k_{cat}/K_m$). Likewise, in the subsequent step, the adding enzyme MurF utilizes both D-Ala-D-lactate and D-Ala-D-Ala as substrates with similar efficiencies. The $Zn^{2+}$-dependent dipeptidase VanX selectively and efficiently catalyzes the hydrolysis of the D-Ala-D-Ala dipeptide pool, shunting its entry away from the vancomycin-sensitive biosynthetic pathway.

## Preparation

Currently there is no commercial source for VanX, as the pathogenic nature of the vancomycin-resistant *Enterococcus faecium* host most likely precludes large-scale production and isolation from this organism. However, two recombinant sources have been developed from the isolated plasmid transposon Tn1546 bearing the *van*X gene. Courvalin and coworkers have utilized *E. coli* harboring a pUC-based expression vector for the purpose of assaying enzymatic activity in crude lysates (Reynolds *et al.*, 1994). Using *E. coli* bearing a pET-based (Novagen) expression vector, Walsh and colleagues reported the overproduction and purification to homogeneity of milligram quantities of the enzyme suitable for mechanistic studies (Wu *et al.*, 1995) (see Appendix II for full names and addresses of suppliers).

## Biological Aspects

Vancomycin acts by disrupting cell wall biosynthesis, binding tightly to the D-Ala-D-Ala portion of the lipid-PP-disaccharide-pentapeptide, the substrate molecule for the $\beta$-lactam-sensitive transglycosylase/transpeptidase (Reynolds, 1989). This transpeptidase adds pentapeptide units to the growing peptidoglycan, then cross-links peptides within and between peptide strands, buttressing the cell wall. Vancomycin binding physically occludes the disaccharyl pentapeptide substrate from the glycosylase/peptidase, resulting in lack of disaccharyl pentapeptide attachment and therefore an insufficient buildup of cross-links. As a result, the mechanical stability of the peptidoglycan layer is reduced and subsequently renders the bacterium susceptible to lysis by osmotic shock.

As stated previously, resistance to vancomycin is conferred by a plasmid-borne transposon encoding the five *van* genes. All five encoded proteins have since been overexpressed in *E. coli*, purified, and characterized by Walsh and coworkers (Bugg *et al.*, 1991a,b; Wright & Walsh, 1992; Walsh, 1993; Wu *et al.*, 1995). Their functions, summarized in Figure 503.2, reflect an elegant and ultimately simple solution to engender resistance. VanS is a sensor kinase and VanR is a response-regulator protein; these gene products comprise a two-component transcriptional-activation system that can turn on transcription of *van*H, A and X to induce the biosynthesis of the altered cell-wall intermediates. VanH, A and X act in collaboration to ensure almost exclusive production of an alternative peptidoglycan biosynthesis intermediate, D-Ala-D-lactate, possessing a 1000-fold lower affinity for vancomycin (Bugg *et al.*, 1991b).

Global kinetic analysis of VanH, A and X provided the necessary clues for delineating the role of VanX in the antibiotic-resistant cell (Figure 503.2; Wu *et al.*, 1995). The $k_{cat}/K_m$ catalytic efficiency ratio for VanX indicated a preference for D-Ala-D-Ala hydrolysis of $>10^6$; $k_{cat}$ for D-Ala-D-lactate was reduced 10 000-fold compared to the rate constant of D-Ala-D-Ala hydrolysis, and the ester substrate also has an apparent 250-fold lower binding affinity than D-Ala-D-Ala. Because a mix of D-Ala-D-Ala and D-Ala-D-lactate are produced by the enterococcal Ddl (dipeptide) and by VanA (dipeptide and depsipeptide) ligases, the role of VanX is to shunt the entry of newly formed D-Ala-D-Ala away from the pathway for peptidoglycan assembly by catalyzing its degradation. For VanH, A, X-producing bacterium, only D-Ala-D-lactate accumulates as a substrate for D-Ala-D-Ala adding enzyme (MurF for *E. coli*); and thus the peptidoglycan advanced intermediates in these bacteria contain only D-Ala-D-lactate termini.

## Further Reading

For a review, see Walsh *et al.* (1996).

## References

Arthur, M. & Courvalin, P. (1993) Genetics and mechanisms of glycopeptide resistance in enterococci. *Antimicrob. Agents Chemother.* **37**, 1563–1571.

Arthur, M., Molinas, C., Depardieu, F. & Courvalin, P. (1993) Characterization of Tn1546, a Tn3-related transposon conferring glycopeptide resistance by synthesis of depsipeptide peptidoglycan precursors in *Enterococcus faecium* BM4147. *J. Bacteriol.* **175**, 117–127.

Barna, J.C.J. & Williams, D.H. (1984) The structure and mode of action of glycopeptide antibiotics of the vancomycin group. *Annu. Rev. Microbiol.* **38**, 339–357.

Bugg, T.D.H., Dutka-Malen, S., Arthur, M., Courvalin, P. & Walsh, C.T. (1991a) Identification of vancomycin resistance protein VanA as a D-alanine:D-alanine ligase of altered substrate specificity. *Biochemistry* **30**, 2017–2021.

Bugg, T.D.H., Wright, G.D., Dutka-Malen, S., Arthur, M., Courvalin, P. & Walsh, C.T. (1991b) Molecular basis for vancomycin resistance in *Enterococcus faecium* BM4147: biosynthesis of a depsipeptide peptidoglycan precursor by vancomycin resistance proteins VanH and VanA. *Biochemistry* **30**, 10408–10415.

Doi, E., Shibata, D. & Matoba, T. (1981) Modified colorimetric ninhydrin methods for peptidase assay. *Anal. Biochem.* **118**, 173–184.

Messer, J. & Reynolds, P.E. (1992) Modified peptidoglycan precursors produced by glycopeptide-resistant enterococci. *FEMS Microbiol. Lett.* **94**, 195–200.

Reynolds, P.E. (1989) Structure, biochemistry, and mechanism of glycopeptide antibiotics. *Eur. J. Microb. Infect. Dis.* **8**, 943–950.

Reynolds, P.E., Depardieu, F., Dutka-Malen, S., Arthur, M. & Courvalin, P. (1994) Glycopeptide resistance mediated by enterococcal transposon Tn1546 requires production of VanX for hydrolysis of peptidoglycan precursors in *Enterococcus faecium* BM4147. *Mol. Microbiol.* **13**, 1065–1070.

Swartz, M.N. (1994) Hospital-acquired infections: diseases with increasingly limited therapies. *Proc. Natl Acad. Sci. USA* **91**, 2420–2427.

Walsh, C.T. (1993) Vancomycin resistance – decoding the molecular logic. *Science* **261**, 308–309.

Walsh, C.T., Fisher, S.L., Park, I.-S., Prahalad, M. & Wu, Z. (1996) Bacterial resistance to vancomycin: five genes and one missing hydrogen bond tells the story. *Chem. Biol.* **3**, 21–28.

Wright, G.D. & Walsh, C.T. (1992) D-Alanyl-D-alanyl ligases and the molecular mechanism of vancomycin resistance. *Accts Chem. Res.* **25**, 468–473.

Wu, Z. & Walsh, C.T. (1995) Phosphinate analogues of D-, D-dipeptides: Slow-binding inhibition and proteolysis protection of VanX, a D-, D-dipeptidase required for vancomycin resistance in *Enterococcus faecium*. *Proc. Natl Acad. Sci. USA* **92**, 11603–11607.

Wu, Z. & Walsh, C.T. (1996) Dithiol compounds: potent, time-dependent inhibitors of VanX, a zinc-dependent D-, D-dipeptidase required for vancomycin resistance in *Enterococcus faecium. J. Am. Chem. Soc.* **118**, 1785–1786.

Wu, Z., Wright, G. & Walsh, C.T. (1995) Overexpression, purification, and characterization of VanX, a D-, D-dipeptidase which is essential for vancomycin resistance in *Enterococcus faecium* BM4147. *Biochemistry* **34**, 2455–2463.

*Dewey G. McCafferty*
*Harvard Medical School,*
*Department of Biological Chemistry*
*and Molecular Pharmacology,*
*Boston, MA 02115, USA*

*Ivan A. D. Lessard*
*Harvard Medical School,*
*Department of Biological Chemistry*
*and Molecular Pharmacology,*
*Boston, MA 02115, USA*

*Christopher T. Walsh*
*Harvard Medical School,*
*Department of Biological Chemistry*
*and Molecular Pharmacology,*
*Boston, MA 02115, USA*
*Email: walsh@walsh.med.harvard.edu*

# 504. *Carboxypeptidase Taq*

## Databanks

Peptidase classification: clan MX, family M32, MEROPS ID: M32.001
NC-IUBMB enzyme classification: EC 3.4.17.19
Databank codes:

| Species | SwissProt | PIR | EMBL (cDNA) | EMBL (genomic) |
|---|---|---|---|---|
| *Bacillus subtilis* | P50848 | – | L47838 | L77246: multiple entry |
| *Thermus aquaticus* | P42663 | – | D17669 | – |

## Name and History

A thermostable carboxypeptidase was purified and characterized for the first time from *Thermus aquaticus* YT-1, an extremely thermophilic bacterium, and named **carboxypeptidase Taq (CPase Taq)** (Lee *et al.*, 1992). Subsequently the gene encoding CPase Taq was cloned and sequenced (Lee *et al.*, 1994). This enzyme has no obvious sequence similarity to any other carboxypeptidases, but a HEXXH motif, which is a conserved sequence in the active site of neutral zinc metallopeptidases (Jongeneel *et al.*, 1989), was found in this enzyme (Lee *et al.*, 1994, 1996). These findings indicate that CPase Taq is a novel type of zinc-dependent metallocarboxypeptidase and the sole member of an independent family (M32) designated as the carboxypeptidase Taq family (Rawlings & Barrett, 1995).

## Activity and Specificity

The relative activity of CPase Taq with various synthetic peptides as substrates has been tabulated (Lee *et al.*, 1994). CPase Taq hydrolyzes peptides and releases amino acids from the C-terminus with a broad substrate specificity, except for peptides with proline at the C-terminus. Amino acids with large side chains were most readily hydrolyzed. Peptides containing a Pro residue at the second (P1) and the third (P2) position from the C-terminal are equally hydrolyzed. When substrates with a Gly or Glu residue in P1 are compared, those with a Gly residue are more reactive than those with a Glu residue. The order of release of amino acids from the C-terminus of oligopeptides has been examined and it was demonstrated that CPase Taq releases amino acids sequentially from the C-terminal end of the substrate, so this enzyme is truly a carboxypeptidase (Lee *et al.*, 1992).

The hydrolytic activity of CPase Taq toward synthetic peptides is measured by the ninhydrin method of Matheson & Tattrie (1964) or that of Rosen (1957), except for peptides containing a Pro residue at the C-terminus. With such peptides as substrates, the activity is measured by the ninhydrin method of Yaron & Mlynar (1968). The enzyme activity is conveniently measured with 0.5 mM Z-Phe-Tyr as the substrate at 70°C for 30 min in 50 mM Tris-HCl buffer (pH 8.5) containing 0.1 mM $CoCl_2$. Enzyme kinetics are studied by assay in 50 mM *N*-2-hydroxyethylpiperazine-*N'*-3-propanesulfonic acid (HEPPS)-NaOH buffer (pH 8.5) containing 0.1 mM $CoCl_2$ at 60°C, by the method of Rosen (1957).

The optimum pH for activity of CPase Taq is about pH 8. The enzyme is completely stable within the pH range of 7–10. The optimal temperature for activity of the enzyme is 80°C.

The enzyme is stable up to 80°C, but at 90 and 95°C it is unstable. The presence of $Co^{2+}$ in the buffer employed does not affect the heat stability of this enzyme.

In the absence of divalent cations, almost no activity is found. Activity is highest when $Co^{2+}$ (1 mM) is present, whereas $Ca^{2+}$ (1 mM) gives about 50% activity. Other metal ions, including $Zn^{2+}$ (1 mM) have little effect (Lee *et al.*, 1992). Later, it was found that CPase Taq is a zinc-dependent enzyme. Excess zinc ion (1 mM) strongly inhibits the enzyme activity, whereas 0.01 mM $Zn^{2+}$ does not inhibit the reaction at all (Lee *et al.*, 1994). EDTA (1 mM) and 1,10-phenanthroline, both of which are metal-chelating reagents, inhibit CPase Taq. It is slightly affected by sulfhydryl reagents. The enzyme is easily denatured by SDS (0.05%), but is more stable against a low concentration of guanidine hydrochloride. DFP and PMSF have no effect.

## Structural Chemistry

CPase Taq is a monomeric enzyme with an $M_r$ of 56 000 or 58 000 as indicated by SDS-PAGE and gel filtration, respectively (Lee *et al.*, 1992). The gene for CPase Taq encodes a protein of 56 210 Da consisting of 511 amino acid residues. Analysis by ICP-AES (inductively coupled plasma atomic emission spectrometry) shows that the recombinant enzyme contained 1 mol of zinc ion per mol of enzyme protein, but not cobalt or calcium ions, although the enzyme examined was prepared from *Escherichia coli* cells cultivated in a medium containing 1 mM $CoCl_2$.

This enzyme possesses the HEXXH sequence at position 276–280; this is the consensus sequence found in the active site of most zinc-dependent endopeptidases and aminopeptidases. Amino acid substitution experiments using site-directed mutagenesis showed that the zinc ligands are His276 and His280 (Lee *et al.*, 1996). Glu277 is involved not in zinc binding but participates directly in the catalytic function, as in the case of Glu143 of thermolysin (Vallee & Auld, 1990). Replacement of Glu298 or Glu332 with Gln only slightly, if at all, reduces the enzyme activity, although a reduction in the zinc content is observed in both cases. Clearly, these Glu residues are not essential for the enzyme function. Therefore, CPase Taq is the first carboxypeptidase to be discovered that binds zinc at a His-Glu-Xaa-Xaa-His motif.

## Preparation

CPase Taq has been found only in *Thermus aquaticus*. The distribution of this enzyme among other strains or other species has not been examined. The enzyme has been purified 63-fold from cell-free extract. In each of the purification steps, it is necessary to add 1 mM EDTA to the buffers to prevent self-digestion of the enzyme or nonspecific binding of the enzyme to proteins. The gene for CPase Taq has been expressed in *E. coli* (Lee *et al.*, 1994). The expressed enzyme showed excellent thermostability, therefore, it was effectively purified by heat treatment (70°C for 1 h) of the supernatant obtained by centrifugation after disruption of the cells by sonication. Then, successive butyl-Toyopearl and DEAE-Toyopearl chromatography steps yielded a purified preparation of the enzyme protein.

## Biological Aspects

CPase Taq has been reported only in the extremely thermophilic bacterium *Thermus aquaticus*. The location of this enzyme in the cell has not been determined, but the enzyme is probably membrane associated or particle bound. The sequence of the gene for CPase Taq has been determined and there is no prepropeptide. The location of the *CPase T* gene in the physical map of the genome of *Thermus thermophilus* strains HB27 has been determined in preliminary fashion (Tabata & Hoshino, 1996), suggesting that the *CPase T* gene might be common in *Thermus* spp. It seems likely that this enzyme might be a constitutive enzyme in the cell. A hypothetical protein from the ypwA gene of *Bacillus subtilis* (Marburg) strain 168, is homologous to CPase Taq, with 38% sequence identity. The hypothetical open reading frame of YpwA encodes 501 amino acids, and has a conserved HEXXH motif. The YpwA protein may be a metallopeptidase. *Bacillus subtilis* is a mesophilic bacterium, so the protein is expected to be thermolabile.

## Distinguishing Features

No other carboxypeptidases in *Thermus* spp. have been reported. CPase Taq seems to be the only component active against Z-Phe+Tyr in the cytoplasm of *Thermus* spp. An extremely heat-stable carboxypeptidase (Chapter 494) from *Sulfolobus solfataricus* is also reported (Colombo *et al.*, 1995), but it belongs to the peptidase family M40 (Chapter 494) and there is no relationship to CPase Taq.

## Further Reading

For a review, see Lee *et al.* (1992).

## References

Colombo, S., Toietta, G., Zecca, L., Vanoi, M. & Tortora, P. (1995) Molecular cloning, nucleotide sequence, and expression of a carboxypeptidase-encoding gene from the archaebacterium *Sulfolobus solfataricus*. *J. Bacteriol.* **177**, 5561–5566.

Jongeneel, C.V., Bouvier, J. & Bairoch, A. (1989) A unique signature identifies a family of zinc-dependent metallopeptidases. *FEBS Lett.* **242**, 211–214.

Lee, S.-H., Minagawa, E., Taguchi, H., Matsuzawa, H., Ohta, T., Kaminogawa, S. & Yamauchi, K. (1992) Purification and characterization of a thermostable carboxypeptidase (carboxypeptidase Taq) from *Thermus aquaticus* YT-1. *Biosci. Biotechnol. Biochem.* **56**, 1839–1844.

Lee, S.-H., Taguchi, H., Yoshimura, E., Minagawa, E., Matsuzawa, H., Kaminogawa, S., Ohta, T. & Yamauchi, K. (1994) Carboxypeptidase Taq, a thermostable zinc enzyme, from *Thermus aquaticus* YT-1: molecular cloning, sequencing, and expression of the encoding gene in *Escherichia coli*. *Biosci. Biotechnol. Biochem.* **58**, 1490–1495.

Lee, S.-H., Taguchi, H., Yoshimura, E., Minagawa, E., Kaminogawa, E., Ohta, T. & Matsuzawa, H. (1996) The active site of carboxypeptidase Taq possesses the active-site motif His-Glu-X-X-His of zinc-dependent endopeptidases and aminopeptidases. *Protein Eng.* **9**, 467–469.

Matheson, A.T. & Tattrie, B.L. (1964) A modified Yemm and Cocking ninhydrin reagent for peptidase assay. *Can. J. Biochem.* **42**, 95–103.

Rawlings, N.D. & Barrett, A.J. (1995) Evolutionary families of metallopeptidases. *Methods Enzymol.* **248**, 183–228.

Rosen, H. (1957) A modified ninhydrin colorimetric analysis for amino acids. *Arch. Biochem. Biophys.* **67**, 10–15.

Tabata, K. & Hoshino, T. (1996) Mapping of 61 genes on the refined physical map of the chromosome of *Thermus thermophilus* HB27 and comparison of genome organization with that of *T. thermophilus* HB8. *Microbiology* **142**, 401–410.

Vallee, B.L. & Auld, D.S. (1990) Zinc coordination, function,and structure of zinc enzymes and other proteins. *Biochemistry* **29**, 5647–5659.

Yaron, A. & Mlyner, D. (1968) Aminopeptidase-P. *Biochem. Biophys. Res. Commun.* **32**, 658–663.

***Hidemasa Motoshima***
*Research Center,*
*Yotsuba Milk Products Co., Ltd,*
*465-1, Wattsu, Kitahiroshima, Hokkaido, 061-12 Japan*
*Email: motosima@ppp.bekkoame.or.jp*

***Shuichi Kaminogawa***
*Department of Applied Biochemistry,*
*The University of Tokyo,*
*1-1-1 Yayoi, Bunkyouku, Tokyo 113 Japan*
*Email: akamino@hongo.ecc.u-tokyo.ac.jp*

# 505. *O-Sialoglycoprotein endopeptidase*

## Databanks

*Peptidase classification: clan MX, family M22, MEROPS ID: M22.001*
*NC-IUBMB enzyme classification: EC 3.4.24.57*
*ATCC entries: 43270 (Pasteurella haemolytica A1 strain)*
*Databank codes:*

| Species | SwissProt | PIR | EMBL (cDNA) | EMBL (genomic) |
|---|---|---|---|---|
| *Haemophilus influenzae* | P43764 | – | – | U32735: genomic |
| *Mycobacterium leprae* | P37969 | – | U00015 | – |
| *Mycoplasma genitalium* | P47292 | – | U39683 | – |
| *Pasteurella haemolytica* | P36175 | A38108 | U15958 | – |
| Other related sequences | | | | |
|   *Escherichia coli* | P05852 | – | M16194 | U28379: chromosome 68′ |
|   *Haloarcula marismortui* | P36174 | – | X70117 | – |
|   *Homo sapiens* | – | – | T65524 | – |
|   *Saccharomyces cerevisiae* | P36132 | – | Z28263 | – |
|   *Saccharomyces cerevisiae* | P43122 | – | X79380 | – |
|   *Salmonella typhimurium* | – | – | M14427 | – |
|   *Synechocystis* sp. | – | – | M74801 | – |
|   *Synechocystis* sp. | – | – | M74801 | – |

## Name and History

The enzyme *O-sialoglycoprotein endopeptidase*, abbreviated as *glycoprotease*, derives its name from its unique specificity for the proteolytic cleavage of glycoproteins, and to date, all known substrates are glycoproteins rich in sialoglycans *O*-linked to threonine or serine residues. The enzyme was discovered when human erythrocyte glycophorin A was found to be hydrolyzed by culture supernatants from the cattle pathogen, *Pasteurella haemolytica* A1, an organism which is associated with fibrinous pneumonia in cattle (Otulakowski *et al.*, 1983). The enzyme has been found to be secreted by all *P. haemolytica* A-biotypes except A11 (Abdullah *et al.*, 1990, Lee *et al.*, 1994a). Most substrates are cell surface mucin-like molecules, but soluble large molecular mass mucins are not cleaved. No unglycosylated or solely *N*-glycosylated protein has yet been shown to be cleaved by the enzyme. The enzyme has been purified to homogeneity (Abdullah *et al.*, 1992), and the gene for the enzyme has been cloned and expressed in *E. coli* (Abdullah *et al.*, 1991).

## Activity and Specificity

Human glycophorin A, either *in situ* in erythrocyte ghost membrane or free in solution, is cleaved by the enzyme at several peptide bonds, principally Arg31┼Asp32, but extensive degradation occurs at other sites and the peptide bond specificity is apparently broad (Abdullah *et al.*, 1992; Nakada *et al.*, 1993). Cleavage of glycophorin A is most extensive within the heavily glycosylated extracellular domain, which bears 15 *O*-linked glycans and one *N*-linked glycan. Other cell surface *O*-sialoglycoproteins which are cleaved by the glycoprotease include the human hematopoietic stem cell antigen CD34, leukosialin CD43, the hyaluronate receptor CD44, the leukocyte common antigen CD45 (Sutherland *et al.*, 1992), the ligands for L-selectin and P-selectin (Steininger *et al.*, 1992: Norgard *et al.*, 1993), the receptor for interleukin 7, the human tumor cell antigen epitectin (Hu *et al.*, 1994), the human platelet glycoprotein Ib(a) (Yeo & Sutherland, 1995), cranin which is a laminin-binding membrane *O*-sialoglycoprotein (Smalheiser & Kim, 1995), a human vascular adhesion protein VAP-1 (Salmi & Jalkanen, 1996) and the Epstein–Barr viral glycoproteins gp150 and gp350. Cattle κ-casein was thought to be a substrate (Norgard *et al.*, 1993; Mellors & Lo, 1995) but contamination of glycoprotease preparations with proteases that cleave κ-casein may be responsible for this activity. Cattle fetuin, which has potentially three *O*-linked and two *N*-linked glycans, is not cleaved by the enzyme (Cladman *et al.*, 1996). The removal of sialate residues from the susceptible substrates abolishes or greatly reduces proteolysis by the enzyme.

## Structural Chemistry

The gene sequence predicts a 35.2 kDa enzyme of 325 amino acid residues, four cysteine residues, a pI of 5.2, and a putative metal ion-binding site HMXGH which differs from known zinc ion-binding sites (Hooper, 1994). There is no secretion signal sequence or proenzyme precursor. While the enzyme is inhibited by relatively high concentrations of EDTA or 1,10-phenanthroline, it is not inactivated by dialysis nor activated by the addition of metal ions. Zinc ions are inhibitory to the enzyme at micromolar concentrations. There is no inhibition of the enzyme by serine protease inhibitors, thiol protease inhibitors, aspartate protease inhibitors or by the specific inhibitor of thermolysin-like metalloproteases, phosphoramidon.

## Preparation

*Pasteurella haemolytica* A1 cultures, grown to late-log phase in medium RPMI 1640, secrete the enzyme into the culture supernate. If 5% fetal calf serum or equivalent serum albumin is included in the culture medium, the enzyme can be precipitated at pH 4.5 from the diafiltered culture supernate. The resultant precipitate contains many *P. haemolytica* secreted proteins, as well as serum albumin, but the enzyme retains high activity in lyophilized preparations and is free from other bacterial proteases. Chromatographic separation of the enzyme from major contaminants can be done by DEAE-cellulose chromatography of diafiltered and concentrated culture supernates, from cultures grown in the presence of 0.2% CHAPS (Mellors & Lo, 1995; Cladman *et al.*, 1996).

## Biological Aspects

No similar enzyme activity has yet been demonstrated in species other than *P. haemolytica*, however homologous open reading frames have been reported in prokaryotes and eukaryotes: *Haemophilus influenzae*, *Escherichia coli*, *Mycobacterium leprae*, *Salmonella typhimurium*, *Haloarcula marismortui*, *Mycoplasma genitalium*, *Saccharomyces cerevisiae*. The N-terminal domain of a eukaryotic-like protein kinase from the archaeon *Methanococcus vannielii* also shows significant homology with *P. haemolytica* glycoprotease (Smith & King, 1995).

When the *P. haemolytica* gene *gcp*, which codes for the glycoprotease, is overexpressed in *E. coli*, the gene product rGcp is not secreted but accumulates in the periplasm as an inactive form of the enzyme, a misfolded aggregate cross-linked by intermolecular disulfide bridges. A fusion protein denoted rGcp-F, constructed between rGcp and the C-terminal secretion signal of *E. coli* α-hemolysin was specifically secreted into the medium when the α-hemolysin secretion apparatus was supplied in *trans* on a separate plasmid (Lo *et al.*, 1994), however this fusion protein also lacked enzymatic activity. These recombinant forms of the enzyme were used to generate monoclonal antibodies which can be used to quantify the glycoprotease protein and to specifically inhibit the cleavage of glycophorin A and other substrates by the enzyme. The denatured and reduced form of the enzymatically inactive rGcp can be refolded and activated by treatment with the *E. coli* heat shock protein (chaperone) thioredoxin, as well as by mammalian protein disulfide isomerase (Watt, 1996). Another fusion protein was constructed, this time between *E. coli* thioredoxin (Trx) and Gcp. The 47 kDa rTrx-Gcp fusion protein was retained in the *E. coli* cytoplasm in a monomeric and fully reduced form, but again did not show enzyme activity. However, an enzymatically active form of the glycoprotease was recovered after ion-exchange chromatographic isolation of the rTrx-Gcp fusion protein from fresh *E. coli* lysates (Watt, 1996).

The high specificity of the *P. haemolytica* glycoprotease for cell surface *O*-sialoglycoproteins has made the enzyme a useful tool in cell surface antigen studies (Mellors & Sutherland, 1994). The lack of action of the enzyme against protein substrates which do not bear *O*-linked glycans means that cell surface molecules and their roles can be delineated by the use of this enzyme. For example, the use of the enzyme in the immunomagnetic isolation of human bone marrow CD34+ stem cells has been described (Marsh *et al.*, 1992). More recently, the enzyme has been used to demonstrate the role of the P-selectin ligand in the dynamic aspects of neutrophil rolling (Alon *et al.*, 1995). However, there is as yet no clear indication of the role of the enzyme in the pathogenesis of fibrinous pneumonia in cattle. Cattle vaccinated with *P. haemolytica* culture supernate proteins supplemented with the recombinant glycoprotease fusion protein rGcp-F show enhanced protection against experimental challenge with live *P. haemolytica* A1. Vaccinated cattle and many cattle naturally exposed to the pathogen have circulating antibodies which are neutralizing for the glycoprotease enzyme activity (Lee *et al*., 1994b). The *O*-sialoglycoprotein endopeptidase may interfere with cell–cell adhesion or with cytokine–receptor binding, during the development of the host immune response in the cattle lung.

## Distinguishing Features

The *P. haemolytica O*-sialoglycoprotein endopeptidase remains the only known protease which is specific for glycoproteins. Hundreds of protein substrates have been tested, and only a restricted number of *O*-sialoglycoproteins have been found to be susceptible to proteolysis. The enzyme is commercially available from Cedarlane or from Accurate Chemical (see Appendix 2 for full names and addresses of suppliers).

## Further Reading

For a review, see Mellors & Lo (1995).

## References

Alon, R., Hammer, D.A. & Springer, T.A. (1995) Lifetime of the P-selectin-carbohydrate bond and its response to tensile force in hydrodynamic flow. *Nature* **374**, 539–542.

Abdullah, K.M., Lo, R.Y.C. & Mellors, A. (1990) Distribution of glycoprotease activity and the glycoprotease gene among serotypes of *Pasteurella haemolytica*. *Biochem. Soc. Trans.* **18**, 901–903.

Abdullah, K.M., Lo, R.Y.C. & Mellors, A. (1991) Cloning, nucleotide sequence and expression of the *Pasteurella haemolytica* A1 glycoprotease gene. *J. Bacteriol.* **173**, 5997–5603.

Abdullah, K.M., Udoh, E.A., Shewen, P.E. & Mellors, A. (1992) A neutral glycoprotease of *Pasteurella haemolytica* A1 specifically cleaves *O*-sialoglycoproteins. *Infect. Immun.* **60**, 56–62.

Cladman, W.M., Watt, M.-A.V., Dini, J.-P. & Mellors, A. (1996) The *Pasteurella haemolytica O*-sialoglycoprotein endopeptidase is inhibited by zinc ions and does not cleave fetuin. *Biochem. Biophys. Res. Commun.* **220**, 141–146.

Hooper, N.M. (1994) Families of zinc metalloproteases. *FEBS Lett.* **354**, 1–6.

Hu, R.-H., Mellors, A. & Bhavanandan, V.P. (1994) Cleavage of epitectin, a mucin-type sialoglycoprotein, from the surface of human laryngeal carcinoma cells by a glycoprotease from *Pasteurella haemolytica*. *Arch. Biochem. Biophys.* **310**, 300–309.

Lee, C.W., Lo, R.Y.C., Shewen, P.E. & Mellors, A. (1994a) The detection of the sialoglycoprotease gene and assay for sialoglycoprotease activity among isolates of *Pasteurella haemolytica* A1 strains, serotypes A13, A14, T15 and A16. *FEMS Microbiol. Lett.* **121**, 199–206.

Lee, C.W., Shewen, P.E., Cladman, W.M., Conlon, J.A.R.,

Mellors, A. & Lo, R.Y.C. (1994b) Sialoglycoprotease of *Pasteurella haemolytica* A1: detection of anti-sialoglycoprotease activity in sera of calves. *Can. J. Vet. Res.* **58**, 93–98.

Lo, R.Y.C., Watt, M.-A.V., Gyorffy, S. & Mellors, A. (1994) Preparation of recombinant glycoprotease of *Pasteurella haemolytica* A1 utilizing the *Escherichia coli* α-hemolysin secretion system. *FEMS Microbiol. Lett.* **116**, 225–230.

Marsh, J.C.W., Sutherland, D.R., Davidson, J., Mellors, A. & Keating, A. (1992) Retention of progenitor cell functions in CD34+ cells purified using a novel *O*-sialoglycoprotease. *Leukemia* **6**, 926–934.

Mellors, A. & Lo, R.Y.C. (1995) *O*-Sialoglycoprotease from *Pasteurella haemolytica*. *Methods Enzymol.* **248**, 728–740.

Mellors, A. & Sutherland, D.R. (1994) Tools to cleave glycoproteins. *Trends Biotechnol.* **12**, 15–18.

Nakada, H., Inoue, M., Numata, Y., Tanaka, N., Funakoshi, I., Fukui, S., Mellors, A. & Yamashina, I. (1993) Epitopic structure of Tn glycophorin A for an anti-Tn antibody (MLS 128). *Proc. Natl Acad. Sci. USA* **90**, 2495–2499.

Norgard, K.E., Moore, K.L., Diaz, S., Stults, N.L., Ushiyama, S., McEver, R.P., Cummings, R.D., & Varki, A. (1993) Characterization of a specific ligand for P-selectin on myeloid cells. A minor glycoprotein with *O*-linked oligosaccharides. *J. Biol. Chem.* **268**, 12764–12774.

Otulakowski, G.L., Shewen, P.E., Udoh, E.A., Mellors, A. & Wilkie, B.N. (1983) Proteolysis of sialoglycoproteins by *Pasteurella haemolytica* cytotoxic culture supernatant. *Infect. Immun.* **2**, 64–70.

Salmi, M. & Jalkanen, S. (1996) Human vascular adhesion protein 1 (VAP-1) is a unique sialoglycoprotein that mediates carbohydrate-dependent binding of lymphocytes to endothelial cells. *J. Exp. Med.* **183**, 569–579.

Smalheiser, N.R. & Kim, E. (1995) Purification of cranin, a laminin binding protein: identity with dystroglycan and reassessment of its carbohydrate residues. *J. Biol. Chem.* **270**, 15425–15433.

Smith, R.A. & King, K.Y. (1995) Identification of a eukaryotic-like protein kinase gene in *Archaebacteria*. *Protein Sci.* **4**, 126–129.

Steininger, C.N., Eddy, C.A., Leimgruber, R.M., Mellors, A. & Welply, J.K. (1992) The glycoprotease of *Pasteurella haemolytica* A1 eliminates binding of myeloid cells to P-selectin but not to E-selectin. *Biochem. Biophys. Res. Commun.* **188**, 760–766.

Sutherland, D.R., Abdullah, K.M., Cyopick, P. & Mellors, A. (1992) Cleavage of the cell-surface *O*-sialoglycoproteins CD34, CD43, CD44 and CD 45, by a novel glycoprotease from *Pasteurella haemolytica*. *J. Immunol.* **148**, 1458–1464.

Watt, M.-A.V. (1996) Activation of the recombinant *Pasteurella haemolytica O*-sialoglycoprotease. PhD thesis, University of Guelph.

Yeo, E. & Sutherland, D.R. (1995) Selective cleavage of platelet gpIba (CD42b) but not gpIIb/IIIa (CD41/CD61), P-selectin (CD62), or CDw109 with a novel *O*-sialoglycoprotein glycoprotease. In: *Leukocyte Typing (V): White Cell Differentiation Antigens*, vol. 2 (Schlossman, S.F. *et al*., eds). Oxford: Oxford University Press, pp. 1313–1315.

**M**

*Alan Mellors*
*Guelph-Waterloo Centre for Graduate Work in Chemistry,*
*Department of Chemistry and Biochemistry, University of Guelph,*
*Guelph, Ontario N1G 2W1, Canada*
*Email: mellors @chembio.uoguelph.ca*

# 506. β-Lytic metalloendopeptidase

## Databanks

*Peptidase classification: clan MX, family M23, MEROPS ID: M23.001*
*NC-IUBMB enzyme classification: EC 3.4.24.32*
*Databank codes:*

| Species | SwissProt | PIR | EMBL (cDNA) | EMBL (genomic) |
|---|---|---|---|---|
| *Achromobacter lyticus* | P27458 | A37151 | M60896 | – |
| *Lysobacter enzymogenes* | P00801 | A00994 | – | – |

## Name and History

*β-Lytic metalloendopeptidase*, formerly *β-lytic protease*, was discovered over three decades ago in the culture filtrate of *Myxobacter* 495, a soil bacterium initially defined as *Sorangium* sp. (Whitaker, 1965) and currently classified as *Lysobacter enzymogenes*. This organism has a remarkable ability to lyse other bacteria and some soil nematodes, a property by which it was originally selected (Gillespie & Cook, 1965; Katznelson *et al*., 1964). The trivial name *β*-lytic protease stems from the enzyme's ability to cause lysis of bacterial cell walls and differentiates it from *α*-lytic protease (Chapter 82), another major bacteriolytic endopeptidase produced by *Myxobacter* 495 (Whitaker, 1965, 1970). A *β*-lytic protease with amino acid sequence almost identical to that of the *L. enzymogenes* enzyme has recently been identified in the culture medium of *Achromobacter lyticus* (Li *et al*., 1990).

## Activity and Specificity

*L. enzymogenes* β-lytic metalloendopeptidase lyses *Arthrobacter globiformis* and *Micrococcus luteus* (previously *M. lysodeikticus*) cells. Lysis of the former is more efficient (Whitaker, 1965). The enzyme is likely to also lyse *Staphylococcus aureus* cells (Gillespie & Cook, 1965). There is suggestive evidence that in *M. luteus* peptidoglycan, cleavages occur at the *N*-acetylmuramoyl┼Ala and Gly┼(ε-amino)Lys peptide bonds (Tsai *et al*., 1965). Limited casein degradation has also been observed (Whitaker, 1965). The enzyme cleaves the Gly23┼Phe24 and (more slowly) the Val18┼Cya19 bonds in oxidized insulin B chain (Whitaker *et al*., 1965b):

FVNQHLCGSHLVEALYLV┼CGERG┼FFYTPKA

It also degrades elastin-orcein and hydrolyzes the synthetic substrates FA-Gly┼Leu-NH$_2$ (FAGLA) and Z-Gly┼Phe-NH$_2$ but not Z-Gly-Phe (Oza, 1973). From this sparse information it appears the enzyme has restricted specificity, although the determinants of this are not clear. The *Achromobacter* enzyme may lyse *M. luteus* cells (Li *et al*., 1990).

Bacteriolytic activity can be determined with *Arthrobacter globiformis* (Whitaker *et al*., 1965a; Kessler, 1995) or *M. luteus* (for the *A. lyticus* enzyme; Li *et al*., 1990) cells. Peptidase activity of the *Lysobacter* β-lytic protease can be determined with FAGLA (Oza, 1973; Kessler, 1995). pH optimum for bacteriolysis is ∼9 and the assay is performed in buffers of low (0.01–0.025 M) ionic strength. pH optimum for FAGLA hydrolysis is 6.5. FAGLA is available from Sigma, Calbiochem/Novabiochem or Peninsula Laboratories (see Appendix 2 for full names and addresses of suppliers). The enzyme is inhibited by 1,10-phenanthroline and is insensitive to DFP (Whitaker & Roy, 1967; Oza, 1973).

## Structural Chemistry

*L. enzymogenes* β-lytic metalloendopeptidase is a basic 19.1 kDa protein. The enzyme (178 amino acid residues) contains two disulfide bonds between residues 65–111 and 155–168 and one zinc atom per mol that is essential for activity (Damaglou *et al*., 1976). It does not contain the consensus zinc-binding site HEXXH but has a conserved HXH sequence that is implicated in zinc binding (Li *et al*., 1990). The molecular mass of the 179 residue *Achromobacter* enzyme is 19.3 kDa. It is synthesized as a 374 residue preproenzyme with putative 24 and 171 residue signal and pro sequences, respectively. The propeptide shows no homology to propeptides of other proteinases, although, in common with the propeptides of some other bacterial endopeptidases, it contains several clusters of arginine (Li *et al*., 1990). The amino acid sequences of mature *A. lyticus* and *L. enzymogenes* β-lytic metalloendopeptidases are ∼96% identical. The *Lysobacter* enzyme has been crystallized (Cruse & Whitaker, 1975).

## Preparation

A purification procedure for *L. enzymogenes* β-lytic protease that yields approximately 2 g of the purified enzyme has been described (Whitaker, 1967, 1970). The purified enzyme fraction may contain traces of *α*-lytic protease that can be selectively inhibited with DFP. Purification of the *Achromobacter* β-lytic protease (Li *et al*., 1990) involved a reverse phase HPLC step and thus yielded a denatured enzyme.

## Biological Aspects

β-Lytic metalloendopeptidase inhibits growth of sensitive organisms and may potentially serve as an antimicrobial agent (Gillespie & Cook, 1965; Li *et al*., 1990).

## Distinguishing Features

β-Lytic metalloendopeptidase can be distinguished from the α-lytic protease (Chapter 82) by its resistance to DFP (Whitaker, 1967; Oza, 1973).

## Related Peptidases

A bacteriolytic protease (formerly, EC 3.4.99.29) with properties reminiscent of *L. enzymogenes* β-lytic metalloendopeptidase was isolated from the culture media of *Myxobacter* AL-1 and termed *Myxobacter* AL-1 protease (Jackson & Wolfe, 1968; Jackson & Matsueda, 1970). The enzyme (pI ~10) lyses *S. aureus, A. crystallopoietes* and *M. luteus* cells (optimum pH ~9) by a primary attack on the cell wall peptide cross-bridges pentaglycine and L-Ala in *S. aureus* and *A. crystallopoietes*, respectively (Tipper *et al.*, 1967), and hydrolysis of the acetylmuramoyl┼L-Ala linkage in the *M. luteus* peptidoglycan (Katz & Strominger, 1967). Because of differences in molecular mass ($M_r$ of *Myxobacter* AL-1 protease is ~14 000), number of disulfide bonds (1 in *Myxobacter* AL-1 protease as opposed to 2 in the *Lysobacter* enzyme), and zinc contents (0.27 mol per mol protein in AL-1 protease), as well as lack of sequence information on the *Myxobacter* AL-1 protease, the question of whether this protease and β-lytic metalloendopeptidase are mechanistically and evolutionarily related remains open. Two endopeptidases with documented homology to β-lytic metalloendopeptidase, *Pseudomonas aeruginosa* staphylolysin (LasA) and *Aeromonas hydrophila* protease, are described in Chapter 507.

## Further Reading

For a review, see Kessler (1995).

## References

Cruse, W.B.T. & Whitaker, D.R. (1975) Preliminary crystallographic data for beta-lytic protease. *J. Mol. Biol.* **102**, 173–175.

Damaglou, A.P., Allen, L.C. & Whitaker, D.R. (1976) Beta-lytic protease – Myxobacter 495. In: *Atlas of Protein Sequence and Structure*, vol. 5 (suppl. 2) (Dayhoff, M.O., ed.). Washington, DC: National Biomedical Research Foundation, p. 198.

Gillespie, D.C. & Cook, F.D. (1965) Extracellular enzymes from strains of *Sorangium. Canad. J. Microbiol.* **11**, 109–118.

Jackson, R.L. & Matsueda, G.R. (1970) Myxobacter AL-1 protease. *Methods Enzymol.* **19**, 591–599.

Jackson, R.L. & Wolfe, R.S. (1968) Composition, properties, and substrate specificities of *Myxobacter* AL-1 protease. *J. Biol. Chem.* **243**, 879–888.

Katz, W. & Strominger, J.L. (1967) Structure of the cell wall of *Micrococcus lysodeikticus.* II. Study of the structure of the peptides produced after lysis with the *Myxobacterium* enzyme. *Biochemistry* **6**, 930–937.

Katznelson, H., Gillespie, D.C. & Cook, F.D. (1964) Studies on the relationships between nematodes and other soil microorganisms. III. Lytic action of soil myxobacters on certain species of nematodes. *Can. J. Microbiol.* **10**, 699–704.

Kessler, E. (1995) β-Lytic endopeptidases. *Methods Enzymol.* **248**, 740–756.

Li, S.L., Norioka, S. & Sakiyama, F. (1990) Molecular cloning and nucleotide sequence of the β-lytic protease from *Achromobacter lyticus J. Bacteriol.* **172**, 6506–6511.

Oza, N.B. (1973) β-Lytic protease, a neutral sorangiopeptidase. *Int. J. Pept. Protein Res.* **5**, 365–369.

Tipper, D.J., Strominger, J.L. & Ensign, J.C. (1967) Structure of the cell wall of *Staphylococcus aureus*, strain Copenhagen. VII. Mode of action of the bacteriolytic peptidase from *Myxobacter* and the isolation of intact cell wall polysaccharides. *Biochemistry* **6**, 906–920.

Tsai, C.S., Whitaker, D.R. & Jurasek, L. (1965) Lytic enzymes of *Sorangium* sp. Action of the α- and β-lytic proteases on two bacterial mucopeptides. *Can. J. Biochem.* **43**, 1971–1983.

Whitaker, D.R. (1965) Lytic enzymes of *Sorangium* sp. Isolation and enzymatic properties of the α- and β-lytic proteases. *Can. J. Biochem.* **43**, 1935–1954.

Whitaker, D.R. (1967) Simplified procedures for production and isolation of the bacteriolytic proteases of *Sorangium* sp. *Can. J. Biochem.* **45**, 991–993.

Whitaker, D.R. (1970) The α-lytic protease of a *Myxobacterium. Methods Enzymol.* **19**, 599–613.

Whitaker, D.R. & Roy, C. (1967) Concerning the nature of the α- and β-lytic proteases of *Sorangium sp. Can. J. Biochem.* **45**, 911–916.

Whitaker, D.R., Cook, F.D. & Gillespie D.C. (1965a) Lytic enzymes of *Sorangium sp.* Some aspects of enzyme production in submerged culture. *Can. J. Biochem.* **43**, 1927–1933.

Whitaker, D.R., Roy, C., Tsai, C.S. & Jurasek, L. (1965b) Lytic enzymes of *Sorangium sp.* A comparison of the proteolytic properties of the α- and β-lytic proteases. *Can. J. Biochem.* **43**, 1961–1970.

**M**

*Efrat Kessler*
*Maurice & Gabriela Goldschleger Eye Research Institute,*
*Tel-Aviv University Sackler Faculty of Medicine,*
*Sheba Medical Center,*
*Tel-Hashomer 52621, Israel*
*Email: ekessler@post.tau.ac.il*

# 507. *Staphylolysin*

## Databanks

*Peptidase classification: clan MX, family M23, MEROPS ID: M23.002*
*NC-IUBMB enzyme classification: none*
*Databank codes:*

| Species | SwissProt | PIR | EMBL (cDNA) | EMBL (genomic) |
|---|---|---|---|---|
| Fibrinogenolytic endopeptidase | | | | |
| *Aeromonas hydrophila* | – | A46041 | – | – |
| | | | | |
| *Pseudomonas aeruginosa* | P14789 | A46076 | M20982 | – |
| | | | U68175 | |
| | | | X55904 | |

## Name and History

*Staphylolysin* is a new name suggested for *Pseudomonas aeruginosa* **LasA endopeptidase** that denotes its ability to cause lysis of *Staphylococcus aureus* cells. It is based on the finding (Kessler *et al.*, 1993) that staphylolysin (LasA) is identical to the long-recognized **staphylolytic endopeptidase** (previously known as the **bacteriolytic enzyme**) of *P. aeruginosa* (Lache *et al.*, 1969). Staphylolysin was rediscovered through a mutation in a gene (*lasA*) required for full expression of the elastin-degrading (elastolytic) phenotype of *P. aeruginosa* (Ohman *et al.*, 1980). The *lasA* gene was once thought to encode either pseudolysin (Chapter 357) or an enzyme involved in the maturation of pseudolysin (Goldberg & Ohman, 1987; Schad & Iglewski, 1988). Recent studies show that staphylolysin is a secreted endopeptidase that can slowly degrade insoluble elastin, rendering it a better substrate for pseudolysin and other endopeptidases (Peters & Galloway, 1990; Toder *et al.*, 1991; Wolz *et al.*, 1991; Peters *et al.*, 1992). Hence, staphylolysin has been called *Pseudomonas elastase*, but its limited elastolytic power makes this name less appropriate (Peters *et al.*, 1992; Kessler *et al.*, 1997).

## Activity and Specificity

Staphylolysin cleaves peptide bonds within the pentaglycine cross-linking peptides of *S. aureus* peptidoglycan and leads to cell lysis. It also lyses cell walls of *Micrococcus radiodurans* and *Gaffkia tetragena*, which contain di- and triglycine sequences in their interpeptides, respectively (Lache *et al.*, 1969). Synthetic penta- and hexaglycine are cleaved into di- and tri- or triglycine, respectively (Kessler *et al.*, 1993). Oligopeptides containing internal Gly-Gly or Gly-Gly-Gly sequences are also cleaved, but the cleavage sites in these substrates are not defined (Park & Galloway, 1995). Tropoelastin is cleaved preferentially at Gly-Gly+Ala sequences surrounded by apolar residues. The same bonds and possibly Gly-Gly+Phe and Gly-Gly+Gly are apparently cleaved in insoluble elastin (Kessler *et al.*, 1997). β-Casein contains no Gly-Gly sequences and is resistant to staphylolysin (Kessler *et al.*, 1997) even though it has been reported as a substrate (Peters *et al.*, 1992; Park & Galloway, 1995). Staphylolysin specificity seems to be restricted to Gly in the P1 and P2 positions, and Gly, Ala or Phe are preferred at P1′. Apolar residues are favored in flanking sequences.

Staphylolysin activity can be assayed spectrophotometrically by the reduction in turbidity of *S. aureus* cell suspensions in buffers of low ionic strength (Kessler, 1995). Other proteases in the *P. aeruginosa* culture filtrates can enhance the apparent activity of staphylolysin in this assay (Gustin *et al.*, 1996). The optimum pH for staphylolytic activity is 8.5. Staphylolytic activity is inhibited by 1,10-phenanthroline, tetraethylene pentamine, excess $Zn^{2+}$ (0.1 mM) and DTT. EDTA and EGTA are inhibitory only at concentrations higher than 20 mM (Kessler, 1995; Kessler *et al.*, 1997).

## Structural Chemistry

Staphylolysin is a basic (pI 9.24) 19 965 Da protein. It consists of 182 residues and its amino acid sequence is ~40% identical to those of *Achromobacter lyticus* and *Lysobacter enzymogenes* β-lytic endopeptidases (Chapter 506) (Darzins *et al.*, 1990; Kessler *et al.*, 1993; Gustin *et al.*, 1996). Staphylolysin contains four conserved Cys residues that, by analogy to the *Lysobacter* enzyme, are likely to be engaged in disulfide bonds. Staphylolysin also contains one atom of zinc as determined by atomic absorption spectrophotometry (J.K. Gustin, E. Kessler & D.E. Ohman, unpublished results). Although it does not have a classical zinc-binding motif (HEXXH), a conserved HXH sequence has been speculated to be a potential zinc-binding site (where X = Leu121). A His120Ala substitution in this motif results in the secretion of a stable enzymatically inactive protein by *P. aeruginosa* (Gustin *et al.*, 1996). Staphylolysin is synthesized as a 418 residue preproenzyme with a 31 residue (3298 Da) signal peptide and a 205 residue (22 319 Da) N-terminal propeptide (Gustin *et al.*, 1996).

## Preparation

Most *P. aeruginosa* wild-type strains secrete staphylolysin, although it often represents a minor constituent in the culture filtrate (Kessler *et al.*, 1993). An overexpressing *P. aeruginosa* strain (FRD2128[pJKG107]) that secretes about 3-fold more staphylolysin than the parental strain (FRD2) (Gustin *et al.*, 1996) may serve as an improved source for the enzyme.

Staphylolysin has been purified from culture filtrates of several *P. aeruginosa* strains with purification factors ranging from 140 to 520 (Lache *et al.*, 1969; Brito *et al.*, 1989; Peters & Galloway, 1990; Kessler, 1995; Gustin *et al.*, 1996). Purified staphylolysin preparations often contain another protease(s) that is undetectable in SDS gels, but its presence is responsible for $\beta$-casein hydrolytic activity (Kessler *et al.*, 1997).

## Biological Aspects

Biologically and clinically relevant substrates of staphylolysin include *S. aureus* cells and elastin. Staphylolysin appears to play a role in pathogenesis of corneal infections: in a mouse model of *P. aeruginosa* corneal infection, a defined *lasA* mutant causes mild to no disease following infection at doses of $10^8$ colony-forming units per eye in contrast to its virulent parental strain PAO1; application of 5 µg of purified staphylolysin on to scarified mouse corneas produces an acute toxic reaction (Preston *et al.*, 1997). In experimental models of acute (Blackwood *et al.*, 1983) and chronic (Woods *et al.*, 1982) lung infections, bacterial production of staphylolysin increases the virulence of *P. aeruginosa*. Staphylolysin inhibits the growth of *S. aureus* (Perestelo *et al.*, 1985) and thus may play a role in acquiring certain niches, such as the cystic fibrosis lung environment. Its ability to stimulate elastin degradation by other endopeptidases, as demonstrated *in vitro* (Peters & Galloway, 1990; Kessler *et al.*, 1997), suggests that it may contribute to the tissue destruction associated with *P. aeruginosa* infections.

The *lasA* gene encodes staphylolysin and is located at ~55 min on the 75 min *P. aeruginosa* chromosome map (Shortridge *et al.*, 1991). Its transcription is positively regulated by *lasR* (Toder *et al.*, 1991) and *rhlR* (Brint & Ohman, 1995), members of two autoinducer-responsive regulatory systems in *P. aeruginosa* that respond to cell density by a mechanism known as quorum sensing. The processing protease and cellular location of prostaphylolysin processing are not known but there is evidence suggesting that processing does not involve autoproteolysis (Gustin *et al.*, 1996).

## Distinguishing Features

A *lasA* deletion derivative of *P. aeruginosa* shows no detectable staphylolytic activity under standard assay conditions (Gustin *et al.*, 1996). However, another staphylolytic endopeptidase called LasD has been reported in *P. aeruginosa* culture filtrates and is presumed to be a serine protease (Park & Galloway, 1995). Staphylolysin can be differentiated from LasD by its slightly faster migration in SDS-PAGE, pH optimum (8.5 as opposed to 9.5 of LasD), distinct N-terminal sequence and lack of immunological cross-reactivity.

Polyclonal antibodies to staphylolysin and to a 22 residue synthetic peptide (positions 77–98) have been described (Peters & Galloway, 1990; Peters *et al.*, 1992; Kessler

*et al.*, 1993; Gustin *et al.*, 1996), but are not commercially available.

## Related Peptidases

A related zinc metalloendopeptidase, AhP, has been isolated from the culture filtrate of *Aeromonas hydrophila* (Loewy *et al.*, 1993). The sequence of the first 40 residues of AhP is 69% identical with that of staphylolysin. AhP hydrolyzes the Gly-Gly+Ala sequence near the cross-link site in the fibrin $\gamma$-chain. This site in fibrinogen, which contains an unsubstituted lysine residue in position P2′, is resistant. Synthetic fragments mimicking this region in fibrinogen, Ac-Glu-Gly-Gln-Gln-His-His-Leu-Gly-Gly+Ala-Lys(Tfa)Gln-Ala-NH$_2$ and Ac-Leu-Gly-Gly+Ala-Lys(Tfa)Gln-Ala-NH$_2$ (Tfa: trifluoroacetyl) are cleaved readily, but Ac-Leu-Gly+Ala-Lys(Tfa)-Gln-Ala-NH$_2$, with one of the glycines missing, is hydrolyzed 200 times more slowly. Synthetic substrates with the general structure Ac- and Suc-Gly-Gly+Xaa-NH$_2$, where X = Ala, Phe or Nph, are readily cleaved. Gly+Gly cleavages also take place in these substrates, although at a slower rate. Polar residues near the scissile bonds are not well tolerated. Thus, AhP prefers Gly-Gly in the P2 and P1 positions, and nonpolar residues seem to be favored in the P1′ and P2′ positions. This suggests that AhP may have staphylolytic activity. A staphylolytic protease of *A. hydrophila*, which shares several properties with the known $\beta$-lytic endopeptidases, has been described (Coles *et al.*, 1969) and may be the same as AhP. AhP is a basic 19 kDa protein containing one atom of zinc and inhibited by 1,10-phenanthroline. Its optimum pH with synthetic substrates is 7.5.

## Further Reading

For a review, see Kessler (1995).

## References

Blackwood, L.L., Stone, R.M., Iglewski, B.H. & Pennington, J.E. (1983) Evaluation of *Pseudomonas aeruginosa* exotoxin A and elastase as virulence factors in acute lung infection. *Infect. Immun.* **39**, 198–201.

Brint, J.M. & Ohman, D.E. (1995) Synthesis of multiple exoproducts in *Pseudomonas aeruginosa* is under the control of RhlR-RhlI, another set of regulators in strain PAO1 with homology to the autoinducer-responsive LuxR-LuxI family. *J. Bacteriol.* **177**, 7155–7163.

Brito, N., Falcón, M.A., Carnicero, A., Gutiérrez-Navarro, A.M. & Mansito, T.B. (1989) Purification and peptidase activity of a bacteriolytic extracellular enzyme from *Pseudomonas aeruginosa*. *Res. Microbiol.* **140**, 125–137.

Coles, N.W., Gilbo, C.M. & Broad, A.J. (1969) Purification, properties and mechanism of action of a staphylolytic enzyme produced by *Aeromonas hydrophila*. *Biochem. J.* **111**, 7–15.

Darzins, A., Peters, J.E. & Galloway, D.R. (1990) Revised nucleotide sequence of the *lasA* gene from *Pseudomonas aeruginosa* PAO1. *Nucleic Acids Res.* **18**, 6444.

Goldberg, J.B. & Ohman, D.E. (1987) Activation of an elastase precursor by the *lasA* gene product of *Pseudomonas aeruginosa*. *J. Bacteriol.* **169**, 4532–4539.

**M**

Gustin, J.K., Kessler, E. & Ohman, D.E. (1996) A substitution at His-120 in the LasA protease of *Pseudomonas aeruginosa* blocks enzymatic activity without affecting propeptide processing or extracellular secretion. *J. Bacteriol.* **178**, 6608–6617.

Kessler, E. (1995) β-Lytic endopeptidases. *Methods Enzymol.* **248**, 740–756.

Kessler, E., Safrin, M., Olson, J.C. & Ohman, D.E. (1993) Secreted LasA of *Pseudomonas aeruginosa* is a staphylolytic protease. *J. Biol. Chem.* **268**, 7503–7508.

Kessler, E., Safrin, M., Abrams, W.R., Rosenbloom, J. & Ohman, D.E. (1997) Inhibitors and specificity of *Pseudomonas aeruginosa* LasA. *J. Biol. Chem.* **272**, 9884–9889.

Lache, M., Hearn, W.R., Zyskind, J.W., Tipper, D.J. & Strominger, J.L. (1969) Specificity of a bacteriolytic enzyme from *Pseudomonas aeruginosa. J. Bacteriol.* **100**, 254–259.

Loewy, A.G., Santer, U.V., Wieczorek, M., Blodgett, J.K., Jones, S.W. & Cheronis, J.C. (1993) Purification and characterization of a novel zinc-proteinase from cultures of *Aeromonas hydrophila. J. Biol. Chem.* **268**, 9071–9078.

Ohman, D.E., Cryz, S.J. & Iglewski, B.H. (1980) Isolation and characterization of a *Pseudomonas aeruginosa* PAO mutant that produces altered elastase. *J. Bacteriol.* **142**, 836–842.

Park, S.J. & Galloway, D.R. (1995) Purification and characterization of LasD: a second staphylolytic proteinase produced by *Pseudomonas aeruginosa. Mol. Microbiol.* **16**, 263–270.

Perestelo, F.R., Blanco, M.T., Gutierrez-Navarro, A.M. & Falcón, M.A. (1985) Growth inhibition of *Staphylococcus aureus* by a staphylolytic enzyme from *Pseudomonas aeruginosa. Microbios Lett.* **30**, 85–94.

Peters, J.E. & Galloway, D.R. (1990) Purification and characterization of an active fragment of the LasA protein from *Pseudomonas aeruginosa*: enhancement of elastase activity. *J. Bacteriol.* **172**, 2236–2240.

Peters, J.E., Park, S.J., Darzins, A., Freck, L.C., Saulnier, J.M., Wallach, J.M. & Galloway, D.R. (1992) Further studies on *Pseudomonas aeruginosa* LasA: analysis of specificity. *Mol. Microbiol.* **6**, 1155–1162.

Preston, M.J., Seed, P.C., Toder, D.S., Iglewski, B.H., Ohman, D.E., Gustin, J.K., Goldberg, J.B. & Pier, G.B. (1997) Contribution of proteases and LasR to virulence of *Pseudomonas aeruginosa* during corneal infections. *Infect. Immun.* **65**, 3086–3090.

Schad, P.A. & Iglewski, B.H. (1988) Nucleotide sequence and expression in *Escherichia coli* of the *Pseudomonas aeruginosa lasA* gene. *J. Bacteriol.* **170**, 2784–2789.

Shortridge, V.D., Pato, M.L., Vasil, A.I. & Vasil, M.L. (1991) Physical mapping of virulence-associated genes in *Pseudomonas aeruginosa* by transverse alternating-field electrophoresis. *Infect. Immun.* **59**, 3596–3603.

Toder, D.S., Gambello, M.J., Iglewski, B.H. (1991) *Pseudomonas aeruginosa* LasA: a second elastase under the transcriptional control of *lasR. Mol. Microbiol.* **5**, 2003–2010.

Wolz, C., Hellstern, E., Haug, M., Galloway, D.R., Vasil, M.L. & Doring, G. (1991) *Pseudomonas aeruginosa* LasB mutant constructed by insertional mutagenesis reveals elastolytic activity due to alkaline proteinase and the LasA fragment. *Mol. Microbiol.* **5**, 2125–2131.

Woods, D.E., Cryz, S.J., Friedman, R.L. & Iglewski, B.H. (1982) Contribution of toxin A and elastase to virulence of *Pseudomonas aeruginosa* in chronic lung infections of rats. *Infect. Immun.* **36**, 1223–1228.

*Efrat Kessler*
Maurice & Gabriela Goldschleger Eye Research Institute,
Tel-Aviv University Sackler Faculty of Medicine,
Sheba Medical Center, Tel-Hashomer 52621, Israel
Email: ekessler@ccsg.tau.ac.il

*Dennis E. Ohman*
Department of Microbiology & Immunology,
University of Tennessee & Veteran Affairs Medical Center,
Memphis, TN 38163, USA
Email: dohman@utmem1.utmem.edu

# 508. IgA-specific metalloendopeptidase

## Databanks

Peptidase classification: clan MX, family M26, MEROPS ID: M26.001
NC-IUBMB enzyme classification: EC 3.4.24.13
Databank codes:

| Species | SwissProt | PIR | EMBL (cDNA) | EMBL (genomic) |
|---|---|---|---|---|
| *Streptococcus pneumoniae* | – | – | U47687 | – |
| *Streptococcus pneumoniae* | – | – | X94909 | – |
| *Streptococcus sanguis* | – | – | L29504 | – |

## Name and History

IgA proteinases were recognized by identification of unusual fragments of immunoglobulin A (IgA) in cell-free fluids of the human digestive tract (Mehta *et al.*, 1973). It had already been established that antibodies in 'external' human secretions, including those of the gastrointestinal tract, are nearly all of the IgA isotype. During purification of human intestinal IgA for immunological studies, it was noted that the intestine contained intact Fc domains of IgA, the region of an antibody protein that carries out many antibody effector functions once the Fab domains have bound to antigens. At the time, several laboratories had been trying to separate and isolate the various domains of IgA to analyze their respective functions. In the case of IgA, *in vitro* proteolysis with purified trypsin, chymotrypsin and papain all cleaved the hinge region to yield intact, functional Fab domains, but an intact Fc could not be recovered, this fragment having been further proteolyzed to small peptides. Thus, finding intact Fc fragments in the lower gut was unexpected and suggested the presence of one or more enzymes that cleaved IgA, but could not secondarily attack the Fc product.

This cleaving activity was then found in normal human saliva (Plaut *et al.*, 1974), indicating that it was of bacterial origin, not from the pancreas or intestine. This was confirmed by finding IgA-cleaving activity in culture filtrates of many bacteria colonizing or infecting human beings (Kilian & Holmgren, 1981). The enzymes were named *IgA proteinases* (later also *IgA1 proteinases* or *immunoglobulin A*

*proteinases*) because of their remarkable substrate specificity for human IgA1. IgA proteinases are a group of endopeptidases produced by medically important bacteria in the genera *Streptococcus* (Kilian *et al.*, 1979, 1980; Male, 1979), *Neisseria* (Plaut *et al.*, 1975), *Haemophilus* (Kilian *et al.*, 1979; Male, 1979), *Ureaplasma* (Spooner *et al.*, 1992), *Clostridium* (Fujiyama *et al.*, 1986), *Capnocytophaga* (Kilian, 1981; Frandsen *et al.*, 1987) and *Prevotella* (formerly *Bacteroides*) (Frandsen *et al.*, 1987; Mortensen & Kilian, 1984). They differ in catalytic mechanism, with serine-type, metallo-type and sulfhydryl proteases all represented (Mulks & Shober, 1994). The serine-type IgA proteases are discussed in Chapter 87.

*IgA-specific metalloproteinases* are confined to streptococci that colonize the human oral cavity and upper respiratory tract: *Streptococcus sanguis, S. pneumoniae, S. oralis, S. mitis*, and several species of the genus *Capnocytophaga*. Other closely related streptococci cocolonizing these sites are IgA proteinase negative. The complex taxonomy of oral streptococcal species has been advanced by the classification scheme of Kilian *et al.* (1989), which should be consulted.

## Activity and Specificity

All IgA proteinases cleave the glycosylated hinge region of human IgA, a flexible, peptide stretch that lies within the heavy ($\alpha$) polypeptide chain, between the Fab and Fc domains. The hinge segment contains multiple prolines, and

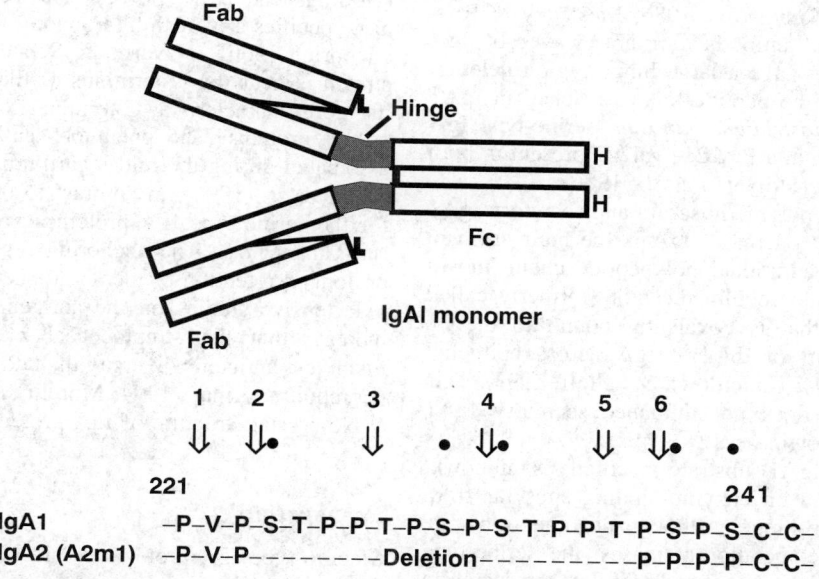

*Figure 508.1*   Cleavage sites of IgA1 proteinases in human IgA. The IgA1 monomer diagram shows heavy (H) and light (L) polypeptide chains linked by S–S bonds (light lines), and the hinge region of the heavy chain. IgA proteinases hydrolyze the hinge to yield intact Fab and Fc domains. The IgA1 hinge is aligned with the commonest allotype of the IgA2 subclass to define the 13 residue deletion in IgA2. Residue numbers are according to Tsuzukida *et al.* (1979). Filled circles indicate IgA1 hinge glycosylation sites. Numbered open arrows show which hinge peptide bond is cleaved by each IgA proteinase: (1) *Clostridium ramosum*, (2) *Capnocytophaga* and *Prevotella* species, (3) *Streptococcus pneumoniae, S. sanguis, S. oralis* and *S. mitis*, (4) *Haemophilus influenzae* enzyme type 1, and *H. aegyptius*, (5) *Neisseria gonorrhoeae, N. meningitidis* and *H. influenzae* type 2 enzymes, and *Ureaplasma urealyticum*, (6) *Neisseria gonorrhoeae* and *N. meningitidis* type 1 enzymes. Enzyme type is discussed in the text. *Clostridium ramosum* proteinase cleaves IgA1 and IgA2 proteins, because the site is preserved in both subclasses.

a tandemly duplicated octapeptide, Thr-Pro-Pro-Thr-Pro-Ser-Pro-Ser, as shown in Figure 508.1. Each IgA proteinase cleaves after only one Pro residue in the hinge and in only one of the duplicated octapeptides. This high degree of substrate specificity mandates use of IgA for assay (Plaut & Bachovchin, 1994), obtained as the purified monoclonal protein from the plasma of patients with multiple myeloma, or as a polyclonal mixture of normal serum or secretory IgA. A method for identifying IgA-cleaving activity in proteins after SDS-PAGE has recently been reported (Poulsen et al., 1996). Among Haemophilus influenzae and Neisseria isolates cleavage site varies depending on the actual strain producing the enzyme, leading to designation of enzyme 'type', as in Figure 508.1. As discussed below, metallo-type IgA proteinases all cleave the same bond, so no type designation is needed.

Metallo-type IgA proteinases identified to date all cleave bond Pro227↓Thr228 (numbered as in Tsuzukida et al., 1979) in the IgA heavy chain (at site 3, in Figure 508.1). This limited IgA1 proteolysis yields large, intact functional domains – two Fab (antigen-binding) fragments per IgA monomer, and one Fc fragment per monomer. The metallo-type IgA proteinases are active over a relatively broad pH range, 5.5–7.5, and the $K_m$ of the S. sanguis protease for human IgA1 is $5.5 \times 10^{-6}$ M (Plaut et al., 1978). Gorilla and chimpanzee IgA, which have only a few scattered differences in the hinge region amino acid sequence compared to human IgA (Kawamura et al., 1992), are cleaved extremely slowly by these enzymes (Cole & Hale, 1991; Qiu et al., 1996). IgA1 and IgA2 are isotypes present in all normal people, and they differ in function and relative concentration at various mucosal sites (Kett et al., 1986; Mestecky & Russell, 1986). Human IgA2 antibodies are not cleaved by IgA proteinases because they have a large hinge region deletion that includes nearly all the peptide bonds attacked in IgA1 (Figure 508.1). Only Clostridium ramosum serine-type IgA proteinase cleaves IgA2, at a Pro221↓Val222 present in both IgA1 and IgA2 proteins (Fujiyama et al., 1986).

Both monomeric and polymeric serum and secretory IgA1 proteins are susceptible. Secretory IgA is the main form in secretions, and has an additional polypeptide chain known as the polymeric immunoglobulin receptor (formerly called secretory component) that is covalently bound to the Fc domain during transport of dimeric IgA across epithelial cells (Neutra et al., 1994; Underdown & Schoff, 1993). This additional polypeptide does not influence secretory IgA1 susceptibility to IgA proteinases.

The basis for human IgA1 substrate specificity is unknown, but activity may depend on enzyme-binding sites far from the cleaved bond in the hinge. Although no other substrate has been found for the metalloproteinases, the serine-type enzymes cleave their own precursor to complete the secretion pathway in gram-negative bacteria (Pohlner et al., 1987; Poulsen et al., 1989). No smaller substrates are available, and no inhibitors active in the submillimolar range have been successfully made for the metallo-type enzymes (see below). Inhibitory antibodies are present in most human serum and secretions including milk (Plaut et al., 1992; Reinholt & Kilian, 1995), but plasma proteinase inhibitors such as $\alpha_2$-macroglobulin and $\alpha_1$-antiprotease do not inhibit these enzymes.

## Structural Chemistry

The S. sanguis and S. pneumoniae IgA proteinases are single-chain proteins of about 200 kDa, and contain no cysteine residues. There is some size heterogeneity of the extracellullar enzyme among S. pneumoniae strains (Poulsen et al., 1996), but no clearly established autocatalytic processing similar to that of the serine-type enzymes (Pohlner et al., 1987).

All cloned iga genes encoding IgA proteinases are in single copy on the bacterial chromosome, and enzyme production is not known to be regulated. Nucleotide and deduced amino acid sequences for the metallo-type Streptococcus sanguis (Gilbert et al., 1991) and S. pneumoniae enzymes (Poulsen et al., 1996; Wani et al., 1996) have been reported, and these are 64–70% identical. In sharp contrast, neither of these sequences has any homology with the iga genes encoding serine-type IgA proteinases (Pohlner et al., 1987; Poulsen et al., 1989), despite the functional equivalence of all the enzymes.

The S. sanguis and S. pneumoniae metallo-type IgA enzymes are most similar in the C-terminal two-thirds of the proteins, and each has an HEMTH metal-binding site motif in this region. Site-specific mutagenesis of this pentapeptide to Phe-Lys-Met-Thr-His in S. sanguis yielded a full-length, inactive protein (Gilbert et al., 1991), defining a critical role for enzyme metal in IgA-cleaving activity. Both S. sanguis and S. pneumoniae enzymes have a proline-rich, tandem-repeat region in the N-terminal third of the protein which differs in length and composition among strains, indicating considerable structural variability. Tandem repeats occur frequently among extracellular and surface proteins of streptococci, and may be important in pathogenesis and/or immune evasion (Hollingshead et al., 1987). The S. pneumoniae sequence also specifies a typical, hydrophobic streptococcal membrane anchoring motif (Navarre & Schneewind, 1994), but this stretch is near the N-terminus of the IgA proteinase, unlike its more characteristic C-terminal location in other streptococcal proteins. The pneumococcal enzyme is tightly cell associated during the mid-logarithmic phase of broth culture (Wani et al., 1996), in contrast to the rapid secretion noted for the serine type IgA proteinases of gram-negative bacteria (Chapter 87). The anchoring segment may contribute to prolonged retention of membrane activity.

Extensive studies on the antigenicity of IgA proteinases indicate that the streptococcal metalloproteinases exhibit much less antigenic diversity than do the serine-type enzymes (Lomholt & Kilian, 1994; Morelli et al., 1994).

No crystal structure of IgA proteinases has been reported.

## Preparation

Metallo-type IgA proteinases are obtained from spent, cell-free culture supernatants of S. pneumoniae or S. sanguis cells that have reached stationary phase in liquid media (Plaut & Wright, 1995). The recombinant enzymes expressed in E. coli accumulate in the periplasmic space (Gilbert et al., 1991).

## Biological Aspects

The independent evolution of several distinct classes of enzymes with a similar biological function points to the

6666

importance of inactivation of IgA1 for colonization by bacterial mucosal pathogens. Yet no specific role for any IgA proteinase in infection has been proven. Locally produced secretory IgA is the principal antibody in human milk, and in secretions that bathe mucosal surfaces of the oral cavity, upper respiratory tract, intestine, colon, genital tract and the conjunctiva (Tomasi, 1970). These are tissue sites where bacteria first encounter the human host, either to become harmless commensals, or to initiate clinical infection. *Streptoccocus pneumoniae*, one of the most important human pathogens, is highly adapted to human beings and causes many infections including otitis media, sinusitis, pneumonia and meningitis. The oral streptococci are implicated in a number of local infections such as periodontal disease and dental caries, and can cause invasive illness such as bacterial endocarditis. The biological significance of IgA proteinases has recently been reviewed (Kilian *et al.*, 1988, 1996).

Secretory IgA consists of antibodies synthesized as IgA dimers and higher polymers by B cells located in the lamina propria of the mucous membranes. These antibodies then reach the secretions by receptor-mediated endocytosis and transcytosis across epithelial cells (Mostov & Deitcher, 1986) and serve many immune functions including prevention of bacterial and viral attachment to host cells, mucosal penetration, virus neutralization, inactivation of toxins (Kilian *et al.*, 1996; Mazanec *et al.*, 1996), and elimination of viruses once they have entered epithelial cells (Mazanec *et al.*, 1996). IgA proteinase activity is, in principle, capable of interfering with most functions of IgA, because cleavage separates the antigen-binding Fab from the effector Fc domains of the molecule (Kilian *et al.*, 1988, 1996; Hajishengallis *et al.*, 1992). Not only can this facilitate bacterial evasion of immunity by establishing a localized zone of immunodeficiency around the microbe (Kilian *et al.*, 1996), but the secretory IgA cleavage products themselves may contribute to immune evasion. For example, the Fab hydrolytic products retain antigenic-binding capacity, and in monomeric form can bind to bacteria, thereby obstructing access of intact antibodies to these antigens (Kilian *et al.*, 1988; Hajishengallis *et al.*, 1992). A complicating factor in defining the role of IgA proteinases in immune evasion is that human secretions contain antibodies that inhibit the enzymes (Plaut *et al.*, 1992; Reinholdt & Kilian, 1995) and these antibodies are themselves secretory IgA. Thus secretory IgA is both antibody and substrate, illustrating the highly complex interactions among these large proteins in secretions bathing human mucous membranes.

It is also possible that IgA proteinases may be able to hydrolyze other host proteins (Pohlner *et al.*, 1995), or may have nonenzymatic properties that contribute to microbial tissue adhesiveness.

## Distinguishing Features

The *iga* genes encoding metallo-type IgA proteinases are not homologous with the serine-type (Chapter 87). However, all IgA proteinases, irrespective of mechanism, are functionally equivalent in cleaving after proline in the human IgA1 hinge.

Attempts to synthesize small, substrate-based inhibitors for the metallo-type IgA proteinases have been unsuccessful, while millimolar inhibitors have been made for the serine-type enzymes (Bachovchin *et al.*, 1990). A large series of proline-based derivatives of ketomethylene, phosphonate, phosphoramidon and hydroxyamine classes of metalloproteinase inhibitors have been ineffective, emphasizing that binding sites outside the hinge may be critical for this metallo-type proteinase activity. This is similar to the requirement for long polypeptide substrates seen in botulinum and tetanus toxins (Chapters 509 and 510), also metallo-type bacterial proteinases (Yamasaki & Baumeister, 1994).

## Further Reading

A comprehensive discussion of the infectious agents that produce IgA proteinases, and the possible role of the enzymes in the infectious process is presented by Kilian *et al.* (1996).

## References

Bachovchin, W.W., Plaut, A.G., Flentke, G.R., Lynch, M. & Kettner, C.A. (1990) Inhibition of IgA1 proteinases from *Neisseria gonorrhoeae* and *Haemophilus influenzae* by peptide prolyl boronic acids. *J. Biol. Chem.* **265**, 3738–3743.

Cole, M.F. & Hale, C.A. (1991) Cleavage of chimpanzee secretory immunoglobulin A by *Haemophilus influenzae* IgA1 protease. *Microb. Pathogen.* **11**, 39–46.

Cover, T.L. & Blaser, M.J. (1992) Purification and characterization of the vacuolating toxin from *Helicobacter pylori*. *J. Biol. Chem.* **267**, 10570–10575.

Frandsen, E.V.G., Reinholdt, J. & Kilian, M. (1987) Enzymatic and antigenic characterization of immunoglobulin A1 proteases from *Bacteroides* and *Capnocytophaga* spp. *Infect. Immun.* **55**, 631–638.

Fujiyama, Y., Iwaki, M. Hodohara, K., Hosoda, S. & Kobayashi, K. (1986) The site of cleavage in human alpha chains of IgA1 and IgA2:A2m(1) allotype paraproteins by the Clostridial IgA protease. *Mol. Immun.* **23**, 147–150.

Gilbert, J.V., Plaut, A.G. & Wright, A. (1991) Analysis of the immunoglobulin A protease gene of *Streptococcus sanguis*. *Infect. Immun.* **59**, 7–17.

Hajishengallis, G., Nikolova, E. & Russell, M.W. (1992) Inhibition of *Streptococcus mutans* adherence to saliva-coated hydroxyapatite by human secretory immunoglobulin A (S-IgA) antibodies to cell surface protein antigen I/II: reversal by IgA1 protease cleavage. *Infect. Immun.* **60**, 5057–5064.

Hollingshead, S.K., Fischetti, V.A. & Scott, J.R. (1987) Size variation in group A streptococcal M protein is generated by homologous recombination between intragenic repeats. *Mol. Gen. Genet.* **207**, 196–203.

Kawamura, S., Naruya, S. & Ueda, S. (1992) Concerted evolution of the primate immunoglobulin alpha-gene through gene conversion. *J. Biol. Chem.* **267**, 7359–7367.

Kett, K., Brandtzaeg, P., Radl, J. & Haajiman, J.J. (1986) Different subclass distribution of IgA-producing cells in human lymphoid organs and various secretory tissues. *J. Immunol.* **136**, 3631–3635.

Kilian, M. (1981) Degradation of immunoglobulin A1, A2 and G by suspected principal periodontal pathogens. *Infect. Immun.* **34**, 757–765.

Kilian, M. & Holmgren, K. (1981) Ecology and nature of IgA1 protease producing Streptococci in the human oral cavity and pharynx. *Infect. Immun.* **31**, 868–873.

Kilian, M., Mestecky, J. & Schrohenloher, R.E. (1979) Pathogenic species of the genus *Haemophilus* and *Streptococcus pneumoniae* produce immunoglobulin A1 protease. *Infect. Immun.* **26**, 143–149.

Kilian, M., Mestecky, J. Kulhavy, R., Tomana, M. & Butler, W.T. (1980) IgA1 proteases from *Haemophilus influenzae*, *Streptococcus pneumoniae*, *Neisseria meningitidis* and *Streptococcus sanguis*: comparative immunochemical studies. *J. Immunol.* **124**, 2596–2600.

Kilian, M., Mestecky, J. & Russell, M.W. (1988) Defense mechanisms involving Fc-dependent functions of immunoglobulins A and their subversion by bacterial immunoglobulin A proteases. *Microbiol. Rev.* **52**, 296–303.

Kilian, M, Mikkelsen, L. & Henrichsen, J. (1989) Taxonomic study of viridans streptococci: description of *Streptococcus gordonii* sp.nov. and emended descriptions of *Streptococcus sanguis* (White and Niven 1946), *Streptococcus oralis* (Bridge and Sneath 1982), and *Streptococcus mitis* (Andrewes and Horder 1906). *Int. J. Syst. Bacteriol.* **39**, 471–484.

Kilian, M., Reinholdt, J., Lomholt, H., Poulsen, K. & Frandsen, E.V.G. (1996) Biological significance of IgA1 proteases in bacterial colonization and pathogenesis: critical evaluation of experimental evidence. *APMIS* **104**, 321–338.

Lomholt, H. & Kilian, M. (1994) Antigenic relationships among immunoglobulin A1 proteases from *Haemophilus*, *Neisseria* and *Streptococcus* species. *Infect. Immun.* **62**, 3178–3183.

Male, C. (1979) Immunoglobulin A1 protease production by *Haemophilus influenzae* and *Streptococcus pneumoniae*. *Infect. Immun.* **26**, 254–261.

Mazanec, M.B., Huang, Y.T., Pimplikar, S.W. & Lamm, M.E. (1996) Mechanisms of inactivation of respiratory viruses by IgA, including intraepithelial neutralization. *Semin. Virol.* **7**, 285–292.

Mehta, S.K., Plaut A.G., Calvanico, N.S. & Tomasi, T.B., Jr. (1973) Human immunoglobulin A: production of an Fc fragment by an enteric microbial proteolytic enzyme. *J. Immunol.* **111**, 1274–1276.

Mestecky, J. & Russell, M.W. (1986) IgA subclasses. *Monogr. Allergy* **19**, 277–301.

Morelli, G., del Valle, J., Lammel, C.J. *et al.* (1994) Immunogenicity and evolutionary variability of epitopes within IgA1 protease from serogroup A *Neisseria meningitidis*. *Mol. Microbiol.* **11**, 175–187.

Mortensen, S.B. & Kilian, M. (1984) Purification and characterization of an immunoglobulin A1 protease from *Bacteroides melaninogenicus*. *Infect. Immun.* **45**, 550–557.

Mostov, K.E. & Deitcher, D.L. (1986) Polymeric immunoglobulin receptor expressed in MDCK cells transcytoses IgA. *Cell* **46**, 613–621.

Mulks, M.H. & Shober, R.J. (1994) Bacterial immunoglobulin A1 proteases. *Methods Enzymol.* **235**, 543–554.

Navarre, W.W. & Schneewind, O. (1994) Proteolytic cleavage and cell wall anchoring at the LPXTG motif of surface proteins in gram-positive bacteria. *Mol. Microbiol.* **14**, 115–121.

Neutra, M.R., Michetti, P. & Kraehenbuhl, J.-P. (1994) Secretory immunoglobulin A: structure, synthesis, and function. In: *Physiology of the Gastrointestinal Tract*, 3rd edn (Johnson, L.R., ed.). New York: Raven, pp. 975–1009.

Plaut, A.G. & Bachovchin, W.W. (1994) IgA-specific prolyl endopeptidases: serine type. *Methods Enzymol.* **244**, 137–151.

Plaut, A.G. & Wright, A. (1995) IgA-specific prolyl endopeptidases (metallo-type). *Methods Enzymol.* **248**, 634–642.

Plaut, A.G., Genco, R.J. & Tomasi, T.B., Jr. (1974) Isolation of an enzyme from *Streptococcus sanguis* which specifically cleaves IgA. *J. Immunol.* **113**, 289–291.

Plaut, A.G., Gilbert, J.V. Artenstein, M.S. & Capra, J.D. (1975) *Neisseria gonorrhoeae* and *N. meningitidis*: production of an extracellular enzyme which cleaves human IgA. *Science* **190**, 1103–1105.

Plaut, A.G., Gilbert, J.V. & Heller, I. (1978) Assay and properties of IgA protease of *Streptococcus sanguis*. *Adv. Exp. Med. Biol.* **107**, 489–495.

Plaut, A.G., Qiu, J., Grundy, F. & Wright, A. (1992) Growth of *Haemophilus influenzae* in human milk: synthesis, distribution, and activity of IgA protease as determined by study of *iga*+ and mutant *iga*− cells. *J. Infect. Dis.* **166**, 43–52.

Pohlner, J., Halter, R., Beyreuther, K. & Meyer, T.F. (1987) Gene structure and extracellular secretion of *Neisseria gonorrhoeae* IgA protease. *Nature* **325**, 458–462.

Pohlner, J., Langenberg, U., Wölk, U., Beck, S.C. & Meyer, T.F. (1995) Uptake and nuclear transport of Neisseria IgA1 protease-associated alpha proteins in human cells. *Mol. Microbiol.* **17**, 1073–1083.

Poulsen, K., Brandt, J., Hjorth, J.P., Thøgersen, H.C. & Kilian, M. (1989) Cloning and sequencing of the immunoglobulin A1 protease gene (*iga*) of *Haemophilus influenzae* serotype b. *Infect. Immun.* **57**, 3097–3105.

Poulsen, K., Reinholdt, J. & Kilian, M. (1996) Characterization of the *Streptococcus pneumoniae* immunoglobulin A1 protease gene (*iga*) and its translation product. *Infect. Immun.* **64**, 3957–3966.

Provence, D.L. & Curtiss, R. (1994) Isolation and characterization of a gene involved in hemagglutination by an avian pathogenic *Escherichia coli* strain. *Infect. Immun.* **62**, 1369–1380.

Qiu, J., Brackee, G.P. & Plaut, A.G. (1996) Analysis of the specificity of bacterial IgA proteases by a comparative study of ape serum IgA as substrates. *Infect. Immun.* **64**, 933–937.

Reinholdt, J. & Kilian, M. (1995) Titration of inhibiting antibodies to bacterial IgA1 proteases in human sera and secretions. In: *Advances in Mucosal Immunology*, part A (Mestecky, J., Russell, M.W., Jackson, S., Michalek, S.M., Tlaskalova, H. & Sterzl, J., eds). New York: Plenum Press, pp. 605–608.

St. Geme, J.W., de la Morena, M.L. & Falkow, S. (1994) A *Haemophilus influenzae* IgA protease-like protein promotes intimate interaction with human epithelial cells. *Mol. Microbiol.* **14**, 217–233.

Spooner, R.K., Russell, W.C. & Thirkell, D. (1992) Characterization of the immunoglobulin A protease of *Ureaplasma urealyticum*. *Infect. Immun.* **60**, 2544–2546.

Tomasi, T.B. (1970) Structure and function of human mucosal antibodies. *Annu. Rev. Med.* **21**, 281–298.

Tsuzukida, Y., Wang, C. & Putnam, F.W. (1979) Structure of the A2m(1) allotype of human IgA – a recombinant molecule. *Proc. Natl Acad. Sci. USA* **76**, 1104–1108.

Underdown, B.J. & Schoff, J.M. (1993) Immunoglobulin A: strategic defense initiative at the mucosal surface. *Annu. Rev. Immunol.* **4**, 389–417.

Wani, J.H., Gilbert, J.V., Plaut, A.G. & Weiser, J.N. (1996) Identification, cloning, and sequencing of the immunoglobulin A1 protease gene of *Streptococcus pneumoniae. Infect. Immun.* **64**, 3967–3974.

Yamasaki, S. & Baumeister, A. (1994) Cleavage of members of the synaptobrevin/VAMP family by types D and F botulinal neurotoxins and tetanus toxin. *J. Biol. Chem.* **269**, 12764–12772.

*Andrew G. Plaut*
*New England Medical Center Hospital and Tufts University School of Medicine,*
*Box 006, 750 Washington Street,*
*Boston, MA 02111, USA*
*Email: aplaut_mib@opal.tufts.edu*

# 509. *Tentoxilysin*

## Databanks

*Peptidase classification: clan MX, family M27, MEROPS ID: M27.001*
*NC-IUBMB enzyme classification: EC 3.4.24.68*
*Databank codes:*

| Species | SwissProt | PIR | EMBL (cDNA) | EMBL (genomic) |
|---|---|---|---|---|
| *Clostridium tetani* | P04958 | A25689 | M12739 X04436 X06214 | – |

## Name and History

Tetanus was first described by Hippocrates of Kos in the fourth century BC. The disease is characterized by an often fatal spastic paralysis with contraction of skeletal muscles that work one against the other (van Heyningen, 1968). Tetanus was thought to be a neurologic syndrome until 1884, when it was demonstrated to be caused by a novel bacterium, *Clostridium tetani* (Carle & Rattone, 1884; Kitasato, 1889). Soon after, it was shown that a single protein produced by this bacterium was solely responsible for all the symptoms of tetanus and it was thereafter named tetanus toxin (Faber, 1890; Tizzoni & Cattani, 1890). Formaldehyde treatment of partially purified preparations of this toxin generates a tetanus toxoid that is a very effective antitetanus vaccine (Middlebrook & Brown, 1995). Together with the diphtheria toxoid, this is a major product of the biotechnology industry and it is currently produced in more than one hundred million doses per year. Tetanus toxin it is not the only protein toxin produced by *C. tetani* and hence, to avoid confusion, it is preferable to use the term ***tetanus neurotoxin***, a name which also refers to its absolute specificity for nerve cells. The sequence of the gene encoding tetanus neurotoxin was published in 1986 (Eisel *et al.*, 1986; Fairweather & Lyness, 1986) and this paved the way to the discovery of its metallopeptidase activity specific for a vesicle-bound cytosolic protein. Jongeneel *et al.* (1989) detected in tetanus toxin an extended motif around an HEXXH sequence that is characteristic of many zinc metallopeptidases, and activity was detected soon after (Schiavo *et al.*, 1992a,b). IUBMB has recommended the name ***tentoxilysin*** to emphasize this proteolytic activity.

## Activity and Specificity

The only known proteolytic substrate of tentoxilysin is a 120-residue protein anchored to the membrane of cell vesicles. This protein has been termed VAMP (vesicle-associated membrane protein) (Trimble *et al.*, 1988), but also synaptobrevin (Baumert *et al.*, 1989) and cellubrevin (McMahon *et al.*, 1993). Several isoforms of this protein have been identified, but only VAMP-1, VAMP-2 and cellubrevin are sensitive to tetanus neurotoxin. They are cleaved at a single peptide bond: Gln76┼Phe77 (numbering of rat VAMP-2), which occurs within a 65 residue region that is highly conserved among isoforms and species. Tentoxilysin does not cleave short peptides encompassing the cleavage site of VAMP: the minimal length of VAMP fragment optimally cleaved by tentoxilysin is VAMP33-94 (Cornille *et al.*, 1994; Foran *et al.*, 1994; Yamasaki *et al.*, 1994a). Such a requirement results from the neurotoxin's recognition of its substrate via a nine-residue-long motif N-terminal to the cleavage site (Rossetto *et al.*, 1994; Pellizzari *et al.*, 1996;

Witcome *et al.*, 1996) and a cluster of positively charged residues (VAMP83–87) (Cornille *et al.*, 1994; Foran *et al.*, 1994; Yamasaki *et al.*, 1994a), in addition to the active-site binding segment to be cleaved.

VAMP proteolysis *in vitro* is assayed by separation of the two VAMP fragments by SDS-PAGE or HPLC followed by their quantitative determination with various methods (Cornille *et al.*, 1994; Foran *et al.*, 1994). The proteolytic activity of tentoxilysin on intact cells, or on cell fragments such as synaptosomes, is assayed by high-sensitivity immunoblotting or by following the disappearance of the VAMP immunofluorescence signal (Osen-Sand *et al.*, 1996; Matteoli *et al.*, 1996).

Tentoxilysin has to be reduced with thiol compounds (e.g. 10 mM DTT for 30 min at 37°C) to display its proteolytic activity. Such reduction is not necessary when the catalytic domain (L chain) alone is used (see sections on Structural Chemistry and Preparation; Weller *et al.*, 1989; De Filippis *et al.*, 1995). The pH optimum of the proteolytic activity is 7.0. 1,10-Phenanthroline and dipicolinic acid are rapid and effective inhibitors of tentoxilysin, whereas EDTA is active only upon prolonged exposure at 37°C. Specific inhibitors of tentoxilysin are not available.

## Structural Chemistry

Tetanus neurotoxin is produced as a single polypeptide chain of 150 kDa (1315 amino acid residues), which folds into three domains. Selective proteolysis of an exposed loop generates within the bacterial culture two polypeptide chains held together by a single interchain disulfide bond and by non-covalent interactions. The heavy chain (H, 100 kDa) consists of two domains: $H_C$ is responsible for the specific binding to synaptic terminals and $H_N$ is involved in the membrane translocation of the light chain in the cytosol. The light chain of tetanus neurotoxin (L, 50 kDa, 447 amino acid residues) is released from the H chain upon reduction and it is responsible for the metallopeptidase activity of the toxin. It binds one atom of zinc via the characteristic HEXXH motif (Schiavo *et al.*, 1992a; Wright *et al.*, 1992; Yamasaki *et al.*, 1994b) and possibly via Tyr242 (Morante *et al.*, 1996), a residue within a 20-residue segment that is highly conserved among clostridial neurotoxins. The zinc atom can be replaced with various metal ions and the Mn(II) and Co(II) derivatives are highly active (Tonello *et al.*, 1997).

## Preparation

Tentoxilysin is isolated from culture supernatants of *Clostridium tetani* by ammonium sulfate precipitation followed by chromatography (Matsuda & Yoneda, 1975; Schiavo & Montecucco, 1995). In old bacterial cultures, some free L chain may be present as a consequence of the digestion of the H chain by bacterial proteases, and it can be purified to homogeneity via HPLC (Ozutsumi *et al.*, 1985; De Filippis *et al.*, 1995). Alternatively, the L chain can be separated from the H chain by isoelectrofocusing under reducing conditions in the presence of 2 M urea (Weller *et al.*, 1989). The metallopeptidase L chain of tetanus neurotoxin has been produced by expression in *E. coli* as a (His)$_6$-tagged protein (Yamasaki *et al.*, 1994b) or as a glutathione-*S*-methyltransferase fusion

protein (F. Tonello *et al.*, unpublished results). Tetanus neurotoxin and its L chain can be purchased from Alomone Laboratories, Calbiochem, List Biologicals and Sigma (see Appendix 2 for full names and addresses of suppliers). Purity should be checked by 13% SDS-PAGE electrophoresis.

## Biological Aspects

Tetanus is acquired by contamination of wounds with spores of toxigenic strains of *Clostridium tetani*, which are almost ubiquitous, but highly enriched in feces from many animals including farmyard animals. Under anaerobic conditions, spores germinate, and tetanus neurotoxin is produced and released upon autolysis. The toxin spreads in the body and binds, via the $H_C$ domain, to an as yet unidentified receptor(s) of the presynaptic terminal of the neuromuscular junction (Halpern & Neale, 1995). After endocytosis, the toxin migrates retroaxonally inside the motor neuron and reaches the spinal cord. The toxin is then released in the intersynaptic space between the motorneuron and the inhibitory interneuron (Renshaw cell). It then enters small synaptic vesicles when they expose their lumen to the outside following fusion with the presynaptic membrane and release of neurotransmitter (Matteoli *et al.*, 1996). Synaptic vesicles are then endocytosed and their lumen is acidified by a vacuolar ATPase proton pump to drive the reuptake of neurotransmitter. Such luminal acidification is essential for intoxication to occur (Williamson & Neale, 1994; Matteoli *et al.*, 1996) because tetanus neurotoxin at low pH changes conformation and becomes hydrophobic (Boquet & Duflot, 1982; Montecucco, 1986; Schiavo *et al.*, 1991). In such a form, the toxin inserts in the lipid bilayer of the vesicle membrane and somehow the $H_N$ domain manages to translocate the L chain into the cytosol (Hoch *et al.*, 1985; Schmid *et al.*, 1993; Montecucco & Schiavo, 1995), with the consequent cleavage of VAMP. Such specific proteolysis is sufficient to cause a prolonged blockade of neuroexocytosis, and this demonstrates the essential role of VAMP in neuroexocytosis.

VAMP is involved in several other events of vesicle fusion with a target membrane which take place in many non-neuronal cells as well (Rothman & Warren, 1994). The toxin cannot access the cytosol of these cells because of lack of receptors, but if the L chain is injected or the cells are permeabilized, or the gene encoding the L chain is inserted in a cell or in a transgenic animal (Anhert-Hilger *et al.*, 1989; Höhne-Zell *et al.*, 1993; Penner *et al.*, 1986; Sweeney *et al.*, 1995), VAMP will be cleaved and the event in which it is involved will consequently be blocked. Thus, tentoxilysin is increasingly used in cell biology to study the involvement of VAMP in cellular phenomena. Because of its retroaxonal transport and trans-synaptic migration, tentoxilysin and its $H_C$ domain are used to map neuronal routes from the peripheral nervous system to the central nervous system by coupling it to horseradish peroxidase or gold particles (Cabot *et al.*, 1991).

Toxicity is tested by intraperitoneal injection of different amounts of toxin. Mouse LD$_{50}$ is 0.2–0.5 ng kg$^{-1}$ and that for humans is estimated to be even lower (Gill, 1982). Tetanus neurotoxin toxicity varies in different animal species and this is due to different binding at the neuromuscular junction and/or to mutations at the site of action of tentoxilysin inside cells. An example of this latter possibility is illustrated by chickens and rats, which are tetanus-resistant species. Their VAMP-1, which is the predominant isoform in the spinal cord, shows the replacement of Gln with a Val in the P1 position and this variant is very slowly hydrolyzed by the toxin (Schiavo *et al.*, 1992b; Patarnello *et al.*, 1993).

The use of this toxin does not present problems to workers vaccinated with tetanus toxoid vaccine. If more than 10 years have elapsed from the last vaccine injection, it is advisable to have a single booster injection. The toxin is very sensitive to oxidants and simple washings with diluted hypochlorite solutions are sufficient to eliminate residual toxin after experiments.

## Distinguishing Features

Tetanus neurotoxin, as well as the other clostridial neurotoxins (Chapter 510), is characterized by a three-domain structure which reflects the different steps of its mechanism of cell intoxication. The metallopeptidase activity is confined in the L chain, whose sequence has no similarity to other metallopeptidases, apart from the presence of the zinc-binding motif. The other major peculiarity of tentoxilysin resides in its specificity for VAMP, which is recognized via multiple interactions with non-sequential VAMP regions.

Polyclonal and monoclonal antibodies, made in a variety of organisms and specific for the three domains of the toxin molecule, have been produced (Middlebrook & Brown, 1995). Some of them can be obtained from national health authorities. Antitetanus neurotoxin-specific human polyclonal antisera are used as passive antitetanus immunotherapy after wounding.

## Further Reading

Montecucco (1995) provides a collection of reviews that cover all aspects of the biology of tetanus and botulinum neurotoxins.

## References

Anhert-Hilger, G., Weller, U., Dauzenroth, M.E., Habermann, E. & Gratzl, M. (1989) The tetanus toxin light chain inhibits exocytosis. *FEBS Lett.* **242**, 245–248.

Baumert, M., Maycox, P.R., Navone, F., De Camilli, P. & Jahn, R. (1989) Synaptobrevin: an integral membrane protein of 18,000 daltons present in small synaptic vesicles of rat brain. *EMBO J.* **8**, 379–384.

Boquet, P. & Duflot, E. (1982) Tetanus toxin fragment forms channels in lipid vesicles at low pH. *Proc. Natl Acad. Sci. USA* **79**, 7614–7618.

Cabot, J.B., Mennone, A., Bogan, N., Carroll, J., Evinger, C. & Erichsen, J.T. (1991) Retrograde, trans-synaptic and transneuronal transport of fragment C of tetanus toxin by sympathetic preganglionic neurons. *Neuroscience* **40**, 805–823

Carle, A. & Rattone, G. (1884) Studio esperimentale sull'eziologia del tetano [Experimental studies of the etiology of tetanus]. *Giorn. Accad. Med. Torino* **32**, 174–179.

Cornille, F., Goudreau, N., Ficheux, D., Niemann, H. & Roques, B.P. (1994) Solid-phase synthesis, conformational analysis and *in vitro* cleavage of synthetic human synaptobrevin II 1–93 by tetanus toxin L chain. *Eur. J. Biochem.* **222**, 173–181.

De Filippis, V., Vangelista, L., Schiavo, G., Tonello, F. & Montecucco, C. (1995) Structural studies on the zinc-endopeptidase light chain tetanus neurotoxin. *Eur. J. Biochem.* **229**, 61–69.

Eisel, U., Jarausch, W., Goretzki, K., Henschen, A., Engels, J., Weller, U., Hudel, M., Habermann, E. & Niemann, H. (1986) Tetanus toxin: primary structure, expression in *E. coli* and homology with botulinum toxins. *EMBO J.* **5**, 2495–2502.

Faber, K. (1890) Die Pathogenie des Tetanus [The pathogenesis of tetanus]. *Berl. klin. Wochenschr.* **27**, 717–720.

Fairweather, N.F. & Lyness, V.A. (1986) The complete nucleotide sequence of tetanus toxin. *Nucleic Acids Res.* **14**, 7809–7812.

Foran, P., Shone, C.C. & Dolly, J.O. (1994) Differences in the protease activities of tetanus and botulinum B toxins revealed by the cleavage of vesicle-associated membrane protein and various sized fragments. *Biochemistry* **31**, 15365–15374.

Gill, D.M. (1982) Bacterial toxins: a table of lethal amounts. *Microbiol. Rev.* **46**, 86–94.

Halpern, J.L. & Neale, E.A. (1995) Neurospecific binding, internalization, and retrograde axonal transport. *Curr. Top. Microbiol. Immunol.* **195**, 221–241.

Hoch, D.H., Romero-Mira, M., Ehrlich, B.E., Finkelstein, A., DasGupta, B.R. & Simpson, L.L. (1985) Channels formed by botulinum, tetanus and diphteria toxin in planar liquid bilayers: relevance to translocation of protein across membranes. *Proc. Natl Acad. Sci. USA* **82**, 1692–1696.

Höhne-Zell, B., Stecher, B. & Gratzl, M. (1993) Functional characterization of the catalytic site of the tetanus toxin light chain using permeabilized adrenal chromaffin cells. *FEBS Lett.* **336**, 175–180.

Jongeneel, C.V., Bouvier, J. & Bairoch, A. (1989) A unique signature identifies a family of zinc-dependent metallopeptidases. *FEBS Letters* **242**, 211–214

Kitasato, S. (1889) Ueber den Tetanus bacillus [On the tetanus bacillus]. *Z. Hyg. Infekt. Kr.* **7**, 225–233.

Matsuda, M. & Yoneda, M. (1975) Isolation and purification of two antigenically active, complementary polypeptide fragments of tetanus neurotoxin. *Infect. Immun.* **12**, 1147–1153.

Matteoli, M., Verderio, C., Rossetto, O., Iezzi, N., Coco, S., Schiavo, G. & Montecucco, C. (1996) Synaptic vesicle endocytosis mediates the entry of tetanus neurotoxin into hippocampal neurons. *Proc. Natl Acad. Sci. USA* **93**, 13310–13315.

McMahon, H.T., Ushkaryov, Y.A., Edelmann, L., Link, E., Binz, T., Niemann, H., Jahn, R. & Südhof, T.C. (1993) Cellubrevin is a ubiquitous tetanus-toxin substrate homologous to a putative synaptic vesicle fusion protein. *Nature* **364**, 346–349.

Middlebrook, J.L. & Brown, J.E. (1995) Immunodiagnosis and immunotherapy of tetanus and botulinum neurotoxins. *Curr. Top. Microbiol. Immunol.* **195**, 89–122.

Montecucco, C. (1986) How do tetanus and botulinum neurotoxins bind to neuronal membranes? *Trends Biochem. Sci.* **II**, 314–317.

Montecucco, C. (ed.) (1995) *Clostridial Neurotoxins. Current Topics in Microbiology and Immunology*, vol. 195. Heidelberg: Springer-Verlag.

Montecucco, C. & Schiavo, G. (1995) Structure and function of tetanus and botulinum neurotoxins. *Q. Rev. Biophys.* **28**, 423–472.

Morante, S., Furenlid, L., Schiavo, G., Tonello, F., Zwilling, R. & Montecucco C. (1996) X-ray absorpion spectroscopy study of zinc coordination in tetanus neurotoxin, astacin, alkaline protease and thermolysin. *Eur. J. Biochem.* **235**, 606–612.

Ozutsumi, K., Sugimoto, N. & Matsuda, M. (1985) Rapid, simplified method for production and purification of tetanus toxin. *Appl. Environ. Microbiol.* **49**, 939–943.

**M**

Osen-Sand, A., Staple, J.K., Naldi, E., Schiavo G., Rossetto, O., Petitpierre, S., Malgaroli, A., Montecucco, C. & Catsicas, S. (1996) Common and distinct fusion proteins in axonal growth and transmitter release. *J. Comp. Neurol.* **367**, 222–234.

Patarnello, T., Bargelloni, L., Rossetto, O., Schiavo, G. & Montecucco, C. (1993) Neurotransmission and secretion. *Nature* **364**, 581–582.

Pellizzari, R., Rossetto, O., Lozzi, L., Giovedì, S., Johonson, E., Shone C.C. & Montecucco, C. (1996) Structural determinants of the specificity for VAMP/synaptobrevin of tetanus and botulinum B and G neurotoxins. *J. Biol. Chem.* **271**, 20353–20358.

Penner, R., Neher, E. & Dreyer, F. (1986) Intracellularly injected tetanus toxin inhibits exocytosis in bovine adrenal chromaffin cells. *Nature* **324**, 76–77.

Rothman, J.E. & Warren, G. (1994) Implications of the SNARE hypothesis for intracellular membrane topology and dynamics. *Curr. Biol.* **4**, 220–233.

Rossetto, O., Schiavo, G. Montecucco, C., Poulain, B., Deloye, F., Lozzi, L. & Shone, C.C. (1994) SNARE motif and neurotoxin recognition. *Nature* **372**, 415–416.

Schiavo, G. & Montecucco, C. (1995) Tetanus and botulism neurotoxins: isolation and assay. *Methods Enzymol.* **248**, 643–652.

Schiavo, G., Demel, R. & Montecucco, C. (1991) On the role of polysialoglycosphingolipids as tetanus toxin receptors: a study with lipid monolayers. *Eur. J. Biochem.* **199**, 705–711.

Schiavo, G., Poulain, B., Rossetto, O., Benfenati, F., Tauc, L. & Montecucco, C. (1992a) Tetanus toxin is a zinc protein and its inhibition of neurotransmitter release and protease activity depend on zinc. *EMBO J.* **11**, 3577–3583.

Schiavo, G., Benfenati, F., Poulain, B., Rossetto, O., Polverino de Laureto, P., DasGupta, B.R. & Montecucco, C. (1992b) Tetanus and botulinum-B neurotoxins block neurotransmitter release by a proteolytic cleavage of synaptobrevin. *Nature* **359**, 832–835.

Schmid, M.F., Robinson, J.P. & DasGupta, B.R. (1993) Direct visualization of botulinum neurotoxin-induced channels in phospholipid vesicles. *Nature* **364**, 827–830.

Sweeney, S.T., Broadie, K., Keane, J., Niemann, H. & O'Kane, C.J. (1995) Targeted expression of tetanus toxin light chain in *Drosophila* specifically eliminates synaptic transmission and causes behavioral defects. *Neuron* **14**, 341–351.

Tizzoni, G. & Cattani, G. (1890) Uber das Tetanusgift [On tetanus toxin]. *Zentralbl. Bakt.* **8**, 69–73.

Tonello, F., Schiavo, G. & Montecucco, C. (1997) Metal substitution of tetanus neurotoxin. *Biochem. J.* **322**, 507–510.

Trimble, W.S., Cowan, D.M. & Scheller, R.H. (1988) VAMP-1: a synaptic vesicle-associated integral membrane protein. *Proc. Natl Acad. Sci. USA* **85**, 4538–4542.

van Heyningen, W.E. (1968) Tetanus. *Sci. Am.* **218**(4), 69–77.

Weller, U., Dauzenroth, M.-E., Meyer Heringdorf, D. & Habermann, E. (1989) Chains and fragments of tetanus toxin. Separation, reassociation and pharmacological properties. *Eur. J. Biochem.* **182**, 649–656.

Williamson, L.C. & Neale, E.A. (1994) Bafilomycin A1 inhibits the action of tetanus toxin in spinal cord neurons in cell culture. *J. Neurochem.* **63**, 2342–2345.

Witcome, M., Rossetto, O., Montecucco, C. & Shone, C.C. (1996) Substrate residues N-terminal to the cleavage site of botulinum type B neurotoxin play a role in determining the specificity of its endopeptidase activity. *FEBS Lett.* **386**, 133–136.

Wright, J.F., Pernollet, M., Reboul, A., Aude, C. & Colomb, M. (1992) Identification and partial characterization of a low affinity metal-binding site in the light chain of tetanus toxin. *J. Biol. Chem.* **267**, 9053–9058.

Yamasaki, S., Hu, Y., Binz, T., Kalkuhl, A., Kurazono, H., Tamura, T., Jahn, R., Kandel, E. & Niemann, H. (1994a) Synaptobrevin/VAMP of *Aplysia californica*: structure and proteolysis by tetanus and botulinal neurotoxins type D and F. *Proc. Natl Acad. Sci. USA* **91**, 4688–4692.

Yamasaki, S., Baumeister, A., Binz, T., Blasi, J., Link, E., Cornille, F., Roques, B., Fykse, E.M., Südhof, T.C., Jahn, R. & Niemann, H. (1994b) Cleavage of members of the synaptobrevin/VAMP family by types D and F botulinal neurotoxins and tetanus toxin. *J. Biol. Chem.* **269**, 12764–12772.

***Cesare Montecucco***
*Centro CNR Biomembrane,*
*Dipartimento di Scienze Biomediche,*
*Università di Padova,*
*Via Trieste 75,*
*35121 Padova, Italy*
*Email: toxin@civ.bio.unipd.it*

# 510. Bontoxilysin

## Databanks

*Peptidase classification: clan MX, family M27, MEROPS ID: M27.002*
*NC-IUBMB enzyme classification: EC 3.4.24.69*

*Databank codes:*

| Species | Type | Strain[a] | SwissProt | PIR | EMBL (cDNA) | EMBL (genomic) |
|---|---|---|---|---|---|---|
| *Clostridium barati* | F | | – | S33411 | X68262 | – |
| *Clostridium botulinum* | A | 62A | P10845 | – | M30196 | |
| | | NCTC2916 | | | X52066 | |
| | | 667Ab | | | X87848 | |
| | | Kyoto f | – | – | X73423 | |
| *Clostridium botulinum* | B | Danish | P10844 | – | M81186 | – |
| | | NCTC7273 | | | Z11934 | |
| | | Eklund 17B | – | | X71343 | |
| *Clostridium botulinum* | C | 6813 | – | – | D49440 | – |
| | | 6813 phage | P18640 | S46431 | X53751 | – |
| | | 468 | | | X72793 | |
| *Clostridium botulinum* | D | phage d-16 | P19321 | – | S49407 | |
| | | BVD/-3 | | | X54254 | |
| *Clostridium botulinum* | E | Beluga | Q00496 | – | X62089 | – |
| | | NCTC11219 | | | X62683 | |
| *Clostridium botulinum* | F | 202F | P30996 | | M92096 | |
| | | – | – | – | L35496 | |
| | | NCTC10281 | – | | X81714 | |
| | | Langeland | – | | S76749 | |
| *Clostridium botulinum* | G | NCFB3012 | – | | X74162 | |
| *Clostridium butyricum* | E | ATCC43755 | P30995 | – | X62088 | |

[a] 'Strain' applies to the organism, not the enzyme, and is only relevant for the EMBL/GenBank entries: the protein sequence databases lump all the variants together under one entry.

## Name and History

Botulism is a neurologic syndrome of vertebrates, characterized by the loss of function of peripheral cholinergic synapses. This functional loss is variable and may be so minimal as to go unnoticed until a generalized flaccid paralysis develops. Such a manifestation accounts for the fact that botulism was first described only at the beginning of the nineteenth century and infant botulism was identified only 20 years ago (Hatheway, 1995). Botulism was shown to be caused by a bacterial toxin by van Ermengen (1897) and the toxin was thereafter named botulinum toxin. Presently, the generally used name is **botulinum neurotoxin** to distinguish it from other toxins released by *Clostridium botulinum*. However, the name recommended by IUBMB is **bontoxilysin**. Seven different bontoxilysins have been identified to date and they are labeled with capital letters from A to G. The sequences of the genes encoding for the seven bontoxilysins and their variants have been recently determined (Minton, 1995). Following the identification of the enzymic activity of tentoxilysin (Chapter 509), they were soon also shown to be metallopeptidases able to enter into the cytosol of neurons. There, they cleave one or two out of three proteins which are essential components of the neuroexocytosis apparatus (Ferro-Novick & Jahn, 1994; Montecucco & Schiavo, 1995).

## Activity and Specificity

Similarly to tentoxilysin, bontoxilysins B, D, F and G cleave specifically VAMP (vesicle-associated membrane protein, also termed synaptobrevin) at different single peptide bonds, as shown in Table 510.1 (Schiavo *et al.*, 1992a, 1993a,b, 1994; McMahon *et al.*, 1993; Binz *et al.*, 1994; Yamasaki *et al.*, 1994a,b). The three VAMP isoforms, VAMP-1, VAMP-2 and cellubrevin, of human and mice are cleaved. VAMPs of other species may carry mutations at the cleavage site or at recognition site(s) which render them neurotoxin insensitive, as is the case for rat and chicken VAMP-1, which are not cleaved by bontoxilysin B (Schiavo *et al.*, 1992a; Patarnello *et al.*, 1993). The assay of the activity of type B bontoxilysin with VAMP fragments has been described in detail (Shone *et al.*, 1993; Shone & Roberts, 1994; Witcome *et al.*, 1996).

At variance with this, bontoxilysins A, C and E cleave SNAP-25 (synaptosomal-associated protein of 25 kDa) at three different sites close to the C-terminus of the protein (see Table 510.1; Blasi *et al.*, 1993a; Schiavo *et al.*, 1993b,c; Binz *et al.*, 1994; Foran *et al.*, 1994; Osen-Sand *et al.*, 1996; Williamson *et al.*, 1996). Assay of the proteolytic activity of types A and E can be performed also with the SNAP-25 [136-206] fragment (Binz *et al.*, 1994), which includes the common recognition motif of clostridial neurotoxins, located in mouse SNAP-25 in the segment 145–153 (Washbourne *et al.*, 1997). Bontoxilysin C does not cleave SNAP-25 *in vitro*, most likely because the four cysteines of recombinant SNAP-25 are not palmitoylated and hence the recombinant substrate cannot bind to membranes.

Bontoxilysin C also cleaves syntaxin, a 30 kDa protein anchored on the cytosolic surface of the plasma membrane and several intracellular membranes (Blasi *et al.*,

M

*Table 510.1*  Target and peptide bond specificities of bontoxilysins

| Bontoxilysin type | Target | Peptide bond cleaved[a] | | | | | | |
|---|---|---|---|---|---|---|---|---|
| | | P3 | P2 | P1 | + | P1' | P2' | P3' |
| A | SNAP-25 | Ala | Asn | Gln | + | Arg | Ala | Thr |
| B | VAMP | Ala | Ser | Gln | + | Phe | Glu | Thr |
| C | syntaxin | Thr | Lys | Lys | + | Ala | Val | Lys |
| C | SNAP-25 | Asn | Gln | Arg | + | Ala | Thr | Lys[b] |
| D | VAMP | Asp | Gln | Lys | + | Leu | Ser | Glu |
| E | SNAP-25 | Ile | Asp | Arg | + | Ile | Met | Gly |
| F | VAMP | Arg | Asp | Gln | + | Lys | Leu | Ser |
| G | VAMP | Thr | Ser | Ala | + | Ala | Lys | Leu |

[a]The cleavages shown are those in VAMP-2, SNAP-25a and syntaxin IA of the rat.
[b]Personal communication of Professor H. Niemann.

1993b; Schiavo *et al.*, 1995). Bontoxilysin C cleaves syntaxin inserted into liposomes, but is totally inactive on the recombinant molecules in solution. Eight isoforms of syntaxin have so far been identified in rats (Hata & Südhof, 1995), and bontoxilysin C cleaves isoforms 1A, 1B, 2 and 3 (Blasi *et al.*, 1993b; Schiavo *et al.*, 1995). The enzymic activity of the bontoxilysins is determined *in vitro* by separation of the two proteolytic fragments by SDS-PAGE or HPLC followed by quantitative determination with various methods (Blasi *et al.*, 1993a; Schiavo *et al.*, 1993a,b; Shone *et al.*, 1993). In intact cells or cell fragments, the activity of bontoxilysins B, D, F and G is assayed as described for tentoxilysins (Chapter 509). Nerve cell intoxication by bontoxilysins A, C and E is assayed by high-sensitivity immunoblotting or by following the disappearance of the SNAP-25 signal with an antibody raised against its C-terminal 12-residue-long segment (Galli *et al.*, 1994; Osen-Sand *et al.*, 1996; Williamson *et al.*, 1996). Type C-induced cleavage of syntaxin in cells is monitored with an antibody raised against the recombinant syntaxin molecule (Foran *et al.*, 1996; Osen-Sand *et al.*, 1996; Williamson *et al.*, 1996; Igarashi *et al.*, 1996).

All seven botulinum neurotoxins have to be reduced with thiol compounds (e.g. 10 mM DTT for 30 min at 37°C) to free the proteolytic activity of their L chains. Reduction is not necessary when the catalytic domain (L chain) alone is used (see sections on Structural Chemistry and Preparation). The pH optimum of the proteolytic activity of bontoxilysin B is around 7.0 (Shone & Roberts, 1994). All the bontoxilysins are rapidly inactivated by 1,10-phenanthroline, whereas EDTA is variably effective with the different toxins. Apart from zinc chelators, no other inhibitors of proteolytic activity of bontoxilysins are known.

## Structural Chemistry

The overall structure of botulinum neurotoxins is closely similar to that of tetanus neurotoxin, being released as single 150 kDa polypeptide chains folded into three distinct domains. Selective proteolysis carried out with trypsin or other proteases cleaves an exposed loop between the first and the second domain and generates the active two-chain toxin with a single interchain disulfide bond. The heavy chain (H, 100 kDa) consists of two domains: $H_C$ is responsible for the specific binding to synaptic terminals and $H_N$ is involved in the membrane translocation of the light chain in the cytosol. The light chains (L, 50 kDa) of the seven botulinum neurotoxins range in length between the 422 residues of type E and the 445 residues of type D. The L chain is endowed with metallopeptidase activity and binds one atom of zinc via a conserved 20 residue long segment (DPhLXLh**HELXHXXH**XLY**G**h, where h indicates aliphatic and X any residue), which includes the HEXXH motif (Schiavo *et al.*, 1992b,c; Wright *et al.*, 1992). There is evidence that the four residues marked in bold are involved in binding the catalytic atom of zinc. Type C neurotoxin binds an additional atom of zinc via as yet unidentified residues (Schiavo *et al.*, 1995).

## Preparation

The preparation of the various bontoxilysins has recently been reviewed (Shone & Tranter, 1995; Schiavo & Montecucco, 1995). All bontoxilysins, with the exception of type G, can be purchased from Alomone Laboratories, Calbiochem, Sigma or Wako (see Appendix 2 for full names and addresses of suppliers). Purity should be checked by 13% SDS-PAGE. Traces of contaminant proteases can be easily removed by immobilized metal-ion affinity chromatography (Rossetto *et al.*, 1992), except for type D, which is not retained by this matrix (Schiavo *et al.*, 1993a). The metallopeptidase domain can be produced in *E. coli* as a (His)_6-tagged protein (Yamasaki *et al.*, 1994b) or as a glutathione-S-methyltransferase fusion protein (F. Tonello *et al.*, unpublished results).

Bontoxilysin A is largely used as a therapeutic agent for dystonias, strabismus and other syndromes, where a depression of neuromuscular junction activity is sought (Jankovic & Hallett, 1994). For such use, it is sold by Allerghan with the product name BOTOX and by Porton with the trade name of Dysport. Moreover, Associated Synapse has received permission to market bontoxilysin A as an orphan drug.

## Biological Aspects

Clostridial spores harboring the genes encoding botulinum neurotoxins are widespread in the environment and may germinate in anaerobic foods, and produce and release the neurotoxins. Eating enough of such poisoned food causes botulism, following transcytosis of botulinum neurotoxin across the intestinal epithelial layer and its spread through the body. Botulinum neurotoxins bind at the unmyelinated presynaptic membrane of motor neurons and of other cholinergic synapses. Botulinum neurotoxin type B binds to the N-terminal portion of synaptotagmin complexed with polysialo-gangliosides (Nishiki *et al.*, 1996). Receptors for the other types have not yet been identified, but different receptors must be implicated to account for the lack of competition among neurotoxin serotypes and for the different sensitivity of animal species. Binding is followed by endocytosis inside vesicles of unknown nature. However, it is clear that the lumen of such vesicles must acidify in order to induce a structural transition in the toxin molecule such that it is able to translocate the L chain into the cytosol, where it displays its metallopeptidase activity. These peptidases impair neurotransmission via the selective cleavage of either VAMP or SNAP-25 and/or syntaxin, three main components of the machinery that mediates docking and fusion of synaptic vesicles with the presynaptic membrane. This results in a flaccid paralysis as well as in alterations of the autonomic nervous system.

Toxicity is usually evaluated by intraperitoneal or intravenous injection of serially diluted neurotoxin solutions in mice. Mouse $LD_{50}$ of BoNT/A and E is in the range of $0.1$–$1\,\mathrm{ng\,kg^{-1}}$ and varies greatly in different animal species (Gill, 1982). These neurotoxins are estimated to be even more potent in humans: the human lethal dose may lie in the range 2000–4000 times the mouse lethal dose. Hence, they have to be handled with care. It may be sufficient to use them in a contained place, to protect hands with rubber gloves and to avoid the use of needles and glass pipettes. Wastes should be collected and treated with diluted hypochlorite solutions that quickly inactivate these toxins. Operators handling large toxin amounts should be vaccinated with the pentavalent toxoid vaccine prepared with toxin serotypes A, B, C, D, E, which can be obtained through the Botulism Laboratory of the Center for Disease Control. The California Department of Health Sciences at Berkeley has recently developed a human-derived botulinum antitoxin type A. A pentavalent horse antiserum is also available through national health services, but its use may give rise to serum-sickness side-effects (Hatheway, 1995; Middlebrook & Brown, 1995). Polyclonal and monoclonal antibodies made in a variety of organisms specific for the different botulinum neurotoxins are available and are used to determine the type of toxin involved in episodes of botulism (Middlebrook & Brown, 1995). Some antibodies are specific for selected domains.

Primary neurons in culture, neuronal-derived cell lines and brain cortex synaptosomes are able to internalize the two-chain neurotoxins A and E (Sanchez-Prieto *et al.*, 1987; Blasi *et al.*, 1993a,b; Schiavo *et al.*, 1993a; Osen-Sand *et al.*, 1996; Williamson *et al.*, 1996). Non-neuronal cells have to be injected or permeabilized to allow the access of the L chain to the cytosol (Penner *et al.*, 1986; Boyd *et al.*, 1995; Sadoul *et al.*, 1995; Foran *et al.*, 1996).

These neurotoxins are increasingly used in cell biology to probe the role of VAMP, SNAP-25 and syntaxin in cell processes including exocytosis, axonal growth, synapse stability and membrane repair (Boyd *et al.*, 1995; Sadoul *et al.*, 1995; Osen-Sand *et al.*, 1996; Igarashi *et al.*, 1996).

After the first report that BoNT/A injection is very effective in the treatment of strabismus (Scott, 1989), the therapeutic use of this neurotoxin has been extended to dystonias and other diseases that benefit from a functional paralysis of the neuromuscular junction (Jancovic & Hallett, 1994). Currently, other botulinum neurotoxin types are undergoing clinical tests.

## Distinguishing Features

The structural organization of botulinum neurotoxins in three domains is a characteristic feature and reflects the different steps of their mechanism of cell intoxication. The metallopeptidase activity is confined to the L chain, whose sequence has no similarity with any known metallopeptidase other than tentoxilysin, apart from the presence of the zinc-binding motif. The other major peculiarity of botulinum neurotoxins is their absolute specificity for VAMP or SNAP-25 or syntaxin. There is clear evidence that they recognize solely these three proteins because only these three proteins contain a nine-residue-long motif characterized by an appropriate spacing of three negative charges and three hydrophobic residues. Binding of the botulinum neurotoxins to their three substrates takes place via an interaction with two nonsequential regions of the substrate: the recognition motif and a segment that includes the peptide bond to be cleaved.

## Further Reading

Montecucco (1995) provides a collection of reviews that cover all aspects of the biology of tetanus and botulinum neurotoxins.

## References

Binz, T., Blasi, J., Yamasaki, S., Baumeister, A., Link, E., Südhof, T.C., Jahn, R. & Niemann, H. (1994) Proteolysis of SNAP-25 by types E and A botulinal neurotoxins. *J. Biol. Chem.* **269**, 1617–1620.

Blasi, J., Chapman, E.R., Link, E., Binz, T., Yamasaki, S., De Camilli, P., Südhof, T., Nieman, H. & Jahn, R. (1993a) Botulinum neurotoxin A selectively cleaves the synaptic protein SNAP-25. *Nature* **365**, 160–163.

Blasi, J., Chapman, E.R., Yamasaki, S., Binz, T., Nieman, H. & Jahn, R. (1993b) Botulinum neurotoxin C blocks neurotransmitter release by means of cleaving HPC-I/syntaxin. *EMBO J.* **12**, 4821–4828.

Boyd, R.S., Duggan, M.J., Shone, C.C. & Foster, K.A. (1995) The effect of botulinum neurotoxins on the release of insulin from insulinoma cell lines HIT-15 and RINm5F. *J. Biol. Chem.* **270**, 18216–18218.

Ferro-Novick, S. & Jahn, R. (1994) Vesicle fusion from yeast to man. *Nature* **370**, 191–193.

Foran, P., Shone, C.C. & Dolly, J.O. (1994) Differences in the protease activities of tetanus and botulinum B toxins revealed by the cleavage of vesicle-associated membrane protein and various sized fragments. *Biochemistry* **31**, 15365–15374.

M

Galli, T., Chilcote, T., Mundigl, O., Binz, T., Niemann, H. & De Camilli, P. (1994) Tetanus toxin-mediated cleavage of cellubrevin impairs exocytosis of transferrin receptor-containing vesicles in CHO cells. *J. Cell. Biol.* **125**, 1015–1024.

Gill, D.M. (1982) Bacterial toxins: a table of lethal amounts. *Microbiol. Rev.* **46**, 86–94.

Hata, Y. & Südhof, T.C. (1995) A novel ubiquitous form of Munc-18 interacts with multiple syntaxins. Use of the yeast two-hybrid system to study interactions between proteins involved in membrane traffic. *J. Biol. Chem.* **270**, 13022–13028.

Hatheway, C.L. (1995) Botulism: the present status of the disease. *Curr. Top. Microbiol. Immunol.* **195**, 55–76.

Igarashi, M., Kozaki, S., Terakawa, S., Kawano, S., Ide, C. & Komiya, Y. (1996) Growth cone collapse and inhibition of neurite growth by botulinum neurotoxin C: a t-SNARE is involved in axonal growth. *J. Cell Biol.* **134**, 205–215.

Jancovic, J. & Hallett, M. (eds) (1994) *Therapy with Botulinum Toxin.* New York: Marcel Dekker.

McMahon, H.T., Ushkaryov, Y.A., Edelmann, L., Link, E., Binz, T., Niemann, H., Jahn, R. & Südhof, T.C. (1993) Cellubrevin is a ubiquitous tetanus-toxin substrate homologous to a putative synaptic vesicle fusion protein. *Nature* **364**, 346–349.

Middlebrook, J.L. & Brown, J.E. (1995) Immunodiagnosis and immunotherapy of tetanus and botulinum neurotoxins. *Curr. Top. Microbiol. Immunol.* **195**, 89–122.

Minton, N.P. (1995) Molecular genetics of clostridial neurotoxins. *Curr. Top. Microbiol. Immunol.* **195**, 161–194.

Montecucco, C. (ed.) (1995) *Clostridial Neurotoxins. Current Topics in Microbiology and Immunology,* vol. 195. Heidelberg: Springer-Verlag.

Montecucco, C. & Schiavo, G. (1995) Structure and function of tetanus and botulinum neurotoxins. *Q. Rev. Biophys.* **28**, 423–472.

Nishiki, T., Tokuyama, Y., Kamata, Y., Nemoto, Y., Yoshida, A., Sato, K., Sekiguchi, M., Takahashi, M. & Kozaki, S. (1996) The high-affinity binding of *Clostridium botulinum* type B neurotoxin to synaptotagmin II associated with gangliosides GT1b/GD1a. *FEBS Lett.* **378**, 253–257.

Osen-Sand, A., Staple, J.K., Naldi, E., Schiavo, G., Rossetto, O., Petitpierre, S., Malgaroli, A., Montecucco, C. & Catsicas, S. (1996) Common and distinct fusion-proteins in axonal growth and transmitter release. *J. Comp. Neurol.* **367**, 222–234.

Patarnello, T., Bargelloni, L., Rossetto, O., Schiavo, G. & Montecucco, C. (1993) Neurotransmission and secretion. *Nature* **364**, 581–582.

Penner, R., Neher, E. & Dreyer, F. (1986) Intracellularly injected tetanus toxin inhibits exocytosis in bovine adrenal chromaffin cells. *Nature* **324**, 76–77.

Rossetto, O., Schiavo, G., Polverino de Laureto, P., Fabbiani, S. & Montecucco, C. (1992) Surface topography of histidine residues of tetanus toxin probed by immobilized-metal-ion affinity chromatography. *Biochem. J.* **285**, 9–12.

Sadoul, K., Lang, J., Montecucco, C., Weller, U., Catsicas, S., Wollheim, C. & Halban, P. (1995) SNAP-25 is expressed in islets of Lagerhans and is involved in insulin release. *J. Cell Biol.* **128**, 1019–1028.

Sanchez-Prieto, J., Shira, T.S., Evans, D., Ashton, A., Dolly, J.O. & Nicholls, D. (1987) Botulinum toxin A blocks glutamate exocytosis from guinea-pig cerebral cortical synaptosomes. *Eur. J. Biochem.* **165**, 675–681.

Schiavo, G. & Montecucco, C. (1995) Tetanus and botulism neurotoxins: isolation and assay. *Methods Enzymol.* **248**, 643–652.

Schiavo, G., Benfenati, F., Poulain, B., Rossetto, O., Polverino de Laureto, P., DasGupta, B.R. & Montecucco, C. (1992a) Tetanus and botulinum-B neurotoxins block neurotransmitter release by a proteolytic cleavage of synaptobrevin. *Nature* **359**, 832–835.

Schiavo, G., Rossetto, O., Santucci, A., DasGupta, B.R. & Montecucco, C. (1992b) Botulinum neurotoxins are zinc proteins. *J. Biol. Chem.* **267**, 23479–23483.

Schiavo, G., Poulain, B., Rossetto, O., Benfenati, F., Tauc, L. & Montecucco, C. (1992c) Tetanus toxin is a zinc protein and its inhibition of neurotransmitter release and protease activity depends on zinc. *EMBO J.* **11**, 3577–3583.

Schiavo, G., Shone, C.C., Rossetto, O., Alexandre, F.C.G. & Montecucco, C. (1993a) Botulinum neurotoxin serotype F is a zinc endopeptidase specific for VAMP/synaptobrevin. *J. Biol. Chem.* **268**, 11516–11519.

Schiavo, G., Rossetto, O., Catsicas, S., Polverino de Laureto, P., DasGupta, B.R., Benfenati, F. & Montecucco, C. (1993b) Identification of the nerve-terminal targets of botulinum neurotoxins serotype A, D and E. *J. Biol. Chem.* **268**, 23784–23787.

Schiavo, G., Santucci, A., DasGupta, B.R., Mehta, P.P., Jontes, J., Benfenati, F., Wilson, M.C. & Montecucco, C. (1993c) Botulinum neurotoxins serotype A and E cleave SNAP-25 at distinct COOH-terminal peptide bonds. *FEBS Lett.* **335**, 99–103.

Schiavo G., Malizio, C., Trimble, W.S., Polverino de Laureto, P., Milan, G., Sugiyama, H., Johnson, E.A. & Montecucco, C. (1994) Botulinum G neurotoxin cleaves VAMP/synaptobrevin at a single Ala/Ala peptide bond. *J. Biol. Chem.* **269**, 20312–20216.

Schiavo, G., Shone, C.C., Bennett, M.K., Sheller, R.H. & Montecucco, C. (1995) Botulinum neurotoxin type C cleaves a single Lys-Ala bond within the carboxyl-terminal region of syntaxins. *J. Biol. Chem.* **270**, 10566–10570.

Scott, A.B. (1989) Clostridial toxin as therapeutic agents. In: *Botulinum Neurotoxin and Tetanus Toxin* (Simpson, L.L., ed.). New York: Academic Press, pp. 399–412.

Shone, C.C. & Roberts, A.K. (1994) Peptide substrate specificity and properties of the zinc-endopeptidase activity of botulinum type B neurotoxin. *Eur. J. Biochem.* **225**, 263–270.

Shone, C.C. & Tranter, H.S. (1995) Growth of clostridia and preparation of their neurotoxins. *Curr. Top. Microbiol. Immunol.* **195**, 143–160.

Shone, C.C., Quinn, C.P., Wait, R., Hallis, B., Fooks, S.G. & Hambleton, P. (1993) Proteolytic cleavage of synthetic fragments of vesicle-associated membrane protein, isoform-2 by botulinum type B neurotoxin. *Eur. J. Biochem.* **217**, 965–971.

van Ermengen, E. (1897) Über ein neuen anaeroben Bacillus und seine Beziehungen zum Botulismus [About a new anaerobic Bacillus and its relation to botulism]. *Z. Hyg. Infektkr.* **26**, 1–56.

Washbourne, O., Pellizzari, R., Baldini, G., Wilson, M.C. & Montecucco, C. (1997) Botulinum neurotoxin types A and E require the SNARE motif in SNAP-25 for proteolysis. *FEBS Lett.* **418**, 1–5.

Williamson, L.C., Halpern, J.L., Montecucco, C., Brown, J.E. & Neale, E.A. (1996) Clostridial neurotoxin and substrate proteolysis in intact neurons. *J. Biol. Chem.* **271**, 7694–7699.

Witcome M., Rossetto, O., Montecucco, C. & Shone, C.C. (1996) Substrate residues N-terminal to the cleavage site of botulinum type B neurotoxin play a role in determining the specificity of its endopeptidase activity. *FEBS Lett.* **386**, 133–136.

Wright, J.F., Pernollet, M., Reboul, A., Aude, C. & Colomb, M.G. (1992) Identification and partial characterization of a low affinity metal-binding site in the light chain of tetanus toxin. *J. Biol. Chem.* **267**, 9053–9058.

Yamasaki, S., Hu, Y., Binz, T., Kalkuhl, A., Kurazono, H., Tamura, T., Jahn, R., Kandel, E. & Niemann, H. (1994a) Synaptobrevin/VAMP of *Aplysia californica*: structure and proteolysis by tetanus and botulinal neurotoxins type D and F. *Proc. Natl Acad. Sci. USA* **91**, 4688–4692.

Yamasaki, S., Baumeister, A., Binz, T., Blasi, J., Link, E., Cornille, F., Roques, B., Fykse, E.M., Südhof, T.C., Jahn, R. & Niemann, H. (1994b) Cleavage of members of the synaptobrevin/VAMP family by types D and F botulinal neurotoxins and tetanus toxin. *J. Biol. Chem.* **269**, 12764–12772.

***Cesare Montecucco***
*Centro CNR Biomembrane,*
*Dipartimento di Scienze Biomediche,*
*Università di Padova,*
*Via Trieste 75, 35121 Padova, Italy*
*Email: toxin@civ.bio.unipd.it*

***Fiorella Tonello***
*Centro CNR Biomembrane,*
*Dipartimento di Scienze Biomediche,*
*Università di Padova,*
*Via Trieste 75, 35121 Padova, Italy*
*Email: toxin@civ.bio.unipd.it*

# 511. *Anthrax toxin lethal factor*

## Databanks

Peptidase classification: clan MX, family M34, MEROPS ID: M34.001
NC-IUBMB enzyme classification: none
Databank codes:

| Species | SwissProt | PIR | EMBL (cDNA) | EMBL (genomic) |
|---|---|---|---|---|
| *Bacillus anthracis* | P15917 | JQ0032 | M29081 M30210 | – |

## Name and History

***Anthrax toxin lethal factor*** is one of the three components that are collectively termed anthrax toxin (Leppla, 1995). The combination of protective antigen (82.7 kDa) and lethal factor (90.2 kDa) is designated lethal toxin because it kills certain animals and rapidly lyses macrophages. The combination of protective antigen and edema factor (88.8 kDa), termed edema toxin, inhibits phagocytes by increasing cAMP concentrations to unphysiologic levels. Both lethal factor and edema factor enter cells by binding to furin-activated, receptor-bound protective antigen (Leppla *et al*., 1988). The low pH of endosomes causes membrane insertion of protective antigen, creating a heptameric, protein-translocating channel by which lethal factor and edema factor cross to the cytosol (Leppla, 1995; Milne *et al*., 1994). Edema factor is a calmodulin-dependent adenylate cyclase (Leppla, 1982). Evidence suggests that lethal factor is a metalloprotease (Klimpel *et al*., 1994), but its cellular substrate(s) and mechanism of toxicity are unknown.

## Activity and Specificity

No protein or polypeptide substrate has been identified. Common proteins and peptide substrates are not cleaved. Lethal factor may be highly specific for one or several cytosolic proteins, analogous to the specificity of the clostridial neurotoxins (Chapters 509 and 510) for synaptic vesicle proteins.

## Structural Chemistry

Lethal factor is a soluble, secreted protein of 90.2 kDa, calculated pI 6.1, which contains no cysteines. Three domains are recognized. The sequence of the N-terminal 250 amino acids is similar to the same region of edema factor. This region binds to protective antigen and causes translocation to the cytosol. Residues 282–383 contain five imperfect repeats of 19 amino acids. This region may act as a flexible linker to allow binding of up to seven lethal factor molecules to the heptameric protective antigen channel by relieving steric crowding of their catalytic domains. Residues 400–776 contain the putative catalytic domain, and the sequence HEFGH, residues 686–690, characteristic of a zinc-binding metalloprotease (Klimpel *et al*., 1994). Substitutions at His686, Glu687 or His690 inactivate lethal factor toxicity and decrease its zinc-binding ability.

## Preparation

Lethal factor is usually made from avirulent strains of *B. anthracis* lacking the plasmid encoding the antiphagocytic polyglutamate acid capsule (Leppla, 1991). A mutated *B. anthracis* strain is available that produces no protective antigen or edema factor, but only lethal factor (Pezard *et al*., 1993). Culture supernatants contain about $10\,\mu g\,ml^{-1}$ lethal factor and few other proteins. Purification from culture supernatants is by any of several chromatographic techniques, and routinely yields homogeneous material in milligram amounts (Quinn *et al*., 1988; Leppla, 1991). Lethal factor has also been produced as a recombinant fusion protein containing residues 1–164 of protective antigen joined to lethal factor through a sequence cleaved by factor Xa (Klimpel *et al*., 1994). The toxin components are not commercially available.

### Biological Aspects

Lethal toxin (lethal factor + protective antigen) is the major determinant of *B. anthracis* virulence. This toxin lyses mouse macrophages in 90 min (Friedlander, 1986). The lethality of toxin to mice is blocked by antibodies to tumor necrosis factor and interleukin 1, suggesting that activation and lysis of macrophages causes the shock-like response to lethal toxin (Hanna *et al*., 1993). Further evidence that lethal factor is a protease is that macrophages are protected by certain hydrophobic peptides such as Leu-NH$_2$ (EC$_{50}$ 1 mM), Phe-NH$_2$ (0.2 mM), Leu-CH$_2$Cl (0.1 mM) and bestatin (0.2 mM).

### Distinguishing Features

No other toxin or protease is known to cause similar biological effects. Both polyclonal and monoclonal antibodies have been described (Little *et al*., 1990) but are not commercially available.

### References

Friedlander, A.M. (1986) Macrophages are sensitive to anthrax lethal toxin through an acid-dependent process. *J. Biol. Chem.* **261**, 7123–7126.

Hanna, P.C., Acosta, D. & Collier, R.J. (1993) On the role of macrophages in anthrax. *Proc. Natl Acad. Sci. USA* **90**, 10198–10201.

Klimpel, K.R., Arora, N. & Leppla, S.H. (1994) Anthrax toxin lethal factor contains a zinc metalloprotease consensus sequence which is required for lethal toxin activity. *Mol. Microbiol.* **13**, 1093–1100.

Leppla, S.H. (1982) Anthrax toxin edema factor: a bacterial adenylate cyclase that increases cyclic AMP concentrations of eukaryotic cells. *Proc. Natl Acad. Sci. USA* **79**, 3162–3166.

Leppla, S.H. (1991) Purification and characterization of adenylyl cyclase from *Bacillus anthracis*. *Methods Enzymol.* **195**, 153–168.

Leppla, S. (1995) Anthrax Toxins. In: *Bacterial Toxins and Virulence Factors in Disease. Handbook of Natural Toxins*, vol. 8 (Moss, J., Iglewski, B., Vaughan, M. & Tu, A., eds). New York: Marcel Dekker, pp. 543–572.

Leppla, S.H., Friedlander, A.M. & Cora, E. (1988) Proteolytic activation of anthrax toxin bound to cellular receptors. In: *Bacterial Protein Toxins* (Fehrenbach, F., Alouf, J.E., Falmagne, P., Goebel, W., Jeljaszewicz, J., Jurgen, D. & Rappouli, R., eds). New York: Gustav Fischer, pp. 111–112.

Little, S.F., Leppla, S.H. & Friedlander, A.M. (1990) Production and characterization of monoclonal antibodies against the lethal factor component of *Bacillus anthracis* lethal toxin. *Infect. Immun.* **58**, 1606–1613.

Milne, J.C., Furlong, D., Hanna, P.C., Wall, J.S. & Collier, R.J. (1994) Anthrax protective antigen forms oligomers during intoxication of mammalian cells. *J. Biol. Chem.* **269**, 20607–20612.

Pezard, C., Duflot, E. & Mock, M. (1993) Construction of *Bacillus anthracis* mutant strains producing a single toxin component. *J. Gen. Microbiol.* **139**, 2459–2463.

Quinn, C.P., Shone, C.C., Turnbull, P.C. & Melling, J. (1988) Purification of anthrax-toxin components by high-performance anion-exchange, gel-filtration and hydrophobic-interaction chromatography. *Biochem. J.* **252**, 753–758.

*Stephen H. Leppla*
*National Institute of Dental Research,*
*National Institutes of Health,*
*Bethesda, MD 20892-4350, USA*
*Email: sleppla@irp30.nidr.nih.gov*

# 512. *Penicillolysin*

### Databanks

*Peptidase classification: clan MX, family M35, MEROPS ID: M35.001*
*NC-IUBMB enzyme classification: none*

*Databank codes:*

| Species | SwissProt | PIR | EMBL (cDNA) | EMBL (genomic) |
|---|---|---|---|---|
| *Penicillium citrinum* | P47189 | – | D25535 | – |

## Name and History

Studies of substrate specificity with the oxidized insulin B chain (Ichishima *et al.*, 1991) and bioactive oligopeptides (Yamaguchi *et al.*, 1993) led to the discovery of a new 18 kDa metalloendopeptidase from *Penicillium citrinum*, a fungus from the imperfect group of fungi. Named ***penicillolysin***, this enzyme has a distinct mode of action and a specificity unique among the metalloendopeptidases (Gripon *et al.*, 1980; Hooper, 1994; Morihara *et al.*, 1968; Morihara, 1974; Vallee & Auld, 1990). It contains 1 mol of zinc per mol of protein, which is catalytically essential.

## Activity and Specificity

The most susceptible bond in the oxidized B chain of insulin is Tyr16↓Leu17; additional cleavages are found at Glu13↓Ala14 and Ala14↓Leu15 (Yamaguchi *et al.*, 1993). The cleavage specificity of the enzyme is distinct from those of other metalloendopeptidases: neutral protease II from *P. roqueforti* (Chapter 513) (Gripon *et al.*, 1980), neutral proteinase I from *Aspergillus oryzae* (Morihara, 1974) and *A. sojae* (Sekine, 1976) (Chapter 513), neutral proteinase II from *A. sojae* (Chapter 513) (Sekine, 1976) and thermolysin (Chapter 351) (Matsubara *et al.*, 1966).

A summary of penicillolysin activity on a series of oligopeptides is presented by Yamaguchi *et al.* (1993). The enzyme degrades a variety of peptides which possess various amino acids at the P1 position, including proline, arginine and hydrophobic amino acids such as tryptophan, phenylalanine, tyrosine and leucine. It most rapidly degrades substance P and dynorphin $A_{1-13}$. A characteristic feature of penicillolysin is the hydrolysis of Pro↓Xaa (where Xaa is Gln, Lys, Leu or Arg) bonds of substance P, dynorphin $A_{1-13}$, neurotensin and chicken brain peptapeptide:

| | |
|---|---|
| Substance P: | RPKP↓QQFF↓G↓LM-NH₂ |
| Dynorphin $A_{1-13}$: | YGGFLR↓RIRP↓KLK |
| Neurotensin: | <QL↓YENKP↓R↓RPYIL |
| Chicken brain pentapeptide: | LP↓LRF-NH₂ |

Such Pro↓Xaa-cleaving activity is also found for prolyl oligopeptidase (Chapter 125), a serine proteinase from vertebrates, plants and *Flavobacterium*.

Penicillolysin recognizes a structural feature in dynorphin A and neurotensin. The enzyme shows a high affinity towards the Arg6↓Arg7 bond in dynorphin $A_{1-13}$ and Arg8↓Arg9 bond in neurotensin. The preferential cleavage by penicillolysin of bonds with hydrophobic amino acid residues at the P1 position was observed for Phe8↓Gly9 in substance P, Leu2↓Tyr3 in neurotensin, Phe5↓Ser6 in bradykinin, Trp3↓Ser4 and Tyr5↓Gly6 in luteinizing hormone releasing hormone, Trp10↓Gly11, Tyr2↓Ser3 and Phe7↓Arg8 in α-melanotropin, and Leu5↓Arg6 and Phe4↓Leu5 bonds in α-neoendorphin. Penicillolysin shows no preferential cleavage specificity for bonds with hydrophobic amino acid residues at position P1′, and its specificity is thus different from that of microbial metalloproteinases such as thermolysin and *Aspergillus oryzae* neutral protease I. It has no ability to hydrolyze angiotensin I, angiotensin II or cholecystokinin (CCK) octapeptide₂₆₋₃₃. The specificity of penicillolysin for oligopeptides is different from those of microbial metalloendopeptidases (Matsubara *et al.*, 1966; Morihara *et al.*, 1968; Morihara, 1974; Sekine, 1976; Gripon *et al.*, 1980).

The pH optimum is about 7 for herring clupeine, ovalbumin or casein (Ichishima *et al.*, 1991; Yamaguchi *et al.*, 1993). The rate of clupeine hydrolysis is 60-fold greater than that for casein. The specific activities of penicillolysin for clupeine and casein are $3.0 \times 10^{-1}$ and $5.2 \times 10^{-3}$ kat kg⁻¹ protein at pH 7.0, respectively. When zinc is removed, the enzyme is completely inactive, and readdition of zinc restores the dual activities towards clupeine and casein. Depending on the casein substrate, the cobalt-penicillolysin could be up to 1.6 times more active than the native zinc enzyme. On the other hand, in clupeine hydrolysis, the cobalt enzyme is about 70% as active as the native enzyme. Thus, replacement of zinc-penicillolysin with cobalt markedly decreases the activity towards clupeine while it increases it towards casein (Yamaguchi *et al.*, 1993). Similar strong action on protamine and histone was reported for *P. roqueforti* neutral protease ('acid' metalloprotease) (Gripon *et al.*, 1980) and neutral proteinase II from *A. sojae* (Sekine, 1973). Basic proteins such as protamine and histone are cleaved by neutral proteases from *P. roqueforti* and *A. sojae* with better yield than casein or hemoglobin. The enzyme has no detectable activities for Z-Gly-Leu-NH₂ or Z-His-Leu-NH₂, for fluorogenic peptidyl-aminomethylcoumarins or for chromogenic *p*-nitroanilides.

## Structural Chemistry

Penicillolysin is a single-chain protein of 177 amino acid residues with relative molecular mass of 18 529 and pI 9.6, which contains three disulfide bonds (Yamaguchi *et al.*, 1993). It contains 1 mol of zinc per mol of enzyme (Ichishima *et al.*, 1991; Yamaguchi *et al.*, 1993). When zinc is removed by EDTA, the enzyme is completely inactive and readdition of zinc or cobalt restores activity (Yamaguchi *et al.*, 1993). A marked pH dependence on inactivation of the enzyme by EDTA is observed; the enzyme is completely inactivated by EDTA at acidic pH below 5, but the inactivation is slight at an alkaline pH of 8.0.

The deduced amino acid sequence of penicillolysin from cDNA cloned in the plasmid pHMP3 shows a unique primary structure (Matsumoto *et al.*, 1994). From N-terminal amino acid sequencing, it was deduced that penicillolysin is synthesized as a precursor consisting of 351 amino acids, with a 19-residue signal peptide (as predicted by the weight-matrix approach of von Heijne (1986)), and a 155-residue propart. Mature penicillolysin has 177 amino acids and a calculated relative molecular mass of 18 525. In the primary structure of the prepro-penicillolysin, there are two potential N-linked glycosylation sites, Asn7-Ala-Ser and Asn123-Val-The (Matsumoto *et al.*, 1994). There is a zinc-binding motif, **HEXXH**132 (Matsumoto *et al.*, 1994); the third zinc ligand is not known.

Comparison of the penicillolysin amino acid sequence with that of other known metalloendopeptidases (Hooper, 1994; Vallee & Auld, 1990) shows that the primary structure of penicillolysin has only a low degree of sequence identity with that of thermolysin (Chapter 351) (Matsubara *et al.*, 1966), while it has 68% identity with that of neutral proteinase II from *A. oryzae* (Tatsumi *et al.*, 1991, 1994) (Chapter 513). The N-terminal sequences of penicillolysin and a 19 000 Da protease from *P. roqueforti* are 57% identical over 35 residues (Gripon *et al.*, 1980).

According to the CD spectra of penicillolysin, α helix and β structures constitute 19% and 58%, respectively (Yamaguchi *et al.*, 1993), while the α helix and β structure contents of the apoenzyme without zinc are about 9% and 61%. This shows that about 50% of the α helix is destroyed in the conformational change from native to apoenzyme. The CD curve of $Co^{2+}$-reconstituted enzyme is shown by Yamaguchi *et al.* (1993): the contents of the α helix and β structure were about 20% and 58%, respectively.

## Preparation

A crude enzyme preparation is commercially available as 'protease B' of *P. citrinum* produced by Amano Pharmaceutical (see Appendix 2 for full names and addresses of suppliers) (Ichishima *et al.*, 1991; Yamaguchi *et al.*, 1993). The recombinant enzyme of penicillolysin has been expressed in *Saccharomyces cerevisiae* cells (E. Ichishima *et al.*, unpublished results).

## Biological Aspects

Penicillolysin is an extracellular proteinase of the fungus *P. citrinum* (Ichishima *et al.*, 1991; Yamaguchi *et al.*, 1993). It appears to be an important digestive enzyme for fungal nutrition.

## Distinguishing Features

Penicillolysin contains 1 mol of zinc per mol of enzyme and no calcium (Yamaguchi *et al.*, 1993). The enzyme is heat labile and its activity is completely destroyed at 60°C for 10 min. For the native enzyme, the zinc atom is indispensable to both catalytic activity and structural function. In contrast, neutral proteinases II from *A. oryzae* (Tatsumi *et al.*, 1991, 1994) and *A. sojae* (Sekine, 1972) are heat resistant at 100°C for 10 min. Neutral proteinase II from *A. sojae* contains 1 mol of zinc and 2 mol of calcium per mol with a relative molecular mass of 19 800 (Sekine, 1972). Metal determination confirmed that *P. roqueforti* and *P. caseicolum* proteases are metalloproteases, each containing one atom of zinc per molecule, which is essential for enzymatic activity (Gripon *et al.*, 1980). The amounts of calcium of the inactive apoenzymes of *P. roqueforti* and *P. caseicolum* remain unchanged after dialysis. These last two enzymes are discussed further in Chapter 513. A heat-labile serine proteinase from *P. citrinum* has also been purified (Yamamoto *et al.*, 1993).

## Related Peptidases

*A. fumigatus* (Monod *et al.*, 1993) and *A. flavus* (Rhodes *et al.*, 1990), two causative agents of invasive aspergillosis, each produce 23 kDa metalloproteinases (MEP20) with pI 5.5 and pI 7.6, respectively. Cloning and characterization of the cDNAs and genes (*mep20*) encoding homologous metalloproteinases from these two agents were reported by Ramesh & Sirakova (1995). *A. flavus* MEP20 has 58 and 64% sequence identity with *A. oryzae* neutral protease II (Ramesh & Sirakova, 1995) and *Penicillium citrinum* penicillolysin (Matsumoto *et al.*, 1994), respectively. *A. fumigatus* MEP20 has 60% identity with penicillolysin (Matsumoto *et al.*, 1994) and with the neutral protease II from *A. oryzae* (Tatsumi *et al.*, 1991). These two 23 kDa proteases are discussed further in Chapter 513.

## References

Gripon, J.C., Auberger, B. & Lenoir, J. (1980) Metalloproteases from *Penicillium caseicolum* and *P. roqueforti*: comparison of specificity and chemical characterization. *Int. J. Biochem.* **12**, 451–455.

Hooper, N.M. (1994) Families of zinc metalloproteases. *FEBS Lett.* **354**, 1–6.

Ichishima, E., Yamaguchi, M., Yano, H., Yamagata, Y. & Hirano, K. (1991) Specificity of a new metalloproteinase from *Penicillium citrinum*. *Agric. Biol. Chem.* **55**, 2191–2193.

Matsubara, H., Sasaki, R., Single, A. & Jukes, T.H. (1966) Specific nature of hydrolysis of insulin and tobacco mosaic virus protein by thermolysin. *Arch. Biochem. Biophys.* **115**, 324–331.

Matsumoto, K., Yamaguchi, M. & Ichishima, E. (1994) Molecular and nucleotide sequence of the complementary DNA for penicillolysin gene, *plnC*, an 18 kDa metalloendopeptidase gene from *Penicillium citrinum*. *Biochim. Biophys. Acta* **1218**, 469–472.

Monod, M., Paris, S., Sanglard, D., Jaton-Ogay, K., Bille, J. & Latge, J.P. (1993) Isolation and characterization of a secreted metalloprotease of *Aspergillus fumigatus*. *Infect. Immun.* **61**, 4099–4104.

Morihara, K. (1974) Comparative specificity of microbial proteases. *Adv. Enzymol.* **41**, 179–243.

Morihara, K., Tsuzuki, H. & Oka, T. (1968) Comparison of the specificities of various neutral proteinases from microorganisms. *Arch. Biochem. Biophys.* **123**, 572–588.

Ramesh, M.V. & Sirakova, T.D. (1995) Cloning and characterization of the cDNAs and genes (*mep20*) encoding homologous metalloproteinases from *Aspergillus flavus* and *A. fumigatus*. *Gene* **165**, 121–125.

Rhodes, J., Amlung, T.W. & Miller, M.S. (1990) Isolation and characterization of an elastolytic proteinase from *Aspergillus flavus*. *Infect. Immun.* **58**, 2529–2534.

Sekine, H. (1972) Some properties of neutral proteinases I and II of *Aspergillus sojae* as zinc-containing metalloenzyme. *Agric. Biol. Chem.* **36**, 2143–2150.

Sekine, H. (1973) Neutral proteinase II of *Aspergillus sojae*. An enzyme specifically active on protamine and histone. *Agric. Biol.*

Chem. **37**, 1765–1767.

Sekine, H. (1976) Neutral proteinases I and II of *Aspergillus sojae*. Action on various substrates. *Agric. Biol. Chem.* **40**, 29–31.

Tatsumi, H., Murakami, S., Tsuji, R.F., Ishida, Y., Murakami, K., Masaki, A., Kawaba, H., Arimura, H., Nakano, K. & Motai, H. (1991) Cloning and expression in yeast of a cDNA clone encoding *Aspergillus oryzae* neutral protease II, a unique metalloprotease. *Mol. Gen. Genet.* **228**, 97–103.

Tatsumi, H., Ikegawa, K., Murakami, S., Kawabe, H., Nakano, E. & Motai, H. (1994) Elucidation of the thermal stability of the neutral proteinase II from *Aspergillus oryzae*. *Biochim. Biophys. Acta* **1208**, 179–185.

Vallee, B.L. & Auld, D.S. (1990) Zinc coordination, function, and structure of zinc enzymes and other proteins. *Biochemistry* **29**, 5647–5659.

von Heijne, G. (1986) A new method for predicting signal sequence cleavage site. *Nucleic Acids Res.* **14**, 4683–4690.

Yamaguchi, M., Hanzawa, S., Hirano, K., Yamagata, Y. & Ichishima, E. (1993) Specificity and molecular properties of penicillolysin, a new metalloproteinase from *Penicillium citrinum*. *Phytochemistry* **33**, 1317–1321.

***Eiji Ichishima***
*Laboratory of Molecular Enzymology,*
*Department of Applied Biological Chemistry,*
*Faculty of Agriculture,*
*Tohoku University,*
*1-1, Tsutsumidori-Amamiyamachi,*
*Aoba-ku, Sendai 981, Japan*
*Email: ichisima@t.soka.ac.jp*

# 513. *Deuterolysin*

## Databanks

*Peptidase classification: clan MX, family M35, MEROPS ID: M35.002*
*NC-IUBMB enzyme classification: EC 3.4.24.39*
*Databank codes:*

| Species | SwissProt | PIR | EMBL (cDNA) | EMBL (genomic) |
|---|---|---|---|---|
| *Aspergillus flavus* | P46073 | | L37524 | – |
| *Aspergillus oryzae* | P46076 | S16547 S47562 | S53810 | – |
| *Sartorya fumigata* | – | – | U24146 | – |

## Name and History

Proteases secreted from *Aspergillus sojae* and *A. oryzae*, which are closely related in taxonomy, play an important role in producing the distinctive taste of soy sauce by hydrolyzing soybean proteins. The molds secrete two kinds of metalloproteases, neutral proteinase I and **neutral proteinase II** (**NpII**) (Sekine, 1972a; Nakadai *et al.*, 1973a,b). The neutral proteinase I has properties similar to those of *Bacillus thermoproteolyticus* thermolysin (Chapter 351), an extensively characterized metalloprotease, while NpII has quite different substrate specificity, molecular mass and thermal stability (Nakadai *et al.*, 1973a,b; Sekine, 1972a,b,c, 1973a,b, 1976). Gripon and colleagues characterized metalloproteases from *Penicillium caseicolum* and *P. roqueforti*, and suggested that these metalloproteases and NpII should be grouped under the term 'acid metalloproteases' (Gripon & Hermier, 1974; Gripon *et al.*, 1980). These metalloproteases were classified as **deuterolysin** by the IUBMB in 1992.

## Activity and Specificity

Deuterolysin shows high activity on basic nuclear proteins such as histone and protamine but has low activities on substrates routinely used in the laboratory, such as casein, hemoglobin, albumin and gelatin (Sekine, 1973a, 1976; Gripon & Hermier, 1974; Gripon *et al.*, 1980). Deuterolysin does not cleave any disubstituted dipeptides tested (Gripon & Hermier, 1974; Gripon *et al.*, 1980; Nakadai *et al.*, 1973b; Sekine, 1976), but it cleaves oxidized insulin B chain with a broad specificity (Gripon & Hermier, 1974; Sekine, 1976; Gripon *et al.*, 1980):

*A. sojae*:

F+VN+Q+HLCG+S+H+L+VE+A+LY+LVCGER+GFFY+TPKA

*P. roqueforti*:

F+VN+Q+HLCG+SHLV+E+A+LY+LVCGERGFFY+TPKA

It is likely that deuterolysin requires a minimal chain length to act. The specificity is distinctly different from that of other proteases, but the rules governing the specificity remain difficult to define.

The pH optimum is 5.5–6.0 on casein (Sekine, 1972b; Nakadai *et al.*, 1973b; Gripon & Hermier, 1974; Gripon *et al.*, 1980). Assays are made with casein (Sekine, 1972a; Nakadai *et al.*, 1973b), protamine sulfate (Tatsumi *et al.*, 1994) or a fluorescent substrate Boc-Val-Leu-Lys-NHMec (Tatsumi *et al.*, 1991), which is available from Peptide Institute (see Appendix 2 for full names and addresses of suppliers).

Deuterolysin contains one atom of zinc per molecule which is essential for activity (Sekine, 1972c; Gripon *et al.*, 1980). $Zn^{2+}$, $Co^{2+}$ and $Mn^{2+}$, in that order, reactivate EDTA-treated apo-deuterolysin (Sekine, 1972c; Gripon & Hermier, 1974).

Deuterolysin is inhibited by chelators but DFP, sulfhydryl reagents, DAN and phosphoramidon are not inhibitory (Sekine, 1972b; Gripon & Hermier, 1974; Gripon *et al.*, 1980). EDTA is effective only at acidic pH (Sekine, 1972c).

## Structural Chemistry

Deuterolysin is a single-chain protein. Among the deuterolysins only the gene (cDNA) for NpII from *A. oryzae* (NpII-O) has been isolated (Tatsumi *et al.*, 1991). According to the nucleotide sequence, NpII-O consists of 177 amino acids and has an $M_r$ of 19018. It contains six Cys residues in three intramolecular disulfide bonds, Cys6-Cys78, Cys85-Cys103 and Cys117-Cys177 (Tatsumi *et al.*, 1994). The two His residues in the sequence -His-Glu-Phe-Thr-His-Ala- are presumed to be zinc ligands. The $M_r$ (around 20000) and the amino acid composition of other deuterolysins are similar to those of NpII-O (Sekine, 1973b; Gripon & Hermier, 1974; Gripon *et al.*, 1980). Deuterolysin does not contain carbohydrates (Sekine, 1973b; Tatsumi *et al.*, 1991). The pI is around 4 (Sekine, 1972a; Tatsumi *et al.*, 1991).

## Preparation

Deuterolysin has so far been found in certain molds belonging to the genera *Aspergillus* and *Penicillium*. Deuterolysin may exist in other microorganisms but has not yet been identified, possibly because of low activity on normal substrates. Deuterolysin has been purified to homogeneity (Sekine, 1972a; Nakadai *et al.*, 1973b; Gripon & Hermier, 1974; Gripon *et al.*, 1980). Recombinant NpII-O was expressed from the yeast *Saccharomyces cerevisiae* as a secreted form (Tatsumi *et al.*, 1991) and some mutant NpII-O forms were produced using this expression system (Tatsumi *et al.*, 1994).

The biological role of deuterolysin is not clear, but it is likely that it enables organisms to utilize substrates in their environment.

According to the nucleotide sequence of NpII-O, it has a prepro region consisting of 175 amino acids at the N-terminus of a mature region consisting of 177 amino acids. When a plasmid designed to express the prepro-cDNA was introduced into *S. cerevisiae* and the transformant was cultured in YPD medium (2% glucose, 2% polypeptone, 1% yeast extract), it secreted a pro-NpII-O. However, in a culture of the same medium containing 0.2 mM $ZnCl_2$, it secreted mature NpII-O. This observation suggests that autoproteolytic activity effected the processing of the pro region (Tatsumi *et al.*, 1991).

Deuterolysin shows unique thermal stability. It is most unstable after 10 min at about 65–75°C, but it regains stability beyond this temperature, and it is relatively stable at 100°C (Sekine, 1972b; Gripon & Hermier, 1974; Tatsumi *et al.*, 1994). This phenomenon is attributed to unique properties of deuterolysin in regard to reversible thermal unfolding and autoproteolysis (Tatsumi *et al.*, 1994).

## Distinguishing Features

Deuterolysin shows high activities on basic nuclear protein substrates, but has low activities on substrates routinely used in the laboratory. Its $M_r$ is around 20 000. Deuterolysin has a unique thermal stability that passes through a minimum around 70°C and recovers at 100°C.

## Related Peptidases

Recently, genes for metalloproteases from *P. citrinum*, *A. flavus* and *A. fumigatus* have been isolated and their primary structures were found to be homologous to that of NpII-O (Matsumoto *et al.*, 1994; Ramesh *et al.*, 1995). The specificity of the metalloprotease (penicillolysin) from *P. citrinum* (Chapter 512) appears to be distinct from that of deuterolysin (Ichishima *et al.*, 1991). However, the *A. flavus* and *A. fumigatus* (*Sartorya fumigata*) 23 kDa proteases appear to belong here in the deuterolysin group and their cDNA data are included in the table above. Further details of this 23 kDa protease activity may be found in Chapter 514 under Related Peptidases.

## References

Gripon, J.-C. & Hermier, J. (1974) Le système protéolytique de *Penicillium roqueforti*. III. Purification, propriétés et spécificité d'une protéase inhibée par l'E.D.T.A [The proteolytic system of *Penicillium roqueforti*. III. Purification, properties and specificity of a protease inhibited by EDTA]. *Biochimie (Paris)* **56**, 1324–1332.

Gripon, J.C., Auberger, B. & Lenoir, J. (1980) Metalloproteases from *Penicillium caseicolum* and *P. roqueforti*: comparison of specificity and chemical characterization. *Int. J. Biochem.* **12**, 451–455.

Ichishima, E., Yamaguchi, M., Yano, H., Yamagata, Y. & Hirano, K. (1991) Specificity of a new metalloproteinase from *Penicillium citrinum*. *Agric. Biol. Chem.* **55**, 2191–2193.

Matsumoto, K., Yamaguchi, M. & Ichishima, E. (1994) Molecular cloning and nucleotide sequence of the complementary DNA for penicillolysin gene, *plnC*, an 18 kDa metalloendopeptidase gene from *Penicillium citrinum*. *Biochim. Biophys. Acta* **1218**, 469–472.

Nakadai, T., Nasuno, S. & Iguchi, N. (1973a) Purification and properties of neutral proteinase I from *Aspergillus oryzae*. *Agric. Biol. Chem.* **37**, 2695–2701.

Nakadai, T., Nasuno, S. & Iguchi, N. (1973b) Purification and properties of neutral proteinase II from *Aspergillus oryzae*. *Agric. Biol. Chem.* **37**, 2703–2708.

Ramesh, M.V., Sirakova, T.D. & Kolattukudy, P.E. (1995) Cloning and characterization of the cDNAs and genes (*mep20*) encoding homologous metalloproteinases from *Aspergillus flavus* and *A. fumigatus*. *Gene* **165**, 121–125.

Sekine, H. (1972a) Neutral proteinases I and II of *Aspergillus sojae*. Isolation in homogeneous form. *Agric. Biol. Chem.* **36**, 198–206.

Sekine, H. (1972b) Neutral proteinases I and II of *Aspergillus sojae*. Some enzymatic properties. *Agric. Biol. Chem.* **36**, 207–216.

Sekine, H. (1972c) Some properties of neutral proteinases I and II of *Aspergillus sojae* as zinc-containing metalloenzymes. *Agric. Biol. Chem.* **36**, 2143–2150.

Sekine, H. (1973a) Neutral proteinases II of *Aspergillus sojae*: an enzyme specifically active on protamine and histone. *Agric. Biol. Chem.* **37**, 1765–1767.

Sekine, H. (1973b) Neutral proteinases I and II of *Aspergillus sojae*: some physicochemical properties and amino acid composition. *Agric. Biol. Chem.* **37**, 1945–1952.

Sekine, H. (1976) Neutral proteinases I and II of *Aspergillus sojae*. Action on various substrates. *Agric. Biol. Chem.* **40**, 703–709.

Tatsumi, H., Murakami, S., Tsuji, R.F., Ishida, Y., Murakami, K., Masaki, A., Kawabe, H., Arimura, H., Nakano, E. & Motai, H. (1991) Cloning and expression in yeast of a cDNA clone encoding *Aspergillus oryzae* neutral protease II, a unique metalloprotease. *Mol. Gen. Genet.* **228**, 97–103.

Tatsumi, H., Ikegaya, K., Murakami, S., Kawabe, H., Nakano, E. & Motai, H. (1994) Elucidation of the thermal stability of the neutral proteinase II from *Aspergillus oryzae*. *Biochim. Biophys. Acta* **1208**, 179–185.

**M**

***Hiroki Tatsumi***
*Research & Development Division,*
*Kikkoman Corporation,*
*399 Noda, Noda City, Chiba 278, Japan*

# 514. *Fungalysin*

## Databanks

*Peptidase classification: clan MA, family M36, MEROPS ID: M36.001*
*NC-IUBMB enzyme classification: none*
*Databank codes:*

| Species | Strain/ isolate | SwissProt | PIR | EMBL (cDNA) | EMBL (genomic) |
|---|---|---|---|---|---|
| *Sartorya fumigata* | 13 | P46074 | – | – | L29566: complete gene |
| *Sartorya fumigata* | delta 18 | P46075 | – | – | Z30424: complete gene |

## Name and History

Neutral metalloproteinase activity is produced by a variety of fungal species including several species of *Aspergillus* such as *A. fumigatus (Sartorya fumigata)* (Monod *et al*., 1993; Markaryan *et al*., 1994), *A. flavus* (Rhodes *et al*., 1990), *A. oryzae* (neutral proteinase II) (Tatsumi *et al*., 1991) and *Penicillium citrinum* (penicillolysin) (Matsumoto *et al*., 1994). Two distinct sizes are found. Thus, *A. fumigatus* contains both a 42 kDa (Monod *et al*., 1993; Markaryan *et al*., 1994) and a 23 kDa protease, also found in *A. flavus* (Ramesh *et al*., 1995). The smaller protease of 23 kDa is discussed in Chapter 513 and penicillolysin in Chapter 512. The present chapter deals with the 42 kDa, for which the name *fungalysin* has been coined. Earlier names such as *Aspergillus fumigatus metalloproteinase* or *elastase* were unsatisfactory in that they did not distinguish between the two sizes of proteases.

## Activity and Specificity

Fungalysin readily hydrolyzes laminins. It also hydrolyzes elastin at orygin about 60% of the rate observed with the serine proteinase oryzin of *A. fumigatus* (Chapter 105) but hydrolyzes collagen less efficiently. The substrate specificity is similar to that of thermolysin (Chapter 351). The specificity is for the amino side of hydrophobic residues with bulky side chains (Markaryan *et al*., 1994). The cleavage positions in the oxidized B chain of insulin are:

FVNQHLCGSH┼LVEA┼LY┼LVCGERG┼F┼FYTPKA

The enzyme releases fluorescent product(s) from Abz-Ala-Ala-Phe-Phe-NHPhNO$_2$ and this is a useful substrate for a spectrofluorometric assay of this proteinase. Hydrolysis of this substrate probably involves cleavage of the Phe┼Phe bond and does not cause release of *p*-nitroaniline. The enzyme does not hydrolyze Z-Ala-Ala-Leu-NHPhNO$_2$, Suc-Ala-Ala-Pro-Leu-NHPhNO$_2$, or Z-Phe-Arg-NHMec.

The pH optimum of this proteinase is 7.5–8.0. Maximal proteinase activity is obtained at 60°C and at this temperature the decrease in enzyme activity becomes measurable after 20 min. It is inhibited by EDTA, 1,10-phenanthroline and phosphoramidon and not inhibited by PMSF, antipain, leupeptin, chymostatin and pepstatin. The enzyme that is inactivated by removal of the metal with chelators can be fully reactivated by the addition of Zn$^{2+}$ and partially reactivated by Co$^{2+}$.

## Structural Chemistry

The enzyme is a single-chain protein of mass 42 kDa. Comparison of the amino acid sequence of this proteinase with those of other metalloproteinases show that fungalysin contains motif 1 (**HEYTH**) with the two His residues involved in zinc binding and the active-site Glu residue, motif 2 (**YALESGGMGEGWSD**) containing the zinc-binding Glu and the active-site Tyr and Asp residues, and motif 3 (**TYTSVNSLNAVHAIGTVWASILY**) containing the active-site His (Sirakova *et al*., 1994; see also Alignment 350.1). It is not a glycoprotein.

## Preparation

Fungalysin, being a secreted protein, is purified about 320-fold from the culture medium of *A. fumigatus* grown on elastin. Bacitracin-Sepharose 4B affinity chromatography is followed by Sephadex G-75 gel filtration (Markaryan *et al*., 1994). This method can be scaled up. Recombinant protein expressed in *E. coli* at 30°C showed the same properties as the native enzyme and catalyzed elastin hydrolysis (Sirakova *et al*., 1994). This method can also be scaled up.

## Biological Aspects

Fungalysin is produced when the fungus is grown on protein substrates or the insoluble material from lung tissue. It is secreted into the host tissue when the fungus invades the lungs of immunocompromised animals (Markaryan *et al*., 1994). This proteinase probably helps to degrade proteinaceous structural barriers in the host during invasion of host lung. Both cDNA and the gene have been cloned and sequenced. The cDNA is composed of an open reading frame of 1899 bp preceded by 27 bp 5′ and 97 bp 3′ untranslated sequences. The transcript in the fungus comprises 2.4 kb. The genomic sequence has four introns (Jaton-Ogay *et al.*, 1992; Sirakova *et al.*, 1994). The primary translation product of the transcript would encode a 634 amino acid protein containing a signal sequence followed by a 227 amino acid pro sequence and a 389 amino acid mature enzyme with a calculated $M_r$ of 42 106. Recombinant propeptide binds the mature fungalysin and selectively inhibits it with a $K_i$ of $3 \times 10^{-9}$ M, whereas this propeptide does not inhibit the 23 kDa protease from *A. flavus* (Chapter 513) or thermolysin (Markaryan *et al.*, 1996) (Chapter 351). The very long propeptide probably keeps the proteinase inactive until the mature protein is secreted. The presence of $Zn^{2+}$ in the culture medium stimulates the production of fungalysin by *A. fumigatus*.

## Related Peptidases

Fungalysin was not found in *A. flavus* (Ramesh *et al.*, 1995). Instead a smaller metalloproteinase was found to be a major proteinase produced by *A. flavus* grown on elastin. This enzyme was first purified to homogeneity as a 23 kDa protein in 5–10% yield (Rhodes *et al.*, 1990). It is not a glycoprotein; its pH optimum is 7.0–8.0; it is inhibited by a variety of divalent metal ions and strongly inhibited by 1,10-phenanthroline. The cDNA and gene have been cloned (Ramesh *et al.*, 1995), and the enzyme is classed with deuterolysin (Chapter 513) in family M35. Its pro region contains a 54 bp segment with an intron character that is lacking in the corresponding genes in *A. fumigatus* (Ramesh *et al.*, 1995), *A. oryzae* (Tatsumi *et al.*, 1991) or *Pencillium citrinum* (Matsumoto *et al.*, 1994). Antibodies prepared against this enzyme show weak cross-reactivity with the corresponding enzyme of *A. fumigatus*. *A. oryzae* has a similar gene and produces a similar proteinase (Tatsumi *et al.*, 1991). Disruption of the gene for the serine proteinase oryzin (Chapter 105) in *A. flavus* by homologous recombination with hygromycin resistance gene as a marker totally abolishes production of serine proteinase and causes a compensatory increase in the levels of the small metalloprotease transcript, protein and enzyme activity (Ramesh & Kolattukudy, 1996).

## References

Jaton-Ogay, K., Suter, M., Crameri, R., Falchetto, R., Faith, A. & Monod, M. (1992) Nucleotide sequence of a genomic and a cDNA clone encoding an extracellular alkaline proteinase of *Aspergillus fumigatus*. *FEMS Microbiol. Lett.* **92**, 163–168.

Markaryan, A., Morozova, I., Hongshi, Y. & Kolattukudy, P.E. (1994) Purification and characterization of an elastinolytic metalloprotease from *Aspergillus fumigatus* and immunoelectron microscopic evidence of secretion of this enzyme by the fungus invading murine lung. *Infect. Immun.* **62**, 2149–2157.

Markaryan, A., Lee, J.-D., Sirakova, T.D. & Kolattukudy, P.E. (1996) Specific inhibition of mature fungal serine and metalloproteinases by their propeptides. *J. Bacteriol.* **178**, 2211–2215.

Matsumoto, K., Yamaguchi, M. & Ichishima, E. (1994) Molecular cloning and nucleotide sequence of the complementary DNA for penicillolysin gene *pinC*, and 18 kDa metalloendopeptidase gene from *Penicillium citrinum*. *Biochim. Biophys. Acta* **1218**, 469–472.

Monod, M., Paris, S., Sanglard, D., Jaton-Ogay, K., Bille, J. & Latgé, J.P. (1993) Isolation and characterization of a secreted metalloprotease of *Aspergillus fumigatus*. *Infect. Immun.* **61**, 4099–4104.

Ramesh, M.V. & Kolattukudy, P.E. (1996) Disruption of the serine proteinase gene (*sep*) in *Aspergillus flavus* leads to a compensatory increase in the expression of a metalloproteinase gene (*mep20*). *J. Bacteriol.* **173**, 3899–3907.

Ramesh, M.V., Sirakova, T.D. & Kolattukudy, P.E. (1995) Cloning and characterization of the cDNAs and genes (*mep20*) encoding homologous metalloproteinases from *Aspergillus flavus* and *A. fumigatus*. *Gene* **165**, 121–125.

Rhodes, J.C., Amlung, T.W. & Miller, M.S. (1990) Isolation and characterization of an elastinolytic proteinase from *Aspergillus flavus*. *Infect. Immun.* **58**, 2529–2534.

Sirakova, T.D., Markaryan, A. & Kolattukudy, P.E. (1994) Molecular cloning and sequencing of the cDNA and gene for a novel elastinolytic metalloproteinase from *Aspergillus fumigatus* and its expression in *Escherichia coli*. *Infect. Immun.* **62**, 4208–4218.

Tatsumi, H., Murakami, S., Tsuji, R.F., Ishida, K., Murakami, A., Masaki, H., Kawabe, H., Arimura, H., Nakano, E. & Motai, H. (1991) Cloning and expression in yeast of a cDNA clone encoding *Aspergillus oryzae* neutral protease II, a unique metalloprotease. *Mol. Gen. Genet.* **228**, 97–103.

*P.E. Kolattukudy*
*The Ohio State University,*
*Departments of Biochemistry and Medical Biochemistry,*
*The Neurobiotechnology Center,*
*206 Rightmire Hall,*
*1060 Carmack Drive,*
*Columbus, OH 43210, USA*
*Email: kolattukudy.2@osu.edu*

*Tatiana D. Sirakova*
*The Ohio State University,*
*The Neurobiotechnology Center,*
*206 Rightmire Hall,*
*1060 Carmack Drive,*
*Columbus, OH 43210, USA*
*Email: sirakova.1@osu.edu*

**M**

# 515. *Lysostaphin*

## Databanks

*Peptidase classification: clan MX, family M37, MEROPS ID: M37.001*
*NC-IUBMB enzyme classification: EC 3.4.24.75*
*Chemical Abstracts Service registry number: 0911-93-2*
*Databank codes:*

| Species | SwissProt | PIR | EMBL (cDNA) | EMBL (genomic) |
|---|---|---|---|---|
| *Staphylococcus simulans* | P10547 | A25881 | M15686 | – |
| *Staphylococcus staphylolyticus* | P10548 | S01079 | X06121 | – |

## Name and History

During a period in which antibiotic-resistant staphylococci were producing the first real need for alternative antistaphylococcal agents, Schindler and Schuhardt isolated from a single known strain of *Staphylococcus simulans* a lytic substance specific for staphylococci, but not for various other gram-positive and negative bacteria (Schindler & Schuhardt, 1964, 1965). Based on these observations, the substance was named **lysostaphin**. Lysostaphin has been tested as a novel antistaphylococcal agent (Zygmunt & Tavormina, 1972) and also as a reagent for typing staphylococci (Huber & Huber, 1989; Severance *et al.*, 1980). One of its more important functions, however, is in the research laboratory where it is frequently used to lyse staphylococcal cell walls for the liberation of intracellular enzymes, nucleic acids, and cell membrane and surface components. The importance of lysostaphin in staphylococci-related studies is underlined by the fact that many staphylococci are resistant to conventional lysis methods.

## Activity and Specificity

Lysostaphin cleaves staphylococcal peptidoglycans in general but is directed specifically to *Staphylococcus aureus* target cells. In lysing staphylococci, the key enzymatic reaction of lysostaphin is the specific cleavage of the Gly┼Gly bond in the pentaglycine subunit of the cell wall peptidoglycan (Iversen & Grov, 1973). The specificity for the pentaglycine domain is high in that staphylococcal strains with mutations in this region are no longer susceptible to lysostaphin (Maidhof *et al.*, 1991; de Jonge *et al.*, 1993). Lysostaphin also hydrolyzes glycine-rich proteins such as insoluble elastin (Park *et al.*, 1995). Both activities have a pH optimum in the neutral range and are inhibited by agents that would chelate the endogenous zinc atoms (1,10-phenanthroline and EDTA) as well as by exogenously added zinc. Binding of lysostaphin to elastin, however, is not affected by exogenously added zinc, suggesting that zinc interferes directly with the catalytic activity of the enzyme.

## Structural Chemistry

Lysostaphin is synthesized as a preproenzyme of $M_r$ 42 000 with a signal peptide that is cleaved intracellularly and an N-terminal propeptide removed extracellularly to yield the mature 27 kDa enzyme (Heinrich *et al.*, 1987; Recsei *et al.*, 1987). The sequence element that is responsible for binding to the cell wall of *Staphylococcus aureus* is at the C-terminus of the molecule (Baba & Schneewind, 1996). Lysostaphin contains 1 mol of zinc per mol of protein, but the zinc coordination site has not been elucidated. Like other elastolytic enzymes such as human leukocyte elastase (Chapter 15) and pancreatic elastase (Chapters 11 and 12), mature lysostaphin has a basic pI of 9.7. The genes for lysostaphin and lysostaphin resistance reside on plasmid pACK1 (Heath *et al.*, 1989; DeHart *et al.*, 1995).

## Preparation

Conventional purification of lysostaphin from the culture supernatant of *Staphylococcus simulans* involves ammonium sulfate precipitation, ion-exchange chromatography (DEAE cellulose), and an additional ammonium sulfate precipitation step (Schindler & Schuhardt, 1965). This purified material is commercially available (Sigma; see Appendix 2 for full names and addresses of suppliers), but in addition to the endopeptidase activity of lysostaphin, this preparation has minute endo-β-N-acetylglucosaminidase and amidase contamination (Malatesta *et al.*, 1992). These contaminants can be removed easily with reverse-phase HPLC. More recently, a single-step purification procedure has been described that utilizes adsorption of lysostaphin on to bacterial cells of lysostaphin-resistant *Staphylococcus aureus* mutants (Chan, 1996). A recombinant form of lysostaphin has recently been made available by a commercial source (Sigma).

## Biological Aspects

The strain that secretes lysostaphin (*Staphylococcus simulans* biovar. *staphylolyticus*) is not susceptible to autolysis since the resistance gene is coexpressed by the same plasmid (Heath *et al.*, 1989). Recsei *et al.* (1987) have suggested that lysostaphin may be involved in cell wall remodeling during bacterial cell division. This hypothesis, however, has been questioned since the plasmid containing the lysostaphin gene is not necessary for growth in the producing strain. Since elastolytic activities of microbial pathogens are generally associated with virulence, lysostaphin or a lysostaphin-like enzyme may play an important role in staphylococcal disease pathogenesis. It is interesting to note that while the staphylolytic enzyme lysostaphin is elastolytic, the elastolytic enzyme human leukocyte elastase (Chapter 15) (Janoff & Blondin, 1973) is not staphylolytic. However, the *Pseudomonas las*A gene product staphylolysin (Chapter 507) (Kessler *et al*., 1993) is both staphylolytic and also a weak elastase.

## Related Peptidases

Potential zinc coordination sites for lysostaphin have been suggested in Chapter 497 (see Alignment 497.3 on the CD-ROM). Interestingly, $\beta$-lactamase is encoded by the same plasmid that contains the lysostaphin gene.

## Further Reading

The protein sequence and motif structure are described in Recsei *et al*. (1987). The article by Park *et al*. (1995) thoroughly covers background and known activities of the enzyme, including degradation of mammalian proteins.

## References

Baba, T. & Schneewind, O. (1996) Target cell specificity of a bacteriocin molecule – a C-terminal signal directs lysostaphin to the cell wall of *Staphylococcus aureus*. *EMBO J.* **15**, 4789–4797.

Chan, E.C. (1996) Expression and purification of recombinant lysostaphin in *Escherichia coli*. *Biotechnol. Lett.* **18**, 833–838.

de Jonge, B.L.M., Sidow, T., Chang, Y.S., Labischinski, H., Berger-Bachi, B., Gage, D.A. & Tomasz, A. (1993) Altered muropeptide composition in *Staphylococcus aureus* strains with an inactivated *femA* locus. *J. Bacteriol.* **175**, 2779–2782.

DeHart, H.P., Heath, H.E., Heath, L.S., LeBlanc, P.A. & Sloan, G.L. (1995) The lysostaphin endopeptidase resistance gene (*epr*) specifies modification of peptidoglycan cross bridges in *Staphylococcus simulans* and *Staphylococcus aureus*. *Appl. Environ. Microbiol.* **61**, 1475–1479.

Heath, H.E., Heath, L.S., Nitterauer, J.D., Rose, K.E. & Sloan, G.L. (1989) Plasmid-encoded lysostaphin endopeptidase resistance of *Staphylococcus simulans* biovar *staphylolyticus*. *Biochem. Biophys. Res. Commun.* **160**, 1106–1109.

Heinrich, P., Rosenstein, R., Böhmer, M., Sonner, P. & Götz, F. (1987) The molecular organization of the lysostaphin gene and its sequences repeated in tandem. *Mol. Gen. Genet.* **209**, 563–569.

Huber, M.M. & Huber, T.W. (1989) Susceptibility of methicillin-resistant *Staphylococcus aureus* to lysostaphin. *J. Clin. Microbiol.* **27**, 1122–1124.

Iversen, O. & Grov, A. (1973) Studies on lysostaphin. Separation and characterization of three enzymes. *Eur. J. Biochem.* **38**, 293–300.

Janoff, A. & Blondin, J. (1973) The effect of human granulocyte elastase on bacterial suspensions. *Lab. Invest.* **29**, 454–457.

Kessler, E., Safrin, M., Olson, J.C. & Ohman, D.E. (1993) Secreted lasA of *Pseudomonas aeruginosa* is a staphylolytic protease. *J. Biol. Chem.* **268**, 7503–7508.

Maidhof, H., Reinicke, B., Blümel, P., Berger-Bächi, B. & Labischinski, H. (1991) *femA*, which encodes a factor essential for expression of methicillin resistance, affects glycine content of peptidoglycan in methicillin-resistant and methicillin-susceptible *Staphylococcus aureus* strains. *J. Bacteriol.* **173**, 3507–3513.

Malatesta, M.L., Heath, H.E., LeBlanc, P.A. & Sloan, G.L. (1992) EGTA inhibition of DNase activity in commercial lysostaphin preparations. *Biotechniques* **12**, 71–72.

Park, P.W., Senior, R.M., Griffin, G.L., Broekelmann, T.J., Mudd, M.S. & Mecham, R.P. (1995) Binding and degradation of elastin by the staphylolytic enzyme lysostaphin. *Int. J. Biochem. Cell Biol.* **27**, 139–146.

Recsei, P.A., Gruss, A.D. & Novick, R.P. (1987) Cloning, sequence, and expression of the lysostaphin gene from *Staphylococcus simulans*. *Proc. Natl Acad. Sci. USA* **84**, 1127–1131.

Schindler, C.A. & Schuhardt, V.T. (1964) Lysostaphin: A new bacteriolytic agent for the *Staphylococcus*. *Proc. Natl Acad. Sci. USA* **51**, 414–421.

Schindler, C.A. & Schuhardt, V.T. (1965) Purification and properties of lysostaphin: a lytic agent for *Staphylococcus aureus*. *Biochim. Biophys. Acta* **97**, 242–250.

Severance, P.J., Kauffman, C.A. & Sheagren, J.N. (1980) Rapid identification of *Staphylococcus aureus* by using lysostaphin sensitivity. *J. Clin. Microbiol.* **11**, 724–727.

Vallee, B.L. & Auld, D.S. (1992) Active zinc binding sites of zinc metalloenzymes. *Matrix* **suppl. 1**, 5–19.

Zygmunt, W.A. & Tavormina, P.A. (1972) Lysostaphin: model for a specific enzymatic approach to infectious disease. *Prog. Drug Res.* **16**, 309–333.

**M**

*Pyong Woo Park*
*Harvard Medical School,*
*300 Longwood Ave., Enders-9,*
*Boston, MA 02115, USA*
*Email: park_p@a1.tch.harvard.edu*

*Robert P. Mecham*
*Department of Cell Biology & Physiology,*
*Washington University School of Medicine,*
*660 S. Euclid Ave., Box 8228,*
*St. Louis, MO 63110, USA*
*Email: bmecham@cellbio.wustl.edu*

# 516. FtsH protease

## Databanks

Peptidase classification: clan MX, family M41, MEROPS ID: M41.001
NC-IUBMB enzyme classification: none
Databank codes:

| Species | SwissProt | PIR | EMBL (cDNA) | EMBL (genomic) |
|---|---|---|---|---|
| *Bacillus subtilis* | P37476 | – | D26185 | – |
| *Escherichia coli* | P28691 | S35109 | M83138 | U18997: chromosomal region 67.4–76.0' |
| | | | U01376 | |
| *Haemophilus influenzae* | P45219 | – | U32824 | – |
| *Helicobacter pylori* | – | – | U59625 | – |
| *Lactococcus lactis* | P46469 | – | X67015 | – |
| | | | X69123 | |
| *Mycoplasma genitalium* | P47695 | – | U39732 | – |
| *Mycoplasma pneumoniae* | – | – | Z32663 | – |
| *Synechocystis* sp. | – | – | – | D90904: genome section 6 of 27 |

## Name and History

*Escherichia coli ftsH* has been identified genetically as mutations causing diverse cellular phenotypes. A temperature-sensitive mutation, *ftsH1*, (filamentation temperature-sensitive) was originally described as causing a cell-division defect at 42°C (Santos & Almeida, 1975), but more recently the original mutant strain proved to carry another mutation in the *ftsI* gene that is mainly responsible for the cell division phenotype (Begg *et al.*, 1992), whereas the *ftsH1* mutation itself is responsible for the loss of viability at high temperature. Other mutations in *ftsH* include the *hflB29* (high frequency lysogenization for bacteriophage λ) mutation (Herman *et al.*, 1993), the *tolZ21* (colicin-tolerant) mutation (Qu *et al.*, 1996), and *std* (stop transfer defective) mutations which cause abnormal periplasmic localization of an alkaline phosphatase (PhoA) mature moiety that has been attached to a normally cytoplasmic domain of SecY, a multispanning membrane protein (Akiyama *et al.*, 1994a,b). This protein is called either **FtsH protease** or **HflB protease**.

## Activity and Specificity

FtsH has an ATP-dependent protease activity against $\sigma^{32}$ and SecY proteins. However, the specificity of its peptide bond cleavage reaction has not been elucidated. In the absence of substrate proteins, FtsH hydrolyzes ATP (about 230 nmol min$^{-1}$ mg$^{-1}$ protein) with an apparent $K_m$ value for ATP of about 80 μM (Tomoyasu *et al.*, 1995). *In vitro* degradation of $\sigma^{32}$ and SecY by FtsH depends on ATP and less on CTP. Nonhydrolyzable ATP analogs, as well as GTP, are totally ineffective (Tomoyasu *et al.*, 1995; Akiyama *et al.*, 1996), indicating that hydrolysis of ATP is required for proteolysis. Some degradation of SecY was also observed in the presence of UTP (Akiyama *et al.*, 1996). Zn$^{2+}$ and Mn$^{2+}$ stimulate *in vitro* degradation of $\sigma^{32}$ (Tomoyasu *et al.*, 1995). Proteolytic activity of FtsH is inhibited by chelators such as 1,10-phenanthroline and EDTA, as well as by vanadate, whereas it is insensitive to *N*-ethylmaleimide, azide, KNO$_3$ and PMSF (Tomoyasu *et al.*, 1995). The temperature sensitivity of the *ftsH1* mutant is suppressed by inclusion of Fe$^{2+}$, Ni$^{2+}$, Mn$^{2+}$ or Co$^{2+}$ in the medium (Herman *et al.*, 1995a). These results suggest that FtsH is a metalloprotease.

## Structural Chemistry

The nucleotide sequence predicts that *E. coli* FtsH is a 70.7 kDa protein with 633 amino acid residues (Tomoyasu *et al.* 1993b). An alterative initiation codon (GTT) which extends the open reading frame by two amino acid residues at the N-terminus also exists. FtsH is an integral cytoplasmic membrane protein with two transmembrane segments near the N-terminus and a large cytoplasmic domain (Tomoyasu *et al.*, 1993a). The cytoplasmic domain includes a segment that is homologous to the members of an ATPase family, called AAA (**A**TPase **a**ssociated with diverse cellular **a**ctivity) (Confalonieri & Duguet, 1995) (Chapter 517). Proteins of this family share a homologous region of about 200 amino acid residues including a set of Walker A and B motifs and highly conserved 'AAA signature' sequences. Mutational substitution of Asn for the conserved Lys or Thr in Walker A motif abolished both ATPase and proteolytic activities of FtsH (Akiyama *et al.*, 1996). Another motif A-like sequence is located N-terminally and outside the AAA region, but its role in the FtsH functions is obscure (Akiyama *et al.*, 1996). The cytoplasmic domain of FtsH contains an HEXXH metalloprotease zinc-binding motif (Tomoyasu *et al.*, 1995). The importance of this motif for the proteolytic (but not ATPase) activity has been suggested by mutational studies (Akiyama *et al.*, 1996; Qu *et al.* 1996). The quaternary structure of FtsH is quite complex. Homo-oligomeric interaction mediated by the N-terminal region was documented (Akiyama *et al.*, 1995). FtsH also forms a high molecular mass complex in the membrane, which includes the HflK and HflC membrane proteins and probably several other unidentified proteins (Akiyama *et al.*, 1995; Kihara *et al.*, 1996). The FtsH–HflKC association appears to be stimulated by ATP (Kihara *et al.*, 1996).

## Preparation

The *E. coli* FtsH protein has been overproduced, solubilized with detergent NP40 and purified to near homogeneity by ion-exchange and gel-filtration chromatographies (Tomoyasu *et al.*, 1995). More conveniently, wild-type and mutant forms of FtsH were tagged with hexahistidine at the C-terminus, overproduced and purified by immobilized nickel affinity chromatography. The histidine tag did not seem to affect the ATPase and the proteolytic activities (Akiyama *et al.*, 1996).

## Biological Aspects

The heat-shock factor $\sigma^{32}$ is unstable *in vivo*, and its degradation is catalyzed by FtsH (Herman *et al.*, 1995b; Tomoyasu *et al.*, 1995). Another soluble protein, λ cII, is also suggested to be a substrate of FtsH *in vivo* (Herman *et al.*, 1993). The cII protein is unstable and critical for lysogenization of λ (Banuett *et al.*, 1986). Thus, FtsH plays a key role in proteolytic quantity control over certain regulatory molecules. Another physiological role of FtsH is in the proteolytic quality control over some integral membrane protein complexes. SecY is a subunit of the SecYEG protein translocase in the cytoplasmic membrane. It is rapidly degraded (half-life, about 2 min) when it fails to associate with SecE. This degradation, mediated by FtsH, acts to eliminate potentially harmful incomplete assemblies of SecY (Kihara *et al.*, 1995). These important roles of FtsH may well explain the fact that FtsH is the only *E. coli* protease that is essential for viability. In addition, FtsH might have roles that are not ascribable to its proteolytic activity, since a loss of its function leads to an altered topology of a SecY–PhoA fusion protein. Also, *ftsH* mutants are partially defective in export of normally secreted proteins (Akiyama *et al.*, 1994a,b). These phenotypes are suppressible by overproduction of some molecular chaperones (Shirai *et al.*, 1996). FtsH could have a chaperone-like function.

Expression of the *ftsJ-ftsH* operon is directed by two promoters, one constitutive and the other subject to heat shock regulation by $\sigma^{32}$ (Herman *et al.*, 1995b). The *Bacillus subtilis* FtsH protein is also induced by heat or osmotic shock (Deurling *et al.*, 1995). The proteolytic activity of FtsH may be regulated: *in vivo* and *in vitro* studies suggest that activity is negatively modulated by association with the HflKC protein (Kihara *et al.*, 1996). HflKC itself was suggested to be a protease acting against the cII protein (Cheng *et al.*, 1988), but this has been questioned (Kihara *et al.*, 1996).

## Distinguishing Features

*E. coli* has several other ATP-dependent proteolytic systems, but FtsH is unique in that it is membrane bound and essential for growth. Polyclonal antibodies have been raised against peptides corresponding to the AAA signature region of FtsH (Tomoyasu *et al.*, 1993a).

## Related Peptidases

Yta10/Yta12 proteins are close homologs of FtsH and located in the mitochondrial inner membrane of *Saccharomyces cerevisiae* (Chapter 517). They form a complex of about 850 kDa, and participate in degradation of some mitochondrially encoded membrane proteins. Like FtsH, Yta10/Yta12 has been proposed to possess chaperone-like activity required for assembly of the ATP synthase (Arlt *et al.*, 1996).

## Further Reading

Tomoyasu *et al.* (1995) provide a recent overview of this enzyme and its function.

## References

Akiyama, Y., Ogura, T. & Ito, K. (1994a) Involvement of FtsH in protein assembly into and through the membrane. I. Mutations that reduce retention efficiency of a cytoplasmic reporter. *J. Biol. Chem.* **269**, 5218–5224.

Akiyama, Y., Shirai, Y. & Ito, K. (1994b) Involvement of FtsH in protein assembly into and through the membrane. II. Dominant mutations affecting FtsH functions. *J. Biol. Chem.* **269**, 5225–5229.

Akiyama, Y., Yoshihisa, T. & Ito, K. (1995) FtsH, a membrane-bound ATPase, forms a complex in the cytoplasmic membrane of *Escherichia coli. J. Biol. Chem.* **270**, 23485–23490.

Akiyama, Y., Kihara, A., Tokuda, H. & Ito, K. (1996) FtsH (HflB) is an ATP-dependent protease selectively acting on SecY and some other membrane proteins. *J. Biol. Chem.* **271**, 31196–31201.

Arlt, H., Tauer, R., Feldmann, H., Neupert, W. & Langer, T. (1996) The YTA10-12 complex, an AAA protease with chaperone-like activity in the inner membrane of mitochondria. *Cell* **85**, 875–885.

Banuett, F., Hoyt, M.A., McFarlane, L., Echols, H. & Herskowitz, I. (1986) *hflB*, a new *Escherichia coli* locus regulating lysogeny and the level of bacteriophage lambda cII protein. *J. Mol. Biol.* **187**, 213–224.

Begg, K.J., Tomoyasu, T., Donachie, W.D., Khattar, M., Niki, H., Yamanaka, K., Hiraga, S. & Ogura, T. (1992) *Escherichia coli* mutant Y16 is a double mutant carrying thermosensitive *ftsH* and *ftsI* mutations. *J. Bacteriol.* **174**, 2416–2417.

Cheng, H.H., Muhlrad, P.J., Hoyt, M.A. & Echols, H. (1988) Cleavage of the cII protein of phage 1 by purified HflA protease: control of the switch between lysis and lysogeny. *Proc. Natl Acad. Sci. USA* **85**, 7882–7886.

Confalonieri, F. & Duguet, M. (1995) A 200-amino acid ATPase module in search of a basic function. *BioEssays* **17**, 639–650.

Deurling, E., Paeslack, B. & Schumann, W. (1995) The *ftsH* gene of *Bacillus subtilis* is transiently induced after osmotic and temperature upshift. *J. Bacteriol.* **177**, 4105–4112.

Herman, C., Ogura, T., Tomoyasu, T., Hiraga, S., Akiyama, Y., Ito, K., Thomas, R., D'Ari, R. & Bouloc, P. (1993) Cell growth and λ

M

phage development controlled by the same essential *Escherichia coli* gene, *ftsH/hflB*. *Proc. Natl Acad. Sci. USA* **90**, 10861–10865.

Herman, C., Lecat, S., D'Ari, R. & Bouloc, P. (1995a) Regulation of the heat-shock response depends on divalent metal ions in an *hflB* mutant of *Escherichia coli*. *Mol. Microbiol.* **18**, 247–255.

Herman, C., Thévenet, D., D'Ari, R. & Bouloc, P. (1995b) FtsH/HflB controls the heat shock response in *Escherichia coli* via σ³² proteolysis. *Proc. Natl Acad. Sci. USA* **92**, 3516–3520.

Kihara, A., Akiyama, Y. & Ito, K. (1995) FtsH is required for proteolytic elimination of uncomplexed forms of SecY, an essential protein translocase subunit. *Proc. Natl Acad. Sci. USA* **92**, 4532–4536.

Kihara, A., Akiyama, Y. & Ito, K. (1996) A protease complex in the *Escherichia coli* plasma membrane: HflKC (HflA) forms a complex with FtsH (HflB), regulating its proteolytic activity against SecY. *EMBO J.* **15**, 6122–6131.

Qu, J.-N., Makino, S., Adachi, H., Koyama, Y., Akiyama, Y., Ito, K., Tomoyasu, T., Ogura, T. & Matsuzawa, H. (1996) The *tolZ* gene of *Escherichia coli* is identified as the *ftsH* gene. *J. Bacteriol.* **178**, 3457–3461.

Santos, D. & Almeida, D.F. (1975) Isolation and characterization of a new temperature-sensitive cell division mutant of *Escherichia coli* K-12. *J. Bacteriol.* **142**, 1502–1507.

Shirai, Y., Akiyama, Y. & Ito, K. (1996) Suppression of *ftsH* mutant phenotypes by overproduction of molecular chaperones. *J. Bacteriol.* **178**, 1141–1145.

Tomoyasu, T., Yamanaka, K., Murata, K., Suzaki, T., Bouloc, P., Kato, A., Niki, H., Hiraga, S. & Ogura, T. (1993a) Topology and subcellular localization of FtsH protein in *Escherichia coli*. *J. Bacteriol.* **175**, 1352–1357.

Tomoyasu, T., Yuki, T., Morimura, S., Mori, H., Yamanaka, K., Niki, H., Hiraga, S. & Ogura, T. (1993b) The *Escherichia coli* FtsH protein is a prokaryotic member of a protein family of putative ATPases involved in membrane functions, cell cycle control, and gene expression. *J. Bacteriol.* **175**, 1344–1351.

Tomoyasu, T., Gamer, J., Bukau, J., Kanemori, M., Mori, H., Rutman, A.J., Oppenheim, A.B., Yura, T., Yamanaka, K., Niki, H., Hiraga, S. & Ogura, T. (1995) *Escherichia coli* FtsH is a membrane-bound, ATP-dependent protease which degrades the heat-shock transcription factor σ³². *EMBO J.* **14**, 2551–2560.

*Yoshinori Akiyama*
*Department of Cell Biology,*
*Institute for Virus Research,*
*Kyoto University,*
*Kyoto 606-01, Japan*
*Email: yakiyama@kyoto-u.ac.jp*

*Koreaki Ito*
*Department of Cell Biology,*
*Institute for Virus Research,*
*Kyoto University,*
*Kyoto 606-01, Japan*
*Email: kito@kyoto-u.ac.jp*

# 517. Mitochondrial m- and i-AAA proteases

## Databanks

*Peptidase classification: clan MX, family M41, MEROPS ID: M41.003*
*NC-IUBMB enzyme classification: none*
*Databank codes:*

| Species | Gene | SwissProt | PIR | EMBL (cDNA) | EMBL (genomic) |
|---|---|---|---|---|---|
| *Caenorhabditis elegans* | M03c11.5 | – | – | – | Z49128: cosmid M03C11 |
| *Saccharomyces cerevisiae* | AFG3 | P39925 | S46611 | X76643 X81066 | U18778: cosmids 9537, 9581, 9495 and 9867 |
| *Saccharomyces cerevisiae* | RCA1 | P40341 | S54465 | U09358 X81068 | Z49259: chromosome XIII cosmid 9582 |
| *Saccharomyces cerevisiae* | YME1 | P32795 | S46608 | D16332 L14616 X81067 | Z49274: chromosome XVI cosmid 9367 |
| *Schistosoma mansoni* | – | P46508 | – | Z29947 | – |

## Name and History

*YME1* (for **y**east **m**itochondrial **e**scape), also variously called *OSD1* and *YTA11*, was isolated from *Saccharomyces cerevisiae* in three independent mutant screens. An assay that allows the detection of DNA escape from mitochondria to the nucleus revealed that mutations in *YME1* caused a high rate of mitochondrial DNA escape to the nucleus when compared to wild-type yeast (Thorsness & Fox, 1993). *OSD1* was isolated as one of three independent loci that, when mutated, created a defect in the degradation of unassembled cytochrome oxidase subunit II (Nakai *et al.*, 1995). Both *YME1* and *OSD1* were cloned based on the inability of the respective mutations (*yme1-1*, *osd1-1*) to grow on nonfermentable carbon sources at 37°C (Thorsness *et al.*, 1993; Nakai *et al.*, 1995). *RCA1* (for **r**espiratory **c**hain **a**ssembly), also called *YTA12*, was cloned via complementation of a respiration-defective mutant of the G25 complementation group in *S. cerevisiae*. G25 mutants are grossly deficient in mitochondrial respiratory and ATPase complexes (Tzagoloff *et al.*, 1994). *AFG3* (for **A**TPase **f**amily **g**ene), also known as *YTA10*, was cloned based on its complementation of a temperature-sensitive respiratory-deficient mutation (Guélin *et al.*, 1994). All three genes were isolated as members of a novel gene family in a PCR-based screen. The genes were designated *YTAs* (for **y**east **T**at-binding **a**nalogs) based on their sequence homology with *TBP1*, a human immunodeficiency **T**at-**b**inding **p**rotein (Schnall *et al.*, 1994), but are better characterized as members of the 'AAA family' (for **A**TPases **a**ssociated with a variety of cellular **a**ctivities) (Arlt *et al.*, 1996). The proteases corresponding to these genes are designated **Yme1p** or **i-AAA protease** and **Afg3p** and **Rca1p** or **m-AAA protease**.

## Activity and Specificity

Cells lacking Yme1p stabilize nonassembled cytochrome oxidase subunit II in *cox4* mutants, suggesting that this subunit is a putative substrate for Yme1p (Nakai *et al.*, 1995; Pearce & Sherman, 1995; Weber *et al.*, 1996). Moreover, degradation of a hybrid fusion protein [Su9(1–66)-pCOXII(1–74)-DHFR], containing the mitochondrial pre sequence of subunit 9 of the ATPase of *Neurospora crassa*, the first 74 amino acids of cytochrome oxidase subunit II and mouse dihydrofolate reductase, occurs in the intermembrane space (Leonhard *et al.*, 1996). Degradation of the hybrid protein resulted in the formation of several subfragments, with cleavage of the hybrid protein occurring at the outer surface of the inner membrane. Cleavage of the hybrid protein occurred near residue 40, at the N-terminal end of the transmembrane domain of the cytochrome oxidase subunit II portion of the hybrid protein (Leonhard *et al.*, 1996). Degradation of the hybrid protein substrate was inhibited in mitochondria that had been depleted of ATP in the intermembrane space via addition of carboxyatractyloside and apyrase and in cells lacking Yme1p (Leonhard *et al.*, 1996). This proteolytic activity has been termed 'i-AAA protease' due to its location in the intermembrane space of mitochondria.

The Su9(1–66)-pCOXII(1–74)-DHFR hybrid fusion protein was also cleaved by an activity in the mitochondrial matrix referred to as 'm-AAA protease' (Leonhard *et al.*,

1996). This proteolytic complex is composed of Rca1p and Afg3p. *In vitro* labeling of isolated mitochondria revealed that the m-AAA protease complex is directly or indirectly involved in the degradation of a number of mitochondrially encoded gene products, including cytochrome *b*, the cytochrome *c* oxidase subunits I and III, and the $F_0$-ATP synthase subunits 6, 8 and 9 (Guélin *et al.*, 1996). *In vitro* synthesis of mitochondrial proteins in the presence of puromycin produces incompletely synthesized polypeptides. Rapid proteolysis of these peptides is dependent upon m-AAA protease (Pajic *et al.*, 1994; Arlt *et al.*, 1996).

Afg3p-mediated proteolysis can be inhibited with the membrane-permeable chelating agent 1,10-phenanthroline and can be almost completely restored via addition of $Fe^{2+}$, $Co^{2+}$ or $Mn^{2+}$ and partially restored with $Zn^{2+}$, whereas addition of $Mg^{2+}$ and $Ca^{2+}$ had no effect (Nakai *et al.*, 1994; Pajic *et al.*, 1994; Yasuhara *et al.*, 1994; Leonhard *et al.*, 1996). Afg3p-mediated proteolysis can also be inhibited with the cysteine-modifying agent *N*-methylmaleimide. A cysteine residue (at position 565) in close proximity to the HEXXH motif may be responsible for the sensitivity to thiol-blocking agents (Pajic *et al.*, 1994).

In addition to proteolytic activity, the m-AAA protease also displays chaperone activity specific for the assembly of ATP synthase and respiratory chain complexes in the mitochondrial inner membrane (Arlt *et al.*, 1996; Rep *et al.*, 1996). Cells lacking Rca1p and Afg3p fail to utilize nonfermentable carbon sources. However, the respiratory function of a *rca1/yta12 afg3/yta10* double mutant strain was restored upon expression of proteolytically inactive mutant forms of Rca1p and Afg3p. It has thus been concluded that m-AAA protease plays a role in assembly of respiratory chain complexes and ATP synthase. The respiratory-growth defect found in yeast lacking m-AAA protease can be corrected by overexpression of Oxa1p, which is required for assembly of the cytochrome *c* oxidase and ATP synthase (Altamura *et al.*, 1996), providing additional evidence for the notion that the Rca1p-Afg3p complex fulfills important functions in assembly of mitochondrial membrane complexes.

## Structural Chemistry

Yme1p, Rca1p and Afg3p form a subfamily of the AAA family of proteins. All AAA family members exhibit a highly conserved domain of 300 amino acids containing specialized forms of ATPase A and B boxes (GPPGTGKT and DE[L/I]D, respectively) and a highly homologous region of 63 amino acid residues, referred to as the DEXD domain (Schnall *et al.*, 1994). Yme1p, Rca1p and Afg3p share extensive sequence similarities with each other and with the FtsH protein of *E. coli* (Tomoyasu *et al.*, 1993) (Chapter 516). All these proteins are similar in size, are integral membrane proteins and contain a sequence motif (HEXXH) characteristic of metallopeptidases in clans MA and MB.

The predicted molecular mass of Yme1p, as deduced from the DNA sequence, is 81 768 Da with a pI of 6.89. Yme1p is embedded in the inner mitochondrial membrane and is resistant to extraction by alkaline carbonate or 6 M urea (Nakai *et al.*, 1995; Leonhard *et al.*, 1996; Weber *et al.*,

**M**

1996). The C-terminus, which comprises the bulk of Yme1p, is located in the intermembrane space and Yme1p is part of an 850 kDa complex (Leonhard *et al.*, 1996).

Rca1p protein consists of 785 amino acid residues with a calculated molecular mass of 88 412 and a pI of 6.12. Afg3p contains 741 amino acids with a predicted molecular mass of 82 180 and a pI of 9.18. Both Rca1p and Afg3p contain two hydrophobic sequences of sufficient length to form membrane spanning $\alpha$ helices and both proteins can be released only with solubilizing detergents, indicating that they are integrated in the inner mitochondrial membrane (Pajic *et al.*, 1994; Tzagoloff *et al.*, 1994; Paul & Tzagoloff, 1995). The N- as well as the longer C-termini of Afg3p (Pajic *et al.*, 1994) and Rca1p (Arlt *et al.*, 1996) protrude into the mitochondrial matrix space, thus placing the bulk of the polypeptide chain in the mitochondrial matrix. This is similar to the membrane topology proposed for the FtsH protein in *E. coli* (Tomoyasu *et al.*, 1993) (Chapter 516). Rca1p and Afg3p form a complex of about 850 kDa (*m*-AAA protease) in which both proteins are present in equimolar amounts. Formation of this complex requires the presence of nucleotides (ATP-gS, AMP-PNP, ADP, ATP or GTP) but not the hydrolysis of ATP. However, ATP hydrolysis is essential for enzymatic activity of the complex (Arlt *et al.*, 1996). Assembly of Rca1p and Afg3p into a Rca1p–Afg3p complex is crucial for proteolytic activity. Both subunits of this complex exert metal and ATP-dependent peptidase activity (Arlt *et al.*, 1996).

## Preparation

No purification scheme for Yme1p, Afg3p or Rca1p has been reported.

## Biological Aspects

This family of proteases has only been found in the mitochondria of the yeast *Saccharomyces cerevisiae*. Related proteases have been found in *Escherichia coli* (*ftsH*; Tomoyasu *et al.*, 1995) (Chapter 516) and *Lactococcus lactis* (*tma1*; Nilsson *et al.*, 1994). The pleiotropic phenotypes resulting from loss of any one of these mitochondrial proteases highlights the fundamental importance of their activities to the assembly and function of mitochondrial compartments. The *m*-AAA protease complex, composed of Afg3p and Rca1p, has an important chaperone function in the assembly of mitochondrial inner-membrane protein complexes (Arlt *et al.*, 1996; Rep *et al.*, 1996). This function is dependent upon the inherent ATPase activity of Afg3p and Rca1p. Proteins that cannot be assembled into higher-order complexes due to misfolding, blockages in complex formation, or molar excess are subjected to proteolysis. A similar chaperone-like function has not been demonstrated for Yme1p, although loss of Yme1p activity has pleiotropic consequences for yeast. Phenotypes of *yme1* mutant yeast include an increased escape of DNA from mitochondria to the nucleus, inability to utilize nonfermentable carbon sources for growth at 37°C, slow growth on rich glucose media at 14°C (Thorsness *et al.*, 1993), slow growth in the complete absence of mitochondrial DNA (Weber *et al.*, 1995) and mitochondrial compartments with altered morphology (Campbell *et al.*, 1994).

## Further Reading

For a review, see Langer *et al.* (1995).

## References

Altamura, N., Capitanio, N., Bonnefoy, N., Papa, S. & Dujardin, G. (1996) The *Saccharomyces cerevisiae OXA1* gene is required for the correct assembly of cytochrome *c* oxidase and oligomycin-sensitive ATP synthase. *FEBS Lett.* **382**, 111–115.

Arlt, H., Tauer, R., Feldmann, H., Neupert, W. & Langer, T. (1996) The *YTA10-12* complex, an AAA protease with chaperone-like activity in the inner membrane of mitochondria. *Cell* **85**, 875–885.

Campbell, C.L., Tanaka, N., White, K.H. & Thorsness, P.E. (1994) Mitochondrial morphological and functional defects in yeast caused by *yme1* are suppressed by mutation of a 26S protease subunit homologue. *Mol. Biol. Cell* **5**, 899–905.

Guélin, E., Rep, M. & Grivell, L.A. (1994) Sequence of the *AFG3* gene encoding a new member of the FtsH/Yme1/Tma subfamily of the AAA-protein family. *Yeast* **10**, 1389–1394.

Guélin, E., Rep, M. & Grivell, L.A. (1996) Afg3p, a mitochondrial ATP-dependent metalloprotease, is involved in the degradation of mitochondrially-encoded Cox1, Cox3, Cob, Su6, Su8, and Su9 subunits of the inner membrane complexes III, IV and V. *FEBS Lett.* **381**, 42–46.

Langer, T., Pajic, A., Wagner, I. & Neupert, W. (1995) Proteolytic breakdown of membrane-associated polypeptides in mitochondria of *Saccharomyces cerevisiae*. *Methods Enzymol.* **260**, 495–503.

Leonhard, K., Herrmann, J.M., Stuart, R.A., Mannhaupt, G., Neupert, W. & Langer, T. (1996) AAA proteases with catalytic sites on opposite membrane surfaces comprise a proteolytic system for the ATP-dependent degradation of inner membrane proteins in mitochondria. *EMBO J.* **15**, 4218–4229.

Nakai, T., Mera, Y., Toshimasa, Y. & Ohashi, A. (1994) Divalent metal ion-dependent degradation of unassembled subunits 2 and 3 of cytochrome oxidase. *J. Biochem.* **116**, 752–758.

Nakai, T., Yasuhara, T., Fujiki, T. & Ohashi, A. (1995) Multiple genes, including a member of the AAA family, are essential for degradation of unassembled subunit 2 of cytochrome *c* oxidase in yeast mitochondria. *Mol. Cell. Biol.* **15**, 4441–4452.

Nilsson, D., Lauridsen, A.A., Tomoyasu, T. & Ogura, T. (1994) A *Lactococcus lactis* gene encodes a membrane protein with putative ATPase activity that is homologous to the essential *Escherichia coli ftsH* gene product. *Microbiology* **140**, 2601–2610.

Pajic, A., Tauer, R., Feldmann, H., Neupert, W. & Langer, T. (1994) Yta10p is required for the ATP-dependent degradation of polypeptides in the inner membrane of mitochondria. *FEBS Lett.* **353**, 201–206.

Paul, M.F. & Tzagoloff, A. (1995) Mutations in *RCA1* and *AFG3* inhibit F1-ATPase assembly in *Saccharomyces cerevisiae*. *FEBS Lett.* **373**, 66–70.

Pearce, D.A. & Sherman, F. (1995) Degradation of cytochrome oxidase subunits in mutants of yeast lacking cytochrome *c* and suppression of the degradation by mutation of *yme1*. *J. Biol. Chem.* **270**, 20879–20882.

Rep, M., van Dijl, J.M., Suda, K., Schatz, G., Grivell, L.A. & Suzuki, C.K. (1996) Promotion of mitochondrial membrane complex assembly by a proteolytically inactive yeast Lon. *Science* **274**, 103–107.

Schnall, R., Mannhaupt, G., Stucka, R., Tauer, R., Ehnle, S., Schwarzlose, C., Vetter, I. & Feldmann, H. (1994) Identification of a set of yeast genes coding for a novel family of putative ATPases with high similarity to constituents of the 26S protease complex. *Yeast* **10**, 1141–1155.

Thorsness, P.E. & Fox, T.D. (1993) Nuclear mutations in *Saccharomyces cerevisiae* that affect the escape of DNA from mitochondria to the nucleus. *Genetics* **134**, 21–28.

Thorsness, P.E., White, K.H. & Fox, T.D. (1993) Inactivation of *YME1*, a gene coding a member of the *SEC18, PAS1, CDC48* family of putative ATPases, causes increased escape of DNA from mitochondria in *Saccharomyces cerevisiae*. *Mol. Cell. Biol.* **13**, 5418–5426.

Tomoyasu, T., Yamanaka, K., Murata, K., Suzaki, T., Bouloc, P., Kato, A., Niki, H., Hiraga, S. & Ogura, T. (1993) Topology and subcellular localization of FtsH protein in *Escherichia coli*. *J. Bacteriol.* **175**, 1352–1357.

Tomoyasu, T., Gamer, J., Bukau, B., Kanemori, M., Mori, H., Rutman, A.J., Oppenheim, A.B., Yura, T., Yamanaka, K., Niki, H., Hiraga, S. & Ogura, T. (1995) *Escherichia coli* FtsH is a membrane-bound, ATP-dependent protease which degrades the heat shock transcription factor $\sigma$32. *EMBO J.* **14**, 2551–2560.

Tzagoloff, A., Yue, J., Jang, J. & Paul, M.F. (1994) A new member of a family of ATPases is essential for assembly of mitochondrial respiratory chain and ATP synthesis complexes in *Saccharomyces cerevisiae*. *J. Biol. Chem.* **269**, 26144–26151.

Weber, E.R., Rooks, R.S., Shafer, K.S., Chase, J.W. & Thorsness, P.E. (1995) Mutations in the mitochondrial ATP synthase gamma subunit suppress a slow-growth phenotype of *yme1* yeast lacking mitochondrial DNA. *Genetics* **140**, 435–442.

Weber, E.R., Hanekamp, T. & Thorsness, P.E. (1996) Biochemical and functional analysis of the *YME1* gene product, an ATP and zinc-dependent mitochondrial protease from *S. cerevisiae*. *Mol. Biol. Cell* **7**, 307–317.

Yasuhara, T., Mera, Y., Nakai, T. & Ohashi, A. (1994) ATP-dependent proteolysis in yeast mitochondria. *J. Biochem.* **115**, 1166–1171.

***Theodor Hanekamp***
*Department of Molecular Biology,*
*University of Wyoming,*
*Laramie, WY 82071-3944, USA*
*Email: hanekamp@uwyo.edu*

***Peter E. Thorsness***
*Department of Molecular Biology,*
*University of Wyoming,*
*Laramie, WY 82071-3944, USA*
*Email: thorsnes@uwyo.edu*

# *518. Cytophagalysin*

## Databanks

*Peptidase classification: clan MX, family M43, MEROPS ID: M43.001*
*NC-IUBMB enzyme classification: none*
*Databank codes:*

| Species | SwissProt | PIR | EMBL (cDNA) | EMBL (genomic) |
|---------|-----------|-----|-------------|----------------|
| *Cytophaga* sp. | – | – | D50600 | – |

## Name and History

A strain of bacteria that secretes collagenase was isolated and identified as *Cytophaga* sp. L43-1 in 1991 (Kojima, 1991). Later, the enzyme was purified from the culture supernatant of this strain (Sasagawa *et al.*, 1993) and then the gene of the enzyme (*cog*) was cloned and sequenced (Sasagawa *et al.*, 1996). The name **cytophagalysin** denotes the collagenase originally obtained from *Cytophaga* sp. L43-1.

## Activity and Specificity

Both insoluble and acid-soluble collagens and gelatin are digested by cytophagalysin. The enzyme does not hydrolyze three synthetic substrates of other bacterial collagenases: Pz-Pro-Leu-Gly-Pro-Arg, Z-Gly-Pro-Gly-Gly-Pro-Ala and Z-Gly-Pro-Leu-Gly-Pro (Sasagawa *et al.*, 1993). The collagenases from *Vibrio alginolyticus* (Chapter 367) and *Clostridium histolyticum* (Chapter 368) cleave $\beta$-casein (Gilles & Keil, 1976), and cytophagalysin also does so; it cleaves at Pro63$\downarrow$Gly64 and at Met102$\downarrow$Ala103 (Sasagawa *et al.*, 1993).

The optimal pH of the enzyme is 7.5 using acid-soluble collagen as a substrate. By pH 6.0 and 8.5, the activity decreases by 50%. The optimal temperature is 30°C using gelatin as a substrate.

Assays can be made using insoluble collagen, acid-soluble collagen, and gelatin as substrates as described by Sasagawa *et al.* (1993). Collagens (insoluble type I from Achilles' tendon of cattle and acid-soluble type III from calf skin) are available from Sigma (see Appendix 2 for full names and addresses of suppliers).

EDTA (0.1 mM) inhibits the activity of cytophagalysin completely. DFP inhibits the activity partially (70%) at 0.1 mM and completely at 1 mM. Iodoacetamide (1 mM) or soybean trypsin inhibitor (0.1 mg ml$^{-1}$) have no inhibitory activity (Sasagawa *et al.*, 1993). When acid-soluble collagen, gelatin and $\beta$-casein are used as substrates, similar pH and heat stabilities of cytophagalysin and also similar effects of inhibitors on the activity are observed. Therefore, the enzymatic reaction of cytophagalysin with these various substrates collagen, gelatin and $\beta$-casein seems to be catalyzed by the same active center in cytophagalysin.

## Structural Chemistry

Cytophagalysin is a single-chain protein of about 120 kDa, pI 4.8. The structural gene of the enzyme was cloned and sequenced (Sasagawa *et al.*, 1996). In the deduced amino acid sequence, the putative zinc-binding motif, His-Glu-Phe-Gly-His, is found. The two His residues in this motif may be zinc ligands.

## Preparation

Cytophagalysin has been purified 16.5-fold from the culture fluid of *Cytophaga* sp. L43-1. The gene of the enzyme has been expressed in *Escherichia coli* (Sasagawa *et al.*, 1996).

## Biological Aspects

A *cog* gene encoding cytophagalysin has been cloned, and the nucleotides sequenced. The structural gene of *cog* consists of 3846 bp and encodes a polypeptide consisting of 1282 amino acid residues with a predicted molecular mass of 130 kDa which is synthesized as a precursor. The *cog* gene is expressed in *E. coli* using the *lac* promoter and ribosomal binding sequence in plasmid vector pUC119 or pKK223-3, but not its own putative promoter and Shine–Dalgarno sequence (Sasagawa *et al.*, 1996).

## Distinguishing Features

Although cytophagalysin is similar to collagenase from *Vibrio* (Chapter 367) (Gilles & Keil, 1976) in properties including cleavage sites in $\beta$-casein, it does not hydrolyze small peptides (Pz-PLGPR, Z-GPGGPA and Z-GPLGP)

unlike the latter enzyme. Also, cytophagalysin is unrelated in sequence to the *Vibrio* enzyme (Takeuchi *et al.*, 1992; Sasagawa *et al.*, 1996). The collagenase from *Clostridium histolyticum* (Chapter 368) cleaves $\beta$-casein at the position of Ser57↓Leu58 and Ser124↓Leu125, whereas cytophagolysin cleaves at Pro63↓Gly64 and Met102↓Ala103.

## Related Peptidases

Bacterial collagenases have been obtained from a variety of microorganisms such as *Clostridium histolyticum* (Chapter 368) (Seifter & Harper, 1970), *Vibrio alginolyticus* (Chapter 367) (Reid *et al.*, 1980), *Vibrio* B-30 (Chapter 368) (Merkel & Dreisbach, 1978), *Pseudomonas marinoglutinosa* (Chapter 368) (Hanada *et al.*, 1973), *Streptomyces* sp. (Chapter 368) (Endo *et al.*, 1987) and *Cytophaga* sp. (Sasagawa *et al.*, 1993).

## Further Reading

For a review, see Sasagawa *et al.* (1996).

## References

Endo, A., Murakawa, S., Shimizu, H. & Shiraishi, Y. (1987) Purification and properties of collagenase from a *Streptomyces* species. *J. Biochem.* **102**, 163–170.

Gilles, A.M. & Keil, B. (1976) Cleavage of $\beta$-casein by collagenases from *Achromobacter iophagus* and *Clostridium histolyticum*. *FEBS Lett.* **65**, 369–372.

Hanada, K., Mizutani, T., Yamagishi, M., Tsuji, H., Misaki, T. & Sawada, J. (1973) The isolation of collagenase and its enzymological and physico-chemical properties. *Agric. Biol. Chem.* **37**, 1771–1781.

Kojima, H. (1991) Methods for the preparation of collagenase. Jpn Kokai Tokkyo Koho No. 3–91478.

Merkel, J.R. & Dreisbach, J.H. (1978) Purification and characterization of a marine bacterial collagenase. *Biochemistry* **17**, 2857–2863.

Reid, G.C., Woods, D.R. & Robb, F.T. (1980) Peptone induction and rifampicin-insensitive collagenase production by *Vibrio alginolyticus. J. Bacteriol.* **142**, 447–454.

Sasagawa, Y., Kamio, Y., Matsubara, Y., Suzuki, K., Kojima, H. & Izaki, K. (1993) Purification and properties of collagenase from *Cytophaga* sp. L43–1 strain. *Biosci. Biotechnol. Biochem.* **57**, 1894–1898.

Sasagawa, Y., Izaki, K., Matsubara, Y., Suzuki, K., Kojima, H. & Kamio, Y. (1996) Molecular cloning and sequence of the gene encoding the collagenase from *Cytophaga* sp. L43–1 strain. *Biosci. Biotechnol. Biochem.* **59**, 2068–2073.

Seifter, S. & Harper, E. (1970) Collagenases. *Methods Enzymol.* **19**, 613–635.

Takeuchi, H., Shibano, Y., Morihara, K., Fukushima, J., Inami, S.,

Keil, B., Gilles, M., Kawamoto, S. & Okuda, K. (1992) Structural gene and complete amino acid sequence of *Vibrio alginolyticus* collagenase. *Biochem. J.* **281**, 703–708.

***Yoshiyuki Kamio***
*Department of Applied Biological Chemistry,*
*Faculty of Agriculture,*
*Tohoku University,*
*1-1 Amamiya-machi, Tsutsumi-dori, Aoba-ku, Sendai*
*981, Japan*
*Email: ykamio@biochem.tohoku.ac.jp*

***Yoshikiyo Sasagawa***
*Biochemical Laboratory, Nitto Boseki Co., Ltd.,*
*Koriyama 963, Japan*

M

# 519. Introduction: unsequenced metallopeptidases

About one tenth of the metallopeptidases that have been characterized in sufficient detail to merit inclusion in the present volume are still of unknown sequence, or sequence data became available too late to allow the chapter to be placed with those on other members of the family. In the latter category is *peptidyl-Lys metalloendopeptidase* (Chapter 537). Fungal metalloendopeptidases cleaving on the carboxyl side of Lys residues have been known for many years, but only very recently have amino acid sequences of the enzyme from two species of fungus been reported (Nonaka *et al.*, 1997). These show distant relationship to penicillolysin and deuterolysin, in family M35.

In the absence of any structure-based classification we have arranged these chapters according to the sub-subclasses on the IUBMB enzyme list. Amongst the unsequenced aminopeptidases are *tripeptide aminopeptidase* (Chapter 520), *clostridial aminopeptidase* (Chapter 521), *X-Trp aminopeptidase* (Chapter 522) and *tryptophanyl aminopeptidase* (Chapter 523). Also included are seven chapters on as yet unsequenced dipeptidases: *X-Arg dipeptidase* (Chapter 524), *Pro-X dipeptidase* (Chapter 525), *Met-X dipeptidase* (Chapter 526), *cytosol nonspecific dipeptidase* (Chapter 527), *β-Ala-His dipeptidase* (Chapter 528), *X-His dipeptidase* (Chapter 529) and *lysosomal dipeptidase* (Chapter 530).

Two of the unsequenced metallopeptidases release C-terminal dipeptides, either from such bioactive peptides as atriopeptin II (*peptidyl-dipeptidase B*) (Chapter 531) or from C-terminal tetra-proline sequences: -Pro-Pro+Pro-Pro (*peptidyl-dipeptidase* (*Streptomyces*)) (Chapter 532).

There are four unsequenced carboxypeptidases to mention: *alanine carboxypeptidase* (Chapter 533), *membrane Pro-X carboxypeptidase* (Chapter 534), *carboxypeptidase G₃* (Chapter 535) and *mitochondrial carboxypeptidase* (Chapter 536).

Finally, there are six articles on metalloendopeptidases for which amino acid sequences are not yet (or only recently) available. These are *peptidyl-Lys metalloendopeptidase* (Chapter 537) *aureolysin* (Chapter 538), *peptidyl-Asp metalloendopeptidase* (Chapter 539), *magnolysin* (Chapter 540), *dactilysin* (Chapter 541) and *dynorphin-converting enzyme* (Chapter 542).

## Reference

Nonaka, T., Dohmae, N., Hashimoto, Y. & Takio, K. (1997) Amino acid sequences of metalloendopeptidases (MEP) specific for acyl-lysin bonds from *Grifolia frondosa* and *Pleurotus ostreatus* fruiting bodies. *J. Biol. Chem.* **267**, 30032–30039.

# 520. Tripeptide aminopeptidase

## Databanks

*Peptidase classification: clan MX, family M99, MEROPS ID: M9A.002*
*NC-IUBMB enzyme classification: EC 3.4.11.7*
*Chemical Abstracts Service registry number: 9074-83-3*
*Databank codes: no sequence data available*

## Name and History

In studies conducted on peptidases of pig intestinal mucosa many years ago, an enzyme that hydrolyzes Leu+Gly-Gly to yield leucine and glycylglycine was demonstrated. This enzyme was classified as a **glycine imidoendopeptidase** (Smith & Bergmann, 1944). The occurrence of this peptidase activity in serum was considered to be related to its liberation into the circulating body fluids during the course of the rapid turnover of lymphoid cells. The peptidase partially purified from crude saline extracts of calf thymus was termed **lymphopeptidase** (Fruton *et al.*, 1948); however, the substrate specificity of more highly purified enzyme preparations, and also the activity in extracts of several animal tissues, justifies the assignment of the better term **tripeptidase** to replace the designation lymphopeptidase (Ellis & Fruton, 1951). Afterwards the enzyme was classified into the xx-aminopeptide amino acid hydrolase category as **aminotripeptidase** (EC 3.4.1.3 ) in the first report of the Commission on Enzyme Nomenclature in 1961, and then transferred to the entry **tripeptide aminopeptidase** (EC 3.4.11.4) in 1964.

## Activity and Specificity

The enzyme acts as an aminopeptidase capable of removing N-terminal neutral amino acid residues from tripeptide substrates having an L-configuration for the first two amino acids. Tripeptides with a charged N-terminal residue such as lysine or glutamic acid are poor substrates or not hydrolyzed at all (Chenoweth *et al.*, 1973a; Doumeng & Maroux, 1979; Sachs & Marks, 1982). The enzyme has no action on dipeptides, dipeptide amides, acyltripeptides, tripeptide amides, or tetrapeptides (Smith, 1955); also, peptides lacking a hydrogen at the N-terminal peptide bond, such as Gly-L-Pro-Gly are not susceptible to hydrolysis (Chenoweth *et al.*, 1973a).

Several assay methods can be employed to measure the hydrolysis of peptides by this enzyme: spectrophotometric measurement of the decrease in absorbance at 230 nm (Doumeng & Maroux, 1979), determination of the nature and amount of released amino acids with an amino acid analyzer (Doumeng & Maroux, 1979), automated fluorimetric determination of released amino acid coupled with *o*-phthaldialdehyde (Sachs & Marks, 1982), and detection of cleavage products by HPLC at 210 nm (Sachs & Marks, 1982; Strauch & Miller, 1983; Hiraoka & Harada, 1993). A coupling reaction of the liberated N-terminal amino acid with a specific enzyme can also be applied, e.g. Leu enzymatically formed from Leu+Gly-Gly is reacted with leucine dehydrogenase in the presence of $NAD^+$, and the increased absorbance at 340 nm is monitored (Kanda *et al.*, 1984a). Alternatively, the absorbance at 530 nm is monitored using the assay mixture of L-amino acid oxidase, horseradish peroxidase and *o*-dianisidine (Hermsdorf, 1978; Hayashi & Oshima, 1980).

The pH optimum of the purified enzyme varies from 7.5 to 8.5 depending upon the source. Among chelating agents 1,10-phenanthroline is a potent inhibitor of brain (Sachs & Marks, 1982) and kidney (Doumeng & Maroux, 1979) enzymes, but EDTA shows a weak inhibitory effect on both enzymes. Bestatin shows a strong inhibitory effect on several purified preparations. The inhibition is competitive, and the $K_i$ value is estimated to be $5 \times 10^{-7}$ M for the monkey brain enzyme (Hayashi & Oshima, 1980). Among divalent metal ions, $Zn^{2+}$ (0.1–1.0 mM) shows strong inhibition, whereas others, e.g. $Mn^{2+}$, $Co^{2+}$, $Cd^{2+}$, $Pb^{2+}$, $Ca^{2+}$ and $Hg^{2+}$, exhibit different degrees of inhibition (Hayashi & Oshima, 1980; Sachs & Marks, 1982; Chenoweth *et al.*, 1973a; Hiraoka & Harada, 1993). Captopril, a mercaptoprolyl agent known to inhibit angiotensin-converting enzyme (Chapter 359), was found to effect reversible and competitive inhibition (Sachs & Marks, 1982). No activators are required for the enzyme. The amino acid composition of various purified enzymes has been reported and compared (Hayashi & Oshima, 1980).

## Structural Chemistry

Tripeptide aminopeptidase purified from pig kidney is composed of monomeric and dimeric forms (Chenoweth *et al.*, 1973b). Based on sedimentation-equilibrium studies, the $M_r$ values of the two purified forms are calculated to be 71 100 and 142 000, respectively. The zinc content of the pig kidney enzyme has not yet been stoichiometrically determined, but the enzyme purified from rabbit intestinal mucosa possesses 1 mol zinc per relative molecular mass of 50 000 (Doumeng

& Maroux, 1979). For the purified enzyme from rat brain, the activity is decreased after dialysis with EDTA, but the activity is not restored upon addition of $Zn^{2+}$ (Sachs & Marks, 1982). Therefore, the role of this metal ion in the purified enzymes is unclear. The pI is 4.92 to 5.2 for pig kidney enzyme (Chenoweth *et al.*, 1973b; Khilji *et al.*, 1979) and 7.0 for dental follicle enzyme (Hiraoka & Harada, 1993).

## Preparation

Several organs have been used for purification of the enzyme, including calf thymus (Ellis & Fruton, 1951), pig kidney (1100-fold; Chenoweth *et al.*, 1973a), rabbit intestinal mucosa (630-fold; Doumeng & Maroux, 1979), monkey brain (3429-fold; Hayashi & Oshima, 1980), rat brain (686-fold; Sachs & Marks, 1982) and cattle dental follicle (793-fold; Hiraoka & Harada, 1993).

## Biological Aspects

Tripeptide aminopeptidase is distributed in various human tissues, with the highest activity observed in liver and lymphocytes. The highest specific activity of the enzyme is observed in the soluble fraction prepared from liver, and 65% of the total activity is recovered in this fraction. Elevation of the enzyme activity is observed in the sera of patients with liver disorders, leukemia and autoimmune diseases (Kanda *et al.*, 1984b). Clinical usefulness of serum tripeptide aminopeptidase activity in diagnosing liver diseases has been reported (Kanda *et al.*, 1987).

In terms of biological function, this enzyme has been postulated to hydrolyze the Pro+Xaa-Yaa-type peptides crossing the brush border membrane (Doumeng & Maroux, 1979). The possible role of activity of the enzyme towards biologically active peptides, e.g. chemotactic factor (Met-Leu-Tyr), liver growth factor (Gly-His-Lys) and central nervous system tripeptide (Thr-Val-Leu), has also been discussed (Sachs & Marks, 1982).

## Related Peptidases

Aminotripeptidase (peptidase T) activity has been isolated from *Escherichia coli* strain AJ005 (Hermsdorf, 1978) and *Salmonella typhimurium* (Strauch & Miller, 1983). The activity pattern of the enzyme from *Lactococcus lactis* subsp. *cremoris* AM2 toward different tripeptides was studied comparatively (Bacon *et al.*, 1991). The relationship of the bacterial peptidases to the mammalian enzyme is not clear at this time; these bacterial enzymes are considered separately in Chapter 485.

## References

Bacon, C.L., Ni Fhaolain, I., Jennings, P.V. & O'Cuinn, G. (1991) Aminotripeptidase activity in *Lactoccocus lactis* subsp *cremoris* AM2. *Biochem. Soc. Trans.* **19**, 35S.

Chenoweth, D., Mitchel, R.E.J. & Smith, E.L. (1973a) Aminotripeptidase of swine kidney. I. Isolation and characterization of three different forms, utility of the enzyme in sequence work. *J. Biol. Chem.* **248**, 1672–1683.

Chenoweth, D., Brown, D.M., Valenzuela, M.A. & Smith, E.L. (1973b) Aminotripeptidase of swine kidney II. Amino acid composition and molecular weight. *J. Biol. Chem.* **248**, 1684–1686.

Doumeng, C. & Maroux, S. (1979) Aminotripeptidase, a cytosol enzyme from rabbit intestinal mucosa. *Biochem. J.* **177**, 801–808.

Ellis, D. & Fruton, J.S. (1951) On the proteolytic enzymes of animal tissues. IX. Calf thymus tripeptidase. *J. Biol. Chem.* **191**, 153–159.

Fruton, J.S., Smith, V.A. & Driscoll, P.E. (1948) On the proteolytic enzymes of animal tissues. VII. A peptidase of calf thymus. *J. Biol. Chem.* **173**, 457–469.

Hayashi, M. & Oshima, K. (1980) Isolation and characterization of aminopeptidase from monkey brain. *J. Biochem.* **87**, 1403–1411.

Hermsdorf, C.L. (1978) Tripeptide-specific aminopeptidase from *Escherichia coli* AJ005. *Biochemistry* **17**, 3370–3376.

Hiraoka, B.Y. & Harada M. (1993) Purification and characterization of tripeptide aminopeptidase from bovine dental follicles. *Mol. Cell Biochem.* **129**, 87–92.

Kanda, S., Sudo, K. & Kanno, T. (1984a) A specific kinetic assay for tripeptide aminopeptidase in serum. *Clin. Chem.* **30**, 843–846.

Kanda, S., Maekawa, M., Kohno, H., Sudo, K., Hishiki, S.,

Nakamura, S. & Kanno, T. (1984b) Examination of the subcellular distribution of tripeptide aminopeptidase and evaluation of its clinical usefulness in human serum. *Clin. Biochem.* **17**, 253–257.

Kanda, S., Nakamura, S. & Kanno, T. (1987) Clinical usefulness of serum tripeptide aminopeptidase activity in diagnosing liver diseases. *Clin. Biochem.* **20**, 53–56.

Khilji, M.A., Akrawi, A.F. & Bailey, G.S. (1979) Purification and partial characterization of a bovine kidney aminotripeptidase (capable of cleaving prolyl-glycylglycine). *Mol. Cell Biochem.* **23**, 45–52.

Sachs, L. & Marks, N. (1982) A highly specific aminotripeptidase of rat brain cytosol. *Biochim. Biophys. Acta* **706**, 229–238.

Smith, E.L. (1955) Aminotripeptidase (tripeptidase). *Methods Enzymol.* **2**, 83–87.

Smith, E.L. & Bergmann, M. (1944) The peptidases of intestinal mucosa. *J. Biol. Chem.* **153**, 627–651.

Strauch, K.L. & Miller, C.G. (1983) Isolation and characterization of *Salmonella typhimurium* mutants lacking a tripeptidase (peptidase T). *J. Bacteriol.* **154**, 763–771.

*Minoru Harada*
*Department of Oral Biochemistry,*
*Matsumoto Dental College,*
*Shiojiri, Nagano 399-07, Japan*
*Email: minoharada@mdc.ac.jp*

# 521. *Clostridial aminopeptidase*

## Databanks

*Peptidase classification: clan MX, family M99, MEROPS ID: M9A.005*
*NC-IUBMB enzyme classification: EC 3.4.11.13*
*Databank codes: no sequence data available*

## Name and History

This enzyme was isolated by Kessler & Yaron (1973) from the culture medium of *Clostridium histolyticum*. They were searching for enzymes that could cleave Pro residues from various substrates such as (Pro-Gly-Pro)$_n$, (Pro-Gly-Pro)$_n$-OMe, Dnp-(Pro-Gly-Pro)$_n$ and poly-L-proline (average $M_r$ 6000). The ability of the enzyme to rapidly cleave N-terminal Pro from unblocked peptides distinguished it from the usual leucyl aminopeptidases. However, the enzyme could not remove the terminal Pro from poly-L-Pro. The name *clostridial aminopeptidase* was applied to this novel activity.

## Activity and Specificity

The enzyme is strictly an aminopeptidase and has no endopeptidase action on polypeptides or proteins. The amino group of the N-terminal amino acid must be free. Di-, tri- and polypeptide substrates are cleaved and the optimum size

accommodated by the active center is four residues. The peptidase readily releases N-terminal proline and hydroxyproline, as well as all other natural amino acids, except when the penultimate (P1′) residue is Pro. The residues cleaved most rapidly are hydrophobic ones; the larger the hydrophobic side chain the smaller the $K_m$ value. The $k_{cat}$ values are highest with Ala and Pro. Charged amino acids are cleaved slowly. Compounds such as Ala-amide and Ala-NHNap are cleaved, but not Ala-NHPhNO$_2$ (Kessler & Yaron, 1976a,b).

The enzyme is readily assayed (once purified) by incubation with Pro-Gly-Pro for 30 min at pH 8.6 and 40°C. The reaction mixture is then boiled with ninhydrin and the red color measured at 520 nm (Kessler & Yaron, 1976b).

It has been proposed to use clostridial aminopeptidase in amino acid sequencing, to degrade proline-containing polypeptides sequentially from the N-terminus by use of columns containing a combination of clostridial aminopeptidase and X-Pro aminopeptidase (Chapter 479). The enzymes are coupled to glass beads or cellulose and digestion of angiotensin II, bradykinin, substance P, etc. have been demonstrated. The amino acids emerge from the column in order, permitting determination of peptide sequences (Fleminger & Yaron, 1983, 1987).

In free or bound state, the aminopeptidase is metal dependent and no activity is seen in the absence of metal ions (Fleminger & Yaron, 1984). $Mn^{2+}$ and, to a lesser extent, $Co^{2+}$ can activate the enzyme. Lower activity is produced by $Ni^{2+}$ and $Cd^{2+}$, while $Ca^{2+}$, $Ba^{2+}$, $Sr^{2+}$ and $Mg^{2+}$ have no effect. In the presence of magnesium, $Cu^{2+}$ (millimolar) and $Zn^{2+}$ ($10\,\mu M$) inhibit the enzyme. Complete inhibition is produced by 1.7 mM $p$-mercuribenzoate; DFP has no effect (Kessler & Yaron, 1976a). Competitive inhibition is produced by intramolecularly quenched fluorescent peptides developed as substrates for leucyl aminopeptidase (Chapter 473). Thus, Lys(Abz)-Phe-ONBzl (in which ONBzl is $p$-nitrobenzyloxy) inhibits the digestion of Leu-NHPhNO$_2$ with a $K_i$ of $3\,\mu M$ (Carmel *et al.*, 1977).

## Structural Chemistry

The primary sequence of clostridial aminopeptidase is not known. The $M_r$ determined by molecular sieving is 340 000. This hexamer is dissociated by SDS into subunits of 56 000 Da (Kessler & Yaron, 1976a,b).

## Preparation

The enzyme is readily purified from the culture filtrate of *Clostridium histolyticum*. Most of the enzyme is cell associated, so mild sonication of the cells would probably yield higher amounts of enzyme. EDTA is added to 1 mM and the preparation is passed through DEAE-cellulose, Sephadex G-150 and a second DEAE-cellulose column. An overall purification of 51-fold results in a preparation that is homogeneous by gel electrophoresis and by immunodiffusion and immunoelectrophoresis with rabbit antibodies raised against crude medium. Exposure to zinc should be avoided.

The enzyme may be stored frozen in EDTA for one year, but may not be lyophilized (Kessler & Yaron, 1976a,b).

## Biological Aspects

*Clostridium histolyticum* is known for its ability to penetrate the extracellular matrix. It has a potent collagenase (Chapter 368) that degrades collagen to small peptides. It may be speculated that the aminopeptidase evolved, in part, to hydrolyze these proline-rich products as part of the organism's nutritional requirements.

## Distinguishing Features

The enzyme is readily distinguished from leucyl aminopeptidase by its sensitivity to inhibition by zinc and by the fact that it is inhibited by Lys(Abz)-Phe-ONBzl, a good substrate for leucyl aminopeptidase.

## References

Carmel, S., Kessler, E. & Yaron, A. (1977) Intramolecularly-quenched fluorescent peptides as fluorogenic substrates of leucine aminopeptidases and inhibitors of clostridial aminopeptidase. *Eur. J. Biochem.* **73**, 617–625.

Fleminger, G. & Yaron, A. (1983) Sequential hydrolysis of proline-containing peptidase with immobilized aminopeptidases. *Biochim. Biophys. Acta* **743**, 437–446.

Fleminger, G. & Yaron, A. (1984) Soluble and immobilized clostridial aminopeptidase and aminopeptidase P as metal-requiring enzymes. *Biochim. Biophys. Acta* **789**, 245–256.

Fleminger, G. & Yaron, A. (1987) Application of immobilized aminopeptidases to the sequential hydrolysis of proline-containing peptides. *Methods Enzymol.* **136**, 170–178.

Kessler, E. & Yaron, A. (1973) A novel aminopeptidase from *Clostridium histolyticum. Biochem. Biophys. Res. Commun.* **50**, 405–412.

Kessler, E. & Yaron, A. (1976a) Extracellular aminopeptidase from *Clostridium histolyticum. Eur. J. Biochem.* **63**, 271–287.

Kessler, E. & Yaron, A. (1976b) Extracellular aminopeptidase from *Clostridium histolyticum. Methods Enzymol.* **45**, 544–552.

*J. Fred Woessner*
*Department of Biochemistry and Molecular Biology,*
*University of Miami School of Medicine,*
*Miami, FL 33101, USA*
*Email: fwoessne@mednet.med.miami.edu*

# 522. X-Trp aminopeptidase

## Databanks

*Peptidase classification: clan MX, family M99, MEROPS ID: M9A.007*
*NC-IUBMB enzyme classification: EC 3.4.11.16*
*Chemical Abstracts Service registry number: 137010-33-4*
*Databank codes: no sequence data available*

## Name and History

*X-Trp aminopeptidase* was discovered by affinity chromatography using a monoclonal antibody generated to pig kidney microvillar membranes (Gee & Kenny, 1985). A long search for hydrolyzable substrates was rewarded when it was found to cleave Leu↓Trp. The observation that the enzyme prefers short peptides with an aromatic residue in the P1' position, with Glu↓Trp being the best substrate, led to the name *aminopeptidase W* (W being the single-letter abbreviation for Trp).

## Activity and Specificity

X-Trp aminopeptidase hydrolyzes short peptides and exhibits maximal rates towards dipeptides (Gee & Kenny, 1987). Dipeptides with Trp, Phe or Tyr in the P1' position are rapidly hydrolyzed. The requirement in the P1 position is not stringent. The enzyme does not hydrolyze N-blocked peptides. Activity is usually monitored with suitable dipeptide substrates such as Glu↓Trp (Gee & Kenny, 1987) or Asp↓Phe-$NH_2$ (Tieku & Hooper, 1992). A coupled fluorometric assay with Glu↓Trp and glutamate dehydrogenase has been described (Jackson *et al.*, 1988).

The activity of X-Trp aminopeptidase is markedly influenced by ionic conditions; the highest activity is observed in 0.1 M Tris–HCl, pH 8.0 (Gee & Kenny, 1987). Phosphate ions are strongly inhibitory. The chelating agents EDTA and 1,10-phenanthroline are without inhibitory effect up to a concentration of 10 mM (Gee & Kenny, 1987). The enzyme is inhibited by the zinc aminopeptidase inhibitors amastatin, bestatin and probestin (Tieku & Hooper, 1992). Selective inhibition of X-Trp aminopeptidase relative to the other mammalian zinc aminopeptidases can be obtained with the angiotensin-converting enzyme inhibitors rentiapril and zofenoprilat (Tieku & Hooper, 1992).

## Structural Chemistry

In the pig, X-Trp aminopeptidase is a 130 kDa integral membrane glycoprotein which associates as a noncovalent dimer (Gee & Kenny, 1985). The enzyme can be solubilized from the membrane by papain treatment. Deglycosylation with *N*-glycanase reduced the mass to 90 kDa and the enzyme was shown to contain 1.2 atoms of zinc per subunit (Gee & Kenny, 1987).

## Preparation

The enzyme has only been purified from pig kidney by immunoaffinity chromatography with a monoclonal antibody as in the original isolation (Gee & Kenny, 1985).

## Biological Aspects

The location of the enzyme in the microvillar membrane of the intestine and kidney proximal tubules is consistent with a role in the metabolism of dietary peptides and peptides filtered by the glomerulus, respectively (Gee & Kenny, 1985). X-Trp aminopeptidase may play a minor role in the intestinal metabolism of the artificial sweetener aspartame (Asp↓PheOMe; Hooper *et al.*, 1994). In the peripheral nervous system the enzyme has been localized to both myelinated and unmyelinated neurons (Barnes *et al.*, 1991) and may be involved in neuropeptide metabolism.

## Distinguishing Features

X-Trp aminopeptidase has an overlapping substrate specificity and inhibitor profile with the related mammalian membrane-bound zinc aminopeptidases membrane alanyl aminopeptidase (Chapter 337) and glutamyl aminopeptidase (Chapter 339), but can be distinguished from them by its inhibition by rentiapril and zofenoprilat (Tieku & Hooper, 1992).

## Further Reading

For a general review of ectopeptidases which includes X-Trp aminopeptidase, see Hooper (1993).

## References

Barnes, K., Bourne, A., Cook, P.A., Turner, A.J. & Kenny, A.J. (1991) Membrane peptidases in the peripheral nervous system of the pig: their localisation by immunohistochemistry at light and electron microscopic levels. *Neuroscience* **44**, 245–261.

Gee, N.S. & Kenny, A.J. (1985) Proteins of the kidney microvillar membrane. The 130 kDa protein in pig kidney recognised by monoclonal antibody GK5C1 is an ectoenzyme with aminopeptidase activity. *Biochem. J.* **230**, 753–764.

Gee, N.S. & Kenny, A.J. (1987) Proteins of the kidney microvillar membrane. Enzymic and molecular properties of aminopeptidase W. *Biochem. J.* **246**, 97–102.

Hooper, N.M. (1993) Ectopeptidases. In: *Biological Barriers to Protein Delivery* (Audus, K.L. & Raub, T.J., eds). New York: Plenum Press, pp. 23–50.

Hooper, N.M., Hesp, R.J. & Tieku, S. (1994) Metabolism of aspartame by human and pig intestinal microvillar peptidases. *Biochem. J.* **298**, 635–639.

Jackson, M.C., Choudry, Y., Bourne, A., Woodley, J.F. & Kenny, A.J. (1988) A fluorimetric assay for aminopeptidase W. *Biochem. J.* **253**, 299–302.

Tieku, S. & Hooper, N.M. (1992) Inhibition of aminopeptidases N, A and W. A re-evaluation of the actions of bestatin and inhibitors of angiotensin converting enzyme. *Biochem. Pharmacol.* **44**, 1725–1730.

*Nigel M. Hooper*
*Department of Biochemistry and Molecular Biology,*
*The University of Leeds,*
*Leeds LS2 9JT, UK*
*Email: n.m.hooper@leeds.ac.uk*

# 523. *Tryptophanyl aminopeptidase*

## Databanks

*Peptidase classification: clan MX, family M99, MEROPS ID: M9A.008*
*NC-IUBMB enzyme classification: EC 3.4.11.17*
*Databank codes: no sequence data available*

## Name and History

In the process of developing a new enzymatic method of L-tryptophan production, a potent L-tryptophanamide-hydrolyzing activity in a strain of *Trichosporon cutaneum* was found through a survey of some hundreds of microbial strains (Iwayama *et al.*, 1983). Production of L-tryptophan from chemically synthesized DL-tryptophanamide has been accomplished by growing cells, resting cells or an immobilized cell system. A cell-free extract of the organism may also be utilized for this purpose. The enzyme is useful for this manufacturing process because it is not inhibited by high levels of the product. The paper by Iwayama *et al.* (1983) is the only report on the purification and properties of *tryptophanyl aminopeptidase*.

## Activity and Specificity

Assay of enzyme activity is performed by two sequential reactions: L-tryptophanamide is first hydrolyzed to L-tryptophan by L-tryptophan aminopeptidase and then the liberated L-tryptophan is further broken down into indole, pyruvate and ammonia by tryptophanase, an enzyme that scarcely cleaves L-tryptophanamide. Ehrlich reagent is added to produce a color with the indole; the reagent is prepared by dissolving 1 g of *p*-dimethyl-aminobenzaldehyde in 20 ml of concentrated HCl. The total volume is adjusted to 100 ml with ethanol. Hydrolysis of various amides of amino acids and various peptides can be examined by coupling with glutamate dehydrogenase in the presence of NADPH, followed by color development with ninhydrin (Iwayama *et al.*, 1983).

The optimum pH of the enzyme is 9.0–9.5 when determined with the crystalline enzyme and L-tryptophanamide as the substrate. There is no appreciable loss of enzyme activity when the enzyme solution is held at pH 7.5–8.5 for 20 h at room temperature. Hydrolysis of L-tryptophanamide proceeds most rapidly at 40–45°C at pH 7.5. However, the enzyme is stable on heating up to 55°C for 10 min; heating the enzyme above 60°C causes rapid inactivation. The enzyme specifically requires $Mn^{2+}$ for its activity; cobalt ions produce only 25% as much activation. Other metal ions have no effect. $Mn^{2+}$ higher than 2.5 mM should be added to the reaction mixture to obtain full enzymatic activity. Reagents such as $\alpha, \alpha'$-dipyridyl and *N*-ethylmaleimide appear to be potent inhibitors of the enzyme.

L-Tryptophanamide is the most rapidly hydrolyzed substrate. L-Phenylalaninamide is also hydrolyzed (47% as rapidly as L-tryptophanamide). Other substrates such as L-leucinamide (26%), L-tyrosinamide (22%), L-valinamide (15%), L-alaninamide (14%), L-methioninamide (14%) and L-serinamide (7%) are hydrolyzed by the enzyme. Hydrolysis of L-glutamine (18%), L-asparagine (22%) and L-citrulline (22%) is also catalyzed by the enzyme. D-Tryptophanamide, the sole D-isomer tested, is not hydrolyzed at all. Peptides having an L-tryptophan residue as the N-terminal moiety such as Trp┼Leu (94% to that of L-tryptophanamide), Trp┼Ala (83%), Trp┼Gly (55%), Trp┼Phe (81%) and Trp┼Tyr (44%) are preferentially hydrolyzed with some exceptions such as Trp┼Trp (25%) or Trp┼Glu (18%). Alanyl-, glycyl- or glutamyl-dipeptides seem to be less susceptible to the enzyme. From such data, it can be concluded that the enzyme has a relatively high affinity toward peptides having an L-tryptophan residue as the N-terminal moiety. Together with the results for aminoacylamides, the enzyme is concluded to differ in substrate specificity from leucyl aminopeptidase (Chapter 473), although both enzymes hydrolyze aminoacylamides as well as dipeptides.

## Structural Chemistry

The $M_r$ of the enzyme is 270 000 by gel filtration; the enzyme is dissociated into four identical subunits having $M_r$ 68 000 upon SDS-PAGE. The pI of the enzyme is 4.7 as determined by isoelectrofocusing.

## Preparation

*Trichosporon cutaneum* IFO 0173 is available from the Institute for Fermentation (see Appendix 2 for full names and addresses of suppliers). The culture medium consists of 5% glycerol, 5% corn steep liquor, 1% mineral mixture and 1% 1 M potassium phosphate buffer, pH 7.0, in tap water. The microbial strain is precultured on 1 liter of the medium by shaking at 30°C for 2 days and then transferred to a 50 liter jar fermenter containing 25 liters of the medium. Cultivation is carried out at 30°C under vigorous aeration for 2 days until the microbial growth enters the early exponential phase. Cells are then harvested, washed, disrupted in a French press and by sonication. After centrifugation to remove intact cells, the cell-free extract is precipitated by adding ammonium sulfate to 65% saturation, pH 7.5. The precipitate is removed by centrifugation at 10 000 × *g* for 20 min and ammonium sulfate is added to 85% saturation. The precipitate containing enzyme activity is collected and dialyzed against 50 mM Tris–HCl, pH 7.5, containing 5 mM ME and 50 mM KCl.

The clarified enzyme solution is chromatographed on DEAE-cellulose column in dialysis buffer, washed with buffer and eluted with buffer containing 0.1 M KCl. The enzyme is concentrated by ammonium sulfate and applied to a hydroxyapatite column. The final step is isoelectric focusing on an LKB column between pH 4 and 6. The enzyme is recovered at its isoelectric point between pH 4.5 and 4.9; purification is 1600-fold with a recovery of 23%. The purified enzyme could be crystallized from ammonium sulfate to produce hexagonal crystals. The crystals showed no increase in specific activity indicating purity of the preparation. Full details of the preparation may be found in Iwayama *et al*. (1983).

## Biological Aspects

L-Tryptophan aminopeptidase from *T. cutaneum* is an N-terminal exopeptidase with an unusual substrate specificity, highly specific to L-tryptophanamide and L-tryptophanyl dipeptides. The enzyme is localized to the cytosol fraction by conventional subcellular fractionation. But a fair amount of enzyme activity is still found in the precipitated cells, most of which remain unbroken as judged by light microscopy.

## Distinguishing Features

The absolute requirement of a divalent cation for enzyme activity, with $Mn^{2+}$ being the best, is an important characteristic of the enzyme. Of the comparable enzymes activated by $Mn^{2+}$, the enzyme from *T. cutaneum* is more similar to X-Pro aminopeptidase (Chapter 479) (Yaron & Mlynar,

1968) than leucyl aminopeptidase (Chapter 473) (Hanson & Frohne, 1976) or the aminopeptidase from *Clostridium histolyticum* (Chapter 521) (Kessler & Yaron, 1976) in its $Mn^{2+}$ requirement. However, the substrate specificities and inhibitor sensitivities are different for these four enzymes.

The most interesting character of the enzyme could be its unusual substrate specificity: the preferential hydrolysis of L-tryptophanamide and L-tryptophanyl dipeptides at the N-terminal moiety. An increasing number of N-terminal exopeptidases have been reported from various origins, but there has been no other report on an enzyme which hydrolyzes L-tryptophanamide and L-tryptophanyl peptides so specifically as the enzyme from *T. cutaneum*.

## References

Hanson, H. & Frohne, M. (1976) Crystalline leucine aminopeptidase from lens (α-aminoacyl-peptide hydrolyase; EC 3.4.11.1). *Methods Enzymol.* **45**, 504–521.

Iwayama, A., Kimura, T., Adachi, O. & Ameyama, M. (1983) Crystallization and characterization of a novel aminopeptidase from *Trichosporon cutaneum*. *Agric. Biol. Chem.* **47**, 2483–2493.

Kessler, E. & Yaron, A. (1976) Extracellular aminopeptidase from *Clostridium histolyticum*. *Methods Enzymol.* **45**, 544–552.

Yaron, A. & Mlynar, D. (1968) Aminopeptidase-P. *Biochem. Biophys. Res. Commun.* **32**, 658–663.

*Osao Adachi*
*Laboratory of Applied Microbiology,*
*Department of Biological Chemistry,*
*Faculty of Agriculture, Yamaguchi University,*
*1677-1 Yoshida, Yamaguchi 753, Japan*
*Email: osao@agr.yamaguchi-u.ac.jp*

# 524. X-Arg dipeptidase

## Databanks

*Peptidase classification: clan MX, family M99, MEROPS ID: M9B.001*
*NC-IUBMB enzyme classification: EC 3.4.13.4*
*Chemical Abstracts Service registry number: 37288-72-5*
*Databank codes: no sequence data available*

## Name and History

$N^{\alpha}$-(γ-aminobutyryl)lysine is a compound found in mammalian brain. Although presumably synthesized by carnosine synthase (EC 6.3.2.11), it is not hydrolyzed by X-His dipeptidase (EC 3.4.13.3) (Chapter 529). An *$N^{\alpha}$-(γ-aminobutyryl)-lysine hydrolase* was purified 62-fold from pig kidney by Kumon *et al*. (1970). The name *X-Arg dipeptidase* is recommended by IUBMB.

## Activity and Specificity

Enzyme activity was determined following incubation of substrate and enzyme in a Tris–HCl buffer, pH 7.0. After termination of the reaction with 10% trichloracetic acid, released amino acids were measured with an amino acid analyzer. The enzyme generally hydrolyzed dipeptides in which a basic amino acid was present at the C-terminus. Of the substrates tested, highest activity was found with $N^\alpha$-($\gamma$-aminobutyryl)arginine followed by $N^\alpha$-($\gamma$-amino-butyryl)lysine. $N^\varepsilon$-($\gamma$-aminobutyryl)lysine was not hydrolyzed. Dipeptides in which the N-$\alpha$-amino group was blocked, such as $N^\alpha$-carbobenzoxy-leucyl-arginine, were not hydrolyzed. Dipeptides in which the C-terminus was blocked, such as $N^\alpha$-($\gamma$-aminobutyryl)lysine amide, were also not hydrolyzed. The enzyme was differentiated from carboxypeptidase B (Chapter 455) by comparison of the rate of cleavage of benzoyl-glycyl-lysine versus $N^\alpha$-($\gamma$-aminobutyryl)lysine.

## Preparation

A 62-fold purification from the supernatant of an homogenate of 10 kg of hog kidney was achieved by ammonium sulfate and acetone fractionation and hydroxyapatite chromatography. In the initial step, the enzyme precipitates at 35% ammonium sulfate saturation and is solubilized in a buffer containing 0.4% sodium deoxycholate.

## Biological Aspects

The enzyme is widely distributed. In the rat, highest activity is present in skeletal muscle, whereas in the rabbit, highest activity is present in the kidney. It is possible that the enzyme plays a wider role than hydrolysis of $N^\alpha$-($\gamma$-aminobutyryl)lysine, perhaps acting in the general hydrolysis of dipeptides containing C-terminal lysine or arginine.

## Reference

Kumon, A., Matsuoka, Y., Kakimoto, Y., Nakajima, T. & Sano, I. (1970) A peptidase that hydrolyzes $N^\alpha$-($\gamma$-aminobutyryl)lysine. *Biochim. Biophys. Acta* **200**, 466–474.

*Sherwin Wilk*
*Department of Pharmacology,*
*Mount Sinai School of Medicine,*
*One Gustave l. Levy Place,*
*New York, NY 10029, USA*
*Email: S_Wilk@smtplink.mssm.edu*

# 525. *Pro-X dipeptidase*

## Databanks

*Peptidase classification: clan MX, family M99, MEROPS ID: M9B.003*
*NC-IUBMB enzyme classification: EC 3.4.13.8*
*Databank codes: no sequence data available*

## Name and History

This activity was described by Sarid *et al*. (1962). At the time of writing it is listed by IUBMB as **Pro-X dipeptidase**, in preference to earlier names that included *prolyl dipeptidase*, *iminodipeptidase*, *prolinase* and *prolylglycine dipeptidase*. It was described in the 1992 EC listing as showing preferential hydrolysis of Pro┼Xaa dipeptides and also acting on hydroxyprolyl analogs. But the Comments section noted that prolyl dipeptides are also hydrolyzed by cytosol nonspecific dipeptidase, EC 3.4.13.18. A series of experiments beginning as early as 1971 and culminating in the work of Lenney (1990a) established that this enzyme, was in fact, identical to the **cytosolic nonspecific dipeptidase**. The same enzyme is also responsible for the activities referred to as **human tissue carnosinase**, **mouse Mn-dependent carnosinase** and **rat brain β-Ala-Arg hydrolase** (Lenney, 1990b). A second pig kidney prolinase described by Mayer & Nordwig (1973) and cited in the EC entry is much larger (300 kDa versus 100 kDa) and is insensitive to PCMB. The status of this second peptidase is unclear at this time.

It is recommended that the current EC entry for Pro-Xaa dipeptidase be discontinued and that the enzyme be included at EC 3.4.13.18 under the heading cytosol nonspecific dipeptidase. The reader of this *Handbook* is, therefore, referred to Chapter 527 for a full description of the enzyme and its history.

## References

Lenney, J.F. (1990a) Human cytosolic carnosinase: evidence of identity with prolinase, a non-specific dipeptidase. *Biol. Chem. Hoppe-Seyler* **371**, 167–171.

Lenney, J.F. (1990b) Separation and characterization of two carnosine-splitting cytosolic dipeptidases from hog kidney (carnosinase and non-specific dipeptidase). *Biol. Chem. Hoppe-Seyler* **371**, 433–440.

Mayer, H. & Nordwig, A. (1973) The cleavage of prolyl peptides by kidney peptidases. Detection of a new peptidase capable of removing N-terminal proline. *Z. Physiol. Chem.* **354**, 371–379.

Sarid, S., Berger, A. & Katchalski, E. (1962) Proline iminopeptidase. II. Purification and comparison with iminopeptidase (prolinase). *J. Biol. Chem.* **237**, 2207–2212.

*J. Fred Woessner*
*Department of Biochemistry and Molecular Biology,*
*University of Miami School of Medicine,*
*PO Box 106960,*
*Miami, FL 33101, USA*
*Email: fwoessne@mednet.med.miami.edu*

# 526. *Met-X dipeptidase*

## Databanks

*Peptidase classification: clan MX, family M99, MEROPS ID: M9B.004*
*NC-IUBMB enzyme classification: EC 3.4.13.12*
*Chemical Abstracts Service registry number: 37341-91-6*
*Databank codes: no sequence data available*

## Name and History

This enzyme was first reported in cell lysates of *Escherichia coli* strain B by Brown & Krall (1971) during a search for peptidases that could remove the N-terminal Met residue of bacterial proteins once these had undergone a deformylase reaction. The enzyme was purified (Brown, 1973) and given the name **dipeptidase M**, based on facile cleavage of certain methionyl dipeptides and failure to cleave any peptides of length greater than two residues. The name **methionyl dipeptidase** has also been used, but the NC IUBMB adopted the name **Met-X dipeptidase** when the enzyme was added to the 1975 Supplement. The hypothesis that this enzyme could be the methionine-releasing peptidase was subsequently abandoned with the discovery of methionyl aminopeptidase (Chapter 476).

## Activity and Specificity

The enzyme is strictly a dipeptidase and has no aminopeptidase action on tripeptides. The amino group of the N-terminal amino acid must be free. Peptides such as Met+Ala and Met+Ser are cleaved most rapidly; Met+Leu, Met+Gly and Met+Thr are cleaved at about 60% of the maximal rate. Met+Glu, Met+Val and Met+Ile are cleaved at 10% of the maximal rate. Peptides with Ala, Ser, Thr, Leu and Gly N-terminal to -Ala are also cleaved at significant rates (Brown, 1973). A similar enzyme from *E. coli* strain K-12 (Simmonds, 1972) cleaves Met+Met, Met+Leu and Met+Gly optimally; has less activity on Leu+Met, Leu+Gly and Lys+Lys; and has almost no action on Leu-Leu.

The enzyme is unstable in the absence of metal ions and reducing agents. $Mn^{2+}$ is the best cation for maintaining stability and is the only divalent cation to restore activity following EDTA treatment. Serum albumin is also added to stabilize the enzyme during assay. Activity is completely blocked by iodoacetic acid or PCMB; the latter action can be completely reversed with 2 mM 2-ME (Brown, 1973). Simmons *et al.* (1976) similarly found rapid loss of activity upon storage with complete restoration by 15 mM 2-ME. They also noted that the enzyme from strain AJ005, when assayed with $Co^{2+}$ in the absence of Tris, digested Lys-Lys twice as rapidly as in the $Tris-Mn^{2+}$ system. It therefore seems possible that the $Co^{2+}$-activated Lys-Lys dipeptidase reported by Sussman & Gilvarg (1970) might correspond to the Met-X dipeptidase.

There is a broad pH range between 5.5. and 9.5, with the optimum lying between 7.8 and 9.0. Assay of the enzyme activity is performed with 1 mM Met-Ala as substrate in Tris buffer, pH 7.4, for 10 min at 37°C. Serum albumin (200 μg ml$^{-1}$), 1 mM 2-ME and 2.5 mM $MnCl_2$ stabilize and activate the enzyme. The reaction is stopped with HCl and color is developed with trinitrobenzenesulfonic acid (Brown, 1973).

## Structural Chemistry

The primary sequence of Met-X dipeptidase is unknown. The $M_r$ determined by molecular sieving is approximately 94 000 and the enzyme comprises two identical subunits of $M_r$ 47 000 (Brown, 1973).

## Preparation

The procedure of Brown (1973) is to disrupt 2 kg of *E. coli* strain B cells (ATCC 11303) and centrifuge at $16\,000 \times g$ for 20 min. Manganese acetate, DNAase and RNAase are added overnight. Ammonium sulfate fractionation at 50–75% saturation is followed by KCl-gradient elution from a DEAE-cellulose column. Precipitation of inactive protein at pH 5.5 is followed by QAE-Sephadex and DEAE-Sephadex chromatographies to provide a 450-fold purification. The product is pure as judged by three criteria.

## Biological Aspects

This peptidase is believed to contribute to the amino acid nutrition of *E. coli*. Simmons *et al.* (1976) discuss a possible role in providing lysine and phenylalanine, even though dipeptides with these residues are cleaved relatively slowly.

## Distinguishing Features

The major dipeptidase of *E. coli*, at least in certain strains including B, appears to be the cytosolic nonspecific dipeptidase (EC 3.4.13.18) which is widely distributed in the bacterial, plant and animal kingdoms. The size of the nonspecific dipeptidase is similar to that of Met-X dipeptidase: two subunits of $M_r\,53\,000$. However, this is a zinc metalloenzyme that is inhibited by reducing agents such as 2-ME, because of their chelating effect. On this basis, Hayman *et al.* (1974) conclude that the nonspecific dipeptidase is distinct from the Met-X dipeptidase, which is activated by ME.

There are also similarities of Met-X dipeptidase to *E. coli* X-His dipeptidase (pepD) (Chapter 487). Hermsdorf *et al.* (1979) found that K-12 mutants lacking the *pepD* gene lost dipeptidase activity of the type reported earlier by Simmonds *et al.* (1976); however, the recent characterization of the cloned X-His dipeptidase (Schroeder *et al.*, 1994) indicates a number of distinct properties including inhibition by $Mn^{2+}$. Unfortunately, although 108 peptide substrates were tested, only three of these match substrates tested with Met-X dipeptidase. Met-Ala was digested 7 times faster than Ala-Phe by X-His dipeptidase, while the rates were almost equal for Met-X dipeptidase.

The fact that 20 years have passed without further report on Met-X dipeptidase and that no cDNA sequences or bacterial mutants have been reported leaves the status of this peptidase under a cloud of uncertainty.

## Related Peptidases

As mentioned above, the $Co^{2+}$-activated Lys-Lys dipeptidase of *E. coli* reported by Sussman & Gilvarg (1970) might possibly be the same enzyme as Met-X dipeptidase. A very similar enzyme has also been reported from strain AJ005 by Ota (1986). Finally, Johnson & Brown (1974) have purified a Met-X dipeptidase with very similar specificity to the *E. coli* enzyme from *Neurospora crassa*. There is not sufficient evidence to make a definitive identification of these enzymes with *E. coli* Met-X dipeptidase.

## References

Brown, J.L. (1973) Purification and properties of dipeptidase M from *Escherichia coli* B. *J. Biol. Chem.* **248**, 409–416.

Brown, J.L. & Krall, J.F. (1971) Studies on the substrate specificity of an *Escherichia coli* peptidase. *Biochem. Biophys. Res. Commun.* **42**, 390–397.

Hayman, S., Gatmaitan, J.S. & Patterson, E.K. (1974) The relationship of extrinsic and intrinsic metal ions to the specificity of a dipeptidase from *Escherichia coli* B. *Biochemistry* **13**, 4486–4493.

Hermsdorf, C.L., Simmonds, S. & Saunders, A. (1979) Soluble di- and aminopeptidases in *Escherichia coli* K-12. *Int. J. Pept. Protein Res.* **13**, 146–151.

Johnson, G.L. & Brown, J.L. (1974) Partial purification and characterization of two peptidases from *Neurospora crassa*. *Biochim. Biophys. Acta* **370**, 530–540.

Ota, A. (1986) Purification and properties of dipeptidase from *Escherichia coli* AJ005. *Mol. Cell. Biochem.* **71**, 87–93.

Schroeder, U., Henrich, B., Fink, J. & Plapp, R. (1994) Peptidase D of *Escherichia coli* K-12, a metallopeptidase of low substrate specificity. *FEMS Microbiol. Lett.* **123**, 153–160.

Simmonds, S. (1972) Peptidase activity and peptide metabolism in *Escherichia coli*. In: *Peptide Transport in Bacteria and Mammalian Gut*. Ciba Foundation Symposium 4. Amsterdam: Elsevier, pp. 43–57.

Simmonds, S., Szeto, K.S. & Fletterick, C.G. (1976) Soluble tri- and dipeptidases in *Escherichia coli* K-12. *Biochemistry* **15**, 261–271.

Sussman, A.J. & Gilvarg, C. (1970) Peptidases in *Escherichia coli* K-12 capable of cleaving lysine homopeptides. *J. Biol. Chem.* **245**, 6518–6524.

*J. Fred Woessner*
*Department of Biochemistry and Molecular Biology,*
*University of Miami School of Medicine,*
*Miami, FL 33101, USA*
*Email: fwoessne@mednet.med.miami.edu*

# 527. *Cytosol nonspecific dipeptidase*

## Databanks

*Peptidase classification: clan MX, family M99, MEROPS ID: M9B.005*
*NC-IUBMB enzyme classification: EC 3.4.13.18*
*Chemical Abstracts Service registry number: 9032-23-9 + 9025-31-4*
*Databank codes: no sequence data available*

## Name and History

It has long been known, especially from the pioneering work of Smith (1948) and coworkers, that dipeptides are rapidly degraded by dipeptidases, a group of enzymes that hydrolyze only dipeptides with free amino and carboxyl groups. These peptidases, distinctly different from aminopeptidases, are found in the soluble fractions of almost all mammalian tissue preparations, notably the kidney and intestines. Initial studies with a limited number of dipeptides indicated that there are various enzymes which exhibit a rather narrow range of specificity, but subsequent studies with highly purified enzyme preparations from various tissues provided strong evidence that the *glycyl-leucine dipeptidase* (Smith, 1948) actually exhibits a broad substrate specificity (Traniello & Vescia, 1964; Hayman & Patterson, 1971; Das & Radhakrishnan, 1973; Norén *et al*., 1973; Piggott & Fottrell, 1975) and hence this enzyme was termed *cytosol nonspecific dipeptidase.* With these purified enzyme preparations hydrolysis of Pro-Ala was also noted, but this activity was generally attributed to a contaminating enzyme known as *prolinase* (also described as *prolylglycine dipeptidase* or *iminodipeptidase*), although it has been proposed that the two activities may be due to a single enzyme (Hayman & Patterson, 1971). This latter interpretation was supported by the observation that prolinases purified from pig kidney (Sarid *et al*., 1962; Mayer & Nordwig, 1973) and cattle kidney (Akrawi & Bailey, 1976) also hydrolyzed nonspecific dipeptidase substrates. Priestman & Butterworth (1985) provided firm evidence that both activities are due to a single enzyme. Furthermore, although hydrolysis of $\beta$-Ala-Gly was not observed for the purified dipeptide preparation (Hayman & Patterson, 1971; Norén *et al*., 1973), studies on the degradation of carnosine ($\beta$-Ala-His) revealed that this dipeptide is not only hydrolyzed by carnosinase ($\beta$-Ala-His dipeptidase; Chapter 528) but also by another peptidase (Margolis *et al*., 1979; Lenney *et al*., 1985; Kunze *et al*., 1986) which exhibits properties almost identical to the cytosol nonspecific dipeptidase and prolinase. Subsequent studies by Lenney (1990) provided further evidence that the *'manganese-dependent carnosinase'* the carnosine-degrading *'β-Ala-Arg hydrolase'* and *'human tissue carnosinase'* are actually identical to prolinase, the cytosol nonspecific dipeptidase.

## Activity and Specificity

As the name 'nonspecific dipeptidase' implies, this enzyme is a true dipeptidase that hydrolyzes dipeptides only (or pseudodipeptides such as carnosine) with free amino and carboxyl groups. The hydrolysis of the substrates varies widely depending on the conditions of the assay. Of the dipeptidase substrates Gly+Leu gave the highest activity and of the prolinase substrates Pro+Leu had the highest activity; it was hydrolyzed at half the rate of Gly-Leu (Priestman & Butterworth, 1985). Generally, substrates that are hydrolyzed most rapidly are those with a bulky, aliphatic C-terminal residue and a small side-chain N-terminal residue. The $K_m$ values against the various dipeptides range between 0.5 and 2 mM. With prolinase substrates such as Pro+Ala, enzyme activity can be conveniently assayed by following fluorometrically the liberation of the $\alpha$-amino acid by use of *o*-phthaldialdehyde (Butterworth & Priestman, 1982). With nonspecific dipeptidase substrates such as Gly+Leu or Gly+Phe the liberated Leu or Phe can be determined spectrophotometrically by the L-amino acid oxidase/peroxidase-coupled oxidation of *o*-dianisidine (Butterworth & Priestman, 1982). Carnosinase activity has been measured either by radiometric assays (Margolis *et al*., 1979; Kunze *et al*., 1986) or by following the liberation of free histidine fluorometrically after reaction with *o*-phthaldialdehyde (Lenney *et al*., 1985). Although the optimal assay conditions have not been determined systematically, the accumulated data indicate that buffers based on Tris, HEPES, *N*-ethylmorpholine and Na-barbital are more suitable, while phosphate buffers seem to exhibit inhibitory effects, probably due to their metal-binding capacity. Depending on the buffer used, the optimal pH values range between 8.2 and 9.3.

The enzyme is stabilized and activated by DTT (2 mM) or 2-ME and is strongly activated by $Mn^{2+}$ at an optimal concentration of about 120 µM (higher concentrations of $Mn^{2+}$ and other bivalent metal ions especially $Ca^{2+}$, $Ni^{2+}$, $Zn^{2+}$, $Cd^{2+}$ and $Co^{2+}$ are inhibitory). Conversely, the enzyme is inhibited by thiol-reactive substances such as *p*-chloromercuriphenyl sulfonic acid, *N*-ethylmaleimide or 2-iodoacetamide, as well as by chelating agents such as EDTA or 1,10-phenanthroline. The enzyme is extremely sensitive to bestatin inhibition, being completely inhibited in the nanomolar range (Reith & Neidle, 1979; Lenney *et al*., 1985; Kunze *et al*., 1986), but is unaffected by other peptidase inhibitors of microbial origin such as antipain, chymostatin, leupeptin or phosphoramidon.

## Structural Chemistry

The reported relative molecular masses as estimated by gel filtration range between 70 000 and 100 000 (Lenney, 1990;

Priestman & Butterworth, 1985). Values around 85 000 have been estimated by most investigators with the exception of Mayer & Nordwig (1973) who reported an $M_r$ of 300 000 but also indicated that the high molecular mass could be due to the formation of aggregates. Since the enzyme has not been purified to homogeneity, the molecular mass has not been determined by SDS-PAGE. Although the studies with metal chelators strongly suggest that the enzyme is a metalloenzyme, the present data with impure enzyme preparations are not conclusive. By column chromatofocusing and isoelectric focusing pI values ranging between 5.05 and 5.4 have been reported by various investigators with the exceptions of Mayer & Nordwig (1973) and Akrawi & Bailey (1976) who estimated a pI value of 4.3.

## Preparation

Cytosol nonspecific dipeptidase is widely distributed and found in almost every tissue tested (Butterworth & Priestman, 1982) with only a 10-fold difference in the specific activities. Apart from heart muscle, all tissues have moderate activity, with kidney being particularly active. Significant activities are also found in white blood cells and even in red blood cells. The enzyme has been partially purified from various sources (about 500-fold from mouse ascites tumor by Hayman & Patterson, 1971; about 340-fold from pig intestinal mucosa by Norén *et al.*, 1973; and about 500-fold from human kidney by Priestman & Butterworth, 1985). Purification to electrophoretic homogeneity has not yet been achieved.

Several studies indicate that the enzyme exists in two forms which can be partially separated by anion-exchange chromatography (Piggott & Fottrell, 1975; Butterworth & Priestman, 1982; Lenney, 1990). It has been suggested (Lenney, 1990) but not yet demonstrated, that the two forms result from an autolytic conversion process.

Earlier reports have always indicated that the enzyme is very unstable, especially during purification. As noted by Priestman & Butterworth (1985), the enzyme is very unstable only at low protein concentrations (less than 0.01 mg ml$^{-1}$). Stability is maintained at higher protein concentrations and even highly purified enzyme preparations are stable for 9 months at 4°C in the presence of serum albumin.

## Biological Aspects

Cytosol nonspecific dipeptidase may be important for hydrolyzing dipeptides which diffuse through the lysosomal membrane when proteins are digested by lysosomal enzymes. It is thought that during collagen breakdown dipeptides of the type Pro-X are hydrolyzed by 'prolinase' and dipeptides of the type X-Pro by prolidase (X-Pro dipeptidase, Chapter 475). In contrast to prolidase deficiency, however, no deficiency has been reported for the prolinase activity of the cytosol nonspecific dipeptidase and thus the biological function of this enzyme remains unknown.

## Distinguishing Features

Dipeptidases with similar, although not identical properties have been characterized from some microorganisms such as *E. coli* (Patterson, 1976) and yeast (Röhm, 1974), while dipeptidases from other microorganisms, e.g. *Mycobacterium phlei* (Plancot & Han, 1972) or *Streptococcus thermophilus* (Rabier & Desmazeand, 1973) are less similar.

## References

Akrawi, A.F. & Bailey, G.S. (1976) Purification and specificity of prolyl dipeptidase from bovine kidney. *Biochim. Biophys. Acta* **422**, 170–178.

Butterworth, J. & Priestman, D.A. (1982) Fluorimetric assay of prolinase and partial characterization in cultured skin fibroblasts. *Clin. Chim. Acta* **122**, 51–60.

Das, M. & Radhakrishnan, A.N. (1973) Glycyl-L-leucine hydrolase, a versatile 'master' dipeptidase from monkey small intestine. *Biochem. J.* **135**, 609–615.

Hayman, S. & Patterson, E.K. (1971) Purification and properties of a mouse ascites tumor dipeptidase, a metalloenzyme. *J. Biol. Chem.* **246**, 660–669.

Kunze, N., Kleinkauf, H. & Bauer, K. (1986) Characterization of two carnosine-degrading enzymes from rat brain. Partial purification and characterization of a carnosinase and a $\beta$-alanyl-arginine hydrolase. *Eur. J. Biochem.* **160**, 605–613.

Lenney, J.F. (1990) Human cytosolic carnosinase. Evidence of identity with prolinase, a non-specific dipeptidase. *Biol. Chem. Hoppe-Seyler* **371**, 161–171.

Lenney, J.F., Peppers, S.C., Kucera-Orallo, C.M. & George, R.P. (1985) Characterization of human tissue carnosinase. *Biochem. J.* **228**, 653–660.

Margolis, F.L., Grillo, M., Brown, C.E., Williams, T.H., Pitcher, R. & Elgar, G.J. (1979) Enzymatic and immunological evidence for two forms of carnosinase in the mouse. *Biochim. Biophys. Acta* **570**, 311–323.

Mayer, H. & Nordwig, A. (1973) The cleavage of prolyl peptides by kidney peptidases. Purification of iminodipeptidase (prolinase). *Z. Physiol. Chem.* **354**, 371–379.

Norén, O., Sjöström, H. & Josefsson, L. (1973) Studies on a soluble dipeptidase from pig intestinal mucosa. *Biochim. Biophys. Acta* **327**, 446–456.

Patterson, E.K. (1976) A dipeptidase from *Escherichia coli* B. *Methods Enzymol.* **45**, 377–386.

Piggott, C.O. & Fottrell, P.F. (1975) Purification and characterization from guinea-pig intestinal mucosa of two peptide hydrolases which preferentially hydrolyze dipeptides. *Biochim. Biophys. Acta* **391**, 403–409.

Plancot, M.-T. & Han, K.-K. (1972) Purification and characterization of an intracellular dipeptidase from *Mycobacterium phlei*. *Eur. J. Biochem.* **28**, 327–333.

Priestman, D.A. & Butterworth, J. (1985) Prolinase and non-specific dipeptidase of human kidney. *Biochem. J.* **231**, 689–694.

Rabier, D. & Desmazeaud, M.J. (1973) Inventaire des différentes activités peptidasiques intracellulaires de *Streptococcus thermophilus* [Inventory of the various intracellular peptidase activities in *Streptococcus thermophilus*]. *Biochimie* **55**, 389–404.

Reith, M.E.A. & Neidle, A. (1979) The isolation of two dipeptide hydrolases from mouse brain cytosol. *Biochem. Biophys. Res. Commun.* **90**, 794–800.

Röhm, K.-H. (1974) Properties of a highly purified dipeptidase (EC 3.4.13.?) from brewer's yeast. *Z. Physiol. Chem.* **355**, 675–686.

M

Sarid, S., Berger, A. & Katchalski, E. (1962) Proline iminopeptidase. II. Purification and comparison with iminodipeptidase (prolinase). *J. Biol. Chem.* **237**, 2207–2212.

Smith, E.L. (1948) Studies on dipeptidases. II. Some properties of
the glycyl-L-leucine dipeptidases of animal tissues. *J. Biol. Chem.*
**176**, 9–19.

Traniello, S. & Vescia, A. (1964) Properties of a dipeptidase from
swine kidney. *Arch. Biochem. Biophys.* **105**, 465–469.

*Karl Bauer*
*Max-Planck-Institut für experimentelle Endokrinologie,*
*Feodor-Lynen-Strasse 7,*
*D-30625 Hannover, Germany*
*Email: 106001,2503@compuserve.com*

# 528. *β-Ala-His dipeptidase*

## Databanks

*Peptidase classification: clan MX, family M99, MEROPS ID: M9B.006*
*NC-IUBMB enzyme classification: EC 3.4.13.20*
*Databank codes: no sequence data available*

## Name and History

The first observation of the hydrolysis of carnosine (β-Ala-
His) and anserine (β-Ala-1-methyl-His) by human serum was
made by Perry *et al.* (1968). **Serum carnosinase** was the
name used to describe the inferred enzyme activity and was
derived from the name of the distinct pig kidney carnosinase
(see Chapter 529), discovered by Hanson & Smith (1949).
Serum carnosinase is the most commonly used name, but **β-
Ala-His dipeptidase** has been recommended by the IUBMB.
A radiometric assay for β-Ala-His dipeptidase (as serum
carnosinase) was developed by van Munster *et al.* (1970).
A simplified and more sensitive fluorometric assay was subsequently developed (Murphey *et al.*, 1972). Lenney *et al.*
(1982) increased the assay sensitivity 10-fold and established a number of distinctions between serum carnosinase
(β-Ala-His dipeptidase), tissue carnosinase (X-His dipeptidase) (Chapter 529) and hog kidney carnosinase (nonspecific
cytosolic dipeptidase) (Chapter 527). The observed distinctions indicated that the enzymes were different gene products.
β-Ala-His dipeptidase has been purified to apparent homogeneity and structurally characterized (Jackson *et al.*, 1991)
and shown to be present in the blood sera of higher primates only, in contrast to the carnosinase activities of cytosol
nonspecific dipeptidase (Chapter 527) and X-His dipeptidase
(Chapter 529) which are found in the tissues of many mammals. β-Ala-His dipeptidase is the only human enzyme capable of hydrolyzing homocarnosine (γ-aminobutyryl-His, or
GABA-His). Thus, β-Ala-His dipeptidase presents a very
interesting evolutionary history.

## Activity and Specificity

Substrate specificity includes a preference for His in the
P1′ position (Jackson *et al.*, 1991). Relative to its rate
of hydrolysis of β-Ala┼His (100), rates on other dipeptides are reported to be as follows: β-Ala┼1-methyl-His
88, Ala┼His 38, Gly┼His 37, Gly┼Leu 21, β-Ala┼Phe 18,
Ala┼Leu 13, Ala┼Ala 11, GABA┼His 11, β-Ala┼Ala 10,
Phe┼Ala 8, Ala┼Tyr 8, Ser┼His 8, Leu┼Leu 8, Gly┼Gly 7.
The enzyme also hydrolyzes two tripeptides Gly┼His-Gly 13
and Gly┼His-Lys 7, suggesting a degree of aminopeptidase
activity. Peptides that are not hydrolyzed include β-Ala-Tyr,
β-Ala-Gly, β-Ala-Lys, Ala-Phe, Ala-Val, His-Ala, His-Gly,
His-Ser, Pro-Ala, Pro-Leu, γ-Glu-His, Leu-Arg, Gly-Gly-
Gly, Ac-carnosine, Ac-Met, Ac-His, Leu-NHNap, and β-Ala-
NHNap.

The pH optimum for activity against carnosine is about
8.0–8.5 (Lenney *et al.*, 1982). The most sensitive assay
method was developed by Lenney *et al.* (1982) and utilizes the fluorescence method for the determination of histidine described by Ambrose *et al.* (1969). Activity against
carnosine is enhanced in the presence of 0.8 mM $Cd^{2+}$,
($K_m = 4.0$ mM). For activity against homocarnosine (GABA-
His) the pH optimum is 7.6, with maximum activity in the
presence of 0.4 mM $Co^{2+}$ ($K_m = 0.4$ mM), indicating a significantly higher affinity for homocarnosine than for carnosine.
Carnosine and homocarnosine are competitive substrates. The
enzyme is stabilized in the presence of 0.25 mM $Mn^{2+}$, but
unlike other carnosinases, is not dependent upon $Mn^{2+}$ for
activity.

Carnosine hydrolysis is inhibited by 1.0 mM DTT, however PCMB, chloroacetate and PMSF have no effect on activity, indicating that the enzyme does not have essential serine or sulfhydryl residues in its active center (Lenney *et al.*, 1982). Its inactivation by DTT is probably due to chelation of a metal ion.

The substrates carnosine and homocarnosine are available from Sigma (see Appendix 2 for full names and addresses of suppliers).

## Structural Chemistry

β-Ala-His dipeptidase is a homodimer with a native $M_r$ of 160 000 (Lenney *et al.*, 1982). SDS-PAGE of the purified enzyme under reducing conditions reveals a single band with an $M_r$ of 75 000 (Jackson *et al.*, 1991). It has a pI of 4.4 and is glycosylated (approximately 18% w/w).

## Preparation

Human blood plasma has been the only source used to isolate β-Ala-His dipeptidase. It has been purified to apparent homogeneity (approximately 18 000-fold) in a four-step procedure described by Jackson *et al.* (1991). However, purification of active enzyme was not achieved, due to a marked instability as purity was neared. After 5000-fold purification, only approximately 7.5% of the original activity remained. Immunoaffinity purification of the enzyme using a rabbit polyclonal antiserum (Jackson & Lenney, 1992) did not result in purification of an active form.

## Biological Aspects

β-Ala-His dipeptidase is found predominantly in the blood sera of higher primates (Jackson *et al.*, 1991), including chimpanzee, gibbon, orangutan and pygmy chimpanzee. Only trace amounts were found in three monkey species tested. Activity was not detected in the sera of dog, horse, calf, hog, rat, mouse, rabbit, baboon, guinea pig, armadillo or Chinese hamster. However, the golden hamster possesses a distinct serum enzyme capable of hydrolyzing carnosine.

Children less than one year of age possess only a trace amount of β-Ala-His dipeptidase in their sera. The serum enzyme concentration gradually rises thereafter until approximately 15 years of age, when it reaches adult levels (van Munster *et al.*, 1970, Lenney *et al.*, 1982). It is synthesized predominantly in the brain and reaches the serum via the cerebrospinal fluid (CSF) (Jackson *et al.*, 1991, 1994). Using a rabbit polyclonal antiserum (Jackson & Lenney, 1992), raised against pure but inactive β-Ala-His dipeptidase, immunohistochemical studies localized the enzyme in certain neuronal layers of the human eye and also in certain neuronal tracts of the cerebrum, hippocampus, cerebellum and the olfactory bulb (Jackson *et al.*, 1994). A trace amount of the enzyme was also localized in hepatocytes. This may be a secondary synthesis site.

The central nervous system (CNS) localization of this enzyme provides a possible molecular basis for early clinical studies associating the enzyme with neurological function. Perry *et al.* (1967) reported on two patients with carnosinemia associated with progressive neurological disease developing into severe psychomotor retardation. It was later shown (Perry *et al.*, 1968) that these patients were deficient in β-Ala-His dipeptidase activity. In addition, certain urea cycle defects have been associated with low plasma enzyme concentrations and in some cases attendant hypercarnosinemia (Burgess *et al.*, 1975). In phenylketonuria, van Sande *et al.* (1970) reported that of 12 untreated cases, three had elevated homocarnosine concentrations in CFS, implying a β-Ala-His dipeptidase deficiency. However, this could not be confirmed in later studies (Lunde *et al.*, 1982). Duane & Peters (1988) reported that in patients with chronic alcoholic myopathy the activity of β-Ala-His dipeptidase was inversely proportional to the degree of type II muscle fiber atrophy. In 1994 the same group of researchers demonstrated that β-Ala-His dipeptidase activity was significantly reduced in patients suffering from stroke (Wassif *et al.*, 1994). More recently, Butterworth *et al.* (1996) have demonstrated a significant correlation between the serum neuron specific enolase/β-Ala-His dipeptidase ratio and the clinical outcome from a stroke.

The physiological function of β-Ala-His dipeptidase has not been definitively proven. However, homocarnosine is a likely substrate in the CNS for a number of reasons. Among the histidine-containing dipeptides, homocarnosine has by far the highest concentration in human brain (Kish *et al.*, 1979) and is only found in the CNS (Abraham *et al.*, 1962; Pisano *et al.*, 1961), with no anserine present and only a relatively small amount of carnosine in the olfactory bulb. It should be noted that β-Ala-His dipeptidase has a significantly higher affinity for homocarnosine than for carnosine (10-fold lower $K_m$) and is the only human enzyme capable of hydrolyzing homocarnosine. Homocarnosine has been localized in the retina (Margolis & Grillo, 1984; Jackson *et al.*, 1994), either colocalizing with, or in, cells immediately adjacent to those in which β-Ala-His dipeptidase has been localized (Jackson *et al.*, 1994). These authors demonstrated a similar colocalization in other regions of the human CNS. Further evidence to support the hypothesis of homocarnosine as a physiological substrate is the observation that in patients suffering from homocarnosinosis (Gjessing & Sjaastad, 1974; Sjaastad *et al.*, 1976), where CSF homocarnosine concentrations are 20 times the normal level, there is a lack of β-Ala-His dipeptidase activity (Lenney *et al.*, 1983; Jackson & Lenney, 1992), progressive dementia, spastic paraplegia and an abnormal retinal pigmentation. Considering all of this evidence, Jackson *et al.* (1994) postulated that β-Ala-His dipeptidase is part of a novel metabolic pathway for the release of the inhibitory neurotransmitter GABA from homocarnosine, separate from glutamate decarboxylation, the predominant means of GABA production.

## Distinguishing Features

There has been considerable confusion in the literature, with the tacit assumption that β-Ala-His dipeptidase, X-His dipeptidase (Chapter 529) and nonspecific cytosolic dipeptidase (Chapter 527) are the same enzyme. Notable differences include the observation that unlike the other two enzymes, β-Ala-His dipeptidase is able to hydrolyze homocarnosine and anserine, is not activated, but inhibited in the presence of DTT and has a different molecular mass, pH optimum, metal ion activation and tissue localization. Unlike X-His dipeptidase (Lenney, 1990), it is not a prolinase: it cannot hydrolyze Pro-Leu.

## Further Reading

Excellent reviews of human carnosinases and homocarnosinosis can be found in Gjessing (1985) and Gjessing *et al.* (1990). See Lenney *et al.* (1982) for the first full characterization of human β-Ala-His dipeptidase and Jackson *et al.* (1994) for distribution of β-Ala-His dipeptidase and homocarnosine in the human CNS.

## References

Abraham, D., Pisano, J.J. & Udenfriend, S. (1962) The distribution of homocarnosine in mammals. *Arch. Biochem. Biophys.* **99**, 210–213.

Ambrose, J.A, Crimm, A., Burton, J., Paullin, K. & Ross, C. (1969) Fluorimetric determination of histidine. *Clin. Chem.* **15**, 361–366.

Burgess, E.A., Oberholzer, V.G., Palmer, T. & Levin, B. (1975) Plasma carnosinase in patients deficient with urea cycle defects. *Clin. Chim. Acta* **61**, 215–218.

Butterworth, R.J., Wassif, W.S., Sherwood, R.A., Gerges, A., Poyser, K.H., Garthwaite, J., Peters, T.J. & Bath, P.M.W. (1996) Serum neuron specific enolase, carnosinase, and their ratio in acute stroke. An enzymatic test for predicting outcome? *Stroke* **27**, 2064–2068.

Duane, P. & Peters, T.J. (1988) Serum carnosinase activities in patients with alcoholic skeletal muscle myopathy. *Clin. Sci.* **75**, 185–190.

Gjessing, L.R. (ed.) (1985) Homocarnosinosis: symposium in honor of James F. Lenney. *J. Oslo City Hosp.* **35**, 21–46.

Gjessing, L.R. & Sjaastad, O. (1974) Homocarnosinosis, a new metabolic disorder associated with spasticity and mental retardation. *Lancet* **ii**, 1028.

Gjessing, L.R., Lunde, H.A., Mokrid, L., Lenney, J.F. & Sjaastad, O. (1990) Inborn errors of carnosine and homocarnosine metabolism. *J. Neural Transm. Suppl.* **29**, 91–106.

Hanson, H.T. & Smith, E.L. (1949) Carnosinase: an enzyme of swine kidney. *J. Biol. Chem.* **179**, 789–801.

Jackson, M.C. & Lenney, J.F. (1992) Homocarnosinosis patients and great apes have a serum protein that cross-reacts with human serum carnosinase. *Clin. Chim. Acta* **205**, 109–116.

Jackson, M.C., Kucera, C.M. & Lenney, J.F. (1991) Purification and properties of human serum carnosinase. *Clin. Chim. Acta* **196**, 193–206.

Jackson, M.C., Scollard, D.M., Mack, R.J. & Lenney, J.F. (1994) Localization of a novel pathway for the liberation of GABA in the human CNS. *Brain Res. Bull.* **33**, 379–385.

Kish, S.J., Perry, T.L. & Hansen, S. (1979) Regional distribution of homocarnosine, homocarnosine-carnosine synthetase and homocarnosinase in human brain. *J. Neurochem.* **32**, 1629–1636.

Lenney, J.F. (1990) Human cytosolic carnosinase: evidence for identity with prolinase, a non-specific dipeptidase. *Biol. Chem. Hoppe-Seyler* **371**, 167–171.

Lenney, J.F., George, R.P., Weiss, A.M., Kucera, C.M., Chan, P.W.H. & Rinzler, G.S. (1982) Human serum carnosinase: characterization, distinction from cellular carnosinase and activation by cadmium. *Clin. Chim. Acta* **123**, 221–231.

Lenney, J.F., Peppers, S.C., Kucera, C.M. & Sjaastad, O. (1983) Homocarnosinosis: lack of serum carnosinase is the defect probably responsible for elevated brain and CSF homocarnosine. *Clin. Chim. Acta* **132**, 157–165.

Lunde, H., Sjaastad, O. & Gjessing, L. (1982) Homocarnosinosis: hypercarnosinuria. *J. Neurochem.* **38**, 242–245.

Margolis, F.L. & Grillo, M. (1984) Carnosine, homocarnosine and anserine in vertebrate retinas. *Neurochem. Int.* **6**, 207–209.

Murphey, W.H., Patchen, L. & Lindmark, D.G. (1972) Carnosinase: a fluorimetric assay and demonstration of two electrophoretic forms in human tissue extracts. *Clin. Chim. Acta* **41**, 309–314.

Perry, T.L., Hansen, S., Tischler, B., Bunting, R. & Berry, K. (1967) Carnosinemia: a new metabolic disorder associated with neurological disease and mental defect. *N. Engl. J. Med.* **277**, 1219–1227.

Perry, T.L., Hansen, S. & Love, D.L. (1968) Serum carnosinase deficiency in carnosinemia. *Lancet* **i**, 1229–1230.

Pisano, J.J., Wilson, J.D., Cohen, L., Abraham, D. & Udenfriend, S. (1961) Isolation of γ amino-butyrylhistidine (homocarnosine) from brain. *J. Biol. Chem.* **236**, 499–502.

Sjaastad, O., Berstad, J., Gjesdahl, P. & Gjessing, L. (1976) Homocarnosinosis 2. A familial metabolic disorder associated with spastic paraplegia, progressive mental deficiency and retinal pigmentation. *Acta Neurol. Scand.* **52**, 275–290.

van Munster, P.J.J., Trijbels, J.M.F., van Heeswijk, P.J., Schut-Jansen, B. & Moerkerk, C. (1970) A new sensitive method for the determination of serum carnosinase activity using L-carnosine-[1-$^{14}$C]β-alanyl as substrate. *Clin. Chim. Acta* **42**, 309–314.

van Sande, M., Mardens, Y., Adriaenssens, K. & Lowenthal, A. (1970) The free amino acids in cerebrospinal fluid. *J. Neurochem.* **17**, 125–136.

Wassif, W.S., Sherwood, R.A., Amir, A., Idowu, B., Summers, B., Leigh, N. & Peters, T.J. (1994) Serum carnosinase activities in central nervous system disorders. *Clin. Chim. Acta* **225**, 57–64.

Dedicated in honor of Dr James F. Lenney and his tireless pursuit of the truth.

*Mel C. Jackson*
*Hawaii Agriculture Research Center,*
*99–196 Aiea Heights Drive,*
*Aiea. HI 96701, USA*
*Email: mjackson@harc-hspa.com*

# 529. *X-His dipeptidase*

## Databanks

*Peptidase classification: clan MX, family M99, MEROPS ID: M9B.006*
*NC-IUBMB enzyme classification: EC 3.4.13.3*
*Chemical Abstracts Service registry number: 9027-21-8*
*Databank codes: no sequence data available*

## Name and History

Carnosine, the first peptide isolated in homogeneous form from natural material (Liebig's meat extract) in 1900, was subsequently identified as the pseudodipeptide $\beta$-Ala-His. This peptide and structurally related compounds ($\omega$-aminoacyl amino acids such as anserine ($\beta$-alanyl-1-methyl-histidine) and homocarnosine ($\gamma$-aminobutyryl-histidine)) are major constituents of excitable tissues (muscle and brain). Hanson & Smith (1949) first described a carnosine-degrading enzyme from pig kidney, an enzyme that was inhibited by cysteine and sulfide but was strongly activated and stabilized by manganese. Rosenberg (1960a,b) purified this enzyme, termed carnosinase, and studied in detail the effects of metal ions on activity and stability. Subsequent studies by various investigators provided clear evidence for the existence of two carnosine-degrading enzymes. One enzyme termed 'manganese-dependent carnosinase' by Margolis *et al*. (1979, 1983), 'human cytosolic carnosinase' by Lenney *et al*. (1985) and $\beta$-Ala-Arg-hydrolase by Kunze *et al*. (1986) actually is not a carnosine-specific enzyme but an enzyme with broad substrate specificity which exhibits characteristics identical to the cytosol nonspecific dipeptidase (Chapter 527). In the view of the present author, the term **carnosinase** should be retained for the enzyme which has been described as '*homocarnosinase*' by Lenney *et al*. (1977), as '*manganese-independent carnosinase*' by Margolis *et al*. (1979, 1983) and as 'carnosinase' by Kunze *et al*. (1986). The name **aminoacylhistidine dipeptidase** has also been used for this enzyme and the name recommended by the NC-IUBMB is **X-His dipeptidase**. Dependent on the assay conditions used by the different laboratories, considerable differences in various parameters are reported for these carnosinases from various tissues of several species. It remains to be investigated whether these differences might be attributed to distinct isoenzymes since major differences are also observed between different tissues of the same species (Margolis *et al*., 1979, 1983).

## Activity and Specificity

Enzyme activity is always determined with carnosine as substrate, preferably in Tris–HCl or HEPES buffer, pH 7.4–8.0. The radiometric assays described by Margolis *et al*. (1979) and by Kunze *et al*. (1986) follow the degradation of radiolabeled carnosine using ion-exchange chromatography to separate the substrate from the radiolabeled product $\beta$-alanine. Murphey *et al*. (1972) and Lenney *et al*. (1982) used unlabeled carnosine and followed the liberation of histidine fluorometrically after reaction with *o*-phthalaldehyde.

With regard to the substrate specificity there are considerable differences. For the enzyme purified from mouse kidney, Margolis *et al*. (1983) reported that anserine is a poor substrate and homocarnosine is not a substrate. The enzyme also cleaved Gly╂His and Ala╂His at reduced rate but not His-$\beta$-Ala or His-Gly. In contrast, Lenney *et al*. (1977) reported that their enzyme purified from pig kidney had a narrow specificity when tested in the absence of metal ions and a broad specificity when assayed in the presence of cobalt ions (Lenney *et al*., 1977; Lenney, 1990). According to their reports, the enzyme was activated by $Co^{2+}$ and $Mn^{2+}$. Margolis *et al*. (1983) reported that $Mn^{2+}$ is inhibitory at all but very high carnosine concentrations, resulting in a shift of the $K_m$ for carnosine from 60 $\mu$M to 2.5 mM and an accompanying increase in $V$ of about 50%. This phenomenon was not observed with the enzyme preparation from rat brain (Kunze *et al*., 1986). The brain enzyme exhibits a low $K_m$ value (20 $\mu$M) towards carnosine, degrades anserine at reduced rate but does not hydrolyze homocarnosine. This brain enzyme is inhibited by $Mn^{2+}$ as well as by all other transition metal ions tested.

With regard to the chemical characteristics, the different enzymes exhibit similar properties: carnosinase activities are strongly inhibited by chelating agents such as EDTA, 1,10-phenanthroline and phosphate, and, in agreement with the original report by Hanson & Smith (1949), by DTT, ME and sulfide. Interestingly, however, these enzymes are also inhibited by PCMB. Carnosinase activities are not affected by serine protease inhibitors such as DFP and are also not sensitive to 0.2 $\mu$M bestatin and other microbial peptidase inhibitors.

## Structural Chemistry

Molecular weights of 57 000 and 58 000 were estimated by SDS-PAGE for the kidney carnosinases and for the brain enzyme an $M_r$ of 85 000 was determined by gel filtration. Cross-linking experiments suggest that mouse kidney carnosinase is a dimeric protein (Margolis *et al*., 1983).

## Preparation

The kidney is by far the richest source of carnosinase activities. Significant activities are also found in uterus and, depending on the assay conditions used, in liver and brain whereas in other tissues carnosinase activities are rather low or not detectable. Carnosinase was obtained in homogeneous form from pig kidney (purified 1340-fold; Lenney *et al.*, 1977) and from mouse kidney (purified 480-fold; Margolis *et al.*, 1983). Antibodies generated against mouse kidney carnosinase precipitated carnosinase activity from extracts of kidney, uterus or olfactory mucosa but not from extracts of brain, spleen or liver (Margolis *et al.*, 1979). Cytochemical studies with these antibodies revealed carnosinase to be localized in the proximal tubules of the kidneys and in glandular cells of several tissues such as uterine epithelium and secretory epithelium of the nasal cavity (Margolis *et al.*, 1983).

## Biological Aspects

The physiological role of these enzymes remains largely obscure since the tissue distribution differs from that of the dipeptides. Homocarnosine in the brain is not degraded by brain carnosinase and in skeletal muscle carnosine-degrading activities are almost undetectable. Kidney carnosinase may be important in the hydrolysis of (dietary) carnosine circulating in the bloodstream. In contrast to serum carnosinase (Chapter 528), there are no reports on clinical conditions that could be attributed to cellular carnosinase deficiency.

## Distinguishing Features

According to Lenney *et al.* (1982) serum carnosinase, or *β*-Ala-His dipeptidase (Chapter 528) is entirely different from the tissue carnosinase. The carnosinase activities in tissue extracts can be measured in the presence of $0.2\,\mu$M bestatin, which inhibits the degradation of carnosine by the cytosol nonspecific dipeptidase (Chapter 527) completely without affecting carnosinase activity.

## References

Hanson, H.T. & Smith, E.L. (1949) Carnosinase: an enzyme of swine kidney. *J. Biol. Chem.* **179**, 789–801.

Kunze, N., Kleinkauf, H. & Bauer, K. (1986) Characterization of two carnosine-degrading enzymes from rat brain. Partial purification and characterization of a carnosinase and a *β*-alanyl-arginine hydrolase. *Eur. J. Biochem.* **160**, 605–613.

Lenney, J.F. (1990) Human cytosolic carnosinase. Evidence of identity with prolinase, a non-specific dipeptidase. *Biol. Chem. Hoppe-Seyler* **371**, 161–171.

Lenney, J.F., Kan, S.-C., Sin, K. & Sugiyama, G.H. (1977) Homocarnosine: a hog kidney dipeptidase with a broader specificity than carnosinase. *Arch. Biochem. Biophys.* **184**, 257–266.

Lenney, J.F., George, R.P., Weiss, A.M., Kucera, C.M., Chan, P.W.H. & Rinzler, G.S. (1982) Human serum carnosinase: characterization, distinction from cellular carnosinase, and activation by cadmium. *Clin. Chim. Acta* **123**, 221–231.

Lenney, J.F., Peppers, S.C., Kucera-Orallo, C.M. & George, R.P. (1985) Characterization of human tissue carnosinase. *Biochem. J.* **228**, 653–660.

Margolis, F.L., Grillo, M., Brown, C.E., Williams, T.H., Pitcher, R. & Elgar, G.J. (1979) Enzymatic and immunological evidence for two forms of carnosinase in the mouse. *Biochim. Biophys. Acta* **570**, 311–323.

Margolis, F.L., Grillo, M., Grannot-Reisfeld, N. & Farbinan, A.I. (1983) Purification, characterization and immunocytochemical localization of mouse kidney carnosinase. *Biochim. Biophys. Acta* **744**, 237–248.

Murphey, W.H., Patchen, L. & Lindmark, D.G. (1972) Carnosinase: a fluorometric assay and demonstration of two electrophoretic forms in human tissue extracts. *Clin. Chim. Acta* **42**, 309–314.

Rosenberg, A. (1960a) Purification and some properties of carnosinase of swine kidney. *Arch. Biochem. Biophys.* **88**, 83–93.

Rosenberg, A. (1960b) The activation of carnosinase by divalent metal ions. *Biochim. Biophys. Acta* **45**, 297–316.

*Karl Bauer*
*Max-Planck-Institut für experimentelle Endokrinologie,*
*Feodor-Lynen-Strasse 7,*
*D-30625 Hannover, Germany*
*Email: 106001,2503@compuserve.com*

# 530. *Lysosomal dipeptidase*

## Databanks

*Peptidase classification: clan MX, family M99, MEROPS ID: M9B.007*
*NC-IUBMB enzyme classification: none*
*Databank codes: no sequence data available*

## Name and History

This enzyme was originally detected as a contaminating activity when highly purified preparations of dipeptidyl-peptidase I (DPP I) (Chapter 214) were employed in time-course studies designed to establish the extent of dipeptide release from polypeptide hormones such as pig α-corticotropin (ACTH) and glucagon. Following their release from α-corticotropin, Ser-Tyr and Ser-Met underwent a partial breakdown during extended periods of digestion (McDonald *et al.*, 1969b). Similarly, Thr-Phe and Thr-Ser were attacked following their appearance in glucagon and secretion digests (McDonald *et al.*, 1969a). Because Ser-Met was especially sensitive to hydrolysis, the responsible enzyme was named *Ser-Met dipeptidase* (McDonald *et al.*, 1972b). Five years later it was renamed *lysosomal dipeptidase* (*LDP*) as a consequence of an effort to organize and name some mammalian exopeptidases in accordance with their subcellular localization (McDonald & Schwabe, 1977). Later, in response to evidence suggesting that lysosomes may harbor yet another dipeptidase (Bouma *et al.*, 1976), the first was dubbed *lysosomal dipeptidase I* (McDonald & Barrett, 1986). But now, in the absence of additional studies providing unequivocal support for the existence of a second lysosomal dipeptidase, the Roman numeral designation is herewith being abandoned. IUBMB has not yet assigned an EC number to lysosomal dipeptidase.

Lysosomal dipeptidase closely resembles a zinc-activated dipeptidase detected in thyroid extracts from cattle by Loughlin & Trikojus (1964). Although they provisionally named the responsible enzyme *cysteinylglycinase*, they later noted that the partially purified enzyme hydrolyzed Leu-Tyr more rapidly. (The name cysteinylglycinase is also used for a manganese-activated dipeptidase discussed in Chapter 338.) Another dipeptidase activity that may be attributable to LDP is that of *Leu-Gly dipeptidase*. It occurs in rat liver lysosomes (tritosomes), and, like LDP, is acid stable, and has no sulfhydryl requirement. On the other hand, its pH 6.5 optimum is one unit higher, it is only partially inhibited by EDTA, and activation by zinc could not be demonstrated (Bouma *et al.*, 1976).

## Activity and Specificity

Characteristic of a true dipeptidase, LDP activity is restricted to the hydrolysis of dipeptides possessing unsubstituted (free) N- and C-termini. Although Ser+Met is readily hydrolyzed, the enzyme shows no action on Ser-Met-NH$_2$, Z-Ser-Met, Ser-Met-Glu or Ser-NHNap. Other susceptible dipeptides include Ser+Ala, Ser+Tyr, Ser+Phe, Ser+Leu, Ser+His, Ser+Ser, Ala+Met, Ala+Phe, Gly+Met, Met+Ser, Met+Phe, Thr+Phe, His+Phe, Arg+Phe and Asp+Phe. Little or no action is exhibited on Gly-Phe, Gly-Gly, Gly-Leu, Lys-Lys, Ser-Pro, Phe-Ser and Pro-Phe (McDonald *et al.*, 1972b). The preparation of LDP employed in the foregoing specificity survey, conducted at pH 5.5, was derived from beef spleen and was known to contain both

dipeptidyl-peptidase I (Chapter 214) and lysosomal Pro-X carboxypeptidase (Chapter 137), but their activities were effectively eliminated with *p*-chloromercuriphenyl sulfonate (0.5 mM) and DFP (1 mM), respectively. Neither inhibitor had any effect on LDP. LDP was also present in similar DPP I preparations derived from rat liver (McDonald *et al.*, 1972b).

The dipeptidase as 'cysteinyltyrosinase' purified (60-fold) from cattle thyroid glands hydrolyzes Cys+Tyr optimally at pH 5.3. Other dipeptides readily hydrolyzed include Ala+Ala, Ala+Gly, Glu+Tyr, Leu+Ala, Leu+Gly, Leu+Leu, Leu+Phe, Leu+Tyr (hydrolyzed faster than Cys-Tyr), Leu+Trp, Met+Met and Tyr+Leu. Lower rates occur on Gly+Leu, Gly+Phe, Gly+Tyr, Val+Val and Gly+Asn, but little or no action is seen on Gly-Gly, Lys-Gly, Gly-Pro, Pro-Gly, Gly-Leu-Tyr, Gly-Phe-NH$_2$, Ac-Phe-Tyr, Leu-NH$_2$ and Leu-NHNapOMe (Loughlin & Trikojus, 1964).

The pH optimum for Ser-Met hydrolysis by LDP derived from cattle spleen or rat liver is about 5.5 (McDonald *et al.*, 1972b). The partially purified dipeptidase from cattle thyroid acts optimally at about pH 5.3 (Loughlin & Trikojus, 1964). Highly purified LDP from human thyroid tissue hydrolyzes Ser-Met actively at about pH 5.5, and effectively cleaves T$_4$-Gln (T$_4$ = 3,5,3,5-tetraiodothyronyl-) following its liberation from thyroglobulin, at pH 4.5, by cathepsin B purified from human thyroid tissue (Dunn *et al.*, 1996).

The activity of LDP can be determined using a colorimetric assay procedure that employs Ser-Met as the substrate in a sodium cacodylate or acetate buffer, pH 5.5. (Citrate buffers are inhibitory.) Zinc acetate (0.3 mM) may be included to ensure full activity (McDonald *et al.*, 1972b; Dunn *et al.*, 1996). The liberated amino acids are quantified colorimetrically by employing a trinitrobenzenesulfonic acid reagent (Okuyama & Satake, 1960) to which copper is added to suppress reaction with the substrate and other peptides, thus greatly increasing the sensitivity of the assay. The addition of copper to the reagent also makes it possible to assay for the dipeptidase in fractions containing Ampholine (McDonald *et al.*, 1972b).

Lysosomal dipeptidase is strongly inhibited by chelating agents such as EDTA and 1,10-phenanthroline. Citrate buffers and metals such as Hg$^{2+}$, Cu$^{2+}$, Fe$^{2+}$ and Fe$^{2+}$ are also inhibitory. *p*-Chloromercuriphenyl sulfonate (1 mM), a thiol-blocking reagent, and DFP (1 mM) have no effect (McDonald *et al.*, 1972b).

## Structural Chemistry

Although the molecular structure of native lysosomal dipeptidase has yet to be firmly established, a recent study suggests that the dipeptidase of human thyroid is a dimeric glycoprotein with a relative molecular mass ($M_r$) of about 100 000 (Dunn *et al.*, 1996). The native enzyme appears to be a metalloprotein wherein Zn$^{2+}$ appears to best satisfy its catalytic activity, and to a decreasing degree Mg$^{2+}$, Mn$^{2+}$ and Co$^{2+}$. The metal of the native enzyme appears to be very firmly bound since the activity of the EDTA-inhibited enzyme can

**M**

be fully restored by dialysis as well as by the addition of zinc acetate (McDonald *et al.*, 1972b). LDP of cattle spleen is a stable enzyme with a pI of about 5.4. The fact that it copurifies with dipeptidyl-peptidase I indicates that it tolerates tissue autolysis at pH 3.5 (37°C) and heat treatment at 65°C (McDonald *et al.*, 1972a). It appears that the S1 subsite on the dipeptidase is capable of accommodating residues with a large hydrophobic side chain as evidenced by its ability to cleave T$_4$-Gln.

## Preparation

As a lysosomal peptidase, LDP is most probably present in the extract of any lysosome-rich tissue. Although LDP characteristically copurifies with DPP I when the latter is purified from cattle spleen or rat liver, it is possible finally to separate them by isoelectric focusing in a pH 5–7 Ampholine gradient (McDonald *et al.*, 1972a,b). LDP present in extracts of human thyroid tissue has recently been purified to near-homogeneity by employing a series of column chromatographies that utilize hydroxyapatite, molecular exclusion, lectin affinity and high-performance anion and cation exchangers. Such preparations are free of contamination by other protease activities (Dunn *et al.*, 1996).

## Biological Aspects

LDP has been shown by differential and density-equilibrium centrifugation to have a lysosomal localization in cattle spleen (McDonald *et al.*, 1972b) and in human thyroid (Dunn *et al.*, 1991, 1996). Its lysosomal distribution in the tissues, and its selective action on dipeptides, strongly suggests a role for this enzyme in the final stages of intracellular protein degradation. The broad specificity of LDP and its action over a range of acidic pH is well suited to the task of cleaving the many dipeptides successively released from polypeptides by dipeptidyl-peptidases I and II (Chapter 138), and by the peptidyl-dipeptidase activity of cathepsin B (Chapter 209). It has been reported that the complex of endopeptidases and exopeptidases present in isolated rat liver lysosomes is capable of extensively degrading globin substrates. About 30% of the globin residues appear as dipeptides resistant to further hydrolysis, and 40% appear as free amino acids (Coffey & de Duve, 1968). Most of the free amino acids probably arise through the action of LDP on dipeptide fragments. Since it is generally accepted that the lysosomal membrane is permeable to dipeptides and free amino acids, but not tripeptides and larger oligopeptides (Lloyd & Foster, 1986), dipeptides resistant to the action of LDP would be expected to penetrate the lysosomal membrane and undergo hydrolysis in the cytosol, thus returning all the residues of the globin substrate to the cytosolic amino acid pool.

The proteolytic processing of thyroglobulin within the lysosomes of the secretary epithelium of thyroid follicles has been under investigation for many years (De Robertis, 1941). Of special interest are the lysosomal proteolytic enzymes responsible for releasing the hormone thyroxine or T$_4$ (3,5,3,5-tetraiodothyronine) from this 660 kDa glycoprotein. It has recently been shown that T$_4$-Gln, a dipeptide corresponding to residues 5 and 6 of thyroglobulin, which represents the most important hormone-forming site in all mammalian species so far examined (Roe *et al.*, 1989), is excised by the successive endopeptidase and exopeptidase actions of cathepsin B (Chapter 209) purified from human thyroid tissue. This dipeptide is then cleaved by purified thyroidal lysosomal dipeptidase to release the biologically active hormone (Dunn *et al.*, 1996).

## Distinguishing Features

The enzyme has a lysosomal localization and catalyzes an EDTA-sensitive hydrolysis of Ser-Met at acidic pH; this activity is unaffected by sulfhydryl reagents and inhibitors of the serine- and aspartic-type proteases.

## References

Bouma, J.M.W., Scheper, A., Duursma, A. & Gruber, M. (1976) Localization and some properties of lysosomal dipeptidases in rat liver. *Biochim. Biophys. Acta* **444**, 853–862.

Coffey, J.W. & de Duve, C. (1968) Digestive activity of lysosomes. The digestion of proteins by extracts of rat liver lysosomes. *J. Biol. Chem.* **243**, 3255–3263.

De Robertis, E. (1941) Proteolytic enzyme activity of colloid extracted from single follicles of the rat thyroid. *Anat. Rec.* **80**, 219–231.

Dunn, A.D., Crutchfield, H.E. & Dunn, J.T. (1991) Proteolytic processing of thyroglobulin by extracts of thyroid lysosomes. *Endocrinology* **128**, 3073–3080.

Dunn, A.D., Myers, H.E. & Dunn, J.T. (1996) The combined action of two thyroidal proteases releases T$_4$ from the dominant hormone-forming site of thyroglobulin. *Endocrinology* **137**, 3279–3285.

Lloyd, J.B. & Foster, S. (1986) The lysosomal membrane. *Trends Biochem. Sci.* **11**, 365–368.

Loughlin, R.E. & Trikojus, V.M. (1964) A metal-dependent peptidase from thyroid glands. *Biochim. Biophys. Acta* **92**, 529–542.

McDonald, J.K. & Barrett, A.J. (1986) Lysosomal dipeptidase I. In: *Proteases: A Glossary and Bibliography*, vol. 2: *The Exopeptidases*. London: Academic Press, pp. 298–300.

McDonald, J.K. & Schwabe, C. (1977) Intracellular exopeptidases. In: *Proteinases in Mammalian Cells and Tissue* (Barrett, A.J., ed.). Amsterdam: North-Holland Publishing, pp. 311–391 (see pp. 346–348).

McDonald, J.K., Callahan, P.X., Zeitman, B.B. & Ellis, S. (1969a) Inactivation and degradation of glucagon by dipeptidyl aminopeptidase I (cathepsin C) of rat liver. Including a comparative study of secretin degradation. *J. Biol. Chem.* **244**, 6199–6208.

McDonald, J.K., Zeitman, B.B., Reilly, T.J. & Ellis, S. (1969b) New observations on the substrate specificity of cathepsin C (dipeptidyl aminopeptidase I). Including the degradation of α-corticotropin and other peptide hormones. *J. Biol. Chem.* **244**, 2693–2709.

McDonald, J.K., Callahan, P.X. & Ellis, S. (1972a) Preparation and specificity of dipeptidyl aminopeptidase I. *Methods Enzymol.* **25B**, 272–281.

McDonald, J.K., Zeitman, B.B. & Ellis, S. (1972b) Detection of a lysosomal carboxypeptidase and a lysosomal dipeptidase in highly purified dipeptidylamino peptidase I (cathepsin C) and the elimination of their activities from preparations used to sequence peptides. *Biochem. Biophys. Res. Commun.* **46**, 62–70.

Okuyama, T. & Satake, K. (1960) The preparation and properties of 2,4,6-trinitrophenylamino acids and peptides. *J. Biochem.* **47**, 454–466.

Roe, M.T., Anderson, P.C., Dunn, A.D. & Dunn, J.T. (1989) The hormonogenic sites of turtle thyroglobulin and their homology with those of mammals. *Endocrinology* **124**, 1327–1332.

*J. Ken McDonald*
*Department of Biochemistry and Molecular Biology,*
*Medical University of South Carolina,*
*171 Ashley Avenue,*
*Charleston, SC 29425, USA*

# 531. *Peptidyl-dipeptidase B*

## Databanks

*Peptidase classification: clan MX, family M99, MEROPS ID: M9D.001*
*NC-IUBMB enzyme classification: EC 3.4.15.4*
*Databank codes: no sequence data available*

## Name and History

An enzyme termed **dipeptidyl carboxy hydrolase** was discovered in cattle atrial tissue by Harris & Wilson (1984). Subsequent studies showed that the enzyme, as **dipeptidyl carboxyhydrolase**, was similar, but not identical to peptidyl-dipeptidase A (Chapter 359). The name **atrial peptide convertase** has also been suggested (Harris & Wilson, 1985a), but that recommended by NC-IUBMB is **peptidyl-dipeptidase B**.

## Activity and Specificity

Peptidyl-dipeptidase B (PDB) has been only partially purified, which means that the data for activity must be regarded as somewhat preliminary. That said, it was found that PDB was active on Bz-Gly┼His-Leu, a standard substrate of peptidyl-dipeptidase A, but the enzyme also removed the C-terminal dipeptide, Phe-Arg, from atriopeptin II (to form atriopeptin I) and from Bz-Gly-Ser┼Phe-Arg, an analog of the C-terminus of atriopeptin II. Atriopeptin I was not further degraded, probably because of the presence of a disulfide bond (Harris & Wilson, 1985a,b).

When the hydrolysis of Bz-Gly┼His-Leu and Bz-Gly-Ser┼Phe-Arg by PDB and peptidyl-dipeptidase A were compared, it was found that PDB exhibited selectivity for Bz-Gly-Ser┼Phe-Arg and peptidyl-dipeptidase A for Bz-Gly┼His-Leu (Harris & Wilson, 1985b). A continuous fluorometric assay with Abz-Gly-Ser┼Phe(NO₂)-Arg was described by Soler & Harris (1988). PDB was found to act also as a peptidyl-tripeptidase, releasing the C-terminal tripeptide from atriopeptin III, -Cys-Asn-Ser┼Phe-Arg-Tyr (Soler & Harris, 1989).

PDB was completely inhibited by EDTA, 1,10-phenanthroline, DTT and 2-ME, leaving little doubt that it is a metallopeptidase. After inactivation by EDTA, the enzyme could be reactivated by $Co^{2+}$, $Zn^{2+}$ and $Mn^{2+}$, but not by $Cu^{2+}$, $Mg^{2+}$, $Ca^{2+}$ or $Cd^{2+}$ (Soler & Harris, 1989). The enzyme was also inhibited by D-2-methyl-3-mercaptopropanoyl-L-Pro (captopril), 3-mercaptopropanoyl-L-Pro, 2-D-methylsuccinyl-L-Pro, and bradykinin potentiating factor, all inhibitors of peptidyl-dipeptidase A. PDB differs from peptidyl-dipeptidase A, however, in having no requirement for chloride ions, a slightly lower pH optimum (pH 7.3), and much lower sensitivity to inhibition by D-Cys-L-Pro (Harris & Wilson, 1984).

## Structural Chemistry

The amino acid sequence of PDB is unknown. The $M_r$ is estimated as 240 000 from ultracentrifugation (Harris & Wilson, 1984).

## Preparation

PDB has not been purified to homogeneity, but was obtained in partially purified form after extraction from cattle atrial tissue with Triton X-100 (Harris & Wilson, 1984, 1985a,b). Tissue (400 g) was homogenized three times in buffer containing 0.4 M NaCl to remove soluble proteins, and then twice in 2% Triton X-100 to solubilize the atrial enzyme. The enzyme was then partially purified by chromatography on DEAE-cellulose, hydroxyapatite and Sephadex G-200.

It should be mentioned that attempts to detect PDB in cattle atrial tissue in the present author's laboratory were unsuccessful, despite kind help from Dr R.B. Harris (P.M. Dando & A.J. Barrett, unpublished results), so the strain of cattle or other factors as yet unidentified may be critical.

M

## Biological Aspects

The activity of PDB in forming atriopeptin I from atriopeptin II and atriopeptin III suggested that the enzyme functions in this way *in vivo*, so that the heart can be considered an endocrine organ (Soler & Harris, 1989). The enzyme was present in the microsomal fraction of cattle atrium, and specifically in sucrose density gradient fractions enriched for atrial granules. Ventricular tissue, which does not contain the atrial peptides, also did not contain PDB, consistent with the possibility that the processing of atriopeptin II is a biological function of PDB.

## Distinguishing Features

The peptidase that would be most likely to be confused with PDB is peptidyl-dipeptidase A (angiotensin-converting enzyme), which is also a membrane-associated metallopeptidase, with generally similar substrate specificity and sensitivity to inhibitors. Only PDB was found to act on atriopeptin II, however, and the enzyme also differed in showing no chloride ion dependence for activity, as well as lesser sensitivity to inhibition by D-Cys-L-Pro. Whether PDB has any relationship to a captopril-sensitive peptidyl-dipeptidase that acts on [Met⁵]-enkephalin+Arg-Phe (Ronai *et al.*, 1995) remains to be established.

## References

Harris, R.B. & Wilson, I.B. (1984) Atrial tissue contains a metallo-dipeptidyl carboxyhydrolase not present in ventricular tissue: partial purification and characterization. *Arch. Biochem. Biophys.* **233**, 667–675.

Harris, R.B. & Wilson, I.B. (1985a) Conversion of atriopeptin II to atriopeptin I by atrial dipeptidyl carboxy hydrolase. *Peptides* **6**, 393–396.

Harris, R.B. & Wilson, I.B. (1985b) Comparison of hydrolysis of atriopeptin II stand-in substrate by atrial dipeptidyl carboxyhydrolase and angiotension I-converting enzyme. *Int. J. Pept. Protein Res.* **26**, 78–82.

Ronai, A.Z., Feher, E., Botyanszki, J., Hepp, J., Magyar, A. & Medzihradszky, K. (1995) Substrate- and inhibitor-specificity of a non-endothelial enzyme which forms [Met5]-enkephalin from [Met5]-enkephalin-Arg6,Phe7 in isolated rabbit ear artery: pharmacological characterization. *Neuropeptides* **28**, 137–145.

Soler, D.F. & Harris, R.B. (1988) Continuous fluorogenic substrates for atrial dipeptidyl carboxyhydrolase. Importance of Ser in the P1 position. *Int. J. Pept. Protein Res.* **32**, 35–40.

Soler, D.F. & Harris, R.B. (1989) Atrial dipeptidyl carboxyhydrolase is a zinc-metallo proteinase which possesses tripeptidyl carboxyhydrolase activity. *Peptides* **10**, 63–68.

*Alan J. Barrett*
*MRC Peptidase Laboratory, The Babraham Institute,*
*Cambridge CB2 4AT, UK*
*Email: alan.barrett@bbscr.ac.uk*

# 532. Peptidyl-dipeptidase (Streptomyces)

## Databanks

*Peptidase classification: clan MX, family M99, MEROPS ID: M9D.002*
*NC-IUBMB enzyme classification: none*
*Databank codes: no sequence data available*

## Name and History

Several peptidases specific for proline residues have been isolated (Walter *et al.*, 1980; Mentlein, 1988), but no peptidase that efficiently hydrolyzes peptides containing two or more proline residues in their C-termini had previously been described. However, a novel proline-specific peptidase, ***Streptomyces peptidyl-dipeptidase***, was recently detected and purified (Miyoshi *et al.*, 1992).

## Activity and Specificity

The enzyme removes Pro-Pro from the C-terminals of proline-containing peptides such as Boc-Pro-Pro+Pro-Pro and Leu-Pro-Pro-Pro+Pro-Pro (Miyoshi *et al.*, 1992). The peptides with proline, 4-hydroxyproline or 3,4-dehydroproline at the P2' position are good substrates, while those with pipecolic acid, D-proline, or other usual amino acids at the P2' position are scarcely hydrolyzed. The peptides with proline,

3,4-dehydroproline, pipecolic acid or *N*-methylalanine at the P1′ position are well hydrolyzed, while those with 4-hydroxyproline or D-proline at the P1′ position are not hydrolyzed. Thus, the enzyme has high specificity for proline or some imino acids at subsites S2′ and S1′ and some specificity at S1 (Maruyama *et al.*, 1993). Besides hydrolyzing proline-containing peptides, the enzyme can catalyze the synthesis of Boc-Pro-Pro-Pro-Pro using Boc-Pro-Pro as an acidic component and Pro-Pro as a basic component (Maruyama *et al.*, 1993). The pH optimum of the enzyme is about 7.6. The enzyme is inhibited by EDTA, 1,10-phenanthroline or DTT. Angiotensin I-converting enzyme inhibitors such as captopril (Ondetti *et al.*, 1977) and Leu-Pro-Pro (Maruyama *et al.*, 1989) are effective inhibitors (Miyoshi *et al.*, 1992).

## Structural Chemistry

The molecular mass of the enzyme is 70 kDa by SDS-PAGE and 65 kDa by gel filtration (Miyoshi *et al.*, 1992).

## Preparation

The enzyme has been purified 730-fold from the culture broth of a microorganism, *Streptomyces* sp., and a culture of this strain has been deposited at the Patent Microorganism Depository, National Institute of Bioscience and Human Technology, Japan, as FERM TU-212 (Miyoshi *et al.*, 1992).

## References

Maruyama, S., Miyoshi, S., Kaneko, T. & Tanaka, H. (1989) Angiotensin I-converting enzyme inhibitory activities of synthetic peptides related to the tandem repeated sequence of a maize endosperm protein. *Agric. Biol. Chem.* **53**, 1077–1081.

Maruyama, S., Miyoshi, S., Nomura, G., Suzuki, M., Tanaka, H. & Maeda, H. (1993) Specificity for various imino-acid-residues of a proline-specific dipeptidylcarboxypeptidase from a *Streptomyces* species. *Biochim. Biophys. Acta* **1162**, 72–76.

Mentlein, R. (1988) Proline residues in the maturation and degradation of peptide hormones and neuropeptides. *FEBS Lett.* **234**, 251–256.

Miyoshi, S., Nomura, G., Suzuki, M., Fukui, F., Tanaka, H. & Maruyama, S. (1992) Purification and characterization of a novel dipeptidyl carboxypeptidase from a *Streptomyces* species. *J. Biochem.* **112**, 253–257.

Ondetti, M.A., Rubin, B. & Cushman D.W. (1977) Design of specific inhibitors of angiotensin-converting enzyme: new class of orally active antihypertensive agents. *Science* **196**, 441–444.

Walter, R., Simmons, W.H. & Yoshimoto, T. (1980) Proline specific endo- and exopeptidases. *Mol. Cell. Biochem.* **30**, 111–127.

*Sususmu Maruyama*
*National Institute of Bioscience and Human-Technology,*
*Agency of Industrial Science and Technology,*
*1-1 Higashi, Tsukuba, Ibaraki 305, Japan*
*Email: smaruyama@ccmail.nibh.go.jp*

# 533. Alanine carboxypeptidase

## Databanks

*Peptidase classification: clan MX, family M99, MEROPS ID: M9E.002*
*NC-IUBMB enzyme classification: EC 3.4.17.6*
*Chemical Abstracts Service registry number: 37288-70-3*
*Databank codes: no sequence data available*

## Name and History

Characterization of enzymes hydrolyzing folate analogs led Levy & Goldman (1969) to synthesize other such compounds for screening soil cultures. The folate analog 4-amino-4-deoxypteroyl alanine and subsequently 4-aminobenzoyl alanine were synthesized. Miyagawa *et al.* (1986) sought to isolate enzymes related to hippuryl hydrolase (*N*-benzoylglycine hydrolase). *N*-Benzoyl-L-alanine was used to isolate an enzyme from the soil microorganism *Corynebacterium equi* H-7. The name **alanine carboxypeptidase** has been adopted by the Enzyme Commission to describe these enzymes isolated from soil bacteria. The enzyme from *Corynebacterium equi* H-7 has also been referred to as **N-benzoyl-L-alanine amidohydrolase** (Miyagawa *et al.*, 1986).

## Activity and Specificity

The enzyme isolated by Levy & Goldman (1969) cleaved substrates of the general formula RCO┼L-Ala, with a preference for an aromatic residue at R. No hydrolysis occurred when the carboxyl group of alanine was blocked. The enzyme isolated by Miyagawa *et al*. (1986) hydrolyzed Bz+Ala > Bz+Gly > Bz+ aminobutyrate. In contrast to the enzyme described by Levy & Goldman, the enzyme isolated by Miyagawa *et al*. (1986) did not hydrolyze Ac-Ala or Z-Ala.

The enzyme isolated by Levy & Goldman (1969) has an absolute requirement for $Zn^{2+}$. The enzyme isolated by Miyagawa *et al*. (1986) is completely inhibited by metal chelators and heavy metal ions. It is activated by $Co^{2+}$ and stabilized by $Cl^-$.

## Structural Chemistry

The $M_r$ of the enzyme isolated by Miyagawa *et al*. (1986) is 230 000. This protein gives a single band on SDS-PAGE with an $M_r$ of 40 000 and a pI of 4.6.

## Preparation

Levy & Goldman (1969) purified their enzyme more than 100-fold to near homogeneity by classical techniques. Miyagawa *et al*. (1986) purified their enzyme 323-fold to apparent homogeneity by classical techniques, i.e. ammonium sulfate fractionation, and sequential chromatography on DEAE-cellulose, DEAE-Sephadex A50 and hydroxyapatite.

## Biological Aspects

Various species of microorganisms degrade hippuric acid to benzoic acid and glycine. The enzyme described by Miyagawa *et al*. (1986) behaves as an atypical hippuryl hydrolase preferring Bz-Ala over Bz-Gly.

## Distinguishing Features

Although both enzymes release C-terminal alanine from selected substrates, they are not identical. The enzyme described by Miyagawa *et al*. (1986) can be considered as an atypical hippuryl hydrolase whereas the enzyme of Levy & Goldman (1969) can be considered as an acyl-alanine hydrolase.

## References

Levy, C.C. & Goldman, P. (1969) Bacterial peptidases. II. An enzyme specific for *N*-acyl linkages to alanine. *J. Biol. Chem.* **244**, 4467–4472.

Miyagawa, E., Harada, T. & Motoki, Y. (1986) Purification and properties of *N*-benzoyl-L-alanine amidohydrolase from *Corynebacterium equi*. *Agric. Biol. Chem.* **50**, 1527–1531.

*Sherwin Wilk*
*Department of Pharmacology,*
*Mount Sinai School of Medicine,*
*One Gustave I. Levy Place,*
*New York, NY 10029, USA*
*Email: S_Wilk@smtplink.mssm.edu*

# 534. Membrane Pro-X carboxypeptidase

## Databanks

*Peptidase classification: clan MX, family M99, MEROPS ID: M9E.004*
*NC-IUBMB enzyme classification: EC 3.4.17.6*
*Chemical Abstracts Service registry number: 38288-70-3*
*Databank codes: no sequence data available*

## Name and History

The study of Dehm & Nordwig (1970) on the enzyme involved in the final step of the breakdown of collagen, a protein especially rich in proline and endowed with an unusually high proportion of Xaa-Pro-Yaa sequences, provided the background which led to the discovery of a novel carboxypeptidase in a pig kidney microsomal fraction. Because this enzyme was the first intracellular, nonlysosomal carboxypeptidase, these investigators called it **microsomal carboxypeptidase**. They also suggested the name **carboxypeptidase P** in order to indicate the hydrolysis of proline-containing peptides by a C-terminal-specific peptidase. However, the same name was also used for a serine-type proline carboxypeptidase from *Penicillium* (Yokoyama *et al.*, 1974). To avoid confusion, the IUBMB recommended that the name be changed to **membrane Pro-X carboxypeptidase**; the first EC entry for the enzyme was made in 1992.

## Activity and Specificity

The enzyme catalyzes the hydrolysis of C-terminal amino acid residues from peptides bearing a penultimate (P1) proline residue. The purified kidney enzyme can hydrolyze peptide bonds such as Pro┼Phe, Pro┼Met, Pro┼Thr, Pro┼Ala, Pro┼Ile and Pro┼Gly, whereas Pro-Pro, Pro-Hyp, Hyp-Gly and Pro-Leu bonds are not recognized as substrates (Dehm & Nordwig, 1970; Hedeager-Sørensen & Kenny, 1985). The enzyme is not strictly specific for imino acids since nonproline amino acid-containing peptides such as Z-Ala┼Tyr, Z-Gly┼Tyr, Z-Ala┼Phe and Z-Gly┼Phe can also be hydrolyzed at a measurable rate. Alanine can substitute for proline in the penultimate position such as Z-Ala┼Met; and especially the tetrapeptide Ala-Ala-Ala┼Ala is a relatively good substrate, although the tripeptide Ala-Ala-Ala is resistant to hydrolysis (Hedeager-Sørensen & Kenny, 1985). Although the peptide chain length might appear to be a factor for the expression of enzyme action, the enzyme also exhibits dipeptidase activity toward Pro┼Ala and Pro┼Phe (Dehm & Nordwig, 1970). The highest rate of hydrolysis was observed with the peptide substrate Val-Ala-Ala┼Phe, which was 4-fold greater than that for Z-Pro┼Phe or Z-Pro┼Met (Hedeager-Sørensen & Kenny, 1985).

The enzyme assay is carried out with peptides blocked with a benzyloxycarbonyl group. The method for determination of the activity is based upon amino acid analysis measurement of the released C-terminal amino acid from Z-Pro┼Ala or Z-Gly-Pro┼Ala (Dehm & Nordwig, 1970). The enzyme may also be assayed by following the release of methionine from Z-Pro┼Met as substrate in a coupled assay system utilizing L-amino acid oxidase (Kenny *et al.*, 1977; Booth *et al.*, 1979; Hedeager-Sørensen & Kenny, 1985). An HPLC method utilizing the same synthetic substrate, Z-Pro┼Ala, as used for the assay of lysosomal Pro-X carboxypeptidase (Chapter 137) (Suzawa *et al.*, 1995) may also be applied. An HPLC analysis was employed to measure the hydrolyzing activity of the same biologically active peptides (Hedeager-Sørensen & Kenny, 1985; Huneau *et al.*, 1994; Bouras *et al.*, 1995). The assay system typically includes 1 mM $MnCl_2$.

The enzyme has a pH optimum of 7.75 and is stimulated 5- to 10-fold by 1 mM $MnCl_2$ (Dehm & Nordwig, 1970; Hedeager-Sørensen & Kenny, 1985). Both EDTA (0.5 mM) and 1,10-phenanthroline (10 mM) inhibit all activity (Hedeager-Sørensen & Kenny, 1985).

## Structural Chemistry

Membrane Pro-X carboxypeptidase has an $M_r$ of 240 000 (Dehm & Nordwig, 1970) and is a dimer of subunits of $M_r$ 135 000 (Hedeager-Sørensen & Kenny, 1985). The chemical analysis data indicate the enzyme to be a glycoprotein containing 33.2% (w/w) carbohydrate, with each subunit having one zinc atom (Hedeager-Sørensen & Kenny, 1985).

## Preparation

Membrane Pro-X carboxypeptidase was solubilized and purified 274-fold from a pig kidney microsomal fraction by toluene-trypsin treatment (Dehm & Nordwig, 1970), and about 600-fold from kidney microvillus membranes by either Triton X-100 or papain treatment (Hedeager-Sørensen & Kenny, 1985).

## Biological Aspects

After the first report of detection of membrane Pro-X carboxypeptidase in pig kidney microsomes by Dehm & Nordwig (1970), Kenny *et al.* (1977) found the enzyme to be localized in the microvillus membrane of the kidney. The enzyme constitutes about 1.5% of the kidney microvillus protein (Hedeager-Sørensen & Kenny, 1985). Auricchio *et al.* (1978) reported that the enzyme is also present in rabbit intestinal brush border preparations. This enzyme was identified as membrane Pro-X carboxypeptidase; however, it was suggested to differ from the kidney enzyme in some important respects such as substrate specificity, requirement of divalent cations, inhibitory effect of chemicals and optimum pH (Bouras *et al.*, 1995; Erickson *et al.*, 1989). The enzyme in kidney microvillar membrane or intestinal brush border membrane may play a role in the terminal phase of digestion and absorption of proline-containing peptides (Dehm & Nordwig, 1970; Kenny *et al.*, 1977; Yoshioka *et al.*, 1988; Huneau *et al.*, 1994).

Membrane Pro-X carboxypeptidase cleaves the C-terminal Pro┼Phe bond in angiotensins II (Asp-Arg-Val-Tyr-Ile-His-Pro┼Phe) and III (des-Asp[1]-angiotensin II). It has also been suggested to hydrolyse enterostatin, a powerfully anorectic pentapeptide that is produced by the tryptic cleavage of procolipase in the small intestine. The cleavage is Ala-Pro-Gly-Pro┼Arg in human, chicken and rabbit and Val-Pro-Asp-Pro┼Arg in pig and cattle (Huneau *et al.*, 1994; Bouras *et al.*, 1995).

## Distinguishing Features

Membrane Pro-X carboxypeptidase is distinguishable from serine-type carboxypeptidases (Chapter 132) and lysosomal Pro-X carboxypeptidase (Chapter 137) by its stability to heating and by its pH optimum. Unlike the serine enzymes, it has not been shown to be inhibited by DFP.

## Further Reading

Vanhoof *et al.* (1995) gives a useful summary of the structure and function of proline-containing biologically important peptides.

## References

Auricchio, S., Greco, L., DeVizia, B. & Buonocore, V. (1978) Dipeptidylaminopeptidase and carboxypeptidase activities of the brush border of rabbit small intestine. *Gastroenterology* **75**, 1073–1079.

Booth, A.G., Hubbard, L.M.L. & Kenny, A.J. (1979) Proteins of the kidney microvillar membrane. Immunoelectrophoretic analysis of the membrane hydrolases: identification and resolution of the detergent- and proteinase-solubilized forms. *Biochem. J.* **179**, 397–405.

Bouras, M., Huneau, J.F., Luengo, C., Erlanson-Albertsson, C. & Tomé, D. (1995) Metabolism of enterostatin in rat intestine, brain membranes, and serum: differential involvement of proline-specific peptidases. *Peptides* **16**, 399–405.

Dehm, P. & Nordwig, A. (1970) The cleavage of prolyl peptides by kidney peptidases. Isolation of a microsomal carboxypeptidase from swine kidney. *Eur. J. Biochem.* **17**, 372–377.

Erickson, R.H., Song, I.-S., Yoshioka, M., Gulli, R., Miura, S. & Kim, Y.S. (1989) Identification of proline-specific carboxypeptidase localized to brush border membrane of rat small intestine and its possible role in protein digestion. *Dig. Dis. Sci.* **34**, 400–406.

Hedeager-Sørensen, S. & Kenny, A.J. (1985) Proteins of the microvillar membrane. Purification and properties of carboxypeptidase P from pig kidneys. *Biochem. J.* **229**, 251–257.

Huneau, J.F., Erlanson-Albertsson, C., Beauvallet, C. & Tomé, D. (1994) The *in vitro* intestinal absorption of enterostatin is limited by brush-border membrane peptidases. *Regul. Pept.* **54**, 495–503.

Kenny, A.J., Booth, A.G. & Macnair, R.D.C. (1977) Peptidases of kidney microvillus membrane. *Acta Biol. Med. Germ.* **36**, 1575–1585.

Suzawa, Y., Hiraoka, B.Y., Harada, M. & Deguchi, T. (1995) High-performance liquid chromatographic determination of prolylcarboxypeptidase activity in monkey kidney. *J. Chromatogr. B* **670**, 152–156.

Vanhoof, G., Goossens, F., DeMeester, I., Hendriks, D. & Scharpé, S. (1995) Proline motifs in peptides and their biological processing. (review) *FASEB J.* **9**, 736–744.

Yokoyama, S., Oobayashi, A., Tanabe, O., Sugawara, S., Araki, E. & Ichishima, E. (1974) Production and some properties of a new types of acid carboxypeptidase of *Penicillium* molds. *Appl. Microbiol.* **27**, 953–960.

Yoshioka, M., Erickson, R.H. & Kim, Y.S. (1988) Digestion and assimilation of proline-containing peptides by rat intestinal brush border membrane carboxypeptidases. *J. Clin. Invest.* **81**, 1090–1095.

*B. Yukihiro Hiraoka*
*Department of Oral Biochemistry,*
*School of Dentistry,*
*Matsumoto Dental University,*
*Shiojiri, Nagano, 399-07, Japan*
*Email: byh@po.mdu.ac.jp*

# 535. *Carboxypeptidase G₃*

## Databanks

*Peptidase classification: clan MX, family M99, MEROPS ID: M9E.007*
*NC-IUBMB enzyme classification: EC 3.4.17.11*
*Databank codes: no sequence data available*

## Name and History

An enzyme useful for preparing semisynthetic polymyxins was found to be produced by a *Pseudomonas* sp. strain, M-6-3 (closely related to *Pseudomonas acidovorans*). This enzyme, tentatively named polymyxin acylase, is the first enzyme known to remove the acyl moiety from an acyl peptide without affecting the peptide moiety (Kimura *et al.*, 1987, 1989). During purification of the deacylase from the polymyxin acylase-producing strain M-27, which is similar to the M-6-3 strain, another new enzyme that hydrolyzed only *N*-fatty acyl-Glu or -Asp, but not polymyxins, was found. It seemed to be a kind of carboxypeptidase because it acted on C-terminal amino acids in peptides (Yasuda *et al.*, 1992). The enzyme resembles the known enzyme carboxypeptidases G₁ (Chapter 483) (McCullough *et al.*, 1971) and G₂ (Chapter 483) (Sherwood *et al.*,

1985) in its glutamate-releasing activity. However, it acts not only on the L-form but also on the D-form of acidic amino acids and shows affinity for the long-chain fatty acyl group but not the benzoyl group. Because this enzyme differs from carboxypeptidase G₁ and G₂, it was named ***carboxypeptidase G₃***.

## Activity and Specificity

*N*-Octanoyl-amino acids are useful substrates for this enzyme. Of 11 kinds of *N*-octanoyl-amino acids, *N*-octanoyl-ɪDL-Glu and -ɪDL-Asp are hydrolyzed, while the other *N*-octanoyl-amino acids are only very slightly or not at all hydrolyzed. The most susceptible chain length among the acyl groups is C10 (decanoyl group) as shown in Table 535.1.

*Table 535.1* Relative activities of carboxypeptidase $G_3$ for acyl amino acids

| Substrate | Relative activity (%) |
|---|---|
| Ac-(C2)-DL-Glu | 38 |
| Butyroyl-(C4)-DL-Glu | 42 |
| Hexanoyl-(C6)-DL-Glu | 81 |
| Octanoyl-(C8)-DL-Glu | 100 |
| Decanoyl-(C10)-DL-Glu | 111 |
| Dodecanoyl-(C12)-DL-Glu | 50 |
| Tetradecanoyl-(C14)-DL-Glu | 39 |
| Hexadecanoyl-(C16)-DL-Glu | 27 |
| Z-DL-Glu | 29 |
| Bz-DL-Glu | 6 |
| Octanoyl-(C8)-DL-Asp | 118 |
| Octanoyl-(C8)-Gly | 0.1 |
| Octanoyl-(C8)-DL-Ala | 0.1 |
| Octanoyl-(C8)-DL-Val | 2 |
| Octanoyl-(C8)-DL-Leu | 3 |
| Octanoyl-(C8)-DL-Ser | 3 |
| Octanoyl-(C8)-DL-Thr | 0.1 |
| Octanoyl-(C8)-DL-Met | 1 |
| Octanoyl-(C8)-DL-Phe | 1 |
| Octanoyl-(C8)-6-aminohexanoic acid | 0.1 |

The enzyme activity for octanoyl-(C8)-DL-Glu, corresponding to $33\,\mu mol\,min^{-1}\,mg^{-1}$, was taken as 100%.

*Table 535.2* Substrate specificity of carboxypeptidase $G_3$ for peptides, folic acid and its related compounds

| Substrate | Activity ($\mu mol\,min^{-1}\,mg^{-1}$) |
|---|---|
| Octanoyl-Gly-L-Glu | 23.4 |
| Z-L-Tyr-L-Glu | 12.5 |
| L-Glu-L-Glu | 8.8 |
| Folic acid | 1.4 |
| Dihydrofolic acid | 0.6 |
| Methotrexate | 0.4 |
| Aminopterin | 0.2 |
| 5-Methyltetrahydrofolic acid | 0.04 |
| Leucovolin | 0.02 |

L-Glu-L-Glu, folic acid and its analog, methotrexate, are susceptible to hydrolysis and the enzyme acts on C-terminal amino acids in peptides such as *N*-octanoyl-Gly+L-Glu or Z-L-Tyr+L-Glu, but not on C-terminal proline, as shown in Table 535.2.

Also, although the optical specificity (D/L ratio) of the enzyme is different for each substrate, it acts not only on the L-form but also on the D-form of acidic amino acids.

Carboxypeptidase or acyl-amino acid acylase activities were assayed from the ninhydrin color spot of liberated amino acids by thin-layer chromatography, and activities for high molecular weight peptides were measured using gel-permeation chromatography (Yasuda *et al.*, 1992). The pH optimum of carboxypeptidase $G_3$ is 8.0 for *N*-octanoyl-DL-Glu. Enzyme activity falls to 29% in the presence of 0.1 mM EGTA and to 35% in the presence of 0.1 mM 1,10-phenanthroline. The inactivated enzyme is activated by addition of $Zn^{2+}$ and slightly activated by $Cu^{2+}$, $Mn^{2+}$ and $Co^{2+}$.

## Structural Chemistry

Carboxypeptidase $G_3$ consists of four subunits of identical molecular mass of 15 kDa. The pI is estimated to be 7.2. The enzyme is suspected to be a zinc-dependent metallopeptidase. The amino acid sequence and gene for carboxypeptidase $G_3$ are not known yet.

## Preparation

Carboxypeptidase $G_3$ was found in a strain of *Pseudomonas* sp. M-27 isolated from a soil sample from Nishinomiya, Japan. The cell-free extract, solubilized by the detergent activity of colistin sulfate, has been purified to a homogeneous state (Yasuda *et al.*, 1992).

## Biological Aspects

The enzyme exists on the cell membrane of *Pseudomonas* sp. bacteria, and seems to be a constitutive enzyme. The enzyme shows *in vitro* antitumor activity. Carboxypeptidase $G_3$ very slightly inhibits the growth of murine leukemia, but shows remarkable antitumor activities against human carcinoma KB and PC-3 cells among ten kinds of human carcinoma cell lines (Yasuda *et al.*, 1994). Its antitumor activity mechanism seems to be based on the degradation of folic acid.

## Distinguishing Features

Unlike carboxypeptidases G (Goldman & Levy, 1967), $G_1$ (McCullough *et al.*, 1971) and $G_2$ (Sherwood *et al.*, 1985), the carboxypeptidase $G_3$ acts not only on the L-form of acidic amino acids, but also on the D-form, and slightly hydrolyzes Bz-DL-Glu (Yasuda *et al.*, 1992). However, some properties such as C-terminal glutamate-releasing activity and the requirement for $Zn^{2+}$ are quite similar to those of the other enzymes. The molecular mass of carboxypeptidase $G_3$ is 60 kDa, consisting of four identical subunits, while those of carboxypeptidase $G_1$ and $G_2$ are 92 kDa and 83 kDa, respectively, with two identical subunits each. Carboxypeptidase $G_3$ also differs from the other enzymes in enzymatic properties such as the optimal pH for methotrexate and the pI. Other bacteria with peptidases reported to cleave the amino bond of folic acid are *Alcaligenes faecalis* (McNutt, 1963) and *Flavobacterium polyglutamicum* (Pratt *et al.*, 1968), but their enzymatic properties have not been described in detail.

Carboxypeptidase $G_1$ can inhibit the growth of L5178Y and L1210 murine leukemia (Bertino *et al.*, 1971; Chabner *et al.*, 1971), and carboxypeptidase $G_2$ exhibits antitumor activity against Walker carcinoma in rats (Sherwood *et al.*, 1985). Carboxypeptidase $G_3$ has shown *in vitro* antitumor activities against human carcinoma KB and PC-3.

## References

Bertino, J.R., O'Brien, P. & McCullough, J.L. (1971) Inhibition of growth of leukemia cells by enzymic folate depletion. *Science* **172**, 161–162.

M

Chabner, B.A., McCullough, J.L. & Bertino, J.R. (1971) Carboxypeptidase G₁: a new approach to anti-folate chemotherapy. *Proc. Am. Assoc. Cancer Rec.* **12**, 88.

Goldman, P. & Levy, C.C. (1967) Carboxypeptidase G: purification and properties. *Proc. Natl Acad. Sci. USA* **58**, 1299–1306.

Kimura, Y. & Yasuda, N. (1989) Polymyxin acylase: purification and characterization, with special reference to broad substrate specificity. *Agric. Biol. Chem.* **53**, 497–504.

Kimura, Y., Matsunaga, H., Yasuda, N., Tatsuki, T. & Suzuki, T. (1987) Polymyxin acylase: a new enzyme for preparing starting materials for semisynthetic polymyxin antibiotics. *Agric. Biol. Chem.* **51**, 1617–1623.

McCullough, J.L., Chabner, B.A. & Bertino, J.R. (1971) Purification and properties of carboxypeptidase G₁. *J. Biol. Chem.* **246**, 7207–7213.

McNutt, W.S. (1963) The enzymic deamination and amide cleavage of folic acid. *Arch. Biochem. Biophys.* **101**, 1–6.

Pratt, A.G., Crawford, E.J. & Friedkin, M. (1968) The hydrolysis of mono-, di-, and triglutamate derivatives of folic acid with bacterial enzymes. *J. Biol. Chem.* **243**, 6367–6372.

Sherwood, R.F., Melton, R.G. Alwan, S.M. & Hughes, P. (1985) Purification and properties of carboxypeptidase G₂ from *Pseudomonas* sp. strain RS-16. *Eur. J. Biochem.* **148**, 447–453.

Yasuda, N., Kaneko, M. & Kimura, Y. (1992) Isolation, purification, and characterization of a new enzyme from *Pseudomonas* sp. M-27, carboxypeptidase G₃. *Biosci. Biotechnol. Biochem.* **56**, 1536–1540.

Yasuda, N., Shimo, Y., Kimura, Y. & Sasaki, T. (1994) *In vitro* antitumor activity of carboxypeptidase G₃ from *Pseudomonas* sp. M-27. *Bull. Mukogawa Women's Univ. Nat. Sci.* **42**, 63–66.

*Yukio Kimura*
*Noriko Yasuda,*
*Laboratory of Biochemistry,*
*Faculty of Phamaceutical Sciences,*
*Mukogawa Women's University,*
*11-68 Koshien Kyuban-cho,*
*Nishinomiya 663, Japan*
*Email: met10947@miet.mukogawa-u.ac.jp*

# 536. *Mitochondrial carboxypeptidase*

## Databanks

*Peptidase classification: clan MX, family M99, MEROPS ID: M9F.003*
*NC-IUBMB enzyme classification: none*
*Databank codes: no sequence data available*

## Name and History

The history of this mitochondrial carboxypeptidase goes back to results from our laboratory demonstrating that the exopeptidase activity of rat liver mitochondria is mainly located in the matrix fraction (Duque-Magalhães & Régnier, 1986) and that most of such exoproteolytic activity is abolished by chelating agents (Duque-Magalhães, 1989). The mitochondrial matrix of rat liver possesses a number of carboxypeptidases. The most significant of them, *mitochondrial carboxypeptidase III (mCP III)*, was purified to apparent homogeneity (Figueiredo & Duque-Magalhães, 1994). The name mCP III was used because the enzyme represents the third major peak of carboxypeptidase activity isolated from the mitochondrial matrix fraction after the most decisive step of purification by metal chelate affinity chromatography.

Although other metalloproteases have been described as associated with the matrix fraction of eukaryotic mitochondria (Yasuhara & Ohashi, 1987) and two processing metalloendopeptidases have already been purified from the matrix of rat liver mitochondria (Miura *et al.*, 1982; Kalousek *et al.*, 1992) (Chapters 377 and 469) this is the first clearly identified mitochondrial carboxypeptidase.

## Activity and Specificity

mCP III is clearly a carboxypeptidase: it releases the C-terminal residue from N-blocked dipeptides with preference for Z-Phe+Ala. The measurable release of a C-terminal alanine is 2-fold faster than that of any other amino acid residue. On the other hand, the rate of hydrolysis seems to be enhanced by the presence of a penultimate (P1) phenylalanine residue. In contrast the enzyme does not act on peptides containing a penultimate proline residue. The enzyme shows no degrading activity towards [$^{125}$I]casein.

mCP III is a typical metallopeptidase, inhibited by 1 mM 1,10-phenanthroline and reactivated in a concentration-dependent manner by zinc or cobalt. The best reactivation is achieved with 50 μM $Zn^{2+}$. Higher concentrations of zinc are inhibitory. The pH optimum is 7.6.

The enzyme is very labile and is stabilized by $Ca^{2+}$. The activity towards N-blocked peptides can be assayed by direct spectrophotometry (Feder, 1968) or by secondary methods such as quantification of released amino acids with ninhydrin plus hydrindantin (Moore, 1968).

## Structural Chemistry

The active enzyme has a molecular mass of 120 kDa and is constituted by two polypeptidic chains, of 63 and 56 kDa, not linked by disulfide bonds. The exact structure of these two subunits still awaits elucidation.

## Preparation

The mCP III has only been extracted from the matrix fraction of rat liver mitochondria where it exists in minute amounts, so large-scale methods of purification are needed. Two affinity approaches have been used: immobilized metal affinity chromatography in a matrix loaded with $Zn^{2+}$ or $Cu^{2+}$ through the chelating compound iminodiacetic acid (Porath & Olin, 1983) or affinity chromatography with potato carboxypeptidase inhibitor (PCI), based on the tight binding of PCI to metallocarboxypeptidases through direct interaction with the active site (Molina *et al.*, 1994).

## Biological Aspects

mCP III is clearly localized in the inner core of rat liver mitochondria, based on isolation from the mitochondrial matrix fraction free of contaminants ((Duque-Magalhães, 1979; Duque-Magalhães & Régnier, 1982). This novel metallocarboxypeptidase may play an important role in intramitochondrial proteolysis.

## References

Duque-Magalhães, M.C. (1979) On a neutral proteolytic system in rat liver mitochondria. *FEBS Lett.* **105**, 317–320.

Duque-Magalhães, M.C. (1989) Proteolytic enzymes in mitochondria. In: *Cell Biology Review* vol. 2 (Knecht, E. & Grisolia, S., eds). New York: Springer, pp. 475–490.

Duque-Magalhães, M.C. & Régnier, P. (1982) Study on the localization of proteases of mitochondrial origin. *Biochimie* **64**, 907–913.

Duque-Magalhães, M.C. & Régnier, P. (1986) Discrimination of distinct proteases at the four structural levels of rat liver mitochondria. *Biochem. J.* **233**, 283–286.

Feder, J. (1968) A spectrophotometric assay for neutral protease. *Biochem. Biophys. Res. Commun.* **32**, 326–331.

Figueiredo, E. & Duque-Magalhães, M.C. (1994) Identification, purification and partial characterization of a carboxypeptidase from the matrix of rat liver mitochondria: a novel metalloenzyme. *Biochem. J.* **300**, 15–19.

Kalousek, F., Isaya, G. & Rosenberg, L.E. (1992) Rat liver mitochondrial intermediate peptidase (MIP): purification and initial characterization. *EMBO J.* **11**, 2803–2809.

Miura, S., Mori, M., Amaya, Y. & Tatibana, M. (1982) A mitochondrial protease that cleaves the precursor of ornithine carbamoyltransferase. Purification and properties. *Eur. J. Biochem.* **122**, 641–647.

Molina, M.A., Marino, C., Oliva, B., Aviles, F.X. & Querol, E. (1994) C-tail Val is a key residue for stabilization of the complex between PCI and carboxypeptidase A. *J. Biol. Chem.* **269**, 21467–21478.

Moore, S. (1968) Amino acid analysis aqueous DMSO as solvent for hydrindantin in the ninhydrin reaction. *J. Biol. Chem.* **243**, 6281–6283.

Porath, J. & Olin, B. (1983) Immobilized metal ion affinity chromatography of biomaterials. *Biochemistry* **22**, 1621–1628.

Yasuhara, T. & Ohashi, A. (1987) New chelator-sensitive proteases in matrix of yeast mitochondria. *Biochem. Biophys. Res. Commun.* **144**, 277–283.

*Maria Conceição Duque-Magalhães*
*Instituto Gulbenkian de Ciencia,*
*2781 Oeiras, Codex, Portugal*
*Email: cduque@isa.utl.pt*

# 537. *Peptidyl-Lys metalloendopeptidase*

## Databanks

*Peptidase classification: clan MX, family M35, MEROPS ID: M35.004*
*NC-IUBMB enzyme classification: EC 3.4.24.20*
*Chemical Abstracts Service registry number: 72561-05-8; 65979-41-1*
*Databank codes: no sequence data available*

## Name and History

The name **peptidyl-Lys metalloendopeptidase** is based on the substrate specificity of the enzyme. The first enzyme with this specificity documented was **Myxobacter AL-1 protease II** (*My*MEP) which was isolated as a crystalline protein with no cell wall lytic activity from the culture fluid of *Myxobacter* strain AL-1 during CM-cellulose chromatography (Wingard *et al.*, 1972). At about the same time, **Armillaria mellea protease** (*Am*MEP) extracted from caps of the basidiomycete *Armillaria mellea* was patented by I.C.I. Ltd (Walton *et al.*, 1972). Although these enzymes are valuable tools for protein structure studies because of their rather strict unique substrate specificity, they have never been made commercially available. Recently, two enzymes, **GfMEP** (**thermostable lysine-specific metalloendopeptidase**) and **PoMEP** (**ProB**), were reported; these were extracted from the fruiting bodies of the basidiomycetes *Grifola frondosa* (Nonaka *et al.*, 1995) and *Pleurotus ostreatus* (Dohmae *et al.*, 1995), respectively.

## Activity and Specificity

Peptidyl-Lys metalloendopeptidases cleave preferentially acyl-lysine bonds in peptides and proteins including Pro↓Lys bonds. They do not release terminal lysine residues. *S*-Aminoethylcysteine, a lysine homolog, is also recognized as a substrate (Doonan & Fahmy, 1975). With *Am*MEP, Pro (Shipolini *et al.*, 1974) or Asp (Doonan *et al.*, 1975) residues at the P2' position prevented cleavage and a Glu residue adjacent to Lys (P1 or P2') restricted cleavage. Although *Am*MEP shows high substrate specificity, some Arg↓Ile/Leu bond cleavage was reported (Doonan *et al.*, 1975). While *Gf*- and *Po*MEP cleave polyarginine (Dohmae *et al.*, 1995; Nonaka *et al.*, 1995), no peptide bond other than Xaa↓Lys was cleaved in protein digests. *Gf*MEP cleaved Val-Glu21↓Lys22-Gly in horse cytochrome *c* but failed to cleave exactly the same tetrapeptide sequence, Val-Glu3-Lys4-Gly (Nonaka *et al.*, 1995), indicating that the enzyme has a substrate recognition site that extends beyond P2–P2'.

Because there are no convenient synthetic substrates, all assays are carried out with Azocoll (Wingard *et al.*, 1972; Lewis *et al.*, 1978) or azocasein (Starkey, 1977). While *Po*MEP exhibits a slightly acidic pH optimum, others have neutral to alkaline pH optima (*Po*MEP, 5.6; *Am*MEP, 7–7.5; *My*MEP, 8.5–9; *Gf*MEP, 9.5). One zinc atom per molecule is an essential component; activities are inhibited by chelating agents such as EDTA and 1,10-phenanthroline. DTT also inhibits *Gf*MEP (78% inhibition at 1 mM). Heavy metal ions

Table 537.1   pH and temperature stabilities of MEP enzymes

| Enzyme | pH | Time | Temperature (°C) | Recovery |
|---|---|---|---|---|
| *My*MEP | 5–10 | 18 h | 4 | 100 |
| *Gf*MEP | 5–10 | 18 h | 4 | 100 |
| *Po*MEP | 4–9 | 18 h | 4 | 100 |
| *My*MEP | 7.0 | 24 h | 50 | 90 |
| *Gf*MEP | 7.2 | 3 h | 80 | 60 |
| *Po*MEP | 7.0 | 15 min | 50 | 90 |

inhibit *My*MEP ($Hg^{2+}$, 36% inhibition at 0.1 mM), *Gf*MEP ($Hg^{2+}$, 68% at 0.1 mM; $Co^{2+}$, 41% at 1 mM; $Cu^{2+}$, 31% at 1 mM) and *Po*MEP (at 1 mM, 90% inhibition with $Hg^{2+}$ or $Ag^+$ and 70% inhibition with $Cu^{2+}$). Apoenzymes can be prepared by treatment of *Gf*- or *Po*MEP with 10 mM EDTA followed by dialysis. Activities are restored by the addition of divalent metal ions (*Gf*MEP at 0.1 mM metal ion, >200% with $Mn^{2+}$, 140% with $Zn^{2+}$, 130% with $Ca^{2+}$ or 100% with $Co^{2+}$; *Po*MEP at 1 mM metal ion, 100% with $Zn^{2+}$, 80% with $Co^{2+}$ or 30% with $Mn^{2+}$).

Enzymes in this group are generally pH stable and temperature stable as shown in Table 537.1. They are also resistant to denaturants. *Po*MEP retains 96% activity after 5 h in 5 M urea at 37°C and pH 5 or 56% activity after 5 h in 2 M guanidine-HCl under the same conditions. *Gf*MEP remains fully active in 4 M urea.

Inhibition by various amino acids and their derivatives have been reported. *Am*MEP is inhibited strongly by lysine ($K_i$ 1 mM), *n*-propylamine ($K_i$ 8 mM) and *S*-aminoethylcysteine ($K_i$ 12.7 mM), but weakly by ornithine ($K_i$ 140 mM) and arginine ($K_i$ 170 mM). This inhibition spectrum is in accord with the specificity of the enzyme, suggesting a specificity based on a rather rigid substrate pocket (Lewis *et al.*, 1978). On the other hand, at a concentration of 5 mM D,L-homoarginine (85%), agmatine (81%) and L-arginine (70%) inhibit *Gf*MEP more strongly than L-lysine ethylester (66%) and L-lysine (42%; Nonaka *et al.*, 1995). This result may indicate the presence of an acidic subsite in the substrate-binding site of *Gf*MEP, in good agreement with the larger substrate recognition site described above.

## Structural Chemistry

The reported isoelectric points are pH 7.5 (*Gf*MEP) and 8.4 (*Po*MEP). These enzymes contain one zinc atom per molecule (*Am-*, *Gf-* and *Po*MEP) as an essential component. Two enzymes, *My-* and *Am*MEP, are reported to be composed of 157 and 154 amino acids with minimum molecular masses of 16 600 and 16 650 Da, respectively (Wingard *et al.*, 1972; Lewis *et al.*, 1978). Primary structures of the two mushroom enzymes have been determined (Nonaka *et al.*, 1997). They are homologous proteins of 167 (*Gf* MEP) and 168 (*Po*MEP) amino acid residues with 61.3% identity and molecular masses are calculated to be 18 040 and 17 921 Da, respectively. They have two disulfide bonds and Thr42 of *Gf* MEP is partially mannosylated. They are found to share sequence homology (21–26% sequence identity) with deuterolysins (Chapter 513) and with penicillolysin (Matsumoto *et al.*, 1994) (Chapter 512), *Aspergillus oryzae* neutral protease II (Tatsumi *et al.*, 1991) (Chapter 513) and the 23 kDa metalloproteinases of *Aspergillus flavus* and *Sartorya fumigata* (Ramesh *et al.*, 1995). Because deuterolysins are expressed as precursor proteins, the mushroom, as well as *Myxobacter*, enzymes are likely to be the proteolytically processed mature forms of precursor proteins.

## Preparation

*My*MEP (Wingard *et al.*, 1972) was purified from the culture medium using ammonium sulfate precipitation or column chromatography in 11.4% yield. *Am*MEP (Walton *et al.*, 1972; Lewis *et al.*, 1978) was isolated from *A. mellea* caps in a yield of 35.1% (486-fold) and was homogeneous on PAGE at pH 4.3 and on SDS-PAGE in the presence and absence of 2-ME. *Gf-* (Nonaka *et al.*, 1995) and *Po*MEP (Dohmae *et al.*, 1995) were purified 212- and 34.8-fold from homogenates of edible mushrooms, *Grifola frondosa* and *Pleurotus ostreatus* in yields of 24.7 and 8.3%, respectively, by conventional purification methods.

## Biological Aspects

The enzymes are found exclusively in fruiting bodies of higher basidiomycetes, except for *My*MEP. Although *My*MEP is secreted into culture media, possibly to extract nutrients from extracellular proteins, the function of the mushroom enzymes internally is not clear. It is not known yet whether the enzyme is found also in mycelia. Unlike many serine proteinases, no intrinsic inhibitor nor inhibitory propeptide is likely to exist since the enzyme is fully active in the homogenate (Dohmae *et al.*, 1995).

*Gf* MEP, but not *Po*MEP, binds to cellulose, curdlan ($\beta$-1,3-glucan) and chitin (Nonaka *et al.*, 1995). No activity could be detected in the breakthrough fraction when *Gf* MEP was loaded onto a cellulose column at pH 5.0. Buffers containing 1.2 M KCl, 20 mM EDTA, 4 M urea or 0.5 M cellobiose failed to elute bound enzyme. This enzyme can be eluted quantitatively with a pH 8.5 buffer containing 1.2 M KCl. *Gf* MEP is fully active in the presence of cellulose. However, *Gf* MEP passes through a Sephadex G-25 column at pH 5.0, indicating that the enzyme lacks affinity for dextran, $\alpha$-1,6-linked glucose polymer. Its strong interaction with both $\beta$-1,3-glucan and chitin, major polysaccharides constituting the fungus cell wall, may provide an indication of the cellular localization and function of this enzyme.

## Distinguishing Features

Enzymes of this group may become an indispensable tool for structural studies of proteins owing to their strict specificity, high heat stability, resistance to denaturing agents, activity over a wide pH range, and ease of inhibition. When used in the digestion of an N-terminally blocked protein, they are likely to generate an N-terminal peptide with no positive charge which will not be expected to bind to a cation-exchange column. *Gf* MEP has already been used successfully to obtain the N-terminally blocked peptide of GAL80, a yeast negative regulatory protein (Yun *et al.*, 1991). Mass spectrometry indicates that many of the peptides generated have positive charges localized to the N-termini and are expected to produce only N-terminal sequence ions upon fragmentation by collision-induced dissociation or postsource decay, making interpretation of the spectrum simple.

## References

Dohmae, N., Hayashi, K., Miki, K., Tsumuraya, Y. & Hashimoto, Y. (1995) Purification and characterization of intracellular proteinases in *Pleurotus ostreatus* fruiting bodies. *Biosci. Biotechnol. Biochem.* **59**, 2047–2080.

Doonan, S. & Fahmy, H.M.A. (1975) Specific enzymic cleavage of polypeptides at cysteine residues. *Eur. J. Biochem.* **56**, 421–426.

Doonan, S., Doonan, H.J., Hanford, R., Vernon, C.A., Walker, J.M., Airoldi, L.P. da S., Bossa, F., Barra, D., Carloni, M., Fasella, P. & Riva, F. (1975) The primary structure of aspartate aminotransferase from pig heart muscle. *Biochem. J.* **149**, 497–506.

Lewis, W.G., Basford, J.M. & Walton, P.L. (1978) Specificity and inhibition studies of *Armillaria mellea* protease. *Biochim. Biophys. Acta* **522**, 551–560.

Matsumoto, K., Yamaguchi, M. & Ichishima, E. (1994) Molecular cloning and nucleotide sequence of the complementary DNA for penicillolysin gene, *plnC*, an 18 kDa metalloendopeptidase gene from *Penicillium citrinum*. *Biochim. Biophys. Acta* **1218**, 469–472.

Nonaka, T., Ishikawa, H., Tsumuraya, Y., Hashimoto, Y., Dohmae, N. & Takio, K. (1995) Characterization of a thermostable lysine-specific metalloendopeptidase from the fruiting bodies of a basidiomycete, *Grifola frondosa*. *J. Biochem. (Tokyo)* **118**, 1014–1020.

Nonaka, T., Dohmae, N., Tsumuraya, Y., Hashimoto, Y. & Takio, K. (1997) Amino acid sequences of metalloendopeptidases specific for acyl-lysine bonds from *Grifola frondosa* and *Pleurotus ostreatus* fruiting bodies. *J. Biol. Chem.* **272**, 30032–30039.

Ramesh, M.V., Sirakova, T.D. & Kolattukudy, P.E. (1995) Cloning and characterization of the cDNAs and genes (mep20) encoding homologous metalloproteinases from *Aspergillus flavus* and *A. fumigatus*. *Gene* **165**, 121–125.

Shipolini, R.A., Calleweart, G.L., Cottrell, R.C. & Vernon, C.A. (1974) The amino-acid sequence and carbohydrate content of phospholipase A$_2$ from bee venom. *Eur. J. Biochem.* **48**, 465–476.

Starkey, P.M. (1977) Elastase and cathepsin G; the serine proteinases of human neutrophil leucocytes and spleen. In: *Proteinases in Mammalian Cells and Tissues* (Barrett, A.J., ed.). Amsterdam: Elsevier/North-Holland Biomedical Press, pp. 57–89.

Tatsumi, H., Murakami, S., Tsuji, R.F., Ishida, Y., Murakami, K., Masaki, A., Kawabe, H., Arimura, H., Nakano, E. & Motai, H. (1991) Cloning and expression in yeast of a cDNA clone encoding *Aspergillus oryzae* neutral protease II, a unique metalloprotease. *Mol. Gen. Genet.* **228**, 97–103.

M

Walton, P.L., Turner, R.W. & Broadbent, D. (1972) British Patent No. 1263956.

Wingard, M., Matsueda, G. & Wolfe, R.S. (1972) Myxobacter AL-1 protease II: specific peptide bond cleavage on the amino side of lysine. *J. Bacteriol.* **112**, 940–949.

Yun, S.-J., Hiraoka, Y., Nishizawa, M., Takio, K., Titani, K., Nogi, Y. & Fukasawa, T. (1991) Purification and characterization of the yeast negative regulatory protein GAL80. *J. Biol. Chem.* **266**, 693–697.

*Koji Takio*
*Division of Biomolecular Characterization,*
*The Institute of Physical and Chemical Research (RIKEN),*
*2-1, Hirosawa, Wako, Saitama, Japan*
*Email: takio@postman.riken.go.jp*

# 538. Aureolysin

### Databanks

*Peptidase classification: clan MX, family M99, MEROPS ID: M9G.017*
*NC-IUBMB enzyme classification: EC 3.4.24.29*
*Databank codes: no sequence data available*

### Name and History

The production of an extracellular metalloproteinase by *Staphylococcus aureus* strain V8 was first documented by Arvidson *et al*. (1972) who found an EDTA-sensitive proteolytic activity in bacterial culture supernatants. The enzyme, initially confused with staphylokinase, was purified to homogeneity by several groups either from V8 or M139 strains and referred to as **protease III** (Arvidson, 1973) or simply **staphylococcal metalloprotease** (Saheb, 1976; Hasche *et al*., 1977; Drapeau, 1978). The name **aureolysin** was recommended by the NC-IUBMB in 1992.

### Activity and Specificity

Specificity of aureolysin was extensively studied using horse liver alcohol dehydrogenase (Björklind & Jörnvall, 1974) and oxidized B chain of insulin as substrates (Saheb, 1976; Drapeau, 1978) and was found to be similar to that of thermolysin (Chapter 351) with preference for hydrophobic P1′ residues (leucine, valine, tyrosine, isoleucine and phenylalanine) and alanine:

FVNQH↓LCGSH↓LVEA↓LY↓LVCGERG↓FF↓YTPKA

The same specificity is also observed on substrates with potential pathological effect, including $\alpha_1$-antichymotrypsin (Potempa *et al*., 1991a), $\alpha_1$-proteinase inhibitor (Potempa *et al*., 1986) and other serpins (Potempa *et al*., 1988a, 1991b, 1995), each of which is inactivated by limited proteolysis of an Xaa↓Leu peptide bond in the inhibitor reactive site loop.

The pH optimum of aureolysin is neutral, with both casein and FA-Gly-Leu-NH$_2$ (FAGLA) (Sigma; see Appendix 2 for full names and addresses of suppliers) as substrates (Arvidson, 1973; Drapeau, 1978). Activity is not affected by typical inhibitors/activators of serine, cysteine and aspartic proteinases but is inhibited by EDTA, 1,10-phenanthroline (Drapeau, 1978) and $\alpha_2$-macroglobulin. The last inhibitor can also be used as an active-site titrant (Potempa *et al*., 1995). Assays are most conveniently made with FAGLA or casein and are described by Arvidson (1983).

### Structural Chemistry

Aureolysin is a single-chain protein of pI about 4.6 and molecular mass 28 kDa or 38 kDa as determined by gel filtration/sedimentation equilibrium studies (Arvidson, 1973; Saheb, 1976) and SDS-PAGE (Drapeau, 1978), respectively. The discrepancy is not simply the result of proteolytic removal of a 10 kDa peptide because both forms have an identical relative amino acid composition; they completely lack cysteine residues. Zinc is required for aureolysin catalytic activity, but it can be substituted with cobalt producing a proteinase that is more active than native enzyme (Drapeau, 1978). The native protein also contains 2–3 mol of calcium per mol of the active enzyme (Saheb, 1976). Calcium ions stabilize the structure and their removal by chelation irreversibly inactivates the enzyme, apparently due to conformational changes in tertiary structure (Wasylewski *et al*., 1976). The apometalloproteinase is extremely sensitive to proteolytic degradation (Potempa *et al*., 1989).

## Preparation

The richest source of aureolysin is strain V8 (available from the ATCC), where this proteinase constitutes at least 50% of the total proteinase activity in the culture fluid. The purification procedure is simple and includes ammonium sulfate and acetone precipitation, as well as repeated ion-exchange chromatography on DEAE-cellulose. The yield may be up to 50 mg of pure enzyme per liter of medium (Drapeau, 1978).

## Biological Aspects

Aureolysin seems to have an important function in activating the staphylococcal serine glutamyl endopeptidase (V8 protease; Chapter 79). Mutants which do not produce this metalloproteinase accumulate the inactive precursor of V8 protease (Drapeau, 1978). Assuming a similar mechanism of activation in all *S. aureus* strains, aureolysin would be produced by the same strains as the serine proteinase, i.e. by at least 67% of the strains (Björklind & Arvidson, 1977). The enzyme is also responsible for staphylokinase conversion from a form of pI 6.7 to a form of pI 5.7, both *in vivo* and *in vitro* (Arvidson, 1983).

The role of aureolysin in the pathophysiology of *S. aureus* diseases is unknown and the enzyme is not considered as a virulence factor. In spite of this, there is some indication that the metalloproteinase can indirectly modulate an inflammatory reaction. The enzyme is responsible for the pseudocoagulase activity of coagulase-negative *S. aureus* strains, apparently through direct proteolytic activation of prothrombin in human plasma (Wegrzynowicz *et al.*, 1980). On the other hand, inactivation of plasma proteinase inhibitors ($\alpha_1$-antichymotrypsin and $\alpha_1$-proteinase inhibitor) (Potempa *et al.*, 1986, 1991a) may result in the loss of control of neutrophil proteinases (elastase and cathepsin G) and, ultimately, abnormal connective tissue degradation. In addition, aureolysin may modulate immunogenic reactions since it affects stimulation of both T and B lymphocytes by polyclonal activators and inhibits immunoglobulin production by lymphocytes in culture (Prokesova *et al.*, 1991).

## Distinguishing Features

At the time of this review, the primary and tertiary structure of aureolysin is not available and, therefore, its relation to metalloproteinases from coagulase-negative staphylococci including metalloelastase from *S. epidermidis* (Teufel & Götz, 1993) and lysostaphin (Chapter 515) from *S. simulans* (Recsei *et al.*, 1987) is unknown. A feature distinguishing it from many other bacterial metalloproteinases is a lack of elastinolytic activity (Potempa *et al.*, 1988b).

## Further Reading

An extensive review of all aspects of bacterial metalloproteinases, including aureolysin, is given in Häse & Finkelstein (1993).

## References

Arvidson, S. (1973) Studies on extracellular proteolytic enzymes from *Staphylococcus aureus*. II. Isolation and characterization of an EDTA-sensitive protease. *Biochim. Biophys Acta* **302**, 149–157.

Arvidson, S.O. (1983) Extracellular enzymes from *Staphylococcus aureus*. In: *Staphylococci and Staphylococcal Infections* (Easmon, C.S.F. & Adlam, C., eds). London: Academic Press, pp. 745–808.

Arvidson, S., Holme, T. & Lindholm, B. (1972) The formation of extracellular proteolytic enzymes by *Staphylococcus aureus*. *Acta Pathol. Microbiol. Scand. Sect. B* **80**, 835–844.

Björklind, A. & Arvidson, S. (1977) Occurrence of an extracellular serinoproteinase among *S. aureus* strains. *Acta Pathol. Microbiol. Scand. Sect. B* **85**, 277–280.

Björklind, A. & Jörnvall, H. (1974) Substrate specificity of three different extracellular proteolytic enzymes from *Staphylococcus aureus*. *Biochim. Biophys. Acta* **370**, 542–529.

Drapeau, G.R. (1978) Role of a metalloprotease in activation of the precursor of staphylococcal protease. *J. Bacteriol.* **136**, 607–613.

Hasche, K.D., Schaeg, W., Blobel, H. & Bruckler, J. (1977) Purification of protease from *Staphylococcus aureus*. *Zbl. Bakt. Parasitenk. Hyg. Abt. I. Orig. A* **238**, 300–309.

Häse, C.C. & Finkelstein, R.A. (1993) Bacterial extracellular zinc-containing metallo-proteinases. *Microbiol. Rev.* **57**, 823–837.

Potempa, J., Watorek, W. & Travis, J. (1986) The inactivation of human $\alpha$-1-proteinase inhibitor by proteinases from *Staphylococcus aureus*. *J. Biol. Chem.* **261**, 14330–14334.

Potempa, J., Dubin, A., Watorek, W. & Travis, J. (1988a) An elastase inhibitor from equine leukocyte cytosol belongs to the serpin superfamily. Further characterization and amino acid sequence of the reactive center. *J. Biol. Chem.* **263**, 7364–7369.

Potempa, J., Dubin, A., Korzus, G. & Travis, J. (1988b) Degradation of elastin by a cysteine proteinase from *Staphylococcus aureus*. *J. Biol. Chem.* **263**, 2664–2667.

Potempa, J., Porwit-Bobr, Z. & Travis, J. (1989) Stabilization vs. degradation of *Staphylococcus aureus* metalloproteinase. *Biochim. Biophys. Acta* **993**, 301–304.

Potempa, J., Fedak, D., Dubin, A., Mast, A. & Travis, J. (1991a) Proteolytic inactivation of $\alpha$-1-anti-chymotrypsin. Sites of cleavage and generation of chemotactic activity. *J. Biol. Chem.* **266**, 21482–21487.

Potempa, J., Wunderlich, J.K. & Travis, J. (1991b) Comparative properties of three functionally different but structurally related serpins variants from horse plasma. *Biochem. J.* **274**, 465–471.

Potempa, J., Enghild, J.J. & Travis, J. (1995) The primary elastase inhibitor (elastasin) and trypsin inhibitor (contrapsin) in the goat are serpins related to human $\alpha$-1-anti-chymotrypsin. *Biochem. J.* **306**, 191–197.

Prokesova, L., Porwit-Bobr, Z., Baran, K., Potempa, J., Pospisil, M. & John, C. (1991) Effect of metalloproteinase from *Staphylococcus aureus* on *in vitro* stimulation of human lymphocytes. *Immunol. Lett.* **27**, 225–230.

Recsei, P.A., Gruss, A.D. & Novick, R.P. (1987) Cloning, sequence, and expression of the lysostaphin gene from *Staphylococcus simulans*. *Proc. Natl Acad. Sci. USA* **84**, 1127–1131.

Saheb, S.A. (1976) Purification et caractérisation d'une protéase extracellulaire de *Staphylococcus aureus* inhibée par l'E.D.T.A [Purification and characterization of an extracellular protease of *Staphylococcus aureus* inhibited by EDTA]. *Biochimie* **58**, 793–804.

Teufel, P. & Götz, F. (1993) Characterization of an extracellular metalloproteinase with elastase activity from *Staphylococcus epidermidis*. *J. Bacteriol.* **175**, 4218–4224.

M

Wasylewski, Z., Stryjewski, W., Wasniowska, A., Potempa, J. & Baran, K. (1976) Effect of calcium binding on conformational changes of staphylococcal metalloproteinase measured by means of intrinsic protein fluorescence. *Biochim. Biophys. Acta* **871**, 177–181.

Wegrzynowicz, Z., Heczko, P.B., Drapeau, G., Jeljaszewicz, J. & Pulverer, G. (1980) Prothrombin activation by a metalloproteinase from *Staphylococcus aureus. J. Clin. Microbiol.* **12**, 138–139.

***Jan Potempa***
*Jagiellonian University,*
*Institute of Molecular Biology,*
*Al. Mickiewicza 3,*
*31-120 Krakow, Poland*
*Email: potempa@mol.uj.edu.pl*

***Adam Dubin***
*Jagiellonian University,*
*Institute of Molecular Biology,*
*Al. Mickiewicza 3,*
*31-120 Krakow, Poland*
*Email: dubin@mol.uj.edu.pl*

***James Travis***
*University of Georgia,*
*Department of Biochemistry,*
*Athens, GA 30602, USA*
*Email: jtravis@uga.cc.uga.edu*

# 539. *Peptidyl-Asp metalloendopeptidase*

## Databanks

*Peptidase classification: clan MX, family M99, MEROPS ID: M9G.020*
*NC-IUBMB enzyme classification: EC 3.4.24.33*
*Databank codes: no sequence data available*

## Name and History

***Peptidyl-Asp metalloendopeptidase*** was first isolated from the culture supernatant of *Pseudomonas fragi*. This microorganism was thought to produce only a single extracellular proteinase (Porzio & Pearson, 1975). However, during the isolation of the extracellular proteinase from a derepressed mutant of *P. fragi* ATCC 4973 which produces 40 times higher proteinase levels, it became clear that *P. fragi* in fact produces more than one proteinase species (Noreau & Drapeau, 1979). A proteinase isolated from this mutant strain was found to exhibit an unusual cleavage specificity. Soon after its discovery the enzyme became available commercially under the name ***endoproteinase Asp-N***, based on its cleavage specificity. Another synonym is ***X-Asp metalloendopeptidase***.

## Activity and Specificity

Peptidyl-Asp metalloendopeptidase specifically cleaves peptide bonds at the N-terminal side of either aspartic acid (Xaa┼Asp) or cysteic acid residues (Xaa┼Cya) (Drapeau, 1980; Ponstingl *et al.*, 1986). The activity of peptidyl-Asp metalloendopeptidase is determined by general proteinase assays using proteins as substrates. No assay procedure using a specific chromogenic substrate has yet been described (Hagmann *et al.*, 1995). To check for cleavage specificity, Fischer

*et al.* (1988) digested hormone peptides, e.g. glucagon:

$$HSQGTFTS┼DYSKYL┼DSRRAQ┼DFVQWLMNT$$

which was cleaved into four fragments. Smaller synthetic peptides may also be used (Tarentino *et al.*, 1995). Cleavage at the N-terminal side of glutamic acid residues can occur under certain conditions (Ingrosso *et al.*, 1989; Tatez *et al.*, 1990) but aspartyl residues are kinetically preferred (Tschakert, 1989; Geuss *et al.*, 1990).

Due to its limited specificity the protease is useful in the digestion of peptides and proteins prior to amino acid sequence analysis (Ponstingl *et al.*, 1986). Peptidyl-Asp metalloendoproteinase may also be used to detect glycosylation sites within a protein molecule (Leonard *et al.*, 1990).

The pH optimum is in the range of pH 7.0–8.5 (Drapeau, 1980; U. Geuss, unpublished results). There is no significant difference when various buffer systems are used. Activity is completely inhibited by chelating substances like 1,10-phenanthroline, EDTA or EGTA. The 1,10-phenanthroline-inhibited enzyme cannot be reactivated by $Zn^{2+}$. No inhibition is observed with the serine protease inhibitor PMSF. Inhibition by $\alpha_2$-macroglobulin occurrs at a molar ratio of inhibitor:protease of about 18:1 (Tschakert, 1989).

## Structural Chemistry

Only partial amino acid sequences are available at present and are shown in Figure 539.1. The sequence fragment spanning

```
DIATDSSTSP   YAYGHGYRYE   PATGwRTIMA   YNCTRSCPRL   NYWSNPNISY   digp
DCATGYYSFA   HEIGHLQsar
DNQRVLVNTK   ATIAAFr
ELARYETTNY   TESGSF
DSIHTSRNTY   TAA
ESNQGYVNSN   VGI
DTDLARFRGT   S
```

*Figure 539.1* Partial amino acid sequences of peptidyl-Asp metalloendopeptidase. Amino acids in lower case are tentative.

20 amino acids contains a HEXXH pattern which is a potential zinc-binding site, but no significant match has been found between these fragments and any other sequence in the public databases.

Peptidyl-Asp metalloendopeptidase is a single-chain protein with a relative molecular mass of 24 440 as determined by laser-desorption mass spectrometry (M. Hagmann, unpublished results). On SDS gels a single band with a relative molecular mass of 27 000 is found.

## Preparation

Purification of peptidyl-Asp metalloendoproteinase from a *Pseudomonas fragi* mutant is described by Noreau & Drapeau (1979). However, for its application in protein sequencing, peptidyl-Asp metalloendopeptidase may have to be further purified by ion-exchange chromatography. In highly purified form, for use in protein sequencing, the enzyme is available commercially as endoproteinase Asp-N from several suppliers.

## Distinguishing Features

The glycosylated metalloendopeptidase flavastacin found in *Flavobacterium meningosepticum* has similar specificity (Tarentino *et al.*, 1995) (Chapter 412). However peptidyl-Asp metalloendopeptidase and flavastacin differ in their molecular masses (24 vs. 40 kDa).

## References

Drapeau, G.R. (1980) Substrate specificity of a proteolytic enzyme isolated from a mutant of *Pseudomonas fragi*. *J. Biol. Chem.* **255**, 839–840.

Fischer, S., Geuss, U., Schäffer, M., Kresse, G.-B. & Drapeau, G.R. (1988) A new commercially available endoproteinase which cleaves specifically at the amino terminal side of aspartic acid. *J. Protein Chem.* **7**, 225–226.

Geuss, U., Schäffer, M., Tschakert, J. & Kresse, G.-B. (1990) Characterization of glutamyl cleavage activity of endoproteinase Asp-N sequencing grade. *J. Protein Chem.* **9**, 299–300.

Hagmann, M.-L., Geuss, U., Fischer, S. & Kresse, G.-B. (1995) Peptidyl-Asp metalloendopeptidase. *Methods Enzymol.* **248**, 782–787.

Ingrosso, D., Fowler, A.V., Bleibaum, J. & Clarke, S. (1989) Specificity of endoproteinase Asp-N (*Pseudomonas fragi*): cleavage at glutamyl residues in two proteins. *Biochem. Biophys. Res. Commun.* **162**, 1528–1524.

Leonard, C.K., Spellman, M.W., Riddle, L., Harris, R.J., Thomas, J.N. & Gregory, T.J. (1990). Assignment of intrachain disulfide bonds and characterization of potential glycosylation sites of the type I recombinant human immunodeficiency virus envelope glycoprotein (gp120) expressed in Chinese hamster ovary cells. *J. Biol. Chem.* **265**, 10373–10382.

Noreau, J. & Drapeau, G.R (1979) Isolation and properties of the protease from the wild-type and mutant stains of *Pseudomonas fragi. J. Bacteriol.* **140**, 911–916.

Ponstingl, H., Maier, G., Little, M. & Krauhs, E. (1986) Use of a metalloproteinase specific for the amino side of Asp in protein sequencing. In: *Advanced Methods in Protein Microsequence Analysis* (Wittmann-Liebold, B., ed.). Berlin and Heidelberg: Springer, pp. 316–319.

Porzio, M.A. & Pearson, A.M. (1975) Isolation of an extracellular neutral protease from *Pseudomonas fragi. Biochim. Biophys. Acta* **384**, 235–241.

Tarentino, A.L., Quinones, G., Grimwood, B.G., Hauer, C.R. & Plummer, T.H., Jr. (1995) Molecular cloning and sequence analysis of flavastacin: an O-glycosylated procaryotic zinc metalloendopeptidase. *Arch. Biochem. Biophys.* **319**, 281–285

Tetaz, T., Morrison, J.R., Andreou, J. & Fidge, N.H. (1990) Relaxed specificity of endoproteinase Asp-N: this enzyme cleaves at peptide bonds N-terminal to glutamate as well as aspartate and cysteic acid residues. *Biochem. Int.* **22**, 561–566.

Tschakert, J. (1989) Charakterisierung der Endoprotease Asp N [Characterization of endoprotease Asp N]. Diploma thesis, University of Munich.

***Marie-Luise Hagmann***
*Boehringer Mannheim GmbH,*
*Nonnenwald 2,*
*D-82377 Penzberg, Germany*
*Email: marie-luise_hagmann@bmg.boehringer-mannheim.com*

# 540. *Magnolysin*

## Databanks

*Peptidase classification: clan MX, family M99, MEROPS ID: M9G.025*
*NC-IUBMB enzyme classification: EC 3.4.24.62*
*Databank codes: no sequence data available*

## Name and History

The search for processing endoproteases that cleave peptide hormone precursors at basic amino acid doublets (Rholam *et al.*, 1986; Cohen, 1987) has led some authors to use as substrates synthetic peptides that either reproduce or mimic the prohormone primary structure around the cleavage site(s) (reviewed in Rholam & Cohen, 1994). In the primary structure of pro-ocytocin-neurophysin (pro-TO/Np), the N-terminal nonapeptide ocytocin is separated from the C-terminal neurophysin by a cleavage motif, Gly10-Lys11-Arg12 (Land *et al.*, 1983). There is evidence for a precursor processing pathway (Camier *et al.*, 1985) in which the primary cleavage occurs at the Arg12┼Ala13 bond (Ala13 is the N-terminal amino acid of the ocytocin-binding protein, neurophysin) (Rholam *et al.*, 1982; Rose *et al.*, 1996). These observations have led to the synthesis of a large number of peptides related to the [1–20]-N-terminal sequence of pro-TO/Np (Clamagirand *et al.*, 1986; Créminon *et al.*, 1988; Plevrakis *et al.*, 1989; Brakch *et al.*, 1993; Rholam & Cohen, 1994). They were used to identify, monitor and characterize an endoprotease first called ***pro-ocytocin convertase*** because of its ability to perform the Arg12┼Ala13 cleavage in the pro-TO/Np(1–20) peptide (Cohen *et al.*, 1995) as well as in the full-length, semisynthetic (Brakch *et al.*, 1989) or natural (Camier *et al.*, 1991) pro-TO/Np. This ***pro-ocytocin/neurophysin-converting enzyme*** has recently been named ***magnolysin***, in reference to the nuclei of the magnocellular system of the hypothalamus, the origin of the ocytocin-producing neurons which end in the neurohypophysis. It is proposed as a candidate processing enzyme for the corresponding prohormone.

## Activity and Specificity

Under certain conditions, magnolysin *in vitro* produces cleavage of synthetic peptide substrates derived from the reference pro-TO/Np(1–20) on the C-terminal side of basic doublets (Rholam & Cohen, 1994). In the light of data obtained with these synthetic peptides, the selective recognition of precursor dibasic processing sites by magnolysin appeared to be governed by a certain number of structural parameters (Rholam & Cohen, 1994):

1. The integrity of the pair of basic amino acids is essential for substrate recognition by the enzyme: peptides bearing Nle or D-Arg at subsite P1 or containing Nle or D-Lys at subsite P2 remained uncleaved (Créminon *et al.*, 1988; Plevrakis *et al.*, 1989).

2. The substrate bearing the Lys-Arg doublet was hydrolyzed with the best efficiency ($K_m$ 100 μM, V 260 pmol h$^{-1}$; $V/K_m$ 2.60 pmol h$^{-1}$ μM$^{-1}$) although other types of paired basic amino acids were also accepted (Créminon *et al.*, 1988).

3. A minimal length of eight residues (surrounding the dibasic motif) is necessary to allow for substrate binding to the active enzyme site: the forms of the reference peptide substrate (Cys1 → Arg20), either truncated on the N-terminal (Lys11 → Arg20) or on the C-terminal (Cys1 → Ala13) side of the Lys11-Arg12 doublet, remained uncleaved (Créminon *et al.*, 1988; Plevrakis *et al.*, 1989; Brakch *et al.*, 1989). Moreover, the flexibility of the peptide segment around the dibasic site is essential for its adaptability to the active site (Brakch *et al.*, 1989).

4. The sequence of four residues immediately next to the N-terminal of the Lys11-Arg12 doublet participates in a key fashion in providing a β-turn structure (Brakch *et al.*, 1989; Rholam *et al.*, 1990; Paolillo *et al.*, 1992). This structural motif can be interchanged by another β-turn promoting sequence without any major change in the cleavage efficiency (Brakch *et al.*, 1993). Examples of values are $K_m$ 150 μM, V 720 pmol h$^{-1}$; $V/K_m$ 4.8 pmol h$^{-1}$ μM$^{-1}$ for peptide Pro7 → Leu15 and $K_m$ 430 μM$^{-1}$, V 1420 pmol h$^{-1}$, $V/K_m$ 3.3 pmol h$^{-1}$ mM$^{-1}$ for peptide Asn7 → Leu15.

5. The dibasic prohormone sequences adopting either an α-helix or a β-sheet conformation are recognized by magnolysin but with a significantly lower affinity (Brakch *et al.*, 1993; $K_m$ 6450 μM for peptide Ile7 → Leu15 and $K_m$ 4760 μM for peptide Ala3 → Arg20).

6. The P1′ residue of precursor dibasic processing sites is also an important feature in the discrimination between *in vivo* cleaved and uncleaved dibasic sites by magnolysin (Rholam *et al.*, 1995). Indeed, substitution of Ala13 at position P1′, by the α-carbon branched side chain residues (Leu, Val, Ile and Thr) in peptide Pro7 → Leu15, totally abolished the recognition of substrates by magnolysin.

7. Finally, magnolysin cleaves the pro-ocytocin-neurophysin, either obtained by semisynthesis (Brakch *et al.*, 1989) or purified from producing organs (Camier *et al.*, 1991), at the C-terminal of the Lys11-Arg12 doublet. Other peptide substrates, representing either the entire prohormone sequence (proneuropeptide Y) or fragments of the precursor (proglucagon, proneurotensin and prodynorphin) were also hydrolyzed at the dibasic moieties which are known to be processed *in vivo* (Brakch *et al.*, 1993).

Assays of activity are conveniently performed by using peptide Cys1 → Arg20 as the substrate and measuring the production of either peptide Cys1 → Arg12 or Ala13 → Arg20 (Clamagirand *et al.*, 1986; Cohen *et al.*, 1995). The optimal pH for enzyme activity is around 7.0 (Clamagirand *et al.*, 1986; Plevrakis *et al.*, 1989). The apparent pI of the protein is around 6.9 (Plevrakis *et al.*, 1989).

Magnolysin is significantly inhibited by millimolar concentrations of EDTA and EGTA (Clamagirand *et al.*, 1987a,b). It is also partially sensitive to thiol-blocking reagents such as *p*-(chloromercuri)-benzenesulfonic acid and PCMB, suggesting either a direct or indirect involvement of thiol group(s) in the endoproteolytic reaction (Plevrakis *et al.*, 1989). Serine protease inhibitors (PMSF, aprotinin) or aspartic protease inhibitors (pepstatin) had no effect (Clamagirand *et al.*, 1987a,b). Finally, the D-Arg12 derivative of pro-TO/Np(1–20) is an inhibitor of the protease (Plevrakis *et al.*, 1989; $K_i$ 30 μM).

## Structural Chemistry

The enzyme was obtained in an homogeneous form as a 55–65 kDa monomeric chain (Plevrakis *et al.*, 1989).

## Preparation

Magnolysin was isolated in four main purification steps from extracts of enriched preparations of secretory granules from either cattle neurohypophysis hypothalamus (Clamagirand *et al.*, 1987a,b; Plevrakis *et al.*, 1989) or corpus luteum (Plevrakis *et al.*, 1989). Optimal purification was achieved by using, additionally, affinity chromatography on D-Arg12-[Cys1 → Arg20] peptide conjugated to an HMD-Ultrogel (Plevrakis *et al.*, 1989).

## Biological Aspects

Although it performs an adequate endoproteolytic cleavage *in vitro*, there is no direct evidence to date that magnolysin is indeed the pro-TO/Np-processing enzyme. Because of its colocalization with the precursor and its maturation products in the cattle neurohypophyseal tracts (Clamagirand *et al.*, 1986) as well as in the cattle (Clamagirand *et al.*, 1987a,b) and human ovary (Plevrakis *et al.*, 1990; Guillou *et al.*, 1992, 1994), it is tempting to postulate such a role. Generation of fully active, C-terminally amidated ocytocin implies the successive action of an endoprotease, a carboxypeptidase B and the amidating enzyme (Clamagirand *et al.*, 1986, 1987a,b). It is noteworthy that cattle granulosa cells (Camier *et al.*, 1991) accumulate TO-Gly, the precursor for amidated ocytocin.

## Distinguishing Features

In addition to pro-ocytocin-neurophysin convertase, a number of proteases involved in the processing of protein precursors at some single basic residues and/or selective pairs of basic amino acids have been described. They include the serine/subtilisin-like enzymes, furin (Chapter 117) and the proprotein convertases (Chapters 118, 119, 120, 121, 122, 123) (Steiner *et al.*, 1992), representatives from the cysteine (Chapter 263) (Hook *et al.*, 1996) and aspartic protease classes (yapsin 3, Chapter 306) (Egel-Mitani *et al.*, 1990) and recently a new dibasic endoprotease (nardilysin, Chapter 468) (Pierotti *et al.*, 1994) related to the zinc metalloendopeptidase family M16 (Rawlings & Barrett, 1995). Despite the fact that selective proteolysis of inactive precursors is a general mechanism, some substantial differences in properties of these dibasic specific endoproteases were observed. For example, yapsin 3 prefers substrates with basic residues in the P2, P1 and P2′ positions (Ledgerwood *et al.*, 1996). In contrast, furin and PC3 of the $Ca^{2+}$-dependent serine protease family S8 were characterized by a preference for substrates with a minimal consensus sequence Arg-Xaa-Xaa-Arg (Denaults & Leduc, 1996). Finally, nardilysin cleaves on the N-terminal side of Arg in basic doublets (Gluschankof *et al.*, 1987; Chesneau *et al.*, 1994). Accordingly, magnolysin appears as a prototype of a putative new type of proprotein convertases.

## References

Brakch, N., Boussetta, H., Rholam, M. & Cohen, P. (1989) Processing endoprotease recognizes a structural feature at the cleavage site of peptide prohormones: the pro-ocytocin/neurophysin model. *J. Biol. Chem.* **264**, 15912–15916.

Brakch, N., Rholam, M., Boussetta, H. & Cohen, P. (1993) Role of β-turn in proteolytic processing of peptide hormone precursors at dibasic sites. *Biochemistry* **32**, 4925–4930.

Camier, M., Barre, N. & Cohen, P. (1985) Hypothalamic biosynthesis and transport of neurophysins and their precursors to the rat brain stem. *Brain Res.* **334**, 1–8.

Camier, M., Benveniste, D. Barre, N., Brakch, N. & Cohen, P. (1991) Synthesis and processing of pro-ocytocin in bovine corpus luteum and granulosa cells. *Mol. Cell. Endocrinol.* **77**, 141–147.

Chesneau, V., Pierotti, A., Barre, N., Créminon, C, Tougard, C. & Cohen, P. (1994) Isolation and characterization of a dibasic selective metalloendopeptidase from rat testes which cleaves on the amino-terminus of arginine residues. *J. Biol. Chem.* **269**, 2056–2061.

Clamagirand, Ch., Camier, M., Boussetta, H., Fahy, C., Morel, A., Nicolas, P. & Cohen., P. (1986) An endopeptidase associated with bovine neurohypophysis secretory granules cleaves proocytocin/neurophysin peptide at paired basic residues. *Biochem. Biophys. Res. Commun.* **134**, 1190–1196.

Clamagirand, C., Camier, M., Fahy, C., Clavreul, C., Créminon, C. & Cohen, P. (1987a) C-terminally extended ocytocin and proocytocin/neurophysin peptide converting enzyme in bovine corpus luteum. *Biochem. Biophys. Res. Commun.* **143**, 789–796.

Clamagirand, C., Créminon, C., Boussetta, H., Fahy, C., Nicolas, P. & Cohen, P. (1987b) Partial purification and functional properties of an endroprotease from bovine neurosecretory granules cleaving pro-ocytocin/neurophysin peptides at the basic amino acid doublet. *Biochemistry* **26**, 6018–6023.

Cohen, P. (1987) Proteolytic events in the post-translational processing of polypeptide hormone precursors. *Biochimie* **59**, 87–89.

Cohen, P., Rholam, M. & Boussetta, H. (1995) Methods for the identification of neuropeptide processing pathways. In: *Methods in Neurosciences*, vol. 23 (Smith, I.A., ed.). New York: Academic Press, pp. 155–193.

M

Créminon, C., Rholam, M., Boussetta, H., Marrakchi, N. & Cohen, P. (1988) Synthetic peptide substrates as models to study a pro-ocytocin/neurophysin converting enzyme. *J. Chromatogr.* **440**, 439–448 (Special Issue for the Jubilee of Professor E. Lederer).

Denaults, J.-B. & Leduc, R. (1996) Furin/PACE/SPC1: a convertase involved in exocytic and endocytic processing of precursor proteins. *FEBS Lett.* **379**, 113–116.

Egel-Mitani, M., Flygenring, H. & Hansen, M.T. (1990) A novel aspartyl protease allowing *KEX2*-independent *Mf α* propheromone processing in yeast. *Yeast* **6**, 127–137.

Gluschankof, P., Gomez, S., Morel, A. & Cohen, P. (1987) Enzymes that process somatostatin precursors. A novel endoprotease that cleaves before the arginine-lysine doublet is involved in somatostatin-28 convertase activity of the rat brain cortex. *J. Biol. Chem.* **262**, 9515–9520.

Guillou, M., Barre, N., Bussenot, I., Plevrakis, I. & Clamagirand, C. (1992) COOH-terminally-extented processing forms of oxytocin in human ovary. *Mol. Cell. Endocrinol.* **88**, 233–238.

Guillou, M., Camier, M. & Clamagirand, C. (1994) Evidence for the presence of pro-ocytocin/neurophysin converting enzyme in the human ovary. *J. Endocrinol.* **142**, 345–352.

Hook, V.Y.H., Schiller, M.R. & Azaryan, A.V. (1996) The processing proteases prohormone thiol protease, PC1/3 and PC2, and 70-kDa aspartic proteinase show preferences among proenkephalin, proneuropeptide Y, and proopiomelanocortin substrates. *Arch. Biochem. Biophys.* **328**, 107–114.

Land, H., Grez, M., Ruppert, S., Schmale, H., Rehbein, M., Richter, D. & Schütz, G. (1983) Deducted amino acid sequence from the bovine oxytocin-neurophysin I precursor cDNA. *Nature (Lond.)* **302**, 343–344.

Ledgerwood, E.C., Brennan, S.O., Cawley, N.X., Loh, Y.P. & George, P.M. (1996) Yeast aspartic protease 3 (Yap3) prefers substrates with basic residues in the $P_2$, $P_1$ and $P'_2$ positions. *FEBS Lett.* **383**, 67–71.

Paolillo, L., Simonetti, M., N. Brakch, N., D'Auria, G., Saviano, M., Dettin, M., Rholam, M., Scatturin, A., Di Bello, C. & Cohen, P. (1992) Evidence for the presence of a secondary structure at the dibasic processing site of prohormone: the pro-ocytocin model. *EMBO J.* **11**, 2399–2405.

Pierotti, A., Prat, A.,Chesneau, V., Gaudoux, F., Lesney, A.M.,

Foulon, T. & Cohen, P. (1994) N-Arginine dibasic convertase, a metalloendopeptidase as a prototype of a class of processing enzymes. *Proc. Natl Acad. Sci. USA* **91**, 6078–6082.

Plevrakis, I., Créminon, C., Clamagirand, C., Brakch, N., Rholam, M. & Cohen. P. (1989) Pro-ocytocin/neurophysin convertase from bovine neurohypophysis and corpus luteum secretory granules: complete purification, structure-function relationship and competitive inhibitor. *Biochemistry* **28**, 2705–2710.

Plevrakis, I., Clamagirand, C. & Pontonnier, G. (1990) Oxytocin biosynthesis in serum-free cultures of human granulosa cells. *J. Endocrinol.* **124**, R5–R8.

Rawlings, N.D. & Barrett. A.J. (1995) Evolutionary families of peptidases. *Methods Enzymol.* **248**, 183–228.

Rholam, M. & Cohen, P. (1994) Strategies and techniques of processing enzyme characterization: techniques for the determination of prohormone conformation and its role in processing. In: *Neuroprotocols: A Companion to Methods in Neurosciences*, vol. 8 (Beinfeld, MC., ed.). New York: Academic Press, pp. 130–143.

Rholam, M., Nicolas, P. & Cohen, P. (1982) Binding of neurohypophyseal peptides to neurophysin dimer promotes the formation of compact and spherical complexes. *Biochemistry* **21**, 4968–4973.

Rholam, M., Nicolas, P. & Cohen, P. (1986) Precursors for peptide hormones share common secondary structure forming features at the proteolytic processing sites. *FEBS Lett.* **207**, 1–6.

Rholam, M., Cohen, P., Brakch, N., Paolillo, L., Scatturin, A. & Di Bello, C. (1990) Evidence for $β$-turn structure in model peptides reproducing pro-ocytocin/neurophysin proteolytic processing site. *Biochem. Biophys. Res. Commun.* **168**, 1066–1073.

Rholam, M., Brakch, N. Germain, D., Thomas, D., Fahy, C., Boussetta, H., Boileau, G. & Cohen, P. (1995) Role of amino acid sequences flanking dibasic cleavage sites in precursor proteolytic processing. The importance of the first residue C-terminal of the cleavage site. *Eur. J. Biochem.* **227**, 707–714.

Rose, J., Wu, C.-K., Hsiao, C.-D., Breslow, E. & Wang, B.-C. (1996) Crystal structure of the neurophysin-oxytocin complex. *Nature Struct. Biol.* **3**, 163–169.

Steiner, D.F., Smeekens, S.P., Ohagi, S. & Chan, S.J. (1992) The new enzymology of precursor processing endoproteases. *J. Biol. Chem.* **267**, 23435–23438.

*Mohamed Rholam*
*Laboratoire de Biochimie des Signaux Régulateurs*
*Cellulaires et Moléculaires,*
*Unité de Recherche Associée au Centre National de la*
*Recherche Scientifique 1682,*
*Université Pierre et Marie Curie,*
*96 Boulevard Raspail,*
*F-75006 Paris, France*
*Email: rholam@infobiogen.fr*

*Christine Clamagirand*
*Laboratoire de Biochimie des Signaux Régulateurs*
*Cellulaires et Moléculaires,*
*Unité de Recherche Associée au Centre National de la*
*Recherche Scientifique 1682,*
*Université Pierre et Marie Curie,*
*96 Boulevard Raspail,*
*F-75006 Paris, France*
*Email: clamagirand@infobiogen.fr*

*Paul Cohen*
*Laboratoire de Biochimie des Signaux Régulateurs Cellulaires*
*et Moléculaires,*
*Unité de Recherche Associée au Centre National de la*
*Recherche Scientifique 1682,*
*Université Pierre et Marie Curie,*
*96 Boulevard Raspail,*
*F-75006 Paris, France*
*Email: pcohen@infobiogen.fr*

# 541. Dactylysin

## Databanks

*Peptidase classification: clan MX, family M99, MEROPS ID: M9G.026*
*NC-IUBMB enzyme classification: EC 3.4.24.60*
*Databank codes: no sequence data available*

## Name and History

The original observation (Kuks *et al.*, 1989) that several prohormonal sequences produced in the secretions of *Xenopus laevis* skin possess a potential maturation site, highly conserved around a consensus RXVR╈G motif, with the typical basic residue at position P4 of the cleavage site, has led to the synthesis of a peptide substrate (called Kermit): DVDERDVR╈GFASFL-NH₂. This tetradecapeptide was used to monitor the Arg╈Gly proteolytic processing by the so-called RXVRG-endoprotease (Kuks *et al.*, 1989). This enzyme has turned out to be, most probably, the prohormone convertase furin (Chapter 117). During the purification however, the peptide was found to be cleaved by a copurifying endoproteolytic activity able to cleave the Ser╈Phe bond. The enzyme responsible for this reaction was first called *peptide hormone inactivating enzyme (PHIE)* (Carvalho *et al.*, 1992; Joudiou *et al.*, 1993) in reference to its ability to perform inactivating cleavage(s) in a number of hormonal peptides by selective hydrolysis of peptide bond(s) at hydrophobic bulky amino acid residues. It was then named *dactylysin* from dactylètre du Cap (South African frog: *Xenopus laevis*).

## Activity and Specificity

This endoprotease exhibits a marked preference for peptide substrates at least six amino acid long and produces cleavage with hydrophobic, bulky, residues Trp, Tyr, Phe, Ile, Leu, Val in P1′. Moreover it was observed on several substrates that there is also a preference for a hydrophobic amino acid in P1. This type of cleavage was clearly and unambiguously identified when peptides of the atrial natriuretic factor (ANF), substance P, enkephalins and dynorphins series, just to cite a few, were used as substrates for the endoprotease (Joudiou *et al.*, 1993). For example:

    [21–28]ANF:  LGCNS╈FRY
    Substance P:  RPKQQF╈FG╈LM-NH₂
    [Leu⁵,Arg⁶]-enkephalin:  YGGF╈LR
    Neurokinin A:  HKTDSFVG╈LM-NH₂

Indeed, dactylysin cleaves the peptide link Ser╈Phe in Kermit as well as Phe╈Phe in the interdisulfide bridge motif of somatostatin-14 ($K_m$ 18 µM) or Phe╈Leu in dynorphin A$_{1-16}$ ($K_m$ 50 µM). This thermolysin-like cleavage selectivity was analyzed in more detail on various peptides of the ANF, bradykinin and dynorphin families. The following cleavages were identified (Joudiou *et al.*, 1993):

- Ser╈Phe in ANF-derived synthetic peptides at least six residues long. A $K_m$ of 50 µM was measured for [5–25]ANF;
- Phe4╈Leu5 ($K_m$ 11 µM and 79 µM) for [Arg⁰, Leu⁵, Arg⁶]- and [Leu⁵, Arg⁶]-enkephalins respectively;
- Gly9╈Leu10 in substance P ($K_m$ 20 µM) and related peptides;
- Tyr4╈Ile5 in angiotensin II ($K_m$ 63 µM);
- Gly4╈Phe5 in bradykinin ($K_m$ 41 µM).

Additionally, neurokinin A and some $\beta$-amyloid derived peptides were cleaved at the Gly╈Leu bond ($K_m > 125$ µM) or else were inhibitors of the activity when tested on substance P. Finally, PGLa, the antimicrobial peptide of *X. laevis* skin secretions, was predominantly cleaved at the Lys12╈Ile13 bond ($K_m$ 28 µM).

Dactylysin is a metallopeptidase and its activity is strongly inhibited by divalent ion-chelating agents like EDTA, EGTA and particularly 1,10-phenanthroline (1 µM concentration produced 98% inhibition of the enzyme activity) whereas 10 µM DTT produced 70% inhibition of activity when tested on Kermit. Other cysteine protease, serine protease and aspartic peptidase inhibitors were inactive (Joudiou *et al.*, 1993).

## Structural Chemistry

Whereas the activity was clearly associated with high molecular mass fractions and described as a 90–100 kDa enzyme (Carvalho *et al.*, 1992), more recent observations would suggest that the activity can be dissociated to a much smaller species, perhaps a subunit, of 14 kDa which still retains the endoproteolytic activity.

## Preparation

Dactylysin is purified from exudates of *X. laevis* skin generated by the granular glands of this amphibian following an epinephrine stress. A 2000-fold purification is achieved by a four-step fractionation including successively ion-exchange column, hydrophobic chromatographies and gel filtrations (Joudiou *et al.*, 1993).

## Biological Aspects

Since the antimicrobial peptides produced by the *X. laevis* skin glands possess in their primary structures at least one potential cleavage site, it is tempting to hypothesize that dactylysin inactivates these amphipathic peptides by selective hydrolysis. Moreover, the *in vitro* selectivity of cleavages produced by dactylysin on a number of neuropeptides suggests that a related enzyme may be produced in mammals, including humans. This is supported by observations on human neuroblastoma cell lines (Delporte *et al*., 1992; Melino *et al*., 1996).

## Distinguishing Features

Because of its specificity for the N-terminus of bulky hydrophobic residues, dactylysin resembles a number of metallo-endopeptidases including neprilysin (Chapter 362), and meprin (Chapter 406). It is distinguished from these enzymes by the following features (Joudiou *et al*., 1993): dactylysin does not cleave the pentapeptide [Leu[5]]- and [Met[5]]-enkephalins, which in fact, are both inhibitors when activity is tested towards substance P. Phosphoramidon (1 μM concentration), a potent inhibitor of neprilysin, produces only 41% inhibition of dactylysin. Dactylysin is not inhibited by TIMPs (tissue inhibitors of matrix metalloproteinases).

Dactylysin is not inhibited by 1 mM captopril, a strong inhibitor of angiotensin-converting enzyme (ACE, Chapter 359): moreover different preferential cleavages are produced by these enzymes in substance P (Gly9∤Leu10 for dactylysin versus Phe8-Gly9 for ACE). This Gly9-Leu10 cleavage is also different from those generated by a number of peptidases called 'substance P-degrading metalloproteases' (Lee *et al*., 1981; Nyberg *et al*., 1984; Endo *et al*., 1988).

## References

Carvalho, K. de M., Joudiou, C., Boussetta, H., Leseney, A.M. & Cohen, P. (1992) A peptide-hormone-inactivating endopeptidase (PHIE) in *Xenopus laevis* skin secretion. *Proc. Natl Acad. Sci. USA* **89**, 84–88.

Delporte, C., Carvalho, K. de M., Leseney, A.M., Winand, J., Christophe, J. & Cohen, P. (1992) A new metallo-endopeptidase from human neuroblastoma NB-OK1 cells which inactivates atrial natriuretic peptide by selective cleavage at the Ser[123]-Phe[124] bond. *Biochem. Biophys. Res. Commun.* **182**, 158–164.

Endo, S., Yokosawa, H. & Ishii, S.I. (1988) Purification and characterization of a substance P-degrading endopeptidase from rat brain. *J. Biochem.* **104**, 999–1006.

Joudiou, C., Carvalho, K.de M., Camarao, G., Boussetta, H. & Cohen, P. (1993) Characterization of the thermolysin-like cleavage of biologically active peptides by *Xenopus laevis* peptide hormone inactivating enzyme, PHIE. *Biochemistry* **32**, 5959–5966.

Kuks, P., Creminon, C., Leseney, A.M., Bourdais, J., Morel, A. & Cohen, P. (1989) *Xenopus laevis* skin Arg-Xaa-Val-Arg-Gly endoprotease: a highly specific protease cleaving after a single arginine of a consensus sequence of peptide hormone precursors. *J. Biol. Chem.* **264**, 14609–14612.

Lee, C.M., Sandberg, B.E.B., Hanley, M.R. & Iversen, L.L. (1981) Purification and characterisation of a membrane-bound substance-P-degrading enzyme from human brain. *Eur. J. Biochem.* **114**, 315–327.

Melino, G., Draoui, M., Bernardini, S., Bellincampi, L., Reichert, U. & Cohen, P. (1996) Regulation by retinoic acid of insulin-degrading enzyme (IDE) and of a related endoprotease in human neuroblastoma cell lines. *Cell Growth Differ.* **7**, 787–796.

Nyberg, F., Le Greves, P., Sundqvist, C. & Terenius, L. (1984) Characterization of substance P (1–7) and (1–8) generating enzyme in human cerebrospinal fluid. *Biochem. Biophys. Res. Commun.* **125**, 244–250.

*Carine Joudiou*
*Laboratoire de Biochimie des Signaux Régulateurs*
*Cellulaires et Moléculaires,*
*Unité de Recherche Associée au Centre National de la*
*Recherche Scientifique No. 1682,*
*Université Pierre et Marie Curie,*
*96 Blvd. Raspail,*
*F-75006 Paris, France*

*Christine Clamagirand*
*Laboratoire de Biochimie des Signaux Régulateurs*
*Cellulaires et Moléculaires,*
*Unité de Recherche Associée au Centre National de la*
*Recherche Scientifique No. 1682,*
*Université Pierre et Marie Curie,*
*96 Blvd. Raspail,*
*F-75006 Paris, France*

*Paul Cohen*
*Laboratoire de Biochimie des Signaux Régulateurs Cellulaires*
*et Moléculaires,*
*Unité de Recherche Associée au Centre National de la*
*Recherche Scientifique No. 1682,*
*Université Pierre et Marie Curie,*
*96 Blvd. Raspail,*
*F-75006 Paris, France*
*Email: pcohen@infobiogen.fr*

# 542. Dynorphin-converting enzyme

## Databanks

*Peptidase classification: clan MX, family M99, MEROPS ID: M9G.030*
*NC-IUBMB enzyme classification: none*
*Databank codes: no sequence data available*

## Name and History

Dynorphins (Dyn) are a class of potent opioid peptides with wide distribution in the nervous system. They are synthesized as large proteins containing biologically active peptides flanked by dibasic and monobasic processing signals. A monobasic processing enzyme capable of converting dynorphin B-29 to dynorphin B-13 by a cleavage at a monobasic site was first characterized in the Triton X-100 extract of rat brain membranes (Devi & Goldstein, 1984) and recently purified to homogeneity from the cattle neurointermediate lobe (Berman *et al.*, 1995). This enzyme was designated, ***dynorphin-converting enzyme (DCE)*** since it converts a $\kappa$-opioid receptor active peptide (Dyn B-29) to another $\kappa$-opioid receptor active peptide (Dyn B-13). It should be noted that this monobasic processing dynorphin-converting enzyme is different from the dibasic processing dynorphin-A converting enzyme that generates Leu-enkephalin-Arg[6] from dynorphin A and B (Silberring *et al.*, 1992). These enzymes also differ in their physicochemical properties and protease inhibitory profiles (Berman *et al.*, 1995). The dibasic processing metallopeptidase may be nardilysin (Chapter 468).

## Activity and Specificity

The monobasic processing DCE cleaves Dyn B-29 at Thr↓Arg14 and recognizes peptides that fit the consensus for monobasic processing (Berman *et al.*, 1995). The consensus for recognition by the monobasic processing enzyme is hypothesized to require basic residues at the −3, −5 or −7 (P4, P6, P8) positions, and not to tolerate an aliphatic residues at the +1 (P1′) position, or an aromatic residue at the −1 (P2) position (Devi, 1991).

The pH optimum for DCE is about 7.0. The activity is completely inhibited by 1 mM 1,10-phenanthroline and partially inhibited by 5 mM EDTA or EGTA. The activity is stimulated with high concentrations of $MgCl_2$ and $MnCl_2$ but not by $CoCl_2$ and $ZnCl_2$ (Berman *et al.*, 1995).

The enzyme activity is measured by a radioimmunoassay using an antibody specific to Dyn B-13 (Devi & Goldstein, 1984). An assay using an intramolecularly quenched fluorescent peptide substrate, Dyn B-29-(6–16) [Abz-Arg-Arg-Gln-Phe-Lys-Val-Val-Thr↓Arg-Ser-Gln-*N*-(2,4-dinitrophenyl)ethylenediamine] has been useful for rapid characterization of the enzyme. An assay using matrix-assisted laser desorption ionization time-of-flight mass spectrometry has also been useful for the identification of the peptide products. This and other assays are described in Berman *et al.* (1995).

## Structural Chemistry

DCE exhibits a pI of about 5.1 (Berman *et al.*, 1995). It runs as a 180 kDa protein under nondenaturing electrophoresis conditions and as single band of 54 kDa under denaturing electrophoresis conditions. Increasing the reducing conditions does not result in further changes in the size of this protein (Berman *et al.*, 1995).

## Preparation

DCE has been purified to homogeneity from the cattle neurointermediate lobe of the pituitary (Berman *et al.*, 1995) using phenyl-Sepharose chromatography, preparative isoelectrofocusing, nondenaturing electrophoresis and FPLC with Mono-Q. The enzyme has also been partially purified (about 2500-fold) from the anterior pituitary (Devi *et al.*, 1991). Other sources from which DCE has been partially purified include rat brain and ileum (Devi, 1992) and cell lines AtT-20 (Devi, 1992), GH4C1 (Greco *et al.*, 1992), and BRL 2A (Petanceska *et al.*, 1993).

## Biological Aspects

The enzyme is present in most tissues, but is particularly abundant in rat brain and pituitary (Devi, 1992). In tissue homogenates, the activity is equally distributed between soluble and membrane-bound forms (L. Ageyeva, Y. Berman & L.A. Devi, unpublished results). Subcellular fractionation studies have shown that DCE is present in the neuropeptide-containing secretory vesicles in cattle pituitary (Devi *et al.*, 1991). DCE is also found in the functional secretory compartment in endocrine cell lines such as AtT-20 cells (Devi, 1992) and GH4C1 cells (Greco *et al.*, 1991). Agents that modulate the levels of endogenous peptides or secretory vesicles also modulate DCE activity, suggesting that DCE is coregulated with the components of the secretory vesicles. Treatment of cultured astrocytes or C6 glioma cells with nitric oxide-generating agents causes substantial reduction in DCE activity, suggesting that DCE is regulated by nitric oxide (Devi *et al.*, 1994).

It is likely that DCE is involved in the generation of many neuropeptide and peptide hormones since it seems to be widely distributed (Devi, 1993; Berman *et al*., 1994) and has a fairly wide substrate specificity (Devi, 1992; Devi & Goldstein, 1986).

## Distinguishing Features

DCE processes on the N-terminal side of Arg in peptide substrates that fit the consensus for monobasic processing. DCE is a 54 kDa neutral thiol-sensitive metalloprotease with a pI of 5.1. It is distinct from the prohormone thiol protease (Chapter 263) (Krieger & Hook, 1991) and from all other thiol-sensitive proteinases in its monobasic cleavage site specificity, physiochemical properties and sensitivity to specific inhibitors (Berman *et al*., 1995).

## References

Berman, Y., Rattan, A., Carr, K. & Devi, L. (1994) Regional distribution of neuropeptide processing enzymes in rat brain. *Biochimie* **76**, 245–250.

Berman, Y.L., Juliano, L. & Devi, L. (1995) Purification and characterization of a dynorphin processing endoprotease. *J. Biol. Chem.* **270**, 23845–23850.

Devi, L. (1991) Consensus sequence for processing of peptide precursors at monobasic sites. *FEBS Lett.* **280**, 189–194.

Devi, L. (1992) Secretion and regulation of a neuropeptide processing enzyme in AtT-20 cells. *Endocrinology* **131**, 1931–1935.

Devi, L. (1993) Tissue distribution of a dynorphin processing endopeptidase. *Endocrinology* **132**, 1139–1144.

Devi, L. & Goldstein, A. (1984) Dynorphin converting enzyme with unusual specificity from rat brain. *Proc. Natl Acad. Sci. USA* **81**, 1892–1896.

Devi, L. & Goldstein, A. (1986) Opioid peptides as inhibitors of leumorphin (dynorphin B-29) converting enzyme activity. *Peptides* **7**, 87–90.

Devi, L., Gupta, P. & Fricker, L. (1991) Subcellular localization, partial purification, and characterization of a dynorphin processing endopeptidase from bovine pituitary. *J. Neurochem.* **56**, 320–329.

Devi, L., Petanceska, S., Liu, R., Arbabha, B., Bansinath, M. & Garg, U. (1994) Regulation of neuropeptide-processing enzymes by nitric oxide in cultured astrocytes. *J. Neurochem.* **62**, 2387–2393.

Greco, L., Daly, L., Kim, S. & Devi, L. (1992) A dynorphin processing endoprotease in the lactotrophic rat anterior pituitary derived cell line, GH4C1. *Neuroendocrinology* **55**, 351–356.

Krieger, T.J. & Hook, V.Y.H. (1991) Purification and characterization of a novel thiol protease involved in processing the enkephalin precursor. *J. Biol. Chem.* **266**, 8376–8383.

Petanceska, S., Zikherman, J., Fricker, L.D. & Devi, L. (1993) Processing of prodynorphin in BRL-3A cells, a rat liver-derived cell line: implications for the specificity of neuropeptide processing enzymes. *Mol. Cell Endocrinol.* **94**, 37–45.

Silberring, J., Castello, M.E. & Nyberg, F. (1992) Characterization of dynorphin A-converting enzyme in human spinal cord. *J. Biol. Chem.* **267**, 21324–21328.

**Lakshmi A. Devi**
*Department of Pharmacology,*
*NYU School of Medicine,*
*New York, NY 11016, USA*
*Email: Lakshmi.Devi@Med.Nyu.Edu*

# UNCLASSIFIED PEPTIDASES

# 543. Introduction: peptidases of unknown catalytic type

## Databanks

| Species | SwissProt | PIR | EMBL (cDNA) | EMBL (genomic) |
|---|---|---|---|---|
| **Family U3** | | | | |
| Endopeptidase GPR (Chapter 559) | | | | |
| **Family U4** | | | | |
| Sporulation factor SpoIIGA | | | | |
| *Bacillus subtilis* | P13801 | A29812 | M57606 | – |
| | | JT0495 | X17344 | |
| | | S08224 | | |
| *Bacillus thuringiensis* | P26767 | – | X56697 | – |
| *Clostridium acetobutylicum* | – | – | Z23079 | – |
| **Family U6** | | | | |
| Murein endopeptidase MepA (Chapter 556) | | | | |
| **Family U7** | | | | |
| Protein C | | | | |
| Bacteriophage lambda | P03711 | – | – | – |
| Bacteriophage P21 | P36273 | – | M81255 | – |
| Protease IV (Chapter 562) | | | | |
| sohB gene product | | | | |
| *Buchnera aphidicola* | – | – | U09185 | – |
| *Escherichia coli* | P24213 | – | M73320 | – |
| *Haemophilus influenzae* | P45315 | – | U32841 | – |
| **Family U9** | | | | |
| Bacteriophage T4 prohead endopeptidase (Chapter 563) | | | | |
| **Family U12** | | | | |
| Type IV prepilin peptidase (Chapter 560) | | | | |
| **Family U26** | | | | |
| D-Ala-D-Ala carboxypeptidase vanY (Chapter 554) | | | | |
| **Family U28** | | | | |
| Dipeptidase E (Chapter 544) | | | | |
| **Family U29** | | | | |
| Aphthovirus and cardiovirus 2A autolytic sequence (Chapter 566) | | | | |
| **Family U32** | | | | |
| Microbial collagenase | | | | |
| *Porphyromonas gingivalis* | P33437 | A41881 | M60404 | – |
| Others | | | | |
| *Escherichia coli* | P45527 | – | – | U18997: chromosomal region 67.4–76.0' |
| *Haemophilus influenzae* | P44700 | – | U32725 | – |
| **Family U34** | | | | |
| Dipeptidase DA (Chapter 545) | | | | |
| **Family U35** | | | | |
| Prohead proteinase | | | | |
| Bacteriophage HK97 | – | – | U18319 | – |
| **Family U36** | | | | |
| Acidic endopeptidase | | | | |
| *Myxococcus xanthus* | – | – | X75892 | – |
| **Family U39** | | | | |
| Hepatitis C virus endopeptidase 2 (Chapter 567) | | | | |

U

*continued overleaf*

| Species | SwissProt | PIR | EMBL (cDNA) | EMBL (genomic) |
|---|---|---|---|---|
| **Family U40** | | | | |
| Murein endopeptidase P5 (Chapter 557) | | | | |
| **Family U42** | | | | |
| HycI endopeptidase (Chapter 561) | | | | |
| **Family U43** | | | | |
| Nsp Vp4 endopeptidase | | | | |
| Avian infectious bursal disease virus | | | | |
| strain Australian 002-73 | P08364 | A24382 | X03993 | – |
| strain CV-1 | P15480 | A35353 | – | D00867: genome segment X16107: genome segment |
| strain E | P29802 | PQ0283 | D10065 | – |
| strain OH | – | A40569 | M66722 | – |
| strain PBG-98 | P25220 | JQ0944 | D00868 | – |
| strain STC | P22351 | JS0360 | D00499 | – |
| strain 52/70 | P25219 | JQ0941 | D00869 | – |
| Infectious pancreatic necrosis virus endopeptidase (Chapter 564) | | | | |
| **Family U44** | | | | |
| Pestivirus N$^{pro}$ endopeptidase (Chapter 565) | | | | |
| **Family U45** | | | | |
| Sporulation factor SpoIVFB putative endopeptidase | | | | |
| *Bacillus subtilis* | P26937 | S18438 | X59528 | – |
| **Family U46** | | | | |
| PfpI protease (Chapter 558) | | | | |

There are a number of peptidases for which the catalytic type remains to be established. Those for which amino acid sequences are known can be grouped into families, which have been given names that begin with the letter 'U' to signify peptidase **u**nclassified with regard to catalytic mechanism, and are completed with a number assigned arbitrarily as the families were recognized. When the biochemical data become available and the catalytic type is discovered for a family, then the family will be renamed for the appropriate catalytic type. Because this has already happened on a number of occasions, the list of families is more interrupted than are those for other catalytic types. The databanks table shows 20 families of unclassified peptidases, 15 of which are the subjects of individual chapters, and 5 of which have no separate chapter, and are dealt with briefly only in this Introduction. There are also 11 chapters on peptidases that cannot be placed in peptidase families. For 8 of them, this is because amino acid sequences are not known, and for the other 3 it is because the type of reaction catalysed is not a peptidase reaction according to the strictest definitions, but nevertheless is likely to be of interest to the reader of the *Handbook*.

We have not been able to arrange this heterogeneous group of peptidases in any very structured way. The family numbers, being arbitrary, do not provide a rational sequence, and no clans are recognizable amongst the unclassified peptidases. Because the arrangement of chapters here is potentially confusing, a brief overview may be helpful, before we introduce the section more fully. We start with a series of chapter on miscellaneous exopeptidases: several dipeptidases, a carboxypeptidase, and omega peptidases. Then we have a group of 5 chapters on exopeptidases and endopeptidases that are linked by the fact that all the enzymes participate in the metabolism of bacterial cell walls. This is followed by

a series of 12 chapters on enzymes that hydrolyse proteins, first from bacteria and then from viruses. And the last two chapters describe proteolytic proteins that have not normally been thought of as peptidases at all. This sequence may be described more fully as follows.

*Family U28* contains **dipeptidase E** (Chapter 544), which is known not only from bacteria such as *Escherichia coli*, but also from *Xenopus*. The bacterial enzyme is cytoplasmic, and preferentially cleaves Asp┼Pro dipeptides, and its activity is regulated by the cyclic AMP receptor (Conlin *et al*., 1994). The dipeptidase has been shown not to be a metallopeptidase, being unaffected by EDTA. The *Xenopus* homolog is one of 35 genes upregulated by thyroid hormone during metamorphosis (Brown *et al*., 1996). Absence of signal peptide, transmembrane domains or *N*-glycosylation sites implies that the protein is cytoplasmic.

*Family U34* contains **dipeptidase DA** (Chapter 545) from *Lactobacillus*. This is a general dipeptidase, optimally active at pH 6 and 55°C; it is inhibited by *p*-hydroxymercuribenzoate and can be reactivated by DTT, but is not inhibited by EDTA, suggesting that it may be a cysteine-type enzyme.

A number of peptidases of unknown catalytic type cannot yet be assigned to families because no amino acid sequence data are available. These include three dipeptidases, **Glu-Glu dipeptidase** (Chapter 546), **non-stereospecific dipeptidase** (Chapter 547), **X-methyl-His dipeptidase** (Chapter 548), and also **tubulinyl-Tyr carboxypeptidase** (Chapter 549). The group also includes two omega peptidases: **N-formyl-methionyl-peptidase** (Chapter 550) and **β-aspartyl-peptidase** (Chapter 551).

**Destabilase** (Chapter 552) hydrolyzes the secondary amide bonds between the side-chains of Glu and Lys that are formed by factor XIII in the cross-linking of fibrin.

An account of *LD-dipeptidase* (Chapter 553) leads in to the sequence of chapters on peptidases involved in the metabolism of bacterial cell walls. Families *U6*, *U26* and *U40* contain peptidases involved in bacterial cell wall lysis and turnover. Bacterial cell walls are complex polymers of amino sugars and amino acids, and chains of alternating *N*-acetylglucosamine and *N*-acetylmuramic acid units are cross-linked by short peptides. The structure of the linking peptide differs between bacterial species, but in *Escherichia coli* the important cross-link is between *meso*-diaminopimelic acid and D-alanine. It is this bond that is broken by the murein endopeptidases from *E. coli* (family U6; Chapter 556) and bacteriophage phi6 (family U40; Chapter 557). Both endopeptidases are penicillin insensitive, unlike the DD-endopeptidase PBP7 (Chapter 143) from family S11 and penicillin-binding protein 4 (Chapter 148) from family S13. Antibiotic resistance (to vancomycin) led also to the discovery of *Enterococcus* D-Ala-D-Ala carboxypeptidase vanY from family U26 (Chapter 554), which processes the precursor of the link peptide, removing a C-terminal D-alanine. This reaction can be performed by a number of other peptidases from different bacterial species, including serine-type peptidases such as D-Ala-D-Ala carboxypeptidase PBP5 (Chapter 142) from family S11, *Streptomyces* R61 D-Ala-D-Ala carboxypeptidase (Chapter 145) from family S12, and metallopeptidases such as *Streptomyces albus* zinc D-Ala-D-Ala carboxypeptidase (Chapter 463) from family M15.

*Family U32* contains a microbial collagenase from *Porphyromonas*, the product of the *prtC* gene. The enzyme cleaves the Pz-Pro-Leu-Gly-Pro peptide that is a substrate for clostridial collagenase (Chapter 368) and thimet oligopeptidase (Chapter 371), as well as soluble type I collagen. The collagenase can be inhibited with *p*-hydroxymercuribenzoate and EDTA (Takahashi *et al.*, 1991), and calcium has been shown to enhance activity (Kato *et al.*, 1992). Homologs are also known from *Escherichia coli* and *Haemophilus influenzae*.

*Family U36* contains an acidic endopeptidase from the soil saprophytic actinomycete *Myxococcus*. The enzyme is secreted during vegetative growth, and is active under acidic conditions (pH 5.9). It cleaves a Phe105↓Met106 bond in κ-casein that leads to clotting of casein and is also cleaved by chymosin (Chapter 274). Because of this, the *Myxococcus* proteinase has been described as 'chymosin-like' (Lucas *et al.*, 1994), but it does not contain either of the sequence motifs conserved in aspartic endopeptidases of clan AA (Chapter 309). The endopeptidase is synthesized as a preproprotein.

*Families U3, U4* and *U45* include endopeptidases involved in sporulation in *Bacillus* species. Sporulation is controlled by a number of σ factors, principally $\sigma^E$, $\sigma^F$, $\sigma^G$ and $\sigma^K$. The sporulation process is divided into stages I–VII, and the stage number is included in the name of any gene that is active during that particular stage. It is only during stages II and IV that proteolytic processing of transcription factor precursors occurs.

During stage II of the sporulation process, in which the septum is completed, the precursor of transcription factor $\sigma^E$ is cleaved near the N-terminus to release the transcription factor and a 27 residue peptide (Haldenwang, 1995). It is thought that the membrane-bound spoIIGA proteinase (family U4) is responsible for this cleavage. The presence of an Asp-Ser-Gly motif in spoIIGA proteinase has led to the speculation that it is an aspartic endopeptidase acting at neutral pH (Peters & Haldenwang, 1991). The spoIIGA endopeptidase can be activated by transcription factor $\sigma^F$, or by an extracellular signal from the secreted spoIIR protein. The spoIIGA protein has an extracellular domain that acts as a receptor for the spoIIR protein, and binding of spoIIR switches on the peptidase activity (Hofmeister *et al.*, 1995). Transcription factor $\sigma^F$ is also activated during stage II, but by phosphorylation and not proteolytic processing; its activator is the membrane-bound serine phosphatase spoIIE. Transcription factor $\sigma^F$ brings about synthesis of the GPR endopeptidase (family U3; Chapter 559), which autoprocesses itself to release a 16 residue propeptide during stage II (Sanchez-Salas & Setlow, 1993).

During stage IV of sporulation, a peptidoglycan layer surrounds the protoplast. It is during this stage that transcription factor $\sigma^E$ has its effect, upregulating several genes, including that of the membrane-bound spoIVFB endopeptidase (family U45). The spoIVFB endopeptidase is responsible for releasing transcription factor $\sigma^K$ from its precursor. It is possible that spoIVFB endopeptidase is a metallopeptidase, because it possesses an 'HEXXH' motif, although located in one of the four predicted transmembrane domains. Transcription factor $\sigma^K$ is active during the latter stages of sporulation, in which endopeptidases are not known to be involved.

Finally, when the spores are produced and begin to germinate, the gpr endopeptidase has its effect. A further autolytic cleavage takes place to remove the N-terminal Leu, and the gpr endopeptidase is active in the degradation of small, acid-soluble spore proteins that are not bound to DNA (Sanchez-Salas & Setlow, 1993; Setlow & Setlow, 1995).

*Family U12* includes type IV prepilin peptidase (Chapter 560). Pili are hair-like structures that occur on the cell surfaces of bacteria, and each is assembled from one or more protein subunits known as pilins. Type IV pili are found in gram-negative pathogens, including *Neisseria gonorrhoeae* and *Pseudomonas aeruginosa*, and are thought to be responsible for attaching the organism to the surface of host epithelial cells. The pilins are secreted to the periplasm as precursors with special leader peptides at the N-terminus that are removed by the type IV prepilin leader peptidase. These leader peptides are 6–8 residues long and rich in charged amino acids, quite unlike the leader peptides removed by signal peptidase I (Chapter 153). All mature type IV pilins have a methylated N-terminal Phe residue. The type IV prepilin leader peptidase is located in the inner membrane, and cleavage and methylation of the pilin precursors appears to occur on the cytoplasmic face of the membranes. The type IV prepilin leader peptidase cleaves Gly↓Phe bonds, and is also responsible for the methylation of the new N-terminal Phe. It had been thought the enzyme was a cysteine-type peptidase, because site-directed mutagenesis had implicated four Cys residues (Strom *et al.*, 1993); accordingly, we had previously named the family C20 (Rawlings & Barrett, 1994). However, a homolog from *Xanthomonas* lacks any of the conserved Cys residues and yet is functional (Hu *et al.*, 1995). For this reason, this family of peptidases has been reclassified as of unknown catalytic type.

*Family U42* includes the hydrogenase activator from *Escherichia coli* (Chapter 561). Hydrogenases are nickel-containing, multisubunit enzymes. The subunits are synthesized as precursors that are activated by processing at the C-terminus. In *E. coli*, the gene encoding the processing endopeptidase has been identified and sequenced, and is part of the *hyc* operon that also controls expression of the large subunit gene (*hycE*) and has been named *hycI* (Rossmann *et al.*, 1995). The hydrogenase activator is not affected by typical inhibitors of serine or metallopeptidases. Homologs have been identified in a number of other bacteria, and the sequence alignment (Alignment 543.1 on CD-ROM) shows conservation of two aspartyl and a histidine residue. This suggests that the hydrogenase activator is either an aspartic-type peptidase acting at neutral pH, or a metallopeptidase.

The thermophilic archaean *Pyrococcus* produces at least two endopeptidases: the subtilisin homolog pyrolysin (Chapter 112), and protease I, which is included in *family U46* (Chapter 558). In keeping with its origin in an extreme thermophile, protease I is maximally active at 95°C. The active enzyme is a homohexamer of 19 kDa subunits, although a homotrimeric form also exists. A number of homologs are known, which are mainly from bacteria, but the human DJ-1 protein and the B0432.2 protein from *Caenorhabditis* are distantly related. Most of these homologs have not been biochemically characterized, and none is known to be a peptidase. The *Pyrococcus* enzyme has inhibition characteristics of a serine peptidase, but the only completely conserved residues in the family of proteins that might be catalytically important are Glu15 and Cys100 (numbered according to protease I). The homolog from *Burkholderia cepacia* has a long N-terminal extension relative to the other sequences.

*Family U7* includes protease IV (Chapter 562), an endopeptidase from *Escherichia coli* that contributes to the degradation of signal peptides released by the action of signal peptidase I (Chapter 153). The sohB protein is a homolog of protease IV that when overexpressed can compensate for the absence of protease Do (Chapter 83) in *E. coli* mutants lacking this activity (Baird *et al.*, 1991).

Another homolog of protease IV is protein C from bacteriophage λ. The sequence similarity is in the C-terminal half of protease IV, and although this may represent a mosaic relationship not including peptidase units, there is evidence that protein C may be an endopeptidase involved in the activation of protein B, one of the components of the bacteriophage prohead, and degradation of the scaffold protein around which the prohead is built (Kochan & Murialdo, 1983).

In addition to family U7, *families U9* and *U35* also contain endopeptidases involved in phage prohead assembly. The bacteriophage T4 prohead endopeptidase of family U9 (Chapter 563) is responsible for the proteolytic activation of all but one of the prohead proteins, and, like bacteriophage λ protein C, the prohead endopeptidase also degrades the scaffold protein. For both bacteriophage λ and bacteriophage T4, the scaffold protein is a product of the same gene as the endopeptidase, but transcription begins at an alternative start codon. The bacteriophage HK97 endopeptidase from family U35 is required for processing of the capsid protein precursor (Duda *et al.*, 1995).

*Families U29, U39, U43* and *U44* contain polyprotein-processing endopeptidases from RNA viruses. Family U29 includes endopeptidase 2A from cardioviruses (Chapter 566). Cardioviruses possess a picornain 3C (Chapter 239) that

processes most of the sites in the viral nonstructural polyprotein, but do not have a picornain 2A (Chapter 241). The 2A/2B cleavage is not performed by picornain 3C, but cleavage does occur at a Gly+Pro bond to release a P1–2A product that is subsequently cleaved by picornain 3C. This cleavage appears to be mediated by the 150 residue 2A protein itself. The aphthoviruses, such as foot-and-mouth disease virus, also have a picornain 3C and lack a picornain 2A. Some of the cleavages performed by picornain 2A in other viruses are done by the L-peptidase (Chapter 226), but not the release of the 2A protein. In aphthoviruses, the 2A protein is a 19 residue polypeptide. Surprisingly, given the size of the polypeptide, it appears to act as an autolytic oligopeptide. Expression systems for cardiovirus polyproteins have been engineered, with the finding that only the C-terminal 19 amino acids from encephalomyocarditis virus 2A protein are required for cleavage between the 2A and 2B proteins (Donnelly *et al.*, 1997).

The hepatitis C polyprotein is processed by virally encoded endopeptidases as well as by host endopeptidases. Most of the cleavages are performed by the NS3 polyprotein peptidase (Chapter 89), which is a serine-type endopeptidase with a trypsin-like fold. Cleavage of the NS2/3 site at a Leu+Ala bond is mediated by the second virally encoded endopeptidase, endopeptidase 2, often termed the 'NS2-3 endopeptidase', not because of the cleavage site but because the endopeptidase consists of the NS2 protein and one-third of the NS3 protein (Chapter 567). Endopeptidase 2 is included in family U39. Site-directed mutagenesis had identified His952 and Cys993 as the catalytic dyad (Grakoui *et al.*, 1993), and we had previously considered the enzyme to be a cysteine-type endopeptidase in family C18 (Rawlings & Barrett, 1994). However, evidence that the NS2-3 endopeptidase is zinc dependent, including inhibition with EDTA and increased activity with the addition of zinc (Hijikata *et al.*, 1993), has cast doubt on the identification of the catalytic type. For this reason, the NS2-3 endopeptidase family has been renamed U39 until the situation becomes clearer.

Birnaviruses are double-stranded RNA viruses. The genomes of infectious pancreatic necrosis virus and the avian infectious bursal disease virus include at least three open reading frames, one of which encodes a polyprotein for the structural proteins. Within the polyprotein there is a processing endopeptidase known as nonstructural protein VP4 (Chapter 564). The endopeptidase is in family U43. There are no conserved cysteines or aspartates in the family, but there are three conserved serines, two glutamates and a histidine (see Alignment 543.2 on CD-ROM), suggesting that the endopeptidase is most probably of serine-type.

Pestiviruses possess two polyprotein-processing endopeptidases, the serine-type NS2-3 endopeptidase (Chapter 91), which is the more general processing activity, and the p20 endopeptidase, which is the N-terminal protein and releases itself from the polyprotein by autolytic cleavage of a Cys+Ser bond. The p20 endopeptidase is included in family U44 (Chapter 565).

We conclude the series of chapters on proteolytic enzymes of unknown catalytic type with accounts of two systems in which peptide bonds are broken, but not by agents that would normally be thought of as proteolytic enzymes at all. One of these is *kedarcidin* (Chapter 568), representative of a group of antitumor chromoproteins with proteolytic activity, and the second is the class of *self-splicing proteins* (Chapter 569), the members of which are able to eject an internal polypeptide sequence, and in the process, religate the N- and C-terminal flanking sequences.

## References

Baird, L., Lipinska, B., Raina, S. & Georgopoulos, C. (1991) Identification of the *Escherichia coli sohB* gene, a multicopy suppressor of the Htra (DegP) null phenotype. *J. Bacteriol.* **173**, 5763–5770.

Brown, D.D., Wang, Z., Furlow, J.D., Kanamori, A., Schwartzman, R.A., Remo, B.F. & Pinder, A. (1996) The thyroid hormone-induced tail resorption program during *Xenopus laevis* metamorphosis, *Proc. Natl. Acad. Sci. USA* **93**, 1924–1929.

Conlin, C.A., Håkensson, K., Liljas, A. & Miller, C.G. (1994) Cloning and nucleotide sequence of the cyclic AMP receptor protein-regulated *Salmonella typhimurium pepE* gene and crystallization of its product, an α-aspartyl dipeptidase. *J. Bacteriol.* **176**, 166–172.

Donnelly, M.L.L., Gani, D., Flint, M., Monaghan, S. & Ryan, M.D. (1997) The cleavage activities of aphthovirus and cardiovirus 2A proteins. *J. Gen. Virol.* **78**, 13–21.

Duda, R.L., Martincic, K. & Hendrix, R.W. (1995) Genetic basis of bacteriophage HK97 prohead assembly. *J. Mol. Biol.* **247**, 636–647.

Grakoui, A., McCourt, D.W., Wychowski, C., Feinstone, S.M. & Rice, C.M. (1993) A second hepatitis C virus-encoded proteinase. *Proc. Natl Acad. Sci. USA* **90**, 10583–10587.

Haldenwang, W.G. (1995) The sigma-factors of *Bacillus subtilis*. *Microbiol. Rev.* **59**, 1–30.

Hijikata, M., Mizushima, H., Akagi, T., Mori, S., Kakiuchi, N., Kato, N., Tanaka, T., Kimura, K. & Shimotohno, K. (1993) Two distinct proteinase activities required for the processing of a putative nonstructural precursor protein of hepatitis C virus. *J. Virol.* **67**, 4665–4675.

Hofmeister, A.E.M., Londoño-Vallejo, A., Harry, E., Stragier, P. & Losick, R. (1995) Extracellular signal protein triggering the proteolytic activation of a developmental transcription factor in B-subtilis. *Cell* **83**, 219–226.

Hu, N.T., Lee, P.F. & Chen, C.H. (1995) The type IV pre-pilin leader peptidase of *Xanthomonas campestris* pv *campestris* is functional without conserved cysteine residues. *Mol. Microbiol.* **18**, 769–777.

Kato, T., Takahashi, N. & Kuramitsu, H.K. (1992) Sequence analysis and characterization of the *Porphyromonas gingivalis prtC* gene, which expresses a novel collagenase activity. *J. Bacteriol.* **174**, 3889–3895.

Kochan, J. & Murialdo, H. (1983) Early intermediates in bacteriophage lambda prohead assembly. II. Identification of biologically active intermediates. *Virology* **131**, 100–115.

Lucas, N., Mazaud-Aujard, C., Bremaud, L., Cenatiempo, Y. & Julien, R. (1994) Protein purification, gene cloning and sequencing of an acidic endoprotease from *Myxococcus xanthus* DK101. *Eur. J. Biochem.* **222**, 247–254.

Peters, H.K., III & Haldenwang, W.G. (1991) Synthesis and fractionation properties of spoIIGA, a protein essential for Pro-σ$^E$ processing in *Bacillus subtilis*. *J. Bacteriol.* **173**, 7821–7827.

Rawlings, N.D. & Barrett, A.J. (1994) Families of cysteine peptidases. *Methods Enzymol.* **244**, 461–486.

Rossmann, R., Maier, T., Lottspeich, F. & Bock, A. (1995) Characterisation of a protease from *Escherichia coli* involved in hydrogenase maturation. *Eur. J. Biochem.* **227**, 545–550.

Sanchez-Salas, J.-L. & Setlow, P. (1993) Proteolytic processing of the protease which initiates degradation of small, acid-soluble proteins during germination of *Bacillus subtilis* spores. *J. Bacteriol.* **175**, 2568–2577.

Setlow, B. & Setlow, P. (1995) Binding to DNA protects α/β-type, small, acid-soluble spore proteins of *Bacillus* and *Clostridium* species against digestion by their specific protease as well as by other proteases. *J. Bacteriol.* **177**, 4149–4151.

Strom, M.S., Bergman, P. & Lory, S. (1993) Identification of active-site cysteines in the conserved domain of PilD, the bifunctional type IV pilin leader peptidase/*N*-methyltransferase of *Pseudomonas aeruginosa*. *J. Biol. Chem.* **268**, 15788–15794.

Takahashi, N., Kato, T. & Kuramitsu, H.K. (1991) Isolation and preliminary characterization of the *Porphyromonas gingivalis prtC* gene expressing collagenase activity. *FEMS Microbiol. Lett.* **84**, 135–138.

# 544. *Dipeptidase E*

## *Databanks*

*Peptidase classification: clan UX, family U28, MEROPS ID: U28.001*
*NC-IUBMB enzyme classification: none*
*Databank codes:*

| Species | SwissProt | PIR | EMBL (cDNA) | EMBL (genomic) |
|---|---|---|---|---|
| *Escherichia coli* | P32666 | – | – | U00006: chromosomal region from 89.2 to 92.8′ |
| *Haemophilus influenzae* | P44766 | – | U32740 | – |
| *Salmonella typhimurium* | P36936 | – | U01246 | – |
| *Xenopus laevis* | – | – | U37377 | |

## Name and History

A leucine auxotroph of *Salmonella typhimurium* can use almost any leucine-containing dipeptide as a source of leucine (Carter & Miller, 1984). But a mutant strain lacking the broad-specificity aminopeptidases lysylaminopeptidase (Chapter 345), leucylaminopeptidase (Chapter 474) and aminopeptidase B, the broad-specificity X-His dipeptidase (Chapter 487), and the X-Pro-specific aminopeptidase (Chapter 479) and dipeptidase (Chapter 478), will not grow when supplied with most leucine-containing dipeptides as leucine sources. Indeed, a screen of Leu-containing dipeptides led to only one peptide, Asp-Leu, which supported growth. Analysis of extracts of such a mutant strain led to the discovery that *S. typhimurium* contains three distinct enzymes that hydrolyze Asp-Xaa dipeptides. Only one of these, **dipeptidase E (PepE)**, has been characterized. This is also known as **aspartyl dipeptidase**.

## Activity and Specificity

Dipeptidase E catalyzes the hydrolysis of Asp$\downarrow$Xaa dipeptides. It does not attack peptides with N-terminal Glu, Asn or Gln, nor does it cleave isoaspartyl peptides. A free carboxyl group is not absolutely required since Asp$\downarrow$Phe-NH$_2$ and Asp$\downarrow$Phe-OMe are hydrolyzed somewhat more slowly than peptides with free C-termini. No peptide larger than a C-blocked dipeptide is known to be a substrate. Asp$\downarrow$NHPhNO$_2$ is hydrolyzed and is a convenient substrate for routine assay (Conlin *et al.*, 1994). The enzyme can be separated from the other Asp-Xaa hydrolyzing enzymes in crude extracts by nondenaturing PAGE, and detected *in situ* using an activity stain with Asp-Leu as substrate (Carter & Miller, 1984). PepE activity is maximal at about pH 7.0. The enzyme is not inhibited by DFP, PMSF, dipicolinic acid or EDTA (Carter & Miller, 1984; Conlin *et al.*, 1994).

## Structural Chemistry

The nucleotide sequence of *pepE* predicts a protein of 229 amino acids (24.8 kDa). The enzyme is active as a monomer. Alignment of the sequences of the four known PepE proteins reveals three conserved Ser residues, four conserved His residues, and two conserved Asp residues. One of the serines (Ser120, *S. typhimurium* numbering) lies in a conserved Gly-Xaa-Ser-Yaa-Gly sequence. The presence downstream from Ser120 of conserved Asp (Asp135) and His (His157) residues, both lying in regions of strong conservation, suggest that PepE may be a serine peptidase despite the fact that classical serine peptidase inhibitors do not seem to affect it.

## Preparation

The enzyme has been purified from strains carrying a single copy of the gene (Carter & Miller, 1984) and from strains carrying the *pepE* gene on a high copy number plasmid (Conlin *et al.*, 1994). Crystallization of the protein has been reported (Conlin *et al.*, 1994).

## Biological Aspects

PepE is localized to the cytoplasm of the bacterial cell. Transcription of *pepE* is regulated by the cAMP receptor protein, suggesting that the enzyme might be involved in producing amino acids that can be used as carbon sources (Conlin *et al.*, 1994). Although aspartate cannot serve as a carbon source for either *E. coli* or *S. typhimurium*, it is a precursor to several other amino acids, and its availability might spare carbon under starvation conditions. The *pepE* gene is dispensable for growth in standard media. In addition to the *pepE* genes identified in *S. typhimurium*, *E. coli*, and *H. influenzae*, a *Xenopus laevis* gene which encodes a protein with substantial similarity (41% identity, 60% similarity) to *pepE* has been found. This gene is upregulated in response to thyroid hormone during tail resorption in *Xenopus* metamorphosis (Wang & Brown, 1993). This gene can be expressed in *S. typhimurium* and its product hydrolyzes Asp-X peptides (L.-Y. Li & C.G. Miller, unpublished results).

## Distinguishing Features

To our knowledge, no other peptidase with a rigid specificity for N-terminal aspartate has been reported. At least one other peptidase (peptidase B) present in extracts of *S. typhimurium* and *E. coli* hydrolyzes Asp-Leu but this enzyme is a member of the leucyl aminopeptidase family (Z. Mathew & C.G. Miller, unpublished results) and is completely inhibited by EDTA. PepE does not hydrolyze isoaspartyl peptides (Carter & Miller, 1984) which are hydrolyzed by at least two distinct activities present in extracts of *E. coli* (Haley, 1968; Gary & Clarke, 1995; Chapters 486 and 502).

## Further Reading

Carter & Miller (1984) describe the first characterization of PepE and the isolation of *pepE* mutants. Conlin *et al.* (1994) report the sequence of the gene, additional characterization and crystallization of its product.

### References

Carter, T.H. & Miller, C.G. (1984) Aspartate-specific peptidases in *Salmonella typhimurium*: mutants deficient in peptidase E. *J. Bacteriol.* **159**, 453–459.

Conlin, C.A., Haakensson, K., Liljas, A. & Miller, C.G. (1994) Cloning and nucleotide sequence of the cyclic AMP receptor protein-regulated *Salmonella tyuphimurium pepE* gene and crystallization of its product, an α-aspartyl dipeptidase. *J. Bacteriol.* **176**, 166–172.

Gary, J.D. & Clarke, S. (1995) Purification and characterization of an isoaspartyl dipeptidase from *Escherichia coli*. *J. Biol. Chem.* **270**, 4076–4087.

Haley, E.E. (1968) Purification and properties of a β-aspartyl peptidase from *Escherichia coli*. *J. Biol. Chem.* **243**, 5748–5752.

Wang, Z. & Brown, D.D. (1993) Thyroid hormone-induced gene expression program for amphibian tail resorption. *J. Biol. Chem.* **268**, 16270–16278.

*Charles G. Miller*
*Department of Microbiology,*
*B103 Chemical and Life Sciences Laboratory MC-110,*
*University of Illnois at Champagne-Urbana,*
*Urbana, IL 61801, USA*
*Email: charlesm@uiuc.edu*

# 545. Dipeptidase DA

*Peptidase classification: clan UX, family U34, MEROPS ID: U34.001*
*NC-IUBMB enzyme classification: none*
*Databank codes:*

| Species | SwissProt | PIR | EMBL (cDNA) | EMBL (genomic) |
|---|---|---|---|---|
| *Lactobacillus helveticus* | – | – | U34257<br>Z38063 | – |
| *Lactobacillus sake* | – | – | X98238 | – |

## Name and History

Plasmids from a genomic library of *Lactobacillus helveticus* CNRZ32 constructed in *Escherichia coli* DH5$\alpha$ (Nowakowski *et al*., 1993) were electroporated into *E. coli* CM89, and the resulting transformants were screened for dipeptidase activities using the substrates Leu-Leu, Leu-Leu-$NH_2$, Pro-Leu and Met-Pro. Of the peptidase-positive strains identified, one hydrolyzed Leu-Leu, but did not have any apparent activity on the other three dipeptides. The plasmid carried by this strain, designated pSUW10, was found to encode a 5.8 kbp *L. helveticus* chromosomal insert. The peptidase activity encoded by pSUW10 was designated ***DPI***. Later, this designation was changed to ***pepDA***, indicating this was the first dipeptidase genetically identified in *L. helveticus* (Dudley *et al*., 1996). This dipeptidase was also purified by another group, which designated it ***PepD*** (Vesanto *et al*., 1996).

## Activity and Specificity

Substrate specificity studies of PepDA have been performed using both cell-free extracts of *E. coli* CM89 (pSUW10) (Nowakowski *et al*., 1993) and purified enzyme (Vesanto *et al*., 1996). Assays with cell-free extract were performed in 0.5 M *N*-(2-hydroxyethyl)piperazine-*N'*-(2-ethanesulfonic acid), pH 7.5, using a coupled L-amino acid oxidase assay (Tan & Konings, 1990). Assays with purified protein were performed in 50 mM 2-(*N*-morpholino)ethanesulfonic acid (MES), pH 6.0 at 55°C with a Cd-ninhydrin reagent (Doi *et al*., 1981). Both studies have shown that PepDA is active on a variety of dipeptides other than those containing proline. While the enzyme was able to hydrolyze Leu+Leu and Phe+Leu, no activity was detected when either Leu-Leu-$NH_2$ or Phe-Leu-$NH_2$ was used as substrate. This suggests that PepDA only hydrolyzes peptides containing a free C-terminus. This enzyme has not been found to hydrolyze the tripeptides Leu-Leu-Leu or Val-Gly-Gly, and has no detectable activity on any amino acid-$NHPhNO_2$ substrate tested. Therefore, our current knowledge of the substrate specificity of PepDA suggests the enzyme is active only on dipeptides.

Inhibitor studies have been performed on PepDA using the above conditions for the purified enzyme. Chemical agents were added to the purified enzyme to a final concentration of either 0.1 or 1.0 mM and incubated at room temperature for 30 min prior to assaying with Leu-Leu. At either concentration of inhibitor, 2-ME, DTT, 1,10-phenanthroline and EDTA were found to stimulate PepDA activity by 5–97%, except for 0.1 mM 2-ME which reduced activity by 50%. Additionally, 0.1 and 1 mM PMSF inhibited PepDA by 9 and 37% respectively. However, no PepDA activity was detected after preincubation with either concentration of *para*-hydroxymercuribenzoate, while this inhibition could be restored with DTT. These results suggest that a sulfhydryl group is important for PepDA activity.

## Structural Chemistry

The molecular mass of PepDA was estimated to be 420 kDa by gel filtration chromatography (Vesanto *et al*., 1996). As the deduced amino acid sequence of PepDA derived from the nucleotide sequence suggests a monomeric mass of 53.5 kDa, it appears that the native enzyme contains eight subunits.

## Preparation

The *pepDA* gene from *L. helveticus* 53/7 was overexpressed in *E. coli* JM105, and the enzyme was subsequently purified to SDS-PAGE homogeneity (Vesanto *et al*., 1996).

## Biological Aspects

Nucleotide sequencing of *pepDA* from *Lactobacillus helveticus* CNRZ32 identified a 1422 bp open reading frame (ORF) which could encode a polypeptide of 53.5 kDa (Dudley *et al*., 1996). Preceding the ORF was a putative ribosome-binding site which resembles the ribosome-binding site of other *L. helveticus* sequences. Downstream of *pepDA* is a putative *rho*-independent transcriptional terminator with a $\Delta G$ of $-19.8\ \mathrm{kcal\ mol^{-1}}$. The deduced amino acid sequence of *pepDA* is 99% identical (473 of 474 amino acids) to the *pepD* sequence from *L. helveticus* 53/7, and also demonstrates high identity to an unidentified ORF from *Lactobacillus sake* (van den Berg *et al*., 1996). The best alignment between the deduced amino acid sequence of *pepDA* and the *L. sake* ORF is between amino acids 119–305 and 118–304 respectively, where the two sequences have 43% identity and 65% similarity to one another.

U

Additionally, southern hybridization experiments were performed using a probe synthesized from an internal portion of *pepDA* (Dudley *et al.*, 1996). When the hybridization was performed at 42°C in the presence of 10% formamide, hybridization was detected with total DNA isolated from two strains of *L. helveticus* and one of two strains of *Lactobacillus delbrueckii* subsp. *bulgaricus*. No hybridization was detected with total DNA from a second strain of *L. delbrueckii* subsp. *bulgaricus* or strains of *Lactobacillus casei, Leuconostoc, Lactococcus, Pediococcus* or *Streptococcus thermophilus*. This result suggests PepDA is not present in most lactic acid bacteria, or the identity between *pepDA* and its homologs is too low to be detected by this technique.

## Distinguishing Features

While a number of dipeptidases have been purified and characterized from lactic acid bacteria (Kunji *et al.*, 1996), PepDA is the first which was not found to be inhibited by metal chelating agents such as EDTA or 1,10-phenanthroline. The inhibition studies of Vesanto *et al.* suggest that a cysteine residue may be involved in catalysis.

## Further Reading

The papers of Dudley *et al.* (1996), Kunji *et al.* (1996) and Vesanto *et al.* (1996) are recommended.

## References

Doi, E., Shibata, D. & Matoba, T. (1981) Modified colorimetric ninhydrin methods for peptidase assay. *Anal. Biochem.* **118**, 173–184.

Dudley, E.G., Husgen, A.C., He, W. & Steele, J.L. (1996) Sequencing, distribution, and inactivation of the dipeptidase A gene (*pepDA*) from *Lactobacillus helveticus* CNRZ32. *J. Bacteriol.* **178**, 701–704.

Kunji, E.R.S., Mierau, I., Hagting, A., Poolman B. & Konings, W.N. (1996) The proteolytic systems of lactic acid bacteria. *Antonie van Leeuwenhoek* **70**, 187–221.

Nowakowski, C.M., Bhowmik, T.K. & Steele, J.L. (1993) Cloning of peptidase genes from *Lactobacillus helveticus* CNRZ32. *Appl. Microbiol. Biotechnol.* **39**, 204–210.

Tan, P.S.T. & Konings, W.M. (1990) Purification and characterization of an aminopeptidase from *Lactococcus lactis* subsp. *cremoris* Wg2. *Appl. Environ. Microbiol.* **56**, 526–532.

van den Berg, D.J.C., Toonen, M.Y., Robijn, G.W., Kamerling, J.P., Vliegenthart, J.F.G., van der Swaluw, C.D.M., Ledoboer, A.M. & Verrips, C.T. (1996) Isolation, characterization, and disruption of the putative exopolysaccharide gene cluster from *Lactobacillus sake* 0-1. GenBank accession X98238.

Vesanto, E., Peltoniemi, K., Purtsi, T., Steele, J.L. & Palva, A. (1996) Molecular characterization, over-expression and purification of a novel dipeptidase from *Lactobacillus helveticus*. *Appl. Microbiol. Biotechnol.* **45**, 638–645.

*Edward G. Dudley*
*Department of Bacteriology,*
*University of Wisconsin-Madison,*
*Madison, WI 53706, USA*

*James L. Steele*
*Department of Food Science,*
*University of Wisconsin-Madison,*
*Madison, WI 53706, USA*
*Email: jlsteele@facstaff.wisc.edu*

# 546. Glu-Glu dipeptidase

## Databanks

*Peptidase classification: clan UX, family U99, MEROPS ID: U9B.003*
*NC-IUBMB enzyme classification: EC 3.4.13.7*
*Databank codes: no sequence data available*

## Name and History

In an effort to produce pteroic acid microbiologically, a soil organism of the genus *Flavobacterium* (ATCC 25012) was employed that is capable of utilizing naturally occurring folic acid and its polyglutamate forms as a sole source of carbon and nitrogen (derived from the liberated glutamate residues). Pteroic acid, which is poorly utilized, accumulates in the growth medium. An investigation of the intracellular enzymes capable of degrading the pteroylpolyglutamate molecule resulted in the discovery of an enzyme that catalyzes the hydrolysis of $\alpha$-glutamylglutamate (Glu+Glu), but shows little or no action on folic acid (pteroylmonoglutamate) or its poly-$\gamma$-glutamate derivatives (Pratt *et al.*, 1968). Based on the selective affinity of the enzyme for Glu-Glu, IUB recommended, in 1972, the name **α-glutamyl-glutamate dipeptidase**, and assigned an EC number (3.4.13.7). In 1992, IUBMB recommended the abbreviated name **Glu-Glu dipeptidase**.

## Activity and Specificity

This enzyme catalyzes the hydrolysis of the $\alpha$-glutamyl bond in $\alpha$-glutamylglutamate (Glu+Glu). No action occurs if the N-terminus of the dipeptide is substituted, as shown by a lack of activity on pteroyl-$\alpha$-diglutamate. Pteroylmonoglutamate (folic acid) is also resistant. The $\gamma$-glutamyl bond in the dipeptide $\gamma$-glutamylglutamate is not cleaved, and only a trace of activity is seen on pteroyl-$\gamma$-diglutamate. Reactions are performed at pH 7.4 and 37°C.

It is unclear whether the Glu-Glu substrate must have a free C-terminal $\alpha$-carboxyl group, a specificity characteristic that is required for classification as a dipeptidase. Questions also remain as to whether the enzyme's specificity extends to other $\alpha$-glutamyl dipeptides ($\alpha$-Glu-Xaa) and acidic dipeptides such as Asp-Asp.

The assay of dipeptidase activity is done by combining enzyme and Glu-Glu in a Tris–HCl buffer, pH 7.4. Samples (10 µl) taken at timed intervals are applied to Whatman No. 1 filter paper. The paper is reacted with ninhydrin, the colored spots eluted into ethanolic copper sulfate solution, and the absorbance determined at 520 nm.

Effects of inhibitors and activators have not been reported.

## Structural Chemistry

Not yet established.

## Preparation

Cultured cells of a *Flavobacterium* species (ATCC 25012) are harvested, ruptured by sonication or high pressure, and centrifuged to obtain a supernatant fluid. Enzyme activities are fractionated by gradient elution chromatography on DEAE-cellulose, and finally by gel filtration on Sephadex G-200.

## Biological Aspects

Glu-Glu dipeptidase apparently participates in a pathway that enables *Flavobacterium* to utilize pteroylpolyglutamates as a sole source of carbon and nitrogen.

## Distinguishing Features

No distinguishing features are yet recognized, apart from its affinity for unsubstituted $\alpha$-glutamylglutamate.

## Reference

Pratt, A.G., Crawford, E.J. & Friedkin, M. (1968) The hydrolysis of mono-, di-, and triglutamate derivatives of folic acid with bacterial enzymes. *J. Biol. Chem.* **243**, 6367–6372.

*J. Ken McDonald*
*Department of Biochemistry and Molecular Biology,*
*Medical University of South Carolina,*
*171 Ashley Avenue,*
*Charleston, SC 29425, USA*

# 547. *Non-stereospecific dipeptidase*

## Databanks

*Peptidase classification: clan UX, family U99, MEROPS ID: U9B.004*
*NC-IUBMB enzyme classification: EC 3.4.13.17*
*Databank codes: no sequence data available*

## Name and History

When free D-aspartic acid was identified in the brain and other parts of the nervous system of *Octopus vulgaris* Lam. (D'Aniello & Giuditta, 1977), and in other cephalopods as well (D'Aniello & Giuditta, 1978), an effort was made to locate a peptidase in these tissues that is capable of catalyzing the hydrolysis of peptide bonds formed by D-amino acids.

Such a peptidase was identified in the squid, *Loligo vulgaris*, and named **peptidyl-D-amino acid hydrolase** (D'Aniello & Strazzullo, 1984). In 1992, IUBMB recommended the name **non-stereospecific dipeptidase**, indicating (a) that the enzyme acts without preference on peptide bonds involving D- or L-amino acids, and (b) that its specificity is directed primarily, if not exclusively, against unsubstituted dipeptides.

U

## Activity and Specificity

Rates of hydrolysis of L-Ala+D-Ala and L-Ala+L-Ala are the highest reported and are essentially equal. A free N-terminus is required, as indicated by a lack of activity on Ac-L-Ala-L-Ala, Bz-Gly-L-Phe and Bz-Gly-L-Arg. Compared to the rate seen on L-Ala-D-Ala (set to 100%), relative rates displayed on Gly-D-Ala and Gly-L-Ala are about 80%, and on D-Leu-D-Leu and D-Leu-L-Leu, about 70%. Specificity for dipeptides appears to be broad, judging from rates (ranging from 10 to 90%) exhibited on Gly-Gly, Gly-D-Ser, Gly-D-Leu, Gly-D-Phe, Gly-D-Asp, L-Tyr-D-Arg and D-Leu-L-Tyr. Negligible rates, however, occur on D-Ala-D-Ala and D-Ala-L-Ala. Unlike many other peptidases, non-stereospecific dipeptidase does not hydrolyze amino acid amides, esters or other derivatives.

Maximal rates of hydrolysis of dipeptides occur at pH 8.0, with notable activity seen over a pH range of 7.2–8.8.

Although dipeptides are hydrolyzed at especially high rates, activity may not be limited to dipeptides. Detectable rates occur on some tripeptides containing D- or L-amino acids, but these are trivial: 0 to 2% compared to the rate of L-Ala-D-Ala hydrolysis. Examples include such tripeptides as Gly-Gly-D-Leu, Gly-Gly-L-Leu, D-Ala-D-Ala-D-Ala, L-Ala-L-Ala-L-Ala, L-Ala-D-Ala-L-Ala, D-Ala-Gly-Gly, L-Leu-Gly-Gly, Gly-Gly-Gly, L-Ala-L-Val-L-Leu and L-Leu-L-Ala-L-Pro. A relative rate of 30% seen on D-Leu-Gly-Gly is exceptional.

No action occurs on oligopeptides such as tetra-Gly, penta-Gly and hexa-Gly, but other oligopeptides are attacked. They include pentapeptides such as [Leu[5]]enkephalin (Tyr-Gly-Gly-Phe-Leu) and [Met[5]]enkephalin. Although some internal bonds are cleaved in these substrates, degradation is primarily through a carboxypeptidase attack. [D-Ala[2],Met[5]]enkephalin, which is also reduced to free amino acids, is similarly degraded sequentially from the C-terminus. No action occurs on large polypeptides such as glucagon and insulin, or on the proteins hemoglobin, albumin and casein.

The carboxypeptidase-like activity displayed on certain oligopeptides raises questions regarding the possible presence of a contaminating activity.

Activity is typically determined with a colorimetric hydrazone procedure that measures the amount of D-alanine liberated from Gly-D-Ala in a reaction mixture containing enzyme and a Tris-HCl buffer, pH 8.0. D-Amino acid oxidase and catalase are also included for the purpose of converting free D-alanine to a ketoacid. Timed reactions are stopped by the addition of 2,4-dinitrophenylhydrazine in 1 N HCl. The absorbance of the resulting hydrazone solution is determined at 445 nm, and converted to D-alanine concentration by comparison with a standard curve.

Sensitivity to inhibitors is not well understood. Exposure to 50 mM EDTA has no effect on activity, nor do metals such as $Co^{2+}$, $Mn^{2+}$, $Mg^{2+}$, $Zn^{2+}$, $Cu^{2+}$ and $Ca^{2+}$ at concentrations up to 1 mM; $Pb^{2+}$ (1 mM) inhibits only 5–10%. Dissociating agents such as guanidinium chloride (50 mM) and urea (3 M) have no effect on activity.

## Structural Chemistry

The native enzyme is a 140 kDa heterodimer with a pI of 6.1. It is composed of 106 kDa and 36 kDa subunits. The native enzyme is comprised of 1065 amino acid residues. The most abundant residues are aspartic acid/asparagine, glutamic acid/glutamine, glycine, and valine. The enzyme contains 7.12% aromatic, 2.2% sulfur and 13.9% heterocyclic residues. Half-cystines account for 7.3 residues per mol of enzyme. Less than 0.3% carbohydrate is detectable as glucose.

## Preparation

The cecal sac contents from 6–8 specimens of adult squid (*Loligo vulgaris* Lam.) are dialyzed against Tris–HCl buffer, pH 8.0, and centrifuged to remove insoluble material. Four purification steps involving ion-exchange and molecular-exclusion chromatographies are employed to achieve a 70-fold purification with a 17% yield. Analysis by PAGE, at pH 8.8 and 4.5, shows a single band of protein.

## Biological Aspects

High concentrations of D-aspartic acid exist in the brain of *Octopus vulgaris* (D'Aniello & Giuditta, 1977), in squid axoplasm and in other regions of the nervous system of *O. vulgaris, Loligo vulgaris*, and the cuttlefish *Sepia officinalis* (D'Aniello & Giuditta, 1978). Since racemases have not been found in these tissues, it has been postulated that free D-aspartic acid found in nerve tissue of cephalopods is the product of the breakdown of D-amino acid peptides by non-stereospecific dipeptidase. Although the cecal sac of the above-mentioned cephalopods is a rich source of the enzyme, it is also present in the nerve tissue (central brain and optic lobes) of these invertebrates.

Significant quantities of D-alanine have been detected in the muscle and hepatopancreas of several crustacean species, such as *Eriphia spinifrons* (crab), *Palinurus vulgaris* (lobster) and *Parapenaeus longirostris* (shrimp) (D'Anniello & Giuditta, 1980). Since crustaceans are the usual food of cephalopods, and since non-stereospecific dipeptidase is abundant in the digestive tissues (hepatopancreas and intestine) of these mollusks, this peptidase may serve to hydrolyze exogenous D-amino acid peptides present in the crustacean tissues.

## Distinguishing Features

The ability to hydrolyze L-Ala-D-Ala and L-Ala-L-Ala at near-equal rates at pH 8 is quite distinctive. The specificity of the bacterial LD-dipeptidase (Chapter 553) is similar in some respects.

## References

D'Aniello, A. & Giuditta, A. (1977), Identification of D-aspartic acid in the brain of *Octopus vulgaris* Lam. *J. Neurochem.* **29**, 1053–1057.

D'Aniello, A. & Giuditta, A. (1978), Presence of D-aspartic acid in squid axoplasm and in other regions of the cephalopod nervous system. *J. Neurochem.* **31**, 1107–1108.

D'Aniello, A. & Giuditta, A. (1980), Presence of D-alanine in crustacean muscle and hepatopancreas. *Comp. Biochem. Physiol.* **66B**, 319–322.

D'Aniello, A. & Strazzullo, L. (1984), Peptidyl-D-Amino acid hydrolase from *Loligo vulgaris* Lam. Purification and characterization. *J. Biol. Chem.* **259**, 4237–4243.

**J. Ken McDonald**
*Department of Biochemistry and Molecular Biology,*
*Medical University of South Carolina,*
*171 Ashley Avenue,*
*Charleston, SC 29425, USA*

# 548. *X-Methyl-His dipeptidase*

## Databanks

*Peptidase classification: clan UX, family U99, MEROPS ID: U9B.002*
*NC-IUBMB enzyme classification: EC 3.4.13.5*
*Databank codes: no sequence data available*

## Name and History

This enzyme was originally detected in the skeletal muscle of cod fish (*Gadus callarias*) and was named **anserinase** (Jones, 1954) because it cleaved anserine ($\beta$-alanyl+1-methylhistidine), a dipeptide that is present at high levels in codling skeletal muscle, which lacks the related dipeptide carnosine ($\beta$-Ala-His). The name **aminoacyl-methylhistidine dipeptidase** (EC 3.4.3.4) was later recommended by the IUB. In a separate line of investigation, an enzyme was found in the brain of skipjack tuna (*Katsuwonus pelamis*) that catalyzed the deacetylation of *N*-acetylhistidine (Baslow & Lenney, 1967). The authors referred to the tuna enzyme as **$\alpha$-N-acetyl-L-histidine amidohydrolase**, but in 1972 the IUB recommended the name **acetylhistidine deacetylase** (EC 3.5.1.34). A subsequent comparative study (Lenney *et al.*, 1978) of the deacetylase of tuna brain and the dipeptidase of cod brain and muscle furnished evidence that these activities were attributable to a single enzyme. In 1992, the IUBMB recommended that the enzyme be named **X-methyl-His dipeptidase** (3.4.13.5). A systematic name is **aminoacyl-pros-methyl-L-histidine hydrolase**.

## Activity and Specificity

The enzyme acts at neutral pH to cleave certain dipeptides that are resistant to most other dipeptidases; these are dipeptides in which the free amino group is on the $\beta$-carbon, as in anserine and carnosine (Jones, 1956), or on the $\gamma$-carbon, as in homocarnosine (Lenney *et al.*, 1978). The enzyme also shows deacetylase activity that is very specific for $\alpha$-*N*-acetyl-L-histidine (Baslow & Lenney, 1967) and thus is quite unlike that of aminoacylase (EC 3.5.1.14), which hydrolyzes a range of acetylated amino acids and also dipeptides (Kördel & Schneider, 1975), but has no action on carnosine or homocarnosine (Lenney *et al.*, 1978).

Only impure enzyme preparations have so far been used in studies of the substrate specificity of X-methyl-His dipeptidase, but the enzyme from brain of cod and tuna is reported also to hydrolyze $\alpha$-linked dipeptides. These include Gly+His and Gly+Leu (both cod and tuna enzyme: Lenney *et al.*, 1978), and Ala+Gly, Ala+His, Leu+Tyr, Leu+Gly (cod enzyme: Jones, 1956; Lenney *et al.*, 1978).

Activity may be assayed with anserine at pH 7.3 (Jones, 1955) or *N*-acetylhistidine at pH 6.5 (Lenney *et al.*, 1978), with ninhydrin to detect the liberated amino acid (Moore, 1968). The deacetylation of *N*-acetylhistidine can also be followed by use of the Pauly color reagent to quantify the unchanged substrate spectrophotometrically at 510 nm (Baslow & Lenney, 1967).

X-Methyl-His dipeptidase probably contains one or more thiol groups that are essential for activity, as well as an essential zinc ion. Activity is commonly increased by the presence of $ZnSO_4$ (0.16 mM) and DTT (0.5 mM), whereas PCMB and $HgCl_2$ are strongly inhibitory. Phosphate buffers and higher levels of thiol compounds inhibit, possibly by complexing the metal ion (Lenney *et al.*, 1978). Activity is lost upon exposure to EDTA, but can then be restored with

**U**

1 mM $Zn^{2+}$ or $Co^{2+}$, but not by $Mg^{2+}$ or $Mn^{2+}$ (Lenney *et al.*, 1978).

## Structural Chemistry

Little is known about the structure of X-methyl-His dipeptidase. The molecular mass of the enzyme from tuna brain was estimated to be about 120 kDa by molecular exclusion chromatography in a buffer that contained Brij-35 (0.05%) and $ZnSO_4$ (0.1 mM). The native enzyme probably occurs as a zinc metalloprotein (Lenney *et al.*, 1978) containing one or more essential thiol groups.

## Preparation

Although the enzyme was first detected in the skeletal muscle of codling and was partially purified from that source (Jones, 1955), it was later discovered that brain from both cod and tuna, as well as ocular fluid (both aqueous and vitreous humors) from the latter, are rich sources of the enzyme (Lenney *et al.*, 1978). In the tuna fish, moderate activity occurs in liver and intestine, with only traces in stomach, ovary, spleen, white muscle and gallbladder. Red muscle, pancreas and heart lack detectable activity (Lenney *et al.*, 1978). The enzyme has been partially purified from skipjack tuna brain (87-fold) and ocular fluid (58-fold), and from cod brain (30-fold) by Lenney *et al.* (1978).

Activity in crude extracts is stable to freeze-drying and acetone powder preparation (Jones, 1954), but the partially purified enzyme is unstable to freezing and storage in solution at 4°C, activity decreasing by 10–20% per week. Activity is very unstable below pH 4.0 and above pH 8.5 (Lenney *et al.*, 1978).

## Biological Aspects

Anserine (but not carnosine) is present at high levels in the skeletal muscle of certain fish (e.g. tuna, salmon, swordfish), and is believed to serve as an endogenous substrate for the anserine-splitting activity of X-methyl-His dipeptidase present in those tissues (Jones, 1955; Lenney *et al.*, 1978). This may serve to recycle $\beta$-alanine and 1-methylhistidine into the metabolic pool.

The lens and retina of the eye of poikilothermic vertebrates are known to contain high levels of *N*-acetyl-L-histidine (Baslow, 1965). The lens is capable of generating this derivative from histidine and acetyl-coenzyme A (Baslow, 1966), but lacks *N*-acetylhistidine deacetylase activity. The deacetylase is abundant in the ocular fluids, however, and Baslow (1967) described an ocular fluid-lens histidine cycle, which mediates the active transport of histidine from the ocular fluid into the lens, against a concentration gradient, where it is acetylated. The acetylhistidine then diffuses out of the lens into the ocular fluid, where it is immediately hydrolyzed by X-methyl-His dipeptidase. Histidine may serve to transport zinc into the lens, which contains a high concentration of this metal. Histidine binds zinc more strongly than does *N*-acetylhistidine.

As far as is known, neither *N*-acetylhistidine nor X-methyl-His dipeptidase is present in mammalian tissues (Jones, 1955; Baslow, 1965; Baslow & Lenney, 1967).

## Distinguishing Features

X-Methyl-His dipeptidase is distinctive in displaying both dipeptidase activity that includes the hydrolysis of dipeptides having the free amino group on the $\beta$-carbon, and deacetylase activity that is essentially limited to *N*-acetylhistidine.

### References

Baslow, M.H. (1965) Neurosine, its identification with *N*-acetyl-L-histidine and distribution in aquatic vertebrates. *Zoologica* **50**, 63–66.

Baslow, M.H. (1966) *N*-Acetyl-L-histidine synthetase activity from the brain of the killifish, *Fundulus heteroclitus*. *Brain Res.* **3**, 210–213.

Baslow, M.H. (1967) *N*-Acetyl-L-histidine metabolism in the fish eye: evidence for ocular fluid-lens L-histidine recycling. *Exp. Eye Res.* **6**, 336–342.

Baslow, M.H. & Lenney, J.F. (1967) α-*N*-Acetyl-L-histidine amidohydrolase activity from the brain of the skipjack tuna *Katsuwonus pelamis*. *Can. J. Biochem.* **45**, 337–340.

Jones, N.R. (1954) Enzymatic cleavage of anserine in skeletal muscle of codling (*Gadus callarias*). *Biochem. J.* **57**, XXIV (abstract).

Jones, N.R. (1955) The free amino acids of fish. 1-Methylhistidine and β-alanine liberation by skeletal muscle anserinase of codling (*Gadus callarias*). *Biochem. J.* **60**, 81–87.

Jones, N.R. (1956) Anserinase and other dipeptidase activity in skeletal muscle of codling (*Gadus callarias*). *Biochem. J.* **64**, 20p (abstract).

Kördel, W. & Schneider, F. (1975) The pH dependence of the peptidase activity of aminoacylase. *Z. Physiol. Chem.* **356**, 915–920.

Lenney, J.F., Baslow, M.H. & Sugiyama, G.H. (1978) Similarity of tuna *N*-acetylhistidine deacetylase and cod fish anserinase. *Comp. Biochem. Physiol. B. Comp. Biochem.* **61**, 253–258.

Moore, S. (1968) Amino acid analysis: aqueous dimethyl sulfoxide as solvent for the ninhydrin reaction. *J. Biol. Chem.* **243**, 6281–6283.

*J. Ken McDonald*
*Department of Biochemistry and Molecular Biology,*
*Medical University of South Carolina,*
*171 Ashley Avenue,*
*Charleston, SC 29425, USA*

# 549. *Tubulinyl-Tyr carboxypeptidase*

## Databanks

*Peptidase classification: clan UX, family U99, MEROPS ID: U9E.002*
*NC-IUBMB enzyme classification: EC 3.4.17.17*
*Databank codes: no sequence data available*

## Name and History

The discovery of the tubulin tyrosination cycle was made serendipitously during a study of the content of tRNAs in rat brain extracts (Barra *et al.*, 1973). Under conditions where no protein synthesis was expected, $^{14}$C-labeled tyrosine became incorporated into trichloroacetic acid (TCA)-insoluble material. ***Tubulinyl-Tyr carboxypeptidase***, more commonly known as ***tubulin carboxypeptidase (TCP)***, but also referred to as either ***carboxypeptidase tubulin (CP-T)*** or ***tubulinyl-tyrosine carboxypeptidase***, was first described in 1977 as a distinct enzyme activity that post-translationally modified the α subunit of tubulin (Hallak *et al.*, 1977). CP-T is no longer an appropriate term for this enzyme, as a bacterial carboxypeptidase is named carboxypeptidase T (Chapter 456). Soon after the 1977 publication appeared, Argaraña *et al.* (1978) characterized the activity more completely and demonstrated that it was distinct from the reverse reaction of the tyrosinating enzyme.

## Activity and Specificity

Tubulinyl-Tyr carboxypeptidase cleaves the C-terminal peptide bond from α-tubulin sequences that end with either -Glu┼Tyr or -Glu┼Phe (Barra *et al.*, 1973). The most commonly used assays for TCP require the enzymatic tyrosination of brain tubulin for use as substrate with partially purified tubulin tyrosine ligase and [$^{14}$C]tyrosine (Barra *et al.*, 1973; Raybin & Flavin, 1977). The radiolabeled substrate (usually in unpolymerized form) is incubated with the enzyme-containing solution for a fixed period, and the enzyme activity is then calculated from the amount of TCA-soluble radioactivity present (Argaraña *et al.*, 1978). However, modifications of this assay have been introduced which increase its sensitivity and specificity (see below). The reaction seems specific for α-tubulin, because other proteins (aldolase) or peptides that possess similar C-termini fail to compete for the activity of TCP on tubulin (Kumar & Flavin, 1981). It is also known that the reaction does not progress further along the polypeptide chain, because the product of the reaction, identified by HPLC analysis, corresponds only to [$^{14}$C]tyrosine (Webster *et al.*, 1992). Although the exact sequence requirements for detyrosination are not known, it has been reported that neither β-tubulin nor the N-terminal portion of α-tubulin is required (Weizetfel *et al.*, 1989). Finally, the specific activity of the enzyme is significantly higher (2.5–10-fold) on polymerized than on unpolymerized tubulin substrate (Thompson *et al.*, 1979; Kumar & Flavin, 1981; Arce & Barra, 1983; Wehland & Weber, 1987; Webster *et al.*, 1992), suggesting that the conformational changes that occur in tubulin during microtubule assembly contribute favorably to either the binding or the activity of TCP. For this reason, an assay for TCP that uses taxol-stabilized microtubules as substrate has been developed (Webster *et al.*, 1992; Webster & Oxford, 1996). Using the modified assay we have determined that the pH optimum for activity is 7.5 (using a MOPS/HEPES buffer), and that either $Ca^{2+}$ or $Mg^{2+}$, over the concentration range from 2 to 10 mM, stimulates activity. Higher concentrations of these ions, as well as submillimolar levels of sulfhydryl-modifying compounds, inhibit activity (Webster & Oxford, 1996). No specific activators or inhibitors of TCP have been described.

## Structural Chemistry

Because TCP has not yet been purified to homogeneity, little is known about its structure, molecular mass or pI. One report suggests that TCP has a basic pI of ~9.0 (Smania *et al.*, 1992), which corresponds well with chromatofocusing experiments performed by the author (unpublished results). TCP is heat labile (90°C, 5 min, unpublished result) and (unlike pancreatic carboxypeptidase A) is probably not a metallopeptidase, since activity is unaffected by EDTA (Argaraña *et al.*, 1978). In addition, while metallopeptidases often lose activity after dialysis (due to the loss of the metal ion) and are stimulated by the addition of metal ions, TCP activity is retained after dialysis and is strongly inhibited by $Zn^{2+}$ and $Cu^{2+}$ and, to a lesser extent, $Co^{2+}$ (Webster & Oxford, 1996).

## Preparation

Although several attempts have been made to purify TCP (Argaraña *et al.*, 1980; Martensen, 1980; Flavin & Murofushi, 1984), none has resulted in more than a moderate enrichment, due in part to the assay used. Mammalian brain has proven to be the richest and most convenient source of the enzyme. Ample starting material for partial purification can be obtained from 6–8 cattle brains. TCP specific activity is roughly equal in white and gray matter, while cerebellum has somewhat less. Typical specific activities for TCP in crude brain extract range from 6 to 12 pmol min$^{-1}$ mg$^{-1}$ (Webster *et al.*, 1992). Partial purification of TCP has been reported by two different protocols (Argaraña *et al.*, 1980; Flavin & Murofushi, 1984), but these procedures were developed using the older, less specific assay, and must therefore be viewed with some caution.

## Biological Aspects

Results obtained by use of antibodies specific for detyrosinated tubulin have indicated that TCP is present in most eukaryotic organisms, ranging from trypanosomes (Stieger *et al.*, 1984) to mammalian cells (Gundersen & Bulinski, 1986), whereas it is absent from *Schizosaccharomyces pombe* (Alfa & Hyams, 1991). TCP activities (measured directly) are highest in the developing brain ($\sim$40 pmol min$^{-1}$ mg$^{-1}$ from P2 rat brain), while the measured levels from liver, cardiac and skeletal muscle tissues range from 0.04 to 1 pmol min$^{-1}$ mg$^{-1}$ (unpublished results of the author). PC-12 cells also possess high levels of activity, which increase after the cells are induced to differentiate (Webster *et al.*, 1992).

The function of this or any other post-translational modification of tubulin is currently not well understood. Modified microtubules have been correlated with a stable subset in many cell lines (Kreis, 1987; Webster *et al.*, 1987), but detyrosination does not confer stability on microtubules *in vivo* (Khawaja *et al.*, 1988; Webster *et al.*, 1990). One function of detyrosinated microtubules may be to organize the array of intermediate filaments within cells (Gurland & Gundersen, 1995). No diseases are known to be caused by aberrant TCP activities, although correlations have been made between alterations in the tyrosination cycle and experimentally induced phenylketonuria (Rodriguez & Borisy, 1979) and Chédiak–Higashi syndrome (Nath *et al.*, 1982).

## Distinguishing Features

The activity of TCP can be mimicked *in vitro* by pancreatic carboxypeptidase A (CPA; Chapter 451) in that CPA can detyrosinate either assembled or unassembled tubulin with similar kinetics (Kumar & Flavin, 1981). However, significant differences between the two enzymes have been noted. For example, agents that inhibit CPA significantly (EDTA, DTT, potato-derived inhibitor, 8-hydroxyquinoline sulfonic acid) have little or no effect on TCP activity (Webster *et al.*, 1992). Further, substrate analogs that interfere directly and potently with CPA, including DL-benzylsuccinic acid ($K_i \sim 10^{-6}$ M; Byers & Wolfenden, 1972) and ZAA$^P$(O)F ($K_i \sim 10^{-12}$ M; Hanson *et al.*, 1989), only begin to inhibit TCP at $\sim 10^{-2}$ M concentrations (Webster & Oxford, 1996; unpublished results of the author). Finally, antisera raised against CPA show no apparent cross-reactivity with TCP (unpublished results of the author). Therefore, TCP should be considered a distinct enzyme from CPA. Lysosomal carboxypeptidase A (Chapter 133) can also be distinguished from TCP, as the pH optimum for TCP is 7.5 while the optimum for the lysosomal enzyme is more acidic. The activities of other neutral proteases can be distinguished from that of TCP by use of the modified assay (Webster *et al.*, 1992).

## Further Reading

Barra *et al.* (1973) describe the serendipitous discovery of the tubulin tyrosination cycle, Kumar & Flavin (1981) report a good, initial biochemical characterization of the enzyme, and Webster *et al.* (1992) describe an improved assay for TCP that permits more accurate and sensitive detection of activity.

## References

Alfa, C.E. & Hyams, J.S. (1991) Microtubules in the fission yeast *Schizosaccharomyces pombe* contain only the tyrosinated form of α-tubulin. *Cell Motility and the Cytoskeleton* **18**, 86–93.

Arce, C.A. & Barra, H.S. (1983) Association of tubulinyl-tyrosine carboxypeptidase with microtubules. *FEBS Lett.* **157**, 75–78.

Argaraña, C.E., Barra, H.S. & Caputto, R. (1978) Release of [14C] tyrosine from tubulinyl-[14C] tyrosine by brain extract. Separation of a carboxypeptidase from tubulin-tyrosine ligase. *Mol. Cell. Biochem.* **19**, 17–21.

Argaraña, C.E., Barra, H.S. & Caputto, R. (1980) Tubulinyl-tyrosine carboxypeptidase from chicken brain: properties and partial purification. *J. Neurochem.* **34**, 114–118.

Barra, H.S., Rodriguez, J.A., Arce, C.A. & Caputto, R. (1973) A soluble preparation from rat brain that incorporates into its own proteins [14C]arginine by a ribonuclease-sensitive system and [14C]tyrosine by a ribonuclease-insensitive system. *J. Neurochem.* **20**, 97–108.

Byers, L.D. & Wolfenden, R. (1972) A potent reversible inhibitor of carboxypeptidase A. *J. Biol. Chem.* **247**, 606–608.

Flavin, M. & Murofushi, H. (1984) Tyrosine incorporation in tubulin. *Methods Enzymol.* **106**, 223–237.

Gundersen, G.G. & Bulinski, J.C. (1986) Distribution of tyrosinated and nontyrosinated alpha-tubulin during mitosis. *J. Cell Biol.* **102**, 1118–1122.

Gurland, G. & Gundersen, G.G. (1995) Stable detyrosinated microtubules function to localize vimentin intermediate filaments in fibroblasts. *J. Cell Biol.* **131**, 1275–1290.

Hallak, M.E., Rodriguez, J.A., Barra, H.S. & Caputto, R. (1977) Release of tyrosine from tyrosinated tubulin. Some common factors that affect this process and the assembly of tubulin. *FEBS Lett.* **73**, 147–150.

Hanson, J.E., Kaplan, A.P. & Bartlett, P.A. (1989) Phosphonate analogs of carboxypeptidase A substrates are potent transition-state analogue inhibitors. *Biochemistry* **28**, 6294–6305.

Khawaja, S., Gundersen, G.G. & Bulinski, J.C. (1988) Enhanced stability of microtubules enriched in detyrosinated tubulin is not a direct function of detyrosination level. *J. Cell Biol.* **106**, 141–150.

Kreis, T. (1987) Microtubules containing detyrosinated tubulin are less dynamic. *EMBO J.* **6**, 2597–2606.

Kumar, N. & Flavin, M. (1981) Preferential action of a brain detyrosinolating carboxypeptidase on polymerized tubulin. *J. Biol. Chem.* **256**, 7678–7686.

Martensen, T. (1980) Preparation and properties of enzymes catalyzing tubulin tyrosinolation and detyrosinolation. *Fedn Proc.* **39**, 2161.

Nath, J., Flavin, M. & Gallin, J.I. (1982) Tubulin tyrosinolation in human polymorphonuclear leukocytes: Studies in normal subjects and in patients with the Chédiak–Higashi syndrome. *J. Cell Biol.* **95**, 519–526.

Raybin, D. & Flavin, M. (1977) Enzyme which specifically adds tyrosine to the α chain of tubulin. *Biochemistry* **16**, 2189–2194.

Rodriguez, J.A. & Borisy, G.G. (1979) Experimental phenylketonuria: Replacement of carboxyl terminal tyrosine by phenylalanine in infant rat brain tubulin. *Science* **206**, 463–465.

Smania, A.M., Argaraña, C.E., Weizetfel, J.C. & Barra, H.S. (1992) Immunodetection of tubulin carboxypeptidase activity on nitrocellulose membrane after gel electrophoresis and blotting. *Biochem. Int.* **28**, 921–928.

Stieger, J., Wyler, T. & Seebeck, T. (1984) Partial purification and characterization of microtubular protein from *Trypanosoma brucei. J. Biol. Chem.* **259**, 4596–4602.

Thompson, W.C., Deanin, G.G. & Gordon, M.W. (1979) Intact microtubules are required for rapid turnover of carboxyl-terminal tyrosine of α-tubulin in cell cultures. *Proc. Natl Acad. Sci. USA* **76**, 1318–1322.

Webster, D.R. & Oxford, M.G. (1996) Regulation of cytoplasmic tubulin carboxypeptidase activity *in vitro* by cations and sulfhydryl-modifying compounds. *J. Cell. Biochem.* **60**, 424–436.

Webster, D.R., Gundersen, G.G., Bulinski, J.C. & Borisy, G.G. (1987) Differential turnover of tyrosinated and detyrosinated microtubules. *Proc. Natl Acad. Sci. USA* **84**, 9040–9044.

Webster, D.R., Wehland, J., Weber, K. & Borisy, G.G. (1990) Detyrosination of alpha tubulin does not stabilize microtubules in vivo. *J. Cell Biol.* **111**, 113–122 (and printer's corrections on pages 1325–1326).

Webster, D.R., Modesti, N. & Bulinski, J.C. (1992) Regulation of cytoplasmic tubulin carboxypeptidase activity during neural and muscle differentiation: characterization using a microtubule-based assay. *Biochemistry* **31**, 5849–5856.

Wehland, J. & Weber, K. (1987) Turnover of the carboxy-terminal tyrosine of α-tubulin and means of reaching elevated levels of detyrosination in living cells. *J. Cell Sci.* **88**, 185–203.

Weizetfel, J.C., Argaraña, C.E., Beltramo, D.M. & Barra, H.S. (1989) The integrity of tubulin molecule is not required for the activity of tubulin carboxypeptidase. *Biochem. Biophys. Res. Commun.* **159**, 770–776.

*Daniel R. Webster*
*Department of Cell Biology and Biochemistry,*
*Texas Tech University H.S.C.,*
*3601 4th Street,*
*Lubbock, TX 79430, USA*
*Email: cbbdrw@wpoffice.net.ttuhsc.edu*

# 550. *N-Formylmethionyl peptidase*

*Peptidase classification: clan UX, family U99, MEROPS ID: U9F.002*
*NC-IUBMB enzyme classification: EC 3.4.19.7*
*Databank codes: no sequence data available*

## Name and History

The addition of formylmethionyl-β-naphthylamide substrate to cell-free homogenates of rat mucosa led to the discovery of an *N*-acyl methionyl aminopeptidase (Sherriff *et al.*, 1992). This enzyme was active against formyl and acetyl methionyl peptides. It was described as an ***N*-formylmethionine aminopeptidase** because the biologically important substrates for the intestinal enzyme are believed to be *N*-formylmethionyl oligopeptides arising from N-terminal peptide sequences of bacterial precursor proteins. The name recommended by NC-IUBMB for this omega peptidase is ***N*-formylmethionyl-peptidase**.

## Activity and Specificity

The enzyme was initially characterized with formyl-amino acid-β-naphthylamide derivatives. Activity was demonstrated with fMet┼NHNap but not with fAla-, fVal-, fLeu-, fAsp-, fSer-, fArg- or fPhe-NHNap, indicating that fMet in position 1 was essential. Both acetyl and formyl methionyl di- and tripeptides were hydrolyzed. The fMet┼Leu-Phe analog fNle-Leu-Phe was also hydrolyzed. The enzyme exhibited a pH optimum between 7.5 and 9.0 and was unaffected by the addition of 2-ME (5 mM), DTT (20 mM), azide (5 mM) and EDTA (4 mM). The enzyme was markedly inhibited by the addition of $Cu^{2+}$ and $Cd^{2+}$ and activity was abolished by $Hg^{2+}$. The enzyme was activated by $Cl^-$ with maximal activation at 1 M.

## Structural Chemistry

Little is known about the structural chemistry of the fMet-peptidase. The purified enzyme has a native molecular mass of 320–340 kDa, and on SDS-PAGE runs as a major polypeptide of 83 kDa, consistent with a native enzyme comprised of four subunits. The enzyme was not retained on wheatgerm agglutinin- or concanavalin A-Sepharose.

## Preparation

The formylmethionyl peptidase was purified from rat intestinal mucosa homogenates by $CaCl_2$ precipitation, Accell anion exchange, Mono-Q anion exchange and FPLC gel filtration. The chromatography of the enzyme was followed colorimetrically with fMet-NHNap as substrate. The enzyme was purified 2300-fold with a recovery of 12.1% of activity.

## Biological Aspects

Formyl-peptides secreted by bacteria of the intestinal lumen are implicated in intestinal inflammatory disorders (Chadwick *et al.*, 1988; Broom *et al.*, 1992). Many *N*-formylmethionyl oligopeptides are potent bioactive agents, recruiting neutrophils (Chadwick *et al.*, 1988) through activation of leukocyte formylpeptide receptors (Showell *et al.*, 1976) and eliciting inflammatory responses. The rat intestinal *N*-fMet-peptidase is active against both synthetic and naturally occurring biologically active formyl peptides (Sherriff *et al.*, 1992), indicating that this enzyme may play a prominent role in the defense of the host against such intestinal inflammatory agents. The rat mucosal enzyme was present throughout the intestine in normal and germ-free animals. Similar activity was also detected in human intestinal tissue.

## Related Peptidases

Several enzymes with *N*-acyl-aminoacyl peptidase activities have been purified from sheep (Witheiler &.Wilson, 1972), rabbit (Yoshida & Lin, 1972), human erythrocytes (Jones & Manning, 1988), rat liver (Suda *et al.*, 1980) and skeletal muscle (Radhakrishna & Wold, 1986). The reported specificities suggest there are at least two different enzymes. The first can be described as an α-*N*-acyl-aminoacyl peptide hydrolase; this enzyme has a broad specificity for acyl groups (acetyl > formyl > propionyl > butyl) and for the first amino acid (Kobayashi & Smith, 1987; Jones & Manning, 1988). Alanine is the preferred substrate but glycine, serine, methionine and phenylalanine are all cleaved. This is a serine peptidase of family S9, described in Chapter 130. The second enzyme cleaves acetyl or formyl methionine from short peptides but has no activity against other N-terminally blocked peptides (Suda *et al.*, 1980; Radhakrishna & Wold,

1986). These characteristics are similar to those of the *N*-fMet-peptidase from rat intestinal mucosa.

## References

Broom, M.F., Sherriff, R.M., Dunster, D. & Chadwick, V.S. (1992) Identification of formyl Met-Leu-Phe in culture filtrates of *Helicobactor pylori*. *Microbios* **72**, 239–245.

Chadwick, V.S., Mellor, D.M., Myers, D.B., Selden, A.C., Keshavarzian, A., Broom, M.F. & Hobson, C.H. (1988) Production of peptides inducing chemotaxis and lysosomal enzyme release in human neutrophils by intestinal bacteria *in vitro* and *in vivo*. *Scand. J. Gastroenterol.* **23**, 121–128.

Jones, W.M. & Manning, J.M. (1988) Substrate specificity of an acylaminopeptidase that catalyzes the cleavage of the blocked amino termini of peptides. *Biochim. Biophys. Acta* **953**, 357–560.

Kobayashi, K & Smith, J.A. (1987) Acyl-peptide hydrolase from rat liver. Characterization of enzyme reaction. *J. Biol. Chem.* **262**, 11435–11445.

Radhakrishna, G. & Wold, F. (1986) Rabbit muscle extracts catalyze the specific removal of *N*-acetylmethionine from acetylated peptides. *J. Biol. Chem.* **261**, 9572–9575.

Sherriff, R.M., Broom, M.F. & Chadwick, V.S. (1992) Isolation and purification of N-formyl methionine aminopeptidase from rat intestine. *Biochim. Biophys. Acta* **1119**, 275–280.

Showell, H.J., Freer, R.J., Zigmond, S.H., Schiffman, E., Aswanikumar, S., Corcoran, B. & Becker, E.L. (1976) The structure-activity relations of synthetic peptides as chemotactic factors and inducers of lysosomal enzyme secretion for neutrophils. *J. Exp. Med.* **143**, 1154–1169.

Suda, H., Yamamoto, K., Aoyagi, T. & Umezawa, H. (1980) Purification and properties of *N*-formylmethionine aminopeptidase from rat liver. *Biochim. Biophys. Acta* **616**, 60–67.

Witheiler, J. & Wilson, D.B. (1972) The purification and characterization of a novel peptidase from sheep red cells. *J. Biol. Chem.* **247**, 2217–2221.

Yoshida, A. & Lin, M. (1972) NH₂-Terminal formylmethionine and NH₂-terminal methionine-cleaving enzymes in rabbits. *J. Biol. Chem.* **247**, 952–957.

*Murray Broom*
*Molecular Biology Unit,*
*Biochemistry Department,*
*University of Otago,*
*Dunedin, New Zealand*
*Email: Murray.Broom@stonebow.otago.ac.nz*

# 551. β-Aspartyl-peptidase

## Databanks

*Peptidase classification: clan UX, family U99, MEROPS ID: U9F.003*
*NC-IUBMB enzyme classification: EC 3.4.19.5*
*Databank codes: no sequence data available*

## Name and History

Studies by Dorer *et al*. (1968) revealed extensive metabolism of β-aspartyl peptides after their administration to rats. These findings stimulated a search for the enzyme with hydrolytic activity toward β-aspartyl peptides. A partially purified preparation of an enzyme termed **β-aspartyl-peptidase** was obtained from the cytosolic fraction of a rat liver homogenate. A similar activity was partially purified from rat kidney.

## Activity and Specificity

Hydrolysis of β-aspartyl peptides was determined by following the release of glycine from β-aspartylglycine by a ninhydrin method. The product was separated from the substrate by sequential chromatography on columns of Dowex 50X4 and Dowex 2X8.

Cleavage of β-aspartylglycine proceeds optimally at pH 7.5–8.0 in sodium phosphate buffer and between pH 7 and 9 in Tris–HCl buffer. A wide series of β-aspartyl peptides is cleaved with highest activity toward β-Asp╪Gly and β-Asp╪Gly-Gly. Some hydrolysis of other peptides was attributed to the presence of contaminating enzymes in the preparation, but there was little or no activity toward Ac-Gly, β-Ala-Gly or γ-Glu-Leu.

## Preparation

The enzyme was partially purified from a $105\,000 \times g$ supernatant of rat liver homogenate. An initial 40–55% ammonium sulfate fractionationation step was followed by a 5 min heat inactivation at 60°C. The resulting supernatant was subjected to sequential chromatography on columns of CM-cellulose and hydroxyapatite.

## Biological Aspects

The hydrolysis of β-Asp-Gly was detected in homogenates of rat kidney, brain, lung, skeletal muscle and heart muscle in addition to rat liver. The rat kidney enzyme was also partially purified and exhibits a similar specificity to the rat liver enzyme. It is likely that this enzyme is responsible for the reported *in vivo* metabolism of β-aspartyl peptides.

## Distinguishing Features

The enzyme is inactivated when dialyzed against Tris–HCl. Addition of NaCl to the dialyzed enzyme does not restore activity, but if NaCl is added to the Tris–HCl buffer prior to dialysis, inactivation is prevented. Maximal enzymatic activity requires either $Na^+$ or $K^+$.

The β-aspartyl-peptidase, which is so far known only from mammalian sources (Haley, 1970), should be distinguished from an enzyme hydrolyzing β-aspartyl dipeptides that was partially purified from *E. coli* by Haley (1968) and was shown more recently (Gary & Clarke, 1995) to be a zinc metallopeptidase. This enzyme does not hydrolyze tripeptides and is therefore called β-aspartyl-dipeptidase or isoaspartyl-dipeptidase (Chapter 502).

## References

Dorer, F.E., Haley, E.E. & Buchanan, D.L. (1968) The hydrolysis of -aspartylpeptides in rat tissue. *Arch. Biochem. Biophys.* **127**, 490–495.

Gary, J.D. & Clarke, S. (1995) Purification and characterization of an isoaspartyl dipeptidase from *Escherichia coli*. *J. Biol. Chem.* **270**, 4076–4087.

Haley, E.E. (1968) Purification and properties of a β-aspartyl peptidase from *Escherichia coli*. *J. Biol. Chem.* **243**, 5748–5752.

Haley, E.E. (1970) β-Aspartyl peptidase from rat liver. *Methods Enzymol.* **19**, 737–741.

*Sherwin Wilk*
*Department of Pharmacology,*
*Mount Sinai School of Medicine, One Gustave I. Levy Place,*
*New York, NY 10029, USA*
*Email: S_Wilk@smtplink.mssm.edu*

# 552. Destabilase

## Databanks

Peptidase classification: none
NC-IUBMB enzyme classification: none
Databank codes:

| Species | SwissProt | PIR | EMBL (cDNA) | EMBL (genomic) |
|---|---|---|---|---|
| Destabilase I | | | | |
| *Hirudo medicinalis* | – | – | U24122 | – |

*continued overleaf*

| Species | SwissProt | PIR | EMBL (cDNA) | EMBL (genomic) |
|---|---|---|---|---|
| Destabilase II *Hirudo medicinalis* | – | – | U24121 | – |
| F22A3.6 protein *Caenorhabditis elegans* | – | – | – | U41547: cosmid F22A3 |

## Name and History

The name **destabilase** was given to a highly specific enzyme capable of solubilizing stabilized insoluble fibrin cross-linked by $\varepsilon$-($\gamma$-Glu)-Lys isopeptide bonds. The enzyme was first discovered in salivary gland secretion of the medicinal leech *Hirudo medicinalis* (Baskova & Nikonov, 1985). Later it was isolated from water extracts of the medicinal leech.

## Activity and Specificity

Destabilase does not catalyze hydrolysis of peptide bonds in fibrinogen, fibrin or other proteins. However, it splits $\varepsilon$-($\gamma$-Glu)-Lys isopeptide bonds in cross-linked fibrin and in its fragment, D-D dimer, which also contains such isopeptide bonds linking two D-monomers. The identity of the N-terminal amino acid sequences of $\gamma$ chains prior to and after the incubation of D-D dimers, and the absence of the new N-termini confirm the splitting of isopeptide bonds by destabilase (Zavalova *et al*., 1993). A weak amidolytic activity of destabilase was detected using $\gamma$-Glu-NHPhNO$_2$ as substrate (Baskova *et al*., 1990).

The catalytic activity of destabilase was estimated using PAGE, measuring the progress of D-D dimer monomerization during incubation with the enzyme at 37°C, pH 7.4. The incubation of destabilase in the presence of SDS or other detergents was accompanied by the loss of specific D-D isopeptidase activity and acquisition of pronounced proteolytic activity towards D-D dimer and other proteins (Zavalova *et al*., 1994). The enzyme is inhibited by PMSF and to some extent also by iodoacetamide.

## Structural Chemistry

Destabilase ($M_r$ 12 300) is composed of a single peptide chain of 115 amino acid residues of known sequence, including 14 cysteine residues (Fradkov *et al*., 1996).

## Preparation

Destabilase has been purified 2500-fold from the whole aqueous extract of the leech. Nucleic acids were removed from the extract by streptomycin sulfate precipitation, and the supernatant was fractionated with ammonium sulfate (40–60% saturation). The resulting protein precipitate was dissolved in Tris–HCl buffer, pH 7.5, and further fractionated by ion-exchange chromatography on DEAE-Toyopearl and chromatography on CM-Toyopearl. The eluted protein was finally purified by gel filtration through Superose-12 HR (Zavalova *et al*., 1996).

The enzyme was successfully expressed in *Spodoptera frugiperda* cells with the baculovirus system, as confirmed by detection of the destabilase activity in the cell extracts (Zavalova *et al*., 1996).

## Biological Aspects

Three full-length cDNAs (*Ds1, Ds2* and *Ds3*) were cloned and sequenced. They form the destabilase gene family that includes three different proteins. The protein purified from the secretion of the medicinal leech is encoded by *Ds3* form cDNA. All the three cDNAs contained open reading frames (of 345 bp length for *Ds1* and *Ds3*, and 348 bp for *Ds2* form) The deduced amino acid sequences consisted of 135 residues for *Ds1* and *Ds3*, and 136 residues for *Ds2*, including signal peptides of 20 amino acids. Although the signal peptide sequence of *Ds2* differed from those of *Ds1* and *Ds3*, all of them possessed the major features for protein transport into the endoplasmic reticulum. The mature proteins encoded by corresponding mRNAs are remarkable in their high cysteine contents (14 out of 115 or 116 amino acids), the positions of the Cys residues being the same in all three deduced amino acid sequences. A computer analysis of the proteins revealed three very hydrophobic and conservative regions at positions 14–20, 53–65 and 97–105 of the mature polypeptides. We suggest that these are directly involved in the catalytic mechanism of the enzyme (Zavalova *et al*., 1996; Fradkov *et al*., 1996).

## Distinguishing Features

Rabbit polyclonal antibodies have been produced against the *Ds2* and *Ds3* forms of destabilase (unpublished results of the authors).

## References

Baskova, I.P. & Nikonov, G.I. (1985) Destabilase – an enzyme of the medicinal leech salivary gland secretion hydrolyses isopeptide bonds in stabilized fibrin. *Biokhimia (USSR)* **50**, 424–431.

Baskova, I.P., Nikonov, G.I., Zavalova, L. & Larionova, N.I. (1990) Kinetics of hydrolysis of Glu-pNA by destabilase, the enzyme from the leech *Hirudo medicinalis*. *Biokhimia (USSR)* **55**, 674–679.

Fradkov, A., Berezhnoy, S., Barsova, E., Zavalova, L., Lukyanov, S., Baskova, I. & Sverdlov, E. (1996) Enzyme from the medicinal leech (*Hirudo medicinalis*) that splits endo-$\varepsilon$-($\gamma$-Glu)-Lys isopeptide bonds: cDNA cloning and protein primary structure. *FEBS Lett.* **390**, 145–148.

Zavalova, L.L., Kuzina, E.V., Levina, N.B. & Baskova, I.P. (1993) Monomerization of fragment D-D by destabilase from the medicinal leech does not alter the N-terminal sequence of the gamma chain. *Thromb. Res.* **71**, 241–244.

Zavalova, L.L., Kuzina, E.V. & Baskova, I.P. (1994) The appearance of non-specific proteolytic activity in destabilase, ε-(γ-Glu)-Lys isopeptidase, in the presence of SDS. *Biokhimia (USSR)* **59**, 905–910.

Zavalova, L., Lukyanov, S., Baskova, I., Snezhkov, E., Akopov, S., Berezhnoy, S., Bogdanova, E., Barsova, E. & Sverdlov, E. (1996) Genes from the medicinal leech (*Hirudo medicinalis*) coding for unusual enzymes that specifically cleave endo-ε-(γ-Glu)-Lys isopeptide bonds and help to dissolve blood clots. *Mol. Gen. Genet.* **253**, 20–25.

*Ludmila L. Zavalova*
*Shemyakin Institute*
*of Bioorganic Chemistry,*
*Miklukho-Maklaya Str., 16/10,*
*117871 Moscow, Russia*
*Email: leech@humge.siobc.ras.ru*

*Isolda P. Baskova*
*Laboratory of Blood Coagulation,*
*Biological Department,*
*Moscow State University,*
*119899 Moscow, Russia*

*Eugene D. Sverdlov*
*Shemyakin Institute*
*of Bioorganic Chemistry,*
*Miklukho-Miklukho-Maklaya Str., 16/10,*
*117871 Moscow, Russia*

# 553. LD-Dipeptidase

## Databanks

*Peptidase classification: clan UX, family U99, MEROPS ID: U9B.004*
*NC-IUBMB enzyme classification: none*
*Databank codes: no sequence data available*

## Name and History

In *Bacillus sphaericus* a peptidase was found to hydrolyze dipeptides with a free N-terminal L-amino acid and a free C-terminal D-amino acid (or Gly). In consequence, the enzyme was termed **LD-dipeptidase** (Vacheron *et al.*, 1978).

## Activity and Specificity

Assays were made with either commercial dipeptides or labeled dipeptides prepared by condensing the Boc-L-amino acid with the D-amino acid methyl ester (Ciroussel *et al.*, 1990). In addition, the dipeptide *ms*-A$_2$pm(L)-D-[$^{14}$C]Ala (*ms*-A$_2$pm: *meso*-diaminopimelate) was derived from a peptidoglycan precursor (Guinand *et al.*, 1974; Arminjon *et al.*, 1977). With radioactive peptides, the reaction products were separated by cellulose thin-layer chromatography and quantified by radioactivity measurements. With unlabeled peptides, free amino acids were detected by a colorimetric assay (Setlow, 1975).

Specific activity (μmol min$^{-1}$ mg$^{-1}$) decreases through a range of substrates in the order: L-Lys┼Gly (15.3), L-Lys┼D-Ala (11), *ms*-A$_2$pm(L)┼D-Ala (8), L-Ala┼D-Ala (7), L-Leu┼Gly (5), L-Lys┼D-Glu (3.5), L-Ala┼Gly (2.3) and L-Ala┼D-Glu (0.85). Thus LD-dipeptidase has a wide range of specificity. Moreover, its acts on positively or negatively charged dipeptides such as L-Ala-D-Glu or L-Lys-D-Ala. Activity is stimulated by 10$^{-3}$ M Co$^{2+}$; the pH optimum is about 8.

## Structural Chemistry

LD-Dipeptidase is a single-chain protein of molecular mass 38 kDa derived from SDS-PAGE (Ciroussel *et al.*, 1990). Sequence data are not presently available.

## Preparation

The enzyme has been purified 250-fold from the sporulation medium of *Bacillus sphaericus* (Ciroussel *et al.*, 1990).

## Biological Aspects

The D-amino acids are essentially the constituents of the bacterial peptidoglycan. The LD-dipeptidase is located in the mother cell and the spore cytoplasm. Its function is presumably to salvage the amino acids from the dipeptides *ms*-A$_2$pm-D-Ala, L-Lys-D-Ala and L-Ala-D-Glu released from the peptidoglycan by the coordinated action of several other peptidoglycan hydrolases (see Chapters 265 and 457). It is different from the L-lysyl-D-alanine carboxypeptidase whose activity on tetrapeptides and glycotetrapeptides is discussed in Guinand *et al.* (1974), Arminjon *et al.* (1977), Tipper *et al.* (1977) and Vacheron *et al.* (1977) (see also Chapter 555). The specificity has some similarity to that of non-stereospecific dipeptidase isolated from molluscan tissues (Chapter 547).

U

## References

Arminjon, F., Guinand, M., Vacheron, M.J. & Michel, G. (1977) Specificity profiles of the membrane-bound γ-D-glutamyl-(L)*meso*-diaminopimelate endopeptidase and LD-carboxy peptidase from *Bacillus sphaericus* 9602. *Eur. J. Biochem.* **73**, 557–565.

Ciroussel, F., Vacheron, M.J., Guinand, M. & Michel, G. (1990) Purification and properties of an LD-dipeptidase from *Bacillus sphaericus*. *Int. J. Biochem.* **22**, 525–532.

Guinand, M., Michel, G. & Tipper D.J. (1974) Appearance of a γ-D-glutamyl-(L)*meso*-diaminopimelate peptidoglycan hydrolase during sporulation in *Bacillus sphaericus. J. Bacteriol.* **120**, 173–184.

Setlow, P. (1975) Protease and peptidase activities in growing and sporulating cells and dormant spores of *Bacillus megaterium. J.* *Bacteriol.* **122**, 642–649.

Tipper, D.J., Pratt, I., Guinand, M., Holt, S.C. & Linnett, P.E. (1977) Control of peptidoglycan synthesis during sporulation in *Bacillus sphaericus. Microbiology* **3**, 50–68.

Vacheron, M.J., Guinand, M., Arminjon, F. & Michel, G. (1977) Mise en évidence et séparation d'une γ-D-glutamyl-(L)*meso*-diaminopimelate endopeptidase et d'une LD-carboxypeptidase exocellulaires à partir de *Bacillus sphaericus* 9602. *Biochimie* **59**, 15–21.

Vacheron, M.J., Guinand, M. & Michel, G. (1978) Mise en évidence au cours de la sporulation de *Bacillus sphaericus* d'une activité di-peptidasique sur des substrats constituants du peptidoglycane. *FEMS Microbiol. Lett.* **3**, 71–75.

*Micheline Guinand*
*Laboratoire de Biochimie Analytique,*
*Université Claude Bernard Lyon 1,*
*43 Boulevard du 11 Novembre 1918,*
*F-69622 Villeurbanne Cedex, France*

# 554. D-Ala-D-Ala *carboxypeptidase VanY*

## Databanks

*Peptidase classification: clan UX, family U26, MEROPS ID: U26.001*
*NC-IUBMB enzyme classification: none*
*Databank codes:*

| Species | SwissProt | PIR | EMBL (cDNA) | EMBL (genomic) |
|---|---|---|---|---|
| *Enterococcus faecalis* | – | – | U35369 | – |
| *Enterococcus faecium* | P37711 | JC1427 | M90647 | – |
| | | | M97297 | |

## Name and History

Resistance to the glycopeptide antibiotics vancomycin and teicoplanin in isolates of enterococci had been demonstrated to be correlated with increased DD-carboxypeptidase activity (Al-Obeid *et al.*, 1990; Gutmann *et al.*, 1992). Cloning of the five genes located on plasmid pIP816 which are necessary and sufficient for high-level vancomycin resistance in *Enterococcus faecium* BM4147 followed by the determination of the function of the corresponding proteins did not reveal a protein with DD-carboxypeptidase activity (see Chapter 503). However, further cloning of additional clustered genes revealed an open reading frame termed *vanY*, so named because it was adjacent (3′) to the gene encoding the dipeptidase VanX (Chapter 503) (Arthur *et al.*, 1992). Functional analysis revealed that VanY is a DD-carboxypeptidase (Wright *et al.*, 1992).

## Activity and Specificity

VanY has been demonstrated to cleave the C-terminal D-Ala from *N,N*-diacetyl-L-Lys-D-Ala+D-Ala and the peptidoglycan precursors UDP-MurNAc-L-Ala-D-Glu-mDAP-D-Ala+D-Ala and UDP-MurNAc-L-Ala-D-Glu-mDAP-D-Ala+D-Lac (in which UDP-MurNAc is undecaprenylpyrophosphatidyl-*N*-acetyl-muramoyl) (Wright *et al.*, 1992). The enzyme is insensitive to β-lactam antibiotics and does not possess β-lactamase or cross-linking transpeptidase activities.

VanY activity was assayed by titration with *o*-phthaldial-dehyde for fluorometric assay of enzymatically released D-Ala and by an HPLC assay which monitors conversion of UDP-MurNAc-L-Ala-D-Glu-mDAP-D-Ala+D-Ala (or D-Lac) to the tetrapeptide UDP-MurNAc-L-Ala-D-Glu-mDAP-D-Ala (Wright *et al*., 1992). Alternatively, the release of D-Ala can be measured by amino acid analysis (Gutmann *et al*., 1992).

## Structural Chemistry

The 909 bp *vanY* gene encodes a 303 residue protein of 34 980 Da. The predicted amino acid sequence does not contain the SXXK motif common to serine-type DD-carboxypeptidases (Chapter 141), consistent with the lack of reactivity with $\beta$-lactam antibiotics. There is also no significant sequence similarity with the zinc-dependent DD-carboxypeptidase (Chapter 463) from *Streptomyces albus*. Thus VanY appears to belong to a novel class of DD-carboxypeptidases.

## Preparation

The N-terminus of VanY is composed primarily of hydrophobic residues (MKKLFFLLLLLFLIYLGYDY), concordant with the observation that much of the activity can be released by detergent extraction of membranes obtained from *Escherichia coli* expressing plasmid-borne VanY (Wright *et al*., 1992). Membrane extracts of *E. coli* expressing VanY show a protein of approximately 35 kDa that is not apparent in cytosolic fractions. Purification to homogeneity has not been reported.

## Biological Aspects

The association of inducible DD-carboxypeptidase activity with vancomycin resistance in enterococci suggested an intimate relationship between this activity and drug resistance (Al-Obeid *et al*., 1990; Gutmann *et al*., 1992). The realization that VanY was not required for high-level vancomycin resistance but that the *vanY* gene was located immediately downstream from *vanX* and under control of the positively regulated PvanH promoter (Arthur *et al*., 1993; Holman *et al*., 1994) suggested a more subtle association between VanY and the vancomycin resistance phenotype. The presence of VanY does increase the tolerance to vancomycin under conditions where significant D-Ala is present in the medium or when the accumulation of peptidoglycan precursors terminating in D-Ala-D-Ala reaches a critical level (Arthur *et al*., 1994). These results demonstrate that the presence of VanY serves to contribute to high-level vancomycin resistance in only a minor fashion, but can play a role in drug resistance under some conditions. The demonstration that all the vancomycin resistance genes, including *vanY*, are located on transposable elements (e.g. Tn*1546*) (Arthur *et al*., 1993),

suggests the real possibility that VanY may eventually be found in a context where its role in drug resistance is more significant.

## Distinguishing Features

The inhibitor-sensitivities VanY DD-carboxypeptidase are distinct from those of the serine- and zinc-dependent DD-carboxypeptidases.

## Further Reading

For reviews of vancomycin resistance and the roles of VanY see Walsh *et al*. (1996) and Arthur & Courvalin (1993).

## References

Al-Obeid, S., Collatz, E. & Gutmann, L. (1990) Mechanism of resistance to vancomycin in *Enterococcus faecium* D366 and *Enterococcus faecalis* A256. *Antimicrob. Agents Chemother.* **34**, 252–256.

Arthur, M. & Courvalin, P. (1993) Genetics and mechanisms of glycopeptide resistance in enterococci. *Antimicrob. Agents Chemother.* **37**, 1563–1571.

Arthur, M., Molinas, C. & Courvalin, P. (1992) Sequence of the *vanY* gene required for production of a vancomycin-inducible DD-carboxypeptidase in *Enterococcus faecium* BM4147. *Gene* **120**, 111–114.

Arthur, M., Molinas, C., Depardieu, F. & Courvalin, P. (1993) Characterization of Tn*1546*, an Tn*3*-related transposon conferring glycopeptide resistance by synthesis of depsipeptide peptidoglycan precursors in in *Enterococcus faecium* BM4147. *J. Bacteriol.* **175**, 117–127.

Arthur, M., Depardieu, F., Snaith, H.H., Reynolds, P.P. & Courvalin, P. (1994) Contribution of VanY DD-carboxypeptidase to glycopeptide resistance in *Enterococcus faecalis* by hydrolysis of peptidoglycan precursors. *Antimicrob. Agents Chemother.* **38**, 1899–1903.

Gutmann, L., Billot-Klein, D., Al-Obeid, S., Klare, I., Francoual, S., Collatz, E. & van Heijenoort, J. (1992) Inducible carboxypeptidase activity in vancomycin-resistant enterococci. *Antimicrob. Agents Chemother.* **36**, 77–80.

Holman, T.T., Wu, Z., Wanner, B.B. & Walsh, C.C. (1994) Identification of the DNA binding site for phosphorylated VanR protein required for vancomycin resistance in *Enterococccus faecium*. *Biochemistry* **33**, 4625–4631.

Walsh, C.C., Fisher, S.S., Park, I.-S., Prohalad, M. & Wu, Z. (1996) Bacterial resistance to vancomycin: five genes and one missing hydrogen bond tell the story. *Chem. Biol.* **3**, 21–28.

Wright, G.G., Molinas, C., Arthur, M., Courvalin, P. & Walsh, C.T. (1992) Characterization of VanY, a DD-carboxypeptidase from vancomycin-resistant *Enterococcus faecium* BM4147. *Antimicrob. Agents Chemother.* **36**, 1514–1518.

***G.D. Wright***
*Department of Biochemistry,*
*McMaster University,*
*1200 Main St. W.,*
*Hamilton, Ontario, L8N 3Z5 Canada*
*Email: wrightge@fhs.csu.mcmaster.ca*

# 555. *Murein tetrapeptide* LD-*carboxypeptidases*

## Databanks

*Peptidase classification: clan UX, family U99, MEROPS ID: U9E.003*
*NC-IUBMB enzyme classification: EC 3.4.17.13*
*Databank codes: no sequence data available*

## Name and History

Several bacterial muramoyl-carboxypeptidases have been described. These enzymes cleave the peptide bond between the dibasic amino acid (either L-lysine or *meso*-diaminopimelic acid, *m*-$A_2$pm) and the C-terminal D-alanine in the tetrapeptide moiety of the murein (peptidoglycan) or in the activated precursors, the undecaprenyl pyrophosphate (UDP) muropeptides. The tetrapeptide is formed by the action of the penicillin-sensitive DD-carboxypeptidases of serine (Chapter 141) or zinc (Chapter 463) type, which are known to specifically hydrolyze the terminal D-Ala┼D-Ala peptide bond of the pentapeptide side chains present in newly synthesized murein (Rogers *et al.*, 1980; Ghuysen & Hakenbeck, 1994). The tetrapeptides can be further degraded to tripeptide side chains in a reaction in which an L-amino acyl┼D-amino acid bond is split. This hydrolysis is catalyzed by carboxypeptidases, first described as **carboxypeptidase II** (Izaki & Strominger, 1968) and later renamed LD-**carboxypeptidase**. With the biochemical characterization of several LD-carboxypeptidases from different bacteria, it became clear that at least two types of this specificity exist, one acting on the activated murein precursors (activity I) and one acting on nascent murein (activity II) (Metz *et al.*, 1986b).

Some of the enzymes possessing LD-carboxypeptidase activity are not only able to hydrolyze the peptide bond, but in an amino acid exchange reaction they can also replace the terminal D-Ala with another D-amino acid, therefore acting as LD-transpeptidases. It is possible that the carboxypeptidase activity found with these enzymes is only the result of a side reaction in which a water molecule rather than an amino acid functions as the acceptor (Caparros *et al.*, 1993).

Despite the fact that the presence of LD-carboxypeptidases has been demonstrated in a number of bacteria and that an important function during growth and division has been proposed for this class of bacterial carboxypeptidases (Beck & Park, 1976; Begg *et al.*, 1990; Höltje, 1996) they remain relatively poorly characterized. Several different proteins possessing LD-carboxypeptidase activity have been isolated, but for none of them could the gene be identified and no sequences have been obtained.

## Activity and Specificity

The LD-carboxypeptidases act on the tetrapeptide moieties of a nucleotide-activated precursor of murein synthesis such as UDP-*N*-acetylmuramoyl-L-Ala-D-Glu-*m*-$A_2$pm┼D-Ala, the lipid-linked precursor (undecaprenylpyrophosphatidyl-*N*-acetyl-muramoyl-L-Ala-D-Glu-*m*-$A_2$pm┼D-Ala), or the tetrapeptide found in the murein sacculus (-L-Ala-D-Glu-*m*-$A_2$pm┼D-Ala). In addition, the breakdown products of the murein, the muropeptides containing the L-Ala-D-Glu-*m*-$A_2$pm┼D-Ala peptidyl side chains, are also cleaved by some of the LD-carboxypeptidases. In the case where *meso*-diaminopimelic acid rather than L-Lys is present as the dibasic amino acid, the peptide bond hydrolyzed is the one that links the carboxyl group of the L-center of the *meso*-diaminopimelic acid with the amino group of the D-Ala.

The differences in substrate requirement are reflected in the biochemical properties of the known activities. At least six enzymes have been described and all of them differ in certain aspects. An LD-carboxypeptidase from *Proteus vulgaris* ($M_r$ 32 000, pH optimum 8.9) acts on UDP-*N*-acetylmuramoyl-tetrapeptide; other substrates have not been tested (Rousset *et al.*, 1982). The enzyme from *Streptococcus faecalis* (pH optimum 6) does not recognize the activated precursor, but acts on degradation products of the sacculus (*N*-acetylmuramoyl-tetrapeptide); its activity on whole sacculi has not been tested, but it is able to catalyse the amino acid exchange reaction (Coyette *et al.*, 1974). The LD-carboxypeptidase isolated from *Bacillus megaterium* ($M_r$ 60 000, pH optimum 8) catalyzing the exchange reaction, accepts whole sacculi as well as murein degradation products as substrate, but is not active on the murein precursors (DasGupta & Fan, 1979). LD-carboxypeptidase activity has also been demonstrated in membrane preparations of *Gaffkya homari* (Hammes & Seidel, 1978a,b).

In *Escherichia coli*, several LD-carboxypeptidases have been described. Izaki & Strominger (1968) used an enriched fraction of the protein (called carboxypeptidase II) for an initial determination of its properties. Activity on UDP-*N*-acetylmuramoyl-tetrapeptide and on breakdown products could be demonstrated. Purification of an enzyme acting on UDP-*N*-acetylmuramoyl-tetrapeptide demonstrated the presence of a 43 kDa protein, pH optimum 8, which is able to dimerize (Beck & Park, 1977). Experimental evidence for two types of LD-carboxypeptidases was obtained by Metz *et al.* (1986a) and the terms carboxypeptidase I (for enzymes acting on murein precursors) and carboxypeptidase II (for enzymes acting on murein sacculi or murein degradation products) were introduced. These authors purified an enzyme ($M_r$ 12 000, pH optimum 8.5) which possesses type I activity

(Metz *et al*., 1986b). Another enzyme, purified by affinity chromatography on nocardicin A (Ursinus *et al*., 1992) possesses type I activity; later it could be shown to act also on soluble murein degradation products, while not being active on whole sacculi (Leguina *et al*., 1994).

No one set of reaction conditions can be described for assay of the activities of all of the LD-carboxypeptidases because of their different properties. For the preparation of substrate also different methods have been employed. A useful method for the isolation of UDP-*N*-acetylmuramoyl-tetrapeptide employs HPLC (Kohlrausch & Höltje, 1991). This precursor can be radiolabeled and the reaction products can be separated by paper chromatography (Beck & Park, 1976). Alternatively, an HPLC system as described by Ursinus *et al*. (1992) can be used to quantitate substrate and reaction products. When using murein sacculi or murein degradation products as substrate, the methods used by Leguina and coworkers (1994) are recommended.

A method using $S$-[$^3$H]methyl-D-cysteine has been established to specifically assay for LD-transpeptidase activity (Caparros *et al*., 1992).

Interestingly, different inhibitors of LD-carboxypeptidase activity have been described. While penicillin G, a potent inhibitor of DD-carboxypeptidases, has no effect on this class of enzymes, some other $\beta$-lactam antibiotics do inhibit enzymatic activity (Hammes & Seidel, 1978a,b). Thienamycin, one of the penem antibiotics, inhibits the 12 kDa enzyme from *E. coli* (Metz *et al*., 1986b) and an enzyme found in *G. homari*. Nocardicin A, a $\beta$-lactam containing a D-amino acid, has been shown to be inhibitory for two of the *E. coli* enzymes (Metz *et al*., 1986a; Ursinus *et al*., 1992). D-Amino acids (e.g. D-methionine) have not only been shown to be substrates for the amino acid exchange reaction (see above), but are also inhibitory for different LD-carboxypeptidases (Hammes, 1978; Caparros *et al*., 1993).

## Structural Chemistry

For most of the known LD-carboxypeptidases only an initial biochemical characterization has been performed, so knowledge of structural features is limited. Besides the molecular masses of the above-mentioned proteins (*P. vulgaris*, 32 kDa; *B. megaterium*, 60 kDa; *E. coli*, 12 kDa, 32 kDa, 43 kDa) only the isoelectric point (pI 5.5) of the 32 kDa protein from *E. coli* is known (Ursinus *et al*., 1992).

## Preparation

Various methods have been described for purifying the LD-carboxypeptidases from different bacterial species. The only protein for which an essentially homogeneous preparation (purification factor of 1000) could be obtained is the 32 kDa LD-carboxypeptidase from *E. coli* (Ursinus *et al*., 1992).

## Biological Aspects

Although present in several bacteria (*Proteus vulgaris, Streptococcus faecalis, Bacillus megaterium, Gaffkya homari, Escherichia coli*), LD-carboxypeptidase activity could not be detected in *Caulobacter crescentus* (Markiewicz *et al*., 1983).

The different substrate specificities (i.e. acting on murein precursors or acting on whole murein sacculi or their degradation products) suggest that different enzymes with distinct functions exist. It has been speculated that enzymes acting on the cytoplasmic intermediate may have a function in controlling the cell cycle (Beck & Park, 1976, 1977; Höltje, 1996). It is proposed that by providing the cells with tripeptide precursor molecules, the preferred acceptor for the transpeptidation leading to septum formation, cell division is triggered (Begg *et al*., 1990).

The formation of tripeptide side chains in the growing murein sacculus of *E. coli* has been shown to be catalyzed by a periplasmic enzyme. This conversion of tetrapeptides to tripeptides is a late event during maturation of the murein as shown by pulse-chase experiments (Glauner & Höltje, 1990).

## Further Reading

The articles of Ghuysen & Hakenbeck (1994) and Rogers *et al*. (1980) are recommended.

## References

Beck, B.D. & Park, J.T. (1976) Activity of three murein hydrolases during the cell division cycle of *Escherichia coli* K-12 as measured in toluene-treated cells. *J. Bacteriol.* **126**, 1250–1260.

Beck, B.D. & Park, J.T. (1977) Basis for the observed fluctuation of carboxypeptidase II activity during the cell cycle in BUG 6, a temperature-sensitive division mutant of *Escherichia coli*. *J. Bacteriol.* **130**, 1292–1302.

Begg, K.J., Takasuga, A., Edwards, D.H., Dewar, S.J., Spratt, B.G., Adachi, H., Ohta, T., Matsuzawa, H. & Donachie, W.D. (1990) The balance between different peptidoglycan precursors determines whether *Escherichia coli* cells will elongate or divide. *J. Bacteriol.* **172**, 6697–6703.

Caparros, M., Aran, V. & de Pedro, M.A. (1992) Incorporation of $S$-[$^3$H]methyl-D-cysteine into the peptidoglycan of ether-treated cells of *Escherichia coli*. *FEMS Microbiol. Lett.* **72**, 139–146.

Caparros, M., Quintela, J.C. & de Pedro, M.A. (1993) Amino acids as useful tools in the study of murein metabolism in *Escherichia coli*. In: *Bacterial Growth and Lysis – Metabolism and Structure of the Bacterial Sacculus*, vol. 65 (de Pedro, M.A., Höltje, J.-V. & Löffelhardt, W., eds). New York-London: Plenum, pp. 147–160.

Coyette, J., Perkins, H.R., Polacheck, I., Shockman, G.D. & Ghuysen, J.-M. (1974) Membrane-bound DD-carboxypeptidase and LD-transpeptidase of *Streptococcus faecalis* ATCC 9790. *Eur. J. Biochem.* **44**, 459–468.

DasGupta, H. & Fan, D.P. (1979) Purification and characterization of a carboxypeptidase-transpeptidase of *Bacillus megaterium* acting on the tetrapeptide moiety of the peptidoglycan. *J. Biol. Chem.* **254**, 5672–5683.

Glauner, B. & Höltje, J.-V. (1990) Growth pattern of the murein sacculus of *Escherichia coli*. *J. Biol. Chem.* **265**, 18988–18996.

Ghuysen, J.-M. & Hakenbeck, R. (1994) Bacterial cell wall. In: *New Comprehensive Biochemistry*, vol. 27 (Neuberger, A. & van Deenen, L.L.M., eds). Amsterdam: Elsevier.

Hammes, W.P. (1978) The LD-carboxypeptidase activity in *Gaffkya homari*. The target of the action of D-amino acids or glycine on the formation of wall-bound peptidoglycan. *Eur. J. Biochem.* **91**, 501–507.

Hammes, W.P. & Seidel, H. (1978a) The activities in vitro of DD-carboxypeptidase and LD-carboxypeptidase of *Gaffkya homari* during biosynthesis of peptidoglycan. *Eur. J. Biochem.* **84**, 141–147.

Hammes, W.P. & Seidel, H. (1978b) The LD-carboxypeptidase activity in *Gaffkya homari*. The target of the action of certain beta–lactam antibiotics on the formation of wall-bound peptidoglycan. *Eur. J. Biochem.* **91**, 509–515.

Höltje, J.-V. (1996) A hypothetical holoenzyme involved in the replication of the murein sacculus of *Escherichia coli*. *Microbiology* **142**, 1911–1918.

Izaki, K. & Strominger, J.L. (1968) Biosynthesis of the peptidoglycan of bacterial cell walls. XIV. Purification and properties of two D-alanine carboxypeptidases from *Escherichia coli*. *J. Biol. Chem.* **243**, 3193–3201.

Kohlrausch, U. & Höltje, J.-V. (1991) One-step purification procedure for UDP-*N*-acetylmuramyl-peptide murein precursors from *Bacillus cereus*. *FEMS Microbiol. Lett.* **78**, 253–258.

Leguina, J.I., Quintela, J.C. & de Pedro, M.A. (1994) Substrate specificity of *Escherichia coli* LD-carboxypeptidase on biosynthetically modified muropeptides. *FEBS Lett.* **339**, 249–252.

Markiewicz, Z., Glauner, B. & Schwarz, U. (1983) Murein structure and lack of DD- and LD-carboxypeptidase activities in *Caulobacter crescentus*. *J. Bacteriol.* **156**, 649–655.

Metz, R., Henning, S. & Hammes, W.P. (1986a) LD-carboxypeptidase activity in *Escherichia coli*. I. The LD-carboxypeptidase activity in ether treated cells. *Arch. Microbiol.* **144**, 175–180.

Metz, R., Henning, S. & Hammes, W.P. (1986b) LD-carboxypeptidase activity in *Escherichia coli*. II. Isolation, purification and characterization of the enzyme from *E. coli* K12. *Arch. Microbiol.* **144**, 181–186.

Rogers, H.J., Perkins, H.R. & Ward, J.B. (1980) *Microbial Cell Walls and Membranes*. London: Chapman & Hall.

Rousset, A., Nguyen-Disteche, M., Minck, R. & Ghuysen, J.-M. (1982) Penicillin-binding proteins and carboxypeptidase/transpeptidase activities in *Proteus vulgaris* P18 and its penicillin-induced stable L-forms. *J. Bacteriol.* **152**, 1042–1048.

Ursinus, A., Steinhaus, H. & Höltje, J.-V. (1992) Purification of a nocardicin A-sensitive LD-carboxypeptidase from *Escherichia coli* by affinity chromatography. *J. Bacteriol.* **174**, 441–446.

*Markus F. Templin*
*Abteilung Biochemie,*
*Max-Planck-Institut für Entwicklungsbiologie,*
*Spemannstrasse 35,*
*72076 Tübingen, Germany*
*Email: mt@mpib-tuebingen.mpg.de*

*Joachim-Volker Höltje*
*Abteilung Biochemie,*
*Max-Planck-Institut für Entwicklungsbiologie,*
*Spemannstrasse 35,*
*72076 Tübingen, Germany*
*Email: joho@gen.mpib-tuebingen.mpg.de*

# 556. Murein endopeptidase MepA

## Databanks

*Peptidase classification: clan UX, family U6, MEROPS ID: U06.001*
*NC-IUBMB enzyme classification: none*
*Databank codes:*

| Species | SwissProt | PIR | EMBL (cDNA) | EMBL (genomic) |
|---|---|---|---|---|
| *Brucella abortus* | – | – | U21919 | – |
| *Escherichia coli* | P14007 | S08345 | X16909 | – |
| *Haemophilus influenzae* | P44566 | H64053 | U32705 | U32705: complete genome section 20 of 163 |

## Name and History

During the purification of the penicillin-sensitive D-alanine carboxypeptidase 1B that also shows DD-endopeptidase activity (PBP4; Chapter 148) a novel species of *murein endopeptidase (MepA)*, insensitive to the action of β-lactam antibiotics, was identified in *Escherichia coli*. This enzyme, independently discovered by two different laboratories (Tomioka & Matsuhashi, 1978; Keck & Schwarz, 1979), is able to degrade high molecular mass murein (peptidoglycan) sacculi completely into uncross-linked murein, and thus represents a potential enzyme for autolysis of the bacterium.

## Activity and Specificity

The penicillin-insensitive endopeptidase hydrolyzes the high molecular mass murein sacculus by splitting the peptide bonds interlinking the peptidyl moieties of neighboring glycan strands in the macromolecular murein net (for a detailed discussion of the structure of the murein see Rogers *et al.*, 1980; Glauner *et al.*, 1988; Ghuysen & Hakenbeck, 1994). These peptide cross-bridges in the murein are formed by two types of connections. One of these is a DD-peptide bond between the $\alpha$-carboxyl group of the D-Ala residue of one peptidyl group and the $\varepsilon$-amino group of the *meso*-2,6-diaminopimelic acid of a second peptide moiety. The second type of bond connecting neighboring glycan strands is an LD-peptide bond connecting two *meso*-diaminopimelic acids. Both of these bonds are recognized and split: L-Ala-D-Glu-*meso*-diaminopimelic acid+D-Ala-*meso*-diaminopimelic acid-D-Glu-L-Ala; L-Ala-D-Glu-*meso*-diaminopimelic acid+*meso*-diaminopimelic acid-D-Glu-L-Ala). Therefore MepA possesses both DD- and LD-endopeptidase activity (unpublished results of the author).

MepA has a pH optimum of about 6, and requires the presence of divalent cations ($Mg^{2+}$, $Ca^{2+}$, $Co^{2+}$) for optimal activity.

Since the endopeptidase acts on high molecular mass sacculi, activity can be measured by following the formation of soluble murein degradation products (muropeptides). A sensitive assay uses radioactively labeled murein sacculi (Romeis & Höltje, 1994). Alternatively, isolated oligomeric muropeptides such as bis(disaccharide-tetrapeptide) formed by the action of lysozyme on bacterial cell walls is used as substrate, and the formation of monomeric reaction products is monitored by paper chromatography (Tomioka & Matsuhashi, 1978; Keck & Schwarz, 1979) or more conveniently by HPLC (Glauner & Schwarz, 1988). The addition of penicillin G to the reaction mixture inactivates the two other known murein endopeptidases PBP4 (Chapter 148) and PBP7 (Chapter 143), making the assay specific for the penicillin-insensitive enzyme.

The endopeptidase is strongly inhibited by DNA and polynucleotides, with single-stranded polymers being more effective than double-stranded structures (Tomioka & Matsuhashi, 1978; Keck & Schwarz, 1979).

## Structural Chemistry

The murein endopeptidase MepA is a soluble protein of about 28.4 kDa, pI 8.0. It is synthesized as a preprotein and gets processed upon export to the periplasm.

Besides the similarity to the two obvious homologs in *Haemophilus influenzae* and *Brucella abortus* no significant similarity to other proteins in the databases can be detected.

## Preparation

A purification procedure from crude *E. coli* cell extracts involving three different chromatography steps has been described (Keck & Schwarz, 1979). Using this procedure, a highly enriched preparation (purification factor about 20 000) can be obtained. Since the enzymatically active protein is produced from a high copy number plasmid containing the *mepA* gene, cells carrying this construct can be used for an efficient purification procedure (Keck *et al.*, 1990).

## Biological Aspects

An extensive screen of a mutant library for clones with reduced endopeptidase activity led to the identification of two mutant genes, *mepA* and *mepB*. These gene loci are found at 50.4 min and 90 min, respectively, on the map of the *E. coli* chromosome. The cloning of *mepA* showed that this locus represents the structural gene of the murein endopeptidase; the function of *mepB* remains unclear. This work and the fact that a deletion mutant of *mepA* is viable, show that this murein endopeptidase is not an essential enzyme (unpublished results of the author).

## Distinguishing Features

Like the other known murein endopeptidases, PBP4 (Chapter 148) and PBP7 (Chapter 143), the enzyme splits the DD-bond in the murein peptide cross-bridges. The fact that MepA is penicillin insensitive and that it is capable of hydrolyzing the stereochemically different LD-linkage between the peptide moieties in the murein, clearly distinguishes this enzyme from the serine peptidases PBP4 and PBP7.

## Further Reading

Fuller accounts of the murein endopeptidase have been provided by Ghuysen & Hakenbeck (1994) and Rogers *et al.* (1980).

## References

Glauner, B. & Schwarz, U. (1988) Investigation of murein structure and metabolism by high pressure liquid chromatography. In: *Modern Microbiological Methods*, vol. III: *Bacterial Cell Surface Techniques* (Hancock, I.C. & Poxton, I.R., eds). Chichester: John Wiley & Sons, pp. 158–171.

Glauner, B., Höltje, J.V. & Schwarz, U. (1988) The composition of the murein of *Escherichia coli*. *J. Biol. Chem.* **263**, 10088–10095.

Ghuysen, J.-M. & Hakenbeck, R. (1994) Bacterial cell wall. In: *New Comprehensive Biochemistry*, vol. 27 (Neuberger, A. & van Deenen, L.L.M., eds). Amsterdam: Elsevier.

Keck, W. & Schwarz, U. (1979) *Escherichia coli* murein-DD-endopeptidase insensitive to $\beta$-lactam antibiotics. *J. Bacteriol.* **139**, 770–774.

Keck, W., van Leeuwen, A.M., Huber, M. & Goodell, E.W. (1990) Cloning and characterisation of *mepA*, the structural gene of the penicillin-insensitive murein hydrolase from *Escherichia coli*. *Mol. Microbiol.* **4**, 209–219.

Rogers, H.J., Perkins, H.R. & Ward, J.B. (1980) *Microbial Cell Walls and Membranes*. London: Chapman & Hall.

Romeis, T. & Höltje, J.-V. (1994) Penicillin-binding protein7/8 of *Escherichia coli* is a DD-endopeptidase. *Eur. J. Biochem.* **224**, 597–604.

Tomioka, S. & Matsuhashi, M. (1978) Purification of penicillin-insensitive DD-endopeptidase: a new cell wall peptidoglycan-hydrolyzing enzyme in *Escherichia coli*, and its inhibition by deoxyribonucleic acids. *Biochem. Biophys. Res. Commun.* **84**, 978–984.

*Markus F. Templin*
*Abteilung Biochemie,*
*Max-Planck-Institut für Entwicklungsbiologie,*
*Spemannstrasse 35,*
*72076 Tübingen, Germany*
*Email: mt@mpib-tuebingen.mpg.de*

*Joachim-Volker Höltje*
*Abteilung Biochemie,*
*Max-Planck-Institut für Entwicklungsbiologie,*
*Spemannstrasse 35,*
*72076 Tübingen, Germany*
*Email: joho@gen.mpib-tuebingen.mpg.de*

# 557. *Murein endopeptidase P5*

## Databanks

*Peptidase classification: clan UX, family U40, MEROPS ID: U40.001*
*NC-IUBMB enzyme classification: none*
*Databank codes:*

| Species | SwissProt | PIR | EMBL (cDNA) | EMBL (genomic) |
|---|---|---|---|---|
| Bacteriophage phi6 | P07582 | D23368 | M12921 | – |

## Name and History

Early studies on the protein composition of phage phi6 by use of SDS-PAGE resulted in the identification of ten different polypeptide chains as components of purified virus particles. These proteins were designated P1–P10 according to their mobilities in electrophoresis, the slowest moving protein being P1, and the fastest P10 (Sinclair *et al.*, 1975). Further studies on phi6 revealed the presence of a lytic enzyme activity associated with purified virions, as demonstrated by a spot test on chloroform-treated bacterial lawns (Kakitani *et al.*, 1978). Studies carried out with several phage mutants showed protein P5 to be a component of the bacteriophage lysin (Mindich & Lehman, 1979). Iba *et al.* (1979) partially purified the phi6 lytic enzyme. The instability of that preparation, however, precluded further studies on the enzymatic activity. More recently, P5 has been purified to homogeneity and its activity characterized (Caldentey & Bamford, 1992).

## Activity and Specificity

Protein P5 is active against the cell walls of several gram-negative bacteria but no activity against gram-positive species has been detected. The enzyme appears to cleave the cell wall peptide bridge formed by *meso*-2,6-diaminopimelic acid and D-Ala. Lytic activity can be measured by the ability of protein P5 preparations to decrease the optical density of a suspension of chloroform-treated bacterial cells. The pH optimum of the endopeptidase is about 8.5 and the assay system typically includes Triton X-100 (0.1%) and NaCl (up to 150 mM). Further increase of the salt concentration or addition of divalent cations, such as $Ca^{2+}$ and $Mg^{2+}$, have an inhibitory effect on the enzymatic activity of P5. The enzyme is inactivated at temperatures above 20°C (Caldentey & Bamford, 1992).

## Structural Chemistry

Phi6 protein P5 is a single-chain, basic protein of approximately 24 kDa with an estimated pI of about 10. The enzyme does not contain any Cys residues.

## Preparation

Protein P5 is located between the nucleocapsid and the envelope of phage phi6, in a loose contact with the nucleocapsid surface protein P8 (Hantula & Bamford, 1988). The enzyme has been isolated from purified phi6 particles that had been disrupted with butylated hydroxytoluene and Triton X-114, and purified by chromatography on CM-agarose columns (Caldentey & Bamford, 1992).

## Biological Aspects

The genome of phage phi6 is composed of three dsRNA segments, the gene encoding P5 being located in the smaller genomic fragment (McGraw *et al*., 1986). The synthesis of an additional protein, P11, with an apparent molecular mass of about 25 kDa, is closely associated with the synthesis of P5, both proteins being similarly affected by mutations (Sinclair *et al*., 1976). Since the N-terminal sequence of P5 is consistent with the earliest possible start for gene 5 (Caldentey & Bamford, 1992), protein P11 could either be the result of a read-through of gene 5 or a modified form of P5. No precursor–product relationship between the two proteins is observed *in vivo* and the relative amounts of P11 produced depend on host and temperature (Sinclair *et al*., 1975).

In addition to its role in host cell lysis late in infection, protein P5 appears to contribute to infection of the bacterial cell by phi6 by locally digesting the cell wall after the phage membrane has fused with the host cell outer membrane. The infectivity of phi6 has a temperature sensitivity similar to that of P5, and electron microscopical observations indicate that heat-treated phage particles are capable of adsorption and membrane fusion *in vivo*, but cannot penetrate the peptidoglycan layer (Bamford *et al*., 1987).

Early observations suggested that protein P10 might also be involved in host cell lysis (Mindich & Lehman, 1979), and the effects of P10 on cell lysis were found to be host dependent. Recent studies confirm the involvement of P10 in host cell lysis, although the protein seems not to be required for infection (Johnson & Mindich, 1994). In contrast to P5, the absence of P10 can be compensated *in vitro* by the addition of few drops of chloroform, suggesting that P10 may assist the endopeptidase by creating lesions in the bacterial membrane in a holin-like manner. This, however, has yet to be confirmed.

## Further Reading

The article of Caldentey & Bamford (1992) is recommended.

## References

Bamford, D.H., Romantschuk, M. & Somerharju, P.J. (1987) Membrane fusion in prokaryotes: bacteriophage phi6 membrane fuses with the *Pseudomonas syringae* outer membrane. *EMBO J.* **6**, 1467–1473.

Caldentey, J. & Bamford, D.H. (1992) The lytic enzyme of the *Pseudomonas* phage phi6. Purification and biochemical characterization. *Biochim. Biophys. Acta* **1159**, 44–50.

Hantula, J. & Bamford, D.H. (1988) Chemical crosslinking of bacteriophage phi6 nucleocapsid proteins. *Virology* **165**, 482–488.

Iba, H., Nanno, M. & Okada, Y. (1979) Identification and partial purification of a lytic enzyme in the bacteriophage phi6 virion. *FEBS Lett.* **103**, 234–237.

Johnson III, M.D. & Mindich, L (1994) Isolation and characterization of nonsense mutations in gene 10 of bacteriophage phi6. *J. Virol.* **68**, 2331–2338.

Kakitani, H., Emori, Y., Iba, H. & Okada, Y. (1978) Lytic enzyme activity associated with double-stranded RNA bacteriophage phi6. *Proc. Japan Acad. Ser. A* **54**, 337–340.

McGraw, T., Mindich, L. & Frangione, B. (1986) Nucleotide sequence of the small double-stranded RNA segment of bacteriophage phi6: novel mechanism of natural translation control. *J. Virol.* **58**, 142–151.

Mindich, L. & Lehman, J. (1979). Cell wall lysin as a component of the bacteriophage phi6 virion. *J. Virol.* **30**, 489–496.

Sinclair, J.F., Tzagoloff, A., Levine, D. & Mindich, L. (1975) Proteins of bacteriophage phi6. *J. Virol.* **16**, 685–695.

Sinclair, J.F., Cohen, J. & Mindich, L. (1976) The isolation of suppressible nonsense mutants of bacteriophage phi6. *Virology* **75**, 198–208.

*Javier Caldentey*
*Institute of Biotechnology,*
*PO Box 56 (Viikinkaari 9),*
*00014 University of Helsinki, Finland*
*Email: javier.caldentey@helsinki.fi*

# 558. *PfpI protease (Pyrococcus furiosus)*

## Databanks

*Peptidase classification: clan UX, family U46, MEROPS ID: U46.001*
*NC-IUBMB enzyme classification: none*
*Databank codes:*

| Species | Gene | SwissProt | PIR | EMBL (cDNA) | EMBL (genomic) |
|---|---|---|---|---|---|
| PfpI endopeptidase | | | | | |
| *Pyrococcus furiosus* | *pfpI* | – | JC6003 | U57642 | – |

*continued overleaf*

| Species | Gene | SwissProt | PIR | EMBL (cDNA) | EMBL (genomic) |
|---|---|---|---|---|---|
| **Others homologs** | | | | | |
| *Bacillus subtilis* | *yraA* | – | – | – | X92868: 23.9 kb fragment |
| *Burkholderia cepacia* | – | – | – | U41162 | – |
| *Escherichia coli* | *thiJ* | Q46948 | – | U34923 | – |
| *Escherichia coli* | *yhbo* | P45470 | – | – | U18997: chromosomal region 67.4–76.0' |
| *Homo sapiens* | – | – | – | D61380 | – |
| *Methanococcus jannaschii* | *MJ0967* | – | – | – | U67540: genome section 82 of 150 |
| *Mycoplasma pneumoniae* | – | – | – | – | AE000015: genome section 15 of 53 |
| *Staphylococcus aureus* | *orf1* | – | – | L19300 | – |
| *Synechocystis* sp. | – | – | – | – | D90908: genome section 10 of 27 |

## Name and History

*Pyrococcus furiosus protease 1 (PfpI protease)* is one of the most prominent intracellular proteases in *Pyrococcus furiosus*, and consequently was among the first proteases to be discovered in this organism. Previously, PfpI was named *S66*, due to the fact that after heating at 95°C in 1% SDS for 24 h, it ran at approximately 66 kDa on a denaturing gel (Blumentals *et al.*, 1990). It was later noted that there were several forms of the enzyme, and that 66 kDa was not an accurate size, so the name was changed to PfpI to avoid confusion.

## Activity and Specificity

PfpI shows specificity towards basic and bulky, hydrophobic P1 amino acid residues in peptide substrates (Halio *et al.*, 1997). To date, this endopeptidase is most active toward the artificial substrate Suc-Ala-Ala-Phe↓NHMec (Halio *et al.*, 1997). Using this substrate, the pH optimum is 6.3, and the temperature optimum is 86°C (Halio *et al.*, 1997). Large proteins such as azocasein and gelatin are also degraded by PfpI, but the specificity of bond cleavage in these proteins has not yet been studied.

Assays are commonly made with fluorogenic substrates such as Suc-Ala-Ala-Phe-NHMec in 50 mM sodium phosphate buffer, pH 7. Endpoint assays can be done as described by Halio *et al.* (1997), with microtiter plates and fluorometric analysis. Proteolytic bands are often identified on gelatin overlay gels (Blumentals *et al.*, 1990) or 0.1% gelatin zymograms (Novex; see Appendix 2 for full names and addresses of suppliers) as well.

PfpI activity on an overlay gel appears to be inhibited by PMSF and DFP, but not EDTA or iodoacetic acid (Blumentals *et al.*, 1990). The protease is resistant to SDS, urea and other common denaturants, and retains some proteolytic activity in the presence of 2-ME and DTT.

## Structural Chemistry

The monomer for PfpI contains 166 amino acid residues, and has a molecular mass of 18.8 kDa. Two active forms of PfpI have been purified to date, a hexamer ($124 \pm 6$ kDa on gel filtration) and a trimer ($59 \pm 3$ kDa on gel filtration) (Halio *et al.*, 1997). Since these proteins are not denatured fully by SDS, molecular mass determination is not accurate. These forms have apparent molecular masses of 86 and 66 kDa,

respectively, on denaturing gels. There is evidence from western blots and zymogram gels that there may be an even larger active assembly of PfpI of over 200 kDa (Blumentals *et al.*, 1990; Halio *et al.*, 1996; unpublished results of the authors). The pI calculated from the amino acid sequence of the 18.8 kDa subunit is 6.1, whereas the experimentally determined pI values of the hexamer and trimer are 6.1 and 3.7–3.9, respectively. If all asparagine and glutamine residues are changed to their corresponding acidic residues, the pI of the monomer is calculated to be 4.8. At high temperatures it is possible that the protein is deamidated, especially in its trimeric form since more surface area is left unprotected, which would explain the calculated pI difference from the predicted value (Halio *et al.*, 1997). Upon long exposures to heat (24–48 h), the predominant band on an overlay gel is, in fact, the smaller form (Halio *et al.*, 1997).

Genetic analysis of *pfpI* did not show any common structural motifs (Halio *et al.*, 1996). However, many homologous protein sequences have been found (see Databanks table), and an alignment of these suggests that there is a sequence consensus around a Ser, a His, and two Asp residues. This would be consistent with the possibility that PfpI is a serine protease, as is indicated by the inhibitor studies.

## Preparation

PfpI is not commercially available. For most studies it has been purified from its native source, *P. furiosus* (DSM 3638), which is grown anaerobically at 98°C in an artificial seawater-based medium supplemented with maltose or peptides and sulfur (Fiala & Stetter, 1986). Purification of two intracellular functional forms (trimer and hexamer) is described by Halio *et al.* (1997), with a 200-fold purification and a 2% yield.

The gene for PfpI (*pfpI*) has been expressed in *Escherichia coli* using the T7 promoter and infection with λ phage CE6 (Novagen). Vector constructs are often unstable, and toxicity has been a problem in producing recombinant protease. Expression levels were best (1 mg per 100 ml culture) when a histidine tag was used to make a fusion protein, which was then purified using an affinity column (Halio *et al.*, 1996).

## Biological Aspects

The biological function of PfpI is still under investigation, but it is a predominant protease in *P. furiosus* cell extracts and has a 33 h half-life at 98°C after heating in 1% SDS (Blumentals *et al*., 1990). Continuous culture experiments showed that under nutrient-limiting conditions, *P. furiosus* produced more of PfpI (trimer) at lower dilution rates (Snowden *et al*., 1992). In batch experiments it was shown that total proteolytic activity of the cell increases at early log phase and late log phase, when grown on peptides as the sole carbon source (Snowden *et al*., 1992). It is possible that this protein may have many roles, as *P. furiosus* is an evolutionarily primitive organism (Woese *et al*., 1990). The archaea are not known to cause disease so pathogenicity can probably be ruled out.

Only ten nucleotides were found upstream of the start of translation, so it is unlikely that any pre or pro regions are associated with PfpI for targeting or assembly. Two strongly conserved box A archaean promoter sequences, TTATA at −180 and TTTAAA at −116 nucleotides 5′ to the initiation codon, were identified, along with another possible promoter sequence, TTAAA, found at −36 and −63. The sequence downstream of the TGA stop codon is thymidine-rich and could function in transcription termination (Halio *et al*., 1996).

## Distinguishing Features

PfpI has been shown to have immunological relationships with ClpP from *E. coli*, and both archaean and eukaryotic proteasomes (Chapters 169, 172). However, the archaean proteasome has also been purified from *P. furiosus* (Bauer *et al*., 1997). It also appears that PfpI has homologs in all three domains of life, and its sequence is well conserved (see Table 559.1), which suggests that this enzyme is important for cellular function or survival. However, it should be noted that all of the homologous proteins are of unknown function.

## Further Reading

General references on high-temperature enzymes, and summaries of known hyperthermophilic proteases may be found in Bauer *et al*. (1996) and Lauschner & Antranikian (1995).

## References

Bauer, M.W., Halio, S.B. & Kelly, R.M. (1996) Proteases and glycosyl hydrolases from hyperthermophilic microorganisms. *Adv. Protein Chem.* **48**, 271–310.

Bauer, M.W., Bauer, S.H. & Kelly, R.M. (1997) Purification and characterization of a proteasome from the hyperthermophilic archaeon *Pyrococcus furiosus*. *Appl. Environ. Microbiol.* **63**, 1160–1164.

Blumentals, I.I., Robinson, A.S. & Kelly, R.M. (1990) Characterization of sodium dodecyl sulfate-resistant proteolytic activity in the hyperthermophilic archaebacterium *Pyrococcus furiosus*. *Appl. Environ. Microbiol.* **56**, 1992–1998.

Fiala, G. & Stetter, K.O. (1986) *Pyrococcus furiosus* sp. nov. represents a novel genus of marine heterotrophic archaebacteria growing optimally at 100°C. *Arch. Microbiol.* **145**, 56–60.

Halio, S.B., Blumentals, I.I., Short, S.A., Merrill, B.M. & Kelly, R.M. (1996) Sequence, expression in *Escherichia coli*, and analysis of the gene encoding a novel intracellular protease (PfpI) from the hyperthermophilic archaeon *Pyrococcus furiosus*. *J. Bacteriol.* **178**, 2605–2612.

Halio, S.B., Bauer, M.W., Mukund, S., Adams, M.W.W. & Kelly, R.M. (1997) Purification and characterization of two functional forms of intracellular protease PfpI from the hyperthermophilic archaeon *Pyrococcus furiosus*. *Appl. Environ. Microbiol.* **63**, 289–295.

Leuschner, C. & Antranikian, G. (1995) Heat-stable enzymes from extremely thermophilic and hyperthermophilic microorganisms. *World J. Microbiol. Biotechnol.* **11**, 95–114.

Snowden, L.J., Blumentals, I.I. & Kelly, R.M. (1992) Regulation of proteolytic activity in the hyperthermophile *Pyrococcus furiosus*. *Appl. Environ. Microbiol.* **58**, 1134–1140.

Woese, C.R., Kandler, O. & Wheelis, M.L. (1990) Towards a natural system of organisms: proposal for the domains archaea, bacteria, and eucarya. *Proc. Natl Acad. Sci. USA* **87**, 4576–4579.

*P.M. Hicks*
*Department of Chemical Engineering,*
*North Carolina State University, Box 7905,*
*Raleigh, NC 27695, USA*
*Email: pameese@eos.ncsu.edu*

*R.M. Kelly*
*Department of Chemical Engineering,*
*North Carolina State University, Box 7905,*
*Raleigh, NC 27695, USA*
*Email: rmkelly@eos.ncsu.edu*

# 559. *Endopeptidase GPR*

## Databanks

*Peptidase classification: clan UX, family U3, MEROPS ID: U03.001*
*NC-IUBMB enzyme classification: none*

*Databank codes:*

| Species | SwissProt | PIR | EMBL (cDNA) | EMBL (genomic) |
|---|---|---|---|---|
| *Bacillus megaterium* | P22321 | A39198 | M55262 | – |
| *Bacillus subtilis* | P22322 | B39198 | D17650 | D84432: genome 283 kb fragment |
| | | | M55263 | |

## Name and History

The discovery that 10–20% of the protein of dormant spores of *Bacillus* species is rapidly degraded during spore germination led to the identification of **germination proteinase (GPR)** as the endopeptidase that initiates this proteolysis (Setlow, 1976, 1988). Disruption of the gene coding for GPR (*gpr*) led to spores in which protein degradation during spore germination was greatly slowed, indicating the predominant role for GPR in this proteolysis (Sanchez-Salas *et al*., 1992).

## Activity and Specificity

GPR cleaves internal bonds in proteins. Major substrates *in vivo* are the small acid-soluble proteins (SASPs) degraded during spore germination (Setlow, 1988). Following cleavage of SASPs by GPR *in vivo* the resulting oligopeptides are degraded to amino acids by other peptidases, generating a large supply of amino acids for the germinating spore (Setlow, 1976, 1988). The GPR zymogen also autoprocesses to the active enzyme (Sanchez-Salas & Setlow, 1993; Illades-Aguiar & Setlow, 1994a,b). SASPs and the GPR zymogen are also cleaved *in vitro*. The complete rules governing GPR specificity are unclear, but acidic residues (preferentially Glu) are at P1, hydrophobic residues at P1', Ala at P2', a variety of residues at P3' and another acidic residue at P4' (Carillo-Martinez & Setlow, 1994; Illades-Aguiar & Setlow, 1994a; Sanchez-Salas *et al*., 1992; Setlow, 1988). Examples of cleavage sites are: SASP-α of *B. subtilis*, Met-Lys-Leu-Glu⊦Ile-Ala-Ser-Glu-Phe; SASP-γ of *B. subtilis*, Phe-Gly-Thr-Glu⊦Phe-Ala-Ser-Glu-Thr, and *B. megaterium* GPR zymogen, Val-Arg-Thr-Asp⊦Leu-Ala-Val-Glu-Ala (Sanchez-Salas & Setlow, 1993; Setlow, 1988).

The enzyme is routinely assayed with a mixture of SASPs from *B. megaterium* spores as substrate, endopeptidase action on SASPs being amplified with an aminopeptidase also partially purified from spores, and generation of free amino groups being quantitated with ninhydrin (Setlow, 1976). The enzyme can also be assayed with purified SASPs, cleavage being monitored by PAGE at low pH (Setlow, 1976). $K_m$ values determined with purified SASPs are below 14 µM (Dignam & Setlow, 1980). The enzyme is active on small (tetra- to hepta-) peptides containing a cleavage site, with $V$ values similar to those on SASPs, but the $K_m$ values for these peptides are 20 mM or more (Dignam & Setlow, 1980).

The pH optimum of the enzyme is about 7.5 (Setlow, 1976). The enzyme is rather unstable but is stabilized by glycerol (20%) and $Ca^{2+}$ (5 mM) (Setlow, 1976). The enzyme is not inhibited by EDTA, 1,10-phenanthroline, DFP, *N*-ethylmaleimide or iodoacetamide, or by inhibitors of aspartic proteases, and the enzyme lacks typical signature sequences of serine, cysteine, metallo or aspartic proteases (Setlow,

1976; Sussman & Setlow, 1991; B. Setlow & P. Setlow, unpublished results).

## Structural Chemistry

GPR is a homotetramer of 40 kDa subunits and is active only as the tetramer (Loshon & Setlow, 1982). The zymogen is also a homotetramer of approximately 43 kDa subunits as measured on SDS-PAGE (Loshon *et al*., 1982). The protein shows no sequence similarity to any protein in current databases (Setlow, 1994).

## Preparation

The enzyme has been purified to homogeneity (about 2000-fold) from germinated spores of *B. megaterium* (Loshon & Setlow, 1982). The recombinant *B. megaterium* enzymes (both zymogen and active enzyme) have been expressed in *Escherichia coli* as soluble proteins and purified on a scale of tens of milligrams (Illades-Aguiar & Setlow, 1994a; Sanchez-Salas & Setlow, 1993).

## Biological Aspects

The enzyme is synthesized only during sporulation within the developing spore, and as a zymogen whose sequence appears identical to that encoded in the *gpr* gene (Loshon *et al*., 1982; Sanchez-Salas & Setlow, 1993; Illades-Aguiar & Setlow, 1994a). Approximately 2 h after its synthesis the zymogen autoprocesses to the active enzyme by intramolecular removal of 15 (*B. megaterium*) or 16 (*B. subtilis*) N-terminal residues and undergoes a conformational change during this process (Loshon *et al*., 1982; Sanchez-Salas & Setlow, 1993; Illades-Aguiar & Setlow, 1994a). The physiological stimuli for the autoprocessing include the acidification and dehydration of the forespore and its accumulation of a large depot of pyridine-2,6-dicarboxylic acid (Illades-Aguiar & Setlow, 1994b). Although potentially active GPR is generated by this autoprocessing reaction, it does not act on its SASP substrates in sporulation or in the dormant spore, probably because of the low water content of the forespore and spore (Illades-Aguiar & Setlow, 1994b). However, the spore rehydrates early in the germination process, allowing GPR to act on SASPs. Eventually GPR itself is degraded by the ATP-dependent proteolytic system of the cell (Loshon & Setlow, 1982).

The only substrates identified to date for endopeptidase GPR *in vivo* (other than the GPR zymogen) are the SASPs. These proteins are DNA-binding proteins that protect dormant spore DNA from damage, but also inhibit transcription (Setlow, 1988, 1994, 1995). Consequently, spores of *gpr*-negative mutants do not rapidly proceed to vegetative growth following spore germination, as transcription is inhibited by the undegraded SASPs (Sanchez-Salas *et al.*, 1992). This phenotype of *gpr* mutants can be suppressed by also inactivating genes coding for most SASPs (Sanchez-Salas *et al.*, 1992). Although SASPs are the only GPR substrates identified *in vivo*, high levels of the active enzyme are deleterious to *E. coli* suggesting that proteins in addition to SASPs can be degraded (Illades-Aguiar & Setlow, 1994a).

## Distinguishing Features

Although many proteases can hydrolyze SASPs, no others are known with the strict cleavage specificity of GPR. Polyclonal antisera have been raised against the active and zymogen forms of *B. megaterium* GPR, and these show no cross-reaction with proteins in growing cells of *B. megaterium* or in *E. coli* (Loshon & Setlow, 1982; Sanchez-Salas & Setlow, 1993). The antiserum is not commercially available.

## Further Reading

Recent reviews include those of Setlow (1994, 1995).

## References

Carillo-Martinez, Y. & Setlow, P. (1994) Properties of *Bacillus subtilis* small, acid-soluble, spore proteins with changes in the sequence recognized by their specific protease. *J. Bacteriol.* **176**, 5357-5363.

Dignam, S.S. & Setlow, P. (1980) *Bacillus megaterium* spore protease: action of the enzyme on peptides containing the amino acid sequence cleaved *in vivo*. *J. Biol. Chem.* **255**, 8408-8412.

Illades-Aguiar, B. & Setlow, P. (1994a) Studies of the processing of the protease which initiates degradation of small, acid-soluble proteins during germination of spores of *Bacillus* species. *J. Bacteriol.* **176**, 2788-2795.

Illades-Aguiar, B. & Setlow, P. (1994b) Autoprocessing of the protease that degrades small, acid-soluble proteins of spores of *Bacillus* species is triggered by low pH, dehydration and dipicolinic acid. *J. Bacteriol.* **176**, 7032-7037.

Loshon, C.A. & Setlow, P. (1982) *Bacillus megaterium* spore protease: purification, radioimmunoassay, and analysis of antigen level and localization during growth, sporulation and spore germination. *J. Bacteriol.* **150**, 303-311.

Loshon, C.A., Swerdlow, B.M. & Setlow, P. (1982) *Bacillus megaterium* spore protease: synthesis and processing of precursor forms during sporulation and germination. *J. Biol. Chem.* **257**, 10838-10845.

Sanchez-Salas, J.-L. & Setlow, P. (1993) Proteolytic processing of the protease which initiates degradation of small, acid-soluble, proteins during germination of *Bacillus subtilis* spores. *J. Bacteriol.* **175**, 2568-2577.

Sanchez-Salas, J.-L., Santiago-Lara, M.L., Setlow, B., Sussman, M.D. & Setlow, P. (1992) Properties of mutants of *Bacillus megaterium* and *Bacillus subtilis* which lack the protease that degrades small, acid-soluble proteins during spore germination. *J. Bacteriol.* **174**, 807-814.

Setlow, P. (1976) Purification and properties of a specific proteolytic enzyme present in spores of *Bacillus megaterium*. *J. Biol. Chem.* **251**, 7853-7862.

Setlow, P. (1988) Small acid-soluble, spore proteins of *Bacillus* species: structure, synthesis, genetics, function and degradation. *Annu. Rev. Microbiol.* **42**, 319-338.

Setlow, P. (1994) Mechanisms which contribute to the long-term survival of spores of *Bacillus* species. *J. Appl. Bacteriol.* **176** (Symposium Supplement), 49S-60S.

Setlow, P. (1995) Mechanisms for the prevention of damage to the DNA in spores of *Bacillus* species. *Annu. Rev. Microbiol.* **49**, 29-54.

Sussman, M.D. & Setlow, P. (1991) Cloning, nucleotide sequence, and regulation of the *Bacillus subtilis gpr* gene which codes for the protease that initiates degradation of small, acid-soluble, proteins during spore germination. *J. Bacteriol.* **173**, 293-300.

*Peter Setlow*
*Biochemistry Department,*
*University of Connecticut Health Center,*
*Farmington, CT 06030, USA*
*Email: setlow@sun.uchc.edu*

# 560. Type IV prepilin peptidase

U

## Databanks

*Peptidase classification: clan UX, family U12, MEROPS ID: U12.001*
*NC-IUBMB enzyme classification: none*

*Databank codes:*

| Species | SwissProt | PIR | EMBL (cDNA) | EMBL (genomic) |
|---|---|---|---|---|
| *Aeromonas hydrophila* | P45794 | – | U20255 | – |
| *Bacillus subtilis* | P15378 | – | M30805 | |
| *Dichelobacter nodosus* | – | – | U17138 | |
| *Erwinia carotovora* | P31712 | – | X70049 | – |
| *Erwinia chrysanthemi* | P31711 | – | L02214 | |
| *Escherichia coli* | P25960 | – | L28106 | U18997: chromosomal region 67.4–76.0' |
| | | | M27176 | |
| *Escherichia coli* | – | – | – | X62169: plasmid R721 |
| *Escherichia coli* | – | – | Z34464 | |
| *Haemophilus influenzae* | P44620 | – | – | U32715: genomic |
| *Klebsiella pneumoniae* | P15754 | – | – | – |
| *Neisseria gonorrhoeae* | P33566 | S32915 | U32588 | – |
| *Pseudomonas aeruginosa* | P22610 | A39131 | M32066 | – |
| | | | M61096 | |
| *Pseudomonas putida* | P36642 | – | X74276 | |
| *Synechocystis* sp. | – | – | – | D90899: genome section 1 of 27 |
| *Vibrio cholerae* | P27717 | A40582 | M74708 | – |
| *Vibrio cholerae* | – | – | L25661 | – |
| *Vibrio vulnificus* | – | – | U48808 | – |
| *Xanthomonas campestris* | – | – | U12432 | – |

## Name and History

Type IV pili (fimbriae), implicated as important cell-associated virulence determinants, are produced by a large number of gram-negative pathogens of humans, animals and plants (Tennent & Mattick, 1994). These pili are composed of repeating subunits which are synthesized as a precursor (type IV prepilin) that is subsequently cleaved by a specific endopeptidase (type IV prepilin peptidase) prior to assembly into pili. Nunn & Lory (1991) first showed that the *Pseudomonas aeruginosa* type IV prepilin peptidase is encoded by *pilD* (*xcpA*) and can process the type IV prepilin, PilA (*pil* denotes a pilus-related gene). Homologous genes have been identified in many other bacterial species. The cleavage of type IV prepilin is carried out by the products of *tcpJ* (*tcp* stands for **t**oxin **c**oregulated **p**ilus) in *Vibrio cholerae* (Kaufman *et al.*, 1991), *pilD*$_{ng}$ of *Neisseria gonorrhoeae* and other *Neisseria* spp. (Lauer *et al.*, 1993; Dupuy & Pugsley, 1994), *fimP* (*fim* stands for **fim**briae) in *Dichelobacter nodosus* (Johnston *et al.*, 1995) and *bfpP* (where *bfp* denotes **b**undle-**f**orming **p**ili) in *Escherichia coli* EPEC (Zhang *et al.*, 1994). The machinery required for biogenesis of type IV pili (which has been studied most extensively in *P. aeruginosa* and *V. cholorae*) is shared by diverse bacterial systems, including those involved in secretion of virulence factors by gram-negative bacteria, biogenesis of bacteriophages, natural competence in gram-positive and gram-negative bacteria and DNA transfer (Hobbs & Mattick, 1993). These systems have in common several sets of related genes including those encoding proteins homologous to the type IV prepilin and type IV prepilin peptidase. In *Klebsiella oxytoca*, the type IV prepilin-like peptidase, PulO, is involved in secretion of pullulanase, a lipoprotein that degrades complex starches (Pugsley & Dupuy, 1992). The products of the *xpsO* and *outO* genes are required in *Xanthomonas campestris* (Hu *et al.*,

1995) and in *Erwinia chrysanthemi* (Lindeberg & Collmer, 1992) respectively for extracellular secretion of degradative enzymes. The only known homolog of type IV prepilin-like peptidases not associated with extracellular protein secretion is *B. subtilis comC*, a protein required for the competence state and DNA uptake (Mohan *et al.*, 1989; Dubnau, 1991). **Type IV prepilin peptidases** and **type IV prepilin-like peptidases** are homologous over their entire amino acid sequences, and in some but not all cases are interchangeable *in vivo* (de Groot *et al.*, 1991; Dupuy *et al.*, 1992).

## Activity and Specificity

The maturation of type IV prepilin and related proteins that are required for extracellular protein secretion and DNA uptake involves two consecutive post-translational modifications, i.e. removal of the 'leader sequence and then N-methylation of the N-terminal residue of the mature polypeptide. The leader sequences of these proteins resemble those of other secretory proteins. However, they are processed by the type IV prepilin peptidase at an atypical site, in most cases releasing only a short, basic leader peptide of 5–8 amino acids. Longer leader peptides have been reported, such as those of TcpA of *V. cholerae* (25 amino acids), BfpA of *E. coli* EPEC (13 amino acids) and PilV of *P. aeruginosa* (14 amino acids) (Kaufman *et al.*, 1991; Donnenberg *et al.*, 1992; Alm & Mattick, 1995). The consensus sequence of the cleavage site is the Gly↓Phe (or Gly↓Met) peptide bond within the sequence $Gly^{-1}$↓(Phe/Met)-Thr-Leu-(Ile/Leu)-$Glu^{+5}$. Mutational analysis showed that the Gly at the −1 (P1) position is absolutely required for complete type IV prepilin processing (Koomey *et al.*, 1991; Strom & Lory, 1991; Chiang *et al.*, 1995). Substitution of Ala for Gly at this position resulted in partial processing, suggesting that an amino acid with a small side chain is preferred at this position. Mutations that eliminate the positively charged residues within the leader sequence prevent processing and type IV pilus assembly in *P. aeruginosa* (Strom & Lory, 1991). However, the mutant type IV prepilin is still exported to the cytoplasmic membrane, indicating that cleavage is not required for export (Plasloske *et al.*, 1988). Although PulO has been shown to cleave the gonococcal type IV prepilin (Dupuy *et al.*, 1992), PilD/XcpA (of *P. aeruginosa*) and ComC (of *B. subtilis*) are unable to process prePulG (a type IV prepilin-like protein of *K. oxytoca*), indicating that recognition of the consensus cleavage site may depend on other sequences in the mature part of the polypeptide (Pugsley & Dupuy, 1992). For cleavage of the leader peptide from the *P. aeruginosa* substrate *in vitro*, a $K_m$ of approximately $650\,\mu M$ and a turnover rate ($k_{cat}$) of $180\,min^{-1}$ were measured (Strom & Lory, 1992; Strom *et al.*, 1994). Similar kinetic data were obtained for processing of *N. gonorrhoeae* type IV prepilin by type IV prepilin peptidase of *P. aeruginosa*. This shows that the differences in length and net charge of the *N. gonorrhoeae* prepilin leader sequence do not affect the overall rate of cleavage by PilD.

Strom *et al.* (1993a) showed that *P. aeruginosa* type IV prepilin peptidase, in addition to cleaving the leader peptide, also catalyzes the *N*-methylation of the first amino acid of the mature pilin. The methyl donor is *S*-adenosyl-L-methionine. Based on mutagenesis studies the Glu residue at position +5 of the type IV prepilin gene is essential for efficient methylation (Strom & Lory, 1991). In *N. gonorrhoeae*, *N*-methylation and the conserved Glu residue at position +5 may play a role as a mechanism for recognition and 'registration' of the N-terminal helix of type IV pilin during pilus assembly (Parge *et al.*, 1995). Other residues (excluding Gly and Asp) can replace the Phe and still be methylated, suggesting that the side chain of Phe is not required for substrate recognition by the methylation activity of type IV prepilin peptidase (Strom & Lory, 1991). Moreover, the methyltransferase activity can be competitively inhibited by *S*-adenosyl-methionine analogs without affecting peptidase activity. These data suggest that cleavage and methylation involved two different active sites. Other bacterial type IV prepilin peptidases that are homologs of PilD, including TcpJ in *V. cholerae* (Shaw & Taylor, 1990; Kaufman *et al.*, 1991) and PulO in *K. oxytoca* (Pugsley, 1993a), have *N*-methyltransferase activity as well.

## Structural Chemistry

Type IV prepilin peptidases are located in the cytoplasmic membrane, but have large, relatively hydrophilic domains in the cytoplasm, and these cytoplasmic domains are the parts of the proteins that are most highly conserved in the type IV prepilin peptidase family (Strom *et al.*, 1993b). The peptidases have molecular masses ranging from 24.8 kDa (PulO) to 32 kDa (PilD/XcpA). Theoretical pI values range from 5.31 (PulO) to 9.13 (XpsO). Sequence alignment of type IV prepilin peptidases from different bacterial species reveals the presence of two pairs of conserved cysteines (Cys-Xaa-Xaa-Cys). These motifs are located in the cytoplasmic domain. Replacement of these Cys residues with Ser or Gly results in a large decrease in both peptidase and *N*-methyltransferase activities *in vitro*. However, the processing of prepilin *in vivo* is only slightly affected in these mutants (Strom *et al.*,

1993b), suggesting that other parts of type IV prepilin peptidase, including segments within the membrane could participate in proteolysis and methylation. These data indicate that the *N*-methyltransferase and the leader peptidase activities are dependent, at least in part, on overlapping domains of the protein, including the shared Cys residues (Strom *et al.*, 1993b). This view has recently been challenged, however, by Hu *et al.* (1995), who have characterized a type IV prepilin-like peptidase from *X. campestris* (XpsO) that lacks the conserved cysteine residues. Moreover, the *xpsO* gene is able to complement a *pilD* mutant of *P. aeruginosa*. However, this insertional inactivation mutant of *pilD* has the first pair of Cys residues (Koga *et al.*, 1993). The XpsO of *X. campestris* may represent a special class of type IV prepilin peptidase.

## Preparation

Because of the inherent difficulties in the purification of integral membrane proteins by conventional methods, the purification of type IV prepilin peptidase from *P. aeruginosa* is accomplished by detergent solubilization of the membranes and immunoaffinity chromatography (Strom *et al.*, 1994). This method remains the method of choice for obtaining type IV prepilin-like peptidase in pure form.

## Biological Aspects

It is now clear that, in a large range of gram-negative and gram-positive bacterial species, type IV prepilin peptidase and its homologs play a central role not only in type IV pilus biogenesis but also in the export of extracellular proteins across cell membranes, and the import of DNA (Hobbs & Mattick, 1993). The post-translation modifications of type IV prepilin by the type IV prepilin peptidase are essential for assembly of subunits of pilin into pili (Nunn & Lory, 1991). Indeed, mutations in type IV prepilin peptidase genes lead to periplasmic accumulation of prepilin (Nunn *et al.*, 1990). The machinery of protein secretion and DNA uptake during transformation includes type IV prepilin-like proteins that are substrates of the type IV prepilin-like peptidases. These homologs undergo similar post-translation modifications to type IV pilin. However, in *K. oxytoca* the role of the processing of these proteins by PulO is not clear. Indeed, multimers of PulG are observed even in the absence of processing of pre-PulG (Pugsley, 1996) whereas in *N. gonorrhoeae* unprocessed type IV prepilin is not assembled (Koomey *et al.*, 1991). The similarity and the role of proteins required for extracellular secretion and DNA uptake to those involved in biogenesis of type IV pili, suggests that they might be assembled to form oligomeric structures akin to pili or even more complex structures. In general terms, these particular structures seem designed to connect the cytoplasmic membrane to the outer surface of the cell, forming a gate or passage for the exit or entry of macromolecules. Recently, Alm *et al.* (1996) identified two additional genes (*pilV* and *pilE*) that have type IV prepilin-like leader sequences and are required for pilus biogenesis. They proposed that type IV pili may be assembled via a pathway analogous to that used for complexes involved in macromolecular trafficking.

## Distinguishing Features

One of the more interesting properties of the type IV prepilin peptidase is its bifunctionality. This protease not only cleaves the leader sequence but also carries out N-terminal methylation of the mature pilin. This peptidase differs from the two other known signal peptidases, LepB (Chapter 153), the major signal peptidase, and LspA (Chapter 333), the lipoprotein signal peptidase (Dalbey, 1991; Munoa *et al.*, 1991), in that type IV prepilin peptidase cleaves signal sequences on the cytoplasmic side of the membrane and appears to be Sec independent. Indeed, TcpJ has been shown to cleave prepilin in the presence of sodium azide (which inhibits SecA protein of the general secretory pathway) (Kaufman *et al.*, 1991). Type IV prepilin peptidase shares no significant homology with other known peptidases, although slight similarity to thiol methyltransferase has been noted (Strom *et al.*, 1993b). This enzyme may represent a new class of thiol proteases.

## Further Reading

For reviews, see Pugsley (1993b) and Hobbs & Mattick (1993).

## References

Alm, R.A. & Mattick, J.S. (1995) Identification of a gene *pilV*, required for type 4 fimbrial biogenesis in *Pseudomonas aeruginosa*, whose product possesses a pre-pilin-like leader sequence. *Mol. Microbiol.* **16**, 485–496.

Alm, R.A., Hallinan, J.P., Watson, A.A. & Mattick, J.S. (1996) Fimbrial biogenesis genes of *Pseudomonas aeruginosa*: *pilW* and *pilX* increase the similarity of type 4 fimbriae to the GSP protein-secretion systems and *pilY1* encodes a gonococcal PilC homologue. *Mol. Microbiol.* **22**, 161–173.

Chiang, S.L., Taylor, R.K., Koomey, M. & Mekalanos, J.J. (1995) Single amino acid substitutions in the N-terminus of *Vibrio cholerae* TcpA affect colonization, autoagglutination, and serum resistance. *Mol. Microbiol.* **17**, 1133–1142.

Dalbey, R.E. (1991) Leader peptidase. *Mol. Microbiol.* **5**, 2855–2860.

de Groot, A., Filloux, A. & Tommassen, J. (1991) Conservation of *xcp* genes involved in the two-step protein secretion process, in different *Pseudomonas species* and other Gram-negative bacteria. *Mol. Gen. Genet.* **229**, 278–284.

Donnenberg, M.S., Giròn, J.A., Nataro, J.P. & Kaper, J.B. (1992) A plasmid-encoded type IV fimbrial gene of enteropathogenic *Escherichia coli* associated with localized adherence. *Mol. Microbiol.* **6**, 3427–3437.

Dubnau, D. (1991) The regulation of genetic competence in *Bacillus subtilis*. *Mol. Microbiol.* **5**, 11–18.

Dupuy, B. & Pugsley, A.P. (1994) Type IV prepilin peptidase gene of *Neisseria gonorrhoeae* MS11: presence of a related gene in other piliated and non piliated *Neisseria* strains. *J. Bacteriol.* **176**, 1323–1331.

Dupuy, B., Taha, M.-K., Possot, O., Marchal, C. & Pugsley, A.P. (1992) PulO, a component of the pullulanase secretion pathway of *Klebsiella oxytoca*, correctly and efficiently processes gonococcal type IV prepilin in *Escherichia coli*. *Mol. Microbiol.* **6**, 1887–1894.

Hobbs, M. & Mattick, J.S. (1993) Common components in the assembly of type 4 fimbriae, DNA transfer systems, filamentous phage and protein secretion apparatus: a general system for the formation of surface-associated protein complexes. *Mol. Microbiol.* **10**, 233–243.

Hu, N.-T., Lee, P.-F. & Chen, C. (1995) The type IV pre-pilin leader peptidase of *Xanthomonas campestris* pv. *campestris* is functional without conserved cysteine residues. *Mol. Microbiol.* **18**, 769–777.

Johnston, J.L., Billington, S.J., Haring, V. & Rood, J.I. (1995) Identification of fimbrial assembly genes from *Dichelobacter nodusus* – evidence that *fimP* encodes the type IV prepilin peptidase. *Gene* **161**, 21–26.

Kaufman, M.R., Seyer, J.M. & Taylor, R.K. (1991) Processing of TCP pilin by TcpJ typifies a common step intrinsic to a newly recognized pathway of extracellular protein secretion by Gram-negative bacteria. *Genes Dev.* **5**, 1834–1846.

Koga, T., Ishimoto, K. & Lory, S. (1993) Genetic and functional characterization of the gene cluster specifying expression of *Pseudomonas aeruginosa*. *Infect. Immun.* **61**, 1371–1377.

Koomey, M., Bergstrom, S., Blake, M. & Swanson, J. (1991) Pilin expression and processing in pilus mutants of *Neisseria gonorrhoeae*: critical role of Gly$^{-1}$ in assembly. *Mol. Microbiol.* **5**, 279–287.

Lauer, P., Albertson, N.H. & Koomey, M. (1993) Conservation of genes encoding components of a type IV pilus assembly/two-step protein export pathway in *Neisseria gonorrhoeae*. *Mol. Microbiol.* **8**, 357–368.

Lindeberg, M. & Collmer, A. (1992) Analysis of eight *out* genes in a cluster required for pectic enzyme secretion by *Erwinia chrysanthemi*: sequence comparison with secretion genes from other Gram-negative bacteria. *J. Bacteriol.* **174**, 7385–7395.

Mohan, S., Aghion, J., Guillen, N. & Dubnau, D. (1989) Molecular cloning and characterization of *comC*, a late competence gene of *Bacillus subtilis*. *J. Bacteriol.* **171**, 6043–6051.

Munoa, F.J., Miller, K.W., Beers, R., Graham, M. & Wu, H.C. (1991) Membrane topology of *Escherichia coli* prolipoprotein signal peptidase (signal peptidase II). *J. Biol. Chem.* **266**, 17667–17672.

Nunn, D. & Lory, S. (1991) Product of the *Pseudomonas aeruginosa* gene *pilD* is a prepilin leader peptidase. *Proc. Natl Acad. Sci. USA* **88**, 3281–3285.

Nunn, D., Bergman, S. & Lory, S. (1990) Products of three accessory genes, *pilB*, *pilC*, and *pilD*, are required for biogenesis of *Pseudomonas aeruginosa* pili. *J. Bacteriol.* **172**, 2911–2919.

Parge, B.L., Forest, K.T., Hickey, M.J., Christensen, D.A., Getzoff, E.D. & Tainer, J.A. (1995) Structure of the fiber-forming protein pilin at 2.6 Å resolution. *Nature* **378**, 32–38.

Plasloske, B.L., Carpenter, M.R., Frost, L.S., Finlay, B.B. & Paranchych, W. (1988) The expression of *Pseudomonas aeruginosa* PAK pilin gene mutants in *Escherichia coli*. *Mol. Microbiol.* **2**, 185–195.

Pugsley, A.P. (1993a) Processing and methylation of PulG, a pilin-like component of the general secretory pathway of *Klebsiella oxytoca*. *Mol. Microbiol.* **9**, 295–308.

Pugsley, A.P. (1993b) The complete general secretory pathway in Gram-negative bacteria. *Microbiol. Rev.* **57**, 50–108.

Pugsley, A.P. (1996) Multimers of the presursor of a type IV pilin-like component of the general secretory pathway are unrelated to pili. *Mol. Microbiol.* **20**, 1235–1245.

Pugsley, A.P. & Dupuy, D. (1992) An enzyme with type IV pre-pilin peptidase activity is required to process components of the general extracellular protein secretion pathway of *Klebsiella oxytoca*. *Mol. Microbiol.* **6**, 751–760.

Shaw, C.E. & Taylor, R.K. (1990) *Vibrio cholerae* 0395 *tcpA* pilin gene sequence and comparison of predicted protein structural features to those of type IV pilins. *Infect. Immun.* **58**, 3042–3049.

Strom, M.S. & Lory, S. (1991) Amino acid substitutions in pilin of *Pseudomonas aeruginosa*. Effect on leader peptide cleavage, amino-terminal methylation, and pilus assembly. *J. Biol. Chem.* **266**, 1656–1664.

Strom, M.S. & Lory, S. (1992) Kinetics and sequence specificity of processing of prepilin by PilD, the type IV leader peptidase of *Pseudomonas aeruginosa*. *J. Bacteriol.* **174**, 7345–7351.

Strom, M.S., Nunn, D.N. & Lory, S. (1993a) A single bifunctional enzyme, PilD, catalyzes cleavage and N-methylation of proteins belonging to the type IV pilin family. *Proc. Natl Acad. Sci. USA* **90**, 2404–2408.

Strom, M.S., Bergman, P. & Lory, S. (1993b) Identification of active-site cysteines in the conserved domain of PilD, the bifunctional type IV pilin leader peptidase/*N*-methyltransferase of *Pseudomonas aeruginosa*. *J. Biol. Chem.* **268**, 15788–15794.

Strom, M.S., Nunn, D.N. & Lory, S. (1994) Posttranslational processing of type IV prepilin and homologs by PilD of *Pseudomonas aeruginosa*. *Methods Enzymol.* **235**, 527–540.

Tennent, J. & Mattick, J.S. (1994) Type IV fimbria. In: *Fimbriae: Aspects of Adhesion, Genetics, Biogenesis and Vaccines* (Klemm, P., ed.). Boca Raton, Florida: CRC Press, pp. 127–146.

Zhang, H.-Z., Lory, S. & Donnenberg, M.S. (1994) A plasmid-encoded prepilin peptidase gene from Enteropathogenic *Escherichia coli*. *J. Bacteriol.* **176**, 6885–6891.

*Bruno Dupuy*
Unité de Génétique Moléculaire Bactérienne,
Institut Pasteur,
28, Rue du Dr Roux,
75724 Paris Cedex 15, France
Email: bdupuy@pasteur.fr

*Muhamed-Kheir Taha*
Unité des Neisseria,
Institut Pasteur,
28, Rue du Dr Roux,
75724 Paris Cedex 15, France
Email: mktaha@pasteur.fr

# 561. *Hydrogenase maturation endopeptidase*

## Databanks

Peptidase classification: clan UX, family U42, MEROPS ID: U42.001
NC-IUBMB enzyme classification: none
Databank codes:

| Species | Gene | SwissProt | PIR | EMBL (cDNA) | EMBL (genomic) |
|---|---|---|---|---|---|
| *Alcaligenes eutrophus* | *hoxM* | P31909 | – | M96433 | – |
| *Azotobacter vinelandii* | *hoxM* | P40591 | A44915 M80522 | L23970 | – |
| *Bradyrhizobium japonicum* | *hupD* | – | – | L24446 | – |
| *Escherichia coli* | *hycI* | – | – | X17506 | – |
| *Escherichia coli* | *hyaD* | P19930 | JV0075 | M34825 | – |
| *Escherichia coli* | *hybD* | P37182 | – | U09177 | – |
| *Methanobacterium thermoautotrophicum* | *frhD* | P19497 | B35620 | J02914 | – |
| *Rhizobium leguminosarum* | *hupD* | P27649 | – | X52974 | – |
| *Rhodobacter capsulatus* | *hupD* | Q03004 | S25686 S32941 | Z15089 | – |
| *Thiocarpa roseopersicina* | *hupD* | – | – | L22980 | – |
| *Wolinella succinogenes* | *orf4* | P31876 | S22407 | X65198 | – |

U

## Name and History

[NiFe]Hydrogenases consist of subunits of 30–35 kDa and of 60–65 kDa, depending on the organism. The large subunit carries the [NiFe]metallocenter, and it is synthesized in a precursor form and proteolytically processed to the form present in the mature enzyme. Three different approaches showed that processing takes place at the C-terminus at a conserved site. First, electrospray mass spectrometry of the large subunit (HoxG) of the hydrogenase from *Azotobacter vinelandii* indicated that cleavage occurred after a conserved His at the C-terminus (Gollin *et al*., 1992). An identical cleavage site was determined by amino acid sequencing of the tryptic C-terminal peptide of the mature large subunit of the *Desulfovibrio gigas* hydrogenase (Menon *et al*., 1993) and by determining the total sequence of the atypical small subunit corresponding to the C-terminal end of the $F_{420}$-nonreducing [NiFeSe]hydrogenase from *Methanococcus voltae*. The sequence terminated C-terminally to the conserved His although the genetically derived sequence was 18 amino acids longer (Sorgenfrei *et al*., 1993). Finally, Rossmann *et al*. (1994) in an *in vitro* system showed the C-terminal removal of a 3.85 kDa peptide from the large subunit of hydrogenase 3 from *Escherichia coli*. The processing activity was purified and shown to be the product of the promoter-distal gene (*hycI*) of the *hyc* operon from *E. coli* which codes for the components of the formate hydrogen lyase system (Sauter *et al*., 1992; Rossmann *et al*., 1995). **Hydrogenase maturation endopeptidases** have now been identifed from many bacteria.

## Activity and Specificity

HycI endopeptidase cleaves an internal -Arg+Met- bond, three residues C-terminal to the nickel-binding motif 2 (Asp-Pro-Cys-Xaa-Xaa-Cys-Xaa-Xaa-Arg) and it removes an intact 32 amino acid peptide. Cleavage takes place only when the precursor of the large subunit contains nickel. It is as yet unknown whether HycI endopeptidase directly interacts with nickel or whether coordination of the metal causes the precursor to adopt a conformation compatible with cleavage. The P1 Arg residue corresponds to the conserved His residue (see above) present in most other hydrogenase large subunits. It can be replaced by a His residue without loss of activity (unpublished results of the author). The P1′ Met residue is also not strictly conserved in other hydrogenases, but must be a hydrophobic residue (Ile, Val, Ala) (Rossmann *et al*., 1995). Peptidase activity is optimal between pH 7.0 and 7.5. Activity is not inhibited by PMSF, benzamidine or EDTA.

## Structural Chemistry

HycI endopeptidase is a monomeric protein with a molecular mass of 17 kDa, containing no metal ions. There are no primary structure motifs characteristic of serine, cysteine or metalloproteases. A comparison of sequences of hydrogenase maturation proteases (derived from the nucleic acid sequences), however, indicates the existence of a conserved Asn, an Asp and a His residue (see Alignment 543.1 on the CD-ROM). HycI lacks the N-terminal Met residue.

## Preparation

HycI can be purified (4000-fold) from *Escherichia coli* cells grown under fermentative conditions. Recombinant enzyme can be obtained in milligram amounts. Prerequisites for overexpression of the gene carried on a plasmid are that the GTG initiation codon of *hycI* is changed to the canonical ATG, and that a tightly controlled expression system is used.

## Biological Aspects

The function of HycI endopeptidase is the formation of active hydrogenase 3 in *E. coli*. It is thought (Rossmann *et al*., 1995; Binder *et al*., 1996) that the large subunit in its precursor form has an open conformation able to coordinate the nickel ligand; once nickel has been inserted, the C-terminal processing catalyzed by HycI endopeptidase initiates a conformational change, burying the metallocenter inside the large subunit where it has been located by X-ray crystallography (Volbeda *et al*., 1995).

Homologs of the *hycI* gene have been identified in each organism possessing [NiFe]hydrogenases. The sequence conservation of the gene products, with the exception of the Asn, Asp and His residues mentioned above, is low, paralleling the low conservation of sequence in the C-terminal propeptide that is cleaved off (Rossmann *et al*., 1995). This may be the biochemical basis for the strict substrate specificity of the endopeptidase. For example, inactivating the *hycI* gene in *E. coli* leads to the maturation defect solely of hydrogenase 3. Active isoenzymes 1 and 2 are still formed since the operons coding for their structural proteins also contain a gene homologous to *hycI*. The exact molecular basis of the substrate specificity, however, is still unresolved. All genes homologous to *hycI* are coexpressed with the respective hydrogenase structural genes, which guarantees that the endopeptidase is synthesized together with its substrate.

## Distinguishing Features

Neither the primary structure nor the inhibitory pattern allows the classification of HycI into the serine, cysteine, aspartic or metallo peptidase groups. The involvement in hydrogenase maturation is distinctive (Maier & Böck, 1996a).

## Further Reading

For a review, see Maier & Böck (1996b).

## References

Binder, U., Maier, T. & Böck, A. (1996) Nickel incorporation into hydrogenase 3 from *Escherichia coli* requires the precursor form of the large subunit. *Arch. Microbiol.* **165**, 69–72.

Gollin, D.J., Mortenson, L.E. & Robson, R.L. (1992) Carboxyl-terminal processing may be essential for production of active NiFe hydrogenase in *Azotobacter vinelandii*. *FEBS Lett.* **309**, 371–375.

Maier, T. & Böck, A. (1996a) Generation of active [NiFe] hydrogenase *in vitro* from a nickel-free precursor form. *Biochemistry* **35**, 10089–10093.

Maier, T. & Böck, A. (1996b) Nickel incorporation into hydrogenases. In: *Mechanisms of Metallocenter Assembly* (Hausinger,

R.P., Eichhorn, G.L. & Marzilli, L.G., eds). New York: VCH Publishers Inc., pp. 173–192.

Menon, N.K., Robbins, J., Vartanian, M.D., Patil, D., Peck, H.D., Menon, A.L., Robson, R.L. & Przybyla, A.E. (1993) Carboxyterminal processing of the large subunit of (NiFe) hydrogenases. *FEBS Lett.* **331**, 91–95.

Rossmann, R., Sauter, M., Lottspeich, F. & Böck, A. (1994) Maturation of the large subunit (HycE) of hydrogenase 3 of *Escherichia coli* requires nickel incorporation followed by C-terminal processing at Arg537. *Eur. J. Biochem.* **220**, 377–384.

Rossmann, R., Maier, T., Lottspeich, F. & Böck, A. (1995) Characterisation of a protease from *Escherichia coli* involved in

hydrogenase maturation. *Eur. J. Biochem.* **227**, 545–550.

Sauter, M., Böhm, R. & Böck, A. (1992) Mutational analysis of the operon (*hyd*) determining hydrogenase 3 formation in *Escherichia coli. Mol. Microbiol.* **6**, 1523–1532.

Sorgenfrei, O., Linder, D., Karas, M. & Klein, A. (1993) A novel very small subunit of a selenium containing [NiFe] hydrogenase of *Methanococcus voltae* is posttranslationally processed by cleavage at a defined position. *Eur. J. Biochem.* **213**, 1355–1358.

Volbeda, A., Charon, M.H., Piras, C., Hatchikian, E.C., Frey, M. & Fontecilla-Camps, J.C. (1995) Crystal structure of the nickel-iron-hydrogenase from *Desulfovibrio gigas. Nature* **373**, 580–587.

*August Böck*
*Lehrstuhl für Mikrobiologie,*
*Universität München,*
*Maria-Ward-Strasse 1a,*
*D-80638 München, Germany*

# *562. Protease IV*

## *Databanks*

*Peptidase classification: clan UX, family U7, MEROPS ID: U07.001*
*NC-IUBMB enzyme classification: none*
*Databank codes:*

| Species | SwissProt | PIR | EMBL (cDNA) | EMBL (genomic) |
|---------|-----------|-----|-------------|----------------|
| *Campylobacter jejuni* | – | – | U38524 | – |
| *Escherichia coli* | P08395 | A24813 | M13359 | – |
| *Haemophilus influenzae* | P14181 | – | M27280 | – |
| *Haemophilus influenzae* | P45243 | – | U32829 | – |
| *Leptospira borgpetersenii* | – | – | L27482 | – |
| *Methanococcus jannaschii* | – | – | – | U67512: genome section 54 of 150 |
| *Mycobacterium tuberculosis* | – | – | – | Z84395: complete cosmid Y210 |
| *Mycoplasma capricolum* | P43044 | – | D14982 | – |
| *Mycoplasma hominis* | P43052 | – | X77529 | – |
| *Synechocystis* sp. | – | – | – | D64000: genome section 19 of 27 |
| *Synechocystis* sp. | – | – | – | D90908: genome section 10 of 27 |

## *Name and History*

**Protease IV** was first described by Pacaud (1982a,b) as one of two membrane-associated *Escherichia coli* activities able to hydrolyze N-blocked amino acid *p*-nitrophenyl esters. Protease IV is associated with the cytoplasmic membrane and prefers substrates with aliphatic side chains (Ala, Val, Leu) whereas protease V is associated with the outer membrane and preferentially hydrolyzes the Phe-containing substrate. Considerable overlap in specificity between the two enzymes was reported (Pacaud, 1982b). Interest in the enzyme was stimulated by the finding of Ichihara *et al.* (1984) that protease IV is able to degrade the lipoprotein signal peptide *in vitro*. Since the signal peptide was attacked only after its release from the lipoprotein precursor, protease IV is a ***signal***

*peptide peptidase* and its gene was designated *sppA* (Suzuki *et al.*, 1987). It should be noted that *sppA* encodes a substantially larger protein (67 kDa monomer) (Ichihara *et al.*, 1986) than reported by Pacaud (1982b) for her purified protease IV (34 kDa). Ichihara *et al.* (1986) suggest that their protease IV is the same activity studied by Pacaud, but that the major protein in her purified preparation was a lower molecular mass contaminant. For the purposes of the present article we shall assume that the two activities are the same, but this assumption may not be correct. Another 'protease IV' described by Regnier (1981) is clearly different from Pacaud's protease IV.

## Activity and Specificity

Protease IV hydrolyzes the N-blocked (Z- or Boc-) *p*-nitrophenyl esters of Gly, Ala, Phe, Val, Leu and (more slowly) Trp (Pacaud, 1982b). Z-Val+NHPhNO$_2$ has been used as substrate for the assay of protease IV (Pacaud, 1982b; Ichihara *et al.*, 1984). Protease IV does not hydrolyze casein although it appears to hydrolyze some of the proteins present in a detergent extract of *E. coli* membranes (Pacaud, 1982b; Palmer & St. John, 1987). Analysis of the peptide products of the protease IV-catalyzed hydrolysis of the 20 amino acid lipoprotein signal peptide suggests that protease IV requires neither a free C- nor N-terminus, prefers hydrophobic amino acids on either side of the scissile bond and will not cleave a peptide containing fewer than six amino acids. The enzyme is most active at pH 7.2–7.6. Pacaud (1982b) found protease IV to be inhibited by DFP (100%, 1 mM, 1 h), PMSF, and Ac-Ala-Phe-CH$_2$Cl, but not by EDTA, *p*-aminobenzamidine, PCMB, iodoacetate, Tos-Lys-CH$_2$Cl, or 2-ME. Ichihara *et al.* (1984) reported inhibition by antipain, leupeptin, chymostatin and elastatinal, although the reversibility of the inhibition by antipain seemed to vary with different preparations of the inhibitor (Ichihara *et al.*, 1986).

## Structural Chemistry

The *sppA* gene encodes a 67 kDa (618 amino acid) protein and the active enzyme (268 kDa) is believed to be a tetramer. Comparisons of deduced amino acid sequences show that the *E. coli* enzyme is closely related to a protein (68 kDa, 615 amino acids) from *H. influenzae*. A somewhat smaller *E. coli* protein SohB (349 amino acids) related to these enzymes is encoded by the *sohB* gene (Baird *et al.*, 1991). Several other proteins, from the cyanobacterium *Synechocystis*, the spirochete *Leptospira borgpetersenii*, and the archaeon *Methanococcus janaschii* are clearly related to protease IV. All of these predicted proteins are substantially smaller than protease IV (204–285 amino acids), however, and show clear similarities to a region of *E. coli* protease IV beginning at about amino acid 370 and extending to about amino acid 525. This region includes three conserved Ser residues (comparing *E. coli* and *H. influenzae* protease IVs, SohB, the three proteins for *Synechocystis*, *Leptospira* and *Methanococcus*).

## Preparation

Protease IV has been purified from strains of *E. coli* carrying the *sppA* gene on a high copy number plasmid (Ichihara, 1986) and from a strain carrying only a singly copy of the gene (Pacaud, 1982b).

## Biological Aspects

Based on *in vitro* studies, protease IV of *E. coli* has been implicated in the degradation of signal peptides (Ichihara *et al.*, 1984, 1986; Suzuki *et al.*, 1987; Novak & Dev, 1988) after they are released from precursor proteins. Such peptides are known to be so rapidly degraded that they are essentially undetectable *in vivo*. Mutants lacking protease IV show no phenotype that indicates signal peptide degradation is defective, however, and it is not clear exactly what role protease IV plays in the cell.

## Distinguishing Features

Protease IV can be distinguished from protease V, another *E. coli* membrane-associated activity by its ability to hydrolyze Z-Val-OPhNO$_2$. It differs from protease VI, yet another *E. coli* membrane-associated enzyme, in its resistance to inhibition by benzamidine and *p*-aminobenzamidine (Palmer & St. John, 1987).

## Further Reading

The initial characterization of protease IV is described by Pacaud (1982b). Its identification as a signal peptide peptidase and the cloning and sequencing of *sppA* are reported in the papers of Ichihara *et al.* (1984, 1986). Additional data concerning the role of protease IV as a signal peptide peptidase and the possible involvement of other enzymes in this process are presented by Novak & Dev (1988).

## References

Baird, L., Lipinska, B., Raina, S. & Georgopoulos, C. (1991) Identification of the *Escherichia coli sohB* gene, a multicopy suppressor of the HtrA (DegP) null phenotype. *J. Bacteriol.* **173**, 5763–5770.

Ichihara, S., Beppu, N. & Mizushima, S. (1984) Protease IV, a cytoplasmic membrane protein of *Escherichia coli*, has signal peptide peptidase activity. *J. Biol. Chem.* **259**, 9853–9857.

Ichihara, S., Suzuki, T., Suzuki, M. & Mizushima, S. (1986) Molecular cloning and sequencing of the *sppA* gene and characterization of the encoded protease IV, a signal peptide peptidase, of *Escherichia coli*. *J. Biol. Chem.* **261**, 9405–9411.

Novak, P. & Dev, I.K. (1988) Degradation of a signal peptide by protease IV and oligopeptidase A. *J. Bacteriol.* **170**, 5067–5075.

Pacaud, M. (1982a) Identification and localization of two membrane-bound esterases from *Escherichia coli*. *J. Bacteriol.* **149**, 6–14.

Pacaud, M. (1982b) Purification and characterization of two novel proteolytic enzymes in membranes of *Escherichia coli*. Protease IV and protease V. *J. Biol. Chem.* **257**, 4333–4339.

Palmer, S.M. & St. John, A.C. (1987) Characterization of a membrane-associated serine protease in *Escherichia coli*. *J. Bacteriol.* **169**, 1474–1479.

Regnier, P. (1981) The purification of protease IV of *E. coli* and the demonstration that it is an endoproteolytic enzyme. *Biochem. Biophys. Res. Commun.* **99**, 1369–1376.

Suzuki, T., Itoh, A., Ichihara, S. & Mizushima, S. (1987) Characterization of the *sppA* gene coding for protease IV, a signal peptide peptidase of *Escherichia coli. J. Bacteriol.* **169**, 2523–2528.

*Charles G. Miller*
*Department of Microbiology,*
*B103 Chemical and Life Sciences Laboratory MC-110,*
*University of Illinois at Champagne-Urbana,*
*Urbana, IL 61801, USA*
*Email: charlesm@uiuc.edu*

# 563. *Bacteriophage T4 prohead endopeptidase*

## Databanks

*Peptidase classification: clan UX, family U9, MEROPS ID: U09.001*
*NC-IUBMB enzyme classification: none*
*Databank codes:*

| Species | SwissProt | PIR | EMBL (cDNA) | EMBL (genomic) |
|---|---|---|---|---|
| Bacteriophage T4 | P06807 | JF0025 | J02512 M15359 | – |

## Name and History

Bacteriophages are double-stranded DNA viruses that infect bacteria, and bacteriophage T4 infects *Escherichia coli*. Infection is usually followed by incorporation of phage DNA into the host cell genome as a prophage (lysogeny). Subsequently, the prophage is activated and phage proteins and DNA are synthesized by the host. Assembly of phage proteins into a nucleic acid-free prohead then takes place. The proteins assemble around a central scaffold protein. Cleavage of the capsid proteins confers stability while degradation of the scaffold permits phage DNA to enter the prohead. In bacteriophage T4, the prohead is bound to the bacterial cell membrane, and the mature phage is able to lyse the host cell and escape into the medium.

The phage T4 prohead protease has been described as the key enzyme in the morphopoietic pathway of the phage prohead (Hintermann & Kuhn, 1992). Cleavage of the phage proteins during head assembly was reported in 1970 by Dickson *et al.* (1970), Hosoda & Kone (1970), Kellenberger & Kellenberger-van der Kamp (1970) and Laemmli (1970) (who reported degradation of the scaffold protein, P22). Two capsid proteins, four assembly core components and the B1 protein are cleaved (Showe *et al.*, 1976a). Onorato & Showe (1975) demonstrated that all these proteins could be cleaved *in vitro* by the product of gene 21. Showe *et al.* (1976b) were able to isolate the product of gene 21 and show that it was also able to degrade itself. It is an autolytically activated form of the protein that is the active peptidase. The nucleotide sequence of gene 21 was determined by Keller & Bickle (1986).

Hintermann & Kuhn (1992) found that gene 21 encodes two proteins, confirming the observation of Keller & Bickle (1986) that a second ATG codon (Met45) in the sequence could act as an alternative initiation site.

The **T4 prohead proteinase** has also been called the **T4 prehead proteinase** (**T4PPase**), and **P21*** (Showe *et al.*, 1976a).

## Activity and Specificity

The only substrates known for the T4 prohead proteinase are the T4 prohead proteins. *In vivo*, assembly into a prohead structure is a prerequisite for cleavage of these proteins, but the individual proteins can be cleaved by the enzyme in *in*

**U**

*vitro* assays. The proteins cleaved include the capsid proteins P22 and P23, core proteins IPI, IPII and IPIII, and B1 protein. Cleavage of P22, IPII, IPIII and P68 is at Glu╪Ala bonds, but other proteins with Glu-Ala bonds are not cleaved (Showe *et al.*, 1976a). Isobe *et al.* (1976) compared sites at which cleavage occurs with those at which it does not, and described a motif for cleavage: Xaa-Xbb-Glu╪Ala-Xcc-Xaa, in which Xaa residues are hydrophobic (Leu, Ile or Val), Xbb may be limited to Thr or Ala, and Xcc is hydrophilic. Not all Glu╪Ala bonds cleaved fit this consensus well, however, and one exception is in the minor protein gp68 (Keller *et al.*, 1985). Moreover, cleavage at Glu╪Gly bonds was observed in P23 (Showe *et al.*, 1976a). The scaffold protein, P22, has been found to be degraded to at least 30 peptides (Showe *et al.*, 1976a). The T4 prohead proteinase precursor contains three Glu-Gly bonds and two Glu-Ala bonds, cleavage of which could account for its autoactivation and self-destruction.

The standard assay for T4 prohead proteinase is done in 1 M Tris–HCl, pH 7.6. The substrate is a mixture of radiolabeled proteins in which P22 is labeled only with $^{14}$C, and other proteins are labeled with both $^{14}$C and $^{3}$H (Onorato & Showe, 1975; Showe *et al.*, 1976a). After incubation for 20 min at 37°C, the reaction is stopped by the addition of 10% trichloroacetic acid and serum albumin. After standing for 10 min, the samples are centrifuged, and radioactivity of a sample of the trichloroacetic acid supernatant is determined. At least a 2-fold excess of $^{14}$C released over the $^{14}$C/$^{3}$H ratio in the substrate mixture is expected (Onorato & Showe, 1975; Showe *et al.*, 1976a). The endopeptidase is active over a wide pH range, 6.5–9.5, with an optimum at pH 8.5 (Showe *et al.*, 1976a).

The T4 prohead proteinase has been shown to be unaffected by DFP, Tos-Phe-CH$_2$Cl, HgCl$_2$ and iodoacetate, and not to be activated by 2-ME or cysteine; it has been concluded that the enzyme is not a serine- or cysteine-type peptidase. Activity was enhanced by MnCl$_2$, but not by any other divalent cation, and neither EDTA nor EGTA had any effect on activity (Showe *et al.*, 1976a).

## Structural Chemistry

The $M_r$ of the precursor of the prohead endopeptidase is predicted from the deduced sequence to be 23 251, or possibly rather less, according to the start codon used (Keller & Bickle, 1986). Autocatalytic activation is known to be required, and the active enzyme has an $M_r$ of 18.5 kDa in SDS-PAGE (Showe *et al.*, 1976b).

The amino acid sequence of the proteinase reveals no relationship to any other protein. The proteinase has not been crystallized.

## Preparation

In the purification scheme of Showe *et al.* (1976a), infected colonies of *E. coli* were homogenized in 50 mM potassium phosphate, pH 6.0, containing DNAase. Cell debris was removed by centrifugation at 170 000 × *g* for 4 h. The pellet was re-extracted twice with the phosphate buffer. The combined supernatants were precipitated with saturated ammonium sulfate (pH 6.0) for 2 h. The pellet was resuspended in 0.05 M potassium phosphate (pH 6.0), and applied to a

DEAE-cellulose column. The column was washed with a linear gradient from 0.1 M to 0.3 M NaCl, and the proteinase was eluted at 0.25 M NaCl. The peak fractions were pooled, dialyzed against 5 mM sodium citrate containing 0.1 mM MnCl$_2$ (which is essential to maintain activity), and applied to a Sephadex G100 column. The proteinase was eluted in the void volume. Peak fractions were lyophilized, and applied to a second DEAE-cellulose column. Activity was eluted with steps of increasing NaCl concentration, and peaks of activity obtained at 0.1 M and 0.12 M were combined as T4PPase I. Peaks of activity at 0.08 M and 0.14 M NaCl were rechromatographed on DEAE-cellulose to give T4PPase II. T4PPase I and II were found to be identical in all respects.

## Biological Aspects

The prohead proteinase is synthesized as a precursor that autoactivates and is subsequently degraded, at least partially by autolysis. About 100 precursor molecules are incorporated centrally into the phage prohead (Van Driel *et al.*, 1980). The enzyme is responsible for processing several capsid protein precursors, which is a prerequisite for stability of the bacteriophage prohead (Showe *et al.*, 1976a; Isobe *et al.*, 1976). It has been shown that the polypeptide removed from the N-terminus of internal protein IPIII acts as a targeting signal for incorporation into the prohead (Mullaney & Black, 1996). The proteinase also degrades the scaffold protein P22, around which the prohead is assembled. Degradation permits entry of phage DNA into the prohead (Laemmli, 1970). Cleavage of the prohead proteins and entry of the phage DNA are steps required for phage maturation, after which lysis of the host cells and infection of other cells will take place.

## Related Peptidases

The T4 prohead proteinase shows no homology to any other protein. Protein C from bacteriophage λ is homologous to protease IV (Chapter 562) of *Escherichia coli*, implying that it may be a peptidase. Protein C is known to be located at the center of the prohead and necessary for the processing of protein B to its active form. It may also be responsible for degrading the scaffold protein (Kochan & Murialdo, 1983), which is a product of the same gene at an alternate initiation site.

## Further Reading

The catalytic activity of the prohead proteinase was detected in 1976 by Showe *et al.* (1976a) and Isobe *et al.* (1976). Showe *et al.* (1976a,b) went on to purify and characterize the endopeptidase. The sequence was determined by Keller & Bickle (1986). Hintermann & Kuhn (1992) demonstrated that gene *21* also encoded the scaffold protein. Muller *et al.* (1994) demonstrated that structural changes and lattice expansion accompanied proteolytic cleavage of the prohead proteins. Mullaney & Black (1996) were able to show that the prohead proteinase removes a capsid targeting sequence (CTS) from the N-terminus of internal protein III (IPIII), and that engineering the CTS to the N-terminus of a fragment of luciferase permits the protein to be packaged and processed in the capsid.

## References

Dickson, R.C., Barnes, S.L. & Eiserling, F.A. (1970) Structural proteins of bacteriophage T4. *J. Mol. Biol.* **53**, 461–473.

Hintermann, E. & Kuhn, A. (1992) Bacteriophage T4 gene 21 encodes two proteins essential for phage maturation. *Virology* **189**, 474–482.

Hosoda, J. & Cone, R. (1970) Analysis of T4 phage proteins. I. Conversion of precursor proteins into lower molecular weight peptides during normal capsid formation. *Proc. Natl Acad. Sci. USA* **66**, 1275–1281.

Isobe, T., Black, L.W. & Tsugita, A. (1976) Protein cleavage during virus assembly: a novel specificity of assembly dependent cleavage in bacteriophage T4. *Proc. Natl Acad. Sci. USA* **73**, 4205–4209.

Kellenberger, E. & Kellenberger-van der Kamp, C. (1970) On a modification of the gene product *P23* according to its use as subunit of either normal capsids of phage T4 or of polyheads. *FEBS Lett.* **8**, 140–144.

Keller, B. & Bickle, T.A. (1986) The nucleotide sequence of gene *21* of bacteriophage T4 coding for the prohead protease. *Gene* **49**, 245–251.

Keller, B., Kellenberger, E., Bickle, T.A. & Tsugita, A. (1985) Determination of the cleavage site of the phage T4 prohead protease in gene product 68. Influence of protein secondary structure on cleavage specificity. *J. Mol. Biol.* **186**, 66

Kochan, J. & Murialdo, H. (1983) Early intermediates in bacteriophage lambda prohead assembly. II. Identification of biologically active intermediates. *Virology* **131**, 100–115.

Laemmli, U.K. (1970) Cleavage of structural proteins during the assembly of the head of bacteriophage T4. *Nature* **227**, 680.

Mullaney, J.M. & Black, L.W. (1996) Capsid targeting sequence targets foreign proteins into bacteriophage T4 and permits proteolytic processing. *J. Mol. Biol.* **261**, 372–385.

Muller, M., Mesyanzhinov, V.V. & Aebi, U. (1994) *In vitro* maturation of prehead-like bacteriophage T4 polyheads – structural changes accompanying proteolytic cleavage and lattice expansion. *J. Struct. Biol.* **112**, 199–215.

Onorato, L. & Showe, M.K. (1975) Gene *21* protein-dependent proteolysis *in vitro* of purified gene *22* product of bacteriophage T4. *J. Mol. Biol.* **92**, 395–412.

Showe, M.K., Isobe, E. & Onorato, L. (1976a) Bacteriophage T4 prehead proteinase. I. Purification and properties of a bacteriophage enzyme which cleaves capsid protein precursors. *J. Mol. Biol.* **107**, 35–54.

Showe, M.K., Isobe, E. & Onorato, L. (1976b) Bacteriophage T4 prehead proteinase. II. Its cleavage from the product of gene 21 and regulation in phage-infected cells. *J. Mol. Biol.* **107**, 55–69.

Van Driel, R., Traub, F. & Showe, M.K. (1980) Probable localization of the bacteriophage T4 prehead proteinase zymogen in the center of the prehead core. *J. Virol.* **36**, 220–223.

*Neil D. Rawlings*
*MRC Peptidase Laboratory,*
*The Babraham Institute,*
*Cambridge, Cambs CB2 4 AT, UK*
*Email: neil.rawlings@bbsrc.ac.uk*

# 564. Infectious pancreatic necrosis virus endopeptidase

## Databanks

*Peptidase classification: clan UX, family U43, MEROPS ID: U43.001*
*NC-IUBMB enzyme classification: none*
*Databank codes:*

| Species | Strain | SwissProt | PIR | EMBL (cDNA) | EMBL (genomic) |
|---|---|---|---|---|---|
| Infectious pancreatic necrosis virus | Jasper | P05844 | A23599 | X04124 | – |
| Infectious pancreatic necrosis virus | NI | P22495 | B34148 | D00701 | – |

## Name and History

The genome segment A of the birnavirus, infectious pancreatic necrosis virus (IPNV), contains a large open reading frame of 2900–2903 nucleotides (Duncan & Dobos, 1986; Haverstein *et al*., 1990; Mason, 1992; Heppell *et al*., 1993; Chung *et al*., 1994; Tseng *et al*., 1996). The primary translation product of this segment is a polyprotein with the gene order pVP2–NS–VP3. It is subsequently cleaved into its three component viral proteins, pVP2 (major capsid protein precursor), NS (protease), and VP3 (minor capsid protein) (Duncan & Dobos, 1986; Haverstein *et al*., 1990; Mason, 1992) (Figure 564.1A). The precise borders for each of these genes have not yet been determined. The processing of the precursor polyprotein into its component parts has been attributed to the activity of the NS protein (Duncan *et al*., 1987; Manning *et al*., 1990) whose amino acid sequence is unique and unlike other proteases (Mason, 1992).

## Activity and Specificity

Fine structure mapping of the coding region for the NS protease was conducted by *in vitro* translation of 5′ and 3′ truncated and deleted mRNAs and comparison of the products produced to that produced by full-length mRNA (Duncan *et al*., 1987; Manning & Leong, 1990; Manning *et al*., 1990). Although these studies defined a region where the protease cleaves the polyprotein (see Figure 564.1A), the exact cleavage sites at either end of NS have not been determined. Amino acid sequence analysis from the N-terminal of VP3 indicated that the terminus was blocked (Mason, 1992).

The NS proteolytic activity has been localized through plasmid deletion analysis and site-directed mutagenesis to a region spanning approximately 740 nt. This sequence encodes amino acids 455–721 in the polyprotein (Mason, 1992). The deletion analysis was used to identify critical regions in NS by either removing a proteolytic active site or by deleting a region critical to the native folding of NS. C-terminal truncation of NS by 30–35 amino acid residues abolished proteolytic cleavage at the upstream pVP2/NS junction (Duncan *et al*., 1987; Nagy *et al*., 1987; Manning & Leong, 1990; Mason, 1992). Similarly, N-terminal truncation of NS by 80–85 amino acid residues resulted in the loss of cleavage at the NS/VP3 junction (Manning *et al*., 1990; Mason, 1992). When the deletion at the N-terminus was confined to a region of 15–20 amino acid residues, the cleavage at the NS/VP3 junction proceeded as normal (Mason, 1992; L. Perez & J.C. Leong, unpublished results). In all cases, the regions surrounding the putative pVP2/NS and NS/VP3 cleavage sites were required. The viral protease did not work in *trans* and antibody to NS did not block *in vitro* cleavage of the polyprotein (Manning *et al*., 1990; Magyar & Dobos, 1994). These results are in sharp contrast to those obtained with other virus-encoded proteases.

Site-specific mutations were introduced into the cDNA clone of the viral genome segment A in order to identify the active site of the protease and its proteolytic processing sites. Possible cleavage sites were identified on the basis of

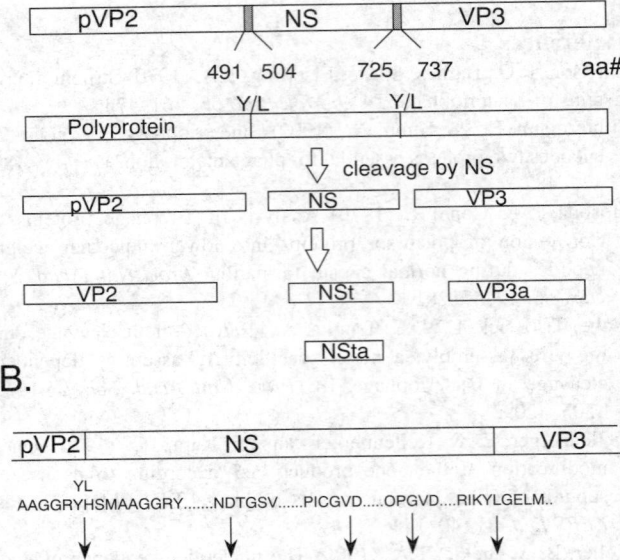

*Figure 564.1* (A) Schematic representation of the gene arrangement of genome segment A of IPNV. The bars represent the different proteins encoded by the A segment. The amino acid residues number 491–504 and 725–737 contain the Tyr-Leu dipeptides that may be the cleavage sites for the NS protease. The initial cleavage by NS to yield pVP2, NS and VP3 occurs very rapidly. The further processing of pVP2 to VP2 occurs at its C-terminus, but the exact cleavage site is not known. VP3a is a truncated form of VP3 and whether this processing occurs at the N- or C-terminus of VP3 is unknown. The same holds true for NSt and NSta. (B) Site-directed mutagenesis at the putative catalytic site and cleavage sites of IPNV NS coding region. The bar represents the NS coding region with the putative cleavage sites represented by a vertical line at the pVP2/NS and NS/VP3 junctions. Amino acids are shown by single-letter code. The arrows mark the five mutations introduced by site-directed mutagenesis. The dipeptide Tyr-His is found instead of Tyr-Leu at position 491 in the Sp strain of IPNV (Mason, 1992).

conserved amino acid residues and the predicted molecular masses of the IPNV proteins. Those sites are indicated on Figure 564.1A. The only dipeptide occurring at either side of the NS sequence for several different IPNV strains is a Tyr-Leu dipeptide; the same dipeptide brackets the VP4 region of infectious bursal disease virus, a birnavirus that infects poultry (Hudson *et al*., 1986; Duncan *et al*., 1987). Site-directed mutagenesis in combination with *in vitro* transcription and translation revealed that changing the Try-Leu at the NS/VP3 junction to Pro-Gly resulted in a great reduction in the proteolytic cleavage at the NS/VP3 junction (Mason, 1992). These results suggested that Tyr-Leu is either the processing site of the NS/VP3 junction or that the Pro-Gly introduced a secondary structural change in the protein that interfered

*Table 564.1* Effect of proteinase inhibitors on the IPNV-NS proteinase

| Inhibitor | Proteinase class specificity | Inhibitor concentration tested | Inhibition |
|---|---|---|---|
| Tos-Lys-CH$_2$Cl | Serine | 37 µg ml$^{-1}$ | None |
| Tos-Phe-CH$_2$Cl | Serine | 70 µg ml$^{-1}$ | None |
| EDTA | Metallo | 100 µM | None |
| Pepstatin A | Aspartic | 0.7 µg ml$^{-1}$ | None |
| H261[a] | Aspartic | 2 µM | None |
| Iodoacetamide | Cysteine | 10 µM | None |
| N-Ethylmaleimide | Cysteine | 10 µM | None |
| ZnSO$_4$ | Cysteine | 20 µM | None |

From Mason (1992).

[a]H261 is an aspartic proteinase inhibitor kindly provided by Dr John Kay, University of Cardiff, Wales, UK. Its structure is Boc-His-Pro-Phe-His-Leu[CHOH-CH$_2$]Val-Ile-His.

with normal processing. A similar change in Tyr-Leu at the pVP2/NS junction did not inhibit cleavage at that site.

Computer alignment of the NS sequence with those of other known proteinases uncovered no clear sequence similarities. However, the NS proteinase sequence alignments can be manipulated so that sequence similarities with the aspartic and serine-like cysteine proteinases are discernible. In this case, a canonical Asp-Thr-Gly sequence generally present in aspartic endopeptidases (Chapter 270) was observed at position 585. When this was mutated to Val-Thr-Gly (Figure 564.1B), no effect on the proteolytic processing of NS was observed. Similarity alignments identified a Cys residue at 669 as a potential catalytic site should NS be a serine-like cysteine proteinase. When this was changed to Ser, no reduction in proteolytic activity was observed. Duncan and coworkers also identified a slow processing mutant of NS which had a Glu instead of Gly residue at position 566 and an Arg instead of Gly at position 691 (Duncan *et al.*, 1987; Nagy *et al.*, 1987). The Gly at 691 was changed to Arg by site-specific mutagenesis (Mason, 1992). No effect on the proteolytic processing was observed. Unfortunately, the change from Gly to Glu at position 566 was not made.

A number of proteinase inhibitors were used in an *in vitro* translation system to determine whether the NS protease could be classified according to its sensitivity profile. None of the eight inhibitors used (Table 564.1) altered the processing pattern. The combined results of the inhibitor studies and site-directed mutagenesis experiments indicate that NS may belong to novel class of viral proteinase (Mason, 1992).

## *Preparation*

NS has been prepared from virus purified and then dissociated by treatment with guanidine hydrochloride and 2-ME. The dissociated proteins were separated on a Sepharose G-200 column with 6 M guanidine/2-ME elution buffer. The NS protein has been expressed in *Escherichia coli*, but it is inactive (Manning *et al.*, 1990; Mason, 1992).

## *Biological Aspects*

The enzyme is an autocatalytic protease that cleaves a large polyprotein synthesized by the large genome segment A of IPNV. IPNV belongs to the virus family Birnaviridae, which is characterized as an icosahedral virus containing two double-stranded RNA genome segments. The large genome segment A encodes the major capsid protein pVP2, NS (protease), and VP3 (minor capsid protein). There is an additional 17 kDa protein encoded on a different reading frame in A as well. The smaller segment B encodes the virion RNA-dependent RNA polymerase.

Several investigations have shown that the NS proteinase participates in the cleavage of the A segment-encoded polyprotein, but its role in the processing of pVP2 to VP2 is unclear and the processing of NS to NSt and VP3 to VP3a is also unknown (see Figure 564.1A).

## References

Chung, H.-K., Lee, S.-H., Lee, H.-H., Lee, D.-S. & Kim, Y.-S. (1994) Nucleotide sequence analysis of the VP2-NS-VP3 gene of infectious pancreatic necrosis virus DRT strain. *Mol. Cells* **4**, 349–354.

Duncan, R. & Dobos, P. (1986) The nucleotide sequence of infectious pancreatic necrosis virus (IPNV) dsRNA segment A reveals one large ORF encoding a precursor polyprotein. *Nucleic Acids Res.* **14**, 5934.

Duncan, R., Nagy, E., Krell, P.J. & Dobos, P. (1987) Synthesis of the infectious pancreatic necrosis virus polyprotein, detection of a virus-encoded protease, and the fine structure mapping of the genome segment A coding regions. *J. Virol.* **61**, 3655–3664.

Havarstein, L.S., Kalland, K.H., Christie, K.E. & Endresen, C. (1990) Sequence of the large double-stranded RNA segment of the N1 strain of infectious pancreatic necrosis virus: a comparison with other Birnaviridae. *J. Gen. Virol.* **71**, 299–308.

Heppell, J., Berthiaume, L., Corbin, F., Tarrab, E., Lecomte, J. & Arella, M. (1993) Comparison of amino acid sequences deduced from a cDNA fragment obtained from infectious pancreatic necrosis virus (IPNV) strains of different serotypes. *Virology* **195**, 84.

Heppell, J., Tarrab, E., Lecomte, J., Berthiaume, L. & Arella, M. (1995) Strain variability and localization of important epitopes on the major structural protein (VP2) of infectious pancreatic necrosis virus. *Virology* **214**, 40–49.

Hudson, P.J., McKern, N.M., Power, B.E. & Azad, A.A. (1986) Genomic structure of the large RNA segment of infectious bursal disease virus. *Nucleic Acids Res.* **14**, 5001–5012.

Magyar, G. & Dobos, P. (1994) Expression of infectious pancreatic necrosis virus polyprotein and VP1 in insect cells and the detection of the polyprotein in purified virus. *Virology* **198**, 437–445.

Manning, D.S. & Leong, J. C. (1990) Expression in *Escherichia coli* of the large genomic segment of infectious pancreatic necrosis virus. *Virology* **179**, 16–25.

Manning, D.S., Mason, C.L. & Leong, J.C. (1990) Cell-free translational analysis of the processing of infectious pancreatic necrosis virus polyprotein. *Virology* **179**, 9–15.

Mason, C.L. (1992) Molecular characterization of the proteins and RNA dependent RNA polymerase of infectious pancreatic necrosis virus, a fish birnavirus. PhD thesis, Oregon State University.

**U**

Nagy, E., Duncan, R., Krell, P.J. & Dobos, P. (1987) Mapping of the large RNA genome segment of infectious pancreatic necrosis virus by hybrid arrested translation. *Virology* **158**, 211–217.

Tseng, C.C., Lo, C.F. & Kou, G.H. (1996) Establishment and characterization of a IPNV SP strain persistent infection cell line. Gen Bank Accession No. U56907.

***Jo-Ann C. Leong***
*Department of Microbiology,*
*Oregon State University,*
*Corvallis, OR 97331-3804, USA*
*Email: leongj@ccmail.orst.edu*

***Carla Mason***
*Department of Microbiology,*
*Oregon State University,*
*Corvallis, OR 97331-3804, USA*

# 565. *Pestivirus N$^{pro}$ endopeptidase*

## Databanks

*Peptidase classification: clan UX, family U44, MEROPS ID: U44.001*
*NC-IUBMB enzyme classification: none*
*Databank codes:*

| Species | Strain/isolate | SwissProt | PIR | EMBL (cDNA) | EMBL (genomic) |
|---|---|---|---|---|---|
| Border disease virus | Clover Lane | – | – | U17143 | – |
| | 59386 | – | – | U17147 | – |
| Cattle viral diarrhea virus | NADL | P19711 | A29198 | M62430 | M31182: complete genome |
| | SD-1 | Q01499 | A44217 | – | M96751: complete polyprotein |
| | Pe515CP | – | – | M96640 | – |
| Hog cholera virus | Alfort | P19712 | A34037 | – | J04358: complete genome |
| | Brescia | P21530 | A35317 | – | M31768: complete genome |
| | CAP | – | – | – | X96550: complete genome |
| | Glentorf | – | – | – | U45478: complete genome |

## Name and History

Pestiviruses (family Flaviviridae, genus *Pestivirus*) comprise a group of small enveloped RNA viruses responsible for diseases in cattle (bovine viral diarrhea virus, BVDV), pigs (classical swine fever virus, CSFV) and sheep (border disease virus, BDV) (Thiel *et al.*, 1996). The positive-stranded genomic RNA with a length of usually 12.3 kb contains one large open reading frame and is translated into a polyprotein that is cleaved co- and post-translationally by host cell- and two virally-encoded proteases. One of these viral proteases is the pestivirus NS2–3/NS3 serine peptidase (Chapter 91). The second, the subject of the present chapter, is the *N-terminal protease*, for which the designation *N$^{pro}$* has been proposed by the Flaviviridae study group of the International Committee on the Taxonomy of Viruses (as an acronym for **N**-terminal **pro**tease) (Stark *et al.*, 1993). The observed molecular mass of 20–23 kDa led to the earlier designations *p20* or *p23* (Collett *et al.*, 1988; Rümenapf *et al.*, 1991). N$^{pro}$ was initially considered to be the viral capsid protein (Collett *et al.*, 1989; Rümenapf *et al.*, 1991). However, the viral capsid protein is represented by a 14 kDa protein (p14, C) that is located on the polyprotein just after N$^{pro}$ (Thiel *et al.*, 1991). The first indication that N$^{pro}$ might be an autoprotease was provided by the observation that a precursor of the protein could not be demonstrated regardless of whether pulse-chase experiments or *in vitro* translation of viral RNA were performed (Collett *et al.*, 1991; Rümenapf *et al.*, 1991; Thiel *et al.*, 1991). *In vitro* translation of the N$^{pro}$-C coding region showed that a proteolytic activity responsible for cleavage between both proteins resides within N$^{pro}$ (Wiskerchen *et al.*, 1991; Stark *et al.*, 1993). *In vitro* mutagenesis experiments and deletion experiments with BVDV N$^{pro}$ (Wiskerchen *et al.*, 1991) as well as *in vitro* translation of C-terminally truncated CSFV cDNA constructs encompassing the N$^{pro}$-C encoding region (Stark *et al.*, 1993) supported the hypothesis that the first protein translated by the pestiviral RNA is an autoprotease.

## Activity and Specificity

N$^{pro}$ is responsible for its own release from the nascent polyprotein, thereby generating the N-terminus of the following capsid protein C. The cleavage site was determined by N-terminal sequencing of CSFV protein C; cleavage occurs between Cys168+Ser169 (Stark *et al.*, 1993). Comparison of the cleavage site with sequences of other pestiviruses suggested a conserved processing site at which P1 is always Cys. In addition, the residues in the P8–P2 positions are highly conserved: -Cys-Pro-Leu-Trp-Val-Thr-Ser-Cys+Ser (Stark *et al.*, 1993). P1′ is occupied by Ser, although Ala and Gly can apparently be accepted by N<sup>pro</sup> (Wiskerchen *et al.*, 1991; Meyers *et al.*, 1992; Tautz *et al.*, 1994). According to sequence comparisons, proteolytic activity of N<sup>pro</sup> seems to be largely independent of sequences downstream of P1′. This conclusion is drawn from the analysis of some cytopathic BVDV strains (see below) (Meyers *et al.*, 1992; Tautz *et al.*, 1994).

Several studies of pestivirus gene expression suggest that N<sup>pro</sup> is involved only in a *cis* cleavage between its own C-terminus and protein C (Wiskerchen *et al.*, 1991). It has been reported that no *trans* activity of N<sup>pro</sup> could be detected (Wiskerchen *et al.*, 1991).

There are no published reports on inhibition of N<sup>pro</sup> by proteinase inhibitors for the different protease classes.

### Structural Chemistry

N<sup>pro</sup> is a single-chain, nonglycosylated protein usually with a length of 168 amino acids and a calculated pI of 9.0. However, it can be concluded from the analysis of a duplicated N<sup>pro</sup> sequence within the genome of a cytopathogenic BVDV strain (BVDV Pe515CP) that residues corresponding to amino acids 1–14 of the protease are not required for autoproteolytic activity (Meyers *et al.*, 1992). On the other hand, N-terminal extension of N<sup>pro</sup> is also possible without affecting proteolytic activity; this has been shown by expression of N<sup>pro</sup> as a fusion protein in *Escherichia coli* (Stark *et al.*, 1993) as well as detection of viral fusion proteins containing a duplicated N<sup>pro</sup> (Meyers *et al.*, 1992).

Computer-assisted analyses of pestiviral N<sup>pro</sup> sequences did not result in identification of motifs typical for proteases (Wiskerchen *et al.*, 1991). Since the experimental data strongly supported the presence of a protease within N<sup>pro</sup>, the pestiviral sequences were reinspected for potential catalytically active amino acid residues applying less stringent criteria. As a result, two hypotheses concerning the amino acid residues of the putative active center of the protease together with classification of N<sup>pro</sup> have been postulated:

1. *Serine-type peptidase*: Wiskerchen *et al.* (1991) provided experimental evidence that His49 is essential for activity of the endopeptidase. The surroundings of a Ser residue at position 124 appeared to be similar to the catalytic center Ser of serine-type peptidases (pestivirus motif Gly-Ser-Asp-Gly versus serine-type proteinase motif Gly-Asp-Ser-Gly). However, an exchange of Ser124 to Ala affected proteolytic activity of N<sup>pro</sup> only slightly. Looking at pestiviral sequences which were available at that time, the Gly-Ser-Asp-Gly motif appeared to be conserved (Wiskerchen *et al.*, 1991). However, using all the sequences which are available now it can be shown that two pestivirus strains do not contain the Gly-**Ser**-Asp-Gly sequence; CSFV strains Glentorf and CAP contain instead

the sequence Gly-**Asn**-Asp-Gly. It is therefore now considered unlikely that N<sup>pro</sup> is a serine-type peptidase.

2. *Cysteine-type peptidase*: Sequence comparisons indicated some common features between N<sup>pro</sup> and a group of virus-encoded papain-like cysteine proteases of clan CC (Chapter 225) (Gorbalenya *et al.*, 1991; Snijder *et al.*, 1992; Baker *et al.*, 1993; Agranovsky *et al.*, 1994; Roberts & Belsham, 1995): (a) the spacing of the two proposed catalytic residues (Cys69 and His130) is similar to the spacing observed for catalytic cysteine and histidine residues in clan CC, (b) several papain-like autoproteases of positive-stranded RNA viruses are located at the N-termini of their polyproteins and mediate a single cleavage at their own C-termini, (c) cleavage occurs within a certain distance range downstream of the active His (Stark *et al.*, 1993). However, experimental data supporting the suggested relationship of N<sup>pro</sup> to cysteine peptidases are not available.

### Preparation

N<sup>pro</sup> has been expressed in a proteolytically active form using different approaches including *in vitro* translation of viral or *in vitro* transcribed RNA, transient expression in mammalian cells, expression by vaccinia virus/pestivirus recombinants and expression in a prokaryotic system (Rümenapf *et al.*, 1991; Wiskerchen *et al.*, 1991; Stark *et al.*, 1993). Purification of N<sup>pro</sup> has not been reported.

### Biological Aspects

The strategy of gene expression used by pestiviruses implies that N<sup>pro</sup> can be found in every infected cell. The protease is expected to be located in the cytosol, maybe in association with membranes.

With regard to function of the nonstructural protein N<sup>pro</sup>, it appears unlikely that its only purpose is generation of the N-terminus of capsid protein C. Interestingly, polyproteins of the other members of the family Flaviviridae, namely flaviviruses and hepatitis C virus, start with the capsid protein C. Maintenance of N<sup>pro</sup>-encoding sequences during pestiviral evolution suggests functional importance in virus replication. However, the function is not known.

As already mentioned, the cleavage performed by N<sup>pro</sup> results in generation of the N-terminus of the following capsid protein C. In a few BVDV isolates, N<sup>pro</sup> mediates a second *cis* cleavage within the polyprotein due to a complex recombination event (Meyers *et al.*, 1992); such viruses are cytopathic and their occurrence is linked to a lethal disease, mucosal disease, in cattle (Thiel *et al.*, 1996). In the polyprotein encoded by these viruses, a duplicated N<sup>pro</sup> is juxtaposed to a viral nonstructural protein the N-terminus of which is generated in a *cis* cleavage performed by the second copy of N<sup>pro</sup> (Meyers *et al.*, 1992).

### Distinguishing Features

Rabbit polyclonal antisera against N<sup>pro</sup> have been prepared against bacterial fusion proteins as well as synthetic peptides (Collett *et al.*, 1988; Wiskerchen *et al.*, 1991; Stark *et al.*,

1993). Monoclonal antibodies against N^pro have not been described.

### Further Reading

The expression of N^pro by *in vitro* translation and transient expression in mammalian cells, and the mutation of amino acid residues important for proteolysis are described by Wiskerchen *et al*. (1991). Stark *et al*. (1993) describe expression of N^pro as a bacterial fusion protein, determination of the cleavage site, and sequence alignment to papain-like viral proteases.

### References

Agranovsky, A.A., Koonin, E.V., Boyko, V.P., Maiss, E., Frötschl, R., Lunina, N.A. & Atabekov, J.G. (1994) Beet yellows closterovirus: complete genome structure and identification of a leader papain-like thiol protease. *Virology* **198**, 311–324.

Baker, S.C., Yokomori, K., Dong, S., Carlisle, R., Gorbalenya, A.E., Koonin, E.V. & Lai, M.M.C. (1993) Identification of the catalytic sites of a papain-like cysteine proteinase of murine coronavirus. *J. Virol.* **67**, 6056–6063.

Collett, M.S., Larson, R., Belzer, S. & Retzel, E. (1988) Proteins encoded by bovine viral diarrhea virus: The genome organization of a pestivirus. *Virology* **165**, 200–208.

Collett, M.S., Moennig, V. & Horzinek, M.C. (1989) Recent advances in pestivirus research. *J. Gen. Virol.* **70**, 253–266.

Collett, M.S., Wiskerchen, M., Welniak, E. & Belzer, S.K. (1991) Bovine viral diarrhea virus genomic organization. *Arch. Virol.* **suppl. 3**, 19–27.

Gorbalenya, A.E., Koonin, E.V. & Lai, M.M. (1991) Putative papain-related thiol proteases of positive-strand RNA viruses. Identification of rubi- and aphtovirus proteases and delineation of a novel conserved domain associated with proteases of rubi-, alpha- and coronaviruses. *FEBS Lett.* **288**, 201–205.

Meyers, G., Tautz, N., Stark, R., Brownlie, J., Dubovi, E.J., Collett, M.S. & Thiel, H.-J. (1992) Rearrangement of viral sequences in cytopathogenic pestiviruses. *Virology* **191**, 368–386.

Roberts, P.J. & Belsham, G.J. (1995) Identification of critical amino acids within the foot-and-mouth disease virus leader protein, a cysteine protease. *Virology* **213**, 140–146.

Rümenapf, T., Stark, R., Meyers, G. & Thiel, H.-J. (1991) Structural proteins of hog cholera virus expressed by vaccinia virus: further characterization and induction of protective immunity. *J. Virol.* **65**, 589–597.

Snijder, E.J., Wassenaar, A.L.M. & Spaan, W.J.M. (1992) The 5′ end of the equine arteritis virus replicase gene encodes a papainlike cysteine protease. *J. Virol.* **66**, 7040–7048.

Stark, R., Meyers, G., Rhmenapf, T. & Thiel, H.-J. (1993) Processing of pestivirus polyprotein: cleavage site between autoprotease and nucleocapsid protein of classical swine fever virus. *J. Virol.* **67**, 7088–7095.

Tautz, N., Thiel, H.-J., Dubovi, E.J. & Meyers, G. (1994) Pathogenesis of mucosal disease: a cytopathogenic pestivirus generated by an internal deletion. *J. Virol.* **68**, 3289–3297.

Thiel, H.-J., Stark, R., Weiland, E., Rümenapf, T. & Meyers, G. (1991) Hog cholera virus: molecular composition of virions from a pestivirus. *J. Virol.* **65**, 4705–4712.

Thiel, H.-J., Plagemann, P.G.W. & Moennig, V. (1996) Pestiviruses. In: *Fields Virology*, 3rd edn (Fields, B.N., Knipe, D.M. *et al.*, eds). Philadelphia: Lippincott–Raven Publishers, pp. 1059–1073.

Wiskerchen, M., Belzer, S.K. & Collett, M.S. (1991) Pestivirus gene expression: the first protein product of the bovine viral diarrhea virus large open reading frame, p20, possesses proteolytic activity. *J. Virol.* **65**, 4508–4514.

*Robert Stark*
*Institut für Virologie (FB Veterinärmedizin),*
*Justus-Liebig-Universität Giessen,*
*Frankfurter Strasse 107, 35392*
*Giessen, Germany*
*Email: robert.stark@vetmed.uni-giessen.de*

*Tillmann Rümenapf*
*Institut für Virologie (FB Veterinärmedizin),*
*Justus-Liebig-Universität Giessen,*
*Frankfurter Strasse 107, 35392*
*Giessen, Germany*
*Email: till.h.ruemenapf@vetmed.uni-giessen.de*

*Heinz-Jürgen Thiel*
*Institut für Virologie (FB Veterinärmedizin),*
*Justus-Liebig-Universität Giessen,*
*Frankfurter Strasse 107, 35392*
*Giessen, Germany*
*Email: heinz-juergen.thiel@vetmed.uni-giessen.de*

# 566. Aphthovirus and cardiovirus 2A autolytic sequence

### Databanks

*Peptidase classification: clan UX, family U29, MEROPS ID: U29.001*
*NC-IUBMB enzyme classification: none*

*Databank codes:*

| Species | Strain | SwissProt | PIR | EMBL (genomic) |
|---|---|---|---|---|
| Encephalomyocarditis virus | | P03304 | A03906 | X00463: polyprotein gene |
| Encephalomyocarditis virus | EMC-B | P17593 | B31473 | M22457: complete genome |
| Encephalomyocarditis virus | EMC-D | P17594 | A31473 | M22458: complete genome |
| Equine rhinovirus | 1 | – | – | X96870: complete genome |
| Equine rhinovirus | 2 | – | – | X96871: complete genome |
| Foot-and-mouth disease virus | A10-61 | P03306 | A93508 | X00429: complete genome |
| | A22/500 | P49303 | – | X74812: complete genome |
| Theiler's murine encephalomyelitis virus | Bean 8386 | P08544 | A29535 | M16020 |
| Theiler's murine encephalomyelitis virus | GDVII | P08545 | | M14703: polymerase and proteinase genes 3' end<br>M20562: complete genome |
| Theiler's murine encephalomyelitis virus | DA | P13899 | A31228 | M20301 |

## Name and History

Picornaviruses encode all of their proteins in the form of a single, long, open reading frame (ORF). Full-length translation products are not observed, however, due to co- and post-translational autoproteolytic processing ('primary' and 'secondary' processing respectively). The organization of the genome and the polyproteins is similar across the various groups within the picornavirus family and standard nomenclature has been adopted to refer to individual proteins and polyprotein domains of the disparate viruses. Entero- and rhinovirus 2A picornains (Chapter 241) are cysteine endopeptidases of some 153 amino acids. The equivalent region of cardiovirus polyproteins also encodes a protein of similar size (ca. 143 amino acids), but with no sequence similarity to the 2A picornains. In contrast, the 2A regions of the aphtho- or foot-and-mouth disease (FMD) viruses and the recently sequenced equine rhinoviruses (ERV) 1 and 2 are very much shorter. The length was originally reported as 16 amino acids, but now seems to be 18 amino acids (Donnelly *et al.*, 1997). The C-terminal region of cardiovirus 2A proteins is very similar in sequence to that of FMD 2A (Figure 566.1).

## Activity and Specificity

In both the cardio- and aphthoviruses the primary polyprotein cleavage in this region occurs at the C-terminus of the 2A region (unlike 2A picornains which cleave N-terminally). The cleavage occurs at a conserved -Gly+Pro- bond (Robertson *et al.*, 1985; Palmenberg *et al.*, 1992) and has all the characteristics of a primary cleavage, being extremely rapid (no precursors are detected spanning this site), insensitive to dilution when translated *in vitro*, and resistant to proteinase inhibitors. Analyses of recombinant FMD and artificially constructed polyproteins containing two reporter proteins flanking 2A in a single ORF (Ryan *et al.*, 1991; Ryan & Drew, 1994) showed: (a) the cleavage activity resided within a defined minimal 13 amino acids sequence, although longer versions produced higher levels of cleavage, and (b) the uncleaved translation product was not a precursor form, since cleavage occurred cotranslationally or not at all. Almost complete cleavage activity was observed with a 24 amino acids version (Donnelly *et al.*, 1997). The C-terminal regions of the cardiovirus 2A proteins also possess this activity (Palmenberg *et al.*, 1992; Donnelly *et al.*, 1997). The peptidase activity of this sequence appears to be entirely intramolecular (in *cis*), as we have failed to demonstrate activity in *trans* using a number of experimental approaches. Site-directed mutations of this sequence show all mutations to be deleterious or, much more commonly, to abolish activity altogether (unpublished results of the authors; Hahn & Palmenberg, 1996). These studies indicate that the activity requires a number of key residues that are completely conserved among the picornavirus sequences at the base of a helical structure (Donnelly *et al.*, 1997). The -Gly+Pro- dipeptide at which cleavage occurs must be maintained for activity.

```
EMC    YAGYFADLLIHDIETNPG+P
TME    HADYYRQRLIHDVETNPG+P
FMD    TLNFDLLKLAGDVESNPG+P
ERV-1  CTNFSLLKLAGDVESNPG+P
ERV-2  ATNFSLLKLAGDVELNPG+P
```

*Figure 566.1* Amino acid sequences of 2A regions of encephalomyocarditis virus (EMC), Theiler's murine encephalomyelitis virus (TME), foot-and-mouth disease virus (FMD), and equine rhinoviruses (ERV-1, ERV-2), showing the site of cleavage.

## Structural Chemistry

The 2A sequence has been used to construct a range of self-processing artificial polyproteins in coordinated expression studies and we (and others) have found the 2A region to be active with a variety of sequences flanking the 2A oligopeptide sequence (Ryan *et al.*, 1991; Ryan & Drew, 1994; Percy *et al.*, 1994; Precious *et al.*, 1995; Santa-Cruz *et al.*, 1996). The 2A sequence is inserted at the desired site of cleavage such that the single ORF is maintained. The N-terminal cleavage product possesses a C-terminal extension of the 2A sequence, and the C-terminal cleavage product has an N-terminal Pro residue.

Although synthetic peptides corresponding to these sequences may be induced to form secondary structures, we have not been able to demonstrate any peptide cleavage. Others have reported cleavage of the synthetic tetrapeptide Asn-Pro-Gly-Pro, although at a site (Asn-Pro+Gly-Pro) different from that observed *in vivo* (see above) (Palmenberg, 1990). Algorithms predicting protein secondary structure and dynamic molecular modeling all predict a helical structure 'capped' by a tight-turn (-NPGP-) at the C-terminus of the 2A sequence. The putative helix is predicted to be amphipathic with polar/charged residues aligning along one face of the helix. In our model, the turn at the bottom of the helix serves to orient the scissile bond to the nucleophile, which is predicted to be not a side-chain group but a main-chain carbonyl group at the base of the helix. The helical structure could serve, therefore, to provide a dipole moment altering the electrochemical environment at the base of this 'capped' helix.

### Biological Aspects

The 2A/2B cleavage of the virus polyprotein serves to physically separate the domains responsible for encapsidation and for the RNA replicative functional of the nascent polyprotein. This cleavage is an essential primary step in polyprotein processing and therefore in the replication of the virus. As a target for antiviral agents it might disappointingly be intractable, since the cleavage is cotranslational, occurs extremely rapidly, and, as our structural predictions indicate, occurs in a highly unusual and inaccessible environment quite unlike a 'classical' proteinase active-site pocket or cleft.

### References

Donnelly, M.L.L., Gani, D., Flint, M., Monaghan, S. & Ryan, M.D. (1997) The cleavage activities of aphthovirus and cardiovirus 2A proteins. *J. Gen. Virol.* **78**, 13–21.

Hahn, H. & Palmenberg, A.C. (1996) Mutational analysis of the encephalomyocarditis virus primary cleavage. *J. Virol.* **70**, 6870–6875.

Palmenberg, A.C. (1990) Proteolytic processing of picornaviral polyprotein. *Annu. Rev. Microbiol.* **44**, 603–623.

Palmenberg, A.C., Parks, G.D., Hall, D.J., Ingraham, R.H., Seng, T.W. & Pallai, P.V. (1992) Proteolytic processing of the cardioviral P2 region: primary 2A/2B cleavage in clone-derived precursors. *Virology* **190**, 754–762.

Percy, N., Barclay, W.S., Garcia-Sastre, A. & Palese, P. (1994) Expression of a foreign protein by influenza A virus. *J. Virol.* **68**, 4486–4492.

Precious, B., Young, D.F., Bermingham, A., Fearns, R., Ryan, M.D. & Randall, R.E. (1995) Inducible expression of the P, V, and NP genes of the paramyxovirus simian virus 5 in cell lines and an examination of the NP-P and NP-V interactions. *J. Virol.* **69**, 8001–8010.

Robertson, B.H., Grubman, M.J., Weddell, G.N., Moore, D.M., Welsh, J.D., Fischer, T., Dowbenko, D.J., Yansura, D.G., Small, B. & Kleid, D.G. (1985) Nucleotide and amino acid sequence coding for polypeptides of foot-and-mouth disease virus type A12. *J. Virol.* **54**, 651–660.

Ryan, M.D. & Drew, J. (1994) Foot-and-mouth disease virus 2A oligopeptide mediated cleavage of an artificial polyprotein. *EMBO J.* **13**, 928–933.

Ryan, M.D., King, A.M.Q. & Thomas, G.P. (1991) Cleavage of foot-and-mouth disease virus polyprotein is mediated by residues located within a 19 amino acid sequence. *J. Gen. Virol.* **72**, 2727–2732.

Santa-Cruz, S., Chapman, S., Roberts, A.G., Roberts, J.M., Prior, D.A.M. & Oparka, K.J. (1996) Assembly and movement of a plant virus carrying a green fluorescent protein overcoat. *Proc. Natl Acad. Sci. USA* **93**, 6286–6290.

*Martin D. Ryan*
*Department of Biochemistry,*
*University of St. Andrews,*
*St. Andrews, Fife, Scotland*
*Email martin.ryan@st-and.ac.uk*

*Michelle L.L. Donnelly*
*School of Chemistry,*
*University of St. Andrews,*
*St. Andrews, Fife, Scotland*

*David Gani*
*School of Chemistry,*
*University of St. Andrews,*
*St. Andrews, Fife, Scotland*

# 567. *Hepatitis C virus endopeptidase 2*

### Databanks

*Peptidase classification: clan UX, family U39, MEROPS ID: U39.001*
*NC-IUBMB enzyme classification: none*
*Databank codes:*

| Species | Isolate | SwissProt | PIR | EMBL (genomic) |
|---|---|---|---|---|
| Hepatitis C virus type 1a[a] | HCV-1 | P26664 | A39166 | M62321: complete genome |
| Hepatitis C virus type 1b[a] | HCV-J | P26662 | A39253 | D90208: complete genome |

| Species | Isolate | SwissProt | PIR | EMBL (genomic) |
|---|---|---|---|---|
| Hepatitis C virus type 1c | HC-G9 | – | – | D14853: complete genome |
| Hepatitis C virus type 2a | HC-J6 | P26660 | JQ1303 | D00944: complete genome |
| Hepatitis C virus type 2b | HC-J8 | P26661 | A40250 | D10988: complete genome |
| Hepatitis C virus type 2c | BEBE1 | – | – | D50409: complete genome |
| Hepatitis C virus type 3a[a] | NZL1 | – | – | D17763: complete genome |
| Hepatitis C virus type 3b | HCV-Tr | – | – | D49374: complete genome |
| Hepatitis C virus type 11a(6)[+] | JK046 | – | – | D63822: complete genome |
| Hepatitis C virus | BK | P26663 | A38465 | M58335: complete genome |
| Hepatitis C virus | H | P27958 | A36814 | M67463: complete genome |
| Hepatitis C virus | HC-JT | Q00269 | A45573 | D11168: complete genome |
| Hepatitis C virus | Taiwan | P29846 | A40244 | M84754: 5′ end of genome |
| Hepatitis G virus | | – | – | U45966: complete genome |
| Hepatitis GB virus A | | – | – | U22303: polyprotein gene |
| Hepatitis GB virus B | | – | – | U22304: complete genome |
| Hepatitis GB virus C | | – | – | U36380: polyprotein gene |

[a]Additional cloned isolates are available for these genotypes; see Bukh *et al*. (1995) for a discussion of HC genotypes and quasispecies, and Tokita *et al*. (1996) for an additional listing of isolate designations and accession numbers.

*Notes:* The hepatitis GB viruses listed above show similarity with the hepatitis C virus NS2/3 proteins over all but the first 60 residues. Because the NS3 protein contains a helicase, it is possible that the relationship between the hepatitis C and hepatitis GB NS2/3 proteins is that of mosaic proteins, with the catalytic domain of the hepatitis C NS2/3 protein restricted to the first 60 residues. So small a catalytic domain is unlikely, however.

## Name and History

The activity responsible for cleavage between nonstructural proteins 2 and 3 (NS2 and NS3) of the hepatitis C virus (HC) polyprotein, known variously as **hepatitis C endopeptidase 2, NS2-3 proteinase, NS2 proteinase** or **Cpro-1**, was first revealed during studies of polyprotein-processing events mediated by an HC-encoded serine proteinase located in NS3, referred to in this *Handbook* as hepatitis C polyprotein peptidase (Chapter 89). Sequence analysis of the HC NS3 region and alignments with other members of the flavivirus family and endopeptidases of the trypsin family led to the suggestion that His1083, Asp1107 and Ser1165 (numbers refer to amino acid positions in HC isolates of genotype 1a and 1b) in the NS3 region form a catalytic triad typical of serine proteinases (Bazan & Fletterick, 1989, 1990; Gorbalenya *et al*., 1989; Chambers *et al*., 1990; Miller & Purcell, 1990). Mutation of the putative active-site nucleophile Ser1165 abolished cleavage at four downstream sites in the polyprotein, but had little or no effect on cleavage at the NS2/NS3 site (Leu1026↓Ala1027), hereafter referred to as the 2/3 site, indicating that a distinct proteinase was responsible for scission of the 2/3 site (Grakoui *et al*., 1993b; Hijikata *et al*., 1993a). The discovery that approximately 130 amino acids upstream and 180 downstream of the 2/3 site were required for cleavage, coupled with the observation that certain mutations in NS2 prevented or inhibited 2/3 cleavage suggested that cleavage at this site is catalyzed by a virus-encoded proteolytic activity residing in the NS2–NS3 region of the polyprotein (Grakoui *et al*., 1993b; Hijikata *et al*., 1993a).

## Activity and Specificity

Cleavage at the 2/3 site is thought to occur in *cis* on the basis of the following observations: (a) processing at this junction seems to be very rapid, since no precursors to this cleavage event are detected after a short pulse (15–20 min)

and the cleavage products do not accumulate during chase periods of up to 6 h, as monitored by immunoprecipitation of NS3 (Lin *et al*., 1994; Tanji *et al*., 1994); (b) 2/3 cleavage is very efficient when the NS2 and NS3 regions are expressed as part of the same polypeptide in mammalian cells or in the rabbit reticulocyte experiments of Hijikata *et al*. (1993a), but (c) apparent *trans* cleavage occurs inefficiently in mammalian cells (Grakoui *et al*., 1993b; Hijikata *et al*., 1993a; Reed *et al*., 1995) and has never been reported in cell-free systems. Analysis of cleavage-site specificity by amino acid substitutions at several positions near the 2/3 cleavage site have suggested that the overall conformation of the NS2–NS3 region may have a greater influence on specificity than individual amino acid side chains at the cleavage site, since most single amino acid mutations were tolerated by the proteinase, even at the P1 and P1′ positions (Hirowatari *et al*., 1993; Reed *et al*., 1995). Only mutations with a high potential for structural disruption, such as proline substitutions, multiple amino acid substitutions or deletions significantly inhibited 2/3 cleavage (Reed *et al*., 1995). Since no homology was found between the NS2 or NS3 regions and known proteinases, the type of proteolytic mechanism used to achieve 2/3 cleavage is unknown. However, the observation that 2/3 cleavage is stimulated by the addition of $ZnCl_2$ and inhibited by EDTA has led to the suggestion that HC endopeptidase 2 is a metalloproteinase (Hijikata *et al*., 1993a). However, no 'HEXXH' or 'HXXEH' motif (present in many metallopeptidases) is to be found in the HC NS2–NS3 region. Another possibility worth considering is that zinc plays a structural or regulatory role rather than participating directly in catalysis. Clarification of the role of zinc in 2/3 cleavage awaits further biochemical and structural studies.

## Structural Chemistry

Three proteinase activities are required for the release of individual NS2 and NS3 proteins from the HC polyprotein:

host signal peptidase (Chapter 156) for production of the N-terminus of NS2, HC endopeptidase 2 for cleavage at the NS2–NS3 junction, and the hepatitis C polyprotein peptidase for production of the C-terminus of NS3. The N- and C-terminal boundaries of the full-length NS2 and NS3 proteins produced by the combined action of these three proteinases are Leu810 and Leu1026 for NS2 (Grakoui *et al.*, 1993b; Mizushima *et al.*, 1994) and Ala1027 and Thr1657 (Grakoui *et al.*, 1993a,b) for NS3, and the molecular masses of the two proteins are approximately 21 and 70 kDa, respectively. Evidence for signal peptidase-mediated processing at the N-terminus of NS2 includes the presence of a hydrophobic putative signal peptide immediately preceding the start of the NS2 region in the HC polyprotein (Grakoui *et al.*, 1993b; Mizushima *et al.*, 1994), the observation that an Ala809Arg mutation at the P1 position abolishes cleavage (Mizushima *et al.*, 1994), consistent with the specificity of signal peptidase, the membrane dependence of cleavage (Grakoui *et al.*, 1993b; Hijikata *et al.*, 1993a; Mizushima *et al.*, 1994; Santolini *et al.*, 1995), and the partial restoration of NS2 production in the presence of trypsinized membranes by the addition of the SRP receptor $\alpha$ subunit (Santolini *et al.*, 1995). A signal peptide located just upstream of the NS2 region would be expected to direct the translocation of its N-terminus into the endoplasmic reticulum (ER) lumen where signal peptidase cleavage occurs. Since NS2 remains membrane associated after extraction with urea, it is predicted to be an integral membrane protein with one or more membrane-spanning domains (Santolini *et al.*, 1995). The topology of the C-terminus of NS2 has not been clearly determined, although potential glycosylation sites introduced at amino acids 983 and 933 were glycosylated, suggesting that those residues were located in the ER lumen (Santolini *et al.*, 1995). Neither has the precise localization of NS3 been defined with respect to the ER membrane, although it has been shown to associate with membranes, perhaps through NS4A, another hydrophobic protein located just downstream from the C-terminus of NS3 (Hijikata *et al.*, 1993b; Satoh *et al.*, 1995). In any case, proper membrane topology and signal peptidase processing of the NS2–NS3 region may affect 2/3 cleavage or vice versa.

The minimum region required for cleavage at the 2/3 site has been mapped to residue 898 on the N-terminal side (Hijikata *et al.*, 1993a) and 1207 on the C-terminal side (Grakoui *et al.*, 1993b), although the activity of a single construct extending from amino acid 898 to 1207 has never been formally tested. This region includes most of NS2 and the N-terminal one-third of NS3, which overlaps with the serine proteinase domain of the hepatitis C polyprotein peptidase that is also located in the N-terminal one-third of NS3. Two mutations in NS2, His952Ala and Cys993Ala, have been identified that abolish cleavage at the 2/3 site while having little or no effect on downstream cleavages catalyzed by hepatitis C polyprotein peptidase (Grakoui *et al.*, 1993b; Hijikata *et al.*, 1993a). This result is reminiscent of that obtained for the Ser1165Ala mutation, which abolished four cleavages in the polyprotein without affecting 2/3 cleavage (Grakoui *et al.*, 1993b; Hijikata *et al.*, 1993a). Other mutations that inhibited cleavage at the 2/3 site to various extents were Cys922Ala and Glu972Gln in the NS2 region, and Cys1123Ala, Cys1125Ala and Cys1171Ala in the NS3 region (Hijikata *et al.*, 1993a). These latter three

mutations also inhibited viral serine proteinase-dependent cleavages (Hijikata *et al.*, 1993a). Interestingly, crystal structures recently obtained for the hepatitis C polyprotein peptidase (Kim *et al.*, 1996; Love *et al.*, 1996) and biochemical studies (DeFrancesco *et al.*, 1996) have indicated that these three residues are involved in zinc binding. The function of this zinc atom, which appears to be bound to NS3 in equimolar quantities, is still unclear, but it may have a structural role in folding or stabilizing NS3. Whether this zinc atom is involved in catalysis by HC endopeptidase 2 is unknown. Mutation of His1175, which may interact indirectly with zinc, also slightly inhibited both HC proteinases (Grakoui *et al.*, 1993b; Hijikata *et al.*, 1993a). Due to its highly hydrophobic nature, native NS2 has not been purified either alone or linked to NS3, so the crystal structure for HC endopeptidase 2 has yet to be determined.

## Preparation

Processing at the 2/3 site has been observed following expression of the NS2–NS3 region in cell-free translation systems (D'Souza *et al.*, 1994; Grakoui *et al.*, 1993b; Hijikata *et al.*, 1993a; Santolini *et al.*, 1995), in *E. coli* (Grakoui *et al.*, 1993b), in insect cells infected with baculovirus recombinants (Hirowatari *et al.*, 1993), and in transiently transfected mammalian cells (Grakoui *et al.*, 1993b; Hijikata *et al.*, 1993a; Reed *et al.*, 1995), but HC endopeptidase 2 has never been purified for a number of reasons, including the hydrophobicity of the NS2 region and the apparently autocatalytic nature of 2/3 cleavage. In all expression systems tested, HC endopeptidase 2 is cleaved into individual NS2 and NS3 components. Although NS2 and NS3 may be capable of association after their release from the polyprotein to form a proteinase capable of *trans* cleavage, such *trans*-active complexes have never been reported. Possibly an uncleaved NS2–NS3 polypeptide could be purified by introducing mutations at the cleavage site which prevent recognition of that site without disrupting the activity of HC endopeptidase 2, although the observed tolerance of the proteinase to cleavage-site mutations increases the likelihood that mutations in the cleavage site that abolish 2/3 cleavage will also affect proteinase activity.

## Biological Significance

The reason that HC employs three different proteinases for processing of its polyprotein is not clear, but the potential to regulate the three types of cleavage events independently could have important ramifications for many steps in the viral life cycle, such as RNA replication, packaging of the viral genome, virion assembly, or budding and release of infectious particles. The NS2–NS3 region is highly conserved among divergent isolates of HC, and His952 and Cys993, two residues essential for HC 2/3 cleavage, are conserved in the GB agents, as well as the N-terminal one-third of NS3 (Muerhoff *et al.*, 1995). This conservation of the NS2–NS3 region among HC isolates and of essential amino acids in NS2, in addition to the N-terminal region of NS3 among HC and the GB agents, suggests that 2/3 cleavage may be an essential step in the propagation of both types of viruses. However, no published reports on 2/3 processing are currently available for the GB agents, and the importance of 2/3 cleavage in HC replication has yet to be established.

## Distinguishing Features

As indicated above, HC is a member of the family Flaviviridae, which contains three genera, the flaviviruses, the pestiviruses and a provisional genus known as hepaciviruses, of which HC is currently the only member. Cleavages at the 2B/3 site, as it is called in flaviviruses, and the 2/3 site of pestiviruses seem to be catalyzed by mechanisms distinct from that of HC endopeptidase 2. The flavivirus 2B/3 site is cleaved by a viral serine proteinase (flavivirin) consisting of a tightly associated NS2B–NS3 heterodimer (see Chapter 88 and Rice, 1996, for reviews). The cleavage specificity as well as the catalytic mechanism of flavivirin seems to differ from that of HC, since flavivirin prefers to cleave after a pair of basic amino acids, while HC endopeptidase 2 seems to prefer hydrophobic amino acids at its cleavage site (Hirowatari *et al.*, 1993; Reed *et al.*, 1995). Moreover, the tight association observed between flavivirus NS2B and NS3 has not been observed for HC NS2 and NS3, although the possibility that HC NS2 and NS3 interact has not been excluded (Reed *et al.*, 1995).

The pestivirus bovine viral diarrhea virus (BVDV) has evolved a unique mechanism for cleavage at the 2/3 site which seems to be related to its pathogenicity (see Chapter 565 and Meyers & Thiel, 1996, for reviews of pestivirus processing). BVDV may exist as either cytopathic or noncytopathic biotypes, with the most obvious difference between them at the molecular level being the production of fully processed NS3 by the cytopathic but not the noncytopathic isolates. The genomes of the cytopathic isolates contain various cellular insertions and viral duplications and/or rearrangements that result in cleavage of the NS2–NS3 region and the ensuing production of NS3. The mechanisms by which these alterations cause cleavage at the 2/3 site differ and in some cases are unknown. In some cytopathic isolates, cleavage resulting in the production of NS3 is accomplished by the insertion of cellular ubiquitin sequences, which serve as a substrate for cellular ubiquitin C-terminal hydrolase (Chapter 222). In another case, an autocatalytic viral proteinase located at the extreme N-terminus of the BVDV polyprotein that is normally responsible for cleavage at its own C-terminus is duplicated and inserted so that it cleaves to produce the N-terminus of NS3. Although these alterations in the viral genome result in cleavage of the NS2–NS3 region, some uncleaved NS2–NS3 polypeptides are also present, in contrast to HC for which cleavage at the 2/3 site usually occurs quickly and efficiently. The pestivirus classical swine fever virus (CSFV) resembles BVDV in its production of a persistent NS2–NS3 precursor in addition to cleaved NS3, but in CSFV this partial cleavage occurs by an unknown mechanism that does not involve drastic genomic alterations. The genomes of all three of these virus groups are small and compact, and tight regulation of polyprotein processing is likely to be essential for the viability of each. However, additional experiments are needed to verify this supposition and to identify similarities and disparities in processing of the NS2–NS3 and NS2B–NS3 regions of the three groups and the relations of these features to various stages of the viral life cycles.

## Further Reading

A full, recent review is that of Rice (1996).

## References

Bazan, J.F. & Fletterick, R.J. (1989) Detection of a trypsin-like serine protease domain in flaviviruses and pestiviruses. *Virology* **171**, 637–639.

Bazan, J.F. & Fletterick, R.J. (1990) Structural and catalytic models of trypsin-like viral proteases. *Semin. Virol.* **1**, 311–322.

Bukh, J., Miller, R.H. & Purcell, R.H. (1995) Genetic heterogeneity of hepatitis C virus: quasispecies and genotypes. *Sem. Liver Dis.* **15**, 41–63.

Chambers, T.J., Hahn, C.S., Galler, R. & Rice, C.M. (1990) Flavivirus genome organization, expression, and replication. *Annu. Rev. Microbiol.* **44**, 649–688.

DeFrancesco, R., Urbani, A., Nardi, M.C., Tomei, L., Steinkuhler, C. & Tramontano, A. (1996) A zinc binding site in viral serine proteinases. *Biochemistry* **35**, 13282–13287.

D'Souza, E.D., O'Sullivan, E., Amphlett, E.M., Rowlands, D.J., Sangar, D.V. & Clarke, B.E. (1994) Analysis of NS3-mediated processing of the hepatitis C virus non-structural region *in vitro*. *J. Gen. Virol.* **75**, 3469–3476.

Gorbalenya, A.E., Donchenko, A.P., Koonin, E.V. & Blinov, V.M. (1989) N-terminal domains of putative helicases of flavi- and pestiviruses may be serine proteases. *Nucleic Acids Res.* **17**, 3889–3897.

Grakoui, A., McCourt, D.W., Wychowski, C., Feinstone, S.M. & Rice, C.M. (1993a) Characterization of the hepatitis C virus-encoded serine proteinase: determination of proteinase-dependent polyprotein cleavage sites. *J. Virol.* **67**, 2832–2843.

Grakoui, A., McCourt, D.W., Wychowski, C., Feinstone, S.M. & Rice, C.M. (1993b) A second hepatitis C virus-encoded proteinase. *Proc. Natl Acad. Sci. USA* **90**, 10583–10587.

Hijikata, M., Mizushima, H., Akagi, T., Mori, S., Kakiuchi, N., Kato, N., Tanaka, T., Kimura, K. & Shimotohno, K. (1993a) Two distinct proteinase activities required for the processing of a putative nonstructural precursor protein of hepatitis C virus. *J. Virol.* **67**, 4665–4675.

Hijikata, M., Mizushima, H., Tanji, Y., Komoda, Y., Hirowatari, Y., Akagi, T., Kato, N., Kimura, K. & Shimotohno, K. (1993b) Proteolytic processing and membrane association of putative nonstructural proteins of hepatitis C virus. *Proc. Natl Acad. Sci. USA* **90**, 10773–10777.

Hirowatari, Y., Hijikata, M., Tanji, Y., Nyunoya, H., Mizushima, H., Kimura, K., Tanaka, T., Kato, N. & Shimotohno, K. (1993) Two proteinase activities in HCV polypeptide expressed in insect cells using baculovirus vector. *Arch. Virol.* **133**, 349–356.

Kim, J.L., Morgenstern, K.A., Lin, C., Fox, T., Dwyer, M.D., Landro, J.A., Chambers, S.P., Markland, W., Lepre, C.A., O'Malley, E.T., Harbeson, S.L., Rice, C.M., Murcko, M.A., Caron, P.R. & Thomson, J.A. (1996) Crystal structure of the hepatitis C virus NS3 protease domain complexed with a synthetic NS4A cofactor peptide. *Cell* **87**, 343–355.

Lin, C., Prágai, B., Grakoui, A., Xu, J. & Rice, C.M. (1994) The hepatitis C virus NS3 serine proteinase: *trans* processing requirements and cleavage kinetics. *J. Virol.* **68**, 8147–8157.

Love, R.A., Parge, H., Wickersham, J.A., Hostomsky, Z., Habuka, N., Moomaw, E.W., Adachi, T. & Hostomska, Z. (1996) The crystal structure of hepatitis C virus NS3 proteinase reveals a trypsin-like fold and a structural zinc binding site. *Cell* **87**, 331–342.

Meyers, G. & Thiel, H.-J. (1996) Molecular characterization of pestiviruses. *Adv. Virus Res.* **47**, 53–118.

Miller, R.H. & Purcell, R.H. (1990) Hepatitis C virus shares amino acid sequence similarity with pestiviruses and flaviviruses as well as members of two plant virus supergroups. *Proc. Natl Acad. Sci. USA* **87**, 2057–2061.

Mizushima, H., Hijikata, M., Tanji, Y., Kimura, K. & Shimotohno, K. (1994) Analysis of N-terminal processing of hepatitis C virus nonstructural protein 2. *J. Virol.* **68**, 2731–2734.

Muerhoff, A.S., Leary, T.P., Simons, J.N., Pilotmatias, T.J., Dawson, G.J., Erker, J.C., Chalmers, M.L., Schlauder, G.G., Desai, S.M. & Mushahwar, I.K. (1995) Genomic organization of GB viruses A and B: two new members of the *Flaviviridae* associated with GB agent hepatitis. *J. Virol.* **69**, 5621–5630.

Reed, K.E., Grakoui, A. & Rice, C.M. (1995) The hepatitis C virus NS2-3 autoproteinase: cleavage site mutagenesis and requirements for bimolecular cleavage. *J. Virol.* **69**, 4127–4136.

Rice, C.M. (1996) Flaviviridae: the viruses and their replication. In: *Fields Virology* (Fields, B.N., Knipe, D.M. & Howley, P.M., eds). New York: Raven Press, pp. 931–960.

Santolini, E., Pacini, L., Fipaldini, C., Migliaccio, G. & Monica, N. (1995) The NS2 protein of hepatitis C virus is a transmembrane polypeptide. *J. Virol.* **69**, 7461–7471.

Satoh, S., Tanji, Y., Hijikata, M., Kimura, K. & Shimotohno, K. (1995) The N-terminal region of hepatitis C virus nonstructural protein 3 (NS3) is essential for stable complex formation with NS4A. *J. Virol.* **69**, 4255–4260.

Tanji, Y., Hijikata, M., Hirowatari, Y. & Shimotohno, K. (1994) Hepatitis C virus polyprotein processing: kinetics and mutagenic analysis of serine proteinase-dependent cleavage. *J. Virol.* **68**, 8418–8422.

Tokita, H., Okamoto, H., Iizuka, H., Kishimoto, J., Tsuda, F., Lesmana, L.A., Miyakawa, Y. & Mayumi, M. (1996) Hepatitis C virus variants from Jakarta, Indonesia classifiable into novel genotypes in the second (2e and 2f), tenth (10a) and eleventh (11a) genetic groups. *J. Gen. Virol.* **77**, 293–301.

*Karen E. Reed*
*Department of Molecular Microbiology,*
*Washington University School of Medicine,*
*660 S. Euclid Avenue,*
*St Louis, MO 63110, USA*

*Charles M. Rice*
*Department of Molecular Microbiology,*
*Washington University School of Medicine,*
*660 S. Euclid Avenue,*
*St Louis, MO 63110, USA*
*Email: Rice@Borcim.wustl.edu*

# 568. Kedarcidin chromoprotein and apoprotein

*Peptidase classification: clan UX, family U99, MEROPS ID: U9G.025*
*NC-IUBMB enzyme classification: none*
*Databank codes: sequence data are given below*

## Name and History

In the search for new DNA-damaging antitumor drug substances from fermentation sources, natural product chemists at Bristol-Myers Squibb discovered a novel cytotoxic antitumor antibiotic agent, kedarcidin, produced by an actinomycete. The principal bioassays used to guide the purification of kedarcidin were the *Escherichia coli* SOS chromotest, which is a sensitive assay to detect DNA-damaging materials, *in vitro* cytotoxicity testing using a human colon tumor (HCT 116) cell line, and the *in vivo* P388 murine leukemia model. Kedarcidin proved to be a new enediyne-containing chromoprotein. Much of the research on these protein complexes has led to the general view that the biological activity is due to the chromophore component, and that the protein is little more than a stabilizing carrier for the highly labile nine-membered rings of the chromophores. Our results suggest that the kedarcidin protein may have a more specific function, however.

## Preparation and Structural Chemistry

Kedarcidin was produced by an unknown actinomycete strain L585-6 (ATCC 53650) isolated from a soil sample collected in the Maharastra State in India. The production titers of kedarcidin in shake flask and fermenter cultures are $1300 \, \mu g \, ml^{-1}$ and $1050 \, \mu g \, ml^{-1}$, respectively (Lam *et al.*, 1991). Kedarcidin was found to be highly acidic, with a pI of 3.65, and it was recovered from broth filtrate by adsorption on a QAE anion exchanger. The purification of kedarcidin was accomplished by gel-filtration and ion-exchange chromatography.

Structural studies demonstrated that kedarcidin belongs to the family of enediyne chromoproteins (Hofstead *et al.*, 1992; Leet *et al.*, 1993). These chromoproteins include kedarcidin, neocarzinostatin, actinoxanthin, maduropeptin and C1027 (Hofstead *et al.*, 1992; Khoklov *et al.*, 1969; Hanada *et al.*, 1991; Otani *et al.*, 1988; Zein *et al.*, 1995b). Similarly to the previously isolated chromoproteins, kedarcidin consists

ASAAVSVSPA  TGLADGATVT  VSASGFATST  SATALQCAIL  ADGRGACNVA
EFHDFSLSGG  EGTTSVVVRR  SFTGYVMPDG  PEVGAVDCDT  APGGCQIVVG
GNTGEYGNAA  ISFG

*Figure 568.1*  Amino acid sequence of the kedarcidin apoprotein.

of a highly labile, nine-membered, enediyne-containing chromophore, noncovalently associated with a highly acidic, stabilizing polypeptide. The chromophore is a solvent-extractable molecule. HPLC with ultraviolet (UV) or mass spectroscopic monitoring showed the presence of three different chromophores in varying ratios and molecular masses. The major chromophore, of molecular mass 1029, has been fully characterized.

The amino acid sequence of the kedarcidin apoprotein (114 amino acids) was determined by automated Edman degradation of unmodified kedarcidin and of enzymatically derived peptide fragments of *S*-pyridylethylated kedarcidin (Hofstead *et al.*, 1992). Depending on fermentation conditions, the kedarcidin complex varies with respect to the amino acid composition of the apoprotein, and amino acid sequencing revealed the presence of three variants of the kedarcidin apoprotein. The major variant polypeptide consists of 114 amino acid residues (Figure 568.1) and is the one used in our studies. The two minor variants lacked one or two N-terminal residues, respectively, as compared to the major form.

The solution conformation of the kedarcidin apoprotein has been characterized by heteronuclear multidimensional NMR spectroscopy. The overall fold of apokedarcidin is well defined; it is composed of an immunoglobulin-like, seven-stranded, antiparallel $\beta$ barrel, and a subdomain containing two antiparallel $\beta$ ribbons (Constantine *et al.*, 1994). Very similar tertiary structures have been reported previously for the related proteins neocarzinostatin, macromomycin and actinoxanthin (Gao, 1992; Adjaj *et al.*, 1992; Van Roey & Beerman, 1989; Pletnev *et al.*, 1982). The solution structure of the kedarcidin chromoprotein could not be obtained, due to the difficulty of producing a 1:1 apoprotein to chromophore complex, however.

The ratio of apoprotein to chromophore in the complex varied from 1:1 to 18:1 as shown by a quantitative UV method of analysis (Leet *et al.*, 1993). The difficulty in consistently recovering a kedarcidin complex with a constant apoprotein-to-chromophore ratio hampered our efforts in pursuing the holoantibiotic as a drug lead. For availability reasons, the complex, chromophore and apoprotein used in our studies came from a lot that had a ratio of 16:1 of apoprotein to chromophore.

## Biological Aspects and Proteolytic Activity

It has been suggested that the antitumor activity of all enediynes involves DNA cleavage, a process about which there is substantial *in vitro* chemical and structural information. As had been shown previously for neocarzinostatin, DNA experiments with both the isolated chromophore and the kedarcidin chromoprotein demonstrate that DNA cleavage is primarily due to the chromophore (Zein *et al.*,

1993b). However, cytotoxicity assays with several lots of the chromoprotein complex, with apoprotein to chromophore ratios ranging from 1:1 to 16:1, showed that irrespective of the lot used, the chromoprotein exhibits equivalent cytotoxicity to that of the purified chromophore. This is intriguing, since if all the activity of the complex were solely due to the chromophore, then different $IC_{50}$ values should have been observed for the different lots. These observations implied that the apoprotein must be contributing actively to the activity of the holomolecule and prompted us to investigate an additional role for the kedarcidin polypeptide beyond that of an inert chromophore stabilizer.

### Activity on Cells

In the course of investigating a new role for the kedarcidin apoprotein, the effects of the chromophore, the complex, and the kedarcidin apoprotein on macromolecular synthesis in HCT 116 cells were examined and compared. This was achieved by a 2 h treatment of the cells with each of the three kedarcidin species followed by labeling of the cells with [$^3$H]thymidine, [$^3$H]uridine and [$^3$H]leucine. The cellular incorporation of the tritiated molecules was measured by scintillation counting and compared to controls. As expected for any cytotoxic agent, the kedarcidin chromophore inhibited synthesis of DNA, RNA and protein. The kedarcidin apoprotein, free from any detectable chromophore, had no measurable effect in this assay. However, the chromoprotein complex inhibited DNA synthesis at nanomolar concentrations, similarly to the chromophore, but did not have any effect on synthesis of RNA or cellular protein. Premixing the chromophore with the apoprotein prior to treatment of the cells yielded identical results to those obtained with the complex, i.e. the DNA synthesis was inhibited after a 2 h treatment, but RNA and protein synthesis were unaffected (W. Solomon & N. Zein, unpublished results). These observations indicated that the apoprotein may play a role in addition to that of a carrier/stabilizer.

### Proteolytic Activity

Because the ultimate effect of the chromophore in cells is believed to be on DNA, and because of the acidic nature of the apoprotein, we decided to examine possible interactions of the complex, the chromophore and the kedarcidin apoprotein with histones. Histones are highly basic proteins that form the DNA scaffold in mammalian cells, and significantly, one of the histones, histone H1, holds the nucleosomes together (Grunstein, 1992). To reach the DNA intact, the chromophore must be protected by the protein, but then it must be brought into contact with the DNA in the chromatin, perhaps with the help of the apoprotein. The implication of this hypothesis was that the apoprotein might have an effect on histones.

(A)  A-E-K-T-P-V-K-K-K†A-A-K-K-P-A†G-A-R-R-K-A-S-G-P-NH₂

(B)  A-E-K-T-P-V-K-K-K†A-A-K-K-P-A-G-A-R†R†K†A†S-G-P-NH₂

(C)  A-E-K-T-P-V-K-K†K†A-A-K-K-P-A-G-A-R-R-K†A-S-G-P-NH₂

(D)  A-E-K-T-P-V-K-K-K-A-A-K-K-P-A-G-A-R-R-K†A-S-G-P-NH₂

*Figure 568.2* Cleavage of a synthetic, 24 amino acid peptide. The peptide represents a basic region of histone H1, and the cleavage sites shown are those by (A) kedarcidin chromoprotein, (B) kedarcidin apoprotein, (C) neocarzinostatin chromoprotein, and (D) maduropeptin chromoprotein.

To test our hypothesis, carefully purified kedarcidin apoprotein was reacted with total calf thymus histones *in vitro*. SDS-PAGE analyses demonstrated that the apoprotein cleaved the histones into low molecular mass peptides. Incubation of each individual calf thymus histone (H1, H2A, H2B, H3 and H4) with the apoprotein showed that all histones were cleaved. The relative susceptibility to cleavage was: H1 > H2B = H3 = H2A > H4. Histone H1 is richest in Lys, followed by H2A, H2B, H3 and H4. The results suggested that the kedarcidin apoprotein possesses proteolytic activity similar to that of an endopeptidase, since it is able to cleave histones into small peptides.

The presence of the chromophore in the complex does not influence histone cleavage. Reacting the naked kedarcidin chromophore with histones for 15 h however, caused the formation of high molecular mass aggregates, but only at relatively high levels of chromophore. Histone H4, richest in Arg, was the most sensitive to this effect of the enediyne (Zein *et al.*, 1993a).

### Specificity of Proteolytic Cleavage

To assess the selectivity of the apoprotein towards histones, several less basic proteins were reacted with the kedarcidin apoprotein and chromoprotein. These proteins include 3':5' cyclic AMP-dependent protein kinase, prostatic phosphatase, calf brain tubulin, apo- and holo-transferrin and HCT 116 cell membrane protein extract. None of these proteins was cleaved suggesting that, *in vitro*, the apoprotein and chromoprotein are selective in the proteins they cleave. Histones which are most opposite in net charge to the highly acidic apoprotein are damaged most readily. Histone H1, richest in Lys, is the most susceptible to cleavage (Zein *et al.*, 1993a).

To allow further investigation of the proteolytic cleavage, a 24 amino acid peptide amide was synthesized. This was AEKTPVKKKAAKKPAGARRKASGP-NH₂, and it represents a basic region of histone H1. The peptide was incubated separately with similar concentrations of the kedarcidin chromoprotein and the kedarcidin apoprotein, and the reaction products were analyzed by liquid chromatography and mass spectrometry. The results (Figure 568.2) indicate that the chromoprotein cleaves the peptide at specific sites and that the kedarcidin apoprotein is less specific than the kedarcidin chromoprotein. Specifically, the kedarcidin chromoprotein cleaved the peptide primarily at two sites, Lys†Ala and Ala†Gly, whereas the apoprotein cleaved at five sites (which did not include the Ala-Gly site sensitive to the chromoprotein). The difference in proteolytic specificity between the apo- and the chromo- forms of kedarcidin is intriguing. In the absence of NMR data and a crystal structure of the kedarcidin chromoprotein, it is difficult to offer a definitive interpretation

of this result, but it is possible that the chromophore induces a conformational change in the protein such that the structure of the proteolytic active site is slightly different from that of the apoprotein (Zein *et al.*, 1995a).

### Serine Protease?

Since the cleavage sites identified for apo- and chromo-kedarcidin resemble those found for certain serine proteases, we tested the effect of common serine protease inhibitors on the cleavage of histone H1. Leupeptin, antipain, aprotinin, DFP and 4-(2-aminoethyl)-benzenesulfonyl fluoride HCl all inhibited the cleavage reactions. In contrast, pepstatin, an aspartic protease inhibitor, had no effect. We do not yet know the location of the protease active site, so no definitive interpretation of these observations is yet possible. However, these preliminary results suggest that kedarcidin protein acts as a serine protease (Zein *et al.*, 1993a).

### Related Peptidases and Distinguishing Features

Neocarzinostatin apoprotein and maduropeptin also cleave calf thymus histones, and as is the case with kedarcidin, H1 is the preferred substrate. In addition, incubation of either chromoprotein with the model peptide results in peptide bond cleavages that vary according to the chromoprotein used (Figure 568.2) (Zein *et al.*, 1993a, 1995a). However, the action of neither neocarzinostatin nor maduropeptin is inhibited by the above-mentioned serine protease inhibitors.

### References

Adjaj, E., Quiniou, E., Mispelter, J., Favaudon, V. & Lhoste, J.M. (1992) The seven-stranded β-barrel structure of apo-neocarzinostatin as compared to the immunoglobulin domain. *Biochimie* **74**, 853–858.

Constantine, K.L., Colson, K.L., Wittekind, M., Friedrichs, M.S., Zein, N., Tuttle, J., Langley, D.R., Leet, J.E., Schroeder, D.R., Lam, K.S., Farmer II, B.T., Metzler, W.J., Bruccoleri, R.E. & Mueller, L. (1994) Sequential ¹H, ¹³C, and ¹⁵N NMR assignments and solution conformation of apokedarcidin. *Biochemistry* **33**, 11438–11452.

Gao, X. (1992) Three-dimensional solution structure of apo-neocarzinostatin. *J. Mol. Biol.* **225**, 125–135.

Grunstein, M. (1992) Histones as regulators of genes. *Sci. Am.* 68–74B.

Hanada M., Ohkuma, H., Yonemoto, T., Tomita, K., Ohbayashi, M., Kamei, H., Konishi, M., Kawagushi, H. & Forenza, S. (1991) Maduropeptin, a complex of new macromolecular antitumor antibiotics. *J. Antibiot.* **44**, 403–414.

Hofstead, S.J., Matson, J.A., Malacko, A.R. & Marquardt, H. (1992) Kedarcidin, a new chromoprotein antitumor antibiotic II. Isolation, purification and physico-chemical properties. *J. Antibiot.* **45**, 1250–1254.

Khoklov, A.S., Cherches, B.Z., Reshetov, P.D., Smirnova, G.M., Sorokina, I.B., Prokoptzeva, T.A., Koloditskaya, T.A., Smirnov, V.V., Navashin, S.M. & Fomina, I.P. (1969) Physico-chemical and biological studies on actinoxanthin, an antibiotic from *Actinomyces globisporus* 1131. *J. Antibiot.* **22**, 541–544.

Lam, K.S., Hesler, G.A., Gustavson, D.R., Crosswell, A.R., Veitch, J.M. & Forenza, S. (1991) Kedarcidin, a new chromo-protein antitumor antibiotic I. Taxonomy of producing organism, fermentation and biological activity. *J. Antibiot.* **44**, 472–478.

Leet, J.E., Schroeder, D.R., Langley, D.R., Colson, K.L., Huang, S., Klohr, S.E., Lee, M.S., Golik, J., Hofstead, S.J., Doyle, T.W. & Matson, J.A. (1993) Chemistry and structure elucidation of the kedarcidin chromophore. *J. Am. Chem. Soc.* **115**, 8432–8443.

Otani, T., Minami, Y., Marunaka, T., Zhang, R. & Xie, M.-Y. (1988) A new macromolecular antitumor antibiotic, C-1027. II. Isolation and physico-chemical properties. *J. Antibiot.* **41**, 1580–1585.

Pletnev, V.Z., Kuzin, A.P., Trakhanov, S.D. & Kostetsky, P.V. (1982) Three-dimensional structure of actinoxanthin IV at 2.5 Å

resolution. *Biopolymers* **21**, 287–300.

Van Roey, P. & Beerman, T.A. (1989) Crystal structure analysis of auromomycin apoprotein (macromomycin) shows importance of protein side chains to chromophore binding selectivity. *Proc. Natl Acad. Sci. USA* **86**, 6587–6591.

Zein, N., Casazza, A.M., Doyle, T.W., Leet, J.E., Schroeder, D.R., Solomon, W. & Nadler, S.G. (1993a) Selective proteolytic activity of the antitumor agent Kedarcidin. *Proc. Natl Acad. Sci. USA* **90**, 8009–8012.

Zein, N., Colson, K.L., Leet, J.E., Schroeder, D.S., Solomon, W., Doyle, T.W. & Casazza, A.M. (1993b) Kedarcidin chromophore: an enediyne that cleaves DNA in a sequence-specific manner. *Proc. Natl Acad. Sci. USA* **90**, 2822–2826.

Zein, N., Reiss, P., Bernatowicz, M. & Bolgar, M. (1995a) The pro-teolytic specificity of the natural enediyne-containing chromopro-teins is unique to each chromoprotein. *Chem. Biol.* **2**, 451–455.

Zein, N., Solomon, W., Colson, K.L. & Schroeder, D. (1995b) Maduropeptin: a novel antitumor chromoprotein with selective protease activity and DNA cleaving properties. *Biochemistry* **34**, 11591–11597.

*Nada Zein*
*Oncology Drug Discovery,*
*Bristol-Myers Squibb K2123E,*
*Princeton, NJ 08558, USA*
*Email: zein@bms.com*

# 569. *Self-splicing proteins*

## Databanks

*Peptidase classification: none*
*NC-IUBMB enzyme classification: none*
*Databank codes: It is not within the scope of the Handbook to tabulate the many known sequences of self-splicing proteins, but these are to be found in the articles cited.*

## Name and History

The purpose of the present chapter is to describe a form of protein splicing that is a self-catalyzed, post-translational event involving the excision of an intervening protein sequence (termed an 'intein') from a protein precursor and the coupled formation of a peptide bond between the two flanking protein domains (termed 'exteins'; see Figure 569.1). Excision of the intein involves the cleavage of a peptide bond at each of the two extein/intein junctions, coupled to a reaction involving the formation of a peptide bond between the N- and C-exteins (Cooper & Stevens, 1995).

Although two peptide bonds are cleaved within the pre-cursor during splicing, proteins of this class are not con-sidered proteolytic enzymes in the normal sense, since the two resulting proteins, the intein and the ligated exteins, are distinct proteins that to date have shown no further pro-teolytic activity. Thus, the present chapter is intended to highlight a novel mechanism of peptide bond cleavage dis-tinct from any of those mediated by conventional proteolytic enzymes.

## Activity, Specificity and Structural Chemistry

The first described example of protein splicing involved the *TFP1* gene of *S. cerevisiae*, which encodes the ATP-binding subunit of the V-ATPase (Kane *et al.*, 1990). Upon sequencing the gene, it was found that the continuous open reading frame predicted a protein much larger than expected for the described protein. By comparison to homologous ATP-binding subunits, the flanking domains of the *TFP1*

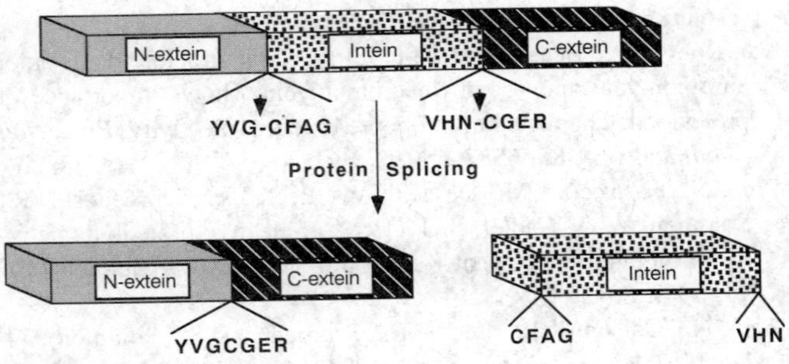

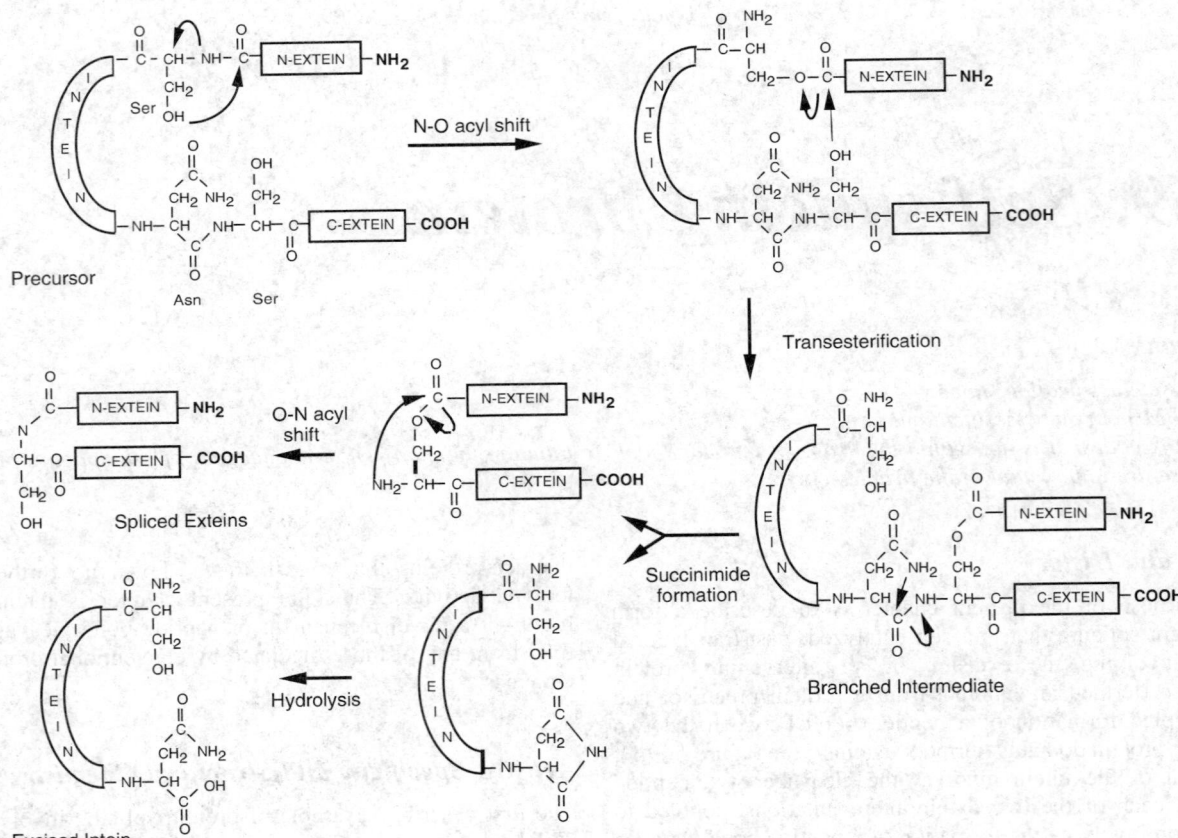

*Figure 569.1* Protein splicing of the *S. cerevisiae* Tfp1p precursor. The amino acid sequences shown are the predicted residues at the precursor splice junctions, the known tryptic peptide spanning the extein/extein spliced junction, and the known N-terminal and C-terminal sequences of the purified intein (Cooper *et al.*, 1993, 1995).

encoded protein were predicted to encode the ATP-binding subunit of the V-ATPase, but these two domains were separated by an inframe insertion of 454 codons. From this initial discovery it was ascertained that protein splicing was responsible, and the elucidation of the mechanism progressed in several stages. First, the information required for protein

splicing was found to be contained within the intein sequence, as demonstrated by the inframe insertion of the intein-encoding DNA into a different gene to create a new artificial precursor that underwent protein splicing (Cooper *et al.*, 1993). Next, mutagenesis of the intein sequence, particularly at the junctions, identified key amino acid residues crucial to the splicing reaction, particularly the Cys, Ser or Thr residues at the splice junctions as well as the Asn residue N-terminal to the intein/C-extein junction (Davis *et al.*, 1992; Hirata & Anraku, 1992; Cooper *et al.*, 1993).

A major advance was provided by an *in vitro* splicing system. Native unspliced precursors are difficult to isolate as the splicing reaction proceeds swiftly. However, an artificial construct was generated using an intein from an extreme thermophile. When expressed in *E. coli* at low temperatures the precursor failed to splice and permitted the purification of the precursor, which subsequently spliced *in vitro* upon warming (Xu *et al.*, 1993). Using this system, a branched splicing intermediate was identified that contained two N-terminal polypeptide chains that corresponded to the N-termini of both the intein and the N-extein (Xu *et al.*, 1993). Further analysis of this branched intermediate showed that the conserved Asn residue at the intein/C-extein junction formed an intramolecular succinimide ring, resulting in cleavage of the appropriate

*Figure 569.2* The protein splicing model. It is postulated that an N-O shift at the N-extein/intein junction, followed by transesterification, produces the branched intermediate. This is resolved by the Asn residue at the intein/C-extein junction forming a succinimide ring that cleaves the adjacent peptide bond. The exteins are now linked by an ester bond that is transformed into a peptide bond by an O-N shift. The succinimide at the C-terminus of the excised intein is hydrolyzed to give Asn (Shao *et al.*, 1996).

peptide bond (Xu *et al.*, 1994). A number of models were proposed to account for the key amino acids, the branched intermediate and the role of the succinimide ring (Xu *et al.*, 1994; Clarke, 1994), but the evidence to date favors the model of Shao *et al.* (1996) as shown in Figure 569.2. Although Cys, Ser or Thr may exist at the splice junctions, the example is shown with Ser. The reaction is proposed to start with an N-O acyl shift that abolishes the peptide bond at the N-extein/intein junction, and is followed by a transesterification step to produce the branched intermediate. Peptide bond cleavage occurs at the intein/C-extein junction due to the adjacent Asn residue forming an intramolecular succinimide in which the Asn side chain amino group attacks the Asn peptide carbonyl. Formation of the succinimide releases the intein, while the exteins remain joined by an ester linkage that is transformed into a peptide bond by an O-N shift to give the mature spliced product. It is proposed that the tertiary structure of the folded intein within the precursor brings the splice junctions into close proximity to react, and that splicing may occur prior to complete translation of the C-extein.

## Preparation, Biological Aspects and Distinguishing Features

Proteins that undergo splicing have been found in archaea, prokaryotes and eukaryotes. The genome of the archaean *Methanococcus jannaschii* has recently been sequenced (Bult *et al.*, 1996) and sequence comparisons to other inteins have suggested the presence of eight protein precursors predicted to undergo protein splicing. The spliced exteins from these organisms encode a number of proteins including DNA polymerases, ATPase subunits and the RecA protein. Due to the broad range of products resulting from protein splicing there is no template for purification of the inteins or spliced exteins.

Most surprising was the discovery that inteins are mobile genetic elements. Inteins as a class of proteins have homology to each other and to endonucleases encoded by group I self-splicing *introns*. Four inteins have so far been shown to be highly specific endonucleases and one spliced intein has been shown to confer genetic mobility upon the DNA sequence that encodes it (Gimble & Thorner, 1992). Inteins comprise an extremely interesting class of elements that are genetically mobile and in which the insertional nature of the intein into the host genome is kept phenotypically silent from the host by protein splicing.

## Further Reading

The review of Cooper & Stevens (1995) is recommended.

## References

Bult, C.J., White, O., Olsen, G.J. *et al.* (1996) Complete genome sequence of the methanogenic Archaeon, *Methanococcus jannaschii. Science,* **273,** 1058–1072.

Clarke, N.D. (1994) A proposed mechanism for the self-splicing of proteins. *Proc. Natl Acad. Sci. USA* **91,** 11084–11088.

Cooper, A.A. & Stevens, T.H. (1995) Protein splicing: self-splicing of genetically mobile elements at the protein level. *Trends Biochem. Sci.* **20,** 351–356.

Cooper, A.A., Chen, Y.J., Lindorfer, M.A. & Stevens, T.H. (1993) Protein splicing of the yeast *TFP1* intervening protein sequence: a model for self-excision. *EMBO J.* **12,** 2575–2583.

Davis, E.O., Jenner, P.J., Brooks, P.C., Colston, M.J. & Sedgewick, S.G. (1992) Protein splicing in the maturation of *M. tuberculosis* recA protein: a mechanism for tolerating a novel class of intervening sequence. *Cell* **71,** 201–210.

Gimble, F.S. & Thorner, J. (1992) Homing of a DNA endonuclease gene by meiotic gene conversion in *Saccharomyces cerevisiae. Nature* **357,** 301–306.

Hirata, R. & Anraku, Y. (1992) Mutations at the putative junction sites of the yeast *VMA1* protein, the catalytic subunit of the vacuolar membrane H(+)-ATPase, inhibit its processing by protein splicing. *Biochem. Biophys. Res. Commun.* **188,** 40–47.

Kane, P.M., Yamashiro, C.T., Wolczyk, D.F., Goebl, M., Neff, N. & Stevens, T.H.. (1990) Protein splicing converts the yeast *TFP1* gene product to the 69 kDa subunit of the vacuolar H$^+$-adenosine triphosphatase. *Science* **250,** 651–657.

Shao, Y., Xu, M.Q. & Paulus, H. (1996) Protein splicing: evidence for an N-O acyl rearrangement as the initial step in the splicing process. *Biochemistry* **35,** 3810–3815.

Xu, M.Q., Southworth, M.W., Mersha, F.B., Hornstra, L.J. & Perler, F.B. (1993) *In vitro* protein splicing of purified precursor and the identification of a branched intermediate. *Cell* **75,** 1371–1377.

Xu, M.Q., Comb, D.G., Paulus, H., Noren, C.J., Shao, Y. & Perler, F.B. (1994) Protein splicing: an analysis of the branched intermediate and its resolution by succinimide formation. *EMBO J.* **13,** 5517–5522.

*Antony Cooper*
*Division of Cell Biology & Biophysics,*
*University of Missouri-Kansas City,*
*Kansas City, MO 64110, USA*
*Email: coopera@cctr.umkc.edu*

# *Appendix 1. Organisms mentioned in the Handbook*

Few readers will be familiar with all of the scientific names of the 750 or so species of organism mentioned in the *Handbook*, so the list below is provided to show what type of organism each species is, and whether it has a common English name. The taxonomy of viruses is very different from that of other organisms, so the list is presented in two parts: Viruses and Other organisms.

## *Viruses*

| | |
|---|---|
| African green monkey immunodeficiency virus | ss-RNA virus |
| AKV murine leukemia virus | ss-RNA virus |
| Andean potato mottle virus | ss-RNA virus |
| apple chlorotic leaf spot virus | ss-RNA virus |
| apple stem grooving virus | ss-RNA virus |
| apple stem pitting virus | ss-RNA virus |
| Aura virus | ss-RNA virus |
| *Autographa californica* nuclear polyhedrosis virus | ds-DNA virus |
| avian adenovirus 1 | ds-DNA virus |
| avian infectious bronchitis virus | ss-RNA virus |
| avian infectious bursal disease virus | ds-RNA virus |
| avian myeloblastosis-associated virus | ss-RNA virus |
| avian reticuloendotheliosis virus | ss-RNA virus |
| baboon endogenous virus | ss-RNA virus |
| bacteriophage HK97 | ds-DNA virus |
| bacteriophage lambda | ds-DNA virus |
| bacteriophage P21 | ds-DNA virus |
| bacteriophage P22 | ds-DNA virus |
| bacteriophage P42D | ds-DNA virus |
| bacteriophage PA-2 | ds-DNA virus |
| bacteriophage Phi-6 | ds-RNA virus |
| bacteriophage T3 | ds-DNA virus |
| bacteriophage T4 | ds-DNA virus |
| bacteriophage T7 | ds-DNA virus |
| barley yellow mosaic virus | ss-RNA virus |
| Barmah Forest virus | ss-RNA virus |
| bean common mosaic virus | ss-RNA virus |
| bean yellow mosaic virus | ss-RNA virus |
| beet necrotic yellow vein virus | ss-RNA virus |
| black beetle virus | ss-RNA virus |
| blueberry scorch virus | ss-RNA virus |
| *Bombyx mori* nuclear polyhedrosis virus | ds-DNA virus |
| Boolarra virus | ss-RNA virus |
| border disease virus | ss-RNA virus |
| brome streak mosaic rymovirus | ss-RNA virus |
| caprine arthritis-encephalitis virus | ss-RNA virus |
| carnation etched ring virus | ds-DNA virus |
| cassava vein mosaic virus | Retroid virus |
| cattle adenovirus 3 | ds-DNA virus |
| cattle adenovirus 7 | ds-DNA virus |
| cattle enterovirus | ss-RNA virus |
| cattle herpesvirus 1 | ds-DNA virus |
| cattle immunodeficiency virus | ss-RNA virus |
| cattle leukemia virus | ss-RNA virus |
| cattle viral diarrhea virus | ss-RNA virus |
| cauliflower mosaic virus | ds-DNA virus |
| cherry capillovirus A | ss-RNA virus |
| chimpanzee immunodeficiency virus | ss-RNA virus |
| *Choristoneura fumiferana* nuclear polyhedrosis virus | ds-DNA virus |
| citrus tatter leaf virus | ss-RNA virus |
| citrus tristeza virus | ss-RNA virus |
| cocksfoot mottle virus | ss-RNA virus |
| *Commelina* yellow mottle virus | ds-RNA virus |
| cowpea mosaic virus | ss-RNA virus |
| cowpea severe mosaic virus | ss-RNA virus |
| coxsackievirus a9 | ss-RNA virus |
| coxsackievirus a21 | ss-RNA virus |
| coxsackievirus a23 | ss-RNA virus |
| coxsackievirus a24 | ss-RNA virus |
| coxsackievirus b1 | ss-RNA virus |
| coxsackievirus b3 | ss-RNA virus |
| coxsackievirus b4 | ss-RNA virus |
| coxsackievirus b5 | ss-RNA virus |
| *Cryphonectria* hypovirus | ds-RNA virus |
| *Dendrobates vertrimaculatus* retrovirus | ss-RNA virus |
| dengue virus type 2 | ss-RNA virus |
| dengue virus type 4 | ss-RNA virus |
| dog mastadenovirus c1 | ds-DNA virus |
| eastern equine encephalitis virus | ss-RNA virus |
| eastern equine encephalomyelitis virus | ss-RNA virus |
| echovirus 9 | ss-RNA virus |
| echovirus 11 (strain Gregory) | ss-RNA virus |
| eggplant mosaic virus | ss-RNA virus |
| encephalomyocarditis virus | ss-RNA virus |
| enterovirus 71 | ss-RNA virus |
| Epstein–Barr virus | ds-DNA virus |
| equine arteritis virus | ss-RNA virus |
| equine herpesvirus type 1 | ds-DNA virus |
| equine herpesvirus type 2 | ds-DNA virus |
| equine infectious anemia virus | ss-RNA virus |
| equine rhinovirus type 1 | ss-RNA virus |
| equine rhinovirus type 2 | ss-RNA virus |
| *Erysimum* latent virus | ss-RNA virus |
| European brown hare syndrome virus | ss-RNA virus |
| feline calicivirus | ss-RNA virus |

| | | | |
|---|---|---|---|
| feline immunodeficiency virus | ss-RNA virus | Middleburg virus | ss-RNA virus |
| feline leukemia virus | ss-RNA virus | Moloney murine leukemia virus | ss-RNA virus |
| figwort mosaic virus | ds-DNA virus | mouse intracisternal A-particle | ss-RNA virus |
| Flock House virus | ss-RNA virus | mouse mammary tumor virus | ss-RNA virus |
| foot-and-mouth disease virus | ss-RNA virus | murine hepatitis virus | ss-RNA virus |
| fowlpox virus | ds-DNA virus | Murray Valley encephalitis virus | ss-RNA virus |
| Friend murine leukemia virus | ss-RNA virus | Nodamura virus | ss-RNA virus |
| garlic latent virus | ss-RNA virus | Norwalk calicivirus | ss-RNA virus |
| garlic yellow streak virus | ss-RNA virus | O'nyong-nyong virus | ss-RNA virus |
| gibbon leukemia virus | ss-RNA virus | *Ononis* yellow mosaic virus | ss-RNA virus |
| grapevine fanleaf virus | ss-RNA virus | *Ornithogalum* mosaic virus | Nonclassified virus |
| hamster intracisternal A-particle | ss-RNA virus | ovine adenovirus | ds-DNA virus |
| hepatitis A virus | ss-RNA virus | ovine lentivirus | ss-RNA virus |
| hepatitis C virus | ss-RNA virus | papaya ringspot virus | ss-RNA virus |
| hepatitis C virus type 3b | ss-RNA virus | parsnip yellow fleck virus | ss-RNA virus |
| hepatitis C virus type 6a | ss-RNA virus | pea seed-borne mosaic virus | ss-RNA virus |
| hepatitis E virus | ss-RNA virus | peanut chlorotic streak virus | ds-DNA virus |
| hepatitis G virus | ss-RNA virus | peanut mottle virus | ss-RNA virus |
| hepatitis GB virus A | ss-RNA virus | peanut stripe virus | ss-RNA virus |
| hepatitis GB virus B | ss-RNA virus | pepper mottle virus | ss-RNA virus |
| hepatitis GB virus C | ss-RNA virus | pestivirus type 1 | ss-RNA virus |
| herpes simplex virus | ds-DNA virus | pig adenovirus type 3 | ds-DNA virus |
| herpes simplex virus type 2 | ds-DNA virus | pig vesicular disease virus | ss-RNA virus |
| herpesvirus saimiri | ds-DNA virus | pig vesicular exanthema virus | ss-RNA virus |
| hog cholera virus | ss-RNA virus | plum pox potyvirus | ss-RNA virus |
| human adenovirus type 2 | ds-DNA virus | poliovirus type 1 | ss-RNA virus |
| human adenovirus type 3 | ds-DNA virus | poliovirus type 2 | ss-RNA virus |
| human adenovirus type 4 | ds-DNA virus | poliovirus type 3 | ss-RNA virus |
| human adenovirus type 5 | ds-DNA virus | porcine transmissible gastroenteritis virus | ss-RNA virus |
| human adenovirus type 12 | ds-DNA virus | potato leaf roll luteovirus | ss-RNA virus |
| human adenovirus type 40 | ds-DNA virus | potato virus C | ss-RNA virus |
| human adenovirus type 41 | ds-DNA virus | potato virus M | ss-RNA virus |
| human coronavirus | ss-RNA virus | potato virus T | ss-RNA virus |
| human cytomegalovirus | ds-DNA virus | potato virus Y | ss-RNA virus |
| human enterovirus 70 | ss-RNA virus | pseudorabies virus | ds-DNA virus |
| human herpesvirus type 6 | ds-DNA virus | rabbit hemorrhagic disease virus | ss-RNA virus |
| human immunodeficiency virus type 1 | ss-RNA virus | radiation murine leukemia virus | ss-RNA virus |
| human immunodeficiency virus type 2 | ss-RNA virus | rat sarcoma virus | ss-RNA virus |
| human rhinovirus | ss-RNA virus | red clover mottling virus | ss-RNA virus |
| human rhinovirus 1b | ss-RNA virus | rice tungro bacilliform virus | ds-DNA virus |
| human rhinovirus 2 | ss-RNA virus | Ross River virus | ss-RNA virus |
| human rhinovirus 14 | ss-RNA virus | Rous avian sarcoma virus | ss-RNA virus |
| human rhinovirus 89 | ss-RNA virus | rubella virus | ss-RNA virus |
| human spumaretrovirus | ss-RNA virus | ryegrass mosaic virus | ss-RNA virus |
| human T cell leukemia virus | ss-RNA virus | San Miguel sea lion virus | ss-RNA virus |
| Hungarian grapevine chrome mosaic virus | ss-RNA virus | Semliki Forest virus | ss-RNA virus |
| ictalurid herpesvirus 1 | ds-DNA virus | sheep pulmonary adenomatosis virus | ss-RNA virus |
| infectious laryngotracheitis virus | ds-DNA virus | simian cytomegalovirus | ds-DNA virus |
| infectious pancreatic necrosis virus | ds-RNA virus | simian foamy virus (type 1) | ss-RNA virus |
| Japanese encephalitis virus | ss-RNA virus | simian foamy virus (type 3) | ss-RNA virus |
| Johnson grass mosaic virus | ss-RNA virus | simian Mason–Pfizer virus | ss-RNA virus |
| *Kennedya* yellow mosaic virus | ss-RNA virus | simian retrovirus | ss-RNA virus |
| Kunjin virus | ss-RNA virus | Sindbis virus | ss-RNA virus |
| lactate dehydrogenase-elevating virus | ss-RNA virus | Sindbis-like virus | ss-RNA virus |
| Lelystad virus | ss-RNA virus | skunk calicivirus | ss-RNA virus |
| lettuce mosaic virus | ss-RNA virus | sooty mangabey immunodeficiency virus | ss-RNA virus |
| macaque immunodeficiency virus | ss-RNA virus | | |
| mandrill immunodeficiency virus | ss-RNA virus | Southampton virus | ss-RNA virus |
| mastadenovirus mus1 | ds-DNA virus | soyabean chlorotic mottle virus | ds-DNA virus |
| Mengo virus | ss-RNA virus | soyabean mosaic virus | ss-RNA virus |

| | |
|---|---|
| squirrel monkey retrovirus | ss-RNA virus |
| strawberry vein banding virus | ds-DNA virus |
| sugar beet yellow virus | ss-RNA virus |
| sugarcane bacilliform virus | ds-DNA virus |
| sweet potato feathery mottle virus | ss-RNA virus |
| Theiler's murine encephalomyelitis virus | ss-RNA virus |
| tick-borne encephalitis virus | ss-RNA virus |
| tobacco etch virus | ss-RNA virus |
| tobacco vein mottling virus | ss-RNA virus |
| tomato black ring virus | ss-RNA virus |
| turkey lymphoproliferative disease virus | ss-RNA virus |
| turnip mosaic virus | ss-RNA virus |
| turnip yellow mosaic virus | ss-RNA virus |
| vaccinia virus | ds-DNA virus |
| varicella-zoster virus | ds-DNA virus |
| Venezuelan equine encephalitis virus | ss-RNA virus |
| Visna lentivirus | ss-RNA virus |
| watermelon mosaic virus II | ss-RNA virus |
| West Nile virus | ss-RNA virus |
| western equine encephalitis virus | ss-RNA virus |
| western equine encephalomyelitis virus | ss-RNA virus |
| yam mosaic virus | ss-RNA virus |
| yellow fever virus | ss-RNA virus |
| zucchini yellow mosaic virus | ss-RNA virus |

## Other organisms

| | |
|---|---|
| *Absibia zychae* | Fungi |
| *Acetobacter turbidans* | Eubacteria |
| *Achromobacter lyticus* | Eubacteria |
| *Acinetobacter calcoaceticus* | Eubacteria |
| *Acipenser transmontanus* (white sturgeon) | Animalia: Teleostei |
| *Acremonium chrysogenum* | Fungi |
| *Actinidia deliciosa* (kiwi fruit) | Viridiplantae: Magnoliopsida |
| *Actinobacillus actinomycetemcomitans* | Eubacteria |
| *Actinobacillus pleuropneumoniae* | Eubacteria |
| *Actinomadura R39* | Eubacteria |
| *Aedes aegypti* (Egyptian mosquito) | Animalia: Insecta |
| *Aeromonas hydrophila* | Eubacteria |
| *Aeromonas salmonicida* | Eubacteria |
| *Aeromonas sobria* | Eubacteria |
| *Agkistrodon bilineatus* (tropical moccasin) | Animalia: Reptilia |
| *Agkistrodon contortrix* (southern copperhead) | Animalia: Reptilia |
| *Agrobacterium rhizogenes* | Eubacteria |
| *Agrobacterium tumefaciens* | Eubacteria |
| *Alcaligenes eutrophus* | Eubacteria |
| *Alicyclobacillus acidocaldarius* | Eubacteria |
| *Alnus glutinosa* (alder) | Viridiplantae: Magnoliopsida |
| *Alteromonas* sp. | Eubacteria |
| *Anabaena variabilis* | Eubacteria |
| *Ananas comosus* (pineapple) | Viridiplantae: Magnoliopsida |
| *Anas platyrhynchos* (mallard) | Animalia: Aves |

| | |
|---|---|
| *Ancyclostoma caninum* | Animalia: Nematoda |
| *Anisakis simplex* | Animalia: Nematoda |
| *Anopheles gambiae* | Animalia: Insecta |
| *Anopheles quadrimaculatus* | Animalia: Insecta |
| *Aplysia californica* | Animalia: Mollusca |
| *Arabidopsis thaliana* | Viridiplantae: Magnoliopsida |
| *Arenicola marina* | Animalia: Annelida |
| *Arthrobacter viscosus* | Eubacteria |
| *Arthrobotrys oligospora* | Fungi |
| *Ascaris lumbricoides* | Animalia: Nematoda |
| *Asclepias syriaca* (milkweed) | Viridiplantae: Magnoliopsida |
| *Aspergillus awamori* | Fungi |
| *Aspergillus flavus* | Fungi |
| *Aspergillus fumigatus* | Fungi |
| *Aspergillus niger* | Fungi |
| *Aspergillus oryzae* | Fungi |
| *Aspergillus phoenicis* | Fungi |
| *Astacus astacus* (broad-fingered crayfish) | Animalia: Crustacea |
| *Astacus fluviatilis* (crayfish) | Animalia: Crustacea |
| *Azospirillum brasilense* | Eubacteria |
| *Azotobacter vinelandii* | Eubacteria |
| *Bacillus alcalophilus* | Eubacteria |
| *Bacillus amyloliquefaciens* | Eubacteria |
| *Bacillus anthracis* | Eubacteria |
| *Bacillus caldolyticus* | Eubacteria |
| *Bacillus cereus* | Eubacteria |
| *Bacillus coagulans* | Eubacteria |
| *Bacillus lentus* | Eubacteria |
| *Bacillus licheniformis* | Eubacteria |
| *Bacillus megaterium* | Eubacteria |
| *Bacillus natto* | Eubacteria |
| *Bacillus pumilus* | Eubacteria |
| *Bacillus smithii* | Eubacteria |
| *Bacillus* sp. | Eubacteria |
| *Bacillus sphaericus* | Eubacteria |
| *Bacillus stearothermophilus* | Eubacteria |
| *Bacillus subtilis* | Eubacteria |
| *Bacillus thermoproteolyticus* | Eubacteria |
| *Bacillus thuringiensis* | Eubacteria |
| *Bacteroides fragilis* | Eubacteria |
| *Balaenoptera acutorostrata* (minke whale) | Animalia: Mammalia |
| *Bartonella bacilliformis* | Eubacteria |
| *Bartonella henselae* | Eubacteria |
| *Beauveria bassiana* | Fungi |
| *Bitis gabonica* (Gaboon viper) | Animalia: Reptilia |
| *Blastocladiella emersonii* | Fungi |
| *Blattella germanica* | Animalia: Insecta |
| *Bombyx mori* (silk moth) | Animalia: Insecta |
| *Boophilus microplus* (southern cattle tick) | Animalia: Arachnida |
| *Bordetella pertussis* | Eubacteria |
| *Borrelia burgdorferi* | Eubacteria |

| | |
|---|---|
| *Bos taurus* (cattle) | Animalia: Mammalia |
| *Bothrops atrox* | Animalia: Reptilia |
| *Bothrops jararaca* (jararaca) | Animalia: Reptilia |
| *Botryllus schlosseri* | Animalia: Tunicata |
| *Bradyrhizobium japonicum* | Eubacteria |
| *Branchiostoma californiensis* | Animalia |
| *Brassica napus* (oil-seed rape) | Viridiplantae: Magnoliopsida |
| *Brassica oleracea* (cauliflower) | Viridiplantae: Magnoliopsida |
| *Brevibacillus brevis* | Eubacteria |
| *Brevundimonas diminuta* | Eubacteria |
| *Brucella abortus* | Eubacteria |
| *Buchnera aphidicola* | Eubacteria |
| *Burkholderia cepacia* | Eubacteria |
| *Caenorhabditis briggsae* | Animalia: Nematoda |
| *Caenorhabditis elegans* | Animalia: Nematoda |
| *Caenorhabditis vulgaris* | Animalia: Nematoda |
| *Calloselasma rhodostoma* (Malayan pit viper) | Animalia: Reptilia |
| *Calotropis* sp. | Viridiplantae: Magnoliopsida |
| *Campylobacter jejuni* | Eubacteria |
| *Canavalia ensiformis* (jack bean) | Viridiplantae: Magnoliopsida |
| *Candida albicans* | Fungi |
| *Candida parapsilosis* | Fungi |
| *Candida tropicalis* | Fungi |
| *Canis familiaris* (dog) | Animalia: Mammalia |
| *Capra hircus* (goat) | Animalia: Mammalia |
| *Capsicum annuum* | Viridiplantae: Magnoliopsida |
| *Carica candamarcensis* | Viridiplantae: Magnoliopsida |
| *Carica papaya* (papaya) | Viridiplantae: Magnoliopsida |
| *Carnobacterium divergens* | Eubacteria |
| *Carnobacterium piscicola* | Eubacteria |
| *Caulobacter crescentus* | Eubacteria |
| *Cavia porcellus* (guinea pig) | Animalia: Mammalia |
| *Celuca pugilator* (sand fiddler crab) | Animalia: Crustacea |
| *Centaurea calcitrapa* | Viridiplantae: Magnoliopsida |
| *Cerastes vipera* (Sahara sand viper) | Animalia: Reptilia |
| *Chionoecetes opilio* (crab-beetle) | Animalia: Crustacea |
| *Chlamydia trachomatis* | Eubacteria |
| *Chlamydomonas eugametos* | Viridiplantae: Chlorophyta |
| *Chlamydomonas reinhardtii* | Viridiplantae: Chlorophyta |
| *Choristoneura fumiferana* | Animalia: Insecta |
| *Chryseobacterium meningosepticum* | Eubacteria |

| | |
|---|---|
| *Cicer arietinum* (chickpea) | Viridiplantae: Magnoliopsida |
| *Citrobacter freundii* | Eubacteria |
| *Citrus sinensis* (orange) | Viridiplantae: Magnoliopsida |
| *Clostridium acetobutylicum* | Eubacteria |
| *Clostridium barati* | Eubacteria |
| *Clostridium botulinum* | Eubacteria |
| *Clostridium butyricum* | Eubacteria |
| *Clostridium histolyticum* | Eubacteria |
| *Clostridium perfringens* | Eubacteria |
| *Clostridium tetani* | Eubacteria |
| *Clostridium thermocellum* | Eubacteria |
| *Coccidioides immitis* | Fungi |
| *Cochliobolus carbonum* | Fungi |
| *Corynebacterium ammoniagenes* | Eubacteria |
| *Corynebacterium pseudotuberculosis* | Eubacteria |
| *Coturnix coturnix* (Japanese quail) | Animalia: Aves |
| *Crithidia fasciculata* | Protozoa |
| *Crotalus adamanteus* (eastern diamondback rattlesnake) | Animalia: Reptilia |
| *Crotalus atrox* (western diamondback rattlesnake) | Animalia: Reptilia |
| *Crotalus basiliscus* | Animalia: Reptilia |
| *Crotalus ruber* (red rattlesnake) | Animalia: Reptilia |
| *Cryphonectria parasitica* (chestnut blight fungus) | Fungi |
| *Cucumis melo* (oriental pickling melon) | Viridiplantae: Magnoliopsida |
| *Culex pipiens* | Animalia: Insecta |
| *Culex quinquefasciatus* | Animalia: Insecta |
| *Cyanophora paradoxa* | Viridiplantae: Thallobionta |
| *Cynara cardunculus* | Viridiplantae: Magnoliopsida |
| *Cynops pyrrhogaster* (Japanese newt) | Animalia: Amphibia |
| *Cyprinus carpio* (carp) | Animalia: Teleostei |
| *Cytophaga* sp. | Eubacteria |
| *Daboia russellii* (Russell's viper) | Animalia: Reptilia |
| *Danio rerio* (zebra fish) | Animalia: Teleostei |
| *Deinagkistrodon acutus* (sharp-nosed viper) | Animalia: Reptilia |
| *Dermasterias imbricata* | Animalia: Echinodermata |
| *Dermatophagoides farinae* | Animalia: Arachnida |
| *Dermatophagoides microceras* | Animalia: Arachnida |
| *Dermatophagoides pteronyssinus* | Animalia: Arachnida |
| *Desmodus rotundus* (vampire bat) | Animalia: Mammalia |
| *Dianthus caryophyllus* (carnation) | Viridiplantae: Magnoliopsida |
| *Dichelobacter nodosus* | Eubacteria |
| *Dictyostelium discoideum* | Protozoa |
| *Diheterospora chlamydosporia* | Fungi |
| *Dissostichus mawsoni* | Animalia: Teleostei |
| *Drosophila buzzatii* | Animalia: Insecta |
| *Drosophila erecta* | Animalia: Insecta |

| | |
|---|---|
| *Drosophila melanogaster* | Animalia: Insecta |
| *Drosophila simulans* | Animalia: Insecta |
| *Echis carinatus* (saw scaled viper) | Animalia: Reptilia |
| *Echis pyramidum* (carpet viper) | Animalia: Reptilia |
| *Eimeria acervulina* | Protozoa |
| *Eimeria bovis* | Protozoa |
| *Emericella nidulans* | Fungi |
| *Entamoeba dispar* | Protozoa |
| *Entamoeba histolyticum* | Protozoa |
| *Entamoeba invadens* | Protozoa |
| *Enterobacter aerogenes* | Eubacteria |
| *Enterobacter cloacae* | Eubacteria |
| *Enterococcus faecalis* | Eubacteria |
| *Enterococcus faecium* | Eubacteria |
| *Enterococcus hirae* | Eubacteria |
| *Epifagus virginiana* (beechdrops) | Viridiplantae: Magnoliopsida |
| *Eptatretus stoutii* | Animalia |
| *Equus caballus* (horse) | Animalia: Mammalia |
| *Erwinia amylovora* | Eubacteria |
| *Erwinia carotovora* | Eubacteria |
| *Erwinia chrysanthemi* | Eubacteria |
| *Escherichia coli* | Eubacteria |
| *Euroglyphus maynei* | Animalia: Arachnida |
| *Fasciola hepatica* | Animalia: Platyhelminthes |
| *Felis catus* (cat) | Animalia: Mammalia |
| *Ficus glabrata* (fig) | Viridiplantae: Magnoliopsida |
| *Filobasidiella neoformans* | Fungi |
| *Flavobacterium* sp. | Eubacteria |
| *Fugu rubripes* | Animalia: Teleostei |
| *Fusarium oxysporum* | Fungi |
| *Fusarium* sp. | Fungi |
| *Gadus morhua* (Atlantic cod) | Animalia: Teleostei |
| *Gallus gallus* (chicken) | Animalia: Aves |
| *Gecko gecko* | Animalia: Reptilia |
| *Giardia intestinalis* | Archezoa: Diplomonadida |
| *Glomerella cingulata* | Fungi |
| *Gloydius halys* | Animalia: Reptilia |
| *Glycine max* (soyabean) | Viridiplantae: Magnoliopsida |
| *Haematobia irritans* | Animalia: Insecta |
| *Haemonchus contortus* | Animalia: Nematoda |
| *Haemonchus* sp. | Animalia: Nematoda |
| *Haemophilus influenzae* | Eubacteria |
| *Hafnia alvei* | Eubacteria |
| *Haliotis rufescens* | Animalia: Mollusca |
| *Haloarcula marismortui* | Archaea |
| *Halobacterium mediterranei* | Archaea |
| *Helicobacter pylori* | Eubacteria |
| *Heliothis virescens* | Animalia: Insecta |
| *Heloderma horridum* (Mexican beaded lizard) | Animalia: Reptilia |
| *Hemerocallis* sp. (day lily) | Viridiplantae: Liliopsida |
| *Herdmania momus* | Animalia |
| *Hirudo medicinalis* (medicinal leech) | Animalia: Annelida |
| *Homarus americanus* (American lobster) | Animalia: Crustacea |
| *Homo sapiens* (human) | Animalia: Mammalia |
| *Hordeum vulgare* (barley) | Viridiplantae: Liliopsida |
| *Hydra attenuata* | Animalia: Cnidaria |
| *Hydra vulgaris* | Animalia: Cnidaria |
| *Hypoderma lineatum* | Animalia: Insecta |
| *Klebsiella pneumoniae* | Eubacteria |
| *Kluyvera citrophila* | Eubacteria |
| *Kluyveromyces lactis* | Fungi |
| *Lachesis muta* (bushmaster snake) | Animalia: Reptilia |
| *Lactobacillus casei* | Eubacteria |
| *Lactobacillus delbrueckii* | Eubacteria |
| *Lactobacillus helveticus* | Eubacteria |
| *Lactobacillus leichmannii* | Eubacteria |
| *Lactobacillus paracasei* | Eubacteria |
| *Lactobacillus plantarum* | Eubacteria |
| *Lactobacillus sake* | Eubacteria |
| *Lactobacillus* sp. | Eubacteria |
| *Lactococcus lactis* | Eubacteria |
| *Lampetra japonica* (Japanese lamprey) | Animalia: Agnatha |
| *Legionella longbeachae* | Eubacteria |
| *Legionella pneumophila* | Eubacteria |
| *Leishmania amazonensis* | Protozoa |
| *Leishmania chagasi* | Protozoa |
| *Leishmania donovani* | Protozoa |
| *Leishmania guyanensis* | Protozoa |
| *Leishmania major* | Protozoa |
| *Leishmania mexicana* | Protozoa |
| *Leishmania pifanoi* | Protozoa |
| *Leptospira borgpetersenii* | Eubacteria |
| *Leuconostoc gelidum* | Eubacteria |
| *Leuconostoc mesenteroides* | Eubacteria |
| *Lilium longifolium* (trumpet lily) | Viridiplantae: Liliopsida |
| *Limulus polyphemus* (Atlantic horseshoe crab) | Animalia: Merostomata |
| *Listeria grayi* | Eubacteria |
| *Listeria ivanovii* | Eubacteria |
| *Listeria monocytogenes* | Eubacteria |
| *Listeria seeligeri* | Eubacteria |
| *Listeria welshimeri* | Eubacteria |
| *Lonomia achelous* | Animalia: Insecta |
| *Lophius americanus* (anglerfish) | Animalia: Teleostei |
| *Lucilia cuprina* (Australian sheep blowfly) | Animalia: Insecta |
| *Lumbricus rubellus* | Animalia: Annelida |
| *Lupinus albus* (white lupin) | Viridiplantae: Magnoliopsida |
| *Lupinus angustifolius* (narrow-leaved blue lupin) | Viridiplantae: Magnoliopsida |
| *Lupinus arboreus* (tree lupin) | Viridiplantae: Magnoliopsida |

| | |
|---|---|
| *Lycopersicon esculentum* (tomato) | Viridiplantae: Magnoliopsida |
| *Lymnaea stagnalis* | Animalia: Mollusca |
| *Lysobacter enzymogenes* | Eubacteria |
| *Macaca fascicularis* (crab-eating macaque) | Animalia: Mammalia |
| *Macaca fuscata* (Japanese macaque) | Animalia: Mammalia |
| *Macaca mulatta* (rhesus monkey) | Animalia: Mammalia |
| *Macrovipera lebetina* | Animalia: Reptilia |
| *Malbranchea sulfurea* | Fungi |
| *Manduca sexta* (tobacco hawkmoth) | Animalia: Insecta |
| *Marchantia polymorpha* | Viridiplantae: Bryophyta |
| *Mastomys natalensis* (African soft-furred rat) | Animalia: Mammalia |
| *Medicago sativa* | Viridiplantae: Magnoliopsida |
| *Megaderma lyra* (false vampire bat) | Animalia: Mammalia |
| *Meriones unguiculatus* (Mongolian gerbil) | Animalia: Mammalia |
| *Mesembryanthemum crystallinum* | Viridiplantae: Magnoliopsida |
| *Mesocricetus auratus* (golden hamster) | Animalia: Mammalia |
| *Mesocricetus longicaudatus* (long-tailed hamster) | Animalia: Mammalia |
| *Metarhizium anisopliae* | Fungi |
| *Methanobacterium thermoautotrophicum* | Archaea |
| *Methanococcus jannaschii* | Archaea |
| *Methanosarcina thermophila* | Archaea |
| *Methanothermus fervidus* | Archaea |
| *Methylbacterium extorquens* | Eubacteria |
| *Micrococcus luteus* | Eubacteria |
| *Moraxella lacunata* | Eubacteria |
| *Mus musculus* (house mouse) | Animalia: Mammalia |
| *Mycobacterium leprae* | Eubacteria |
| *Mycobacterium paratuberculosis* | Eubacteria |
| *Mycobacterium tuberculosis* | Eubacteria |
| *Mycoplasma capricolum* | Eubacteria |
| *Mycoplasma genitalium* | Eubacteria |
| *Mycoplasma hominis* | Eubacteria |
| *Mycoplasma pneumoniae* | Eubacteria |
| *Mycoplasma salivarium* | Eubacteria |
| *Myxococcus xanthus* | Eubacteria |
| *Naegleria fowleri* | Protozoa |
| *Natrialba asiatica* | Archaea |
| *Neisseria flavescens* | Eubacteria |
| *Neisseria gonorrhaea* | Eubacteria |
| *Neisseria gonorrhoeae* | Eubacteria |
| *Neisseria meningitidis* | Eubacteria |
| *Nephrops norvegicus* (Norwegian lobster) | Animalia: Crustacea |
| *Neurospora crassa* | Fungi |
| *Nicotiana otophora* (tobacco) | Viridiplantae: Magnoliopsida |
| *Ochrobactrum anthropi* | Eubacteria |
| *Odontella sinensis* | Viridiplantae: Thallobionta |
| *Oenothera organensis* (evening primrose) | Viridiplantae: Magnoliopsida |
| *Onchocerca volvulus* | Animalia: Nematoda |
| *Oncorhynchus mykiss* (rainbow trout) | Animalia: Teleostei |
| *Ophiostoma piceae* | Fungi |
| *Oryctolagus cuniculus* (rabbit) | Animalia: Mammalia |
| *Oryza sativa* (rice) | Viridiplantae: Liliopsida |
| *Oryzias latipes* | Animalia: Teleostei |
| *Ostertagia ostertagi* | Animalia: Nematoda |
| *Ostertagia* sp. | Animalia: Nematoda |
| *Ovis aries* (sheep) | Animalia: Mammalia |
| *Ovophis okinavensis* (hime-habu) | Animalia: Reptilia |
| *Paecilomyces lilacinus* | Fungi |
| *Paenibacillus polymyxa* | Eubacteria |
| *Papio cynocephalus* (yellow baboon) | Animalia: Mammalia |
| *Papio hamadryas* (hamadryas baboon) | Animalia: Mammalia |
| *Paracentrotus lividus* | Animalia: Echinodermata |
| *Paracoccus denitrificans* | Eubacteria |
| *Paragonimus westermani* | Animalia: Platyhelminthes |
| *Paralithodes camtschatica* (Kamchatka crab) | Animalia: Crustacea |
| *Paramecium tetraurelia* | Protozoa |
| *Paranotothenia magellanica* | Animalia: Teleostei |
| *Pasteurella haemolytica* | Eubacteria |
| *Pediococcus acidilactici* | Eubacteria |
| *Penaeus monodon* | Animalia: Crustacea |
| *Penaeus vannamei* | Animalia: Crustacea |
| *Penicillium citrinum* | Fungi |
| *Penicillium janthinellum* | Fungi |
| *Petromyzon marinus* (sea lamprey) | Animalia: Agnatha |
| *Petroselinum crispum* (parsley) | Viridiplantae: Magnoliopsida |
| *Petunia hybrida* (petunia) | Viridiplantae: Magnoliopsida |
| *Phalaenopsis* sp. | Viridiplantae: Liliopsida |
| *Phaseolus vulgaris* (French bean) | Viridiplantae: Magnoliopsida |
| *Phormidium laminosum* | Eubacteria |
| *Physarum polycephalum* | Protozoa |
| *Pichia pastoris* | Fungi |
| *Pinus contorta* (shore pine) | Viridiplantae: Pinophyta |
| *Pinus thunbergiana* (green pine) | Viridiplantae: Pinophyta |

| | |
|---|---|
| *Pisum sativum* (pea) | Viridiplantae: Magnoliopsida |
| *Plasmodium berghei* | Protozoa |
| *Plasmodium cynomolgi* | Protozoa |
| *Plasmodium falciparum* | Protozoa |
| *Plasmodium fragile* | Protozoa |
| *Plasmodium gallinaceum* | Protozoa |
| *Plasmodium malariae* | Protozoa |
| *Plasmodium ovale* | Protozoa |
| *Plasmodium reichenowi* | Protozoa |
| *Plasmodium vinckei* | Protozoa |
| *Plasmodium vivax* | Protozoa |
| *Pleuronectes platessa* (plaice) | Animalia: Teleostei |
| *Polyporus tulipiferae* | Fungi |
| *Pontastacus leptodactylus* (narrow-fingered crayfish) | Animalia: Crustacea |
| *Porphyra purpurea* | Rhodophyta |
| *Porphyromonas gingivalis* | Eubacteria |
| *Praomys natalensis* | Animalia: Mammalia |
| *Proteus mirabilis* | Eubacteria |
| *Proteus vulgaris* | Eubacteria |
| *Protopterus aethiopicus* (marbled lungfish) | Animalia: Dipnoi |
| *Providencia rettgeri* | Eubacteria |
| *Pseudomonas aeruginosa* | Eubacteria |
| *Pseudomonas fluorescens* | Eubacteria |
| *Pseudomonas putida* | Eubacteria |
| *Pseudomonas* sp. | Eubacteria |
| *Pseudomonas* sp. 101 | Eubacteria |
| *Pseudotsuga menziesii* (Douglas fir) | Viridiplantae: Pinophyta |
| *Pyrobaculum aerophilum* | Archaea |
| *Pyrococcus furiosus* | Archaea |
| *Ralstonia eutropha* | Eubacteria |
| *Rana catesbeiana* (bull frog) | Animalia: Amphibia |
| *Rana esculenta* (edible frog) | Animalia: Amphibia |
| *Rarobacter faecitabidus* | Eubacteria |
| *Rattus norvegicus* (brown rat) | Animalia: Mammalia |
| *Rattus rattus* (black rat) | Animalia: Mammalia |
| *Renibacterium salmoninarum* | Eubacteria |
| *Rhizobium etli* | Eubacteria |
| *Rhizobium leguminosarum* | Eubacteria |
| *Rhizomucor miehei* | Fungi |
| *Rhizomucor pusillus* | Fungi |
| *Rhizopus microsporus* | Fungi |
| *Rhizopus niveus* | Fungi |
| *Rhodobacter capsulatus* | Eubacteria |
| *Rhodococcus erythropolis* | Eubacteria |
| *Rhodococcus* sp. | Eubacteria |
| *Ricinus communis* (castor bean) | Viridiplantae: Magnoliopsida |
| *Rickettsia prowazekii* | Eubacteria |
| *Rickettsia tsutsugamushi* | Eubacteria |
| *Saccharomyces cerevisiae* (baker's yeast) | Fungi |

| | |
|---|---|
| *Saccharomyces kluyveri* | Fungi |
| *Saccharomycopsis fibuligera* | Fungi |
| *Saccharopolyspora erythraea* | Eubacteria |
| *Salmo salar* (Atlantic salmon) | Animalia: Teleostei |
| *Salmonella typhimurium* | Eubacteria |
| *Sarcophaga bullata* | Animalia: Insecta |
| *Sarcophaga peregrina* | Animalia: Insecta |
| *Schistosoma japonicum* | Animalia: Platyhelminthes |
| *Schistosoma mansoni* | Animalia: Platyhelminthes |
| *Schizophyllum commune* | Fungi |
| *Schizosaccharomyces pombe* (fission yeast) | Fungi |
| *Scytalidium lignicolum* | Fungi |
| *Serratia marcescens* | Eubacteria |
| *Serratia* sp. | Eubacteria |
| *Shigella flexneri* | Eubacteria |
| *Simulium vittatum* | Animalia: Insecta |
| *Sinorhizobium meliloti* | Eubacteria |
| *Solanum tuberosum* (potato) | Viridiplantae: Magnoliopsida |
| *Spinacia oleracea* (spinach) | Viridiplantae: Magnoliopsida |
| *Spirometra mansonoides* | Animalia: Platyhelminthes |
| *Spodoptera frugiperda* (fall armyworm) | Animalia: Insecta |
| *Squalus acanthias* (spiny dogfish) | Animalia: Chondrichthyes |
| *Staphylococcus aureus* | Eubacteria |
| *Staphylococcus carnosus* | Eubacteria |
| *Staphylococcus epidermidis* | Eubacteria |
| *Staphylococcus hyicus* | Eubacteria |
| *Staphylococcus simulans* | Eubacteria |
| *Staphylococcus staphylolyticus* | Eubacteria |
| *Staphylothermus marinus* | Archaea |
| *Stenotrophomonas maltophilia* | Eubacteria |
| *Streptococcus agalactiae* | Eubacteria |
| *Streptococcus gordonii* | Eubacteria |
| *Streptococcus pneumoniae* | Eubacteria |
| *Streptococcus pyogenes* | Eubacteria |
| *Streptococcus salivarius* | Eubacteria |
| *Streptococcus sanguis* | Eubacteria |
| *Streptococcus thermophilus* | Eubacteria |
| *Streptomyces albogriseolus* | Eubacteria |
| *Streptomyces albus* | Eubacteria |
| *Streptomyces anulatus* | Eubacteria |
| *Streptomyces cacaoi* | Eubacteria |
| *Streptomyces caespitosus* | Eubacteria |
| *Streptomyces capreolus* | Eubacteria |
| *Streptomyces coelicolor* | Eubacteria |
| *Streptomyces exfoliatus* | Eubacteria |
| *Streptomyces fradiae* | Eubacteria |
| *Streptomyces glaucescens* | Eubacteria |
| *Streptomyces griseus* | Eubacteria |
| *Streptomyces* sp. K15 | Eubacteria |
| *Streptomyces lactamdurans* | Eubacteria |
| *Streptomyces lincolnensis* | Eubacteria |
| *Streptomyces lividans* | Eubacteria |

| | |
|---|---|
| *Streptomyces* sp. | Eubacteria |
| *Strongylocentrotus purpuratus* | Animalia: Echinodermata |
| *Strongyloides ratti* | Animalia: Nematoda |
| *Sulfolobus acidocaldarius* | Archaea |
| *Sulfolobus solfataricus* | Archaea |
| *Sus scrofa* (pig) | Animalia: Mammalia |
| *Synechococcus* sp. | Eubacteria |
| *Synechocystis* sp. | Eubacteria |
| *Tachypleus gigas* (Singapore horseshoe crab) | Animalia: Merostomata |
| *Tachypleus tridentatus* (Japanese horseshoe crab) | Animalia: Merostomata |
| *Tetrahymena thermophila* | Protozoa |
| *Thaumatococcus daniellii* | Viridiplantae: Liliopsida |
| *Theileria annulata* | Protozoa |
| *Theileria parva* | Protozoa |
| *Thermoactinomyces* sp. | Eubacteria |
| *Thermoactinomyces vulgaris* | Eubacteria |
| *Thermoplasma acidophilum* | Archaea |
| *Thermus aquaticus* | Eubacteria |
| *Thermus* sp. | Eubacteria |
| *Thermus thermophilus* | Eubacteria |
| *Thiocarpa roseopersicina* | Eubacteria |
| *Thunnus thynnus* (Northern Pacific bluefin tuna) | Animalia: Teleostei |
| *Toxocara canis* | Animalia: Nematoda |
| *Treponema denticola* | Eubacteria |
| *Treponema pallidum* | Eubacteria |
| *Trichoderma harzianum* | Fungi |
| *Trichomonas foetus* | Protozoa |
| *Trichomonas vaginalis* | Archezoa: Parabasalidea |
| *Trichoplusia ni* (cabbage looper moth) | Animalia: Insecta |
| *Trimeresurus flavoviridis* (habu snake) | Animalia: Reptilia |
| *Trimeresurus gramineus* (green habu snake) | Animalia: Reptilia |
| *Trimeresurus mucrosquamatus* (Taiwan habu snake) | Animalia: Reptilia |
| *Trimeresurus stejnegeri* (Chinese green tree viper) | Animalia: Reptilia |
| *Triticum aestivum* (wheat) | Viridiplantae: Liliopsida |
| *Tritirachium album* | Fungi |
| *Tritrichomonas foetus* | Archezoa: Parabasalidea |
| *Trypanosoma brucei* | Protozoa |
| *Trypanosoma congolense* | Protozoa |
| *Trypanosoma cruzi* | Protozoa |
| *Trypanosoma rangeli* | Protozoa |
| *Urechis caupo* | Animalia: Xenopneusta |
| *Ursus americanus* (black bear) | Animalia: Mammalia |
| *Vespa crabro* (European hornet) | Animalia: Insecta |
| *Vespa orientalis* (oriental hornet) | Animalia: Insecta |
| *Vibrio alginolyticus* | Eubacteria |
| *Vibrio anguillarum* | Eubacteria |
| *Vibrio cholerae* | Eubacteria |
| *Vibrio metschnikovii* | Eubacteria |
| *Vibrio mimicus* | Eubacteria |
| *Vibrio parahaemolyticus* | Eubacteria |
| *Vibrio proteolyticus* | Eubacteria |
| *Vibrio vulnificus* | Eubacteria |
| *Vicia faba* (broad bean) | Viridiplantae: Magnoliopsida |
| *Vicia sativa* (spring vetch) | Viridiplantae: Magnoliopsida |
| *Vigna aconitifolia* (moth bean) | Viridiplantae: Magnoliopsida |
| *Vigna mungo* (mung bean) | Viridiplantae: Magnoliopsida |
| *Vigna radiata* | Viridiplantae: Magnoliopsida |
| *Vigna unguiculata* | Viridiplantae: Magnoliopsida |
| *Wolinella succinogenes* | Eubacteria |
| *Xanthomonas campestris* | Eubacteria |
| *Xanthomonas* sp. T-22 | Eubacteria |
| *Xenopus laevis* (African clawed frog) | Animalia: Amphibia |
| *Yarrowia lipolytica* | Fungi |
| *Yersinia enterocolitica* | Eubacteria |
| *Yersinia pestis* | Eubacteria |
| *Zea mays* (maize) | Viridiplantae: Liliopsida |
| *Zinnia elegans* (zinnia) | Viridiplantae: Magnoliopsida |

# *Appendix 2. Suppliers cited in the Handbook*

**Abbott Laboratories**, 200 Abbott Park Rd, Abbott Park, IL 60064, USA [http://www.abbott.com]

**Accurate Chemical and Scientific Corp.**, 300 Shamer Drive, Westbury, NY 11590, USA [http://www.accurate-assi-leeches.com/accurate/index.html]

**Advance Biofactures Corp.**, see Biospecifics Technologies Corp.

**Affinity Research Products**, Exeter, UK [sheppardpw@affinity-re.com]

**Allergan Inc.**, P.O. Box 19534, Irvine, CA 92623-9534, USA [http://www.allergan.com/contact/index.htm]

**Alomone Laboratories Ltd.**, Shatner Center 3, PO Box 4287, Jerusalem 91042, Israel [Fax: (972) 2-525233]

**Amano Pharmaceutical Co., Ltd.**, Nishiki 1-2-7, Naka-ku, Nagoya-shi, Aichi-ken, Japan [http://www.inter.co.jp/fmf/index.html]

**American Diagnostica Inc.**, 222 Railroad Ave., P.O. Box 1165, Greenwich, CT 06836-1165, USA [http://www.aacc.org/]

**American Peptide Company**, 777 E. Everlyn Avenue, Sunnyvale, CA 94086, USA [http://www.americanpeptide.com/ci.htm]

**American Type Culture Collection**, 12301 Parklawn Drive, Rockville, MD 20852, USA [http://www.atcc.org]

**Amersham North America**, 2636 South Clearbrook Drive, Arlington Heights, Illinois 60005 USA [http://www.amersham.com/life/index.html]

**Athens Research and Technology**, P.O. Box 5494, Athens, GA 30604, USA [http://www.athensresearch.com/proteins.html]

**Bachem California Inc.**, 3132 Kashiwa Street, Torrance, CA 90505, USA [Fax: (310) 530-1571]

**Bachem AG**, Haupstrasse 144, Ch-4416 Bubendorf, Switzerland [http://www.bachem.com]

**Bio-Ass**, Fritz-Winter-Strasse 32, D-86911 Diessen, Germany

**Biodesign International**, 105 York Street, Kennebunk, Maine 04043, USA [http://www.biodesign.com]

**Biospecifics Technologies Corp.**, 35 Wilbur Street, Lynbrook, NY 11563, USA [Fax: (516) 593 7000]

**Biozyme Laboratories International Ltd.**, 9939 Hibert Street, Suite 101, San Diego, CA 92131-1029, USA [http://www.42.com/biochemicals/]

**Boehringer Mannheim GmbH**, 9115 Hague Road, P.O. Box 50414, Indianapolis, Indiana 46250-0414, USA [Fax: (317) 576-7317, *http://www.biochem.boehringer-mannheim.com*; E-mail: biochemts_us@bmc.boehringer-mannheim.com]

**CAB International**, Wallingford, Oxfordshire OX10 8DE, UK

**Calbiochem-Novabiochem Corp.**, 10394 Pacific Center Court, San Diego, CA 92121, USA [http://www.calbiochem.com]

**Cambio Ltd.**, 34 Newnham Road, Cambridge, CB3 9EY, UK [http://www.cambio.co.uk]

**Cayman Chemical Company**, 690 KMS Place, Ann Arbor, MI 48108, USA [http://www.caymenchem.com/]

**Cedarlane Laboratories**, RR2, Hornby, Ontario L0P 1E0, Canada [http://www.cedarlanelabs.com/what.htm]

**Chemicon International Inc.**, 28835 Single Oak Drive, Temecula, CA 92590, USA [Fax: (909) 676-9209; Email: *custserv@chemcon.attmail.com*, http://www.chemicon.com]

**Chr. Hansens Biosystems**, DK-2970 Hørsholm, Denmark [http://www.chrhansen.com]

**Chromogenix AB**, Taljegardsgatan 3, S-431 53, Mölndal, Sweden [http://www.chromogenix.com]

**Cosmo Bio. Co., Ltd.**, 2-2-2- Toyo, Koto-tu, Tokyo 135, Japan [http://www.globe.or.jp/cosmobio]

**Coulter Corporation**, 601 Coulter Way Hialeah, FL 33012-0145, USA [http://www.coulter.com]

**DAKO Corporation**, 6392 Via Real, Carpinteria, CA 93013 USA [Fax: (805) 566-6688; *http://www.dream-catcher.com/dako/index.html*]

**Diagnostic Products Corp.**, 5700 W. 96th St., Los Angeles, CA 90045, USA [Fax: (213) 642-0192]

**Diagnostica Stago**, 9 rue des Freres Chausson, 92600 Asnieres-sur-Seine, France [http://www.stago.fr/]

**Difco Laboratories**, P.O. Box 331058, Detroit, MI 48201, USA [Fax: (410) 316-4060]

**Dupont NEN Research Products**, 549 Albany Street, Boston, MA 02118 USA [http://www.nenlifesci.com/]

**Elastin Products Company Inc.**, PO Box 568, Owensville, MI 65066, USA [Fax: (573) 437-4632]

**Enzyme Research Products Laboratories Inc**, 300 N. Michigan St., #103, South Bend, IN 46601, USA [Fax: (886) 239-4924]

**Enzyme Systems Products**, 488 Lindberg, Livermore, CA 94550, USA [http://www.enzymesys.com]

**Eurogenetics N.C.**, Transportstraat 4, B-3980 Tessenderlo, Belgium [Fax: (32) 13 67 24 94]

**European Collection of Animal Cell Cultures**, PWLS Centre, Porton Down, Salisbury SP4 0JG, UK [Fax: (01980) 610316]

**Fluka Chemika-BioChemika**, Industriestrasse 25, CH-9471 Buchs, Switzerland [http://www.fluka.sial.com]

**Fuji Chemical Industries Inc.**, 530 Chokeiji, Takaoka, Toyama 933, Japan [http://www.fuji-chemi.co.jp]

**Grunenthal GMBH**, Steinfeldstrasse 2, 5190 Stolberg, Germany

**Hematologic Technologies Inc**, Pinewood Plaza, P.O. Box 1-21, Essex Junction, VT 05452, USA [Fax: (802) 878 1776]

**IBF Biotechniques Sepracor**, IBF Technics, 35 ave Jaen-Jaures, F-92390, Villanueve La Gareen, France [Fax: 1 27 92 2655] **also**: 7151 Columbia Gateway Drive, Columbia, MD 21046, USA [Fax: (301) 290-1509]

**ICN Biomedicals Inc., see ICN Pharmaceuticals, Inc.**

**ICN Pharmaceuticals, Inc.**, 3300 Hyland Avenue, P.O. Box 5023, Costa Mesa, CA 92626, USA [http://www.icnpharm.com]

**ICN ImmunoBiomedicals, see ICN Pharmaceuticals, Inc**

**Institute for Fermentation**, 17-85, Juso-honmachi 2-chome, Yodogawa-ku, Osaka 532, Japan

**Invitrogen Corporation**, 1600 Faraday Avenue, Carlsbad, CA 92008, USA [http://www.invitrogen.com]

**Kabi Pharmacia Diagnostics**, 800 Centennial Ave., Piscataway, NJ 08854 [Fax: (908) 457-8010]

**KabiVitrum**, 160 Industrial Drive, Franklin, OH 45005, USA [Fax: (513) 746 9855]

**Kamiya Biomedical Co.**, P.O. Box 6067, Thousand Oaks, CA 91359, USA [Fax (818) 706-8564]

**Kikkoman Co., Ltd.**, Research & Development Division, 399 Noda, Noda-City, Chiba Prefecture, 278, Japan [http://www.kikkoman.co.jp]

**Knoll Laboratories**, 3000 Continental Dr., North Mt. Olive, NJ 07828, USA [http://www.knoll.basf.de]

**Latoxan**, A.P. 1724, F-05150, Rosans, France

**Life Technologies**, Corporate Headquarters, 9800 Medical Center, PO Box 6482, Gaithersburg, MD 20849-6482, USA [http://www.lifetech.com]

**List Biological Labs Inc.**, 501-B Vandell Way, Campbell, CA 95008, USA [Fax: (408) 866-6364]

**Miami Serpentarium Laboratories**, 34879 Washington Loop Road, Punta Gorda, FL 33982, USA [Fax: (813) 639-1811]

**Miles Laboratories**, Elkhart, IN, USA, see **Boehringer Mannheim Corp.**,

**Merck & Co., Inc.**, Rahway, NJ 07065, USA [http://www.merck.com]

**Molecular Genetic Resources, Inc.**, 6201 Johns Road, Suite 8, Tampa, FL 33634, USA [Fax: (813) 881 1589]

**Molecular Probes Inc.**, 4849 Pitchford Avenue, PO Box 22010 Eugene, OR 97402-0414, USA [http://www.probes.com]

**MONOzyme**, Agern Allé, 2970 Høsholm, Denmark

**MyoGenics, Inc.**, Cambridge, MA, USA (see also ProScript, Inc.)

**Nagase Biochemicals Ltd.**, 5-1, Nihonbashi-Kobunacho, Chuo-ku, Tokyo 103, Japan [Fax: 03 3665 3118]

**Neuroprobe**, PO Box 400, Cabin John, MD 20818, USA

**Novagen, Inc.**, 597 Science Drive, Madison, WI, 537111, USA [Tel: (800) 526-7319; http://www.novagen.com]

**Novex**, 4202 Sorrrento Valley Blvd., San Diego, CA 92121, USA [http://www.novex.com]

**Novocastra Laboratories Ltd.**, 24 Claremont Place, Newcastle upon Tyne NE2 4AA, UK [Fax: 0191 222 8687]

**Novus Molecular Inc.**, see **Chemicon Intl. Ltd.**,

**Oncogene Research Products**, 84 Rogers Street, Cambridge, MA 02142, USA [htpp://www.apoptosis.com]

**Oncogene Science, Inc.**, 106 Charles Lindbergh Blvd., Uniondale, NY 11553-3649, USA [http://www.oncogene.com]

**Peninsula Laboratories, Inc.**, 611 Taylor Way, Belmont, California 94002-4041, USA [http://www.penlabs.com]

**Pentapharm AG Ltd.**, Engelstrasse 109, CH-4002 Basel, Switzerland [http://www.pentapharm.ch/]

**Peptides International**, P.O. Box 24658, Louisville, KY 40224, USA [http://www.pepnet.com]

**Peptide Institute, Inc.**, 4-1-2 Ina, Minoh-Shi, Osaka 562, Japan [Fax: 81 (727) 29-4124]

**Pharmacia Biotech Inc.**, 800 Centennial Ave., Piscataway, NJ 08855-1327, USA [htpp://www.biotech.pharmacia.se]

**PharMingen**, 10975 Torreyana Road, San Diego, CA 92121, USA [http://www.pharmingen.com]

**Porton**, Salisbury, Wiltshire, UK, [http://www.dera.gov.uk/dera.htm]

**ProScript, Inc** (formerly MyoGenics, Inc., Cambridge, MA) 38 Sidney Street, Cambridge, MA 02139-4135, USA [Tel: (617) 374-1477]

**Qiagen Inc.**, 9600 De Soto Avenue, Chatsworth, CA 91311, USA [Fax: (800) 426-8157]

**RBI Research Biochemicals International**, One Strathmore Road, Natick, MA 01760-24447, USA [http://www.callrbi.com]

**Research Plus Inc.**, P.O. Box 324, Bayonne, NJ 07002, USA [Fax: (201) 823-9590]

**Santa Cruz Biotechnology Inc.**, 2161 Delaware Ave., Santa Cruz, CA 95060, USA [http://www.scbt.com]

**Dr B. Schircks Laboratories**, Buechstrasse 10, CH-8645 Jona, Switzerland

**Seikagaku Corporation**, Tokyo Yakugyo Bldg., 2-1-5 Nihonbashi-honcho, Chuo-ku, Tokyo 103-0023, Japan [http://www.seikagaku.com.jp]

**Seisin Pharmaceutical**, Seisin Enterprise Co., Ltd., Nippon Brunswick Bldg., 5-27-7 Sendagaya, Shiboya-ku, Tokyo 161, Japan [Fax: 3 350 5793]

**Serono Laboratories S.A.**, 15 Bis-Chemin Desmines, CH-1202, Geneva, Switzerland [http://www.serono-usa.com]

**Serva Biochemicals**, P.O. Box 1531, Paramus, NJ 07653-1531 [Fax: 81 75 241 5208]

**Sigma Chemical Company**, PO Box 14508, St Louis, MO 63178-9916, USA [htpp://www.sigma.sial.com]

**Takara Shuzo Co., Ltd**, Bio-Medical Group, 3-4-1 Seta, Otsu 520-21, Japan [Fax: 81 75 241 5208]

**Takeda Chemical Industries, Ltd.**, 1-2, Doshomachi 4-chome, Chuo-ku, Osaka 541, Japan [htpp://www.medic.mie-u.ac.jp/takeda/takeda.html]

**TCS Biologicals**, Botolph Claydon, Buckingham, Bucks, MK18 2LR, UK

**Tosoh Co., Ltd.**, Hayakawa 2743, Ayaseshi, Kanagawa Prefecture 252, Japan

**Toxin Technology Inc.**, 7165 Curtiss Ave., Sarasota, FL34231, USA [htpp://www.toxintechnology.com]

**Transduction Laboratories**, 133 Venture Court, Suite 5, Lexington, KY 40511-2600, USA [htpp://www.translab.com/]

**Transformation Research Inc.**, P.O. Box 2411, Framingham, MA 01701, USA

**Twyford Laboratories Ltd.**, Park Royal Brewery, London NW10 7RR, UK

**Twyford Pharmaceutical Services Ltd.**, D-6700 Ludswigshafen am Rhein, Postfach 21 08 05, Germany

**Upstate Biotechnology Inc.**, 199 Saranac Avenue, Lake Placid, NY 12946, USA [htpp://www.biosignals.com/]

**Wako Pure Chemical Industries**, 3-1-2, Doshomachi, Chuo-ku, Osaka, Japan [Fax: +81-6-201-5964]

**Wako Chemicals USA Inc.**, 1600 Bellwood Road, Richmond, Virginia 23237, USA [Fax: 804 271-7791]

**Worthington Biochemical Corp.**, 730 Vassar Ave., NJ 08701, USA [htpp://www.worthington-biochem.com/]

**Yeast Genetics Stock Center**, Department of Molecular and Cellular Biology, University of California, 305 Donner Lab., Berkeley, CA 94720, USA

# Index

Page numbers in *italic* refer to illustrations and tables; **bold** page numbers indicate a main discussion.

ISBN 0120793709